MOTOR
TRUCK REPAIR MANUAL

36th Edition

First Printing

Louis C. Forier, SAE
Editorial Director

Michael J. Kromida, SAE
Editor

Dan Irizarry, SAE • Warren Schildknecht, SAE
Senior Editors

Michael E. Pallien, SAE • Mark E. Flynn, SAE
Associate Editors

Daniel E. Doku • Patrick Peyton, SAE • Robert R. Savasta
Derek K. Long, SAE • Mark Sharrock • Russell Macrae
Denver Steele • John R. Lypen • James M. Garripoli
Assistant Editors

Rose-Ellen Lorber-Termaat
Editorial Assistant

Published by

MOTOR

224 West 57th St., New York, N.Y. 10019

Library of Congress Catalog Number: 75-643228

ISBN 0-87851-663-1

DECIMAL & MILLIMETER EQUIVALENTS

INCH	INCH	MM
1/64	.015625	.397
1/32	.03125	.794
3/64	.046875	1.191
1/16	.0625	1.587
5/64	.078125	1.984
3/32	.09375	2.381
7/64	.109375	2.778
1/8	.125	3.175
9/64	.140625	3.572
5/32	.15625	3.969
11/64	.171875	4.366
3/16	.1875	4.762
13/64	.203125	5.159
7/32	.21875	5.556
15/64	.234375	5.953
1/4	.25	6.350
17/64	.265625	6.747
9/32	.28125	7.144
19/64	.296875	7.541
5/16	.3125	7.937
21/64	.328125	8.334
11/32	.34375	8.731

INCH	INCH	MM
23/64	.359375	9.128
3/8	.375	9.525
25/64	.390625	9.922
13/32	.40625	10.319
27/64	.421875	10.716
7/16	.4375	11.113
29/64	.453125	11.509
15/32	.46875	11.906
31/64	.484375	12.303
1/2	.5	12.700
33/64	.515625	13.097
17/32	.53125	13.494
35/64	.546875	13.890
9/16	.5625	14.287
37/64	.578125	14.684
19/32	.59375	15.081
39/64	.609375	15.478
5/8	.625	15.875
41/64	.640625	16.272
21/32	.65625	16.669
43/64	.671875	17.065

INCH	INCH	MM
11/16	.6875	17.462
45/64	.703125	17.859
23/32	.71875	18.265
47/64	.734375	18.653
3/4	.75	19.050
49/64	.765625	19.447
25/32	.78125	19.884
51/64	.796875	20.240
13/16	.8125	20.637
53/64	.828125	21.034
27/32	.84375	21.431
55/64	.859375	21.828
7/8	.875	22.225
57/64	.890625	22.622
29/32	.90625	23.019
59/64	.921875	23.415
15/16	.9375	23.812
61/64	.953125	24.209
31/32	.96875	24.606
63/64	.984375	25.003
1		25.400

Special Service Tools

Throughout this manual references are made and illustrations may depict the use of special tools required to perform certain jobs. These special tools can generally be ordered through the dealers of the make vehicle being serviced. It is also suggested that you check with local automotive supply firms as they also supply tools manufactured by other firms that will assist in the performance of these jobs. The vehicle manufacturers special tools are supplied by:

American Motors & General Motors .... Service Tool Division
Kent-Moore Corporation
29784 Little Mack
Roseville, Michigan 48066

Chrysler Corp. .... Miller Special Tools
A Division of Utica Tool Co.
32615 Park Lane
Garden City, Michigan 48135

Ford Motor Co. .... Owatonna Tool Company
Owatonna, Minnesota 55060

INDEX

This Edition Covers Mechanical Specifications and Service Procedures on 1974–83 Models

TRUCK SECTION

GENERAL SERVICE SECTION

AXLES, DRIVING

BRAKES

CLUTCHES

COOLING SYSTEM

ELECTRICAL

FUEL & EXHAUST

STEERING

TRANSMISSIONS, *Automatic*

TRANSMISSIONS, *Manual Shift*

STOCK & FARM TRACTOR ENGINE SPECIFICATIONS

LIGHT DUTY DIESEL ENGINE REPAIR SECTION*

*Follows page 1508

TUNE UP SERVICE

Engine tune-up has become increasingly important to the modern automotive engine with its vastly improved power and performance. With the emission control systems, improved electrical systems and other advances in design, engines have become more sensitive to usage and operating conditions, all of which have a decided effect on power and performance. Therefore, it is important that this service be performed on the engine at the recommended interval, or more often if conditions warrant.

In addition to the servicing of spark plugs, ignition points and condenser, a proper tune up includes a number of tests to check the condition of the engine and its related systems and uncover sources of future problems.

TUNE UP PROCEDURE

Since a quality tune up is dependent upon the proper operation of a number of systems, we have listed here in a logical sequence, the steps to be followed.

1. Diagnosis. This to consist of a compression test, cylinder balance test, oscilloscope check, manifold vacuum test, charging circuit and cranking voltage test.
2. Service spark plugs.
3. Service ignition system, secondary wiring, distributor and coil.
4. Check and service battery and charging system.
5. Service manifold heat valve, if used.
6. Service carburetor, linkage, fuel and air filters.

Once these mechanical checks have been performed the tune up can be finalized. The final steps are:

1. Setting the dwell.
2. Setting slow idle, idle fuel mixture, choke and ignition timing.
3. Adjusting the fast idle.
4. Checking ignition output and secondary resistance.
5. Road test.

DIAGNOSIS & TESTING

Before a satisfactory tune up can be performed, the existing condition of the engine and its related systems must be determined. A tune up should not be attempted if tests indicate internal engine problems such as burnt valves, worn rings, blown head gasket, etc., until such conditions have been corrected.

Oscilloscope Test

Although oscilloscopes differ in many ways, they all display a light or "trace" on a screen which measures the voltage present at a given point and time. As the ignition system operates, its voltage creates a pattern on the screen. This pattern, when read in accordance with the manufacturers instructions for the particular unit, indicates the condition of the entire ignition system.

Compression Test

CAUTION: Should it become necessary to turn on the ignition switch to crank the engine under the hood for a compression test or for any other reason, it is highly recommended that the high tension cable be removed from the coil tower or from the distributor cap. This is necessary because on many vehicles the ignition and starter circuits are such that the ignition switch and/or starter solenoid may be damaged if this precaution is not taken. Another reason for removing the cable is to prevent the engine from starting accidentally.

The engine cannot be tuned to develop maximum power and smooth performance unless reasonably high and uniform compression pressure is obtained in each cylinder. The compression in each cylinder should, therefore, be tested before any other tune-up operations are performed.

1. Remove any foreign matter from around spark plugs by blowing out plug area with compressed air. Then loosen all plugs one turn.
2. Start engine and accelerate to a fast idle to blow out loosened carbon. Stop engine and remove spark plugs. *Cleaning out carbon in this manner is important in preventing false compression readings due to particles of carbon lodged under the valves.*
3. Remove air cleaner and block throttle and choke in wide open position.
4. Insert compression gauge firmly in spark plug opening and crank engine through at least four or five compression strokes to obtain highest possible reading.
5. Test and record compression of each cylinder and compare the maximum and minimum readings obtained. Refer to the tune-up charts in the truck chapters and determine if the minimum reading is within limits as indicated by the percentages in the chart in Fig. 1.
6. If one or more cylinders read low, inject about a tablespoon of engine oil on top of pistons in the low reading cylinders. Crank engine several times and recheck compression.
7. If compression comes up but does not reach normal it indicates worn piston rings. If compression does not improve, valves are sticking or seating poorly. If two adjacent cylinders indicate low compression and injecting oil does not improve the condition, the cause may be a head gasket leak between the cylinders.

Manifold Vacuum Test

Manifold vacuum is affected by carburetor adjustment, valve timing, ignition timing, condition of valves, cylinder compression, condition of positive crankcase ventilation system, and leakage of manifold, carburetor, carburetor spacer or cylinder head gaskets.

Because abnormal gauge readings may in-

MAXIMUM PSI	MINIMUM PSI		MAXIMUM PSI	MINIMUM PSI		MAXIMUM PSI	MINIMUM PSI	
	75%	80%		75%	80%		75%	80%
134	101	107	174	131	139	214	160	171
136	102	109	176	132	141	216	162	173
138	104	110	178	133	142	218	163	174
140	105	112	180	135	144	220	165	176
142	107	114	182	136	146	222	166	178
144	108	115	184	138	147	224	168	179
146	110	117	186	140	149	226	169	181
148	111	118	188	141	150	228	171	182
150	113	120	190	142	152	230	172	184
152	114	122	192	144	154	232	174	186
154	115	123	194	145	155	234	175	187
156	117	125	196	147	157	236	177	189
158	118	126	198	148	158	238	178	190
160	120	128	200	150	160	240	180	192
162	121	130	202	151	162	242	181	194
164	123	131	204	153	163	244	183	195
166	124	133	206	154	165	246	184	197
168	126	134	208	156	166	248	186	198
170	127	136	210	157	168	250	187	200
172	129	138	212	158	170			

Fig. 1 Compression pressure limit chart

dicate that more than one of the above factors are at fault, use care in analyzing an abnormal reading. For example, if the vacuum is low, the correction of one item may increase the vacuum enough so as to indicate that the trouble has been corrected. It is important, therefore, that each cause of an abnormal reading be investigated and further tests conducted, where necessary, in order to arrive at the correct diagnosis of the trouble. To check manifold vacuum, proceed as follows:

1. Operate engine for 30 minutes minimum at a fast idle speed to be sure engine is at normal operating temperature.
2. Connect an accurate vacuum gauge to the intake manifold vacuum fitting.
3. Operate engine at recommended idle speed with transmission selector lever in neutral.
4. Check vacuum reading on gauge.

Test Conclusions

NORMAL READING: 18 inches or more. Allowance should be made for the effect of altitude on the gauge reading. Engine vacuum will decrease with an increase in altitude.

LOW & STEADY: Loss of power in all cylinders possibly caused by late ignition or valve timing, or loss of compression due to leakage around piston rings.

VERY LOW: Intake manifold, carburetor, carburetor spacer or cylinder head gasket leak.

NEEDLE FLUCTUATES STEADILY AS SPEED INCREASES: A partial or complete loss of power in one or more cylinders caused by a leaking valve, cylinder head or intake manifold gasket, a defect in ignition system, or a weak valve spring.

GRADUAL DROP IN READING AT IDLE SPEED: Excessive back pressure in exhaust system.

INTERMITTENT FLUCTUATION: An occasional loss of power possibly caused by a defect in ignition system or a sticking valve.

SLOW FLUCTUATION OR DRIFTING OF NEEDLE: Improper idle mixture adjustment or carburetor, carburetor spacer or intake manifold gasket leak or crankcase ventilation system restricted.

Cylinder Balance Test

CAUTION: When performing engine diagnosis on vehicles equipped with catalytic converters, a "Cylinder Balance Test", Fig. 2, is *not* to be performed as damage to the catalytic converter will result. The alternate method of shorting out one cylinder at a time and noting the rpm drop of each individual cylinder can be used, but test should be performed as rapidly as possible.

It is sometimes difficult to locate a weak cylinder especially in an eight cylinder engine. A compression test, for example, will not locate a leaky intake manifold, a valve not opening properly due to a worn camshaft, or a defective spark plug.

With the cylinder balance test, the power output of one cylinder may be checked against another, using a set of grounding leads, Fig. 2. When the power of each cylinder is not equal, the engine will lose power and run roughly. The cylinder balance test is as follows:

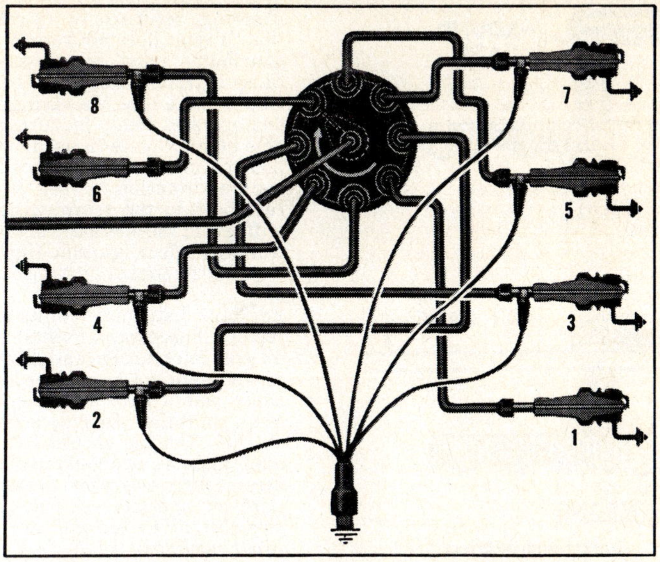

Fig. 2 Cylinder balance test connections. The firing order in this example is 1-8-4-3-6-5-7-2. Therefore, the cylinders to be tested together are 1-6, 8-5, 4-7, 3-2, using the grounding leads as shown.

1. Connect a tachometer and vacuum gauge.
2. Start engine and run it at fast idle.
3. Ground large clip of grounding leads and connect individual leads to all spark plugs *except the pair being tested* (see Fig. 1).
4. Divide the firing order in half and write down the first half over the second half. The cylinders to be tested together appear one over the other:

Firing Order	Pairs Tested
1-8-4-3-6-5-7-2	1-6, 8-5, 4-7, 3-2
1-2-7-8-4-5-6-3	1-4, 2-5, 7-6, 8-3
1-5-4-8-6-3-7-2	1-6, 5-3, 4-7, 8-2
1-5-4-2-6-3-7-8	1-6, 5-3, 4-7, 2-8
1-8-7-3-6-5-4-2	1-6, 8-5, 7-4, 3-2
1-6-5-4-3-2	1-4, 6-3, 5-2
1-5-3-6-2-4	1-6, 5-2, 3-4
1-4-5-2-3-6	1-2, 4-3, 5-6
1-3-4-2	1-4, 3-2

5. Operate engine on each pair of cylinders in turn and note engine rpm and manifold vacuum for each pair. A variation of more than one inch of vacuum or 40 rpm between pairs of cylinders being tested indicates that the cylinders are off balance.
6. To isolate one weak cylinder, short out one bank of cylinders at a time. The bank giving the lower readings will include the weak cylinder.

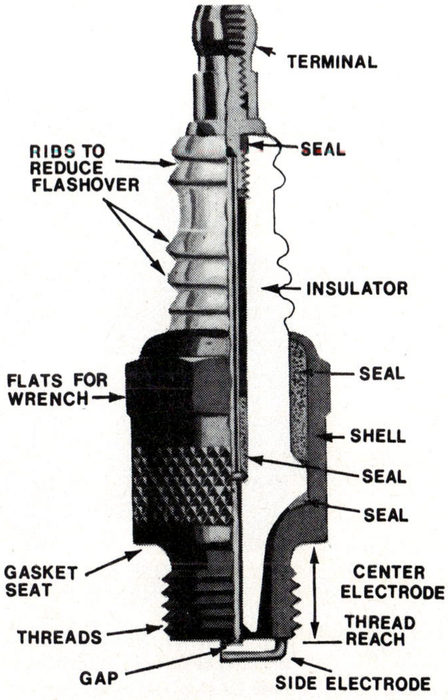

Fig. 3 Spark plug details

Spark Plugs

1. Examine firing ends of plugs for evidence of oil fouling, gas fouling, burned or overheated condition. *Oil fouling is usually identified by wet, sludgy deposits caused by excessive oil consumption. Gas fouling is identified by dry, black, fluffy deposits caused by incomplete combustion. Burned or overheated spark plugs are identified by white, burned or blistered insulator nose and badly burned electrodes. Improper fuel, insufficient cooling or improper ignition timing normally are the cause. Normal conditions are usually identified by white powdery deposits or rusty-brown to grayish-tan powdery deposits.*
2. Clean plugs with a suitable sand blast cleaner following the manufacturers instructions.
3. Remove carbon and other deposits from threads with a stiff wire brush.
4. Dress electrodes with a small file to secure flat, parallel surfaces on both center and side electrodes, Fig. 3.
5. Use a round wire gauge to check the gap, Fig. 4, and adjust by bending the side (never center) electrode to the proper specifications as shown in the *Tune Up Charts* in the vehicle chapters.
6. If gaskets are used, place new ones on plugs and torque them.

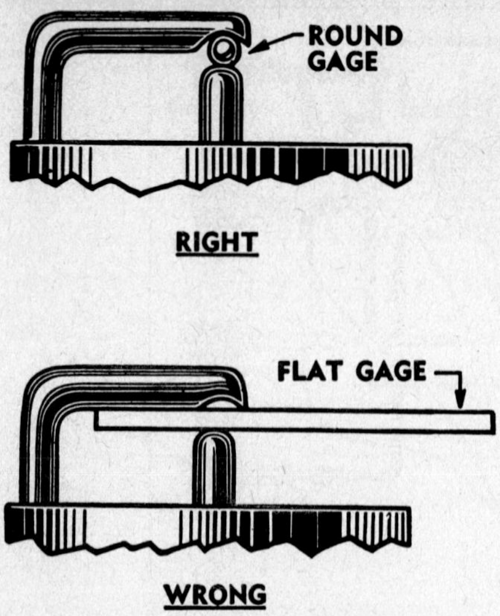

Fig. 4 Correct and incorrect spark plug gauges

CAUTION: Improper installation of spark plugs is one of the greatest single causes of unsatisfactory spark plug performance. Improper installation is the result of one or more of the following practices: 1) Installation of plugs with insufficient torque to fully seat the gasket; 2) excessive torque which changes gap settings; 3) installation of plugs on dirty gasket seal; 4) installation of plugs to corroded spark plug hole threads.

Failure to install plugs properly will cause them to operate at excessively high temperatures and result in reduced operating life under mild operation or complete destruction under severe operation where the intense heat cannot be dissipated rapidly enough.

Always remove carbon deposits in hole threads before installing plugs. When corrosion is present in threads, normal torque is not sufficient to compress the plug gasket (if used) and early failure from overheating will result.

Always use a new gasket (if required) and wipe seats in head clean. The gasket must be fully compressed on clean seats to complete heat transfer and provide a gas tight seal in the cylinder. For this reason as well as the necessity of maintaining correct plug gap, the use of correct torque is extremely important during installation.

Ignition System Service

1. Replace brittle or damaged spark plug wires. Install all wires to proper spark plug. The *Tune Up Charts* in the truck chapters give firing orders should wires become mixed.
2. Tighten all ignition system connections.
3. Replace or repair any wires that are frayed, loose or damaged, Fig. 5.
4. Remove distributor cap, clean and inspect for cracks, carbon tracks and burned or corroded terminals. Replace cap if necessary.
5. Clean rotor and inspect for damage or deterioration. Replace rotor if necessary.
6. Check distributor centrifugal advance mechanism (if used) by turning distributor rotor in direction of running rotation as far as possible, then release rotor to see if springs return it to its retarded position. If rotor does not return readily, the distributor must be disassembled and cause of trouble corrected.
7. Check to see that the vacuum spark control operates freely by turning the movable breaker plate (if used) or distributor housing in a direction opposite to that of running rotation to see if the spring returns it to the retarded position. Any stiffness in the operation of the spark control will affect ignition timing. Correct any interference or binding condition noted.
8. Examine distributor points and clean or replace if necessary. Points with an overall gray color and only slight roughness or pitting need not be replaced.
9. Dirty points should be cleaned with a clean point file. Use only a few strokes of the file. The file should not be used on other metals and should not be allowed to become dirty or greasy. *Never use emergy cloth or sandpaper to clean points since particles will embed and cause arcing and rapid burning of points.* Do not attempt to remove all roughness nor dress the point surface down smooth. Merely remove scale or dirt.
10. Replace points that are badly burned or pitted. Where burned or badly pitted points are encountered, the ignition system and engine should be checked to determine the cause of the trouble so it can be eliminated. Unless the condition causing point burning is corrected, new points will provide no better service than the old points. See *Standard Distributor* chapter for an analysis of point burning or pitting, and for proper installation of points.

Battery & Cables

Inspect for signs of corrosion on battery, cables and surrounding area, loose or broken carriers, cracked or bulged cases, dirt and acid, electrolyte leakage and low electrolyte level. Fill cells to proper level with colorless, odorless drinking water.

The top of the battery should be clean and the battery hold-down bolts properly tightened.

For best results when cleaning batteries, wash first with a dilute ammonia or soda solution to neutralize any acid present and then flush off with clean water. Care must be taken to keep vent plugs tight so that the neutralizing solution does not enter the battery. The holdown bolts should be kept tight enough to prevent the battery from shaking around in its holder, but not so tight that the battery case will be placed under a severe strain.

To insure good contact, the battery cables should be tight on the battery posts. Oil battery terminal felt washer. If the battery posts or cable terminals are corroded, the cables should be cleaned separately with a soda solution and a wire brush. After cleaning and before installing the clamps, apply a thin coating of vaseline to the posts and cable clamps to help retard corrosion.

If the battery has remained undercharged, check for a loose generator belt, defective generator, high resistance in charging circuit, oxidized voltage regulator contact points, or a low voltage setting.

If the battery has been using too much water, the voltage regulator setting is too high.

Fan Belt & Generator

1. Inspect fan belt condition and check adjustment.
2. Inspect generator commutator and brushes for cleanliness or wear. The commutator should be cleaned if dirty and brushes should be replaced if worn down to less than half their original length. *The commutator may be cleaned by holding a strip of No. 00 sandpaper or a cleaning stone against it while the generator is operating.*
3. Replace or repair frayed or broken generator wires and tighten all wire connections.

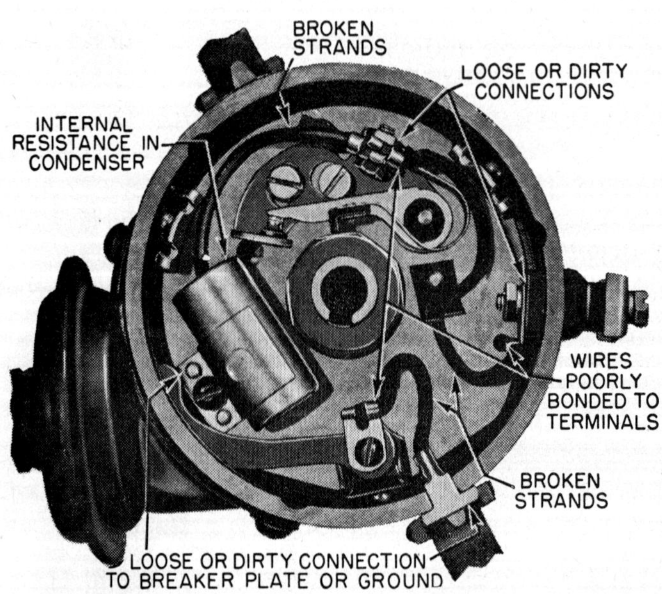

Fig. 5 What to look for when checking for high resistance in the primary circuit of the distributor. In addition to the points indicated, look for external circuit high resistance at ignition switch terminals, ammeter terminals, coil terminals and broken or poorly insulated wires in this circuit.

4. Lubricate generator by filling hinge cap oilers with light engine oil.

Ignition Timing

The use of a timing light, Fig. 6, is recommended for checking and setting ignition timing. This setting is critical especially on units equipped with emission control systems.

Some engines have a provision for monolithic method of ignition timing utilizing a timing receptacle designed to accept an electronic probe. The receptacle is usually mounted at the front of the engine and the electronic probe is connected to electronic equipment which reads out the engine timing. These engines can also be timed using a timing light.

NOTE: When setting timing, be sure idle speed is set lower than speed at which centrifugal advance begins. See *Distributor Specifications Chart* in truck chapters.

NOTE: The timing light should be connected to the proper spark plug lead by the use of an adapter. The boots around the connections should not be pierced to connect the light as this can cause spark arcing and misfiring.

Lacking a timing light, the timing can be set with the engine stopped by using a jumper light. Be sure to use a light bulb that corresponds with the vehicle voltage, Fig. 7.

1. Rotate engine until No. 1 cylinder is positioned at the specified timing mark.
2. Connect jumper light between distributor ignition terminal and ground.
3. Turn on ignition switch.
4. Loosen distributor and turn it in the direction of normal rotation until the points just close (light out). Then slowly turn the distributor in the opposite direction just to the exact point that the light goes on. Tighten distributor in this position.

CARBURETION

Since carburetion is dependent in several ways on both compression and ignition, it should always be checked last when tuning an engine. See the car chapter for adjustments for the unit you are interested in.

Before adjusting the carburetor, consider the factors outlined below and which definitely affect engine performance.

Performance Complaints

Flooding, stumble on acceleration or other performance complaints are in many instances caused by the presence of dirt, water or other foreign matter in the carburetor. To aid in diagnosing the cause of the complaint, the carburetor should be carefully removed from the engine without draining the fuel from the bowl. The contents of the fuel bowl may then be examined for contamination as the carburetor is disassembled.

Check the fuel in the bowl for contamination by dirt, water, gum or other foreign matter. A magnet moved through the fuel in the bowl will pick up and identify any iron oxide dust that may have caused intake needle and seat leakage.

Inspect gasketed surfaces between body and air horn. Small nicks or burrs should be smoothed down to eliminate air or fuel leakage. On carburetors having a vacuum piston, be especially particular when inspecting the top surface of the inner wall of the bowl around the vacuum piston passage. A poor seal at this location may contribute to a "cutting out" on turns complaint.

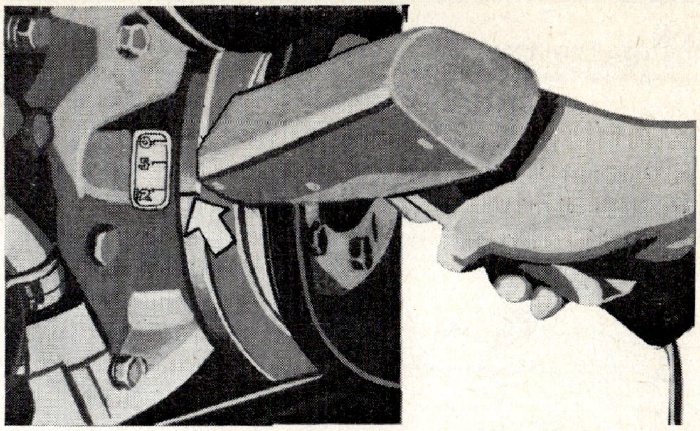

Fig. 6 Checking ignition timing with timing light

Fill the carburetor bowl with clean fuel before installing on manifold. This will help prevent dirt trapped in the fuel system from being dislodged by the free flow of fuel as the carburetor is primed. The operation of the floats and intake needle and seats may be checked under pressure if a fuel pump is used at the bench to fill the carburetor bowl. Operate the throttle several times and visually check the discharge from pump jets.

Poor Mileage and Engine Loading Complaints

Cases of poor mileage and engine loading may be due in many instances to sluggish choke valve opening during cold driveaway, caused by insufficient vacuum in choke housing, a plugged or restricted heat pipe or inlet in choke cover. To check for this condition, have engine warm and running at slow idle. Remove choke heat pipe and hold a finger over the heat inlet hole (hole is on choke housing on some carburetors). If there is little or no vacuum pull on the finger, check the choke housing for gasket leaks or plugged vacuum passages. If these are OK, check choke vacuum passages in carburetor between choke housing and manifold.

Dirty or Rusty Choke Housing

In cases where it is found that the interior of the choke housing is dirty, gummed or rusty while the carburetor itself is comparatively clean, look for a puncture or eroded manifold heat tube (if one is used).

Manifold Heat Control Valve

An engine equipped with a manifold heat control valve can operate with the valve stuck in either the open or closed position, Fig. 8. Because of this, an inoperative valve is frequently overlooked at vehicle lubrication or tune-up.

A valve stuck in the "heat-off" position can result in slow warm up, deposits in combustion chamber, carburetor icing, flat spots during acceleration, low gas mileage and spark plug fouling.

A valve stuck in the "heat-on" position can result in power loss, engine knocking, sticking or burned valves and spark plug burning.

To prevent the possibility of a stuck valve, check and lubricate the valve each time the vehicle is lubricated or tuned-up. Check the operation of the valve manually. To lubricate the valve, place a few drops of penetrating oil on the valve shaft where it passes through the manifold. Then move the valve up and down a few times to work the oil in. *Do not use engine oil to lubricate the valve as it will leave a residue which hampers valve operation.*

Idle Speed & Misture Adjustments

Less Emission Controls

Start the engine and run it until it reaches normal operating temperature, at which time the choke valve should be wide open with the idle speed screw resting on lowest step of fast idle cam. Slowly turn the idle mixture screw a little at a time until the engine shows a tendency to hesitate and stall. Then turn it to the left until the engine runs smoothly. Continue turning the screw to the left until the fuel mixture is rich and the engine starts to "lope" or "gallop". Finally turn the screw to the right until the engine runs evenly.

On two-barrel carburetors, there is an idle mixture screw for each barrel. On four-barrel carburetors there are also two idle mixture screws on the primary side of the carburetor. When idle ports are provided on the secondary side, they are non-adjustable or they are rendered non-functional by being blocked with gaskets.

When adjusting the idle mixture on two and four-barrel carburetors, adjust one screw at a time until the engine runs smoothly as directed above. Then adjust the other screw in like manner.

After the idle mixture has been adjusted, it is recommended that a tachometer (engine speed indicator) be used to set the slow idle speed to the rpm indicated in the *Tune Up* table in the truck chapter.

With Emission Controls

1. With engine at operating temperature, set parking brake and block drive wheels.
2. Make sure choke valve is wide open.
3. On C.C.S. equipped vehicles, see that the air cleaner thermostatic valve is open.
4. On carburetors so equipped, hold hot idle compensator hole closed with eraser on pencil.
5. Turn air conditioner off or on according to directions given in *Tune Up Charts* in truck chapters.
6. Set idle mixture screw(s) for maximum idle rpm.
7. Adjust speed screw (or idle stop solenoid screw on C.C.S.) to obtain the specified

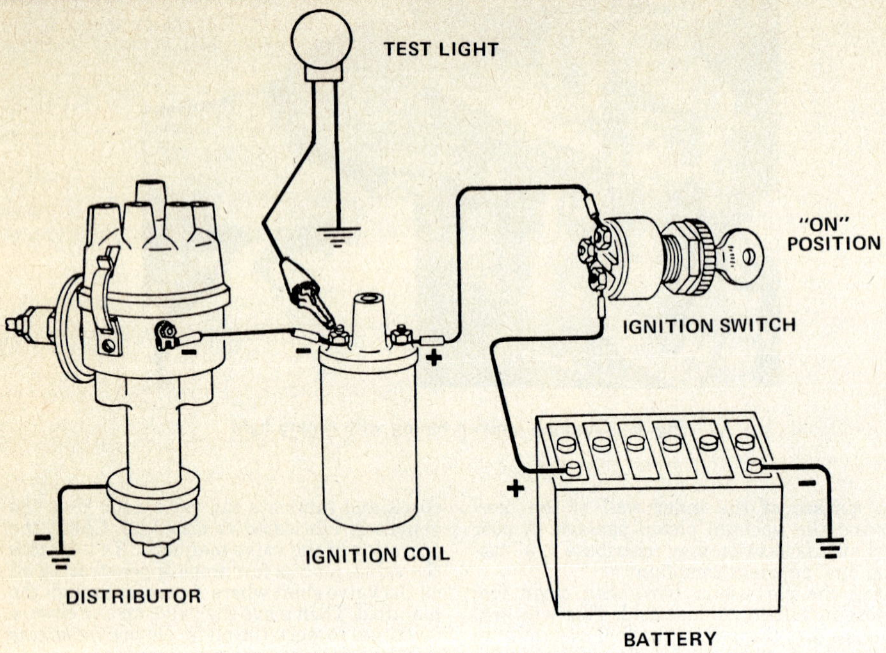

Fig. 7 Jumper light circuit for static ignition timing

rpm in Drive or Neutral as specified.

8. Set ignition timing according to specifications with vacuum advance line disconnected and hole in manifold plugged.
9. Adjust mixture screw IN to obtain a 20 rpm drop (lean roll).
10. Adjust mixture screw OUT 1/4 turn.
11. Repeat Steps 9 and 10 for second mixture screw (2 and 4 barrel carbs).
12. Readjust speed screw (or solenoid screw) if necessary to obtain specified rpm.
13. On C.C.S. with idle solenoid stop on carburetor, electrically disconnect solenoid and adjust carburetor idle speed screw to obtain 400 rpm in neutral, then reconnect wire to solenoid.

NOTE: Exact instructions for each C.C.S. equipped engine transmission combination is given for this Lean Roll (low idle) speed method on a decal permanently affixed to the vicinity of the radiator support.

Carburetor Flange

Check the flange for looseness on the manifold. If one of the flange nuts is loose as little as one-half turn, a sufficient amount of air will enter the intake manifold below the throttle plate to destroy engine idle and all engine performance.

If a tight fit cannot be obtained by tightening the nuts, install a new gasket but be sure that all the old gasket material has been removed.

Throttle Linkage

If the throttle linkage is adjusted so that the accelerator pedal will strike the floor board before the throttle plate is wide open, it will result in low top speed.

Fuel Lines

A restriction of the fuel line will result in an apparent vapor lock action or a definite cut-off of gasoline. This can generally be corrected by blowing out the line with compressed air. In some cases, it may be necessary to replace the line.

Fuel Pump

The pump should be tested to make sure that it will draw an adequate supply of fuel from the tank and deliver it to the carburetor under all conditions of operation. If the pump functions inefficiently, proper adjustment and operation of the carburetor is impossible because the fuel will not be maintained at the prescribed level in the idle passages and main discharge jet (or jets) of the carburetor under all operating conditions.

Fuel Tank

The fuel tank should not be overlooked as a possible source of trouble with carburetion. A shortage of fuel at the fuel pump or carburetor may be caused by material obstructing the mouth of the feed pipe in the tank, or by a restriction of the air vents in the filler cap and neck.

An unusual amount of dirt, water or gum in the fuel filter indicates that the tank is contaminated with these substances, which should be cleaned out to prevent future failure of the pump or carburetor.

Intake Manifold Leaks

Leakage of air into the intake manifold at any point will affect carburetion and general engine performance. Air may leak into the manifold through the joints at the carburetor or cylinder head, cracks in the manifold, cracks or poor connections in the windshield wiper or windshield washer hose lines, or the connections of any accessories which may be connected to the manifold. All such joints should be tested for leaks.

To test the intake manifold for leaks, apply oil from an oil can along the gasket joints with the engine idling. An air leak is indicated when oil is drawn past the gaskets by the suction of the engine. Tighten the nuts or cap screws holding the manifold to the engine and retest for leaks. If tightening fails to stop the leaks, replace the manifold gaskets. If the new gaskets fail to stop the leaks, carefully inspect the manifold for cracks and test any suspicious area with oil.

Air Cleaner

An air cleaner with a dirty element, or with oil that is dirty, too heavy, or too high in the sump, will restrict the air flow through the carburetor and cause a rich mixture at high speeds. In such a condition the air cleaner likewise will not properly remove dirt from the air, and the dirt entering the engine will cause rapid formation of carbon, sticking valves, and wear of piston rings and cylinder bores.

Automatic Choke

The choke mechanism must be inspected and cleaned to make sure it is operating freely. Sluggish action or sticking of the choke will cause excessive fuel consumption, poor performance during warm-up, and possibly hard starting.

The choke thermostat should be set in accordance with the average air temperature as well as the volatility of the fuel being used. It is desirable to have the thermostat set as lean as operating conditions permit in order to avoid an overrich mixture during engine warm-up.

Choke Thermostat

If necessary to adjust the choke more than two marks from the specified setting, either rich or lean, it indicates that the thermostat spring may be bent or has lost its tension.

PERFORMANCE TEST

After an engine has been tuned up, the car

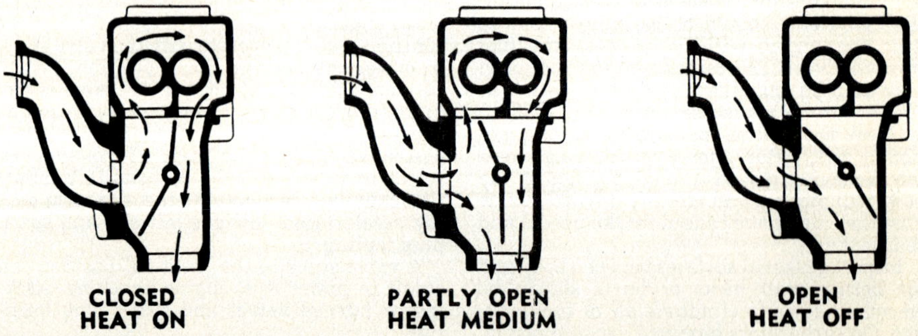

Fig. 8 Operation of a typical manifold heat control valve

should be given a thorough and systematic road test to make certain that engine power and performance are up to standard under all operating conditions. The gasoline used in making the test must be of good quality and proper octane rating in order to obtain the performance described in the following tests.

Engine Warm Up

On vehicles with automatic chokes, a cold engine should operate on fast idle for two to five minutes, depending upon air temperature.

At 32 deg. F. the fast idle cam should move to the slow idle position in approximately 1/2 to 3/4 mile of driving. At higher temperatures, it should move to the slow idle position in a correspondingly shorter distance.

If the engine loads excessively or runs rich on warm up due to a rich choke setting, excessive fuel consumption, carbon formation, and spark plug fouling will result.

An engine which is adjusted for smooth idling in cold outside temperature will not idle smoothly for any length of time in a warm building, since the required carburetor adjustment will cause richness of mixture in the warmer atmosphere.

Gradual Acceleration Test

Starting at idling speed in high gear, gradually open the throttle to increase smoothly the speed of the car through the entire range. Note any roughness, flat spots, or surging in engine performance during acceleration, and the speed at which the unusual condition occurs.

Roughness or poor performance at speeds below 20 m.p.h. indicates improper carburetor idle adjustment, restriction in idle passages in carburetor, tight valve lash or sticking valves or faulty ignition.

Faulty ignition usually causes a more pronounced roughness than imperfect compression or carburetion.

Roughness or poor performance at speeds above 20 m.p.h. indicates restriction or improper settings in the high speed circuit of the carburetor, or faulty ignition.

Wide Open Throttle Acceleration Test

With the car running at idling speed in high gear, quickly press the accelerator pedal to the floor and hold it there, meanwhile noting the performance of the engine as the car is accelerated. Repeat the acceleration test, starting at different constant speeds throughout speed range of car. The car should accelerate smoothly without hesitation, spitting, or leading of the engine.

A hesitation, spitting, or a flat spot indicates that the accelerating pump is not discharging sufficient gasoline into the engine. Sluggishness or loading indicates that the accelerating pump is adjusted too rich.

Constant Speed Test

Hold the truck speed constant at various points through the speed range and note engine performance. The engine should operate smoothly without hesitation or surging under load at all constant speeds.

At some point between 15 and 22 m.p.h. with the car rolling along on a level road or slight upgrade, a slight leanness, surging or missing may be detected. Depressing or releasing the accelerator pedal slightly will eliminate this condition and no attempt should be made to correct it by altering carburetion or ignition. This condition seldom appears in the normal operation of the vehicle.

Spark Knock

Light detonation or spark knock will occur when operating with part throttle on a hard pull, even through the ignition is properly timed and high octane fuel is used. Light detonation also will occur when accelerating with fully opened throttle on a hard pull. These operating conditions are normal and no attempt should be made to eliminate light detonation by retarding the ignition timing.

If regular or low octane fuel is used, detonation will probably be excessive with the standard ignition timing, and it may be necessary to retard the timing, which will reduce fuel economy and over-all performance.

Extreme heavy detonation is injurious to any automobile engine. A car driven continuously under conditions and fuels which produce heavy detonation will overheat and lose power, with the possibility of damage to pistons and bearings.

IGNITION COILS & RESISTORS

IGNITION COILS

If poor ignition performance is obtained and the coil is suspected, it may be tested and the car or it may be removed for the test.

Ignition coils are often condemned when the trouble is actually in the ignition switch. A completely defective ignition switch will produce an open primary circuit, giving the same indications as if the coil were completely dead. A partly defective ignition switch will cause a weak spark. Both of these conditions are often blamed on the coil.

By cutting the ignition switch out of the circuit, it can easily be determined whether or not the coil is defective or whether fault lies with ignition switch.

In the absence of any testing equipment a simple check of an ignition coil can be made as follows: Turn on ignition switch with breaker points closed. Remove the high tension cable from the center socket of the distributor cap and hold it 1/4" to 3/8" away from a clean spot on the engine. If the coil and other units connected to it are in good condition a spark should jump from the wire to the engine. If not, use a jumper wire from the distributor terminal to the engine; if the primary is in good condition a spark will occur.

All ignition coils with metal containers can be tested for grounded windings by placing one test clip on a clean part of the metal container and touching the other clip to the primary and high tension terminals. If the lamp lights or tiny sparks appear at the points of contact, the windings are grounded and the coil should be replaced.

Coil Polarity

Most coils are marked positive and negative at the primary terminals. When installing or connecting a coil be sure to make the connections as shown in Figs. 1 and 2. A reversal of this polarity may affect the performance of the engine (or the radio).

If perchance the coil is not marked as to its polarity, it can be checked by holding any high tension wire about 1/4" away from its spark plug terminal with the engine running. Insert the point of wooden lead pencil between the spark plug and the wire, Fig. 3. If the spark flares and has a slight orange tinge on

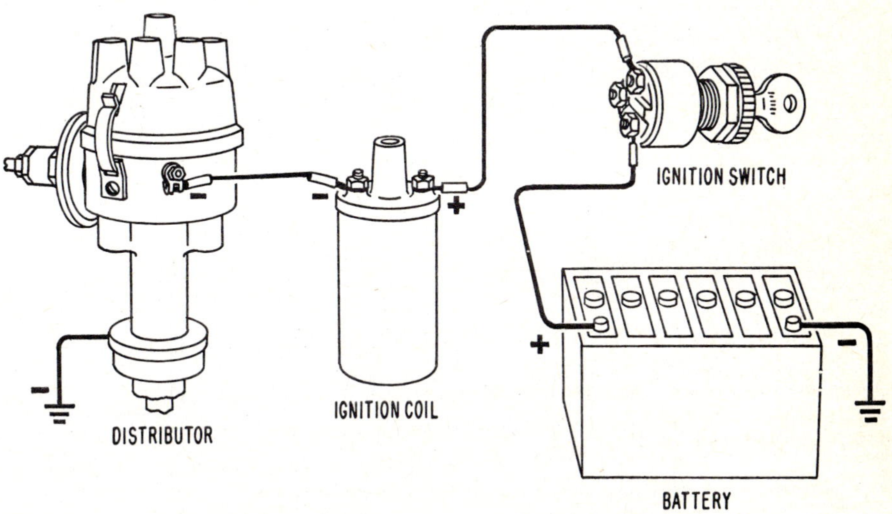

Fig. 1 Wiring connections for coil with negative grounded system

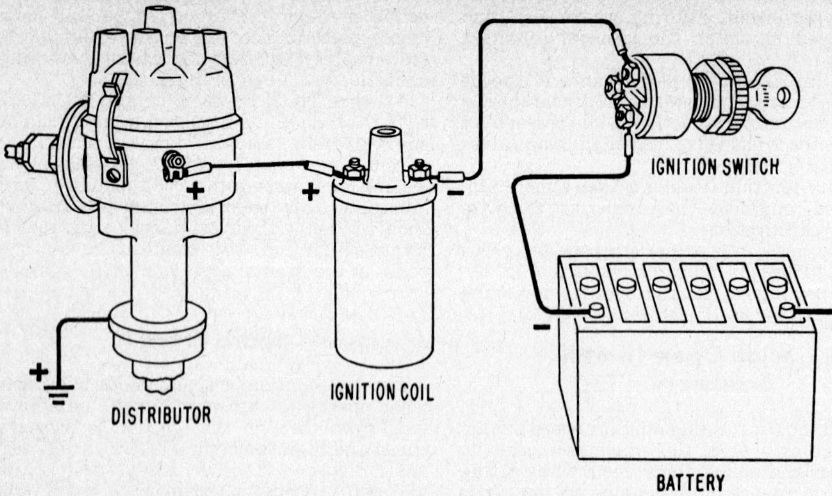

Fig. 2 Wiring connections for coil with positive grounded system

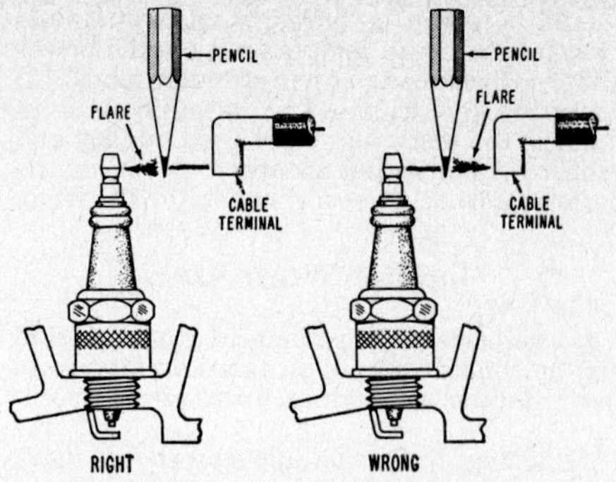

Fig. 3 Checking coil polarity

the spark plug side of the pencil, polarity is correct. If the spark flares on the cable side, coil connections should be reversed.

IGNITION RESISTORS

The purpose of the resistor is to prolong the service life of the distributor breaker points.

Block Type Ballast Resistors

Its basic purpose is to allow full battery voltage to the ignition coil during engine starting, and to reduce battery voltage to the coil when the engine is running. The higher voltage during starts means easier starts. But sustained high voltage to the breaker points can cause point failure. The reduced voltage during engine operation increases breaker point life.

These resistors are normally very dependable, but if one fails it can be one of the most difficult of all ignition system malfunctions to check out and diagnose. This type of resistor can fail in several ways. The resistance wire can separate, with a result similar to that caused by an open switch. An open resistor means that no current reaches the coil and the engine cannot operate. It is possible for the resistance wire to warp or bend enough to touch the side of the case and when this happens the engine may continue to run, but the overall performance would be poor.

Resistors can change value. Any creeping change in resistance value of a ballast resistor is invariably an increase in resistance. This means that coil output to the spark plugs is reduced proportionately. If a ballast resistor is slowly increasing in value the engine could gradually deliver less and less horsepower, particularly under high load conditions. An unsuspecting mechanic could unsuccessfully try to get the engine back to where it will deliver acceptable power output by changing spark plugs, adjusting timing, etc. Replacement of the faulty resistor is the only cure in this case.

To check a ballast resistor, replace it with one of known good quality. Then road-test the vehicle for improved performance.

It is important to remember that new spark plugs can temporarily mask the need for resistor replacement because new plugs require less voltage to fire. This is why a new-plug tune-up may prove satisfactory for a time. But if the ballast resistor is faulty, eventually the engine will misfire under load.

In Auto-Lite, Prestolite and Chrysler systems the resistor consists of an ordinary resistance wire that is sensitive to temperature, Fig. 4. The wire has a lower resistance value when cold than when hot. When the ignition is first turned on, more current will flow through the primary windings of the coil for a very short time until the resistor heats up.

In Delco-Remy and Ford systems, Figs. 5 and 6, the resistor is of the constant temperature type; that is, it is not affected by temperature and its resistance is approximately the same when cold as when hot. However, a feature is employed which shorts out the resistor while the engine is being cranked by the starter and automatically puts the resistance back in the coil circuit as soon as the starter switch is released. This is accomplished by bypassing the resistor through the starter solenoid. The solenoid has an additional terminal from which a wire runs directly to the coil.

On Delco-Remy systems, the resistor is bypassed by means of a "finger" inside the solenoid switch housing which is attached to the

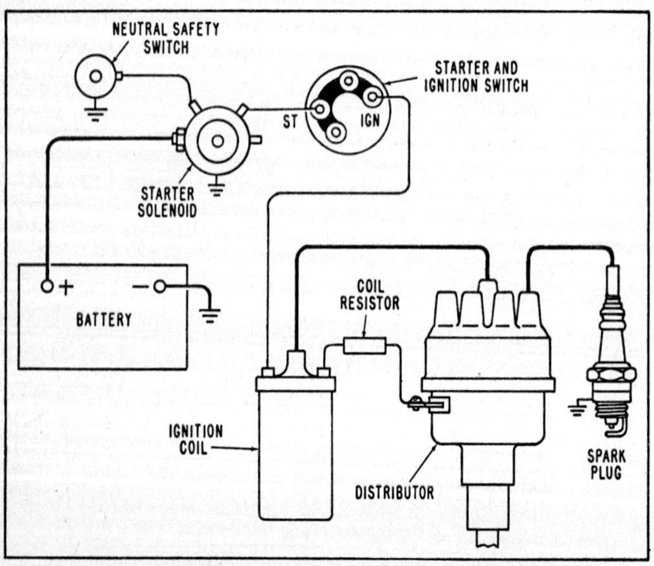

Fig. 4 Auto-Lite and Chrysler ignition circuit diagram with a temperature sensitive block-type resistor

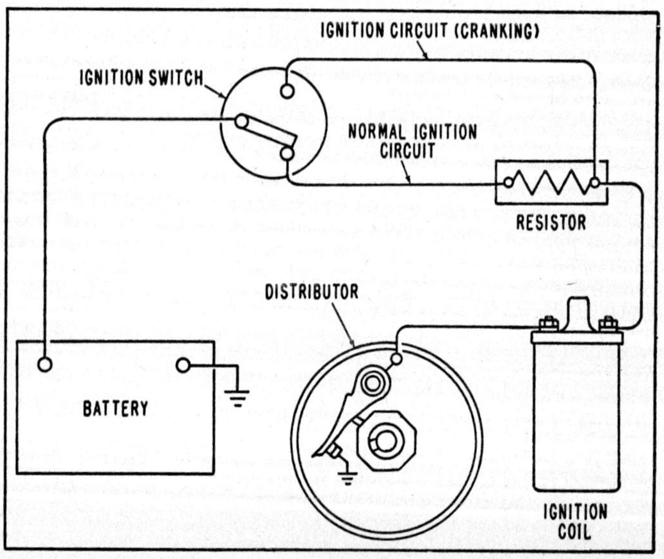

Fig. 5 Delco-Remy ignition circuit diagram with a constant temperature block-type resistor

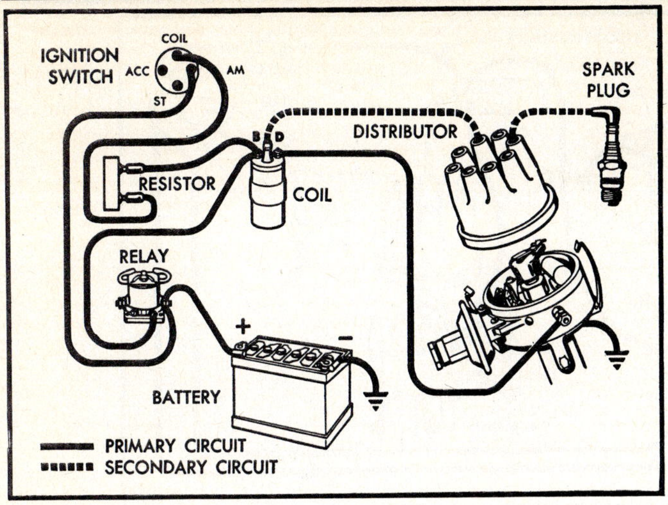

Fig. 6 Ford ignition circuit diagram with a constant temperature block-type resistor

additional switch terminal.

On Ford systems, Fig. 6, the resistor is by-passed through a terminal on the starter relay which is connected directly to the positive terminal of the coil.

NOTE: If the engine fires when the ignition switch is turned on but quits when the switch is released to its running position, it indicates that the resistor is defective and must be replaced.

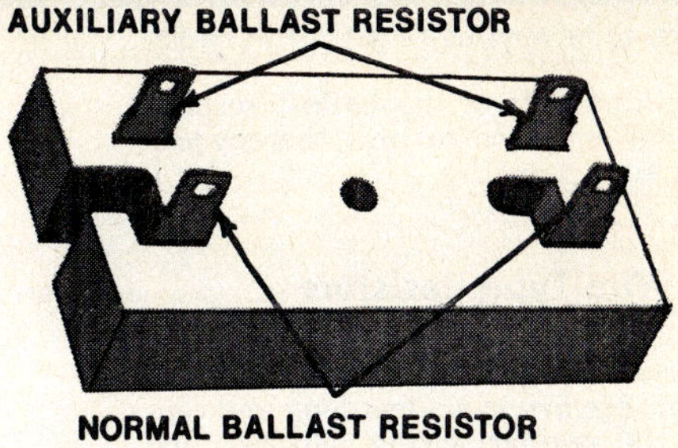

Fig. 7 Dual ballast resistor. Chrysler Corp. electronic ignition system

A dual ballast resistor, Fig. 7, is used on Chrysler Corp. vehicles equipped with electronic ignition. The normal side of the resistor is a compensating resistance in the primary circuit. At low engine speeds, current is maintained for a longer period of time in this side of the resistor, causing the unit to heat up, in turn increasing resistance. This increased resistance reduces the primary circuit voltage, protecting the ignition coil from high voltage at low engine speeds.

As engine speed increases, the period of time in which current is maintained in the normal side of the resistor is shorter, causing the unit to cool, in turn decreasing resistance. This decreased resistance permits the primary circuit voltage to increase for high speed operation.

During engine start, the normal side of the ballast resistor is by-passed, allowing full bat-

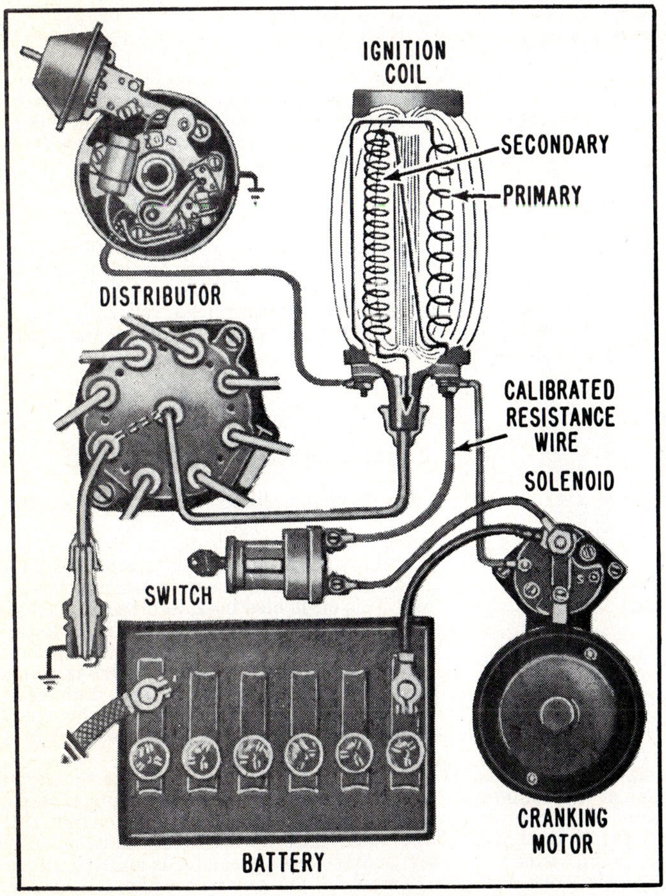

Fig. 8 Delco-Remy ignition circuit diagram with a resistance wire connected to a two-terminal ignition switch

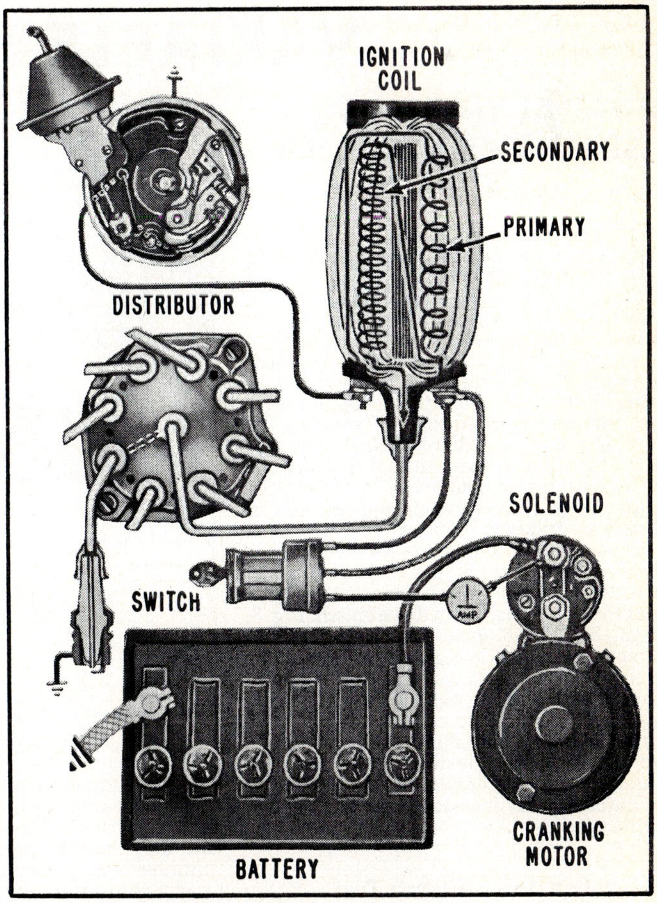

Fig. 9 Delco-Remy ignition circuit diagram with a resistance wire connected to a three-terminal ignition switch

tery voltage to be applied to the primary circuit.

The auxiliary side of the ballast resistor limits voltage to the control unit, thereby protecting the unit.

Wire Type Resistors

The special resistance wires used with 12-volt systems are five to six feet long and contained in the regular wiring harness. The wire is made of stainless steel or special alloy, plastic-coated and covered with a glass braid. There is relatively small temperature rise and the resistance wire is switched out of the circuit for starting and back in again for running.

With the Delco-Remy circuit shown in Fig. 8, the resistance wire is by-passed by means of a "finger" inside the starter solenoid housing which is attached to the additional switch terminal.

The Delco-Remy and Ford circuit diagrams shown in Figs. 9 and 10 have the resistance wire connected to a separate terminal on the ignition switch. When the ignition switch is released to its running position after the engine fires, the resistance wire is put back into the circuit.

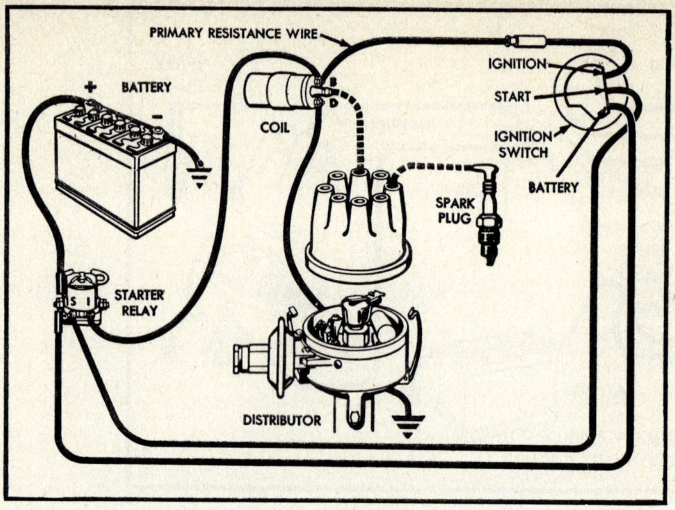

Fig. 10 Ford ignition circuit diagram with a resistance wire connected to a three-terminal ignition switch

NOTE: If the engine fires when the ignition switch is turned on but quits when the switch is released to its running position, it indicates that the resistance wire has lost its continuity or there is a bad connection at the resistor terminals. If the wire is defective, it must be replaced. To do this, first identify the terminal to which it is attached to the ignition switch (corresponding to the position of the ignition key when the engine is running). Then note the color code of the wire and disconnect it at both ends. Connect the new wire to the terminals and tape it to the main wiring harness. Finally, cut off the exposed sections of the defective wire.

CAUTION: Do not attempt to operate the engine for an extended length of time with the resistor shorted out by means of a jumper wire as the breaker points will burn up in short order.

VARIABLE SPEED FANS

BI-METAL THERMOSTATIC SPRING CONTROLLED TYPE

The fan drive clutch, Fig. 1, is a fluid coupling containing silicone oil. Fan speed is regulated by the torque-carrying capacity of the silicone oil. The more silicone oil in the coupling the greater the fan speed, and the less silicone oil the slower the fan speed.

Two types of fan drive clutches are in use. On one, Fig. 2, a bi-metallic strip and control piston on the front of the fluid coupling regulates the amount of silicone oil entering the coupling. The bi-metallic strip bows outward with a decrease in surrounding temperature and allows a piston to move outward. The piston opens a valve regulating the flow of silicone oil into the coupling from a reserve chamber. The silicone oil is returned to the reserve chamber through a bleed hole when the valve is closed.

On the other type of fan drive clutch, Fig. 3, a heat-sensitive, bi-metal spring connected to an opening plate brings about a similar result. Both units cause the fan speed to increase with a rise in temperature and to decrease as the temperature goes down.

In some cases a Flex-Fan is used instead of a Fan Drive Clutch. Flexible blades vary the volume of air being drawn through the radiator, automatically increasing the pitch at low engine speeds.

Fan Drive Clutch Test

Fig. 1 Typical variable-speed fan installed

CAUTION: Do not operate the engine until the fan has been first checked for possible cracks and separations.

Run the engine at a fast idle speed (1000 rpm) until normal operating temperature is reached. This process can be speeded up by blocking off the front of the radiator with cardboard. Regardless of temperatures, the unit must be operated for at least five minutes immediately before being tested.

Stop the engine and, using a glove or a cloth to protect the hand, immediately check the effort required to turn the fan. If considerable effort is required, it can be assumed that the coupling is operating satisfactorily. If very little effort is required to turn the fan, it is an indication that the coupling is not operating properly and should be replaced.

If the clutch fan is the coiled bi-metal spring type, it may be tested while the vehicle is being driven. To check, disconnect the bi-metal spring, Fig. 4, and rotate 90° counter clockwise. This disables the temperature-controlled free-wheeling feature and the clutch performs like a conventional fan. If this cures the overheating condition, replace the clutch fan.

ENGINE OIL CONTROLLED TYPE

The fluid fan clutch is a type of variable speed fan which operates off the engine oil supply. Engine oil is supplied to the drive assembly and torus housing to provide lubrication of the clutch assembly seals and bearings. This oil is bled back into the crankcase at all times from the port on the front lower part of the torus housing. Under cold operating conditions the fan control valve is open, Fig. 5, allowing oil to return to the crankcase from the rear lower part of the torus housing, at this time the fluid fan clutch is in an idle circuit, Fig. 6.

As the engine reaches operating temperature (190°–192° F.), the fan control valve closes, shutting off the oil return line from the lower rear part of the torus housing. The engine oil collects in the torus housing as air is vented to the engine crankcase. As the torus housing fills with engine oil the fan will rotate faster until it is engaged in the coupled or operating cycle, Fig. 5.

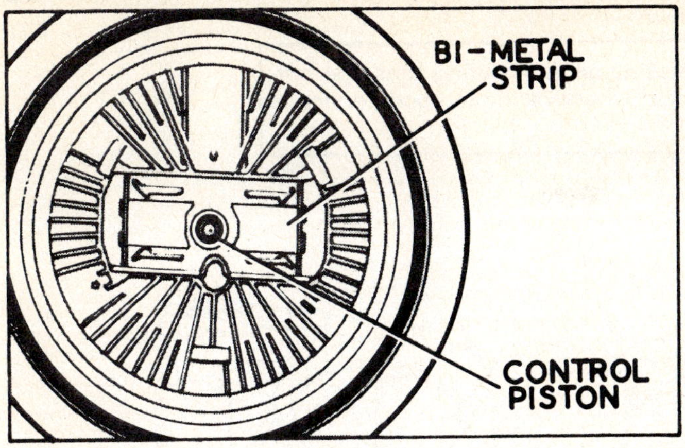

Fig. 2 Variable-speed fan with flat bi-metal thermostatic spring

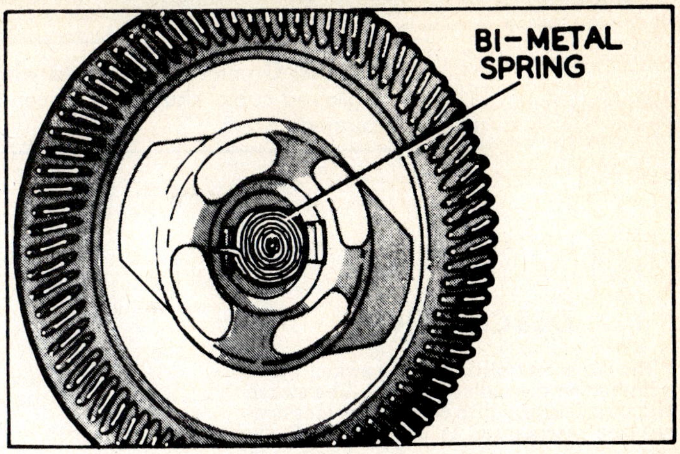

Fig. 3 Variable-speed fan with coiled bi-metal thermostatic spring

To prevent excessive oil pressure in the torus housing the fan control valve has an integral pressure relief valve to limit oil pressure to 23–27 psi.

SERVICE PROCEDURE

Bi-Metal Thermostatic Spring Type

CAUTION: To prevent silicone fluid from draining into fan drive bearing, do not store or place drive unit on bench with rear of shaft pointing downward.

The removal procedure for either type of fan clutch assembly is generally the same for all cars. Merely unfasten the unit from the water pump and remove the assembly from the car.

The type of unit shown in Fig. 2 may be partially disassembled for inspection and cleaning. Take off the capscrews that hold the assembly together and separate the fan from the drive clutch. Next remove the metal strip on the front by pushing one end of it toward the fan clutch body so it clears the retaining bracket. Then push the strip to the side so that its opposite end will spring out of place. Now remove the small control piston underneath it.

Check the piston for free movement of the coupling device. If the piston sticks, clean it with emery cloth. If the bi-metal strip is damaged, replace the entire unit. These strips are not interchangeable.

When reassembling, install the control piston so that the projection on the end of it will contact the metal strip. Then install the metal strip with any identification numerals or letters facing the clutch. After reassembly, clean the clutch drive with a cloth soaked in solvent. Avoid dipping the clutch assembly in any type of liquid. Install the assembly in the reverse order of removal.

The coil spring type of fan clutch cannot be disassembled, serviced or repaired. If it does not function properly it must be replaced with a new unit.

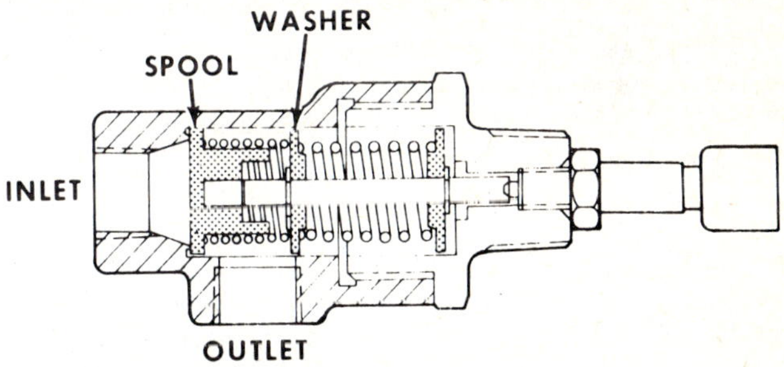

Fig. 5 Fluid fan clutch control valve

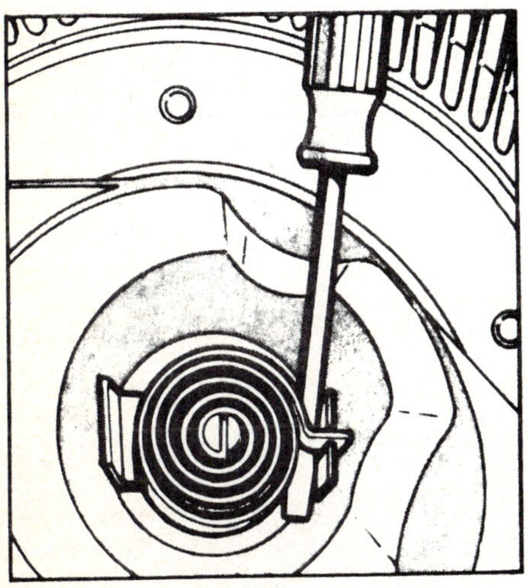

Fig. 4 Disconnecting bi-metal spring

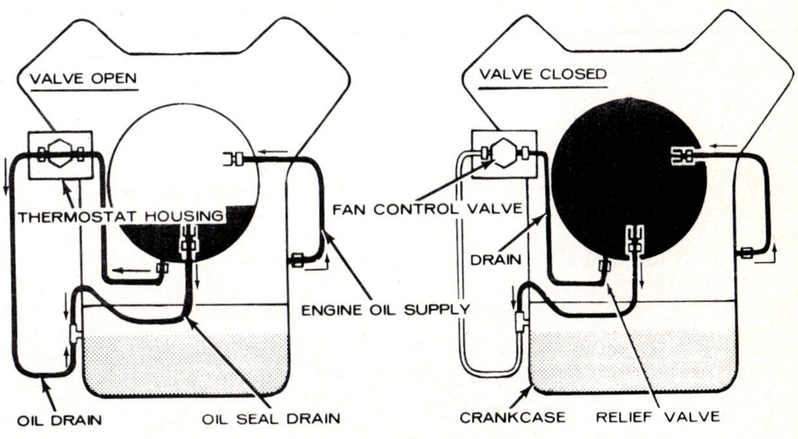

Fig. 6 Fluid fan clutch schematic.

ELECTRIC ENGINE COOLING FANS

CAUTION: On models equipped with electric engine cooling fans, the battery ground cable should be disconnected whenever underhood service is performed.

DESCRIPTION

The fan is controlled by a fan switch which is located on the radiator, Fig. 1. The switch will automatically turn on when coolant temperature reaches 193 to 207 degrees F. On models with A/C, when the A/C system is in operation, the fan motor will operate continually regardless of engine coolant temperature. When the ignition switch is turned off the fan motor will stop operating.

Radiator Fan Switch

The radiator fan switch is located on left hand radiator tank. The switch is normally open and incorporates a bimetal disc which pushes the plunger when coolant temperature reaches approximately 200 degrees F. If the fan motor turns on and off at the appropriate temperature, the switch is operating properly.

To check switch continuity, drain coolant until level is below switch. The switch can be viewed by looking downward through the radiator filler neck. Disconnect electrical connector from switch and remove switch from radiator. Dip switch into an oil bath which has achieved a temperature of 208 degrees F or higher and check switch for continuity using a test lamp or ohmmeter. If continuity is not indicated, replace switch.

Electric Fan Motor

Diagnosis

Disconnect wire connector from fan motor terminal, then connect a 14 guage jumper wire from battery to fan motor terminal. If fan motor does not operate properly, replace fan motor.

Fan Motor, Replace

1. Disconnect wire connectors from fan motor and fan switch, then drain cooling system.
2. Disconnect upper and lower radiator hoses from radiator. On models equipped with automatic transmission, disconnect fluid cooling hoses from radiator, if necessary.
3. Remove upper fan shroud attaching screws, then lift shroud upward and out of bottom shroud retaining clips, separating shroud from radiator.

NOTE: When removing shroud from radiator, use care not to damage fan or radiator cooling fins.

4. Remove fan motor to shroud attaching screws, then while supporting fan motor and shaft, remove fan motor retaining clip and fan motor.
5. Reverse procedure to install.

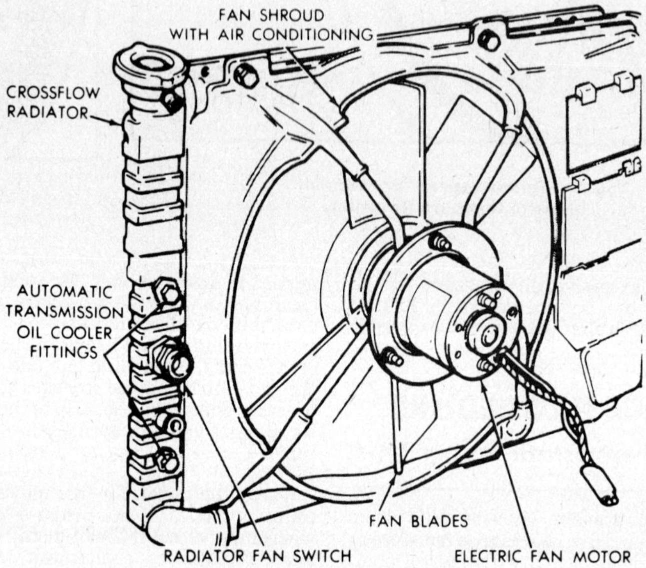

Fig. 1 Electric engine cooling fan

Electric Fan Motor Relay

The fan motor relay is used on models equipped with A/C. The relay is located on the left hand shock absorber housing. If radiator fan switch and fan motor test results are satisfactory, but fan motor will not operate the fan motor relay is then suspected. After replacing relay, disconnect wire connector from radiator fan switch and connect a 14 gauge jumper wire between wire connector terminals. Place ignition switch in the Accessory position. Fan motor should operate.

ALTERNATORS

INDEX

INTRODUCTION

Alternators under discussion here are of the modern type with built-in silicon diode rectifiers. Under normal operating conditions, these alternators have a rating of 30 to 150 amperes, depending upon the requirements of the vehicle, and usually deliver 5 to 10 amperes at curb idle speed.

Alternators are composed of the same functional parts as the conventional D.C. generator but they operate differently. The field is called a rotor and is the turning portion of the unit. A generating part, called a stator, is the stationary member, comparable to the armature in a D.C. generator. The regulator, similar to those used in a D.C. system, regulates the output of the alternator-rectifier system.

The power source of the system is the alternator. Current is transmitted from the field terminal of the regulator through a slip ring to the field coil and back to the ground through another slip ring. The strength of the field regulates the output of the alternating current. This alternating current is then transmitted from the alternator to the rectifier where it is converted to direct current.

These alternators employ a three-phase stator winding in which the phase windings are electrically 120 degrees apart. The rotor consists of a field coil encased between interleaved sections producing a magnetic field with alternate north and south poles. By rotating the rotor inside the stator, the alternating current is induced in the stator windings. This alternating current is rectified (changed to D.C.) by silicon diode rectifiers and brought out to the output terminal of the alternator.

Diode Rectifiers

Six silicon diode rectifiers are used and act as electrical one-way valves. Three of the diodes have ground polarity and are pressed or screwed into a heat sink which is grounded. The other three diodes (ungrounded) are pressed or screwed into and insulated from the end head; these diodes are connected to the alternator output terminal.

Since the diodes have a high resistance to the flow of current in one direction and a low resistance in the opposite direction, they may be connected in a manner that allows current to flow from the alternator to the battery in the low resistance direction. The high resistance in the opposite direction prevents the flow of current from the battery to the alternator. Because of this feature no circuit breaker is required between alternator and battery.

SERVICE PRECAUTIONS

1. Be certain that the battery polarity of the system is known so that the battery is connected to the proper ground. *Reversed battery polarity will damage rectifiers and regulator.*
2. If a booster battery is used for starting the vehicle, it must be connected to the vehicle battery properly to prevent damage to rectifiers and regulator. Negative cable from booster battery to negative terminal on vehicle battery, and positive booster cable to positive terminal.
3. When a fast charger is used to charge a vehicle battery, the vehicle battery cables should be disconnected *unless the charger is equipped with a special Alternator Protector,* in which case the vehicle battery cables need not be disconnected. Also the fast charger should never be used to start an engine as damage to rectifiers will result.
4. Lead connections to grounded rectifiers on Auto-Lite and Chrysler units should never be soldered as the excessive heat may damage the rectifiers.
5. Unless the system includes a load relay or field relay, grounding the alternator output terminal will damage the alternator and/or circuits. This is true even when the system is not in operation since no circuit breaker is used and the battery is applied to the alternator output terminal at all times. The field or load relay acts as a circuit breaker in that it is controlled by the ignition switch.
6. When adjusting the voltage regulator, do not short the adjusting tool to the regulator base as the regulator may be damaged. The tool should be insulated by taping or by installing a plastic sleeve.
7. Before making any "on vehicle" tests of alternator or regulator, the battery should be checked and the circuit inspected for faulty wiring or insulation, loose or corroded connections and poor ground circuits.
8. Check alternator belt tension to be sure belt is tight enough to prevent slipping under load.
9. The ignition switch should be off and the battery ground cable disconnected before making any test connections to prevent damage to the system.
10. The vehicle battery must be fully charged or a fully charged battery may be installed for test purposes.

Chrysler Alternators

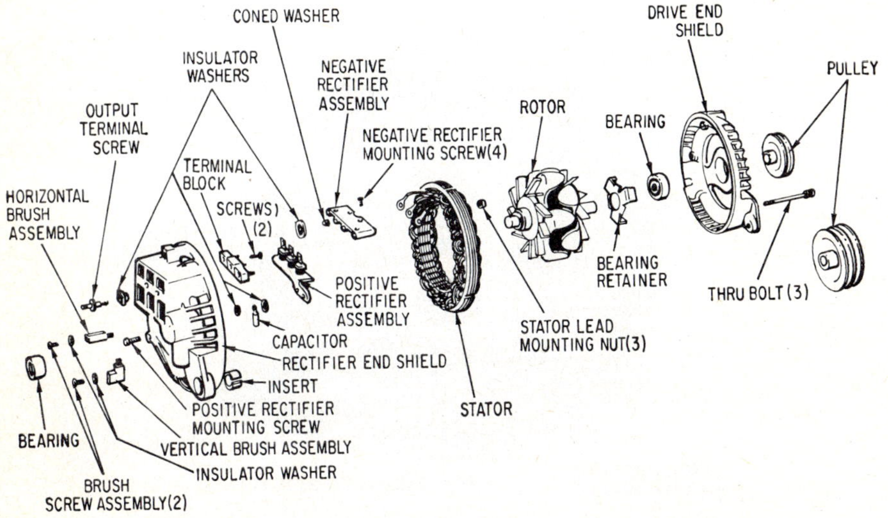

Fig. C1 Isolated field alternator disassembled. 1974–78 (Typical)

The main components of the alternator are the rotor, stator, diodes, end shields and drive pulley. Direct current is available at the output "BAT" terminal. The function of the voltage regulator is to limit output voltage. This is accomplished by controlling the current flow in the rotor field coil, in turn controlling the strength of the rotor magnetic field. The electronic voltage regulator is a sealed, non-adjustable unit, Fig. C1 through C5.

TESTING SYSTEM ON VEHICLE

CHARGING CIRCUIT RESISTANCE TEST

1. Disconnect battery ground cable. Disconnect "Batt" lead at the alternator.
2. Complete test connections as per Fig. C6.
3. Connect battery ground cable, start engine and operate at idle.
4. Adjust engine speed and carbon pile to obtain 20 amps in the circuit and check

voltmeter reading. Reading should not exceed .7 volts. If a voltage drop is indicated, inspect, clean and tighten all connections in the circuit. A voltage drop test at each connection can be performed to isolate the trouble.

CURRENT OUTPUT TEST

1. Disconnect battery ground cable, complete test connections as per Fig. C7 and start engine and operate at idle. *Immediately after starting, reduce engine speed to idle.*
2. Adjust the carbon pile and engine speed in increments until a speed of 1250 rpm and 15 volts on all except 100 and 117 amp units. On 100 and 117 amp units, adjust carbon pile and engine speed in increments until a speed 900 rpm and 13 volts is obtained.

CAUTION: While increasing speed, do not allow voltage to exceed 16 volts.

3. Check ammeter reading. Output current should be within specifications.

VOLTAGE REGULATOR TEST

NOTE: Battery must be fully charged for test to be accurate.

1. Complete test connections as per Fig. C8.
2. Start and operate engine at 1250 rpm with all lights and accessories turned off. Voltage should be 13.9–14.6 if temperature at the regulator is 80 degrees F. and 13.3–14.0 at 140 degrees F.
3. It is normal for the car ammeter to show an immediate charge then gradually return to normal position.
4. If voltage is above, fluctuating or below limits, check for good voltage regulator ground, and voltage drop between regulator cover and body on low voltage scale of voltmeter.
5. Turn off ignition switch and disconnect regulator connector.
6. Turn on ignition switch but do not start engine. Check for battery voltage at the wiring harness terminal connected to the blue and green leads.

NOTE: Disconnect wiring harness from regulator when checking the leads.

7. Turn off ignition switch. If voltage is not present at either lead, the problem is in the vehicle wiring or alternator field circuit. *Use care to avoid bending the terminals with voltmeter probe.*
8. If Steps 4–7 tested satisfactorily, change voltage regulator and repeat Step 2.

BENCH TESTS

If the alternator performance does not meet current output specification limits, it will have to be removed and disassembled for further tests and servicing.

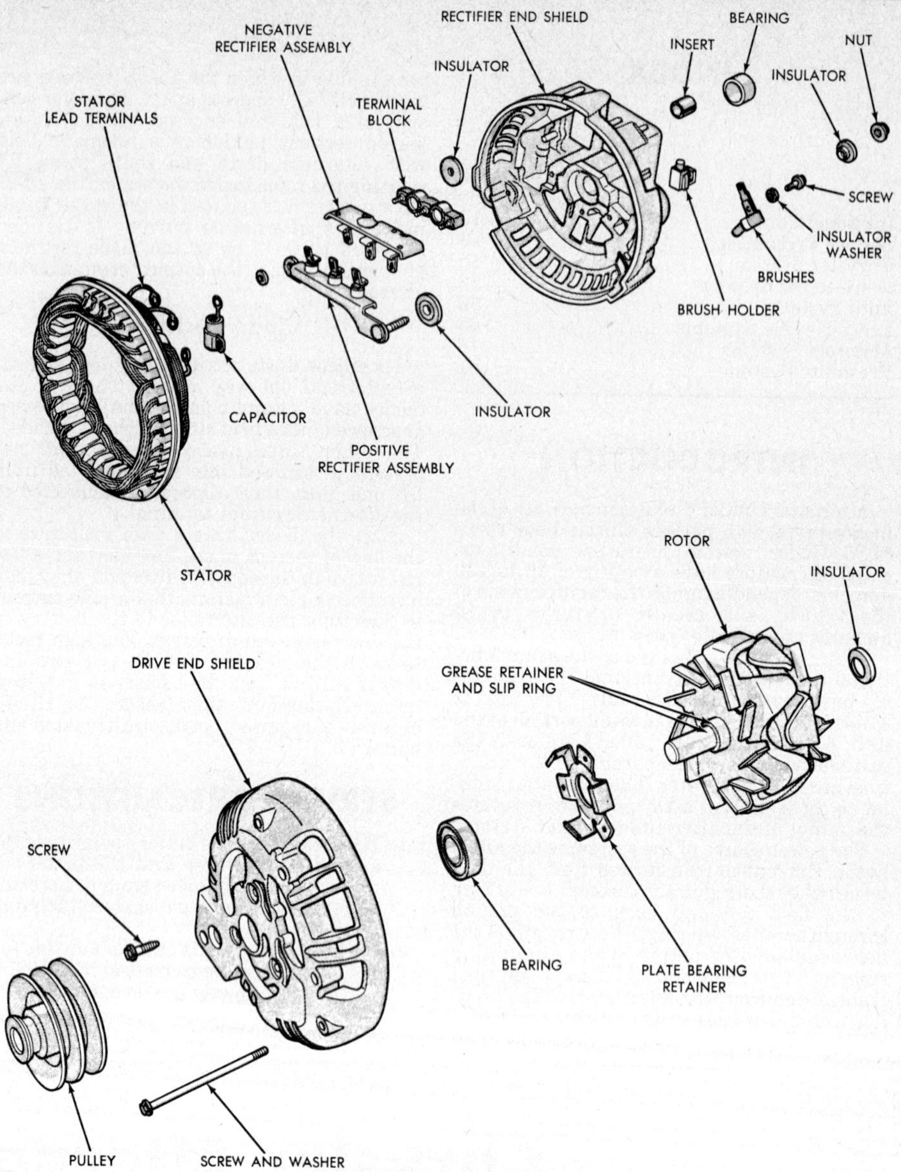

Fig. C2 Isolated field alternator disassembled. 1979–83

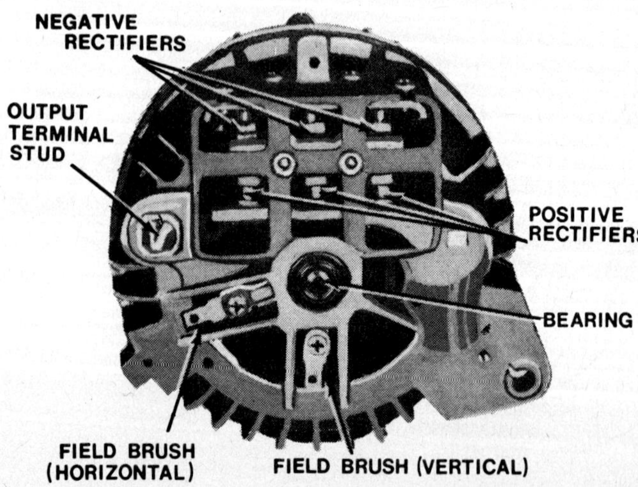

Fig. C3 Isolated field alternator. 1974–83

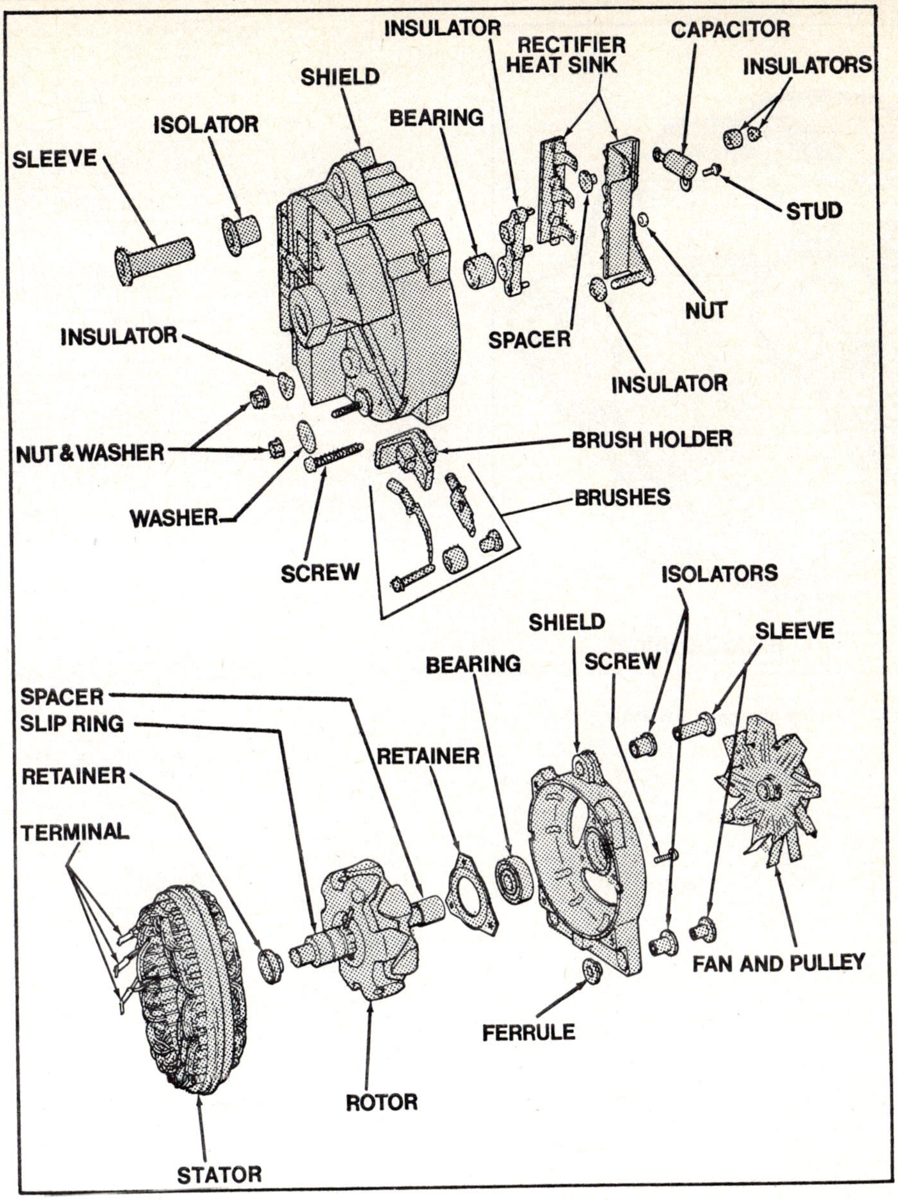

Fig. C4 100 & 117 amp alternator disassembled

To remove the alternator, disconnect the battery ground cable and the leads at the alternator. Then unfasten and remove the alternator from the vehicle.

Field Coil Draw

1974–76

1. Connect jumper wire between one alternator field terminal and the positive terminal of a fully charged battery.
2. Connect test ammeter positive lead to the other alternator field terminal and the ammeter negative lead to the negative battery terminal.
3. Slowly rotate alternator by hand. Field current draw at 12 volts should be 2.5–3.7 amps.
4. A low rotor coil draw is an indication of high resistance in the field coil circuit, (brushes, slip rings or rotor coil). A high rotor coil draw indicates shorted rotor coil or grounded rotor.

1977–83

1. On units except 100 and 117 amp, place alternator on a insulated surface.
2. On all units, connect a jumper wire between one alternator field terminal and the negative terminal of a fully charged battery.
3. Connect test ammeter positive lead to the other alternator field terminal and the ammeter negative lead to the positive battery terminal.
4. On units except 100 and 117 amp, connect a jumper wire between alternator end shield and negative terminal of battery.
5. On all units, slowly rotate rotor by hand and note ammeter reading. On units except 100 and 117 amp, field current at 12 volts should be 4.5 to 6.5 amps. On 100 and 117 amp units, field current should be 4.75 to 6.0 amps.
6. A low rotor coil draw is an indication of high resistance in the field coil circuit, (brushes, slip rings or rotor coil). A high rotor coil draw indicates shorted rotor coil or grounded rotor. No reading indicates an open rotor or defective brushes.

ALTERNATOR REPAIRS EXCEPT 100 & 117 AMP UNITS

Disassembly

To prevent possible damage to the brush assemblies, they should be removed before disassembling the alternator. Both brushes are insulated and mounted in plastic holders.

1. Remove both brush screws, insulating nylon washers and remove brush assemblies, Fig. C1 & C2.

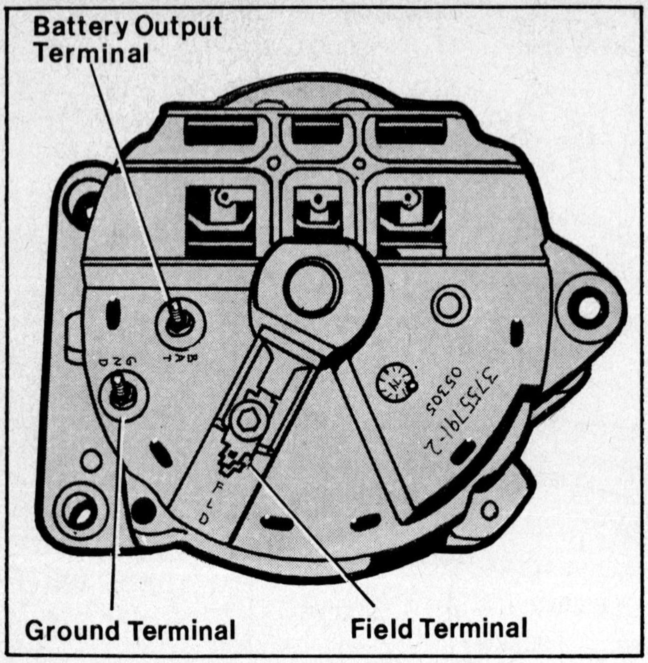

Fig. C5 100 & 117 amp alternator assembly

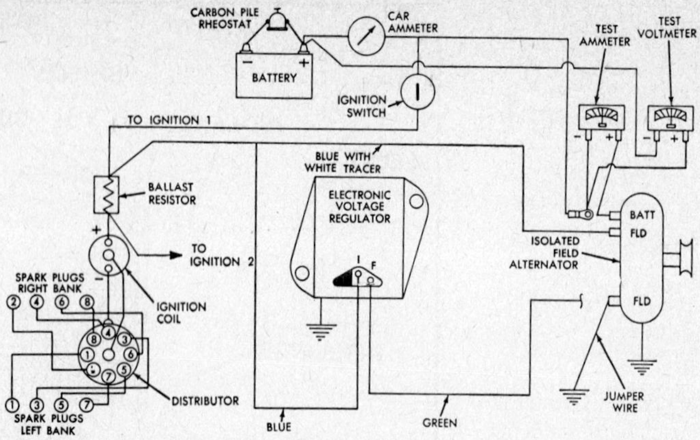

Fig. C6 Charging circuit resistance test with electronic regulator (Typical)

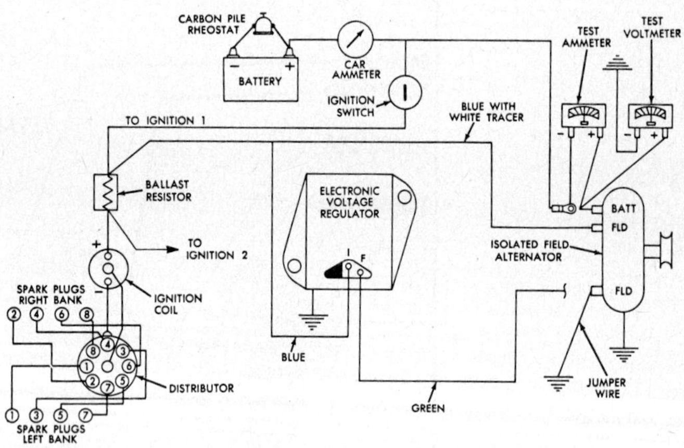

Fig. C7 Current output test with electronic regulator (Typical)

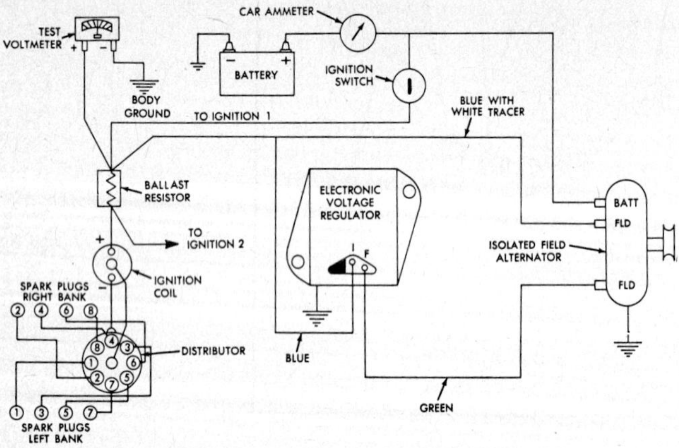

Fig. C8 Electronic voltage regulator test (Typical)

Fig. C9 Separating drive end shield from stator

NOTE: The stator is laminated; do not burr it or the end shield.

2. Remove through bolts and pry between stator and drive end shield with a screwdriver. Carefully separate drive end shield, pulley and rotor from stator and diode rectifier shield, Fig. C9.
3. The pulley is an interference fit on the rotor shaft; therefore, a suitable puller must be used to remove it, Fig. C10.
4. Pry drive end bearing spring retainer from end shield with a screwdriver, Fig. C11.
5. Support end shield and tap rotor shaft with a plastic hammer to separate rotor from end shield.
6. The drive end ball bearing is an interference fit with the rotor shaft; therefore, a

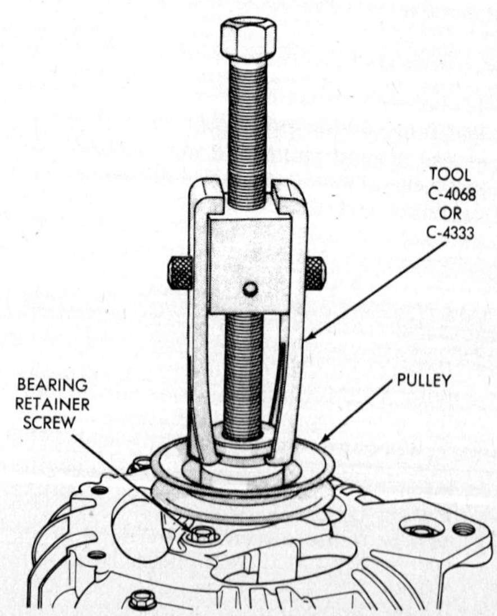

Fig. C10 Removing pulley

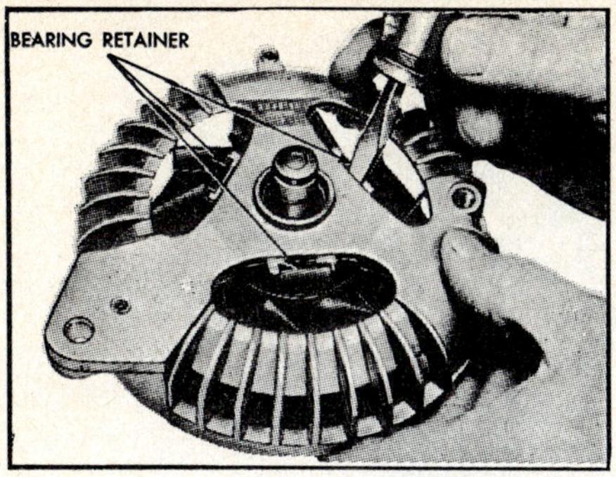

Fig. C11 Disengaging bearing retainer from end shield

suitable puller must be used to remove it, Fig. C12.

7. To remove rectifiers and heat sinks, loosen screws securing negative rectifier and heat sink asembly to end shield, remove the two outer screws and lift assembly from end shield. Remove nuts securing positive rectifier and heat sink assembly to insulated terminals in end shield. Then, remove capacitor ground screw and lift insulated washer, capacitor and positive rectifier and heat sink assembly from end shield.
8. The needle roller bearing in the rectifier end shield is a press fit. If it is necessary to remove the rectifier end frame needle bearing, protect the end shield by supporting the shield when pressing out the bearing as shown in Fig. C13.

Testing Diode Rectifiers

A special Rectifier Tester Tool C-3829 provides a quick, simple and accurate method to test the rectifiers without the necessity of disconnecting the soldered rectifier leads. This instrument is commercially available and full instructions for its use are provided. Lacking this tool, the rectifiers may be tested with a 12 volt battery and a test lamp having a No. 67 bulb. The procedure is as follows:

1. Remove nuts securing stator windings, positive and negative rectifier straps to terminal block. Remove stator winding terminals and pry stator from end shield.

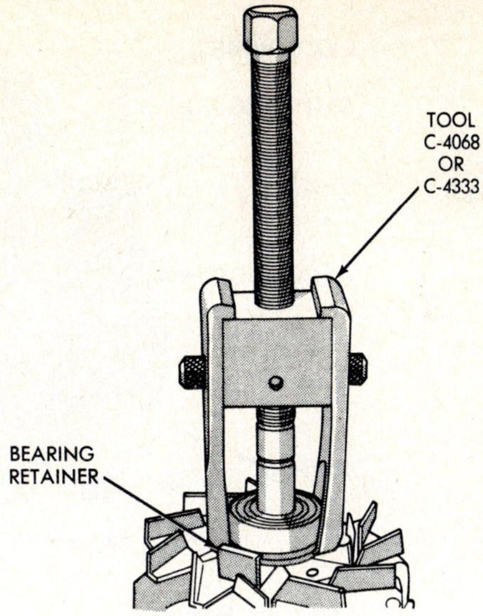

Fig. C12 Removing bearing from rotor shaft

2. Connect one side of test lamp to positive battery post and the other side of the test lamp to a test probe. Connect another test probe to the negative battery post, Fig. C14.
3. Contact heat sink with one probe and strap on top of rectifier with the other probe.
4. Reverse position of probes. If test lamp lights in one direction only, the rectifier is satisfactory. If test lamp lights in both directions, the rectifier is shorted. If test lamp lights in neither direction, the rectifier is open.

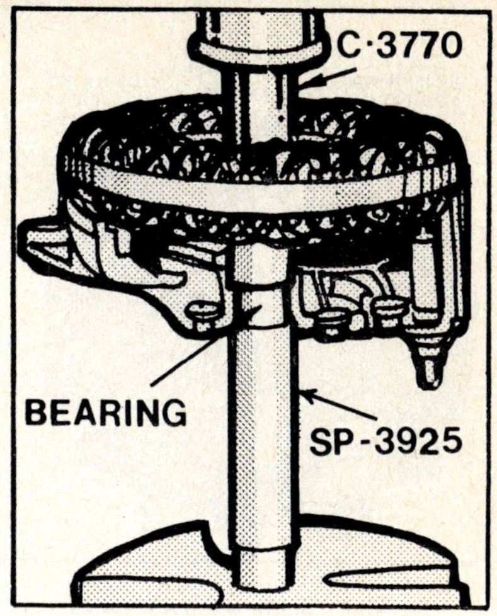

Fig. C13 Removing diode end shield bearing

NOTE: *Possible cause of an open or a blown rectifier is a faulty capacitor or a battery that has been installed on reverse polarity. If the battery is installed properly and the rectifiers are open, test the capacitor capacity, which should be .50 microfarad plus or minus 20%.*

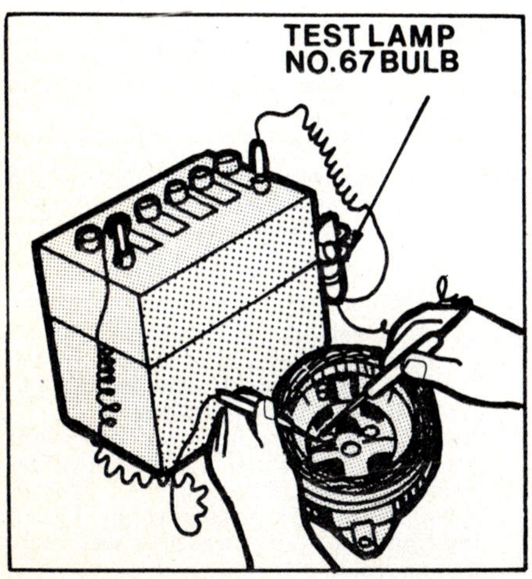

Fig. C14 Testing diodes with a test lamp

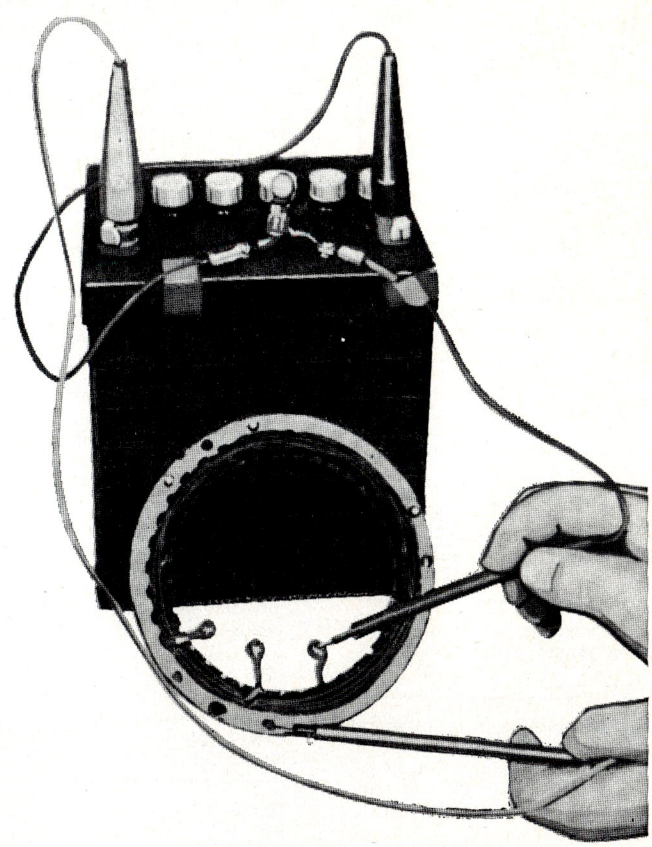

Fig. C15 Testing stator for grounds

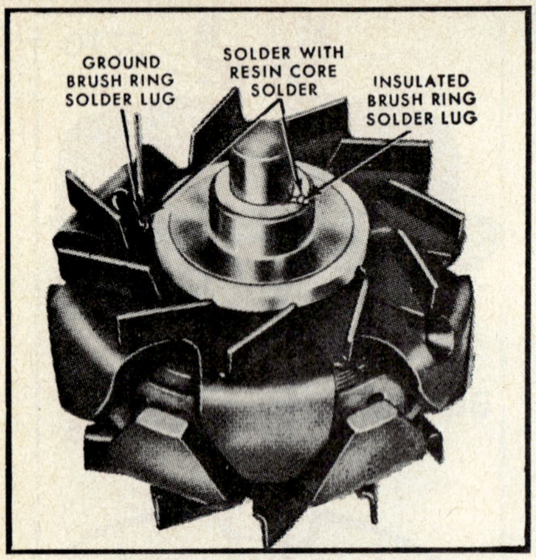

Fig. C16 Soldering points with slip ring installed

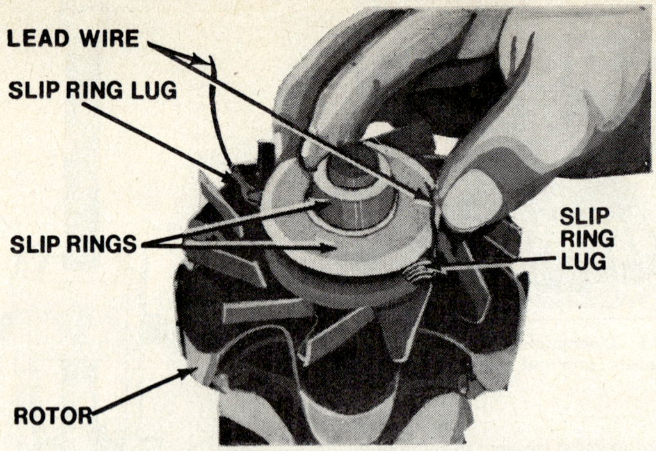

Fig. C18 Aligning slip ring with field lead wires

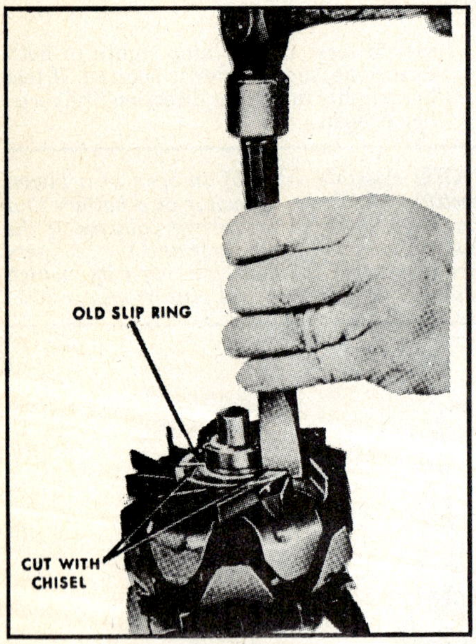

Fig. C17 Cutting old slip rings for removal

Testing Stator

1. Separate stator from end shields.
2. Using a 12 volt test lamp, Fig. C15, test stator for grounds. Contact one test probe to any pin on stator frame and the other to each stator lead. If lamp lights, stator is grounded.

NOTE: Remove varnish from stator frame pin to ensure proper electrical connection.

3. Use a 12-volt test lamp to test stator for continuity. Contact one stator lead with one probe and the remaining two leads with the other probe, Fig. C12. If test lamp does not light, the stator has an open circuit.
4. Install new stator if one tested is defective.

Testing Rotor

The rotor may be tested electrically for grounded, open or shorted field coils as follows:

Grounded Field Coil Test: Connect an ohmmeter between each slip ring and the rotor shaft. The ohmmeter should indicate infinite resistance. If reading is zero or higher, rotor is grounded.

Open or shorted Field Coil Test: Connect an ohmmeter between the slip rings. A resistance of 3 to 4 ohms on 1974–75 units and 1.5 to 2 ohms on 1976–83 units at room temperature indicates rotor is satisfactory. A reading of 4 to 6 ohms on 1974–75 units and 2.5 to 3 ohms on 1976–83 units, indicates the alternator was operated at a high underhood temperature, however, rotor is still satisfactory. Resistance below 3 ohms on 1974–75 units, below 1.5 ohms on 1976–83 units, indicates a shorted field coil. Resistances above 6.5 ohms on 1974–75 units, 3.5 ohms on 1976–83 units, indicates a high resistance in the field coils and further testing or rotor replacement is required.

Replacing Slip Rings

NOTE: On 1975–83 units, slip rings and rotor are serviced as an assembly.

1. On 1974 units, remove rotor plastic grease retainer.
2. Unsolder field coil leads at solder lugs, Fig. C32.
3. Cut through copper of both slip rings at opposite points with a chisel, Fig. C17.
4. Break insulator and remove old ring.
5. Clean away dirt and particles of old slip ring from rotor.
6. Scrape ends of field coil lead wires clean for good electrical contact.
7. Position field coil wires aside and place new slip ring on rotor shaft so slip ring lugs are properly positioned for field coil wire connections, Fig. C18.
8. Place installing tool over rotor shaft and with a suitable press, press slip ring onto shaft, Fig. C19.

NOTE: With slip ring bottomed against fan, the field lead wire should clear the access hole, fan and pole piece.

9. Tin field coil lead wires, then coil each lead wire around slip ring lug with first wrap against lug shoulder and winding outward. Solder leads with resin core solder.
10. Test slip rings for ground with a 110 volt test lamp by touching one test lead prod to rotor pole shoe and remaining prod to slip rings. The lamp should not light. If lamp lights, slip rings are shorted to ground, possibly due to a grounded insulated field lead when installing slip ring.
11. If rotor is not grounded, lightly clean slip ring surfaces with No. 00 sandpaper and assemble to alternator.
12. Position grease retainer gasket and retainer on rotor shaft and press retainer on shaft, Fig. C20. Retainer is properly positioned when inner bore of installer tool bottoms on rotor shaft.

Alternator, Assemble

1. Press grease retainer onto rotor shaft using tool No. C-3921, Fig. C20. The grease retainer is properly positioned when the inner bore of the installed tool bottoms on the rotor shaft.
2. Install diode end shield bearing, Fig. C21.
3. Install drive end bearing in end shield with bearing retainer plate to hold bearing in position. Place assembly on rotor shaft and press into position, Fig. C22.

NOTE: If a metal spacer is provided with replacement rotor to ball bearings, position this spacer on pulley end of rotor shaft first.

4. Press pulley onto rotor shaft until it contacts inner race of bearing.

NOTE: Do not exceed 6800 lbs.

5. Install output terminal stud and insulator through end shield. Then place positive heat sink assembly over studs, guiding rectifier straps over studs.
6. Place capacitor terminal over capacitator end stud and install capacitor shoulder insulator. Ground the capacitor bracket to end shield with a metal screw. Install and tighten positive heat sink lockwashers and nuts.
7. Slide negative rectifier and heat sink assembly into place, position straps on

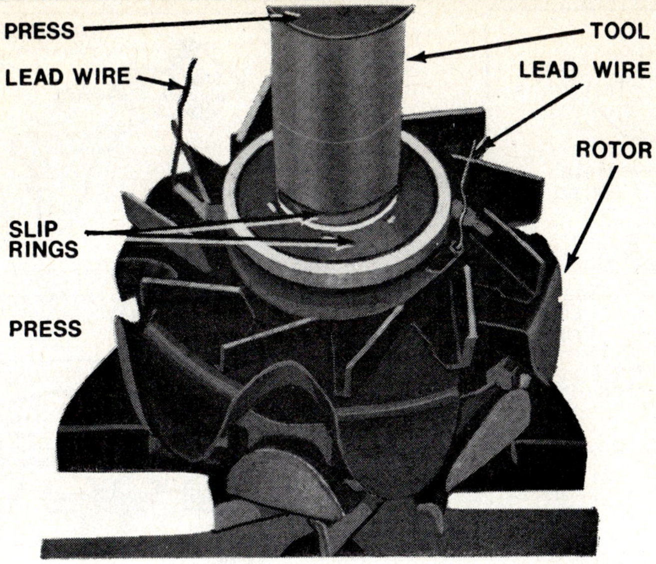

Fig. C19 Installing slip ring

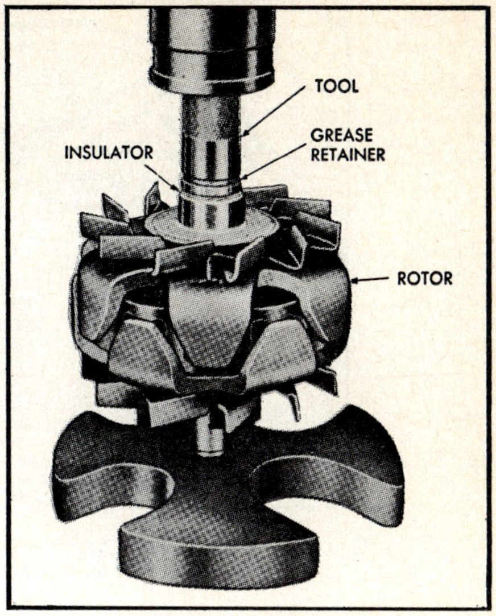

Fig. C20 Installing bearing grease retainer

terminal block studs, then install and tighten attaching screws.

8. Position stator on diode end shield and install stator leads on terminal block. Press stator pins into end shield and install winding terminal nuts.
9. Position rotor and drive end shield on stator and diode end shield.
10. Align through bolt holes in stator, diode end shield and drive end shield.
11. Compress stator and both end shields by hand and install through bolts, washers and nuts.
12. Install field brush into vertical and horizontal holders. Place an insulating washer on each field brush terminal and install lock-washers and attaching screws.

NOTE: Ensure brushes are not grounded.

13. Rotate pulley slowly by hand to be sure rotor fans do not touch diodes, capacitor lead and stator connections.
14. Install alternator and adjust drive belt.
15. Connect leads to alternator.
16. Connect battery ground cable.
17. Start and operate engine and observe alternator operation.
18. If necessary, test current output and regulator voltage setting.

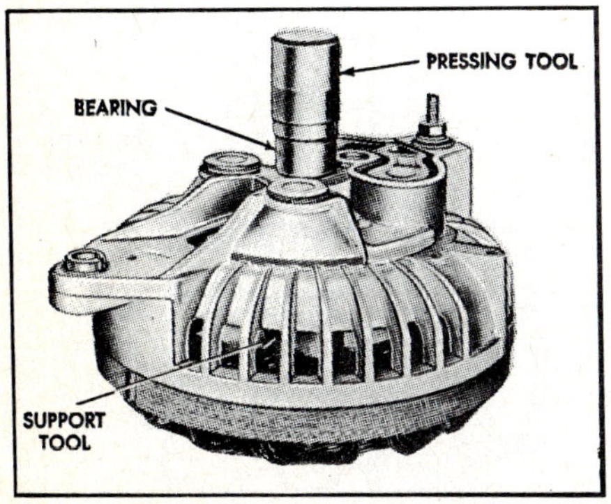

Fig. C21 Installing diode end shield bearing

100 & 117 AMP ALTERNATOR REPAIRS

Disassembly & Testing

Separating End Shields

1. Remove brush holder screw and insulating washer, then lift brush holder from end shield.

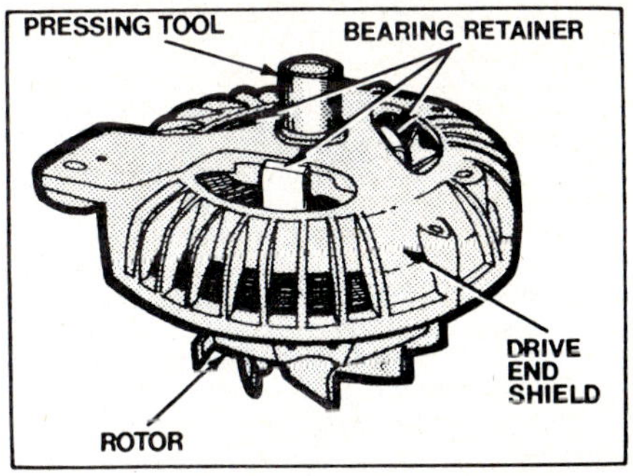

Fig. C22 Installing drive end shield & bearing

Fig. C23 Separating end shields 100 & 117 amp unit

2. Remove the through bolts, then using a screwdriver, pry between the stator and end shield in the slot provided to separate end shields, Fig. C23.

Rectifier Testing

1. Remove stator winding leads to terminal block stud nuts, Fig. C24.
2. Lift stator winding leads and pry stator from end shield.
3. Using a 12 volt battery and a test lamp equipped with a #67 bulb, test rectifiers as follows:
   A. Connect one test probe to rectifier heat sink and the other test probe to the metal strap on top of rectifier, Fig. C25. Reverse the probes.
   B. If test lamp lights in one direction and does not light in the other, rectifier is satisfactory. If test lamp lights in both directions, rectifier is shorted. If test lamp does not light in either direction, rectifier is open.

Rectifier & Heat Sink Assembly Removal

1. Remove nut and insulator securing positive heat sink assembly to end shield stud.

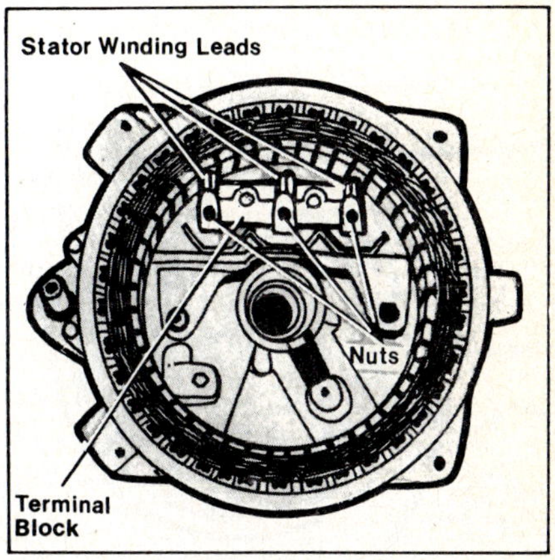

Fig. C24 Removing stator winding leads. 100 & 117 amp unit

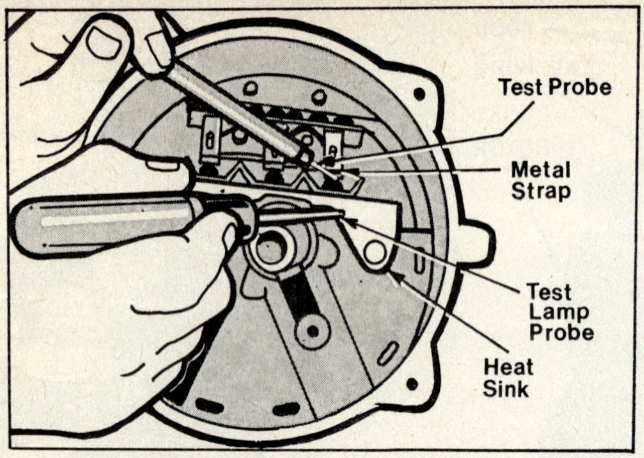

Fig. C25 Testing rectifiers with a test lamp. 100 & 117 amp unit

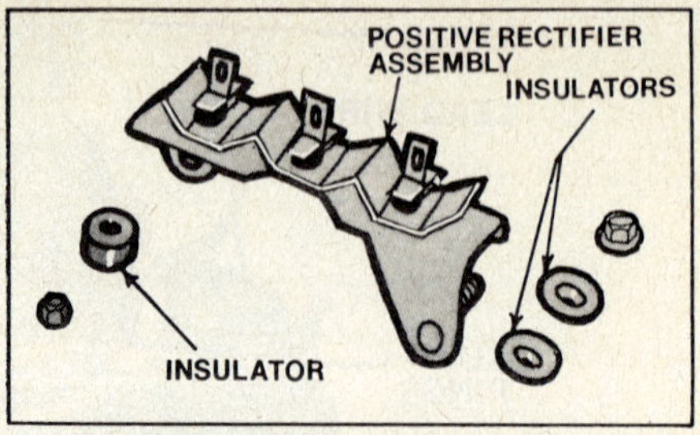

Fig. C26 Positive rectifier assembly. 100 & 117 amp unit

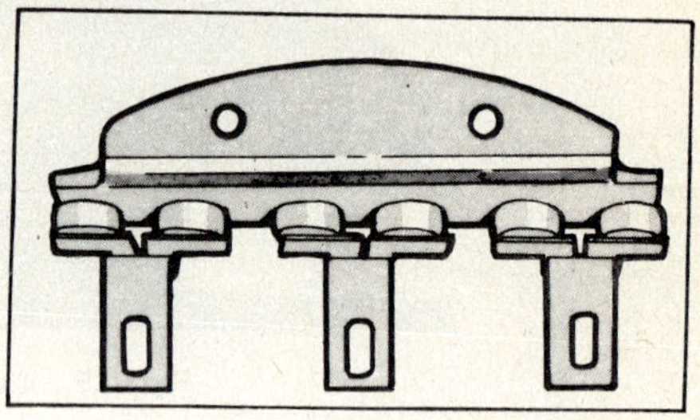

Fig. C27 Negative rectifier assembly. 100 & 117 amp unit

2. Remove capacitor attaching screw.
3. Remove nut and insulator securing positive heat sink assembly stud to end shield, then remove positive heat sink assembly, Fig. C26, noting location of the insulators.
4. Remove screws securing negative heat sink assembly to end shield, then the negative heat sink assembly, Fig. C27.
5. Remove terminal block, then the capacitor and insulator.

Stator Testing

1. Contact one test lamp probe to outer diameter of stator frame and the other probe to each of the stator lead terminals, one at a time, Fig. C28.
2. If test lamp lights, the stator lead is grounded, requiring replacement.

NOTE: The stator windings are Delta Wound, therefore the windings cannot be tested for opens or shorts using a test lamp. If the stator is not grounded, and all other electrical circuits and alternator components test satisfactory, the stator may be open or shorted.

Pulley & Bearing Removal

1. On 100 amp unit alternators, remove pulley with tool C-4333, Fig. C29. On 117 amp unit alternators remove pulley with tool C-4467.
2. Remove bearing retainer to drive end shield attaching screws.
3. Support end shield and using a mallet, tap rotor from end shield.
4. Remove bearing with a suitable puller, Fig. C30.
5. If necessary to remove needle roller bearing in rectifier end shield, use tool C-4330, Fig. C31.

Rotor Testing

Grounded Field Coil Test: Connect a test lamp between each slip ring and the rotor shaft, Fig. C32. If test lamp lights, the rotor is grounded, requiring replacement.

Open Field Coil Test: Connect a test lamp between the slip rings, Fig. C33. If test lamp does not light, the rotor is open, requiring replacement.

Shorted Field Coil Test: Connect an ohmmeter between the slip rings, Fig. C33. If reading is below 1.7 ohms, the rotor is shorted.

High Resistance Test: with an ohmmeter connected across the slip rings, reading should be between 1.7 and 2.1 ohms at 80° F. If not, replace rotor.

Assembly

1. Press grease retainer onto rotor shaft, Fig. C31 (Omit tool C-4330-3).
2. Place rectifier end shield bearing on base of tool C-4330-1, Fig. C31 (Omit tool C-4330-3), then place rectifier end shield on top of bearing. Using tool C-4330-2, press end shield onto bearing until end shield contacts press base.
3. Install drive end bearing and retainer in drive end shield.
4. Position bearing and drive end shield on rotor and while supporting base of rotor shaft, press end shield onto shaft. Ensure rotor spacer is in place before pressing bearing and end shield on shaft.

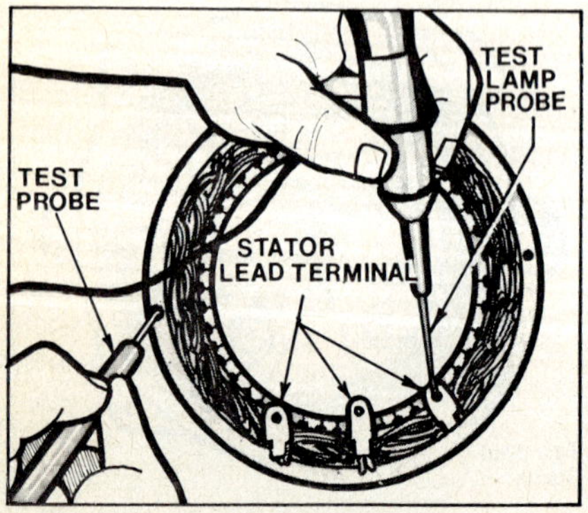

Fig. C28 Testing stator winding. 100 & 117 amp unit

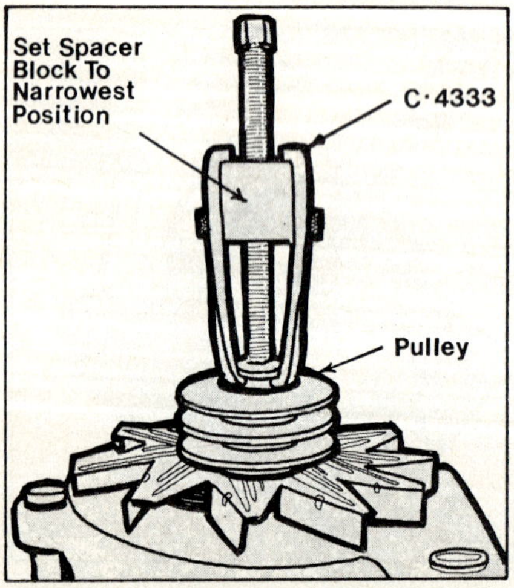

Fig. C29 Removing pulley. 100 & 117 amp unit

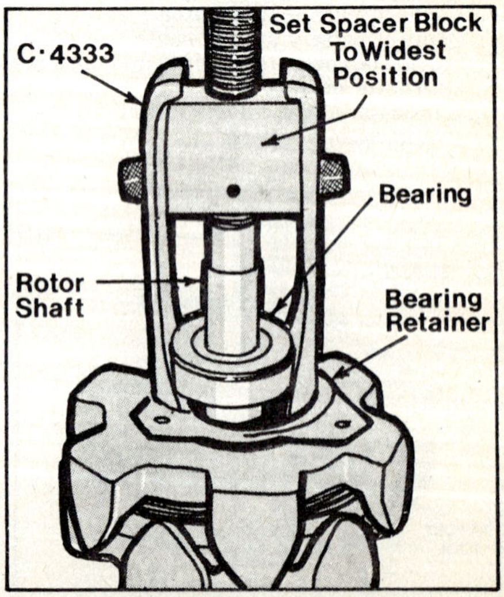

Fig. C30 Removing bearing from rotor shaft. 100 & 117 amp unit

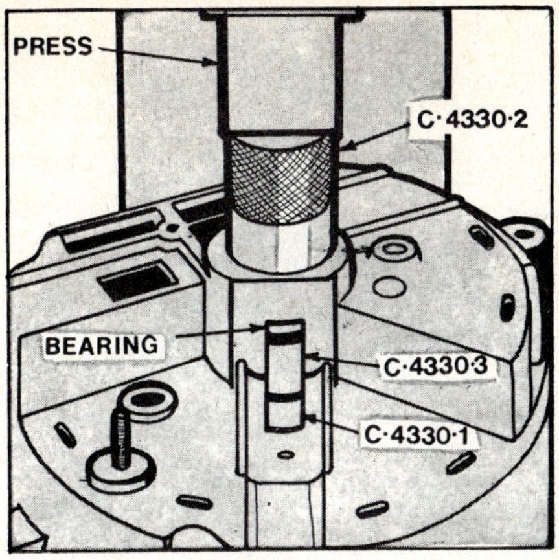

Fig. C31 Removing rectifier end shield bearing. 100 & 117 amp unit

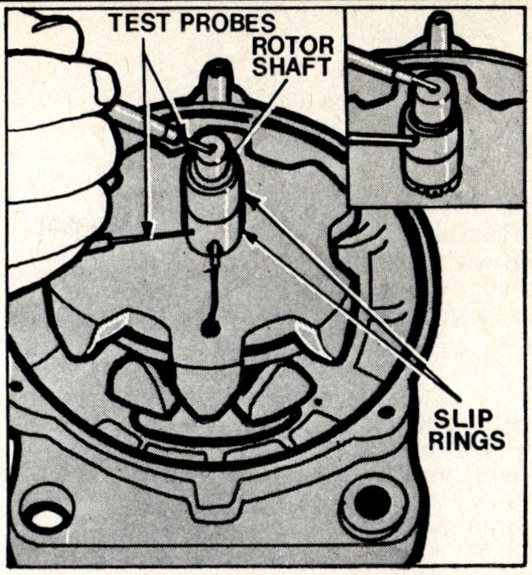

Fig. C32 Testing rotor for grounds. 100 & 117 amp unit

5. Press pulley onto rotor shaft until it contacts inner race of bearing.

NOTE: Do not exceed 6800 lbs.

6. Place insulator and capacitor on positive heat sink mounting stud, then install capacitor mounting screw.
7. Place terminal block into position in rectifier end shield and install mounting screws.
8. Place negative heat sink into position, ensuring metal straps are properly located over studs on terminal block, then install negative heat sink mounting screws.
9. Place insulator over positive heat sink stud and install positive heat sink assembly into position in end shield, ensuring metal straps are properly located over terminal block studs. From inside end shield, place insulator on positive heat sink mounting stud, then install mounting nut. From outside end shield, place insulator on positive heat sink stud, then install mounting nut.
10. Place stator over rectifier end shield and install terminals on terminal block. Then, press stator pins into end shield and install terminal nuts.

NOTE: Route leads to avoid contact with rotor or sharp edge of negative heat sink.

11. Place rotor and drive end shield assembly over stator and rectifier end shield assembly, aligning bolt holes. Compress stator and both end shields, then install and torque through bolts to 40–60 inch lbs.
12. Install field brushes into brush holder with long terminal on bottom and the short terminal on top, Fig. C34. Then, install insulators and mounting screw.
13. Place brush holder on end shield, ensure it is properly seated and tighten mounting screw.
14. Slowly rotate pulley to ensure rotor poles do not contact stator winding leads.

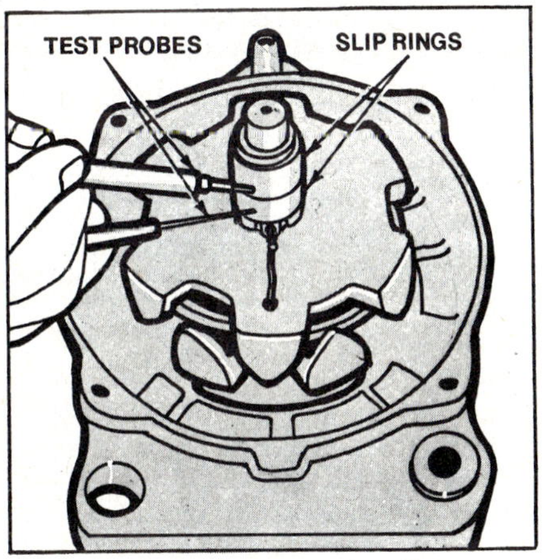

Fig. C33 Testing rotor for open or short circuits. 100 & 117 amp unit

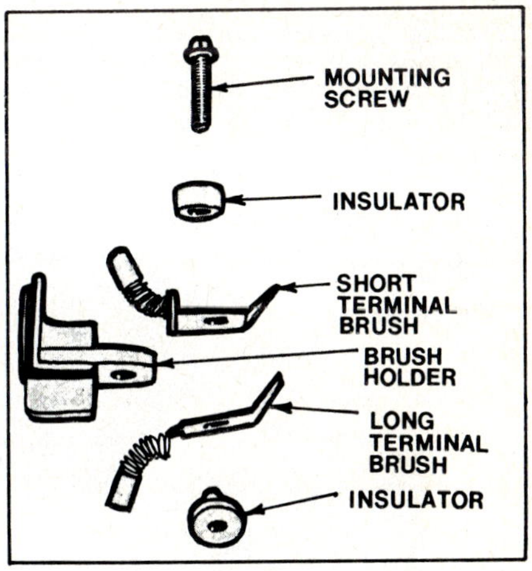

Fig. C34 Assembling field brushes. 100 & 117 amp unit

Delco-Remy "Delcotron"

DESCRIPTION

Figs. 1 through 4 illustrate the four types of Delcotron alternators used on automotive truck applications. All models have the rotor mounted on ball or roller bearings, and each bearing has a supply of lubrication to provide long periods of service. Current to the coil winding mounted on the rotor is supplied through brushes riding on smooth slip rings.

All models develop three-phase A.C. voltage which is then rectified to a single D.C. voltage available at the output terminals of the alternator. All models are also designed to provide an output at engine idle, the amount depending on the application. They all are "self-limiting" in their maximum output, which occurs as the magnetic field produced by the current in the stator windings opposes in polarity and approaches in value the magnetic field provided by the rotor as the output increases. This causes the alternator to limit its own output to a maximum value.

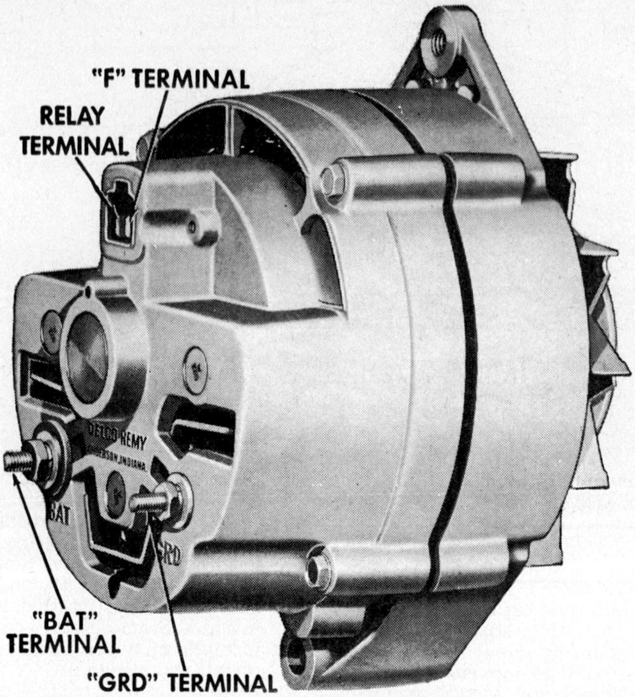

Fig. 1 Delcotron 5.5" 1D Series

OUTPUT TEST ON VEHICLE

Systems with Two-Unit or Transistorized Regulator

If evidence of an undercharged battery exists, test alternator output to determine if alternator is capable of producing its normal rated output.

1. Disconnect negative cable from battery.
2. Remove wire from "BAT" terminal on alternator and connect an ammeter between wire and "BAT" terminal on alternator, Fig. 5.
3. Install voltmeter between "BAT" terminal and "GRD" terminal on alternator.
4. Pull latch on regulator upward to disengage latch from connector lug, then pull wiring harness connector from regulator. With jumper lead, make connection between "F" terminal socket and #3 or "V" socket, Fig. 6.

CAUTION: With wiring connected as shown in Fig. 7, the voltage regulator is taken out of the circuit and causes the field to be energized by full battery voltage. Do not permit voltage to exceed setting specified for regulator. Refer to "Voltage" chart later on in this section.

NOTE: Instead of using a carbon pile, the headlights and heater switch can be turned on.

5. Connect tachometer to engine; connect battery cable, then start engine. Adjust engine speed and carbon pile (if used) or with accessories to provide rated voltage. Ratings for each generator are given in the Truck Chapters. If alternator does not produce current within its rated capacity, it can be considered defective and in need of repair.
6. Remove instruments and jumper lead, then make connections at alternator and regulator.
7. If no defects are discovered by the foregoing tests, yet the battery remains undercharged, the cause is probably a low voltage regulator setting.

Fig. 2 Delcotron 6.2" 2D Series

Systems with Full Transistor Regulator

Two methods may be used for checking alternator output on these systems: Method using a special adapter (J-21600) installed in harness connector at voltage regulator or method not requiring use of an adapter.

Using Special Adapter (J-21600)

1. Connect voltmeter to adapter as shown in View A, Fig. 8.
2. Turn on ignition switch.
3. Operate alternator at specified speed and check for rated output. Load battery with

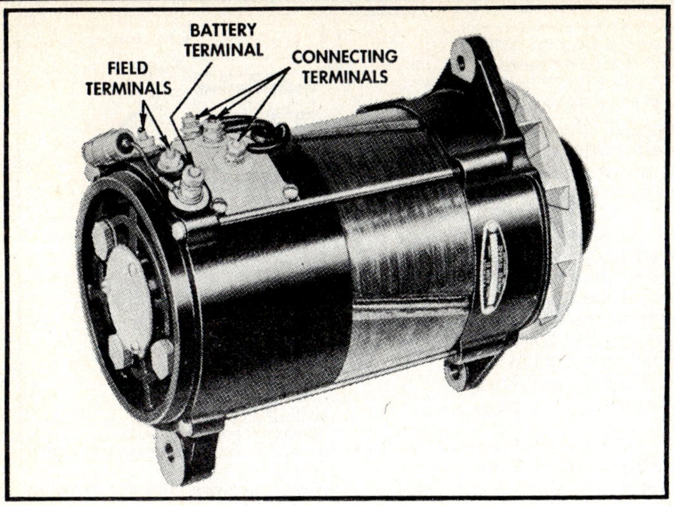

Fig. 3 Delcotron Series 3D

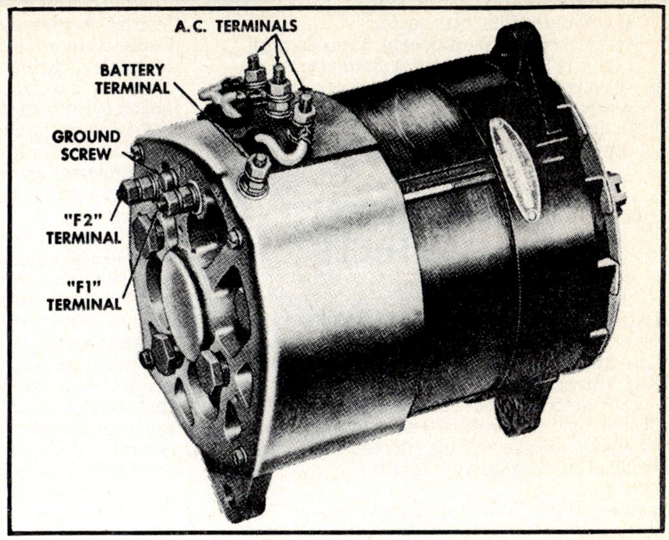

Fig. 4 Delcotron Series 4D

carbon pile or with accessories if needed to obtain rated output. If alternator does not provide rated output, replace or repair alternator.

Without Using Adapter

1. Separate harness connector at regulator.
2. Install voltmeter between alternator "BAT" terminal and base of alternator (to ground).
3. Connect a jumper lead between alternator "F" or "F2" terminal.
4. Load battery with carbon pile or accessories (headlight and heater switch on).
5. Start engine and operate at specified speed and check for rated output.
6. If rated output is not obtained, replace or repair alternator.

REGULATOR UNIT FIELD RELAY TEST ON VEHICLE

Systems with Two-Unit or Transistorized Regulator

If charge indicator shows no charge, the regulator field relay or possibly the alternator is at fault. To determine which, proceed as follows:

1. Make connections to regulator and connector terminals as shown in Fig. 9.
2. Start engine and operate at idle speed. Observe voltmeter reading. If reading is 5 volts or over, and indicator shows no charge, the field relay is defective and must be checked.
3. If voltmeter reading is below 5 volts, trouble is in the alternator.
4. The field relay closing voltage can be checked on the vehicle as follows (refer to Fig. 10 which shows hookup of test equipment):
   a. Connect a 100-150 ohm variable resistor and a voltmeter to adapter as shown. If a 0–50 ohms variable resistor unit is used, it will be necessary to add a 15 and a 115 ohm resistor in same series to provide required resistance.
   b. Turn resistor to the open or "full resistance" position. Leave ignition switch off.
   c. Slowly decrease resistance and note closing voltage. Voltage should be 1.5 to 2.7 for a two-unit regulator, or 2.3 to 3.5 volts for a transistorized regulator. If necessary, adjust as directed under applicable regulator section explained later.

FIELD RELAY CIRCUIT TEST

Separate Unit Type

Using Special Adapter J-21600

1. Connect voltmeter to adapter as shown in View A, Fig. 8.
2. Make sure ignition switch is off.
3. If voltmeter indicates battery voltage, the field relay contacts are stuck closed (replace relay). If reading is zero, proceed as follows:
4. Turn ignition switch on.
5. If voltmeter indicates battery voltage, relay is operating properly. Trouble is most likely in alternator field circuit.
6. If voltmeter indicates zero, check for excessive resistance or an open in wiring or in the ignition switch. If wiring and switch are satisfactory, replace relay.

Without Use of Special Adapter

1. Connect a voltmeter to No. 1 terminal of field relay and to ground.
2. Make sure ignition switch is off.
3. If voltmeter indicates battery voltage the relay points are stuck closed (replace relay).
4. If voltmeter indicates zero, turn ignition switch on.

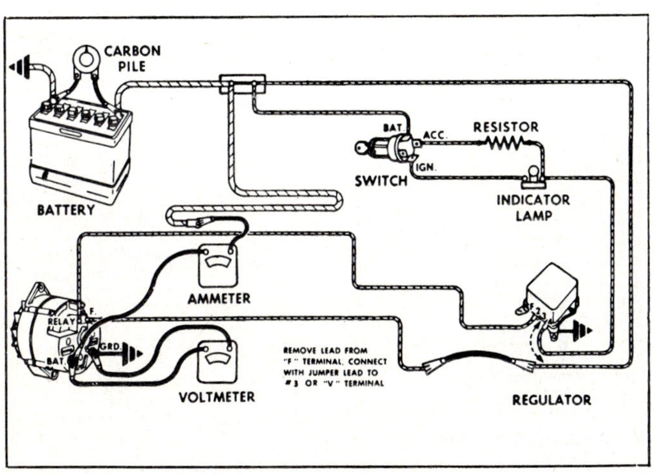

Fig. 5 Connections for testing alternator output with two-unit or transistorized regulator

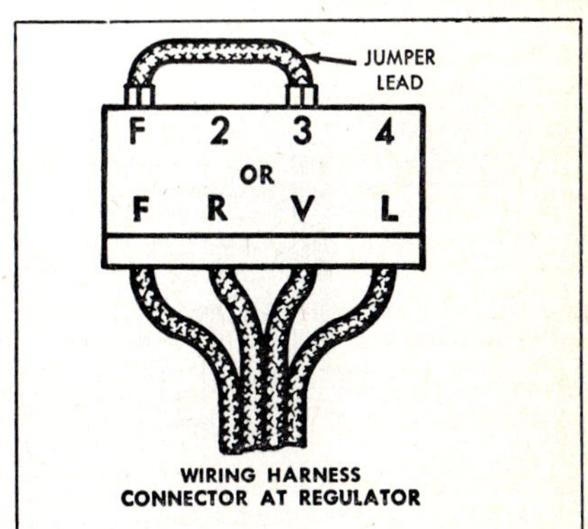

Fig. 6 Alternator output test on systems with two-unit or transistorized regulator. Jumper leads connected at regulator

5. If voltmeter indicates battery voltage, the relay is operating properly. Trouble most likely is in the alternator field circuit.
6. If voltmeter indicates zero, check for excessive resistance or an open in wiring or in ignition switch. If wiring and switch are satisfactory, replace field relay.

REGULATOR VOLTAGE TEST ON VEHICLE

The voltage at which the regulator operates varies with changes in ambient temperature. The ambient temperature is the temperature of air at a distance of 1/4" from the regulator. The full transistor regulator as used with 85 to 130 ampere systems is not noticeably affected by temperature changes. To check and adjust voltage setting, proceed as follows. *Do not remove regulator cover.*

Systems With Two-Unit Regulator

1. Refer to Fig. 11 which shows all test equipment connected into system. Fig. 12 shows use of special adapter and jumper lead at regulator.
2. Connect an ammeter and a 1/4 ohm resistor with a rating of 25 watts or more in series in the circuit at the "BAT" terminal on the alternator, Fig. 11. In the event the battery is discharged, the 1/4 ohm resistor will limit the alternator output to 10 amperes or less which is required when checking and adjusting the voltage setting.
3. Install special adapter as shown in Fig. 11. Use a 25 ohm 25 watt variable resistor in series with the alternator field windings at the regulator "F" terminal, and connect a jumper lead from the #3 adapter lead to the alternator "BAT" terminal as shown. Connect a voltmeter from #3 adapter lead to ground as shown. Turn the resistor to the closed or "no resistance" position.
4. Secure a thermometer close to regulator to establish operating temperature. With all accessories and lights off, start and operate engine for 15 minutes at 1500 engine rpm.
5. After warm-up, cycle alternator as follows:

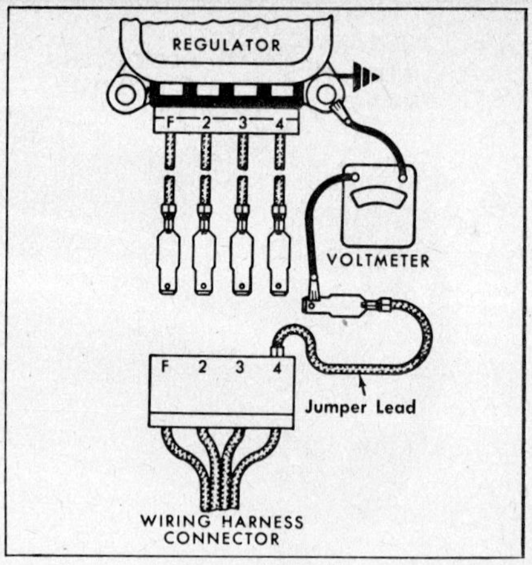

Fig. 7 Testing resistance wire on 1D Series

a. Turn variable resistor to "off" or "full resistance" position.
b. Disconnect lead at #4 terminal of harness connector momentarily, then reconnect lead.
c. Return variable resistor to the closed or "no resistance" position.
d. Bring engine speed up to approximately 2500 rpm and note voltage setting. Refer to voltage chart, Fig. 15, and specifications in Truck Chapters.

NOTE: The regulator unit should be operating on the upper or shorting contacts. If it will not operate on the upper contacts, the battery is in extreme state of discharge, and must be at least partially charged before proceeding with test.

6. To prevent accidental grounding and consequent damage to internal regulator parts when removing or installing regulator cover, perform the following steps in order listed.
   a. Disconnect #4 lead at harness connector.
   b. Disconnect jumper lead at alternator "BAT" terminal.

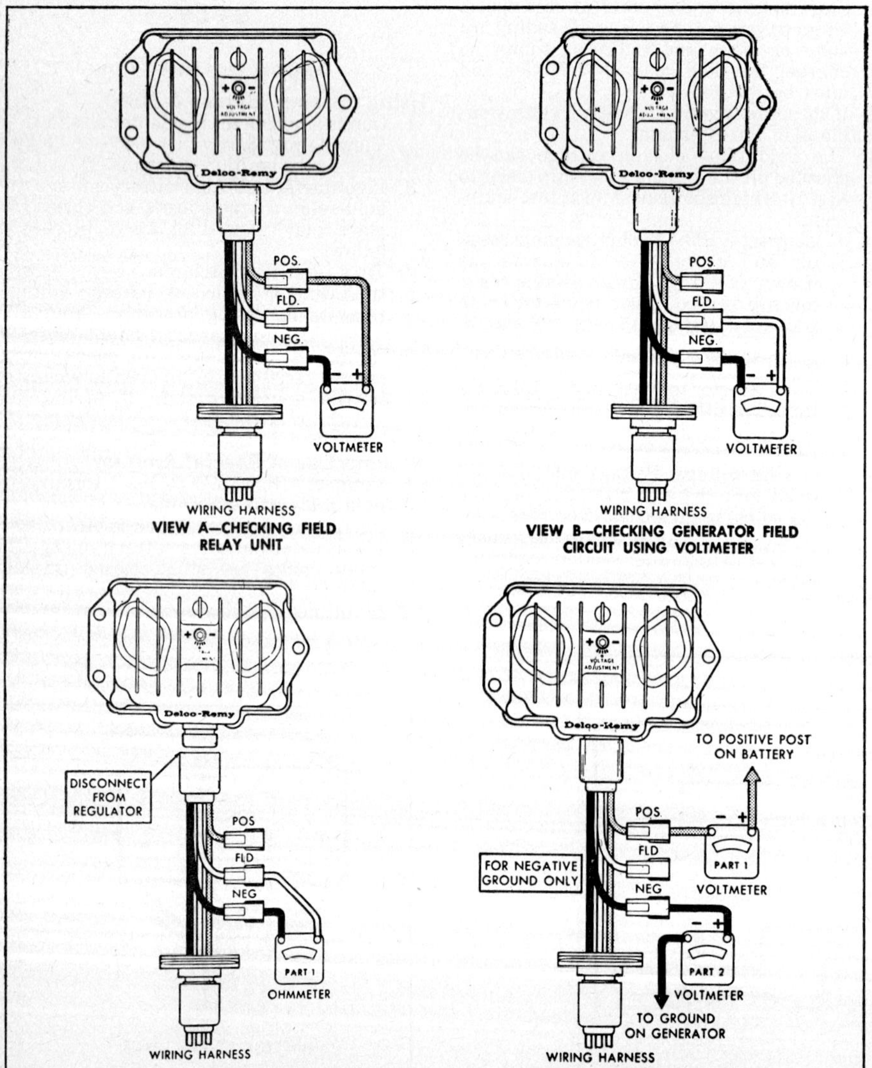

Fig. 8 Using adapter (J-21600) to check alternator output on Series 3D

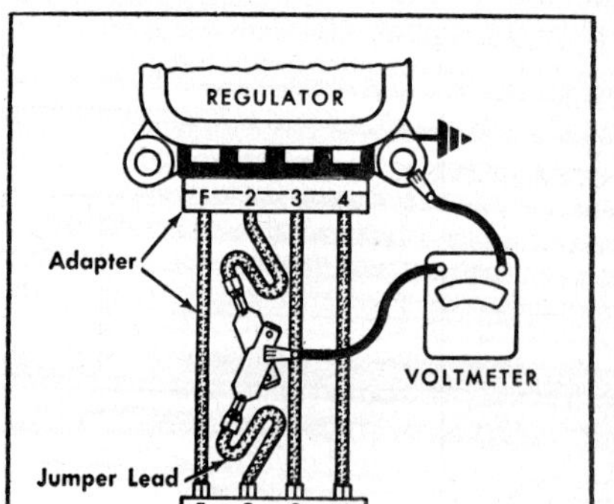

Fig. 9 Testing regulator field relay on two-unit or transistorized regulator

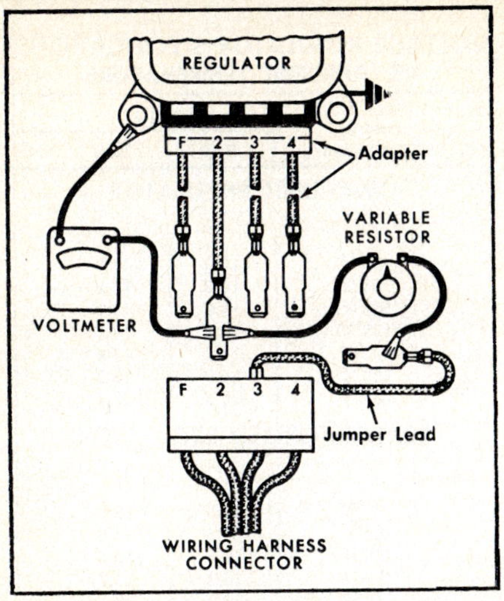

Fig. 10 Testing field relay closing voltage on two-unit or transistorized regulator

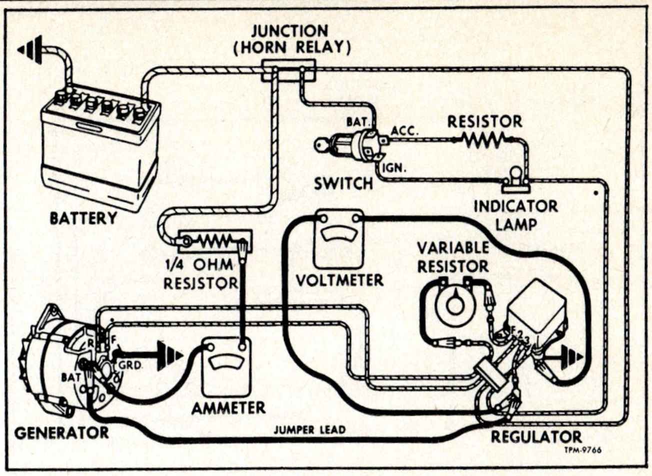

Fig. 11 Connections for testing closing voltage on two-unit regulators

c. Remove regulator cover.
d. Reconnect jumper lead to alternator "BAT" terminal.
e. Connect #4 lead to harness connector.

7. To adjust voltage setting, turn adjusting screw as shown in Fig. 13.

CAUTION: Always turn screw clockwise to make final setting to insure spring holder being against head of adjusting screw. If necessary, pry holder up against screw head before turning screw clockwise.

8. After making setting, cycle alternator again as outlined previously in step 5.
9. Operate engine at approximately 2500 rpm and note voltage setting. Readjust if necessary.
10. Check voltage setting while operating on lower set of contacts as follows:
    a. Slowly increase resistance of variable resistor with engine operating at 2500 rpm until regulator begins to operate on lower sets of contacts. Then note change in voltage reading. The upper set voltage should be 0.1 to 0.3 volts higher.

NOTE: The most desirable method for determining that the regulator is operating on the lower set of contacts when the cover is installed is to use earphones (if available) connected across the regulator "F" terminal to ground. As the variable resistor is turned, and operation changes from the upper set of contacts to the lower set, the earphone sound will fade away and stop completely and then return when the lower set of contacts begin to operate. The alternate method is visual observation, but this is less desirable because the cover must be removed which affects temperature stabilization.

    b. If turning the variable resistor does not cause the regulator to operate on the lower set of contacts, return the variable resistor to the "no resistance" position. Turn the carbon pile to slightly load the battery, and then adjust the variable resistor to cause the regulator to operate on the lower set of contacts. Usually, turning on the vehicle headlights can substitute for the carbon pile.
    c. The difference in voltage between the operation of the upper set of contacts and the lower set is increased by slightly increasing the air gap between the armature and center of core, and decreased by slightly decreasing the air gap, using the Nylon adjusting nut. This adjustment can be made while the regulator is operating. If necessary to make this air gap adjustment, recheck the voltage setting of both sets of contacts. Always make final voltage test after regulator cover is installed, referring to step 6 for precautions when removing and installing cover.

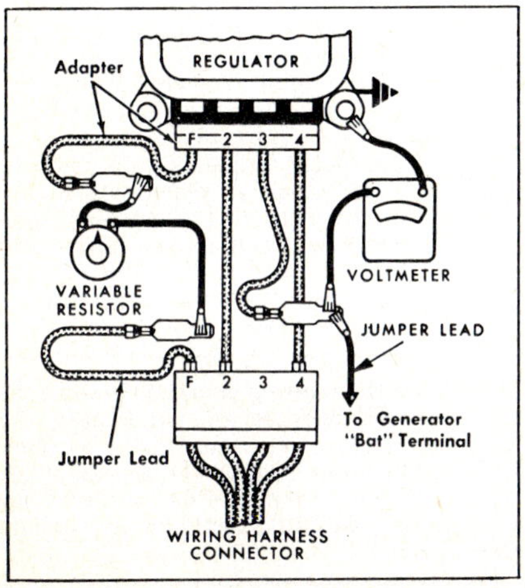

Fig. 12 Testing voltage setting on two-unit regulator

Systems with Transistorized Regulator

1. Refer to chart, Fig. 15. Then make connections as shown in Fig. 14.
2. With switch on, operate engine at 1275 rpm for 15 minutes. *Leave cover on regulator.*
3. After the warm-up, cycle the alternator by turning off the ignition switch, stopping the alternator and temporarily disconnecting the jumper lead from the alternator. Then reconnect the jumper lead, re-run engine at 1275 rpm and note voltage reading.
4. To adjust setting, remove cover and turn adjusting screw shown in Fig. 16.

CAUTION: To avoid grounds, turn off switch and disconnect jumper lead when replacing cover. Always make final setting by turning screw clockwise. After turning screw, pry holder up against screw head, then turn clockwise to make setting.

5. Recycle alternator as explained in step 3, then turn on switch, operate engine at 1275 rpm and note voltage setting. Readjust if necessary.
6. Always recycle alternator before reading final voltage setting on voltmeter.

NOTE: When decreasing voltage, the alternator output may decrease to below the 3 amperes minimum allowable, or even to no output. If this happens, load battery with carbon pile rheostat connected across its posts, or turn on electrical accessories in order to obtain 3 to 30 amperes output required when adjusting the voltage setting. The transistor and diodes can be checked as explained later under "Transistorized Voltage Regulator."

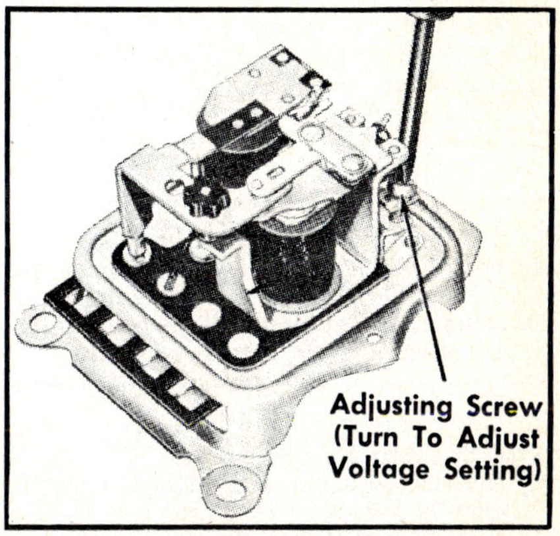

Fig. 13 Adjusting voltage setting on two-unit regulator

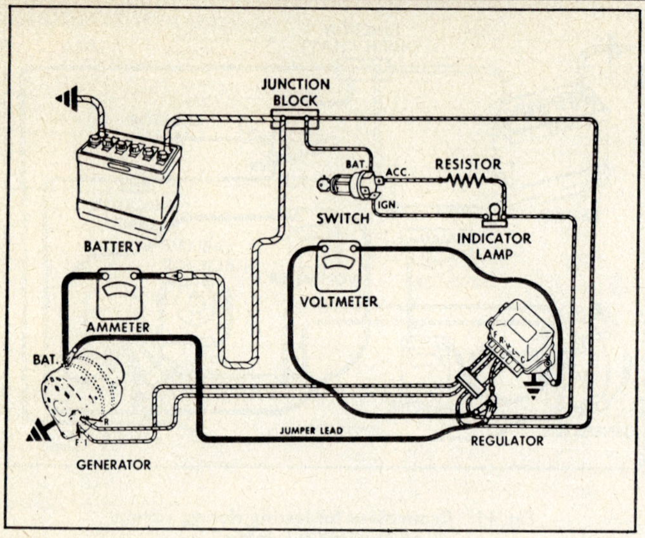

Fig. 14 Connections for testing voltage setting on transistorized regulator

VOLTAGE REGULATOR SPECIFICATIONS VS. AMBIENT TEMPERATURE		
Regulator Ambient Temperature	Voltage	
	High	Low
TWO-UNIT REGULATOR		
65°F.	15.0	13.9
85°F.	14.8	13.8
105°F.	14.6	13.7
125°F.	14.4	13.5
145°F.	14.2	13.4
165°F.	14.0	13.2
185°F.	13.9	13.1
TRANSISTORIZED REGULATOR		
65°F.	14.5	13.6
85°F.	14.5	13.4
105°F.	14.3	13.3
125°F.	14.2	13.2
145°F.	14	13
165°F.	13.8	12.8
FULL TRANSISTOR REGULATOR		
65°F.	14.9	14.1
85°F.	14.7	13.9
105°F.	14.5	13.7
125°F.	14.3	13.5
145°F.	14.2	13.4

Fig. 15 Voltage chart for various type regulators

Series 2D Systems with Full Transistor Regulator

1. Make connections as shown in View B, Part 1, Fig. 17, and record voltage drop.

CAUTION: Do not leave jumper lead connected longer than five minutes.

2. Make connections as shown in View B, Part 2, Fig. 17, and record voltage drop.
3. Add voltage in step 3A to voltage in step 3B. If total voltage is above 2.5 volts, check system wiring for high resistance.
4. If total voltage in step 3C is below .25 volt, make connections as shown in View C, Fig. 17. Then with switch on, operate engine at 1275 rpm for 15 minutes. *Leave cover on regulator.* Place a thermometer 1/4" from regulator cover and compare voltage given in chart, Fig. 15. If voltage is not within specified range, replace regulator. If voltage is within specified range, remove plug and turn slotted adjusting button inside regulator, Fig. 18.
5. For an undercharged battery, raise voltage setting by turning one notch (clockwise). Then check for an improved battery condition after a service period of reasonable length. After two notches in each direction there is a positive stop.
6. For overcharged battery, lower voltage setting by turning one notch (counterclockwise). Then check for an improved battery condition after a service period of reasonable length.
7. If the regulator cannot be adjusted to a value within the specified range, replace the regulator.

NOTE: If repeated regulator failures are experienced on the vehicle, but no defects are found, a shorted, grounded or open alternator field winding, or grounded leads of an intermittent nature should be suspected.

85 to 130 Amp. Systems

1. Connect voltmeter to adapter as shown in View B, Fig. 8.
2. Turn all accessories off.
3. Operate alternator at approximately 3000 rpm (1200 engine rpm).
4. The alternator output should be at least 10 amperes below the rated output for this check. For example, if rated output is 85 amps the voltage setting output should be 75 amps or less.
5. To adjust voltage setting, see Fig. 18. If regulator cannot be adjusted to a value within the specified range, replace or repair the regulator.

NOTE: If repeated regulator failures are experienced on the vehicle, but no defects are found, a shorted, grounded or open alternator field winding, or grounded leads of an intermittent nature should be suspected.

TAILORING VOLTAGE SETTING

On Vehicle

Proper setting is obtained when the battery remains fully charged with a minimum use of water. If no circuit defects are found, yet the battery remains undercharged, raise the setting by 0.3 volt, then check the battery over a period of time to see if improvements were achieved. Reset regulator if necessary. If the system is overcharging, lower the setting by 0.3 volt, then check the battery over a period of time.

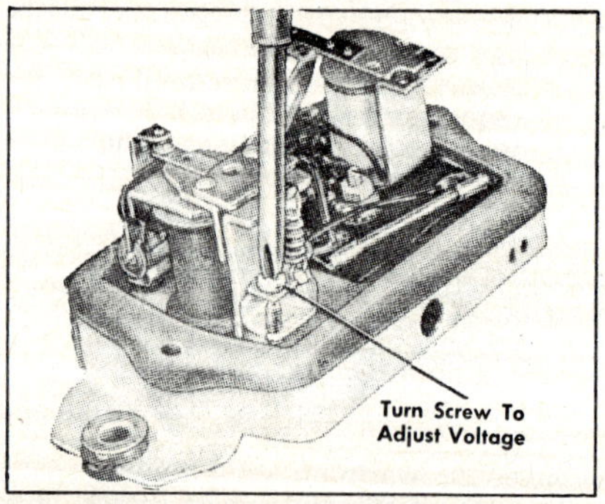

Fig. 16 Adjusting voltage setting on transistorized regulator

ALTERNATOR FIELD CIRCUIT CHECK

Series 3D & 4D Systems

1. Connect a voltmeter to adapter as shown in View B, Fig. 8.
2. Turn on ignition switch.
3. If voltmeter indicates 1 or 2 volts less than battery voltage, proceed to check alternator output explained previously.
4. If voltmeter indicates zero volts, the regulator is defective and must be replaced.

CAUTION: The regulator defect may have been caused by a defective alternator field. Check the field as follows before installing the new regulator.

5. Turn off ignition switch and disconnect battery ground strap.
6. Disconnect adapter from regulator.
7. Connect an ohmmeter to the adapter as shown in View C, Fig. 8.
8. If the ohmmeter indicates high, there is an open or excessive resistance in the field winding or in wiring between regulator positive terminal and alternator "F1" terminal (Lead E, Fig. 19).
9. If ohmmeter indicates low the winding is shorted or grounded.

NOTE: Since the reading is taken through the adapter, leads, brushes and slip rings, the ohmmeter reading on a good field winding will be slightly higher than the specified value. This is because the specified values is for an ohmmeter reading directly across the slip rings.

10. Disconnect ohmmeter and reconnect battery ground strap.

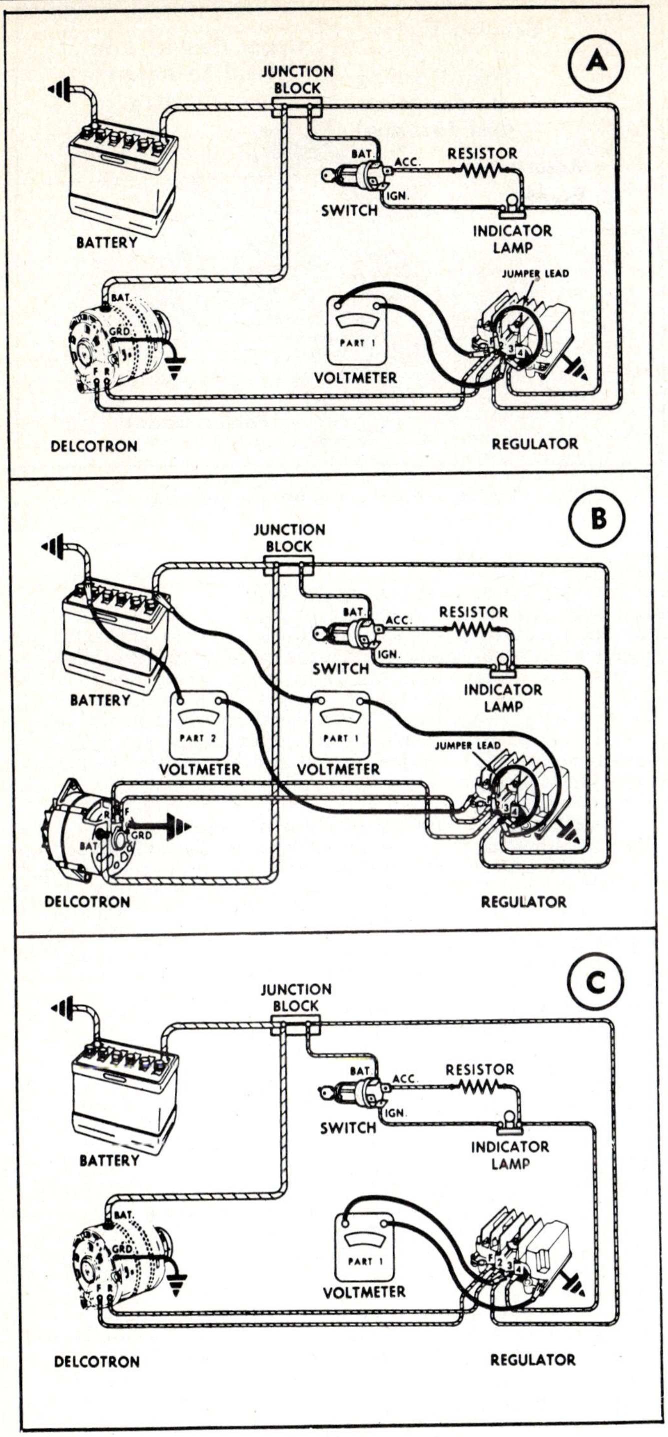

Fig. 17 Connections for testing regulator voltage setting on vehicle with full transistor regulator

11. If the voltmeter indicates battery voltage, the regulator is shorted and must be replaced, or the alternator field winding is open or grounded. Check the field winding as outlined above.
12. To check the regulator, connect a voltmeter as shown in View A, Fig. 8, and operate engine at moderate speed. If the voltage is uncontrolled and increases with speed to values above the specified setting range, replace or repair the regulator.

EXCESSIVE RESISTANCE TEST

Series 3D & 4D Systems

Excessive resistance in the sensing circuit, consisting of the wiring and voltage regulator, can cause an overcharged battery. If the trouble is not battery overcharge, proceed with the "Regulator Voltage Test" explained previously. Otherwise, proceed as follows:

1. Connect a voltmeter as shown in Part 1 and Part 2 in view D, Fig. 8.
2. Turn on ignition switch but do not start engine.
3. If the two voltmeter readings total more than 0.3 volt, check for excessive resistance in leads B, C and D, Fig. 19, which can cause an overcharged battery. If these leads are satisfactory, the field relay contacts may have excessive resistance. In this event, replace field relay unit.

VOLTAGE REGULATOR ADJUSTMENTS

CAUTION: Before removing a voltage regula-

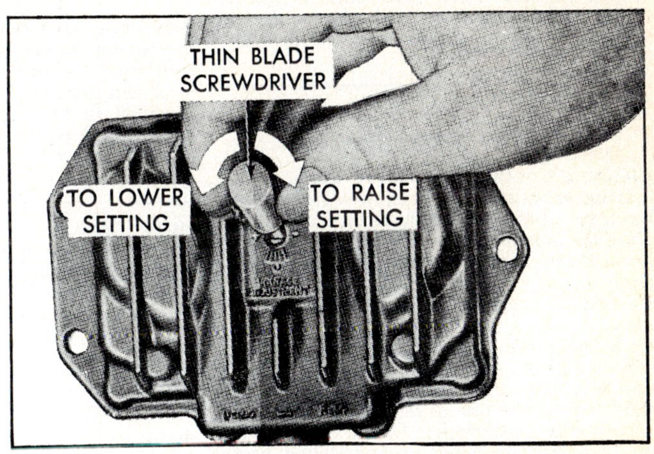

Fig. 18 Adjusting voltage setting on full transistor regulator

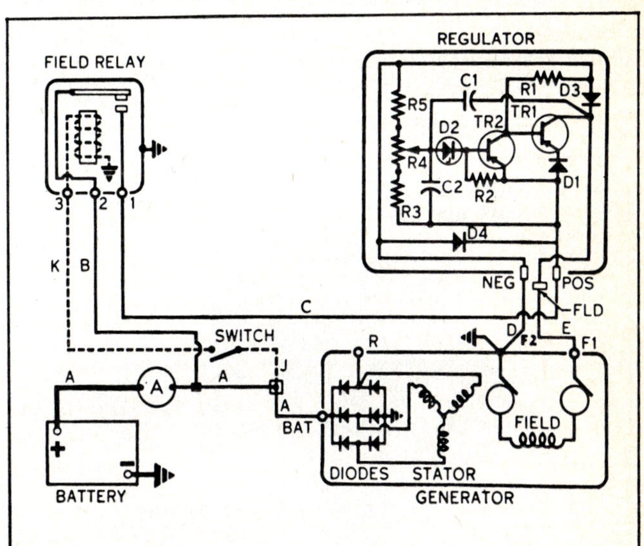

Fig. 19 Schematic diagram (typical) of Series 3D & 4D

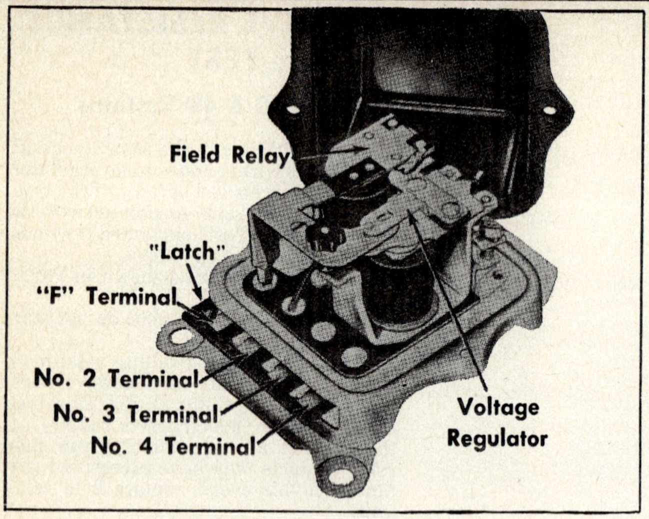

Fig. 20 Two-unit voltage regulator

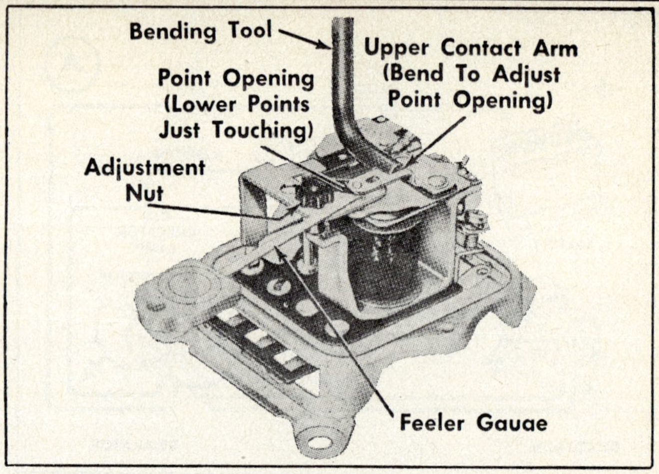

Fig. 21 Adjusting contact point opening

tor from a vehicle, be sure to disconnect one battery cable and leave it disconnected until after the unit is installed on the vehicle.

NOTE: In making the following adjustments, refer to the "Alternator Specifications" charts in the vehicle chapters for the proper feeler gauges to use to obtain the specified settings.

Two-Unit Regulators

Cleaning Contact Points, Fig. 20

If the points are dirty or oxidized, they must be cleaned before any adjustments are made since the cleaning may change the mechanical settings. The upper contacts may develop slight cavities. These surfaces should be cleaned to the bare metal, using #400 silicone carbide paper (or equivalent) folded over, then pulled back and forth between contacts.

It is not necessary to remove the cavities entirely. The lower contacts are of softer material; use a piece of cloth with solvent to remove any discoloration. *Do not use abrasive on these points.* After contacts have been cleaned, they should be washed with a solvent (non-toxic type) to remove any foreign material.

Adjusting Contact Point Opening

With the lower contacts touching, measure the opening between the upper contacts as shown in Fig. 21. Point opening should be as specified in the "Alternator Specifications" chart in the vehicle chapters. Adjust by bending the upper contact arm as shown, being careful not to bend the hinge.

Adjusting Air Gap

Measure the air gap between armature and core when the lower contacts are just touching as shown in Fig. 22. If the clearance is not as specified in the "Alternator Specifications" chart in vehicle chapters, turn the adjusting nut located on the contact support as required.

NOTE: Only an approximate air gap setting should be made by the feeler gauge method. The final setting must be whatever is required to obtain the specified difference in voltage between the upper and lower contacts. Instructions for making the final setting are explained previously under "Regulator Voltage Test On Vehicle."

Field Relay Unit

Cleaning Contact Points

If points appear to be pitted or burned, clean them with a thin cut file. Remove only enough material to clean points. Never use emergy cloth or sandpaper for this operation.

Adjusting Air Gap

Referring to Fig. 23, measure clearance between armature and core and exert just enough pressure on the armature to allow it to touch the feeler gauge. The gauge should be the thickness of the specified clearance in "Alternator Specifications" chart in the vehicle chapters. The contact points should just close at this time. Adjust by bending the flat contact support spring.

Adjusting Contact Point Gap

Referring to Fig. 24, measure gap between the contact points with the armature in its normal at rest position. If the gap is not according to the clearance listed in the "Alternator Specifications" chart in the vehicle chapter, adjust be bending the relay heel iron.

Adjusting Closing Voltage

To adjust the closing voltage, connect a 100–150-ohm variable resistor with an "OFF" position between the positive post of a 12-volt battery and the No. 2 terminal of the regulator.

Connect a jumper lead from the regulator base to the battery negative cable as a ground. Connect a voltmeter between the regulator No. 2 terminal and regulator base. Slowly decrease the resistance, noting the voltage at which the relay closes. Adjust the voltage to specifications by bending the armature heel

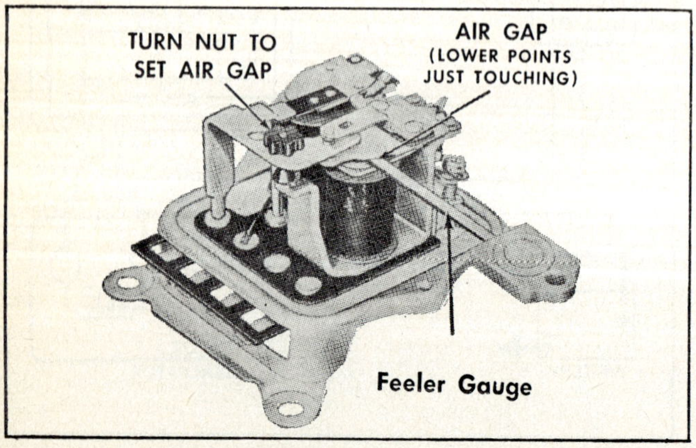

Fig. 22 Adjusting regulator air gap

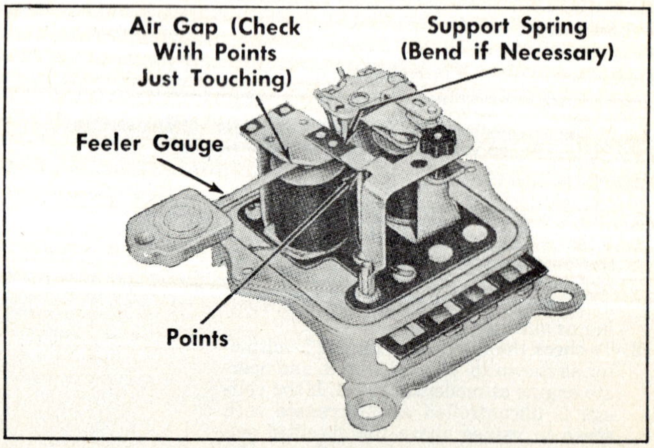

Fig. 23 Adjusting field relay air gap

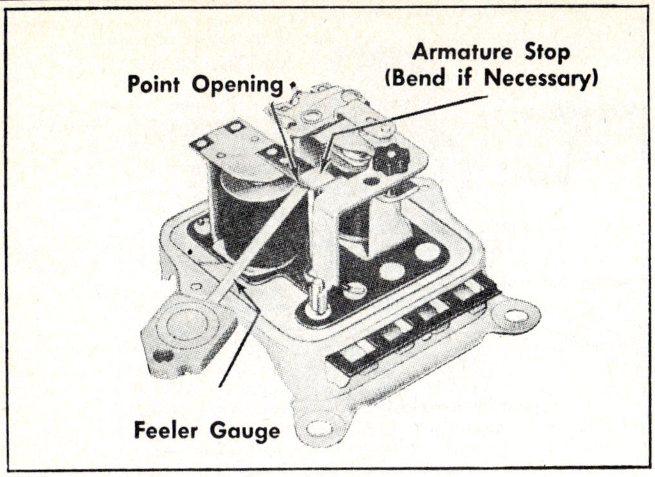

Fig. 24 Adjusting field relay contact points

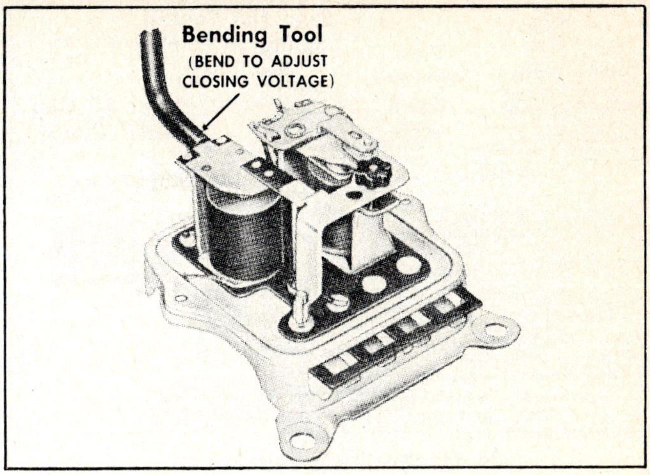

Fig. 25 Adjusting field relay closing voltage

iron shown in Fig. 25 (see "Alternator Specifications" in vehicle chapters).

Transistorized Regulator

When removing or installing the regulator, be sure to have one of the battery cables disconnected. After removing the regulator, remove the cover and inspect the various components, connections, and contact points for signs of damage, Fig. 26. When making adjustments, see the "Alternator Specifications" charts in the vehicle chapters.

Cleaning Contact Points

If the points are dirty or oxidized, they must be cleaned before any adjustments are made since the cleaning may change the mechanical settings. The contact surfaces may oxidize and develop a slight cavity. These surfaces should be cleaned to the bare metal, using a riffle file as shown in Fig. 27. It may be necessary to remove the point support as shown to provide file clearance. Do not file points excessively.

Adjusting Air Gap

Measure the gap between the armature and core, Fig. 28. Push down on the armature (not the flat spring that contains the movable point). The points should just touch when the armature bottoms on the proper feeler gauge. If the air gap is not as specified, loosen the contact bracket mounting screws and adjust the bracket, up or down as required. Make sure points are squarely aligned, then tighten screws and recheck air gap.

The regulator limiting voltage is adjusted after the unit is installed on the vehicle. See instructions for making final setting under "Voltage Regulator Check."

Field Relay

Adjusting Air Gap

Measure the clearance with a feeler gauge of the specified thickness between armature and core. Press down on the armature; the contact points should just touch when the armature bottoms. If the clearance is not as specified, adjust by carefully bending the flat contact support spring, Fig. 29.

Adjusting Point Opening

The point opening is measured in the open or at rest position. The opening should be as

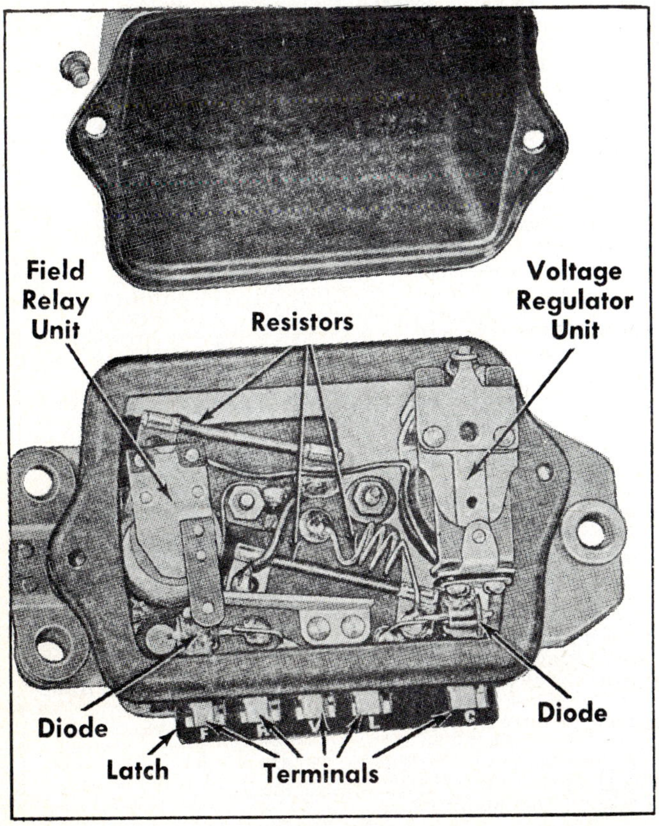

Fig. 26 Transistorized voltage regulator with cover removed

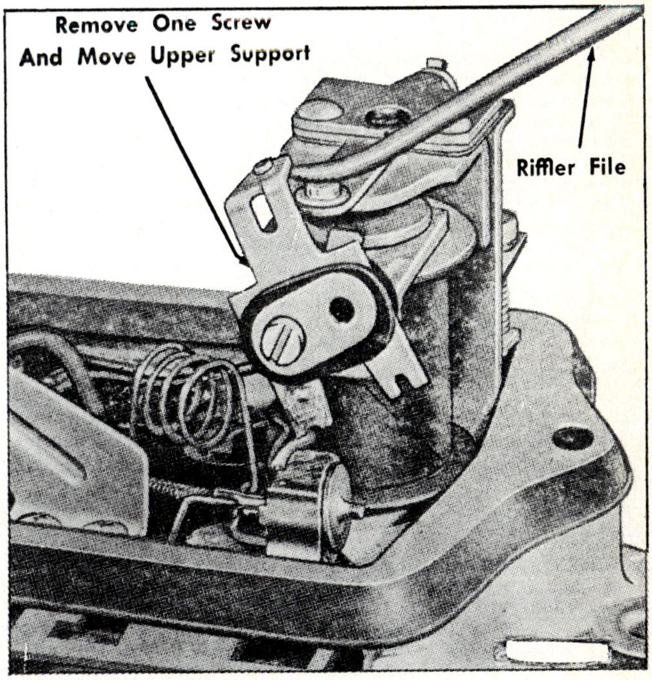

Fig. 27 Cleaning regulator contact points

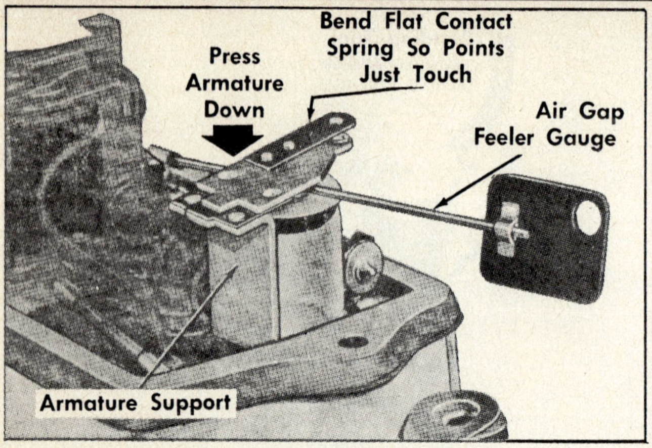

Fig. 28 Adjusting regulator air gap

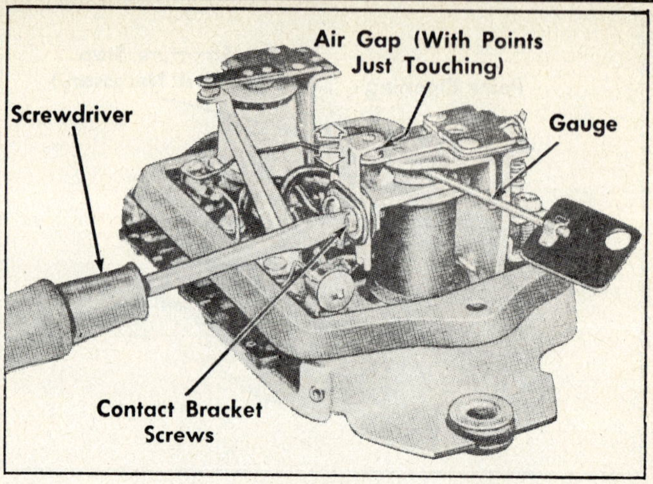

Fig. 29 Checking field relay air gap

specified and may be adjusted by bending the heel iron in the same manner as when adjusting the closing voltage, Fig. 30. This setting is in general a reference point and is not critical. If the air gap and closing voltage are correct, the relay will operate satisfactorily even though the point opening may not be exactly according to specifications.

Adjusting Closing Voltage

Connect a 100–150 ohm variable resistor with an "OFF" position between the positive post of a 12-volt battery and the "R" terminal of the regulator. Connect a jumper lead from the regulator base to the battery negative post as a ground. Connect a voltmeter to the "R" terminal and the regulator base, then slowly decrease the resistance, noting the voltage at which the relay closes. Adjust to the specified voltage by bending the heel iron, Fig. 30.

Transistor and Diode Check

Referring to Fig. 31, the transistor can be checked by connecting a jumper lead between the regulator "V" terminal and the positive post of a 12-volt battery. Use another jumper lead to ground the regulator frame to the negative battery post.

Connect a voltmeter between the "F" terminal and ground, manually close the field relay, and check the meter reading. Meter should indicate 8 or more volts. A reading under 8 volts indicates an open transistor that must be replaced.

With the field relay still held closed, manually open the voltage regulator and note the voltage. If the voltage is more than 2 volts, the transistor is shorted and must be replaced.

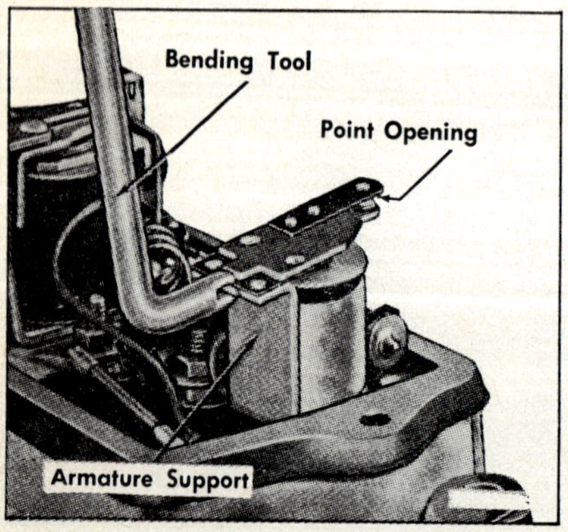

Fig. 30 Adjusting field relay closing voltage

Disconnect the test leads.

If the transistor is open or shorted, replace the unit by removing the two retaining nuts and screws, then unsolder the connections (upper view, Fig. 31). When installing the new transistor, use rosin core solder (60% tin, 40% lead). Limit the soldering time to avoid damage to the transistor.

If the voltage setting was found to be high or low during the diagnosis, or a failed transistor was found, check the two regulator diodes, Fig. 32. It is necessary to unsolder the diode leads before testing. Using the test lamp as shown, check the current flow through the field diode, then reverse test lamp leads. Test lamp should light in one direction only. If lamp lights in both directions, the diode is defective and must be replaced. Check the suppression diode in the same manner.

To replace a diode, melt the solder from around its mounting and at lead connection. When installing the new diode, limit soldering time to a minimum as excessive heat may damage the diode.

NOTE: After performing the foregoing operations, install the unit in the vehicle and adjust the voltage setting as outlined previously under "Voltage Regulator Check."

Full Transistor Regulator

Four-Terminal Type, Fig. 33

This type regulator is composed of transistors, diodes, capacitors, and resistors. These components form a completely static electrical unit containing no moving parts. A combination field and light relay is also part of the entire assembly.

The function of the regulator in the charging circuit is to limit alternator voltage to a pre-set value by controlling the alternator field current.

Regulator Repairs

With the regulator removed from the vehicle, remove the cover and inspect the various components, connections, and field relay contact points for signs of damage.

If the field relay contact points are dirty or oxidized, they should be cleaned. The contact surfaces may oxidize and develop a slight cavity. These surfaces should be cleaned to the bare metal, using a riffler file. Do not file excessively.

If the voltage regulator cannot be adjusted to specifications listed in the "Alternator Specifications" charts in the vehicle chapters, replace the regulator assembly.

Three-Terminal Plug Type, Fig. 34

Like the four-terminal type, the regulator is an assembly composed principally of transistors, diodes, capacitors, and resistors. These components form a completely static electrical unit containing no moving parts.

A separate field relay is included in the control system to disconnect the regulator feed (positive) terminal from the battery circuit

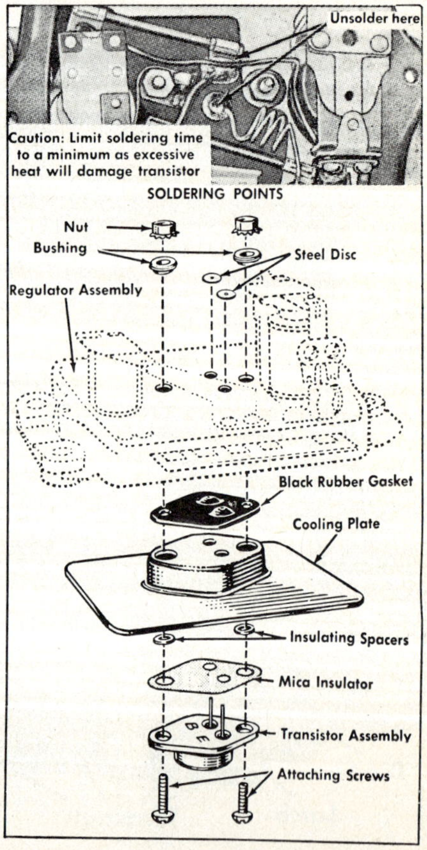

Fig. 31 Transistorized regulator installation views

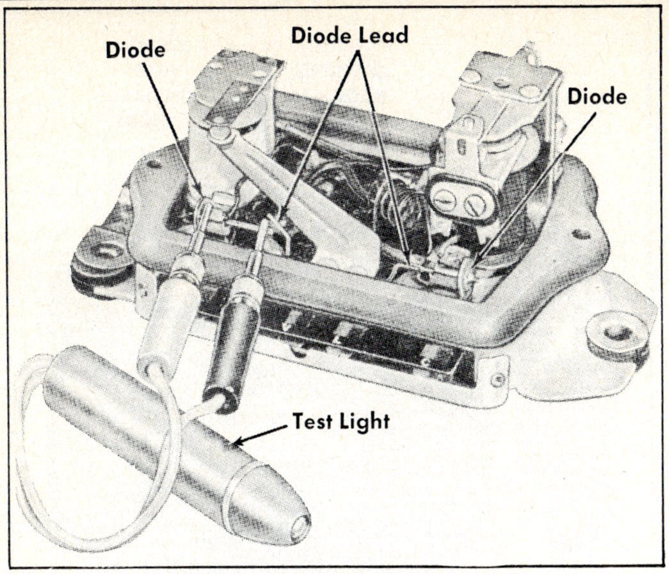

Fig. 32 Checking diodes with test lamp

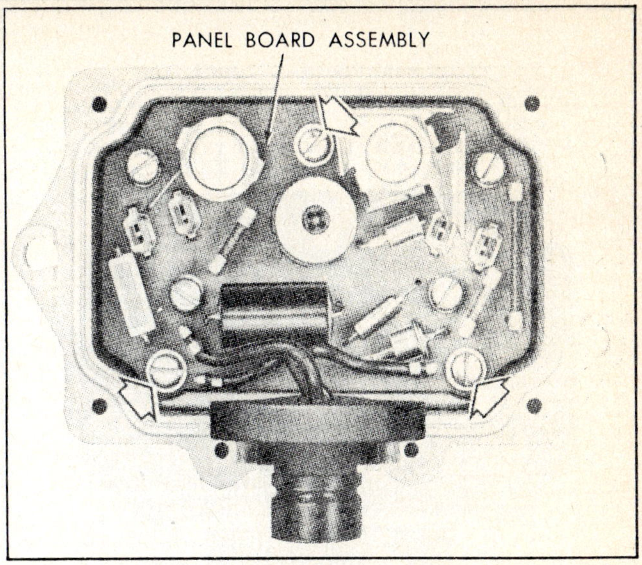

Fig. 35 Regulator with bottom plate removed

when the ignition switch is in the "OFF" position. Supression of radio interference is accomplished by a capacitor on the alternator.

CAUTION: The field wire terminal must not be grounded or flashed when the regulator is connected into the circuit as instant damage to the transistors will occur.

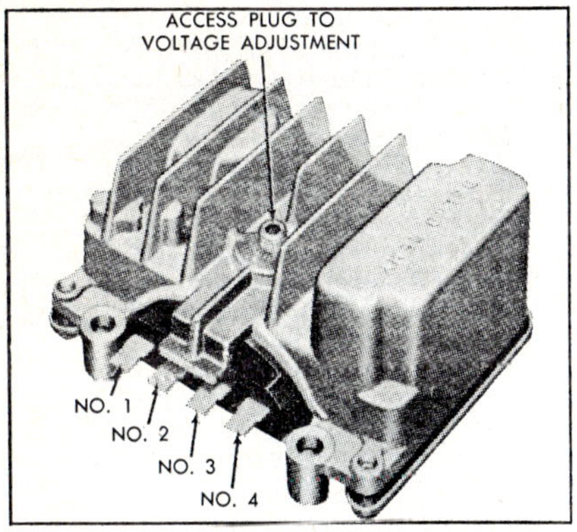

Fig. 33 Full transistor type voltage regulator with four terminals

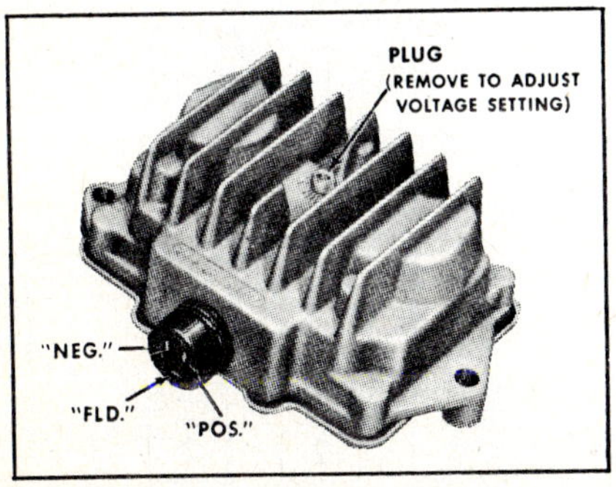

Fig. 34 Full transistor type voltage regulator with three-terminal plug

Analyzing System Troubles

If the trouble is located in the alternator during test procedures outlined previously, refer to the applicable alternator section for corrective procedures.

When analyzing the system, make sure all the connections between the battery, junction block, and alternator are clean and tight. Then remove the wiring harness connector from the regulator and connect adapter (J-21600) between wiring harness connector and the regulator.

Regulator Repairs

1. Remove bottom plate from regulator. Remove three panel board attaching screws identified by arrows, Fig. 35, and lift assembly from housing.
2. To aid in reassembly, note or make any identifying marks on the two transistors and their respective locations on the panel board and heat sink assembly, Fig. 36.
3. Note the insulators between the transistors and the heat sink, and the insulators separating the heat sink from the panel board, Fig. 36.
4. Remove the transistor attaching screws, and separate transistors and heat sink from panel board.

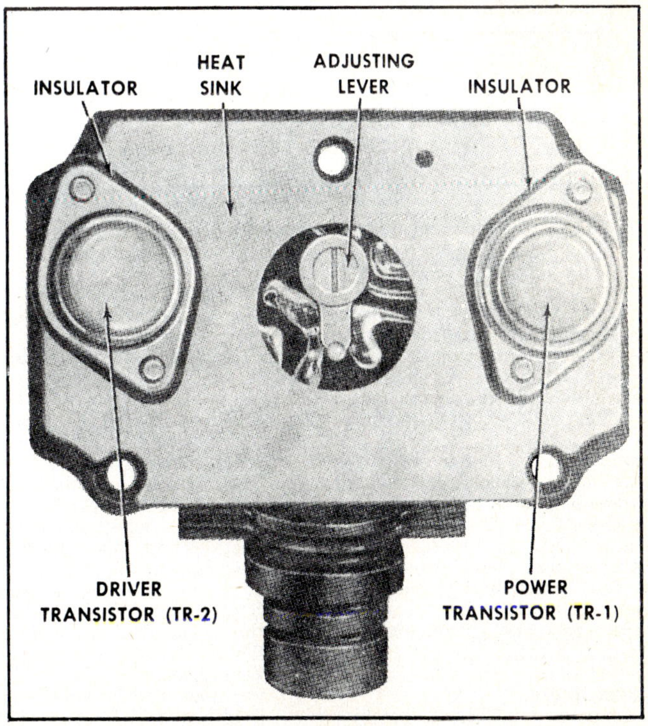

Fig. 36 Top side of regulator panel board

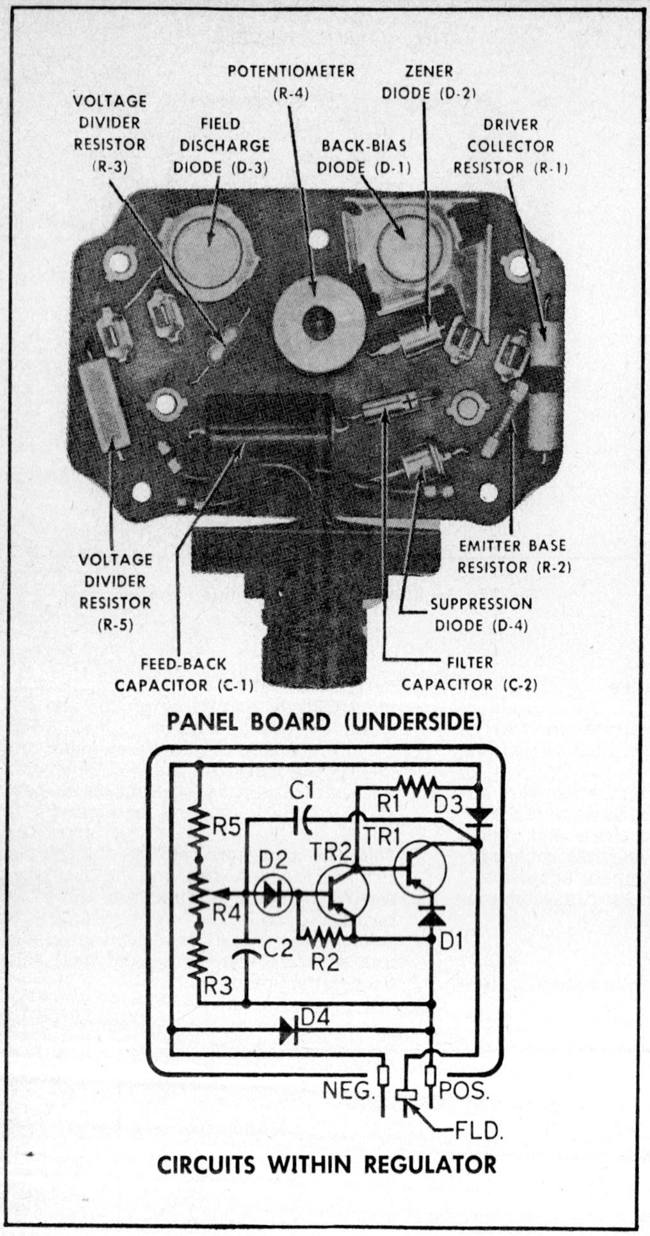

Fig. 37 Underside of regulator panel board

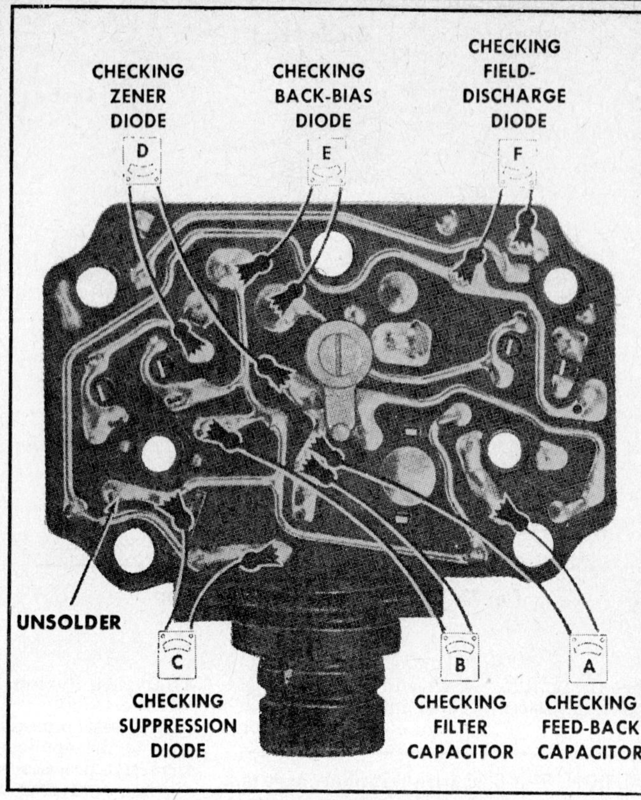

Fig. 38 Checking regulator circuits

5. With the transistors separated from the assembly, an ohmmeter may be used to check the transistors and components on the panel board for defects. An ohmmeter having a 1½ volt cell is recommended. The low range scale on the ohmmeter should be read.
6. If a component part on the panel board is found to be faulty, it should be replaced before proceeding with the remaining checks. A 25-watt soldering gun is recommended, and a 60% tin 40% lead solder should be used when re-soldering. Avoid excessive heat which may damage the panel board. Chip away any epoxy involved, and apply new epoxy, which is commercially available.

Checking Panel Board

In order to check the panel board assembly, it is necessary to unsolder the emitter-base resistor at the location shown in Fig. 38. In all the following checks, connect the ohmmeter as shown and then reverse the ohmmeter leads to obtain two readings. Refer to Figs. 36, 37, 38.

1. Feed-Back Capacitor (C1), Part A, Fig. 38: If both readings are zero, the capacitor is defective. Visually inspect for open soldered connections and broken leads.
2. Filter Capacitor (C2), Part B, Fig. 38: If both readings are zero, the capacitor is defective. Inspect for open soldered connections and broken leads. To assemble a new capacitor properly, note the location of the "+" identifying mark in Fig. 37.
3. Suppression Diode (C4), Part C, Fig. 38: If the two readings are identical the diode is faulty.
4. Zener Diode (D2), Part D, Fig. 38: Replace the diode if both readings are zero, if both readings are infinite, or if both readings are identical.
5. Back-Bias Diode (D1) Part E, Fig. 38: Replace diode if both readings are zero, if both readings are infinite or if both readings are identical.
6. Field Discharge Diode (D3), Part F, Fig. 38: Replace the diode if both readings are zero, if both readings are infinite, or if both readings are identical.
7. Driver-Collector Resistor, Part A, Fig. 39: If both readings are infinite the resistor is open.
8. Voltage-Divider Resistor (R3), Part B, Fig. 39: If one reading is infinite or nearly so, or if both readings are infinite or nearly so the resistor is open.
9. Voltage-Divider Resistor (R5), Part C, Fig. 39: If one reading is infinite or nearly so, or if both readings are infinite or nearly so, the resistor is open.
10. Potentiometer, Parts D and E, Fig. 39: If

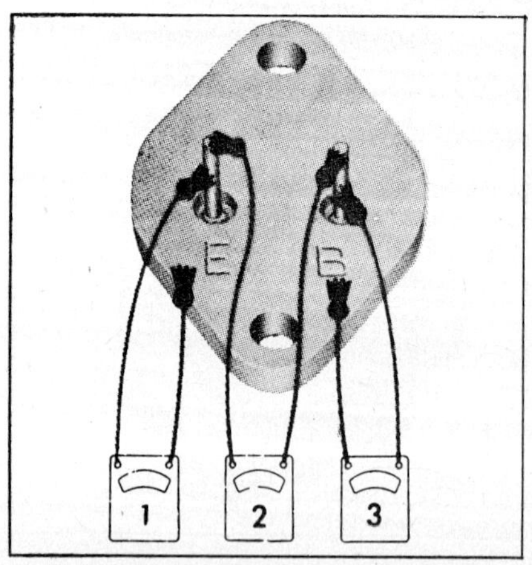

Fig. 40 Checking transistor for shorts

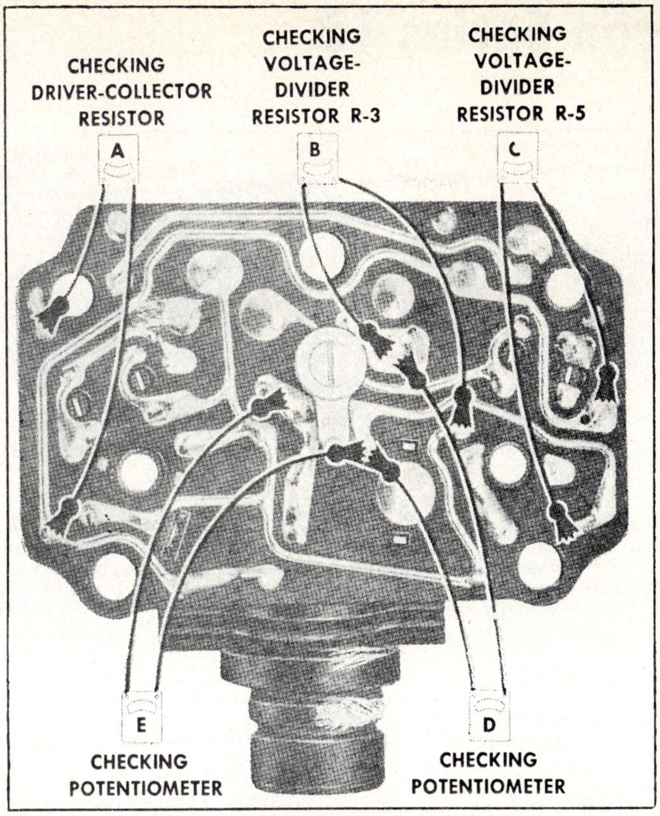

Fig. 39 Checking regulator circuits

one reading is infinite or nearly so in Part D, the potentiometer is open. If both readings are infinite in Part E, the potentiometer is open.

NOTE: When installing a new potentiometer, locate the adjusting lever in a vertical position, Fig. 36. Turn the potentiometer resistance adjustment to the middle position, then use a soldering iron to melt the adjusting lever into the potentiometer.

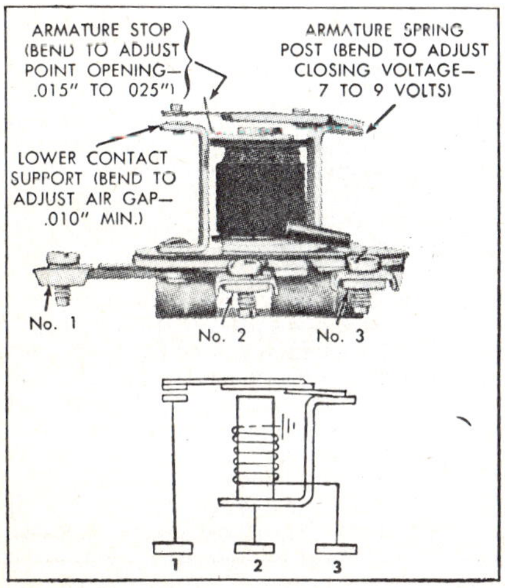

Fig. 42 Separate field relay unit

11. Emitter-Base Resistor: Since this resistor has been unsoldered from the panel board at one end, merely connect an ohmmeter across the resistor; an infinite reading indicates an open, in which case replace the resistor.
12. Driver and Power Transistors, Fig. 40: If both readings in Step 1 are zero or if both readings are very low or indentical, the transistor is shorted. Similarly, if both readings in Step 2 or in Step 3 are zero or very low and identical, the transistor is shorted.
13. Driver and Power Transistors (TR1 and 2), Fig. 41: If both readings in Step 1 are infinite or if both readings are very high and identical, the transistor is open. Similarly, if both readings in Step 2 are infinite or very high and identical, the transistor is open.

Reassembly and Final Check

During assembly, coat with silicone grease both sides of the flat insulators used between the transistors and heat sink, and also the heat sink on the side on which the transistors are mounted. The silicone grease increases heat condition.

Field Relay Unit

When used, the separate field relay unit completes the battery circuit to the voltage regulator, which in turn energizes the alternator field windings when the ignition switch is on. The procedure for checking the field relay on the vehicle is explained previously under "Field Relay Unit Test."

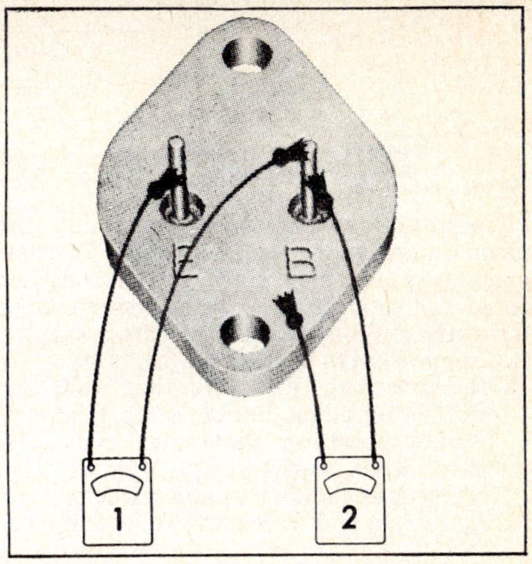

Fig. 41 Checking transistor for open

Relay Adjustments

Remove cover from relay, Fig. 42. Before attempting adjustment of the relay, make sure points are clean. Clean contacts with a thin, fine-cut contact file if pitted or burned.

Air Gap, Fig. 42

Disconnect wire from No. 2 terminal on relay. With contact points just touching, measure air gap (.010" minimum) between armature and center of coil. If necessary, adjust air gap by bending the lower point support.

Point Opening, Fig. 42

Measure point opening with armature up against stop. Correct dimension is .015 to .025". If necessary, adjust by bending armature stop.

Closing Voltage, Fig. 43

Connect an accurate reading voltmeter parallel with the relay operating circuit (No. 3 terminal to ground). Connect a variable resistance unit in series with the operating circuit (at No. 3 terminal). Turn ignition switch on. Adjust variable resistance until relay points close and note voltmeter reading. Points should close between 7 and 9 volts. Adjustment is made by bending armature spring post to increase or decrease spring tension on armature, Fig. 42.

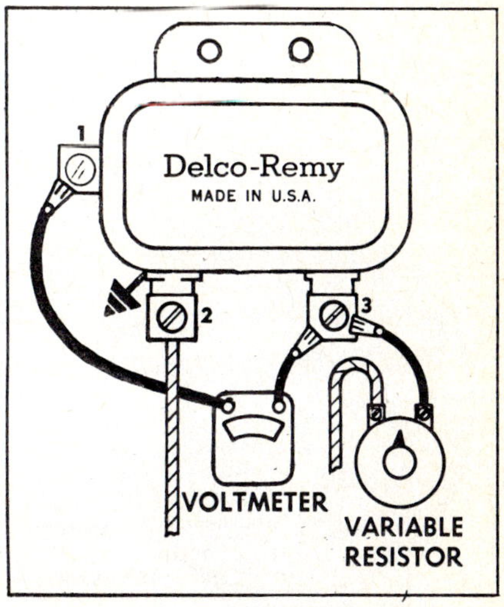

Fig. 43 Checking relay operation

Delcotron Series 1D

DISASSEMBLY

The alternator may be disassembled into major components shown in Fig. 1. After bench tests have been made, the slip ring end frame components may be disassembled if necessary to replace defective parts, Fig. 2.

1. Remove drive pulley and fan, Fig. 3.
2. Remove four thru-bolts that hold end frames together. Scribe a mark on end frames and stator frame to aid in locating parts when reassembling. Pry between stator frame and drive end frame to separate drive end frame and rotor from stator.

NOTE: Separation must be made between drive end frame and stator frame due to stator windings being attached to slip ring end frame assembly. Seal opening at end frame bearing with pressure-sensitive tape to prevent entry of dirt and other foreign material. Also tape bearing contact surface on end of rotor shaft to protect shaft from accidental damage. Do not use friction tape which could leave a gummy deposit on shaft.

3. If any lubricant from front end of rotor shaft has come in contact with brushes, wipe brushes clean. A nontoxic solvent may be used.
4. Remove three stator leads from terminals, Fig. 4. Remove stator from slip ring and end frame assembly.
5. Remove drive end frame and bearing assembly from rotor shaft. Note location of spacers at either end of bearing. There is one long spacer and one short spacer.

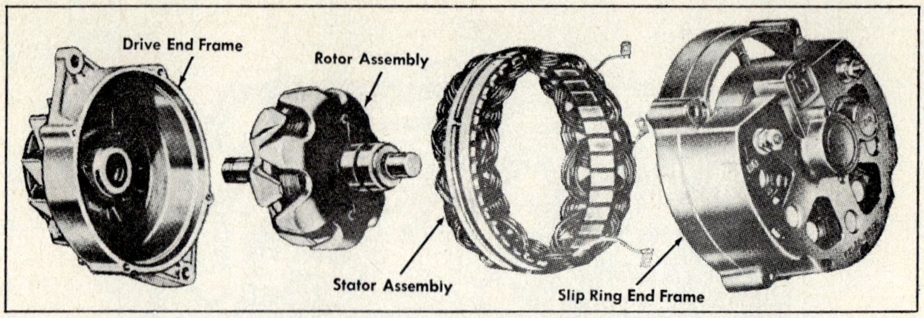

Fig. 1 Delcotron Series 1D alternator major components disassembled

ELECTRICAL TESTS

Rotor Checks, Fig. 5

Checking for Ground with Test Lamp
Touch one test lamp lead to one of the slip rings and other lead to rotor pole or shaft. If lamp lights, windings are grounded and rotor assembly must be replaced.

Checking for Ground with Ohmmeter
Make test in same manner as with test lamp and observe reading. If reading is low indicating low resistance, the winding is grounded and rotor assembly must be replaced.

Checking for Open Circuit
Touch leads from lamp or ohmmeter to each of the slip rings. If lamp fails to light or if ohmmeter reading is high (infinite), the winding is open.

Checking for Shorted Rotor Winding
Either of two methods may be used to check rotor winding for shorts, 1) ammeter and battery method and 2) ohmmeter method.

1. Connect a 12-volt battery and ammeter in series with the two slip rings. Note ammeter reading. If reading is over 2.6 amperes, there is a short circuit in winding.
2. Connect ohmmeter leads to the two slip rings and note resistance reading. If resistance is less than 4.6 ohms, there is a short circuit in rotor winding.

NOTE: If an ohmmeter is not available, the resistance can be calculated by dividing the voltage by the current (amperes), using values from step 1 above.

If there are no defects found in the rotor assembly and alternator does not produce rated output when checked as outlined in "Alternator Output Test," trouble is in either the stator assembly or in the rectifying diodes.

Stator Checks, Fig. 6

1. Connect one test lamp lead or ohmmeter to any one of the three stator leads, and touch the other lead to the stator frame. If lamp lights or if ohmmeter reading is low, stator windings are grounded.
2. Connect test lamp leads or ohmmeter leads successively between pairs of stator winding leads. If lamp fails to light or if ohmmeter reading is high, an open circuit in stator windings is indicated.

NOTE: A short circuit in stator windings is difficult to locate without laboratory test equipment due to low resistance of the windings. However, if alternator output is below specified rating and none of the other tests show defects in other units, shorted stator windings are indicated. When defective stator windings are found, the complete stator assembly must be replaced.

Diode Checks, Fig. 7

Each of the six diodes may be checked electrically for a shorted or open condition, using either of the methods described below. Stator leads must be disconnected and lead from diode being tested must not be in contact with lead from other diode.

Checking Diodes with Ohmmeter
Use lowest range scale and a 1½ volt cell in ohmmeter. Connect one lead from ohmmeter to diode lead and other lead to diode case and note reading on ohmmeter. Reverse ohmmeter lead connections and note reading. If both readings are very low or if both readings are very high, the diode is defective. A good diode will show one low reading and one high reading. Check each diode in same manner.

NOTE: All diodes, if in good condition, will show one low reading and one high reading. But the respective reading at diodes in the heat sink will be opposite from those in the end frame assembly.

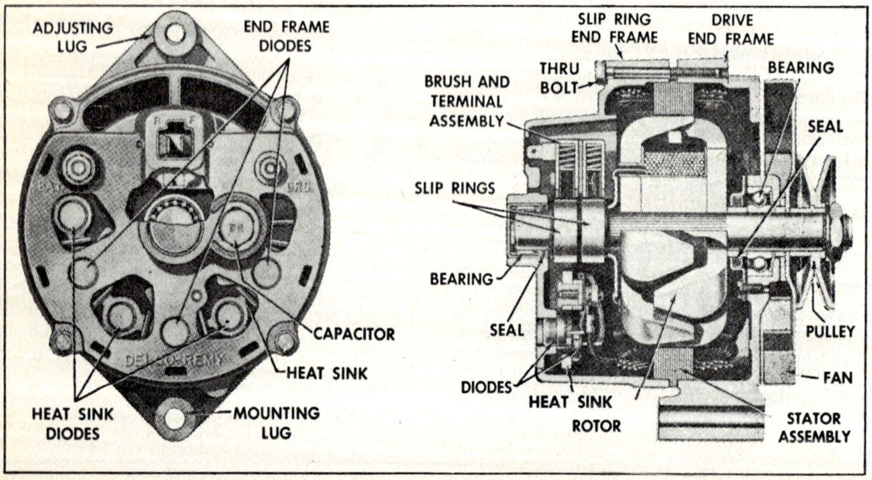

Fig. 2 Delcotron Series 1D alternator

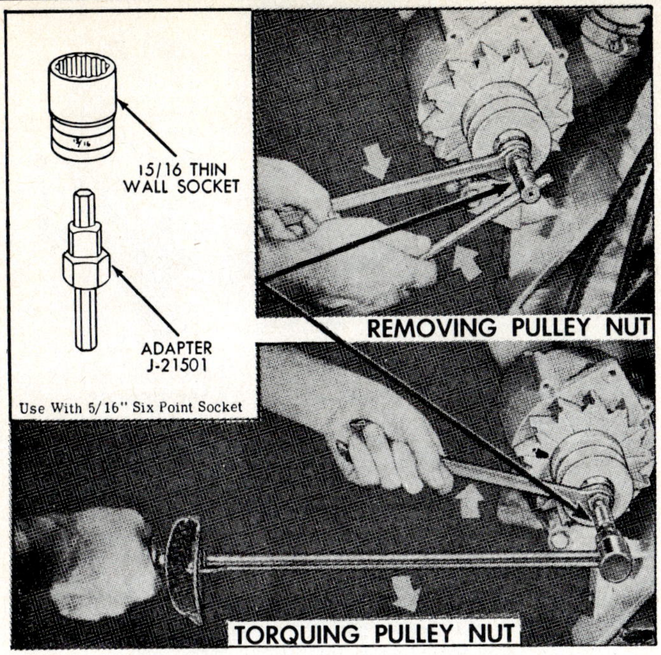

Fig. 3 Replacing pulley nut

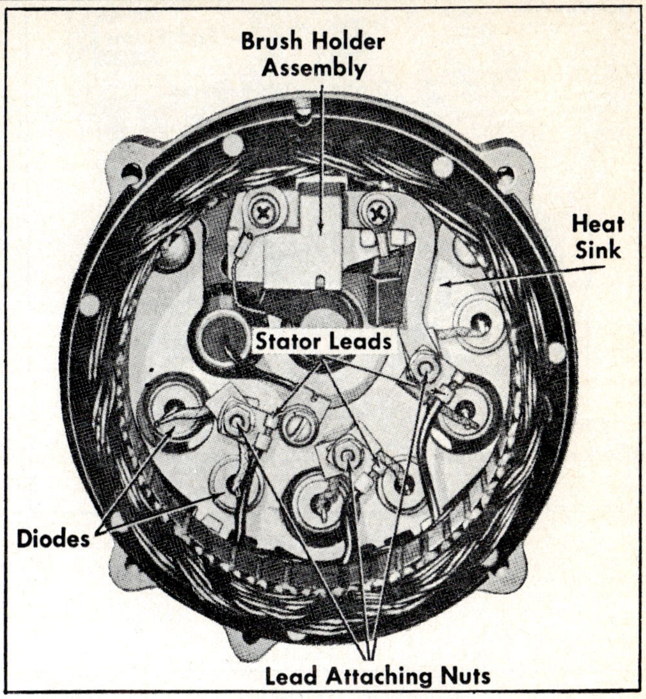

Fig. 4 Stator leads at junction studs

Checking Diodes with Test Lamp

An alternate method of checking diodes is to use a test lamp of not more than 12 volts in place of an ohmmeter. *Do not use 110-volt test lamp to check diodes.*

With stator leads disconnected, connect test lamp leads across each diode in same manner as given for the ohmmeter method. First check for current flow in one direction, then in reverse direction. If the test lamp lights in both checks, or fails to light in both checks, the diode is defective. When checking a good diode, the test lamp will light only when connected in one direction.

Special Testers

Special testers are available which operate without disconnecting stator. When using this type of equipment, follow manufacturer's instructions.

Capacitor Check, Fig. 7

The capacitor is pressed into retainer in the alternator end frame and capacitor lead is connected to the heat sink.

Since the capacitor rating is only 0.5 microfarads, it is impractical to test the capacitance with conventional equipment. However, when the alternator is disassembled for overhaul, a check should be made to be sure the insulation between elements is not permitting a current leak to ground.

Check can be made with a 12-volt test lamp or a 1½ volt ohmmeter. Connect one test lead to capacitor lead and other lead to condenser case. Wiggle the "pig tail" lead while observing ohmmeter or test lamp. If capacitor is good, the ohmmeter will show infinite (high) resistance or a test lamp will not glow during test.

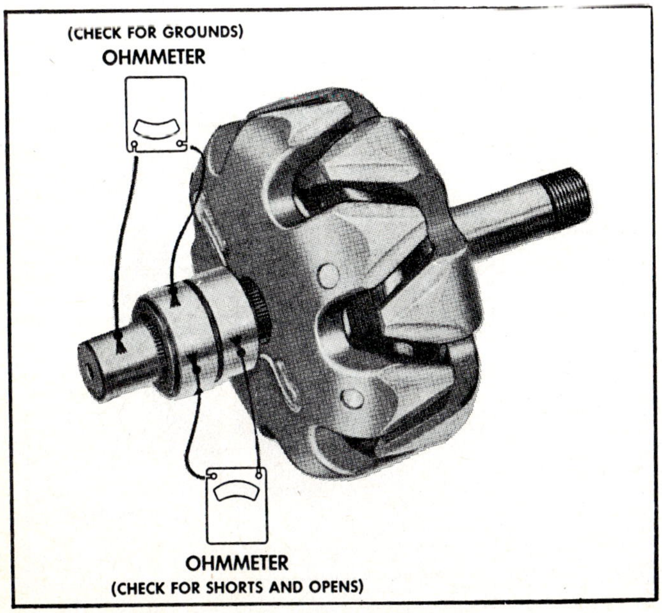

Fig. 5 Checking for grounded, shorted or open circuit in rotor

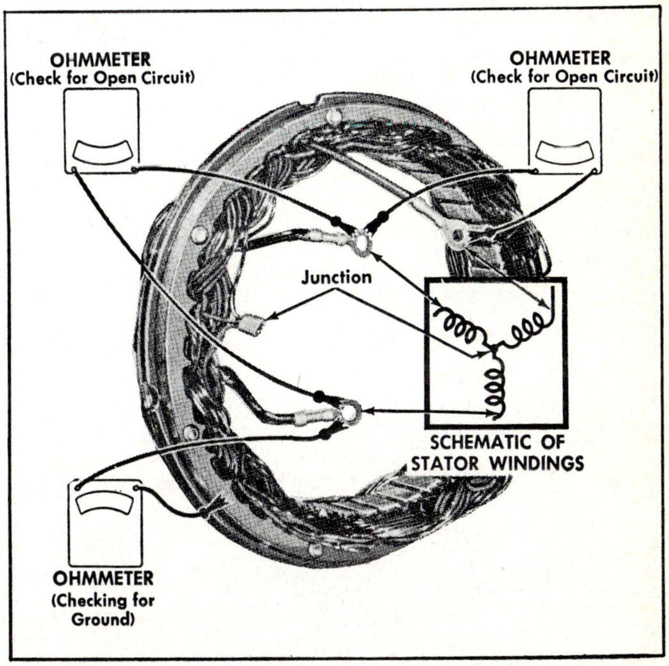

Fig. 6 Checking for grounded or open circuit in stator or windings

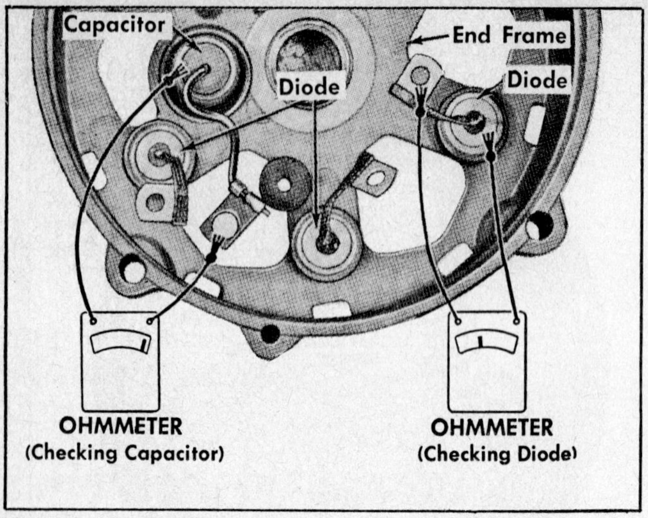

Fig. 7 Testing diode and capacitor

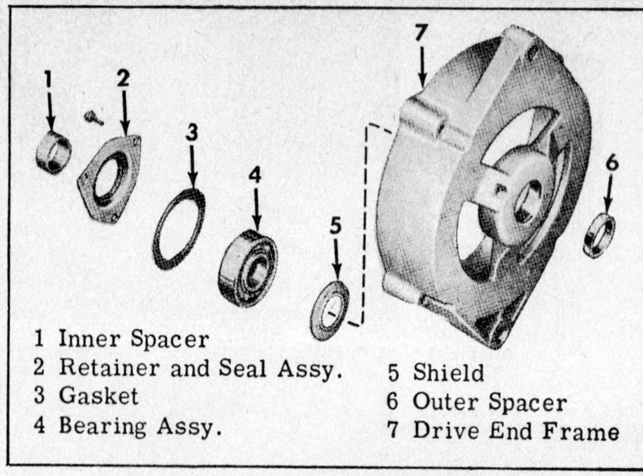

Fig. 8 Drive end frame and bearing components

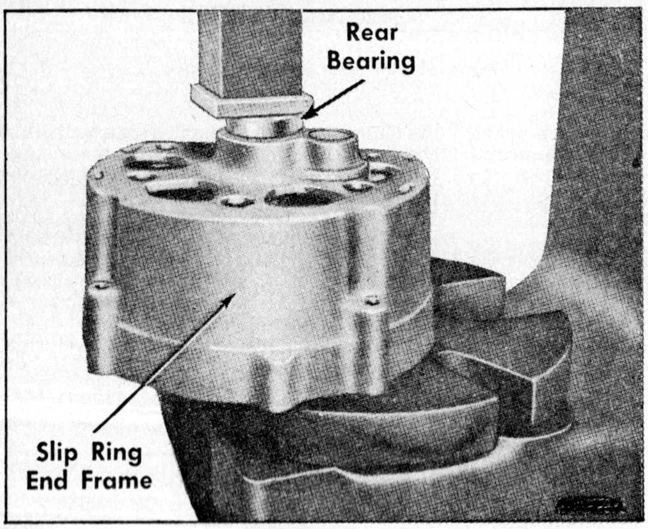

Fig. 9 Installing bearing in slip ring end frame

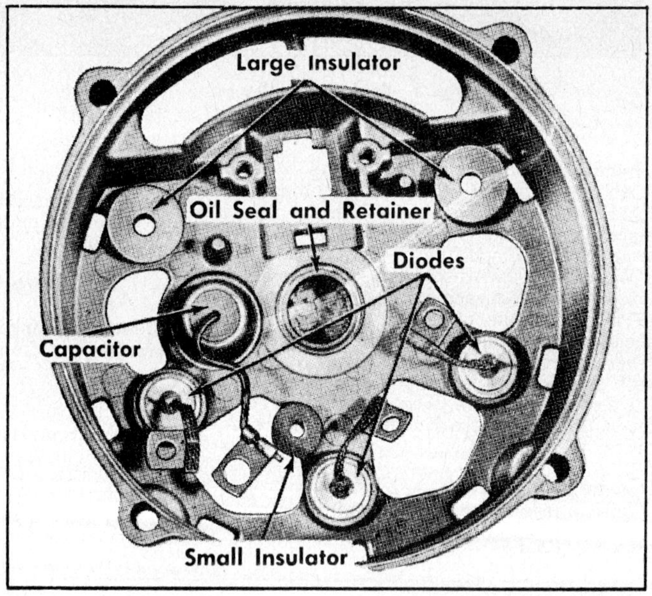

Fig. 11 Slip ring end frame with heat sink removed

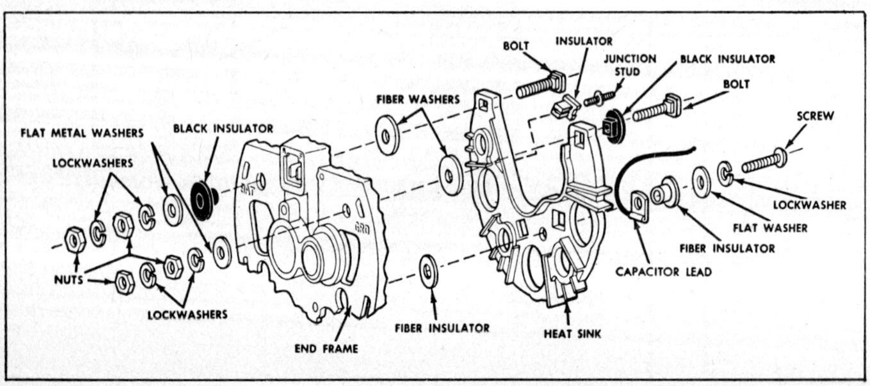

Fig. 10 Heat sink and associated parts disassembled

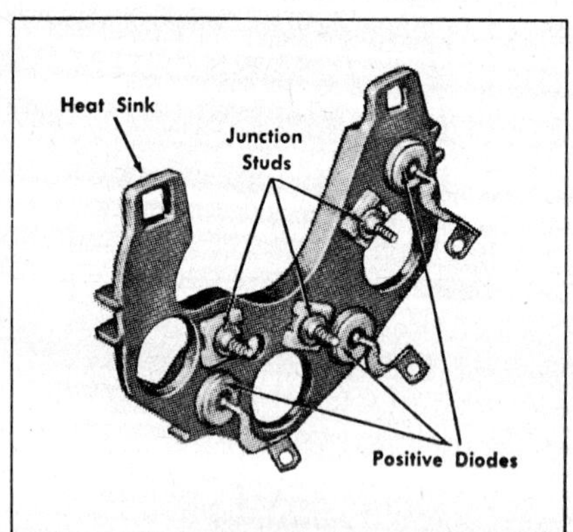

Fig. 12 Heat sink with positive diodes and junction studs installed

Any capacitor that allows current leakage as indicated by low ohmmeter reading must be replaced.

INSPECTION & REPAIRS

Inspection

1. Check drive end frame and bearing assembly for condition of bearing, felt seal, and retainer. Examine end frame for cracks and for damaged threads.
2. Rotor assembly should be inspected for scoring at slip rings and at surfaces contacted by shaft bearings. If shaft is damaged, a new rotor assembly should be installed.
3. Inspect brushes, brush springs and brush holder attached to slip ring end frame assembly. If brushes are worn to half of their original length or are damaged they must be replaced. Visually inspect brush springs; if distorted, worn or broken, replace with new springs.
4. Inspect rotor shaft bearing in slip ring end frame. If there is evidence that bearing has been run without lubricant, or if bearing is not in good condition, it must be replaced.

Repairs

Drive End Frame Bearing, Fig. 8

1. Remove retainer and seal from end frame (3 screws), then press bearing out of frame. Remove bearing shield from bearing recess in frame.
2. Place shield in recess and press bearing into place. Lubricant cavity in bearing and recess must not be packed solid, but should be only 1⁄4 filled.
3. Install bearing retainer and seal.

Slip Ring End Frame Bearing, Fig. 9

1. To remove bearing, use arbor press and a tube or shaft that just fits inside bore without interference. Support end frame and press from outer side to inner side of end frame.
2. Use a flat plate to press new bearing into end frame. Press from outer side of frame while supporting end frame. Use extreme care to avoid cocking bearing or placing undue strain on bearing. Outer (rear) end of bearing must be pressed in flush with outer surface of end frame.
3. Saturate felt seal with SAE 20 engine oil and install seal and steel retainer at inner end of bearing.

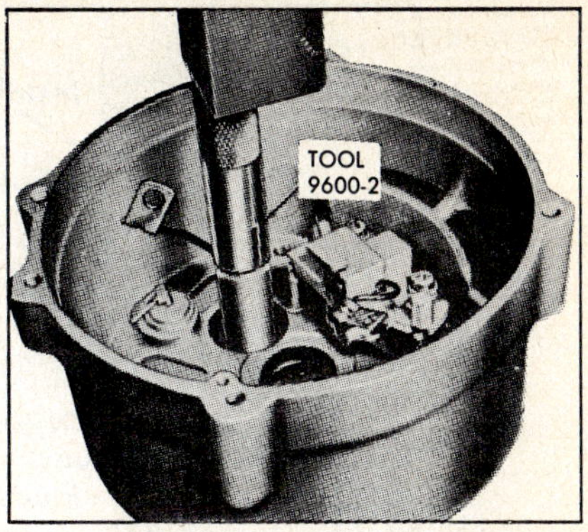

Fig. 13 Installing capacitor in end frame

Heat Sink, Fig. 10

Heat sink and positive diodes can be replaced as an assembly after alternator is separated into major components. The following procedure covers replacing heat sink assembly on slip ring end frame:

1. If brush holder has not been removed from end frame, loosen screw that holds strap at brush holder so lower end of strap can be disengaged from junction terminal without bending strap. Also disengage negative diode leads from the three junction terminals if not previously done.
2. Remove "BAT" and "GRD" terminals with respective washers and insulators from end frame and heat sink.
3. Remove heat sink from end frame and the three insulating washers shown in Fig. 11.
4. With diodes and junction terminals installed in heat sink, place insulating washers in slip ring end frame, Fig. 11, then locate heat sink on washers in end frame. Be sure leads on diodes in end frame protrude through holes in heat sink. Assemble "BAT" and "GRD" termi-

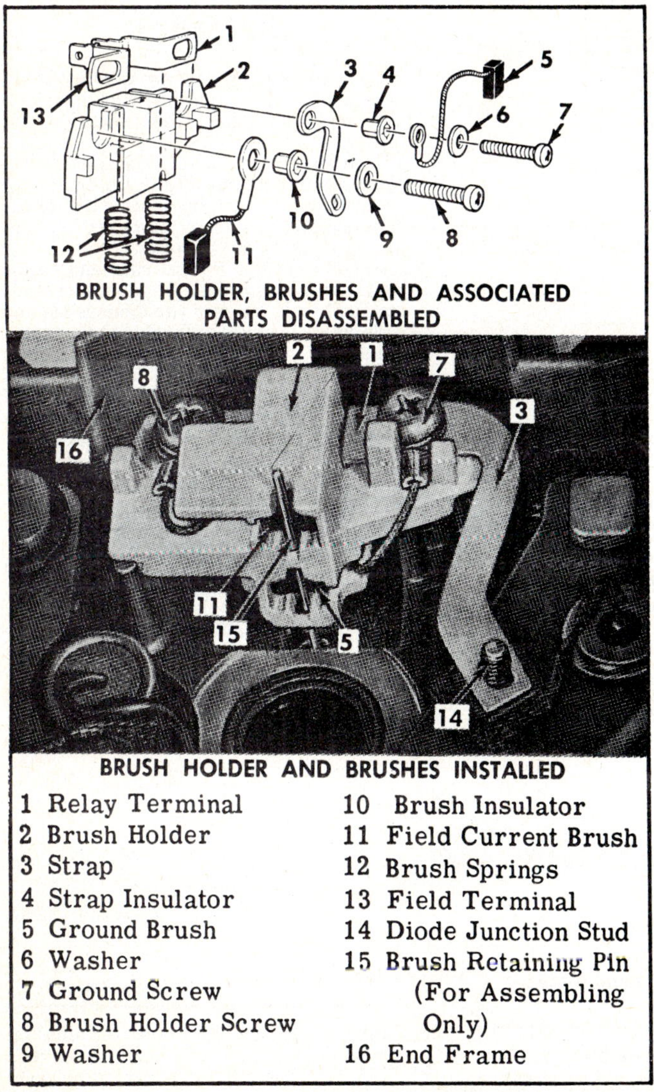

1 Relay Terminal	10 Brush Insulator
2 Brush Holder	11 Field Current Brush
3 Strap	12 Brush Springs
4 Strap Insulator	13 Field Terminal
5 Ground Brush	14 Diode Junction Stud
6 Washer	15 Brush Retaining Pin (For Assembling Only)
7 Ground Screw	16 End Frame
8 Brush Holder Screw	
9 Washer	

Fig. 14 Brush and brush holder installation

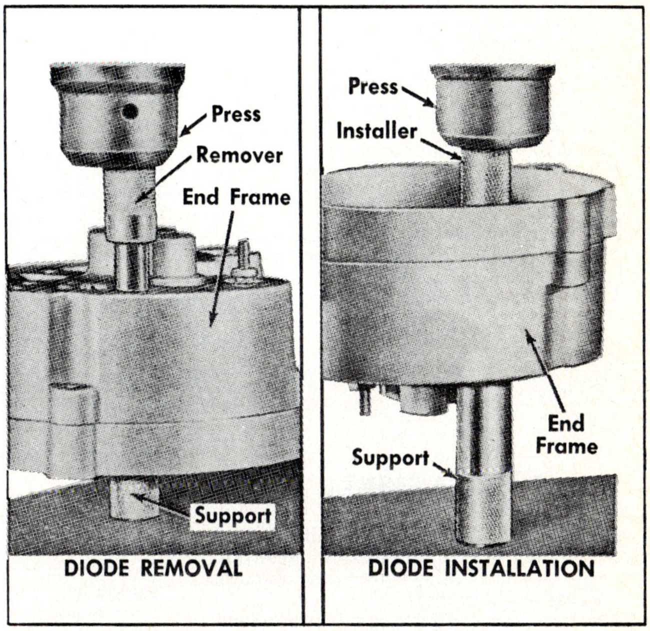

Fig. 15 Replacing diodes in end frame

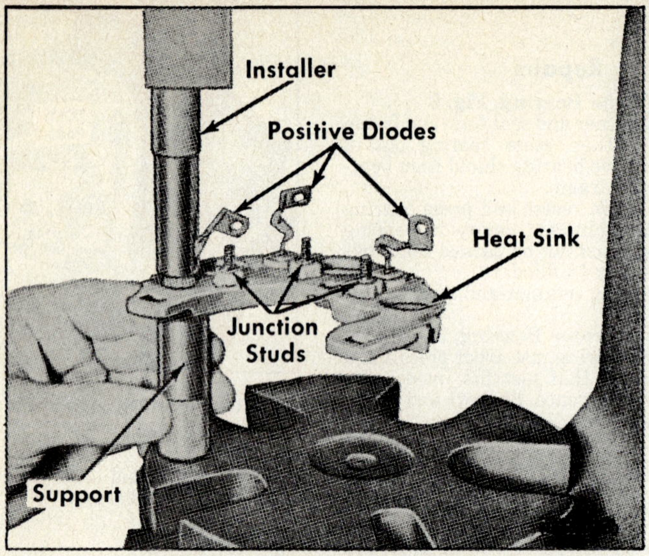

Fig. 16 Installing diodes in heat sink

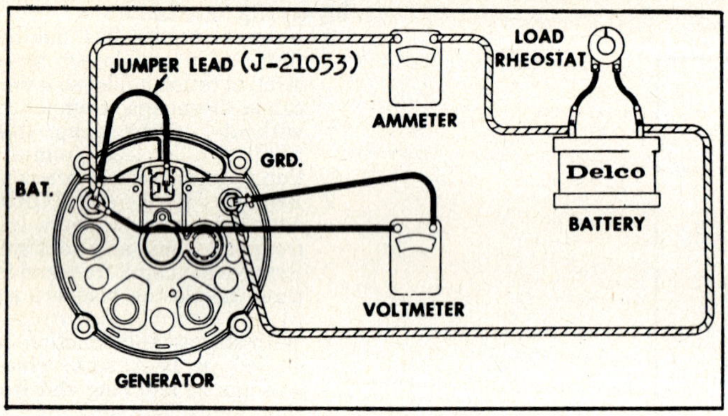

Fig. 18 Connections for bench-testing alternator

nal bolts, insulators and washers in the order shown in Fig. 10. Place lower end of strap on junction stud and tighten screw at brush holder.

5. Attach capacitor lead with screw, insulator, flat washer and lock washer, making sure insulating washer between heat sink and end frame is in place. Place diode leads in pairs on the three junction studs. *Do not install nuts on studs until stator is being assembled to slip ring end frame.*

Diode Junction Stud and Insulators

Diode junction studs and insulators are available for service replacement. Stud and insulator are shown in Figs. 10 and 12. To remove studs, support heat sink at flat side and press studs out of insulators. With pliers, remove insulators that are retained in heat sink by lugs.

Rotor Slip Ring Recondition, Fig. 17

Slip rings on rotor shaft may be cleaned with 400 grit or finer polishing cloth. Spin rotor in lathe to clean and polish slip rings.

Slip rings that are rough or out-of-round should be trued in a lathe to .002" total indicator reading. *Slip ring end of rotor shaft must not be gripped in lathe jaws.* Chuck rotor shaft only at pulley end.

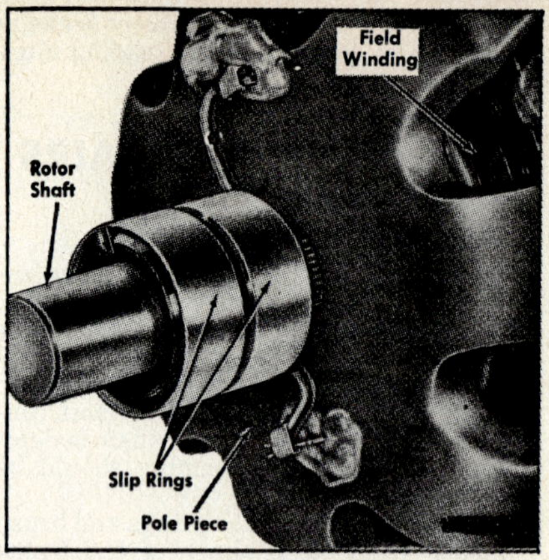

Fig. 17 Rotor and slip ring assembly

Remove only enough material to make slip rings smooth and round. After lathe operation, finish slip ring surface with 400 grit or finer polishing cloth and blow away all dust.

ASSEMBLY

1. Place inner spacer on end of rotor shaft. Place drive end frame and bearing assembly on rotor shaft, then install outer spacer.
2. Place fan on rotor shaft, then install pulley, washer and nut, Fig. 3. Hold nut with wrench while turning shaft with torque wrench and hex driver in socket in shaft. Torque nut to 40–50 ft-lbs.
3. Assemble stator to slip ring end frame with stator leads engaging junction studs on heat sink. Be sure two diode leads and one stator lead are connected to each junction stud. Install lock washer and nut on each stud. Tighten "BAT" terminal stud to 20–25 inch-lbs torque. Tighten "GRD" terminal stud to 15–20 inch-lbs torque.
4. With slip ring brushes in respective holders and retained with pin or wire as shown in Fig. 14, place slip ring end frame assembly over rotor assembly. Refer to marks made at disassembly, align marks, then install through-bolts.
5. Pull out brush retaining wire or pin to allow brushes to contact with slip rings.

Bench Test After Assembly

Fig. 18 shows wiring connections for making alternator output test on bench. Refer to charts in Truck Chapters for specifications.

Delcotron Series 2D

DISASSEMBLE

1. Referring to Figs. 1 and 2, remove two nuts and washers from sockets and terminals. Remove relay wire, then remove socket and terminals.
2. Remove two capacitor and brush holder mounting screws and separate capacitor from brush holder. Lift brush holder and brushes out of cavity in end frame.

NOTE: Capacitor lead is attached to heat sink inside end frame. Avoid unnecessary strain on capacitor lead and relay lead. If brush springs remain inside alternator, remove them and place with brushes.

3. Remove nut from field relay terminal on brush holder to remove insulated brush.
4. Remove three slip ring end frame bolts and washers that secure end frame to stator shell. Remove three special nuts from

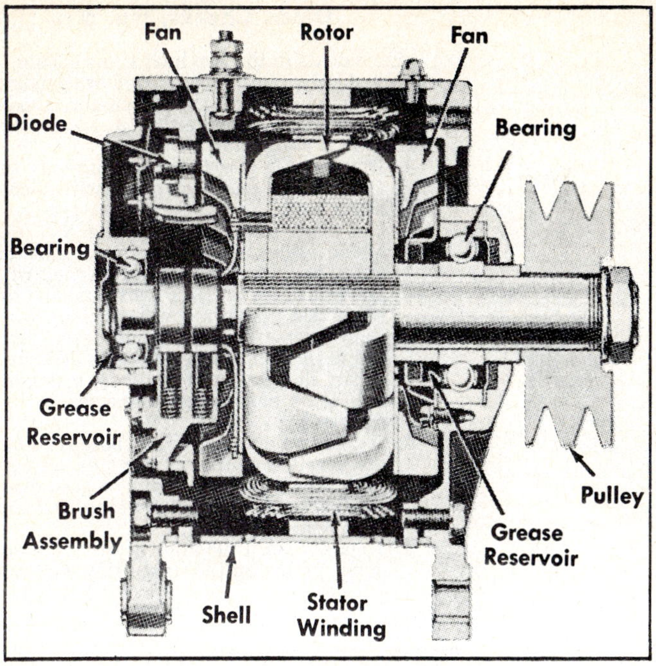

Fig. 1 Delcotron Series 2D

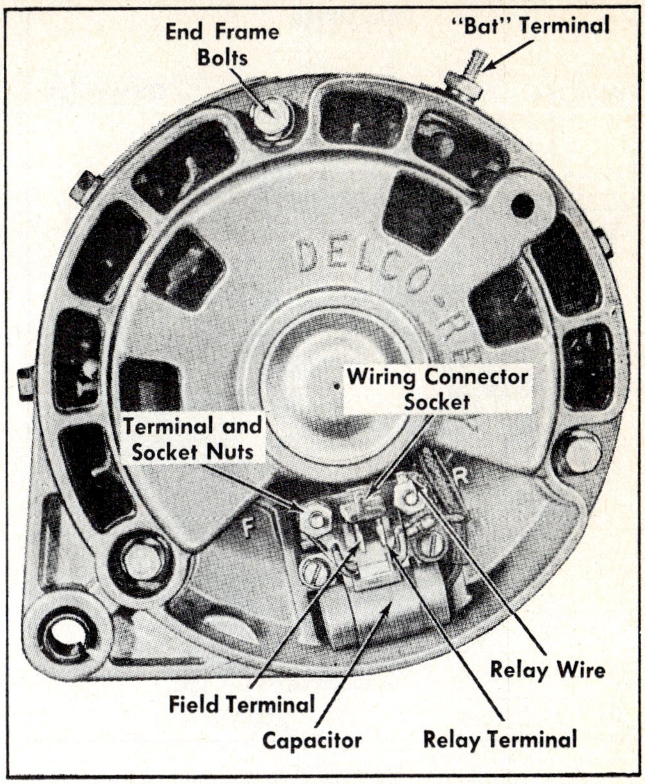

Fig. 2 Rear view of series 2D alternator

slots in stator shell.

5. Pry between stator shell and slip ring end frame to force end frame off rotor shaft rear bearing. Do not cock frame excessively as it is pried off. Fig. 9 shows slip ring end frame removed from alternator. Note that capacitor and relay leads must pass through opening in end frame as end frame is removed and must remain with stator assembly.
6. Remove three bolts, washers and special nuts that attach drive end frame to stator shell. Separate stator and shell from drive end frame and rotor.
7. Remove pulley and Woodruff key and slide end frame and two spacers off shaft. Further disassembly is not required unless inspection and tests indicate necessity for parts replacement.

ELECTRICAL TESTS

Rotor Checks, Fig. 3

Checking for Ground

When using a test lamp, touch one test lamp lead to one slip ring and other lead to rotor pole or shaft. If lamp lights, winding is grounded and rotor must be replaced.

When using an ohmmeter, make test in same manner as with test lamp and observe reading. If reading is low, indicating low resistance, the winding is grounded and rotor must be replaced.

Checking for Open Circuit

Touch leads from test lamp or ohmmeter to each of the slip rings. If lamp fails to light, or ohmmeter reading is high (infinite) the winding is open.

Checking for Shorted Winding

Either of two methods may be used to check rotor winding for shorts.

1. Ammeter and battery method: Connect 12-volt battery and ammeter in series with the two slip rings and note ammeter reading. If reading is over 4.4 amperes, these is a short circuit in winding.
2. Ohmmeter Method: Connect ohmmeter leads to the two slip rings and note the resistance reading. If the resistance is less than 2.7 ohms, there is a short circuit in rotor winding.

NOTE: If an ohmmeter is not available the resistance can be calculated by dividing the voltage by the current (amperes), using values from Step 1 above.

If there are no defects found in the rotor assembly and the alternator does not produce its rated output when checked as outlined in "Alternator Output Test," the trouble is either in the stator assembly or in the rectifying diodes.

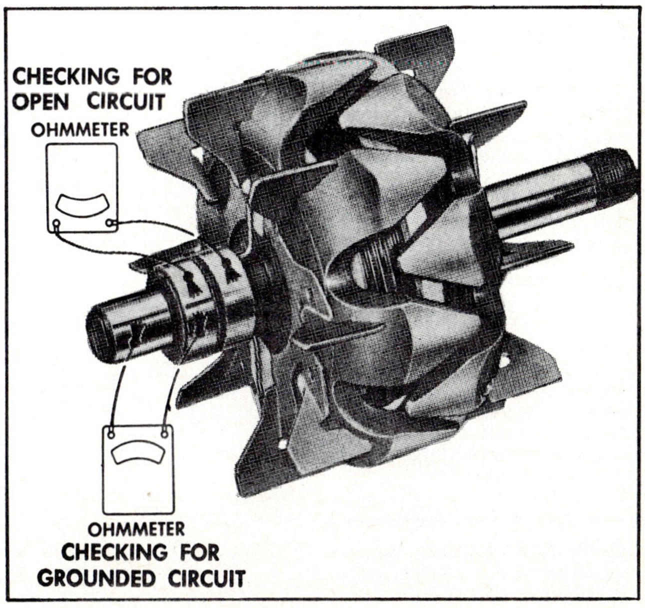

Fig. 3 Checking rotor for grounded, shorted or open circuit

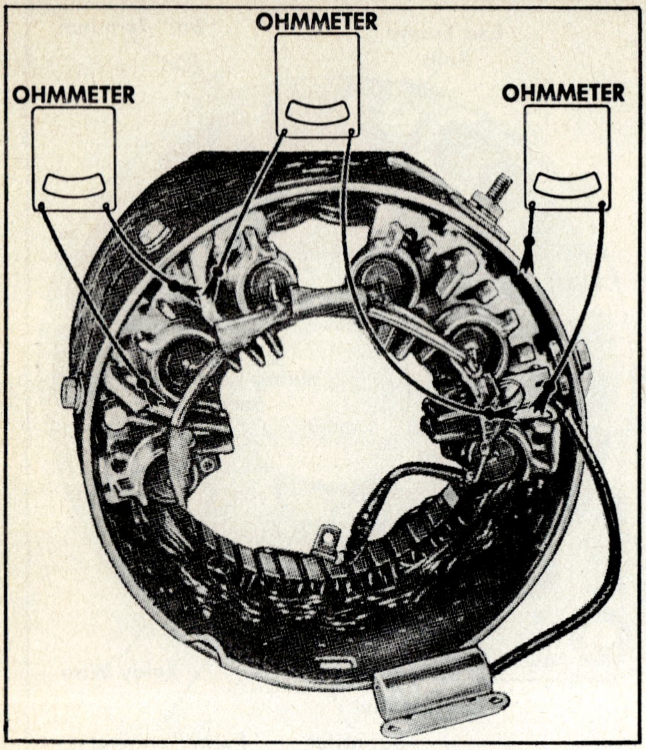

Fig. 4 Checking for grounded or open circuit in stator

Stator Checks, Fig. 4

CAUTION: Do not use test lamps of more than 12 volts. Higher voltage, such as 110-volt test lamp, will immediately destroy a diode if prods should accidentally contact a diode lead when checking the stator.

Due to the low resistance of the stator winding, it is not practical to check stator windings for short circuit without laboratory equipment. However, if all other electrical checks show units to be in good condition, but alternator fails to supply rated output, shorted stator windings are indicated.

1. Scrape epoxy coating from lead connections. Unsolder stator leads, using a minimum amount of heat to avoid damage to diodes to which stator leads are connected. An alternate method is to cut stator leads between coil and diodes. Leads must be spliced after making checks.
2. Check for grounded stator windings by attaching one lead from ohmmeter or test lamp to stator shell, and other lead to any one of the stator winding leads. If the test lamp glows or if ohmmeter shows very low reading, the windings are grounded.
3. To check for open circuits in stator windings, connect ohmmeter leads or test lamp leads successively between the pairs of stator leads. If test lamp fails to light, or if ohmmeter reading is very high (infinite) an open circuit is indicated.

NOTE: If these checks show that the stator is defective, a complete stator and shell assembly must be installed when reassembling the alternator.

Diode Checks, Fig. 5

Each diode in the two heat sinks should be checked individually for shorted or open condition when overhauling alternator. Diodes can be checked using either a 12-volt test lamp or an ohmmeter.

1. Disassemble alternator leaving only the shell, stator and heat sinks with diodes assembled.
2. Cut stator leads as shown, Fig. 5.
3. Using either a test lamp or ohmmeter, check No. 1 positive diode, starting with one test probe to "BAT" terminal and the other test probe to the light relay wire, then reverse this connection. If test lamp lights in both checks, diode is defective. When checking a good diode, test lamp will light only when connected in one direction.
4. If using an ohmmeter, and both readings are very low or if both readings are very high, the diode is defective. A good diode will indicate one low reading and one high reading.
5. To check No. 1 negative diode, place one of the probes on the light relay wire and

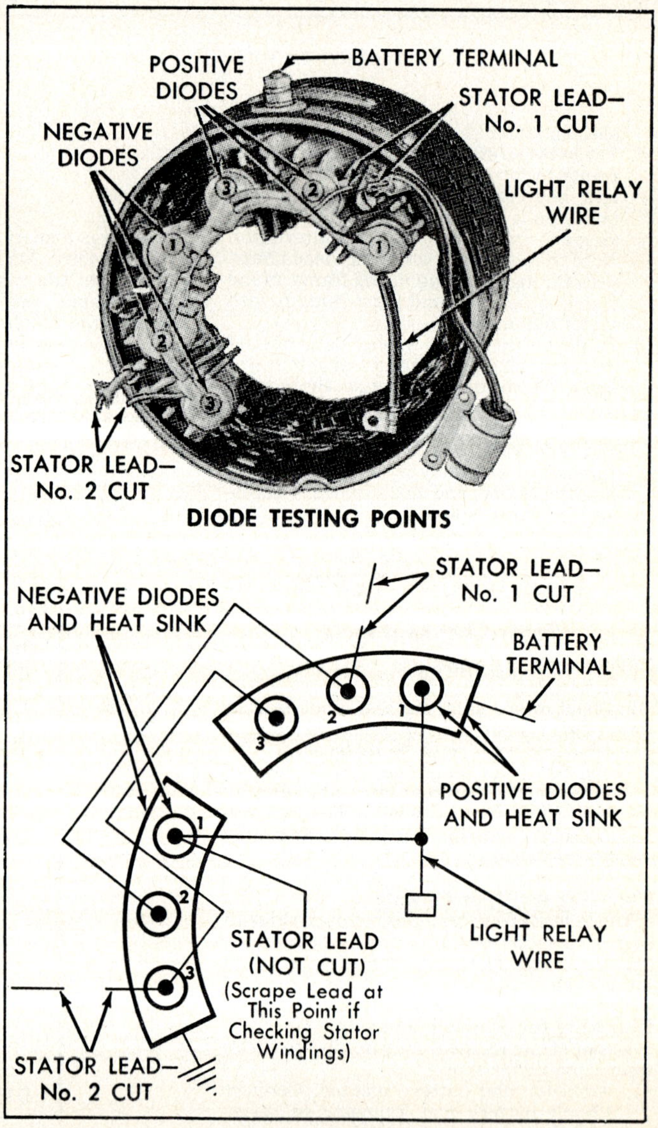

Fig. 5 Testing diodes

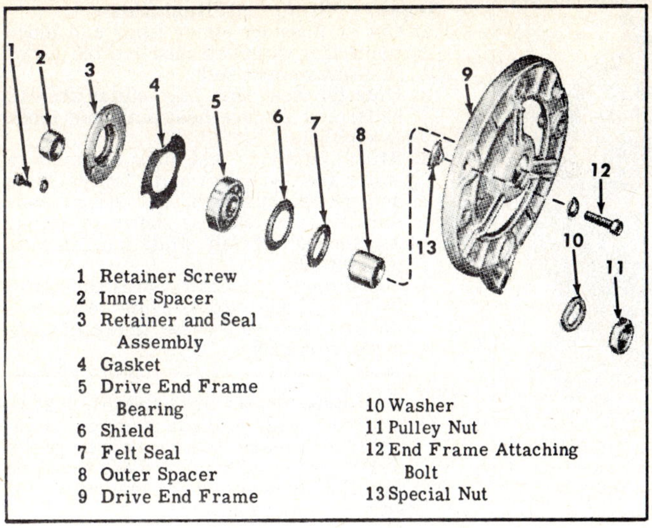

Fig. 6 **Drive end frame and bearing components**

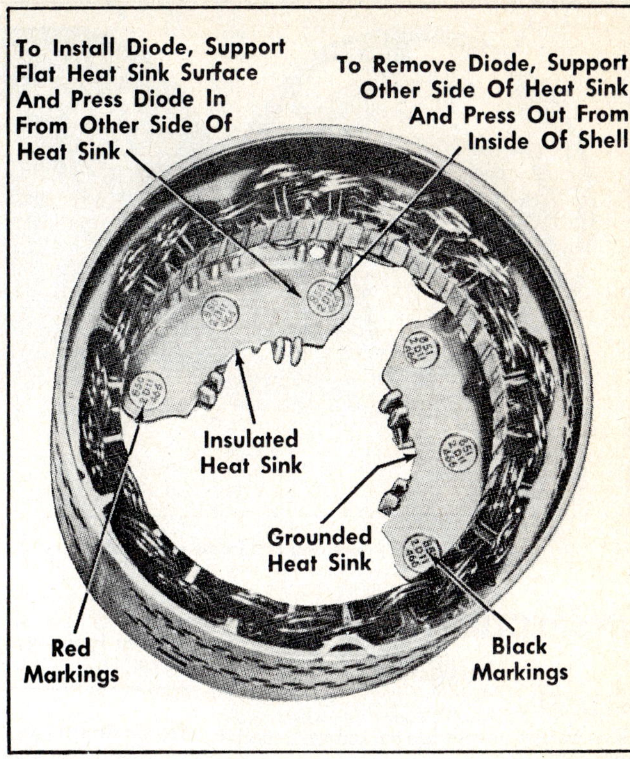

Fig. 8 **Diode replacement**

the other probe on the negative heat sink. Reverse this connection to check this diode.

6. To check No. 2 positive diode, touch one probe to the No. 1 cut stator lead which comes from the diode (not lead that goes to stator) and the other probe to ground on stator shell. Reverse this connection to determine state of this diode.
7. To check No. 2 negative diode, place one probe to the No. 1 cut stator lead and ground the other probe on the negative heat sink. Then reverse connections to determine state of this diode.
8. To check No. 3 positive diode, place one probe on No. 2 cut stator lead (diode end) and ground other probe on positive heat sink. Then reverse connection to determine state of this diode.
9. To check No. 3 negative diode, place a probe on the No. 2 stator lead (diode end) and other probe on negative heat sink. Reverse these connections to determine state of this diode.

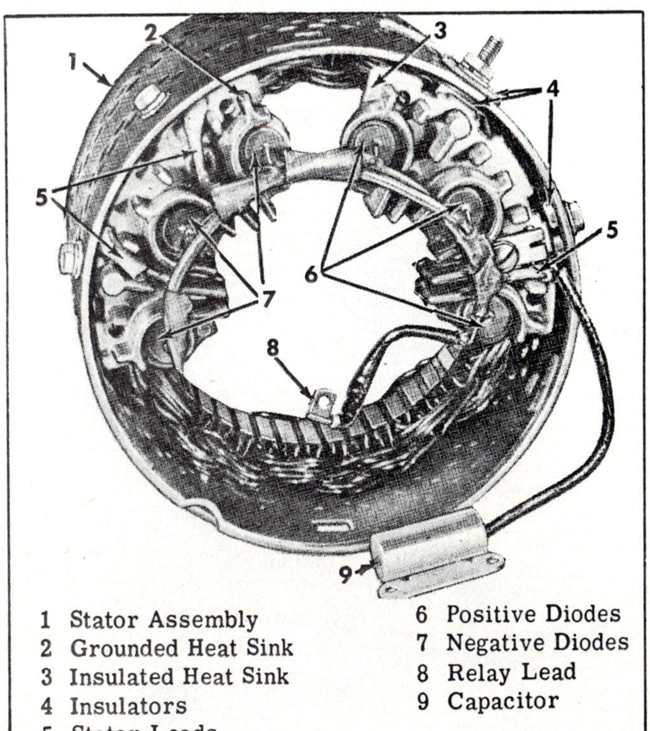

Fig. 7 **Heat sinks installed in rotor shell**

NOTE: If all six diodes check as being good, then the stator should be checked. In order to check the stator the third lead from the stator must also be cut. This will make it unnecessary to disturb the epoxy and solder at the No. 1 negative diode. Proceed to check the stator as outlined above.

If one or more diodes were found to be defective, replace diodes as directed under "Repairs."

Special Testers

Special testers are available which operate without disconnecting the stator. When using this type of equipment, follow the manufacturer's instructions.

Capacitor Check, Fig. 2

The capacitor is mounted to the alternator end frame and capacitor lead is connected to insulated heat sink.

Since the capacitor rating is only 0.5 microfarads, it is impractical to test the capacitance with conventional equipment. However, when the alternator is disassembled for overhaul a check should be made to be sure the insulation between elements is not permitting a current leak to ground.

Check can be made with a 12-volt test lamp or a 1½ volt ohmmeter. Connect one test lead to capacitance lead and other lead to capacitor case, Fig. 7. Wiggle "pig tail" lead while observing ohmmeter or test lamp. If capacitor is good, the ohmmeter will show infinite (high) resistance or a test lamp will not glow during test. Capacitor should be replaced if it does not measure up to this test.

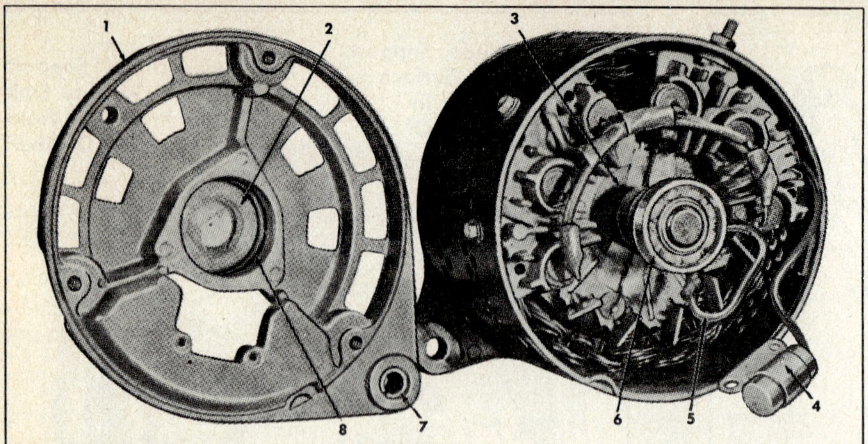

1 Slip Ring End Frame
2 Bearing Lubricant
3 Rotor
4 Capacitor
5 Lead (To Relay Terminal)
6 Bearing Assembly
7 Bushing
8 Bearing Seal Ring

Fig. 9 Slip ring end frame prepared for installation

INSPECTION & REPAIRS

Inspection

Clean all parts except bearings which must not be washed as they are packed with special lubricant.

1. Check drive end frame and bearing assembly for condition of bearing, felt seal and retainer, Fig. 6. Examine end frame for cracks and for other damage.
2. Inspect rotor shaft rear bearing. If there is evidence that the bearing has been run without lubricant, or if bearing is not in good condition, it must be replaced.
3. Rotor assembly should be inspected for scoring at slip rings, Fig. 3. If shaft is damaged a new rotor assembly should be installed at assembly.
4. Inspect brushes, brush springs and brush holder. If brushes are worn to half their original length or are damaged they must be replaced. Visually inspect brush springs; if distorted, worn or broken, replace with new springs.

Repairs

Drive End Frame Bearing, Fig. 6

1. Remove retainer and gasket from end frame, then press out bearing. Remove bearing shield and felt seal from bearing recess.
2. Place shield and seal in recess and press bearing into place. Lubricant cavity in bearing and recess must not be packed solid with lubricant but should be only one-half filled, otherwise bearing will overheat.
3. Install bearing retainer, seal and gasket.

Slip Ring Reconditioning

If slip rings on rotor shaft are scored or dirty, they may be cleaned with 400 grit or finer polishing cloth. Spin rotor in lathe to clean and polish slip rings.

Slip rings that are rough or out-of-round should be trued in a lathe to .002" total indicator reading. Remove only enough material to make slip rings smooth and round. After lathe operation, finish slip ring surface as directed above and blow away all dust.

CAUTION: Brush noise due to flat spots on slip rings may be present if slip rings are cleaned by hand without spinning rotor assembly.

Rotor Shaft Rear Bearing

1. Use a suitable puller or arbor press to remove bearing.
2. With shielded side of bearing toward slip rings, press bearing squarely onto rotor shaft until rear end of shaft protrudes .08 to .10 inch through bearing inner race.

Heat Sink Replacement

The two heat sinks, Fig. 7, are available with diodes for service replacement. The following operations are performed after alternator is disassembled into major components.

Negative (Grounded) Heat Sink

1. Cut or unsolder stator leads and diode connecting wires attached to three diodes in negative heat sink.
2. Unfasten heat sink from stator shell (2 bolts) and remove heat sink and diode assembly.
3. Attach heat sink and diode assembly to stator shell with two bolts, lock washers and special bushings. Two stator leads must fit in notches as shown in Fig. 7. Tighten bolts firmly while holding nuts with wrench.
4. Solder diode connecting wires and stator leads to diode stems, using rosin core (60% tin, 40% lead) solder, then coat diode leads with epoxy.

CAUTION: Do not use excessive heat in soldering leads to diode stems as excessively high temperature may damage the diode. Also avoid bending diode stems, as internal damage to diode may occur, resulting in diode failure.

5. Tape leads and connecting wires as shown in Fig. 7 to prevent wire movement which could eventually wear away insulation.

Positive (Insulated) Heat Sink

1. Detach capacitor from heat sink.
2. Cut or unsolder one stator lead, relay lead and three diode connecting wires which are soldered to diodes.
3. Remove nuts and washers from battery terminal at outer side of stator shell. Remove bolt, lock washer and special bushing which retains heat sink.
4. Remove heat sink and diode assembly and two insulators that are used to prevent contact between heat sink and stator shell.
5. To install, locate battery terminal insulator at inner side of stator shell with projecting flange piloted in terminal hole in shell. Set heat sink mounting bolt insula-

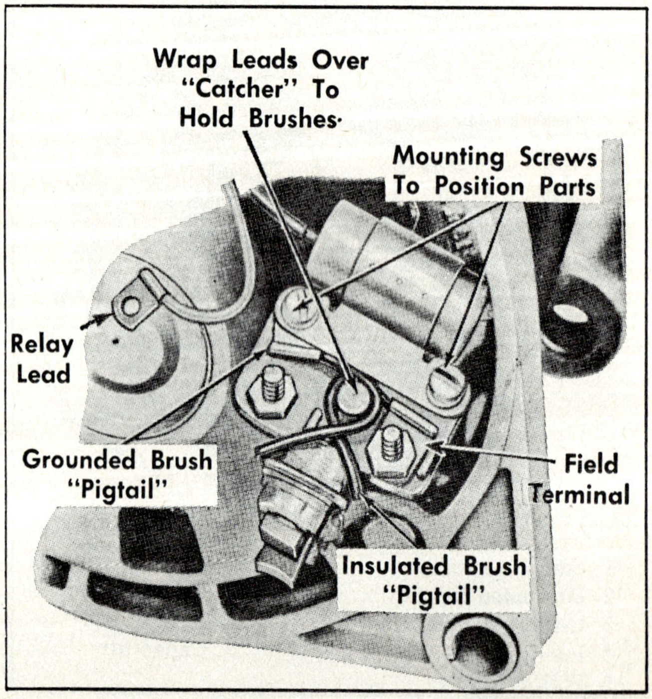

Fig. 10 Brushes and brush holder installation

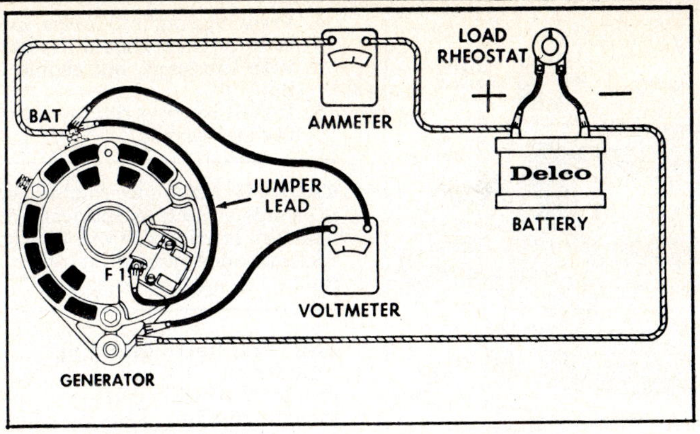

Fig. 11 Equipment and wiring connections for testing alternator on bench

tor in notch in heat sink, then position heat sink in stator shell.

6. Install mounitng bolt with lock washer and special bushing through stator shell and insulator and place nut on inner end of bolt at insulator.
7. Insert battery terminal (bolt) from inside stator shell, through heat sink and insulating washer.
8. Assemble flat fiber washer, steel washer, lock washer and nut on outer end of terminal. Tighten nuts firmly.
9. Referring to Fig. 7, solder diode connecting wires, one stator lead and relay lead to diode stems. Observe the caution previously mentioned to prevent diode damage due to heat or bending diode stems.
10. Tape leads and connecting wires to prevent wire movement, and attach capacitor lead to heat sink.

Diode Replacement, Fig. 8

With proper tools, any of the diodes may be replaced with heat sinks either removed or installed.

NOTE: Negative diodes have black markings on the case and positive diodes have red markings on the case. Note that the heat sink connected to the alternator output terminal is insulated from the shell and contains the positive diodes. The other heat sink is grounded to the shell and contains the negative diodes.

To replace a defective diode, scrape away epoxy coating, then cut only the the leads connected to that particular diode stem. Use diagonal cutters to clip the leads as close to the diode stem as possible. Then remove and install diodes as directed in Fig. 8. After diode is installed, resolder leads to diode stem. Limit soldering time to a minimum so excessive heat will not damage diode.

ASSEMBLE

Follow steps in sequence to assemble the alternator major components, referring to Fig. 1 for identification of various parts except as otherwise indicated.

1. Place inner (short) spacer on pulley end of rotor shaft, then install drive end frame and bearing assembly on rotor shaft with bearing inner race in contact with spacer.
2. Place outer (long) spacer on rotor shaft, then install pulley key and pulley. Retain pulley with spring washer and self-locking nut. Tighten pulley nut to 60 ft-lbs torque.
3. Place stator and shell assembly over rotor and into place at drive end frame. Dimples in stator shell serve to properly locate end frames on stator shell. While holding T-shaped nut in opening in stator shell, install drive end frame bolt and lock washer. Before tightening bolt, install remaining two bolts, lock washers and special nuts, then tighten the three bolts evenly and solidly.
4. Place rear bearing seal ring in groove in slip ring end frame and add recommended lubricant to cavity in bearing. Do not fill reservoir more than half full. Overfilling will cause bearing seal to be pushed out of place or bearing to become overheated.
5. Move slip ring end frame into position with recess in frame aligned with dimple in stator shell. Guide capacitor and relay lead out through brush holder opening in end frame. With slip ring end frame positioned on stator shell, assemble bolts, lock washers and special nuts to secure end frame to stator shell. Tighten bolts evenly and solidly.
6. Referring to Fig. 10, attach insulated alternator brush lead to brush holder with terminal bolt, nut and lock washer. Compress brush spring into brush slot, insert brush against spring and wrap brush spring over "catcher" as shown.
7. Insert two brush holder mounting screws and lock washers through holes in capacitor bracket, and place grounded brush lead on one screw as shown.
8. Insert mounting screws in holes in brush holder, then assemble brush spring and grounded brush in slot in brush holder and wrap lead around "catcher" to hold brush in place until brush holder is installed.
9. Move brush holder and capacitor into place and screw the two mounting screws into threads in end frame.
10. Disengage brush leads from "catcher" to permit brushes to move into contact with slip rings. Install wire connector socket and blade type terminals over terminal screws, attach relay lead on terminal screw as shown in Fig. 2, and install retaining nuts.

Test After Assembly

Fig. 11 shows wiring connections for making alternator output test on bench. Refer to chart in truck chapter for specifications.

CAUTION: When making battery connections, always connect negative terminal to alternator "GRD" terminal. Reversing battery connections will damage alternator.

Delcotron Series 3D

DISASSEMBLY

1. Separate terminal plate from terminals, then remove terminal plate and insulators.
2. Remove two large screws that attach brush holders onto slip ring end frame through window in frame side after removing plates.
3. Disconnect two lead clips that connect the two field terminals to brush holders.
4. Remove four thru-bolts.
5. Tap slip ring end frame lightly to achieve initial separation, then pry it from stator with two screwdrivers located between frames at opposite points.
6. Separate drive end frame from stator in same manner.
7. Referring to Fig. 2, remove shaft nut and washer, then remove pulley.
8. Use an arbor press to force rotor shaft from drive end frame.
9. To remove drive end bearing, detach bearing retainer plate and force bearing from housing.
10. Use a puller to remove slip ring end bearing and collar from shaft.

ELECTRICAL TESTS

Rotor Checks

Checking for Ground

Touch one lead of a 110-volt test lamp to one of the slip rings and the other lead to rotor shaft or laminations. If lamp lights, field windings are grounded.

Checking for Open Circuit

Touch leads from a 110-volt test lamp to each of the slip rings. If lamp fails to light, the windings are open.

Checking for Shorted Windings

Connect a 12-volt battery and ammeter in series with the two slip rings. If ammeter reading is over 6.6 amperes, there is a short circuit in windings.

NOTE: If there are no defects found in the

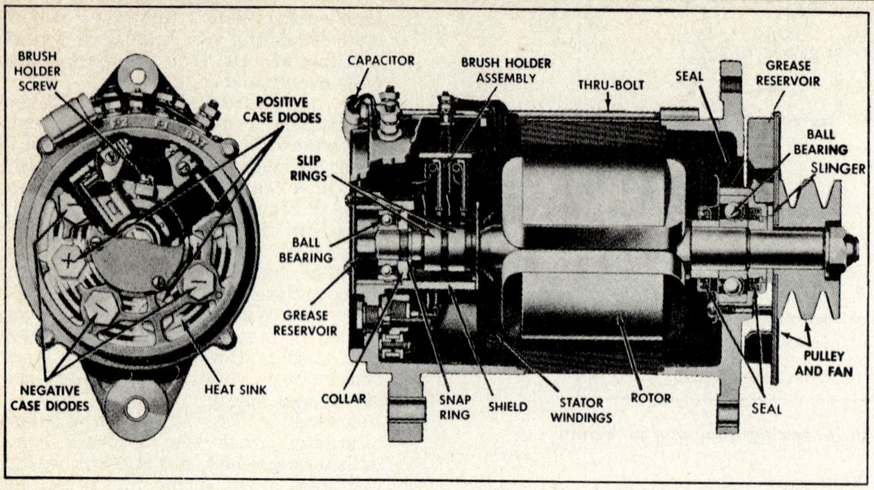

Fig. 1 Delcotron Series 3D

rotor and alternator does not produce its rated output when checked as outlined under "Alternator Output Test", the trouble is in either the stator or in the rectifying diodes.

Stator Checks

1. If a 110-volt test lamp lights when connected from any terminal in the stator windings to the stator frame, the windings are grounded. If the lamp fails to light when successively connected between each pair of stator leads, the windings are open.
2. A short circuit in the stator windings is difficult to locate without laboratory test equipment due to the low resistance of the windings. However, if all other electrical checks are normal and the alternator fails to supply its rated output, shorted stator windings are indicated.

Diode Checks

As shown in Fig. 1, six threaded-type diodes are mounted in the slip ring end frame. Three positive diodes are mounted in the slip ring end frame, and three negative diodes are mounted into a bracket or heat sink located in the frame. The heat sink is attached to but insulated from the slip ring end frame and is attached to the "BAT" terminal on the alternator. Note in Fig. 1 that all diodes are marked either with a + or − on the hexagonal head to identify the polarity of the case.

Diodes can be checked with an ohmmeter as shown in Fig. 3. The ohmmeter leads may be touched to each diode as shown without disconnecting the diodes or removing them from either the end frame or heat sink. Use an ohmmeter with a 1½ volt cell and a scale on which a 300-ohm value is at or near mid-scale. To determine the cell voltage of the ohmmeter, select the scale on the ohmmeter that is to be used, then connect the ohmmeter leads of a voltmeter. The voltmeter will indicate the ohmmeter cell voltage. *Do not use high voltage to test diodes, such as 110-volt test lamp.*

To check each diode, connect ohmmeter one lead to the diode case, and the other ohmmeter lead to the diode lead. Note ohmmeter reading. Then reverse ohmmeter lead connections and note the reading. Ohmmeter readings may vary considerably when checking different diodes. However, if both readings are 300 ohms or less, the diode is defective. Also if both readings are greater than 300 ohms the diode is defective. A good diode will give one very low and one very high reading. When checking each diode, push and then pull on the diode lead slightly to determine if there are loose connections within the diode.

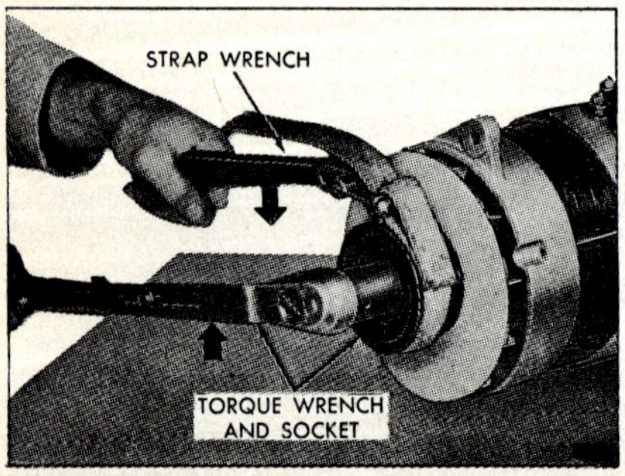

Fig. 2 Replacing alternator pulley nut

INSPECTION & REPAIRS

Inspection

1. Clean all parts except the bearings, which should not be washed as they are packed with special lubricant.
2. Check drive end frame and bearing for condition of bearing and retainer. Examine end frame for cracks or other damage.
3. Inspect rotor shaft rear bearing. If there is evidence bearing has been run without lubricant, or if bearing is not in good condition, it must be replaced.
4. Rotor should be inspected for scoring at slip rings. If shaft is damaged, a new rotor should be installed at assembly.

Repairs

Drive End Frame Bearing

1. Remove four screws attaching retainer and seal to end frame.
2. Remove retainer and gasket, then press bearing out of end frame. Remove lip-type seal from bearing recess in end frame and discard.
3. Install seal in recess in end frame with lip toward reservoir.
4. Press bearing into recess. Lubricant cavity in bearing should not be packed solid with lubricant, but should be only one-half full.
5. Install bearing retainer and seal assembly and gasket.

CAUTION: Make sure shielded side of bearing is on side away from lubricant reservoir. Satisfactory bearing life will be obtained only if recommended lubrication procedures are followed.

Slip Ring Reconditioning

Slip rings that are rough or out-of-round should be trued in a lathe to .002″ maximum indicator reading. Remove only enough material to make the rings smooth and round. Finish with 400 grit polishing cloth and blow away all dust.

Slip rings that must be replaced can be removed from the shaft with a gear puller after the leads have been unsoldered and the bearing, collar and snap ring have been removed. The new assembly should be pressed on with a sleeve that just fits over the shaft. This will apply all the pressure to the inner slip ring collar and prevent damage to the outer slip ring. Only pure tin solder should be used when reconnecting the field leads. Make sure the soldered connections are secure. New slip rings must be turned in a lathe to a smooth finish with .002″ maximum indicator reading. Finish with 400 grit or finer polishing cloth.

Diode Replacement

Negative diodes have right-hand threads, and positive diodes have left-hand threads.

1. To replace any diode mounted in the end frame, disconnect diode lead from terminal, then remove the defective diode. For ease in removing the defective diode it is recommended that both the diode and end frame be heated in an oven to 150°F, or briefly immersed in hot water at a temperature just below the boiling point.
2. Before installing the new diode, lightly coat the threads with silicone grease or light engine oil. With the diode and end frame at room temperature, install the diode and torque to 160–190 inch-pounds. Since diodes use pipe threads, the depth of penetration may vary from one diode to another. Reconnect the diode lead to the terminal.

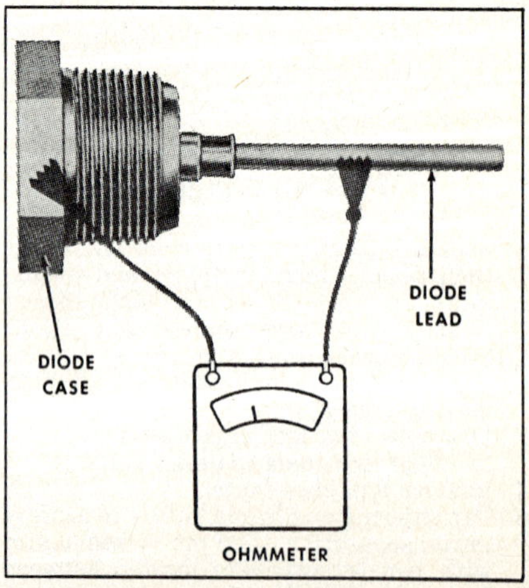

Fig. 3 Checking diodes with ohmmeter

3. To replace any diode mounted in the heat sink, first disconnect all three heat sink diode leads from the three terminals. Then remove the heat sink from the end frame and follow the procedure outlined for removing and installing a new one.

ASSEMBLY

1. Improvise a collar-type tool that just fits over the shaft to press the slip ring end frame bearing and collar with snap ring over shaft. Install collar so shoulder is against bearing inner race.
2. Install drive end frame bearing as outlined previously under "Repairs".
3. Place stator over rotor and into drive end frame. Notches in stator shell serve to properly locate end frame on stator shell.
4. Move slip ring end frame into position on stator shell and install four thru-bolts. Tighten bolts evenly and solidly.
5. Through opening in end frame, install brush and holder assembly and attach to slip ring end frame with two screws.
6. Tighten terminal plate over terminals, then assemble terminal plate and insulators to slip ring end frame.
7. Install terminal nuts and washers on three terminals located in terminal plate.

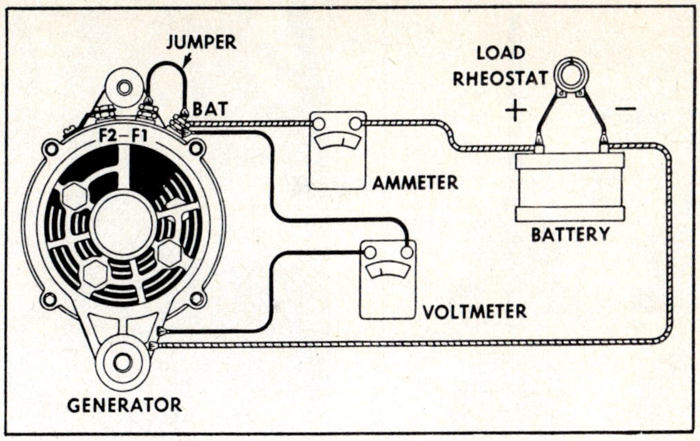

Fig. 4 Wiring connections for checking alternator output on bench

Bench Testing Alternator

Fig. 4 shows the wiring connections for making alternator output test on bench. Connect the jumper lead to the "BAT" and "F1" terminals as shown, operate at specified speed and check for rated output as listed in the "Alternator Specifications" chart in Truck Chapters.

When making battery connections, always connect negative terminal to alternator "GRD" terminal. Reversing battery connections will damage alternator.

Delcotron Series 4D

DISASSEMBLY

1. Referring to Fig. 1, remove four thru-bolts that attach slip ring end frame to drive end frame. Insert a thru-bolt into slip ring end frame and lift brushes off slip rings.
2. Remove drive end frame and rotor assembly. Remove nut and washer from rotor shaft, then use a suitable pulley to remove pulley and fan assembly.
3. When removing slip ring end frame bearing, care should be taken to press the shaft out against the inner bearing race. Because of the small space between the slip ring and the bearing, it may be necessary to secure the outer race of the bearing on this end of the rotor in order to remove it from the shaft. If this is necessary, the bearing may be ruined during removal, in which case a new bearing must be installed. When installing the bearing over the shaft, be sure to press against the inner race only.
4. Disconnect leads from three A.C. terminals, then separate slip ring end frame from stator assembly.

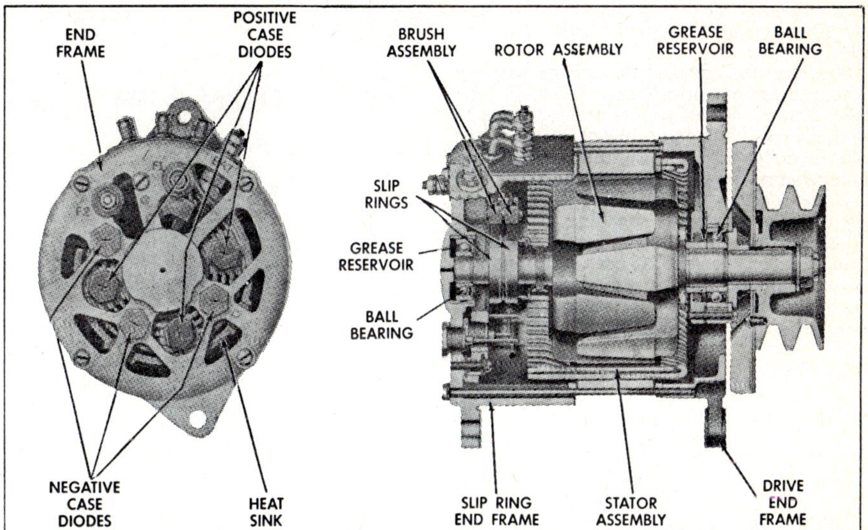

Fig. 1 Delcotron Series 4D

ELECTRICAL CHECKS

Rotor Checks

Checking for Grounds

Touch one lead of a 110-volt test lamp to one of the slip rings and the other lead to the rotor shaft or laminations. If the lamp lights, the field windings are grounded.

Checking for Open Circuit

Touch the leads from a 110-volt test lamp to each slip ring. If the lamp fails to light, the windings are open.

Checking for Shorted Windings

Connect a 12-volt battery and ammeter in series with the two slip rings. If ammeter reading is above 2.4 amperes, there is a short circuit in the winding.

Stator Checks

The stator windings may be checked for grounded or open windings with a 110-volt test lamp. Due to the low resistance of the stator windings, it is not practical to check stator windings for short circuits without laboratory test equipment. However, if all other electrical checks show units to be in good condition, but alternator fails to supply rated output, shorted stator windings are indicated.

1. Check for grounded stator windings by attaching one lead from test lamp to stator frame and the other lead to any one of the stator winding leads. If test lamp lights, the windings are grounded.
2. Check for open circuits in stator windings by connecting test lamp leads successively between each pair of stator terminals. If lamp fails to light, the windings are open.

Diode Checks

With the drive end frame and rotor re-

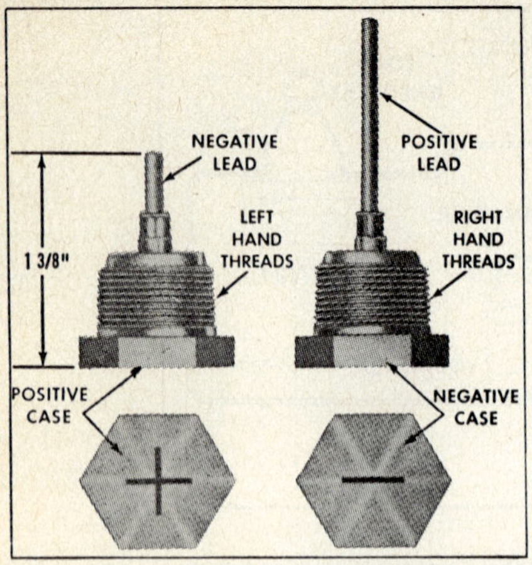

Fig. 2 Diode polarity markings

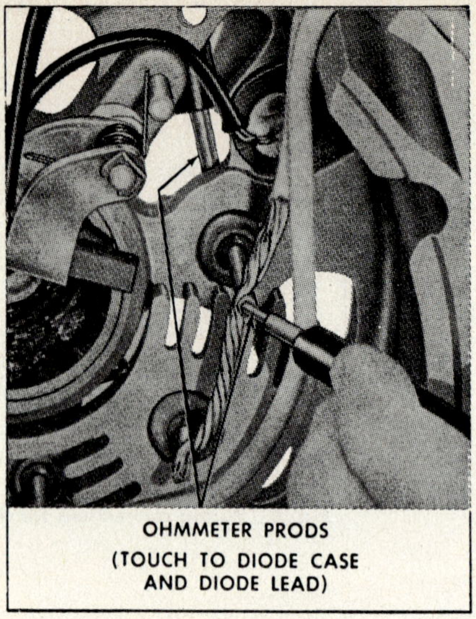

Fig. 3 Checking diodes

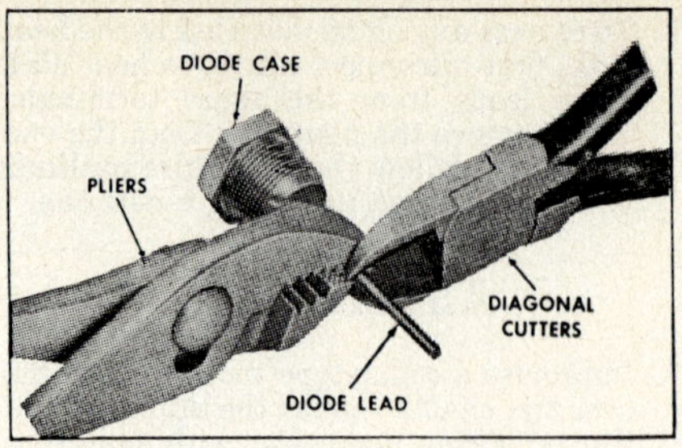

Fig. 4 Cutting diode leads

moved, the slip ring end frame can readily be separated from the stator by disconnecting the leads from the three terminals at top of alternator.

As shown in Fig. 2, there are six diodes mounted in the slip ring end frame. Three positive diodes are mounted in the slip ring end frame, and three negative diodes are mounted into a heat sink which is attached to but insulated from the slip ring end frame. The "BAT" terminal on the alternator is attached to the insulated heat sink.

All diodes are marked either with a + or − on the diode head to identify the polarity of the case. Diodes with a negative case have positive polarity leads whereas positive case diodes have negative polarity leads as shown in Fig. 2.

Diodes can be checked for defects with an ohmmeter as shown in Fig. 3. To check each diode, connect one of the ohmmeter leads to the diode case, and the other ohmmeter lead to the diode lead. Note the ohmmeter reading. Then reverse the ohmmeter connections and note the reading. Ohmmeter readings may vary considerably when checking diodes, but if both readings are 300 ohms or less, the diode is defective. Also, if both readings are greater than 300 ohms, the diode is defective. A good diode will give one low and one high reading.

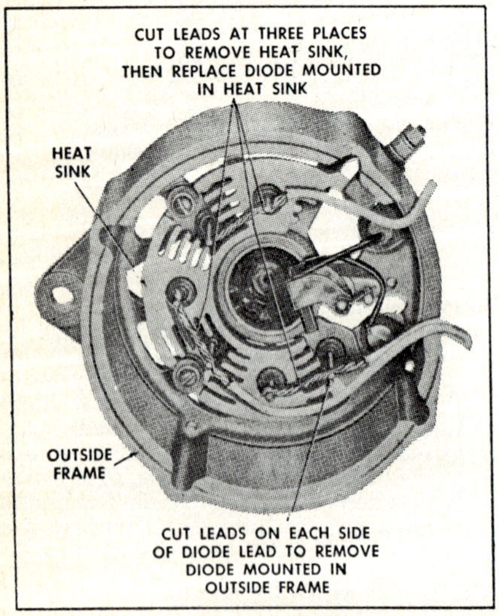

Fig. 5 Replacing diodes

When checking each diode, push and then pull diode lead slightly on the the diode lead to check for loose connections. Use an ohmmeter with a 1½ volt cell and a scale on which the 300 ohm value can be accurately read. To determine the cell voltage of the ohmmeter, select the scale on the ohmmeter which is to be used, then connect the ohmmeter leads to the leads of the voltmeter. The voltmeter will indicate the ohmmeter cell voltage. *Do not use high voltage to test diodes, such as a 110-volt test lamp.*

INSPECTION & REPAIRS

Inspection

1. Clean all parts except bearings which should not be washed as they are packed with special lubricant.
2. Check drive end frame and bearing for condition of bearing, felt seal and retainer. Examine end frame for cracks or other damage.
3. Inspect rotor shaft rear bearing. If there is evidence that bearing has been run without lubricant or if bearing is not in good condition, it must be replaced.
4. The rotor should be inspected for scoring at slip rings. If shaft is damaged, a new rotor assembly should be installed at assembly.
5. Inspect brushes, brush holders and brush springs. If brushes are worn to one-half their original length, they must be replaced.

Repairs

Slip Ring Reconditioning

1. Slip rings that are rough or out-of-round should be trued in a lathe to .002" maximum indicator reading. Remove only enough material to make the rings smooth and round. Finish with 400 grain or finer polishing cloth and blow away all dust.
2. Slip rings that must be replaced can be removed from the shaft with a gear puller after the leads have been unsoldered and the shaft bearing has been removed. The new assembly should be pressed on with a sleeve that just fits over the shaft. This will apply all the pressure to the inner slip ring collar and prevent damage to the outer slip ring. Only pure tin solder should be used when reconnecting the field leads. Make sure the soldered connections are secure. New slip rings must be turned in a lathe to a smooth finish with .002" maximum indicator reading. Finish with 400 grain or finer polishing cloth.

Diode Replacement

Negative diodes have right-hand threads and positive diodes have left-hand threads.

The diode in the service package has a long diode lead. If the diode is to be assembled into the heat sink, it is necessary to cut off the diode lead to an overall length of 1⅜", Fig. 2. To cut off the diode lead, hold the diode lead with ordinary pliers or vise-grip pliers, and use diagonal cutters, Fig. 4, or a hacksaw.

CAUTION: Do not grip diode case when cutting the lead as this will place stress between the diode case and the diode lead, and will damage the diode internally.

1. To replace a diode that is mounted in the outside frame, use diagonal cutters to clip the flexible leads on each side of the diode lead. Clip the flexible leads as close to the diode lead as possible, Fig. 5. For ease in removing the defective diode, it is recommended that the diode and outside frame be heated in an oven to 150°F. or briefly immersed in hot water just below the boiling point.

Fig. 6 Checking brush spring tension

2. Before installing the new diode, lightly coat the threads with silicone grease or light engine oil. With the diode and outside frame at room temperature, install the diode and torque to 160–190 inch-pounds. Since diodes use pipe threads, the depth of penetration may vary from one diode to another.
3. Place the single diode lead clip furnished in service package over the diode lead, and place the ends of the flexible leads into the clip. (The two identical clips are not used when installing a diode in the outside frame.) Crimp the clip tightly over the flexible leads, and liberally solder leads to the clip, and the clip to the diode lead.

CAUTION: Use only 60% tin, 40% lead solder, or other solder with melting point of 360°F., or above. Do not hold the soldering iron on the leads any longer than necessary as excessive heat may damage the diode.

4. To replace a diode mounted in the heat sink, it is necessary to remove the heat sink from the end frame. This is accomplished by clipping with diagonal cutters the flexible lead as close to the diode lead as possible. Also clip the flexible leads midway between the other two pairs of diodes, Fig. 5. Then remove the "BAT" terminal and heat sink attaching screws, and the long leads from the nylon connector.
5. Complete the operation as given for the diodes in the outside frame.

Brush Replacement

The extent of brush wear can be determined by comparison with a new brush. If brushes are worn over half-way, they should be replaced. Replacement is accomplished by installing a new holder, brushes, leads, and terminal assembly. Note that there are both right- and left-hand holders and springs. Brush holders should be toward the right of their pivot pin as viewed from the outside of the rectifier end frame.

Brush spring tension should be checked as shown in Fig. 6. The spring tension scale should be hooked under the brush. The reading should be taken with the brush in the same position encountered when riding on the slip ring. A ten-ounce pull should register on the scale. Excessive tension will cause rapid wear, whereas low tension results in erratic alternator output and burning of slip rings. Defective springs cannot be adjusted, and therefore must be replaced.

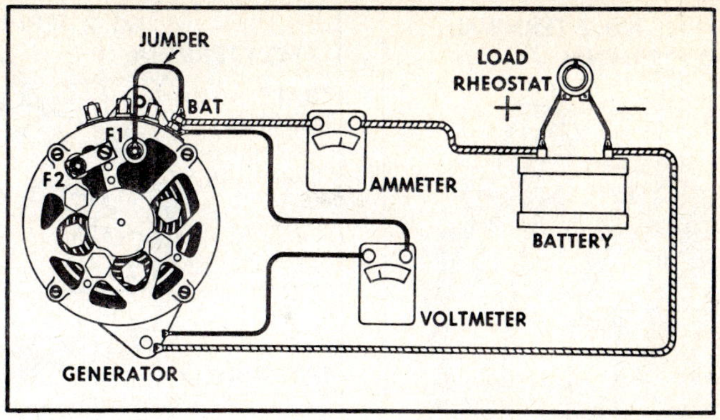

Fig. 7 Wiring connections for checking alternator output on bench

ASSEMBLY

1. Half-fill each reservoir with recommended lubricant.
2. Place inner spacer on pulley end of rotor shaft, then install drive end frame and bearing assembly on rotor shaft with bearing inner race in contact with spacer.
3. Place outer spacer on rotor shaft, then install pulley key and pulley. Retain pulley with self-locking nut and torque nut to 60 ft-lbs.
4. Place stator over rotor and into drive end frame.
5. Fill slip ring end frame bearing cavity half-full of approved lubricant.
6. Insert a thru-bolt through slip ring end frame to hold brushes off slip rings while moving slip ring end frame into position with stator shell. With slip ring end frame positioned on stator shell, install and tighten the four thru-bolts.
7. Connect leads to the three A.C. terminals.

Bench Test

Fig. 7 shows wiring connections for making alternator output test on bench. Refer to "Alternator Specifications" chart in Truck Chapters for specifications. While operating alternator at specified speed, adjust load rheostat to specified voltage. Ammeter will show alternator output which must be within range specified.

When making battery connections, connect negative terminal to alternator "GRD" terminal. Reversing battery connections will damage alternator.

Delcotron 10, 12, 15 & 27 SI (Type 100) Integral Charging Systems

DESCRIPTION

These units, Figs. 1, 2, 3 and 4, feature a solid state regulator mounted inside the alternator slip ring end frame along with the brush holder assembly. All regulator components are enclosed in a solid mold with no need or provision for adjustment of the regulator. A rectifier bridge, containing six diodes and connected to the stator windings, changes A.C. voltage to D.C. voltage which is available at the output terminal. Generator field current is supplied through a diode trio which is also connected to the stator windings. The diodes and rectifiers are protected by a capacitor which is also mounted in the end frame.

NOTE: Some 1975–83 units incorporate a resistor in the warning indicator circuit. Fig. 5.

No maintenance or adjustments of any kind are required on this unit.

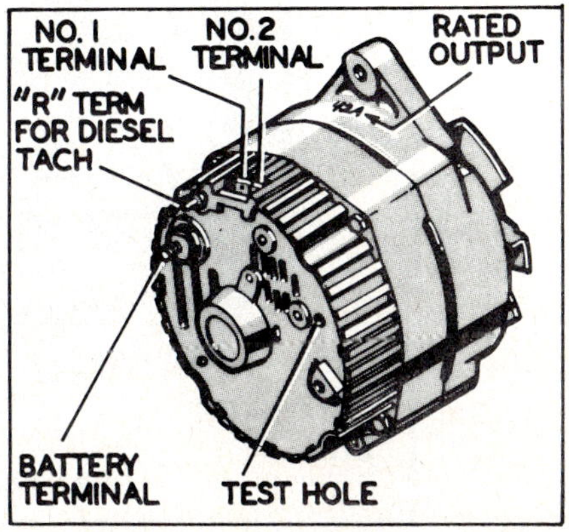

Fig. 1 Delcotron type 10 SI alternator (Typical)

TROUBLE SHOOTING

Undercharged Battery

1. Disconnect battery ground cable.
2. Disconnect wire at "BAT" terminal of alternator, connect ammeter, positive lead to "BAT" terminal and negative lead to wire.
3. Connect battery ground cable.
4. Turn on all accessories, then connect a carbon pile regulator across battery.
5. Operate engine at moderate speed, adjust carbon pile regulator to obtain maximum current output.
6. If ammeter reading is within 10 amps of rated output, alternator is not at fault.

NOTE: Alternator rated output is

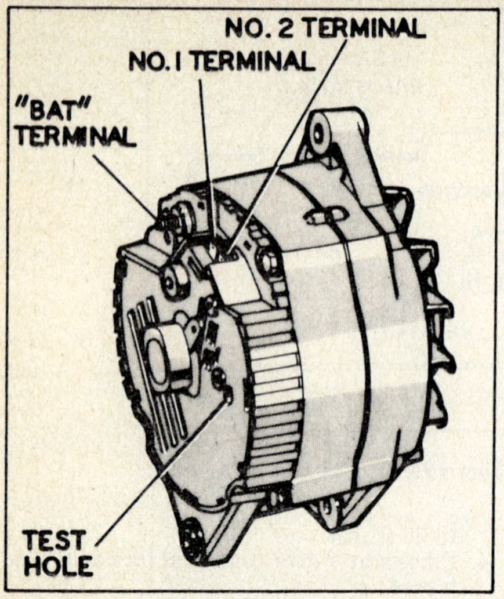

Fig. 2 Delcotron type 27 SI alternator

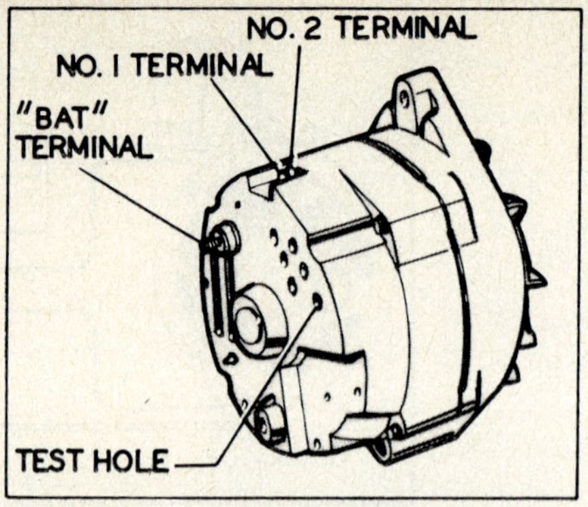

Fig. 3 Delcotron type 15 SI alternator

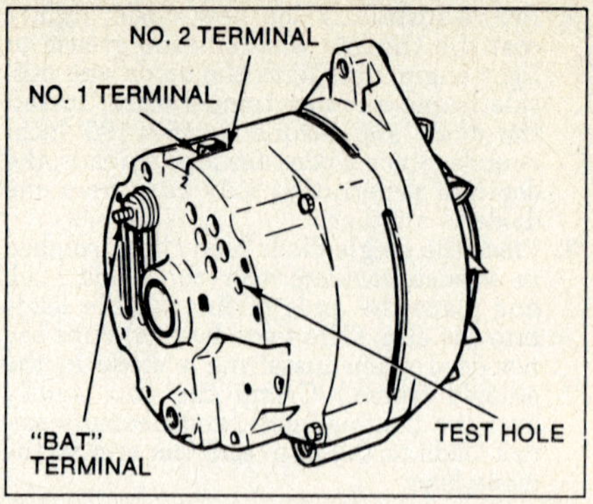

Fig. 4 Delcotron type 12 SI alternator

stamped on alternator frame.

7. If ammeter reading is not within 10 amps of rated output, ground field winding by inserting screw driver in end frame hole, contacting tab. Fig. 6.

NOTE: Do not insert screw driver deeper than one inch, tab is usually located within ¾ inch of casing surface.

8. If reading is within 10 amps of rated output, regulator must be replaced. If reading is not within limits, check field winding, diode trio, rectifier bridge and stator.
9. Turn off all accessories and disconnect ammeter and carbon pile regulator.

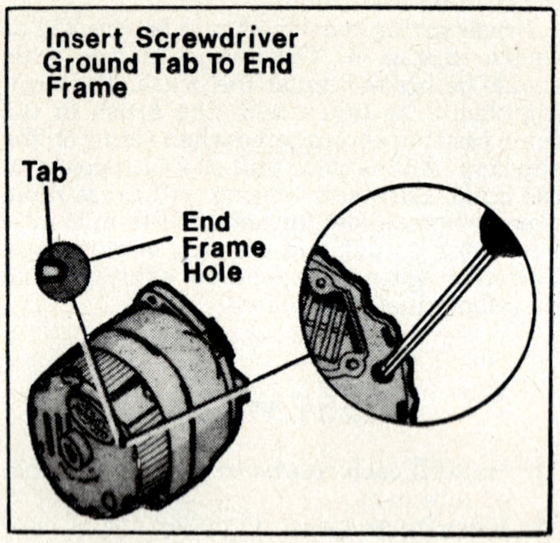

Fig. 6 Grounding field winding

Overcharging Battery

1. Remove alternator from vehicle and separate end frames as outlined under "Alternator Disassembly."
2. Check field winding, if shorted replace rotor and regulator.
3. Connect ohmmeter from brush clip to end frame, set meter on low scale and note reading. Fig. 7.
4. Reverse leads, if both readings are zero remove screw from brush clip and inspect sleeve and insulator.
5. If sleeve and insulator are in good condition, then regulator is at fault and must be replaced.

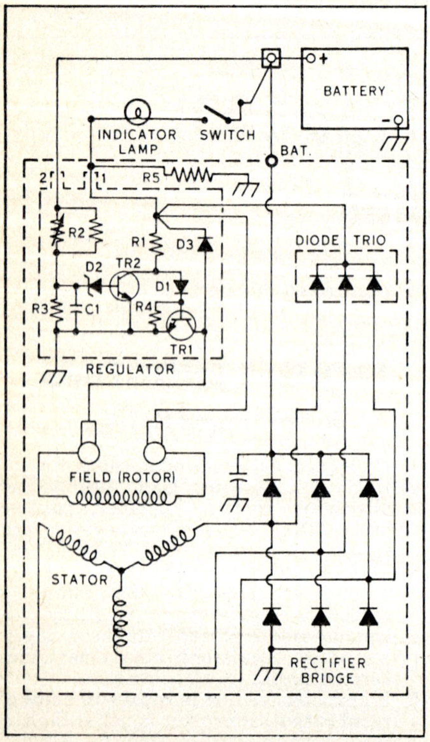

Fig. 5 Wiring diagram of charging circuit

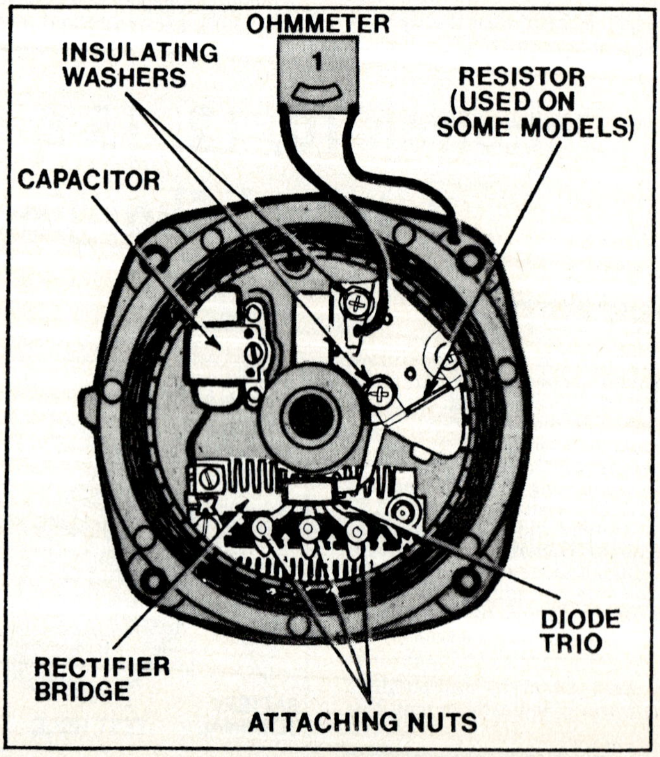

Fig. 7 Testing brush clip

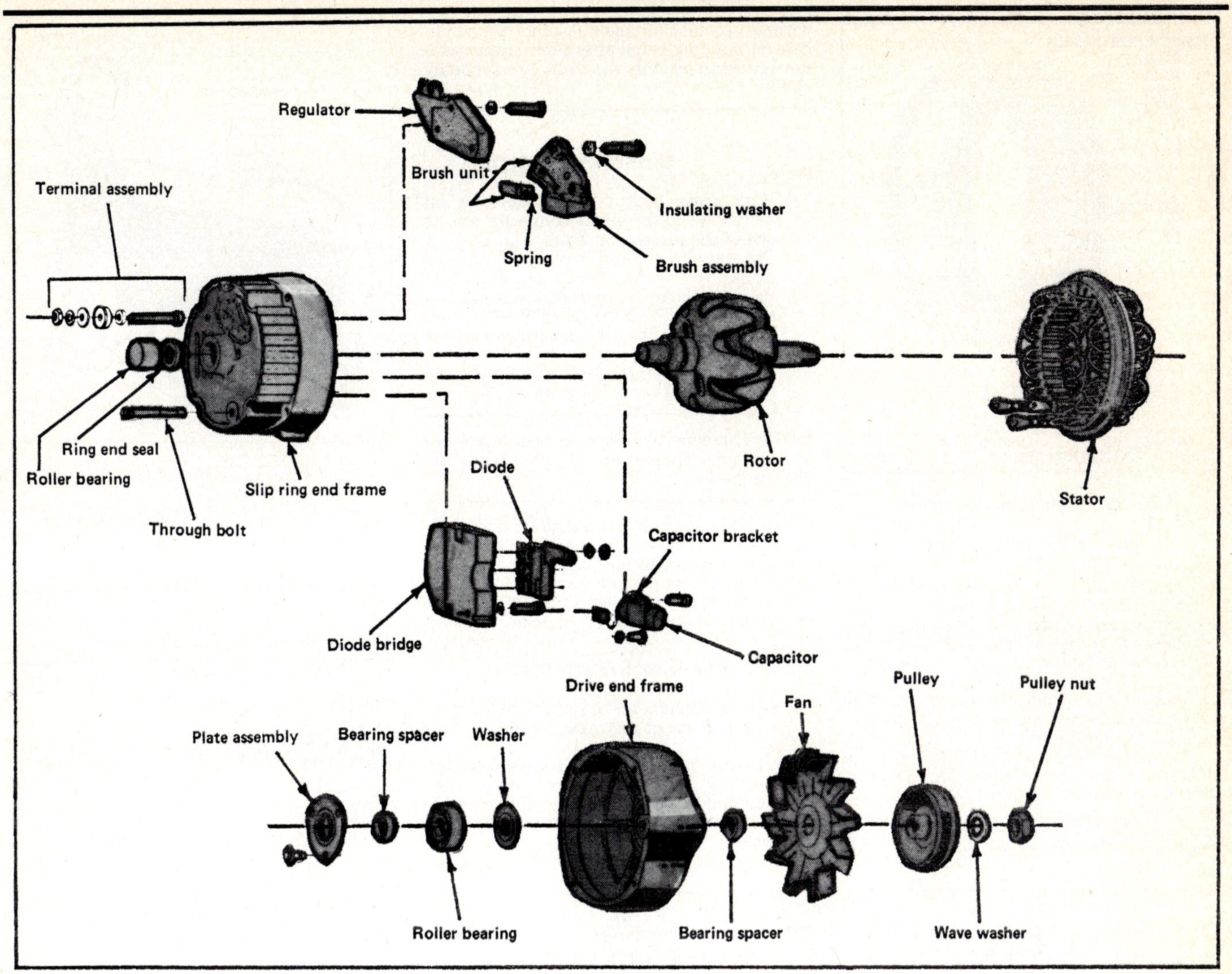

Fig. 8 Alternator disassembled

ALTERNATOR DISASSEMBLY

NOTE: When pressing bearings or seals from end frames, support frames from inside.

1. Scribe mark across end frames and stator ring so parts can be installed in same position.
2. Remove four through bolts, then using screw driver in stator slot pry end frames apart, Fig. 8.

NOTE: Brushes may fall from holders and become contaminated with bearing grease, if so they must be cleaned prior to assembly.

3. Place tape over slip ring end frame bearing and shaft at slip ring end.
4. Remove nut, washer, pulley, fan and collar from rotor shaft, then slide drive end frame from shaft.
5. Remove bearing, retainer and seal from drive end frame.
6. Remove attaching bolts, then pry stator from slip ring end frame.
7. On units except 27 SI, remove ground screw and battery terminal stud nut from rectifier bridge, then remove rectifier bridge, terminal stud and insulating washer from end frame. On 27 SI units, remove two ground screws, connector strap screw and battery stud nut, then remove rectifier bridge, connector, terminal stud and insulating washers from end frame.

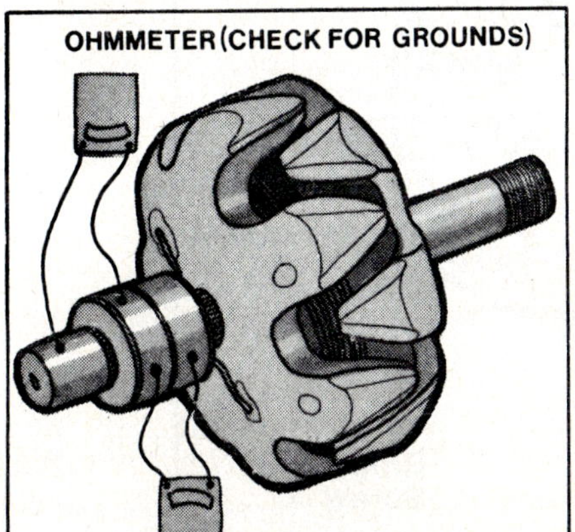

Fig. 9 Testing rotor and slip rings

8. Remove resistor (if equipped), brush holder and regulator.
9. Remove bearing and seal from slip ring end-frame.

BENCH TESTS

Rotor & Slip Ring Test

NOTE: Ohmmeter must be at low scale setting during this test.

1. Inspect rotor for wear or damage.
2. Touch ohmmeter leads to slip rings. Fig. 9.
3. If reading is high an open circuit exists in windings.
4. If reading below 2.5 ohms is obtained winding is shorted.
5. If a reading above 3 ohms is obtained excessive resistance exists in windings.
6. Connect one ohmmeter lead to rotor shaft and touch slip rings with other lead, if any reading is obtained there is a ground in the circuit. Fig. 9.

NOTE: If any of the above problems are present the rotor assembly must be replaced.

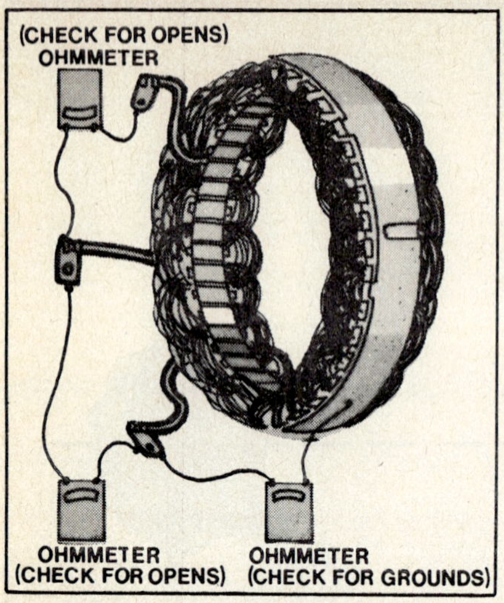

Fig. 10 Testing stator winding

Stator Winding Test

1. Inspect stator for discolored windings, loose connections and damage.
2. Connect an ohmmeter from stator lead to frame, if any reading is obtained windings are grounded. Fig. 10.
3. On 10 SI units, connect ohmmeter between stator leads, if reading is high when connected between each pair of leads, an open circuit exists in windings. The stator windings on 15 SI and 27 SI units cannot be checked for open circuits.

NOTE: Shorted windings are difficult to locate without special equipment. If other tests indicate normal, but rated alternator output cannot be obtained, the windings are probably shorted.

Diode Trio

1. With diode unit removed, connect an ohmmeter to the single connector and to one of the three connectors, Fig. 11.
2. Observe the reading. Reverse ohmmeter leads.
3. Reading should be high with one connection and low with the other. If both readings are the same, unit must be replaced.
4. Repeat between the single connector and each of the three connectors.

NOTE: There are two diode units differeing in appearance. These are completely interchangeable.

The diode unit can be checked for a grounded brush lead while still installed in the end frame by connecting an ohmmeter from the brush lead clip to the end frame as in Steps 1 and 2 above. If both readings are zero, check for a grounded brush or brush lead.

Rectifier Bridge Test

1. Connect ohmmeter to the grounded heat sink and one of the three terminals, Fig. 12.
2. Observe the reading then reverse leads.
3. Reading should be high with one connection and low with the other. If both readings are the same, unit must be replaced.
4. Repeat test for each of the other terminals.

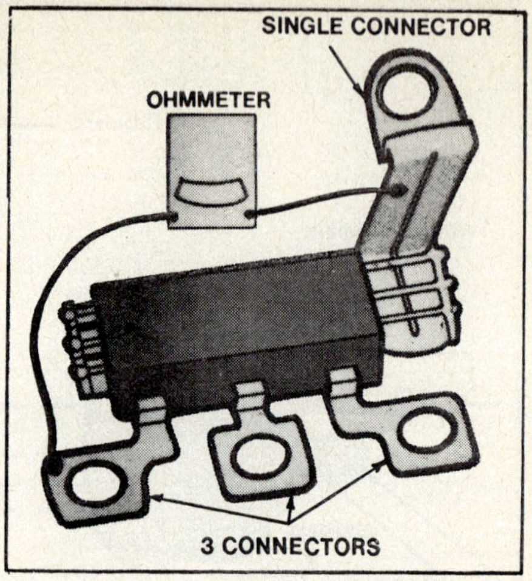

Fig. 11 Testing diode trio

Voltage Regulator/Brush Lead Test

Connect an ohmmeter from the brush lead clip to the end frame, note reading, then reverse connections. If both readings are zero, either the brush lead clip is grounded or the regulator is defective.

ALTERNATOR ASSEMBLY

NOTE: When pressing bearings or seals, end frames must be supported from inside.

1. Lightly lubricate seal and position on slip ring end frame with lip facing toward rotor. Fig. 8.
2. Press seal part way into housing.

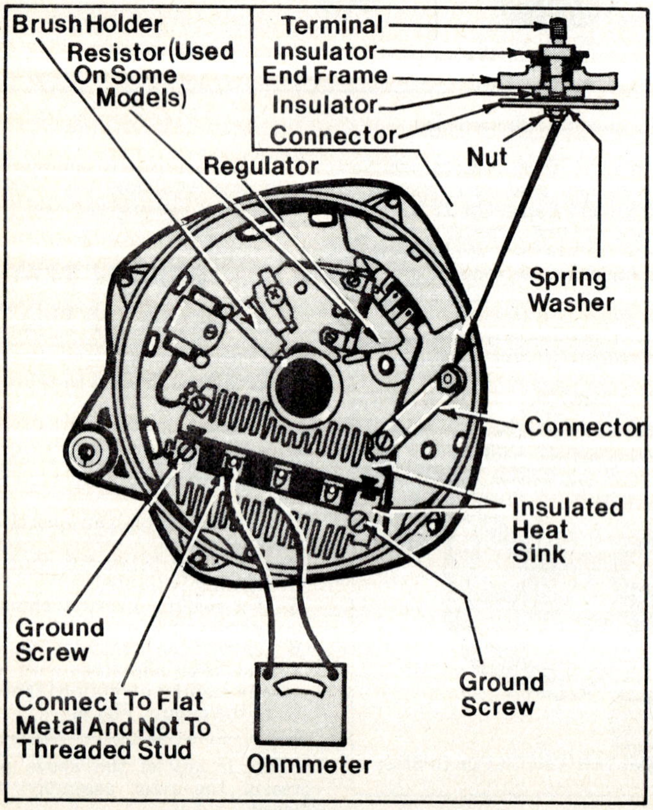

Fig. 12 Testing rectifier bridge diodes

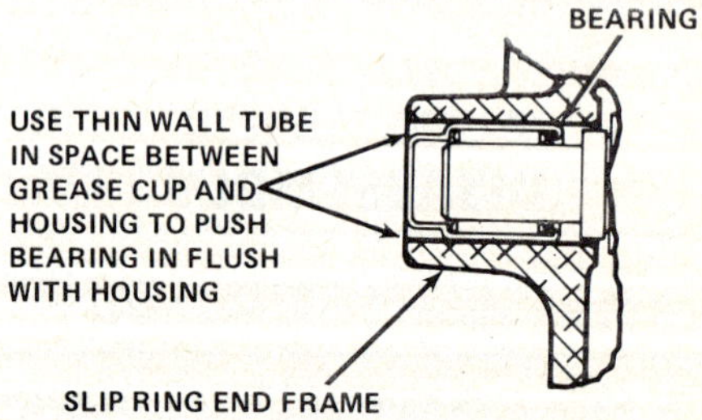

Fig. 13 Slip ring end frame bearing installation. 15 SI units

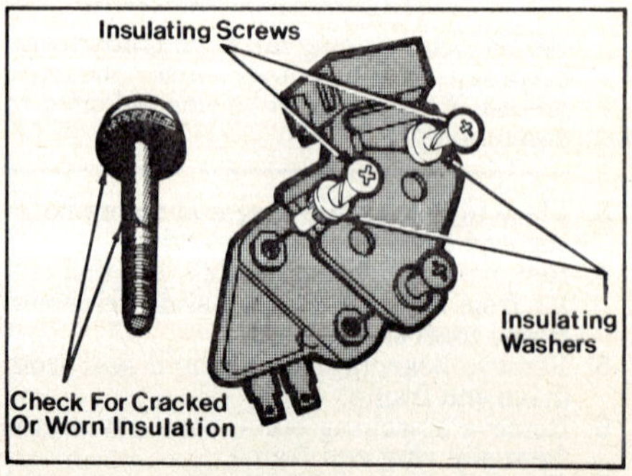

Fig. 14 Brush holder regulator installation

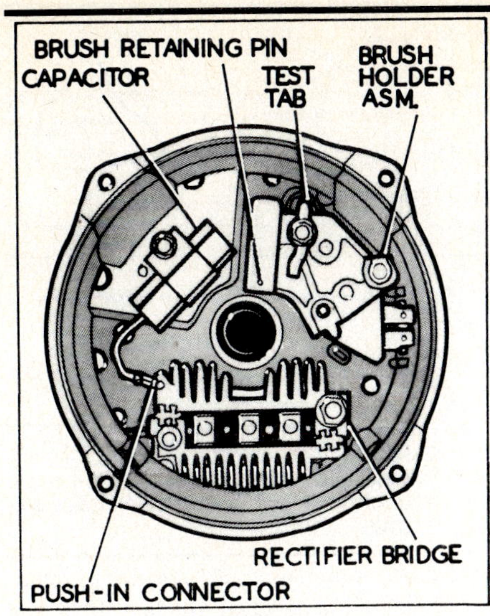

Fig. 15 Slip ring end frame. 15 SI units

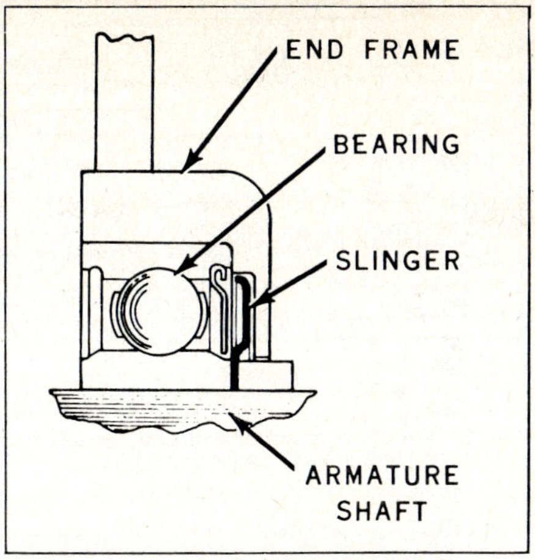

Fig. 16 Drive end frame bearing and slinger installed

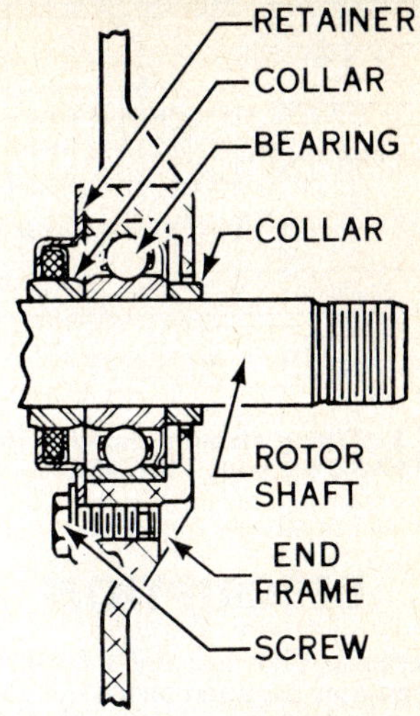

Fig. 17 Drive end frame bearing installation. 15 SI units

3. On 10 SI and 27 SI units, position bearing and end plug on slip ring end frame, then press bearing and plug in until flush with end frame. On 15 SI units, refer to Fig. 13 when installing slip ring end frame bearing.
4. Place regulator in end frame, install brushes and springs in brush holder, use pin to hold brushes in compressed position.

NOTE: Insulating washers are installed under two of the attaching screws, Fig. 14.

5. On units except 27 SI, install insulating washer, battery terminal stud, rectifier bridge and ground screw, then secure bridge to stud with nut and tighten ground screw and nut. On 27 SI units, install rectifier bridge to end frame using two ground screws, then install terminal, insulators and connector and tighten screw and nut.
6. Install diode trio, ensure current only flows one way through single connector.
7. Install capacitor.

NOTE: On 15 SI units, the capacitor lead uses a push clip connector to attach the rectifier bridge instead on a screw, Fig. 15.

8. Install stator, check the three leads for continuity, ensure stator is not grounded against case or holder.
9. On 10 SI and 27 SI units, position slinger on drive end frame, then press ball bearing into end frame, Fig. 16. On 15 SI units, refer to Fig. 17 when installing drive end frame bearing.
10. Fill seal cavity ¼ full with special alternator lubricant, then install retainer.
11. Install rotor in drive end frame, then install collar, fan, pulley, washer and nut.
12. Align scribe marks on end frames and stator plate, install through bolts and remove brush retaining pins.

Delcotron 25 SI Integral Charging System

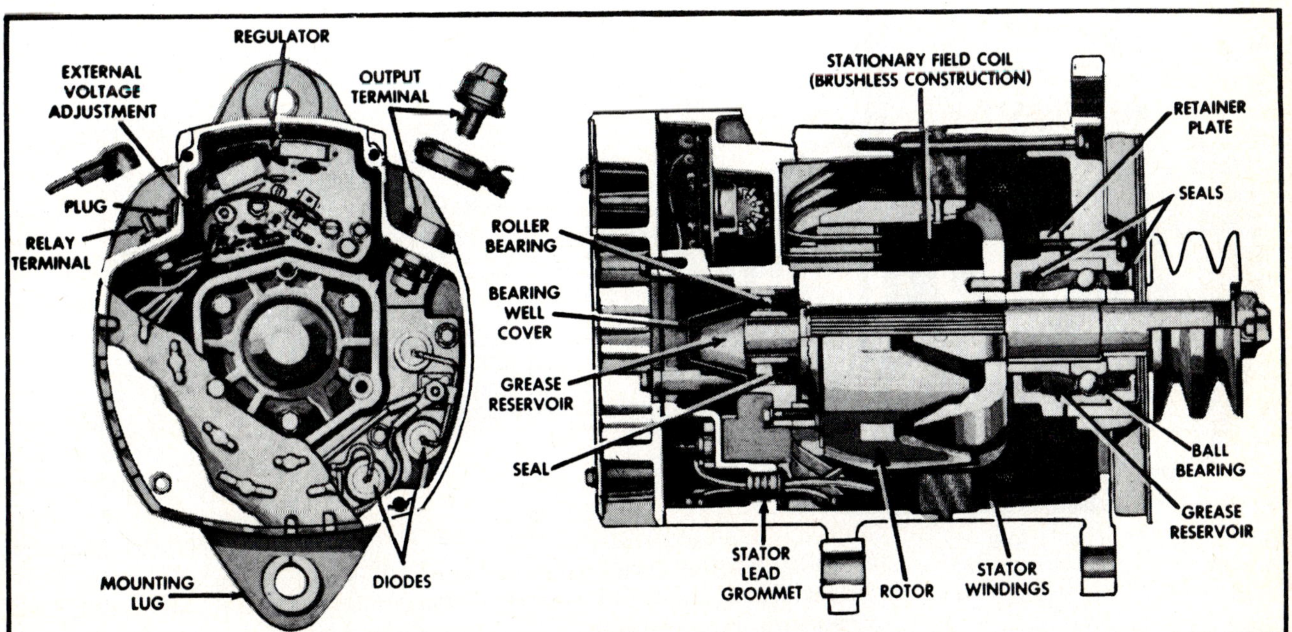

Fig. 1 Delcotron 25 SI Integral Charging System

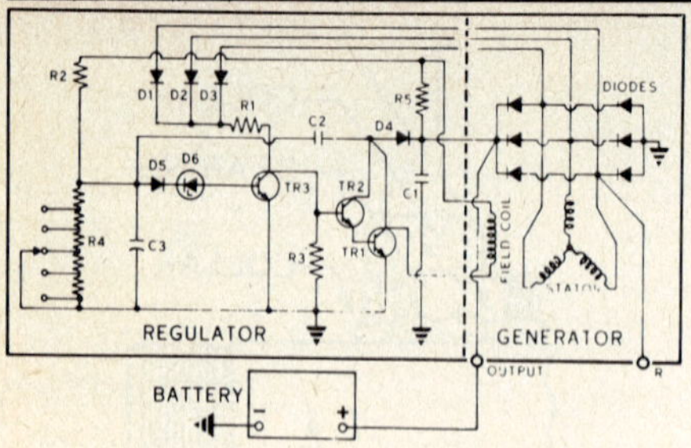

Fig. 2 Wiring diagram charging circuit. 1974–75 & early 1976 25 SI units

DESCRIPTION

These units, Figs. 1, 2 and 3, feature a solid state regulator mounted inside the alternator slip ring end frame, along with the brush holder assembly. All regulator components are enclosed in a solid mold with a provision for adjustment of the regulator. A rectifier bridge, containing six diodes and connected to the stator windings, changes A.C. voltage to D.C. voltage which is available at the output terminal. Generator field current is supplied through a diode trio which is also connected to the stator windings. The diodes and rectifiers are protected by a capacitor.

ALTERNATOR OUTPUT TEST

1. Install test equipment, Fig. 4.
2. With the carbon pile turned "Off", run engine at moderate speed and note voltage reading. Voltage should be 13 to 15 volts.

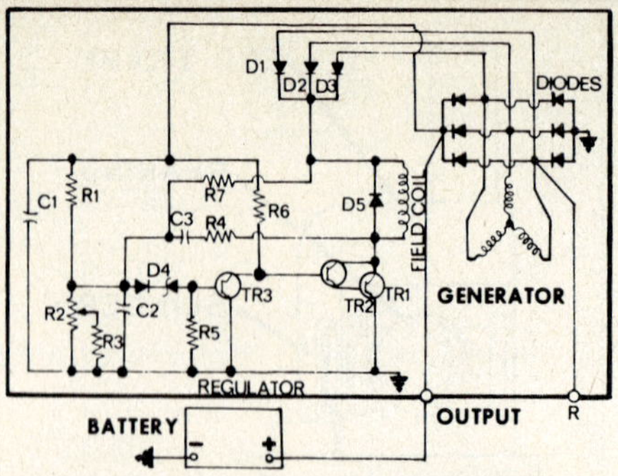

Fig. 3 Wiring diagram of charging circuit. Late 1976 & 1977–83 25 SI units

3. If voltage reading is above 15 volts, check regulator lead for proper connection. If satisfactory, check field coil for shorts. If field coil is shorted, replace field coil and regulator.
4. If voltage reading in step 2 is less than 13 volts increase engine speed and adjust carbon pile to obtain maximum current output. If maximum current is within 10 amp of specified rating, the voltage regulator setting may be changed as follows.
   a. On 1974–75 and early 1976 units, remove pipe plug from internal charging system, Fig. 1. On late 1976 and 1977–83 units, remove end cover to expose voltage adjustment potentiometer, Fig. 5.
   b. Rotate adjusting screw one or two notches clockwise to raise voltage setting.

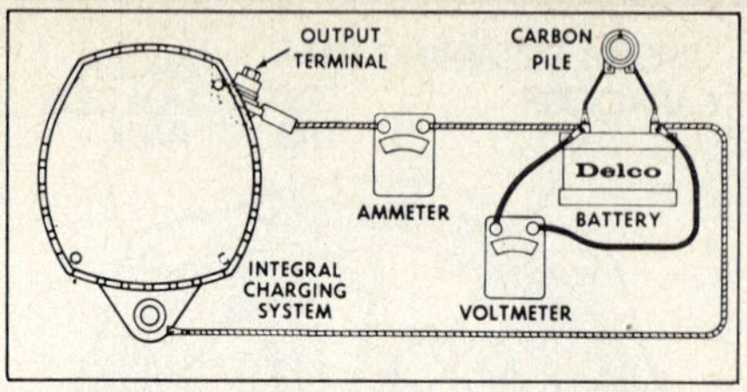

Fig. 4 Alternator output test connections

NOTE: For an overcharged battery condition, rotate adjusting screw counterclockwise to lower the voltage setting.

   c. On 1974–75 and early 1976 units, install pipe plug. On late 1976 and 1977–83 units, install end cover.
5. If maximum current output in step 4 is not within 10 per cent of specified rating and is less than 5 amps, check field coil and if open, replace only the field coil. If field coil is grounded or shorted, replace both the field coil and the regulator. If field coil tests satisfactory, replace regulator.
6. If maximum current output is 5 amps or more in step 4, but is less than specified rating, check field coil, stator and diodes as outlined under "Bench Tests".

ALTERNATOR, DISASSEMBLE, FIG. 6

1. Remove end cover, plate and gasket.
2. If field coil, stator or end frame bearings are to be replaced, remove drive end nut, washer, pulley, fan, slinger and collar from rotor shaft. Scribe an alignment mark across end frame halves, then remove four through bolts and separate rectifier end frame from drive end frame by prying in stator slot.

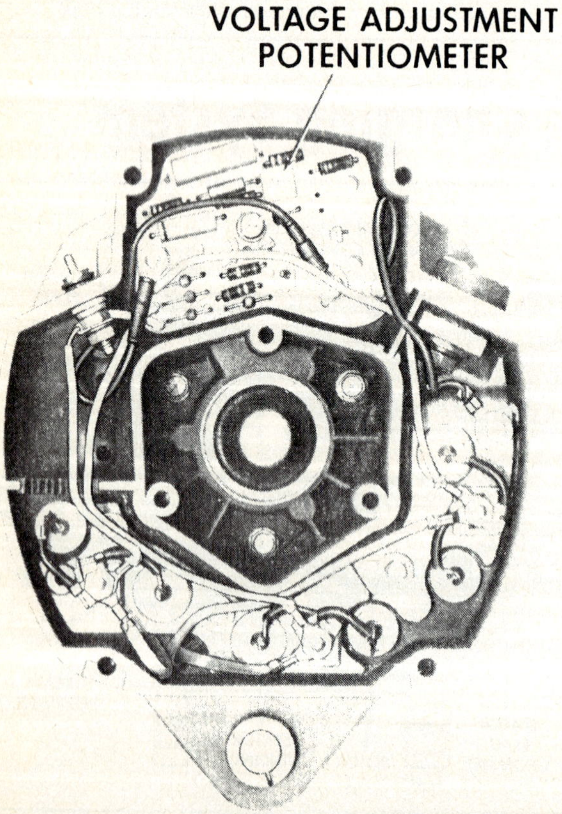

Fig. 5 Voltage regulator adjustment. Late 1976 & 1977–83 25 SI units

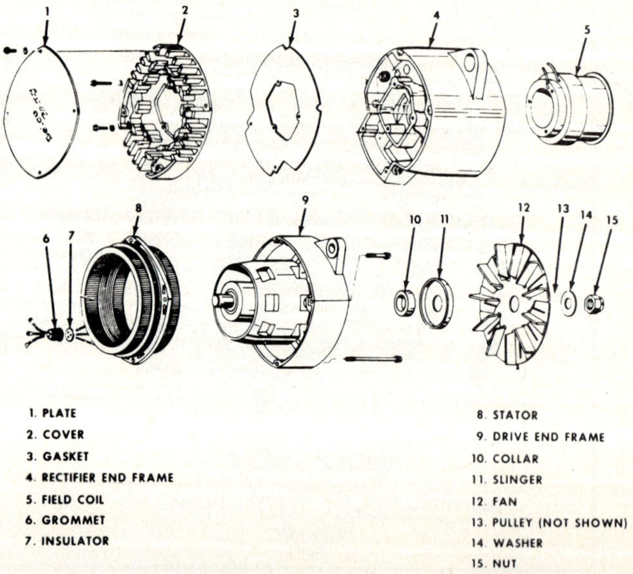

Fig. 6 Disassembled view of Delcotron 25 SI alternator

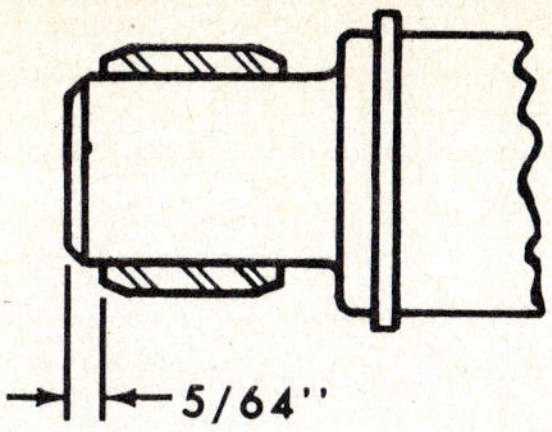

Fig. 7 Bearing inner race location

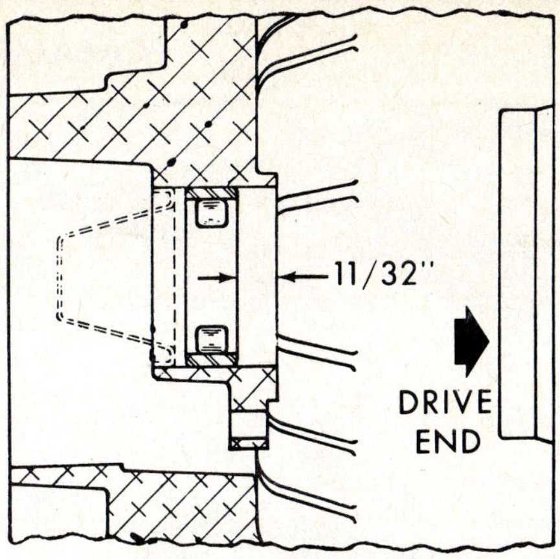

Fig. 8 Bearing location

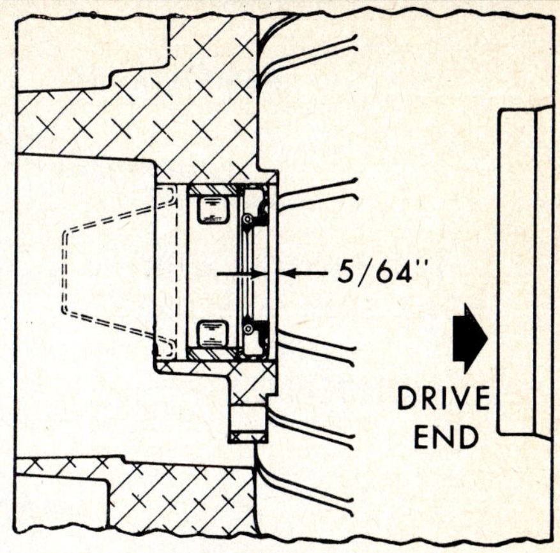

Fig. 9 Seal location

Voltage Regulator, Replace

1. Disconnect three identically colored voltage regulator leads from R terminal and heat sink.
2. Remove two regulator attaching screws, then disconnect red regulator lead from rectifier bridge heat sink, if equipped.
3. Disconnect field coil leads from regulator, then remove regulator from end frame.
4. Reverse procedure to install.

Field Coil, Replace

1. Remove three field coil attaching screws, then disconnect field coil leads from voltage regulator.
2. Remove field coil from end frame, then pull leads and grommet through hole in end frame.
3. Reverse procedure to install. Apply silicone grease to grommet before installing.

Stator, Replace

1. Disconnect stator leads from heat sink terminals.
2. Using a screwdriver, carefully pry stator frame from end frame. Do not pull stator from end frame.
3. Using a screwdriver, carefully drive stator lead grommet in direction of stator removal, from end frame bore, then remove stator from end frame.
4. Reverse procedure to install. Apply silicone grease to stator lead grommet before installing. Position stator on end frame so that through bolt holes are aligned.

Rectifier Diodes, Replace

Heat Sink Diode Replacement

1. Disconnect all lead wires from heat sink and junction terminals.
2. Remove heat sink attaching screws and output terminal.

NOTE: Round insulators are located on attaching screws and flat insulators are located under the heat sink.

---

3. Remove heat sink from end frame.
4. Support heat sink at terminal side of diode using tool No. J-9717-2, then press diode from heat sink using tool No. J-9717-1.
5. With heat sink supported on a suitable tool, press diode in using tool No. J-9600-2 until shoulder is flush.
6. Reverse procedure noted in steps 1

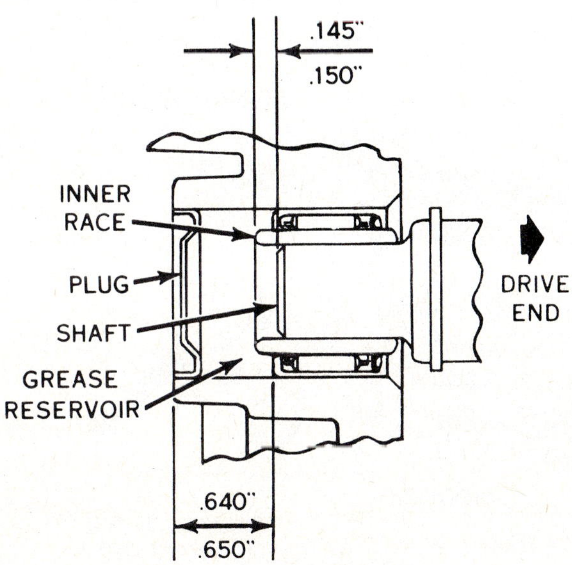

Fig. 10 Rectifier end frame bearing installation dimensions. Late 1976 & 1977–83 25 SI units

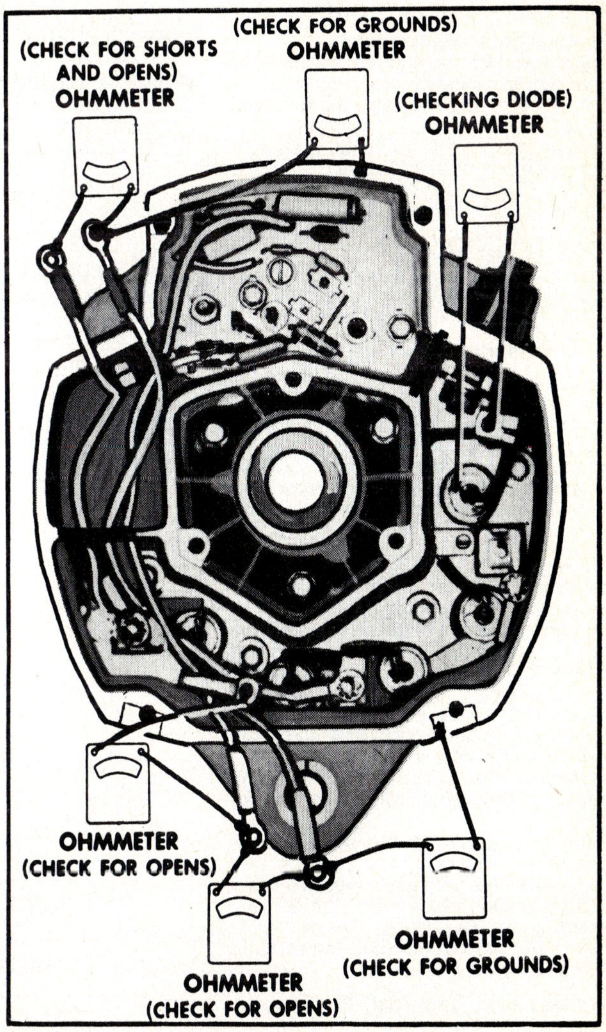

Fig. 11 Bench test connections

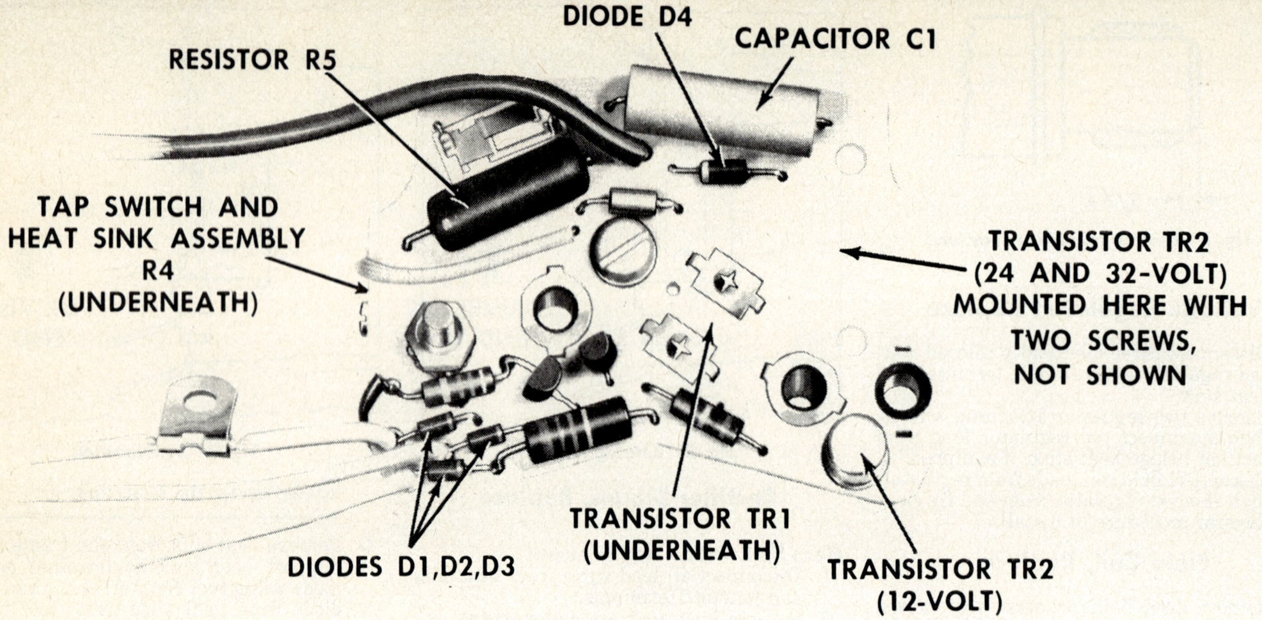

Fig. 12 Regulator panel board serviceable components. 1974–75 & early 1976 25 SI units

through 3 to install heat sink. Apply silicone grease to both sides of flat insulators before installing under heat sink. Torque output terminal nut to 75 ft. lbs.

End Frame Diode Replacement

1. Separate end frames and remove stator as described previously.
2. Remove positive diode heat sink from end frame.
3. Remove lead wire from diode, then drill center of diode with a 19/64 inch drill to a depth of 1/4 inch. Tap diode using a 3/8-16 tap, then insert threaded shaft of slide hammer J-6125-1 into diode and remove diode from end frame.
4. Support end frame on a flat surface, then press diode into end frame using tool No. J-9600-1 until shoulder is flush.

Drive End Frame Bearing, Replace

Removal

1. Remove four bearing retainer attaching screws, then press rotor from end frame.
2. Using a suitable puller, remove bearing from rotor shaft, then remove collar gasket and bearing retainer from rotor.
3. Remove grease seals from bearing retainer and end frame.

Installation

1. Install seals in bearing retainer and end frame with seal lip facing bearing.
2. Fill each reservoir between bearing and seal to the 3/4 full level with lubricant 1948791 or equivalent. Arrange lubricant so that a portion will be in contact with the bearing after assembly. Add lubricant to each seal lip and cavity between lip and steel cage of each seal.
3. Lubricate collar, then install bearing retainer, gasket and collar onto rotor shaft.
4. Using a metal tube against bearing inner race, press bearing onto rotor shaft.
5. With end frame properly supported using a suitable tool, press rotor and bearing assembly into end frame.
6. Install and tighten four bearing retainer attaching screws.

Rectifier End Bearing, Replace

REMOVAL

1. Pull inner race from shaft.
2. Remove old seal and bearing from housing.

INSTALLATION

1974–75 & Early 1976 Units

1. Press inner race onto shaft to dimension shown in Fig. 7.
2. To aid bearing installation, heat bearing in an oven to 200–300°F. Then, press against bearing outer race to install in proper location, Fig. 8.
3. Fill bearing well cover approximately 3/4 full with lubricant, P/N 1948791. A portion of the lubricant should contact bearing after assembly. Press bearing well cover into housing.
4. Lubricate seal lip and fill cavity between rubber lip and steel case of seal with lubricant. Then, install seal with lip facing toward bearing to specified dimension, Fig. 9.

Late 1976 & 1977–82 Units

1. Press bearing inner race onto rotor shaft to dimension shown in Fig. 10.
2. Using tool Nos. J-29033-1 and J-8092, press bearing outer race with integral se facing toward drive end into housing to dimension shown in Fig. 10.

NOTE: Support opposite side of end frame housing on a suitable tool or metal tube when pressing in bearing outer race.

3. Press plug into housing, then fill reservoir to the 1/2 full level with lubricant 1948791 or equivalent. Arrange lubricant so that a portion will contact bearing when assembled.

BENCH TESTS

Field Coil Tests

1. Disconnect field coil leads from regulator

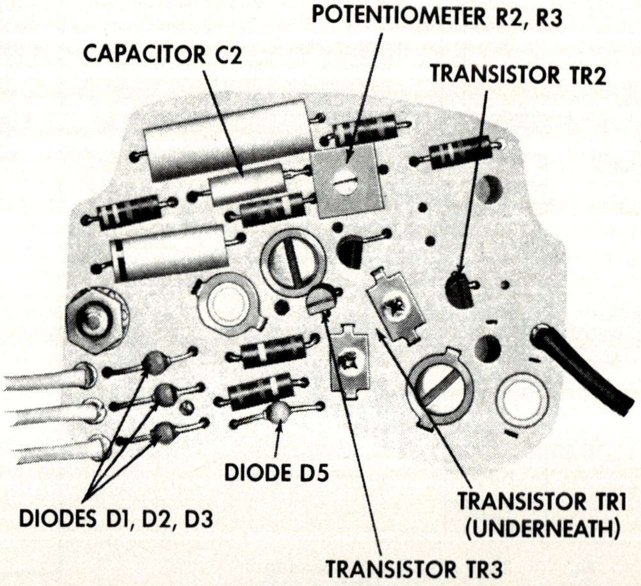

Fig. 13 Regulator panel board serviceable components. Late 1976 & 1977–83 25 SI units

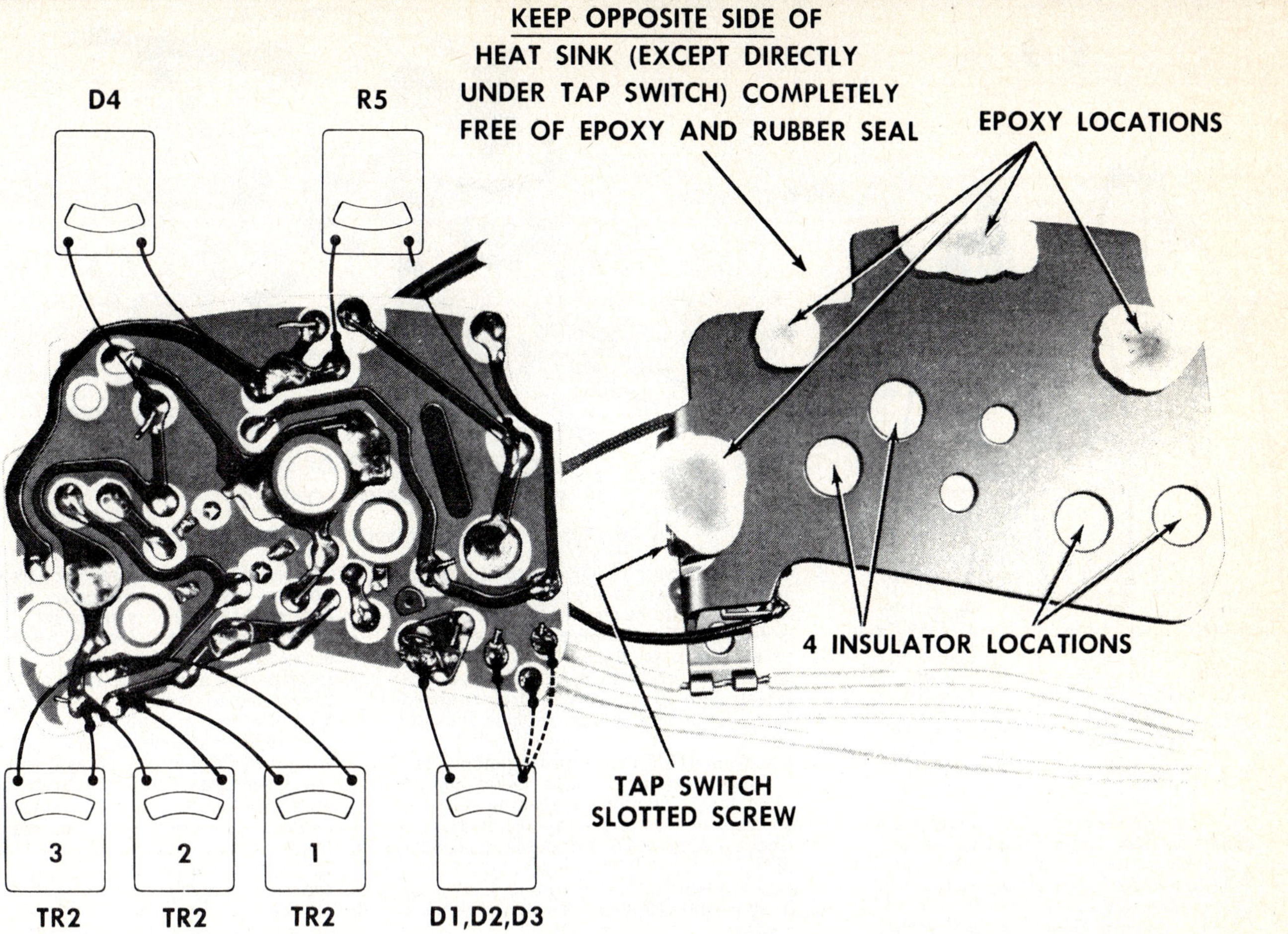

Fig. 14 Checking regulator components. 1974–75 & early 1976 25 SI units

and diode and stator leads from heat sink terminals.

2. To check for grounds, connect one ohmmeter lead to field coil lead and the other ohmmeter lead to the end frame, Fig. 11. If ohmmeter reading is low, the field coil is grounded.
3. To check for open circuits, connect ohmmeter between two field coil leads, Fig. 11. If an infinite reading is obtained, the field coil is open.
4. To check for short circuits, connect a battery and ammeter in series with the field coil and note ammeter reading. Ammeter reading should be 4.1 to 4.5 amps. at 80 degrees F. An ammeter reading above 4.5 amps., indicates a shorted winding. This check may also be performed by connecting an ohmmeter to the field coil, Fig. 11. Resistance reading should be 2.67 to 2.93 ohms. If resistance is below the specified value, the winding is shorted.

Diode & Stator Tests

1. Remove three stud nuts, three regulator leads, three stator leads, six diode leads and R terminal lead from stud. With ohmmeter set at low scale range, connect one lead to diode lead and other lead to alternator case, Fig. 11, and note reading. Reverse leads, one reading should be high and the other low. If both readings are the same the diode must be replaced.
2. To check for grounded stator windings, connect an ohmmeter from any stator lead to alternator case, Fig. 11, if a low reading is obtained stator windings are grounded. To check stator windings for an open circuit, connect an ohmmeter between each pair of stator leads, Fig. 11, if reading is high the windings are opened. To check stator windings for a short circuit special equipment must be used. If all other checks are normal but alternator fails to supply rated output stator windings may be shorted.

Regulator Tests & Repairs

NOTE: The regulator must be removed from the alternator before tests and repairs can be performed. Serviceable components of the circuit board are shown in Figs. 12 and 13.

1. Remove screw, and transistor TR1, then using a screwdriver, carefully pry panel board and heat sink apart.
2. Inspect printed circuit for improperly soldered joints and damaged components.

1974–75 & Early 1976 Units

Perform the following checks using a 1½ volt ohmmeter. Using a sharp instrument, carefully scratch through transparent coating over soldered joints to make ohmmeter connections. When performing ohmmeter checks, reverse leads to obtain two readings. When replacing components, always use a 50 watt soldering iron.

1. Check tap switch and heat sink assembly as follows:
   a. Using a screwdriver, rotate slotted screw to 5 positions and if screw is loose, replace assembly. Assemblies that incorporate the use of a brass screw and nut should be replaced with assemblies having an aluminum screw and no attaching nut whether the screw is loose or not.
2. Using an ohmmeter, check resistor R5, Fig. 14. If either reading is above 1 ohm, replace resistor.
3. Check transistor TR2 as shown in Fig. 14. With ohmmeter connected as shown in steps 1 and 2 Fig. 14, one low and one high reading should be obtained. With ohmmeter connected as shown in step 3 Fig. 14, both readings should be very high. If reading differs from the above, replace transistor.

NOTE: The replacement transistor may be a small black transistor with a red dot and flat side. When installing the diode, the flat side should face diodes D1, D2 and D3, Fig. 12.

4. Check diode D4 as shown in Fig. 14. One high and one low reading should be obtained, if not, replace diode.
5. Check diodes D1, D2 and D3 separately, Fig. 14. One high and one low reading should be obtained, if not, replace diode being checked.
6. Check transistor TR1 as shown in Fig. 15. With ohmmeter connected as shown in step 1 Fig. 15, both readings should be

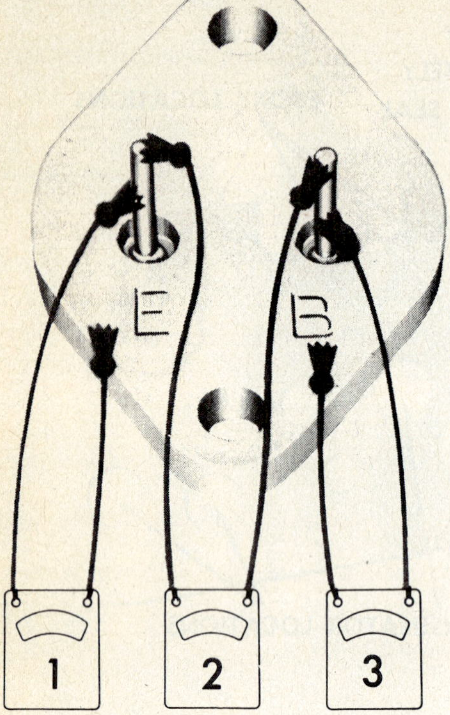

Fig. 15 Checking regulator transistor TR1

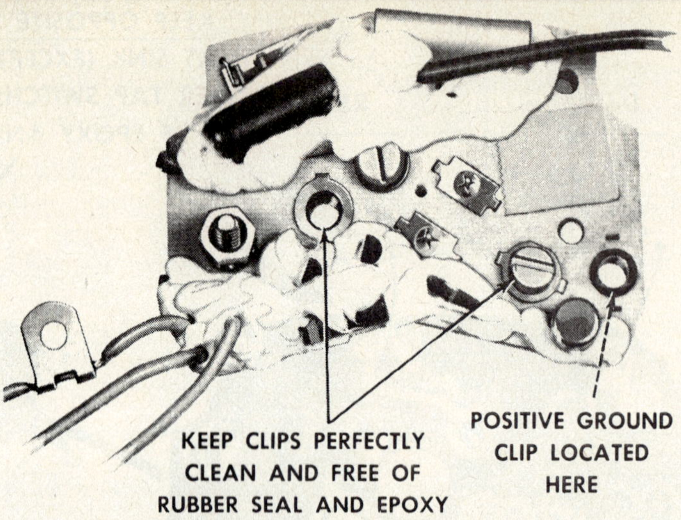

Fig. 16 Applying RTV rubber sealer to circuit board components

very high. With ohmmeter connected as shown in steps 2 and 3, Fig. 15, one high and one low reading should be obtained. If readings differ from those described, replace transistor.

7. Check capacitor C1 for broken or damaged leads.
8. If no defects have been indicated, but regulator fails to operate properly, the regulator assembly must be replaced.
9. Reassemble regulator as follows:
   a. If heat sink is to be reused, burn away epoxy separating circuit board from heat sink using a 50 watt soldering iron. On original or replacement heat sink, apply epoxy at four locations indicated in Fig. 14. Keep opposite side of heat sink free of epoxy.
   b. Using four insulators, assemble heat sink, circuit board and transistor TR1. Apply silicone grease to both sides of mica insulator located between heat sink and transistor.
   c. Apply Dow 732 RTV silicone rubber sealer or equivalent around components, Fig. 16. Use care not to apply sealer to metal clips.
10. Test regulator for proper operation and replace regulator if defective.

Late 1976 & 1977–82 Units

Perform the following checks using a 1½ volt ohmmeter set at low reading scale. Carefully scratch transparent coating over solder to make ohmmeter connections. When replacing soldered components, only use a 50 watt soldering iron.

1. Check transistor TR2 by connecting ohmmeter leads as shown in steps 1, 2 and 3 Fig. 17. With ohmmeter connected as shown in steps 1 and 3, Fig. 17, ohmmeter reading should indicate 5 to 50 ohms. If reading is above 50 ohms or zero ohms, replace transistor. With ohmmeter connected as shown in step 2, Fig. 17, ohmmeter reading should indicate a very high reading; if not replace resistor.
2. Check transistor TR3 by connecting ohmmeter as shown in Fig. 17. A very high reading should be obtained, if not replace resistor. With ohmmeter leads connected as shown in steps 2 and 3, Fig. 17, ohmmeter reading should be approximately 5 to 50 ohms. If reading is above 50 ohms or zero ohms, replace transistor.
3. Check diode D5 by connecting ohmmeter as shown in Fig. 17. Ohmmeter should indicate 5 to 50 ohms. If reading is above 50 ohms or zero ohms, replace diode.
4. Check diode trio D1, D2 and D3 as shown in Fig. 17. Each of the diodes should indicate a reading of 5 to 50 ohms. If reading is above 50 ohms or zero ohms, replace diode.
5. Check capacitor C2 by connecting ohmmeter as shown in Fig. 17. Ohmmeter should indicate a high reading. If reading is zero, replace capacitor.
6. Change ohmmeter setting to the RX10 scale, then check potentiometer with ohmmeter connected as shown in Fig. 17. While rotating potentiometer screw, note ohmmeter reading. Ohmmeter should deflect slightly; if not, replace R2.
7. Change ohmmeter setting to the RX1 scale, then check transistor TR1 as shown in steps 1, 2 and 3 Fig. 15. With ohmmeter connected as shown in step 1 Fig. 15, a very low reading should be indicated; if not, replace transistor. With ohmmeter connected as shown in steps 2 and 3, Fig. 15, a reading of approximately 5 to 50 ohms should be obtained. If reading is above 50 or zero ohms, replace transistor.
8. If no defects have been indicated and regulator is not operating properly, replace the regulator assembly.
9. Reassemble regulator as follows:
   a. Apply silicone grease to both sides of mica under transistor TR1.
   b. Apply Dow 732 RTV silicone rubber sealer around components as shown in Fig. 16. Use care not to apply sealer to metal clips or rubber seal.
10. Test regulator for proper operation and replace if found to be defective.

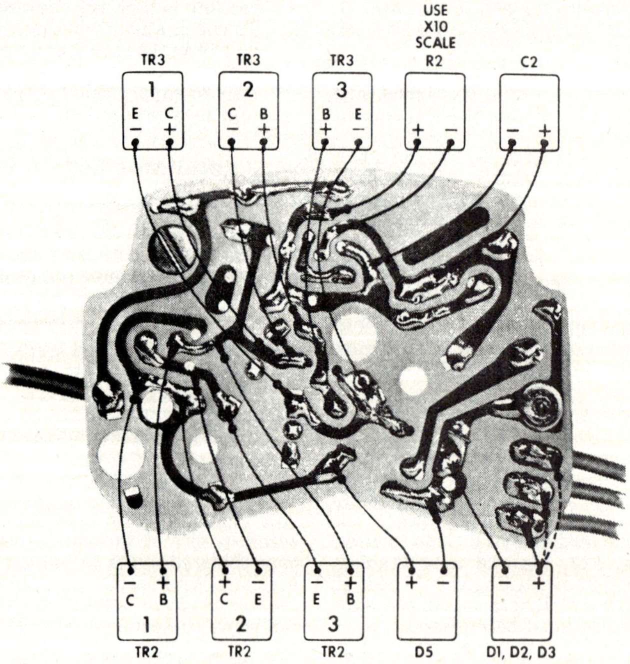

Fig. 17 Checking regulator components. Late 1976 & 1977–83 25 SI units

ALTERNATOR, ASSEMBLE

1. Align through bolt holes, then assemble end frames and install through bolts.
2. Install collar, slinger, fan, pulley, washer and nut on rotor shaft. While using tool No. J-21501 to hold rotor shaft in position, tighten pulley nut to 75 ft. lbs.
3. Install plate, cover and gasket to rectifier end frame, then tighten cover attaching screws.

Delcotron 27 or 51 SI (Type 200) Integral Charging Systems

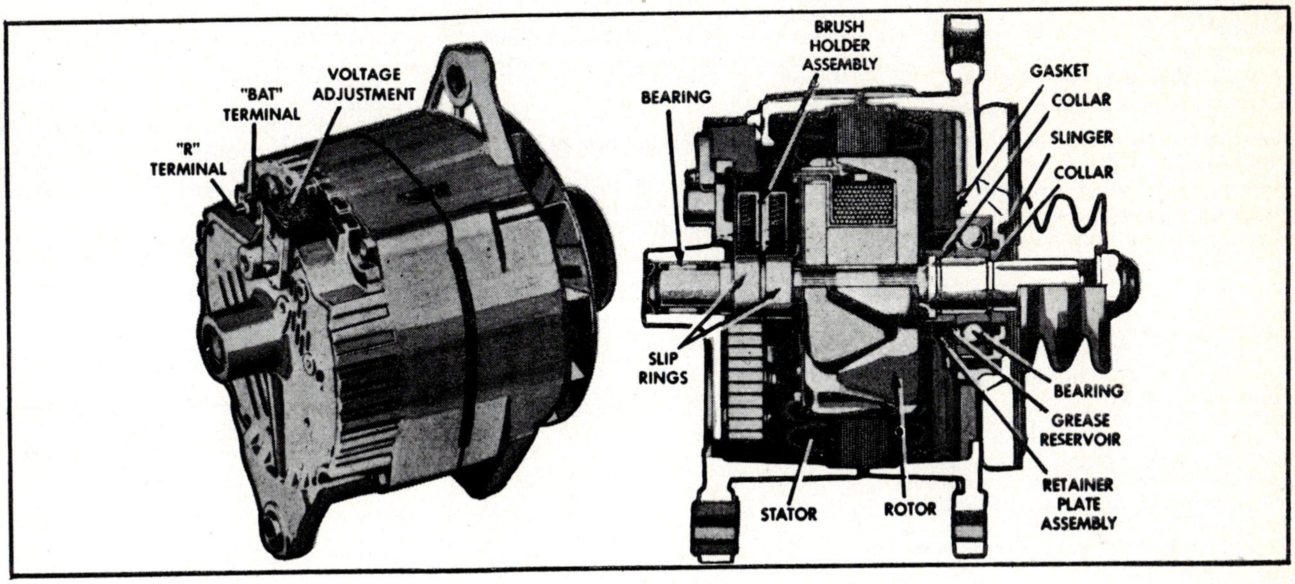

Fig. 1 Delcotron 27 & 51 SI (Type 200) Integral Charging System

DESCRIPTION

This unit, Figs. 1 and 1A, features a solid state regulator mounted inside the alternator slip ring end frame, along with the brush holder assembly. All regulator components are enclosed in a solid mold with a provision for adjustment of the regulator. A rectifier bridge, containing six diodes and connected to the stator windings, changes A.C. voltage to D.C. voltage which is available at the output terminal. Generator field current is supplied through a diode trio which is also connected to the stator windings. The diodes and rectifiers are protected by a capacitor which is also mounted in the end frame.

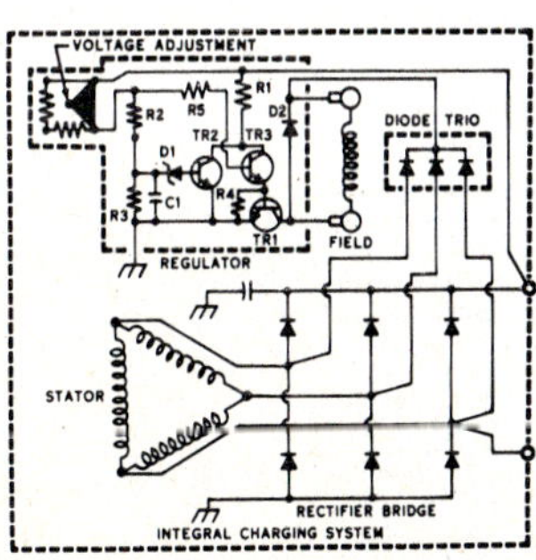

Fig. 1A Wiring diagram of charging circuit

TROUBLE SHOOTING

Undercharged Battery

1. Check alternator drive belt tension and adjust if necessary. Also, check charging system for proper electrical connections.
2. Ensure battery is fully charged at this time.
3. Connect a voltmeter between the alternator "Bat" terminal and the ground, voltmeter check No. 1, Fig. 2. No voltage reading indicates an open circuit between the voltmeter connection and battery.
4. If battery voltage is obtained in step 3, disconnect battery ground cable. Connect voltmeter between battery terminals, voltmeter check No. 2, Fig. 2. Also, connect an ammeter between the battery terminals.
5. Connect battery ground cable.
6. Turn on all accessories, then connect a carbon pile regulator across battery.
7. Operate engine at moderate speed, adjust carbon pile regulator to obtain maximum current output.
8. If ammeter reading is within 10 amps of rated output, alternator is not at fault.

NOTE: Alternator rated output is stamped on alternator frame.

9. If ammeter reading is not within 10 amps of rated output, ground field winding by

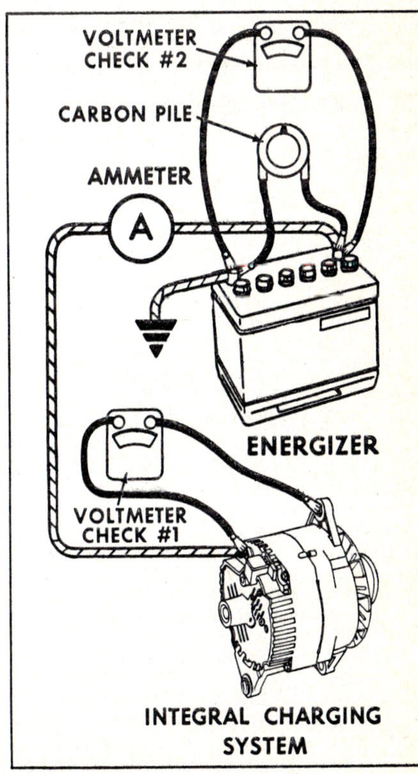

Fig. 2 System test connections

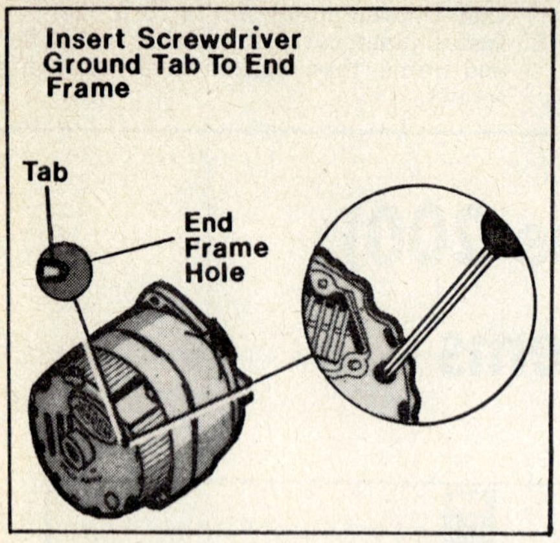

Fig. 3 Grounding field windings

inserting screw driver in end frame hole, contacting tab. Fig. 3.

NOTE: Do not insert screw driver deeper than one inch, tab is usually located within ¾ inch of casing surface.

10. If reading is within 10 amps of rated output, regulator must be replaced. If reading is not within limits, check field winding, diode trio, rectifier bridge and stator.
11. Turn off all accessories and disconnect ammeter and carbon pile regulator.

The voltage setting of this alternator may be adjusted to correct the undercharged condition. Remove voltage adjusting cap from alternator and raise the setting by rotating the cap in 90° increments. Note that in Fig. 4 the cap is set for medium high voltage. With position 2 aligned with the arrow, the setting is medium low. The "Lo" position is the lowest regulator setting and position "Hi" is the highest regular setting. After adjusting the regulator setting, check for improved battery condition after a reasonable length of time in service.

Overcharged Battery

1. Ensure battery is fully charged and connect a voltmeter between the alternator "Bat" terminal and the ground, voltmeter check No. 1, Fig. 2.
2. Start engine and with all accessories turned on, increase engine speed to obtain maximum voltage reading.
3. If maximum voltage is above 15 volts, the alternator is defective.

ALTERNATOR DISASSEMBLY

1. Remove through bolts from drive end frame, Fig. 4A.
2. Separate slip ring end frame and stator assembly from drive end frame and rotor assembly.
3. Remove stator lead attaching nuts and separate stator from end frame.
4. Place a piece of pressure sensitive tape over bearing and shaft.
5. Inspect all leads for burned connections or opens. Check brushes for excessive wear and springs for distortion or discoloration.
6. During servicing and reassembly, hold brushes and springs in holder using a pin or toothpick inserted through end frame.

Drive End Bearing, Replace

1. Remove shaft nut, pulley, fan and slinger.
2. Press rotor from end frame.
3. Remove retainer plate screws, retainer plate assembly, gasket and collar.
4. Press bearing from end frame.
5. Remove retainer plate and replace felt washer. Then, install retainer plate.
6. Press new bearing into end frame with the sealed side facing away from grease reservoir.
7. Fill grease reservoir approximately ½ full with a suitable lubricant, P/N 1948791 or equivalent.
8. Install gasket and new retainer plate assembly in end frame, then with collar on shaft, press rotor into end frame.
9. Install collar, slinger, fan, pulley and shaft nut. Torque shaft nut to 70-80 ft. lbs.

Slip Ring Service

If the slip rings are dirty they may be cleaned with No. 400 silicon carbide paper and finish polished with crocus cloth. Spin the rotor in a lathe, or otherwise spin the rotor, and hold the polishing cloth against the slip rings until they are clean.

CAUTION: The rotor must be rotated in order that the slip rings will be cleaned evenly. Cleaning the slip rings by hand without spinning the rotor may result in flat spots on the slip rings, causing brush noise.

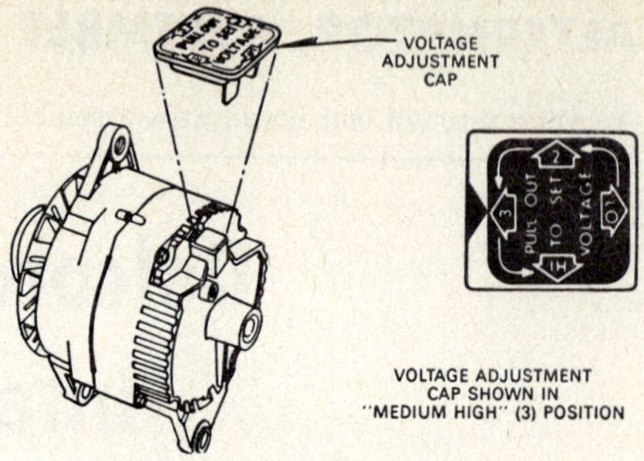

Fig. 4 Voltage adjusting cap

Slip rings that are rough or out-of-round should be trued in a lathe to .002" maximum runout as indicated on a dial gauge. Remove only enough material to make the rings smooth and round. Finish polish with crocus cloth and blow away all dust.

BENCH TESTS

Rotor & Slip Ring Test

NOTE: Ohmmeter must be at low scale setting during this test.

1. Inspect rotor for wear or damage.
2. Touch ohmmeter leads to slip rings. Fig.

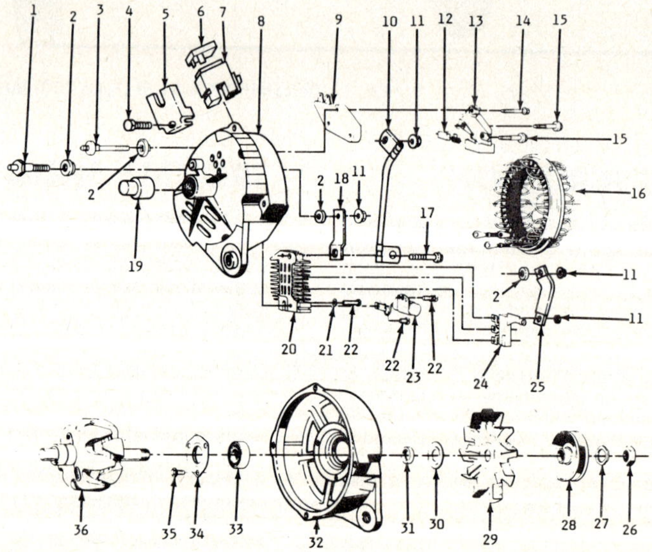

1. Output Terminal
2. Insulator
3. "R" Terminal
4. Screw
5. Clip
6. Cap (Connector)
7. Connector Body
8. Slip Ring End Frame
9. 3 Terminal Regulator
10. Regulator Connector
11. Nut
12. Brushes
13. Brush Holder
14. Ground Screw
15. Insulated Screw
16. Stator
17. Screw
18. Output Terminal Connector
19. Bearing
20. Rectifier Bridge
21. Washer
22. Screw
23. Capacitor
24. Diode Trio
25. "R" Terminal Connector
26. Nut
27. Washer (Wave)
28. Pulley
29. Fan
30. Slinger
31. Collar (Spacer)
32. Drive End Frame
33. Bearing
34. Retainer
35. Screw
36. Rotor

Fig. 4A Disassembled view of 27 & 51 SI Type 200 alternator

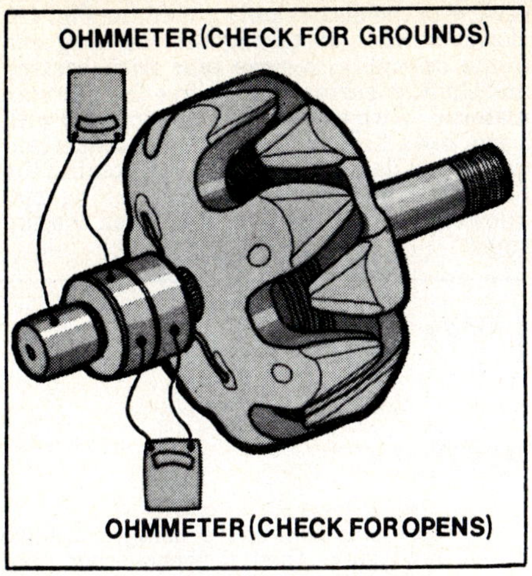

Fig. 5 Testing rotor & slip rings

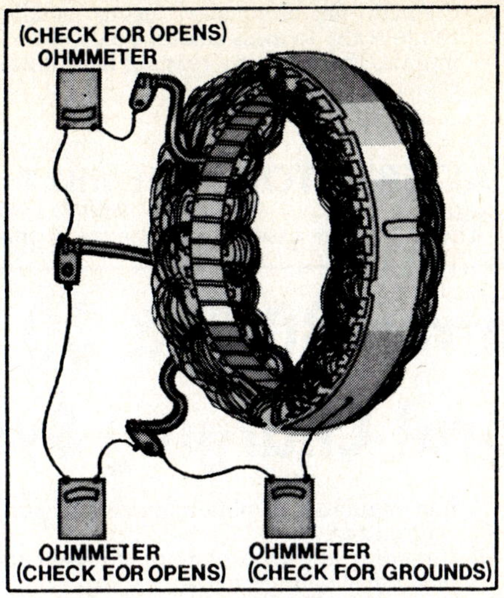

Fig. 6 Testing stator windings

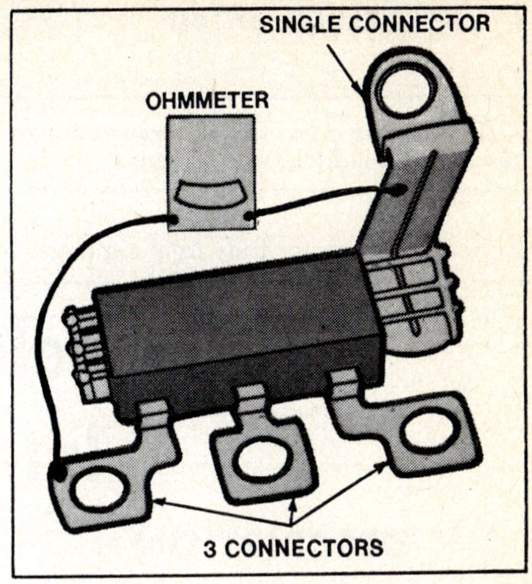

Fig. 7 Testing diode trio

5.
3. If no reading is obtained an open circuit exist in windings.
4. If reading below 3 ohms is obtained winding is shorted.
5. If a reading above 3 ohms is obtained excessive resistance exists in windings.
6. Connect one ohmmeter lead to rotor shaft and touch slip rings with other lead, if any reading is obtained there is a ground in the circuit. Fig. 5.

NOTE: If any of the above problems are present the rotor assembly must be replaced.

Stator Winding Test

1. Inspect stator for discolored windings, loose connections and damage.
2. Connect an ohmmeter from stator lead to frame, if any reading is obtained windings are grounded. Fig. 6.
3. Connect ohmmeter between stator leads, if reading is high when connected between each pair of leads, an open circuit exists in windings.

NOTE: Shorted windings are difficult to locate without special equipment. If other test indicate normal, but rated alternator output cannot be obtained the windings are probably shorted.

Diode Trio

1. With diode unit removed, connect an ohmmeter to the single connector and to one of the three connectors. Fig. 7.
2. Observe the reading. Reverse ohmmeter leads.
3. Reading should be high with one connection and low with the other. If both readings are the same, unit must be replaced.
4. Repeat between the single connector and each of the three connectors.

NOTE: There are two diode units differing in appearance. These are completely interchangeable.

The diode unit can be checked for a grounded brush lead while still installed in the end frame by connecting an ohmmeter from the brush lead clip to the end frame as in Steps 1 and 2 above. If both readings are zero, check for a grounded brush or brush lead.

Rectifier Bridge Test

1. Connect ohmmeter to the grounded heat sink and one of the three terminals. Fig. 8.
2. Observe the reading then reverse leads.
3. Reading should be high with one connection and low with the other. If both readings are the same, unit must be replaced.
4. Repeat test for each of the other terminals.

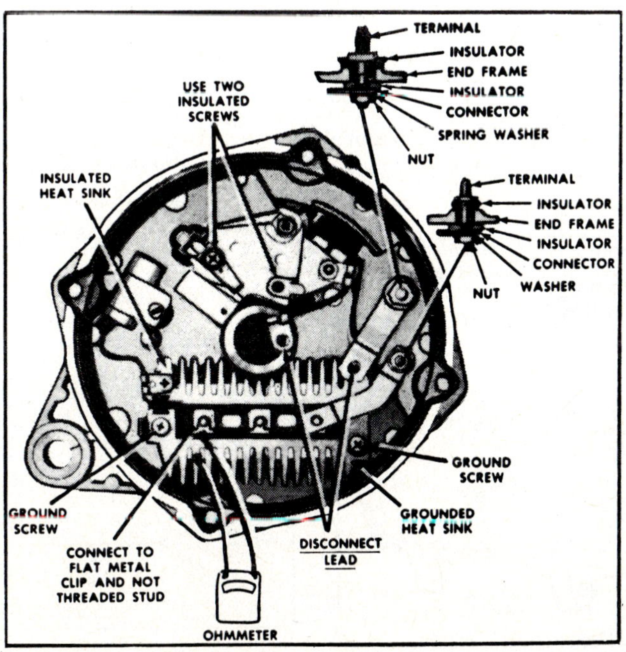

Fig. 8 Testing rectifier bridge

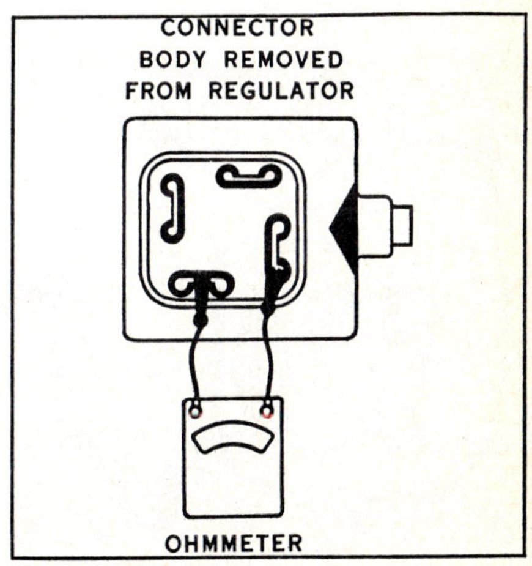

Fig. 9 Testing connector body

Regulator Test

NOTE: This test need only be performed if an overcharged condition was present.

1. Remove connector body from regulator.
2. Connect an ohmmeter, using middle range scale, to each adjacent pair of terminals, Fig. 9. If any reading is infinite, replace connector body.

ALTERNATOR ASSEMBLY

The alternator assembly procedure is the reverse of the disassembly procedure, Fig. 4A. However when installing the slip ring end frame assembly, remove tape from bearing and shaft, ensuring shaft is free from foreign material. With the brushes held in place with a pin, install shaft into end frame, using caution not to damage seal. After tightening the through bolts, remove brush retaining pin, allowing brushes to fall into place on the slip rings.

Delcotron 30 SI & 30 SI/TR Integral Charging System

DESCRIPTION

These units, Figs. 1, 2 and 2A feature a solid state regulator mounted inside the rectifier end frame housing, Figs. 3, 4 and 5. Regulator voltage setting can be adjusted externally by repositioning a voltage adjustment cap. The 30 SI/TR unit is a standard 30 SI with a transformer-rectifier (TR unit) mounted on the end frame, Fig. 5. The TR unit provides a separate voltage to charge a cranking battery, which eliminates the need for a series-parallel switch.

TROUBLE SHOOTING

Undercharged Battery

30 SI

1. Check alternator drive belt tension and adjust if necessary. Also check charging system for proper electrical connections.
2. Ensure battery is fully charged at this time.
3. Connect a voltmeter between alternator "Batt" terminal and ground. No voltage reading indicates an open circuit between voltmeter connection and battery.
4. If battery voltage is obtained in step 3, disconnect battery ground cable. Connect an ammeter between "Batt" terminal of alternator and battery positive post, then connect battery ground cable.
5. Install a carbon pile regulator across battery, then operate engine at a moderate speed.
6. Turn on all accessories and adjust carbon pile regulator to obtain maximum current output.
7. If reading is within 10 amps of rated output, adjust voltage adjustment cap as outlined under "Voltage Setting".
8. If reading is not within 10 amps of rated output, the alternator must be disassembled and checked as outlined under "Alternator Disassembly".

30 SI/TR

NOTE: Do not allow leads or terminals to touch ground.

1. Remove TR unit attaching screws, then pull unit away from alternator to expose lead connections.
2. Disconnect three transformer leads from mounting studs on rectifier bridge then reinstall nuts.
3. Disconnect remaining TR unit lead from rectifier bridge heat sink, then reinstall screw.
4. Check alternator as outlined under 30 SI procedure.
5. If no defect is found, check TR unit rectifier bridge as follows:
   a. Connect one lead of ohmmeter to heat sink and other lead to terminal or flat metal connector, if equipped, Fig. 6.
   b. Note reading, then reverse leads using the same heat sink and terminal or metal connector.
   c. One reading should be high and the other low.
   d. If both readings are the same, replace rectifier bridge.
   e. Repeat test on remaining two terminal or metal connectors using the same heat sink, then between other heat sink and each of the three terminals.
6. If no defects are found, install TR unit on alternator.
7. Disconnect lead from TR unit output terminal, then connect ammeter between terminal and lead.

NOTE: Do not allow terminal to touch ground.

8. Operate engine at moderate speed to obtain maximum output.
9. If specific gravity of cranking battery is 1.200 or below, charging rate of alternator should be 5 amps or higher. If charging rate is below 5 amps, replace TR unit transformer.

Overcharged Battery

NOTE: The TR unit on 30 SI/TR alternators does not require testing for this condition.

1. Operate engine at moderate speed to obtain maximum voltage output, ensure all accessories are turned off.
2. If voltage exceeds 15 volts on a 12 volt system, 30 volts on 24 volt system or 39 volts on a 32 volt system, the alternator must be removed and disassembled and tested as outlined under "Alternator Disassembly".
3. If voltage does not exceed the above values, adjust voltage adjustment cap as outlined under "Voltage Setting".

ALTERNATOR DISASSEMBLY

1. Remove attaching screws and end plate, Fig. 6A.
2. Remove fan and pully, then remove four through bolts.

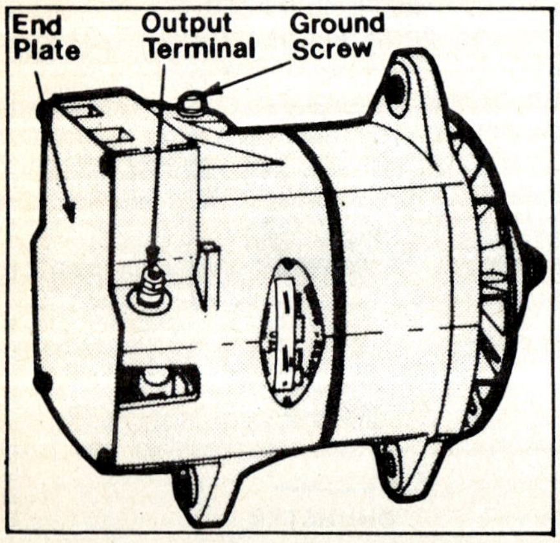

Fig. 1 Delcotron 30 SI alternator

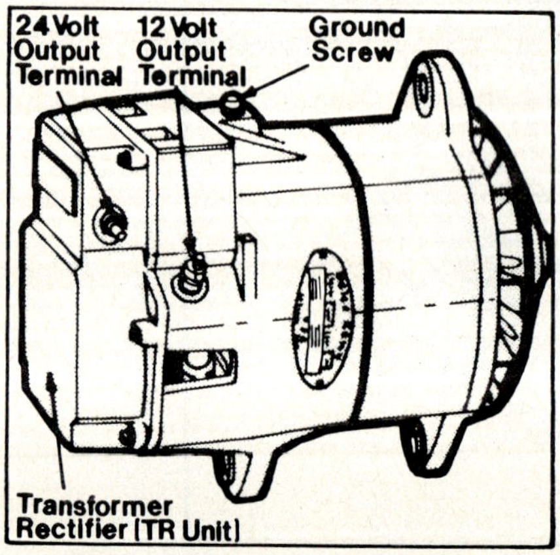

Fig. 2 Delcotron 30 SI/TR alternator

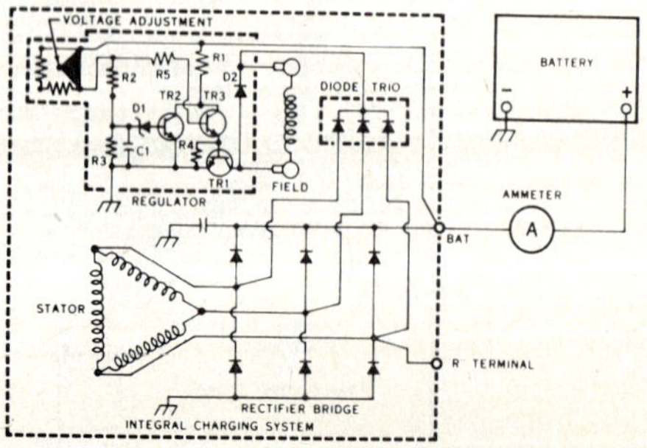

Fig. 2A Wiring diagram of 30 SI charging circuit

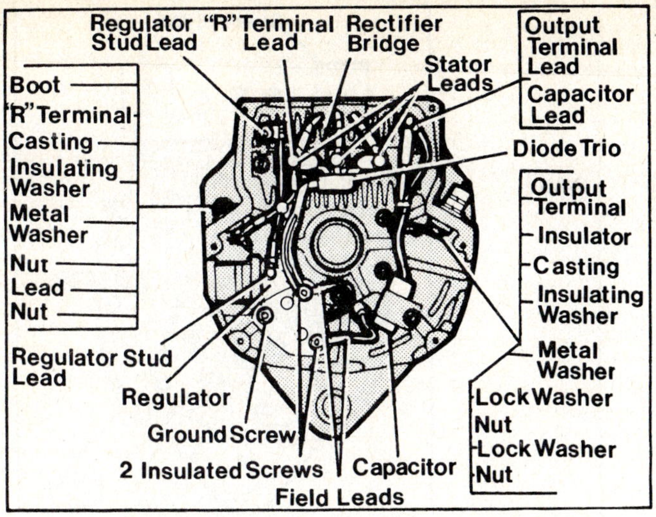

Fig. 3 Delcotron 30 SI negative ground unit

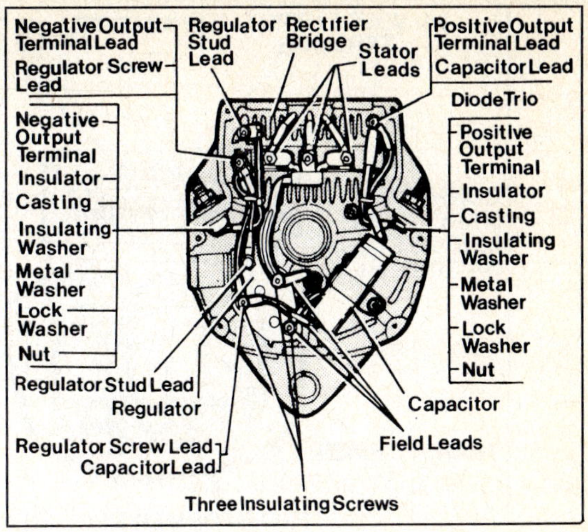

Fig. 4 Delcotron 30 SI insulated unit

3. Separate drive end frame and rotor from rectifier end frame and stator.
4. Press rotor from end frame, then remove inner collar from end frame and outer collar from shaft.

BENCH TEST

Rectifier Bridge Check

Connect ohmmeter between heat sink and one of three terminals or clips, step 1 Fig. 7. If rectifier bridge is constructed with flat metal clips, connect ohmmeter leads to metal clips not threaded studs. Note ohmmeter reading then reverse ohmmeter leads using the same heat sink and terminal or clip. One reading should be high and the other low. If both readings are the same and the rectifier bridge should be replaced. Repeat test on remaining two terminals or metal clips, then check remaining rectifier bridge.

Field Coil Checks

Connect one ohmmeter to field coil lead and other lead to end frame, step 2 Fig. 7. If ohmmeter reading is low, field coil is grounded.

Connect ohmmeter between field coil leads, step 3 Fig. 7. If ohmmeter is high (infinite), field coil is open.

Connect battery and ammeter in series with field coil. Note ammeter reading and refer to Alternator and Regulator Specification in truck chapter. An ammeter reading above specified value indicates shorted windings.

Diode Trio Check

Using an ohmmeter having a 1½ volt cell and set at low reading scale, connect one lead to the single connector and other lead to one of three connectors, Fig. 8. Note reading then reverse leads using the same two connectors. If both readings are the same, replace diode trio. One reading should be high and the other low. Use single connector and repeat test on remaining two connectors.

Stator Check

To check for grounded stator winding, connect one ohmmeter lead to any stator lead and other lead to metal frame, step 4 Fig. 7. Reading should be infinite, if not replace stator. Stator windings cannot be checked for opens or short circuits unless special test equipment is used. If regulator checks good, but alternator fails to supply rated output and stator windings are badly discolored, replace stator.

Regulator Check

The regulator cannot be checked with an ohmmeter. A special regulator tester must be used.

REPAIRS

Bearings, Replace

To replace drive end frame bearing, remove four retainer plate attaching screws, retainer plate and gasket. Push on inner race to remove bearing, then remove seals from end frame and retainer. Press new seals in with lip facing toward bearing. Press bearing in

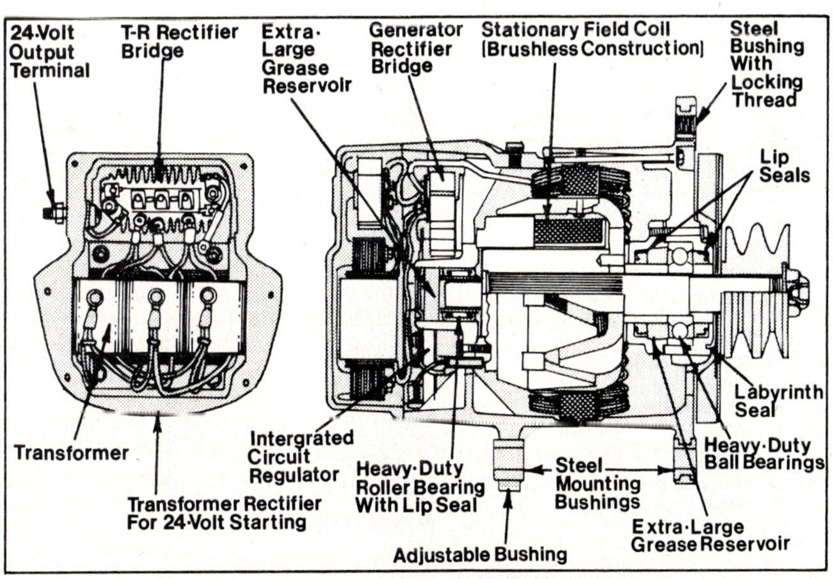

Fig. 5 Cross sectional view. Delcotron 30 SI/TR

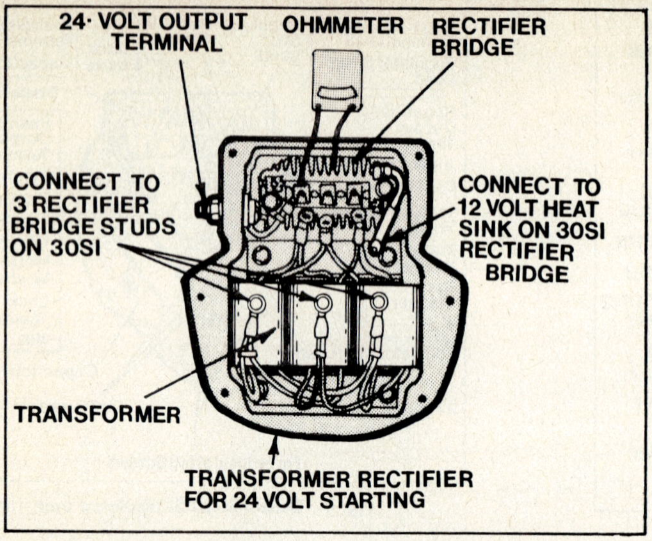

Fig. 6 Testing TR unit rectifier bridge diodes

against outer race, then fill cavity half full with lubricant No. 1948791 or equivalent. Install retainer and four attaching bolts.

To replace rectifier end frame bearing, pull inner race from shaft and bearing from end frame. Install inner race on shaft and bearing in end frame with seal facing away from grease reservoir. Fill reservoir half full with lubricant No. 1948791 or equivalent.

ALTERNATOR ASSEMBLE

1. Install inner collar on shaft and outer collar on end frame under seal next to bearing, Fig. 6A.
2. While supporting outer collar, press rotor into drive end frame.
3. Assemble drive end frame and rotor to rectifier end frame and stator, then install four through bolts.
4. Install fan and pulley on rotor shaft.
5. Install end plate and attaching screws.

VOLTAGE SETTING

Remove adjusting cap and rotate it 1/4 turn in the desired direction, then reinstall cap, Fig. 9. There are four settings, LO-lowest voltage setting, 2-medium low voltage, 3-medium high voltage and HI-highest voltage setting. Alternator voltage should be 13 to 15 volts on a 12 volt system, 26 to 30 volts on a 24 volt system and 33 to 39 volts on a 32 volt system.

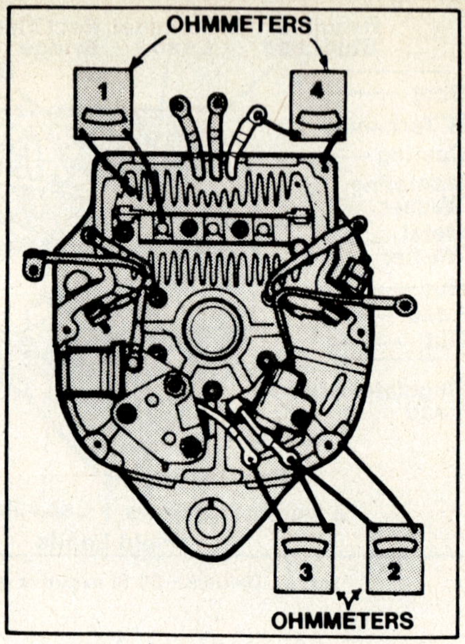

Fig. 7 Bench test connections

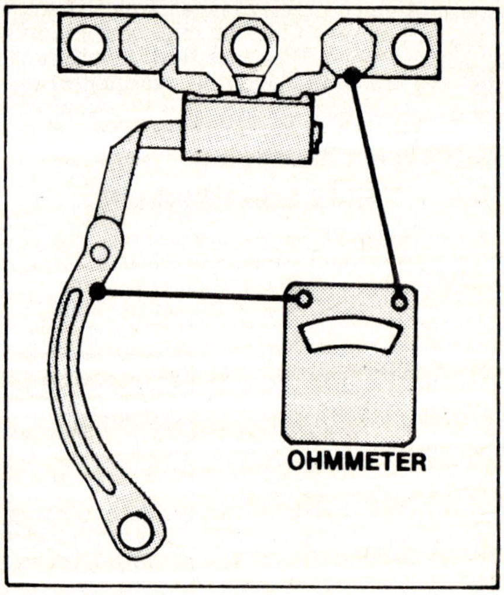

Fig. 8 Testing diode trio

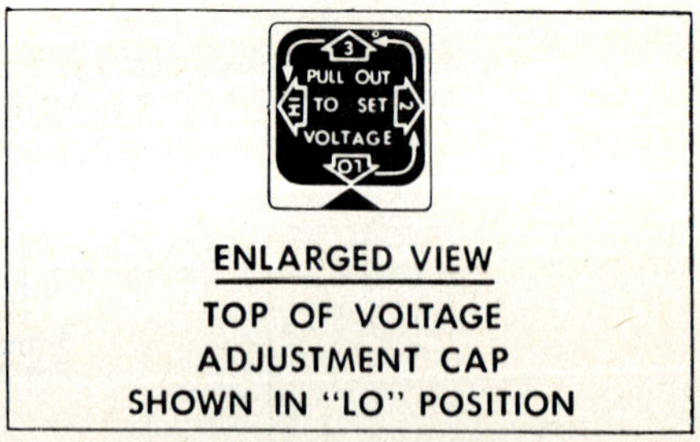

Fig. 9 Voltage adjusting cap

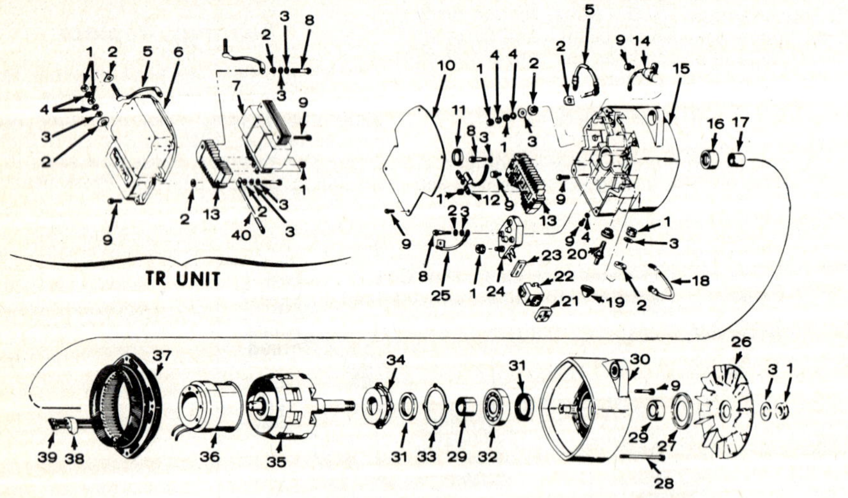

1. NUT
2. WASHER (INSUL.)
3. WASHER (STL.)
4. WASHER (SPRING)
5. 24—V OUTPUT LEAD
6. TR HOUSING
7. TRANSFORMER
8. SCREW (INSUL.)
9. SCREW
10. PLATE (30—SI ONLY)
11. PLUG
12. DIODE TRIO
13. RECTIFIER BRIDGE
14. CAPACITOR
15. HOUSING (RECTIFIER END)
16. BEARING (OUTER)
17. BEARING (INNER)
18. "R" TERM. LEAD
19. CAP ("R" TERM.)
20. "R" TERM. STUD)
21. CAP
22. CONNECTOR BODY
23. BOOT
24. VOLTAGE REGULATOR
25. REGULATOR LEAD
26. FAN
27. SLINGER
28. THRU—BOLT
29. COLLAR
30. HOUSING (DRIVE END)
31. SEAL
32. BEARING
33. GASKET
34. RETAINER
35. ROTOR
36. FIELD COIL
37. STATOR
38. GROMMET
39. CLIP
40. BRIDGE LEAD

Fig. 6A Disassembled view of 30 SI & 30 SI/TR alternator

Delcotron 40 SI Integral Charging System

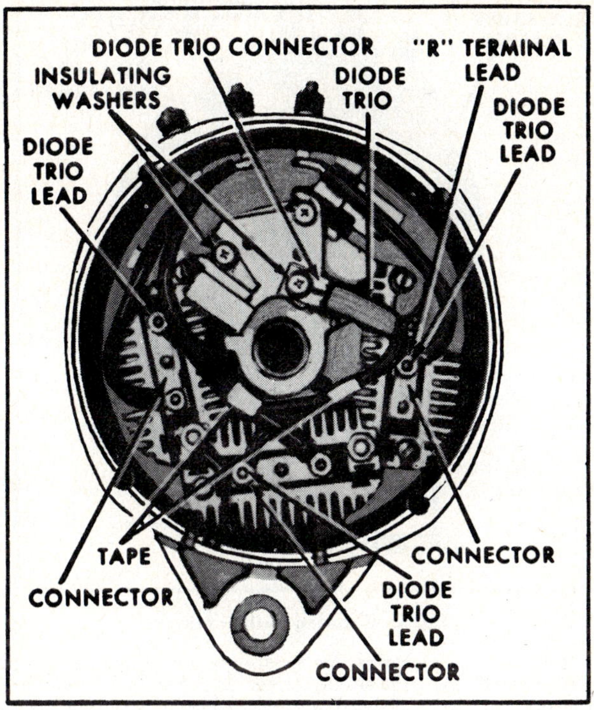

Fig. 1 Slip ring end frame components

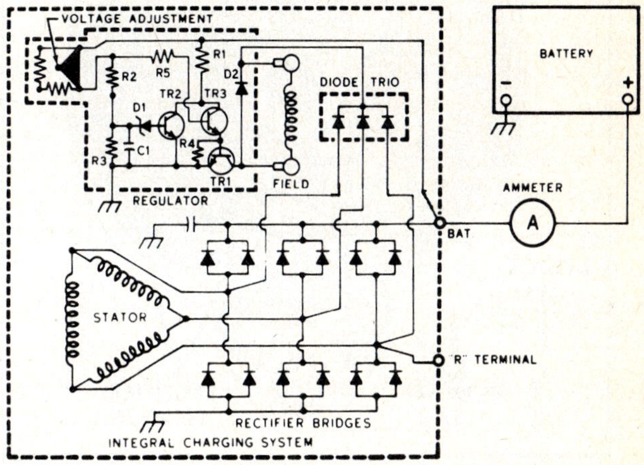

Fig. 1A Wiring diagram of 40 SI charging circuit

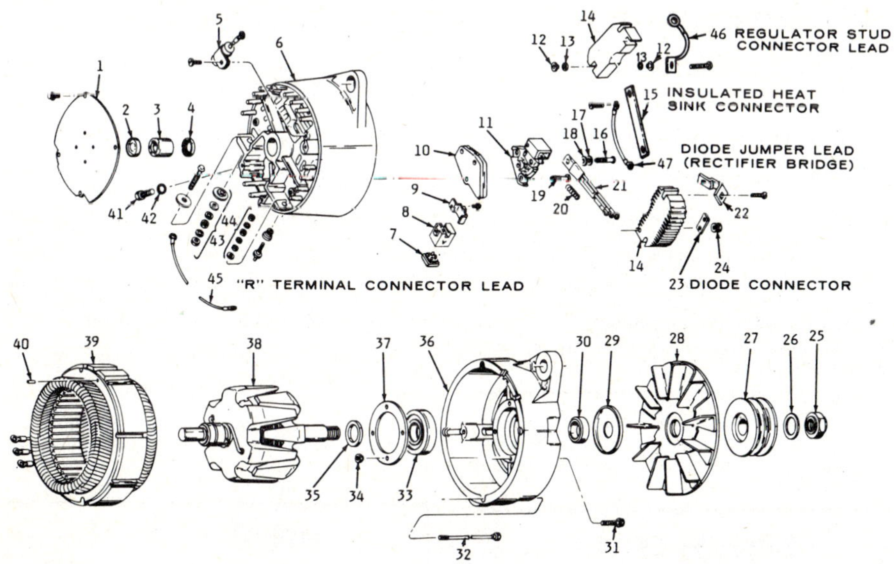

1 Cover Plate	13 Washer (4)	25 Nut (Rotor)	37 Retainer
2 Plug	14 Rectifier Bridge	26 Washer	38 Rotor
3 Bearing	15 Connector	27 Pulley	39 Stator
4 Seal	16 Insulated Screw (2)	28 Fan	40 Pin
5 Capacitor	17 Insulator	29 Slinger	41 Ground Screw
6 Slip Ring End Frame	18 Washer	30 Collar	42 Washer
7 Cap	19 Screw	31 Screw (4)	43 Batt. Terminal
8 Connector	20 Spring (2)	32 Thru-Bolt (4)	44 "R" Terminal
9 Clip	21 Diode Trio	33 Bearing	45 Green Wire
10 Regulator	22 Strap (Batt Conn)	34 Nut (4)	46 Red Wire
11 Brush Holder	23 Strap (2)	35 Collar	47 Black Wire
12 Insulator (4)	24 Nut (6)	36 Drive Housing	

Fig. 1B Disassembled view of 40 SI alternator

DESCRIPTION

This unit features a solid state regulator mounted inside the alternator slip ring end frame, Figs. 1 and 1A, along with brush holder assembly. The voltage setting can be adjusted by repositioning the adjustment cap. The 100 amp model incorporates two rectifier bridges, each containing 6 diodes and connected to the stator winding, changes A.C. voltage to D.C. voltage which is available at the output terminal. The 145 amp model incorporates three rectifier bridges, each containing 6 diodes. Some models incorporate three A.C. terminals located on top of alternator. Alternator field current is supplied through a diode trio which is also connected to the stator windings. The diodes and rectifiers are protected by a capacitor which is also mounted in the end frame.

TROUBLE SHOOTING

Undercharged Battery

1. Disconnect battery ground cable.
2. Connect an ammeter between BAT terminal of alternator and battery positive post and reconnect battery ground cable.
3. Install a carbon pile regulator across battery, then operate engine at a moderate speed.
4. Turn on all accessories and adjust carbon pile regulator to obtain maximum current output.
5. If reading is within 10 amps of rated output, adjust voltage adjustment cap as described under "Voltage Setting".

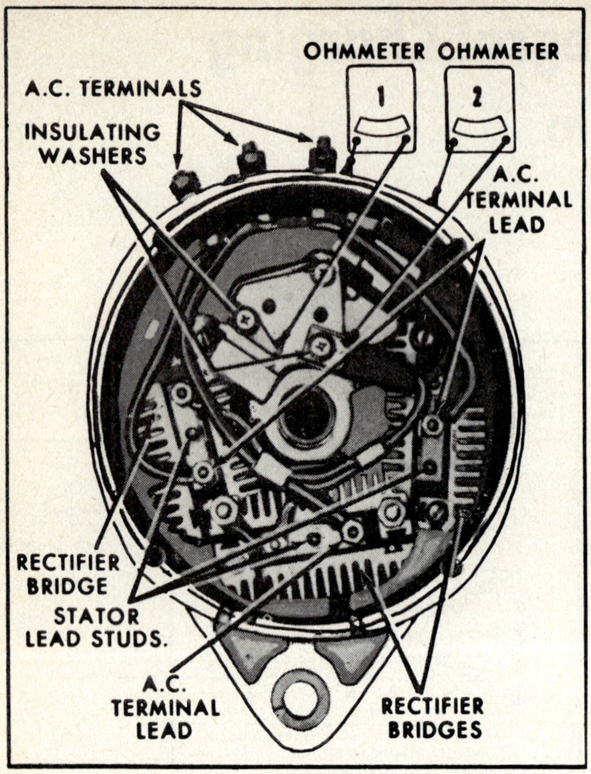

Fig. 2 Voltage regulator and brush lead test

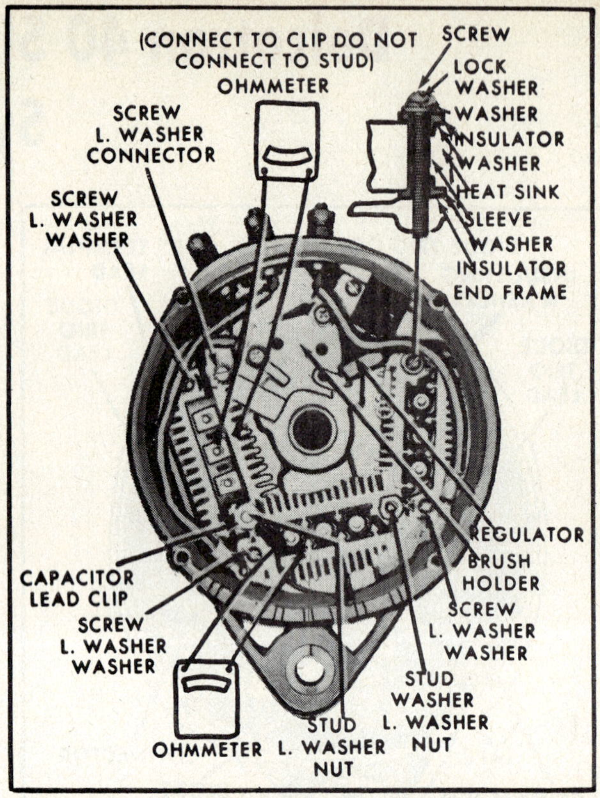

Fig. 5 Checking rectifier bridge

6. If reading is not within 10 amps of rated output, the alternator must be disassembled and checked, as described under "Alternator Disassembly".

Overcharged Battery

1. Connect a voltmeter between alternator BAT terminal and ground.
2. With all accessories turned off, start engine and increase speed until maximum voltage reading is obtained.
3. If voltage reading exceeds 15 volts, the alternator must be disassembled and checked, as described under "Alternator Disassembly".
4. If voltage reading does not exceed 15 volts, adjust voltage adjusting cap as described under "Voltage Setting".

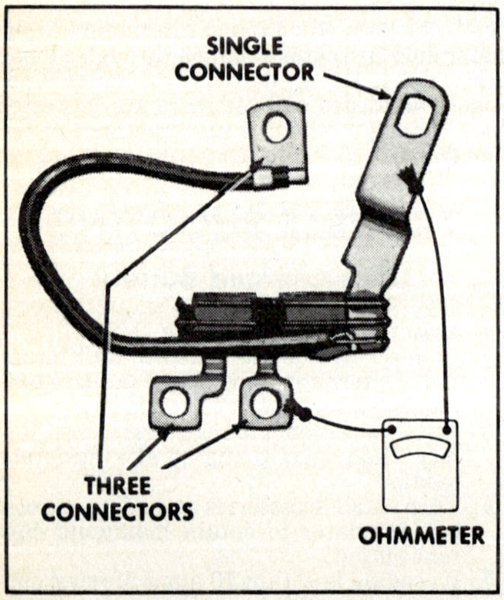

Fig. 3 Checking diode trio. 100 amp units

ALTERNATOR DISASSEMBLY

1. With alternator secured in a vise, remove cover plate from slip ring end frame, Fig. 1B.
2. Using an allen wrench to hold rotor shaft in place remove pulley nut, pulley, fan, slinger and collar.
3. Remove four through bolts, then using a screw driver in stator slot separate slip ring end frame and stator assembly from drive end frame and rotor assembly.
4. Remove three stator lead to rectifier bridge attaching nuts, then separate stator from slip ring end frame.
5. Place tape over slip ring end frame bearing and shaft.

NOTE: Brushes may fall from holders and become contaminated with bearing grease, if so they must be cleaned prior to assembly.

BENCH TESTS

Voltage Regulator/Brush Lead Test

Connect an ohmmeter from the brush lead clip to the end frame, note reading, then reverse connections. If both readings are zero, either the brush lead clip is grounded or the regulator is defective, Fig. 2.

Diode Trio Test

Using an ohmmeter having $1\frac{1}{2}$ volt cell and set at low reading scale, connect one lead to the single connector and the other lead to one of three connectors, Figs. 3 & 4. Note reading, then reverse leads using the same two connectors. If both readings are the same replace the diode trio. One reading should be high and the other low. Using single connector repeat test on the other two connectors.

Rectifier Bridge Test

Connect one lead of ohmmeter to grounded heat sink and other lead to terminal or flat metal connector, if equipped, Fig. 5. Note reading then reverse leads of the same heat sink and terminal or metal connector. One reading should be high and the other low. If both readings are the same the rectifier bridge must be replaced. Repeat test on the remaining two terminals or metal connectors, then check remaining rectifier bridges.

Rotor Field Winding Test

To check if winding is open, connect one ohmmeter to each slip ring, Fig. 6. If reading

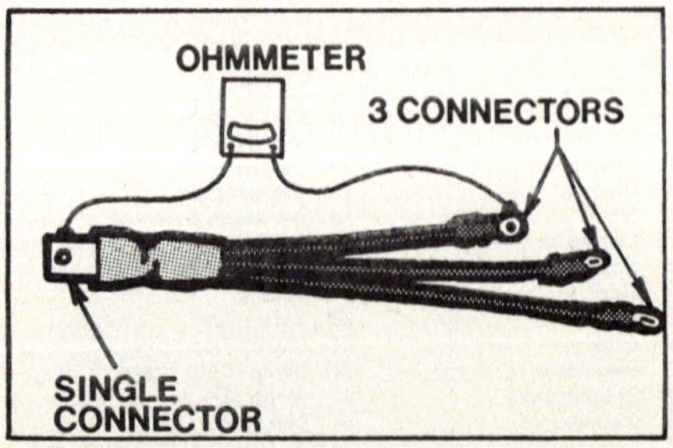

Fig. 4 Checking diode trio. 145 amp units

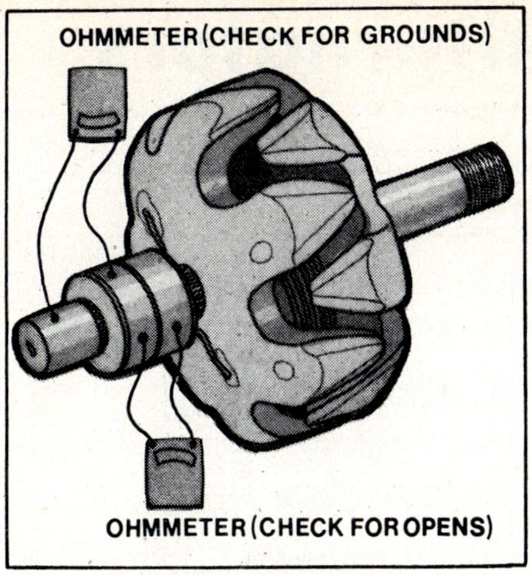

Fig. 6 Checking rotor

is high then the winding is open.

To check for a short circuit or excessive resistance connect an ammeter and battery in series with edges of two slip rings. The ammeter reading should be 4.0 to 4.5 amps at 80° F. If reading is above 4.5 amps the winding is shorted and the rotor and regulator must be replaced.

Rotor may also be checked for shorts by connecting an one ohmmeter lead to each slip ring. Ohmmeter reading should be 2.75 to 3.0 ohms. If reading is below 2.75 ohms the rotor is shorted.

Stator Test

To check for grounded stator winding, connect one ohmmeter lead to any stator lead and the other lead to metal frame, Fig. 7. If ohmmeter reading is low the stator windings are grounded. Stator windings can not be checked for opens or short circuits unless special test equiment is used. If alternator fails to supply rated output, but will supply at least 10 amperes, the stator windings may be shorted.

Regulator Test

If all other checks fail to show any defects and an overcharging condition exists, remove connector body from regulator and check as follows. With ohmmeter set to middle scale range, connect leads to each adjacent pair of terminals until four readings are obtained Fig. 8. If any reading is excessively high replace the connector body. If all previous checks are satisfactory but maximum current output is less than 10 amps or maximum voltage is over 15 volts, replace the regulator.

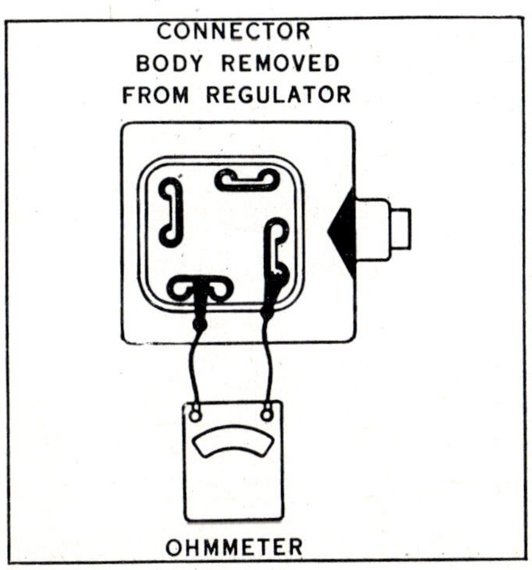

Fig. 8 Checking regulator connector body

VOLTAGE SETTING

Remove adjusting cap and rotate it 1/4 turn in the desired direction, then reinstall cap, Fig. 9. There are four settings, LO—lowest voltage setting, 2—medium low voltage, 3—medium high voltage and HI—highest voltage setting. Alternator voltage should be 13 to 15 volts.

REPAIRS

Bearing Replace

The drive end frame bearing is sealed on both sides and cannot be lubricated. To replace bearing, press rotor from end frame, remove retainer plate and press bearing from end frame. Use a socket that just fits over outer race to press new bearing into end frame.

The slip ring end frame bearing cannot be relubricated, the bearing must be replaced. Using a suitable socket press bearing from end frame. Position new bearing on outside of end frame, then using a suitable socket press bearing in to specified dimension, Fig. 10. Fill plug with lubricant No. 1948791 or equivalent, ensure grease reservoir will be half full when plug is pressed into end frame and that lubricant will contact bearing when plug is installed. Press plug in until flush with end frame. Lubricate seal lip, then position seal on end frame with lip facing toward bearing, press seal in to specified dimension, Fig. 10.

Slip Ring Service

If the slip rings are dirty they may be cleaned with No. 400 silicon carbide paper and finish polished with crocus cloth. Spin the rotor in lathe, or otherwise spin the rotor, and until they are clean.

CAUTION: The rotor must be rotated in order that the slip rings will be cleaned evenly.

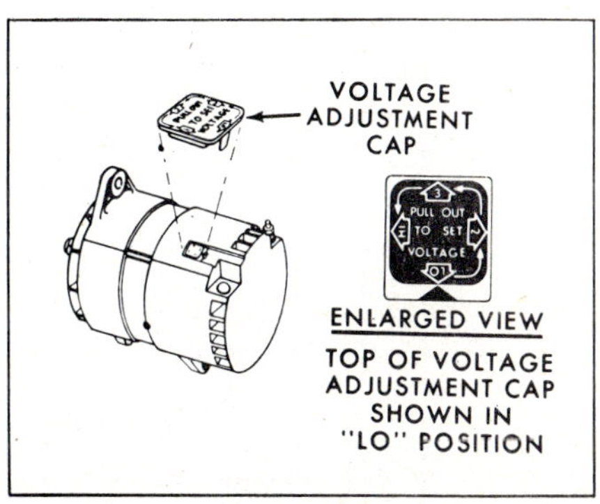

Fig. 9 Adjusting voltage setting

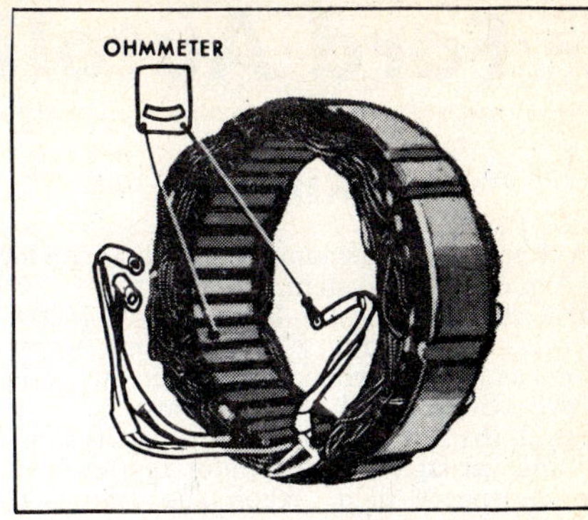

Fig. 7 Checking stator winding

Hold the polishing cloth against the slip rings. Cleaning the slip rings by hand without spinning the rotor may result in flat spots on the slip rings, causing brush noise.

Slip rings that are rough or out-of-round should be trued in a lathe to .002" maximum runout as indicated on a dial gauge. Remove only enough material to make the rings smooth and round. Finish polish with crocus cloth and blow away all dust.

ALTERNATOR ASSEMBLY

1. Remove tape from slip ring end frame bearing and shaft, Fig. 1B.
2. Use pin to hold brushes in compressed position.
3. Install three stator leads on rectifier bridges.
4. Install drive end frame and rotor assembly onto stator and slip ring end frame, use care when installing shaft into slip ring end frame to avoid damaging seal.
5. Install four through bolts.
6. Install collar, slinger, fan, pulley and pulley nut onto rotor shaft, torque pulley nut to 75 ft. lbs.
7. Remove pin holding brushes, then install cover plate onto slip ring and frame.

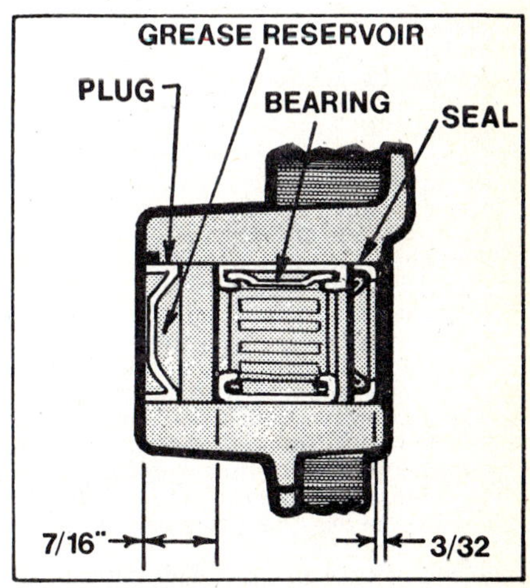

Fig. 10 Installing slip ring end frame bearing assembly

Ford Autolite/Motorcraft Alternators

DESCRIPTION

A charge indicator lamp or ammeter can be used in charging system.

If a charge indicator lamp is used in the charging system, Fig. F1, the system operation is as follows: when the ignition switch is turned ON, a small electrical current flows through the lamp filament (turning the lamp on) and through the alternator regulator to the alternator field. When the engine is started, the alternator field rotates and produces a voltage in the stator winding. When the voltage at the alternator stator terminal reaches about 3 volts, the regulator field relay closes. This puts the same voltage potential on both sides of the charge indicator lamp causing it to go out. When the field relay has closed, current passes through the regulator A terminal and is metered to the alternator field.

If an ammeter is used in the charging system, Fig. F2, the regulator I terminal and the alternator stator terminal are not used. When the ignition switch is turned ON, the field relay closes and electrical current passes through the regulator A terminal and is metered to the alternator field. When the engine is started, the alternator field rotates causing the alternator to operate.

NOTE: The ammeter indicates current flow into (charge) or out of (discharge) the vehicle battery.

Some 1978 and all 1979–83 Ford vehicles are equipped with new electronic voltage regulators, Figs. F3 and F4. These solid state regulators are used in conjunction with other new components in the charging system such as an alternator with a higher field current requirement, a warning indicator lamp shunt resistor (500 ohms) and a new wiring harness with a new regulator connector. When replacing system components, note the following precautions:

1. Always use the proper alternator in the system. If the new 1978–83 alternator is installed on previous model systems, it will destroy the electro-mechanical regulator. If the older model alternator is used on the new system, it will have a reduced output.
2. Do not use an electro-mechanical regulator in the new system since the wiring harness connector will not index properly with this type of regulator.
3. The new electronic regulators are color coded for proper installation. The black color coded unit is installed in systems equipped with a warning indicator lamp. The blue color coded regulator is installed in systems equipped with an ammeter.
4. The new systems use a 500 ohm resistor on the rear of the instrument cluster on vehicles equipped with a warning indicator lamp. Do not replace this resistor with the 15 ohm resistance wire used on previous systems.

On the new systems with an indicator lamp, closing the ignition switch energizes the warning lamp and turns on the regulator output stage. The alternator receives maximum field current and is ready to generate an output voltage. As the alternator rotor speed increases, the output and stator terminal voltages increase from zero to the system regulation level determined by the regulator setting.

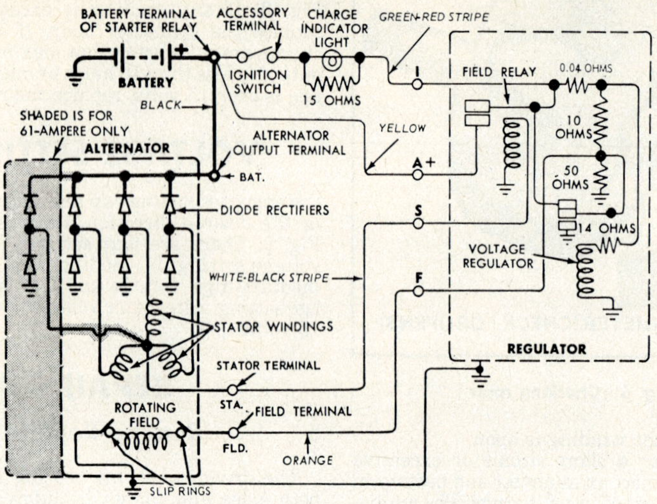

Fig. F1 Indicator light rear terminal alternator charging system (Typical)

When the ignition switch is turned off, the solid state relay circuit turns the output stage off, interrupting current flow through the regulator so there is not a current drain on the battery.

On vehicles equipped with an ammeter, the operating principle is similar.

NOTE: The ammeter indicates current flow into (charge) or out of (discharge) the vehicle battery.

SYSTEM TESTING
In-Vehicle Tests
1974–83 Units

VOLTMETER TEST

NOTE: *All lights and electrical systems in the off position, parking brake applied, transmission in neutral and a charged battery (at least 1.200 specific gravity).*

---

1. Connect the negative lead of the voltmeter to the negative battery cable clamp (not bolt or nut).
2. Connect the positive lead of the voltmeter to the positive battery cable clamp (not bolt or nut).
3. Record the battery voltage reading shown on the voltmeter scale.
4. Connect the red lead of a tachometer to the distributor terminal of the coil and the black tachometer lead to a good ground.
5. Then, start and operate the engine at approximately 1500 rpm. With no other electrical load (foot off brake pedal and car doors closed), the voltmeter reading should increase but not exceed (2 volts) above the first recorded battery voltage reading. The reading should be taken

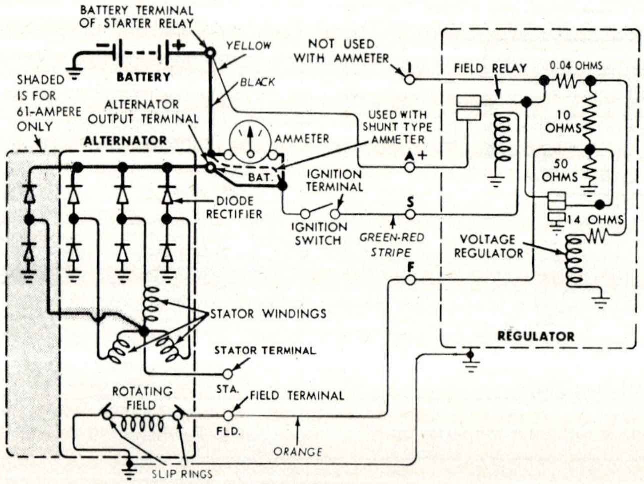

Fig. F2 Ammeter rear terminal alternator charging system (Typical)

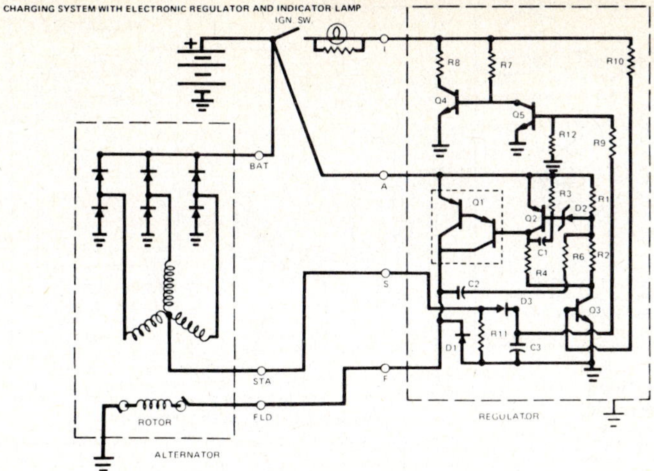

Fig. F3 Indicator light charging system with electronic voltage regulator

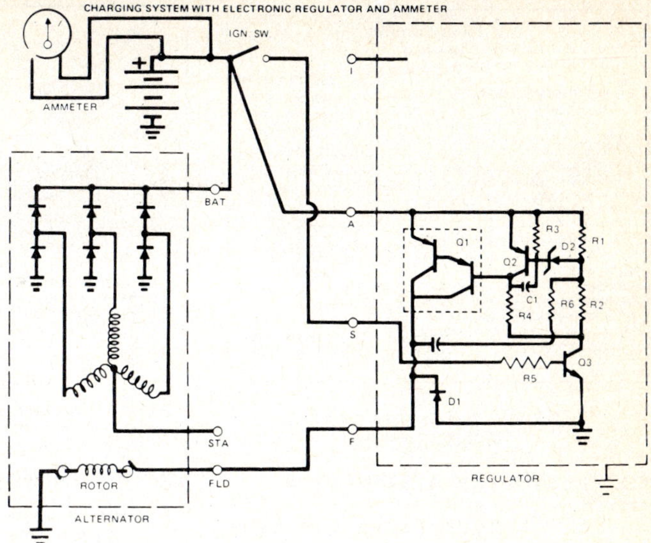

Fig. F4 Ammeter charging system with electronic voltage regulator

when the voltmeter needles stops moving.

6. With the engine running, turn on the heater and/or air conditioner blower motor to high speed and headlights to high beam.
7. Increase the engine speed to 2000 rpm. The voltmeter should indicate a minimum reading of 0.5 volts above the battery voltage, Fig. F5.

NOTE: *If the above tests indicate proper voltage readings, the charging system is operating normally. Proceed to "Test Results" if a problem still exists.*

TEST RESULTS

1. If voltmeter reading indicates 2 volts over battery voltage (over voltage), proceed as follows:
   a. Stop the engine and check the ground connections between the regulator and alternator and/or regulator to engine. Clean and tighten connections securely and repeat the *Voltmeter Test Procedures*.
   b. If *over voltage* condition still exists, disconnect the regulator wiring plug from the regulator and repeat the *Voltmeter Test Procedures*.
   c. If *over voltage* still exists with the regulator wiring plug disconnected, repair the short in the wiring harness between the alternator and regulator. Then, replace the regulator and connect the regulator wiring plug to the regulator and repeat the *Voltmeter Test Procedures*.

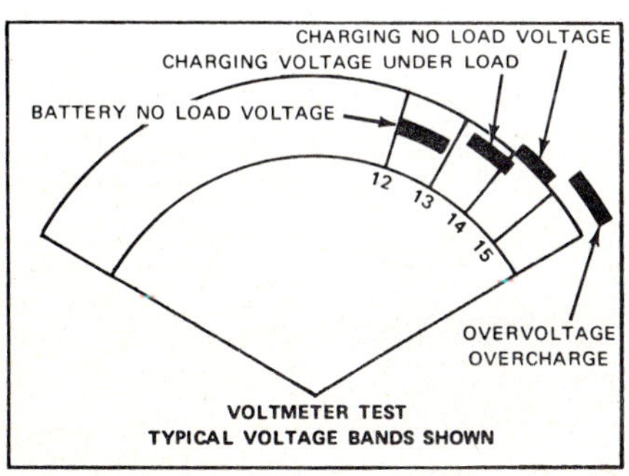

Fig. F5 Voltmeter test scale

2. On 1975–77 units, if voltmeter does not indicate more than ½ volt above battery voltage, proceed as follows:
   a. Check for presence of battery voltage at alternator BAT terminal and the regulator plug A terminal, Fig. 7. Repair the wiring if no voltage is present at these terminals, and repeat the *Voltmeter Test Procedures*.
   b. If voltmeter reading does not increase ½ volt above battery voltage, proceed to next step.
   c. Before performing other tests, the field circuit (regulator plug to alternator) must be checked for a grounding condition. If the field circuit is grounded and the jumper wire is used as a check at the regulator wiring plug from the A to F terminals, Fig. F6, excessive current will cause heat damage to the regulator wiring plug terminals and may burn the jumper wire, Fig. F6. Also, if the field circuit was grounded, the connector wire inside the regulator will be burned open and an under voltage condition will result.
   d. The field circuit should be checked with the regulator wiring plug disconnected and an ohmmeter connected from the *F* terminal of the regulator wiring plug to the battery ground. The ohmmeter should indicate between 4 and 250 ohms, Fig. F7.
   e. A check for the regulator burned-open wire is made by connecting an ohmmeter from the *I* to *F* terminals of the regulator, Fig. F8. The reading should indicate *O* (no resistance). If the reading indicates approximately *10 ohms,* the connector wire inside the regulator is burned open. *The field circuit grounded condition must be found and repaired before installing a new regulator.*
3. On 1978–83 units, if voltmeter does not indicate more than ½ volt above battery voltage, proceed as follows:
   a. Disconnect voltage regulator wire connector and connect an ohmmeter between wire connector F terminal and ground. Ohmmeter reading should indicate more than 3 ohms. If reading is less than 3 ohms, repair grounded field circuit and repeat Voltmeter Test procedure.
   b. If ohmmeter reading is more than 3 ohms, connect a jumper wire between voltage regulator wire connector terminals A and F, Fig. F9, then repeat Voltmeter Test procedure. If voltmeter reading is now more than ½ volt above battery voltage, the voltage regulator or wiring is defective, refer Regulator Test.
   c. If voltmeter still indicates less than ½ volt, disconnect jumper wire from voltage regulator wire connected and leave connector disconnected from regulator. Connect a jumper wire between alternator FLD and BAT terminals, Figs. F10 and F11, then repeat Voltmeter Test procedures.
   d. If voltmeter reading now indicates ½ volt or more above battery voltage, repair alternator to regulator wiring harness.
   e. If voltmeter still indicates less than ½ volt above battery voltage, stop engine and move voltmeter positive lead to alternator BAT terminal.
   f. If voltmeter now indicates battery voltage, the alternator should be removed, inspected and repaired. If zero volts is indicated, repair BAT terminal wiring.

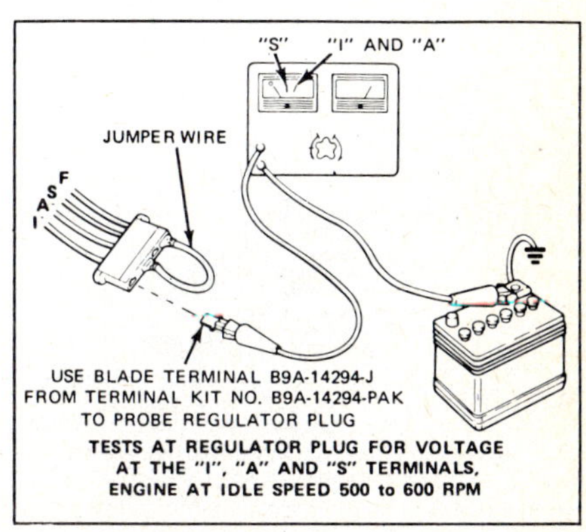

Fig. F6 Regulator plug voltage test

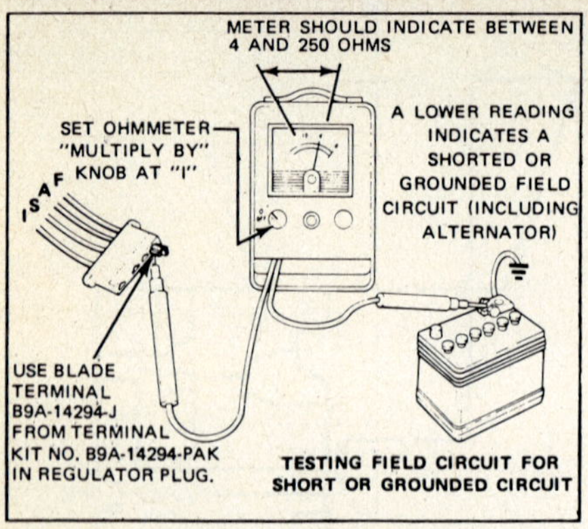

Fig. F7 Testing field circuit with Ohmmeter

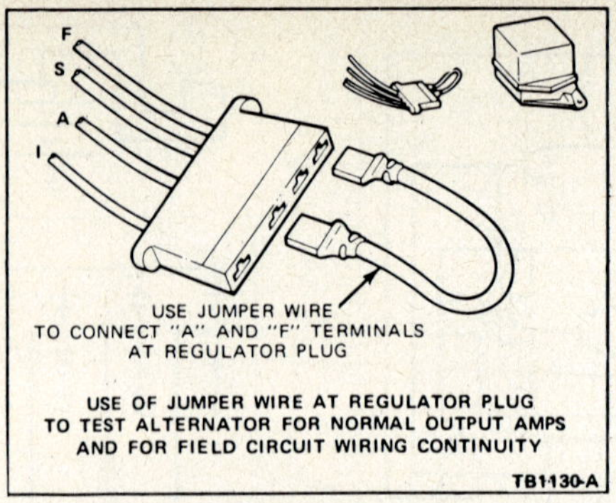

Fig. F9 Regulator plug. Jumper wire connection

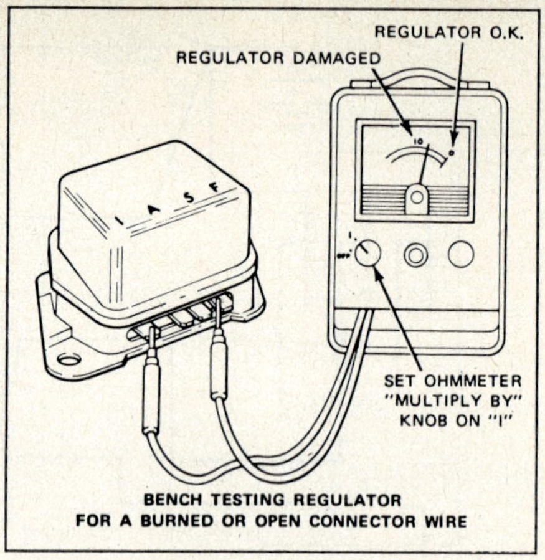

Fig. F8 Testing regulator for a burned or open connector wire

FIELD CIRCUIT & ALTERNATOR TESTS

1. If the field circuit is satisfactory, disconnect the regulator wiring plug at the regulator and connect the jumper wire from *A* to the *F* terminals on the regulator wiring plug, Fig. F9.
2. Repeat the *Voltmeter Test Procedures*.
3. If the *Voltmeter Test Procedures* still indicate a problem of under voltage, remove the jumper wire at the regulator plug and leave the plug disconnected from the regulator, Figs. F10 and F11. Connect a jumper wire to the *FLD* and *BAT* terminals on the alternator, Figs. F10 and F11.
4. Repeat the *Voltmeter Test Procedures*.
5. If the *Voltmeter Test* are now satisfactory, repair the wiring harness from the alternator to the regulator. Then, remove the jumper wire at the alternator and connect the regulator wiring plug to the regulator.
6. Repeat the *Voltmeter Test Procedures,* to be sure the charging system is operating normally.
7. If the *Voltmeter Test* results still indicate under voltage, repair or replace the alternator. With the jumper wire removed, connect the wiring to the alternator and regulator.
8. Repeat the *Voltmeter Test Procedures*.

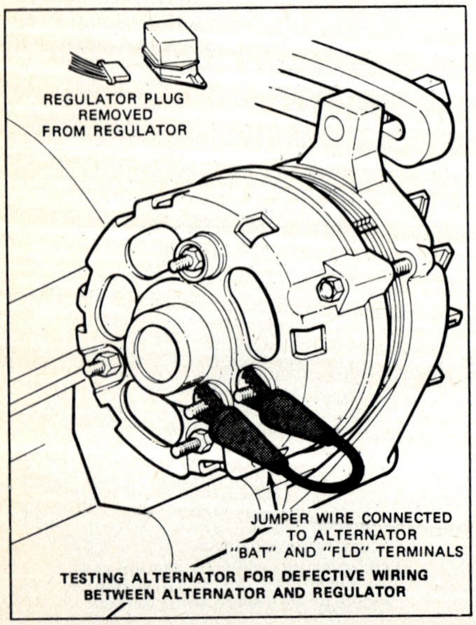

Fig. F10 Rear terminal alternator. Jumper wire connection

REGULATOR TESTS

S Circuit Test—With Ammeter

1. Connect the positive lead to the voltmeter to the S terminal of the regulator wiring plug Fig. F6. Turn the ignition switch to the ON position. *Do not start the engine.*
2. The voltmeter reading should indicate battery voltage.
3. If there is no voltage reading, disconnect the positive voltmeter lead from the positive battery clamp and repair the S wire lead from the ignition switch to the regulator wiring plug.
4. Connect the positive voltmeter lead to the positive battery cable terminal and repeat the *Voltmeter Test Procedures.*

S & I Circuit Test—With Indictor Light

1. With the engine idling, connect the positive lead of the voltmeter to the S terminal and then to the *I* terminal of the regulator wiring plug, Fig. F6. The voltage of the S circuit should read approximately ½ of the circuit.
2. If no voltage is present, repair the alternator or the wiring circuit at fault. Reconnect the positive voltmeter lead to the positive battery cable terminal and repeat the *Voltmeter Test Procedures.*
3. If the above tests are satisfactory, install a new regulator.
4. Then, remove the jumper wire from the regulator wiring plug and connect the wiring plug to the regulator. Repeat the *Voltmeter Test Procedures.*

Bench Tests

REAR TERMINAL ALTERNATOR

Field Open Or Short Circuit Test

The first part of this test will determine if the alternator portion of the field coil system, consisting of the field coil, the field coil slip rings and the field coil brush assembly is satisfactory. The second part of the test will indicate (in case of a field coil system malfunction), which of the above items is causing the malfunction.

Set ohmmeter "Multiple By" knob at 1 and calibrate meter.

Contact one ohmmeter probe to alternator field terminal and other probe to ground terminal. Spin alternator pulley. The ohmmeter reading should be between 2.4 and 200 ohms and should fluctuate while pulley is turning. An infinite reading (no meter movement) indicates an open brush lead, worn or stuck brushes, or faulty rotor assembly. An ohmmeter reading less than 2.4 ohms indicates grounded brush assembly, grounded field terminal or faulty rotor.

If the test shows that the field is shorted or open, determine if the field brush assembly or slip rings are at fault.

Disassemble unit front housing and rotor from the rear housing and stator, check the resistance of the rotor with ohmmeter. Set the ohmmeter multiply-by knob at 1 and calibrate the ohmmeter as indicated inside the ohmmeter cover.

Contact each ohmmeter probe to a slip ring. The resistance should be 3.5 to 5 ohms. A higher reading indicates a damaged slip ring soldered connection or a broken wire. A lower reading indicates a shorted wire or slip ring assembly.

Contact one ohmmeter probe to a slip ring and the other probe to the rotor shaft. The resistance should be infinite. Any reading other than infinite indicates a short to ground. Inspect the slip ring soldered terminals to make certain that they are not bent and touching the shaft or the excess solder is not grounding the rotor coil.

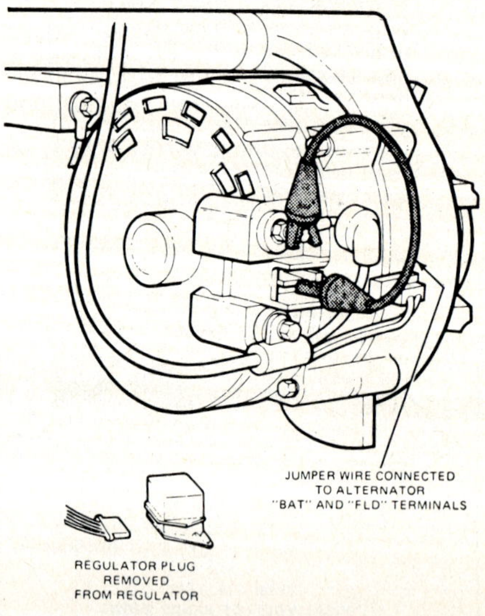

Fig. F11 Side terminal alternator. Jumper wire connection

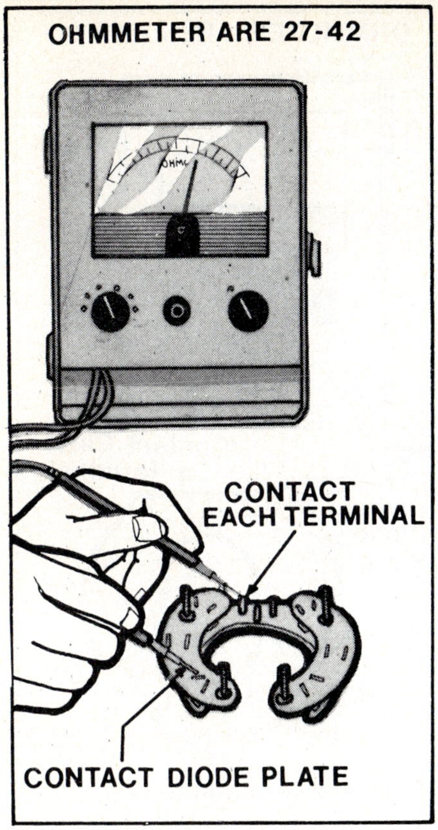

Fig. F12 Diode test. 38, 42 & 35 amp alternators

If the rotor checks indicate that it is in proper operating condition but the overall test indicates trouble, the brushes or brush assembly are the cause.

Diode Tests

Disassemble the alternator. Disconnect the rectifier assembly from the stator and connect leads, Fig. F12 or F13. To test one set of diodes, contact one probe to the diode plate and contact each of the three stator lead terminals with the other probe. Reverse the probes and repeat the test. Test the other set of diodes in the same way. On 61-ampere alternators, test the two additional diodes, Fig. F14.

All 6 tests (8 tests on 61-ampere alternator) should show a low reading of approximately 60 ohms in one direction and an infinite reading (no needle movement) with the probes reversed.

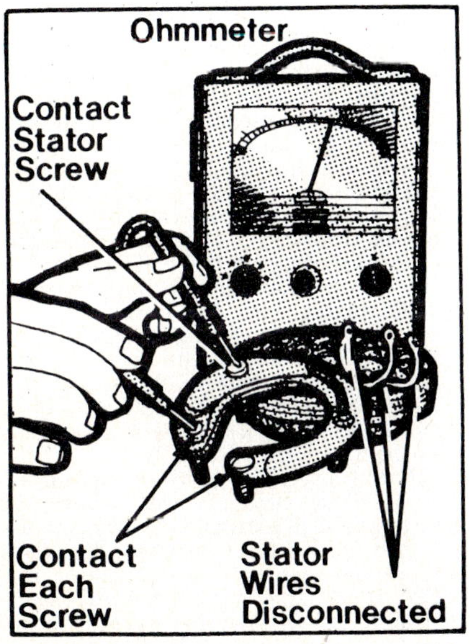

Fig. F14 Booster plate diode test 61-amp alternator

Open Or Grounded Stator Coil Tests

These tests are made to determine if the stator coil is operating properly. Disassemble the stator from the alternator and rectifier assembly.

Open Stator Test

Set ohmmeter multiple-by knob at 1. Connect the ohmmeter probes between each pair of stator leads. If the ohmmeter does not show equal readings between each pair of stator leads, the stator is open and must be replaced.

Grounded Stator Test

Connect the ohmmeter probes to one of the stator leads and to the stator laminated core. Be sure that the probe makes a good electrical connection with the stator core. The metal should show an infinite reading (no meter movement). If the meter does not indicate an infinite reading (needle moves), the stator winding is shorted to the core and must be replaced. Repeat this test for each of the stator leads.

SIDE TERMINAL ALTERNATOR

Rectifier Short or Grounded and Stator Grounded Test

Set ohmmeter Multiply By knob at 10, and calibrate meter.

Contact one ohmmeter probe to the alternator BAT terminal, Fig. F15, the other probe to the STA terminal (rear blade terminal). Then, reverse the ohmmeter probes and repeat the test. A reading of about 60 ohms should be obtained in one direction and no needle movement with the probes reversed. A reading in both directions indicates a bad positive diode, a grounded positive diode plate or a grounded BAT terminal.

Perform the same test using the STA and GND (ground) terminals of the alternator. A reading in both directions indicates either a bad negative diode, a grounded stator winding, a grounded stator terminal, a grounded positive diode plate, or a grounded BAT terminal.

Infinite readings (no needle movement) in all four probe positions in the preceeding tests indicates an open STA terminal lead connection inside the alternator.

Field Open or Short Circuit Test

Set the ohmmeter Multiply By knob at 1 and calibrate meter.

Contact the alternator field terminal with one probe and the ground terminal with the other probe, Fig. F16. Then, spin the alternator pulley. The ohmmeter reading should be between 4 and 200 ohms, and should fluctuate while the pulley is turning. An infinite reading (no meter movement) indicates an open brush lead, worn or stuck brushes, or a bad rotor assembly. An ohmmeter reading less than 4 ohms indicates a grounded brush assembly, a grounded field terminal or a bad rotor.

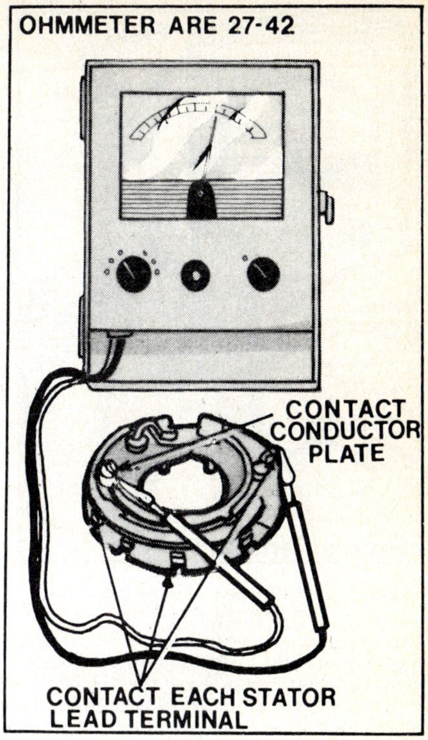

Fig. F13 Diode test. 65-amp rear terminal alternator

Diode Test

Remove the rectifier assembly from the alternator. Set the ohmmeter Multiply By knob at 10 and calibrate meter.

To test one set of diodes contact one probe to the terminal bolt, Fig. F17 and contact each of

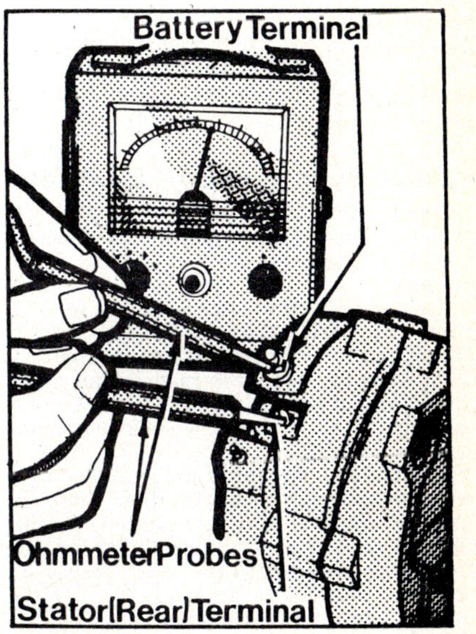

Fig. F15 Rectifier short or grounded and stator grounded test

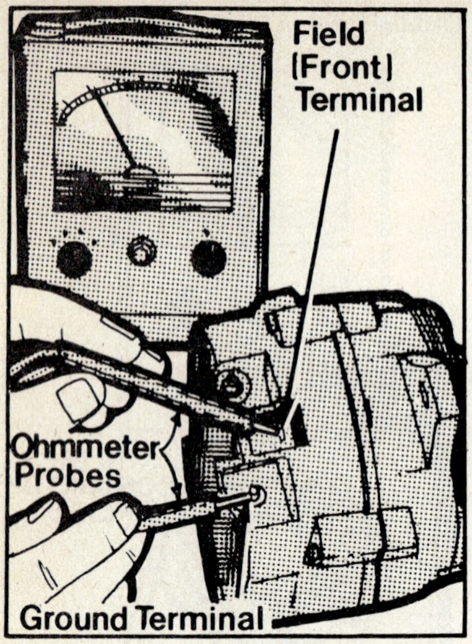

Fig. F16 Field open or short circuit test

the three stator lead terminals with the other probe. Reverse the probes and repeat the test. All diodes should show a low reading of about 60 ohms in one direction, and an infinite reading (no needle movement) with the probes reversed. Repeat the preceding tests for the other set of diodes except that the other terminal screw is used.

If the meter readings are not as specified, replace the rectifier assembly.

Stator Coil Open or Grounded Test

Disassemble the stator from the alternator.

Set ohmmeter Multiply By knob at 1, and calibrate meter. Connect the ohmmeter probes between each pair of stator leads (3 different ways). The ohmmeter must show equal readings for each pair of stator leads. Replace the stator if the readings are not the same.

Set ohmmeter Multiply By knob at 1000. Connect the ohmmeter probes to one of the stator leads and to the stator laminated core. Be sure that the probe makes a good electrical connection with the stator core. The meter should show an infinite reading (no meter movement). If the meter does not indicate an infinite reading (needle moves), the stator winding is shorted to the core and must be replaced. Repeat this test for each stator lead.

Rotor Open or Short Circuit Test

Disassemble the front housing and rotor from the rear housing and stator. Set the ohmmeter. Multiply By knob at 1 and calibrate meter.

Contact each ohmmeter probe to a rotor slip ring. The meter reading should be 3 to 5½ ohms on 1974-77 units, 2.4 to 4.9 ohms on 1978-83 units. A higher reading indicates a damaged slip ring solder connection or a broken wire. A lower reading indicates a shorted wire or slip ring.

Contact one ohmmeter probe to a slip ring and the other probe to the rotor shaft. The meter reading should be infinite (no deflection). A reading other than infinite indicates the rotor is shorted to the shaft. Inspect the slip ring soldered terminals to be sure they are not bent and touching the rotor shaft, or that excess solder is grounding the rotor coil connections to the shaft. Replace the rotor if it is shorted and cannot be repaired.

REGULATOR TESTS

1974–83 UNITS

The alternator must be adjusted within specification, and the charging system electrical connections must be clean and tight before testing the regulator.

1974–83 Regulator Test

For test procedures, refer to In Vehicle Tests, 1974–83 units, Regulator Tests.

Calibration Test Procedure 1974–78 Electro-Mechanical Regulator

Connect the voltmeter positive lead to the battery positive terminal, and the negative lead to the battery negative terminal. Turn off all electrical loads. Then, check and record the voltmeter reading. Connect the red lead of a tachometer to the distributor terminal of the coil and the block tachometer lead to a good ground.

Place the transmission shift lever in the neutral or park position and start the engine. Increase the engine speed to 1800–2200 rpm for 2 or 3 minutes. Check and record the voltmeter reading, it should be 1 to 2 volts higher than the first reading. If the reading is less than 1 volt or greater than 2½ volts, replace the voltage regulator. If the reading is between 1 and 2 volts, turn on the headlights and heater blower. The voltage should not decrease more than ½ volt. Replace the regulator if the voltage drop is greater than ½ volt.

1974–78 ELECTRO-MECHANICAL & 1978–83 ELECTRONIC REGULATORS

The electro-mechanical and electronic regulator is factory calibrated and sealed and is not to be adjusted. If the regulator is not calibrated within the specified limits as listed in truck chapters, it must be replaced.

1974–78 TRANSISTORIZED REGULATOR

The only adjustment of this regulator is the voltage limiter adjustment Fig. F18. This adjustment is made with regulator at normal operating temperature. Remove the regulator cover and using a fiber rod turn voltage adjusting screw clockwise to raise voltage setting or counter-clockwise to lower voltage setting. Refer to the "Alternator & Regulator Specifications" as listed in the individual truck chapters for proper voltage setting.

ALTERNATOR REPAIRS Rear Terminal Alternator Exc. 65, 70, 85 & 90 Ampere Units

NOTE: Use a 100 watt soldering iron.

Disassembly

1. Mark both end housings and the stator with a scribe mark for assembly, Fig. F19.
2. Remove the three housing through bolts.
3. Separate the front housing and rotor from the stator and rear housing.
4. Remove all the nuts and insulators from the rear housing and remove the rear housing from the stator and rectifier assembly.
5. Remove the brush holder mounting screws and remove the holder, brushes, brush springs, insultor and terminal.
6. If replacement is necessary, press the bearing from the rear housing, supporting the housing on the inner boss.
7. If the rectifier assembly is being replaced, unsolder the stator leads from the printed-circuit board terminals, and separate the stator from the rectifier assembly.
8. Original production alternators will have one of three types of rectifier assembly circuit boards, Figs. F20 and F21; one has the circuit board spaced away from the diode plates with the diodes exposed. Another type is a single circuit board with built-in diodes. The third type circuit board has built-in diodes with an additional booster diode plate containing two diodes. This circuit board is used only in the 61-ampere alternator.

If the alternator rectifier has an exposed diode circuit board, remove the screws from the rectifier by rotating the bolt heads ¼ turn clockwise to unlock them and then remove the screws, Fig. F20. Push the stator terminal screw straight out on a rectifier with the diodes built into the circuit board, Fig. F21. Avoid turning the screw while removing to make certain that the straight knurl

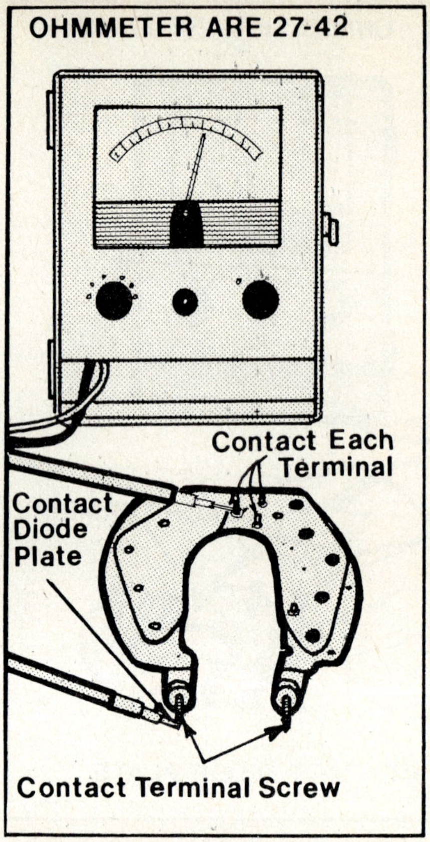

Fig. F17 Side terminal alternator diode test

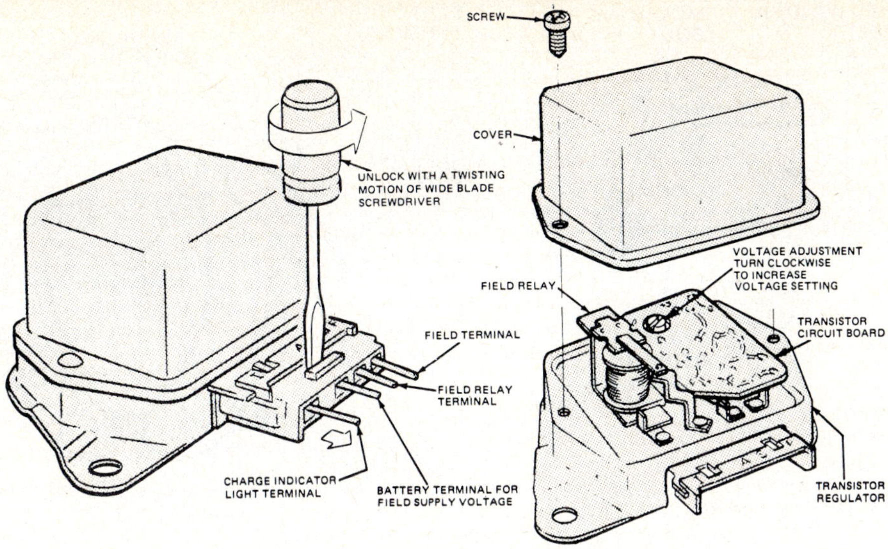

Fig. F18 Transistorized voltage regulator. 1974–78 units

will engage the insulators when installing. Do not remove the grounded screw, Fig. F22.

On 61-ampere alternator rectifier, press the stator terminal screw from the circuit board, Fig. F23. When the terminal screw has moved about 1/4 inch, remove the nut from the end of the screw and lift the screw from the circuit board.

NOTE: Do not twist the screw in the circuit board.

---

9. Remove the drive pulley nut, Fig. F24. Then, pull the lockwasher, pulley, fan, fan spacer, front housing and rotor stop from the rotor shaft.
10. Remove the three screws that hold the front end bearing retainer, and remove the retainer. If the bearing is damaged or has lost its lubricant, support the housing close to the bearing boss and press out the old bearing from the housing.
11. Perform a diode test and a field open or short circuit test.

Cleaning and Inspection

1. The rotor, stator, diodes and bearings are not to be cleaned with solvents. These parts are to be wiped off with a clean cloth. Cleaning solvent may cause damage to electrical parts or contaminate the bearing internal lubricant. Wash all other parts with solvent and dry them.
2. Rotate front bearing on drive shaft. Check for any scraping noise, looseness or roughness that would indicate that the bearing is excessively worn. As the bearing is being rotated, look for any lubricant leakage. If any of these conditions exist, replace the bearing.
3. Place the rear end bearing on the slip ring end of the shaft and rotate the bearing on the shaft. Make the same check for wear or damage as for the front bearing.
4. Check the housings for cracks. Check the front housing for stripped threads in the mounting holes. Replace defective housings.
5. Pulleys that have been removed and installed several times may have to be replaced because of the increase bore diameter. A pulley is not suitable for reuse if more than 1/4 of the shaft length will enter the pulley bore with light pressure. Replace any pulley that is bent out of shape. After installing the pulley, check for clearance between the fins and the alternator drive end housing.
6. Check all the wire leads on both stator and rotor for loose soldered connections and for burned insulation. Resolder poor connections and replace parts that show burned insulation.
7. Check slip rings for damaged insulation. Check the slip rings for runout as shown in Fig. F25. If the slip rings are more than .0005″ out of round, take a light cut (minimum diameter limit 1.22″) from the face of the rings to true them up. If the slip rings are badly damaged, the entire rotor will have to be replaced as they are serviced only as a complete assembly.
8. Replace the terminal spacer block assembly if the neutral terminal is loose. Replace any parts that are burned or cracked. Replace brushes that are worn to less than 5/16″ in length. Replace the brush spring if it has less than 7 to 12 ounces tension.

Assembly

1. The rotor, stator and bearings must not be cleaned with solvent. Wipe these parts off with a soft cloth.
2. Press the front bearing in the front housing bearing boss (put pressure on the out-

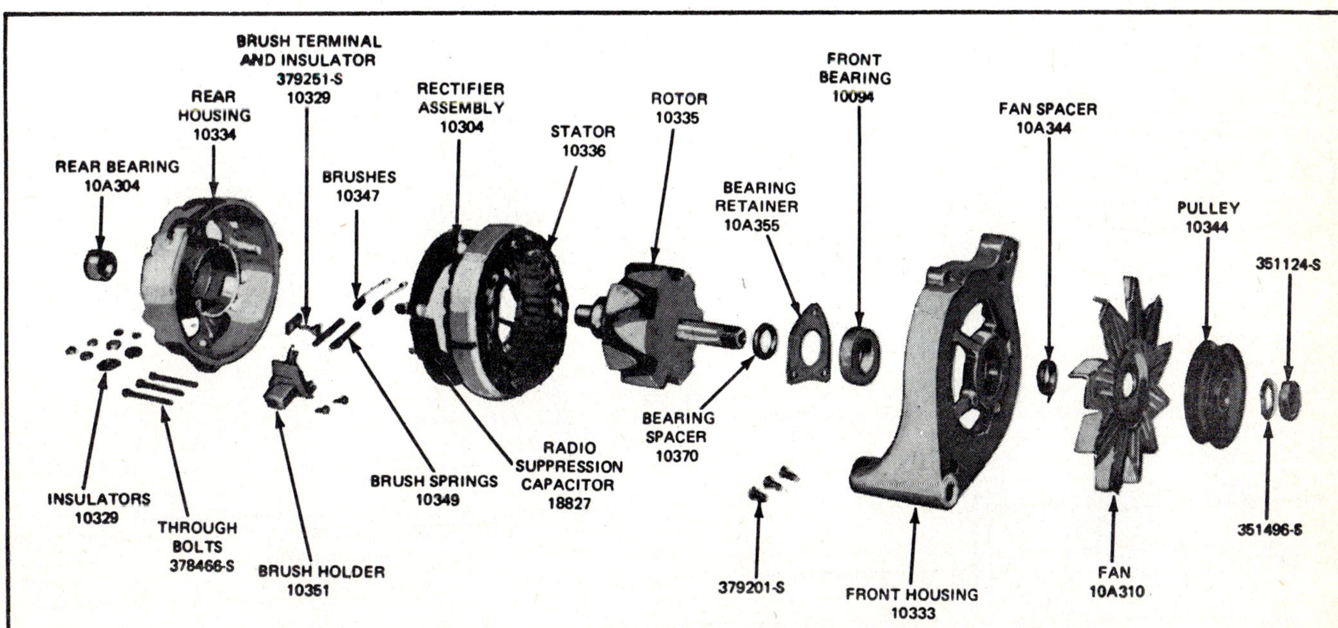

Fig. F19 Disassembled rear terminal alternator (Typical)

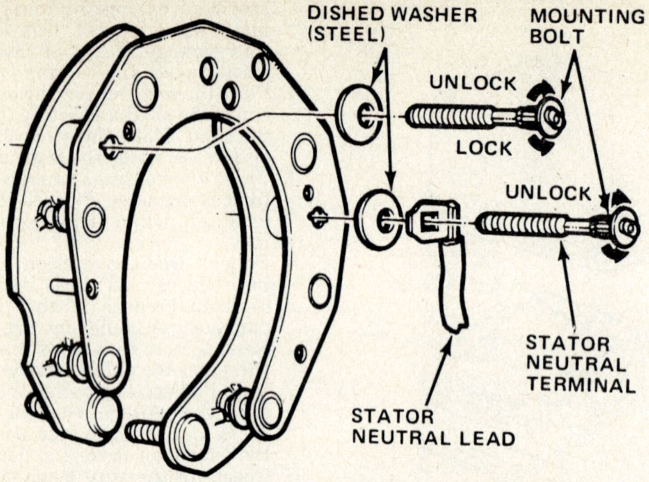

Fig. F20 Rectifier assembly with exposed diodes

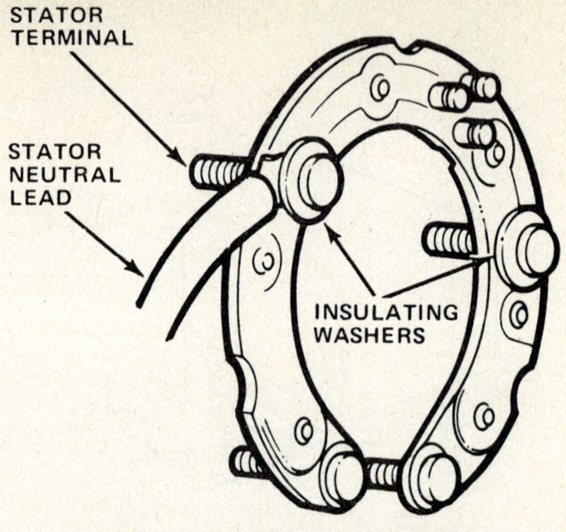

Fig. F21 Rectifier assembly with built-in diodes

er race only), and install the bearing retainer, Fig. F19.

3. If the stop-ring on the rotor drive shaft was damaged, install a new stop-ring. Push the new ring on the shaft and into the groove.

NOTE: Do not open the ring with snap ring pliers as permanent damage will result.

---

4. Position the rotor stop on the drive shaft with the recessed side against the stop-ring.
5. Position the front housing, fan spacer, fan, pulley and lock washer on the drive shaft and install the retaining nut. Torque the retaining nut, Fig. F24, to 60–100 ft lbs.
6. If the rear housing bearing was removed, support the housing on the inner boss and press in a new bearing flush with the outer end surface.
7. Place the brush springs, brushes, brush terminal and terminal insulator in the brush holder and hold the brushes in position by inserting a piece of stiff wire in the brush holder, Fig. F26.
8. Position the brush holder assembly in the rear housing and install the mounting screws. Position the brush leads in the brush holder, Fig. F27.
9. Wrap the three stator winding leads around the circuit board terminals and solder them. Position the stator neutral lead eyelet on the stator terminal screw and install the screw in the rectifier assembly, Fig. F28.
10. For a rectifier with the diodes exposed insert the special screws through the wire lug, dished washers and circuit board, Fig. F20. Turn them 1/4 turn counterclockwise to lock them. For single circuit boards with built in diodes, insert the screws, straight through the wire lug, insulating washer and rectifier into the insulator, Fig. F22.

NOTE: The dished washers are to be used only on the circuit board with exposed diodes, Fig. F20. If they are used on the single circuit board, a short circuit will occur. A flat insulating washer is to be used between the stator terminal and the board when a single circuit board is used, Fig. F22.

---

11. For a rectifier with a booster diode plate (61 Ampere Alternator only), proceed as follows:
   a. Position the stator wire terminal on the stator terminal screw and position the screw into the rectifier. Position the square insulator over the screw

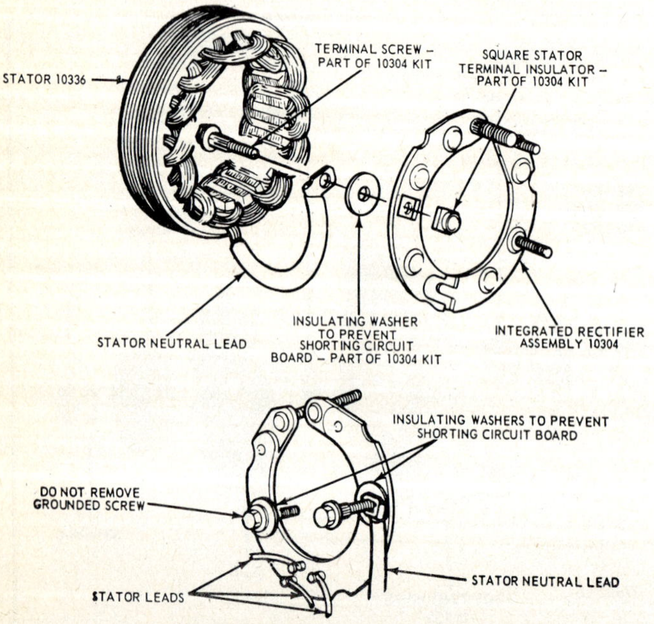

Fig. F22 Stator terminal installation. Integral rectifier circuit board

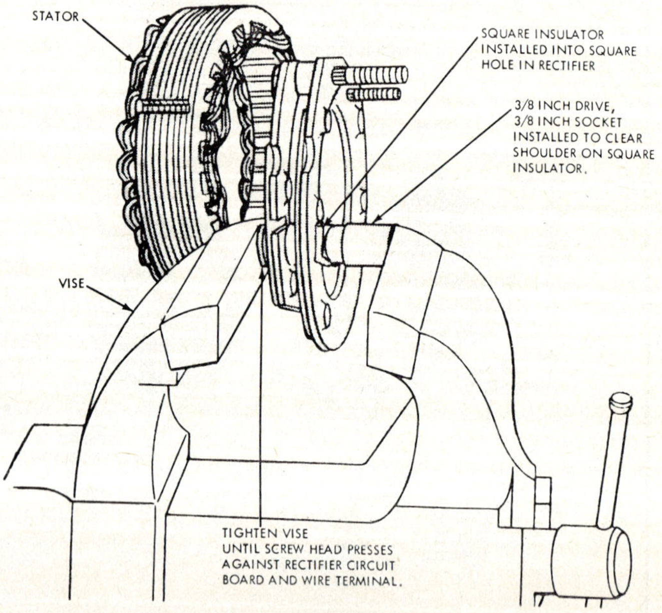

Fig. F23 Stator terminal screw removal. 61-amp alternator

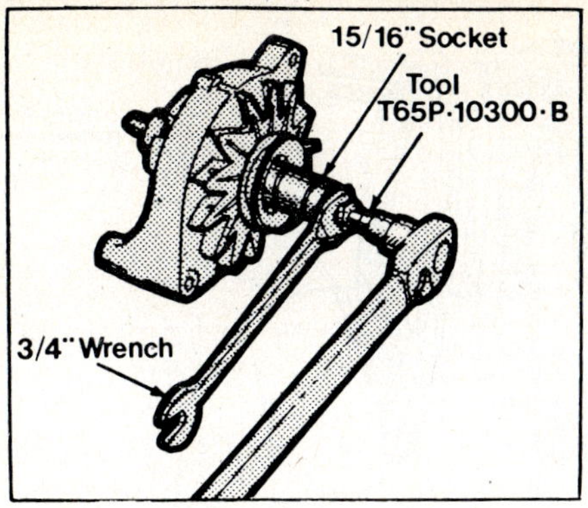

Fig. F24 Typical pulley removal

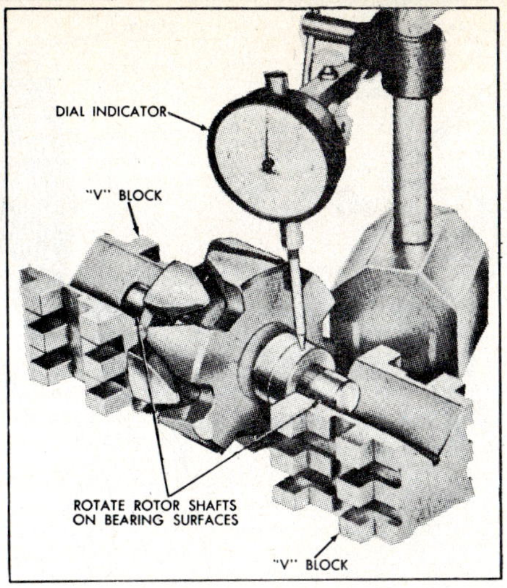

Fig. F25 Slip ring runout check

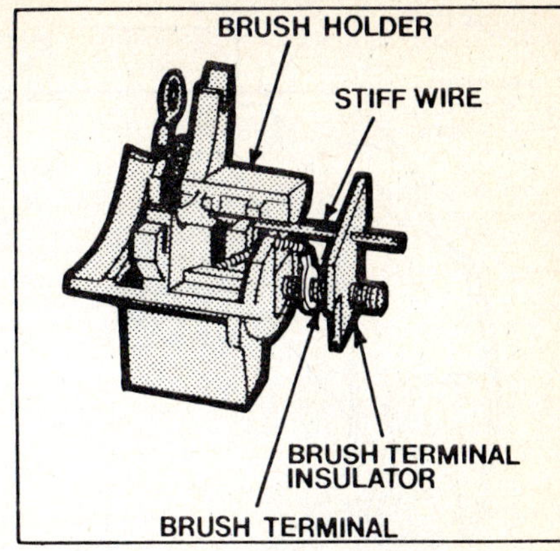

Fig. F26 Brush holder assembly

and into the square hole in the rectifier, Fig. F29.

b. Rotate the terminal screw until it locks in position. Then, press the screw in finger tight.

c. Position the stator wire, Fig. F30. Press the terminal screw into the rectifier and insulator, Fig. F31.

12. Position the radio noise suppression capacitor on the rectifier terminals. On the circuit board with exposed diodes, install the STA and BAT terminal insulators,
Fig. F28. On the single circuit board, position the square stator-terminal insulator in the square hole in the rectifier assembly, Fig. F22. Position the BAT terminal insulator, Fig. F32.
Position the stator and rectifier assembly in the rear housing. Make certain that all terminal insulators are seated properly in the recesses, Fig. F28. Position the STA (black), BAT (red) and FLD (orange) insulators on the terminal bolts, and install the retaining nuts, Fig. F33.
13. Wipe the rear end bearing surface of the rotor shaft with a clean lint-free rag.
14. Position the rear housing and stator assembly over the rotor and align the scribe marks made during disassembly. Seat the machined portion of the stator core into the step in both end housings. Install the housing through bolts. Remove the brush retracting wire, and put a daub of waterproof cement over the hole to seal it.

Fig. F27 Typical brush lead positions

Rear Terminal 65, 70, 85 & 90-Ampere Alternator

NOTE: Use a 200 watt soldering iron.

Disassembly

1. Remove the brush holder and cover assembly from the rear end housing. Fig. F34.
2. Mark both end housings and the stator with a scribe mark for assembly.
3. Remove the three housing through bolts.
4. Separate the front housing and rotor from the stator and rear housing.
5. Remove the drive pulley nut, lock-washer, flat washer, pulley, fan, fan spacer and rotor from the front housing, Fig. F24.
6. Remove the three screws that hold the front bearing retainer, and remove the retainer. If the bearing is damaged or has lost its lubricant, support the housing close to the bearing boss and press out the bearing from the housing.
7. Remove all the nut and washer assemblies and insulators from the rear housing and remove the rear housing from the stator and rectifier assembly.

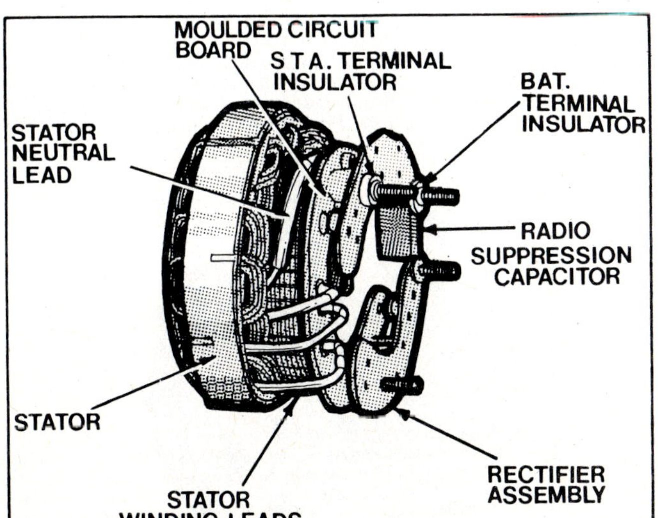

Fig. F28 Stator lead connections. Except 61-amp alternator

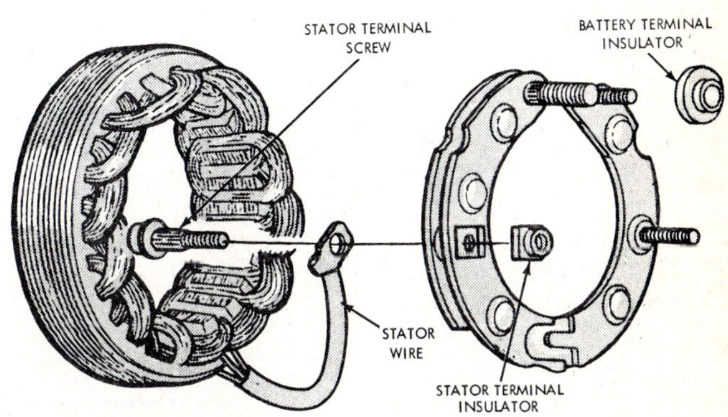

Fig. F29 Stator and rectifier assembly. 61-amp alternator

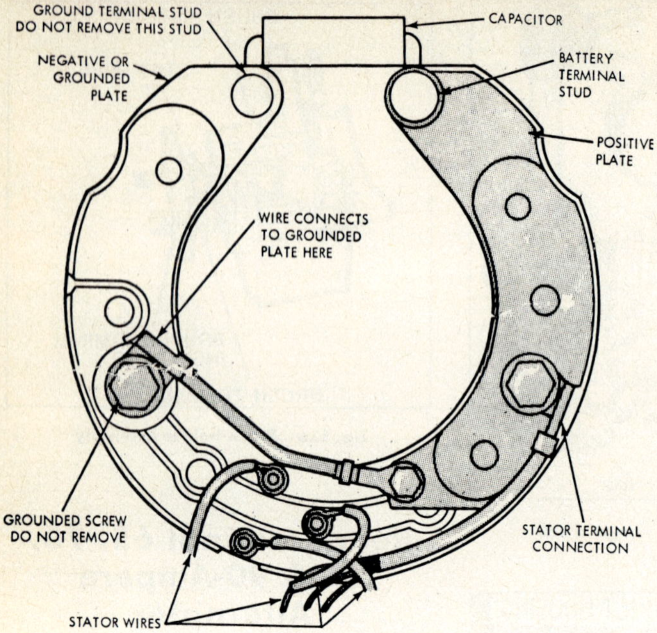

Fig. F30 Rectifier terminal locations. 61-amp alternator

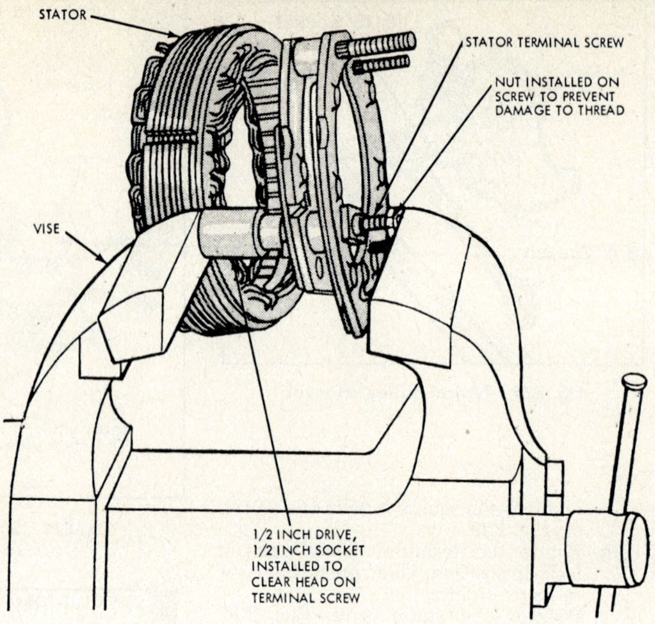

Fig. F31 Stator terminal screw installation. 61-amp alternator

8. If replacement is necessary, press the bearing from the rear housing, supporting the housing on the inner boss.
9. Unsolder the three stator leads from the rectifier assembly, and separate the stator from the assembly.
10. Perform a diode test and an open and grounded stator coil test.

Assembly

NOTE: "Cleaning and Inspection" procedures, are same as procedures used previously in this chapter.

1. If the front bearing is being replaced, press the new bearing in the bearing boss putting pressure on the outer race only. Then, install the bearing retainer, and tighten the retainer screws until the tips of the retainer touch the housing.
2. Position the rectifier assembly to the stator, wrap the three stator leads around the diode plate terminals and solder them, Fig. F35.
3. If the rear housing bearing was removed, press in a new bearing from the inside of the housing until the bearing is flush with the outer end surface. Put pressure on the bearing outer race only.
4. Install the BAT-GRD insulator, Fig. F35, and position the stator and rectifier assembly in the rear housing.
5. Install the STA (purple) and BAT (red) terminal insulators in the terminal bolts and install the nut and washer assemblies.

NOTE: Make certain that the shoulders on all insulators both inside and outside of the housing are seated properly before tightening the nuts.

6. Position the front housing over the rotor and install the fan spacer, fan, pulley, flat and lock washers and nut on the rotor shaft, Fig. F24.
7. Wipe the rear bearing surface of the rotor shaft with a clean lint free rag.
8. Position the rotor with the front housing into the stator and rear housing assembly, and align the scribe marks made during disassembly. Seat the machined portion of the stator core into the step in both housing, and install the through bolts.
9. Hold the brushes in position by inserting a stiff wire in the brush holder, Figs. F36 and F37.
10. Position the brush holder assembly into the rear housing and install the three mounting screws. Remove the brush retracting wire and put a daub of waterproof cement over the hole to seal hole.

Brush Replacement

1. Remove the brush holder and cover assembly from the rear housing.
2. Remove the terminal bolts from the brush holder and cover assembly, and remove the brush assemblies.
3. Position the new brush terminals on the terminal bolts and assemble the terminals, bolts, brush holder washers and nuts, Figs. F36 and F37. The insulating washer mounts under the FLD terminal nut. The entire brush and cover assembly is also available for service.
4. Depress the brush springs in the brush holder cavities and insert the brushes on top of the springs. Hold the brushes in position by inserting a stiff wire in the brush holder as shown in Figs. F36 and F37. Position the brush leads, Figs. F36

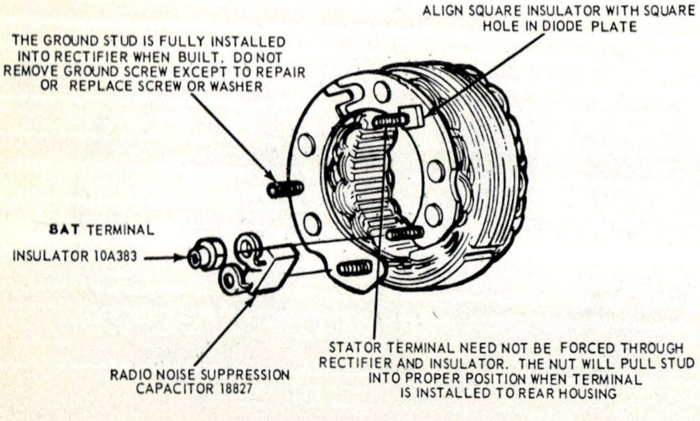

Fig. F32 Fiber-glass circuit board terminal insulators

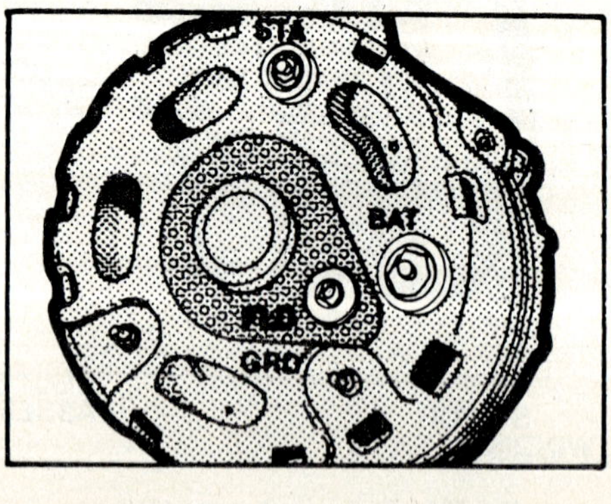

Fig. F33 Alternator terminal locations

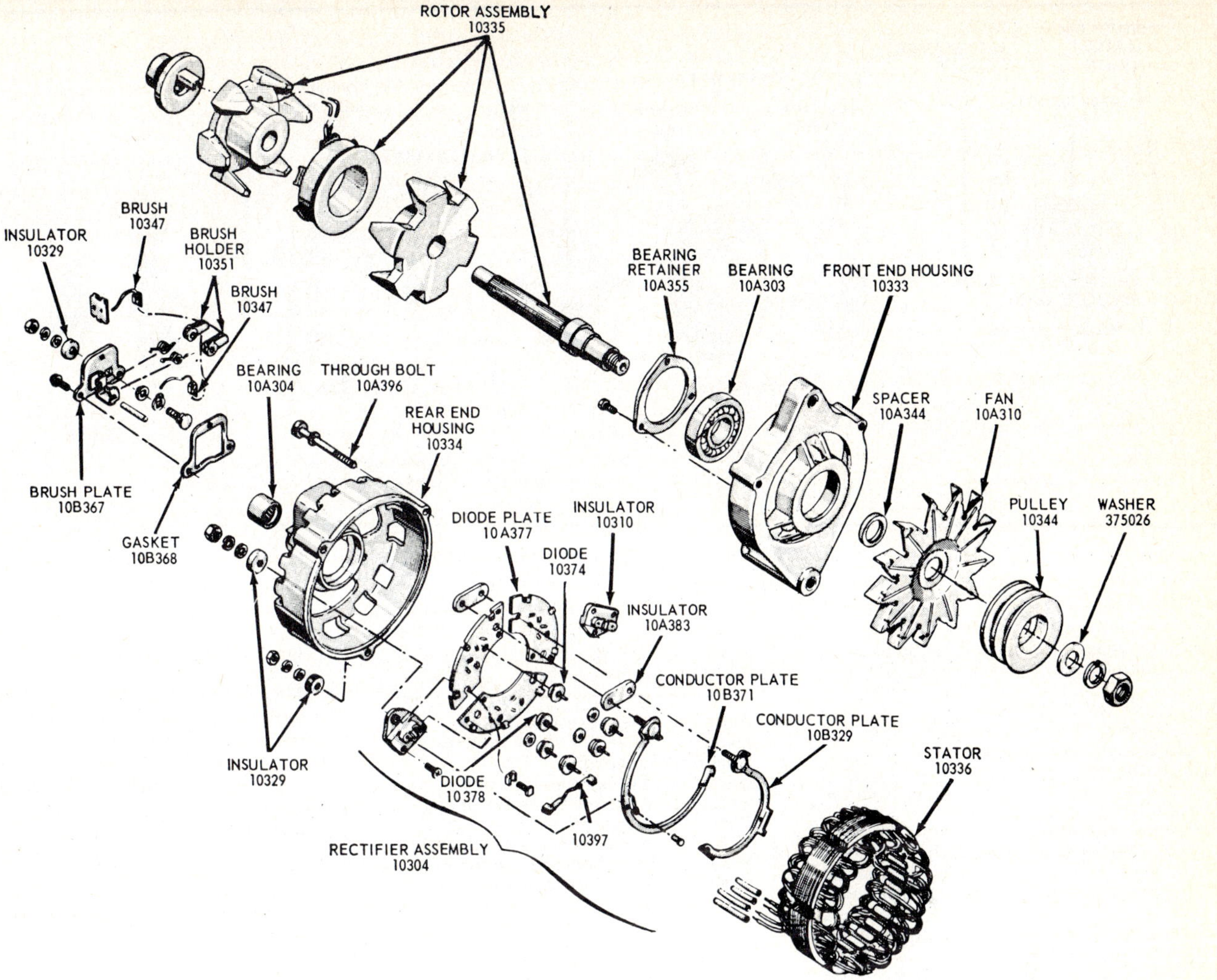

Fig. F34 Disassembled 65-amp rear terminal alternator (Typical)

and F37.

5. Install the brush holder and cover assembly to the rear housing. Remove the brush retracting wire and put a daub of waterproof cement over the hole to seal it.

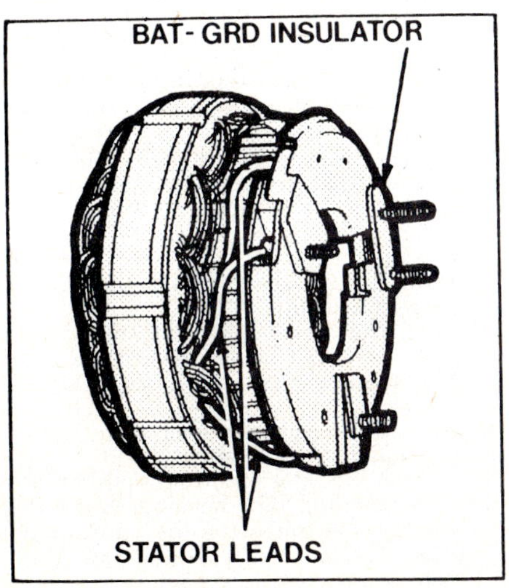

Fig. F35 Stator lead connections

SIDE TERMINAL ALTERNATOR

Disassembly

NOTE: Use a 200 watt soldering iron.

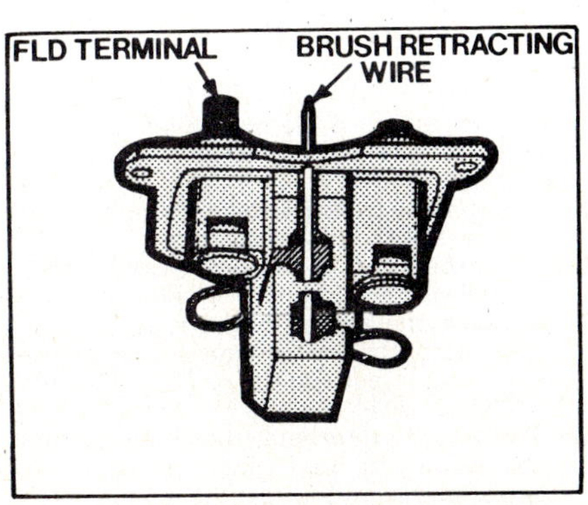

Fig. F36 Field brush assembly. 1974–75

1. Mark both end housings and the stator with a scribe mark for use during assembly, Fig. F38.
2. Remove the four housing through bolts, and separate the front housing and rotor from the rear housing and stator. Slots

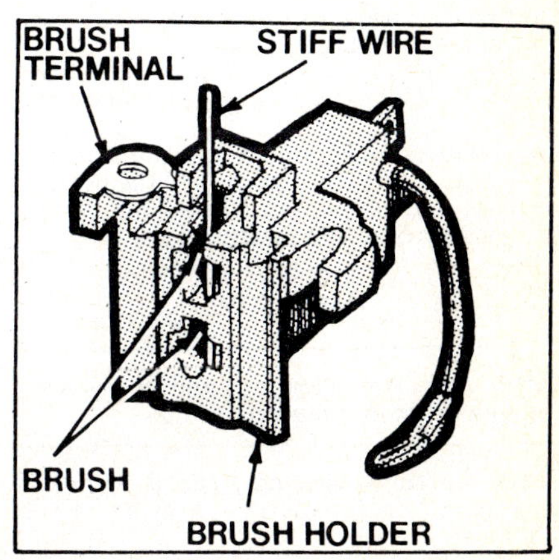

Fig. F37 Brush holder assembly. 1976–83

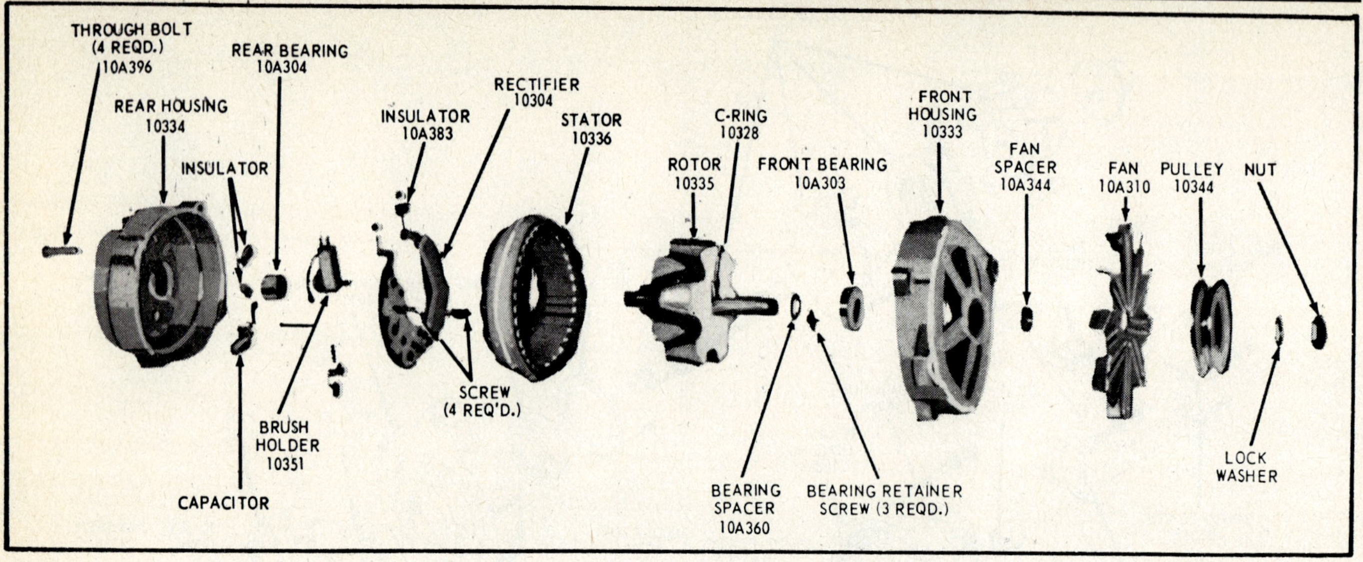

Fig. F38 Side terminal alternator disassembled (typical)

are provided in the front housing to aid in disassembly. *Do not separate the rear housing from the stator at this time.*

3. Remove the drive pulley nut, Fig. F24. Remove the lockwasher, pulley, fan and fan spacer from the rotor shaft.
4. Pull the rotor and shaft from the front housing, and remove the spacer from the rotor shaft, Fig. F38.
5. Remove three screws retaining the bearing to the front housing. If the bearing is damaged or has lost its lubricant, remove the bearing from the housing. To remove the bearing, support the housing close to the bearing boss and press the bearing from the housing.
6. Unsolder and disengage the three stator leads from the rectifier, Fig. F39.
7. Lift the stator from the rear housing.
8. Unsolder and disengage the brush holder lead from the rectifier.
9. Remove the screw attaching the capacitor lead to the rectifier.
10. Remove four screws attaching the rectifier to the rear housing, Fig. F39.
11. Remove the two terminal nuts and insulator from outside the housing, and remove the rectifier from the housing.
12. Remove two screws attaching the brush holder to the housing and remove the brushes and holder.
13. Remove sealing compound from rear housing and brush holder.
14. Remove one screw attaching the capacitor to the rear housing and remove the capacitor.
15. If bearing replacement is necessary, support the rear housing close to the bearing boss and press the bearing out of the housing from the inside.

Assembly

NOTE: Refer to "Cleaning and Inspection" procedures before reassembly.

1. If the front housing bearing is being replaced, press the new bearing in the housing.

NOTE: *Put pressure on the bearing outer race only.* Then, install the bearing retaining screws.

2. Place the inner spacer on the rotor shaft and insert the rotor shaft into the front housing and bearing.
3. Install the fan spacer, fan, pulley, lockwasher and nut on the rotor shaft, Fig. F24. Torque nut to 60–100 ft. lbs.
4. If the rear bearing is being replaced, press a new bearing in from inside the housing until it is flush with the boss outer surface.
5. Position the brush terminal on the brush holder, Fig. F37. Install the springs and brushes in the brush holder, and insert a piece of stiff wire to hold the brushes in place, Fig. F37.
6. Position the brush holder in the rear housing and install the attaching screws. Push the brush holder toward the rotor shaft opening and tighten the brush holder attaching screws.
7. Position the capacitor to the rear housing and install the attaching screw.
8. Place the two cup shaped (rectifier) insulators on the bosses inside the housing, Fig. F40.
9. Place the insulator on the BAT (large) terminal of the rectifier, and position the rectifier in the rear housing. Place the outside insulator on the BAT terminal, and install the nuts on the BAT and GRD terminals *finger tight.*
10. Install but do not tighten the four rectifier attaching screws.
11. Tighten the BAT and GRD terminal nuts on the outside of the rear housing. Then, tighten the four rectifier attaching screws.
12. Position the capacitor lead to the rectifier and install the attaching screw.
13. Press the brush holder lead on the rectifier attaching screws.
14. Position the stator in the rear housing and align the scribe marks. Press the three stator leads on the rectifier pins and solder securely, Fig. F39.
15. Position the rotor and front housing into the stator and rear housing, Align the scribe marks and install the four through bolts. Tighten two opposing bolts and then the two remaining bolts.
16. Spin the fan and pulley to be sure nothing is binding within the alternator.
17. Remove the wire retracting the brushes, and place a daub of waterproof cement over the hole to seal it.

Brush Replacement

Removal

1. Mark both end housings and the stator with a scribe mark for use during assembly.
2. Remove the four housing through bolts, and separate the front housing and rotor from the rear housing and stator. Slots are provided in the front housing to aid in disassembly.

NOTE: *Do not separate the rear housing and stator.*

3. Unsolder and disengage the brush holder lead from the rectifier.
4. Remove the two brush holder attaching screws and lift the brush holder from the rear housing.
5. Remove the brushes from the brush holder.

Installation

1. Insert the brushes into the brush holder and position the terminal on the brush holder.
2. Depress the brushes and insert a 1½ inch piece of stiff wire, Fig. F37, to hold the brushes in the retracted position.
3. Position the brush holder to the rear housing, inserting the wire used to retract the brushes through the hole in the rear housing.
4. Install the brush holder attaching screws. Push the brush holder toward the rotor shaft opening and tighten the attaching screws.
5. Press the brush holder lead on the rectifier pin and solder securely.
6. Position the rotor and front housing into the stator and rear housing. Align the scribe marks and install the four through bolts. Tighten two opposing bolts and then the two remaining bolts.
7. Spin the fan and pulley to be sure nothing is binding within the alternator.

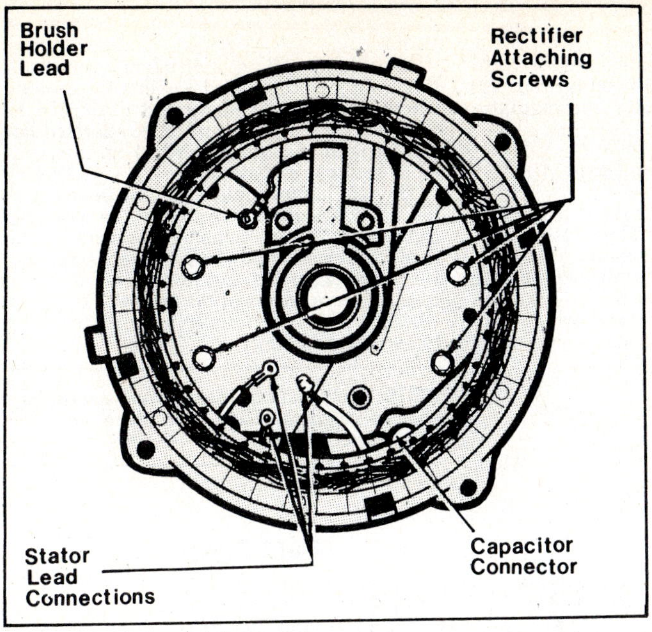

Fig. F39 Stator lead connections

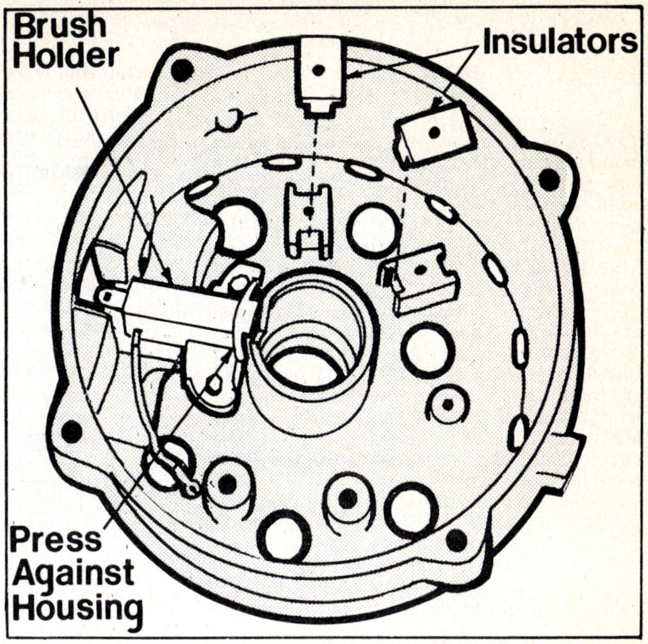

Fig. F40 Brush holder and rectifier insulators installed

8. Remove the wire retracting the brushes, and place a daub of waterproof cement over the hole to seal it.

Rectifier Replacement

Removal

1. Mark both end housings and the stator with a scribe mark for use during assembly, Fig. F38.
2. Remove the four housing through bolts, and separate the front housing and rotor from the rear housing and stator. Slots are provided in the front housing to aid in disassembly.

NOTE: *Do not separate the rear housing and stator at this time.*

3. Unsolder and disengage the three stator leads from the rectifier, Fig. F39.
4. Unsolder and disengage the brush holder lead from the rectifier.
5. Remove the screw attaching the capacitor lead to the rectifier.
6. Remove four screws attaching the rectifier to the rear housing, Fig. F39.
7. Remove the terminal nuts and insulator from outside the housing, and remove the rectifier from the housing.

Installation

1. Insert a piece of wire through the hole in the rear housing to hold the brushes in the retracted position.
2. Place the two cup shaped (rectifier) insulators on the bosses inside the housing, Fig. F40.
3. Place the insulator on the BAT (large terminal) of the rectifier, and position the rectifier in the rear housing. Place the outside insulator on the BAT terminal, and install the nuts on the BAT and GRD terminals finger tight.
4. Install but do not tighten the four rectifier attaching screws.
5. Tighten the BAT and GRD terminal nuts on the outside of the rear housing. Then, tighten the four rectifier attaching screws.
6. Position the capacitor lead to the rectifier and install the attaching screw.
7. Press the brush holder lead on the rectifier pin and solder securely, Fig. F40.
8. Position the stator in the rear housing and align the scribe marks. Press the three stator leads on the rectifier pins and solder securely, Fig. F39.
9. Position the rotor and front housing into the stator and rear housing. Align the scribe marks and install the four through bolts. Partially tighten all four through bolts. Then, tighten two opposing bolts and then the two remaining bolts.
10. Spin the fan and pulley to be sure nothing is binding within the alternator.
11. Remove the wire retracting the brushes in the brush holder, and place a daub of waterproof cement over the hole in the rear housing to seal it.

Cleaning & Inspection Procedures

1. The rotor, stator, and bearings must not be cleaned with solvent. Wipe these parts off with a clean cloth.
2. Rotate the front bearing on the drive end of the rotor drive shaft. Check for any scraping noise, looseness or roughness that will indicate that the bearing is excessively worn. Look for excessive lubricant leakage. If any of these conditions exist, replace the bearing.
3. Inspect the rotor shaft at the rear bearing surface for roughness or severe chatter marks. Replace the rotor assembly if the shaft is not smooth.
4. Place the rear end bearing on the slipring end of the shaft and rotate the bearing on the shaft. Make the same check for noise, looseness or roughness as was made for the front bearing. Inspect the rollers and cage for damage. Replace the bearing if these conditions exist, or if the lubricant is lost or contaminated.
5. Check the pulley and fan for excessive looseness on the rotor shaft. Replace any pulley or fan that is loose or bent out of shape. Check the rotor shaft for stripped or damaged threads. Inspect the hex hole in the end of the shaft for damage.
6. Check both the front and rear housing for cracks. Check the front housings for stripped threads in the mounting gear. Replace defective housings.
7. Check all wire leads on both the stator and rotor assemblies for loose soldered connections, and for burned insulation. Resolder poor connections. Replace parts that show burned insulation.
8. Check the slip rings for nicks and surface roughness. If the slip rings are badly damaged, the entire rotor will have to be replaced, as it is serviced as a complete assembly.
9. Replace any parts that are burned or cracked. Replace brushes and brush springs that are not to specification.

Motorola Alternators

1974–75

The electrical circuit, Fig. M1, of these units differs from previous units in that a field diode assembly is used. Also, on alternators used in 1974–75 Jeep vehicles equipped with a four barrel carburetor, an extra terminal is used on the rear housing which provides about seven volts of alternating current to the heating element of the electric assisted choke.

The field diode (diode trio) assembly incorporates three diodes mounted on a circuit board or on some 1974 & 1975 units, a potted type diode trio is used, Fig. M2. The input lead are connected to the stator windings in parallel with the positive diodes. The diode output leads are connected to a metal grommet in the circuit board and is secured to the insulated regulator terminal.

A portion of the alternating current and voltage developed in the stator windings is rectified by the field diode assembly. This voltage is sensed by the voltage regulator to provide current to the field windings. Fig. M3 shows the charging circuit.

Voltage Regulator

The voltage regulator is an electrical switching device sealed at the factory, requiring no adjustments. It senses the voltage appearing at the regulator terminal of the alternator and supplies the necessary field current for maintaining the system voltage at the output terminal.

TESTING SYSTEM IN VEHICLE

Alternator Output Test 1974–75

1. Connect a voltmeter to the battery, start engine and turn on headlamps.
2. Run engine for two minutes at 1000 RPM and observe voltmeter. If voltage remains above 13 volts, alternator and regulator are satisfactory.

Rectifier Diode Tests

Any commercial in-circuit diode tester will suffice to make the test. Follow Test Equipment Manufacturer's instructions.

Check diodes individually after the diodes have been disconnected from the stator. A shorted stator coil or shorted insulating washers or sleeves on positive diodes would make diodes appear to be shorted.

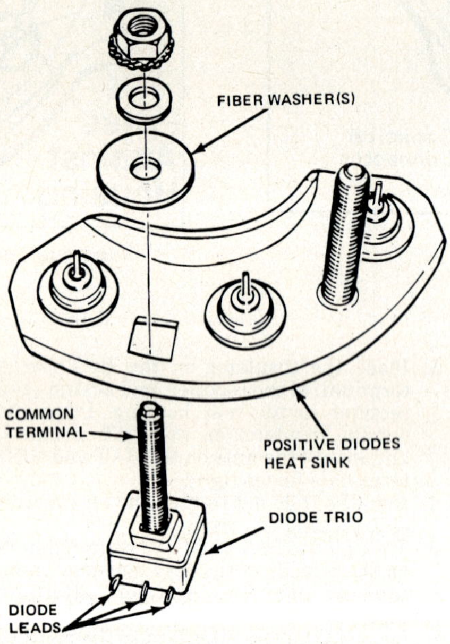

Fig. M2 Potted type diode trio. 1974–75

A test lamp will not indicate an open condition unless all three diodes of either assembly are open. However, a shorted diode can be detected. The test is not 100% effective but can be used if so desired when an in-circuit diode tester is not available.

The test lamp should light in one direction but not in the other direction. If the test lamp lights in both directions, one or more of the diodes of the assembly being tested is shorted. If the test lamp does not light in either direction, *all three diodes in the assembly are open.* Check diodes individually after disassembly to ascertain findings.

NOTE: *A shorted stator coil would appear as a shorted negative diode. Also check stator for shorts after disassembly.*

Field Diode Test

1. Using a voltmeter, connect positive lead to alternator output terminal and negative lead to regulator terminal.
2. Start and run engine at idle speed. The voltmeter should then read .6 volt or less. If reading is over .6 volt, replace field diode assembly.

Voltage-Regulator Test 1974–75

1. Connect a voltmeter to the battery, start engine and turn on headlamps.
2. Run engine at 1000 RPM for several minutes to establish voltage regulator operating temperature.
3. Voltage should be within 13.1 to 14.3 volts when regulator temperature is between 100°F and 150°F.

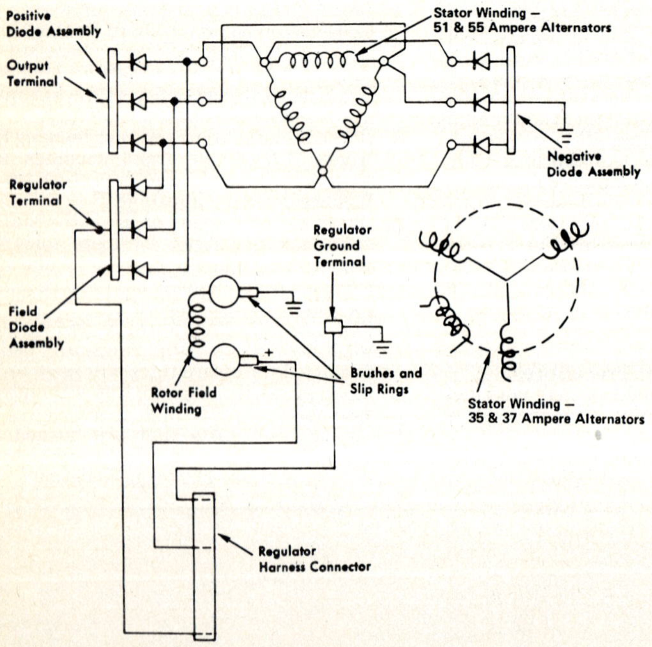

Fig. M1 Alternator circuit, 1974–75, 35, 37, 51, 55 & 62 amp.

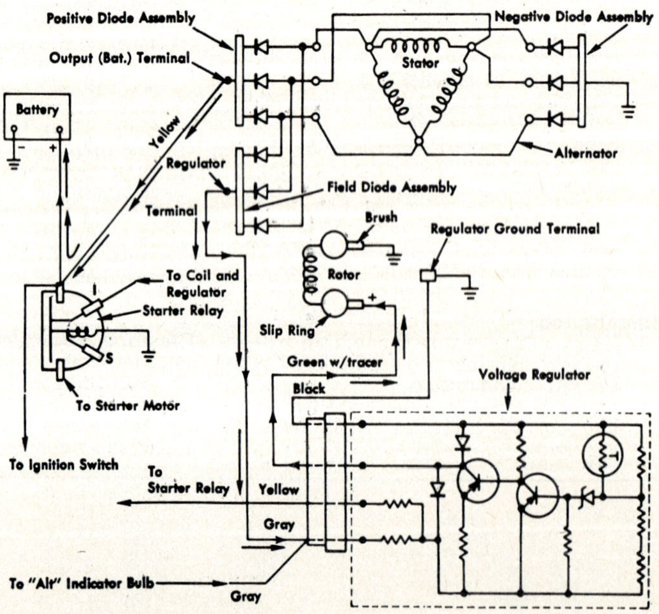

Fig. M3 Charging circuit diagram, 1974–75 (typical)

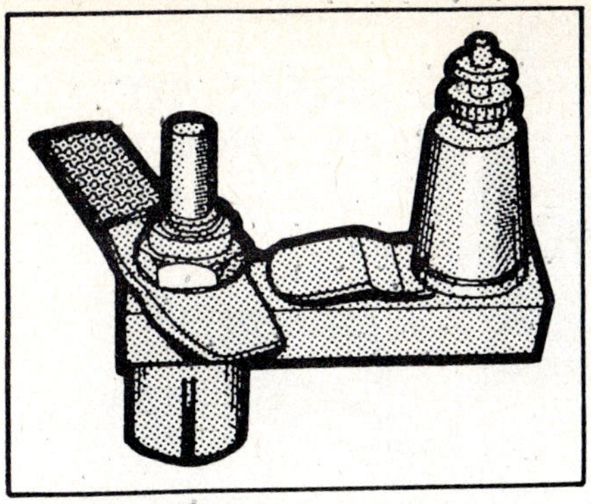

Fig. M5 Battery post adapter tool which provides a convenient method for connecting ammeter leads of volt-ammeter tester to charging system

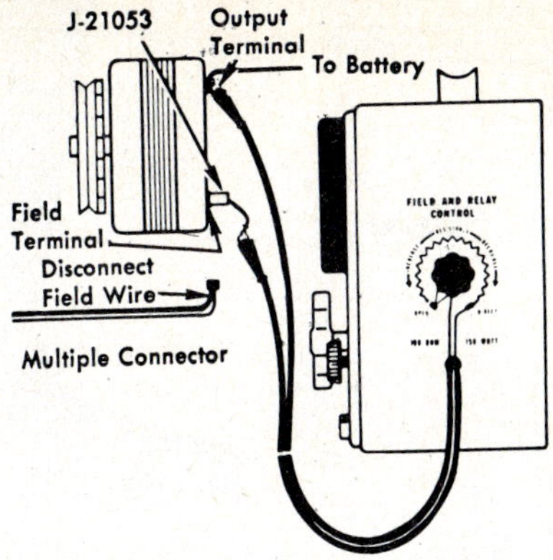

Fig. M6 Alternator output test with external field control

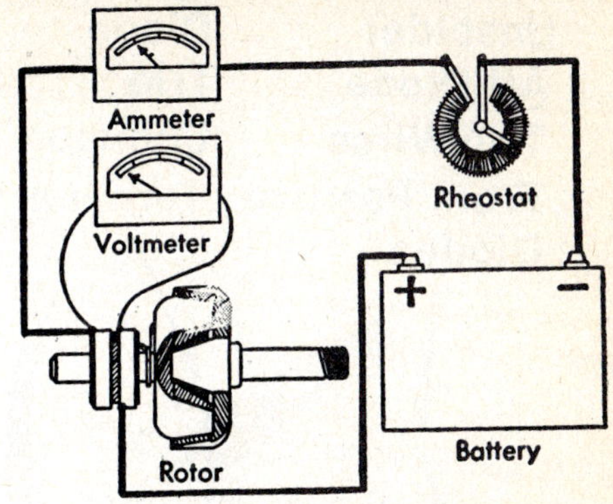

Fig. M8 Field coil test

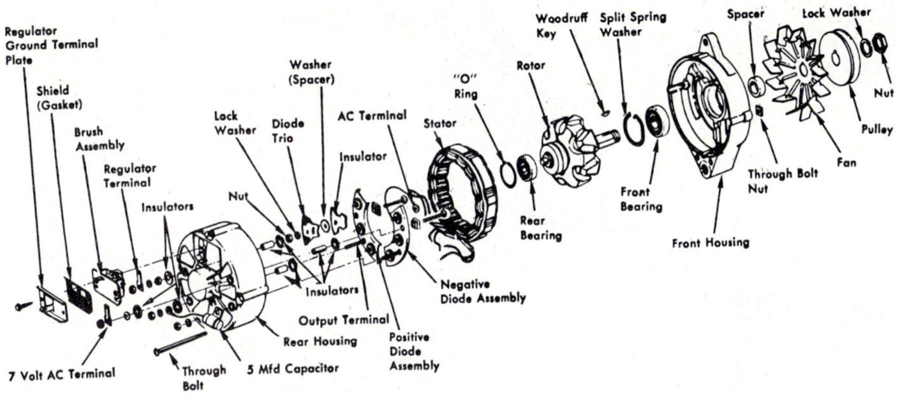

Fig. M7 Alternator disassembled. 1974–75 NOTE: Model shown has AC terminal for 1974–75 electric choke

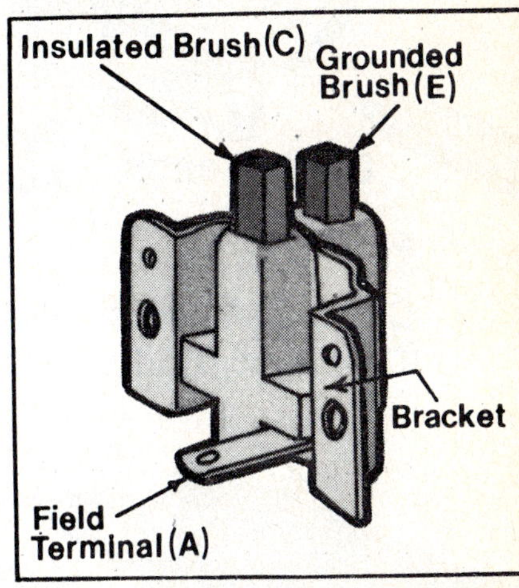

Fig. M9 Brush assembly text

ALTERNATOR REPAIRS

Disassembly, Fig. M7

Brush Assembly

The brush assembly can be removed in most cases with the alternator on the vehicle. The spring clip is bent back so that the field terminal plug can be removed. Remove the two self-tapping screws, field plug retainer spring and cover. Pull brush assembly straight out far enough to clear locating pins, then lift brush assembly out. The complete brush assembly is available for replacement.

Rear Housing

Remove the 4 through bolts and nuts. Carefully separate the rear housing and stator from the front housing by using 2 small screwdrivers and prying the stator from the front housing at 2 opposing slots where the "through bolts" are removed. Do not burr the stator core which would make assembly difficult.

CAUTION: *Do not insert screwdriver blade deeper than 1/16" to avoid damaging stator winding.*

Stator and Diode Assembly

Do not unsolder stator-to-diode wire junction. Remove stator and diode as an assembly. Avoid bending stator wire at junction holding positive and negative diode assembly from housing.

Remove 4 lock nuts and insulating washers. The insulating washers and nylon sleeves are used to insulate the positive plate studs from the housing. With the 4 nuts removed, the stator can be separated from the rear housing by hand.

Diode Replacement

To replace field diode assembly, remove nylon sleeve, insulating washer and hold-down nut from regulator stud. Unsolder field diode wires from positive diodes. *When replacing positive or negative diodes, make note of diode assembly to stator connections, and make sure replacement diode assembly connections are the same. The positive diode assembly has red markings, the negative black markings.*

In soldering and unsoldering leads from diodes, grasp the diode lead with pliers between the diode and stator lead to be removed. This will give better heat dissipation and protect the diode. Do not exert excessive stress on diode lead.

Rotor

The rotor should only require removal from the front housing if there is a defect in the field coil itself or in the front bearing. Front and rear bearings are permanently sealed, self-lubricating type. If the front housing must be removed from the rotor, use a two-jaw puller to remove the pulley. The split spring washer must be loosened with snap ring pliers through the opening in the front housing. Remove the washer only after the housing is removed. The rotor and front bearing can be removed from the front housing by tapping the rotor shaft slightly.

NOTE: *Make certain that the split spring washer has been removed from its groove before attempting to remove the front housing from the bearing.*

Alternator Bench Tests

Field Coil Test

The rotor should be tested for grounds and for shorted turns in the winding. The ground test is made with test probes connected in series with a 110 volt test lamp. Place one test probe on the slip ring and the other probe on the rotor core. If the bulb lights the rotor is grounded.

To test for shorted turns, check rotor field current drawn as shown in Fig. M8. Slowly reduce resistance of rheostat to zero. With full battery voltage applied to the field coil, the field current should be as listed in Alternator Specifications tables in front of Truck chapters. Excessive current draw indicates shorted turn in field winding.

Brush Insulation Test

Connect an ohmmeter or a test lamp to the field terminal and bracket. Resistance should be high (infinite) or test lamp should not light. If resistance is low or if test lamp lights, brush assembly is shorted and must be replaced.

Brush Continuity Test

Connect an ohmmeter to field terminal and brush. Use an alligator clip to assure good contact to brush, test points "A" and "C" in Fig. M9.

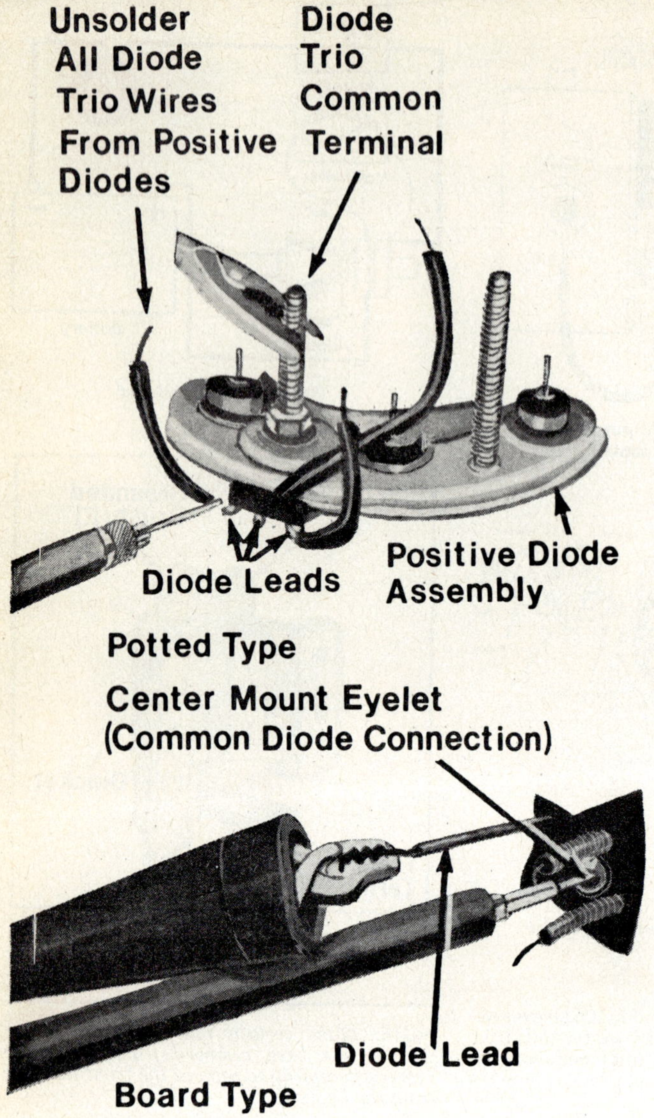

Fig. M10 Diode trio bench test. 1974–75 alternators

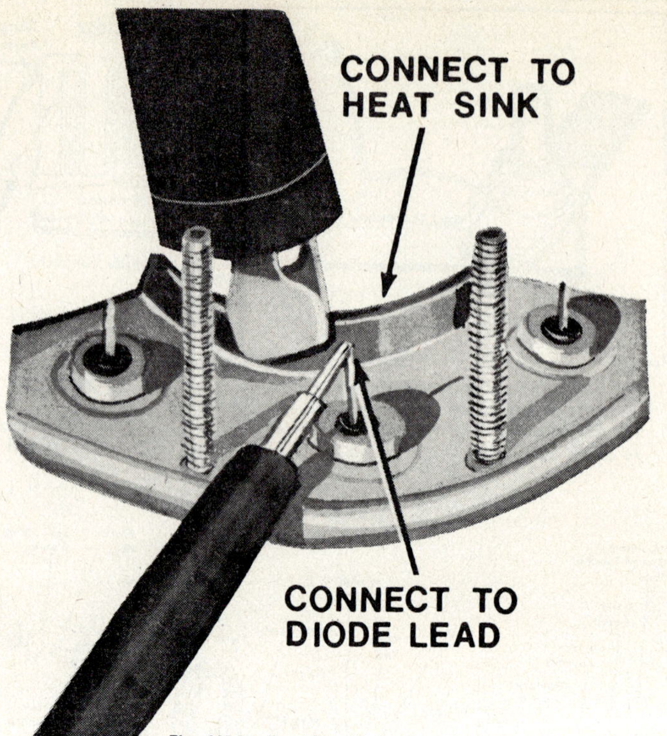

Fig. M11 Rectifier diodes bench test. 1974–75 alternators

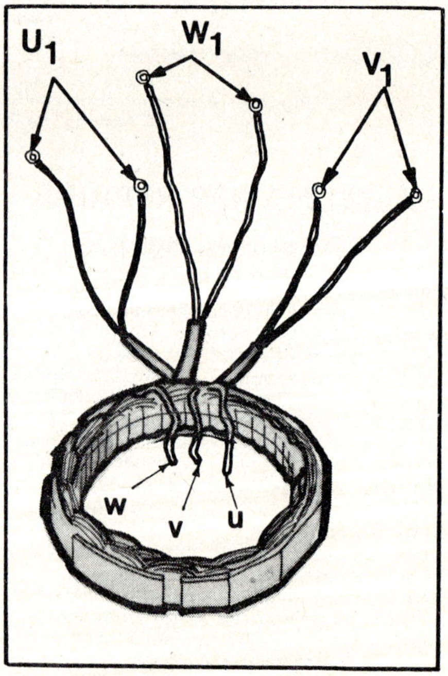

Fig. M12 Stator coil shorts and continuity tests

CAUTION: *Do not chip brush.*

Resistance reading should be zero. Move brush and brush lead wire to make certain that brush lead wire connections are not intermittent. Resistance reading should not vary when brush and lead wire are being moved around. Connect ohmmeter to bracket and grounded brush, test points "E" and "D", Fig. M9. Resistance reading should be zero.

1974–75 Diode Tests

Diode Trio: Unsolder diode trio leads and using a 12 volt meter which draws a one amp maximum load, connect test leads providing current path, Fig. M10. Hold this connection for at least two minutes, then reverse test leads immediately. Test each diode in this manner. If current flows in both directions, neither direction or flows intermittently, the diode trio must be replaced.

Rectifier Diodes: The same method of testing is used for the rectifier diodes as for the diode trio, Fig. M11. However, a 15 amp load is used when testing the rectifier diodes.

NOTE: A diode can be tested for opens or shorts using an ohmmeter. Reason for applying an electrical load to the diodes, in turn generating heat, is to detect intermittent diode failures.

Stator in-Circuit Test

When making the in-circuit stator leakage test, some consideration must be given to the rectifier diodes that are connected to the stator winding. The negative diode assembly will conduct in one direction when properly polarized. A shorted diode in the negative diode assembly would make the stator appear to be shorted. For this reason, the rectifier diode plate assembly and stator must be checked individually after alternator has been disassembled if the problem is localized to the stator.

CAUTION: *Use a special diode continuity light or a DC test lamp. Do not use a 120 volt test lamp as diodes will be damaged.*

1. Connect the test lamp to a diode terminal of the negative assembly and ground terminal.
2. Reverse test probes. The lamp should light in one direction but not in the other.
3. If the test lamp does not light in either direction, this indicates that all three rectifiers in the negative diode assembly are open.

4. If the test lamp lights in both directions, the stator winding is shorted to stator or one of the negative diodes is shorted.
5. Check stator again when it is disassembled from diode assemblies.
6. With alternator disassembled, connect an ohmmeter or test lamp probes to one of the diode terminals and to stator.
7. Resistance reading should be infinite or test lamp should not light.
8. If resistance reading is not infinite or test lamp lights, high leakage or a short exists between stator winding and stator. In either case, stator should be replaced.

Stator Coil Shorts Test

1. This test checks for shorts between stator coil windings. The winding junctions must be separated as shown in Fig. M12. An ohmmeter or test lamp may be used.
2. Connect one of the test probes to test point "U" and the other to test point "V" and then to test point "W". Resistance should be infinite or test lamp should not light.
3. Connect test probes to test V and W. Resistance should be infinite or test lamp should not light. In either test, if the resistance reading is not infinite or the test lamp lights, high leakage or a short exists between stator windings. Stator should be replaced.

Stator Continuity Test

1. Measure resistance of each winding in stator between test points U and U1, V and V1, W and W1, Fig. M12. Resistance should be a fraction of an ohm (approximately .1 Ohm). An extremely accurate instrument would be necessary to ascertain short turns. Only an open condition can be detected with a commercial type ohmmeter.
2. If the alternator has been disassembled because of an electrical malfunction, replace stator only after all other components have been checked and found to be satisfactory.

Assemble Alternator

1. Clean bearing and inside of bearing hub of front housing. Support front housing and, using a suitable driver, apply sufficient pressure to outside race of bearing to seat bearing.
2. Insert split spring washer hub of front housing, seating washer into groove of hub.

NOTE: *Do not use a screwdriver or any small object to compress washer that can slip off and damage bearing seal. Make certain that split spring washer has been installed prior to assembling front housing and rotor.*

3. Use sufficient pressure to seat front bearing against shoulder on rotor shaft. The bearing drive tool must fit the inner race of bearing.
4. Install fan and pulley.
5. Use a 7/16" socket to fit inside race of near bearing and apply sufficient pressure to drive bearing against shoulder of rotor shaft.
6. Assemble front and rear housing.
7. Make certain that rear bearing is properly seated in rear housing hub and that diode wires are properly dressed so that rotor will not contact diode wires.
8. Align stator slots with rear housing through bolt holes, then align front housing through bolt holes with respect to rear housing.

NOTE: *The position of the brush assembly and belt adjusting screw boss must be in the same relative position to each other.*

9. Spin rotor to make certain that rotor is not contacting diode wires install through bolts.
10. Before mounting isolation diode, make certain that positive rectifier diode plate has been properly insulated from housing.
11. Install brush assembly, cover, and field plug retainer spring.

Prestolite Alternators

TESTING SYSTEM IN VEHICLE

Charging Circuit Resistance Test

1. Make connections as shown in Fig. P1.
2. Adjust engine speed and electrical load to obtain amperes in the charging circuit. The voltage drop in this circuit should not exceed .2 volt. If more, locate and correct cause of high resistance before proceeding with tests.
3. With the same operating conditions as in Step 2, connect voltmeter leads from ground battery terminal to alternator frame, and from alternator to regular base. In neither case should the voltage reading exceed .04 volt. If more, locate and correct cause of high resistance ground connections.

Alternator Output Test

1. Make test connections as shown in Fig. P2. *Be sure rheostat is "Off" before connecting leads.*
2. Connect a tachometer, start engine and adjusts its speed to 1750 rpm. Adjust rheostat to obtain 14.2 volts and observe ammeter reading. It should be as specified for the alternator being serviced.
3. If the alternator fails to reach rated output, it should be removed from the vehicle for repairs. A slightly low ammeter reading may indicate an open rectifier while a considerably lower reading may indicate a shorted rectifier.
4. If the output is only slightly low, a temperature check may be made to determine whether a rectifier is open. With the alternator operating as for the output test, place a bulbtype thermometer (250° scale) on the base of each rectifier heat sink for the rectifier bank being checked.
5. The rectifier temperature will normally be several degrees higher than the heat sink temperature. If the rectifier is open its temperature will be the same as that of the heat sink in which it is located.
6. If the rectifier is shorted the temperature of the entire heat sink in which it is located will be abnormally high.

Voltage Regulator Test

The first part of this test checks the operating voltage of the regulator when operating on the upper contract. Make test connections

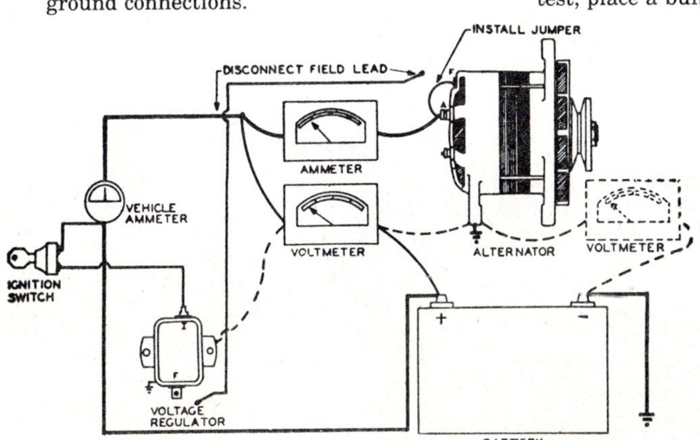

Fig. P1 Meter connections for testing voltage drop in charging circuit

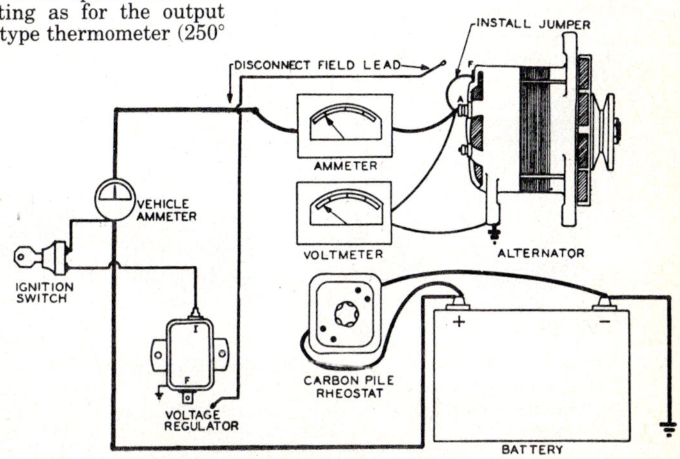

Fig. P2 Meter connections for testing alternator output

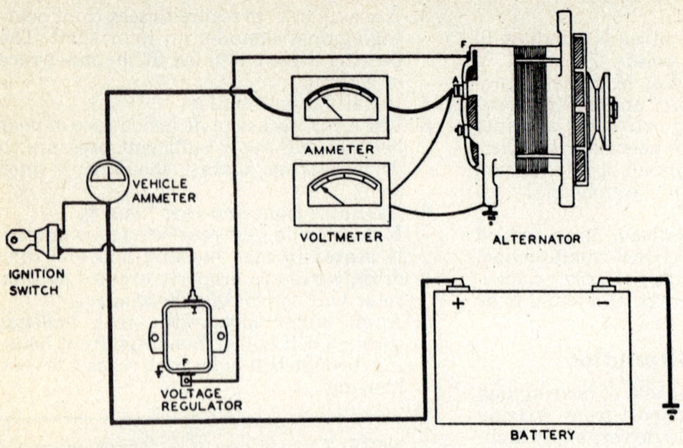

Fig. P3 Meter connections for testing voltage regulator

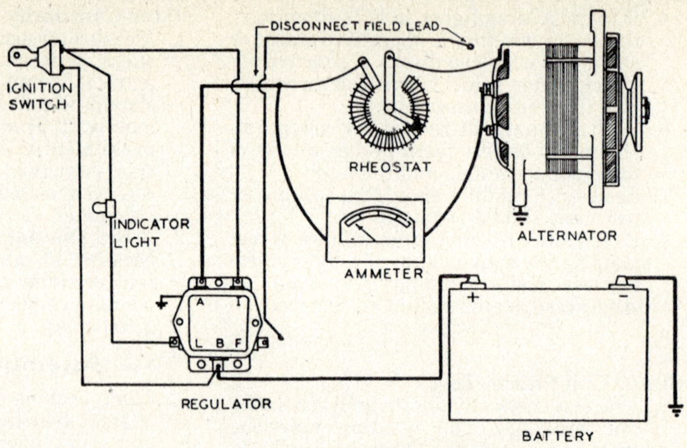

Fig. P4 Meter connections for testing indicator light relay operation

as shown in Fig. P3. *Be sure the ignition switch is off when connecting the field lead. Grounding of the field circuit while the ignition switch is on will damage the regulator.*

1. To test voltage regulator setting, start engine and adjust its speed to 750 rpm. Turn on lights and accessories to obtain a 10-ampere charge rate. Operate the system at this speed and load for 15 minutes to normalize the temperature. (This regulator is temperature compensated.)
2. Cycle the system by stopping and restarting the engine; then note the voltmeter reading. If seriously out of adjustment a rough setting may be made. The final setting is not made until the "spread" between operation on the upper contact and operation on the lower contact is established. This value is determined in the next part of the test which checks voltage when the regulator is operating on the lower contact.
3. Test connections remain as for the previous part of the test. Increase engine speed to 1500 rpm and turn off all lights and accessories. Voltage should increase and amperage should decrease. The "spread" in voltage between this reading and the reading noted in the first part of the test should be from .1 to .3 volt.
4. If the "spread" is greater or less than specified, remove the regulator cover and adjust by loosening the stationary contact support screw and moving the support up or down. Raising the support will increase voltage "spread", and vice-versa.
5. Replace the cover and reduce engine speed to 750 rpm and repeat the first part of the test with the regulator operating on the upper contact. If the voltage setting is not within specifications, remove the cover and adjust by bending the lower spring hanger. Replace the cover and cycle after each trial adjustment to obtain accurate readings.

Indicator Light Relay Test

1. Make meter connections as shown in Fig. P4. *If the vehicle indicator light cannot be seen while making this test, connect a 12-volt test lamp, using a No. 57 bulb between the "L" and "B" terminals of the regulator.*
2. Turn the rheostat to the "resistance in" position and operate the engine at approximately 800 rpm. Slowly cut out resistance and observe the ammeter reading when the vehicle indicator or test light goes out. It should be 4 to 7 amperes.
3. To adjust relay contact opening amperage, remove the cover and bend the lower armature spring hanger up or down. Increasing the spring tension raises the setting, and vice versa. Replace the cover and recheck.

BENCH TESTS

If the voltage regulator is unstable or cannot be adjusted to specifications in the above test, remove and regulator from the vehicle for further tests and adjustments.

Voltage Regulator Contact Gap

Contact gap is checked by placing a .010" gauge between the lower movable contact and the lower stationary contact. Adjust by bending the upper stationary contact up or down. Be sure that proper contact alignment is maintained.

Voltage Regulator Air Gap

To check the voltage regulator armature air gap, connect a No. 57 bulb in series with a 12-volt battery between the regulator field terminal and base.

Place a .034" round wire gauge between the armature and core on the side of the brass stop rivet nearest the center of the core head.

The lower movable contact should barely touch the lower stationary contact when the armature is pressed down against the gauge and the lamp should light. With a .038" gauge between the armature and core the light should go out.

To adjust to the above specifications, loosen the stationary bracket attaching screw and move the bracket up or down. *This is a preliminary adjustment only as the air gap may be*

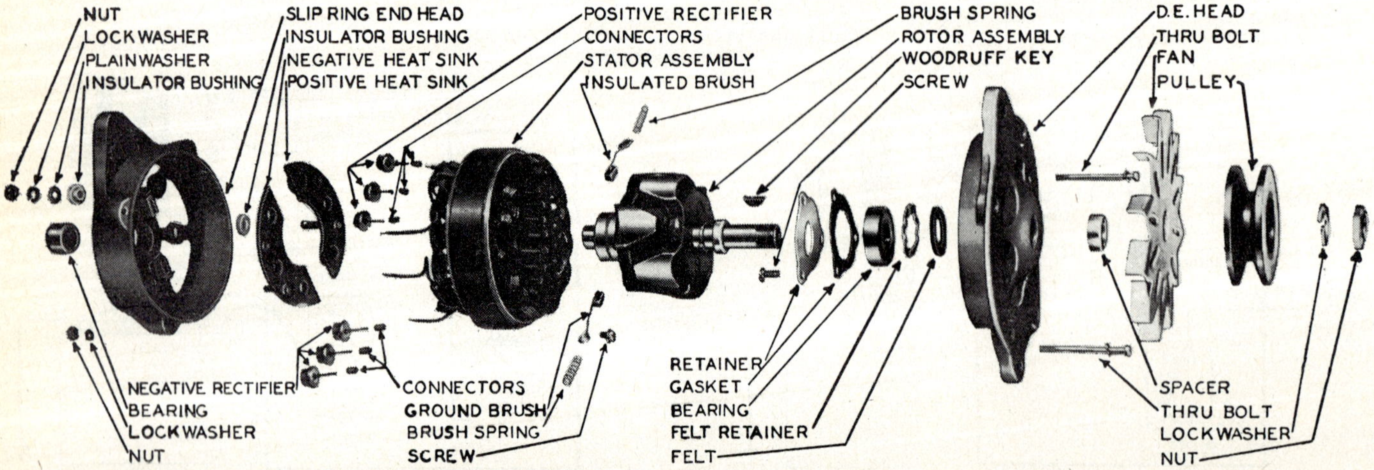

Fig. P5 Exploded view of a Prestolite alternator

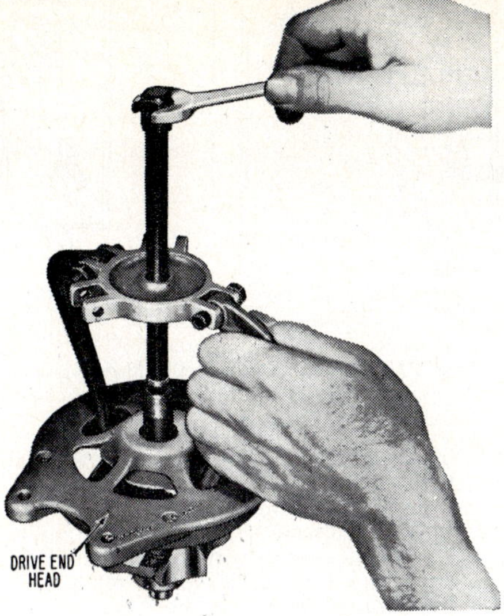

Fig. P6 Removing drive end head and bearing

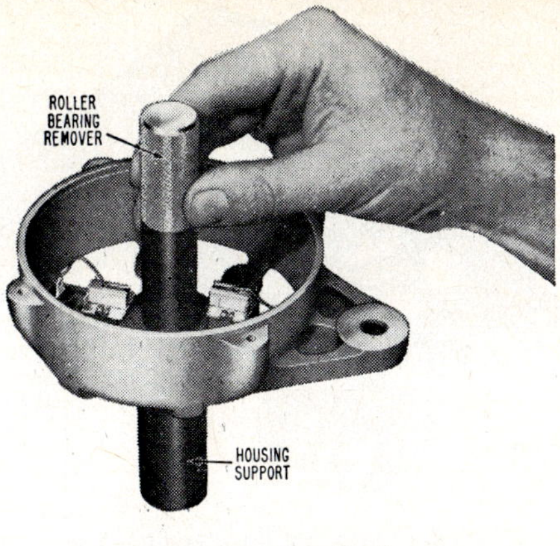

Fig. P7 Removing roller bearing from slip ring end head

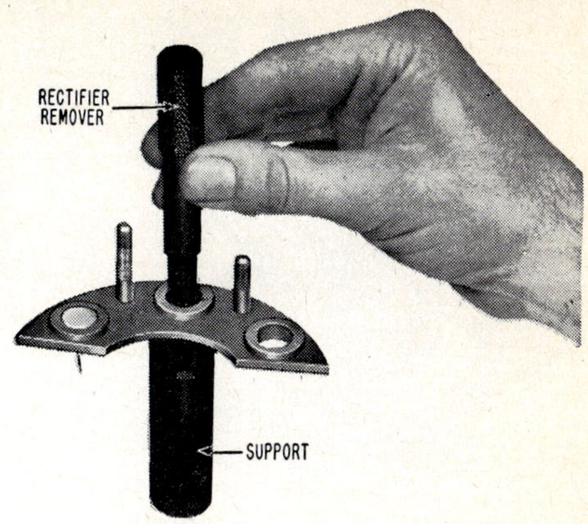

Fig. P8 Removing rectifiers from heat sink

changed when voltage "spread" is established as outlined previously.

Servicing Regulator Contacts

If the contacts are rough or oxidized, they may be cleaned with an American Swiss No. 6 equalling cut file. After filing, the contacts should be cleaned with a strip of linen tape saturated with a few drops of lighter fluid and drawn between the contacts. Then repeat with a dry strip of tape to remove fluid.

ALTERNATOR REPAIRS

When repairing an alternator, Fig. P5, complete disassembly may not be required. In some cases it will be necessary only to perform those operations that are required to repair or replace the defective part. However, the following material covers the complete overhaul procedure.

Disassembly

1. Remove through bolts and tap lightly on end heads to separate them from stator.
2. Remove drive end head and rotor.
3. Remove nuts, etc., from rectifier bracket studs. Separate slip ring end from stator.
4. Both brushes are located in slip ring end housing and should be replaced if worn to 5/16" or less. Remove brushes, being careful that the brush springs are not lost when brushes are removed from holders.
5. Remove pulley and fan.
6. Remove drive end head and bearing, Fig. P6.
7. Remove three scews from retainer plate and press bearing out of drive end head, using tool shown in Fig. P13.
8. Remove roller bearing from slip ring end head, Fig. P7.
9. If rectifiers must be removed, cut rectifier wire as near to crimpled sleeve as possible. Then remove rectifiers, Fig. P8.

Inspection

1. When disassembled, all parts should be wiped clean and inspected for wear, distortion or signs of overheating or mechanical interference.
2. Stator windings and leads should be examined for insulation failures or defects. A shorted phase winding or rectifier will normally be evidenced by discoloration.
3. The stator can be checked for shorted windings with an internal-external growler. The test is made with the stator leads disconnected from rectifiers.
4. The stator can be checked for grounded windings with a 110-volt test lamp and test probes. *Do not make this test with rectifiers connected to stator leads.*
5. Test rectifiers as described below.
6. The rotor should be tested for grounds and for shorted turns in the winding. The ground test is made with the test probes connected in series with a 110-volt test lamp as shown in Fig. P9. If bulb lights, rotor winding is grounded. To test for shorted turns, check rotor field current draw as shown in Fig. P10. Using the rheostat, adjust voltage to 10 volts and read field current draw on ammeter. Reading should be 2.34 to 2.42 amps at room temperature (70°F). Excessive current draw indicates shorted turns in field winding.

Assembly

1. Press rectifiers in heat sink, Fig. P11. *Rectifiers are identified by red markings on positive and black markings on negative.* Strip the insulation back about 1/4" on stator leads and reconnect them to rectifiers. The connector sleeves should be crimped on rectifier leads. *Do not solder these connections as the excessive heat may damage rectifiers.*
2. Install roller bearing, Fig. P12. Enclosed end of bearing should be flush with outer surface of end frame when installed.
3. Install felt retainer and crinkle washer and press bearing into drive end frame.

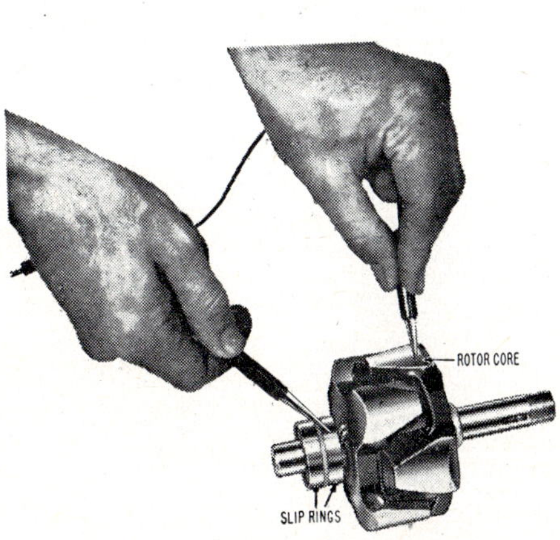

Fig. P9 Testing rotor for grounds

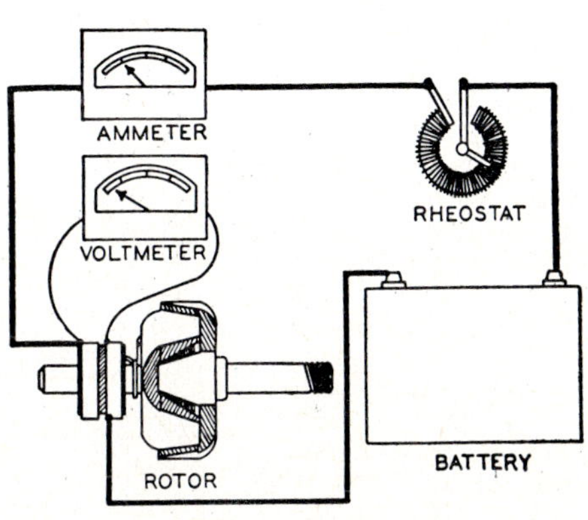

Fig. P10 Meter connections for testing rotor field current draw

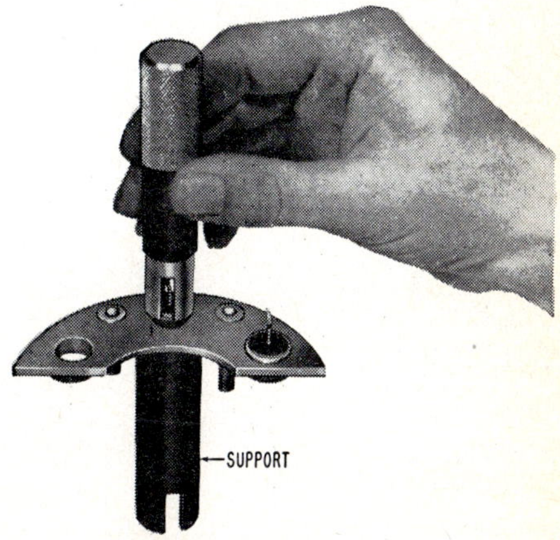

Fig. P11 Installing rectifiers in heat sink

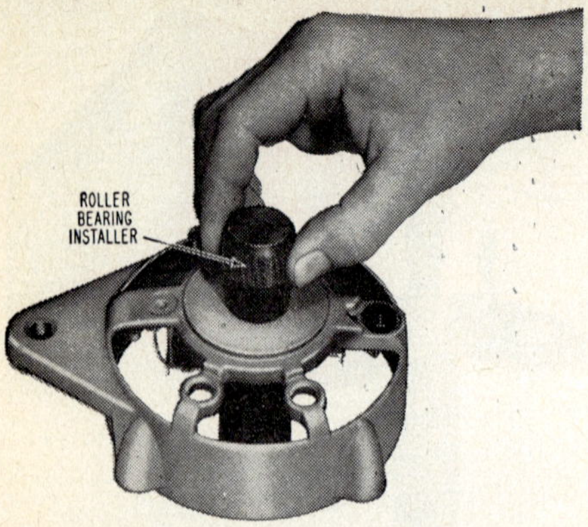

Fig. P12 Installing roller bearing

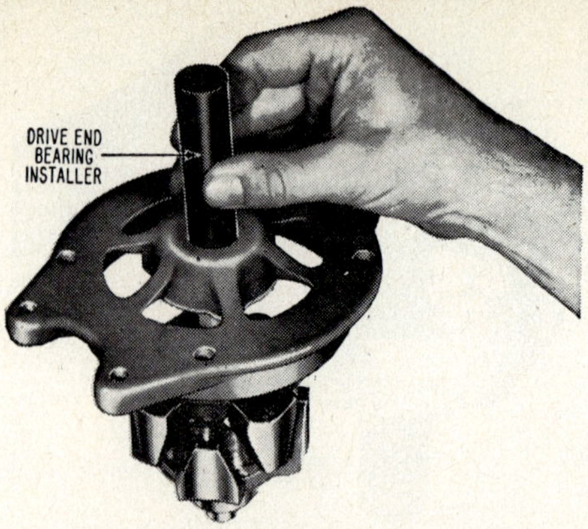

Fig. P13 Installing drive end head and bearing

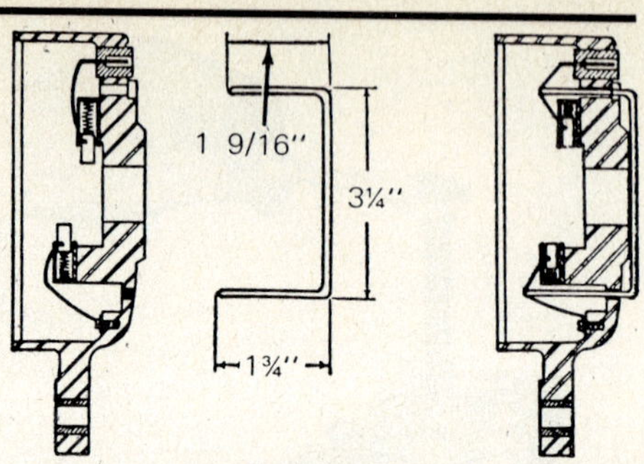

Fig. P14 Brush spreader clip. Right view shows clip installed

Use a flat block 2″ square so that pressure is exerted on outer race. Replace retainer plate and gasket on end frame. Make sure snap ring and retainer are in place on rotor shaft and press end frame on, Fig. P13. Press bearing down against snap ring retainer.

4. Replace spacer on shaft and install Woodruff key, fan and pulley.
5. Install brush springs and brushes. Eyelet on ground brush lead is fastened to end frame with a screw. The blade terminal on the insulated brush lead is pushed into the slot in field terminal insulator bushing. A tab on blade terminal snaps into insulator bushing. If damaged, this bushing may be replaced.
6. To spread brushes so they will clear slip rings when end head is installed, a wire slip can be made and used as shown in Fig. P14. This clip can be fashioned from a coat hanger or 6″ welding rod cut and bent to the dimensions shown. File a "V" groove in ends of wire clip so that it will contact brush leads when inserted through holes in end head.
7. Install negative rectifier heat sink in slip ring end head and install lockwashers and nuts. Install insulator bushings on positive heat sink studs and install heat sink in end head. Install outer insulator bushings, washers and nuts.
8. Manually press stator in position on slip ring end head and install assembled drive end head and install assembled drive end head and rotor. Make sure that through bolt holes line up on the two end heads. Install and tighten through bolts.
9. Remove brush spreader clip and make sure brush leads do not drag on rotor and that rotor turns freely when rotated by hand. Test alternator when assembled.

TESTING RECTIFIERS

There are two methods of testing diode rectifiers. One method requires the use of Tool C-3829. The other is to make the tests with a No. 57 bulb test lamp. The advantage of the former over the latter is that it is not necessary to perform the time consuming job of separating the soldered leads. (The instrument is available through the Miller Mfg. Co., Alliance, Ohio.)

If the current output test indicates that the alternator is not operating within specifications, remove the alternator from the vehicle and proceed as follows:

1. Perform a field coil draw test to provide an indication of the alternator field.
2. Open the alternator by first removing the brushes and then the alternator through bolts. Separate the drive end housing from rectifier end housing.
3. Inspect alternator components, paying special attention to the condition of the slip rings for indications of oil and for being burnt or worn. Check the brushes for evidences of sticking in the holder or housing.
4. Inspect the bearing surfaces of the rotor shaft at the rectifier end and the bearing. Make sure rectifier leads are in good condition, especially at the connections. The insulation should not be worn or broken.

Testing Rectifiers With Test Lamp

1. With alternator disassembled and rectifier leads disconnected, proceed as follows:
2. With a No. 57 test lamp, connect one side of the lamp to the positive battery post and the other side of the lamp to a test probe. Connect another test probe to the negative battery post.
3. Touch the outer case of the rectifier with one probe and the other probe to the wire in the center of the rectifier.
4. Reverse the probes.
5. If the test lamp lights in one direction the rectifier is good. If the lamp lights in both directions the rectifier is shorted. If the lamp does not light in either direction the rectifier is open.

Leece-Neville Alternators

SERIES 6000, 6200

These alternators, Fig. L1, are the 40 and 50 ampere, three-phase type. The slip rings and brushes, which carry field current to the rotor coil, are enclosed for protection from abrasive dust yet are readily accessible for inspection and servicing. One brush is grounded to simplify the circuit.

The rectifier cells, or diodes, are mounted in the slip ring end housing and are internally connected to the stator windings. One end of each of the three stator windings, or phases, is connected to a positive and a negative diode. The other ends of the stator windings are connected together forming a "Y" type connection. The three negative and the three positive diodes are pressed (in sets) into two plates called "heat sinks". These multiple-purpose heat sinks serve as mountings for the diodes, as radiation plates for heat dissipation, and also as current conductors.

The two-element regulator used with these alternators consists of a voltage regulator and a load relay which is connected to the auxiliary terminal on the ignition switch, Fig. L2. The load relay, controlled by the ignition switch, functions as a reverse current relay. When the ignition switch is turned on, the relay contacts close to energize the field coil in the alternator.

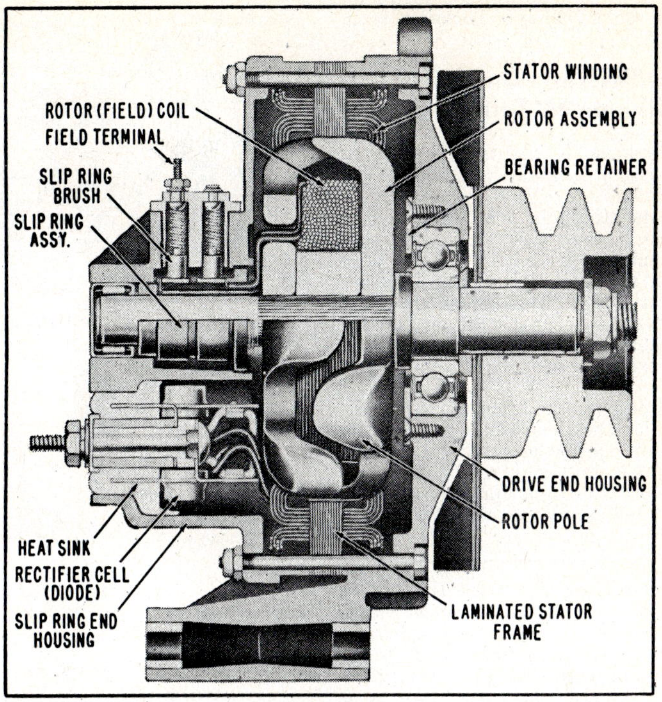

Fig. L1 Sectional view of Leece-Neville 6000 Series alternator

Testing System In Vehicle

Low or No Output

The field circuit must be closed in order to energize the rotor coil before the alternator can generate current.

A quick check of the field circuit can be made (with engine shut off) by turning on ignition switch and disconnecting the field terminal lead at the voltage regulator and then momentarily striking the FLD terminal with it.

If a medium spark is obtained, field circuit is closed. If spark is very light, a poor brush contact at the slip rings, or a poor connection, is indicated. If no spark is obtained, short or broken brushes, collapsed brush springs, poor or broken solder connections at the slip rings, or an open in the rotor coil, is indicated. A very heavy spark would indicate a short in the rotor coil. (Reconnect FLD terminal lead after making this check.)

If field circuit is found to be closed, alternator can be checked for output as follows:

With the engine at idle speed, remove the "F" lead from the alternator and attach a jumper wire from the "F" terminal to the "B" terminal. This jumper connection will take the regulator out of the circuit and allow the alternator to operate at full field strength.

If the alternator shows a high rate of charge when running with full field strength, the alternator is functioning and the problem lies with the regulator, system wiring on battery.

If the alternator shows a very low or no rate of charge when running with full field strength, the alternator will have to be removed from the engine and checked as outlined below if the system wiring and battery are good.

Bench Test

1. Remove pulley, fan and spacer from shaft. Remove brushes. Remove through bolts holding unit together and remove drive end housing and rotor assembly.
2. To test the rotor without removing it from end housing, measure the resistance of the coil by placing the prods of an ohmmeter on the slip rings. If the meter reads from 3.8 to 4.2 ohms the rotor is good. If very little resistance is read, it would indicate a possible shorted coil circuit. Check for loose or broken wire at the slip rings. If no resistance is read, the coil is open. Should no loose or broken wires be detected at the slip rings, the short or open is within the rotor coils and the rotor is beyond repair.
3. Upon removal of the brushes it is found that the brush springs are collapsed, the rotor has a shorted coil and the rotor should be replaced.
4. Check the slip ring brushes or springs. If the brushes or springs are cracked, broken or burned, they should be replaced. Brushes worn to a length less than 3/16" are too short and must be replaced.
5. If the rotor checks good, the stator and rectifier sections should be tested. This can be accomplished without removing the stator and rectifier sections from the slip ring end housing.

Rectifier Tests On Units Having An "N" Terminal

Place the positive prod of an ohmmeter on the "N" terminal at the slip ring end housing and the negative prod on the "B" terminal. With the prods in this position, the meter will read approximately 4 to 5 ohms and indicate that this is the positive section rectifiers. With the prods reversed, the meter needle will read infinity (excessive resistance) which will indicate that the rectifiers are good. Color coding on the bottom of rectifiers is red.

Place the negative prod of the ohmmeter on the "N" terminal and the positive prod on the "GRD" terminal. The ohmmeter will read approximately 4 to 5 ohms and indicate that

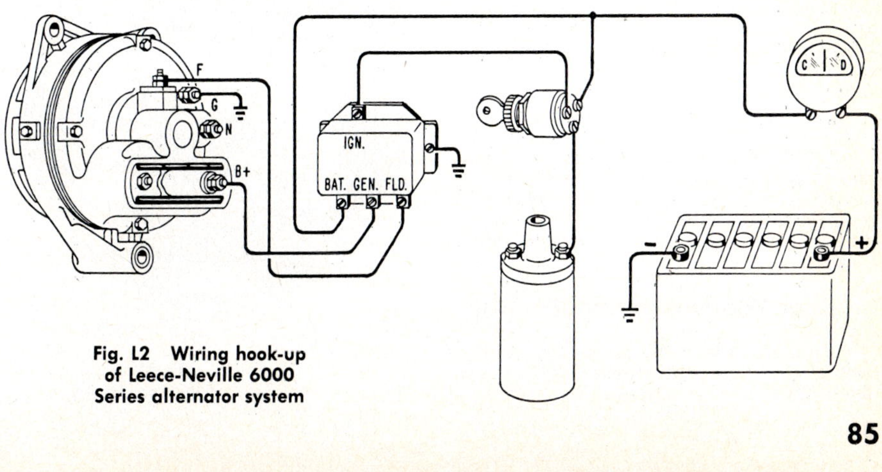

Fig. L2 Wiring hook-up of Leece-Neville 6000 Series alternator system

this is the negative ground secton. Reversing the position of the prods will make the meter read infinity. Color coding on bottom of rectifiers is black.

Rectifier Tests On Units Without "N" Terminal

Since there is no "N" (neutral) terminal, it will be necessary to make the tests at the rectifier sections.

Place the negative prod of an ohmmeter on any one of the three lead connections between the stator and rectifier cells. Place the positive prod on the heat sink of the insulated section of rectifier. The meter should read 4 to 5 ohms. Reverse the prod position and the meter should read infinity and indicate that this is the positive or grounded section of rectifiers.

Place the positive prod on any one of the lead connections between stator and rectifier cells and the negative prod on the opposite heat sink. The meter should read 4 to 5 ohms. Reverse the prod positions and the meter will read infinity indicating that this is the negative rectifier section.

NOTE: *When making the above tests, if the meter readings do not fall in the category mentioned above, one or more of the rectifiers in the section are not operating properly and the section should be replaced. When replacing rectifiers, be certain that the correct section is grounded.*

Alternate Rectifier Test

If an ohmmeter is not available a #57 bulb may be used to test the silicon rectifiers. This test is basically the same except that the bulb and battery replaces the ohmmeter.

If the bulb lights in one direction only the rectifiers are good.

If the bulb lights in both directions one or more of the cells in the section are shorted and the section found bad should be replaced.

If the bulb does not light in either direction one or more of the cells in the section are open and the section should be replaced.

Rectifier Section, Replace

To replace one or both rectifier sections, remove the nuts, etc. from the "B" terminal and the bolt holding the rectifier sections opposite the "B" terminal, and in the case of negative ground units, the "N" terminal. Remove stator and rectifier section from slip ring housing and unsolder the stator leads from the rectifier section or sections to be replaced.

When resoldering the rectifier sections to the stator leads, care must be exercised to apply only enough heat to insure a good connection. Overheating may damage the rectifier cells.

Stator Ground Test

To ground test the stator it will be necessary to unsolder the AC stator leads from the rectifiers. Check each stator phase for grounds to the stator core. A 110 volt test lamp is used for this test. No circuit should be present.

Stator Winding Continuity Test

With the test lamp check the continuity of each of the three stator phases. Each phase should show a closed circuit.

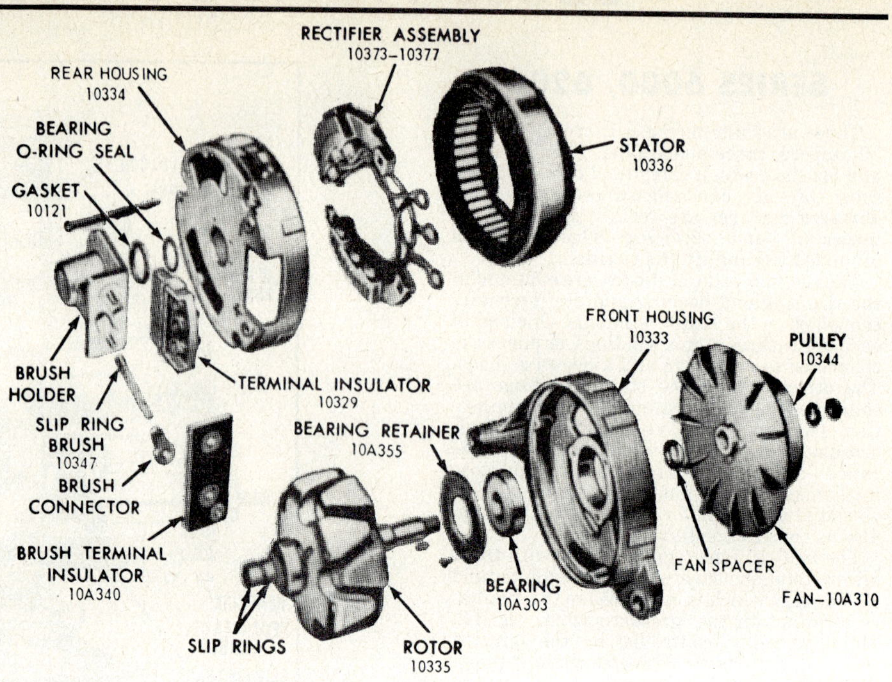

Fig. L3 1974–83 70, 90, 105 & 1979–83 130 amp. alternator disassembled (typical)

Disassembly & Inspection

After the foregoing tests are completed and the part or parts found inoperative are replaced, the following inspection and cleaning of parts should be done.

Rotor

If the rotor has seen considerable service and the slip rings appear to be worn, they should be replaced. Check shaft threads. Press rotor from housing using an arbor press.

To disassemble rotor, unsolder both field coil leads from the slip rings and, using a small puller, carefully pull off the rings. In some cases it will be impossible to save the old slip rings due to the tight fit on the shaft.

To clean the rotor it may be washed with a brush dipped in a cleaning solvent or paint thinner. Rinse with another brush dipped in unleaded gasoline or kerosene and then wipe with a dry cloth or blow dry with compressed air. *Do not immerse complete rotor in th cleaning fluid.*

Stator

After visual inspection finds the stator free of broken or cracked insulation or other damage which would cause failure, the stator can be cleaned in the same manner described for the rotor.

Brushes, Bearings, Etc.

Replace worn or broken brushes, insulation washers, etc. and inspect all tapped holes for good threads. It is recommended that bearings be replaced at time of overhaul.

Remove bearing retainer screws and bearing retainer from drive end housing. Tap or press out bearing. With bearing removed,

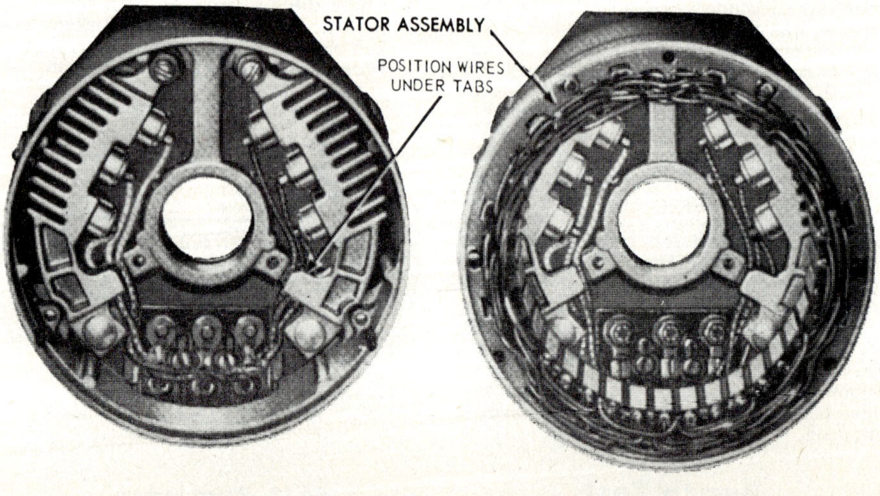

Fig. L4 Slip ring end housing, rectifier & stator assembly, 1974–83 70, 90, 105 & 1979–83 130 amp. alternator

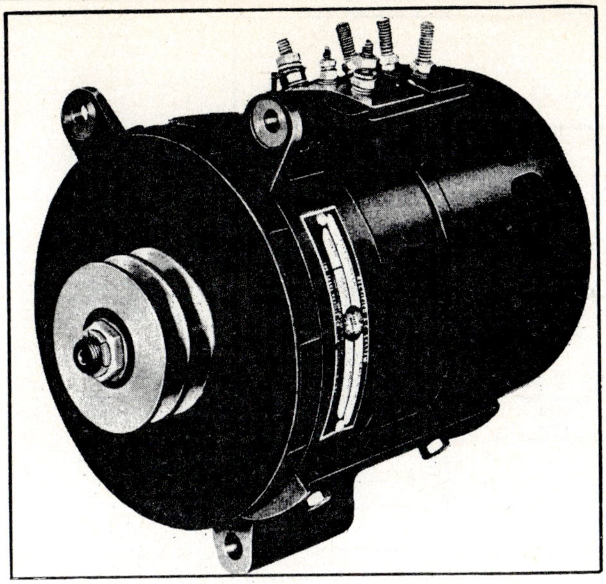

Fig. L5 SI type Leece-Neville alternator

housing can be immersed in cleaning solvent or paint thinner for cleaning.

The slip ring end housing bearing needs no special tool for removal or replacement. It may be tapped or pressed out. However, a piece of fiber or similar material should be placed over the new bearing when installing to prevent damage.

Reassembly

The alternator should be assembled in the reverse order of disassembly. Care should be exercised not to damage the housing pilots and stator windings. When completely assembled, spin the rotor by hand for free rotation check.

Reinstall the alternator on the vehicle and run it "full field" as outlined under Alternator System Check.

Regulator, Adjust

Load Relay

To check the load relay, connect leads from a 12 volt battery to the ground and ignition terminals on the regulator, with a variable resistor in series with one lead. Connect a voltmeter across the "IGN" and "GRD" terminals.

Start with all the resistance cut in and then gradually decrease it to allow the voltmeter reading to rise. The load relay should close at 5.8 to 6.2 volts.

To raise closing voltage, increase spring tension by bending lower arm of the spring bracket. To lower closing voltage decrease spring tension.

Voltage Regulator

To check the voltage regulator remove the cover. Connect one voltmeter lead to the regulator "BAT" terminal and the other lead to the regulator or alternator ground. Then connect a pair of earphones of not less than 1,000 ohms resistance to the "FLD" and "GEN" regulator terminals to hear the operation of the regulator contacts. *After each of the three steps in the test, open the ignition switch to bring the alternator speed to zero.*

1. Raise alternator speed slowly and listen for the vibration indicating the start of regulation on the upper contacts. This should start within the range of 13.9 and 14.3 volts. To increase the voltage, bend the arm of the spring bracket down. To lower the voltage, bend the arm up.
2. Slowly increase alternator speed past the start of regulation until the vibration ceases. Continue until the voltage reaches its maximum value, just before the lower set of contacts starts to vibrate. The maximum should be 14.7 and is adjusted as outlined in Step 1.
3. Increase alternator speed slowly past maximum voltage until the regulator armature vibrates on the lower contacts. The operating voltage should be 13.9 to 14.3 volts. To increase the voltage, increase the armature core gap; reduce the core gap to lower the voltage. The gap is adjusted by loosening the locking screw on the contact block to raise or lower the block.

90, 1974–83 70, 90, 105 & 1979–83 130 AMP. ALTERNATORS

System Testing

Testing procedures for these alternators are same as the procedures outlined in the "Ford Alternator" section of this chapter.

Alternator Repairs

Disassembly

1. Remove pulley nut, pulley, fan, shaft key and spacer.
2. Remove brushes and terminal insulator, then brush holder assembly from housing, Fig. L3.
3. Remove through bolts and separate brush end housing and stator from alternator.
4. Remove AC terminal nuts and stator from end housing.
5. Remove rotor from drive end housing with a gear puller or a suitable press, then front bearing retainer and bearing from housing.
6. Unsolder field leads from slip rings and remove slip rings and bearing from rotor shaft.
7. Remove rectifier assembly bolts, terminals and insulators, then the rectifier assembly and stator terminal insulator.

Assembly

1. Press bearing onto slip ring end of rotor shaft, putting pressure on bearing inner race. Heat slip rings, then carefully press slip rings onto rotor shaft and solder field leads to slip rings.
2. Press bearing into drive end housing, putting pressure on bearing outer race.
3. Press drive end housing and bearing onto rotor shaft, putting pressure on bearing inner race.
4. Install stator insulator and rectifier insulators. Place rectifier assemblies into housing and install mounting screws and terminals, ensuring rectifiers are insulated from end frame and wires are under tabs extending from heat sinks therefore preventing interference with rotor, Fig. L4.
5. Install stator and align through bolt holes with end housing. Place stator terminals over rectifier terminals and install nuts, Fig. L4.
6. Place brush end housing and stator assembly over rotor and install and tighten through bolts.
7. Install brush holder with "O" ring between holder and frame, place brushes and springs into holder with extruded portion of brush connectors against terminal screw shoulders. Hold brush connectors in position and install terminal insulator.
8. Install spacer, shaft key, fan, puller and nut.

SERIES SI

The alternator shown in Fig. L5 is a self-rectifying system integral type unit, having a built-in solid state regulator and interconnecting wiring, Figs. L6, L7 and L8.

The major parts consist of the rotor, stator, drive end and slip ring end housings, and the regulator assembly. The stator is composed of a large number of windings assembled on the inside of a laminated core attached to the alternator frame. As the rotor turns, two brushes mounted behind the slip ring end housing, carry current through the slip rings to the field coil mounted on the rotor shaft and to the regulator. The alternating current produced, is rectified by six silicon rectifier elements mounted on the slip ring end housing which is internally connected to the positive and negative terminal screws of the regulator. Six external terminals (G+, lamp, G− and three A.C. terminals), are provided for connections to the system of the vehicle.

The three A.C. terminals are internally connected to the rectifier heat sinks which in turn are connected to the stator winding transformers, mobile power units and load relay rectifiers. These terminals can also be used to test the silicon rectifiers when the alternator is installed on the vehicle.

One slip ring brush is internally connected to the "F" terminal on the stator and the terminal of the other brush is grounded.

Bench Test

Alternator Output Test

1. Strap unit in test stand and make connec-

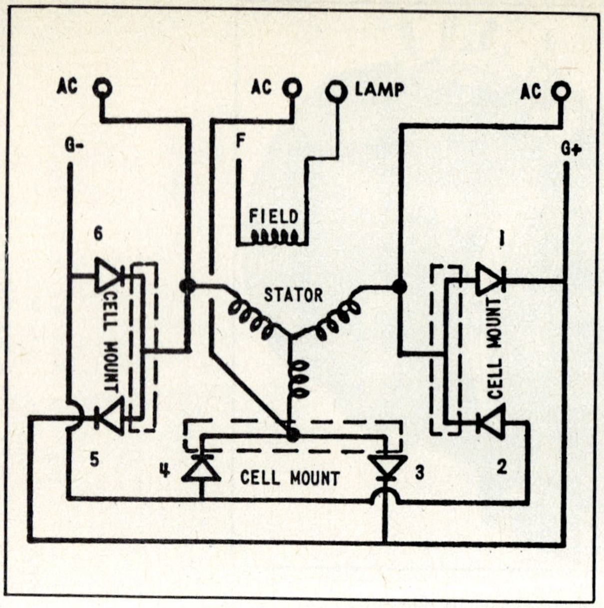

Fig. L6 Internal wiring diagram

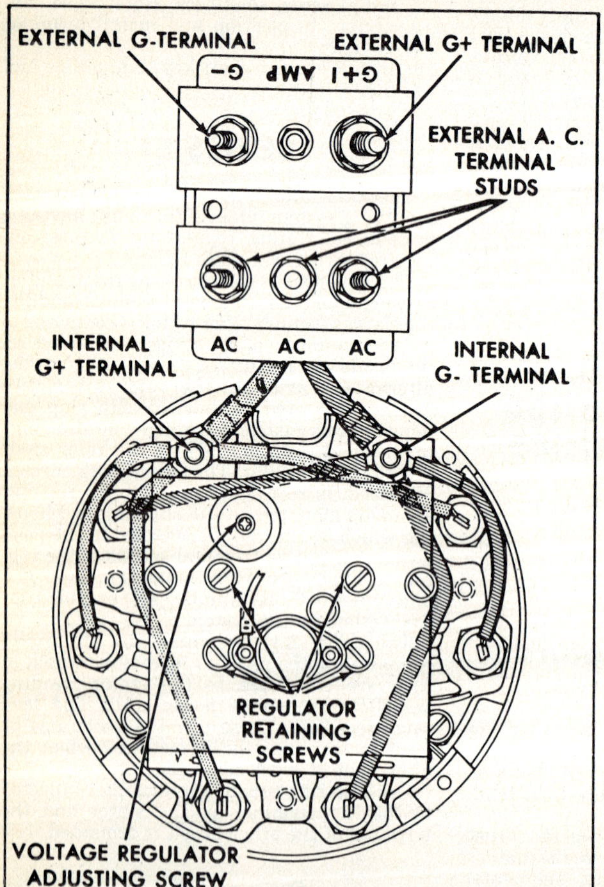

Fig. L7 Terminals & wiring connections

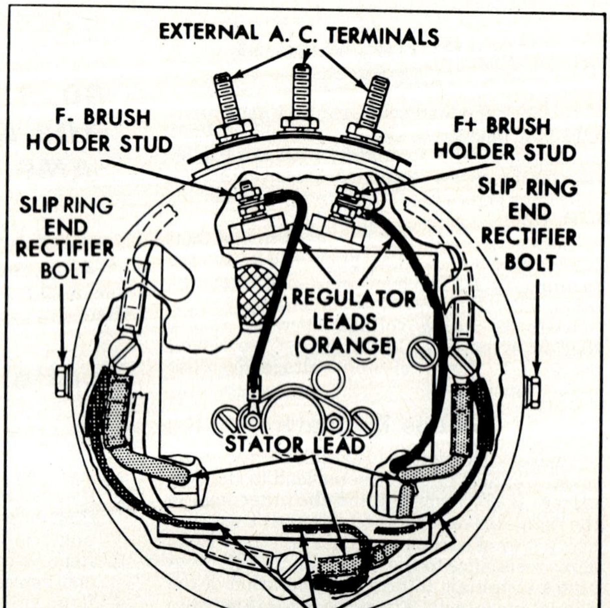

Fig. L8 Regulator terminals & wiring

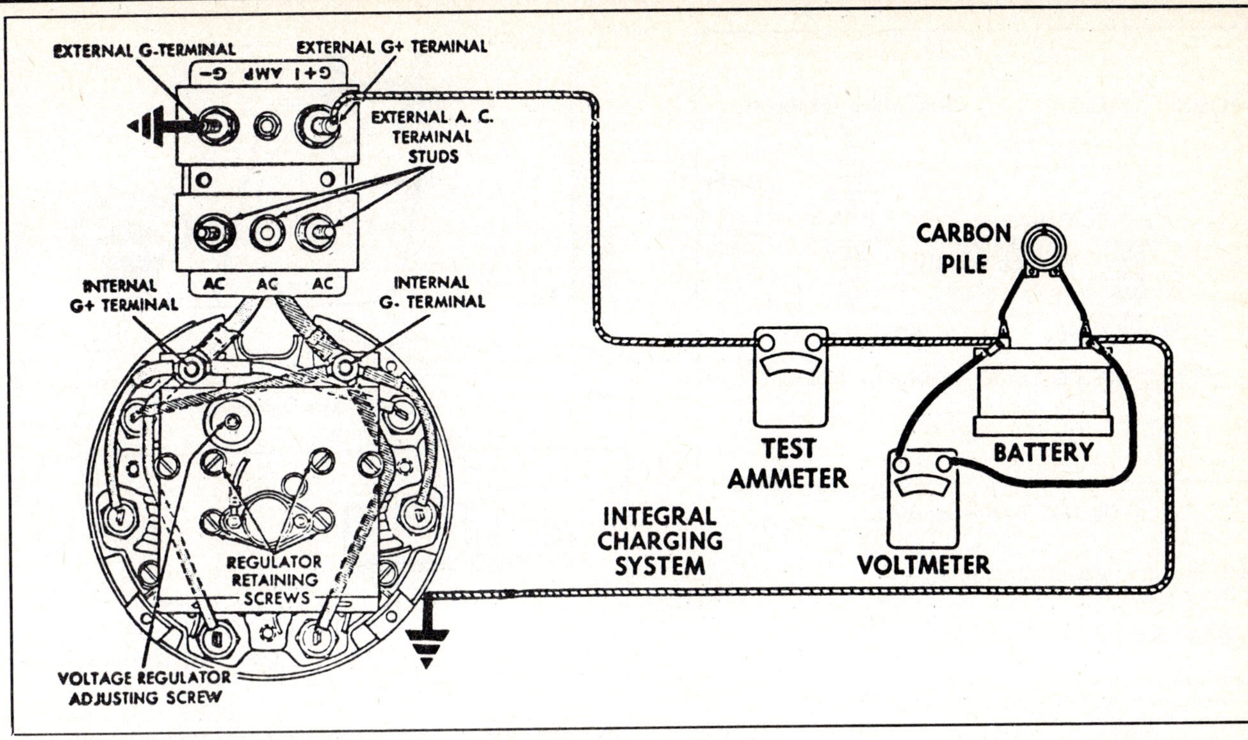

Fig. L9 Output test connections

tions as shown in Fig. L9.

NOTE: Leave carbon pile in "OFF" position or disconnected. Also use a fully charged battery.

2. Connect carbon pile and operate alternator to 2500 rpm. Adjust carbon pile to obtain maximum current output.
3. If ampere reading is within 10% of rated 105 amp system, the charging system is not defective.
4. If reading is not within 10% proceed as follows: Operate alternator at 2000 rpm and adjust carbon pile to place 10 amp load on system.
5. Adjust voltage regulator setting until voltmeter reads 13.9 to 14.1 volts.

NOTE: A hole is provided in the end cover to adjust the regulator. Turn clockwise to increase voltage or counterclockwise to decrease.

6. Repeat steps 1 and 2 to check for normal operation. If unit remains abnormal, make Regulator Voltage Test, otherwise install unit in vehicle and make final adjustment.
7. Install voltmeter at battery. Run engine at 800 rpm. Voltmeter should read 13.5 to 14.1. Adjust regulator if necessary.

Regulator Voltage Test

This test can be made on or off the vehicle and will determine whether the alternator or regulator is at fault when there is low output.

1. Make connections as shown in Fig. L9.
2. Loosen bolts and remove slip ring end rectifier cover.
3. Disconnect field lead from (F−) brush holder stud.
4. Connect external ground jumper wire to (F−) brush holder stud.

NOTE: Unit is now full-fielded and will have no voltage control. Do not exceed 1700 alternator rpm or 500-600 engine rpm.

5. Install suitable ammeter in series at G+ external terminal and voltmeter across battery terminals.

NOTE: Remove all external load from battery by turning off carbon pile.

6. Connect battery and run at 1500 alternator rpm. If there is no positive ammeter reading and battery voltage does not show rise proceed to step 7.
7. Connect a jumper from G+ terminal to F− brush holder after removing existing wires.

NOTE: Leave ground jumper connected.

8. Repeat steps 5 and 6. If voltage increases, replace regulator. If it does not increase, replace alternator.

Rectifier Test

Make note of wiring to each rectifier. With wiring disconnected at terminals, make individual rectifier test as shown in Fig. L10. The results are as follows.

1. If the bulb lights in only one direction, the rectifier is good.
2. If bulb lights in both directions, rectifier is shorted.

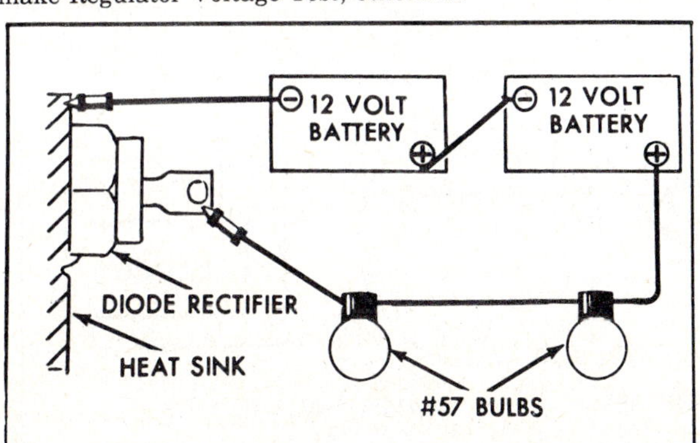

Fig. L10 Rectifier test

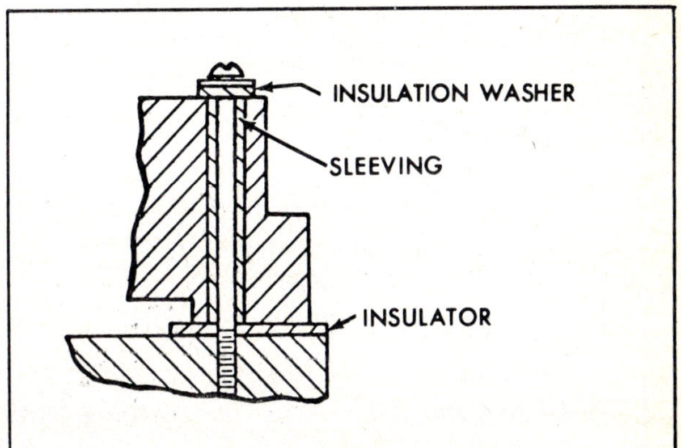

Fig. L11 Rectifier mount insulation

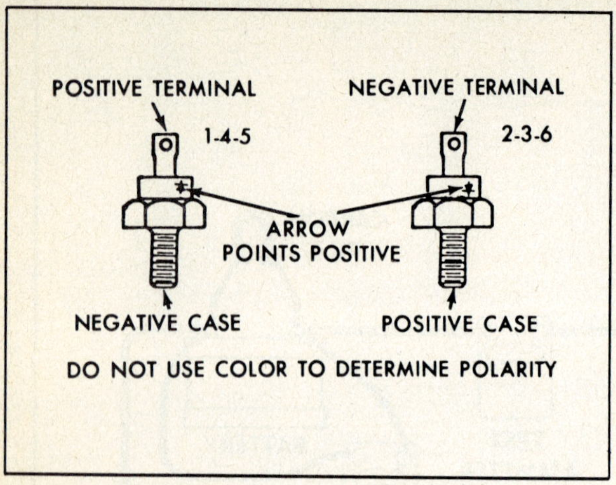

Fig. L12 Rectifier markings

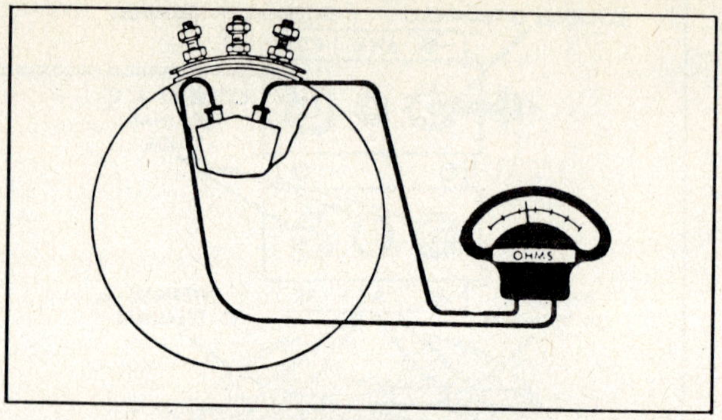

Fig. L14 Rotor coil resistance test

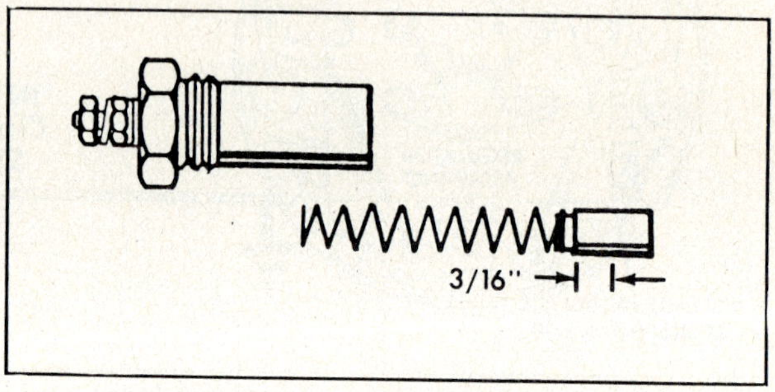

Fig. L15 Minimum serviceable brush length

3. If bulb does not light in either direction, rectifier is open.

Rectifier Mount Ground Test

1. With a 110 volt test lamp, check each of the three cell mounts for grounds to the alternator frame. If a ground is present between mount and frame check the insulator under the mount and the insulation of the attaching screws, Figs. L11 and L12.

Stator Ground Test

Remove external grounding jumper if used. Check each stator phase for grounds to the frame, Fig. L13. A 110 volt test lamp is used. There should be no circuit present.

Stator Winding Continuity Test

With an ohmmeter or test lamp, check the continuity of each of the three stator phases, Fig. L13. Each phase should show a closed circuit or indicate a low resistance, the test lamp should light.

Stator Phase Test

With the alternator on test block, connect a 12 volt battery to the field circuit and run the alternator at 800 rpm. Connect an AC voltmeter or test lamp of the same voltage as the system, across the three A.C. terminals in turn. Voltage or lamp brilliancy should be the same across phases 1-2, 2-3, 1-3. A pronounced difference indicates shorted or grounded stator windings.

Rotor Coil Test

1. Remove bolts attaching cover to slip ring end frame and remove cover.
2. Referring to Fig. L14, remove nuts, washers and leads from F+ and F− brush terminals. Remove brush holders and brushes with springs.
3. To check for ground, place one ohmmeter or test lamp prod on case and the other test lead through brush holder opening on slip ring. If ohmmeter reads high or test lamp lights, rotor coil is open.
4. To check for an open circuit, insert ohmmeter prods into brush holder openings and make direct contact with each of the two slip rings. If ohmmeter reading is very low or test lamp does not light, rotor coil circuit is open.
5. Test rotor coil for short by setting ohmmeter on direct scale and place prods on each slip ring. Ohmmeter reading should be from 4.9 to 5.3 ohms. If lower than 4.9 ohms, coil is shorted or has a loose or poor solder connection.

NOTE: If brush springs were collapsed, the rotor may have shorted turns within the coil.

Slip Ring Brushes

Check the brushes and springs. If cracked,

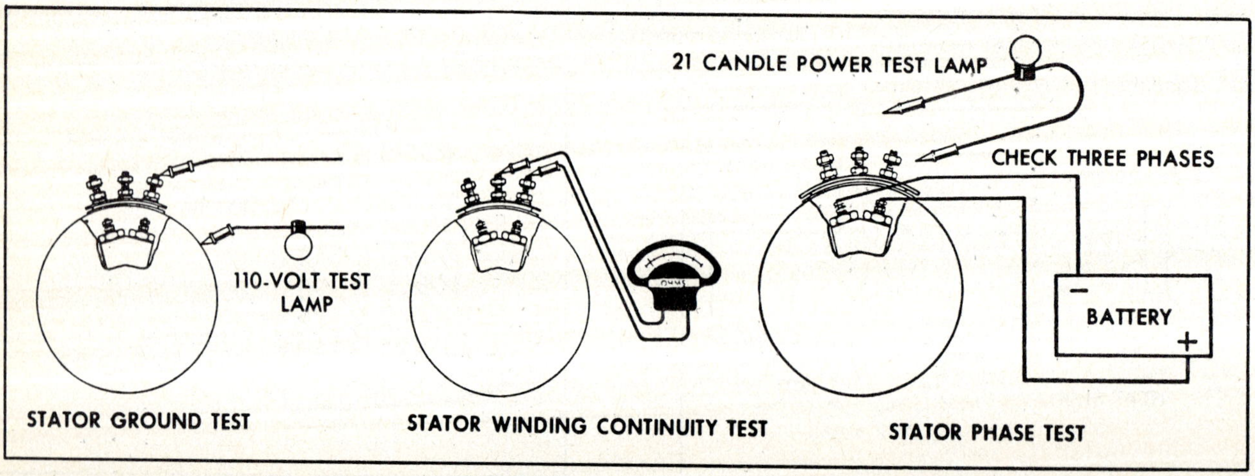

Fig. L13 Stator checks

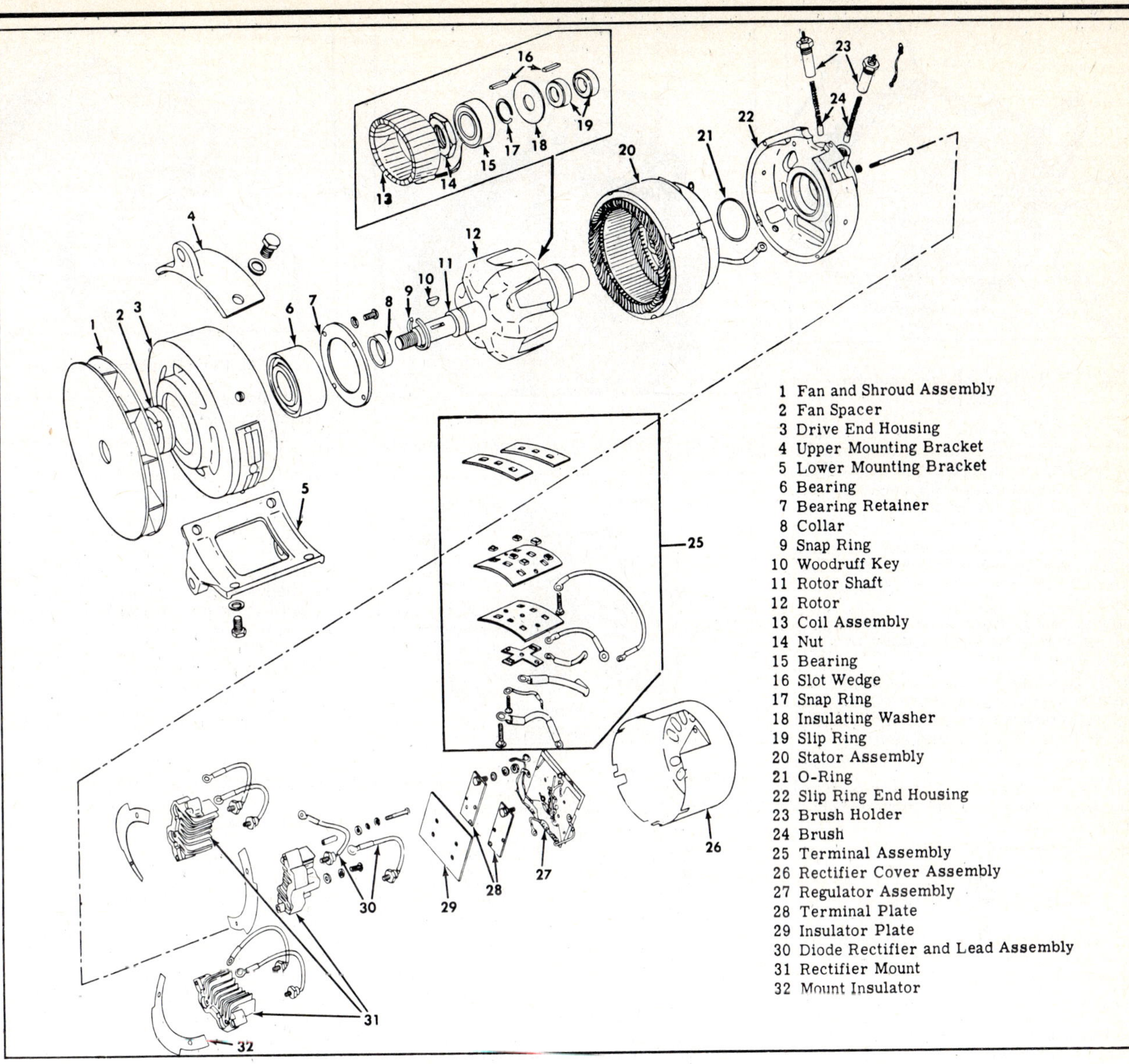

Fig. L16 Type SI alternator exploded

broken or burned, they should be replaced. Brushes worn to less than 3/16" must be replaced, Fig. L15.

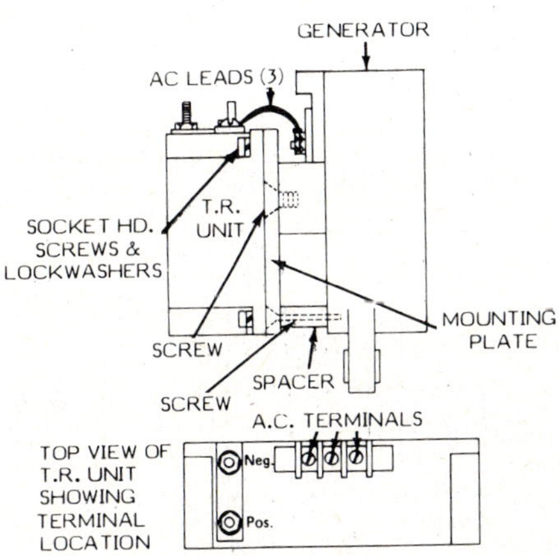

Fig. L17A Alternator with Transformer/Rectifier (T/R) unit installed

Alternator Disassembly

Refer to Fig. L16 for exploded view of alternator.

1. Remove mounting brackets.
2. Place a strap wrench or equivalent around alternator drive pulley and remove shaft nut. Using a puller, remove pulley from rotor shaft. Remove fan, shaft key and fan spacer.
3. Remove slip ring end cover. Remove screws in terminal block mounting plate. Raise terminal block to clear brush holders.
4. Make note of wiring on brush holders and remove brush holders and brushes.
5. Make note of wiring on heat sinks and

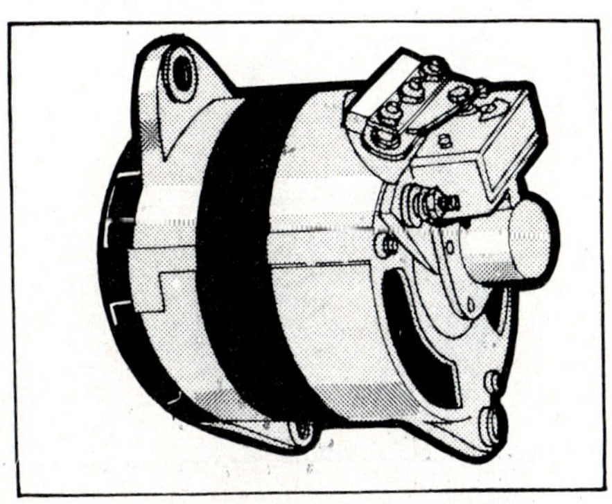

Fig. L17 Alternator assembly

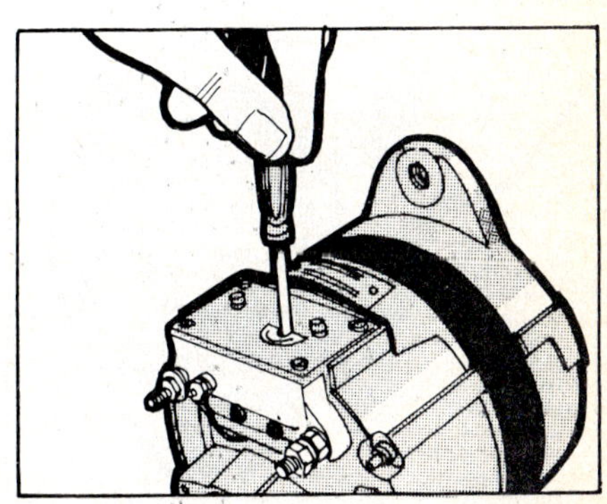

Fig. L18 Voltage regulator adjustment

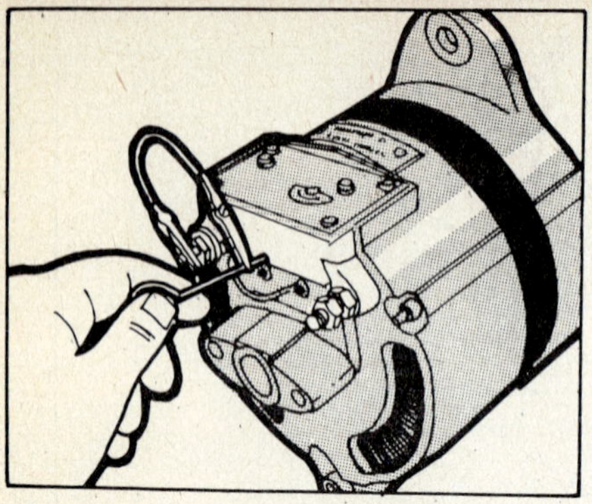

Fig. L19 Testing voltage regulator & diode trio

disconnect A.C. leads.

6. With a socket, take out four screws holding housings and stator and pull off the slip ring end housing . If necessary tap around the housing with a rawhide mallet to free the pilot surfaces. Remove the composition dust shield under the circular bus bars by removing one screw and loosening the two remaining screws to reveal the rotor shaft.
7. Remove stator from drive end housing and rotor. It may be necessary to use a rawhide mallet.
8. Press drive end housing from shaft.
9. Remove bearing retainer screws and bearing retainer from drive end housing. Tap or press out drive end bearing.

NOTE: If the rotor bearings, slip rings and coil are in good condition, further disassembly is unnecessary.

Rotor Disassembly

1. Unsolder both field coil leads from slip rings and using a small puller, carefully pull off the rings.
2. Take of the insulation washer and snap ring in front of the bearing and pull bearing from shaft. Save wedge removed from the lead slot under bearing and remove nut.
3. Loosen field coil lead clip and free the coil lead.
4. Holding rotor in vise, tap off slip ring end rotor half. Remove square key.
5. Remove rotor coil.
6. Using a press, push shaft out of hub and remaining rotor half.
7. Remove snap ring and square key.

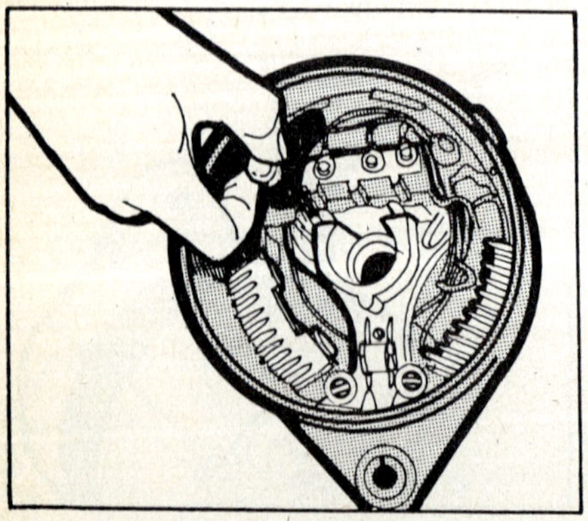

Fig. L23 Removing capacitor

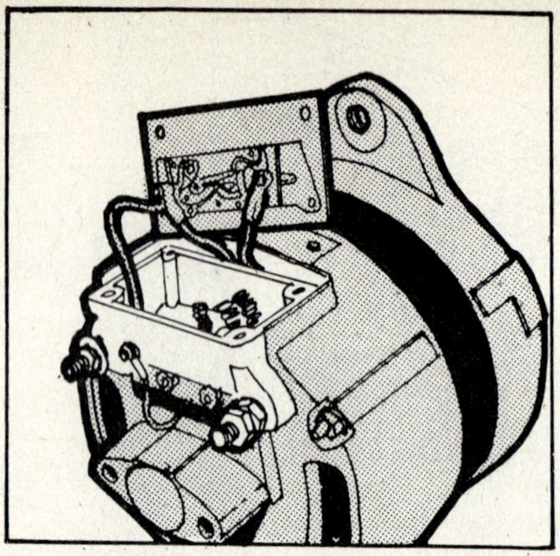

Fig. L20 Removing brush & spring assemblies

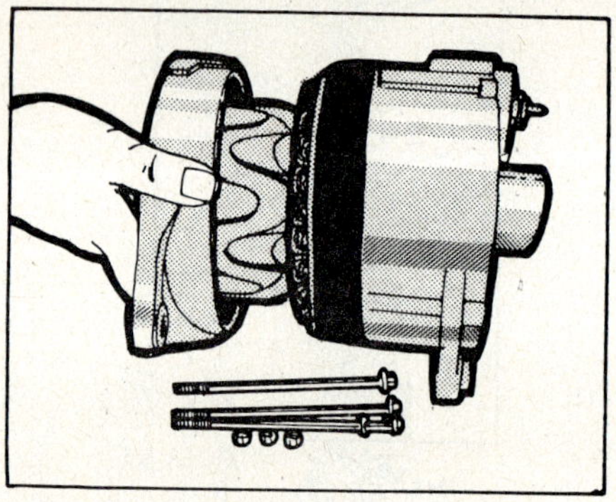

Fig. L22 Separating drive end housing from stator

NOTE: Keep the rotor halves together in marked pairs, to eliminate rebalancing of the assembled rotor.

SERIES 2300JB, 2500JB, 2600JB, 2601JB, 2700JB & 2805JB

These alternators are 14 volt self load limiting units incorporating a fully adjustable integral solid state voltage regulator Fig. L17. Six silicon diodes, mounted in heat sinks, convert alternating current from the delta wound stator into direct current. A capacitor, connected between the heat sinks, assists in suppressing transient voltage spikes which may cause diode damage and failure. The brushes and voltage regulator, installed in a water-

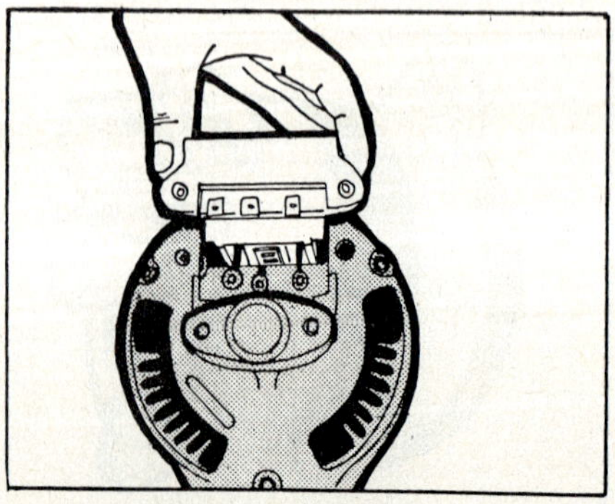

Fig. L24 Removing terminal stud insulating bushings

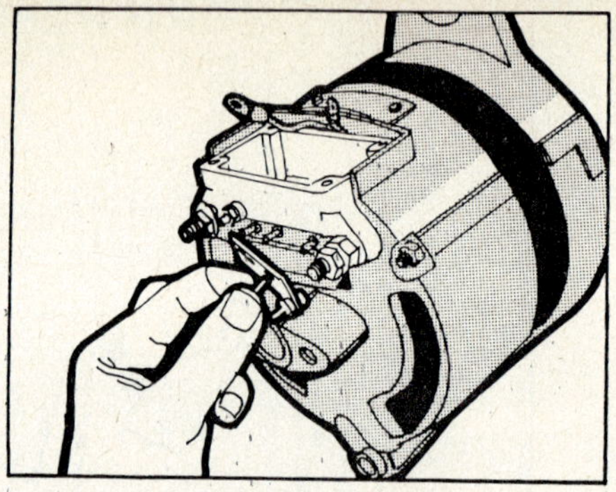

Fig. L21 Removing diode trio

proof housing, may be replaced or inspected without disassembling the unit. An external relay terminal is provided for charge light relays or other accessories which may require current from such a source. These units incorporate ungrounded output terminals and may be utilized on either positive or negative ground systems.

Some units may be equipped with a Transformer/Rectifier, Fig. L17A, unit which is installed on the slip ring end frame. This unit is used to charge the cranking battery on 12 volt vehicles with a 24 volt cranking system. This unit requires no voltage adjustment as voltage is controlled indirectly by the alternator unit.

Trouble Shooting

To determine if problem lies in alternator or regulator, connect a voltmeter across battery and note reading. Start engine, if voltmeter reading rises excessively, the charging system may be defective or require adjustment. With engine operating at approximately 1000 RPM, remove voltage regulator cap and attempt to bring voltage within the 13.6 to 14.2 volt range by turning the adjusting screw, Fig. L18. If voltage cannot be lowered in this manner, the regulator is probably defective and should be replaced.

If alternator voltage cannot be raised, either alternator, regulator or diode trio are defective. Connect one jumper wire lead to alternator negative output terminal and other lead to a piece of stiff wire approximately 1½ inches long Fig. L19. Insert stiff piece of wire into small hole in end of brush holder so it firmly contacts the outer brush holder so it firmly contacts the outer brush terminal. Operate engine at fast idle, if voltmeter reading now rises, the alternator is not at fault and diode trio should be checked as outlined under Bench Test. If diode trio checks out, the regulator should be replaced.

To check alternator output across each phase, construct a test light from a two filament sealed beam unit connected in such a manner that filaments are in parallel. If test lamp is noticeably dimmer on one or two

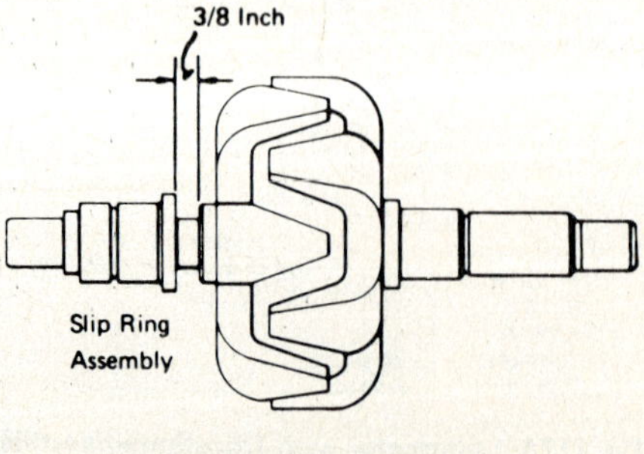

Fig. L25 Slip ring installation

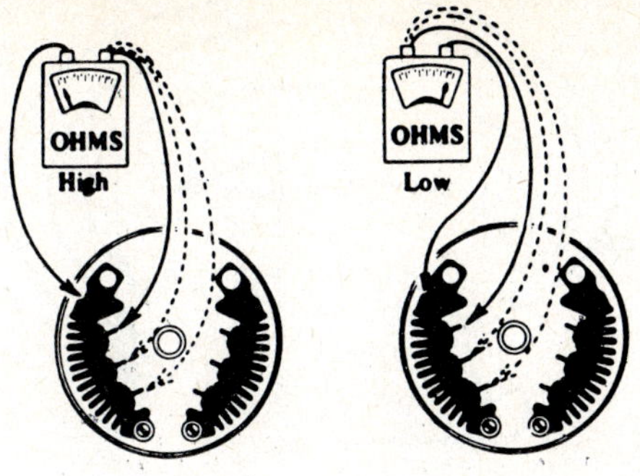

Fig. L26 Positive heat sink tests

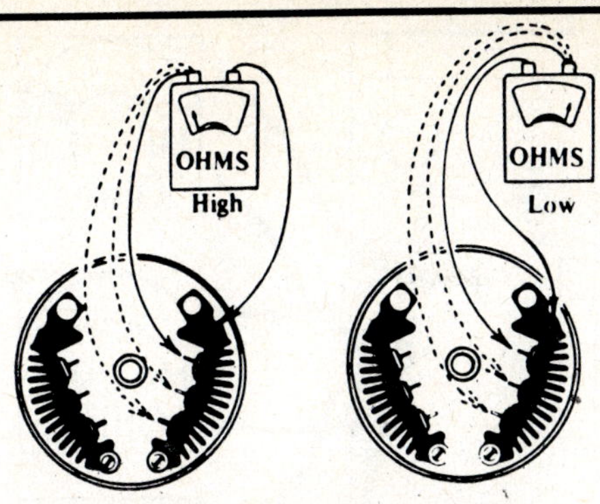

Fig. L27 Negative heat sink tests

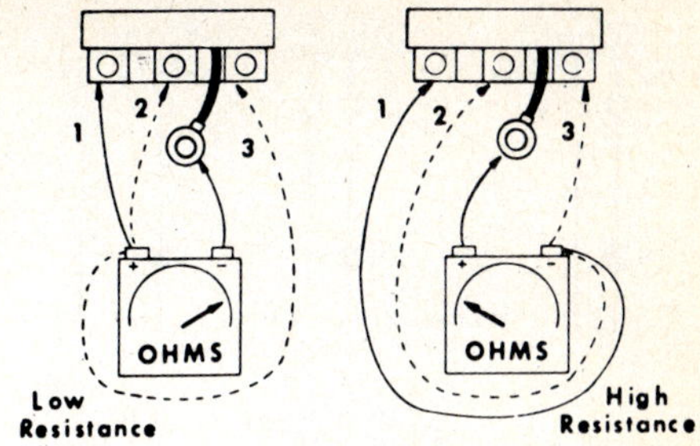

Fig. L28 Diode trio tests

phases, a defective diode trio, stator or power diode is indicated.

Alternator Disassembly

On units equipped with a Transformer/Rectifier (T/R) unit, the T/R unit must be removed before disassembling the alternator. To remove the T/R unit, disconnect A.C. and ground leads from alternator, then remove T/R unit to mounting bracket attaching screws and remove T/R unit. Remove mounting bracket from alternator.

1. Remove nut, pulley and fan.

NOTE: Use a suitable puller to remove pulley to avoid damaging shaft and threads.

2. Remove four regulator attaching screws, then lift regulator and disconnect red and black leads.
3. Disconnect lead from diode trio to terminal on regulator housing. Loosen inner nut, which will allow blue regulator lead to be withdrawn from under terminal screw, then remove regulator.
4. Remove brush and spring assemblies from housing, Fig. L20.
5. Remove three attaching nuts and lift diode trio from AC terminal studs, Fig. L21.
6. Remove three through bolts, then separate rotor and drive end housing from stator and slip ring end housing, Fig. L22.

NOTE: Ensure stator remains attached to slip ring end housing when separating to avoid damaging stator leads.

7. Remove three nuts attaching stator leads to terminals and remove stator.
8. Remove positive and negative output terminals, note location of red and black regulator leads on heat sinks.
9. Remove hex head attaching screws, then remove capacitor connected between heat sinks, Fig. L23.
10. Remove regulator housing, note location of gasket which seals brush compartment.
11. Remove terminal stud insulating bushings from housing, Fig. L24.
12. Remove two screws, lock washers, guard washers and insulating washers which retain lower portion of heat sinks.
13. Remove heat sinks, note location of insulating washers and bushings.
14. Pry metal dust cover from housing.

Slip Ring, Replace

1. Unsolder slip ring leads and position them clear of rotor coil eyelets.
2. Using a suitable puller, remove slip ring assembly from rotor shaft.
3. Clean rotor shaft, then apply a small amount of "Loctite Grade A" on shaft in area where slip rings are to be installed.
4. Position slip rings on shaft so that leads are aligned with rotor coil eyelets.
5. Carefully press slip ring onto shaft until there is a clearance of 3/8 inch between rotor and inner edge of slip ring, Fig. L25.

Bench Test

Heat Sink

To test positive heat sink, connect ohmmeter positive lead to heat sink and contact each of the three diode terminals with negative lead, Fig. L26. A high ohmmeter reading should be obtained, if any of the three diodes shows low resistance, the diode is shorted. Reverse ohmmeter leads and repeat test. A low ohmmeter reading should be obtained, if any of the three diodes show high resistance, the diode is open.

To test negative heat sink, connect ohmmeter negative lead to heat sink and contact each of the three diode terminals with positive lead, Fig. L27. If a resistance is obtained, the diode is shorted. Reverse ohmmeter leads and repeat test. If a high resistance is obtained, the diode is open.

NOTE: If a shorted or open diode is detected, the entire heat sink assembly should be replaced.

Diode Trio

Connect ohmmeter negative lead to diode trio output lead and positive lead to each of three copper terminal pads, Fig. L28. A low ohmmeter reading should be obtained. Reverse ohmmeter leads and repeat test. A high ohmmeter reading should be obtained. Any diode trio which does not conform to the above test should be replaced.

Capacitor

Connect ohmmeter between capacitor terminals. A low reading would indicate a shorted or leaking capacitor, Fig. L29.

Rotor

Connect one ohmmeter lead to rotor shaft and other lead to slip ring, Fig. L30. If ohmmeter reading is other than infinite, the rotor coil is grounded and the rotor should be replaced.

Connect ohmmeter between two slip rings and note reading, Fig. L31. Reading should be 4.4 to 4.8 ohms for 2300JB models manufactured prior to date code 7406, 2.5 to 2.8 ohms for 2300JB models manufactured after date code 7406. Reading should be 3.7 to 4.1 ohms for 2600JB models manufactured before date code 7423, 2.3 to 2.7 ohms for 2600JB models manufactured after date code 7423. Reading for 2700JB models should be 1.9 to 2.3 ohms. If resistance readings differ significantly from these figures, replace rotor.

NOTE: Before replacing rotor, ensure soldier joints at slip ring leads are in good condition.

Stator

Using a 115-220 volt test lamp, check stator

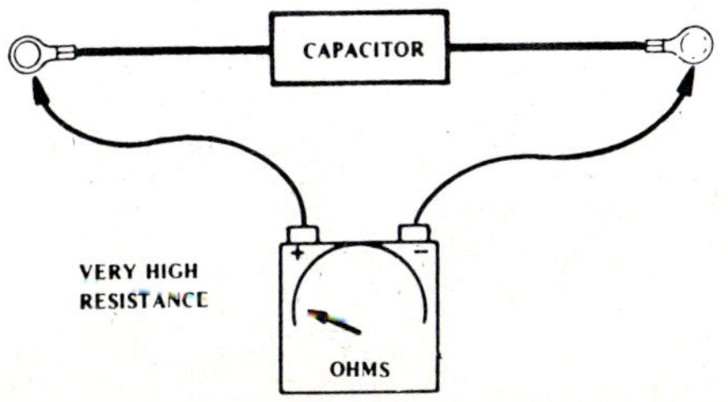

Fig. L29 Testing capacitor for short

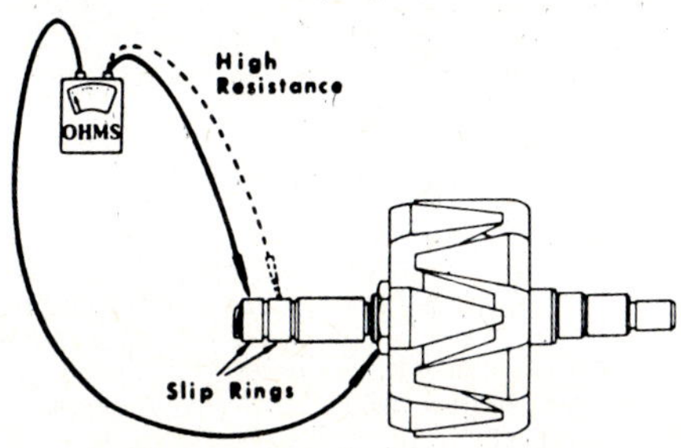

Fig. L30 Rotor coil ground test

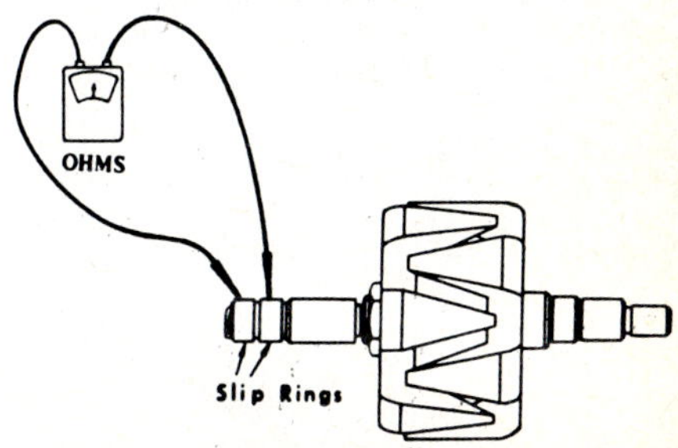

Fig. L31 Rotor coil resistance test

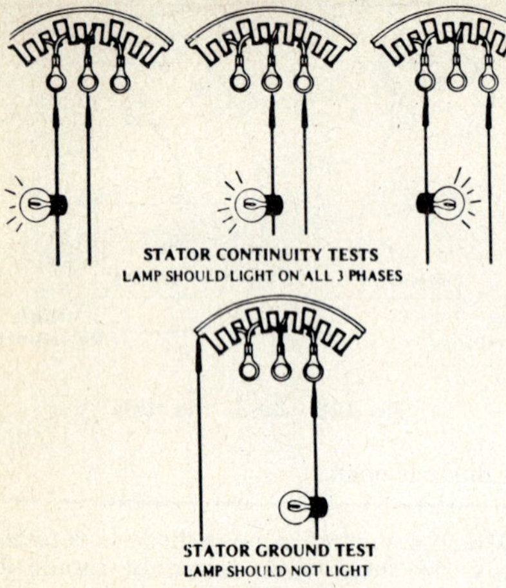

Fig. L32 Testing stator

for continuity between phases, Fig. L32. Test lamp should light on all three phases. To test stator for grounds, contact one stator lead against stator core and other lead to each of three stator leads, Fig. L32. Test lamp should not light. Stator that appear overheated or have charred insulation should be replaced regardless of test results.

Transformer/Rectifier (T/R) Unit

Transformer Primary Continuity Check
Connect an ohmmeter between each of the three A.C. leads, Fig. L32A. All three readings should indicate a low resistance reading. If a infinite reading between any two leads would indicate an open winding and the unit should be replaced.

Ground Test
An infinite ohmmeter reading should be obtained between any of the five terminals and the housing, Fig. L32B. If a reading other than infinite is obtained, there is an internal short and the unit should be replaced.

Rectifier Test
With ohmmeter positive lead connected to positive terminal and ohmmeter negative lead connected to negative terminal, a high reading should be obtained, Fig. L32C. When the ohmmeter leads are reversed, a low resistance should be obtained. If both reading are the same, the rectifier is either opened or shorted and should be replaced.

Alternator, Assemble

1. Install heat sinks on slip ring end housing, ensure upper and lower insulating washers are properly positioned. Install but do not tighten attaching screws, lock

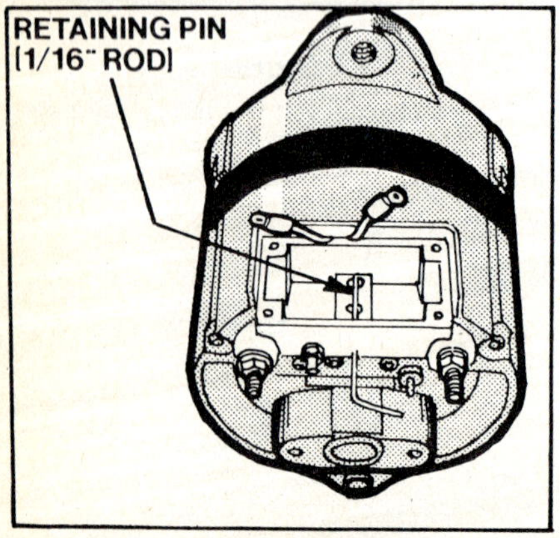

Fig. L33 Brush installation

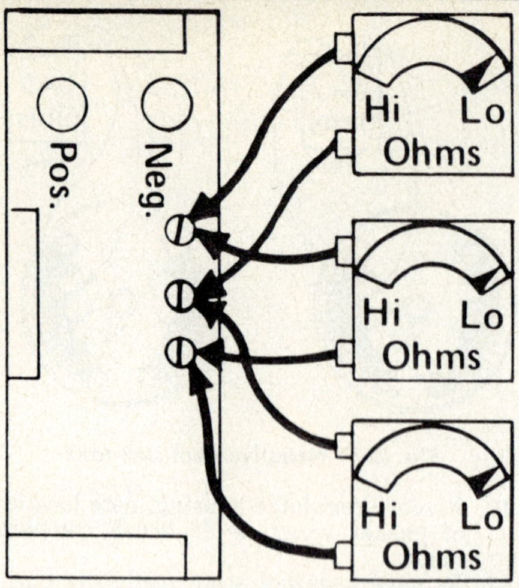

Fig. L32A T/R unit transformer primary continuity check

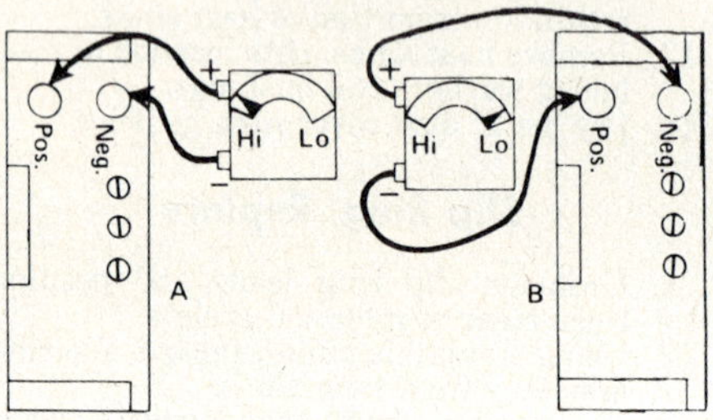

Fig. L32C T/R unit rectifier test

washers, guard washers and insulating washers.
2. Install terminal bolts and regulator lead wires.

NOTE: Red lead goes to positive heat sink and black lead to negative heat sink.

3. Install insulating bushings on each terminal bolt, then install regulator housing. Ensure red and black regulator leads are properly routed through cutaway section of end frame, then install and tighten terminal bolt nuts.
4. Tighten lower heat sink attaching screws, then install capacitor.
5. Install stator leads on terminals, then align stator on end frame and install two through bolts to hold stator position.
6. Press drive end housing and bearing onto rotor shaft.

NOTE: Use a sleeve around shaft and press on inner race to prevent damaging bearing.

7. Install rotor and drive end housing on stator and slip ring end housing, then install through bolts and torque to 50 to 60 in. lbs.
8. Apply a small amount of SRI 2 grease in housing, then carefully press metal dust cap into place.
9. Install diode trio.
10. Insert outer brush and spring assembly into housing, then compress brush spring. Insert a pin through rear of housing to hold brushes in compressed position, Fig. L33. Install remaining brush in same manner.
11. Connect red and black leads to regulator, then attach blue regulator lead to diode

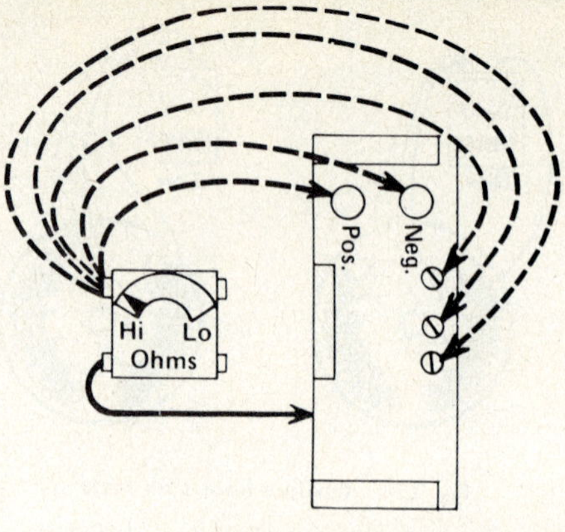

Fig. L32B T/R unit ground test

terminal. Position regulator on end frame housing and loosely install attaching screws, then remove brush retaining pins and tighten attaching screws.
12. Install diode trio lead and attaching nut.
13. Install spacer, key, fan, pulley and nut. Torque nut to 70-80 ft. lbs.

NOTE: The shaft end play designed into this unit will vary from .004-.012 inch.

Regulator Adjustment

With battery in fully charged condition and all accessories turned off, connect voltmeter across battery. Remove cap from adjusting screw, then start engine and operate at approximately 1000 RPM. Turn adjusting screw counter clockwise to lower charging voltage and clockwise to raise charging voltage, Fig. L18. The ideal voltage setting will be a value which maintains a fully charged without resulting in an excessive usage of battery water. After adjusting replace adjusting screw cap.

NOTE: Do not force adjusting screw past its stop as damage may result.

Regulator, Replace

1. Remove four screws and carefully lift regulator from housing, Fig. L20.
2. Disconnect red and black lead from regulator noting their position to aid in assembly.
3. Remove lead from diode trio to regulator terminal. Loosen inner nut which will allow blue regulator lead to be withdrawn under head of terminal screw.
4. Remove regulator, then lift brush and spring assemblies from housing.

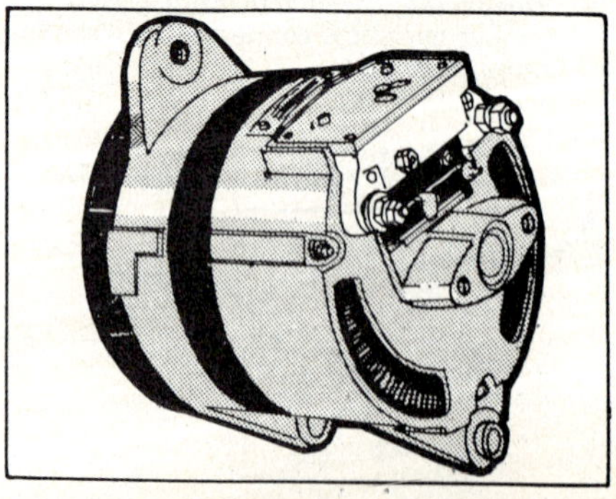

Fig. L34 Alternator assembly

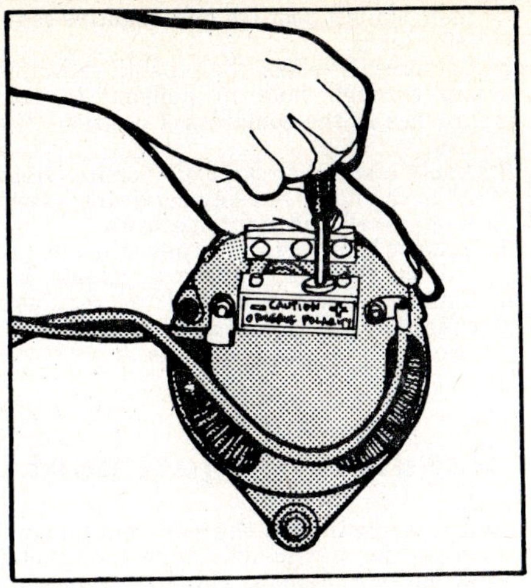

Fig. L35 Voltage regulator adjustment

5. Position brush and spring assembly in housing, then compress brush springs. Insert pin through rear of housing to hold brushes in compressed position. Fig. L33.
6. Connect red and black leads to regulator, then install blue regulator lead on diode terminal screw and tighten screw.
7. Position regulator on housing and loosely install attaching bolts.
8. Remove brush retaining pins, then tighten regulator attaching bolts.
9. Connect diode trio lead to regulator terminal, then install protective caps on terminal screws.

SERIES 3425JA, 4425JA & 4625JA

These alternators are self load limiting units incorporating a fully adjustable integral solid state voltage regulator, Fig. L34. Six silicon diodes, mounted in heat sinks, convert alternating current from delta would stator into direct current. A capacitor, connected between heat sinks, assist in suppressing transient voltage spikes which may cause diode damage and failure. The brushes and voltage regulator, installed in a waterproof housing, may be replaced or inspected without disassembling the unit. An external relay terminal is provided for chargelight relays on negative ground systems only. These units incorporate ungrounded output terminals and may be utilized on either positive or negative ground systems.

Trouble Shooting

To determine if problem lies in alternator or

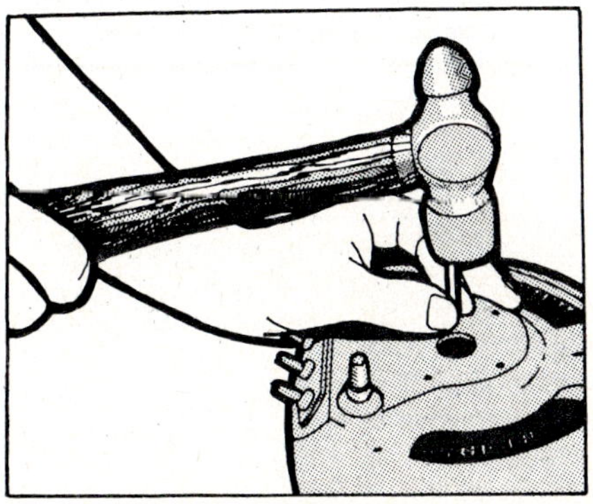

Fig. L38 Removing seal

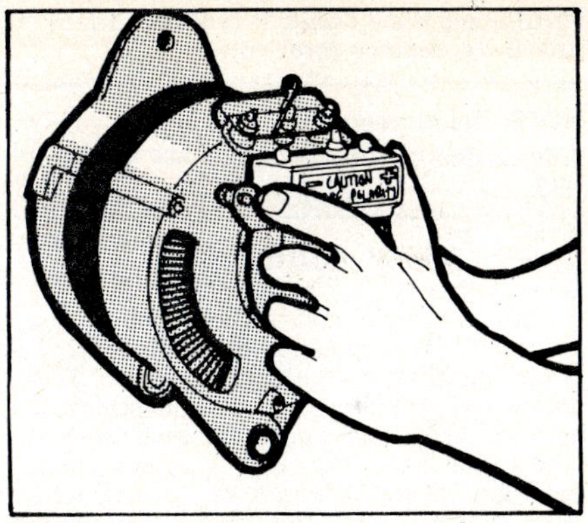

Fig. L36 Remove brush housing assembly

regulator, connect a voltmeter across battery and note reading. Start engine. If voltmeter reading rises excessively, the charging system may be defective or require adjustment. With engine operating at approximately 1000 RPM, remove voltage regulator cap and attempt to bring voltage within the 13.6 to 14.2 volt range by turning the adjusting screw, Fig. L35. If voltage cannot be lowered in this manner, the regulator is probably defective and should be replaced.

If alternator voltage cannot be raised, either alternator, regulator or diode trio are defective. Connect one jumper wire lead to alternator negative output terminal and other lead to a piece of stiff wire approximately 1½ inches long. Insert stiff piece of wire into small hole in end of brush holder so it firmly contacts the outer brush terminal. Operate engine at fast idle. If voltmeter reading now rises, the alternator is not at fault and diode trio should be checked as outlined under Bench Test. If diode trio checks out, the regulator should be replaced.

To check alternator output across each phase, construct a test light from a two filament sealed beam unit connected in such a manner that filaments are in parallel. If test lamp is noticeably dimmer on one or two phases, a defective diode trio, stator or power diode is indicated.

Alternator Disassembly

1. Remove nut, washer, pulley, fan, key and spacer.

NOTE: Use a suitable puller to remove pulley to prevent damage to shaft and threads.

---

2. Remove diode trio lead from top of regulator, then remove nuts from positive and negative output terminals.
3. Remove four regulator attaching screws, then carefully remove regulator and brush holder assembly, Fig. L36.
4. Remove three attaching nuts and lift diode trio from AC terminal studs, Fig. L37.
5. Remove through bolts, then separate rotor and drive end housing from stator and slip ring housing.

NOTE: Ensure stator remains attached to slip ring and housing when separating to avoid damaging stator leads.

---

6. Remove three nuts attaching stator leads to terminals, then remove stator.
7. Remove positive and negative output terminals.

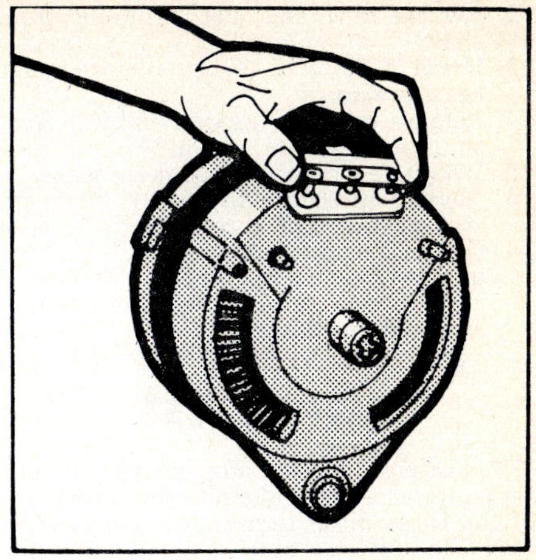

Fig. L37 Remove diode trio from AC terminal studs

8. Remove hex head attaching screws, then remove capacitor connected between heat sinks.
9. Remove two screws, lock washers, guard washers and bushings securing lower end of heat sinks, then remove heat sinks.
10. If slip ring end housing bearings are to be replaced, remove seal, Fig. L38, then using a suitable puller, remove bearing.
11. Press drive end housing bearings are to be replaced, remove four screws and bearing retainer, then press bearing and seal from housing.

Slip Ring Replace

1. Straighten slip ring tabs, then unsolder both connections.
2. Using a suitable puller, remove slip ring assembly.
3. Clean rotor shaft and apply a small amount of Loctite Grade A on shaft in area where slip rings are to be installed.
4. Position slip ring assembly on shaft aligning slot in slip ring with slot in shaft.
5. Carefully press slip ring assembly on shaft until the assembly is against the inner bearing race.
6. Solder rotor coil leads to slip ring tabs, then bend tabs to allow for clearance in brush-regulator housing.
7. If necessary, truse up slip rings in a lathe. Maximum slip ring run out is .002 inch.

Rotor Coil Replace

1. Remove slip ring assembly as outlined under Slip Rings Replace.

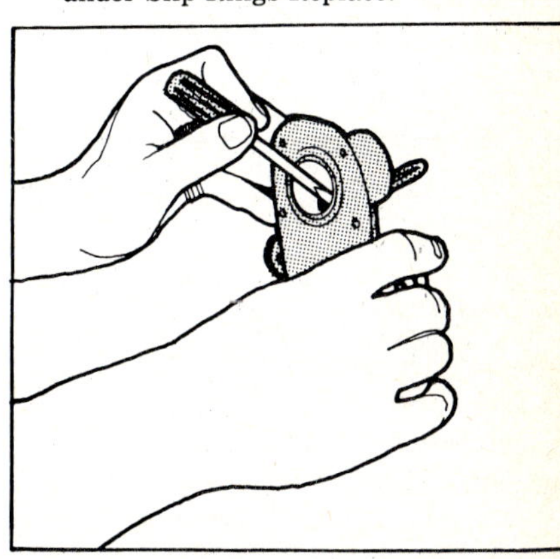

Fig. L39 Brush installation.

2. Remove nut and slot insulator from shaft.
3. Using a suitable puller, remove inner bearing race.
4. Remove rotor halves, coil and hub from shaft and separate.
5. Install new rotor coil and dress leads.
6. Reverse procedure to install. When installing nut coat threads with "Loctite Grade A" and torque to 110 ft. lbs. Press on inner bearing race until a clearance of 9/64 inch is obtained between inner edge of race and nut threads.

BENCH TEST

To test positive heat sink, connect ohmmeter positive lead to heat sink and contact each of the three diode terminals with negative lead Fig. L26. A high ohmmeter reading should be obtained, if any of the three diodes shows low resistance, the diode is shorted. Reverse ohmmeter leads and repeat test. A low ohmmeter reading should be obtained, if any of the three diode shows high resistance, the diode is open.

To test negative heat sink, connect ohmmeter negative lead to heat sink and contact each of the three diode terminals with positive lead Fig. L27. If a low resistance is obtained, the diode is shorted. Reverse ohmmeter leads and repeat test. If a high resistance is obtained, the diode is open.

NOTE: If a shorted or open diode is detected, the entire heat sink assembly should be replaced.

Diode Trio

Connect ohmmeter negative lead to diode trio output lead and positive lead to each of three copper terminal pads Fig. L28. A low ohmmeter reading should be obtained. Reverse ohmmeter leads and repeat test. A high ohmmeter reading should be obtained. Any diode to which does not conform to the above test should be replaced.

Capacitor

Connect ohmmeter between capacitor terminals Fig. L29. A low reading would indicate a shorted or leaking capacitor.

Rotor

Connect one ohmmeter to rotor shaft and other lead to slip ring Fig. L30. If ohmmeter reading is other than infinite, the rotor coil is grounded and should be replaced.

Connect ohmmeter between slip rings and note reading Fig. L31. Ohmmeter reading should be 3.0-3.3 ohms, if reading differs significantly, replace rotor coil.

NOTE: Before replacing rotor coil, ensure soldered joints at slip ring leads are good condition.

Stator

Using a 115-220 volt test lamp, check stator for continuity between phases, Fig. L32. Test lamp should light on all three phases. To test stator for grounds, contact one stator lead against stator core and other lead to each of thre stator leads, Fig. L32. Test lamp should not light. Stator that appear overheated or have charred insulation should be replaced regardless of test results.

Alternator Assemble

1. Press drive end frame bearing into housing with seal facing outward, if removed.

NOTE: When pressing bearing in, apply force to outer race to avoid damaging bearing.

2. Press new bearing seal into bearing retainer, then fill cavity of seal with SRI-2 grease or equivalent.
3. Install retainer, seal and four attaching screws on drive housing, then lubricate bearing seal with grease.
4. Press housing and bearing assembly onto rotor shaft.

NOTE: Use a sleeve around shaft so that force will be exerted on inner race of bearing.

5. Install heat sinks, ensure upper and lower insulating washers and bushing are properly positioned.
6. Install capacitor between heat sinks.
7. Press in rear slip ring housing seal until flush with rear of housing, then fill seal cavity with SRI-2 grease or equivalent.
8. Pack new bearing with grease, then press into slip ring end housing.
9. Pack cavity with grease, then press seal into housing with metal lip facing inside of housing.
10. Position stator on slip ring end housing, then connect leads and install nuts.
11. Install protective cap over slip rings, then assemble drive end housing and rotor to stator and slip ring housing.
12. Ensure both housings are aligned, then install through bolts and torque nuts to 50-60 in. lbs.
13. Install diode trio on AC terminals.
14. Install regulator and brushes on housing, then connect positive and negative regulator leads.
15. Compress brushes, then install retaining pin through hole in housing to hold brushes in the compressed position, Fig. L39.
16. Apply a small amount of "Loctite Grade A" to regulator housing retaining screws then install and tighten screws.
17. Remove brush retaining pins, then install nuts on alternator output terminals.
18. Connect diode trio lead to regulator charge light terminal and output leads to alternator output terminals.

Regulator Adjustment

With battery in fully charged condition and all accessories turned off, connect voltmeter across battery. Remove cap from adjusting screw, then start engine and operate at approximately 1000 RPM. Turn adjusting screw counter clockwise to lower charging voltage and clockwise to raise charging voltage, Fig. L35. The ideal voltage setting will be a value which maintains a fully charged battery without resulting in an excessive usage of battery water. After adjusting replace adjusting screw cap.

NOTE: Do not force adjusting screw past its stop as damage may result.

Regulator Replace

1. Remove diode trio lead from regulator charge light terminal.
2. Remove nuts from positive and negative output terminals.
3. Remove four attaching screws and regulator and brush holder assembly.
4. Remove nuts securing regulator positive and negative leads.
5. Install regulator and brushes on housing, then connect positive and negative regulator leads.
6. Compress brushes, then install retaining pin through hole in housing to hold brushes in the compressed position, Fig. L39.
7. Apply a small amount of "Loctite Grade A" to regulator housing retaining screws then install and tighten screws.
8. Remove brush retaining pins, then install nuts on alternator output terminals.
9. Connect diode trio lead to regulator charge light terminal and output leads to alternator output terminals.
10. Momentarily flash field by connecting a jumper wire between charge light terminal and positive output terminal.

STARTING SWITCHES

Magnetic and solenoid switches are designed to perform mechanical jobs electromagnetically such as closing a heavy circuit or shifting the starter drive pinion with the engine flywheel ring gear for cranking. switches of this type consist basically of contacts and a winding (or windings) around a hollow cylinder containing a movable core or plunger. When the winding (or windings) is energized by the battery through an external control circuit the plunger is pulled inward, producing the necessary mechanical movement.

MAGNETIC SWITCHES

Figs. 1 and 2 illustrate two typical Delco-Remy switches. The switch shown in Fig. 1 is not designed for disassembly and must be replaced if defective.

In the switch shown in Fig. 2 the terminals are assembled into a molded terminal ring which is held in place on the switch case by the cover and screws. Gaskets on both sides of the ring seal contact the compartment as a protection against moisture and dirt. The winding assembly is not removable from the case on this unit although the contact disk, plunger and plunger return spring can be removed after the cover is taken off.

Fig. 3 is a heavy duty magnetic switch. It is completely serviceable and easy to disassemble and assemble. To disassemble, remove the four terminal plate nuts and washers and take off the terminal plate assembly. The contact disk may be removed by taking off the castellated nut. It is necessary to remove the spring and washers on the plunger rod only when the rod needs to be disassembled. To remove the plunger, unscrew the large metal cover, take out the cotter pin in the plunger shaft, remove the spring retainer washer and spring, and withdraw the plunger. The winding and switch case is an integral assembly. The only parts that can be removed are the switch terminals. Before removing the switch terminals, the winding leads must be unsoldered from the terminal studs. Whenever the switch is disassembled, upon reassembly, locate the contact disk properly by turning the castellated nut in or out as required to obtain the dimension shown in Fig. 3 between the contact disk and edge of housing.

NOTE: On vehicles with overrunning clutch starting motors, a magnetic switch is normally used to shift the drive pinion into mesh and to close the starter circuit. There are two variations of three-terminal switches and also one type with four terminals. Any of these switches may be manufactured with either a grounded or insulated base. When installing a switch that is not marked "grounded base" or "insulated base", it must be checked out as follows, using a battery and test lamp connected in series.

THREE TERMINAL SWITCHES

1. If the switch has a grounded base the test lamp will light when connected between the starter ("S") terminal or ignition ("I") terminal and switch mounting bracket.
2. If the switch has an insulated base the test lamp will light when connected between the "S" and "I" terminals and either one of the 5/16" threaded studs.
3. If the test lamp fails to light when connected between the "S" and "I" terminals and any external part of the switch, disassemble the switch to determine whether the base is grounded or insulated.

FOUR TERMINAL SWITCHES

1. If the switch has a grounded base the test lamp will light when connected between the "S" terminal and switch mounting bracket.
2. If the switch has an insulated base the test lamp will light when connected between the "S" and "I" terminals.
3. If the test lamp fails to light when connected between the "S" terminal and any other external part of the switch, disassemble the switch to determine whether it has an insulated or grounded base.

SOLENOID SWITCHES

The solenoid switch on a cranking motor not only closes the circuit between the battery and the cranking motor but also shifts the drive pinion into mesh with the engine flywheel ring gear. This is done by means of a linkage between the solenoid switch plunger and the shaft lever on the cranking motor. Some linkages are adjustable whole others are not. The linkage is not adjustable on the type shown in Fig. 4 but adjustment of the entire assembly is made by moving the switch on the motor frame.

Fig. 4 shows two views of a solenoid switch used on vehicles with 12-volt systems. Like other solenoid switches, this type is energized by the battery through a separate starting switch. Note, however, that the switch includes an additional small terminal and contact finger. This terminal has no functional duty in relation to the switch, but is used to complete a special ignition circuit during the cranking cycle only. When the solenoid is in cranking position, the finger touches the contact disk and provides a direct circuit between the battery and ignition coil.

Fig. 5 is an exploded view of the 12-volt solenoid switch shown in Fig. 4. When reassembling the switch the contact finger should be adjusted to touch the contact disk before the disk makes contact with the main switch terminals. There should be 1/16" to 3/32" clearance between the contact disk and the main terminals when the finger touches.

Fig. 6 is a wiring circuit of a typical solenoid switch. There are two windings in the solenoid: a pull-in winding (shown as dashes) and a hold-in winding (shown dotted). Both windings are energized when the external control switch is closed. They produce a magnetic field which pulls the plunger in so that the drive pinion is shifted into mesh, and the main contacts in the solenoid switch are closed to connect the battery directly to the cranking motor. Closing the main switch contacts shorts out the pull-in winding since this winding is connected across the main contacts. The magnetism produced by the hold-in winding is sufficient to hold the plunger in, and shorting out the pull-in winding reduces drain on the battery. When the control switch is open, it disconnects the hold-in winding from the battery. When the hold-in winding is disconnected from the battery, the shift lever spring withdraws the plunger from the solenoid, opening the solenoid switch contacts and at the same time withdrawing the drive pinion from mesh. Proper operation of the switch depends on maintaining a definite balance between the magnetic strength of the pull-in and hold-in windings.

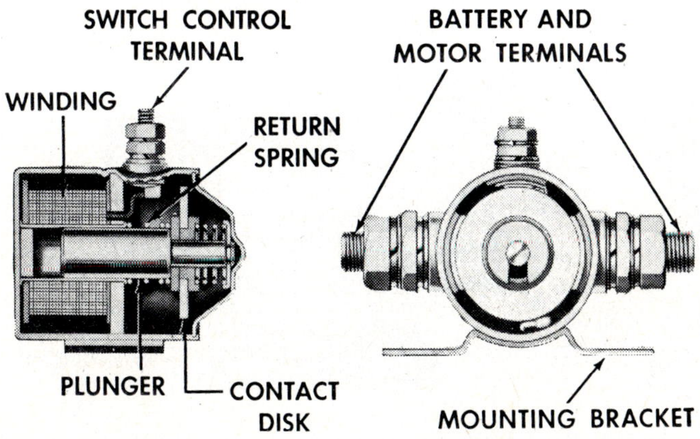

Fig. 1 End and sectional views of a typical magnetic switch

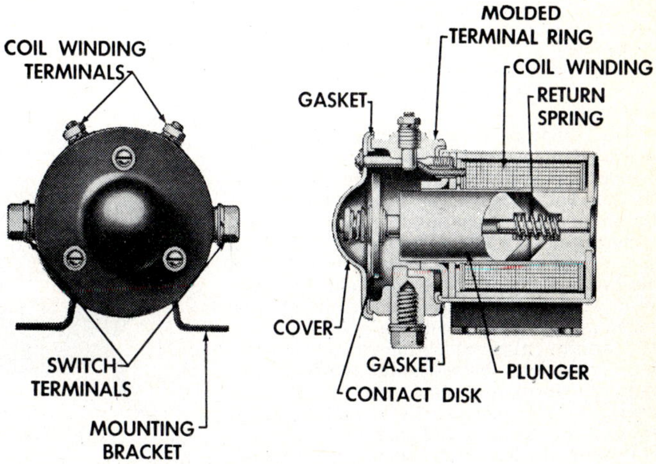

Fig. 2 End and sectional views of a sealed type magnetic switch which uses gaskets to seal the contact compartment.

This balance is established in the design by the size of the wire and the number of turns specified. *An open circuit in the hold-in winding or attempts to crank with a discharged battery will cause the switch to chatter.*

To disassemble the solenoid, remove nuts, washers and insulators from the switch terminal and battery terminal. Remove cover and take out the contact disk assembly.

When the solenoid has been removed from the starter motor for repair or replacement, the linkage must be adjusted to provide the correct pinion clearance or pinion travel when the solenoid is remounted on the motor. Some solenoids equipped with relays have an adjustable plunger stud but others must be moved on the motor frame to adjust pinion travel. When it is not feasible to tabulate the adjustment specifications for all starters likely to be found on trucks, the object of the adjustment is to prevent the drive from slamming against the drive housing.

Solenoid on Dyer Drive Starting Motor

Fig. 7 is an exploded view of this type of solenoid switch. Disassembly of the switch is made by removing the screws or nuts attaching the terminal plate assembly. The contact disk can be detached from the plunger and rod assembly by removing the castellated nut. The contact disk may sometimes require removal so that its face can be wire-brushed to remove corrosion.

On reassembly, the castellated nut should be tightened down so that the face of the disk is 1 1/32" below the edge of the housing with the plunger in the retracted position. After assembly is completed, fill the bushing in the terminal plate through which the lead passes (where present) with sealing material to prevent entrance of dirt.

DELCO-REMY SERIES-PARALLEL SWITCHES

On Diesel and similar heavy-duty engines, cranking requirements are high and consequently it is desirable to use higher voltage cranking motors so that adequate cranking performance will be obtained. The series-parallel switch makes it possible to use two 12-volt batteries which are connected in parallel for normal operating conditions after the engine is started, but which are connected in series by means of the series-parallel switch to provide 24 volts for the cranking motor. Likewise, two 6-volt batteries can be connected either in parallel or in series to provide a 6-12 volt system.

Fig. 8 shows the construction of a solenoid-operated series-parallel switch. Mechanically-operated series-parallel switches are also used. In both types, however, the switch incorporates a heavy copper contact disk and heavy tungsten-faced main terminals which resist the effects of the arcs that occur when the circuits are broken. The main cranking current is carried through these contacts and terminals. In addition there are contacts to the terminal plate which completes the parallel connections between the batteries for normal operation and also complete the connections that energize the cranking motor solenoid in the cranking position.

Fig. 9 is the wiring circuit of a (mechanically-operated) series-parallel system during normal operation of the vehicle. The main charging circuit is shown as a solid line. The current from the generator splits at the −A terminal, half of it going to the left-hand battery (dashed lines) and the other half going to the right-hand battery (dotted lines). Note that there are two circuit breakers or fuses in the right-hand battery circuit. Note also that an optional ammeter, as indicated, may be located in either of the two positions. The position in the −B to cranking motor circuit is preferable, however, since this circuit carries only the charging current while the other circuit (+A to "ground") must also carry the solenoid energizing current.

An ammeter sufficiently heavy to carry the solenoid current is generally unsatisfactory for reading small charging currents. In either position, the ammeter will indicate how much current is entering the right-hand battery, and if this is subtracted from the total shown in the main ammeter, the difference will be that which is entering the left-hand battery.

Fig. 10 illustrates the circuits during the cranking action when the series-parallel switch main contacts are closed (note that this diagram shows a solenoid-operated series-parallel switch). The series connections between the two batteries and the cranking motor are shown as a solid line, while the cranking motor solenoid circuit is shown as a dashed line.

As the switch is operated the solenoid plunger (or push rod in a mechanically-operated switch) strikes a small plunger in the contact plate assembly, causing the small contacts to be actuated so that the parallel connections between the two batteries are broken. Then the main contact disk connects the two main switch terminals (−A and +B); and as the small plunger completes its travel it connects the cranking solenoid through ground through the SW terminal (this circuit is shown as a dashed line). Now the cranking motor solenoid operates and closes its contacts to complete the series connections through the cranking motor so that the motor operates.

After cranking has been accomplished and the series-parallel switch is released, the two batteries again become connected in parallel to provide 12-volt operation of the equipment.

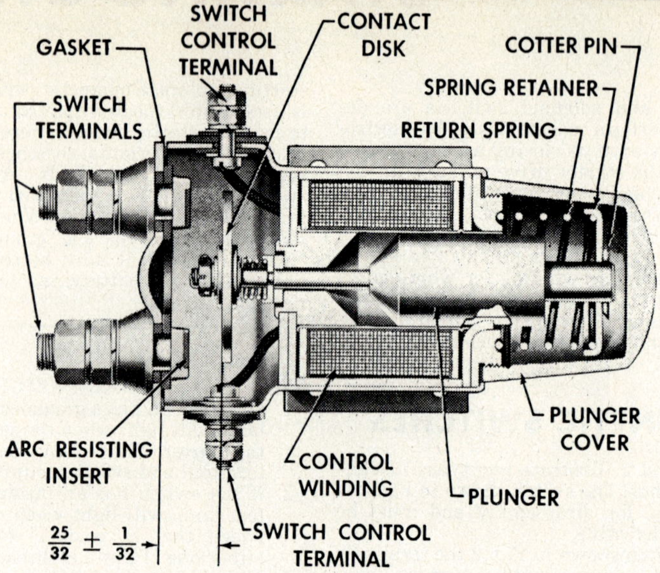

Fig. 3 Sectional view of heavy duty sealed type magnetic switch. To adjust the position of the contact disk, turn the nut on the disk in or out as required to obtained the dimension shown.

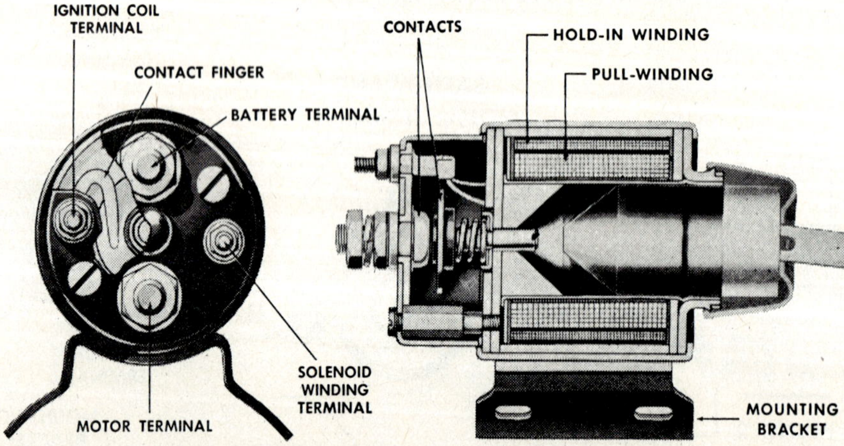

Fig. 4 End and sectional views of solenoid switch. The additional terminal and contact finger are used on 12-volt passenger car applications

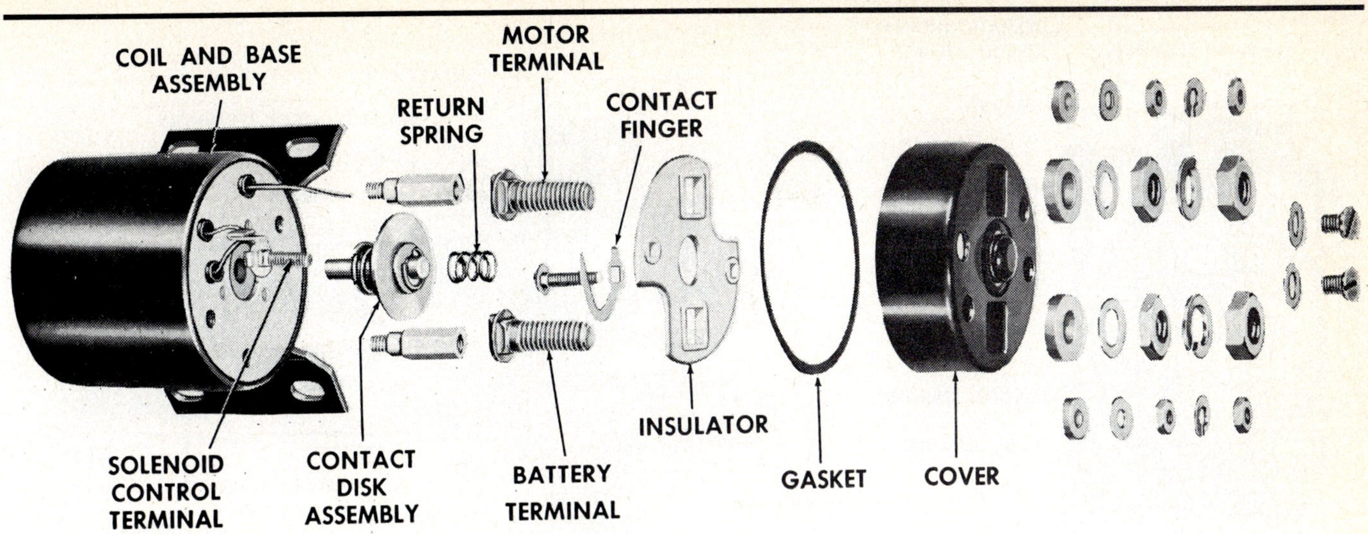

Fig. 5 Exploded view of solenoid switch shown in Fig. 4

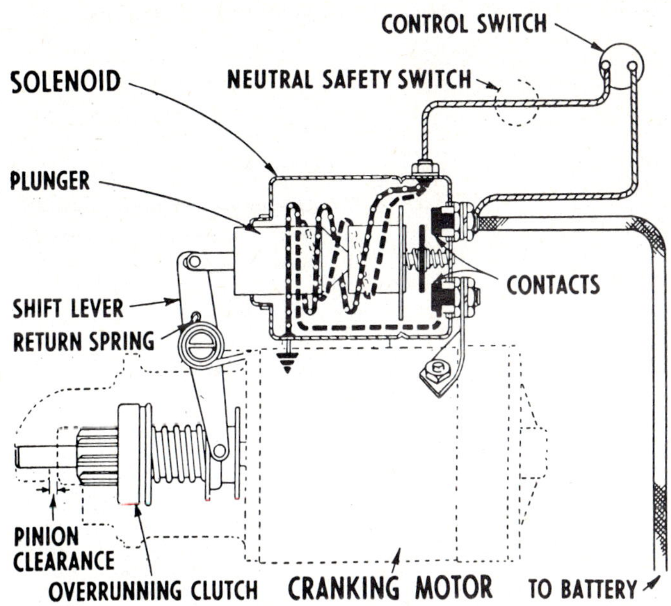

Fig. 6 Wiring circuit of typical solenoid switch

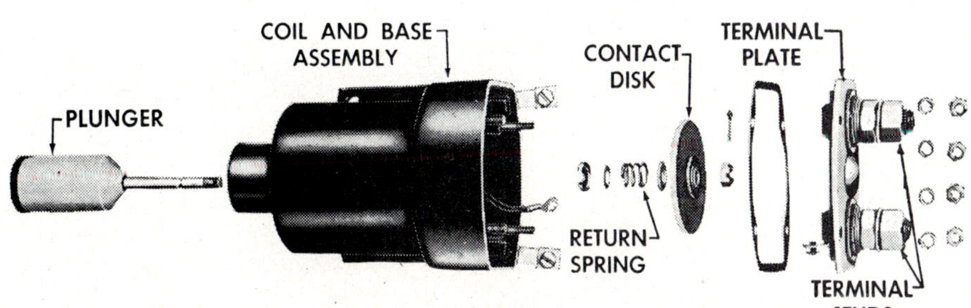

Fig. 7 Exploded view of heavy duty type solenoid used with Dyer cranking motor

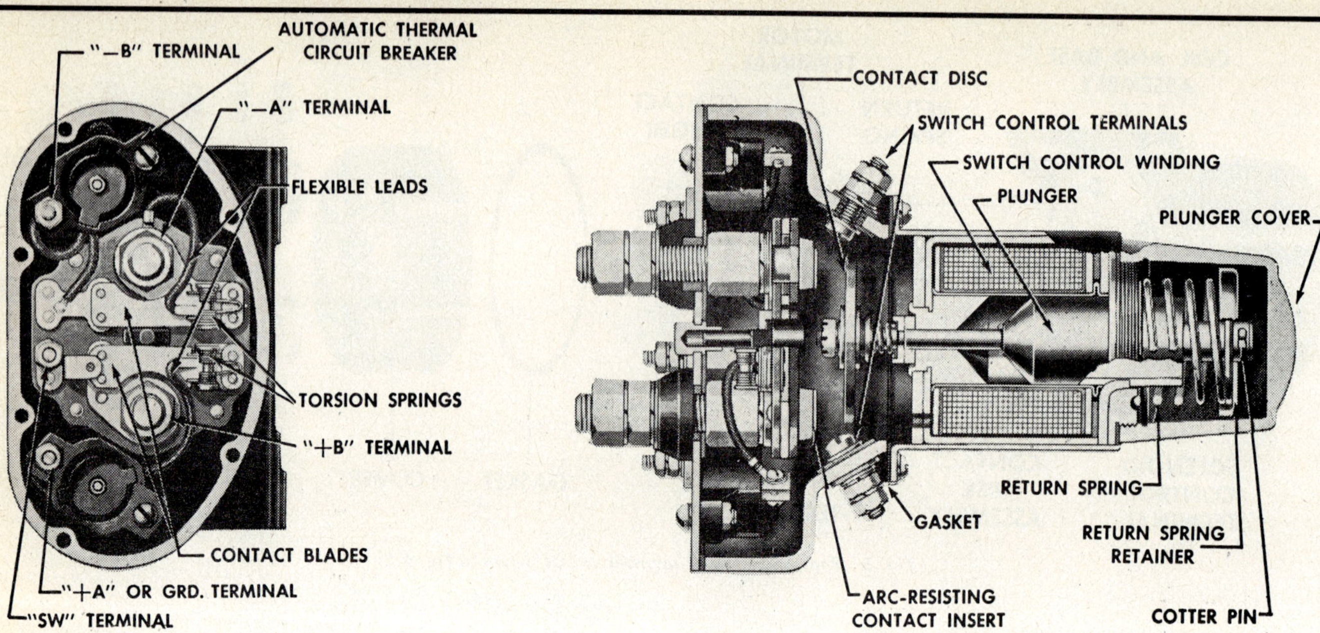

Fig. 8 End and sectional views of a Delco-Remy solenoid-operated series-parallel switch. In the end view the cover has been removed so the terminal plate assembly can be seen

In the mechanically-operated switch, Fig. 9, a quick-break mechanism, consisting of a pair of triggers and a cam, causes the contact disk to be snapped away from the stationary contacts very quickly so that there is a very small amount of arcing.

Fig. 11 illustrates a solenoid-operated series-parallel switch used in conjunction with a cranking motor magnetic type switch. Figs. 12 and 13 are exploded views of the mechanical switch, while Fig. 14 is an exploded view of the solenoid-operated switch. These pictures can be used as guides whenever disassembly is required.

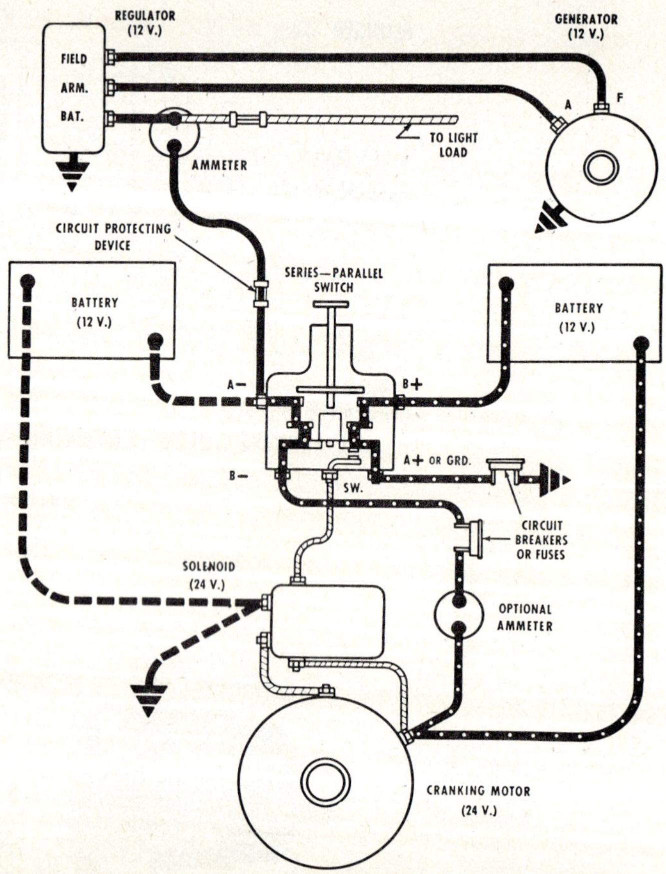

Fig. 9 Wiring diagram of a mechanically-operated series-parallel switch and cranking motor solenoid. This diagram shows the series-parallel switch completing the parallel connections between the batteries for normal operation of the vehicle's electrical equipment on 12 volts.

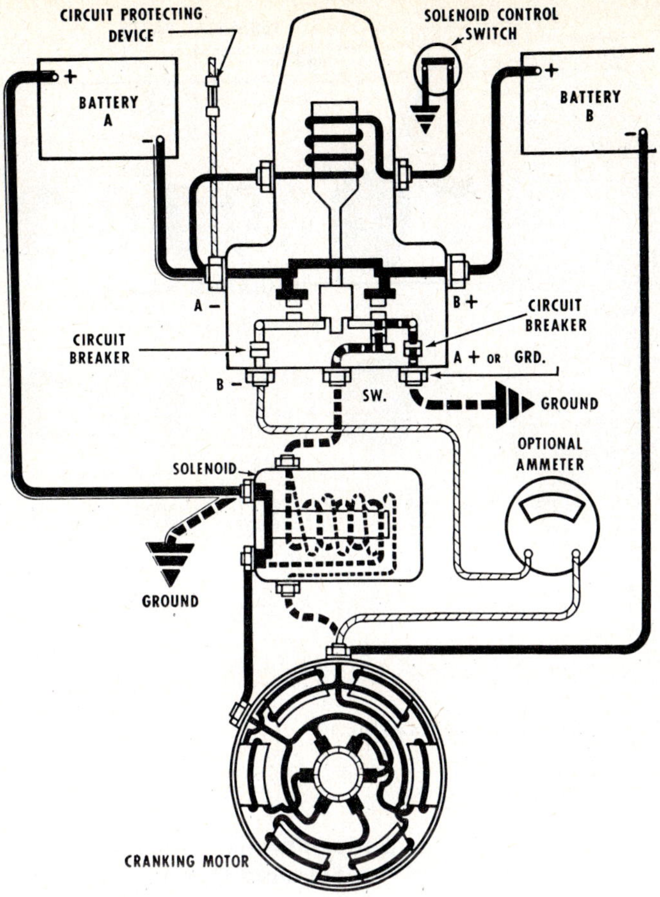

Fig. 10 This diagram shows a (solenoid-operated) series-parallel switch completing the series connections between batteries so 24 volts are imposed on cranking motor.

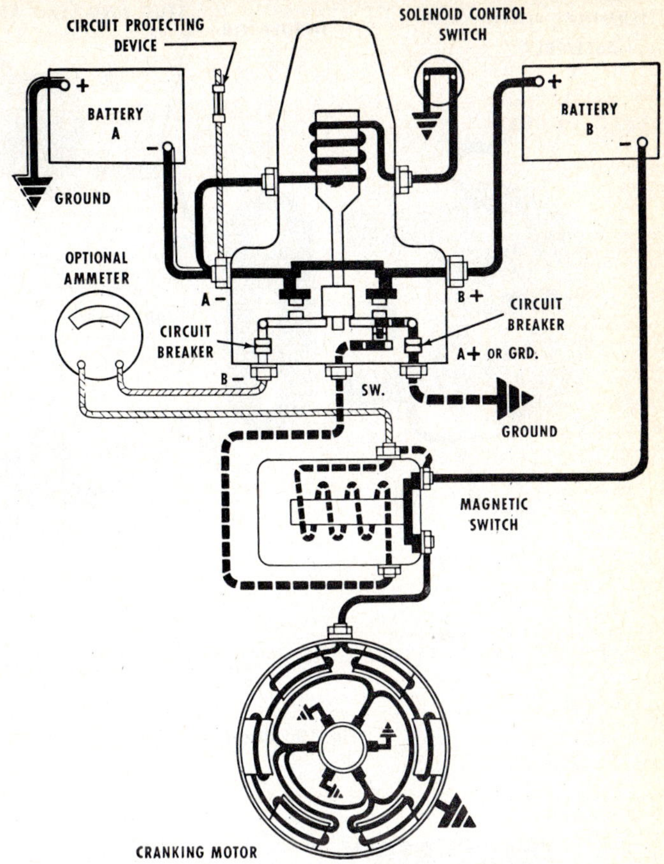

Fig. 11 Same as Fig. 10 except with magnetic type starter switch

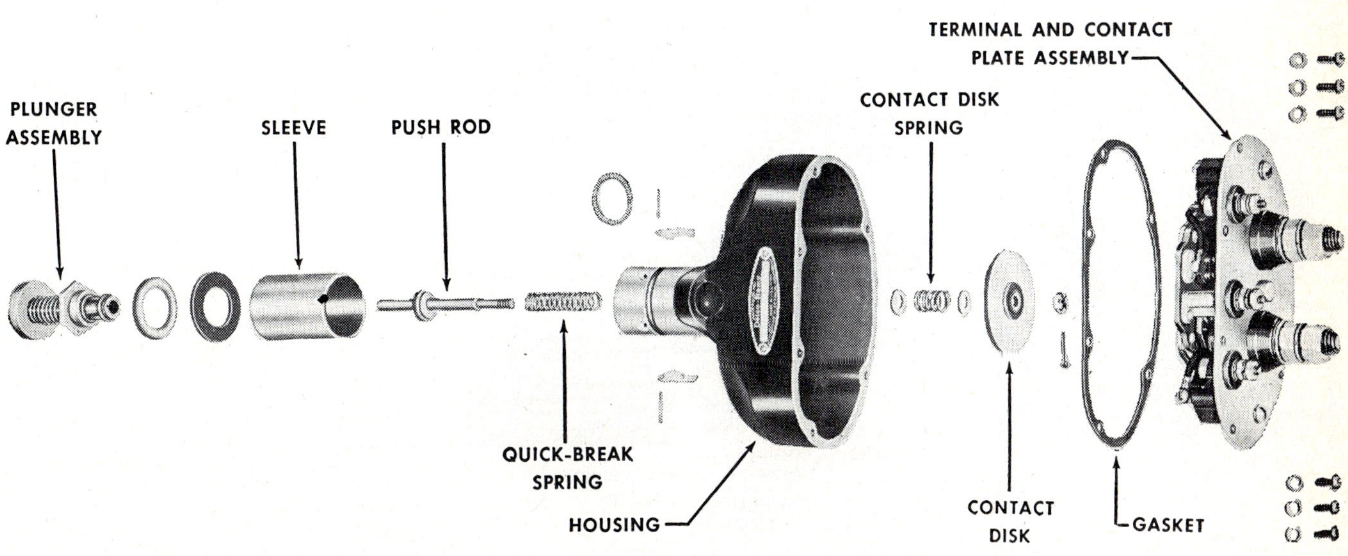

Fig. 12 Disassembled view of mechanically-operated series-parallel switch

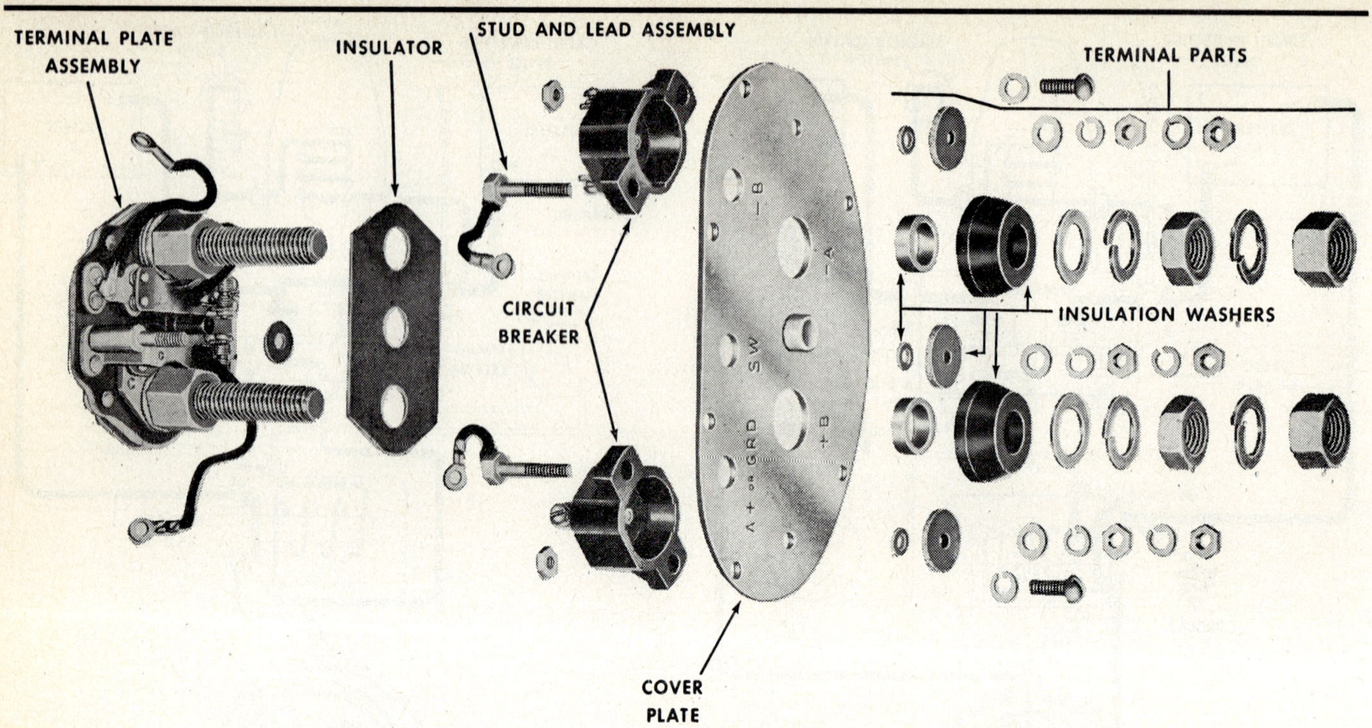

Fig. 13 Disassembled view of terminal plate assembly of switch shown in Fig. 12

DELCO-REMY SPLASHPROOF SERIES-PARALLEL & COMBINED SERIES-PARALLEL & MAGNETIC SWITCHES

Fig. 15 illustrates a low weight series-parallel switch and Fig. 16 pictures the combined series-parallel and magnetic switch of the same type construction. The switch shown in Fig. 15 can be used with either a solenoid or separate magnetic switch controlled starting motor. The switch shown in Fig. 16 is only used with a magnetic switch controlled starting motor and incorporates the magnetic switch with the series-parallel switch into the same splashproof container.

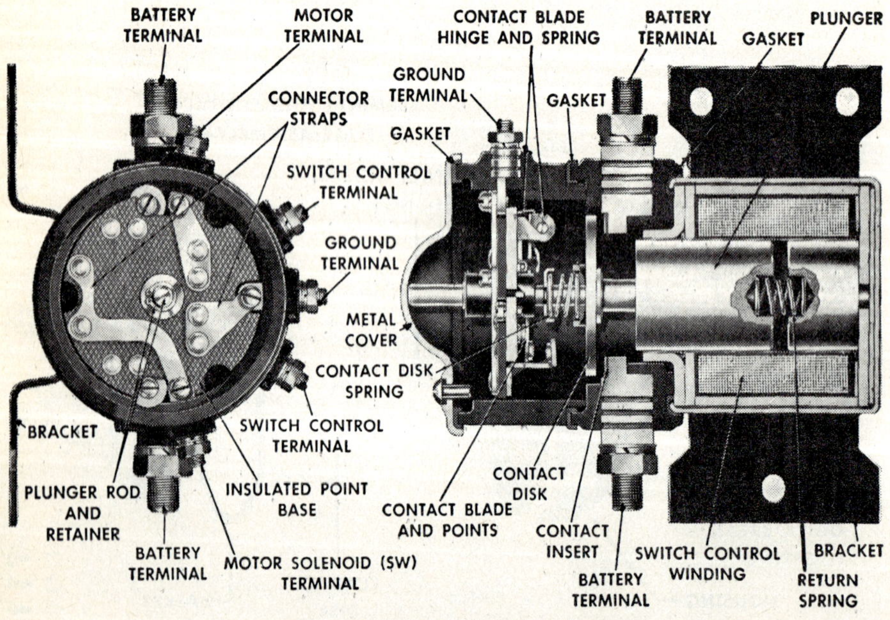

Fig. 15 End and sectional view of Delco-Remy splashproof solenoid operated series-parallel switch

Series-Parallel Switch Used With Solenoid Operated Motor

Fig. 17 is a schematic wiring circuit of the series-parallel system in connection with the solenoid operated motor during the cranking operation. The sequence that takes place as the switch closes is as follows:

As the starting switch is closed, the solenoid coil within the series-parallel switch is energized, creating sufficient magnetic force to attract the series-parallel switch plunger. Movement of the plunger then closes the two main switch terminals and connects the two batteries in series with the motor. At the same time, the starter solenoid coil circuit is completed by a set of points mechanically closed by the series-parallel switch plunger. This completes the battery-to-starting motor circuit and allows cranking to take place.

After cranking has been accomplished and the starting switch is released, the two batteries again become connected in parallel with the series-parallel switch in its "at rest" position. This allows operation of the rest of the vehicle's electrical equipment at a system voltage of 12 volts. Note that there are two circuit breakers or fuses in the circuit of the "B" battery. These should be used in the location shown. The fuse capacity should be 50% of the generator rated output.

Series-Parallel Switch Used With Magnetic Switch Operated Motor

Fig. 18 is the schematic wiring circuit of this system. The position of the switch is that found in normal operation of the vehicle. The current from the generator divides at the terminal of the series-parallel switch, half of it going to the "A" battery, with the other half of the current going to the "B" battery through the series-parallel switch. Note that an optional ammeter, as indicated, may be located

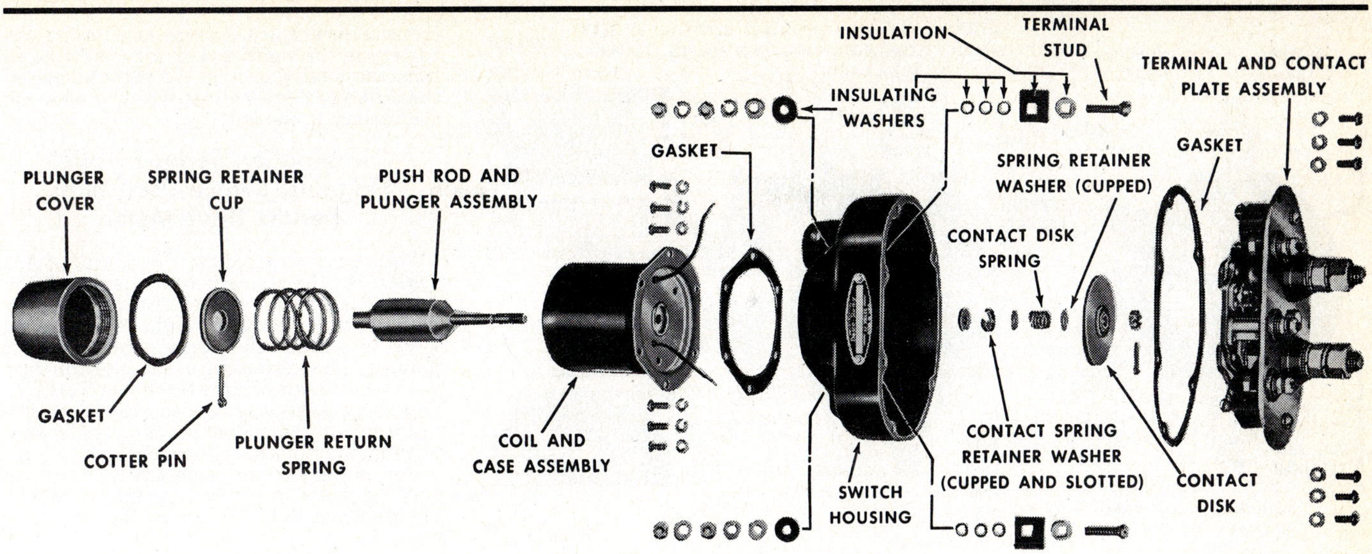

Fig. 14 Disassembled view of solenoid-operated series-parallel switch

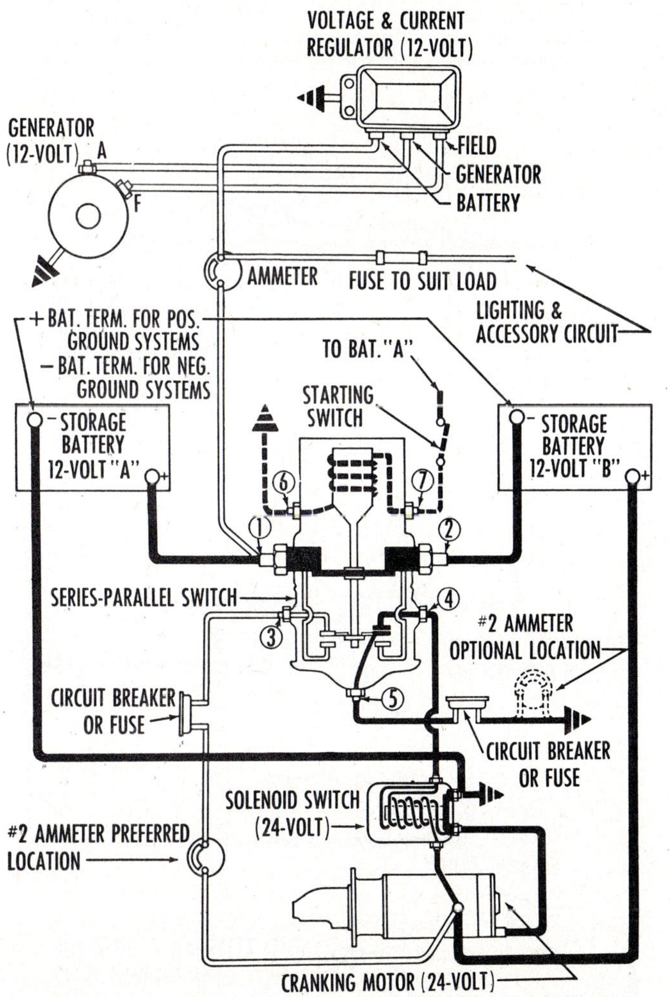

Fig. 17 Circuit diagram of series-parallel system with switch completing series connections between batteries for 24-volt starting motor operation. The motor is operated and controlled by a solenoid.

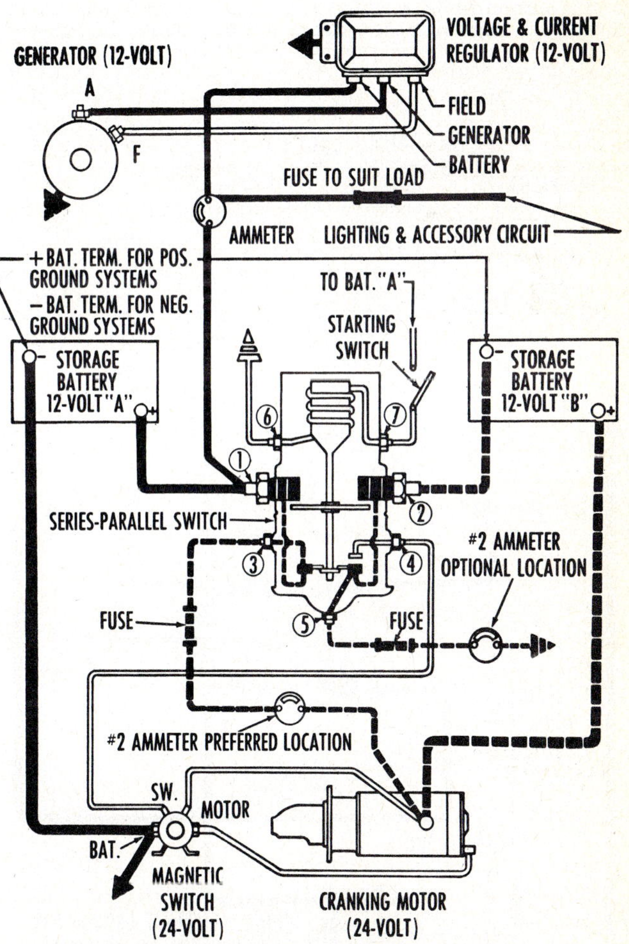

Fig. 18 Circuit diagram of series-parallel system with switch completing parallel connections between batteries for normal operation of vehicle electrical equipment at 12 volts. The starting motor is operated and controlled by a separately mounted magnetic switch.

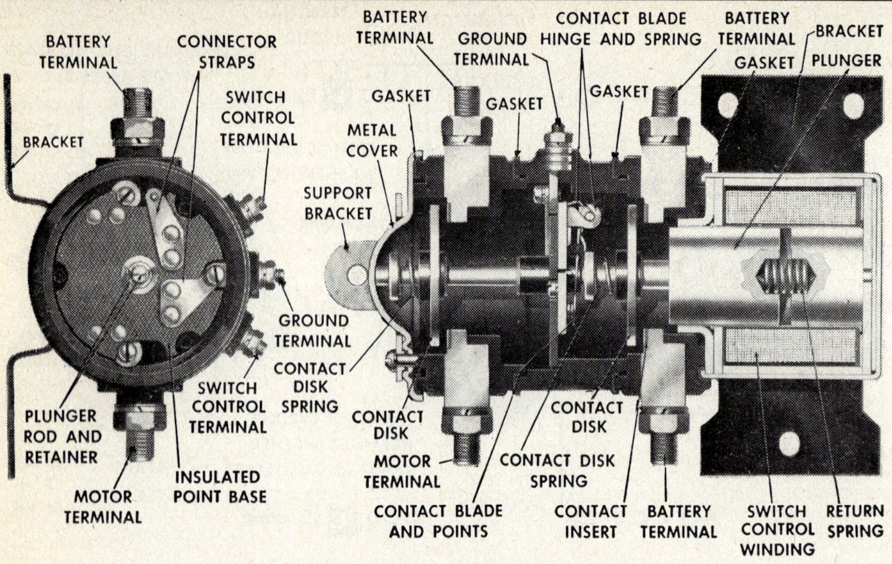

Fig. 16 End and sectional view of Delco-Remy splashproof solenoid operated combined series-parallel and magnetic switch

in either of two positions. The preferred location is shown since this circuit carries only the charging current while the other circuit must also carry the magnetic switch energizing current. An ammeter sufficiently heavy to carry the magnetic switch current is generally unsatisfactory for reading small charging currents. In either position, the ammeter will indicate how much current is entering the battery to the right, and if this is subtracted from the total shown on the main ammeter, the difference will be that which is entering the battery to the left.

Combination Series-Parallel Magnetic Switch Used With Bendix Drive Motor

Fig. 19 is a schematic wiring circuit of this system. As the starting switch is closed, the solenoid coil within the series-parallel switch is energized, creating sufficient magnetic force to attract the series-parallel switch plunger. The movement of the plunger then joins the two sets of large terminals. This connects both batteries in series with the cranking motor in order that 24 volts can be supplied for starting motor operation.

Fig. 20 shows the same switch after the starting switch is released and the series-parallel switch is in its "at rest" or charging position.

WIRING INSTALLATION FOR SERIES-PARALLEL SWITCHES

Due to the high voltage of the starting circuit and the great amount of power available from the two batteries, it is essential that every precaution be taken to avoid short cir-

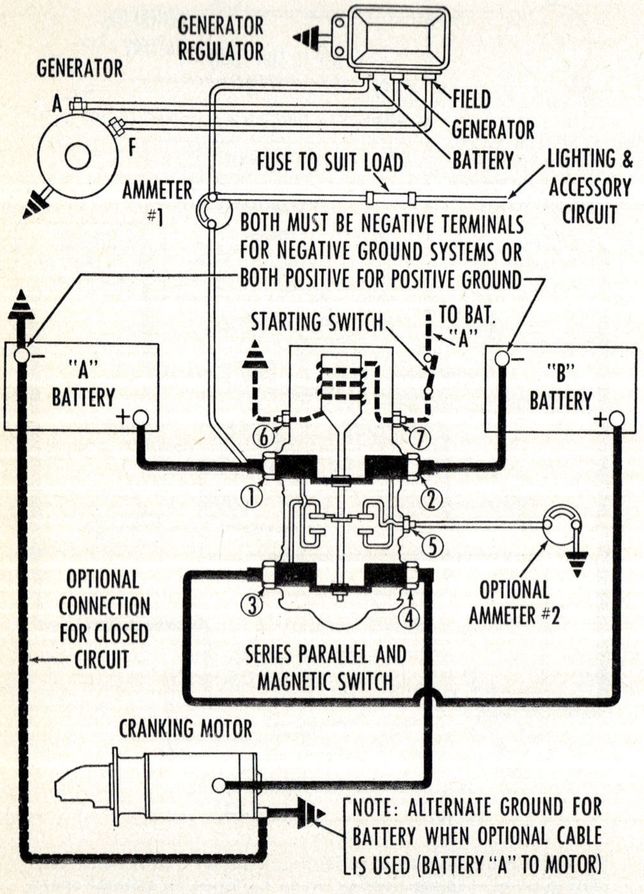

Fig. 19 Circuit diagram of combination series-parallel and magnetic switch completing series connections between batteries for 24-volt starting motor operation

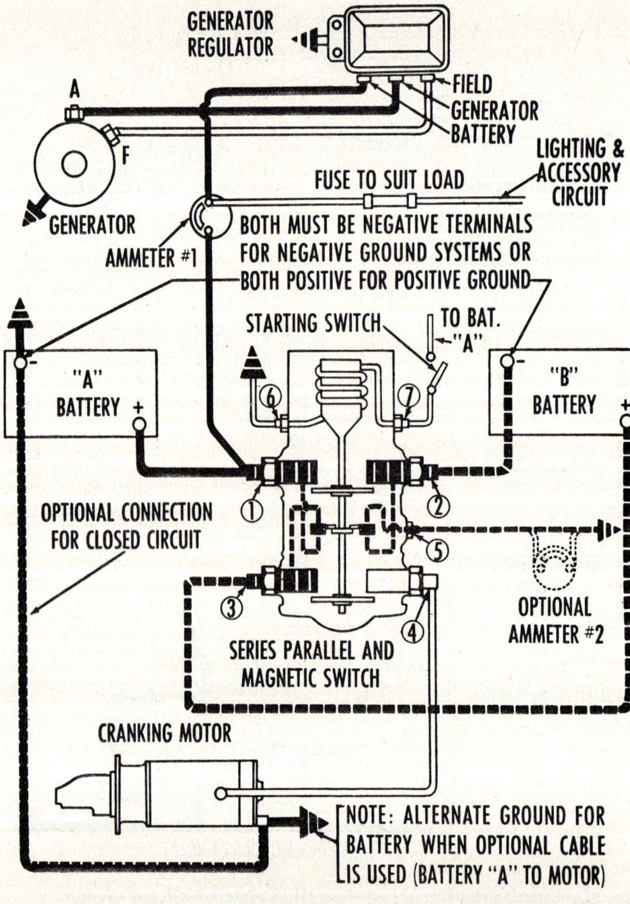

Fig. 20 Circuit diagram of combination series-parallel and magnetic switch completing parallel connections between batteries for normal operation of vehicle electrical equipment at 12 volts.

cuits or grounds. All wires should be of sufficient size to carry the electrical load to which they are subjected without overheating. Stranded wire and cable should be used throughout to reduce the possibility of breakage due to vibration. All connections should be clean and tight and all terminal clips should be soldered to wires or cables.

Only resin flux should be used to solder electrical connections. Acid flux should never be used. All wires should be adequately insulated and supported at enough points to prevent movement and consequent chafing through the insulation. It is desirable that all terminals and clips, which are ordinarily left exposed, be protected by insulation. Thus, rubber boots, rubber tape or friction tape and shellac should be applied to cover all exposed terminals and clips. This will prevent accidental grounding of an exposed terminal which could cause serious damage to the system.

BATTERY CARE

For the best performance of the electrical system in which the series-parallel type switch is used, it is recommended that the "A" and "B" batteries be of the same capacity, size, age and manufacturer. This will assure both batteries of receiving approximately equal charging rates and the danger of overcharging one battery and undercharging the other will be reduced.

Since the electrical loads on most vehicles are tapped off of one battery, it is recommended that periodically the two batteries should be switched in position. This will lessen the possibility of one battery becoming cycled with resultant positive plate shed which can cause unequal battery capacities.

USE OF GLOW PLUGS FOR STARTING

On some applications, glow plus are used to provide some initial heating so that starting can be facilitated. The heater coil and glow plug circuit should be connected to the +B terminal and a quick-break push button switch installed between the SW terminal of the series-parallel switch and the SW terminal of the cranking motor solenoid. Closing the series-parallel switch supplies current to the heater coil and glow plugs. After the proper heating interval, cranking is then accomplished by closing the quick-break push button switch in the cranking motor-solenoid circuit.

STARTING MOTORS

INDEX

STARTING TROUBLE CHECK-OUT

When trouble develops in the starting motor circuit, and the starter cranks the engine slowly or not at all, several preliminary checks can be made to determine whether the trouble lies in the battery, in the starter, in the wiring between them, or elsewhere. Many conditions besides defects in the starter itself can result in poor cranking performance.

To make a quick check of the starter system, turn on the headlights. They should burn with normal brilliance. If they do not, the battery may be run down and it should be checked with a hydrometer.

If the battery is in a charged condition so that the lights burn brightly, operate the starting motor. Any one of three things will happen to the lights: (1) They will go out, (2) dim considerably or (3) stay bright without any cranking action taking place.

If Lights Go Out

If the lights go out as the starter switch is closed, it indicates that there is a poor connection between the battery and starting motor. This poor connection will most often be found at the battery terminals. Correction is made by removing the cable clamps from the terminals, cleaning the terminals and clamps, replacing the clamps and tightening them securely. A coating or corrosion inhibitor (vaseline will do) may be applied to the clamps and terminals to retard the formation of corrosion.

If Lights Dim

If the lights dim considerably as the starter switch is closed and the starter operates slowly or not at all, the battery may be run down, or there may be some mechanical condition in the engine or starting motor that is throwing a heavy burden on the starting motor. This imposes a high discharge rate on the battery which causes noticeable dimming of the lights.

Check the battery with a hydrometer. If it is charged, the trouble probably lies in either the engine or starting motor itself. In the engine, tight bearings or pistons or heavy oil place an added burden on the starting motor.

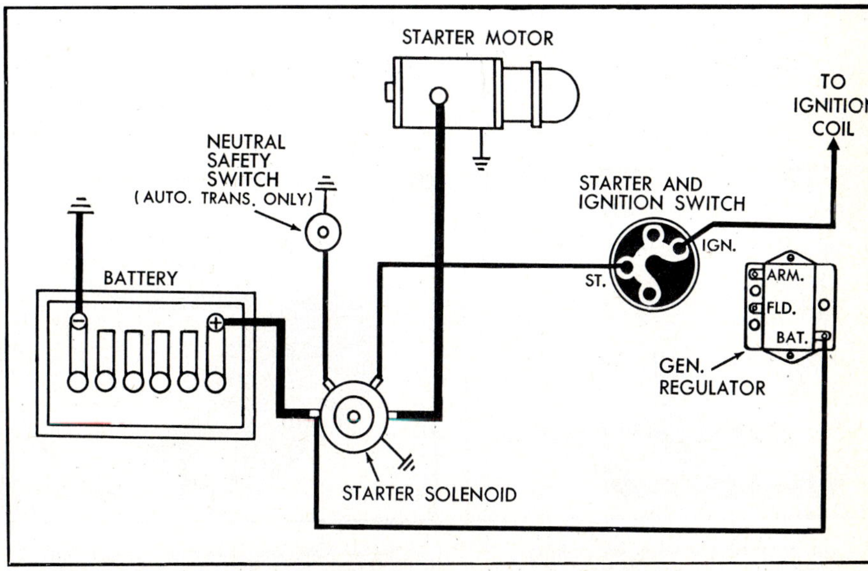

Fig. 1 Wiring diagram of a typical starter circuit

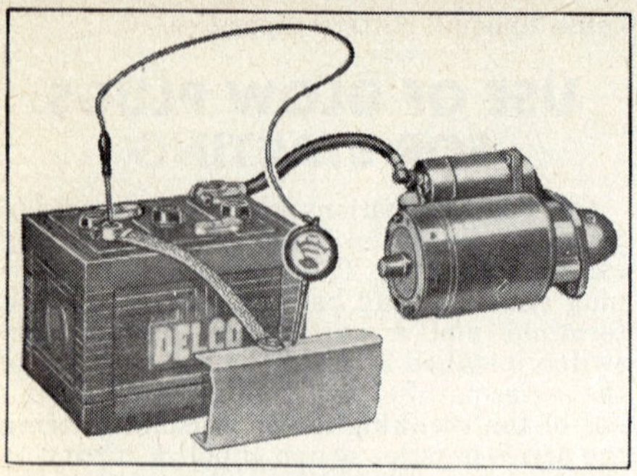

Fig. 2 Checking voltage drop between vehicle frame and grounded battery terminal post

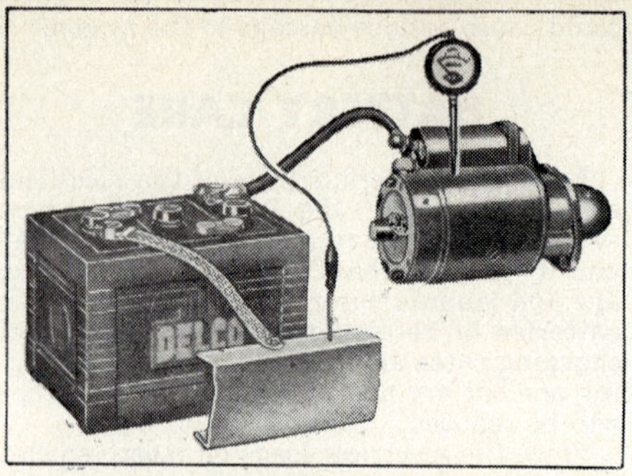

Fig. 3 Checking voltage drop between vehicle frame and starter field frame

Fig. 4 Checking voltage drop between ungrounded battery terminal post and battery terminal on solenoid

Low temperatures also hamper starting motor performance since it thickens engine oil and make the engine considerably harder to crank and start. Also, a battery is less efficient at low temperatures.

In the starting motor, a bent armature, loose pole shoe screws or worn bearings, any of which may allow the armature to drag, will reduce cranking performance and increase current draw.

In addition, more serious internal damage is sometimes found. Thrown armature windings or commutator bars, which sometimes occur on over-running clutch drive starting motors, are usually caused by excessive overrunning after starting. This is the result of such conditions as the driver keeping the starting switch closed too long after the engine has started, the driver opening the throttle too wide in starting, or improper carburetor fast idle adjustment. Any of these subject the overrunning clutch to extra strain so it tends to seize, spinning the armature at high speed with resulting armature damage.

Another cause of failure may be engine backfire during cranking which may result, among other things, from ignition timing being too far advanced.

To avoid such failures, the driver should pause a few seconds after a false start to make sure the engine has come completely to rest before another start is attempted. In addition, the ignition timing should be reset if engine backfiring has caused the trouble.

Lights Stay Bright, No Cranking Action

This condition indicates an open circuit at some point, in the starter itself, the starter switch or control circuit. The solenoid control circuit can be eliminated momentarily by placing a heavy jumper lead across the solenoid main terminals to see if the starter will operate. This connects the starter directly to the battery and, if it operates, it indicates that the control circuit is not functioning normally. The wiring and control units must be checked to locate the trouble, Fig. 1.

If the starter does not operate with the jumper attached, it will probably have to be removed from the engine so it can be examined in detail.

Checking Circuit With Voltmeter

Excessive resistance in the circuit between the battery and starter will reduce cranking performance. The resistance can be checked by using a voltmeter to measure voltage drop in the circuits while the starter is operated. There are three checks to be made:

1. Voltage drop between car frame and grounded battery terminal post (not cable clamp), Fig. 2.
2. Voltage drop between car frame and starting motor field frame, Fig. 3.
3. Voltage drop between insulated battery terminal post and starting motor terminal stud (or the battery terminal stud or the battery terminal stud of the solenoid), Fig. 4.

Each of these checks should show no more than one-tenth (0.1) volt drop when the starting motor is cranking for more than 30 seconds at a time to avoid overheating.

If excessive voltage drop is found in any of these circuits, make correction by disconnecting the cables, cleaning the connections carefully, and then reconnecting the cables firmly in place. A coating of vaseline on the battery cables and terminal clamps will retard corrosion.

NOTE: On some vehicles, extra long battery cables may be required due to the location of the battery and starter. This may result in somewhat higher voltage drop than the above recommended 0.1 volt. The only means of determining the normal voltage drop in such cases is to check several of these vehicles. Then when the voltage drop is well above the normal figure for all cars checked, abnormal resistance will be indicated and correction can be made as already explained.

STARTING MOTOR SERVICE

To obtain full performance data on a starting motor or to determine the cause of abnormal operation, the starting motor should be submitted to a no-load test and torque test. These tests are best performed on a starter bench tester with the starter mounted on it.

From a practical standpoint, however, a simple torque test may be made quickly with the starter in the car. Make sure the battery is

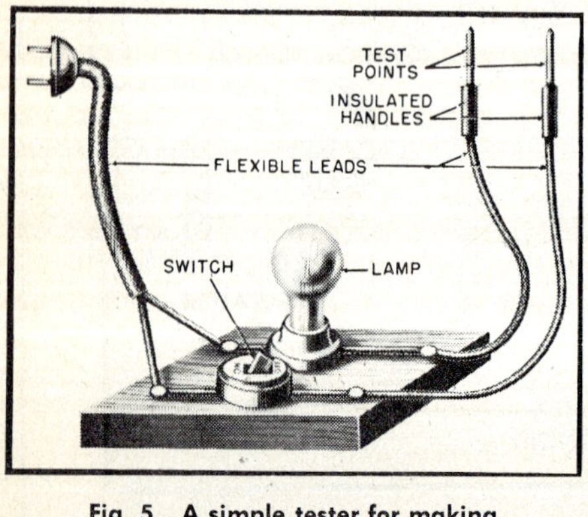

Fig. 5 A simple tester for making continuity and ground tests on armature and field windings

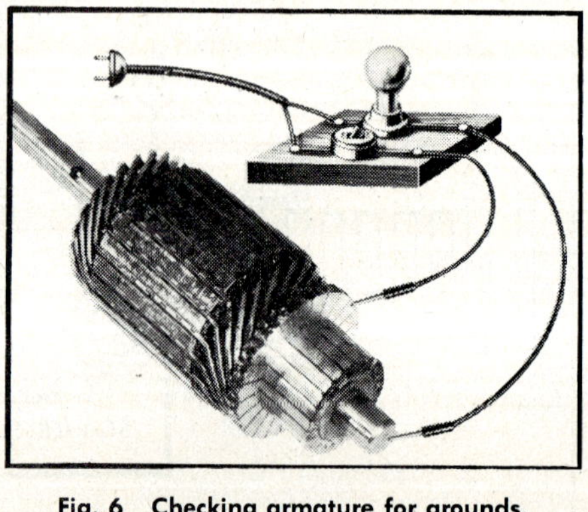

Fig. 6 Checking armature for grounds. If lamp lights armature is grounded and should be replaced

Fig. 7 Measuring commutator runout with a dial indicator. Mount shaft in V blocks and rotate commutator. If runout exceeds .003", commutator should be turned in a lathe to make it concentric.

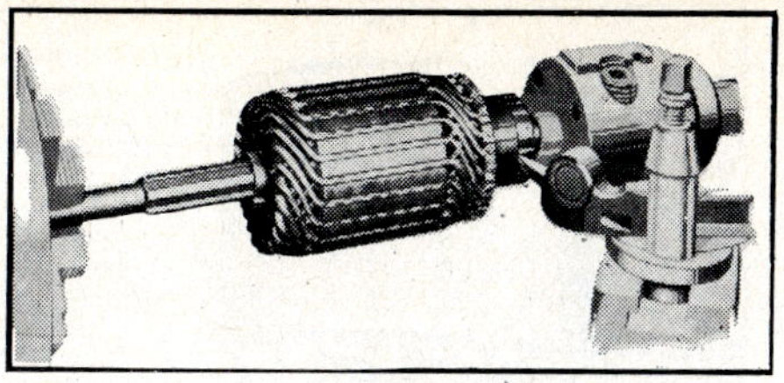

Fig. 8 Turning commutator in a lathe. Take light cuts until worn or bad spots are removed. Then remove burrs with No. 00 sandpaper.

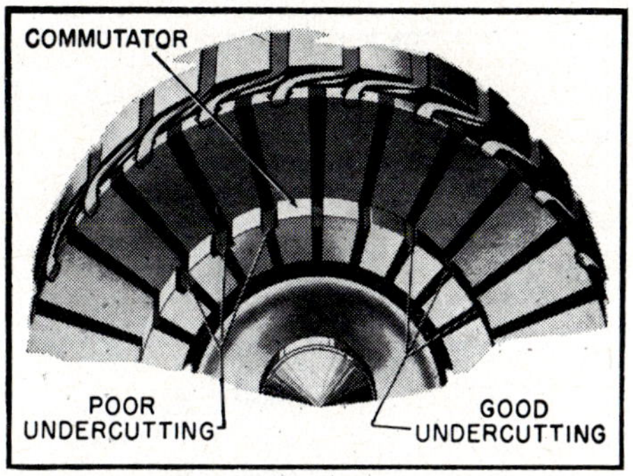

Fig. 9 Good undercutting should be .002" wider than mica insulation, 1/64" deep and exactly centered so that there are no burrs on the mica.

fully charged and that the starter circuit wires and terminals are in good condition. Then operate the starter to see if the engine turns over normally. If it does not, the torque developed is below standard and the starter should be removed for further checking.

Remove starter from engine and disassemble as outlined in the following text and make the tests as suggested in Figs. 5 through 11.

Fig. 10 Checking armature for short circuit. As armature is rotated by hand, steel strip (hacksaw blade) will vibrate if short circuit exists.

CHRYSLER DIRECT DRIVE STARTER

This Chrysler built starting motor, Fig. 12, is a four coil assembly with an over-running clutch type drive and a solenoid shift-type switch mounted on the motor. The brush holders are riveted to a separate brush plate and are not serviced individually. Brush replacement can be made by removing the commutator bearing end head.

Disassembly

1. Remove through bolts and tap commutator end head from field frame.
2. Remove thrust washers from armature shaft.

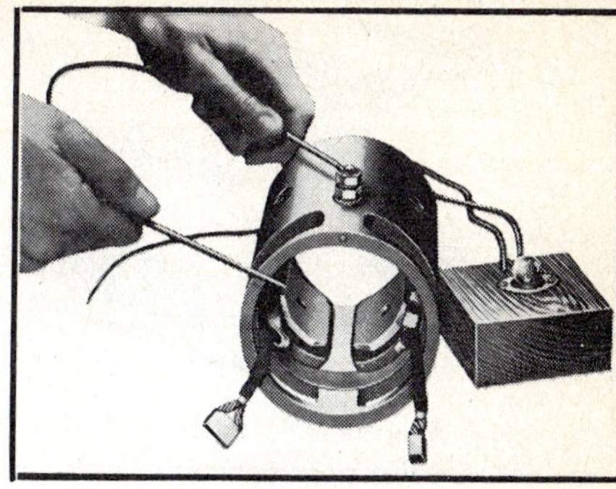

Fig. 11 Testing field coils for grounds. Lamp will light if a ground is present

3. Lift brush holder springs and remove brushes from holders.
4. Remove brush plate.
5. Disconnect field leads at solenoid connector.
6. Unfasten and remove solenoid and boot assembly.
7. Drive out over-running clutch shift fork pivot pin.
8. Remove drive end pinion housing and spacer washer.
9. Note position of shifter fork on starter and remove fork.
10. Slide over-running clutch pinion gear toward commutator end of armature. Drive stop retainer toward clutch pinion gear to expose snap ring and remove snap ring.
11. Slide clutch drive from armature shaft.
12. If it is necessary to replace the field coils, remove screw that holds ground brushes and raise brushes with the terminal and shunt wire up and away from field frame. Remove pole shoe screws and take out field coils.

Reassembly

1. Lubricate armature shaft and splines

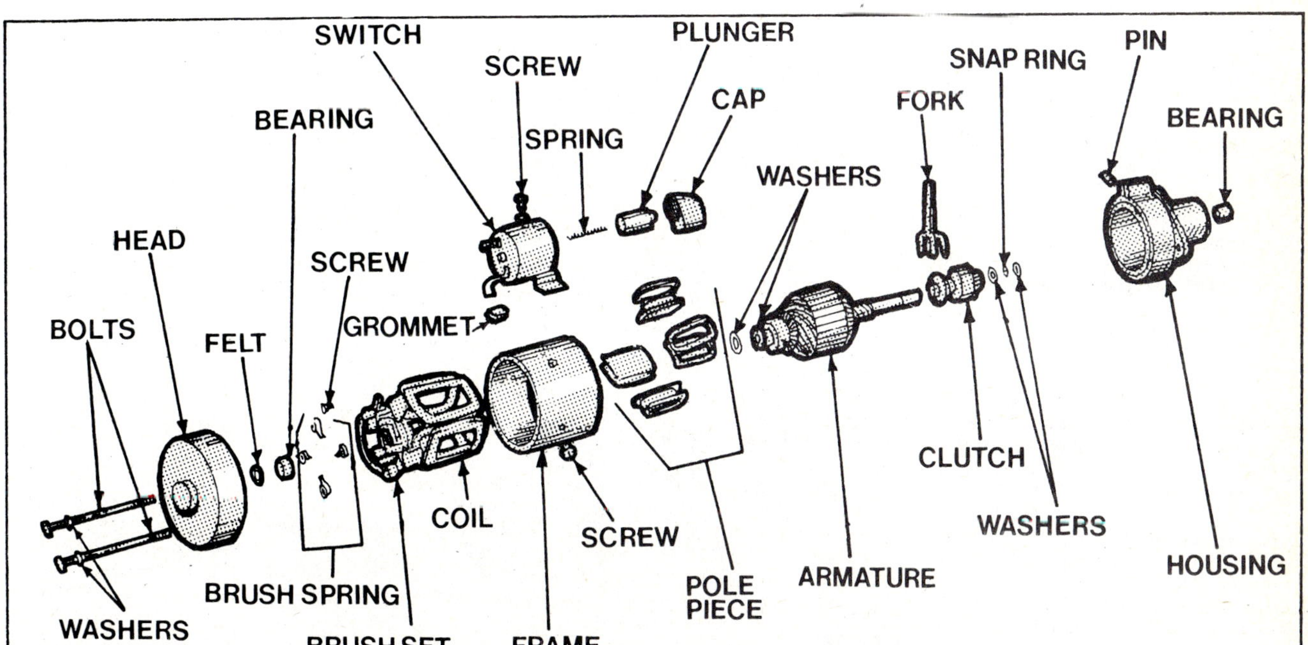

Fig. 12 Chrysler built direct drive starter

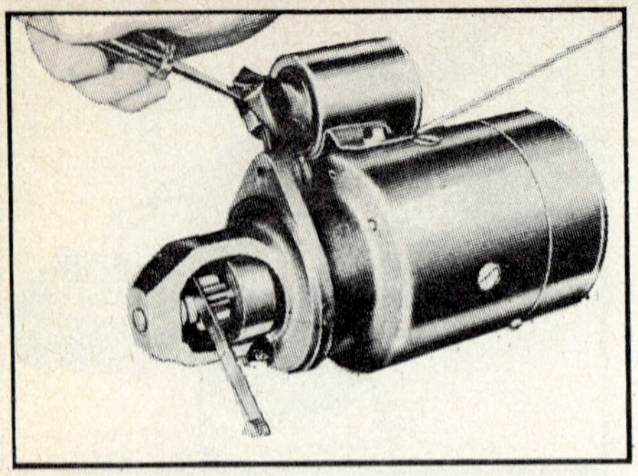

Fig. 13 Clearance between end of pinion and pin stop should be 1/8" with plunger seated and pinion pushed toward commutator.

with SAE 10W or 30W rust preventive oil.

2. Install starter drive, stop collar (retainer), lock ring and spacer washer.
3. Install shifter fork over starter drive spring retainer washer with narrow leg of fork toward commutator. *If fork is not positioned properly, starter gear travel will be restricted, causing a lockup in the clutch mechanism.*
4. Install drive end (pinion) housing on armature shaft, indexing the shift fork with slot in drive end of housing.
5. Install shift fork pivot pin.
6. Install armature with clutch drive, shifter fork and pinion housing. Slide armature into field frame until pinion housing indexes with slot in field frame.
7. Install solenoid and boot assembly and tighten bolts securely.
8. Install ground brushes.
9. Connect field coil leads at solenoid connector.
10. Install brush holder ring, indexing tang of ring in hole of field frame.
11. Position brushes in brush holders. *Be sure field coil lead wires are properly enclosed behind brush holder ring and that they do not interfere with brush operation.*
12. Install thrust washer on commutator end of armature shaft to obtain .010" minimum end play.
13. Install commutator end head.
14. Install through bolts and tighten securely.

Adjusting Pinion Clearance

1. Place starter in vise with soft jaws and tighten vise enough to hold starter. *Place a wedge or screwdriver between bottom of solenoid and starter frame to eliminate all deflection in solenoid when making pinion clearance check.*
2. Push in on solenoid plunger link, Fig. 13 (not fork lever) until plunger bottoms.
3. Measure clearance between end of pinion and pin stop with plunger seated and pinion pushed toward commutator end. Clearance should be 1/8". Adjust by loosening solenoid attaching screws and move solenoid fore and aft as required.
4. Test starter operation for free running and install on engine.

CHRYSLER REDUCTION GEAR STARTER

This reduction gear starting motor, Fig. 14, has an armature-to-engine crankshaft ratio of 45 to 1; a 2 to 1 or 3½ to 1 reduction gear set is built into the motor assembly. The starter utilizes a solenoid shift. The housing of the solenoid is integral with the starter drive end housing.

Fig. 15 Terminal screw replacement, 1973 units

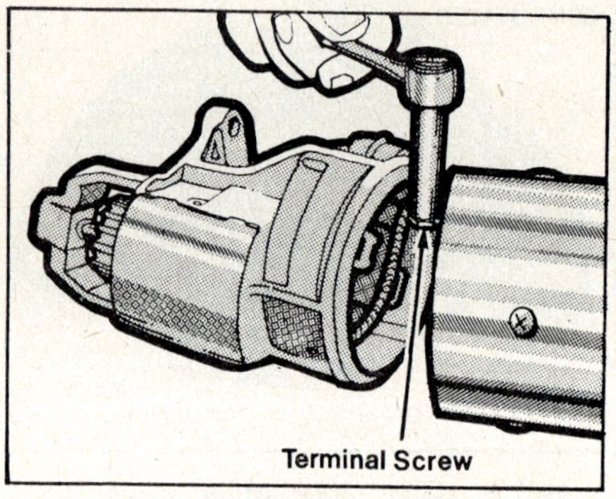

Fig. 16 Terminal screw replacement, 1974–83 units

Disassembly

1. Place gear housing of starter in a vise with soft jaws. *Use vise as a support fixture only; do not clamp.*

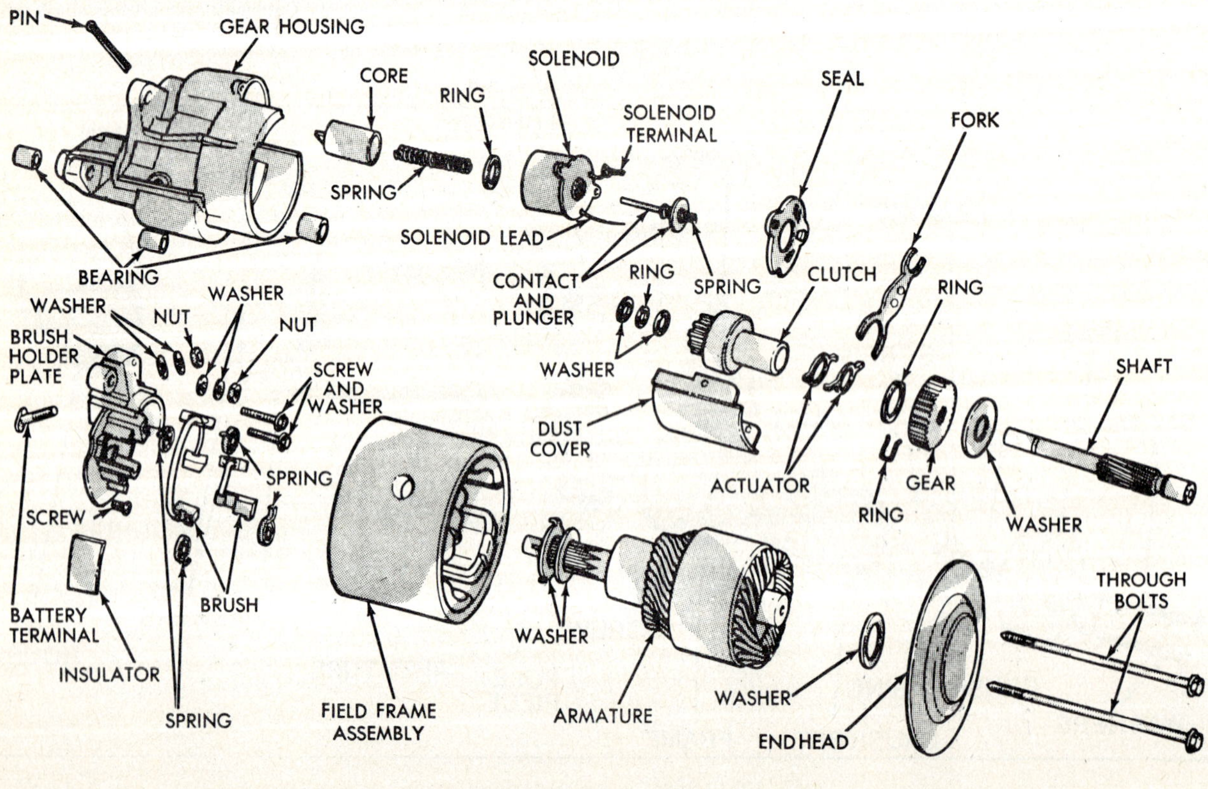

Fig. 14 Chrysler built reduction gear starter

Fig. 17 Unwinding solenoid lead wire, all units (Typical)

2. Remove through bolts and starter end head assembly.
3. On 1973 units, carefully pull armature up and out of gear housing, and starter frame and field assembly. Remove steel and fiber thrust washer. *The wire of the shunt field coil is soldered to the brush terminal. One pair of brushes are connected to this terminal. The other pair of brushes is attached to the series field coils by means of a terminal screw. Carefully pull the frame and field assembly up just enough to expose the terminal screw and the solder connection of the shunt field at the brush terminal. Place two wood blocks between starter frame and gear housing, Fig. 15, to facilitate removal of terminal screw and unsoldering of shunt field wire at brush terminal.*
4. On 1974–83 units, carefully remove armature, then pull field frame assembly from gear housing to expose terminal screw.
5. On all units, support terminal screw with a finger, then remove terminal screw, Figs. 15 and 16.
6. On 1973 units:
   a. Unwrap shunt field coil lead from starter brush terminal.

NOTE: The starter brush holder plate with the starter brush terminal, contact and brushes is serviced as an assembly.

   b. Unwrap solenoid lead wire and unwind wire from starter brush terminal, Fig. 17.
7. On 1974–83 units:
   a. Remove field frame assembly.

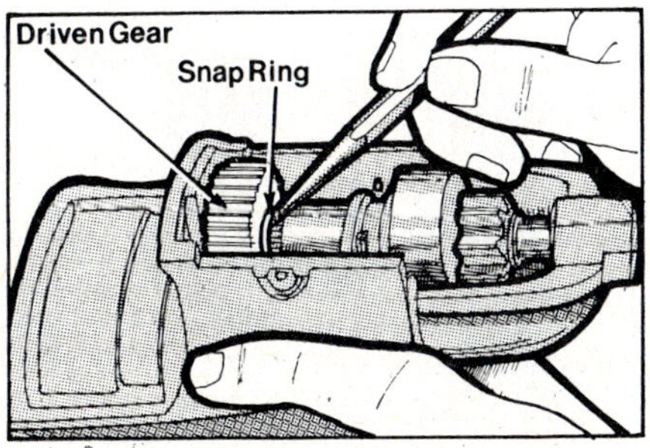

Fig. 20 Driven gear snap ring replacement, all units (Typical)

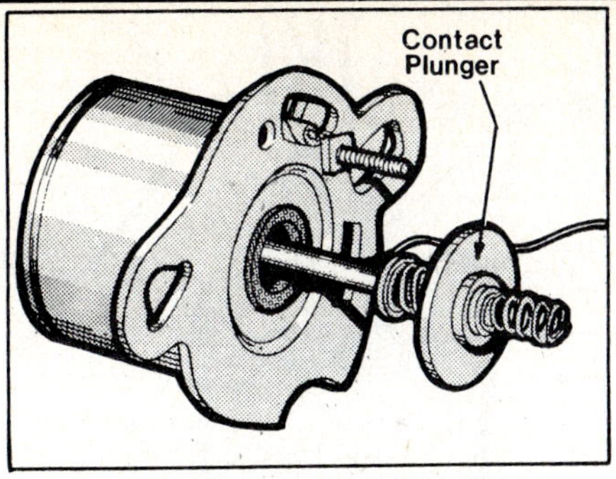

Fig. 18 Solenoid contact & plunger, all units (Typical)

   b. Remove nuts attaching solenoid and brush holder plate assembly to gear housing, then the solenoid and brush plate assembly.
8. On all units, remove nut, steel washer and insulating washer from solenoid terminal.
9. On 1973 units, straighten solenoid wire and remove brush holder plate with brushes and solenoid as an assembly. Remove solenoid from gear housing well.
10. On 1974–83 units, unwind solenoid lead wire from brush terminal, Fig. 17, and remove screws securing solenoid to brush plate, then the solenoid from brush plate.
11. On all units, remove nut from battery terminal on brush plate, then the battery terminal.

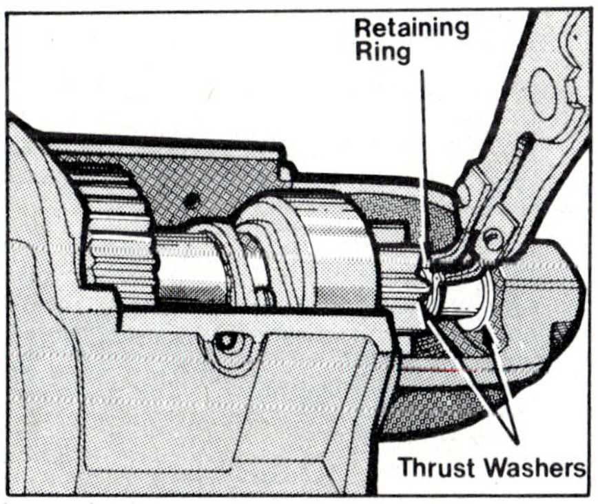

Fig. 21 Pinion shaft retaining ring replacement, 1973 units

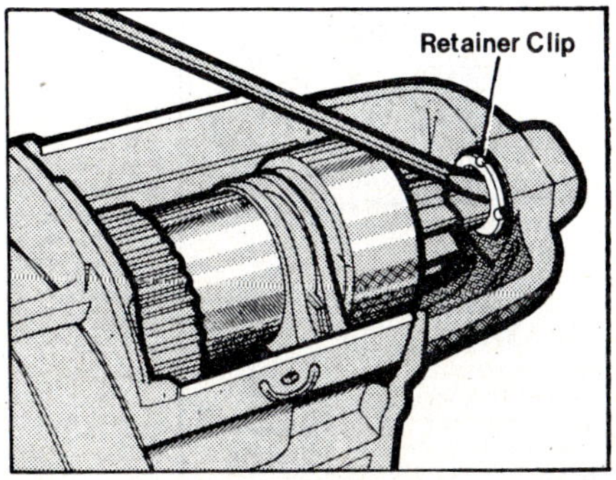

Fig. 22 Pinion shaft "C" clip replacement, 1974–83 units

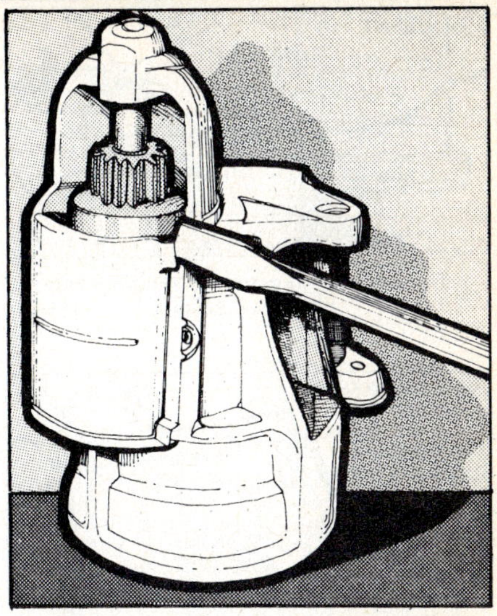

Fig. 19 Dust cover removal, all units (Typical)

12. Remove solenoid contact and plunger assembly from solenoid, Fig. 18, then the return spring from the solenoid moving core.
13. Remove dust cover from gear housing, Fig. 19.
14. Release retainer clip positioning driven gear on pinion shaft, Fig. 20.

NOTE: The retainer clip is under tension. Therefore, it is recommended that a cloth be placed over the retainer clip when released, preventing it from springing away.

15. On 1973 units, release retainer ring at front of pinion shaft, Fig. 21. On 1974–83 units, remove pinion shaft "C" clip, Fig. 22.

NOTE: On 1973 units, do not spread retainer ring greater than the outside diameter of the pinion shaft since lock ring damage may result.

16. On all units, push pinion shaft toward rear of housing, Fig. 23 and remove retainer ring and thrust washers, clutch and pinion assembly, with the two shift fork nylon actuators as an assembly, Fig. 24.
17. Remove driven gear and thrust washer.
18. Pull shifting fork forward and remove solenoid moving core, Fig. 25.
19. Remove shifting fork retainer pin, Fig. 26, and remove clutch shifting fork assembly.

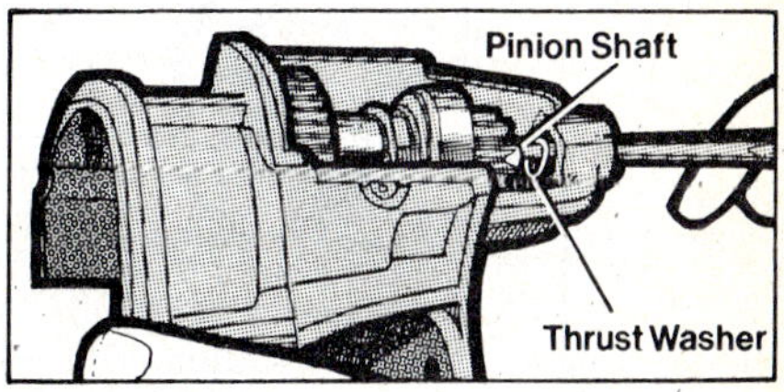

Fig. 23 Removing pinion shaft

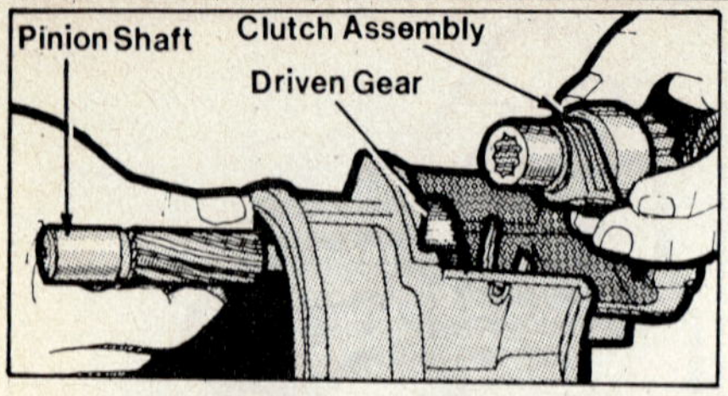

Fig. 24 Clutch assembly replacement, all units (Typical)

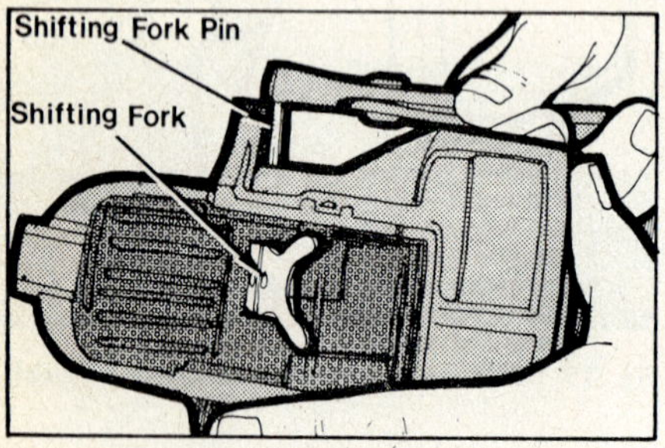

Fig. 26 Shifter fork pin replacement, all units (Typical)

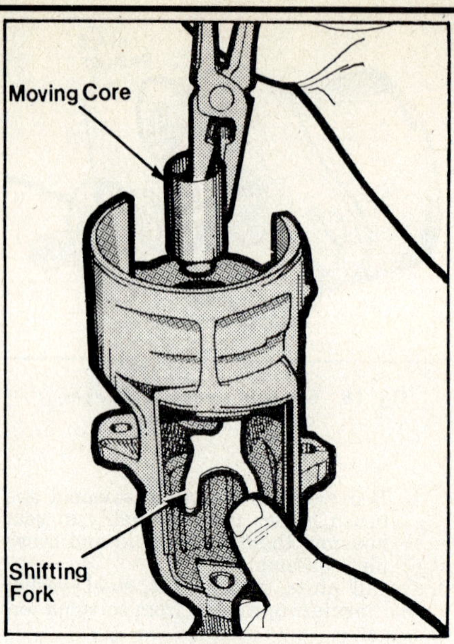

Fig. 25 Solenoid core replacement, all units (Typical)

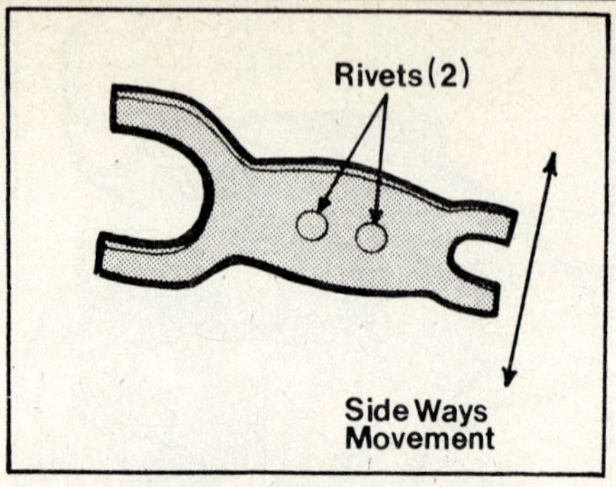

Fig. 27 Shifter fork assembly, all units (Typical)

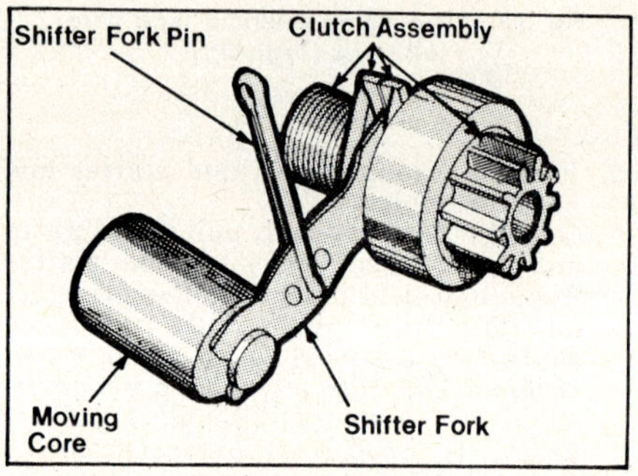

Fig. 28 Shifter fork & clutch assembly, all units (Typical)

Reassembly

NOTE: *The shifter fork consists of two spring steel plates assembled with two rivets, Fig. 27. There should be about 1/16" side movement to insure proper pinion gear engagement. Lubricate between plates sparingly with SAE 10 engine oil.*

1. Position shift fork in drive housing and install fork retaining pin, Fig. 26. *One tip of pin should be straight, the other tip should be bent at a 15 degree angle away from housing. Fork and pin should operate freely after bending tip of pin.*
2. Install solenoid moving core and engage shifting fork, Fig. 25.
3. Enter pinion shaft in drive housing and install friction washer and driven gear.
4. Install clutch and pinion assembly, Fig. 24, thrust washer, retaining ring, and thrust washer.
5. Complete installation of pinion shaft, engaging fork with clutch actuators, Fig. 28. *Friction washer must be positioned on shoulder of splines of pinion shaft before driven gear is positioned.*
6. Install driven gear snap ring, Fig. 20, then the pinion shaft retaining ring or "C" clip, Figs. 21 and 22.
7. Install starter solenoid return spring into movable core bore.

NOTE: Inspect starter solenoid switch contacting washer. If top of washer is burned, disassemble contact switch plunger assembly and reverse the washer.

8. Install solenoid contact plunger assembly into solenoid, Fig. 18. On 1973 units, reform the double wire to permit proper entry of terminal stud into brush holder with the double wires curved around the contactor. The contactor must not touch the double wires when the solenoid is energized after assembly. On 1974–83 units, ensure contact spring is properly positioned on shaft of solenoid contact plunger assembly.
9. On all units, install battery terminal stud in brush holder.

NOTE: Inspect contacts in brush holder. If contacts are badly burned, replace brush holder with brushes and contacts as an assembly.

10. Position seal on brush holder plate.
11. Install solenoid lead wire through hole in brush holder, Fig. 29 then the solenoid stud, insulating washer, flat washer and nut.

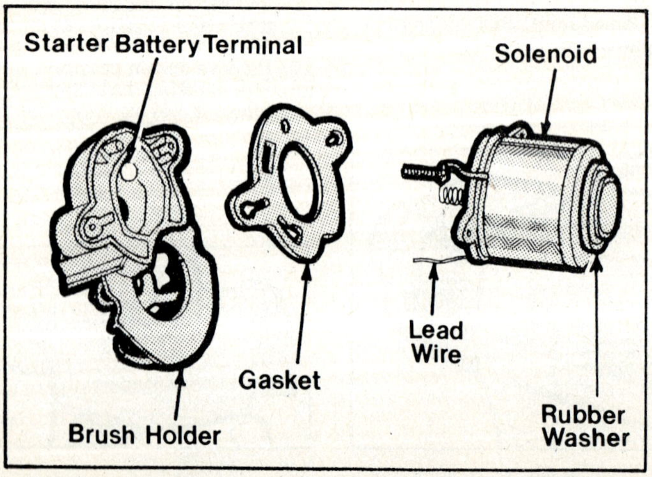

Fig. 29 Solenoid to brush holder plate assembly, all units (Typical)

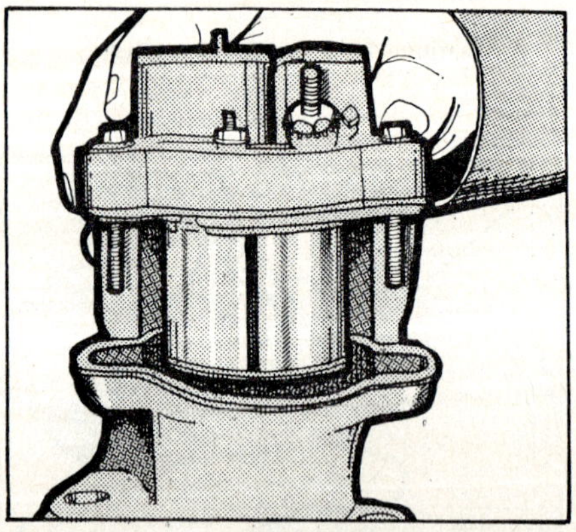

Fig. 30 Solenoid & brush installation, all units (Typical)

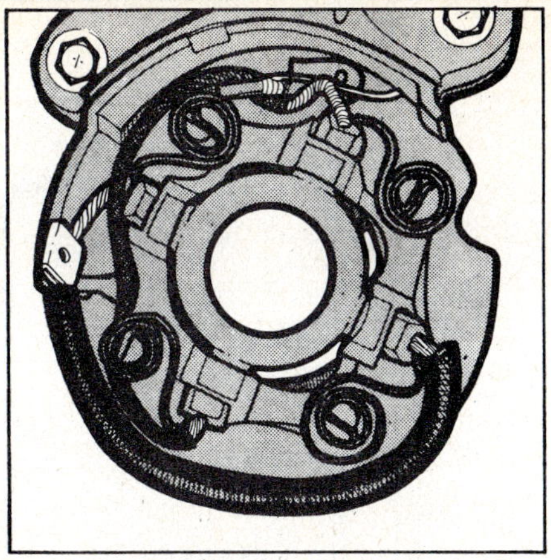

Fig. 31 Installation of brushes & armature thrust washer, all units (Typical)

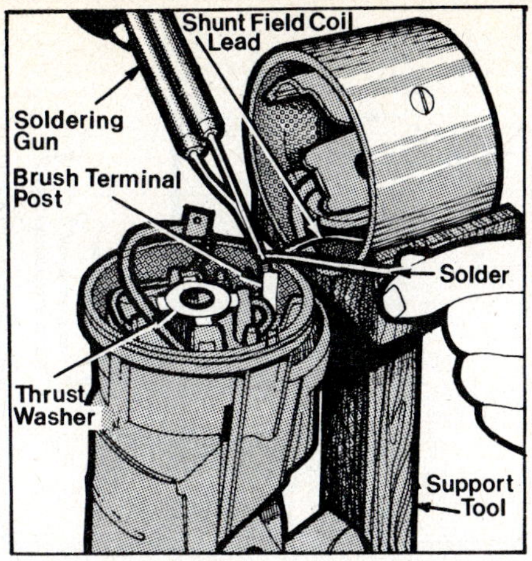

Fig. 32 Soldering shunt coil lead wire, 1970–73 units

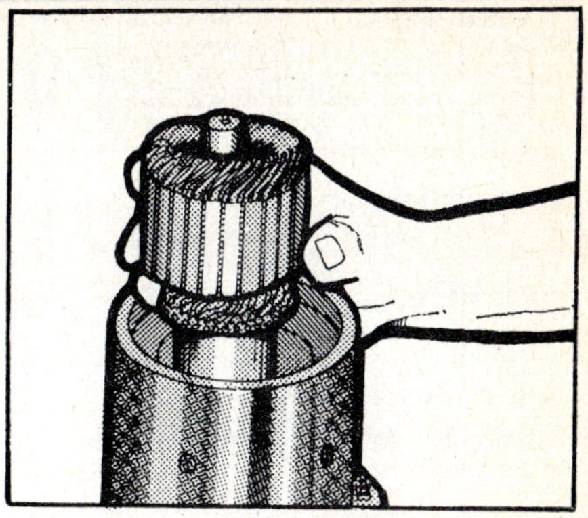

Fig. 33 Armature installation, all units (Typical)

12. Wrap solenoid lead wire around brush terminal post, Fig. 17, and solder with a high temperature resin core solder and resin flux.
13. Install brush holder attaching screws.
14. Install solenoid coil and brush plate assembly into gear housing bore and position brush plate assembly into starter gear housing, Fig. 30. Then install and tighten housing attaching nuts.
15. Install brushes with armature thrust washer, Fig. 31. This holds brushes out and facilitates proper armature installation.
16. On 1973 units, solder shunt coil lead wire to starter brush terminal, Fig. 32.
17. On all units, install brush terminal screws, Figs. 15 and 16.
18. Position field frame on gear housing and install armature into field frame and starter gear housing, Fig. 33, carefully engaging splines of shaft with reduction gear by rotating armature slightly to engage splines.
19. Install thrust washer on armature shaft. On 1973 units, install steel washer on armature shaft.
20. On all units, install starter end head assembly, then the through bolts.

DELCO-REMY STANDARD DUTY ENCLOSED SHIFT LEVER TYPE MOTOR

Except 1978–83 V8-350 Diesel

This type of starting motor, Fig. 34, has the solenoid shift lever mechanism and the solenoid plunger enclosed in the drive housing, thus protecting them from exposure to road dirt, icing conditions and splash. They have an extruded field frame and an overrunning clutch type of drive. The overrunning clutch is operated by a solenoid switch mounted to a flange on the drive housing.

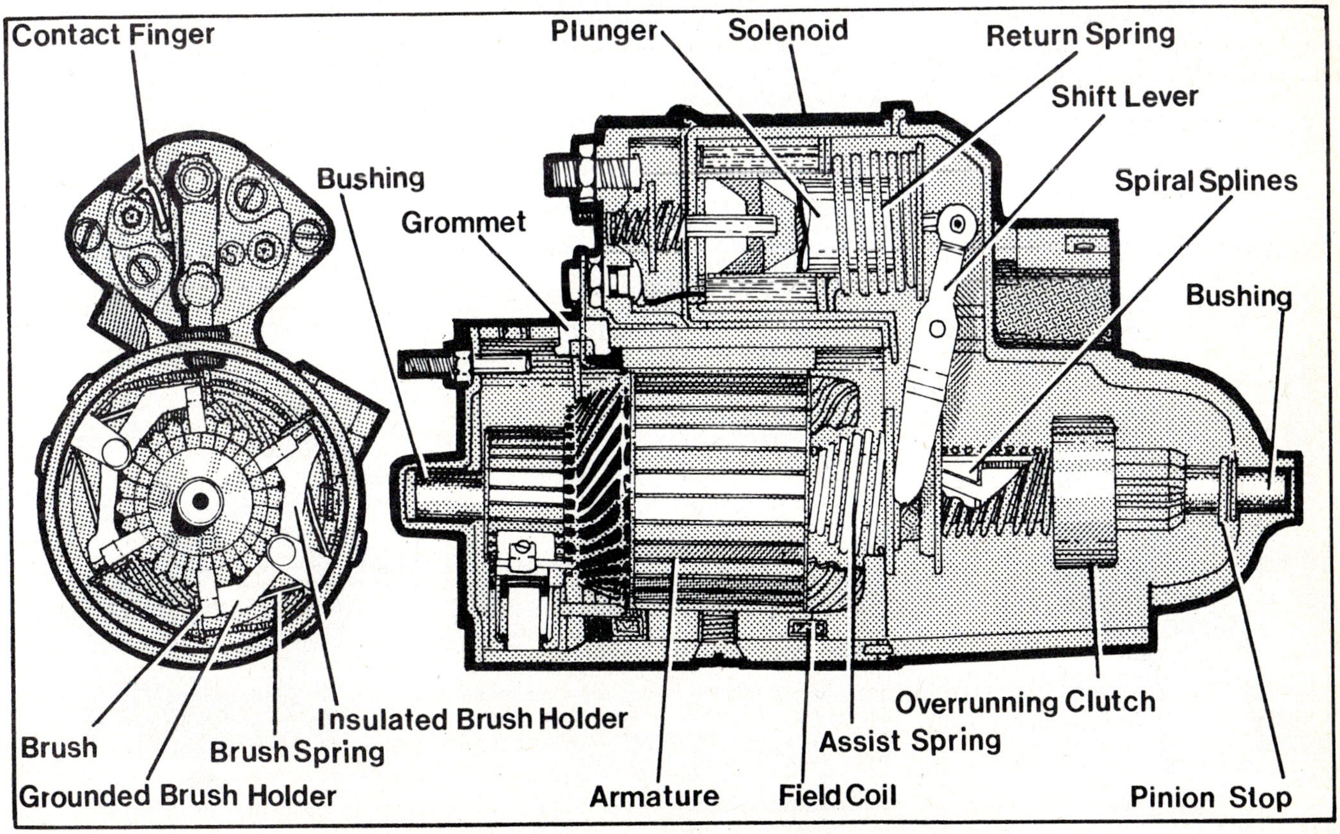

Fig. 34 Delco-Remy Standard Duty starter with enclosed shift lever (Typical)

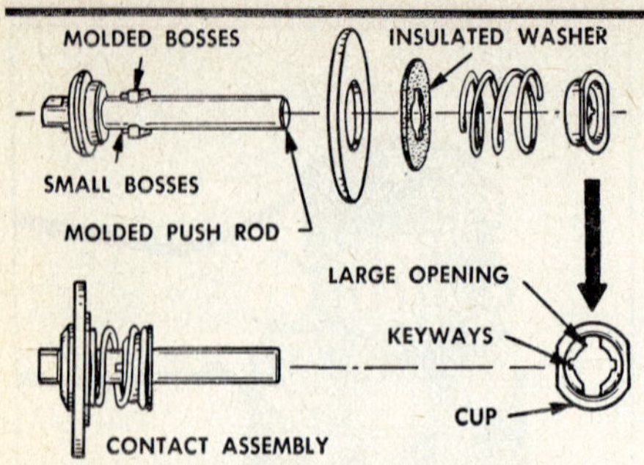

Fig. 35 Solenoid contact assembly

Solenoid

The solenoid is attached to the drive end housing by two screws. The angle of the nose of the plunger provides a greater bearing area between the plunger and core tube. A molded push rod, Fig. 35, is assembled in the contact assembly. A shoulder molded on the push rod and a cup that can easily be assembled to the rod and locked into position over two molded bosses holds the contact assembly in place.

To disassemble the cup from the push rod, push in on the metal cup and rotate 1/4 turn so the molded bosses on the rod are in line with openings in the cup, then slide the metal cup off the rod.

To assemble the metal cup on the rod, locate the parts on the rod as shown and align the large openings in the cup with the molded bosses on the rod): then push in on the cup and rotate it 1/4 turn so the small bosses on the rod fall into the keyways of the cup.

Solenoid Terminals

The terminals of the solenoid are assembled in a molded cover. Some solenoids have an additional small terminal which is identified with the letter "R". To this terminal is attached a small metal finger which makes contact with a disc inside the solenoid when it is energized. On the vehicle, this terminal is connected to the battery side of the ignition coil. The purpose of this is to short out the ignition resistor during cranking and thereby provide high ignition coil output for starting the engine.

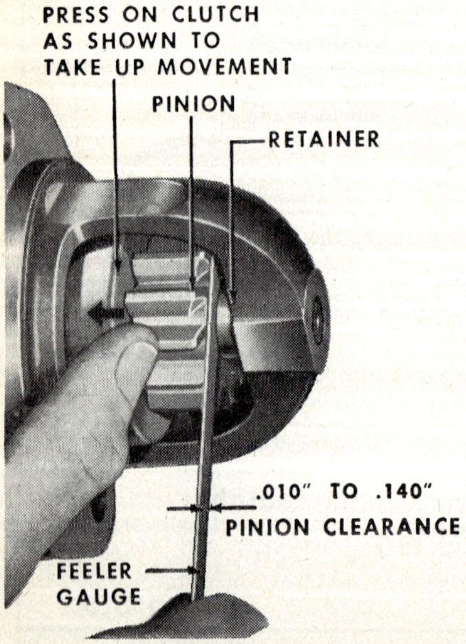

Fig. 38 Checking pinion clearance

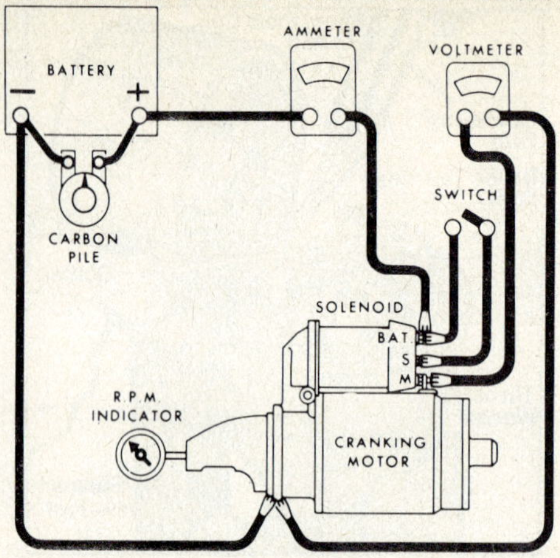

Fig. 36 Connections for checking free speed of motor

Maintenance

Most motors of this type have graphite and oil impregnated bronze bearings which ordinarily require no added lubrication except at time of overhaul when a few drops of light engine oil should be placed on each bearing before reassembly.

Motors provided with hinge cap oilers should have 8–10 drops of light engine oil every 5000 miles, or every 300 hours of operation. Since the motor and brushes cannot be inspected without disassembling the unit, there is no service that can be performed with the unit assembled on the vehicle.

Starting Motor Test

Free Speed Check—With the circuit connected as shown in Fig. 36, use a tachometer to measure armature revolutions per minute. Failure of the motor to perform to specifications may be due to tight or dry bearings, or high resistance connections.

Pinion Clearance—There is no provision for adjusting pinion clearance on this type motor. When the shift lever mechanism is correctly assembled, the pinion clearance should fall within the limits of .010 to .140". When the clearance is not within these limits, it may indicate excessive wear of the solenoid linkage or shift lever yoke buttons.

Pinion clearance should be checked after the motor has been disassembled and reassembled. To check, disconnect motor field coil connector from solenoid terminal and insulate end. Connect one battery lead to solenoid switch terminal and the other lead to the solenoid frame, Fig. 37. Using a jumper lead connected to the solenoid motor terminal, momentarily flash the lead to the solenoid frame. This will shift the pinion into the cranking position until the battery is disconnected.

After energizing the solenoid with the clutch shifted toward the pinion stop retainer, push the pinion back toward the commutator end as far as possible to take up any slack movement; then check the clearance with feeler gauge, Fig. 38.

Disassembling Motor

NOTE: The 5 MT series starting motor will be used on some 1978–82 models, Fig. 39. On this type starting motor, the field coils and pole shoes are permanently bonded to the motor frame. The frame and field coils must be replaced as an assembly.

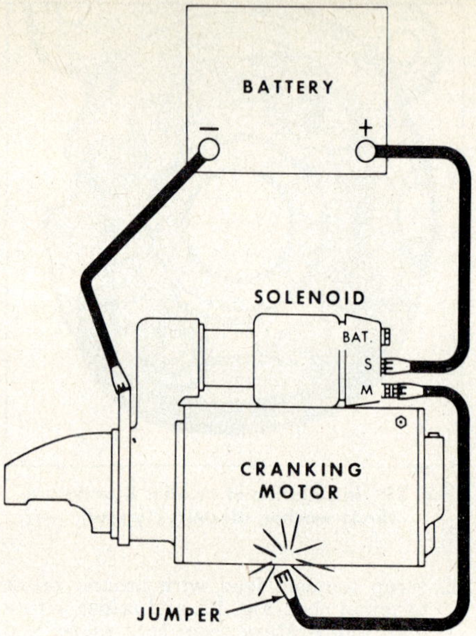

Fig. 37 Connections for checking pinion clearance

Normally the motor should be disassembled only so far as necessary to repair or replace defective parts.

1. Disconnect field coil connectors from solenoid "motor" terminal.
2. Remove thru bolts.
3. Remove commutator end frame and field frame assembly.
4. Remove armature assembly from drive housing. On some models it may be necessary to remove solenoid and shift lever assembly from the drive housing before removing the armature assembly.

IMPORTANT: *When solenoid is installed, apply sealing compound between field frame and solenoid flange* (Fig. 40).

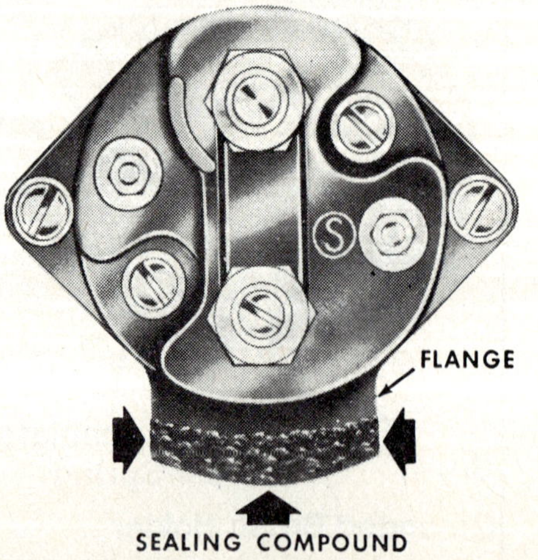

Fig. 40 Sealing solenoid housing to frame

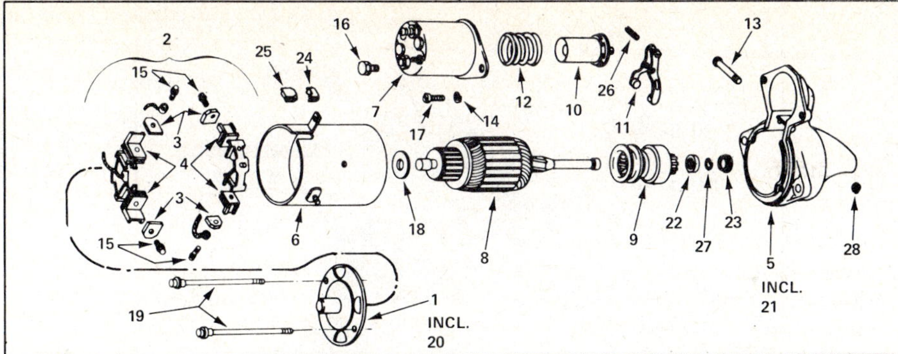

		19. THRU BOLT
1. FRAME–COMMUTATOR END	10. PLUNGER	20. BUSHING–COMMUTATOR END
2. BRUSH AND HOLDER PKG.	11. SHIFT LEVER	21. BUSHING–DRIVE END
3. BRUSH	12. PLUNGER RETURN SPRINGER	22. PINION STOP COLLAR
4. BRUSH HOLDER	13. SHIFT LEVER SHAFT	23. THRUST COLLAR
5. HOUSING–DRIVE END	14. LOCK WASHER	24. GROMMET
6. FRAME AND FIELD ASM.	15. SCREW–BRUSH ATTACHING	25. GROMMET
7. SOLENOID SWITCH	16. SCREW–FIELD LEAD TO SWITCH	26. PLUNGER PIN
8. ARMATURE	17. SCREW–SWITCH ATTACHING	27. PINION STOP RETAINER RING
9. DRIVE ASM.	18. WASHER–BRAKE	28. LEVER SHAFT RETAINING RING

Fig. 39 Disassembled view of Delco-Remy 5MT series starting motor

5. Remove overrunning clutch from armature shaft as follows:
   a. Slide thrust collar off end of armature shaft, Fig. 41.
   b. Slide a deep socket or pipe coupling of suitable size over shaft so end of coupling or cylinder butts against edge of retainer. Tap end of coupling with hammer, driving retainer toward armature and off snap ring, Fig 42.
   c. Remove snap ring from groove in shaft. If snap ring is too badly distorted during removal, use a new one when reassembling the clutch.
   d. Slide retainer, clutch and assist spring from armature shaft.

Reassembling Motor, Figs. 39 & 45

1. Lubricate drive end and splines of armature shaft with SAE 10 oil. *If heavier oil is used it may cause failure to mesh at low temperatures.*
2. Place "assist" spring on drive end of shaft next to armature, with small end against lamination stack.
3. Slide clutch assembly onto armature shaft with pinion outward.
4. Slide retainer onto shaft with cupped surface facing end of shaft.
5. Stand armature on end on wood surface with commutator down. Position snap ring on upper end of shaft and hold in place with a block of wood. Hit wood block with a hammer forcing snap ring over end of shaft, Fig. 43. Slide snap ring into groove, squeezing it to ensure a good fit in groove.
6. Assemble thrust collar on shaft with shoulder next to snap ring.
7. Position retainer and thrust collar next to snap ring. With clutch pressed against assist spring, for clearance next to retainer, use two pairs of pliers at the same time (one pair on either side of shaft) to grip retainer and thrust collar. Then squeeze until snap ring is forced into retainer, Fig. 44.
8. Place 4 or 5 drops of SAE 10 oil in drive housing bushing. Make sure thrust collar is in place against snap ring and retainer; then slide armature and clutch assembly into place in drive housing.
9. Attach solenoid and shift lever assembly to drive housing. Be sure lever buttons are located between sides of clutch collar.
10. Position field frame over armature, *applying sealing compound between frame and solenoid flange, Fig. 40.* Position frame against drive housing, using care to prevent damage to brushes.
11. Place 4 or 5 drops of SAE 10 oil in bushing in commutator end frame. Make sure leather brake washer is on armature

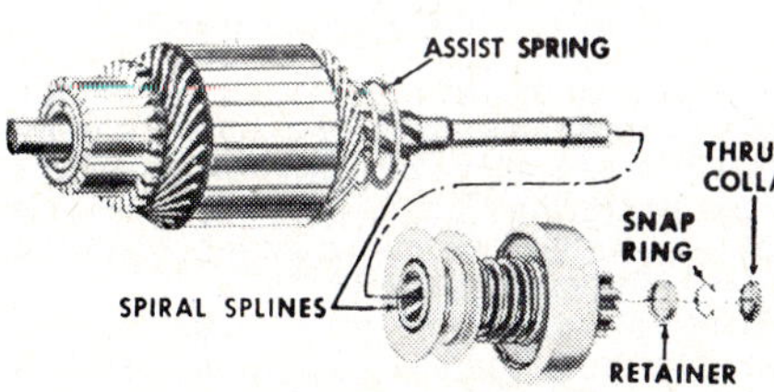

Fig. 41 View of armature and overrunning clutch

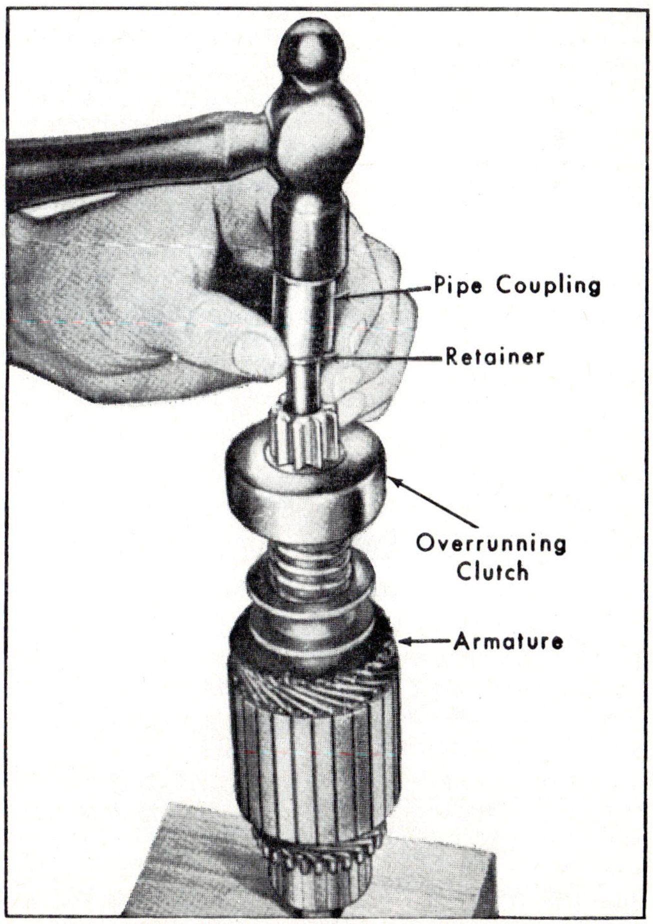

Fig. 42 Removing over-running clutch snap ring retainer

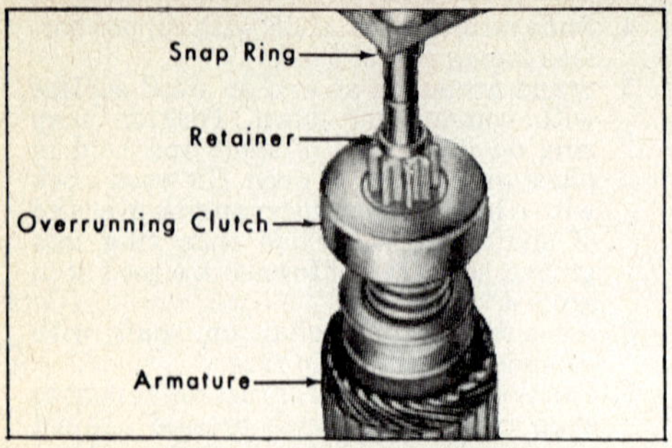

Fig. 43 Installing snap ring onto armature shaft

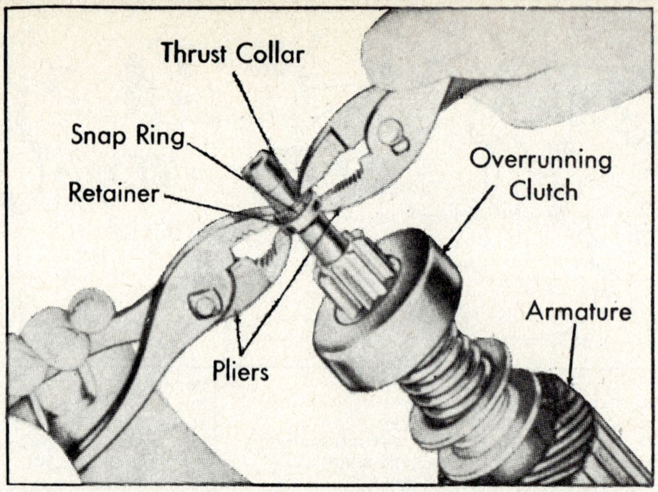

Fig. 44 Installing snap ring into retainer

shaft; then slide commutator end frame onto shaft.
12. Install thru bolts and tighten securely.
13. Reconnect field coil connectors to solenoid "motor" terminal.

Cleaning

The overrunning clutch, armature and fields should not be cleaned in any degreasing tank, or with grease dissolving solvents, since these would dissolve the lubricant in the clutch mechanism and would damage the insulation in the armature and field coils. It is suggested that all parts, except the clutch, be cleaned with oleum spirits and a brush. The clutch can be wiped with a clean cloth.

If the commutator is dirty it may be cleaned with No. 00 sandpaper; *never use emery cloth for this purpose.*

1978–83 V8-350 Diesel

Disassemble

1. Remove screw from field coil connector and solenoid mounting screws, then rotate solenoid 90° and remove solenoid with plunger return spring, Fig. 46.
2. Remove two through bolts, then commutator end frame with washer.
3. Remove end frame assembly from drive gear housing.
4. Remove shift lever pivot bolt.
5. Remove center bearing screws, then remove drive gear housing from armature shaft. Shift lever and plunger assembly, should fall from starter clutch.
6. Remove thrust washer or collar from armature shaft.
7. Position a 5/8 in. deep socket over shaft against retainer, then tap socket to move retainer off snap ring.
8. Remove snap ring from groove in shaft, then remove retainer, clutch assembly, fiber washer and center bearing from armature shaft.
9. Remove roll pin and separate shift lever and plunger.
10. Remove brush pivot pin and brush spring, then replace brushes as necessary.

Assemble

1. Lubricate drive end of armature with lubricant 1960954 or equivalent.
2. Install center bearing with bearing facing toward armature winding, then install fiber washer on armature shaft, Fig. 49.
3. Position clutch assembly on armature shaft with pinion facing away from armature.
4. Position retainer on armature shaft with cupped side facing end of shaft.
5. Install snap ring into groove on armature shaft, then install thrust washer.
6. Using two pliers, grip retainer and thrust washer or collar and squeeze until snap ring is forced into retainer and is held securely in groove on armature shaft.
7. Lubricate drive end housing bushing with lubricant 1960954 or equivalent.
8. Engage shift lever yoke with clutch and slide complete assembly into drive gear housing.
9. Install center bearing attaching screws and shift lever pivot bolt.
10. Install solenoid on drive gear housing.
11. Apply sealer No. 1050026 or equivalent to solenoid flange where field frame contacts solenoid.
12. Position field frame against drive gear housing on alignment pin, using care to prevent damage to brushes.
13. Lubricate commutator end frame bushing with lubricant 1960954 or equivalent.
14. Install washer on armature shaft and slide end frame onto shaft, then install and tighten through bolts.
15. Connect field coil connector to solenoid terminal, then check pinion clearance.

Pinion Clearance Check

1. Disconnect field coil connector from solenoid terminal and insulate carefully.
2. Connect one 12 volt battery lead to solenoid switch terminal and other lead to starter motor frame.
3. Flash a jumper lead momentarily from solenoid motor terminal to starter motor frame. This will shift pinion into cranking position until battery is disconnected.
4. Push pinion back as far as possible to take up any movement, then check pinion clearance using a feeler gauge. Pinion clearance should be .010–.140 in. If clearance is not within limits, check for improper installation or worn parts and replace as necessary.

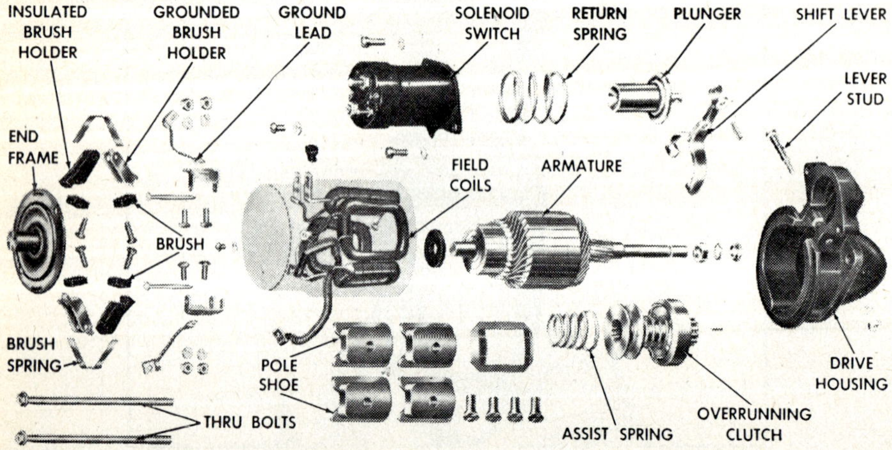

Fig. 45 Disassembled view of 10MT series starting motor (Typical)

DELCO-REMY HEAVY DUTY ENCLOSED SHIFT LEVER TYPE STARTER

In this type motor, Figs. 47, 48 and 49, the nose housing can be rotated to obtain a number of different solenoid positions with respect to the mounting flange, which is a feature that makes these motors adaptable to a wide variety of mounting applications.

Positive lubrication is provided to the bronze bushing located in the commutator end frame, in the lever housing, and in the nose housing by an oil saturated wick that projects through each bushing and contacts the arma-

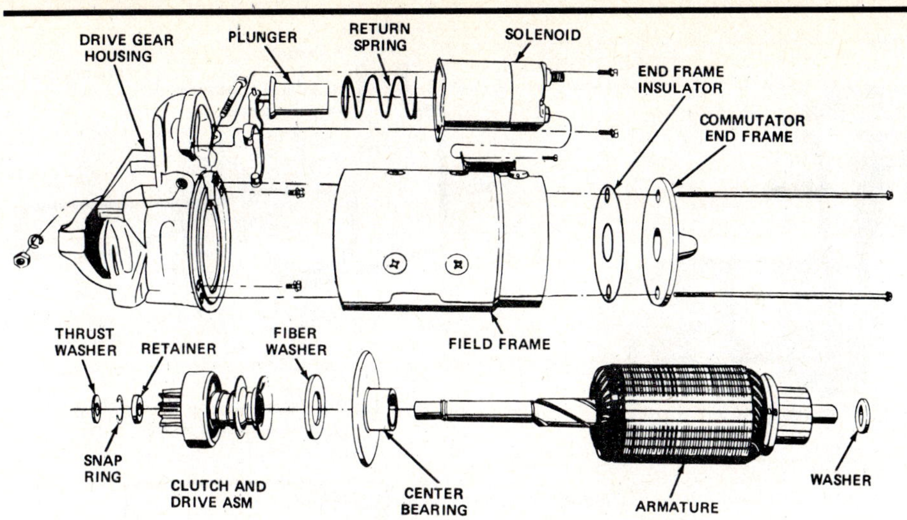

Fig. 46 Disassembled view of Delco-Remy 25MT series starting motor

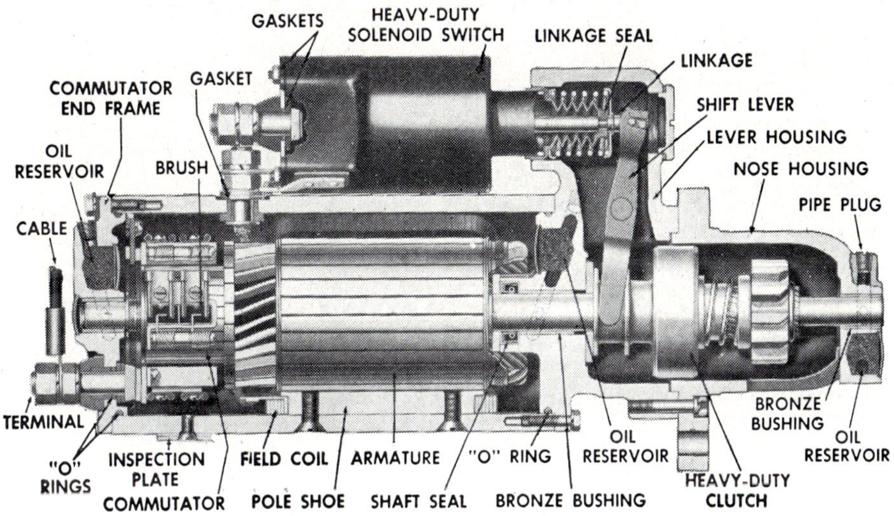

Fig. 47 Delco-Remy heavy duty enclosed shift lever type motor

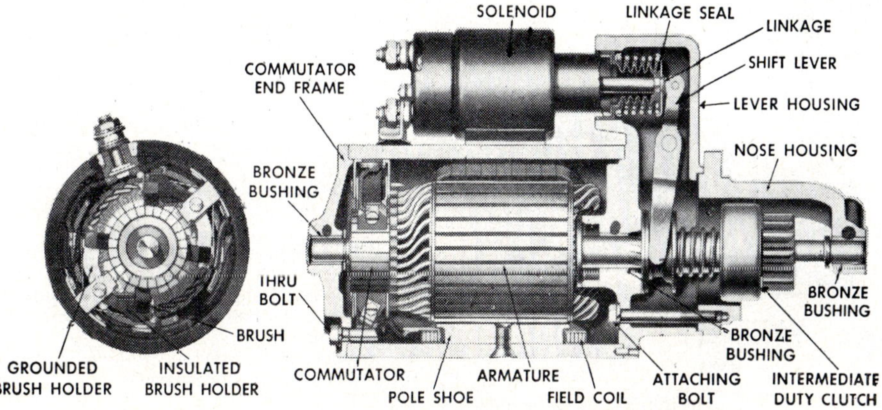

Fig. 48 Delco-Remy intermediate duty enclosed shift lever type motor (with through bolts)

ture shaft. Oil can be added to each wick by removing a pipe plug which is accessible on the outside of the motor.

Available as an optional feature is a waste-filled oil reservoir for each wick which makes available a larger supply of oil thereby extending the time required between lubrication periods. Another optional feature is "O" rings which can be added to resist entry of dirt and moisture into the entire motor assembly.

Many models feature a seal between the shaft and lever housing, and all models have a rubber boot or linkage seal over the solenoid plunger. The seal and the boot, when used together, prevent entry of transmission oil into the motor main frame and solenoid case, allowing the motor to be used on wet clutch applications.

Overrunning Clutches

As shown in Figs. 47, 48 and 49, two kinds of clutches, heavy duty or intermediate duty overrunning type, may be used with these motors. The clutches may be either the sprag type or the four-roll type. Both types are moved into mesh with the ring gear by the action of the solenoid. Once engaged, the clutch will not disengage during intermittent engine firing, which prevents damage to pinion and ring gear teeth. The pinion remains engaged until starting is assured and the solenoid circuit is interrupted.

Maintenance

Under normal operating conditions, no maintenance will be required between engine overhaul periods. At time of engine overhaul, motors without oil reservoirs and "O" rings, and "long life" motors incorporating "O" rings and oil reservoirs should be disassembled, inspected, cleaned and tested.

Adjustable Nose Housing

Two methods are employed to attach the nose housing to the lever housing. As shown in Fig. 47, one method attaches the nose housing to the lever housing by six bolts located around the outside of the housing. To relocate the housing, it is only necessary to remove the bolts, rotate the housing to the desired position, reinstall the bolts and torque them to 13–17 ft. lb. during reassembly.

In the second method, Fig. 48, the lever housing and commutator end frame are held to the field frame by thru bolts extending from the commutator end frame to threaded holes in the lever housing. The nose housing is held to the lever housing by internal attaching bolts. With this arrangement, it is necessary to partially disassemble the motor to provide access to the attaching bolts when relocating the nose housing.

To accomplish this, remove the electrical connector and the screws attaching the solenoid assembly to the field frame and then remove the thru bolts from the commutator end frame. Separate the field frame from the remaining assembly, and pull the armature away from the lever housing until the pinion stop rests against the clutch pinion. This will clear the nose housing attaching bolts so they can be removed, permitting relocation of the nose housing. During reassembly, torque the nose housing attaching bolts to 11–15 ft. lb.

Starting Motor Tests

On intermediate duty motors with through bolts, make test connections as shown in Figs. 36 and 37. On intermediate motors with-

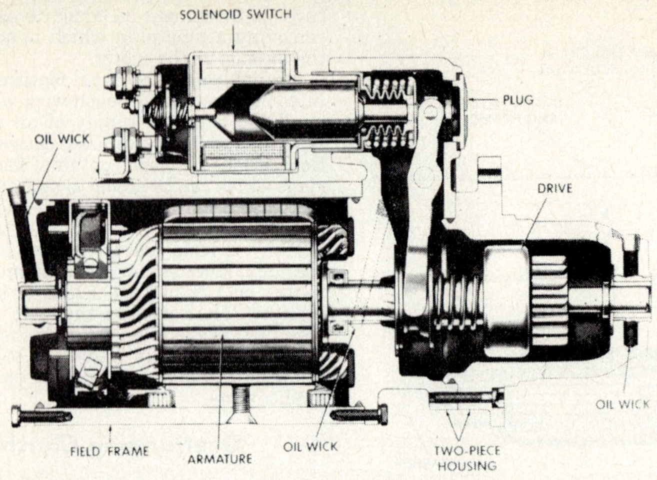

Fig. 49 Delco-Remy intermediate duty enclosed shift lever type motor (without through bolts)

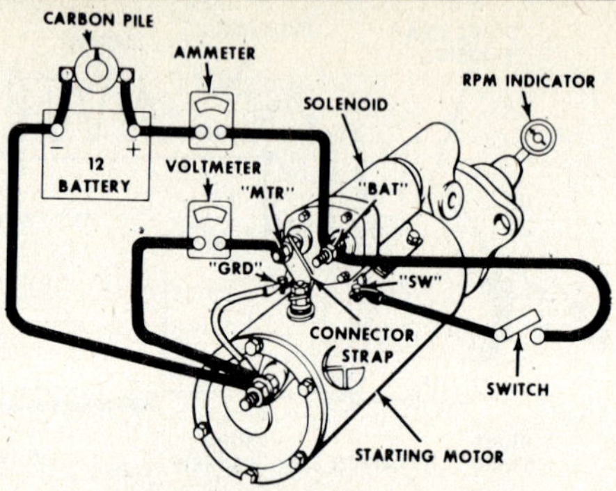

Fig. 50 Connections for checking free speed of motor

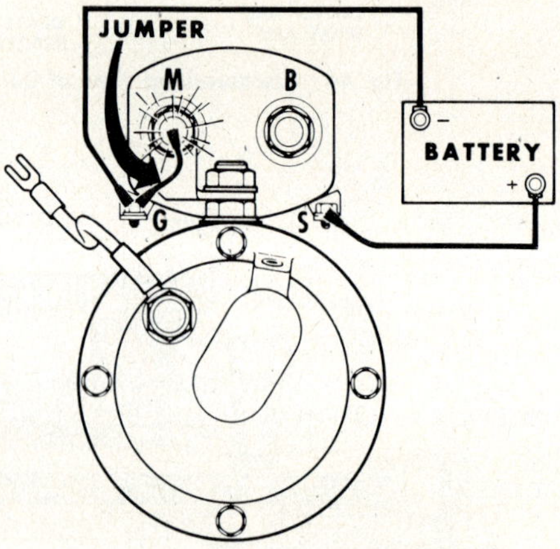

Fig. 51 Connections for checking pinion clearance

out through bolts and heavy duty motors, make test connections shown in Figs. 50 and 51.

Free Speed Test, Figs. 36 and 50—Failure of the motor to perform to specifications may be due to tight or dry bearings, or high resistance connections.

Pinion Clearance: There are no provisions for adjusting pinion clearance on motors using the intermediate duty clutch. However, this should be checked after reassembly of motor to make sure the clearance is within specifications. On intermediate duty motors with through bolts, make test connections as shown in Fig. 37. On intermediate duty motors without through bolts, make test connections shown in Fig. 51. After energizing the solenoid with the clutch shifted toward the pinion stop retainer, push the pinion back toward the commutator end as far as possible to take up any slack movement, then check the clearance between pinion and pinion stop retainer with a feeler gauge, Fig. 52. If clearance is not within specified limits, check the solenoid linkage and shift lever yoke buttons for excessive wear.

The pinion clearance for motors using the heavy duty clutch is checked in the same manner as for intermediate duty clutches. The clearance for these motors, however, is measured as indicated in Fig. 53. The clearance is adjustable and is accomplished by removing the plug on the lever housing and turning the nut on the plunger rod inside the housing. Turn the nut clockwise to decrease clearance, and vice versa.

Disassembling Motor

Note the relative position of the solenoid, lever housing and nose housing so the motor can be reassembled in the same manner.

INTERMEDIATE DUTY MOTOR WITH THROUGH BOLTS

1979–83 Units

1. Remove through bolts, then remove end frame.
2. Remove spacer from commutator end of armature shaft, then separate field frame from lever housing assembly. Use care not to damage solenoid boot.
3. Disconnect field coil connector from solenoid motor terminal, then remove solenoid to end frame attaching screws and remove solenoid.
4. Remove nose housing attaching bolts, then separate nose housing from lever housing.
5. To remove drive assembly from lever housing, proceed as follows:
   a. Using a metal cylinder and a hammer, drive collar away from retainer ring, then remove retainer ring from armature shaft.
   b. Remove armature shaft from lever housing and separate driver assembly from shift lever fingers.
6. Remove solenoid plunger from lever housing:
   a. Remove expansion plugs from housing by piercing the plugs and prying out, Fig. 54.
   b. Using needle nose pliers, remove lever pin retainer clip, then remove pin and solenoid plunger from housing.
7. Disassemble shift lever from housing by piercing one of the small expansion plugs in lever housing. Drive shift lever pin out of housing, then remove shift lever from housing.
8. Inspect brake washer in lever housing and replace as necessary.

1974–78 Units

1. Disconnect field coil connector from solenoid motor terminal, and remove solenoid

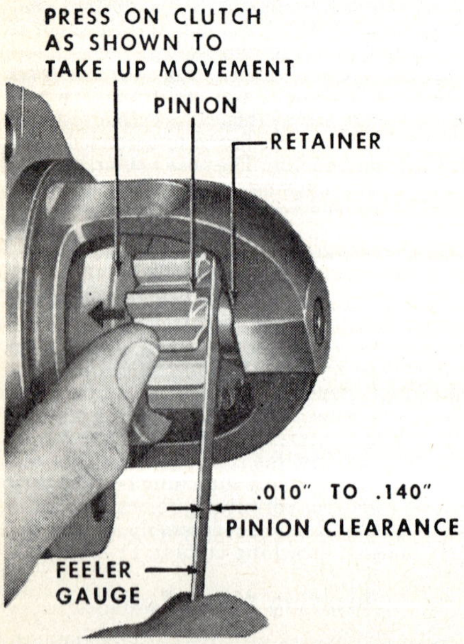

Fig. 52 Checking pinion clearance on intermediate duty motor

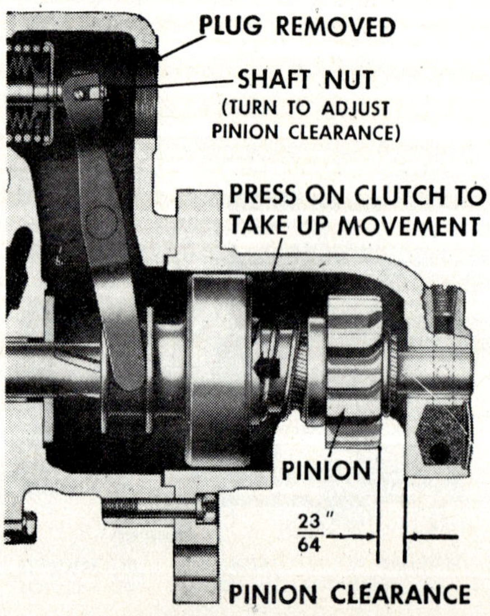

Fig. 53 Checking pinion clearance on heavy duty motor

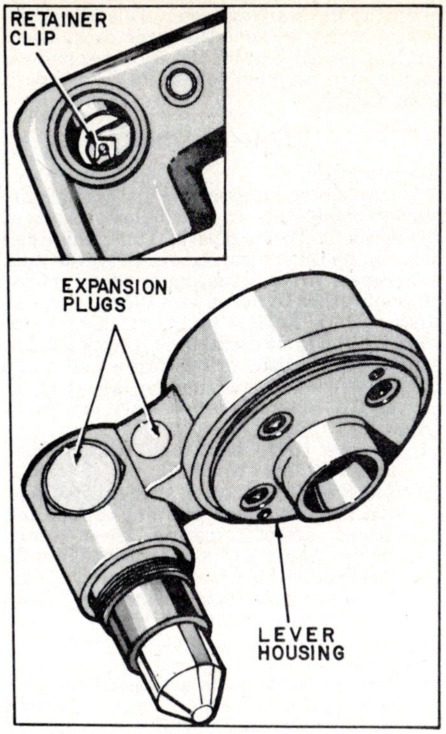

Fig. 54 Lever housing expansion plugs & shift lever retainer. Intermediate duty motor with through bolts

mounting screws.
2. Remove thru bolts.
3. Remove commutator end frame from field frame and field frame from lever housing.
4. Remove nose housing attaching bolts and separate nose housing from lever housing.
5. Slide a deep socket or pipe coupling of suitable size over shaft so end of coupling or cylinder butts against edge of retainer. Tap end of coupling with a hammer, driving retainer toward armature and off snap ring.
6. Remove snap ring from groove in shaft, using pliers or other suitable tool. If snap ring is too badly distorted during removal, use a new one when reassembling clutch.
7. Remove armature and clutch from lever housing.
8. Separate solenoid from lever housing.

INTERMEDIATE DUTY MOTORS WITHOUT THROUGH BOLTS

1. Remove screws attaching to nose housing to lever housing, then separate nose housing from motor.
2. Remove solenoid access plug from lever housing.
3. Remove bolts attaching lever housing to field frame.
4. Disconnect field coil lead from solenoid motor terminal.
5. Remove solenoid to frame attaching bolts.
6. Remove end frame and spacer from armature shaft.
7. Separate lever housing assembly from field frame.
8. Remove solenoid from lever housing. Use care not to damage solenoid boot.
9. Remove adjustment nut from solenoid plunger shaft.

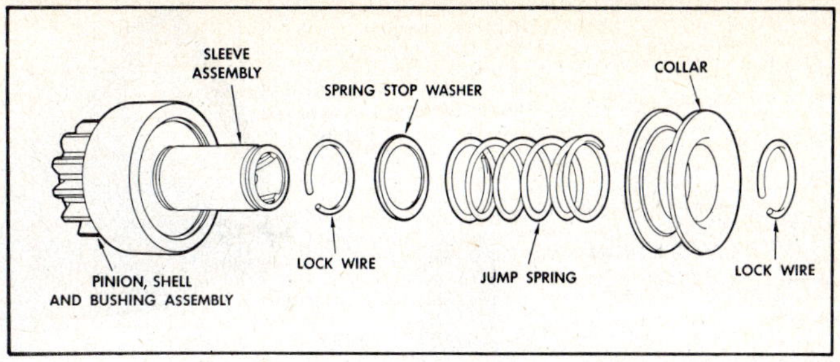

Fig. 55 Disassembled view of early type intermediate duty sprag clutch drive

10. Remove armature and drive assembly as follows:
    a. Using a metal cylinder and a hammer, drive stop collar away from retainer ring, then remove retainer ring from armature shaft.
    b. Remove armature shaft from lever housing and separate drive assembly from shift lever fingers.
11. Remove snap ring from shift lever retainer pin, then drive retainer pin from housing and remove shift lever.
12. Inspect brake washer in lever housing and replace as necessary.

HEAVY DUTY MOTORS

1. Disconnect field coil connector from solenoid motor terminal and lead from solenoid ground terminal.
2. On motors having brush inspection plates, remove plates and then brush lead screws. This will disconnect the field leads from the brush holders.
3. Remove attaching bolts and separate commutator end frame from field frame.
4. Separate nose housing and field frame from lever housing by removing attaching bolts.
5. Remove armature and clutch assembly from lever housing.
6. Separate solenoid from lever housing by pulling apart.

Intermediate Duty Sprag Clutch

1. Remove lock wire and collar from sleeve, Fig. 55 and 56.
2. Remove jump spring, spring stop washer and second lock wire.
3. Remove retainer ring and large washers. Do not remove sleeve or sprags from shell.
4. Lubricate sprags, Fig. 57, and saturate the felt washer with SAE 5W-20 oil. *Heavier oil must not be used.*
5. Reassemble washers and retaining ring.
6. Assemble lock wire in groove next to sprag; then assemble stop washer.
7. Assemble jump spring, collar and other lock wire.
8. Lubricate armature shaft with a few drops of SAE 10 oil before sprag clutch is reinstalled.

Heavy Duty Sprag Clutch

1. Remove cupped pinion stop and split washer, Figs. 58, 59 and 60. In removing cupped pinion stop, it will probably be damaged to the extent that a new one will be required on reassembly.
2. Remove pinion, spring and retainer cups.
3. Remove retainer ring and washers. *Do not remove collar sleeve and shell, or sprags from shell.*
4. Lubricate sprags and saturate felt washer with SAE 5W-20 oil—no heavier.
5. Reassemble washers and retainer ring.
6. Lubricate spiral spline on collar sleeve and shell with SAE 10 oil.
7. Assemble retainer cups and spring.
8. Assemble pinion with collar side toward spring, Figs. 58, 59 and 60.
9. Install new cupped pinion stop and split washer as shown. Stake cupped pinion stop in about 10 places with a round-nosed tool.

Cleaning

Follow cleaning instructions outlined for standard duty motors.

Reassembling Motor

The reassembly procedure for each type of motor is the reverse of disassembly. On motors using a snap ring and retainer on the shaft as a pinion stop, the ring and retainer can be assembled as follows:

With the retainer placed over the shaft with the cupped surface facing the end of the shaft, force the ring over the shaft with a light hammer blow and then slide the ring down into the groove. To force the retainer over the snap ring, place a suitable washer over the shaft and squeeze with pliers. Remove the washer, Fig. 61.

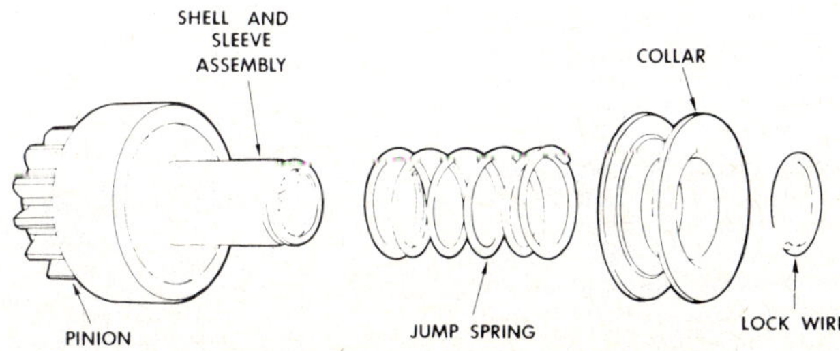

Fig. 56 Disassembled view of late type intermediate duty sprag clutch drive

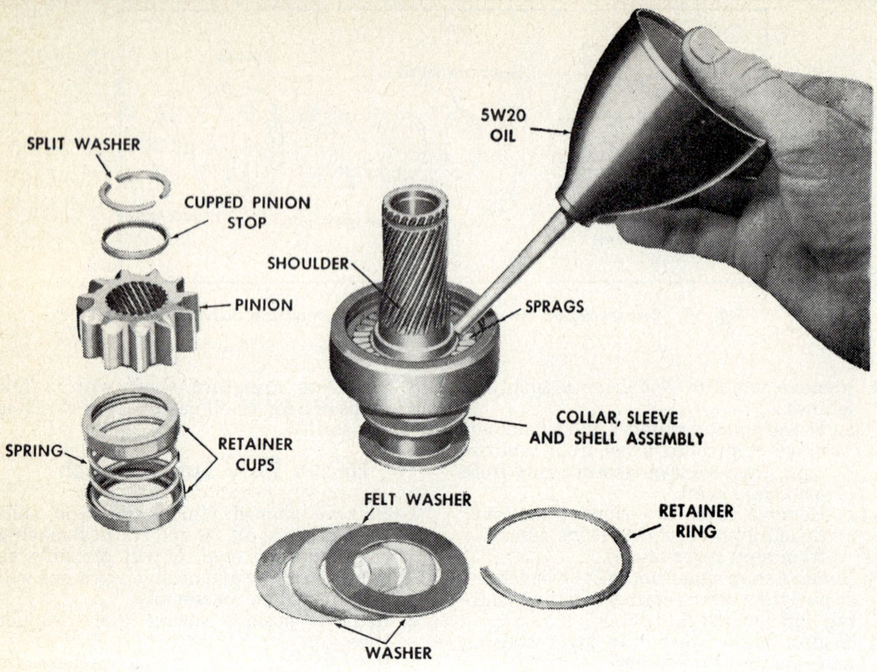

Fig. 57 Lubricating clutch sprags

To reassemble the end frame having eight brushes onto the field frame, pull the armature out of the field frame just far enough to permit the brushes to be placed over the commutator, Fig. 62. Then push the commutator end frame and the armature back against the field frame.

Lubrication

All wicks and reservoirs should be saturated with SAE 10 oil, and the splines underneath the clutch should be lubricated with a light coat of SAE 10 oil. *Heavier oil may cause failure to mesh at low temperatures.*

FORD STARTER WITH INTEGRAL POSITIVE ENGAGEMENT DRIVE

This type starting motor, Figs. 63 and 64 is a four pole, series parallel unit with a positive engagement drive built into the starter. The drive mechanism is engaged with the flywheel by lever action before the motor is energized.

When the ignition switch is turned on to the start position, the starter relay is energized and supplies current to the motor. The current flows through one field coil and a set of contact points to ground. The magnetic field given off by the field coil pulls the movable pole, which is part of the lever, downward to its seat. When the pole is pulled down, the lever moves the drive assembly into the engine flywheel, Fig. 65.

When the movable pole is seated, it functions as a normal field pole and opens the contact points. With the points open, current flows through the starter field coils, energizing the starter. At the same time, current also flows through a holding coil to hold the movable pole in its seated position.

When the ignition switch is released from the start position, the starter relay opens the circuit to the starting motor. This allows the return spring to force the lever back, disengaging the drive from the flywheel and returning the movable pole to its normal position, Fig. 66.

1974–77 Units

Disassembly

It may not be necessary to disassemble the starter completely to accomplish repair or replacement of certain parts. Thus, before disassembling the motor, remove the cover band and starter drive actuating lever cover. Examine brushes to make sure they are free in their holders. Replace brushes if defective or worn beyond their useful limit. Check the tension of each brush spring with a pull scale. Spring tension should not be less than 45 ounces. If disassembly is necessary, proceed as follows:

1. Remove cover band and starter drive actuating lever cover.
2. Remove through bolts, starter drive gear housing, drive gear retaining clip cup and starter drive actuating lever return spring. Some units incorporate needle bearings at the drive end. If needle bearings need not be replaced, insert a dummy shaft into the housing while removing the armature. This will prevent loss of bearings.
3. Remove pivot pin retaining starter gear actuating lever and remove lever and

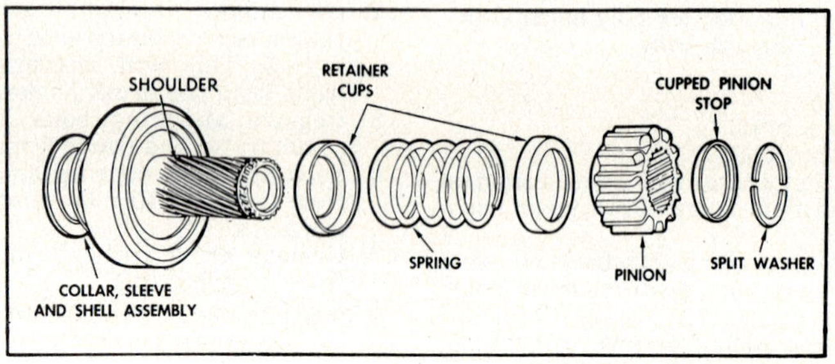

Fig. 58 Disassembled view of early type heavy duty sprag clutch drive

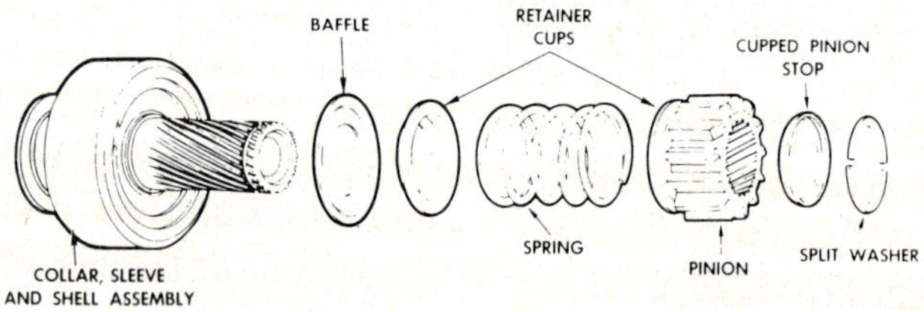

Fig. 59 Disassembled view of late type heavy duty sprag clutch drive

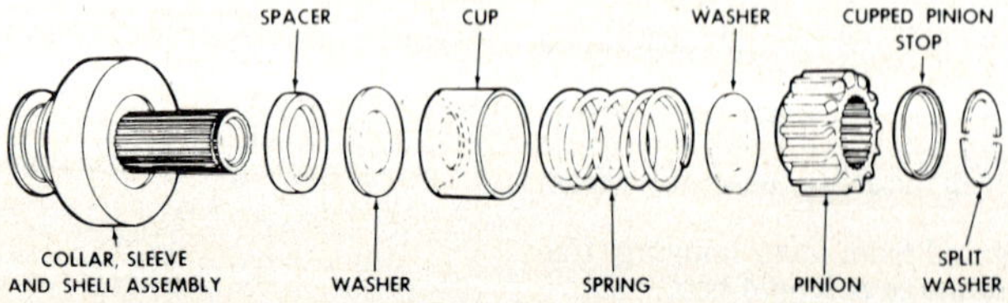

Fig. 60 Disassembled view of DR-250 clutch drive

armature.
4. Remove and discard spring clip retaining starter drive gear to end of armature shaft, and remove drive gear.
5. Remove commutator brushes from holders and remove brush end plate.
6. Remove two screws retaining ground brushes to frame, Fig. 67.
7. On the field coil that operates the drive actuating lever, bend tab up on field retainer and remove retainer.
8. Remove field coil retainer screws, Fig. 68. Unsolder field coil leads from terminal screw, and remove pole shoes and coils from frame.
9. Remove starter terminal nut and related parts. Remove any excess solder from terminal slot.

Reassembly

1. Install starter terminal, insulator, washers and retaining nut in frame, Fig. 67. Be sure to position slot in screw perpendicular to frame end surface.
2. Install field coils and pole pieces. As pole shoe screws are tightened, strike frame several sharp blows with a soft-faced hammer to seat and align pole shoes, then stake the screws.
3. Install solenoid coil retainer and bend tabs to retain tabs to frame.
4. Solder field coils and solenoid wire to starter terminal, using rosin core solder.
5. Check for continuity and grounds in the assembled coils.
6. Position solenoid coil ground terminal over ground screw hole nearest starter terminal.
7. Position ground brushes to starter frame and install retaining screws, Fig. 67.
8. Position starter brush end plate to frame with end plate boss in frame slot.
9. Install drive gear to armature shaft and install a new retaining spring clip.
10. Position fiber thrust washer on commutator end of armature shaft and install armature in frame.
11. Install starter drive actuating lever to frame and starter drive, and install pivot pin.
12. Position actuating lever return spring and drive gear housing to frame and install through bolts. Do not pinch brush leads between brush plate and frame.
13. Install brushes in holders, being sure to center brush springs on brushes.
14. Position drive gear actuating lever cover on starter and install brush cover band.
15. Check starter for free running and install on engine.

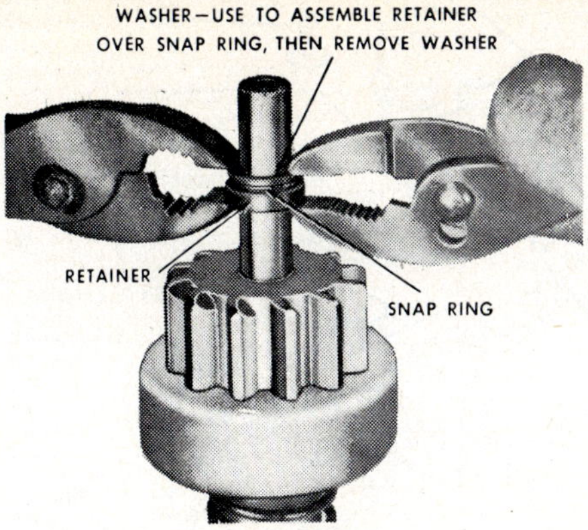

Fig. 61 Forcing retainer over snap ring

Fig. 62 Assembling end frame with eight brushes to field frame

1978–83 Units

Disassemble, Fig. 64

1. Remove cover screw, cover, through bolts, starter drive end housing and starter drive plunger lever return spring.
2. Remove plunger lever retaining pin, then remove lever and armature.
3. Remove stop ring retainer from armature shaft, then remove stop ring and starter drive gear assembly.
4. Remove brush end plate and insulator assembly.
5. Remove brushes from brush holder, then remove brush holder. Note position of brush holder to end terminal.
6. Remove two screws retaining ground brushes to frame.
7. Bend up edges of sleeve inserted in frame, then remove sleeve and retainer.
8. Detach field coil ground wire from copper tab on frame.
9. Using tool No. 10044-A, remove three coil retaining screws, Fig. 68. Cut field coil connection at switch post lead and pole shoes and coils from frame.
10. Cut positive brush leads from field coils, as close to field connection as possible.

Assemble, Fig. 64

1. Position pole pieces and coils in frame, then install retaining screws using tool No. 10044-A, Fig. 68. As pole shoes are tightened, strike frame several times with a soft faced mallet to seat and align pole shoes, then stake screws.
2. Install plunger coil sleeve and retainer, then bend tabs to retain coils to frame.
3. Position grommet on end terminal, then insert terminal and grommet into notch on frame.
4. Solder field coil to starter terminal post strap.

NOTE: Use 300 watt soldering iron and rosin core solder.

5. Check coils for continuity and grounds.
6. Position brushes to starter frame and install retaining screws.
7. Apply a thin coat of Lubriplate 777 or equivalent to armature shaft splines.
8. Install drive gear assembly on armature shaft, then install stop ring and stop ring retainer.
9. Install armature into starter frame.
10. Position starter drive gear plunger to frame and starter drive assembly then install pivot assembly. Fill end housing bearing bore approximately 1/4 full with grease, then position drive end housing to frame.
11. Install brush holder, then insert brushes into holder and install brush springs.

NOTE: Ensure positive brush leads are properly positioned in slots on brush holder.

12. Install brush end plate. Ensure brush end plate insulator is properly positioned.
13. Install through bolts, then install starter drive plunger lever cover and tighten retaining screw.

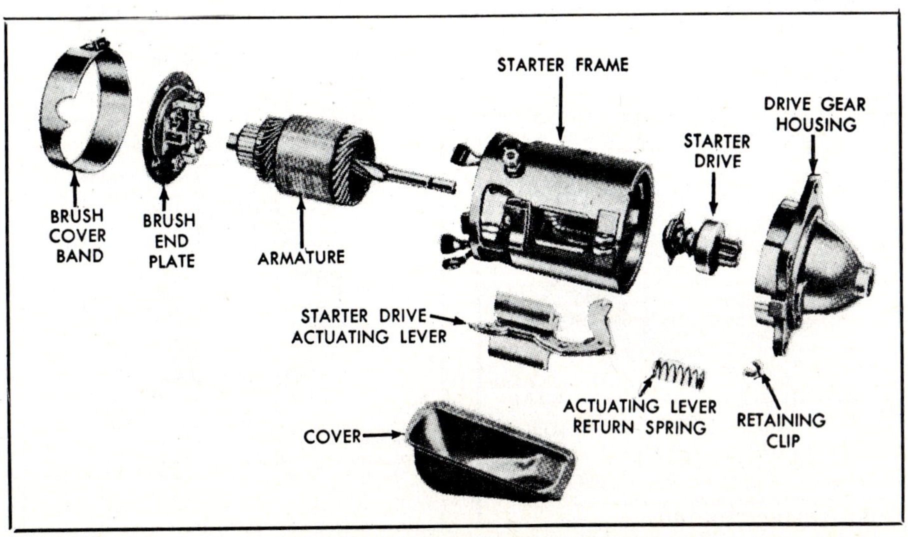

Fig. 63 Ford starter with an integral positive engagement drive 1974–77

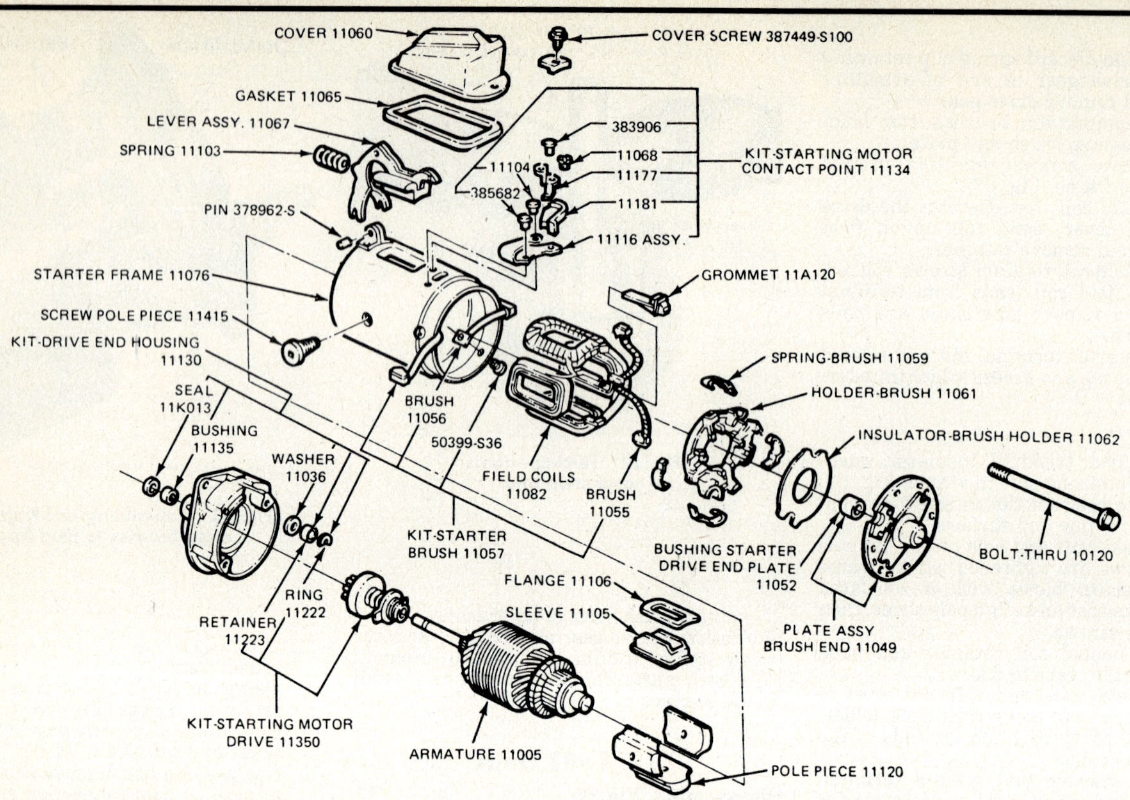

Fig. 64 Ford Motorcraft positive engagement starting motor. 1978–83

FORD MOTORCRAFT SOLENOID ACTUATED STARTER

Description

The solenoid assembly, in this unit, Fig. 69, is mounted to a flange on the starter drive housing which encloses the entire shift lever and solenoid plunger mechanism. The solenoid incorporated a pull-in winding and a hold-in winding.

Operation

As the solenoid is energized, it shifts the starting motor pinion into mesh with the engine flywheel ring gear.

At the same time, the solenoid contacts are closed and battery current flows to the motor, turning it and the engine.

After the engine starts, the starter drive is disengaged when the ignition switch is returned from the start position to the run position and the solenoid spring pushes the shift lever back, disengaging the starter drive from the flywheel ring gear.

The starting motor is protected by an overrunning clutch built into the starter drive.

Disassembly, Fig. 69

1. Disconnect the copper strap from the starter terminal of the solenoid, remove the retaining screws and remove solenoid.
2. Loosen retaining screw and slide brush cover band back on frame.
3. Remove commutator brushes from holders. Hold each spring away from the brush with a hook while sliding brush from holder.
4. Remove through bolts and separate end plates and frame.
5. Remove solenoid plunger and shift fork assembly.
6. Remove armature and drive assembly from frame. Remove drive stop ring and slide drive assembly from shaft. Remove fiber thrust washer from commutator end of shaft.
7. Remove drive stop ring retainer from shaft.

Reassembly, Fig. 69

1. Install drive assembly on shaft and in-

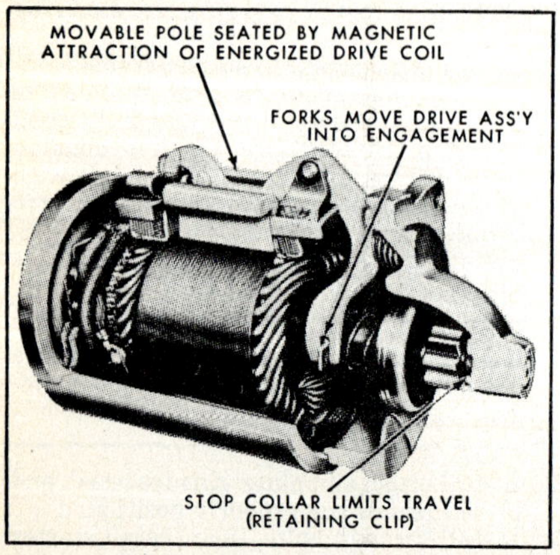

Fig. 65 Starter drive engaged

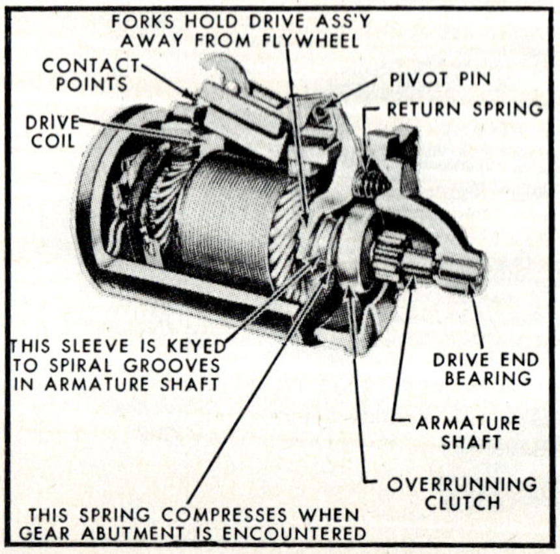

Fig. 66 Starter drive disengaged

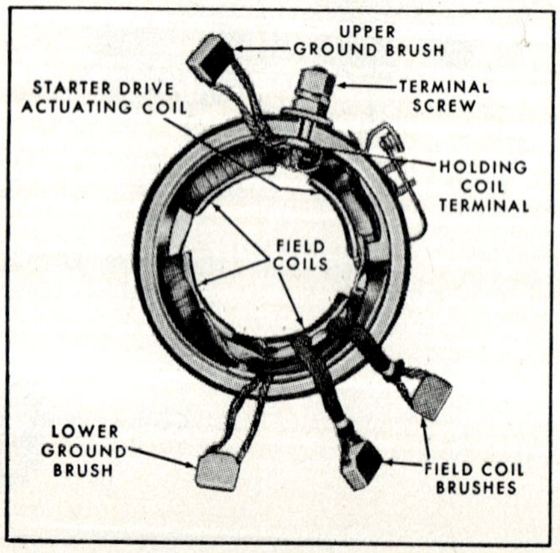

Fig. 67 Field coil assembly

stall new stop ring.
2. Install solenoid plunger and shift fork.
3. Place new retainer in drive housing and install armature and drive in housing. Be sure shift lever tangs properly engage drive assembly.
4. Install fiber washer on commutator end of shaft and position frame to drive housing, being sure to index frame and drive housing correctly.
5. Install brush plate assembly being sure to index it properly, install through bolts and tighten to 55-75 in. lbs.
6. Install brushes by pulling each spring away from holder with a hook to allow entry of the brush. Center the brush springs on the brushes. Press insulated brush leads away from all other components to prevent possible shorts.
7. Install rubber gasket and solenoid.
8. Connect copper strap to starter terminal of solenoid.
9. Position cover band and tighten retaining screw.
10. Connect starter to battery and check operation.

PRESTOLITE STARTER W/POSITORK DRIVE

This heavy duty starter is solenoid actuated and has an armature shaft center bearing. The indexing type Positork drive assures complete drive pinion engagement before the motor begins to rotate, thereby minimizing flywheel ring gear and drive pinion wear and also protects the starter from over-speeding.

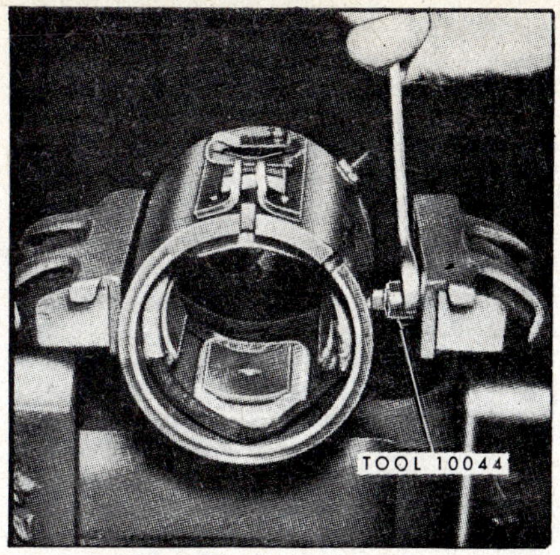

Fig. 68 Removing field coil pole shoe screws

Disassembly, Fig. 70

1. Remove plug and gasket from intermediate housing, then remove the nut from the end of the moving core shaft.

NOTE: To ease removal of adjusting nut, an adapter should be fabricated, Fig. 71. Drill a 3/16 inch hole through a piece of 3/8 inch square and 1½ inches long stock. Install a 3/8 drive ½ inch socket onto nut, then install adapter onto socket. Insert a 3/32 inch allen wrench through adapter to hold the moving core shaft. While holding allen wrench, turn adapter with a ½ inch wrench.

---

2. Disconnect solenoid lead from the front motor terminal, then remove the nut and lockwasher from the motor terminal stud. Remove the four solenoid retaining screws and remove the solenoid.
3. Remove the nut, lockwasher and washers from the commutator end head, then remove the attaching bolts or screws and remove the commutator.
4. Remove the intermediate housing to frame and field assembly bolts then remove the frame and field assembly.
5. Remove the pipe plugs to expose the shift lever pin, then using a small drift or punch, drive the shift lever pin out of the intermediate housing.
6. Scribe or punch aligning marks on intermediate housing and pinion housing, then remove the attaching bolts and separate the two housings. Slide the intermediate housing and drive assembly off the armature.

Assembly

To assemble starter, reverse disassembly procedure while observing the following precautions:
1. When installing new bearings, always use the proper arbor to obtain correct fit. Remove felt oil wick from pinion housing reservoir before installing bearing. After bearings are installed, saturate felt wick with SAE 20 oil, then install felt wick and fill reservoir with SAE 20 oil.

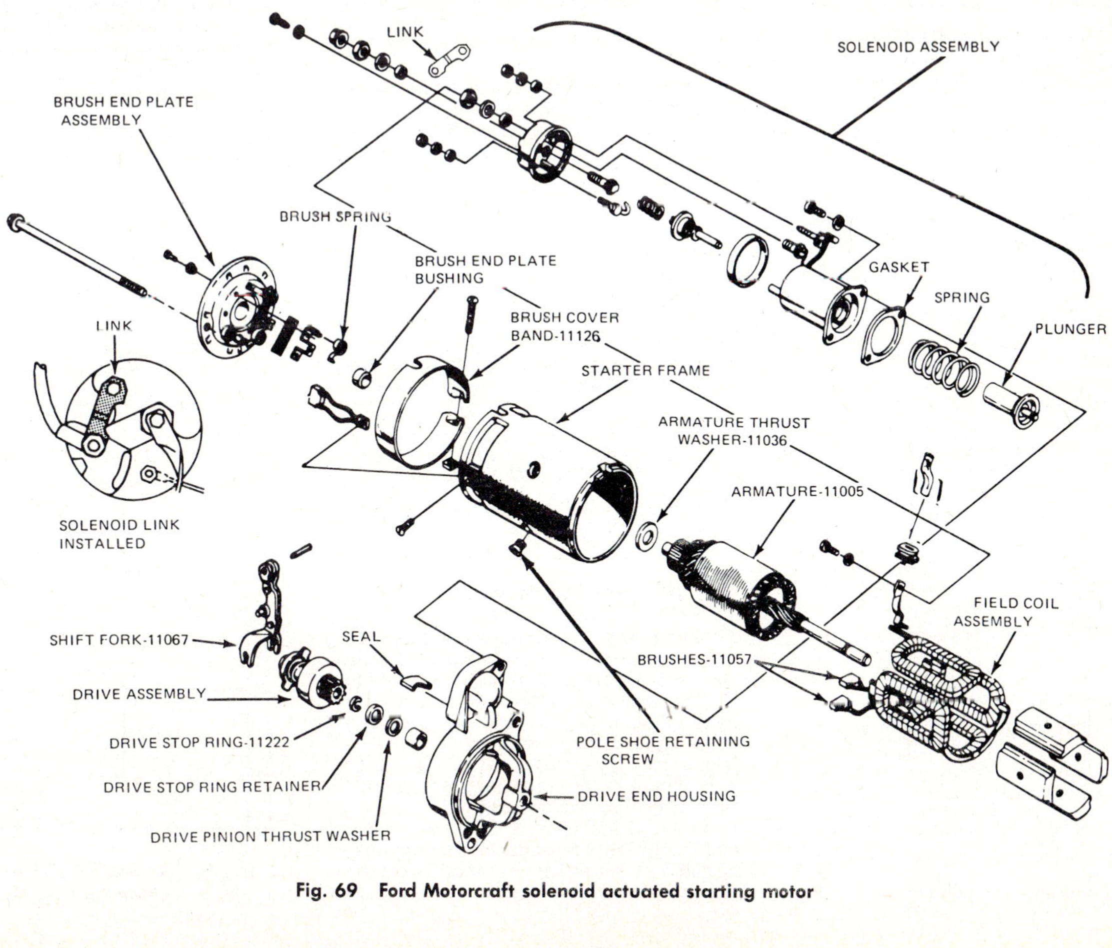

Fig. 69 Ford Motorcraft solenoid actuated starting motor

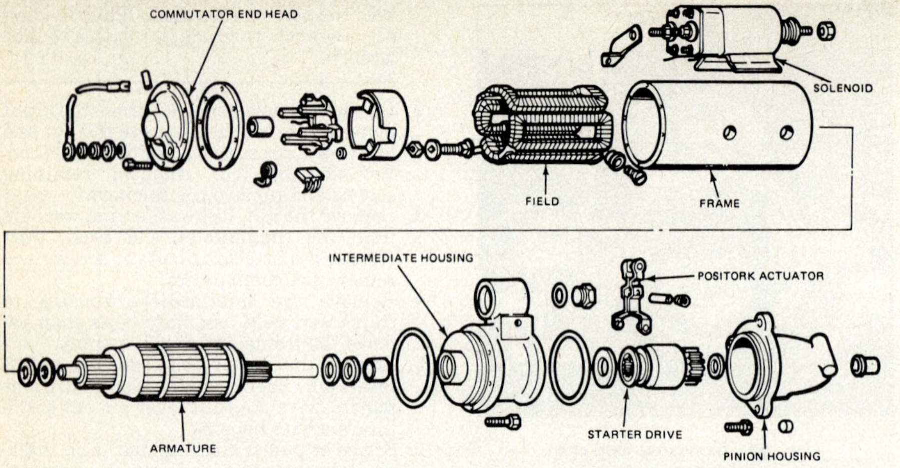

Fig. 70 Prestolite starter with Positork drive

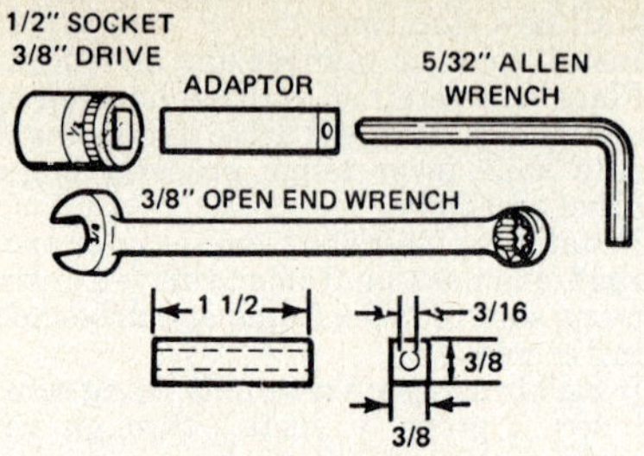

Fig. 71 Fabrication dimensions for adapter tool

2. Apply a light coat of grease to all O-ring seals prior to installation to prevent damaging seals during motor assembly.
3. Any sealing components that are worn or were damaged during removal, should be replaced.
4. Lubricate bearing surfaces of armature with SAE 10 oil. Lubricate shaft splines under drive assembly with low temperature EP grease.
5. When assembling intermediate and pinion housings, torque allen head retaining bolts to 8 ft. lbs.
6. After assembly, coat the insulation of the two terminal studs with sealing compound.

NOTE: Make sure to keep sealer off contact surfaces of terminal studs and nuts.

Inspection & Testing

Armature

1. Check armature for excessive wear, indications of excessive arcing, rough bearing surfaces on the shaft or damaged shaft splines.
2. If commutator is rough, it should be turned down. If commutator was originally undercut, undercut mica to a depth of 1/32 inch below commutator segments.
3. Check armature for shorts or grounds as described previously.

Field Coils

Check field coils for grounds as described previously. Assure that brushes or coil terminals are not grounded while checking field coils.

Brush Holder

Check brush holder assembly for grounds by touching one probe of test light to the holder and the other probe to each brush. If test light goes on, replace brush holder assembly.

Positork Drive

Clean drive assembly with solvent and wipe drive. Do not immerse drive assembly in solvent as this will wash away the internal lubrication which cannot be replaced. Check drive splines for wear or damage. Check to make sure that drive moves freely on armature shaft splines.

Pinion Housing

Inspect housing for cracks. Clean any paint or rust off the mounting flange.

Intermediate Housing

Inspect oil seal and bearing for wear. Replace seal if it is worn, or is hard and brittle.

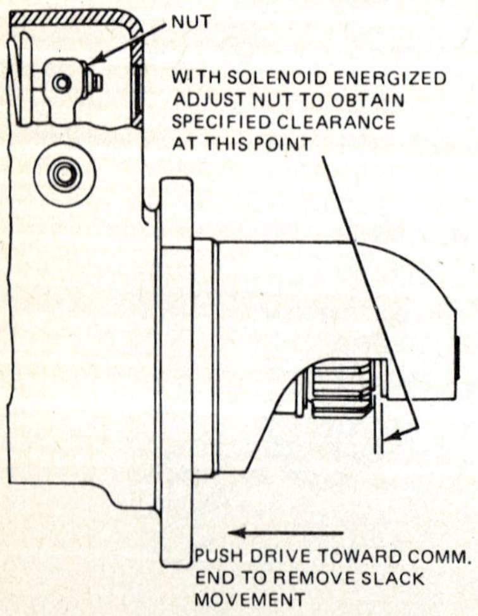

Fig. 72 Starter drive adjustment

Armature End Play

Adjust to obtain .005-.030 inch end play, by adding or removing thrust washers on the commutator end of the armature shaft.

Pinion Adjustment

1. Apply 12 volts to 24 volt solenoids, then momentarily connect a jumper lead from the motor terminal stud of the solenoid to the terminal stud on the commutator end head. This will shift the solenoid and drive into cranking position.
2. Disconnect the jumper wire. The drive will remain in the cranking position until the battery is disconnected.
3. Push the drive toward the commutator end to eliminate any backlash in the linkage and measure the clearance between the outside edge of the pinion and nose housing, Fig. 72. If the clearance is not within .020-.050 inch, adjust by turning the adjusting nut in or out as required.
4. Fabricate interference block, Fig. 73.

CAUTION: Due to the high amount of current passing through the series windings, the following tests should be made as quickly as possible.

5. Place the 63/64 inch side of the interference block against the drive gear, Fig. 74, then connect a 12 volt battery with a test light in series to 12 volt solenoids, or a 24- volt battery and test light in series to 24 volt solenoids.
6. If the test light goes on, the solenoid has been assembled incorrectly. Remove the solenoid cover and check the contact component assembly.
7. If the test light does not go on, connect a carbon pile and ohmeter to the circuit.
8. Place the 1/2 inch side of the interference block against the drive gear, Fig. 74, and adjust the voltage with the carbon pile. The test light must go on before 16 volts on 24 volt solenoids or 8 volts on 12 volts solenoids are reached. If the light does not go on as specified, turn the adjusting nut out until the light goes on as specified.
9. After adjusting, replace the plug and washer in the shift linkage cover.
10. Check all starters for 80 ampere draw at 11 volts. If current draw is excessive, check end play and bearing alignment. Normally several taps with a rawhide mallet will correct bearing misalignment.

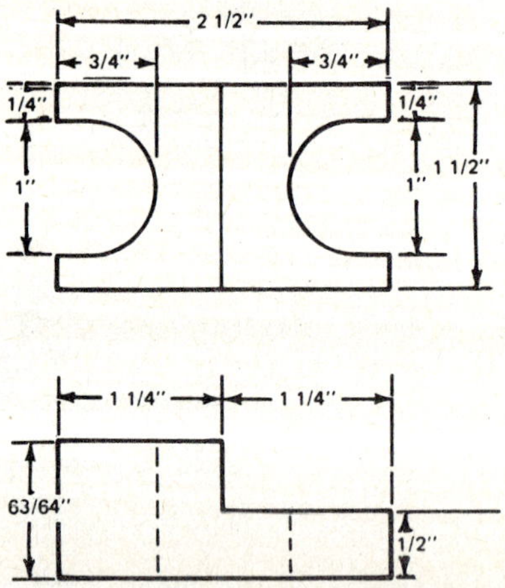

Fig. 73 Fabrication dimensions for interference block

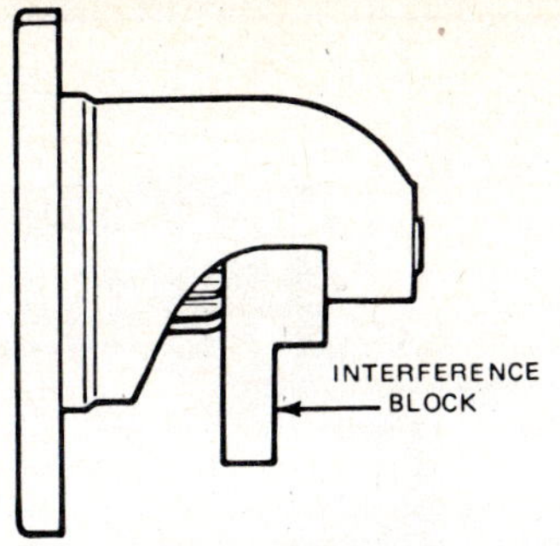

Fig. 74 Interference block installed

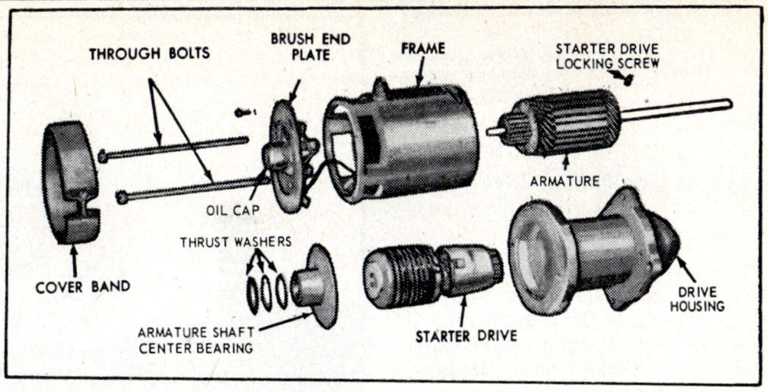

Fig. 75 Prestolite starter with inertia drive

PRESTOLITE STARTER W/INERTIA DRIVE

This heavy duty starter has an armature shaft center bearing and an inertia drive with an overrunning clutch protecting starter from overspeed.

Disassembly, Fig. 75

1. Remove cover band and brushes.
2. Remove through bolts then the end plate retaining screws.
3. Pull out armature, brush end plate and drive housing from frame.
4. Remove terminal from frame.
5. Remove field pole retaining screws and slide field poles out of frame.
6. If field brushes are to be replaced, unsolder the field brushes and solder the new brushes onto the field coil. If ground brushes are to be replaced, remove the rivets and rivet the new brushes onto the end plate.
7. To remove the starter drive, compress the drive spring slightly, then remove the locking screw and slide the drive off the shaft.

Assembly, Fig. 75

1. Position thrust washers, center bearing and starter drive on the shaft and install the locking screw.
2. Install armature assembly into drive housing so that slot center bearing support plate engages pin in drive housing and bearing is firmly seated in housing.
3. Install field coils, pole shoes and retaining screws. As the retaining screws are being tightened, tap the frame several times with a soft-face hammer to align and seat the pole shoes.
4. Install terminal, brush end plate and the six retaining screws.
5. Slide armature and drive housing into frame with frame dowel engaging the hole in the drive housing and install the through bolts.
6. Install the brushes making sure that springs are centered on the brushes and that insulated brushes are not grounded.
7. Install cover band, oil felt and oil cap. Add several drops of oil into oil cap opening.

STARTER DRIVE SERVICE

Starter drives fall into one or the other of two basic groups, the type that uses the principle of the over-running clutch, Figs. 76 and 77, and the Bendix, which uses the spinning nut principle, Figs. 78 and 79.

Starter drive troubles are easy to diagnose and they usually cannot be confused with ordinary starter difficulties. If the starter does not turn over at all or if it drags, look for trouble in the starter or electrical supply system. Concentrate on the starter drive or ring gear if the starter is noisy, if it turns but does not engage the engine, or if the starter won't disengage after the engine is started. After the starter is removed, the trouble can usually be located with a quick inspection.

Worn or chipped ring gear or starter pinion are the usual causes of noisy operation. Before replacing either or both of these parts, try to find out what caused the damage. With the Bendix type drive incomplete engagement of the pinion with the ring gear is a common cause of tooth damage. The wrong pinion clearance on starter drives of the overrunning clutch type leads to poor meshing of the pinion and ring gear and too rapid tooth wear.

A less common cause of noise with either type of drive is a bent starter armature shaft. When this shaft is bent, the pinion gear alternately binds and then only partly meshes with the ring gear. Most manufacturers specify a maximum of .003" radial run-out on the armature shaft.

When Clutch Drive Fails

The over-running clutch type drive seldom becomes worn that they fail to engage since they are directly activated by a fork and lever, Fig. 76. The only thing that is likely to happen is that, once engaged, it will not turn the engine because the clutch itself is worn out. A much more frequent difficulty and one that rapidly wears ring gear and teeth is partial engagement. Proper meshing of the pinion is controlled by the end clearance between the pinion gear and the starter housing or the pinion stop, if one is used.

The clearance is set with the starter off the vehicle and with the drive in the engaged position. While it is not feasible to list the clearances for all starters used on all trucks, the idea is to have enough clearance so that the drive does not slam against the end of the drive housing or pinion stop when it is engaged.

To check the clearance, supply current to the starter solenoid with the electrical connection between starter and solenoid removed. Supplying current to the solenoid but not the starter will prevent the starter from rotating during the test. Take out all slack by pushing lightly on the starter drive clutch housing while inserting the proper thickness feeler gauge between pinion and housing or pinion

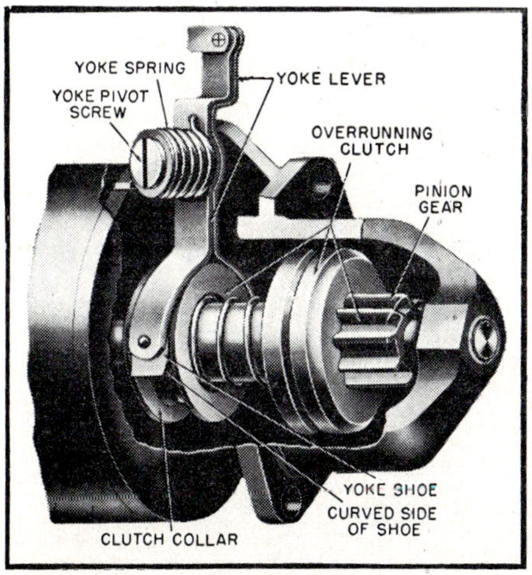

Fig. 76 Overrunning clutch drive. When assembling make sure curved sides of yoke shoes are toward gear end of clutch. Reversed yoke shoes can cause improper meshing of pinion.

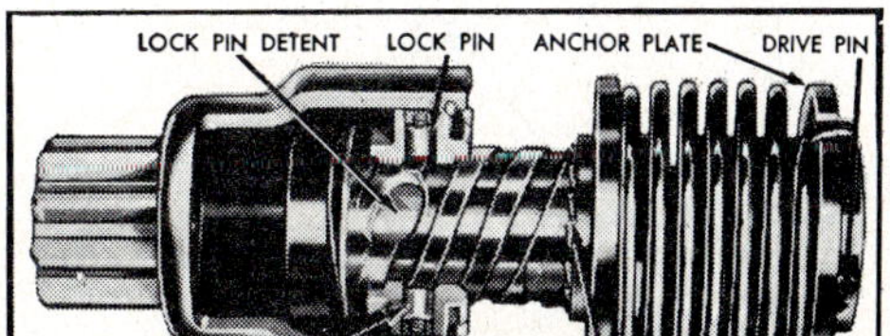

Fig. 77 Folo-Thru starter drive with Bendix spring

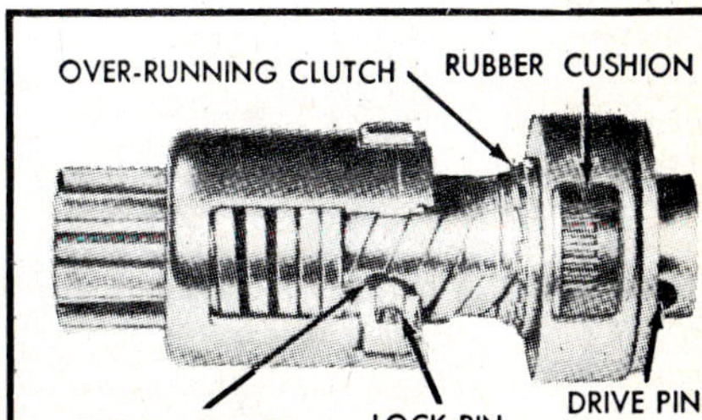

Fig. 78 Folo-Thru starter drive without Bendix spring

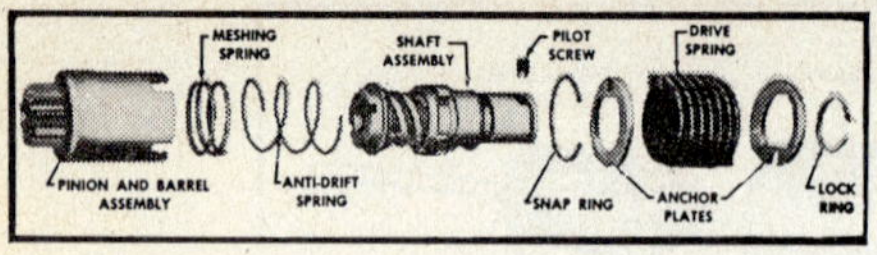

Fig. 79 Barrel type Bendix drive

stop, Fig. 80.

On a number of starters, the solenoids are completely enclosed in the starter housing and the pinion clearance is not adjustable. If the clearance is not correct, the starter must be disassembled and checked for excessive wear of solenoid linkage, shift lever mechanism, or improper assembly of parts.

On starters where the solenoid is exposed, the clearance can be adjusted either by loosening the screws holding the solenoid to the starter and moving the solenoid forward and backward, Fig. 81, or by screwing or unscrewing the link attached to the solenoid plunger, Fig. 82.

Failure of the over-running clutch drive to disengage is usually caused by binding between the armature shaft and the drive. If the drive, particularly the clutch, shows signs of overheating, it indicates that it is not disengaging immediately after the engine starts. If the clutch is forced to overrun too long, it overheats and turns a bluish color. For the cause of the binding, look for rust or gum between the armature shaft and the drive, or for burred splines. Excess oil on the drive will lead to gumming, and inadequate air circulation in the flywheel housing will cause rust.

Over-running clutch drives cannot be overhauled in the field so they must be replaced. In cleaning, never soak them in a solvent because the solvent may enter the clutch and dissolve the sealed-in lubricant. Wipe them off lightly with kerosene and lubricate them sparingly with SAE 10 or 10W oil.

When Bendix Drive Fails

When a Bendix drive doesn't engage the cause usually is one of three things: either the drive spring is broken, one of the drive spring bolts has sheared off, or the screwshaft threads won't allow the pinion to travel toward the flywheel. In the first two cases, remove the drive by unscrewing the set screw under the last coil of the drive spring and replace the broken parts. Gummed or rusty screwshaft threads are fairly common causes of Bendix drive failure and are easily cleaned with a little kerosene or steel wool, depending on the trouble. Here again, as in the cause of overrunning clutch drives, use light oil sparingly, and be sure the flywheel housing has adequate ventilation. There is usually a breather hole in the bottom of the flywheel housing which should be kept open.

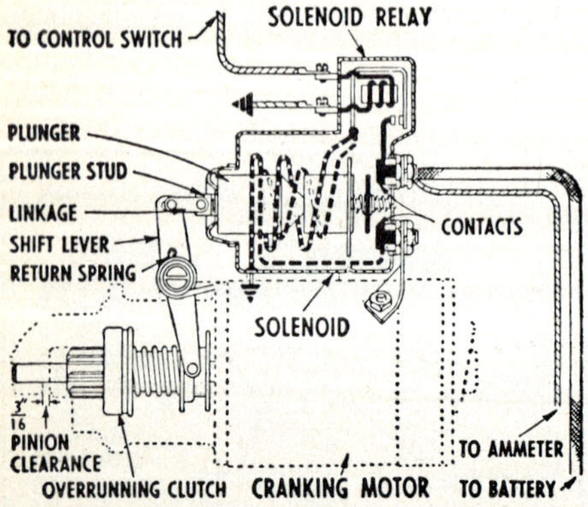

Fig. 82 Overrunning clutch motor equipped with adjustable plunger stud for adjusting pinion clearance

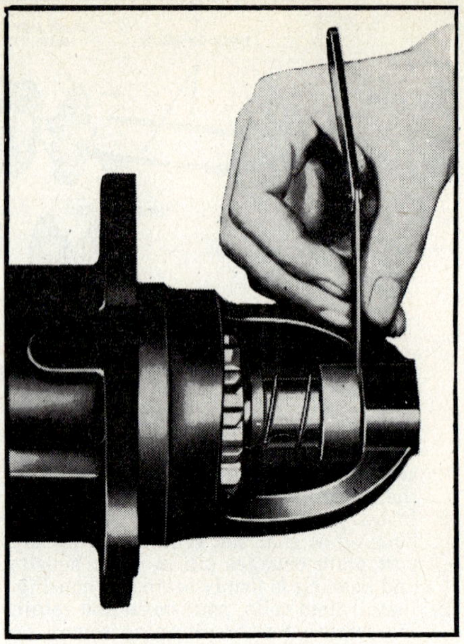

Fig. 80 Measuring overrunning clutch drive stop clearance. Do not compress anti-drift spring as this will give an incorrect clearance. If clearance is not present there is danger of the drive housing being broken as gear or collar slams back against it.

The failure of a Bendix drive to disengage or to mesh properly is most often caused by gummed or rusty screwshaft threads. When this is not true, look for mechanical failure within the drive itself.

Bendix Folo-Thru Drive

This type of drive, Figs. 77 and 78, is in wide use on late model starters. It incorporates a device that keeps the pinion engaged to the flywheel until the engine reaches a specified rpm. When replacing one of these drives, be sure that you have the correct drive for the vehicle. The drives are rated differently and the correct one must be used for the vehicle being serviced. The Folo-Thru, incidentally, is not supposed to be repaired in the field because of the danger of incorrectly assembling the carefully calibrated springs in the pinion head.

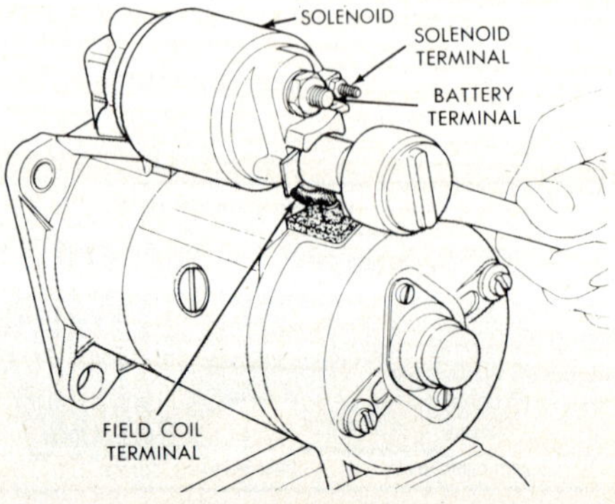

Fig. 83 Disconnecting coil wire from solenoid (typical)

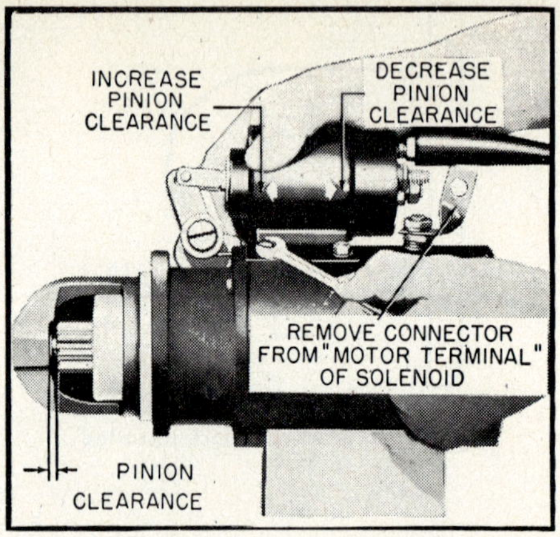

Fig. 81 Adjusting pinion clearance on overrunning clutch motor with an exposed solenoid having a non-adjustable plunger stud

BOSCH STARTER MOTORS

1. Remove nut from solenoid field coil terminal, then remove field coil strap from terminal, Fig. 83.
2. Remove solenoid attaching screws, then work solenoid from shift fork and remove solenoid.
3. Remove starter motor end shield bearing cap attaching screws and remove end shield cap, C-washer and bearing washer, Fig. 84.
4. Remove starter motor through bolts, then remove starter motor brush end shield.
5. Remove two starter motor brushes by prying the brush retaining springs back, then remove brush plate, Fig. 85.
6. Slide field frame from armature, then remove armature and clutch assembly from drive end housing.
7. Remove rubber seal and clutch lever bolt from drive end frame, then remove clutch lever.

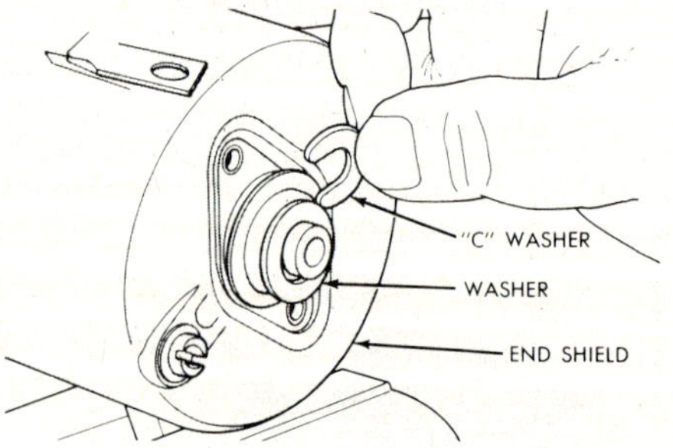

Fig. 84 "C" washer & bearing washer removal

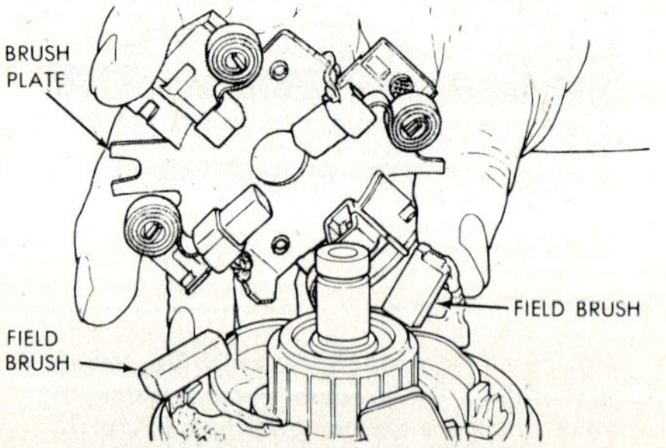

Fig. 85 Brush plate removal

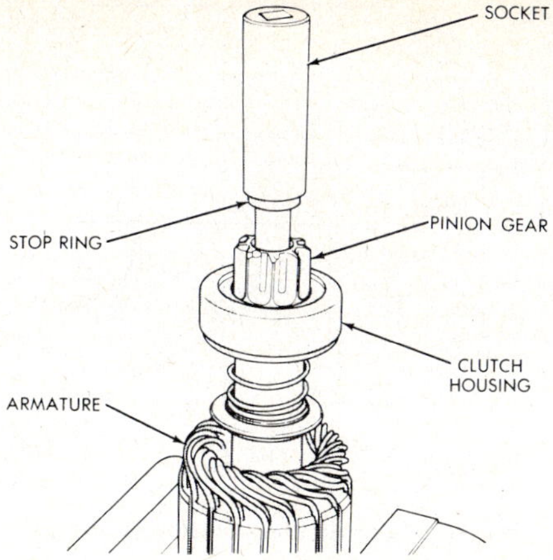

Fig. 86 Pressing stop collar from snap ring. Bosch units except models with A-412 man. transaxle

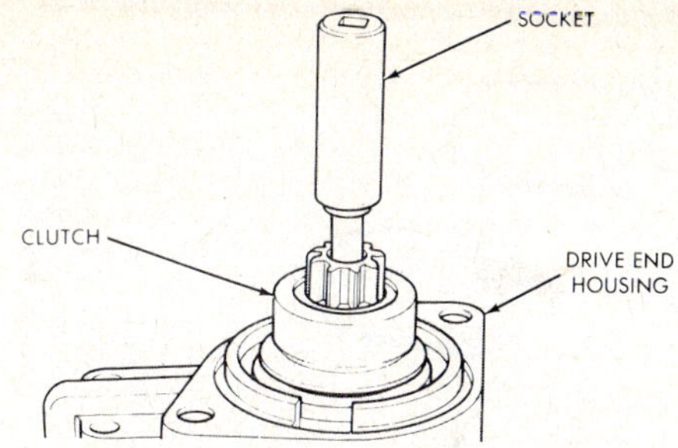

Fig. 87 Pressing stop collar from snap ring. Bosch units for A-412 man. transaxle

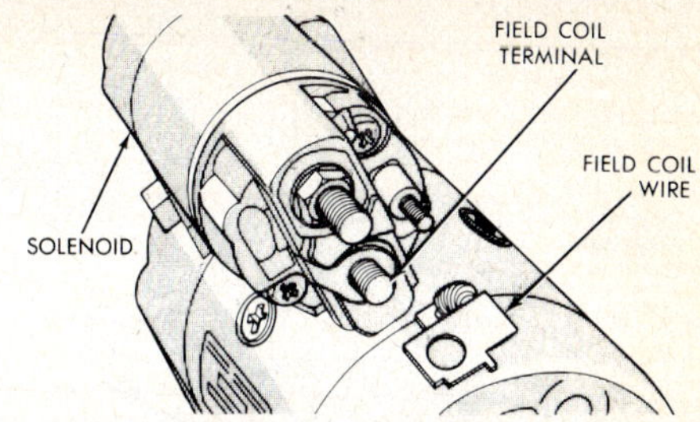

Fig. 88 Field coil terminal identification

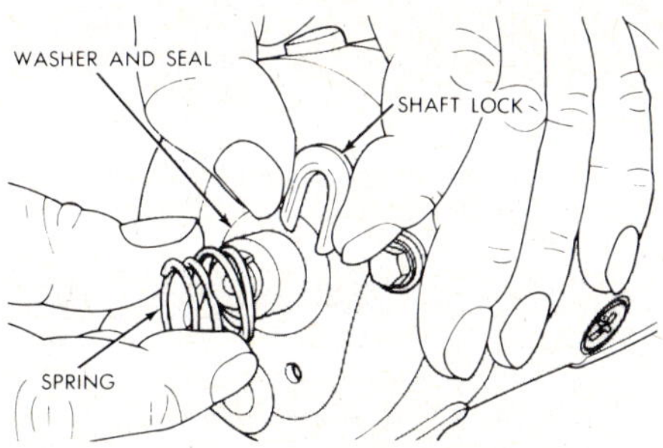

Fig. 89 Armature assembly removal

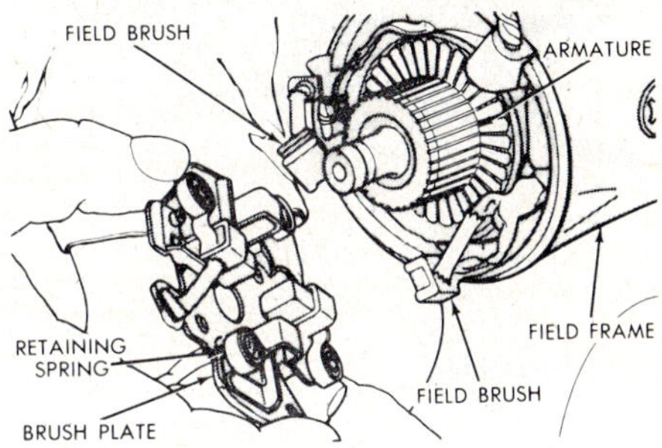

Fig. 90 Brush plate removal

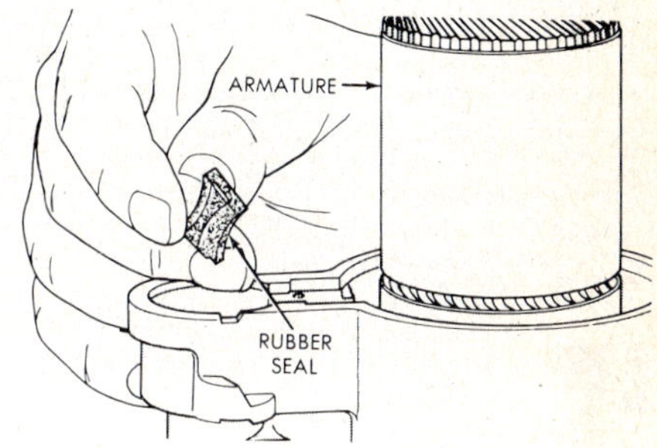

Fig. 91 Rubber seal removal

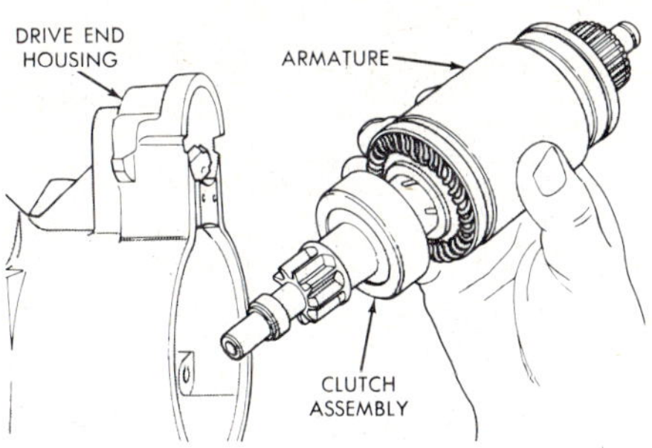

Fig. 92 Armature & clutch assembly

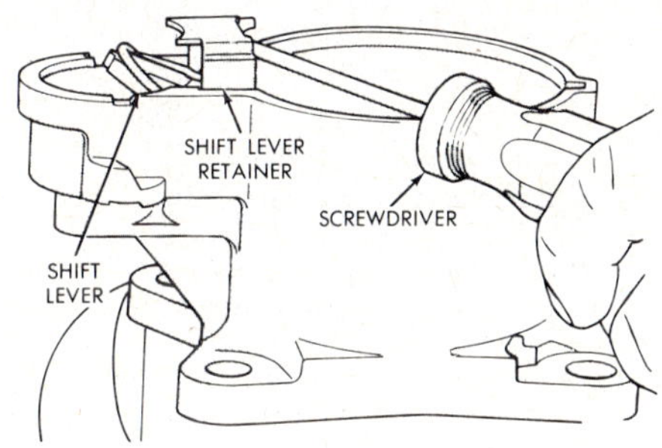

Fig. 93 Clutch shift lever retainer

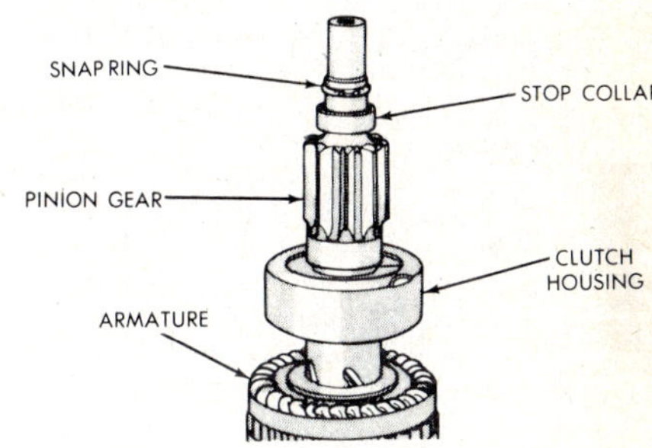

Fig. 94 Stop collar, snap ring & clutch assembly

8. Position armature in a suitable vise, then using a socket, press collar from pinion gear stop snap ring, Fig. 86. Remove snap ring from armature shaft using snap ring pliers.
9. Remove stop collar and clutch assembly from armature shaft, Fig. 87.
10. Reverse procedure to assemble.

NIPPONDENSO STARTER MOTORS

1. Remove nut from solenoid field coil terminal, then remove field coil strap, Fig. 88.
2. Remove nuts attaching solenoid to starter housing, then work solenoid from shift fork and remove solenoid.
3. Remove attaching screws end shield bearing cap, then remove end shield cap, C-washer, bearing spring and bearing washer, Fig. 89.
4. Remove starter motor through bolts, then remove starter motor brush end shield.
5. Pry back on brush spring and remove field brushes, then remove brush plate, Fig. 90.
6. Remove field frame from armature, then remove rubber seal from drive end housing, Fig. 91.
7. Remove armature and clutch assembly from drive end frame, Fig. 92.
8. Remove clutch lever retainer from drive end frame, then remove clutch lever, Fig. 93.
9. Press stop collar from snap ring, then remove snap ring, stop collar and clutch, Fig. 94.
10. Reverse procedure to assemble.

HITACHI STARTING MOTORS

1. Disconnect field lead from "M" terminal of solenoid, Fig. 95.
2. Remove solenoid attaching bolts and separate plunger shaft from shift lever, Fig. 96.
3. Pry off dust cover and remove snap ring washer.
4. Remove the two through bolts.
5. Remove two screws and rear cover assembly, Fig. 97.
6. Raise brush springs and pull out brushes, then remove brush holder assembly, Fig. 98.
7. Separate yoke assembly and gear case.
8. Remove armature and shift lever.
9. Remove drive assembly retainer and snap ring.
10. If field coils are to be removed, remove the four pole shoe retaining screws.
11. Check commutator surface for roughness or burning. Turn commutator down on a lathe if necessary to clean up. The commutator can be turned down to a minimum of 1.497 in. diameter.
12. Measure undercut of insulator from surface of commutator. Depth should be from .0197 to .0315 inch.
13. Check insulation between commutator coils and core using a growler. If growler light turns on, the coils are poorly insulated and should be repaired, if possible, or the armature replaced.
14. Rotate armature slowly on growler while holding strip of steel over each segment. If strip of steel is pulled or vibrates, the

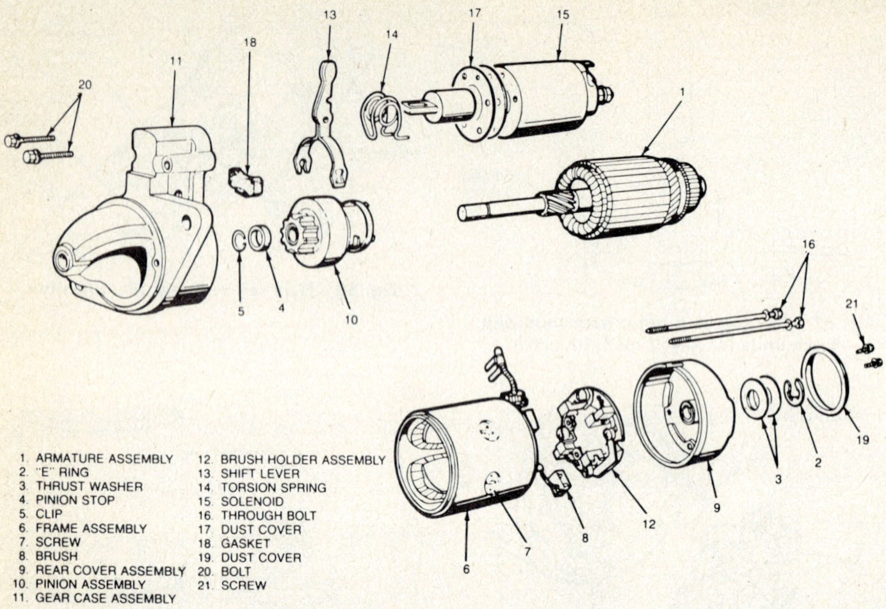

Fig. 95 Hitachi starter, disassembled

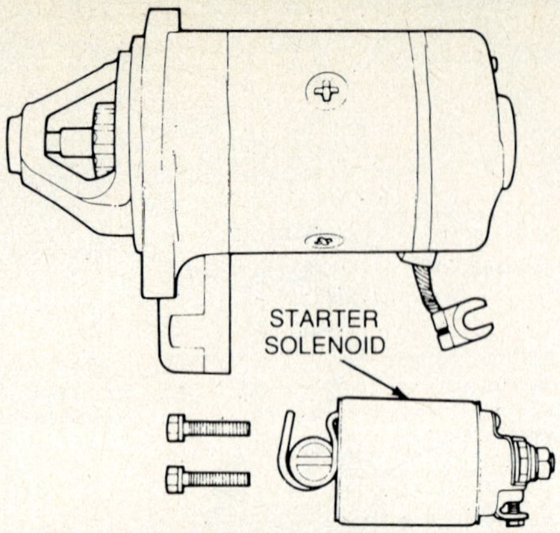

Fig. 96 Solenoid removal, Hitachi starter

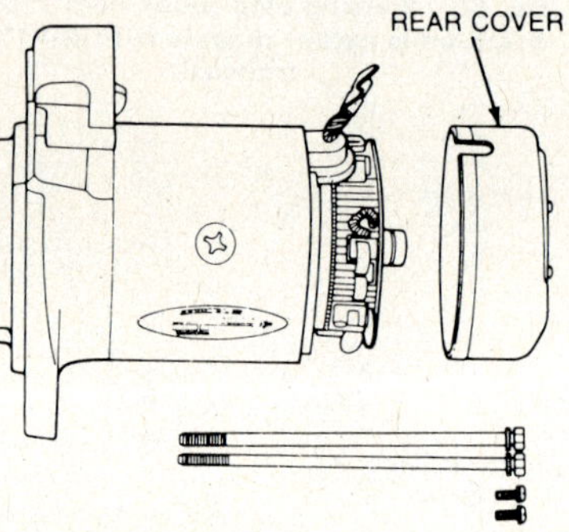

Fig. 97 Rear cover removal

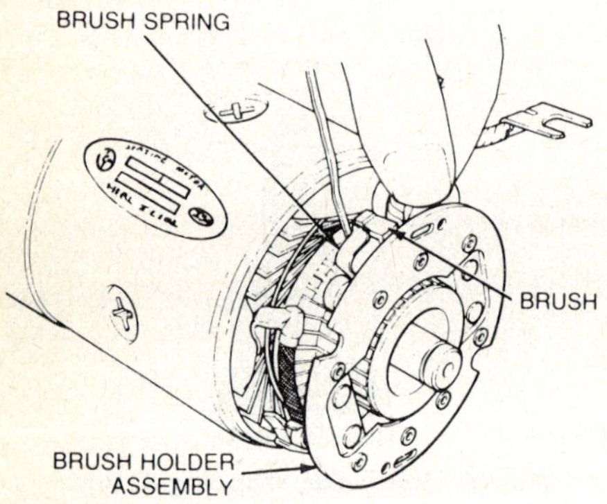

Fig. 98 Removing brush holder assembly

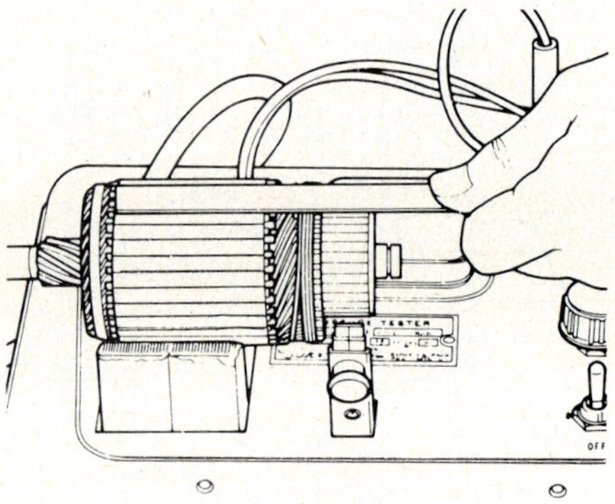

Fig. 99 Using a growler to test armature for shorts

coils are shorted and the armature should be repaired, if possible, or replaced, Fig. 99.

15. Make a continuity test between the commutator segments using the growler. If growler light does not come on when tester leads are connected across the segments, the coils are open and the armature should be repaired if possible or replaced, Fig. 100.
16. Using a pair of "V" blocks and a dial indicator, check armature shaft for runout. If runout exceeds .0031 in., correct or replace the armature, Fig. 101.
17. Check shaft and bearings for wear. Maximum allowable clearance between shaft and bearings is .0079 in.
18. With an ohmmeter check resistance between the field coils and yoke. Insulation resistance should be 1 ohm or more. If poor insulation is noted, remove the pole shoes one at a time to locate grounded parts and correct as necessary.
19. Check brushes for wear, cracks or broken lead and replace if found to be defective. Minimum usable brush length is .472 inch.
20. Inspect brush springs for rusting, distortion, weakening or breakage and replace as necessary. Spring tension should be 3.53 lbs.
21. Clean brush holder to remove carbon and check for insulation between insulated brush holder and plate using an ohmmeter. If poor insulation is indicated, replace brush holder assembly.
22. Remove cover and check solenoid points for fouling or roughness and dress with an oil stone if necessary.
23. Make a continuity test between the "C" and "M" terminals and between the "C" terminal and ground using a circuit tester. If an open circuit is indicated, replace solenoid.
24. The solenoid is satisfactory when the plunger is pulled in as 12 volts is applied between terminals "C" and "M" and returns smoothly as power is cut off.
25. Inspect pinion gear for wear, clutch for slipping and sliding surface in pinion. Replace pinion and clutch assembly, if necessary.
26. Reverse procedure to assemble, noting the following:
    a. Apply lubricant to the bearings and sliding surfaces.

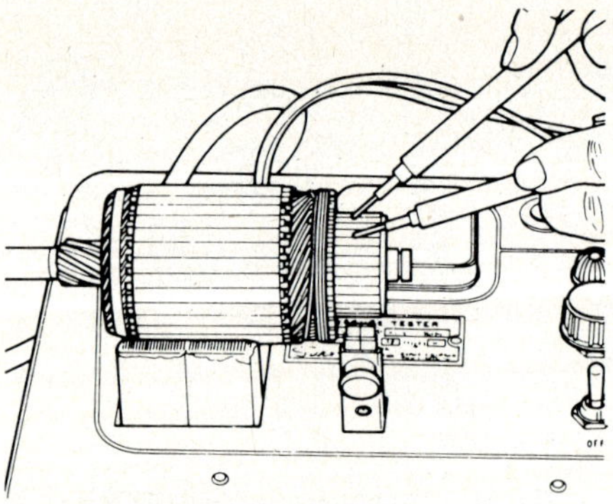

Fig. 100 Using a growler to test armature for opens

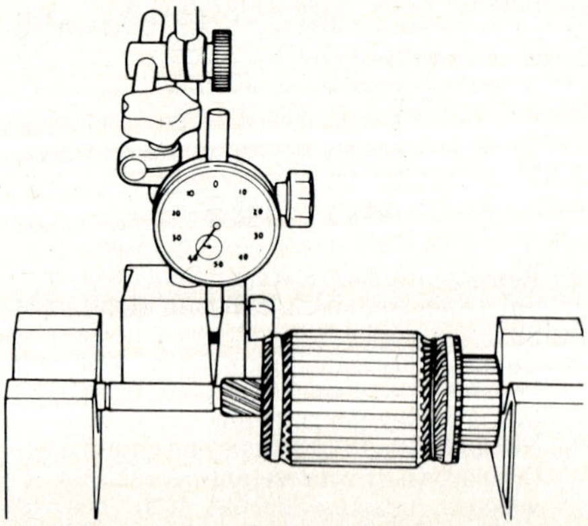

Fig. 101 Checking armature shaft for runout

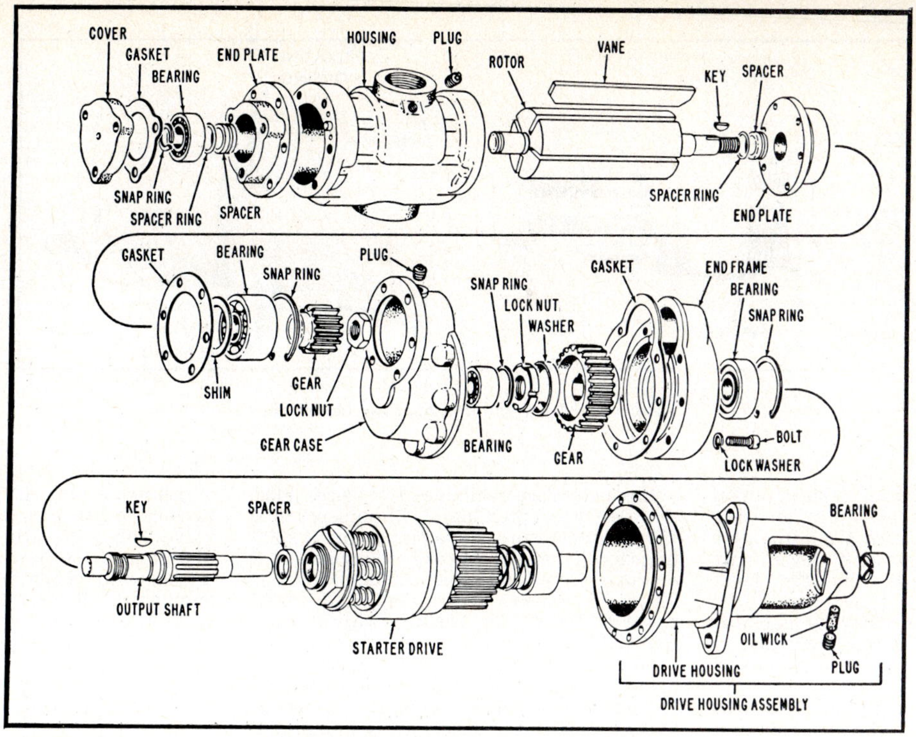

Fig. 102 "Startaire" air cranking motor

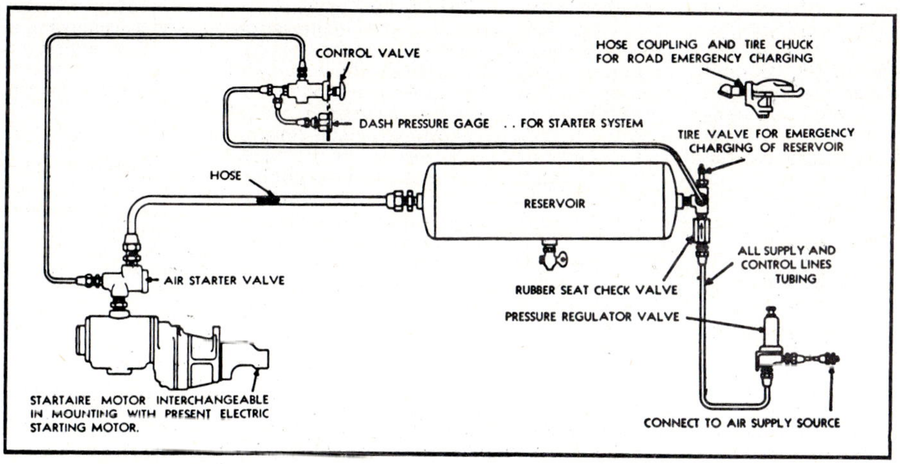

Fig. 103 Typical air starter installation

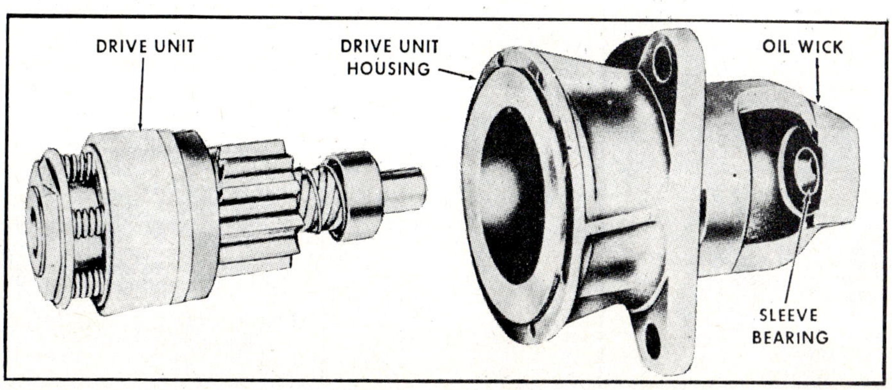

Fig. 102 "Startaire" air cranking Fig. 103 Typical air starter Fig. 104 Drive end and housing. Model 17

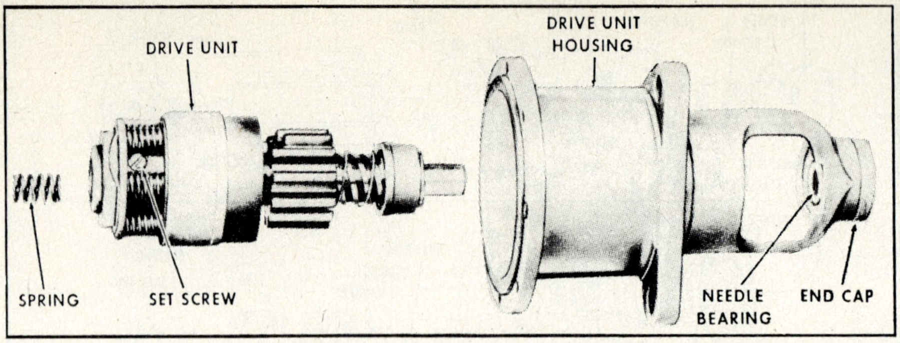

Fig. 105 Drive end and housing. Model 30

b. Set shift lever on lever guide of pinion assembly as the armature is being installed. Do not turn armature after installation since shift lever may become dislodged from the lever guide.
c. Install dust cover and yoke assembly to gear case by aligning the groove in the yoke with corresponding projection.
d. When installing brush and retainer assembly, do not scratch commutator surface.
e. Install adjustment plate and torsion spring to solenoid, then assemble plunger over shift lever with torsion spring in intermediate groove of shift lever and install the two retaining bolts.
f. Apply 12 volts between the "C" and "M" terminals and measure clearance between end of pinion and the retainer. Clearance should be from .011 to .098 inch. To adjust clearance, the solenoid must be removed to replace adjustment plates.

AIR STARTING MOTOR

The principle components of the "Startaire" assembly are a multi-vane type air rotor, a reduction gear assembly, and a friction clutch type drive. Compressed air in the correct volume and pressure range is used to provide the necessary cranking power. Fig. 102 shows an exploded view of an air cranking motor.

The motors are available in two sizes. Model 17 delivers approximately 17 horsepower at 100 psi air pressure with a pinion rpm of 2300. In general, model 17 cranks all engines up to approximately 1100 cubic inch displacement. Model 30 delivers approximately 30 horsepower at 100 psi air pressure with a pinion rpm of 2000. It is normally used on engines of 1100 cubic inch displacement or more.

Fig. 103 shows the various devices used in a "Startaire" installation. A pressure protection valve is used in the air supply line to protect the basic air system on the vehicle against complete loss of air during starting operation. A rubber seated single check valve permits maintenance of the Startaire system as a self-contained system in which leakage can be minimized. A control valve and air pressure gauge registering air pressures in the starting system are installed close to the driver's position. An air starting valve which opens or closes the air line between the starter reservoir and motor, as controlled by the control valve, is mounted on or close to the motor as possible. The motor is installed on the engine in the usual location provided for starting motors. One or two reservoirs are installed in the system to provide a sufficient volume of compressed air to the system. A tire valve, tire chuck and hose coupling are provided to replenish the system from an outside source should it be required.

Operation

For operation, 100 psi or more should register on the air starter system air gauge. Before starting the engine, be sure that all engine adjustments and prior starting procedures are carried out. Depress the knob on the control valve. This action allows air to turn the motor and engage the flywheel. The control valve knob should be kept fully depressed until the engine fires. Release the control valve immediately when the engine fires to allow the drive gear to disengage the flywheel.

Preventive Maintenance

1. Daily or as often as necessary, drain the reservoir to be sure clean dry air is delivered to the motor.
2. Monthly or about every 5000 miles of service, remove pipe plug or grease fitting at air inlet port or air inlet fitting at the motor and inject ½ pint of kerosene into the rotor assembly before operating the motor.

CAUTION: *Do not over-lubricate. Overlubrication or use of lubricants of heavy consistency can cause improper operation. Do not perform the above flushing operation indoors as the kerosene will be discharged at the exhaust outlet with the first start. Also be certain that there is no open flame in the vicinity to ignite the exhaust kerosene.*

---

3. Check to see that the motor mounting is secure to the flywheel housing. Check all air lines and valves in the system for leakage.

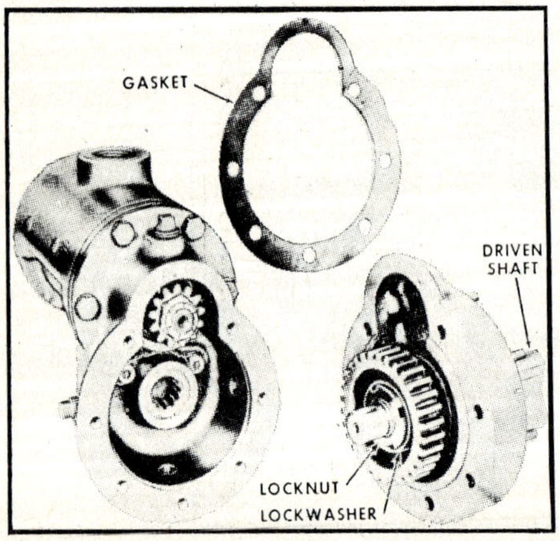

Fig. 106 Cover, shaft and gear. Model 17

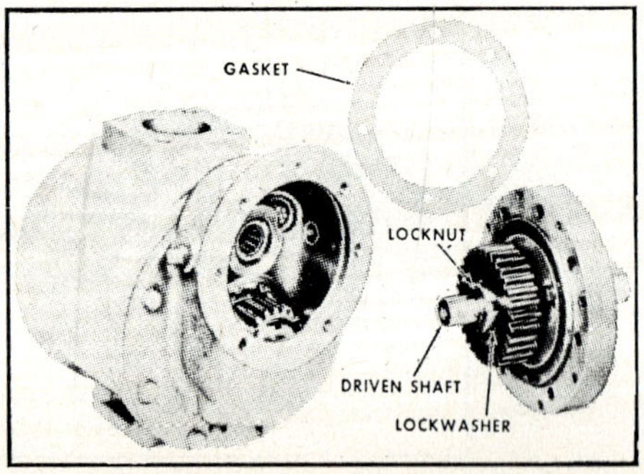

Fig. 107 Cover, shaft and gear. Model 30

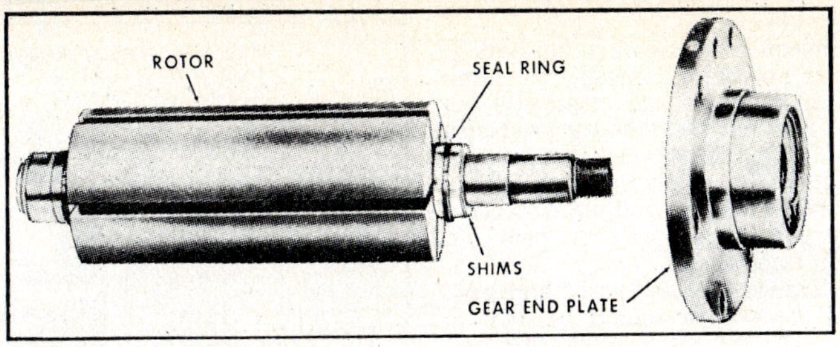

Fig. 108 Rotor and gear end plate. Model 17

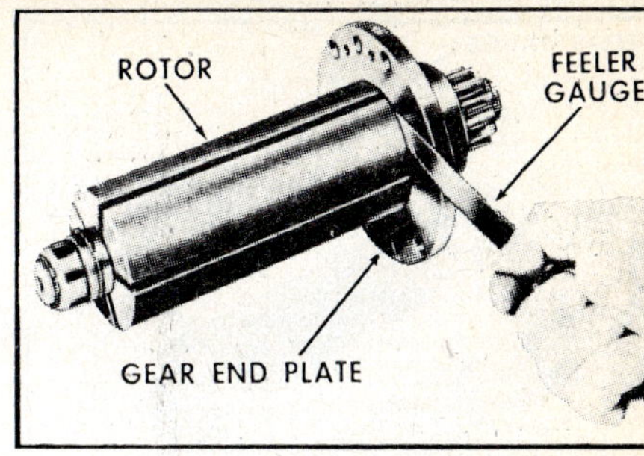

Fig. 109 Checking clearance between rotor and end plate. Model 17

4. Test the operation of the motor. The unit should turn over the engine quickly and at high revolutions per minute. Be sure that air pressure and volume are correct for motor requirements in making this test.
5. To make a leakage test, first see that the system is fully charged and the control valve is in fully released position. Then check to be sure there is no measurable leakage at the exhaust port of motor, at control valve knob, at tire valve, or at other possible leakage points in the system.
6. Periodically check all other devices, fittings and air lines to be sure they are tight.

Remove & Install Motor

1. Disconnect air line at control valve on Model 17 or air line at motor on Model 30.
2. On Model 17, remove motor control valve.
3. Unfasten motor housing from flywheel housing and lift out motor.
4. Reverse removal procedure to install.

Disassemble Motor

1. After covering inlet and exhaust ports, clean the exterior of the motor.
2. The original position of the mating parts should be marked. *This is important because if the rotor housing is turned end for end when reassembled, the motor will rotate in the opposite direction, the pinion will not engage and the motor could rotate at dangerous speeds.*
3. Remove drive unit housing.
4. Remove drive unit by pulling off splines of driven shaft, Fig. 104. If necessary to replace sleeve bearing in drive housing, remove oil wick and press out bearing.
5. When disassembling Model 30, remove set screw in clutch to permit removal of drive unit from driven shaft, Fig. 105. This unit contains a spring at the end of the driven shaft. In Model 30, if it is necessary to replace the needle bearing in drive housing, drive off end cap and press out bearing.
6. Remove spacer from driven shaft on Model 17.
7. Remove capscrews from cover and remove cover, driven shaft and gear, Figs. 106 and 107.
8. Release lockwasher on driven gear and remove lock nut. Use puller to remove gear from driven shaft.
9. For Model 17, remove key and press driven shaft out of bearing. Remove snap ring and press bearing from gear case cover.
10. For Model 30, remove capscrews from bearing retainer plate and press bearing and shaft from cover after which press bearing from driven shaft.
11. Remove capscrews securing gear case to main housing, including the two capscrews inside the case.
12. Before removing gear case on Model 17, remove lock wire inside case. Then remove gear.
13. Remove snap ring from roller bearing and press out bearing which is a sliding fit in case.

Removing Model 17 Rotor

1. Remove gear end plate and rotor from housing. Dowels in main case position end plate correctly.
2. When removing rotor from main housing, inner race of roller bearing remains on rotor shaft. Do not remove it unless bearing is to be replaced.
3. Remove rotor vanes to protect them from damage.
4. Remove lock nut and take off drive gear from rotor shaft with the aid of a gear puller.
5. Remove key from rotor shaft and press rotor from bearing and gear end plate, using an arbor press.
6. Rotor and shaft are serviced as a unit. *A spacer is pressed on each rotor shaft. Do not remove spacer unless replacement is required. If rotor is to be reassembled using the same parts, the shims located between the spacer on the gear end of the rotor shaft and bearing inner race can be reused if not damaged.* Remove snap ring in gear end plate. Bearing is a light press fit in plate.
7. Remove plain end plate cover.
8. Remove plain end plate from housing and press out bearing.

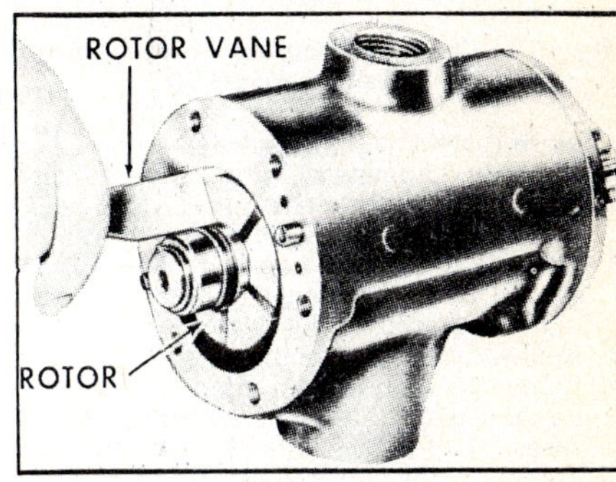

Fig. 110 Installing vanes. Model 17

Removing Model 30 Rotor

1. After removing drive gear from rotor shaft with a gear puller, remove gear end plate and bearing from rotor shaft. There are dowels on end plate for positioning

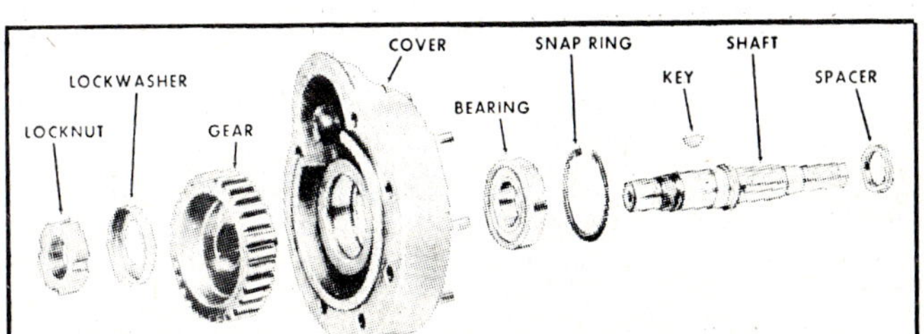

Fig. 111 Cover, driven shaft and gear parts. Model 17

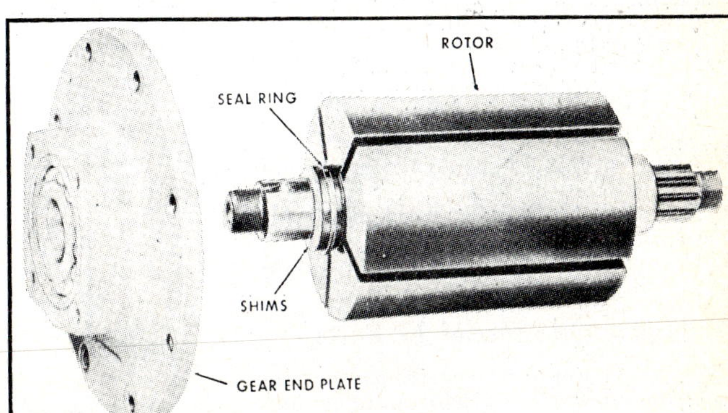

Fig. 112 Rotor and gear end plate. Model 30

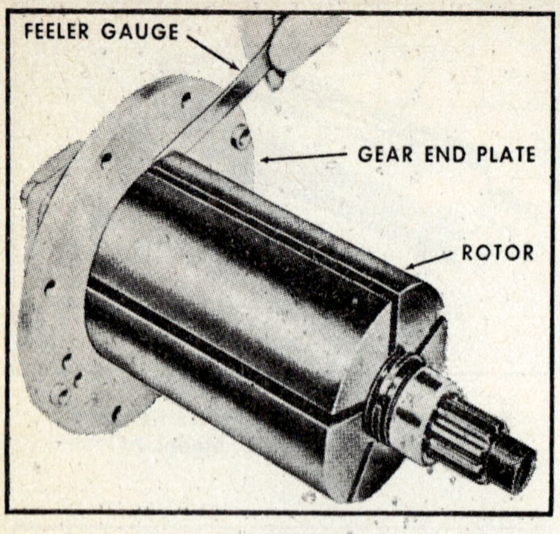

Fig. 113 Checking clearance between rotor and end plate. Model 30

plate correctly.
2. Remove snap ring holding roller bearing in gear end plate. Bearing is a light press fit in plate and can easily be removed.
3. Bearing inner race normally remains on shaft. Do not remove it unless a new bearing is required.
4. Remove plain end plate cover.
5. Remove end plate from main housing and lift end plate and rotor as a unit out of case.
6. Remove rotor vanes to protect them from damage. Release lockwasher and remove luck nut. Carefully press rotor from end plate or use a gear puller to remove end plate from rotor. Bearing is a light press fit in end plate. *Do not misplace or damage shims between rotor shaft shoulder and bearing inner race because the same shims can be used again for reassembly if the same parts are to be reused. The rotor and shaft are serviced as a unit only.*

Inspection of Parts

1. Clean and inspect all parts. Any broken or excessively worn parts must be replaced with new ones.
2. Inspect rotor, housing and end plates for cracks, abrasions, wear spots or build-up of foreign material. Cracked castings must be replaced. Rough or worn spots can sometimes be smoothed out with fine emery cloth. Deposits of gum, carbon, or other foreign material must be removed. Wash and dry all parts.
3. Rotor vanes should be inspected for cracks and wear and should normally be replaced.
4. Gears should be replaced if excessively worn or if teeth are broken.
5. For Model 30 the gear splines should be a light press fit on shaft splines. If fit is loose, replace parts.
6. Broken seal rings must be replaced.
7. Seal ring bores in end plates should be checked for excessive wear or roughness.
8. Seal ring grooves should be checked for visible wear.
9. Bearings should be checked for visible wear and replaced if necessary.
10. New gaskets should be used when reassembling.

Reassemble Model 17 Rotor

1. Install spacers and seal rings on rotor shaft. Press roller bearing on rotor shaft and install snap ring. Press bearing into gear end plate and replace bearing lock wire.
2. Place shims on gear end of rotor shaft. If same parts are being reused, use same shims if undamaged, as originally removed. If new parts are being used, use .015" of shims for initial installation, subject to adjustment as given below.
3. Press bearing and gear end plate on rotor shaft, Fig. 108. Install key, drive gear and locknut on rotor shaft.
4. Check clearance between rotor and gear end plate. Clearance should be .002 to .003", Fig. 109. Shims of .002, .003 and .005" are available for adjustment.
5. After obtaining proper clearance, install rotor and gear end plate in main housing.
6. Install vanes in rotor through open end of housing, Fig. 110. Install bearing in plain end plate and apply end plate to main housing, using care to avoid breaking seal ring.
7. Install capscrews and, using a new gasket, fill cover half full of ball and roller bearing grease and install cover.
8. Install bearing and snap ring in gear case. Use a new gasket and install gear case. After installing capscrews, insert lock wire.
9. Press bearing on driven shaft with its seal facing spline end of shaft. Install key, driven gear, lockwasher and lock nut.
10. Install driven shaft, bearing and gear assembly in cover and install snap ring, Fig. 111. Before assembling cover to gear case, place about 6 ounces of general purpose grease in gear case. Use a new gasket and install cover, shaft and gear in case.
11. Install spacer over driven shaft and install drive unit on shaft.
12. If the sleeve bearing in end of drive unit has been removed, press new bearing into housing flush with end face and finish inside diameter to .751-.752" concentric with drive housing pilot diameter. If finishing tools are not available and bearing is worn, replace drive housing assembly.
13. Install oil wick after soaking it in SAE No. 10 engine oil. Install drive unit to gear case cover.

Reassemble Model 30 Rotor

1. Install bearing in plain end plate, and seal rings on rotor shaft.
2. Install bearing on rotor shaft.
3. Install spacer shims on plain end rotor shaft, Fig. 112. If new parts are being used, start with about .009" shim thickness for initial installation subject to adjustment as given below.
4. Press bearing and plain end plate over rotor shaft, using care to avoid breaking seal ring. Install lockwasher and lock nut. Do not flatten lockwasher until .003" end clearance between rotor and plain end plate is obtained.
5. Install new gasket between small cover and end plate and install cover temporarily. Check clearance as shown in Fig. 113. Adjust as necessary to obtain .003" clearance between rotor shaft shoulder and bearing inner race. Shims of .002, .003 and .005" thicknesses are available for adjustment.
6. After clearance is obtained, remove small cover and fill it half full with bearing grease and reinstall cover.
7. Install rotor and plain end plate on main housing. Install rotor vanes through open gear end of housing, Fig. 114. Press bearing into gear end plate and install snap ring.
8. Install gear end plate to main housing, using care to avoid breaking seal ring.
9. Install gear on rotor shaft. Install new lockwasher and pull gear into position and install lock nut.
10. Press roller bearing in gear, install washer and snap ring, Fig. 115.
11. Using new gasket between end plate and gear case, install gear case.
12. Press bearing on driven shaft with its seal facing drive unit. Install bearing and shaft in cover, and bearing retainer plate, Fig. 115.
13. Press driven gear over splined end of shaft and install lockwasher and lock nut.
14. Place about 8 ounces of general purpose grease in gear case and install case. Using a new gasket, install cover and gear sub-assembly to gear case.
15. Install two keys and drive unit on driven shaft after installing spring between end of driven shaft and drive unit. Install and tighten set screw in drive unit. Install needle bearing in housing flush with drive face. Fill end cap half full with bearing grease before pressing into place.

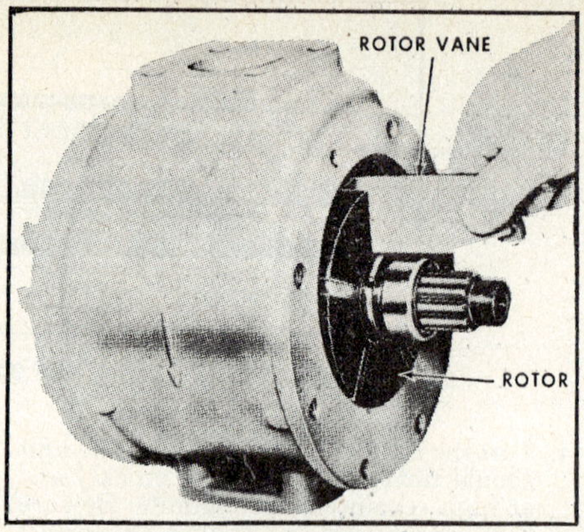

Fig. 114 Installing vanes. Model 30

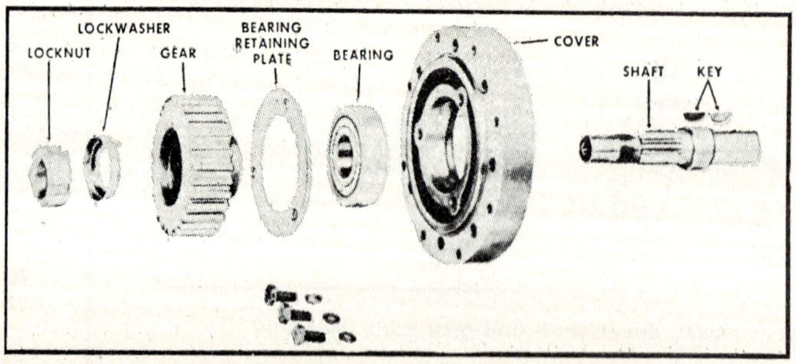

Fig. 115 Cover, driven shaft and gear parts. Model 30

MECHANICAL FUEL PUMP

Fig. 1 illustrates a schematic drawing of a typical fuel system in which a combination fuel and vacuum pump is incorporated. Figs. 2 and 3 illustrate two typical diaphragm fuel pumps. Fig. 4 differs in that the pump has a vacuum booster section. The booster section has nothing to do with the fuel system except that it is operated by the pump arm.

Fuel Pump Operation

During the first or suction stroke, the rotation of the eccentric on the camshaft operates the pump operating arm which pulls the lever and diaphragm downward against the pressure of the diaphragm spring, producing a suction in the pump chamber. The suction holds the outlet valve closed and pulls the inlet valve open, making the fuel flow from the supply tank through the inlet up through the inlet valve into the pump chamber. During the return stroke, the diaphragm is forced up by the diaphragm spring, the inlet valve closes and the outlet valve is forced open, allowing the fuel to flow through the outlet to the carburetor.

Vacuum Section Operation

The vacuum section of combination pumps operates the windshield wipers at almost constant speed. The rotation of the camshaft eccentric in this type pump also operates the vacuum booster section by actuating the pump arm which pushes a link and bellows diaphragm downward, expelling air in the vacuum chamber through its exhaust valve out into the intake manifold of the engine. On the return stroke of the pump arm, the diaphragm is moved upward, producing a suction in the vacuum chamber. This suction operates the vacuum section and draws air through the inlet passage from the windshield wiper.

Fuel Pump Performance

It is essential that the fuel pump deliver sufficient fuel to supply the requirements of the engine under all operating conditions and that it maintain sufficient pressure in the line between the pump and carburetor to keep the fuel from boiling and to prevent vapor lock.

Excessive fuel pump pressure holds the carburetor float needle valve off its seat, causing high gasoline level in the float chamber which in turn increases gasoline consumption.

The pump usually delivers a minimum of ten gallons of gasoline per hour at top engine speeds, under an operating pressure of from 2 to 6 psi (depending on the installation). The highest operating pressure will be attained at idling speed and the lowest at top speed.

Fuel Pump Tests

The fuel pump can be tested on the car with a pressure gauge, a hose and a pint measuring can. With this equipment, it is possible to check the fuel pump to see if it is delivering the proper amount of gasoline at the correct pressure.

Pressure Test

To make the pressure test, disconnect the fuel pipe at the carburetor inlet and attach the pressure gauge and hose between the carburetor inlet and the disconnected fuel pipe, Fig. 5. Take the pressure reading with the engine running. The pressure should be within the limits permissible for the particular vehicle installation. The pressure should remain constant or return very slowly to zero when the engine is stopped.

Capacity Test

To make this test, connect the hose so the pump will deliver gasoline into the pint measure held at carburetor level. Run the engine at idle speed and note the time it takes to fill the measure. On the average it should take from 20 to 30 seconds, depending on the pump being tested.

When Pressure Is Low

Low pressure indicates extreme wear on one part, small wear on all parts, ruptured diaphragm, dirty valve or gummy valve seat.

Wear in the pump usually occurs at the rocker arm pivot pin and on the contacting surfaces of the rocker arm and links. Due to the leverage design, wear at these points is multiplied five times in the movement of the diaphragm. It is apparent, therefore, that very little wear will materially reduce the stroke of the diaphragm. The worn parts must

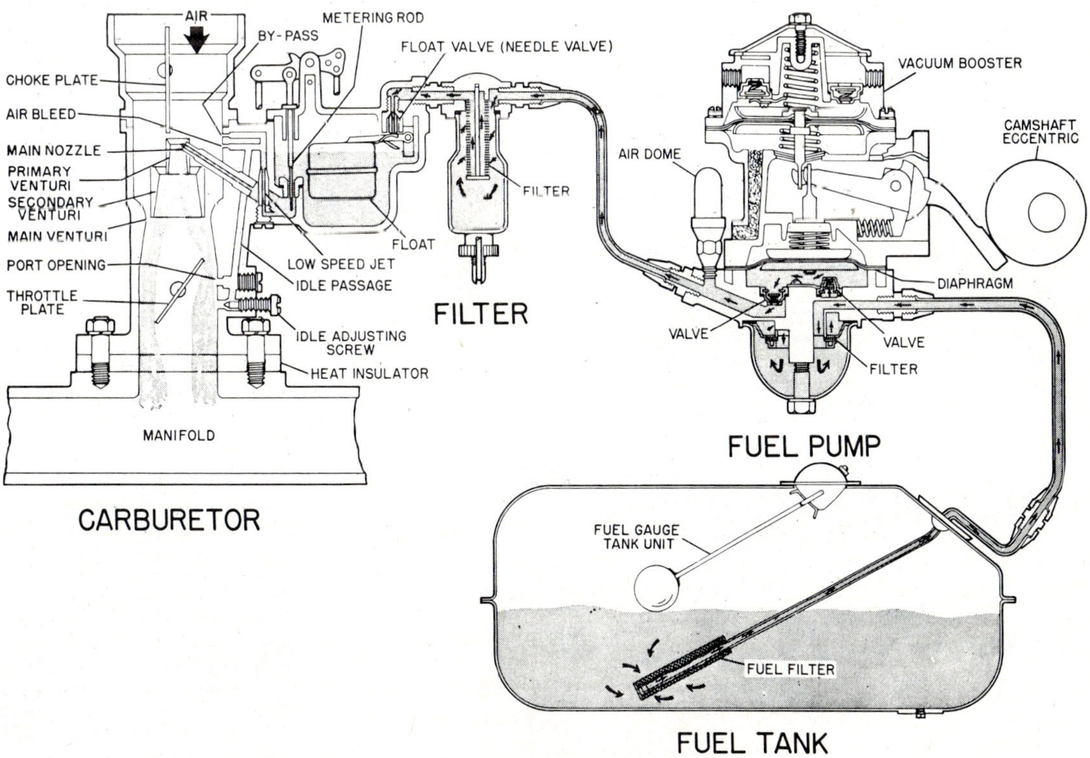

Fig. 1 Diagram of fuel system in operation

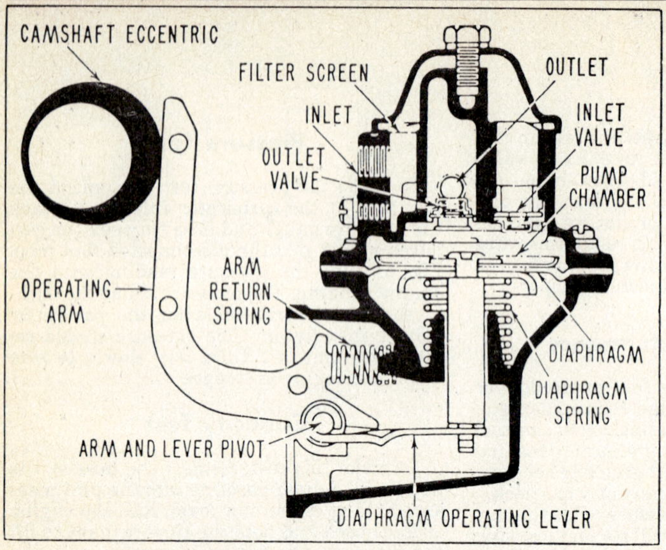

Fig. 2 Fuel pump with built-in fuel filter

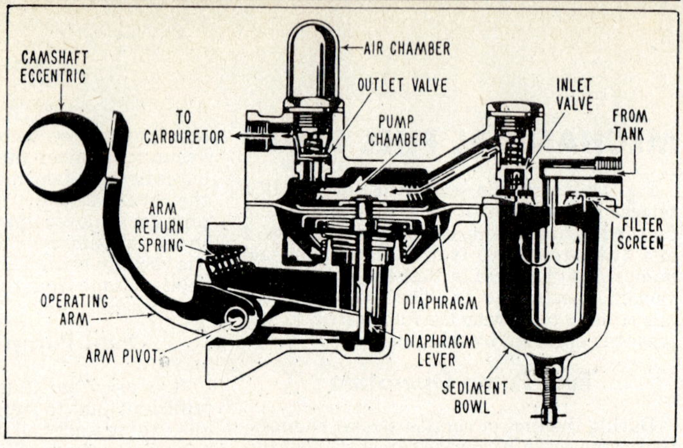

Fig. 3 Fuel pump with separate fuel filter

be replaced for a satisfactory correction.

The diaphragm pull rod has an oil seal around it which prevents the hot oil vapors from the crankcase coming in contact with the diaphragm. If this seal is damaged, the oil vapors have a tendency to shorten the life of the diaphragm.

The first three conditions—extreme wear on one part, small wear on all parts and ruptured diaphragm—are brought about by usage, while dirty and/or poor fuel is usually the cause of valve trouble.

When Pressure Is High

High pressure is caused by a tight diaphragm, fuel between diaphragm layers, diaphragm spring too strong, pump link frozen to rocker arm.

A tight diaphragm will stretch slightly on the down stroke. As the pump operates, the diaphragm will rebound on the up stroke beyond its normal position, much as a stretched rubber band when it is suddenly released. This rebound will cause a higher than normal pressure in the pump chamber.

A loose diaphragm retainer nut or poor riveting on the diaphragm assembly may allow fuel to seep between the diaphragm layers. This will cause a bulge in the diaphragm and have the same effect as a diaphragm that is too tight.

A diaphragm spring that is too strong also causes a high pressure, for the diaphragm will operate longer before pressure of the fuel on the diaphragm will overcome the diaphragm spring.

On a combination pump there are times when the operating parts may become badly corroded and the links freeze to the rocker arm. In this condition the pump operates continually, resulting in a very high pressure and a flooding carburetor.

The remedy for all these conditions is to remove the pump for replacement or repair, using a repair kit.

When Capacity Is Low

A pump is extra efficient and will never starve the engine when it supplies fuel equal to or above the capacity of the pump. The pressure, of course, must be within specifications.

Low capacity is usually caused by an air leak at the intake pipe at these points: fuel pipe fitting at pump, bowl flange or diaphragm flange, fuel bowl. (It is assumed that the conditions of too little fuel have already been checked and the fuel pump is the cause of the difficulty.)

An air leak at the fuel pipe fittings indicate either poor installation of pump or a defective fitting. The fitting should be tightened or replaced.

A leak at the diaphragm flange may be caused by a warped cover casting, loose diaphragm cover screws or foreign material be-

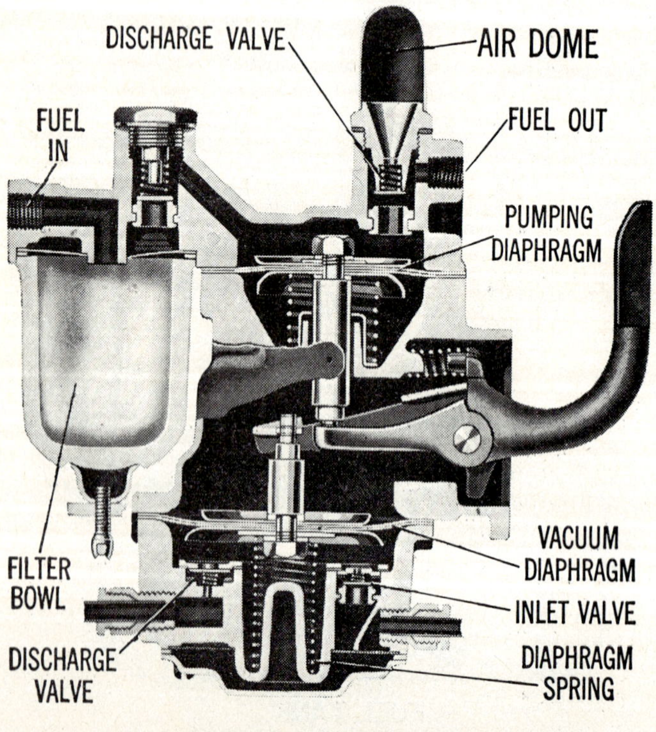

Fig. 4 Fuel pump with fuel filter and vacuum booster for windshield wiper operation

Fig. 5 Testing fuel pump pressure

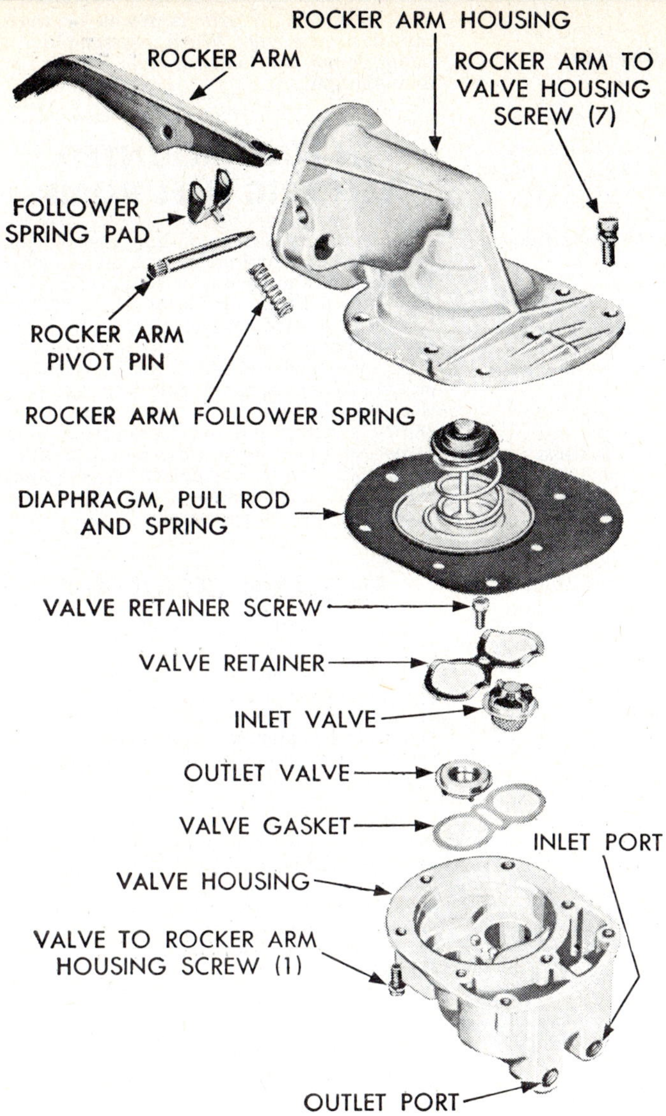

Fig. 6 Auto-Lite fuel pump

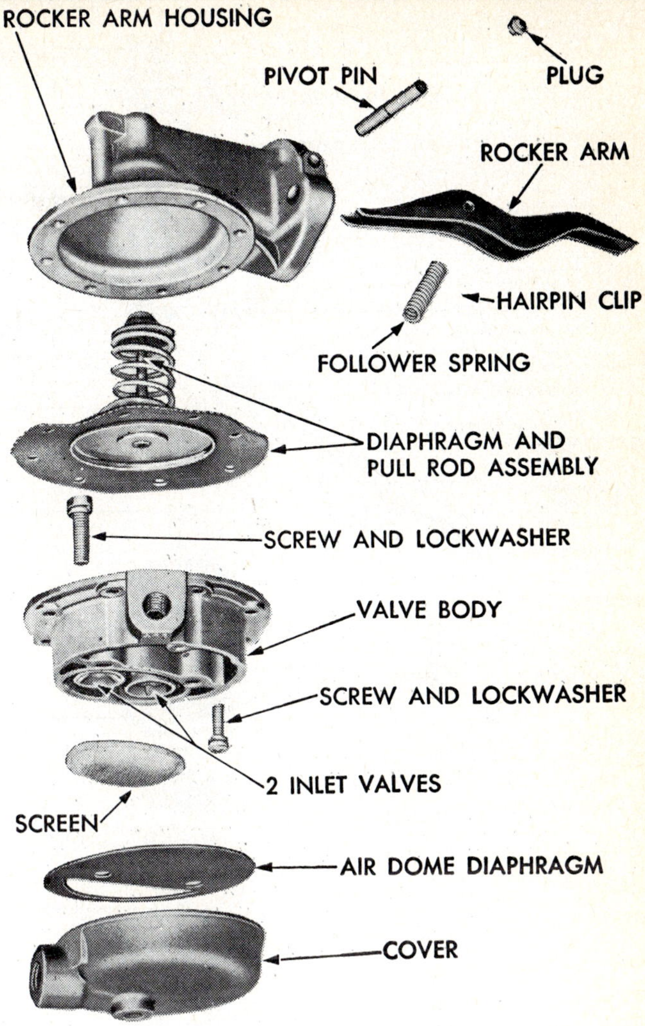

Fig. 7 Carter fuel pump

tween cover casting and diaphragm.

A leak at the bowl flange of the cover casting can usually be corrected by the installation of an extra gasket. A warped top cover indicates that the pump must be replaced.

A chipped glass or bent metal bowl may cause a leak at the bowl flange as may a defective gasket or foreign material between gasket and bowl or cover casting. A chipped glass bowl must be replaced while a dented metal bowl can be straightened.

Vacuum Pump Troubles

To assist the manifold vacuum to operate the windshield wiper at a uniform rate under any engine load is the only function of the vacuum pump. Of course, "uniform rate" infers a wet windshield and not one covered with snow or ice. Failure to do the above indicates difficulty in either the vacuum system of the pump, windshield wiper motor, or tubes and connections.

Symptoms of trouble in the vacuum section show up in four ways: oil consumption, slow windshield wiper action, poor idle, noise.

In some cases it has been found that an engine which has given very good oil mileage suddenly appears to be using oil. Upon investigation, it will often be found that the vacuum booster has a ruptured diaphragm and is drawing oil fumes from the crankcase into the intake manifold. This can be checked by removing the cover of the vacuum section.

When the windshield wiper slows down excessively under engine load, it usually is an indication of a ruptured vacuum diaphragm or defective valve action in the vacuum pump. This condition may not be discovered immediately as the windshield wipers may not be used for long intervals.

Oil will be evidenced in the cover casting recesses if the diaphragm is ruptured. The pump should be removed and the diaphragm replaced.

On some vehicles the engine will idle very poorly when the vacuum diaphragm is ruptured. This is true when the tube from the vacuum pump is connected to one end of the intake manifold. The air leak through the vacuum section will give those cylinders on the end a lean mixture which results in a miss or poor idle. In many cases, this leads one to believe the valves of the engine are sticking; but if they are ground, the miss or poor idle remains.

This condition can be checked by removing the vacuum pump tube at the manifold and plugging the hole. If the miss or rough idle disappears, the trouble is an air leak through the pump or tube connections. The pump should be removed and repaired or the connections tightened to eliminate this trouble.

Sometimes a combination pump will give off a peculiar grunting sound on idle. In some cases this can be remedied by stuffing curled horse hair into the pump breather.

Vacuum Pump Test

With a combination fuel and vacuum pump the windshield wiper should operate at 80 to 100 strokes per minute through all ranges of car speed and load. The windshield should be wet when the test is made, otherwise the action will be slow.

Checking With Vacuum Gauge

To check the vacuum section, disconnect both inlet and outlet tubes and attach a vacuum gauge to the inlet (side that goes to windshield wiper). It is assumed that the engine, windshield wiper motor and blade, and connecting tubing have been checked and are in satisfactory condition.

Read the gauge when the engine is running at 100 rpm (about 20 mph). It should read

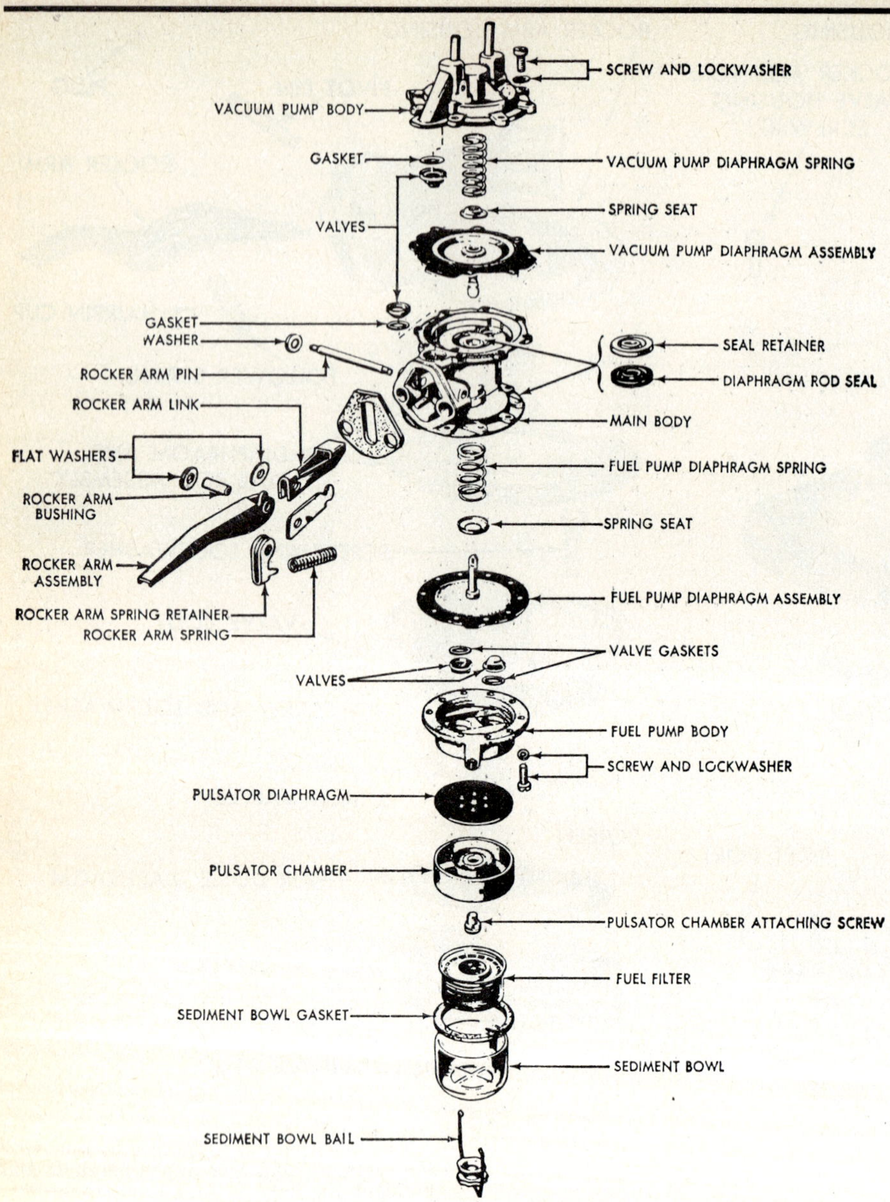

Fig. 8 A.C. combination pump with four vacuum valves and built-in fuel filter

from 7 to 12 inches of vacuum on a normal pump. If the reading is less than 7, the pump should be removed and repaired or replaced. When making this test, the tube to the manifold should be plugged and the pump outlet should always be open or damage may result to the mechanism.

Checking Without Gauge

Disconnect the outlet tube (to manifold) from the pump and plug the end. Then operate the engine from idle through slow acceleration to about 40 mph. If the wiper starts operating at about 15 mph and reaches full speed at about 40 mph, the vacuum section is okay. If it does not operate, it may be the windshield wiper motor. This can be checked by connecting the intake manifold directly to the windshield wiper tube. Then slowly accelerate the engine from idle to about 25 mph. The wiper should operate at full speed. If it does not it can be assumed that the wiper motor or tubing is defective.

Fuel Pump Service

NOTE: Most modern fuel pumps are sealed, non-serviceable units. These pumps cannot be repaired and must be replaced when defective.

Illustrated are representative fuel pumps among the many models that have been produced. Before disassembling any pump, scribe a mark across the housings in such a manner that they may be reassembled with inlet and outlet fitting holes in correct location.

When disassembled, clean all parts except diaphragms) in solvent and blow dry with compressed air. Examine the diaphragm for cracks, torn screw holes or ruptures. If deteriorated, install new diaphragm and pull rod assembly. Check the strainer screen and if found to be corroded or clogged, install a new screen. Check the rocker arm for wear or scoring on that portion which contacts the camshaft eccentric. If arm is scored or worn, install a new one. When reassembling a pump, do not use shellac or other adhesive on a diaphragm.

IN-LINE MOUNTED ELECTRIC FUEL PUMP

Some trucks use an in-line positive displacement type fuel pump with a wet motor. Lubrication and cooling is provided by fuel flow through the pump. The pump is protected by a filter located between the fuel tank and the fuel pump.

On some models a primer switch is located on the instrument panel. This delivers a positive supply of fuel to the carburetor before engine starting and reduces starting time. Some units also incorporate an oil pressure safety switch which will de-activate the pump when the engine has stopped or in the event of excessively low oil pressure.

IN-TANK MOUNTED ELECTRIC FUEL PUMPS

Some trucks use an in-tank, positive displacement electric fuel pump. This pump has an enclosed, dry motor driving the impeller and in some applications, a magnetic coupling between the pump motor and impeller is incorporated to protect the motor in event of impeller seizure.

The pump is actuated by a primer switch on the dash panel before engine starting is attempted. This delivers a positive supply of fuel to the carburetor. Some units incorporate an oil pressure safety switch which de-activates the pump when the engine has stopped or in the event of excessively low oil pressure.

AUTO PULSE ELECTRIC FUEL PUMP

This bellows type pump, Fig. 10, has a magnetic activating coil in the base with a drive spring through the center of the coil. A pressure screw is incorporated to increase or reduce the drive spring pressure to control output pressures. The armature and contacts are located above the coil. The drive spring presses against a vertical shaft to which the upper contacts and one-piece metal bellows are attached. The bellows, in addition to performing the pumping action, acts as a seal between the pumping chamber and the electric mechanism. The pumping chamber is above the electric mechanism and consists of an inlet and outlet valve through which the fuel flows. A screen is located in the top, or dome, of the pump.

Operation

On the intake stroke the coil is energized, extending the bellows and compressing the drive spring. The bellows then returns to its normal position in the pressure stroke, pushing fuel through the outlet valve. The contacts are open during this interval. At the end of the pressure stroke, the contacts close, energizing the coil and drawing the armature and bellows downward on the intake stroke. Electrical contact is made for only a short time during the intake stroke. This cycle is repeated as long as fuel is required.

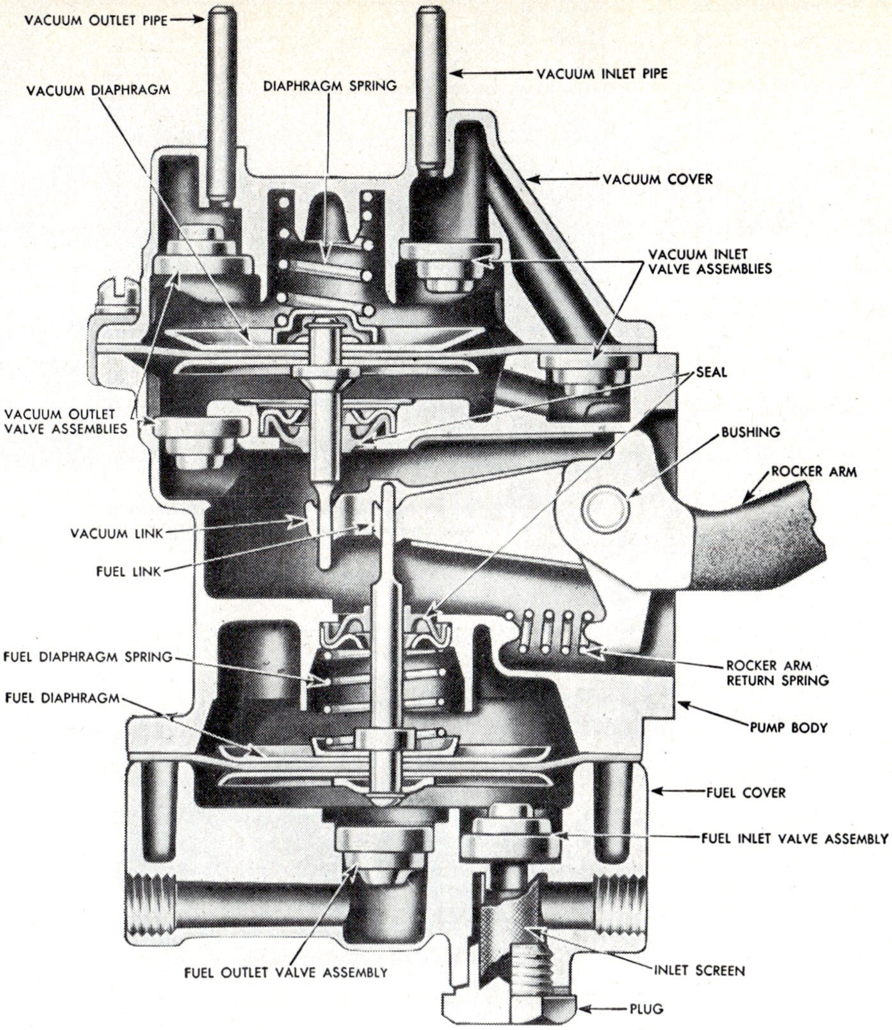

Fig. 9 A.C. combination pump with four vacuum valves

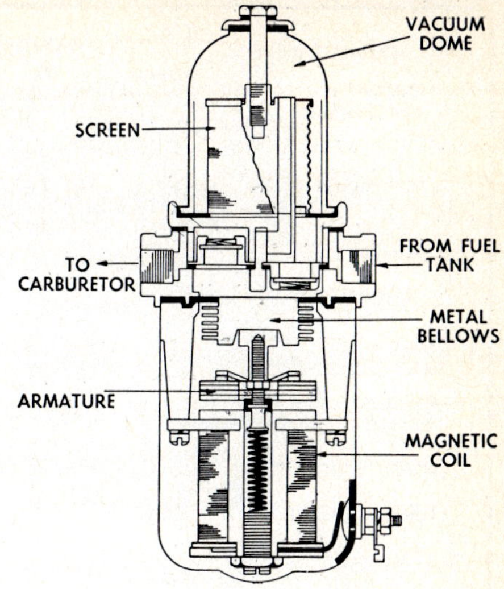

Fig. 10 Auto Pulse electric fuel pump

Trouble Shooting

Disconnect the fuel line at the carburetor and turn on the ignition switch. If ample fuel flows, the pump is okay. If the pump races but fuel does not flow in sufficient volume, the cause is an air leak in the suction portion of the fuel line. If fuel does not flow at all, inspect the fuel lines for restrictions or check the electrical connections for a shorted or open circuit. If the pump still does not operate, remove the pump and replace or repair as necessary.

Disassembly

1. Take off the filter bowl and screen.
2. Remove two screws attaching filter base.
3. The electrical unit of the pump may be exposed by removing the lower cover.
4. To separate the coil from the rest of pump, remove the pressure nut screw. Then loosen the drive screw nut below the bellows stud.
5. Insert a thin screwdriver and loosen the drive screw.
6. Remove the coil mounting screws.
7. Separate the coil from the armature by removing the armature mounting screws.
8. The bellows are removed by taking out the four bellows ring screws from the pump base.
9. Remove bellows ring, bellows, gasket and air seal gasket.

Inspection

Wash the filter and bowl in solvent and blow dry with compressed air. The valves may be resurfaced with crocus cloth. Clean the faces of the pump base and bellows ring with a soft, dry cloth. Replace all defective parts.

Assembly

Reverse the disassembly procedure to reassemble the pump, noting the following: Always use a new bellows when rebuilding. Make sure the inlet and outlet valves are not tight in their chambers. Use new gaskets throughout the assembly.

TURBOCHARGERS

A turbocharger is essentially an exhaust driven blower, the purpose of which is to increase engine power by supplying compressed air to the combustion chambers, permitting greater fuel consumption at an efficient air-fuel ratio. The turbocharger is not connected to the engine power train; therefore, it is not a horsepower parasite during operation. The unit automatically responds to engine load demands without any control connections between turbocharger and engine.

In outward appearance, turbochargers are made in various shapes but they are of two general types. In one type the unit consists basically of a compressor (main) housing and a turbine housing, Fig. 1. The other type consists primarily of a compressor housing, a turbine housing and a separate bearing housing. This design is shown mounted on a diesel engine in Fig. 2.

In both types the turbine housing encloses the turbine wheel and provides an exhaust outlet and inlet. The turbine housing also incorporates a nozzle ring which directs the exhaust gases to the turbine wheel. The nozzle ring is designed to increase the velocity of the exhaust gas before it enters the turbine wheel. The hot gases, expanding through the turbine housing, rotate the turbine wheel, thus producing power to drive the compressor.

The compressor housing in both types encloses the compressor wheel and provides an air inlet and outlet. Intake air from the engine air cleaner enters the compressor housing, is centrifugally compressed by the action of the wheel, and is supplied to the intake manifold of the engine.

The turbocharger shown in Figs. 1 and 3 is provided with a semi-floating sleeve bearing which is lubricated with oil from the engine lubrication system. A piston ring type oil seal prevents leakage at the turbine end of the shaft, while a mechanical-face type seal is usually used at the compressor end.

The turbocharger shown in Figs. 2 and 4 is lubricated by oil from the engine lubricating system also. Passages in the main housing direct oil under pressure to the bearings. The oil circulating through the bearing housing acts as a heat barrier between the hot turbine and compressor, and helps carry away heat generated in the bearings. From the bearing housing, the oil returns by gravity to the engine oil sump. Seal rings are used at each end of the shaft to prevent lubricating oil from entering the turbine or compressor.

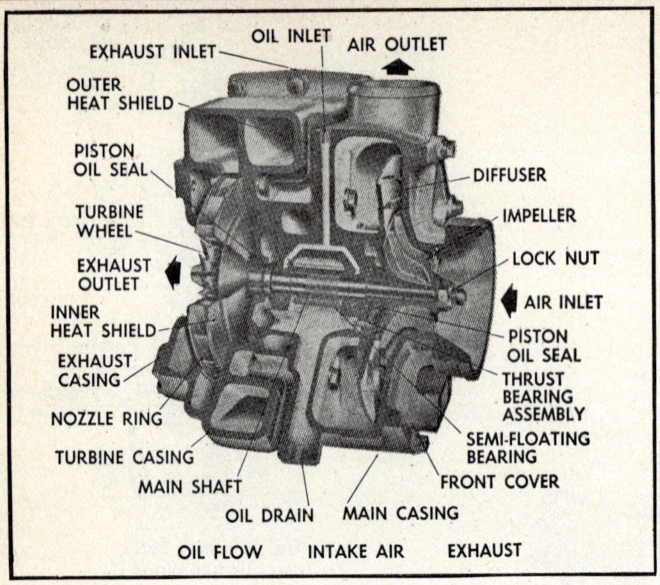

Fig. 1 Cutaway view of a Cummins turbocharger without a separate bearing housing

Fig. 2 Turbocharger with separate bearing housing mounted on a diesel engine

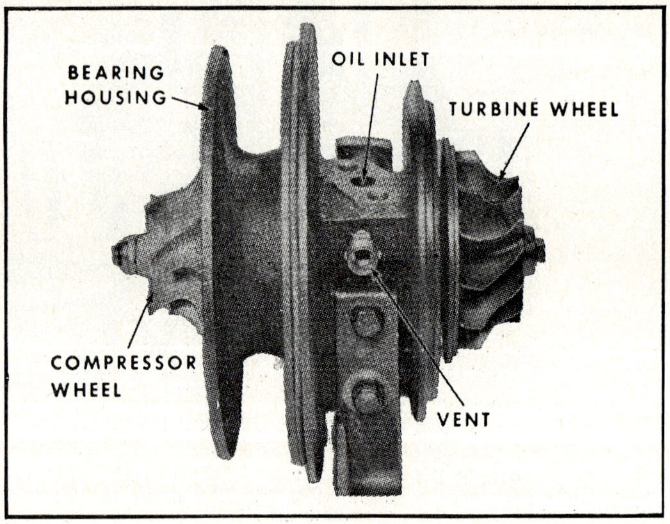

Fig. 4 Turbocharger with separate bearing housing

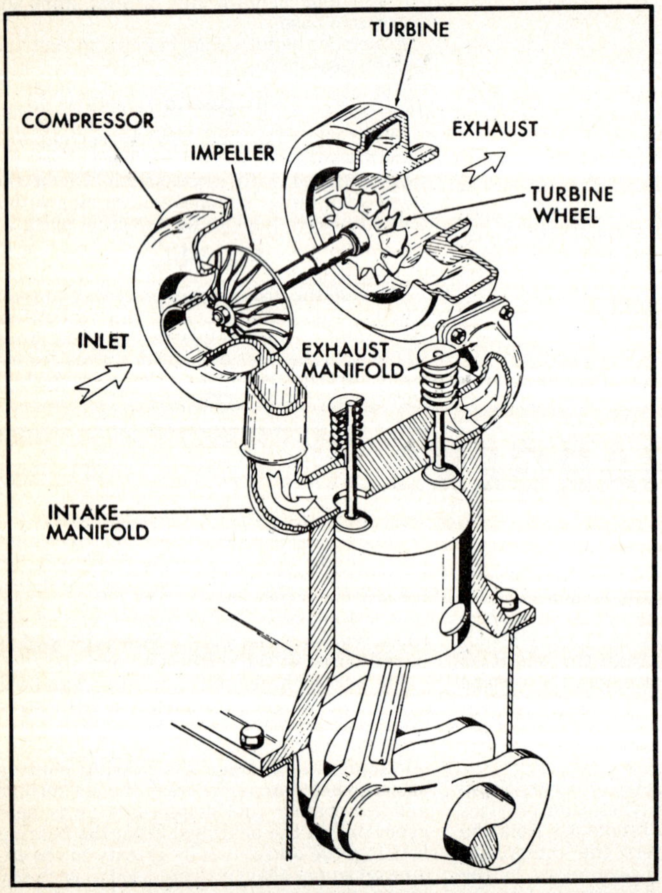

Fig. 3 Turbocharged gasoline engine

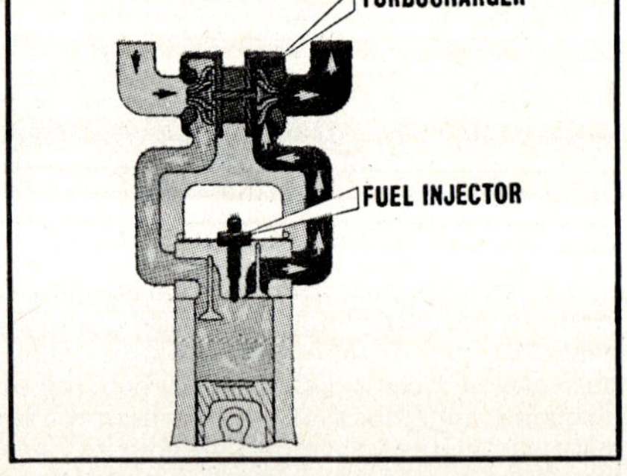

Fig. 5 Turbocharged diesel engine

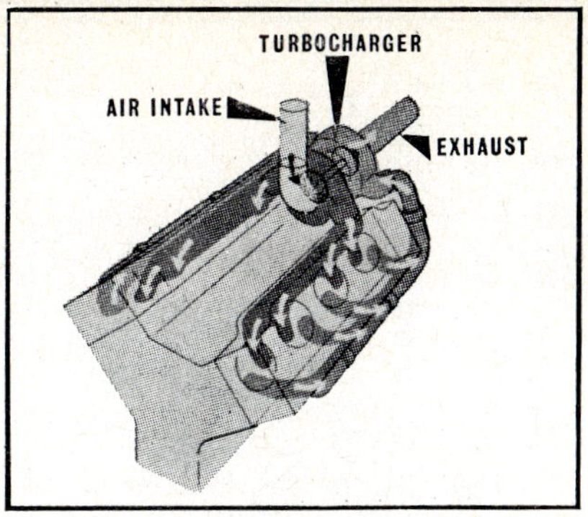

Fig. 6 Air and exhaust flow in turbocharged engine

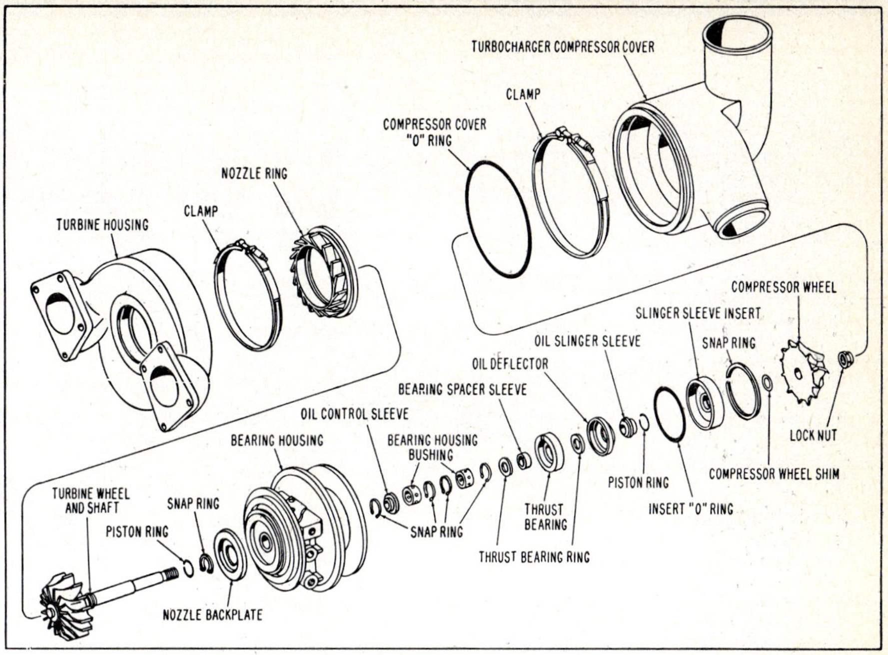

Fig. 7 Exploded view of a turbocharger with separate bearing housing

OPERATION

In a normal breathing engine (one that is neither turbocharged nor supercharged) air enters the engine at atmospheric pressure, mixes with a specified amount of fuel, and is burned in the combustion chamber, producing a certain amount of power.

On a turbocharged diesel engine, Fig. 5, the turbocharger supplies air under pressure to the intake manifold. Thus a greater amount of air enters the combustion chamber. The fuel injection system is calibrated to inject the correct amount of fuel for the increased volume of air. In a gasoline engine, Fig. 3, the principle is the same except that the carburetor is calibrated to provide the correct amount of fuel for the increased volume of air.

The turbocharger, in addition to exhaust volume and velocity, depends upon exhaust heat. Under loaded conditions, at maximum fuel delivery, the turbocharger becomes very efficient and supplies the increased air volume needed to support proper combustion.

Since the turbine speed is governed by the exhaust energy of the engine at any speed, the turbocharger delivers the correct volume of air at any throttle position. Also, less air resistance at higher elevations allows the turbocharger to spin faster and maintain correct air delivery, thereby avoiding the power loss and excessive smoking that occurs on a normal breathing engine and supercharged engines at high altitudes.

Summary: Engine exhaust is directed into the turbine housing. As the exhaust passes through the nozzle ring vanes, it gains velocity and strikes the turbine wheel vanes. This action spins the turbine wheel shaft to which the compressor wheel is attached. As the compressor wheel turns, it draws air from the air cleaner into the compressor housing, compresses it and forces it into the intake manifold. This provides a greater volume of air in the combustion chamber. Because a greater volume of air is delivered to the combustion chamber, a greater amount of fuel is required to obtain the correct air-fuel mixture. This increased consumption of fuel, at the correct air-fuel ratio, results in increased power output.

Service Precautions

NOTE: Regardless of their rugged appearance, turbochargers are to be handled with extreme care. The internal parts are manufactured to close tolerances and are sensitive to the accumulation of dirt, nicks and scratches. The turbine wheel, compressor impeller, shaft and related parts form a balanced assembly which rotates at very high speeds (65,000 rpm and higher in many applications). Bending of the turbine or compressor vanes or removal of any vane material during servicing could cause a serious out-of-balance condition which could quickly destroy the turbocharger. *Therefore, any major repairs to the unit should be performed by a specialty shop having the necessary special tools and equipment to do the work in accordance with the manufacturer's instructions and specifications.*

Because of the greater air flow requirements of a turbocharged engine, it is extremely important that the air intake system be kept clean. The air cleaner must be serviced at recommended intervals to avoid power loss due to air cleaner restriction and to prevent possible entry of dirt into the turbocharger and engine.

Turbocharging an engine is more than merely bolting on the turbocharger and driving away. The installation must be carefully engineered to provide the best performance and to avoid "over-charging" which would be destructive. Also, the engine must be designed to handle the greater air and fuel flow, higher pressures and temperatures, and increased torque and power output.

MAINTENANCE

General Handling

1. Prevent the entrance of foreign material by covering or plugging all openings into the turbocharger when the unit has been removed from the engine. Covers must remain in place until the unit is installed and all connections are secured.
2. All connections to the turbocharger (manifolds and piping) must be clean and free of foreign material, since serious damage to the turbocharger or engine could result. All connections must be air-tight.
3. Exhaust stacks of extra long length and other fixtures must not be rigidly attached to the turbocharger.
4. For initial running when installing a new or rebuilt turbocharger or after the engine has been shut down for 30 days or more or when engine oil filters have been changed, it is recommended that 4 or 5 ounces of oil (same type and grade as used in the engine) be injected into the oil inlet opening in the turbocharger. This will provide initial lubrication for the turbocharger bearings until normal engine lubrication is established.
5. If engine oil becomes contaminated by water, the turbocharger must be drained to prevent sludge formation. It is recommended that after an engine overhaul, an oil inlet filter be installed and used during the first 5 to 25 hours of operation and then removed, as this filter can clog with carbon that is normally suspended in the oil of a diesel engine.
6. It is imperative that the air cleaner be serviced at recommended intervals because of the oil carry-over and power losses that can be incurred with a restricted air cleaner.

NOTE: Air flow requirements for turbocharged diesel engines are considerably greater than for a non-turbocharged engine of the same size running at the same speed. Air inlet accessories must minimize the restriction of this higher air flow and maintain performance of the turbocharger unit.

7. The engine crankcase breather should be cleaned periodically to be sure there is no restriction.
8. During normal operation, the turbocharger should be free from vibration or unusual noises.
9. The exhaust stack should be covered to prevent water from entering and damaging the turbine during shutdown periods or when unit is being transported.
10. Periodic inspection of the compressor wheel should be made to check for soft carbon deposits, damaged blades, interference or excessive end play.

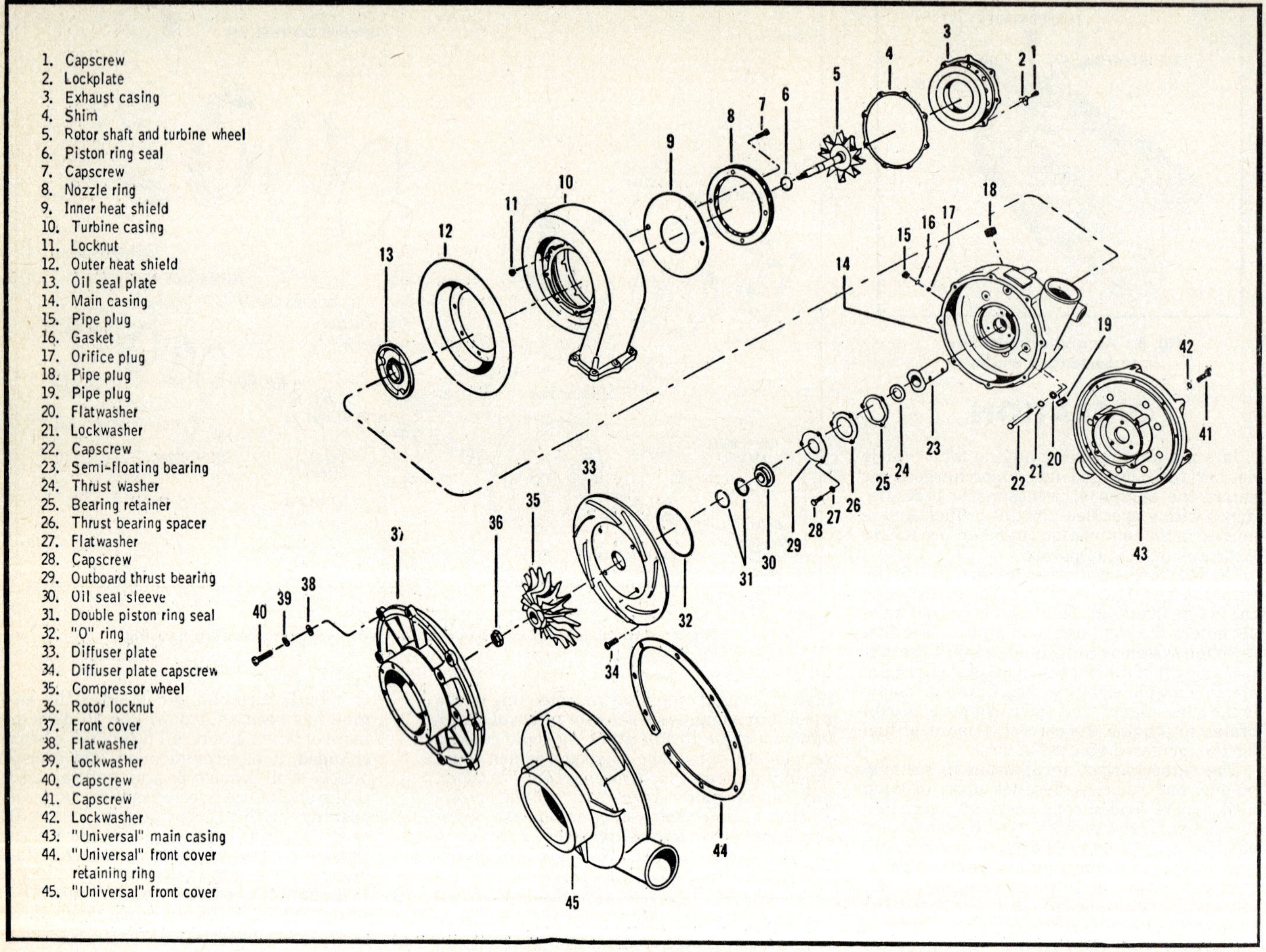

Fig. 8 Exploded view of a turbocharger without a separate bearing housing

LPG CARBURETION

INTRODUCTION

LPG (liquid petroleum gas) is an ideal engine fuel similar to gasoline. Both are a product of crude oil but they differ in that LPG is more volatile; that is it wants to boil or become a vapor at ordinary atmospheric temperatures. When used in an internal combustion engine, LPG is converted from a liquid in the fuel tank to a dry gas before reaching the combustion chamber. By good carburetion and perfect mixing of dry gas with air, uniform distribution throughout the manifold is achieved. With gasoline, or wet fuel, this uniform mixture and distribution is not attained.

With complete burning at a rate slower than gasoline, LPG makes for smoothness of operation because of prolonged power impulses on the power stroke and more uniform bearing pressures.

Being a dry gas, crankcase dilution is minimized and likewise all ill effects such as thinning of the lubricating oil and washing the lubricating oil from the cylinder walls with resultant cylinder wear are eliminated. Lubricating oil can be safely used much longer than on gasoline due to the elimination of carbon deposits, dirt, acid, and water content in the lubricating oil.

Harmful carbon within the engine is eliminated. There is no exhaust smoke or obnoxious odors. The time interval between engine shut-downs for overhaul is prolonged. Rings, valves, cylinders, and all moving parts lubricated from the crankcase last longer, resulting in a very substantial reduction in cost of keeping equipment in operation.

DESCRIPTION

Butane and propane gases, captured at the refinery and liquified by being put under pressure, are blended to a desired mixture and pumped as a liquid under pressure into specially constructed heavy steel tanks mounted on the truck. Tank pressures will vary from 20 to 175 psi (pounds per square inch) depending on the butane-propane mixture and atmospheric temperature. By reason of this tank pressure, no fuel pump is necessary. L. P. Gas tanks are never filled 100% full as 10% to 20% space must be allowed above the liquid for expansion.

Fig. 1 shows the flow diagram of the system.

Filtering

From the tank the liquid is piped to a liquid filter where scale from tank and fittings, and dirt or other solids are removed. From the filter, the L. P. Gas goes into the regulating unit. Final filtering is accomplished through a fine metal screen inserted in the fuel inlet connection of the regulator to further eliminate passage of undersirable foreign matter into the regulator.

Primary or High Pressure Regulation

LP6 is dropped from tank pressure (20–

175 psi) down to a regulated 1 to 10 psi. By use of the oil-controlled pressure regulator, the pressure is maintained at 1 psi with engine stopped and 10 psi when engine is running.

Vaporization

This function is performed by the vaporizer coils heated by hot water circulated by the engine. The high pressure valve would freeze up under the rapid expansion of gases were it not for the application of heat. This is accomplished by taking hot water from the water manifold, directing it first around the expansion valve, then around the heating coils to vaporize the expanded fuel, then returning it to the intake side of the water pump.

In starting a cold engine, there is sufficient differential in temperature between the extremely cold expanding butane and the parts surrounding the valve to provide latent heat for initial vaporization. However, it is desirable, especially in cold weather, to take vapor from the supply tank for starting and engine warm-up.

Final Pressure Regulation

The final regulator in the regulating unit further reduces the pressure of the completely vaporized butane from its 1 to 10 psi to slightly below atmospheric pressure. This final or "metering" regulator is actuated by engine vacuum or "suction" by the intake stroke which opens the final fuel regulator sufficiently to allow enough fuel to pass along to the carburetor to run the engine according to throttle opening. It actually regulates the supply of gas and shuts off the supply when the engine is stopped.

Idling System

The final regulator also contains the idling system. Even though most fuel used for idling enters the carburetor through the main gas orifice, additional and supplemental fuel for idling is taken from the final regulator and by-passed to a point in the carburetor directly above the throttle disc. The idle adjustment is in this line. This source of idling fuel is independent of the main gas supply to the carburetor and is controlled by an entirely different system. The two sources combine to produce the correct idling mixture.

Carburetion

From the regulating unit, the L. P. Gas is piped to the carburetor for final mixing and proportioning of the fuel with air. The carburetor is closely related to the final regulator of the regulating unit and with the two in correct coordination, air-fuel ratios are automatically controlled for greatest power, economy and flexibility.

The carburetor has an "easy-starting" feature wherein by pulling the choke entirely shut, a separate air-gas mixing device is put into operation to produce a pre-determined starting mixture. A separate adjustment is provided on the carburetor for adjusting the starting mixture. When the choke is released, the starting device is rendered inoperative and the main carburetor air-fuel orifices come into play. Therefore, the choke must be entirely *closed* for starting and entirely *open* for running.

A fuel economizer, actuated by manifold vacuum, is also utilized to reduce fuel consumption when operating at less than 80% full power.

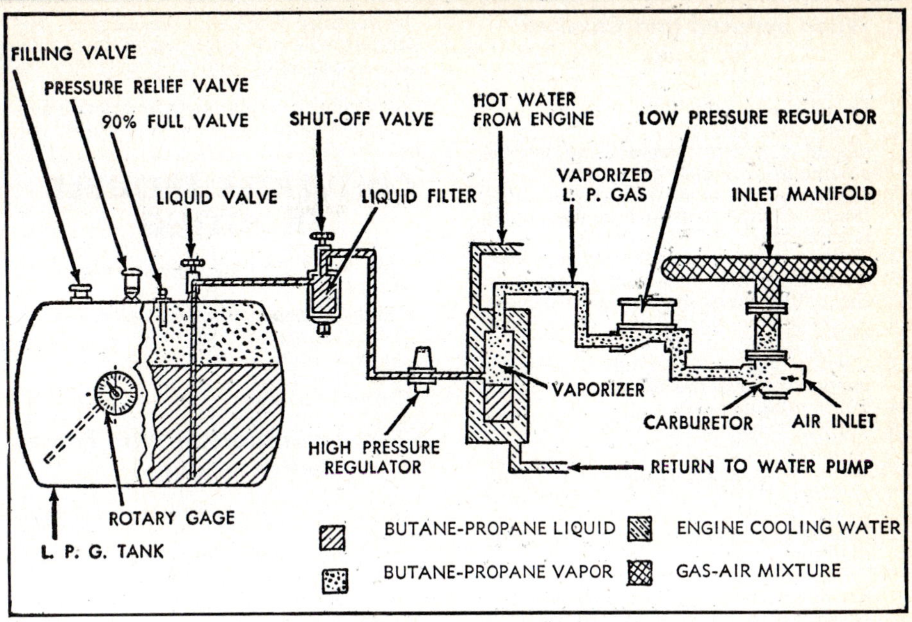

Fig. 1 Fuel diagram of LPG (liquid petroleum gas) fuel system

LPG SAFETY MEASURES

When Servicing Truck

1. To avoid accumulation of gas-air mixtures in and around a truck caused by undetected leaks, select a place where there is good air circulation.
2. Keep as far as possible from hot dip tanks, hot water cleaners, steam cleaners, welders or any device operating with an open flame or which may produce sparks such as cutting, welding, grinding, chiseling or any such similar operation.
3. If the truck is placed over a pit or near one, a fan should be put in the pit for ventilation as the gas is heavier than air and can easily accumulate in a pit with no air circulation. Nearby open stairways, or elevator shafts leading to basements should be provided with air circulation to keep gas out, or basements should be ventilated.
4. Absolutely no smoking.

Preparing Truck for Service

1. Before taking the truck into the shop for service, exhaust all fuel from the system by shutting off the main valves at the fuel tanks and allowing the engine to run until it stops. Then turn off the ignition.
2. Before towing a disabled vehicle into the shop, shut the valve at the tank and bleed the fuel system of all fuel and gas.
3. Never bring a vehicle into the shop without first checking the fuel system and fuel tanks for leaks.

Before Starting Repairs

1. Remember that the fuel system is pressurized.
2. Make certain the tank valves are tightly closed and all fuel exhausted from the lines before starting any repair work on the fuel system.
3. Make certain there are no leaks before doing any repair work on any part of the vehicle.
4. Place a fire extinguisher close by so it can be quickly put to use.
5. It is advisable to place Danger and No Smoking signs on each side of the vehicle as well as warn other people who might wander into the shop.

When LPG Gets on Fire

LPG should be allowed to burn until, if possible, the source of fuel is shut off. A fire extinguished before this is accomplished can and may result in dangerous accumulations of gas which might result in a more serious flash or explosion.

When Servicing Fuel System

1. Replace worn or defective fittings.
2. Treat all threaded connections with an insoluble lubricant such as Permatex or aviation gasket maker.
3. Check entire system for leaks after connecting it up.
4. LPG is treated with rather an offensive strong compound or odorant as a warning agent to indicate even small quantities of gas leakage.

Checking For Leaks

1. Detection of the peculiar odor indicates a possible leak. If detected, do not turn on or off any lights or electrical switches on any device or vehicle. Immediately tow vehicle outdoors until leak is located and danger eliminated.
2. Frosting of connections indicates leakage at that point.
3. Normally there should be a slight film of frost on the hex headed plug and boss just above the inlet fuel line connection to the regulator.
4. Never use open flame to check for leakage.
5. Brush a lather of soap on with a soft brush which will indicate the presence of leaks.
6. All rubber hose regardless of length used anywhere in the fuel line or piping system should be very carefully inspected for loose connections or any sign of checking, cracking, etc. At the least sign of injury or deterioration replace immediately.

When Serviced and Checked for Leaks

1. Do not start engine first.
2. Raise hood and allow air to circulate around engine to drive out any possible accumulation of gas.
3. Check fuel in tank with rotary gauge located on tank to make sure fuel supply is available.
4. To start engine open vapor valve *slowly*.
5. Turn on ignition and start engine with choke fully closed and throttle 1/4 open.
6. After engine starts open choke completely for normal operation.
7. Operate engine at fast idle until water temperature reaches 125°.
8. Turn *on* liquid valve.
9. Turn *off* vapor valve.

Precautions

1. LPG is not intended for any use other than to run the engine.
2. Do not use it as a substitute for solvents to clean parts, seats, cushions, trim, etc.
3. Do not use it in place of compressed air to inflate tires, blow out lines, or any other object or device which can be operated by air.
4. Do not perform any repair work on tanks. Have the necessary tank work performed by qualified firms who service such containers who are informed as to regulations, inspection and test after any repairs are made.
5. Never check fuel tanks without first shutting off engine.

CONVERTER TROUBLE SHOOTING

No Fuel At Converter Inlet

1. Check voltage at filter fuelock coil.
2. Defective filter fuelock coil.
3. Plugged filter.
4. LPG tank empty.
5. Crushed fuel line.

No Fuel To Carburetor With Primer Switch Actuated

1. Primary orifice in primary regulator plugged.
2. Missing primary spring in primary regulator.
3. Secondary regulator valve stuck in closed position.
4. Secondary regulator levers not adjusted properly.
5. Secondary regulator primary plunger not opening valve.
6. Defective primer switch.
7. Defective primer coil.
8. Check primer wiring for proper connections.

Fuel Leaking Through Primary Regulator

1. Dirty or defective valve seat.
2. Broken diaphragm.
3. Distorted valve lever.
4. Improperly installed spring or washers.

Fuel Leaking Through Secondary Regulator

1. Dirt on valve seat.
2. Missing spring.
3. Improperly adjusted levers.
4. Primer plunger too long.
5. Actuated primer.

Freezing

NOTE: Repeated freezing may loosen back cover screws or cause cover distortion.

1. Converter mounted too high.
2. Converter hose fittings too small or restricted by internal corrosion.
3. Internal fuel leak within converter.

LOCKING DIFFERENTIALS

The main purpose of these differentials is that they supply a greater percentage of torque to the wheel with better traction and still furnishes to the wheel with poorer traction as much torque as it is able to absorb under the circumstances.

In the conventional differential the wheel easiest to turn receives the power. Therefore, when traction conditions under the rear wheels are not the same, the driving force is limited by the wheel with the poorer traction (easiest to turn) even though one wheel is on good traction surface.

These locking differentials do not permit shock loads or full engine torque to be transmitted to one axle shaft. They provide power to both rear wheels and maintain the differential action that is necessary when the vehicle is turning a corner to permit the outer wheel to turn faster than the inner wheel.

NOTE: "On car" type wheel balancers are not recommended for use on rear wheels of cars equipped with a locking differential. The rear wheel will drive if in contact with the ground or with a block even though the opposite wheel is raised. *However, this type of wheel balancer may be used by removing the wheel opposite of the one being balanced with the complete assembly raised.*

SPICER POWR-LOK

Fig. 1 shows the cone clutch type whereas Fig. 2 illustrates the plate clutch type. Each assembly consists of two case halves, two cross

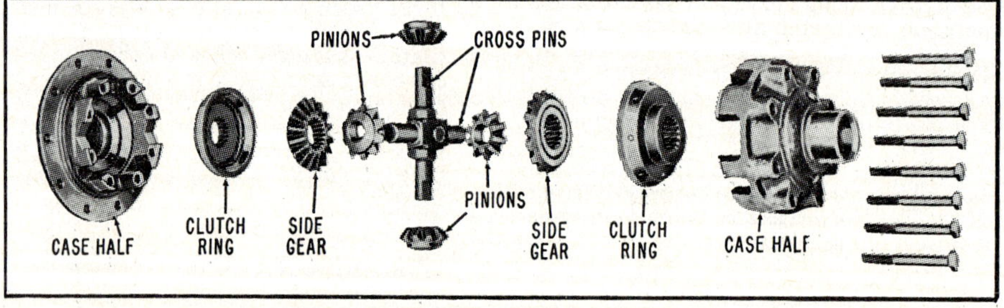

Fig. 1 Cone clutch type of POWR-LOK differential

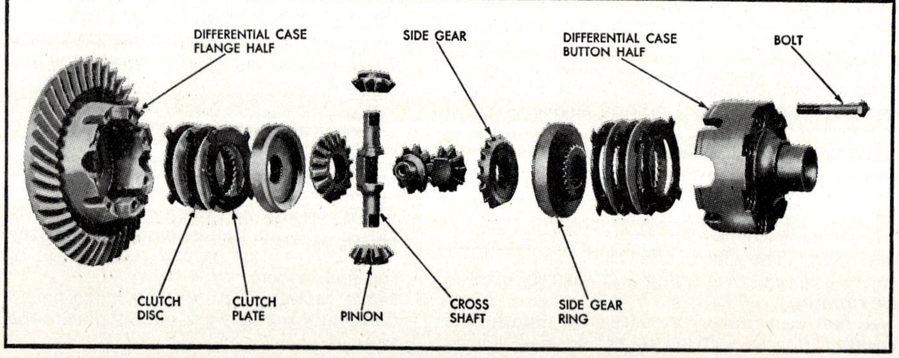

Fig. 2 Plate clutch type of POWR-LOK differential

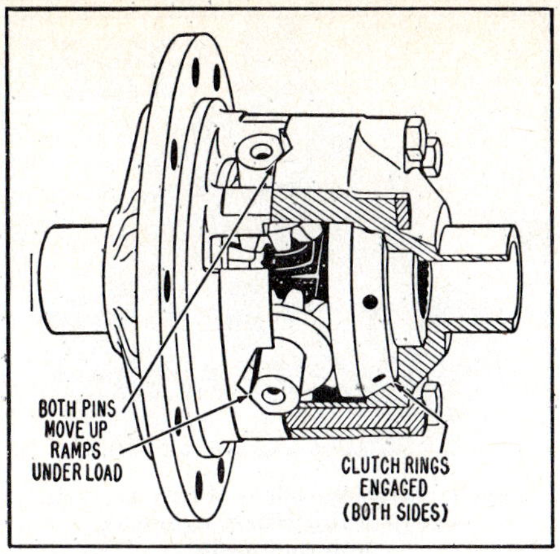

Fig 3. Differential action when moving straight ahead

pins held together by a thrust block, differential pinions, side gears, and clutch rings (Fig. 1) or clutch plates (Fig. 2).

The cross pins are made with a movable joint at the center to permit each one to move independently while in continuous engagement. The pin ends are made in the form of a V, and a similar V is machined in each case half to provide a ramped cam surface.

In Fig. 1, the clutch ring is fitted over each bevel side gear and mates with a cone surface machined in each half of the differential case.

In Fig. 2, the clutch plates operate between the flat surface on the clutch retainer and the machined surface in each half of the differential case.

Operation

As the vehicle is put in motion, the driving force moves the cross pin up the ramp of the cam surface, applying a load to the clutches and restricts turning of the differential through the friction of the clutches on the mating surfaces in the differential case, Fig. 3. This provides a torque ratio between the axle shafts which is based on the amount of friction in the differential and the amount of load that is being applied to the differential.

When turning a corner this process is, in effect, partially reversed, Fig. 4. The differential gears become a planetary gear set, with the gear on the inside of the curve becoming the fixed gear of the planetary. The outer gear of the planetary overruns as the outside wheel on the curve has a further distance to travel. With the outer gear overrunning and the inner gear fixed, the differential pinions are caused to rotate, but since they are restricted by the fixed gear, they must first move the cross pins back down the cam surface, relieving the thrust load of the clutch on its mating surface in the differential case. Thus when turning a corner, the differential, for all practical purposes, is similar to a conventional differential and the wheels are free to rotate at different speeds.

With locking differentials, certain characteristics must be taken into consideration. Under average driving conditions, such as straight ahead, reverse, extreme right and left turns, the operation will be the same as with the conventional differential. However, a little more backlash may be noticed due to the lateral movement of the differential cross pins. In addition, a slight chatter may occur under surge torque with one wheel on a slippery surface. These conditions are considered normal.

CAUTION: Do not attempt to spin a wheel under power with a jack under only one side; both wheels must be clear of the ground. If one wheel remains in contact with the ground, there is a possibility that when spinning the other wheel, friction of the clutch rings or clutch disc may set the vehicle in motion.

Servicing Cone Clutch Type

This type differential, Fig. 1, must be replaced as a unit as individual parts are not available. It is furnished less ring gear and side bearings.

Servicing Plate Clutch Type

Before disassembling the differential, scribe a line across the differential case halves for alignment upon reassembly. Then, after removing the attaching bolts, separate the case halves and remove the differential parts.

Note the relative position of the friction plates, friction discs and Belleville spring plates (if used) so that they may be reassembled in the proper sequence, Figs. 5, 6 and 7.

Inspect all parts. See that there are no worn, cracked or distorted clutch plates. All parts must be free of nicks, burrs or any imperfections that will reduce the efficient operation of the unit.

1. Assemble differential as suggested in Figs. 5, 6 and 7.
2. Make sure the side gear retainer will rotate with a slight drag when in the case. Repeat for opposite side.
3. Install side gear on ring gear flange half.
4. With ring gear flange half in a vertical position, install one shaft and gears, making certain that notch in shaft is up.
5. Install side gear in other half of case.
6. Hold the remaining case half through the bearing trunnion and install it on the ring gear flange half, being sure scribed mark made previously is in alignment.
7. Insert axle shafts to align side gear ring

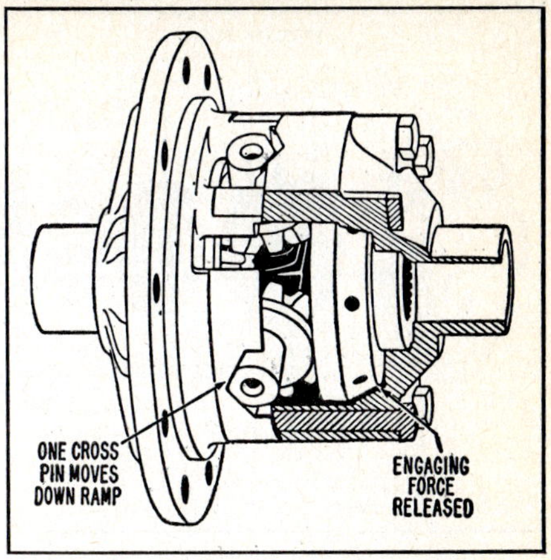

Fig. 4 Differential action when turning a corner

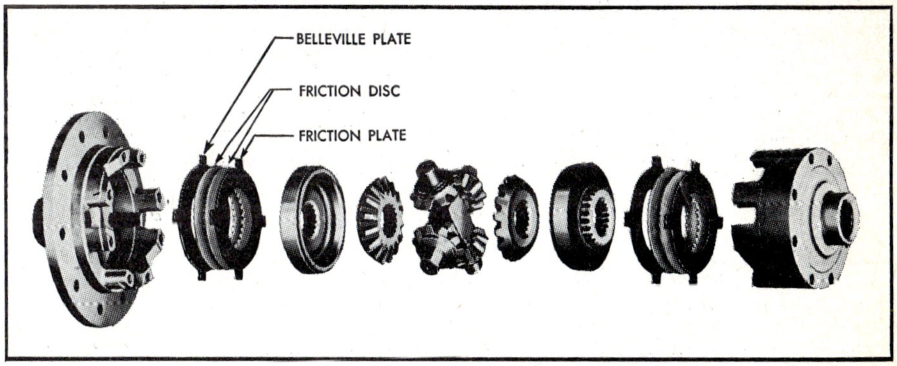

Fig. 5 Plate clutch type of POWR-LOK differential (4 stack unit) with Belleville spring plates at outside of stacks

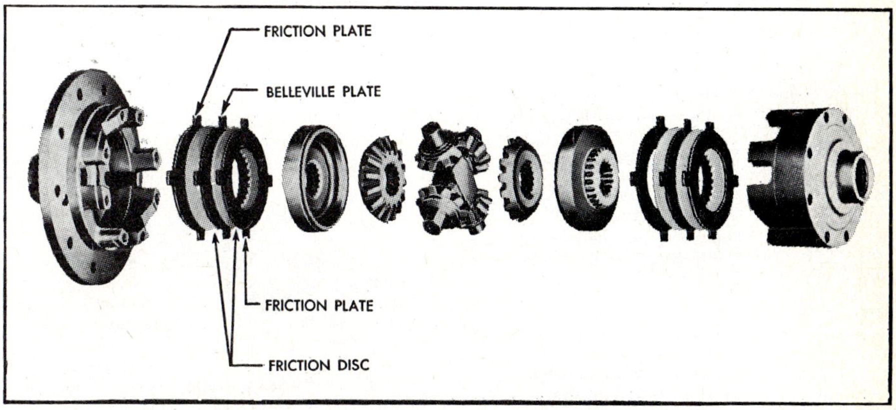

Fig. 6 Plate clutch type of POWR-LOK differential (5 stack unit) with Belleville spring plates in center of stacks

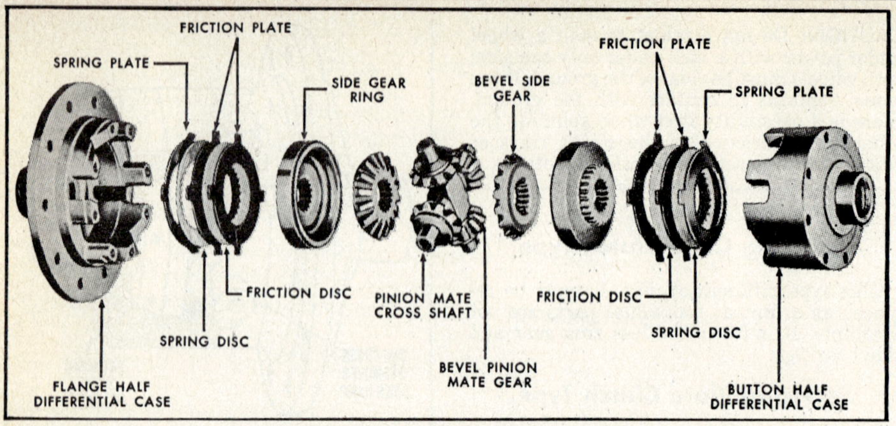

Fig. 7 Plate clutch type of POWR-LOK differential (5 stack unit) with Belleville spring plates at outside of stacks

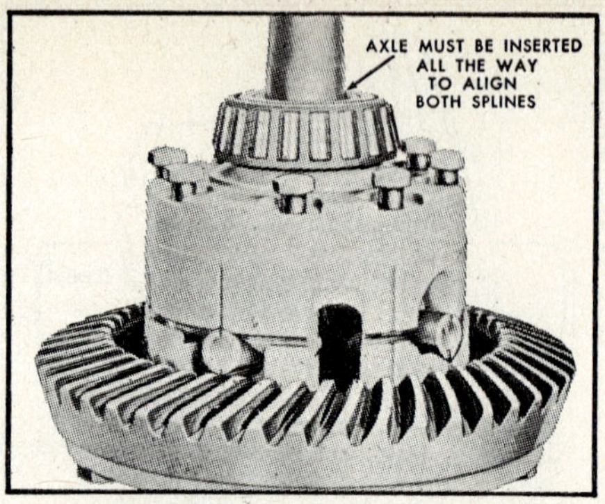

Fig. 8 Aligning splines of side gear and side gear ring before tightening attaching bolts

with bevel side gear, Fig. 8. If this is not done there may be some difficulty when installing axle shafts during final assembly. Tighten attaching bolts evenly to 35–45 ft-lb.

8. Check clearance between pinion mate shaft and the V of the case, Fig. 9. Do this by placing shim stock or feeler gauges on both ends of the same shaft and on opposite sides of the V. The clearance should be approximately .015″.

Installation Note

Make sure the spline end of the axle shaft does not interfere with the pinion mate shafts. This can be checked with a steel tape by measuring from the bottom of the axle shaft bearing bore to the pinion mate shafts. Then measure the axle shafts from the corresponding point of the bearing to the end of the spline. The minimum clearance required is 1/8″. Grind off the spline end of the axle shaft if it is too long. Check the other axle shaft in the same manner.

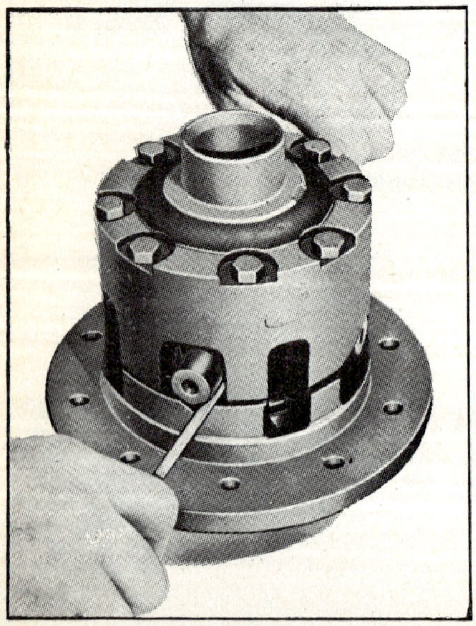

Fig. 9 Checking clearance between pinion mate shaft and V in case. Feeler gauges must be inserted on both ends of same shaft and on opposite sides of V

CAUTION: Use rear axle lubricant of the type specially recommended for locking differentials. Failure to use the proper lubricant will result in chattering or other serious malfunction.

FORD EXC. TRACTION-LOK

Disassembly

1. Referring to Fig. 10, mark one differential bearing cap and the mating support to help position the parts properly during assembly of the carrier.
2. Remove bearing adjusting nuts and lift differential assembly out of carrier.
3. Remove differential bearings.
4. Loosen *alternate* drive gear attaching bolts evenly to release the spring pressure between differential case and cover.
5. Remove differential case cover and then the drive gear.
6. Remove clutch plates.
7. Remove two Belleville springs.
8. Remove clutch hub, side gear and thrust washer.
9. Drive out differential pinion shaft lock pin.
10. Drive out differential pinion shaft. Then remove pinion gears and thrust washers.

Assembly

1. Lubricate all parts during assembly. Referring to Fig. 10, place side gear and thrust washer in differential case.
2. With soft-faced hammer, drive pinion shaft into case only far enough to retain a pinion thrust washer and pinion gear.
3. Place a second pinion and thrust washer in position, and drive pinion shaft into place. *Carefully line up pinion shaft retainer holes.*
4. Place second pinion and thrust washer in position. Drive pinion shaft into place. Install pinion shaft retainer. *The retainer must not extend beyond the machined surface of the case.*
5. Insert two 2″ × 7/16″ bolts through differential flange and thread them 3 or 4 turns into the drive gear as a guide in aligning drive gear bolt holes. Press or tap drive gear into position.
6. Clamp differential case in a vise. Install differential side gear on pinion gears. Place clutch hub on side gear and place thrust washer on hub.
7. Insert two 3/16 × 2″ dowel pins into differential case. Place a steel plate on differen-

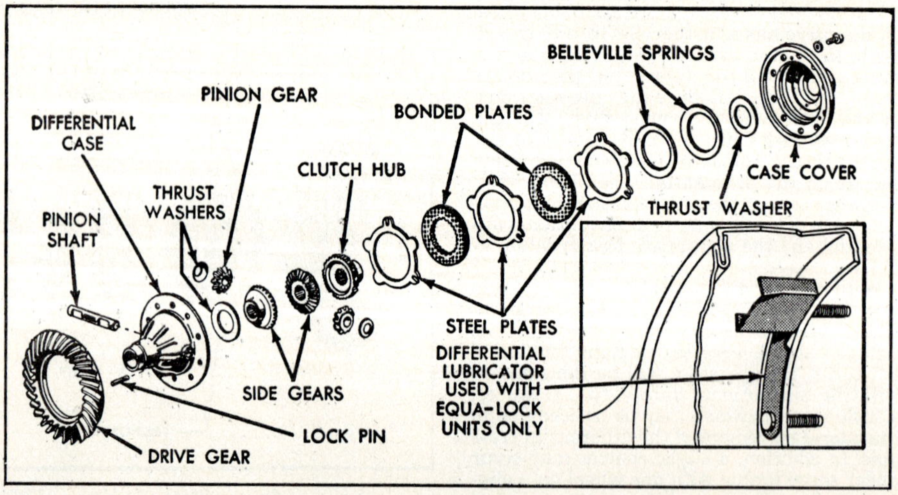

Fig. 10 Ford differential, exc. Traction-Lok

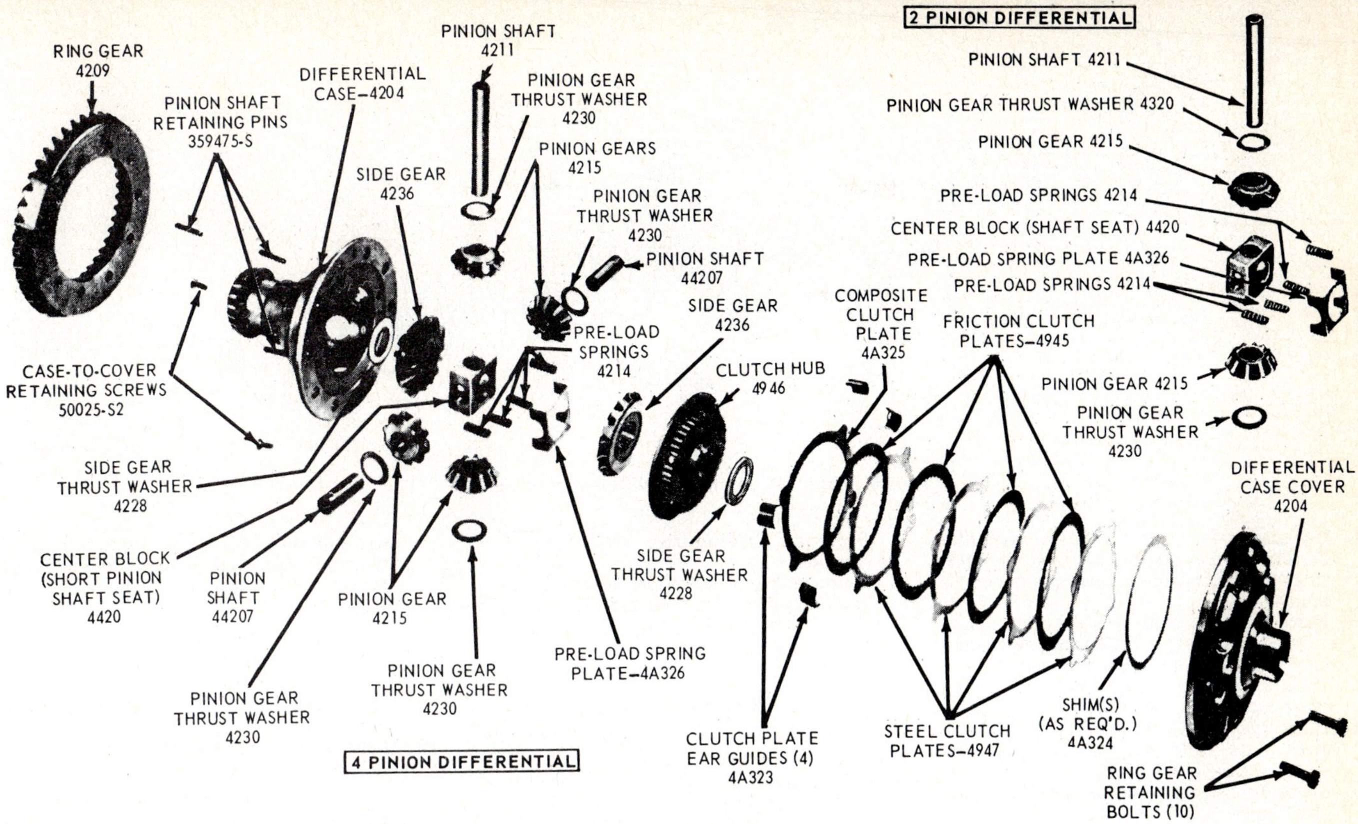

Fig. 11 Ford Traction-Lok differential

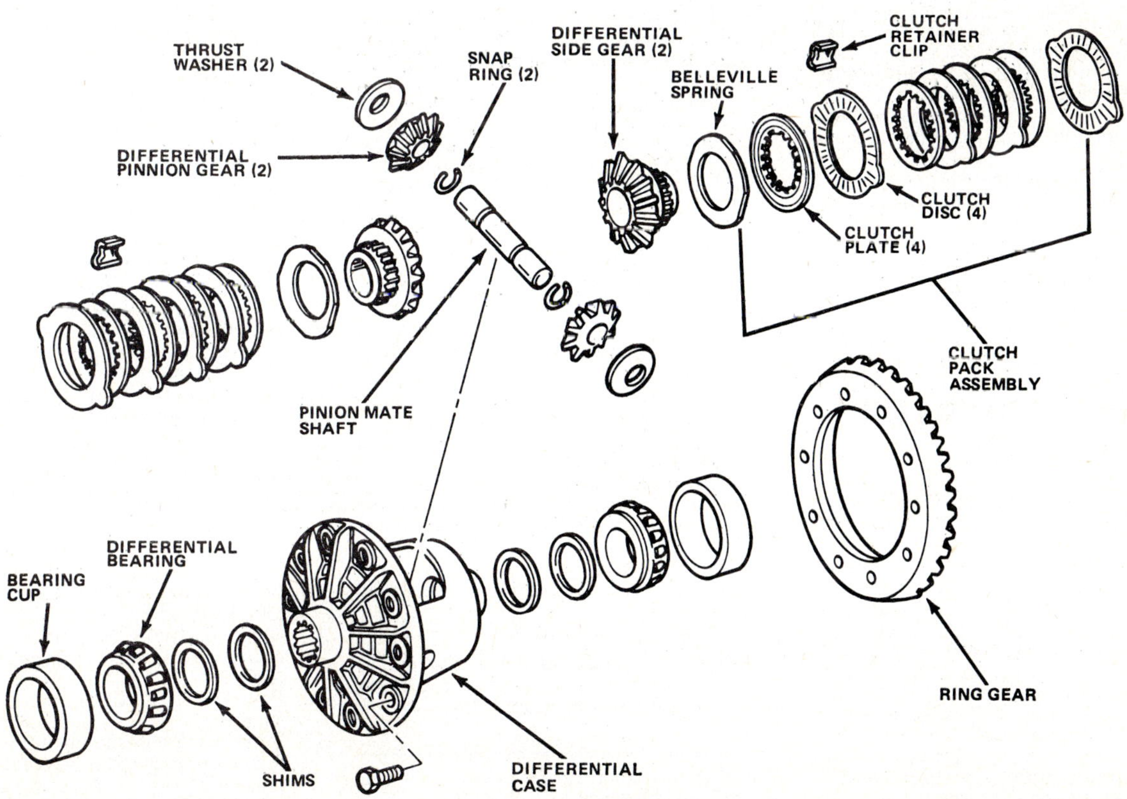

Fig. 12 Dana Trac-Lok locking differential, exc. Model 70 (Typical)

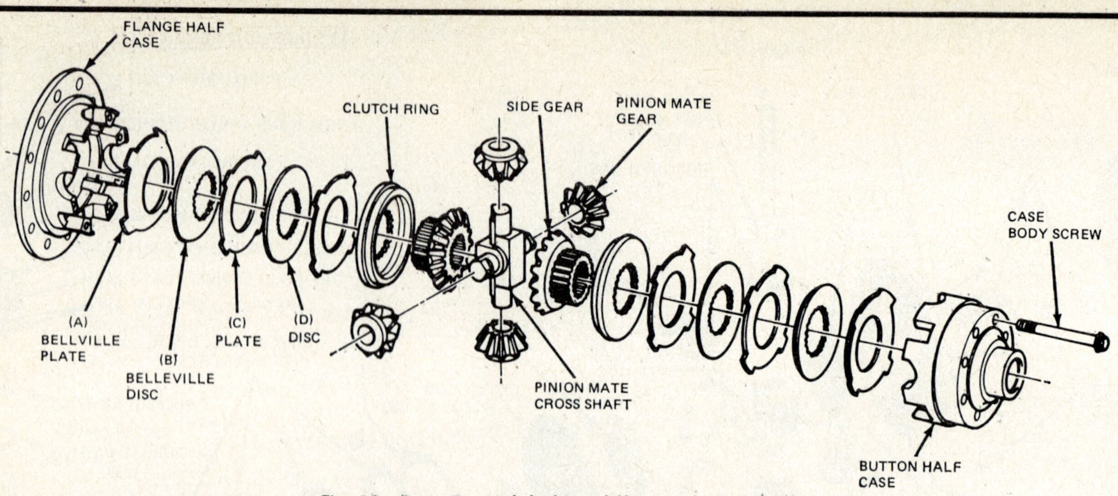

Fig. 13 Dana Trac-Lok locking differential, Model 70

tial case so that slots in dog ears straddle dowel pins. *Lubricate all differential parts so that an accurate torque check can be made.*

8. Place a bonded plate on steel plate. Make sure bonded plate inner spline teeth properly engage hub spline. Assemble remaining plates: a steel plate, a bonded plate and lastly a steel plate.
9. Place two Belleville springs on the top steel plate. Both springs are assembled with concave sides down.
10. Place differential case cover on case and start bolts.
11. Tighten bolts evenly and alternately across the diameter of drive gear. As the bolts are tightened the Belleville springs are compressed and the case and cover are pulled together.
12. Torque the bolts to 65–75 ft-lbs.
13. Check torque required to rotate one side gear while the other side gear is held. Ignore the torque required to start the side gear turning. The torque required to keep it moving steadily should be between 155 and 195 ft-lbs. If not within these limits, check for improper assembly.
14. Install differential in carrier.

CAUTION: Use rear axle lubricant of the type specially recommended for locking differentials. Failure to use proper lubricant will result in chattering or other serious malfunction.

FORD TRACTION-LOK

Disassembly

1. Referring to Fig. 11, remove ring gear-to-differential case bolts. Remove ring gear.
2. With case in press or using two bolts and nuts through ring gear mounting holes to compress case halves and overcome preload spring tension, loosen two Allen or Phillips head screws until only a few threads of the screws are still engaged then remove case from press and tap cover to spring it loose. Remove screws.
3. With cover facing down, remove case, preload spring plate and springs.
4. From cover, remove side gear, clutch guides, hub and clutch discs.
5. Drive pinion shaft lock pins from case.
6. With brass drift, drive long pinion shaft from case, driving from end opposite lock pin hole.
7. Remove two short pinion shafts by driving each shaft from center outward. Lift out center block, pinion gears, thrust washers and side gear.
8. Remove differential bearings.

Reassembly

1. With case mounted in a soft jawed vise, place a side gear in case.
2. Install pinion gears, center block and pinion shafts. Position center block so long shaft is driven through rough side and short shaft through machined side. Install lock pins.
3. Install pre-load springs and position preload plate over springs, straddling center block over narrower machined width.
4. Install selective shim in cover.
5. Starting with a composite plate, install clutch plates alternately. New plates should be soaked in gear lubricant before installation.
6. Install four guides and side gear.
7. Place two halves in press, compress springs and install two Allen or Phillips head screws and tighten evenly.
8. Install ring gear and tighten evenly to 65–80 ft-lbs.

DANA TRAC-LOK

Exc. Model 70

Disassembly, Fig. 12

1. Mount one axle shaft in a vise with the splined end facing upward. Do not allow end of shaft to extend more than 2¾ inch above top of vise. This prevents the shaft from fully entering the side gear.
2. Mount differential case on axle shaft with ring gear bolt heads facing upward.
3. Remove ring gear bolts.
4. Loosen ring gear with a mallet, remove case from axle shaft, then the ring gear.
5. Remount differential case on axle shaft.
6. Remove snap rings from pinion mate shaft using two screwdrivers to disengage snap rings.

NOTE: On model 60 units, the pinion mate shaft is retained by a roll pin. Use a 3/16 inch diameter punch to remove roll pin.

7. Remove pinion mate shaft with a suitable drift and hammer.

NOTE: The gear rotating tool J-23781 is required to perform the following steps. The tool consists of three parts: gear rotating tool, forcing screw and step plate.

8. Install step plate in lower differential side gear.
9. Position pawl end of gear rotating tool on step plate.
10. Insert forcing screw through top of case and thread into gear rotating tool.

NOTE: Apply a small amount of grease to centering hole in step plate and oil forcing screw threads before using forcing screw.

11. Center forcing screw in step plate and tighten screw to move differential side gears away from pinion gears.
12. Remove pinion gear thrust washers with a .030 feeler gauge or shim stock. Insert gauge or shim stock between washer and case and withdraw thrust washer and gauge.

NOTE: Momentarily, loosen the forcing screw. This step is very important to relieve the pressure of the clutch pack.

13. Tighten forcing screw until movement of pinion gears is observed.
14. Insert pawl end of gear rotating tool between teeth of one differential side gear. Pull handle of tool to rotate side and pinion gears. Remove the pinion gears as gears appear in case opening.

NOTE: It may be necessary to adjust the tension applied on the belleville springs by the forcing screw before the gears can be rotated in the case.

15. Retain upper side gear and clutch pack in case by holding bottom of rotating tool while removing forcing screw. Remove rotating tool, upper side gear and clutch pack.
16. Remove differential case from axle shaft.
17. Invert case with flange or ring gear side facing upward and remove step plate tool, lower side gear and clutch pack.
18. Remove retainer clips from both clutch packs to permit separation of plates and discs.

Assembly, Fig. 12

1. Lubricate differential gear teeth, thrust faces, splines and the clutch discs and plates.
2. Assemble clutch packs placing discs and plates in original position. Then, install retainer clips.
3. Install clutch packs on differential side gears, then install one assembly in case.

NOTE: Ensure clutch packs remain assembled on side gear splines and the retainer clips are fully seated in case pockets.

---

4. Mount case assembly on axle shaft while clutch packs in place. Ensure that the side gear and axle shaft splines are properly aligned when mounting case on shaft. Also, ensure that the clutch packs are properly assembled in case.
5. Install step plate tool in side gear. Apply a small amount of grease to centering hole of step plate.
6. Install remaining clutch pack and side gear.

NOTE: Ensure clutch packs remain assembled on side gear splines and the retainer clips are fully seated in case pockets.

---

7. Position gear rotating tool in upper side gear.
8. Hold side gear and gear rotating tool in position and insert forcing screw through top of case and thread into gear rotating tool.
9. Install both differential pinion gears into case, ensuring that gear bores are aligned.
10. Tighten forcing screw to compress belleville springs and provide clearance between teeth of pinion and side gears.
11. Place pinion gears in case and insert rotating tool pawl between side gear teeth. Rotate side gears by pulling on tool handle and install pinion gears.

NOTE: If the side gears will not rotate, adjust belleville spring load. Loosen or tighten forcing screw to permit side gear rotating.

---

12. Rotate side gears until shaft bores are aligned with case bores.
13. Lubricate pinion gear thrust washers.
14. Tighten or loosen forcing screw to permit thrust washer installation. Install thrust washers using a small screwdriver to guide washers into position. Ensure that the shaft bores in the thrust washers and gears are properly aligned with case bores.
15. Remove forcing screw, gear rotating tool and step plate.
16. Lubricate pinion mate shaft and seat shaft in case using a hammer. Ensure that the snap ring grooves in shaft are exposed to permit snap ring installation.
17. Install pinion mate shaft snap rings.

NOTE: On models 60 units, align the shaft and shaft retaining pin bore and case pin bore. Drive the shaft into position and install retaining pin. If the case is mounted in a vise with the machined side of the ring gear flange facing upward, use a $\frac{5}{16}$ inch diameter punch to install the retaining pin. Seat the pin until the punch bottoms in case bore. If the case is mounted in a vise with the machined side of the ring gear flange facing downward, place a piece of tape around a $\frac{3}{16}$ inch diameter punch approximately $1\frac{3}{4}$ inch from end of punch. Install retaining pin until edge of tape is flush with pin bore.

---

18. Remove case from axle shaft.
19. Place ring gear on case and install ring gear bolts finger tight.
20. Remount case on axle shaft and tighten ring gear bolts.

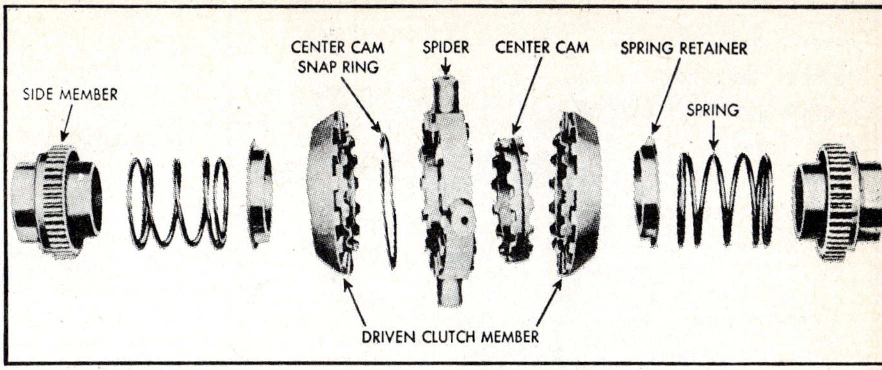

Fig. 14 NoSPIN locking differential

Model 70

Disassembly, Fig. 13

1. Scribe a mark between the differential case halves to be used during assembly. Also, mark the pinion mate shafts and corresponding ramps for assembly in original locations.
2. Clamp differential assembly in a soft jawed vise.
3. Loosen but do not remove the case half retaining bolts.
4. Place assembly on bench with the ring gear half down and remove case half retaining screws.
5. Remove the cover half of the case.
6. Remove upper mate shaft, side gear, side gear ring and clutch pack. Retain parts with cover half of case so they can be installed in original position.
7. Remove lower mate shaft, side gear, side gear ring and clutch pack from drive gear case half.

Assembly, Fig. 13

1. Place side gear ring from ring gear half of

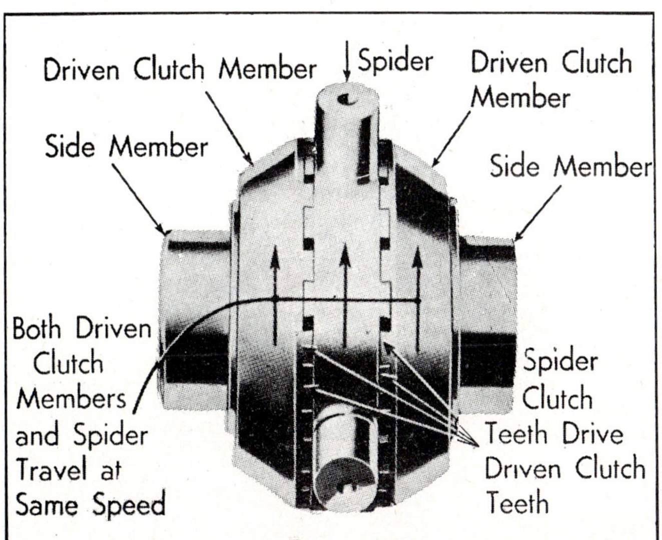

Fig. 15 NoSPIN differential action when moving straight ahead

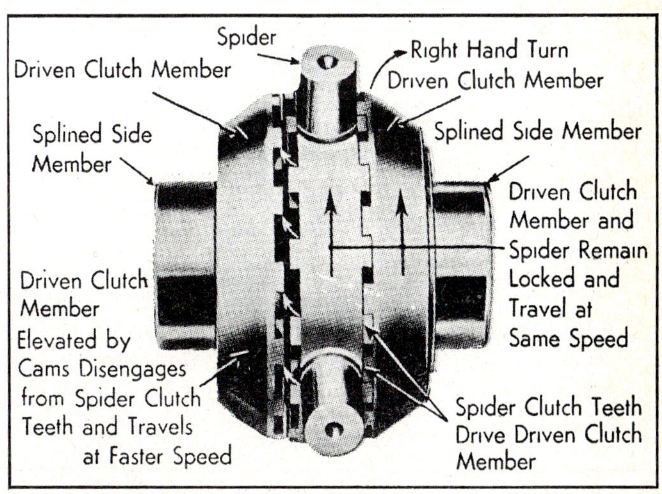

Fig. 16 NoSPIN differential action when making a forward right turn

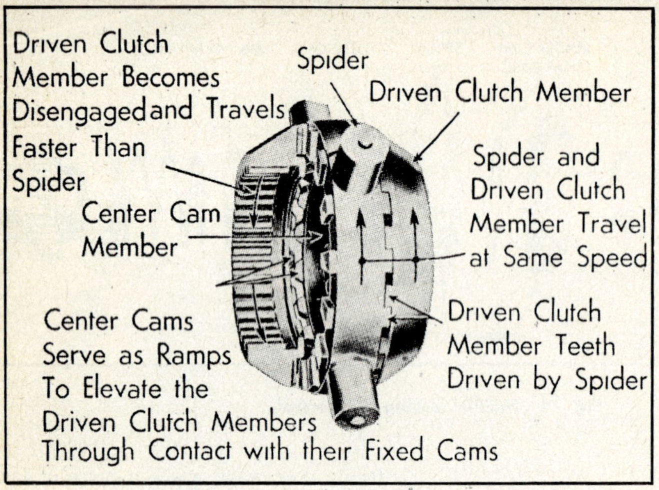

Fig. 17 **Cross sectional view of NoSPIN differential when making a forward right turn**

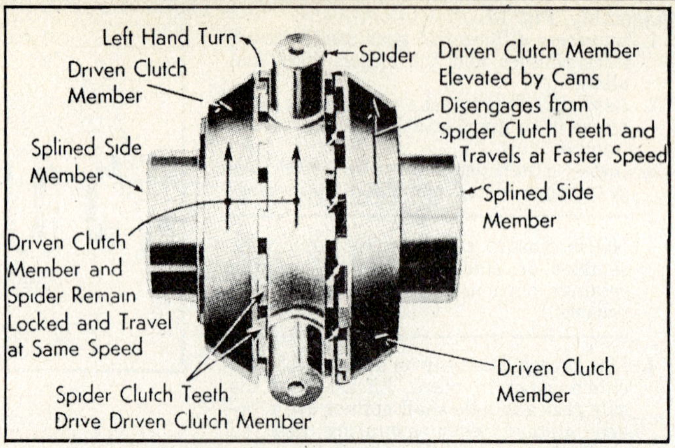

Fig. 18 **NoSPIN differential action when making a forward left turn**

case on a pinion flange or other suitable fixture so the case is approximately 4 inches above bench.

2. Lubricate clutch plates and assemble clutch pack on side gear ring.
3. Place ring gear side of case over clutch pack and side gear ring. Ensure that the clutch plate lugs enter the slots in the case and that the case bottoms on the clutch pack.
4. Invert the assembly. Hold assembly together while inverting.
5. Install the ring gear case half side gear in side gear ring.
6. Install axle shaft spacer in cross shaft.
7. Install the ring gear case half pinion mate shaft and pinions on the side gear ring.
8. Install the cover half pinion mate shaft and pinions.
9. Place side gear on pinions, then the side gear ring on side gear.
10. Assemble clutch pack on side gear ring, aligning clutch pack lugs.
11. Install cover half of case over assembly, aligning the marks made during disassembly.
12. Install the case half retaining bolts, turning the bolts to engage a few threads only.
13. Insert axle shafts into assembly and align the splines of the side gears and side gear rings. With the axle shafts in position, torque case half retaining bolts to 35–45 ft. lbs. up to 1978. For 1979–82 torque case half retaining bolts to 65–70 ft. lbs.
14. Remove axle shafts.
15. If the differential has been assembled properly, each pinion mate cross shaft should be tight on its ramp. However, if clearance is present, it should not exceed .010 inch and be equal at all four cross shaft ends.

NoSPIN DIFFERENTIAL

Fig. 14 illustrates the components of this unit while Figs. 15 through 20 show how it works.

NOTE: Before disassembling the differential case, insert a bolt through the center of the NoSPIN unit (axle shaft openings) with a flat washer on each end against the side members, Fig. 21. Thread a nut on the bolt against the flat washer finger tight. This will prevent possible injury caused by the unit flying apart due to the spring pressure within itself during disassembly of the differential case.

Disassembly

1. Remove nuts from differential case and lift out NoSPIN unit. Mount unit in a small press, Fig. 22.
2. Apply enough pressure on head of bolt to release spring pressure against nut.
3. Remove nut and flat washer by reaching underneath the press.
4. Slowly release press and allow unit to disassemble itself until spring pressure is fully released.
5. Remove unit from press and remove parts. The center cam may be removed

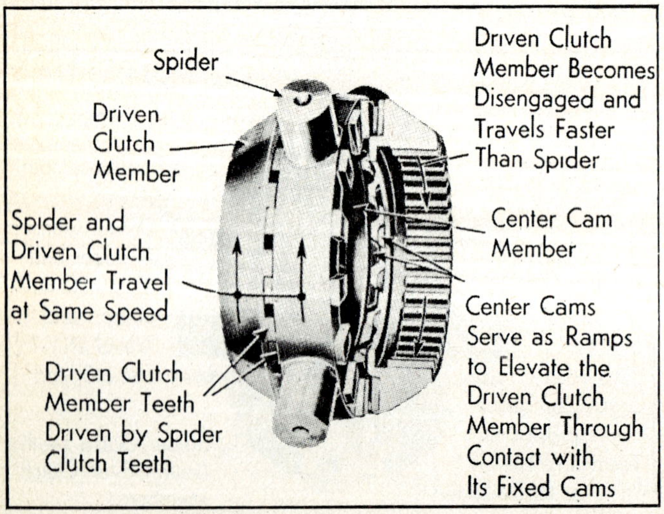

Fig. 19 **Cross sectional view of NoSPIN differential when making a forward left turn**

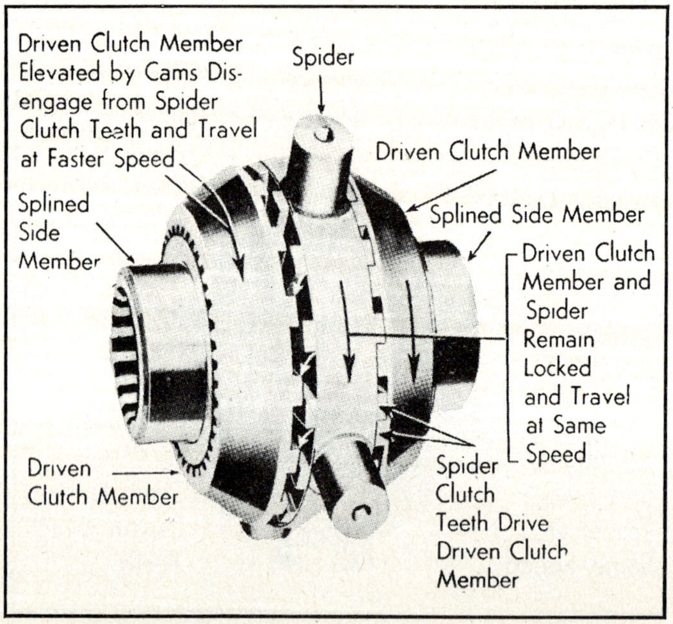

Fig. 20 **NoSPIN differential when making a rearward right turn**

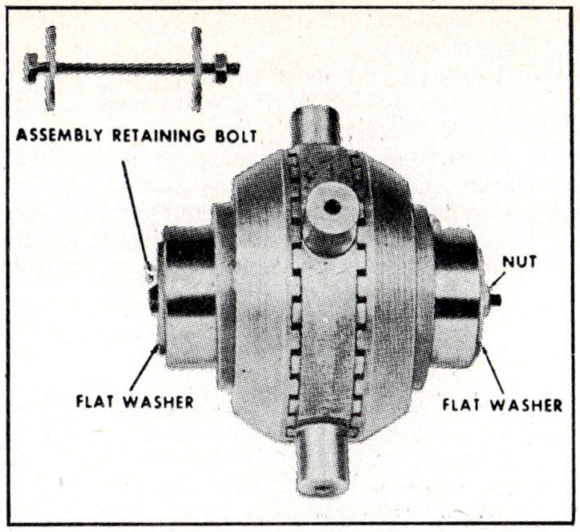

Fig. 21 Inserting assembly retaining bolt

from the spider by expanding the snap ring with small screwdrivers or wedges. However, it is not necessary to remove the center cam from the spider even if only one part is damaged since the center cam, snap ring and spider are not serviced separately—only as an assembly.

6. Wash all parts with cleaning solvent. Inspect all mating surfaces and teeth for possible wear or damage. Replace all worn or damaged parts upon reassembly. Lubricate all parts with SAE-30 oil during reassembly.

Reassembly

1. Place side member upright and install spring on same.
2. Place spring retainer on spring with flange end toward side member.
3. Install driven clutch member on spring retainer.
4. Place spider on driven clutch member, indexing teeth of same.
5. Install other driven clutch member, spring retainer, spring and side member on spider.
6. Insert a bolt through the center of the NoSPIN unit with a flat washer against the side member, and mount unit in press, Fig. 22.
7. Compress springs by pressing on head of bolt and index splines of side members with those of driven clutch members. *Keep entire unit aligned in press to prevent it from kicking out while springs are being compressed.*
8. Compress unit until side member splines are completely indexed and flush with driven clutch member.
9. Install flat washer on bolt against side member and thread nut on bolt finger tight so it has the appearance of Fig. 21 when removed from press.

TRACTION EQUALIZER

Cartridge Type

Fig. 23 is a cross-sectional view of this device. It consists of an internally splined cartridge, Fig. 24. One end mates with the externally splined differential case extension. The shell is loaded with a seris of specially treated clutch plates which are splined alternately on the inside and outside diameters to match the internal splines of the cartridge and the external splines on the lengthened axle shaft splines. A series of Belleville spring washers maintain a constant pressure on the clutch plates. Oil scoops automatically pick up the standard axle lubricant.

Operation

During a straight forward drive, the drive torque is transmitted to the wheels in the same manner as with a conventional rear axle. Likewise, straight rearward driving employs the same principles as does the conventional rear axle.

When making a turn, differential action is required in order to permit the outside wheel to travel a greater distance, and faster, than the inside wheel. While differentiation takes place within the differential case, the outer wheel ground traction overpowers the clutch pack friction, thereby causing the clutch plates to rotate. This action takes place constantly in both units mounted on the axle shafts as the vehicle deviates minutely from a straight forward or straight rearward drive.

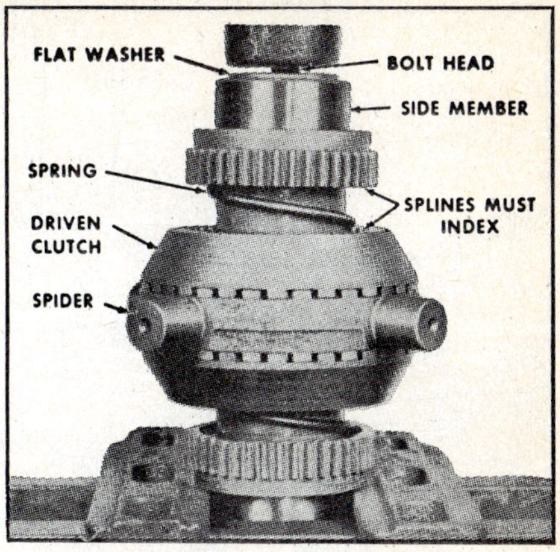

Fig. 22 Release of spring pressure

Differential Removal

1. Remove plug from bottom of axle housing and drain lubricant.
2. Back off and remove Traction Equalizer retaining set screws from housing, Fig. 25. *If only axle shafts are to be removed, do not disturb cartridge set screws.*
3. Remove axle shaft drive nuts and washers.
4. Hold a short drift firmly against center of axle shaft flange *between raised circular lugs* and strike drift a sharp blow to loosen tapered dowels. *Do not strike raised circular lugs with hammer as the lugs are liable to shatter and explode. Do not pry shaft loose. Pry bars, chisels and wedges will damage hub, shaft, flange and oil seals.*
5. Remove dowels and pull out axle shafts.
6. If cartridge remains on differential case splines when axle shaft is removed, insert long hooked rod through housing sleeve. Hook cartridge on inside, pull off differential case splines and leave in housing.
7. Disconnect universal joint at pinion shaft.
8. Disconnect power unit (if equipped).
9. Remove carrier-to-housing stud nuts and washers, leaving the two top nuts partial-

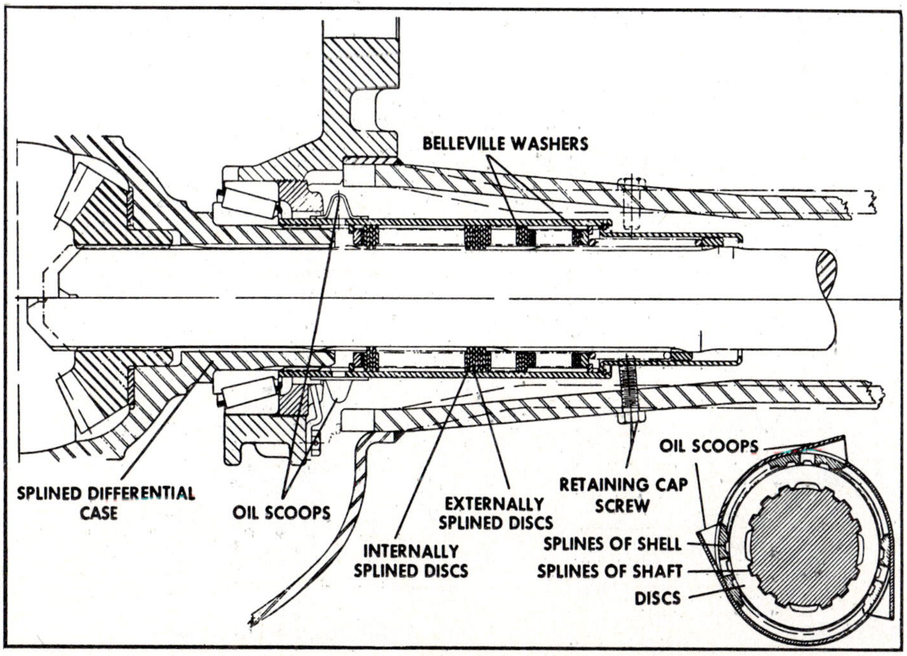

Fig. 23 Cross sectional view of Traction Equalizer differential

Fig. 24 Traction equalizer cartridge

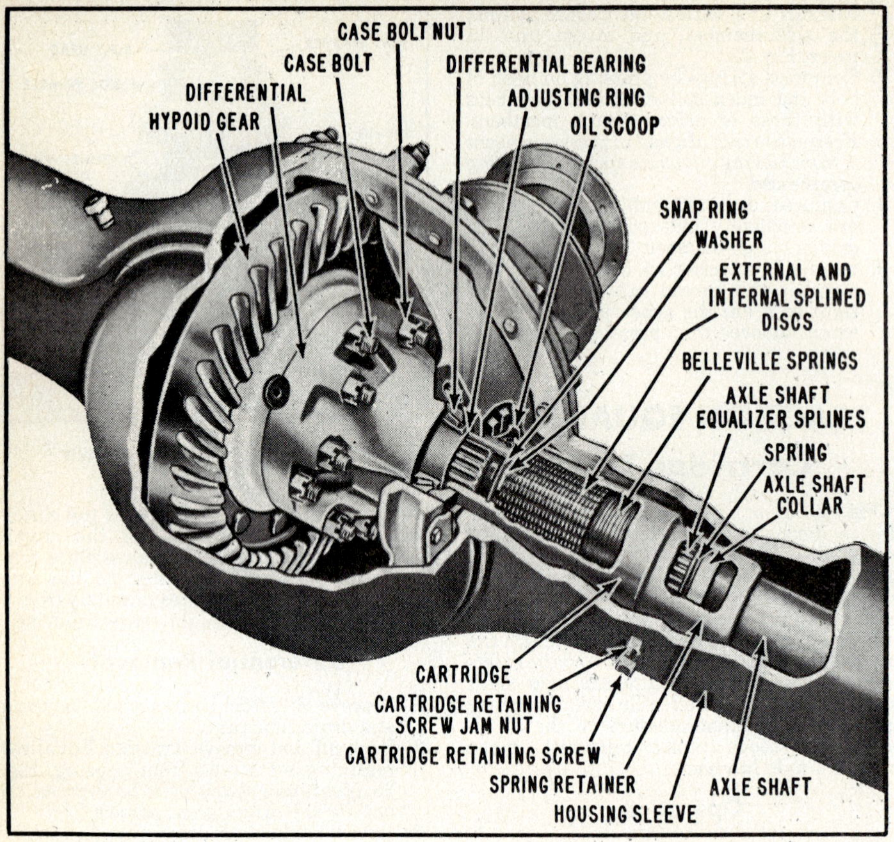

Fig. 25 Cutaway view of rear axle is showing Traction Equalizer unit

ly screwed on studs to prevent carrier from falling.

10. Break carrier loose from housing with a rawhide mallet (remove tapered dowels where used).
11. To remove carrier from a front mounted axle, place a roller jack under carrier, remove the two top nuts and work carrier free. To remove carrier from a top mounted axle, use a chain fall and puller screws in holes provided.

Disassembly

1. Place cartridge assembly under a press with oil scoop end down. A slight press pressure on the retainer assembly in the cartridge will compress the spring and permit removal of the outer snap ring.
2. Remove snap ring. Release press pressure and remove retainer, spring and axle collar.
3. Replace cartridge under press, oil scoop end down. Insert a mandrel with flange end up. Apply a slight pressure to compress Belleville springs and remove retaining snap ring.
4. Release press pressure and remove mandrel. Invert cartridge assembly and place over a post or bar to keep internal parts in order. Hold internal parts down with a sleeve (or equivalent tool) and lift off cartridge shell.
5. Clean and inspect shell and snap ring at oil scoop end.

Inspection & Repair

1. Wire internal parts together in original order and clean in kerosene or diesel fuel oil (not gasoline). Dry immediately after cleaning, using lintless, absorbent paper towels or wiping rags free of abrasive

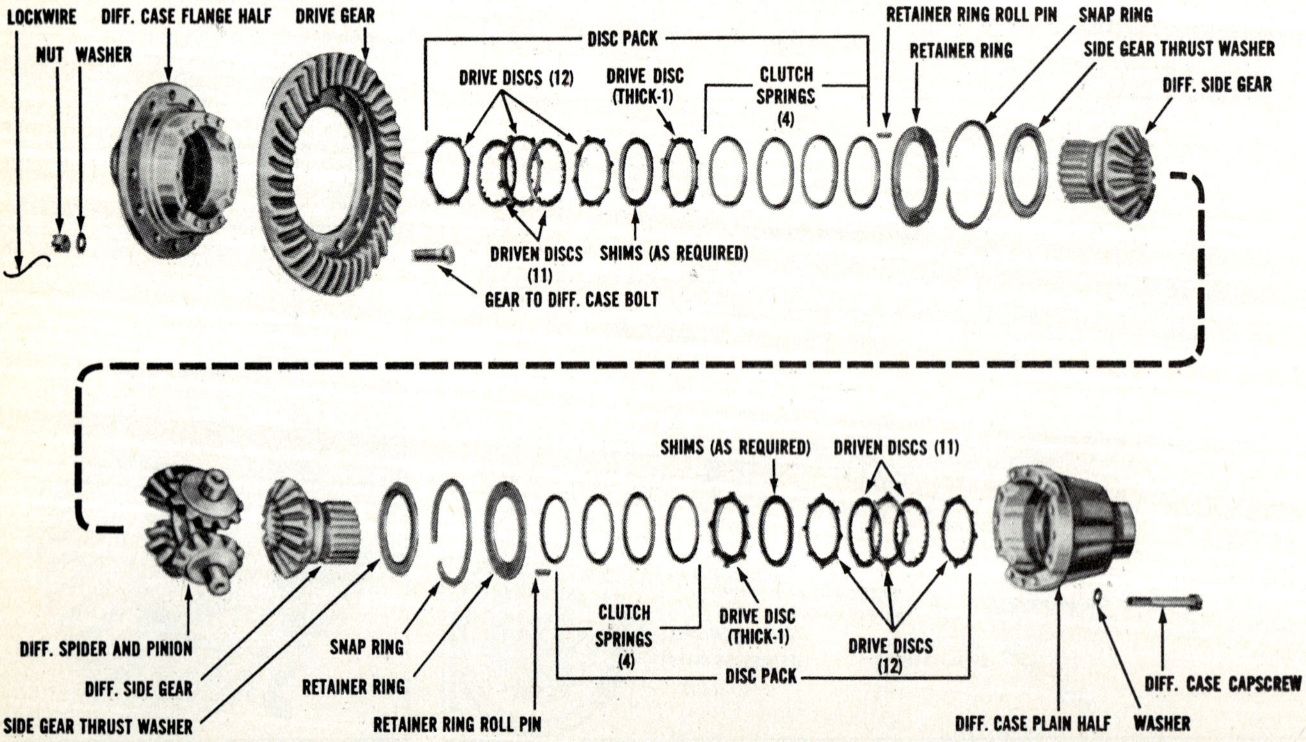

Fig. 26 Exploded view of self contained Traction Equalizer

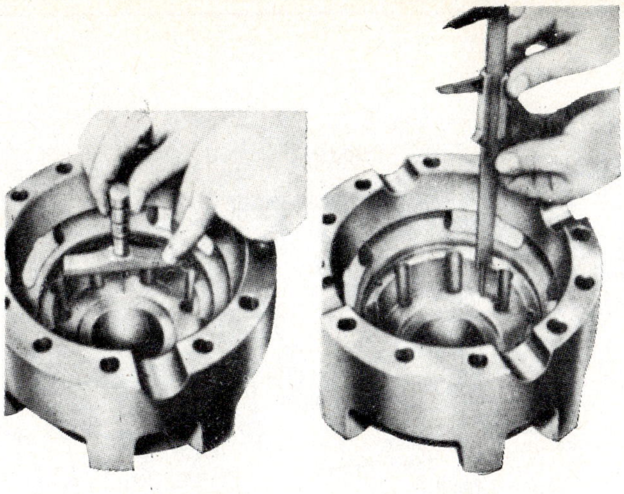

Fig. 27 Measuring depth of clutch bore in case half

material.
2. Replace dry internal parts on post and remove wire.
3. Remove any accumulation of grit, dirt or gum from housing bowl and sleeve. Clean housing with solvent and dry with compressed air.
4. Inspect housing for cracks, loose studs, nicks or burrs at machined surfaces. Remove nicks or burrs with a stone or file.
5. Make all necessary repairs or parts replacement before installing drive unit and cartridge in housing.

Reassembly

1. Install snap ring at oil scoop end (if removed).
2. Place cartridge shell with oil scoop end down over mandrel.
3. Starting with a thick washer and following with discs and Belleville springs, take internal parts from post, one at a time. Inspect and re-oil with axle shaft lubricant, install in shell and over mandrel. *Discs are to be installed alternately starting with an externally splined disc. Belleville springs are to be installed in alternating groups of four starting with concave side down toward last disc which is to be an externally splined disc. If discs bind in cartridge, tap into position with a screwdriver; do not force down as they will be cocked.*
4. Carefully remove mandrel, being careful not to misalign splines, and insert mandrel at top of cartridge with flange end up.
5. Place cartridge under press and apply a slight pressure to compress Belleville springs and install snap ring.
6. Release press pressure and mount a dial indicator so that its stem is against the top of the mandrel flange.
7. Apply a slight pressure and read dial indicator for movement required to compress discs and springs. Compression should be .120″ to .160″. If compression is more than desired, add one or more splined discs, being sure last disc is externally splined. (Discs are about .040″ thick.)
8. Release press pressure and remove mandrel. Install axle shaft collar, spring and retainer with chamfered edge of collar up.
9. Compress spring with slight pressure and install snap ring.
10. Insert axle shafts in housing and carefully mount cartridges on splined ends of shafts through housing bowl opening. Withdraw shafts and cartridges from housing just enough to allow clearance for carrier installation.
11. Install carrier in housing.
12. Complete assembly of one axle shaft in housing. Engage shaft with splines in side gear. Install taper dowels, lockwashers and nuts. Restrain pinion shaft and rotate hub or wheel until cartridge engages splines on differential case. Install other axle shaft in same manner.
13. Install housing drain plug and insert lubricant.
14. Insert cartridge retaining set screw and jam nut to prevent loss of lubricant.
15. Drive vehicle in a circle to permit second cartridge to engage splines on differential case.
16. Turn each cartridge set screw until it touches spring retainer, then back off screw 1½ to 2 turns and tighten jam nut to 30–35 ft. lbs.

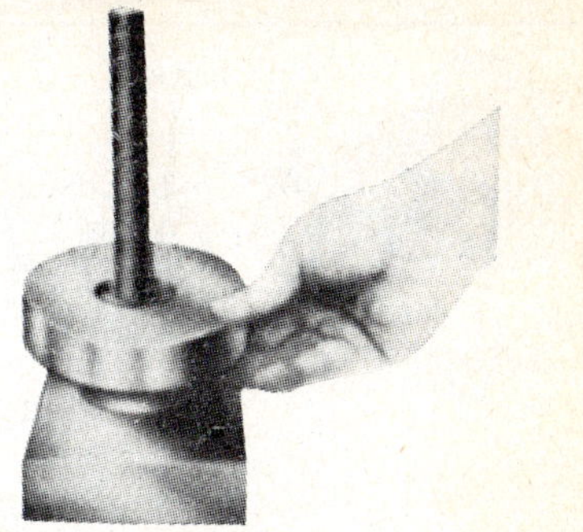

Fig. 28 Disc pack platform in place

Self Contained Type

When servicing this unit, Fig. 26, a clamping fixture must be used to permit the tension to be slowly released when the snap ring is removed. Both halves of the unit are serviced in the same manner.

Inspection

Parts having ground and polished surfaces such as gears, bearings, bearing journals and differential case halves, should be cleaned in a suitable solvent such as kerosene or diesel fuel oil. Do not use gasoline or any alkaline or hot tank solution.

Inspect all parts for pitted, scored or worn surfaces. Thrust washers must be replaced in sets. The use of a combination of old and new parts will result in premature failure.

Assembly

Preload Check

1. Using a depth micrometer, measure

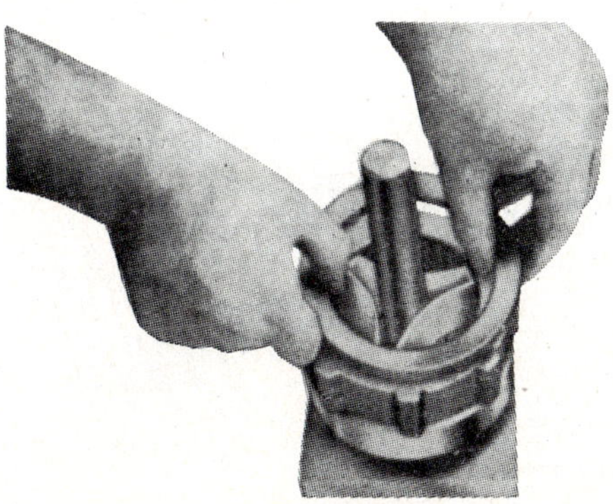

Fig. 29 Installing Belleville springs, concave side up

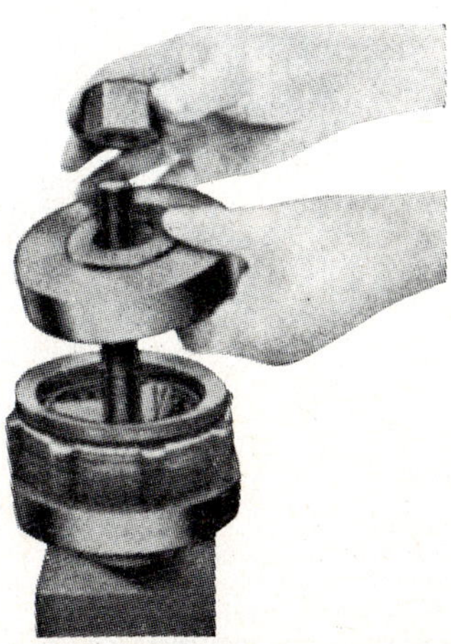

Fig. 30 Installing nut to compress pack

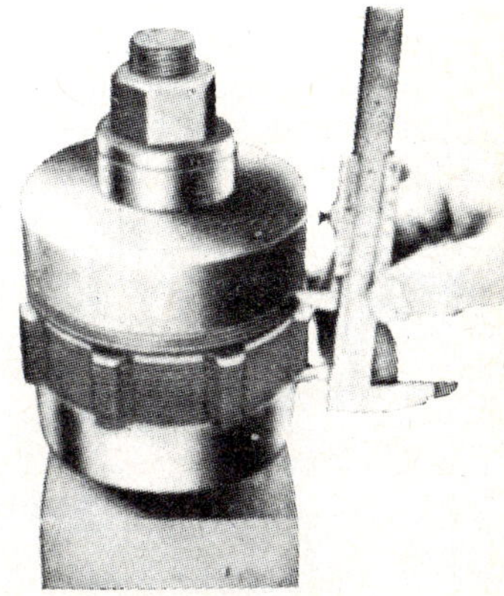

Fig. 31 Measuring height of pack

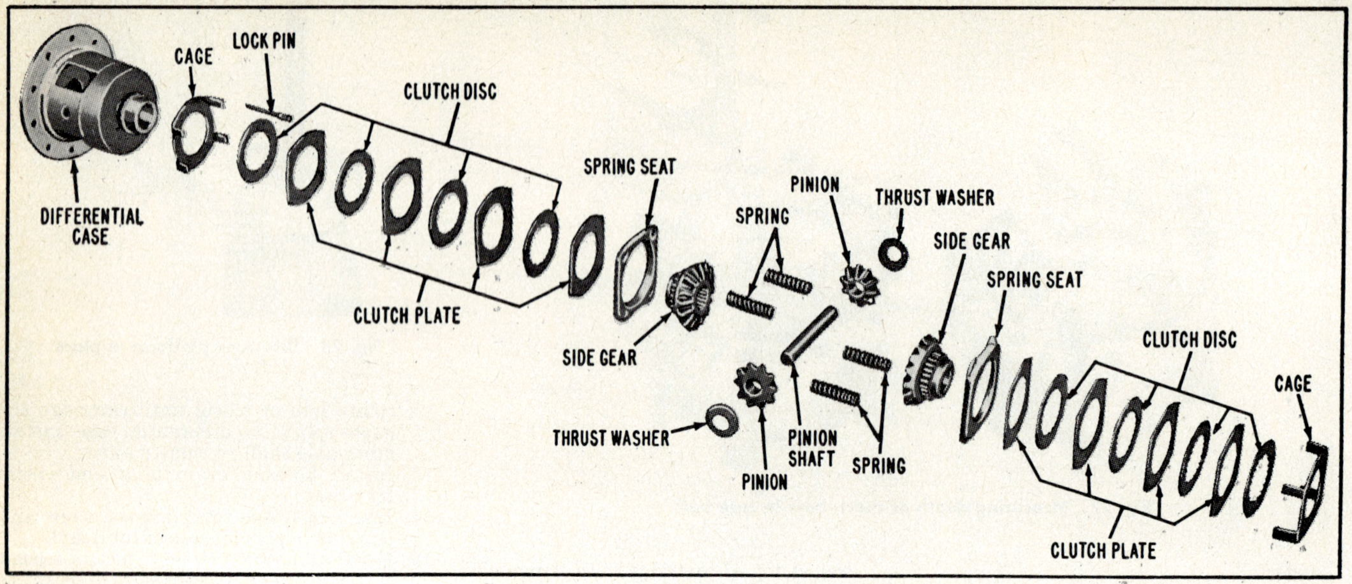

Fig. 32 Spring-loaded "Positraction" locking differential disassembled

depth of clutch bore in case half, Fig. 27, and record measurement.

2. Using a fixture as shown in Fig. 27, position the 1" thick spacer over the locating washer of fixture.
3. Place disc pack on platform. Omit shims at this time.
4. Place drive disc over disc pack.
5. Position four Belleville springs, concave side up, over thick disc, Fig. 28.
6. Place pressure plate, washer and nut over disc pack and tighten nut to 110–120 ft. lbs. to compress pack.
7. Using a Vernier gauge, measure height of pack, Fig. 30. Measure between top of disc pack platform (1" spacer) and bottom of pressure plate. Record measurement.
8. Subtract pack height obtained in step 7 from bore depth obtained in step 1. A shim pack of .004–.009" less than the difference of these two measurements should be used. If the desired shim pack exceeds .040", one of the old drive may be used as part of the shim pack.
9. Remove nut washer and pressure plate from fixture. Remove complete clutch pack and set aside for later use.
10. The unit is now ready for assembly.

DANA POSITRACTION

Disassembly, Fig. 32

1. Remove ring gear and install spring tool shown in Fig. 33, making sure spring jaws are securely indexed between spring coils.
2. Compress tool until spring is compressed enough to slide out of retainer seats. Repeat on remaining springs.
3. Remove differential pinions by rotating side gear with axle shaft until they move around to large opening in case. Pry clutch pack away from pinion until pinion drops out. Remove pinion thrust washers.
4. Remove clutch packs and side gears.

Inspection

1. Clean and inspect side gear thrust surfaces, pinion gear thrust surfaces and thrust washers for excessive wear and replace as required.
2. Inspect clutch pack for excessive wear or or overheating and, if necessary, replace as a unit.

Reassembly

1. Place clutch pack cages into case with side tabs up and flat sides parallel with flat sides in case.
2. Place aluminum spring seat plate over side gear.
3. Alternately place clutch plate and clutch disc on side gear until assembly of pack is complete. Make sure clutch discs index with teeth in side gear.

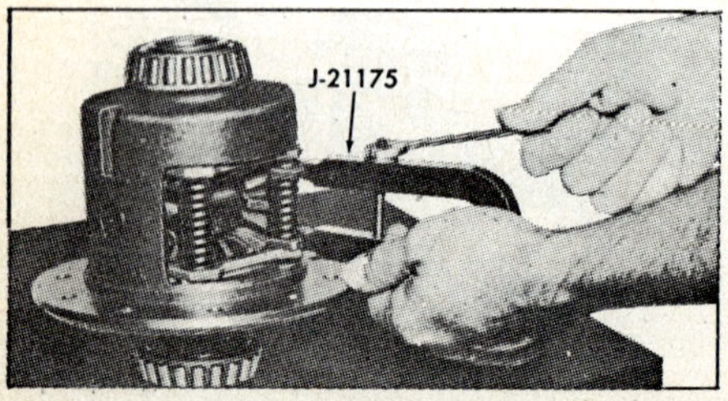

Fig. 33 Spring removal

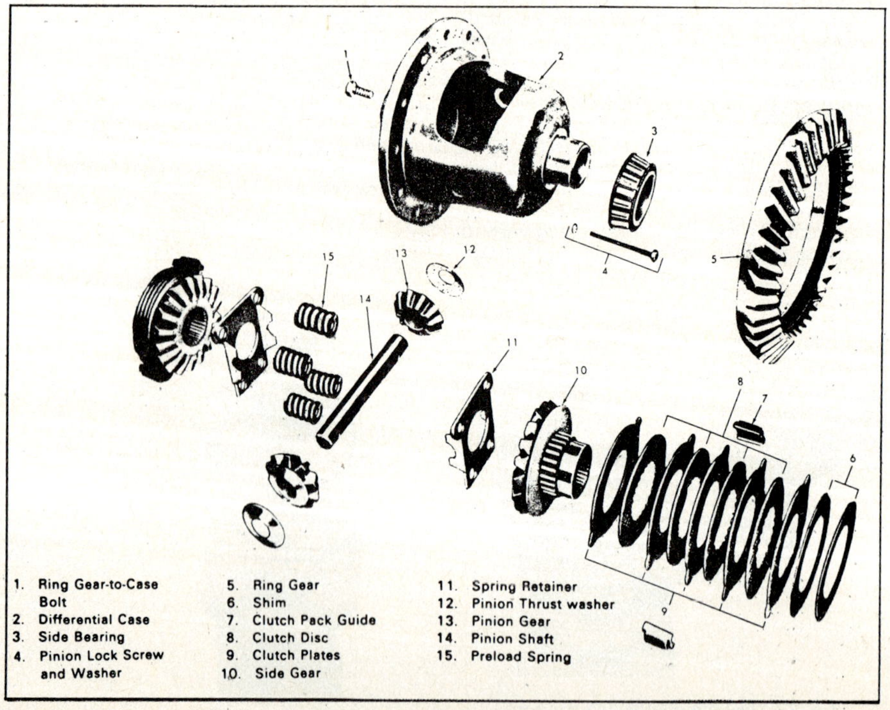

Fig. 34 Eaton locking differential

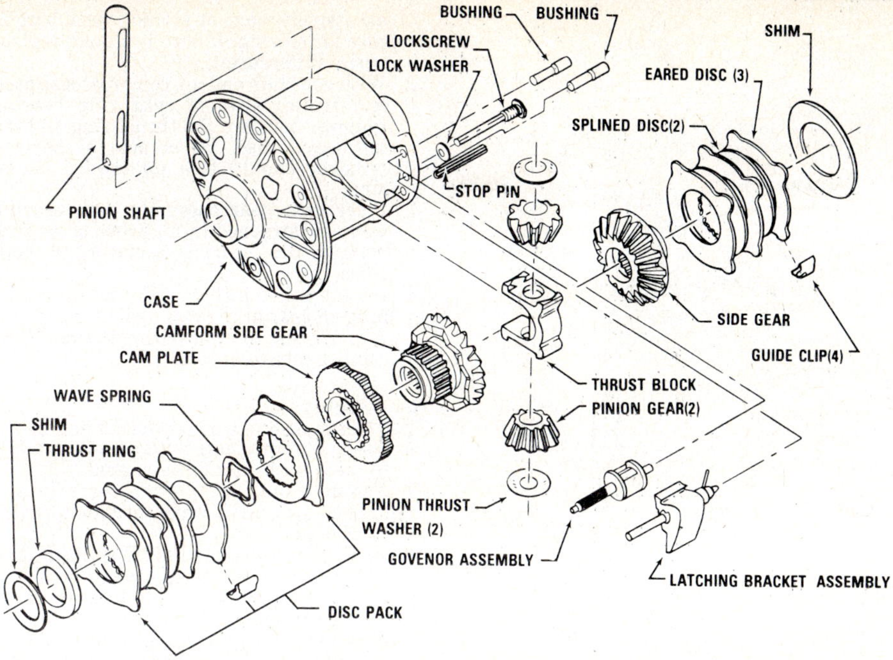

Fig. 35 Eaton positive locking differential. 1981–82 ½ ton unit

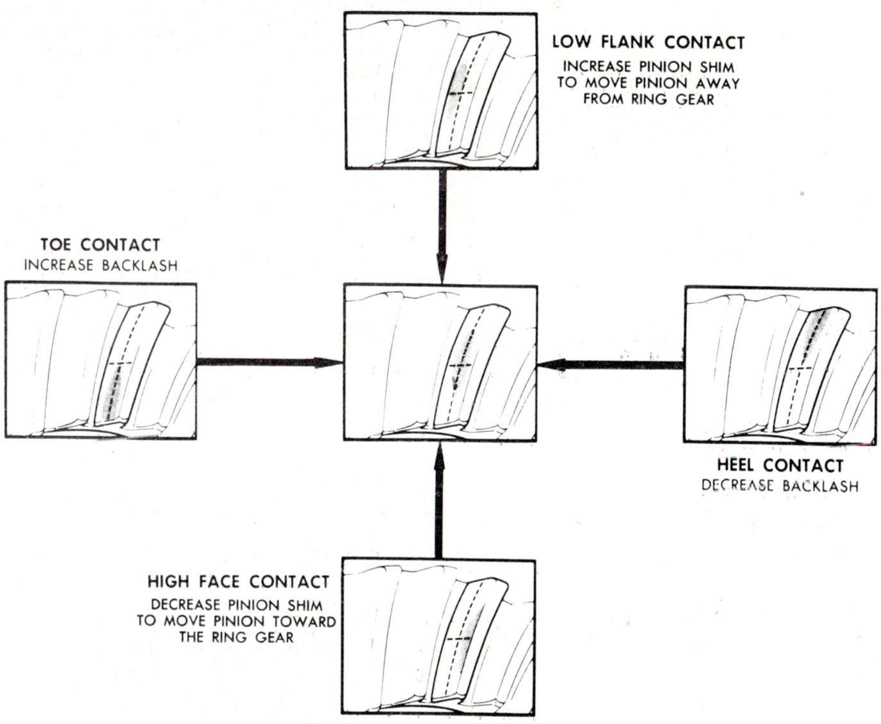

Fig. 36 Gear teeth contact pattern check

4. Install clutch pack and side gear assemblies in case. Tilt case on its side to prevent side gears from falling out.
5. Place pinion gear between side gears through large opening in case. Rotate side gears and install second pinion directly opposite first. Install pinion thrust washers.
6. Place clutch spring in jaws of tool shown in Fig. 33. Compress spring to about 1.90″ and carefully insert into spring seats. Release spring and remove tool. Repeat for remaining springs.
7. Install ring gear.
8. Place differential in carrier.
9. Test operation of differential as outlined at the beginning of this chapter.

EATON

1973–83

1. Remove pinion shaft lock screw and pull pinion shaft from case, Fig. 34.
2. Remove preload spring retainer and springs.
3. Rotate side gears until pinions are in open area of case and pick out pinions and thrust washers.
4. Remove a side gear, clutch pack and shims, noting shim location in case to aid in reassembly. Remove side gear clutch pack and shims from opposite side. If side gear or clutch pack cannot be removed readily, drive it out with a brass drift.
5. Remove clutch plate guides and separate shims and clutch plates from side gears.

NOTE: Keep clutch plates in their original location in clutch pack.

Inspection

1. If pinion shaft, pinions or side gears are excessively scored, pitted or worn, replace parts affected.
2. Inspect clutch plates for scored, worn, cracked or distorted condition. If any of these conditions exist, new clutch plates must be installed.

Reassembly

1. Alternately position nine clutch plates on side gear, starting and ending with a plate with external lugs.
2. Install two clutch guides over clutch plate lugs.
3. Install same shims which were removed or an equal amount on clutch plate.
4. Repeat above steps on other clutch pack.
5. Install one side gear with clutch pack and shims in case.
6. Position two pinion gears and thrust washers on side gear and install pinion shaft.
7. Compress clutch pack by inserting a screwdriver or wedge between side gear and pinion shaft.
8. Install a suitable dial indicator with contact button against pinion gear.
9. Rotate pinion gear. Clearance should be .001″ to .006″.
10. If clearance is more than .006″, add shims between clutch pack and case. If clearance is less than .001″, remove shims. A .002″ shim will change clearance about .001″. Recheck clearance after adding or subtracting shims.
11. Remove side gear and repeat procedure with remaining clutch pack on opposite side of case.
12. Remove pinion shaft, pinions and thrust washers.
13. Install remaining side gear and clutch pack with correct shims in case.
14. Place pinion gears on side gears and rotate into correct position.
15. Compress preload springs and drive preload retainer and springs between side gears.
16. Insert thrust washers behind pinion gears.
17. Install pinion shaft and retain with lock bolt, tightening lock bolt to 15–25 ft-lbs.
18. Check side gear splined hole to be certain it is in line with hole in preload spring retainer. Spring retainer can be moved slightly to correct alignment.

EATON POSITIVE LOCKING DIFFERENTIAL

1981–83 ½ Ton Unit, Fig. 35

1. Note the position of the governor and latching bracket assembly. Remove the ring gear and side bearings.

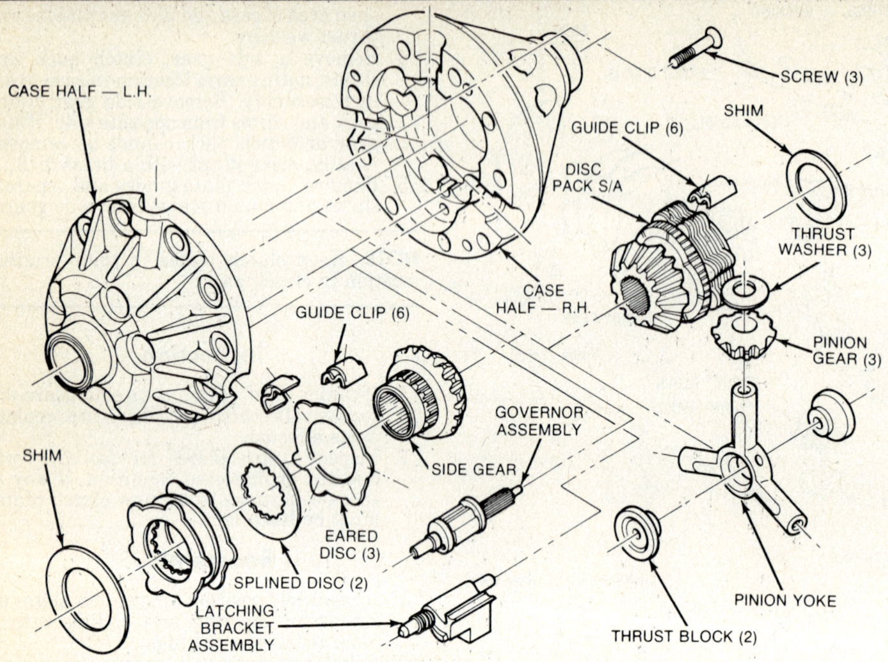

Fig. 37 Eaton positive locking differential. 1981–83 1 & 3/4 ton unit

2. Using bushing puller tool No. J-26252, remove governor assembly and latching bracket by pulling the retaining bushings as shown in Fig. 34C. Pull the latching bracket spring out of the way while pulling the governor assembly bushing to prevent damage. Remove the stop pin by driving through the case with a drive pin punch.
3. Remove lock screw and pinion shaft, and roll out differential pinion gear.
4. Remove thrust block and pinion thrust washers.
5. Remove the cam gear, disc pack assembly and disc pack guide clips.
6. Remove bell disc pack assembly and shim.
7. Clean and inspect all parts. Replace parts as needed. The following parts are serviceable components: governor assembly and latching bracket with spring and stop pin, cam plate, clutch disc pack with guides [flanged end only], pinion cross shaft, pinion gears, pinion thrust washers, reaction block, thrust ring, flange end shim. Attempts to service other components may disturb critical clearances. If reaction block or flange and shim must be replaced, the original pieces must be carefully measured for thickness and replaced with a piece of the same size.

Cam/Clutch Service

Disassembly

If cam plate or clutch discs must be replaced, the cam gear sub-assembly must be serviced as follows:

1. Measure and record overall length of gear assembly (front face of gear to back side of thrust ring, including shim). This dimension will be needed to reassemble unit if thrust ring is replaced. Do not replace thrust ring unless necessary. If ring is excessively worn or scored, check bore in case for scoring. If bore is scored, replace entire differential.
2. With gear hub end up, compress disc pack and install jaws of split ring bearing remover between the thrust ring and the top eared disc. Bevel side of bearing remover should face up toward thrust ring.
3. Place cam gear assembly with bearing remover attached in an arbor press supporting the bearing remover on both sides.
4. Install a 1½ inch to 1¾ inch diameter plug on gear hub. Press against plug with ram of press to remove thrust ring. Keep components in proper order.

Reassembly

1. Place gear on bench with hub end up.
2. Assemble cam plate with cam form down to mate with cam form on gear.
3. Assemble onto cam plate: 2 eared discs, 1 splined disc, and 1 wave spring alternately, Fig 35.
4. Assemble onto gear hub: 2 splined discs and 3 eared discs alternately, Fig 35.
5. Locate cam gear assembly in arbor press with hub end up. Place thrust ring on gear hub and press to shoulder making sure that ring is square with hub. Compress disc pack by pushing down on the discs to keep the splined discs from becoming wedged between the thrust ring and gear shoulder while pressing the subassembly together.

Reassembly

1. Install 4 clutch pack guide clips on the ears of the cam gear clutch pack using grease for retention.
2. Install cam gear assembly and original shim in flange end of case. If a new thrust ring has been pressed on the cam gear, reshimming may be necessary. Measure the overall length of the new cam gear assembly, including the shim and compare to the dimension obtained in Step 1 of cam gear service section. If the length of the new assembly varies by more than .003 inch, larger or smaller than the original dimension, reshimming will be necessary.

NOTE: Incorrect shimming will result in differential failure.

---

3. Lock an axle shaft in vise. Mount the differential case over the end of the axle shaft engaging the spline of the side gear with the shaft. Grease the two pinion gear thrust washers and locate them in their power position. Assemble on to bell end gear hub 2 splined discs and 3 eared discs alternately, Fig. 34C. Begin and end with an eared disc. Install 4 small clutch pack guide clips on the ears of the bell end clutch pack using grease for retention. Install case with original shim.
4. Insert one pinion gear through the small window opening in the case while at the same time inserting the reaction block and other pinion gear through the large window opening. Rotate the two pinion gears and thrust block 90° so as to position the reaction block with the open side toward the small window opening in the case. Be sure the two pinion gears and thrust washers are in their proper location.

NOTE: Reaction block must be replaced with a block of the identical thickness.

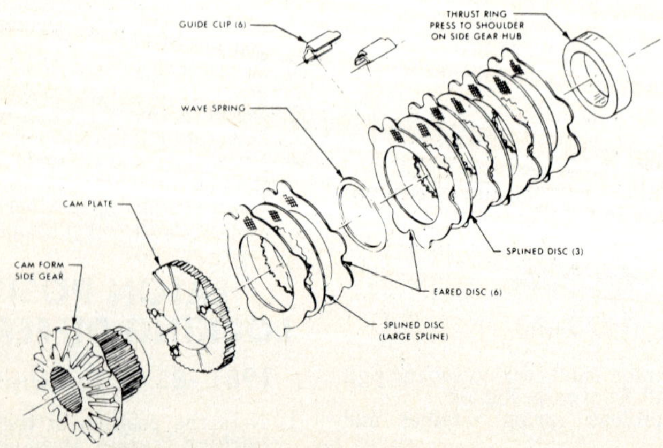

Fig. 38 Cam plate and clutch discs assembly

Incorrect block thickness may result in differential failure.

5. Install shaft and lock screw.
6. Insert the governor assembly and latching bracket into case. Place the straight end of the latching bracket spring over and onto the outside of the engagement shaft to preload the latching bracket against the governor assembly. Press bushing and 1/4 inch stop pin into case. Press bushing for governor assembly into case to give .004 to .020 inch shaft end play. Press bushing for latching bracket into case to remove end play. Press stop pin flush with top of case. For the latching bracket use bushing with the tapered hole. The bushing for the governor assembly has a straight hole.
7. Install ring gear and side bearings.
8. Place differential unit in carrier and adjust ring gear and pinion backlash and gear tooth pattern.

Gear Tooth Pattern Check, Fig. 36

1. Wipe oil out of carrier and carefully clean each tooth of the ring gear.
2. Use gear marking compound and apply this mixture to ring gear teeth. When properly aligned, the area of tooth contact will be clearly visible after load is applied.
3. Tighten bearing cap bolts to 55 ft. lbs.
4. Expand brake shoes until a torque of 20–30 ft. lbs. is required to turn the pinion. A test made without loading the gears will not give a satisfactory pattern. Turn companion flange with wrench so that ring gear rotates one full revolution then reverse rotation so that ring gear rotates one revolution in opposite direction.

Ring Gear and Pinion Adjustment

1. Loosen right hand adjusting nut and tighten left hand adjusting nut until all backlash is removed between ring gear and pinion.
2. Back off left adjusting nut approximately two notches to a point where notch in nut is aligned with lock.
3. Tighten right adjusting nut firmly to force differential and planet assembly in solid contact with left adjusting nut. Loosen right adjusting nut until it is free of its bearing, then retighten snugly against bearing. Tighten right nut from one to two notches more to a locking position. This method provides for the proper pre-load on the bearings.
4. Mount a dial indicator on the housing and check the backlash between ring gear and pinion. Backlash should be from .005 inch to .008 inch. If backlash is less than .005 inch, loosen left nut one notch and tighten right nut one notch. If backlash is more than .008 inch loosen right adjusting nut one notch and tighten left nut one notch. When backlash has been adjusted, tighten bearing cap bolts securely. Recheck backlash and then install adjuster locks.

1981–83 3/4 & 1 Ton Unit, Fig. 37

1. Remove ring bolts and, using a soft drift and a hammer, tap ring gear off the case.
2. Remove three screws on front face of ring gear flange.
3. Set unit on the right side case half and gently pry apart at yoke hole locations.
4. Remove left side case half. To prevent side gear from falling out, hold thumb against inside of gear hub while separating case halves.
5. Clean and inspect all parts. The following are serviceable parts for this differential: governor assembly and latching bracket with spring, cam plate, clutch disc pack with wave spring and guide clips, pinion yoke, pinion gears, pinion thrust washers, reaction blocks, thrust ring, right side shim and left side guide clips. Attempts to service other components will disturb critical clearances and may result in differential failure. If reaction blocks or right side shim must be replaced, the original pieces must be carefully measured for thickness and replaced with a piece of the same size.

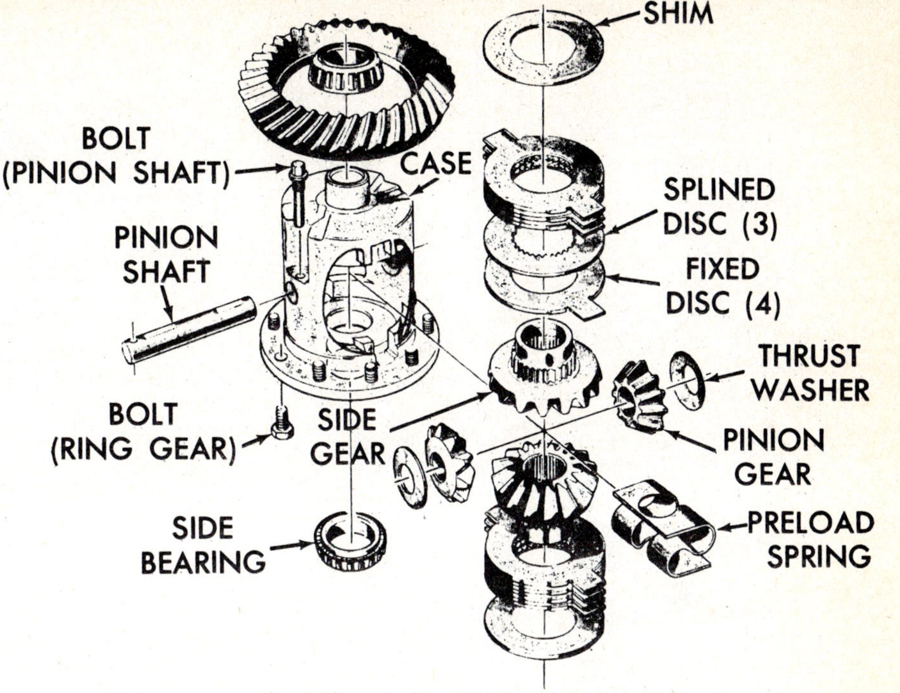

Fig. 39 GM Corp. locking differential disassembled

Cam/Clutch Service

If cam plate or clutch discs must be replaced, the cam gear sub-assembly must be serviced as follows.

1. Measure and record overall length of gear assembly [front face of gear to back side of thrust ring, including shim]. This dimension will be needed to reassemble unit if thrust ring is replaced. Do not replace thrust ring unless necessary. If ring is excessively worn or scored, check bore in case for scoring. If bore is scored, replace entire differential.
2. With gear hub end up, compress disc pack and install jaws of split ring bearing remover between the thrust ring and the top eared disc. Bevel side of bearing remover should face up toward thrust ring.
3. Place cam gear assembly with bearing remover attached in an arbor press supporting the bearing remover on both sides.
4. Install a 1 1/2 inch to 1 3/4 inch diameter plug J8107-4 on gear hub. Press against plug with ram of press to remove thrust ring. Keep all components in proper order.

Reassembly

1. Place gear on bench with hub end up.
2. Assemble cam plate with cam form down to mate with cam form on gear.
3. Assemble onto cam plate 2 eared discs, 1 splined disc and 1 wave spring alternately, Fig. 38.
4. Assemble onto gear hub 4 eared discs and 3 splined discs alternately, Fig. 38.
5. Locate cam gear assembly in arbor press with hub end up. Place thrust ring on gear hub and press to shoulder making sure that ring is square with hub. Compress disc pack by pushing down on the discs to keep the splined discs from becoming wedged between the thrust ring and gear shoulder while pressing the sub-assembly together.

Reassembly

1. Install 6 clutch pack guide clips on the ears of the cam gear clutch pack using grease for retention.
2. Install cam gear assembly and original shim in right side case half. If a new thrust ring has been pressed on the cam gear, reshimming may be necessary. Measure the overall length of the cam gear assembly including the shim and compare to the dimension obtained in Step 1 of cam gear service section. If the length of the new assembly varies by more than .003 inch than the original dimension, reshimming will be necessary.

NOTE: Incorrect shimming will result in differential failure.

3. Install right side reaction block on gear face with button side of block facing up. Use original block unless severely scored or worn. If new block is used, determine face-to-face thickness of original block

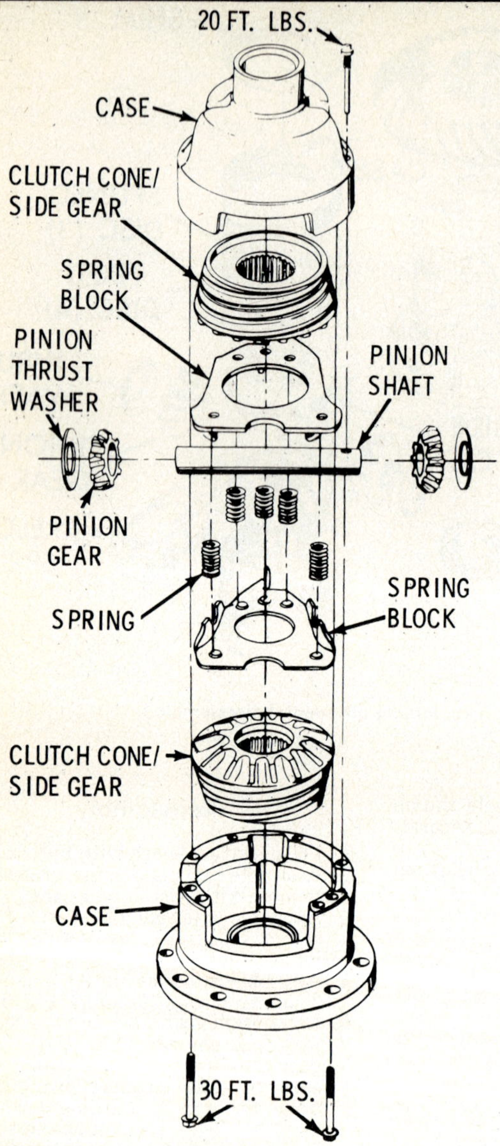

Fig. 40 Borg Warner cone type locking differential disassembled

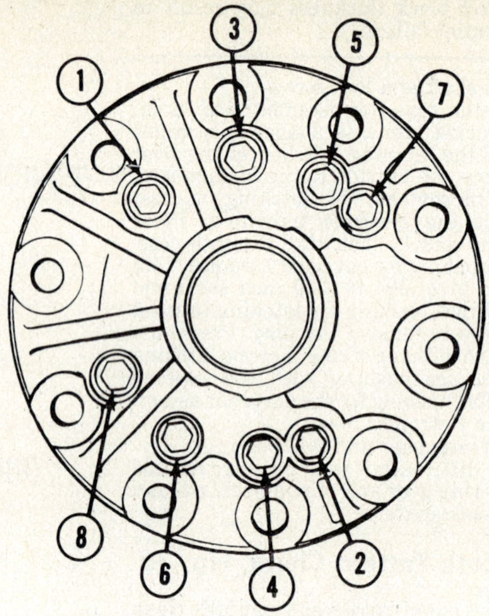

Fig. 41 Differential case bolt torque sequence

with calipers and replace with a service block of the same thickness.

NOTE: Incorrect block thickness will disturb critical clearances and may result in differential failure.

---

4. Preassemble pinions and pinion thrust washers on pinion yoke. Index yoke to proper position and install in housing with center over button end of reaction block. A light tap on the yoke may be needed to seat it in the housing.
5. Install left side reaction block on the yoke with flange side up. Use original block unless severely scored or worn. If new block is used, determine face-to-face thickness of original block with calipers and replace with a service block of the same thickness.

NOTE: Right and left side reaction blocks may not be the same thickness. If blocks are broken or cannot be accurately measured for thickness, the entire differential must be replaced. Incorrect block thickness will disturb critical clearances and may result in differential failure.

---

6. Install governor assembly and latching bracket in their respective locations. Place straight end of latching bracket spring over and to the outside of the governor shaft to preload the latching bracket against the governor assembly.
7. Install the original 3 eared and 2 splined clutch plates on the left side gear alternately, starting and ending with an eared disc.

NOTE: Original discs must be used. Attempts to install other than original discs will result in differential failure.

---

8. Install 6 clutch pack guide clips on left side disc pack assembly using grease for retention.
9. Install original shim in left side case half.

NOTE: Original shim must be used. Attempts to use other than original shims will disturb critical clearances and may result in differential failure. Slight polishing or scuffing is normal.

---

10. Remove disc pack from side gear and carefully lower disc pack assembly into left side case half. Make sure car guide clips are in their proper position.
11. Install side gear in left side case half. Rotate gear to engage spline with splines on discs.
12. Being careful not to dislodge side gear assembly, lower the left hand case assembly onto the right hand case. Index left hand case so holes in housing will line up for the governor assembly and latching bracket shafts.
13. Turn the entire unit over and install 3 screws.
14. Lock an axle shaft in a vise in a vertical position. Install differential assembly on axle shaft engaging the spline of the axle with a side gear. Rotate unit slowly. A short shaft held in pinion yoke will aid in rotating. The unit should turn smoothly without binding or locking up. Differential will lock if turned rapidly—50 mph or greater.
15. Install ring gear and side bearings.
16. Place differential in carrier and adjust ring gear to pinion backlash and gear tooth pattern. Use same procedure as for the Eaton 8⅞ Ring Gear to perform these checks.

NOTE: Use standard differential lubricant. DO NOT use lubricants specifically designed for "limited slip differentials".

GM CORP.

Disassembly, Fig. 39

1. Remove ring gear and side bearings, then remove pinion shaft.
2. Using a brass drift drive the preload spring from the case.
3. Support an axle shaft in a vise, and slide the case into the shaft, then turn the case to remove both pinions and thrust washers.
4. Remove the case from the axle shaft, then remove both side gears, clutch packs and shims. Mark the gears, clutch packs and shims for reinstallation in same position.

Inspection

Check the clutch plates and discs for wear and signs of overheating. Check the gear teeth for signs of wear. Replace parts as needed. The force required to compress the preload spring to 15/16 inch. should be 300 ± 30 lbs.

Reassembly

1. Lubricate the clutch discs and plates with limited slip lubricant.
2. Alternately position clutch plates and discs on a side gear, beginning and ending with a clutch plate.
3. Position the side gear, clutch pack and original shim into the case.
4. Install both pinion gears and thrust washers into the case, and install the pinion shaft.
5. Install the case onto an axle shaft supported in a vise.
6. Insert a screwdriver between the pinion shaft and the face of the side gear. Force the screwdriver in until the clutch pack is compressed.
7. Check the backlash between the side gear and pinion gears. If backlash does not fall into the range of .005 inch. to .008 inch., adjust the shim dimension as required. Increasing shim thickness will decrease backlash; decreasing shim thickness will increase backlash. Service shims are available from .070 inch. to .122 inch. in increments of .004 inch.
8. Remove the pinion shaft, pinion gears, side gear, clutch pack and shim from the case.
9. Install the opposite gear, clutch pack and original shim into the opposite side of the case. Place both pinion gears and thrust washers into place, and install the pinion shaft.
10. Follow the procedure in steps 5, 6, and 7 to determine the proper shim dimension.
11. When proper shim dimensions have been determined, remove pinion gears and pinion shaft and install both side gears, shims and clutch packs into case.
12. Mount the case onto the axle shaft locked in a vise. Place both pinions and thrust washers into position 180° apart and carefully "roll in" by turning the case on the shaft. A large "C" clamp may be used to apply slight compression against pinion gears to aid the "rolling in" procedure.
13. Tap the preload spring into place with a hammer.
14. Install the pinion shaft and lock screw.
15. Install the side bearings and ring gear using the procedure outlined for conventional units.
16. Place the differential unit in the carrier and adjust ring gear and pinion backlash, and gear tooth pattern.

BORG WARNER CONE TYPE

Disassembly, Fig. 40

1. If removing ring gear, clamp case in a vise so jaws are 90° to pinion shaft holes and remove ten ring gear retaining bolts.
2. Partially install two bolts on opposite sides of ring gear.
3. Remove ring gear from case by alternately tapping on bolts. Do not pry between case and ring gear.
4. Remove differential case half attaching bolts.
5. Lift cap half of case from flange half. Remove clutch cone/side gears, spring blocks, preload springs, pinion gears and shaft. Mark each clutch cone/side gear and pinion gear so they can be reinstalled in their original position.
6. Clean and inspect all parts and replace as necessary. Slight grooves or scratches, indicating passage of foreign material, are permissable and normal. If case or clutch cone/side gear are damaged, it is necessary to replace case assembly. All other parts are serviceable.

Assembly

1. Install proper cone/gear assembly into position in cap half of case.
2. Place one spring block in position over gear face, in alignment with pinion gear shaft grooves. Install pinion shaft, pinion gears and thrust washers into cap half or differential case in such a manner that pinion shaft retaining dowel can be inserted through pinion gear shaft into differential case. Be certain that pinion gears are installed in their original location.
3. Insert five springs into spring block that is already installed in case, then place second spring block over springs.
4. Install second cone/gear assembly face down on spring block so that gear will mesh with pinion gears.
5. Install flange half of differential case over cone, insert case bolts finger tight. Tighten bolts one turn at a time in sequence shown in Fig. 41. Then torque case bolts to 30 ft. lbs.
6. If ring gear was removed, position it so that holes line up with case holes.
7. Lubricate NEW attaching bolts with clean engine oil and install.
8. Tighten bolts evenly and alternately around the case. When all bolts are snug, torque evenly and alternately to 120 ft. lbs.

EATON DRIVING AXLES

INDEX

LUBRICANT CAPACITIES

Lubricant capacities are given as a guide only. The capacities of two similar axles in the same series may vary considerably due to design changes and the vehicle manufacturers' installation. The actual service capacity may be determined by carefully measuring the amount of lubricant necessary to fill the assembly to the correct level and measuring the lubricant again when it is drained. The vehicle should be on a level floor when this inspection is made.

Single Reduction Axles

Model	Forged Housing Capacity Pts.	Fabricated Housing Capacity Pts.
1315	19	19
1614-5, 1618-9	24	25
1790A-1A	29	26
1792-3	29	26
1862-3, 1880-1	29	26
1890-1, 2, 3	29	26
1910-1	34	—
1918-9	34	—
2010-1	38	—
2018-9	38	—

EATON DRIVING AXLES

Single Reduction Axles—Cont.

Model	Forged Housing Capacity Pts.	Fabricated Housing Capacity Pts.
17101, 121	31	—
17120	29	—
18101, 121	33½	—
18301	33	—
19121	35½	—
22121	29	—
23121	34	—
23421	—	—
26121	—	—

Two Speed Axles

Model	Forged Housing Capacity Pts.	Fabricated Housing Capacity Pts.
1350	19	19
13600-2	19	19
13800-2	19	19
15201	17½	—
16220	20	—
16221	24	—
16244	23	—
16500-1	24	25
16600-1	24	25
16802-3	24	25
17201, 221	31	—
17202, 222	22	—
17220	29	—
17500-1	29	26
17800-1	29	26
18201, 221	33½	—
18202	—	—
18222	28½	—
18500-1, 2, 3	29	26
18800-1, 2, 3	29	26
19201	34	—
19202	29½	—
19221	36	—
19500-1, 2, 3	34	—
19800-1	34	—
20500-1, 2, 3	38	—
20800-1	38	—
22221	29	—
22500-1	35	—
23221	34	—
26221	34	—

Planetary Double Reduction Axles

Model	Forged Housing Capacity Pts.	Fabricated Housing Capacity Pts.
350	19	19
3800, 3802	19	19
6600-1	24	25
6802-3	24	25
7800-1	29	26
8802-3	29	26
9502-3	34	—
9800-1	34	—
0502-3	38	—
0800-1	38	—
2500-1	35	—
17301	—	—
18302, 322	28½	—
45AR	38	—
35327	—	—

Tandem Drive Axles

Model	Forged Housing Capacity Pts.	Fabricated Housing Capacity Pts.
22MF, MR	12	—
22M (Power Divider)	9	—
28MF, MR	17	—
28M (Power Divider)	9	—
30DS, DT, DP	30	—

Tandem Drive Axles—Cont.

Model	Forged Housing Capacity Pts.	Fabricated Housing Capacity Pts.
30DSC, DTC, DPC	28	—
30D-3	30	—
30RS, RT, RP	27	—
30R-3	27	—
32MF	29	—
32M (Power Divider)	3	—
32MR	32	—
34DS, DT, DP	29	—
34DSC	32½	—
34DSE	32	—
34 DTC, DTE	32	—
34DPC, DPE	32	—
34D-3, C	29	—
34D-3E	24	—
34RS, RT, RP	32	—
34R-3	32	—
36MF	24	—
36M (Power Divider)	3	—
36MR	24	—
38DPC	35½	—
38 DS, DT, DP	29	—
38DSC	34	—
38DSE	31	—
38DTC, DPE	31	—
38D-3	29	—
38D-3C, D	27	—
38RS, RT, RP	32	—
38R-3	32	—
42D-3	31	—
42DT, DP	31	—
42DPB, DPD	29	—
42RT, RP	29	—
44DS	43	—
44DS, (Power Divider)	2	—
44DSC, DTC	20	—
50DP	24	—
50MF (In Rear Filler Hole)	20	—
50MF (In Front Filler Hole)	2	—
50M (Power Divider)	6	—
50MR (In Rear Filler Hole)	20	—
50MR (In Front Filler Hole)	2	—
DS340	29	—
DS340 (Power Divider)	2	—
DT340	—	—
DS-341 (Front)	33	—
DS-341 (Rear)	30.5	—
DS-341 (Power Divider)	2	—
DP380	—	—
DS380	29	—
DS380	—	—
DT380	—	—
DS-381 (Front)	33	—
DS-381 (Rear)	30.5	—
DS-381 (Power Divider)	2	—
DP400-P	—	—
DS400-P (Front)	29	—
DS400-P (Rear)	32	—
DT400-P	—	—
DS-401-P	—	—
DP440-P	—	—
DS440-P (Front)	43	—
DS440-P (Rear)	34	—
DT440-P	—	—
DP480-P	—	—
DS480-P (Front)	43	—
DS480-P (Rear)	34	—
DT480-P	—	—
DP580-P	—	—
DP650-P	—	—

TORQUE SETTINGS

Thread Size	Ft Lbs.
1/4–20	18–25
5/16–18	18–25
3/8–16	35–45
3/8–24	25–35
7/16–14	50–70 (1)
7/16–14	35–50 (2)
7/16–20	50–70
1/2–13	80–100
1/2–20	80–100
9/16–12	100–125
9/16–18	125–150 (3)
9/16–18	95–125 (4)
9/16–18	150–175 (5)
5/8–11	150–175
5/8–18	135–160 (6)
5/8–18	150–175 (7)
5/8–18	160–185 (8)
11/16–11	175–200
3/4–10	200–225
13/16–10	225–250
7/8– 9	300–350
1–20	225–350
1 1/8–18	325–450 (11)
1 1/8–18	500–700 (10)
1 1/8–18	175–200 (9)
1 1/4–12	400–600 (13)
1 1/4–12	175–200 (12)
1 1/4–18	400–600
1 1/2–18	175–200 (15)
1 1/2–18	500–700 (14)
1 3/4–12	600–800 (17)
1 3/4–12	175–200 (16)
2–12	700–900

(1)—Differential case.
(2)—Inspection cover to axle housing. Also oil collector drum to differential case on D Series forward axles.
(3)—Gear support case on two speed and planetary double reduction axles.
(4)—Ring gear to case on M Series axles.
(5)—Bolt locations other than specified in notes (3) and (4).
(6)—Ring gear to differential case on single reduction axles.
(7)—Herringbone gear to differential case on M Series forward axles, and gear case support on two-speed and planetary double reduction axles.
(8)—Differential case to axle housing on two-speed, plantetary double reduction, D Series forward and rear axles, and M Series forward axles.
(9)—Drive pinion nut on two speed and planetary double reduction axles.
(10)—Input and output shaft nuts on M Series axles.
(11)—Bolt locations other than specified in notes (9) and (10).
(12)—Output shaft nut on D Series forward axles.
(13)—Bolt locations other than specified in note (12).
(14)—Drive pinion nut on D Series forward axle and M Series rearward axle.
(15)—Bolt locations other than specified in note (14).
(16)—Input shaft nut on M Series forward axle.
(17)—Bolt locations other than specified in note (16).

Axle Adjustments

PROCEDURE

Measuring ring gear to drive pinion backlash and checking tooth contact pattern before disassembly, plus "punch marking" components for related position are recommended procedures for easier adjustment of axle after reassembly. If major axle parts are replaced, these identifying marks and reference measurements may not necessarily provide correct adjustment. However, usual return of parts to the identifying marks will serve as a reference to position parts as near as possible to correct adjustment.

The following is an outline of axle adjustment procedure and includes information on when adjustment should be performed and references to detailed instructions:

1. *Power Divider Input and Output Shaft Bearing Clearances.* These adjustments are made during reassembly of power divider. Refer to individual axle instructions.
2. *Drive Pinion Bearing Preload.* This adjustment is made before installing drive pinion (see below).
3. *Drive Pinion Position Adjustment.* Initial or preliminary adjustment is made during reassembly (see below).
4. *Differential Bearing Preload.* This adjustment is made in combination with ring gear and drive pinion adjustments (see below).
5. *Ring Gear and Drive Pinion Backlash.* Adjustments are given below.
6. *Ring Gear Tooth Contact.* Check ring gear tooth contact when adjusting ring gear and drive pinion (see below).
7. *Ring Gear Thrust Block Clearance.* This adjustment is made after drive pinion adjustments are completed and before differential carrier is installed in axle housing. Refer to individual axle instructions.
8. *Wheel Bearing Adjustments.* Procedure is given further on in this chapter.

DRIVE PINION BEARING PRELOAD

If the original pinion spacer can be used in reassembly, this should provide a close to correct adjustment of pinion bearing preload. If original spacer cannot be used start with spacer thickness recommended in Fig. 1.

Fig. 2 Checking drive pinion bearing preload with torque wrench (pinion in vise)

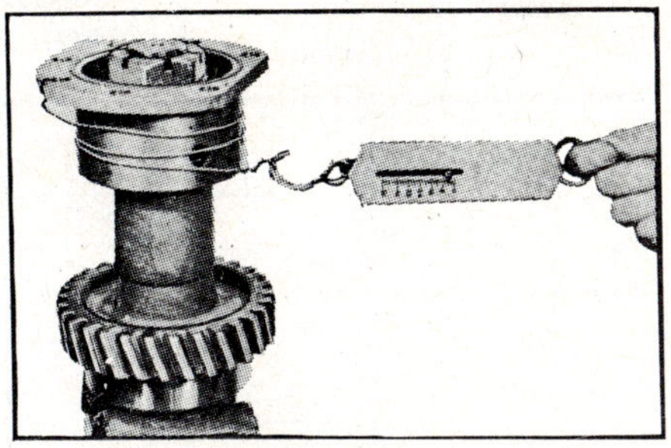

Fig. 3 Checking drive pinion bearing preload with spring scale for Tandem Drive Axles (pinion in vise)

With drive pinion, bearings, bearing cage and other related parts (oil seal and retainer removed) assembled and lubricated, measure and adjust pinion bearing preload.

Torque Wrench Method

1. Mount an adapter on pinion bearing cage, Fig. 2.
2. Hold drive pinion in a vise or arbor press. If an arbor press is used, apply approximately 5 tons pressure to bearings.
3. Rotate bearing cage by hand several revolutions to assure normal bearing contact. Then rotate bearing cage with a torque wrench and note torque reading as cage is rotated.
4. Correct bearing preload is indicated by a torque of 15 to 35 inch lbs. If incorrect, adjust by changing pinion bearing spacer. Use a thicker spacer to decrease preload or a thinner one to increase preload. *Do not use shim stock to increase thickness of spacer; always use correct size spacer.*

Spring Scale Method

1. Wrap a soft wire around bearing cage, Figs. 3 and 4.
2. Hold drive pinion in a vise or arbor press. If a press is used, apply approximately 5 tons pressure to bearings.
3. Pull scale on a line tangent to the cage outside diameter as shown with the scale attached to the wire and note scale reading.
4. To calculate torque from scale reading measure cage radius (see Fig. 1). Then multiply radius times scale reading. Example: With cage radius $2^{1/2}$" and scale pull 12 lbs., torque would be $12 \times 2^{1/2}$ or 30 in. lbs.
5. Correct bearing preload is indicated by a torque of 15 to 35 inch lbs. If incorrect, use a thicker spacer to decrease preload and a thinner one to increase preload. *Do not use shim stock to increase thickness of spacer; always use correct size spacer.*

DRIVE PINION POSITION

Drive pinion position in relation to ring gear is adjusted by adding or removing shims between pinion bearing cage and differential carrier or power divider case.

To assist in adjustment of ring gear and pinion, pinion should be positioned as near as possible to the correct position during reassembly. This can be accomplished as follows:

If the original ring gear and pinion are used in reassembly, use same size and quantity of shims removed during disassembly. If a new ring gear and pinion are used in reassembly, use shim pack recommended in Fig. 1.

The drive pinion position may also be set with a depth gauge. Fig. 1 also shows the basic dimensions used for gauge setting.

RING GEAR & DRIVE PINION

Except Model 42M Tandem

1. With differential bearings lubricated and bearing cap screws tightened *finger tight,* adjust ring gear and pinion as follows, referring to Figs. 5 and 6 for tooth contact pattern identification. Also refer to Fig. 7 for tooth nomenclature.

NOTE: *Adjustment of bearing adjusters to change ring gear position is affected by the relative mounting positions of the ring gear and drive pinion. The following instructions are for axles with ring gear on left-hand side of drive pinion. Left and right-hand instructions in parentheses (left-hand) (right-hand) are for axles with ring gear on right-hand side of drive pinion.*

---

2. Tighten differential bearing adjusters to correctly seat bearing cups.
3. Loosen right-hand (or left-hand) bearing adjuster until it is very loose. Tighten left-hand (or right-hand) bearing adjuster until backlash is removed between ring gear and drive pinion. Tightening left-hand (or right-hand) adjuster moves ring gear toward drive pinion.
4. Tighten right-hand (or left-hand) bearing

Fig. 4 Checking drive pinion bearing preload with spring scale (pinion in press)

adjuster until snug, then continue to tighten adjuster two additional notches. This will normally position ring gear within the specified .006 to .016" backlash. This adjustment will also preload differential bearings. Tightening right-hand (or left-hand) adjuster moves ring gear away from drive pinion.

5. Tighten differential bearing capscrews to correct torque, then measure ring gear movement with a dial indicator, Fig. 8, to determine specific backlash.
6. Paint ring gear teeth and check tooth contact pattern. Compare pattern with those illustrated in Figs. 5 and 6. If contact pattern similar to "A" is obtained, adjustment is correct.
7. If Pattern "B" is indicated, loosen left-hand (or right-hand) bearing adjuster and tighten right-hand (or left-hand) adjuster equal amounts. Recheck tooth contact pattern. If pattern is not corrected, continue to move bearing adjusters (the same number of notches) until proper lengthwise pattern is obtained.
8. If Pattern "C" is indicated, loosen right-hand (or left-hand) bearing adjuster and tighten left-hand (or right-hand) adjuster equal amounts. Recheck tooth pattern. If pattern is not corrected, continue to turn bearing adjusted equal amounts until proper lengthwise bearing pattern is obtained.
9. If Pattern "D" or "E" is indicated, adjustment of drive pinion position is necessary. Add or remove pinion bearing cage shims to move drive pinion as pattern indicates. When drive pinion position is changed, the adjustment may affect lengthwise bearing. If necessary, readjust backlash for proper lengthwise bearing. Recheck tooth contact pattern.
10. When correct tooth pattern "A" is obtained, install adjuster locks and cotter pins. Lock wire differential bearing capscrews.

		Dimensions for Drive Pinion Bearing Preload Adjustment		Dimensions for Drive Pinion Position Adjustment	
	AXLE MODELS	PINION BEARING SPACER	PINION BEARING CAGE RADIUS	PINION BEARING CAGE SHIM PACK	PINION DEPTH GAUGE SETTING
SINGLE REDUCTION AXLES	1315	0.528	$2\frac{1}{8}$	0.046	3.28125
	1614	0.637	$2\frac{1}{2}$	0.043	3.5625
	1618	0.528	$2\frac{1}{8}$	0.030	4.125
	1790A, 1862	0.639	$2\frac{1}{2}$	0.030 + 0.125 Spacer	4.2187
	1792, 1892	0.190	$2\frac{13}{16}$	0.040 + 0.125 Spacer	4.0625
	1910, 2010	0.185	$3\frac{1}{8}$	0.040	4.600
2-SPEED AXLES	1350, 13800	0.530	$2\frac{1}{8}$	0.040	3.6244
	13600	0.330	$2\frac{23}{32}$	0.045	4.28125
	13602	0.330	$2\frac{23}{32}$	0.165	4.28125
	13802	0.527	$2\frac{1}{8}$	0.040 + 0.250 Spacer *	3.6244
	16500	0.530	$2\frac{1}{8}$	0.040	4.125
	16600	0.331	3	0.056	4.8125
	16802	0.528	$2\frac{1}{8}$	0.043	4.125
	17500, 18500	For replacement, use 17800 or 18802 Gear Set plus 0.250 in. Bearing Cage Spacer			4.4062
	17800, 18802	0.638	$2\frac{1}{2}$	0.037	4.4062
	19502, 20502	0.188	$2\frac{13}{16}$	0.040	4.7812
	22501	0.185	$3\frac{1}{8}$	0.040	5.4375
PLANETARY DOUBLE REDUCTION AXLES	45AR, 0502, 9502	0.188	$2\frac{13}{16}$	0.040	4.7812
	2501	0.185	$3\frac{1}{8}$	0.040	5.4375
	3800	0.530	$2\frac{1}{8}$	0.040	3.6244
	3802	0.527	$2\frac{1}{8}$	0.040 + 0.250 Spacer †	3.6244
	6600	0.331	3	0.056	4.8125
	6802	0.528	$2\frac{1}{8}$	0.043	4.125
	7800, 8802	0.638	$2\frac{1}{2}$	0.037	4.4062
TANDEM DRIVE AXLES	22MF	0.530	$2\frac{1}{8}$	0.050	3.253
	22MR	0.530	$2\frac{1}{8}$	0.040	3.253
	28MF	0.530	$2\frac{1}{8}$	0.050	3.4725
	28MR	0.640	$2\frac{7}{16}$	0.040	3.4725
	30DS, 30DT, 30DP	0.638	$2\frac{1}{2}$	0.040	4.125
	30RS	0.528	$2\frac{1}{8}$	0.030	4.125
	30RT, 30RP	0.528	$2\frac{1}{8}$	0.043	4.125
	32MF	0.637	$2\frac{15}{32}$	0.050	4.2187
	32MR	0.639	$2\frac{1}{2}$	0.030	4.2187
	34DS, 34DT, 34DP	0.642	$2\frac{13}{16}$	0.056	4.4062
	34RS	0.639	$2\frac{1}{2}$	0.030 + 0.125 Spacer	4.2187
	34RT, 34RP	0.638	$2\frac{1}{2}$	0.037	4.4062
	36MF	0.638	$2\frac{15}{32}$	0.060	4.875
	36MR	0.177	$3\frac{1}{8}$	0.040	4.875
	38DS, 38DT, 38DP	0.642	$2\frac{13}{16}$	0.056	4.4062
	38RS	0.639	$2\frac{1}{2}$	0.030 + 0.125 Spacer	4.2187
	38RP, 38RT	0.638	$2\frac{1}{2}$	0.037	4.4062
	42DT, 42DP	0.642	$2\frac{13}{16}$	0.056	4.4062
	42RT, 42RP	0.638	$2\frac{1}{2}$	0.037	4.4062
	42MF	0.638	$2\frac{15}{32}$	0.060	2.8125
	42MR	0.175	$3\frac{3}{8}$	0.060	2.8125

* spacer not used on all model 13802 axles.   † spacer not used on all model 3802 axles.

Fig. 1 Eaton axle drive pinion adjustment specifications

Model 42M Tandem Axle

For easier adjustment, it is recommended that the differential assembly and power divider assembly be removed from the differential carrier. In the following instructions, it is assumed that only the countershaft assembly and bearings are installed in differential carrier. Refer to Fig. 5 for tooth contact pattern and Fig. 7 for tooth nomenclature.

1. To adjust countershaft bearings, turn left-hand bearing adjuster until two or three threads are engaged. Tighten right-hand bearing cage until end play is removed from countershaft bearings.
2. To preload countershaft bearings, continue to tighten right-hand bearing cage until resistance is felt when rotating countershaft. *This preload adjustment can be maintained during ring gear adjustment by turning bearing adjuster and cage equal amounts. Preload should be 10 to 22 inch lbs.*
3. Install power divider assembly.
4. To adjust ring gear backlash, loosen right-hand bearing cage and tighten left-hand bearing adjuster (the same number of turns) until .006 to .016" backlash is obtained. Measure backlash with dial indicator, Fig. 8.
5. Paint ring gear teeth and check tooth contact pattern. Compare pattern with those shown in Fig. 5.
6. If Pattern "B" is indicated, loosen left-hand bearing adjuster and tighten right-hand bearing cage equal amounts. Recheck tooth pattern. If pattern is not corrected, continue to move bearing cage and adjuster (the same number of notches) until proper lengthwise bearing pattern is obtained.
7. If Pattern "C" is indicated, loosen right-hand bearing cage and tighten left-hand bearing adjuster equal amounts. Recheck tooth pattern. If pattern is not corrected, continue to turn bearing cage and adjuster equal amounts until proper lengthwise bearing pattern is obtained.
8. If Pattern "D" or "E" is indicated, adjustment of drive pinion is necessary. Add or remove pinion bearing cage shims to move drive pinion as pattern indicates. When drive pinion position is changed, the adjustment may affect lengthwise bearing. If necessary, readjust backlash for proper lengthwise bearing. Recheck tooth contact pattern.
9. When Pattern "A" is obtained, install countershaft bearing covers, gaskets and capscrews. Install differential assembly.

WHEEL BEARING ADJUSTMENT

Wheel bearings should be adjusted to provide .001 to .010" end play.

1. Remove outer nut and lockwasher.
2. While turning wheel in both directions, tighten inner nut until a slight bind on nut is present.
3. Back off inner nut 1/3 turn to permit wheel to rotate freely.
4. Install lockwasher and outer nut, and lock both nuts in this position.

The ring gear tooth contact patterns illustrated are approximate shapes. Actual contact may vary, however the same general shape should be obtained. When adjusting gear sets that have been in service, tooth contact may vary because of wear. To obtain best results, strive to obtain a pattern coinciding with original patterns.

Pattern "A"
CORRECT TOOTH CONTACT

Correct adjustment is obtained when pattern of tooth bearing *(both lengthwise and profile)* appear as shown.

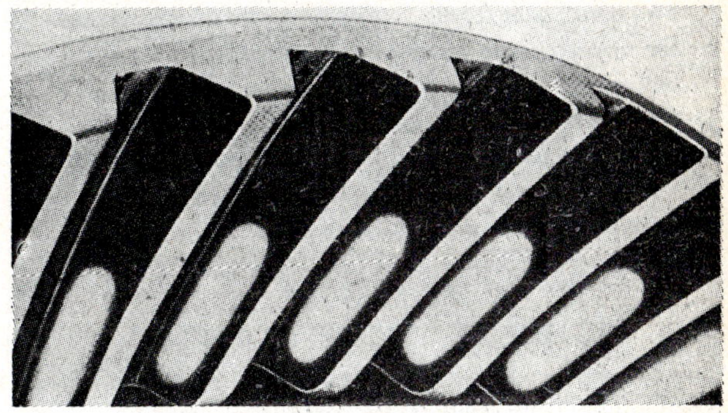

Pattern "A"

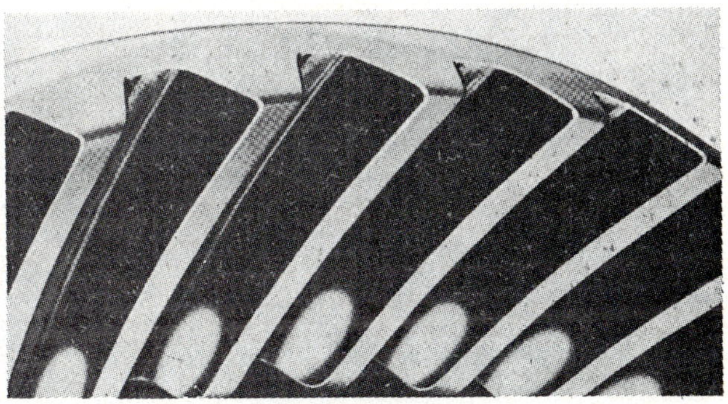

Pattern "B"
CONCENTRATED BEARING AT TOE

Not enough backlash . . . move ring gear *away* from drive pinion to increase lengthwise bearing. This may change the profile bearing to some extent and an adjustment of the pinion may be required.

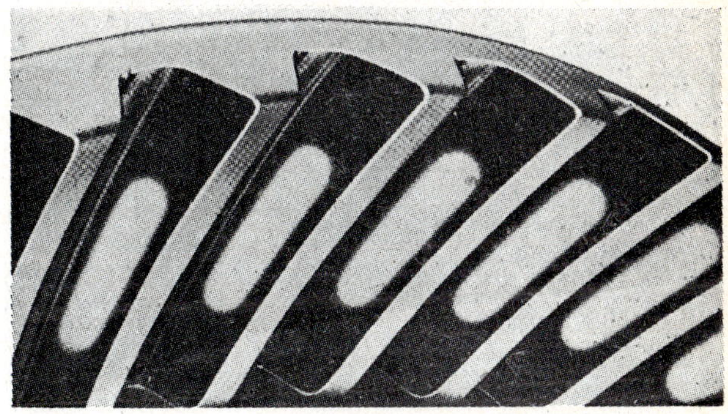

Pattern "C"
CONCENTRATED BEARING AT HEEL

Too much backlash . . . move ring gear *toward* drive pinion to obtain correct lengthwise bearing. This may change the profile bearing to some extent and an adjustment of the pinion may be required.

Pattern "D"
PROFILE BEARING HIGH

Pinion is out too far . . . remove shims to move drive pinion toward ring gear. Then move ring gear away from pinion to obtain correct lengthwise bearing.

Pattern "E"
PROFILE BEARING LOW

Pinion is in too far . . . add shims to move drive pinion away from ring gear. Then move ring gear toward pinion to obtain correct lengthwise bearing.

Fig. 5 Correct and incorrect tooth contact patterns for all models except 22M, 28M, 36M Tandem Axles

The ring gear tooth contact patterns illustrated are approximate shapes. Actual contact may vary, however the same general shape should be obtained. When adjusting gear sets that have been in service, tooth contact may vary because of wear. To obtain best results, strive to obtain a pattern coinciding with original patterns.

Pattern "A"
CORRECT TOOTH CONTACT

Correct adjustment is obtained when pattern of tooth bearing *(both lengthwise and profile)* appear as shown.

Pattern "A"

Pattern "B"
CONCENTRATED BEARING AT TOE

Not enough backlash . . . move ring gear *away* from drive pinion to increase lengthwise bearing. This may change the profile bearing to some extent and an adjustment of the pinion may be required.

Pattern "C"
CONCENTRATED BEARING AT HEEL

Too much backlash . . . move ring gear *toward* drive pinion to obtain correct lengthwise bearing. This may change the profile bearing to some extent and an adjustment of the pinion may be required.

Pattern "D"
PROFILE BEARING HIGH

Pinion is out too far . . . remove shims to move drive pinion toward ring gear. Then move ring gear away from pinion to obtain correct lengthwise bearing.

Pattern "E"
PROFILE BEARING LOW

Pinion is in too far . . . add shims to move drive pinion away from ring gear. Then move ring gear toward pinion to obtain correct lengthwise bearing.

Fig. 6 Correct and incorrect tooth contact pattern for Tandem Axle Models 22MF, 22MR, 28MF, 28MR, 36MF, 36MR

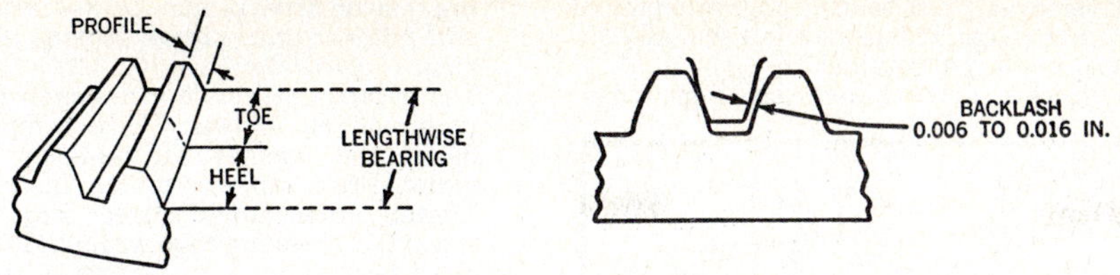

Ring Gear Tooth Nomenclature

check tooth contact pattern

Tooth contact pattern consists of the lengthwise bearing and profile bearing. The lengthwise bearing is the bearing along the tooth of the ring gear. The profile bearing is the bearing up and down the tooth (See fig. 206). In determining correct ring gear tooth contact, these two types of bearings must be considered separately to obtain proper results in combination.

With differential bearings adjusted correctly, paint at least ten ring gear teeth with red lead or prussian blue. Turn ring gear by hand a few revolutions in both directions to obtain impressions (or patterns) of tooth contact. Compare contact patterns with those shown in figure 208 or 209. If tooth contact patterns are not correct, move ring gear and/or drive pinion as necessary to adjust for correct pattern. Keep in mind that ring gear movement affects lengthwise bearing and that drive pinion movement affects the profile bearing.

Adjust to obtain correct tooth contact on drive side of teeth. Under this condition, the coast side tooth contact is usually satisfactory.

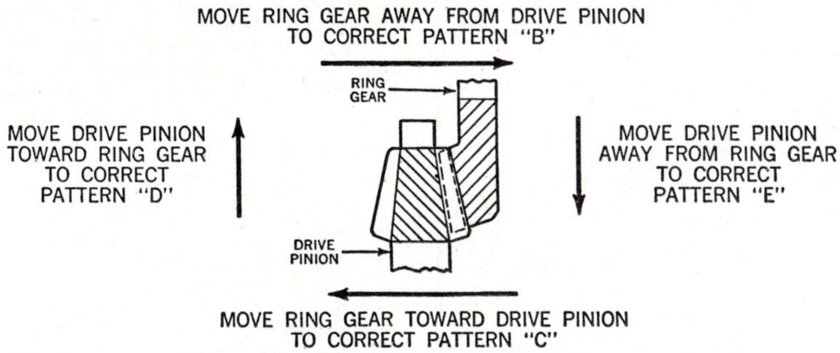

Fig. 7 Ring gear and drive pinion movement to obtain correct tooth contact pattern

Fig. 8 Measuring ring gear backlash with dial indicator

Eaton Single Reduction Axles

DISASSEMBLE

Differential Carrier, Fig. 9

1. Drain lubricant.
2. Disconnect propeller shaft and remove axle shafts.
3. Unfasten and remove differential carrier from axle housing and discard gasket.
4. Unscrew pipe plug, then remove spring and oil distributor (if used).
5. Remove ring gear thrust block from end of adjusting screw after removing locknut and screw (if used).
6. Punch mark differential bearing caps and carrier for correct location on reassembly. Also punch mark differential bearing adjusters for correct location of adjuster locks.
7. Remove differential bearing caps and adjusters. Tilt ring gear away from pinion and lift out differential assembly.
8. Remove pinion bearing cage-to-carrier capscrews. Drive pinion assembly out of carrier, being careful not to allow assembly to fall on flange. *Note size and quantity of shims under bearing cage flange for easier adjustment during reassembly.*

Drive Pinion

1. Hold companion flange in vise, then remove cotter pin, nut and flat washer.
2. Drive pinion out of flange.
3. Remove pinion bearings cage assembly. Remove oil seal retainer, flange flat washer (if used) and outer bearing cone from cage. Discard bearing cage cork seal.
4. If replacement is necessary, remove bearing cups from cage, using suitable puller. Remove oil seal and felt seal from seal retainer.
5. Remove bearing spacer washer (if used) and spacer from pinion.
6. Remove inner bearing cone and pilot bearing from pinion, using suitable pullers.

Differential

1. Punch mark differential cases for correct location in reassembly. Separate case halves and lift off plain half.
2. Lift out differential spider, pinions and thrust washers.
3. Remove bearing cones from differential cases with a split-type puller.
4. To remove ring gear from flanged differential, drill off the flat head of rivets (on ring gear side) and drive out rivets. Some ring gears are fastened with bolts and locknuts.

REASSEMBLE

Differential, Fig. 10

1. Rivet (cold) ring gear to flanged differential case, using a press. Compress rivets from 44 to 48 tons on Model 1315 and from 47–50 on all other models. If bolts and locknuts are used, tighten to correct torque. Lubricate internal parts of differential during reassembly.
2. Press bearing cones on differential cases.
3. Assemble spider, pinions and thrust washers in flanged differential case. Then tap plain differential case into place, aligning punch marks made during disassembly. Install capscrews, bolts and nuts. Tighten nuts and capscrews to correct torque and lockwire capscrews and nuts.

Drive Pinion, Fig. 11

1. Press pilot bearing on drive pinion and stake in four places as shown in Fig. 12.
2. Press inner bearing cone on pinion. Place spacer and washer (if used) on pinion.
3. If removed, press bearing cups into pinion bearing cage. Install new felt seal and oil seal in oil seal retainer.
4. Lubricate drive pinion bearings, then place pinion bearing cage and outer bearing cone on drive pinion. *Do not install oil seal retainer until pinion bearing adjustment is completed.*
5. Temporarily assemble companion flange to pinion as follows: Place companion flange flat washer (if used) on drive pinion. Press flange on pinion. Install and tighten pinion nut to correct torque. Rotate pinion bearing cage while tightening nut. Check pinion bearing preload adjustment as outlined previously.
6. With pinion bearings correctly adjusted, remove companion flange. Install new cork seal in groove of pinion bearing cage. Install oil seal retainer, then reinstall companion flange, flat washer and pinion nut. Tighten nut to correct torque and install cotter pin.

Differential Carrier

1. Install same quantity and size shims that were removed in disassembly on pinion bearing cage, then install drive pinion assembly in carrier. Install and tighten capscrews to correct torque.
2. Thread thrust block adjusting screw (if used) far enough into carrier to mount thrust block. Lubricate and install thrust block on end of adjusting screw.
3. Place differential assembly in carrier and install differential bearing cups, adjusters and caps. Install and finger tighten differential bearing capscrews. Tighten adjusters to correctly seat differential bearings. Adjust ring gear and drive pinion as outlined previously.
4. With ring gear and pinion correctly adjusted, and differential bearing capscrews tightened to correct torque, install cotter pins to secure adjuster locks. Lock wire capscrews.
5. Install oil distributor (if used), spring and pipe plug.
6. Adjust thrust block clearance as follows: Tighten adjusting screw to position thrust block against ring gear. Loosen adjusting screw 1/8–1/4 turn to obtain .010 to .015″ clearance between thrust block and ring gear. Tighten locknut. Bend ear of lock against nut to secure adjustment.

Install Carrier In Housing

1. Using a new gasket, install carrier in axle housing. Install capscrews and tighten to correct torque.
2. Install axle shaft and connect propeller shaft. Fill housing with correct lubricant.

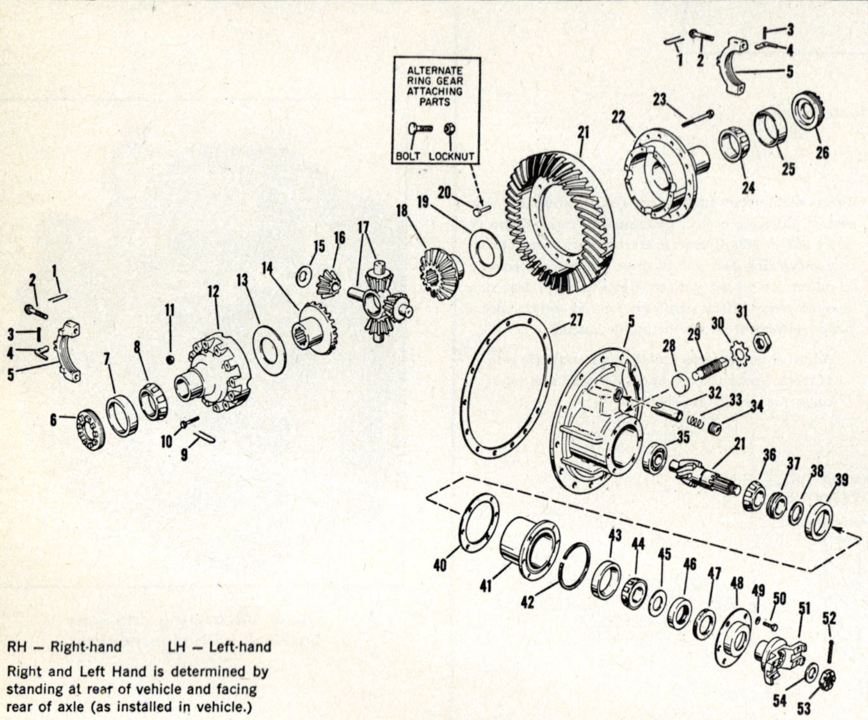

1. Lockwire
2. Cap screw
3. Cotter pin
4. Adjuster lock
5. Differential carrier and bearing caps (matched parts)
6. Differential bearing adjuster (RH)
7. Differential bearing cup (RH)
8. Differential bearing cone (RH)
9. Lockwire
10. Cap screw
11. Nut
12. Differential case (plain half)
13. Side gear thrust washer (RH)
14. Side gear (RH)
15. Side pinion thrust washer
16. Side pinion
17. Spider
18. Side gear (LH)
19. Side gear thrust washer (LH)
20. Rivet
21. Ring gear and drive pinion (matched set)
22. Differential case (flanged half)
23. Bolt
24. Differential bearing cone (LH)
25. Differential bearing cup (LH)
26. Differential bearing adjuster (LH)
27. Carrier gasket
28. Ring gear thrust block
29. Adjusting screw
30. Lock
31. Locknut
32. Oil distributor
33. Spring
34. Pipe plug
35. Pinion pilot bearing
36. Pinion bearing cone (inner)
37. Pinion bearing spacer
38. Pinion bearing spacer washer
39. Pinion bearing cup (inner)
40. Pinion bearing cage shims
41. Pinion bearing cage
42. Cork seal
43. Pinion bearing cup (outer)
44. Pinion bearing cone (outer)
45. Companion flange flat washer
46. Oil seal
47. Felt seal
48. Oil seal retainer
49. Lockwasher
50. Cap screw
51. Companion flange
52. Cotter pin
53. Pinion nut
54. Flat washer

Fig. 9 Disassembled view of Eaton Single Reduction Axle Differential Carrier

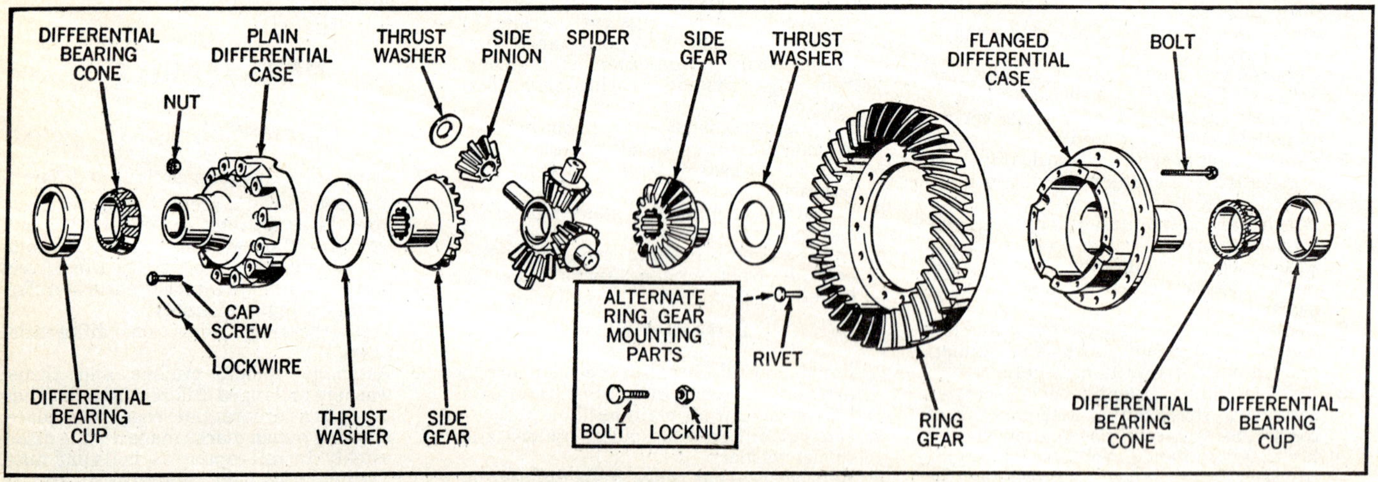

Fig. 10 Disassembled view of differential

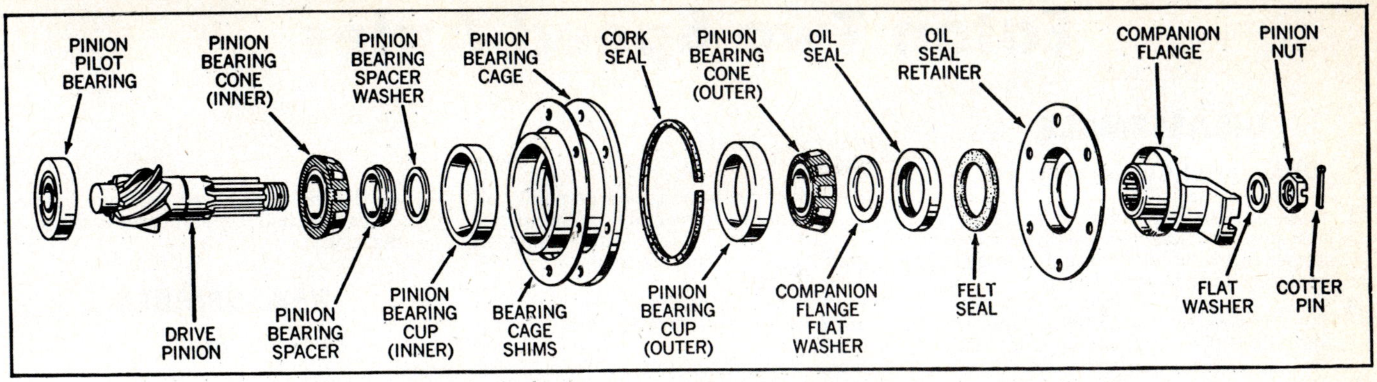

Fig. 11 Disassembled view of drive pinion. Companion flange flat washers and pinion bearing spacer washers are not used on all axles

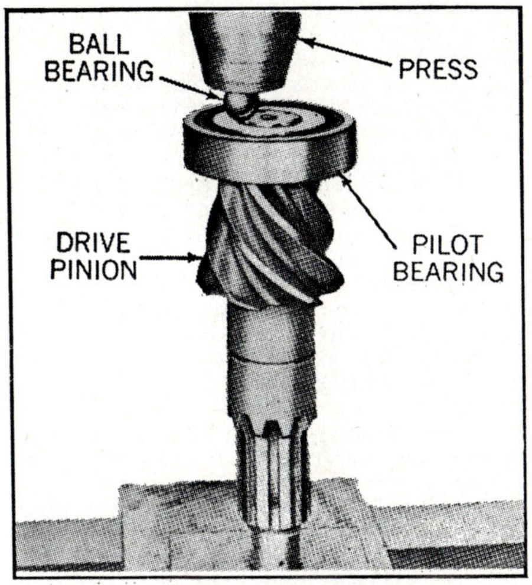

Fig. 12 Staking drive pinion pilot bearing (stake in four places)

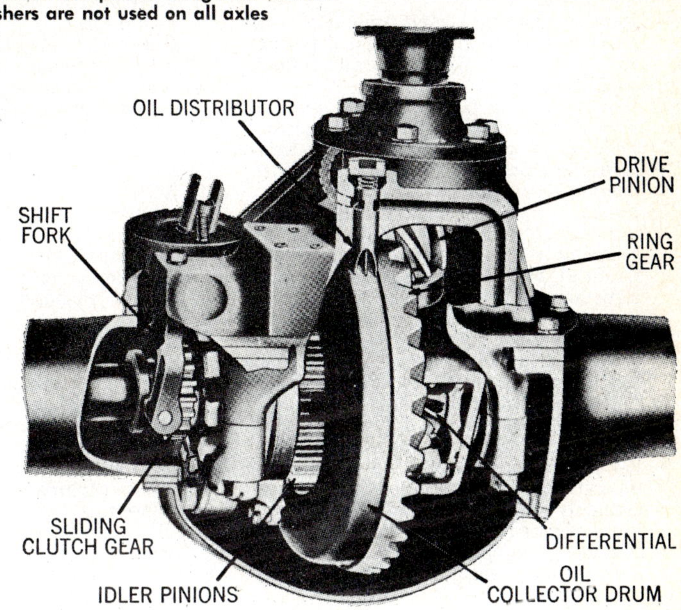

Fig. 13 Cutaway view of Eaton Two Speed Axle assembly

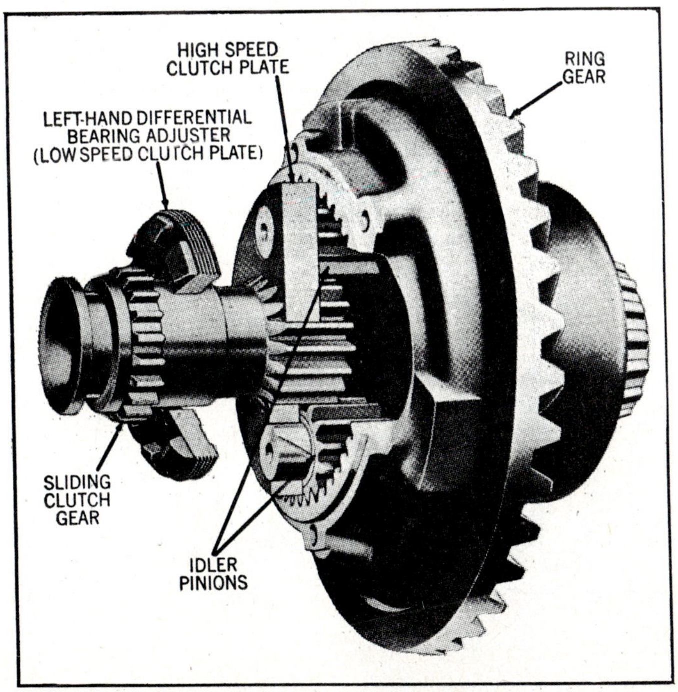

Fig. 14 Sliding clutch gear position for high axle ratio

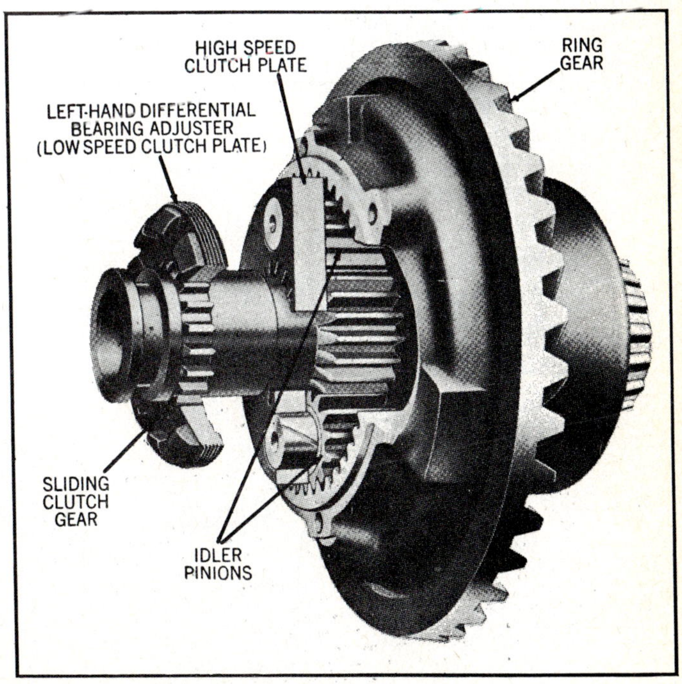

Fig. 15 Sliding clutch gear position for low axle ratio

Eaton Two Speed Axles

DISASSEMBLE

Differential Carrier, Fig. 16

1. Place axle in low range. Remove shift unit nuts and washers. Disconnect electric wires, air tubing or vacuum tubing as necessary to remove shift unit.
2. Drain lubricant. Disconnect propeller shaft and remove axle shafts. Unfasten and lift differential carrier from axle housing. Discard carrier gasket.
3. Unscrew pipe plug and remove spring and oil distributor.
4. Remove seal and spring from shift fork. Remove expansion plugs and drive out shift fork shaft. Disengage and remove shift fork.
5. Remove sliding clutch gear. *If preliminary inspection reveals that ring gear and drive pinion are satisfactory, check backlash and tooth pattern for easier adjustment in reassembly.*
6. Punch mark right-hand differential bearing adjuster for location of lock in reassembly. Cut lockwire and remove differential bearing cap screws. Remove right-hand differential bearing cap, bearing adjuster and lock.
7. Cut lockwire and remove left-hand differential bearing cap, bearing adjuster and lock as an assembly to assure correct position of ring gear during reassembly.
8. Tilt up planetary end of differential, then lift out differential assembly.
9. Remove pinion bearing cage-to-differential carrier capscrews. Then drive pinion assembly out of carrier, being careful not to allow assembly to fall on flange. *Note size and quantity of shims on bearing cage flange* for easier adjustment in reassembly.

Drive Pinion

1. Hold companion flange in vise and, after removing cotter pin, nut and flat washer, drive pinion out of flange.
2. Remove pinion bearing cage assembly, oil seal retainer assembly (flanged or pressed-in type) flange flat washer (if used) and outer bearing cone from pinion bearing cage.
3. Discard pinion bearing cage cork seal (used only with flanged type oil seal retainer).
4. Remove oil seal and felt seal from oil seal retainer. If replacement is necessary, remove bearing cups from cage, using suitable puller.
5. Remove pinion bearing spacer washer (if used) and spacer from drive pinion.
6. Remove inner bearing cone and pilot bearing, using pullers.

Differential

1. Remove locknuts and bolts holding gear support cases together. Tap alternately on opposite sides of ring gear with a soft-faced hammer until ring gear is free of flange on support case.
2. Lift off left-hand support case, oil collector drum and thrust washer.
3. Lift off ring gear.
4. Pry off high-speed clutch plate, then remove idler pinions and pins.
5. Lift out differential assembly. Remove support case thrust washer.
6. Punch mark differential cases for correct location in reassembly. Cut lockwires and remove capscrews, then lift off right-hand differential case.
7. Lift out long-hub side gear and thrust washer.
8. Lift out spider, pinions and thrust washers. Remove short-hub side gear and thrust washer from left-hand differential case.
9. To remove differential bearing cones, place pilot punch in holes provided in gear case support. Tap on bearing cone inner race, alternately through each hole until cone is removed. Remove bearing cones from both support cases in the same manner.

REASSEMBLE

Differential, Fig. 17

1. Press bearing cones on gear support case.
2. Place thrust washer and short-hub side gear in left-hand differential case. Assemble side pinions and thrust washers to spider, then place this assembly in left-hand differential case.
3. Place long-hub side gear with thrust washer on side pinions. Place right-hand differential case over differential gears, aligning punch marks on both cases. Then install capscrews and tighten to correct torque. Lockwire capscrews.
4. Place right-hand support case thrust washer and differential assembly in right-hand support case. Install spider and pinions. Place high speed clutch plate on idler pinions. *Make certain that chamfered teeth of clutch plate are positioned toward idler pinions.*
5. Mount ring gear, oil collector drum, thrust washer and left-hand support case on right-hand support case. *During assembly of these parts, temporarily place two bolts in mounting holes. This procedure will assure bolt hole alignment as these parts are assembled into final position. Secure assembly with bolts and locknuts. Temporarily install sliding clutch gear and rotate idler pinions to make sure parts are assembled correctly. Tighten locknuts to correct torque.*

Drive Pinion, Fig. 18

1. Press pinion pilot bearing on drive

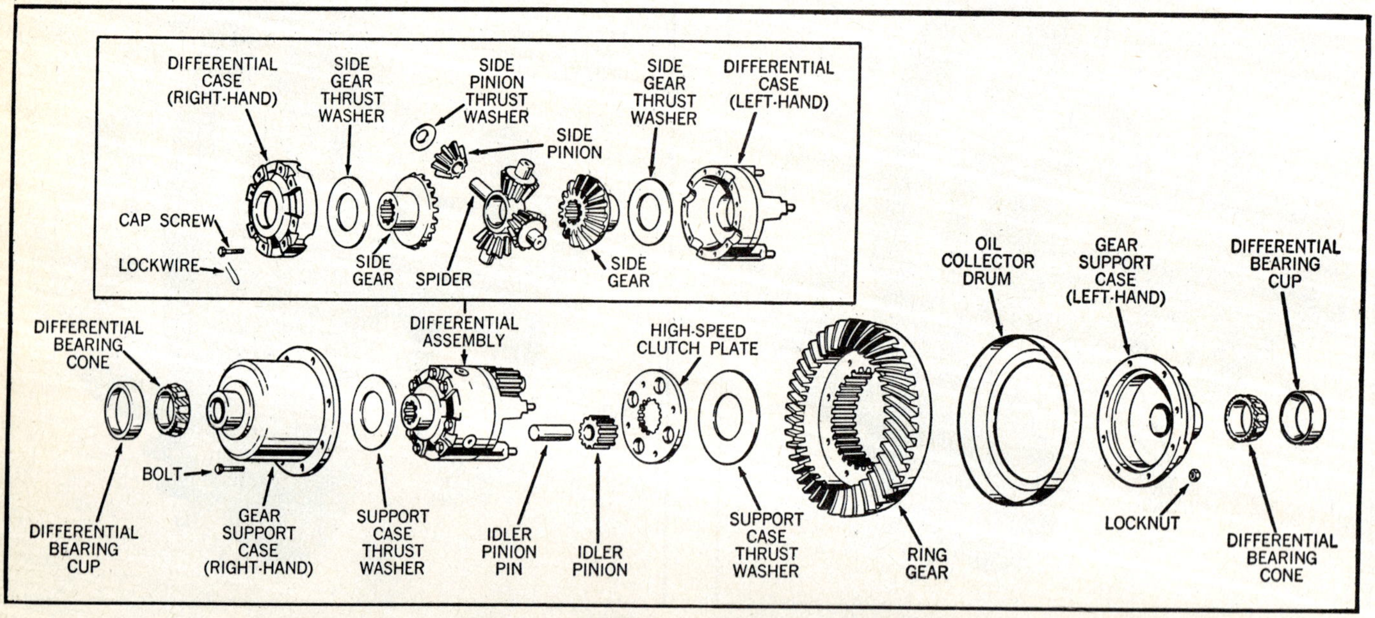

Fig. 17 Disassembled view of differential assembly

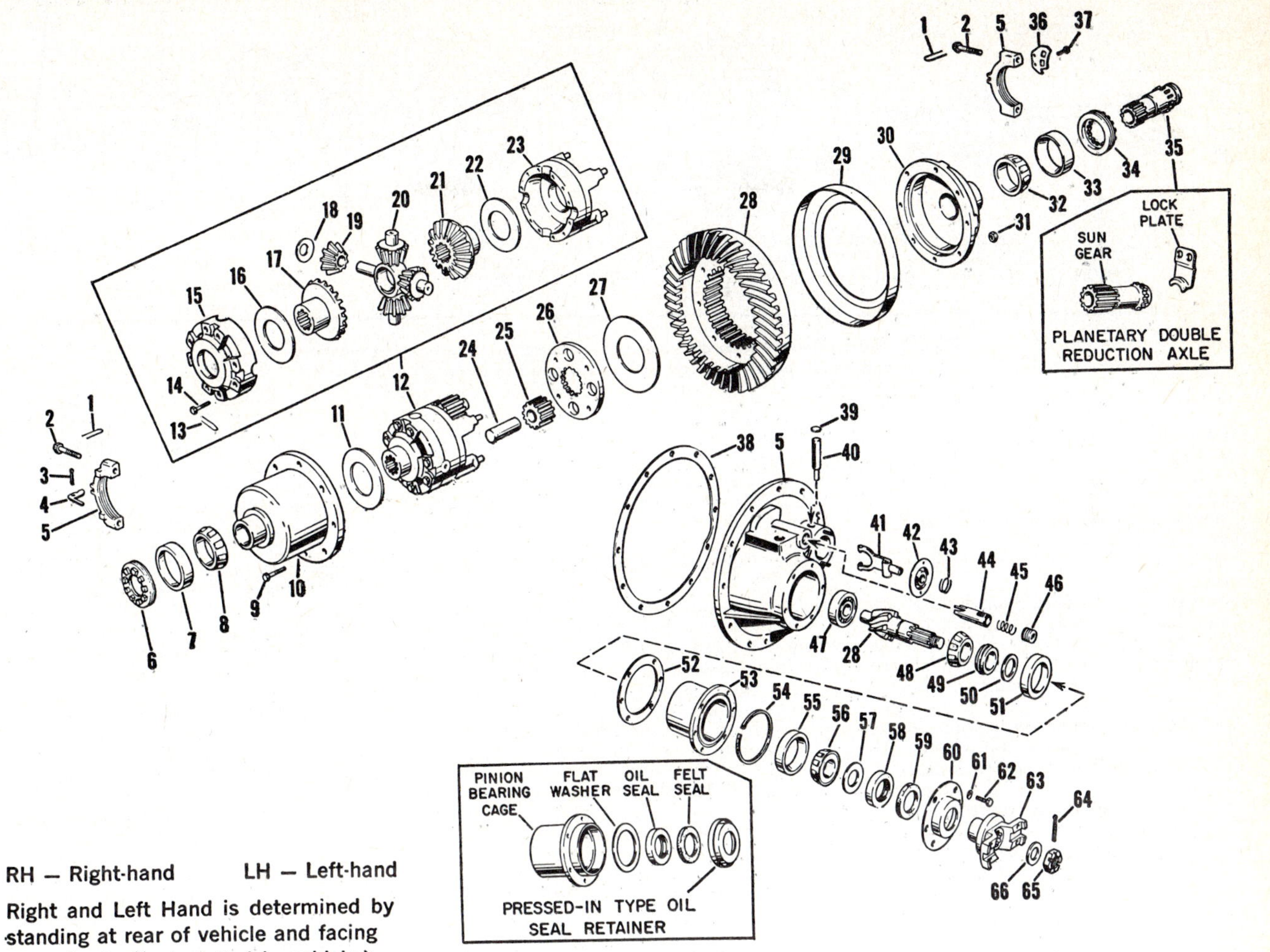

RH — Right-hand LH — Left-hand

Right and Left Hand is determined by standing at rear of vehicle and facing rear of axle (as installed in vehicle.)

1. Lockwire
2. Cap screw
3. Cotter pin
4. Adjuster lock (RH)
5. Differential carrier and bearing caps (matched parts)
6. Differential bearing adjuster (RH)
7. Differential bearing cup (RH)
8. Differential bearing cone (RH)
9. Bolt
10. Gear support case (RH half)
11. Support case thrust washer (RH)
12. Differential assembly
13. Lockwire
14. Cap screw
15. Differential case (RH half)
16. Side gear thrust washer (RH)
17. Side gear (long-hub, RH)
18. Side pinion thrust washer
19. Side pinion
20. Spider
21. Side gear (short hub, LH)
22. Side gear thrust washer (LH)
23. Differential case (LH half)
24. Idler pinion pin
25. Idler pinion
26. High-speed clutch plate
27. Support case thrust washer (LH)
28. Ring gear and drive pinion (matched set)
29. Oil collector drum
30. Gear support case (LH half)
31. Locknut
32. Differential bearing cone (LH)
33. Differential bearing cup (LH)
34. Differential bearing adjuster (LH) (low-speed clutch plate is integral part of adjuster)
35. Sliding clutch gear
36. Adjuster lock (LH)
37. Cap screw
38. Carrier gasket
39. Expansion plug
40. Shift fork shaft
41. Shift fork
42. Shift fork seal
43. Shift fork seal spring
44. Oil distributor
45. Spring
46. Pipe plug
47. Pinion pilot bearing
48. Pinion bearing cone (inner)
49. Pinion bearing spacer
50. Pinion bearing spacer washer
51. Pinion bearing cup (inner)
52. Pinion bearing cage shims
53. Pinion bearing cage
54. Cork seal (used with flanged-type oil seal retainer)
55. Pinion bearing cup (outer)
56. Pinion bearing cone (outer)
57. Companion flange flat washer
58. Oil seal
59. Felt seal
60. Oil seal retainer (flanged type) (See insert for components of pressed-in type oil seal retainer)
61. Lockwasher
62. Cap screw
63. Companion flange
64. Cotter pin
65. Pinion nut
66. Flat washer

Fig. 16 Disassembled view of differential carrier for Two Speed Axle

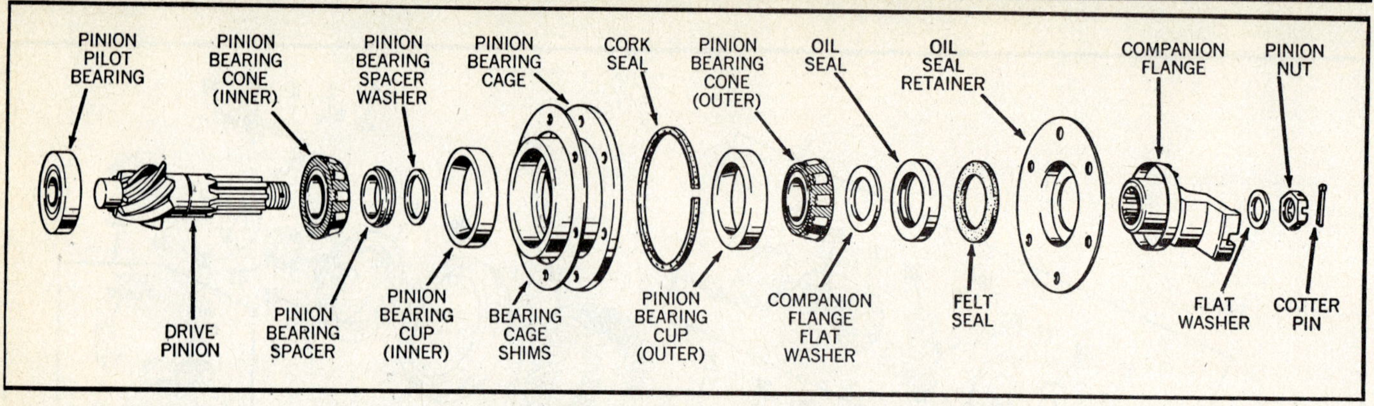

Fig. 18 Disassembled view of drive pinion. Companion flange flat washers and pinion bearing spacer washers are not used on all axles

pinion, then stake in place as shown in Fig. 12.
2. Press inner bearing cone on drive pinion.
3. If removed, press bearing cups in bearing cage.
4. Install new felt seal and oil seal in retainer.
5. Lubricate drive pinion bearings. Place pinion bearing spacer, spacer washer (if used), bearing cage and outer bearing cone on drive pinion. *Do not install oil seal retainer until pinion bearing adjustment is completed.*
6. Temporarily assemble companion flange as follows: Place flange flat washer (if used) on drive pinion. Press flange on pinion and install flat washer and pinion nut, tightening nut to correct torque. *Rotate bearing cage while tightening pinion nut.* Check pinion bearing preload adjustment as outlined previously.
7. To install oil seal retainer assembly (after drive pinion bearings are correctly adjusted) remove nut, flat washer and companion flange. Install oil seal retainer assembly *(pressed-in type mounts in recess of bearing cage with flat washer positioned between cage recess and oil seal retainer. Flanged-type is attached with same capscrews that fasten bearing cage).* Install companion flange, flat washer and nut, tightening nut to correct torque.

Differential Carrier

1. Using same quantity and size of shims that were removed during disassembly on pinion bearing cage flange, install drive pinion assembly in carrier. Install bearing cage capscrews and lockwashers, tightening capscrews to correct torque. Lubricate differential bearings during reassembly.
2. Place differential assembly in carrier and install differential bearing cups. Install assembled left-hand differential bearing cap, adjuster and lock. Install right-hand differential bearing cap and adjuster. Install and finger tighten differential bearing capscrews. Tighten right-hand bearing adjuster to seat differential bearings. Adjust ring gear and drive pinion as outlined previously.
3. With ring gear and drive pinion correctly adjusted, tighten differential bearing cap screws to correct torque. Then install cotter pins to secure locks and lockwire capscrews.
4. Install sliding clutch gear.
5. Install and engage shift fork with sliding clutch gear. Install shift fork shaft and expansion plugs.
6. Place seal on shift fork with bottom hole (stamped on seal) over lower stud. Install spring on seal.
7. Install oil distributor, spring and pipe plug in carrier.

Install Carrier In Housing

1. Using a new gasket, install differential carrier in axle housing. Install and tighten capscrews to correct torque. Install axle shafts and connect propeller shaft. Fill housing with correct lubricant.
2. Install electric shift unit or air-torsion spring unit on carrier, engaging swivel with slot in shift fork shaft. Install mounting nuts and washers. For electric shift, connect electric wires. For air-torsion spring shift, connect air line to unit.

Eaton Planetary Double Reduction Axles

This axle assembly, Fig. 19, is designed to provide a compact unit with heavy-duty load capacity. This is accomplished by incorporating a rugged ring gear and drive pinion set and a durable planetary unit in a single axle housing.

Differential Carrier Service

Remove the differential carrier from the axle housing as outlined for the two speed axle. Procedures for disassembly are the same as for the two speed axle. However, the Planetary Double Reduction Axle does not use a shift mechanism.

A sun gear is used in place of the sliding clutch gear. To remove the sun gear, remove the retainer which holds the sun gear in position, then remove sun gear.

Fig. 19 Power flow through Planetary Double Reduction Axle. This unit is similar to the two speed unit but it does not have a shift unit

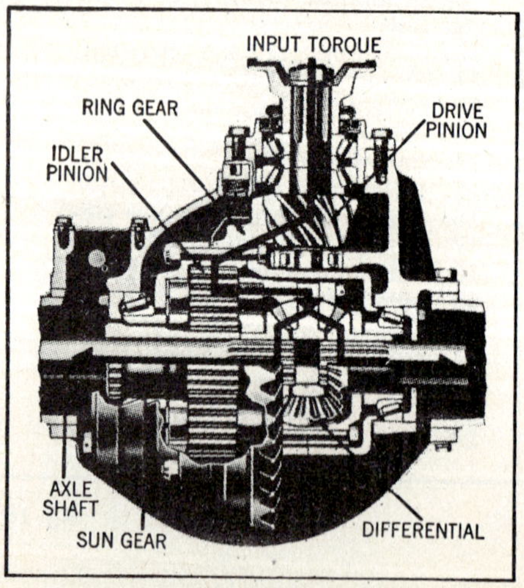

Eaton "D" Series Tandem Drive Axles

The Series "D" driving unit consists of two axles coupled by a power divider. Three different combinations are available:

1. A single reduction forward axle with power divider, and a single reduction rearward axle.
2. A two-speed forward axle with power divider, and a two-speed rearward axle.
3. A planetary double reduction forward axle with power divider, and a planetary double reduction rearward axle.

Disassemble

It is recommended, but not necessary, that the axle housing cover be removed to replace the differential carrier and power divider assembly.

1. Drain lubricant, disconnect propeller shafts at companion flanges and remove axle shafts.
2. Remove companion flange from output shaft.
3. Remove axle housing cover and discard gasket.
4. To disassemble axle housing cover, remove oil seal from output shaft opening, using a suitable puller. Remove snap ring from output shaft opening, then press rear bearing from axle housing cover. *Rear bearing sleeve is mounted on output shaft.*
5. Disconnect lockout shift lever connections.
6. If axle is "Two-Speed", remove shift units, referring to "Two-Speed Axle" Section.
7. After removing stud nuts and lockwashers, lift and pull differential carrier and power divider out of axle housing.

Carrier & Power Divider, Figs. 20, 21, 22

1. If preliminary inspection reveals that ring gear and drive pinion are satisfactory, check backlash and tooth contact pattern for easier adjustment in reassembly.
2. Remove carrier cover from carrier.
3. Lift inter-axle differential off output shaft side gear.
4. Before disassembling inter-axle differential, punch mark cases for correct position during reassembly. Cut lockwire and remove nuts and bolts. Separate cases and remove thrust washers, side pinions, bushings and spider.

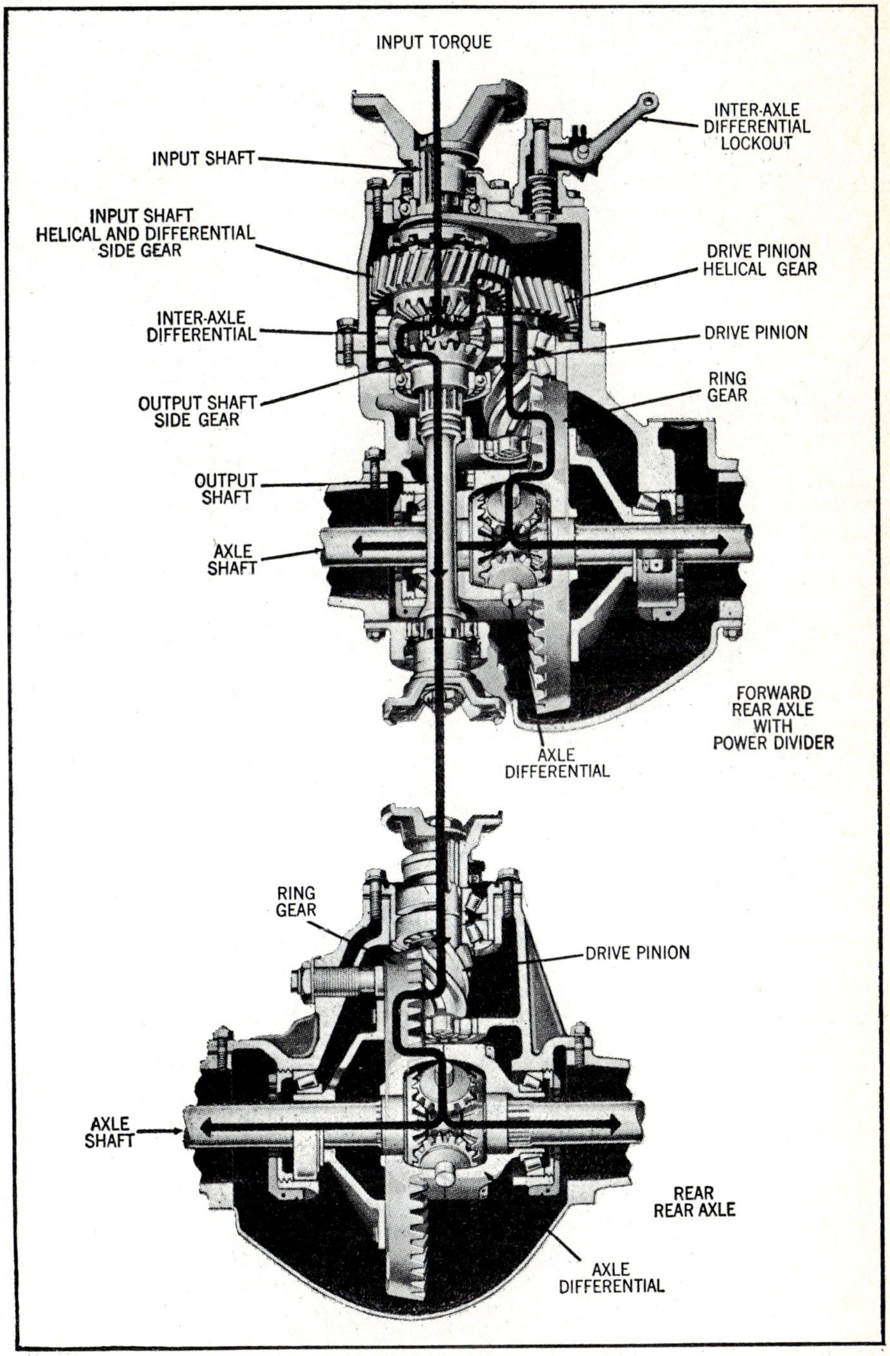

Fig. 20 Power flow through Single Reduction "D" Series Tandem Drive Axle

Output Shaft

1. Drive output shaft from differential carrier, driving at companion flange end of shaft.
2. To disassemble output shaft, remove snap ring, then lift off side gear assembly. Discard output shaft "O" rings. If replacement is necessary, remove bushing from bore of output shaft.
3. To remove output shaft front bearing from side gear, place side gear assembly in press with split-type puller plates under front bearing, then press side gear out of front bearing.

Input Shaft

1. Remove snap ring from input shaft, then lift off helical side gear, thrust washer and "D" washer. If replacement is necessary, press bushings from bore of side gear.
2. Remove companion flange from input shaft.
3. Remove input shaft bearing cover and shims. *Note size and quantity of shims for easier adjustment during reassembly.* Remove oil seal and felt seal from bearing cover.
4. Drive input shaft out of differential carrier cover.
5. To remove bearing from input shaft, place assembly in a press with threaded end of input shaft upward. Then press input shaft out of bearing.

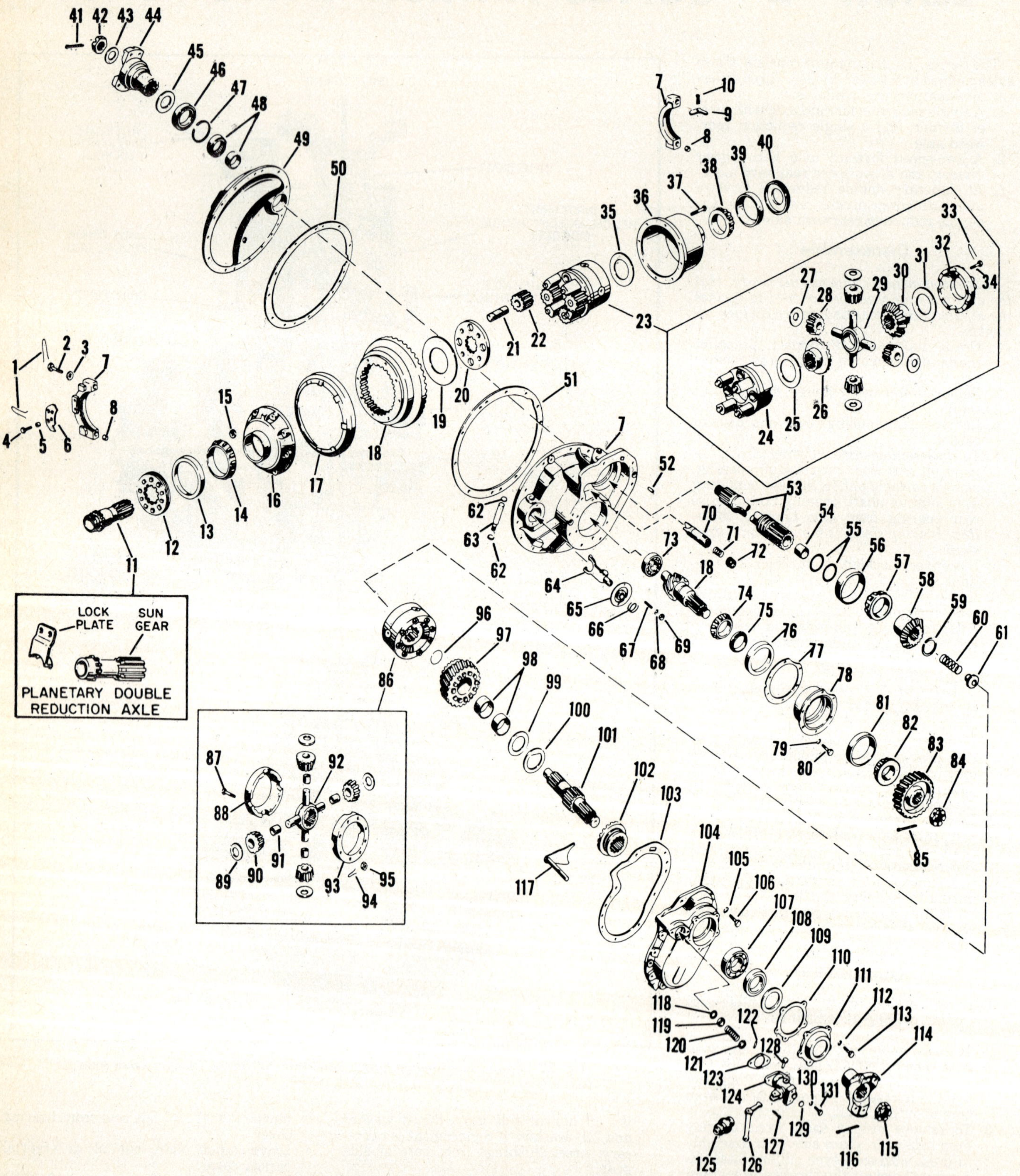

Fig. 21 Disassembled view of forward rear axle for Two-Speed and Planetary Double Reduction "D" Series Tandem Drive Axles (See next page for legend)

Forward Rear Axle for 2-Speed and Planetary Double Reduction "D" Series Tandem Drive Axles

RH — Right-hand LH — Left-hand

Right and Left Hand is determined by standing at rear of vehicle and facing rear of axle (as installed in vehicle)

1. Lockwire
2. Cap screw
3. Flat washer
4. Cap screw
5. Dowel bushing
6. Adjuster lock (RH)
7. Differential carrier and bearing caps (matched parts)
8. Dowel bushing
9. Adjuster lock (LH)
10. Cotter pin
11. Sliding clutch gear
12. Differential bearing adjuster (RH) (low speed clutch plate is integral part of adjuster)
13. Differential bearing cup (RH)
14. Differential bearing cone (RH)
15. Locknut
16. Gear support case (RH half)
17. Oil collector drum
18. Ring gear and drive pinion (matched set)
19. Support case thrust washer (RH)
20. High speed clutch plate
21. Idler pinion pins
22. Idler pinions
23. Differential asembly
24. Differential case (RH half)
25. Side gear thrust washer (RH)
26. Side gear (short hub, RH)
27. Side pinion thrust washer
28. Side pinion
29. Spider
30. Side gear (long hub, LH)
31. Side gear thrust washer (LH)
32. Differential case (LH half)
33. Lockwire
34. Cap screw
35. Support case thrust washer (LH)
36. Gear support case (LH half)
37. Bolt
38. Differential bearing cone (LH)
39. Differential bearing cup (LH)
40. Differential bearing adjuster (LH)
41. Cotter pin
42. Output shaft nut
43. Flat washer
44. Output companion flange
45. Rear bearing retainer washer
46. Oil seal
47. Rear bearing snap ring
48. Output shaft rear bearing and sleeve (matched parts)
49. Axle housing cover
50. Cover gasket
51. Carrier gasket
52. Dowel pins
53. Output shaft
54. Output shaft bushing
55. Output shaft "O" rings
56. Output shaft front bearing cup
57. Output shaft front bearing cone
58. Output shaft side gear
59. Output shaft snap ring
60. Output shaft holdout spring
61. Output shaft holdout thrust washer
62. Expansion plug
63. Shift fork shaft
64. Shift fork
65. Shift fork seal
66. Seal spring
67. Stud
68. Lockwasher
69. Stud nut
70. Oil distributor
71. Spring
72. Pipe plug
73. Pinion pilot bearing
74. Pinion bearing cone (inner)
75. Pinion bearing spacer
76. Pinion bearing cup (inner)
77. Pinion bearing cage shims
78. Pinion bearing cage
79. Lockwasher
80. Cap screw
81. Pinion bearing cup (outer)
82. Pinion bearing cone (outer)
83. Pinion helical gear
84. Pinion nut
85. Cotter pin
86. Inter-axle differential assembly
87. Bolt
88. Inter-axle differential case (female half)
89. Side pinion thrust washer
90. Side pinion
91. Side pinion bushing
92. Spider
93. Inter-axle differential case (male half)
94. Lockwire
95. Nut
96. Input shaft snap ring
97. Helical and differential side gear
98. Helical and differential side gear bushings
99. Helical and differential side gear thrust washer
100. Helical and differential side gear "D" washer
101. Input shaft
102. Input shaft sliding clutch
103. Differential carrier cover gasket
104. Differential carrier cover
105. Lockwasher
106. Cap screw
107. Input shaft bearing
108. Oil seal
109. Felt seal
110. Bearing cover shims
111. Bearing cover
112. Lockwasher
113. Cap screw
114. Input companion flange
115. Input shaft nut
116. Cotter pin
117. Shift fork and push rod
118. Felt seal
119. Seal retainer
120. Compression spring
121. Retainer washer
122. Retainer washer pin
123. Shift lever bracket gasket
124. Shift lever bracket
125. Shift lever boot
126. Shift lever
127. Cotter pin
128. Clevis pin
129. Expansion plug
130. Lockwasher
131. Cap screw

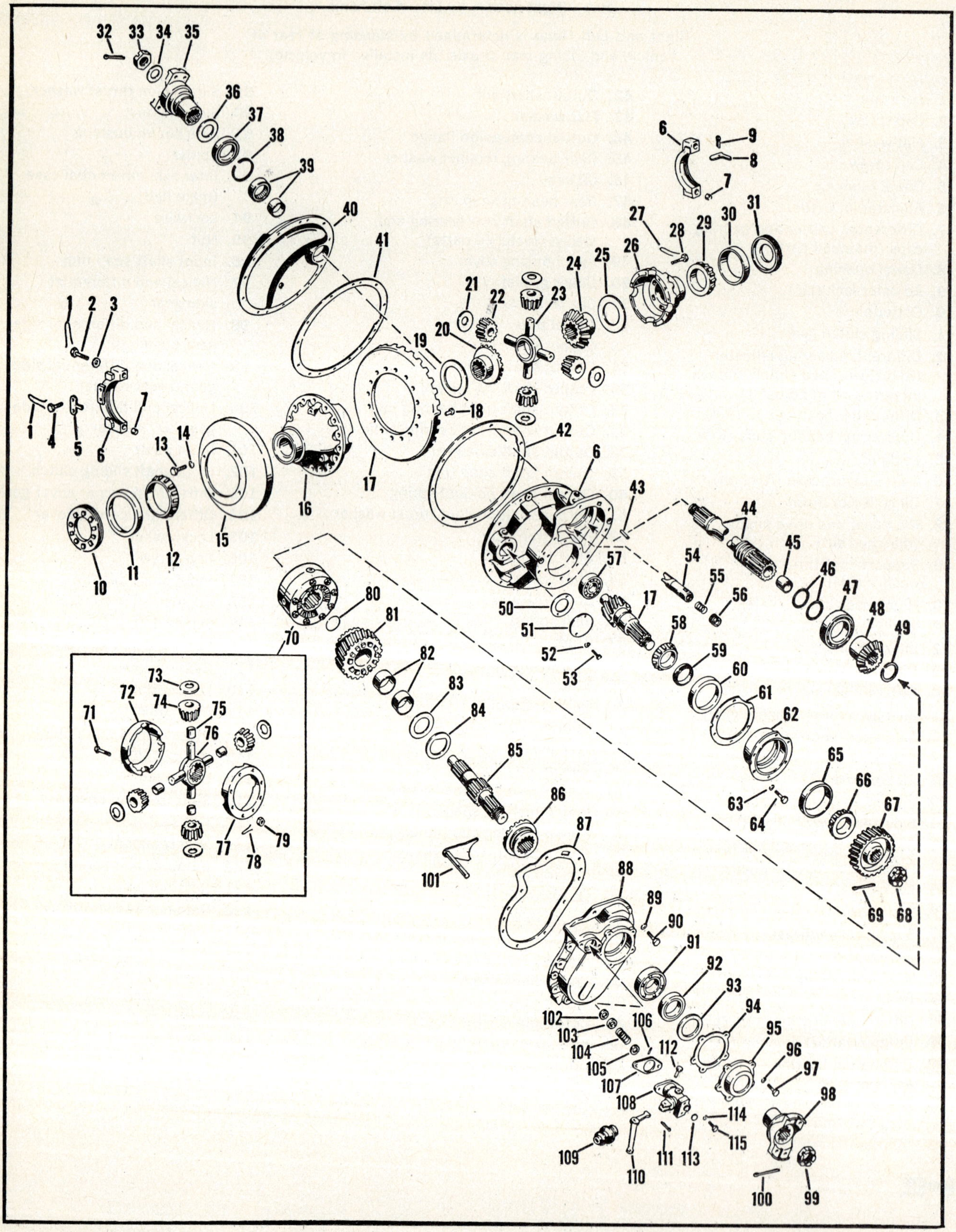

Fig. 22 Disassembled view of forward rear axle for Single Reduction "D" Series Tandem Drive Axle (See next page for legend)

RH – Right-hand LH – Left-hand

Right and Left Hand is determined by standing at rear of vehicle and facing rear of axle (as installed in vehicle)

1. Lockwire
2. Cap screw
3. Flat washer
4. Cap screw
5. Adjuster lock (RH)
6. Differential carrier and bearing caps (matched parts)
7. Dowel bushing
8. Adjuster lock (LH)
9. Cotter pin
10. Differential bearing adjuster
11. Differential bearing cup (RH)
12. Differential bearing cone (RH)
13. Cap screw
14. Lockwasher
15. Oil collector drum
16. Differential case (RH half)
17. Ring gear and drive pinion (matched set)
18. Rivet
19. Side gear thrust washer (RH)
20. Side gear (RH)
21. Side pinion thrust washer
22. Side pinion
23. Spider
24. Side gear (LH)
25. Side gear thrust washer (LH)
26. Differential case (LH half)
27. Lockwire
28. Cap screw
29. Differential bearing cone (LH)
30. Differential bearing cup (LH)
31. Differential bearing adjuster (LH)
32. Cotter pin
33. Output shaft nut
34. Flat washer
35. Output companion flange
36. Rear bearing retainer washer
37. Oil seal
38. Rear bearing snap ring
39. Output shaft rear bearing and sleeve (matched parts)
40. Axle housing cover
41. Cover gasket
42. Carrier gasket
43. Dowel pins
44. Output shaft
45. Output shaft bushing
46. Output shaft "O" rings
47. Output shaft front bearing
48. Output shaft side gear
49. Output shaft snap ring
50. Hole cover gasket
51. Hole cover
52. Lockwasher
53. Cap screw
54. Oil distributor
55. Spring
56. Pipe plug
57. Pinion pilot bearing
58. Pinion bearing cone (inner)
59. Pinion bearing spacer
60. Pinion bearing cup (inner)
61. Pinion bearing cage shims
62. Pinion bearing cage
63. Lockwasher
64. Cap screw
65. Pinion bearing cup (outer)
66. Pinion bearing cone (outer)
67. Pinion helical gear
68. Pinion nut
69. Cotter pin
70. Inter-axle differential assembly
71. Bolt
72. Inter-axle differential case (female half)
73. Side pinion thrust washer
74. Side pinion
75. Side pinion bushing
76. Spider
77. Inter-axle differential case (male half)
78. Lockwire
79. Nut
80. Input shaft snap ring
81. Helical and differential side gear
82. Helical and differential side gear bushings
83. Helical and differential side gear thrust washer
84. Helical and differential side gear "D" washer
85. Input shaft
86. Input shaft sliding clutch
87. Differential carrier cover gasket
88. Differential carrier cover
89. Lockwasher
90. Cap screw
91. Input shaft bearing
92. Oil seal
93. Felt seal
94. Bearing cover shims
95. Bearing cover
96. Lockwasher
97. Cap screw
98. Input companion flange
99. Input shaft nut
100. Cotter pin
101. Shift fork and push rod
102. Felt seal
103. Seal retainer
104. Compression spring
105. Retainer washer
106. Retainer washer pin
107. Shift lever bracket gasket
108. Shift lever bracket
109. Shift lever boot
110. Shift lever
111. Cotter pin
112. Clevis pin
113. Expansion plug
114. Lockwasher
115. Cap screw

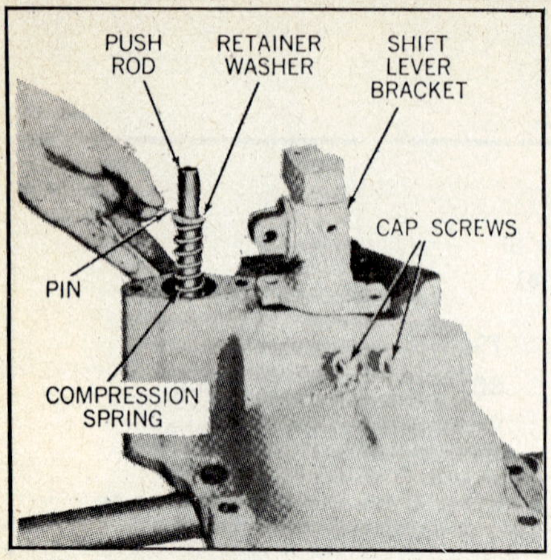

Fig. 23 Vacuum-operated lockout details

Lockout Mechanism (Vacuum Type)

Either vacuum-operated or air-operated lockout may be used on "D" Series Axles. The air-operated type is covered further on. The following instructions are for the vacuum-operated lockout.

1. Pull boot off lockout shift lever. Remove clevis pin and remove shift lever.
2. Referring to Fig. 23, remove shift lever bracket. Compress compression spring, then remove pin, retainer washer and compression spring from shift fork and push rod.
3. Lift input shaft sliding clutch and shift fork and push rod assembly from differential carrier cover.

Final Disassembly

1. Before removing differential assembly, remove cotter pin and loosen drive pinion helical gear nut.
2. Be sure to remove pipe plug, spring and oil distributor from differential carrier before attempting to remove differential assembly. *Refer to separate sections of this chapter for instructions covering the three types of axles used with the "D" Series combination.*
3. Removing and disassembling the drive pinion is covered in the separate sections of this chapter for the three axle types used, with the following variations:
   a. The drive pinions for the individual axles are equipped with companion flanges. For this type installation, the pinion bearing cage capscrews are accessible to permit removal of pinion with flange assembled.
   b. The drive pinion for "D" Series Tandem Axles are equipped with a helical gear, Fig. 24, which must be removed to gain access to pinion bearing cage capscrews. Therefore, to remove drive pinion, remove nut previously loosened before differential removal.
   c. Mount bearing puller on helical gear, gripping gear with puller legs and placing puller screw against pinion. Then pull gear off pinion. Remove capscrews from bearing cage and then remove drive pinion assembly.

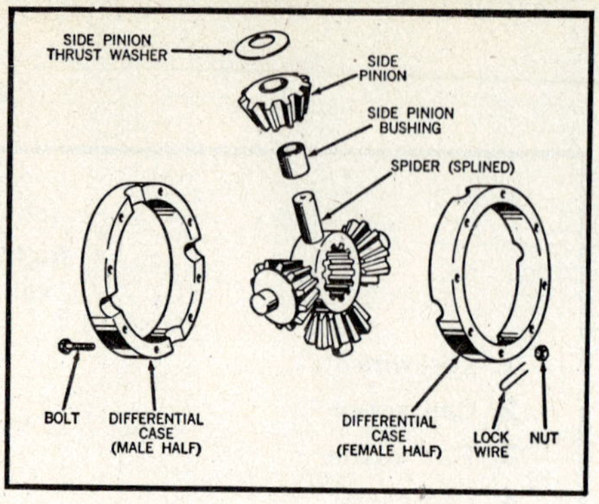

Fig. 26 Disassembled view of inter-axle differential

REASSEMBLE

Drive Pinion

Refer to separate sections of this chapter for instructions covering the three types of axles except for the following variation:

The drive pinions for the standard axle units are equipped with companion flanges. The pinion bearing cage capscrews are accessible to permit installation of pinion with companion flange installed.

The drive pinion for the "D" Series Tandem Axle units are equipped with a helical gear, Fig. 24, and the bearing cage capscrews cannot be installed with the helical gear attached to the drive pinion. Install the pinion in these

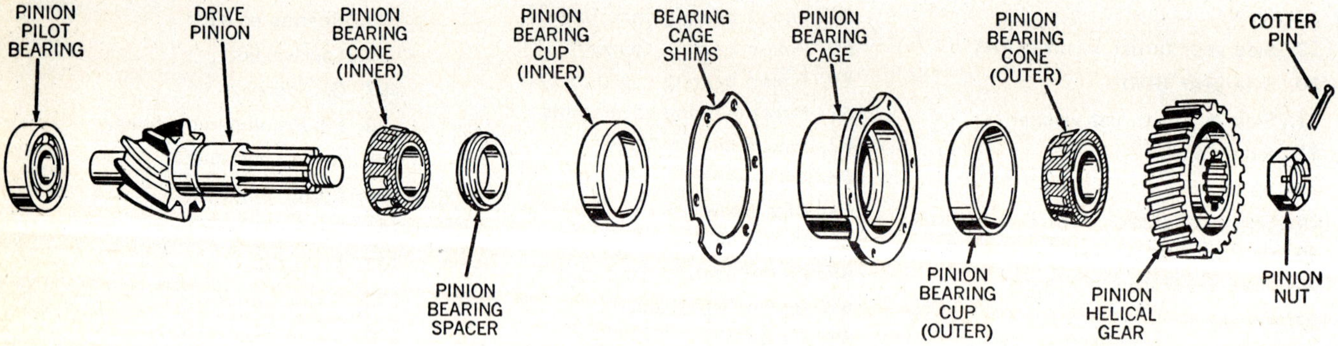

Fig. 24 Disassembled view of "D" Series drive pinion

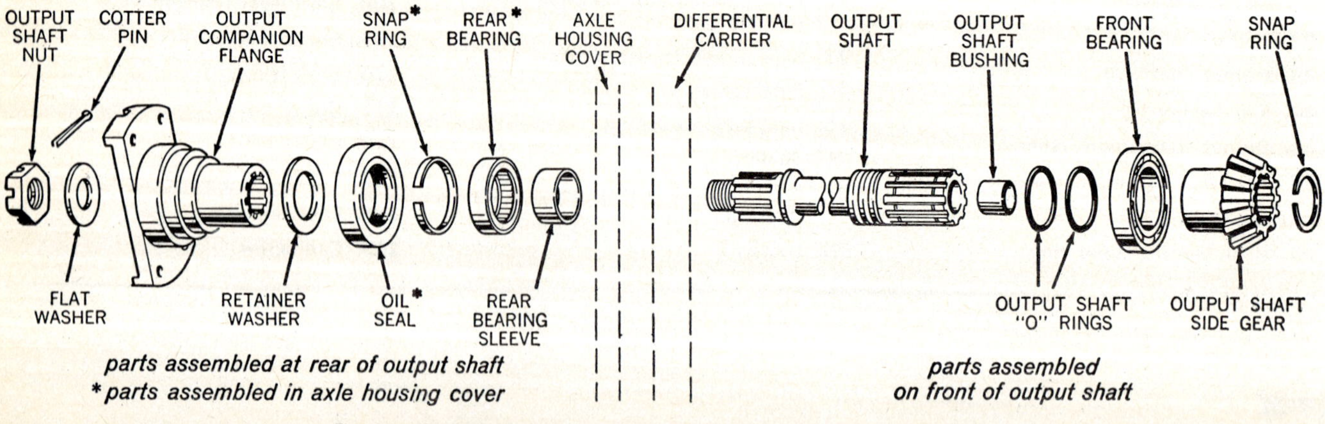

Fig. 25 Disassembled view of output shaft and axle housing cover parts

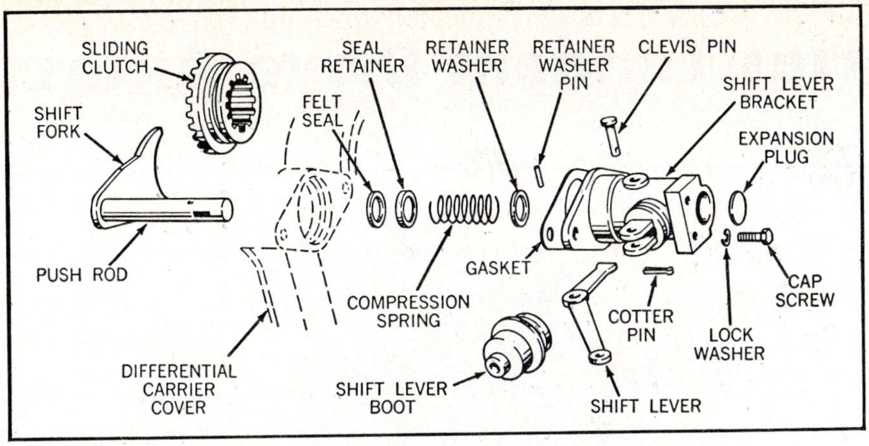

Fig. 27 Disassembled view of vacuum-operated lockout mechanism

axles with the helical gear removed. After pinion is installed, place helical gear on pinion. Install and tighten pinion nut to correct torque, then install cotter pin.

With drive pinion and differential assembled in differential carrier and correctly adjusted, install power divider components as follows:

Output Shaft

1. If removed, press bushing in bore of output shaft.
2. Press front bearing on output shaft side gear. Place side gear assembly on output shaft and install snap ring.
3. Install two "O" rings in grooves of output shaft.
4. Lubricate output shaft "O" rings, front bearing and bushing, then install output shaft assembly in differential carrier by tapping front bearing into bearing bore of carrier.

Inter-Axle Differential, Fig. 26

1. Install bushings, side pinions and thrust washers on journals of differential spider. Lubricate all parts during reassembly.
2. Place spider assembly in male differential case. Install female case, aligning punch marks made during disassembly, and secure with bolts and nuts. Tighten nuts to correct torque and install lockwire.
3. Place inter-axle differential on output shaft side gear. Position differential with nuts away from output shaft side gear.

Lockout Mechanism (Vacuum Type), Fig. 27

Instructions for the air-operated type lockout is covered further on.

1. Install felt seal and seal retainer in recess in carrier cover. Engage lockout shift fork with groove in input shaft sliding clutch. Place this assembly in differential carrier cover.
2. Place compression spring and retainer washer on push rod, depress spring and install retainer washer pin. Coat spring and retainer washer with grease.
3. Install new gasket and shift lever bracket on differential carrier cover and install capscrews and lockwashers. Tighten screws to correct torque. If removed, install grease fitting in bracket.
4. Place lockout shift lever in bracket, engaging slot in push rod. Install clevis pin and cotter pin.
5. Place boot over shift lever and secure boot to bracket.

Input Shaft, Fig. 28

1. Press bearing on threaded end of input shaft.
2. Place input shaft assembly in differential carrier cover, engaging sliding clutch.
3. Install felt seal and oil seal in input shaft bearing cover. Using same size and quantity of shims removed during disassembly, install shims and bearing cover on carrier cover. Install capscrews finger tight, with lockwashers. *To check bearing adjustment or determine thickness of shims required, install bearing cover on carrier cover without shims. Install and tighten capscrews finger tight. Then, using a feeler gauge, measure clearance between bearing cover and carrier cover. This clearance measurement plus .003" would equal the thickness of shims required for correct bearing clearance.*
4. Install input companion flange and nut on input shaft. Tighten nut to correct torque and install cotter pin. Tighten input shaft bearing cover capscrews to correct torque.
5. If removed, press two bushings in bore of helical side gear. Place "D" washer, thrust washer and helical side gear assembly on input shaft. Install snap ring on input shaft to secure these parts.

Differential Carrier Cover

1. Place new cover gasket on differential carrier.
2. Place carrier cover on differential carrier, aligning dowel pins with holes in cover. *During reassembly of cover to carrier, slowly rotate input shaft to engage its splines with splines in bore of inter-axle differential spider.*
3. Carefully install two cover capscrews, watching for possible binding between cover and carrier. If cover assembles to carrier without bind, install remaining capscrews and tighten to correct torque.
4. After cover is assembled to carrier, check operation of inter-axle differential. Hold output shaft stationary and rotate input shaft companion flange. If the assembly differentiates, the unit is correctly assembled.

Carrier & Power Divider

1. Place new gasket on carrier. Install carrier and power divider assembly into axle housing. Install and tighten stud nuts to correct torque.
2. Assemble axle housing cover, Fig. 25, as follows: Press output shaft rear bearing in cover. Install snap ring to secure bearing. Press oil seal in same opening in axle housing cover.
3. Place new gasket on axle housing and install axle housing cover, taking care not to damage oil seal. Install stud nuts and lockwashers and tighten nuts to correct torque.
4. Install rear bearing sleeve in output shaft. Install retainer washer, output companion flange, flat washer and nut. Tighten nut to correct torque and install cotter pin.
5. Install axle shafts and fill axle with correct lubricant. Connect propeller shafts. On Two-Speed Axles, install shift units. Make lockout shift lever connections.

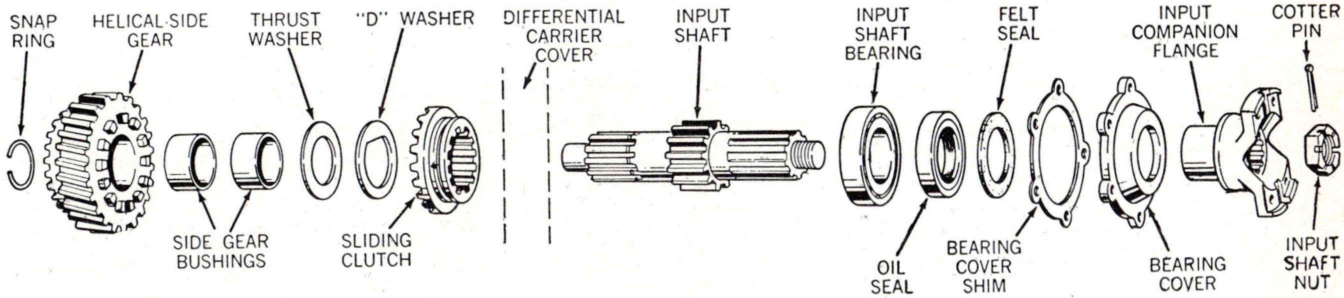

Fig. 28 Disassembled view of input shaft

Eaton "M" Series Tandem Drive Axles

MODELS 22M, 28M, 32M, 36M

This series of axles consist of two axle units coupled by a single propeller shaft and a power divider unit. The forward rear axle on Models 22M, 28M, 32M and 36M are single reduction units combined with a power divider. The rearward rear axle is also a single reduction unit similar to the standard Eaton Single Reduction Axle, Fig. 29.

Model 42M consists of three units: Two knee-mounted double reduction axle units and a power divider. The power divider is mounted on the forward rear axle. An inter-axle propeller shaft connects the rearward rear axle to the power divider.

NOTE: *Inasmuch as the rearward rear axle is similar to the standard Eaton Single Reduction Axle, refer to that section of this chapter for service procedures. The following material covers the procedures on the forward rear axle.*

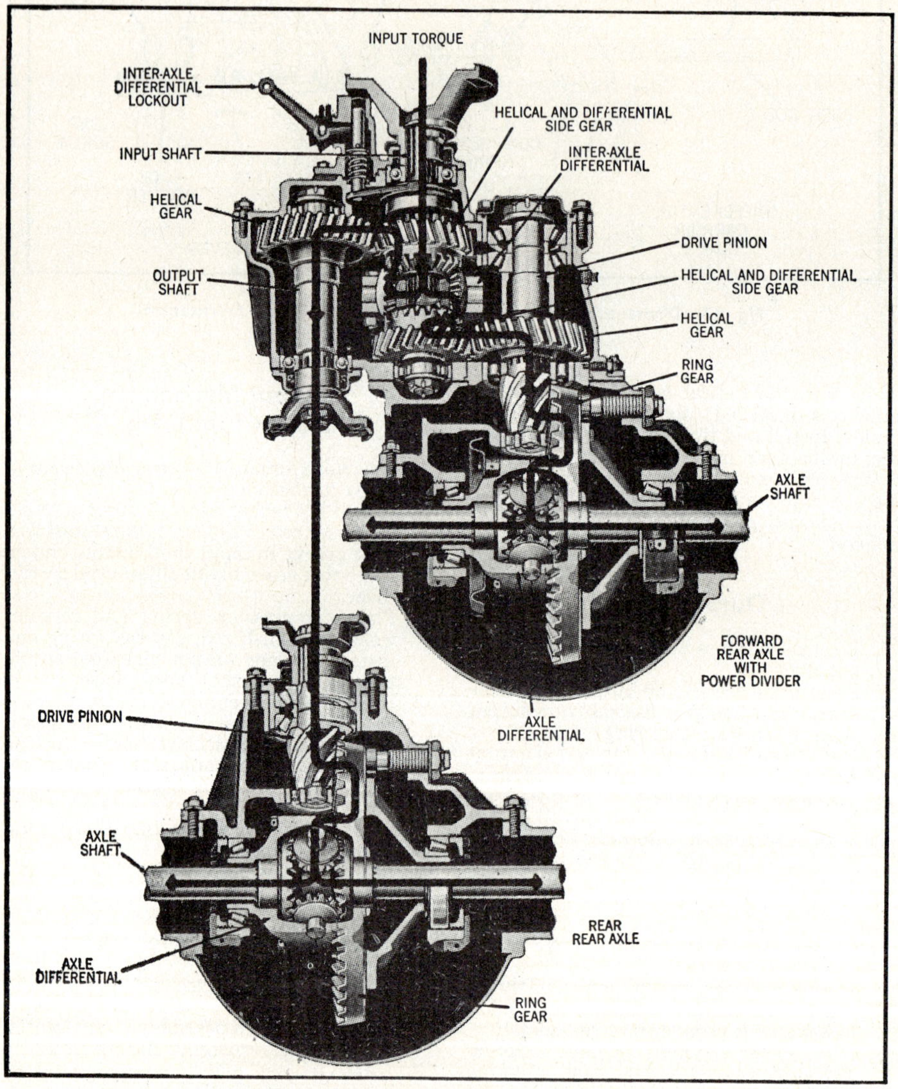

Fig. 29 Power flow through M Series Tandem Drive Axles

DISASSEMBLE

Carrier & Power Divider

1. Drain lubricant. Disconnect propeller shafts and remove axle shafts.
2. Disconnect lockout shift lever.
3. Unfasten and remove differential carrier and tower divider assembly from axle housing.
4. If preliminary inspection reveals that ring gear and drive pinion are satisfactory, check backlash with dial indicator and take a tooth contact pattern for reference and easier adjustment in reassembly.
5. Remove pinion bearing cage cover. Remove and discard cork seal. Hold input and output companion flanges and loosen (but do not remove) pinion nut.
6. Hold input companion flange and loosen (but do not remove) flange nut.
7. Hold output companion flange and loosen (but do not remove) flange nut.
8. Punch mark differential bearing caps and carrier for correct location in reassembly. Also punch mark differential bearing adjusters for correct location of adjuster locks, Fig. 30.
9. Cut lockwires and remove differential bearing cap screws, caps, adjusters, locks and bearing cups from both sides of differential.
10. On Models 22M, 28M and 36M, remove differential assembly from carrier.
11. On Model 32M, straighten ear of lock and loosen nut. Loosen thrust block adjusting screw to move thrust block away from ring gear and permit removal of differential assembly. Remove oil collector drum attaching capscrews. Tilt differential to move ring gear away from pinion and lift differential assembly and oil collector drum from carrier.
12. Unfasten and remove power divider from carrier. Discard divider case gasket. *On Models 22M, 28M and 32M, the differential must be removed from the carrier before removing the power divider. On Model 36M, the power divider may be removed with the differential assembled in the carrier.*

36M Power Divider

The 36M power divider is the same as Model 42M axle. Refer to 42M instructions given further on.

22M, 28M, 32M, Power Divider

NOTE: *Reference to "front" indicates the input companion flange side of power divider. "Rear" indicates differential carrier side.*

1. Mount power divider in a repair stand that will permit rotation of the entire assembly for easier handling during reassembly.
2. Remove input flange nut and flange.
3. Loosen (but do not remove) input shaft front bearing cover, Fig. 31.
4. Remove nuts securing power divider case cover to case. Then, using a suitable puller, grip front bearing cover with puller legs and position puller screw against input shaft. Pull bearing cover and case cover assembly from case.
5. Remove loosened output flange nut, flat washer and flange.
6. Remove output shaft rear bearing cover. Remove oil seal and felt seal from bearing cover.
7. Working at rear end of output shaft, drive shaft out of rear bearing. Lift output shaft assembly from divider case, then remove rear bearing from divider case.
8. Remove inner bearing sleeve set screw and lockwasher from divider case.
9. Remove loosened drive pinion nut and tongued washer. Drive pinion assembly out rear of divider case. Remove outer bearing cone from bearing cage.

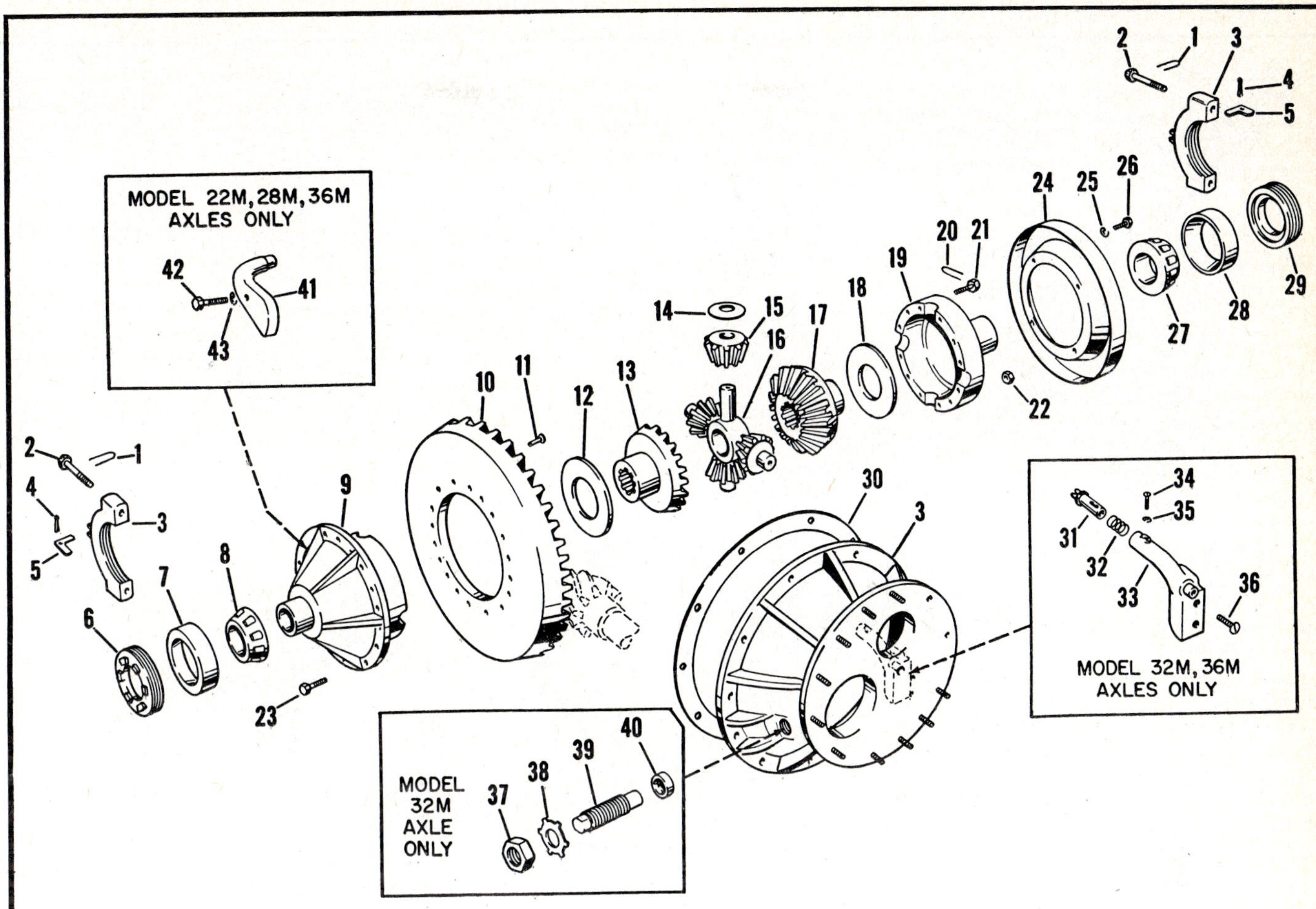

RH — Right-hand   LH — Left-hand   Right and Left Hand is determined by standing at rear of vehicle and facing rear of axle (as installed in vehicle).

1. Lockwire
2. Cap screws
3. Differential carrier and bearing caps (matched parts)
4. Cotter pin
5. Adjuster lock
6. Differential bearing adjuster (RH)
7. Differential bearing cup (RH)
8. Differential bearing cone (RH)
9. Differential case (flanged half)
10. Ring gear and drive pinion (matched set)
11. Ring gear rivet
12. Side gear thrust washer (RH)
13. Side gear (RH)
14. Side pinion thrust washer
15. Side pinion
16. Spider
17. Side gear (LH)
18. Side gear thrust washer (LH)
19. Differential case (plain half)
20. Lockwire (32M, 36M)
21. Cap screw (32M, 36M)
22. Locknut (22M, 28M)
23. Bolt (22M, 28M)
24. Oil collector drum (32M, 36M)
25. Lockwasher (32M)
26. Oil collector drum cap screw (32M)
27. Differential bearing cone (LH)
28. Differential bearing cup (LH)
29. Differential bearing adjuster (LH)
30. Carrier gasket
31. Differential carrier oil feeder (32M, 36M)
32. Compression spring (32M, 36M)
33. Oil feeder bracket (32M, 36M)
34. Locating screw (32M, 36M)
35. Lockwasher (32M, 36M)
36. Machine screw (32M, 36M)
37. Nut (32M)
38. Lock (32M)
39. Adjusting screw (32M)
40. Ring gear thrust block (32M)
41. Differential oil lubricator (22M, 28M, 36M)
42. Cap screw (22M, 28M, 36M)
43. Lockwasher (22M, 28M, 36M)

Fig. 30 Disassembled view of differential carrier for Models 22M, 28M, 36M

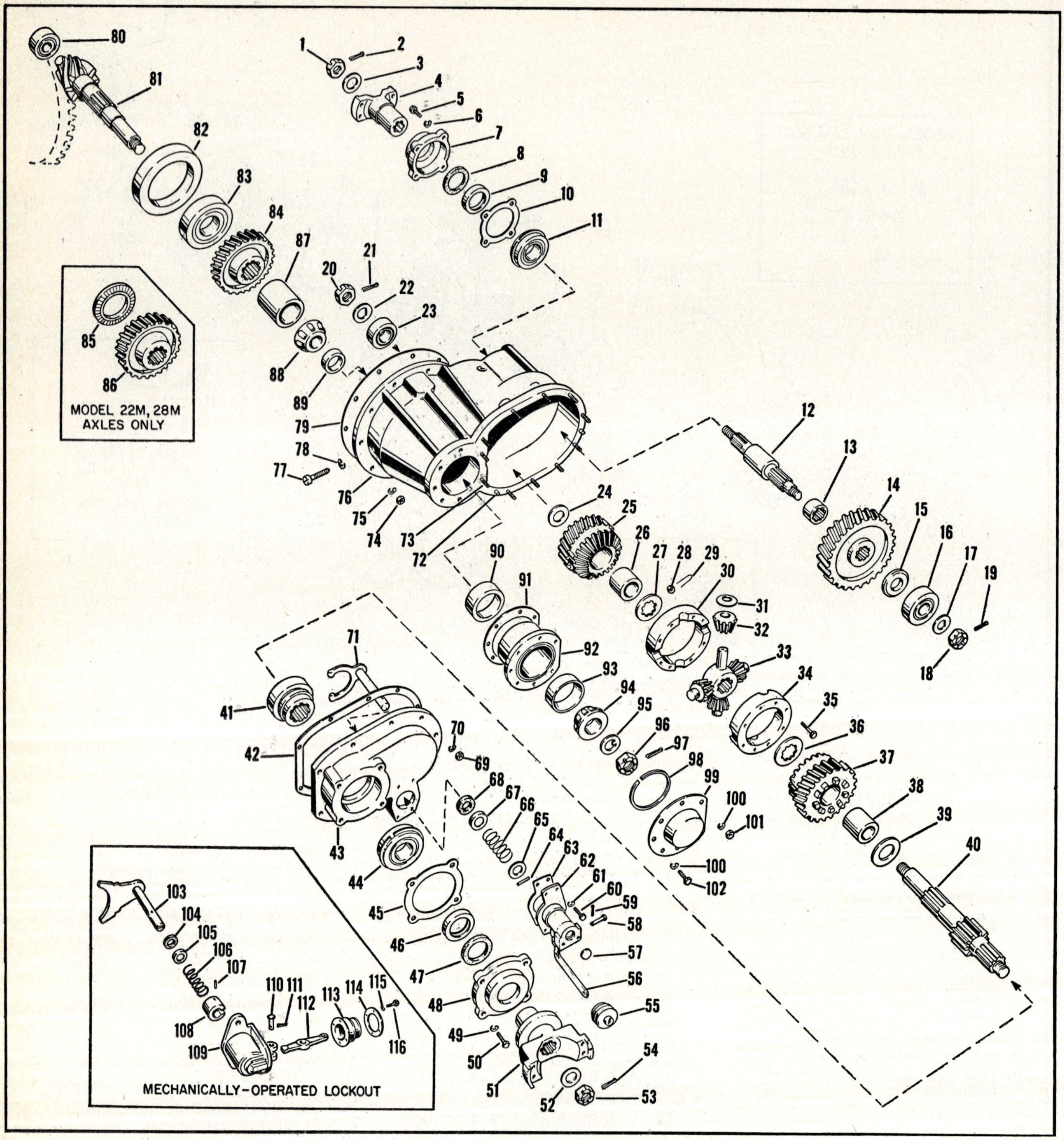

Fig. 31 Disassembled view of power divider for Models 22M, 28M, 32M (See next page for legend)

Reference to "FRONT" indicates the input companion flange side of power divider. "REAR" indicates the differential carrier side.

1. Flange nut
2. Cotter pin
3. Flat washer
4. Output companion flange
5. Cap screw
6. Lockwasher
7. Output shaft bearing cover (rear)
8. Felt seal
9. Oil seal
10. Bearing cover gasket
11. Output shaft bearing (rear)
12. Output shaft
13. Helical gear spacer
14. Output shaft helical gear
15. Front bearing spacer washer
16. Output shaft bearing (front)
17. "D" washer (32M)
18. Output shaft nut (front)
19. Cotter pin
20. Input shaft nut (inner)
21. Cotter pin
22. "D" washer
23. Input shaft bearing (rear)
24. Rear bearing spacer washer
25. Helical and differential side gear (rear)
26. Helical and differential side gear bushing (rear, short)
27. Differential spider splined washer (rear)
28. Nut
29. Lockwire
30. Inter-axle differential case (male half, rear)
31. Side pinion thrust washer
32. Side pinion
33. Spider
34. Inter-axle differential case (female half, front)
35. Bolt
36. Differential spider splined washer (front)
37. Helical and differential side gear (front)
38. Helical and differential side gear bushing (front, long)
39. Input shaft spacer washer
40. Input shaft
41. Sliding clutch
42. Power divider case cover gasket
43. Power divider case cover
44. Input shaft bearing (front)
45. Bearing cover gasket
46. Oil seal
47. Felt seal
48. Input shaft bearing cover (front)
49. Lockwasher
50. Cap screw
51. Input companion flange
52. Flat washer
53. Flange nut
54. Cotter pin
55. Shift lever boot
56. Shift lever
57. Expansion plug
58. Shift lever clevis pin
59. Cotter pin
60. Cap screw
61. Lockwasher
62. Shift lever bracket
63. Shift lever bracket gasket
64. Retainer washer pin
65. Retainer washer
66. Compression spring
67. Seal retainer
68. Felt seal
69. Stud nut
70. Lockwasher
71. Shift fork and push rod
72. Dowel pins
73. Stud
74. Stud nut
75. Lockwasher
76. Power divider case
77. Bearing sleeve set screw
78. Lockwasher
79. Power divider case gasket
80. Pinion pilot bearing
81. Drive pinion and ring gear (matched set)
82. Inner bearing sleeve
83. Pinion bearing (inner)
84. Helical gear (32M)
85. Helical gear seal (22M, 28M)
86. Helical gear (22M, 28M)
87. Pinion spacer (long)
88. Pinion bearing cone (inner)
89. Pinion bearing spacer
90. Pinion bearing cup (inner)
91. Shims
92. Pinion bearing cage
93. Pinion bearing cup (outer)
94. Pinion bearing cone (outer)
95. Tongued washer
96. Pinion nut
97. Cotter pin
98. Cork seal
99. Pinion bearing cage cover
100. Lockwasher
101. Stud nut
102. Cap screw

LEGEND FOR INSERT

(Mechanically-Operated Lockout)

103. Shift fork and push rod
104. Felt seal
105. Seal retainer
106. Compression spring
107. Rod bushing pin
108. Rod bushing
109. Shift lever bracket
110. Shift lever clevis pin
111. Cotter pin
112. Shift lever
113. Shift lever boot
114. Boot retainer
115. Lockwasher
116. Cap screw

10. Remove pinion bearing cage and shims from divider case. *Note quantity and size of shims under cage flange for easier adjustment during reassembly.* If replacement is necessary, remove bearing cups from cage.
11. Remove cotter pin from input shaft inner nut. Temporarily place companion flange on input shaft. Hold flange and remove nut and "D" washer from input shaft.
12. Place divider case in press with front helical gear supported, then press on rear of input shaft until shaft end clears rear bearing. Remove input shaft assembly from divider case, and also rear bearing.
13. To disassemble input shaft, lift rear bearing spacer washer and rear helical side gear off input shaft. Place input shaft in press with splined end down and spacer washer resting on press plate. Then press input shaft out of component parts.
14. To disassemble inter-axle differential, punch mark differential cases for location on reassembly. Cut lockwire and remove nuts and bolts holding cases together. Separate cases, then remove spider, side pinions and thrust washers.
15. To disassemble drive pinion, remove sleeve from inner bearing. Place drive pinion in arbor press with teeth down and disassemble in two steps as follows: (a) Place press support plates under helical gear; then press pinion out of bearing spacer, inner bearing cone, long pinion spacer and helical gear. (b) Place bearing puller plates between pinion teeth and inner race of inner bearing and press pinion out of inner bearing. Remove pilot bearing from pinion, using press or suitable puller.
16. To disassemble output shaft, remove cotter pin from output shaft front nut. Temporarily install companion flange on rear end of output shaft. Hold flange and loosen output shaft front nut. Remove nut and "D" washer ("D" washer used on 32M only). Place output shaft in press with splined end down and helical gear resting on press plate. Press shaft through front bearing, spacer washer and helical gear. Remove helical gear spacer.

Power Divider Case Cover

NOTE: *Either vacuum-operated, air-operated or mechanically-operated lockout may be used on "M" Series axles. The air-operated lockout is covered further on. The following instructions are for vacuum-operated type. Mechanically-operated lockout varies slightly from the vacuum type (refer to insert in Fig. 31 for components).*

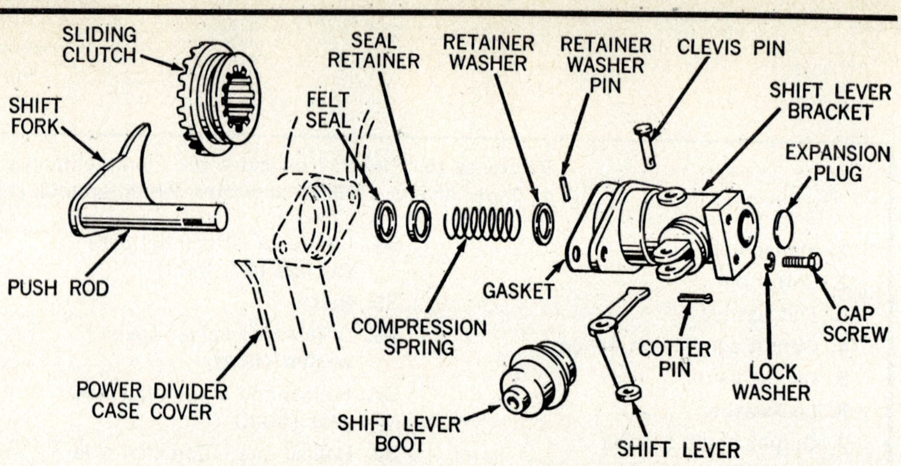

Fig. 32 Disassembled view of vacuum-operated lockout

1. Remove loosened capscrews, lockwashers and front bearing cover from divider case cover. Remove oil seal and felt seal from bearing cover.
2. Remove sliding clutch and input shaft front bearing from case cover.
3. Push back shift lever boot. Remove clevis pin and shift lever. Unfasten and remove shift lever bracket from case cover.
4. Depress compression spring. Then remove pin, retainer washer and compression spring from push rod.
5. Withdraw shift fork and push rod from case cover. Remove seal retainer and felt seal from case cover.

Axle Differential, Models 22M, 28M

1. Punch mark differential cases for correct location in reassembly. Remove locknuts and bolts and lift off plain differential case.
2. Remove differential components from ring gear side of case.
3. Remove capscrew, lockwasher and lubricator from flanged differential case.
4. Remove bearing cones from differential cases, using split-type bearing puller.
5. To remove ring gear from flanged differential case, drill off the flat head of rivets (on ring gear side) and drive out rivets.

32M Axle Differential

1. Punch mark differential cases for correct location in reassembly. Cut lockwire and remove differential case cap screws. Lift off plain differential case.
2. Remove differential components ring gear side of differential case.
3. Remove bearing cones from differential case, using split-type bearing puller. To remove ring gear from flanged differential case, drill off flat head of rivets (on ring gear side) and remove rivets.

36M Axle Differential

1. Punch mark differential cases for correct location in reassembly. Cut lockwire and remove differential case capscrews. Lift off oil collector drum and plain differential case.
2. Remove differential components from flanged differential case.
3. Remove capscrew, lockwasher and differential lubricator from flanged differential case.
4. Remove bearing cones from differential cases, using split-type bearing puller. To remove ring gear from flanged differential case, drill off flat head of rivets (on ring gear side) and remove rivets.

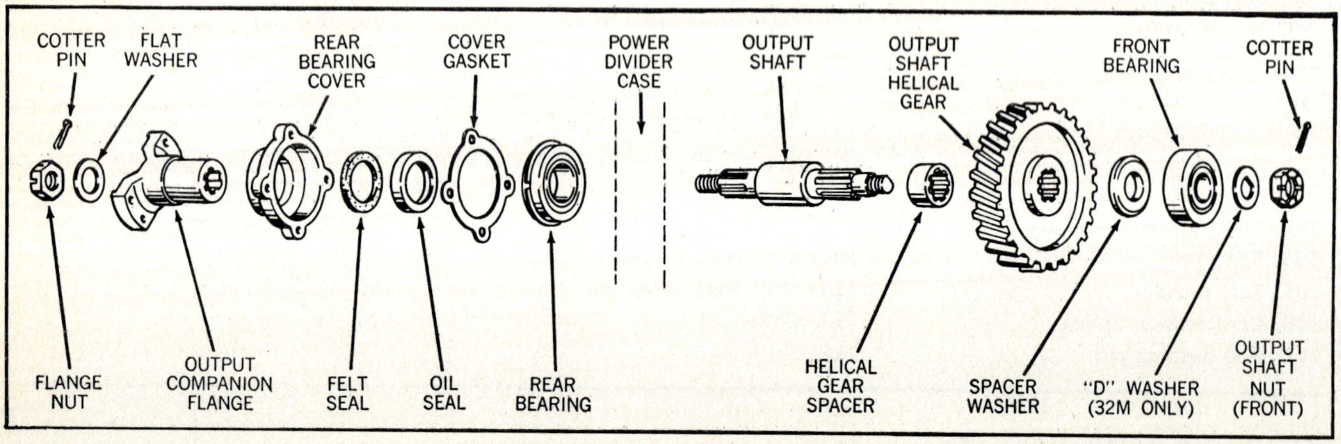

Fig. 33 Disassembled view of output shaft

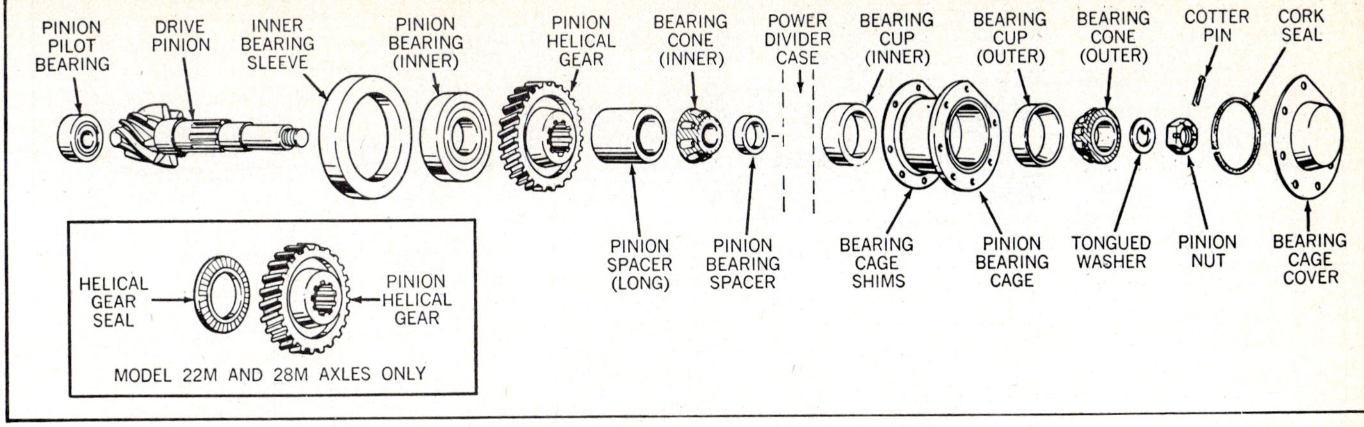

Fig. 34 Disassembled view of drive pinion

REASSEMBLE

36M Power Divider

NOTE: *The power divider assembly for Model 36M is the same as 42M. Refer to 42M instructions further on for procedures.*

22M, 28M, 32M, Power Divider

NOTE: *Partially reassemble shafts and case covers before reassembling power divider case.*

Power Divider Case Cover (Vacuum Type)

NOTE: *The following instructions are for the vacuum-type lockout. Mechanically operated lockout varies slightly from the vacuum type (refer to insert in Fig. 31). The air-operated type is covered further on. A disassembled view of the vacuum-type lockout is shown in Fig. 32.*

1. Install input shaft front bearing in divider case cover.
2. Place felt seal and seal retainer in recess in case cover. Insert push rod through case cover opening. Place compression spring and retainer washer on push rod, compress spring and install retainer washer pin. Coat spring and retainer washer with grease.
3. Install shift lever bracket and gasket over push rod and install capscrews and lockwashers.
4. Position shift lever on bracket, engaging end of lever with slot in push rod. Install clevis and cotter pins to connect lever to bracket.
5. Place boot on shift lever and bracket.
6. Depress shift lever and position sliding clutch in case cover engaging shift fork in clutch grooves.

Output Shaft, Fig. 33

1. Press helical gear spacer on front end of output shaft. Then press helical gear, spacer washer and front bearing on output shaft.
2. Temporarily place companion flange on rear end of output shaft. Place output shaft in vise with rear end down, gripping flange in vise jaws.
3. Install "D" washer (32M only) on output shaft. Install and tighten output shaft front nut to correct torque. Install cotter pin.

Drive Pinion, Fig. 34

1. Press pilot bearing on drive pinion and stake in four places (see Fig. 12).
2. Lubricate and press inner bearing on drive pinion, positioning bevel side of bearing toward pinion teeth.
3. On Models 22M and 28M, install oil seal in helical gear, positioning segment springs inside. Press helical gear on drive pinion, placing seal towards inner bearing. On Model 32M, press helical gear on pinion with short hub toward inner bearing.
4. Press long spacer on pinion. Press inner bearing cone on pinion.
5. If removed, press bearing cups in pinion bearing cage.
6. Lubricate pinion bearings. Temporarily place bearing cage, spacer, outer bearing cone, tongued washer and nut on pinion. Tighten nut to correct torque. Rotate pinion bearing cage while tightening nut. Check pinion bearing preload adjustment as already explained at the beginning of this chapter.
7. With pinion bearings correctly adjuted, remove temporarily installed nut, tongued washer, outer bearing cone and bearing cage from pinion.

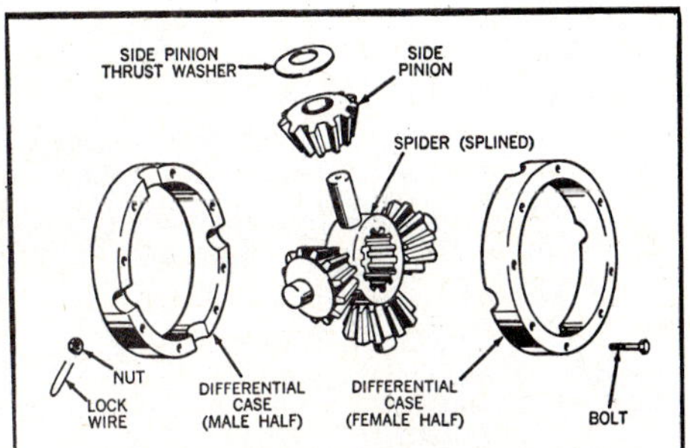

Fig. 35 Disassembled view of inter-axle differential

Inter-Axle Differential, Fig. 35

1. Lubricate parts during reassembly. Place side pinions and thrust washers on spider journals. Assemble spider assembly and case halves, aligning punch marks made during disassembly. Install bolts and nuts to secure case halves. Tighten nuts to correct torque and lockwire.

Input Shaft, Fig. 36

1. Temporarily install flange nut on front end of input shaft to protect threads during assembly. Place input in press, with shaft supported at nut and install parts on shaft as follows:
2. Place spacer washer on shaft.
3. Press front helical side gear bushing on shaft. Lubricate bushing, then install front helical side gear over bushing. *Bushing and gear must rest firmly against spacer washer.*
4. Place front spider splined washer on shaft. Place inter-axle differential on shaft, positioning bolt heads toward companion flange. Place rear spider splined washer on shaft, *making certain splines are engaged with input shaft splines.*
5. Press rear helical side gear bushing on shaft. Lubricate bushing, then install rear helical side gear over bushing. *Bushing and gear must rest firmly against splined washer.*
6. Place rear bearing spacer washer on shaft.
7. Temporarily place a "dummy sleeve" (1½" I.D. × 1" long), "D" washer and input shaft inner nut on shaft. Tighten nut to correct torque, then rotate shaft parts by hand. Gears and interaxle differential should rotate freely without bind. After testing, remove temporarily installed parts.

Power Divider

1. Temporarily install companion flange on input shaft. Place input shaft in vise, with front end down and gripping flange in vise jaws.
2. Place power divider case, front side down, over input shaft. With rear bearing spacer washer on input shaft, install rear bearing in divider case bore and on input shaft. Install "D" washer and inner nut on shaft. Tighten nut to correct torque and install cotter pin. Install output shaft rear bearing in divider case. Lubricate bearing. With new felt seal and oil seal

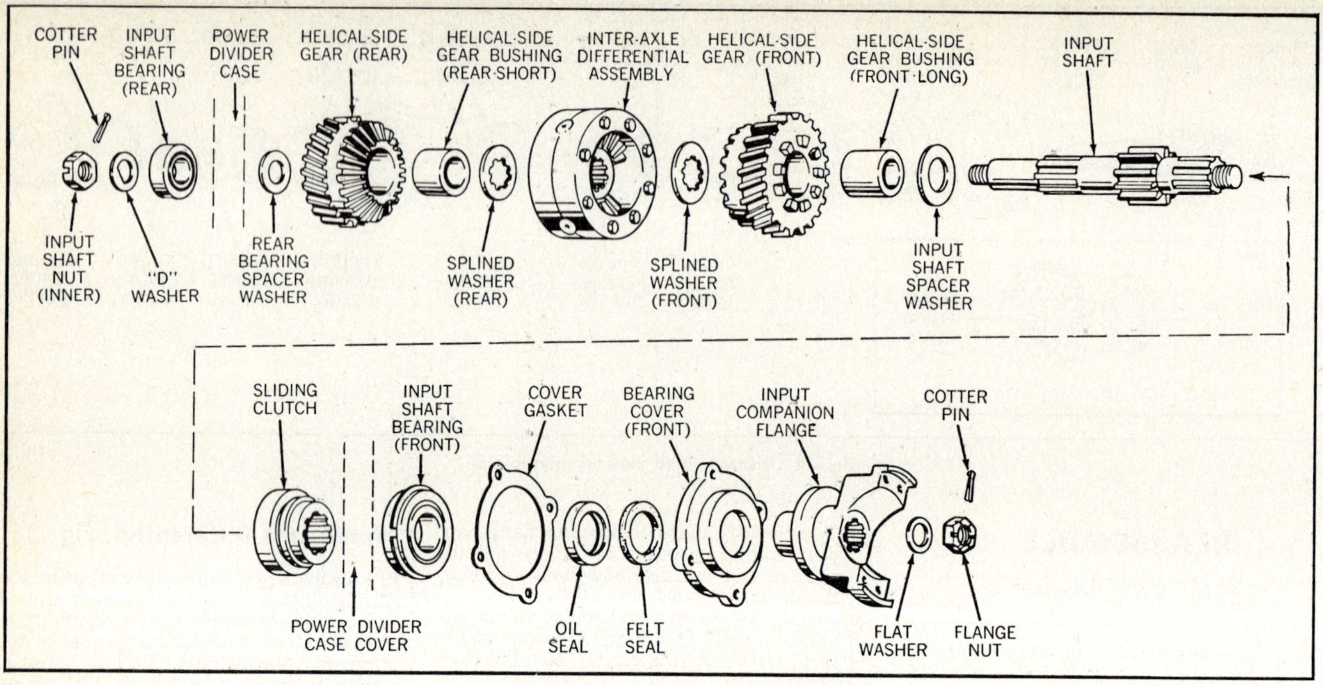

Fig. 36 Disassembled view of input shaft

installed in rear bearing cover, install gasket and bearing cover on divider case. Install cover capscrews and lockwashers.
3. Remove divider case and input shaft assembly from vise.
4. Position divider case with front side up. Place output shaft in case, taking care not to damage rear bearing oil seal.
5. Place gasket and case cover on divider case. Make certain sliding clutch engages splines of input shaft. Fasten cover to case and tighten stud nuts to correct torque.
6. Install felt seal and oil seal in input shaft front bearing cover. Place gasket and bearing cover on case cover. Install capscrews and lockwashers but do not tighten until companion flange is installed.
7. Install companion flange flat washer and nut. Tighten nut to correct torque and insert cotter pin. Now tighten front bearing cover capscrews to correct torque.
8. Install same quantity and size shims that were removed in disassembly on pinion bearing cage. Place shims and cage in divider case.
9. Insert drive pinion through opposite side of divider case until inner bearing cone is seated in inner bearing cup. Then place outer bearing cone, tongued washer and nut on drive pinion. Tighten nut to correct torque and insert cotter pin.
10. Install new cork seal in cage groove. Install cage cover and tighten nuts and capscrews to correct torque.
11. Insert bearing sleeve in divider case and over pinion inner bearing. Align sleeve hole with hole in case, then install set screw and lockwasher.
12. Loosen capscrews holding output shaft bearing cover. Install companion flange, flat washer and nut on output shaft. Tighten nut to correct torque and insert cotter pin. Tighten bearing cover capscrews to correct torque.

22M, 28M Axle Differential, Fig. 37

1. Rivet (cold) ring gear to flanged differential case, placing flat side head of rivets on ring gear side. Compress rivets to 34–38 tons on Model 22M or 40–42 tons on 28M.
2. Press bearing cones on differential cases. If removed, install oil lubricator in flanged differential case and secure with capscrew and lockwasher.
3. Lubricate parts during assembly. Place thrust washer and right-hand side gear in flanged differential case. Assemble side pinions and thrust washers to spider, then place this assembly on flanged differential case. Place left-hand side gear and thrust washer on side pinions.
4. Tap plain differential into place, aligning previously made punch marks. Draw case halves together with four bolts and lock

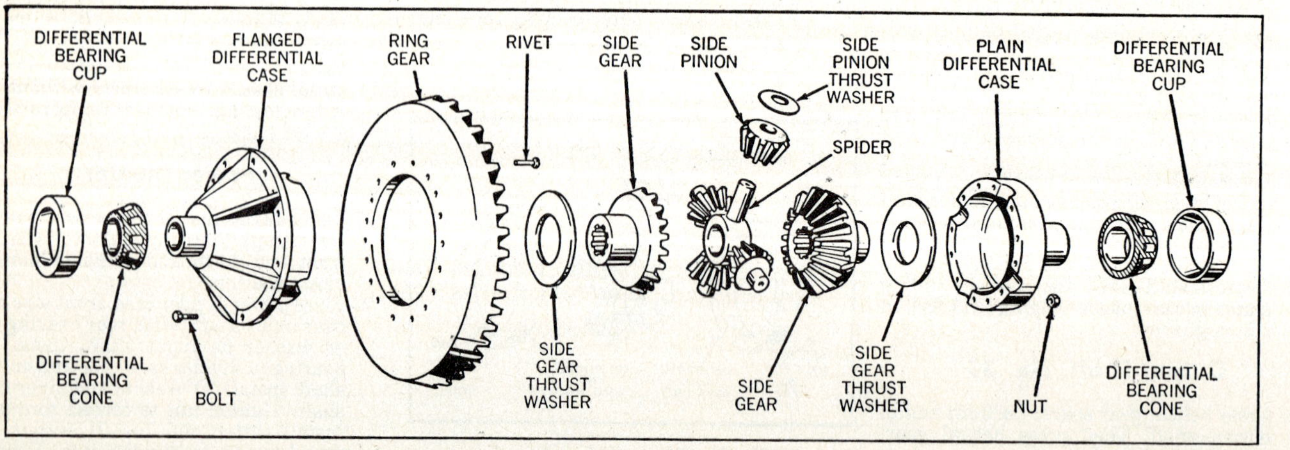

Fig. 37 Disassembled view of axle differential for Models 22M, 28M

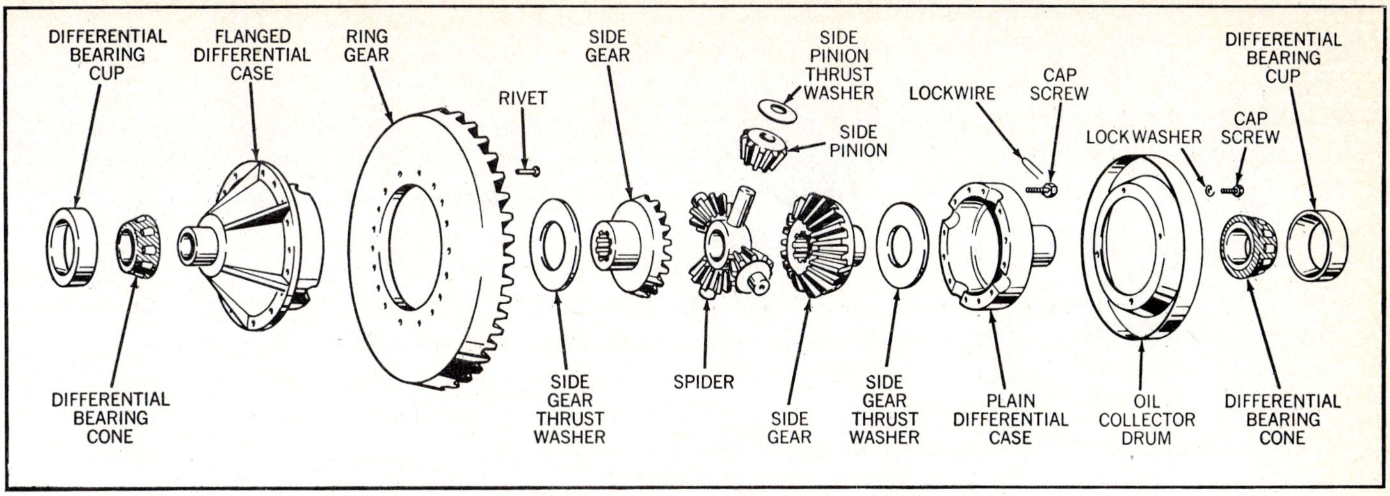

Fig. 38 Disassembled view of axle differential for Models 32M, 36M

nuts. Check assembly for free rotation. Install remaining bolts and locknuts and tighten all nuts to correct torque.

32M Axle Differential, Fig. 38

1. Rivet ring gear to flanged differential case, compressing rivets to 47–50 tons.
2. Press bearing cones on differential case. Lubricate parts during assembly.
3. Assemble differential parts in flanged differential case.
4. Tap plain differential case into place, aligning previously made punch marks. Do not install oil collector drum at this time, but assemble drum to case after differential is installed in carrier.
5. Install four large tapped head capscrews, equally spaced, to draw case halves together. Check assembly for free rotation. Install eight remaining capscrews and tighten all to correct torque. Lock with wire.

36M Axle Differential, Fig. 38

1. Rivet ring gear to flanged differential case, compressing to 39–41 tons.
2. Press bearing cones on differential cases. If removed, install oil lubricator in flanged differential case.
3. Assemble differential parts in flanged differential case. Then tap plain differential case half into place, aligning previously made punch marks.
4. Place oil collector drum on plain differential case. Install four capscrews to draw case halves together. Check assembly for free rotation. Install remaining capscrews and tighten all to correct torque.

Final Assembly

1. Using a new gasket, place power divider on differential carrier. Install and tighten stud nuts to correct torque.
2. On Models 22M, 28M and 36M, install differential in carrier as follows: Lubricate differential bearings. Place differential in carrier and install differential bearing cups, adjusters and caps, aligning punch marks previously made on caps. Install and finger-tighten bearing capscrews. Tighten adjusters to correctly seat bearings. Adjust ring gear and drive pinion as per instructions at the beginning of this chapter. When correctly adjusted, tighten bearing capscrews to correct torque. Install cotter pins to secure adjuster locks and lock capscrews with wire.
3. On Model 32M, install differential in carrier as follows:
   a. Lubricate differential bearings. Place oil collector drum on left-hand side of differential, but do not install capscrews. Place differential and drum in carrier and install bearing cups, adjusters and caps. Install and finger-tighten bearing cap screws.
   b. Secure oil collector drum to plain differential with four capscrews and lockwashers.
   c. Tighten bearing adjusters to seat bearings. Adjust ring gear and pinion as described at the beginning of this chapter.
   d. With ring gear and pinion correctly adjusted, tighten bearing capscrews to correct torque. Install cotter pins to secure adjuster locks and lockwire capscrews.
   e. Tighten thrust block adjusting screw to position thrust block against ring gear. Loosen adjusting screw (1/8 to 1/4 turn) to obtain .010–.015" clearance between thrust block and ring gear. Tighten nut and bend ear of lock to secure adjustment.

Install Assembly In Axle Housing

1. Using a new gasket, install differential and power divider assembly in axle housing. Install capscrews or stud nuts and lockwashers. Tighten screws or nuts to correct torque.
2. Connect lockout shift lever.
3. Install axle shafts and connect propeller shafts.
4. Fill axle with correct lubricant.

MODEL 42M

Rearward Rear Axle

Instructions for the rearward rear axle are essentially the same as the forward rear axle except for removal of the power divider and drive pinion.

The rearward rear axle is not equipped with a power divider. The drive pinion for the rearward rear axle is mounted in a bearing cage at the front side of the differential carrier. The drive pinion can be removed and installed as a complete assembly, consisting of pinion, bearings, cage and companion flange.

Disassembly and reassembly of the differential assembly and countershaft are the same as instructions for the forward rear axle which follows:

DISASSEMBLE FORWARD REAR AXLE

1. Drain lubricant. Disconnect propeller shafts and remove axle shafts.
2. Disconnect lockout shift lever.
3. Unfasten and remove carrier and power divider from axle housing. *If preliminary inspection reveals that ring gear and drive pinion are satisfactory, check backlash with dial indicator and take a tooth contact pattern for reference and easier adjustment during reassembly.*
4. Remove pinion cage cover. Remove and discard cork seal. Hold input and output companion flanges and loosen (but do not remove) pinion nut.
5. Hold input flange and loosen (but do not remove) flange nut.
6. Hold output flange and loosen (but do not remove) flange nut.
7. Unfasten and lift power divider from differential carrier.

Differential Carrier, Fig. 39

1. Punch mark differential bearing caps and carrier for correct location during reassembly.
2. Remove bearing caps and lift differential out of carrier. Remove outer bearing races from differential cases.
3. Unscrew oil distributor plug, spring and distributor.
4. Punch mark countershaft bearing covers and carrier for correct location during reassembly. Remove covers from both sides.
5. Punch mark countershaft bearing cage (right-hand), bearing adjuster (left-hand) and differential carrier for correct location on reassembly. Then remove bearing cage and adjuster from carrier. Remove cage seal.
6. Remove oil reservoir from carrier.
7. Position herringbone pinion halfway through right-hand bearing cage opening, then tilt countershaft to position ring gear at top and lift out countershaft.

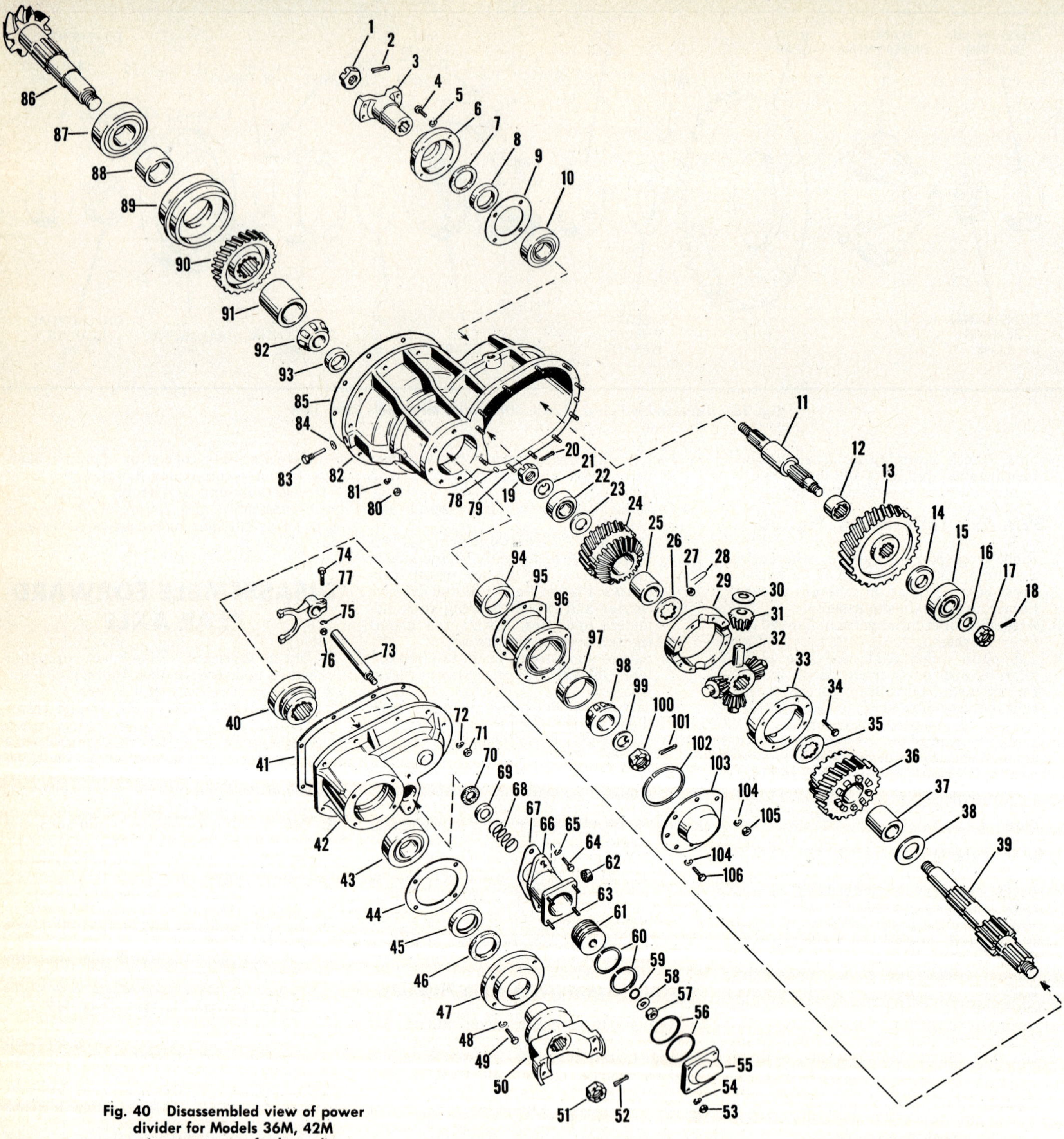

Fig. 40 Disassembled view of power divider for Models 36M, 42M
(see next page for legend)

Differential

1. If match marks are not present, make marks to assure correct position of each case half in reassembly.
2. Unfasten and separate herringbone gear and case halves.
3. Lift herringbone gear and left-hand case off right-hand case and remove differential components.
4. Place side gear in vise and drive left-hand case from herringbone gear and case support.
5. Remove snap rings from case halves.
6. Remove bearing inner race from case halves using a split-type bearing puller.

Countershaft

1. Remove capscrews and oil collector disc.
2. If match marks are not present, make marks to assure correct position of ring gear on countershaft flange.
3. Remove ring gear from countershaft.
4. Remove bearing cones from countershaft using a split-type bearing puller.

Power Divider, Fig. 40

1. Remove input flange. Loosen (do not remove) capscrew on input shaft front bearing cover.
2. Remove divider cover to case stud nuts. Using suitable bearing puller, grip front bearing cover with puller legs and puller screw against input shaft. Pull bearing cover and case cover from case.
3. Remove output flange.
4. Working at rear end of output shaft, drive shaft end out of rear bearing. Lift output

Reference to "FRONT" indicates the input companion flange side of power divider. "REAR" indicates the differential carrier side.

1. Flange nut
2. Cotter pin
3. Output companion flange
4. Cap screw
5. Lockwasher
6. Output shaft bearing cover (rear)
7. Felt seal
8. Oil seal
9. Bearing cover shim
10. Output shaft bearing (rear)
11. Output shaft
12. Helical gear spacer
13. Output shaft helical gear
14. Front bearing spacer washer
15. Output shaft bearing (front)
16. Tongued washer
17. Output shaft nut (front)
18. Cotter pin
19. Input shaft nut (inner)
20. Cotter pin
21. Tongued washer
22. Input shaft bearing (rear)
23. Rear bearing spacer washer
24. Helical and differential side gear (rear)
25. Helical and differential side gear bushing (rear, short)
26. Differential spider splined washer (rear)
27. Nut
28. Lockwire
29. Inter-axle differential case (male half, rear)
30. Side pinion thrust washer
31. Side pinion
32. Spider
33. Inter-axle differential case (female half, front)
34. Bolt
35. Differential spider splined washer (front)
36. Helical and differential side gear (front)
37. Helical and differential side gear bushing (front, long)
38. Input shaft spacer washer
39. Input shaft
40. Sliding clutch
41. Power divider case cover gasket
42. Power divider case cover
43. Input shaft bearing (front)
44. Bearing cover shim
45. Oil seal
46. Felt seal
47. Input shaft bearing cover (front)
48. Lockwasher
49. Cap screw
50. Input companion flange
51. Flange nut
52. Cotter pin
53. Stud nut
54. Lockwasher
55. Shift cylinder cover
56. Shift cylinder cover grommets
57. Push rod nut
58. Flat washer
59. Push rod grommet
60. Piston felt oilers
61. Shift cylinder piston
62. Shift cylinder body filter
63. Cover stud
64. Cap screw
65. Lockwasher
66. Shift cylinder body
67. Shift cylinder body gasket
68. Compression spring
69. Seal retainer
70. Felt seal
71. Stud nut
72. Lockwasher
73. Push rod
74. Bolt
75. Lockwasher
76. Nut
77. Shift fork
78. Dowel
79. Stud
80. Stud nut
81. Lockwasher
82. Power divider case
83. Bearing sleeve set screw
84. Lockwasher
85. Power divider case gasket
86. Drive pinion and ring gear (matched set)
87. Pinion bearing (inner)
88. Pinion inner bearing spacer (42M only)
89. Inner bearing sleeve
90. Helical gear
91. Pinion spacer (long)
92. Pinion bearing cone (inner)
93. Pinion bearing spacer
94. Pinion bearing cup (inner)
95. Shims
96. Pinion bearing cage
97. Pinion bearing cup (outer)
98. Pinion bearing cone (outer)
99. Tongued washer
100. Pinion nut
101. Cotter pin
102. Cork seal
103. Pinion bearing cage cover
104. Lockwasher
105. Stud nut
106. Cap screw

shaft from divider case, then remove bearing from case.

5. Remove inner bearing sleeve set screw from divider case.
6. Remove loosened pinion nut and tongued washer. Drive pinion out rear of divider case. Remove outer bearing cone from bearing cage.
7. Remove pinion bearing cage and shims from front side of divider case. *Note size and quantity of shims removed for easier adjustment in reassembly.* If replacement is necessary remove bearing cups from cage.
8. Remove input shaft assembly. This can easily be accomplished by temporarily installing companion flange and nut on end of shaft. Use flange to lift assembly from case.

Input Shaft & Inter-Axle Differential

1. Place input shaft in vise, gripping companion flange. Loosen and remove inner nut and tongued washer.
2. Place input shaft in press, with front helical side gear down and input shaft front spacer washer resting on press plates. Press shaft out of component parts.
3. Punch mark differential cases for correct location in reassembly. Cut lockwire and remove bolts holding cases together. Separate case halves and remove differential components.

Drive Pinion

1. On Model 36M, remove sleeve from inner

RH – Right-hand LH – Left-hand
Right and Left Hand is determined by standing at rear of vehicle and facing rear of axle (as installed in vehicle)

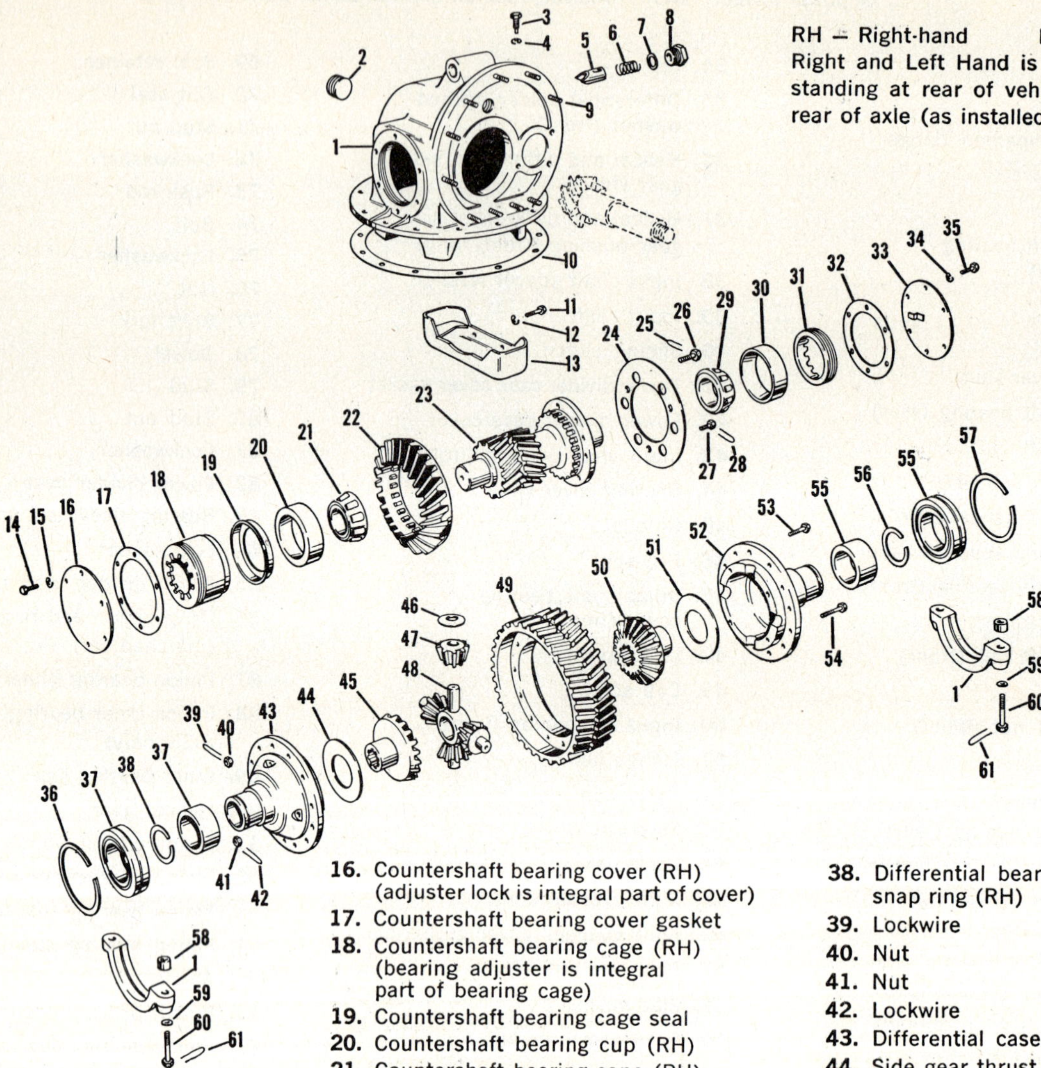

1. Differential carrier and bearing caps (matched parts)
2. Pipe plug
3. Cap screw
4. Lockwasher
5. Oil distributor
6. Oil distributor spring
7. Oil distributor plug gasket
8. Oil distributor plug
9. Carrier stud
10. Carrier gasket
11. Cap screw
12. Lockwasher
13. Oil reservoir
14. Cap screw
15. Lockwasher
16. Countershaft bearing cover (RH) (adjuster lock is integral part of cover)
17. Countershaft bearing cover gasket
18. Countershaft bearing cage (RH) (bearing adjuster is integral part of bearing cage)
19. Countershaft bearing cage seal
20. Countershaft bearing cup (RH)
21. Countershaft bearing cone (RH)
22. Ring gear and drive pinion (matched set)
23. Countershaft (herringbone pinion is integral part of countershaft)
24. Oil collector disc
25. Lockwire
26. Cap screw
27. Cap screw
28. Lockwire
29. Countershaft bearing cone (LH)
30. Countershaft bearing cup (LH)
31. Countershaft bearing adjuster (LH)
32. Countershaft bearing cover gasket
33. Countershaft bearing cover (LH) (adjuster lock is integral part of cover)
34. Lockwasher
35. Cap screw
36. Differential bearing outer race snap ring (RH)
37. Differential bearing inner and outer race (RH) (matched parts)
38. Differential bearing inner race snap ring (RH)
39. Lockwire
40. Nut
41. Nut
42. Lockwire
43. Differential case (RH)
44. Side gear thrust washer (RH)
45. Side gear (RH)
46. Side pinion thrust washer
47. Side pinion
48. Spider
49. Herringbone gear
50. Side gear (LH)
51. Side gear thrust washer (LH)
52. Differential case (LH)
53. Bolt
54. Bolt
55. Differential bearing inner and outer race (LH) (matched parts)
56. Differential bearing inner race snap ring (LH)
57. Differential bearing outer race snap ring (LH)
58. Dowel bushing
59. Flat washer
60. Cap screw
61. Lockwire

Fig. 39 Disassembled view of differential carrier for Model 42M

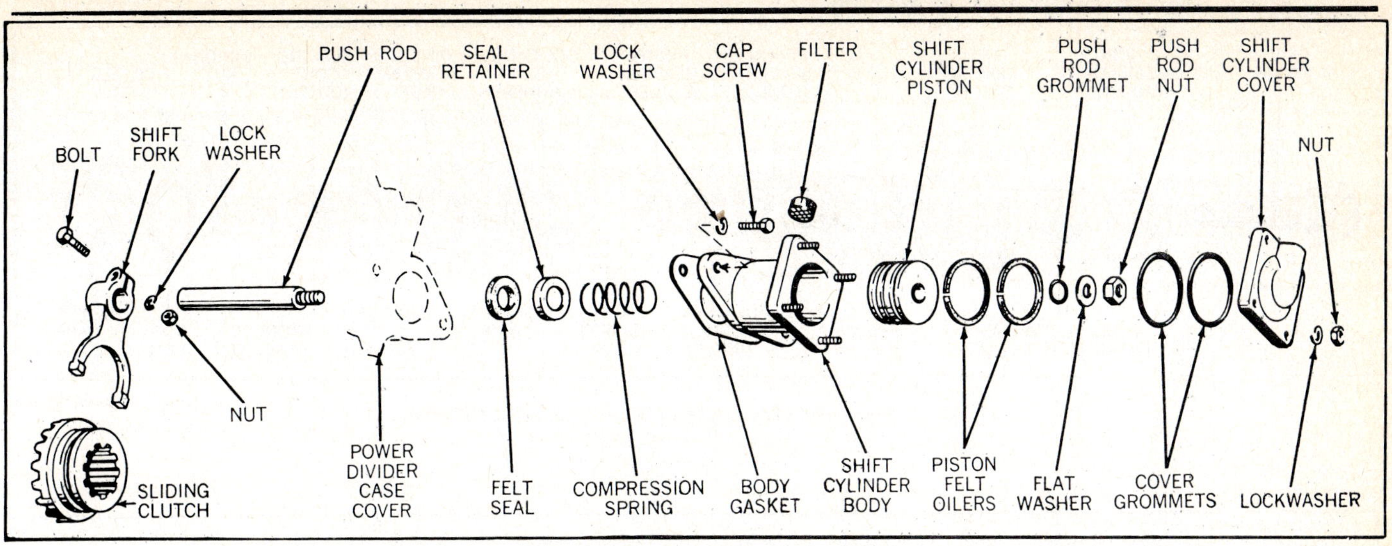

Fig. 41 Disassembled view of air-operated lockout mechanism

bearing. Place pinion in press with pinion teeth down. Then first press pinion out of bearing spacer, inner bearing cone, long spacer and helical gear. Then press pinion out of inner bearing.

2. On Model 42M, remove pinion bearing spacer. Then first press pinion out of inner bearing cone, long spacer, helical gear and inner bearing sleeve. Then press pinion out of inner bearing spacer and inner bearing.

Output Shaft

1. Remove cotter pin from output shaft front nut. Temporarily place companion flange on shaft. Then remove front nut, tongued washer and flange.
2. Place output shaft in press with rear end of shaft down and helical gear resting on press plates. Press shaft out of front bearing, spacer washer and helical gear. Remove helical gear spacer.

Divider Case Cover (Air-Operated Lockout)

Either air-operated or mechanically-operated lockout may be used on 36M and 42M axles. Refer to insert in Fig. 31 for components of mechanically-operated lockout. The following instructions are for air-operated lockout.

1. Remove cover from shift cylinder body. Discard grommets. Remove nut, flat washer and grommet from push rod.
2. Remove shift cylinder body from case cover.
3. Remove compression spring from push rod. Remove sliding clutch and shift fork from case cover.
4. Remove front bearing cover and shims. Note size and quantity of shims removed for easier adjustment during reassembly. Remove oil seal and felt seal from bearing cover. Remove input shaft bearing from case cover.

REASSEMBLE

Divider Case Cover (Air Operated), Fig. 41

1. Press front input shaft bearing in divider case cover.
2. Place felt seal and seal retainer in recess in case cover. Assemble push rod to shift fork and secure with nut, bolt and lock-washer.
3. Assemble shift fork and sliding clutch to case cover as follows: Partially insert push rod in case cover, engaging sliding clutch with shift fork, then push this assembly into correct mounting position in case cover.
4. Place compression spring, gasket and shift cylinder body over push rod and onto case cover. Secure body with capscrews and lockwashers. *Piston felt oilers should be soaked in SAE-30 oil for one hour prior to assembly.*
5. Install felt oilers on piston and insert piston assembly in body and on push rod. Place grommet on push rod, then install flat washer and nut.
6. Install grommets on cylinder cover, then install cover, nuts and lockwashers on shift cylinder body.

Output Shaft, Fig. 42

1. Press helical gear spacer on front end of shaft. Then press helical gear, bearing spacer washer and front bearing on shaft.
2. Temporarily place companion flange on rear end of shaft. Place shaft in vise with rear end down, gripping flange in vise jaws.
3. Install tongued washer on shaft. Install and tighten shaft front nut to correct torque. Install cotter pin.

Drive Pinion, Fig. 43

1. Lubricate and press inner bearing on pinion, positioning bevel side of bearing toward pinion teeth.
2. On Model 42M, press inner bearing spac-

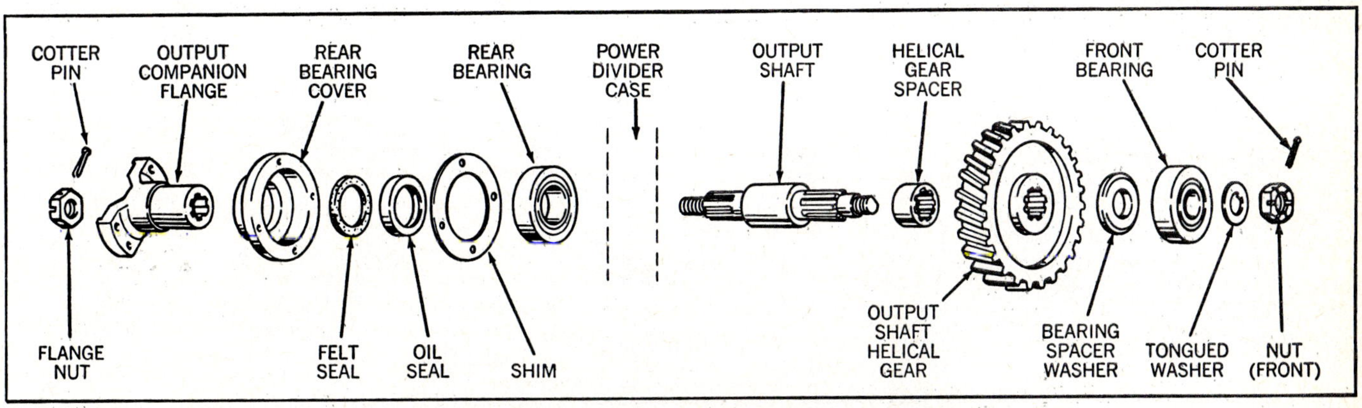

Fig. 42 Disassembled view of output shaft for Models 36M, 42M

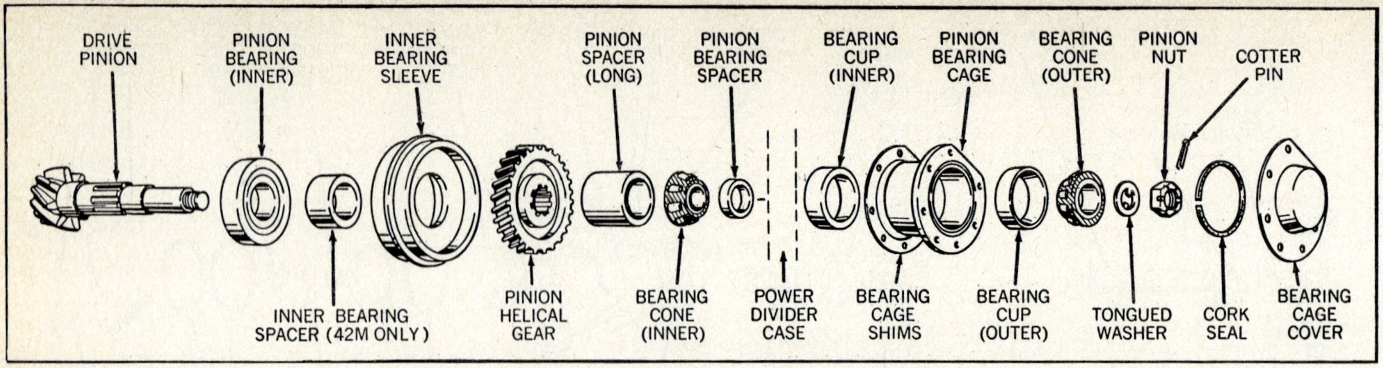

Fig. 43 Disassembled view of drive pinion for Models 36M, 42M

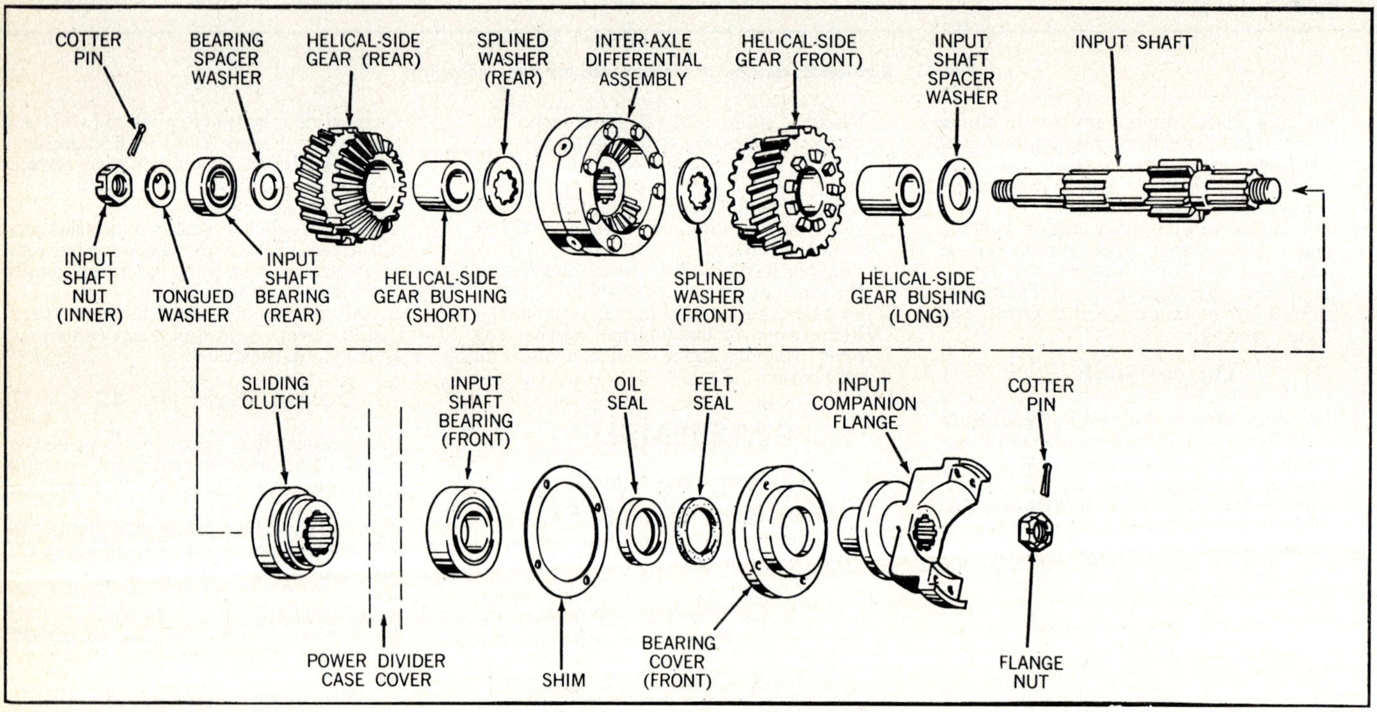

Fig. 45 Disassembled view of input shaft for Models 36M, 42M

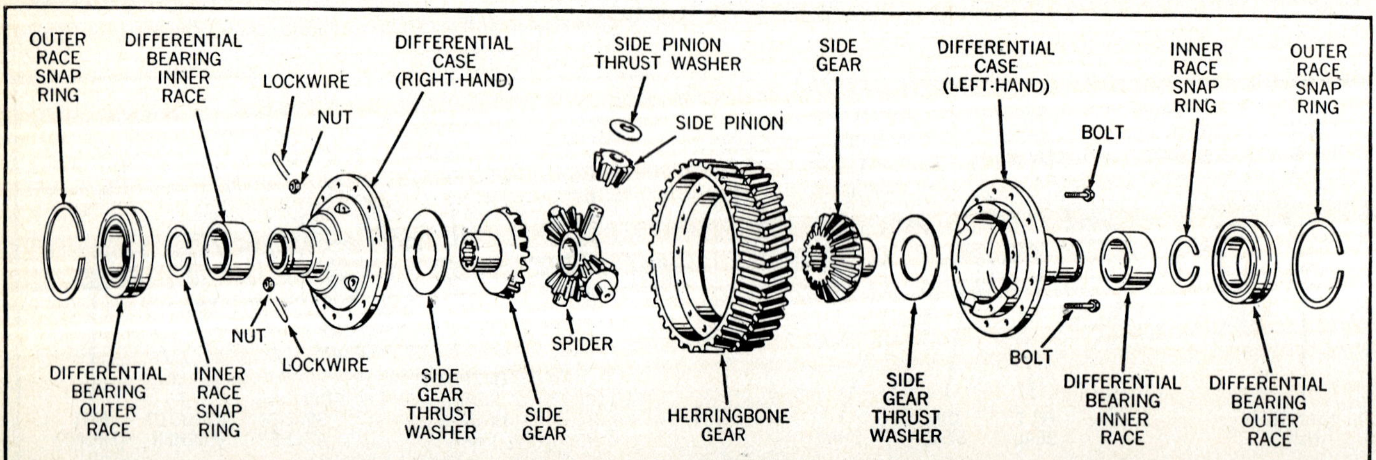

Fig. 46 Disassembled view of axle differential for Model 42M

er and inner bearing sleeve on pinion. *On Model 36M, install inner bearing sleeve when pinion is assembled to power divider case (spacer is not used).*

3. Press helical gear in pinion splines, making certain long hub is against inner bearing spacer (42M). On Model 36M, long hub is against inner bearing.
4. Press long pinion spacer and inner bearing cone on pinion.
5. If removed, press pinion bearing cups n bearing cage.
6. Lubricate pinion bearings. Temporarily place bearing cage, bearing spacer, outer bearing cone, tongued washer and nut on pinion. Tighten nut to correct torque, rotating bearing cage while tightening nut. Check pinion bearing preload as instructed at the beginning of this chapter.
7. With pinion bearing preload correctly adjusted, remove temporarily installed parts from pinion.

Inter-Axle Differential, Fig. 44

Lubricate parts during reassembly. Place side pinions and thrust washers on spider journals. Assemble spider assembly and case halves, aligning previously made punch marks. Secure case halves with bolts and nuts, tightening nuts to correct torque. Lock nuts with wire.

Input Shaft, Fig. 45

1. Temporarily install flange nut on front end of shaft to protect threads during assembly. Place shaft in a press, with shaft supported at nut and install parts on shaft as follows:
2. Place spacer washer on shaft. Press long helical side gear bushing on shaft. Lubricate bushing, then install front helical side gear on bushing. *Bushing and gear must rest firmly against spacer washer.*
3. Place front spider splined washer on shaft. Place inter-axle differential on shaft, positioning bolt heads toward companion flange. Place rear spider splined washer on shaft. *Make certain washer splines are engaged with shaft splines.*
4. Press short helical side gear bushing on shaft. Lubricate bushing, then install rear helical side gear over bushing. *Bushing and gear must rest firmly against splined washer.*
5. Place rear bearing spacer washer on shaft. Temporarily place a dummy sleeve (1½" I.D. × 1" long), tongued washer and shaft inner nut on shaft. Tighten nut to

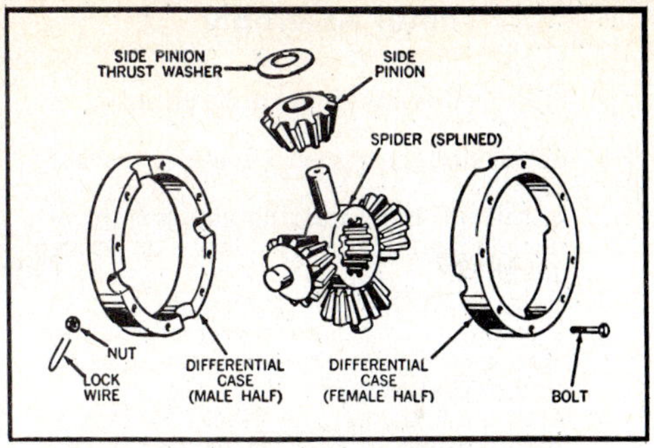

Fig. 44 Disassembled view of inter-axle differential for Models 36M, 42M

correct torque. Rotate shaft parts by hand; they should rotate freely without binding.

6. After testing assembly for free movement, remove temporarily installed parts. Then press rear bearing on shaft and install tongued washer and inner nut. Tighten nut to correct torque and install cotter pin.

Power Divider, Fig. 40

1. Position divider case with rear side up.
2. Lubricate and install output shaft rear bearing in divider case.
3. Install felt seal and oil seal in output shaft rear bearing cover. Using same size and quantity of shims removed at disassembly, install shims and bearing cover on divider case. Install (but do not tighten) cover capscrews and lockwashers until companion flange is installed.
4. *To check bearing clearance to determine thickness of shims required, install bearing cover on divider case without shims. Install and finger-tighten cover capscrews. Then, using a feeler gauge, measure clearance between bearing cover and divider case. This clearance measurement plus .003" would equal the thickness of shims required for correct bearing clearance.*
5. Rotate divider case to position front side up. Insert assembled input shaft, engaging rear bearing in bore of divider case.
6. Insert assembled output shaft in divider case, inserting end of shaft through rear bearing and taking care not to damage oil seal.
7. Using new case cover gasket, place divider case cover on divider case, engaging sliding clutch splines with input shaft splines. Install stud nuts and lockwashers. Tighten stud nuts to correct torque.
8. Install felt seal and oil seal in input shaft front bearing cover.
9. Using same size and quantity of shims removed in disassembly, install shims and bearing cover on divider case cover. Install capscrews and lockwashers but do not tighten until companion flange is installed. Check bearing clearance as explained for output shaft bearings above.
10. With front bearing cover capscrews loose, install companion flange on input shaft and secure with flange nut. Tighten nut to correct torque and install cotter pin. Tighten front bearing cover capscrews to correct torque.
11. Install same size and quantity of shims removed as disassembly on pinion bearing cage, then install cage on divider case. Temporarily install cage capscrews to retain cage during pinion installation.
12. With divider case rear side up, and with output shaft rear bearing cover capscrews loose, install companion flange on output shaft, taking care not to damage oil seal. Install and tighten flange nut to correct torque and install cotter pin. Tighten rear bearing cover capscrews to correct torque.
13. Lubricate pinion bearings. Install partially assembled pinion in divider case.
14. On Model 42M, position inner bearing sleeve in case bore and end of pinion through inner bearing cup. Align sleeve with set screw hole in case and install set screw and lockwasher.
15. Install pinion bearing spacer, outer pinion bearing cone, tongued washer and pinion nut. Tighten nut to correct torque and install cotter pin.
16. Remove temporarily installed bearing cage capscrews. Install new cork seal in cage groove. Install cover and tighten nuts and capscrews to correct torque.
17. On Model 36M, place inner bearing sleeve on pinion inner bearing and in divider case. Align sleeve with set screw hole in case and install set screw and lockwasher.

Differential, Fig. 46

1. Press bearing inner races on differential case halves and install snap rings.
2. Lubricate parts during reassembly.
3. Place herringbone gear on left-hand differential case. Place thrust washer and side gear in left-hand case. Assemble side pinions and thrust washers to spider, then place this assembly in left-hand case. Place right-hand side gear on side pinions.

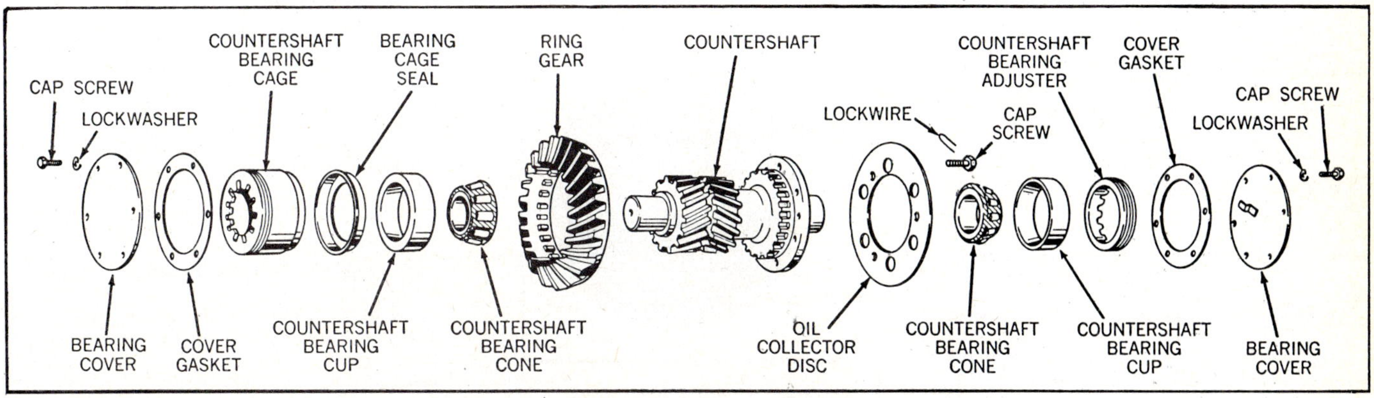

Fig. 47 Disassembled view of countershaft for Model 42M

4. Place side gear thrust washer on right-hand side gear. Position right-hand case over left-hand case, aligning matching marks on each case and, using bolts as guides, drive right-hand case into herringbone gear.
5. Install bolts with heads positioned on left-hand side and tighten to correct torque. *The four oversize bolts are mounted midway between spider arms.* Lock nuts with wire.

Countershaft, Fig. 47

1. Place ring gear on countershaft, aligning punch marks.
2. Place oil collector disc on opposite side of countershaft flange.
3. Install and tighten capscrews to correct torque. Lock screws with wire.
4. Press bearing cones on end of countershaft.

Final Assembly

1. Place countershaft in differential carrier.
2. Lubricate bearings. Install bearing cups.
3. Install left-hand bearing adjuster in carrier, engaging only two or three threads.
4. Install right-hand bearing cage.
5. Tighten bearing cage to preload countershaft bearings as explained at the beginning of this chapter.
6. With new gasket in place on carrier, install power divider on carrier. Install and tighten lockwashers and nuts to correct torque.
7. Adjust ring gear and pinion for proper tooth contact (explained earlier).
8. After correct adjustment, install countershaft bearing cover gaskets, bearing covers, capscrews and lockwashers. Tighten screws to correct torque.
9. Place oil reservoir in carrier.
10. Lubricate bearing outer races. If removed, place snap rings on differential bearings. Place outer races on inner races, then install in carrier. Install bearing caps, making certain that snap rings rest in grooves in carrier and caps.
11. Install bearing capscrews and lockwashers and tighten to correct torque. Lock screws with wire.
12. Insert oil distributor in carrier and install capscrew and lockwasher. Install spring, gasket and plug.
13. Install differential carrier and power divider in axle housing.
14. Connect lockout shift lever.
15. Install axle shafts and connect propeller shafts.
16. Fill axle with correct lubricant.

Eaton Three-Speed Tandem Axles

DESCRIPTION

The Eaton Three-Speed Tandem Drive Axle consists of two planetary-type, two-speed axle units, coupled with a power divider and an inter-axle differential.

One control switch, mounted on the gear shift lever, effects the speed changes. This manually-operated switch controls operation of an Air-Torsion Spring Shift System which, in turn, shifts the axles from one ratio to next.

Service instructions on the shift system are given below. Axle instructions are covered in the Eaton Two-Speed Axle Section of this chapter.

Control Switch

The control switch, Fig. 48, is a three-position switch mounted on the gear shift lever and is manually operated by the vehicle driver. The switch controls current flow through the electrical system to operate the speedometer adapters and solenoid valves.

Speedometer Adapters

As shown in Fig. 49, two speedometer adapters are mounted in the system and compensate for drive shaft speed variations between low, intermediate and high axle range. In the low range the adapters are unexcited. In intermediate range one adapter is energized to 1:1 ratio while the other adapter is unexcited. In high range both adapters are energized to 1:1 ratio. The foregoing arrangement is accomplished by connecting each adapter in parallel to one of each of the two solenoid valves in the system.

NOTE: *For systems that use the front wheel drive speedometer, the speedometer adapters are not required in the electrical system.*

Wiring Harness & Circuit Breaker

Electric wiring in the system is a completely assembled unit including a circuit breaker. Individual wires in the harness are identifed by various colors, Fig. 49. The circuit breaker is connected to the wire leading from the power source. If a short circuit occurs, the circuit breaker will open and cut off electrical current to the system.

Solenoid Valves

Two solenoid valves provide the link between the electrical control system and the air-torsion spring shift units. When the solenoids are energized, air is allowed to travel to the air shift units and thus shift the axles. When the solenoid valves are de-energized, by movement of control switch, air pressure supply is shut off allowing the air-torsion spring shift unit to bleed air through the solenoid valve exhaust port. In units with an electric lockout, an additional solenoid valve controls air pressure supply to operate the lockout unit mounted on the power divider.

Lockout With Low Pressure Switch

A low pressure switch is included in the shift system. This switch is mounted in the lockout valve and prevents the use of the inter-axle differential lockout except when the axles are in low ratio. When the lockout is engaged this switch immediately allows the axles to go to low ratio, no matter what range they were in previously. With lockout engaged, the axles cannot be shifted.

Fig. 48 Eaton three speed axle control switch

Lockout With Electric Switches

The electric type lockout consists of two electric switches and a solenoid valve, Fig. 49. One switch is located in the vehicle cab and the other is mounted on the forward rear axle shift unit. This type system permits lockout to engage in low range only. When the shift unit is in intermediate and high range position, the electric switch at the shift unit is open. This prevents completion of the electrical circuit to the lockout solenoid valve.

Air-Torsion Spring Shift Unit

This unit is mechanically connected to the axle shift fork and shifts the axle. The unit includes an air chamber and a torsion spring drive assembly. A diaphragm, operated by air pressure, moves a push rod. The end of the push rod connects to a spring winding lever. This lever is part of the torsion spring drive assembly, which acts to move the shift fork and change axle range as described under *System Operation.*

SYSTEM OPERATION

Operation In High Range

When control switch is moved to "HI" position, current to the two solenoid valves allows air to pass from its supply through the valves and be applied to the air shift diaphragm on both forward and rearward driving axles.

Diaphragm movement operates the push rod on each unit which, in turn, moves the spring winding levers. The spring winding levers increase the load placed on the torsion springs. When this occurs, the axles are ready to shift to high range. When torque on the gears is relieved by closing and opening the throttle, or declutching, the shift is completed.

Operation In Intermediate Range

When the control switch is moved to the "INT" position, the solenoid valve shuts off the air supply to the rearward rear axle shift unit, and the solenoid valve for the forward rear axle remains open, allowing the forward axle shift unit to retain its air. Thus, when

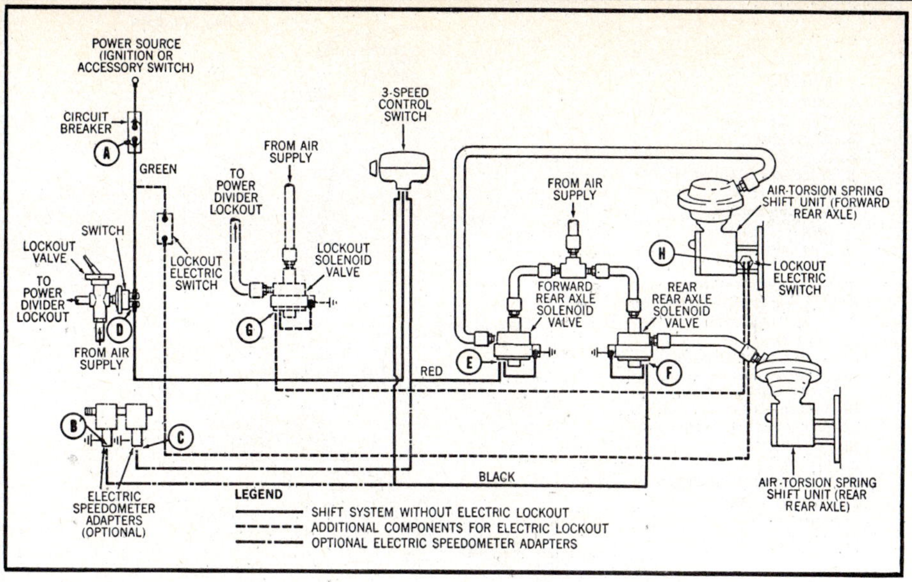

Fig. 49 Eaton air-torsion spring shift system diagram

torque on the gears is relieved by closing and opening the throttle, or declutching, the rearward rear axle is shifted to low range and the forward rear axle remains in high range. The shift is now completed to intermediate.

Operation In Low Range

When the control switch is moved to the "LO" position, both solenoid valves shut off the air supply. Air pressure in the air-torsion spring shift units bleeds back through the solenoid valves and air lines. Pressure on diaphragm is released. Push rod moves toward diaphragm and moves spring winding lever. Additional load is placed on torsion spring, providing a condition ready for axle to shift to low range. When torque on gears is relieved by closing and opening throttle, or declutching, the shift to low range is completed.

TROUBLE SHOOTING

If the axle will not shift, test system as follows, using a 12-volt test lamp at the terminal references indicated in Fig. 49.

Circuit Breaker Operation

Disconnect circuit breaker lead wire ("A"). Connect test lamp across terminal "A" and ground on vehicle frame. Turn on ignition (or diesel accessory switch) and observe test lamp operation:

If lamp lights and stays on continuously, circuit breaker and lead wire are satisfactory.

If lamp does not light, check for poor electrical connections or broken lead wire. If these are satisfactory, circuit breaker is faulty.

If lamp does not light immediately then starts to flash on and off, the circuit breaker is faulty.

Control Switch & Wiring Harness

1. Check harness for damage or worn insulation that may cause a ground connection, especially where harness passes through cab floor. Check for short circuits between wire terminals.
2. To determine condition of control switch, it is recommended that a new switch be temporarily installed. However, if a new switch is not available, check its operation with the test lamp as follows:

NOTE: *Under actual installation conditions, lead wires to speedometer adapters could be transposed without affecting operation. However, in the following tests it is assumed that the black and red wires are connected as shown in Fig. 49.*

---

3. Disconnect wires at speedometer adapter terminals "B" and "C" and at solenoid terminals "E" and "F". Turn ignition on and alternately connect test lamp leads to disconnected lead wires. Operate control switch and observe test lamp for following conditions:
4. When test lamp is connected to speedometer adapter lead wire "B", lamp should light when control switch is in "HI" position. Lamp should go out with control switch in "INT" or "LO" position.
5. When test lamp is connected to lead wire ("C"), lamp should light in "HI" or "INT" position. Lamp should go out in "LO" switch position.
6. When test lamp is connected to solenoid valve lead wire "E", lamp should light in "INT" and "HI" positions of control switch. Lamp should go out when control switch is in "LO" position.
7. When test lamp is connected to solenoid lead wire "F", lamp should light in "HI" position, and should go out when control switch is in "INT" or "LO".
8. If test indications are correct in the foregoing tests, current supply is correct to speedometer adapters and solenoid valves.
9. If lamp does not light correctly, trouble may be a short circuit in wiring harness or control switch, or a wire in harness may be broken.

Solenoid Valve Operation

1. Disconnect solenoid lead wires at terminals "E", "F" and "G". Also disconnect air lines leading to respective shift unit.
2. Install an air pressure gauge in air line opening in solenoid valve.
3. Apply power supply (vehicle voltage) to solenoid terminal and observe pressure gauge reading. Operating pressure should approximate reservoir pressure.
4. If gauge indicates approximate reservoir pressure, solenoid operation is satisfactory. If gauge indicates low or no pressure, solenoid is faulty.

Air Torsion Spring Shift Unit Operation

If electrical system is satisfactory and axle does not shift correctly, trouble may be caused by a faulty spring shift unit which should be disassembled and repaired.

Speedometer Adapter Operation

Before condemning speedometer adapter operation, make certain that ground connection is satisfactory. If testing is required, proceed as follows:

1. Disconnect lead wire at terminal "B".
2. Connect test lamp to disconnected wire terminal and ground on vehicle frame.
3. Turn ignition switch on and observe test lamp. Test lamp should light in high range and should go out in intermediate and low ranges.
4. To test other speedometer adapter, disconnect wire at terminal "C".
5. Connect test lamp to disconnected wire and ground on vehicle frame.
6. Lamp should light in intermediate and high ranges, and should go out in low range.
7. If lamp indications are correct, current supply to adapters is correct.
8. If lamp indications are correct and axle shifts normally but speedometer does not operate properly, replace faulty adapter.
9. If test lamp indications are not correct, trouble is in wiring harness or control switch.

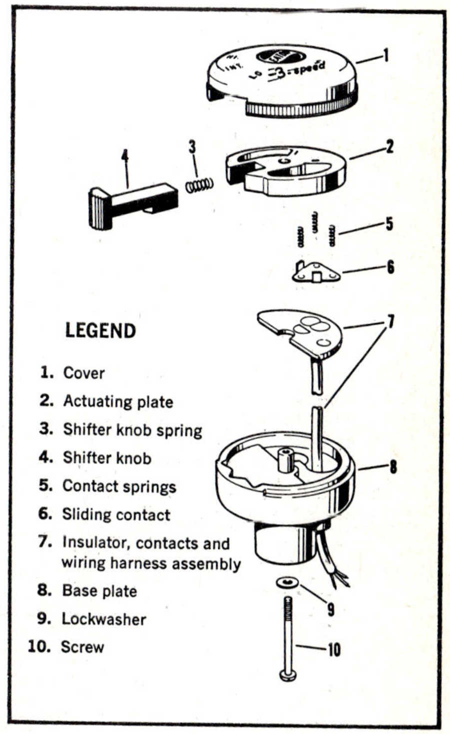

Fig. 50 Control switch disassembled

Low Pressure Switch Operation

1. Place control switch in "HI".
2. Turn on ignition switch, then engage inter-axle differential lockout.
3. Air should bleed from shift system and axles should shift to low range.
4. If this does not occur, check low pressure switch circuit with test lamp as follows:
5. Disconnect wire at terminal "D".
6. Connect test lamp to terminal "D" and ground on vehicle frame.
7. Turn on ignition switch.
8. Place lockout valve in engaged and disengaged positions, and observe test lamp.
9. In disengaged position, lamp should light. If not, check for poor electrical connection or broken lead wire. If wire and connections are satisfactory, low pressure switch is faulty.
10. In the engaged position, lamp should not light. If it does, low pressure switch is faulty.

Electric Switch At Shift Unit

1. Disconnect wire at terminal "H".
2. Connect test lamp to terminal "H" and ground.
3. Close manual electrical lockout switch and operate control switch in all positions.
4. Test lamp should light in "LO" and should not light in "INT" and "HI".
5. Replace faulty switch.

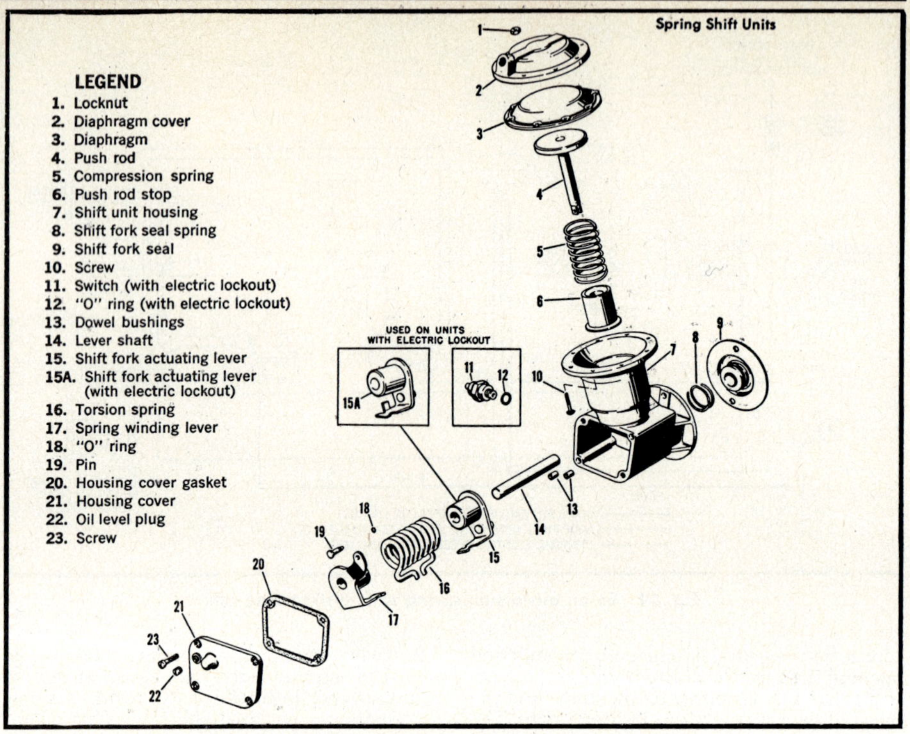

Fig. 51 Air torsion spring shift unit disassembled

UNIT REPAIRS

Control Switch

1. Referring to Fig. 50, disconnect wires at end of switch wiring harness, and unscrew switch from gear shift lever.
2. Remove screw from mounting side of switch and lift off over, actuating plate, shifter knob, knob spring, contact springs and contact.
3. Remove insulator, contacts and wiring harness from base plate.

REASSEMBLE

1. Pass wire harness through base plate, positioning insulator in recessed area in base plate.
2. Lightly coat contacts and pivot hub (at center of base plate) with non-melting silicone grease.
3. Hold actuating plate (top side down) in one hand and install three springs and sliding contact in correct mounting position. Place base plate over actuating plate with other hand.
4. Hold actuating and base plates together with one hand and install spring and shifter knob.
5. Continue to hold actuating plate in mounting position and install cover. Fasten cover to base plate with screw and lock washer.
6. Thread switch onto shift lever and connect wiring harness (see Fig. 49).

Air Torsion Spring Shift Unit

1. Referring to Fig. 51, disconnect air line connection from shift lever cover (when equipped with electric locknut disconnect wires at switch).
2. Unfasten and lift shift unit off studs.
3. Remove shift fork seal and spring.

NOTE: *Shift unit need not be completely disassembled to replace torsion spring drive assembly. Air pressure applied to opening in diaphragm cover will move push rod enough to remove or install pin which connects push rod to spring winding lever.*

DISASSEMBLE

1. Remove housing cover and drain lubricant.
2. Remove diaphragm and cover from housing.
3. When equipped with electric lockout, unscrew switch and O-ring from housing.
4. Place housing in a vise and compress spring until pin which connects push rod and spring winding lever together, can be removed.
5. Remove pin and O-ring.
6. Carefully open vise until tension on compression spring is completely relieved.
7. Remove housing from vise and remove push rod, compression spring and push rod stop from housing.
8. Remove lever shaft and lift torsion spring drive from housing. Do not disassemble these parts unless replacement is necessary.

REASSEMBLE

1. Place torsion spring drive in housing. Insert lever shaft through torsion spring drive and into hole provided in housing.
2. Place push rod stop, compression spring and push rod in housing.
3. Place housing in vise. Close vise to compress spring until pin can be installed to connect push rod to spring winding lever.
4. Install O-ring on pin, then insert pin in openings provided in spring winding lever and push rod end.
5. Remove housing from vise.
6. Place shift fork seal on studs and differential carrier. Install spring on seal.
7. Place shift unit on mounting studs and make certain shift fork actuating lever engages slot in shift fork. Install stud nuts and lock washers.
8. Place diaphragm and cover on housing and secure with screws and lock nuts. Tighten lock nuts to 50–60 inch-lbs. After initial torquing, retighten until 50–60 inch-lbs is permanently maintained.
9. Install dowel bushings (if removed), gasket and housing cover. Secure cover and torque screws to 35–40 inch-lbs.
10. When equipped with electric lockout, install switch and O-ring in housing, and connect wiring to switch.
11. Connect air line to shift unit cover.

Lubrication

1. Use SAE 10 engine oil all year.
2. Each 10,000 miles or three months, remove level plug in cover and check oil level. Level should be even with bottom of filler hole.
3. At least twice a year remove housing cover and drain oil. Wash parts thoroughly and air dry. Reinstall cover and remove level plug. Fill through oil hole to bottom of filler hole.

SPICER DRIVING AXLES

SINGLE SPEED UNITS

Spicer axles are of the integral carrier, hypoid gear type and are available in two and four pinion models, Figs. 1 and 2. Pinion and differential bearing preload and ring gear and pinion backlash adjustments are made through use of shim packs. Axle assembly and axle shaft replace procedures are covered in the applicable truck chapters. The following service procedures apply to both front and rear axle carriers.

Differential Carrier, Disassemble

1. Remove axle shafts as outlined in applicable truck chapter. Axle shafts may be pulled out far enough to clear differential side gears.
2. Drain lubricant and remove rear cover.
3. Make sure differential side bearing caps and axle housing are marked, then remove the side bearing caps.
4. Pry differential from housing, Fig. 3.
5. Remove side bearing cups.
6. Pull off side bearing and adjusting shims, tagging the shims for identification on reassembly.
7. Unfasten ring gear from case.
8. Drive out differential pinion shaft pin and pull out the shaft, pinions and side gears.
9. Hold companion flange from turning and remove flange nut.
10. Pull flange from pinion shaft.
11. Remove pinion from carrier by tapping on front end with soft hammer.
12. Remove pinion shaft bearing oil seal and bearings from carrier, keeping separate the shim pack at each bearing.

Pinion & Bearings, Replace

If the original ring gear and pinion are being used in the original carrier, use the original shim packs at each bearing, Fig. 4. If a new pinion or differential carrier is installed, note the markings on the end of the pinion gear and on the differential carrier to obtain the correct thickness of shimming to be used with these parts from your supplier. The shims behind the rear bearing establishes the correct pinion depth.

1. Press the rear pinion bearing cup in the housing with the proper thickness of shims. Press the rear pinion bearing on the shaft.
2. Install the front bearing cup and shims, and front bearing.
3. Install the companion flange and, while holding the flange from turning, tighten the nut to a torque load of 200–220 ft. lbs.

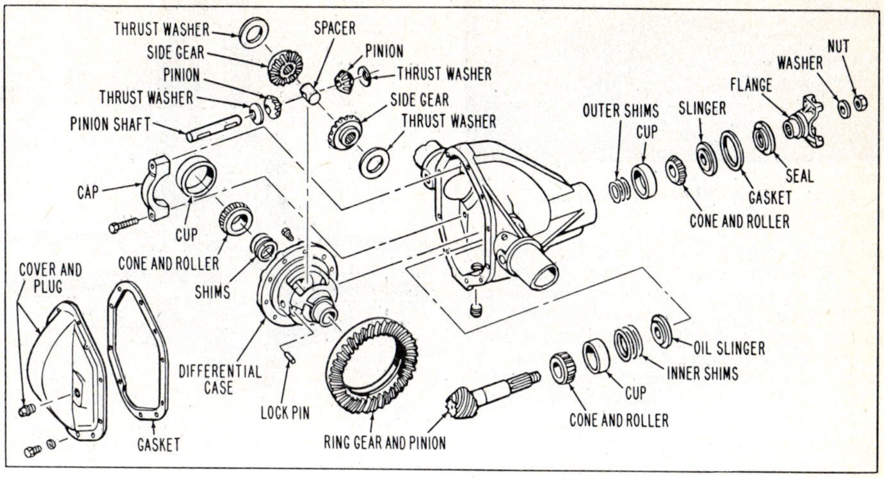

Fig. 1 Spicer 2 pinion driving axle

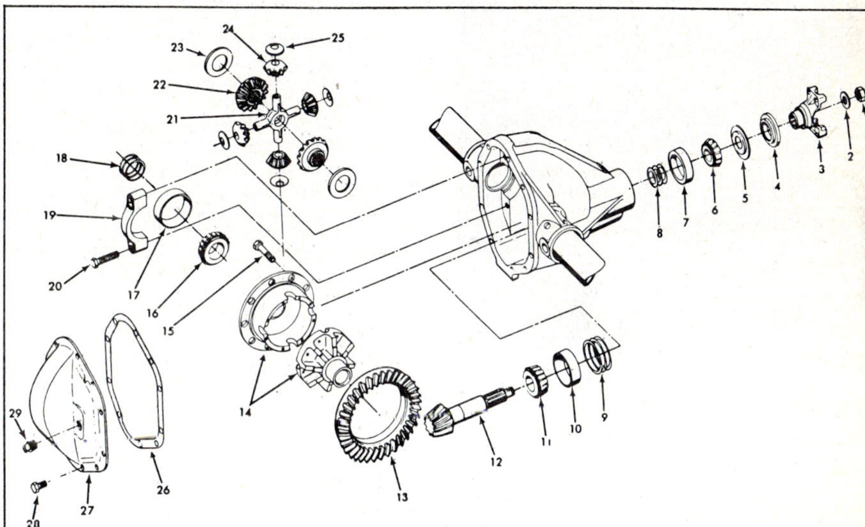

1. Pinion Nut
2. Washer
3. Companion Flange
4. Oil Seal
5. Oil Slinger
6. Pinion Front Bearing
7. Front Bearing Cup
8. Preload Shim Pack
9. Pinion Depth Shim Pack
10. Rear Bearing Cup
11. Pinion Rear Bearing
12. Drive Pinion
13. Ring Gear
14. Differential Case
15. Ring Gear Bolt
16. Differential Side Bearing
17. Side Bearing Cup
18. Side Bearing Adjusting Shims
19. Bearing Cap
20. Bearing Cap Bolt
21. Differential Spider
22. Differential Side Gear
23. Washer
24. Pinion Gear
25. Washer
26. Gasket
27. Cover
28. Cover Screw
29. Drain Plug

Fig. 2 Spicer 4 pinion driving axle

SPICER SINGLE SPEED DRIVING AXLE SPECIFICATIONS

Model	Ring Gear & Pinion Backlash		Pinion Bearing Preload			Differential Bearing Preload		
	Method	Adjustment	Method	New Bearings Inch-Lbs.	Used Bearings Inch-Lbs.	Method	New Bearings Inch	Used Bearings Inch
27	Shims	.005–.010	Shims	10–25	—	Shims	.015	.015
30	Shims	.005–.009	Shims	15–35	10–20	Shims	.015	.015
44	Shims	.005–.010	Shims	20–40	10–20	Shims	.015	.015
53	Shims	.005–.010	Shims	10–25	—	Shims	.015	.015
60, 61	Shims	.005–.009	Shims	20–40	10–20	Shims	.015	.015
70	Shims	.005–.010	Shims	20–40	10–20	Shims	.015	.015

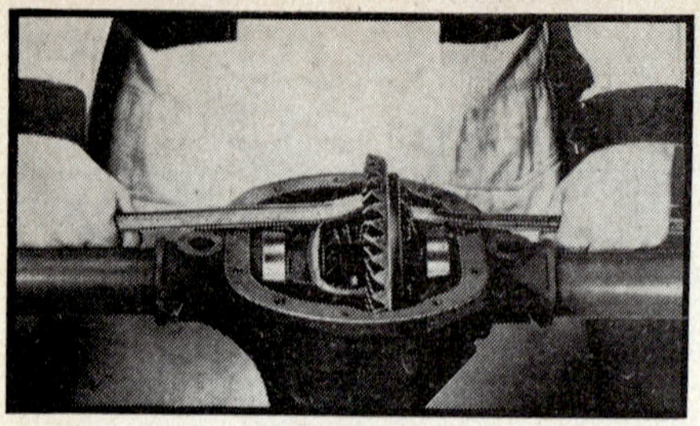

Fig. 3 Prying differential from housing

4. Check pinion bearing preload with an inch-pound torque wrench. Measure amount of torque required to turn pinion. If preload is not within specifications, add or remove shims from behind front bearing to obtain correct preload.
5. Remove the companion flange and install a new oil seal (well soaked). Reinstall the companion flange and tighten the nut to a torque load of 200–220 lb. ft.

Fig. 6 Installing differential in housing

Fig. 7 Ring gear and pinion backlash adjustment

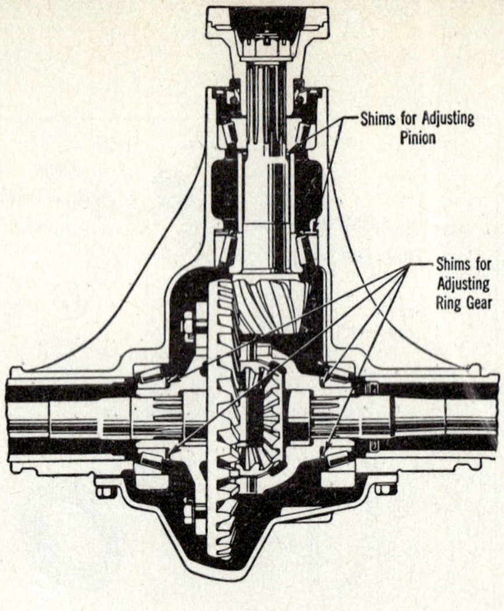

Fig. 4 Differential and pinion bearing adjustments

Ring Gear, Replace

1. Install guide pins in every other hole in the ring gear. These pins can be made from 1½" long capscrews with heads cut off the ends slotted.
2. Make sure back face of ring gear and face of case are free of dirt and burrs and slip gear over pilot diameter of case.
3. Install every other ring gear bolt. Draw them up evenly and snugly so ring gear face is flush with face of case.
4. Remove guide pins and install remaining bolts.

Differential Carrier, Assemble

The differential bearings are adjusted by shims, Fig. 4. These shims also establish the ring gear position with the pinion. Therefore, backlash must be checked whenever a bearing adjustment is made.

The correct bearing adjustment is one which will provide a .015–.020" pinch fit when the differential unit is assembled into the carrier. To make the adjustment, install the bearing cones without shims and place the assembly in the housing with the bearing cups. Force the unit to one side and check the clearance between the bearing cup and differential case with a feeler gauge, Fig. 5. When the clearance is determined, select shims of this amount plus .015–.020" extra to establish the proper pre-load.

Remove the differential bearings again and divide the shims into two packs of equal thickness and install on each side and replace the

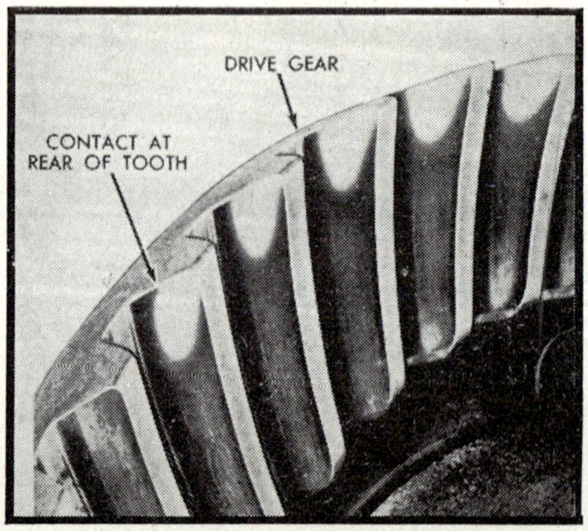

Fig. 8 Short heel contact. Move drive gear toward pinion

Fig. 5 Checking clearance between bearing cup and differential case

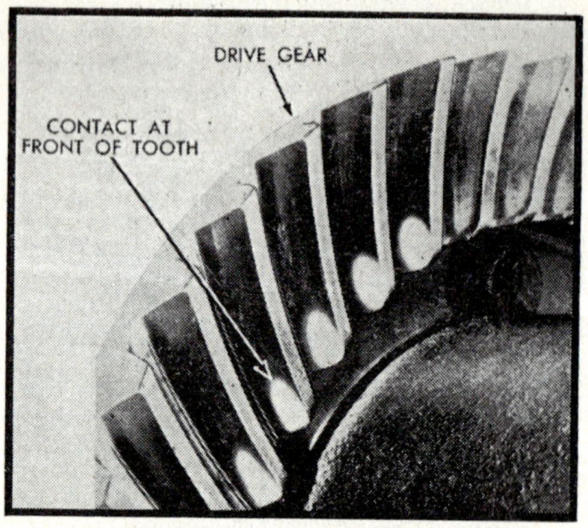

Fig. 9 Short toe contact. Move drive gear away from pinion

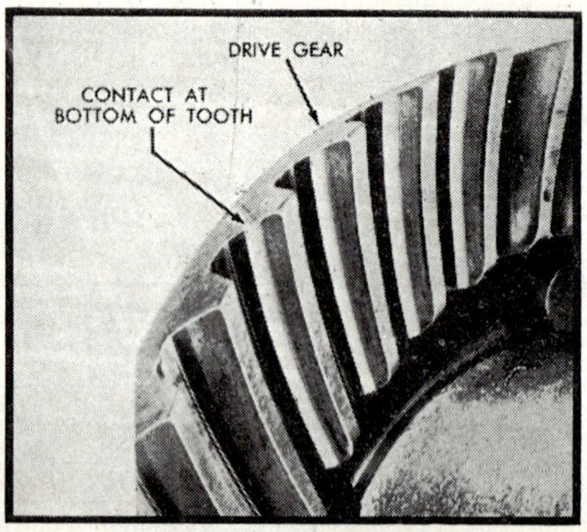

Fig. 10 Contact too low and narrow. Pinion should be moved away from center of axle

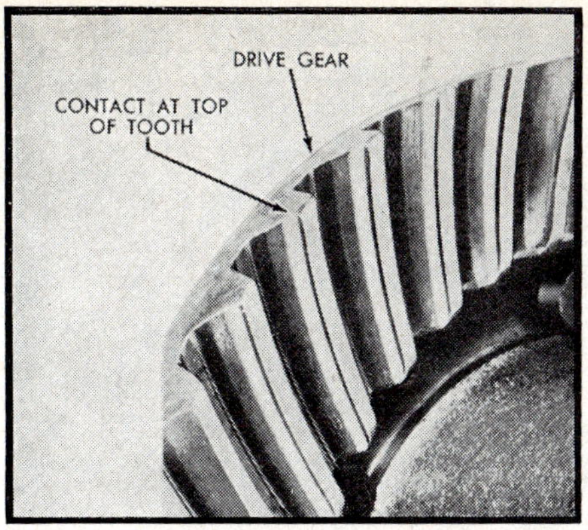

Fig. 11 Contact too high and narrow. Pinion should be moved toward center of axle

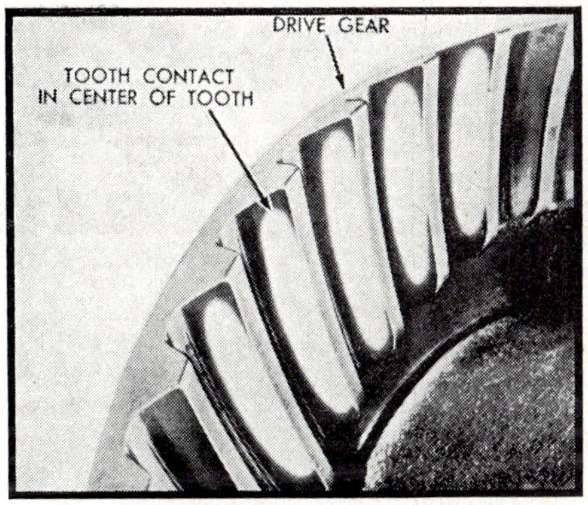

Fig. 12 Correct tooth contact

bearings. Reinstall the unit in the carrier. This operation is made easier by cocking the bearing cups slightly when the differential is placed in the housing and then tapping them lightly with a mallet, Fig. 6. However, when installing the differential in the housing, be sure the ring gear teeth mesh with the pinion teeth before tapping the bearings in place.

After the bearing cups are firmly in place, install the bearing caps. The bearing caps and gasket surface of the housing are marked with a horizontal numeral and on the other side by a vertical numeral. The position of the numerals should correspond when reinstalling the bearing caps.

Ring Gear & Pinion Backlash, Adjust

Mount a backlash gauge indicator on the carrier and start checking for the correct backlash between the ring gear and pinion. If the backlash is not within specifications, it will be necessary to change the arrangement of the shims back of the bearings. Make corrections in backlash according to Fig. 7, bearing in mind that shims removed from one side must be installed on the opposite side so that the total shim thickness of the right and left side will remain unchanged and the bearing adjustment undisturbed.

Gear Tooth Contact Pattern

Allowable variations in the carrier or drive pinion may cause the pinion to be too far in or out even when shimmed properly. Thus, the tooth contact must be tested and corrected as necessary or the gears may be noisy.

Paint the ring gear teeth with a light coating of red lead, white lead or prussian blue. Revolve the gears and observe the contact, referring to Figs. 8 to 12.

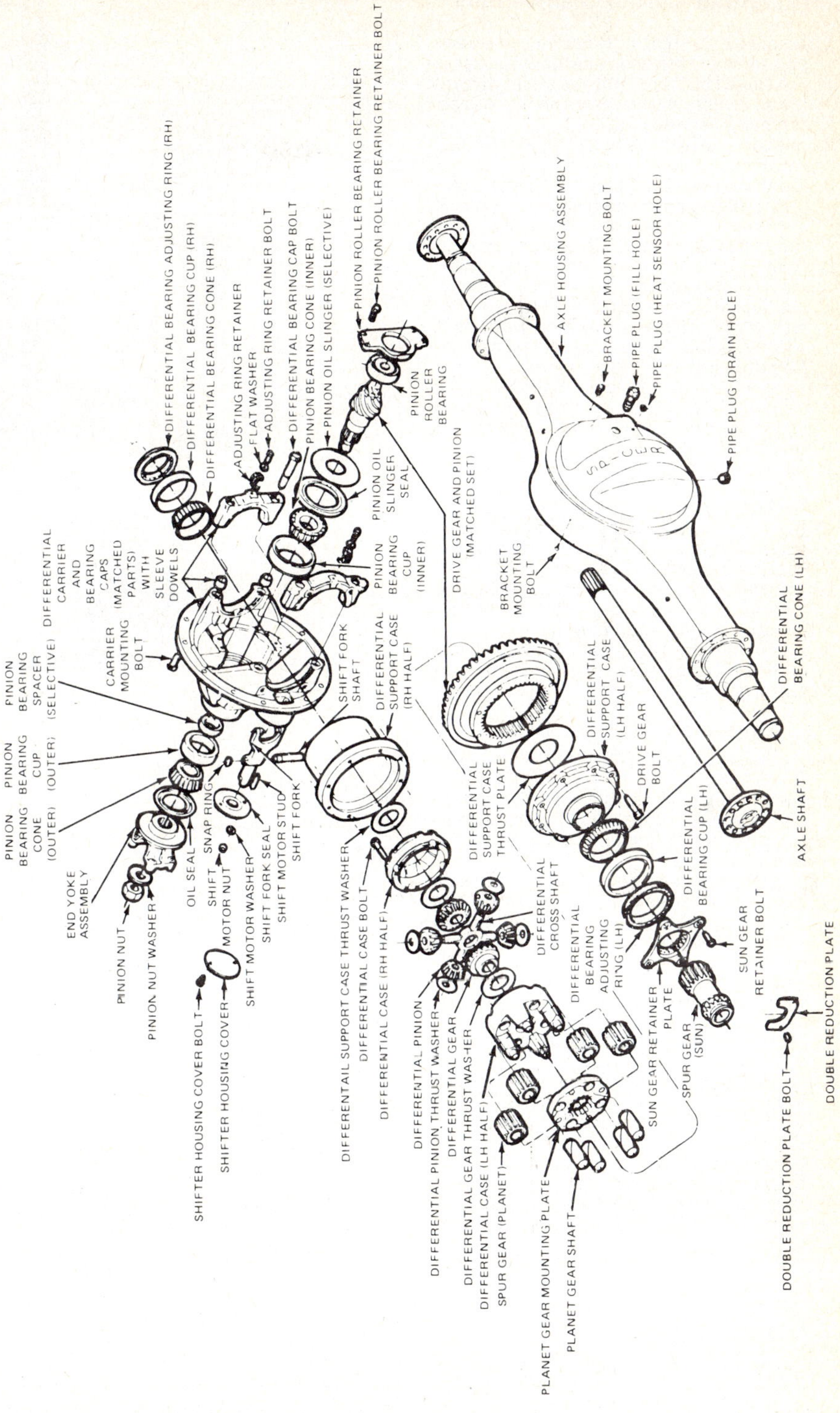

Fig. 13 Spicer two speed double reduction axle

TWO-SPEED DOUBLE REDUCTION AXLE

The Spicer two-speed double reduction axle, Fig. 13, consists of a full floating banjo type axle housing member with a bolt-in differential carrier assembly. The differential carrier assembly consists of a primary hypoid gear set for high range operation, and a secondary planetary type spur gear set, for low range operation.

Range changes are made with a control button on the gear shift lever. Actuating the button causes the planetary sun gear to act as a sliding clutch member and move in and out of engagement with two clutching plates to complete range changes.

Differential Carrier Disassembly

1. Remove axle shaft nuts and/or cap screws and remove both axle shafts. If gasket is positioned between hub and shaft, discard it and replace with a new one when reassembling carrier.
2. Support carrier with suitable jack and remove.
3. Clean carrier-to-housing RTV gasket material from carrier and housing mounting surfaces.
4. Place carrier in suitable holding fixture.
5. Remove shift fork boot from carrier housing. Remove shift fork retaining ring and using a drift and hammer, remove shift fork shaft. Retrieve retaining ring so it will not be lost in the differential carrier. Remove shift fork from carrier.
6. Remove the 4 sun gear retainer plate screws and sun gear retainer plate. Pull sliding clutch sun gear out of differential carrier.
7. Remove bearing adjusting ring retainers, 4 differential bearing cap bolts, both adjusting rings, and both bearing caps. Using tool T70P-4067-A, differential bearing nut wrench or equivalent to remove bearing adjusting ring.
8. Lift ring gear and support case subassembly out of carrier using caution to avoid damage to ring and pinion gears.
9. Mark right and left support case halves in order to ensure reassembly in same relative position.

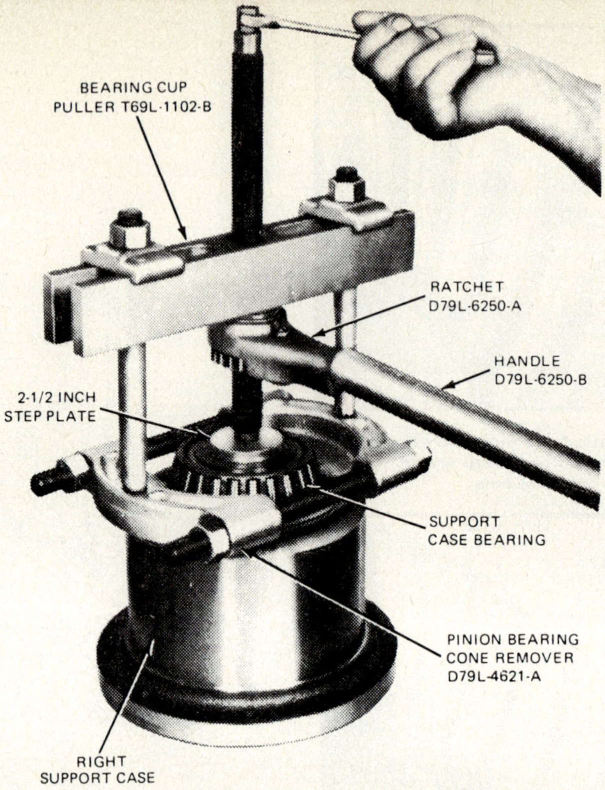

Fig. 14 Removing right support case bearing

Ring Gear Support Case Disassembly

1. Clamp ring gear and support case assembly in soft-jaw vise. Remove drive gear screws, lift off left support case flange and thrust washer.
2. Tap the ring gear alternately on each side with a soft hammer to loosen it from the right support case. Always replace cap screws when removing ring gear from support case.
3. Lift right support case and thrust washer from differential and planetary gear assembly.
4. Mark left and right sides of differential case for proper reassembly.
5. Separate right and left differential cases. Remove differential side gears and thrust washers, differential spider gears, their thrust washers and differential spider.
6. Remove planetary gear mounting plate from left side differential case by tapping with soft hammer if necessary and remove planetary pins and planetary gears.
7. If the right differential support case bearing requires replacement, it can be removed with the special tools shown in Fig. 14.
8. If the left differential support case bearing requires replacement, the same tools can be used with the substitution of a 3¾ inch step plate.

Drive Pinion Disassembly

1. Remove 4 pilot bearing retainer cap screws from carrier. Lift pinion roller bearing retainer cap from the pilot bearing and carrier housing.
2. Hold end yoke or flange using tool T57T-4851-B and remove pinion nut and washer.
3. Remove end yoke using puller T65L-4851-B or equivalent. Remove pinion with plastic hammer using care to prevent pinion from falling on ground.

NOTE: A selective spacer for pinion bearing preload is inserted between the pinion bearing near the spline end of the pinion. The spacer should be retained for possible use during reassembly.

4. Remove inner oil seal from carrier and discard. Install new seal when reassembling.
5. If necessary, the inner pinion bearing cone can be removed with tool D79L-4621-A or equivalent.
6. Remove selective inner pinion oil slinger and retain for possible use in reassembly.
7. Remove pinion pilot bearing assembly with tool D79L-4621-A or equivalent.
8. Pull out pinion seal with puller D78P-1225-B or equivalent. Discard seal and replace with new one during reassembly. Remove outer pinion bearing cone.
9. Remove outer pinion bearing cup with puller D78P-1225-B or equivalent.

CAUTION: Do not nick carrier bore.

10. Remove inner bearing cup with puller D78P-1225-B or equivalent.

Differential Case Assembly

1. Install the 4 planetary pins in holes of differential case. Place planetary gears in position on planetary pins after applying small amount of hypoid lubricant to pins.
2. Position planetary clutch mounting plate on planetary gear pins with chamfered teeth facing planetary gears. Tap plate in place with soft hammer if necessary.
3. Assemble thrust washers to differential side gears. Apply hypoid lubricant on side gear hubs and between side gears and thrust washers.
4. Assemble side gears, pinion mate gears, cross-shaft, and pinion mate washers into differential case after greasing cross-shaft and washers.
5. Assemble right half of differential case, making sure scribe marks align.
6. Assemble and torque body bolts evenly to 110–125 ft. lbs.

Ring Gear Support Case Assembly

1. Use bearing installation tool D80T-4200-E (17,500 lb. Axle) or D80T-4200-F (18,500 lb. Axle) or equivalent to seat right support case bearing cone.
2. Repeat step 1 for left support case bearing cone.
3. Place right differential support case in vise with bearing side facing down. Position differential case-to-gear support case thrust washer in case after lightly greasing both sides of washer.
4. Place differential assembly in gear support case with planetary end facing upward. Place ring gear over differential assembly with the hypoid teeth on the ring gear facing downward, and mesh the planetary gears with the internal teeth on the ring gear.
5. Align bolt holes in ring gear with bolt holes in gear support case. Apply oil to each side of the planetary clutch plate thrust washer and position washer on clutch plate.

NOTE: Using a stone, remove any burrs or nicks from clutch plate top surface prior to positioning thrust washer.

MM	Inches	MM	Inches
1.321	.052	1.778	.070
1.397	.055	1.854	.073
1.473	.058	1.930	.076
1.549	.061	2.007	.079
1.626	.064	2.083	.082
1.702	.067		

NOTE: The inner pinion oil slinger is a selective slinger that is used to determine the position of the pinion (mounting distance) during reassembly. The slinger is available in the following thicknesses. Measure the slinger with a micrometer before assembly to assure correct thickness.

Fig. 15 Slinger-spacer selection chart

MM	Inches	MM	Inches
17.780	.700	18.085	.712
17.805	.701	18.110	.713
17.831	.702	18.136	.714
17.856	.703	18.161	.715
17.882	.704	18.186	.716
17.907	.705	18.212	.717
17.932	.706	18.237	.718
17.958	.707	18.263	.719
17.983	.708	18.288	.720
18.009	.709	18.313	.721
18.034	.710	18.339	.722
18.059	.711	18.364	.723

Fig. 16 Preload spacer chart

6. Place left gear support case flange on ring gear and align bolt holes in ring gear with both gear support cases.
7. Assemble and torque ring gear bolts to 160–180 ft. lbs.

Pinion Bearing Cup Assembly

1. Assemble inner pinion bearing cup into carrier using tool D80T-4200-D (18,500 lb. Axle) or D80T-4200-E (17,500 lb. Axle) or equivalent.
2. Assemble outer pinion bearing cup into carrier using tool D80T-4200-C (18,500 lb. Axle) or D80T-4200-D (17,500 lb. Axle) or equivalent.

NOTE: Both inner and outer pinion bearings can be installed simultaneously using draw-bar T75T-1175-A and discs D80T-4000-D.

Pinion Depth Adjustment

Pinion depth is determined by placing a combination oil slinger and spacer between the pinion shaft gear and inner bearing cone.

1. Install depth gauge tool in carrier housing. For 17,500 lb. Axle, use spacer D80T-4020-R71, gauge disc D80T-4020-R61, screw D80T-4020-R69, aligning adapter D80T-4020-R60, and aligning plate D80T-4020-R70. For 18,500 lb. Axle, use spacer D80T-4020-R71, gauge disc D80T-4020-R61, screw D80T-4020-R69, aligning adapter D80T-4020-R63, and aligning plate D80T-4020-R70. Place bearing to be installed in axle on tool when determining pinion depth.
2. Assemble dummy bearings (D80T-4020-R67 and D80T-4020-R68 for 17,500 lb. axle and D80T-4020-R65 and D80T-4020-R66 for 18,500 lb. axle) on gauge tube (D80T-4020-R64). If necessary, coat gauge tube with engine oil to ease installation.
3. Install dummy bearings and gauge tube into axle carrier housing. Install bearing caps and torque bolts to 220–250 ft. lbs for 17,500 lb. axle and 290–350 ft. lbs for 18,500 lb. axle.
4. Determine correct shim usage by placing feeler gauges between gauge block and gauge tube. When slight drag is felt between gauge and tube, read feeler gauge.
5. Compare feeler gauge reading with number stamped on bottom of pinion. If pinion is marked +5, 0.005 inch must be subtracted from feeler gauge reading. If pinion is stamped −5, 0.005 inch must be added to feeler gauge reading.
6. After determining correct shim size refer to chart in Fig. 15. Select shims to the closest available size.
7. Remove pinion depth gauge tools from axle carrier housing and remove pinion from carrier.
8. Install correct size combination oil slinger-spacer on drive pinion. Place rear pinion bearing onto pinion and press into place with special tools axle bearing seal plate T75L-1165-B or equivalent and D80T-4200-B (17,500 lb. axle) or D80T-4200-C (18,500 lb. axle).

Pinion Bearing Preload

NOTE: To prevent damage to seals do not install pinion shaft seals until after pinion bearing preload has been determined.

1. Install drive pinion assembly into carrier housing with spacer tool D80T-4020-R71. Assemble outer pinion bearing cone, yoke washer, yoke and nut onto the pinion. Torque nut to 700–900 ft. lbs.
2. Rotate pinion with torque wrench. Correct preload with new bearings installed should be 15–25 inch lbs.
3. If pinion bearing preload is not to specifications install spacers until preload of 15–25 ft. lbs is obtained. Refer to chart in Fig. 16. Remove spacer tool.
4. Remove drive pinion, outer pinion bearing cone, yoke washer, yoke and nut from carrier housing.

Drive Pinion Assembly

1. Install pilot bearing on drive pinion and press bearing onto pinion.
2. Install inner pinion oil seal into carrier with tabs locking into oil channel. Make sure flat side of seal is seated against bottom of bearing cup bore so that lip side will be against pinion oil slinger.
3. Apply light coat of hypoid lubricant to lips of new pinion seal and assemble it into housing, using installation tool D80T-4200-G or equivalent.
4. Check yoke oil slinger for wear. If wear on sleeve from seal lip is detectable with fingernail, replace sleeve. Install slinger by driving into place on yoke with special tool D80T-4200-E for 17,500 lb. axle or D80T-4200-F for 18,500 lb. axle.
5. Install drive pinion into carrier housing.
6. Turn carrier over and install preload spacers.
7. Install end yoke and washer with new pinion nut. Torque pinion nut to 700–900 ft. lbs while holding yoke with tool T78P-4851-A or equivalent.
8. Install pilot bearing retainer cap, tapping lightly with soft hammer if necessary.
9. Assemble retainer cap screws and torque to 110–125 ft. lbs.

Ring Gear Support Case Assembly Installation

1. Install ring gear and support case assembly into carrier housing. Use care to avoid nicking teeth of ring gear and pinion during assembly. Check that differential bearing cups are in proper position with respect to bearing cones.
2. Install differential bearing cups and assemble bearing cap screws finger tight.
3. Install bearing adjusters. Tighten adjuster on ring gear side (left) until there is zero backlash between the ring gear and pinion.
4. Tighten bearing adjuster on opposite (right) side until adjuster has made contact with bearing cup and cup is seated. From this point, continue to tighten the adjuster four or five holes to obtain specified backlash and preload differential bearings.
5. Check ring gear and pinion backlash in four positions around the ring with a dial indicator. Backlash tolerance is 0.008–0.012 inch.
6. Check tooth contact pattern according to procedure under single speed axle section.
7. After proper backlash has been established torque differential bearing caps to 290–350 ft. lbs. Recheck ring gear and pinion tooth contact pattern.
8. Install adjuster ring retainers and screws. Torque the retainer screws to 30–40 ft. lbs.
9. Install sliding clutch sun gear.
10. Install the sun gear retainer plate and 4 bolts. Torque bolts to 200–250 ft. lbs.
11. Position shift fork on sliding clutch gear after sliding fork into carrier housing. Insert shift fork shaft through hole in differential carrier into shift fork and install shift fork retaining ring. Fill shift fork shaft hole in carrier with RTV to prevent lubricant seepage and install shift fork boot.

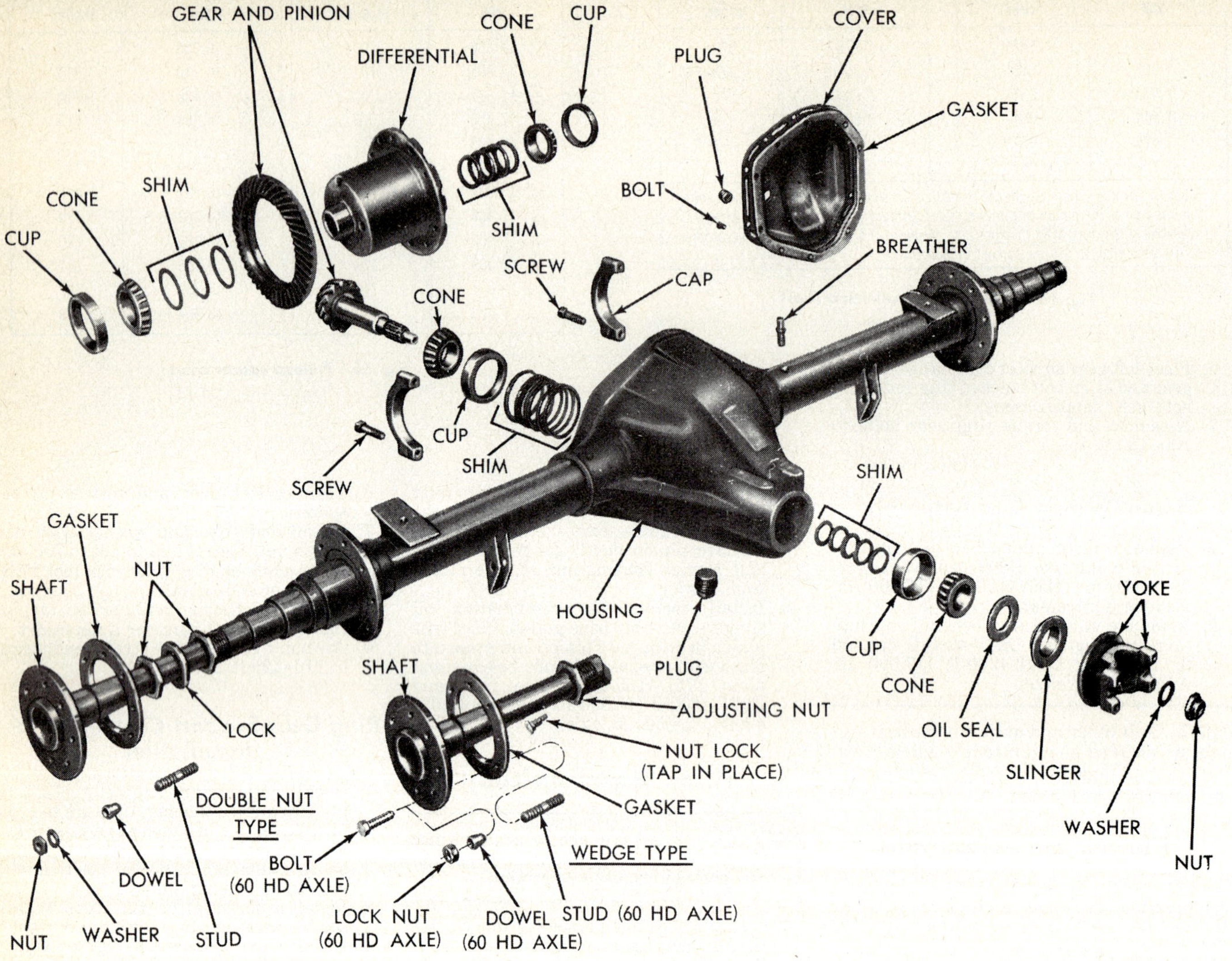

Fig. 17 Spicer 60 & 60 HD Rear Axle

Differential Carrier Installation

1. Clean inside of carrier housing thoroughly. Stone surface if necessary to remove burrs or nicks.
2. Apply 1/4 inch bead of RTV gasket material or equivalent to carrier mounting surface laying it on inside and outside of all carrier mounting screw holes.
3. Thread two studs into housing to serve as carrier assembly alignment guides. This procedure avoids damage to RTV seal from realigning carrier bolt holes.
4. With carrier supported on roller jack, assemble carrier to housing and install carrier bolts, replacing two guide studs with mounting bolts. Torque bolts alternately to 180–230 ft. lbs. Allow one hour cure time for RTV sealant prior to adding hypoid lubricant or operating vehicle.
5. Insert both axle shafts, using new gaskets and install axle nuts or cap screws.
6. Reinstall shift unit.
7. Replace drain plug after cleaning magnet, plug thread and surrounding area. Refill axle housing to proper level with hypoid lubricant. Replace filler hole plug after cleaning magnet, plug thread and surrounding area.

60 & 60 HD

Differential Carrier, Disassembly Fig. 17

1. Remove carrier cover screws and cover, then drain lubricant.
2. Measure drive gear face runout as follows:
   a. Mount a dial indicator so plunger contacts rear face of ring gear.
   b. Zero dial indicator.
   c. Rotate drive gear through several complete revolutions and read dial indicator. Mark drive gear and differential case at point of maximum runout. The marking of differential case will be helpful later in checking differential case runout. Total indicator readings in excess of .006 inch might indicate possible loose drive gear or damaged differential case.
3. Check the clearance between the differential bearing cap and bearing cup. Insert a .003 inch feeler gage between cap and cup. The gage should not enter between cap and cup. If so, the bearing cup may have turned in carrier, causing excessive wear.

NOTE: Identifying letters are stamped on bearing caps and face of carrier housing seal surface. Letters are stamped on left side in horizontal position while letters on right side are stamped in vertical position. Always match identifying letters for proper reassembly.

4. Remove differential bearing caps and locate spreader tool W-129 with dowel pins seated in locating holes of axle housing. Turn tool screw finger tight at this time.
5. Install pilot stud on left side of axle housing. Attach a dial indicator with the plunger contacting the opposite side of the axle housing.
6. Tighten spreader tool nut sufficiently enough to obtain .015 inch movement of dial indicator to permit removal of differential case and ring gear assembly.

NOTE: Do not spread over .002 inch as it will result in permanent damage to carrier casting.

TIMKEN (Rockwell) DRIVING AXLES

INDEX

LUBRICANT CAPACITIES

Lubricant capacities are given as a guide only. The capacities of two similar axles of the same series may vary considerably due to design changes and the vehicle manufacturer's installation. The actual service capacity may be determined by carefully measuring the amount of lubricant necessary to fill the assembly to the correct level and measuring the lubricant again when it is drained. The vehicle should be on a level floor when this inspection is made.

Single Axles

Model	Oil, Pints
A-150	5½
B-100	10
B-140	12
B-150	3½
C-100	12½
D-100	12½
D-140	13
E-100	15
E-105	12½
E-150	9
E-300	13①
E-350, 370	22①
F-30	6
F-35	7
F-37	7
F-38	10
F-46	10
F-50	10
F-53	12
F-54	11
F-56	14
F-58	15
F-75	9
F-77	10
F-100	13
F-106	13½
F-140	14
F-141	18
F-147	14
F-200	12
F-223	16
F-233, 234, 235	23
F-300	16
F-337	24
F-340	16①
F-400	16
F-409	28
F-501	10
F-544	10
F-551, 552	11
F-580, 583	15
F-2090	12
F-3100	16
F-3110	26
F-3200	22
F-4700	40
F-4710	32
F-7900	40
F-7910	32
FS-4711	32
FDS-75	13
FDS-750	7
FDS-1600	23
FDS-1800	35
G-161	21
G-340	24①
G-341	22①
G-361	21①
H-100	20
H-140	21
H-141	18
H-150	11
H-162	20
H-170	26½
H-171	27
H-172	27
H-200	28①
H-240	22①
H-262	23
H-300	26①
H-340	22①
H-350, 360, 370	24①
H-362	26
L-100	23
L-140	24
L-162	24
L-172	—
L-200	31①
L-240	22①
L-300	29①
L-340	22①
L-345	33
L-350	24①
L-362	26
L-370	32①
LT-200	31①
LT-300	29①
QT-140	24
QT-200	31①
QT-230	44③
QT-240	34①
QT-300	29①
QT-330	44③
QT-340	32①
Q-100	31
Q-145	—
Q-200	34①
Q-245	34①
Q-246A	36
Q-300	32①
Q-345	32①
Q-346A	34
Q-350	34①
Q-370	34①
Q-380	36①
Q-390	36①
RL-170	—
RT-240	32①
RT-340	32①
RT-341A	34
R-100	30
R-140, 160	28
R-155	—
R-163	34
R-170	43
R-171B	44½
R-172	—
R-200	36①
R-230	36①
R-230③	45①
R-255	—
R-270	—
R-300	34①
R-330	35①
R-330③	44①
R-390	60①
R-2090	10
R-3100	20
S-170	—
S-200	38①
S-300	39①
U-170	—
U-200	38①
U-240	—
U-240	—
U-270	—
U-300	39①
U-340	—
W-170	—
W-270	—

Tandem Axles

Model	Oil, Pints
SDHD	16②
SDHR	16
SFDD	28
SFD-DD-4640	—
SFHD	17②
SFHR	16½
SHHD	26②
SHHR	26
SLHD	32½②
SLHR	32
SQHD	34②
SQHP	40②
SQHR	31
SQTT-335	44①
SRD-DD	—
SHRD	37②
SR-170	—
SR-270	—
SRHR	36
SSHD	32②
SSHR	28
SRT-235	45①
SRT-335	44①
STHD	—
STD-DD	—
SLD	28
SLDD	28②
SQD	22
SQDD	22②
SRD	22
SRDD	22②
STD	28
STDD	30
STDDA	30
SUDD	46
SUDDA	46
SUHD	—
SUD-DD	—
SU-170	—
SU-270	—
SW-170	—
SW-270	—
4600, SFD	28
4600, SFDD	28②
4700, SFD	28
4700, SFDD	28②
75, SFD	16
157, SFD	9
353, SD	24
375, SFD	23
450, SFD	36
454, SD	26

TIMKEN (Rockwell) DRIVING AXLES

Model	Oil, Pints
456, SW	28
457, SW	28
460, SW	28
460, SFD	29
SQW	33
SQW	40③
3010, SD	19
3010, SW	14
3013, SW	23
3020, SD	31
3020, SFD	31
3020, SFDD	31③
3020, SW	28
3022, SW	27
472, SD	28
473, SD	28
700, SBD	7
760, SBD	5
1000, SBD	8
1055, SBD	19
1500, SBD	12
1555, SBD	26
3000, SD	19
3000, SW	19
3002, SW	17
3456, SW	24
3458, SW	33

①—Add one pint of lubricant to pinion cage when new or reconditioned drive unit is installed.
②—Add two pints of lubricant to inter-axle differential housing when new or reconditioned drive unit is installed in addition to specified amount of lubricant in housing.
③—Housing cover 6½" deep overall.

A new clearance for the ring gear thrust screw has been established for models that are not equipped with a thrust block and adjusting screw. The new clearance is .025–.035 inch. Refer to Fig. A. Proceed as follows:

1. Install thrust screw into side of drive unit carrier until it bottoms out on ring gear.
2. Back out thrust screw ½ turn. Tighten lock nut to 150–190 ft. lbs.

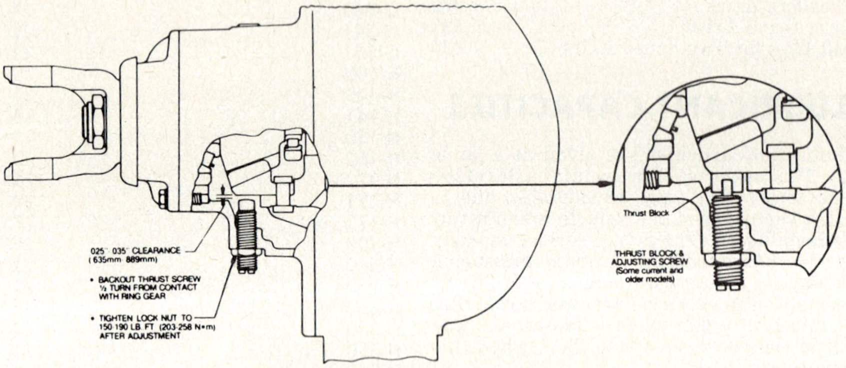

Fig. A Ring gear thrust screw clearance adjustment. Single & tandem axles with thrust screw

Single Reduction Axles

DISASSEMBLE

Remove Carrier From Housing

1. Remove plug from bottom of housing and drain lubricant.
2. Remove axle shaft drive stud nuts and lockwashers.
3. Rap axle shafts sharply in center of flange to free dowels. Remove taper dowels and pull out axle shafts. *Prying shafts loose will damage hubs and oil seals.*
4. Disconnect universal at pinion shaft.
5. Remove carrier-to-housing stud nuts and washers. Loosen two top nuts and leave on studs to prevent carrier from falling.
6. Break carrier loose from axle housing with rawhide mallet.
7. Remove top nuts and washers and work carrier free, using puller screws in holes provided. A small pinch bar may be used to straighten carrier in housing bore. However, the end must be rounded to prevent indenting carrier flange. A roller jack may be used to facilitate removal of carrier.

Disassemble Carrier, Fig. 1

1. To remove differential and gear assembly, cut lock wire. Remove capscrews and adjusting nut locks.
2. Center punch one differential carrier leg and bearing cap to identify for proper reassembly.
3. Remove bearing cap stud nuts or capscrews, bearing caps and adjusting nuts.
4. Loosen jam nut and back off thrust block

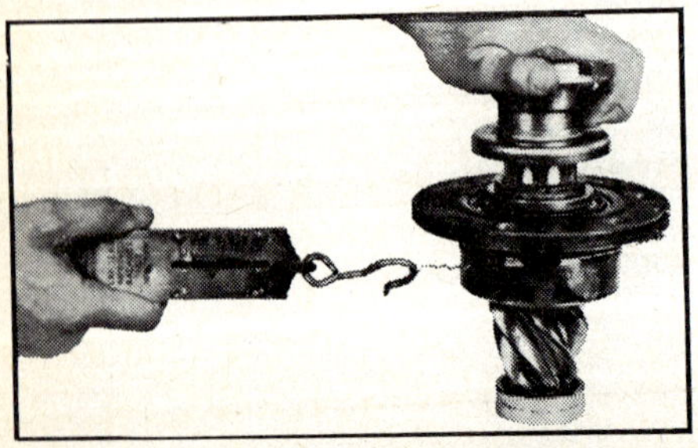

Fig. 2 Measuring pinion bearing preload

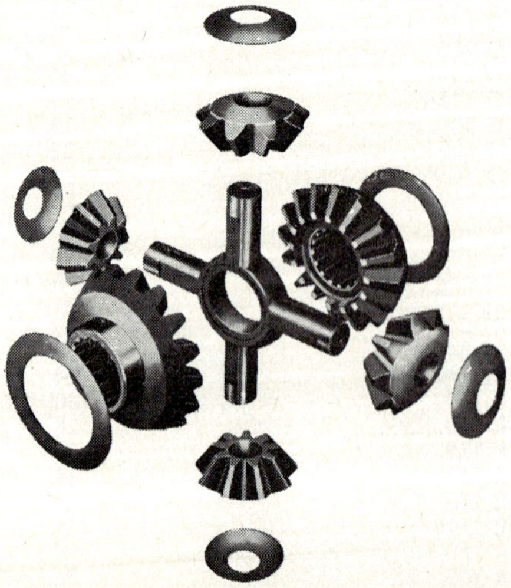

Fig. 3 Differential pinion and side gear assembly

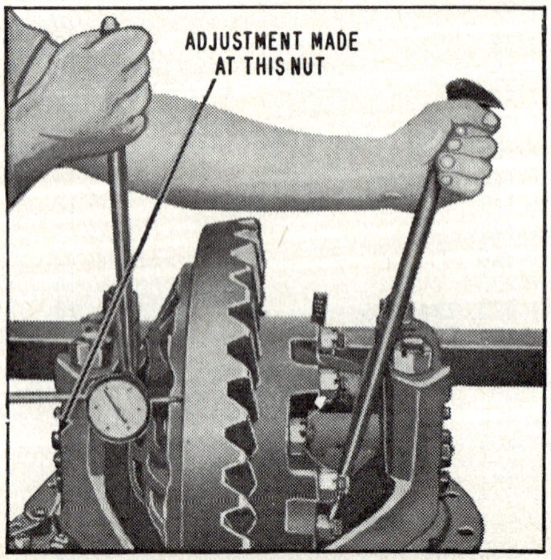

Fig. 4 Adjusting differential bearing preload

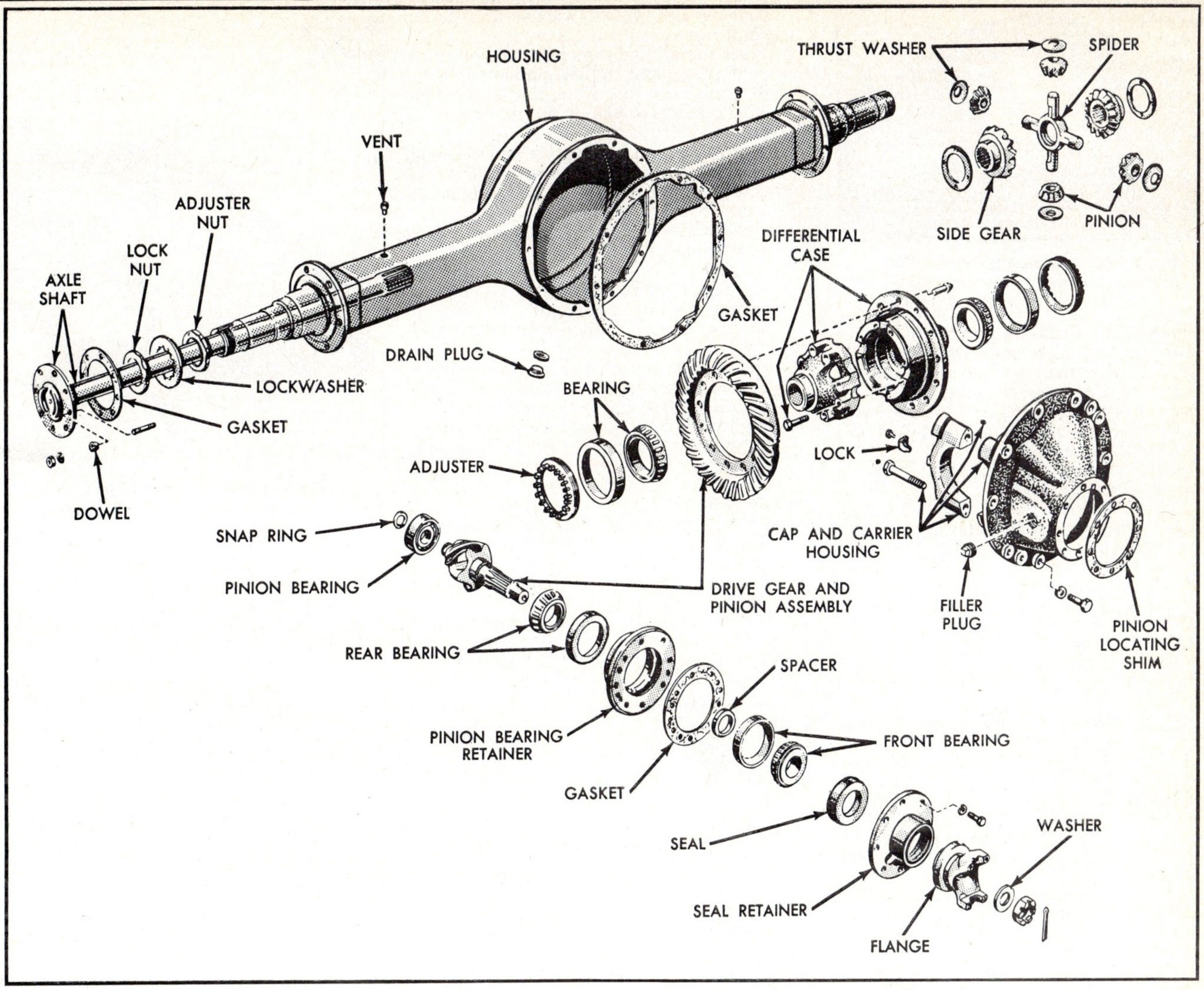

Fig. 1 Timken (Rockwell) single reduction axle (typical)

adjusting screw.
5. Lift out differential and gear assembly.
6. Remove thrust block from inside of carrier housing.

Disassemble Differential & Gear

1. If original identification marks are not clear, mark differential case halves with a punch or chisel for correct alignment upon reassembly.
2. Cut lock wire, remove bolts and separate case halves.
3. Remove spider, pinions, side gears and thrust washers.
4. To remove ring gear rivets, carefully center punch rivets in center of head. Use a drill 1/32" smaller than body of rivet to drill through head. Then press out rivets.
5. If necessary to remove differential bearings, remove with a suitable puller.

Removing Pinion & Cage

1. Hold flange or yoke with suitable tool and remove pinion shaft nut and washer.
2. Remove flange or yoke with a suitable puller. *Driving flange off will cause runout.*
3. Remove pinion cage stud nuts or capscrews.
4. Remove bearing cover and oil seal assembly.
5. Remove bearing cage, using puller screws in holes provided. *The use of a pinch bar will damage shims. Driving pinion from inner end with a drift will damage bearing lock ring groove.*
6. Wire shim pack together to facilitate adjustment on reassembly.

Disassemble Pinion & Cage

Both splined and tapered pinion shafts are used in single reduction carriers. Where the tapered shaft is used, the thrust bearings are adjusted by means of adjusting and lock nuts. On the splined shaft, this adjustment is secured with a selective spacer or spacer combination.

Splined Shaft

1. Tap shaft out of cage with a soft mallet and press shaft from cage.
2. Remove outer bearing from cage.
3. Remove spacer or spacer combination from pinion shaft.
4. If necessary to replace rear thrust bearing or radial bearing, remove with suitable puller.
5. Remove oil seal from bearing cover.

Tapered Shaft

1. Straighten lock washer, and remove lock nut, washer, adjusting nut and thrust washer.
2. Tap pinion out of cage with a soft mallet or press shaft from cage.
3. Remove bearing from cage.
4. Remove bearings from shaft with suitable puller if necessary.
5. Remove oil seal from bearing cover.

REPAIRS

1. Replace all worn or damaged parts. Hex nuts with rounded corners, all lock washers, oil seals and gaskets should be replaced.
2. Remove nicks, mars or burrs from machined or ground surfaces. Threads must be clean and free to obtain accurate adjustment and correct torque. A fine mill file or India stone is suitable for this purpose. Studs must be tight prior to reassembling of parts.
3. All bronze bushed differential pinions should be ball burnished after bushing

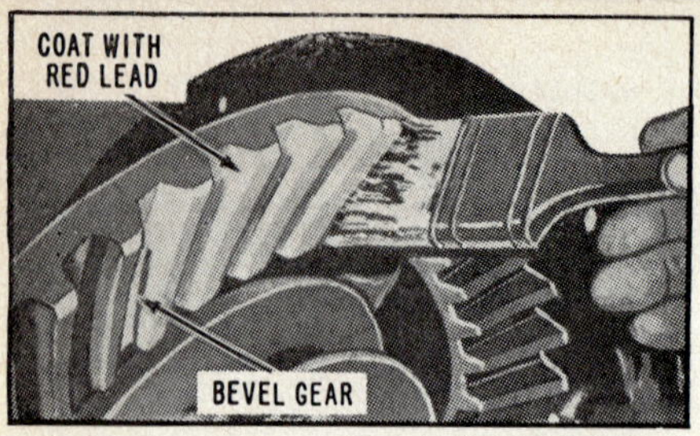

Fig. 5 Painting ring gear teeth for obtaining tooth contact impressions

installation. Install bushing with a small stepped drift. The small O.D. should be .010" smaller than the bushing burnished I.D. and $1\frac{1}{2}$ times bushing length. Always install bushings so end is even with I.D. chamfer or about $\frac{1}{16}$" below the spherical surface.

4. When assembling components, use a press where possible.
5. Tighten all nuts to the specified torque (see torque limits following service instructions). Use soft iron locking wire to prevent possibility of wire breakage.
6. The burrs, caused by lockwashers, at the spot face of stud holes of cages and covers should be removed to assure easy reassembly of these parts.

REASSEMBLY OF CARRIER

Pinion & Cage

Splined Shaft

1. Press rear thrust and radial bearings firmly against pinion shoulders with a suitable sleeve that will bear only on bearing inner race.
2. Install radial bearing lock ring and squeeze ring into pinion shaft groove with pliers.
3. If new cups are to be installed, press firmly against pinion bearing cage shoulders.
4. Lubricate bearings and cups with light machine oil.
5. Insert pinion and bearing assembly in pinion cage and position spacer or spacer combination over pinion shaft.
6. Press front bearing firmly against spacer.
7. Rotate cage several revolutions to assure normal bearing contact.
8. While in press under pressure, check bearing preload torque. Wrap soft wire around cage and pull on a horizontal line with a spring scale, Fig. 2. If a press is not available, the pinion nut may be tightened to the correct torque and preload checked.
9. The correct pressures and torque for checking pinion bearing preload are as follows:

Thread Size	Pressure	Torque, Ft. Lbs.
1" × 20	6 tons	300–400
$1\frac{1}{4}$" × 18	11 tons	800–1100
$1\frac{1}{2}$" × 12	14 tons	800–1100
$1\frac{1}{2}$" × 18	14 tons	800–1100
$1\frac{1}{3}$" × 12	14 tons	800–1100

10. Use rotating torque, not starting torque. If rotating torque is not within 5 to 15 pound inches, use thinner spacer to increase or thicker spacer to decrease preload. Assuming pinion cage diameter to be 6", the radius would be 3" and with 5 pounds pull would equal 15 pound inches of preload torque.
11. Press flange or yoke against forward bearing and install washer and pinion shaft nut.
12. Place pinion and cage over carrier studs, hold flange and tighten pinion shaft nut to correct torque. Flange must be held with a suitable tool or fixture to tighten nut.
13. Recheck pinion bearing preload torque. If rotating torque is not within 5 to 15 pound inches, repeat the foregoing procedure.
14. Hold flange and remove pinion shaft nut and flange.
15. Lubricate pinion shaft oil seal and cover outer edge of seal body with a non-hardening sealing compound. Press seal against cover shoulder with a seal driver.
16. Install new gasket and bearing cover.
17. Press flange against forward bearing and install washer and pinion shaft nut.
18. Tighten to correct torque and install cotter key. *Do not back off nut to align cotter key holes.*

Tapered Shaft

1. Press rear thrust and radial bearings firmly against pinion shaft shoulder.
2. Install radial bearing lock ring and squeeze ring into pinion shaft groove with pliers.
3. If new cups are to be installed, press firmly against pinion cage shoulders.
4. Lubricate bearings and cups with light machine oil.
5. Install forward bearing, thrust washer and adjusting nut.
6. Install new lockwasher and lock nut.
7. Adjust pinion bearing preload to 5 to 15 pound inches with lock nut tightened securely against washer. *Lock nut must be tight to secure correct preload.*
8. Bend lock washer when correct preload has been obtained.
9. Lubricate pinion shaft oil seal and cover outer edge of seal body with a non-hardening sealing compound. Press seal against cover shoulder with a seal driver.
10. Install new gasket and bearing cover. Cover should be carefully installed to prevent cutting seal on keyway.
11. Install key, press flange on taper and install washer and pinion shaft nut.
12. Tighten to correct torque and install cotter key. *Do not back off nut to align cotter key holes.*

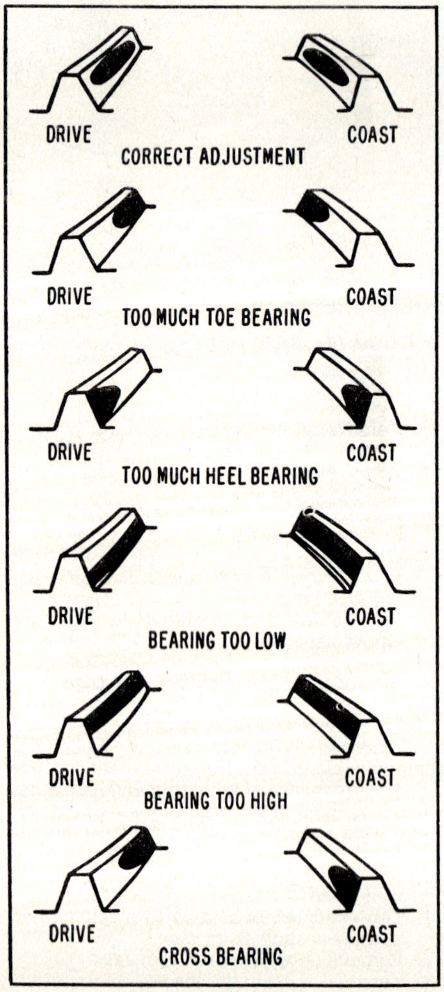

Fig. 6 Location, size and shape of tooth contacts

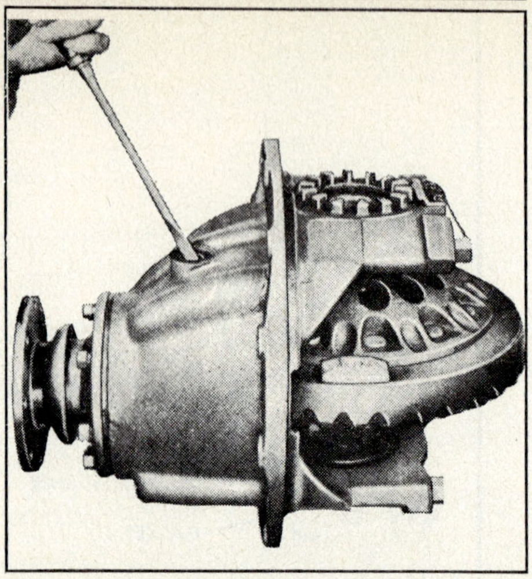

Fig. 7 Showing thrust block placed on back face of ring gear

Install Pinion & Cage

1. Install correct shim pack. Locate thin shims on both sides for maximum sealing ability.
2. Position pinion and cage assembly over studs and tap into position with soft mallet.
3. Install lock washers and stud nuts or capscrews. Tighten to correct torque.

Assemble Differential & Gear

1. Rivet ring gear to case half with new rivets. Rivets should not be heated, but always upset cold. When the correct rivet is used, the head being formed will be at least $\frac{1}{8}$" larger in diameter than the rivet hole. The head will then be approximately the same height as the preformed head. Excessive pressure will cause distortion of the case holes and result in gear eccentricity. Bolts are available for service replacement of rivets.
2. Lubricate differential case inner walls and all component parts with axle lubricant.
3. Position thrust washer and side gear in ring gear and case half, Fig. 3.
4. Place spider with pinions and thrust washers in position.

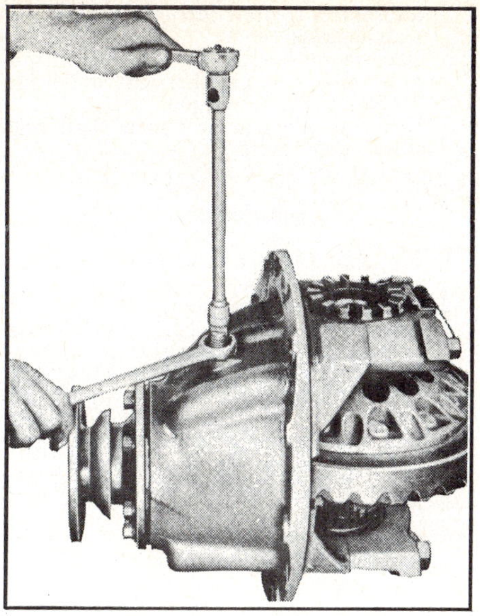

Fig. 8 Adjusting clearance between thrust block and back face of ring gear

5. Install component side gear and thrust washer.
6. Align mating marks and position component case half. Draw assembly together with four bolts or capscrews equally spaced.
7. Check assembly for free rotation of differential gears and correct if necessary.
8. Install remaining bolts and capscrews. Tighten to correct torque and lock with wire.
9. If bearings are to be replaced, press squarely and firmly on differential case halves.

Install Bearing Cups In Carrier Leg Bores

1. Temporarily install bearing cups or threaded adjusting rings and bearing caps. Tighten capscrews to proper torque.
2. The bearing cups must be a hand push fit in the bores, otherwise the bores must be re-worked with a scraper or some emery cloth until a hand push fit is obtained. Use a blued bearing cup as a gauge and check the fits as work progresses.

Install Differential & Gear

1. After checking related parts, coat differential bearing cones and cups with rear axle lubricant.
2. Place cups over assembled bearing cones and position differential assembly in carrier.
3. Insert bearing adjusting nuts and turn hand-tight against bearing cups.
4. Install bearing caps in correct location as marked and tap lightly in position. *If bearing caps do not position properly, adjusting nuts may be cross-threaded. Remove caps and reposition the adjusting nuts. Forcing caps into position will result in irreparable damage to carrier housing or bearing caps.*
5. Install flat washers where used and stud nuts or capscrews. Tighten to correct torque.

Adjusting Differential Bearing Preload

1. Using dial indicator at back face of ring gear, Fig. 4, loosen bearing adjusting nut *on side opposite gear only* enough to notice end play on dial indicator.
2. Tighten same adjusting nut only enough to obtain zero end play.
3. Check gear for runout. If runout exceeds .008″, remove differential and check for cause.
4. Tighten adjusting nuts *one notch each* from zero end play to preload differential bearings.

Check Ring Gear Backlash

If original gears are reused the backlash should be the same as before disassembly. For new gears the new backlash should be from .006″ to .012″. Adjust backlash by moving ring gear only. This is done by backing off one adjusting ring and advancing the opposite ring the same amount.

Check Gear Tooth Contact

Temporarily bolt up the pinion and cage to the differential carrier and coat the ring gear teeth with oiled red lead, Fig. 5. When the pinion is rotated, the red lead is squeezed away by the contact of the teeth, leaving bare areas the exact size, shape and location of the contacts. If these contacts are not acceptable, shims must be added to or removed from the shim pack located between the pinion cage flange and differential carrier housing. It may be necessary to make several attempts at the

CAP SCREWS AND STUD NUTS

LOCATION	DIAMETER	NO. THREADS	TORQUE — LB. FT. MIN.	TORQUE — LB. FT. MAX.
Pinion cage	3/8″	16	27	35
	1/2″	13	81	104
	9/16″	12	116	149
	5/8″	11	160	205
Carrier to housing	7/16″	14	53	67
	1/2″	20	92	118
	5/8″	18	185	235
	3/4″	16	320	415
Differential case bolt	1/2″	20	92	118
	9/16″	18	130	167
	5/8″	18	185	235
	3/4″	16	320	415
Pinion shaft	7/8″	20	175	250
	1″	20	300	400
	1 1/4″	18	700	900
	1 1/2″	18	800	1100
	1 1/2″	12	800	1100
	1 3/4″	12	800	1100
Differential bearing adjusting nut lock	5/16″	18	16	20
	1/2″	13	81	104

DIFFERENTIAL BEARING CAP
CAP SCREWS OR STUD NUTS
(Earlier Axle Models Without Hardened Washers)

CAP SCREW OR STUD NUT DIAMETER	CAP SCREW OR COARSE STUD THREAD	STUD NUT OR FINE THREAD	TORQUE — LB. FT. MIN.	TORQUE — LB. FT. MAX.
5/8″	11	18	130	170
3/4″	10	16	230	300
7/8″	9	14	345	440
7/8″	14	14	380	485
1″	14	14	380	485

DIFFERENTIAL BEARING CAP
CAP SCREWS OR STUD NUTS
(Later Axle Models Employing Hardened Washers)

CAP SCREW OR STUD NUT DIAMETER	CAP SCREW OR COARSE STUD THREAD	STUD NUT OR FINE THREAD	TORQUE — LB. FT. MIN.	TORQUE — LB. FT. MAX.
5/8″	11	18	160	205
3/4″	10	16	290	370
7/8″	9	14	470	595
7/8″	14	14	510	655
1″	14	14	580	745

Fig. 9 Torque settings for Timken (Rockwell) single reduction axles. Specifications apply to parts coated with machine oil. For dry (or "as received") parts, increase torque 10%. For parts coated with multi-purpose gear oil, decrease torque 10%. Nuts on studs to use same torque as for driving the stud

right selection of shims to obtain the correct tooth contact, Fig. 6.

Sharper tooth contact impressions may be obtained by applying a small amount of resistance to the gear with a flat steel bar and using a wrench to rotate the pinion. When making adjustments, check the drive side of the ring gear teeth. Coast side contact should be automatically correct when drive side contact is correct. As a rule coating about 12 teeth is sufficient for checking purposes.

Install Thrust Block

1. Remove thrust block adjusting screw and lock nut.
2. Place thrust block on rear face of ring gear, Fig. 7, and rotate gear until hole in thrust block is aligned with adjusting screw hole.
3. Install adjusting screw and lock nut and tighten screw enough to locate thrust block firmly against back face of ring gear.
4. To secure correct adjustment of .010″ to .015″ clearance Fig. 8, loosen adjusting nut 1/4 turn and lock securely with nut.
5. Recheck to assure minimum clearance of .010″ during full rotation of ring gear.

Install Carrier In Housing

1. Install new gasket over housing studs.
2. Roll carrier into position on roller jack. Start carrier into housing with four flat washers and nuts equally spaced. *Do not drive carrier into housing with a hammer at the carrier stud flange. The flange may easily be distorted and cause severe oil leakage.*
3. Install lock washers and stud nuts on any studs under carrier housing offsets. *It is impossible to start these nuts after carrier is drawn into housing.*
4. Tighten the four nuts over flat washers alternately to draw carrier squarely into axle housing.
5. Remove nuts and flat washers. Install taper dowels, lock washers and stud nuts. Tighten to correct torque.
6. Connect universal at pinion shaft and install axle shafts.

Lubrication

1. Fill axle housing to correct level with specified lubricant.
2. Lubricate universal joint.
3. Jack up both rear wheels and operate vehicle in high transmission gear at approximately 25 mph for five minutes to assure satisfactory lubrication of all parts. *Do not operate with one wheel jacked up. Operation in this manner will result in over-heating the differential spider with resultant galling or shearing of spider pins.*

Double Reduction Axle

DISASSEMBLE

Remove Carrier From Housing

Front Mounted Carrier

The procedures for removing the differential carrier from the axle housing and removing the differential and drive gear from the carrier is the same as outlined for the single reduction axle. During the procedure which follows, refer to Figs. 1 and 2.

Top Mounted Carrier

1. Drain lubricant from housing.
2. Remove axle shaft stud nuts, lock washers and tapered dowels.

NOTE: To loosen the dowels, use a brass drift and hammer driving against the center of the axle shaft head, inside the circular driving lugs. Do not strike the circular driving lugs since the lugs may splinter or shatter.

3. Remove axle shafts and the front and rear propeller shafts.
4. Remove carrier to housing stud nuts and washers.
5. Loosen carrier from housing with a mallet and remove tapered dowels.
6. Pull carrier from housing with suitable lifting equipment.

Differential Case & Drive Gear

1. If original identification marks are not clear, center punch case halves for correct alignment on reassembly.
2. Cut lock wire, remove long bolts and separate case halves.
3. Remove spider, pinions, side gears and thrust washers.
4. Remove short bolts and separate gear from case half.
5. Remove differential bearings with suitable puller if necessary.
6. Remove drive gear rivets as directed for the Single Reduction Axle.

Pinion & Cage Assembly

1. Remove cage stud nuts or capscrews and lockwashers. Lift out cage assembly. *If cage is not free, tap it loose, using a soft drift on inner face of pinion or use puller screws in holes provided.*
2. Wire shim pack together to facilitate adjustment on reassembly.
3. Place pinion cage over carrier studs or secure with three capscrews. Hold flange and remove pinion shaft nut.
4. Press pinion shaft out of flange or yoke and cage.
5. Remove adjusting spacers or shims.
6. If necessary to renew the pinion or rear bearing, remove bearing with suitable puller.
7. Press front bearing and pinion shaft oil seal from cage. On assemblies where bearing cover is used, remove seal assembly from cover.

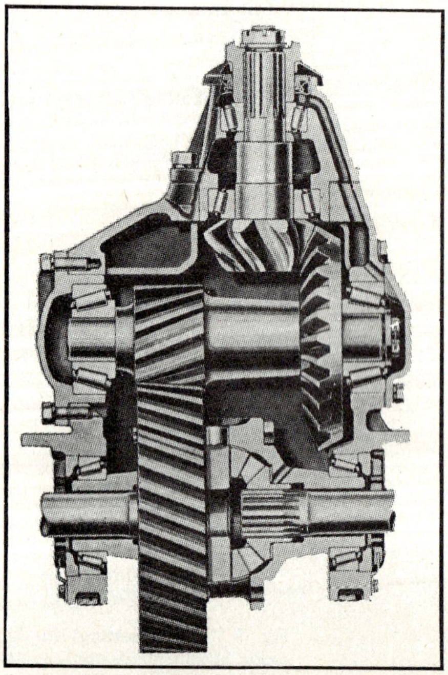

Fig. 1 Cutaway view of Timken (Rockwell) Double Reduction Axle (typical)

Cross Shaft Assembly

1. Remove stud nuts or capscrews and lockwashers from cross shaft bearing cage (side opposite drive gear).
2. Force off bearing cage with small pinch bar between back of gear and carrier housing (puller screw holes are also provided).
3. Wire shim pack together to facilitate adjustment on reassembly.
4. Remove cross shaft assembly.
5. Cut lock wire, remove capscrews and bearing retainer plate from gear end of shaft.
6. Press shaft from gear and bearing.
7. If necessary to replace bearing, remove bearing from opposite end of cross shaft with a suitable puller.
8. If necessary to replace bearing cups, remove cross shaft bearing cover or cage (drive gear side). Wire shim pack together to facilitate adjustment on reassembly.

REPAIRS

Precautions when making repairs on this axle are the same as outlined for the Single Reduction axle.

REASSEMBLY

Cross Shaft

1. Insert key and start shaft into drive gear in line with keyway or splines.
2. Press shaft squarely into drive gear. Gear must be firmly against cross shaft shoulder. *To facilitate installation, drive gear may be heated in oil from 200 to 250 degrees F.* Install capscrews where used.
3. Press bearing firmly against cross shaft shoulder on spur gear end.
4. Install bearing retainer plate and capscrews on drive gear end. Tighten to

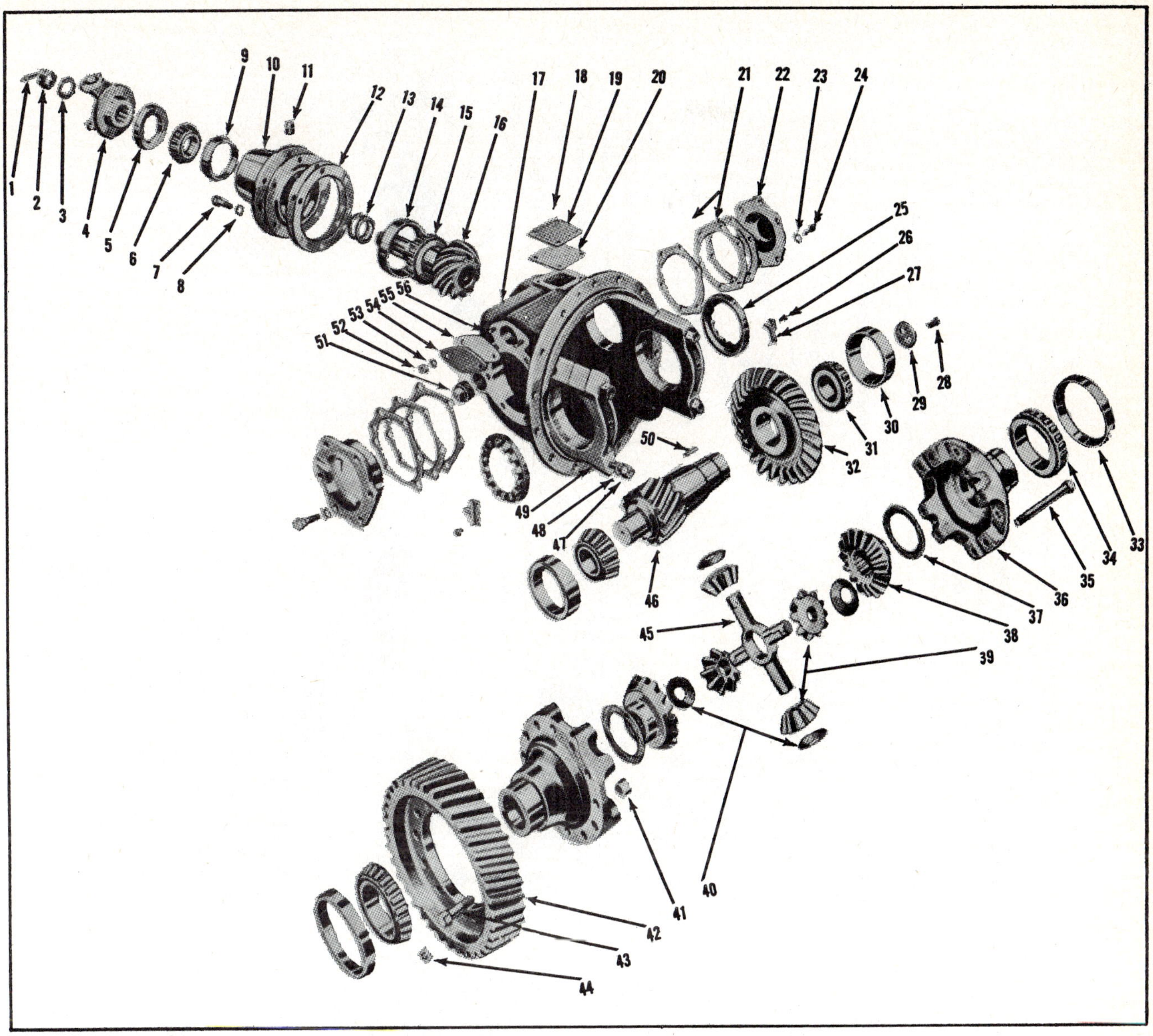

1. Cotter key
2. Yoke nut
3. Washer
4. Yoke and slinger assembly
5. Oil seal
6. Drive pinion outer bearing
7. Cap screw
8. Lock washer
9. Drive pinion outer bearing cup
10. Drive pinion bearing cage
11. Oil fill plug
12. Drive pinion shims
13. Drive pinion bearing spacer
14. Drive pinion inner bearing cup
15. Drive pinion inner bearing
16. Drive pinion
17. Differential carrier housing
18. Cap screw
19. Inspection hole cover
20. Gasket
21. Cross shaft cover shims
22. Cross shaft cover
23. Lock washer
24. Cap screw
25. Differential bearing adjusting nut
26. Cap screw
27. Differential bearing adjusting nut lock
28. Cap screw
29. Cross shaft bearing retainer
30. Cross shaft bearing cup
31. Cross shaft bearing
32. Drive gear
33. Differential bearing cup
34. Differential bearing
35. Differential case bolt
36. Differential case
37. Differential side gear thrust washer
38. Differential side gear
39. Differential pinion
40. Differential pinion thrust washer
41. Gear to case bolt nut
42. Helical spur gear
43. Gear to case bolt
44. Differential case bolt nut
45. Differential spider
46. Cross shaft (integral with spur pinion)
47. Differential bearing cap cap screw
48. Washer
49. Differential bearing cap
50. Square key
51. Oil level and fill plug
52. Stud nut
53. Lock washer
54. Shift shaft hole cover
55. Gasket
56. Stud

Fig. 2 Disassembled view of front mounted Timken (Rockwell) Double Reduction Axle

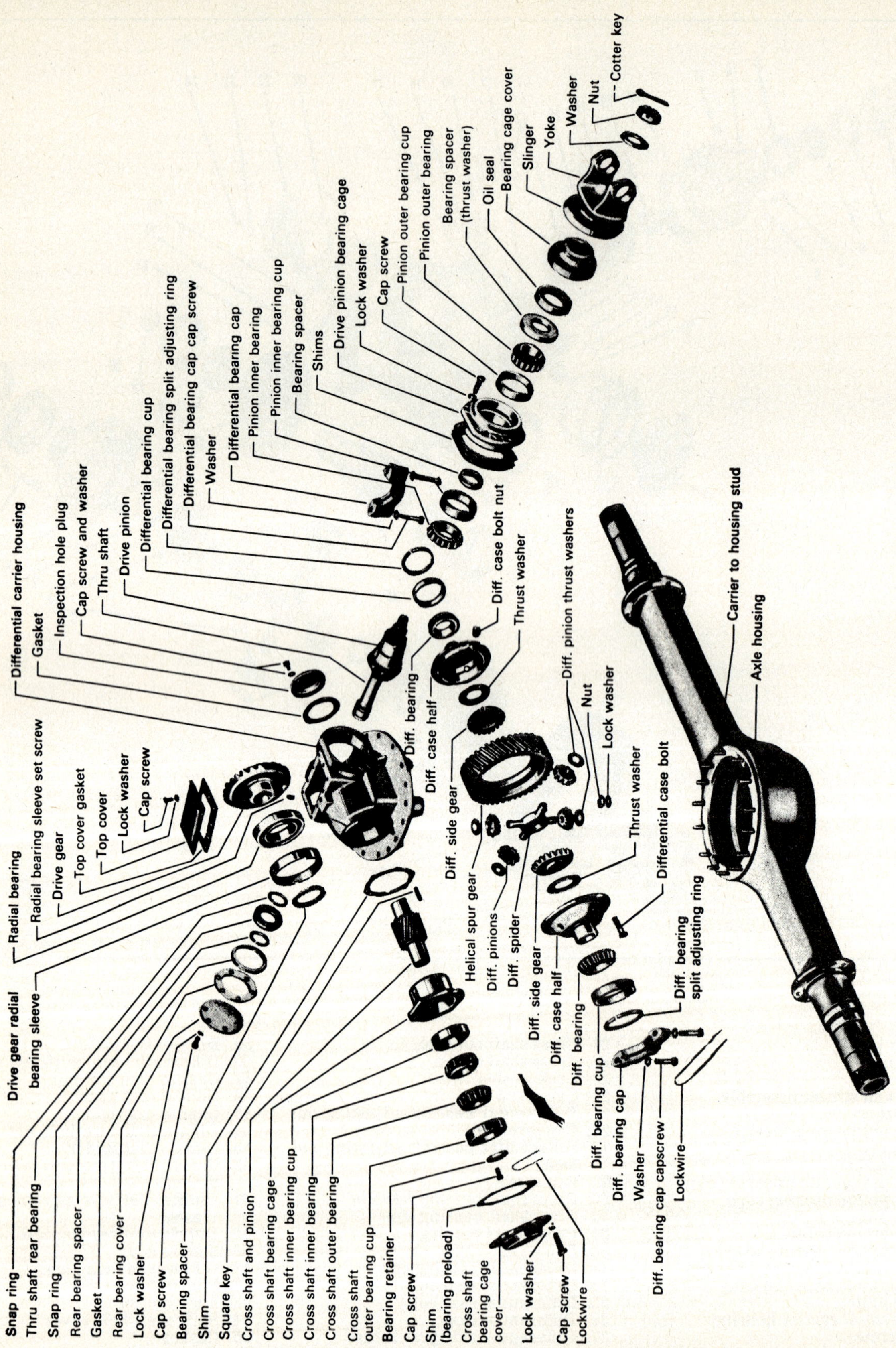

Fig. 3 Disassembled view of top mounted Timken (Rockwell) Double Reduction Axle, Series STD & STDD

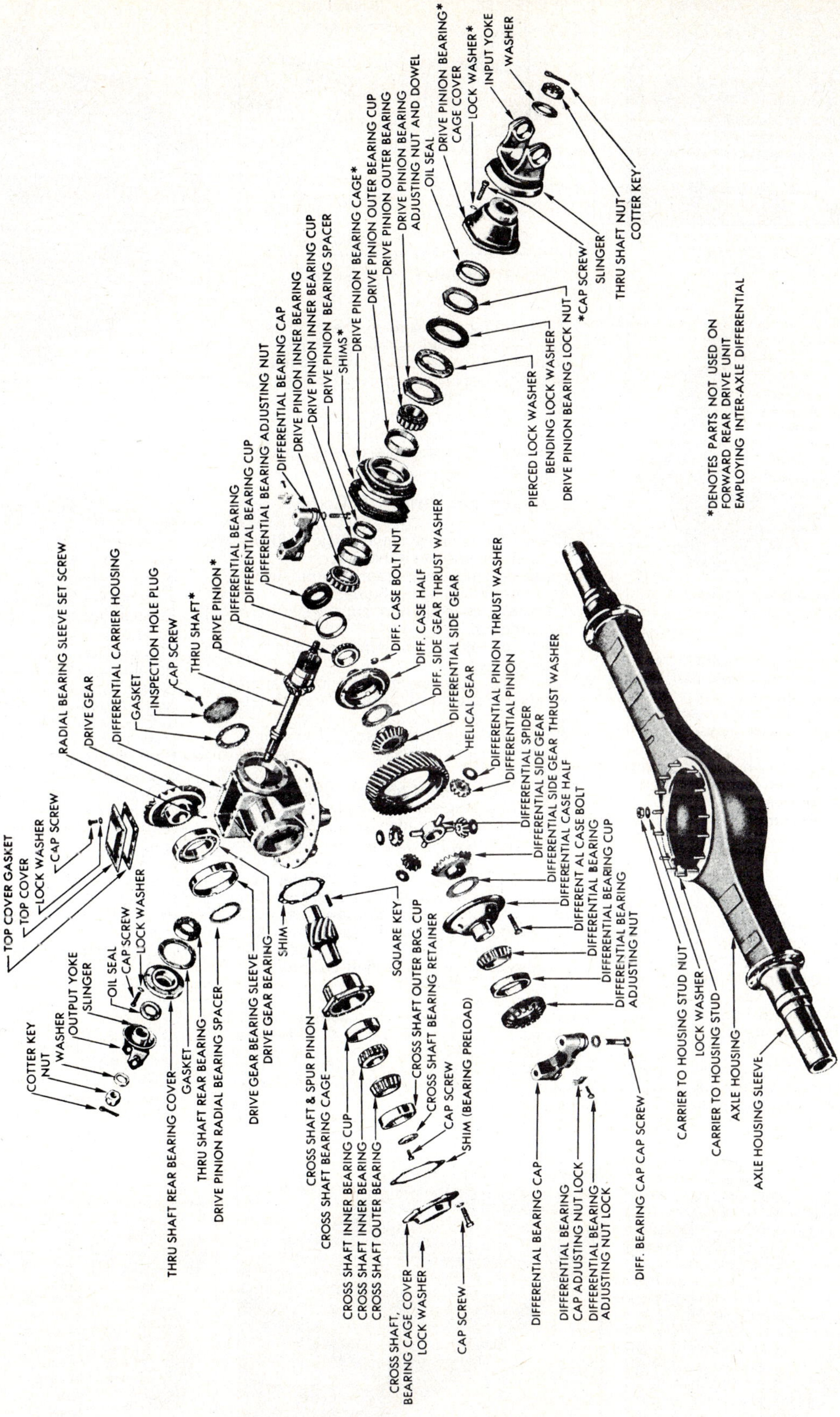

Fig. 3A Disassembled view of top mounted Timken (Rockwell) Double Reduction Axle, Series SRDD, SFDD, SUD & SUDD

Fig. 4 Checking preload torque on cross shaft bearing

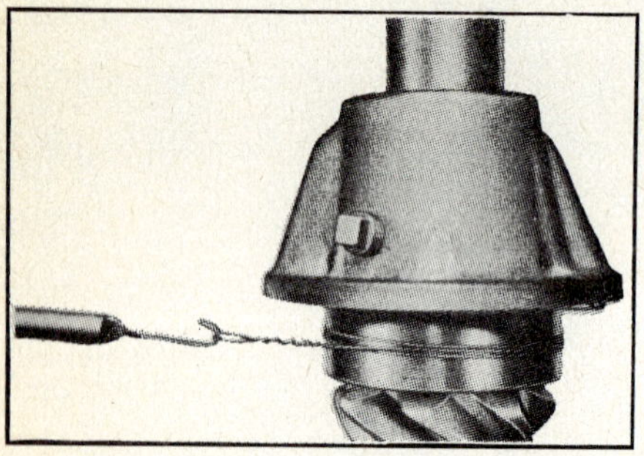

Fig. 4A Checking preload torque on pinion bearing

correct torque and lock with soft wire.

5. Lubricate cross shaft bearings and cups with light machine oil.
6. If bearing cover or cage on drive gear side has been removed, replace using the correct shim pack. Tighten stud nuts or capscrews to correct torque. If cover has not been removed, retighten stud nuts or capscrews to correct torque.
7. Ease cross shaft assembly past differential bearing supports and position in bearing cup.
8. Press bearing firmly against drive gear, using a suitable sleeve.
9. Install correct shim pack and start cross shaft bearing cage into carrier housing.
10. Tap bearing cage into position with soft mallet. Install lockwashers and stud nuts or capscrews and tighten to correct torque.
11. Rotate assembly several revolutions before checking bearing preload to assure full bearing contact.
12. Check cross shaft bearing preload torque. Wrap soft wire around pinion and pull on a horizontal line with a spring scale, Fig. 4. Assuming spur gear diameter to be 4″, the radius would equal 2″. Thus, five pounds pull on the scale would equal 10 pound-inches bearing preload torque. Use rotating, not starting torque.
13. To obtain correct preload torque of 5 to 15 pound-inches, add or remove shims from pack under bearing cage on side opposite bevel gear.

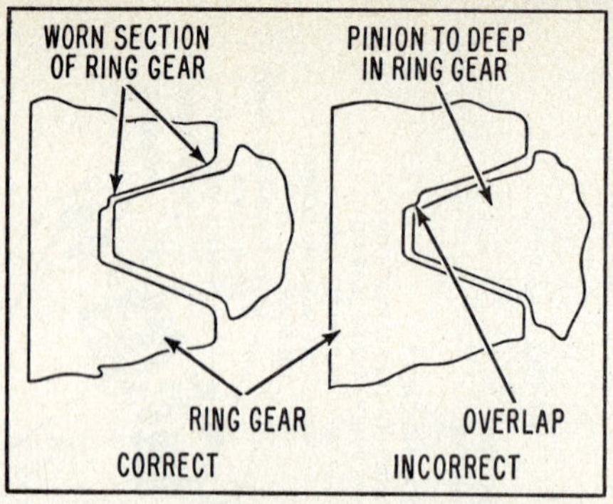

Fig. 5 Adjusting drive gear lash

Assemble Pinion & Cage

1. Press rear bearing squarely and firmly against pinion gear shoulder.
2. Press bearing cups squarely and firmly against pinion cage shoulder.
3. Lubricate bearing and cups with light machine oil.
4. Insert pinion and bearing assembly in pinion cage.
5. Install selective spacer over pinion shaft with bevel side toward pinion shaft shoulder.
6. Press front bearing squarely and firmly against selective spacer with a suitable sleeve.
7. Rotate cage several revolutions to assure normal bearing contact.
8. While in press under pressure, check pinion bearing preload torque. Wrap soft wire around pinion cage and pull on a horizontal line with a spring scale, Fig. 4A. If a press is not available, the preload may be checked with the flange or yoke installed and the nut tightened to the correct torque. The correct pressures or torques for checking pinion bearing preload are as follows:

Thread Size	Pressure	Torque, Lb. Ft.
1″ × 20	6 tons	300–400
1¼″ × 18	11 tons	700–900
1½″ × 12	14 tons	800–1100
1½″ × 18	14 tons	800–1100
1⅓″ × 12	14 tons	800–1100

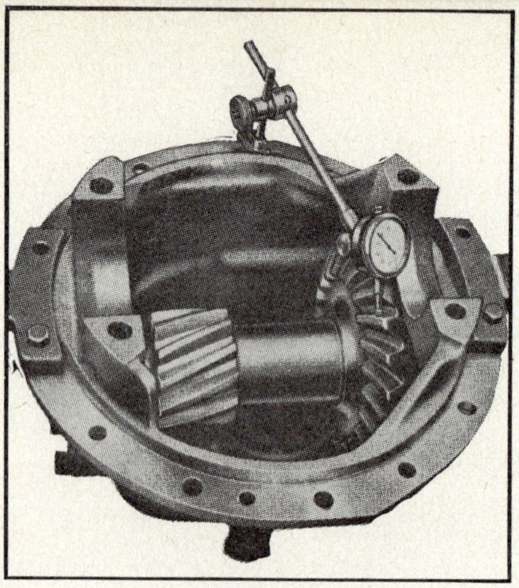

Fig. 6 Checking drive gear lash with dial indicator

9. Assuming pinion cage diameter to be 6″ the radius would be 3″. Four pounds pull on the scale would equal 12 pound-inches bearing preload torque. Use rotating, not starting torque. If rotating torque is not within 5 to 15 pound-inches, use a thinner spacer to increase or a thicker spacer to decrease bearing preload.
10. Press flange or yoke firmly against pinion forward bearing.
11. Place pinion and cage assembly over carrier studs or secure with capscrews. Hold flange and tighten pinion shaft nut to correct torque. If rotating torque is not within 5 to 15 pound-inches, repeat foregoing procedure.
12. Hold flange and remove pinion shaft nut and yoke or flange.
13. Lubricate pinion shaft oil seal. Coat outer edge of seal body with a non-hardening sealing compound and press seal firmly against cage or cover shoulder. Install bearing cover where used with capscrews.
14. Reinstate flange or yoke. Tighten pinion shaft nut to correct torque.

Install Pinion & Cage

1. The oil passage holes in carrier housing, gasket, shim pack and pinion bearing cage must be aligned.
2. Install cross shaft assembly as outlined previously.
3. Install correct shim pack.
4. Install pinion cage assembly, lockwashers and stud nuts or capscrews. Tighten to correct torque.

Fig. 7 Satisfactory tooth contact (gears unloaded)

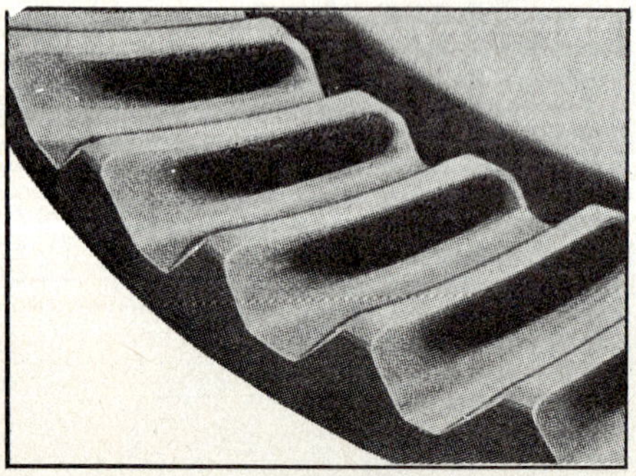

Fig. 8 Satisfactory tooth contact (gears loaded)

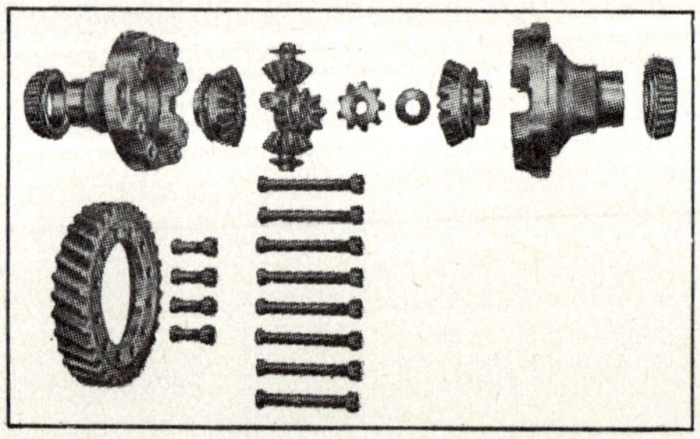

Fig. 9 Differential and gear assembly

WRENCH TORQUE FOR PINION SHAFT NUTS			
DIAMETER OR SIZE	THREADS PER INCH	TORQUE—LB. FT.	
		MIN.	MAX.
7/8"	20	175	250
1"	20	300	400
1 1/4"	18	700	900
1 1/2"	12	800	1100
1 1/2"	18	800	1100
1 3/4"	12	800	1100

DIFFERENTIAL BEARING CAP CAP SCREWS OR STUD NUTS (Earlier Axle Models Without Hardened Washers)				
CAP SCREW OR STUD NUT DIAMETER	CAP SCREW OR COARSE STUD THREAD	STUD NUT OR FINE THREAD	TORQUE—LB. FT.	
			MIN.	MAX.
5/8"	11	18	126	140
3/4"	10	16	225	250
7/8"	9	14	330	370
7/8"	14	14	375	415
1"	14	14	375	415

DIFFERENTIAL BEARING CAP CAP SCREWS OR STUD NUTS (Later Axle Models Employing Hardened Washers)				
CAP SCREW OR STUD NUT DIAMETER	CAP SCREW OR COARSE STUD THREAD	STUD NUT OR FINE THREAD	TORQUE—LB. FT.	
			MIN.	MAX.
5/8"	11	18	160	180
3/4"	10	16	290	320
7/8"	9	14	470	520
7/8"	14	14	510	570
1"	14	14	570	630

Fig. 10 Torque specifications for Timken Double Reduction Axles

Checking Drive Gear Lash

Generally, if original gears are being installed, red leading of teeth will not indicate the same contact as new gears and can be misleading. Gears that have been in service for long periods form running contacts due to wear of teeth. Therefore, the thickness of the original shim pack plus approximately .015" additional shim stock should be maintained to check gear lash. In the event that gear lash is in excess of maximum tolerance, as stated under gear adjustment, reduce gear lash only in the amount that will avoid overlap of the worn tooth section, Fig. 5. Bevel gear lash can only be reduced to a point of maintaining smooth rotation of bevel gears.

Smoothness or roughness can be noted by rotating the bevel gear. If a slight overlap, Fig. 5, takes place at the worn tooth section, rotation will be rough. Generally with the original gears, tone should be satisfactory. Fig. 6 shows how a dial indicator is used to check bevel gear lash.

Drive Gear Adjustment for Correct Tooth Contact

Checking tooth contact is accomplished by means of oiled red lead applied lightly to the drive gear teeth. When the pinion is rotated, the red lead is squeezed away by the contact of the teeth, leaving bare areas the exact size, shape and location of the contacts.

With adjustments properly made (pinion at correct depth and backlash set at .010") the contacts shown in Figs. 7 and 8 will be procured. The area of contact favors the toe and is centered between the top and bottom of the tooth.

The hand-rolled pattern shown in Fig. 7, will result in a pattern centered in the length of the tooth when the gears are under load, shown in Fig. 8. The loaded pattern will be almost full length and the top of the pattern will approach the top of the gear tooth.

After the correct contacts shown have been established with a backlash of .010", open the backlash to measure .020 to .026."

Recheck Cross Shaft Bearing Preload

1. Remove pinion and cage, keeping shim pack intact.
2. Tighten cross shaft bearing cage and cover stud nuts or capscrews to the correct torque.
3. Recheck cross shaft bearing preload torque.
4. Add or remove shims from pack under cross shaft bearing cage on the side opposite drive gear only as necessary to obtain correct bearing preload torque.
5. Replace pinion cage and install lockwashers, nuts or capscrews.
6. Tighten bearing cage stud nuts to correct torque.
7. Reinstall pinion and cage over shim pack determined by correct gear adjustment.
8. Install lockwashers and stud nuts and tighten to correct torque.

Drive Gear Backlash Adjustment

1. Adjust gear backlash to .020 to .026" (regardless of what markings might be etched on the gear set) by transposing cross shaft bearing cage or cover shims.
2. To move gear away from pinion, remove shims from pack under cross shaft bearing cage on opposite side of gear and add shims of equal thickness to pack under cross shaft bearing cover or cage on gear side. Shims should be transposed in this manner to maintain established preload.
3. To move gear toward pinion, remove shims from pack from under cross shaft bearing cover or cage on the gear side and add shims of equal thickness to pack under cross shaft bearing cage on the opposite side.

Assemble Differential & Spur Gear

When new spur gears or a new differential case is installed, the differential case holes must be line-reamed with the gear in order to assemble, using correct size bolts or rivets.

Type Using Long and Short Bolts, Fig. 9

1. Join spur gear to either differential case half with short bolts.
2. Lubricate differential case inner walls and all components parts with axle lubricant.
3. Install thrust washer and side gear in one of the case halves. Place spider with pinions and thrust washers in position. Install other side gear and thrust washer.
4. Check assembly for free rotation and correct if necessary.
5. Install remaining bolts and tighten to correct torque.
6. Install lock wire.
7. Press differential bearings squarely and firmly on differential case halves.

Type Using Rivets or Short Bolts

1. Lubricate differential case inner walls and all component parts with axle lubricant.
2. Install thrust washer and side gear in one of the differential case halves.
3. Position spur gear on case half.

Location	Diameter	Threads Per Inch	Torque Lb. Ft. Min.-Max.
Carrier to Housing	7/16"	14	52-58
	7/16"	20	52-58
	1/2"	13	82-91
	1/2"	20	82-91
	5/8"	11	160-180
	5/8"	18	160-180
Pinion Cage	3/8"	16	33-37
	7/16"	14	52-58
	7/16"	20	52-58
	1/2"	13	82-91
	1/2"	20	82-91
	9/16"	12	116-129
	9/16"	18	116-129
	5/8"	11	160-180
Cross Shaft Bearing Cage and Cover	1/2"	13	82-91
	1/2"	20	82-91
	9/16"	12	116-129
	9/16"	18	116-129
	5/8"	11	160-180
Cross Shaft Bearing Lock	7/16"	14	42-45
	9/16"	12	92-101
Differential Bolts	3/8"	16	33-37
	7/16"	14	52-58
	1/2"	20	92-102
	9/16"	18	130-145
	5/8"	18	185-205
	3/4"	16	320-360
Adjusting Nut Lock	5/16"	18	15-17
Inspection Cover	3/8"	16	26-29
Shift Unit (Mounting)	3/8"	16	26-29
Shift Unit Lock Nut, Set Screw and Clamp Screw	3/8"	24	30-33
	7/16"	20	30-33
Shift Unit Travel Limiting Screws	1/2"	13	55-60
	5/8"	11	30-35

Fig. 11 Torque specifications for capscrews or stud nuts on front-mounted Timken (Rockwell) Double Reduction Axles. Torques given apply to parts coated with machine oil. For dry (or "as received") parts, increase torque 10%. For parts coated with multi-purpose gear oil, decrease torque 10%. Nuts on studs to use same torque as for driving the stud

LOCATION	DIAMETER	NO. THREADS	TORQUE LB. FT. MIN.	TORQUE LB. FT. MAX.
INTER-AXLE DIFFERENTIAL GROUP				
Case (Input) Shaft Nut	1-3/4 1-3/4	12 20	300	400
Cover Capscrews	7/16	14	60	77
Housing to Carrier Capscrews	1/2	13	93	120
Case Bolts and Nuts	1/2	20	105	135
Shift Yoke Adjusting Screw Jam Nut	3/8	24	43	56
Shift Yoke Bolt and Nut	3/8	24	43	56
SHIFT UNIT AND HOUSING GROUP				
Housing Capscrews and Studs	3/8	16	27	35
Shift Shaft Stopscrew Nut	1/2	20	75	96
DRIVE PINION GROUP				
Bearing Adjusting Nut	3-3/8	12	800	1000
Bearing Jam Nut	3-3/8	12	1000	1200
THROUGH SHAFT GROUP				
Rear Bearing Cage Capscrews	3/8	16	38	49
Rear Bearing Cage Cover Capscrews	3/8	16	38	49
Rear (Output) Nut	1-1/2 1-3/4	18 12	300	400
CROSS SHAFT GROUP				
Cover (L.H.) Capscrews	3/8	16	27	35
Bearing Cage Cover (R.H.) Capscrews	1/2 9/16	13 12	66 130	85 170
Bearing Retainer (Washer) Capscrews	7/16 9/16	14 12	60 130	77 170
MAIN DIFFERENTIAL AND CARRIER GROUP				
Bearing Adjusting Nut Lock Capscrews	5/16	18	16	20
Bearing Cap Capscrews	3/4 7/8	10 9	290 480	320 520
Case and Gear to Case Bolts and Nuts	9/16	18	148	190
Top Cover Capscrews	3/8	16	38	49

Fig. 12 Torque specifications for capscrews or stud nuts on top mounted Timken (Rockwell) Double Reduction Axles

NOTE: Torques given apply to parts coated with machine oil. For dry (or "as received") parts, increase torque 10%. For parts coated with multi-purpose gear oil, decrease torque 10%. Nuts on studs to use same torque as for driving the stud

4. Place spider with pinions and thrust washers in position and install other side gear and thrust washer.
5. Align mating marks and position other case half.
6. Draw either bolted or riveted assembly together with four bolts equally spaced and check for free rotation of differential gears.
7. Install remaining bolts or rivets together as required. Install rivets as directed for the Single Reduction Axle.

Final Assembly & Installation

The installation of bearing cups in carrier leg bores, of differential assembly in carrier and of carrier in axle housing is the same as outlined for the Single Reduction Axle.

Two Speed Double Reduction Axles

Fig. 1 shows the various types of Timken (Rockwell) two speed double reduction drive axles. The first reduction is through a hypoid pinion and gear. The hypoid pinion and gear set operates in conjunction with either of two sets of helical gears and pinions of different ratios. This second reduction is selective between high and low ratio.

The hypoid pinion is mounted on two tapered roller bearings in a pinion cage. The hypoid gear is locked on the cross shaft by a press fit and key. The cross shaft is mounted on tapered roller bearings and carries two free-rolling helical pinions. Both helical pinions engage helical gears attached to the differential, which is mounted on tapered roller bearings.

To establish and maintain differential bearing preload, these axles may incorporate any one of the three types of carrier leg and cap machining shown in Figs. 2, 3 and 4.

FRONT MOUNTED DRIVE UNIT

TOP MOUNTED DRIVE UNIT

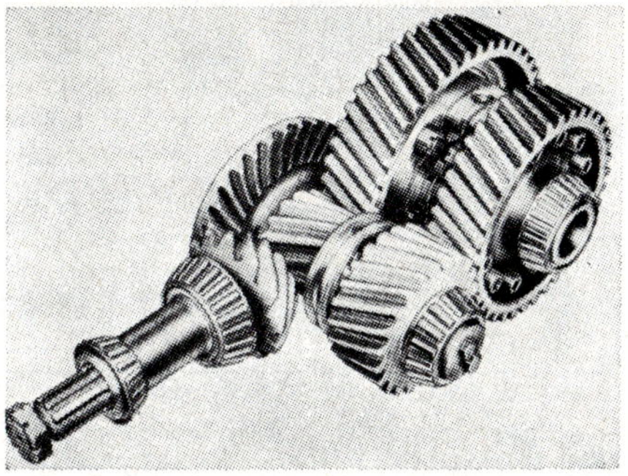

FRONT MOUNTED GEAR TRAIN

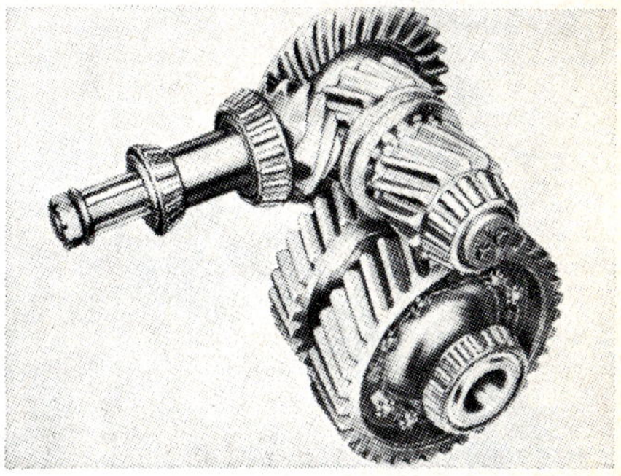

TOP MOUNTED GEAR TRAIN

Fig. 1 Timken (Rockwell) two-speed double reduction axles

REMOVING CARRIER FROM AXLE HOUSING

Front Mounted Type

1. Drain lubricant from housing.
2. Remove axle shafts.
3. Disconnect U-joint at pinion shaft.
4. Disconnect power shift unit.
5. Remove carrier-to-housing stud nuts, leaving two top nuts partially screwed on studs to prevent carrier from falling.
6. Break carrier loose from axle housing with a soft mallet. Remove tapered dowels if present.
7. Place a roller jack under carrier, remove two top nuts and work carrier free, using puller screws in holes provided. A small pinch bar may be used to straighten carrier in housing bore. However, be sure the end is rounded to prevent indenting carrier flange circle.

Top Mounted Type

1. Perform Steps 1 through 5 as outlined above.
2. Then break carrier loose from housing with soft mallet and remove tapered dowels. If necessary, back out studs.
3. Pull carrier straight out of housing with chain falls, boom, "A" frame or puller screws. A small pinch bar may be used to straighten carrier in housing bore. However, be sure end of punch is rounded to prevent indenting carrier flange circle.

DISASSEMBLE, FIG. 5

1. Center punch one carrier leg and bearing cap to identify for proper reassembly.
2. Remove capscrews and adjusting ring locks (where used).
3. Remove bearing cap, screws, bearing cups and adjusters.
4. Lift out differential and gear.
5. If original identification marks are not clear, center punch differential case halves for correct alignment upon reassembly.
6. Separate case halves and remove differential components.
7. Remove short bolts or rivets and separate

Fig. 2 Type with threaded adjusting rings on both sides. This type positions differential as well as preloading bearings by moving threaded rings in or out

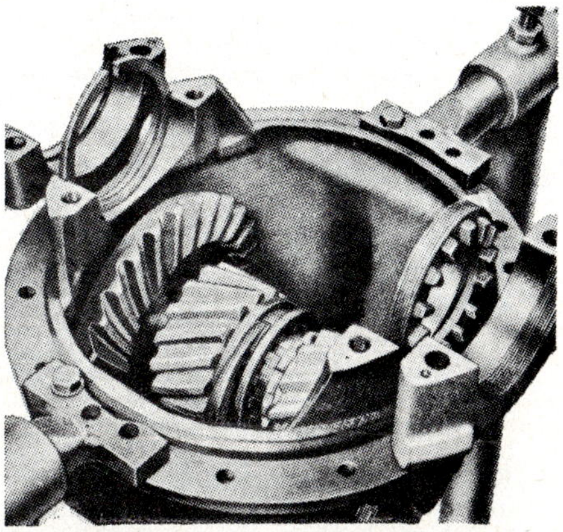

Fig. 3 Type with threaded ring on one side and split ring on opposite side. Split ring and groove locates differential whereas threaded ring is moved in or out to preload bearings

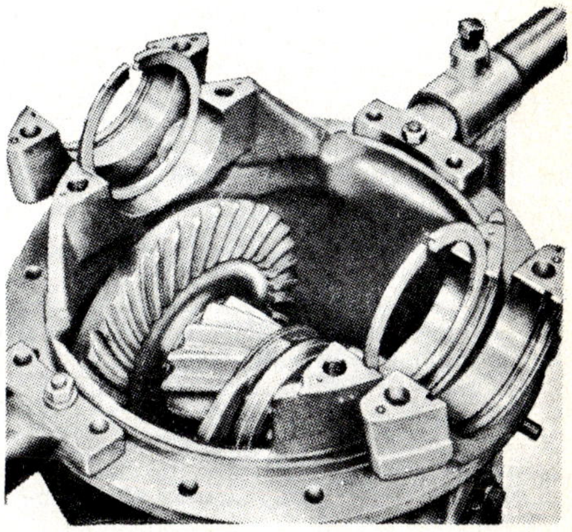

Fig. 4 Type with split rings on both sides. Bearing preload is established by using various thickness split rings. Differential positioning is automatic

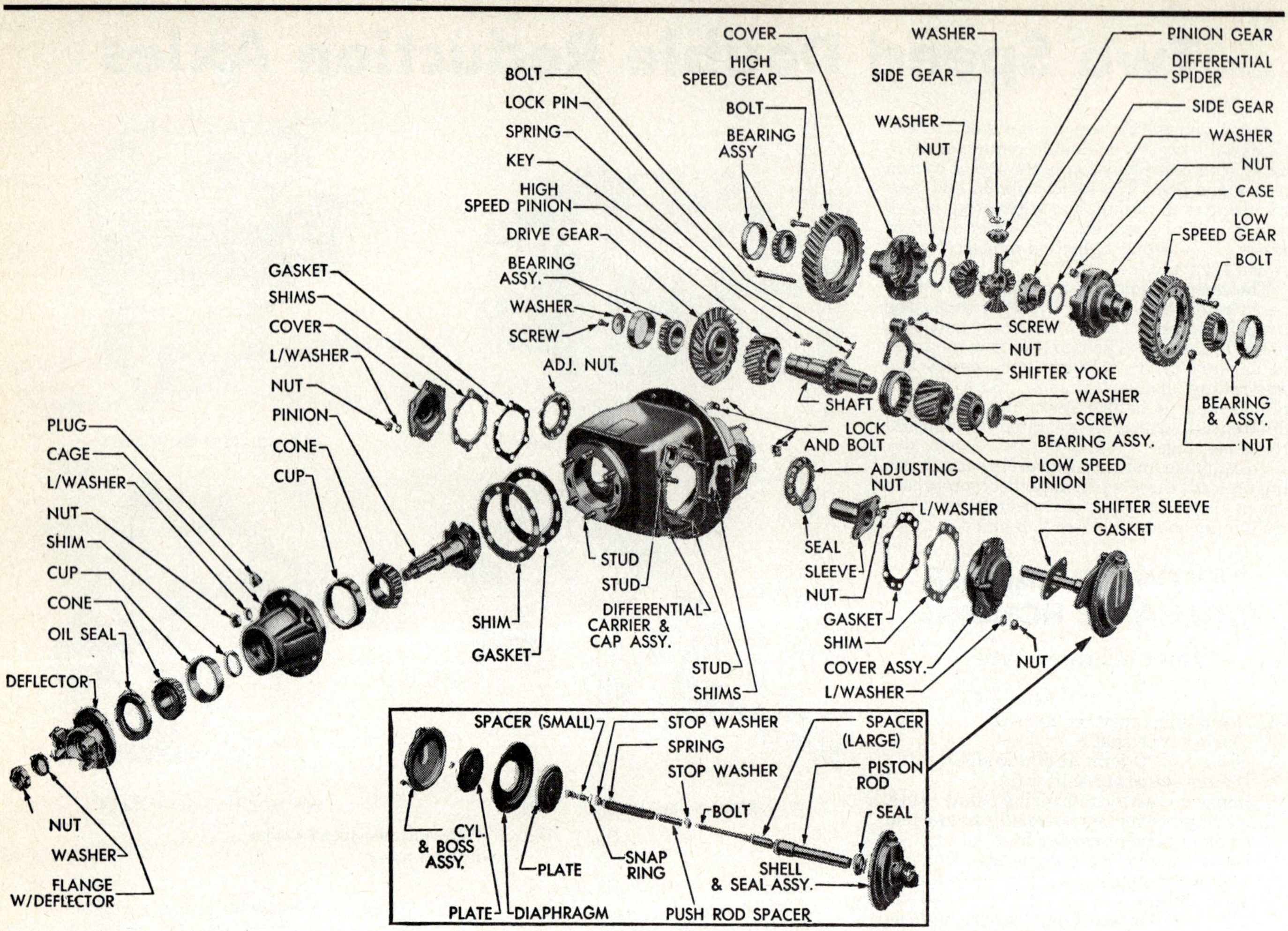

Fig. 5 Exploded view of Timken (Rockwell) two speed double reduction axle of the front mounted type with vacuum shift unit

gears from case when gear or case is to be replaced. Remove differential bearings with suitable puller when desired.

8. Hold flange and remove pinion shaft nut. Remove attaching capscrews and lift out pinion cage. If cage is not free, tap loose with a soft drift on inner face of pinion or use puller screws in holes provided.
9. Wire pinion cage shims together to facilitate adjustment on reassembly.
10. Press pinion shaft out of flange and cage. Remove adjusting spacers.
11. If necessary to replace pinion or rear bearing, remove bearing with suitable puller.
12. Press front bearing and pinion shaft oil seal from cage.

Remove Side Mounted Shift Unit

1. Remove lockwire at fork locking screw.
2. Loosen lock nut on shift fork lock screw and remove screw and nut.
3. Remove two shift unit stud nuts and washers.
4. Remove shift unit and shaft assembly and lift out shift fork.
5. Tap sleeve from carrier with a soft mallet.
6. Wire shim pack to keep intact for proper adjustment on reassembly.

Remove Front Mounted Shift Unit

1. Remove four shift unit adapter screws and washers.
2. Remove shift unit and adapter assembly from carrier.

Remove Cross Shaft

1. Remove capscrews from cross shaft bearing cage (side opposite hypoid gear).
2. Force out bearing cage with small pinch bar between back of hypoid gear and carrier housing (puller screw holes also provided).
3. Wire shim pack together to facilitate adjustment on reassembly.
4. On models with front mounted shift units, remove lockwire and set screw holding shift fork on shift shaft. Slide shaft out and remove fork from inside housing.
5. Thread out cross shaft assembly. Some wide ratio carriers will require removal of high speed pinion and bearing from cross shaft before removing cross shaft from carrier. On these units, proceed as follows:
6. Raise cross shaft through cage bore. Remove bearing retaining plate. Use puller to remove bearing. Remove high speed helical pinion from cross shaft. Thread out cross shaft. *Some wide range carriers will permit using cross shaft to that high speed pinion and bearing can be pulled off together.*

Disassemble Cross Shaft

1. Remove capscrews and retaining plates.
2. Press cross shaft from hypoid gear and bearing. Lift off helical pinion and remove shift collar, poppets and springs.
3. Replace shift collar and press shaft through remaining helical pinion and bearing.
4. If necessary to replace cross shaft bearing cups, remove cross shaft bearing cover (hypoid gear side). Wire shim pack together to facilitate adjustment on reassembly.
5. Remove cups from cages with suitable puller. Cup in cover side of carrier can be removed by tapping.

Service Notes

1. Replace all worn or damaged parts. Hex nuts with rounded corners, all lock washers, oil seals and gaskets should be replaced.
2. Remove nicks, mars or burrs from machined or ground surfaces. Threads must be clean and free to obtain accurate adjustment and correct torque. A fine mill file or India stone is suitable for this purpose. Studs must be tight prior to reassembling parts.
3. All bronze bushed differential pinions should be ball burnished after bushing

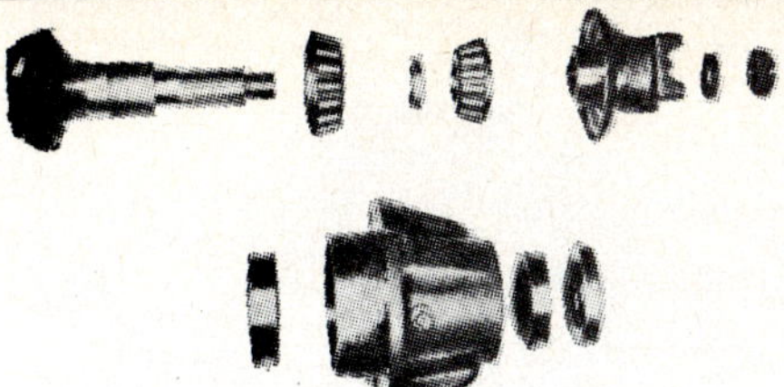

Fig. 6 Hypoid pinion and cage assembly

installation. Install bushing with small stepped drift. The small O.D. should be .010" smaller than the bushing burnished I.D. and 1½ times bushing length. Always install bushings so end is even with the I.D. chamfer or about 1/16" below the spherical surface.

4. Burrs, caused by lock washers at the spot face of stud holes of cages and covers should be removed to assure easy reassembly of these parts.

REASSEMBLE CARRIER

Hypoid Pinion & Cage

1. Refer to Fig. 6. Press rear bearing squarely and firmly against pinion shaft shoulder.
2. Press bearing cups squarely and firmly against pinion cage shoulder.
3. Lubricate bearing and cups with light machine oil.
4. Insert pinion and bearing in cage.
5. Install selective single or combination spacers over pinion shaft.
6. Install single spacer with O.D. bevel toward outer bearing cone. Install two combination spacers with I.D. bevels opposite each other and large flat faces together. *Spacers must be installed in this manner to prevent interference with outer bearing cone roller cage.*
7. Press front bearing squarely and firmly against selective spacer with suitable sleeve.
8. Rotate cage several revolutions to assure normal bearing contact.
9. While in press under pressure, check pinion bearing preload torque. Wrap strong cord around pinion cage and pull on a horizontal line with a spring scale, Fig. 7.
10. If a press is not available, pinion nut may be tightened to the correct torque and preload checked as follows:

Pinion Shaft Thread Size	Nut Torque Ft. Lbs.	Equivalent Pressure In Tons
1" × 20	300–400	6
1¼" × 18	700–900	11
1½" × 12	800–1100	14
1½" × 18	800–1100	14
1¾" × 12	800–1100	14
2" × 16	800–1100	14
2⅛" × 12	800–1100	14

Example: Assuming pinion cage diameter to be 6", the radius would be 3", and with 4 pounds pull on scale would equal 12 inch pounds bearing preload torque.

11. Read rotating pounds pull on scale (not starting pounds pull). If rotating torque is not within 5 to 15 inch pounds, use a thinner spacer to increase or a thicker spacer to decrease preload torque.
12. Press flange or yoke firmly against forward pinion bearing.
13. Use carrier as a convenient holding fixture for pinion and cage assembly. Hold flange and tighten shaft nut to correct torque.
14. Recheck pinion bearing preload torque. If rotating torque is not within 5 to 15 inch pounds repeat foregoing procedure.
15. Hold flange and remove pinion shaft nut and yoke or flange.
16. Lubricate pinion shaft oil seal. With suitable sleeve, press seal firmly against bearing cage shoulder.
17. Reinstall yoke or flange and tighten nut to correct torque.
18. Remove pinion and cage assembly from carrier.

Cross Shaft Assembly

1. Refer to Figs. 8 and 9. Lubricate inner bearing surfaces on helical pinions with axle lubricant.
2. Position high or low speed helical pinion (depending on type of carrier) on cross shaft with splined row of teeth toward cross shaft teeth.
3. Install key and start cross shaft and pinion in hypoid gear in line with keyway.
4. Press shaft squarely into hypoid gear. *Cross shaft of all wide range ratio carriers and a number of conventional ratio carriers employ no hypoid gear shoulder. Extreme care must be exercised to assure pinion end play of .010" minimum.*
5. Check helical pinion end play with feeler gauge, Fig. 10. End play must be .010" minimum to .026" maximum. *On conventional ratio carriers where definite gear stop is provided by cross shaft shoulder, it is still necessary to have this end play. To facilitate installation, hypoid gear may be heated in oil to 200–250°.*
6. Press bearing firmly against hypoid gear using suitable sleeve. Install bearing retaining plate and capscrews.
7. Install poppets and springs and coat with axle lubricant.
8. Align all three tapered shift collar splines with poppets and position with shift collar marked "low side" toward low speed helical pinion. Due to tooth spacing, some shift collars can be assembled in one position only. *Some wide range ratio carriers will require installing cross shaft in carrier before assembly of high speed helical pinion and bearing on cross shaft. On these units skip Steps 9 to 12.*
9. Install opposite helical pinion (high or low speed depending on carrier) with splined teeth toward cross shaft teeth.
10. Press cross shaft bearing squarely and firmly against cross shaft shoulder using suitable sleeve.
11. Check low speed pinion end play which must be a minimum of .010". *Less than .010" may result in seizure.*
12. Install bearing retaining plates and capscrews.

Install Cross Shaft In Carrier

Gaskets at the pinion cage, cross shaft covers

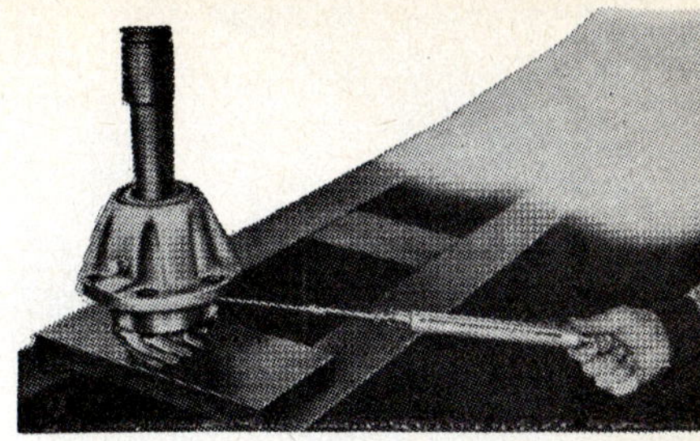

Fig. 7 Checking pinion bearing preload

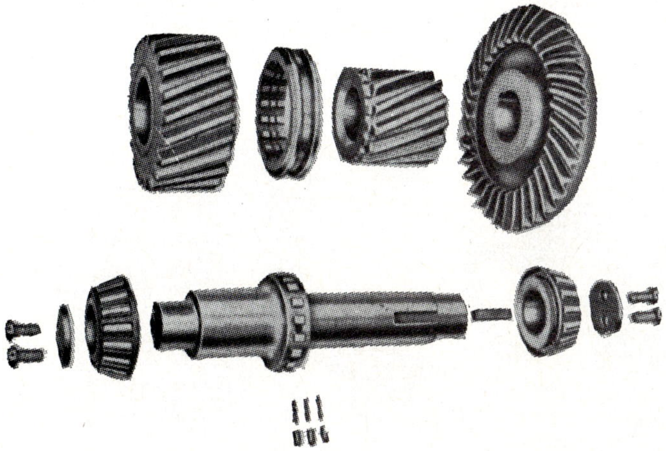

Fig. 8 Wide range ratio and some conventional ratio cross shaft assembly

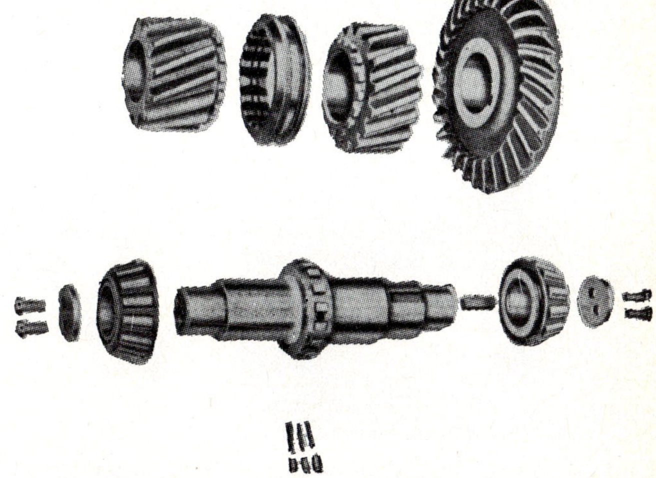

Fig. 9 Most conventional ratio cross shaft assemblies

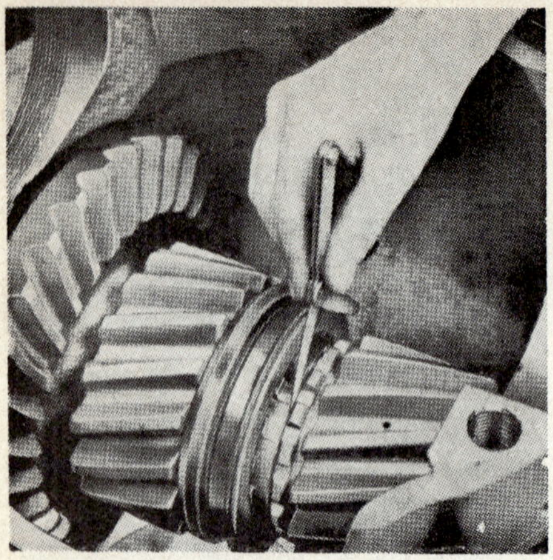

Fig. 10 Checking helical pinion end play with feeler gauge

Fig. 11 Checking cross shaft bearing preload

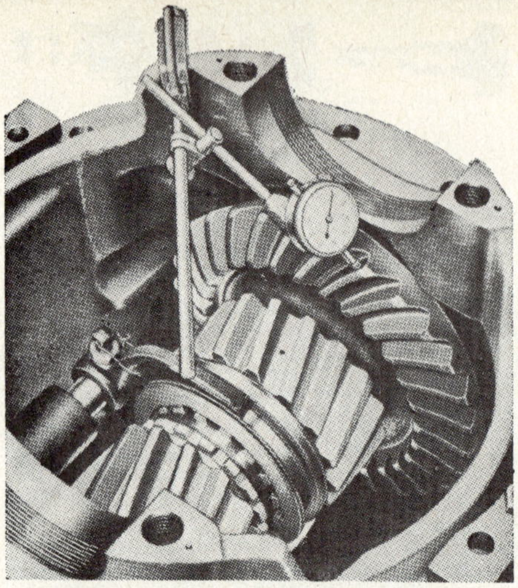

Fig. 12 Using dial indicator to determine gear lash

and cages have not been used in units produced for several years. Therefore, when assembling a drive unit formerly using gaskets, add approximately .015" shim stock to original shim pack to maintain correct bearing preload and gear adjustment. Thin shims should be located on both sides of pack to obtain maximum sealing ability.

Wide Range Ratio Carriers

On wide range ratio carriers requiring assembly of high speed helical pinion and bearings after cross shaft is in the carrier, proceed as follows:

1. Lubricate cross shaft bearings and cups with light machine oil.
2. Install cross shaft assembly (minus high speed helical pinion) into carrier with cage or cover of hypoid side removed.
3. Install high speed pinion and set up in a press so that bearing retaining plate at hypoid side of cross shaft may be rested on a support at cover or cage opening.
4. Press bearing opposite hypoid gear onto shaft. Install bearing retaining plate and capscrews.
5. Install bearing cage or cover with shim pack on hypoid side of carrier. Replace capscrews and lock washers, then tighten to proper torque. *Should a press not be available, cross shaft bearing opposite gear side may be installed with a suitable driver placed against the inner race of bearing providing gear side of shaft is blocked up against bearing retainer. Do not drive or press bearings onto cross shaft with opposite bearing resting on cage or in cup. To do so will cause brinelling of the cup and roller surfaces.*

All Other Carriers

On carriers permitting complete assembly of cross shaft before threading into carrier housing, proceed as follows:

1. Lubricate bearings and cups with light machine oil.
2. If bearing cover or cage on hypoid gear side has been removed, replace it using correct shim pack.
3. Thread cross shaft past differential bearing supports and position in bearing cup.
4. On carriers with front mounted shift units, position shift fork and install shift shaft. Align lock nut hole in fork with shaft indent and tighten lock screw and nut. Wire lock screw to fork.
5. Start bearing cage (side opposite hypoid gear) into carrier housing.
6. Tap bearing cage into position with soft mallet. Install capscrews.

Check Cross Shaft Bearing Preload

1. In checking cross shaft bearing preload torque, always rotate cross shaft and gear assembly several revolutions before checking preload to assure normal bearing contacts.
2. Lock low speed pinion and cross shaft with shift collar. Wrap a strong cord around pinion and pull on a horizontal line with a pound spring scale, Fig. 11. Assuming the pinion diameter to be 4", the radius would be 2", and with a pull of 7 pounds on the scale, the bearing preload would be 14 inch pounds of preload torque.
3. Read rotating pull (not starting pull). If preload torque is not within 5 to 15 inch pounds, add shims under cage (opposite hypoid gear) to decrease or remove shims to increase preload torque.

Install Pinion & Cage

1. Install pinion cage with original shim pack. Install capscrews and tighten.
2. Adjust hypoid gear backlash to .014–.020" (regardless of what markings might be etched on gear set) by transposing cross shaft bearing cage or cover shims, Fig. 12.
3. To move bevel gear away from pinion, remove shims from pack under cross shaft bearing cage on side opposite bevel gear and add shims of equal thickness to pack under cross shaft bearing cover or cage on bevel gear side. *Shims should be transposed in this manner to maintain established preload.*
4. To move bevel gear toward pinion, remove shims from pack under cross shaft bearing cover or cage on bevel gear side and add shims of equal thickness to pack under cross shaft bearing cage on side opposite bevel gear. *The actual backlash*

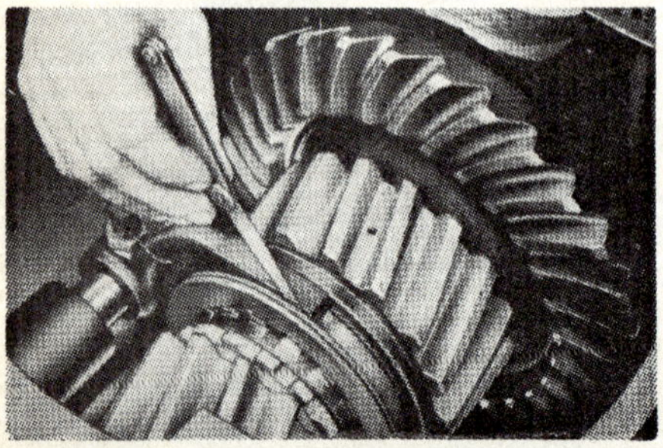

Fig. 13 Checking clearance of shift fork pads in shift collar with feeler gauge

Fig. 14 Differential pinion and side gear assembly

Fig. 15 Checking differential bearing preload

changes approximately .008" for each .010" movement of the gear.

5. Check gear tooth contact for correct adjustment as outlined in the *Single Reduction Axle* section.

Install Shift Units

Power Actuated Side Mounted Type

1. Tap shift unit sleeve into carrier housing over original shim pack. Install lock washers and stud nuts.
2. Install new flange gasket. Hold shift fork in position. Align lock screw holes in shift shaft and slide shift unit and shaft into position.
3. Tighten fork screw and lock nut to 30–35 ft. lbs.
4. Install shift unit lock washers and stud nuts.
5. Check clearance of shift fork pads in shift collar with gauge, Fig. 13. Clearance should not be less than .010" on each side of fork in both high and low speed positions. *When checking clearance, shift collar must be flush with end face of spur pinion in both high and low speed positions.* Add or remove shims from pack under sleeve to obtain correct adjustment.
6. Check operation of shift unit. Ten pounds of air may be used to check vacuum type unit by applying air to opposite side of diaphragm, using vent opening on the single line shift unit.

Power Actuated, Front Mounted Type

1. Check bellcrank buttons and remove any scores at operating faces.
2. Place shift fork, collar and shift unit in high speed position. They must be in the same position to install shift unit.
3. Install new gasket and position shift unit with bellcrank in slot of shift shaft. Install lockwashers and capscrews. Do not disturb Allen set screw or travel positioning screw unless shift fork adjustments are needed. If adjustments are required, proceed as follows:
4. With fork and collar shifted to engage helical pinion next to hypoid gear, adjust Allen screw and lock nut in top of carrier so that fork is centered in collar groove within .005".
5. With fork and collar shifted to engage helical pinion away from hypoid gear, adjust hex head screw and lock nut in cross shaft cage so that fork is centered in collar groove within .005". When checking clearance, shift collar must be flush with end face of spur pinion in both high and low speed positions.
6. Check operation of shift unit in same manner as outlined for side mounted shift unit.

WRENCH TORQUE FOR PINION SHAFT NUTS

DIAMETER OR SIZE	THREADS PER INCH	TORQUE—LB. FT.	
		MIN.	MAX.
7/8"	20	175	250
1"	20	300	400
1¼"	18	700	900
1½"	12	800	1100
1½"	18	800	1100
1¾"	12	800	1100

DIFFERENTIAL BEARING CAP CAP SCREWS OR STUD NUTS (Earlier Axle Models Without Hardened Washers)

CAP SCREW OR STUD NUT DIAMETER	CAP SCREW OR COARSE STUD THREAD	STUD NUT OR FINE THREAD	TORQUE—LB. FT.	
			MIN.	MAX.
5/8"	11	18	130	170
¾"	10	16	230	300
7/8"	9	14	345	440
7/8"	14	14	380	485
1"	14	14	380	485

DIFFERENTIAL BEARING CAP CAP SCREWS OR STUD NUTS (Later Axle Models Employing Hardened Washers)

CAP SCREW OR STUD NUT DIAMETER	CAP SCREW OR COARSE STUD THREAD	STUD NUT OR FINE THREAD	TORQUE—LB. FT.	
			MIN.	MAX.
5/8"	11	18	160	205
¾"	10	16	290	370
7/8"	9	14	470	595
7/8"	14	14	510	655
1"	14	14	580	745

Fig. 16 Torque specifications for capscrews or stud nuts for Timken (Rockwell) Two Speed Double Reduction Axles

NOTE: Torques given apply to parts coated with machine oil. For dry (or "as received") parts, increase torque 10%. For parts coated with multi-purpose gear oil, decrease torque 10%. Nuts on studs to use same torque as for driving the stud.

Assemble Differential

1. Join high and low speed spur gears to their respective case halves with bolts or rivets as desired. *When new spur gears or differential cases are installed, the case holes may require line reaming with the gear in order to assemble correct size bolts or rivets. When rivets are used they should be upset cold (not heated).*
2. Lubricate differential case inner walls and all parts with axle lubricant, Fig. 14.
3. Install thrust washer and side gear in one case half. Place spider with pinions and thrust washers in position. Install component side gear and thrust washer.
4. Align mating marks, position component case half and draw assembly together with four long bolts or capscrews equally spaced.
5. Check assembly for free rotation of differential gears and correct if necessary.
6. Install and tighten remaining bolts and secure with lock wire.
7. Press differential bearings squarely and firmly on case halves.

Install Bearing Cups In Carrier Leg Bores

1. Temporarily install bearing cups, threaded adjusting rings (where used) and bearing caps and tighten capscrews.
2. Bearing cups must be a hand push fit in bores otherwise bores must be re-worked with a scraper or emergy cloth until a hand push fit is obtained. Use a blued bearing cup as a gauge and check fits as work progresses. This applies to all three types of carrier leg bores.
3. On carriers employing adjuster rings, if the rings cannot be turned by hand or with a maximum force of 10 lbs. applied at one end of a 2-ft. wrench (20 ft. lbs. torque) the O.D. should be slightly reduced with a fine mill file or use others that turn more freely. Be sure adjusting rings and threads are free from burrs or nicks.

Install Differential

1. After checking related parts, coat bearings with axle lubricant.

2. Place bearing cups over assembled bearing cones and position differential in carrier.
3. On the threaded adjusting ring type (see Fig. 2) insert rings and turn them hand tight against bearing cups. Position bearing caps in place.
4. On type with one threaded adjusting ring and one split ring (see Fig. 3) position split ring in groove of carrier leg and insert threaded adjusting ring on opposite side. Turn threaded ring hand tight against bearing cup. Install bearing caps.
5. With the split ring type (see Fig. 4) insert one ring in carrier leg groove. Move differential over so that face of bearing cup is held tightly against inserted ring. Install opposite split ring by tapping it into groove with a blunt end drift, tapping on lower I.D. of ring. Install bearing caps.
6. Install and tighten carrier leg screws.

Adjusting Differential Bearing Preload

Type With One or Two Threaded Rings

1. On units employing two threaded rings center helical gears in relation to helical pinions by means of the threaded rings.
2. Using a dial indicator as shown in Fig. 15 and starting with a definite amount of end play adjust differential bearings to zero end play. A pair of pinch bars may be used as shown. Rotate assembly several revolutions each time adjusting rings are moved before checking end play to assure normal bearing contact.
3. Tighten adjusting rings 1¾ to 2½ notches (total of both rings) to correctly preload bearings.
4. Install ring locks and capscrews (where used) and install lock wire.

Type Employing Two Split Rings

1. Temporarily install differential in carrier and center between carrier leg grooves.
2. Insert thin split rings making certain that there is clearance between bearing cup faces and rings (do not install bearing caps).
3. Measure end play as shown in Fig. 15.
4. Remove and measure thickness of thin rings. To the total thickness of the two thin rings add the end play figure plus another .017 to .022" to obtain total thickness of the two thicker rings required to obtain proper bearing preload. Hardened split rings are ground in increments of .005".
5. When proper bearing preload has been established complete the assembly of the differential. Install unit in axle housing in reverse order of removal.

Location	Diameter	Threads Per Inch	Torque Lb. Ft. Min.-Max.
Carrier to Housing	7/16"	14	53-67
	7/16"	20	53-67
	1/2"	13	81-104
	1/2"	20	81-104
	5/8"	11	160-205
	5/8"	18	160-205
Pinion Cage	3/8"	16	33-43
	7/16"	14	53-67
	7/16"	20	53-67
	1/2"	13	81-104
	1/2"	20	81-104
	9/16"	12	116-149
	9/16"	18	116-149
	5/8"	11	160-205
Cross Shaft Bearing Cage and Cover	1/2"	13	81-104
	1/2"	20	81-104
	9/16"	12	116-149
	9/16"	18	116-149
	5/8"	11	160-205
Cross Shaft	7/16"	14	42-54
Bearing Lock	9/16"	12	94-120
Differential Bolts	3/8"	16	33-43
	7/16"	14	53-67
	1/2"	20	92-118
	9/16"	18	130-167
	5/8"	18	185-235
	3/4"	16	320-415
Adjusting Nut Lock	5/16"	18	16-20
Inspection Cover	3/8"	16	27-35
Shift Unit (Mounting)	3/8"	16	27-35
Shift Unit Lock Nut,	3/8"	24	31-39
Set Screw and Clamp Screw	7/16"	20	31-39
Shift Unit Travel	1/2"	13	55-60
Limiting Screws	5/8"	11	30-35

Fig. 17 Torque specifications for capscrews or stud nuts for Timken (Rockwell) Two Speed Double Reduction Axles

NOTE: Torques given apply to parts coated with machine oil. For dry (or "as received") parts, increase torque 10%. For parts coated with multi-purpose gear oil, decrease torque 10%. Nuts on studs to use same torque as for driving the stud.

Single Reduction Thru-Drive Tandem Axles

DESCRIPTION

This type drive, Fig. 1, utilizes a modified conventional single reduction axle for the forward rear axle and a conventional single reduction axle for the rearward axle. A power divider or inter-axle differential is mounted integrally with the forward axle. This type of drive permits a tandem hook-up with a minimum of angles in the drive line.

The pinion bearing preload is adjusted and maintained by a hardened precision spacer between the inner and outer tapered bearings. The differential bearing preload is adjusted and maintained by threaded adjusting rings located in the carrier legs and caps.

The inter-axle differential may be engaged or disengaged by a power actuated shift unit which moves a sliding collar on the through shaft splines. The shift unit is controlled by a selector switch or lever within the cab of the vehicle and may be engaged or disengaged under any normal operating conditions.

The inter-axle differential when engaged (unlocked) divides the engine torque between the forward and rearward axles; when disengaged (locked) it converts the two axles to a through drive type tandem.

DISASSEMBLE

Remove Unit From Housing

1. Disconnect forward and rear propeller shafts.
2. Drain lubricant from axle housing and inter-axle differential.
3. Remove axle shafts as outlined for standard single reduction axles.
4. Remove shift shaft housing assembly.
5. Disassemble and remove shift lever attaching nut, button, lever, cup and spring. The body fit bolt should not be removed.
6. Remove thru-shaft, cage and yoke, Fig. 4. To free cage from case bore it may be necessary to tap yoke with a soft mallet. While thru-shaft assembly is being threaded out, the sliding clutch must be eased along the shaft at the shift lever opening. When thru-shaft clears opening, the clutch may be lifted out.

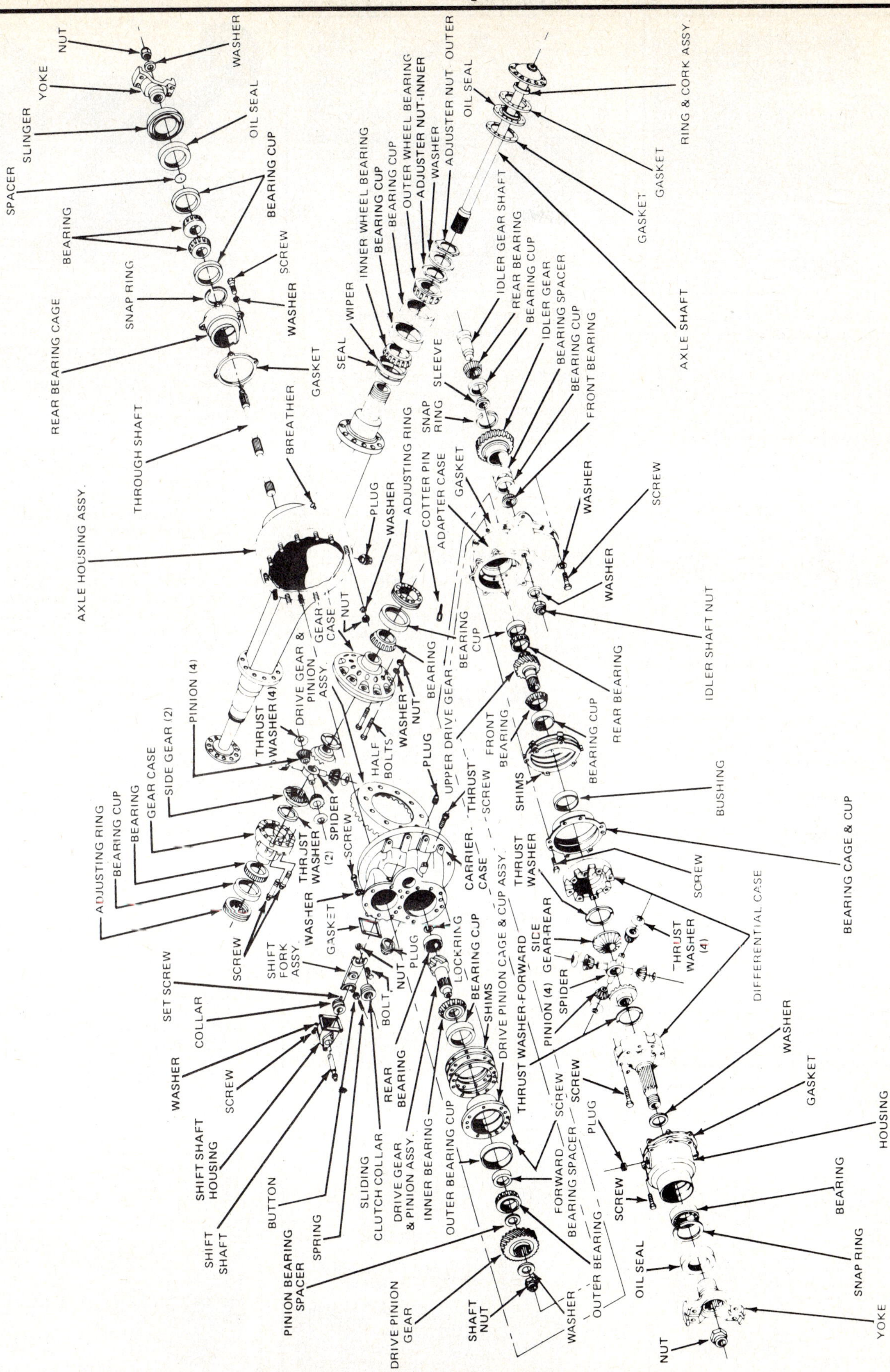

Fig. 1 Type 3 single reduction tandem drive forward axle. This unit differs from Type 2 in that it employs splines instead of dog teeth on the inter-axle differential, as shown in Figs. 2 and 3. Type 1 is similar to Type 2 except that the impeller (oil scoop), oil baffles, etc. used in Type 2 are not employed in Type 1

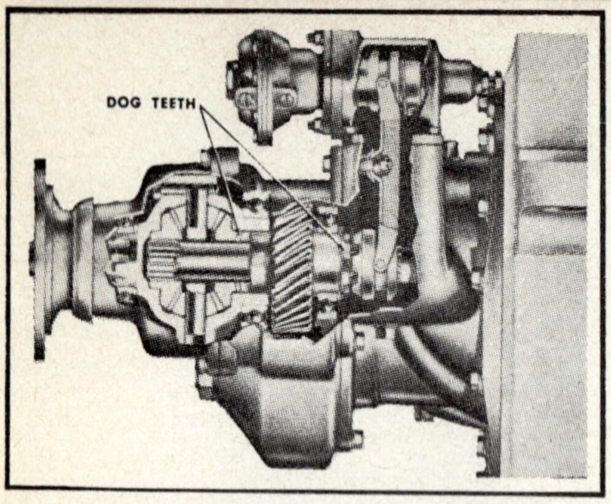

Fig. 2 Types 1 and 2 inter-axle differential shift mechanism

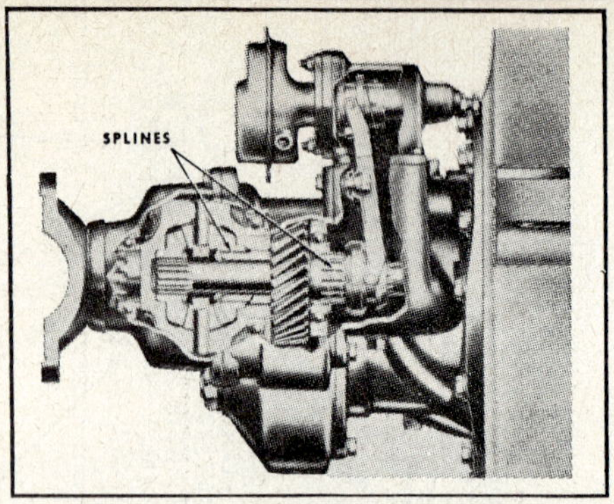

Fig. 3 Type 3 inter-axle differential shifting mechanism

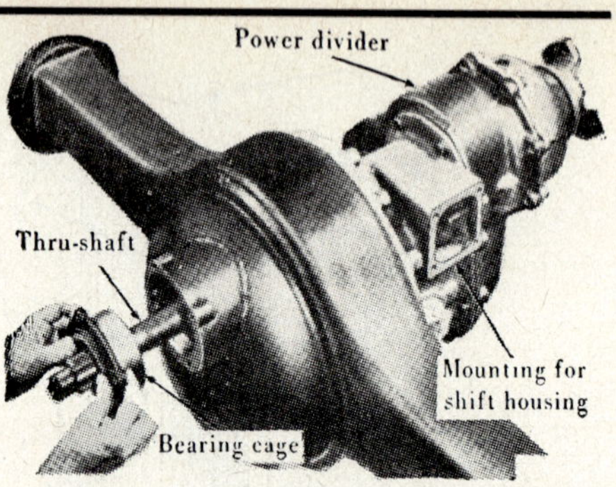

Fig. 4 Removing thru-shaft and cage

7. Remove carrier from axle housing in same manner outlined for the standard single reduction axle.

Disassemble Thru-Shaft

1. After removing nut, use puller to remove yoke.
2. Press thru-shaft from cage, using a suitable sleeve against bearing inner race.
3. On units with ball type bearings, remove cage snap ring, Fig. 5. Then tap radial bearing out of bore from seal end. If seal is damaged or worn, press out with suitable sleeve. If seal is in good condition, do not disturb.

NOTE: On units with roller type bearings, the bearing and cage assembly must be replaced as a pre-assembled unit.

Disassemble Inter-Axle Differential

1. Separate inter-axle differential and adapter case from carrier case, Fig. 6.
2. Unfasten and lift inter-axle differential from adapter case, Fig. 7.
3. Remove input shaft flange or yoke nut.
4. Press inter-axle differential from cover.
5. If original identification marks are not clear, mark differential case halves with a punch for correct alignment on reassembly.
6. Disassemble case halves and remove differential components.
7. On Types 1 and 2, remove radial bearings from differential case halves with a suitable puller, pulling against inner race of bearing. On Type 3, do not remove radial bearing from case unless replacement is necessary.
8. To disassemble adapter case, tap upper drive gear back until gear touches case. Remove spacer and snap ring from bearing outer race. Tap drive gear out of adapter case, Fig. 9. Do not remove radial bearing from gear hub unless replacement is necessary. If necessary, fold back tab locks and remove hub snap ring. Then remove bearing with suitable puller.
9. To disassemble idler gear, remove nut and spacing washer. Remove shaft and slide out gear and bearing. Remove tapered bearings and tap out cups and related parts, Fig. 10.

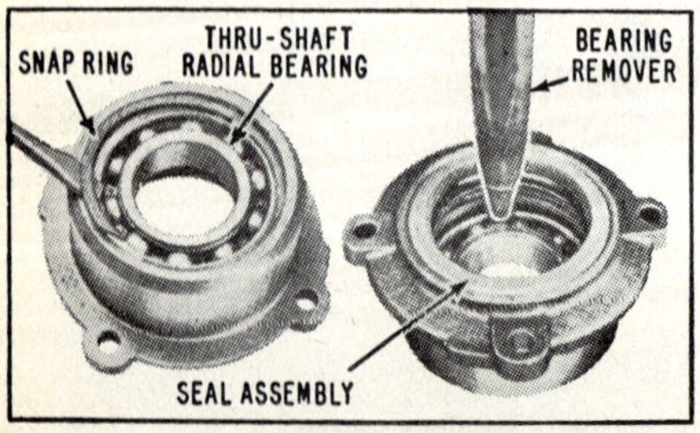

Fig. 5 Removing thru-shaft radial bearing

Main Differential Case, Ring Gear & Drive Pinion

Service procedures on these components are the same as outlined for the standard single reduction axle.

REASSEMBLE

Main Differential Case, Ring Gear & Drive Pinion

Assembly procedure, adjusting pinion bearing and differential bearing preloads, checking ring gear and pinion backlash are the same as described for the standard single reduction axles. Complete the assembly and adjustment of these parts, then proceed with the adapter case and inter-axle as follows:

Assemble Adapter Case

1. Install snap ring and press idler gear inner bearing cup squarely against snap ring.
2. Insert idler gear cup spacing sleeve against opposite side of snap ring.
3. Press idler gear outer bearing cup squarely against spacing sleeve.
4. Position idler gear inner and outer bearings into cups with hardened bearing spacer (or combination of two spacers) between them. *Some of the Type 1 drive units do not incorporate the idler bearing spacer or the idler shaft "O" ring. These parts should be added to any of the units that do not have them.*
5. Slide complete assembly through adapter case drive pinion opening. *On Types 1 and 2, position assembly so that bearings are aligned with adapter case shaft hole. On Type 3, temporarily push the assembly to the far side of the pocket and proceed to install upper drive gear assembly.*
6. On Types 1 and 2, tap idler shaft through idler gear so that inner bearing is squarely against idler shaft shoulder.

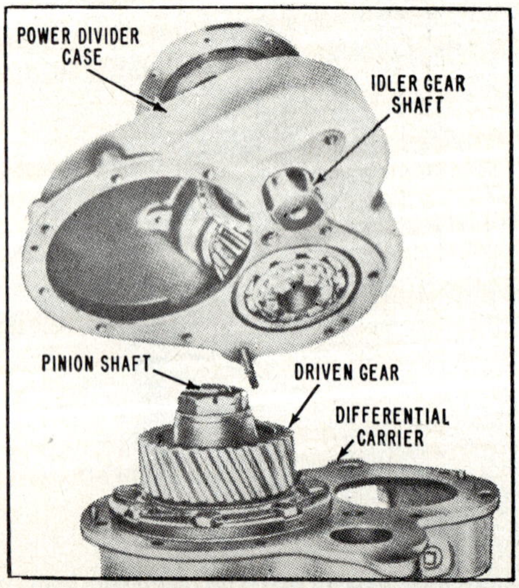

Fig. 6 Lifting power divider from differential carrier

Idler Shaft Bearing Adjustment

On Types 1 and 2, this adjustment is made before assembling the upper drive gear. On Type 3, first assemble the upper drive gear, then make the adjustment as follows:

1. Hold adapter case and idler gear securely in a vise with jaws clamped on flattened end of idler shaft.
2. Install washer and nut and torque nut from 350 to 400 ft. lbs.
3. Measure idler shaft bearing end play with a dial indicator mounted to adapter case proper with stem set against idler gear face. End play should be .001"-.005".
4. If not within these limits, use a thinner or thicker spacer or combination of two spacers as required.
5. After proper end play has been established, remove washer and nut. Insert "O" ring and reinstall washer and nut. Tighten nut from 350 to 400 ft. lbs. and install cotter key.

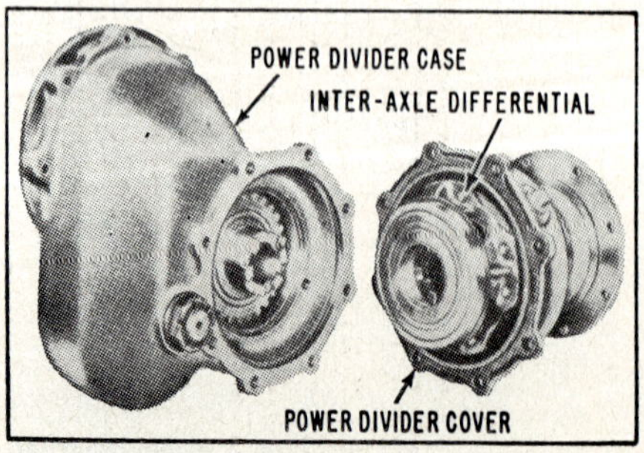

Fig. 7 Separating power divider cover from case

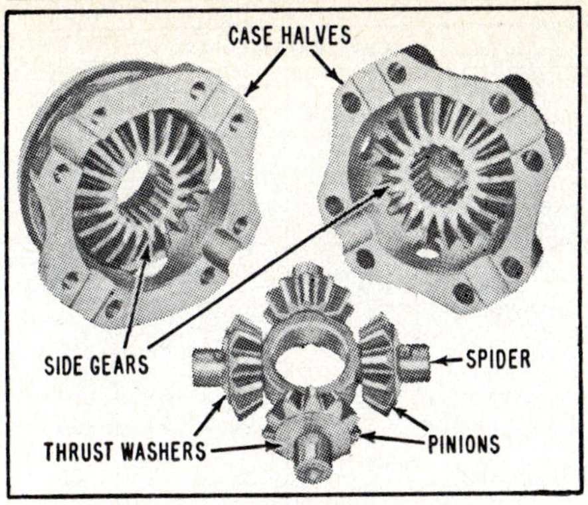

Fig. 8 Inter-axle differential parts

Fig. 9 Drive gear and bearing removed from case

Fig. 10 Idler gear components

Assemble Upper Drive Gear

On Type 3, this gear is assembled before adjusting idler shaft bearing, the procedure for all types being as follows:

1. If rear bearing has been removed for replacement, press new bearing on hub using a suitable sleeve against bearing inner race.
2. Install tabbed lock ring and hub snap ring.
3. With radial bearing outer race snap ring removed, position drive gear in case. Tap assembly through bore until gear touches wall of case.
4. Install radial bearing outer race snap ring in the groove. Tap gear back so that snap ring seats in recess in case.
5. On Types 1 and 2, install spacer over outer race of bearing against snap ring.
6. On Type 3, position spacer over bearing with flat toward pinion cage bore. Position idler assembly so that bearings are lined up with idler shaft hole. Tap idler shaft through gear so that inner bearing is squarely against idler shaft shoulder.

Assemble Inter-Axle Differential

1. Lubricate differential case walls and all components parts with axle lubricant.
2. Position thrust washer and rear side gear into rear case half.
3. Place spider with pinions and thrust washers into position.
4. Install forward side gear and thrust washer.
5. Align mating marks, position forward case half and draw together with case capscrews. Types 2 and 3 employ an impeller over forward case half.
6. Tighten case capscrews to correct torque.
7. Check for free rotation of gears and correct if necessary.
8. On Types 1 and 2, install forward radial bearing spacer and press on forward and rear radial bearings.

Install Inter-Axle Differential

Types 1 and 2

1. If snap ring was removed, reinstall in nose of cover.
2. Install cover-to-case gasket. Position cover over inter-axle differential and tap down with a sleeve until forward radial bearing seats against snap ring.
3. Position cover and differential over upper drive gear hub and tap down until rear radial bearing snap ring seats in adapter case recess. It may be necessary to rotate the assembly slightly to line up the dog teeth.
4. Install capscrews and lockwashers and draw cover and adapter case together. Tighten capscrews to specified torque.
5. Install cover oil seal with suitable driver.

Type 3

1. If cover was disassembled, install forward radial bearing and snap ring.
2. Install spacer on input shaft.
3. Position cover over input shaft and tap down until radial bearing seats against spacer.
4. Install gasket and position cover and differential over upper drive gear hub. Tap assembly into position. It will be necessary to line up the splines of the drive gear with those of the side gear.
5. Install capscrews and lockwashers and torque to specifications.
6. Install cover oil seal with suitable driver.

Assemble Adapter Case & Inter-Axle Differential to Carrier

1. Mount carrier in stand in upright position. Install gasket and place adapter case over carrier. The idler shaft flat must be lined up with corresponding flat in carrier.
2. Install lockwashers and capscrews and torque to specifications.

Final Assembly

1. Install carrier in axle housing in a manner outlined for standard single reduction axles.
2. Install thru-shaft rear radial bearing into cage and lock in place with snap ring.
3. Press bearing and cage on splined end of thru-shaft with suitable sleeve.
4. Install thru-shaft cage oil seal with suitable driver.
5. Enter thru-shaft and cage with new cage gasket into cage bore at rear of axle housing until forward end of shaft is even with shift lever opening.
6. Install sliding clutch collar over forward

CAP SCREWS OR STUD NUTS

LOCATION ON UNIT	DIA. INCH	THDS. NO.	TORQUE MIN.	TORQUE MAX.	LOCATION ON UNIT	DIA. INCH	THDS. NO.	TORQUE MIN.	TORQUE MAX.
Shift Unit	3/8"	24	38	49	Pinion Shaft Nut	1 3/4"	12	800	1100
Carrier to Housing	1/2" 5/8" 3/4"	20 18 16	81 160 325	104 205 420	Diff. Bearing Adj. Lock	5/16"	18	15	17
Shift Shaft Adjusting Screw Lock Nut	1/2"	20	66	85	Through-Shaft Rear Cage	3/8" 7/16"	24 14	38 60	49 77
					Inter-Axle Diff. Case Bolt	7/16" 1/2"	14 20	60 93	77 120
Input Shaft Yoke or Flange Nut	1 3/4"	12	300	400	Input Cage	7/61"	14	60	77
Pinion Cage Cap Screw	1/2"	13	82	91	Pinion Cover	3/8"	24	38	49

REPLACEMENT—DIFFERENTIAL GEAR TO CASE BOLTS

DIFFERENTIAL CARRIER	SERVICE BOLT	NUT	LOCKWIRE	ORIGINAL RIVET	SPOTFACE DIAMETER	TORQUE LBS. FT. MIN.	TORQUE LBS. FT. MAX.
Q—gearing	15X-786	N-210-1	LW-1610	RV-71014	1 3/8"	186	205
R—gearing	15X-659	N-210-1	LW-1610	RV-71016	1 3/8"	186	205

Fig. 11 Torque specifications for Timken (Rockwell) Single Reduction Thru-Drive Tandem Axles, Two Gear Units

NOTE: Torques given apply to parts coated with machine oil. For dry (or "as received") parts, increase torque 10%. For parts coated with multi-purpose gear oil, decrease torque 10%. Nuts on studs to use same torque as for driving the stud.

end of shaft through shift housing opening. Ease shaft into forward side gear of inter-axle differential, while at the same time passing clutch collar onto collar splines.

7. Install thru-shaft cage capscrews and washers and torque to specifications.
8. Install over shift lever bolt the shift lever spring, cup and lever. Lever inner yoke must be properly located in collar groove at this time. Install shift lever button and nut. Tighten nut securely and install cotter key.
9. Position gasket and shift housing against drive unit, making sure that shift lever yoke is properly located in shift collar groove.
10. Install shift housing capscrews and torque to specifications.

Adjusting Shift Shaft

1. Adjust positioning screw at rear of shift housing as follows: With shift shaft moved back its full travel, locking the inter-axle differential, turn adjusting screw in until end of screw touches end of shift shaft.
2. From this point proceed 1 to 1¼ turn more and lock adjusting screw with jam nut. This will allow approximately .012" clearance between the yoke and groove of collar and thus eliminate yoke and collar wear. The shift collar provides a definite stop against housing wall when shifted in the opposite direction.

Install Thru-Shaft Yoke

1. Install yoke on splines with suitable sleeve. Install washer and nut.
2. By using a holder on yoke, tighten yoke nut to proper torque and install cotter key.

BOLT AND STUD NUTS				
LOCATION ON UNIT	DIA. INCH	THDS. NO.	TORQUE — LB. FT.	
			MIN.	MAX.
Shift Unit	⅜"	24	27	35
Carrier to Housing	½" ⅝"	20 18	81 160	104 205
Shift Shaft Adjusting Screw Lock Nut	½"	20	45	50
Through-Shaft Yoke and Input Shaft Yoke Nuts	1" 1¼" 1½"	20 18 18	300 300 300	400 400 400
Differential Case Cap Screws	½" ⅝"	13 18	93 210	120 270
Bearing Cap Cap Screws	⅝" ¾"	11 10	130 230	170 300

CAP SCREWS				
LOCATION ON UNIT	DIA. INCH	THDS. NO.	TORQUE — LB. FT.	
			MIN.	MAX.
Through-Shaft Cage	⅜"	16	27	35
Carrier to Adapter Case	7/16"	14	60	77
Pinion Cage	½"	13	93	120
Inter-Axle Cover	7/16"	14	42	54
Inter-Axle Differential Case	7/16" ½"	14 13	53 93	67 120
Shift Housing	⅜"	16	27	35
Shift Collar Set Screw	⅜"	24	12	16

REPLACEMENT—DIFFERENTIAL GEAR TO CASE BOLTS

TANDEM MODEL CARRIER	SERVICE BOLT	NUT	LOCK WIRE	ORIGINAL RIVET	SPOTFACE DIAMETER	TORQUE — LB. FT.	
						MIN.	MAX.
DHD —	15X-775	N-28-1	LW-1610	RV-7812	1⅛"	105	135
FHD —	15X-775	N-28-1	LW-1610	RV-7812	1⅛"	105	135
†LHD —	15X-666	N-29-1	LW-1610	RV-7913	1⅜"	130	170
†QHD —	15X-666	N-29-1	LW-1610	RV-7913	1⅜"	130	170
‡LHD —	15X-786	N-210-1	LW-1610	RV-71014	1⅜"	185	235
‡QHD —	15X-786	N-210-1	LW-1610	RV-71014	1⅜"	185	235

† Original production had 9/16" rivets and require 9/16" bolts and nuts for service.

‡ Current production has ⅝" rivets and require ⅝" bolts and nuts for service.

Fig. 12 Torque specifications for Timken (Rockwell) Single Reduction Thru-Drive Tandem Axles, Three Gear Units

NOTE: Torques given apply to parts coated with machine oil. For dry (or "as received") parts, increase torque 10%. For parts coated with multi-purpose gear oil, decrease torque 10%. Nuts on studs to use same torque as for driving the stud.

Double Reduction Thru-Drive Tandem Axles

In these units, Fig. 1, tapered roller bearings support the forward and rear ends of the thru-shaft and pinion assembly of spiral bevel gear units. Bearing preload is adjusted and maintained by shim packs between carrier machined surfaces and bearing covers. Yokes or flanges, held in place by thru-shaft nuts, press against the bearing cones or spiral bevel pinion splined to thru-shaft.

The thru-shafts of hypoid gear drive units are supported at the forward end by tapered roller bearings in a cage and at the rear end by a straight roller bearing. Pinion bearing preload is adjusted and maintained by a hardened precision spacer between inner and outer tapered bearings which are held in place on the thru-shaft by separate thru-shaft nuts.

Hypoid gear top-mounted forward drive units may be equipped with an inter-axle differential of the planetary type which divides engine torque between the forward and rear axles. The differential is equipped with a lock-up device which converts the tandem assembly to a thru-drive type tandem. The lock-up is actuated by an air chamber and is controlled in the vehicle cab. It can be engaged and disengaged under all normal operating conditions. Fig. 2 illustrates the early

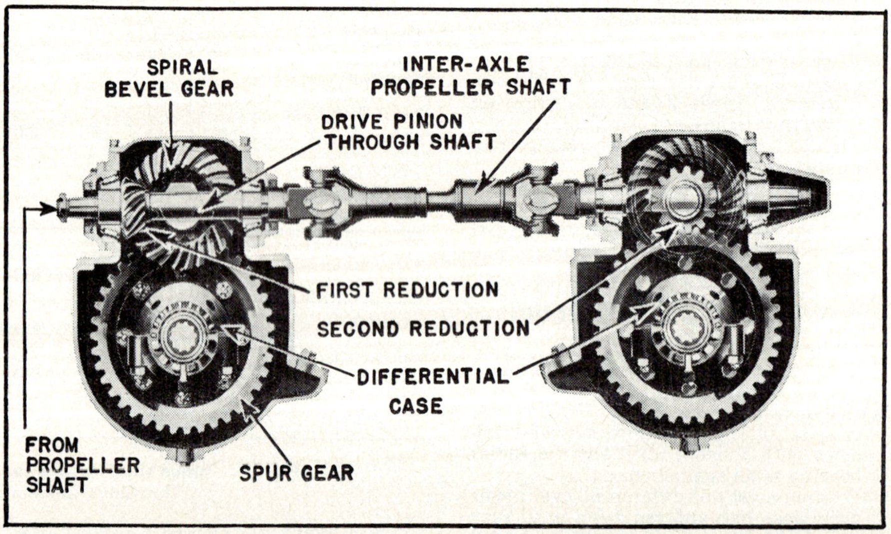

Fig. 1 Timken (Rockwell) double reduction, top-mounted thru-drive axles. Both hypoid and spiral bevel gears are used in the first reduction. Either straight spur or helical gears are used in the second reduction.

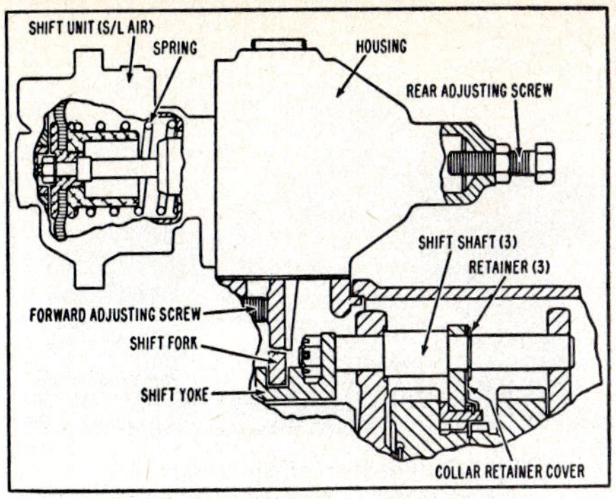

Fig. 2 Shift unit assembly (early models)

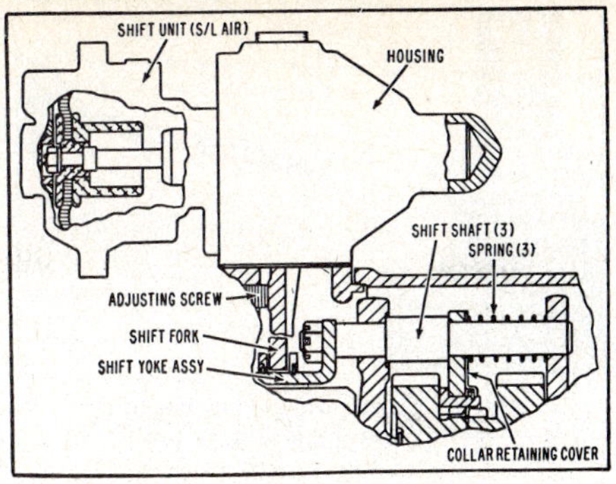

Fig. 3 Shift unit assembly (later models)

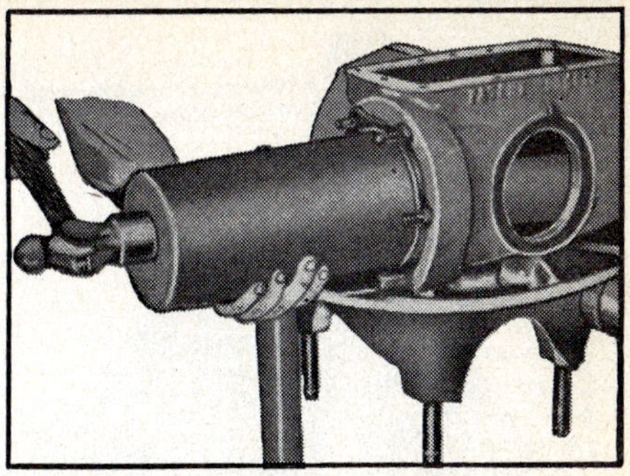

Fig. 4 Installing pinion, bearing and thru-shaft assembly with sleeve shown

design shift unit, while Fig. 3 is the current design.

DISASSEMBLE

During the disassembly process, be sure to keep any shims found with parts intact and identified so that when reassembling, adjustment of unit involved will be facilitated.

1. Drain axle and inter-axle differential housing.
2. Remove axle shafts and carrier from housing.
3. Match mark one differential carrier leg and bearing cap for identification on reassembly.
4. Lift out differential. Before dismantling differential, match mark case halves for proper alignment on reassembly.
5. Remove pinion and thru-shaft on models not equipped with inter-axle differential.
6. If inter-axle differential is used, remove shift unit.
7. Remove inter-axle differential bearing and cover assembly. Hoist inter-axle differential out with a chain fall.
8. Disassemble thru shaft.
9. Disassemble inter-axle differential.
10. Disassemble cross shaft.

REASSEMBLY

Service Note

The thru-shaft, pinion and bearing assembly of the spiral bevel type drive units must be installed in the carrier before the cross shaft assembly is installed. This permits determining the thickness of shim packs that control pinion bearing preload without interference of the cross shaft assembly. The thru-shaft assembly is then removed and the cross shaft assembly installed. Cross shaft bearing preload must be established at this time before final installation of the thru-shaft assembly with previously determined shim packs.

The cross shaft assembly of hypoid gear units must be installed in the carrier first so cross shaft bearing preload can be established without interference of the thru-shaft and pinion assembly. The thru-shaft, pinion, bearing and cage may be assembled at the bench and then installed in the carrier.

Assemble Spiral Bevel Pinion & Thru-Shaft

1. Press tapered bearing on pinion.
2. Coat pinion and thru-shaft splines with heavy grease and press parts together. *Press parts together about 1/4" to 3/8", then relieve press pressure to permit them to realign themselves to prevent distortion and damage. Continue the pressing operation until parts are correctly assembled.*
3. Install pinion, bearing and thru-shaft in drive unit. Install pinion bearing cup with suitable sleeve, Fig. 4.
4. Install spacer (if used) over splined end of thru-shaft and tighten nut on spacer end of thru-shaft.
5. Coat seal with non-hardening sealing compound and install in bearing covers. Lubricate sealing element of seal.
6. Install bearing cover and seal assembly (where used) and original shim pack over splined end of thru-shaft. Tighten cover nuts.
7. Install bearing opposite pinion firmly against shaft shoulder.
8. Install bearing cup in drive unit, rotating shaft to position all bearing rollers.
9. Install bearing cover and seal over original shim pack and tighten stud nuts.

Measuring Thru-Shaft Bearing Preload

1. As shown in Fig. 5, wrap strong cord around thru-shaft and pull on a horizontal line with a pound scale.
2. The preload specification for tapered roller bearings at opposite ends of thru-shaft is zero thru-shaft end play to 5 inch pounds torque.
3. Alter shim pack under bearing cover at end opposite pinion so there is no thru-shaft end play when measured with a dial indicator and so bearing preload torque does not exceed 5 inch pounds.
4. For example, assuming thru-shaft, diameter is $1\frac{1}{2}$", the radius is 3/4" and with 4 pounds pull on scale, preload torque is 3 inch pounds.
5. Read rotating pounds pull (not starting pull). If preload torque is not within specified limits, add shims under bearing cover at end opposite pinion to decrease torque or remove shims at same location to increase preload torque.
6. Complete bearing adjustments before installing flanges or yokes.
7. Remove thru-shaft. Keep both shim pack with their respective covers to maintain established bearing preload.

Assemble Cross Shafts

Check I.D. of sleeve and O.D. of radial bearing and replace either or both of these parts if there is more than .006" clearance between them. When new, these parts have a clearance

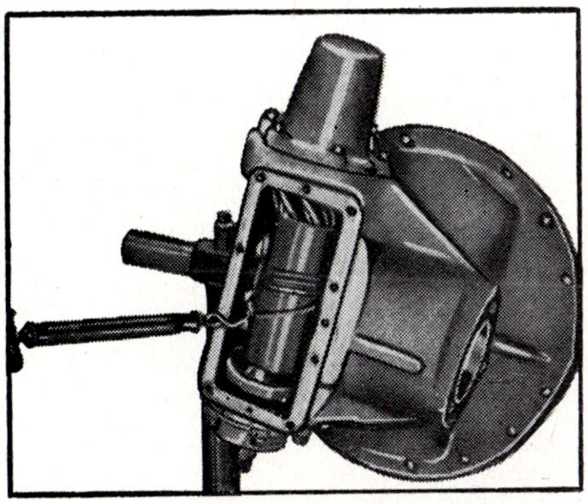

Fig. 5 Measuring thru-shaft bearing preload

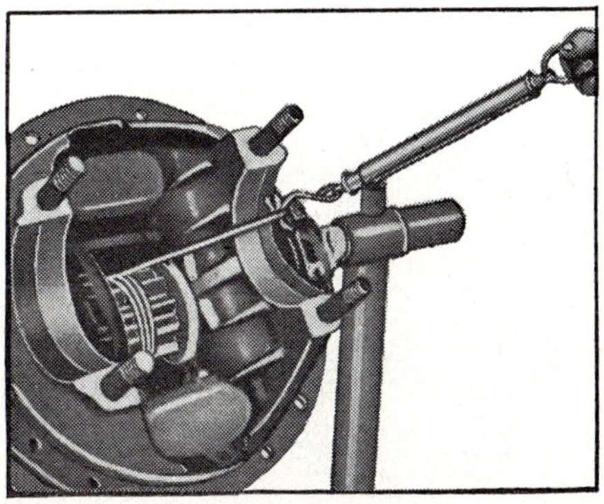

Fig. 6 Measuring Type 1 cross shaft bearing preload

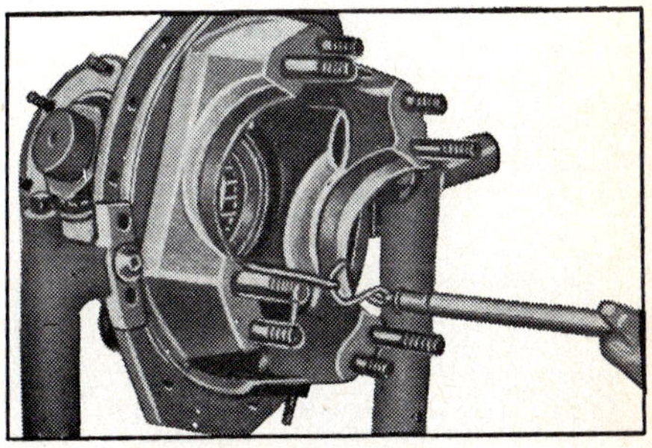

Fig. 7 Measuring Type 2 cross shaft bearing preload

Fig. 8 Install inner bearing on Type 3 cross shaft, using wood block as shown

of .0024" to .004". The radial bearing must be free to float in the sleeve.

Check the I.D. of the bearing bore on older drive units that do not have replaceable sleeves. If the I.D. is more than .006" larger than the bearing O.D., replace carrier and cap assembly with the newer type carrier and cap that incorporates replaceable sleeves.

If sleeve is to be replaced, press new sleeve firmly against housing shoulder. Drill hole for lock screw and remove burrs from sleeve. Install and tighten lock securely or install pin and stake in place.

Assemble Type 1 Cross Shaft

Gear Journal Fit to Cross Shaft

1. Install 6 studs at cross shaft flange. Assemble radial bearing on gear hub with large radius of bearing inner race toward back of gear.
2. Install bearing washer on gear hub against bearing with chamfer of spacer away from bearing. Install spacers with I.D. chamfer so chamfer is next to radius at root of pinion teeth. *A large flat washer is used at this location when O.D. of pinion teeth is smaller than I.D. of bearing.*

Fig. 11 Assemble rear gear against insert with locking teeth up

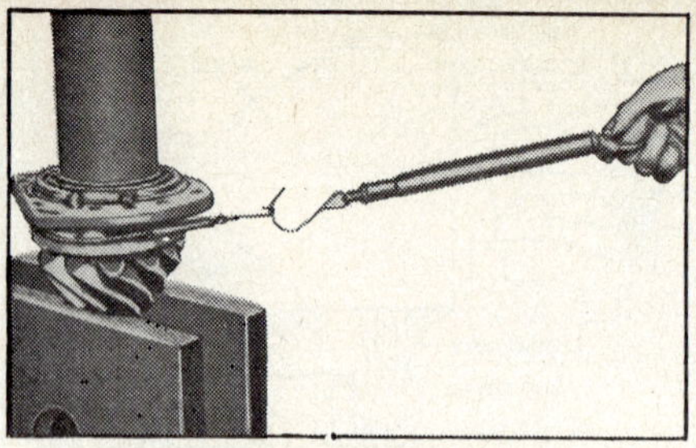

Fig. 9 Measuring pinion bearing preload on hypoid gear unit

3. Coat I.D. of gear with heavy grease. Install gear, bearing and spacer assembly in drive unit sleeve and block up to hold in place.
4. Inspect entering end of cross shaft and remove any nicks or burrs. Coat O.D. of shaft and I.D. of gear with heavy grease.
5. Position housing in press with thru-shaft chamber down and gear supported on suitable sleeve.
6. Align key in cross shaft with key-way in gear hub and press shaft firmly against gear, bearing and spacer. *Continue pressing to exert 10 to 20 tons pressure in excess of that required for secure assembly.*
7. Install tapered bearing cage original shim pack (which controls gear position) over cross shaft cage studs.
8. Apply graphite lubricant to cross shaft tapered bearing journal and assemble inner and outer bearings, using suitable sleeves.
9. Install cage and bearing assembly in drive unit, pressing against outer bearing cone with suitable sleeve. Be sure oil holes in cage are aligned with oil holes in drive unit.
10. Assemble bearing retainer plate with two capscrews and secure with lock wire.
11. Install bearing cage cover over original shim pack (which control tapered bearing preload) over studs next to cage flange.
12. Assemble bearing cage cover, lock washers and stud nuts.
13. Wrap a strong cord around spur pinion, Fig. 6, and pull on a horizontal line on pound scale to measure cross shaft bearing preload torque. Proper preload torque is 5 to 15 inch pounds, which is equivalent to 2½ to 7½ pounds pull on scale.
14. Read rotating pull (not starting pull). If torque is not within limits, add shims between cover and cage to decrease, or remove shims to increase torque.
15. Install thru-shaft and pinion.
16. Adjust pinion and gear to secure correct gear tooth contact and backlash.

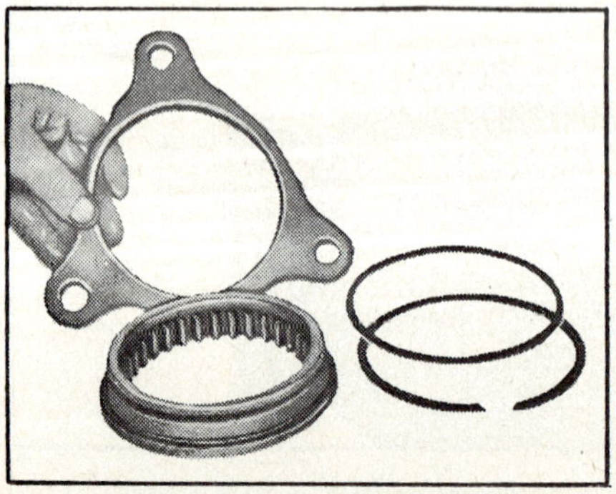

Fig. 12 Assemble yoke to shift collar

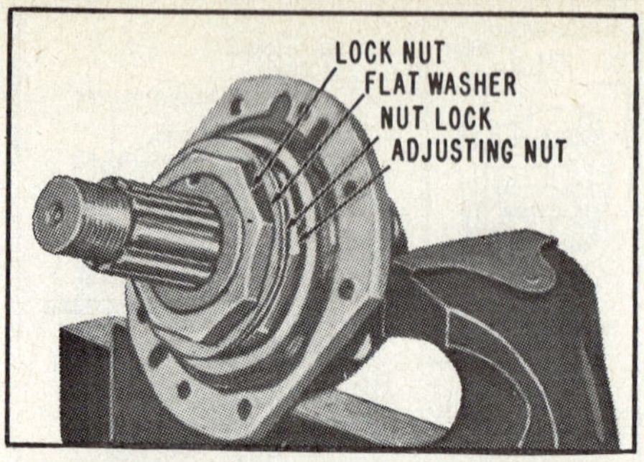

Fig. 10 Position of pinion locking components on hypoid gear units

Assemble Type 2 Cross Shaft

Gear Splined to Hub, Hub Keyed to Cross Shaft

1. Apply graphite to cross shaft tapered bearing journal. Press inner bearing firmly against shaft shoulder with suitable sleeve.
2. Assemble gear to hub, aligning 6 gear-to-hub capscrew holes. Tighten screws securely.
3. Position drive unit on press bed with thru-shaft chamber up, and install radial bearing in drive unit sleeve. Large radius of bearing inner race must be toward bevel gear.
4. Support inner race of radial bearing with sleeve and press gear and hub assembly into bearing.
5. Turn drive unit over in press with thru-shaft chamber down, and support gear, hub and bearing assembly on sleeve with notches in end that fit over capscrew heads.
6. Inspect entering end of cross shaft and remove any nicks or burrs. Coat O.D. of shaft with heavy grease.
7. Align keyway in hub with key in cross shaft and press shaft and inner tapered bearing assembly firmly against hub.
8. Remove drive unit from press and install 6 cross shaft cage studs.
9. If tapered bearing cup has been removed from cross shaft cage, reassemble in press using sleeve to install cup firmly against cage shoulder.
10. Install original shim pack (which controls gear backlash) over studs, and cross shaft cage over shim pack, *aligning oil holes in cage with oil holes in drive unit.*
11. Install outer tapered bearing using suitable sleeve.

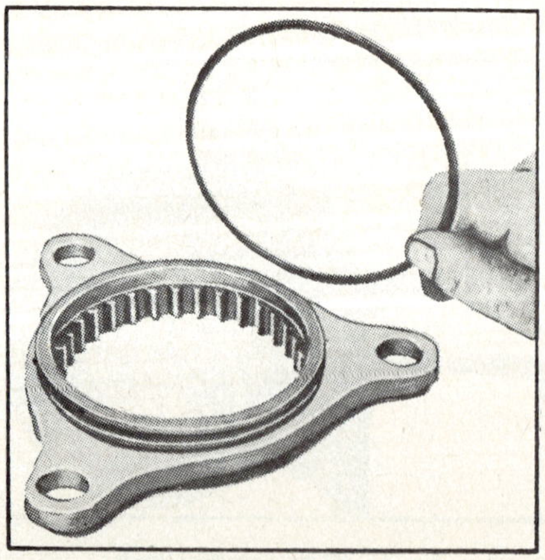

Fig. 13 Assemble spacer over collar yoke

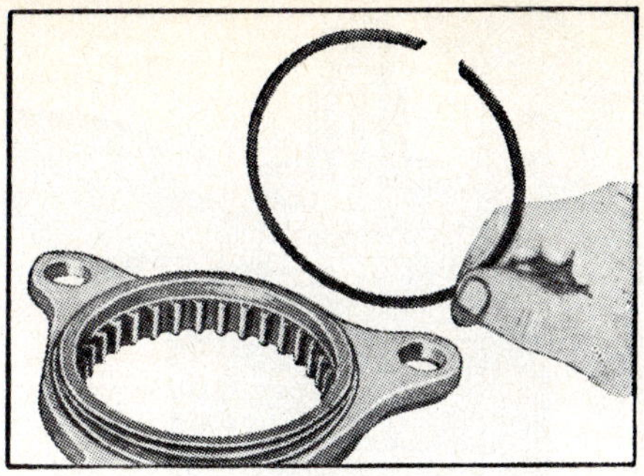

Fig. 14 Install retainer in shift collar groove

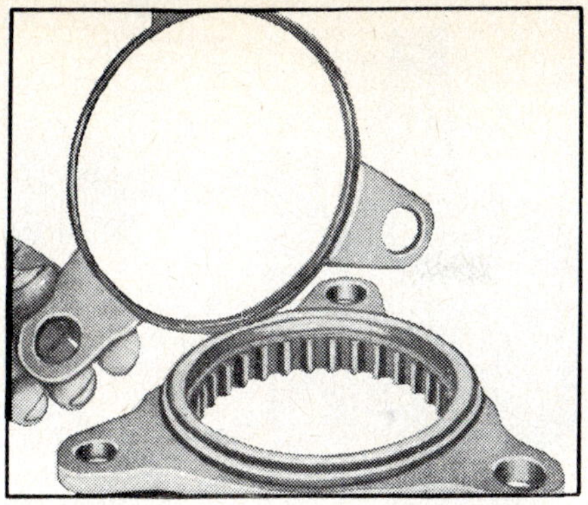

Fig. 15 Position stamped collar retainer cover over retainer

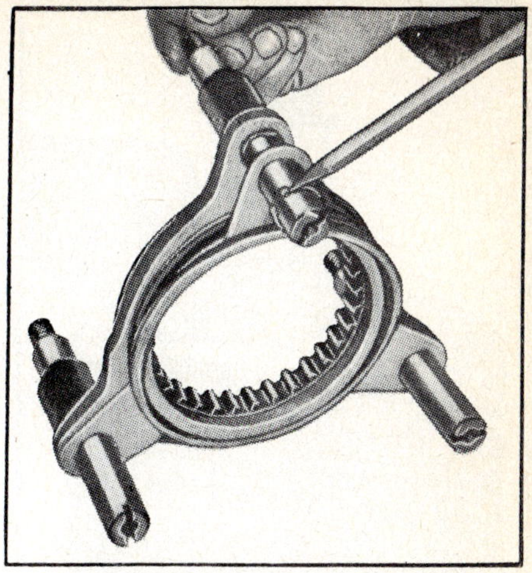

Fig. 16 Assemble shift shafts in collar and yoke and install snap rings in grooves as indicated

12. Install cross shaft bearing adjusting nut. Tighten nut as cross shaft is revolved to position all bearing rollers.
13. Measure cross shaft bearing preload torque by wrapping strong cord around spur pinion, Fig. 7, and pull on a horizontal line with pound scale. Preload torque should be 5 to 15 inch pounds which is equivalent to $2^1/_2$ to $7^1/_2$ pounds pull on scale.
14. Read rotating pull (not starting pull). If torque is not within limits, tighten bearing adjusting nut to increase, or loosen nut to decrease torque. Back off adjusting nut $^1/_4$ to $^1/_3$ turn. Then install nut lock, flat washer and jam nut. Tighten jam nut to 500 ft. lbs. Recheck bearing preload. When correct, bend flat washer over jam nut.
15. Install thru-shaft and pinion.
16. Adjust pinion and gear to secure correct gear tooth contact and backlash.
17. Remove stud nuts and flat washers. Install bearing cage cover gasket, cover, stud nuts and lock washers.

Assemble Type 3 Cross Shaft

Gear Splined to Cross Shaft

1. Install radial bearing in drive unit with large radius of bearing inner race toward bevel gear.
2. Position drive unit in press with thru-shaft chamber up. Coat O.D. of shaft with heavy grease and press parts together about $^1/_4$" to $^3/_8$". Then release press pressure to permit parts to align themselves to prevent distortion and damage. Continue pressing operation until parts are properly assembled.
3. Align gear and cross shaft capscrew holes and assemble gear to shaft. Torque capscrews to 65–85 ft. lbs.
4. Block up gear as shown in Fig. 8. Apply graphite lubricant to cross shaft tapered bearing journal. Install inner bearing firmly against shaft shoulder with suitable sleeve.
5. If tapered bearing cup has been removed, press cup firmly against shaft shoulder.
6. Install original shim pack (which controls gear backlash) over cross shaft cage studs, and cage over shim pack, aligning oil holes in cage with oil holes in drive unit. Install 4 evenly spaced nuts over thick flat washers to hold assembly in place. Tighten nuts securely.
7. Install outer tapered cross shaft bearing with suitable sleeve.
8. Install bearing adjusting nut. Tighten nut as cross shaft is being revolved to correctly position all bearing rollers.
9. Measure cross shaft bearing preload. Wrap strong cord around spur pinion, Fig. 7, and pull on horizontal line with pound scale. Perload torque is 5 to 15 inch pounds which is equivalent to $2^1/_2$ to $7^1/_2$ pounds pull on scale. If preload is not within limits, tighten bearing adjusting nut to increase, or loosen nut to decrease preload torque.
10. Back off adjusting nut $^1/_4$ to $^1/_3$ turn. Install nut lock, flat washer and jam nut. Tighten jam nut to 500 ft. lbs. Recheck bearing preload. When correct, bend flat washer over jam nut.
11. Install gear cover spacers, cover and 3 capscrews. Install pinion and thru-shaft. Adjust pinion and gear for correct gear

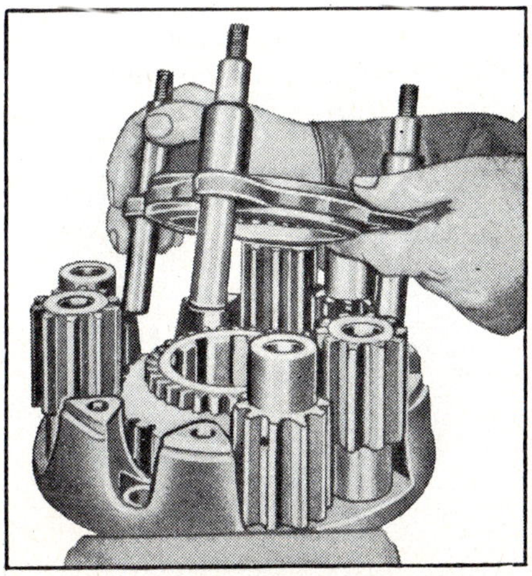

Fig. 17 Install pinions and shift yoke, collar and shaft assemblies, mating internal teeth of collar with teeth of rear side gear

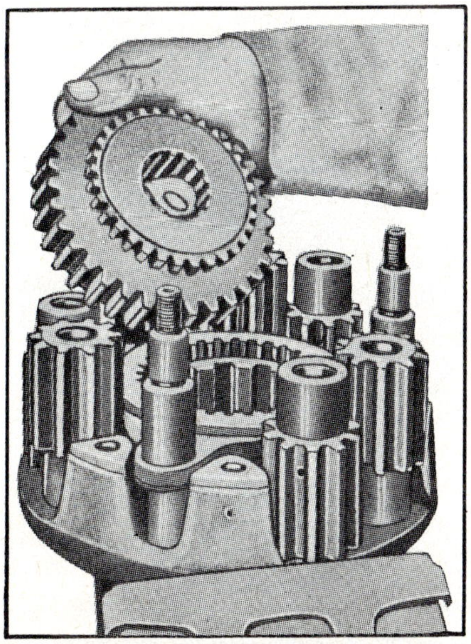

Fig. 18 Install forward differential gear. Turn gears to be sure all parts are correctly assembled and free-moving. Install forward thrust washer

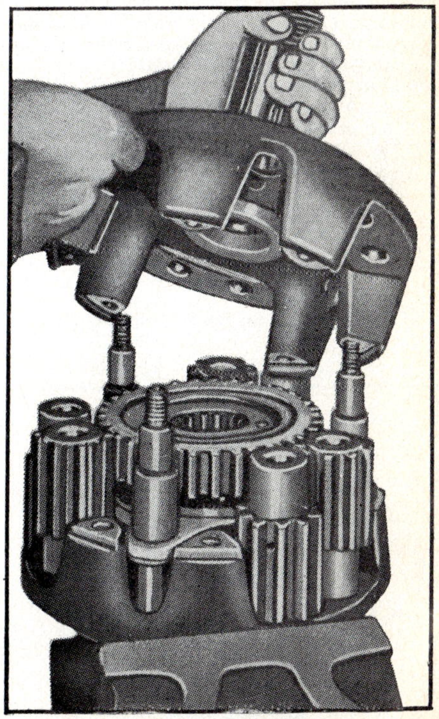

Fig. 19 Install forward case half, being sure to align match marks on case halves

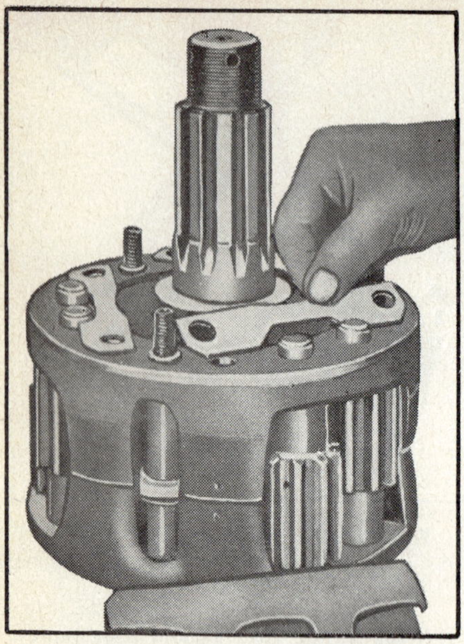

Fig. 20 Install pinion shaft locks

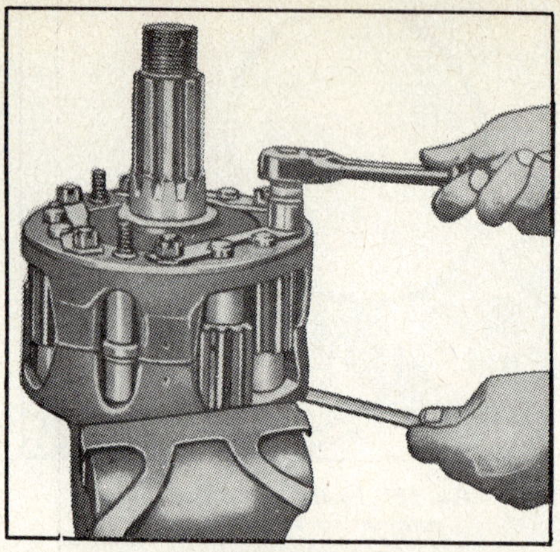

Fig. 21 Fasten case halves together. Check assembly for free movement by rotating pinions

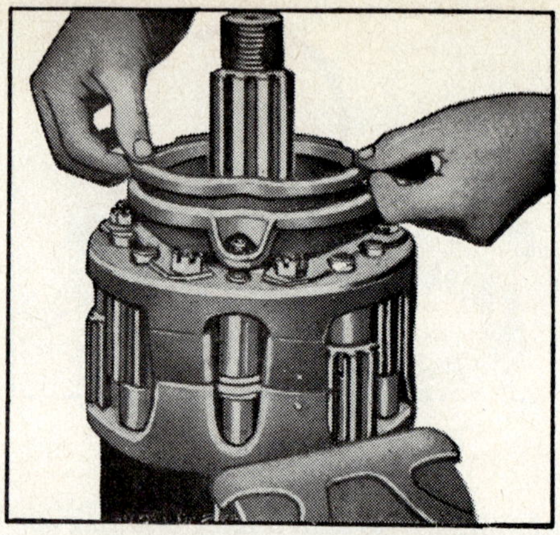

Fig. 22 Install shift collar. Tighten nuts to 65–85 ft. lbs. Be sure shift collar and shafts work up and down freely. If necessary rap assembly at several points on case to free up parts

backlash and tooth contact.

12. Remove stud nuts and flat washers. Install cage cover gasket, cover, nuts and lockwashers.

Install Hypoid Pinion & Thru-Shaft

1. Press pinion bearing cups in cage firmly against shoulder.
2. Coat thru-shaft O.D. and rear bearing inner race I.D. with heavy grease. *Press rear bearing inner race against rear shoulder of thru-shaft of forward carriers and rear carriers that have double splined end shafts. On rear carriers with short single end thru-shafts, install rear bearing inner race retainer, then press race in place and install outer retainer.*
3. Coat pinion and thru-shaft splines with heavy grease. Begin assembly operation in a press and press parts together 1/4" to 3/8". Then relieve press pressure to permit parts to realign themselves to prevent distortion and damage. Continue pressing operation until all parts are correctly assembled.
4. Lubricate all bearing journals with a few drops of engine oil and press inner pinion bearing firmly in place.
5. Position original bearing spacer next to bearing, being sure O.D. chamfer is toward outer bearing.
6. Install cage and cup and outer bearing. Press bearing firmly against selective spacer with suitable sleeve, rotating cage to assure normal bearing contact.
7. Measure pinion bearing preload torque while in press under 9 tons pressure. Wrap strong cord around cage pilot, Fig. 9, and pull on a horizontal line with a pound scale. The preload torque should be 5 to 15 inch lbs. which is equivalent to 2 1/2 to 7 1/2 pounds pull on scale. Use a thinner spacer to increase or a thicker spacer to decrease preload.
8. Remove assembly from press and install and tighten inner nut. Assemble as shown in Fig. 10. When properly tightened, bend lock washer over nut flat.
9. Install thru-shaft rear bearing with suitable sleeve. Check position in forward carrier with cover and seal assembly. In rear carriers with long thru-shaft (both ends splined) check position with cast end cover. In rear carriers with short thru-shaft, use large bearing spacer as a gauge.
10. Loosely install two guide studs in cage capscrew holes to aid reassembly. Position original shim pack over studs so oil holes are aligned. Carefully tap pinion, cage and thru-shaft in place.
11. Install and tighten 4 capscrews over flat washers to hold pinion cage firmly in place while checking hypoid tooth contact. Alter shim packs under cross shaft bearing cage to secure correct tooth contact.
12. Coat pinion oil seal with non-hardening sealing compound and install in cover. Lubricate sealing element.
13. Remove 4 capscrews and guide studs. Coat outer machined surface of cage and inside of cover with gasket cement. Install new gasket on cage flange.
14. Install cover and seal. Apply gasket cement to all covers and mating surfaces and install covers.
15. On rear drive units with thru-shafts splined at both ends, assemble parts as follows: Narrow spacer next to rear bearing inner race. Wide spacer next to narrow spacer. Nut washer, then rear nut.
16. Install forward yokes or flanges. Install narrow spacer at rear end of forward carrier thru-shaft, then yoke or flange.

Assemble Inter-Axle Differential

Assemble inter-axle differential as shown in Figs. 11 to 25. Then proceed with the

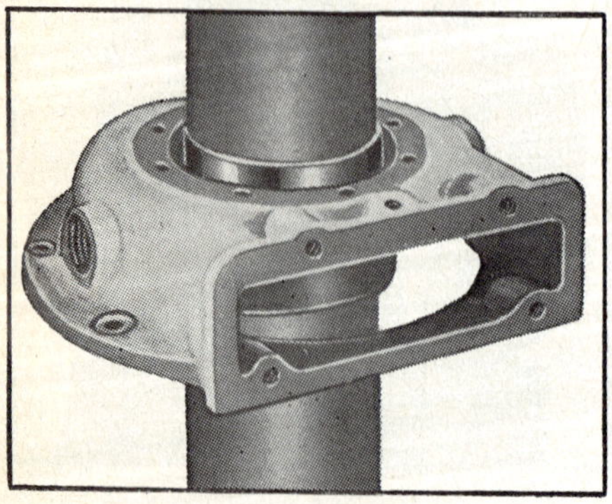

Fig. 23 Install ball bearing in housing cover

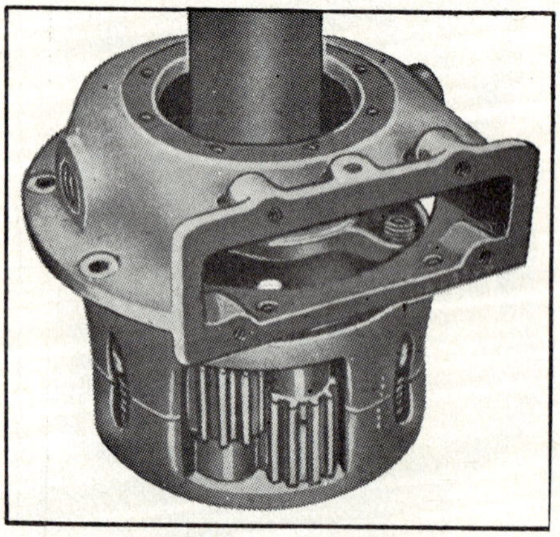

Fig. 24 Assemble cover and bearing assembly to differential. Use a small sleeve over input shaft against bearing inner race. Rotate cover to be sure it clears differential

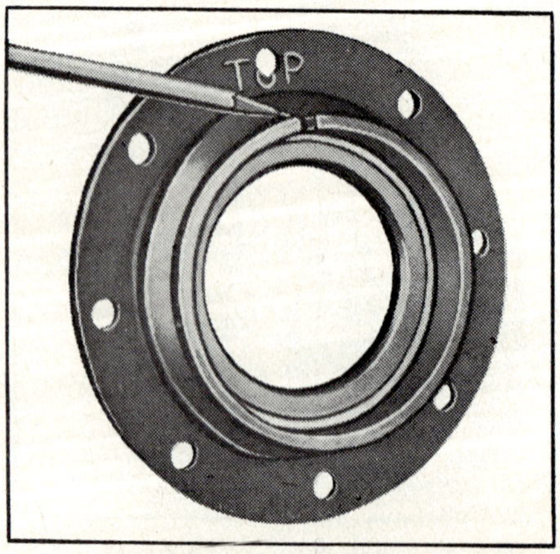

Fig. 25 Install retainer with slot indicated up, aligned with oil hole in cover

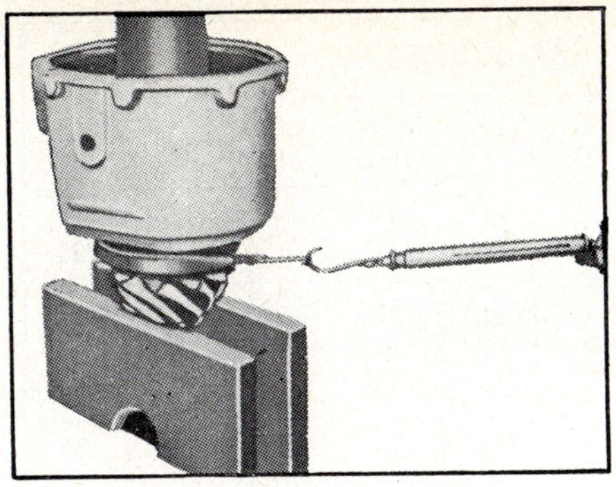

Fig. 26 Measuring pinion bearing preload

assembly as follows:

1. Position new gasket on cover and install bearing and seal retainer.
2. Carefully inspect completed assembly to be sure capscrew ends do not interfere with shift yoke or shift rings.
3. Install yoke or flange, washer and nut.

Assemble Hypoid Pinion & Quill

1. Press pinion bearing cups against housing shoulder. Assemble rear bearing inner race inner retainer. Coat quill O.D. and race I.D. with heavy grease and press race in place. Install outer retainer.
2. Lubricate all bearing journals with a few drops of engine oil and press inner bearing on pinion and quill assembly.
3. Install original spacer on inner bearing with O.D. chamfer of spacer toward outer (forward) bearing.
4. Position housing and cup assembly on pinion. Press bearing firmly against selective spacer with suitable sleeve, rotating housing to assure normal bearing contact.
5. Measure pinion bearing preload while in press under 9 tons pressure. Wrap strong cord around housing pilot, Fig. 26, and pull on a horizontal line with a pound scale. Preload should be 5 to 15 inch pounds which is equivalent to 2½ to 7½ pounds pull on scale.
6. Use a thinner spacer to increase or a thicker spacer to decrease preload. Tighten adjusting nut and recheck preload.
7. Install pinion quill rear bearing in carrier with suitable sleeve. Check location with thru-shaft ball bearing retainer.
8. Loosely install 2 guide studs in carrier if capscrews are used to hold housing in place. Position original shim pack over studs so oil holes are aligned. Carefully tap pinion, housing and gear assembly in place.
9. Install 4 equally spaced capscrews or stud nuts to hold assembly in place while checking hypoid gear tooth contact. Alter shim pack under cross shaft bearing cage and inter-axle differential housing to secure correct tooth contact.
10. Install remaining capscrews or stud nuts and secure with lock wire.

Fig. 28 Adjusting shift fork, using feeler gauge

Install Inter-Axle Differential

1. Turn carrier so pinion and quill assembly is in vertical position and loosely install 3 guide studs in differential housing. Position gasket in place and install differential and cover. Revolve differential back and forth to engage splines of pinion with rear gear. Install capscrews.
2. Revolve differential to be sure it is free turning. If not, remove it and inspect parts.
3. Check position of oil scoop for interference at helix baffle at left inside housing of early models.
4. Install and tighten 2 capscrews at shift lever housing. Secure with lock wire.

Install Shift Unit Housing

1. Hold shift unit in vise and install push rod in housing and fork, Fig. 27. Be sure round button of fork faces front. Assemble push rod in place with sleeve.
2. Place copper gasket over end of push rod and install diaphragm assembly. Hold push rod with screwdriver while starting nut over lock washer. Be sure flats of push rod fit in slot of diaphragm insert as nut is tightened. Tighten nut to 4–6 ft. lbs. Install shift unit cover plug.
3. Attach assembly to cover with 2 capscrews over lock washers. Install 2 tapered dowels, special lock washers and stud nuts.

Adjust Shift Fork (Late Models)

This adjustment is for later models without shift shaft springs that have both forward and rear adjusting screws. Always check shift fork location with feeler gauge, Fig. 28, to be sure there is .010" clearance (minimum) at guide rings or groove.

Lockout Disengaged, Forward Adjusting Screw

1. Back off adjusting screw lock nut so screw is free turning.
2. Shift assembly to disengaged position with power chamber.
3. Turn in adjusting screw until it is finger-tight against fork.
4. Turn in screw ½ revolution more to center fork in groove.
5. Tighten lock nut.

Lockout Engaged, Rear Adjusting Screw

1. Back off adjusting screw lock nut so screw is free turning.
2. Shift assembly to engaged position with power chamber. Be sure shift collar is fully engaged with rear gear.
3. Turn in screw until it is finger-tight against push rod.
4. Turn in screw ½ revolution more to center fork in groove. Tighten lock nut.

Adjust Shift Fork (Early Models)

This adjustment is for early models that have shift shaft springs and only a forward adjusting screw. With lockout disengaged, adjust as follows:

1. Back off adjusting screw lock nut.

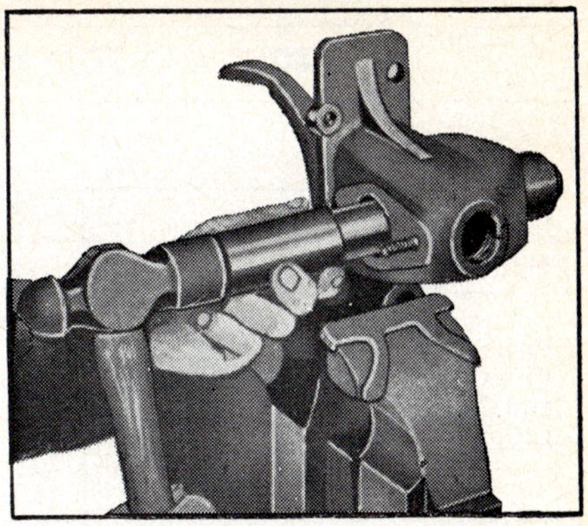

Fig. 27 Installing shift unit push rod

2. Turn screw in finger-tight against fork. This forces fork against inner guide ring.
3. Back off screw ½ revolution to center fork between guide rings.
4. Tighten lock nut.

Install Thru-Shaft

1. Coat seal body and seat with non-hardening sealing compound and install in cover with suitable sleeve. Lubricate sealing element.
2. Install thru-shaft ball bearing firmly against shaft shoulder and assemble shaft and bearing in retainer.
3. Position new gasket over studs and install shaft assembly. Be sure thru-shaft splines mesh with forward gear of inter-axle differential. Do not force in place.
4. Install cover and seal with new gasket. Alternately tighten 4 evenly spaced stud nuts to draw assembly in place and correctly position pinion quill rear bearing. Install and tighten stud nuts.
5. Install thru-shaft yoke or flange.

Tooth Contact & Gear Backlash

Tooth contact may be checked as outlined in the *Single Reduction Axle* section. Spiral bevel gear first reduction units have shim packs between the forward and rear thru-shaft tapered bearing covers and the carrier to control pinion position. Hypoid gear first reduction units have a single shim pack between the pinion cage and carrier or inter-axle differential housing and carrier to control pinion position. A shim pack between the cross shaft bearing cage flange and carrier controls the

Fig. 29 Installing split ring

BOLT AND STUD NUTS

LOCATION	DIAMETER	NO. THREADS	NUT THICKNESS	TORQUE—LB. FT. MIN.	TORQUE—LB. FT. MAX.
Push Rod Nut	5/16″	18		4	6
Shift Unit	3/8″	24		39	42
Pinion, Thru-Shaft and Cross Shaft Bearing Cages and Covers and Inter-Axle Differential Housings	3/8″	24		39	42
	7/16″	20		61	66
	1/2″	20	3/8″	73	80
	1/2″	20	7/16″	94	102
	1/2″	20		94	102
	9/16″	18		132	145
	5/8″	18		186	205
Carrier to Housing	1/2″	20		94	102
	5/8″	18		186	205
	3/4″	16		325	360
Axle Differential Case Bolt	1/2″	20		94	102
	9/16″	18		132	145
	5/8″	18		186	205
	3/4″	16		325	360
Inter-Axle Differential Shift Shaft Nut	1/2″	20		65	85
Inter-Axle Differential Case Bolt Nut	1/2″	20		94	102
Inter-Axle Differential Shift Housing	7/16″	20		61	66
Shift Fork Adjusting Screw Lock Nut	1/2″	20	5/16″	40	55
Shift Fork Set Screw Lock Nut	3/8″	24		30	33
Pinion Shaft, Thru-Shaft and Inter-Axle Differential Case Shaft	1″	20		300	400
	1¼″	18		700	900
	1½″	18		800	1100
	1½″	12		800	1100
	1¾″	12		800	1100
Pinion Bearing Adjusting and Lock Nut	3⅜″	12		1000	1200

CAP SCREWS

LOCATION	DIAMETER	NO. THREADS	TORQUE—LB. FT. MIN.	TORQUE—LB. FT. MAX.
Differential Bearing Adjusting Nut Lock	5/16″	18	15	17
	1/2″	13	85	91
Cross Shaft Bearing Lock	7/16″	14	42	45
	1/2″	20	73	80
	9/16″	12	92	101
Inter-Axle Differential Shift Unit Housing	7/16″	14	54	58
Pinion, Thru-Shaft and Cross Shaft Bearing Cages and Covers and Inter-Axle Differential Housings	3/8″	16	26	29
	1/2″	13	85	91
	9/16″	12	120	129
	5/8″	11	168	180
Inter-Axle Differential Housing Cover and Bearing Retainer	3/8″	24	30	33
	7/16″	14	54	58
Differential Carrier Top, Rear and Side Covers	3/8″	16	26	29
Inter-Axle Differential Shift Fork Set Screw	3/8″	24	30	33

AXLE DIFFERENTIAL BEARING-CAP CAP SCREWS OR STUD NUTS

CAP SCREW OR STUD NUT DIAMETER	CAP SCREW OR COARSE STUD THREAD	STUD NUT OR FINE THREAD	TORQUE—LB. FT. MIN.	TORQUE—LB. FT. MAX.
5/8″	11	18	127	140
3/4″	10	16	230	250
7/8″	9	14	345	370
7/8″	14	14	375	415
1″	14	14	570	630

Fig. 30 Torque specifications for double reduction thru-drive tandem axles. Torques given apply to parts coated with machine oil. For dry (or "as received") parts increase torques 10%. For parts coated with multipurpose gear oil decrease torques 10%. Nuts on studs use same torque as driving the stud

position of the first reduction spiral bevel or hypoid gear in all top-mounted double reduction drive units.

Set backlash to the amount specified on gear O.D. When backlash is not specified on gear set to .006–.012″.

Be sure all bearing preloads are within specified limits before checking tooth contacts. Transpose equal amounts of shims between front and rear pinion bearing covers of spiral bevel drive units to maintain correct bearing preload while adjusting tooth contact. Add or remove shims between pinion cage or inter-axle differential housing of hypoid drive gear units to alter pinion location. Change the location of spiral bevel or hypoid gear by altering the thickness of the shim pack between the cross shaft bearing cage flange and carrier.

In hypoid gearing the actual backlash changes .008″ for each .010″ movement of the gear. In spiral bevel gearing, the actual backlash changes .007″ for each .010″ movement of the gear.

Install Axle Differential

1. Press differential bearings firmly against case shoulders. Coat inside of case and all differential parts with rear axle lubricant.
2. Assemble differential case half and gear. Install side gear thrust washers, side gears and pinion gears, thrust washers and spider.
3. Note case alignment marks and assemble opposite case half. Hold assembly together with 4 bolts and nuts and check for free rotation of parts.
4. Install remaining case bolts so heads are locked by machined relief in case half. Be sure case halves are assembled to gear so there is adequate nut clearance. Check clearance in carrier before completing assembly.

NOTE: Apply Loctite grade 277 to the case bolts. Allow Loctite to cure for 4 hours before adding oil and 24 hours before subjecting axle to heavy duty use.

---

5. When a new gear or differential case is installed, the case holes must be line-reamed with the gear in order to assemble the parts, using the correct size rivets or bolts. Align case halves and hold together with 4 bolts. Line ream the holes and clean parts before assembly.

Install Differential Bearing Cups

1. Temporarily install bearing cups, threaded adjusting rings (when used) securely.
2. Bearing cups must be a hand push fit in bores, otherwise bores must be re-worked with a scraper or emery cloth until a hand push fit is obtained. Use a blued bearing cup as a gauge and check fit as work progresses.

Bearings With Two Threaded Adjusting Nuts

1. Apply rear axle lubricant to bearing cups and cones. Position cups over cones and install in carrier.
2. Insert differential bearing adjusting nuts and turn hand tight against bearing cups.
3. Install bearing caps. Install and tighten stud nuts or capscrews.
4. Alternately loosen one adjusting nut and tighten the other while turning the differential to assure normal bearing contact and to keep bearing cups straight in the bores.
5. With a dial gauge, establish zero end play. Then tighten adjusting nuts 1¾ to 2½ notches (total for both nuts) tight to correctly preload bearings.

6. Install adjusting nut locks and capscrews.

Type With Two Split Rings

1. With differential and bearings installed, insert thin split rings, making certain that there is clearance between bearing cup faces and rings (do not install bearing caps).
2. With a dial indicator, measure end play of differential by shifting assembly back and forth between rings with a pair of small pinch bars placed between carrier legs and spur gears.
3. Remove and measure thickness of rings. To the total thickness of the 2 thin rings add the end play figure plus another .017" to .022" to obtain the total thickness of the 2 thicker rings required to obtain proper bearing preload. Split rings are available in increments of .005".
4. Insert one split ring in carrier leg groove. Move differential over so that face of bearing cup is held tightly against inserted ring. Rings should be positioned in carrier leg grooves so that split portion will locate in center of cap.
5. Install opposite split ring by tapping it into the carrier leg, Fig. 29, tapping on I.D. of ring opposite split portion.
6. Install differential bearing caps and install capscrews.

Timken Worm Drive Axle

This type axle, Fig. 1, incorporates a forged, ground and polished worm of alloy steel and a worm wheel compounded of special phosphor-bronze. Worm drive units are usually incorporated in tandem drive vehicles where high engine horsepower is required to move heavy loads and where chassis design may be a factor limiting drive unit size.

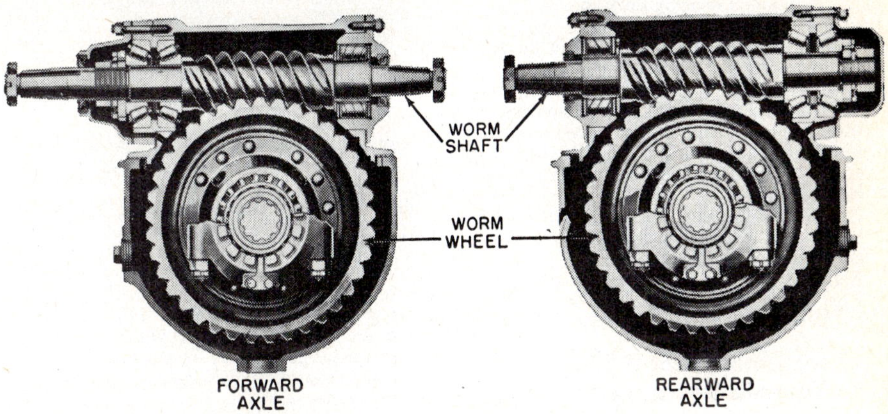

Fig. 1 Timken worm drive tandem axle units. Some units have the tapered bearing cage at the rear of both drive units. The tapered bearings shown are maintained by adjusting nuts. Units of more recent design have selective spacers to establish and maintain tapered bearing adjustment

Remove Drive Unit from Housing

1. Disconnect U-joints at ends of worm shafts. Remove plug at bottom of housing and drain lubricant.
2. Remove axle shafts.
3. Remove attaching stud nuts and break drive unit loose from axle housing with soft mallet and remove tapered dowels. Dowels must be removed before drive unit can be lifted from housing. Back out housing studs if necessary.

DISASSEMBLE DRIVE UNIT

1. Remove flange nuts and stud nuts or capscrews from tapered cage cover. Hold forward yoke or flange or rear drive unit while loosening tapered bearing jam nut.
2. Pull yokes or flanges from worm shaft with suitable puller. Remove key (if used) from worm shaft.
3. Match mark one differential bearing cap and carrier leg to identify at reassembly.
4. Remove all locking wire and worm wheel baffle (if used). Remove differential bearing caps, bearing cups, adjusting nuts and differential assembly.

Disassemble Differential Unit

Two types of differential assemblies are used. One uses bolts and nuts to hold differential case halves to worm wheel; the other uses rivets.

1. To disassemble the bolt-on type, match mark differential case halves and worm wheel so parts can be correctly identified if they are to be reused. Drive bolts from assembly with brass drift and pull assembly apart.
2. To disassemble the riveted-on type, match mark differential case halves and worm wheel for identification on reassembly if parts are to be re-used. Drill out rivets. Tap differential case half free and remove side gear and thrust washer. Turn worm wheel over and repeat operation.
3. Match mark two halves of differential driving ring at matching lugs for reassembly. Some driving rings have a dowel and hole in respective halves to insure correct assembly.
4. Drive or press driving ring, spider, pinion and thrust washer assembly from worm wheel. Then separate halves of driving ring to disassemble spider and pinions.

Remove Worm Shaft

1. Remove worm shaft and tapered bearing

Fig. 2 Removing straight roller bearing outer race lock screw

Fig. 3 Removing oil baffle (when used)

Fig. 4 Installing selective bearing spacers

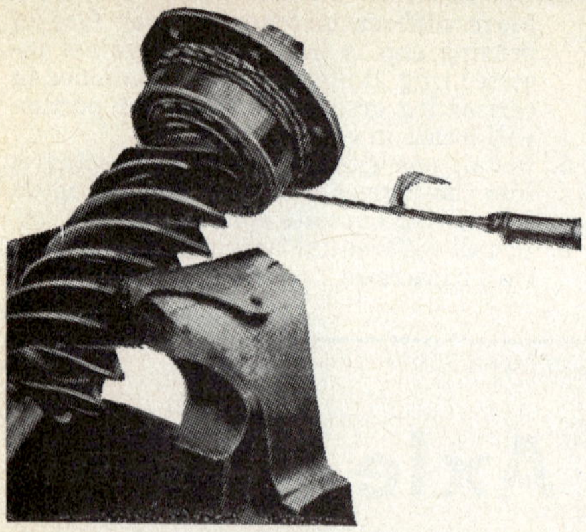

Fig. 5 Measuring worm shaft tapered bearing preload

cage from drive unit by tapping on end opposite tapered bearing cage.

2. Remove roller bearing or self-aligning bearing end of drive unit. Remove bearing cage or straight roller bearing cover and seal assembly. Remove roller bearing spacer (if used), roller bearing, and outer race lock screws. Fig. 2.
3. Carefully tap race from drive unit with brass drift. Remove roller bearing from cage and seal as required.
4. Remove worm shaft oil baffle (if used), Fig. 3. Identify location of one end of baffle to aid in reassembly.

Disassemble Worm Shaft & Tapered Bearing Cage

1. Remove attaching nut and related parts and press shaft out of bearing cage. Do not lose selective spacers that are between bearings.
2. Tap hardened steel spacer from roller bearing end of worm shaft.
3. Remove bearing cups from cage with suitable puller.
4. Remove seals or felts from front and rear covers or cages.

REASSEMBLY

Replace worm wheels that have worn or broken teeth and worm shafts where there is evidence of checked or damaged case hardening.

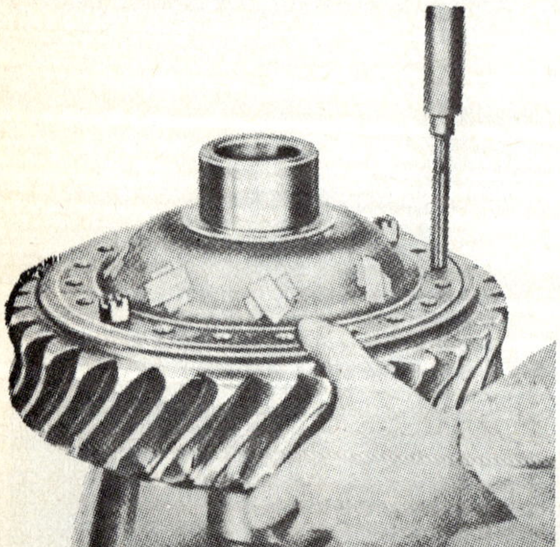

Fig. 8 Differential case holes covered with masking tape to prevent chips from falling into gear assembly when line-reaming case-to-worm wheel holes for bolt or rivet installation

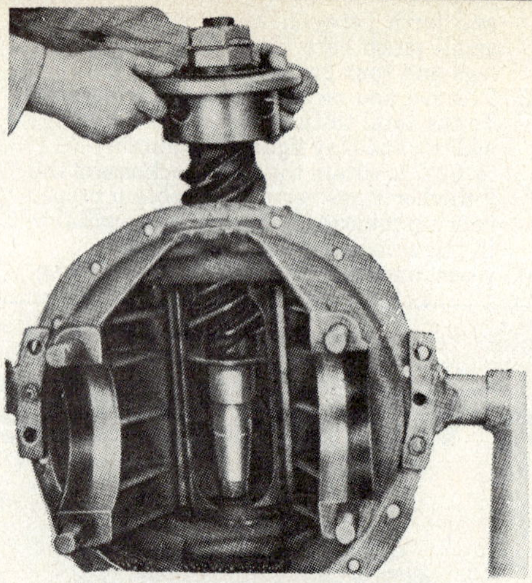

Fig. 6 Installing worm shaft assembly

Inspect differential for pitted, scored or worn thrust surfaces of differential case halves, thrust washers, spider trunnions and differential gears. *Thrust washers must be replaced in sets. The use of a combination of old and new thrust washers will result in premature failure.* Check for wear or damage to differential gear teeth. Inspect spider trunnions for looseness in differential case or driving ring bores.

Worm & Tapered Bearing Cage

1. Press inner and outer tapered bearing cups in cage with suitable sleeve.
2. Position outer tapered bearing on suitable sleeve in press. Place cage and cup on outer bearing.
3. Position original selective spacers in cage on outer bearing, Fig. 4.
4. Press worm shaft into cage and bearing as cage is rotated to assure normal bearing contact. *Begin the assembly operation in the press making sure the parts are properly aligned. Press the parts together about 1/4" to 3/8", then relieve the press pressure to permit them to realign themselves to prevent distortion or damage.* Continue the pressing operation until the parts are correctly assembled.
5. Install and tighten adjusting nut.

Tapered Bearing Preload

As shown in Fig. 5, wrap strong cord around tapered bearing cage three or four times and attach a pound scale to cord end. Pull scale on horizontal line and read pounds pull. Use rotating pounds pull (not starting pull). Tapered bearing preload may be measured while the assembly is in the press under pressure. The following table gives equivalent nut torques and press pressures:

Worm Shaft Thread Size	Specified Nut Torque Ft. Lbs.	Equivalent Press Pressure In Tons
1" × 20	300–400	6
1¼" × 18	700–900	11
1½" × 12	800–1100	14
1½" × 18	800–1100	14
1¾" × 12	800–1100	14
2" × 16	800–1100	14
2⅛" × 12	800–1100	14

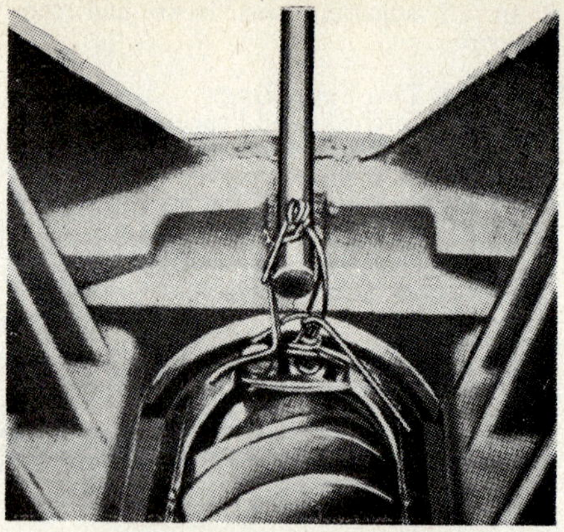

Fig. 7 Wire threaded through oil baffle holes to contract baffle

Compute bearing preload torque by multiplying radius of bearing cage pilot by the number of pounds pull. For example, assume diameter of cage pilot is 6", the radius is 3". If the scale reading is one pound rotating pull, the bearing preload is 3 inch pounds. Tapered bearing preload should be zero worm shaft end play to 5 inch pounds maximum torque. If in excess of 5 inch pounds when nut is correctly tightened, remove nut and outer bearing cone and substitute a thicker spacer. If less than zero worm shaft end play, remove spacer and add one thinner than the one removed. Continue this procedure until correct bearing preload is obtained.

Adjust splined worm shaft tapered bearing preload in the same manner, omitting the seal and cover assembly (where used) until correct preload is obtained.

Position nut lock and washer (where used) over adjusting nut and tighten jam nut. Check tapered bearing preload and readjust if necessary. Carefully tap steel washer on opposite end of worm shaft.

Assemble Worm Shaft In Drive Unit

1. Coat oil seals with non-hardening sealing compound and assemble in cover or cage

Fig. 9 Differential components used with riveted case

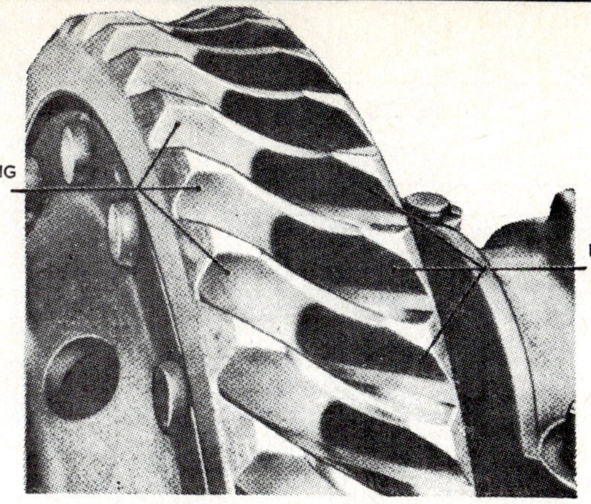

Fig. 10 Satisfactory tooth contact. Starting at leaving side 30 to 60% full tooth length

with suitable sleeve.
2. Install roller bearing outer race (where used) in drive unit with suitable sleeve. Be sure lock screw holes are properly aligned. Install and tighten bearing lock screw.
3. Note identifying location marks previously made on worm shaft oil baffle (new baffles have stenciled installation directions).
4. Install baffle so its tangs engage notches in drive unit. Position new gasket over tapered bearing cage studs.
5. Install worm shaft and bearing cage assembly in drive unit, Fig. 6.
6. Thread two or three strands of wire through oil baffle holes and twist wire to contract baffle, Fig. 7.
7. Tap worm shaft assembly into drive unit so baffle will fit against I.D. of tapered bearing cage. *Cut and remove wire, being sure no small bits of wire drop into drive unit.*
8. Install new gasket over tapered bearing cage flange and install cage cover or cover and seal assembly. Tighten stud nuts over lock washers.
9. Install roller bearing in opposite end of drive unit, or install self-aligning bearing and cage with new gasket. Install roller bearing outer washer and dowel assembly (if used).
10. Position new gasket over cover or cage and install assembly. Tighten stud nuts or capscrews.
11. When aluminum cover and seal assemblies are used, large flat steel washers are located beneath stud nut or capscrew lock washers.

Assemble Worm Wheel & Differential

Bolt-On Type

1. Assemble worm wheel to differential case half, carefully noting case matching marks. A suitable drift can be used in aligning case half and worm wheel holes.
2. Turn assembly over and install side gear thrust washer and side gear in case half.
3. Assemble differential pinions and thrust washers on spider. Install spider and pinion assembly in case half.
4. Install opposite side gear and thrust washer. Install opposite case half, carefully noting case matching marks.
5. Install two 7/16" × 4 1/2" bolts in assembly and tighten nuts. Check assembly for free rotation of gears.
6. *When a anew worm wheel or differential case is installed, the case holes must be line-reamed with the worm wheel holes to properly reassemble, using correct size bolts. Use care to prevent ream chips from falling into assembly. Case holes may be covered with masking tape, Fig. 8, or plugged with stoppers.*
7. Remove two bolts used to hold assembly together.
8. Install differential assembly bolts so heads are locked by case half flange shoulder and safety wire holes in bolts are roughly parallel to worm wheel O.D. Install and tighten nuts. The two case halves are usually identical except for width of bolt circle flange shoulder. Install bolts so that bolt heads will fit snugly against locking shoulder of one case half.
9. Lock bolts in groups of four with soft iron wire. Install differential bearings with suitable sleeve. Lubricate assembly.

Riveted Type

1. Assemble pinion and thrust washers on spider. Assemble spider assembly in half of differential driving ring.
2. Place assembly in worm wheel driving ring slots and assemble other half of driving ring, Fig. 9. Note previously made matching marks (or dowel and its hole) in matching driving ring lugs.
3. Support worm wheel on steel blocks and assemble driving ring and spider assembly to wheel by driving it in wheel slots with a soft hammer. Continue procedure until driving ring assembly is flush with machined face of worm wheel.
4. Install side gear and thrust washer. Assemble differential case half to worm wheel and driving ring assembly, noting differential case matching marks. A suitable drift can be used to aid in aligning case half-to-worm wheel holes.
5. Turn assembly over and install opposite side gear and thrust washer and case half. Note matching marks on case half.
6. Hold assembly together with two 7/16" × 4 1/2" bolts. Check for free rotation of gears.
7. *When a new worm wheel or differential case is installed, the case holes must be line-reamed with worm wheel holes to properly reassemble, using the correct size rivets. Use care to prevent rivet chips from falling into assembly. Case holes may be covered with masking tape, Fig. 8.*
8. Install rivets cold. When correct, rivet head being formed will be at least 1/8" larger in diameter than rivet hole. The formed head should not be more than 1/16" lower than preformed head as excessive pressure may distort assembly and result in unsatisfactory tooth contact.
9. Install differential bearings with suitable sleeve. Lubricate assembly.

Install Differential

1. Note identifying marks on bearing caps and install bearing cups, nuts and caps. Tighten nuts or capscrews. Then remove caps, cups and nuts. Lubricate parts and reassemble again, tightening nuts or capscrews securely.
2. Alternately loosen one adjusting nut and tighten the opposite nut while turning assembly to straighten bearing cups in bores and assure normal bearing contact.
3. Mount dial indicator to contact machine face of worm wheel and check end play. Established a .000" end play (no preload condition in bearings).
4. Tighten adjusting nuts 1 3/4 to 2 1/2 notches tight (total for both nuts) to correctly preload bearings.

Worm Tooth Patterns

The worm wheel teeth have an entering side (where the worm shaft teeth enter) and a leaving side (where the worm shaft teeth leave). The entering sides of the teeth have the flattest angle while the leaving sides are

Fig. 11 Not enough leaving side contact. Move worm wheel to leaving side

Fig. 12 Contact at entering side. Move worm wheel to entering side

almost straight or vertical.

The proper tooth contact during operation under load is approximately 80% of the tooth length, starting at the leaving side of the tooth, progressing toward the entering side of the tooth. *Do not have worm tooth patterns to have contact at the entering side.*

The correct tooth contact, when turned by hand before operation under load, should be 30 to 60% of worm wheel tooth length, starting at the leaving side of the tooth. To establish correct tooth contact, proceed as follows:

1. Apply a thin coating of Prussian blue to 6 or 8 worm wheel teeth.
2. Set the worm to its maximum amount of backlash, using a dial indicator.
3. Rotate worm shaft to wipe blue from worm wheel, indicating tooth contact.
4. Move worm wheel to one side or the other until contact, starting at the leaving side is 30 to 60% of full tooth length, Fig. 10.
5. Always use drive side of teeth when making adjustments as coast side will automatically show a satisfactory pattern when drive side contact is correct. If tooth pattern shows as illustrated in Figs. 11 and 12, adjust as indicated. As a rule it is necessary to move the worm wheel only a few thousandths of an inch to alter tooth contact.
6. To adjust, move differential bearing adjusting nuts about 1/4 turn at a time until proper tooth contact is obtained. Maintain differential bearing preload during adjustment by first loosening the adjusting nut on the side toward which the differential is to be moved and tightening the opposite adjusting nut the same amount. *Make normal wear adjustments at the left hand bearing only. This will maintain correct worm and wheel alignment.*

Final Assembly

1. Tighten differential bearing cap nuts or capscrews. Install nut locks and lock screws and secure with lock wire.
2. Install and tighten yoke and flange nuts.
3. With new gasket, install drive unit in housing. First install lock washer and stud nuts under any drive unit and housing offset. It is impossible to start these nuts after drive unit is tight in housing.
4. Install tapered dowels (where used). Alternately tighten nuts.
5. Connect U-joints at worm shaft. Install axle shafts and connect brake lines.

BOLTS AND STUD NUTS

Location	Dia.	No. Thds.	Torque—Lb. Ft. Min.	Max.
Oil Baffle	5/16″	24	16	18
Carrier to Housing	½″	20	92	102
	⅝″	18	185	205
	¾″	16	320	360
Worm Shaft Covers and Cages	7/16″	20	59	66
	½″	20	92	102
	9/16″	18	130	145
Differential Assembly and Worm Wheel Bolt	½″	20	92	102
Worm Shaft Yoke or Flange Nut	1″	20	300	400
	1¼″	18	700	900
	1½″	12	800	1100
	1½″	18	800	1100
	1¾″	12	800	1100
Worm Shaft Bearing Locknut	1½″	12	800	1100
	2″	16	800	1100
	2⅛″	12	800	1100

CAP SCREWS

Location	Dia.	No. Thds.	Torque—Lb. Ft. Min.	Max.
Adjusting Nut Lock	5/16″	18	15	17
	5/16″	24	16	18
Oil Baffle	⅜″	16	26	29
Carrier to Housing	½″	13	82	91
Roller Bearing Race Lock	½″	13	82	91
Worm Shaft Covers and Cages	7/16″	14	52	58
	½″	13	82	91
	9/16″	12	116	129

DIFFERENTIAL BEARING CAP CAP SCREWS OR STUD NUTS (Earlier Axle Models Without Hardened Washers)

Cap Screw or Stud Nut Diameter	Cap Screw or Coarse Stud Thread	Stud Nut or Fine Thread	Torque—Lb. Ft. Min.	Max.
⅝″	11	18	126	140
¾″	10	16	225	250
⅞″	9	14	330	370
⅞″	14	14	375	415
1″	14	14	375	415

DIFFERENTIAL BEARING CAP CAP SCREWS OR STUD NUTS (Later Axle Models Employing Hardened Washers)

Cap Screw or Stud Nut Diameter	Cap Screw or Coarse Stud Thread	Stud Nut or Fine Thread	Torque—Lb. Ft. Min.	Max.
⅝″	11	18	160	180
¾″	10	16	290	320
⅞″	9	14	470	520
⅞″	14	14	510	570
1″	14	14	570	630

Fig. 13 Torque specifications for Timken worm drive axles

Shift Unit For Two-Speed Axles

ELECTRIC SHIFT UNIT

Description

Basically, the shift unit, Fig. 1, consists of an electric motor, a worm shaft and wheel, an eccentric and a connecting rod. The electric motor drives the worm shaft and wheel; the eccentric is attached to the worm wheel and drives the connecting rod which operates the shift shaft.

The actual shifting is done by a double acting compressing spring contained in the push rod. This spring is mounted on the shift shaft which is attached to the connecting rod. As the connecting rod pushes (or pulls) the shift shaft, the spring cocks and, in turn, pushes or pulls on the push rod and shift collar. When the driving torque is relieved (by releasing the accelerator or clutch) the spring pressure will complete the shift.

Side Mounted Type

Removal

1. Disconnect electric harness at terminals at shift unit cover.
2. Remove attaching screws and filler plug from cover.
3. Screw a threaded tool into the filler plug hole and pull the cover from the housing. The tool is made from a 1/8 × 27 pipe plug welded to the end of a T handle. *Do not pry cover from housing.*
4. Drive roll pin from shift shaft, using tool having a 5/8″ body and a 5/32″ nose section. *If the roll pin is in line with the worm shaft, it will be necessary to change the position of the shift shaft before the pin can be driven out.*
5. To change the position of the shift shaft, turn on the vehicle ignition switch. Shift the unit to the desired position by touch-

1. Shift Motor Assembly
2. Shift Motor Lock Nut
3. Worm Shaft Spring Thrust Washer
4. Worm Shaft Spring
5. Worm Shaft Spring Washer
6. Worm Shaft
7. Worm
8. Worm Shaft Spring Retainer
9. Shift Motor Gasket
10. Shift Housing
11. Shift Housing Cover Gasket
12. Worm Wheel Cap Screw and Lock Washer—Long
13. Worm Wheel
14. Eccentric
15. Eccentric Connecting Rod
16. Eccentric Cover
17. Eccentric Shaft
18. Shift Cover
19. Shift Cover Attaching Screws
20. Shift Cover Filler Plug
21. Electric Switch and Shift Cover Nut
22. Electric Switch and Shift Cover Washer
23. Electric Switch and Shift Cover Bushing
24. Electric Switch and Shift Cover Insulator
25. Electric Switch Assembly
26. Electric Switch Nut
27. Worm Wheel Cap Screw and Lock Washer—Short
28. Ball Bearing
29. Adjusting Screw Jam Nut
30. Adjusting Screw
31. Oil Seal Assembly
32. Snap Ring
33. Roll Pin
34. Shift Shaft Assembly
35. Push Rod
36. Shift Housing to Sleeve Gasket
37. Push Rod Sleeve
38. Push Rod Sleeve Shim
39. "O" Ring

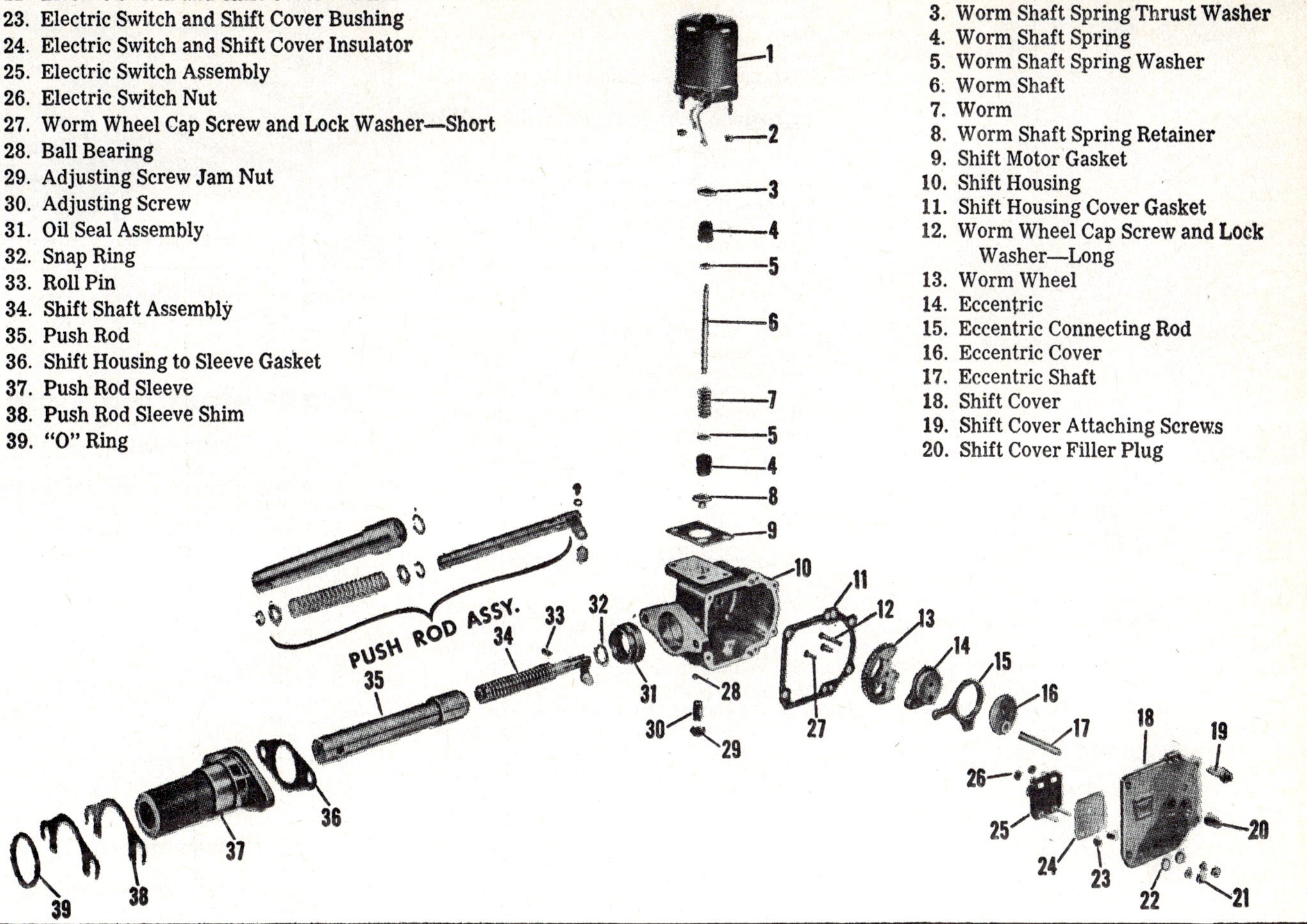

Fig. 1 Electric shift for two speed axles

ing the hot lead of the wiring harness to the left-hand terminal on the shift unit cover. It is not necessary to attach the cover to the housing. *If the shift motor is not operating, disconnect the wires at the inner face of the cover and remove the motor as outlined under Disassembly. Then install a new motor and shift the unit to the desired position. In case a new motor is not available, use a slotted tool that fits over the tang of the worm shaft and turn it until the shift shaft changes position.*

6. Unfasten and remove shift unit. Do not damage oil seal located in flange section.

Disassembly

1. Disconnect wires from terminals on inside face of cover.
2. Remove two stud nuts and star washers that hold motor to housing.
3. Carefully break loose motor and remove it and gasket from housing.
4. Loosen jam nut and remove adjusting screw. *Do not lose ball bearing that rides in recess of adjusting screw.*
5. Remove worm shaft assembly and disassemble it if necessary. *The spring retainer at bottom of worm shaft is pressed on and cannot be removed by hand. However, it can be removed by holding it in a vise and driving the shaft out.*
6. Remove worm wheel and eccentric assembly. To disassemble, hold assembly in vise and remove three capscrews and lockwashers. Drive eccentric shaft out of assembly.

Reassembly

1. Reassemble worm shaft and position it in shift housing.

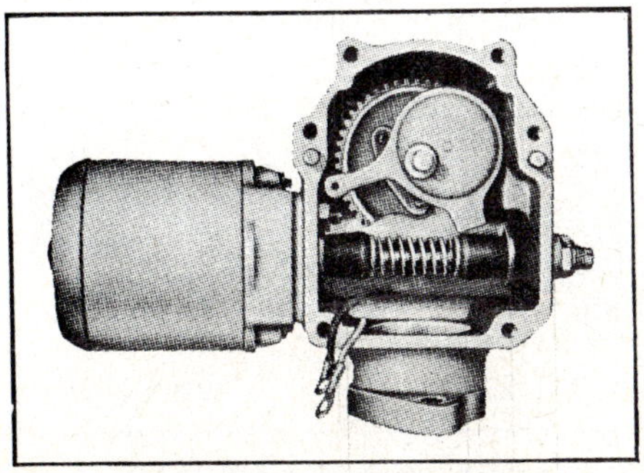

Fig. 2 Worm wheel and eccentric installed

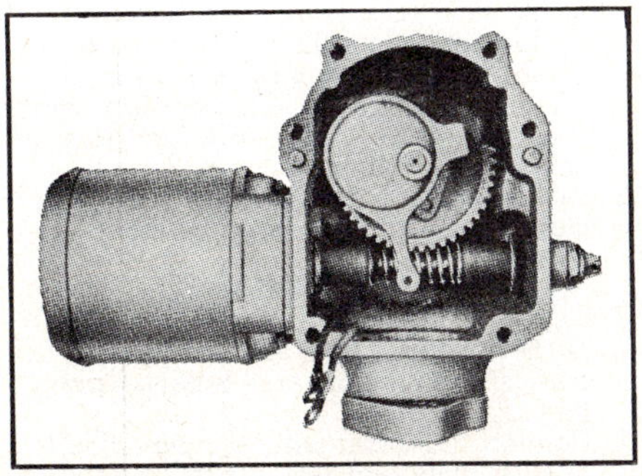

Fig. 3 Shift unit in neutral position

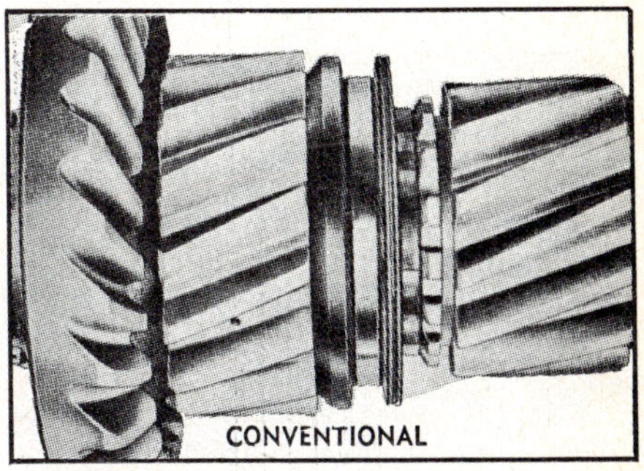

Fig. 4 Location of gears on cross shaft with conventional ratio

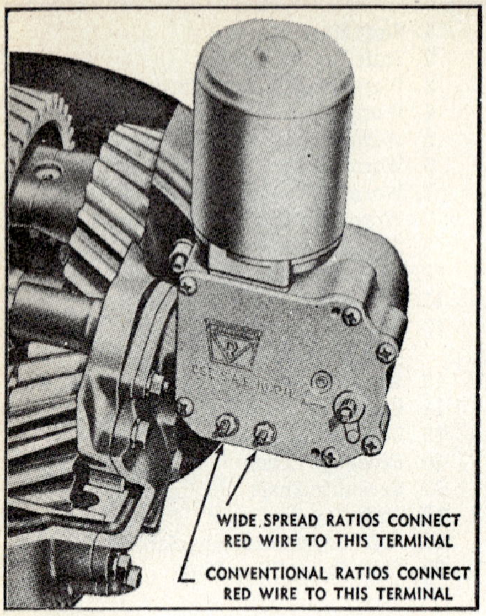

Fig. 5 Electric shift wiring hook-up

2. Install shift motor with gasket to housing. Slot in motor shaft fits over tang of worm shaft.
3. Install and tighten two locking stud nuts that secure motor.
4. Place ball bearing in recess of adjusting screw.
5. Turn adjusting screw in until the ball bearing is snug against bottom of worm shaft. Then back off screw 1/8 turn for proper adjustment. Hold adjusting screw in place and tighten lock nut. Check to see that worm shaft turns freely by hand.
6. Reassemble worm wheel and eccentric. *The eccentric shaft should protrude an equal distance at top and bottom of the assembly.* Install assembly in position shown in Fig. 2.
7. Connect electric leads on inside face of shift cover. Attach red wire to terminal closet to flange of shift unit.

Installation

1. Position shift fork in cross shaft shift collar groove.
2. Slide push rod into shift fork. Line up set screw hole in fork with recess in push rod.
3. Install shift fork set screw and tighten jam nut to 30–33 ft. lbs. torque. Be sure set screw enters recess in push rod.
4. Lock set screw to fork with wire.
5. Remove screws from shift unit cover.

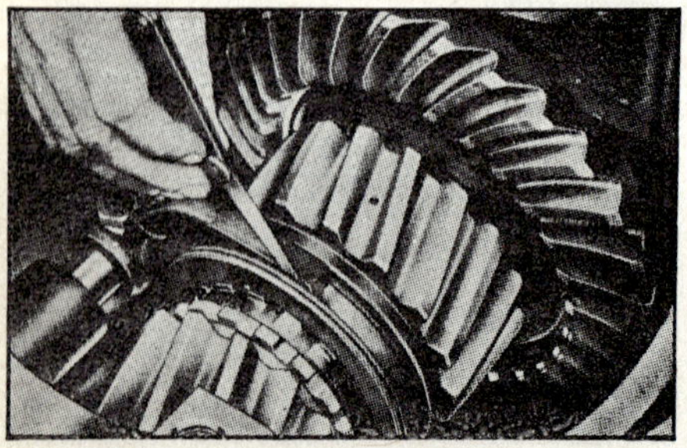

Fig. 7 Checking clearance of shift fork pads in shift collar with feeler gauge

6. Remove cover from housing, using a tool made from a 1/8 × 27 pipe plug welded to the end of a T handle and screwed into filler plug hole. Do not pry cover.
7. Move shift unit to neutral position as shown in Fig. 3.
8. Move cross shaft shift collar to neutral (central) position.
9. Install new shift housing to sleeve gasket.
10. Position shift unit over mounting studs. Do not damage oil seal installed in flange section of shift housing.
11. Move connecting rod into slot in end of shift shaft.
12. Install lockwashers and mounting stud nuts. Tighten nuts to 26–29 ft. lbs. torque.
13. Line up hole in connecting rod with hole in end of shift shaft.
14. When holes are lined up, drive roll pin into place with a suitable driver.
15. Install gasket and cover. *To install cover, shift unit and axle assembly must be in neutral position to avoid damaging the switches.*
16. Fill unit to bottom of filler hole with SAE 10 motor oil. Install filler plug.

Connect Electric System

If a vacuum or air shift unit is being replaced by the electric type, it will be necessary to install an electric hook-up from the cab of the vehicle to the electric shift unit.

If the spur pinions on the cross shaft are located in the conventional manner, with the HI speed (larger) pinion next to the drive gear, Fig. 4, connect the red wire to the terminal indicated in Fig. 5, with the black wire connected to the other terminal.

If the spur pinions are located in the wide range manner, with the LO (smaller) pinion next to the drive gear, Fig. 6, connect the red wire to the terminal indicated in Fig. 5, with the black wire connected to the other terminal.

Check Operation of Shift Unit

1. Turn on ignition switch.
2. Flip switch button in cab to operate shift unit. One man can operate switch while another observes movement of shift collar on cross shaft through pinion cage bore.
3. Check to make sure shift collar completes shift into both HI range and LO range. The engaging teeth on shift collar must completely engage corresponding teeth of spur pinions. Incomplete shifts can cause serious axle failures.

Adjustments

1. Shift collar travel can be equalized by altering the shim pack installed behind flange of push rod sleeve.
2. Check clearance of shift fork pads in shift collar with feeler gauge, Fig. 7. Clearance should not be less than .010" on each side of fork in both HI and LO speed positions. *When checking shift fork clearance, shift collar must be flush with end face of engaged spur pinion.*
3. Add or remove shims from pack under sleeve to obtain correct adjustment.

Final Assembly

1. If above adjustments are satisfactory, install shims, pinion and cage assembly, and capscrews.
2. Connect universal joint at pinion shaft.
3. Load vehicle and test operation of shift unit under actual operating conditions.

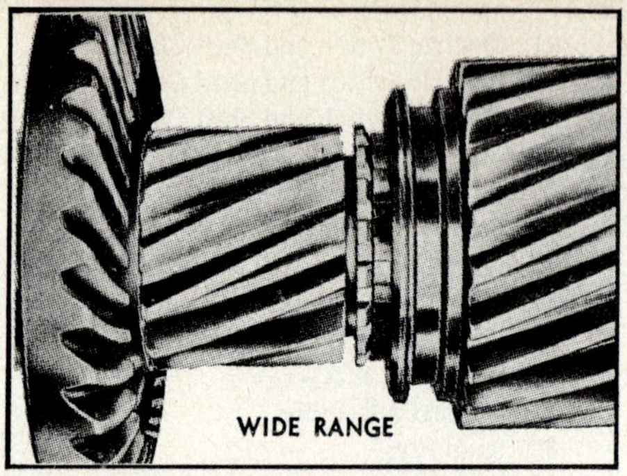

Fig. 6 Location of gears on cross shaft with wide range ratio

Front Mounted Type

Removal

1. Disconnect wires from terminals on shift cover.
2. Unfasten and remove shift unit from carrier. *The push rod must be drawn in toward shift housing before it can be removed from carrier, otherwise it will not clear bore in carrier.*
3. To move push rod to correct position, first turn on ignition switch. Then shift the unit by touching the hot lead of the wiring harness to the right-hand terminal. If shift motor is not working, see procedure under *Change Position of Shift Shaft* in the Side Mounted Type procedure. Do not attempt to force or pry shift unit from carrier.

Disassembly

1. Remove cover attaching screws and filler plug.
2. Screw a threaded tool into filler plug hole and pull off cover. Tool can be made from a 1/8 × 27 pipe plug welded to the end of a T handle. Do not pry off cover.
3. Drive roll pin from shift shaft. *If roll pin is in line with worm shaft, it will be necessary to change position of shift shaft by turning the worm.*
4. Remove two stud nuts and separate shift unit from push rod and bracket assembly.
5. Proceed with disassembly as outlined for the Side Mounted Type.

Reassembly

1. Install worm shaft assembly, shift motor, adjusting screw, and worm wheel and eccentric assembly as outlined for the Side Mounted Type.
2. Place worm shaft and worm wheel in the neutral position (see Fig. 3).
3. Install push rod and bracket assembly.

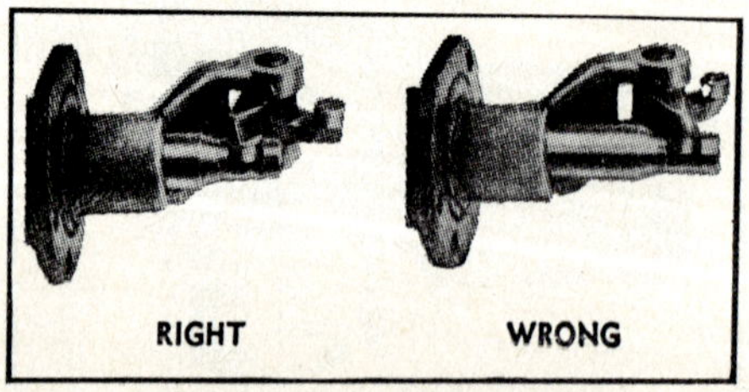

Fig. 8 Positioning push rod to permit entry in carrier bore

Fig. 9 Shift collar travel adjustment. Front mounted type

Tighten nuts to 26–29 ft. lbs. torque.
4. Slide connecting rod into slot at end of shift shaft.
5. Line up hole in connecting rod with hole in shift shaft and install roll pin with suitable driver.
6. Connect wire leads at inside face of shift cover. Attach red wire closest to flange of shift unit.
7. Install cover with new gasket. The shift unit must be in neutral position when cover is installed to avoid damage to switches.

Installation

1. Move cross shaft shift collar to engage spur pinion opposite drive gear.
2. Shift the unit to the corresponding position (with push rod drawn in toward shift unit housing, Fig. 8). Push rod must be in "right" position, otherwise it will not fit in carrier bore. *To shift the unit, use a battery and two leads; ground one lead and touch the other to the right-hand terminal on the cover.*
3. Install new gasket and slide shift unit into carrier bore.
4. Remove inspection cover from top of carrier and look inside to guide push rod bellcrank into slot in shift shaft.
5. Install and tighten attaching capscrews to 26–29 ft. lbs. torque.
6. Fill unit with SAE 10 motor oil to filler plug hole. Install filler plug.

Connect Electric System & Check Operation

Connect wires as outlined for the Side Mounted Type. Check the operation of the shift unit as outlined for the Side Mounted Type, observing the movement of the shift collar through the inspection hole at top of carrier.

Adjustments

Shift collar travel can be adjusted by the stop screws installed in the carrier assembly, Fig. 9. One stop screw is installed in cross shaft bearing cage, opposite the drive gear, and the other is installed in top of the carrier. *Do not disturb settings of stop screws unless shift fork adjustments are required.*

1. If necessary to adjust, first be sure fork and collar are shifted to engage the helical pinion next to the drive gear. Then adjust the stop screw and lock nut in the top of the carrier so that the fork is centered in the collar groove within .005".
2. With fork and collar shifted to engage pinion away from drive gear, adjust stop screw and lock nut in the cross shaft bearing cage so that the fork is centered in the collar groove with .005".
3. When checking shift fork clearance in collar groove, collar must be flush with end face of spur pinion in both the HI and LO speed positions.

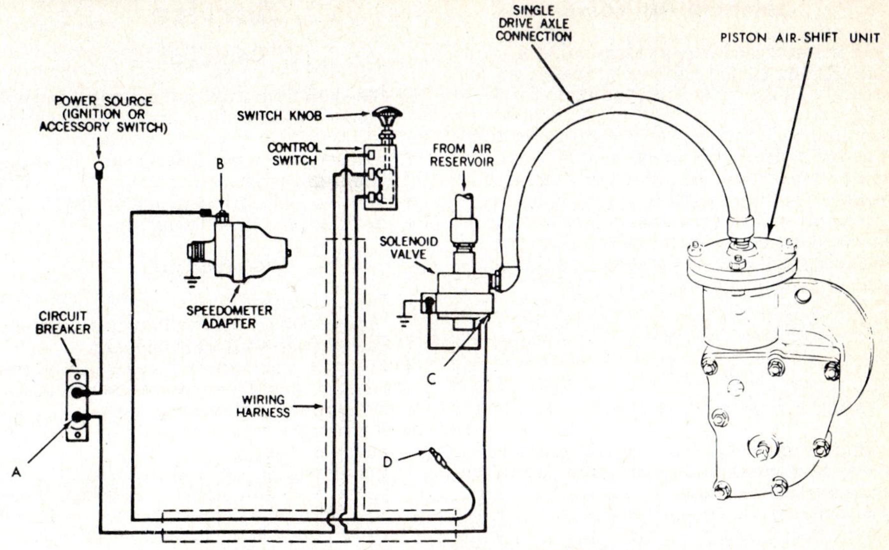

Fig. 10 Two speed air-electric shift control system

AIR-ELECTRIC SHIFT UNIT

DESCRIPTION

The air-electric shift control system consists of a control switch, speedometer adapter, solenoid valve and an air shift unit. The control switch is mounted on the gearshift lever. The switch controls current flow to operate the speedometer adapter and solenoid valve. The speedometer adapter is mounted to the speedometer head and is electrically operated and connected to the control switch. The solenoid valve allows air pressure to enter the shifting unit to shift the axle into high axle range. The air shift unit is mechanically connected to the axle shift fork and shifts the axle into low or high range. The air shift unit incorporates an air chamber, piston, compression spring and mechanical linkage. When air is admitted into the chamber, the piston travels downward against a compression spring, transferring motion through a push rod and an actuating lever to the shift fork, thus shifting the axle into the high range. Air pressure exhaust permits spring to return axle to low range.

Testing

NOTE: Refer to Fig. 10 when testing the system.

Speedometer Adapter

The speedometer adapter is energized in low axle range only. The adapter is grounded through the speedometer mounting or speedometer cable and cable housing. Ensure proper ground is achieved before proceding with this test.

Disconnect lead wire at terminal B, connect a suitable test lamp at B and vehicle ground. Turn ignition switch on. Test lamp should light in low range position of the control switch. Lamp should go out when control switch is placed in the high position. If test lamp indications are correct, current supply to the adapter is correct. If lamp indications are not correct, trouble is in the wiring harness or control switch.

Circuit Breaker

Disconnect circuit breaker lead wire at terminal A and connect test lamp from A to ground. Turn ignition switch on, if lamp stays on, circuit breaker and lead wire are in proper operating condition. If test lamp does not light, check for poor electrical connections or an open lead wire. If connections and lead wire are satisfactory, circuit breaker is inoperative. Also if test lamp does not immediately light but starts to flicker, and a faint clicking can be heard, circuit breaker is inoperative. Remove and replace circuit breaker.

Control Switch & Wiring Harness

If trouble in the control switch or wiring harness is suspected, visually check harness for worn or damaged wire insulation which can cause a ground connection. Check for accidental grounding of wire terminals to the transmission lever. Also check for short circuits between wire terminals. If trouble still persists use the following procedure:

1. Disconnect lead wires at terminals B and C.
2. Turn ignition switch on and alternately connect test lamp leads to the disconnected lead wire.
3. Operate control switch and observe test lamp for the following conditions:
   a. Connect test light to lead wire B, test lamp should light in low range of control switch. Test lamp should go out when control switch is in high range position.
   b. Connect test 1 to lead C, test lamp should light in high range position, test lamp should go out in low range on the control switch. If test 1 indications are correct in steps a and b, current supply to the speedometer adapter and solenoid valve is correct. If test lamp does not go out, trouble may be in the wiring harness or control switch. Remove and replace inoperative components.

Solenoid Air Valve

Disconnect lead wire at terminal C, and air line leading to the shift unit. Install an air pressure gauge in air line opening in the solenoid valve. Apply voltage to solenoid valve terminal and observe pressure gauge reading. If gauge indication is almost that of the reservoir pressure, solenoid valve is operating correctly.

If gauge indication is low or there is no pressure indication, solenoid valve is faulty. If solenoid with voltage applied does not energize, replace entire assembly.

Air Shift Unit

Start engine and build up air pressure in the system, check and repair any leaks in the system. Place control switch from low to high positions. If axle does not shift and air pressure is being applied to the shift unit, shift unit needs repair. Refer to the disassembly procedure to repair this unit.

Service

Removal

1. Bleed air from shift unit cylinder, and disconnect air line at the piston housing cover.
2. Remove nuts and shifting unit from differential carrier.
3. Remove shift fork seal and spring.
4. Reverse procedure to install.

Disassembly

1. Drain lubricant from shift housing, then remove bolts, washers, housing cover and gasket from shift unit housing.
2. Position and secure housing in a soft jawed vice, remove nuts, bolts and piston housing cover from shift unit.
3. Remove "O" ring seal and "O" ring from piston housing.
4. Install tool No. J-8107 or equivalent, over piston bore and clamp onto flange. Position a suitable socket wrench over nut and tighten down on tool until tension is relieved on clevis pin, then remove pin.
5. Unscrew tool from piston bore until compression spring tension is relieved and remove tool.
6. Push and remove piston, push rod and compression spring through top of cylinder.
7. Remove actuating lever from housing.

Cleaning & Inspection

Clean all parts with a suitable solvent and blow dry using compressed air. Inspect all contact surfaces for wear and corrosion. Replace any defective part. Replace all "O" rings and seals.

Assembly

1. Lubricate all "O" rings, felt washers and wipers. And install parts onto push rod.
2. Install actuating lever into housing.
3. Install piston and washer onto push rod and torque nut to 135 inch. lbs. Install push rod, spring and piston into cylinder bore.
4. Install clevis pin on push rod and actuator arm connection, then install housing cover and gasket. Torque screws to 100 inch. lbs.

CLARK TRANSMISSIONS

SERVICE INDEX

LUBE CAPACITY

MODEL	OIL, PINTS
250V, 250VO, 251VO	9
261F	12
262-1V	11½
264V, 264VO	11½
265V, 265VO	11½
267V, 267VO	12
268V, 268VO	6
269V	12
280VO, 282V, 285V	12
280VHD	8
300V, 301V	12
305V, 307V, 308V, 308VO	15
320	—
325V	—
327V	—
380	13
390	8
397V	8
401V, 406V	20
407V, 408V	20
2621V, 2622V	12
2651V	11½
2653V	12
2681V	11½

250 SERIES

Five Speed Synchronized Units, Figs. 1 and 2

When assembling the transmission, use new snap rings, cotter pins, gaskets and oil seals. Lubricate each moving part as it is being assembled, and coat all bushings with a light film of petroleum before installation.

Assembling Mainshaft

1. Clamp mainshaft in vise with front end up and install 3rd gear with internal teeth down and slip on thrust washer.
2. Place lock pin in 4th gear bushing (5th on O.D. units) and position bushing on shaft with lock pin toward upper end. Install gear on bushing with hub upward. Install thrust washer and snap ring.
3. Check end play of 4th gear. If not between .008–.012" replace thrust washer with one of a size that will provide correct end play. Thrust washers are available in four thicknesses ranging from .111" to .125".
4. Remove assembly from vise and install 2nd gear with internal teeth toward front. Install low-reverse gear with fork slot toward front.

Assemble Transmission

1. Install countershaft front bearing snap ring in case, then install new expansion plug in countershaft bore in case. Be sure to expand plug enough to prevent oil leakage.
2. Install countershaft front bearing in case. Install thrust washer on front end of shaft. Tip rear end of countershaft down and lower into case. Push shaft forward and insert it in front bearing.
3. Place rear bearing snap ring on bearing. Position bearing on countershaft with snap ring toward rear and install bearing on shaft.
4. Install reverse idler gear assembly in case with smaller gear toward rear of case. Drive reverse idler shaft into case until slot in shaft is flush with case. Install retainer in slot and secure with lock washer and bolt.
5. Tilt rear of mainshaft downward, inserting end of shaft through rear bore in case. Then lower front end of shaft until it is in line with pilot bearing opening and move assembly forward into position.
6. Place synchronizer on shaft with longer hub forward.
7. Install rear bearing on mainshaft with snap ring groove toward rear. Install snap ring on bearing and tap bearing onto mainshaft until snap ring is seated against case. Install speedometer drive gear spacer and gear on rear of shaft and pilot bearing on front end.
8. Install main drive gear and bearing in case. Tap front end of shaft with soft hammer until snap ring in bearing is seated against case. Position new gasket

of proper thickness on case and install bearing retainer. Be sure that oil holes in retainer and gasket are aligned with hole in case.
9. Place countershaft rear bearing cap and gasket on case and install lock washers and bolts.
10. Install mainshaft rear bearing retainer with new seal and gasket.
11. Install 1st-reverse shift fork in case with fork engaged in sliding gear groove. Insert shift rail through hole in rear of case and through fork. Drive rail into case until edge of lock slot is flush with case. Install rail lock plate and secure with capscrew.
12. Place 1st-reverse rocker arm in case and install pivot bolt and nut.
13. Place new gasket on top of case and with gears in neutral, install gearshift housing, being sure shift forks engage slots in gears.
14. Install emergency brake and U-joint flange. Shift transmission into gear and tighten flange nut. If cotter pin holes do not line up, always tighten nut (not loosen) until cotter pin can be inserted.

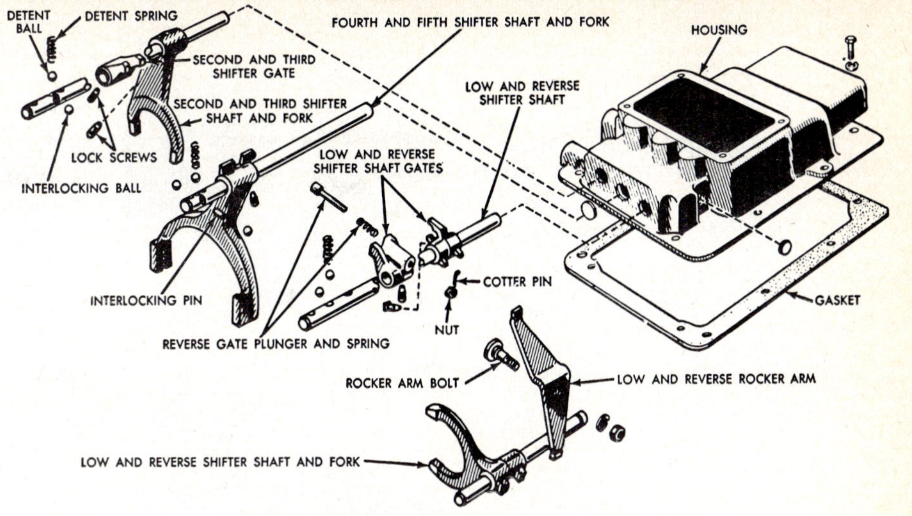

Fig. 1 Clark 250 Series gearshift housing

260, 2600 & 300 SERIES

4 and 5 Speed Synchronized Units, Figs. 3 and 4

When assembling the transmission, use new snap rings, cotter pins, gaskets and oil seals. Lubricate each moving part as it is being assembled, and coat all bushings with a light film of petrolatum before installation.

Assembling Mainshaft

1. Clamp mainshaft in vise with soft jaws with front end up.
2. Position 2nd speed gear on mainshaft with internal teeth facing up.
3. Place 2nd gear retainer washer in position of shaft.
4. Install 2nd gear retaining ring in groove of shaft over locating ring.

Four Speed Models

1. Place 2nd speed synchronizer on mainshaft with counterbore in shift hub down.
2. Position synchronizer support over shaft above synchronizer and install locating washer.
3. With lock pin in place in fluted bushing, install fluted bushing on shaft with lock pin toward upper end. Tap bushing into place.
4. Position 3rd speed gear on mainshaft with toothed hub up.
5. Install 3rd speed gear retaining washer on mainshaft, rotating gear so external teeth on washer mesh with internal teeth of gear.
6. Install snap ring on mainshaft, making sure it is securely locked in place.
7. Check end play on 3rd gear, which must be .004" to .012". Retaining washers of various thicknesses are available to obtain required end play.
8. Remove mainshaft from vise and install 3rd and 4th synchronizer assembly with lip on shift hub toward front.
9. Install 1st and reverse gear on shaft with fork slot toward front.

Five Speed Models

1. Referring to Figs. 3 and 5, place 2nd and 3rd speed synchronizer on mainshaft with counterbore in shift hub down.
2. Position 3rd speed gear on shaft with toothed hub down.
3. Install 3rd speed gear locating washer on shaft over 3rd speed gear.
4. With lock pin in place in fluted bushing, install fluted bushing on mainshaft with lock pin toward upper end. Tap bushing into position.
5. Position 4th speed gear on mainshaft with toothed hub up.
6. Install 4th speed gear retaining washer on splines of mainshaft, rotating gear to align external teeth of washer with internal teeth of gear so washer will drop into place.
7. Install snap ring on shaft so it is securely locked in groove. Check end play on 4th gear, which should be .005" to .012". Retaining washers are available in various thicknesses to obtain required end play.
8. Remove assembly from vise and install 4th and 5th speed synchronizer with lip on shift hub toward the front.
9. Install 1st and reverse speed gear with fork slot toward front.

Assemble Transmission

NOTE: If countershaft front bearing was previously removed and not reinstalled, install bearing as follows, using care not to damage the bearing during installation:

1. Using a suitable sleeve and an arbor press or hammer, press bearing in transmission case from clutch housing end of case. Open end of bearing goes toward rear of case.
2. Press bearing .007" below front face of case.
3. Tap rear end of countershaft down and lower into transmission case, running rear of countershaft through opening for bearing in rear of case far enough so front countershaft can be lowered into position. Push countershaft forward into position in front bearing.
4. Install countershaft rear bearing, being sure snap ring is in groove in outer race of bearing. Press bearing into place on shaft with snap ring toward rear. Press bearing into case until snap ring contacts case.
5. Install snap ring in groove in countershaft to retain bearing, being sure it is well seated in groove.
6. Insert one reverse idler bearing in reverse idler gear; then install bearing spacer and the second bearing in gear.
7. Install reverse idler gear in case with bevel on gear toward rear of case.
8. Insert reverse idler shaft through opening in rear of case and into idler gear. Make sure groove for retainer lock in rear of shaft is at bottom of shaft. Drive shaft in until inner edge of groove is flush with transmission case. Install retainer lock and lock screw.
9. Install mainshaft assembly in case by tilting rear end down and lowering into and through opening in fear of case, which is provided for rear bearing. Lower front end in line with pilot bearing opening. Move mainshaft forward into position, lifting assembly so 2nd gear will clear countershaft oil slinger.
10. Position mainshaft rear bearing over end of shaft with snap ring in outer race of bearing facing the rear.
11. Using a suitable sleeve and hammer, drive bearing on mainshaft and into case until snap ring on bearing contacts case.
12. Slide pilot bearing on front end of mainshaft.
13. Position main drive gear and bearing into opening in front of case so that drive gear engages and meshes with countershaft gear and snap ring on bearing contacts case. It may be necessary to use a suitable sleeve and hammer to drive gear and bearing into position.
14. Install a new gasket on main drive gear bearing cap. Install bearing cap on case and tighten screws securely. *Oil return hole in bearing cap must line up with oil hole in transmission case.*
15. Install countershaft rear bearing cap and new gasket.
16. Install spacer washer and speedometer drive gear on mainshaft.
17. Install mainshaft rear bearing cap and oil seal and new gasket.
18. Install 2nd speed gear oil scraper to transmission case.
19. Install brake drum and flange.
20. Lock transmission in two speeds at once and install companion flange nut, torquing it to 250 ft. lbs. Then advance nut to nearest cotter pin hole and install cotter

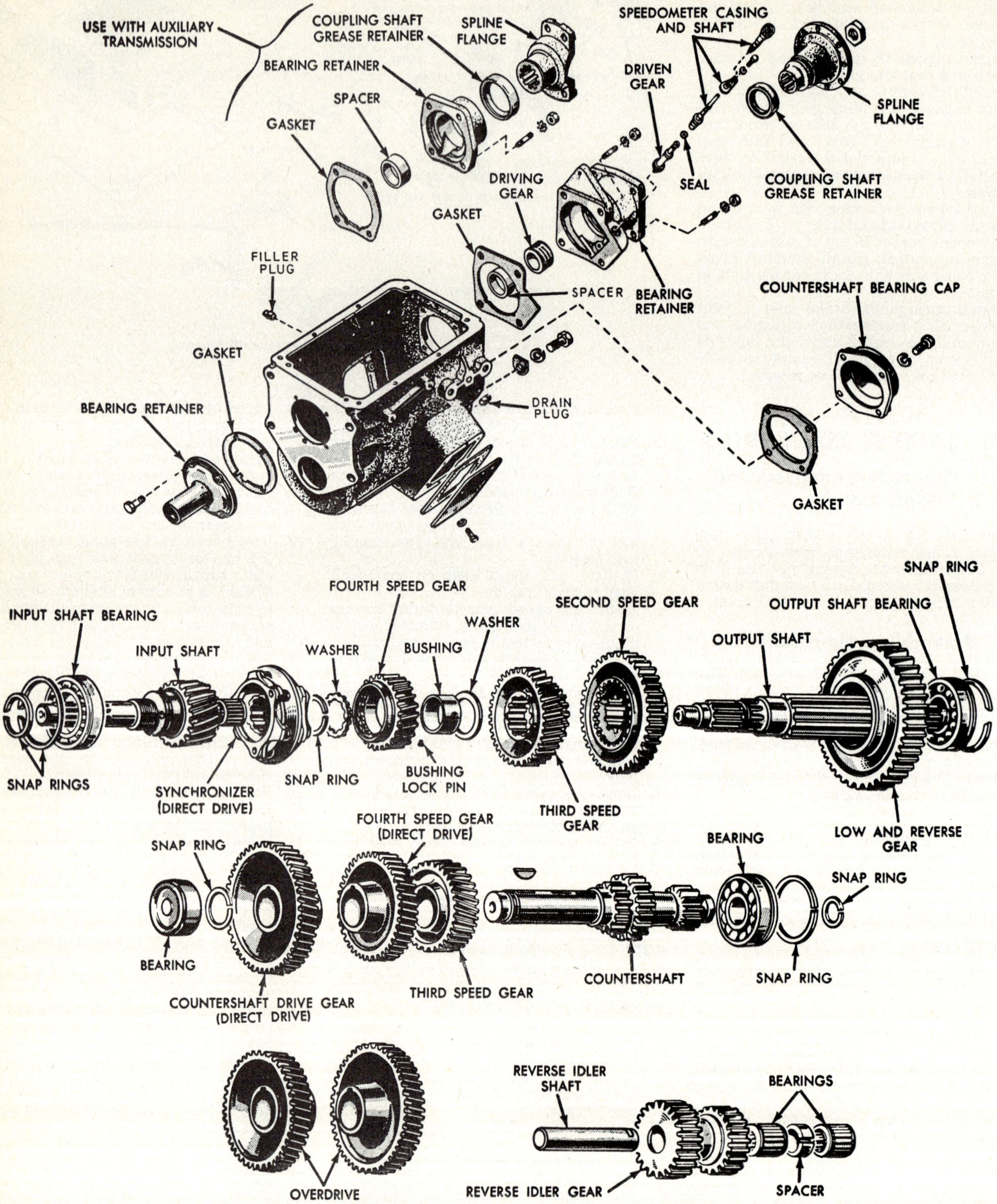

Fig. 2 Clark 250 Series five speed synchronized transmissions

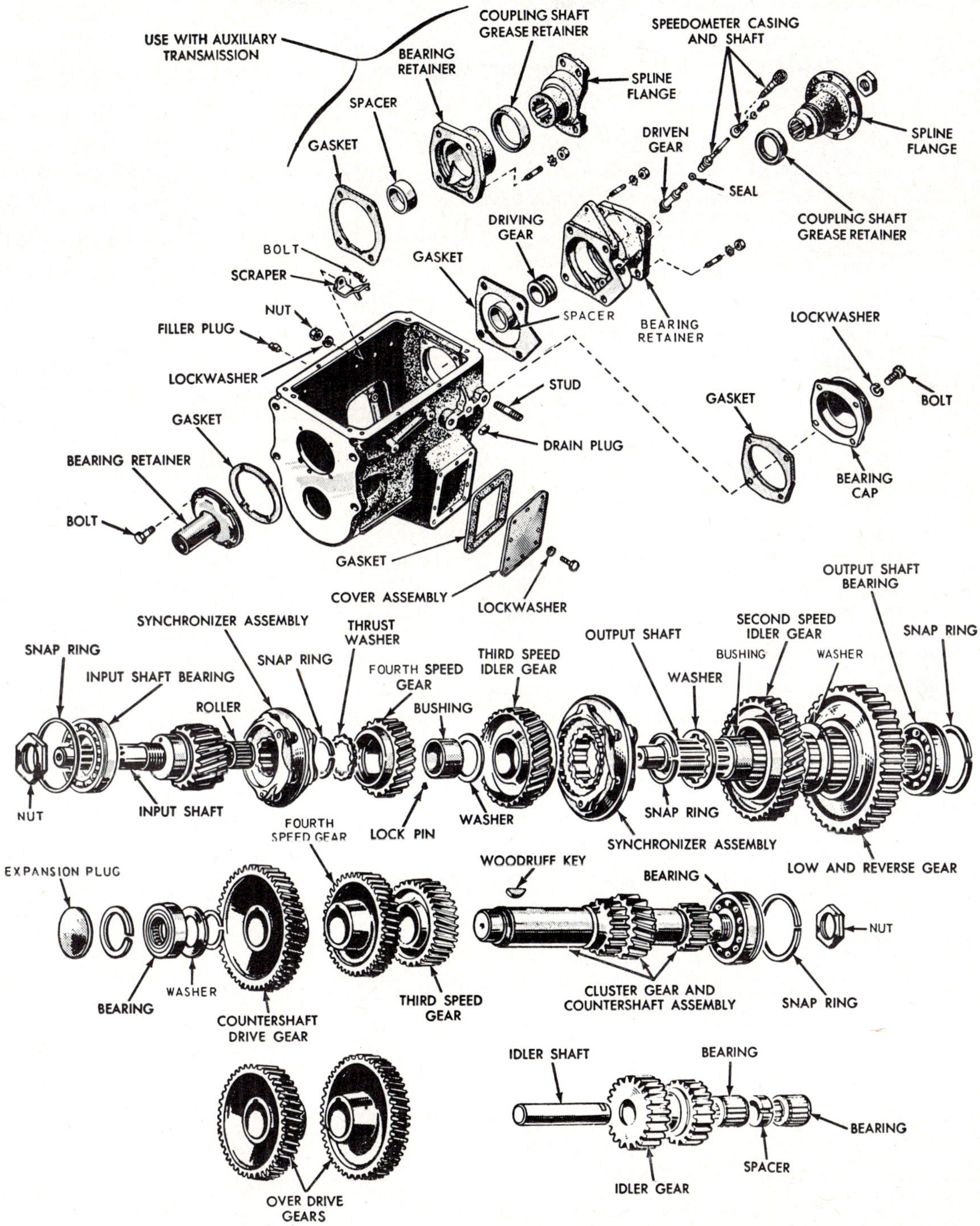

Fig. 3 Clark 260 and 300 Series four and five speed synchronized transmissions

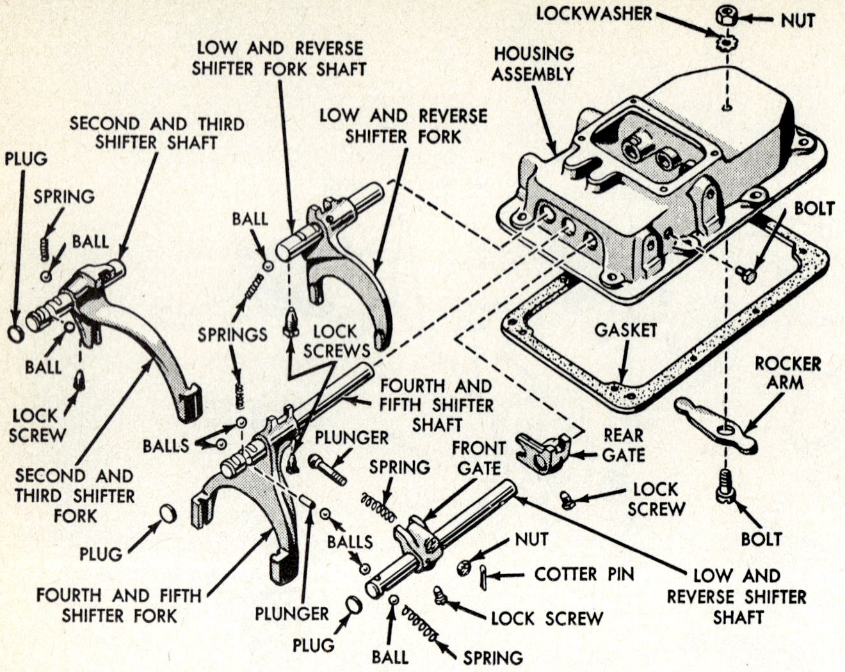

Fig. 4 Clark 260 and 300 Series gearshift housing

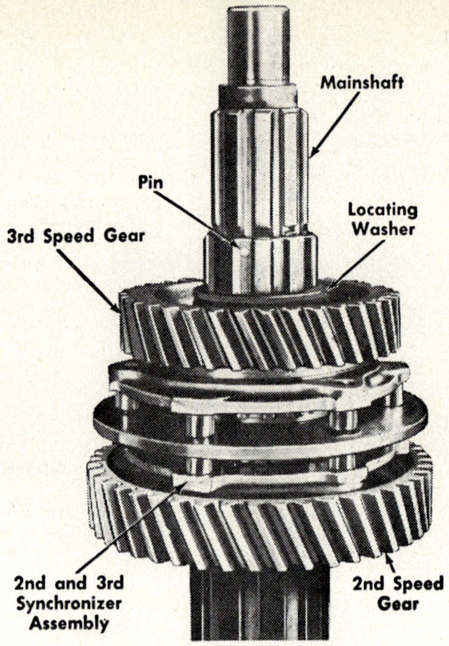

Fig. 5 Clark 260 and 300 Series five speed mainshaft assembly with fourth speed gear removed

pin.
21. Install shift bar housing with new gasket, being sure forks engage slots and rings in gears.
22. Install control cover and new gasket on shift bar housing.
23. Make sure transmission is in neutral by holding brake drum and turning main drive gear.
24. Install speedometer driven gear and adapter, hand brake parts, P.T.O. covers, and clutch housing.

390 & 280 VHD SERIES

Transmission Disassemble, Figs. 5A & 5B

1. Remove remote control or shift tower from shifter housing, then remove shifter housing and back-up light switch.
2. Remove U-joint assembly and drive shaft from parking brake drum.
3. Remove parking brake actuating lever from linkage.
4. Remove spline flange, then remove bolts retaining carrier plate to transmission housing and slide plate with brake shoes and retaining springs off transmission.
5. Lock transmission in two gears, then remove retaining nut and drum from output shaft.
6. Remove output shaft rear bearing retainer and speedometer drive gear.
7. Remove countershaft rear bearing retainer and snap ring.
8. Remove input shaft bearing retainer and pull input shaft out of case.

NOTE: Use care to avoid dropping output shaft pilot bearing rollers into transmission case.

9. Push output shaft rearward until rear bearing is exposed, then using a puller, remove rear bearing.
10. Remove output shaft from case.
11. Remove reverse idler shaft and lift idler gear, bearing and thrust washers from case.
12. Push countershaft rearward until rear bearing is exposed, then using a puller, remove bearing and oil slinger.
13. Lift countershaft from case, then if countershaft front bearing or pilot bearing is to be replaced, remove clutch housing.
14. If pilot bearing is to be replaced, press bearing from case.

NOTE: Do not hammer or drive out bearing, as damage or distortion to the bearing bore will result.

Output Shaft

280 VHD UNITS
Disassembly

1. Remove fourth/fifth synchronizer assembly, fourth/fifth bearing and race and fourth/fifth shift hub sleeve.
2. Remove fourth gear, third gear retaining ring, locating washer and third gear.
3. Remove second/third synchronizer cup and assembly, third gear shift hub sleeve retaining ring, second/third shift hub sleeve and second gear shift hub sleeve retaining ring.
4. Remove second gear and locating washer.
5. Remove output shaft rear main bearing and first/reverse sliding gear. Remove second gear retaining ring.

NOTE: Rear main bearing is removed along with first/reverse sliding gear by using a suitable puller tool on the gear, Fig. 5C. Do not pull on bearing.

Assembly

1. Install second gear retaining ring and first/reverse sliding gear on output shaft.

NOTE: Groove for shift fork in sliding gear must face toward front of shaft.

2. Install rear main bearing.
3. Install second gear locating washer and second speed gear on output shaft with clutching teeth on gear toward front of shaft.
4. Install second gear shift sleeve hub retaining ring and place second/third shift hub sleeve on output shaft.
5. Install third gear shift hub sleeve retaining ring. Install second and third synchronizer and cup on shift hub sleeve.
6. Place third gear and locating washer on output shaft and install retaining ring.
7. Place fourth gear on output shaft with clutching teeth toward front of shaft. Slide fourth/fifth shift hub sleeve onto output shaft with chamfer toward fourth gear.
8. Install fourth/fifth shift hub bearing and race on output shaft.
9. Install fourth gear synchronizer cup, fourth/fifth synchronizer and fifth gear synchronizer cup onto output shaft.

390 UNITS
Disassembly

1. Place output shaft in soft-jawed vise with front end up. Remove fifth speed synchronizer cup and fourth/fifth synchronizer assembly.
2. Remove fourth/fifth shift hub bearing and race, fourth speed synchronizer cup and fourth/fifth shift hub sleeve.
3. Remove fourth gear from output shaft.

NOTE: Use care not to lose third and fourth gear needle rollers when removing gears.

4. Remove third gear retaining ring, locating washer, lock ball (if equipped) and remove third gear from output shaft.
5. Remove third speed synchronizer cup and second/third speed synchronizer assembly.
6. Reverse position of output shaft in vise and remove rear bearing and first gear from output shaft.

NOTE: Output shaft first gear is retained

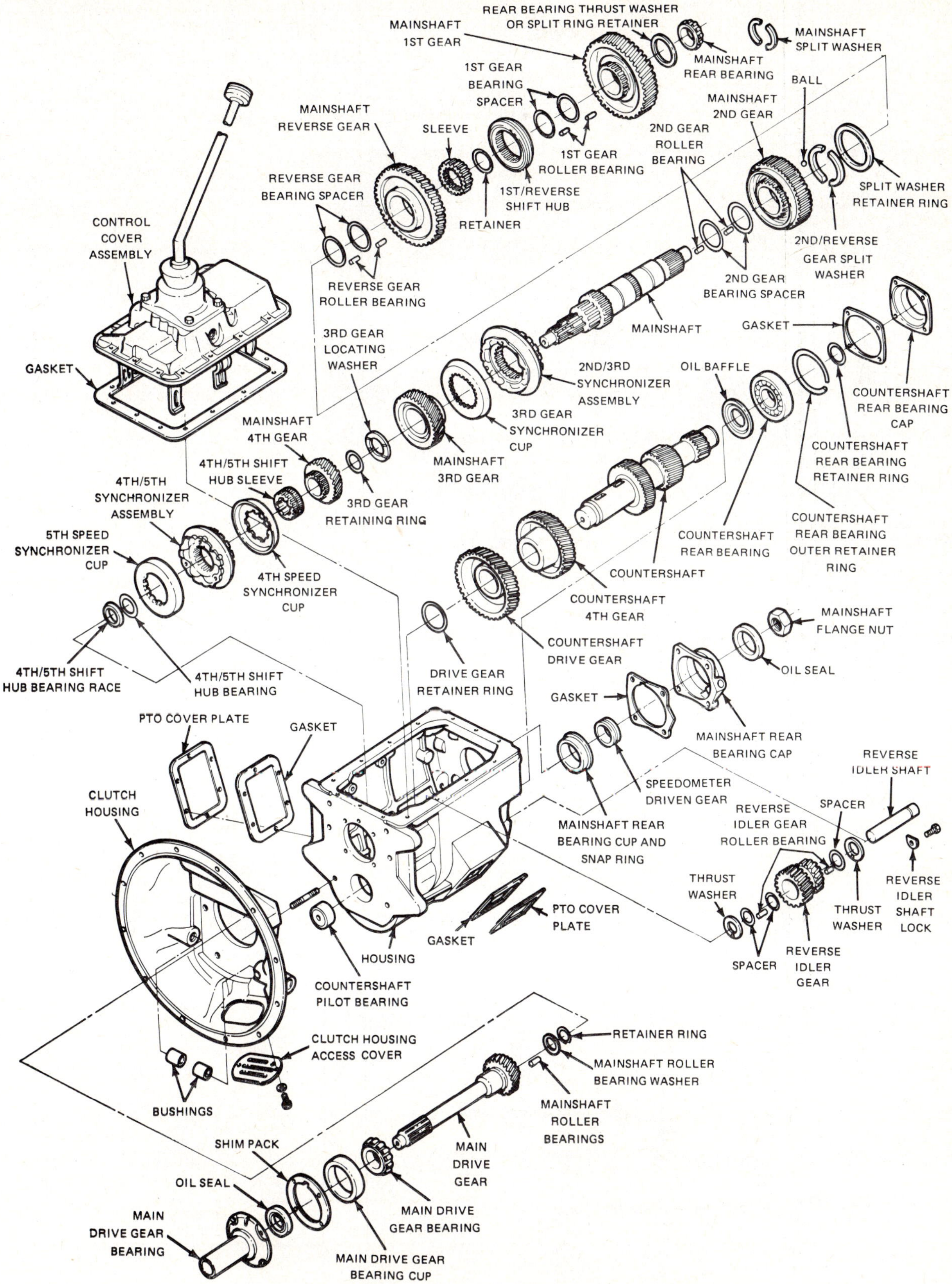

Fig. 5A Clark 390 Series five speed synchronized transmissions

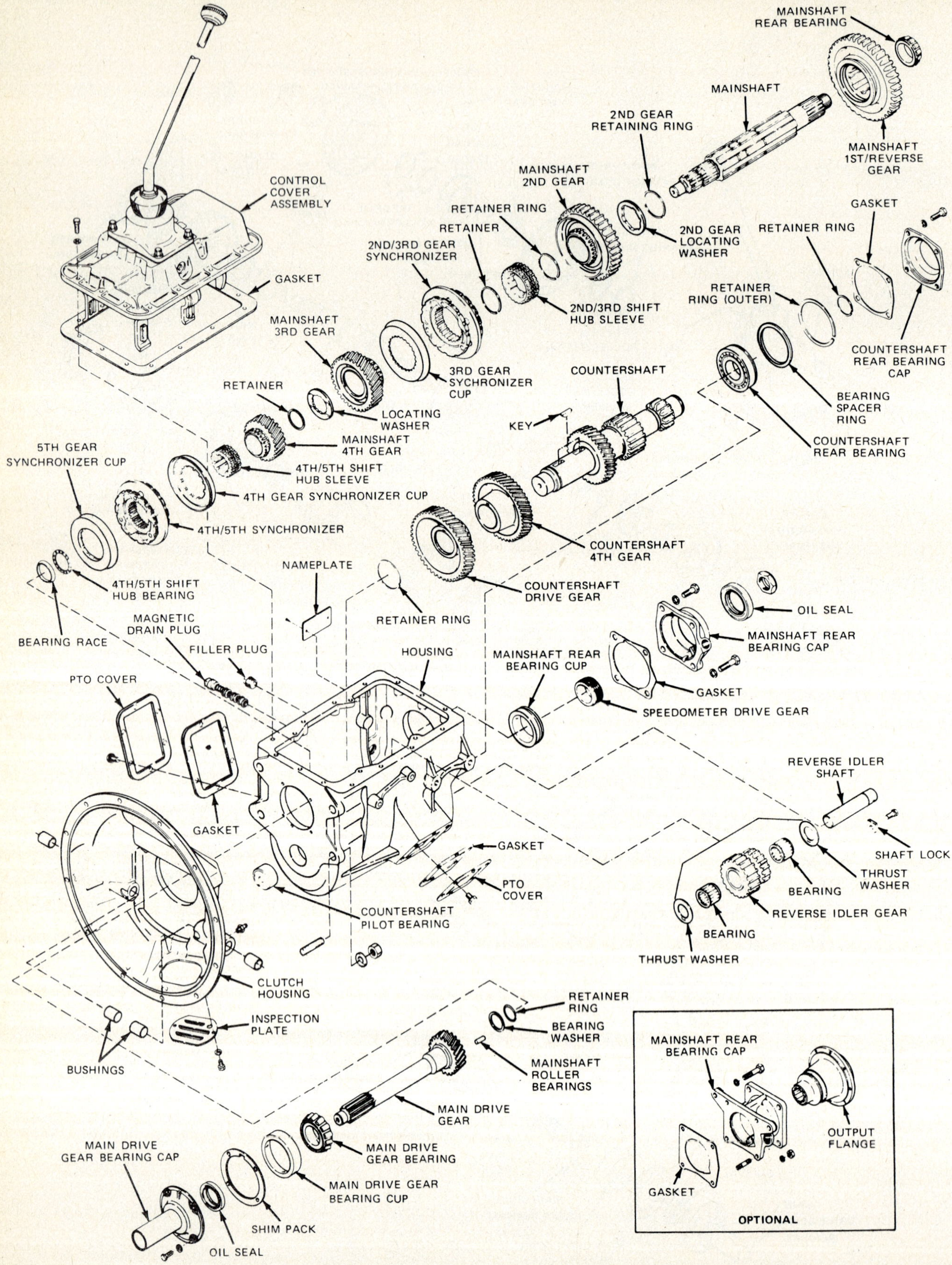

Fig. 5B Clark 28Q VHD Series five speed synchronized transmission

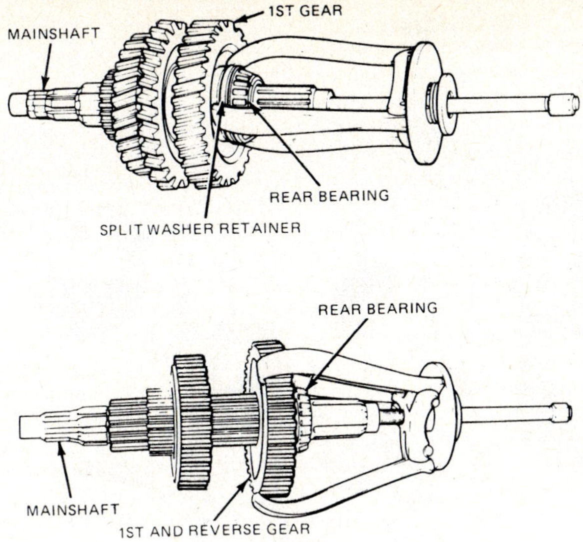

Fig. 5C Mainshaft rear bearing & 1st gear removal

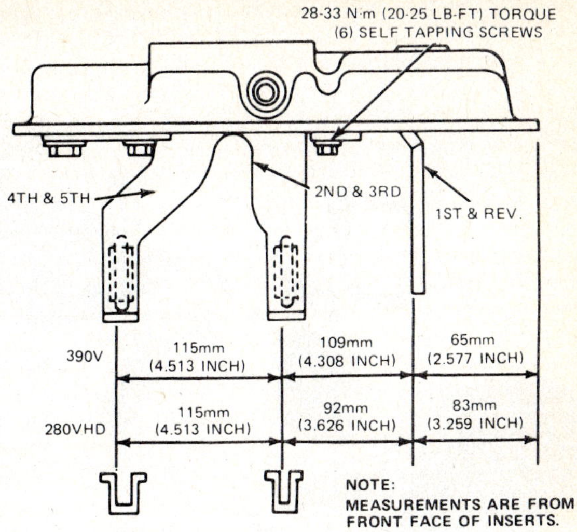

Fig. 5D Shift rail installation

by two different methods. One method uses a thrust washer between rear bearing and first gear. For this arrangement, remove rear bearing along with first gear by pulling on gear with suitable puller tool. The other method uses a split washer and split washer retaining ring. For this arrangement, use a three legged puller tool to pull against split washer retaining ring to remove bearing. Then remove split washer and retaining ring. Do not pull on gear itself. Use care not to lose first gear needle rollers when removing gear.

7. Remove first/reverse shift hub sleeve retainer and remove shift hub and sleeve.
8. Remove reverse gear from output shaft.

NOTE: Use care not to lose reverse gear needle rollers when removing gear.

9. Remove second gear split washer retaining ring, second/reverse gear split washer, locating ball and second gear from output shaft.

NOTE: Use care not to lose locating ball and second gear needle rollers when removing gear.

Assembly

1. Place output shaft with forward end up in a soft jawed vise.
2. Place third gear on output shaft with clutching teeth down, then install the locking washer and snap ring.
3. Turn outer shaft over and install third gear synchronizer ring and second and third synchronizer assembly.
4. With clutching teeth of second gear down, coat inside diameter of gear with high quality heavy grease to retain needle rollers, then install first row of needle rollers.
5. Install bearing spacer, then install the second row of needle rollers and outer bearing spacer.
6. Install second gear, being careful not to catch needle rollers on edge of spline or snap ring groove.
7. Install second gear split washer locating ball and split washer on output shaft, then install retainer ring over split washer.
8. Coat inside of reverse gear with grease and install bearing spacers and bearings.
9. Install reverse gear on output shaft, then position shaft hub sleeve and shift hub and install shift hub sleeve snap ring.
10. Install spacers and bearings in first gear and install first gear on output shaft.
11. Install first gear retaining washer and output shaft bearing.

NOTE: Make certain that bearing is tight against washer. On some units a thrust washer is used instead of a split washer and retainer.

12. Turn output shaft over and install fourth gear with clutching teeth facing upward.
13. Place fourth gear synchronizer ring on clutching teeth of fourth gear, then install fourth and fifth shift hub sleeve on output shaft with sleeve chamfer down.
14. Install fourth and fifth synchronizer and fifth speed synchronizer ring on output shaft.
15. Install thrust bearing and race.

Countershaft

Disassembly

1. Remove drive gear retaining ring and the drive gear from countershaft.
2. Remove key, fourth gear and remaining key from shaft.
3. On 280 VHD units, remove rear bearing inner race retaining ring and the race.

Assembly

1. Install fourth speed gear key in slot on shaft, then place fourth speed gear on shaft with long hub facing toward front.
2. Install countershaft main drive gear key in keyway in shaft.
3. Install countershaft drive gear on shaft with long hub facing toward rear.
4. Install snap ring on countershaft.

NOTE: On 280 VHD units, inner bearing race is assembled with outer bearing components and installed after countershaft is placed in case.

Input Shaft

Disassemble

1. Remove roller bearing from input shaft.
2. Remove output shaft roller bearing retaining ring and washer. Remove output shaft roller bearings from bore of input shaft.

Assemble

1. Press bearing on input shaft.
2. Install needle bearings, washer and snap ring.

Gear Shifter Housing

Disassemble

1. Place shift mechanism in neutral and move fourth/fifth speed shift fork rearward into fourth-speed position. Remove front rail support capscrew and front rail support.
2. Remove interlock tapered pin supports.

NOTE: Position of interlock pins should be noted for reassembly purposes.

3. Remove rear rail support capscrews and rear rail support.
4. Remove first/reverse shift fork and rail assembly, fourth/fifth shift fork and rail assembly and second/third shift fork rail assembly.

NOTE: Do not lose interlock cross pin, interlock tapered pins or mesh lock poppet balls.

5. Remove first/reverse shift rail and four mesh lockpoppet balls and springs.
6. Remove first/reverse rocker arms, reverse-latch plunger spring retaining plug, spring and plunger.

Assemble

1. Place first and reverse rocker arm on pivot pin, then install reverse latch plunger, spring and retaining plug.
2. Install the four poppet springs and poppet balls, noting the first and reverse shift fork rail poppet ball in pocket, Fig. 5D.
3. Align one tapered interlock cross pin with the hole in the first and reverse shaft rail, then position rail on poppet ball with rail in neutral position, noting position of the

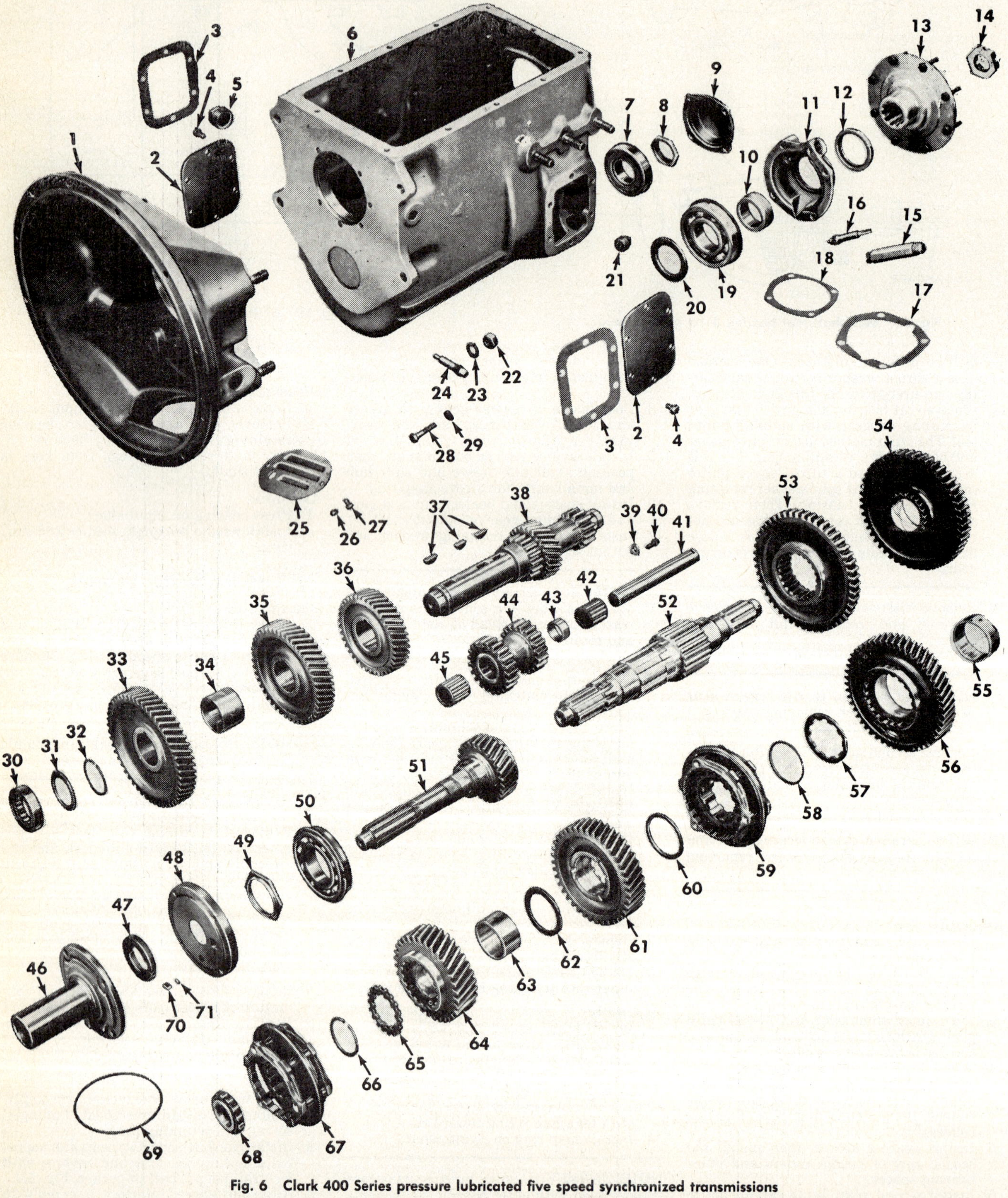

Fig. 6 Clark 400 Series pressure lubricated five speed synchronized transmissions

tapered interlock cross pin in relation to rail.

4. Install interlock cross pin in second and third shift rail, then position rail on poppet ball in neutral position with interlock pin aligned with first interlock tapered pin. Install second interlock tapered pin aligned with interlock cross pin hole.
5. Position fourth and fifth shift fork and rail assembly on poppet ball in neutral position, then raise rear of rail slightly to align second interlock tapered pin with cross hole in rail. Note positions of tapered interlock pins and shift rails.
6. Install first and reverse shift fork and rail assembly on poppet ball in neutral position. Align first and reverse rocker arm in notch at rear of rail.
7. Position rear rail support, then install washers and screws and tighten screws slightly.
8. Install interlock tapered supports, then tap fourth and fifth shift fork rearward to fourth speed position.
9. Position front rail support and install washers and screws. Torque screw to 20–25 ft. lbs.
10. Tap fourth and fifth shift fork rail assembly forward to neutral position.

Transmission Assemble

1. Coat front countershaft needle bearings with grease and install them into front countershaft bore.
2. Tip rear of countershaft down and lower it into case, then push countershaft forward and insert shaft into front bearing.
3. Position countershaft rear bearing oil slinger and bearing on shaft, then drive bearing into bore and install rear bearing retainer.

NOTE: Support countershaft drive gear while driving shaft to prevent damage to front bearing.

4. Coat reverse idler thrust washers with grease and place them into case.
5. Insert bearings in reverse idler gear bore and place gear assembly into transmission case with small gear toward rear of case.
6. Insert reverse idler shaft through hole in case, then through reverse idler gear, and into forward support boss.
7. Drive reverse idler shaft into case until slot in shaft aligns with lock bolt hole, then install retainer and lock bolt and tighten lock bolt securely.
8. Tilt rear of output shaft assembly down and insert end of shaft through output shaft bore in case. Lower front end of output shaft until it is aligned with pilot bearing opening and move assembly forward into position.
9. Insert pilot bearing in input shaft bore.
10. Position input shaft and bearings into forward end of transmission case, then tap front end of shaft until snap ring is seated against case.

NOTE: Make sure that clutching teeth of input shaft gear mesh with fifth speed synchronizer without binding.

11. Position new gasket on input shaft bearing retainer making sure that oil return holes in retainer and gasket are properly aligned with oil return hole in case. Position retainer on case, then install bolts and torque to 15–20 ft. lbs.
12. Place output shaft rear bearing on shaft and drive bearing into bore until snap ring is seated against case.
13. Place countershaft rear bearing cap and gasket on case, then install washers and bolts and torque bolts to 20–25 ft. lbs.
14. Install speedometer drive gear on output shaft and install a new seal on output shaft bearing retainer making sure that seal is correctly installed.
15. Place a new gasket and bearing retainer on case, then install washers and bolts and torque bolts to 60 ft. lbs.
16. Install parking brake flange and torque yoke retaining nut to 400–450 ft. lbs.
17. With transmission in neutral, position gear shifter housing over gears, aligning shift forks with gear shift hubs.

NOTE: If gear shifter housing is in neutral and transmission is in neutral, the transmission drive gear should turn without brake drum or output shaft turning.

18. Install shifter housing bolts and torque to 20–25 ft. lbs. Install remote control or shift tower and torque bolts to 20–25 ft. lbs.

1	Clutch Housing
2	P.T.O. Cover
3	Gasket
4	Cap Screw
5	Filler Plug
6	Transmission Case
7	Countershaft Rear Bearing
8	Rear Bearing Retainer Nut
9	Rear Bearing Cap
10	Speedometer Drive Gear
11	Rear Bearing Cap
12	Rear Bearing Cap Oil Seal
13	Hand Brake Companion Flange
14	Companion Flange Nut
15	Driven Gear Adapter
16	Speedometer Driven Gear
17	Gasket
18	Gasket
19	Mainshaft Rear Bearing
20	Mainshaft 1st Gear Thrust Washer
21	Drain Plug
22	Clutch Housing Stud Nut
23	Lock Washer
24	Oil Gallery Magnetic Plug
25	Clutch Housing Cover
26	Lock Washer
27	Cover Screw
28	Bearing Cap Attaching Screw
29	Lock Washer
30	Countershaft Front Bearing
31	Front Bearing Spacer
32	Countershaft Drive Gear Retainer Ring
33	Countershaft Drive Gear
34	Drive Gear Spacer
35	Countershaft 4th Gear
36	Countershaft 3rd Gear
37	Countershaft Gear Keys
38	Countershaft
39	Reverse Idler Shaft Lock
40	Lock Screw
41	Reverse Idler Shaft
42	Reverse Idler Gear Bearing
43	Bearing Spacer
44	Reverse Idler Gear
45	Reverse Idler Gear Bearing
46	Main Drive Gear Bearing Cap
47	Pressure Rotor
48	Rotor Backing Plate
49	Bearing Retainer Nut
50	Main Drive Gear Bearing
51	Main Drive Gear
52	Mainshaft
53	Mainshaft Reverse Gear
54	Mainshaft 1st Gear and Bushing Assy.
55	Mainshaft 2nd Gear Bushing Sleeve
56	Mainshaft 2nd Gear and Bushing Assy.
57	2nd Gear Retainer Washer - Selective
58	2nd Gear Retainer Snap Ring
59	2nd and 3rd Synchronizer Assy.
60	Mainshaft 3rd Gear Thrust Washer
61	Mainshaft 3rd Gear and Bushing Assy.
62	Mainshaft 4th Gear Locating Washer
63	Mainshaft 4th Gear Bushing Sleeve
64	Mainshaft 4th Gear and Bushing Assy.
65	4th Gear Retainer Washer - Selective
66	4th Gear Retainer Snap Ring
67	4th and 5th Synchronizer Assy.
68	Mainshaft Pilot Bearing
69	Main Drive Gear O-Ring
70	Rotor Vane
71	Rotor Drive Pin
72	Bushing Sleeve Pin

Fig. 6

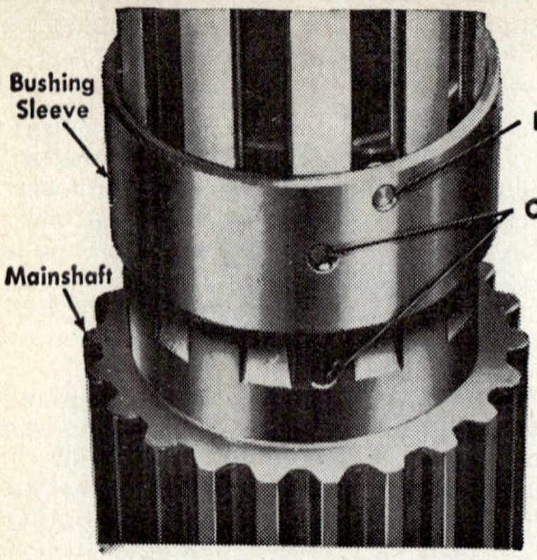

Fig. 7 Clark 400 Series. Installing mainshaft second gear bushing sleeve

400 SERIES

Five Speed Synchronized Units

Because of the increasingly higher engine speeds and extended periods of high gear operation, positive lubrication of the mainshaft is provided in this transmission in the form of a mainshaft pump that supplies mainshaft gear bushings and bearings with adequate oil at all times.

The pump is composed of a rotor which surrounds and is driven by the input shaft, a single free-floating vane confined in the rotor, and a stationary housing provided by the input shaft bearing cap which forms an eccentric bore for the single vane to contact.

Assemble Sub-Assemblies

When assembling the transmission, use new snap rings, cotter pins, gaskets and oil seals. Lubricate each moving part as it is being installed. Coat all bushings with a light coat of petrolatum before installation. Refer to Fig. 6.

Assemble Mainshaft

1. Clamp mainshaft in a soft-jawed vise with front end up.
2. Install mainshaft 2nd gear bushing sleeve and lock pin. Pin must be installed toward pilot end of mainshaft with oil hole in sleeve lined up with oil hole in shaft, Fig. 7.
3. To insure a proper tight stack-up between the 2nd gear bushing sleeve, selective retaining washer and snap ring, proceed as follows:
   a. Before 2nd gear is installed on bushing sleeve, place retaining washer on shaft and hold down securely on bushing sleeve.
   b. While holding snap ring parallel to retaining washer, slip snap ring into ring groove in mainshaft. Select variable thickness retaining washers until one is selected that will allow snap ring to fit securely in groove, Fig. 8.
   c. Remove selected washer and install 2nd gear and bushing over bushing sleeve.
   d. Place selected washer in position; then install snap ring into shaft groove. End play on 2nd gear must be .004" to .010".
4. Place 2nd and 3rd synchronizer with large end of hub toward rear on splines of mainshaft and drop into position. Install 3rd speed gear locating thrust washer on shaft.
5. Position 3rd gear and bushing over mainshaft with toothed hub down and drop into place.
6. Position 4th gear locating washer on shaft.
7. With lock pin in place in 4th gear bushing sleeve, install bushing sleeve on shaft with lock pin toward pilot or front end of shaft. It may be necessary to use a suitable sleeve and hammer to properly install sleeve. *Oil hole in sleeve must line up with oil hole in mainshaft.*
8. To insure proper stack-up between 4th gear locating washer, bushing sleeve, selective retainer washer and snap ring, follow the same procedure outlined for the 2nd gear parts.
9. Install 4th and 5th speed synchronizer with large end of hub toward rear.
10. Press pilot bearing on front end of mainshaft. Remove assembly from press and cover it with clean cloth until ready for installation.

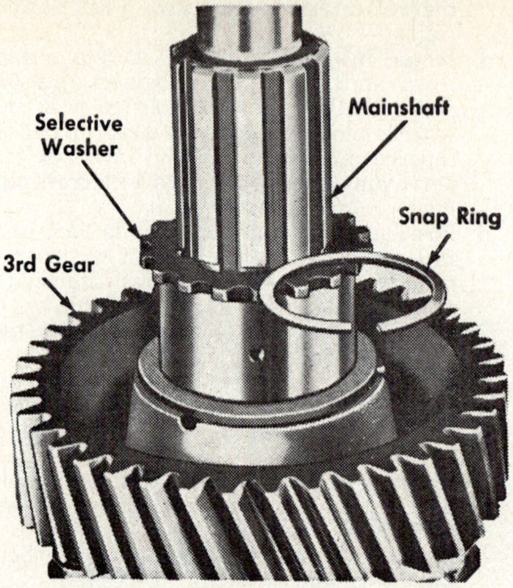

Fig. 8 Clark 400 Series. Selecting fourth gear variable thickness retaining washer

Assemble Transmission

1. Install countershaft from bearing in bore of case, applying pressure on outer race of bearing only.
2. Place countershaft front bearing spacer in position on front end of countershaft.
3. Tip rear of countershaft down and lower assembly into transmission case, running rear end of countershaft through rear bearing opening in case far enough to permit front of countershaft to be lowered into place. Push countershaft forward into proper position.
4. Using hammer and suitable installer sleeve, install countershaft rear bearing. Install rear bearing retainer nut loosely.
5. Insert two roller bearings, separated by spacer, into hub of reverse idler gear.
6. Lower reverse idler gear into position in case. Gear with chamfered teeth goes toward rear of case.
7. Insert reverse idler shaft through rear of case, through reverse idler gear and into forward support boss. The shaft should be driven in until forward face of slot in shaft is flush with rear face of case and slot lined up to permit installation of locking plate.
8. Install reverse idler shaft lock on idler shaft, engaging lock with slot in shaft. Secure lock to shaft with cap screw.
9. Position mainshaft 1st gear and bushing and reverse gear in case.
10. Install mainshaft in case by tilting rear end of assembly down and lowering into and through reverse gear, 1st gear, and opening in rear of case. Lower front end of shaft in line with pilot bearing opening in case and move assembly forward in proper position.
11. Install pressure rotor backing plate on main drive gear.
12. Position pressure rotor drive pin on drive gear.
13. Install pressure rotor on drive gear, with side of rotor marked "Front" toward front of transmission away from backing plate. With rotor drive pin at top and vane slot of rotor at bottom, see that oil inlet hole in rotor is on left side of rotor vane slot, Fig. 9.
14. Position new O-ring seal in groove in pressure rotor backing plate.
15. Install bearing cap on main drive gear, aligning oil inlet and outlet holes between rotor backing plate and drive gear bearing cap. Align word "Top" on cap with "Top" on backing plate.
16. Position main drive gear, bearing cap, pressure rotor, and new gasket into case so the main drive gear engages and meshes with countershaft drive gear. *Oil inlet and outlet holes must line up with oil holes in transmission case and word "Top" on bearing cap toward top of case.* Install four cap screws and tighten securely.
17. Install mainshaft 1st gear thrust washer; then, using a suitable installer sleeve and hammer, install mainshaft rear bearing.
18. Lock transmission gears in two speeds at same time and tighten countershaft rear bearing nut. Draw nut up tight and stake in position.
19. Position speedometer gear on mainshaft.
20. If previously removed, install new mainshaft rear bearing cap oil seal, using a suitable sleeve and hammer. Lip of seal goes toward front of bearing cap.
21. Install mainshaft and countershaft rear bearing caps and new gaskets.
22. With transmission still "locked up", install hand brake companion flange and retainer nut.
23. Install speedometer driven gear and adapter, P.T.O. covers and drain plugs.
24. Install pressure rotor oil gallery magnetic plug in case.
25. Install clutch housing, shift bar housing, control top oil baffle and control top assembly, Fig. 10.
26. Make sure transmission is in neutral by holding brake drum and turning main drive gear.
27. Install hand brake parts.

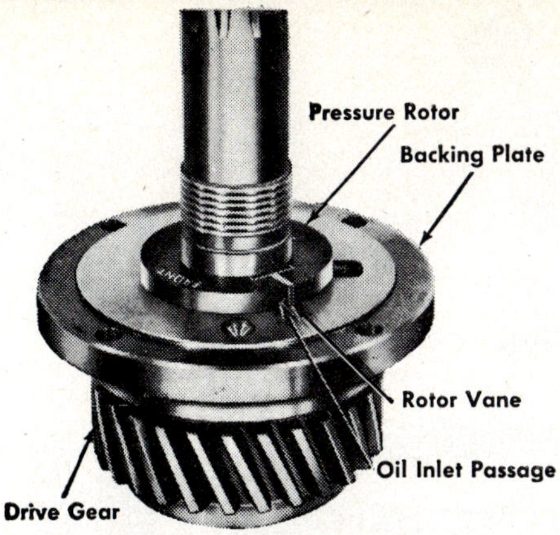

Fig. 9 Clark 400 Series. Oil pressure rotor installed

280 SERIES

Transmission Disassemble, Fig. 11

1. Remove shifter housing.
2. Remove parking brake and drum.
3. Remove mainshaft rear bearing cap and speedometer drive gear.
4. Remove countershaft rear bearing cap and snap ring.
5. Remove main drive gear.
6. Pry mainshaft rearward to expose rear bearing. Remove snap ring and pull off rear bearing.
7. Raise front end of mainshaft and lift assembly out of case.
8. Remove reverse idler lock.
9. Use puller to remove reverse idler gear shaft.
10. Remove reverse idler gear, bearings and two spacers.
11. Pry countershaft rearward to expose rear bearing, then remove snap ring from bearing.
12. Use puller to remove countershaft rear bearing and oil slinger.
13. Raise front end of countershaft and lift assembly out of case.
14. Disassemble mainshaft and countershaft as required.

Mainshaft Assemble, Fig. 12

1. Install 2nd speed gear retainer ring, locating washer and 2nd speed gear on mainshaft with clutch teeth up. Install 2nd gear retainer ring.
2. Install 2–3 shift hub sleeve and sleeve retainer ring.
3. Install 2–3 synchronizer assembly.
4. Install 3rd speed gear with clutch teeth down, then install locating washer.
5. Install 3rd gear retainer ring.
6. Install 4th speed gear with clutch teeth up. Note chamfer on 4–5 shift hub sleeve. Chamber must go down. Install bottom cup of 4–5 synchronizer on 4th gear clutch teeth before installing 4–5 shift hub sleeve. Install sleeve retainer ring.
7. Install 4–5 synchronizer on shift hub sleeve.
8. Turn mainshaft over and install 1st-reverse sliding gear with shift fork slot down.

Transmission Assemble

1. Coat countershaft pilot needle bearings with heavy grease to hold them in place until countershaft is installed.
2. Tip rear of countershaft down and into transmission case. Feed rear of countershaft through rear countershaft bearing bore. Move countershaft forward and into pilot bearing.
3. Position oil slinger and start rear bearing. Drive rear bearing on countershaft and into rear bearing bore.

NOTE: Countershaft drive gear must be supported on each side with a 1/4" flat bar to prevent damage to countershaft pilot bearing. Install bearing retainer ring.

4. Use heavy grease on reverse idler thrust washers to hold them in place.
5. Insert two reverse idler gear bearings in idler gear.
6. Install reverse idler gear, then insert idler shaft through case and idler gear.

NOTE: Idler shaft lock groove must line up with lock bolt hole. Drive shaft into position. Install shaft lock and bolt. Torque bolt to 20–25 ft-lbs.

7. Install mainshaft assembly in case.
8. Assemble pilot bearing in main drive gear as follows: If a new pilot bearing is to be used it comes from the factory with a plastic sleeve. Stand drive gear on end. Set bearing and sleeve over bearing pocket in drive gear. Slide bearing rollers and cage from plastic sleeve into bearing pocket. If old pilot bearing is being used, set rollers in bearing cage and hold in place with a rubber band. Slide bearing rollers and cage from rubber band into bearing pocket.
9. Install main drive gear assembly in case. Clutch teeth must enter 5th speed synchronizer cup without binding.
10. Press oil seal into drive gear bearing cap with lip of seal up. Shellac a new gasket to bearing cap, using care so as not to cover oil return groove in bearing cap. Install bearing cap with oil grooves lined up with oil holes in case. Torque to 20–25 ft-lbs.
11. Install parking brake and drum.
12. With transmission gears in neutral, install shift control assembly. Torque bolts to 20–25 ft-lbs.

320 & 380 SERIES

Transmission Disassemble, Fig. 13

1. Remove shifter housing.
2. Remove parking brake (if used).
3. Pull companion flange from rear end of mainshaft.
4. Remove speedometer driven gear and adapter (if used) from mainshaft rear bearing cap.

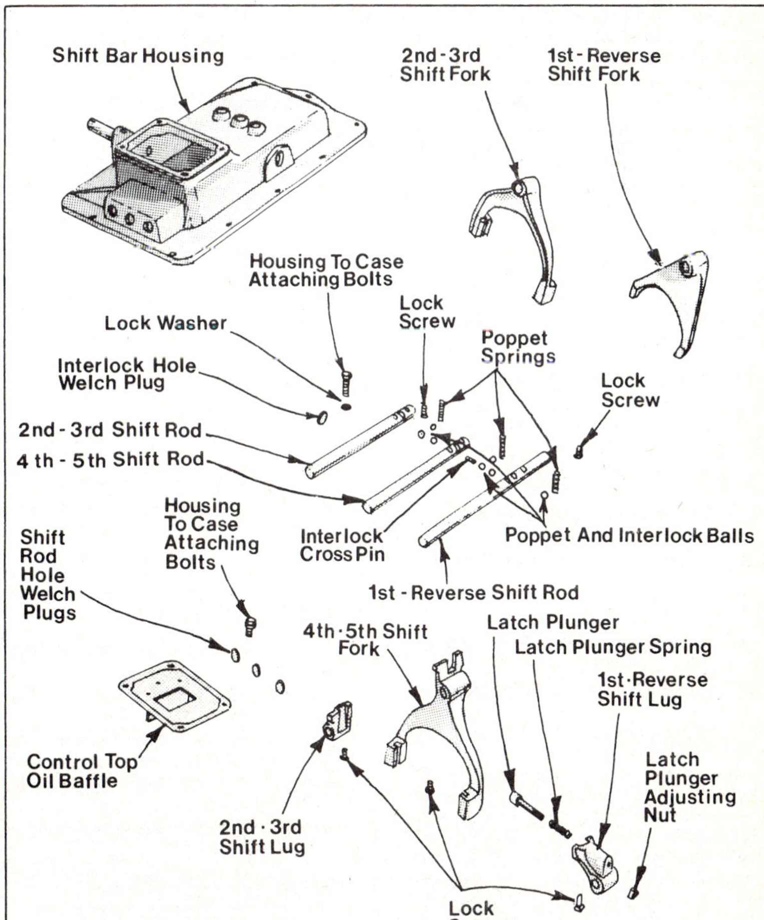

Fig. 10 Clark 400 Series gearshift housing

5. Remove mainshaft and countershaft rear bearing caps.
6. Remove countershaft rear bearing nut.
7. Remove clutch release bearing and support from main drive gear.
8. Remove clutch release yoke.
9. Remove main drive gear bearing cap.
10. Remove main drive gear and bearing from front of transmission case.
11. Remove speedometer drive gear and spacer (if used).
12. Push mainshaft toward rear of case far enough to expose rear bearing, then pull bearing from mainshaft.
13. Tilt front end of mainshaft and lift out through top of case.
14. Remove reverse idler gear and related parts.
15. Push countershaft toward rear far enough to remove snap ring from rear bearing, then pull off bearing.
16. Raise front end of countershaft and lift assembly through top of case.
17. Disassemble mainshaft and countershaft as required.

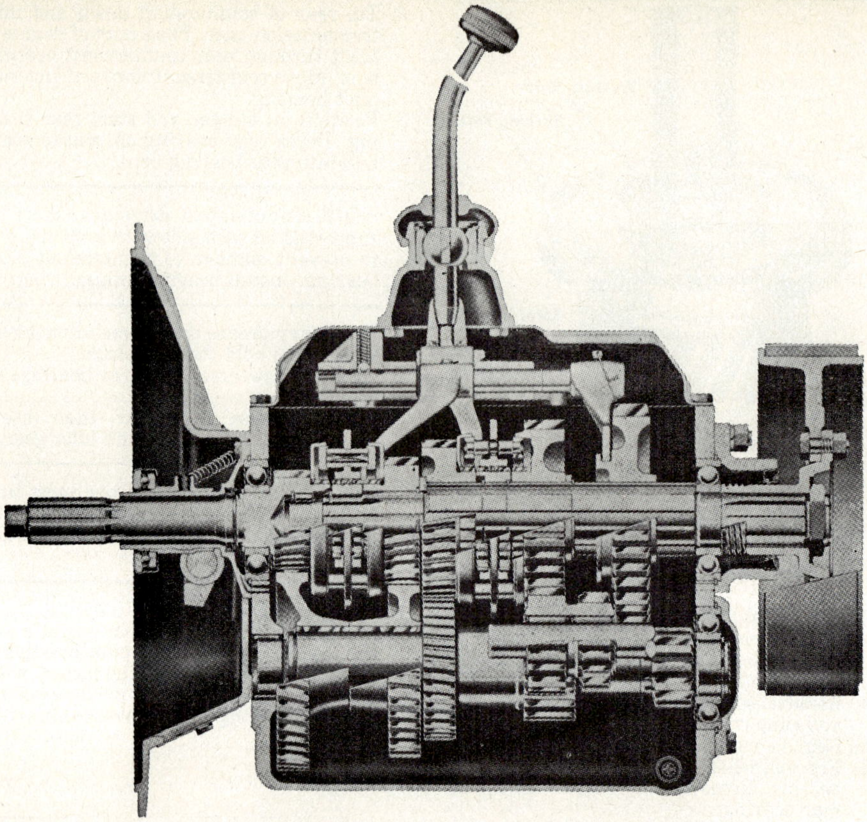

Fig. 11 Sectional view of Clark 280 Series five-speed transmission

Main Drive Gear Assemble, Fig. 14

1. Lubricate main drive gear bearing and shaft. Start bearing straight on shaft of drive gear with bearing snap ring toward pilot end of gear. Support front inner race of bearing and press gear into bearing tight against front face of gear.
2. Install bearing retainer nut or snap ring.
3. If a new pilot bearing is to be used it comes from the factory with a plastic sleeve. Stand drive gear on end. Set bearing and sleeve over bearing pocket in drive gear. Slide bearing rollers and cage from plastic sleeve into bearing pocket. When new bearing is used always install new bearing race on mainshaft. If old pilot bearing is used, set rollers in bearing cage and hold in place with a rubber band. Slide bearing rollers and cage from rubber band into bearing pocket.

Countershaft Assemble, Fig. 14

1. Install Woodruff keys in 3rd and 4th gear slots. Lubricate countershaft and bore of each gear before installing gears on countershaft.
2. Press 3rd speed gear straight on countershaft with long hub of gear toward rear. Align keyway in gear with key and press gear tight against shoulder on shaft.
3. Press 4th speed gear on countershaft in same manner used for 3rd speed gear. Long hub of gear goes toward rear.
4. Install drive gear spacer and key.
5. Press drive gear onto countershaft with long hub toward rear.
6. Install snap ring in countershaft groove.

Mainshaft Assemble, Fig. 14

1. Clamp mainshaft in a vise with front end up.
2. Drop 2nd speed gear locating washer into position over mainshaft.
3. With lock pin in place in 2nd gear bushing sleeve, install bushing sleeve on mainshaft with lock pin toward upper end of mainshaft.
4. To insure proper tight stack-up between 2nd gear locating washer and bushing sleeve, 2nd gear retaining washer and snap ring, the following procedure should be used:
   a. Before 2nd gear is installed on bushing sleeve, place retaining washer on mainshaft and hold down securely against bushing sleeve.
   b. Holding snap ring parallel to retaining washer, slip snap ring into its groove in mainshaft.
   c. Select variable thickness 2nd gear retaining washers until a washer is selected that will assure a tight stack-up through the assembly.
   d. Remove selected washer and install 2nd gear over bushing sleeve.
   e. Place selected washer in position on shaft and install snap ring so it is locked securely in groove. End play of 2nd gear must be a minimum of .006".
5. Place 2-3 speed synchronizer with long end of hub toward rear on splines of mainshaft and drop into position. Install 3rd speed gear locating thrust washer on mainshaft.
6. Position 3rd gear and bushing assembly over mainshaft with toothed hub down and drop into place. Position 4th gear locating washer on mainshaft.

NOTE: When assembling 4th gear bushing sleeve and lock pin assembly, make certain sleeve lock pin is centered in a spline of mainshaft before pressing sleeve into place. Failure to do this may shear head of lock pin and cause bushing seizure, Fig. 15.

7. With lock pin in place in 4th gear bushing sleeve, install bushing sleeve on mainshaft with lock pin toward pilot or front end of mainshaft. It may be necessary to use a suitable sleeve and hammer to install sleeve properly.
8. To insure a tight stack-up between 4th gear locating washer, bushing sleeve, gear selective retaining washer and snap ring, the following procedure must be used:
   a. Before 4th gear is installed on bushing sleeve, place retaining washer on mainshaft and hold down against bushing sleeve.
   b. While holding snap ring parallel to retaining washer, slip snap ring into its groove in mainshaft. Select variable thickness retaining washers until a washer is selected that will assure a tight stack-up through the assembly, Fig. 16.
   c. Remove selected washer and install 4th speed gear and bushing assembly over bushing sleeve.
   d. Place selected washer into position, then install snap ring into groove in mainshaft. End play of 4th speed gear must be a minimum of .006".
9. Remove assembly from vise and install 4-5 speed synchronizer on mainshaft with long end of hub toward rear.
10. Install 1st-reverse gear on mainshaft with shift fork slot toward front.

Transmission Assemble

Countershaft Installation

1. Install countershaft front bearing in case. Apply pressure on outer race of bearing only.
2. Place countershaft front bearing spacer on forward end of countershaft.
3. Tip rear end of countershaft assembly down and lower into case, running rear of countershaft through opening for rear bearing in rear of case far enough so front of countershaft can be lowered into position. Push countershaft forward into position in front bearing.

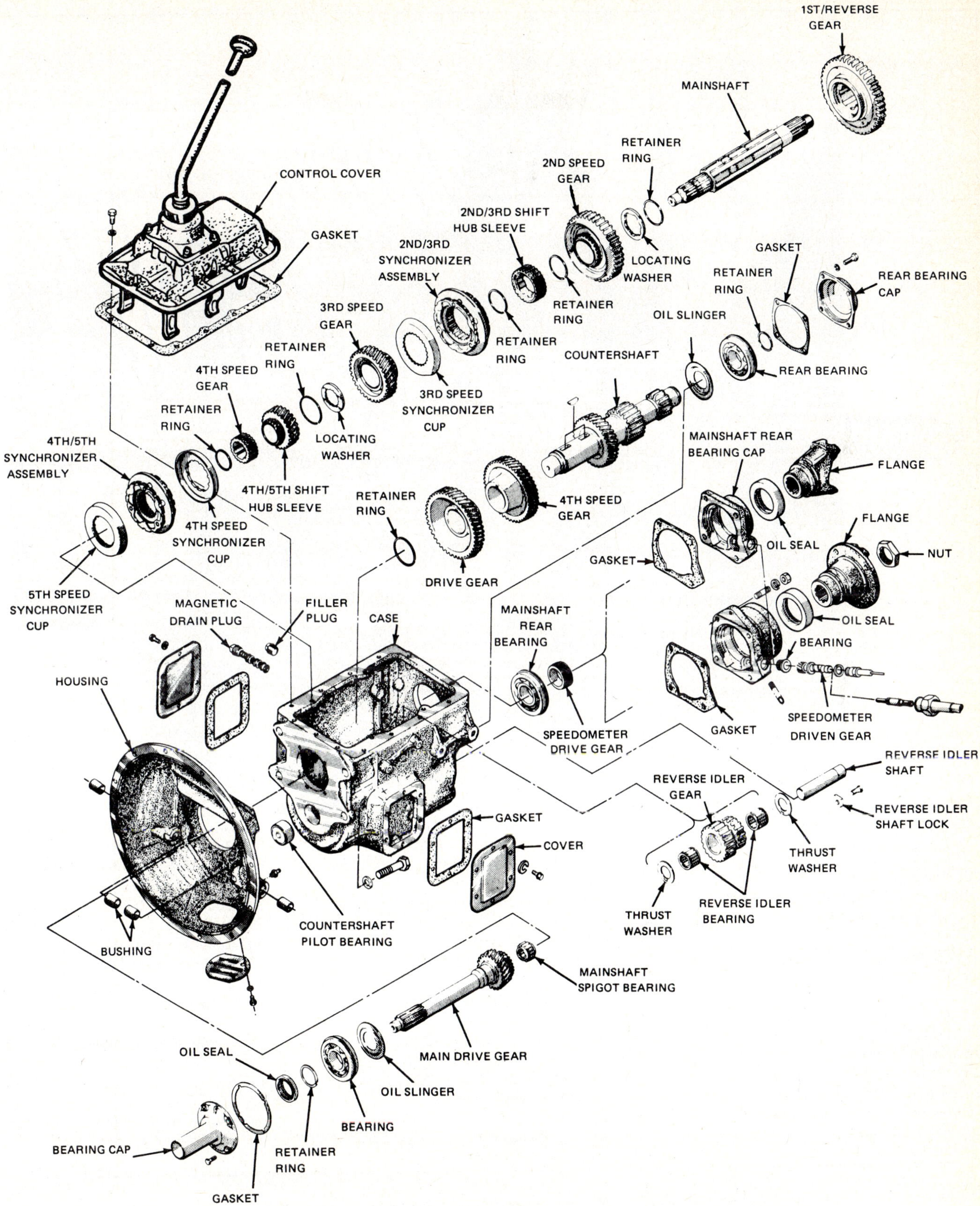

Fig. 12 Exploded view of Clark 280 Series transmission

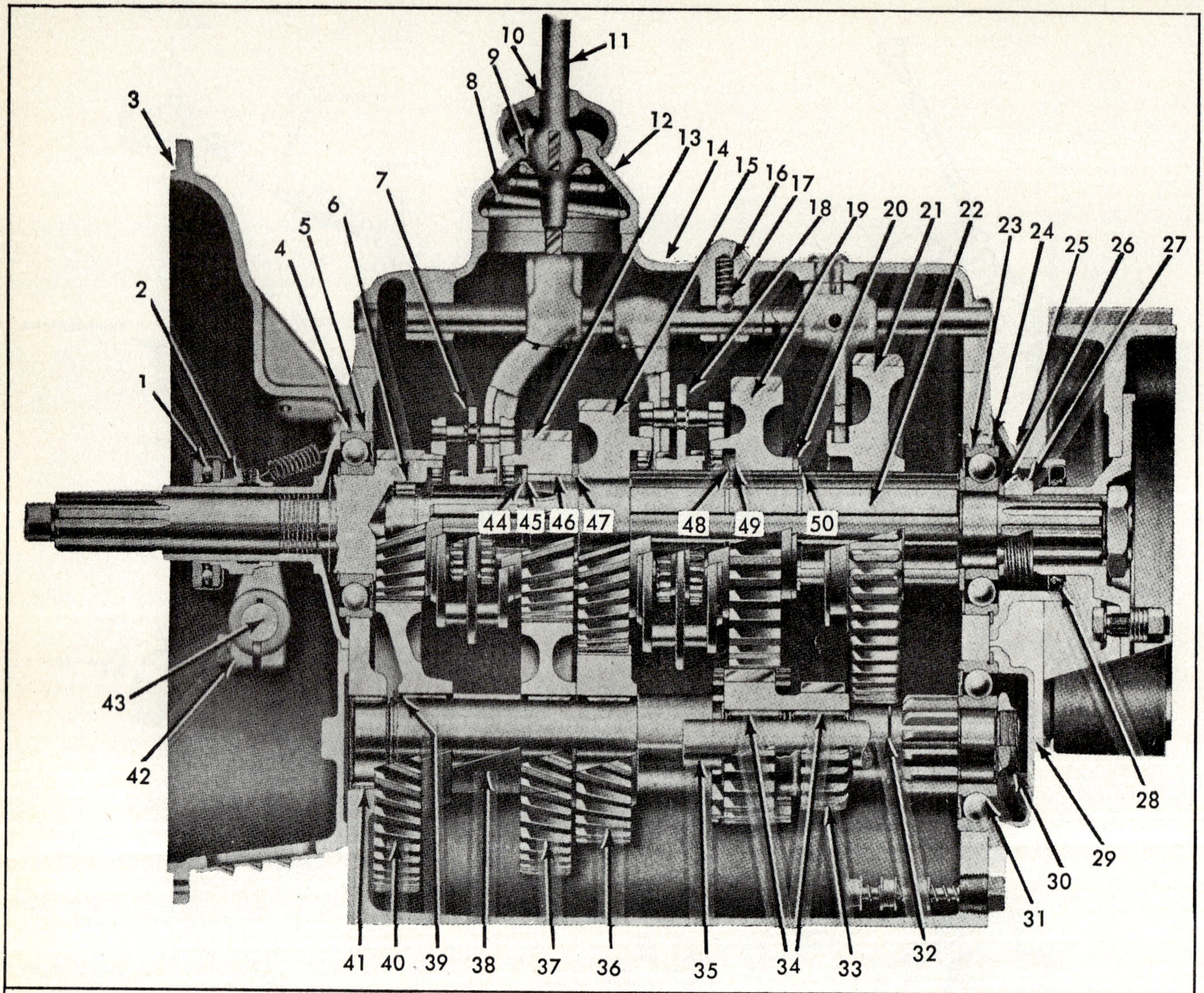

1. Clutch Release Bearing
2. Release Bearing Support
3. Clutch Housing
4. Drive Gear Bearing Retainer Ring
5. Drive Gear Bearing
6. Mainshaft Pilot Bearing
7. 4th and 5th Synchronizer Assy
8. Gearshift Lever Spring
9. Spring Seat Washer
10. Dust Cover
11. Gearshift Lever
12. Control Tower
13. Mainshaft 4th Speed Gear
14. Shift Bar Housing
15. Mainshaft 3rd Speed Gear
16. Poppet Spring
17. Poppet Ball
18. 2nd and 3rd Synchronizer Assy.
19. Mainshaft 2nd Speed Gear
20. 2nd Gear Locating Washer
21. Mainshaft 1st and Reverse Gear
22. Mainshaft
23. Mainshaft Rear Bearing
24. Mainshaft Rear Bearing Snap Ring
25. Mainshaft Rear Bearing Cap
26. Speedometer Drive Gear Spacer
27. Speedometer Drive Gear
28. Rear Bearing Cap Oil Seal
29. Countershaft Rear Bearing Cap
30. Countershaft Rear Bearing Retainer Nut
31. Countershaft Rear Bearing
32. Countershaft
33. Reverse Idler Gear
34. Reverse Idler Gear Bearings
35. Reverse Idler Shaft
36. Countershaft 3rd Speed Gear
37. Countershaft 4th Speed Gear
38. Countershaft Drive Gear Spacer
39. Snap Ring
40. Countershaft Drive Gear
41. Countershaft Front Bearing
42. Clutch Release Yoke
43. Release Yoke Cross Shaft
44. Snap Ring
45. 4th Speed Gear Retainer Washer
46. 4th Speed Gear Bushing Sleeve
47. 4th Speed Gear Locating Washer
48. Snap Ring
49. 2nd Speed Gear Retainer Washer
50. 2nd Speed Gear Snap Ring

Fig. 13 Sectional view of Clark 320 & 380 Series five-speed transmission. Typical

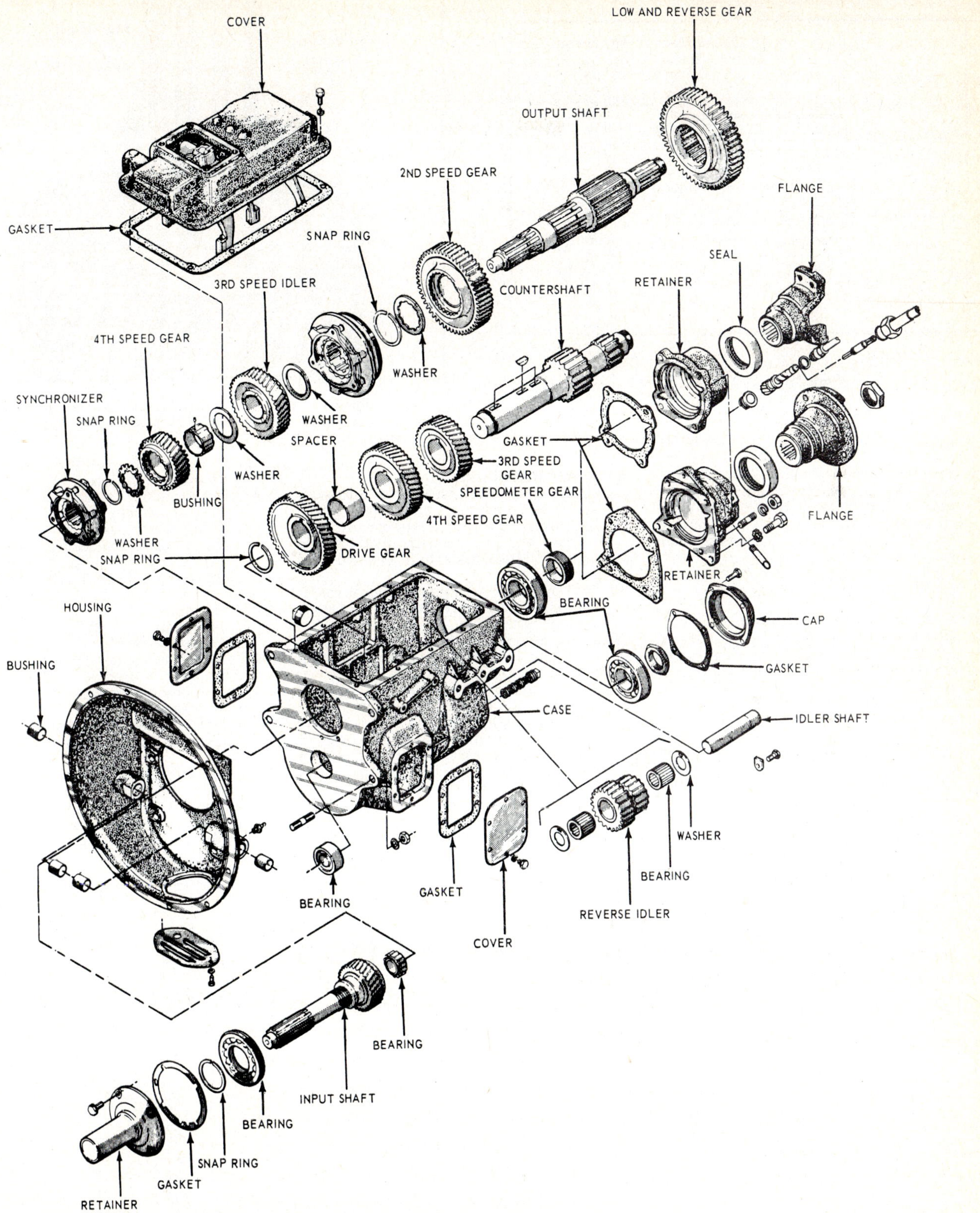

Fig. 14 Exploded view of Clark 320 & 380 Series transmission. Typical

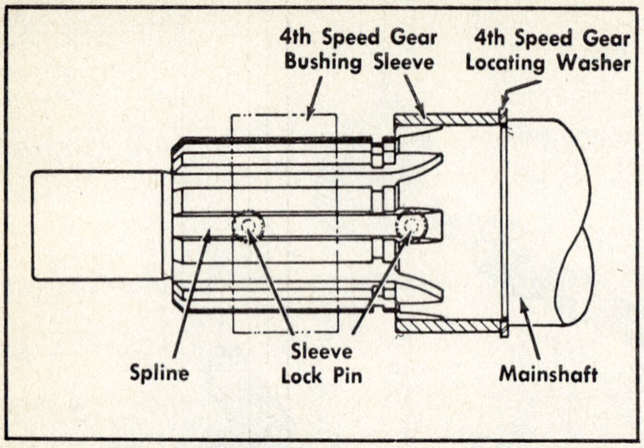

Fig. 15 Sleeve lock properly installed. Clark 320 & 380 Series. Typical

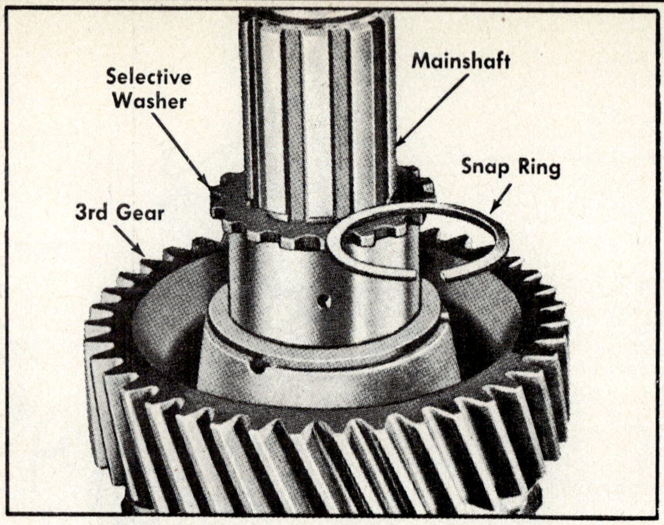

Fig. 16 Selecting 4th speed gear variable thickness retainer washer. Clark 320 & 380 Series. Typical

4. Install countershaft rear bearing. Be sure snap ring is in position in groove of outer race of bearing. Press bearing into place on countershaft with snap ring toward rear. Press bearing into case until snap ring contacts case.
5. Install countershaft rear bearing nut loosely.

Reverse Idler Gear Installation

1. Insert one reverse idler gear bearing in idler gear, then install bearing spacer and second bearing in bore of idler gear.
2. Install reverse idler gear assembly in case with chamfered teeth on gear toward rear of case.
3. Insert reverse idler gear shaft through opening in rear of case, into idler gear and into forward support boss. Make sure groove for retainer lock in rear of idler shaft is at bottom of shaft. Drive shaft in until inner edge of groove is flush with transmission case. Install retainer lock with bolt and lock washer.

Mainshaft Installation

1. Install mainshaft assembly into case by tilting rear end down and lowering into and through opening in rear of case. Lower front end in line with pilot bearing opening. Move mainshaft forward into position.
2. Position rear bearing over end of mainshaft with snap ring in outer race of bearing, facing the rear.
3. Drive mainshaft rear bearing on mainshaft and into case until snap ring on bearing contacts case.
4. Position main drive gear and bearing assembly into front of case so that drive gear engages and meshes with countershaft drive gear and snap ring on bearing contacts case. It may be necessary to use a sleeve and hammer to drive gear and bearing into case.
5. Install main drive gear bearing cap with new gasket on transmission case. Oil return hole in bearing cap must line up with oil hole in case.
6. Install clutch release bearing return spring clip under head of bolt when installing bearing cap.
7. Lock transmission in two gears at the same time to lock it up. Tighten countershaft rear bearing nut. After nut is driven up tight, stake it in position with punch and hammer.
8. Install countershaft rear bearing cap with new gasket.
9. Install speedometer drive gear spacer and drive gear (if used) on mainshaft.
10. Install mainshaft rear bearing cap with new gasket.
11. Install parking brake drum (if used) and companion flange on rear end of mainshaft.
12. With transmission still locked in two speeds, install and torque flange nut to 250 ft-lbs.
13. Place transmission gears in neutral and install shifter housing.

FULLER TRANSMISSIONS

Model	Speeds	Oil, Pts.	Page
3-65 Series	3 Aux.	13	286
5A-33, 330	5	11	283
5B-33, 330	5	11	283
5F-33	5	11	283
5A-43, 430	5	16	283
5A-62, 620	5	18	283
5A-65, 650	5	24	283
5C-65, 650	5	24	283
5C-67, 720	5	24	283
5CA-720	5	24	283
5H-74, 5HA-74	5	12	285
5W-74	5	24	283
5WA-74	5	24	283
4C-75	4 Aux.		289
R46	8	19	270
R96, R960	10	33	277
RT610	10	12	256
RT613	13	16	251
RT906	6	26	251
RT910, RTF910	10	26	251
RT1110	10	25	251
RT6610	12	12	256
RT6613	13	16	251

Model	Speeds	Oil, Pints	Page
RT9509	9	25	251
RT12509	9	25	251
RT12510, RTF12510	10	25	251
RTO910, RTOF910	10	26	251
RTO-913	13	24	251
RT-915, RTO-915	15	28	251
RT-9513	13	27	263
RT-12513	13	27	263
RTF-9513	13	27	263
RTL1157DL	9	29.5	251
RTO958LL	10	27	251
RTO1157DLL	9	27	251
RTO1258LL	10	27	251
RTO-9513	13	27	263
RTO11509	9	25	251
RTOF-9513	13	27	263
RTOO-9513	13	27	263
RTOOF-9513	13	27	263
RTQ9513	13	27	251
T-905, TO-905	5	22	251
T905A	5	22	251

RT, RTF, RTO, RTQ, T, TO Models With Twin Countershafts

DESCRIPTION

Series T transmission is shown in Fig. 1. As shown it is a five speed unit and is similar to the main transmission of the RT Series, Figs. 2, 3, 4. The "O" in RTO and TO units indicate that the transmissions include an overdrive ratio. The "F" in RTF models are of the same construction except for the forward position of the gearshift lever.

The twin countershaft design splits torque equally between the two shafts providing a high torque capacity-to-weight ratio. Because of torque splitting, each gear set carries the load, greatly reducing the face width of each gear.

Another unique design feature is the floating gear principle. The mainshaft gears, when not engaged, "float" between the countershaft gears, eliminating the need for gear sleeves and bushings. All gears are in constant mesh and have spur type teeth.

The 9509 and 12509 models have nine forward speeds and two reverse speeds, consisting of a five speed front section and a two speed auxiliary, or range, section. One ratio in the front section is used only in the low range as a starting gear. The remaining four ratios are used once through the reduction gear range and once through the direct or high, range.

The 910 models have ten forward speeds and two reverse speeds, consisting of a five-speed front section and a two-speed auxiliary or range section. First through fifth speeds are obtained by using the five gear ratios in the front section through the low-speed gear of the auxiliary section. Sixth through tenth speeds are obtained by using the five gear ratios in the front section through the high speed (direct drive) auxiliary section.

The 913 and 9513 models have 13 progressive forward ratios and two reverse, consisting of a five-speed front section, a synchronized two-speed range section, and an overdrive splitter gear. One ratio in the front section is used only in low range as a low-low or starting gear. The remaining ratios (four) in the front section are used once through the reduction (low) gear in the range section and once through direct (high) in the range section. While in high range the four ratios can be split by the overdrive splitter to give eight speeds in high range.

The 915 models have 15 speeds forward and three reverse speeds, consisting of the five-speed front section which is identical to the 910 front section, and a three-speed auxiliary or range section. Both sections are contained in one case, the rear plate being extended to accommodate the auxiliary gears. The 15 speeds are obtained by using the five speeds in the front section through direct drive (high range) through the low speed range gear, and through the hole gear of the auxiliary section for deeper reduction provides the 15 forward speeds.

The range shift between high and low is made automatically by air upon preselection by the driver. The hole gear in the 915 models is engaged by air when selected by the driver.

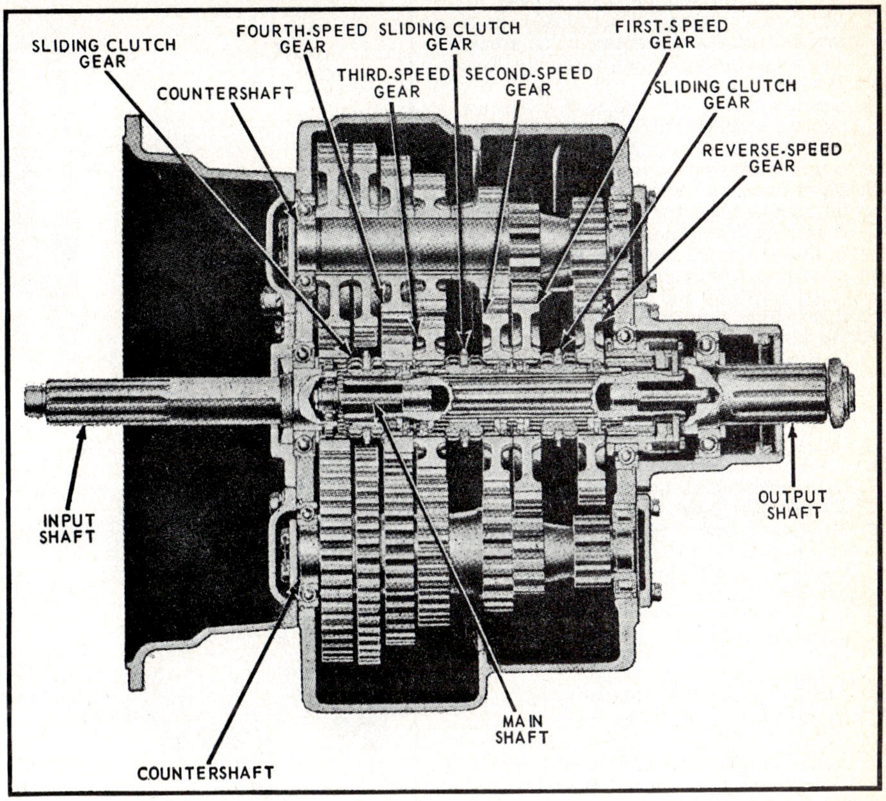

Fig. 1 Sectional view of Fuller T Series twin countershaft transmission

DISASSEMBLE TRANSMISSION

Preparation, Series T

1. Remove bolts that attach shifting bar housing to transmission case. Tap housing with soft hammer to free it from case, then lift from transmission.
2. Lock mainshaft by engaging two gears at one time, then remove companion flange nut and remove flange.
3. Remove attaching screws and pry off rear bearing cover, Fig. 5.
4. Remove speedometer drive gear or spacer and bearing washer from output shaft or from cover.
5. Remove rear bearing from output shaft, Fig. 6. In some instances this bearing will remain in rear cover, in which case, it should be removed from cover.
6. Remove lock wire and the two flat key attaching bolts, Fig. 7. Lift keys from output shaft.
7. Remove outshaft shaft, Fig. 8.
8. Remove clutch release mechanism from flywheel housing. Unfasten and lift flywheel housing from transmission case.

NOTE: At this point the balance of the procedure for disassembling the transmission is the same as given for the compound RT units.

Separation of Main & Auxiliary Units (All Models)

1. Disconnect air hoses at air valve.
2. Remove air filter and air regulator valve assemblies.
3. Remove air valve from adapter plate.
4. Remove alignment sleeve, spring and actuating pin from air valve adapter plate, then remove plate from transmission.
5. With transmission in neutral, remove gearshift housing cover.
6. Lock transmission in two gears and remove U-joint flange.
7. Unfasten auxiliary section from main transmission. Install three puller screws in tapped holes of rear plate and tighten them alternately and evenly to move auxiliary section to rear just enough to break gasket seal.
8. Attach chain hoist to auxiliary section and continue to move it to the rear until free of transmission, Fig. 9.
9. Unfasten and remove clutch housing from transmission.

ALL EXC. 610, 6610, 9513 & 12513 SERIES

Main. Trans. Disassemble

NOTE: All references to any part of the auxiliary sections do not apply, of course, to the five-speed (T Series).

1. Remove snap ring from the inside of the

auxiliary drive gear.
2. Remove mainshaft rear quill support plate from bore of auxiliary drive gear and remove coupling. Remove coupling snap ring.
3. Cut lock wire and remove bolts that attach bearing retainer ring to case. Insert three puller screws in tapped holes of retainer ring and tighten bolts evenly and alternately to pull assembly from case, Fig. 11.
4. Secure auxiliary drive gear in a vise and remove bearing nut (left-hand thread). Then press on drive gear bearing retainer ring and bearing from gear.
5. Move mainshaft reverse gear to rear as far as possible and remove reverse gear snap ring. Then move reverse gear forward and against low-speed gear and into engagement with sliding clutch gear. Remove snap ring and reverse gear splined spacer from mainshaft.
6. Working from inside of case, remove left reverse idler shaft nut and washer.
7. Remove plug, Fig. 12, from reverse idler shaft bore. Using an impact puller, remove reverse idler gear shaft. Remove thrust washer and gear.
8. If necessary, remove bearing inner race from reverse idler shaft, and bearing from reverse idler gear.
9. Move mainshaft to rear as far as possible to separate mainshaft gears from countershaft gears. Tilt front of mainshaft and lift from case.
10. Cut lock wire and remove bolts that attach bearing retainer to front of countershaft. Remove snap ring from rear of countershaft.
11. Using a mallet and soft drift, drive countershaft to rear as far as it will go. This will partially unseat front bearing from shaft and rear bearing from case bore.
12. Use a puller to remove countershaft rear bearing from shaft and case bore.
13. Move countershaft forward until front bearing is clear of case, then pull bearing from shaft. Remove left countershaft bearings in like manner.
14. Remove input shaft bearing retainer bolts and, working from inside of case, tap input shaft forward and remove bearing retainer. Remove snap ring from input shaft bearing.
15. Move input shaft to rear and lift from case.
16. Move front of right countershaft toward center of case and, at the same time, lift it from case. Remove left countershaft in like manner.
17. Working from inside case, remove right reverse idler shaft and nut, then remove component parts in same manner as left-hand assembly.
18. Disassemble shifter housing, mainshaft and countershaft as required.

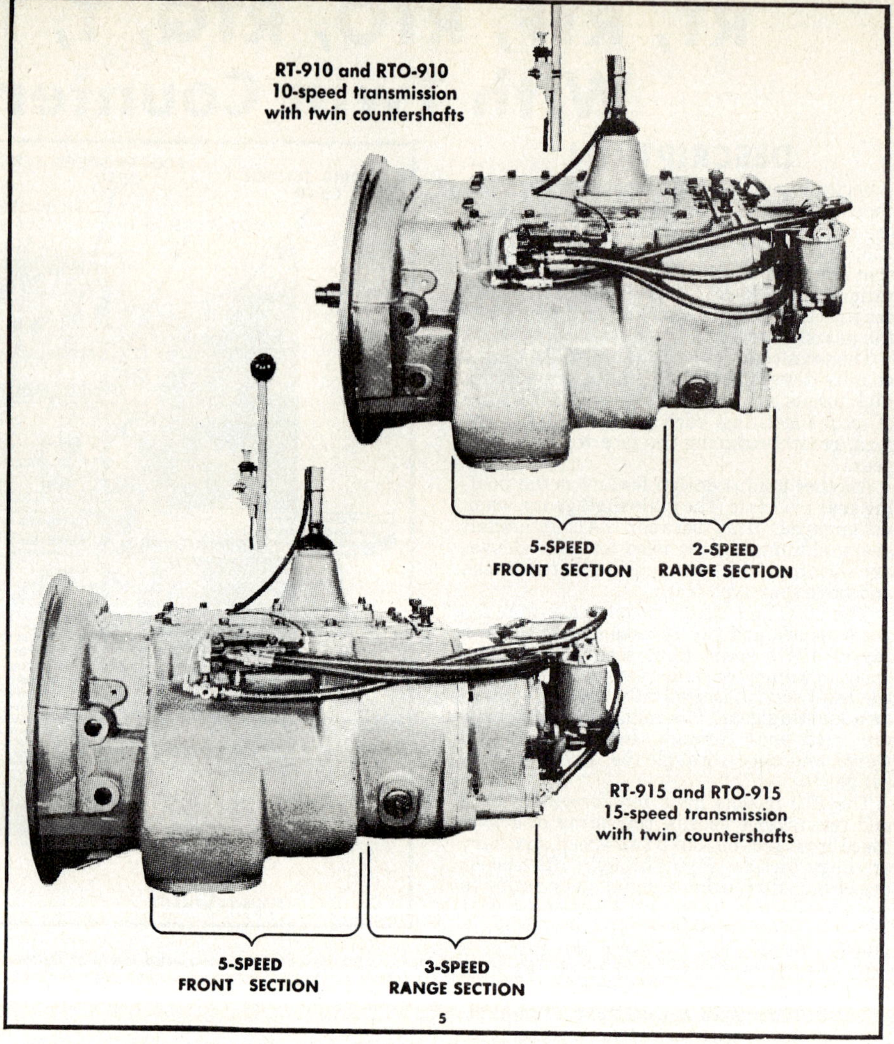

Fig. 2 Exterior view of Fuller RT twin countershaft compound transmissions. The 9513 & 12513 Series is similar to RT-915 and RTO-915 Series

Main Sub-Assemblies, Assemble

Countershaft Assemble, Fig. 13 and 13A

NOTE: Except for the number of teeth on the PTO gears, the countershafts are identical and are assembled in the same manner.

1. Install front washer over rear of countershaft and position keys in keyways.
2. Press countershaft gears on shaft one at a time, starting with the drive gear. The drive, PTO, 3rd and 1st speed gears are

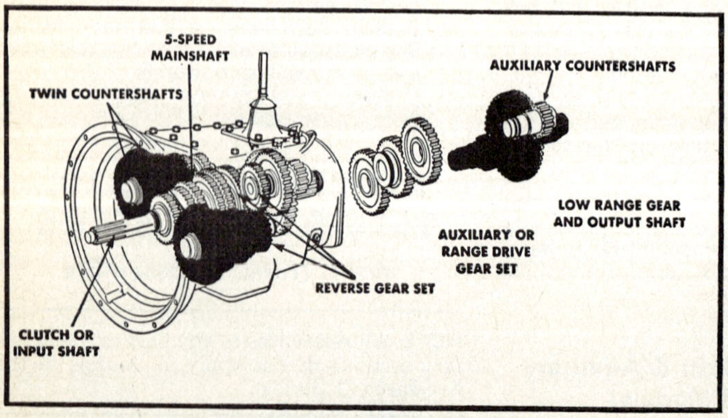

Fig. 3 Expanded view of RT-910 transmission

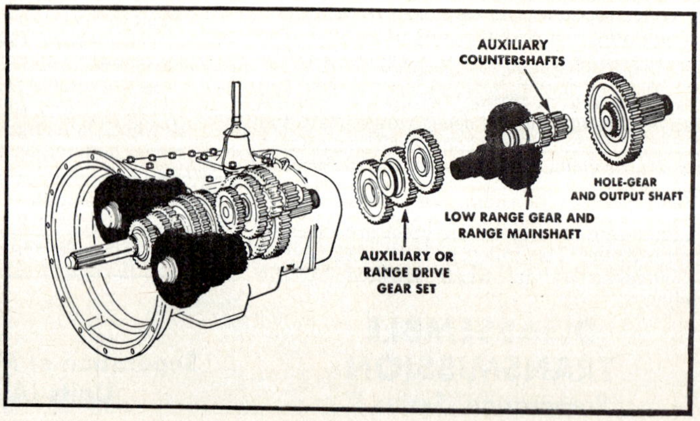

Fig. 4 Expanded view of RT-915 transmission

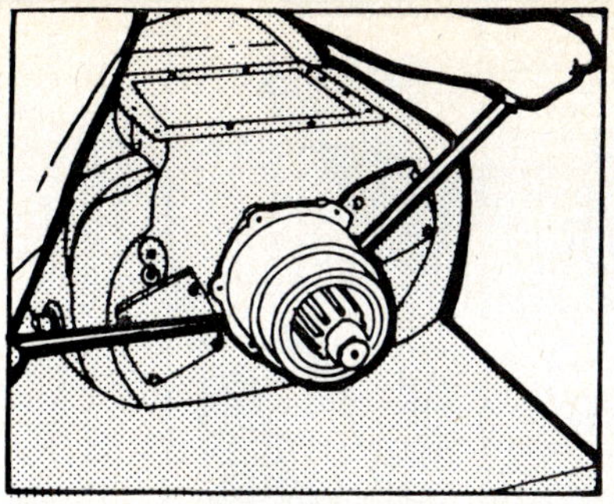

Fig. 5 Removing rear bearing cover. T Series

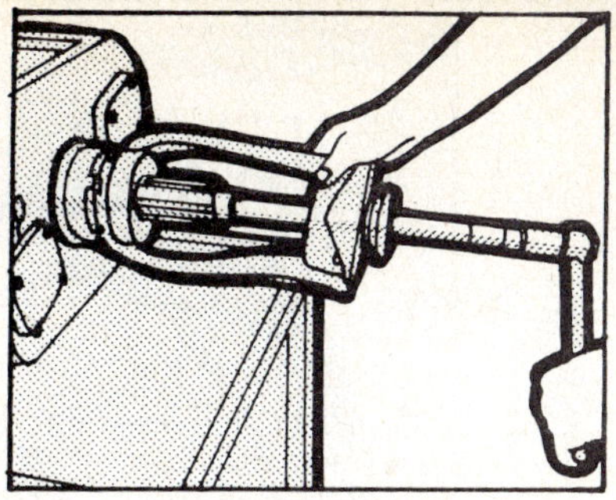

Fig. 6 Removing rear bearing from output shaft. Series T

Fig. 7 Removing output shaft screws. Series T

installed with long hubs to the rear; the 4th, 2nd and reverse gears are installed with long hubs to the front. The left countershaft has a 47 tooth PTO gear; the right side a 45 tooth gear. After installing the PTO gear, mark the assembly either left or right to correspond with the installed PTO gear.

3. Install snap ring on rear of countershaft.

Input Shaft, Assemble, Fig. 14

1. Install snap ring in drive gear if it was removed.
2. Install drive gear on shaft with snap ring toward front.
3. Install drive gear spacer.
4. Press drive gear bearing on shaft with snap ring toward front.
5. Install and tighten bearing attaching nut (left-hand thread). Peen nut in slots milled in shaft.

Mainshaft Assemble, Fig. 15

1. Except for the reverse gear, install snap ring in each gear if it was removed.
2. Place mainshaft in a vise with front end up.
3. Install reverse gear washer (flat side down) in fifth groove from top of mainshaft.
4. From bottom, install long key in keyway in shaft so that forward end is moved through slot of reverse gear washer. Keep snap ring groove in key to the rear and to the outside. Move key upward to engage washer as it is placed on mainshaft.
5. Install 1st-reverse sliding clutch gear on shaft, making sure that keyway is lined up with key in shaft.
6. Install 1st speed gear on shaft with snap ring up. Engage splines of gear with splines of spacer.
7. With snap ring down, install 2nd speed gear on shaft and against 1st gear.
8. Install 2nd gear spacer and washer on shaft (flat sides down). Move key upward through slot in 2nd gear washer until it reaches second groove from top of mainshaft.
9. Align keyway in 2-3 speed sliding clutch with key, and position clutch on mainshaft.
10. Install 3rd gear washer (flat side up) in second groove from top of mainshaft, then move key upward to engage washer.
11. Install 3rd gear spacer with flat side against washer.
12. Install 3rd gear (snap ring up) on shaft and onto splines of spacer.
13. With snap ring down, install 4th speed gear on shaft against 3rd speed gear.
14. With flat side up, install 4th speed gear spacer into 4th gear hub.
15. Install 4th speed gear washer (flat side down) on shaft and move key upward to engage washer.
16. Install 4-5 speed sliding clutch gear, aligning large slot in clutch gear with keyway in shaft.
17. Install front quill bearing if it was removed. When installing the bearing, be sure to use a suitable tool so that the quill is not pushed into the mainshaft. Install front quill snap ring.
18. Remove mainshaft from vise and align snap ring groove in key with snap ring groove in mainshaft.
19. Install reverse gear on mainshaft and move it as far forward as possible into engagement with sliding clutch gear. *Do not install reverse gear spacer and snap ring until mainshaft assembly is installed in transmission.*
20. Install front mainshaft coupling snap ring on shaft and key, securing key to mainshaft.

Gearshift Housing Assemble, Fig. 16

1. Install shifter shaft through rear bores of housing with neutral and detent notches to rear. Keep shafts in neutral position when installed.
2. Place housing in a vise with plunger side up.
3. Install 1st-reverse shifter shaft in bottom bore and through shift fork. Install set screw and secure with lock wire.
4. Install actuating plunger in front opening of housing, and a 3/4" interlock ball in rear opening.
5. Position 2-3 speed shift shaft through shift block and fork. *Be sure to install interlock pin in bore at neutral notch.* Install set screws and secure with lock wire. *Keep shift shaft so that interlock pin remains in vertical position during remainder of assembly. Rotating the shaft*

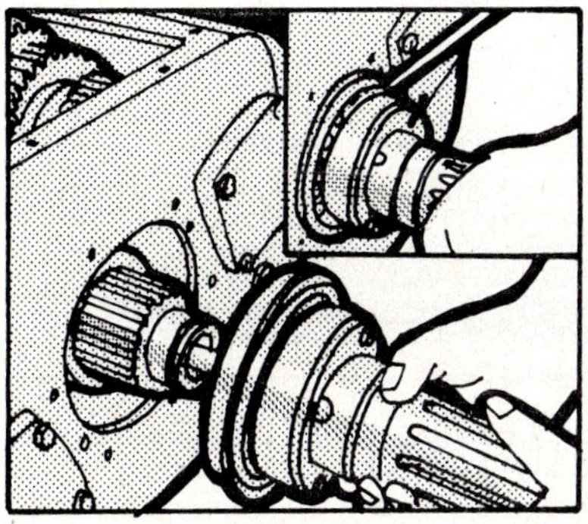

Fig. 8 Removing output shaft. Series T

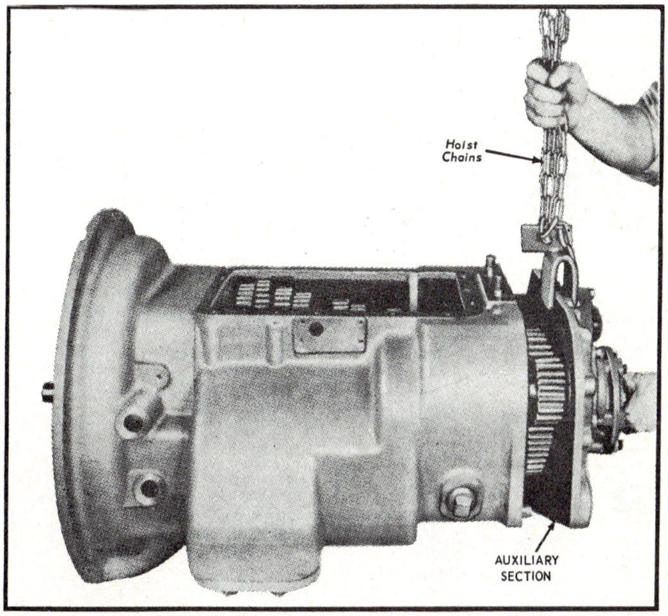

Fig. 9 Separation of auxiliary section. RT-910 shown

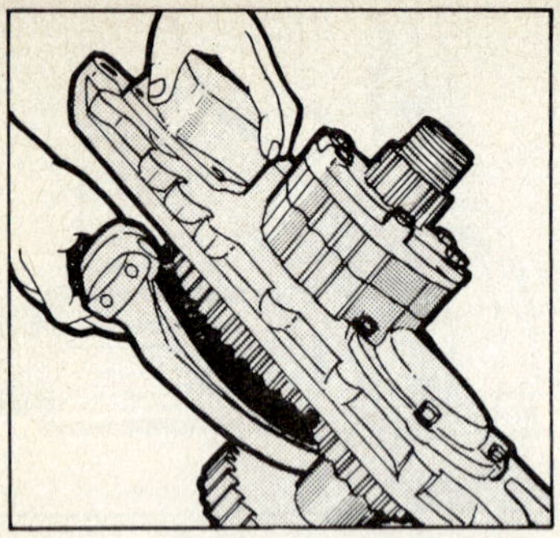

Fig. 10 Removing auxiliary section shift cylinder housing and fork

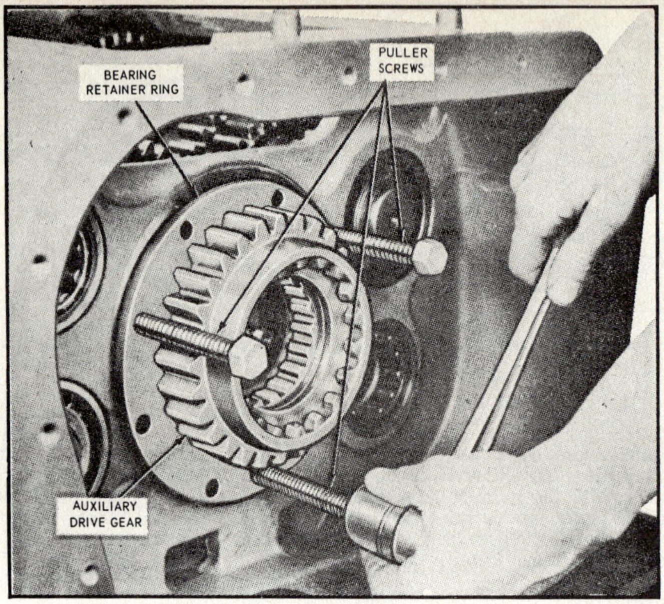

Fig. 11 Removing auxiliary drive gear

will cause interlock pin to jam into tension spring bores.

6. Install 3/4" ball in rear opening in housing.
7. Position 4-5 speed shift shaft in upper bore and through shift block and fork. Install set screws and secure with lock wire.
8. Remove shift housing from vise and place on work bench. Install three detent balls and springs into top of housing.

Auxiliary Section

Disassemble

1. Remove shift cylinder cover.
2. Remove nut from end of shifter shaft. Using puller screws, remove piston from shift cylinder.
3. Remove bolts attaching shift cylinder housing to rear plate.
4. Cut lock wires and remove set screws from shifting fork. If necessary, rotate shift cylinder housing to expose set screws.
5. Push shifter shaft to inside, through fork hub, and out of cylinder housing. Remove shift fork and shift cylinder housing from plate.
6. To remove countershaft gear, remove snap ring and press gear from each shaft.
7. Place synchronizer assembly on work bench. Remove direct (high range) cone synchronizer ring from pins of low-speed synchronizer ring. *Place a cloth over ring during removal to catch the three springs installed in bores of direct cone synchronizer.*
8. Remove sliding clutch gear from pins of low-speed synchronizer.
9. Remove inner bearing spacer from output shaft, Fig. 17.
10. Using front face of low-speed gear as a base, press shaft through gear, washer and bearing. Remove splined and stepped washers from shaft.

Assemble Sub-Assemblies

1. Set output shaft on work bench with threaded end up. Install stepped and splined washers on shaft, Fig. 17. Be sure small diameters are toward threaded end of shaft.
2. Install low-speed gear on shaft (flat side up). Then install cone-shaped washer with flat side against gear.
3. Install bearing on shaft with taper toward threaded end.
4. Install inner bearing spacer on shaft and against bearing.
5. Place low-speed synchronizer ring on work bench with pins up. Then install clutch sliding gear on pins.
6. Install three springs into bores of direct synchronizer ring, Fig. 18, and install direct synchronizer over pins of low-speed synchronizer ring.
7. Apply pressure to direct synchronizer ring to compress springs and fully seat

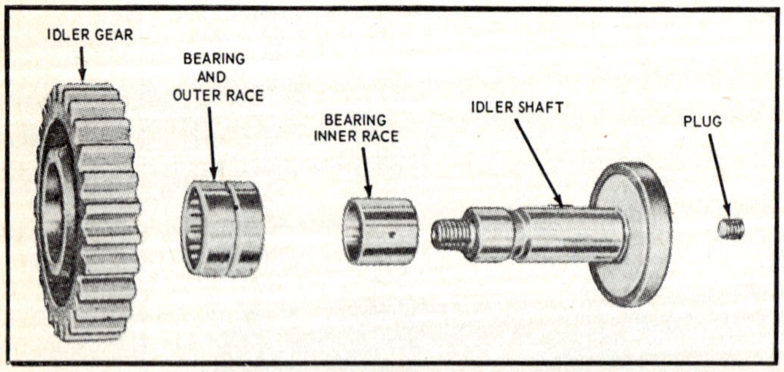

Fig. 12 Reverse idler shaft components, except 9513 & 12513

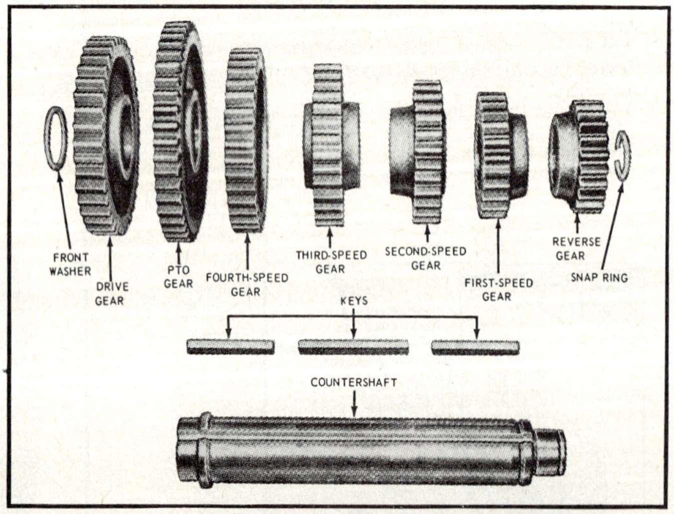

Fig. 13 Main transmission countershaft disassembled, all except 613, 913, 9513 & 12513 Series

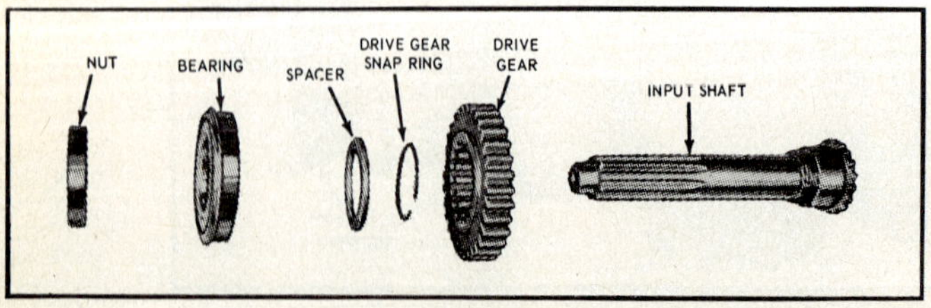

Fig. 14 Main transmission input shaft disassembled

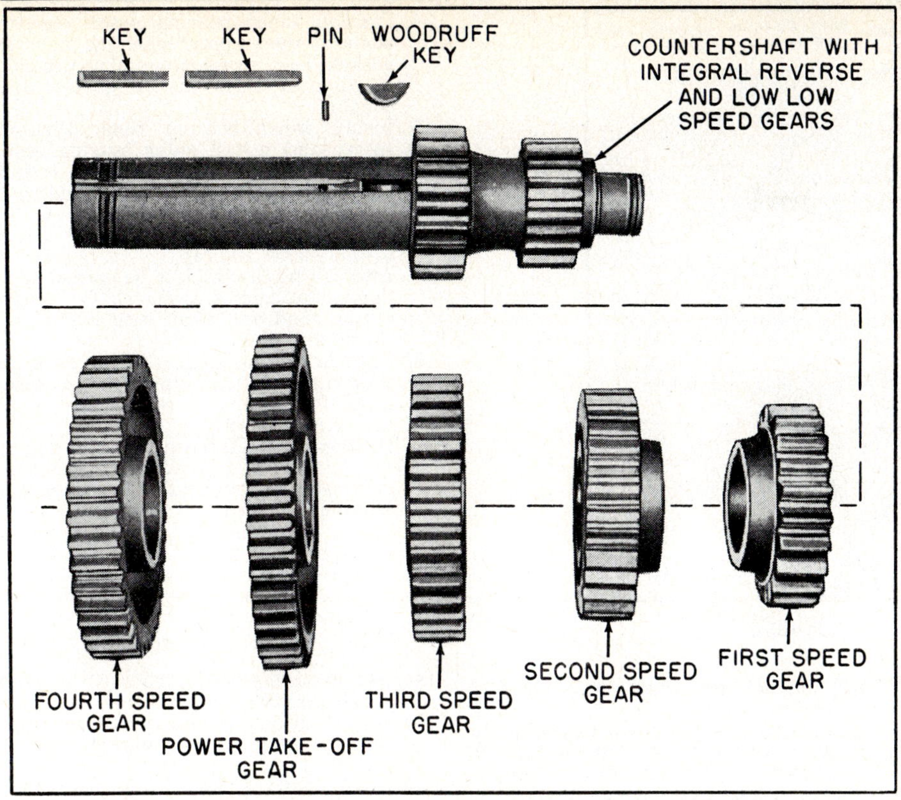

Fig. 13A Main transmission countershaft disassembled. 613, 913, 9513 & 12513 Series

direct synchronizer on pins of low-speed synchronizer.

8. Install Woodruff key on each countershaft and press countershaft into each drive gear. Install snap ring on front end of each shaft.

Assemble Auxiliary Section

1. With rear plate on work bench, place front cup (taper to inside) of rear bearing into rear bore of plate.
2. Place bearing outer spacer on front cup. Place rear cup on spacer and tap the two cups and spacer into rear plate bore.
3. Place rear plate over output shaft. Be sure that the bearing on shaft is seated on rear plate front bearing cup.
4. Install rear bearing on output shaft. Install new seal in rear bearing cover (if it was removed). Position cover to rear plate and install attaching bolts.
5. Install synchronizer assembly on output shaft with low-speed synchronizer ring toward low-speed gear.
6. To install and time auxiliary countershafts, mark gear tooth which is stamped "O" on small diameter low-range gear of each countershaft. This tooth will be aligned with the keyway in the countershaft drive gear. These two teeth, one on each countershaft, must both mesh at the same time with the output shaft low-speed gear and be in directly opposite positions.
7. Mark any two adjacent teeth on the output shaft low-speed gear. Then mark the two adjacent teeth which are directly opposite the first set marked. There should be the same number of teeth between the markings on each side of the gear.
8. Place rear plate assembly on blocking with threaded end of output shaft up. Make sure blocking is also placed under the synchronizer assembly to prevent it from dropping from the output shaft.
9. Place each countershaft into position and place blocking under each shaft. For correct timing, place marked teeth of countershafts into mesh with the low-speed gear. Install rear bearing on each countershaft. Recheck gears to be sure they are correctly timed.
10. Position rear bearing retainer plate on each countershaft and secure them with bolts and lock wire. Install rear bearing covers.
11. Install new O-ring in shift cylinder bore if it was removed.
12. Position shift fork in sliding clutch gear and install shift cylinder into bore of rear plate.
13. Insert threaded end of shift shaft through front of shift cylinder, through shift fork hub and into cylinder bore. Install set screws and secure with lock wire.
14. Install new O-rings on shift cylinder piston if they were removed.
15. Position piston in shift cylinder bore and secure with elastic stop nut.
16. Position cover on shift cylinder (air port to upper left) and secure with bolts.

Main Trans. Assemble

1. Install plug in right reverse idler shaft. Install needle bearings in bore of reverse idler gear if they were removed. Install bearing inner race on reverse idler shaft if it was previously removed.
2. Insert threaded end of idler shaft through right wall of case. As shaft is moved forward, install reverse idler gear and thrust washer. *Be sure needle bearings in gear are seated evenly on shaft inner race before moving shaft forward.* The thrust washer should be positioned between gear and boss, with slot toward gear.
3. Position washer over threaded end of shaft and secure with elastic stop nut.
4. To install and time countershaft assemblies, mark drive gear tooth that aligns with keyway of each countershaft. This tooth will be stamped with an "O." Position left and right countershaft assemblies in case. *Make sure that left countershaft has larger 47-tooth PTO gear.*
5. Mark any two adjacent gear teeth on the input shaft drive gear. Then mark two adjacent teeth which are directly opposite the first set marked.
6. Install input shaft through front bore from inside of case. Move shaft as far forward as possible and install snap ring on input shaft bearing.
7. Using a wood block, center front end of left countershaft in case bore. Then mesh the marked tooth on left countershaft drive gear with the two teeth marked on the input shaft main drive gear.

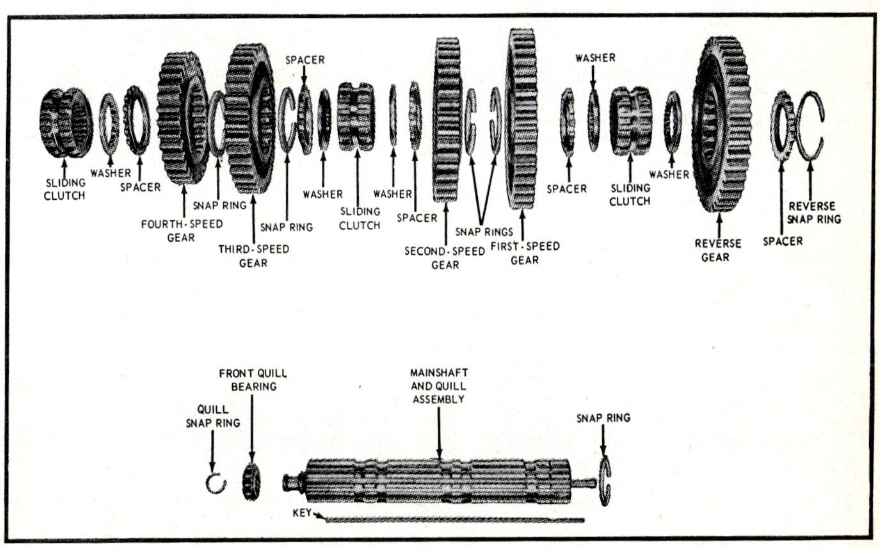

Fig. 15 Main transmission mainshaft disassembled. All except 9513 & 12513

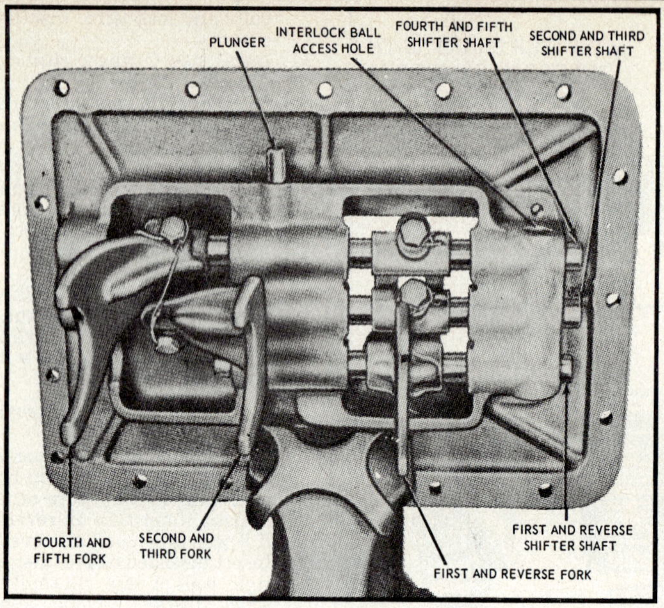

Fig. 16 Gearshift housing assembled

8. With countershaft as far to rear as possible, install rear bearing on shaft and into case bore.
9. Remove wood block and install front bearing on countershaft and into case bore. Install snap ring on rear of countershaft.
10. Position bearing retainer plate on front of countershaft and secure with attaching bolts and lock wire.
11. After installing left countershaft assembly in case, install right countershaft in like manner.
12. Position input shaft bearing retainer to case and secure with attaching bolts.
13. Lower mainshaft assembly in case. Place bar across top of case and install support wire under forward sliding clutch gear, Fig. 19.
14. Align mainshaft assembly, making sure mainshaft gears and mating countershaft gears are properly engaged, and that quill bearing is seated in pocket of main drive gear. Remove support wire and bar. *Keep mainshaft in forward position during installation of left reverse idler gear so that gears will not slip out of mesh.*
15. Install plug in left reverse idler shaft. Install needle bearings in bore of reverse idler gear if they were removed. Install bearing inner race on reverse idler shaft if it was previously removed.
16. Insert threaded end of reverse idler shaft through left wall of case. As shaft is moved forward, install reverse idler gear and thrust washer. *Be sure needle bearings in gear are seated evenly on shaft inner race before moving shaft forward.* The thrust washer should be positioned between gear and boss with slot toward gear. Install washer over threaded end of shaft and install elastic stop nut.
17. Move mainshaft reverse gear to the rear so that it engages reverse idler gears. Then install reverse gear spacer (flat side inward) into hub of reverse gear. Install reverse gear snap ring.
18. If auxiliary drive gear was disassembled, place bearing retainer on gear and press bearing into place. Apply Loctite to threads of drive gear and nut and install nut (left-hand thread).
19. Position auxiliary drive gear assembly in rear case bore and install attaching bolts. Wire bolts in groups of three.
20. Pull mainshaft as far to rear as possible and install mainshaft coupling snap ring. Place coupling gear and rear quill support plate in bore of auxiliary drive gear and install snap ring.
21. Install auxiliary countershaft front bearings into reverse idler shaft bores.

Series T Final Assembly

1. Install countershaft front bearing in case bore. Position front bearing spacer and gasket and install retainer.
2. Position countershaft assembly in case, then tilt case so that rear face is up.
3. Position two wood blocks of equal thickness between drive gear and case.
4. Position inner bearing outer race in adapter, then apply sealer to adapter gasket. Make sure adapter oil return hole is aligned, then install gasket and adapter.
5. Install inner bearing cone. Position spacer and install outer bearing cone. Position outer bearing outer race. Install bearing lock plate and secure bolts with lock wire.
6. Apply sealer to adapter cap gasket and install gasket and cap.
7. After tilting case back to its normal position, install idler gear and shaft. Be sure flat on shaft will align with lock plate. Install lock plate.
8. Position output shaft sliding gears in case, then position output shaft in case. Check for proper gear engagement.
9. Install rear bearing.
10. Position input shaft bearing snap ring in case bore. Position input shaft assembly and install bearing retainer gasket and retainer, using sealer on gasket.
11. Install clutch housing and speedometer drive gear.
12. After installing a new seal in rear bearing retainer, install gasket and retainer, using sealer on gasket.
13. Install parking brake shoe and connect brake linkage.
14. With transmission locked in two gears, install brake drum and flange.
15. Adjust parking brake. Install gearshift housing, using sealer on gasket.

Joining Main & Auxiliary Sections

1. Install clutch housing to transmission case. Using a chain hoist, install auxiliary section to transmission. It may be necessary to rotate drive gear to properly mesh gears. Install and securely tighten attaching bolts.
2. With transmission locked in two gears, install speedometer drive gear and U-joint flange on mainshaft. *If speedometer drive gear is not used, be sure to install replacement spacer of the same width.*
3. Install gearshift housing cover to transmission. Install air valve plate and housing. Install air valve and air regulator valve assemblies, and connect air hoses at air valve.

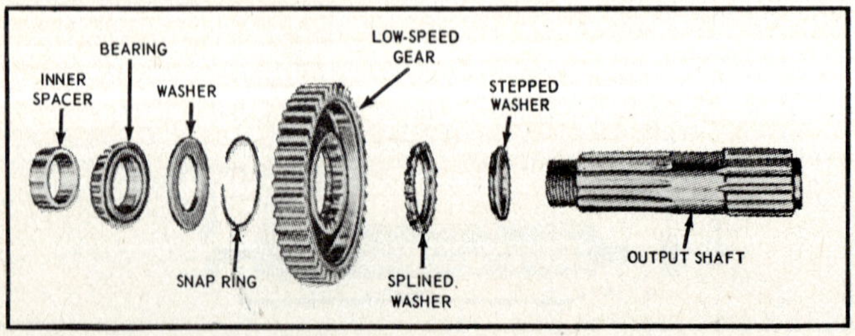

Fig. 17 Auxiliary section output shaft. All except 9513 & 12513

RT610 & RT6610

Separation of Main & Auxiliary Units

1. Disconnect nylon air lines from air valve and remove gear shift lever housing, lever and control valve as an assembly. Then disconnect air line and clamp between air valve and regulator.
2. Remove air regulator and filter assembly, then remove air lines between air valve and air cylinder in auxiliary section.
3. Remove air valve and actuating spring and pin from valve bore, then remove gear shift lever housing or control housing from shift bar housing.
4. Remove shifting bar housing from transmission, then lock transmission by engaging two gears with mainshift sliding clutch gears and remove output shaft nut and flange.
5. Remove auxiliary section attaching screws, then install puller screws in tapped holes of auxiliary section and tighten screws evenly until housing can be removed, Fig. 9.
6. Remove clutch release mechanism and

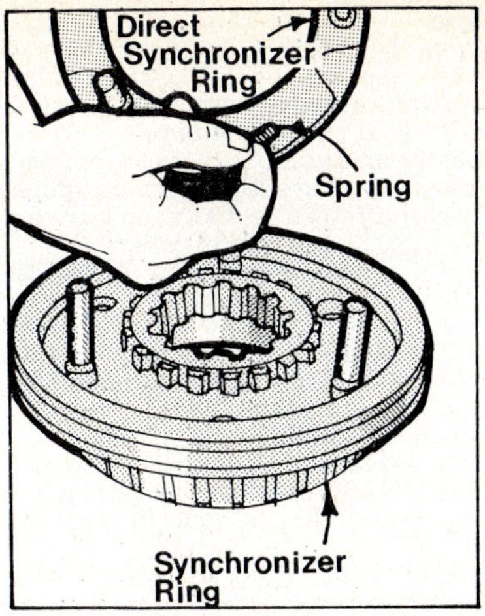

Fig. 18 Installing synchronizer springs

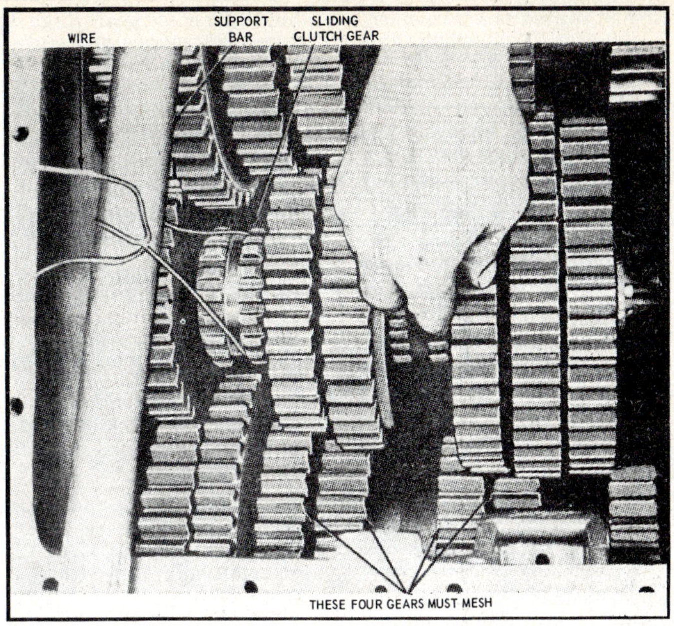

Fig. 19 Aligning mainshaft countershaft gears

clutch housing and remove front gear cover.

Main Trans. Disassemble

1. Remove right countershaft snap ring from rear bearing bore of case, Fig. 20, then using a soft bar and hammer, drive against rear of countershaft, moving assembly forward as far as possible to expose front bearing snap ring groove.

NOTE: If front bearing cannot be exposed, use punch between hub of countershaft drive gear and inner race of bearing to move gear forward on shaft.

2. Install snap ring removed from rear bearing bore into exposed bearing snap ring groove, then loosen capscrews at bearing retaining rings of auxiliary drive gear. This will allow movement to rear of countershaft and mainshaft assemblies.
3. Using soft bar and hammer, drive against front of countershaft and move countershaft assembly rearward as far as possible, thereby moving front bearing forward on shaft. Then, again using soft bar and hammer, drive against countershaft and move countershaft forward as far as possible to unseat bearing from case bore. Using puller, remove bearing.
4. Using a soft curved bar and hammer from inside case, drive rear countershaft bearing rearward far enough to expose snap ring groove in rear bearing. Install snap ring into groove in rear bearing and using puller, remove countershaft rear bearing.
5. Move drive gear and shaft forward as far as possible, then remove bearing retaining snap ring from groove in shaft.
6. Hold shaft in position and tap drive gear forward to unseat bearing from case, then using puller, remove bearing.
7. Remove drive gear washer, and remove snap ring from inside diameter of drive gear, then pull shaft forward and from splines of drive gear.
8. Remove bushing from pocket in clutch shaft, if necessary.
9. Position drive gear rearward and against 4th speed gear, engaging teeth of sliding clutch gear, then remove capscrews and the two bearing retainer rings from case.
10. Remove coupling snap ring from groove in mainshaft located in bore of auxiliary drive gear, then tap against front of mainshaft to move auxiliary drive gear bearing rearward, exposing bearing snap ring.
11. Using pry bars, remove auxiliary drive gear assembly from case bore and from splines of mainshaft, then remove snap ring from groove in auxiliary gear and press or drive bearing from auxiliary drive gear.
12. Using pry bar, remove bearing from idler bore in case, then remove rear idler washer and holder from bore. If necessary bend holder lugs and remove washer from holder.
13. Remove idler gear from case and remove idler bearing from reverse idler gear.
14. Remove front idler washer from case, then block right countershaft to the right as far as possible and lift mainshaft assembly from case.
15. Remove drive gear and 4th-5th sliding clutch from mainshaft, then from inside hub of 4th speed gear, remove snap ring from groove in mainshaft, Fig. 21.
16. Remove 4th and 3rd speed gears and washers, remove 2nd-3rd speed sliding clutch, remove long key from mainshaft slot, remove 2nd and 1st speed gears and washers, remove 1st-reverse sliding clutch and remove reverse gear and washer.
17. Lift right countershaft assembly from case, then remove bearings from left countershaft in same manner in which bearings were removed from right countershaft.

NOTE: Both left and right countershaft assemblies are disassembled in the same manner as their construction is identical.

18. Remove snap ring from front of countershaft, Fig. 22, and press the drive gear then the 4th speed gear from shaft.
19. Using the rear face of PTO gear as a base in press, press the 3rd speed gear, the PTO gear and 2nd speed gear cluster from shaft. If necessary, remove Woodruff key, long key and roll pin from shaft.
20. Using curved pry bar, remove reverse idler gear bearing from idler bore in case, then remove rear idler washer and holder from bore. If necessary, bend holder lugs and remove washer from holder.
21. Remove reverse idler gear from case, then remove idler bearing from reverse idler gear and remove front idler washer from case.

Fig. 20 Removing right countershaft snap ring. RT610 & 6610

Main Trans. Assemble

1. Install small, flat, right reverse idler gear front washer on pin in lower right of case, then set reverse gear in position next to washer, with long hub to the rear.
2. Install needle bearing through rear bore and into hub of reverse gear, then place rear washer in holder with oil slots down and bend lugs to secure washer.
3. Place washer and holder into reverse gear case bore, with oil slots toward gear and holder flange to the rear.
4. Install auxiliary countershaft front bearing into reverse gear case bore to hold reverse idler gear assembly in position.

NOTE: The front portion of the auxiliary countershaft is used as a journal for the reverse idler gear. Countershafts are identical and assembled in the same manner.

5. If removed, install roll pin, long key and Woodruff key in countershaft slots.
6. Press PTO and 2nd speed gear cluster on countershaft with PTO gear to the front, then press 3rd speed gear on shaft with long hub to the rear, 4th speed gear with long hub to the front and drive gear with long hub to the rear.
7. Install snap ring into groove at front of shaft, Fig. 23.
8. On each countershaft drive gear, mark the gear tooth aligned with the keyway in shaft and is stamped with an "O", Fig. 24.
9. First place left countershaft and then right countershaft into place in case. Do not install bearings.
10. Place mainshaft in vise with pilot end down and keeping keyway free for insertion of long key.

NOTE: Each mainshaft gear is held in place by locking the gear splined washer to the mainshaft with key. There is one splined washer for each gear. Splined washers for the reverse, 1st, 2nd and 3rd speed gears are identical. The 4th speed gear splined washer is of smaller diameter.

11. Install 2nd speed gear washer at 2nd speed gear location and insert key from bottom to lock washer in place, then install 2nd speed gear on splined washer with clutching teeth down.
12. Install 1st speed gear on shaft and against 2nd speed gear with clutching teeth up, then insert splined washer in hub of 1st speed gear. Align washer with mainshaft splines and move key up to lock washer in position.
13. Install 1st-reverse gear splined washer, aligning slot in clutch with key, then install reverse gear splined washer, align keyway and lock with key.
14. Reposition assembly in vise with pilot end up.
15. Install 2nd-3rd speed sliding cluth, aligning keyway in sliding clutch with key in shaft.
16. Install 3rd speed gear splined washer and lift key to install washer under pin in key to lock in position.
17. Install 3rd speed gear on mainshaft and slign gear splines with those of washer, Fig. 25.
18. Install 4th speed gear on mainshaft with clutching teeth up and against 3rd speed gear, then install 4th speed gear splined washer on mainshaft and in hub of 4th speed gear.
19. Install snap ring into mainshaft groove which will hold 4th speed gear in position, then install 4th-5th speed sliding clutch on mainshaft and engage it with 4th speed gear.
20. Mark two adjacent teeth of drive gear (5th speed gear) and mark the two adjacent teeth which are directly opposite the set marked. There should be the same number of teeth between markings on each side of the gear.
21. Install drive gear in mainshaft and against 4th speed gear with snap ring in drive gear to the front.
22. Remove assembly from vise and install reverse gear on rear of mainshaft, engaging splines of gear with those of splined washer on mainshaft.
23. Move both countershafts toward case wall as far as possible and position mainshaft assembly in case. Block under front of mainshaft to center it in front of case bore.
24. Press bearing on auxiliary drive gear with snap ring towards gear, then install bearing retaining snap ring in auxiliary drive gear shoulder groove.
25. Center mainshaft in rear bore of case and install auxiliary drive gear on splines of mainshaft, then seat auxiliary drive gear bearing in rear bore of case.
26. Install coupling snap ring in mainshaft groove in hub of auxiliary drive gear, then install the two bearing retainer plates on the case and attach both upper and lower plates using a double locking lug for each plate. Tighten screws and secure by bending locking lugs.
27. If removed, install bushing in pocket of clutch shaft. Install flush with shaft making certain that oil hole in shaft is not plugged.
28. Install clutch shaft in splines of drive gear, moving gear forward against wall of case, then install ring in groove in inside diameter of drive gear.
29. Install the drive gear spacer on shaft and drive gear bearing on shaft and into case bore, then install bearing retaining snap ring in groove in clutch with taper towards the outside.
30. Place left countershaft into mesh mainshaft gears, aligning the marked timing tooth on countershaft drive gear with the two marked timing teeth on main drive gear.
31. Hold left countershaft in position and install front and rear bearings, then center rear of left countershaft in rear bore of case with wood block. Partially install front and rear countershaft bearings, then using a flanged bearing driver, complete seating both front and rear bearings.
32. Install snap ring in groove in rear bore of case, then place right countershaft into mesh with mainshaft gears, aligning the

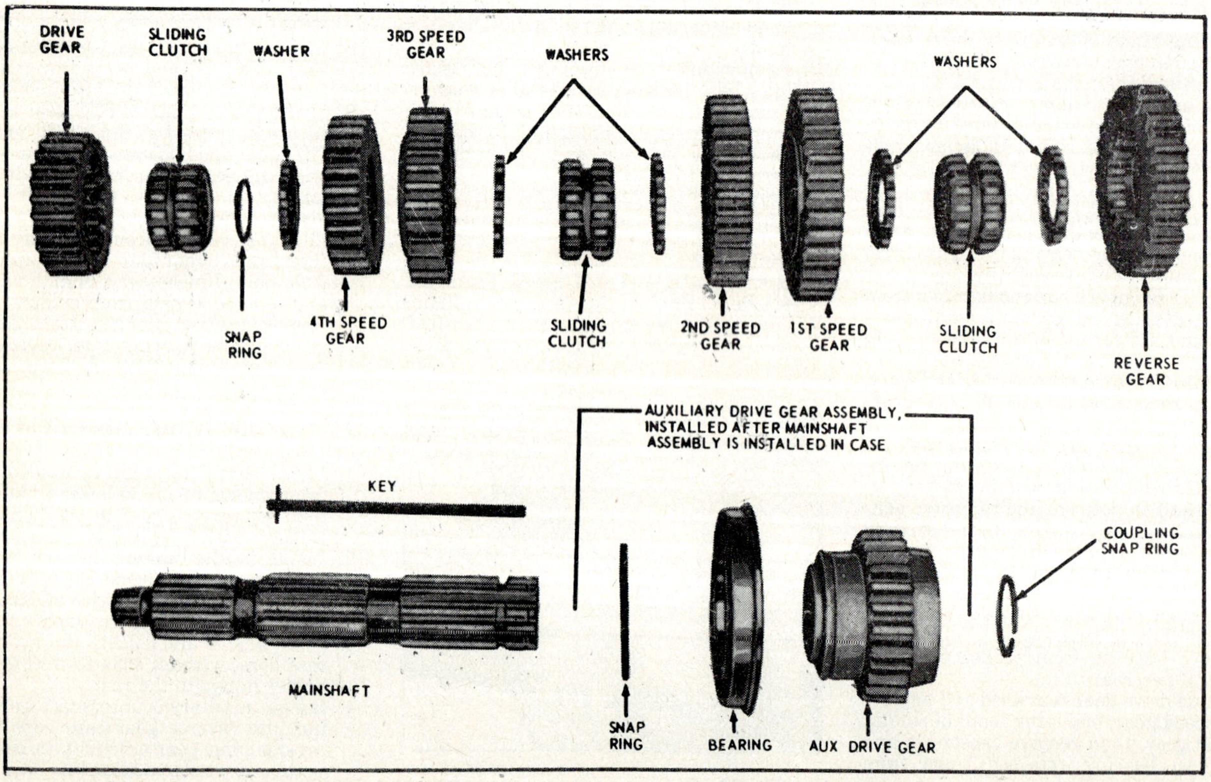

Fig. 21 Mainshaft and auxiliary drive gear. RT610 & 6610

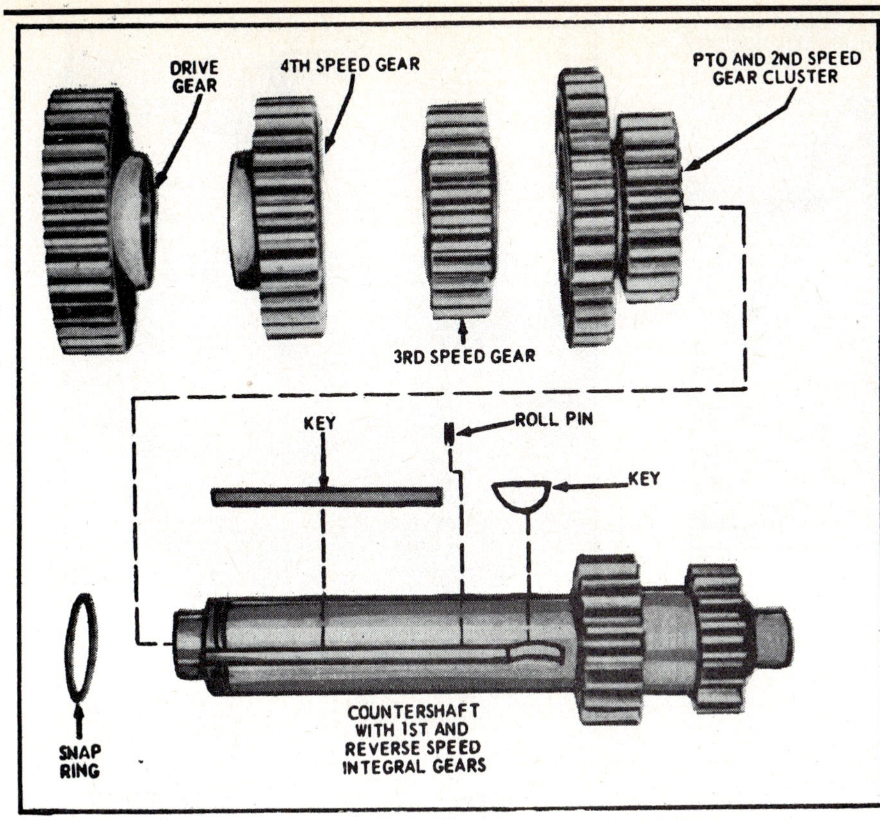

Fig. 22 Front section countershaft. RT610 & 6610

marked tooth on countershaft drive gear with the two marked timing teeth on main drive gear, Fig. 26.

33. Hold right countershaft in position and install front and rear bearings in same manner as installed on left countershaft.
34. Install snap ring in groove in rear bore of case and install left-reverse idler gear front washer on pin in upper left of case, then install needle bearing in hub of reverse idler gear and position reverse idler gear with long hub facing rearward.
35. If removed, place rear washer in holder with oil slot down, and bend lugs to secure washer, then place washer and holder into reverse gear case bore, with oil slots toward gear and holder flange to the rear.
36. Install the auxiliary countershaft front bearing into the reverse idler case bore to hold the reverse idler gear assembly in position.

NOTE: The front portion of the auxiliary countershaft is used as a journal for the reverse idler gear.

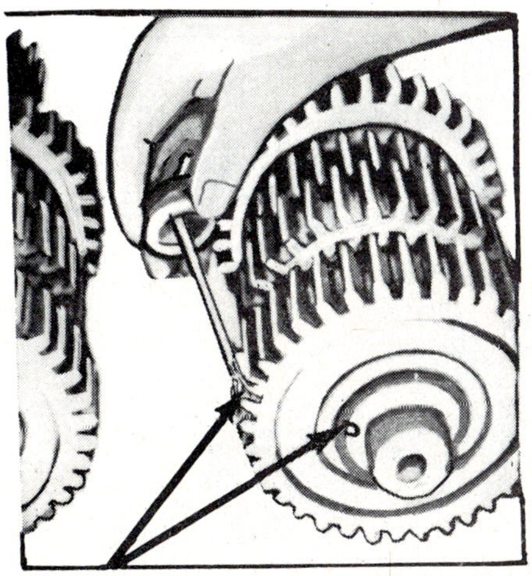

Fig. 24 Marking countershaft drive gear for timing purposes. RT610 & 6610

Auxiliary Section, Disassemble

1. Using puller, remove output shaft bearing, then remove rear bearing cover and if necessary, remove seal from cover.
2. Remove speedometer gear or replacement spacer from output shaft and remove speedometer gear washer.
3. Using a long punch from inside housing, move outer races of countershaft rear bearings rearward approximately 1/2 inch, Fig. 27.
4. Pull countershafts forward and out of housing, then remove bearing inner race from rear of countershaft, Fig. 28.
5. Remove snap ring from groove in front of shaft and press drive gear forward and from shaft. If necessary, remove key from countershaft.
6. Remove air cylinder cover and remove nut and lock washer from air cylinder piston shaft.
7. Using compressed air, remove piston, Fig. 29. If necessary, remove O-ring from outside diameter of piston, then remove copper seal from shaft.
8. Remove synchronizer assembly, shift yoke and piston shaft out of housing as complete assembly.
9. Remove shaft and yoke from synchronizer assembly, then cut lock wire, remove lock screws and remove yoke from shaft.
10. Place larger low range synchronizer ring on bench, then turn and pull upward to remove high range synchronizer from pins.

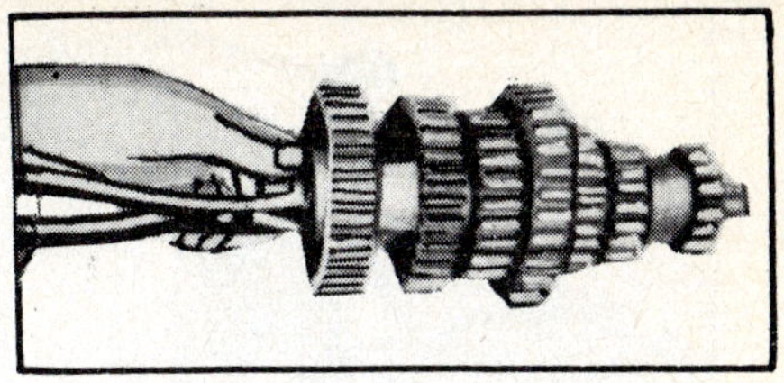

Fig. 23 Installing front countershaft snap ring. RT610 & 6610

NOTE: The three springs located in high range synchronizer housing will be released as high range synchronizer is removed from pins.

11. Remove sliding clutch from low range synchronizer pins.
12. Remove dust seal, air port extension and air cylinder from housing, then remove O-ring from small bore in cylinder, if necessary.
13. Drive or press against rear of tailshaft to move low speed gear and tailshaft assembly forward and from rear bearing, Fig. 30.
14. Remove rear washer and low speed gear from rear of shaft and remove spline washer from hub of low speed gear.
15. Move mainshaft rear bearing rearward and out of housing, then using bearing driver, remove rear countershaft bearings from housing.

Auxiliary Section, Assemble

1. Make certain that magnetic cleaner is installed in recess of auxiliary case.
2. For timing purposes, mark any two adja-

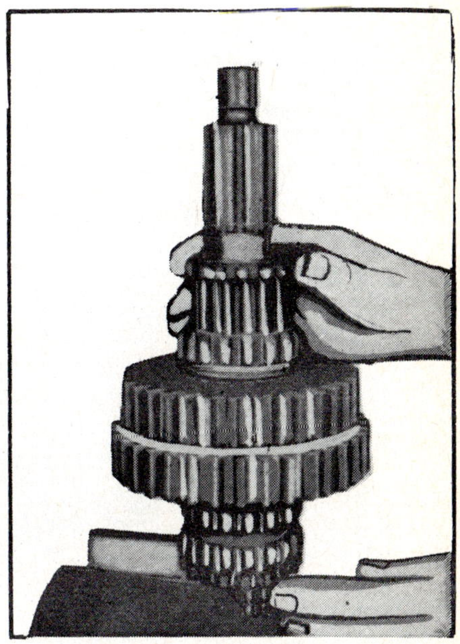

Fig. 25 Assembling mainshaft. RT610 & 6610

Fig. 26 Meshing countershaft and low speed gear for proper timing. RT610 & 6610

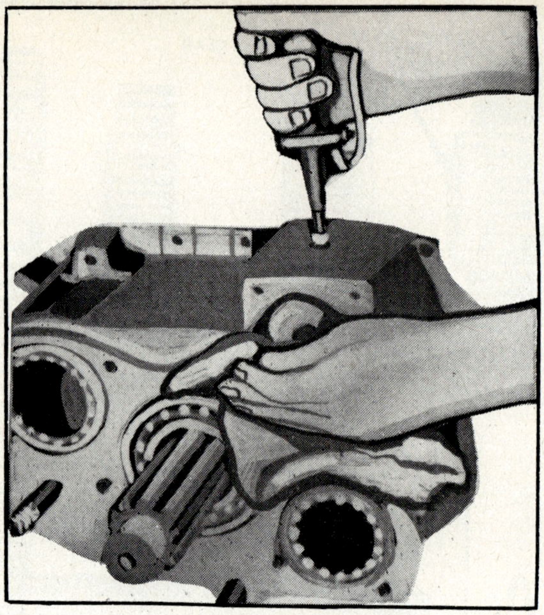

Fig. 29 Removing piston from auxiliary section. RT610 & 6610

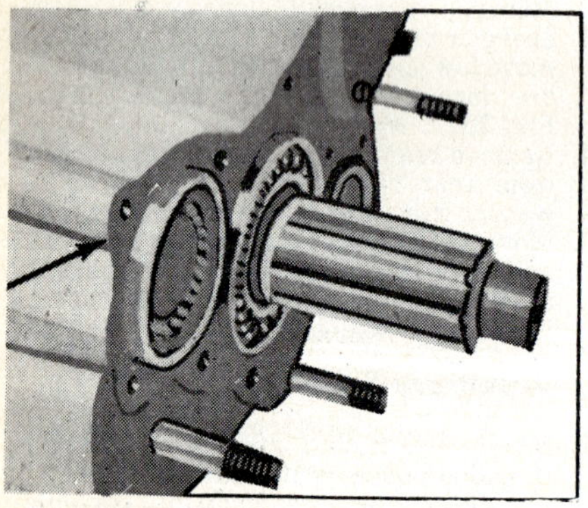

Fig. 27 Moving outer countershaft races rearward. RT610 & 6610

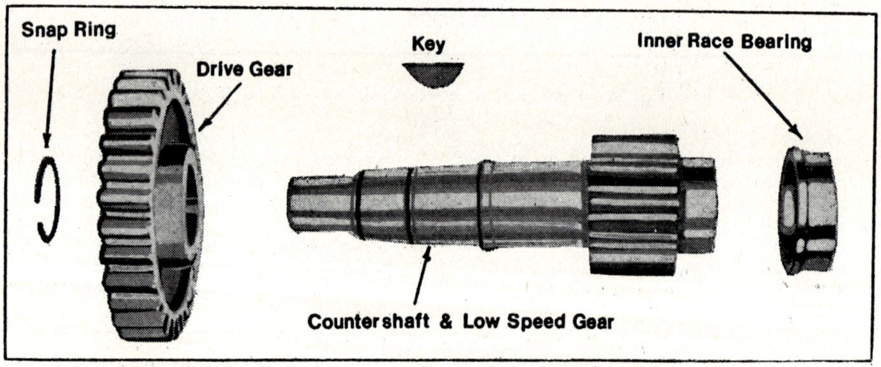

Fig. 28 Auxiliary section countershaft. RT610 & 6610

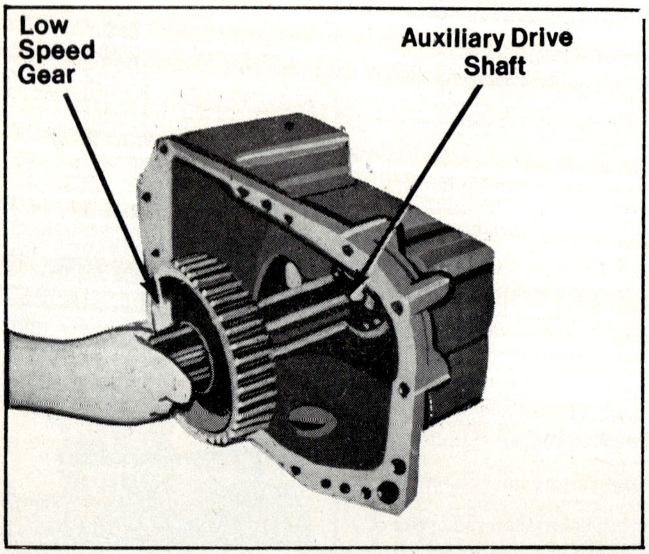

Fig. 30 Removing low speed gear and tailshaft. RT610 & 6610

Fig. 31 Marking timing tooth on low speed gear for timing purposes. RT610 & 6610

cent gear teeth on low speed gear, then mark the two adjacent teeth which are directly opposite the first set marked. There should be the same number of teeth between the markings on each side of the gear.
3. Place splined spacer in hub of low speed gear with shoulder facing rearward, then install low speed gear and spacer over rear of shaft and against shoulder with clutching teeth to the front.
4. Install low speed gear washer on shaft and against gear with chamfer facing rearward.
5. Position tailshaft with forward end down and place auxiliary housing over rear of shaft so that shaft extends through rear bore.
6. Seat mainshaft rear bearing securely on shaft and in bore, then install O-ring into slot of small bore of range shift air cylinder.
7. Install cylinder in housing with small bore in cylinder aligned with air port in top of auxiliary section, then install dust cover and air port extension through housing and into bore of air cylinder.
8. Place shifting yoke on piston shaft with fork facing threaded end, then align slots in shaft with bores in yoke hub and install the two lock screws. Secure with safety wire.
9. Place yoke in slot of sliding clutch with threaded end of piston shaft towards the larger low range synchronizer, then place entire assembly into auxiliary housing and thread splined sliding clutch onto tailshaft and insert piston shaft into cylinder bore.
10. Install copper seal on threaded end of piston shaft and install O-ring on outside diameter of piston.
11. Install lock washer and nut on shaft to secure piston, then install gasket and cylinder cover and torque retaining bolts to 35–45 ft. lbs.
12. With larger low range synchronizer ring facing down with pins up, place sliding clutch with recessed side up on pins of low range synchronizer.
13. Install the three springs into bores in high range synchronizer ring as shown in Fig. 18.
14. Place high range synchronizer ring over pins of low speed synchronizer, seating springs against pins.
15. Apply pressure to high range synchro-

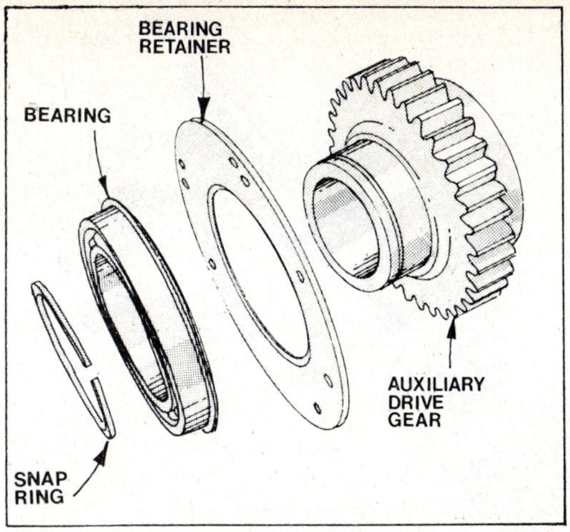

Fig. 32 Auxiliary gear assembly. 9513 & 12513 Series

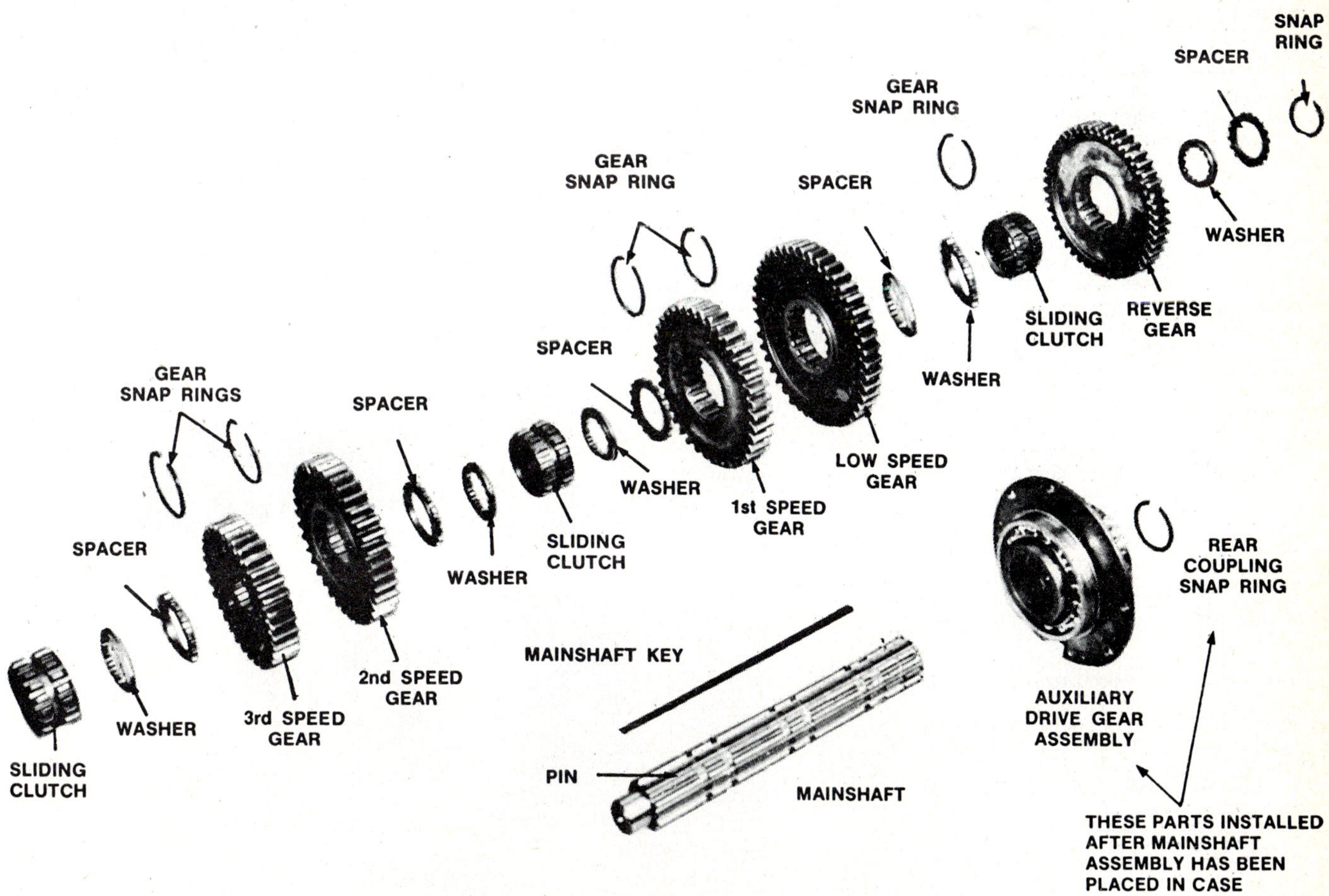

Fig. 33 Mainshaft assembly. 9513 & 12513 Series

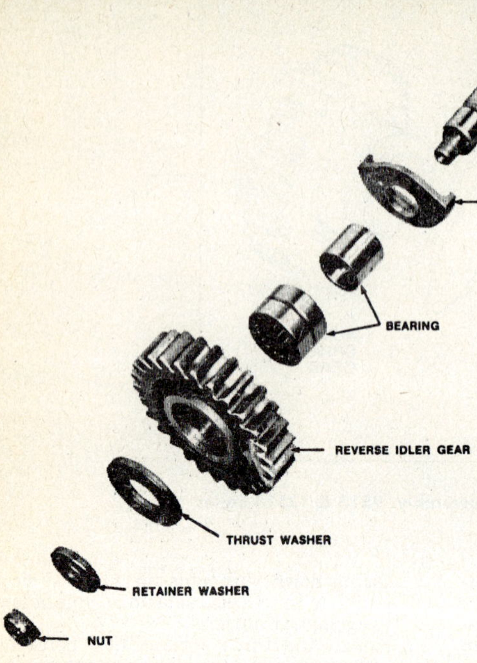

Fig. 34 Reverse idler gear assembly. 9513 & 12513 Series

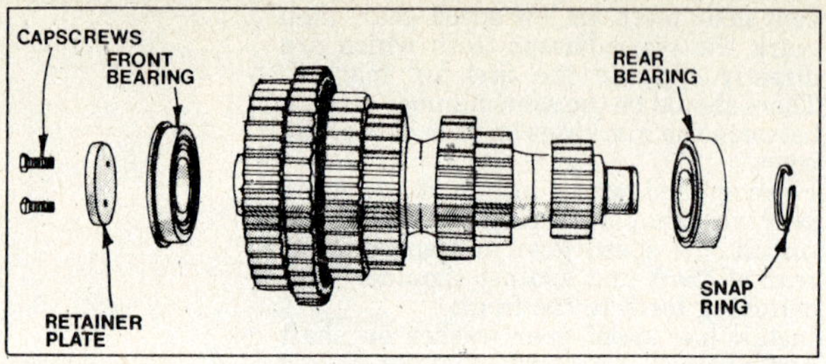

Fig. 35 Countershaft bearing assembly. 9513 & 12513 Series

nizer ring to compress springs and seat pins of low range synchronizer fully into bores of high range synchronizer ring.
16. If removed, install keys into countershaft keyways, then press drive gear onto each countershaft with long hub facing rearward and install snap ring in front of each countershaft.
17. Install bearing inner race on rear of each countershaft.
18. For timing purposes, mark tooth on countershaft low speed gear that is aligned with the drive gear keyway and is stamped with an "O", Fig. 31.
19. Partially seat outer races of countershaft rear bearings and place countershafts into position in auxiliary housing, then as the countershafts are installed, mesh marked low speed gear tooth on each countershaft between each set of marked gear teeth of low speed gear.
20. With countershafts positioned, complete installation of countershaft rear bearings on shaft and into case bore.
21. Install speedometer gear rear washer on tailshaft and against bearing with chamfered inside diameter towards bearing and install speedometer gear or replacement spacer on shaft and against washer. If removed, install oil seal into rear bearing cover with lip of seal facing rearward.
22. Install rear bearing cover with speedometer bore up and to the left, then torque retaining bolts to 35–45 ft. lbs. and install front output shaft bearing.

Joining Main & Auxiliary Sections

1. Referring to Figs. 9, 27 & 28, install drive gear bearing cover and clutch housing, and torque stud nuts to 170–180 ft. lbs.
2. If equipped, install the clutch release mechanism.

NOTE: The two reverse gears and washers must be in perfect alignment with center of case bores as the front of the auxiliary countershafts must be inserted through these parts during auxiliary section installation. A heavy grease will help hold washers in place. Also check rear of front section to make certain that all snap rings have been installed.

3. Install auxiliary section of front section, aligning auxiliary section on dowel pins and the extended portion of auxiliary countershafts with reverse idler gears. The clutch shaft may be rotated to facilitate installation.

NOTE: The auxiliary section can also be installed with front section in a vertical position, by placing clutch housing on blocks and using a chain hoist to lower auxiliary section into position.

4. Install auxiliary section attaching screws, then lock transmission by engaging two gears with mainshaft sliding clutches.
5. Make certain that speedometer gear washer and speedometer gear or replacement spacer are installed on tailshaft, then install companion flange and torque retaining nut to 380–470 ft. lbs.
6. With shift bars and sliding clutches on mainshaft in neutral position, install shifting bar housing on transmission, fitting yokes into yoke slots of corresponding clutch gears. Torque attaching capscrews to 35–45 ft. lbs.
7. Making certain that shifting notches on bars in shifting bar housing are aligned in neutral position, install gear shift lever housing on shifting bar housing, fitting lower end of lever into notches in shifting block and yokes. Install all except rear left attaching screws and torque to 35–45 ft. lbs.
8. If removed, install fittings on air valve and install actuating pin and spring in transmission bore.
9. Make certain that alignment sleeve is installed in air valve.

NOTE: Before installing alignment sleeve, check bore in air valve to make sure the piston is either in a forward or rearward position. Piston can be moved with pencil or compressed air.

10. Install air valve on transmission and torque screws to 15–20 ft. lbs. Actuating plunger in case will fit into alignment sleeve in air valve.
11. Install fittings, air lines, regulator and

Fig. 36 Countershaft gears timing marks

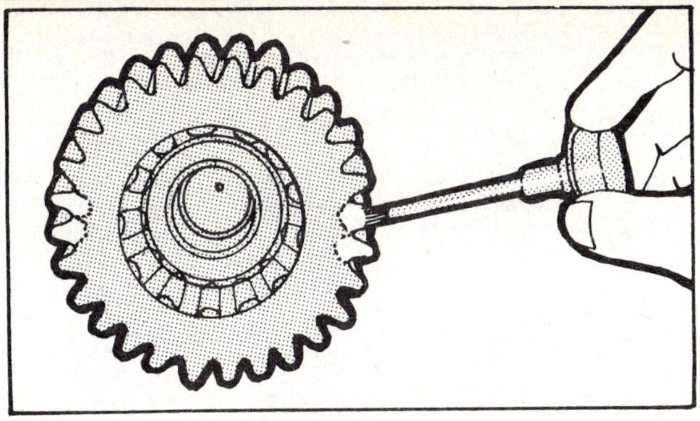

Fig. 37 Drive gear timing marks

Fig. 38 Left countershaft gear and drive gear correctly meshed

filter assembly, gear shift lever housing, lever and control valve. Tighten all attaching nuts and bolts securely.

9513 & 12513 SERIES

Main Trans., Disassemble

1. Remove mainshaft rear coupling snap ring and bearing retainer screws. Remove bearing retainer using three puller screws, Fig. 11, then remove snap ring and press retainer ring and bearing from drive gear, Fig. 32.
2. Remove snap ring from reverse gear, Fig. 33. Move reverse gear into engagement with sliding clutch and remove auxiliary countershaft front bearings from left reverse idler gear bore using an inside jawed puller.
3. Remove nut, washer and plug from idler shaft, pull idler shaft from gear and idler boss, then remove reverse idler gear and thrust washer, Fig. 34.

NOTE: To remove mainshaft assembly, it is necessary to remove right countershaft bearings.

4. Remove snap ring from rear of right countershaft, Fig. 35, and use a soft bar from inside case to move the rear bearing to the rear and off the shaft.
5. Remove the front right bearing retainer plate, Fig. 35. Using a soft bar and mallet, drive the countershaft about a ½ inch to the rear then drive the countershaft as far forward as possible to expose the front bearing snap ring then remove the front bearing using a puller.
6. Place a rubber band around mainshaft to retain key and remove mainshaft.
7. Remove 3rd-4th speed sliding clutch, front coupling snap ring, mainshaft key, reverse gear spacer and washer from mainshaft, Fig. 33.
8. Remove reverse gear and 1st-reverse sliding clutch from mainshaft then remove remaining parts and align gear washers with splines. If necessary remove snap rings from gears.
9. Remove snap ring and retainer plate from left countershaft, then remove front and rear bearings in the same manner as they were removed from the right countershaft.
10. Remove drive gear front bearing cover then tap drive gear outward and remove snap ring, then remove drive gear from inside case.
11. Remove drive gear nut where peened into shaft and remove nut, Fig. 14.
12. Press shaft from drive gear and bearing. If necessary, remove snap ring from drive gear, Fig. 14. Inspect bushing in pocket of input shaft and replace if necessary.

NOTE: Countershaft assemblies are identical except for the number of teeth on PTO gears and are disassembled in the same manner.

13. Remove right and left countershaft assemblies from case, Fig. 13A.
14. Press the 4th, PTO, 3rd and second speed gears off the countershaft.

NOTE: This will require a press of at least a 25 ton capacity. Do not use PTO gear as a base for pressing as its large diameter makes it susceptible to damage.

15. Press 1st speed gear and remove the keys from the countershaft, Fig. 13A.
16. Remove right reverse idler gear bearing and assembly in the same manner as the left reverse idler gear assembly was removed, Fig. 34.

Front Section, Assemble

1. Install plug and cup on reverse idler shaft then install bearing inner race on shaft against cup, Fig. 34. Press needle bearing into reverse idler gear.
2. Place reverse idler gear and thrust washer into case and install shaft into

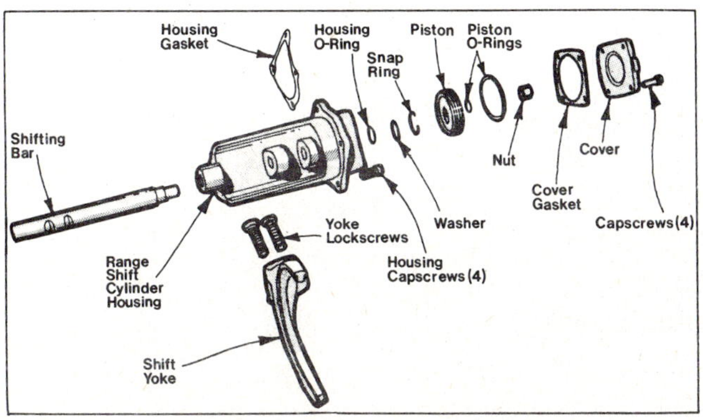

Fig. 39 Range shift cylinder assembly

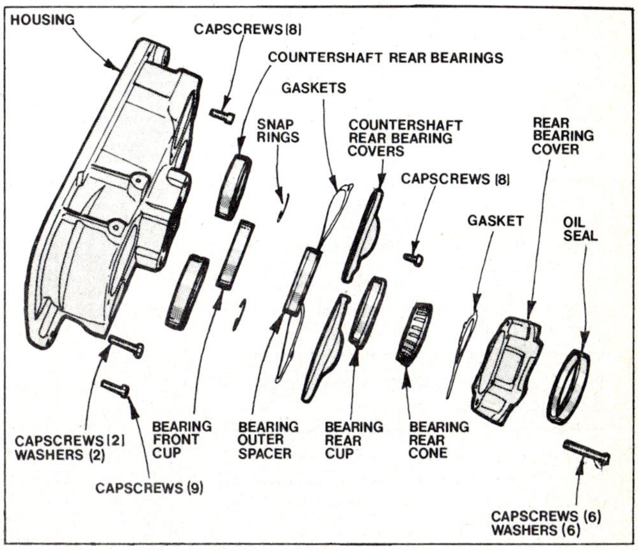

Fig. 40 Auxiliary section bearing assembly

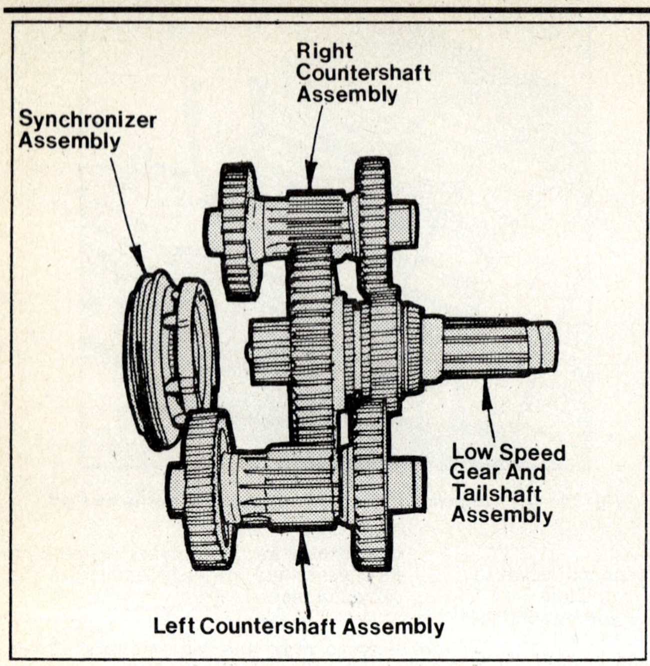

Fig. 41 Auxiliary section gear assembly

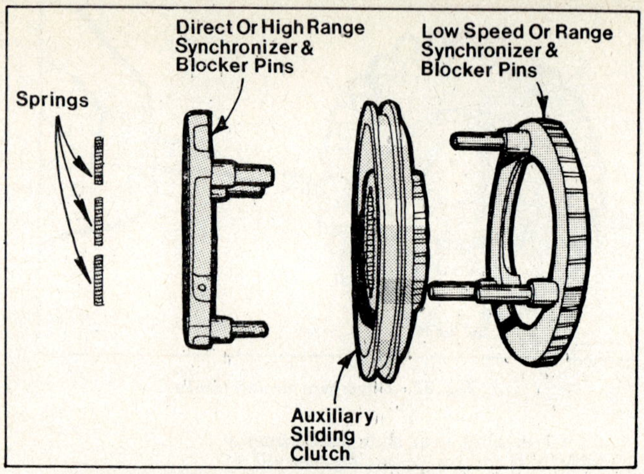

Fig. 42 Synchronizer assembly

bore.

NOTE: Make certain needle bearing in gear seats on inner race evenly before completing installation of shaft.

3. Install stop nut and retainer washer on shaft then install auxiliary countershaft front bearing outer race into reverse idler bore, Fig. 34.
4. Install pin and keys in countershaft then press first speed gear on countershaft with long hub of gear facing front of shaft, Fig. 13A.
5. Press second speed gear with long hub of gear facing rear of shaft followed by the third speed gear with long hub of gear facing front of shaft, Fig. 13A.
6. Press the PTO gear with tapered edge of teeth facing rear of shaft followed by the fourth speed gear with the long hub of gear facing rear of shaft, Fig. 13A.
7. Mark the timing tooth that is aligned with keyway on each countershaft and is stamped with an "O", Fig. 36.
8. Place the left countershaft assembly (47 tooth PTO gear) and the right countershaft assembly (45 tooth PTO gear) into the case. Do not install bearings.
9. Install snap ring into drive gear and install drive gear on input shaft with snap ring facing forward, Fig. 14.
10. Install spacer and press drive gear bearing onto shaft, Fig. 14.
11. Apply Loctite grade AVV to threads of gear nut then install nut and torque to 250–300 ft. lbs., Fig. 14.

NOTE: All threads must be dry and free of dirt and grease.

12. Peen nut into slots of shaft, then for timing purposes mark any two adjacent teeth on drive gear plus the two adjacent teeth directly opposite, Fig. 37.

NOTE: Check to make certain input shaft bushing is in good condition. If replacement is necessary, press bushing flush into shaft. If the three oil holes are restricted by bushing, drill out with a 5/32 inch drill. Radial clearance between bushing and pilot should be .040–.045 inch.

13. Install input shaft from inside of case and install bearing snap ring.
14. Place blocking tool or wooden block in the case bore, to center rear of left countershaft then mesh marked tooth on left countershaft with marked teeth on drive gear, Fig. 38.
15. Install front countershaft bearing, remove blocking and install rear countershaft bearing.
16. Install front bearing retainer plate and safety wire and install snap ring in groove at rear of shaft.

Mainshaft, Assemble

1. Install snap ring in all mainshaft gears except the reverse speed gear and make sure that roll pin is in place in mainshaft near pilot end, Fig. 33.

NOTE: Axial end play is .005–.038 inch for the reverse speed gear and .005–.012 inch for the forward speed gears. Washers are used to obtain the correct limits and are available in six thicknesses as follows:

Axial Clearance (End Play) Washer Table

Part No.	Limits	Color
14274	.248–.250	White
14275	.253–.255	Green
14276	.258–.260	Orange
14277	.263–.265	Purple
14278	.268–.270	Yellow
14279	.273–.275	Black

Always use the low limit washer (14274) in the reverse, 1st and 3rd speed gear positions.

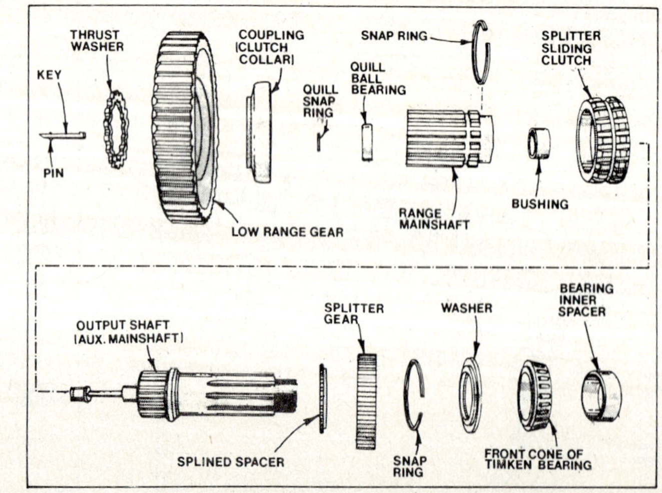

Fig. 43 Tailshaft and low range gear assembly

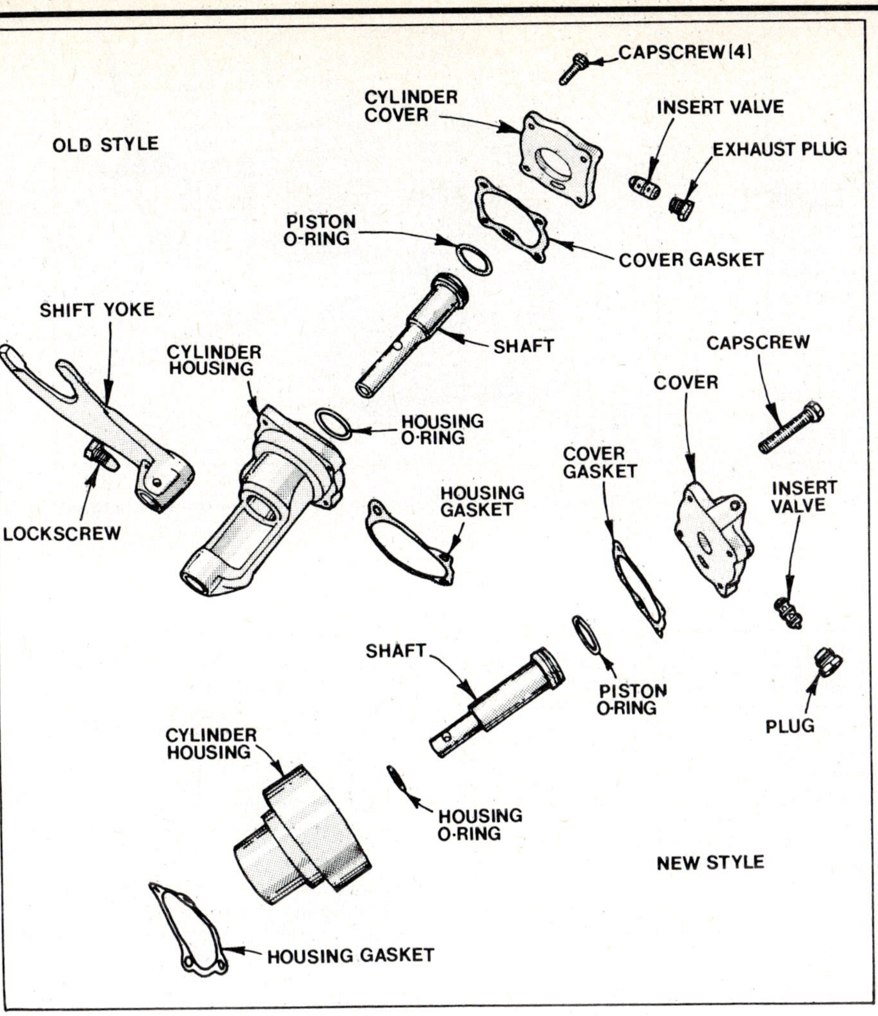

Fig. 44 Splitter shift cylinder assembly

For reference purposes, the gear washers are internally splined and the spacers are externally splined and there is one washer and one spacer for each gear.

2. With mainshaft in a vise, place 3rd speed gear washer (14274) over rear of shaft and down to lowest groove in shaft with flat side up, then turn washer to align internal splines with mainshaft splines and temporarily install key in spline to keep washer aligned.

NOTE: Do not use square cut spline inside washer for locking purposes.

3. Install 3rd speed gear spacer (flat side facing down) followed by 3rd speed gear (clutching teeth facing down).
4. Install 2nd speed gear (clutching teeth facing up) then install spacer into hub of gear.
5. Remove key and install washer, align with splines of mainshaft.

NOTE: Use correct limit washer to obtain specified end play.

6. Install key and the 1st-2nd speed sliding clutch, aligning keyway in sliding clutch with key.
7. Remove key and install 1st speed washer (14274) with flat side up then align splines and reinstall key.
8. Install the 1st speed gear spacer (flat side facing down), 1st speed gear (clutching teeth facing down) and low speed gear (clutching teeth facing up).
9. Install low speed gear spacer in hub of gear (flat side facing up) then remove key and install correct low speed gear washer with flat side facing down to obtain specified clearance, align splines and reinstall key.
10. Install low-reverse sliding clutch, aligning keyway in sliding clutch with key, then install the reverse gear onto sliding clutch, engaging clutch in both low and reverse gear.
11. Remove key and install reverse gear washer (14274) then align splines, reinstall key and install reverse spacer (flat side facing down).
12. Install snap ring in second groove from top with gap of snap ring placed at key location then temporarily place a rubber band on mainshaft to hold key in place.
13. Remove assembly from vise and install 3rd-4th speed sliding clutch.
14. Install assembly into case with rear of mainshaft through rear bearing bore and lower into position, then move mainshaft forward to seat pilot end of shaft in bushing of input shaft and remove rubber band.
15. To time and install the right countershaft assembly, mesh timing tooth on right countershaft with timing teeth of drive gear and mesh countershaft gears with mainshaft gears.
16. Using blocking tool, center rear of countershaft in bore and center rear of mainshaft in bore.

NOTE: Accurate centering of mainshaft is important.

17. Install front and rear countershaft bearings, then install front bearing retainer plate and safety wire. Install rear bearing snap ring.
18. Install plug, cup and bearing inner sleeve on reverse idler shaft then press bearing in bore of reverse idler gear, Fig. 34.
19. With thrust washer and reverse idler gear in position, insert shaft through gear and seat into bore making sure needle bearing is aligned with bearing sleeve then install washer and nut.
20. Seat outer race of auxiliary countershaft front bearing in reverse idler bore, then mesh mainshaft reverse gear with reverse idler gears and install snap ring into mainshaft reverse gear.
21. To reassemble and install auxiliary drive gear, place retainer on auxiliary drive gear, press on bearing and install snap ring in gear shoulder groove, Fig. 32.
22. Seat bearing in bore fitting drive gear

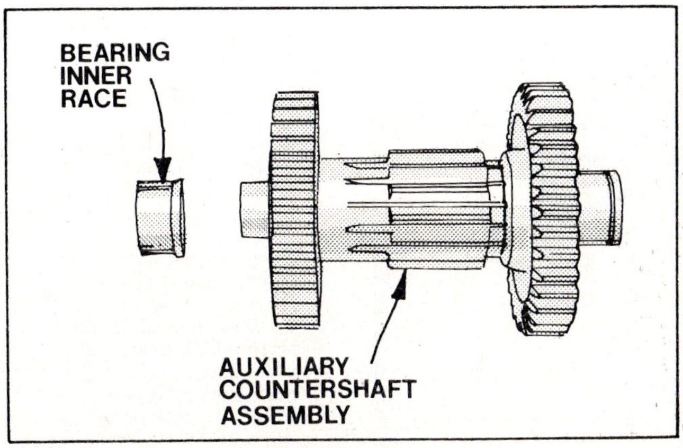

Fig. 45 Auxiliary countershaft assembly

Fig. 46 "O" stamped tooth on overdrive gear marked for proper timing

Fig. 47 Left countershaft gear and splitter gear correctly meshed

with splines of mainshaft, install screws, safety wire, then install rear coupling snap ring.

23. Install front bearing cover and gasket then install clutch housing and torque the four bolts to 70–75 ft. lbs. and the six nuts to 170–185 ft. lbs.
24. Using a chain hoist attach the auxiliary section to the front section.
25. Install speedometer drive gear on flange and install flange on output shaft.

NOTE: If speedometer gear is not used, install a spacer of same width.

26. Lock transmission in two gears and tighten flange nut to 450–500 ft. lbs.

NOTE: Failure to torque and pull flange tightly into place will result in damage to the rear bearing.

Auxiliary Section

Disassemble

1. Remove shift cylinder cover and shift bar stop nut, Fig. 39 then apply a small amount of compressed air through the supply port to remove piston.
2. Remove cylinder retaining screws and remove shift yoke screws.

NOTE: The cylinder may be rotated to expose the screws, Fig. 39.

3. Remove the shift bar from the cylinder, then the cylinder from the housing and the shift yoke from the sliding clutch gear followed by the snap ring, washer and O-ring, Fig. 39.
4. Remove the two countershaft rear bearing covers and the two snapings, Fig. 40.
5. Drive countershafts forward about 1/2 inch then rearward to expose snap rings and remove countershaft bearings, Fig. 40.
6. Separate front ends of countershafts and remove synchronizer assembly. Pull direct synchronizer from blocker pins of low speed synchronizer and remove sliding gear from low speed synchronizer, Figs. 41 and 42.

NOTE: Place a cloth over ring before removal as springs in direct ring will be released, Fig. 42.

7. Remove key from splines of range mainshaft, Fig. 43, then align splines of low speed gear washer with splines of shafts.
8. Remove the low speed gear, washer and clutching collar from shaft, Fig. 43 and then the right countershaft from housing.
9. Remove shift lock screw and splitter gear shift cylinder cover. If necessary remove insert valve cover and valve, Fig. 44.
10. Remove shift shaft, shift yoke, cylinder housing, Fig. 44 and left countershaft assembly, Fig. 47.
11. Drive output shaft assembly out of rear bearing and remove bearing inner spacer from shaft.
12. Press output shaft from splitter gear and bearing front cone and remove snap ring from output shaft quill, Fig. 43.
13. Pull range mainshaft from quill using the sliding clutch as a puller base and remove ball bearing, bushing and snap ring from mainshaft, Fig. 43.
14. Remove rear bearing cover, oil seal, bearing rear cone, cups and outer spacer from housing.
15. Pull bearing inner race from the front of each countershaft, Fig. 45.

NOTE: It may be necessary to start inner race with a pry bar between shoulder of shaft and bearing race.

NOTE: The auxiliary countershaft, drive gear and overdrive gear are one piece and cannot be disassembled.

Reassembly

1. For timing purposes, mark the tooth on each overdrive gear which is stamped with an "O", Fig. 46.
2. Heat countershaft bearing inner race and install it on shaft, seated against shaft shoulder, Fig. 45.
3. Place splitter gear sliding clutch on output shaft with internal splines toward threaded end, Fig. 43.
4. Press range mainshaft bushing into bore so that it is about 1/16 inch below face of shaft, Fig. 36, and install snap ring in splines of range mainshaft, Fig. 43.
5. Install range mainshaft on quill of output shaft, Fig. 43.

NOTE: Radial clearance between quill and bushing is .025–.030 inch.

6. Install quill ball bearing and snap ring into output shaft quill, Fig. 43.

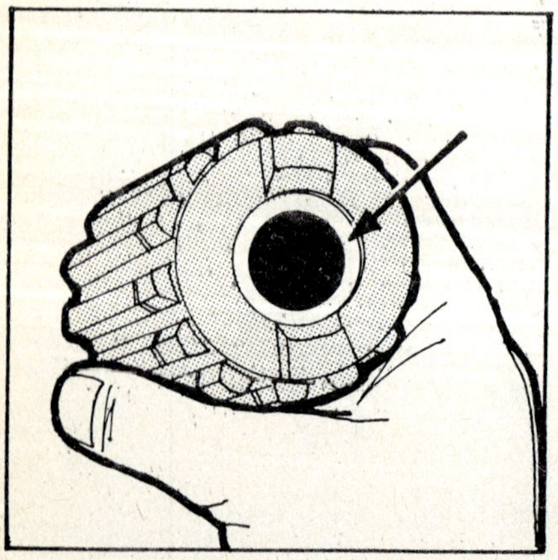

Fig. 48 Bushing pressed in about 1/16 inch below face of mainshaft

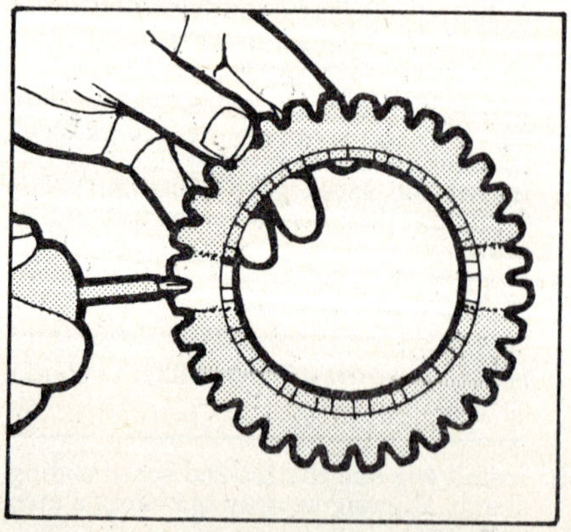

Fig. 49 Splitter gear timing marks

Fig. 50 Right countershaft gear and splitter gear correctly meshed

7. Install snap ring into splitter gear, Fig. 43. For timing purposes mark any two adjacent teeth on the splitter gear, then mark the two adjacent teeth directly opposite, Fig. 49.
8. Place output shaft assembly in a vise with threaded end facing up. Install splined washer with shoulder facing up, splitter gear with clutching teeth facing down and rear washer with flat side facing up, Fig. 43.
9. Heat front cone of Timken bearing and install on shaft against washer, followed by the inner spacer, Fig. 43.

CAUTION: Do not heat bearing over 275° F.

10. Start front cup of bearing (taper facing inside) into rear bore of housing, place outer spacer on front cup and place rear cup onto spacer, then tap lightly and evenly to seat it in the bore, Fig. 40.
11. Place auxiliary housing over output shaft, seating front cup on front cone of bearing then heat rear cone of bearing and install on shaft and into rear cup, Fig. 40.
12. Install oil seal flush into rear bearing cover, Fig. 40, then install cover onto case making sure that output shaft is rearward as far as possible.

NOTE: Use star washer at speed bore location to prevent oil leakage.

13. Position left countershaft in place, meshing marked tooth on countershaft gear between marked teeth on splitter gear, Fig. 47.
14. Install O-ring in splitter cylinder bore then install yoke onto slot of sliding clutch gear. Insert splitter cylinder into housing and place hub of yoke into cylinder, Fig. 44.
15. Install O-ring on piston and shift shaft into cylinder through yoke hub. Install lockscrew and secure with safety wire, Fig. 44.
16. Install splitter insert valve and exhaust screw into cover, Fig. 44, then install cover with exhaust screw facing down and supply port to the right.
17. Position right countershaft in place, meshing marked tooth on countershaft gear between marked teeth on splitter gear, Fig. 50, then install clutching collar on range mainshaft against snap ring with clutching teeth facing the rear, Fig. 43.
18. Install low range gear on mainshaft against clutching collar, Fig. 43 and install splines into hub or low range gear aligning washer splines with shaft splines. Lock low range gear in place by inserting key between splines of shaft and washer, Fig. 51.
19. Place sliding clutch gear on pins of low speed synchronizer and install the three springs in direct synchronizer, Figs. 18 and 42.
20. Place direct synchronizer over pins of low speed blocker pins and compress springs to fully seat direct synchronizer on pins of low speed synchronizer, then install the synchronizer assembly on output shaft.

NOTE: Make certain that marked timing teeth on countershaft gears are aligned with the marked timing teeth on splitter gear.

21. Block against housing or mount housing in a vise and install rear bearings on countershafts and into case bores making certain that bearings are fully against shoulders so that snap ring groove is fully exposed.
22. Install snap ring on each countershaft and install bearing covers, Fig. 40.
23. Install O-ring, washer and snap ring in bore of shift cylinder, Fig. 39.
24. Install the range shift yoke into slot of sliding clutch then install shift cylinder into housing and place hub of shift yoke in place in cylinder.

NOTE: With newer type range cylinders, make certain that long hub of shift yoke is facing the rear plate.

25. Install the shift bar and the two yoke lockscrews, Fig. 39, and safety wire, then install cylinder housing screws with air port facing up.
26. Install O-rings on piston and install piston in cylinder with flat side out, then install nut on shifting bar and cover onto shift cylinder with air port towards upper left side.

Gearshift Housing, Fig. 52

Disassembly

1. After removing springs from housing, remove housing from transmission, invert housing and remove steel balls.
2. Place housing in vise, remove yoke lockscrews, 3rd-4th speed shift bar, 1st-2nd speed shift bar and interlock pin.
3. Remove actuating plunger, low-reverse speed shift bar and interlock balls.

Assembly

1. Install low-reverse shifting bar and yoke then torque lockscrew to 45–55 ft. lbs. and safety wire.
2. Install the actuating plunger and the 3/4 inch interlock ball in the rear boss.
3. Install 1st-2nd speed shifting bar, yoke and block, inserting interlock pin in bore of neutral notch of bar then install lockscrews and safety wire. On "F" models, use short lockscrew in shift block.

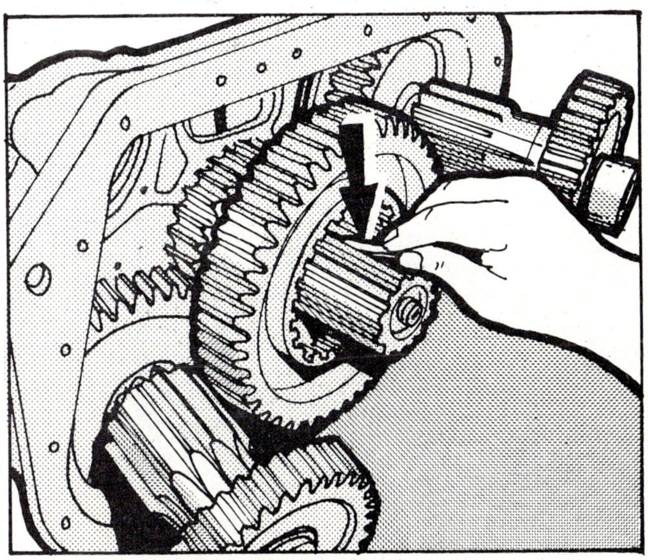

Fig. 51 Low range gear locked in place with key inserted between splines of shaft and washer

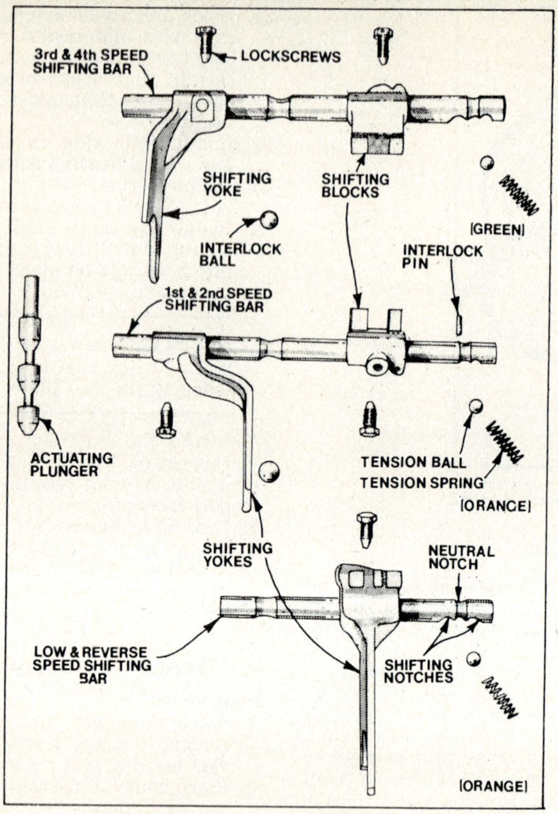

Fig. 52 Shifting bar assembly

4. Install the next 3/4 inch interlock ball.
5. Install 3rd-4th speed shifting bar, yoke, block and lockscrews and safety wire.
6. Install the three tension balls in bores in proper bar locations.
7. Install shifting bar housing on transmission and fit yoke forks into sliding clutch gears and install screws.
8. Install tension springs in bores at bar locations.

NOTE: Install green spring in 3rd-4th speed bar location.

RANGE SHIFT AIR SYSTEM

System Check-Out

The following checks are to be made with normal vehicle air pressure of approximately 60 psi but with the engine off. Refer to Fig. 53, for check points.

Incorrect Hook-Up

1. With gear shift lever in neutral, move control button up and down.
2. If lines are crossed between control valve and air valve the control button will not stay in an up position and transmission will not stay in high range.
3. If lines are crossed between air valve and shift cylinder, transmission range gearing will not correspond with button position. This will have to be tested by starting vehicle and, with control button down, engaging low speed gear.

Air Leaks

1. With gear shift lever in neutral, coat all air lines and fittings with soapy water and check for leaks, moving control button to both positions.
2. If there is a steady leak out control valve exhaust or bottom port, an O-ring is defective in the control valve.
3. If there is a steady leak out breather on air valve, an O-ring is defective in the air valve. This may also indicate a leak in the shift cylinder.

Air Regulator

1. With gear shift lever in neutral, check exhaust port on bottom of air regulator. If there is a steady leak it indicates a ruptured diaphragm, or dirt and rust have clogged the regulator piston in the input port of the seal. Inspect and replace diaphragm if necessary. Clean regulator.
2. Cut off air pressure and install an air gauge in air line at output port of regulator. Bring vehicle air pressure to normal, which should be a steady pressure of 58 to 62 psi. Move range control button up and down; the pressure should drop during each shift and then return quickly to the original reading. If gauge readings do not conform to the above, proceed as follows:
3. If gauge shows full line pressure the diaphragm is ruptured and should be replaced.
4. A pulsating gauge indicates loose lock ring which attaches two sections of regulator.
5. Regulated pressure can be changed by the addition or removal of shims. Add shims to raise air pressure; remove shims to lower the pressure. One shim is equal to

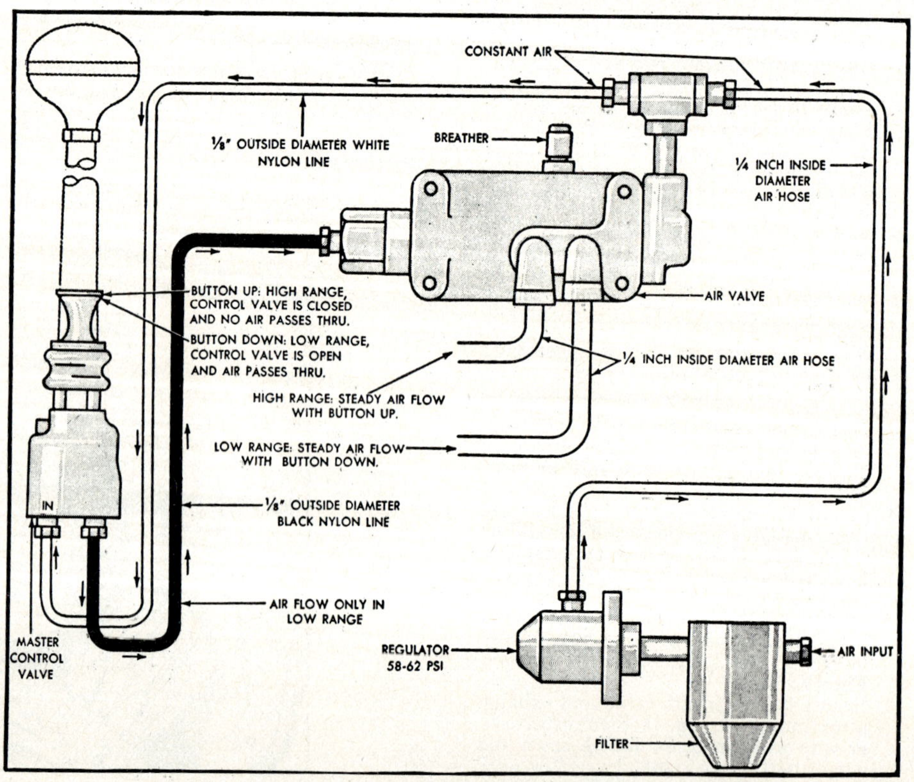

Fig. 53 Transmission air system. RT-910. On RT-915 models the hole gear is engaged by air for deeper reduction

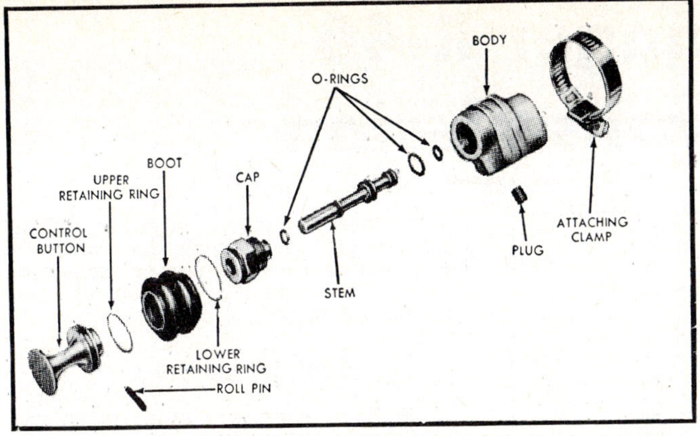

Fig. 54 Range shift control valve disassembled

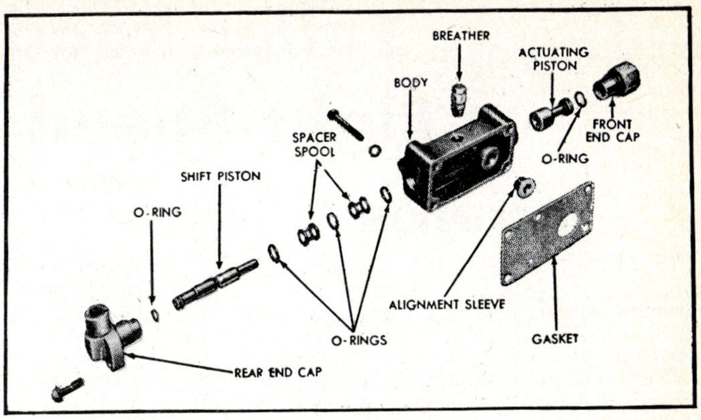

Fig. 55 Air valve disassembled

approximately 2 psi pressure change.

6. Only as a last resort should an adjustment be made with the screw in end cap of regulator. This adjustment has been set for correct operating limits. Any deviation from these limits will in most cases be caused by dirt or worn parts. Adjustment of screw will give only a temporary remedy. Turning screw clockwise raises air pressure; turning screw out lowers air pressure.

Control Valve

1. With gear shift lever in neutral, pull control button up to high range and disconnect 1/8" black nylon air line at air valve.
2. When control button is pushed down, a steady blast of air should flow from the disconnected line. Air will shut off when button is pulled up. This indicates that control valve is operating correctly. Reconnect air line if control valve does not operate correctly, check for leaks, restrictions and defective O-rings.

High Range Check

1. With gear shift lever in neutral, push control button down and disconnect high range air line from shift cylinder cover.
2. Pull control button up. There should be a steady flow of air from high range air line. Push button down to shut off air. If air escapes from high range port in cover with button down, this indicates a defective O-ring on piston in shift cylinder.
3. Make sure vehicle engine is shut off and move gear shift lever to a gear position. Pull control button up; there should be no air at the high range line. Move shift lever to neutral; there should now be a steady flow of air from high range line. Push button down to shut off air and reconnect line.
4. If air system operated incorrectly, it indicates that air valve is defective or that actuating parts in shifting bar housing are jammed or defective.

Low Range Check

1. With gear shift lever in neutral, pull control button up and disconnect low range air line at shift cylinder.
2. Repeat procedure under "High Range Check", reversing the position of control button in order to check low range operation.

Range Shift Cylinder

1. If any of the seals in the range shift cylinder are defective the range shift will be affected. The degree of lost air, or course, will govern the degree of failure, from slow shift to complete failure to shift.
2. Failure to shift into either low or high range indicates a steady leak out breather in air valve in both ranges.
3. Failure to shift into low range is due to pressurizing of transmission. Make sure cylinder bore is clean to prevent damage to piston seal. Use only a very light amount of shellac or Permatex on cover gasket to prevent clogging cylinder. Tighten capscrews securely.

Hole-Gear Air System 915 Units

Air Input Check

1. With gear shift lever in neutral and normal vehicle air pressure, loosen connection at air input (end port) of deep reduction valve (in cab) until it can be determined that there is a constant flow of air at this point. Reconnect line.
2. If there is no air at this point, there is a restriction in the line between deep reduction valve and air valve. Also check to make sure this line is connected to constant supply.

Deep Reduction Valve (In Cab)

1. With deep reduction valve lever to "IN", remove line from deep reduction valve at port in hole-gear shift cylinder; there should be no air at this point.
2. Move deep reduction valve lever to "OUT". There should now be a constant air flow from the line. Move lever to "IN" to shut off air. If the above conditions do not exist, deep reduction valve is faulty or there is a restriction in air line.

Hole-Gear Shift Cylinder

If any of the seals in the hole-gear shift cylinder are defective, the hole-gear shift will be affected. The degree of lost air, of course, will govern the degree of failure, from slow shift to complete failure to shift.

AIR SYSTEM SERVICE

Shift Control Valve, Fig. 54

1. Disconnect two nylon air lines at air valve on transmission. If desired, gear shift lever housing can now be removed from transmission without disconnecting remainder of control valve air lines and without removing control valve from shift lever.
2. Disconnect two nylon air lines at control valve on shift lever.
3. Remove ball grip from shift lever.
4. Loosen mounting clamp and remove control valve and clamp from shift lever.
5. Remove nylon air lines, sheathing and O-ring clamp from shift lever.

Air Regulator & Filter

1. Disconnect and remove air line between air valve and air regulator, including hose clamp.
2. On RT-915 models, remove air line between air regulator and hole gear shift cylinder.
3. Turn out capscrews and remove air regulator and filter assembly from transmission or from vehicle frame if so mounted.
4. Turn large hex nut from threads of air regulator.
5. Turn air regulator from nipple.
6. Remove bracket and nipple from air filter.

Air Valve, Fig. 55

1. Disconnect and remove air lines between air valve and shift cylinder in rear plate of transmission.
2. If necessary, remove fittings from shift cylinder.
3. Turn out four capscrews and remove air valve from adapter plate.
4. Remove alignment sleeve from air valve or from bore in adapter plate.
5. Remove spring and actuator pin from bore in adapter plate.
6. Turn out capscrews and two Allen head screws and remove adapter plate.
7. If necessary, remove fittings from air valve.

Fuller Roadranger R-46 Eight-Speed Transmission

DESCRIPTION

The Fuller R-46 Roadranger 8-speed transmission, Fig. 1, combines a four-speed front (or main) section with a two-speed auxiliary (or rear) section to provide a total of eight forward and two reverse speeds.

The transmission should always be filled through the auxiliary filler plug opening. Openings between the front and rear sections allow the lubricant to seek its proper level in both cases. Fill to the level of the filler opening in the auxiliary case. Overfilling will slow the action of the auxiliary synchronizing clutches, and the extra drag caused may result in damage to the synchronizer discs.

The power flow of the transmission is controlled by manipulation of the gearshift lever and the air-powered range shift control, Fig. 2. The air-powered range shift control system consists of a master valve mounted on the gearshift lever and a slave valve mounted on the shift bar housing on the transmission that are connected by nylon air hose. The air system provides for smooth operation. Each gear shift should be double-clutched.

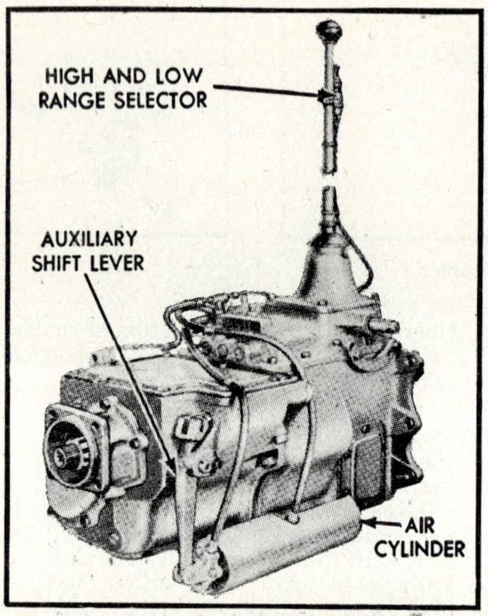

Fig. 1 Roadranger R-46 transmission

TROUBLE SHOOTING

NOTE: Before removing the transmission from the truck for inspection, always check for possible trouble in the clutch, drive shaft, universal joints or rear axle.

Noisy Transmission

Excessive noise may be caused by misalignment due to loose mounting bolts, paint on the clutch housing or transmission faces, flywheel housing misalignment, loose parts, dirt or metal chips in the lubricant, or insufficient lubricant.

Transmission noise may also be caused by worn or damaged parts, which requires the removal of the transmission to replace worn parts.

Transmission Shifts Hard

Check the clutch pedal free travel adjustment and clutch parts. Inspect the transmission linkage for binding caused by bent or worn parts. Hard shifting may also be caused by improper lubricant in the transmission.

Transmission Jumps Out of Gear

Improper shifting may cause the transmission to jump out of gear. Be sure the gears are completely engaged before releasing the clutch pedal.

Check the transmission linkage adjustment and make necessary corrections. Check for excessive end play caused by wear in the shift forks, sliding gear fork grooves, thrust washers, output shaft or countershaft bearings, or clutch pilot bearing or bushing.

Inspect the detent springs in the gear shift housing and replace any that are broken or damaged. Check clutch housing alignment with the engine.

Oil Leakage

Oil leaks may be caused by overfilling the transmission or by using a lubricant that expands and foams while the truck is in operation.

Loose gear shift housing cover screws may allow the lubricant to escape between the housing and transmission case. Check the condition of the bearing retainers and gaskets. See that the transmission vent is open.

Bind in Air Cylinder or Connecting Linkage

With the auxiliary section in high range, relieve the pressure in the air system and disconnect the piston rod of the air cylinder. Slowly move the piston several times through the complete stroke cycle. If either bind or drag is evident, disassemble the cylinder to find the cause. When connecting the piston, check the entire linkage for bind resulting from misalignment.

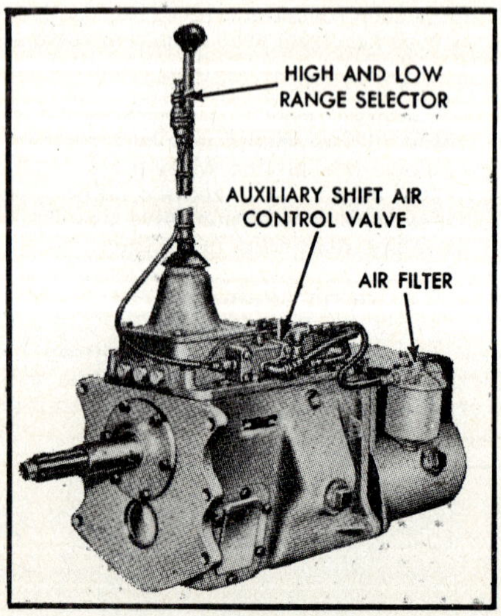

Fig. 2 Roadranger R-46 transmission

Leak in Air Lines

Raise the air pressure in the lines to normal and coat all connections in the system with soapy water to locate leaks. Disassemble the leaky connections, coat them with sealing compound and reassemble. Replace worn, leaky air hose, and repair or replace connectors, lines and hoses which continue to leak.

Leak in Air Cylinder

With the air system at normal line pressure and shift lever in neutral, disconnect the hose at the air cylinder which is not under pressure. Check the open port in the cylinder for a leak past the piston leathers. If no leak is found, connect the hose.

Move the shift lever to neutral and move the range selector handle to its opposite position. At the air cylinder, disconnect the air hose which is *not* now under pressure. Check the open port for a leak past the piston leathers; if no leak is found, connect the hose.

Check for a leak at the seal in the cylinder cap where the piston rod enters the cylinder.

If the above procedures disclose a leak, disassemble the cylinder and check the leathers, barrel, seal and piston rod. Replace any defective parts.

Defective Air Valve

Disconnect air hoses at the cylinder and cap each hose securely. After raising the air pressure to normal, move the shift lever to neutral and raise and lower the range selector handle. With each movement of the handle a short, fast blast of air should be heard at the breather valve. If the air exhaust is not as described, the valve is defective and the "O" rings should be replaced.

With the gear shift lever in neutral and selector handle in low-range position, check the breather valve for an escape of air past the pressure (outside) "O" ring. Move the selector handle to the opposite position and recheck the breather valve for a leak past the pressure "O" ring.

With the engine not running and with normal line pressure, move the shift lever in a geared position and then move the range selector handle to its opposite position. With movement of the selector handle a short, fast blast of air should be heard at the breather valve. After the initial exhaust, check the breather valve for continued escape of air past the pre-exhaust (inside) "O" ring.

Move the shaft lever to neutral to allow the air valve to complete the shift. Again move the shift lever to a geared position and move the selector handle to its opposite position. After the initial exhaust, check the breather valve for continued escape of air past the pre-exhaust "O" ring.

If a leak is found at the breather valve after either a complete range shift or pre-exhaust of the air valve, replace both "O" rings.

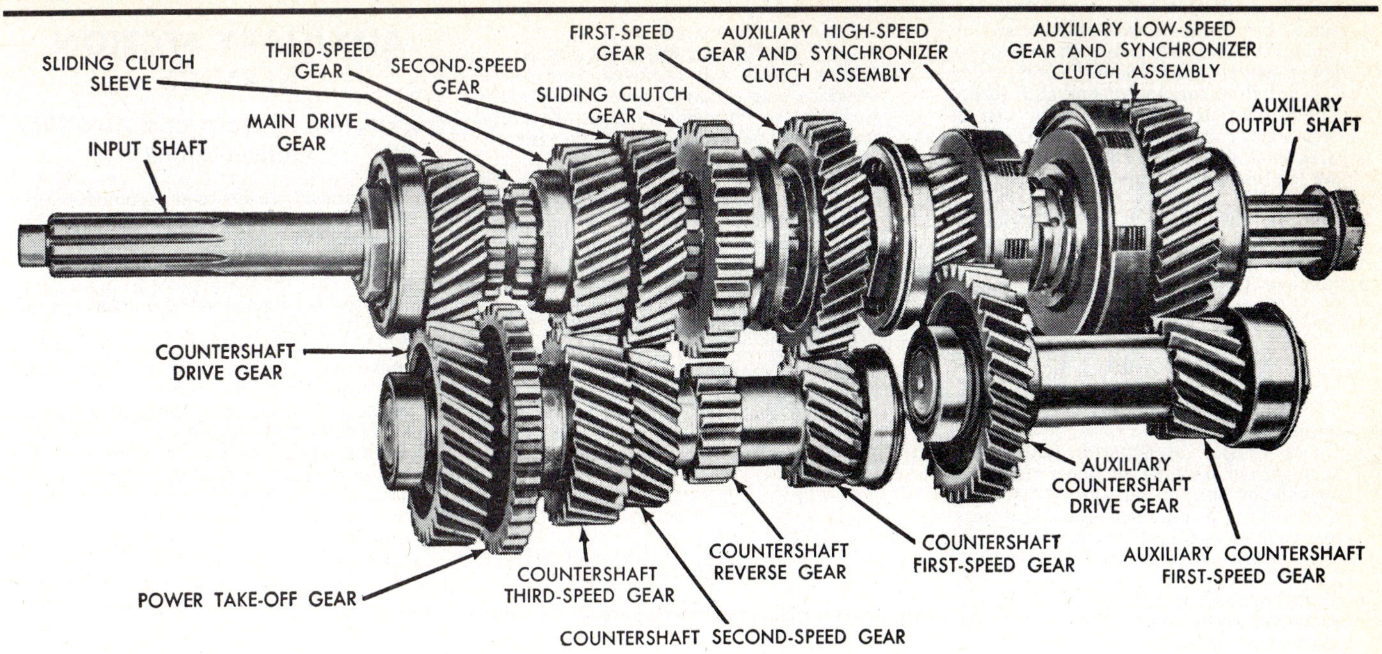

Fig. 3 Roadranger R-46 eight speed transmission

Block in Air System

Raise the air pressure to normal and move the shift lever to neutral. At the air cylinder, disconnect the hose which is not under pressure and move the range selector handle to its opposite position. If no air block exists, a loud, continuous air blast will escape from the disconnected hose. Move the range selector handle to its opposite position and reconnect the loose hose to the cylinder.

Move the range selector handle to its opposite position and proceed as above with the other hose. If this check reveals an air block, it can be located by inspection of the air valve and/or by inspection of the air lines.

TRANSMISSION REMOVAL

The procedure for removing the transmission varies somewhat depending upon the truck in which it is used. In general, however, it is accomplished as follows:

1. Bleed air reservoirs.
2. Drain transmission.
3. Disconnect drive shaft.
4. Set parking brake and remove companion flange nut.
5. Remove parking brake assembly.
6. Disconnect speedometer cable at transmission.
7. Remove parking brake lever.
8. Move selector control cable to high range position and remove cable retaining clamps.
9. Disconnect cable at connector joint and remove cable and cable tube.
10. Remove gear shift lever housing.
11. Disconnect air line from air filter.
12. Disconnnect air lines from air cylinder and remove cylinder from transmission.
13. Remove four lower transmission mounting bolts and install guide studs in the two lower holes.
14. Raise front axle on blocks and position jack under transmission.
15. Install safety chain.
16. Remove transmission support.
17. Remove two upper attaching bolts and remove transmission from truck.

TRANSMISSION INSTALL

Installation of the transmission is the reverse of the removal procedure. However, observe the following:

1. Torque the transmission-to-clutch housing bolts to 120–130 ft. lbs.
2. Torque the brake shoe support bolts to 180–220 ft. lbs.
3. Torque the companion flange nut to 275–350 ft. lbs.
4. Torque the gearshift housing attaching bolts to 35–40 ft. lbs.
5. During installation of the selector cable, eliminate unnecessary bends, and do not bend the cable on a radius of less than 4 inches.
6. On tilt-cab models, route the connector end of the cable along the frame before attempting to adjust the cable. Position the frame clamps on the cable but do not attach the clamps to the frame.
7. Push the selector handle all the way down to the cable ring and shift the air valve into high range position.
8. Partially fill the cable tube with Lubriplate and slide it over the end of the long cable.
9. Attach the connector ends of the two cables and move the cable rearward into its bore in the valve housing.
10. Push the cable casing into the connector tube until it contacts the cable joint. *The selector handle must be held up in the high range position during this operation.*
11. With the cable casing in position against the joint, install the cable clamp and spacer on the adapter. *Tighten the bolts finger-tight.*
12. Position the control cable and clamps on the gear shift lever and tighten the clamp bolts securely. On tilt-cab models do not attach the clamps to the frame at this time.
13. Loosen bolts that attach cable clamp and spacer to adapter.
14. Push selector handle down until there is 1/8" clearance between bottom of selector handle and cable ring, Fig. 5. *This clearance should not exceed 1/8".*
15. Tighten cable clamp bolts to adapter 10–15 ft. lbs., being careful not to pinch cable casing.
16. Check operation of control cable, making certain that air valve shifts freely into both high and low range positions.
17. On tilt-cab models, install frame clamps.
18. Install transmission cover plate and close air reservoir valves.
19. Fill transmission with lubricant, start engine and check for air leaks.

AIR VALVE SERVICE

Removal

1. Remove floor mat and plate.
2. Move selector control cable to high range position and remove cable clamp and spacer from adapter.
3. Disconnect long and short shift cables at connector joint, Fig. 6.
4. Disconnect 3 hoses at air valve and remove valve housing mounting bolts.
5. Remove air valve and plate, Fig. 7.
6. Remove Allen screw which holds adapter to gearshift housing and remove adapter.
7. Remove lockout spring and pin from bore in gearshift housing.

Disassembly

1. Remove 3 air line fittings from valve housing.
2. Remove breather valve, Fig. 8.
3. Remove Allen head pipe plug from center opening in side of valve housing.
4. Remove short control cable only for replacement. Remove Allen set screw from end of lower swivel and pull cable from swivel through opening in front end of valve housing, Fig. 8.
5. Remove upper and lower swivels and

spring by prying outward at lower swivel. Spring tension will free both swivels as lower swivel clears side of housing.
6. Lift out slide bar, spacer and shift fork.
7. Remove welch plug from front end of housing by carefully forcing plug outward with small punch. *Inserts in housing must not be moved or damaged as plug is removed.*
8. Remove valve through welch plug opening by pushing with small screwdriver at shift fork slot.
9. Remove "O" rings from valve.

Inspection

1. Clean and inspect all parts, replacing as necessary.
2. Check tension of actuator and swivel springs, and examine the upper and lower swivels for wear or damage.
3. Check the slide bar, spacer and shift fork for wear or rust, and replace the 4 "O" rings if one or more show wear or damage.
4. Coat the slide bar, spacer, shift fork, lower swivel and contact points with Automatic Transmission Fluid.

Assemble

1. Install "O" rings on valve.
2. Position valve in housing with shift fork slot toward front of housing, and install welch plug.
3. Position slide bar, spacer and shift fork in housing. *Shift fork must engage groove in valve and spacer must be positioned between slide bar and shift fork, Fig. 8.*
4. It short cable was removed from lower swivel, pass new cable completely through swivel. Stake cable securely through set screw hole in swivel. Install set screw.
5. Push connector end of cable through its opening in front of valve housing.
6. Position upper swivel and spring guide and bracket in ends of spring, and position lower swivel in yoke of spring guide.
7. Compress spring until upper swivel enters its depression in slide bar and lower swivel rides on lower track in housing.
8. To provide maximum projection of cable, move slide bar to rear of housing.
9. Install Allen head pipe plug in center opening in side of housing.
10. Install breather valve and 3 air hose fittings.

Installation

1. Install lockout spring and pin in gearshift housing bore.
2. Position adapter and install retaining screw.
3. Position valve plate and air valve, install 4 mounting bolts and connect 3 hoses. Torque bolts to 10–15 ft. lbs.
4. Connect long and short shift cables and pull cable tube toward rear until it bottoms in valve housing.
5. Install cable retaining clamp and spacer on adapter and adjust cable as outlined under *Transmission, Install.*

AUXILIARY SECTION SERVICE

Separation of Main and Auxiliary Transmissions

1. Disconnect 3 air hoses at air control valve and remove the 2 valves-to-cylinder hoses from transmission.
2. Remove air filter and air hose.
3. Remove air valve and plate from air valve adapter. *Do not lose small horseshoe spacer in air valve.*
4. With assembly shifted into neutral, remove main transmission gear shift housing assembly and auxiliary cover.
5. Remove auxiliary rear bearing cover and speedometer drive gear.
6. With a small punch, relieve the auxiliary countershaft rear bearing nut where it is staked to shaft. Shift main transmission into two gears and remove countershaft rear bearing nut. *This nut has a right-hand thread.*
7. Shift main transmission into neutral and auxiliary unit into low range. Remove lock wire from groove at front end of auxiliary high speed synchronizer clutch housing. Remove 3 set screws (in lock wire groove) which secure high speed synchronizer clutch housing to auxiliary drive gear, Fig. 9. Move clutch housing to rear against auxiliary shifting yoke.
8. Remove upper front nuts from two studs inside auxiliary case.
9. Remove 6 nuts retaining auxiliary case to main case.
10. Remove auxiliary case from main case by tapping with a soft mallet.

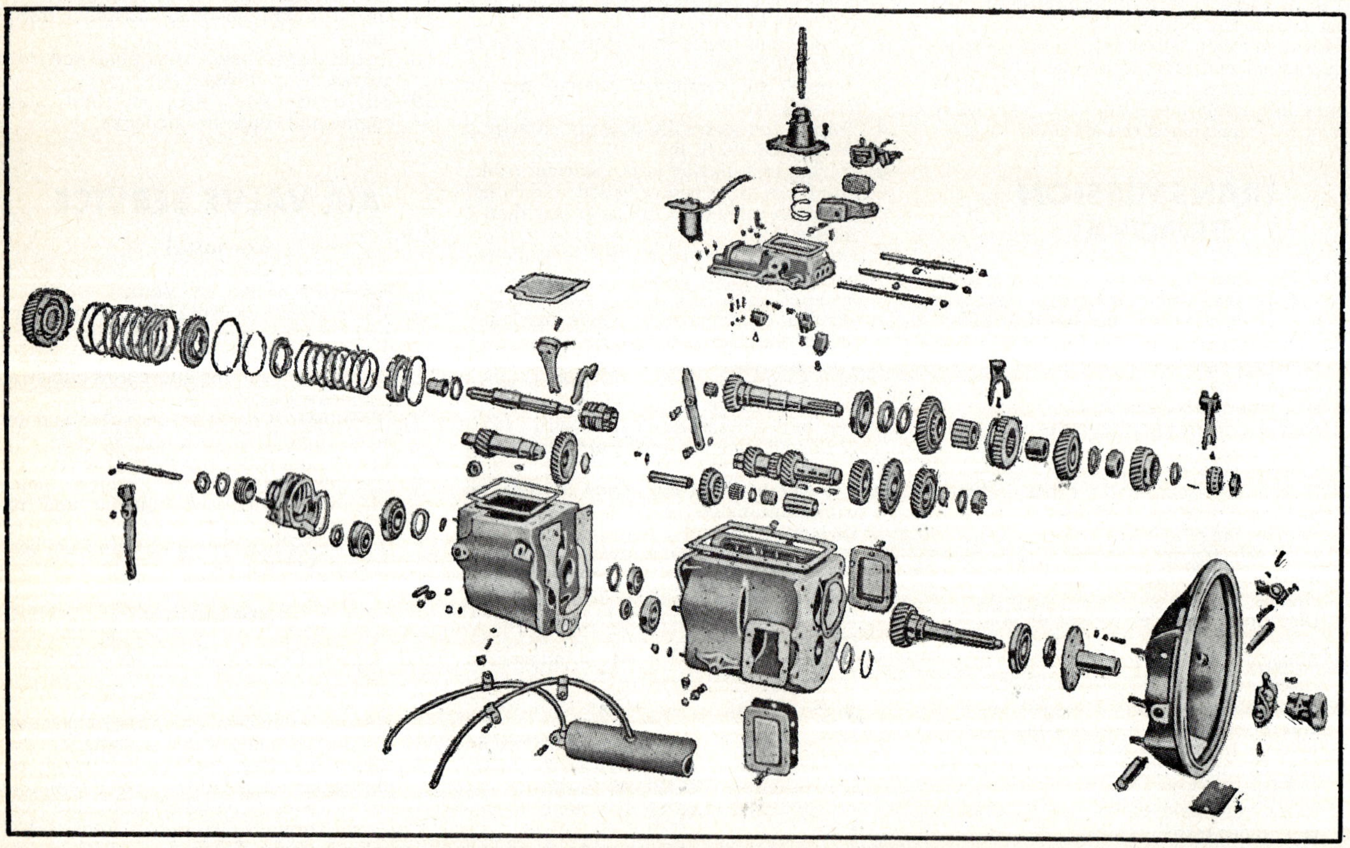

Fig. 4 Roadranger R-46 eight-speed transmission disassembled

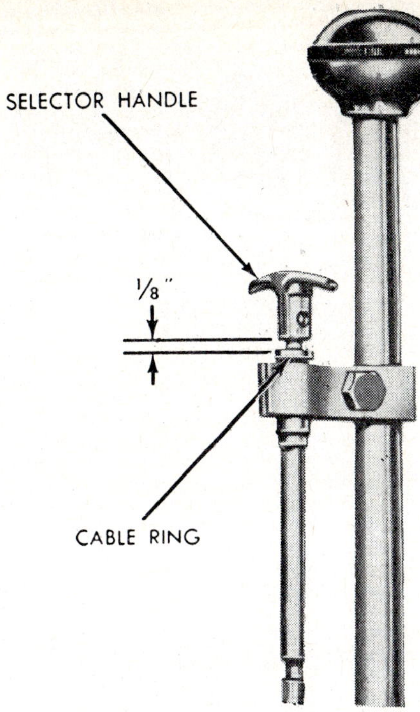

Fig. 5 Selector handle adjustment

Disassemble

1. Remove auxiliary detent spring cap from boss under shift fork shaft, and remove spring and ball.
2. Remove shift lever fulcrum pin lock plate and withdraw pin, using a slide hammer.
3. Remove shift lever.
4. Remove swivel pin from shift shaft.
5. To remove shift fork, cut lock wire and remove shift fork lock screws. Withdraw shift shaft and remove fork and fork shoe.
6. Remove oil seal from shift shaft bore in rear of case.
7. Remove auxiliary output shaft pilot

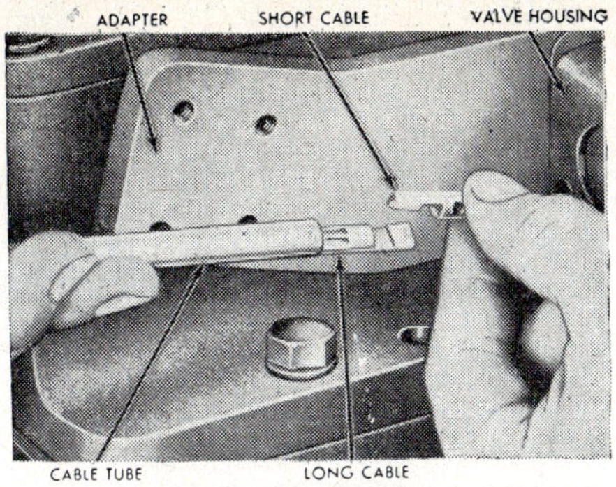

Fig. 6 Long and short shift cable connection

bearing.

8. Using a soft mallet, tap auxiliary output shaft forward out of rear bearing. Remove shaft by forcing it forward out of case through front bearing bore.
9. Remove low speed gear, high speed synchronizer clutch and low speed thrust washer from case.
10. Remove output shaft rear bearing from case by lightly tapping bearing outer race with soft drift.
11. To remove countershaft, tap shaft to rear until rear bearing is out of case. Remove rear bearing from shaft, using a suitable puller.
12. Tilt countershaft and remove it from top of case.
13. Remove front countershaft bearing by tapping lightly on outer race with a soft mallet. If countershaft drive gear is to be replaced, press it off the shaft.
14. To dismantle the output shaft assembly, Fig. 10, first remove the sliding clutch gear.
15. Press the output shaft 5/8" through low speed gear front thrust washer and bushing. *Pressing the shaft more than 5/8" will damage the small Woodruff key and thrust washer.* Move thrust washer forward against shoulder of shaft and press shaft through bushing until key is completely exposed. Remove key and press washer and bushing off shaft.

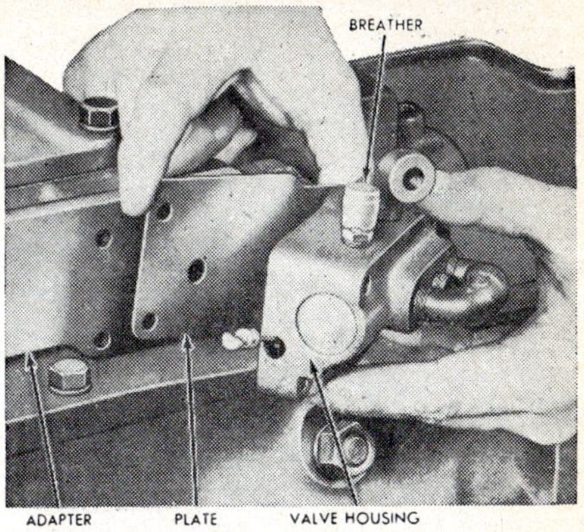

Fig. 7 Air valve and plate

16. Disassembly of the two clutches is the same. At the rim of the clutch, remove the wire stop for the pressure plate retaining snap ring. Remove snap ring, pressure plate and discs from housing.

Inspection

Check the surfaces of all thrust washers. Washers scored and/or reduced in thickness should be replaced. Synchronizer discs scored, burned or warped should be replaced. Bearing covers grooved or showing wear from thrust of adjacent bearings should be replaced. Check oil return threads in bearing covers. If the sealing action of the threads has been destroyed by contact from the input and output shafts, replace the covers.

Assemble

1. Both synchronizer clutches are assembled the same way. Start with an inner disc (projections at inside diameter) and alternately place one inner and one outer disc in housing until 8 discs have been installed, Fig. 11.
2. Install pressure plate and snap ring. *Lock wire hole in rim of gear must be between open ends of snap ring.*
3. Install lock wire in housing rim and bend wire up internally and down externally. *Internal end must not extend above housing.*
4. Press countershaft drive gear on shaft, Fig. 12.

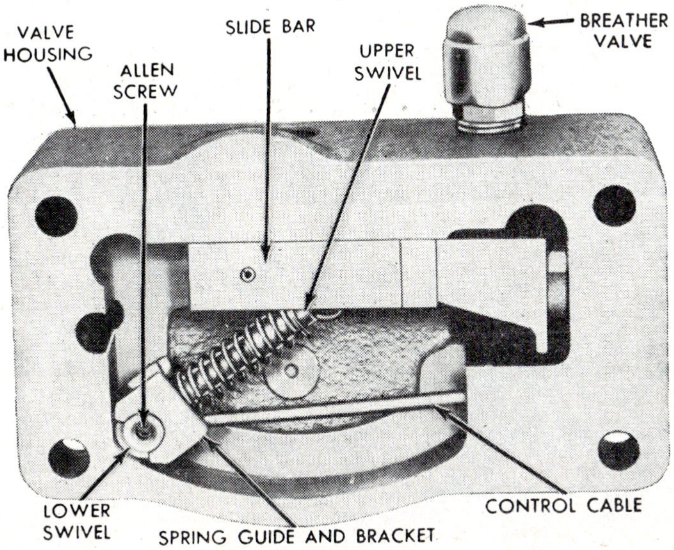

Fig. 8 Air valve

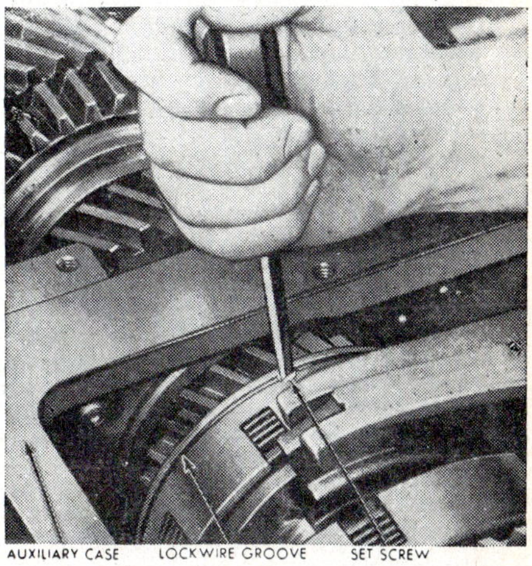

Fig. 9 Removal of one screw from high-speed synchronizer clutch housing

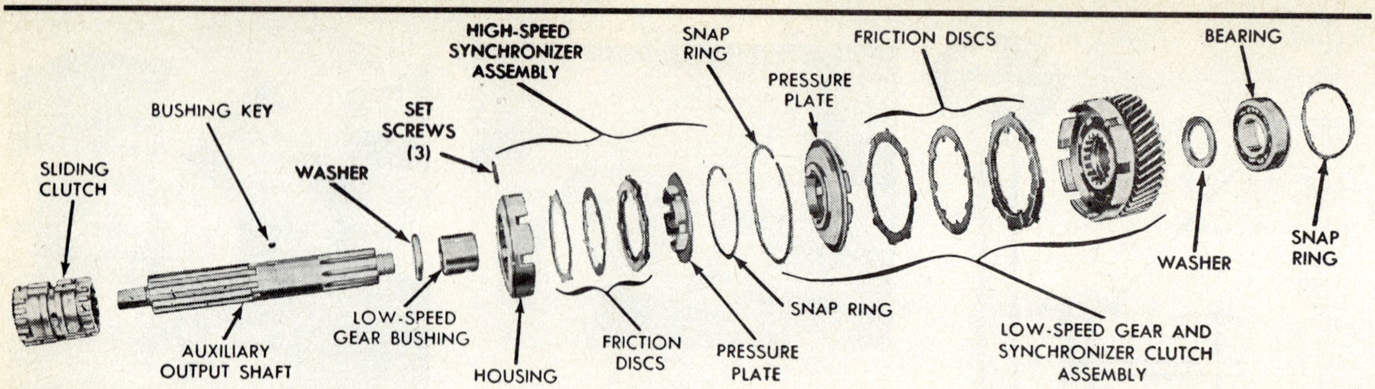

Fig. 10 Auxiliary section output shaft

5. To assemble output shaft, refer to Fig. 10 and proceed as follows:
6. Install low speed gear front thrust washer on rear of shaft *with beveled edge toward front of shaft.*
7. Install Woodruff key and press low speed gear bushing onto shaft until thrust washer is firmly seated against spline shoulder.
8. Install new oil seal in shift shaft bore.
9. Position snap ring on countershaft front bearing and install bearing in case. Tap lightly with soft mallet if necessary.
10. Position countershaft front bearing thrust washer on countershaft and install it in case.
11. Position snap ring on countershaft rear bearing and install bearing, using a suitable bearing driver.
12. Place low speed gear and high speed clutch in case, making sure low gear teeth engage those of low countershaft gear.
13. Install output shaft through front of case, threading it through high speed clutch and low gear until low gear is fully seated on sleeve.
14. Position low gear rear thrust washer on output shaft *with beveled edge toward low gear.*
15. Install snap ring on output shaft rear bearing.
16. Block front of output shaft and install bearing in case, using suitable bearing driver.
17. Install sliding clutch on output shaft through front bearing bore with word "front" toward front end of shaft. Thread sliding clutch through both synchronizer clutches until it engages pocket in low gear. *Interference can be eliminated during installation of sliding clutch by turning synchronizer clutches until lugs on inside diameters line up with grooves in sliding clutch.*
18. Install shoe in shift fork. *Countersunk holes in fork must be up and rib in shoe must face the rear.*
19. Install shoe and fork in sliding clutch.
20. Insert shift shaft in case through fork. End of shaft with milled flats should be to rear. Make sure notches in shaft line up with holes in fork but do not install lock screws.
21. Shift sliding clutch into low speed position and install auxiliary detent spring, ball and cap. Torque cap to 25–35 ft. lbs.
22. Install swivel pin on outer end of shift shaft.
23. Position shift lever on swivel pin. Insert fulcrum pin in case and through shift lever, lining up milled slot in pin with lock screw hole. Drive fulcrum pin into case until slot is flush with case.
24. Install fulcrum pin lock, capscrew and lockwasher, and torque to 20–25 ft. lbs.
25. Install shift fork lock screws and torque to 50–60 ft. lbs.
26. Install safety wires in shift fork screws.

Joining Main and Auxiliary Transmissions

1. Shift auxiliary section into low range. Move high speed clutch to rear against auxiliary shift fork.
2. Install auxiliary output shaft pilot bearing in bore of auxiliary drive gear.
3. Install new gasket between two transmission cases.
4. Attach main and auxiliary cases by fitting auxiliary section over gear which projects from end of mainshaft of main transmission. At the same time, fit auxiliary case over studs which project from rear of main case. The mainshaft rear bearing must be seated in shallow bore in rear of main case before two sections are joined.
5. Install retaining nuts holding two cases together and torque to 65–70 lbs. ft. *Do not overlook the nuts on the two upper studs on the inside of the auxiliary case.*
6. Shift main transmission into two gears and torque auxiliary countershaft rear bearing nut to 250–300 ft. lbs. *This nut has a right-hand thread.* Stake nut at milled slots, and shift transmission into neutral.
7. Move high speed synchronizer clutch forward against auxiliary drive gear.
8. Line up three holes in outside diameter of high speed synchronizer housing with holes in projections on drive gear.
9. Fit high speed clutch on projections of drive gear and move it forward until flush against shoulder of gear, at the same time keeping holes in line.
10. Install three set screws which retain clutch to drive gear. Tighten screws securely and then back them off until the slots are parallel with lock wire grooves in synchronizer housing.
11. Install lock wire in groove at front end of synchronizer housing and through slotted heads of set screws.
12. Install speedometer drive gear.
13. Install auxiliary rear bearing cover with new gasket and torque bolts to 25–35 ft. lbs.
14. Install auxiliary transmission cover and

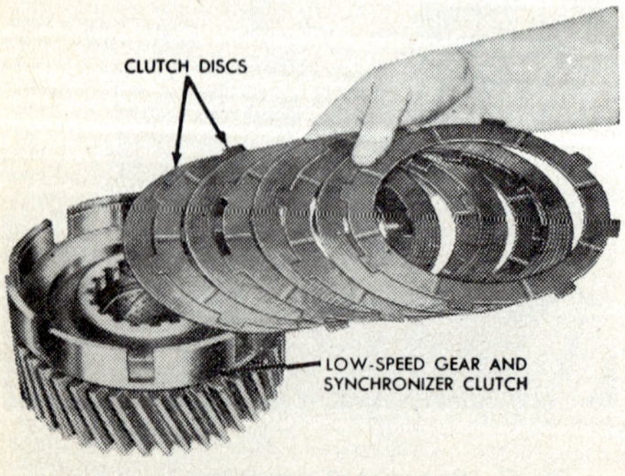

Fig. 11 Positioning synchronizer discs

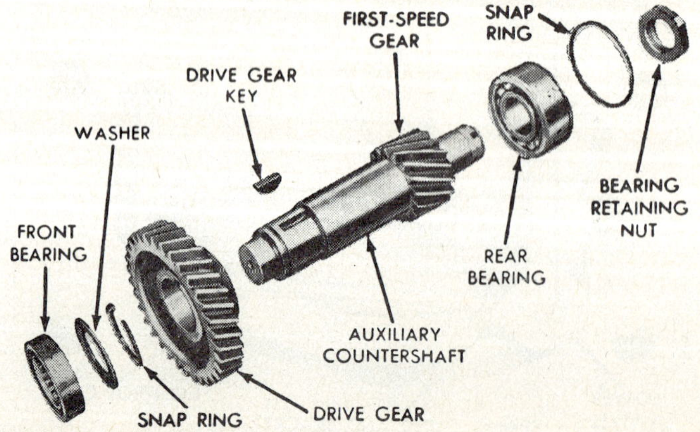

Fig. 12 Auxiliary section countershaft and related parts

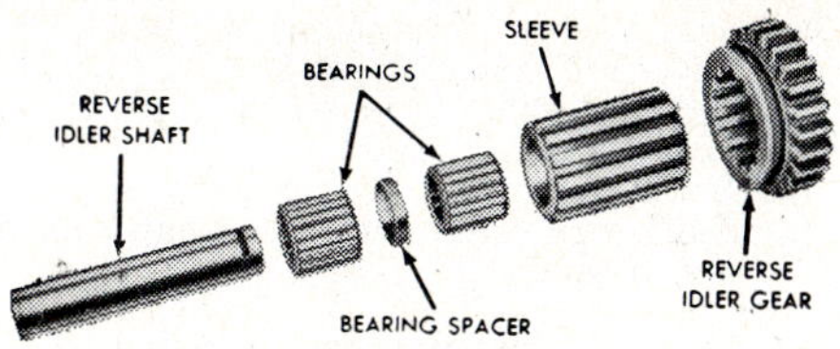

Fig. 13 Reverse idler shaft and related parts

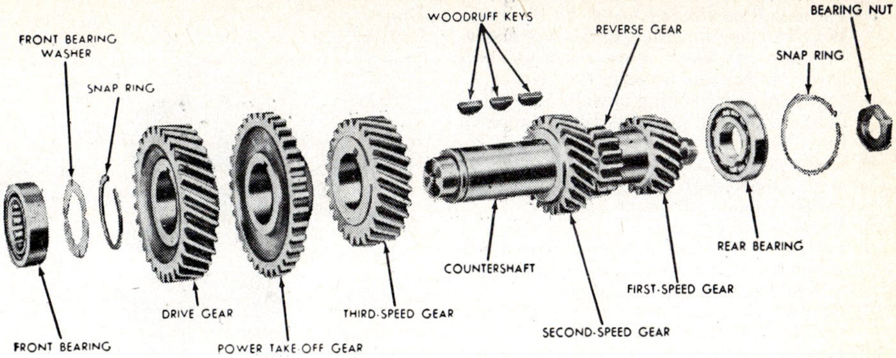

Fig. 14 Countershaft and related parts

gasket and torque bolts to 35–40 ft. lbs.

15. Shift gear shift housing in neutral and install with new gasket. Shift forks must enter grooves in mainshaft sliding gears and reverse shift lever must engage reverse shift gate. Torque gear shift housing bolts to 35–40 ft. lbs.
16. Install air filter.
17. Install air valve and plate on air valve adapter. Torque bolts to 10–15 ft. lbs. Make sure small horseshoe spacer is installed between slide bar and shift fork.
18. Connect air hose to front of air filter and top fitting of air valve.
19. Install two air cylinder hoses and clamps on transmission and connect hoses to air control valve.

4 SPEED MAIN SECTION

NOTE: Separate the main and auxiliary transmission as outlined previously. Then disassemble the main section as follows:

Disassemble

1. Using a small punch, unstake countershaft rear bearing nut. Shift transmission into two gears and remove nut *which has a right-hand thread.*
2. Remove input shaft bearing retainer and pull shaft and bearing out of front of case as an assembly.
3. If input shaft bearing is to be removed, unstake retaining nut and remove the nut, *which has a left-hand thread.* Shift transmission into neutral.
4. To remove mainshaft, shift reverse gear into mesh with countershaft, and shift sliding gear forward into second speed position. Then lift mainshaft from case.
5. Remove nut from reverse shift lever pivot pin on left side of transmission case. Force pin out of case and remove lever and pin.
6. Remove reverse idler shaft retainer through opening in rear of case.
7. Remove shaft by threading a capscrew into the shaft and use a slide hammer to pull shaft from case.
8. Remove splined reverse sleeve with reverse gear, bearings and spacer.
9. Remove reverse idler gear from sleeve.
10. Remove caged bearings and spacer from sleeve, Fig. 13.
11. To remove countershaft, pry it to rear until rear bearing is out of case.
12. Using a suitable bearing puller, remove bearing from countershaft.
13. Lift countershaft out of case.
14. To remove countershaft front bearing, rap front of case sharply with a mallet to jar bearing backward, and remove it from case.
15. To disassemble input shaft, remove bearing nut and press bearing from shaft.
16. If countershaft gears are to be removed, remove snap ring and press gears off shaft one at a time, Fig. 14. Remove Woodruff keys from shaft.

Mainshaft, Disassemble

1. Secure mainshaft assembly in a soft-jawed vise.
2. Remove pilot bearing.
3. Tap sliding clutch off shaft.
4. At pilot end of shaft and between the splines, pry out third speed gear washer locking key. Remove washer and gear, Fig. 15.
5. Position mainshaft in a press, using sliding clutch gear as a base and press mainshaft 1/8" through third speed gear bushing and thrust washer. If a press is not available, pry bushing upward. *Travel of*

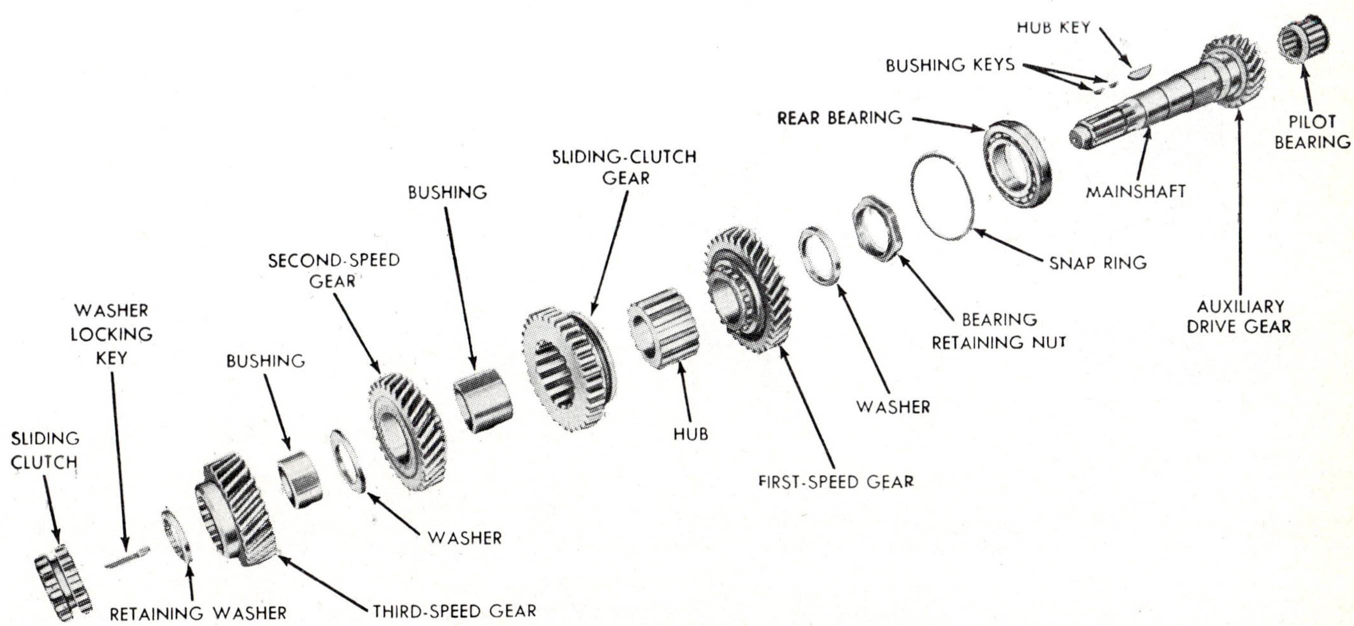

Fig. 15 Transmission mainshaft related parts

mainshaft as limited to 1/8" by the small Woodruff key installed under the sleeve. Pressing the shaft more than 1/8" will shear the key and damage the washer.

6. Pry third gear bushing upward on shaft to expose Woodruff key, Fig. 16, and remove key.
7. Using weight of sliding clutch gear, jar second speed gear, third gear bushing and thrust washer off shaft.
8. Remove sliding gear.
9. Position mainshaft in press. Then press shaft out of first speed gear, sliding gear hub and second gear bushing.
10. Remove Woodruff keys from shaft and remove first gear thrust washer.
11. Unstake and remove mainshaft rear bearing nut, which has a left-hand thread.
12. Press rear bearing from mainshaft.

Mainshaft, Assemble

1. Press mainshaft rear bearing on shaft with snap ring groove toward front and install snap ring.
2. Install gear bearing retaining nut and torque to 300–350 lbs. ft. Stake nut at slots milled into shaft.
3. Install first gear rear thrust washer with beveled edge on the I.D. toward rear of shaft.
4. Install first gear with clutching teeth toward front of shaft, Fig. 15.
5. Install large Woodruff key which prevents movement of sliding clutch gear hub.
6. Position sliding clutch gear hub on shaft with beveled edge at the I.D. toward rear of shaft. Line up keyway with key and press shaft through hub until hub is tight against shoulder of shaft. *When pressing shaft through hub hold first gear against its rear thrust washer.*
7. Install sliding clutch gear on its hub with fork slot to rear.
8. Install small Woodruff key which prevents movement of second gear bushing.

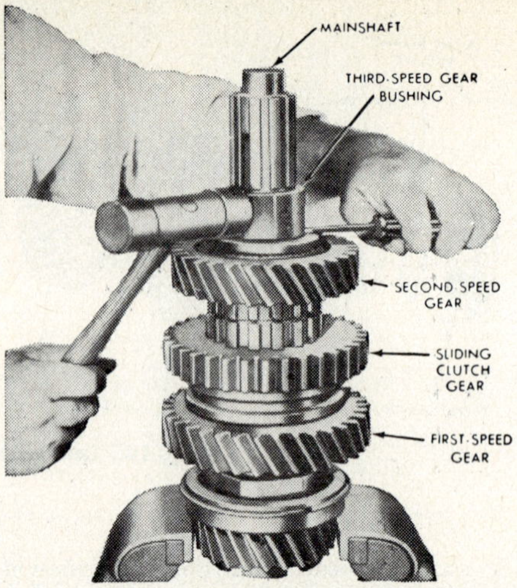

Fig. 16 Gaining access to third speed gear washer locking key

9. Position second gear bushing on shaft, making sure keyway is lined up with Woodruff key.
10. Press or tap bushing into position making sure it fits tightly against hub.
11. Install second gear with clutching teeth toward rear.
12. Install third gear rear thrust washer with beveled edge at the I.D. to rear.
13. Install Woodruff key which prevents movement of third gear sleeve.
14. Line up keyway in third gear bushing with the key, and press or tap bushing into position. Bushing must fit tightly against rear thrust washer.
15. Install third speed gear with internal clutching teeth toward front of shaft.
16. Install third gear retaining washer with beveled edge of lugs at the I.D. to rear. Turn washer in its groove until lugs are under splines on shaft. The thick end of this key must enter slot in washer. Check end clearance between third gear and its thrust washer, which must be .006" to .012". Adjust to specified clearance by selecting a third gear retaining washer of the proper thickness. These washers are available in thicknesses of .253–.255", .258–.260" and .263–.265".
17. Install sliding clutch with word "front" toward front of shaft.
18. Install mainshaft pilot bearing on mainshaft with roller retaining snap ring to rear of shaft.

Main Transmission, Assemble

1. Install countershaft front bearing in its bore.
2. Lower countershaft into case and position front of shaft into front bearing.
3. Install countershaft rear bearing, using a suitable bearing driver.
4. Install countershaft rear bearing retaining nut finger-tight. This nut has a right-hand thread.
5. Install bearings and spacer in splined reverse gear sleeve, Fig. 13.
6. Install reverse gear in splined sleeve. Position gear and sleeve in case *with shift fork groove in gear toward front of transmission.*
7. Insert reverse idler shaft in case and through bearings and spacer. Line up slot in shaft with lock screw hole in case.
8. Drive shaft into case until front edge of milled slot in shaft is flush with case wall.
9. Install reverse idler shaft retainer and torque capscrew to 20–25 lbs. ft.
10. Install reverse shifting lug in bottom or gear end of reverse lever. Insert lug in lever from side opposite bosses. Install snap ring on grooved end of lug.
11. Position reverse shifting lever in case, and guide lug into groove in reverse sliding gear.

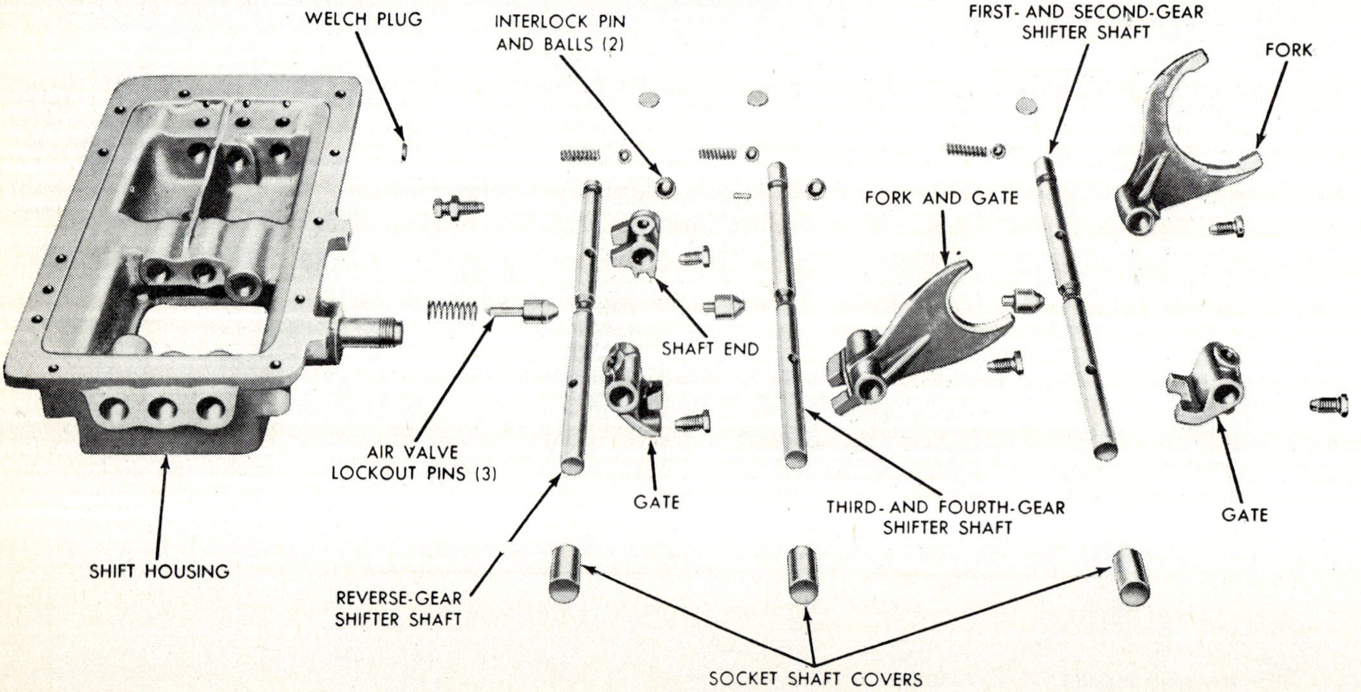

Fig. 17 R-46 gearshift mechanism

12. Line up hole in lever with hole in case and install pivot pin through lever and case. Install washer and nut on pivot pin and torque nut to 100–115 ft. lbs.
13. Shift reverse idler shaft into mesh with countershaft and move mainshaft sliding gear into mesh with second speed gear clutching teeth.
14. Lower mainshaft assembly into case, making sure all mainshaft and countershaft gears properly engage, and that the rear bearing is seated in its shallow bore in rear face of case.
15. Install input shaft, making sure mainshaft pilot bearing enters bore in input shaft. Block against rear of mainshaft as input shaft is installed.
16. Shift transmission into two gears and torque countershaft rear bearing nut to 225–275 ft. lbs. Stake nut in place.
17. If input shaft bearing was removed, install retaining nut and torque to 250–300 ft. lbs. Stake nut in place.
18. Shift transmission into neutral. Install input shaft bearing cover and gasket, making sure oil return hole in cover lines up with hole in case. Torque bolts to 25–35 ft. lbs.

GEARSHIFT MECHANISM

Disassemble, Fig. 17

1. Remove the air valve adapter plate and lock pin and spring from the bore in the air valve mounting surface.
2. Remove the welch plug from the interlock bore on the left side of the shift housing and, with air valve mounting face down, mount housing in a soft-jawed vise.
3. Make sure all shifter shafts are in neutral. Then remove lock wires and lock screws from shifting forks and gates, Fig. 17.
4. Remove welch plugs from rear of shift housing.
5. Remove shifter shafts, top one first, by driving them out front of housing. The detent springs and balls, interlock pin and balls, and intermediate air valve lockout pins must not be lost as shafts are removed. A detent spring and steel ball are depressed by each shaft and will be released from the bore as the shaft clears the hole in the rear hub.

Inspection

Shift forks and gates should be checked for wear. Check forks for alignment. Check lock screws in forks and gates. Replace parts showing wear or misalignment. Retighten and rewire lock screws. Check neutral notches in shafts for wear from interlock balls. Shafts indented at points adjacent to neutral notches should be replaced. Check spring tension on shift lever. Replace tension spring and washer if lever moves too freely. Check shift lever pivot pin and slot in lever for wear. Check auxiliary shift fork and shoe for wear. Replace if necessary.

Assemble

Reverse the order of disassembly to assemble the shift housing. Be sure the interlock pin is in the shaft. Torque shift gate and fork lock screws to 50–60 ft. lbs.

Fuller R-96 & R-960 Roadranger Ten-Speed Transmission

Description

These transmissions, Figs. 1 and 2, consist of a five-speed main (front) section and a two-speed auxiliary (rear) section which provides ten forward and two reverse speeds shifted with one lever. The ten forward speeds are obtained by using the ratios of the five-speed section twice; the first time with the auxiliary section in low gear so that the ratios in the five-speed section are compounded by the auxiliary low gear; the second time with the auxiliary section in high gear. The two reverse speeds are obtained by using reverse in the five-speed section through either high or low gear of the auxiliary section.

Separating Sections

1. Disconnect three air hoses at air control valve. Remove clamps and remove hoses with air cylinder.
2. Remove air filter and air regulator valve (with bracket and hose).
3. Remove air valve housing and its plate from adapter.
4. With transmission in neutral, remove gearshift housing and auxiliary cover.
5. Drain transmission and remove oil filter.
6. Remove brake drum, lock transmission in two gears and remove U-joint flange.
7. Remove parking brake shoe assembly.
8. Shift auxiliary unit in low range.
9. To hold mainshaft of front unit in position during removal of auxiliary section, block large sliding gear forward, Fig. 3. Use a block of wood about two inches wide, and be sure the sliding gear is engaged with the adjacent gear.
10. Secure auxiliary section on a hoist.
11. Remove nuts holding the two sections together. Do not overlook the two upper front nuts that are inside the auxiliary section, Fig. 4.
12. In the right and left sides of the flange at the rear of the main section, locate the two 1/2″ tapped holes. In each of these, start a hardened bolt about 2 1/2″ long, Fig. 5. *Do not try to separate the sections at this time.*
13. Line up one of the two depressions in the O.D. of the clutch disc carrier with the auxiliary countershaft drive gear, Fig. 6, so that there will be clearance when transmission cases are separated.
14. Now separate cases by evenly tightening the bolts started in Step 12. Then remove the bolts.

Fig. 1 Fuller Roadranger R-96 ten-speed transmission

Disassemble Main Trans.

1. Remove wood block (Fig. 3) and, while keeping sliding gear engaged, move mainshaft rearward until it clears input shaft. Remove mainshaft.
2. Remove input shaft bearing retainer bolts. Then, working inside of case, force out input shaft, including bearing, nut and bearing retainer.
3. Remove retainer from input shaft and bearing snap ring from case.
4. Remove reverse idler gear shaft, gear and bearing from case, then remove bearing from gear.
5. Remove bearing retainer washer from rear end of countershaft.
6. Remove countershaft front bearing retainer and bearing spacer.
7. After driving countershaft far enough rearward to expose rear bearing, pull the bearing.
8. After removing the countershaft, remove outer race of countershaft front bearing from case.

Disassemble Sub-Assemblies

Front Unit Mainshaft

1. After removing wire stop from clutch disc carrier, remove pressure plate snap ring from carrier, Fig. 7.
2. Remove pressure plate and friction discs, *keeping discs in order of original assembly.*
3. Remove mainshaft pilot bearing.
4. After removing washer locking key, Fig. 8, turn washer so that it aligns with shaft spline grooves. Remove 4th speed gear and washer (3rd on R-960).
5. Using rear face of 3rd speed gear (2nd on R-960) as a base, press shaft through bushing and gear, Fig. 9. *Do not let shaft drop.* Remove clutch sleeve.
6. Using rear face of 2nd speed gear (1st on R-960) as a base, press shaft through two bushings and gear, Fig. 10.
7. Remove large and small Woodruff keys from shaft.
8. On R-96 only: Using rear face of 1st speed gear as a base, press shaft through bushing, washer and gear until front end of bushing meets lower end of large keyway, Fig. 11. *Pressing the shaft further will damage washer.* Then move gear and washer back to the rear and pry bushing off shaft. Remove Woodruff key now exposed, Fig. 12, then remove washer and two sliding gears.
   a. On R-960 only: Using rear face of sliding gear as a base, press shaft through bushing, washer and gears until front end of bushing meets lower end of large keyway. *Pressing shaft further will damage washer.* Then move gear and washer back to rear with gear

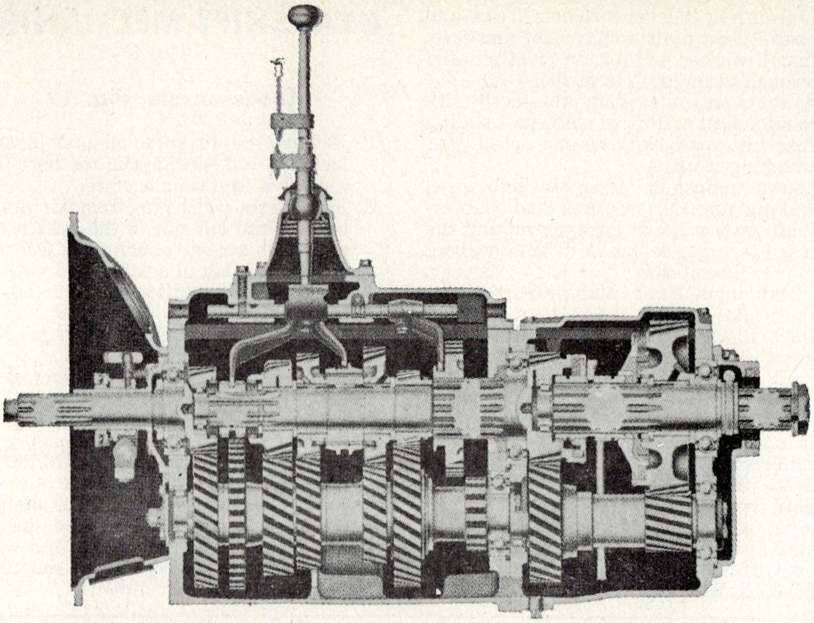

Fig. 2 Fuller Roadranger R-960 ten-speed transmission

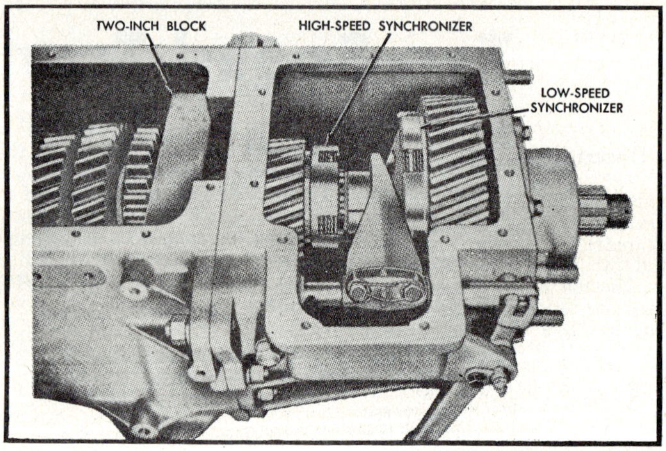

Fig. 3 Preparation for separation of cases

Fig. 4 Auxiliary upper front stud nuts

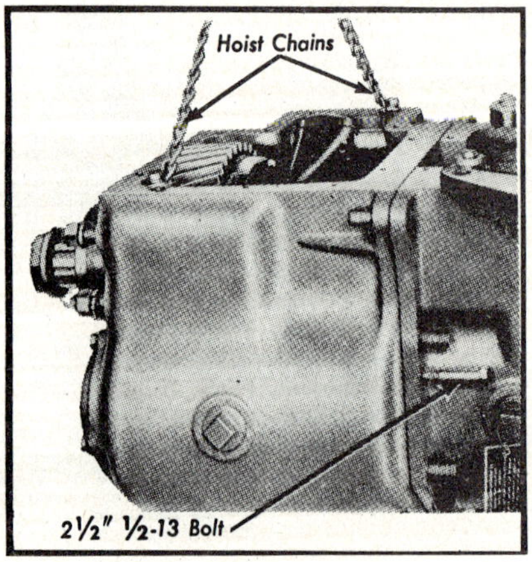

Fig. 5 Separation of cases

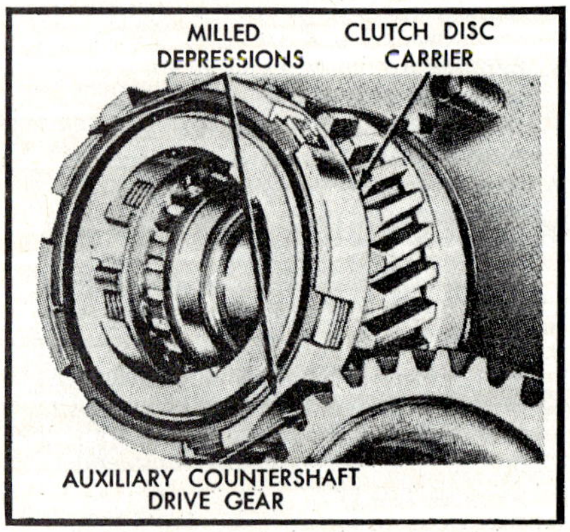

Fig. 6 Clutch disc carrier clearance

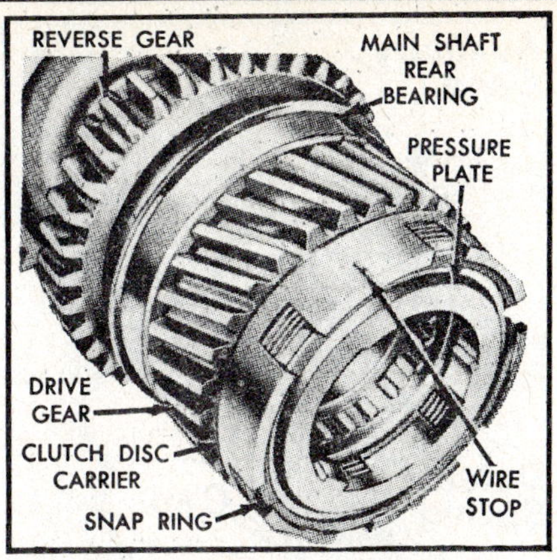

Fig. 7 Rear end of mainshaft

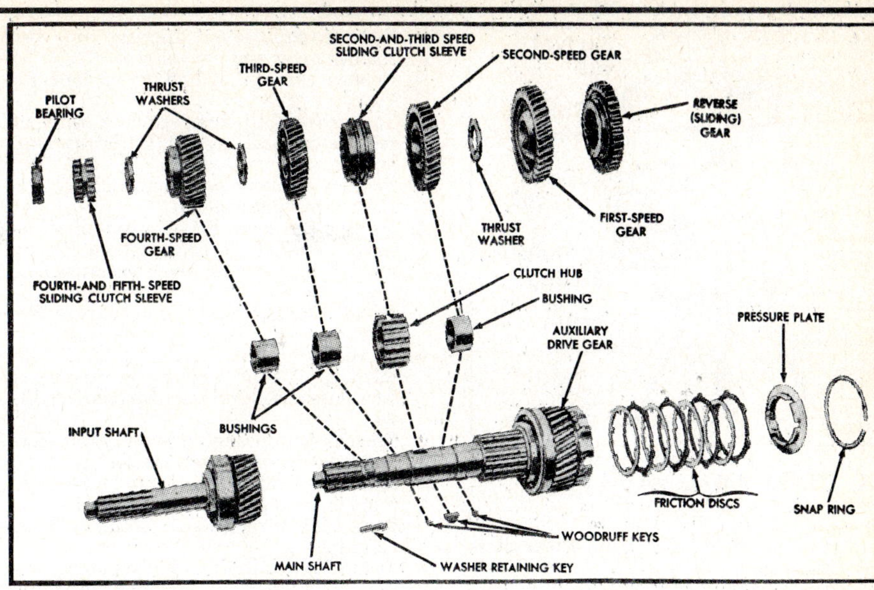

Fig. 8 Mainshaft disassembled

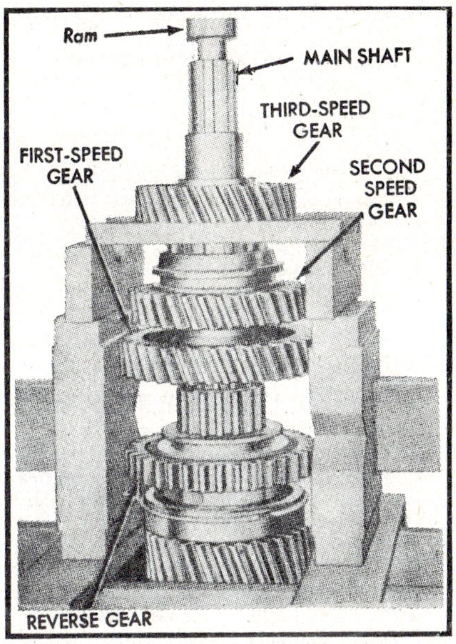

Fig. 9 Removing 4th speed gear bushing

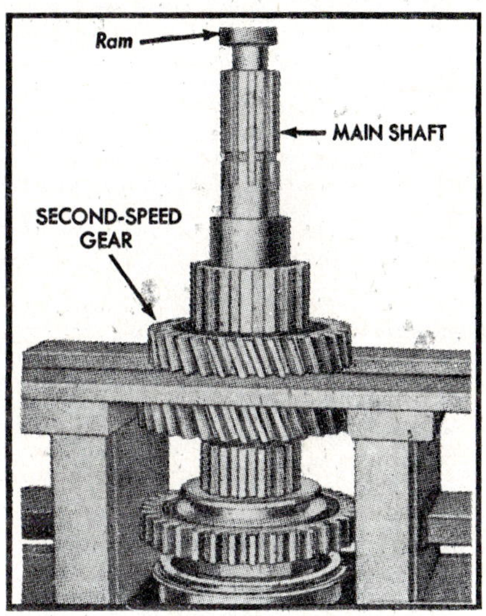

Fig. 10 Removing 3rd speed gear bushing and hub

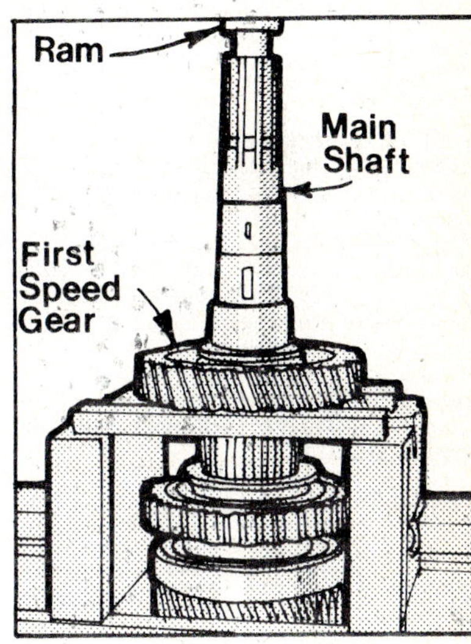

Fig. 11 Removing 2nd speed gear bushing

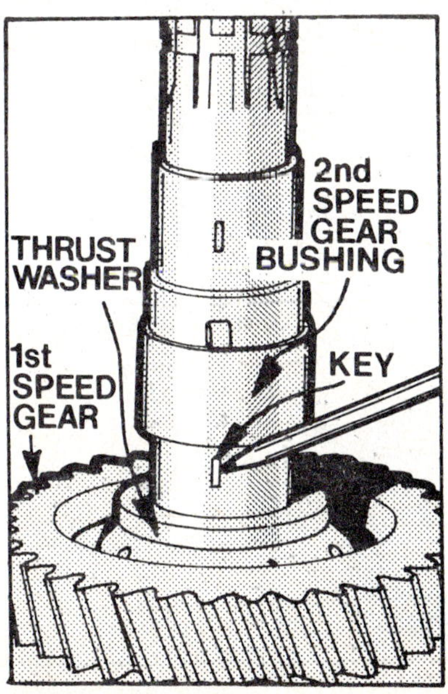

Fig. 12 Second speed gear bushing key

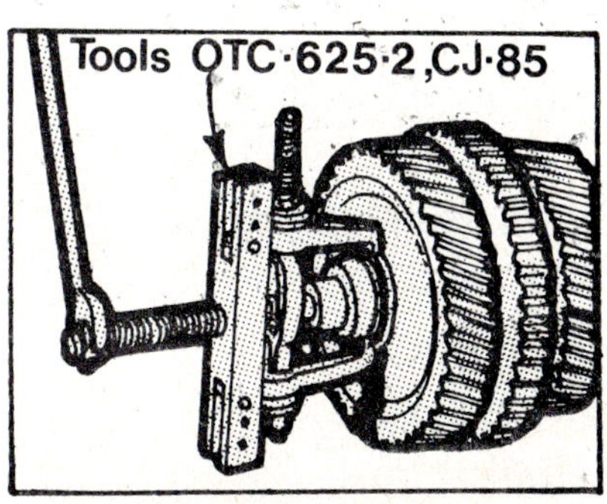

Fig. 13 Removing main countershaft front bearing inner race

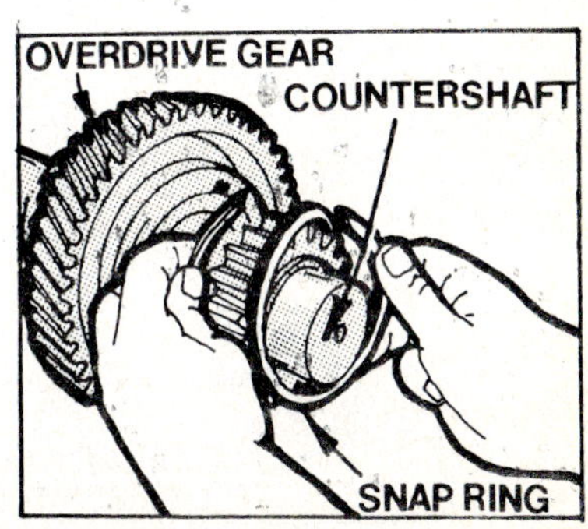

Fig. 14 Spiral snap ring (R-960 only)

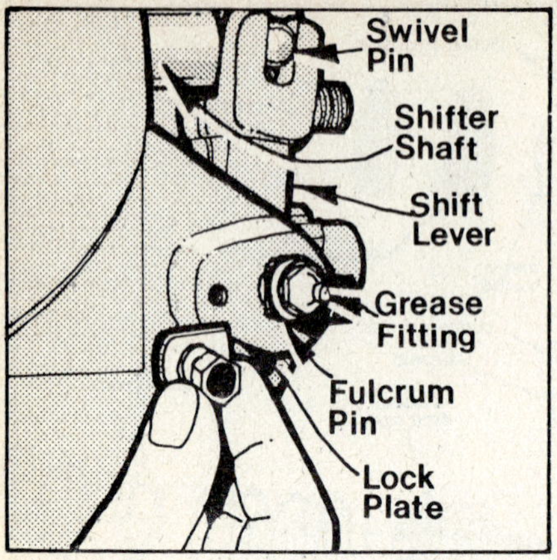

Fig. 15 Fulcrum pin lock plate

against shoulder of shaft. Press shaft through bushing and remove Woodruff key, washer and gears.
9. Mount shaft in soft-jawed vise, relieve peening on bearing nut and remove nut *which has a left-hand thread.*
10. Jar shaft on a block of wood to remove bearing.

Front Unit Input Shaft

Mount shaft in a soft-jawed vise. Relieve peening on bearing nut and remove nut *which has a left-hand thread.* Press or jar bearing front shaft.

Front Unit Countershaft

1. After removing snap ring at front end of shaft, use a puller to remove bearing inner race, Fig. 13.
2. Press off gears *one at a time,* beginning with the drive gear. On R-960, remove spiral type snap ring at rear of overdrive gear before pressing off this gear, Fig. 14. Remove Woodruff keys.

Auxiliary Section

1. Remove shift lever lock plate and grease fitting, Fig. 15.
2. Remove shift lever fulcrum pin, then remove lever and swivel pin.
3. Remove cap, spring and detent ball, Fig. 16.
4. To remove shift fork, cut lock wire and remove fork lock screws. Withdraw shifter shaft and remove fork.
5. Remove bearing retainer and countershaft retaining washer.
6. Using a soft drift, drive output shaft forward through front bearing bore. Remove low-speed gear and rear thrust washer from case, Fig. 17.
7. Remove output shaft rear bearing from case.
8. To remove countershaft, tap shaft rearward until rear bearing is outside its bore. Use a puller to remove bearing, Fig. 18.
9. Remove countershaft assembly from case, and front bearing from its bore.
10. If countershaft or gear is to be replaced, remove snap ring at front of shaft, then press shaft through gear. Remove Woodruff key.
11. To dismantle auxiliary output shaft, position it in a soft-jawed vise and use a puller to remove pilot bearing, Fig. 19.
12. After removing sliding clutch, position shaft in a press. Use low-speed gear thrust washer as a base and press the shaft 1$^3/_8$" through bushing and washer, Fig. 20. *If shaft is pressed further, washer will be damaged.*
13. Move washer against shoulder of shaft and, using front end of bushing as a base, press shaft through bushing, Fig. 21. Remove Woodruff key and thrust washer.
14. To disassemble low-speed gear assembly, first remove wire stop from low-speed clutch disc carrier, then remove pressure plate snap ring.
15. Remove pressure plate and friction discs, *keeping discs in order of original assembly.*

Disassemble Gearshift Housing

1. Carefully remove detent spring retainer, springs and balls from upper side of housing. Remove plug from bore on air valve side of housing.

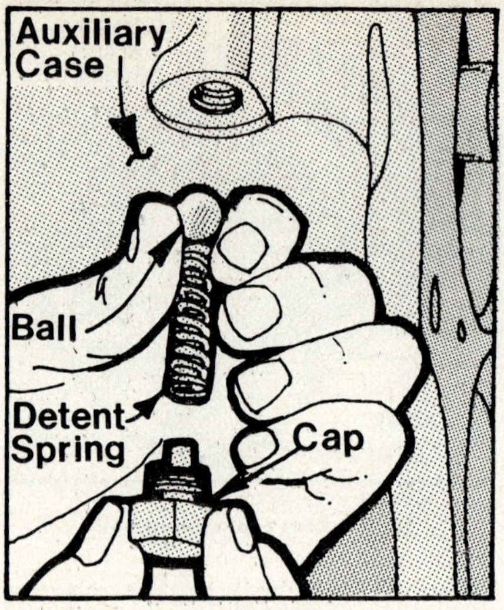

Fig. 16 Auxiliary detent spring, cap and ball

2. Mount air valve side of housing in a vise and, *with all shafts in neutral,* remove set screws from all forks and gates.
3. Remove adapter, Fig. 20, and interlock spring and plunger, Fig. 23.
4. After removing welch plugs from shifter shaft bores, remove shafts, *being sure not to lose interlock plungers and balls as the last shaft is removed,* Fig. 22.

Assemble Gearshift Housing

1. Shifter shafts should be positioned through front of housing with notches to the front. Mount housing in vise with air valve face upward.
2. Position 1st-reverse (overdrive-reverse on R-960) shifter shaft in bottom of bore, through gate and through fork, Fig. 23. Install and safety-wire set screws.
3. Install $^3/_4$" interlock ball in front opening (air valve side) and air valve intermediate lockout plunger in rear opening, Fig. 24. Move shaft to neutral.
4. Position 2nd-3rd (1st-2nd on R-960) shifter shaft in center bore and through shift fork. *Don't forget the lockout pin that must be inserted in its hole at neutral notch.* Install and safety-wire screws.
5. Repeat Step 3 and move shaft to neutral.
6. Position 4th-5th (3rd-4th on R-960) shifter shaft in upper bore and through shift fork. Install and safety-wire retaining screws.

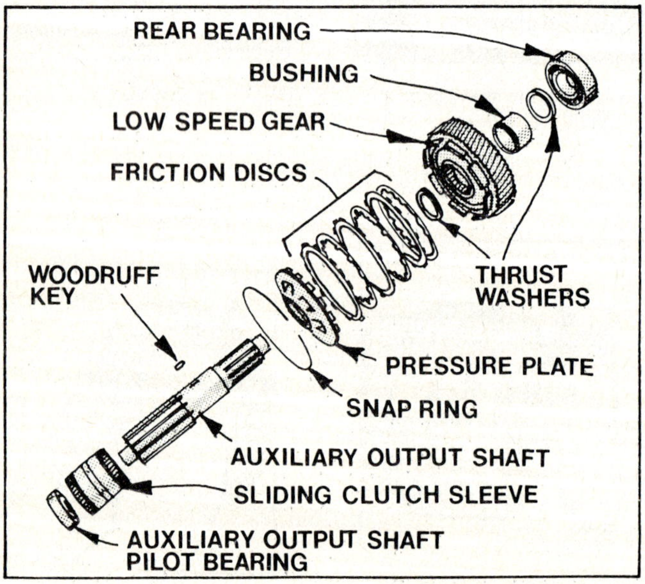

Fig. 17 Auxiliary output shaft (R-960 only)

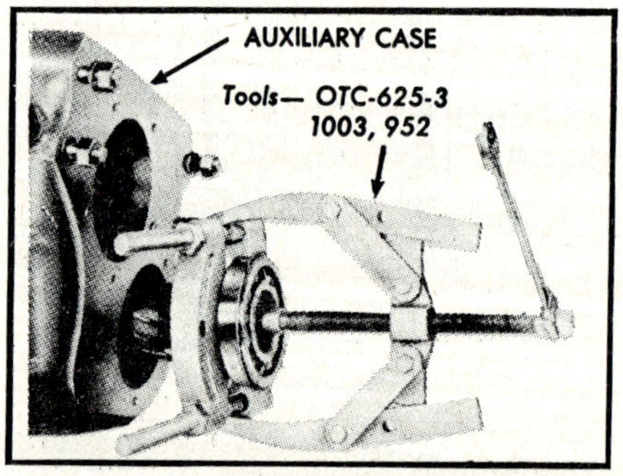

Fig. 18 Removing auxiliary countershaft rear bearing

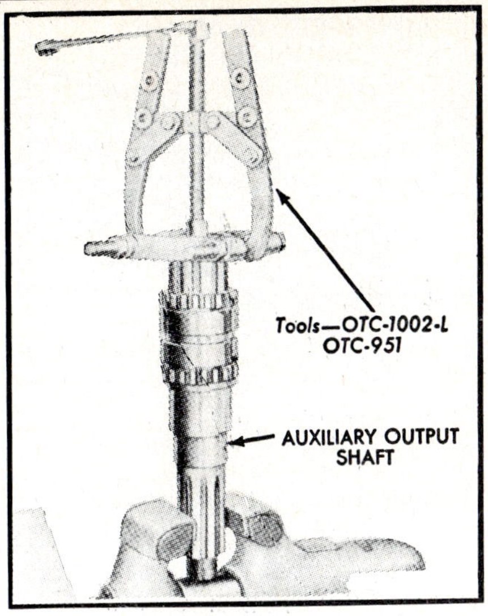

Fig. 19 Removing auxiliary output shaft pilot bearing

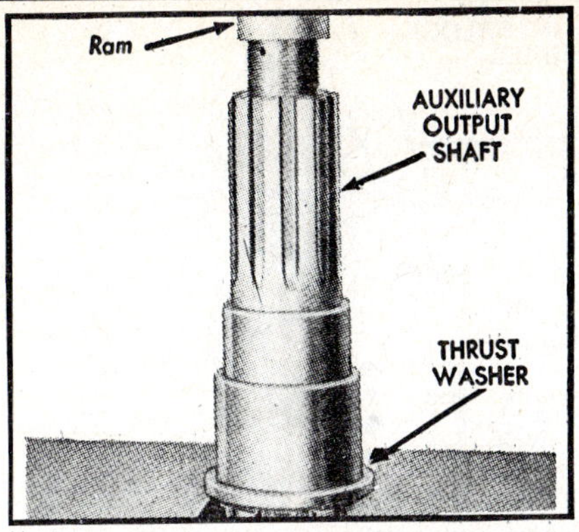

Fig. 20 Removing auxiliary shaft bushing

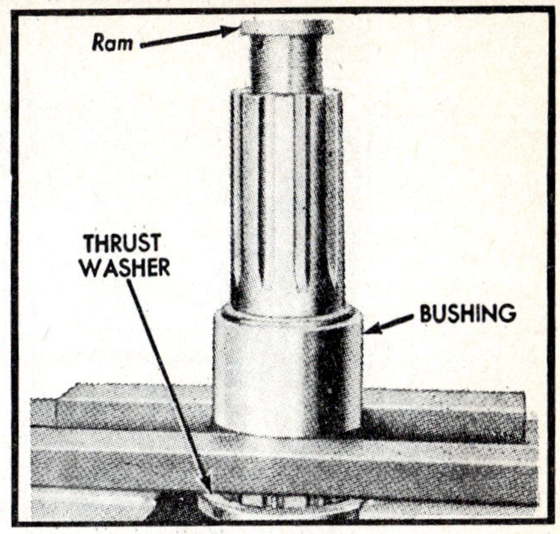

Fig. 21 Pressing auxiliary output shaft through bushing

7. After positioning detent balls and spring in top of housing, install detent spring retainer.
8. Apply sealer to welch plugs and install them in housing. Install plug in bore of air valve side of housing.
9. Install third interlock plunger and its spring, then install adapter.

Auxiliary Section, Assemble

1. Position friction discs in low-speed gear. Start with an inner disc (projections at inside diameter) and alternately one inner and one outer disc in housing until six discs have been positioned, Fig. 25. The last inner disc to be installed will have four bent lugs that should be positioned toward pressure plate.
2. Install pressure plate and snap ring. The lock wire hole in rim of gear must be between open ends of snap ring.
3. Cut a new stop wire and install it in housing rim between open ends of snap ring. Bend wire up internally and down externally. The internal end must not extend above the housing.
4. To assemble output shaft, install low-speed gear front thrust washer with *beveled face to front,* Fig. 26.
5. Position Woodruff key and low-speed gear bushing, and press bushing onto shaft.
6. Position sliding clutch on shaft and install pilot bearing.
7. To install countershaft drive gear, position the key and press gear on shaft. Install snap ring on shaft, Fig. 27.
8. Position countershaft in case, then turn case up on its front surface.
9. Install countershaft rear bearing on shaft and in case bore.
10. Position case in its normally upright position and install inner race of countershaft front bearing. Install outer race.
11. Install rear bearing retainer washer, install and safety-wire bolts.
12. Position output shaft low-speed gear in case, meshing teeth with those on countershaft.
13. Position output shaft through front bearing bore. Move sliding clutch gear through synchronizer assembly into engagement with low-speed gear. Install rear thrust washer with polished face next to gear.
14. Install output shaft rear bearing.
15. Install speedometer drive gear spacer and gear, and bearing retainer.
16. Position auxiliary shift fork and its shaft in case. Install lock screws and wire them securely.
17. With transmission shifted in low, install detent ball, spring and cap.
18. Position swivel pin in end of shifter shaft, then install shift lever and fulcrum pin.
19. Line up fulcrum pin slot with lock screw hole and drive pin into case until inner edge of slot is flush with boss. Install lock plate and grease fitting for the pin.

Assemble Main Trans. Sub-Assemblies

Countershaft, Fig. 28

1. Install Woodruff keys and press gears on, one at a time, over front end of shaft.
2. Install inner race of front bearing on shaft. Install gear retaining snap ring.
3. On R-960, press on overdrive gear (long hub to rear). Install spiral type snap ring.

Input Shaft

1. Press on input shaft bearing with shield toward front of shaft.
2. Install bearing nut (left-hand thread). Tighten nut and peen it into the milled slots.

Mainshaft, Fig. 8

1. Press on rear bearing with snap ring toward front. Install retaining nut (left-hand thread) and peen nut in four places

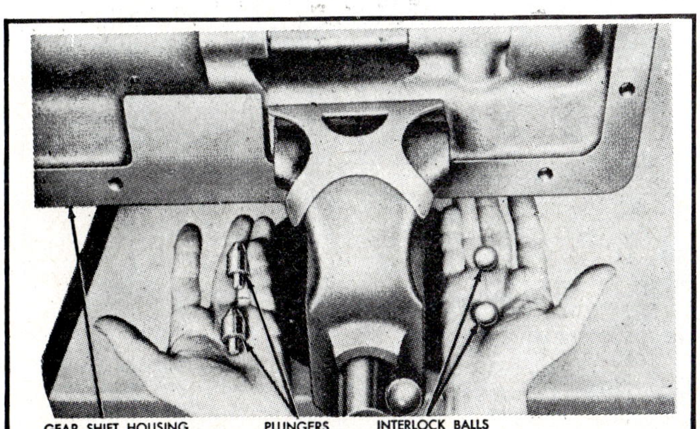

Fig. 22 Interlock plungers and balls

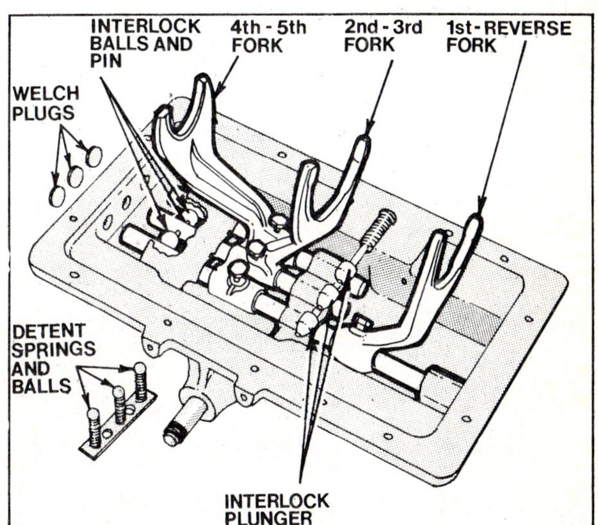

Fig. 23 Shift mechanism

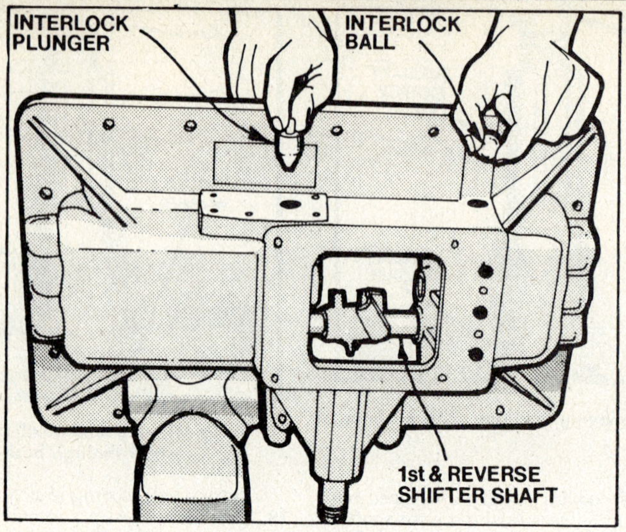

Fig. 24 Interlock ball and plunger installation

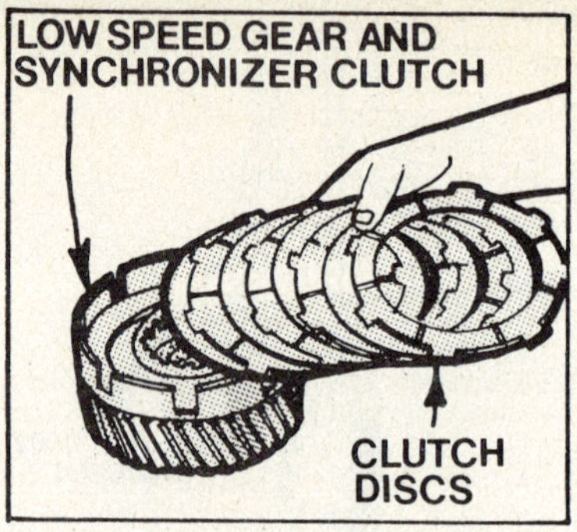

Fig. 25 Positioning synchronizer discs

90° apart.

2. Install sliding gear with fork groove to front.
3. Install 1st speed gear (overdrive gear on R-960) *with clutching teeth to rear.*
4. Install 2nd speed gear washer (1st speed on R-960).
5. Install 2nd speed gear bushing (1st speed on R-960). Bushing must be tight against washer.
6. Install 2nd speed gear (1st speed on R-960) *with clutching teeth to front.*
7. With large Woodruff key installed, press on clutch hub. Hub must be tight against gear.
8. After installing shift collar, the small Woodruff key and its bushing, install 3rd speed gear (2nd speed on R-960) *with clutching teeth to rear.*
9. Install 4th speed gear (3rd speed on R-960) washer, *with ID chamfered edge to rear.*
10. Install 4th speed gear (3rd speed on R-960) bushing and its gear, *with internal clutching teeth to front.*
11. After positioning the internally splined front washer, *with chamfered end of splines to rear,* turn washer in its groove until lugs on its ID are under projecting mainshaft splines.
12. Install washer locking key, making sure that lower end engages washer.
13. Install sliding clutch gear with the word "Front" toward pilot end of shaft. Install pilot bearing with snap ring groove to rear.
14. The high- and low-speed synchronizers are assembled in the same manner. Position high-speed synchronizer discs in their original order in the synchronizer housing. Install pressure plate, retaining snap ring and wire stop.

Assemble Main Trans.

1. Install outer race of countershaft front bearing and spacer.
2. Install countershaft in case and install rear bearing.
4. Install rear bearing retainer.

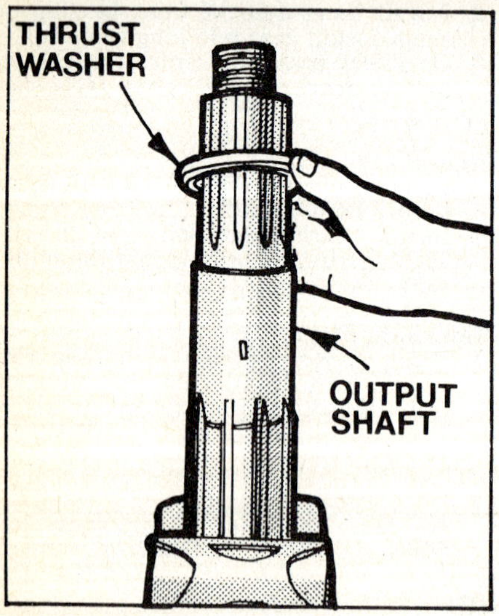

Fig. 26 Thrust washer installation

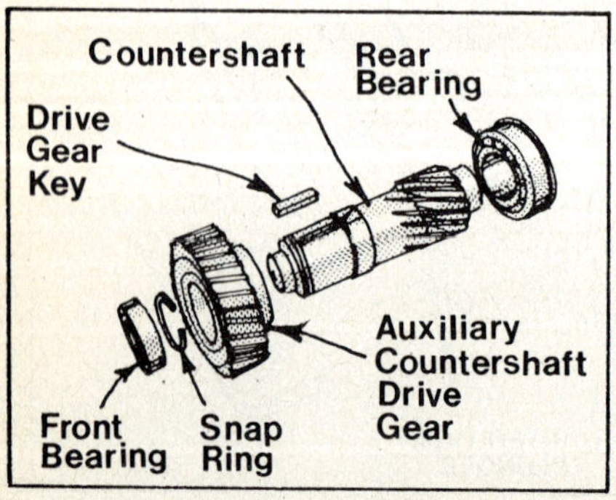

Fig. 27 Auxiliary countershaft

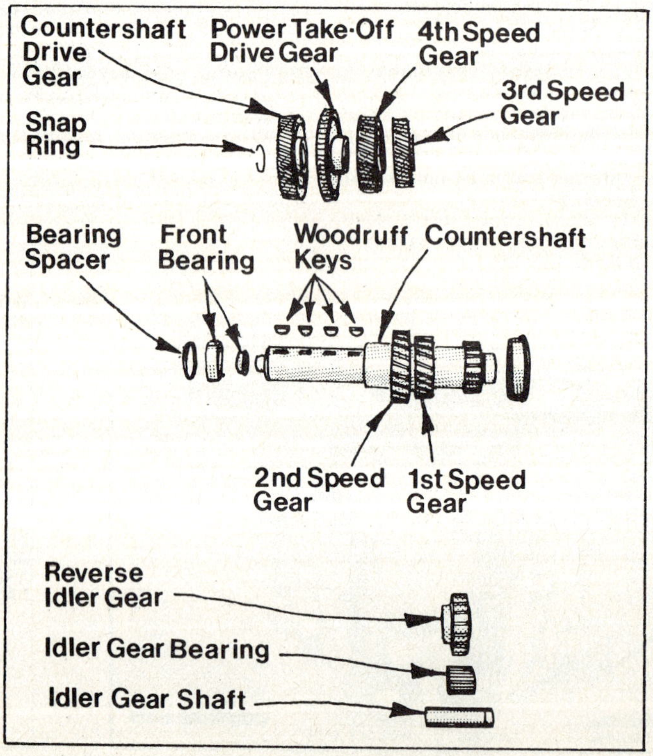

Fig. 28 Main countershaft

5. After positioning bearing in reverse gear, install gear in case *with long hub to front*.
6. Using a soft drift, drive reverse gear shaft into place until front edge of shoulder is flush with rear face of case.
7. To install mainshaft, engage sliding gear with adjacent gear and lower shaft assembly into case. Be sure that corresponding countershaft gears and that rear bearing seats in its bore.
8. Install input shaft bearing retaining ring in its case bore.
9. Block rear end of mainshaft and install input shaft assembly. Install input shaft bearing retainer, making sure oil drain groove in retainer lines up with drain hole in case.

Joining Front & Rear Sections

1. With auxiliary section in low gear and new gasket in place, move auxiliary section into place. Before moving auxiliary all the way forward, be sure that one of the two depressions in the OD of the clutch disc carrier is lined up to provide clearance at the countershaft drive gear. Also line up the ID lugs of the pressure plate with the corresponding grooves in the OD of the auxiliary sliding clutch gear.
2. With case faces flush, install lockwashers and nuts on attaching studs, including the two upper studs on the inside of the auxiliary, Fig. 4.
3. Install auxiliary section cover.
4. After making sure that speedometer gear washer and gear are installed, install parking brake assembly.
5. Install U-joint flange and brake drum.
6. Install air valve plate and housing.
7. Install air filter and air regulator valve assemblies, and connect the three air hoses at the air control valve.
8. Install oil filter and lubricant.

Air Valve Service

For air valve service and trouble shooting data, follow instructions given in the Fuller R-46 transmission section.

Fuller Five-Speed Conventional Transmissions
5-33, 5-43, 5-62, 5-65, 5-72, 5W-74

Disassemble Transmission

Before dismantling a transmission, carefully clean the outside of the case to avoid the possibility of introducing dirt or other foreign material into the unit. This is important because dirt, being an abrasive, is highly detrimental to highly polished parts such as bearings, sleeves and bushings. These, as well as other parts, should be cleaned as removed and protectively wrapped until ready for use.

In removing the control cover, do not force it off the transmission. Forcing may spring the yokes out of alignment and cause partial engagement or gear interference. Follow the directions relative to shifting and the cover can easily be removed. If binding occurs, a slight manipulation or movement to one side or the other will free it.

When disassembling the control cover, all parts should be laid on a clean bench in the same sequence as removed. This procedure will not only simplify reassembling the yokes and bars in their proper places but will also reduce the possibility of omitting the small yoke bar interlock parts. If these parts are left out, the unit will shift into two speeds at the same time.

Unless absolutely necessary, the clutch housing should not be detached from the case. If it is detached, extreme care should be taken when reassembling the housing to maintain the alignment between its machined face and the mainshaft bores in the case; also the concentricity of the mainshaft bores and the flange pilot.

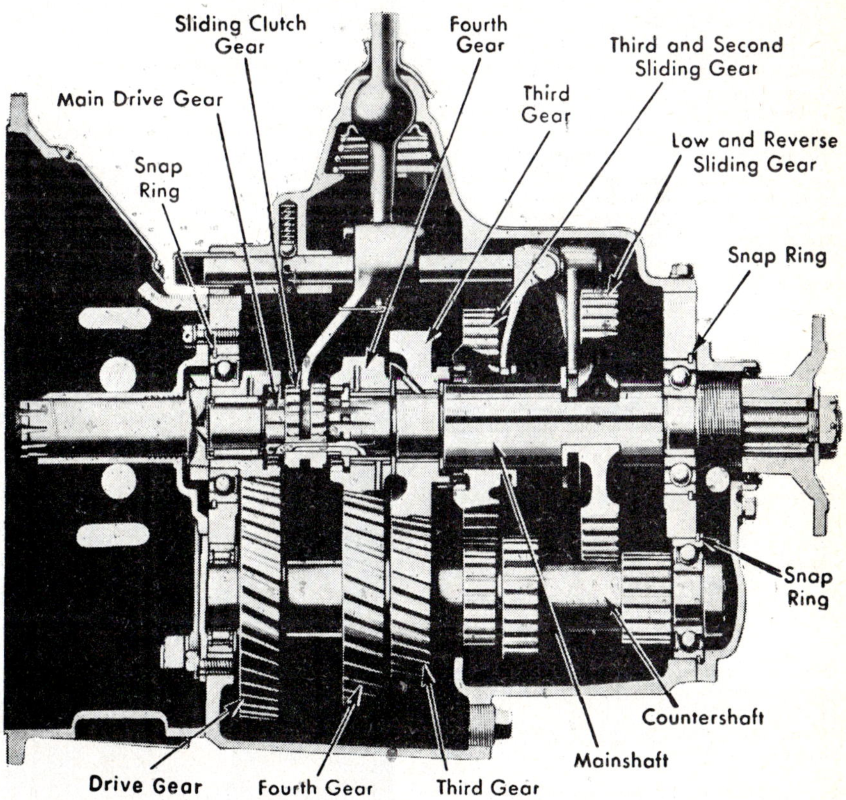

Fig. 1 Fuller five speed transmission. Series 5-33, 330, 5-43, 430

NOTE: The disassembly procedure which follows is based on the assumption that the transmission has been taken from the chassis and that all supplementary equipment such as universal joint companion flanges and brake assemblies have been removed.

Control Cover, Disassemble

1. Shift transmission into 1st speed.
2. Remove hand brake lever.
3. Detach and lift off cover.
4. Place control lever in neutral.
5. Remove shift yoke lock screws.
6. Start with upper bar and drive shift bars from housing. Use care to see that none of the small yoke bar locking parts are lost as bars are removed.
7. To remove shift lever, take off rubber grip, and latch spoon if lever is equipped with latch rod.
8. Force dust cover up and off lever after removing lock screw from those so equipped.
9. Remove tension spring and washer from inside lever housing, and withdraw shift lever.

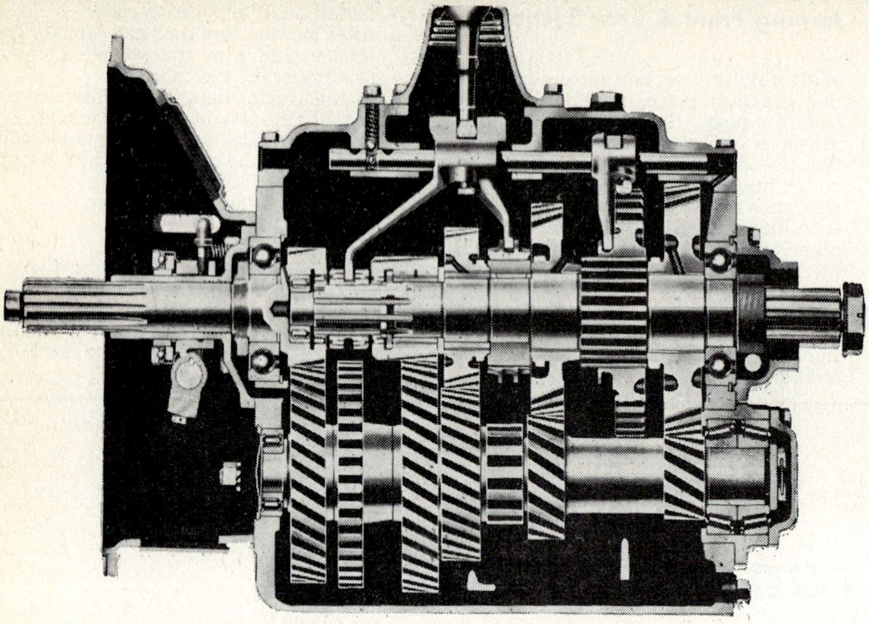

Fig. 2 Fuller five speed transmission. Series 5-72, 720, 5W-74

Mainshaft, Disassemble

1. Remove rear bearing covers from mainshaft and countershaft.
2. Lock gears and remove bearing lock nut from end of countershaft.
3. Pull mainshaft and countershaft assemblies to the rear.
4. Remove mainshaft rear bearing.
5. Tilt mainshaft and remove through top of case, leaving sliding gears inside case.
6. Remove loose sliding gears from case.
7. To remove gears and sleeve from mainshaft, remove pilot bearing and sliding clutch, using the latter to free bearing.
8. Remove mainshaft gear retaining washer key.
9. Rotate mainshaft gear retaining washer until its inside lugs line up with grooves in shaft.
10. Remove mainshaft bushed gears, sleeve and washers, using the rear bushed gear to start the sleeves.

Main Drive Gear, Disassemble

1. Remove clutch release mechanism.
2. Remove drive gear bearing cover.
3. Withdraw gear.

Reverse Gears, Disassemble

1. Remove high speed reverse shifting bar and yoke from inside the case (Models 62, 620, 65, 650, 72, 720).
2. Withdraw shaft or shafts and remove gear or gears from case.

Countershaft, Disassemble

1. Remove countershaft rear bearing.
2. Tilt countershaft and remove through top of case.
3. To disassemble countershaft, remove gear retaining snap ring.
4. Press gears from shaft one at a time.

Reassembly Instructions

With the exception of a few specific points, the general instructions for reassembly are merely the reversal of those for disassembly. The exceptions are as follows and should be given careful attention.

Clearances

1. On mainshaft helical bushed gears, replace worn washers to assist in maintaining original fit.
2. Hold end play of both gears to a minimum of .006″. Check with dial indicator or feeler gauge.

3. Replace bushings in gears having excessive radial clearance.

Assembling Bushed Gears

When assembling either new or rebushed gears on the mainshaft or countershaft, be sure that they move freely after having been assembled in their proper places. Clean all mounting parts and coat with clean oil before replacing gears.

Installing Bushings in Mainshaft Helical Gears

1. Remove old bushing.
2. Clean bore of gear carefully and remove all burrs.
3. Lubricate outside diameter of bushing and inside diameter of gear.
4. Carefully press bushing fully into gear.
5. Face off any projecting end of bushing.
6. Drill oil holes in bushing through holes in gear, making sure all holes are open.
7. Remove all burrs resulting from drilling and break all sharp edges on ends of bushing.

Installing Bushings in Reverse & Countershaft Gears

1. Remove old bushing, clean bore of gear and remove all burrs.
2. Press bushing fully into gear.
3. Spin over the projecting ends of the bushing.
4. Open the ends of the oil grooves with a file.
5. Drill oil holes in bushing through holes in gear, making sure all holes are open.
6. Remove all ridges and burrs resulting from drilling and spinning operations.

Fig. 3 Fuller five speed transmission. Series 5-62, 620, 5-65, 650

U-Joint Flange

The mainshaft is held in its proper place by the universal joint companion flange. The omission of parts between the flange and bearing, or failure to pull the flange tightly into place will allow the mainshaft to move endwise with resultant damage to the pilot bearing, mainshaft and main drive gear.

Sliding Clutches

All sliding clutches should be assembled to the shaft with the counterbore toward the bushed gears. An exception is the sliding clutches in 43 and 430 series, which have no counterbore. They are marked, however on one end with the word "front," and must be assembled with the marked end toward the front of the transmission.

Power Take-Off

When attaching a power take-off, be sure sufficient filler blocks and gaskets are used to prevent the gears from meshing too deeply. If the gears bottom, they will damage not only themselves but also the transmission case.

Fuller Series 5H Five-Speed Transmissions

Disassemble Transmission

1. With transmission in neutral, remove gearshift housing.
2. With transmission locked up in two gears, remove output shaft flange nut.
3. Remove flange and brake drum as a unit, and disconnect parking brake linkage as required so that brake shoe assembly can be removed.
4. Remove output shaft rear bearing retainer and speedometer drive gear. Remove oil seal from retainer.
5. After removing clutch release mechanism, remove six stud nuts (inside clutch housing) and then remove clutch housing.
6. Remove input shaft bearing retainer, then pull out input shaft and bearing assembly. If necessary, use a soft drift from inside of case. Remove bearing retainer snap ring from case bore.
7. Mount input shaft in a vise and relieve bearing nut peening. Remove nut which has a left-hand thread. If bearing is to be replaced, press or jar it from shaft.
8. Remove countershaft rear bearing adapter cap. Remove retainer plate from rear end of countershaft.
9. Remove countershaft rear bearing adapter, using puller screws in holes provided in adapter.
10. Drive output shaft forward until it stops solidly, then drive it to the rear until rear bearing snap ring is exposed. Remove bearing. If bearing is to be replaced, remove the snap ring, Fig. 1.
11. Remove output shaft and gears, except sliding gears. Lift sliding gears out of case.
12. Remove countershaft rear bearings and spacer from retainer.
13. After removing reverse idler shaft retainer, remove shaft, then idler gears and bearings.
14. Move countershaft rearward to clear front bearing, then remove shaft and gears from case.
15. Remove countershaft front bearing retainer, gasket, bearing and spacer.
16. Disassemble output shaft and countershaft as required.

Gearshift Housing, Assemble

1. Referring to Fig. 2, position housing (left-hand side up) in a vise and install rocker arm and pivot.
2. Insert rocker arm shift shaft through webs and gates and install set screws.
3. Insert 2–3 shaft into housing and through its fork. Install set screw.

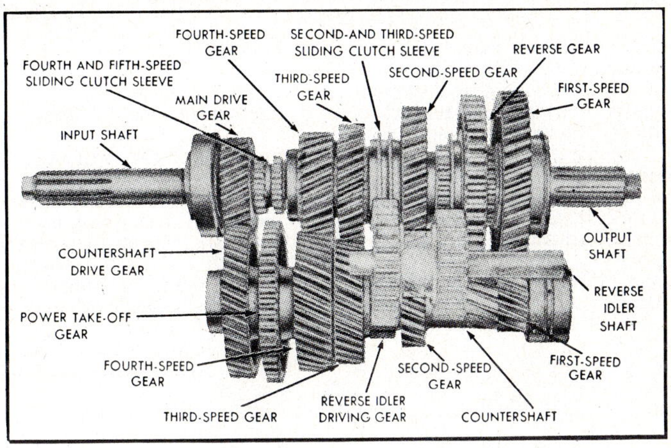

Fig. 1 Gear train of Series 5-74 five-speed transmission

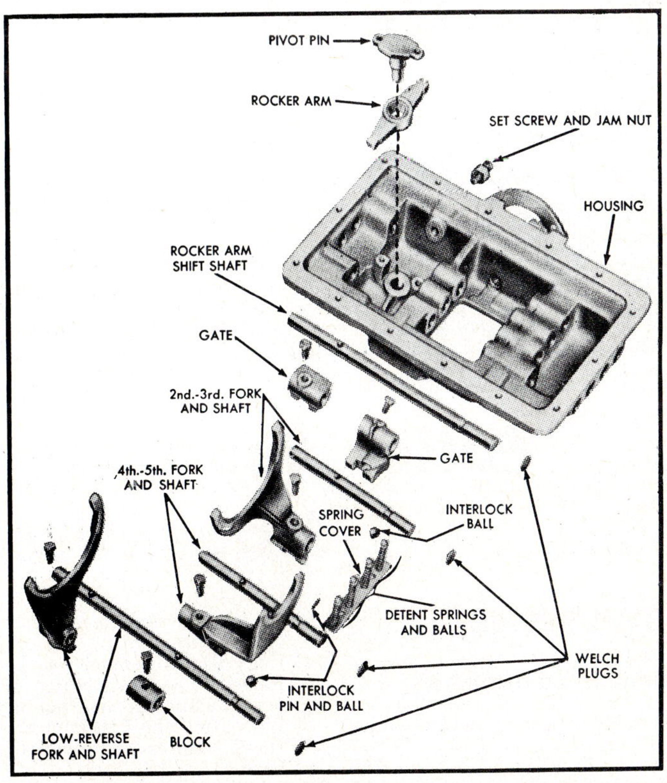

Fig. 2 Gearshift housing disassembled. Series 5-H-74, 740

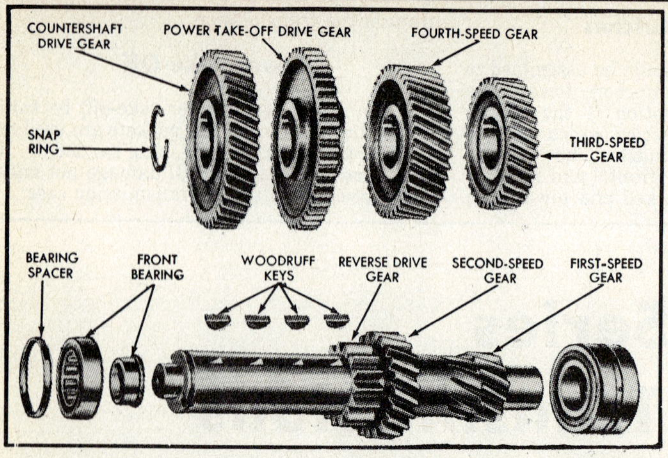

Fig. 3 Countershaft disassembled. Series 5-H-74, 740

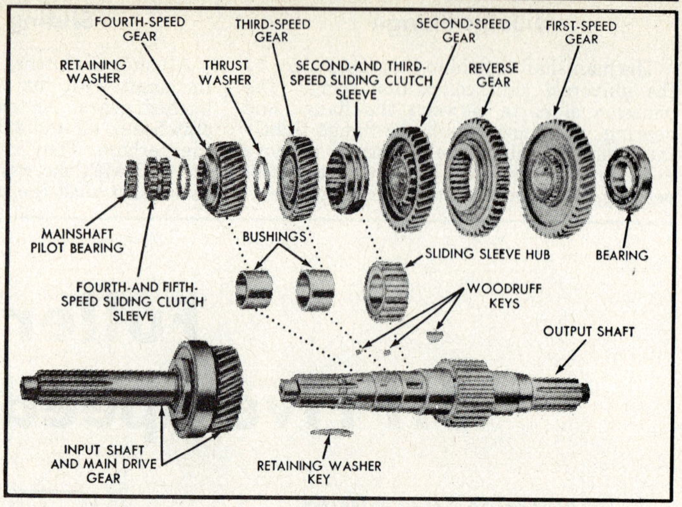

Fig. 4 Main drive line disassembled. Series 5-H-74, 740

4. Position a 3/4" interlock ball in housing opening between 2–3 and 4–5 shafts.
5. Insert 4–5 shaft into housing and through its fork. Insert interlock pin into shaft as neutral notch enters web. Install lock screw.
6. Install remaining 3/4" interlock ball.
7. Insert low-reverse shaft through housing, block and fork. Install set screws.
8. Safety wire all set screws and pivot pin retaining bolts.
9. Install four 1/2" steel balls and detent springs. Be sure all shafts are in neutral position so that balls will fit in their slots.
10. Install detent spring cover and new gasket.
11. Install plugs in shift shaft openings of housing.
12. Start set screw and jam nut and tighten screw until it bottoms. Back off 1/2 turn and tighten jam nut.

Countershaft Assemble, Fig. 3

1. Install Woodruff keys and press gears on one at a time.
2. Press on front bearing inner race and install drive gear snap ring.

Output Shaft Assemble, Fig. 4

1. Install Woodruff keys for sleeve hub and 3rd speed gear bushing.
2. Position 2nd gear (clutching teeth front) on shaft, then press sleeve hub (recessed end to front) tightly against gear.
3. Press 3rd gear bushing on shaft and against sleeve hub. Position clutch sleeve on its hub.
4. Position 3rd gear (clutching teeth to rear) on its bushing.
5. Install thrust washer (chamfer to rear) and install remaining key.
6. Press 4th gear bushing on shaft and position 4th gear (hub forward) on bushing.
7. Install retaining washer and measure clearance between 4th gear and its thrust washer. If clearance is not within the limits of .006 to .012", select the required retaining washer to result in the specified end play.
8. Install retaining washer key and clutch sleeve (recessed end to rear).
9. Install pilot bearing in input shaft gear pocket.
10. If input shaft bearing was removed, install new bearing with shielded side toward front.

Assemble Transmission

1. Install countershaft front bearing in case bore.
2. Position front bearing spacer and gasket and install retainer.
3. Position countershaft in case, then tilt case so that its rear face is up.
4. Position two wood or fiber blocks of equal thickness between drive gear and case.
5. Position inner bearing outer race in adapter, then apply sealer to adapter gasket. Make sure adapter oil return hole is aligned, then install gasket and adapter.
6. Install inner bearing cone.
7. Position spacer and install outer bearing cone. Position outer bearing outer race.
8. Install bearing lock plate and safety wire bolts.
9. Apply sealer to adapter cap gasket and install gasket and cap.
10. After tilting case back to its normal position, install idler gear and shaft. Be sure flat on shaft will align with lock plate. Install lock plate.
11. Position output shaft sliding gears in case, then install output shaft in case. Check for proper gear engagement with countergears.
12. Install output shaft rear bearing.
13. Install input shaft bearing snap ring in case bore.
14. Position input shaft, then install gasket and bearing retainer, using sealer on gasket.
15. Install clutch housing and speedometer drive gear.
16. After installing new seal in rear bearing retainer, install gasket and retainer, using sealer on gasket.
17. Install parking brake shoe assembly and connect brake linkage.
18. With transmission locked in two gears, install brake drum and flange.
19. Adjust parking brake and install gearshift housing, using a new gasket.

Fuller Separate Auxiliary Transmissions

3-65 SERIES

Disassemble Transmission

1. Referring to Fig. 1, remove shifter mechanism.
2. Remove rear cover, speedometer drive gear and washer from output shaft.
3. From inside of case, drive output shaft to the rear, forcing rear bearing out of its seat in case bore.
4. Use puller to remove rear bearing.
5. Remove low-speed clutch gear through output shaft rear bearing bore.
6. Lift output shaft through top of case.
7. Drive countershaft rearward until rear bearing cap has been forced from case.
8. Move countershaft rearward as far as it will go.
9. Remove retaining nuts from input shaft from bearing cover.
10. Drive input shaft backward into case and lift assembly out through top of case.
11. Remove input shaft front bearing cover.
12. Remove front bearing spacer.
13. Place bearing cover in a vise.
14. Drive bearing forward against oil seal, forcing both out through front of case.
15. Disassemble output shaft and counter-

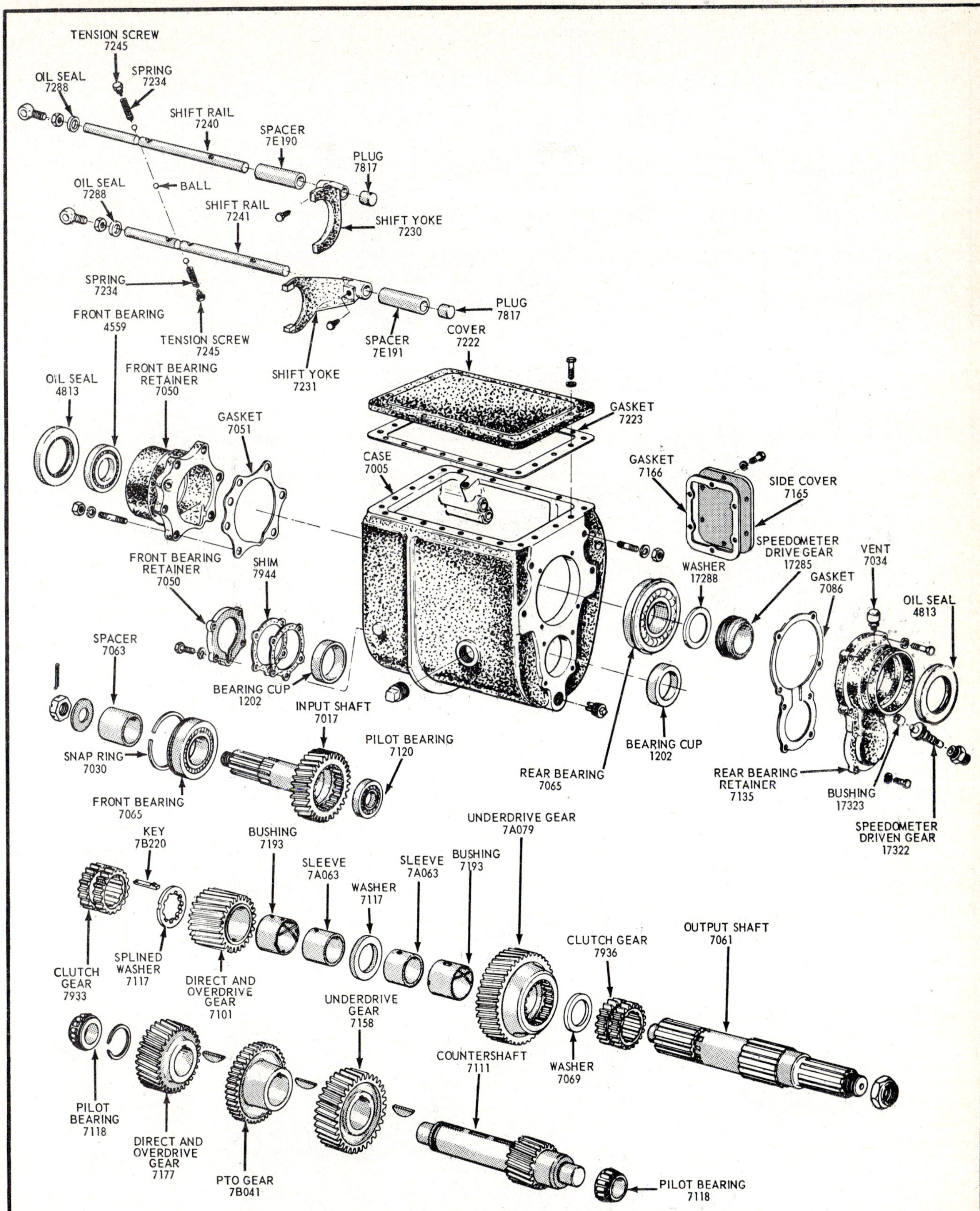

Fig. 1 Exploded view of 3-65 Series three-speed auxiliary transmission

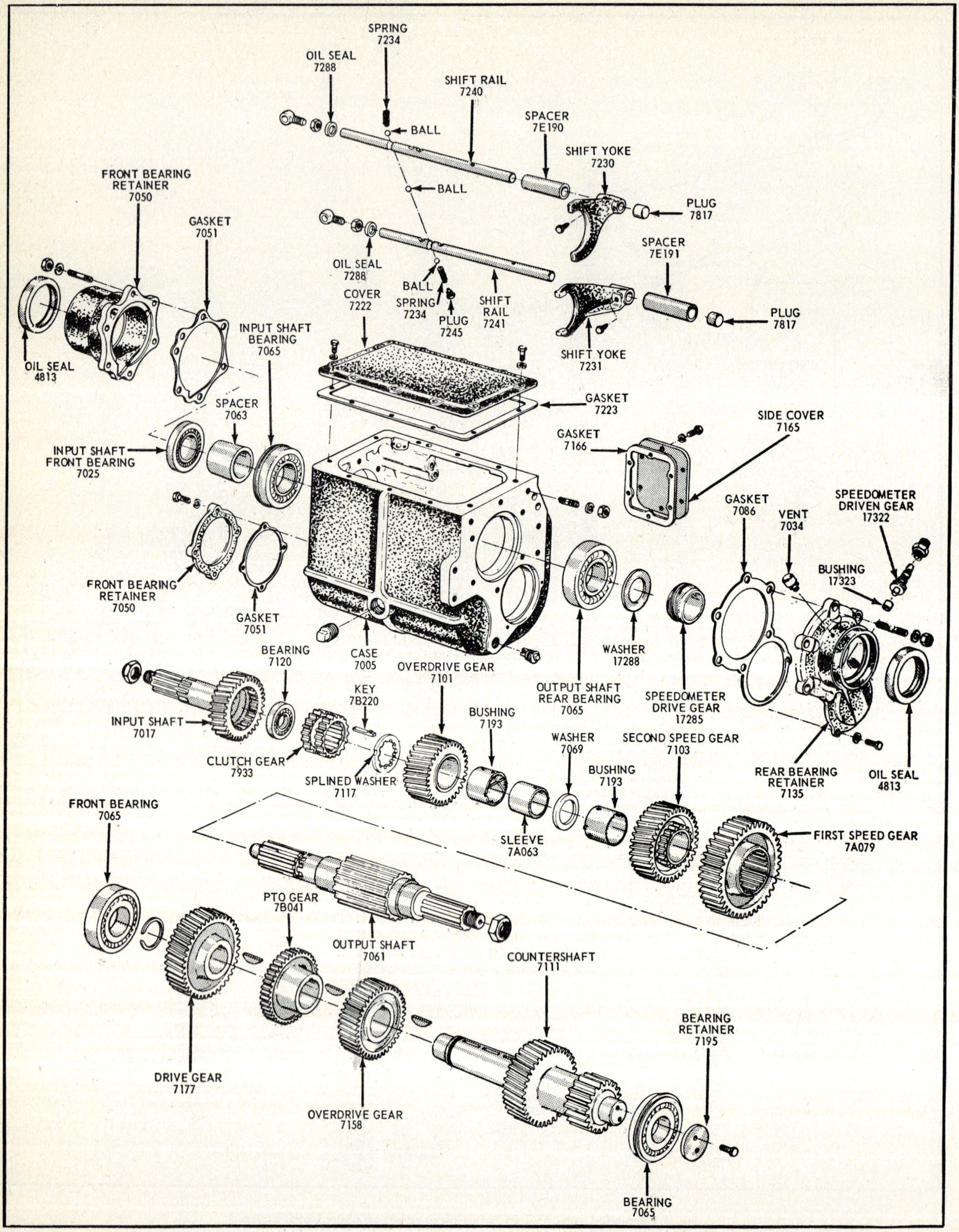

Fig. 2 Exploded view of 4-C-75 four speed auxiliary transmission

shaft as required.

Countershaft Assemble, Fig. 1

1. Insert Woodruff keys in shaft.
2. Press overdrive, PTO and direct drive gears on shaft one at a time. Shaft must center gear bore on chamfered side of gear.
3. Install snap ring on front end of countershaft.
4. Install pilot bearings on each end of countershaft.

Output Shaft Assemble, Fig. 1

1. Place output shaft in vise with front end up.
2. Install underdrive gear rear washer with waffle side facing up.
3. Install underdrive gear bushing.
4. Install underdrive gear on shaft with internal clutching teeth facing down.
5. Position underdrive gear rear thrust washer on shaft.
6. Install overdrive gear.
7. Install overdrive gear splined washer with thrust surface toward gear. Turn washer in its groove until lugs on inside diameter line up with splines on output shaft.
8. Install washer key in keyway between splines of output shaft.
9. Install direct and overdrive clutch gear with counterbore to rear.
10. Install output shaft pilot bearing.

Assemble Transmission

1. Install countershaft in case with rear bearing protruding through rear bore.
2. Remove snap ring from input shaft main bearing.
3. Press input shaft main bearing on shaft, leaving approximately 1/2" space between bearing and drive gear.
4. Install input shaft from inside of case, driving shaft forward until main bearing snap ring groove is clear of case.
5. Install snap ring on main bearing.
6. Block rear of input shaft and, using a sleeve-type bearing installer, drive bearing backward until it seats against drive gear shoulder.
7. Install countershaft front bearing cup in case bore.
8. Move countershaft forward until front bearing seats in bearing cup.
9. Install countershaft rear bearing cup in case bore.
10. Install countershaft rear bearing cover and tighten capscrews.
11. Install countershaft front bearing cover and tighten securely. Then remove bearing cover and measure distance bearing cup protrudes from case. This distance should be .003–.005". Add or subtract shims as required.
12. Remove rear bearing cover.
13. Lower output shaft assembly into case, engaging helical gears of output shaft with those of countershaft.
14. Insert pilot bearing into drive gear bore.
15. Slide underdrive clutch gear onto shaft with chamfered end of clutch facing underdrive gear.
16. Drive output shaft rear bearing into place in case bore.
17. Install speedometer gear washer with chamfered side in, then install speedometer gear.
18. Install output shaft rear bearing cover. Install oil seal in cover.
19. Install input shaft front bearing cover but do not tighten retaining nuts at this time.
20. Install input shaft spacer and outer bearing in cover bore.
21. Tighten input shaft bearing cover nuts.
22. Install oil seal in front bearing cover.
23. Insert direct and overdrive shift bar into lower set of shift bar holes in case. This bar must pass direct and overdrive yoke and spacer. Spacer must be installed to rear of yoke.
24. Install yoke lock screw and wire securely.
25. Insert one 1/2" steel ball in poppet spring hole located between the two shift bars.
26. Be sure direct-overdrive shift bar is in neutral. Insert underdrive shift bar through spacer and shift yoke. Spacer should locate in front of shift yoke. Install shift bar in upper set of holes in case.
27. Install yoke lock screw and wire securely.
28. Shift the bars into neutral. Install poppet balls, springs and plugs in both upper and lower poppet holes.
29. Install top cover and tighten screws securely.

4-C-75 FOUR SPEED

Disassemble Transmission

1. Referring to Fig. 2, remove shifter mechanism.
2. Lock mainshaft in two gears at once. Then remove nuts and take off universal joint companion flanges from input and output shafts.
3. Place transmission gears in neutral.
4. Remove capscrews from rear cover.
5. Move sliding 1st and 2nd speed gears forward on output shaft into engagement with 2nd speed gear on countershaft.
6. Drive output shaft rearward until rear bearing cover is forced approximately 1/4" from case.
7. Place a wood block between rear of case and 1st speed gear.
8. Pry rear bearing cover from case.
9. Remove oil seal from cover.
10. Remove speedometer gear and washer from rear of output shaft.
11. Remove wood block and move output shaft to rear, forcing rear bearing from its seat in case bore.
12. Use a puller to remove output shaft rear bearing.
13. Lift output shaft out through top of case, then remove sliding gear from case.
14. Remove countershaft rear bearing plate.
15. Remove countershaft front bearing cover.
16. From outside of case, drive countershaft rearward until front and rear bearings are forced out of their seats in case bores and 1st speed gear is completely exposed outside the case.
17. Force input shaft rearward until shaft is unseated from bearings.
18. Lift input shaft out of case.
19. Remove front bearing cover.
20. Remove input shaft gear bearing spacer from inside of bearing cover.
21. Remove input shaft front bearing and oil seal from inside of bearing cover.
22. Force input shaft main bearing out through front of case bore.
23. Remove countershaft rear bearing with a puller.
24. Lift countershaft out top of case.
25. Disassemble output shaft and countershaft as required.

Countershaft Assemble, Fig. 2

1. Install Woodruff keys in countershaft keyways.
2. Press overdrive gear on shaft with long shoulder toward rear.
3. Press on PTO gear with long shoulder to the rear.
4. Press drive gear on shaft with long shoulder to the rear.
5. Install snap ring in its groove at front end of shaft.
6. Install front bearing on shaft. Bearing must be flush with end of shaft with stamped face forward.

Output Shaft Assemble, Fig. 2

1. Mount output shaft in a vise with front end up.
2. Install 2nd speed gear on shaft with clutching teeth down.
3. Install overdrive gear rear washer with waffle side down.
4. Install overdrive gear bushing on shaft, fitting bushing tight against washer.
5. Install overdrive gear on bushing, clutching teeth up.
6. Install splined washer in its groove, flat side down.
7. Turn splined washer in its groove until lugs on inside diameter line up with splines of output shaft.
8. Install washer key in its keyway between lugs of splined washer.
9. Install sliding clutch gear on shaft with counterbore toward rear.
10. Install pilot bearing on front of shaft, chamfered inner diameter down.
11. Remove assembly from vise and install sliding gear over rear of shaft with yoke slot facing forward.

Transmission, Assemble

1. Place countershaft assembly in bottom of case, inserting rear of shaft through rear bore of case until 1st gear is completely exposed.
2. Remove snap ring from input shaft main bearing.
3. Press input shaft main bearing on shaft, leaving approximately 1" space between bearing and gear.
4. Install input shaft from inside of case. Be sure front bearing is not cocked in case bore.
5. Move input shaft and bearing forward until bearing snap ring groove is exposed, then install snap ring.
6. Using a sleeve-type bearing installer, force main bearing against shoulder of gear.
7. Install input shaft front bearing cover. Install lockwashers and nuts but do not tighten at this time. Install spacer in bearing cover.
8. Install front bearing on input shaft with shielded side out.
9. Tighten bearing cover nuts securely. Install oil seal in cover.
10. Work countershaft forward and insert front bearing into case bore. At the same time, mesh countershaft drive gear with input shaft drive gear.
11. Install rear bearing in countershaft and into case bore. Bearing must seat against shoulder of shaft.
12. Install rear bearing retainer plate. Tighten capscrews evenly and install lockwire.
13. Install front bearing cover.
14. Move output shaft sliding gear into engagement with 2nd speed gear.

15. Lower output shaft into case, engaging output shaft gears with mating gears on countershaft. At the same time, fit output shaft pilot bearing into drive gear bore.
16. Block front face of 2nd speed gear to prevent output shaft movement. Install output shaft rear bearing in case bore. Bearing should seat against shoulder of shaft with stamped numbers forward.
17. Remove block and install speedometer gear washer and drive gear on rear of shaft.
18. Install rear bearing cover.
19. Install oil seal in rear cover.
20. Install vent in rear bearing cover.
21. Lock output shaft by engaging two gears. Install companion flange on shaft, being sure flange is tight against speedometer gear.
22. Install companion flange to input shaft.
23. Move gears to neutral position.
24. Install direct-overdrive shift bar through shift yoke and long spacer and into lower set of shift bar bores. At the same time, insert shift yoke into yoke slot on clutch gear. Install lock screw and wire securely.
25. Place shift bar in neutral and install steel interlock ball in angled boss at top of case.
26. Install 1st, 2nd shift bar into upper shift bar bore, through short spacer and shift yoke. Keep lower shift bar in neutral to prevent interlock ball from jamming upper shaft bar during installation. Install lock screw and wire securely.
27. With both shift bars in neutral, install lower tension ball, spring and plug on side of case.
28. Install top tension ball and spring in top face of case. Install top cover.

NEW PROCESS TRANSMISSIONS

INDEX & LUBE CAPACITY

Model	Speeds	Oil, Pts.	Page
245, 275	5	12	290
420	4	5½	290
433, 434, 435	4	7	292
445	4	7	298
540, 541, 542	5	10	296
745	3	3¾	300
875, 895	5	8	303
4590	4	8	304
5401, 2, 3, 4	5	10	296
7550, 7590	5	22	306

SERIES 245, 275

Disassemble, Figs. 1, 2

1. Remove nut from output shaft.
2. Remove parking brake, universal joint spline flange and parking brake drum.
3. Remove gearshift housing.
4. Remove power take-off covers and gaskets.
5. Remove speedometer driven gear and bearing from output shaft bearing retainer.
6. Remove output shaft bearing retainer and slide speedometer drive gear and spacer from output shaft.
7. Remove input shaft bearing retainer and pull shaft and bearing out through front of case.
8. Slide synchronizer assembly off output shaft and lift shaft out of case.
9. Pull reverse idler gear shaft out of case and lift out gear.
10. Remove countershaft front bearing retainer, unfasten retainer washer and remove washer.
11. Remove bolts from countershaft rear bearing cap. Tap countershaft toward rear until bearing cap and bearing is forced out of case.
12. Lift out countershaft front bearing.

Disassemble Sub-Assemblies

Input Shaft—Remove lock nut from end of shaft and remove bearing. Remove snap ring from shaft bore and remove output shaft pilot bearings. Remove snap ring from output shaft bearing.

Output Shaft—Slide low and reverse gear and 2nd speed gear from shaft. Remove snap ring and remove 4th and 5th gear synchronizer hub. Slide 4th speed gear (5th gear on overdrive units) and thrust washer from shaft. Remove bushing from gear and pin from bushing.

Slide off 3rd speed gear and thrust washer and remove bearing rollers from gear. Separate synchronizer parts and inspect for wear or damage. Reassemble with new parts as necessary.

Countershaft—Remove thrust washer from front end of countershaft. Press gear cluster off shaft and drive key from shaft slot.

Reverse Idler Gear—Remove thrust washer from bore of gear and remove bearing rollers and spacer.

Assemble Sub-Assemblies

Reverse Idler Gear—Install spacer in gear. Coat bearing rollers with grease and install 31 rollers in each end of gear. Install washer at each end of gear next to rollers.

Countershaft—Install spacer on shaft and key in slot. Press gear cluster on shaft until gear is seated against spacer. Position thrust washer on forward end of shaft.

Output Shaft—Coat 3rd speed gear rollers with lubricant and install rollers in bore of gear. Place thrust washer on shaft and slide gear on shaft with spur gear teeth toward rear.

Install lock pin in 4th speed gear bushing (5th speed gear on overdrive units). Place thrust washer on shaft and install bushing. Install gear on bushing with shorter teeth toward front.

Position clutch gear on shaft and install snap ring in shaft groove. Check 3rd speed gear end play. If not within .006-.008", replace snap ring with one of the proper thickness to give correct end play.

Position 2nd speed gear on shaft with shift fork groove toward rear. Install low and reverse gear on shaft with shift fork groove toward front of shaft. Assemble synchronizer as shown in Fig. 1.

Input Shaft—Install bearing on shaft. Install lock nut and tighten until bearing is firmly seated against gear shoulder. Install snap ring in shaft bearing groove.

Coat pilot bearing rollers with grease, and install in bore of input shaft. Secure bearings with snap ring.

Installation in Case

1. Install snap ring on countershaft front bearing and tap bearing into case.
2. Position countershaft in case and insert forward end in front bearing.
3. Tap countershaft rear bearing and cap into case and onto countershaft. Install bolts and lockwashers.
4. Install countershaft front bearing retaining washer and retainer.
5. Position reverse idler gear in case and tap idler shaft into case until notch in shaft is flush with rear face of case. Install shaft retainer.
6. Place output shaft in case and slip on synchronizer.
7. Install input shaft through front of case. Tap assembly into case until bearing snap ring is seated in recess in case. Install and tighten bearing retainer.
8. Position output shaft into pilot bearing, then install shims on rear of output shaft. Install snap ring on output shaft bearing and tap bearing into case bore until snap ring is seated against rear face of case. Install spacer and speedometer drive gear on output shaft.
9. Install output shaft bearing retainer and tighten securely.
10. Check end play of synchronizer. If not within .040-060", change shims at rear of output shaft until correct end play is obtained.
11. Install speedometer driven gear and bearing, and power take-off covers.
12. Place gears in neutral and install gearshift housing, tightening bolts to 30-40 ft. lbs. torque.
13. Slide parking brake drum onto output shaft and place band and lever on drum. Install bolts to case and torque them to 30-40 ft. lbs.
14. Install parking brake band anchor spring and adjusting screw, then adjust brake.

SERIES 420

Disassemble, Fig. 3

1. Remove nut from companion yoke.
2. Remove hand brake linkage and band.
3. Pull off brake drum and companion yoke.
4. Remove transmission cover.
5. Remove rear bearing retainer and seal assembly, and speedometer drive gear.
6. Remove main drive gear bearing retainer, gear and bearing.
7. Remove snap ring and pull rear bearing from case. Remove shim pack and tie shims together for reassembly.
8. Remove synchronizer snap ring and slide synchronizer gear forward. Then slide

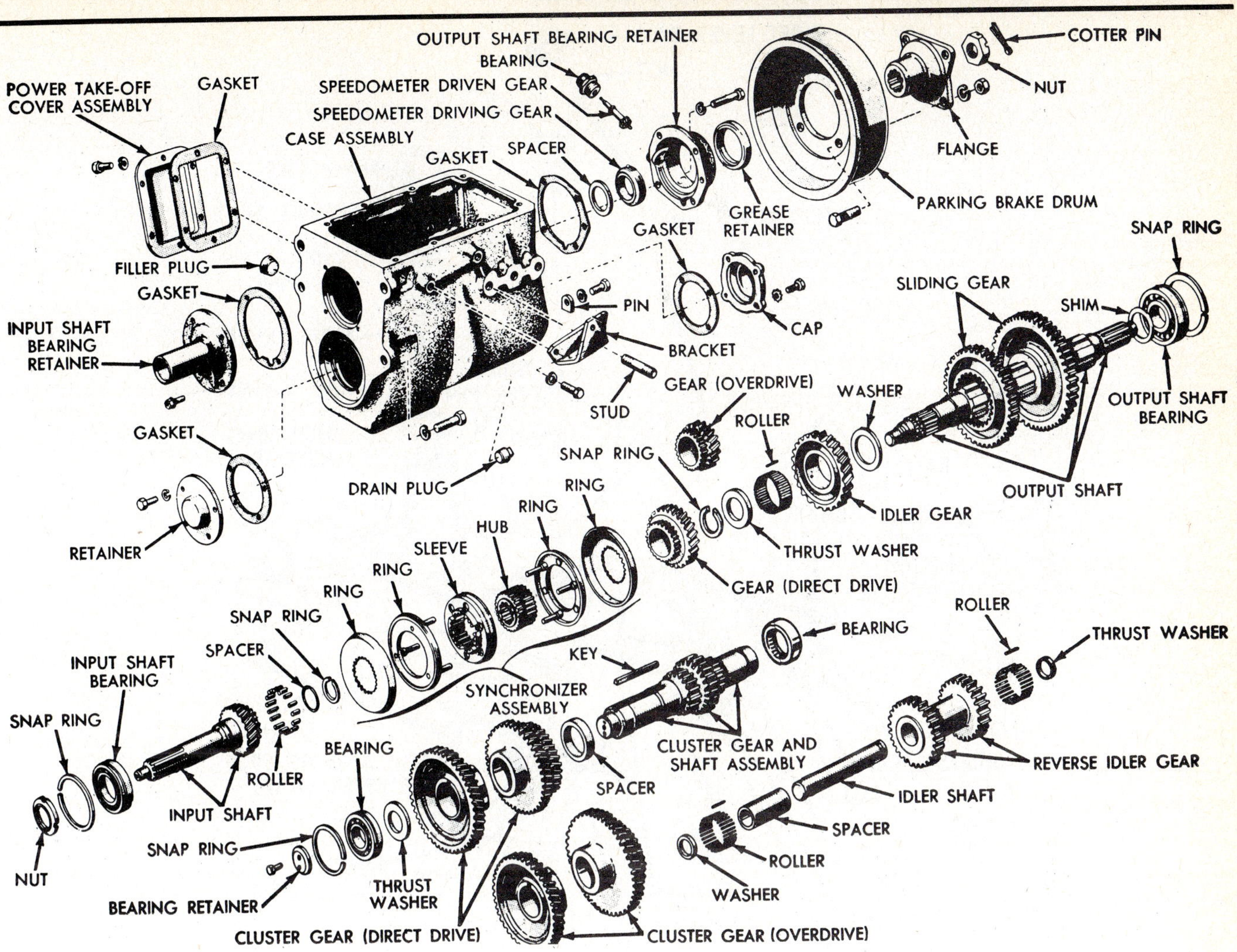

Fig. 1 New Process five speed transmission. Series 245, 275

synchronizer from mainshaft.

9. Lift mainshaft assembly out of case through sliding gears. Then remove sliding gears from case.
10. Remove reverse idler shaft lock plate, pull out shaft and lift gear from case.
11. Remove rear countershaft bearing retainer and bearing. Then remove front countershaft retainer, cap screws and lock plate, and remove bearing.
12. Lift countershaft assembly from case.

Reassembly

Reverse Idler Gear—Install spacer in gear. Coat roller bearings with grease and install a complete set of 31 rollers in each end of the gear. Install a thrust washer in each end of the gear.

Countershaft—Install spacer and drive key in shaft keyway. Remove all burrs from key and shaft and press gear cluster onto shaft until gears bottom on spacer. Then install thrust washer on forward end of shaft.

Mainshaft—Coat third speed gear roller bearings with grease and place in third speed gear. Install thrust washer on mainshaft. Position third speed gear with spur teeth toward rear of shaft and slide gear on shaft.

Install lock pin in fourth speed gear (fifth on overdrive units). Place thrust washer on shaft and install bushing on shaft. Position fourth (or fifth) speed gear with shortest teeth toward front end of shaft. Install gear on bushing.

Install mainshaft synchronizer gear on mainshaft and secure with snap ring in shaft groove. Check snap ring for proper thickness by testing third speed gear end play. Snap rings are available in .002" steps each to obtain desired end play of .006 to .008".

Position second speed gear with shift fork groove toward rear of shaft and slide gear on shaft.

Main Drive Gear—Install bearing on shaft. Install lock nut and tighten securely to seat bearing against gear shoulder. Install snap ring in bearing groove. Lubricate mainshaft pilot bearing and place it in main drive gear bore.

Installation in Case

Countershaft—Install snap ring on countershaft front bearing and tap bearing into case. Lower countershaft in case and insert shaft into front bearing. Install new gasket on rear bearing retainer and tap rear bearing and retainer assembly into case and on countershaft. Install retainer cap screws and lock washers and tighten securely. Install countershaft front bearing lock plate and secure screws with lock wire. Install front bearing retainer, using a new gasket, and tighten securely.

Reverse Idler Gear—Place gear in position and tap shaft through case and gear until notch in shaft is flush with rear face of transmission case. Install shaft lock plate.

Mainshaft & Main Drive Gear—Place low and reverse gear in case. Lower mainshaft into case, passing it through low and reverse gear. Then insert shaft through rear of case far enough to install synchronizer parts one at a time on front end of shaft.

Insert main drive gear in case, tapping it with a soft hammer until the bearing snap ring is seated against case. Install main drive gear bearing retainer, using a new gasket, and tighten securely.

Place synchronizer outer stop ring on main drive gear and position front end of mainshaft into pilot bearing. Then install shim pack (as removed) on rear end of mainshaft.

Install snap ring on mainshaft rear bearing. Tap bearing into case until retainer seats against case. Install speedometer gear on mainshaft. Install mainshaft rear bearing retainer, using new gaskets, and tighten securely.

Install companion yoke and nut, tightening nut from 95–135 ft. lbs. torque. Check gears in neutral and in all speeds for free rotation.

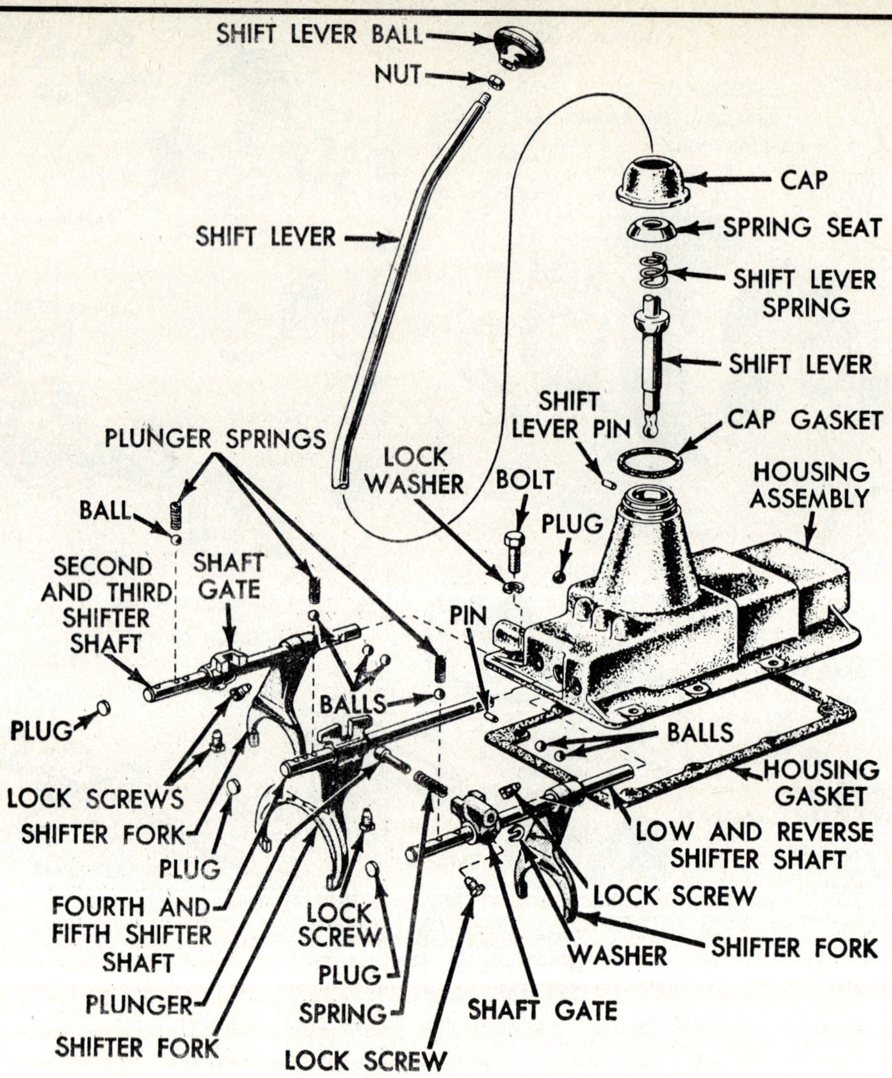

Fig. 2 New Process 245 and 275 gearshift housing

Check synchronizer end play by placing feeler gauges diametrically opposite each other. End play should be .040 to .060"; if not within these limits, adjust by removing or adding shims at the rear of the case to obtain the desired result.

With gears in neutral, install a new gasket and install the transmission cover. Tighten cover screws to 30–40 ft. lbs. torque.

Install hand brake and fill transmission with correct amount of lubricant.

SERIES 433, 434, 435

Transmission Disassemble, Fig. 3A

1. With transmission in neutral, remove gearshift housing.
2. Lock transmission in two gears and remove output flange nut, yoke and brake drum.
3. Remove speedometer drive pinion and mainshaft rear bearing retainer.
4. Remove drive pinion bearing retainer.
5. Rotate drive pinion gear to align space in pinion gear teeth with countershaft drive gear teeth and remove drive pinion gear and tapered roller bearing from transmission.
6. Remove snap ring, washer and pilot roller bearing from recess in drive pinion gear.
7. Using a brass drift placed in front center of mainshaft, drive mainshaft to rear, then when mainshaft rear bearing clears case, remove rear bearing and speedometer gear (with spacer) using a puller.
8. Move mainshaft assembly rearward and tilt front of mainshaft up.
9. Remove roller thrust bearing, synchronizer and stop rings, and mainshaft.
10. Remove reverse idler lock screw and lockplate.
11. Using a brass drift, drive idler gear rearward, then pull shaft and lift reverse idler gear from case.
12. Remove gearing retainer at rear end of countershaft.
13. Tilt cluster gear assembly and work it out of case.
14. Using a suitable driver, remove front bearings from case.

Disassemble Sub-Assemblies

Mainshaft

1. Remove clutch gear snap ring, then remove clutch gear, synchronizer outer stop ring to third speed gear shim(s) and third gear.
2. Using two screwdrivers, remove split lock ring, then remove second speed gear and synchronizer.
3. Remove first and reverse sliding gear.

Reverse Idler Gear

Do not disassemble idler gear. If satisfactory operation is doubtful, replace entire assembly complete with bearings.

Drive Pinion & Bearing Retainer

1. Using a suitable tool, remove roller bearing from pinion shaft.
2. Remove snap ring, washer and pilot rollers from gear bore.
3. Using a puller, remove bearing race from front bearing retainer.
4. Remove pinion shaft seal.

Shifter Housing

1. Remove roll pins from first and second speed shift fork and gate.
2. Push shift rail out through front to force expansion plug out of cover and remove rail, fork and gate.

NOTE: Cover detent ball access hole in cover to prevent ball and spring from flying out as rail clears hole.

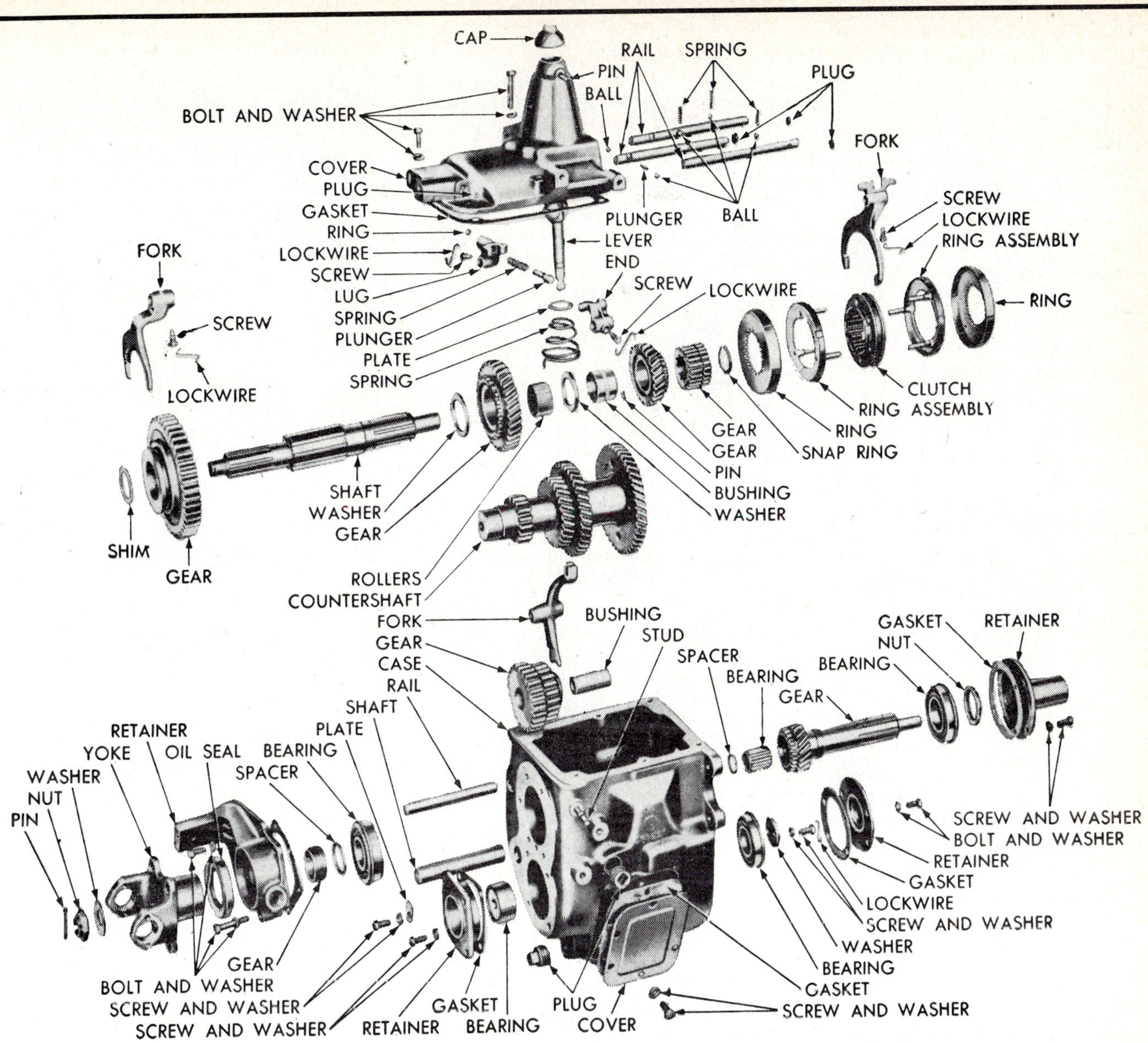

Fig. 3 New Process 420 Series four speed transmissions

3. Remove third and fourth speed rails, then remove reverse rail.
4. Compress reverse gear plunger, then remove retaining clip, plunger and spring from gate.

Assemble Sub-Assemblies

When assembling sub-assemblies, use new expansion plugs, gaskets and seals. Lubricate all parts with clean transmission lubricant. Where grease is recommended, use oil soluble grease. Make certain that oil return passages are not obstructed with grease.

Mainshaft

1. Place mainshaft in soft jawed vise with rear end up and install first and reverse gear making sure that two spline springs (if used) are inside of gear.
2. Reverse mainshaft in vise with forward end up.
3. Assemble second speed synchronizer spring and synchronizer brake to second speed gear, then install snap ring making sure that snap ring tangs are facing away from gear.
4. Slide second speed gear onto front of output shaft making sure that synchronizer brake is toward rear, then install the two piece lock ring and install third speed gear.
5. Install shim(s) between third speed gear and third and fourth speed synchronizer stop ring. Check end play. If end play is not within .050–.070 inch, add or remove shims as necessary to bring end play to specifications.

Drive Pinion & Bearing Retainer

1. Using an arbor press and a wooden block placed on pinion gear, press gear into bearing until it contacts bearing inner race.
2. Coat roller bearings with a light coat of grease to hold rollers in place and insert into pocket of drive pinion.
3. Install washer and snap ring.
4. Press seal into bearing retainer making sure that lip of seal is toward mounting plate.
5. Press bearing race into retainer.

Shifter Housing, Fig. 3B

1. Install spring on reverse gear plunger, then compress spring into reverse shift gate and install retaining clip.
2. Install reverse shift rail into cover and place detent spring and ball in position, then depress ball and slide rail over it.
3. Install gate and fork on rail, then install a new roll pin in gate and fork.
4. With reverse fork in neutral position, install the two interlock plungers in their bores.
5. Install interlock pin in third and fourth speed shift rail, then install rail in same manner as reverse shift rail.
6. Install first and second speed shift rail in same manner as reverse shift rail, making certain that interlock plunger is in

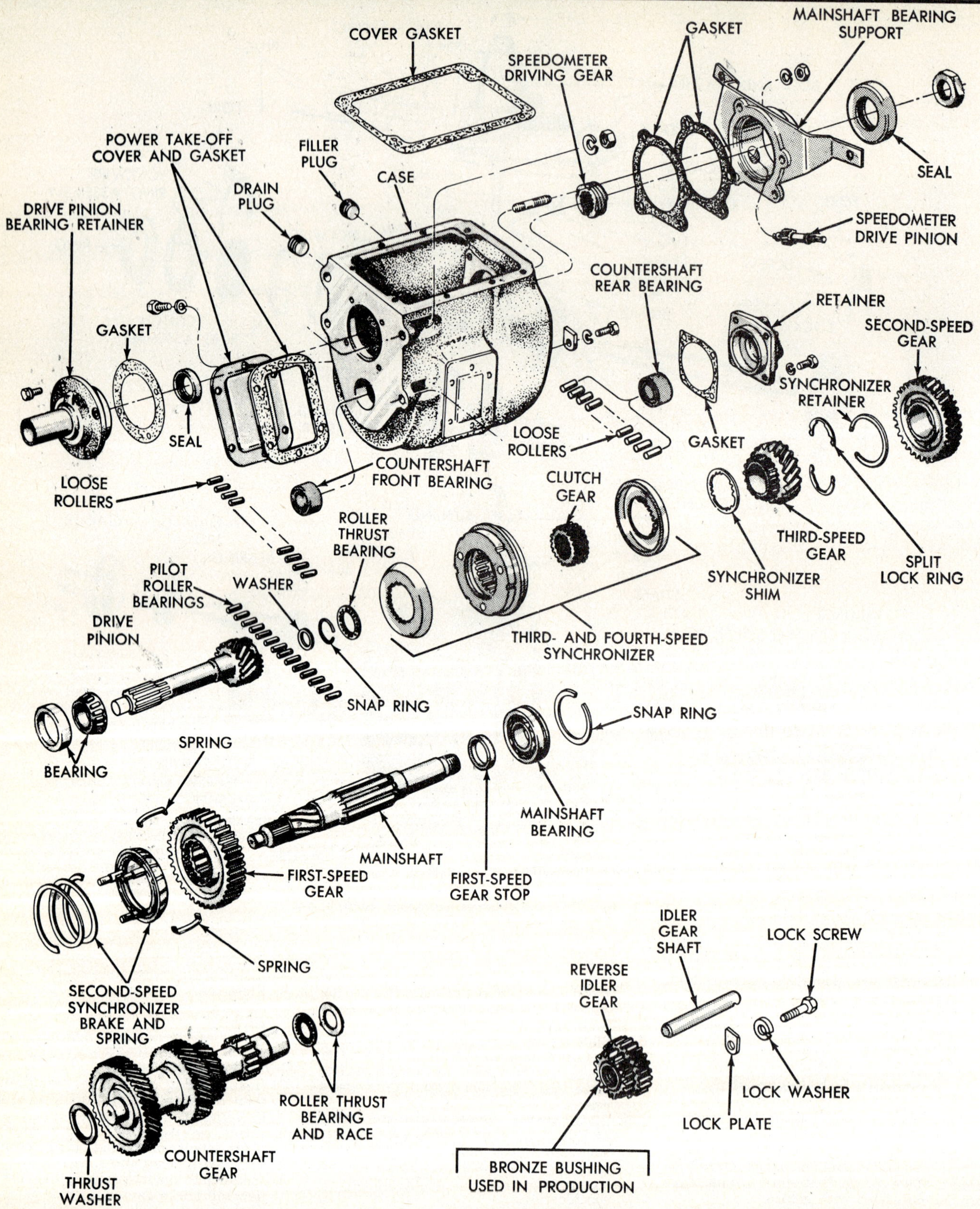

Fig. 3A New Process four speed transmission. Series 433, 434, 435

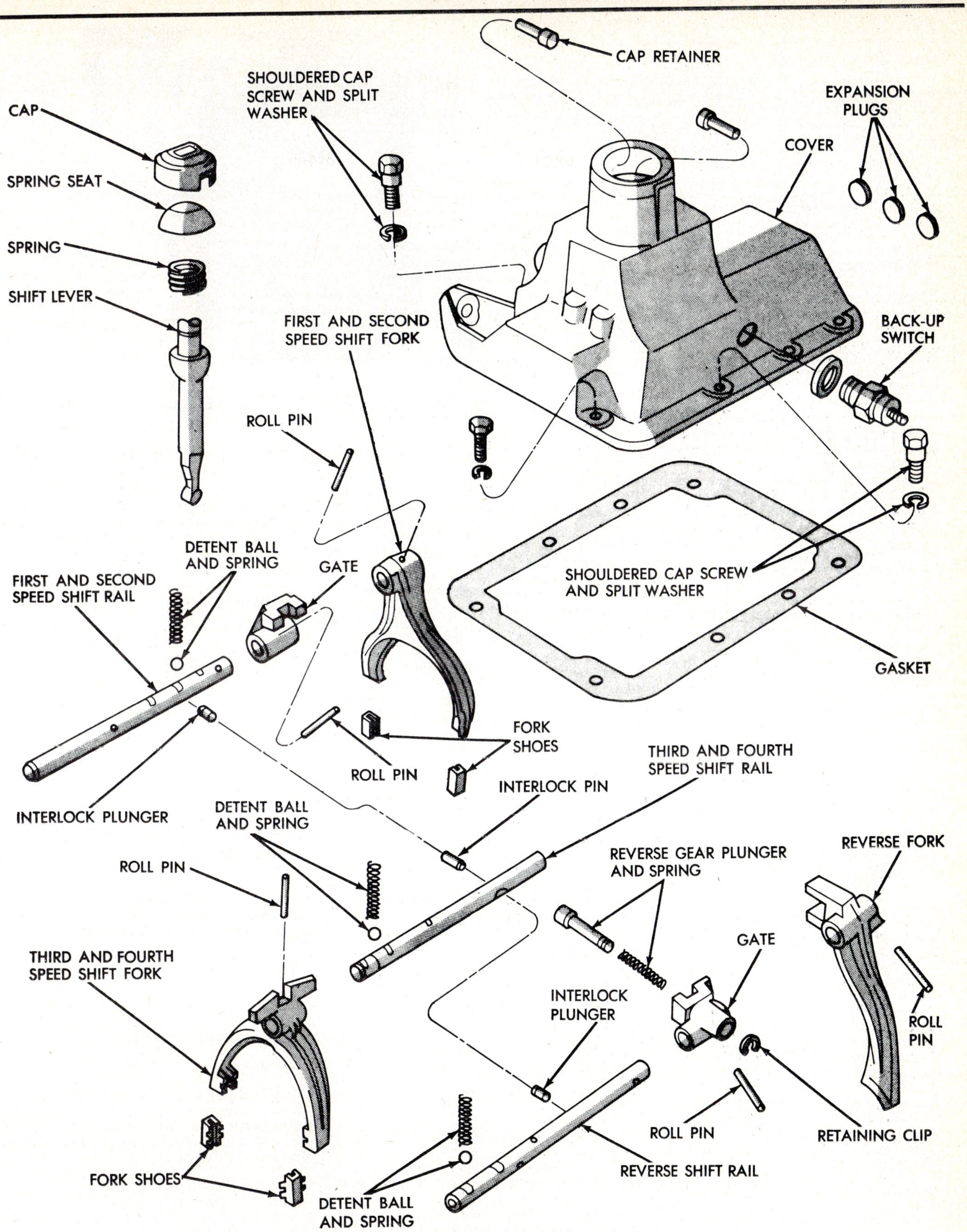

Fig. 3B Shifter housing disassembled. Series 433, 434, 435

place.

NOTE: Check interlocks by shifting reverse shift rail into reverse position. It should be impossible to shift all other rails when this rail is shifted.

7. If installing shift lever, lubricate spherical ball seat and place cap in position.
8. Install back-up light switch, if used.
9. Install new expansion plugs in bores of rail holes and install interlock hole plug.

Transmission Assemble

1. Press front roller bearing into case until cage is flush with front of case and lubricate bearings with a light coat of grease.
2. Position transmission with front of case facing down. If uncaged roller bearings are being re-used, hold rollers in place with a light coat of grease.
3. Lower countershaft assembly into case, placing thrust washer tangs in slots in case and inserting front end of shaft into bearing.
4. Place countershaft rear roller thrust bearing on countershaft and hold in place with a light coat of grease.
5. Align countershaft assembly and install retainer gasket, retainer and bearing, then install and tighten screws.
6. Place reverse idler gear and bearing assembly in case.
7. Align reverse idler shaft so lock plate groove in shaft is properly aligned to install lock plate.
8. Tap reverse idler shaft far enough into case to start reverse gear, then while holding gear in position, tap shaft through case and gear.
9. Install lock plate, washer and screw. Make certain that gear turns freely on shaft.
10. Lower rear end of mainshaft into case, then while holding first speed gear on shaft, maneuver shaft through rear bearing opening.

NOTE: With shaft assembly moved to rear of case, make sure that third and fourth speed synchronizer and preselected shims remain in place.

11. Install roller thrust bearing.
12. Place a wood block between front of case and front of mainshaft.
13. Using a suitable size sleeve against inner race of bearing, install rear bearing onto shaft and into case until snap ring is flush against case.
14. Install drive pinion shaft and bearing assembly making sure pilot rollers remain in place.
15. Install spacer, speedometer gear, gasket and bearing retainer.
16. Place drive pinion bearing retainer over pinion shaft without gasket, then while holding retainer tight against bearing, measure clearance between retainer and case using a feeler gauge. Install a gasket shim pack .010–.015 inch greater than clearance measured between retainer and case to obtain the required .007–.017 inch pinion shaft end play. After retainer bolts are tightened, recheck end play.
17. Check for .050–.070 inch synchronizer end play after all mainshaft components are in place and properly tightened, using two sets of equal size feeler gauges Fig. 3C. If necessary, disassemble mainshaft and change shims to retain end play within specifications.

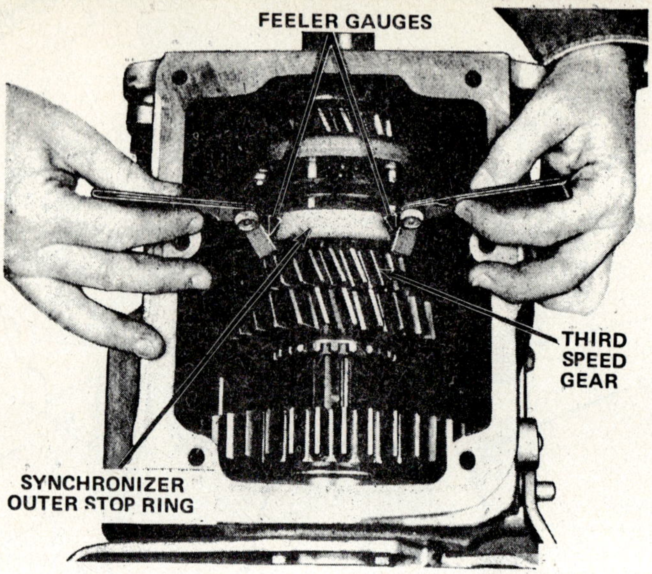

Fig. 3C Checking synchronizer end play. Series 433, 434, 435

18. Install speedometer drive pinion, yoke flange, drum and brake assembly and output flange nut.
19. Shift gears into all speed positions and check for freedom of operation.

NOTE: Before installing shifter housing, fill transmission with lubricant, then oil all exposed transmission parts to avoid damage to transmission at start up.

20. Move gears into neutral position, then lower cover and new gasket over transmission, engaging forks into proper gears.
21. Install one shouldered alignment screw in each of the holes on each side of cover located next to the front hole. Check gears for freedom by shifting gears with a long screwdriver inserted in cover tower.
22. Install remaining cover screws.

SERIES 540, 541, 542, 5400

Disassemble, Fig. 4

1. Remove nut from output shaft.
2. Remove parking brake, universal joint spline flange and parking brake drum.
3. Remove gearshift housing.
4. Remove power take-off covers and gaskets.
5. Remove speedometer driven gear and bearing from output shaft bearing retainer.
6. Remove output shaft bearing retainer and slide speedometer drive gear and spacer from output shaft.
7. Remove input shaft bearing retainer and pull shaft and bearing out through front of case.
8. Remove output shaft bearing, using appropriate puller tool. Slide synchronizer assembly off output shaft and lift shaft out of case.
9. Pull reverse idler gear shaft out of case and lift out gear.
10. Remove countershaft front bearing retainer, unfasten retainer washer and remove washer.
11. Remove bolts from countershaft rear bearing cap. Tap countershaft toward rear of case until bearing cap and bearing is forced out of case.

Disassemble Sub-Assemblies

Clutch Shaft—Remove lock nut from end of shaft and remove bearing. Remove snap ring from shaft bore and remove output shaft pilot bearings. Remove snap ring from output shaft bearing.

Mainshaft—Slide low and reverse gear and 2nd speed gear from shaft. Remove snap ring and remove 4th and 5th gear synchronizer hub. Slide 5th speed gear and thrust washer from shaft. Remove bushing from gear and lock pin from bushing.

Slide off 3rd speed gear and thrust washer and remove bearing rollers from gear. Separate synchronizer parts and inspect for wear or damage. Reassemble with new parts as necessary.

Countershaft—Remove thrust washer from front end of countershaft. Press gear cluster off shaft and drive key from shaft slot.

Reverse Idler Gear—Some models of these transmissions are equipped with a bushing type reverse idler gear, whereas others have roller bearing type gear. Remove thrust washers from bore of gear and remove bearing rollers and spacer (if equipped).

Assemble Sub-Assemblies

Reverse Idler Gear—If roller bearings are used, install spacer in gear. Coat bearing rollers with grease and install 31 rollers in each end of gear. Install washer at each end of gear next to rollers.

Countershaft—Install spacer on shaft and key in slot. Press gear cluster on shaft until gear is seated against spacer. Position thrust washer on forward end of shaft.

Mainshaft—Coat 3rd speed gear rollers with lubricant and install rollers in bore of gear. Place thrust washer on shaft and slide gear on shaft with spur gear teeth toward rear.

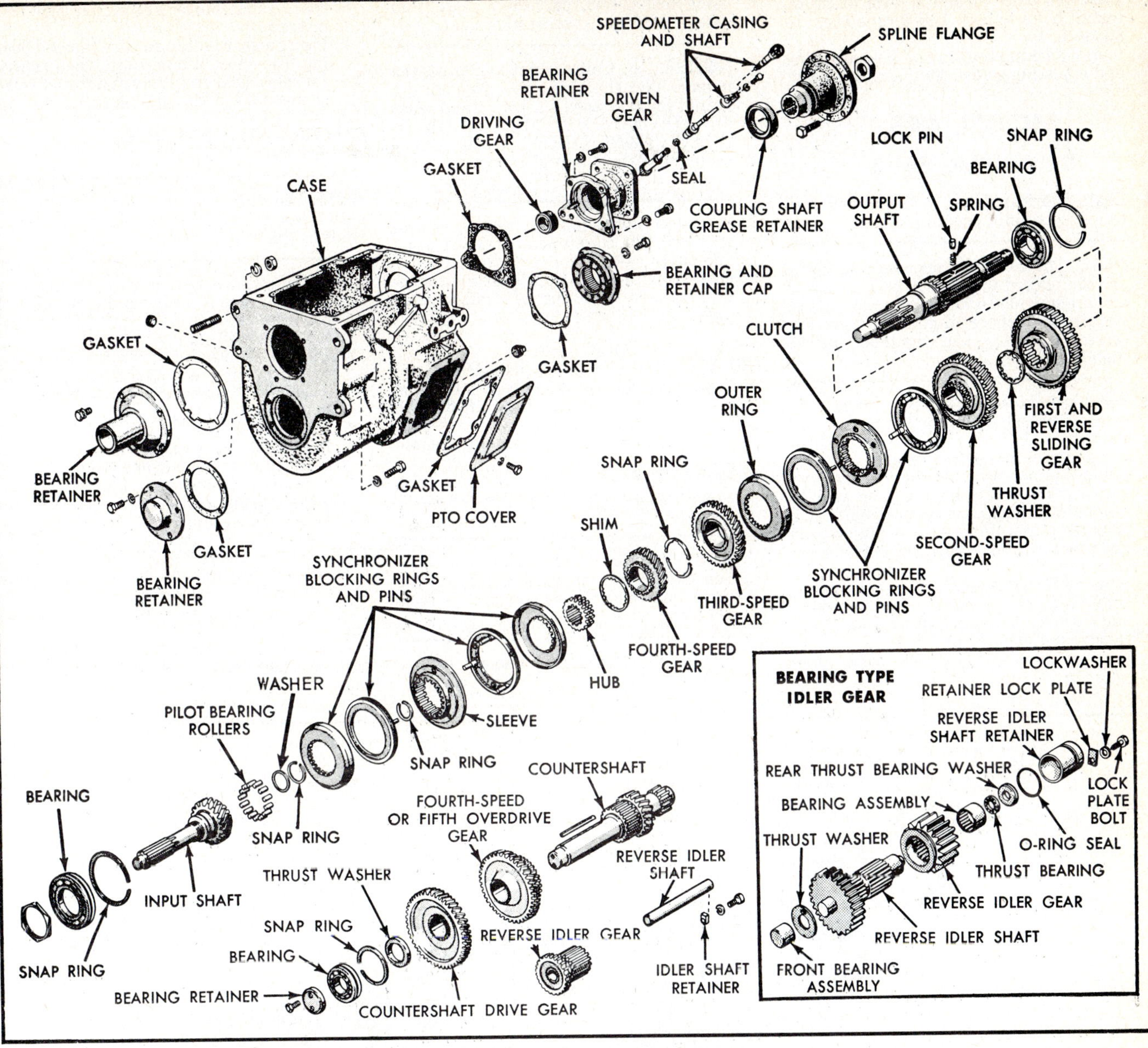

Fig. 4 New Process 540 and 5400 Series five speed transmission

Install lock pin in 5th speed gear bushing. Place thrust washer on shaft and install bushing. Install gear on bushing with shorter teeth toward front.

Position clutch gear on shaft and install snap ring in shaft groove. Check 3rd speed gear end play. If not within .006–.008″, replace snap ring with one of the proper thickness to give correct end play.

Position 2nd speed gear on shaft with shift fork groove toward rear. Install low and reverse gear on shaft with shift fork groove toward front of shaft. Assemble synchronizer as shown in Fig. 4.

Input Shaft—Install bearing on shaft. Install lock nut and tighten until bearing is firmly seated against gear shoulder. Install snap ring in shaft bearing groove.

Coat pilot bearing rollers with grease and install in bore of input shaft. Secure bearings with snap ring.

Installation in Case

1. Install snap ring on countershaft front bearing and tap bearing into case.
2. Position countershaft in case and insert forward end in front bearing.
3. Tap countershaft rear bearing and cap into case and onto countershaft. Install bolts and lockwashers.
4. Install countershaft front bearing retaining washer and retainer.
5. Position reverse idler gear in case and tap idler shaft into case until notch in shaft is flush with rear face of case. Install shaft retainer.
6. Place output shaft in case and slip on synchronizer.
7. Install input shaft through front of case. Tap assembly into case until bearing snap ring is seated in recess in case. Install and tighten bearing retainer.
8. Position output shaft into pilot bearing, then install shims on rear of output shaft. Install snap ring on output shaft bearing and tap bearing into case bore until snap ring is seated against rear face of case. Install spacer and speedometer drive gear on output shaft.
9. Install output shaft bearing retainer and tighten securely.
10. Check end play of synchronizer. If not within .040–.060″, change shims at rear of output shaft until correct end play is obtained.

NOTE: The clearance on new synchronizer assemblies should be .070–.095 inch.

11. Install speedometer driven gear and bearing. Install power take-off covers.
12. Place gears in neutral and install gearshaft housing, tightening bolts to 30–40 lb. ft. torque.
13. Slide parking brake drum onto output

shaft and place band and lever on drum. Install bolts to case and torque them to 30–40 ft. lb.

14. Install parking brake band anchor spring and adjusting screw, then adjust brake.

MODEL 445

Disassemble, Fig. 5

1. Drain lubricant from transmission and remove gearshift cover.

NOTE: The two bolts opposite the tower are shouldered to properly position cover and are provided with lockwashers. Make certain that these alignment bolts are correctly located and tightened before installing cover.

2. Lock transmission in two gears and remove mainshaft nut and yoke.
3. Remove mainshaft rear extension housing and speedometer drive gear.
4. Remove pinion front bearing retainer and gasket, then rotate drive pinion gear to align pinion gear with countershaft drive gear teeth and remove drive pinion gear and tapered roller bearing from transmission, Fig. 6.
5. Remove thrust bearing, then push mainshaft assembly rearward of transmission and tilt front of mainshaft up, Fig. 7, and remove mainshaft from case.
6. Remove reverse idler lock plate, then using tool C-603, pull idler shaft from case and lift out reverse idler gear.

NOTE: If tool C-603 is not available, the reverse idler shaft may be driven out the rear of the case with a hammer and brass drift. DO NOT ATTEMPT TO DRIVE IDLER SHAFT FORWARD, as it is a shouldered shaft and a broken case will result.

7. Remove countershaft rear bearing retainer, then slide countershaft rearward, up and out of case, Fig. 8.
8. Drive countershaft from bearing assembly forward and out of transmission case.

SUB ASSEMBLIES, DISASSEMBLE

1. Place mainshaft vertically in soft jawed vise with front end up, then lift 3rd-4th synchronizer and high speed clutch gear off mainshaft and remove third speed gear.
2. Remove 2nd speed gear snap ring with snap ring pliers and lift off thrust washer, Fig. 9.

NOTE: The thrust washer must be positioned with ground surface toward 2nd speed gear when the mainshaft is reassembled.

3. Remove 2nd speed gear and lift off low speed synchronizer and clutch gear from mainshaft.
4. Using a gear puller, remove tapered bearing from rear of mainshaft, then remove 1st speed gear snap ring and thrust washer and lift off 1st speed gear.

COVER AND SHIFT FORK ASSEMBLY, FIG. 10

NOTE: The shift cover should be disassembled only if it is necessary to replace a shift fork, a shaft or the cover.

1. Using a square type or spiral wound "easy out" mounted in a tap handle, remove roll pin from 1st-2nd speed shift fork and roll pin from shift gate.
2. Push 1st-2nd speed shift rail out of rear of cover, driving out expansion plug and remove shift fork and gate. Be sure to cover detent hole to prevent loss of detent ball and spring.
3. Remove 3rd-4th speed shift rail in same manner as in steps 1 and 2.
4. Compress reverse gear plunger and remove retaining clip. Remove plunger and spring from gate.

CLEANING & INSPECTION

Wash all bearings in clean solvent and check all rollers and races for wear, pitting and spalled areas. Replace parts as necessary.

Check operating gear teeth for wear or pitting on tooth faces. Gears which have been

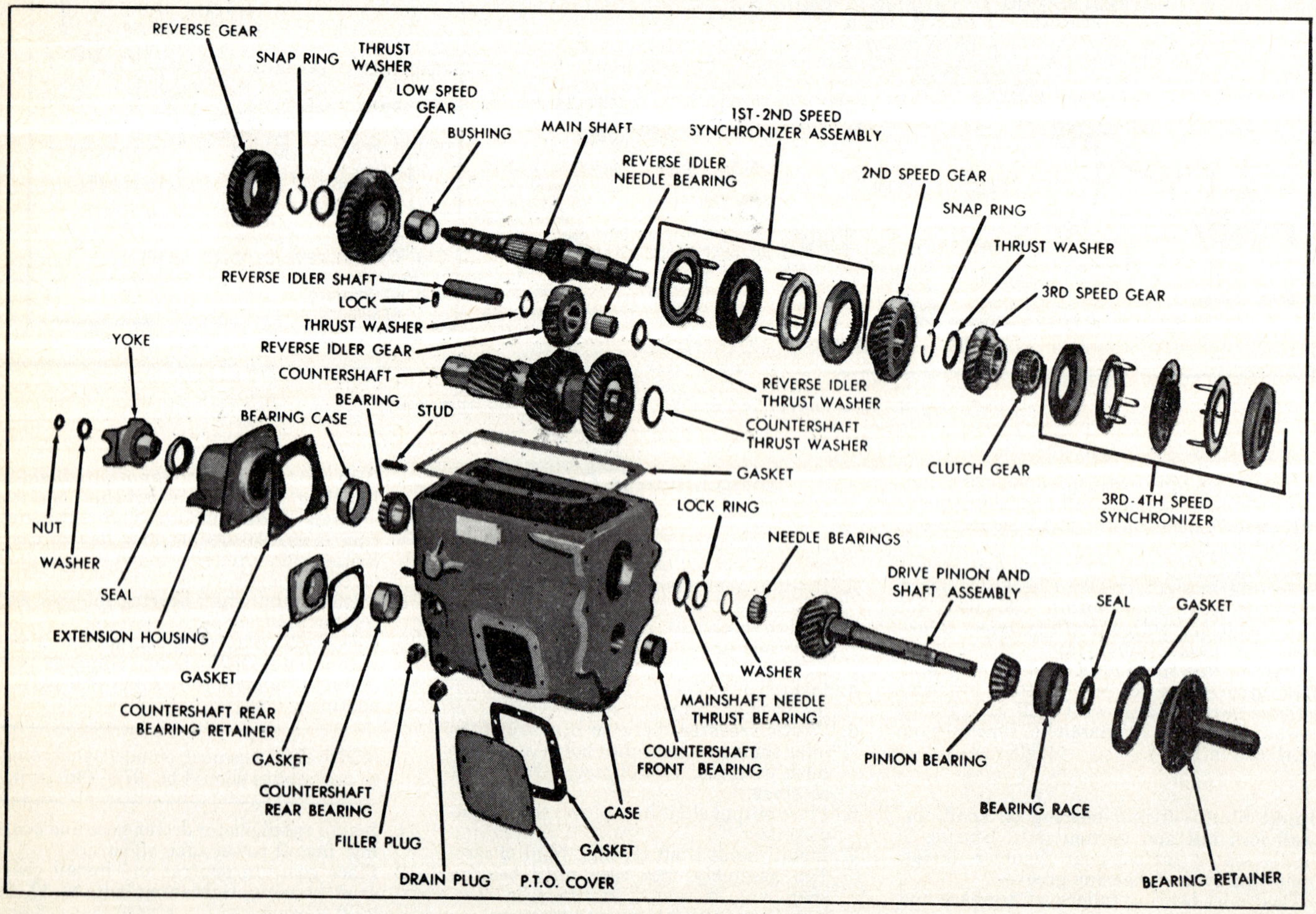

Fig. 5 New Process Model 445 four speed transmission

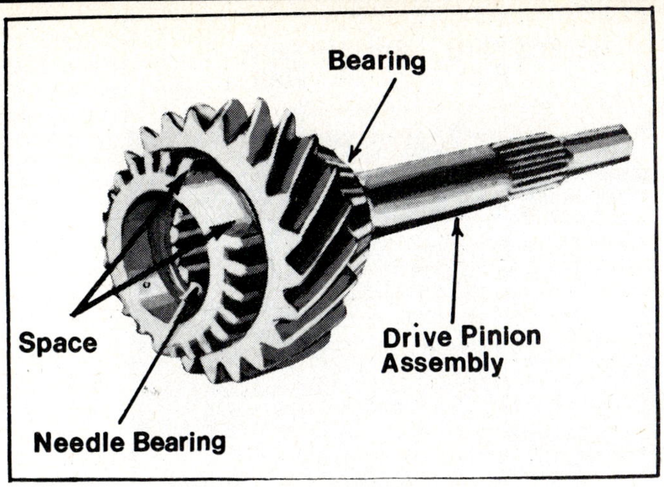

Fig. 6 Drive gear. Model 445

shortened or tapered from clashing during gear shifts should be replaced. If excessive axial gear clearance is found, check snap rings, washers, spacers and gear hubs for excessive wear. Proper axial clearance must be maintained on all mainshaft gears. Replace parts as necessary.

Check splines on mainshaft for excessive wear. If sliding gears, clutch hubs or flanges have been worn into the sides of the splines, the shaft should be replaced.

Check reverse idler gear shaft and rollers for wear and replace as necessary.

Check synchronizer for burrs and uneven or excessive wear at contact surfaces. Check blocker pins for excessive wear or looseness. Check synchronizer contact surfaces on all affected gears for excessive wear.

Check shift bar housing forks and gates for wear at pad and lever slot, check all forks for alignment, check roll pin fit in forks and gates and check neutral notches of shift shafts for wear from interlock balls. Shafts which are indented at points adjacent to neutral notches should be replaced. Replace all worn or damaged parts.

Check all other parts for cracks and fractures. Replace or repair as necessary. Cracked castings may be welded or brazed as long as cracks do not extend into bearing bores or bolting surfaces.

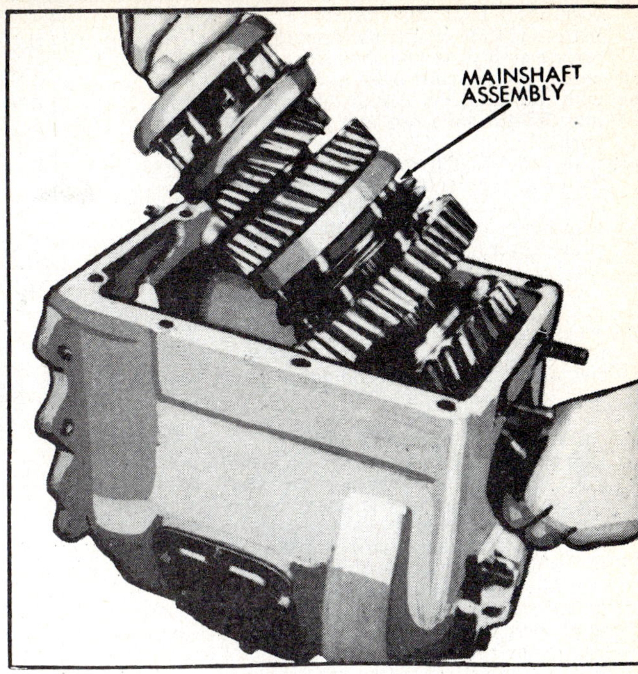

Fig. 7 Removing mainshaft. Model 445

Transmission, Assemble

MAINSHAFT

NOTE: Prior to assembly, lubricate all parts with clean transmission lubricant and replace all expansion plugs, gaskets and seals.

1. Place mainshaft in vertical position with rear end up and lubricate with transmission oil.
2. Slide 1st speed gear over mainshaft with clutch facing down and install thrust washer and snap ring.
3. Install reverse gear over end of mainshaft with fork groove down, then install mainshaft rear bearing and using appropriate size sleeve and tool C-4040, press bearing on inner race.
4. Reverse mainshaft in vise.
5. Install low speed synchronizer and slide 2nd speed gear over mainshaft, then install keyed thrust washer over mainshaft, with ground side facing 2nd gear. Secure with snap ring.
6. Install 3rd speed gear and one shim on mainshaft, then install 3rd-4th synchronizer over mainshaft, making certain that slotted end of clutch gear is positioned toward 3rd speed gear.

GEAR SHIFT COVER, ASSEMBLE, FIG. 10

1. Grease interlock slugs and slide into

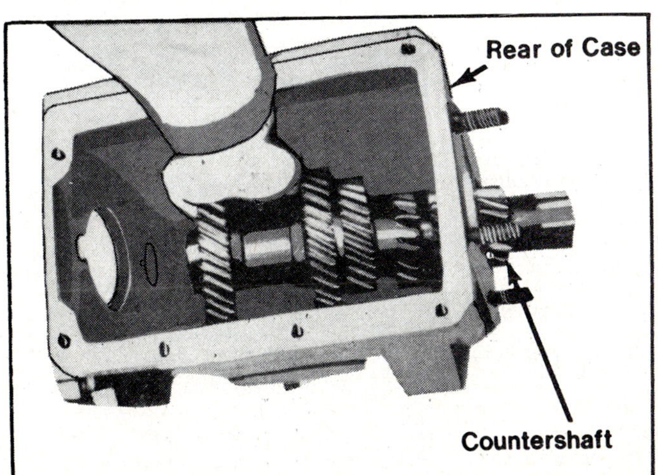

Fig. 8 Removing countershaft. Model 445

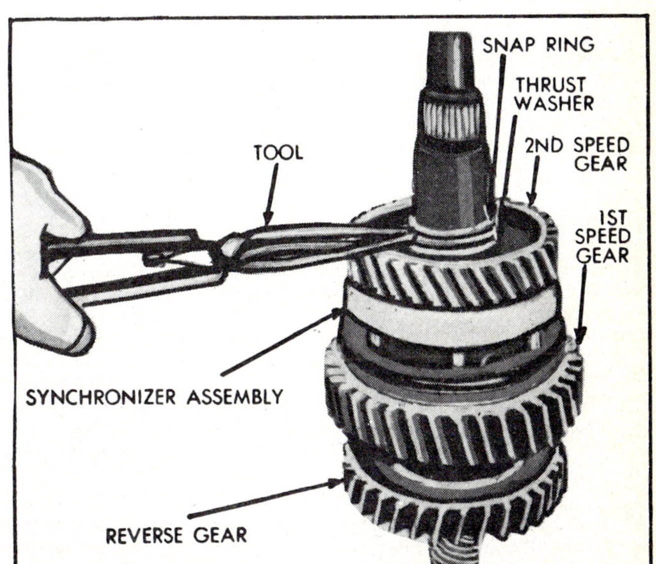

Fig. 9 Removing 2nd-speed gear snap ring. Model 445

openings in shift rail supports.
2. Insert and slide reverse shift rail through reverse shift plate and shift fork.
3. Secure reverse shift plate and shift fork with roll pin, then install interlock pin into 3rd-4th shift rail and secure with grease.
4. Slide 3rd-4th shift rail into rail support from rear of cover, through 3rd-4th speed shift fork and over poppet ball and spring, then secure 3rd-4th speed shift fork to rail with roll pin.
5. Making sure interlock slug is in place, slide 1st-2nd speed shift rail into case through shift fork and shift gate. Hold poppet ball and spring down until shift rail passes and secure 1st-2nd speed shift fork and shift gate with roll pins.

SUB-ASSEMBLY, INSTALLATION

1. Using a 1 3/8 inch socket as driver, install countershaft front bearing in case. Grease needle bearings prior to installation. Hold bearings in place in bearing retainer with socket of appropriate diameter while seating retainer. Drive retainer until flush with front of case.
2. Install tanged thrust washer on countershaft with tangs facing out and install countershaft into case.
3. Install countershaft rear bearing retainer over countershaft rear bearing and using new washer, position retainer with curved segment toward bottom of case.
4. Install reverse idler gear with chamfer toward rear of case, then while holding thrust washers and needle bearings in position, slide reverse idler shaft into case through idler gear. Make certain that shaft lock notch is down and at rear of shaft and install shaft lock and bolt.
5. Place mainshaft vertically in a soft jawed vise with front end up.
6. Mount drive gear on top of mainshaft, then measure clearance between high speed synchronizer and drive gear with two feeler gauges. If clearance is geater than .043–.053 inch, install synchronizer shims between third speed gear and synchronizer brake drum of the thickness required to bring clearance within specifications. After required shim thickness has been determined, remove drive gear from mainshaft.
7. Insert assembled mainshaft into case and place thrust washer over pilot end of mainshaft.
8. Position drive gear so that the cutaway portion of gear is facing down, then slide drive gear into front of case engaging mainshaft pilot in pocket of drive gear.
9. Slide drive gear front bearing retainer over shaft with no gasket or bolts, then install mainshaft rear bearing retainer and tighten retainer bolts.
10. Hold retainer against front of transmission case and measure clearance between front bearing retainer and front of case with feeler gauge. Remove front bearing retainer.
11. Install a gasket pack on front bearing retainer which is .010–.015 inch thicker than the clearance measured in above step. Install front bearing retainer and tighten bolts. This will insure the required .007–.017 inch gear end play.
12. The front synchronizer end play float must be checked before installation of transmission cover. Measure end play float by inserting two feeler gauges diametrically opposite one another between 3rd speed gear and synchronizer stop ring. Accurate measurement can be made only after all mainshaft components are in place and tightened securely. If front synchronizer end play is not within .050–.070 inch, shims should be added or removed as required between 3rd speed gear and synchronizer stop ring.
13. Install yoke and retaining nut on mainshaft, then lock transmission in two gears and torque yoke nut to 125 ft. lbs.
14. Shift transmission into neutral and install cover gasket, then shift transmission and cover into 2nd and carefully lower cover into position. It may be necessary to position reverse gear to permit shift fork to engage groove.
15. Install and tighten cover aligning screws, then install remaining cover screws and tighten all screws evenly and securely.

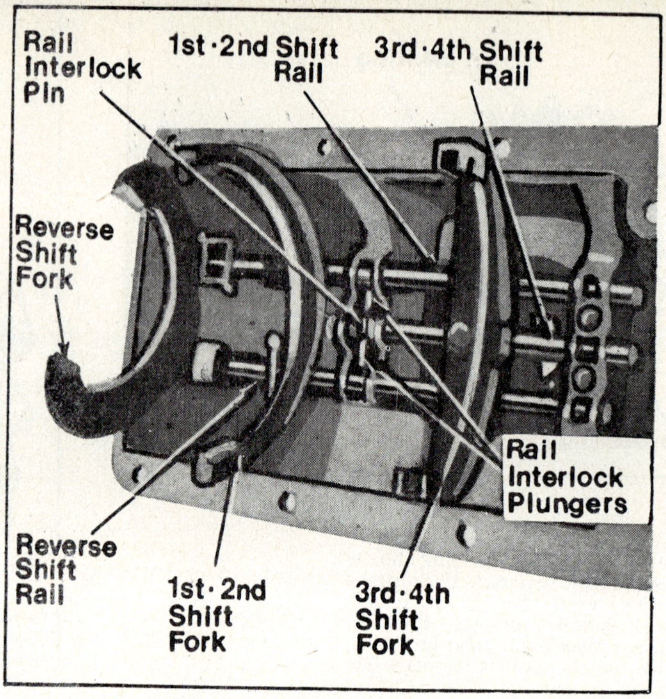

Fig. 10 Cover and shift fork assembly. Model 445

SERIES 745

Disassemble, Fig. 11

1. Remove parking brake drum and flange.
2. Remove speedometer pinion.
3. Remove speedometer drive gear.
4. Remove parking brake.
5. Remove rear bearing seal.
6. Remove transmission cover.
7. Remove drive pinion bearing retainer.
8. When removing drive pinion and bearing from case, slide synchronizer front inner stop ring from clutch teeth on pinion as assembly is being removed from case.
9. Remove snap ring which locks drive pinion bearing on shaft.
10. Press bearing off shaft and remove oil slinger.
11. Remove mainshaft pilot bearing snap ring and remove 15 rollers.
12. Remove seal from retainer.

DISASSEMBLE MAINSHAFT

1. Remove mainshaft rear bearing snap ring.
2. Slide mainshaft and rear bearing to the rear until bearing is out of case.
3. Remove synchronizer from mainshaft.
4. Remove 2–3 shift fork.
5. Remove synchronizer clutch gear snap ring.
6. Remove synchronizer clutch gear 2nd gear and low-reverse gear from mainshaft.

NOTE: *If synchronizer clutch gear cannot be removed easily from the mainshaft, position the low-reverse fork and sliding gear to the rear of case. Then, using a plastic hammer, tap mainshaft back out of synchronizer cluth gear.*

7. Remove mainshaft and bearing out through rear of case, while stripping mainshaft of gears.

COUNTERSHAFT

1. Using a countershaft bearing arbor, drive countershaft toward rear of case until small key can be removed from countershaft.
2. Drive countershaft the remaining way out of the case (toward the rear) keeping arbor tight against end of countershaft to prevent loss of roller bearings.
3. Remove cluster gear and thrust washers.
4. Remove rollers (88), washers (4) and center spacer from cluster gear.

REVERSE IDLER GEAR

1. Using a blunt drift, drive reverse idler shaft toward rear of case far enough to remove key from shaft.
2. Drive shaft all the way out and remove idler gear and bearing assembly.
3. Remove thrust washers and needle bearings (22).

GEARSHIFT MECHANISM

1. Using a small punch, remove low-reverse gear lever shaft tapered lock pin by driving it toward the top of the transmission case.

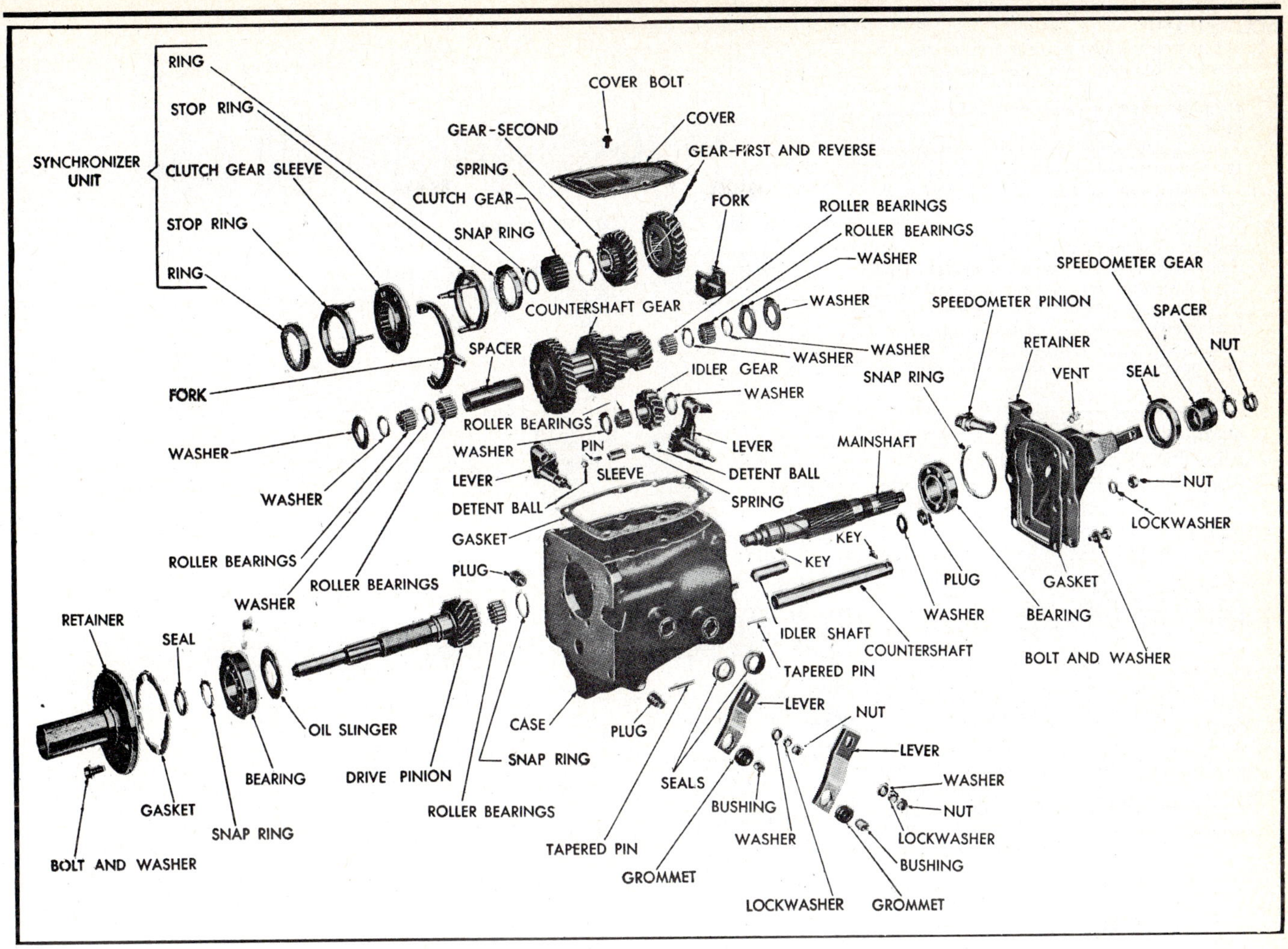

Fig. 11 New Process 745 three speed transmission

2. Remove 2–3 lever pin in same manner.
3. Remove lever shafts from case, taking care not to lose spring-loaded detent balls.
4. Remove interlock sleeve, spring, pin and balls.

Assemble

DRIVE PINION & BEARING

1. Place oil slinger on drive pinion.
2. Place bearing on pinion shaft with outer snap ring away from gear.
3. Press bearing into position so it is seated firmly against oil slinger and pinion gear.
4. Install bearing snap ring on shaft.
5. Coat 15 roller bearings with heavy grease and install them in cavity at rear of drive pinion. Install roller bearing snap ring.

COUNTERSHAFT

1. Place bearing spacer in center of cluster gear bore and use an arbor to assist in assembling roller bearings. Use heavy grease to hold bearings in place and install a row of 22 rollers next to one end of spacer.
2. Place a spacer ring next to rollers and install another row of 22 rollers next to spacer ring.
3. Install another spacer ring at outside end of second row of rollers.
4. At opposite end of cluster gear bore, install two more rows of rollers in same manner.
5. With a small amount of heavy grease to hold it in place, install front thrust washer on arbor at front end of cluster gear, with tabs outward.
6. Install tabbed rear thrust washer on arbor against rear of cluster gear with tabs positioned in grooves provided in cluster gear.
7. Install remaining rear thrust plate on rear thrust washer.
8. Align tabs of front thrust washer vertically to index with notches in transmission case. Position cluster gear and arbor assembly in case. Make sure thrust washers are not dislodged from arbor, and engage thrust washer tabs in case grooves, while sliding assembly into place.
9. Using countershaft, drive arbor forward out of cluster gear and through bore in front of case. Before driving countershaft all the way into case with soft hammer, insert shaft key and continue to drive countershaft forward until it is 1/64″ below rear face of case.

REVERSE IDLER

1. Position arbor in reverse idler gear and, using heavy grease, install 22 roller bearings in gear around an arbor.
2. Place front and rear thrust washers at each end of gear and position assembly in case with chamfered end of gear teeth toward front.
3. Insert reverse idler shaft into bore at rear of case, pushing arbor toward front of transmission.
4. With shaft keyway aligned with recess in case, drive shaft forward, inserting key before keyway is obscured. Continue driving shaft forward until 1/64″ below rear face.

GEARSHIFT MECHANISM

1. Install two lever shaft seals in case.
2. Install low-reverse gear lever shaft in case bore.
3. Install low-reverse gear lever shaft lock pin at top of hole and driving it in firmly to prevent leakage or loss of pin.
4. Place interlock parts in position in case in order: sleeve, ball, spring, pin and ball.
5. Place 2–3 lever shaft in case and, using a suitable tool to depress detent ball against spring tension push lever shaft firmly into position so it prevents ball from escaping.
6. Install 2–3 lever shaft lock pin.
7. Place low-reverse fork in lever shaft with offset toward rear of transmission.

MAINSHAFT

1. While holding low-reverse sliding gear in position in fork, with hub extension to rear, insert mainshaft with rear bearing through rear of case and into sliding gear.
2. Place synchronizer energizer spring and then the inner stop ring on clutch teeth of 2nd speed gear. Install 2nd speed gear on mainshaft.
3. Install synchronizer clutch gear on mainshaft with shoulder to front.
4. Select thickest synchronizer clutch gear snap ring that can be used and install it in mainshaft groove. Make sure ring is bottomed all the way around in groove.
5. Check clearance between clutch gear and 2nd gear. Clearance should be .002" to .016".

NOTE: *End play in excess of .016" may cause 2nd speed gear "jump out".*

6. Hold synchronizer clutch gear sleeve front inner and two outer rings together with pins properly entered in holes in synchronizer gear sleeve and with gear sleeve engaged in groove of 2–3 shift fork, position fork on gear fork shaft.
7. While holding synchronizer parts and fork in position, slide mainshaft forward, placing synchronizer clutch gear into clutch sleeve and at the same time guiding mainshaft rear bearing into case bore.
8. While continuing to hold synchronizer parts in position, tap mainshaft forward until rear bearing bottoms in case bore.
9. Install mainshaft rear bearing snap ring in groove in case bore.

DRIVE PINION

1. Install new seal in retainer until it bottoms on seat of counterbore.
2. Place synchronizer front inner ring in position in front outer ring and guide drive pinion through case bore.
3. Engage splines on rear of pinion with inner stop ring, and tap drive pinion into case until outer snap ring on pinion bearing is against transmission case.
4. Place drive pinion bearing retainer, without gasket, over pinion shaft and against case. While holding retainer with hand pressure against case, measure clearance between retainer and case with a feeler gauge.
5. Select a gasket .003" to .005" thicker than the clearance found. This eliminates end play of front bearing in case and also insures sealing.
6. Install and tighten front bearing retainer bolts to 200 inch-lbs.
7. Install plug or back-up light switch (if equipped) with gasket and tighten securely.

REAR BEARING RETAINER

1. Install new seal in rear bearing retainer.
2. Install rear bearing retainer with new gasket and torque bolts to 50 ft-lbs.
3. Install parking brake assembly.
4. Place speedometer drive gear on main shaft and install parking brake drum and flange assembly. Install washer and nut and torque to 175 ft-lbs.
5. Install speedometer pinion and gear-shift operating levers, tightening nuts to 180 inch-lbs.

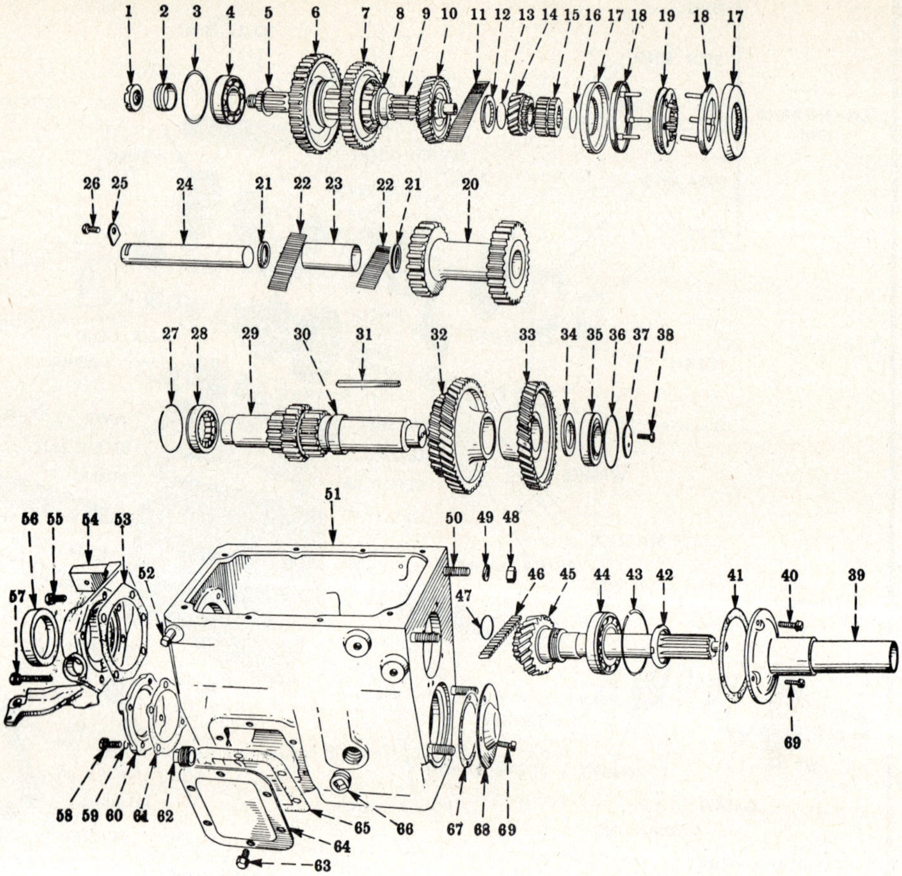

1. Mainshaft nut
2. Speedometer drive gear
3. Snap ring
4. Bearing
5. Shim
6. Low and reverse gear
7. Second and third gear
8. Thrust washer
9. Mainshaft
10. Third speed gear
11. Bearing
12. Thrust washer
13. Snap ring
14. Overdrive gear
15. Synchronizer clutch gear
16. Snap ring
17. Synchronizer ring
18. Synchronizer ring
19. Synchronizer clutch
20. Reverse idler gear
21. Spacer
22. Bearing
23. Spacer
24. Reverse idler shaft
25. Lock
27. Snap ring
28. Bearing
29. Countershaft
30. Spacer
31. Key
32. Third and fourth gear
33. Drive gear
34. Thrust washer
35. Bearing
36. Snap ring
37. Plate
39. Retainer
41. Gasket
43. Snap ring
44. Bearing
45. Main drive gear
46. Bearing
47. Spacer
51. Case
53. Gasket
54. Retainer
56. Seal
60. Bearing cover
61. Gasket
62. Drain plug
64. Cover
65. Gasket
66. Plug
67. Gasket
68. Cover

Fig. 12 New Process five speed transmissions. Models 875, 895

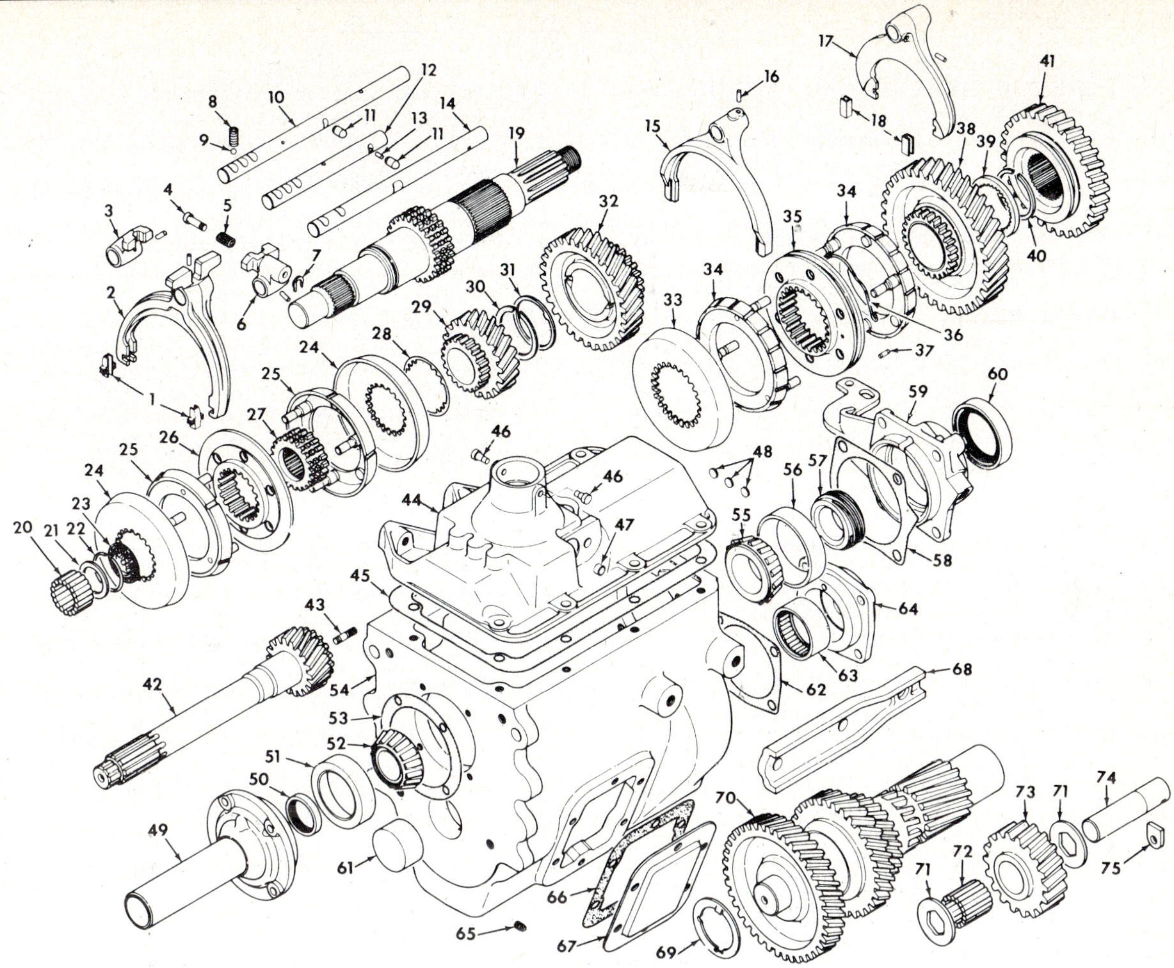

Fig. 12A New Process 4590 Series five speed transmission *(see next page for legend)*

SERIES 875, 895

Disassemble, Fig. 12

1. Drain transmission.
2. Remove hand brake band and drum.
3. Remove transmission control housing.
4. Remove mainshaft rear bearing retainer, noting the number of gaskets present.
5. Remove oil seal and speedometer drive gear.
6. Remove main drive gear bearing retainer, noting the number of gaskets used.
7. Remove main drive gear.
8. Tap mainshaft to rear until rear bearing is free of case.
9. Remove synchronizer.
10. Use puller to remove bearing from rear of mainshaft. Remove shims and note number used as they control floating clearance of synchronizer.
11. Lift mainshaft assembly from case.
12. Remove reverse idler gear shaft from rear of case and lift out reverse idler gear.
13. Remove front and rear countershaft bearing covers, and bearing retainer plate from end of countershaft.
14. Tap front end of countershaft toward rear of case until rear bearing is free of case. Then lift countershaft and gear assembly out of case.

Reassembly

1. Install reverse idler gear assembly in case before installing countershaft gear assembly. This will permit clearance for removal of pilot shaft used to assemble bearings in idler gear.
2. A wood pilot shaft or a discarded idler gear shaft cut to a length of five inches may be used to facilitate assembling bearings in reverse idler gear. Diameter of wood pilot shaft should be 1$^{1}/_{16}$″.
3. A bench assembly of the reverse idler gear and bearings should be made before installing in case. Position pilot shaft and long spacer in idler gear. Insert 31 needle bearings and short spacer between pilot shaft and idler gear. Insert 31 needle bearings and short spacer in same manner at opposite end of idler gear. Position gear and bearing assembly in case. Push shaft flush against pilot shaft, permitting pilot to be removed from gear. Remove pilot shaft from inside of case.
4. The mainshaft third speed gear bearing consists of 50 individual needle bearings. To assemble bearings, position thrust washer and third speed gear on mainshaft. Insert each needle bearing between gear and mainshaft.
5. Select a snap ring of correct thickness that will provide proper running clearance of mainshaft third speed gear. Snap rings are available in three thicknesses (.090″, .093″, .096″) for this purpose.
6. Select a snap ring for securing the clutch gear that will hold the gear tight against the shoulder on mainshaft. Four thicknesses (.087″, .090″, .093″, .096″) are available for this purpose. To check for the thickness required temporarily install the synchronizer clutch gear on the mainshaft before installing the overdrive gear and third speed gear. Insert snap ring in groove on mainshaft and check the gear for tightness.
7. The main drive gear bearing lock nut should be tightened sufficiently to seat bearing against gear shoulder.
8. Like parts of the synchronizer assembly are interchangeable. Preassembling of this unit before installing on the mainshaft is not required since parts may be installed individually.
9. Check and replace all thrust washers showing signs of excessive wear.
10. The mainshaft front and rear bearing retainers fit or end play for their respective bearing is controlled by gaskets. Install bearing retainers and cap screw, check clearance between each retainer and case with a feeler gauge. Then select a gasket or gaskets about .005″ thicker than the clearance between retainer and

Key	Description
1	PLATE, Shift Fork Wear (2)
2	FORK with PLATES, Third and Direct-Speed Shift
3	BLOCK, Low and Second-Speed Shift
4	PLUNGER, Reverse-Speed Shift Block
5	SPRING, Reverse-Speed Shift Block Plunger
6	BLOCK, Reverse-Speed Shift
7	WASHER, Reverse-Speed Plunger
8	SPRING, Shift Bar Poppet Ball (3)
9	BALL, Poppet, 3/8 (3)
10	BAR, Low and Second-Speed Shift
11	PLUNGER, Shift Bar Interlock (2)
12	BAR, Third and Direct-Speed Shift
13	PIN, Shift Bar Interlock
14	BAR, Reverse-Speed Shift
15	FORK, Low and Second-Speed Shift
16	PIN, Shift Fork (5)
17	FORK, Reverse-Speed Shift
18	PLATE, Shift Fork Wear (2)
19	MAINSHAFT
20	BEARING, Mainshaft Roller (14)
21	WASHER, Mainshaft Roller Bearing
22	RING, Mainshaft Snap
23	BEARING, Clutch Gear Thrust
24	RING, Synchronizer Outer (2)
25	RING with PINS, Synchronizer (2)
26	CLUTCH, Sliding
27	GEAR, Synchronizer Clutch
28	SHIM, Synchronizer Adjusting
29	GEAR, Mainshaft Third-Speed
30	RING, Mainshaft Snap
31	WASHER, Mainshaft Thrust
32	GEAR, Mainshaft Second-Speed
33	RING, Synchronizer Outer Stop
34	RING with PINS, Synchronizer (2)
35	CLUTCH with PINS and SPRING, Sliding Rear
36	SPRING, Sliding Clutch (2)
37	PIN, Spring (2)

Key	Description
38	GEAR with BUSHING, Low-Speed
39	WASHER, Mainshaft Thrust
40	RING, Mainshaft Snap
41	GEAR, Mainshaft Reverse
42	GEAR, Main Drive
43	STUD, Transmission Mounting Front (4)
44	HOUSING, Control Lever
45	GASKET, Control Housing
46	PIN, Control Lever Pivot (2)
47	PLUG, Shift Bar Interlock
48	PLUG, Expansion, 3/4 (3)
49	RETAINER, Main Drive Gear Bearing
50	SEAL, Main Drive Gear Oil
51	BEARING, Main Drive Gear Cup
52	BEARING, Main Drive Gear Cone
53	GASKET, Main Drive Gear Brg. Retainer
54	CASE with MAGNET, Transmission
55	BEARING, Mainshaft Cone
56	BEARING, Mainshaft Cup
57	GEAR, Speedometer Drive
58	GASKET, Mainshaft Rear Brg. Retainer
59	RETAINER with SEAL, Mainshaft Rear Bearing
60	SEAL, Mainshaft Bearing Retainer Oil
61	BEARING, Countershaft Front
62	GASKET, Countershaft Rear Brg. Cover
63	BEARING, Countershaft Rear
64	COVER with BEARING, Countershaft
65	PLUG, Drain and Filler (2)
66	GASKET, PTO Opening Cover (2)
67	COVER, PTO Opening (2)
68	BAR, Brake Anchor
69	WASHER, Countershaft Front Thrust
70	GEAR, Countershaft
71	WASHER, Reverse Idler Thrust (2)
72	BEARING, Reverse Idler Needle (21)
73	GEAR with BEARING and WASHERS, Reverse Idler
74	SHAFT, Reverse Idler
75	LOCK, Reverse Idler Gear Shaft

case. Remove retainer, install selected gasket or gaskets and retainer. This will give a satisfactory condition as to bearing end play and sealing of transmission.

SERIES 4590, FIG. 12A

Transmission Disassembly

1. Shift transmission gears into neutral. Remove gear shift cover by lifting straight up.

NOTE: The two bolts opposite tower are shouldered to properly position cover and are provided with lockwashers. Tighten alignment bolts first during installation.

2. Lock transmission in two gears and remove mainshaft nut and yoke.
3. Remove retainer attaching bolts and mainshaft rear bearing retainer and speedometer drive gear.
4. Remove front bearing retainer.
5. Rotate drive gear to align gear flat with countershaft drive gear teeth. Remove drive gear and roller bearing from transmission.
6. Remove thrust bearing and push mainshaft assembly to rear of transmission and tilt front of mainshaft up and remove from transmission case.
7. Remove reverse idler lock screw and lock plate.
8. Using a hammer and brass drift, drive out reverse idler shaft through rear of case.

NOTE: Do not drive idler gear out through front of case as shaft is shouldered and will damage case.

9. Lift out reverse idler gear.
10. Remove countershaft rear bearing retainer and slide countershaft to rear, then up and out of case.
11. Drive countershaft front bearing assembly forward and out of case.

Mainshaft, Disassembly

1. Place mainshaft in soft-jawed vise with front end up.
2. Lift third-fourth synchronizer and high speed clutch gear off mainshaft.
3. Remove third-speed gear.
4. Remove second speed snap ring. Lift off thrust washer.

NOTE: Thrust washer must be positioned with ground surface toward the second speed gear during reassembly.

5. Remove second speed gear and lift off low speed and second speed synchronizer.
6. Reverse mainshaft in vise.
7. Remove rear bearing from mainshaft and lift off reverse sliding gear.
8. Remove first gear snap ring and thrust washer and lift off first speed gear.

Cover & Shift Fork, Disassembly

1. Remove roll pin from first-second speed shift fork. Use a square type or easy out mounted in a tap handle. Remove roll pin from shift gate.
2. Push first-second-speed shift rail out of rear of cover, driving out expansion plug. Remove rail fork and gate.

NOTE: Be sure to cover detent hole to prevent loss of detent ball and spring.

3. Remove third-fourth-speed shift rail and reverse shift rail.
4. Compress reverse gear plunger and remove retaining clip. Remove plunger and spring from gate.

Mainshaft, Reassembly

1. Place mainshaft in soft jawed vise, rear end up.

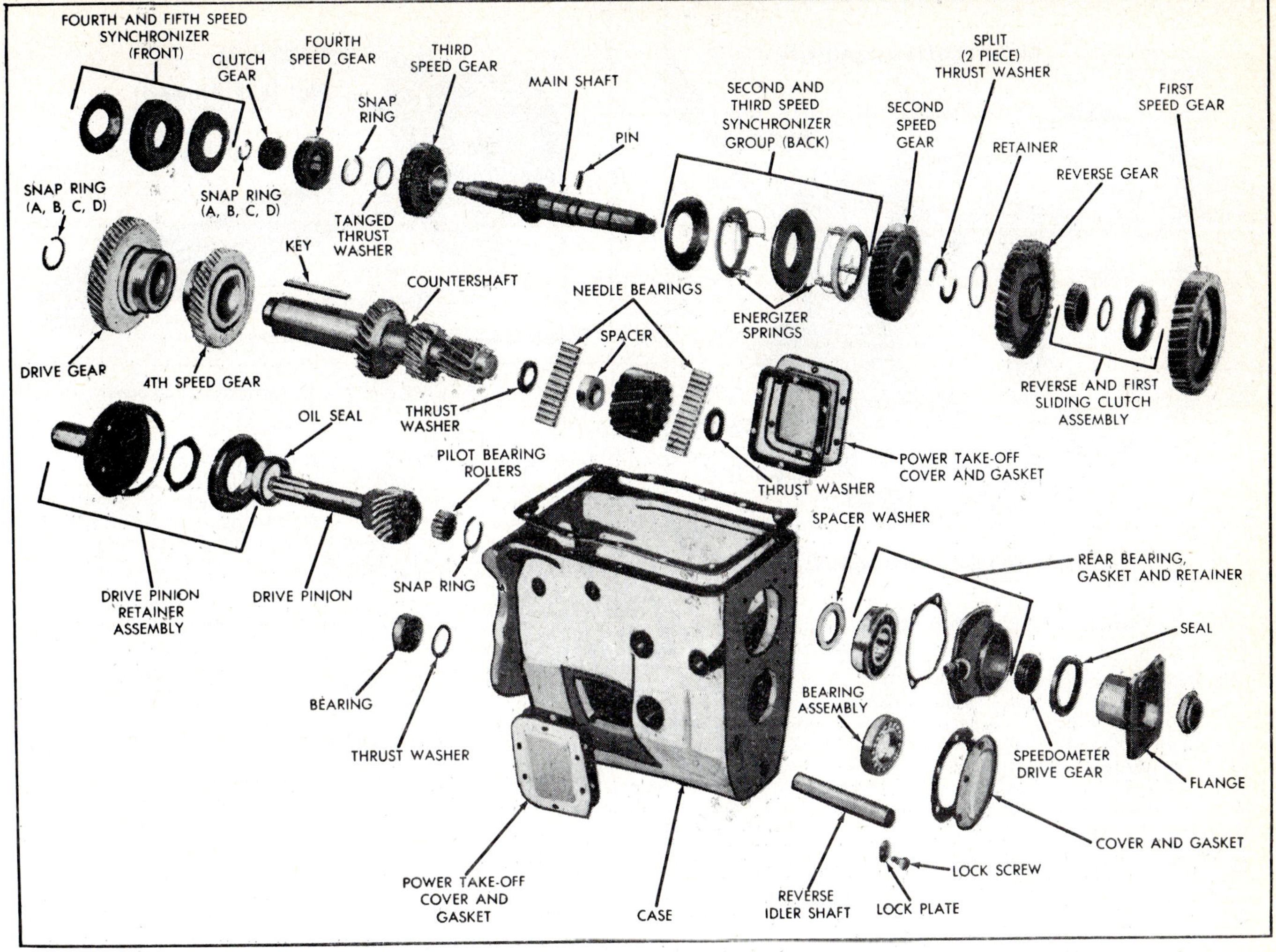

Fig. 13 New Process 7500 Series five speed transmission

2. Slide first-speed gear over mainshaft, clutch gear down, then install thrust washer and snap ring.
3. Slide reverse gear over end of mainshaft, fork groove down.
4. Install main shaft rear bearing on mainshaft while pressing on bearing inner race.
5. Reverse mainshaft in vise.
6. Install low and second speed synchronizer and slide second speed gear over mainshaft.
7. Place keyed thrust washer over mainshaft with ground side toward second gear and install snap ring.
8. Install third speed gear and one shim on mainshaft. Install third-fourth synchronizer over mainshaft with slotted end of clutch gear positioned toward third speed gear.

Transmission Cover, Reassembly

1. Grease interlock slugs and slide into opening in shift rail supports.
2. Insert reverse shift rail from rear of cover and slide into supports.
3. Slide reverse shift rail through reverse shift plate and shift fork and secure with roll pin. Install interlock pin in third-fourth rail.
4. Slide third-fourth-speed shift rail into rail supports from rear of cover. Slide rail through third-fourth-speed shift fork and over poppet ball and spring. Secure with roll pin.
5. Install interlock slug and slide first-second-speed shift rail into case, through shift fork and shift gate. Secure with roll pin.

Installation of Sub-Assemblies

1. Install countershaft front bearing in case.
2. Install tanged thrust washer on countershaft and install countershaft in case.
3. Install countershaft rear bearing retainer over countershaft rear bearing positioning retainer with curved segment toward bottom of case.
4. Install reverse idler gear with gear chamfer to rear of case. Hold thrust washers and needle bearing in position.
5. Slide reverse idler shaft into case and through idler gear.

NOTE: Position shaft lock notch down and to rear of shaft.

6. Install shaft lock and bolt.
7. Place mainshaft in a soft-jawed vise, front end up and mount drive gear on top of mainshaft.
8. Measure clearance between highspeed synchronizer and drive gear using two feeler gauges. If clearance is greater than .043 in.–.053 in. install synchronizer shims between third-speed gear and synchronizer brake drum until clearance is within specifications.
9. Insert assembled mainshaft into case. Place thrust washer over pilot end of mainshaft.
10. Position drive gear so cutaway portion of gear is facing down. Slide into front of case so that mainshaft pilot enters pocket in drive gear.
11. Slip drive gear front bearing retainer over shaft without gasket or capscrews.
12. Install mainshaft rear bearing retainer and torque bolts to 15–25 ft. lbs.
13. Hold front bearing retainer against transmission case and measure clearance. Place a gasket pack on front bearing retainer which is .010 in.–.015 in. thicker than clearance. This will provide required .007–.017 drive gear end play. Install front bearing retainer bolts and torque to 15–25 ft. lbs.
14. Measure end play float on front synchronizer with two feeler gauges diametrically opposed to each other between third-speed gear and synchronizer stop ring. If front synchronizer end play float is not between .050–.070 shims should be added between third-speed gear and synchronizer stop ring.
15. Install yoke and retaining nut on rear of mainshaft. Shift transmission into two gears at once and torque yoke nut to 125 ft. lbs.
16. Shift transmission into neutral and install cover gasket.

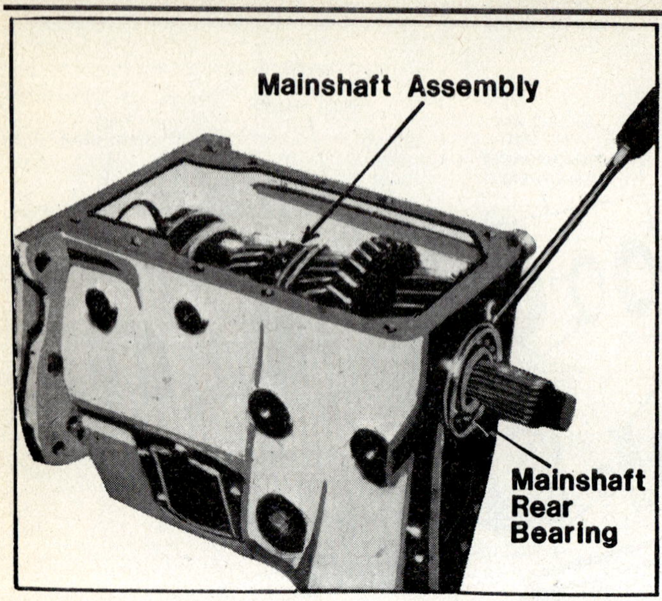

Fig. 13A Removing rear bearing from mainshaft. 7500 Series

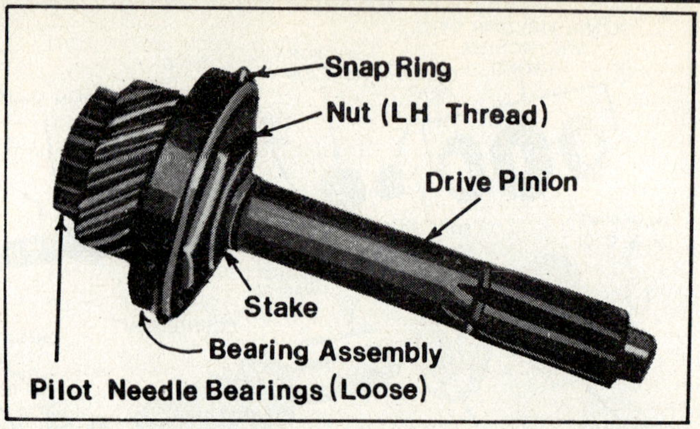

Fig. 14 Drive pinion shaft assembly. 7500 Series

17. Shift transmission and shift cover into second gear.
18. Lower cover into position and install aligning screws finger tight. Install remaining cover screws. Torque all screws to 20–40 ft. lbs.

SERIES 7500

Disassemble

SHIFT COVER & PARKING BRAKE

1. Remove parking brake assembly (if used), then shift transmission into neutral and remove shift cover retaining bolts and shift cover.

NOTE: Second screw from front on each side is shouldered to properly align shift cover.

---

2. Lock transmission in two gears and remove output shaft flange nut, then remove brake drum (if used) and yoke assembly by lightly tapping with a soft hammer.

NOTE: The yoke and drum are balanced as a unit, and unless replacement of parts is required, the yoke and drum should be removed as an assembly.

---

3. Remove brake band assembly bracket (if used), support bolts and lock-washers and remove brake band assembly as a complete unit.

CHECKING FRONT SYNCHRONIZER END PLAY FLOAT

Before removal and disassembly of drive pinion and mainshaft, measure end-play between synchronizer outer stop ring and 4th speed gear. Record reading for reference during assembly. Note deviation from desired .070–.090 inch allowable end play so that correction shims can be installed during assembly. Thick or thin shims are available as required.

DRIVE PINION, BEARING RETAINER & MAINSHAFT

1. Remove drive pinion bearing retainer, Fig. 13.
2. Remove drive pinion gear and ball bearing from transmission. Pull on pinion shaft or rap on face of case with brass hammer.
3. Remove speedometer drive pinion and mainshaft rear bearing retainer, then place a brass drift in front center of mainshaft and drive rearward.
4. Using a puller or two large screwdrivers, Fig. 13A, remove rear bearing from mainshaft, then when mainshaft rear bearing clears case, remove rear bearing and speedometer gear using a puller.
5. Move mainshaft assembly rearward and tilt front of mainshaft up, then while holding 1st-reverse gear and 4th–5th speed synchronizer, lift assembly out of case.

REVERSE IDLER SHAFT & COUNTERSHAFT, FIG. 13

1. Remove reverse idler lock screw and lock plate, then using a brass drift held at an angle, drive idler shaft rearward and pull out.
2. Remove reverse idler gear and thrust washer from case.

NOTE: Loose needle bearings should be replaced with all new needle bearings. Never mix old with new.

NOTE: If excessive side and end play are not found and teeth are not badly worn or chipped, countershaft gears may not require replacement. However, if continued use seems questionable, the countershaft should be removed.

---

3. Remove countershaft rear bearing retainer, gasket and bearing, then tip countershaft upward and remove from case. Remove thrust washers from front end of countershaft.
4. Using a suitable driver from inside of

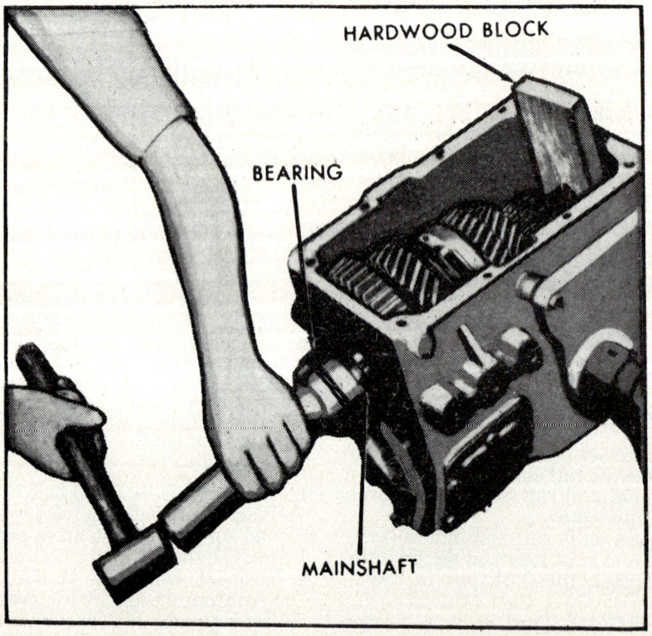

Fig. 14A Installing mainshaft rear bearing. 7500 Series

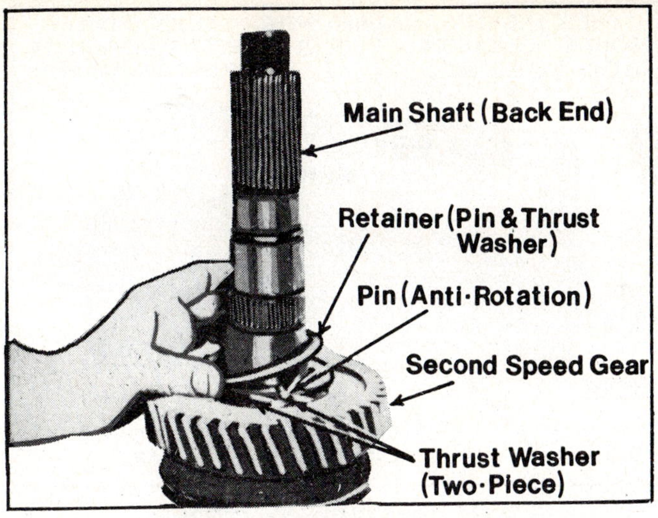

Fig. 15 Installing or removing 2nd-speed gear. 7500 Series

case, remove countershaft front needle bearing by tapping on bearing cage.

NOTE: It is not usually necessary to disassemble countershaft, unless if after inspection, damage or malfunction warrants disassembly.

---

5. Remove snap ring, then place countershaft assembly in a suitable press with blocks supporting drive gear and carefully press shaft.
6. Repeat above procedure supporting 4th speed gear and remove key.

DRIVE PINION, FIG. 14

1. Remove snap ring and washer holding pilot needle bearings in place and remove bearings.
2. Relieve staked area, then remove drive pinion ball bearing retainer nut and remove ball bearing.

NOTE: Retainer nut has left hand thread.

---

3. Remove snap ring from drive pinion ball bearing and seal from drive pinion bearing retainer.

MAINSHAFT, FIG. 15

1. Remove spacer washer and first speed gear from mainshaft, then remove retaining ring and 1st-reverse clutch and clutch gear assembly and reverse gear.
2. Remove 2nd speed gear (retaining) thrust washer assembly. This two piece thrust washer consists of two halves held in position on the mainshaft and prevented from turning by a pin in mainshaft together with a retainer ring.
3. Remove 2nd speed gear, then remove back (2nd–3rd speed) synchronizer assembly. The 2nd–3rd speed clutch is integral with mainshaft.
4. Remove snap ring, then remove front (4th–5th speed) synchronizer assembly, clutch gear and 4th speed gear.
5. Remove retaining snap ring, tanged thrust washer and 3rd speed gear.

SHIFT COVER & FORKS, FIG. 16

NOTE: Shift cover and forks should be disassembled only if rails, poppets or interlock plungers are worn or springs broken allowing end play or simultaneous movement of rails, and if shift forks or cover itself needs replacing.

---

1. Mark each fork and rail for location during assembly and place shift forks and rails in neutral. Mount cover in a soft jawed vise or bench fixture and remove spiral pins from shift forks and shift lugs. Spiral pins may be removed using an "easyout" installed in a tap handle.
2. Drive short, 1st-reverse shift rail rearward and out of transmission cover and remove fork.
3. Remove first-reverse shift rail pivot bolt and nut and remove crossover lever.

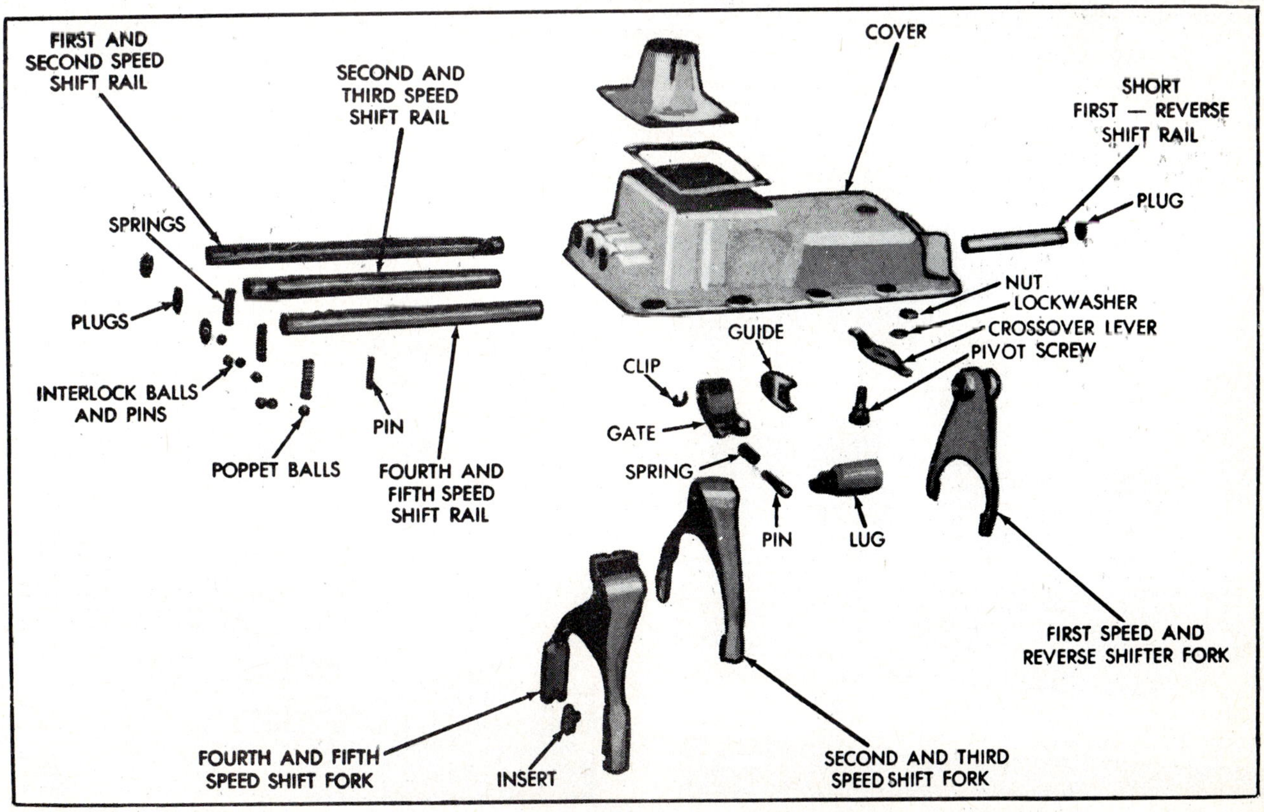

Fig. 16 Shift cover and fork assembly. 7500 Series

4. Drive 4th–5th and 2nd–3rd speed shift rails forward and out of transmission cover, then remove 1st-reverse shift rail in same manner.

NOTE: Place a towel over shift rails while driving shift rails out, to prevent loss of poppet balls and springs.

5. Remove the four interlock balls and pin from cover.

NOTE: Make certain that pin and balls are out by shaking or jarring cover sharply on a block of wood.

Sub-Assembly, Assemble

NOTE: During assembly, use new expansion plugs, gaskets and seals and lubricate all parts with transmission fluid. Where grease is specified, use an oil soluble grease such as Multi-Purpose Grease NLGI Grade 2 or equivalent. Make certain that all oil returns are free of grease or obstructions. Clean and inspect all parts and replace as necessary. Clean and inspect transmission case and wipe all bores and bearing surfaces before reassembly.

Check synchronizers for burrs, uneven or excessive wear at threads and check blocker pins for excessive wear or looseness. Check synchronizer contact surfaces on gears for excessive wear. If replacement is required, replace front (4th–5th speed) synchronizer along with outer stop rings. Replace back (2nd–3rd speed) synchronizer parts as required.

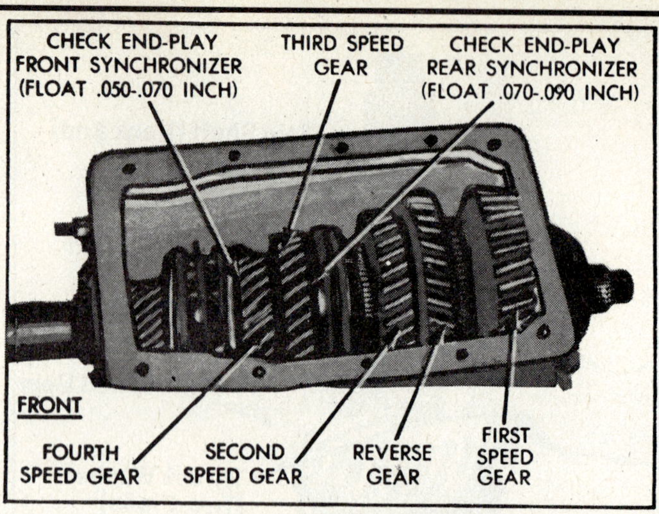

Fig. 17 Checking end-play. 7500 Series

MAINSHAFT FIGS. 13 & 15

1. Place mainshaft with forward end up in a soft jawed vise, then place third speed gear on shaft with clutching teeth facing down and install tanged thrust washer and snap ring.
2. Place 4th speed gear on shaft with clutching teeth up.

NOTE: Referring to dimension recorded during disassembly, select shims which will provide .070–.090 inch end play between 4th speed gear and front synchronizer. Thick and thin shims are available.

3. Place 4th speed synchronizer clutch gear (with oil slots down) on mainshaft. Select a snap ring of greatest thickness (available in four sizes, marked A, B, C or D) to eliminate all clutch gear end play.
4. Remove mainshaft from vise and install 2nd–3rd synchronizer group.

NOTE: Synchronizer sleeve is marked "FRONT" for proper installation.

5. Place second speed gear on shaft and thrust washer retaining pin in hole in mainshaft, then position the two thrust washer halves on mainshaft and install the thrust washer retaining ring with large diameter making contact with 2nd speed gear, Fig. 15.
6. Install reverse gear, then position reverse and 1st speed clutch gear on mainshaft and install retaining snap ring. Select a snap ring of greatest thickness (available in four sizes marked A, B, C or D) to eliminate all clutch gear end play.
7. Position sliding clutch on clutch gear, then install 1st speed gear and spacer on mainshaft.

CHECKING BACK SYNCHRONIZER END-PLAY FLOAT, FIG. 17

Checking end float (.070–.090") at back synchronizer (2nd–3rd speeds) should be performed during assembly of the mainshaft assembly. This can be done by using two equal sized feeler gauges diametrically opposite each other between 3rd speed gear outer stop ring and third speed gear itself.

NOTE: Extreme care should be used to make sure that all synchronizer parts are assembled correctly and parallel and that the feeler gauges are inserted close to the mainshaft and up on the shoulder of 3rd speed gear, otherwise an erroneous reading will result.

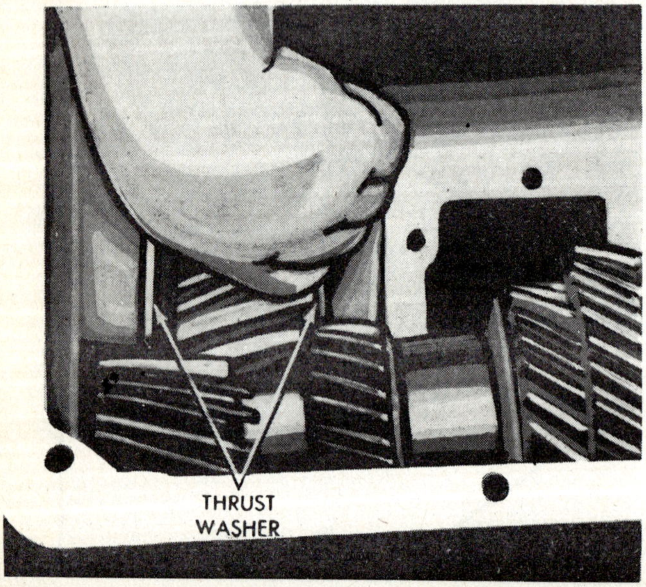

Fig. 18 Reverse idler installation. 7500 Series

Fig. 19 Shift cover and fork installation. 7500 Series

If the end play float is less than .070 inch or more than .090 inch, install new component parts for the synchronizer assembly. This may require new gears if thrust faces are worn, as shims cannot be used at this point.

COUNTERSHAFT, FIG. 13

1. With keys positioned on countershaft, press on gears until properly seated.

NOTE: Make certain that key does not climb out of position as gear is being pressed on.

2. Install snap ring on countershaft. Select a snap ring of greatest thickness (available in four sizes, marked A, B, C or D) to eliminate possible end play.
3. Install washer on countershaft drive gear.

DRIVE PINION, FIG. 14

1. Grease needle bearings and install into pocket drive gear and install washer and snap ring.
2. Press large bearing onto pinion shaft making sure bearing is properly seated, then install bearing retainer nut and tighten securely. Stake in place.

NOTE: Nut has left hand thread.

3. Install snap ring on large bearing, making certain it is properly seated.

SHIFT COVER & FORKS, FIG. 17

1. Push 1st-reverse shift rail into cover bore far enough to permit installation gates, poppet ball springs and roll pins. Move shift rail into neutral position.
2. Coat the four interlock balls and pin with grease, then place two balls in interlock bore, moving both balls toward shift rail and seating ball in neutral notch.
3. Place interlock pin in hole located in 2nd–3rd speed shift rail. Install 2nd–3rd speed shift rail, gate and fork as in step one and move rail into neutral position.
4. Install the two remaining balls into interlock bore, then push 4th–5th speed shift rail into cover bore and install shift fork and move shift rail into neutral position.
5. Position crossover lever in such a manner that the short 1st-reverse shift rail, fork and roll pin can be installed.
6. Position crossover lever and align notches in both long and short 1st-reverse shift rails.
7. Install pivot bolt through corssover lever and transmission cover and install and tighten nut.
8. Install new welch plugs.

Transmission Reassembly

COUNTERSHAFT

1. If removed, install front bearing assembly into countershaft bearing bore.
2. Grease thrust washer and position it into transmission case, then install countershaft front bearing journal into front bearing, seated against thrust washer. Make sure to keep centerline of countershaft aligned with rear bearing bore during installation to prevent damage to countershaft front bearing.
3. Install countershaft rear roller bearing assembly, gasket and cover and tighten screws securely.

NOTE: Check countershaft for .008–.020 inch end play clearance. Clearance can be adjusted by changing countershaft rear bearing cover gasket.

REVERSE IDLER, FIG. 18

1. With new needle bearing held in place on each side of spacer with grease, place reverse idler gear and thrust washers into position in case.
2. Drive shaft through case and gear using a hammer and brass drift. Be sure needles stay in place and lock strap slit in shaft will line up so that lock strap, cap and screw can be installed.
3. Install lock strap on shaft and tighten screw securely.

MAINSHAFT & DRIVE PINION, FIG. 14

1. Carefully lower rear end of mainshaft into case while holding 1st speed gear and spacer washer from slipping off shaft.
2. Using a wooden block at front of mainshaft, drive mainshaft bearing onto shaft and into case, Fig. 14A.
3. Install drive pinion by carefully driving on bearing outer race and forcing it into case while guiding front end of mainshaft into pilot bearing pocket. Make certain bearing is fully seated.
4. Press oil seal into retainer until it is seated. Do not press beyond this point.
5. Install bearing retainer and gasket and tighten retainer screws securely.

MAINSHAFT FLANGE

1. Install speedometer gear on output shaft, then install oil seal into retainer until seal makes contact with its seat. Do not press beyond this point.
2. Install retainer and gasket and tighten retainer screws securely.
3. Position universal joint flange and brake drum (if used) on output shaft.
4. Lock transmission in two gears and install and torque output shaft nut to 125–175 ft. lbs.

CHECKING FRONT SYNCHRONIZER END-PLAY FLOAT

End-play float of .050–.070 inch should be checked just before transmission cover is installed. Follow the same procedure used for initial back synchronizer end play check, except place feeler gauges between 4th speed gear and outer stop pins.

NOTE: If end play float is not within .050–.070 inch, shims should be removed or added between 4th speed gear and outer stop ring of synchronizer.

NOTE: Correct readings can only be obtained after all mainshaft components are properly assembled and tightened.

NOTE: If end-play float is not within .050–.070 inch, shims should be removed or added between 4th speed gear and outer stop ring of synchronizer. Add shims as necessary and recheck clearance.

TRANSMISSION SHIFT COVER, FIG. 19

NOTE: After completing assembly of main case and just before installing cover assembly, recheck rear synchronizer end-play float of .070–.090 inch. If it is not within these limits, install new parts as necessary to correct end-play float.

1. With transmission gears and cover placed in neutral, carefully lower cover into place while guiding shift forks into position.
2. Install shouldered aligning screws and split lockwashers in second hole from front and finger tighten.
3. Install remaining screws and tighten all screws securely after cover is properly seated.
4. Before installing shifter, make certain that transmission shifts through all gears.

SPICER TRANSMISSIONS

SERVICE INDEX

SST Split Torque Units

Compound Units

Auxiliary Units

LUBRICANT CAPACITY BY MODELS

MODEL	PINTS
3152, 3153	10
5000	17
5052, 5252	17
5652	13
5662	17
5752, 5753, 5756	13
5852, 5853	13
6000	18
6052	18
6253	17
6352, 6354	17
6452, 6453, 6454	17
6455, 6456	17
6553	18
6852, 6853, 6854, 6855	17
7016, 7216 (4×4)	28
7352	18
7452, 7453	18
8012, 8212 (4×3)	36
8016, 8216 (4×4)	36
8041, 8045	16
8051, 8055	24
8052, 8054	24
8125, 8125-U	28
8241, 8245	16
8251, 8255	24
8312, 8612	28
8440, 8445	18
8516	36
8552	24
8716	36
10	38
1007-2A	③
1007-2B, C, D	①
1007-3A, B	①
1010-2A	41
1052	②
1062-A, B, C	②
1211-3A	28
1214-2A	①
1252-A	28
1252	②
1253	②
1262-A	②
1263-A	②

Auxiliary Units

MODEL	PINTS
1241	14
5831	4
6041	8
6231	8
7041	11
7231	8
8031, 8035	12
8341, 8345	12
9341	18
9441	12

①—front case 41 pints, rear case 10 pints
②—fill until oil runs out of fill hole
③—front case 38 pints, rear case 10 pints

3000 & 5000 SERIES EXC. CM5000

Five Speed Synchronized Type, Figs. 1 and 2

1. To disassemble remove gearshift housing.
2. Remove brake drum, spline flange, brake shoe, mainshaft rear bearing retainer, speedometer drive gear and spacer.
3. Remove countershaft rear bearing retainer and nut.
4. Remove left side power take-off cover.
5. Use puller to remove reverse idler shaft.
6. With hard wood block against front side of second speed gear, drive mainshaft toward rear until rear bearing is free of case. Remove snap ring and bearing.
7. Slide first speed gear off as mainshaft is removed from case.
8. Remove input shaft from case.
9. Remove countershaft rear bearing.
10. Work countershaft toward rear until it clears front bearing and lift assembly out of case.
11. Remove countershaft front bearing retainer and tap bearing out of case.
12. Remove reverse idler gear.

Assemble Mainshaft

1. Place 2nd speed gear on output shaft with clutch teeth toward front.
2. Install two Woodruff keys in mainshaft. Then press 2nd-3rd synchronizer clutch gear on mainshaft.
3. Place 2nd-3rd speed synchronizer, 3rd gear and 3rd gear sleeve on mainshaft. Press sleeve onto shaft until it bottoms on synchronizer clutch gear. *Third speed gear slots must line up with Woodruff keys in mainshaft.*
4. Install snap ring at front of 3rd gear sleeve.
5. Install 4th gear, thrust washer and snap ring on mainshaft.
6. Install 4th-5th synchronizer.

Assemble Transmission

1. Tap countershaft front bearing into case. Install bearing retainer with new gasket. Line up oil holes in retainer, gasket and case.
2. Place assembled countershaft in case and start shaft into front bearing. Drive rear bearing onto shaft and into case.
3. Install reverse idler and power takeoff cover.
4. Tap input shaft and bearing into case and install mainshaft pilot roller.
5. Install input shaft retainer without gasket and tighten bolts to bottom retainer on bearing snap ring. Check clearance between retainer gasket surface and case. Use gaskets that will prevent oil leakage and any end play. Install retainer with oil hole in line with gasket and case holes.
6. Install low-reverse gear on mainshaft and position assembly in case. Drive mainshaft bearing into position.
7. Install and torque countershaft nut to 500–550 ft. lbs. on all models. Install bearing retainer and torque bolts to 60–80 ft. lbs.
8. Install new seal in mainshaft rear bearing retainer, tightening bolts to 60–80 ft. lbs.
9. Install brake shoe, spline flange and

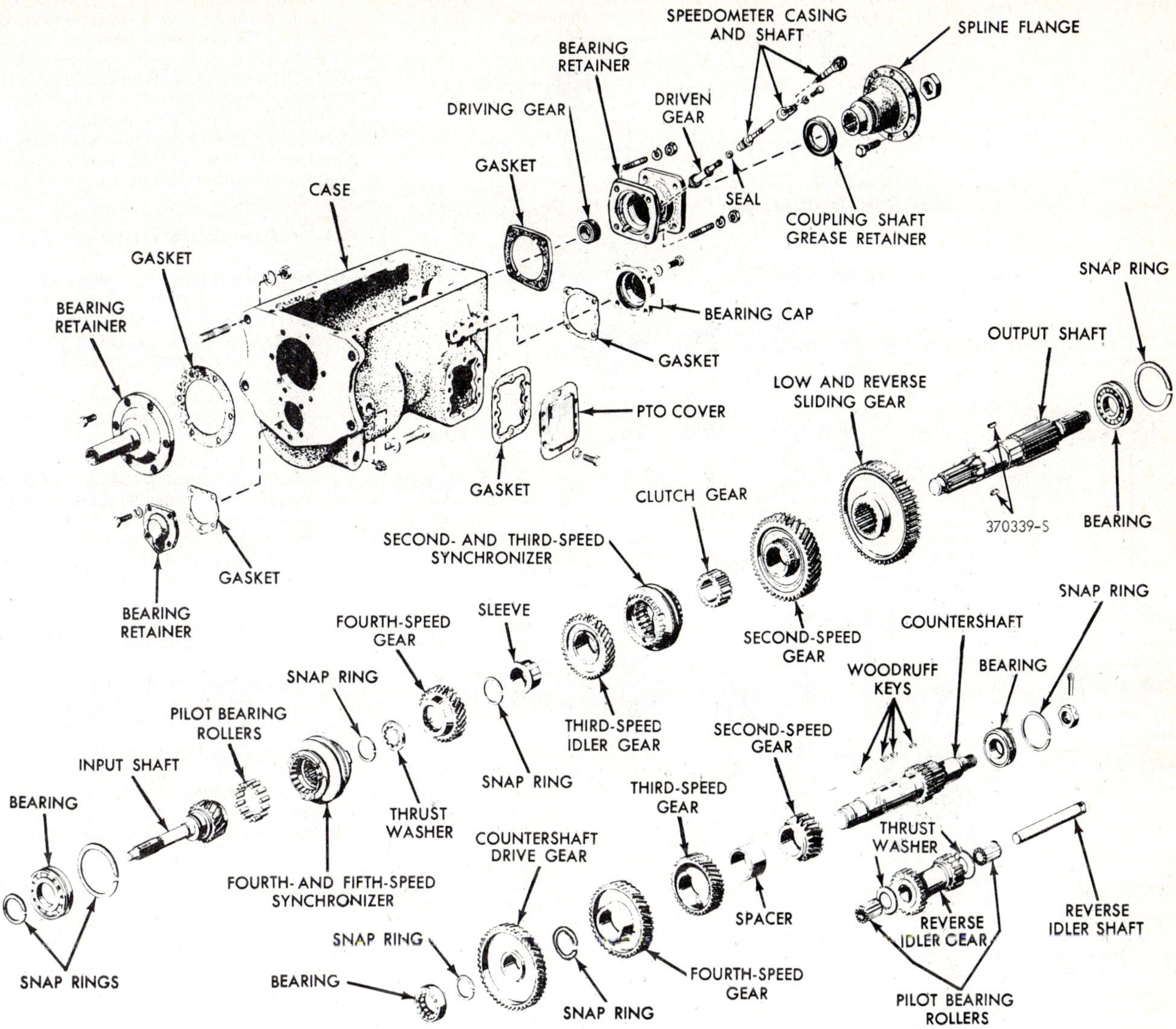

Fig. 1 Spicer 3000 and 5000 Series five speed synchronized transmissions

nut.
10. Shift gears into neutral and install gearshift housing.

6000 SERIES EXC. CM6000

Five Speed Synchronized Type, Figs. 3 and 4

1. To disassemble, remove gearshift housing.
2. Remove brake drum, spline flange and brake shoe.
3. Remove rear bearing retainers from mainshaft and countershaft. Then remove speedometer drive gear and spacer from mainshaft.
4. Remove countershaft rear nut.
5. Remove reverse idler gear shaft.
6. Using a hard wood block against front side of 2nd speed gear, drive mainshaft toward rear until rear bearing clears case. Remove snap ring and bearing from shaft.
7. Slide low-reverse gear from shaft as mainshaft is removed from case. Remove pilot rollers from input shaft.
8. Remove input shaft from case.
9. Remove snap ring and bearing from rear of countershaft. Slide countershaft toward rear of case until shaft clears front bearing and lift assembly out of case.
10. Remove countershaft front bearing retainer and tap bearing out of case.

Assemble Mainshaft

1. Slide 2nd speed gear and synchronizer clutch gear onto shaft. Install clutch gear retaining snap ring.
2. Slide 2nd-3rd synchronizer onto shaft, and press on 3rd speed gear.
3. Install 4th speed gear sleeve and 4th gear. Position thrust washer and install snap ring. Slide 4th-5th speed synchronizer onto shaft. Then slip low-reverse sliding gear on shaft.

Assemble Transmission

1. Tap countershaft front bearing into case. Install retainer with new gasket, lining up oil holes in retainer, gasket and case.
2. Start assembled countershaft into front bearing. Drive rear bearing onto shaft and into case.
3. With bearings in reverse idler gear, install gear in case, lining up flat surface on rear end of idler shaft with position of lock tang on countershaft rear bearing retainer. Drive shaft into position.
4. Position mainshaft assembly in case. With wood block at front end of shaft, drive rear bearing into place.
5. Insert pilot rollers into input shaft, holding rollers with Lubriplate. Last roller must be inserted endwise.
6. Install input shaft and bearing.
7. Lock transmission in two gears and install and tighten countershaft rear bearing nut to 500–550 ft. lbs. on all models.
8. Install countershaft rear bearing retainer.
9. Install new oil seal in mainshaft bearing retainer. Place speedometer gear and spacer on mainshaft. Install bearing re-

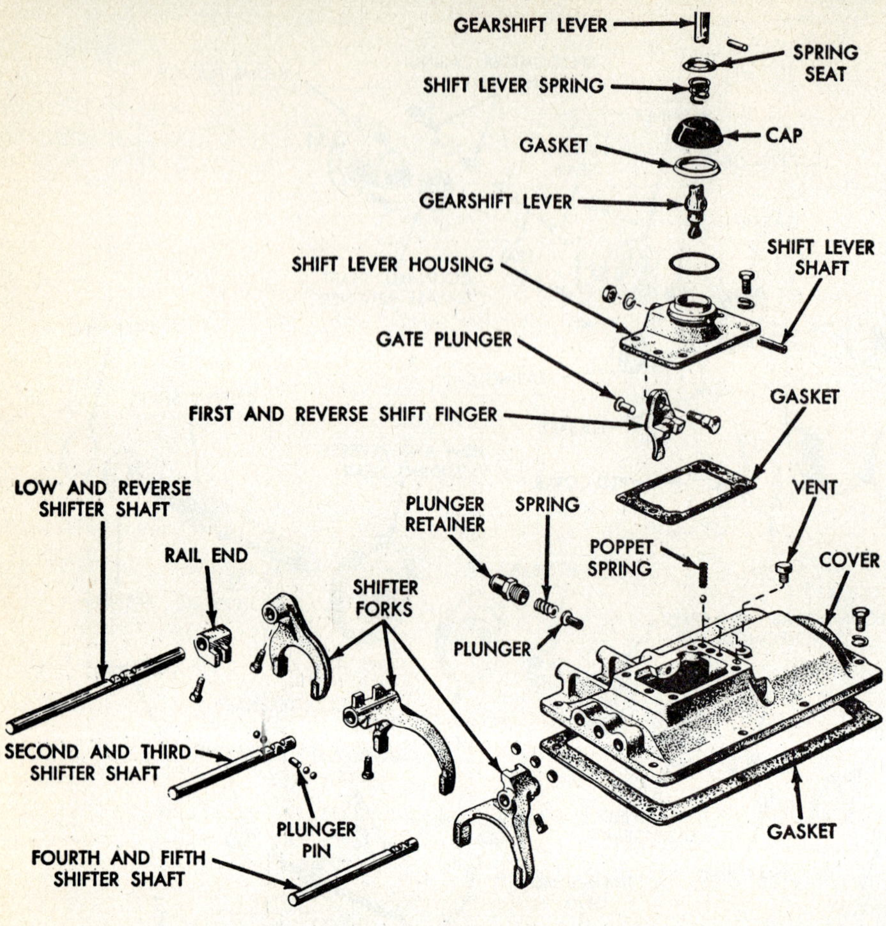

Fig. 2 Spicer 3000 and 5000 gearshift housing

tainer.

10. Install input shaft retainer without gasket and tighten bolts to bottom retainer on bearing snap ring. Check clearance between retainer gasket surface and case, which should be .010–.015″. Select gasket pack that will seal oil and also prevent end play at bearing snap ring.
11. Install input shaft retainer with oil hole in line with gasket and case holes.
12. Install brake shoe assembly.
13. Install U-joint flange and parking brake drum.
14. With transmission in neutral, install gearshift housing.

Oil Pump

The oil pump (when used) is bolted to the inner side of the clutch housing. The pump is fitted with a non-adjustable pressure relief valve consisting of a ball and spring held in place by a pipe plug in the side of the oil pump housing.

When assembling the pump, press the drive gear on the shaft so that the distance from the front end of the shaft to the front side of the gear measures 17/32″. This dimension should be maintained to prevent excessive wear on the side of the gear.

Assemble the oil pump driving gear and shaft in the housing. Press the driven gear shaft in the housing and then on the shaft. Assemble the relief valve and spring and install the pipe plug. Position the housing gasket, being sure the oil holes in the case are not covered by the gasket. Fill the oil pump with gear oil and fasten the assembly in place, being sure the oil holes line up.

5831 SERIES

Three Speed Auxiliary Type, Figs. 5 and 6

1. To disassemble, remove gearshift housing.
2. Remove input shaft U-joint yoke.
3. Remove flange and brake drum from output shaft.
4. Remove brake shoe assembly.
5. Remove rear mounting bracket.
6. Remove output shaft rear bearing retainer and remove speedometer gear from shaft.
7. Remove input shaft bearing retainer and bearing spacer.
8. Remove countershaft rear bearing cap and shims. Tie shims together for use at reassembly.
9. Tap underdrive gear toward rear until bearing clears case. Pull bearing from shaft. Remove underdrive sleeve and hub through rear bearing bore.
10. Tilt front of output shaft upward and slide it out of case.
11. Remove pilot rollers from input shaft. Remove snap ring and tap input shaft and bearing into case.
12. Drive countershaft toward rear until rear bearing cup is out of case. Remove countershaft front bearing cup from case.

Assembly Notes

After countershaft is assembled, measure clearance between rear bearing cap and case. Then select a shim or shims that will provide .001–.003″ end play. When adjustment is satisfactory torque bolts to 45–55 ft. lbs.

When assembling the output shaft, position direct and overdrive hub with face of hub stamped "X" toward front end of shaft. Install underdrive hub with face stamped "X" toward rear of shaft.

Assemble Transmission

1. With gears in neutral and new gasket on case, place gearshift housing on gasket with outer ends of shift rods toward front of transmission and each shift fork engaged in its proper groove, Fig. 6. Torque housing bolts to 45–50 ft. lbs.
2. Install transmission trunnion, making certain it does not bind on bearing retainer.
3. Install U-joint flange and yoke on input shaft and torque nut to 350–400 ft. lbs.
4. Install transmission mounting bracket and torque bolts to 40–45 ft. lbs.
5. Install brake shoe assembly and rear U-joint flange (and drum) and torque nut to 350–400 ft. lbs.

7231 SERIES

Three Speed Auxiliary Type, Figs. 7, 8, 9 & 10

1. Remove gearshift housing.
2. Remove input shaft yoke.
3. Remove trunnion from front bearing retainer.
4. Remove output shaft flange and brake drum, brake shoe and output shaft bearing retainer.
5. Remove countershaft rear bearing cap and shims. Tie shims together for use at reassembly.
6. Remove input shaft assembly.
7. Tap underdrive gear toward rear until rear bearing clears case. Pull bearing off shaft.
8. Remove underdrive sleeve and hub through rear bearing bore and remove output shaft.
9. Drive countershaft toward rear until rear bearing cup clears case. Slide countershaft toward rear and lift out assembly. Remove countershaft front bearing cup.

Assemble Transmission

1. Install countershaft front bearing cup. Lower gear end of shaft into rear bearing bore and position shaft in case. Install rear bearing cup.
2. Position countershaft rear bearing cap without shims. Measure clearance between cap and case. Select shims that will provide .001–.003″ end play, and tighten bolts.
3. Position output shaft in case and slide hub and underdrive sleeve onto shaft. With rear end of output shaft extended beyond rear of case, drive rear bearing onto shaft. Drive bearing on shaft until it firmly seats in bore.
4. Install new oil seal in output shaft bearing retainer. Install speedometer gear on shaft and install bearing retainer, tightening bolts to 45–50 ft. lbs.
5. Install input shaft and bearing retainer as a unit.
6. Install parking brake shoe and tighten

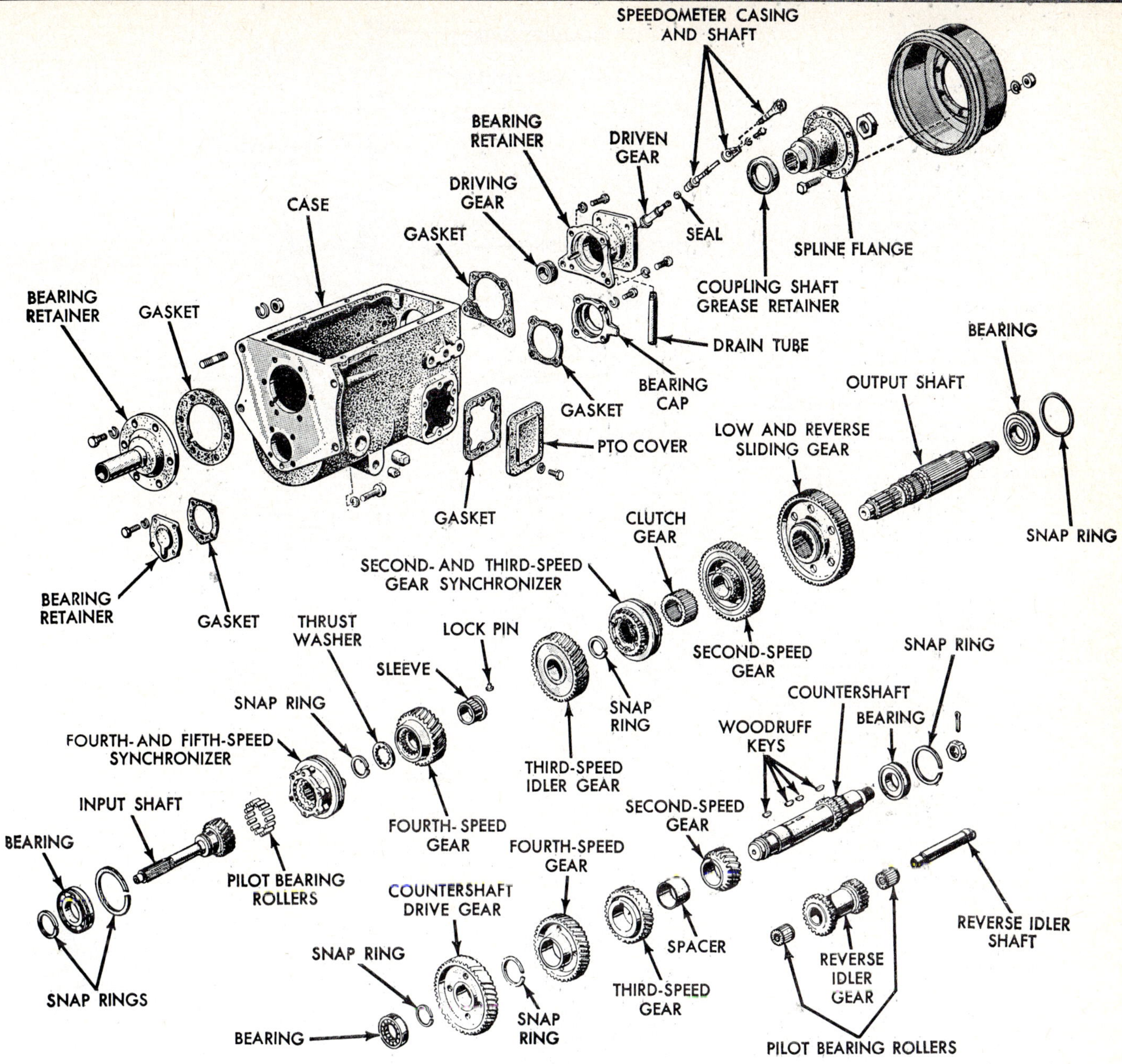

Fig. 3 Spicer 6000 Series five speed synchronized transmission

bolts to 180–220 ft. lbs.

7. Install output shaft flange and tighten nut to 350–400 ft. lbs. on 1970–72 vehicles or 400–450 ft. lbs. on all except 1970–72 vehicles. If brake drum was removed, install it and tighten nuts to 70–75 ft. lbs.
8. Install trunnion but do not tighten bolt. Position input shaft yoke on shaft, install nut and torque to 400–450 ft. lbs. on all vehicles.
9. Install gearshift housing.

7041, 8341, 8345

Four Speed Auxiliary Type, Figs. 11 to 14

1. Remove gearshift housing.
2. Remove flange (and drum) from output shaft.
3. Remove input shaft U-joint flange and dust deflector.
4. Remove trunnion and front bearing retainer cap.
5. Remove lockwashers from retainer cap bolts and reinstall cap. Then after removing bolts from retainer, use puller to remove retainer and front bearing. Remove cap and tap bearing out of retainer.
6. Remove brake shoe, output shaft bearing retainer, speedometer drive gear and 1st gear spacer washer.
7. After sliding input shaft as far forward as possible, tap output shaft forward to start shaft out of its bearing. Then slide output shaft and bearing rearward until there is clearance for a standard bearing puller. Pull bearing from shaft and remove 1st gear thrust washer.
8. Place two 1" thick wood blocks between input shaft and 3rd-4th clutch gear. Press output shaft forward until clutch gear lock ring is exposed. Remove two machined halves and wood blocks.
9. Pull output shaft out through rear of case, noting that there are double rows of bearing rollers in 1st, 2nd and 4th speed gears.
10. Remove input shaft through top of case and pilot bearing from shaft. Remove sleeve and inner bearing race from shaft.
11. Remove countershaft front bearing retainer washer, bearing cap and gasket. Press countershaft toward rear until shaft is freed from front bearing and rear bearing is out of case. Remove bearing

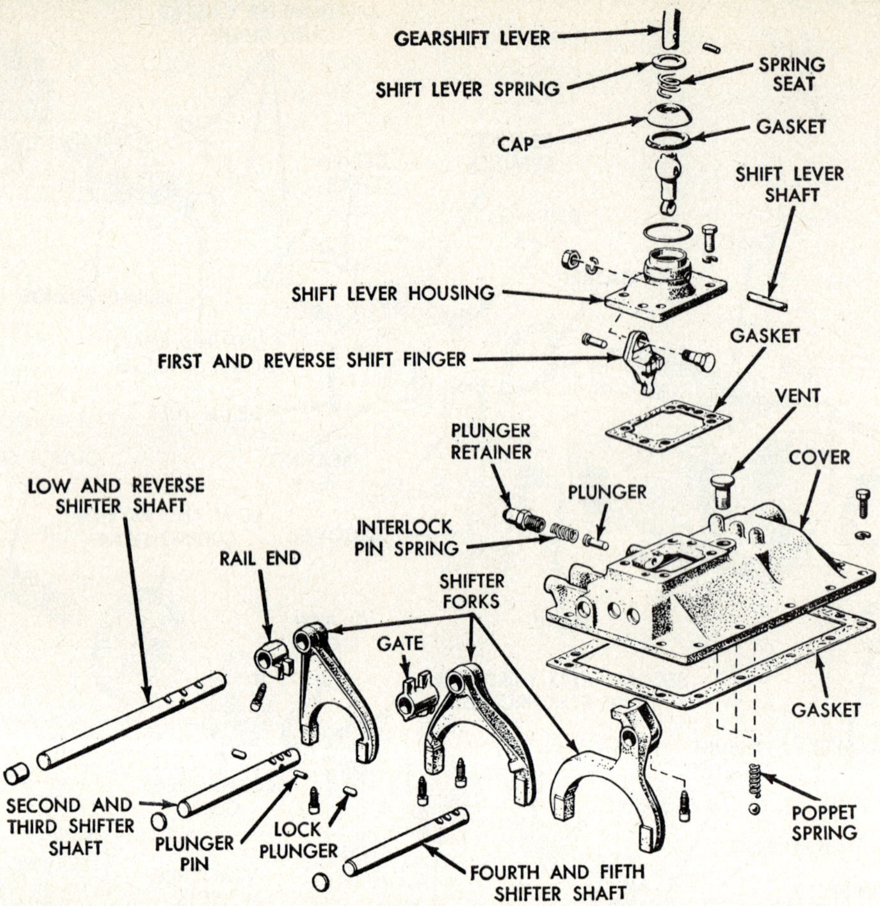

Fig. 4 Spicer 6000 Series gearshift housing

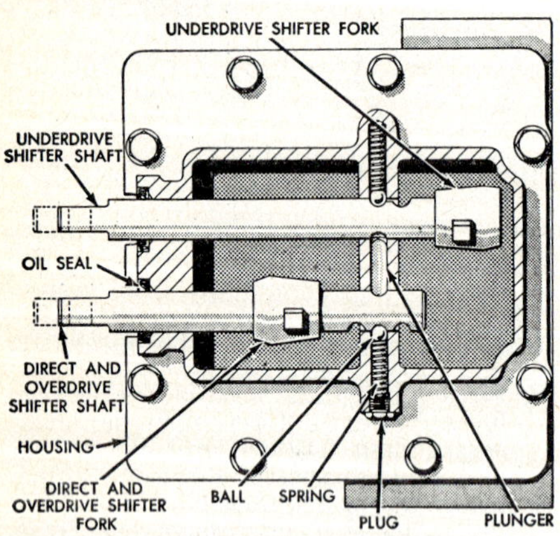

Fig. 6 Spicer 5831 Series gearshift housing

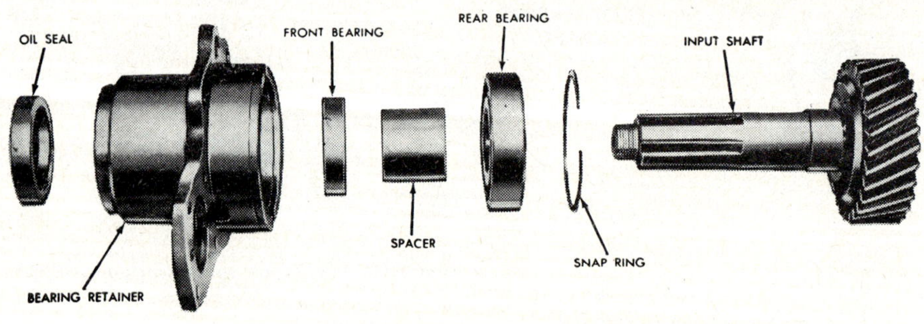

Fig. 7 Series 7231 input shaft parts

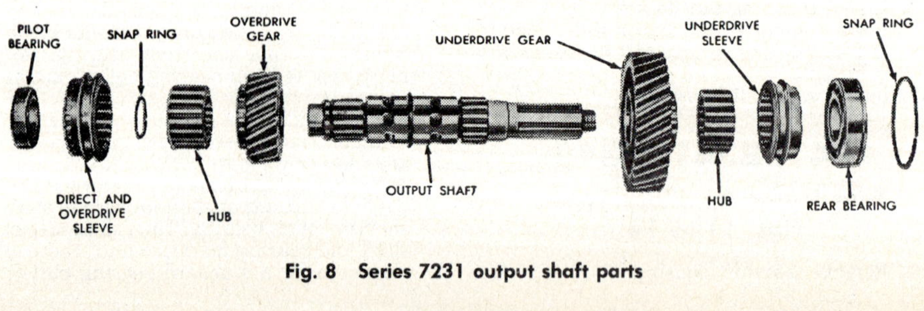

Fig. 8 Series 7231 output shaft parts

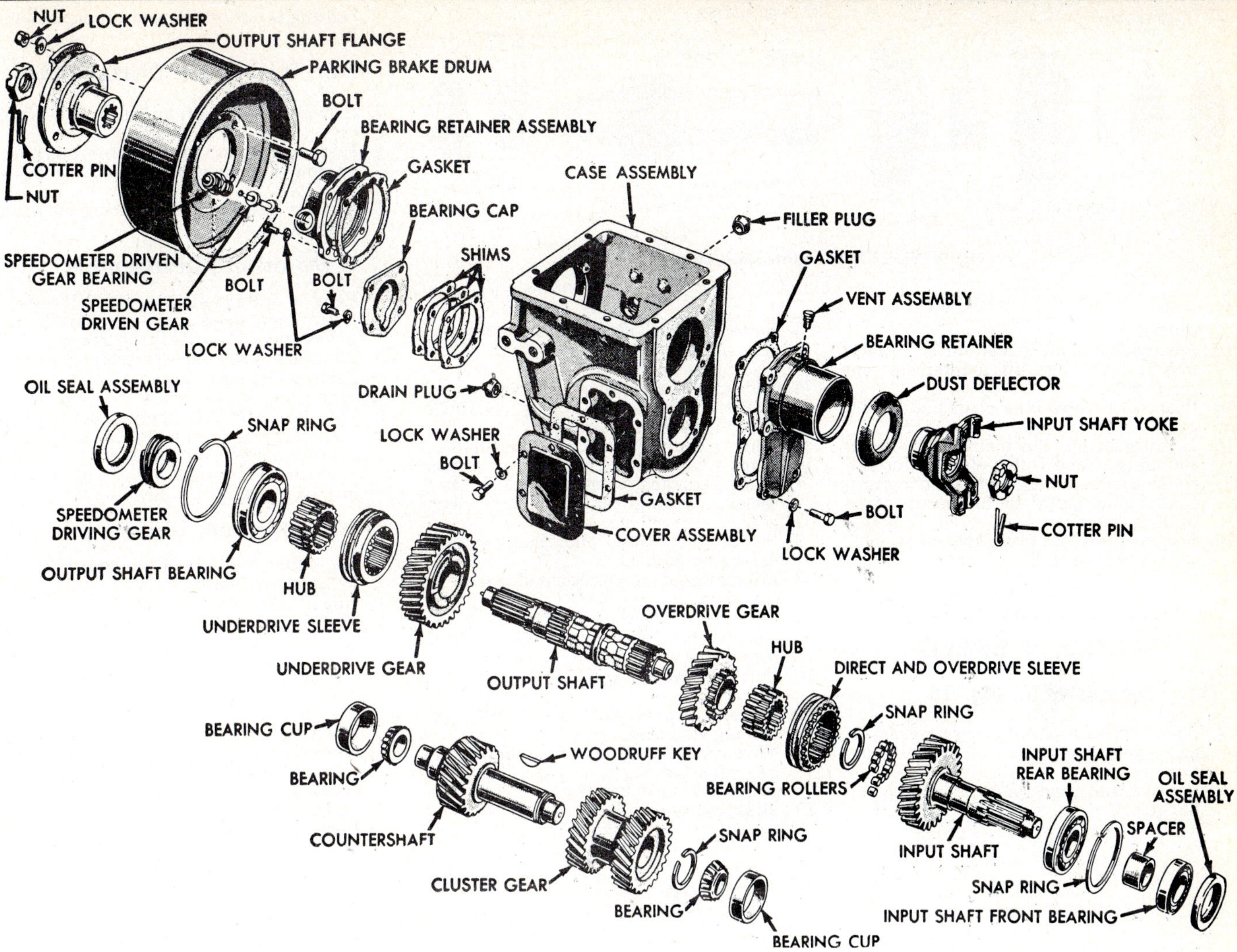

Fig. 5 Spicer 5831 Series three speed auxiliary transmission

from shaft and front bearing from case.
12. Remove countershaft from case, front end first.

Assemble Transmission

1. Assemble and install countershaft in case, being sure oil drain hole in rear bearing cap is at bottom and torque bolts to 45–50 ft. lbs.
2. Assemble and install input shaft in case.
3. Assemble and install output shaft components in case. Then with two 1" thick wood blocks between input shaft and 3rd-4th clutch gear, press output shaft forward until shaft lock ring groove is beyond 3rd-4th clutch gear. Remove wood blocks. Slide 3rd-4th clutch gear forward and make sure that both parts of lock ring will center inside counterbore of clutch gear. Pull output shaft rearward until components are drawn into position.
4. Install output shaft bearing with snap ring groove to rear. Install spacer washer and speedometer drive gear.
5. Install rear bearing retainer and torque bolts to 30–40 ft. lbs. on all 7041 models or 60–80 ft. lbs. on all 8341 and 8345 models. Install brake shoe assembly, output shaft flange and drum and torque nut to the following values: 500–550 ft. lbs.; 1974–80 model 7231, 400–450 ft. lbs. and 1974–80 models 8341 and 8345, 350–420 ft. lbs.
6. Install input shaft bearing retainer and gasket and torque bearing retainer bolts as follows: 1974–80 model 7041, 35–40 ft. lbs.; 1974–80 models 8341 and 8345, 40–50 ft. lbs. Install front bearing into retainer, then install bearing cap and gasket and torque bolts to 25–32 ft. lbs. for small bolts and 60–80 ft. lbs. for large bolts.
7. Install trunnion with finished surface toward front and torque bolt to 50–55 ft. lbs.
8. After positioning dust deflector, install input shaft flange and torque nut to same values as described in step 5 for output flange.
9. After aligning shaft forks with clutch sleeves, install shift cover and gasket and tighten nuts and bolts.

7000 SERIES

Five Speed, Fig. 15

1. Remove transmission cover.
2. Remove universal flange.
3. Unfasten retainer cap from rear of transmission case and remove speedometer drive gear and rear bearing retainer.
4. Pull overdrive shift shaft out through rear and lift out fork.
5. Remove overdrive housing countershaft cover and unscrew lock bolts that retain overdrive gears to countershaft.
6. Reach into overdrive housing and remove bolts which fasten this housing to main case.
7. Pull overdrive housing and gears to rear and force rear bearing from mainshaft.
8. Push mainshaft to rear until intermediate bearing is free of case.
9. Tilt mainshaft and lift assembly out through top.
10. Remove bell housing and main drive gear bearing cap.
11. Unscrew nut from main drive gear shaft and push drive gear into case and lift it out.
12. Remove snap ring from front end of coun-

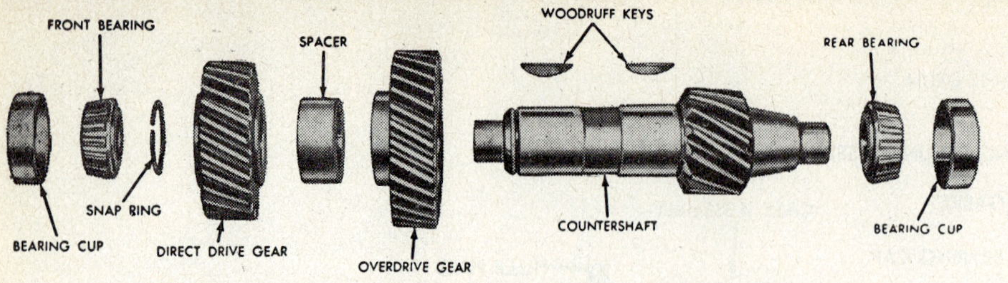

Fig. 9 Series 7231 countershaft parts

tershaft.

13. Push countershaft to rear until bearing is free of case, then lift countershaft out through top.
14. Push out reverse idler shaft and lift out gear.
15. To disassemble the mainshaft, slip off sliding gear and clutch sleeve. Unscrew clutch gear retaining nut and strip mainshaft, being careful not to lose any of the bearing rollers.
16. Reverse the above procedure to assemble the transmission.

8000, 8500, 8700 SERIES, 4 SPEED

Disassemble, Fig. 16

1. Shift transmission into reverse gear, remove cover screws and lift off shift cover.
2. Remove nut and pull off companion flange.
3. Remove mainshaft rear bearing cap and take off speedometer drive gear, spacer and thrust washer.
4. Slide mainshaft to rear far enough to free rear bearing from case and use a suitable puller to remove bearing.
5. Remove clutch release yoke and shafts.
6. Remove cap screws from main drive gear bearing cap and, using a suitable puller, remove bearing cap and main drive gear.
7. Remove mainshaft pilot bearing from drive gear pocket.
8. Lift mainshaft assembly out of case by its front end.
9. Remove countershaft rear bearing cap and the bearing retainer washer from rear end of countershaft.
10. Pull reverse idler gear shaft out with a puller.
11. Remove reverse idler gear, sleeve and bearing assembly.
12. Move countershaft to rear far enough to free rear bearing from case. Then use puller to remove bearing.
13. Slide countershaft to rear until front bearing is out of case and lift out assembly.

Main Drive Gear, Disassemble

1. Remove snap ring from inside of bearing cap.
2. Place bearing cap in arbor press and push bearing and main drive gear out of bearing cap.
3. Remove nut and press bearing off main drive gear.

Mainshaft, Disassemble

1. Remove 3rd and 4th speed clutch collar.
2. Remove low speed gear, being careful not to lose any of the needle bearings.
3. Remove low and 2nd speed clutch gear.
4. Remove 3rd and 4th speed clutch gear snap ring.
5. Place mainshaft assembly in arbor press with rear end down and press mainshaft through gears. Use care to see that none of the needle bearings and 3rd speed gear lock balls are lost as they fall out.

Countershaft, Disassemble

1. Use puller to remove front bearing.
2. Remove countershaft drive gear snap ring.
3. Place assembly in arbor press and press countershaft through gears.

Mainshaft, Assemble

1. Coat inside bore of low speed gear with heavy grease. Then place two rows of needle bearings (72 per row) on inside of low speed gear with spacer between rows.
2. Assemble needle bearings in 2nd and 3rd speed gears in same manner.
3. Place mainshaft in vise with rear end down.
4. Place 2nd speed gear on mainshaft with clutch teeth to rear.
5. Place 3rd speed gear sleeve on mainshaft with flanged end of sleeve toward rear. Line up two notches on inside of sleeve with mainshaft splines. Place two sleeve lock balls in notches of sleeve and press sleeve on mainshaft.
6. Place 3rd speed gear on mainshaft with clutch teeth to front.
7. Place 3rd and 4th speed clutch gear on mainshaft with hub of gear toward front and press in place. Then lock assembly with a new snap ring.
8. Place 3rd and 4th speed clutch collar on clutch gear with longer hub to rear.
9. Take mainshaft from vise and place low and 2nd speed clutch gear with shift fork collar to rear.

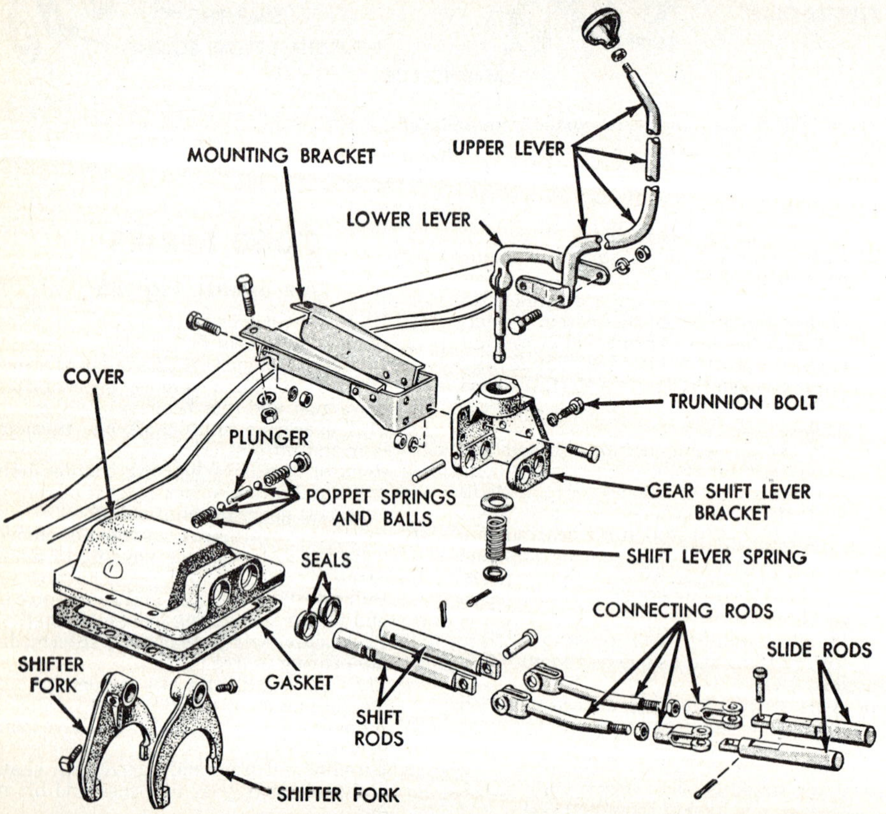

Fig. 10 Spicer 7231 Series gearshift housing

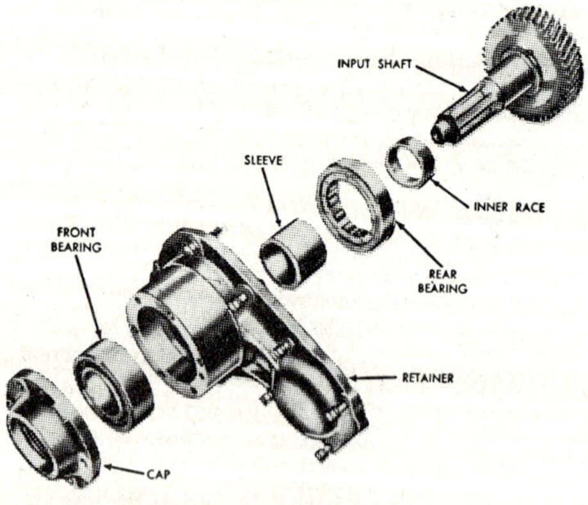

Fig. 11 Spicer 8341 Series input shaft parts

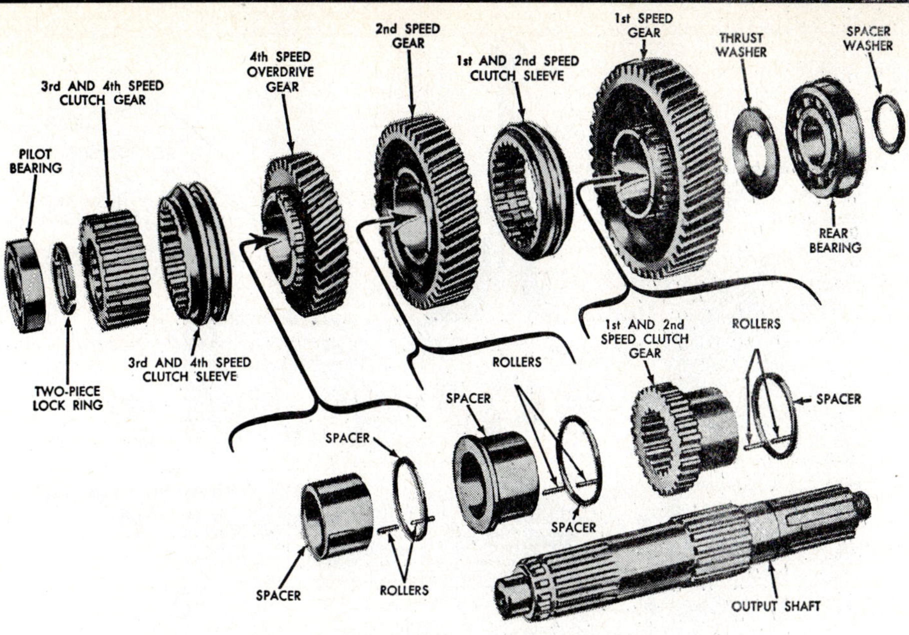

Fig. 12 Spicer 8341 Series output shaft parts

10. Place low speed gear on mainshaft with clutch teeth to front.
11. Grease one side of low speed gear thrust washer and place its greased side against low speed gear.

Transmission, Assemble

Reassembly of the transmission is largely the reverse of the disassembly procedure. However, the following should be noted:

1. Place reverse idler gear and sleeve in case with shift fork collar to rear of case.
2. Make sure flat on rear end of reverse idler shaft is lined up so countershaft rear bearing cap will lock shaft in place.
3. Main drive gear bearing cap is marked "Top" and should be so installed.
4. Be sure mainshaft rear bearing cap and gasket are installed with oil drain holes lined up.

8000, 8500, 8700 SERIES, 5 SPEED

Disassemble, Fig. 17

1. Shift transmission into reverse.
2. After removing cap screws, lift cover and shift forks off transmission.
3. Remove nut and pull off companion flange.
4. Remove mainshaft rear bearing cap.
5. Remove speedometer drive gear or spacer from mainshaft.
6. Slide mainshaft to rear until rear bearing is free of case. Then use a suitable puller to remove bearing.
7. Remove overdrive gear lower shift fork guide shaft and remove shift fork.
8. Remove transmission rear end cover.
9. Remove overdrive gear and bearing spacer from mainshaft.
10. Remove nut from rear end of countershaft and use a suitable puller to remove overdrive sleeve.
11. Remove countershaft overdrive gear and shift collar.
12. Slide mainshaft toward rear to bring center roller bearing out of case far enough to get a puller or "C" clamp on the outside diameter of the bearing. Then tap rear end of mainshaft toward front which will move bearing off shaft.
13. Loosen bearing puller or "C" clamp and slide mainshaft back and get another hold on center mainshaft roller bearing and again tap mainshaft forward as far as possible.
14. Repeat this procedure until center roller bearing is out of case and off mainshaft.
15. Remove clutch release yoke and shafts.
16. Remove cap screws from main drive gear bearing cap and with a suitable puller remove cap and main drive gear out of case, using the two puller screw holes provided.
17. Remove main drive gear pilot bearing from mainshaft.
18. Tilt front end of mainshaft upward and lift assembly out of case.
19. Remove cap screws and take off countershaft rear bearing retainer and thrust washer.

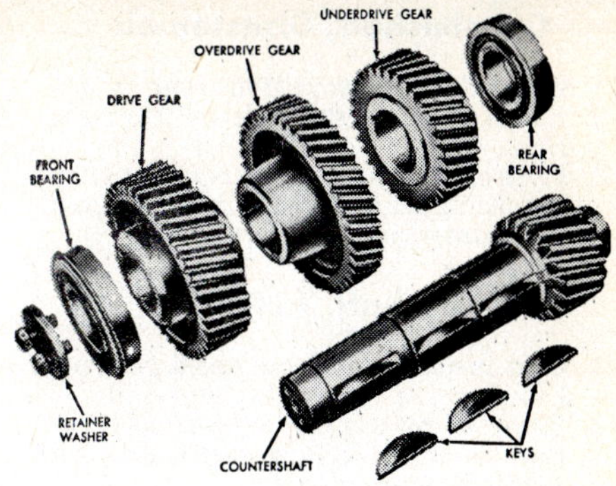

Fig. 13 Spicer 8341 Series countershaft parts

20. Place a suitable puller in threaded hole in reverse idler gear shaft and pull out shaft.
21. Remove reverse idler gear and sleeve and bearing assembly.
22. Move countershaft to the rear until rear bearing is out of case. Then use a puller to remove rear bearing.
23. Slide countershaft toward rear until front end is out of case. Then lift assembly out by its front end.

Main Drive Gear, Disassemble

1. Remove snap ring on inside of bearing cap.
2. Place bearing cap in press and push bearing and main drive gear out of cap.
3. Remove bearing nut and press bearing from main drive gear.

Mainshaft, Disassemble

1. Remove 3rd and 4th speed clutch collar.
2. Remove low speed gear and needle roller bearings, being careful not to lose any of the rollers.
3. Remove low and 2nd speed clutch gear.
4. Remove snap ring from front end of mainshaft and press gears off shaft. Use care to see that needle bearings and 3rd speed gear sleeve lock balls are not lost during the pressing operation.

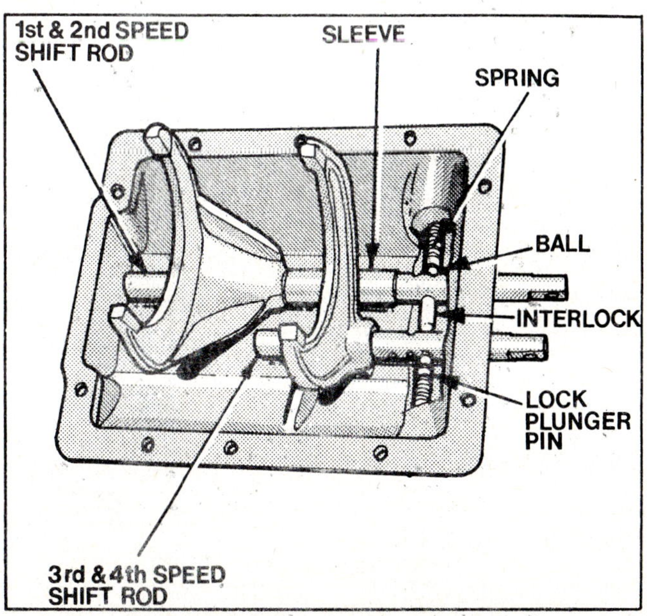

Fig. 14 Spicer 8341 Series gearshift housing

Countershaft, Disassemble

1. Use puller to remove front bearing.
2. Remove countershaft drive gear snap ring.
3. Press off countershaft gears one at a time by placing gears on bed of press and pushing countershaft out of gears.

Mainshaft, Assemble

1. Coat inside bore of low speed gear with heavy grease.
2. Place two rows of needle bearings (71 per row) on inside of low speed gear with spacer between rows.
3. Assemble needle bearings in 2nd and 3rd speed gears in same manner.
4. Place mainshaft in vise with rear end down.
5. Place 2nd speed gear on mainshaft with clutch teeth to rear of shaft.
6. Place 3rd speed gear sleeve on shaft with flanged end toward rear. Line up the two notches on inside of sleeve with mainshaft spline and place two sleeve lock balls in notches of sleeve. Then press sleeve on mainshaft.
7. Place 3rd speed gear on mainshaft sleeve with clutch teeth to front.
8. Press 3rd and 4th speed clutch gear on mainshaft with its hub to front. Then install snap ring in front of clutch gear.
9. Place 3rd and 4th speed clutch collar on clutch gear with longer hub to rear.
10. Remove mainshaft from vise and place low and 2nd speed clutch gear on mainshaft with shift fork collar to rear.
11. Place low speed gear on rear of mainshaft with clutch teeth to front.
12. Grease one side of mainshaft low speed gear thrust washer and place on mainshaft with greased side against low gear.

Transmission, Assemble

Reassembly of the transmission is largely a matter of reversing the disassembly procedure. However, the following should be noted:

1. Shift fork collar of reverse idler gear should be to rear of case.
2. When installing reverse idler gear shaft, make sure flat on rear of shaft is lined up so countershaft rear bearing retainer will lock reverse idler shaft in place.
3. Mainshaft pilot bearing must be assembled on shaft with snap ring in outer race to rear.
4. Main drive gear bearing cap is marked "Top" and it should be so installed.
5. Make sure countershaft overdrive gear clutch collar is installed with hub of collar toward gear. Make sure collar is a free slide fit on spline hub of gear.
6. When installing overdrive shift fork and shaft, place shift fork guide shaft in hole provided in rear of case. Tap it through case and line up shift fork and tap shaft on through hole in center support in main case.

1241 SERIES

4 Speed Auxiliary Unit, Disassemble, Figs. 18 & 19

1. Remove shifter housing, then lock transmission in two gears, remove yoke nuts, using a suitable puller remove companion flanges or yokes.

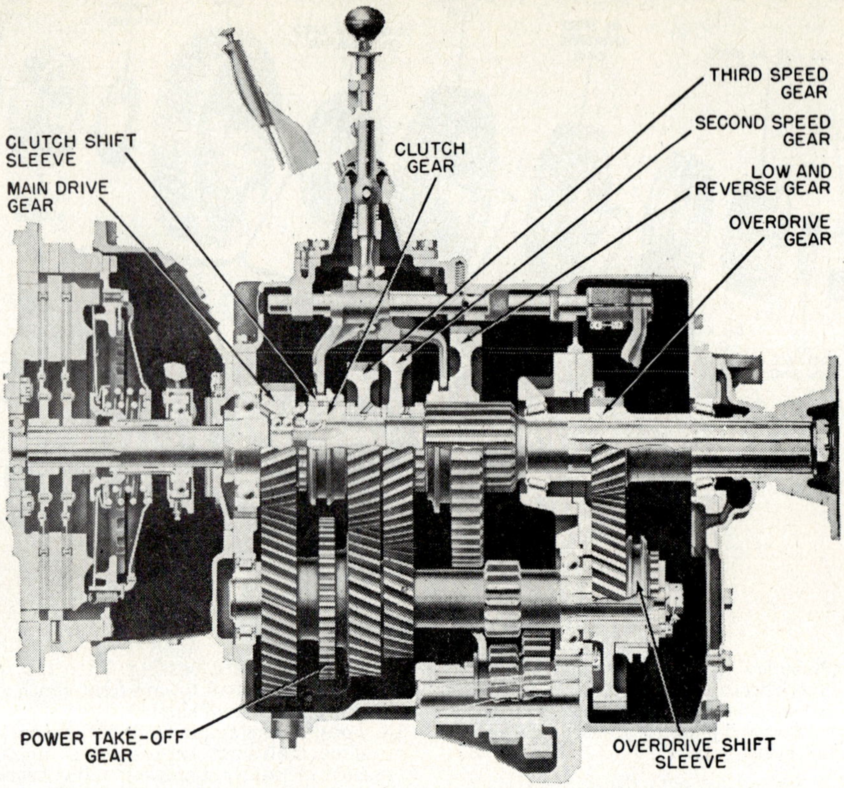

Fig. 15 Spicer Series 7751, 7851 five speed transmissions

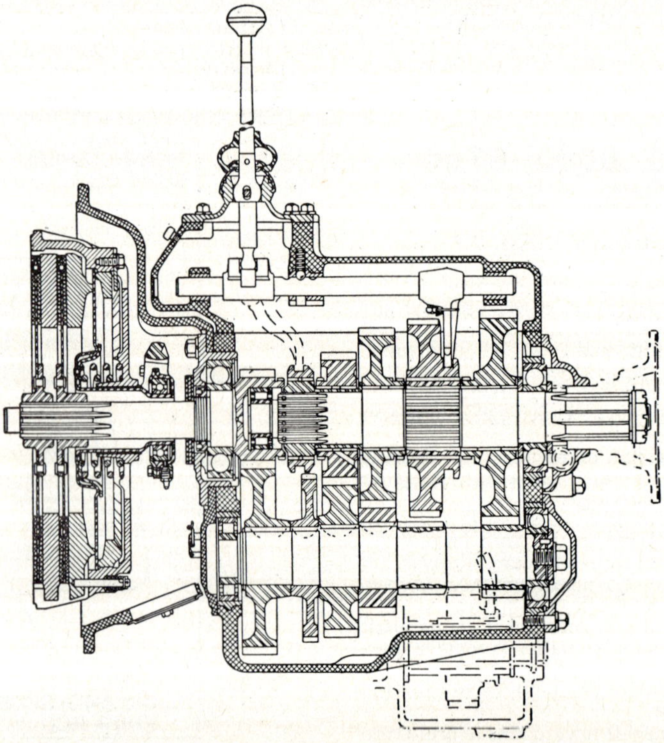

Fig. 16 Spicer 8000, 8500, 8700 Series four speed transmissions

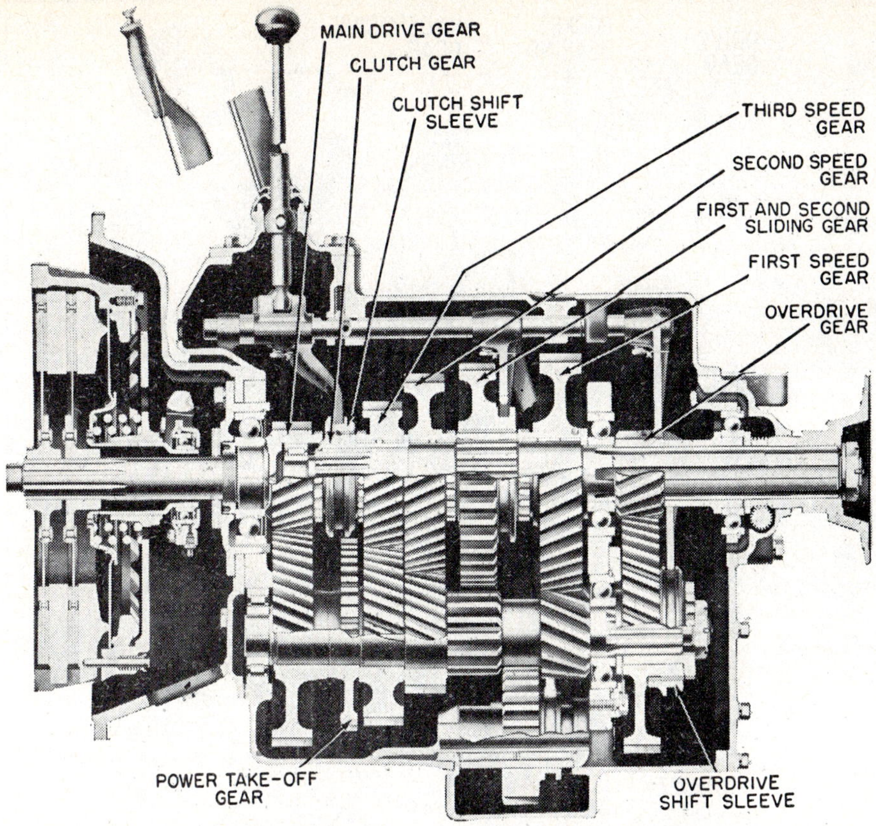

Fig. 17 Spicer 8000, 8500, 8700 Series five speed. Some models have double row ball bearing at rear of mainshaft

NOTE: Do not hammer on these parts as this could cause dimensional distortion and failure of these and other related parts.

2. Remove front transmission hanger (if used) then remove front bearing cap from bearing retainer. Remove front bearing seal only if it requires replacement.
3. Remove drive gear bearing retainer screws. Using a soft hammer, tap trunnion bearing face of drive gear bearing retainer until two pry bars can be inserted between case and retainer. Pry evenly with both bars until bearing retainer and drive gear can be removed.
4. Support bearing retainer on its flange and press drive gear free of bearing retainer, bearings and spacer.
5. Remove bearing and spacer from retainer then using a large screwdriver, pry out drive gear pocket bearing.
6. Remove mainshaft rear bearing caps, both the front and rear countershaft bearing caps then remove bearing retainer plate.
7. Force mainshaft rearward to expose mainshaft bearing snap ring, install puller on snap ring and remove bearing and spacer.
8. Force countershaft sub-assembly forward to expose front bearing snap ring install puller on snap ring and remove bearing.
9. Remove rear bearing, then using a lift strap, move countershaft sub-assembly rearward and remove bearing outer race and roller assembly from inner bearing race and shaft.

NOTE: Do not remove inner race from shaft unless bearing must be replaced.

10. With bearings removed from either or both countershafts, move the countershaft to side of case, then using a lift strap around clutch collar, lift the mainshaft sub-assembly up and out of the case.
11. Remove Lo-Lo gear, thrust washers, snap rings and clutch collar from mainshaft.
12. Remove snap ring located under bore of overdrive gear and slide overdrive gear, thrust washers, snap ring and underdrive gear and its parts off mainshaft then remove the remaining snap ring.

NOTE: All snap rings and thrust washers should be removed from the gear bores at this time for cleaning and inspection.

Countershaft, Disassemble

1. Using lift strap behind head end gear, lift right countershaft sub-assembly from case. Left side countershaft may be removed in the same manner.
2. Support head end with parallel bars as close to hub as possible and press countershaft out of gear.
3. Support underdrive gear with parallel bars as close to hub as possible and press countershaft out of gears.

Countershaft, Reassemble

NOTE: Coat the bores of gears with oil before pressing them onto the countershaft. Install the keys for each gear one at a time as the countershaft is pressed into each gear.

1. Press key closest to countershaft gear into place. Support underdrive gear with long hub facing up and press shaft key into gear.
2. Press second key into place. Support overdrive gear with long hub facing down and press shaft and key until gear seats firmly against face of underdrive gear.
3. Press third key into place. Support direct drive gear with long hub facing up and press shaft and key until gear seats firmly against face of overdrive gear.

NOTE: Tooth timing mark "V" must align with the center of gear keyway and the third key must be slightly under the face of the direct drive shaft.

Mainshaft, Reassemble

NOTE: Prior to assembly, inspect all thrust washers in gear bores for heavy face galling or gear tooth indentations. Replace as necessary.

1. Lubricate thrust washers with #30 engine oil.
2. Install clutch collar over rear end of mainshaft ahead of first snap ring groove at rear of shaft and install snap ring into last groove on shaft.
3. Install Lo-Lo gear with snap ring and thrust washers on rear of shaft and locate internally splined thrust washer against snap ring then install spacer with flanged portion toward face of Lo-Lo gear and shoulder of mainshaft.
4. From the front of the mainshaft install the second snap ring in the groove and install underdrive gear sub-assembly along with snap ring and thrust washers.

NOTE: Internally splined thrust washer should rest against shaft snap ring.

5. Install overdrive gear sub-assembly with snap ring and thrust washers, the gear internal snap ring should face toward the underdrive gear.
6. Place snap ring on last groove on mainshaft, install second clutch collar on front of mainshaft then place assembly in the bore and in mesh with overdrive gear.

Drive Gear Components, Reassembly

1. Press drive gear into the first bearing until race of bearing bottoms against front face of drive gear.
2. Press the pocket bearing firmly against pocket shoulder into the drive gear with the pocket bearing part number visible when pressed in bore.
3. Install spacer followed by the bearing retainer onto the drive gear bearing, then

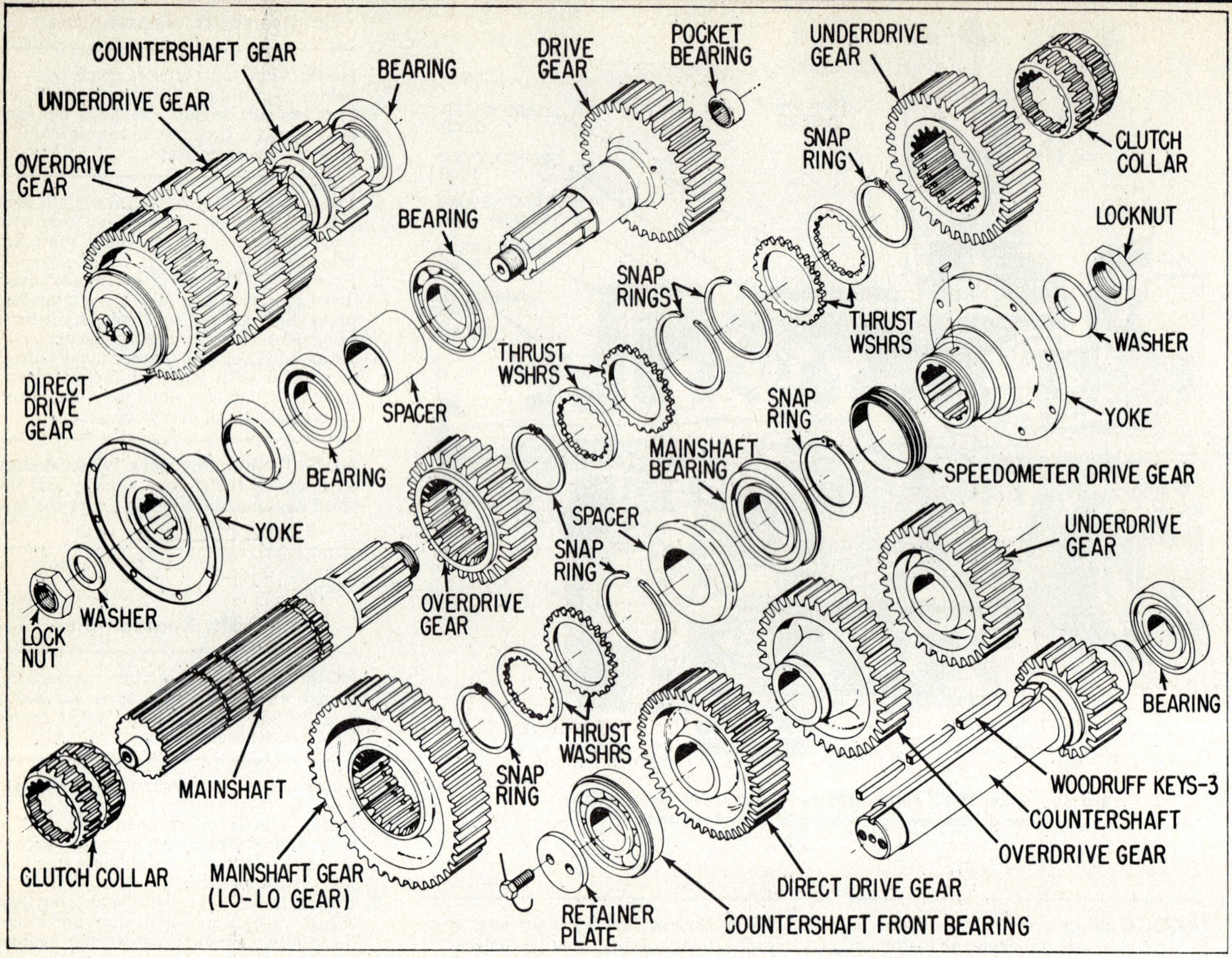

Fig. 18 Gear assembly. Spicer 1241 four speed auxiliary transmission

install the second bearing retainer and bearing and press until it seats firmly against face of spacer and shoulder of bore in the retainer.

4. Assemble front bearing cap and gasket onto bearing retainer and torque screws to 40–50 ft. lbs.

Countershafts, Installation

1. Place either of countershaft sub-assemblies into the left side of case (as viewed from rear of case) with timing mark of head end gear toward center of case.

NOTE: This mark must be aligned with the drive gear timing mark later during reassembly.

2. Place remaining countershaft sub-assembly into the right side of the case with the timing mark on the head end gear facing toward center of case.

NOTE: Do not install front or rear countershaft bearings until the mainshaft sub-assembly is installed in the case.

Mainshaft Sub-Assembly, Installation

1. Using a chain hoist and sling around Lo-Lo underdrive shift collar, lower mainshaft assembly into the approximate center of case rear bore then block mainshaft front stem with a piece of wood or brass plate.
2. Install mainshaft rear bearing on rear of shaft and drive bearing onto shaft and into bore of case then install the bearing snap ring.
3. Install speedometer drive gear (if removed) and rear bearing caps and gaskets then dip screws in sealer and torque to 60–80 ft. lbs.
4. Install front drive gear and bearing cap assembly then dip capscrews in sealer and torque to 60–80 ft. lbs.

NOTE: With all timing gears painted, bring timing teeth of countershaft head end gears, pointing toward center of the case and the two timing teeth aligning with the timing teeth of the countershaft gears.

5. Remove sling from mainshaft clutch collar and disengage shift collar from under overdrive gear and move clutch collar to neutral position on the mainshaft.

Final Assembly

1. Assemble rear bearing inner race onto the countershaft then locate the sling behind the head end gear. Centralize countershaft to case front bore and align timing teeth with the drive gear timing tooth. Install countershaft bearing onto shaft and into case bore, seat bearing snap ring into case bore.
2. Install retaining plate and leave screws loose then lock transmission in two gears by placing clutch collar into Lo-Lo gear bore and direct drive gear bore.
3. Install outer race and roller assembly of countershaft rear bearing onto shaft and into bore of case against shaft shoulder.
4. Install countershaft rear bearing cap and gasket and torque screws to 40–50 ft. lbs., torque retaining plate screws to 60–80 ft. lbs., then install countershaft front bearing cap and gasket and torque screws to 40–50 ft. lbs.
5. Move clutch collars into their neutral positions on the mainshaft, align timing marks on right side countershaft with drive gear timing marks which are parallel to bottom of case, if they were moved out of mesh.

6. Assemble rear bearing inner race to countershaft and using a sling on right countershaft in same manner as before, mesh head and gear timing teeth into drive gear timing tooth.
7. With timing teeth in place, install countershaft bearing onto the front of the right shaft and into bore of case, then install snap ring against face of case and install shaft retaining plate.
8. Install outer race and roller assembly onto shaft and into bore of case and against shoulder of shaft.
9. Install countershaft rear bearing cap and torque screws to 40–50 ft. lbs. then install countershaft front capscrews and torque to 60–80 ft. lbs. and install front bearing cap and torque screws to 40–50 ft. lbs.
10. Move clutch collars back into bore of gears and lock transmission in two gears then place front transmission hanger over front bearing cap.
11. Install end yokes or flanges onto drive gear stem and mainshaft splines, then install washers and locknuts, torque lock nuts to 550–600 ft. lbs.
12. Move clutch collar out of bore of gear and leave it engaged with drive gear, then turn drive gear end yoke or flange turning transmission. If timing teeth have been set correctly, the entire gear train will turn freely. If the unit locks up after several turns, the timing teeth have not been aligned correctly.
13. If transmission locks up, disengage shift collar from drive gear and turn drive gear in reverse rotation until timing marks or paint marks come into view showing alignment or misalignment. If misalignment has occurred it will be necessary to retime the shafts.
14. If timing is correct, move clutch collar to its neutral position on the mainshaft and install shifter housing.

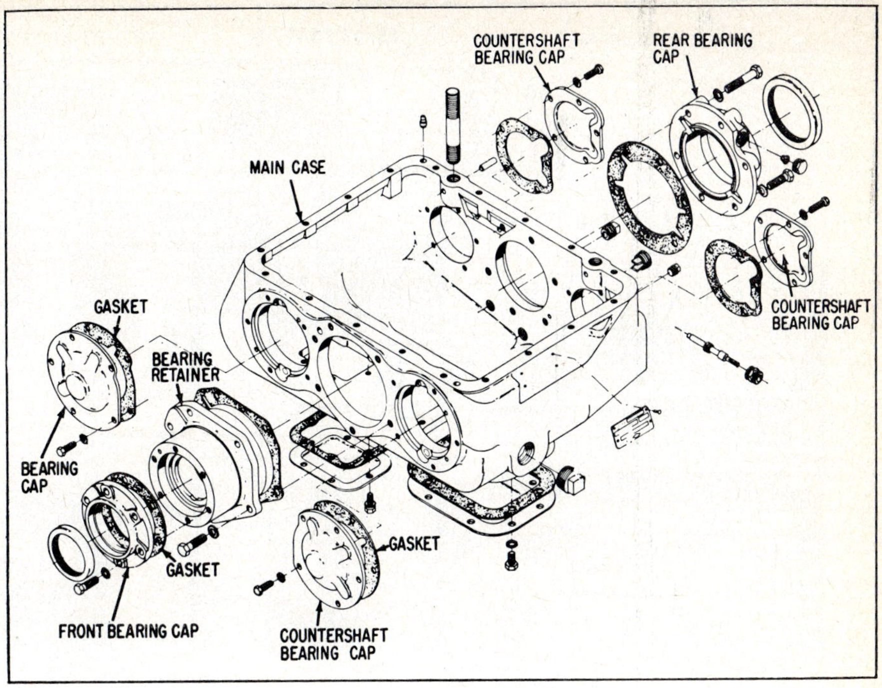

Fig. 19 Case assembly. Spicer 1241 four speed auxiliary transmission

6041 SERIES

4 Speed Auxiliary Disassemble, Fig. 20

1. Remove transmission cover.
2. Remove both propeller shaft flanges.
3. Remove support from main drive gear bearing cap.
4. Remove countershaft rear bearing cap, noting number of shims under cap.
5. Remove mainshaft rear bearing cap, oil seal, and speedometer driven gear in cap. Then remove oil seal and driven gear from cap and sleeve.
6. Remove main drive gear bearing cap with bearings, gear, seal and seal cap.
7. Remove speedometer drive gear.
8. With suitable puller, press mainshaft with front pilot bearing, overdrive gear snap ring, overdrive and direct clutch gear, collar and overdrive gear out through front of case, leaving first speed gear, first and second clutch gear and collar, and second speed gear and sleeve in case. Remove these items from case after removing mainshaft.
9. Remove mainshaft rear bearing.
10. Drive countershaft toward rear far enough to force rear bearing cap out of case. Using snap ring pliers, remove countershaft second speed gear snap ring from groove, slide second gear forward, then lift countershaft assembly out of case.
11. Remove countershaft front bearing cup from case.
12. Remove power take-off covers, and drain and filler plugs from case.
13. Disassemble sub-assemblies as required. If countershaft is to be disassembled, gears should be marked so that correct position will be known when installing.

Countershaft, Assemble

1. Drive gear keys in countershaft slots. Then using arbor press and suitable driver, press countershaft overdrive gear and drive gear into place.
2. Install new snap ring in groove at forward end of drive gear.
3. Press cone on front end of countershaft, making sure that cone seats firmly against countershaft.
4. Install second speed gear front snap ring over end of countershaft and position snap ring in groove in countershaft.
5. Install second gear on countershaft, then position rear snap ring in groove.
6. Press bearing cone on rear end of countershaft, making sure cone seats firmly against countershaft.

Mainshaft, Assemble

1. Place mainshaft in vise with front end up. Then apply heavy gear oil on bearing surfaces.
2. Install overdrive gear and overdrive and direct clutch gear with shoulder toward overdrive gear.
3. Install snap ring in groove at front end of mainshaft.
4. Install overdrive and direct clutch gear collar with external teeth toward front end of mainshaft.
5. Install mainshaft pilot bearing, using suitable bearing replacer.

Main Drive Gear Assemble

1. Apply transmission lubricant to bearings before installing.
2. Press bearing cups into cap and install inner bearing on drive gear. Then install bearing spacer on drive gear.
3. Place drive gear and bearing in arbor press with bearing spacer in place.
4. Apply pressure on front bearing inner race, using a piece of 2″ diameter tubing about 4″ long over end of shaft. Bearing cup should turn readily but not spin on shaft with pressure applied. If bearing adjustment is too loose or too tight, select correct spacer (available in several sizes). After correct spacer has been installed, remove assembly from press.
5. If seal has been removed from seal cap, press new seal in cap.
6. Place washer in position against bearing cone. Then install seal cap with new gasket. Install cap bolts and tighten securely.

Transmission, Assemble

1. Install countershaft front bearing cup in transmission case bore.
2. Lower countershaft assembly in case with rear end of shaft out through rear bearing bore. Then move assembly forward so that front bearing enters cup.
3. Install drive gear bearing cap temporarily with new gasket. Tighten cap screws evenly and firmly. Then slide countershaft second speed gear on shaft and install snap ring in groove ahead of second speed gear.
4. Install rear bearing cup. Using same thickness of shims as was removed at disassembly, install rear bearing cap temporarily. Tighten cap bolts, meanwhile turning countershaft by hand. If any binding of countershaft bearings is noted before cap bolts are fully tightened, remove cap and add shims. If cap bolts are tightened with a shim pack which is too thin, bearings will be damaged. Three sizes of shims (.003″, .010″ and .030″) are available for use in adjusting countershaft bearings. Bearing adjustment is

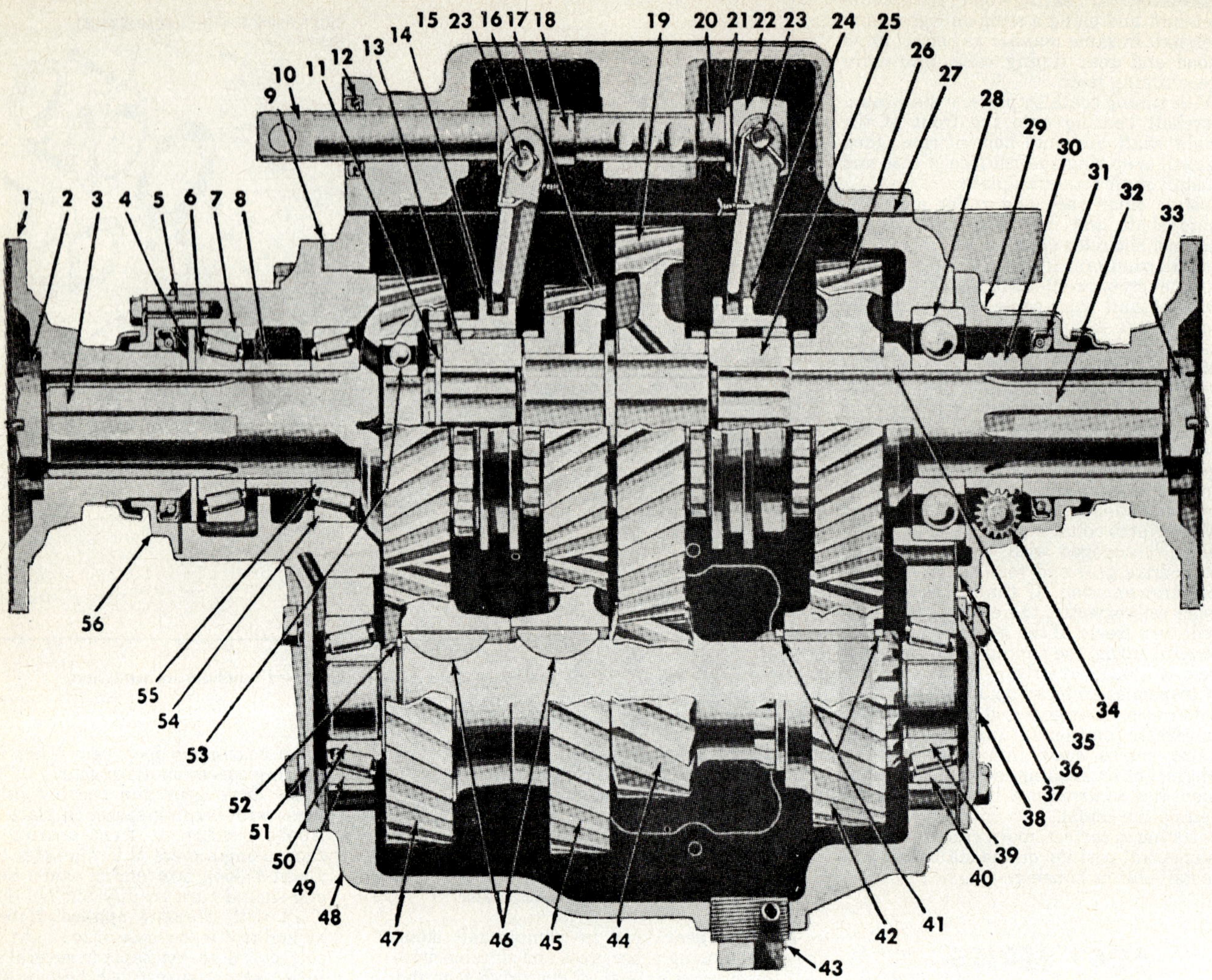

1 Propeller Shaft Flange
2 Main Drive Gear Nut
3 Main Drive Gear
4 Companion Flange Washer
5 Seal Gasket
6 Outer Bearing Cone
7 Outer Bearing Cup
8 Drive Gear Bearing Spacer
9 Bearing Cap Gasket
10 Direct and Overdrive Shift Rod
11 Mainshaft Overdrive Gear Snap Ring
12 Shift Rod Oil Seal
13 Overdrive and Direct Clutch Gear
14 Overdrive and Direct Clutch Collar
15 Shifter Housing (Cover)
16 Direct and Overdrive Shift Fork
17 Mainshaft Overdrive Gear
18 Shift Rod Spacer
19 Mainshaft 1st Speed Gear
20 1st and 2nd Shift Rod
21 Shift Rod Spacer
22 1st and 2nd Shift Fork
23 Shift Fork Screw
24 1st and 2nd Speed Clutch Gear Collar
25 1st and 2nd Speed Clutch Gear
26 Mainshaft 2nd Speed Gear
27 Shifter Housing Gasket
28 Mainshaft Rear Bearing Assembly
29 Mainshaft Rear Retainer Cap Assembly
30 Speedometer Drive Gear
31 Oil Seal
32 Mainshaft
33 Mainshaft Nut
34 Speedometer Driven Gear
35 Mainshaft 2nd Speed Gear Sleeve
36 Retainer Cap Gasket
37 Rear Bearing Cap Adjusting Shims (0. 003, 0. 010 and 0. 030)
38 Countershaft Rear Bearing Cap
39 Countershaft Rear Bearing Cone
40 Countershaft Rear Bearing Cup
41 Countershaft 2nd Speed Gear Snap Rings
42 Countershaft 2nd Speed Gear
43 Drain Plug
44 Countershaft 1st Speed Gear
45 Countershaft Overdrive Gear
46 Countershaft Keys
47 Countershaft Drive Gear
48 Transmission Case
49 Countershaft Front Bearing Cup
50 Countershaft Front Bearing Cone
51 Main Drive Gear Bearing Cap
52 Countershaft Drive Gear Snap Ring
53 Mainshaft Front Bearing
54 Bearing Cup
55 Bearing Cone
56 Seal Cap

Fig. 20 Spicer Series 6041 four speed auxiliary

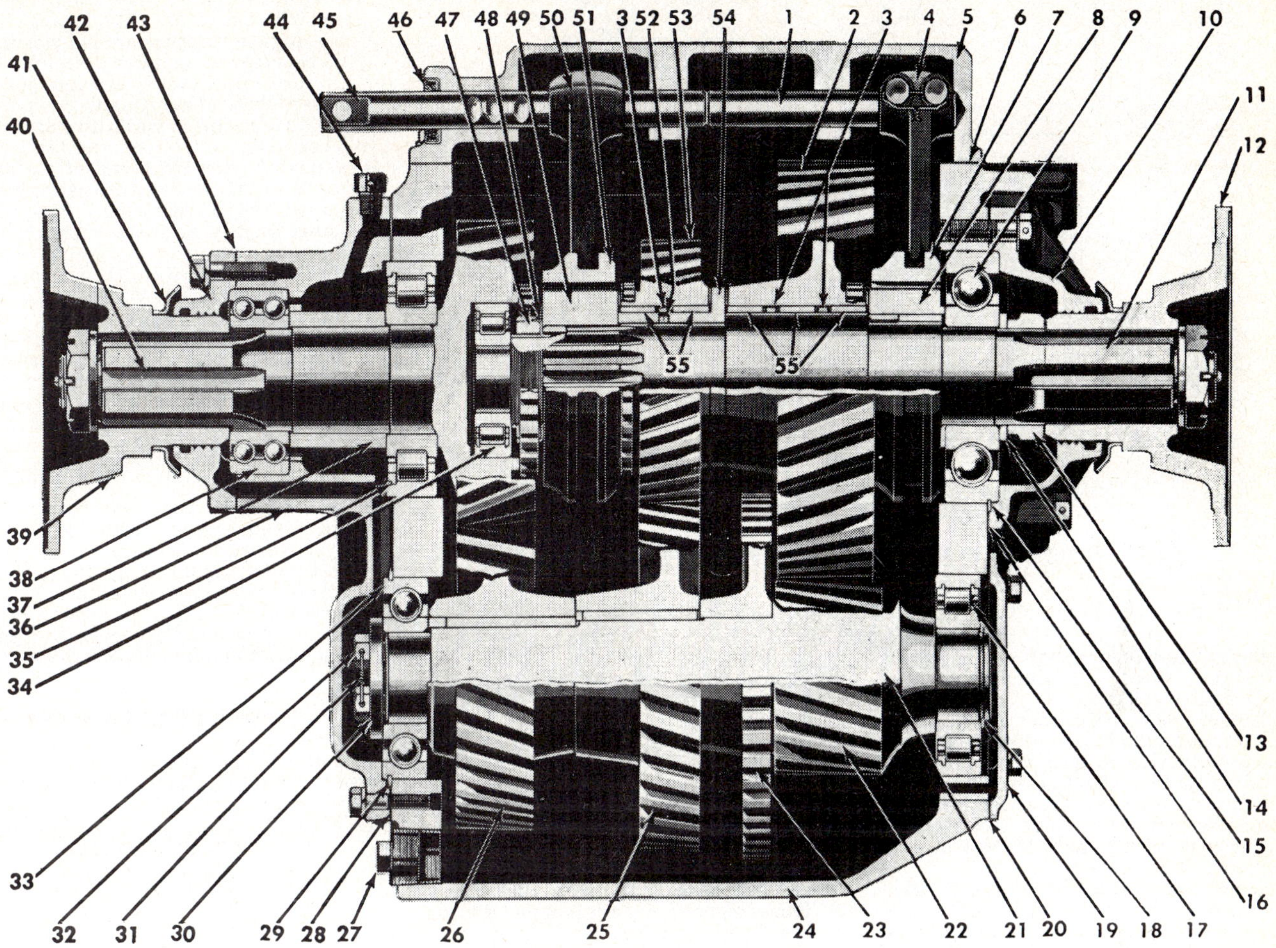

1 Low (1st) Shift Rod
2 Low (1st) Speed Gear
3 Bearing Spacer
4 Low (1st) Speed Shift Fork
5 Cover
6 Cover Gasket
7 Low (1st) Speed Shift Collar
8 Low (1st) Speed Clutch Gear
9 Mainshaft Rear Bearing
10 Mainshaft Rear Bearing Cap
11 Mainshaft
12 Flange
13 Speedometer Drive Gear
14 Spacer Washer
15 Snap Ring
16 Gasket
17 Countershaft Rear Bearing
18 Bearing Snap Ring
19 Countershaft Rear Bearing Cap
20 Cap Gasket
21 Countershaft
22 Low (1st) Speed Gear Teeth (8031-C)
Low (1st) Speed Gear (8031-G, 8035-G)
23 Power Take-Off Drive Gear
24 Case
25 Countershaft Overdrive Gear
26 Countershaft Drive Gear
27 Drain Plug
28 Bearing Cap Gasket
29 Bearing Snap Ring
30 Bearing Retainer Washer
31 Lock Wire
32 Washer Cap Screw
33 Countershaft Front Bearing
34 Drive Gear Pocket Bearing
35 Drive Gear Rear Bearing
36 Drive Gear Bearing Cap
37 Bearing Spacer
38 Drive Gear Front Bearing
39 Propeller Shaft Flange
40 Main Drive Gear
41 Guard
42 Front Bearing Cap
43 Gasket
44 Vent
45 Direct and Overdrive Shift Rod
46 Shift Rod Oil Seal
47 Clutch Gear Lock Nut
48 Lock Washer
49 Clutch Gear
50 Direct and Overdrive Shift Fork
51 Shift Collar
52 Locating Pin
53 Mainshaft Overdrive Gear
54 Sleeve
55 Needle Bearings

TPM-1072

Fig. 21 Spicer Series 8031, 8035 three speed auxiliary transmissions

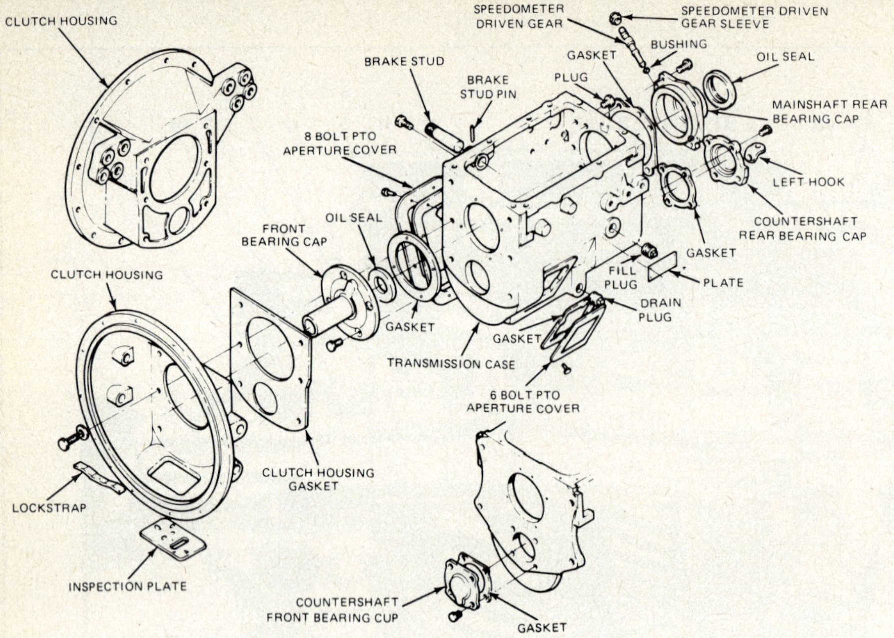

Fig. 22 Transmission case exploded view. CM 5000 & CM 6000 series

correct when countershaft can be turned without evidence of binding and without perceptible end play.

5. Remove main drive gear bearing cap. Lower mainshaft first speed gear, first and second clutch gear and collar, and second speed gear into case. Then guide mainshaft assembly through front of case and through gears. Slide second speed gear sleeve into hub of second gear through rear of case.
6. Using new gasket, install main drive gear bearing cap with main drive gear and bearings on front of case, guiding mainshaft pilot bearing into pocket of main drive gear. Tighten cap bolts securely.
7. Press mainshaft rear bearing onto mainshaft and into case, then place speedometer drive gear on shaft with shoulder toward front of transmission.
8. Install mainshaft rear bearing cap and gasket and tighten capscrews securely.
9. Install propeller shaft flange on mainshaft and tighten nut to a torque of 350–400 ft. lbs.
10. Place transmission front support bracket on trunnion at bearing cap, install front flange and tighten nut to a torque of 350–400 ft. lbs.
11. Install cover, being sure that shift forks enter grooves in shift collars.
12. Install breather in cover, and drain plugs.

8031, 8035 SERIES

3 Speed Auxiliaries Disassemble, Fig. 21

1. Remove transmission cover.
2. Lock mainshaft by placing gears in two speeds at same time, remove nuts and pull off both flanges.
3. Remove front bearing cap and bearing as a unit.
4. Remove washer from front end of countershaft.
5. Remove mainshaft rear bearing cap and pull speedometer drive gear and washer off mainshaft.
6. Move main drive gear forward as far as possible and force mainshaft and gears rearward, thereby forcing mainshaft rear bearing out of case.
7. Use a suitable puller to remove bearing from mainshaft.
8. Slide low speed shift collar off mainshaft and remove through rear of case.
9. Raise front end of mainshaft and lift from case, passing low speed gear teeth through power take-off gear teeth.
10. Move main drive gear rearward, forcing bearing out of bore. Then lift assembly out through top of case.
11. Remove countershaft rear bearing cap. Then push countershaft forward until front bearing is free of case, after which use puller to remove bearing.
12. Move countershaft rearward until rear bearing is out of case. Then remove snap ring and bearing from countershaft. The assembly may then be lifted out of the case by its front end.

Mainshaft, Disassemble

1. Support mainshaft assembly with front side of low speed gear on bed plate of arbor press.
2. Apply pressure on rear of mainshaft to force low speed clutch gear off mainshaft. Use care to prevent loss of needle bearings as mainshaft is pressed out of gear.
3. Grip mainshaft in vise with front end up. Then remove mainshaft pilot bearing from shaft, and remove clutch gear lock nut.
4. Set mainshaft in arbor press with flange on sleeve resting on bed plate. Press on front end of mainshaft to force sleeve with mainshaft overdrive gear and bearings and clutch gear off mainshaft.
5. Slide gear off sleeve, being careful not to lose bearings.

Mainshaft, Assemble

1. Place locating pin in hole in sleeve. Apply heavy grease on surface of sleeve to hold bearings in place. Then stand sleeve on end with flange down and arrange needle bearings and spacer around sleeve. Use 63 bearings in each row. Carefully lower overdrive gear over sleeve and spur teeth toward front of mainshaft.
2. Hold mainshaft in vise with front end up. Then lower overdrive gear and sleeve assembly over front end of mainshaft, carefully guiding head of locating pin into milled slot of mainshaft. Press sleeve flange tightly against mainshaft shoulder.
3. Press direct drive clutch gear on mainshaft up against sleeve. Then install lock washer and nut. Tighten nut securely and bend lock washer against flats on nut.
4. Grip mainshaft in vise with rear end up. Apply heavy transmission grease on mainshaft surface contacted by low speed gear rollers. Then arrange two rows of needle bearings separated by a spacer around mainshaft. Use 63 rollers in each row.
5. Set low speed gear over bearings with spur teeth upward. Drop other spacer over rearward end of mainshaft and insert remaining 63 rollers at gear hub.
6. Install clutch gear on splines at rear of gear with flat side of gear toward front.
7. Install pilot bearing on forward end of mainshaft and install shift collar.

Transmission, Assemble

Reassembly procedure is largely the reverse of disassembly procedure. However, the following should be noted:

1. When installing main drive gear and mainshaft, place main drive gear in place from inside of case, moving the assembly forward as far as possible. Lower mainshaft and gears into case with rear end extending through rear bearing bore. Install low gear shift collar on clutch gear, inserting collar through bearing bore in case.
2. Move mainshaft forward so that mainshaft pilot bearing enters pocket of main drive gear. Then install rear mainshaft bearing, spacer washer and speedometer drive gear.
3. Be sure to install front bearing cap gasket so oil passage is not obstructed.
4. Companion flange nuts should be torque tightened to 350–400 ft. lbs. on 1970–72 models or 400–450 ft. lbs. on all except 1970–72 models.

CM-5000, 6000

Transmission, Disassembly

1. Remove shifter housing from transmission. Use care not to lose poppet balls and springs.
2. Using a suitable tool, engage the 2-3 synchronizer with 2-3 speed gear and the 4-5 synchronizer with the 4-5 speed gear. Lock transmission in the two gears.
3. Remove output flange retaining nut from mainshaft. Remove output flange with a suitable puller.
4. Remove rear main bearing cap to transmission case bolts and the bearing cap, Fig. 22. Inspect bearing cup and oil seal and replace as necessary.
5. Remove speedometer gear from main-

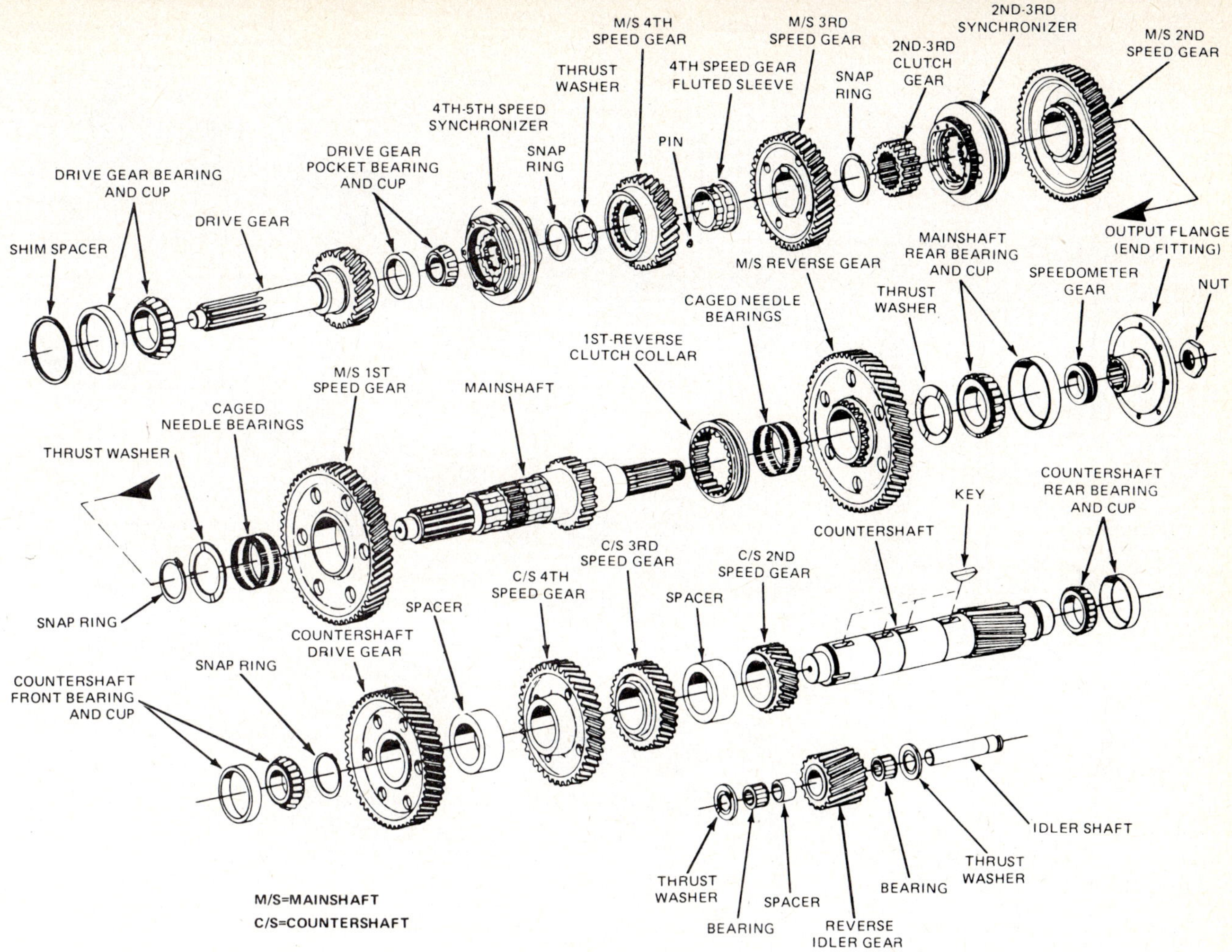

Fig. 23 Gear train assembly exploded view. CM 5000 series

shaft.

6. Remove drive gear front bearing cap bolts and the bearing cap. Inspect bearing cup and oil seal and replace as necessary.

NOTE: A split-ring shim is placed behind the bearing cup to adjust mainshaft end play and should not be replaced unless the end play is not within specifications after assembly. Mainshaft end play should be .003–.008 inch.

---

7. Remove drive gear from case. Use a suitable puller to remove bearing inner race if bearing replacement is required. Inspect drive gear pocket bearing cup for damage. On CM-5000 units if the cup is damaged, the drive gear must be replaced since the cup is machined into the gear. On CM-6000 units the cup may be pressed out and replaced.
8. Remove mainshaft assembly from transmission case.
9. Remove countershaft rear bearing cap.

NOTE: The countershaft may shift toward the rear of transmission, allowing the bearing cup to fall from case bore. Use caution when removing bearing cap to prevent damage to rear bearing or cup.

---

10. Remove reverse gear idler shaft with suitable puller while holding reverse idler gear and bearings inside case. Remove idler gear, bearings, spacer located between caged bearings, and thrust washer from the case.
11. To remove countershaft, carefully pry countershaft toward rear of transmission until rear bearing cup, if not previously removed, is free of case bore, and front bearing is clear of front bearing cup, then remove countershaft from case.
12. Inspect countershaft front bearing cup and if damaged, remove clutch housing and gasket. Then, remove spacer ring and from inside case, tap bearing cup forward with a hammer and drift.

Mainshaft Service

Disassembly, CM5000 Fig. 23

1. Remove rear bearing, thrust washer and reverse gear using suitable puller. Remove needle bearings from end of shaft.
2. Remove 1st and reverse clutch collar from shaft.
3. Remove drive gear pocket bearing and 4th and 5th speed synchronizer using suitable puller.
4. Remove 4th speed gear snap ring, 4th speed gear and thrust washer.
5. Remove 3rd speed gear snap ring, thrust washer and 3rd speed gear. Remove 2nd and 3rd speed synchronizer.
6. Remove 2nd and 3rd speed clutch gear snap ring, then the 2nd and 3rd speed clutch gear using suitable puller. Remove 2nd speed gear from shaft using caution not to lose needle bearings.
7. Remove two keys from mainshaft and three needle bearing spacer rings. Remove 1st speed snap ring, thrust washer and 1st speed gear.

Assembly, CM 5000

1. Lubricate needle bearings with grease, then install into 1st speed gear bore. Position mainshaft vertically and install 1st speed gear onto mainshaft, and ensure that clutch teeth face downward.
2. Install lubricated thrust washer onto mainshaft, then install 1st speed gear snap ring onto mainshaft.
3. Install needle bearing spacer onto rear of mainshaft, then lubricate bore of 2nd speed gear and assemble two rows of needle bearings in bore with spacer between two rows. Position second spacer in clutch teeth end of bore, then install 2nd speed gear onto mainshaft with clutch teeth facing upward.
4. Install keys into mainshaft, then align keyways in 2nd and 3rd speed clutch gear and press clutch gear onto mainshaft using suitable press. Install snap ring.
5. Assemble 2nd and 3rd speed synchronizer onto 2nd and 3rd speed clutch gear.
6. Lubricate mainshaft, then install 3rd

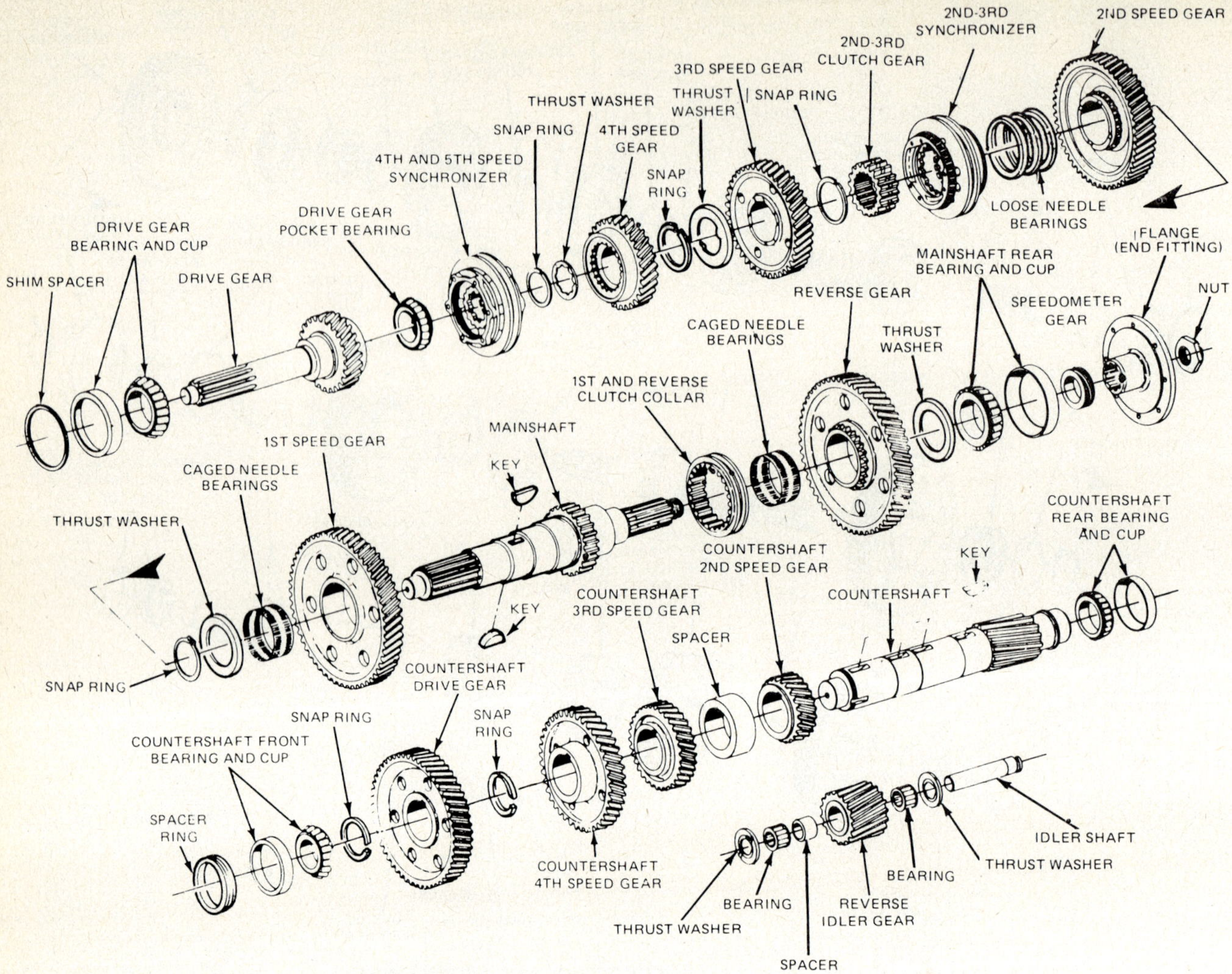

Fig. 24 Gear train assembly exploded view. CM 6000 series

speed gear onto shaft with clutch teeth facing downward. Lubricate thrust washer and install thrust washer and snap ring onto shaft.

7. Lubricate 4th speed gear and install onto mainshaft with hub facing upward. Install lubricated thrust washer, then the snap ring onto shaft.
8. Install 4th and 5th speed synchronizer onto mainshaft with larger brass ring positioned over 4th speed gear hub.
9. Using suitable press, press drive gear pocket bearing onto front of mainshaft until bearing is seated against mainshaft shoulder.
10. Install 1st and reverse clutch collar onto shaft. Lubricate reverse gear needle bearings, then install bearings and reverse gear onto shaft with clutch teeth of reverse gear facing 1st speed gear.
11. Install lubricated thrust washer onto shaft, then press rear bearing onto mainshaft until seated against reverse gear thrust washer.

Disassembly, CM 6000 Fig. 24

1. Remove mainshaft rear bearing, thrust washer and reverse gear using suitable tools. Remove reverse gear bearings and 1st reverse clutch collar.
2. Remove 4th and 5th speed synchronizer and front drive gear pocket bearing using suitable puller.
3. Remove 4th speed gear snap ring, thrust washer and 4th speed gear from mainshaft.
4. Remove 4th speed gear sleeve and 3rd speed gear using suitable puller.
5. Remove 2nd and 3rd speed clutch gear snap ring, 2nd and 3rd speed synchronizer and 2nd and 3rd speed clutch gear. Remove 2nd speed gear from mainshaft.
6. Remove snap ring, thrust washers and 1st speed gear.
7. Remove needle bearings from mainshaft.

Assembly, CM 6000

1. Lubricate needle bearings and install into 1st speed gear.
2. Install 1st speed gear with clutch teeth facing rear of mainshaft.
3. Install lubricated thrust washer, then the snap ring onto mainshaft.
4. Install 2nd speed gear onto mainshaft with clutch teeth facing front of shaft, then install 2nd and 3rd speed clutch gear, 2nd and 3rd speed synchronizer with long hub facing 2nd gear and 2nd and 3rd speed clutch gear snap ring.
5. Install 3rd speed gear with clutch teeth facing rear of shaft. Align pin in bore of 4th speed gear sleeve with spline of mainshaft, then position sleeve in place with flange facing 3rd speed gear.
6. Install 4th speed gear sleeve using suitable tool, then install 4th speed gear, lubricated thrust washer and snap ring onto mainshaft.
7. Install 4th and 5th speed synchronizer onto mainshaft with larger brass ring facing 4th speed gear.
8. Install drive gear pocket bearing onto mainshaft with inner race positioned against mainshaft shoulder.
9. Lubricate needle bearings and install into bore of reverse gear. Install 1st and reverse clutch collar, then install reverse gear onto mainshaft with clutch teeth facing 1st speed gear.
10. Lubricate thrust washer and install onto mainshaft against reverse gear. Install rear bearing using suitable tool.

Countershaft Service

Disassembly, CM 5000 Fig. 23

1. Remove front and rear bearings using suitable puller, then remove drive gear snap ring.
2. Support drive gear close to hub, then press countershaft from gear using suitable press. Remove key from countershaft.
3. Remove 4th speed gear snap ring. Support 3rd speed gear, then press countershaft until free from 3rd speed gear and 4th speed gear. Remove keys from countershaft.

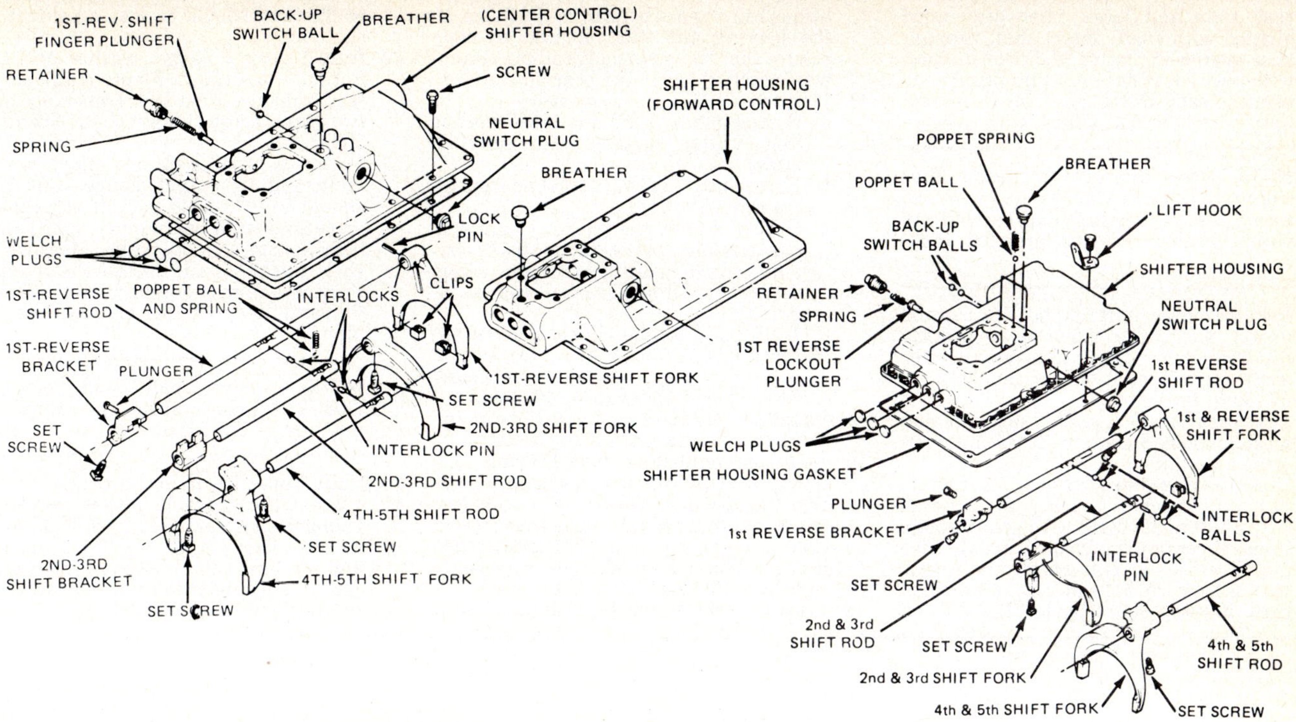

Fig. 25 Shifter housing exploded view. CM 5000 & CM 6000 series

4. Remove spacer, then support 2nd speed gear and press countershaft from gear using suitable press. Remove key from countershaft.

Assembly, CM 5000

1. Install 2nd speed gear key in countershaft, then support hub of 2nd speed gear on suitable press with long hub facing downward and press countershaft into gear. Ensure that key in countershaft aligns with keyway in gear.
2. Install spacer, then the 3rd speed gear key onto countershaft. Install gear onto countershaft as described in previous step. Press shaft until 3rd speed gear seats against spacer.
3. Press countershaft into 4th speed gear using suitable press. Ensure that countershaft key aligns with gear keyway. Install snap ring.
4. Install drive gear key into countershaft, then press shaft into drive gear with long hub facing towards rear of shaft. Ensure that countershaft key and gear keyway align.
5. Install snap ring, then install front and rear bearings using suitable tool.

Disassembly, CM 6000 Fig. 24

1. Remove front and rear bearings using suitable puller, then remove drive gear snap ring.
2. Support drive gear close to hub and press countershaft from gear using suitable press. Remove key and spacer.

NOTE: Spacers used on countershaft are interchangeable.

3. Support 3rd speed gear and press countershaft from 4th speed and 3rd speed gears. Remove keys and spacer from shaft.
4. Support 2nd speed gear and press countershaft from gear using suitable press. Remove key from shaft.

Assembly, CM 6000

1. Install 2nd speed gear key into countershaft, then support hub of 2nd speed gear and press countershaft onto gear using suitable press. Ensure that countershaft key and gear keyway align.
2. Install spacer, then the 3rd speed gear key into shaft. Support hub of 3rd speed gear with long hub facing front of shaft, then press shaft onto gear. Ensure that countershaft key and gear keyway align and that 3rd speed gear seats against spacer.
3. Install 4th speed gear key into countershaft. Support hub of 4th speed gear with long hub facing front of shaft, then press shaft onto gear. Ensure that countershaft key aligns with gear keyway.
4. Install spacer, drive gear key and drive gear onto countershaft and support hub of drive gear with long hub facing rear of shaft. Press shaft onto gear ensuring that countershaft key and gear keyway align. Install snap ring.
5. Install front and rear bearings onto countershaft using suitable tool.

Shifter Housing Service

Disassembly, Fig. 25

1. Remove three poppet springs and balls from top of shifter housing. Invert shifter and position 1st and reverse and 2nd and 3rd shift forks into neutral position.
2. Remove screw from 4th and 5th shift fork. Tap rod forward until end of rod forces welch plug from housing. Use caution not to lose interlock balls. Continue to remove rod, then remove shift fork. Repeat procedure for 2nd and 3rd and 1st and reverse rods. When 1st and reverse shift rod is removed, remove 1st and reverse shift fork and bracket from housing.
3. Remove interlock balls if not done so. Remove back-up switch, then the reverse back-up switch balls.

Assembly

1. Install lockout plunger into 1st and reverse bracket. Lubricate 1st and reverse shift rod and install partially into shifter housing. Align bracket with shift rod and install rod through bracket, bosses and 1st and reverse shift fork. Install screws and torque to 40–50 ft. lbs.
2. Install interlock balls into 2nd and 3rd shift rod and install partially into housing. Install interlock pin and hold in place with suitable grease. Align shift fork with rod and install rod into boss until in detent position. Install screw and torque to 40–50 ft. lbs.
3. Place 2nd and 3rd shift rod in neutral position, then install interlock balls in cross hole of 4th and 5th shift rod boss. Install 4th and 5th shift rod, then install shift fork. Install screw and torque to 40–50 ft. lbs. Install welch plugs and poppet springs and balls.

Assembly

1. Lower countershaft into case and properly position with bearing cups.
2. If removed, install countershaft front bearing cup and spacer ring.
3. Install countershaft front bearing into front cup. Start rear bearing cup into case, then ensuring bearing is properly aligned, drive bearing cup into case bore.
4. Lubricate and install caged pocket bearings with spacer between bearings into reverse idler gear. Coat reverse idler gear thrust washers with grease and position against both ends of gear with tangs facing upward and outward to align with slots in case bore. Hold reverse idler gear assembly inside case with tangs of thrust washers facing upward and insert reverse idler shaft partially into case bore with shoulder and threaded hole facing outward. Align thrust washer tangs with

case slots and lower idler gear until aligned with shaft. Insert shaft through idler gear with flat on shoulder aligned with position of flat on countershaft rear bearing cap. Drive idler shaft inward until bottom of flat is flush with case.

NOTE: When driving the reverse idler shaft into position, it may be necessary to secure the countershaft rear bearing cup with a retainer bolt and large flat washer.

5. Install counter shaft rear bearing cap and gasket. Ensure that flat on side of bearing cap is aligned with flat on reverse idler shaft. Install and torque cap bolts to 60–80 ft. lbs.
6. Lower mainshaft assembly into case and mesh mainshaft gears with mating countershaft gears.
7. Press bearing onto drive gear, if removed. Lubricate pocket bearing at front of mainshaft and install drive gear through front of case onto mainshaft. Install drive gear bearing cap and gasket and ensure that oil return hole in cap, gasket and case is properly aligned. Install and torque bearing cap bolts to 25–32 ft. lbs. If the bearing cup was replaced in cap, ensure that the mainshaft end play shim was installed under the bearing cup.
8. If clutch housing was removed:
   a. Coat clutch housing and transmission case mating surfaces with suitable sealer.
   b. Place clutch housing gasket on front of transmission case.
   c. Ensuring that the gasket clears the countershaft front bearing spacer, install clutch housing onto case using drive gear front bearing cap as a guide.
   d. Install and torque clutch housing attaching bolts to 120–150 ft. lbs. Do not overtorque since mainshaft and countershaft end play will be affected.
9. Install oil seal and bearing cup into mainshaft rear bearing cap, if removed.
10. Lubricate mainshaft rear bearing and install bearing cap and gasket into position. Apply equal pressure to bearing cap to prevent damage to bearing cup. Ensure that oil holes in case, gasket and cap are properly aligned. Install and torque bearing cap bolts to 60–80 ft. lbs.
11. Install speedometer drive gear onto mainshaft. Gear should be seated against the shoulder.
12. Install output flange, washer and lock nut onto mainshaft. Rotate mainshaft to seat bearings, then lock transmission in two gears. Torque output shaft lock nut to 500 ft. lbs.
13. To check mainshaft end play, pry on mainshaft or output flange. End play should be .003–.008 inch. If not, remove drive gear front bearing cap. Remove the bearing cup, then the shim. Measure the shim thickness and install the required thickness shim to obtain correct end play. Reinstall bearing cap and recheck end play.

NOTE: Countershaft end play is .004–.024 inch. However, end play should not require checking.

14. Shift transmission into neutral.
15. Place shifter housing gasket onto top of transmission case. With shift forks in neutral, algn forks with synchronizers and the 1-reverse clutch collar. Assemble shifter housing onto case and torque retaining screws to 25–32 ft. lbs.

Spicer 8125 & 8125-U 12 Speed Units

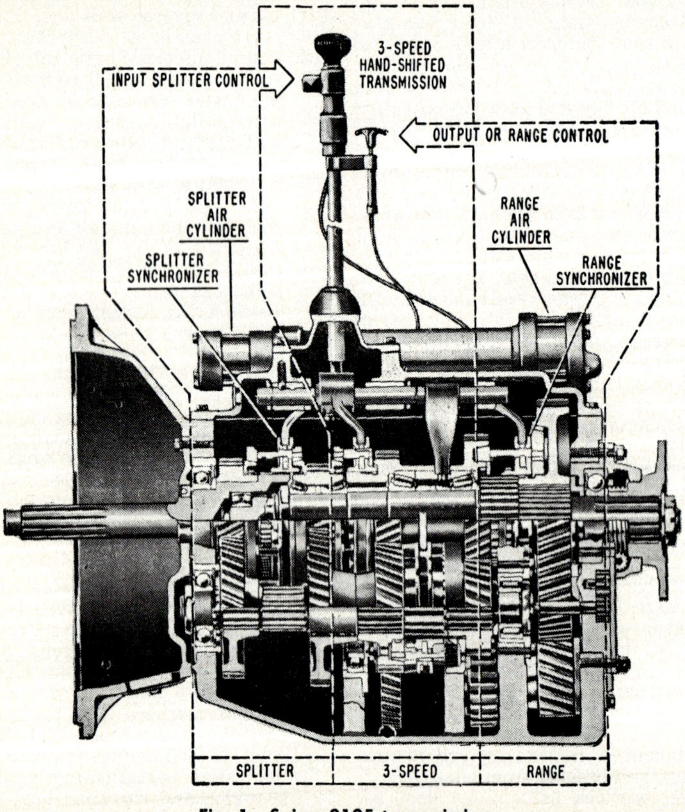

Fig. 1 Spicer 8125 transmission

INDEX

This transmission, Figs. 1 and 2, is made semi-automatic to obtain twelve speeds forward and six reverse with only three shift lever positions. To accomplish the preselected but automatic shifts, air actuated power cylinders built into each end of the shifter housing are used to make the input gear splits and hi-lo range shifts. Air pressure that makes these shifts is admitted and exhausted from opposite ends of the double acting cylinders through a control valve operated by cables attached to the gear shift lever.

Service Bulletin

In an effort to minimize parts mix-up and misapplication, the following information is presented for guidance of parts for both models 8125 and 8125-U transmissions.

In effect there are three types of 12-speed transmissions in use: (1) the fully synchronized model 8125, (2) the desynchronized hand shift model sometimes called "dehorned" model 8125, and (3) the unsynchronized (hand shifts only) model 8125-U. This could be a factory built or a field conversion, using Spicer changeover kits 311329X for No. 1 or 311582X for No. 2 clutch housings.

As noted above the major difference between 8125 and 8125-U is the omission of the hand-shifted synchronizers.

The most critical items involved in parts mix-ups are the high range countershafts with their matching collars and gears, Fig. A. Although these two countershafts look alike and their overall lengths are the same, the splined section for the sliding clutch gear (80-466-18) has been lengthened and the "hopping-guard" is moved rearward on the 8125-U, Fig. B. This change means that clutch gear (80-466-18) and 1-2/7-8 speed gear (80-196-14) *must* be used with (80-30-8) countershaft.

In the high range countershaft synchronizer 310950X or the clutch gear from a cut-up synchronizer and a 1-2/7-8 speed gear 310949X were used with (80-30-8) countershaft, a "walking out" of 3-4/9-10 gear position would probably occur. Also, clutch gear (80-196-14) *cannot* be used with 8125 countershaft (80-30-1).

The 3-4/9-10 speed gear (80-196-8) from a model 8125 can be used with either countershaft if necessary. The tooth pointing is slightly different from the 3-4/9-10 gear (80-195-15) but not enough to prevent its use in an emergency.

In model 8215-U unsynchronized units using the Spicer pull-type clutch, a clutch brake is usually provided. To make room for the clutch brake, the clutch housing and drive gear are made 3/8" longer. The bearing cap is also changed to provide a full face instead of four pads for the brake. When the clutch brake is used it *must* be 311299X. The 8000 Series three-piece clutch brake (125C21 washer and 200C4 disc) is too small on the outside diameter to give effective braking surface on the bearing cap of an 8125-U, Fig. C.

Mainshaft drive gears 310769X and 311300X, Fig. D, differ primarily because of a steel synchronizer cone on one end and the type of tooth pointing on the clutch teeth. If necessary the 8125 gear (310769X) could be used with the 5-6/11-12 clutch gear (80-466-19) in an 8125-U.

The blocker pins in the 5-6/11-12 synchronizer (310706X) may be cut to salvage the sliding clutch gear, Fig. E. This gear, as well as 80-466-19, is symmetrical and may be reversed for longer parts life with either the B or C type units mentioned at the beginning of this bulletin.

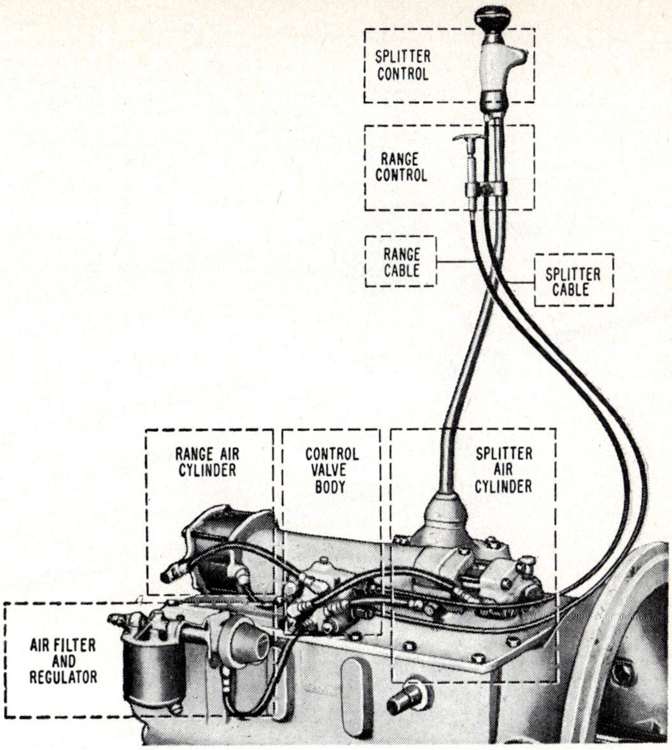

Fig. 2 Spicer 8125 transmission shift control

TRANSMISSION, REMOVE

1. Disconnect linkage and remove forward and reverse shift lever.
2. Drain air reservoir tanks to bleed all air from transmission air system.
3. Disconnect air control valve cable controls by removing capscrew, end plate and sleeves from rear of valve body, Fig. 3.
4. If necessary, push range "T" handle down and rotate splitter control handle clockwise to move both plungers to their rear position. Remove locknuts and rubber rings from end plungers.
5. Loosen upper and lower cable locknuts from front of valve body. Do not remove cable adapter plugs.
6. Grasp cables and pull cable castings and inner wires free of valve body and plungers. Plungers need not be removed from valve body.
7. Place rubber rings inside locknuts and tighten finger tight on plungers. Install sleeves, end plate and capscrew to prevent loss.
8. Remove shift tower connector block and upper shift tower.
9. Disconnect clutch release linkage and remove release shaft.
10. Drain lubricant.
11. Disconnect upper and lower oil filter lines at oil pump and drain. Remove can and cartridge, Fig. 4.
12. Disconnect electrical lead to oil temperature sending unit and remove sending unit to prevent breakage during removal.
13. Remove propeller shaft and parking brake linkage. Disconnect speedometer cable.
14. Disconnect forward and reverse shift lever bracket and linkage at shift rod attaching stud (if equipped).
15. Position suitable jack under transmission and pick up its weight.
16. Remove rear transmission support.
17. Remove all capscrews from bell housing and pull transmission back free of engine. Slide transmission back until pinion shaft clears clutch release bearing housing. Then lower transmission.

TRANSMISSION, INSTALL

1. Use an old pinion shaft or aligning tool to check alignment of clutch discs. Shift transmission into one of the manual gear shift positions. Apply a light coat of lubriplate to splines and entire surface of pinion shaft.
2. Place transmission on a suitable jack and position under truck. Adjust to align transmission with engine.

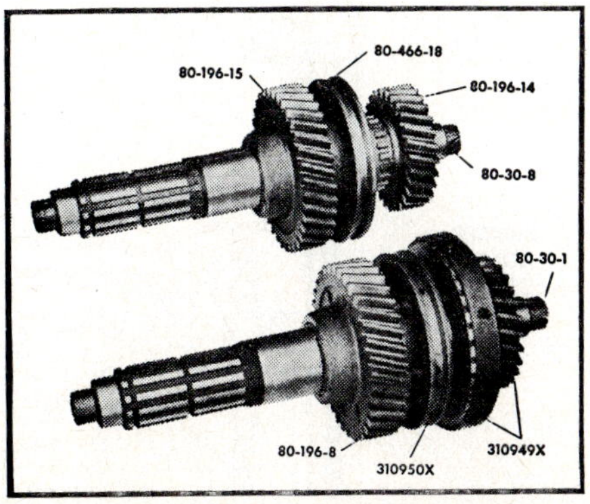

Fig. A Countershafts

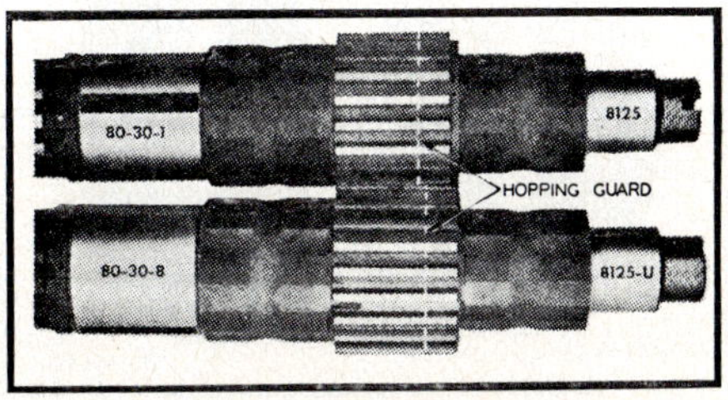

Fig. B Sliding clutch gears

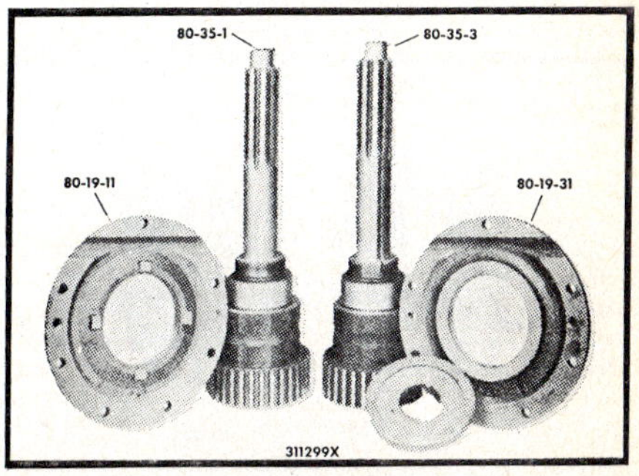

Fig. C Clutch brakes

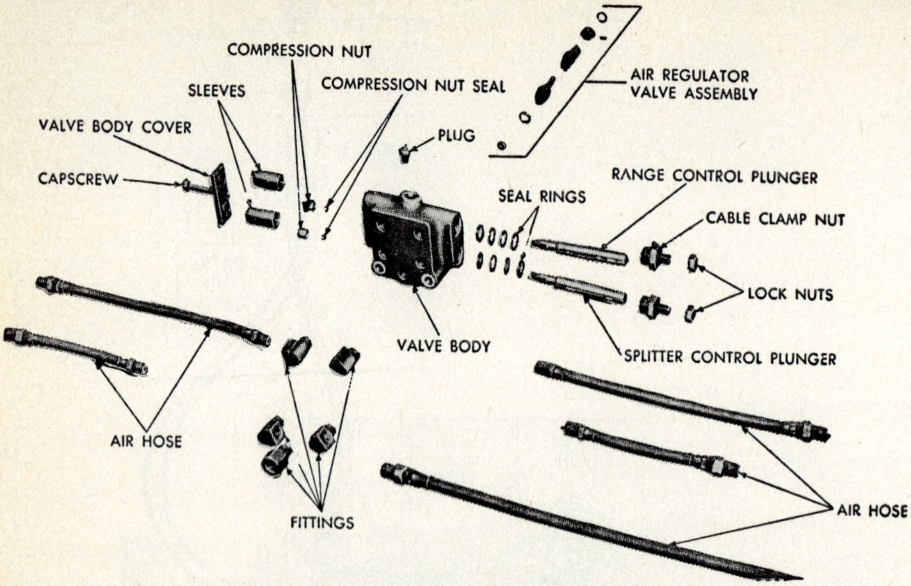

Fig. 3 Air control valve body disassembled

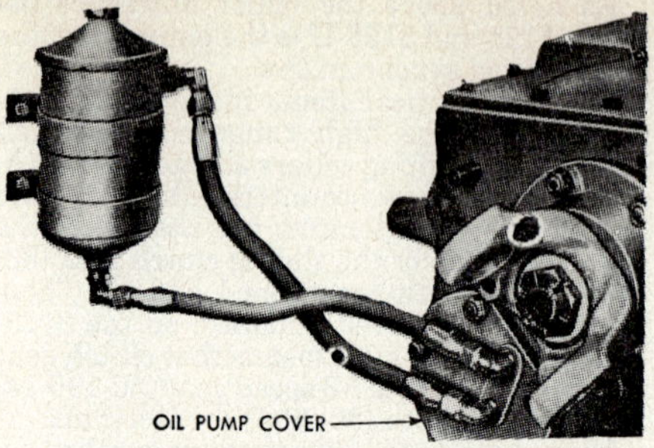

Fig. 4 Transmission oil filter

3. Start pinion gear shaft into clutch release bearing. Enter transmission slowly while checking proper positioning of clutch release bearing housing in clutch release yoke. It may be necessary to rotate end yoke or flange slowly to align splines of pinion shaft with hubs of clutch discs.
4. Enter pilot on bell housing into clutch housing and align capscrews. Start all capscrews and tighten evenly around bell housing. Install rear transmission support.
5. Connect speedometer cable to adapter or driven gear.
6. Install parking brake linkage. Install bracket and connect forward and reverse control cable.
7. Connect propeller shaft.
8. Connect air line to transmission air filter. Close drain cocks on air tanks.
9. Install new filter cartridge. Connect oil lines from filter to pump as shown in Fig. 4. Install temperature sending unit (if removed) and connect lead wire.
10. Install specified type and quantity of transmission oil.
11. Connect clutch release linkage and mount clutch release cylinder (if equipped). Set release bearing clearance to 1/8".
12. Remove capscrew, end plate and sleeves. Pull both plungers to rear and remove locknuts with rubber rings, Fig. 3.
13. Loosen locknuts at front of valve body and insert splitter cable inner wire through lower cable adapter, passing inner wire through plunger.
14. Bottom outer casing of splitter cable in adapter and tighten locknut (190 inch-lbs) to clamp casing. Insert range cable inner wire through upper cable adapter, passing inner wire through plunger. Bottom outer casing in adapter and tighten locknut to 190 inch-lbs.
15. Set range "T" handle so that a 1/4" gap exists between "T" handle and body, Fig. 5. Remove knob, spacer and splitter lever from shift lever stud.

CAUTION: *Take care not to lose small ball bearing in pocket inside splitter lever.*

16. Position inner sleeve so that 1/8" gap exists between bottom of sleeve and shoulder, Fig. 6.
17. At valve body, slide rubber rings over inner wires of splitter and range cables, seating rings against plungers. Assemble locknuts over wires and tighten securely (30 inch-lbs.) to plungers.
18. Assemble splitter control by setting lever over sleeve, noting that inner guide ball registers in spiral groove of sleeve. Install spacer and shift knob. Check operation of splitter control by rotating lever clockwise and counterclockwise. Lever should move easily through a 90 degree arc.

Check range control by pulling up and pushing down on "T" handle. Handle should move freely through 1/2 its travel and should not bottom against body in down position, Fig. 5.

19. Install forward and reverse lever and connect linkage.

DISASSEMBLE TRANS.

1. Disconnect filter-to-cylinder air line at cylinder and remove air filter and bracket.
2. Remove remaining shift housing bolts and use a sling to remove housing. *The transmission must be in Lo-Range and left edge of shifter housing must swing up and to the right, Fig. 7.*
3. Remove shift fork shoes and housing gasket.
4. Remove clutch housing (8 nuts) and release mechanism.
5. Back off parking brake adjustment and lock transmission in two gears.
6. Remove output flange and brake drum.
7. Disconnect parking brake linkage, remove brake shoe assembly, speedometer drive and driven gear, and oil seal.
8. Remove transmission rear support.
9. Remove output shaft rear bearing retainer cap and gasket.
10. Remove Lo-Range countershaft rear bearing cap and gasket.
11. Remove oil pump cover plate and gasket, Fig. 8.
12. Install bolts shown in Fig. 9 into threaded holes in oil pump body and run bolts down evenly until body and gasket are removed from case.
13. Relieve staking on Lo-Range countershaft rear bearing lock nut and remove nut and washer.

Fig. D Main drive gears

Fig. E Synchronizer sleeves

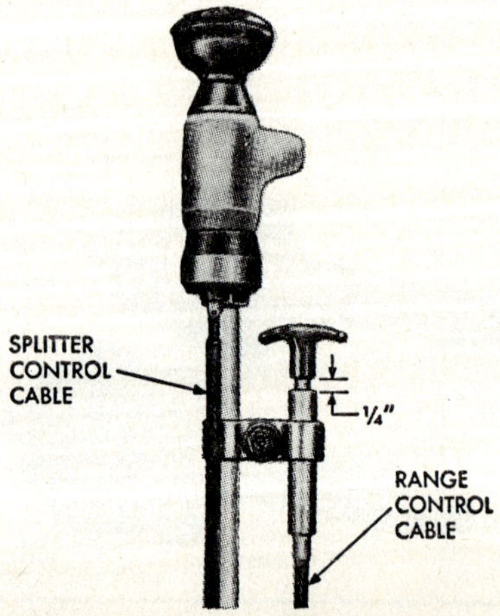

Fig. 5 Range control cable setting

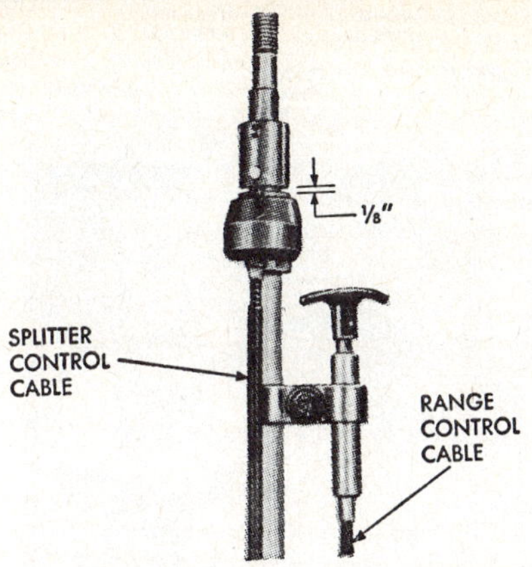

Fig. 6 Splitter control cable setting

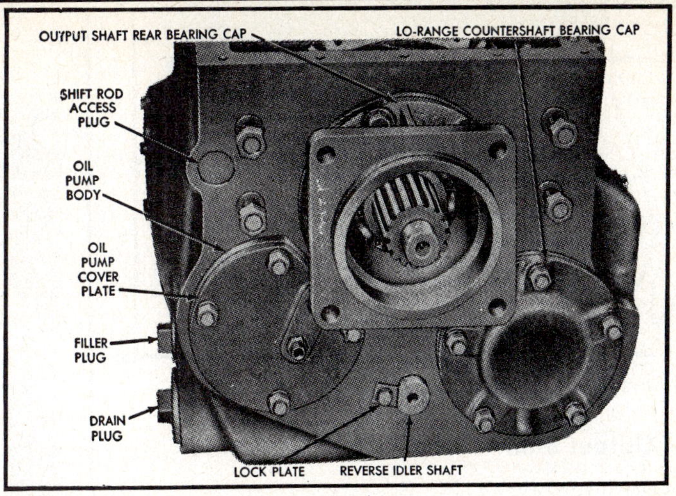

Fig. 8 Transmission rear face

14. Relieve staking on input shaft and Hi-Range front countershaft nut.
15. Remove input shaft nut (left-hand thread) and remove Hi-Range countershaft front nut and washer.
16. Remove input shaft bearing cap nuts.
17. Use a floor crane with lifting sling, support output shaft and cluster gear.
18. Install the two bolts shown in Fig. 10 into the blind holes in input shaft bearing retainer and remove retainer. *Keep input shaft in place against output shaft by tapping with a mallet while pulling retainer.* Remove bearing from retainer.
19. Remove input shaft, thrust washer and front drive gear from shaft, Fig. 11, through front bearing bore and remove input shaft from case.
20. Remove clutch housing gasket.
21. Remove output shaft rear bearing retainer and gasket, using two 7/16-14 puller bolts.
22. Remove output shaft rear bearing spacer from shaft or retainer. *Use a suitable puller as required but do not pull on seal ring. Remove seal rings from spacer.*
23. Remove Lo-Range interlock pin from its boss.
24. Remove output shaft from case. Rock shaft to get Lo-Range gear to clear case.
25. Remove Lo-Range gear and range synchronizer from output shaft.
26. Position output shaft horizontally in a vise (holding shaft splines) with cluster gear resting on bench.
27. Cut lock wire and remove Hi-Range countershaft shift fork set screw.

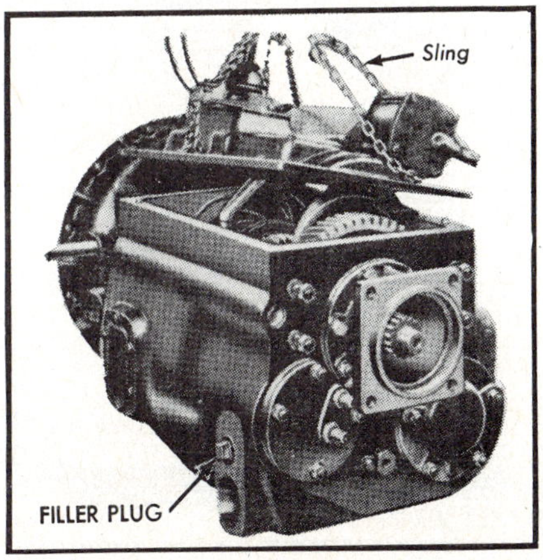

Fig. 7 Removing shift housing

28. Remove welch plug from rear of case by tapping Hi-Range shift rod rearward through hole in case. Remove shift fork as shaft is removed from case.
29. Remove Hi-Range countershaft rear bearing nut.
30. Disengage Hi-Range countershaft front driven gear rear snap ring from its groove and slide front driven gear back against Hi-Range rear driven gear, Fig. 12.
31. Drive Hi-Range countershaft forward until front bearing and retainer are exposed for a suitable puller.
32. Remove Hi-Range countershaft front bearing and retainer, Fig. 13; then remove shaft assembly from case. Remove bearing from retainer.
33. Remove Lo-Range shift rod detent plunger, spring and retainer from outside of case, Fig. 14.
34. After removing shift fork set screw, move Lo-Range shift rod forward out of case and remove fork. Pry shift rod seal from case.
35. Drive Lo-Range countershaft rearward until rear bearing and retainer are exposed. Pull rear bearing, retainer and gasket from shaft. Tap bearing out of retainer.
36. Remove spacer and drive gear from shaft through rear bearing bore.
37. Lift Lo-Range countershaft from case.

Fig. 9 Removing oil pump body

38. Remove reverse idler shaft lock plate and remove shaft, using a drift.
39. Lift reverse idler gear assembly and thrust washers from case and separate the parts, Fig. 15.
40. Remove Hi-Range countershaft rear bearing retainer screws, Fig. 12. Then use two 7/16-14 puller screws to remove retainer from case. Tap bearing out of retainer.
41. Remove Lo-Range countershaft front bearing retainer, Fig. 12, using two 7/16-14 puller screws. Remove retainer from case. Tap bearing out of retainer.
42. If necessary, pry out welch plug above Lo-Range detent mechanism, Fig. 14.
43. Shift case to let interlock pin drop out of case.

DISASSEMBLE SUB-ASSEMBLIES

Oil Pump

1. Drive pin from shaft and remove gears from pump body.
2. Press drive gear from shaft and remove Woodruff key.
3. Drive driven gear shaft from body.
4. Drive driven gear bushing from gear.
5. Press drive shaft bushing from body.

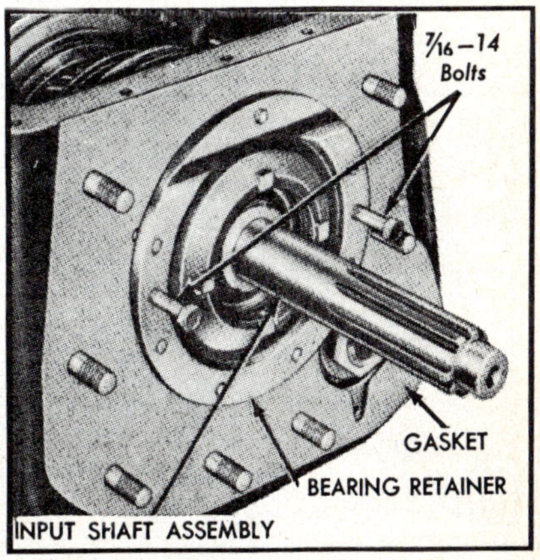

Fig. 10 Removing input shaft bearing retainer

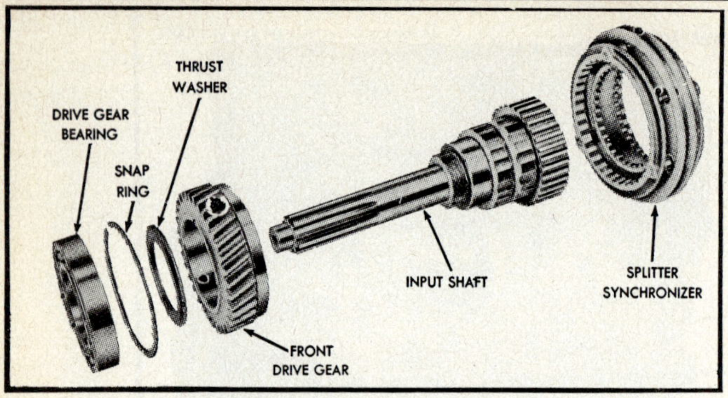

Fig. 11 Input shaft assembly

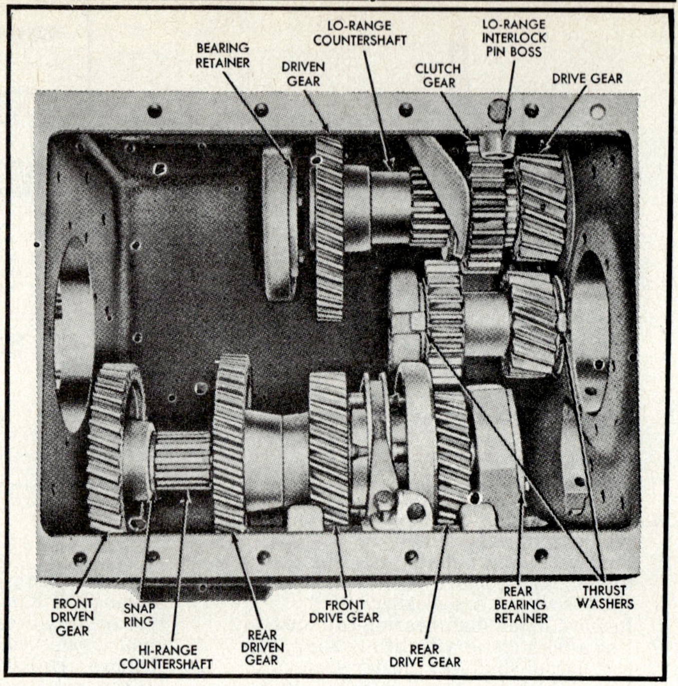

Fig. 12 Transmission with output shaft removed

Output Shaft

1. Tap pilot bearing from shaft.
2. Remove drive gear bearing lock nut.
3. Tap splitter synchronizer forward and remove front bearing cone and rear drive gear. Remove bearing spacers and shim pack. Wire shim pack together.
4. Remove splitter synchronizer from shaft. *Parts for this synchronizer are not serviced.*
5. Remove shaft from vise and position in a press, using rear face of cluster gear as a base. Press output shaft down through rear drive gear bearing cone, spacer and cluster gear, Fig. 16.
6. Remove cluster gear bearing spacers and shim pack from output shaft. Wire shim pack together.
7. Remove cluster gear rear bearing cone Fig. 17.
8. Remove cluster gear bearing snap ring from cluster gear, Fig. 16.
9. Position cluster gear in press with rear face of gear down. Press cluster gear bearing cups and spacers free of cluster gear, Fig. 18.
10. Place range synchronizer on bench with Hi-Range cone up. Lift up to remove Hi-Range cone from clutch gear. *Watch for three poppets and springs, Fig. 19.*
11. With synchronizer on bench (resting on Lo-Range cone) use a small screwdriver to compress poppets back into clutch gear, Fig. 20, and remove gear from Lo-Range cone.
12. Remove oil metering plug from front end of output shaft.

Hi-Range Countershaft

1. Lift off front bearing inner race, thrust washer, rear drive gear and synchronizer assembly. *Parts for synchronizer are not serviced.*
2. Remove snap ring from front end of shaft and remove front driven gear, Fig. 21.
3. Lift off previously disengaged snap ring from countershaft and remove rear driven gear snap ring.
4. Using rear driven gear as a base, press gear from shaft.
5. Remove drive gear Woodruff key from countershaft and slide front drive gear from shaft.

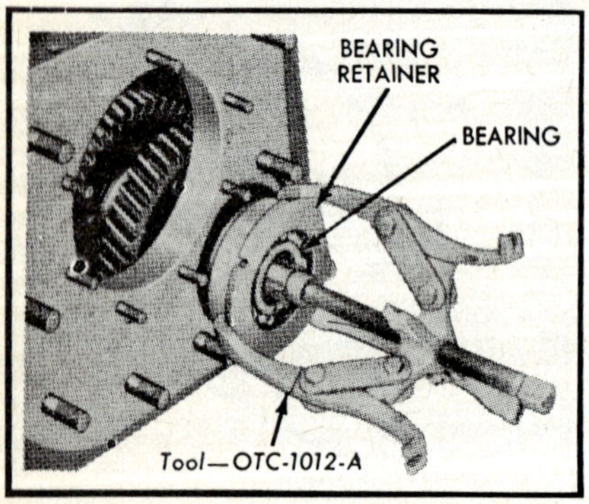

Fig. 13 Removing countershaft front bearing and retainer

Lo-Range Countershaft

1. Remove sliding clutch gear from rear end of shaft, Fig. 22.
2. Remove driven gear snap ring.
3. Press off driven gear and remove Woodruff key.
4. Press off countershaft front bearing inner race, Fig. 23.

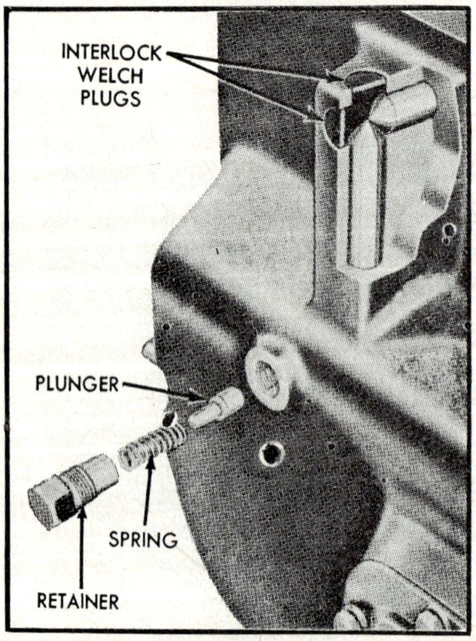

Fig 14 Lo-Range countershaft detent mechanism

ASSEMBLE SUB-ASSEMBLIES

Lo-Range Countershaft, Fig. 22

1. Install Woodruff key and press driven gear on shaft.

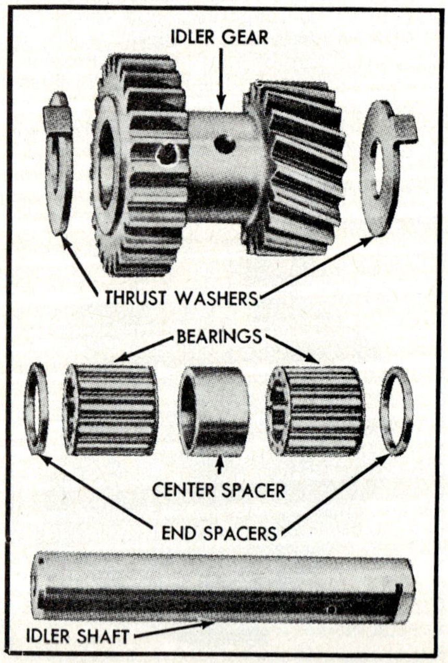

Fig. 15 Idler shaft assembly

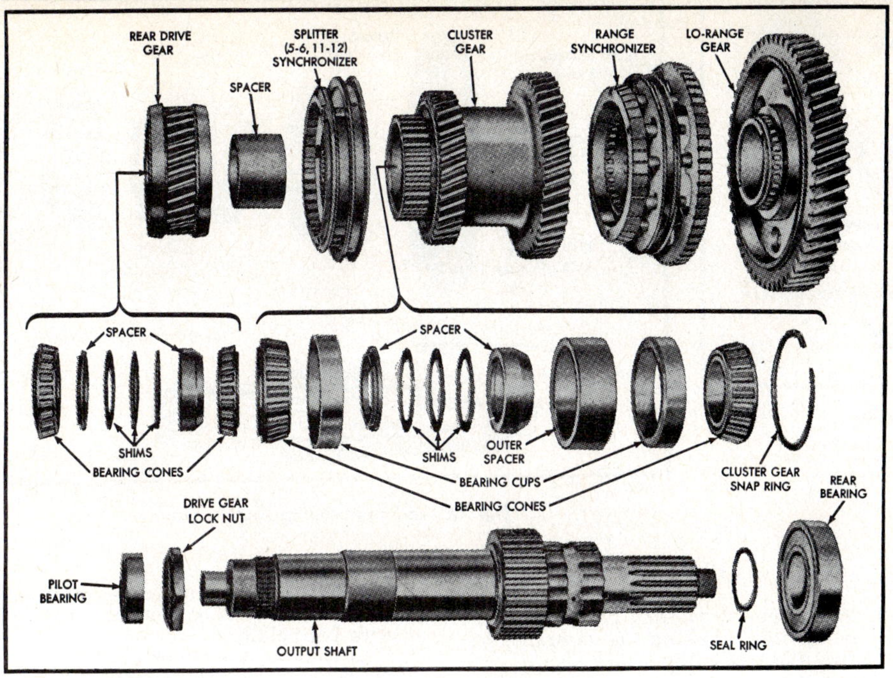

Fig. 16 Output shaft assembly

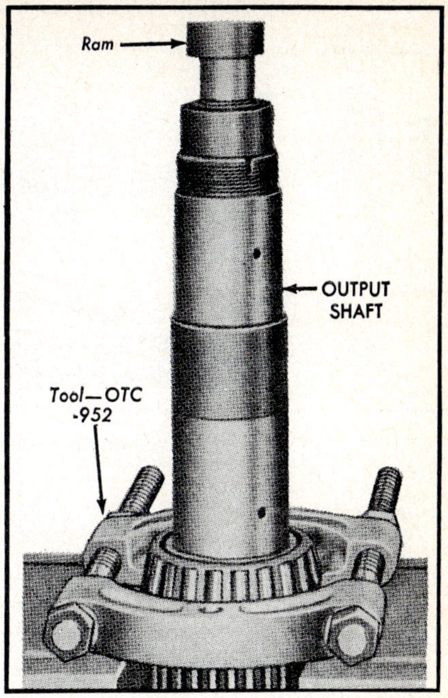

Fig. 17 Removing output shaft cluster gear rear bearing cone

2. Install driven gear snap ring.
3. Press on front bearing inner race.
4. Position sliding clutch gear as shown. Apply Lubriplate to bore of drive gear.

Hi-Range Countershaft, Fig. 21

1. After coating bore of front drive gear with Lubriplate, position gear on shaft.
2. Install Woodruff key and press rear driven gear on shaft.
3. Install rear driven gear snap ring.
4. Position front driven gear rear snap ring over shaft and slide it back against snap ring.
5. Install front driven gear and its snap ring.
6. Apply Lubriplate to front drive gear clutch teeth, the splines of Hi-Range countershaft, the clutch teeth of the synchronizer, the rear drive gear and the fluted bearing surface of shaft.
7. Install synchronizer and rear drive gear.
8. Position thrust washer *with flat side against rear driven gear* on shaft.
9. Tap rear bearing inner race on shaft.

Output Shaft, Fig. 16

1. Install oil metering plug in shaft bore.
2. Position Lo-Range cone on bench with pins up and with poppets and springs in gear. With three spacer tools, position gear on cone. Remove tools, Figs. 19 and 20.
3. Position remaining three poppets and springs in clutch gear. Compress poppets and install Hi-Range cone. Check action of assembled synchronizer.
4. Position cluster gear on press with larger face up, Fig. 24. Position front bearing cone in gear.
5. Position front bearing cup, outer spacer and rear bearing cup. Then press these parts into cluster gear, Fig. 24.
6. Install cluster gear snap ring (tapered side up) in groove inside of cluster gear. Be sure that ring is fully seated.
7. Press cluster gear rear bearing cone on output shaft, Fig. 25.
8. Install remaining cluster gear spacers and shim pack on output shaft.
9. Press cluster gear assembly on output shaft, Fig. 26. Press on front bearing cone to position gear on shaft. *It is essential that end play of the output shaft cluster gear and output shaft drive gear be held to .002–.004" loose.*
10. Install a 5½" length of 3" ID diameter tubing on output shaft, Fig. 27. Install output shaft lock nut.
11. Seat cluster gear bearings by pressing down on cluster gear and rotating it back and forth through about 90 degrees.
12. Mount a dial indicator to check end play, Fig. 27. Make a reference mark where indicator point touches cluster gear and set dial indicator to zero. Pull up on gear

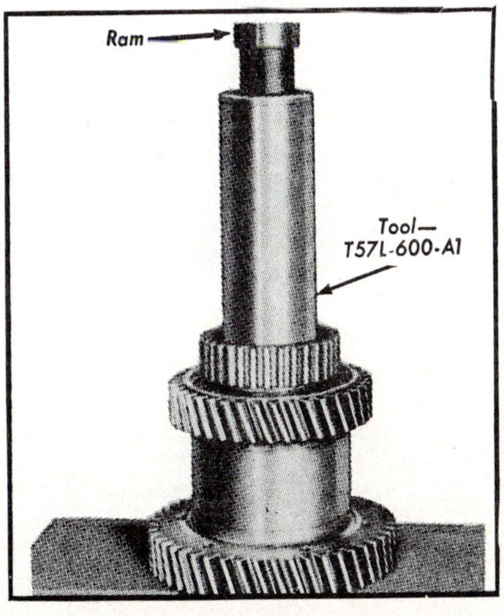

Fig. 18 Removing cluster gear bearings and related parts

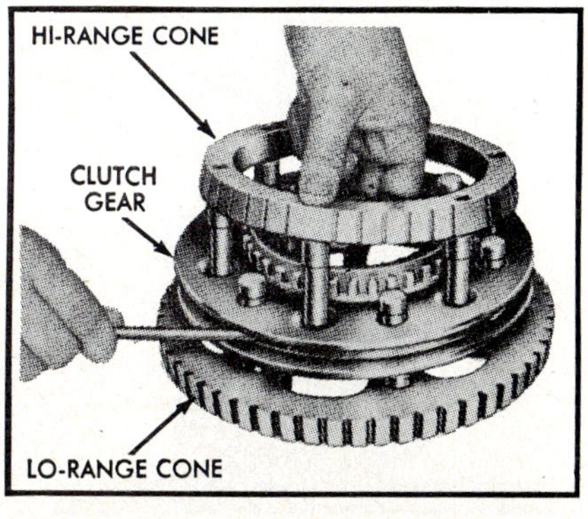

Fig. 19 Removing range synchronizer Hi-Range cone

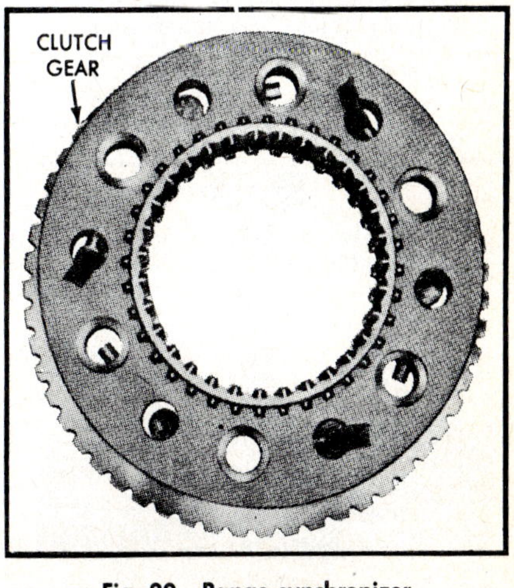

Fig. 20 Range synchronizer cone installation

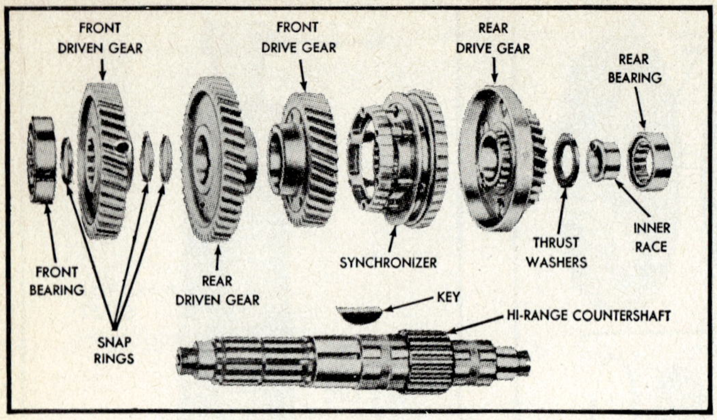

Fig. 21 Hi-Range countershaft assembly

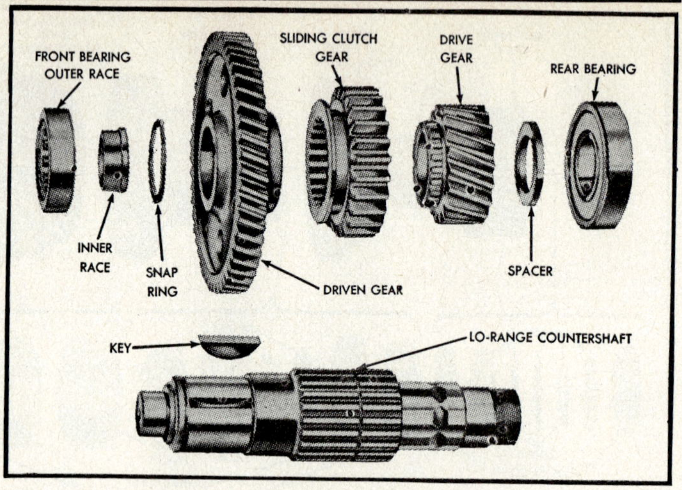

Fig. 22 Lo-Range countershaft assembly

and rotate it back and forth through approximately 90 degrees to seat the other roller bearings.

13. Rotate the cluster gear to line up reference mark with indicator and read indicator while pulling up on cluster gear, Fig. 27. If necessary, remove cluster gear and adjust shim pack to obtain .002–.004″ end play.
14. After securing correct end play adjustment, remove nut and tubing spacer.
15. Install splitter synchronizer on output shaft and position long spacer inside cluster gear.
16. Install rear drive gear rear bearing cone on shaft, Fig. 16. Then position rear drive gear internal spacers and shim pack.
17. Position rear drive gear and install front bearing cone.
18. Install rear drive gear lock nut.
19. Mount a dial indicator and check end play, Fig. 28. Determine end play of output shaft rear drive gear by following a similar procedure to that outlined in Steps 11, 12 and 13. End play should be .002–.004″.
20. Stake drive gear bearing lock nut after final assembly of drive gear, then remove output shaft from vise.
21. Set output shaft sub-assembly on bench with rear end of shaft overhanging edge of bench. Apply Lubriplate to splines and fluted bearing surface of output shaft as well as clutch teeth of synchronizer and gear.
22. Position range synchronizer on shaft with Hi-Range cone toward cluster.
23. Apply Lubriplate to rear clutch teeth of range synchronizer as well as to clutch teeth and pilot of Lo-Range gear.

Oil Pump

1. Press a new drive shaft bushing into oil pump body, allowing 1/16″ clearance between end of bushing and edge of gear pocket.
2. Press new bushing into driven gear.
3. Burnish drive shaft bushing and idler gear bushing with a .5070–.5065″ steel ball.
4. Press a new driven gear shaft into pump body until flush with body.
5. Position driven gear on shaft and check rotational freedom.
6. Tap Woodruff key into drive shaft, then press shaft into gear flush with gear face.
7. Assemble drive shaft and gear assembly into pump body and check rotation of pump gears for freedom. Pump must rotate freely.
8. Press pin through hole in end of drive shaft and equalize pin projection on each side of shaft.
9. If pump rotates freely, lubricate gears with heavy engine oil as a means of priming the pump.

ASSEMBLE TRANSMISSION

1. Drop Lo-Range countershaft vertical interlock pin into case. Install welch plug, Fig. 14.
2. Install Lo-Range countershaft shift rod oil seal in case. Coat bushing and seal with Lubriplate.
3. Coat idler shaft bearings and three spacers with Lubriplate and assemble these parts in reverse idler gear, Fig. 15.
4. Position Lo-Range countershaft front bearing outer race in its retainer.
5. Install two pilot studs (7/16-14) in case to align Lo-Range countershaft front bearing retainer and tap retainer into position. Remove pilot studs.
6. Install bearing retainer bolts and secure with lock wire.
7. Position Hi-Range countershaft rear bearing in its retainer.
8. Install two 7/16-14 pilot studs in case to align Hi-Range countershaft rear bearing retainer and tap retainer into position. Remove pilot studs.
9. Install bearing retainer bolts and secure with lock wire.
10. Apply Lubriplate to reverse idler gear thrust washers and position them in case, Fig. 12.

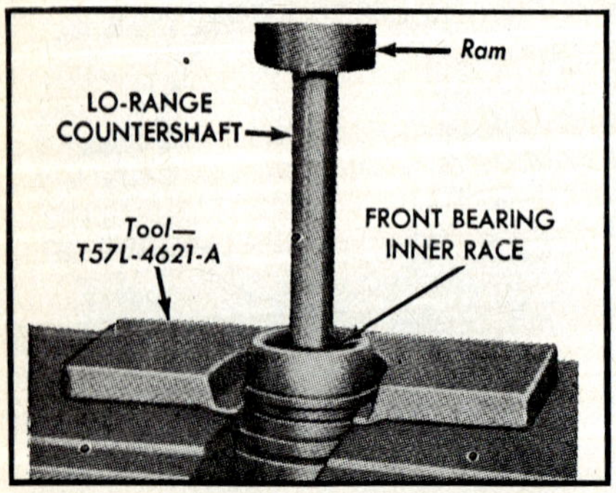

Fig. 23 Removing Lo-Range countershaft front bearing inner race

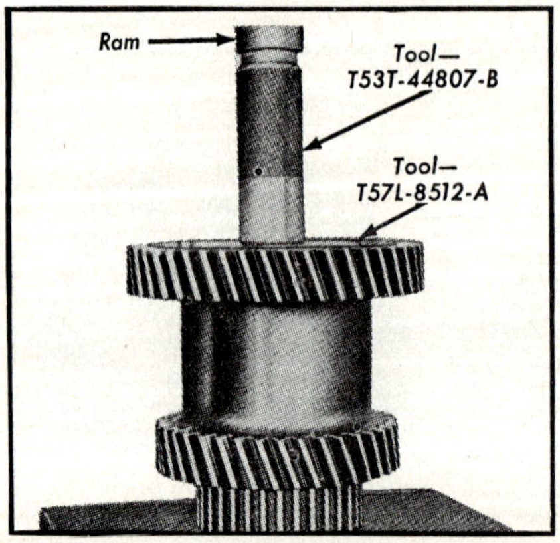

Fig. 24 Assembly of cluster gear

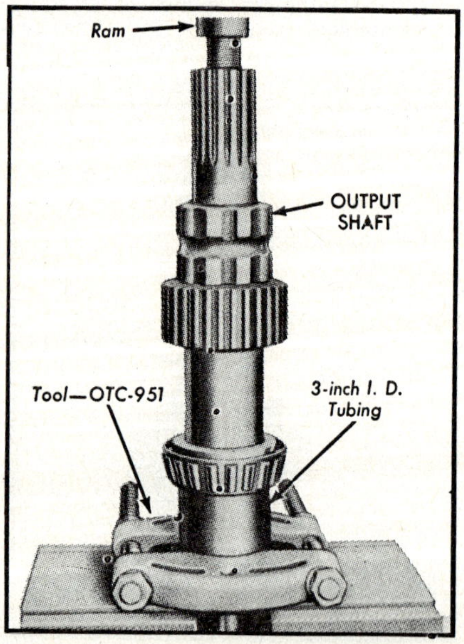

Fig. 25 Cluster gear rear bearing cone installation

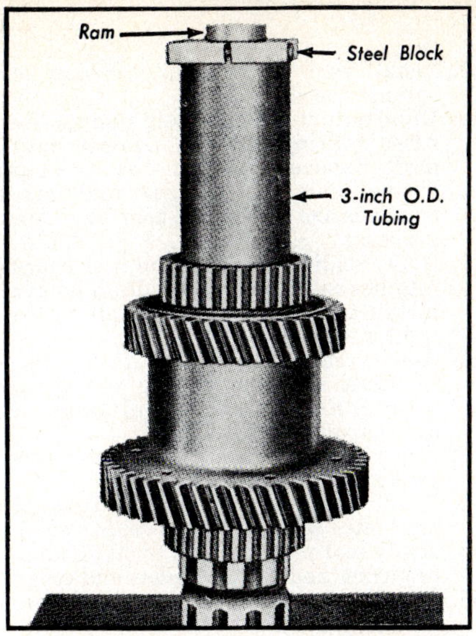

Fig. 26 Installing cluster gear assembly

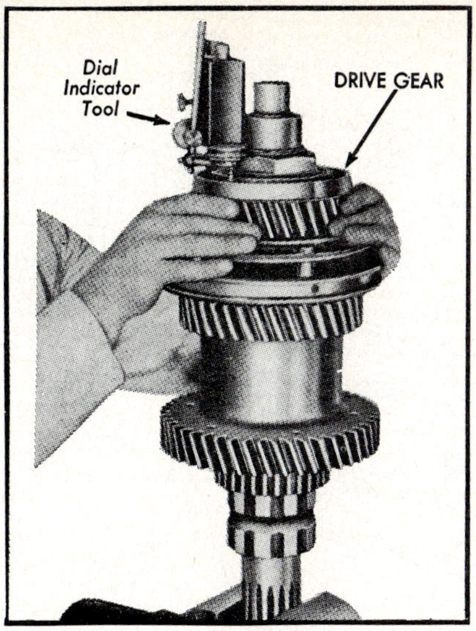

Fig. 27 Checking cluster gear end play

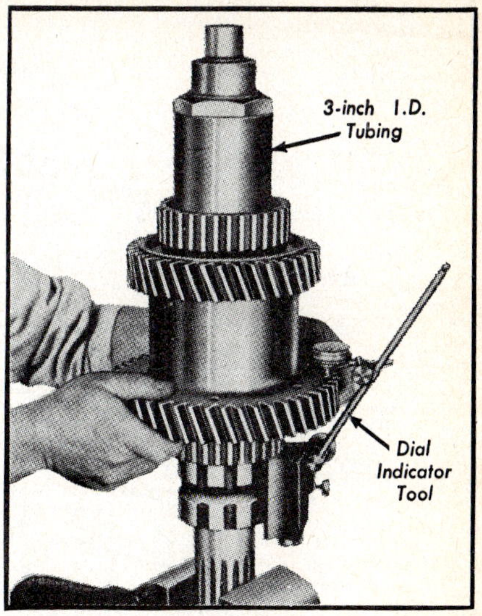

Fig. 28 Checking drive gear end play

11. Install reverse idler gear assembly, Fig. 15, aligning flat at rear of shaft to mate with lock plate. Install lock plate.
12. With Lo-Range countershaft in case, slide drive gear onto shaft.
13. Position gear spacer and apply sealing compound to rear bearing retainer gasket. Position gasket over studs.
14. Tap bearing retainer onto shaft and into case bore. Install bearing.
15. Install flat washer and rear bearing lock nut; *hand-tighten nut.*
16. Start Lo-Range countershaft shift rod from the front through seal and bushing assembly. As rod enters case, tighten shift fork on sliding clutch gear and slide shift rod through fork. Lift interlock pin and push rod into final position.
17. Install set screw and secure with lock wire.
18. Install Lo-Range shift rod detent plunger, spring and retainer, Fig. 14.
19. Position Hi-Range countershaft in case and position front bearing and retainer. Line up the alignment notch with the lock pin and drive front bearing and retainer into case.
20. Install flat washer and nut on front of shaft and *hand-tighten nut.*
21. Tap Hi-Range countershaft front driven gear forward and lock in place with snap ring. Make sure snap ring seats in groove.
22. Position High-Range countershaft shift fork in synchronizer and install shift fork in synchronizer and shift fork rod through rear rod boss, shift fork and front boss.
23. Install fork screw and secure with lock wire.
24. Apply sealing compound to a welch plug and install in shift rod access hole at rear of transmission case.
25. Position output shaft Lo-Range gear (clutch teeth to front) in transmission case. Lower output shaft into transmission case and guide splined end of shaft through Lo-Range gear and rear bearing bore. Align Lo-Range gear with range synchronizer and move output shaft to rear as far as possible.
26. Install output shaft pilot bearing, then align all gears to mesh properly.
27. Install Lo-Range interlock pin in case, Fig. 14.
28. Coat splines and fluted bearing surface of input shaft with Lubriplate and position on output shaft. If input shaft does not enter freely, tap end of shaft lightly. *Do not force assembly because self-aligning bearing may be tipped.*
29. Coat clutching surface of front drive gear with Lubriplate and install gear on output shaft.
30. Install thrust washer on input shaft with flat against front drive gear, Fig. 11.
31. Apply sealing compound on clutch housing gasket and place gasket on front studs.
32. Tap input shaft bearing and retainer into position. It may be necessary to lift input shaft for alignment of bearing retainer stud holes with studs.
33. Install bearing retainer stud nuts.
34. Install input shaft bearing nut *handtight.* Lock transmission in two gears and tighten the following: input shaft nut, Hi-Range countershaft front nut, and Lo-Range countershaft rear nut. Stake these nuts in one place. *Do not use a prick-punch in grooves or threads.*
35. Apply sealing compound to output shaft rear bearing retainer gasket and place gasket on studs. Be sure to align oil return hole in gasket with oil return hole in case.
36. Assemble output shaft bearing retainer components, Fig. 29.

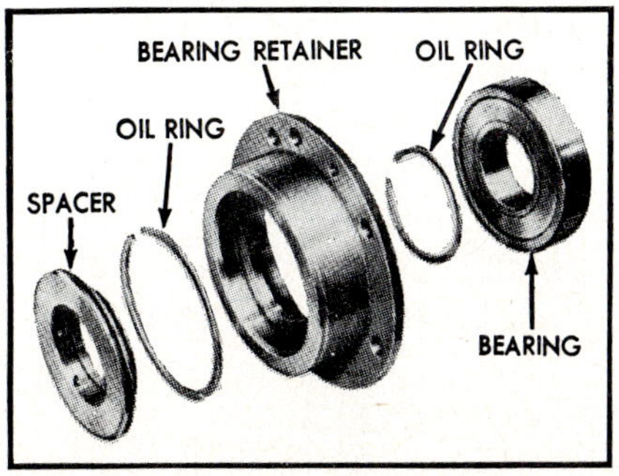

Fig. 29 Output shaft bearing retainer assembly

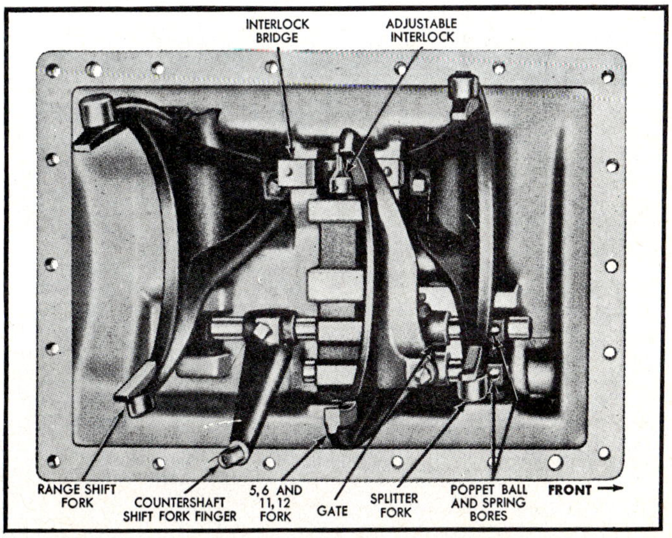

Fig. 31 Assembled gearshift housing

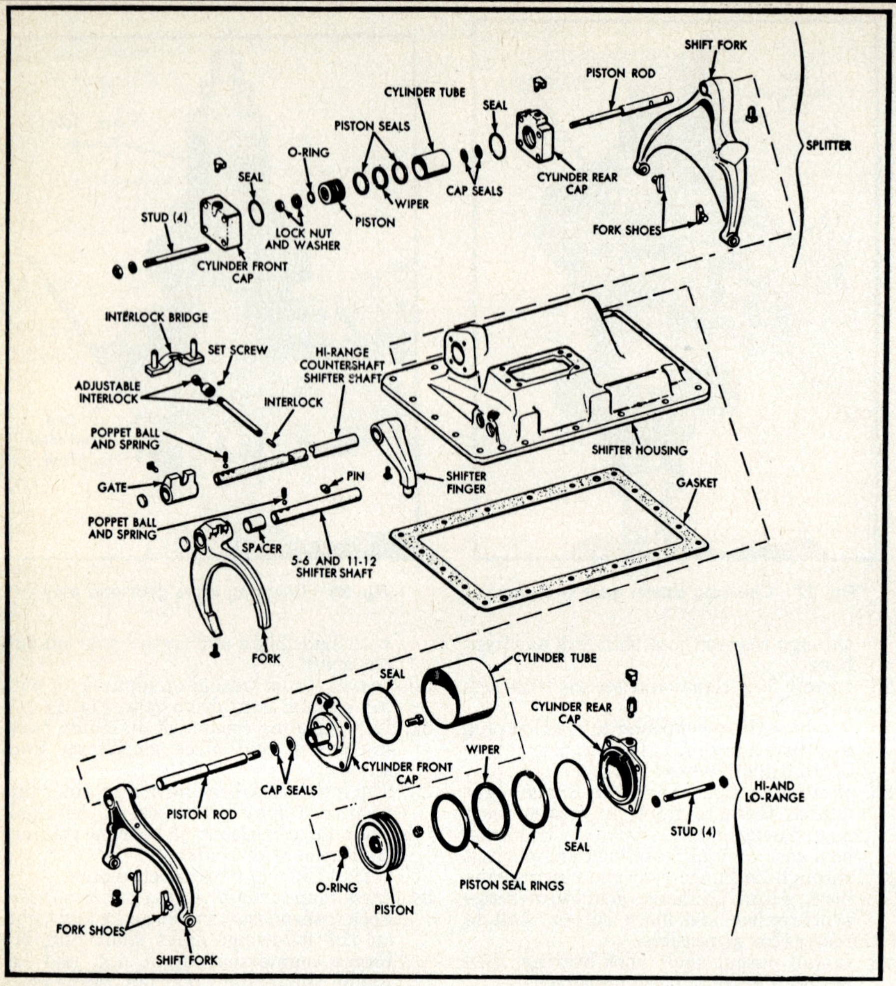

Fig. 30 Gearshift housing disassembled

37. Position output shaft rear bearing retainer over case studs.
38. Block front of output shaft Lo-Range gear to prevent any forward movement of gear.
39. Install seal ring on output shaft and against spacer.
40. Drive output shaft bearing into retainer. After bearing bottoms against output shaft, remove block from case. Use a pry bar between Lo-Range gear and back of transmission case to pull bearing retainer into case.
41. Apply sealing compound to rear bearing retainer cap gasket, align drain hole and install gasket on studs. Install cap and tighten nuts.
42. Apply sealing compound to Lo-Range countershaft rear bearing cap gasket, align drain hole and install gasket. Install cap and tighten nuts.
43. Apply sealing compound to oil pump body gasket and position on studs.
44. Install oil pump, making sure that drive pin mates with Hi-Range countershaft.
45. Apply sealing compound to oil pump cover gasket and install gasket and cover.
46. Install transmission rear support, speedometer gears and oil seal.
47. Install brake shoe assembly on rear retainer cap. Connect parking brake linkage.
48. Lock transmission in two gears and install output shaft flange, brake drum and nut. Shift transmission to neutral and adjust parking brake.
49. Install oil pick-up screen and cap with gasket.
50. Install clutch housing and release mechanism. Apply sealing compound to clutch housing gasket and place gasket on case. Fill transmission (through top opening) with proper amount of lubricant.
51. If disassembled, assemble shift housing as suggested in Figs. 30 and 31. Then position shift housing above transmission. Stick shift fork shoes to forks with grease. Lower shift housing and align all shoes and forks.
52. Install air filter assembly.
53. Install and tighten shift housing bolts.
54. Connect air filter air line to cylinder and install transmission.

Spicer SST Series Exc. 10, 1010-2A, 1020, 1052, 1214, 1252 & 1253

REAR CASE & GEARS

Partial Disassembly

1. Remove capscrews and lockwashers from rear section piston body and remove assembly from case face and off piston and "O" ring. Remove piston body gasket.
2. Remove shift piston locknut from shaft and remove piston, Fig. 1.
3. Lift and place complete transmission unit vertically resting on two 4" × 4" wooden blocks under clutch bell pilot face.
4. Remove 16 capscrews and washers from rear cover.
5. Using a chain hoist, with rope sling slipped through end yoke bores, lift complete rear cover assembly off dowel pins.
6. Remove gasket from either rear case housing or the case cover, Fig. 2.
7. Remove fork rod, fork and stop spacer with shift collar from mainshaft gear.
8. Before removing either countershaft gears with front cones from mating mainshaft gear, pay particular attention to the timing marks or paint marks on web face of gears. Turn head end gear until timing marks align to each other.
9. Pull countershaft gears with front cone straight back to separate from front cup.
10. If it is necessary to remove the mainshaft head end gear, unstake tang of lockwasher from groove of locknut to remove locknut.
11. Use puller to remove pocket bearing from shaft.
12. Slide gear with thrust washer off splines of shaft.
13. Inspect lip of oil seal for damage or wear.

Disassembly for Overhaul of Rear Unit

1. Remove locknut from end of output shaft and remove end yoke or companion flange.
2. Remove output shaft bearing cap and gasket. Inspect lip of oil seal and replace seal if lip wear or damage is found.
3. Pull back on output shaft to expose snap ring on shaft bearing. Use puller on snap ring and pull bearing off shaft. Remove thrust spacer.
4. Remove piston body and gasket from case face and off piston.
5. Remove shift piston locknut and piston from shaft.
6. Remove rear cover from dowel pins by lifting straight up. Remove cover gasket and inspect "O" ring in piston rod hole of cover. Do not remove shift rod welch plug unless necessary.
7. Remove rear cover countershaft bearing caps with shims. Tie shims together and lay aside.
8. Remove cups from countershaft bores of cover. Inspect cups for wear, pitting or abnormal contamination wear.
9. Remove output shaft and gear as an assembly from rear case.
10. Remove piston shift rod with fork and stop spacer as an assembly. Clutch collar will pull away with fork.
11. Before removing countershafts, turn head end gear until timing marks align to each other. Then remove countershafts from rear case.
12. If countershaft bearing cones require replacement, use splitter puller tool to remove cones from countershaft ends.
13. Unstake lockwasher from groove of locknut and remove locknut and washer from end of mainshaft.
14. Use puller and remove pocket bearing from shaft.
15. Slide mainshaft gear and thrust washer from end of shaft.
16. Unfasten and remove rear case and gasket from front unit.
17. Inspect lip of oil seal in rear case centerbore. Also inspect cups in countershaft bores. Replace seal and cups if badly worn or mutilated.
18. Remove air vent and wash with cleaning solution. Replace vent on rear case being careful not to overtighten as this can cause collapse of vent body.
19. Reassembly is done in reverse order of disassembly.

MAIN CASE & GEARS

Disassemble Figs. 3 and 4

1. To disassemble, remove gearshift housing.
2. Remove clutch housing from main case.

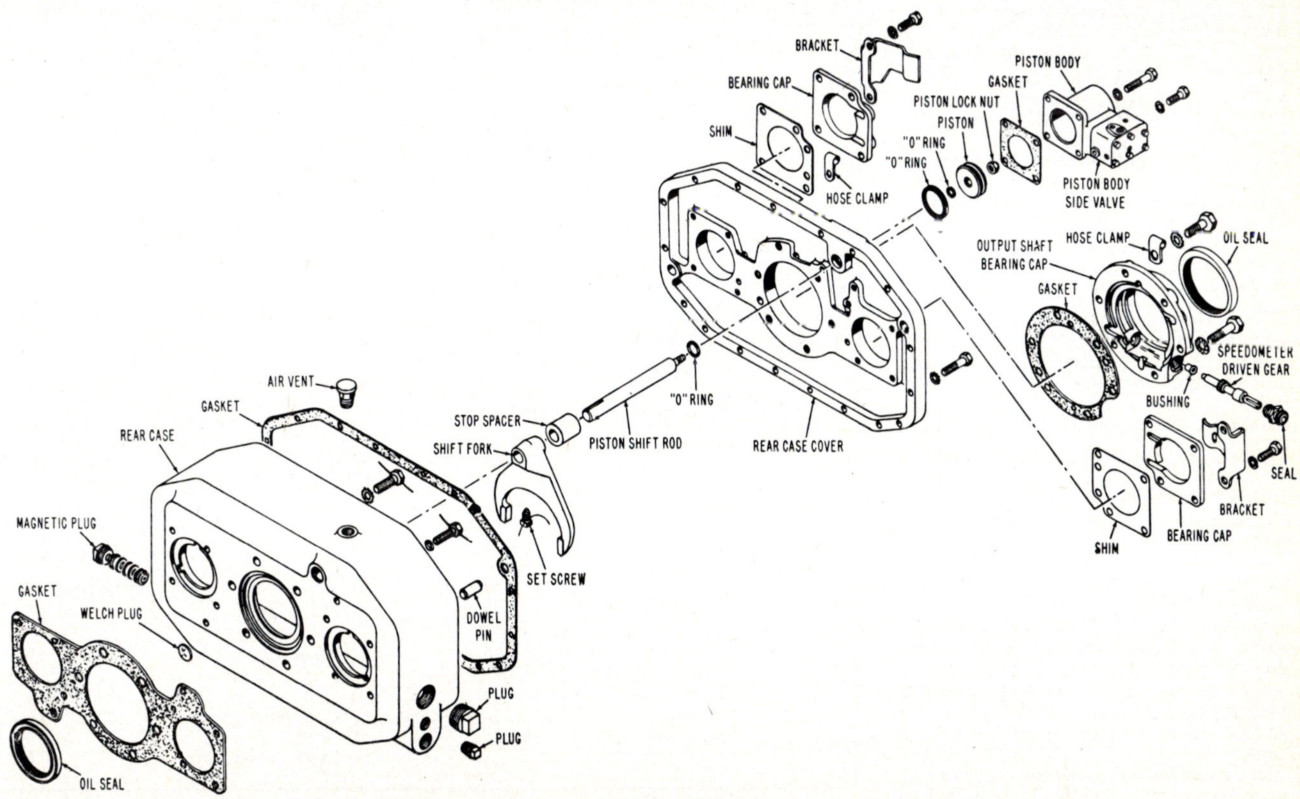

Fig. 1 Rear case cover components

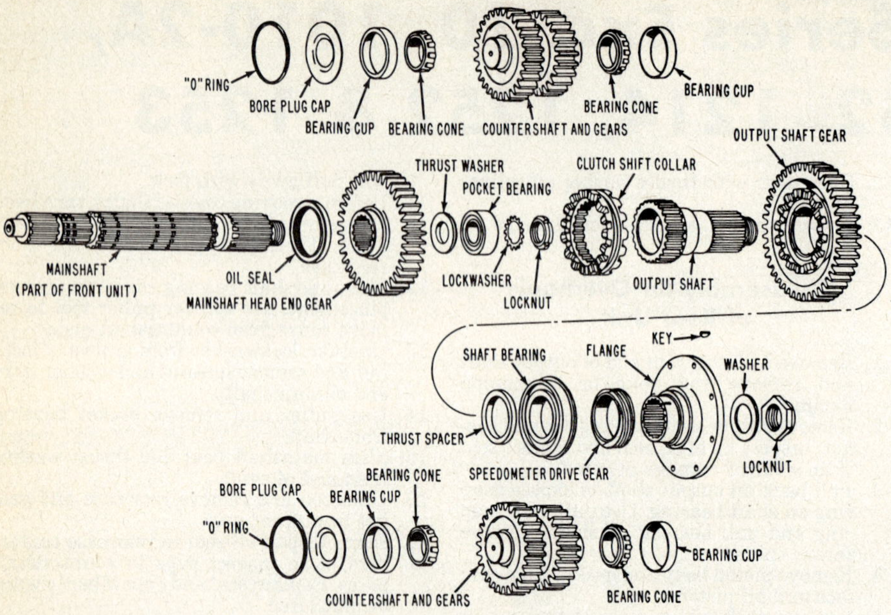

Fig. 2 Rear case gears

Rotate main drive gear until timing marks on back face of drive gear and countershaft drive gears match. Lock transmission into 1st and 6th gears and remove both countershaft front locknuts and mainshaft rear locknut.

3. Using a suitable puller, remove flange and washer.
4. Remove mainshaft rear bearing cap, speedometer gear, bushing and seal.
5. Engage collar into 5th speed gear and use a pry bar against the front face of the collar to force the mainshaft rearward to expose rear bearing snap ring. Install a puller on snap ring and pull bearing from shaft.
6. Remove thrust washer and split rings from shaft.
7. Remove snap ring located in reverse gear bore.
8. Engage 1st-reverse shift collar under reverse gear. Slide reverse gear and collar forward, butting gear against 1st speed gear. Wire or tie both gears together.

NOTE: The countershaft gears will support the mainshaft sub-assembly with mainshaft rear bearing and drive gear sub-assembly removed from the main case.

9. Remove cap screws and lock washers from countershaft rear bearing cap. Separate cap from case.
10. Remove drive gear bearing cap retaining screws and separate cap from case.
11. Remove drive gear sub-assembly from the case bore by pulling forward on the drive gear splined stem.
12. Using a $1^3/_4$" dia. × 6" long soft metal bar, drive backward on the front ends of both countershafts. Force shaft backward to allow bearings to creep forward to expose bearing snap ring. Install suitable puller to bearing snap ring and remove bearings.
13. Move countershaft to expose rear bearing so a bearing split tool can be installed on the front side of the bearing. Remove snap ring from countershaft end. Install puller arms on split tool and remove bearing.
14. Remove upper idler gear shaft and lock ball from case. Leave idler gear in place for later removal after mainshaft is removed.
15. Place a sling or wire around 2nd-3rd shift collar to support mainshaft. Using chain hoist on sling, lift out mainshaft.
16. Remove upper idler gear.
17. Remove right then left side countershafts.
18. Remove lower idler gear.
19. Countershaft, Fig. 5, and mainshaft, Fig. 6, sub-assemblies can now be disassembled.

Assemble Mainshaft, Fig. 6

1. Position mainshaft vertically in vise with output end resting on bed and soft jaws clamping splined area.
2. Install thrust washer on mainshaft to shoulder near output splines.
3. Install snap ring and seat securely in bottom groove of mainshaft.
4. Assemble 1st-reverse clutch collar on shaft and rest collar on snap ring.
5. Install second snap ring in groove closest to 1st-reverse clutch collar.
6. Install thrust washer on shaft.
7. With thrust washer and snap ring assembled into 1st speed gear, assemble gear sub-assembly on shaft with 35° chamfer in gear bore toward clutch collar.
8. Assemble second set of thrust washers and snap ring into second speed gear. Assemble gear sub-assembly on shaft with 35° chamfer up.

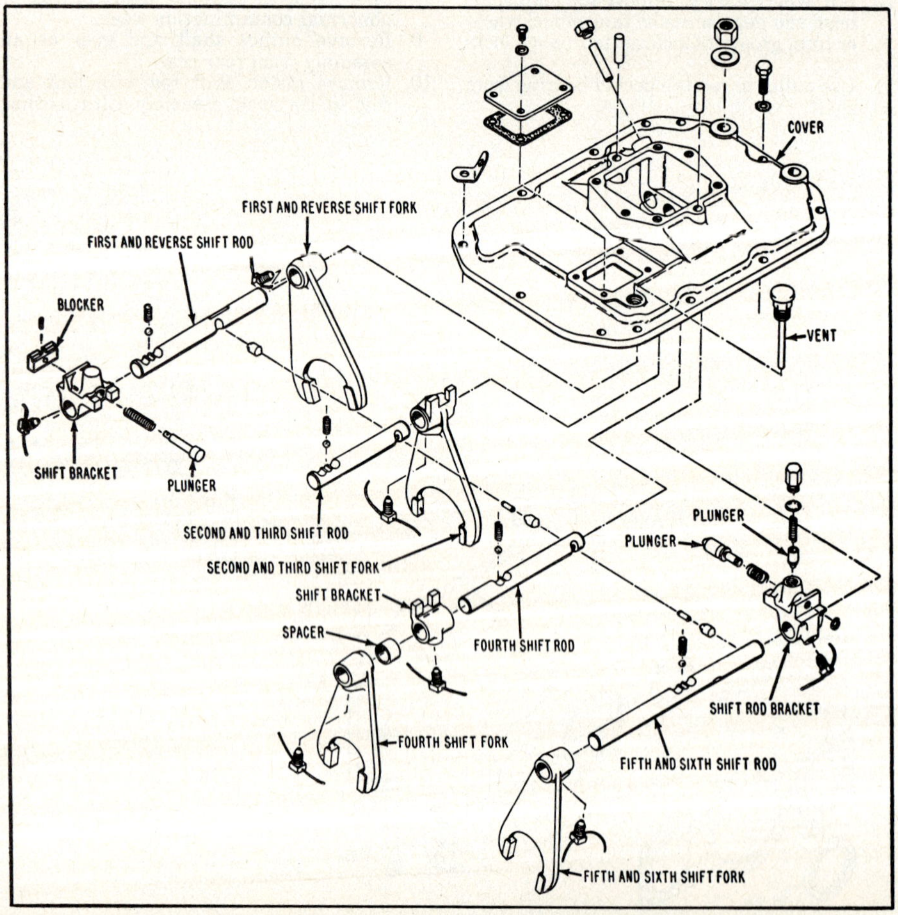

Fig. 3 Spicer SST gearshift housing

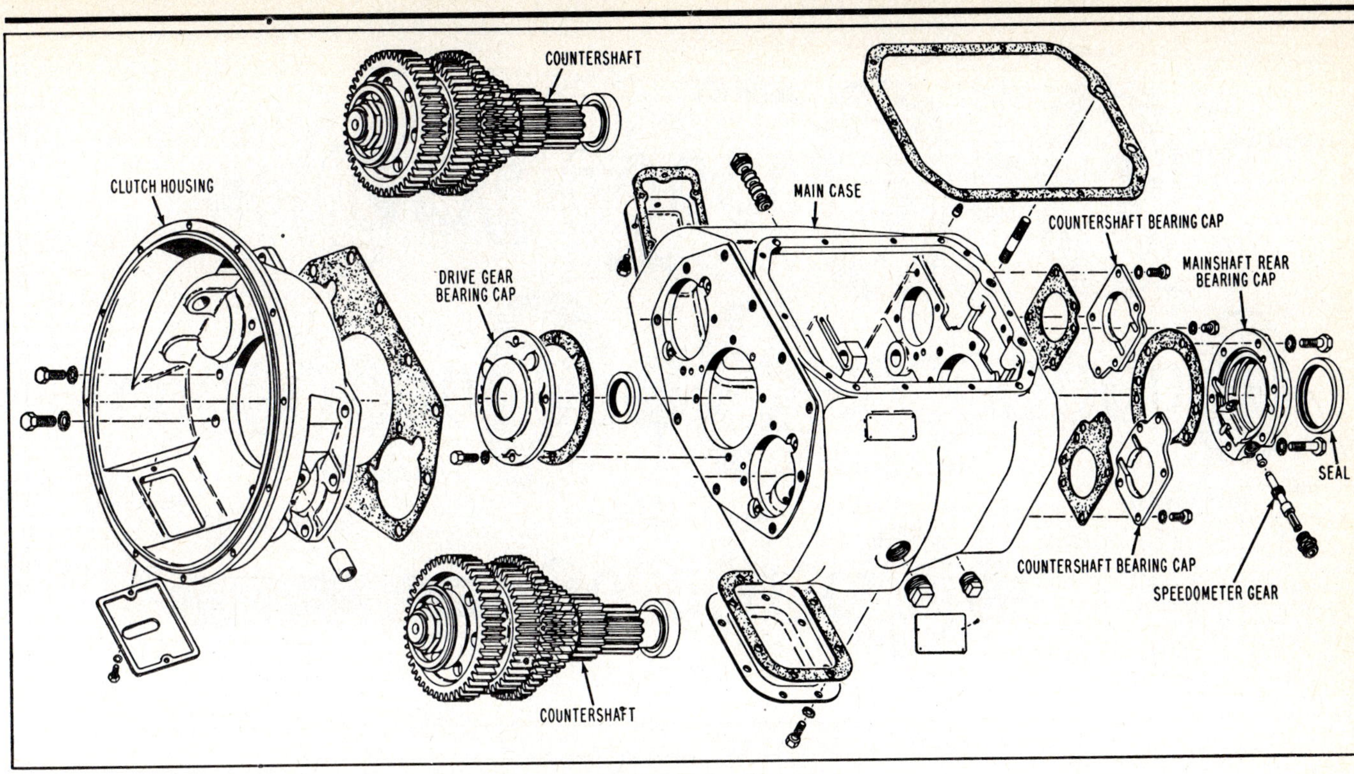

Fig. 4 Spicer SST gear case components

9. Install snap ring.
10. Install 2nd-3rd clutch collar and fourth snap ring. Install thrust washers.
11. Assemble snap ring into 3rd speed gear. Install gear with chamfer down toward clutch collar.
12. Assemble washers and snap ring into 4th speed gear and install gear with chamfer up. Install snap ring.
13. Assemble 4th speed clutch collar on shaft. Install snap ring.
14. Assemble snap rings and thrust washers into 5th speed gear and install gear onto shaft with chamfer up. Install snap ring.
15. Install 5th-6th speed clutch collar on shaft.
16. Remove assembly from vise and place on work bench.
17. Slide 1st-reverse shift collar into bore of 1st gear. Slide thrust washer on end of shaft against snap ring. *Do not install thrust washer or snap ring into bore of reverse gear at this time.*
18. Assemble reverse gear on the rear of the shaft. Slide it forward onto clutch collar. Match gear teeth, butt both gears together and wire the two gears together.
19. The remaining mainshaft parts are assembled after the mainshaft has been placed in the case.

Assemble Transmission

1. Install one of the countershaft sub-assemblies in the case on the left side with timing mark of head and gear toward the center of the case.
2. Assemble two idler shaft bearings and spacer into bore of reverse idler gear.
3. Place one of the reverse idler gears into mesh with countershaft. Do not install idler shaft at this time.
4. Install remaining countershaft into right side of case with head end gear timing mark toward center of case. Install other idler gear, idler gear shaft and lock ball.

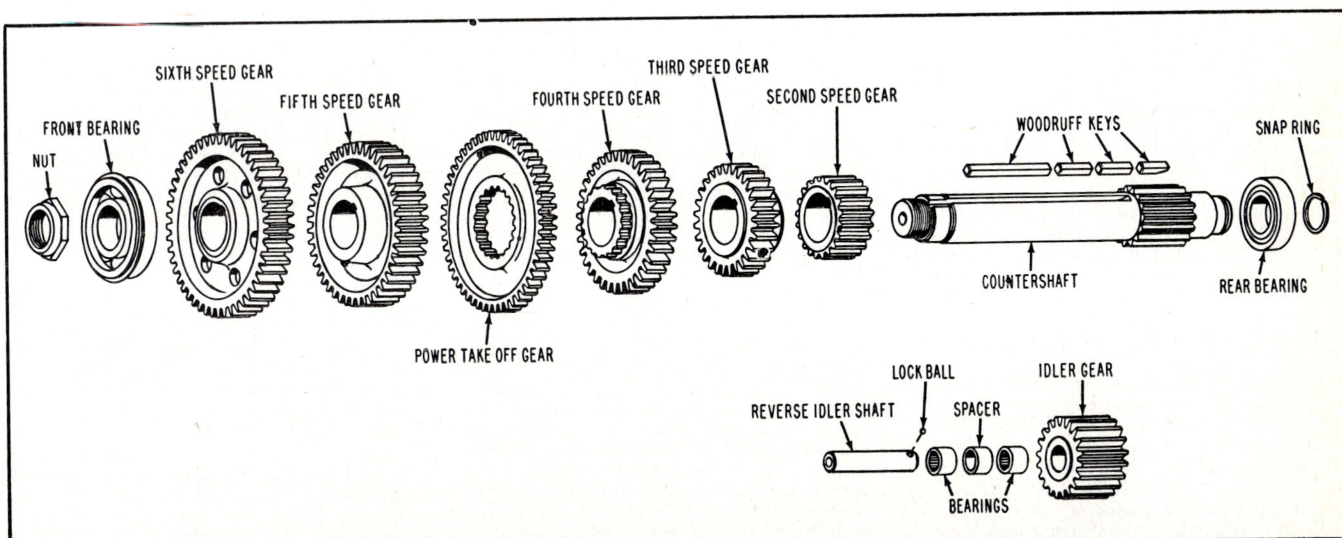

Fig. 5 Spicer SST countershaft. Exploded

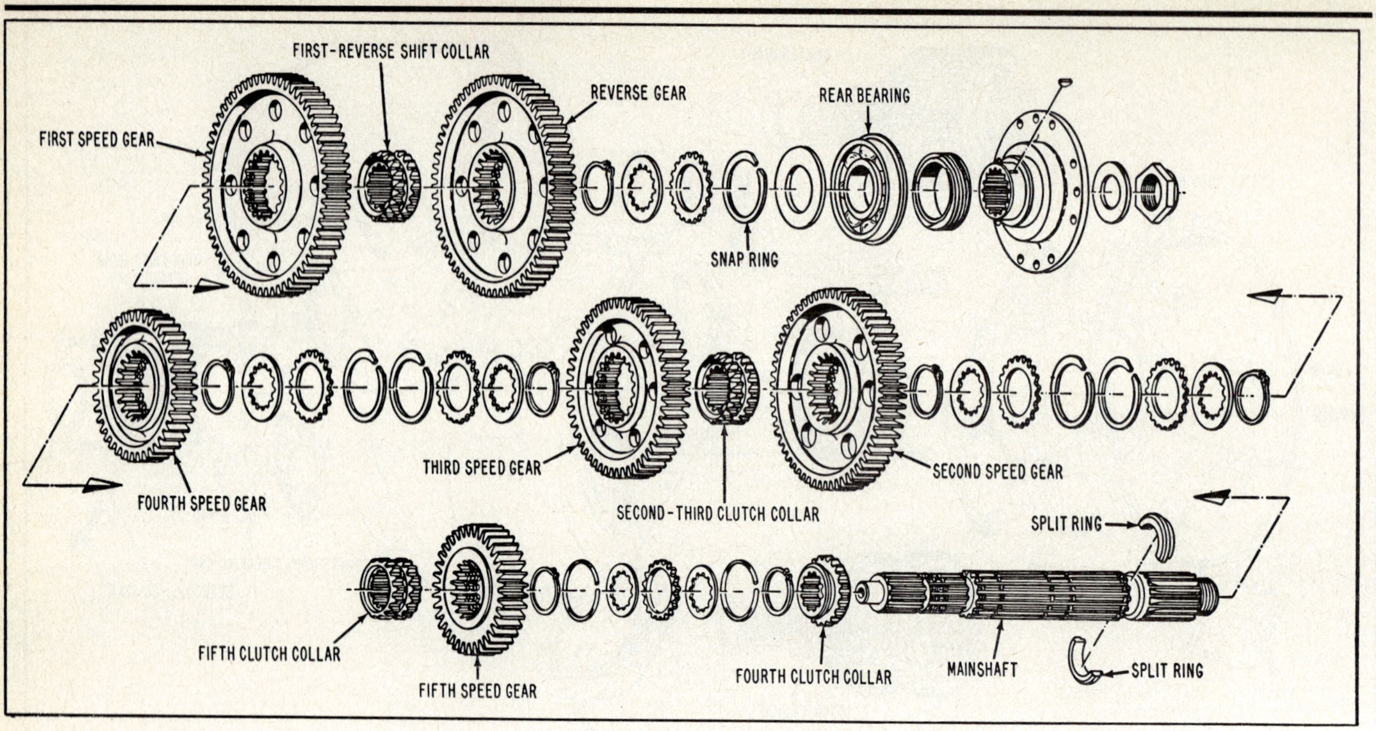

Fig. 6 Spicer SST mainshaft. Exploded

Tap shaft flush with case. Do not install countershaft bearings at this time.

5. Using chain hoist, install mainshaft.
6. Install thrust washers on output end of mainshaft and cut tie wire from 1st-reverse gears.
7. Install split rings with flanges toward gears.
8. Install remaining thrust washers and snap rings on mainshaft.
9. Install drive gear and bearing cap.

NOTE: Mainshaft and both countershafts must be correctly timed. Be sure timing teeth of countershaft head end gears align with timing teeth (two) on mainshaft before completing assembly.

---

10. Install mainshaft rear bearing and remove chain hoist.
11. Install reverse idler shaft into upper reverse idler gear. Install lock ball in shaft and tap shaft end flush with case.
12. During final assembly and tie-up of countershafts, both countershafts must be supported to maintain proper gear timing. Using care not to disturb timing, install countershaft bearings and lock nuts.
13. After all shafts are installed, recheck gear timing and if timing is correct, complete assembly of transmission.

Spicer SST 10, 1010-2A, 1020 & 1214 Units

DISASSEMBLY

1. Using a 15/16 inch socket remove six clutch housing-to-gear case external cap screws and lock washers, Fig. 1. Remove four internal cap screws and lockwashers from inside the clutch housing with a 3/4 inch socket.
2. Using a soft head hammer, tap the clutch housing off the gear case and the OD pilot of the drive gear front bearing cap, Fig. 1. Remove the clutch housing gasket if torn or damaged.
3. Rotate the main drive gear until the timing marks on the outside diameter of the curvic ring of the splitter drive gear align with the timing marks on the back face of the two countershaft head-end drive gears, Fig. 2. Lock the splitter shift clutch collar into the curvic ring of the splitter drive gear.
4. On the mainshaft, shift the clutch collar into the 1st-2nd gear to lock the transmission in two gears, Fig. 2.
5. Remove both countershaft front lock nuts and the yoke lock nut on the mainshaft, Fig. 2.
6. Remove the yoke from the mainshaft, Fig. 2, with a suitable puller. The washer will come off with the yoke.
7. Remove the capscrews and lock washers from the mainshift rear bearing cap with a 3/4 inch socket, Fig. 1. Separate bearing cap from the gear case and gasket. Remove speedometer drive gear and bushing, if used. Remove the seal, if damaged.
8. At the mainshaft, position a pry bar between the splitter shift clutch collar and the 9th-10th gear curvic tooth ring, Fig. 2. Force the mainshaft subassembly rearward to expose the snap ring on the mainshaft rear bearing. Install a suitable puller on the snap ring and pull the bearing off the shaft.
9. Remove thrust washer from the mainshaft, then the split rings, Fig. 2. Remove snap ring located in the bore of the reverse gear. Use a long thin-lipped screw driver to work the snap ring from groove in the gear bore and rearward to remove snap ring from the gear and mainshaft.
10. Pull upper reverse idler gear shaft from gear case, Fig. 3. Recover the idler shaft lock ball. Leave the idler gear in place and force to side of case where it should remain until after the mainshaft subassembly has been removed.
11. On the mainshaft, engage 1st-Reverse clutch collar with reverse gear then slide the reverse gear and collar forward until reverse gear butts against the 1st-2nd speed gear, Fig. 2. Wire or tie both gears together.
12. Remove capscrews and lock washers from both countershaft rear bearing caps and separate the caps from gear case and gaskets, Fig. 1. The air filter and regulator mounting brackets will come off with the bearing caps. Do not remove the connecting nylon line between the air filter and regulator unless damaged. Position the two assemblies with connecting line aside.
13. Remove the capscrews and lockwashers from the drive gear front bearing cap and separate the cap from the gear case and gasket, Fig. 1.
14. Disengage drive gear front bearing snap ring from drive gear and pull the bearing and snap ring forward on drive gear

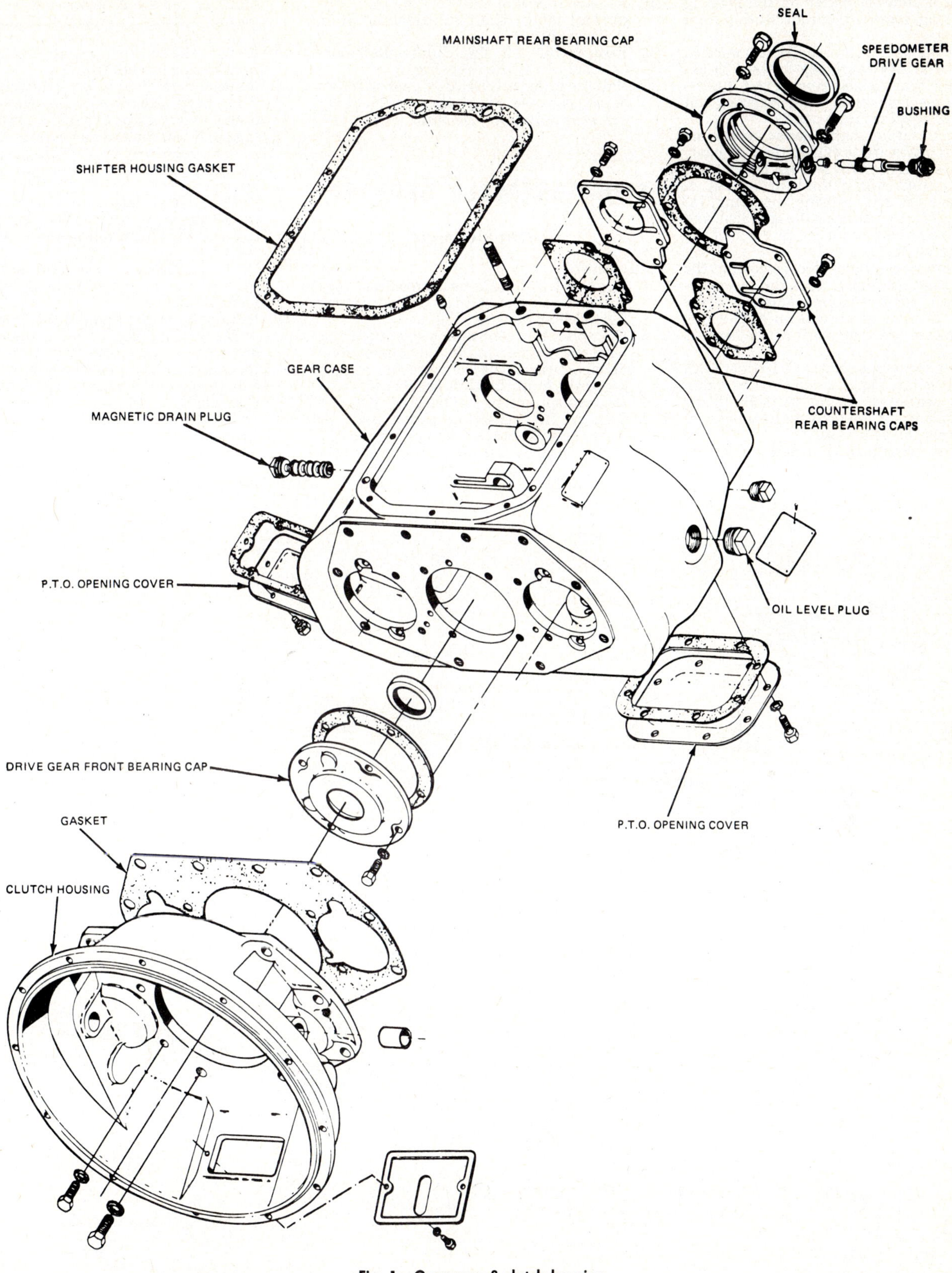

Fig. 1 Gear case & clutch housing

spline stem as far as possible, Fig. 2. This will make the snap ring accessible slightly forward of the gear case face. Using a suitable puller on the bearing snap ring, remove the front bearing from drive gear.

15. Force the two countershafts rearward so front bearings can creep forward to expose the bearing snap rings, Fig. 4. Install puller tool arms on bearing snap ring and remove bearings from shafts.
16. At each countershaft, maneuver the assembly rearward to expose the rear bearing so a bearing split tool can be installed on the front side of the bearing, Fig. 4. Remove the rear bearing snap ring, install puller tool arms on the split tool, and pull the bearing from the rear end of the countershaft.
17. To obtain clearance for removing the mainshaft assembly, force the two countershaft sub-assemblies and the drive gear assembly forward as far as possible. Also, keep the mainshaft as far back as possible.
18. Place a sling rope around 3rd-4th/5th-6th clutch collar, Fig. 2, to support the mainshaft assembly. Using suitable lifting equipment, lift the mainshaft sub-assembly from the gear case.
19. Remove the upper idler gear subassembly, Fig. 3.
20. Remove the drive gear sub-assembly from the case, Fig. 2.
21. Remove the right side (looking from rear of case) countershaft sub-assembly from the case, Fig. 2.
22. Remove the left side countershaft sub-assembly from the case, Fig. 2.
23. Pull the lower reverse idler gear shaft out of the gear case, Fig. 3. Remove idler gear sub-assembly from the case.

SUB-ASSEMBLY SERVICE

Shifter Housing

Disassembly

1. Remove four retaining screws and lock washers and the splitter piston housing from shifter housing, Fig. 5.
2. Clamp splitter control in a soft jawed vise with the splitter piston rod shift bracket facing upward. Cut lock wire and remove set screw from the shift bracket.
3. Remove splitter piston housing cover.
4. Remove the actuating piston and rod from the shift bracket and out of the piston housing by pulling on rod lock nut.

NOTE: If the piston O-rings are damaged by cuts or have flat spots on air sealing surfaces, replace as necessary.

5. Remove the splitter air control valve body from the piston housing.
6. Remove the two valve plungers from valve body. Inspect the O-rings on both plungers for cuts or flat-spot wear on sealing surfaces, replace as necessary.
7. Install required new O-rings on the plungers and lubricate with silicon grease.
8. Reassemble plungers into the valve body, install the valve body on the piston housing with the six retaining screws and lock washers.
9. Place the piston on the actuation piston rod and secure to rod with the lock nut hand tight.
10. Install the piston and rod into the cylindrical housing, Fig. 5. When the rod end protrudes through the front boss, install the shift bracket on the rod with the extended hub of the bracket toward the welch plug.
11. Locate the shift bracket in the proper position on the piston rod, install the set screw, and secure with a lock wire.

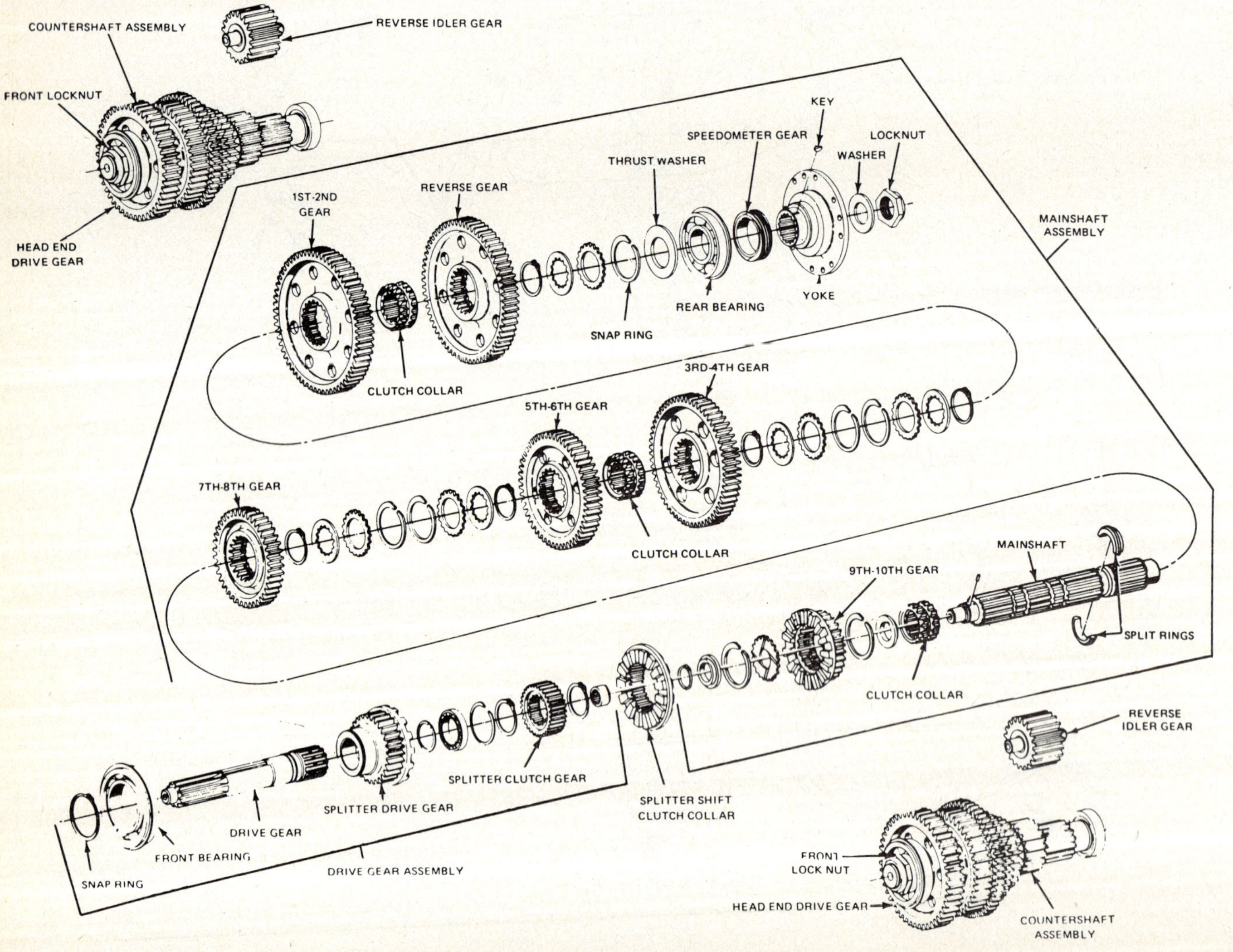

Fig. 2 Sub-assembly installation in transmission case

12. Tighten the piston rod lock nut.
13. Install splitter piston housing cover and gasket, lock washers and retaining screws.
14. Position the air valve sub-assembly aside for later installation on the shifter housing.
15. Remove shoes from the splitter shift fork.
16. Place the edge of the shifter housing in a vise with rods and forks exposed and the front of the housing to the left, or lay the shifter housing on a work bench with forks up and the front of the housing to the left.
17. Cut and remove all wires, then loosen and remove all set screws from shift forks.
18. Turn the splitter shift rod 90 degrees in either direction to preload the poppet detent pin into the recess hole of the center boss on the shifter housing.
19. With a soft drift and hammer, tap the splitter shift rod to the left. As the rod clears housing center boss, remove the detent pin and the poppet spring from the recess hole in the boss. As the rod clears the splitter fork, remove the fork from the rod and the rod from the housing front boss.
20. Tap the 7th-8th/9th-10th shift rod to the left. Remove the bracket as it clears the rod. Remove the poppet ball and spring as the rod clears the front boss. Remove the shift fork as it clears the rod. Then, remove the rod from the front boss.
21. Tap the 3rd-4th/5th-6th shift rod to the left. Remove the poppet ball and spring from the housing boss. Remove the shift rod interlock pin if it should dislodge from the rod during rod movement. Remove the shift fork as it clears the rod, then the rod from the front boss.
22. Remove the (housing boss) interlock pin located internally in the rear boss between the two shift rod (3rd-4th/5th-6th, and 7th-8th/9th-10th) bores.
23. Tap the 1st-2nd/Reverse shift rod to the left and remove the shift fork as it clears the rod. Remove the poppet ball and spring from the housing boss. Remove the bracket sub-assembly as it clears the rod and the rod from the front boss.
24. Remove the second interlock pin located internally in the boss between the two shift rod (3rd-4th/5th-6th and 1st-2nd/Reverse) bores.

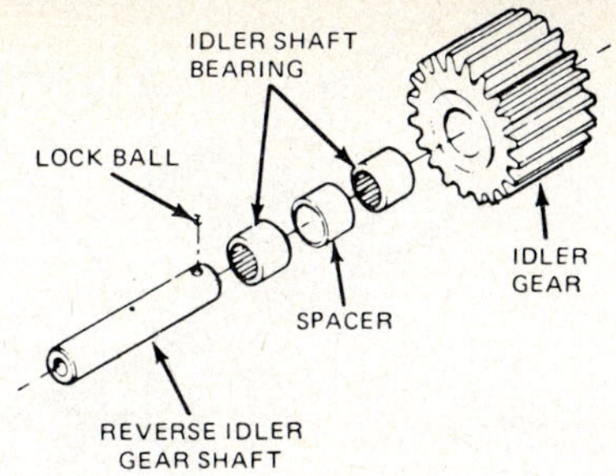

Fig. 3 Idler gear, disassembled

25. To disassemble the 1st-2nd/Reverse shift bracket, clamp the trunion ends in a vise. Shear the block retainer pin using a flathead punch against the small diameter end of the plunger. Strike the plunger end with a sharp blow to shear the pin. Check the block and plunger for excessive wear, and check the spring for broken coils or weak tension.
26. Remove vent from the top surface of the shifter housing.
27. Remove old gaskets and sealing material from the machined surfaces of the shifter housing and top of the gear case.
28. Examine the shift fork shoes for excessive wear. Sharp corners of the shoes should be chamfered slightly with a hand grinder to allow entry of oil between the shoes and shift collar, thereby reducing wear on these surfaces.
29. Check the 1st/reverse blocker pins which are located on the internal side of the shifter housing, Fig. 5. If either pin shows excessive wear or galling, drive it out with a flat head punch. Press a new pin into its hole in the shifter housing flush with the top surface of the housing. This will allow proper length extension to the internal location of the shifter housing.

Assembly

1. If 1st-2nd/reverse shift bracket was disassembled, clamp the trunion ends of the bracket in a vise and insert the plunger and spring into the plunger hole of the bracket, Fig. 5. Pre-load the plunger with a C clamp.
2. Assemble the block to the bracket so the end of the plunger enters the hole in the block. The step on the block must face inward against the block, Fig. 5. Assembling the block outward will cause 1st-2nd/reverse gear block-out.

   Align the plunger pin hole with the block pin hole, and install the lock pin with a .084 inch flat head punch. Tap the C clamp from the plunger and move the plunger in and out to be sure plunger moves freely.
3. Place the shifter housing on a bench with the inside of the housing facing up and the front of the housing to the left.
4. Check all four shift rods to ensure free movement in the boss bores without excessive radial movement. Remove the rods and lightly coat both the rods and the bores in the shifter housing with grease. Install the rods in the bores.
5. Coat interlock pin with heavy grease and insert into the 1st-2nd/reverse shift rod bore through the hole in rear boss.
6. Install and pre-load the poppet spring and ball in the poppet detent bore of 1st-2nd/reverse rod location.
7. To install the 1st-2nd/reverse shift rod and bracket: start the end of the rod, farthest from set screw countersunk holes, through the rear boss of the housing. Assemble bracket on the rod with top boss of the bracket down and the shift gate towards the lower opening. Tap rod sharply to remove the poppet loading tool and slide the rod into the front boss. Assemble the shift fork on the rod with the extended hub of fork to the right. Position the fork and bracket properly on the rod, install the set screws and install the lock wires. Move the shift rod until the poppet ball registers in the neutral detent of the rod.
8. Install and preload the poppet spring and ball in the poppet detent bore of 3rd-4th/5th-6th shift rod location of the boss.
9. To install the 3rd-4th/5th-6th shift rod and fork: coat the small interlock pin, Fig. 5, with heavy grease and insert into the

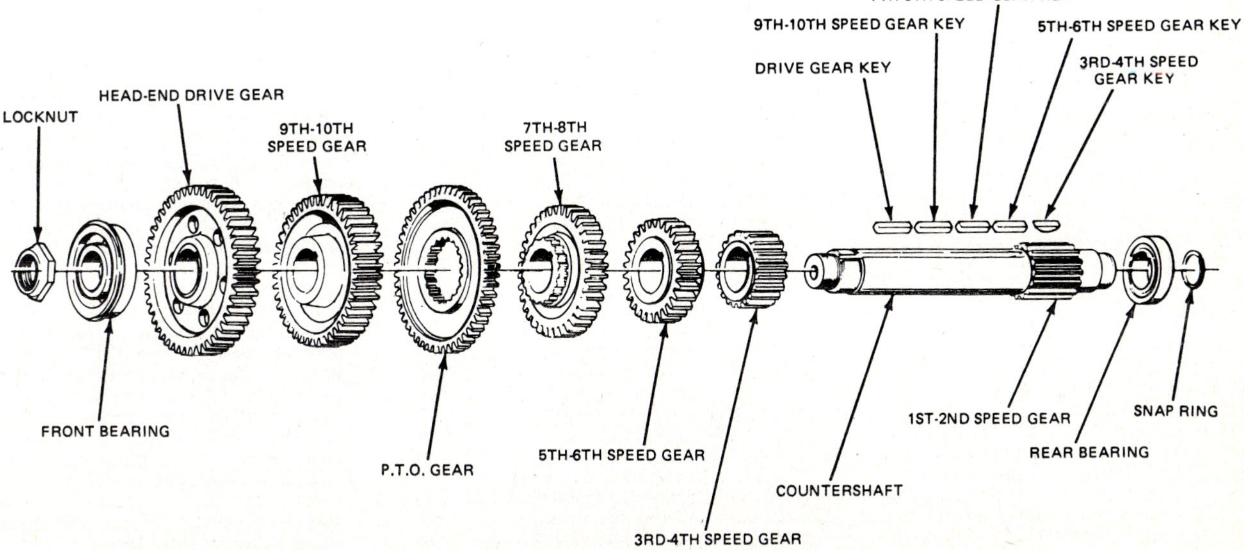

Fig. 4 Countershaft, disassembled

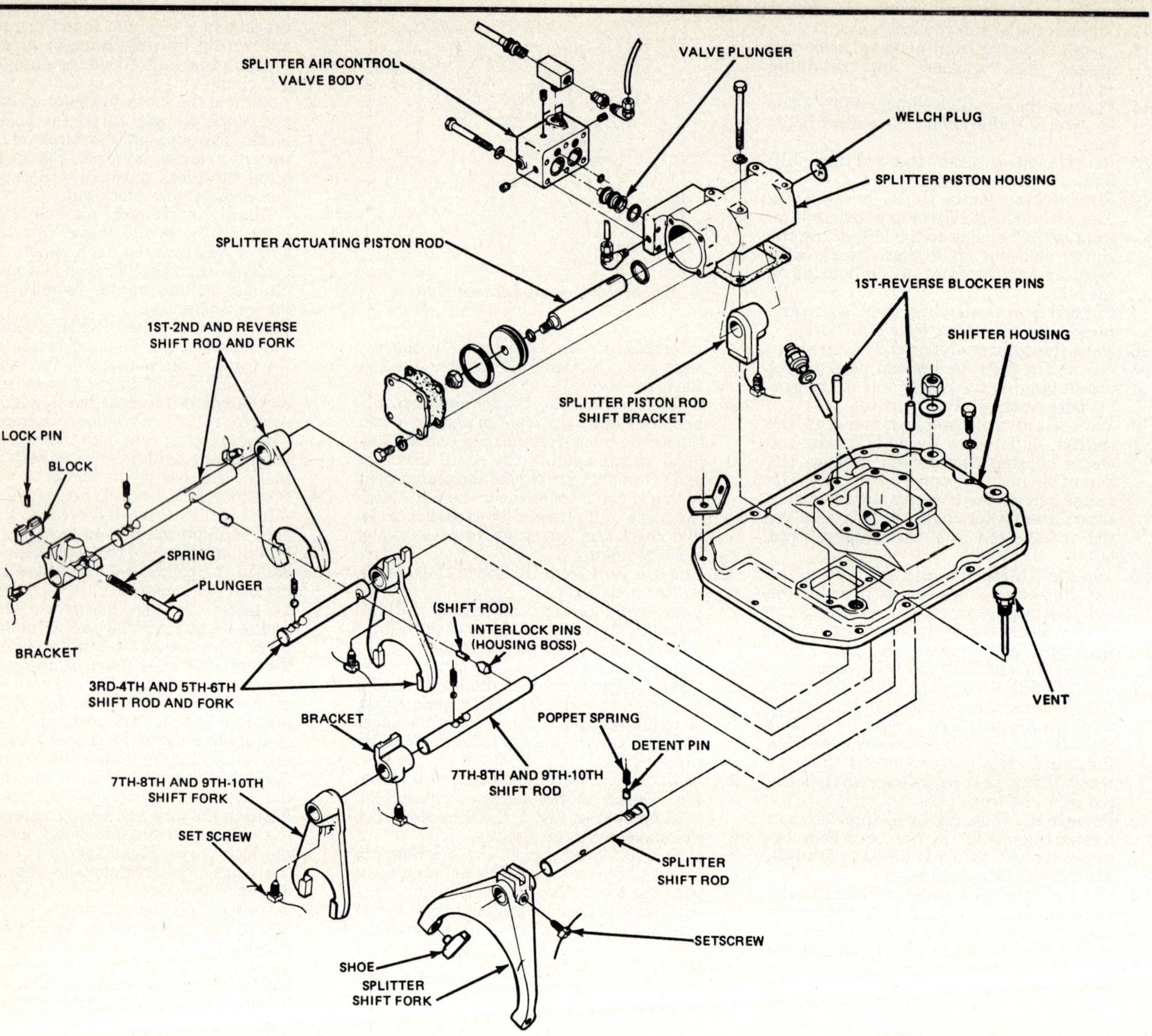

Fig. 5 Shifter housing, disassembled

hole of the shift rod. Enter the rod with interlock pin to the right through the rear boss. Assemble the fork to the rod with fork hub to the left. As the rod enters the front boss, tap it sharply to remove the poppet loading tool. Properly position the shift fork on the rod, install the set screw and secure with the lock wire. Move the rod until the poppet registers in the neutral detent of the rod.

10. Coat the interlock pin with heavy grease and insert into 7th-8th/9th-10th shift rod bore through the hole in the rear boss. Install the poppet spring and ball in the poppet detent bore of 7th-8th/9th-10th shift rod location of the boss.
11. To install the 7th-8th/9th-10th shift rod, bracket and fork: enter the rod end (farthest from interlock detent) into the rear boss. Assemble the bracket on the rod with the shift gate downward. Tap the rod sharply to remove the poppet loading tool in the front boss. Assemble the fork to the rod with fork hub to the left. Properly position the bracket and fork on the rod, install the set screws, and install the lock wires. Move the rod until the poppet registers in the neutral detent of the rod.
12. Install and pre-load the poppet spring and ball in the poppet detent bore of the splitter shift rod location of the boss.
13. To install the splitter shift rod and fork: enter the rod (detent end) into the front boss with detent facing up. Assemble the fork to the shaft with fork extended hub to the right and hub shift gate down. Tap the rod sharply to remove the poppet loading tool. Turn the rod 180 degrees so the poppet pin locates in the rod detent. Properly position the fork on the rod, install the set screw and secure with the lock wire. Move the rod until the poppet registers in the left detent of the rod. This locates the splitter fork in the LO position of splitter-to-transmission gear.
14. Shift 1st-2nd/reverse speed shift fork into gear position and try to shift the other two shift rods (3rd-4th/5th-6th and 7th-8th/9th-10th). If all interlocks are functioning, the rods should be locked in neutral position.
15. Return the 1st-2nd/reverse shift fork to neutral positions and check the movement of each of the other shift rods to the rods move readily and completely into each gear position.
16. Turn the shifter housing assembly over (top up). Install vent in the tapped hole in the top surface of the housing using sealer on the threads, Fig. 5. Tighten vent securely, but be careful not to collapse the body of the vent by over-tightening.
17. Assemble the splitter piston housing and splitter control valve body sub-assembly to the shifter housing with gasket, Fig. 5. As the assembly is mounted to the housing engage the pad on the splitter piston rod shift bracket into the splitter shift fork. Secure the assembly to the housing with four capscrews and lock washers.
18. Coat the trunions of the splitter fork shoes with heavy grease and insert the shoes in the splitter fork.
19. Position complete shifter housing aside for later assembly to gear case.

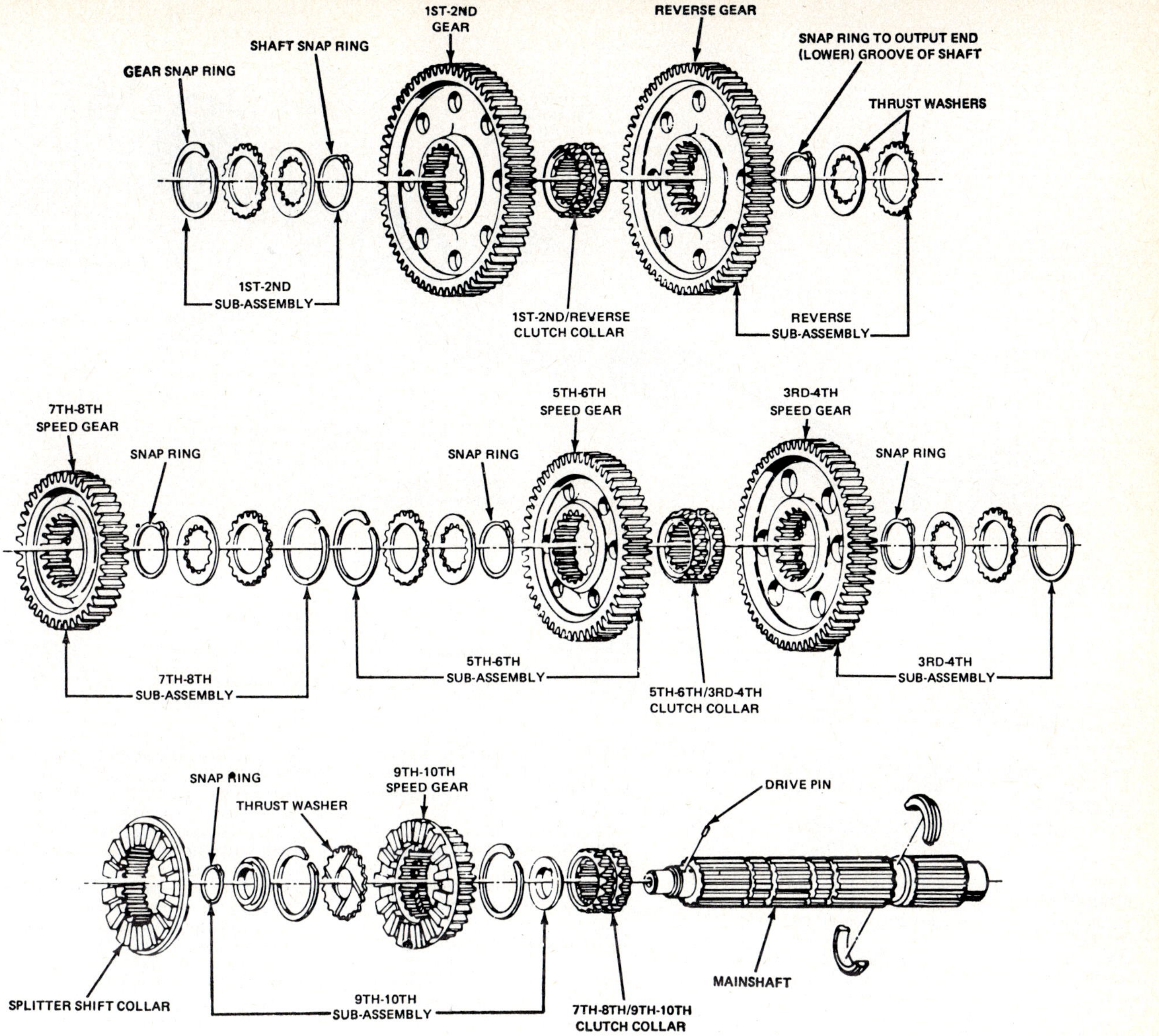

Fig. 6 Mainshaft, disassembled

Mainshaft

Disassembly

1. Remove snap ring from front of mainshaft, and slide the 9th-10th gear sub-assembly off the shaft drive pins, Fig. 6.
2. Remove 7th-8th/9th-10th clutch collar from shaft.
3. Remove the snap ring from the shaft groove, then the 7th-8th speed gear sub-assembly.
4. Remove 5th-6th gear sub-assembly and the snap ring from the shaft groove.
5. Remove 5th-6th/3rd-4th clutch collar from shaft.
6. Remove snap ring from the 3rd-4th speed gear bore, and the 3rd-4th speed gear sub-assembly.
7. Remove the wire used to fasten the 1st-2nd speed gear to the reverse gear when the mainshaft was removed from the gear case.
8. Remove 1st-2nd gear sub-assembly from the shaft and the snap ring from the shaft groove.
9. Remove 1st-2nd/reverse shift collar from shaft.
10. Remove reverse gear sub-assembly from the (rear) output end of the shaft.

Assembly

Before assembly inspect all thrust washers. If heavy face galling or tooth wear is indicated, the thrust washers should be replaced. Lubricate thrust washers with 30 weight engine oil on the thrust faces.

1. Clamp the mainshaft vertically in a vise with the output end splines of the shaft resting on the bed and the soft jaws clamping on the splined area.
2. Install internal tooth thrust washer on the mainshaft and let it rest on the shoulder near the output splines. The external tooth thrust washer is installed after the mainshaft sub-assembly has been installed in the gear case.
3. Install the snap ring in the lower groove of the shaft, Fig. 6. Make sure the ring is seated securely in the groove.
4. Assemble 1st-2nd/reverse clutch collar with either end of collar down. Rest the collar against the snap ring.

NOTE: Clutch collars are identical parts. They can be installed in any of the gear positions. Either end can be installed on the shaft at time of assembly.

---

5. Install 1st-2nd sub-assembly snap ring in the shaft groove closest to the 1st-2nd/reverse clutch collar, Fig. 6.
6. Install the internal tooth snap ring on the shaft to snap ring rests on the snap ring.
7. With both thrust washers and the large snap ring assembled into the 1st-2nd speed gear, install the 1st-2nd sub-assembly on the shaft so the chamfer in the bore of the gear extends down toward the clutch collar.
8. Assemble second set of thrust washers and large snap ring into the 3rd-4th speed gear and install the 3rd-4th sub-assembly on the shaft, chamfer up. Rest 3rd-4th speed gear against the face of the 1st-2nd speed gear. Install third (shaft) snap ring in shaft groove under the bore of the 3rd-4th speed gear.
9. Assemble 3rd-4th/5th-6th clutch collar so clutch collar rests in the bore of 3rd-4th

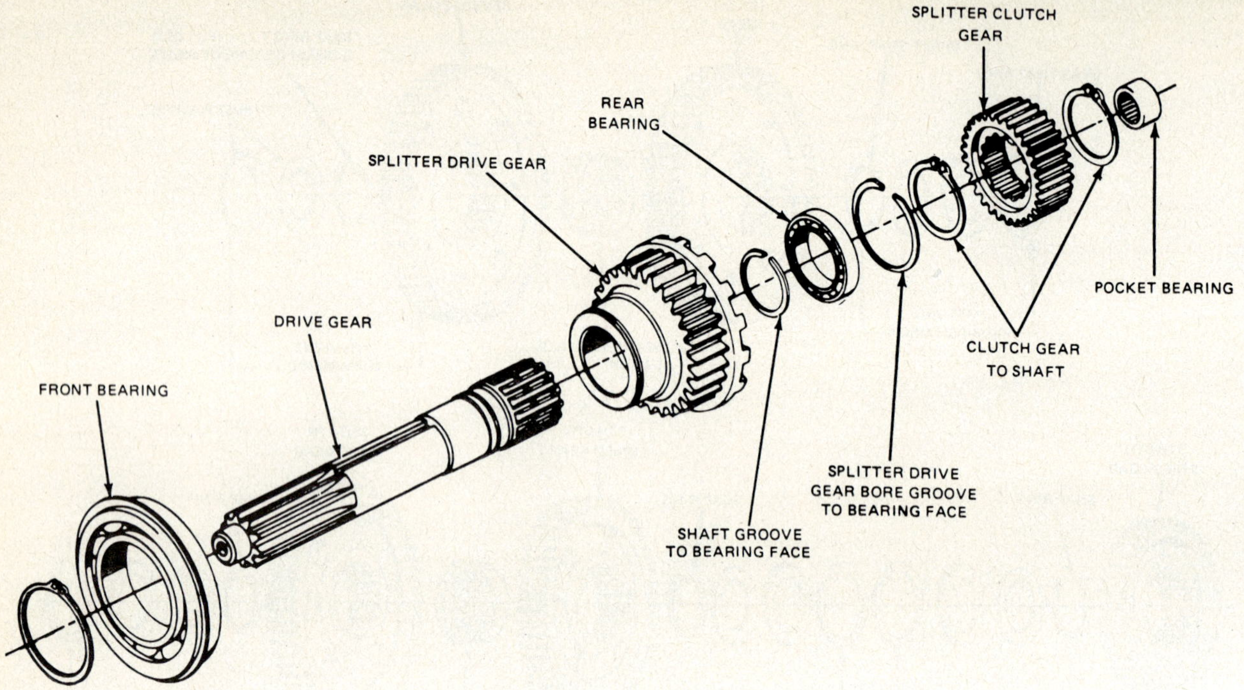

Fig. 7 Drive gear, disassembled

speed gear.
10. Install the fourth (shaft) snap ring in the shaft groove closest to 3rd-4th/5th-6th clutch collar.
11. Install two thrust washers, one internal and one external, on the shaft, resting on the snap ring.
12. Assemble the large snap ring into the 5th-6th speed gear and assemble the gear chamfer in bore down toward the clutch collar.
13. Assemble the fourth set of washers and the large snap ring into the 7th-8th speed gear. Assemble the 7th-8th gear sub-assembly chamfer up and resting against the face of the 5th-6th gear. Install the fifth snap ring in the shaft groove under the bore of the 7th-8th speed gear.
14. Assemble the 7th-8th/9th-10th speed clutch collar on the shaft.
15. If 9th-10th speed gear was disassembled, reassemble one large snap ring in the gear bore groove, install the thrust washer and lock the parts together with the second large snap ring, Fig. 6.
16. If replacement of drive pins on the front of the mainshaft is not necessary, install the 9th-10th sub-assembly on the mainshaft.
17. Lock up all the 9th-10th gear parts on the shaft by installing the small ring in the shaft groove. Install curvic shift collar on the 9th-10th gear curvic ring.
18. Remove the complete assembly from the vise and place it on a work bench.
19. Slide 1st-2nd/reverse clutch collar into the bore of the 1st-2nd speed gear. Also, slide the internal tooth thrust washer on the rear end of the shaft against the snap ring.
20. Do not install the external tooth thrust washer or large snap ring into the bore of the reverse gear at this time. Assemble only the reverse gear on the rear end of the mainshaft. Assemble the reverse gear with the chamfer of clutch teeth toward the clutch collar. Match the O.D. teeth of the clutch collar to the 1st-2nd speed gear internal teeth. Butt both gears together and tie with wire.

The remaining parts, Fig. 2, relative to the reverse gear and rear end of the mainshaft will be assembled to their respective places after the mainshaft sub-assembly has been placed in the gear case.

Countershaft

Use an arbor press with parallel bars to remove all the gears except 9th and 10th speed gear.

Disassembly

1. Support the head end drive gear, Fig. 4, with parallel bars as close to the hub as possible and press the countershaft from gear.
2. Remove 9th-10th speed gear from the countershaft with a suitable puller tool engaged in the lightener holes in the web of the gear.
3. Lift the P.T.O. gear off the splined teeth of the 7th-8th speed gear, Fig. 4.
4. Support the 7th-8th speed gear with parallel bars under the gear teeth, and press the shaft from gear.
5. Support the 3rd-4th speed gear with parallel bars under and as close to the hub as possible. Then press the countershaft out of both 5th-6th and 3rd-4th speed gears.

Assembly

Coat bores of all gears with oil before pressing the gears on the countershaft. The key to each gear should be installed in the shaft keyway one at a time. If the keys become mutilated or burred, use a mill file to align the sides and remove burrs.

1. Press 3rd-4th speed key into place on the countershaft, Fig. 4. Support the 3rd-4th speed gear and set the countershaft and key into position under the arbor. Align the key with the gear keyway and press the shaft and key into the gear. Seat the gear firmly against the face of the 1st-2nd speed gear. If the key crept out from 3rd-4th speed gear face, tap the key back flush with the gear face or slightly under the gear face.
2. Press the 1 9/16 inch, 5th-6th speed gear key into place on the countershaft. Support the 5th-6th speed gear on the bars with the long hub up. Set the shaft into the gear under the press. Align the key with the gear keyway and press the shaft and key into the gear. Set the gear firmly against the face of the 3rd-4th speed gear. If the key crept out of position, tap it back flush with the face of the gear or slightly under the gear face.
3. Press the 2 3/8 inch, 7th-8th speed gear key into place on the countershaft. Support the 7th-8th speed gear on the bars with the splined teeth on the hub down. Set the shaft into the gear under the arbor press. Align the gear keyway with the key and press the shaft and key into the gear. Seat the gear face firmly against the face of the 5th-6th speed gear. Ensure the key is flush with or under the face of the gear.
4. Slide the P.T.O. gear onto the splined teeth of the 7th-8th speed gear with the long hub of the P.T.O. gear against the face of 7th-8th speed gear.
5. Press the 2 7/16 inch, 9th-10th speed gear key into the keyway on the countershaft. Support the 9th-10th speed gears on the bars with the long hub end down. Set the shaft into the gear holding the P.T.O. gear in mesh with its mating gear, Fig. 4. Align the gear keyway with the key and press the shaft and key into the gear. Seat the gear firmly against the face of the 7th-8th speed gear. Ensure the key is flush with or slightly under the face of the gear.
6. Press the head end drive gear key into the keyway on the countershaft. Support the

head end (splitter) drive gear on the bars with the long hub facing upward.

NOTE: Be sure that the timing mark on the tooth web of the gear is in alignment with the center of the gear keyway.

Set the shaft into the gear, align the gear keyway with the key, and press the shaft and key into the gear. Seat the gear face firmly against the face of the 9th-10th speed gear. Ensure the key face is flush with or slightly under the face of the gear.

NOTE: The front and rear bearings will be installed after the countershaft is installed in the gear case.

Drive Gear

Disassembly

1. If the pocket bearing replacement is necessary, use a small puller tool to dislodge it from the pocket of the drive gear, Fig. 7.
2. Remove the clutch gear-to-shaft outer snap ring and the splitter clutch gear from the drive gear splines. Remove the clutch gear-to-shaft outer snap ring.
3. Remove large snap ring from the bore of the splitter drive gear, and force the splitter drive gear from the rear bearing.
4. Remove the shaft groove-to-bearing face snap ring from the drive gear and shaft assembly, Fig. 7. Support the rear bearing on a press bed and press the drive gear and shaft assembly out of the bearing.

Assembly

1. Support rear bearing in an arbor press (inner race of bearing) and press the drive gear shaft assembly into the bearing. Seat the shaft groove-to-bearing face snap ring into its groove, Fig. 7.
2. Assemble shaft and bearing assembly into the bearing bore of the splitter drive gear. Seat the bearing to the bore face of the splitter drive gear and assemble the gear bore groove-to-bearing face snap ring.
3. Install clutch gear-to-shaft inner snap ring in the groove near the bore of the splitter drive gear. Assemble the splitter clutch gear on the shaft and install the second (outer) snap ring on the shaft to secure the clutch gear to the shaft, Fig. 7.
4. Support the drive gear sub-assembly in an arbor press resting the front face of the splitter drive gear on the press bed. Press the pocket bearing into the pocket or small inner bore of the drive gear and shaft assembly. The pocket bearing part number must be visible when pressed into the bore. The opposite end of the bearing is made of softer metal. Press the bearing so bearing is recessed 1/16 inch under the gear face.
5. Do not install the front bearing and snap ring at this time. The drive gear assembly must be installed in the gear case bore from the internal side after the countershaft assemblies have been placed into position on the bottom of the case. The front bearing and snap ring is then installed on the drive gear assembly and into the case bore from the outside. Set the drive gear assembly aside for later installation into the gear case.

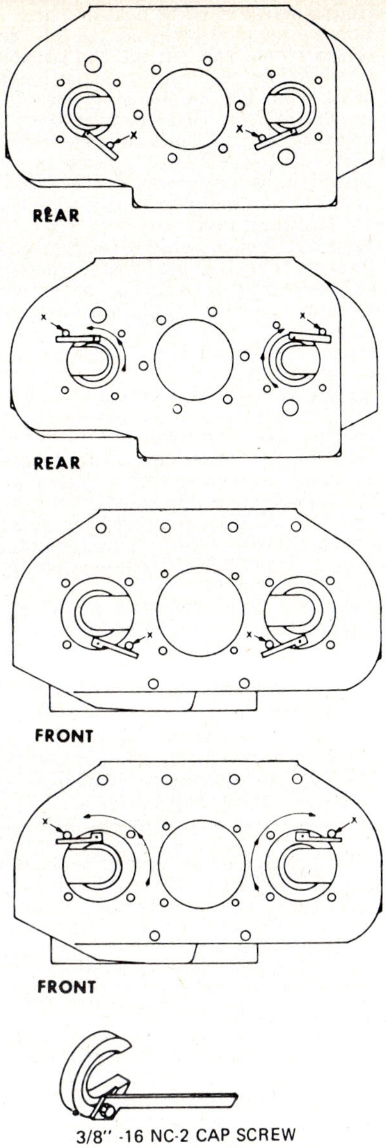

Fig. 8 Positioning pilot tools in countershaft bearing bores

ASSEMBLY

The timing teeth on the head end of the countershafts and on the splitter drive gear, Fig. 2, are designated by a timing mark (letter V). The back face of these three gear teeth should be painted with a bright color before installing any of the sub-assemblies.

The sub-assemblies should be installed in the gear case by performing the following procedures in the specified order.

Countershafts, Reverse Idler Gears, & Drive Gear Assembly Installation

1. Place either of the countershaft subassemblies inside the gear case on the left side with the head-end gear towards the front bore of the case, Fig. 2. Rotate the head-end gear until the timing mark points inward toward the center of the case and the tooth itself is parallel with the bottom of the case.
2. Assemble the two idler shaft bearings with a spacer between the bore of each reverse idler gear, Fig. 3.
3. Position either of the reverse idler gears in the upper bore of the case in mesh with the teeth on the rear end of the countershaft.

NOTE: Do not install the upper idler gear shaft until after the mainshaft is installed.

4. Check if the magnetic drain plug is in place on the right side of the gear case, Fig. 1. If the plug is in the case, remove it so plug will not be damaged during later installation of the lower reverse idler gear.
5. Place the remaining countershaft assembly inside of the gear case on the right side with the head end gear toward the front bore of the case, Fig. 2. Rotate the head end gear until the timing mark points inward toward the center of the case and the timing tooth itself is parallel with the bottom of the case.
6. Position the remaining idler gear to the lower boss location in the gear case. Roll the idler gear on the mating teeth of the countershaft to align the gear bearing bore with the hole in the case. Install the idler shaft into the gear bore, Fig. 3. Install the lock ball into hole in the shaft just before the ball comes into the recess of the gear case. Tap the end of the shaft until it is flush with the face of the case.
7. Do not install the countershaft front and rear bearings at this time. Allow the countershafts to lay free in the bottom of the case.
8. Install the drive gear assembly, Fig. 2, into the gear case center bore from the inside of the case.

Mainshaft Installation

1. Place a sling rope around 3rd-4th/5th-6th clutch collar, Fig. 2, to support the mainshaft assembly. (Using suitable lifting equipment,) lift the mainshaft assembly from the work bench and lower it into the gear case in partial mesh with the countershaft gears. Support the mainshaft in proper position until all bearings and the upper reverse idler shaft have been assembled in the case.
2. Install the external-tooth thrust washer over the output (rear) end of the mainshaft and slide it forward against the internal-tooth thrust washer previously installed during mainshaft assembly, Fig. 2. Force both washers forward against the snap ring already installed on the mainshaft.
3. Unfasten the 1st-2nd speed gear from the reverse gear. Slide the reverse gear rearward into approximate location on the reverse gear bore thrust washers internal and external teeth.
4. Install the large snap ring into the location groove of the bore on the reverse gear, Fig. 2. Use a long thin-lipped screwdriver to work the snap ring into the gear groove.
5. Coat the two split rings, Fig. 2, with heavy grease and install the rings into the recess on the rear of the mainshaft with ring flanges facing the gears.
6. Coat the thick thrust washer with heavy grease and install it on the mainshaft against the flange face of the split rings,

Fig. 2.
7. Install the pilot tool for the mainshaft rear bearing bore. The tool will help align the mainshaft and gears with the drive gear assembly that is already installed in the front center bore of the case.
8. Support the front end of each countershaft with a countershaft front bearing bore tool, Fig. 8.
9. Turn the countershaft head end gears around until the V timing marks point inward toward each other and the timing teeth are parallel with the bottom of the case. Position the two timing marks on the splitter drive gear (curvic ring) where they will match and mate with the timing teeth on the countershaft gears, Fig. 2.
10. Install the snap ring on the drive gear front bearing, and seat the bearing with the snap ring against the face of the gear case, Fig. 2.
11. Install the drive gear front bearing cap with gasket, Fig. 1. Align the oil port holes of the case to the oil return holes on the bearing cap.
12. Dip the cap screws in sealer and attach the bearing cap to the case with cap screws and washers.
13. Remove bearing bore pilot tool from the case center rear bore and mainshaft. Slide the three clutch collars into their neutral positions on the mainshaft, Fig. 2.
14. Start the inner race of the mainshaft rear bearing, Fig. 2, on the mainshaft with the mainshaft rear bearing pilot tool against the bearing face. Use the drive tubing of the tool to drive against the pilot tool and force the bearing onto the shaft and into the bore of the case. Seat the bearing snap ring against the face of the case.
15. Relieve the hoist tension on the sling rope and remove the rope from the mainshaft.
16. Install the shaft into the upper reverse idler gear bore, Fig. 3. Install the shaft lock ball in its hole in the shaft just before the ball enters the recess in the case. Tap the end of the idler shaft flush with the case face.

Final Assembly

1. Support the rear end of each countershaft with a countershaft rear bearing bore pilot tool, Fig. 8. As the pilot tools are inserted into the bores keep the timing teeth in correct mesh with each other.
2. As each front piloting tool is removed and the front bearing installed, both countershafts must be held in proper position relative to the timing teeth mesh and the timing marks must be held in place.
3. Remove the pilot tool from the left countershaft front bore. Install the left countershaft front bearing, Fig. 4. Use the face of the pilot tool on the front end of the bearing to drive the bearing on the shaft and into the case bore. Seat the bearing snap ring to the face of the case.
4. Install the left countershaft front locknut on the shaft and tighten against the bearing face hand tight.
5. Remove the pilot tool from the right countershaft front bore and install the right countershaft front bearing in the same manner as the left countershaft described in Step 3.
6. Remove tools and chain hoist from the countershafts.
7. Install the right countershaft front lock nut on the shaft and tighten against the bearing face hand tight.
8. Lock the transmission in two gears by engaging the splitter shift clutch collar to the curvic ring of the splitter drive gear and by shifting the 1st-2nd; reverse clutch collar into the 1st-2nd speed gear, Fig. 2. Tighten both countershaft front bearing lock nuts.
9. Remove the pilot tool from the left countershaft rear bore and install the countershaft rear bearing.

NOTE: The bearing must be assembled with the snap ring toward the outside of the transmission case. Do not assemble the bearing with the snap ring toward the inside of the transmission.

---

10. Use the pilot tool tubing to drive the bearing on the shaft and into the bore of the case. Install the snap ring on the shaft.
11. Remove pilot tool from the right countershaft rear bore and install the right countershaft rear bearing in the same manner as for the left countershaft described in Steps 9 and 10.
12. If the speedometer gear was removed. install it into the mainshaft rear bearing cap. Install rear bearing cap with the oil return holes of the bearing cap aligning the oil port holes in the case.
13. Dip the bearing cap retaining screws in sealer, install the screws with washers.
14. Assemble both countershaft rear bearing caps with gaskets to the face of the gear case. Secure the caps with retaining screws and washers.
15. Install the end yoke with washer on the mainshaft output splines, Fig. 2.
16. Install the mainshaft locknut, Fig. 2.
17. Move the 1st-2nd/reverse clutch collar into its neutral position, Fig. 2. Leave the splitter shift clutch collar engaged in the splitter drive gear.
18. Roll the gear train by turning the drive gear stem. If tooth timing marks are in their correct positions, the entire gear train will roll freely. If the timing teeth have been set incorrectly or have moved from proper position, the gear train will roll lock up after several turns of the drive gear.
19. If the unit locks up, disengage the shift collar from the splitter drive gear. Turn the drive gear in reverse rotation until the timing marks or painted teeth come into alignment or close mis-match. If mismatch appears, the timing was set incorrectly at the time of assembly or the timing teeth escaped positioning during final tie-up. If this is the case, the shafts must be re-timed.
20. If timing is correct, shift all clutch collars, except the splitter shift collar, into their neutral positions for later installation of the shifter housing assembly, Fig. 2. Move the splitter shift clutch collar into mesh with the 9th-10th speed gear so that it will be in proper position for later assembly with the splitter shift fork, Fig. 5.

Now proceed with installation of the clutch housing and related parts.

Clutch Housing Installation

1. Using the drive gear front bearing cap as a pilot, position the clutch housing gasket to the front face of the gear case, Fig. 1.
2. Set the clutch housing in place using the drive gear bearing cap as a guide.
3. Dip the capscrews in sealer, attach the clutch housing to the gear case with capscrews (six external and eight internal) and washers, Fig. 1.
4. Install clutch release yoke and short shaft to the clutch housing. Assemble the long shaft to the yoke. Align the keyway with the slots in the yoke and assemble the key, lockwashers and capscrews.
5. Replace the P.T.O. opening covers and gaskets if removed.
6. Install magnetic drain plug on right side of gear case and the oil level plug on the left side, Fig. 1.
7. Install shifter housing gasket on the gear case, Fig. 1.
8. Check the shifter housing assembly to see all shift forks are in neutral position, Fig. 5. Install the shifter housing to the gear case so all shift forks engage their pertinent clutch collars.

Spicer SST 1052, 1252 & 1253 Units

DISASSEMBLY

1. Remove clutch housing from main case. Refer to Fig. 1 under "Spicer SST" 10, 1010-2A, 1020 & 1214 Units" under "Disassembly".
2. Remove clutch brake from drive gear stem shaft. On vehicles equipped with wet clutch, remove oil sump if fluid leakage is evident.
3. Rotate main drive gear until gear timing marks on drive gear align with timing marks on countershaft drive gears.
4. Lock sliding shift collars into 1st gear and 5th gear, then using suitable socket remove countershaft lock nuts and rear mainshaft locknut.
5. Using suitable puller, remove mainshaft rear flange and washer.
6. Remove mainshaft rear bearing cap and speedometer drive gear.
7. Position collar into 4th speed gear. Pry against front face of collar to position mainshaft assembly rearward. Remove rear bearing snap ring, then remove rear bearing using suitable puller.
8. Remove thrust washer and split rings. Remove snap ring located in reverse gear.
9. Position 1st and reverse shift collar under reverse gear, then slide reverse gear and collar forward until it contacts 1st speed gear. Tie both gears together.

NOTE: Countershaft gears will support mainshaft sub assembly with mainshaft rear bearing and drive gear sub assembly removed from case.

---

10. Remove countershaft rear bearing caps,

then remove drive gear bearing cap.

11. Remove drive gear sub assembly from case by pulling forward on drive gear stem.
12. Using suitable pry bar, position each countershaft rearward, then remove snap ring and countershaft front bearing.
13. Position countershaft rearward, then remove snap ring and countershaft rear bearing using suitable tool.

NOTE: Ensure that bearing puller tool pulls on inner face of rear bearing. Applying force to outer face of bearing will damage bearing.

14. Using suitable tool, remove upper idler gear shaft from case. Leave idler gear in position for removal after mainshaft sub assembly is removed.
15. Position suitable lifting tool onto 2nd and 3rd shift collar and remove mainshaft sub assembly from case.
16. Remove remaining parts of upper idler gear subassembly.
17. As viewed from rear of case remove right side countershaft subassembly from case.
18. Remove left side countershaft sub assembly from case.
19. Using suitable tool, remove lower idler gear shaft from case, then remove remaining parts of lower idler gear sub assembly.

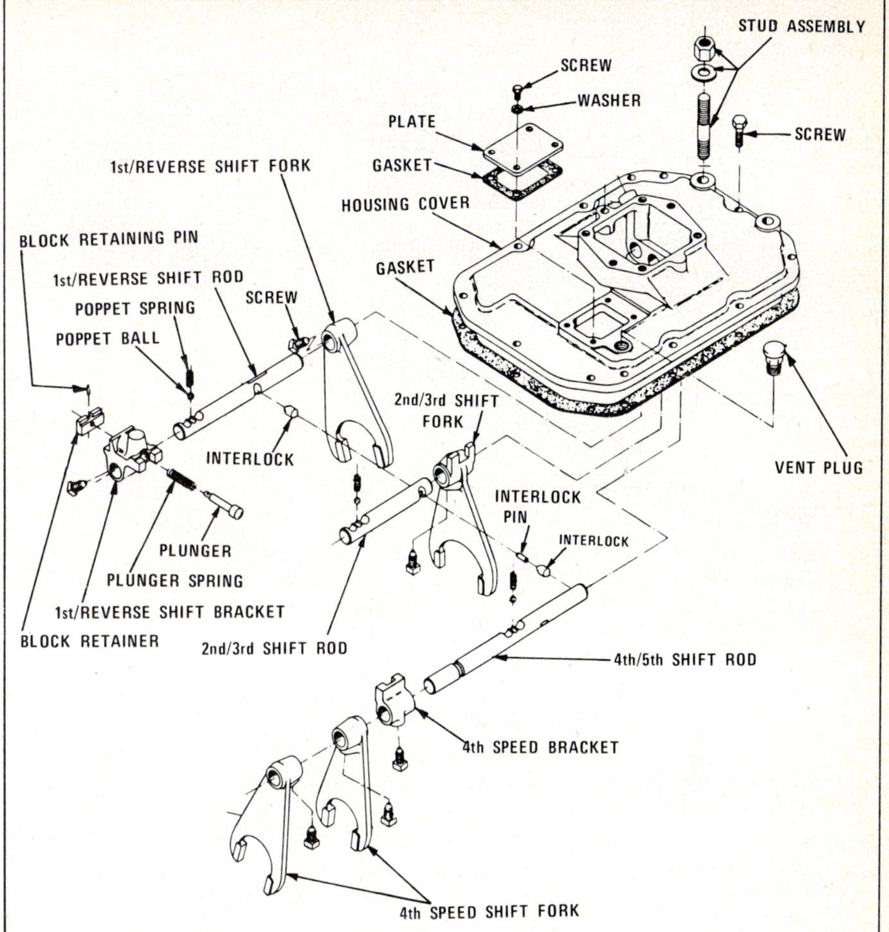

Fig. 1 Shifter housing exploded view

SUB ASSEMBLY SERVICE

Shifter Housing, Fig. 1

Disassembly

1. Place shifter housing in suitable vise, then remove screws from 1st and reverse shift rod bracket and fork.
2. Position 1st and reverse shift rod forward to free from fork and bracket.
3. Remove interlock from cross hole between 1st and reverse shift rods and 2nd and 3rd shift rods.
4. Remove screw from 2nd and 3rd shift fork, then position shift rod forward to free fork.
5. Remove screw from 4th speed bracket and shift fork. Position shift rod forward to free fork and bracket.
6. Remove four poppet springs from shifter housing front boss.
7. Place 1st and reverse shift bracket in suitable vise then remove block retaining pin, plunger and spring.

Assembly

1. Assemble 1st and reverse bracket by placing in suitable vise and installing spring and plunger into access hole. Retain with C clamp to maintain spring preload. Assemble blocker machined step toward shift gate, then place blocker into small diameter end of plunger. Align blocker and plunger pin holes and install .084 inch diameter retaining pin. Install pin until it seats, then remove C clamp from plunger.
2. Lightly grease three shift rods, then assemble in proper positions in housing bosses.
3. Install poppet spring and ball into poppet detent bore located at 2nd from top boss.
4. Install 4th and 5th shift rod and position end that is longest from interlock detent into rear boss on right side of housing.
5. Assemble 4th speed shift rod bracket to rod with shift gate facing downward. Tap shift rod until poppet ball seats in neutral detent of shift rod.
6. Assemble 4th speed shift fork to rod with extended hub facing left. Assemble 5th speed shift fork to rod with extended hub facing right. Assemble bracket and forks into proper position on rod and install screws. Torque screws to 40–50 ft. lbs. and secure with lock wire.
7. Lubricate interlock and assemble into access hole of rear boss through 2nd and 3rd shift rod bore.
8. Install poppet ball and spring into next poppet detent bore.
9. Lubricate interlock pin and insert pin through hole in 2nd and 3rd shift rod. Install rod into rear boss with interlock pin facing right.
10. Assemble 2nd and 3rd shift fork onto rod with extended hub of fork facing left. Install rod into front boss, then assemble shift fork to rod and install screw. Torque screw to 40–50 ft. lbs. and secure with lock wire.
11. Lubricate interlock, then assemble interlock into access hole of rear boss through 1st and reverse shift rod bore. Install poppet ball and spring into poppet detent bore.
12. Install 1st and reverse shift rod with end closest to set screw countersink hole into front boss of housing, then through middle boss of housing. Assemble 1st and reverse bracket onto rod with bracket top boss facing downward and shift gate facing toward tower opening. Slide rod through rear boss and assemble 1st and reverse fork onto rod with extended hub of fork facing right. Assemble shift fork and bracket onto rod and install screws. Torque screws to 40–50 ft. lbs. and secure with lock wire.
13. Shift 1st and reverse speed fork into gear and check that all other rods remain locked in neutral. If any other rod can be shifted, interlock is not operating properly. Position 1st and reverse speed fork in neutral position and check movement of each shift rod.

Mainshaft, Fig. 2

Disassembly

1. Remove 5th speed clutch collar, and on model 1052 only remove 4th speed clutch collar.
2. Remove snap ring, then remove 4th speed gear assembly.
3. Remove 3rd speed gear assembly, then remove snap ring.
4. Remove 2nd and 3rd speed shift collar.
5. Remove snap ring from 2nd speed gear bore, then remove 2nd speed gear assembly.
6. Remove securing wire from 1st and reverse gears, then remove 1st gear assembly. Remove snap ring from shaft groove.
7. Remove 1st and reverse shift collar from shaft.

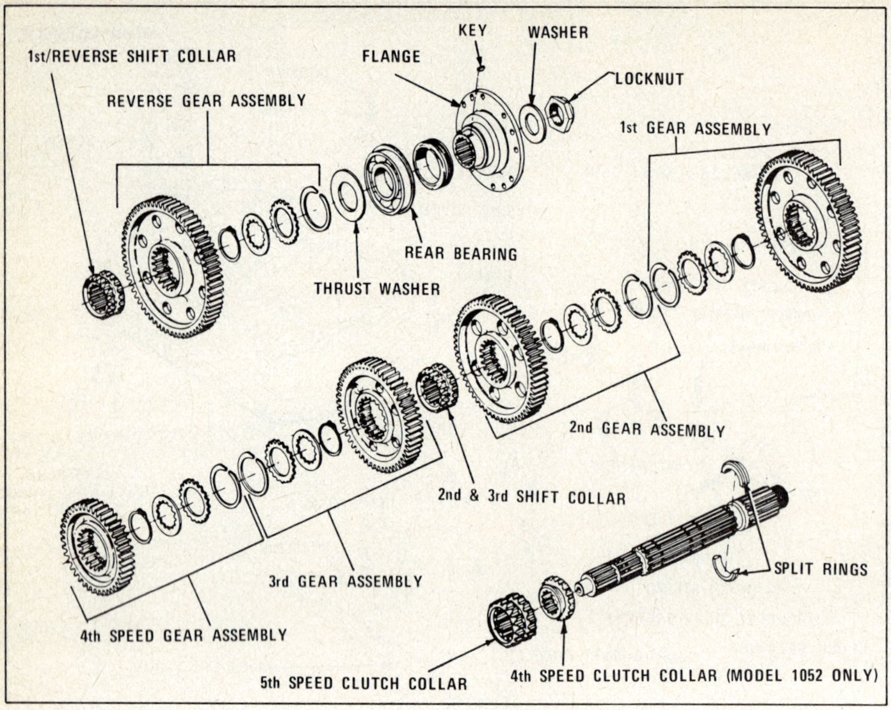

Fig. 2 Mainshaft exploded view

8. Remove reverse speed gear assembly from rear end of shaft.

Assembly
1. Lubricate all thrust washers on thrust face side with engine oil.
2. Place mainshaft in suitable vise and install thrust washer onto mainshaft.
3. Install snap ring onto shaft. Check that snap ring seats securely in bore.
4. Assemble 1st and reverse clutch collar onto shaft and position clutch collar onto snap ring.
5. Install snap ring onto shaft and seat in groove nearest 1st and reverse clutch collar.
6. Install thrust washer onto shaft and position washer on snap ring.
7. Assemble thrust washer and snap ring onto 1st speed gear and install assembly onto mainshaft. Ensure that 35° chamfer in bore of gear faces down, toward clutch collar.
8. Assemble thrust washer and snap ring onto 2nd speed gear and install assembly onto mainshaft with 35° chamfer in bore of gear facing upward. Position 2nd speed gear assembly against face of 1st speed gear. Install snap ring in groove under bore of 2nd speed gear.
9. Assemble 2nd and 3rd speed clutch collar onto mainshaft and position collar in bore of 2nd speed gear.
10. Install snap ring in mainshaft groove closest to 2nd and 3rd clutch collar.
11. Install thrust washers onto shaft and position thrust washers on snap ring.
12. Install snap ring into 3rd speed gear, then install gear onto mainshaft with 35° chamfer in bore of gear facing down towards clutch collar.
13. Assemble thrust washers and snap rings onto 4th speed gear. Install 4th speed gear sub assembly onto mainshaft with 35° chamfer in bore of gear facing upward. Position gear against face of 3rd speed gear.
14. Install snap ring on mainshaft in groove under 4th speed gear.
15. On 1052 models only, install 4th speed clutch collar onto mainshaft with shift fork collar facing upward. Position clutch collar in bore of 4th speed gear.
16. Assemble 5th speed clutch collar onto shaft. Position clutch collar in bore of 5th speed gear.
17. Remove mainshaft from vise, then install 1st and reverse shift collar into 1st gear bore and position thrust washer on end of shaft against snap ring.
18. Install reverse gear from rear of mainshaft only. Position reverse gear onto clutch collar and align outer diameter teeth to 1st speed gear teeth. Secure gears together.

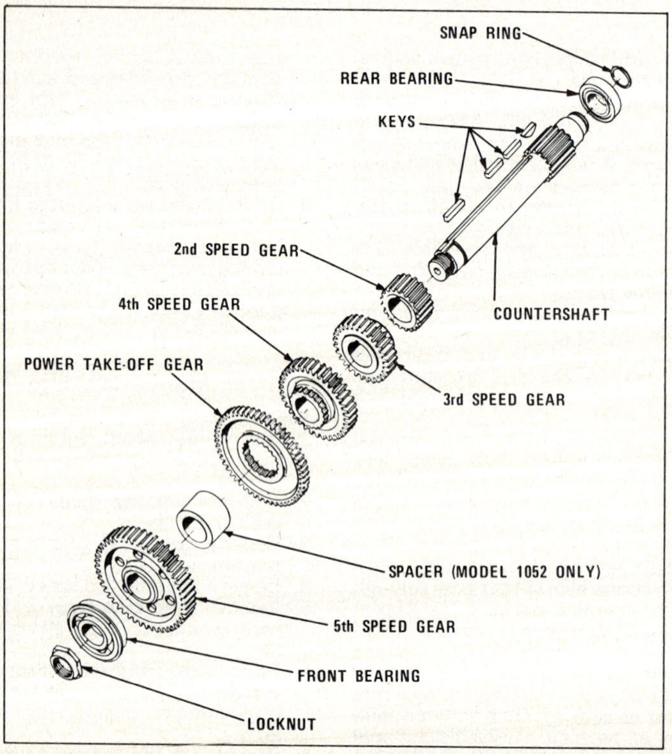

Fig. 3 Countershaft exploded view

Countershaft, Fig. 3

Disassembly
1. With 5th gear properly supported, press countershaft from gear using suitable press.

NOTE: Countershaft keys may remain with gear as shaft is pressed off. Check gear and collect keys as necessary.

2. On model 1052 only remove spacer from shaft.
3. Remove power take-off gear from splined teeth of 4th speed gear.
4. Support 4th speed gear, then press shaft from gear using suitable press.
5. Support 3rd and 2nd speed gears and press shaft from gears using suitable press.

Assembly
1. Lubricate all gear bores with suitable engine oil.
2. Press key into countershaft keyway, then support 2nd speed gear and press shaft and key into gear. Check that keyway aligns with countershaft keyway to prevent damage to key.
3. Press next key into countershaft keyway, then support 3rd speed gear with long hub facing up and press shaft into gear. Check that gear keyway aligns with shaft keyway to prevent damage to key and

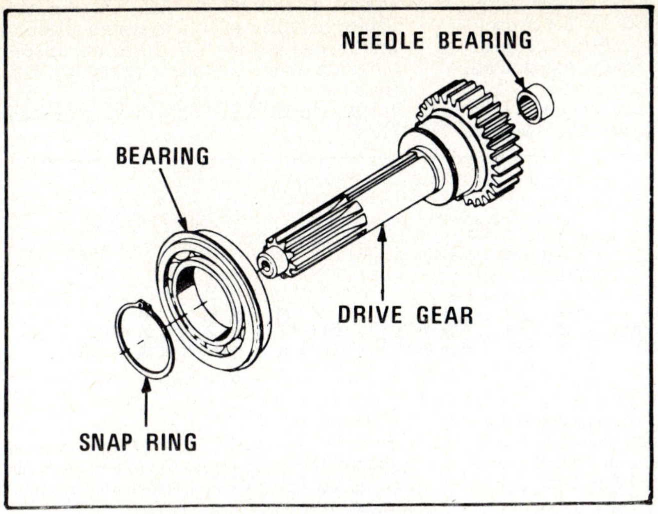

Fig. 4 Drive gear exploded view

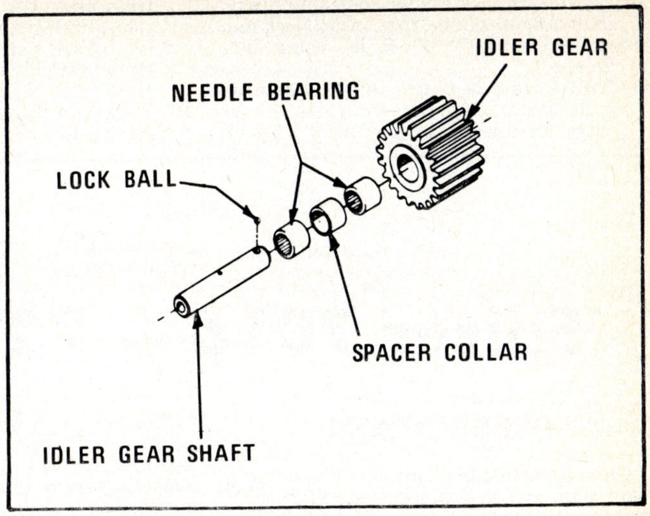

Fig. 5 Reverse idler gear exploded view

that 3rd speed gear face seats firmly against 2nd speed gear face.

4. Press next key into countershaft keyway, then support 4th speed gear with splined teeth on hub facing downward and press shaft onto gear. Check that gear keyway aligns with countershaft keyway to prevent damage to key and that 4th speed gear face seats firmly against 3rd speed gear face.
5. Install power take-off gear onto splined teeth of 4th speed gear with long hub end against gear.
6. On model 1052 install spacer. While holding power take-off in place, seat spacer securely against 4th speed gear face.
7. Support 5th speed gear with long hub facing upward. Check that tooth timing mark on web of gear is aligned with center of gear keyway. Press countershaft onto gear using suitable press. Seat gear face securely against spacer, then check that key face is flush with or slightly under gear face.

Drive Gear, Fig. 4

Disassembly

1. Remove snap ring from drive gear, then support outer bearing race and press drive gear from shaft using suitable press.
2. Remove needle bearing from drive gear.

Assembly

1. Replace drive gear bearing cap seal if necessary.
2. Support drive gear bearing inner race, then press drive gear into bearing using suitable press. Check that drive gear bearing shield is facing upward when bearing is pressed onto gear.
3. Invert drive gear and install snap ring. Install needle bearing into gear using suitable press. Bearing part number must be visible when bearing is pressed into bore. Press bearing until it is recessed .062 inch under gear face.

TRANSMISSION, ASSEMBLY

1. Install either countershaft assembly into case on left side as viewed from rear of case. Ensure that timing mark of gear at head of shaft is positioned toward center of case. This is necessary for proper mating with drive gear "V" timing mark.
2. Assemble two idler shaft bearings with spacer, Fig. 5, into reverse idler gear bores.
3. Install either reverse idler gear assembly and install in upper boss. Check that reverse idler gear meshes with countershaft gear. Do not install reverse idler shaft at this time.

NOTE: Remove magnetic plug from case to avoid damage to plug when lower right side reverse idler gear is installed.

4. Install second countershaft assembly into case. Rotate gear at head of shaft until timing mark is positioned toward center of case. Install reverse idler gear and rotate idler gear on mating tooth of countershaft to determine idler gear bearing bore alignment with case hole. After idler gear bearing bore and case hole are aligned, install reverse idler shaft and ball into case. Do not install countershaft front or rear bearings at this time.
5. Lower mainshaft assembly into case and partially mesh mainshaft gears with countershaft gears.
6. Install suitable pilot tools onto front of case to ease assembly of mainshaft. Refer to Fig. 8 under "Final Assembly" under "Spicer SST 10, 10102A, 1020 & 1214 for tool positioning.
7. Install thrust washer onto output end of mainshaft and position against thrust washer previously assembled during mainshaft assembly.
8. Position both thrust washers against snap ring on rear of mainshaft.
9. Remove tie securing 1st and reverse gears, then position reverse gear rearward into location on reverse gear bore thrust washers.
10. Install snap ring into reverse gear bore, then lubricate split rings and install onto mainshaft with flanges facing gears.
11. Lubricate thrust washer and install onto mainshaft against split ring flange face.
12. Install suitable pilot tool onto rear mainshaft bore. Use pilot tools installed onto front of case to support front of both countershafts, ensuring that timing teeth are kept in proper mesh.
13. Rotate both countershafts until timing teeth on head gears are parallel to bottom of case or pointing toward center of case. Position drive gear timing teeth until they align with countershaft timing teeth.
14. Install drive gear into case until bearing snap ring seats against face of case.
15. Install front drive gear bearing cap and gasket and check that case oil port holes align with bearing cap oil return holes.
16. Install bearing cap onto case and torque retaining bolts to 25–32 ft. lbs.
17. Remove pilot tool from rear of case, then install three shift clutch collars into neutral position on mainshaft.
18. Install rear mainshaft bearing into case, then install snap ring.
19. Install reverse idler shaft into upper reverse idler gear bore. Install idler shaft lock ball into shaft ball hole.
20. Support rear ends of both countershafts with suitable pilot tool, then support each countershaft gear with suitable support tool, ensuring that timing teeth marks remain in alignment.
21. Remove pilot tool from left side front bore, then install countershaft bearing and snap ring.
22. Install countershaft locknut and snug tighten. Remove pilot tool from right side front bore and install countershaft bearing and snap ring. Remove countershaft support tool.
23. Install countershaft locknut and snug tighten, then lock assembly into two gears and torque each locknut to 550–600 ft. lbs. using suitable tool.
24. Remove pilot tool from left side rear end of countershaft and install countershaft bearing, then install snap ring.

NOTE: Bearing must be installed with roller retainer ring on the outside of the case and not on the inside of the case facing transmission.

25. Remove pilot tool from right side gear end of countershaft and install countershaft bearing, then install snap ring. Refer to note in step 24 for bearing installation direction.
26. Install speedometer driven gear into mainshaft rear bearing cap, then install rear cap and gasket. Align oil port holes with bearing cap return hole. Torque bearing cap bolts to 60–80 ft. lbs.
27. Install both countershaft rear bearing caps and gaskets and torque retaining screws to 25–32 ft. lbs.

28. Install end yoke and washer onto mainshaft output splines, then install lock nut and torque to 550–600 ft. lbs. using suitable tool.
29. Position reverse clutch collar in neutral position and rotate gear train assembly to check for freedom of movement. If gear train assembly locks after several rotations, timing teeth marks are not properly aligned. Correct as necessary. If gear train assembly rotates freely, position all clutch collars in neutral position.
30. Install clutch housing and torque 5/8 inch screws to 120–150 ft. lbs. and 1/2 inch screws to 60–80 ft. lbs. On vehicles equipped with wet clutch, install oil sump and torque screws to 7–10 ft. lbs. Install oil sump cover and torque screws to 15 ft. lbs.
31. Install clutch brake onto drive gear stem shaft.

Spicer 7016 & 7216–16-Speed Units

DESCRIPTION

These transmissions consist of a four-speed main unit and a four-speed auxiliary unit joined together in one compact assembly, Fig. 1. The mainshaft in each unit is supported by a single-row roller bearing at the front and a single-row ball bearing at the rear of the shaft. The countershaft in the front unit is supported by a single-row roller bearing at the front; a single-row ball bearing supports the shaft at the rear. The reverse idler gear is supported by two roller-type bearings that are recessed inside the idler gear and revolve on the idler shaft. The countershaft in the rear unit is supported by one single-row tapered roller bearing at each end of the shaft. All gears are helical-mesh type, except the reverse idler and main unit PTO, which are spur type.

Controls for shifting the transmission may be of several types, depending on truck model applications: The one-piece gearshift lever type, the remote prop shaft type, and remote rod and lever type.

The transmission should be overhauled in the sequence outlined unless only one unit is to be worked on or a specific problem has been correctly diagnosed, then complete overhaul is not required. Remove drain plugs from both units and drain before attempting disassembly.

SEPARATING UNITS

1. Remove shifter housing assembly, Figs. 2, 3 and 4.
2. From inside near top of rear case, cut lock wire and remove four capscrews and lockwashers.
3. Support rear unit with a bridge chain as shown in Fig. 5, across top of two internal capscrew holes while performing the next step.
4. From outside of case, remove eight remaining capscrews, three of which are located on filler plug side of main unit and five on the opposite side of the rear unit.
5. Using a pry bar, separate the two units, leveling the rear unit with a chain fall as shown in Fig. 6.

DISASSEMBLE FRONT UNIT

Main Drive Gear, Fig. 7

1. Lock transmission in two gears, then remove countershaft rear bearing nut.

NOTE: The procedure and ease of disassembly for this series transmission varies with the internal gearing and gear ratios provided. If gear ratios are such that the main drive gear is smaller than the case bearing bore, remove drive gear and bearing. If the drive gear is larger than the case bore, it is necessary to remove the mainshaft assembly before the drive gear can be removed from the case.

2. Remove capscrews from main drive gear bearing cap and then use two 7/16" puller screws to remove bearing cap and main drive gear assembly as shown in Fig. 8.

NOTE: Fourteen loose rollers are used for a pilot bearing between drive gear and

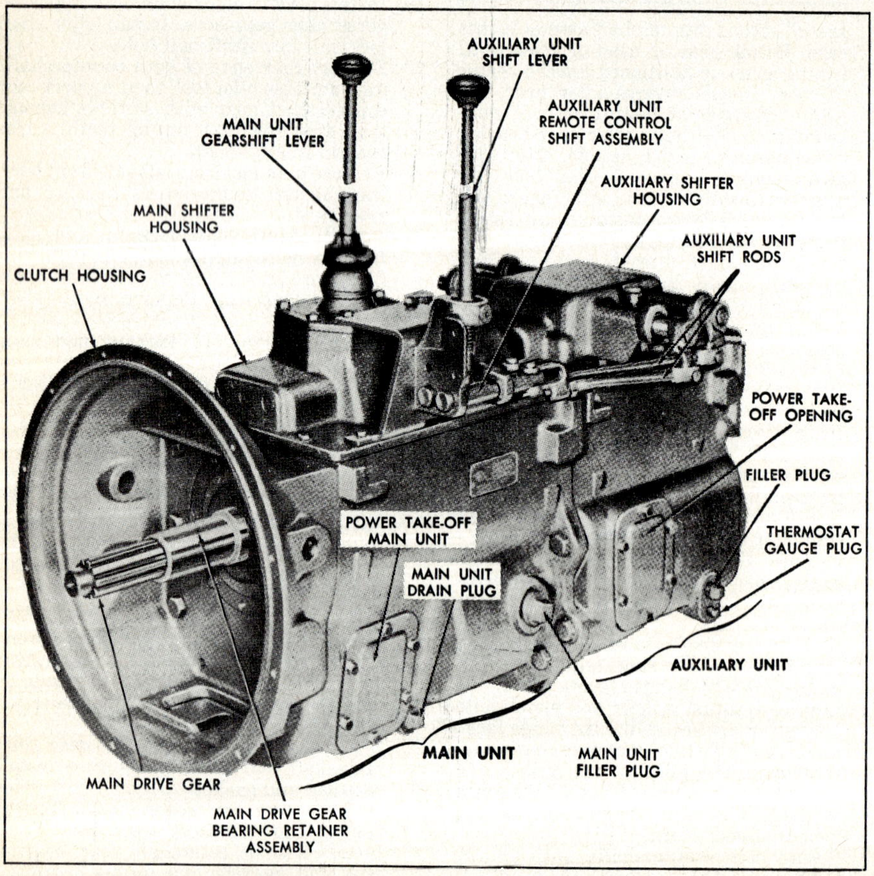

Fig. 1 Exterior view of transmission

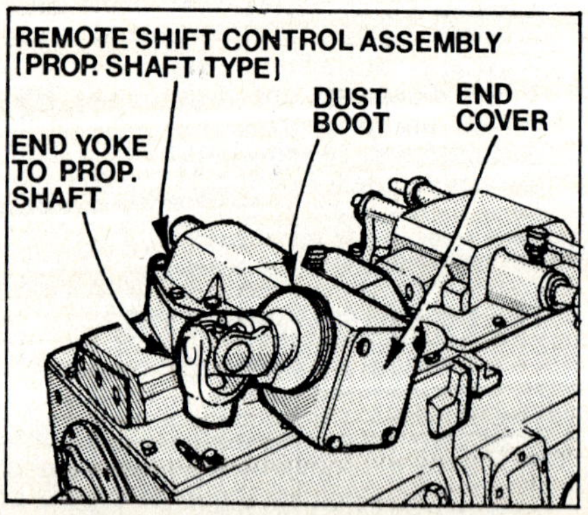

Fig. 2 Prop shaft type remote control

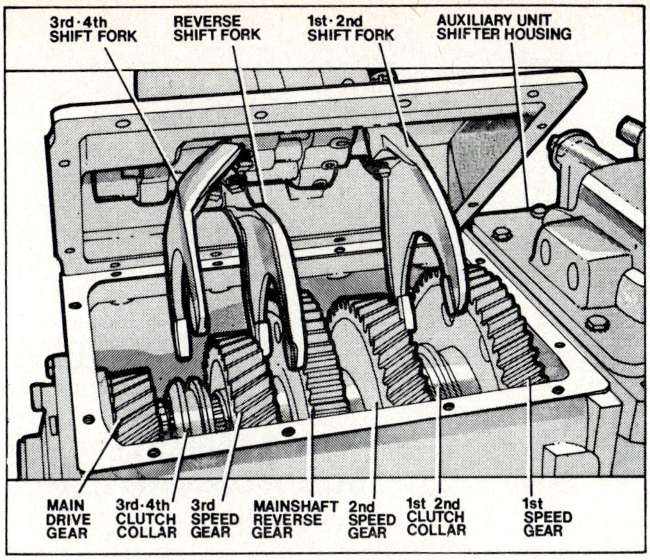

Fig. 3 Removal of front unit shifter housing

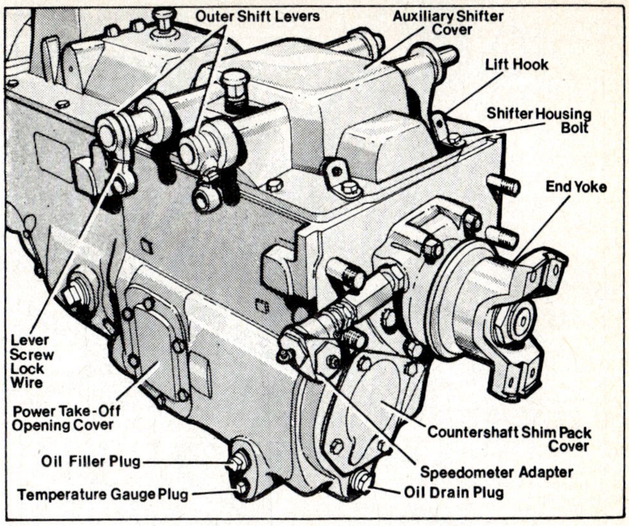

Fig. 4 Rear side view of transmission

mainshaft. Remove rollers from drive gear pocket or from bottom of case if they fell out when gear was removed.

Mainshaft & Gears, Fig. 9

1. Remove auxiliary unit mainshaft front bearing from auxiliary drive gear.
2. Cut lock wire and remove two washer bolts and retainer washer.
3. Remove four bolts from drive gear bearing retainer.
4. Use a suitable puller to remove shaft and gear assembly, Fig. 10. First speed gear may be left in case for easier removal of shaft and gears as shown in Fig. 11.

Reverse Idler Gear, Fig. 12

1. Remove capscrew along with reverse idler gear shaft lock plate. Use a long through-bolt with a 1¾″ × 6″ tube and flat washer to pull reverse idler gear shaft from rear of case.
2. A slide hammer may be used as an option when removing shaft. Remove reverse idler gear sub-assembly as idler shaft pulls free. Remove thrust washers.

Countershaft, Fig. 13

1. Move countershaft assembly toward rear of case until rear bearing is free of case bore. Use a suitable puller to remove bearing from countershaft.
2. With countershaft free of front bearing, lift front end first and remove countershaft assembly from case. Inspect inside of case for lost rollers or reverse idler thrust washers.

DISASSEMBLE REAR UNIT

Mainshaft, Fig. 14

1. Lock transmission in two gears and remove yoke nut. Then use a suitable puller to remove end yoke from mainshaft.
2. Referring to Fig. 15, remove capscrews from mainshaft and countershaft rear bearing caps. Separate bearing caps from gaskets and case, remove shim pack and tie together; then set aside for installation later.
3. Press rear bearing oil seal from cap. Remove speedometer bushing if it needs to be replaced at this time.
4. Remove speedometer drive gear and mainshaft rear bearing thrust washer if loose.
5. Use a soft hammer and tap forward on rear of mainshaft to start bearing off shaft.
6. Slide mainshaft and gears to rear of case as far as possible. Remove rear bearing with suitable puller, clamping on bearing snap ring.
7. Remove mainshaft rear bearing washer.
8. Remove mainshaft direct and overdrive clutch collar from its clutch gear by sliding forward off gear. Remove mainshaft in manner shown in Fig. 11.

Countershaft, Fig. 16

1. Remove snap ring from front of shaft and use a soft drift and hammer to tap countershaft forward, dislodging front bearing cup from case.
2. Remove countershaft rear and front snap rings from grooves, move rings forward on shaft as far as possible, then slide low speed gear forward as far as possible.
3. Use a soft drift to tap countershaft rear-

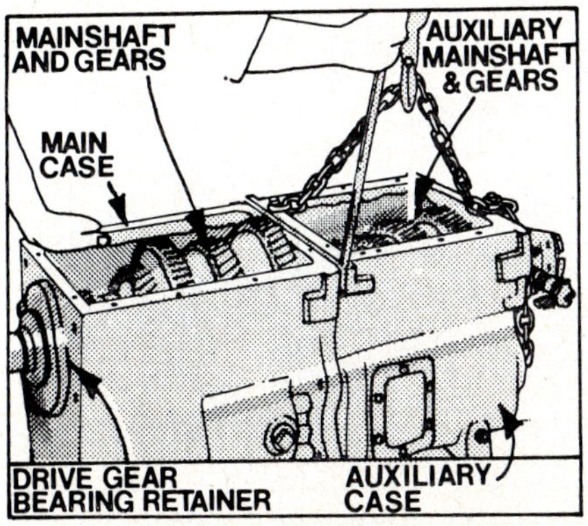

Fig. 5 Separating front and rear units

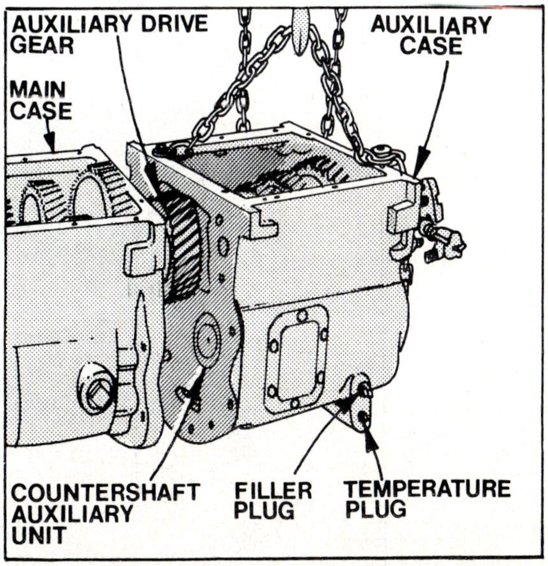

Fig. 6 Front and rear units separated

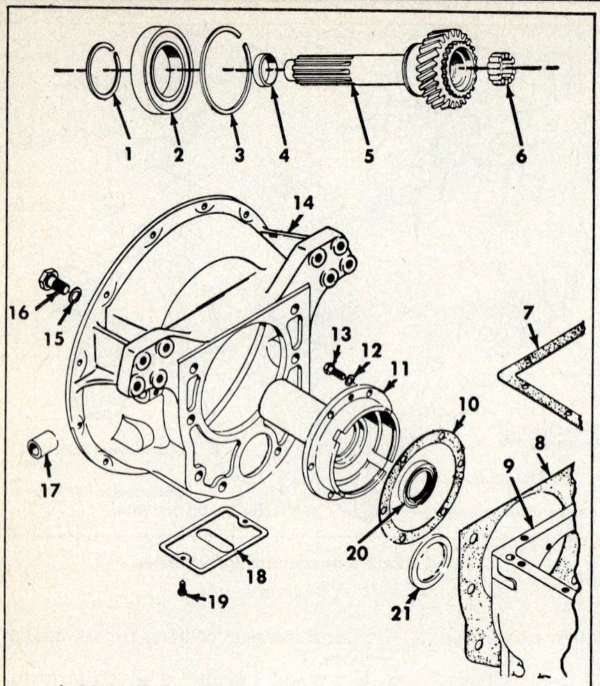

1 Main Drive Gear Bearing Snap Ring
2 Main Drive Gear Bearing
3 Main Drive Gear Bearing Snap Ring Rear
4 Main Drive Gear Wear Sleeve
5 Main Drive Gear
6 Mainshaft Front Bearing Rollers
7 Shifter Housing Gasket
8 Clutch Housing Gasket
9 Main Case
10 Main Drive Gear Bearing Cap Gasket
11 Main Drive Gear Bearing Cap
12 Lock Washer
13 Bolt
14 Clutch Housing
15 Lock Washer
16 Bolt
17 Clutch Housing Bushing
18 Clutch Housing Hand Hole Cover
19 Bolt
20 Main Drive Gear Bearing Cap Oil Seal
21 Countershaft Front Bearing Retainer

Fig. 7 Main drive gear and clutch housing components

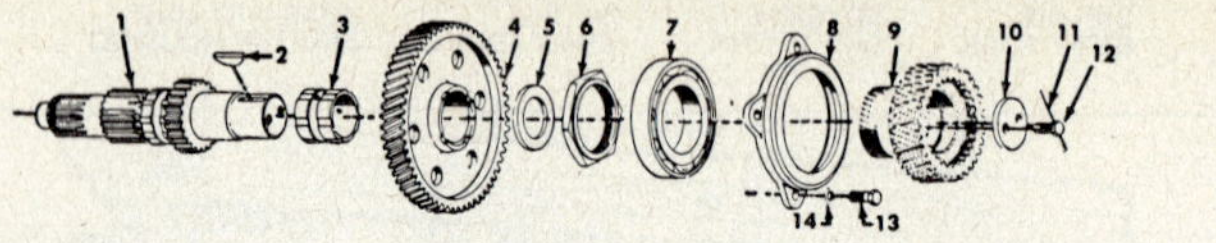

1 Mainshaft
2 Auxiliary Drive Gear Key
3 1st Speed Gear Bushing
4 1st Speed Gear
5 1st Speed Gear Thrust Washer
6 Auxiliary Drive Gear Bearing Retaining Nut
7 Auxiliary Drive Gear Bearing
8 Auxiliary Drive Gear Bearing Retainer
9 Auxiliary Drive Gear
10 Mainshaft Retaining Washer
11 Lock Wire
12 Retaining Washer Bolt
13 Bolt
14 Washer

Fig. 9 Front unit mainshaft and gears

ward to remove rear bearing cup from case. Lift countershaft up and out of case. The 1st speed gear need not be removed unless damaged.

DISASSEMBLE SUB-ASSEMBLIES

Gearshift Housings

Refer to Figs. 17 through 25 and, if necessary, disassemble shifter housing as suggested by the illustrations.

Front Unit Mainshaft, Fig. 26

1. With assembly on bench, remove 3–4 clutch collar by sliding it off clutch gear.
2. Remove snap ring holding 3–4 clutch gear on shaft and remove gear.
3. Remove 1–2 clutch collar and 1st speed gear if not left in case.
4. Remove 1st speed gear bushing from shaft along with keys.
5. Remove 3rd speed gear from its sleeve, also reverse gear.
6. The 3rd speed gear sleeve and reverse clutch gear can be removed by placing shaft and 2nd speed gear in an arbor press. Support gear close to hub to prevent damage, then press shaft from sleeve and clutch gear.
7. Inspect 3rd speed gear sleeve pin for wear and replace if necessary.
8. Unstake drive gear bearing retainer nut (left-hand thread) and remove nut.
9. Place mainshaft drive gear and bearing assembly vertically in an arbor press with front threaded end up; use two 1/4" thick plates to support under bearing retainer. Press drive gear free from bearing and retainer.
10. Press retainer free from bearing.

Front Unit Countershaft, Fig. 13

1. From inside case remove countershaft

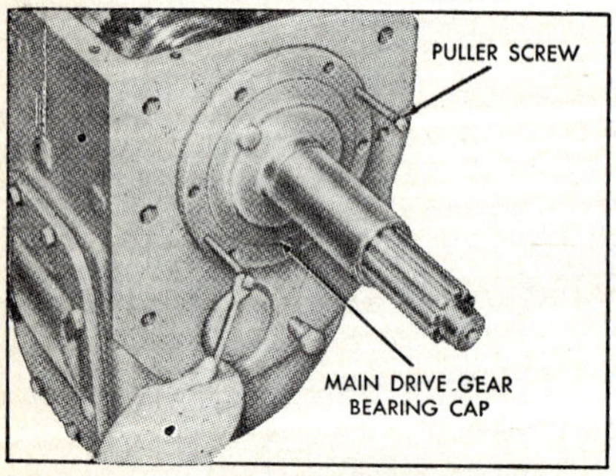

Fig. 8 Removal of main drive gear bearing cap

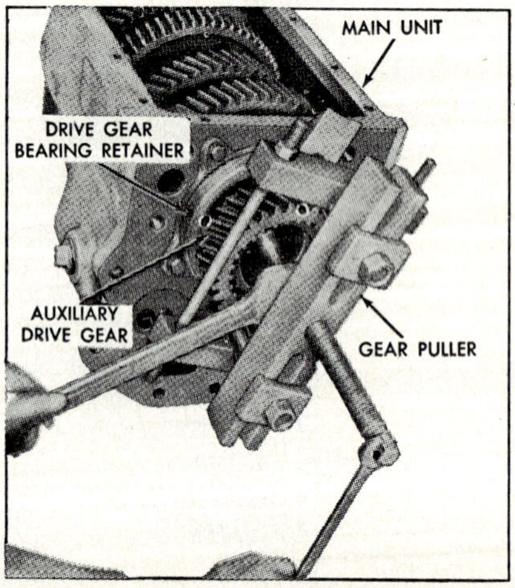

Fig. 10 Removal of auxiliary drive gear

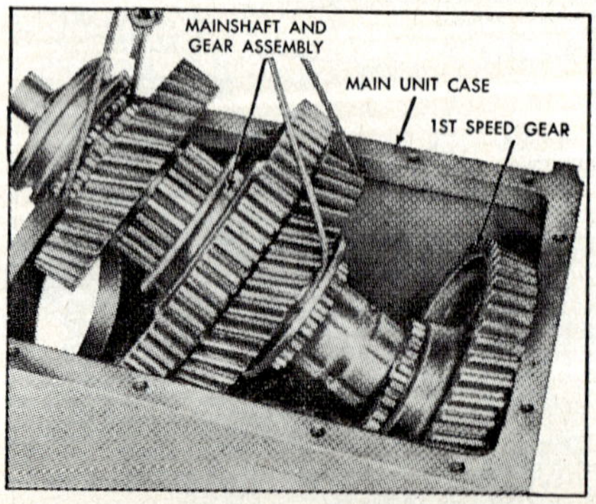

Fig. 11 Removal of mainshaft from main case

front bearing from case bore.
2. Remove drive gear snap ring, 3rd gear snap ring, and 2nd gear snap ring.
3. Press gears off countershaft one at a time by supporting under gear as close to hub as possible. Remove keys one at a time as each gear is pressed from shaft.
4. If inner race of bearing remained on shaft at time of removal from case, remove it at this time. Use puller or drive off with drift tool.

Front Unit Reverse Idler, Fig. 13

1. Remove two sets of idler gear bearings and spacers from bore of idler gear.
2. If spacers are not with gear assembly, retrieve them from bottom of case.
3. Although bearings are identical, it is advisable to install in same location otherwise a different wear pattern will result.
4. Reverse idler gear shaft is larger on outer end; it must be installed in same position.

Rear Unit Mainshaft, Fig. 14

1. Slide 1st speed gear and 1–2 clutch collar off rear of shaft.
2. Remove 1st speed gear bushing if sleeve was not removed with gear.
3. Remove clutch gear snap ring.
4. Using a support under rear of 2nd speed gear, press mainshaft free of direct and overdrive clutch gear and from overdrive gear and sleeve.
5. Remove overdrive gear lock pin from inside overdrive sleeve.

Rear Unit Countershaft, Fig. 16

1. Use a puller to remove front and rear bearing cones.
2. Remove drive gear snap ring and support drive gear underneath with parallel bars as close to hub as possible. Press shaft free of gear.
3. Remove overdrive gear snap ring and support overdrive gear with bar stock on underside, then press countershaft free of gear.
4. Support 2nd speed gear in similar manner and press shaft free of gear.

ASSEMBLE SUB-ASSEMBLIES

Gearshift Housings

Refer to Figs. 17 through 25 and, if disassembled, assemble shifter housing as suggested by the illustration.

Front Unit Countershaft, Fig. 13

IMPORTANT: All countershaft gears should fit tight on the shaft. Since a shrink fit of .0015″ to .003″ is built into new parts, it presents a field assembly problem. If heat is needed to expand gear bores, boiling water, hot oil, steam or heat lamps are satisfactory. Do not exceed 250°F.

CAUTION: Do not use hot plates, acetylene torches or other methods that will turn the steel blue or straw color, and damage the heat-treated gears. If heat is not used, it is advisable to coat the gear bores with white lead to prevent galling or seizing of parts.

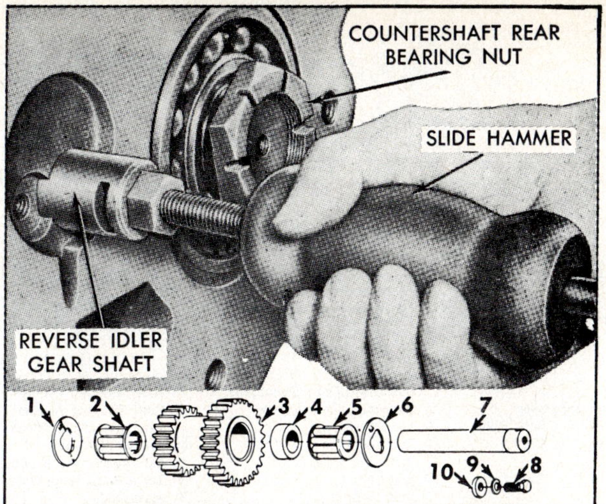

1 Reverse Idler Gear Thrust Washer
2 Reverse Idler Gear Bearing
3 Reverse Idler Gear
4 Reverse Idler Gear Bearing Spacer
5 Reverse Idler Gear Bearing
6 Reverse Idler Gear Thrust Washer
7 Reverse Idler Gear Shaft
8 Reverse Idler Gear Lock Plate Bolt
9 Washer
10 Reverse Idler Gear Lock Plate

T-4205-1

Fig. 12 Removal of reverse idler gear shaft

1. Assemble keys to shaft securely.
2. Using a suitable arbor press, support 2nd speed gear on its hub with chamfer and long end of hub up. Usually two pieces of 5/8″ bar stock will serve this purpose well. Place countershaft in bore of gear. Align key and keyway and press shaft into gear until firmly seated against shoulder. Install 2nd speed gear snap ring.
3. Install 3rd speed gear snap ring on shaft. Using suitable bar stock, support 3rd speed gear as close to center of hub as possible to prevent damage to gear while pressing on shaft. Long end of gear hub should face down with chamfer up. Place shaft in bore and align key with keyway of gear. Press shaft into gear until firmly seated against snap ring face.
4. Support PTO gear in like manner on the short hub end, with chamfer and long end of hub up. Place shaft in chamfered bore, align key with gear keyway and press shaft into gear until seated firmly against 3rd speed gear.
5. Support drive gear on short end of hub with long hub and chamfer up. Place shaft in bore, align key with keyway, and press shaft into gear until firmly seated against PTO gear. Install drive gear snap ring on shaft to lock drive gear to shaft.
6. Press inner race of front roller bearing (flange toward gears) onto shaft until firmly seated against shoulder.

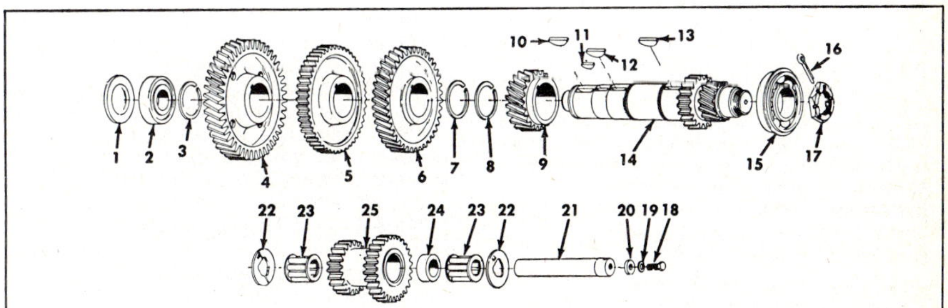

1 Countershaft Front Bearing Retainer
2 Countershaft Front Bearing
3 Countershaft Drive Gear Snap Ring
4 Countershaft Drive Gear
5 Countershaft PTO Gear
6 Countershaft 3rd Speed Gear
7 Countershaft 3rd Speed Gear Snap Ring
8 Countershaft 2nd Speed Gear Snap Ring
9 Countershaft 2nd Speed Gear
10 Countershaft Gear Key
11 Countershaft Gear Key
12 Countershaft Gear Key
13 Countershaft Gear Key
14 Countershaft
15 Countershaft Rear Bearing
16 Countershaft Rear Bearing Nut Cotter Pin
17 Countershaft Rear Bearing Nut
18 Reverse Idler Shaft Lock Bolt
19 Reverse Idler Shaft Lock Washer
20 Reverse Idler Shaft Lock Plate
21 Reverse Idler Gear Shaft
22 Reverse Idler Gear Thrust Washer
23 Reverse Idler Gear Bearing
24 Reverse Idler Gear Bearing Spacer
25 Reverse Idler Gear

T-4217

Fig. 13 Front unit countershaft and reverse idler gears

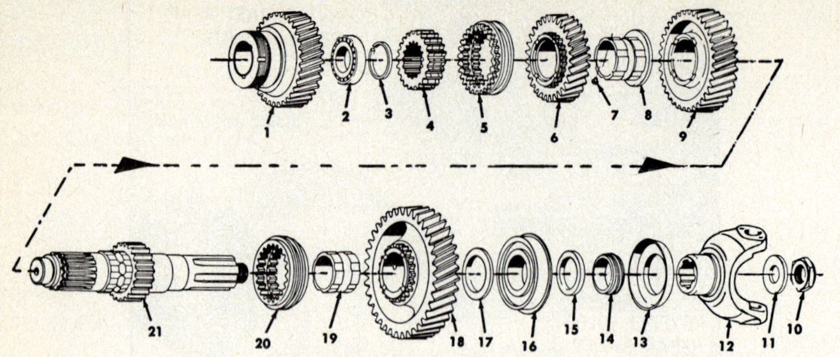

Fig. 14 Auxiliary unit mainshaft and gears

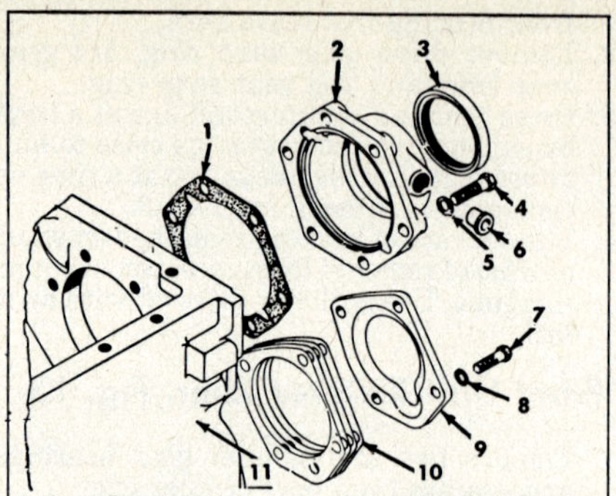

Fig. 15 Auxiliary unit rear bearing caps

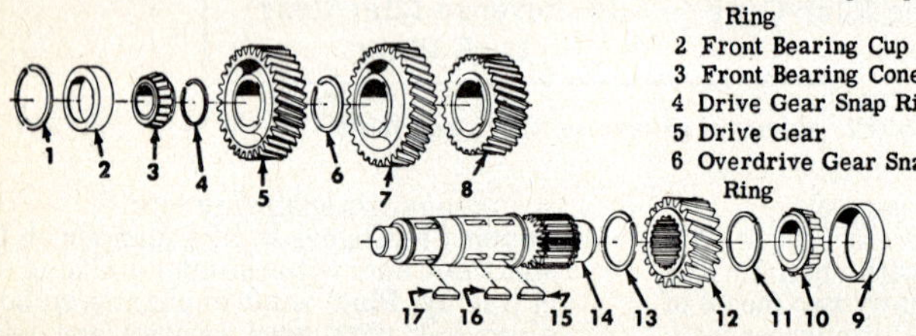

Fig. 16 Auxiliary unit countershaft components

Front Unit Reverse Idler, Fig. 13

1. Hand-pack roller bearings with grease and install inside reverse idler gear at each end with spacer between the two bearings. Place the two thrust washers with assembly for installation with idler gear later.

Front Unit Main Drive Gear, Fig. 26

1. Place drive gear bearing on arbor press, supported underneath with bar stock at shoulder of inner race and with shield of bearing facing up. Press drive gear into inner bearing race until seated firmly against shoulder of drive gear.
2. Install snap ring in groove of drive gear, being sure it is properly seated.
3. Replace wear sleeve at this time if it was removed during disassembly.
4. To install a new bearing cap oil seal, press seal out of cap. Clean cap seal opening. Coat outside diameter of seal with gasket cement and press seal into cap. Lip of seal should face toward bearing. Coat lip of seal with caster oil or equivalent.
5. Place drive gear bearing cap over bearing and press cap into position. Install snap ring in groove provided on inside at rear of cap.
6. Assemble 14 pocket bearing rollers in drive gear. Coat bore of pocket with thick coat of chassis grease to set rollers in place. It is necessary to slide last roller in position endways because of the wedge effect of the other rollers.

Front Unit Mainshaft, Fig. 26

1. Fasten mainshaft vertically in a vise, using soft jaws to clamp on keyway end of shaft.
2. Install 2nd speed gear on shaft with clutch teeth down.
3. Install reverse clutch gear to shaft spline.
4. Install reverse gear to reverse clutch gear spline with fork groove up.
5. Install 3rd speed gear sleeve pin in sleeve. Slide sleeve on shaft, with flange of sleeve down, and head of sleeve pin in proper location in spline of shaft.
6. Install 3rd speed gear on shaft with teeth up and over sleeve.
7. Slide 3-4 clutch gear on shaft.
8. Install 3-4 clutch gear snap ring in groove to secure gear on shaft.
9. Remove mainshaft assembly from vise.
10. Assemble 1-2 clutch collar with long collar toward 2nd gear.
11. Install 1st gear bushing. Assemble 1st gear over bushing with clutch teeth of gear toward (or facing) 2nd gear. Install thrust washer against hub face of gear.
12. Assemble keys to shaft grooves.

NOTE: A rope or wire sling should be attached to the middle section of the mainshaft for easier lifting and positioning, Fig. 11.

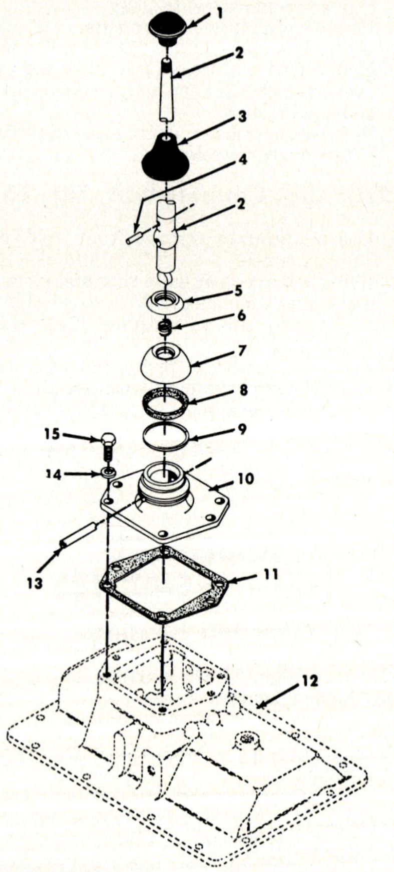

Fig. 17 Front unit shift control tower

13. Press bearing into main drive gear bearing retainer cap. Use a suitable tool to press on outer edge of bearing race.
14. Turn bearing and its retainer over on opposite side on press bed. Support inner race of bearing as close to center as possible to avoid damage to bearing. Press auxiliary drive gear into bearing bore until it seats against inner race of bearing.
15. Remove assembly from press. Install bearing retainer nut on drive gear. Hold gear in suitable manner in a vise and torque nut to 550–600 ft-lbs. Stake nut in three equally spaced places in grooves on drive gear provided for this purpose.

Rear Unit Countershaft, Fig. 16

CAUTION: All countershaft gears should fit tight on the shaft. As a shrink fit of .0015" to .003" is built into new parts, it presents a field assembly problem. If heat is used to expand gear bores, boiling water, hot oil or steam are usually satisfactory. Do not exceed 250°F.

Do not use hot plates, acetylene torches or other methods that would turn steel blue or straw color and damage the heat-treated gears. If heat is not used, it is advisable to coat the gear bores heavily with white lead to prevent galling or seizing of parts.

1. Install keys to shaft, seating them firmly.
2. Support 2nd speed gear in arbor press with long hub down and chamfer up. Set countershaft into position. Align keys with keyways and press shaft into gear until seated firmly against gear hub.
3. Support hub of overdrive gear with long hub down and chamfer up. Set countershaft into position. Align key with keyway and press shaft into gear. Seat shoulder firmly against shoulder of 2nd speed gear.
4. Assemble overdrive gear snap ring to lock 2nd and overdrive gear on shaft.
5. Support drive gear in press with long hub and chamfer up. Set countershaft in position, align key with keyway and press shaft into gear until seated firmly against shoulder.
6. Install drive gear snap ring.
7. Support countershaft in press. Use tubing to press and seat cones of front and rear tapered roller bearings against countershaft shoulders.

Rear Unit Mainshaft, Fig. 14

1. Lubricate all mainshaft free-running gear bearing bores with a light film of grease as gears are assembled to shaft. *Do not plug oil holes with grease.*
2. Position mainshaft vertically in a vise, using soft jaws to clamp on output splines.
3. Install 2nd gear to shaft with clutching teeth down. Insert new lock pin inside fluted overdrive sleeve. Do not press or drive against ground thrust face or flange of sleeve.
4. With flanged end of overdrive sleeve down, align lock pin with mainshaft spline. Push or tap fluted sleeve onto shaft against shoulder on shaft.
5. Install overdrive gear on shaft with clutch teeth up.
6. Install 3–4 clutch gear to shaft. Secure with snap ring. Assemble drive gear pocket bearing to front of shaft. Since bearing fits tight on surface of shaft, it will be necessary to use tubing to drive bearing on. To avoid damage, drive against inner race of bearing. Assemble 3–4 speed clutch collar on clutch gear. *Assemble roller bearing with snap ring in outer race so that snap ring is toward mainshaft.*
7. Turn mainshaft end for end in vise and clamp on 3–4 speed clutch gear.
8. Assemble 1–2 speed clutch collar with long hub down and toward 2nd speed gear.
9. Slide 1st speed gear bushing on shaft and coat fluted area with light grease.
10. Assemble 1st speed gear to shaft sleeve with clutching teeth toward front of mainshaft.

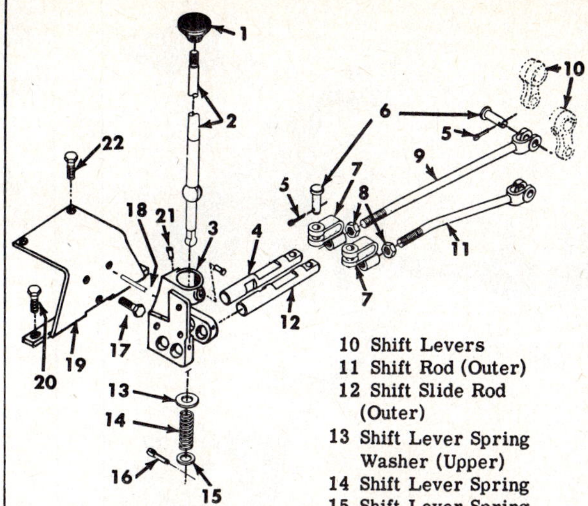

1 Gearshift Lever Knob
2 Gearshift Lever
3 Shift Lever Bracket
4 Shift Slide Rod (Inner)
5 Cotter Pin
6 Clevis Pin
7 Yoke
8 Jam Nut
9 Shift Rod (Inner)
10 Shift Levers
11 Shift Rod (Outer)
12 Shift Slide Rod (Outer)
13 Shift Lever Spring Washer (Upper)
14 Shift Lever Spring
15 Shift Lever Spring Washer (Lower)
16 Cotter Pin
17 Shift Lever Bracket Mounting Bolt
18 Shift Rod Stop Pin
19 Shift Lever Mounting Plate Bracket
20 Mounting Plate Bolt
21 Shift Lever Pivot Pin
22 Mounting Plate Bolt T-4213

Fig. 18 Auxiliary unit shift control and bracket

ASSEMBLE FRONT UNIT

Countershaft

1. Before installation of countershaft and gears in case, check case front bearing retainer. It must be installed in front bearing bore securely. Install outer cup of front bearing into case bore.
2. Lower countershaft assembly into case with rear end through case bore. Align

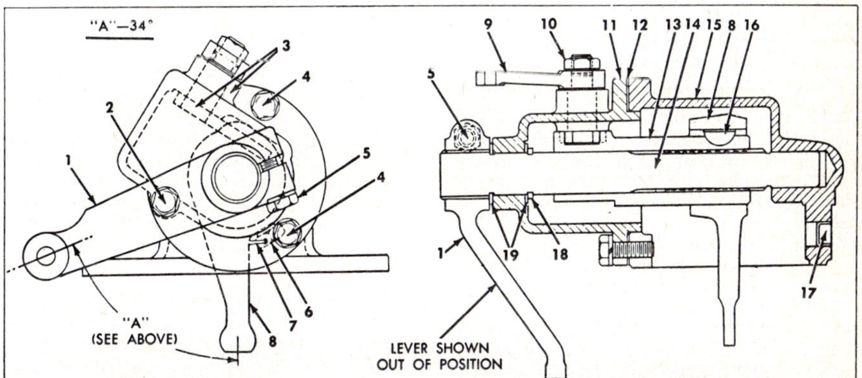

1 Shift Lever
2 End Cover Bolt (Long)
3 Selection Cam and Stud Assy.
4 End Cover Bolt (Short)
5 Shift Lever Bolt
6 Shift Finger Screw
7 Lock Wire
8 Shift Finger
9 Selection Lever
10 Selection Lever Nut
11 End Cover
12 Gasket
13 Selection Lever Shaft
14 Shift Finger Shaft
15 Mechanism Housing
16 Shift Finger Key
17 Housing Plug
18 Snap Ring
19 Thrust Washers

Fig. 19 Lever type remote control

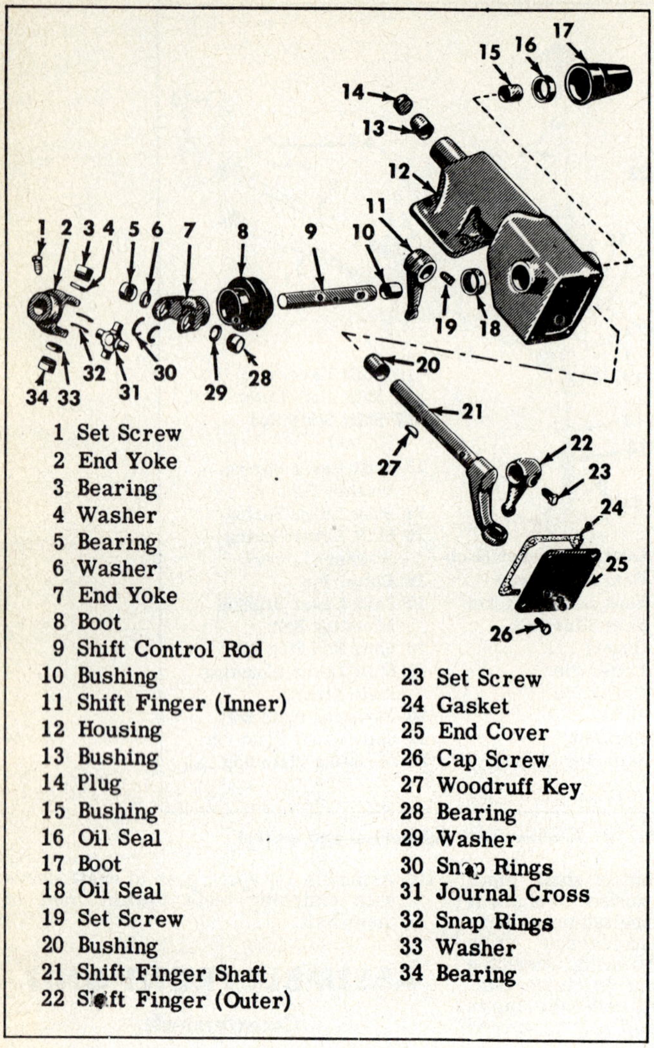

Fig. 20 Prop shaft type remote control

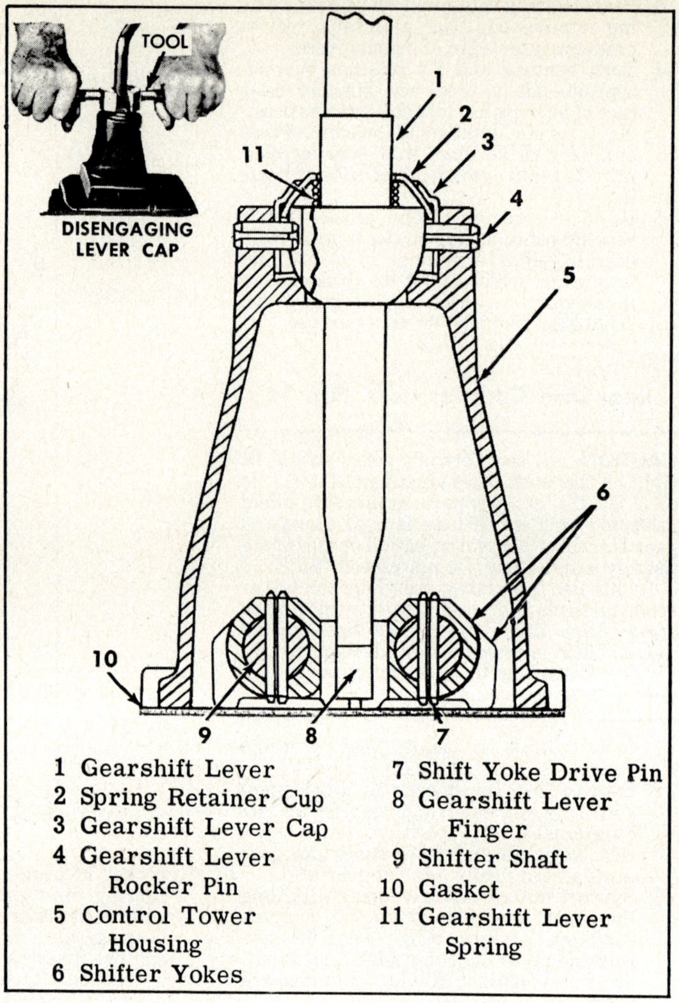

Fig. 21 Auxiliary unit type shift tower

front end of shaft with front bearing. Tap into partial position.

NOTE: To prevent damage to front bearing as rear bearing is being driven on countershaft, place two strips of 3/8" flat stock between countershaft drive gear and front wall of case.

---

3. Position bearing on rear end of countershaft with shield and snap ring of bearing facing toward rear. Align shaft and bearing with case bore. Use tubing to drive bearing until seated against shoulder of shaft.
4. Remove 3/8" strips of flat stock. Tap shaft the rest of the way or until snap ring of rear bearing rests against recessed shoulder of rear bearing bore.
5. Start and turn bearing nut against bearing finger-tight. Nut will be tightened after mainshaft and main drive gear have been assembled in case.

Reverse Idler Gear

1. Position idler gear in case with large gear toward rear. Place thrust washers one at each end of gear. Mesh with countershaft gear teeth and align bore of bearings with bore in case.
2. Insert idler gear shaft (long end first) through opening in rear of case and into idler gear bearings. Align slot on rear of shaft for lock plate and tap into final position.
3. Install lock plate into case and secure with capscrew and lockwasher.

Mainshaft

1. Lift shaft and gear assembly into case by means of overhead hoist and wire and chain sling, Fig. 11. Position gears so they mesh with countershaft gears.
2. Lubricate two keys and mainshaft diameter with grease or white lead for easier assembly of bearing retainer and drive gear, and assemble to shaft.

NOTE: Two methods can be used to press the auxiliary drive gear and bearing retainer assembly on mainshaft and keys. The first method requires the use of a hydraulic press with a minimum press power of 50 tons. If not available, the second method using puller arrangement will be necessary.

---

3. If a hydraulic press is available, a case bore pilot and spacer block is required. The block diameter should be 6½" with a thickness of 1¾".
   a. With mainshaft and gears assembled in case, move entire case to hydraulic press. Before placing unit on press bed, insert spacer block tool into front bore of case. Place case with front end down on press and opening to gears facing press operator.
   b. Place auxiliary drive gear and bearing retainer on end of mainshaft and in line with keys. With a light press, start drive gear and retainer on shaft and keys. Visually align bearing outer race diameter with case bore as assembly is pressed on shaft. Just before bearing retainer flange has bottomed itself against the case surface, adjust retainer flange holes to case holes. Press drive gear and bearing retainer on shaft until flange is flush against face of case.
   c. Relieve press pressure and remove unit from press. Install lock plate to rear of mainshaft and secure with bolts. Torque bolts to 40–45 ft-lbs. Then tie with lock wire, Fig. 27.
   d. With bearing retainer flange holes aligned with holes in case, install and torque bolts to 40–50 ft-lbs.
4. If the puller arrangement method is to be used it will be necessary to make a push spacer to the following dimensions:
   a. Use round steel stock 2 11/16" in diam-

1 Reverse Shift Finger
2 Reverse Shift Plunger Pin
3 Reverse Shift Finger Lock Washer
4 Reverse Shift Finger Lock Washer Nut
5 Breather Cap
6 Reverse Shift Plunger Pin
7 Reverse Shift Finger Poppet Spring
8 Reverse Shift Finger Plunger Retainer
9 Shifter Housing
10 Shifter Housing Bolt
11 Shifter Housing Lock Washer
12 Shifter Housing Gasket
13 Reverse Shift Fork
14 Reverse Shift Rod
15 Shift Fork and Bracket Screw
16 Shift Fork and Bracket Screw Lockwire
17 Shift Rod Interlock
18 Shift Rod Interlock Pin
19 1st & 2nd Shift Fork
20 1st & 2nd Shift Rod
21 3rd & 4th Shift Rod
22 3rd & 4th Shift Fork
23 1st & 2nd Shift Bracket
24 Shift Rod Poppet Ball
25 Shift Rod Poppet Ball Spring
26 Soft Plug
27 Transmission Lift Brackets
28 Shift Rod Hole Plug
29 Reverse Shift Finger Stud

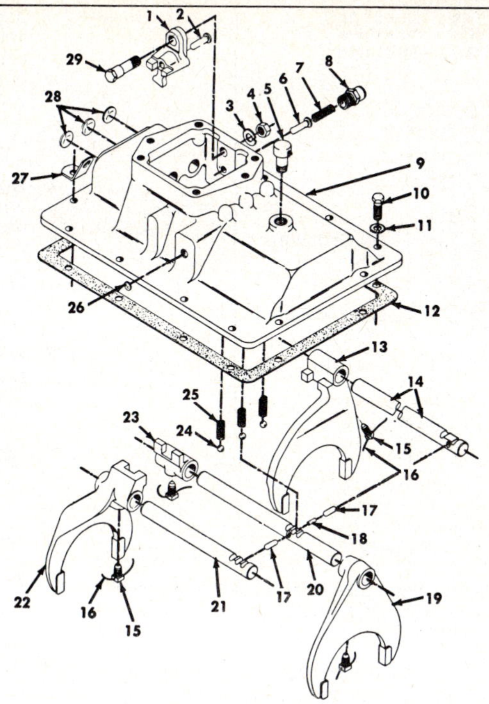

Fig. 22 Main unit shifter housing

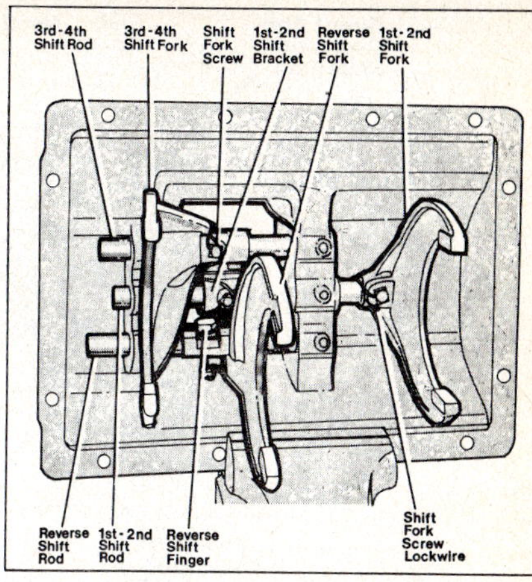

Fig. 23 Front shifter housing

eter and cut to 2 3/8″ long. Drill two 9/16″ holes 5/8″ from center of spacer for each hole. Drill holes completely through stock from one end to the other. These holes must be drilled accurately enough to line up with holes in mainshaft.

b. Secure two hardened bolts (1/2–20 and 4 7/8″ in length) plus two thrust washers to fit bolt diameter, and several flat washers of same hole diameter size. Install spacer washers on bolts and position washer against underside of bolt head. Place each bolt into holes of spacer push tool and through bolt holes in auxiliary drive gear. Start bolts into threaded bolt holes in back end of mainshaft.

c. Progressively tighten each bolt, using the spare washers under bolt heads to obtain full movement of drive gear onto mainshaft and keys. Before bearing retainer bottoms against face of case, align flange holes to case holes. Finish tightening bolts until bearing retainer is flush against case.

d. Remove special bolts and push spacer from shaft. Secure bearing retainer to case and torque bolts to 40–50 ft-lbs. Install lock plate and torque bolts to 40–50 ft-lbs. Install lock wire as shown in Fig. 27.

5. Assemble 3-4 clutch gear collar on clutch gear before assembling main drive gear and bearng cap sub-assembly to transmission.

Main Drive Gear

1. Check drive gear pocket to make sure that all 14 rollers are properly positioned inside pocket.
2. Apply gasket cement to drive gear bearing cap gasket and stick gasket on cap. Apply cement to opposite side of gasket.
3. Install drive gear and bearing cap to front of case. Use a soft head hammer to tap into position and check on pocket rollers to make sure they are still intact.
4. Secure bearing cap to case and torque bolts to 60–80 ft-lbs.
5. Secure case firmly and lock gears, then tighten countershaft rear bearing nut to 400–450 ft-lbs.
6. Place gears in neutral and rotate drive gear to check for free rotation of all gears and shafts.

ASSEMBLE REAR UNIT

Countershaft

1. Lower rear end of countershaft assembly into auxiliary case. Slide end of shaft and 1st speed gear out through rear of case. Lower front of shaft into its approximate position and maintain alignment with a cable, or support underneath by blocking up at countershaft drive gear.
2. Use a soft hammer to tap front and rear roller bearing cups into case and over bearing cones on countershaft. Install snap ring to front bore of case.
3. Assemble rear bearing cap and shim pack to case. Torque capscrews to 35–40 ft-lbs. Check bearing end play. Add or remove shims as required to obtain .001″ to .003″ play. Individual shims are available in various thicknesses.
4. After correct adjustment is made, remove rear bearing cap and coat both sides of

1 Outer Shift Lever Seal
2 Shifter Housing Cap Screw
3 Shifter Housing Cap Screw Lock Washer
4 Housing Stud Nut
5 Shifter Cover Stud
6 Breather
7 Lift Bracket
8 Auxiliary Shifter Housing
9 Cross Shaft Rod
10 Cross Shaft Woodruff Key
11 Outer Shift Lever
12 Outer Shift Lever Set Screw
13 Cross Shaft Rod
14 Outer Shift Lever Bushing
15 Inner Shift Finger
16 Inner Finger Screw Lockwire
17 1st & 2nd Shift Fork
18 Inner Finger Lock Screw
19 Inner Finger Lock Screw Lockwire
20 Shifter Housing Gasket
21 Direct & Overdrive Shift Fork
22 Shift Rod Poppet Spring
23 Shift Rod Poppet Plunger
24 Shift Rod Poppet Plunger Ball
25 Shift Rod Hole Plug
26 Direct & Overdrive Shift Rod
27 1st & 2nd Shift Rod
28 Shift Rod Interlock
29 Shift Rod Interlock Hole Plug

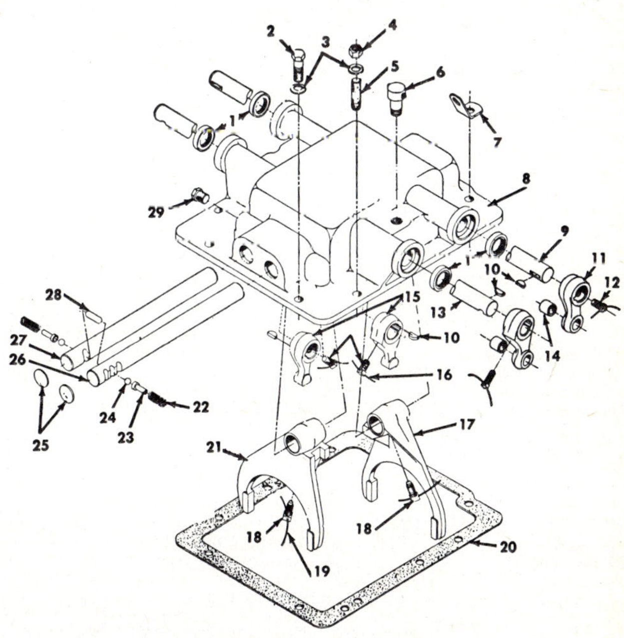

Fig. 24 Auxiliary unit shifter housing

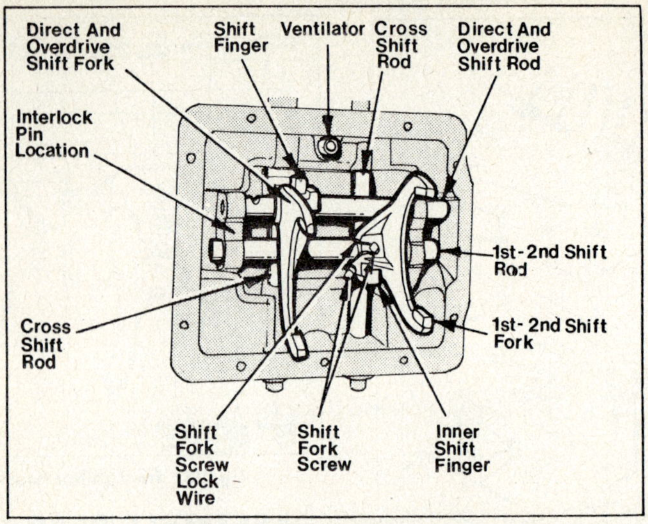

Fig. 25 Auxiliary shifter housing

shim pack with gasket cement. Install bearing cap and torque capscrews to 35–40 ft-lbs.

Mainshaft

1. Lower rear of shaft assembly, Fig. 11, into case and out through rear bearing bore. Lower front of shaft into position and mesh all gears with countershaft gears.
2. Coat thrust face of 1st speed gear washer with light grease and assemble on rear of mainshaft, with flat face toward 1st speed gear.
3. Use a 3/4" piece of stock to block mainshaft across drive gear bearing cap opening at front of case. Position mainshaft rear bearing on shaft with snap ring to rear. Use care to align outer race of bearing with case bore. Drive bearing on shaft with suitable piece of tubing against inner race of bearing until bearing is seated against washer. Remove 3/4" stock and tap bearing into case bore until snap ring seats against case.
4. If necessary, replace speedometer drive gear bushing in rear bearing cap.
5. Apply a light coat of gasket cement to O.D. of rear bearing cap oil seal. Press into cap.
6. Apply gasket cement to rear bearing cap gasket and place gasket on cap. Align oil passages as necessary.
7. Apply cement to other side of gasket. Assemble bearing cap and gasket to rear of case. Dip threads of bolts in gasket cement and install in case with washers. Torque bolts to 60–80 ft-lbs.
8. Assemble end yoke to mainshaft. Install flat washer and nut and torque nut to 400–450 ft-lbs.

ASSEMBLE FRONT & REAR UNITS

1. Install clutch housing to front unit, tightening retaining bolts to 90–95 ft-lbs. Install clutch release parts to clutch housing if previously removed.
2. Position front unit up vertically on clutch housing.
3. Secure rear unit direct and overdrive clutch gear collar to overdrive gear, or insert collar into rear of auxiliary drive gear. Note that clutch teeth of clutch collar point toward auxiliary drive gear.
4. Assemble gasket on front unit rear case surface. Install mainshaft front bearing in auxiliary drive gear bore.
5. Lift rear unit with chain fall safely attached to output yoke and suspend unit vertically over front unit.
6. Carefully lower rear unit into place. Use caution to align rear unit mainshaft with front bearing in rear of auxiliary drive gear of front unit with teeth of countershaft drive gear of auxiliary unit.
7. Bolt two units together, using the four drilled head bolts and lock-washers on upper inside holes of rear unit case.
8. Install remaining bolts, then progressively tighten and pull cases together. Torque all bolts to 120–150 ft-lbs. Secure four inside bolts with lock wire. Tie two side bolts together.
9. Use chain fall to lift entire assembly and lower to a horizontal position.
10. Install shifter housing.

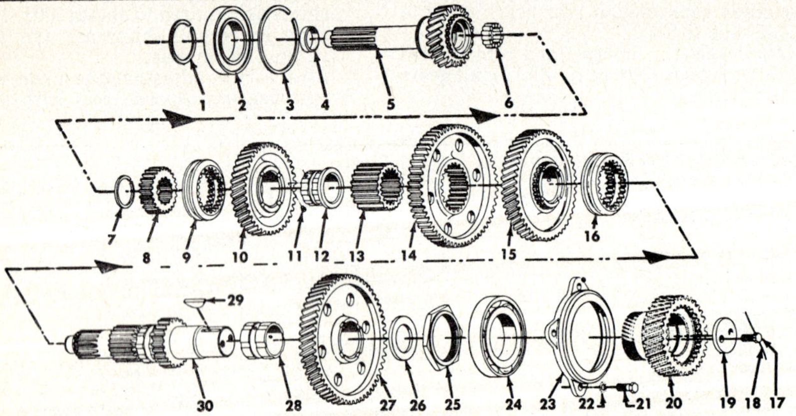

1 Main Drive Gear Bearing Snap Ring (Front)
2 Main Drive Gear Bearing
3 Main Drive Gear Bearing Snap Ring (Rear)
4 Main Drive Gear Wear Sleeve
5 Main Drive Gear
6 Mainshaft Front Bearing Roller
7 Mainshaft 3rd & 4th Clutch Gear Snap Ring
8 Mainshaft 3rd & 4th Clutch Gear
9 Mainshaft 3rd & 4th Clutch Collar
10 Mainshaft 3rd Speed Gear
11 Mainshaft 3rd Speed Gear Sleeve Pin
12 Mainshaft 3rd Speed Gear Sleeve
13 Mainshaft Reverse Clutch Gear
14 Mainshaft Reverse Gear
15 Mainshaft 2nd Speed Gear
16 Mainshaft 1st & 2nd Clutch Collar
17 Mainshaft Retaining Washer Bolt
18 Mainshaft Retaining Washer Bolt Wire
19 Mainshaft Retaining Washer
20 Auxiliary Drive Gear
21 Drive Gear Bearing Retainer Bolt
22 Drive Gear Bearing Retainer Bolt Washer
23 Drive Gear Bearing Retainer
24 Auxiliary Drive Gear Bearing Assembly
25 Auxiliary Drive Gear Bearing Retaining Nut
26 Mainshaft 1st Speed Gear Thrust Washer
27 Mainshaft 1st Speed Gear
28 Mainshaft 1st Speed Gear Bushing
29 Auxiliary Drive Gear Key
30 Mainshaft

Fig. 26 Main unit mainshaft and gears

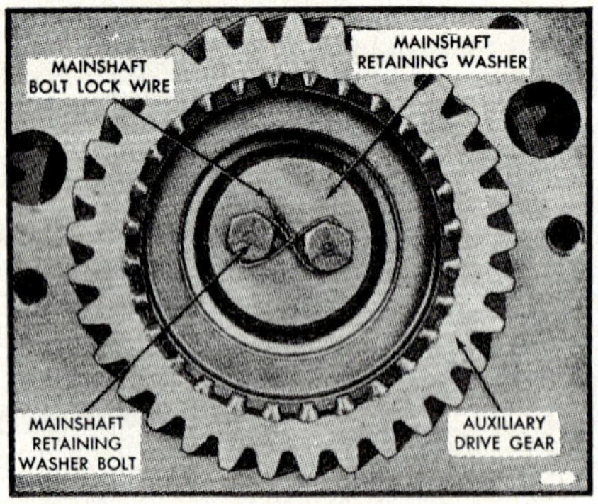

Fig. 27 Auxiliary drive gear and lock plate installed

Spicer 8016 & 8216—16 Speed Units

DESCRIPTION

This unit consists of two 4-speed transmissions joined together with a separate shift lever for each unit. The shift lever for the rear unit is mounted on a bracket bolted to the front unit. The rear unit is operated by means of shift rods from the shift control assembly.

When used on a Tilt Cab model, it is equipped with a remote control cab unit and transmission slave unit, with the rear unit being shifted by means of control cables attached to a shift control assembly in the cab, similar to the controls used with auxiliary transmissions.

Figs. 1 through 5 are exploded views of the various shift control assemblies. The following material covers the disassembly and reassembly of the front and rear gear boxes.

SEPARATING UNITS

1. From inside of rear case, cut lock wire and remove two capscrews.
2. Use a bridge chain across the two internal capscrew holes to support rear unit.
3. Remove six remaining capscrews joining front and rear units from outside case.
4. Using a pry bar, separate the two units, leveling rear unit with a chain fall.

NOTE: As the two units are separated, the mainshaft drive gear bearing retainer, Fig. 6, will pull free of rear case. This allows the integral mainshaft rear drive pinion to pass through the front bearing bore of the rear case and remain with the front unit.

The pocket bearing is the separable type. It may remain intact on the rear mainshaft, stay in the drive pinion pocket or the bearing may separate as the cases are parted.

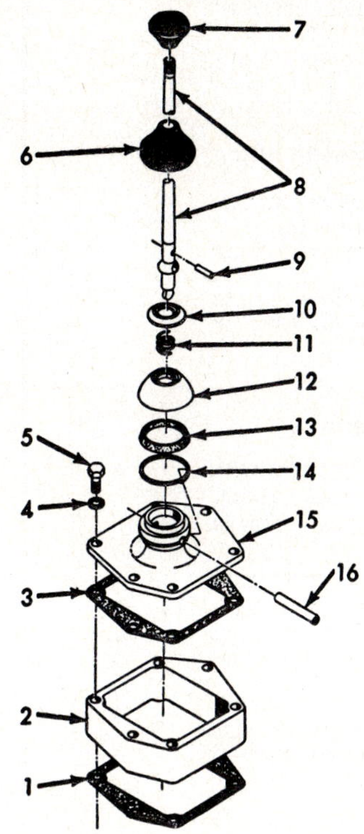

1. Gasket
2. Block
3. Gasket
4. Washer
5. Capscrew
6. Grommet
7. Knob
8. Shift Lever
9. Pin, Collar
10. Collar, Compression
11. Spring
12. Cup, Compression
13. Washer
14. Ring, Snap Rock Shaft
15. Dome, Shift Lever
16. Shaft, Rock

Fig. 1 Front unit overhead control

DISASSEMBLE FRONT UNIT—FIG. 6

Drive Pinion & Clutch, Disassemble

1. Lock transmission in two gears by engaging 3–4 speed clutch collar with drive gear and reverse sliding gear with 2nd speed gear.
2. Remove countershaft rear bearing retaining lockscrew. Remove washer from retaining pin.
3. Remove capscrew attaching *long* clutch release shaft to clutch release yoke. Remove key. Tap release shaft free of yoke and remove through side of clutch housing. Slide yoke and *short* shaft toward inside of housing and remove as a unit.
4. Remove four capscrews from drive pinion bearing cap. Use two 7/16" puller screws to free bearing cap and drive pinion assembly from case bore. Remove as an assembly.
5. Remove snap ring from drive pinion bearing cap and press cap free of bearing.
6. Remove nut from drive pinion.
7. Support outer race of bearing and press drive pinion free of bearing.
8. Remove mainshaft pocket bearing.

Mainshaft, Disassemble

1. Remove front pocket bearing or parts if any remained on mainshaft.
2. Remove rear pocket bearing or parts if any remained while separating the units.
3. Unstake mainshaft 3-4 clutch gear locknut. Using a spanner wrench, remove locknut *which has a left-hand thread.*
4. Slide 3-4 clutch collar from clutch gear.

NOTE: It is impossible to remove the mainshaft from internal gears without dropping any or all of the 154 roller bearings used in the bore of the 1st speed gear. Use caution to prevent loss of bearings as mainshaft is removed from gear.

5. Using a long drift, remove mainshaft by tapping rearward through gears and sleeves, removing each sub-assembly as mainshaft is being tapped through in the following order: (a) 3-4 clutch gear, (b) 3rd gear with sleeve, roller bearings and spacer, (c) 2nd gear with sleeve, bearings and spacer, (d) reverse sliding gear with clutch gear, (e) 1st speed gear with spacer and remaining needle bearings.
6. Remove 1st speed gear thrust washer.
7. Unstake bearing locknut and remove with wrench or hammer and drift. *Nut has a left-hand thread.*
8. Place mainshaft and bearing sub-assembly vertically in arbor press with front end up and press shaft free of bearing.
9. Press bearing retainer free of bearing. Discard two retainer seals.

Reverse Idler Gear, Disassemble

1. Remove reverse idler shaft lock plate.
2. Using a puller bolt and a piece of suitable tubing, remove idler gear shaft.
3. Remove idler gear, two sets of bearings

and sleeve as an assembly.

4. Remove two sets of bearings and gear from sleeve. Remove key.

Countershaft, Disassemble

1. Remove rear bearing retainer capscrew, lockwasher and retainer wahser.
2. Move countershaft sub-assembly toward rear of case until rear bearing is free of case bore. Use a suitable puller to remove bearing from countershaft.
3. Remove countershaft from case by lifting front end up first.
4. Use a suitable puller to remove countershaft front bearing from case bore.
5. Remove countershaft drive gear snap ring.
6. Press gears off countershaft one at a time by supporting behind gears as close to hub as possible.
7. Remove woodruff keys.

Clutch Housing, Remove

NOTE: It is not necessary to remove clutch housing unless it is damaged or if gasket is allowing oil leak.

1. If removal is required, it will be necessary to remove main drive pinion bearing cap if not done previously during disassembly.
2. Remove capscrews from inside clutch housing. *One capscrew is located outside of housing at lower left corner of front case.* Tap housing free of case and remove gasket.

REASSEMBLE FRONT UNIT, FIG. 6

Countershaft, Assemble

NOTE: If woodruff keys or keyways were burred during disassembly, dress with a file before assembly to prevent metal chips from lodging between gear hub faces.

1. Install and seat keys in countershaft.
2. Using a suitable arbor press, support 2nd speed gear on hub with chamfer up. Place countershaft in bore, align key and keyway and press shaft into gear until seated firmly against shoulder.
3. Support 3rd speed gear with chamfer up. Press countershaft in bore, align key and keyway, and press shaft into gear until firmly seated against 2nd speed gear.
4. Support P.T.O. drive on long hub with chamfer up. Place countershaft in bore, align key and keyway, and press shaft into gear until seated firmly against P.T.O. gear.
5. Support countershaft drive gear on short hub and chamfer up. Place countershaft in bore, align key and keyway, and press shaft into gear until seated firmly against P.T.O. gear.
6. Install snap ring on countershaft to lock drive gear in place.
7. Press front roller bearing on countershaft until inner race seats firmly against shoulder on shaft.

Countershaft Installation

1. Lower countershaft assembly into case with rear end through rear bearing bore.

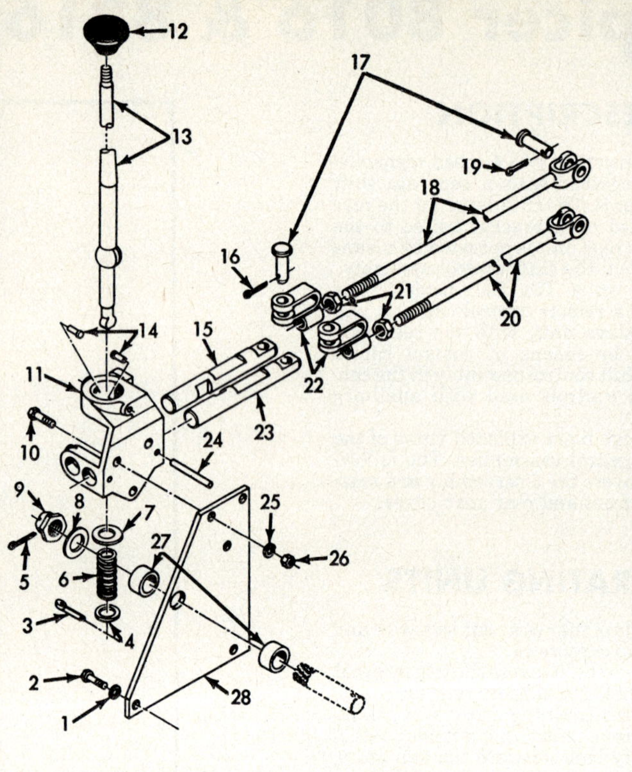

1. Washer	16. Cotter Pin
2. Capscrew	17. Pins Clevis
3. Cotter Pin	18. Rod, Outboard
4. Washer, Spring Lower	19. Cotter Pin
5. Cotter Pin	20. Rod, Inboard
6. Spring	21. Nut, Yoke Jam
7. Washer, Upper	22. Yoke, Offset Clevis
8. Washer, Stud Mounting Plate	23. Slide, Inboard
9. Nut, Mounting Plate Stud	24. Pin, Stop
10. Capscrew	25. Washer
11. Bracket, Shift Lever	26. Nut, Lever Bracket and Plate Screw
12. Knob	27. Sleeve, Bracket Mounting Plate Stud
13. Lever	28. Plate, Shift Lever Bracket
14. Pin, Shift Lever Pivot	
15. Slide, Outboard	

Fig. 2 Rear unit stick control

Align front end of countershaft with front bearing bore and tap into partial position.

NOTE: To prevent damage to front bearing as rear bearing is being driven on countershaft, place two strips of 3/8" flat stock between countershaft drive gear and front wall of case.

2. Position rear bearing on countershaft with shield inward and external snap ring toward rear. Align shaft and bearing with case bore and, using tubing, drive against inner race until bearing is seated on shoulder.
3. Remove steel strips and tap countershaft assembly the remaining way in, or until external snap ring seats against case.
4. Install rear bearing retainer.

Assemble & Install Reverse Idler

1. Install key in sleeve.
2. Install gear in sleeve.

NOTE: The sleeve has been ground to accommodate a hopping guard to prevent "walking" out of reverse. The gear must be assembled on the sleeve with shift fork collar toward rear. Make sure gear slides freely on key.

3. Grease two sets of bearings and install in sleeve bore.
4. Install assembly in case.
5. Insert reverse idler gear shaft (plain end first) through hole in rear of case and through idler gear assembly. Align "flat" on rear of shaft for lock plate and tap into position. Install lock plate.

Assemble Drive Pinion

NOTE: The drive pinion and bearing cap assembly should be installed temporarily to facilitate assembly of mainshaft through the gears stacked inside the case. After mainshaft installation, the drive pinion and bearing cap must be removed to install mainshaft locknut, pocket bearing and gasket.

1. Position drive pinion under arbor press with shield up. Support inner race of bearing and press drive pinion into bearing.
2. Using a vise with soft jaws, clamp on teeth of drive pinion and install bearing locknut. This is a left-hand thread. Tighten locknut 300–350 ft-lbs. and stake in two places.
3. Place bearing cap over drive pinion bearing and press cap into position.
4. Seat snap ring in groove provided inside rear of cap to lock bearing in place.

Install Clutch Housing & Drive Pinion

1. If clutch housing has been removed it must be installed at this time. Using a light coating of gasket cement install gasket on case.
2. Position clutch housing case studs or retain with capscrews and lockwashers. *Do not tighten at this time.*
3. Install drive pinion and bearing cap assembly in bore of clutch housing and transmission case. Bearing cap will align clutch housing bore with transmission case bore. Tighten front bearing cap.
4. Tighten clutch housing capscrews or locknuts, whichever are provided. With cast iron case, tighten capscrews or studs to 90–95 ft-lbs; with aluminum case 70–75 ft-lbs.

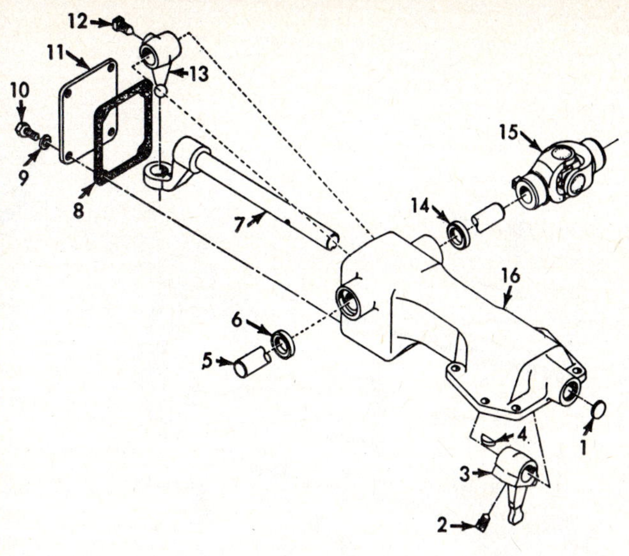

1. Plug
2. Screw, Shift Finger Lock
3. Finger, Inner Shift
4. Key, Shift Finger
5. Rod
6. Seal
7. Rod and Bracket
8. Gasket
9. Washer
10. Capscrew
11. Cover
12. Screw, Shift Finger
13. Finger, Outer Shift
14. Seal
15. "U" Joint
16. Cover, Remote Control

Fig. 3 Front unit remote control

Assemble & Install Mainshaft

NOTE: Due to the one-piece design of the front unit mainshaft and rear drive gear, the shaft must be assembled through the rear bearing retainer bore. Mainshaft installation is accomplished with the front case, front drive pinion and clutch housing set up vertically on two 4×4 wooden blocks.

The temporary installation of drive pinion and front bearing cap assembly holds the vertical stack-up of mainshaft gears and sleeves in their approximate position and alignment inside case during installation of mainshaft.

1. Place 3rd speed gear sleeve on flange end and coat lower one inch and flange with grease. Install one row of 71 needle bearings to sleeve.
2. Install spacer on sleeve and coat upper end of sleeve with grease. Then install second row of needle bearings on sleeve.
3. Install 3rd speed gear on needle and sleeve assembly with clutch teeth up.
4. Install 3-4 clutch gear collar on clutch gear and set in position on 3rd speed gear and sleeve assembly.
5. Engage clutch collar with 3rd speed gear. Grasp sub-assembly securely, turn over and set in position on drive pinion inside case. Allow clutch collar to drop down and engage clutch teeth on drive pinion to hold alignment.
6. Place 2nd speed gear sleeve on a suitable flat plate and coat lower one inch with grease. Install one row of 77 needle bearings on sleeve. Install spacer. Coat upper end with grease. Install second row of 77 needle bearings.
7. With clutch teeth down, install 2nd speed gear over sleeve and needles. Apply an extra heavy coat of grease along ends of needle bearings and face of 2nd speed gear to seal needles in place. Using the plate, turn assembly over quickly and stack in position on 3rd speed gear and sleeve in case.
8. Install 1-2 and reverse clutch gear in 1-2 and reverse sliding gear and stack in position in case on 2nd speed sleeve.

NOTE: Shift collar of sliding gear should be up. Line up assembly by allowng gear to drop down and engage clutch teeth of 2nd speed gear.

9. Place 1st speed gear with clutch teeth facing down into position on gear stack. *Do not install needle bearings at this time.*
10. Position mainshaft drive gear bearing retainer under arbor and press mainshaft drive gear bearing into retainer. Use a suitable tool to permit pressing on outer race of bearing.
11. Turn bearing and retainer assembly over on arbor. Support inner race of bearing and press mainshaft drive gear through bearing until bearing seats against shoulder of mainshaft.
12. Remove mainshaft and bearing sub-assembly from arbor press and install locknut, (left-hand thread). Tighten to 500–550 ft-lbs and stake nut in groove provided.
13. Install two retainer and bearing seals in grooves provided in bearing retainer.
14. Using a mainshaft installing tool, Fig. 10, with a hoist, suspend mainshaft and bearing assembly vertically over rear bearing bore in case.
15. Lower mainshaft slowly into bore of sleeves and clutch gear. Carefully align shaft with splines of 3-4 clutch gear and 1-2 reverse clutch gear by slowly rotating shaft and tapping lightly. When shaft bottoms into position, *mark drive gear position on case,* so mainshaft can be reassembled without difficulty.
16. Use a 2" spacer to hold reverse idler in position to allow mainshaft to rotate.
17. Use caution to hold gear stack in alignment and remove mainshaft out of rear bearing bore.

Install Mainshaft in Gear Stack

1. Stand mainshaft rear drive gear sub-assembly on bench with small end up.
2. Apply a coat of grease to 1st speed gear thrust washer and install on mainshaft, seating it securely on shoulder ahead of locknut.
3. Apply a coat of grease approximately one inch wide to ground diameter of mainshaft above thrust washer. Install one row of 77 needle bearings on mainshaft. Install spacer and seat against end of needle bearings. Apply a coat of grease to the upper shaft area and install the second row of 77 needle bearings on mainshaft. Apply a liberal coat of grease to hold bearing on shaft during assembly.
4. Using a light coat of grease, install gasket on transmission cae.
5. Carefully turn and suspend mainshaft sub-assembly (with bearings intact) vertically with hoist.
6. Position sub-assembly over case and gear stack, and rotate shaft to align reference marks on transmission case and drive

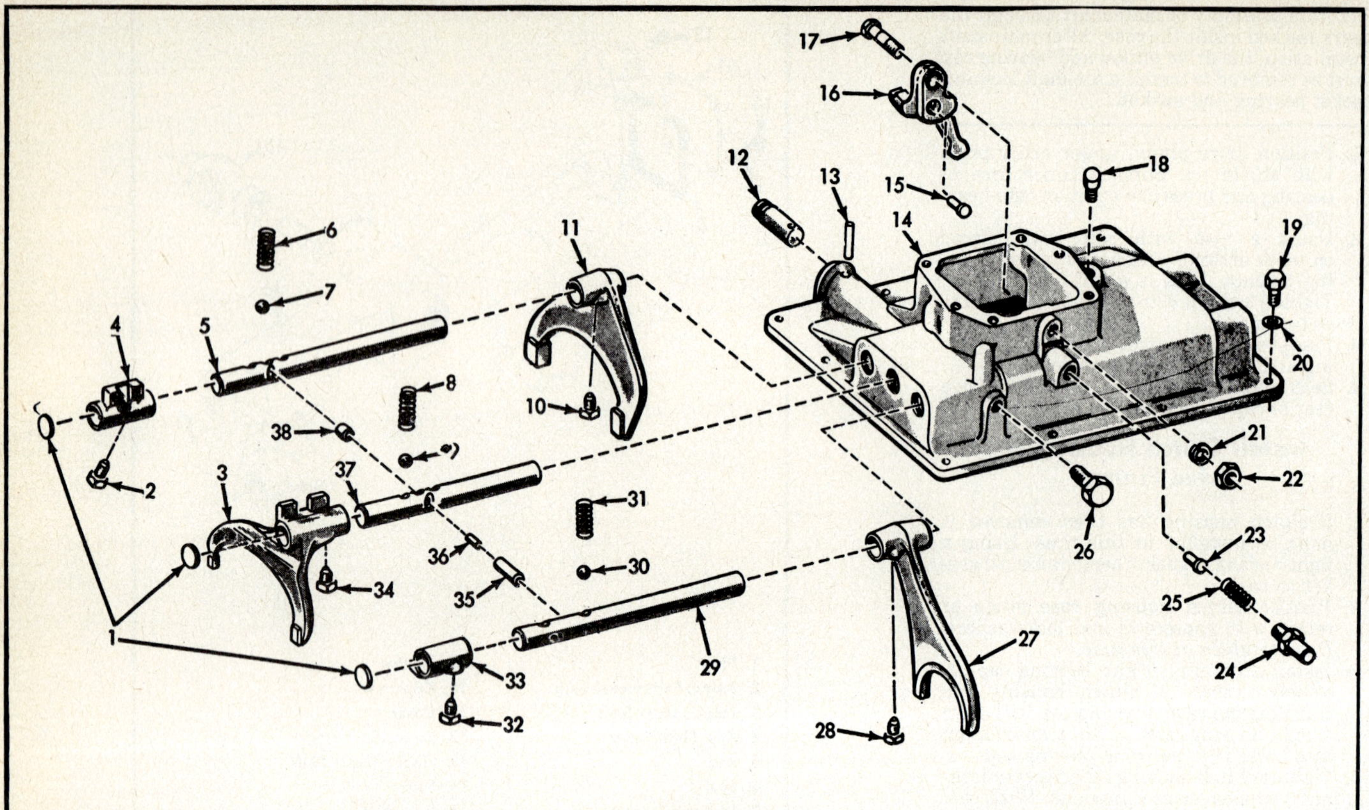

1. Plugs
2. Screw, Fork and Bracket
3. Fork, 3rd-4th Speed
4. Bracket, 1st-2nd Speed
5. Rod, 1st-2nd Speed
6. Spring
7. Ball, Poppet
8. Spring
9. Ball, Poppet
10. Screw
11. Fork 1st-2nd Speed
12. Stud
13. Pin
14. Housing, Shift
15. Pin
16. Finger, Reverse
17. Stud, Reverse Shift Finger
18. Vent
19. Capscrew
20. Washer
21. Washer
22. Nut, Reverse Shift Finger Stud
23. Pin
24. Retainer, Finger Plunger Pin
25. Spring
26. Plug, Interlock Hole
27. Fork, Reverse
28. Screw, Fork
29. Rod, Reverse Shift
30. Ball, Poppet
31. Spring
32. Screw, Fork
33. Bracket, Reverse Shift Rod
34. Screw, Fork and Bracket
35. Interlock, Shift Rod
36. Pin, Interlock
37. Rod, 3rd-4th Speed
38. Interlock, Shift Rod

Fig. 4 Front unit shift housing

gear.

7. Lower mainshaft into position, using caution to align shaft with sleeves and clutch gear splines. The last $1^{1}/_{2}''$ are usually the most difficult of the assembly since mainshaft splines must index and align with the splines of 3-4 speed clutch gear and spline of 1-2 and reverse clutch gear. Relax chain support and tap lightly while rotating back and forth slowly to complete spline assembly.
8. Line up dowel pin, Fig. 9, with notch in mainshaft and drive bearing retainer in case bore.
9. If removed, insert assembled pocket bearing in drive gear end of mainshaft.

DISASSEMBLE REAR UNIT—FIG. 7

Remove & Disassemble Mainshaft

1. Remove pocket bearing from front end of mainshaft. Use a suitable puller if necessary to remove inner race.
2. Lock unit in two gears at once to prevent mainshaft from turning.
3. Remove mainshaft rear flange or end yoke locknut and flat washer.
4. Use a suitable puller to remove flange or yoke.
5. Remove both mainshaft and countershaft rear bearing caps.
6. Press out rear bearing cap seal from mainshaft cap. Remove speedometer bushing if replacement is necessary.
7. Remove speedometer drive gear or spacer and mainshaft rear bearing thrust washer if loose.
8. Use two pieces of hardwood to block 3-4 speed clutch gear against inside of case. Using a brass drift, drive mainshaft forward to expose 3-4 speed clutch gear split lock ring.
9. Remove blocks, split lock ring, 3-4 speed clutch collar and clutch gear.
10. Using a brass drift, drive mainshaft rearward.

NOTE: As mainshaft moves, rear bearing and thrust washer will be forced out of rear case bore. Needle bearings used with 1st speed gear may begin to fall out. Use care to prevent their loss.

---

11. Continue to drive mainshaft rearward and strip off all loose parts.
12. Using a suitable arbor press, support remaining assembly on thrust washer and press shaft out of rear bearing and 1st speed gear thrust washer. If speedometer gear and rear bearing washer could not be removed previously, they will press off with bearing.

Remove & Disassemble Countershaft

1. Remove retainer from front of countershaft.
2. Using a brass drift, drive countershaft rearward to clear front bearing. Rear bearing will be driven out of case.
3. Using a suitable puller, remove rear bearing from countershaft.
4. Lift countershaft assembly from case.
5. Remove front bearing by tapping with soft hammer from inside case.
6. Using an arbor press, remove counter shaft drive gear, 4th and 2nd gears. Care should be used to support each gear as close to hub as possible.

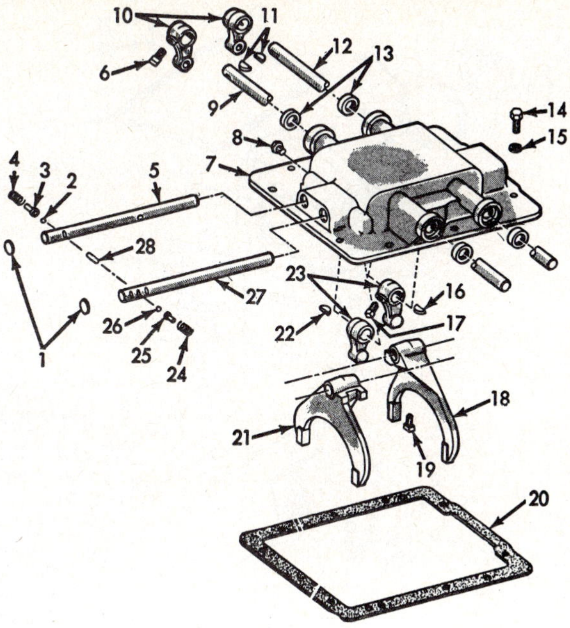

1. Plugs
2. Ball, Poppet
3. Plunger
4. Spring, Poppet
5. Rod, 1st-2nd Speed
6. Screw, Shift Finger
7. Shift Housing
8. Plug, Poppet Hole
9. Rod, 3rd-4th Speed Cross Shift
10. Lever, Cross Shift Outer
11. Keys
12. Rod, 1st-2nd Speed Cross Shift
13. Seals
14. Capscrew
15. Washer
16. Key
17. Screw, Shift Finger
18. Fork 1st-2nd Speed
19. Screw, Set
20. Gasket
21. Fork, 3rd-4th
22. Key
23. Finger, Cross Shift Inner
24. Spring
25. Plunger
26. Ball, Poppet
27. Rod, 3rd-4th Speed
28. Interlock

Fig. 5 Rear unit shift housing

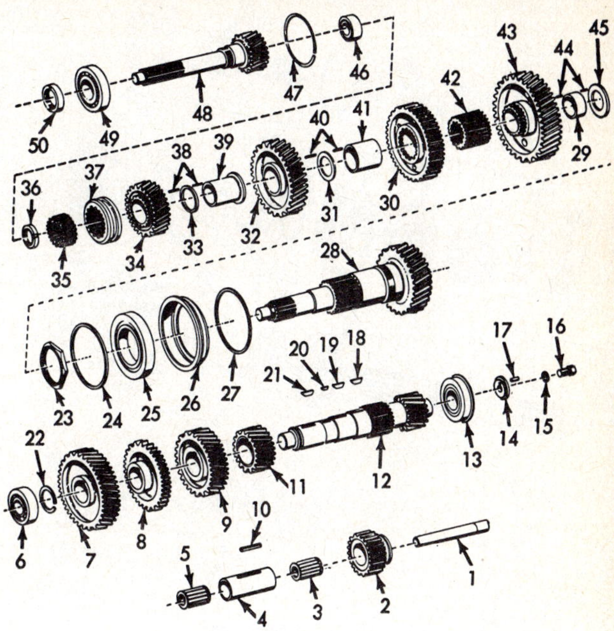

1. Shaft, reverse Idler
2. Gear, Reverse Idler
3. Bearing
4. Sleeve
5. Bearing
6. Bearing
7. Gear, Countershaft Drive
8. Gear P.T.O.
9. Gear, Countershaft 3rd Speed
10. Key
11. Gear Countershaft 2nd Speed
12. Countershaft
13. Bearing, Countershaft Rear
14. Washer, Retainer
15. Washer
16. Capscrew
17. Pin, retainer
18. Key 2nd Speed Gear
19. Key 3rd Speed Gear
20. Key P.T.O.
21. Key, Countershaft Drive
22. Snap Ring
23. Nut
24. Seal
25. Bearing
26. Retainer
27. Seal
28. Mainshaft and Gear
29. Spacer
30. Gear, Reverse Sliding
31. Spacer
32. Gear, Mainshaft 2nd Speed
33. Spacer
34. Gear, Mainshaft 3rd Speed
35. Gear, 3rd-4th Clutch
36. Nut, 3rd-4th Clutch Gear Lock
37. Collar, 3rd-4th Clutch Gear
38. Needle Bearing
39. Sleeve, 3rd Speed Gear
40. Needle Bearings
41. Sleeve, 2nd Speed Gear
42. Gear, 1st-2nd and Reverse Clutch
43. Gear 1st Speed
44. Needle Bearings
45. Washer, Thrust
46. Bearing, Pocket
47. Snap Ring
48. Gear, Drive Pinion
49. Bearing
50. Nut

Fig. 6 Front unit gears and related parts

ASSEMBLE REAR UNIT, FIG. 7

Assemble Countershaft

NOTE: All countershaft gears should fit tight on the shaft. As a shrink fit of .0015" to .003" is built into new parts, it presents a service problem. If heat is used to expand gear bores, boiling water, hot oil or steam are usually satisfactory. Do not exceed 250 deg. F. Do not use flame.

If heat is not used, it is advisable to coat the gear bores heavily with white lead to prevent seizing or galling of parts. If keys or keyways are burred, dress up with a file before assembly.

1. Install keys in countershaft.
2. Support hub of 2nd speed gear in an arbor press with long hub and chamfer up. Place countershaft in position, align key with keyway and press shaft into gear until shoulder seats firmly against gear.
3. Support hub of 4th speed gear in arbor press with long hub down and chamfer up. Place countershaft in position, align key with keyway and press shaft into gear until seated firmly blocking up drive gear.
4. Support hub of countershaft drive gear in arbor with long hub and chamfer up. Press countershaft into gear until seated firmly against 4th speed gear.

NOTE: A two-piece or separable roller bearing is now being used in production transmissions. It is recommended that this two-piece bearing be used as a replacement part.

5. Press inner race of rear bearing on rear of countershaft with lip or flange seated firmly against shoulder of 1st speed gear.

Install Countershaft

1. Lower rear end of countershaft into case with end of shaft and 1st speed gear through rear bearing bore. Lower front of countershaft into its approximate position and maintain alignment with a cable

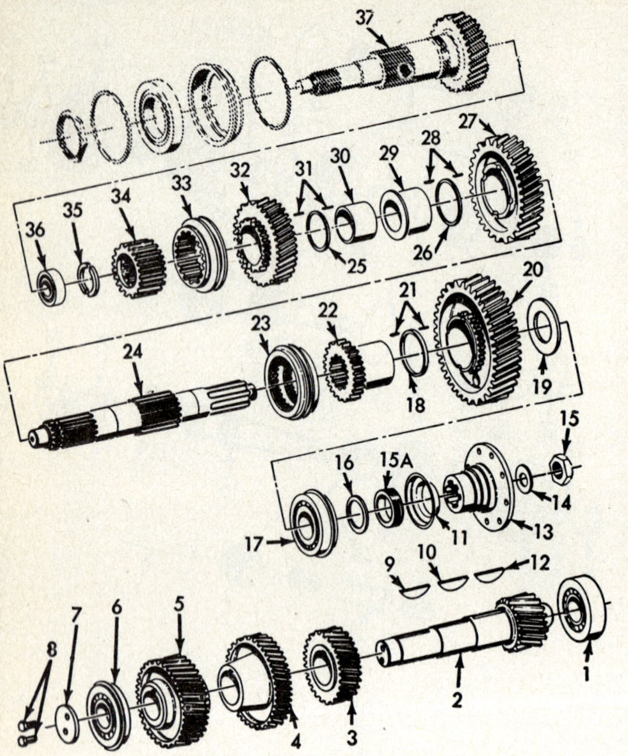

1. Bearing, Countershaft Rear
2. Countershaft
3. Gear, Countershaft 2nd Speed Underdrive
4. Gear, Countershaft 4th Speed Overdrive
5. Gear, Countershaft Drive
6. Bearing, Countershaft Front
7. Washer, retainer
8. Capscrews, Retainer
9. Key, Countershaft Drive Gear
10. Key, Countershaft 4th Speed Gear
11. Finger
12. Key, Countershaft 2nd Speed Gear
13. Flange
14. Washer, Flange Nut
15. Flange Nut
16. Washer
17. Bearing, Mainshaft Rear
18. Spacer
19. Washer, thrust
20. Gear, 1st speed Mainshaft Underdrive
21. Needle Bearings
22. Gear, Mainshaft 1st-2nd Speed Clutch
23. Collar, Mainshaft 1st-2nd Speed Clutch Gear
24. Auxiliary Mainshaft
25. Spacer
26. Spacer
27. Gear, Mainshaft 2nd Speed Underdrive
28. Needle Bearings
29. Sleeve, Mainshaft 2nd Speed Gear
30. Sleeve, Mainshaft 4th Speed Gear
31. Needle Bearings
32. Gear, Mainshaft 4th Speed Overdrive
33. Collar, Mainshaft 3rd-4th Speed Clutch Gear
34. Gear, Mainshaft 3rd-4th Speed Clutch
35. Lock Ring 3rd-4th Speed Clutch Gear
36. Bearing, Drive Pinion Pocket
37. Gear, Drive Pinion

Fig. 7 Rear unit gears and related parts

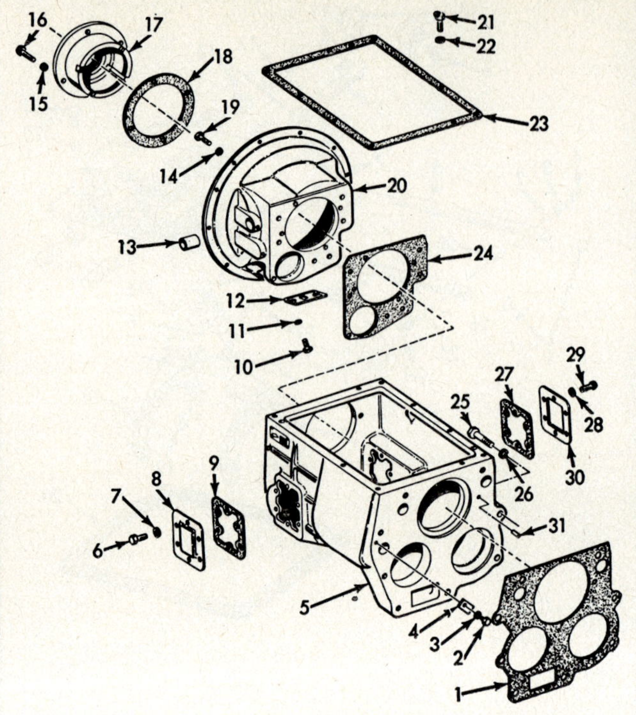

1. Gasket
2. Capscrew, Reverse Idler Lock
3. Washer
4. Plate, Reverse Idler Lock
5. Transmission Case, Front Unit
6. Capscrew
7. Washer
8. Cover, P.T.O. Aperture
9. Gasket P.T.O. Aperture
10. Capscrew
11. Washer
12. Cover, Clutch Housing
13. Bushing, Clutch Release Shaft
14. Washer
15. Washer
16. Capscrew
17. Cap, Front Bearing
18. Gasket
19. Capscrew
20. Housing, Clutch
21. Capscrew
22. Washer
23. Gasket, Shift Housing
24. Gasket, Clutch Housing
25. Capscrew
26. Washer
27. Gasket
28. Washer
29. Capscrew
30. Cover
31. Dowel Pin

Fig. 8 Front unit transmission case

support or by blocking up drive gear.

2. Align outer race of rear bearing (two-piece) with inner race and rear bore and tap flush with case.
3. Using a light coating of gasket cement on gasket, install countershaft rear bearing cap. Secure capscrews dipped in gasket cement. Tighten to 35–40 ft-lbs in cast iron case or 20–25 ft-lbs with aluminum case.
4. Position front bearing on countershaft with external snap ring away from case. Use a 4×4 hard wood block to wedge between rear case wall and 2nd speed gear. Drive against inner race of bearing until seated against hub of drive gear.
5. Secure bearing with retainer washer and capscrews. Tighten capscrews to 45–50 ft-lbs. Secure with lockwire.

Assemble Mainshaft

1. Place 1-2 speed clutch gear on bench with flange down. Apply a light coat of grease approximately one inch wide to ground area above flange. Install one row of 72 needle bearings on clutch gear. Install spacer. Apply a light coat of grease above spacer and install the second row of 72 needle bearings. *Do not plug holes in gears with grease.*
2. With clutch teeth down, install 1st speed gear on clutch gear and over needles and spacer. Set this assembly aside.
3. Coat 2nd speed gear sleeve on thrust face and ground surface with grease and install two rows of 72 needle bearings with spacer in between.
4. Install 2nd speed gear on sleeve and bearings with clutch teeth on gear away from flange. Set this assembly aside.
5. Place 4th speed gear sleeve on a plate and

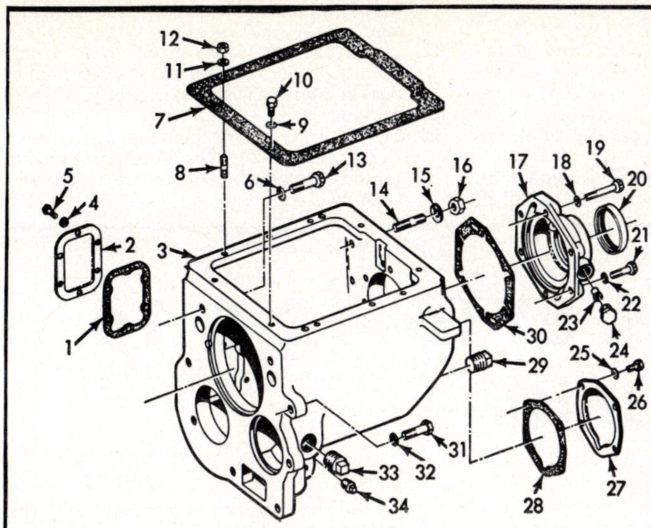

1. Gasket, P.T.O.
2. Cover, P.T.O.
3. Transmission Case, Rear Unit
4. Washer
5. Capscrew
6. Washer
7. Gasket, Shift Housing
8. Stud
9. Washer
10. Capscrew
11. Washer
12. Nut
13. Capscrew
14. Stud
15. Washer
16. Nut
17. Cap, Mainshaft Rear Bearing
18. Washer
19. Capscrew, Rear Bearing Cap
20. Seal, Rear Bearing Cap
21. Capscrew, Rear Bearing Cap
22. Washer
23. Bushing, Speedometer
24. Plug
25. Washer
26. Capscrew
27. Cap, Countershaft Rear Bearing
28. Gasket
29. Plug, Oil Drain
30. Gasket, Mainshaft Rear Bearing Cap
31. Capscrew
32. Washer
33. Plug, Oil Filler
34. Plug, Temperature Indicator

Fig. 9 Rear unit transmission case

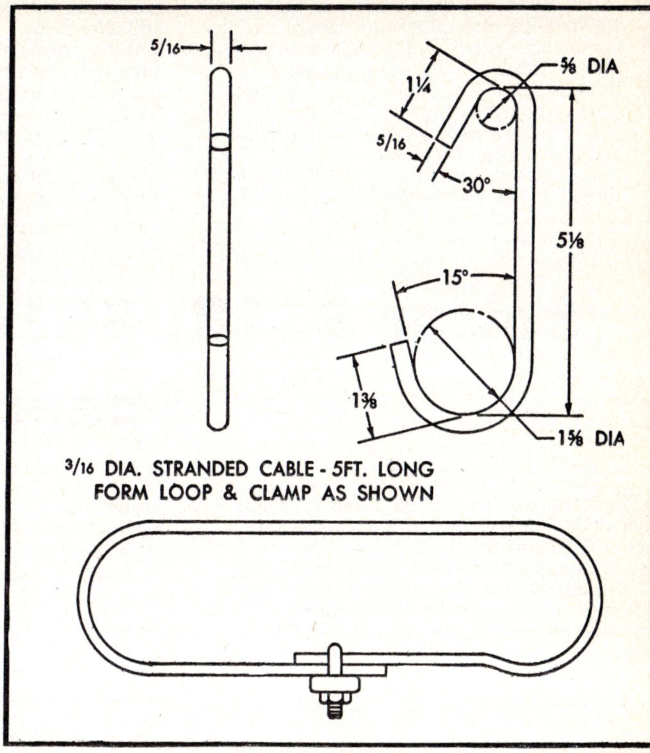

Fig. 10 Dimensions for lifting sling and hook for installing mainshaft

coat lower one inch with grease. Install one row of 62 needle bearings.

6. Install 4th speed gear on sleeve and bearings with clutch teeth of gear up. Set this assembly aside.

Install Mainshaft Sub-Assemblies

1. Position 1st speed gear and sleeve assembly in rear of case with clutch teeth toward front. Mesh with countershaft gear. Install 1-2 clutch collar on 1st speed gear with extended hub toward gear.
2. Place 2nd speed gear sleeve assembly in case with clutching teeth of gear toward 1st speed gear, and mesh with countershaft 2nd speed gear. Move clutch collar to neutral to hold assemblies in relationship.
3. Coat thrust faces and splines of mainshaft with light grease. Enter pilot bearing end through rear bearing bore until mainshaft has passed through 2nd gear sleeve approximately one inch.
4. Place 4th speed gear and sleeve assembly in case with clutch teeth forward or away from 2nd speed gear. Slide mainshaft forward through bore of sleeve.
5. Install 3-4 speed clutch collar on clutch gear. Extended hub of collar goes toward front as does lock ring counterbore in end of gear. Position gear and collar in case and install on mainshaft spline.
6. Block 3-4 clutch gear and tap mainshaft through until locking groove in shaft is beyond clutch gear.
7. Coat two clutch gear lock rings with heavy grease and place them in groove provided.
8. Tap mainshaft rearward and make sure lock rings center inside of counterbore of 3-4 clutch gear.
9. Install 1st speed gear thrust washer on rear of mainshaft next to 1st speed gear.
10. Position mainshaft rear bearing onto mainshaft with external snap ring away from case. Use a 3/4″ thick strip of flat stock to block front of mainshaft against inside of case. Use a pinch bar to roll mainshaft into position to align outer race of bearing with case bore. Using a suitable piece of tubing, drive on inner race until bearing is seated against 1st speed gear thrust washer.
11. Remove 3/4″ block and continue to tap bearing into case bore until snap ring is seated against case.
12. Install rear bearing washer and speedometer gear or spacer on mainshaft.
13. Apply a light coat of gasket cement on OD of rear bearing cap oil seal and press into cap.
14. Using a light coat of gasket cement on cap and gasket, install rear bearing cap. Be sure to align oil passages. Secure with capscrews. Tighten to 45–50 ft-lbs in cast iron case or 25–30 ft-lbs in aluminum case.
15. Install end yoke or flange on mainshaft with pusher tool. *Do not drive flange or yoke on shaft without provisions to keep shaft from moving forward.*
16. Install flat washer and locknut. Lock transmission in two gears and tighten to 350–450 ft-lbs. Install cotter pin.
17. Shift clutch collars back into neutral and make sure all shafts turn freely.
18. The drive pinion pocket bearing is a separable part. Due to the vertical method of joining the front and rear units, the inner race may be pressed on the front end of the rear mainshaft or the outer race may be placed in the "pocket" of the front unit drive gear. *Flanged end of inner race goes toward 3rd speed clutch gear.*

Joining Front & Rear Units

1. Tie 3-4 clutch gear collar to 3rd speed gear, Fig. 6.
2. Pick up rear unit with a chain hoist attached to the output end yoke or flange.
3. Lower rear unit into place, using care to align rear mainshaft with pocket bearing and mainshaft drive gear teeth with countershaft drive gear as the units meet.
4. Bolt cases together using the two drilled head capscrews with star washers on top inside holes in rear case. Install lockwire. Using remaining capscrews on stud nuts with lockwashers to secure cases, tighten

to 95–100 ft-lbs with cast iron case or 70–75 ft-lbs with aluminum case.

5. Use a chain hoist to lift complete assembly and lower to a horizontal position.
6. Remove capscrews and washer from front drive pinion bearing cap. Use two 7/16" puller screws and remove front pinion and bearing cap from case.
7. Install clutch gear locknut (left-hand thread) to front of mainshaft. Lock transmission in two gears and tighten locknut 300–350 ft-lbs.
8. Unlock transmission gears and make sure mainshaft rotates freely before staking locknut to groove provided in mainshaft.
9. Install pocket bearing on end of mainshaft.
10. Using a light coat of gasket cement on bearing cap and gasket, prepare to install drive pinion and bearing cap on transmission.
11. Use a large screwdriver to roll mainshaft up into proper position to aid alignment of pocket bearing with bearing bore of drive pinion.
12. Secure drive pinion bearing cap to case. Tighten to 25–28 ft-lbs. Rotate front drive pinion to make sure all gears and shafts turn freely.

Spicer 8512, 8516 & 8716–12 & 16 Speed Units

These transmissions, Figs. 1 and 2, consist of an 8500 four-speed main transmission coupled with a three-speed auxiliary unit on 8512 and a four-speed auxiliary unit on 8516 and 8716. Inasmuch as the main unit is covered elsewhere in this section, the following material will be confined to the auxiliary units only.

Separating Front & Rear Units

1. Unfasten and lift off shifter housing from rear case.
2. From inside of rear case (near the top), cut lockwire and remove two attaching capscrews.
3. Use a bridge chain across the same two internal capscrew holes to support rear unit as next step is performed.
4. Remove 11 remaining attaching capscrews and bolts from outside of case.
5. Using a pry bar to wedge between front and rear case, separate the two units, leveling the rear unit with a chain fall.

NOTE: As the two cases are separated, the mainshaft drive gear bearing retainer will pull free of the rear case. This allows the integral mainshaft rear drive gear to pass through the front bearing bore of the rear case and remain with the front unit. The pocket bearing is the separable type. It may remain intact on the rear mainshaft, stay in the drive gear pocket or the bearing may separate as the two cases are parted.

Three Speed Rear Unit

Mainshaft Disassemble

1. If necessary, use a suitable puller to remove inner race of pocket bearing from front of mainshaft.
2. Unstake clutch gear lockwasher from locknut at front of mainshaft. If used, pull

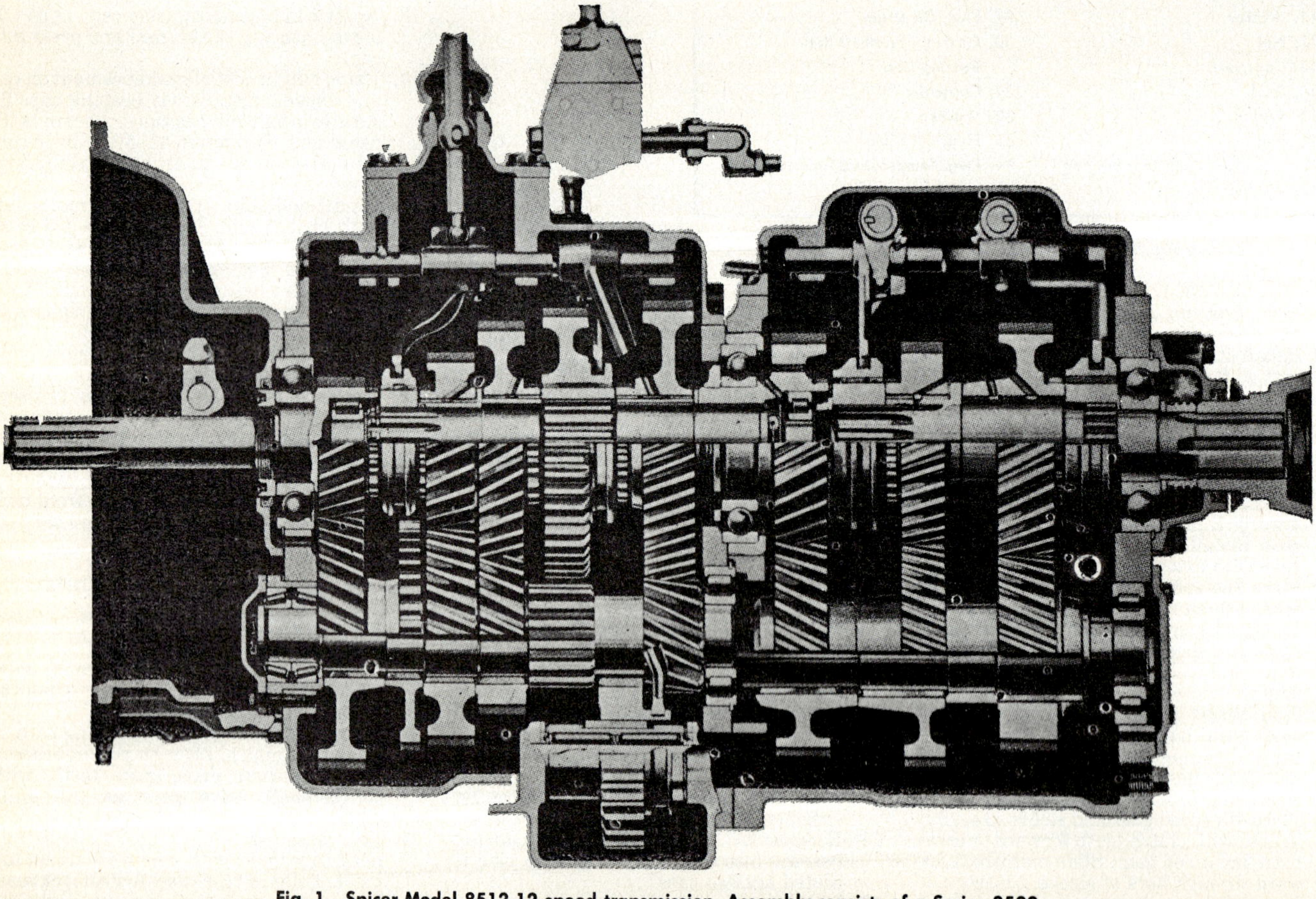

Fig. 1 Spicer Model 8512 12-speed transmission. Assembly consists of a Series 8500 four-speed front unit and a three-speed auxiliary unit

Fig. 2 Spicer Models 8516 and 8716 16-speed transmission. Assembly consists of a Series 8500 four-speed front unit and a four-speed auxiliary unit

cotter pin from rear flange or end yoke locknut.

3. Lock rear unit in two gears by engaging the 2–3 clutch collar with 3rd gear and 1st speed clutch collar with 1st speed gear.
4. Use a piloted tool and impact wrench (or hammer and drift) to loosen locknut. *Locknut has a left-hand thread. Do not remove from shaft at this time.*
5. Remove rear output flange (or end yoke) locknut.
6. Remove flange (or end yoke).
7. Remove mainshaft and countershaft rear bearing caps and gaskets.
8. Remove speedometer drive gear (or spacer) and mainshaft rear bearing thrust washer.
9. Tap or slide mainshaft and gear assembly out rear bearing bore until stopped by 1st speed gear.
10. Remove loosened 2–3 speed clutch gear locknut from front of mainshaft.
11. Remove 2–3 speed clutch gear collar and clutch gear (if slip fit) from front of shaft.
12. Use suitable puller on snap ring to remove ball bearing from rear of mainshaft.
13. Lift remaining assembly, front end first, out top of case.

CAUTION: Do not allow parts to slip off rear of shaft as assembly is removed. Three rows of 62 needle bearing and two spacers contained between 1st speed gear and shaft are loose and can be easily lost.

14. Disassemble mainshaft.

Countershaft Disassembly, Fig. 3

1. Cut lockwire and remove front bearing retaining washer.
2. With brass drift, drive countershaft rearward to clear front bearing, which will remain in case bore. Rear bearing will be driven out of case.
3. Remove snap ring from rear end of shaft and use a suitable puller to remove rear bearing.
4. Lift countershaft assembly out top of case.
5. Remove front bearing by tapping with soft hammer from inside of case.
6. Press countershaft out of gears one at a time by supporting each gear with parallel bars as close to hub as possible.
7. Remove exposed Woodruff keys.
8. Support 1st speed gear and press countershaft out of P.T.O. drive gear spacer at the same time. Remove remaining Woodruff key.

Countershaft, Reassemble, Fig. 3

All countershaft gears should fit tight on the countershaft. As a shrink or interference fit of .0015" to .003" is built into new parts, it presents a field assembly problem.

If heat is used to expand gear bores, boiling water, hot oil or steam are usually satisfactory. Do not exceed 250°F. Do not use hot plates or acetylene torches or any method that will turn the steel blue or straw color and damage the heat treated gears.

If heat is not used, it is advisable to coat the gear bores with white lead to reduce galling or seizing of parts.

When in doubt about which end of the hub to assemble on the shaft first, look for the chamfered end of the bore.

If Woodruff keys or keyways are mutilated or burred during disassembly, clean up with a file before reassembling. This will help prevent metal chips from getting between gear hub faces.

1. Assemble rearmost Woodruff key to countershaft.
2. In a suitable arbor press, support 1st speed gear on hub, with long hub and chamfer up. Place countershaft into position, align key with keyway and press into gear until shoulder on countershaft seats firmly against gear.
3. Assemble P.T.O. drive gear spacer to shaft and seat against gear. Assemble remaining two Woodruff keys to countershaft, seating keys securely.
4. Again, in arbor press, support under hub of 3rd speed gear with long hub down and chamfer up, place countershaft into position, align key with keyway and press into gear until seated against spacer.
5. In a similar manner, support countershaft drive gear with long hub and chamfer up. Press countershaft subassembly into gear until seated against 3rd speed gear.

NOTE: Either one piece or separable bearings will be found as the rear countershaft bearing.

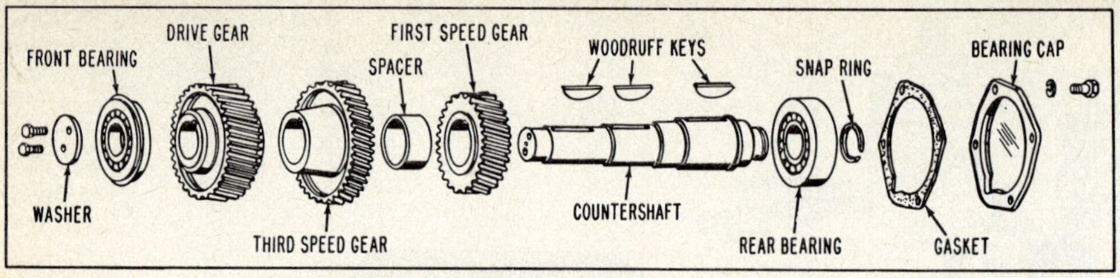

Fig. 3 Countershaft components of the 8512 three-speed rear unit

If a separable bearing is used, press on the inner race with flange seated against shoulder of countershaft. Assemble snap ring to retain bearing.

Installation of Countershaft

1. Lower countershaft sub-assembly into case with rear end of shaft and hub of 1st speed gear out rear bearing bore. Lower front of countershaft into its approximate position. Maintain alignment with a cable support or by blocking under countershaft drive gear.

NOTE: If a one-piece bearing is used, assemble countershaft at this time. This bearing is a tight fit. Block the front end of the countershaft with a 1/2" thick piece of flat steel stock about 8 to 10" long and use tubing to drive on the inner race of rear bearing. Seat bearing firmly on shaft and secure with snap ring.

2. Remove steel plate and move countershaft and rear bearing assembly forward at the same time, tapping outer race of rear bearing into case bore.
3. Use a light coating of gasket cement and assemble countershaft rear bearing cap gasket to bearing cap. Apply gasket cement to case side of gasket and apply to case.
4. Dip capscrews in gasket cement and assemble to case with lockwashers and torque capscrews to 25–32 ft-lbs.
5. Position front bearing on countershaft with snap ring away from face of case. Use a 2 × 4 hard block to wedge between rear case wall and 1st speed gear. Use tubing to drive against inner race of bearing until snap ring seats against front face of case.
6. Lock front bearing to countershaft with retainer washer and capscrews. Torque capscrews to 40–50 ft-lbs and secure with lockwire.

Assemble and Install Mainshaft, Fig. 4

1. Using a suitable arbor press, support extended hub of 1st speed clutch gear and insert mainshaft into spline of clutch gear. Press shoulder of mainshaft flush against clutch gear.
2. Place mainshaft vertically in a vise with soft jaws gripping spline of output end.
3. Coat thrust face of 1st speed clutch gear and about 3/4" of the mainshaft bearing diameter with grease. Assemble one row of 62 rollers to mainshaft. Assemble spacer to mainshaft and slide down into position against end of needle bearings.

NOTE: Due to variations in tolerances and to provide better bearing lubrication, the manufacturer recommends that needle bearings should NOT completely encircle the shaft. Space for approximately one needle bearing should be left unfilled on all Spicer needle bearing applications. This will aid in preventing seizures by allowing easier oil entry and free movement of needle bearings.

If it is more convenient to assemble needle bearings to inside of gears rather than on shaft and sleeve, the method is applicable.

4. Apply a light coat of grease to middle mainshaft bearing area and assemble second row of 62 needle bearings to shaft. In a similar manner, assemble second spacer and third row of 62 needle bearings. Use grease to hold bearings in place.
5. Assemble 1st speed gear over bearings and spacers with clutch teeth down toward clutch gear.
6. Insert new lockpin in sleeve and assemble sleeve on shaft so that pin registers with the one extended spline. If necessary to drive sleeve on shaft, use a tube against end of sleeve only. Seat flange against shoulder of shaft and 1st speed gear.
7. Coat thrust face and about one inch of third speed sleeve with light grease. Assemble one row of 62 needle bearings to sleeve. Assemble spacer over sleeve and down against needle bearings. Apply grease to upper end of sleeve and assemble second row of 62 needle bearings to sleeve.
8. With clutch teeth up, assemble 3rd speed gear over sleeve bearings and spacer.
9. Assemble 3rd speed clutch gear to mainshaft splines with ground end down or toward 3rd speed gear.
10. Assemble 3rd speed gear flat washer, lock washer and lock nut to end of mainshaft. Torque left-hand thread lock nut to 550–600 ft-lbs. Bend tangs of lock washer into slots of nut.
11. Remove mainshaft sub-assembly from vise and install in case with output end of mainshaft through rear bearing bore. Lower front of mainshaft into position and mesh all gears.
12. Maintain approximate alignment of mainshaft by supporting with cable and chain hoist. Assemble 1st speed clutch gear collar to clutch gear, with extended hub toward 1st speed gear.
13. Position rear bearing on mainshaft with snap ring to rear. Use a 3/4" thick flat steel strip to block front of mainshaft against inside of case. With tubing, drive on inner race until bearing seats against 1st speed clutch gear. Use caution to align outer race of bearing with case bore.
14. Assemble rear bearing washer and speedometer gear (or spacer washer) to mainshaft and seat against washer.
15. If necessary, replace speedometer gear bushing in rear bearing cap.
16. Apply a light coat of gasket cement to O.D. of rear bearing cap oil seal and press into cap.
17. Apply gasket cement to mainshaft rear bearing cap gasket and position on cap. Align oil passages.
18. Apply gasket cement to other side of gasket and assemble bearing cap and gasket to rear of case. Dip capscrews in gasket cement and assemble to case with copper washers. Torque capscrews to 60–80 ft-lbs. Secure with lock wire.
19. Assemble end yoke or flange to mainshaft with pusher tool. Do not drive onto mainshaft without provisions to stop shaft from moving forward (see Step 13).
20. Assemble 3rd speed clutch gear collar to clutch gear.
21. Assemble flat washer and lock nut. Lock transmission in two gears and torque flange nut to 400–450 ft-lbs.
22. Shift clutch collars back into neutral and make sure all shafts turn freely.
23. Use pressure type oil can to force lubricant in oil holes and end slots of all floating gears on mainshaft to flush out grease and insure initial lubrication of over-running gears and bearings.
24. Drive gear pocket bearing is a separable part. Due to the vertical method of assembling the rear unit to the front unit, the

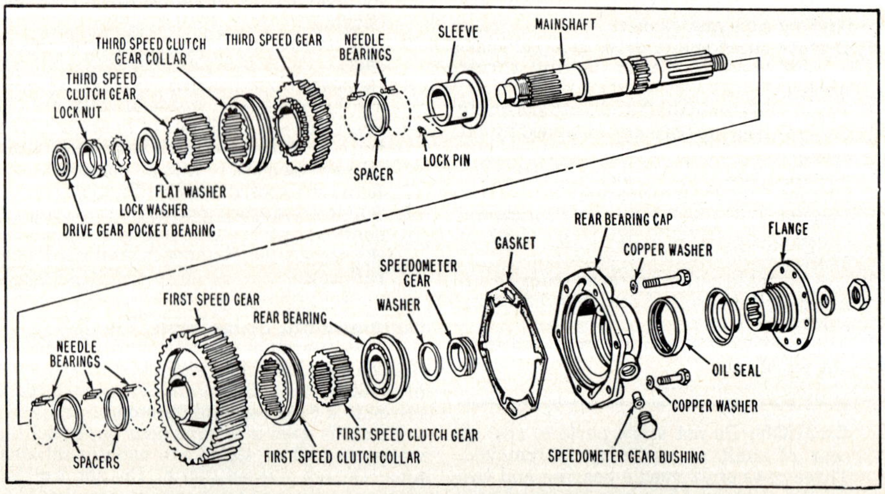

Fig. 4 Mainshaft components of the 8512 three-speed rear unit

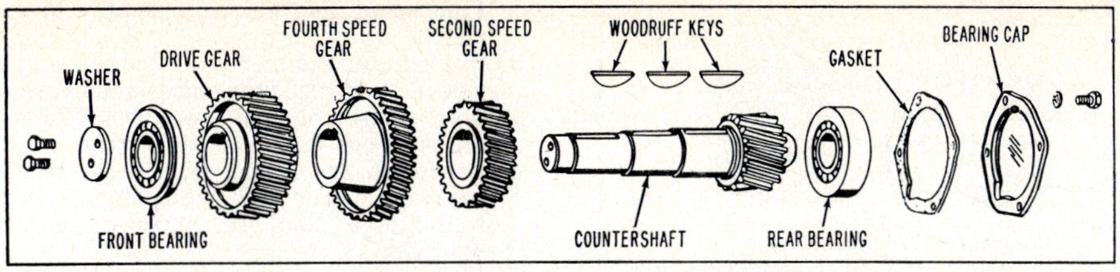

Fig. 5 "A" ratio countershaft components of the 8500 four-speed auxiliary unit

inner race must be pressed on the front end of the rear mainshaft or placed intact in the pocket of the drive gear. Flanged end of inner race assembles toward 3rd speed clutch gear. Use tubing to drive against inner race until seated against shoulder on shaft.

Joining Front and Rear Units

1. Tie the 3–4 clutch gear collar to 3rd speed gear or assemble collar to rear of mainshaft. Long snout of clutch collar points toward mainshaft.
2. Pick up rear unit with a chain hoist attached to output end yoke or companion flange and suspend vertically over front unit.
3. Lower rear unit into place, using caution to align rear mainshaft with pocket bearing of front mainshaft and drive gear teeth with countershaft drive gear.
4. Bolt cases together using two drilled head capscrews and star washers on top inside holes of rear case.
5. Assemble five capscrews with new star washers through flange on righthand side of front case into rear case. Three capscrews and washers assembled from rear into right-hand side of front case.
6. The longest through bolt assembles front to rear in bottom center hole of cases. Two shorter bolts assemble to left of bottom center. Use washer and nuts on all bolts.
7. After starting all capscrews and bolts, progressively tighten and pull cases together. Torque capscrews and nuts to 120–150 ft-lbs. Secure internal capscrews with lockwire.
8. Lower complete assembly in a horizontal position with a chain hoist.

Four Speed Rear Unit

NOTE: The 4 × 4 rear unit is produced with "A" or "B" ratio. This designation will always be found on the name plate as the letter following the model number. For example: 8516-3A or 8716-5B. All "A" models are *deep* Lo-Lo units and all "B" models are *splitter* Lo-Lo units. Only the procedure for removal and installation of the countershafts differ with the "A" and "B" units. Removal and installation of mainshaft is the same for both units.

Mainshaft, Remove and Disassemble

1. Remove shifter housing.
2. Separate rear unit from front unit.
3. Remove pocket bearing from front end of mainshaft. Use suitable puller if necessary to remove inner race.
4. Lock unit in two gears at once and remove mainshaft rear flange or end yoke lock nut. If countershaft is to be removed unstake countershaft lock nut and remove while the gears are locked.
5. Use puller to remove end yoke or flange.
6. Unfasten and remove mainshaft and countershaft rear bearing caps.
7. Press out rear bearing cap oil seal and check for damage. Remove speedometer bushing if it is to be replaced.
8. Remove speedometer drive gear or spacer and mainshaft rear thrust washer if loose.
9. Use two pieces of hard wood to block the 3–4 speed clutch gear against inside of case. With a brass drift, drive mainshaft forward to expose the 3–4 speed clutch gear split lock ring.
10. Remove hard wood blocks and split ring. Remove 3–4 speed clutch collar and clutch gear if it is a slip fit.
11. With brass drift, begin to drive mainshaft rearward.

NOTE: As mainshaft moves, rear bearing and thrust washer will be forced out of rear case bore. Needle bearings used with 1st speed gear may begin to fall out. Use care to prevent their loss.

12. Continue to drive mainshaft rearward. As shaft clears bores, strip the mainshaft of parts.
13. Using a suitable arbor press, support remaining assembly on thrust washer and press shaft out of rear bearing and 1st speed thrust washer. If speedometer gear and rear bearing washer could not be removed previously, they will press off with bearing.

"A" Ratio Countershaft, Disassemble

1. Remove front bearing washer.
2. With brass drift, drive countershaft rearward to clear front bearing which will remain in case bore. Rear bearing will be driven out of case. Use a puller to remove rear bearing from countershaft.
3. Lift countershaft out of case.
4. Remove front bearings by tapping out with soft hammer from inside of case.
5. Press countershaft out of gears one at a time by supporting each gear with parallel bars as close to hub as possible. Woodruff keys need not be removed unless worn or hoose. First speed gear is integral with shaft and cannot be separated.

"B" Ratio Countershaft, Disassemble

1. Remove lock nut and washer.
2. Move countershaft forward to expose snap ring on front bearing. Use suitable puller to remove bearing from countershaft.
3. Block drive gear against moving and tap countershaft rearward until split spacer between drive gear and 4th speed gear can be removed.

NOTE: Early production models had a one-piece bearing assembly. Later production models use a separable bearing which greatly simplifies assembly and disassembly.

4. If a one-piece bearing is used remove roller retaining snap ring from inner race of rear bearing. Remove snap ring from outer race of bearing if present.
5. If snap rings are not exposed, tap countershaft forward to move drive gear back on countershaft splines (where split spacer was removed). Inner race of rear bearing will move forward out of roller cage and outer race will remain in case bore.

NOTE: If old style bearing could not be separated by snap ring removal, use a large chisel as a wedge between face of 1st speed gear and outer race of bearing to start inner race off countershaft. Use drift and hammer to complete removal.

6. With rear of countershaft free of bearing, lift rear end first and remove remaining assembly out through top of case.
7. Slide drive gear forward off countershaft.
8. Press countershaft out of gears one at a time by supporting each gear with parallel bars as close to hub as possible. Woodruff keys need not be removed unless worn or loose. The 1st speed gear is integral with the shaft and cannot be separated.

"A" Ratio Countershaft Assemble, Fig. 5

1. With Woodruff keys in place and seated securely, support hub of 2nd speed gear in an arbor press with long hub and chamfer up. Place countershaft into position and press shaft into gear until shoulder on countershaft seats firmly against gear.
2. Support hub of 4th speed gear with long hub down and chamfer up. Press shaft into 4th speed gear until it is seated firmly against 2nd speed gear.
3. Support hub of drive gear with long hub and chamfer up. Press countershaft into gear until it is seated firmly against 4th speed gear.
4. Press inner race of rear bearing on rear of shaft with lip or flange seated against shoulder of 1st speed gear.

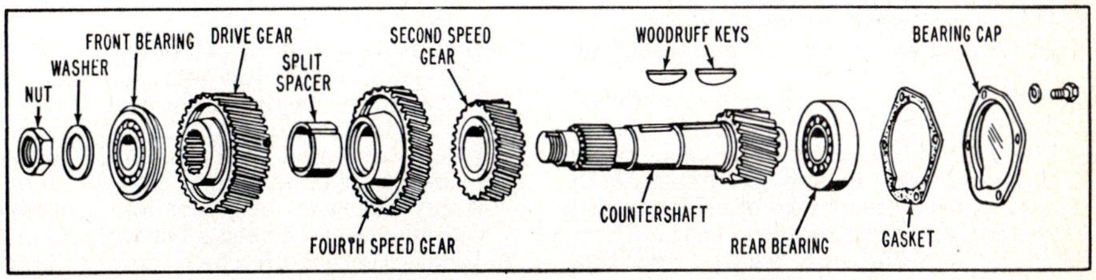

Fig. 6 "B" ratio countershaft components of the 8500 four-speed auxiliary unit

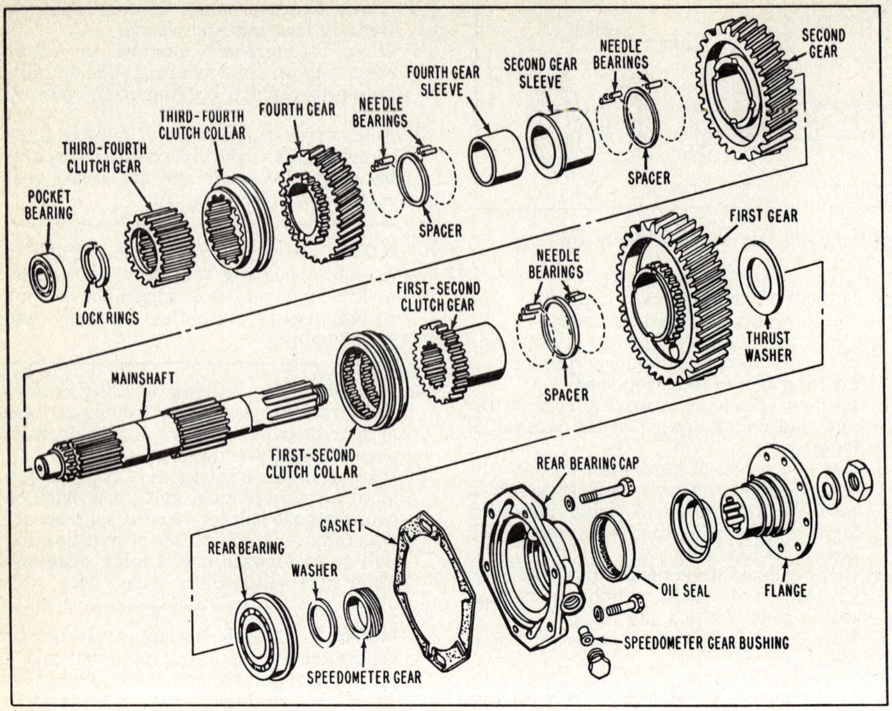

Fig. 7 Mainshaft components of the 8516 and 8716 four-speed auxiliary unit

"B" Ratio Countershaft Assemble, Fig. 6

1. With Woodruff keys in place and seated securely, support hub of 2nd speed gear in an arbor press with long hub and chamfer up. Press countershaft into gear until shoulder on countershaft seats firmly against gear.
2. Support hub of 4th speed gear with long hub down and chamfer up. Press shaft into 4th speed gear until seated firmly against 2nd speed gear.
3. Press inner race of rear bearing on rear of shaft with lip or flange seated against shoulder of 1st speed gear.
4. Assemble drive gear to front of countershaft and gear sub-assembly with long hub of drive gear toward front of countershaft. Slide gear to rear until it butts against 4th speed gear.

"A" Ratio Countershaft Installation

1. Lower rear end of countershaft assembly into case with end of shaft and 1st speed gear through rear bearing bore. Lower front of countershaft into its approximate position and maintain alignment with a cable support or by blocking up drive gear.
2. Align outer race of rear bearing with inner race and rear bore and tap flush with case.
3. Use light coat of gasket cement and assemble gasket to rear bearing cap. Apply cement to other side of gasket and assemble cap and gasket to case.
4. Dip capscrews in gasket cement and install with lockwashers. Torque capscrews to 25–32 ft-lbs.
5. Position front bearing on countershaft with external snap ring away from case. Use a 4 × 4 hard block to wedge between rear case wall and 2nd speed gear. With tubing, drive inner race of bearing until seated against hub of drive gear.
6. Secure bearing with washer and capscrews. Torque to 40–50 ft-lbs and secure with lock wire.

"B" Ratio Countershaft Installation

1. Assemble rear bearing to case and tap outer race in flush with face of case.
2. Use light coat of gasket cement to both sides of gasket and install bearing cap. Dip capscrews in gasket cement, install and torque to 25–32 ft-lbs.
3. Lower front of countershaft into case with end through front case bore. Lower rear end of countershaft into its approximate position and move toward rear until inner race on shaft is inserted in rollers of rear bearing.
4. Use a cable and chain hoist to support countershaft in its correct position. Block countershaft from moving while tapping drive gear forward on splines until stopped by front of case.
5. Assemble split spacer to shaft. Move drive gear back and make sure both halves are seated in counterbores of drive gear and 4th speed gear.
6. Position front bearing on countershaft with external snap ring away from case. Use a 2 × 4 hard block to wedge between rear case and 1st speed gear. Use tubing to drive against inner race of bearing until seated firmly against face of drive gear.
7. Remove block and tap bearing back until snap ring seats against case.
8. Place flat washer against bearing and run lock nut up finger tight. Nut will be tightened after the mainshaft is installed.

Mainshaft Assemble (both ratios) Fig. 7

1. Place 1–2 speed clutch gear on bench with flange down. Apply a light coat of grease about 1" wide to ground area above the flange. Assemble one row of 72 needle bearings to clutch gear. Install spacer. Apply light coat of grease and assemble second row of 72 needle bearings.
2. With clutch teeth down toward flange assemble 1st speed gear to clutch gear and over needle bearings and spacer. Place gear and sleeve assembly to one side.
3. In a similar manner, coat 2nd speed gear sleeve on thrust face and ground surface with grease and apply two rows of 72 needle bearings with spacer between.
4. Assemble 2nd speed gear to sleeve and bearings with clutch teeth of gear away from flange. Place gear and sleeve assembly aside.
5. Place 4th speed gear sleeve on flat plate and coat lower 1" with light grease. Assemble one row of 62 needle bearings to sleeve, add spacer and in like manner, assemble second row of 62 needle bearings.
6. Assemble 4th speed gear to sleeve and bearings with clutch teeth of gear up. Place this assembly aside.

Mainshaft Final Assembly

1. Position 1st speed gear and sleeve in rear of case with clutch teeth toward front of case. Mesh with countershaft gear. Assemble 1–2 speed clutch collar to 1st speed gear with extended hub toward gear.
2. Place 2nd speed gear and sleeve assembly in case with clutching teeth of gear toward 1st speed gear and mesh with countershaft gear. Move clutch collar to neutral to hold assemblies in relationship.
3. Coat thrust faces and splines of mainshaft with light grease. Enter pilot bearing end through rear bearing bore until mainshaft has passed through 2nd gear sleeve about 1".
4. Place 4th speed gear and sleeve in case with clutch teeth forward or away from 2nd speed gear. Slide mainshaft forward through bore of sleeve.
5. Assemble 3–4 speed clutch collar to clutch gear. Extended hub of collar assembles toward front as does lock ring counterbore in end of gear. Position gear and collar in case and assemble to mainshaft spline.
6. Block 3–4 clutch gear and tap mainshaft through until locking groove in mainshaft is beyonid clutch gear.
7. Coat the two clutch gear lock rings with heavy grease and place in groove provided in mainshaft.
8. Tap mainshaft rearward and make sure lock rings center inside of counterbore of 3–4 clutch gear.
9. Assemble 1st speed gear thrust washer on rear of mainshaft next to 1st speed gear.
10. Position mainshaft rear bearing onto mainshaft with external snap ring away from case. Use a 3/4" thick steel strip of flat stock to block front of mainshaft against inside of case. Use a pinch bar to roll mainshaft into position to align outer race of bearing with case bore. With tubing, drive on inner race until bearing is seated against 1st speed gear thrust washer.
11. Remove 3/4" blocking and continue to tap bearing into case bore until snap ring is seated against case.
12. Assemble rear bearing washer and speedometer gear or spacer to mainshaft and seat against rear bearing.
13. If necessary, replace speedometer gear bushing in rear bearing cap.
14. Apply a light coat of gasket cement to O.D. of rear bearing cap oil seal and press into cap.
15. Apply gasket cement to gasket and fasten onto bearing cap. Align oil passages.
16. Apply cement to other side of gasket and assemble bearing cap and gasket to rear

of case. Dip capscrews in gasket cement and install with copper washers. Torque capscrews to 60–80 ft-lbs and secure with lock wire.
17. Assemble end yoke or flange to mainshaft with pusher tool. Do not drive on either piece without provision to stop shaft from moving forward.
18. Assemble flat washer and lock nut. Lock transmission in two gears and torque lock nut to 400–450 ft-lbs.
19. Torque countershaft front lock nut to 550–600 ft-lbs and stake nut in slot in shaft.
20. Shift clutch collars back into neutral and make sure all shafts turn free.
21. Use pressure type oil can to force lubricant down oil holes and end slots of all floating gears on mainshaft to flush out grease and to insure initial lubrication of overrunning gears and bearings.
22. Drive gear pocket bearing is a separable part. Due to the vertical method of reassembling the rear unit to the front unit, the inner race may be pressed on the front end of the rear mainshaft or, with the outer race, be placed in the drive gear pocket. Note that flanged end of inner race assembles toward 3rd speed clutch gear. If necessary, use tubing to drive against inner race until it seats against spline shoulder.
23. Assemble rear unit to front unit in reverse order of disassembly and install shifter housing.

Spicer 9441 Four-Speed Auxiliary Unit

This transmission, Fig. 1, is designed to utilize splash and pump feed lubrication for all internal bearings, bushings, shafts and gears.

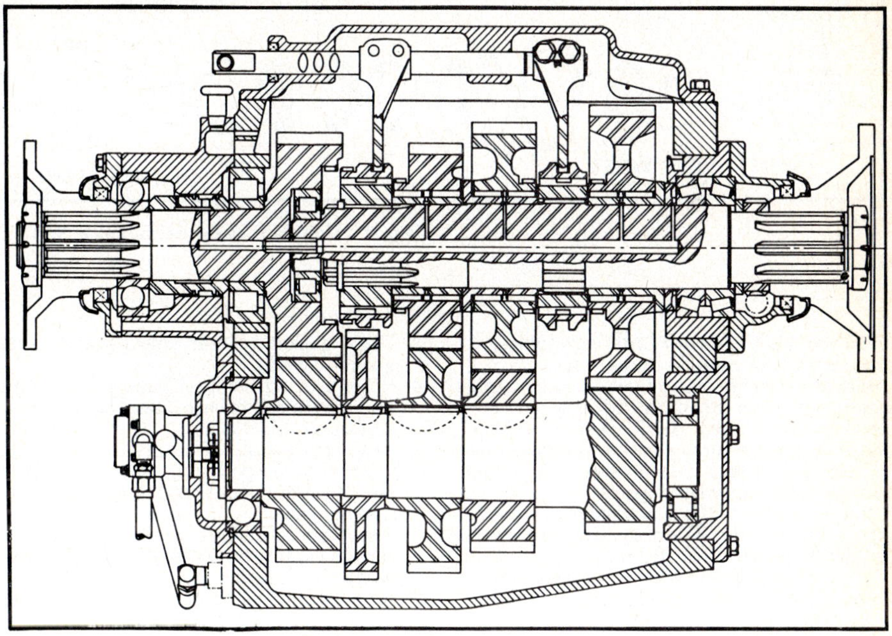

Fig. 1 Sectional view of Spicer Model 9441 four-speed auxiliary transmission

Mainshaft, Disassemble, Fig. 2

1. Remove shifter housing.
2. Lock transmission in two gears.
3. Remove drive gear and mainshaft companion flange (or end yoke) nuts.
4. Use a puller to remove drive gear and mainshaft companion flange or end yokes.
5. Remove front transmission hanger (if used) from drive gear rear bearing cap.
6. Disconnect oil line from front bearing fitting. Disconnect oil lines from fittings on front pump. If it is not necessary to remove the oil lines connected to the case bottom oil screen and filter, leave these oil lines intact.
7. Disconnect oil line from mainshaft rear bearing cap and countershaft rear bearing cap fitting.
8. Separate drive gear front bearing cap from drive gear rear bearing cap.
9. Unfasten drive gear rear bearing cap. Use a soft hammer and tap on trunnion bearing face of drive gear bearing cap until it is separated from case far enough to use two pry bars between case and cap.
10. Pry evenly with both bars until bearing cap and front drive gear bearing are free of drive gear. Remove bearing spacer from drive gear.
11. Tap front bearing free of bearing cap.
12. Unfasten and separate mainshaft rear bearing cap from case. Check and remove speedometer bushing if it is to be replaced.
13. Remove speedometer drive gear (or spacer) and bearing thrust washer from rear of mainshaft, using puller tool.
14. Remove bearing retainer with tapered roller bearing, cups and gasket from case.
15. Use two pry bars to slide mainshaft and gear assembly toward rear of case as far as possible.

NOTE: Some of the needle bearings used with the Lo-Lo gear may fall as the mainshaft moves rearward. Use care not to lose these needles if they are to be used again.

16. Place a 1/2" thick aluminum or brass block between main drive gear and mainshaft direct and overdrive clutch gear.
17. Tap forward on rear of mainshaft until lock rings are exposed from direct and overdrive clutch gear.
18. Remove 1/2" block and remove lock rings.
19. Pull mainshaft out through rear bearing bore, lifting loose parts out through top of case as mainshaft is being withdrawn.

NOTE: If transmission is to be rebuilt using old drive gear and mainshaft, be sure to clean out the shaft center oil reservoir orifices.

20. Before attempting to remove drive gear, countershaft must be removed from front and rear bores by sliding assembly to the rear. This allows drive gear to pass by front face of countershaft power take-off gear.
21. Slide drive gear and inner race off its bearing rearward and lift out through top of case. Remove inner race of bearing. If necessary, remove oil tube from pocket of drive gear.
22. Tap outer race of drive gear roller bearing out of front bearing bore of case. Remove pocket bearing from drive gear.

Countershaft, Disassemble, Fig. 3

1. Remove front bearing cap.
2. Remove rear bearing cap and bearing.
3. Push countershaft rearward and out of front bearing.
4. Lift front end of countershaft up and out of case.
5. Support P.T.O. gear with parallel bars under and as close to hub as possible and press countershaft free from drive gear and P.T.O. gear.
6. Support 2nd speed gear and press countershaft free from 4th and 2nd speed gears. Remove exposed Woodruff keys.

Countershaft, Reasssemble, Fig. 3

CAUTION: All gears should fit tight on countershaft. As a shrink (or interference) fit of

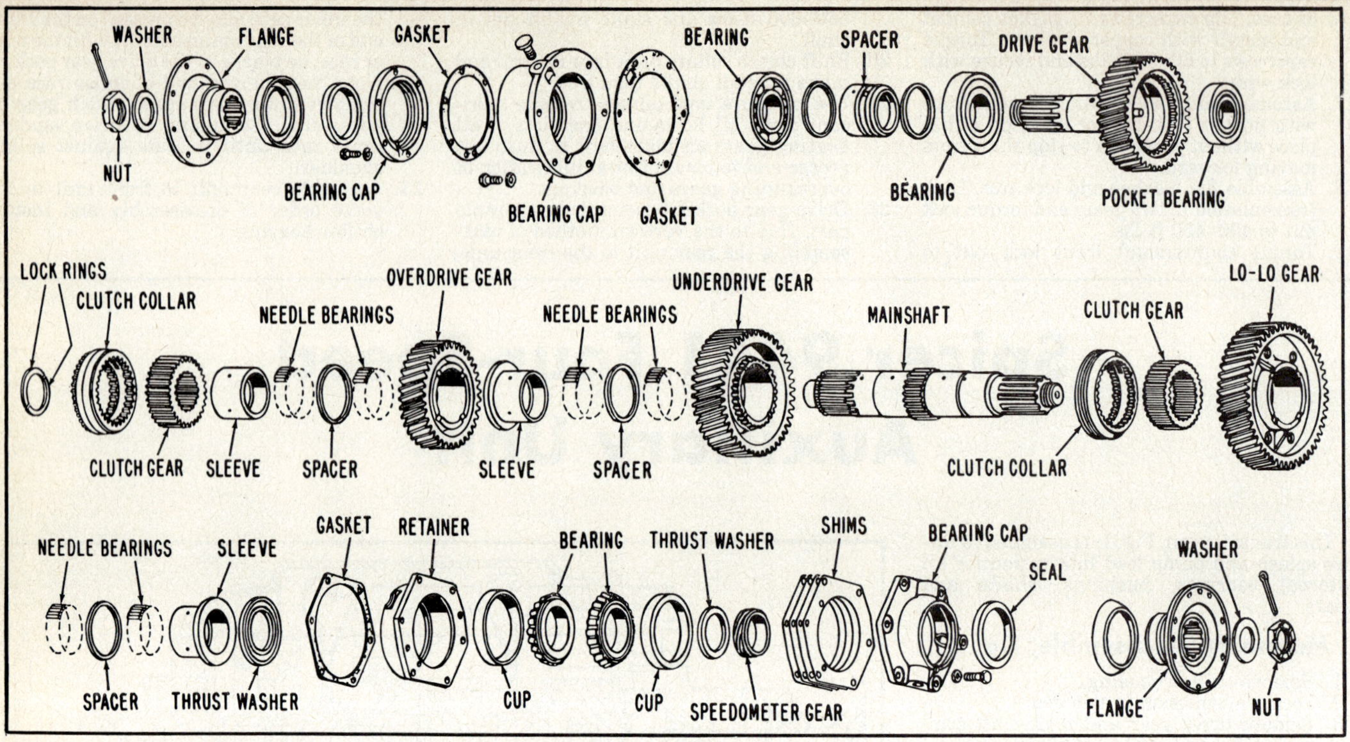

Fig. 2 Mainshaft components. Spicer 9441

.0015" to .003" is built into new parts, it presents a field assembly problem.

If heat is used to expand gear bores, boiling water, hot oil or steam are usually satisfactory. Do not exceed 250°F. Do not use hot plates, acetylene torch or other methods that will turn the steel blue or straw color and damage the heat treated gears.

If heat is not used, it is advisable to coat the gear bores heavily with white lead to prevent galling or seizing of parts.

When in doubt as to which end of the hub to assemble on the shaft first, look for the chamfered end in the bore.

---

1. Assemble all Woodruff keys in countershaft keyways.
2. Support 2nd speed gear on arbor press with chamfer up. Set countershaft into position. Align keys with keyway and press shaft into gear. Seat shoulder firmly against low gear.
3. Support 4th speed gear with long hub down and chamfer up. Align keys with keyway and press shaft into gear firmly against shoulder of 2nd gear.
4. Support P.T.O. gear with long hub up and chamfer up. Align keys with keyway and press shaft into gear firmly against shoulder of 4th speed gear.
5. Support drive gear with long hub down and chamfer up. Align key with keyway and press shaft into gear firmly against shoulder of P.T.O. gear.

Countershaft, Install

1. Assemble bearing to rear of countershaft.
2. Using a rope or wire sling placed in center of countershaft assembly, use hoist to lower assembly in case. Lower rear end of shaft into case and slide small end of shaft out rear bearing bore.
3. Place gasket to rear bearing cap using gasket cement.
4. Slide bearing cap and gasket onto shaft and secure cap to case. Hand tighten capscrews at this time.
5. With front end of countershaft in its approximate position, assemble front bearing onto shaft and seat snap ring against case.
6. Assemble lock plate to end of shaft and torque capscrews to 35–40 ft-lbs. Secure screws with lock wire.
7. Be sure bearing cap is in proper position on rear face of case. Torque capscrews to 35–40 ft-lbs.

CAUTION: Make certain pump coupling is on pump drive shaft, as pump is driven by coupling through slot in countershaft end.

---

8. Install front bearing cap with pump and gasket to case. Line up oil passages in gasket and bearing cap with main case. Torque capscrews to 35–40 ft-lbs.

Main Drive Gear, Assemble

1. Press inner race of bearing with flanged end toward teeth of drive gear. Seat flange of inner race securely against shoulder.
2. Lower front end of drive gear, with inner race of bearing, through top opening and slide it forward through bore of case. Tap outer race of bearing through front of case, using a soft hammer. Position drive gear in outer race of bearing.
3. Install spacer with sealing rings on drive gear shaft.
4. Install drive gear bearing cap with gasket over sealing rings with spacer on case. Line up oil passages in gasket and bearing cap. Secure with lockwashers and capscrews. Install breather cap if removed.
5. Press bearing into bearing cap.
6. Press oil seal in bore of front bearing cap.
7. Assemble front bearing cap and oil seal with gasket to bearing cap. Line up oil passage in gasket and bearing cap. Torque capscrews to 35–40 ft-lbs.
8. Install outer race of bearing into bore of drive gear. Install oil tube into bore of drive gear. This tube must be concentric with bore within .002" total indicator reading.

Mainshaft, Assemble & Install

1. Position case, with gears assembled, in a vertical position with front end down. Lubricate all mainshaft bearing bores with light grease as gears are assembled. Do not plug oil holes with grease.
2. Install collar on clutch gear and set these into clutch teeth of main drive gear. Using heavy grease in 3rd speed (or overdrive) gear, assemble one row of 56 needle bearings in bore. Place spacer in bore and assemble second row of 56 needle bearings in bore of gear.
3. Slide sleeve into needle bearing bore of gear. Place gear assembly with clutch teeth down on top of clutch gear. Use care not to drop or lose any needle bearings.
4. Use heavy grease in bore of 2nd speed gear. Assemble one row of 56 needle bearings in bore. Place spacer in bore and assemble second row of 56 needle bearings.
5. Place a 1/2" thick collar or spacers between overdrive gear and front face of underdrive gear. Slide sleeve into needle bearing bore of gear with flanged end of sleeve away from clutch teeth. Place gear as-

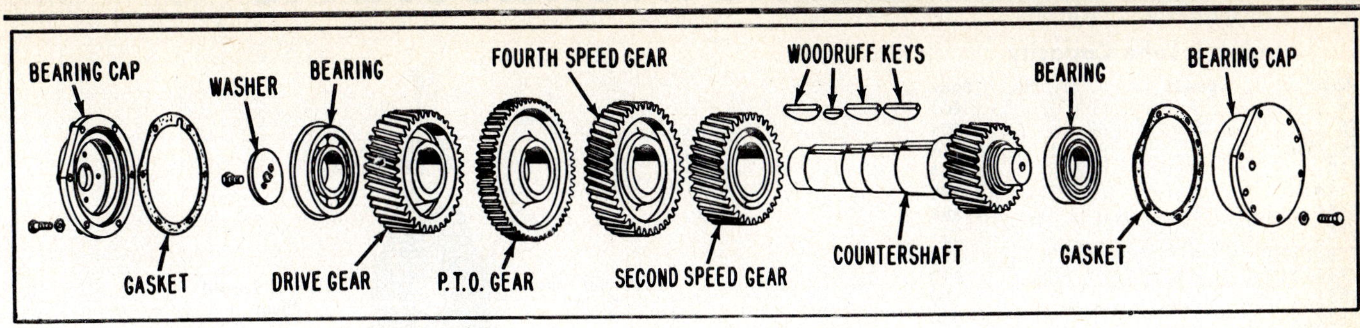

Fig. 3 Countershaft components. Spicer 9441

sembly with clutch teeth up on top of 3rd speed or overdrive gear.

6. Assemble clutch collar to clutch gear and place on top of 2nd speed gear.
7. Use heavy grease in bore of 1st speed gear. Assemble 56 needle bearings in one end of bore. Place spacer in bore and assemble 56 needle bearings on other end of bore.
8. Place 1st speed gear assembly with clutch teeth down on top of clutch gear.
9. Slide mainshaft through all gears. Allow inner race of bearing on front of shaft to seat in rollers of outer race of bearing in drive gear.

CAUTION: Before placing mainshaft into gears, notice drill point mark in back of input spline tooth for alignment teeth of clutch gears. Assemble inner race of bear ing on end of mainshaft to be lowered into drive gear pocket bearing outer race.

10. Assemble 1st speed sleeve, with flange up, into 1st speed gear needle bearing bore. Install thrust washer with flat side against flange face of sleeve.
11. Apply gasket cement to gasket and install on bearing retainer. Assemble bearing retainer to case. Note word "Top" for proper location and line up oil passages.
12. Install front tapered bearing cup in retainer. Use tubing to press bearing cone on mainshaft. Leave rear cone and cup out of bearing retainer at this time as they will be assembled during a later step.
13. Lower entire case assembly to horizontal position. Pull 1/2" collar or spacers from between overdrive gear and underdrive gear.
14. Use pry bars on front face of clutch gear and move gears to rear as far as possible. Tap rear end of mainshaft with a soft hammer until lock ring groove is fully exposed. Use heavy grease to coat faces of lock rings and install in shaft groove.
15. To secure lock rings in place, use a pry bar on rear face of 1st speed gear and force gears forward. This will secure lock rings under clutch gear.
16. Assemble 2nd tapered roller bearing cone on shaft, and 2nd tapered cup into retainer. Assemble spacer and speedometer drive gear (if used) on shaft.
17. If oil seal was removed from rear bearing cap, use gasket cement on O.D. of seal and press into bearing cap. Press in new speedometer bushing (if removed).
18. Assemble rear bearing cap and shim pack to case.

CAUTION: Line up of oil hole on shims with oil hole in retainer is required. Blocking of retainer oil hole by improper shim placement will deprive tapered bearings of lubrication.

19. Secure bearing cap and shim pack and torque capscrews to 35–40 ft-lbs.
20. Assemble end yoke or companion flange and flat washer to rear of shaft. Tighten nut firmly but do not torque at this time.
21. Check end play of mainshaft taper roller bearings. Fasten dial indicator to back face of case with dial indicator shaft on end face of mainshaft. Set dial indicator on "zero".
22. Using body weight and hands, rotate flange or yoke back and forth about 30° while pushing forward. Read indicator until it ceases to change. Leaving indicator as it reads, pull back and twist flange or yoke until reading again ceases to change. Total end play should be .003 to .005". Add or remove shims as required to obtain the desired result.
23. After adjustment is made, remove nut, flange or yoke and rear bearing cap. Coat I.D. and O.D. of shim pack with gasket cement to obtain a good seal. Secure rear cap and torque capscrews to 35–40 ft-lbs.

Final Assembly

1. Lock transmission in two gears.
2. If proper tools are not available, block front or rear face of gears with 3/4" stock. Use tubing to install flange or yoke. Tighten drive gear and mainshaft flange or yoke nuts to 400–450 ft-lbs.
3. Shift clutch collars back into neutral and make sure all shafts turn free.
4. Use a pressure type oil can to force lubricant down the oil holes and end slots of all floating gears on mainshaft to flush out grease and insure initial lubrication of overrunning gears and bearings.

WARNER TRANSMISSIONS

Index & Lube Capacity

①—Capacities vary. Fill to bottom of filler hole.

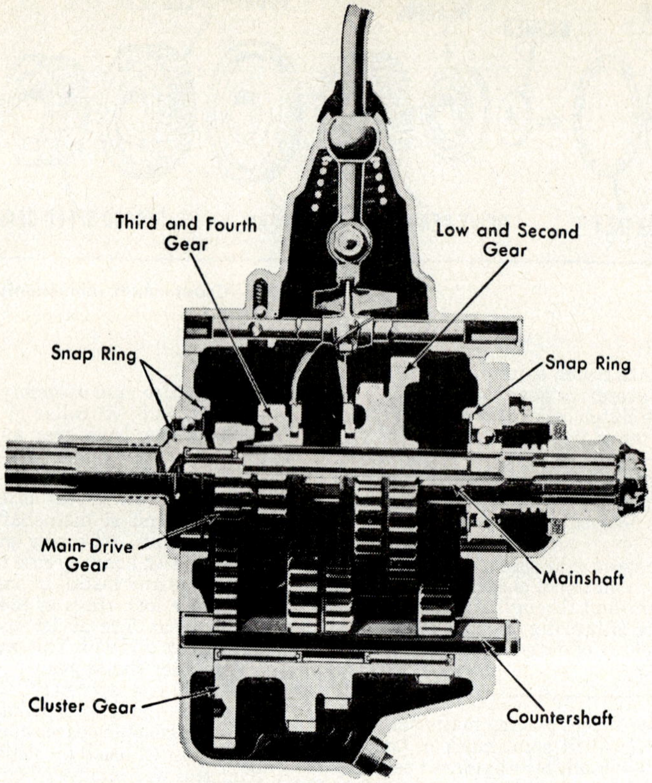

Fig. 1 Warner T9, T9A, T97 Series four speed transmission

T9, T9A & T97 SERIES

Disassemble, Fig. 1

1. Take off the transmission cover.
2. Unscrew the nut from the universal flange.
3. Remove the hand brake mechanism.
4. Remove universal flange.
5. Remove main drive gear bearing retainer and pull drive gear out through front.
6. Take off mainshaft rear bearing retainer oil seal and speedometer drive gear.
7. Before removing the mainshaft, chalkmark the relationship of the mainshaft splines with the mating splines in the sliding gears so that these parts may be assembled in the original position, as the gears are a selective fit on the mainshaft splines.
8. Grasp the mainshaft gears and pull the mainshaft and rear bearing out through the rear. Then lift out the loose gears.
9. Remove the counter shaft lock plate, push the shaft through the rear and lift out the cluster gear.
10. Drive out the reverse shift rail and lift out the fork.
11. Push the reverse idler shaft out through the rear and lift out the gear.

Reassembly

Reverse the order of the above procedure to assemble the unit. And when installing the mainshaft, be sure to match the marked gear splines with the mainshaft splines.

T85 SERIES

Three Speed, Fig. 2

1. To disassemble, remove gearshift housing.
2. Remove mainshaft nut and parking brake.
3. Remove U-joint flange, speedometer gear and spacer.
4. Remove mainshaft and rear bearing retainer rearward out of case as a unit.
5. Remove mainshaft pilot bearing.
6. Remove main drive gear bearing retainer. After removing large snap ring, tap drive gear and bearing back into case, removing it through rear of case.
7. Drive countershaft backward out of case and remove cluster gear assembly.
8. Remove reverse idler gear.
9. To disassemble mainshaft, remove snap ring from its front end. Mark relationship of synchronizer hub and sleeve and slide synchronizer and gears off shaft.

Assemble Transmission

1. Install reverse idler gear and shaft with chamfered teeth ends toward front.
2. Install cluster gear assembly in bottom of case. Tang on front washer fits into slot in case thrust face. Tab on rear steel washer must be up.
3. Install main drive gear bearing and snap ring from front until snap ring bottoms on case.
4. Install drive gear retainer without gasket and finger tighten retainer bolts.
5. Install oil slinger on drive gear shaft with grease and with dished side away from bearing.
6. Install drive gear shaft with oil slinger in case and shaft into bearing.
7. Remove drive gear retainer and install spacer washer and thickest snap ring that will fit. Snap rings are available in thicknesses of .086″ to .103″.
8. Tap drive gear shaft forward out of case as far as gear will allow. Then install mainshaft pilot in bore of drive gear, using grease to hold rollers in place.
9. Install synchronizer stop ring on rear end of drive gear shaft, coating inside of stop ring with grease.

Assemble Mainshaft

1. Press rear bearing on mainshaft. Install rear snap ring into rear bearing retainer housing and tap mainshaft with bearing into retainer. Install thickest snap ring that will fit in front of rear bearing. Snap rings are available in thicknesses of .088″ to .100″.
2. Assemble gears and synchronizer on mainshaft. Grease synchronizer stop ring before assembling, and make sure stop rings are free.
3. Install mainshaft, being sure that one of the pilot bearings is not forced off position and blocks oil hole in mainshaft.
4. Tap main drive gear to rear until snap ring seats on case.

Final Assembly

1. Turn transmission to help align cluster gear to install countershaft.
2. When countershaft is close to end of case, install key and drive it into case.
3. Turn rear bearing retainer into position and torque retainer bolts to 65–75 ft. lbs.
4. Install new gasket on main drive gear bearing retainer, install retainer and torque bolts 25–30 ft. lbs. Select gasket or gaskets that will seal lubricant and prevent end play between snap ring and retainer. Gaskets are available in thicknesses of .10″ to .025″.
5. Install spacer, speedometer drive gear and flange.

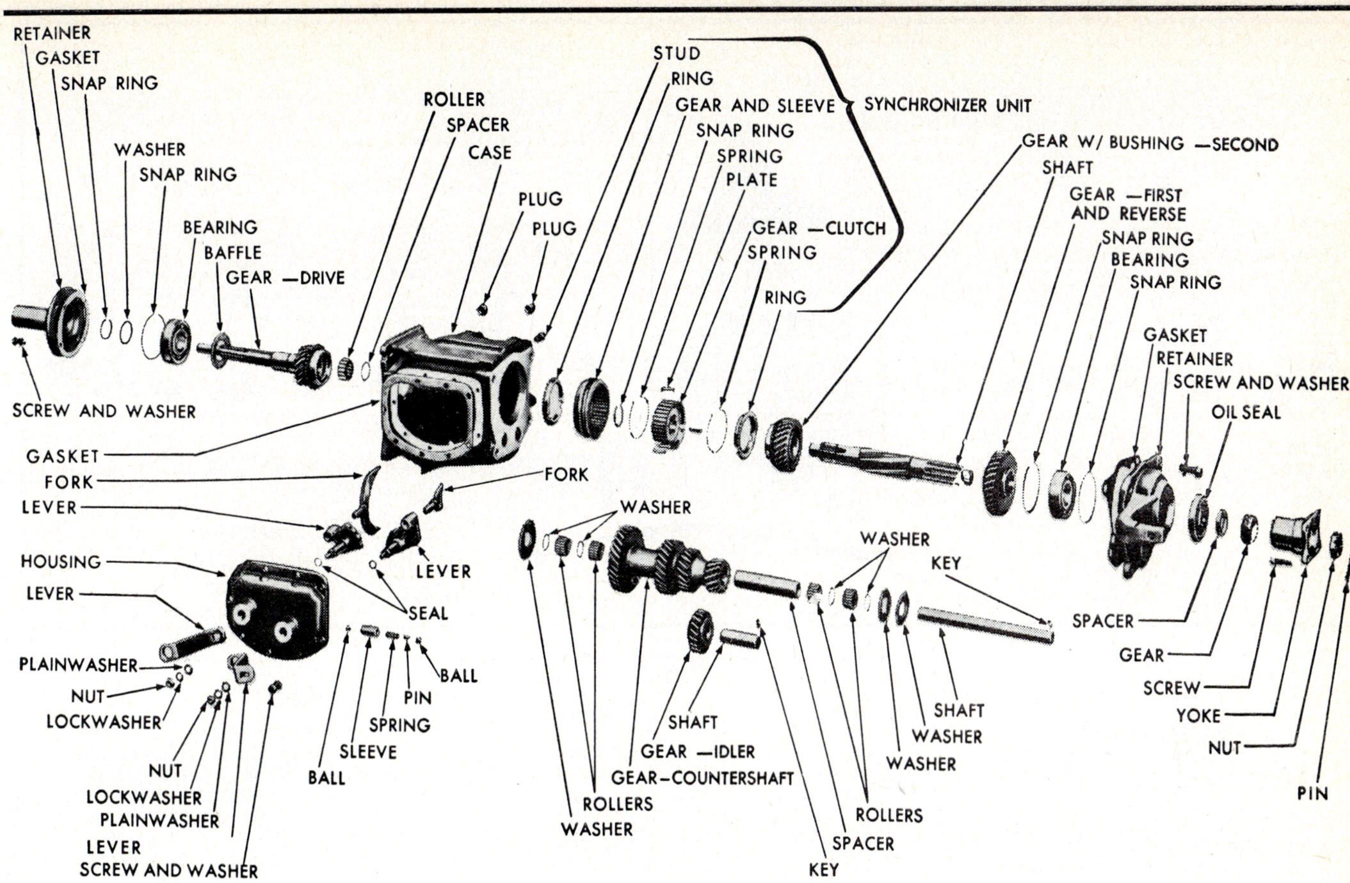

Fig. 2 Warner T85 Series three speed transmission

6. Install parking brake.
7. Shift transmission into neutral. Place 2nd-3rd fork on clutch gear sleeve and place gearshift housing with new gasket on case.

Gearshift Linkage

For any adjustment to bring shift lever to horizontal plane with lever in neutral position, adjust length of 2nd-3rd control rod at its swivel.

The crossover movement of lever should be made without binding or interference. If slots in lever are not aligned, the gearshift tube pin strikes corner of slot. To bring slots in line, adjust low-reverse rod.

T86, T90 SERIES

Three Speed, Fig. 2A

1. Remove shift housing or cover from top of case. On floor shift models, remove shift rails and forks from shift housing.
2. Remove front main drive gear bearing retainer and two socket head screws supporting oil collector inside case, if equipped.
3. Tap front end of countershaft to loosen lock plate, then remove lock plate.
4. With a brass drift, drive countershaft through rear of case and countershaft gear set will drop to bottom of case.
5. Remove mainshaft rear bearing adapter, then mainshaft and gears from case through rear bearing adapter opening.
6. If equipped with oil collector, drive main drive gear into case far enough to remove the collector.
7. Remove main drive gear and on column shift models, remove shift forks.
8. Remove countershaft gear set and thrust washers from bottom of case, then washers, needle bearings and spacer from gear set.
9. With a brass drift, drive reverse idler gear shaft into case, then remove reverse idler gear and shaft.

Assemble Transmission

Reverse above procedure to assemble transmission, observing the following:

1. When replacing either transmission case or transfer case, if used, ensure drilled oil holes on case mating surfaces match and correct gasket is used.
2. Place reverse idler gear and shaft into case, ensuring slot at shaft end is aligned properly to allow lock plate installation.
3. Install spacer, washers and countershaft needle bearings into countershaft gear hub. Place a washer and a set of roller bearings at each end of spacer. Add a set of roller bearings and a washer at each end, completing the assembly. Place countershaft gear set into case, however, do not install countershaft in final position until mainshaft and main drive gear have been installed.
4. Install low and reverse sliding gear on main shaft with shift shoe groove facing front of transmission.
5. To assemble synchronizer unit: Install the two springs in high and intermediate clutch hub with spring tension opposing each other. Install three synchronizer shifting plates into three slots in hub with smooth side of plates facing outward. Hold plates in position then slip second and direct speed clutch sleeve over hub with the long beveled edge toward long part of clutch hub. Install one blocking ring on each side of hub. Place assembly on mainshaft with beveled edge of clutch sleeve toward front end of shaft.
6. Install main drive gear bearing onto main drive gear shaft with tool KF-128A.
7. When installing main shaft, ensure main drive gear needle bearing rollers are correctly positioned.
8. Install front and rear bearing retainers on mainshaft and rotate transmission mainshaft, allowing gears to mesh and permitting alignment and installation of countershaft.
9. To install countershaft gear set: Install large bronze thrust washer at front of case with lip entered in slot of case. Install steel thrust washer at rear of case and partially insert the countershaft into case to hold washer in position. Align slot in countershaft with slot in reverse idler gear shaft, permitting installation of lock plate. Ensure thrust washers are positioned properly, then place bronze thrust washer against rear end of gear and place gear in its running position. Drive countershaft through countershaft gear set and case. When installed, the countershaft gear set should have .012 to .018 inch end play. End play is obtained by the selective thickness of rear steel thrust washers available in thicknesses of .0555 and .0625 inch.

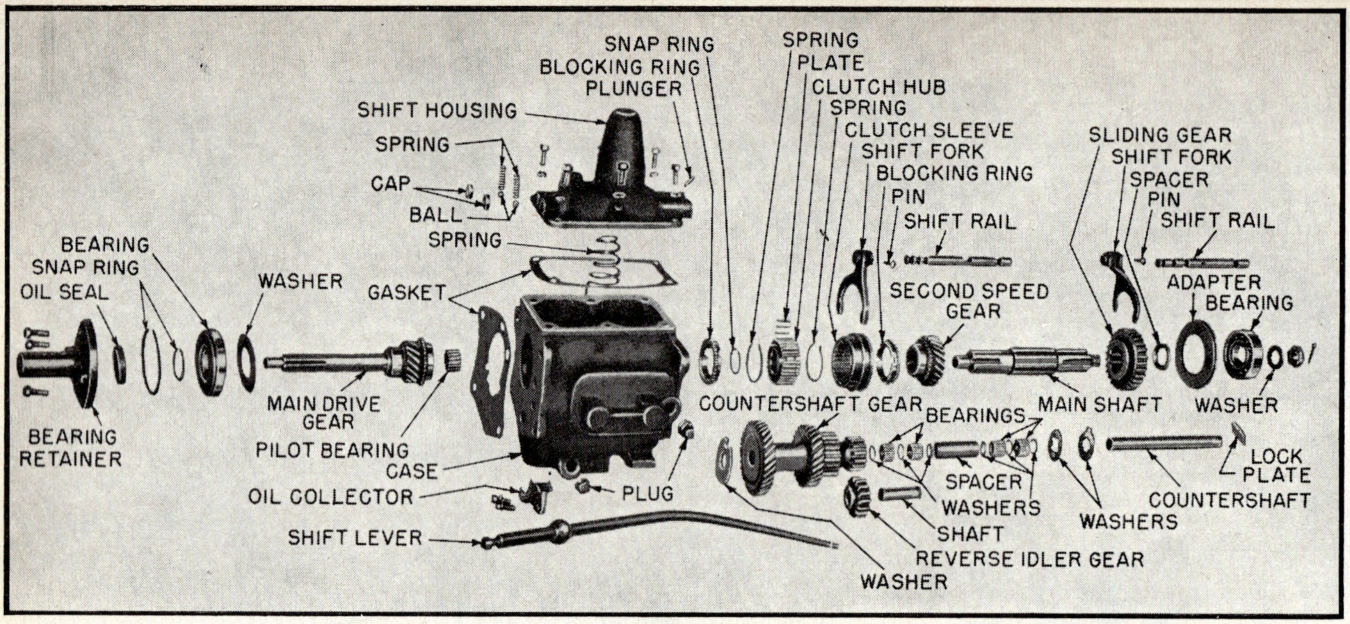

Fig. 2A Warner T86, T90 Series three speed transmission with floor shift

T87 SERIES

Three Speed, Figs. 3 and 4

1. Remove gearshift housing.
2. Remove parking brake.
3. Remove mainshaft bearing retainer.
4. Drive countershaft out through rear of case.
5. Use puller to remove mainshaft bearing.
6. Remove input shaft retainer and tap shaft and bearing as far forward as possible.
7. Remove mainshaft assembly.
8. Remove input shaft. If bearing is to be replaced, assemble with thickest snap ring that will fit.
9. Use puller to remove reverse idler shaft.
10. Remove reverse idler gear and countershaft gear cluster.

Assemble Transmission

1. Install reverse idler gear and shaft.
2. Install cluster gear assembly with front thrust washer tip hooked into groove on inside of case. Check end play (.006-.020″). Change thrust washers as required.
3. Install assembled mainshaft. Place a short 2 × 4 block between input shaft and mainshaft. Slide mainshaft bearing baffle and bearing onto shaft. Baffle must be installed so it does not rub bearing outer race. Drive bearing onto shaft.
4. Install mainshaft pilot rollers in input shaft, holding them with grease.
5. Place front blocking ring on synchronizer and start input shaft bearing in case.
6. Drive mainshaft bearing into case.
7. Tap input shaft and bearing into place. Install retainer without gasket and tighten bolts to bottom retainer on snap ring. With feeler gauge, check clearance between gasket surface and case. Select a gasket or gaskets that will seal oil and eliminate all clearance.
8. Raise cluster gear and insert countershaft. Then install retainer in notches in countershaft and reverse idler shaft. Drive both shafts toward front so that retainer is tight against gasket and torque bolts to 25–37 ft. lbs.
9. Install speedometer gear and spacer, mainshaft bearing retainer and gasket, and parking brake. Tighten output shaft nut to 90–125 ft. lbs.
10. Install gearshift housing.

T89 SERIES

Three Speed, Figs. 5 and 6

1. Remove gearshift housing.
2. Remove parking brake.
3. Tap main drive gear forward as far as possible.
4. Push countershaft out of rear of case.
5. Remove mainshaft assembly.
6. Remove main drive gear; then bearing.
7. Drive reverse idler gear shaft out of rear of case.
8. Lift out countershaft gear cluster.
9. Disassemble mainshaft.

Assemble Transmission

1. Place assembled countershaft gear cluster in case. Tang on front thrust washer fits into slot in case surface. Tab on rear steel washer must be up.
2. Install countershaft. Check end play (.006-.020″). Change thrust washers as required.
3. Install reverse idler gear shaft with chamfered teeth ends to front. Drive shaft in nearly flush with case and insert Woodruff key.
4. Tap input shaft bearing in case until snap ring bottoms in case. Install bearing retainer without gasket and barely tighten retainer capscrews. Lightly coat baffle with grease to hold it in place and install it on input shaft at rear of bearing with dished side away from bearing. Drive shaft into bearing with hard wood block.
5. Remove input shaft bearing retainer and install spacer washer and snap ring on input shaft. Install thickest snap ring that will fit. Then tap input shaft as far forward as possible and install pilot bearing rollers, holding them with grease.
6. Install mainshaft and bearing in bearing retainer casting, using the thickest snap ring that will fit.
7. Assemble mainshaft and secure with snap ring at front of shaft.
8. With new gasket, install mainshaft rear bearing retainer.
9. Tap input shaft bearing toward rear until outer snap ring bottoms on case.
10. Install countershaft gear cluster and insert Woodruff key in countershaft.
11. Torque mainshaft rear bearing retainer bolts to 65–75 ft. lbs.
12. Install and tighten input shaft bearing retainer to 25–30 ft. lbs., using a gasket or gaskets that will seal lubricant and prevent end play between snap ring and retainer.
13. Install parking brake.
14. Install shift housing cover with new gasket.

T96 SERIES

Three Speed, Fig. 6A

1. Remove transmission cover and shift levers.
2. Remove front retainer and gasket and front bearing snap rings.
3. Align notch in clutch shaft with 3rd speed gear and use suitable puller to remove clutch shaft using care not to lose rollers.
4. Remove front bearing with puller.
5. Remove extension case and remove snap rings that retain speedometer drive gear.
6. Remove speedometer gear using care not to lose drive ball.
7. Remove gear bearing snap rings and remove rear bearing with a puller.
8. Move mainshaft to side and remove shift forks.
9. Place front synchronizer in 2nd speed position and remove mainshaft by tilting front of shaft up and lifting through top of

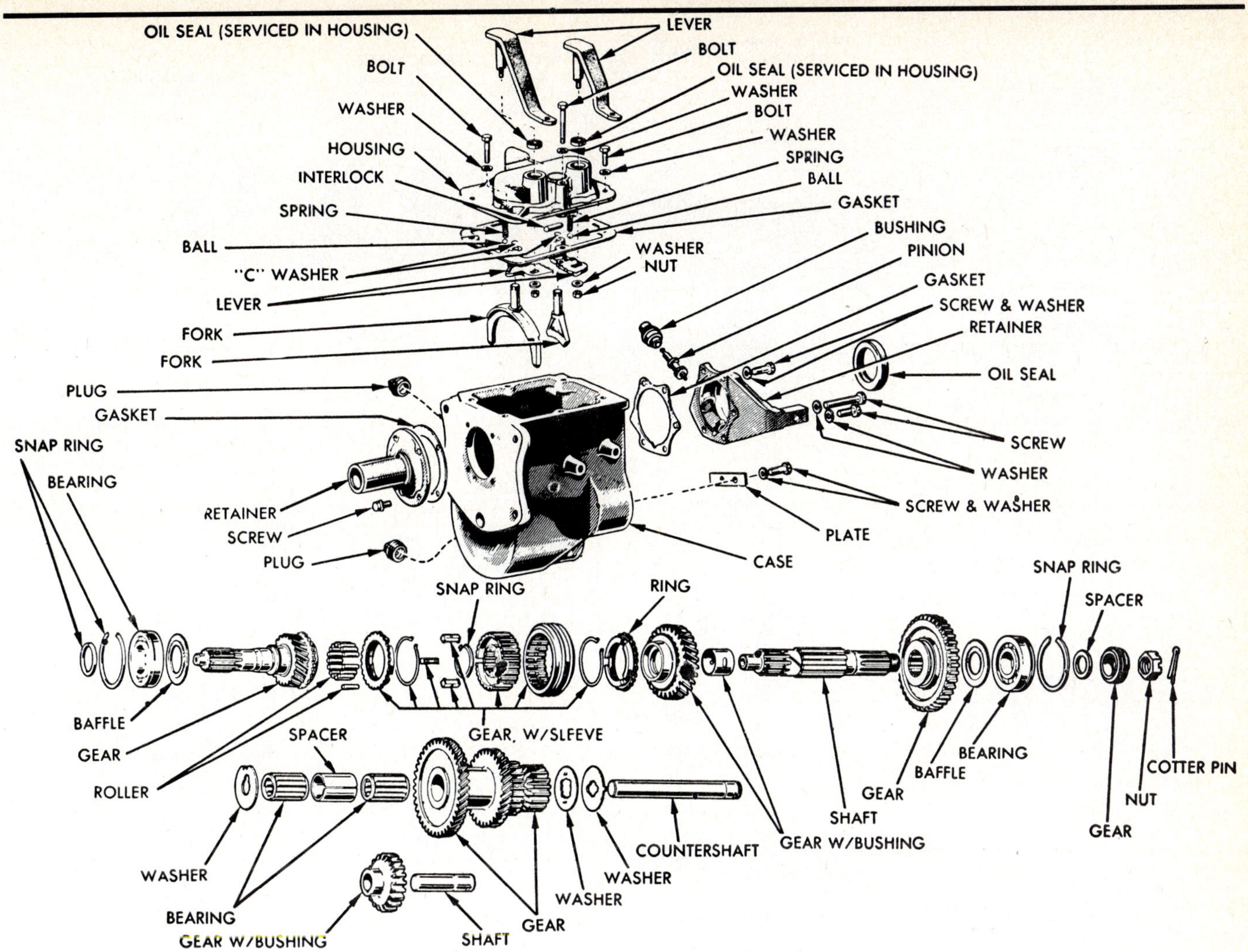

Fig. 3 Warner T87 Series three speed transmission

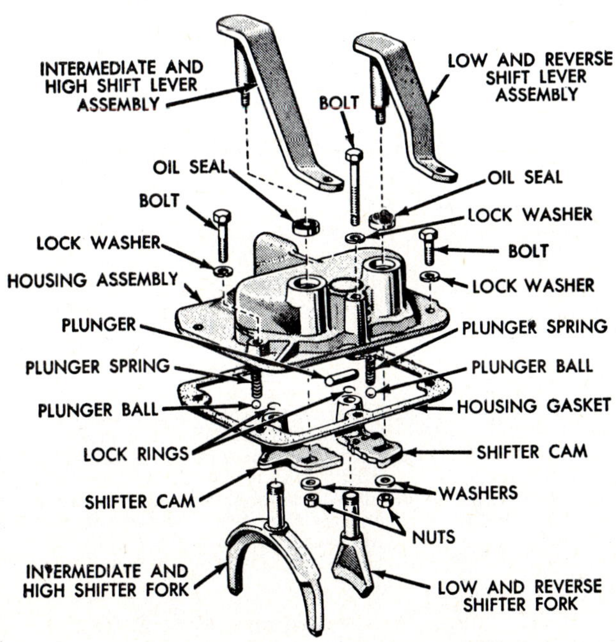

Fig. 4 Warner T87 Series gearshift housing

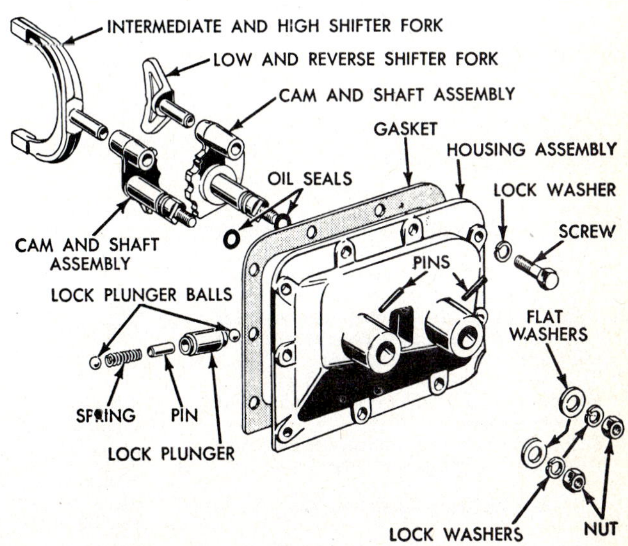

Fig. 6 Warner T89 Series gearshift housing

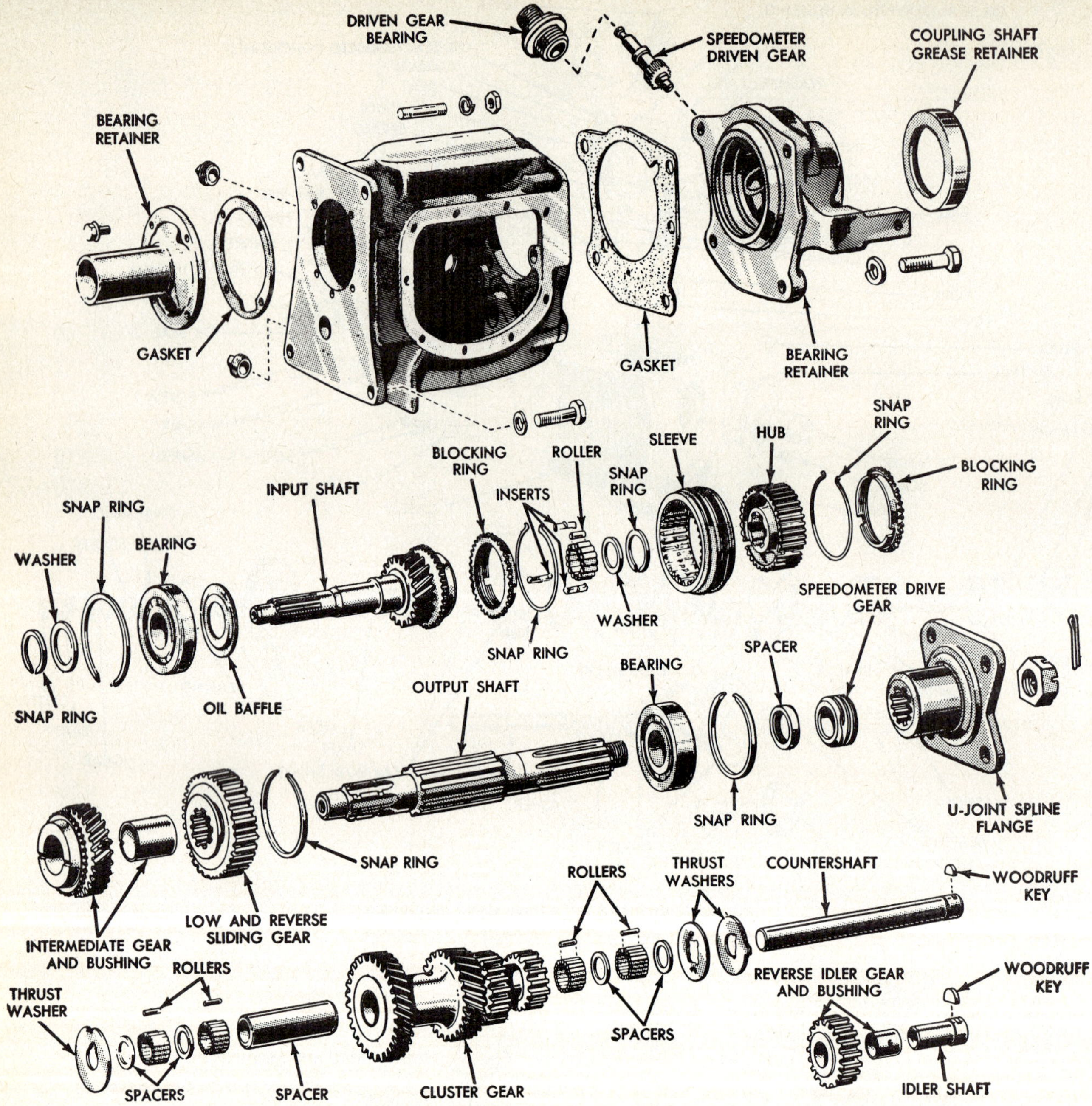

Fig. 5 Warner T89 Series three speed transmission

case.

10. Use a punch to remove roll pins from shift shafts and push shafts into case. Detent assembly may now be removed from case.
11. Using a brass drift drive reverse idler gear shaft out of gear of case and remove idler gear. Do not lose rollers.
12. To retain rollers in countershaft gear, use a dummy shaft to drive countershaft out of rear of case.
13. After disassembling mainshaft carefully inspect all bearings and gear.

Mainshaft Assemble

1. Place low speed and rear synchronizer on mainshaft and install rear synchronizer snap ring.

NOTE: Snap rings are available in select thicknesses. Clearance between first gear and collar on mainshaft must be .003–.012″.

---

2. Place 2nd speed gear and 2nd & 3rd synchronizer on shaft and install snap ring.

NOTE: Snap rings are available in select thicknesses. Clearance between 2nd speed gear and collar on mainshaft must be .003–.010″.

---

3. Slide reverse speed gear on mainshaft and set assembly aside to be installed later.

Countergear

1. Coat bore at each end of counter gear with grease to hold rollers in place.
2. Install dummy shaft in countergear and install spacer, washers and rollers.
3. Place countergear in transmission case and position thrust washers at each end so tabs align with slots in case.
4. Use a plastic mallet to install countershaft.

Reverse Idler Gear

1. Coat bore of idler gear with grease to

retain rollers.
2. Install idler gear in case and position thrust washers.
3. Use plastic mallet to install idler gear shaft.

Shifter Shafts

1. Partially install shifter shafts in transmission case.
2. Align detent assembly with shifter shafts and case stud.
3. Push shift detent assembly and shift shafts into place and install roll pins.

Mainshaft Installation

1. Place front synchronizer in 2nd speed position and place mainshaft in case.
2. Move mainshaft to side and install shift forks by pulling detent lever up and placing forks in the shifting assembly.
3. Position mainshaft assemble in center of case and place rear bearing on mainshaft, drive bearing into position and install snap ring.
4. Install speedometer gear and snap ring.

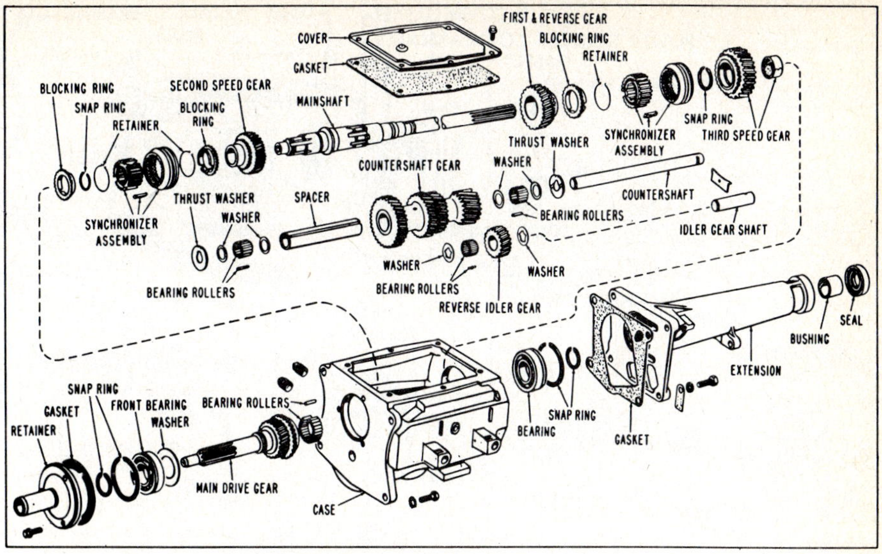

Fig. 6A Warner T96 Series three speed transmission

Final Assembly

1. Install rollers in clutch shaft using grease to retain them.
2. Slide clutch shaft into position through front of case.
3. Install front bearing, snap rings, gasket and retainer.
4. Replace seal if necessary and install extension housing, shift levers and case cover.
5. Fill transmission with lubricant and check operation.

T98 SERIES

Four Speed, Figs. 7 thru 8A

NOTE: Case assembly shown is used on Ford trucks. Jeep vehicles have power take-off located on left side of case.

1. Remove parking brake.
2. Remove mainshaft bearing retainer.
3. Remove countershaft-reverse idler shaft lock.
4. Remove gearshift housing.
5. Remove mainshaft bearing.
6. Remove input shaft bearing.
7. Drive out countershaft.
8. Remove input shaft through front end of case.
9. Pull out reverse idler shaft.
10. Remove reverse gear shifter arm from case.
11. Remove mainshaft assembly through top of case.
12. Remove power take-off cover and roll cluster gear toward top of case until small end of assembly can be pushed through mainshaft bearing bore in case.

Assemble Transmission

1. Position case with power take-off opening downward.
2. Hold thrust washer at front end of inner thrust plate at rear end of cluster gear to prevent rollers from falling out of gear. Position gear in bottom of case and install outer thrust plate.
3. Install reverse idler gear and shaft with slot in end of shaft down at rear face of case.
4. Install countershaft through cluster gear with slot in end of shaft facing downward. Cluster gear end play should be .006–.020″. Change thrust washers as required.
5. Install pilot rollers in input shaft. Place baffle on shaft so it does not rub bearing race. Position input shaft in bore of case, working from inside of case. Install high speed synchronizer blocking ring on input shaft gear.
6. Install mainshaft assembly, inserting forward end in input shaft. Use a stop yoke tool over synchronizer rings to prevent jamming rings.
7. Install input shaft bearing and install thickest snap ring that will fit.
8. Install input shaft and bearing in case and install snap rings on both bearings.
9. Install input shaft retainer without gasket. Tighten bolts enough to bottom retainer on snap ring. Measure clearance between retainer gasket surface and case and select a gasket or gaskets that will seal oil and eliminate clearance. Tighten bolts to 25–30 ft. lbs.
10. Slide second speed synchronizer to rear and install reverse gear shifter arm.
11. Install reverse idler shaft-countershaft retainer and gasket. Torque retainer bolts to 25–37 ft. lbs.
12. Install speedometer gear and spacer, mainshaft bearing retainer and gasket. Make sure longest retainer bolt is in proper hole. Torque bolts to 25–35 ft. lbs.
13. Install gearshift housing and torque bolts to 30–35 ft. lbs.
14. Install U-joint flange and parking brake drum. Torque mainshaft nut to 90–125 ft. lbs.

T16 SERIES

Disassemble, Fig. 9

1. Remove side cover and shift forks.
2. Unfasten attaching screws and rotate extension housing clockwise to expose reverse idler gear shaft.
3. Drive reverse idler shaft and its woodruff key out through rear of case, Fig. 10.
4. Rotate extension housing counter-clockwise to expose countershaft and use a brass drift at front of shaft to drive shaft and its woodruff key out rear of case, Fig. 11.
5. With countergear dropped to bottom of case remove entire mainshaft and extension assembly through rear of case, Fig. 12. Remove mainshaft pilot roller bearings from clutch gear.
6. Expand snap ring in extension and remove extension from rear bearing and mainshaft by tapping on end of mainshaft, Fig. 13.
7. Remove clutch gear bearing retainer, bearing snap ring and washer from mainshaft, Fig. 14.
8. Drive clutch gear through its bearing into case, Fig. 15, and remove bearing by tapping out from inside of case.
9. Remove countergear assembly from case.

Mainshaft Disassembly

1. Remove snap ring from front of mainshaft, Fig. 16.
2. Remove rear bearing snap ring, Fig. 17.
3. Support reverse gear with press plates and press on rear of mainshaft to remove reverse gear, rear bearing, special washer, snap ring and speedometer drive gear from rear of mainshaft.

CAUTION: When pressing rear bearing be sure special washer is clear of snap ring groove. Also, be careful to center gear, bearing, washer and snap ring on mainshaft before attempting to press off speedometer drive gear.

4. Remove low-reverse sliding clutch hub snap ring from mainshaft, Fig. 18, and remove clutch assembly, blocker ring and low gear from mainshaft.

Synchronizer Clutch Keys & Springs

NOTE: The clutch hubs and sliding sleeves are a selected assembly and should be kept together as originally assembled. However, the two keys and springs may be replaced if worn

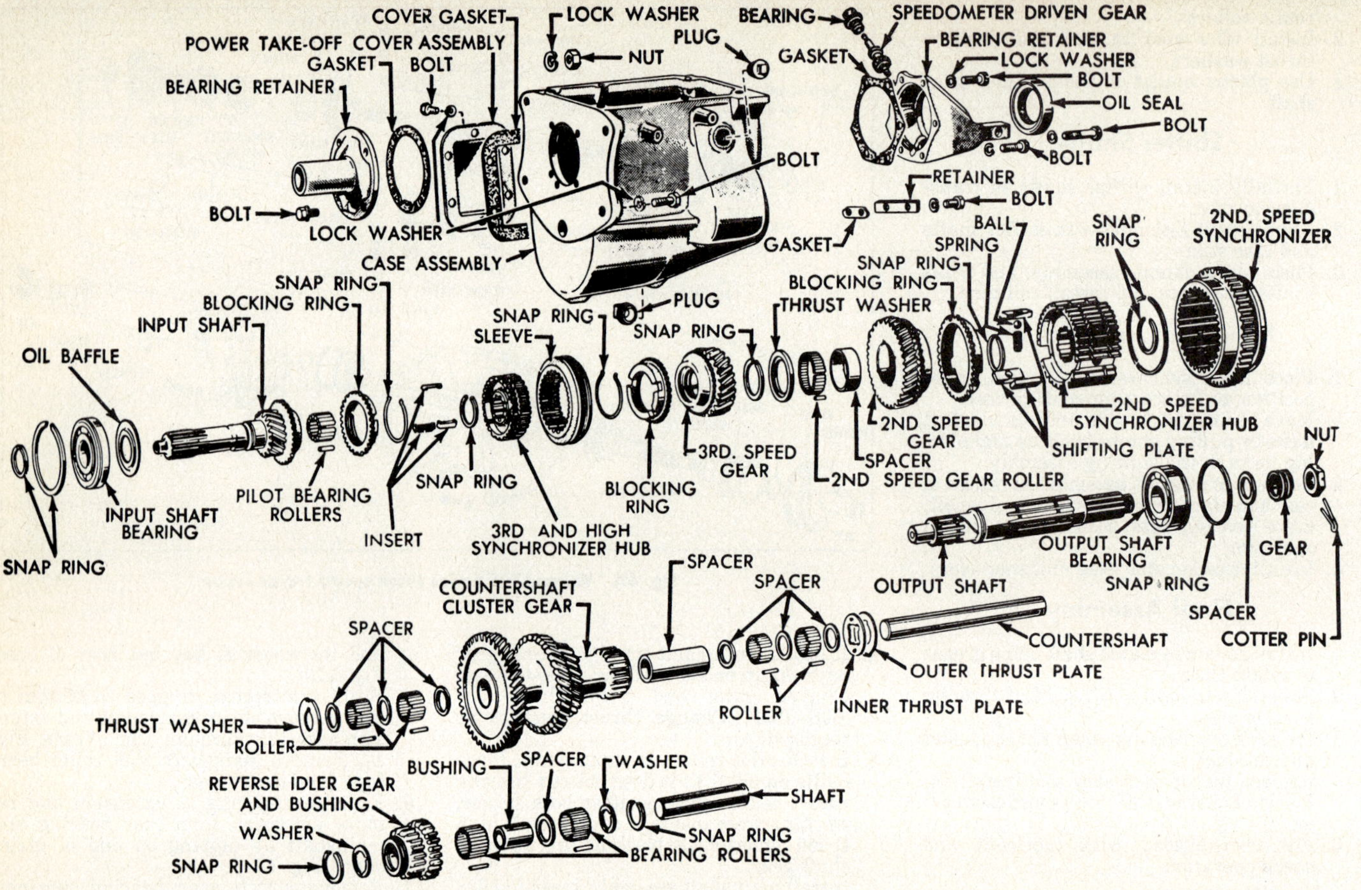

Fig. 7 Warner T98 Series four speed transmission

or broken.

1. If relation of hub and sleeve are not already marked, mark for assembly purposes.
2. Push hub from sliding sleeve. Keys will fall free and springs may easily be removed.
3. Place two springs in position (one on each side of hub) so all three keys are engaged by both springs, Fig. 19. Place keys in position and, holding them in place, slide sleeve onto hub, aligning marks made previously.

Mainshaft Reassemble

1. Referring to Fig. 20, install 1st gear on rear of mainshaft with gear clutching teeth to rear.
2. Install 1st gear blocking ring over 1st gear tapered cone end area with clutch key notches toward rear.
3. Install 1st-reverse sliding clutch assembly over rear of mainshaft, being careful to engage three keys with notches of blocker ring. Properly installed, straightest side of clutch hub and taper of sliding sleeve should both be toward rear of mainshaft (see Fig. 9).
4. Install snap ring in front of 1st-reverse clutch hub, Fig. 18.

NOTE: Snap ring is available in three thicknesses. Use thickest snap ring that will assemble with all parts stacked tight

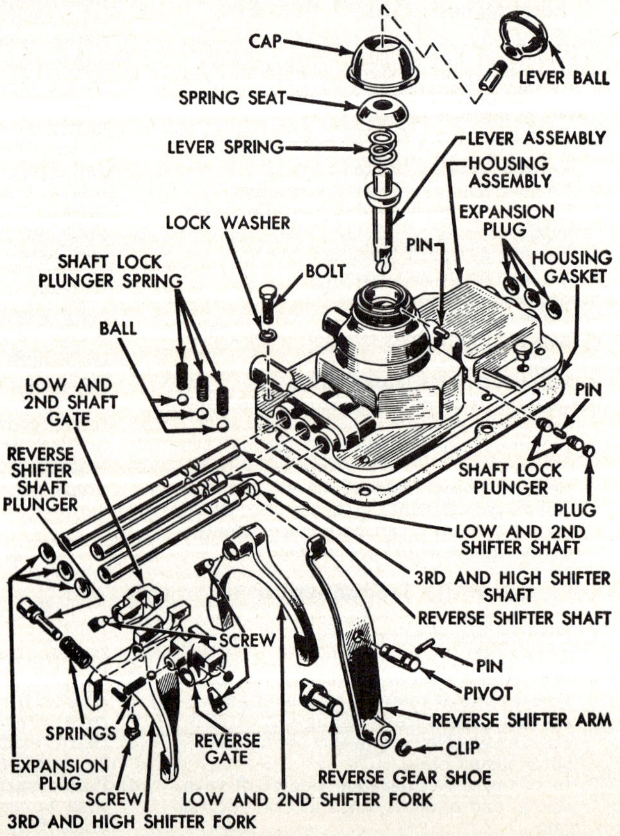

Fig. 8 Warner T-98 Series gear-shift housing

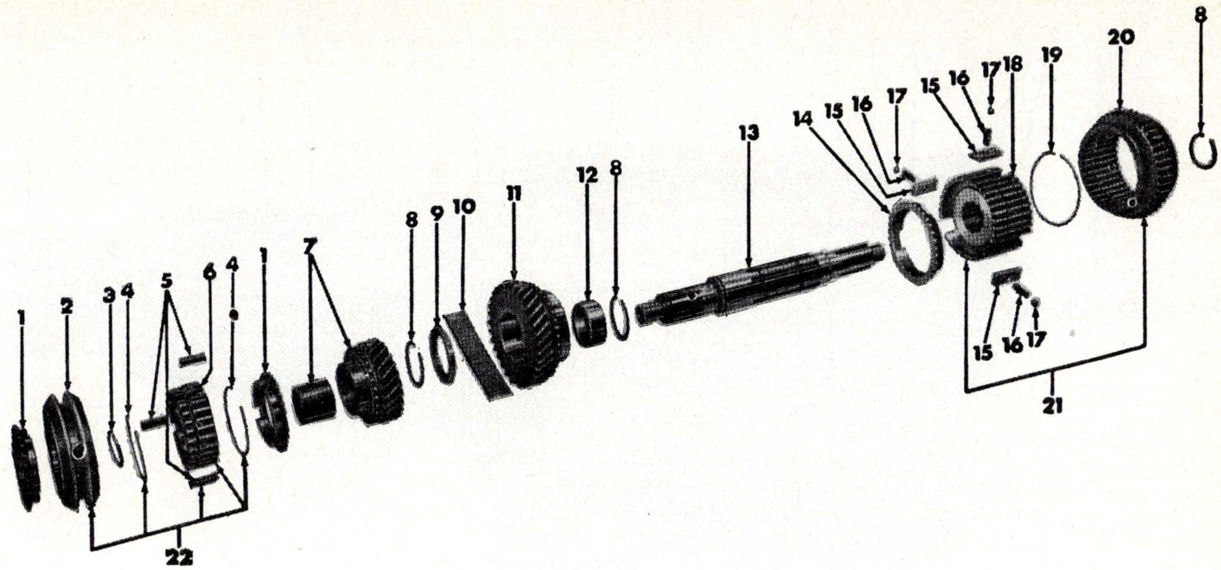

1—Blocking Ring
2—Direct-and-Third Clutch Sleeve
3—Snap Ring
4—Lock Ring
5—Shifting Plate
6—Direct-and-Third Clutch Hub
7—Third-Speed Gear Assembly
8—Snap Ring
9—Thrust Washer
10—Bearing Rollers
11—Second-Speed Gear
12—Spacer
13—Mainshaft
14—Blocking Ring
15—Shifting Plate
16—Poppet Spring
17—Ball
18—Low-and-Second Clutch Hub
19—Retaining Ring
20—Low-and-Second Speed Gear
21—Second-Speed Synchronizer Assembly
22—Direct-and-Third Synchronizer Assembly

Fig. 8A T98 Series mainshaft disassembled

endwise.

5. Install reverse gear over mainshaft with gear clutch teeth toward front.
6. Press rear bearing on mainshaft with its outer race snap ring groove closest to reverse gear.
7. Install rear bearing special washer and snap ring to mainshaft, Fig. 17.

NOTE: This snap ring is available in six thicknesses. Use thickest snap ring that will assemble with all parts stacked tight endwise.

8. Press speedometer drive gear on rear of mainshaft until centered on shaft boss.
9. Install 2nd speed gear over front of mainshaft with gear clutch teeth toward front.
10. Install blocker ring over 2nd gear tapered cone end with clutch key notches toward front.
11. Install 2-3 sliding clutch assembly over front of mainshaft, engaging clutch keys with notches of 2nd gear blocker ring. Properly installed, the straightest side of clutch hub should be toward rear and clutch sliding sleeve taper should be toward front of mainshaft (see Fig. 9).
12. Install 2-3 clutch hub snap ring to mainshaft, Fig. 16.

NOTE: This snap ring is available in four thicknesses. Use thickest snap ring that will assemble with all parts stacked tight endwise.

Assemble Transmission

1. Insert tube spacer and a double row of roller bearings and bearing retainer washers at each end of countergear, using grease to hold them in place, Fig. 22.
2. Place countergear assembly through case rear opening along with a tanged thrust washer (tang away from gear) at each end (large washer at front). Install countershaft and woodruff key from rear of case, Fig. 23.

CAUTION: Be sure countershaft picks up both thrust washers and that washer tangs are aligned with their notches in case.

3. Attach a dial indicator as shown in Fig. 24 and check end play of countergear. If end play is greater than .025" new thrust washers must be installed.
4. Use grease to hold 25 reverse idler gear roller bearings in position and place gear and bearings along with a thrust washer on each end into position inside case so that bevelled edge of gear teeth face toward front of case.
5. Load pilot bearing rollers and spacer into clutch gear with grease to hold them in place and position gear in case. *Do not install clutch gear bearing at this time.*
6. Stand case on end with clutch gear shaft through hole in bench or stand, and place 3rd gear blocker ring over clutch gear.
7. Install mainshaft assembly through rear of case, picking up spacer, pilot bearings and 3rd gear blocker rings, Fig. 25.
8. Install reverse idler gear shaft and woodruff key. *Be sure shaft picks up both thrust washers during installation.*
9. Install extension gasket on rear of case and, using snap ring pliers, expand snap ring, Fig. 13, and install extension over mainshaft and rear bearing. Be sure snap ring has started over bearing, then install and tighten extension case bolts. *Use graphite sealer on the two lower bolts.*
10. Tap on front of clutch gear shaft to force rear bearing snap ring to seat in its groove.

NOTE: This snap ring is available in five thicknesses. Use thickest snap ring that will assemble with all parts stacked tight endwise.

11. Install snap ring in outer race of clutch gear bearing. If snap ring groove is partially inside case opening, tap on inside bearing outer race with a long drift used through side cover opening.

NOTE: If mainshaft does not turn freely, check clutch sliding sleeves for neutral positions and that blocker rings are free on their gear cone surfaces.

12. Install gaskets and clutch gear bearing retainer and lip seal with oil drain passages at bottom and tighten. *Use graphite sealer on threads of retainer bolts.*

NOTE: Install two retainer-to-case gaskets (one .010" and .015") to replace the one .025" production gasket removed.

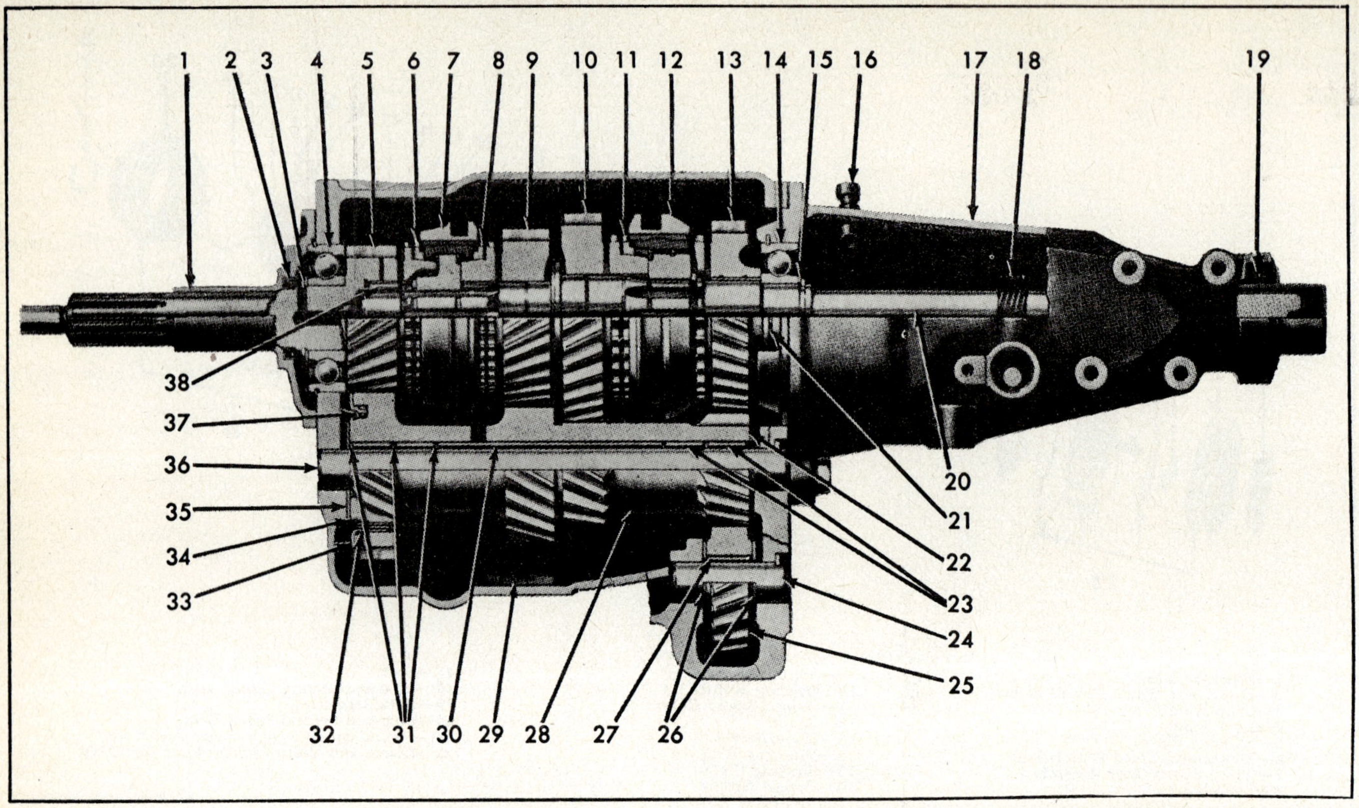

1. Bearing Retainer
2. Lip Oil Seal
3. Snap Ring and Special Washer
4. Clutch Gear Bearing
5. Clutch Gear
6. 3rd Speed Blocker Ring
7. 2-3 Sliding Clutch Sleeve
8. 2nd Speed Blocker Ring
9. Second Speed Gear
10. First Speed Gear
11. 1st Speed Blocker Ring
12. 1st and Reverse Sliding Clutch Sleeve
13. Reverse Gear
14. Mainshaft Rear Bearing
15. Snap Ring and Special Washer
16. Vent
17. Extension
18. Speedo Drive Gear
19. Extension Oil Seal
20. Mainshaft
21. Extension to Bearing Retainer Ring
22. Countergear Thrust Washer
23. Roller Bearings
24. Reverse Idler Shaft
25. Reverse Idler Gear
26. Thrust Washer
27. Roller Bearings
28. Countergear
29. Magnet
30. Tube Spacer
31. Washers
32. Roll Pin
33. Dampener Plate
34. Spiral-Lox Retainer
35. Countergear Thrust Washer
36. Countergear Shaft
37. Dampener Spring
38. Mainshaft Pilot Bearings.

Fig. 9 Warner T16 Series three speed fully synchronized transmission

13. Install shift forks to clutch sleeve grooves with 1st-reverse fork "hump" toward bottom of case, Fig. 26.
14. Install side cover and gasket. *The two rear side cover-to-case bolts have special oil sealing splines and must be used at these two "through" hole locations.*

T10 SERIES

Disassemble, Fig. 27

1. Shift transmission into 2nd gear and remove side cover.
2. Remove front bearing (clutch gear bearing) retainer and gasket.
3. Remove backup lamp switch and plunger from transmission. Drive lock pin from reverse shifter lever boss, Fig. 28 and pull shifter partially out of case to disengage the reverse shifter fork from reverse gear.

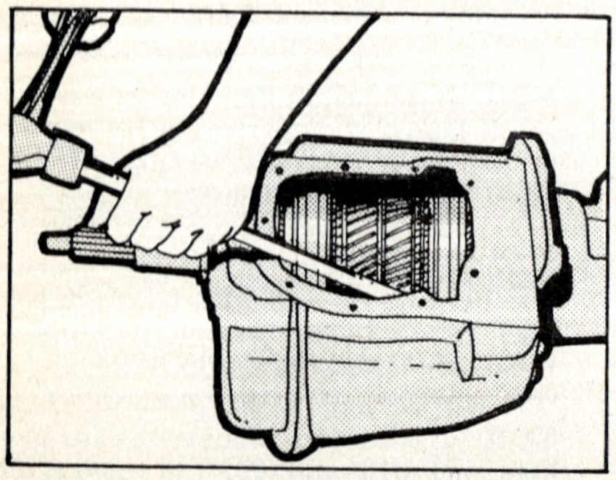

Fig. 10 Removing reverse idler shaft

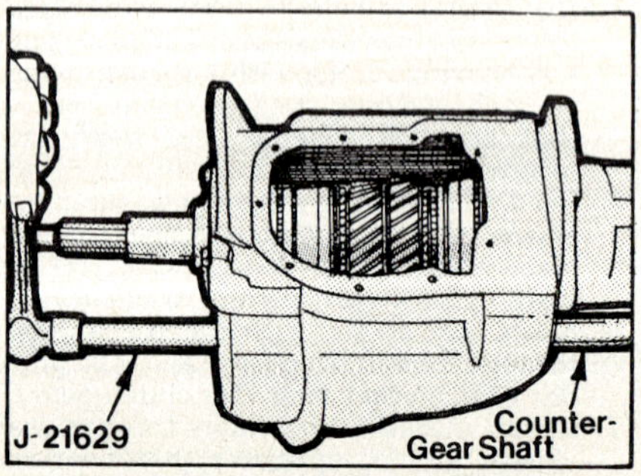

Fig. 11 Removing countershaft

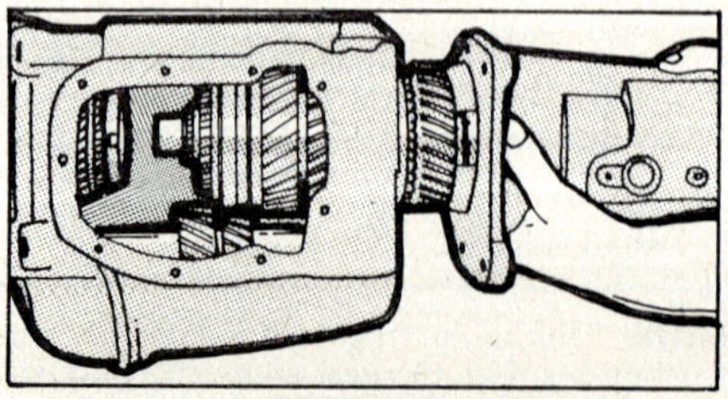

Fig. 12 Removing mainshaft and extension from case

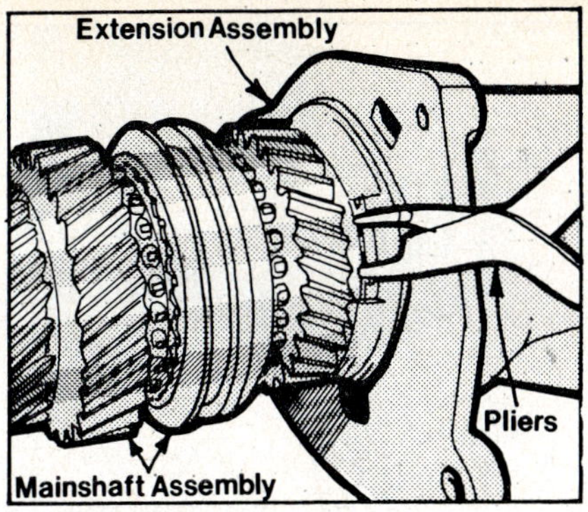

Fig. 13 Extension-to-rear bearing snap ring

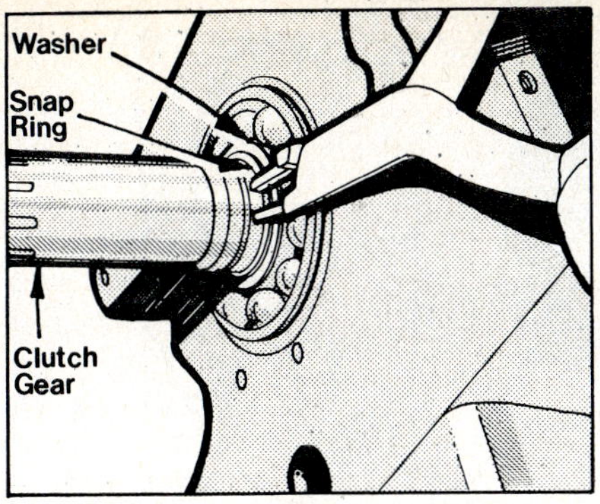

Fig. 14 Clutch gear bearing-to-gear snap ring

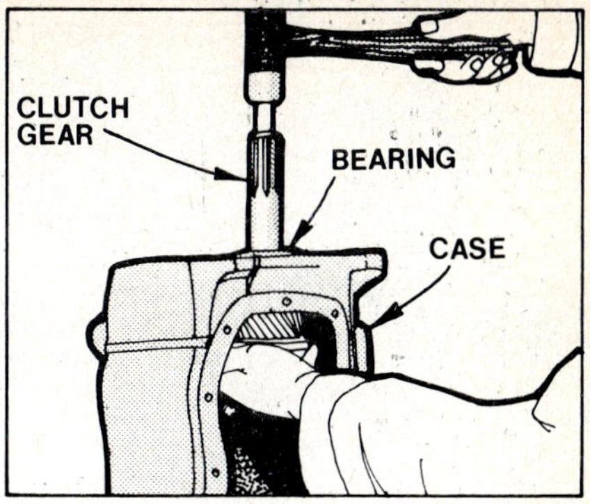

Fig. 15 Removing clutch gear from bearing and case

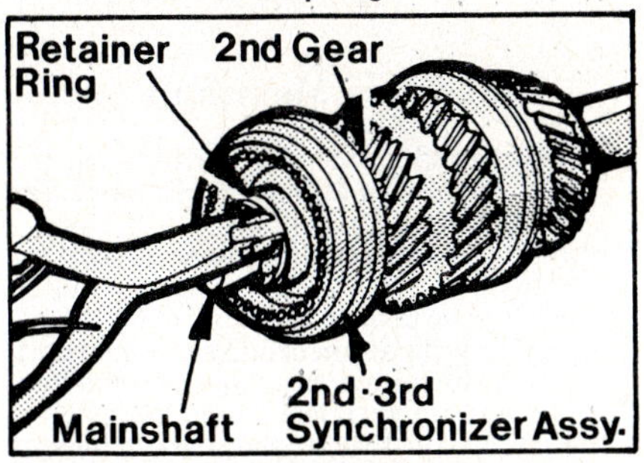

Fig. 16 2nd-3rd clutch hub snap ring

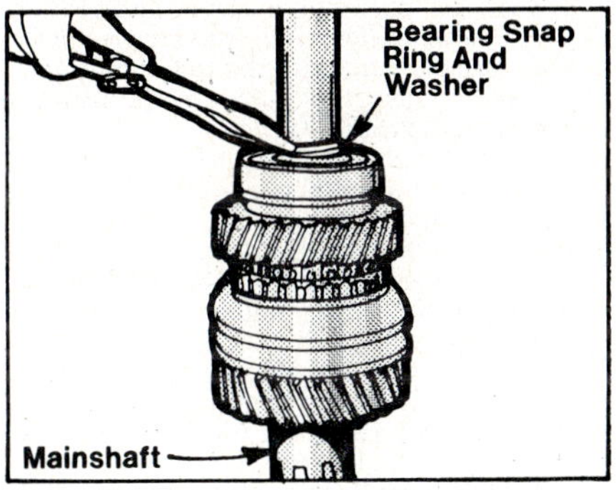

Fig. 17 Rear bearing-to-mainshaft snap ring

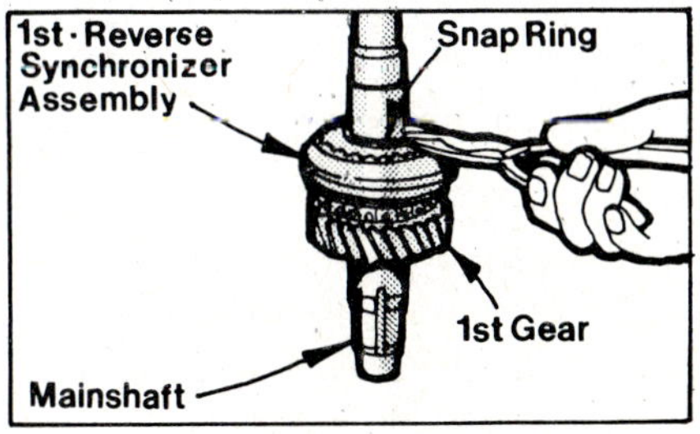

Fig. 18 First-reverse clutch hub snap ring

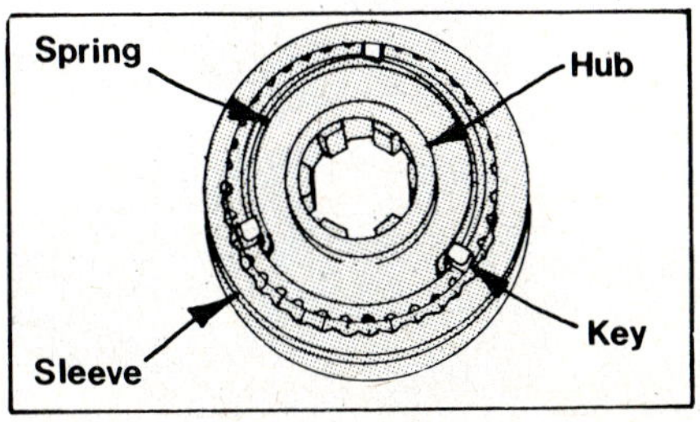

Fig. 19 2nd-3rd synchronizer clutch

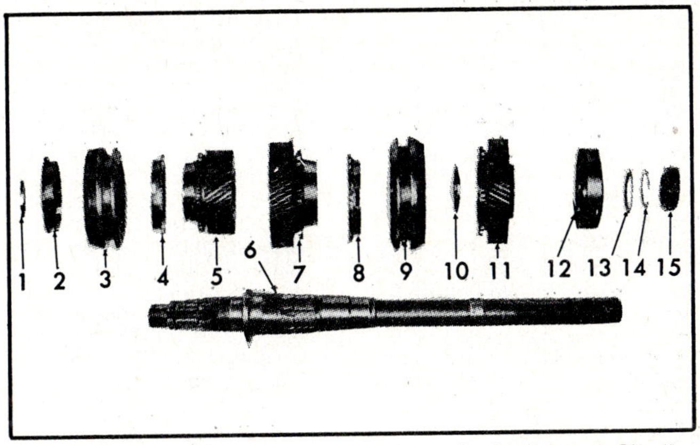

1. Snap Ring
2. 3rd Speed Blocker Ring
3. 2–3 Clutch Assembly
4. 2nd Speed Blocker Ring
5. 2nd Speed Gear
6. Mainshaft
7. 1st Speed Gear
8. 1st Speed Blocker Ring
9. 1st Reverse Clutch Assembly
10. Snap Ring
11. Reverse Gear
12. Rear Bearing
13. Special Washer
14. Snap Ring
15. Speedo Drive Gear

Fig. 20 Mainshaft disassembled

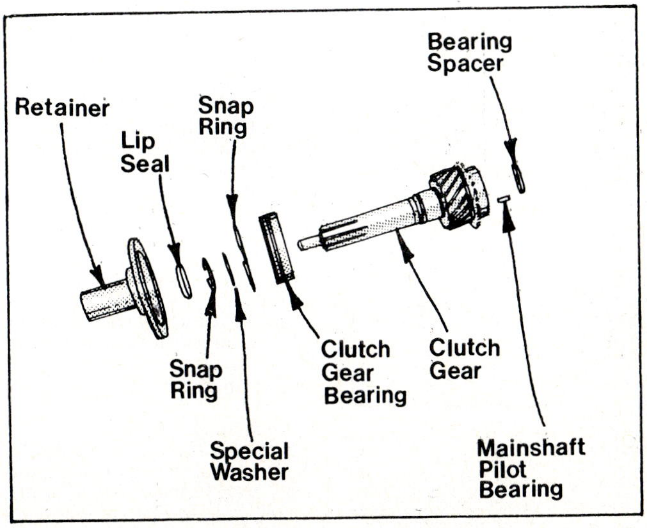

Fig. 21 Clutch gear components

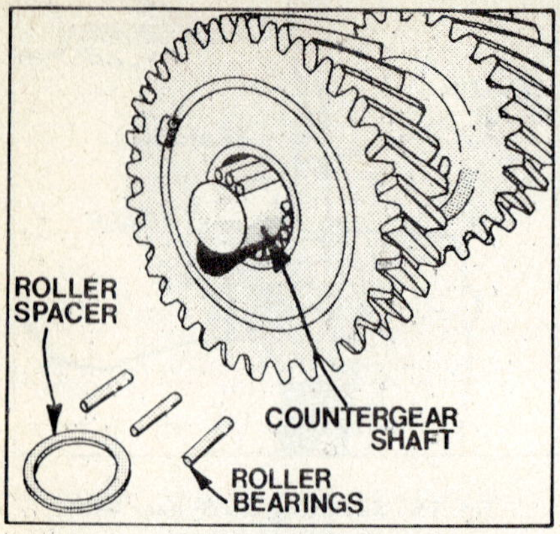

Fig. 22 Loading countershaft bearings

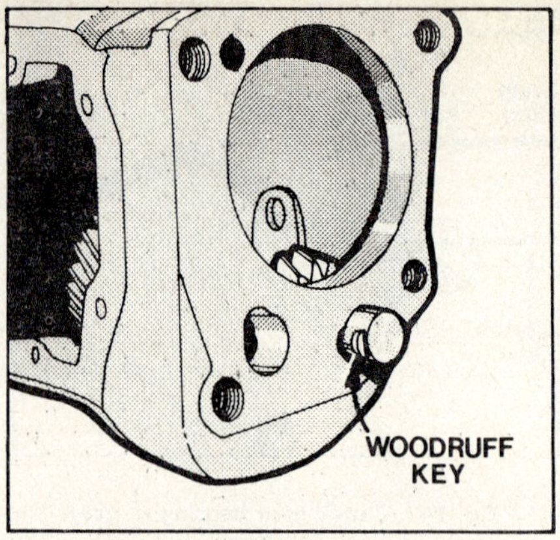

Fig. 23 Installing countershaft

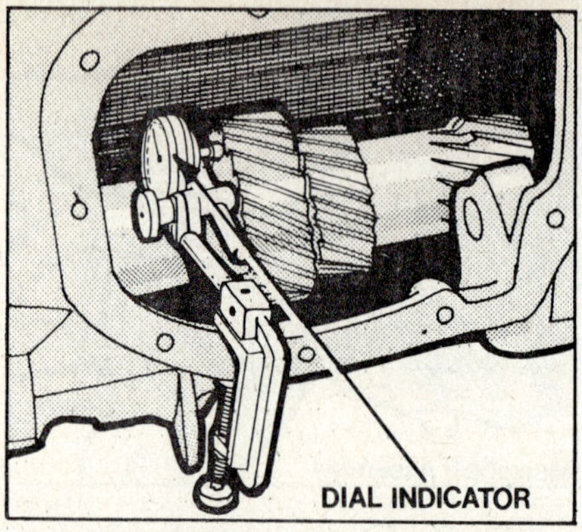

Fig. 24 Checking countergear end play

4. Remove rear extension retaining bolts. Tap extension with a soft hammer in a rearward direction to start removal. Remove extension and gasket. Remove the rearward portion of reverse idler gear from case.
5. Shift transmission into 3rd gear to prevent mainshaft from turning and remove snap ring from rear splines of mainshaft. Using puller, remove speedometer gear, Fig. 29. Remove reverse gear from mainshaft.
6. Remove the self-locking bolt attaching the rear bearing retainer to the transmission case. Remove case to retainer alignment pin with pair of channel locks or vise-grips.
7. Rotate rear bearing retainer until countergear shaft lines up with reverse idler shaft hole in retainer. Using a dummy shaft, drive counter gear shaft out rear of case, Fig. 30, letting countergear drop to bottom of case.

NOTE: Leave dummy shaft inside countergear to retain needle bearings.

8. Slide 3rd-4th gear clutch sleeve forward on hub. Remove mainshaft and rear bearing retainer assembly from case. If 4th gear synchronizer ring has fallen off during removal, remove it from case. The clutch gear roller bearings may also fall into the case. Remove all loose roller bearings from case as well as any rollers retained in clutch gear.
9. Remove front reverse idler gear and thrust washer from case.
10. Remove clutch gear shaft snap ring (do not remove snap ring on outer bearing race). Press clutch gear out of front bearing into case, Fig. 31.
11. From inside case, tap out front main bearing and snap ring.
12. Remove countergear and its tanged thrust washer from case.
13. Remove mainshaft front snap ring, Fig. 32. Slide 3rd and 4th speed clutch assembly, 3rd speed gear, and synchronizing ring from mainshaft.
14. Spread rear bearing retainer snap ring and slide retainer from mainshaft.
15. Remove mainshaft rear snap ring and spacer. Support 1st speed gear using press plates and press rear bearing and 1st gear from mainshaft, Fig. 33.
16. Remove 1st and 2nd speed clutch assembly snap ring. Slide clutch assembly and 2nd speed gear from mainshaft.

Cleaning & Inspection

1. Wash transmission case inside and out with cleaning solvent and inspect for cracks. Inspect front face of case for burrs and, if present, dress them off with a fine cut mill file.
2. Wash front and rear bearings in cleaning solvent. Blow out bearings with compressed air. *Do not allow bearings to spin; turn them slowly by hand. Spinning bearings will damage race and balls.*
3. Make sure bearings are clean, then lubricate them with light engine oil and check them for roughness. Roughness may be determined by turning outer race by hand.
4. All main drive gear and countergear bearing rollers should be inspected closely and replaced if they show wear. Inspect countershaft and replace if necessary. Replace all worn spacers.
5. Inspect all gears and first speed gear bushing (or sleeve) and, if necessary, replace all that are worn or damaged.

Synchronizers

Clutch hubs and sliding sleeves are a selected assembly and should be kept together as originally assembled, but the three keys and two springs may be replaced if worn or broken.

Push hub from sliding sleeve. Keys will fall free and springs may be easily removed. To assemble, place the two springs in position (one on each side of hub) so tanged end of each spring falls into same keyway in hub. Place keys in position and, holding them in place, slide hub into sleeve.

Assemble Mainshaft

1. From rear of mainshaft, assemble the 2nd speed gear (with hub of gear toward rear of shaft).
2. Install 1st-2nd synchronizer clutch assembly to mainshaft (sliding clutch sleeve taper toward the rear, hub to front) together with a synchronizer ring on both sides of the clutch assemblies so the keyways line up with the clutch keys.
3. Install new select fit snap ring (thickest that will assemble) in groove on mainshaft at rear of clutch assembly.
4. Install 1st speed gear (with hub toward front) and supporting inner race, press the rear bearing onto the mainshaft with the snap ring groove toward the front of the transmission, Fig. 34.
5. Install spacer and new select fit snap ring (thickest that will assemble) on mainshaft behind rear bearing.
6. Install the third speed gear (hub to front of transmission) and the 3rd speed gear synchronizer ring (notches to front of transmission).
7. Install the 3rd and 4th speed gear clutch assembly (hub and sliding sleeve) with taper toward the front making sure that the keys in the hub correspond to the notches in the 3rd speed gear synchronizing ring.
8. Install new selective fit snap ring (thickest that will install) in the groove in the mainshaft in front of the 3rd and 4th speed clutch assembly.
9. Install the rear bearing retainer over end of mainshaft. Spread the snap ring in the plate allowing the snap ring to drop around the rear bearing. Release snap ring when it aligns with groove in rear bearing.

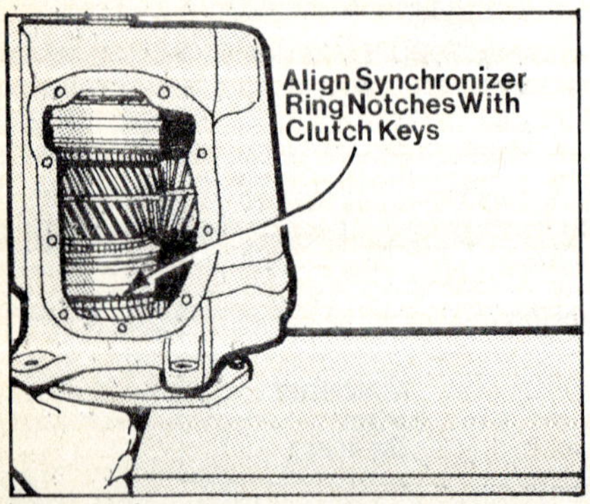

Fig. 25 Installing mainshaft assembly

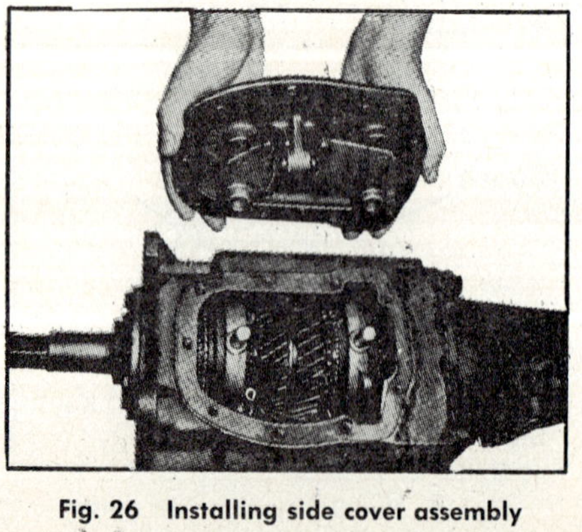

Fig. 26 Installing side cover assembly

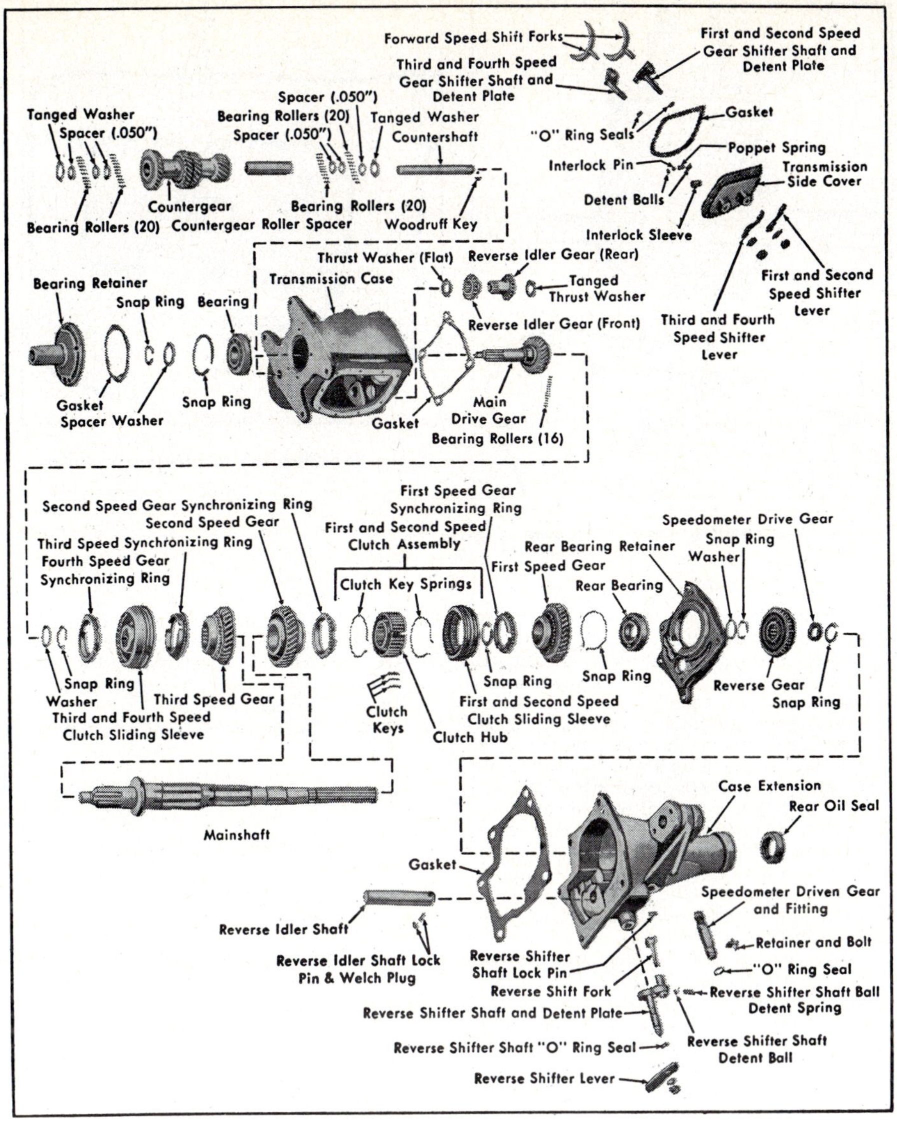

Fig. 27 Exploded view of Warner four speed transmission

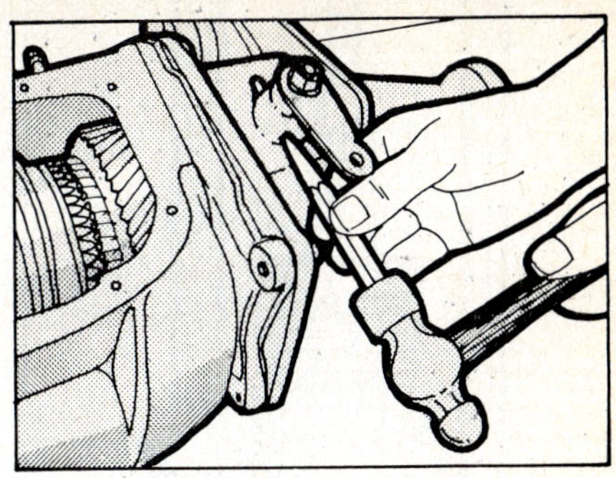

Fig. 28 Removing reverse shifter shaft lock pin

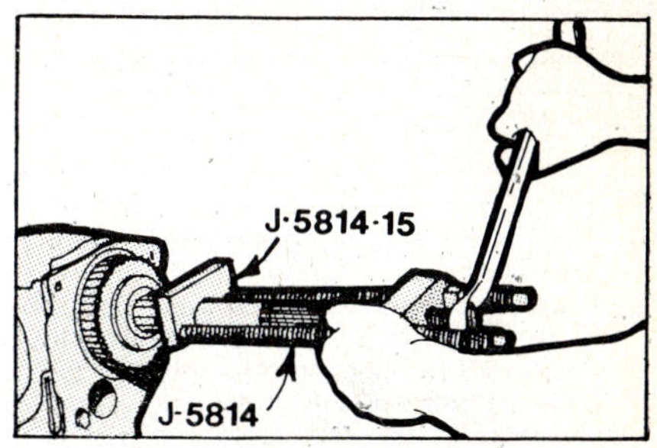

Fig. 29 Removing speedometer gear

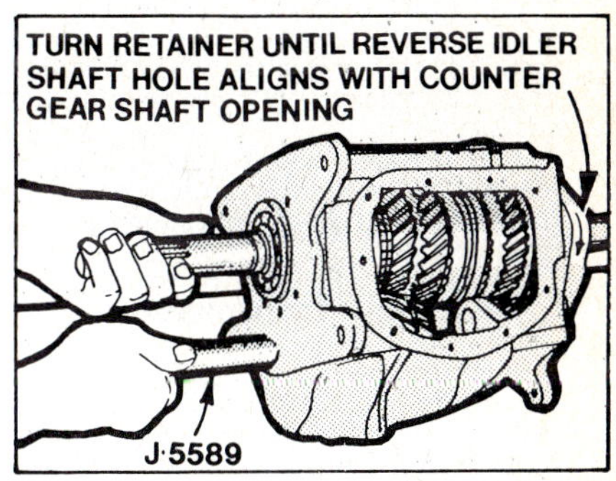

Fig. 30 Removing countergear shaft

Fig. 31 Removing clutch gear

Fig. 32 Removing mainshaft front snap ring

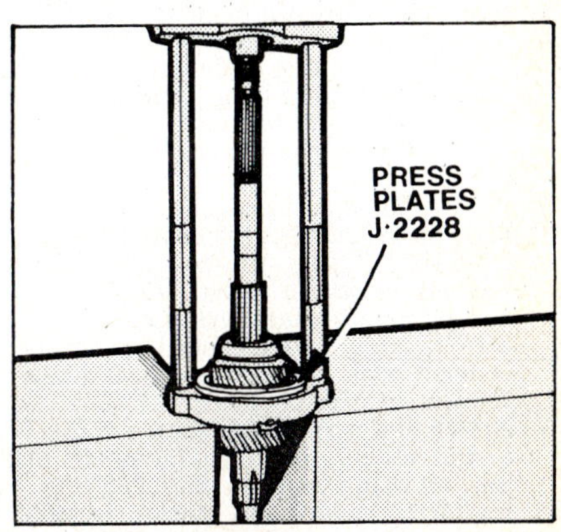

Fig. 33 Removing mainshaft rear bearing

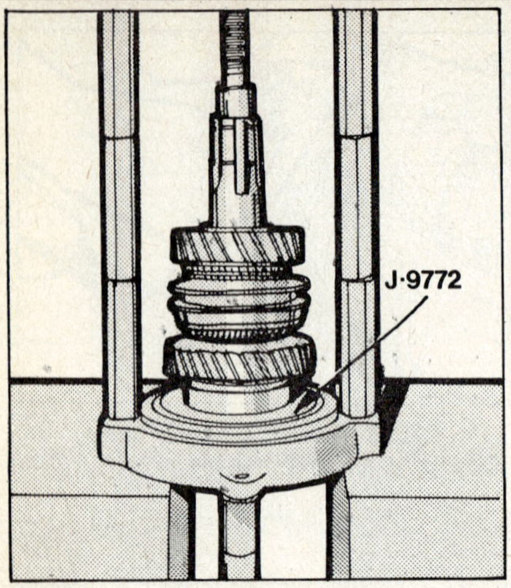

Fig. 34 Installing mainshaft rear bearing

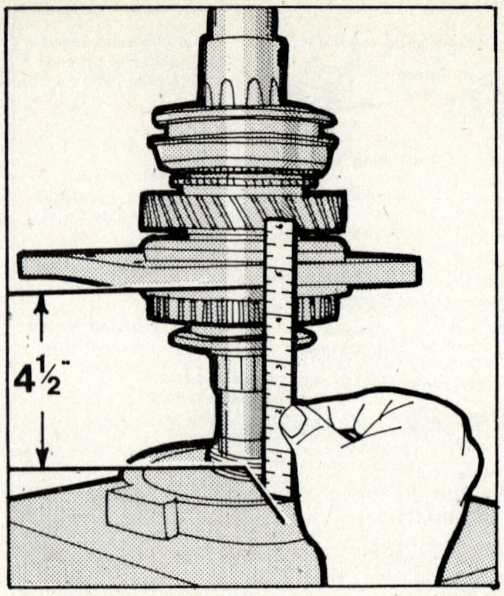

Fig. 35 Installing speedometer gear

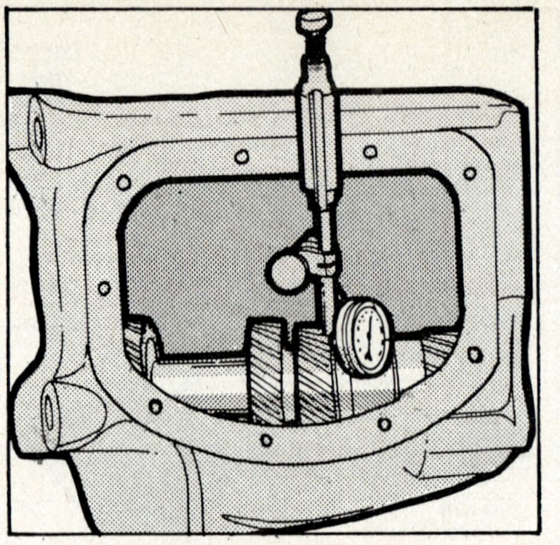

Fig. 36 Checking countergear end play

10. Install the reverse gear (shift collar to rear).
11. Press speedometer drive gear onto the mainshaft using a suitable press plate. Position the speedometer gear to get a measurement of 4½" from the center of the gear to the flat surface of the rear bearing retainer, Fig. 35.

Assemble Countergear

1. Install roller spacer in countergear.
2. Using heavy grease to retain rollers, install 20 rollers in either end of countergear, two spacers, 20 more rollers, then one spacer.
3. Assemble rollers and spacers in the same manner in other end of countergear. Then insert dummy shaft in countergear to retain rollers.

Assemble Transmission

1. Rest case on its side with side cover opening toward you. Place countergear tanged thrust washers in place, retaining them with heavy grease, making sure that tangs are resting in notches in case.
2. Place countergear assembly in bottom of case, making sure that tanged thrust washers are not knocked out of place.
3. Press bearing onto main drive gear with snap ring groove to front.
4. Install spacer washer and selective fit snap ring in groove on gear stem.
5. Install main drive gear assembly through side cover opening and into position in transmission front bore. Tap lightly into place with a soft hammer, if necessary. Place snap ring in groove in front bearing.
6. With transmission resting on its front face, move countergear into mesh with main drive gear, making sure thrust washers remain in place. Install key in end of countershaft and, from front of case, tap or press shaft until end of shaft is flush with rear of case and dummy shaft is displaced.
7. Attach dial indicator as shown in Fig. 36 and check end play of countergear. End play must not be more than .025".
8. Install 14 rollers into main drive gear, using heavy grease to hold them in place. Place gasket in position on front face of rear bearing retainer, using heavy grease to hold it in position.
9. Install 4th gear synchronizing ring on main drive gear with clutch key notches toward rear of case.
10. Position reverse idler gear thrust washer (untanged) on machined face of ear cast in case for reverse idler shaft. Position front reverse idler gear on top of thrust washer, with hub facing toward rear of case.
11. Lower mainshaft assembly into case, making certain that notches on 4th gear synchronizing ring correspond to keys in synchronizer.
12. Install self-locking bolt attaching rear bearing retainer to case.
13. From rear of case, insert rear reverse idler gear, engaging splines with portion of gear within case.
14. Using heavy grease, place gasket into position on rear face of rear bearing retainer.
15. Install remaining tanged thrust washer into place on reverse idler shaft, being sure tang on washer is in notch in idler thrust face of extension.
16. Place two synchronizers in neutral position. *If locking-up of gears is encountered, a small amount of petrolatum may be applied to the 1st speed gear synchronizing ring, enabling it to turn freely on 1st speed gear hub.*
17. Pull reverse shifter shaft to left side of extension and rotate shaft to bring reverse shift fork to extreme forward position in extension. Line up front and rear reverse idler gears, making sure front thrust washer is in place.
18. Start extension into case by carefully inserting reverse idler shaft through reverse idler gears. Slowly push it on shifter shaft until shift fork engages reverse gear shift collar. When fork engages, rotate shifter shaft to move reverse gear rearward, permitting extension to slide onto transmission case.
19. Install extension and retainer to case attaching bolts, and extension to retainer attaching bolts. Use suitable sealer on the lower right attaching bolt.
20. Adjust reverse shifter shaft so that groove in shaft lines up with hole in boss and drive in lock pin from top of boss.
21. Install main drive gear bearing retainer and gasket, being sure oil well lines up with oil outlet hole.
22. Install shift fork in each synchronizer sleeve. With both synchronizers in neutral, install side cover with gasket. Use suitable sealer when installing lower right cover bolt.
23. Install shifter levers, lock washers and nuts.

T14A & T15A SERIES

Disassemble, Fig. 37

1. Remove housing cover and gasket from top of transmission case.
2. Remove nut and flat washer securing transfer case drive gear on the mainshaft. Remove transfer case drive gear, adapter and spacer.
3. Remove main drive gear bearing retainer and gasket.
4. Remove main drive gear and mainshaft bearing snap rings.
5. Using suitable puller, remove main drive gear and mainshaft bearings, Fig. 38.
6. Remove main drive gear from case.

NOTE: T15A unit must be shifted into second gear to permit removal of the mainshaft and gear assembly.

7. Remove mainshaft and gears as an assembly through the case cover opening, Fig. 39.
8. On remote shift units, remove the shifter forks by removing the roll pins from the shift lever shafts and housing. From inside the transmission case, slide the shift levers and interlock assembly toward the outside of the case and remove shifter forks and lever assemblies.
9. Remove lockplate by tapping lightly on the front end of the countershaft and reverse idler shaft. Remove lockplate from the slots in the shafts.
10. Using a suitable drift, drive countershaft rearward out of case. Remove countergear assembly, thrust washers and bearings.
11. Drive reverse idler shaft rearward out of case. Remove idler gear.

Mainshaft, Disassemble

1. Remove clutch hub snap ring and 2-3 synchronizer assembly from the mainshaft.

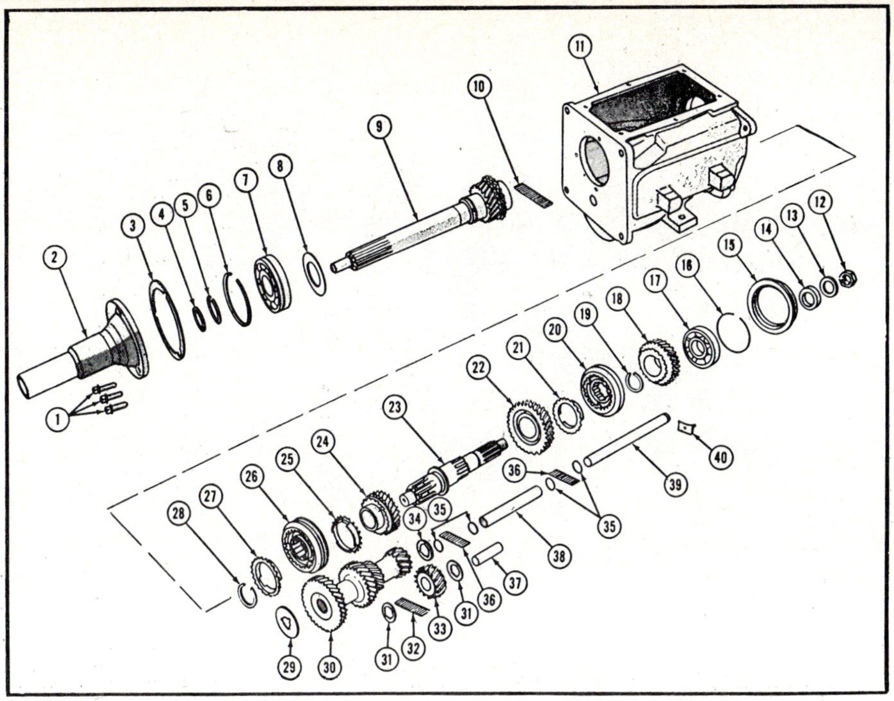

1. Screws
2. Retainer
3. Gasket
4. Seal
5. Snap ring
6. Snap ring
7. Main drive bearing
8. Oil slinger
9. Main drive gear
10. Rollers
11. Case
12. Nut
13. Flatwasher
14. Spacer
15. Bearing adapter
16. Snap ring
17. Mainshaft bearing
18. Reverse gear
19. Snap ring
20. Low synchronizer assembly
21. Blocking ring
22. Low gear
23. Mainshaft
24. Second gear
25. Blocking ring
26. 2-3 synchronizer assembly
27. Blocking ring
28. Snap ring
29. Thrust washer
30. Countergear
31. Washer
32. Rollers
33. Reverse idler gear
34. Thrust washer
35. Spacer washer
36. Rollers
37. Reverse idler shaft
38. Spacer
39. Countershaft
40. Lockplate

Fig. 37 T14A, T15A three speed transmission

2. Remove 2nd speed gear.
3. Remove reverse gear from mainshaft.
4. Remove clutch hub, snap ring, and low synchronizer assembly from the mainshaft.
5. Remove first speed gear from mainshaft.

Transmission, Reassemble

1. Install reverse idler gear assembly, making sure that the slot end of the idler shaft is aligned to receive the lockplate.
2. Assemble countershaft center spacer, bearing spacers and rollers in the countergear. Using a suitable dummy shaft to retain rollers, install gear in case and with thrust washers in place, install countershaft through rear of case with the lockplate slot toward the rear and aligned to receive lockplate.
3. Locate the lockplate in the two shafts and tapping alternately, drive shafts in until lockplate is tight against case.
4. Install mainshaft and gear assembly as a unit through the top cover opening of the transmission case.

NOTE: On remote shift units, the shifter forks and levers must be installed at this time. T15A transmission interlock levers are stamped and must be installed in the respective locations. T14A levers are interchangeable.

---

5. Using a screwdriver, depress interlock lever while installing shift fork into shifter lever and synchronizer clutch sleeve. Be sure poppet spring is properly installed. Install tapered pins securing lever shafts in case.
6. Install main drive gear and oil retainer into case with the cutaway part of the gear downward. Guide main drive gear onto mainshaft using care not to drop the rollers.
7. Install main bearings and snap rings.
8. Install main drive gear bearing retainer. Make sure oil drain holes are aligned.
9. Install transmission cover.

T-331, T-332 SERIES

Transmission Disassembly Figs. 39A & 39B

1. Remove cover and gasket or shift control housing.

NOTE: If shift control housing is to be disassembled proceed as follows:

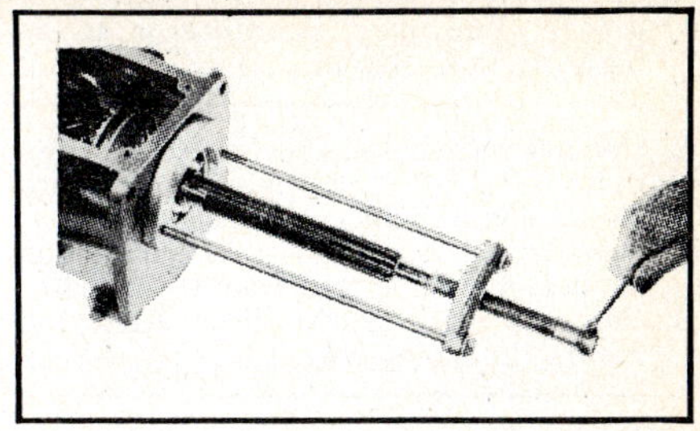

Fig. 38 Removing main drive gear bearing

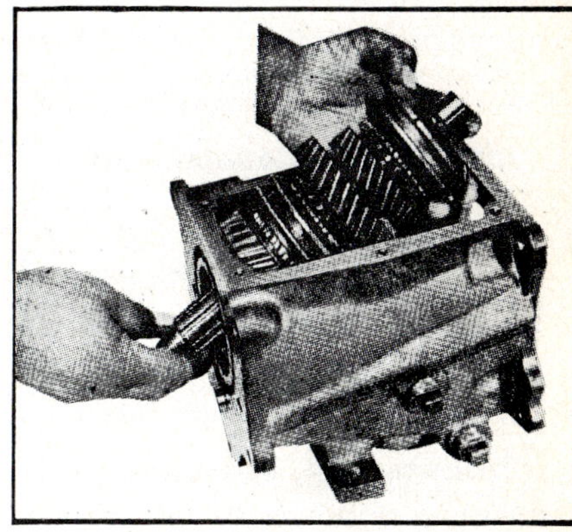

Fig. 39 Removing mainshaft

1. Remove back-up light switch.
2. Remove shift rail sealing plugs from rear of control housing and remove shaft rails.

---

2. Remove main drive gear bearing retainer.

NOTE: When reassembling main drive gear bearing retainer make sure retainer oil drain and gasket oil drain holes align with transmission case oil drain hole.

---

3. Remove main drive gear bearing snap ring and main gear bearing using suitable puller.
4. Remove main drive gear from case.
5. On units with shift control housing, remove extension housing [if applicable], rear bearing, bearing retainer, and speedometer drive gear and gasket. Remove main shaft bearing snap ring and main shaft bearing using suitable puller. On models with side mounted shift lever, place front synchronizer in second speed position. Move mainshaft to side and remove shift forks. Remove shift levers and shift lever interlock lever spring pin.
6. Place front synchronizer in second speed position and remove main shaft assembly by tilting front of assembly up and lifting through top of case.

Mainshaft, Disassembly

1. Remove second and third speed synchronizer snap ring.

NOTE: This snap ring is available in selective thickness. Upon reinstallation, install thickest snap ring that will fit into groove. End play clearance is to be between .004–.020 in.

2. Remove second and third speed synchronizer assembly from mainshaft. Mark sleeve and hub for proper reassembly.
3. Remove second speed gear from main shaft and reverse gear from main shaft.
4. Remove rear synchronizer first and reverse hub snap ring.

NOTE: This snap ring is available in selective thickness. Upon reinstallation, install thickest snap ring that will fit into groove. End play clearance is .005–.020 in. for 1st speed gear and collar on mainshaft.

5. Remove rear synchronizer, synchronizer blocking ring, and first speed gear.

Idler Gear & Countershaft Removal

1. Tap reverse idle gear shaft and countershaft rearward to allow removal of shaft lock plate.
2. Using a brass drift, remove reverse idle gear shaft from case.
3. Using a dummy shaft, drive countershaft from case.

Transmission, Reassembly

Reassembly procedures are the reverse of the disassembly procedures. Use mainshaft pilot end support tool SE-2508 when installing main shaft bearing. When installing main drive gear and oil baffle into case position cutaway portion of the gear downward toward countergear assembly. Bolt torque requirements are as follows: front bearing retainer to case, 10–15 ft. lbs., cover to case, 10–18 ft. lbs., control lever to lever shaft, 15–25 ft. lbs., rear bearing retainer to case, 23–27 ft. lbs., companion flange to mainshaft, 90–120 ft. lbs. and control lever housing bolt, 10–15 ft. lbs.

T18, T18A, T19 & T19D SERIES

Disassemble, Figs. 40, 40A, 41 & 41A

1. Remove gear shift housing, Fig. 41, then the parking brake drum, with the transmission locked in two gears remove U-joint flange and oil seal.
2. Remove speedometer driven gear and bearing assembly.
3. Remove output shaft bearing retainer and speedometer drive gear and spacer, then remove output bearing retainer studs from case, Fig. 40 & 40A.
4. Remove output shaft bearing snap ring, then using suitable tool remove bearing, except on T-19-D. On T-19-D, remove output shaft bearing cup.
5. Remove countershaft and idler shaft retainer and power take-off cover.
6. Remove input shaft bearing snap ring and bearing, except on T-19-D. On T-19-D, remove bearing cup and shims.
7. Remove oil baffle, except on T-19-D. On T-19-D, remove input shaft oil seal from retainer.
8. Remove roll pin from reverse gear shifter arm shaft, then remove shaft from shifter and lift shifter from case.
9. Remove output shaft and gear assembly from case.
10. Using a dummy shaft, drive countershaft out from front of transmission.

Key	Description
1	COVER, Transmission Case
2	GASKET, Transmission Cover to Case
3	MAINSHAFT
4	GEAR, Mainshaft First Speed
5	RING, Synchronizer Blocking
5A	RING, Snap Selective Fit
6	SYNCHRONIZER, Assy., First and Reverse
6A	RING, Snap Selective Fit
7	GEAR, W/Bushing, Mainshaft Reverse
8	BUSHING, Reverse Gear
9	BEARING, Mainshaft Rear
10	RING, Snap, Bearing
11	GEAR, Speedometer Drive
12	NUT, Companion Flange
13	SEAL, Drive Gear Oil
14	RING, Drive Gear Brg Snap Selective Fit
15	RING, Bearing Snap
16	BEARING, Main Drive Gear
17	BAFFLE, Drive Gear Oil
18	GEAR, Main Drive
19	SYNCHRONIZER, Assy. Second and Third
20	GEAR, Mainshaft Second Speed
21	BEARING, Mainshaft Roller
22	WASHER, Countershaft Thrust
22A	SPACER, Countershaft Gear Bearing Short
23	BEARING, Countershaft Roller
24	COUNTERSHAFT
25	SPACER, Countershaft Gear Bearing
26	GEAR, Countershaft Cluster
27	FORK, Shift
28	SPRING, Interlock Lever
29	LEVER, Interlock First and Reverse Speed Shift
30	LEVER, Interlock Second and Third Speed Shift
31	PIN, Interlock Lever
32	LEVER, Shift, Assy.
	WASHER, Flat
	WASHER, Lock
	NUT, Hex.
33	SEAL, Lever Shaft Oil
34	LEVER, Shift, Assy. First and Reverse Speed Shift
35	LEVER, Shift Assy. Second and Third Speed Shift
36	WASHER, Reverse Idler Gear
37	LOCK, Reverse Idler Gear Shaft
38	SHAFT, Reverse Idler Gear
39	GEAR, Reverse Idler
40	BEARING, Reverse Idler Gear Roller
41	RETAINER, Main Drive Gear, W/Bushing
	BOLT, Hex-Hd
	WASHER, Lock
42	GASKET, Main Drive Gear Brg Retainer
43	PLUG, Sq-Hd 3/4 Filler
	PLUG, Magnetic 3/4 Drain
44	PIN, Interlock Lever Spring
45	CASE, Trans, Assy W/Interlock Lever Pin
	STUD, Trans, Case to Bell Housing
	NUT, Hex. 1/2NF
	WASHER, Lock 1/2 Medium
46	GASKET, Rear Bearing Retainer
47	BUSHING, Speedometer Drive Gear
48	RETAINER, W/Bushing and Seal, Rear Bearing
	BOLT, Hex-HD 3/8NC x 1
	BOLT, Hex-HD 3/8NC x 1-1/4
	BOLT, Hex-HD 3/8NC x 1-7/8
	WASHER, Lock 3/8 External
49	SEAL, Mainshaft Oil

Fig. 39A T-331 transmission exploded view

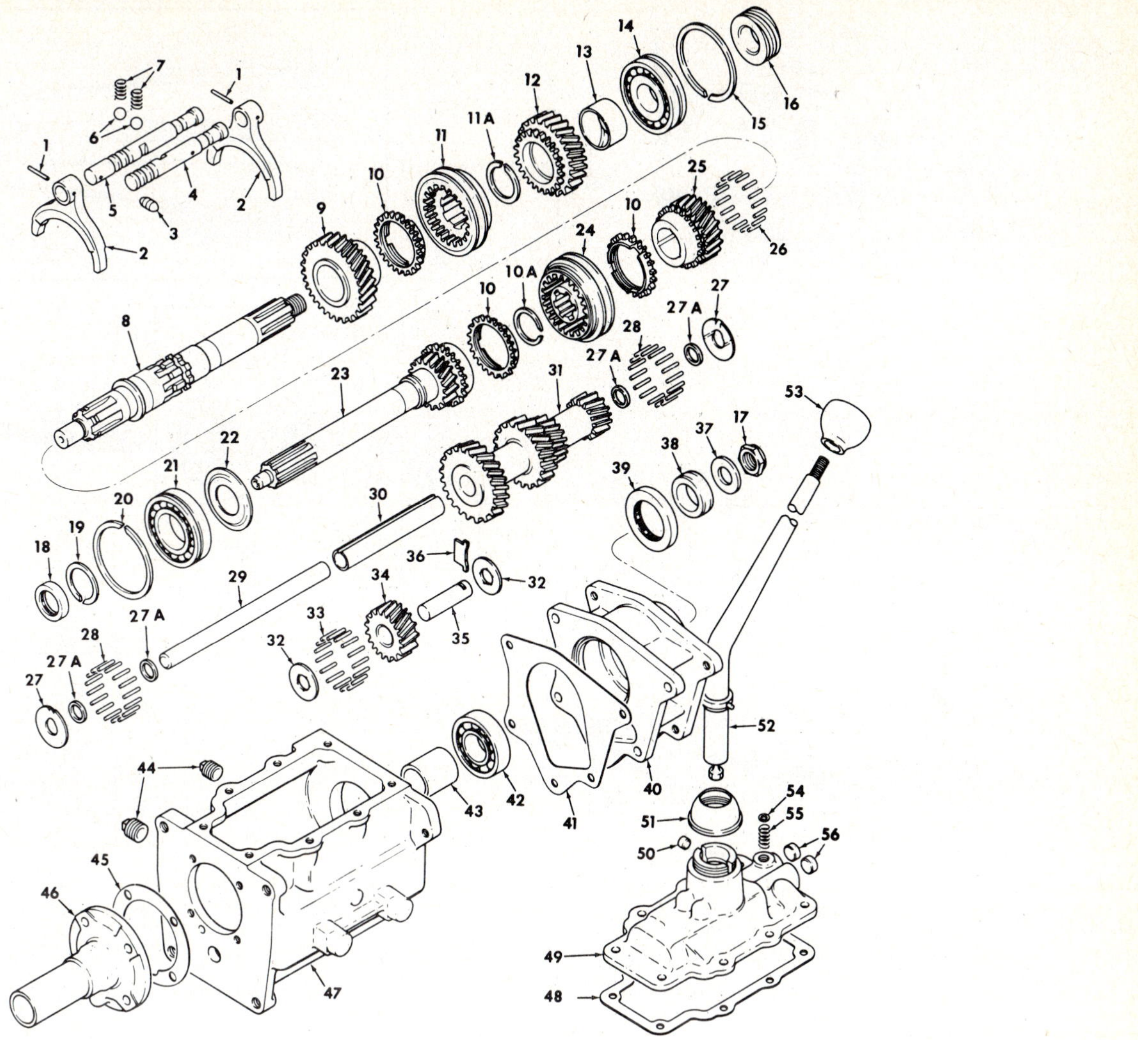

Key	Description
1	PIN, Shift Fork
2	FORK, Shift
3	PLUNGER, Interlock
4	BAR, First and Reverse Speed Shift
5	BAR, Second and Third Speed Shift
6	BALL, Shift Bar Poppet
7	SPRING, Shift Bar Poppet
8	MAINSHAFT
9	GEAR, Mainshaft First Speed
10	RING, Synchronizer Blocking
10A	SNAP, Ring Selective Fit
11	SYNCHRONIZER, First and Reverse Assy.
11A	SNAP, Rings Selective Fit
12	GEAR, W/Bushing, Mainshaft Reverse
13	BUSHING, Reverse Gear
14	BEARING, Mainshaft Rear
15	RING, Snap Bearing
16	GEAR, Speedometer Drive
17	NUT, Companion Flange
18	SEAL, Drive Gear Oil
19	RING, Drive Gear Bearing Snap Selective Fit
20	RING, Bearing Snap
21	BEARING, Main Drive Gear
22	BAFFLE, Drive Gear Oil
23	GEAR, Main Drive
24	SYNCHRONIZER, Second and Third, Assy.
25	GEAR, Mainshaft Second Speed
26	BEARING, Mainshaft Roller
27	WASHER, Countershaft Thrust
27A	SPACER, Countershaft Gear Bearing
28	BEARING, Countershaft Roller
29	COUNTERSHAFT
30	SPACER, Countershaft Gear Bearing
31	GEAR, Countershaft Cluster
32	WASHER, Reverse Idler Gear
33	BEARING, Reverse Idler Gear Roller
34	GEAR, Reverse Idler
35	SHAFT, Reverse Idler Gear
36	LOCK, Reverse Idler Gear Shaft
37	WASHER, Mainshaft Nut
38	SPACER, Mainshaft Bearing
39	SEAL, Mainshaft Oil
40	RETAINER, W/Bushing & Seal, Rear Brg.
41	GASKET, Rear Bearing Retainer
42	BEARING, Mainshaft Rear
43	SPACER, Mainshaft Bearing
44	PLUG, Sq-Hd 3/4 (Filler)
45	GASKET, Main Drive Gear Brg. Retainer
46	RETAINER, W/Bushing, Main Drive Gear
47	CASE, W/Studs, Transmission
	NUT, Hex.
	WASHER, Lock
48	GASKET, Control Housing to Trans. Case
49	HOUSING, Control Lever
	BOLT, Hex-Hd
	WASHER, Lock
50	PLUG, Taper
51	COVER, Control Housing Dust
52	LEVER, Control Assy.
53	HANDLE, Control Lever
54	WASHER, Control Lever Spring
55	SPRING, Control Lever
56	PLUG, Special Cup

Fig. 39B T-332 transmission exploded view

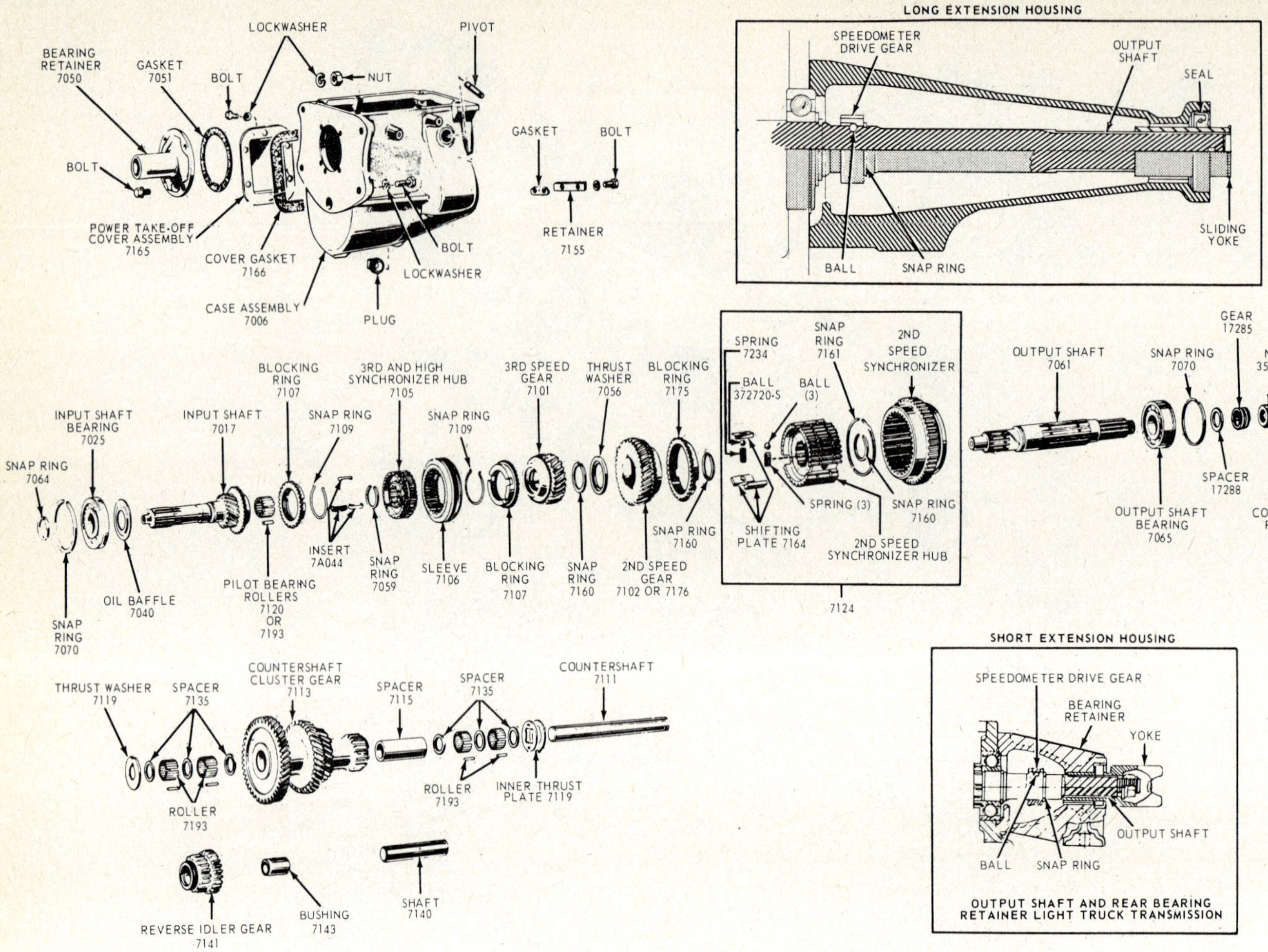

Fig. 40 Exploded view of Warner T18 & T18A Series four speed transmission (typical)

NOTE: Keep dummy shaft in contact with countershaft to avoid dropping rollers.

---

11. Remove input shaft and synchronizer blocking ring, then using a suitable puller remove idler shaft.
12. Remove idler gear and countershaft gear, use care not to lose rollers.

Output Shaft

Disassembly

1. On T-19-D units, remove thrust bearing and race. On all units except T-19-D, remove 3-4 speed synchronizer hub snap ring from output shaft. On all units, slide synchronizer assembly and 3rd gear from shaft.
2. Press reverse gear from output shaft. On T-19-D units, remove bearing cone.
3. Remove 1st speed gear snap ring, then slide gear from shaft.
4. Remove 1st-2nd speed synchronizer snap ring, then slide synchronizer from shaft.
5. Remove snap ring from rear of 2nd speed gear, then remove gear and thrust washer.

Assemble

1. Install 2nd speed gear thrust washer and snap ring on shaft, then hold shaft in vertical position and install 2nd speed gear.
2. Install snap ring at rear of 2nd speed gear, then position blocking ring on gear.
3. Press 1st-2nd speed synchronizer onto shaft and install snap ring.
4. Install 1st speed gear and snap ring on shaft.
5. Press reverse gear onto shaft, then remove shaft from press and install 3rd speed gear and synchronizer blocking ring.
6. Install snap ring with openings staggered at both ends 3rd-4th speed synchronizer hub.
7. Place inserts into synchronizer sleeve and position sleeve on hub.
8. Slide synchronizer assembly onto output shaft and install snap ring at front of synchronizer assembly.

NOTE: Slots in blocking ring must be aligned with synchronizer inserts.

Countershaft Gear

Disassemble

Remove dummy shaft, pilot bearing rollers, bearing spacers and center spacer from countershaft gear.

Assemble

1. Position long bearing spacer into countershaft gear bore, then insert dummy shaft in spacer.
2. Position gear in vertical position, then install spacer and place 22 pilot roller bearings in gear bore.
3. Position bearing spacer on top of rollers, then install 2 more roller bearings and another bearing spacer.
4. Place a large washer against end of countershaft gear to prevent rollers from dropping out then turn assembly over.
5. Install rollers and spacers in other end of countershaft gear.

Reverse Idler Gear

Disassemble

Remove snap ring from end of gear, then remove bearing rollers thrust washers, bearing spacer and bushing.

Assembly

1. Install snap ring in one end of idler gear, then set gear on end with snap ring on bottom.
2. Position thrust washer and bushing in gear bore, then install 37 bearing rollers between bushing and gear bore.

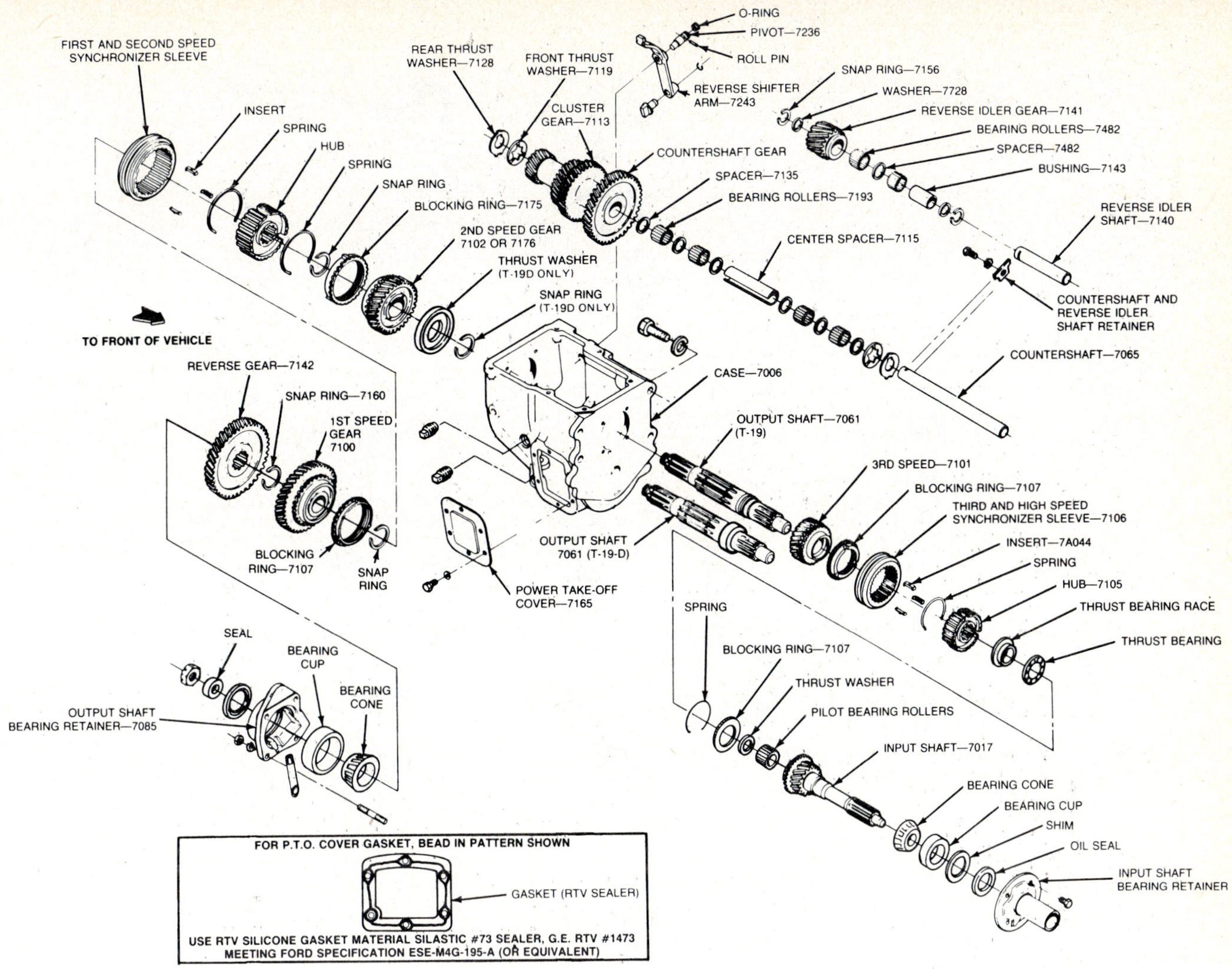

Fig. 40A Exploded view of Warner T19 & T19-D four speed transmission. (Typical)

3. Install spacer on top of rollers then install 37 more bearing rollers.
4. Place remaining thrust washer on rollers and install snap ring.

Assemble Transmission

1. Lubricate all parts with transmission oil, then position countershaft gear assembly in case.
2. Position idler gear assembly in case, then install idler shaft and shifter arm.

NOTE: Position idler shaft so that slot at rear will engage retainer.

3. Drive countershaft in through rear of case forcing dummy shaft out through front.

NOTE: Position countershaft so that slot at rear will engage retainer.

4. Install thrust washers as necessary to obtain countershaft end play of .006 to .020 inch.
5. Install countershaft and idler shaft retainer.
6. Position oil baffle and input shaft pilot rollers so that baffle will not rub on bearing race, then position input shaft in case and install blocking ring.
7. Install output shaft assembly in case.
8. Using a suitable tool drive input shaft bearing onto shaft then install thickest select fit snap ring that will fit on bearing. Snap ring is available in thicknesses of, .117–.119, .120–122, .123–.125 and .127–.129 inch. Install input shaft snap ring.
9. Install output shaft bearing.
10. Install input shaft bearing retainer without gasket and tighten bolts only enough to bottom retainer on bearing snap ring. Measure distance between retainer and case and select a gasket that will seal and also prevent end play retainer and snap ring. Gaskets are available in thicknesses of, .008–.011, .0135–.0165, .018–.022 and .0225–.0275 inch. Install gasket and tighten bolts.
11. Position speedometer drive gear and spacer, then install output shaft bearing retainer and gasket.
12. Install brake drum then lubricate extension housing bushing, seal and U joint flange with ball joint grease.
13. Install U joint flange and tighten bolt.

MODEL T-150

Disassembly, Fig. 42

1. Remove transfer case to transmission attaching bolts, then separate transfer case from transmission.
2. Remove shift control housing, Fig. 43.
3. Move 2-3 synchronizer sleeve forward and the 1st.-reverse sleeve rearward to lock the mainshaft.
4. Remove transfer case gear lock nut, flat washer and drive gear.
5. Move both synchronizers back to the Neutral position.
6. Remove fill plug.
7. Remove countershaft roll pin with a 3/16 inch punch. The roll pin is accessible through fill plug opening.
8. Remove countershaft and access plug using tool J-25232. Remove countershaft from rear of case. Permit countershaft gear to remain at bottom of case after countershaft removal.
9. Punch alignment marks in front bearing cap and transmission case for assembly reference.
10. Remove front bearing cap and gasket.
11. Remove large lock ring from front bearing.

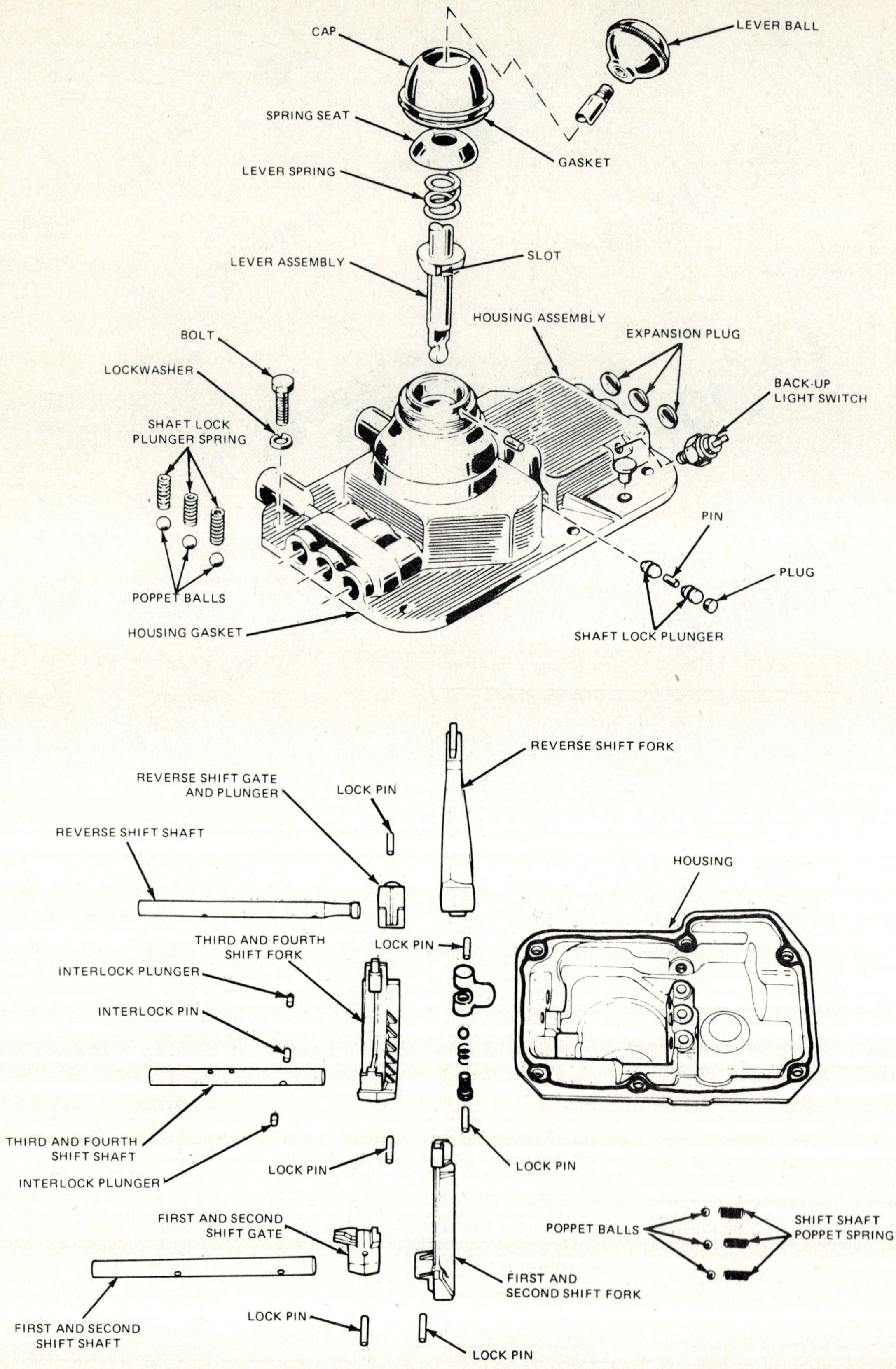

Fig. 41 Gear shift housing. Warner T18 & T18A

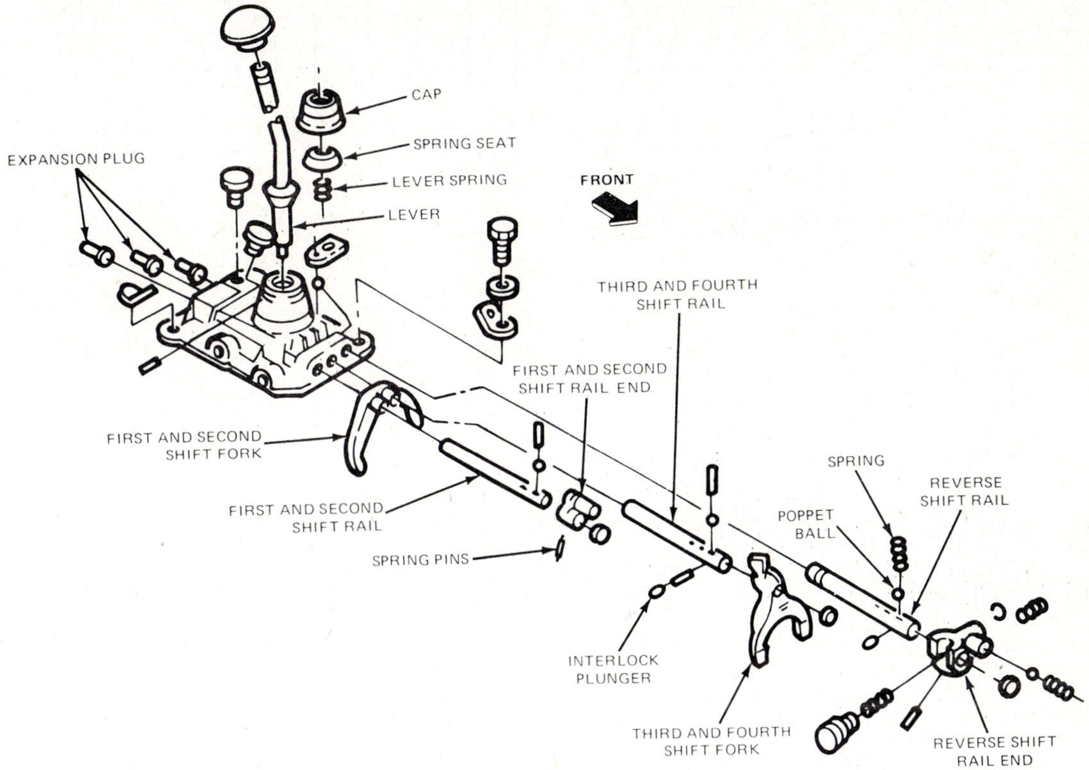

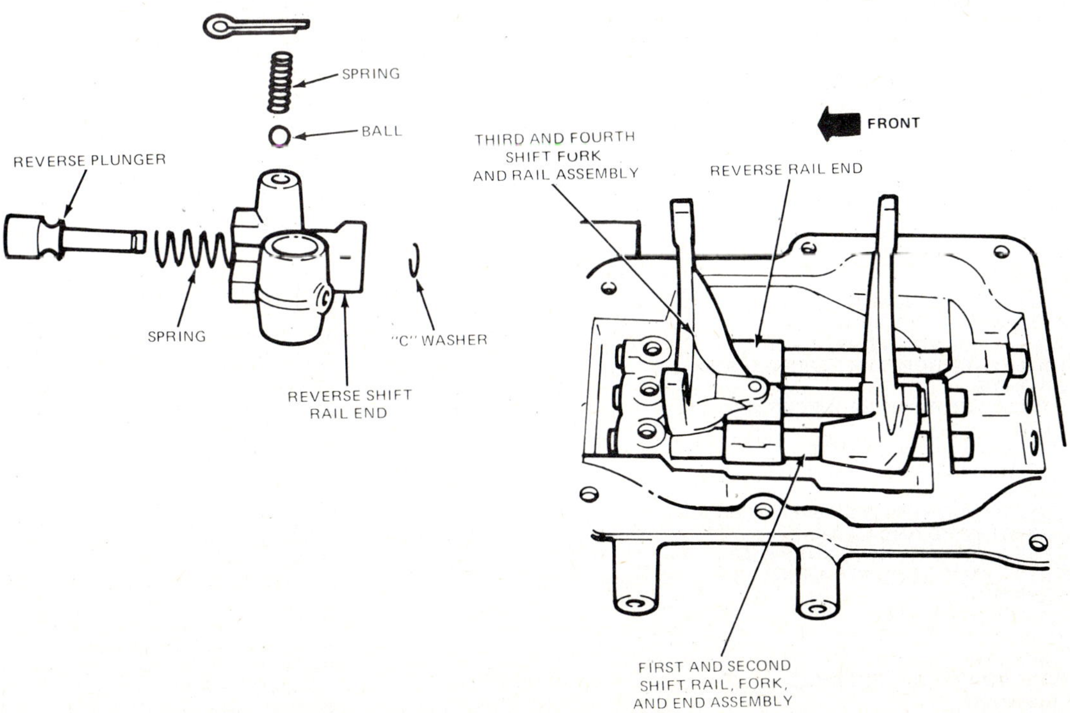

Fig. 41A Gear shift housing (Typical). Warner T19 & T19-D

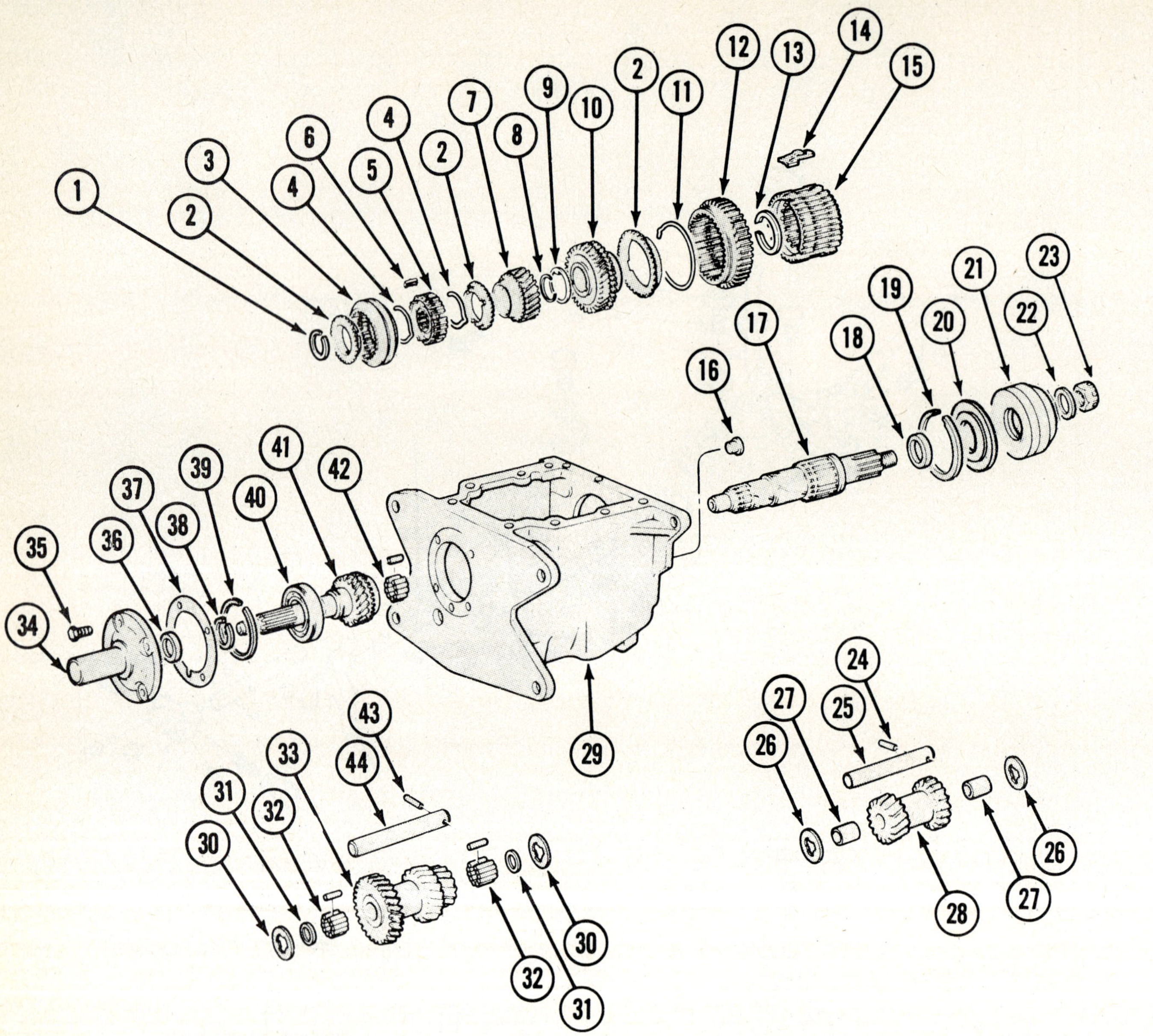

1. MAINSHAFT RETAINING SNAP RING
2. SYNCHRONIZER BLOCKING RINGS (3)
3. SECOND-THIRD SYNCHRONIZER SLEEVE
4. SECOND-THIRD SYNCHRONIZER INSERT SPRING (2)
5. SECOND-THIRD HUB
6. SECOND-THIRD SYNCHRONIZER INSERT (3)
7. SECOND GEAR
8. FIRST GEAR RETAINING SNAP RING
9. FIRST GEAR TABBED THRUST WASHER
10. FIRST GEAR
11. FIRST-REVERSE SYNCHRONIZER INSERT SPRING
12. FIRST-REVERSE SLEEVE AND GEAR
13. FIRST-REVERSE HUB RETAINING SNAP RING
14. FIRST-REVERSE SYNCHRONIZER INSERT (3)
15. FIRST-REVERSE HUB
16. COUNTERSHAFT ACCESS PLUG
17. MAINSHAFT
18. MAINSHAFT SPACER
19. REAR BEARING ADAPTER LOCK RING
20. OIL SLINGER/SPACER
21. REAR BEARING AND ADAPTER ASSEMBLY
22. WASHER
23. LOCKNUT
24. ROLL PIN
25. REVERSE IDLER GEAR SHAFT
26. THRUST WASHER
27. BUSHING (PART OF IDLER GEAR)
28. REVERSE IDLER GEAR
29. TRANSMISSION CASE
30. THRUST WASHER (2)
31. BEARING RETAINER (2)
32. COUNTERSHAFT NEEDLE BEARINGS (50)
33. COUNTERSHAFT GEAR
34. FRONT BEARING CAP
35. BOLT (4)
36. FRONT BEARING CAP OIL SEAL
37. GASKET
38. FRONT BEARING RETAINER SNAP RING
39. FRONT BEARING LOCKRING
40. FRONT BEARING
41. CLUTCH SHAFT
42. MAINSHAFT PILOT ROLLER BEARINGS
43. ROLL PIN
44. COUNTERSHAFT

Fig. 42 Model T150 3 speed transmission, disassembled

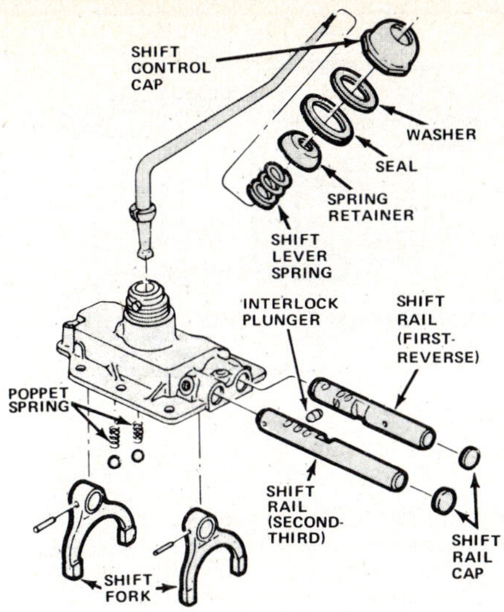

Fig. 43 Shift control housing, disassembled. Model T150

12. Remove clutch shaft and front bearing assembly with tool J-6654-01.
13. Remove 2-3 synchronizer blocking ring from clutch shaft or synchronizer hub.
14. Remove rear bearing and adapter using a suitable drift and hammer.
15. Remove mainshaft and geartrain assembly. Tilt splined end of shaft downward and lift the forward end of the shaft up and out from case.
16. Remove countershaft gear and tool as an assembly.
17. Remove countershaft gear thrust washers, countershaft roll pin and any mainshaft pilot roller bearings which may have fallen into the case during clutch shaft removal.
18. Remove reverse idler gear shaft. Insert a brass drift through clutch shaft bore in front of case and tap shaft until end of shaft with roll pin clears countershaft bore in rear of case, then remove shaft.
19. Remove reverse idler gear and thrust washers.

SUB-ASSEMBLY DISASSEMBLY

Mainshaft

1. Remove retaining snap ring from front of mainshaft.

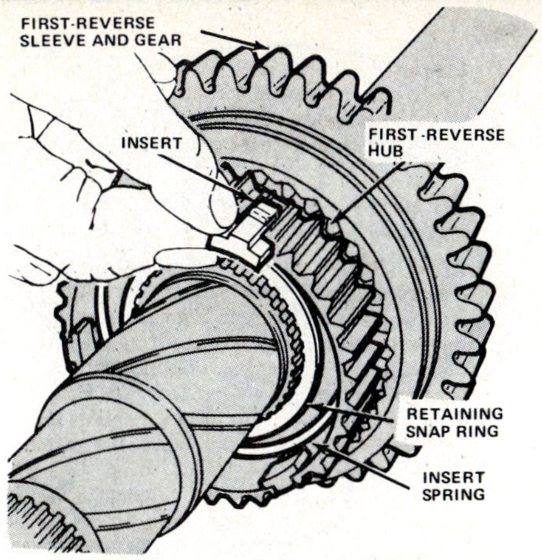

Fig. 44 1st-reverse hub insert

2. Remove 2-3 synchronizer assembly and the second gear. Mark hub and sleeve for assembly reference.

NOTE: Observe position of insert springs and inserts for assembly reference.

3. Remove insert springs from 2-3 synchronizer. Remove the three inserts and separate sleeve from synchronizer hub.
4. Remove snap ring and tabbed thrust washer from shaft.
5. Remove first gear and blocking ring.
6. Remove 1st.-reverse hub retaining snap ring.

NOTE: Observe position of inserts and spring for assembly reference.

7. Remove sleeve and gear, insert spring and three inserts from hub.
8. Remove oil slinger and spacer from rear of mainshaft.
9. Remove hub from output shaft using a suitable press.

NOTE: Do not attempt to hammer the press-fit hub from the shaft since damage to the hub and shaft will occur.

Clutch Shaft

1. Remove front bearing retaining snap ring and any remaining roller bearings.
2. Press front bearing from shaft.

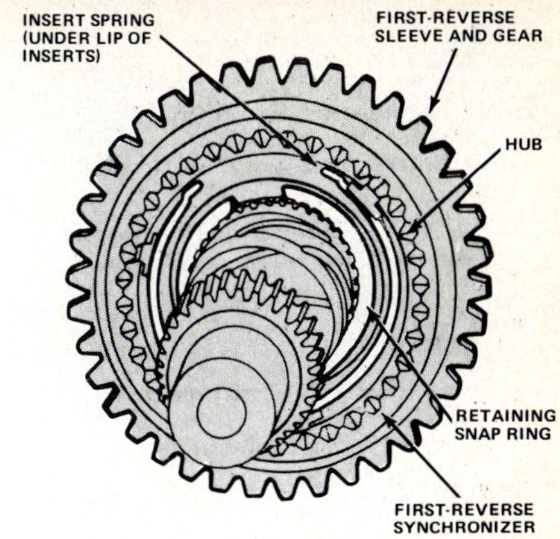

Fig. 45 Snap ring & insert spring positioning in 1st-reverse hub

NOTE: Do not attempt to hammer the bearing from the shaft since damage to the shaft and bearing will occur.

Rear Bearing & Adapter

1. Clamp rear bearing adapter in a soft-jawed vise.
2. Remove rear bearing retaining snap ring.
3. Remove bearing adapter from vise.
4. Press bearing from adapter with a suitable press.
5. Remove bearing adapter lock ring.

Assembly, Fig. 42

1. Lubricate reverse idler gear shaft bore and bushings with transmission lubricant.
2. Coat transmission case reverse idler gear thrust washer surfaces with petroleum jelly, then install thrust washers into case.

NOTE: Ensure that the locating tabs on the thrust washers are engaged into case slots.

3. Install reverse idler gear. Align gear bore, thrust washers and case bores, then install reverse idler gear shaft from rear of case. Align and seat roll pin in shaft in counterbore in rear of case.
4. Measure reverse idler gear end play by inserting a feeler gauge between thrust washer and gear. End play should be

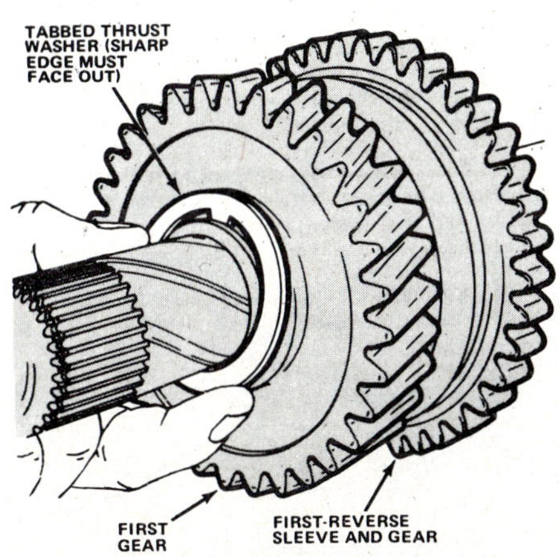

Fig. 46 First gear thrust washer installation

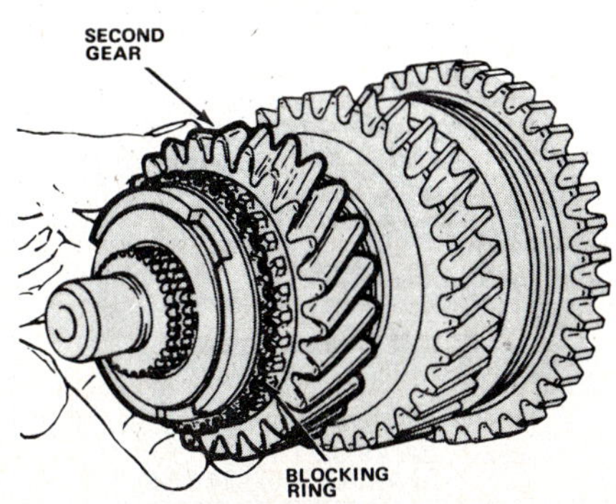

Fig. 47 Second gear installation

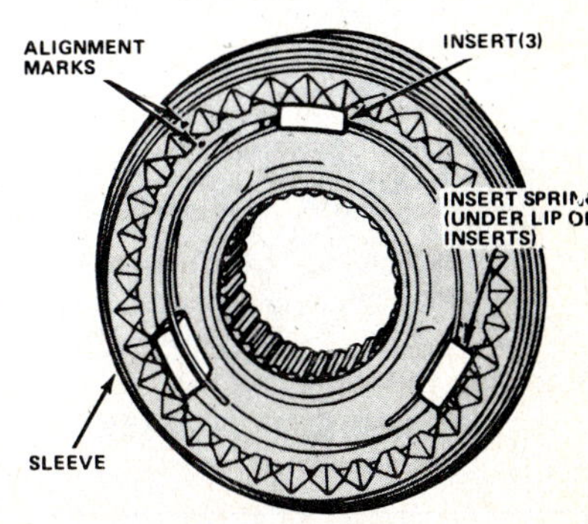

Fig. 48 2nd-3rd synchronizer assembly

Fig. 49 Measuring mainshaft end play

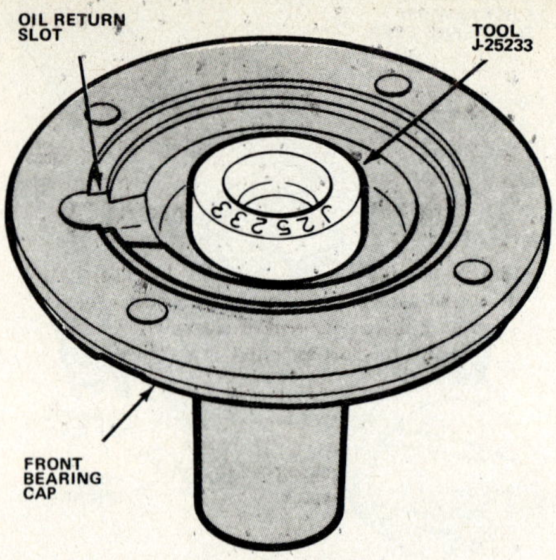

Fig. 50 Front bearing cap oil seal installation

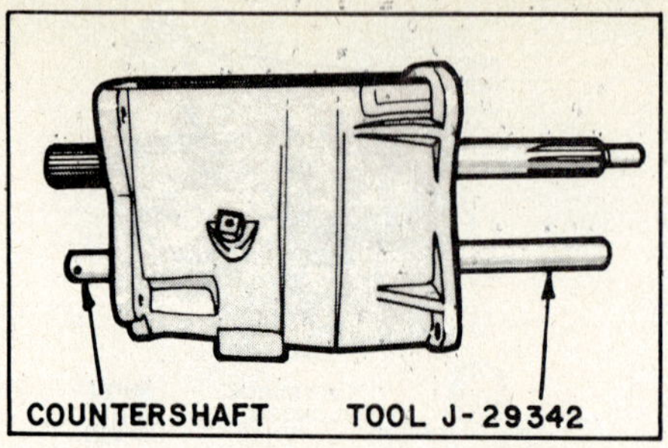

Fig. 51 Countershaft replacemant

.004–.018 inch. If not, replace thrust washers.

5. Coat needle bearings and bearing bores in countershaft gear with petroleum jelly. Insert tool J-25232 in bore of gear and install 25 needle bearings and one retainer in each end of gear.
6. Coat countershaft gear thrust washer surfaces with petroleum and place thrust washers in the case.

NOTE: Ensure that the locating tabs on the thrust washers are engaged into case slots.

7. Insert countershaft into rear case bore far enough to secure the rear thrust washer in position. This prevents the thrust washer from being displaced when countershaft gear is installed.
8. Install countershaft gear but do not install roll pin. Align gear bore, thrust washers and case bores, then install countershaft.

NOTE: Do not remove tool J-25232 completely.

9. Measure countershaft gear end play by inserting a feeler gauge between thrust washer and the countershaft gear. End play should be .004–.018 inch. If not, replace thrust washers. After proper end play is obtained, install arbor tool fully in countershaft gear. Permit gear to remain at bottom of case. Leave countershaft in rear case bore to secure thrust washer in position.

NOTE: The countershaft gear must remain at bottom of case to provide clearance for installation of the mainshaft and clutch shaft assemblies.

10. Coat all mainshaft splines and machined surfaces with transmission lubricant.
11. Start 1st-reverse synchronizer hub onto output shaft splines by hand. The end of the hub with the slots should face toward front of shaft. Use a suitable press to complete installation of the hub, then install retaining snap ring in the most rearward groove in shaft, Fig. 44.

NOTE: Do not attempt to hammer the hub onto shaft since damage to shaft and hub will occur.

12. Coat 1st-reverse hub splines with transmission lubricant.
13. Install 1st-reverse sleeve and gear halfway onto hub. The gear end of the sleeve must face toward rear of shaft. Align sleeve and hub using alignment marks made during disassembly.
14. Install insert springs into 1st-reverse hub. Ensure the spring is bottomed in hub and covers all three insert slots. Position the three "T" shaped inserts into hub with small ends in hub slots, Fig. 44. Push the inserts fully into hub so the inserts seat on the insert spring, then slide 1-reverse sleeve and gear over inserts until inserts engage in sleeve, Fig. 45.
15. Coat bore and blocking ring surface of first gear with transmission lubricant and place first gear blocking ring on tapered surface of gear.
16. Install first gear on mainshaft. Rotate gear until notches in blocking ring engage inserts in 1st-reverse hub, then install tabbed thrust washer, with sharp edge facing outward, and retaining snap on mainshaft, Fig. 46.
17. Coat bore and blocking ring surface of second gear with transmission lubricant and place second gear blocking ring on tapered surface of gear.
18. Install second gear on mainshaft with tapered surface of gear facing toward front of shaft, Fig. 47.
19. Install one insert spring into 2-3 hub. Ensure that the spring covers all three insert slots in hub. Align 2-3 sleeve to hub using alignment marks made during disassembly, and start sleeve onto hub.
20. Place three inserts into hub slots and on top of insert spring, then push sleeve fully onto hub to engage inserts in sleeve, Fig. 48. Install remaining insert spring in the exact same position as the first spring. The ends of both springs must cover same slots in hub and not be staggered.

NOTE: The inserts have a small lip on each end. When properly installed, this lip will fit over the insert spring.

21. Install 2-3 synchronizer assembly on mainshaft. Rotate second gear until notches in blocking ring engages inserts in 2-3 synchronizer assembly.
22. Install retaining snap ring on mainshaft and measure end play between snap ring and 2-3 synchronizer hub with a feeler gauge, Fig. 49. End play should be .004–.014 inch. If not, replace thrust washer and all snap rings on output shaft assembly.

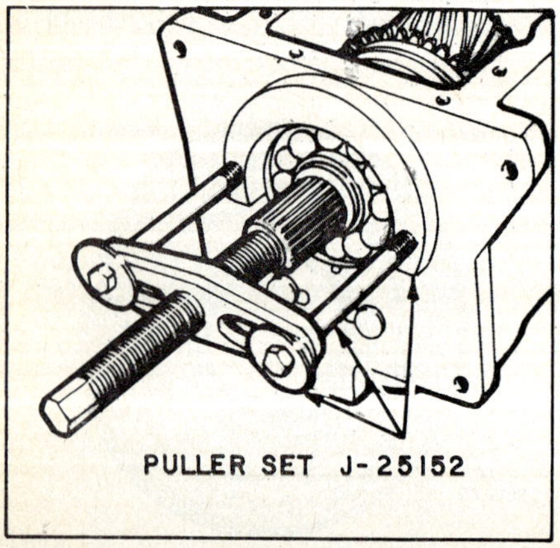

Fig. 52 Rear bearing removal

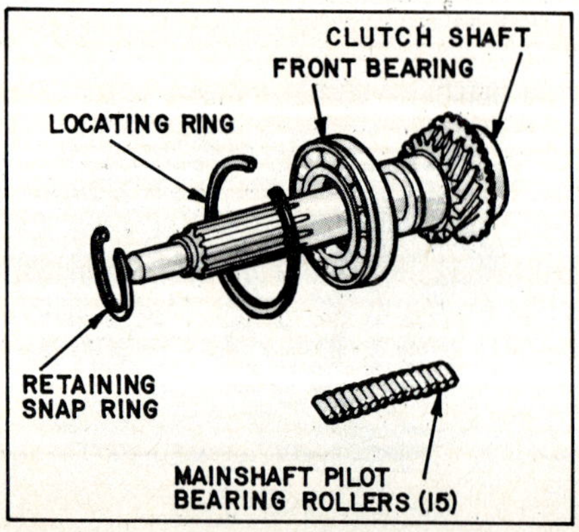

Fig. 53 Clutch shaft & front bearing assembly

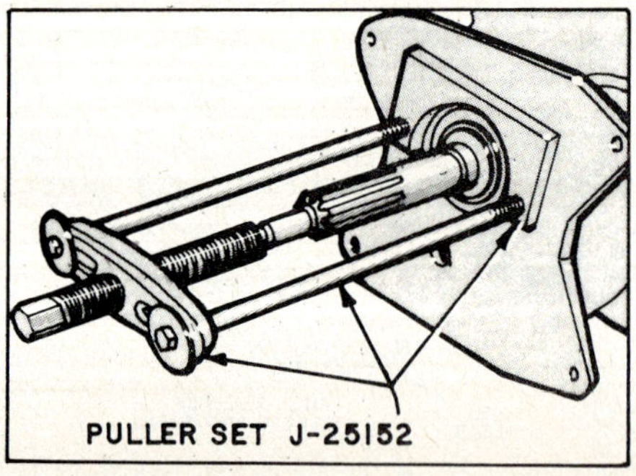

Fig. 54 Clutch shaft & front bearing removal

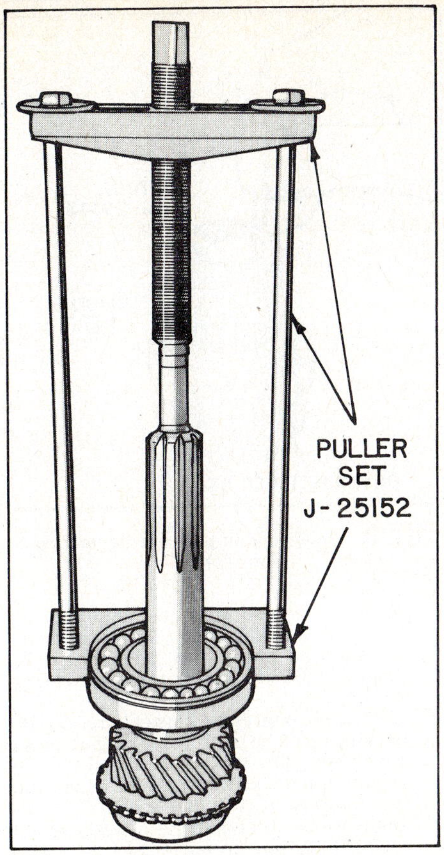

Fig. 55 Removing front bearing from clutch shaft

23. Install spacer and oil slinger on rear of mainshaft.
24. Install mainshaft assembly into case. Ensure 1st.-reverse sleeve and gear is in Neutral (centered) position on hub so gear end of sleeve will clear top of case when output shaft assembly is installed.
25. Press rear bearing into rear bearing adapter using a suitable press.
26. Install rear bearing retaining ring and bearing adapter lock ring in adapter.
27. Support mainshaft assembly and install rear bearing and adapter assembly into case. Use a mallet to seat adapter into case.
28. Press front bearing onto clutch shaft. Install bearing retaining snap ring on clutch shaft and lock ring in front bearing groove.

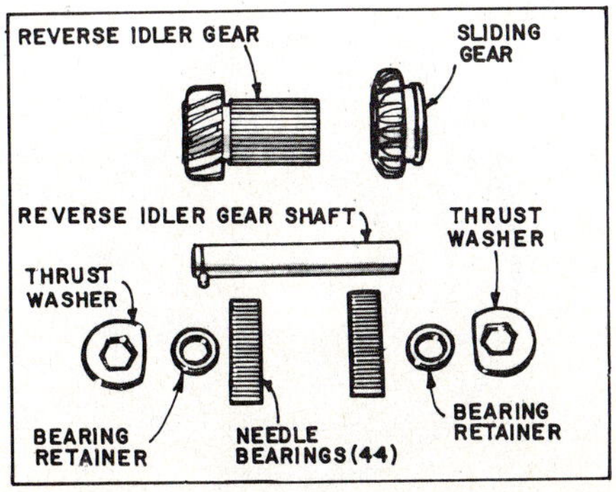

Fig. 57 Reverse idler gear assembly

NOTE: When properly installed, the snap ring groove in the front bearing will be nearest to the front of the clutch shaft.

---

29. Coat bearing bore of clutch shaft with petroleum jelly.
30. Install 15 roller bearings in clutch shaft bore.
31. Coat blocking ring surface of clutch shaft with transmission lubricant and place blocking ring on shaft.
32. Support mainshaft assembly and install clutch shaft through front bearing bore in case. Seat mainshaft pilot in clutch shaft roller bearings and tap front bearing into place with a mallet.
33. Apply a thin film of sealer to front bearing cap gasket and place gasket on case. Ensure that the gasket notch is aligned with the oil return hole in case.
34. Replace front bearing cap oil seal. Remove old seal with a screwdriver and drive new seal into bearing cap with tool J-25233, Fig. 50.
35. Install front bearing cap and torque attaching bolts to 33 ft. lbs.

NOTE: When installing the front bearing cap, align the cap and case index marks. Also, ensure that the cap oil return slot and case oil return hole are aligned.

---

36. Fabricate a wire loop approximately 18 to 20 inches long. Pass the wire loop under the countershaft gear assembly, then raise countershaft gear with loop. Align countershaft gear bore with front thrust washer and countershaft and start countershaft into gear. Align roll pin hole in countershaft and roll pin holes in case, then complete countershaft installation.
37. Install countershaft access plug in rear of case, seating plug with a mallet.
38. Install countershaft roll pin in case. Use a magnet or pair of needlenose pliers to insert pin in case. Use a 1/2 inch diameter punch to seat pin. Install fill plug.
39. Shift synchronizer sleeves into all gear positions to check for proper operation. If clutch shaft and mainshaft binds in Neutral position, check for sticking blocking rings on the first or second speed gear tapers.
40. Shift both synchronizers into gear to prevent gear rotation.
41. Install transfer case drive gear. Install and torque retaining nut to 150 ft. lbs.
42. Shift synchronizers into Neutral position.
43. Attach transmission to transfer case. Torque attaching bolts to 30 ft. lbs.

MODEL T-176

Disassembly

1. Remove transfer case from transmission. Drain lubricant from transmission.
2. Remove shift control housing.

NOTE: Two of the housing attaching bolts are of the dowel pin type. Note location of the bolts during disassembly.

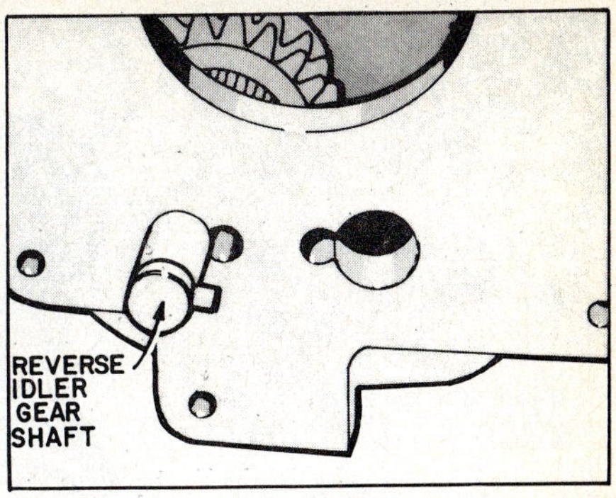

Fig. 56 Reverse idler gear shaft replacement

3. Using arbor tool J-29342, tap countershaft from rear of case, Fig. 51.
4. Remove rear bearing locating ring and snap ring. Using a suitable puller, remove rear bearing from case, Fig. 52.
5. Scribe alignment marks on front bearing cap and transmission case, then remove front bearing cap and gasket and oil seal from cap.
6. Remove locating ring and retaining snap ring from front bearing, Fig. 53.
7. Using suitable puller, remove clutch shaft and front bearing from case, Fig. 54.
8. Remove 3rd-4th gear blocking ring from clutch shaft or synchronizer hub.
9. Press front bearing from clutch shaft, Fig. 55.
10. Remove mainshaft pilot bearing rollers from clutch shaft, Fig. 53.
11. Move 3rd-4th gear synchronizer sleeve rearward to 3rd gear position. Tilt rear end of mainshaft downward and lift front upward and remove mainshaft and geartrain assembly.
12. Remove countershaft gear and arbor tool as an assembly.
13. Remove countershaft gear thrust washers and any fallen mainshaft pilot bearing rollers from the case.
14. Remove reverse gear idler shaft from rear of case. Remove gear assembly and thrust washers, Fig. 56.
15. Remove needle bearings and retainers from gear assembly, Fig. 57. Note position of sliding gear and remove from reverse idler gear.
16. Remove arbor shaft tool from countershaft gear and remove needle bearing and retainers, Fig. 58.

Mainshaft, Disassemble

1. Remove 3rd-4th synchronizer snap ring from front of mainshaft, Fig. 59.

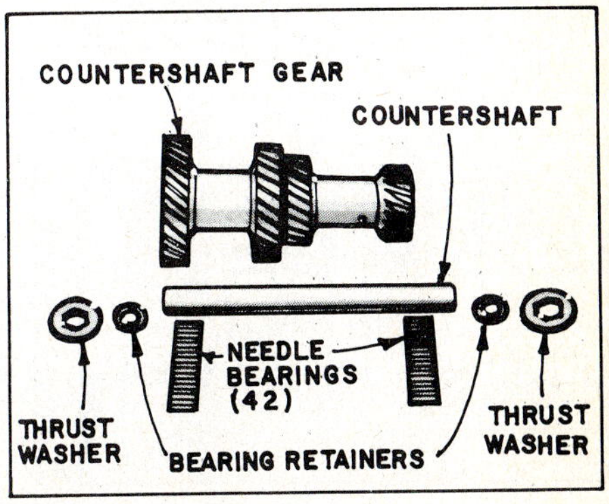

Fig. 58 Countershaft gear assembly

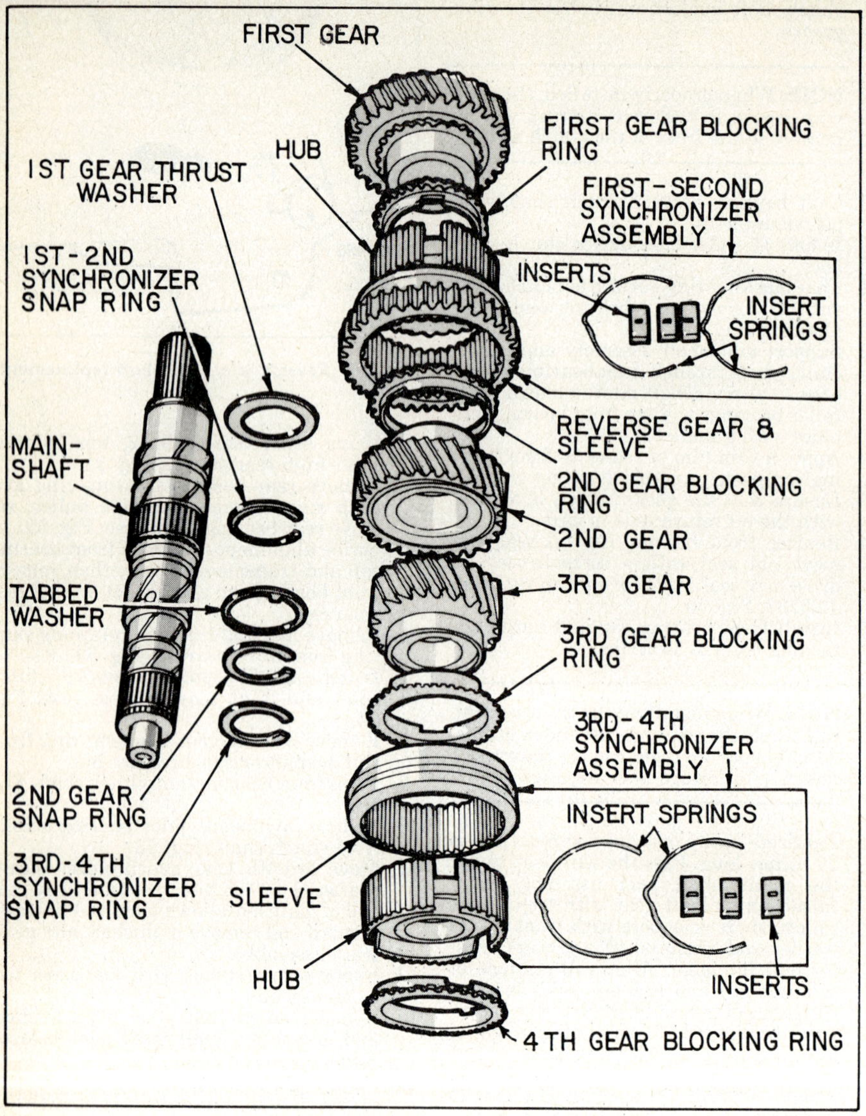

Fig. 59 Mainshaft assembly, Model T176

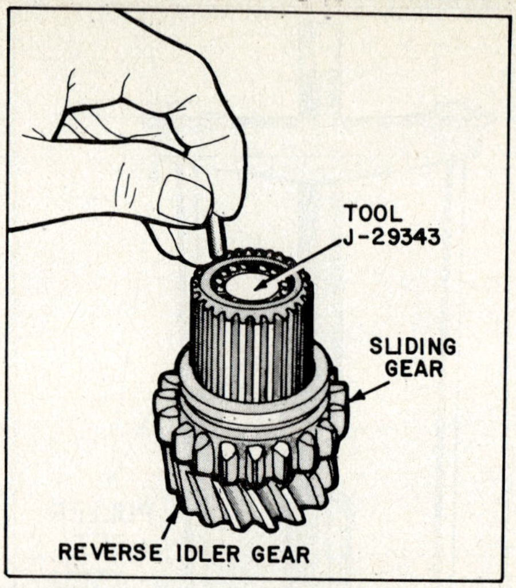

Fig. 60 Reverse idler gear needle bearing installation

2. Remove 3rd-4th assembly from shaft, then slide hub from sleeve. Remove insert springs and three inserts and blocking ring. Note position of insert springs for reassembly.
3. Remove 3rd gear from shaft.
4. Remove 2nd gear snap ring, gear, blocking ring and tabbed washer from shaft, Fig. 59.
5. Remove snap ring from 1st-2nd gear synchronizer hub. Remove hub, reverse gear and sleeve as an assembly. Mark hub and sleeve for assembly reference. Remove insert springs, inserts, sleeve and gear from hub.
6. Remove first gear thrust washer, then the first gear and blocking ring from shaft.

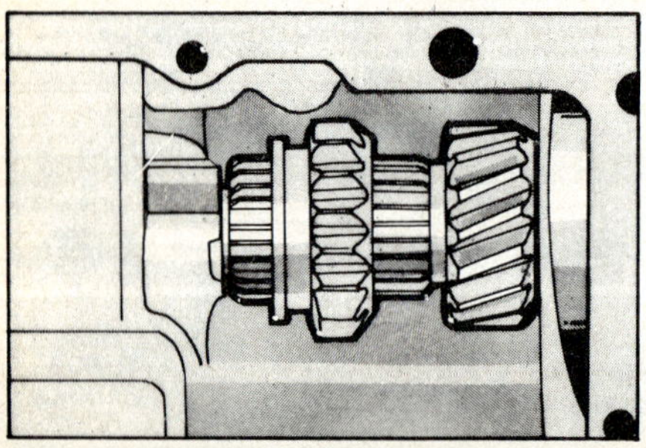

Fig. 61 Reverse idler gear installation

Assembly

1. Lubricate and assemble reverse idler gear and sliding gear, Fig. 57.
2. Install arbor tool J-29343 in reverse idler gear. Install 22 needle bearings and one retainer at each end of gear, Fig. 60.
3. Lubricate reverse idler gear thrust washers with petroleum jelly and install in case.

NOTE: Install thrust washers with flats facing the mainshaft and engage the washer locating tabs into slots of case.

4. Install reverse idler gear assembly into case, Fig. 61. Install reverse idler shaft from rear of case, Fig. 56. Ensure that roll pin seats in case recess.
5. Measure reverse idler gear end play by inserting a feeler gauge between thrust washer and gear. End play should be 0.004 to 0.018 inch. If not, replace thrust washers.
6. Lubricate countershaft gear bore, needle bearings and bearing bores in gear with petroleum jelly. Install arbor tool in bore of gear and install 21 needle bearings and one retainer in each end of gear.
7. Lubricate countershaft gear thrust washers with petroleum jelly and position in case.

NOTE: Install thrust washers with the locating tabs in slots of case.

8. Insert countershaft into rear case bore just enough to hold rear thrust washer in position. This will prevent thrust washer from being displaced during countershaft gear installation.
9. Install countershaft gear. Align gear bore, thrust washers, bores in case and install countershaft partially into case. Ensure arbor tool enters shaft bore at front of case.

NOTE: Do not remove countershaft arbor tool completely.

10. Measure countershaft gear end play with a feeler gauge inserted between gear and thrust washer. End play should be 0.004 to 0.018 inch. If not, replace the thrust washers. After correct end play has been

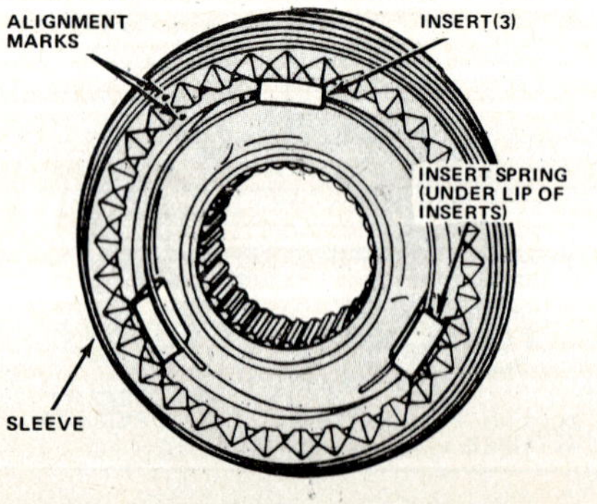

Fig. 62 Synchronizer insert spring installation

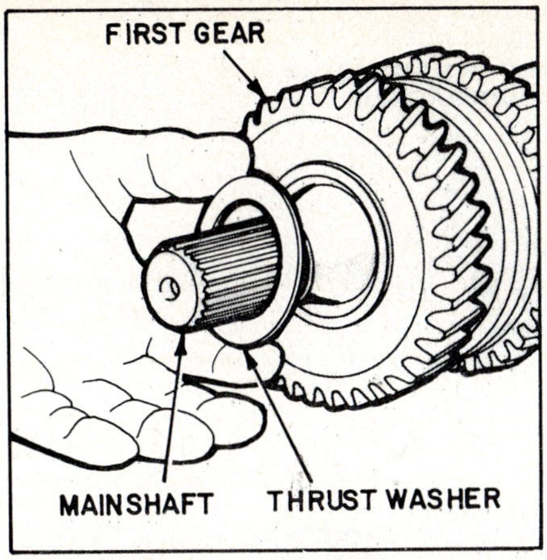

Fig. 63 First gear & thrust washer installation

obtained, reinstall arbor tool in counter shaft gear and allow gear to remain at bottom of case. Leave countershaft in rear case bore to retain thrust washer in position.

NOTE: Countershaft gear must remain at bottom of case to allow clearance for mainshaft and clutch shaft installation.

11. Lubricate mainshaft, synchronizer assemblies and gear bores with transmission lubricant.
12. Assemble 1st-2nd synchronizer hub and reverse gear and sleeve, Fig. 59:
    a. Install gear and sleeve on hub and place assembly flat on bench.
    b. Install the inserts into the hub slots.
    c. Install the insert spring. Position loop-end of spring in one insert, compress spring ends and insert the spring ends under lips of remaining two inserts, Fig. 62.
    d. Turn assembly over and install remaining insert spring as described in previous step. However, install this spring so open end is 180° opposite first spring.
13. Install assembled 1st-2nd synchronizer hub and reverse gear and sleeve on mainshaft, Fig. 59.
14. Install 1st-2nd synchronizer snap ring on mainshaft.
15. Install 1st gear and blocking ring on rear of mainshaft and the first gear thrust washer, Fig. 63.
16. Install new tabbed washer on mainshaft. Ensure that washer tab is seated in mainshaft tab bore, Fig. 64.
17. Install 2nd gear and blocking ring on mainshaft, then a new 2nd gear snap ring.
18. Install 3rd gear and blocking ring on mainshaft.
19. Assemble 3rd-4th synchronizer, Fig. 59:
    a. Install sleeve on synchronizer hub. Align parts using reference marks.
    b. Place assembled hub and sleeve flat on bench.
    c. Install the inserts into the hub slots.
    d. Install insert spring. Position loop-end of spring in one insert, compress spring ends and insert under lips of remaining two inserts, Fig. 62.
    e. Turn assembly over and install remaining insert spring as described in previous step. However, position this spring so open end faces 180° opposite first spring.
20. Install 3rd-4th synchronizer assembly on mainshaft.
21. Install new 3rd-4th synchronizer retaining snap ring on mainshaft and measure end play between hub and snap ring, Fig. 49. End play should be 0.004 to 0.014 inch. If not, replace snap rings and thrust washers.
22. Install mainshaft assembly into case. The synchronizers must be in neutral position so sleeves will clear top of case.
23. Install locating snap ring on front bearing, then install the front bearing partially onto clutch shaft.

NOTE: Do not install bearing completely since the shaft will not clear the countershaft gear during installation.

24. Lubricate bearing bore in clutch shaft and mainshaft roller bearings with petroleum jelly. Install 15 roller bearings in clutch shaft bearing bore.

CAUTION: Do not use chassis grease or other heavy grease in the clutch shaft bore since improper lubrication of the roller bearings may result.

25. Lubricate blocking ring surface of clutch shaft with transmission lubricant and position blocking ring on shaft.
26. Support mainshaft assembly and insert clutch shaft through front bearing bore in case. Seat mainshaft pilot hub in clutch shaft roller bearings and tap front bearing and clutch shaft into case using suitable mallet.
27. Install front bearing cap and tighten bolts finger tight.
28. Position rear bearing on mainshaft. Do not install bearing locating snap ring at this time. Start bearing onto shaft and into case bore using tool J-29345. Remove tool and finish bearing installation using suitable mallet. Install retaining snap ring when bearing is fully seated.

NOTE: To seat bearing on mainshaft, the bearing must be driven into case deeper than the locating snap ring would allow. Therefore, do not install the locating snap ring until after the bearing is fully seated on the shaft and the retaining snap ring is installed.

29. Remove front bearing cap, seat front bearing fully on clutch shaft and install bearing retaining snap ring.
30. Apply a thin film of sealer to front bearing cap gasket and position on case. Ensure that notch in gasket and oil return hole in case are aligned.
31. Remove front bearing cap oil seal using suitable tool. Install new seal using tool J-25233, Fig. 50.
32. Install front bearing cap and torque cap bolts to 12 ft. lbs.
33. Install locating ring on rear bearing. Reseat bearing in case if necessary.
34. Install countershaft as follows:
    a. Turn transmission case on end. Position case at edge of workbench with clutchshaft facing downward. Ensure that countershaft bore in front of case is accessible.
    b. Have an assistant hold case in position.
    c. Align countershaft gear bores with thrust washers and case bores and tap shaft into place.

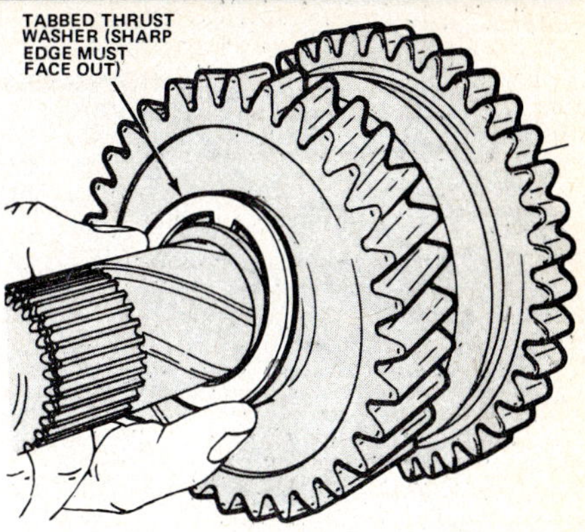

Fig. 64 Tabbed thrust washer installation

NOTE: The arbor tool will be driven through opposite end of case.

CAUTION: Do not damage thrust washers during installation of countershaft. Ensure that the washers, case bores and gear bores are aligned when installing countershaft.

35. Check operation of synchronizer sleeves in all gear positions. If clutch shaft or mainshaft appear to bind in neutral position, check for blocking rings sticking on gears. Use a screwdriver to free any sticking rings.
36. Fill transmission with 3.5 pints of SAE 85W-90 gear lubricant.
37. Install shift control housing with new gasket. Torque housing bolts to 12 ft. lbs.
38. Install transmission on transfer case.

MODEL T4

Transmission, Disassembly, Fig. 65

NOTE: Except for fill plug and gearshift lever attaching bolts, all bolts and threaded holes in the Model T4 Transmission case have metric threads. If replacement bolts are necessary, use only those of the same size and length as the originals.

1. Remove drain plug on transmission and drain lubricant.
2. Using pin punch and hammer, remove roll pin attaching offset lever to shift rail.
3. Remove adapter housing-to-transmission case bolts, then remove housing and offset lever as assembly.

CAUTION: Do not attempt to remove offset lever while the adapter housing is still bolted in place. The lever has a positioning lug engaged in the housing detent plate that prevents moving the lever far enough back for removal.

4. Remove detent ball and spring from offset lever, then remove roll pin from adapter housing or offset lever.
5. Remove and retain countershaft rear thrust bearing and race.
6. Remove transmission cover, shift fork assembly attaching bolts, then transmission cover.

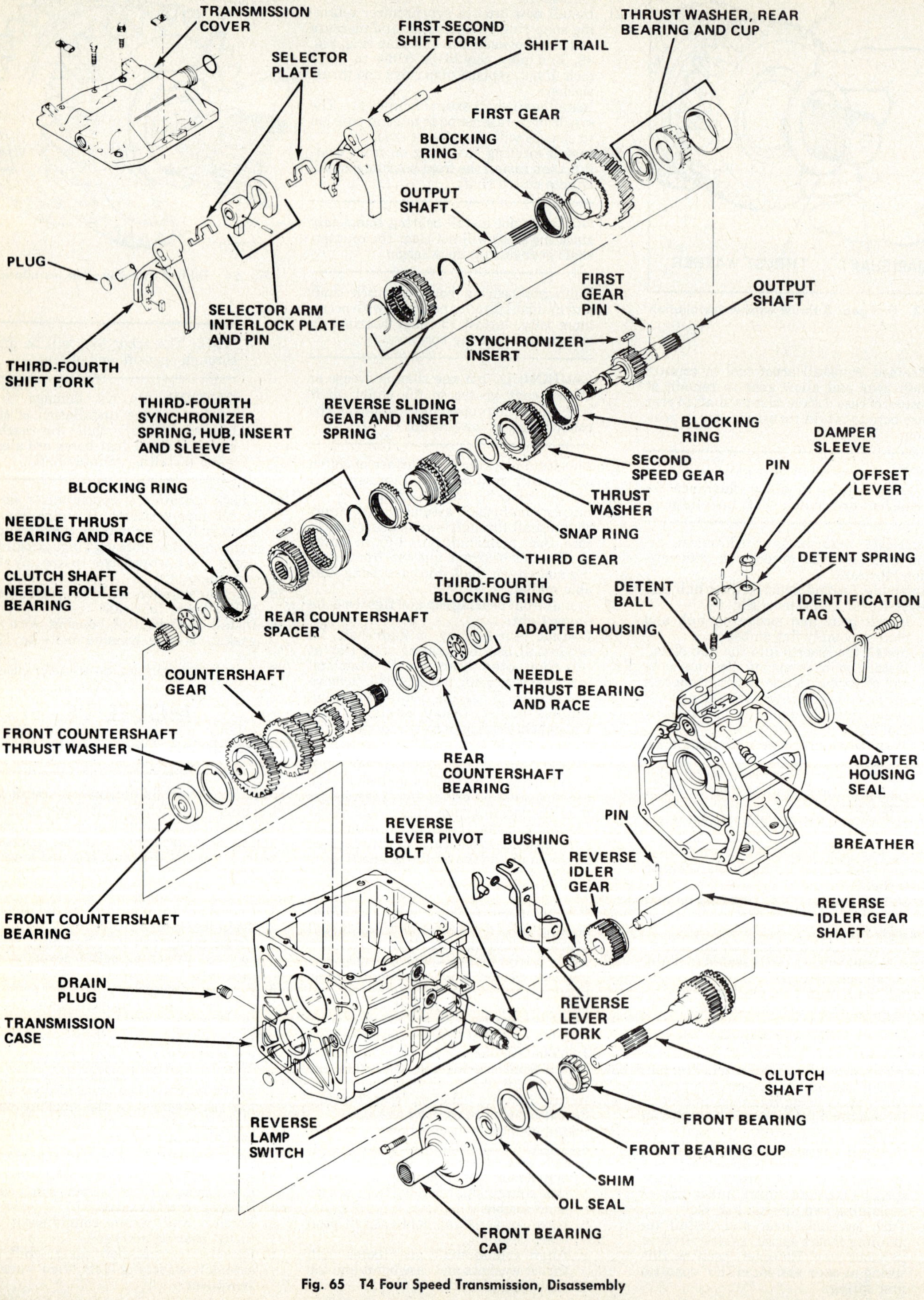

Fig. 65 T4 Four Speed Transmission, Disassembly

NOTE: Two shift control housing cover bolts are dowel-type alignment bolts. Note their location for assembly reference.

7. Remove "C" clip attaching reverse lever to reverse lever pivot bolt, then remove pivot bolt and remove reverse lever and reverse lever fork as assembly.
8. Using center punch, mark position of front bearing cap on transmission, then remove front bearing cap bolts and cap.
9. Remove front bearing race and end play shims from cap, then remove oil seal from cap using screwdriver.
10. Rotate clutch shaft so that flat on gear teeth is facing countershaft, remove shaft, then remove thrust bearing and 15 roller bearings from clutch shaft.
11. Remove output shaft bearing race, tapping front of output shaft with suitable mallet, if necessary.
12. Tilt output shaft assembly upward and remove transmission from case.
13. Using brass drift and arbor press, remove countershaft rear bearing, noting position of bearing for assembly reference. Bearing identification numbers face outward when bearing is installed correctly.
14. Move countershaft rearward, tilt shaft upward and remove from case. Noting position for assembly reference, remove countershaft front thrust washer from case.
15. Remove countershaft rear bearing spacer, then remove reverse idler shaft roll pin using pin punch and hammer.
16. Remove reverse idler shaft and gear, noting position of gear for assembly reference, then remove countershaft front bearing using arbor press.
17. Using removal tool J-29721 and J-22912 or equivalent, remove clutch shaft front bearing, then remove rear adapter housing seal using flat drift and hammer.
18. Remove backup lamp switch from transmission case.

Output Shaft Service

1. Remove thrust bearing washer from front end of output shaft, then scribe alignment marks on third-fourth synchronizer hub and sleeve for assembly reference.
2. Remove third-fourth synchronizer blocking ring, sleeve and hub assembly, noting position of hub and sleeve for assembly reference.
3. Remove third-fourth synchronizer and insert springs, then remove inserts and remove sleeve from hub.
4. Remove third gear from shaft, then remove snap ring retaining second gear on shaft and remove tabbed second gear thrust washer and second gear.
5. Using puller set J-29721 and adapters 293-39 or equivalents, remove output shaft bearing.
6. Remove first gear thrust washer, first gear roll pin, first gear and blocking ring.
7. Scribe alignment marks on first-second gear synchronizer sleeve and output shaft hub for assembly reference and remove insert spring and inserts from first-reverse sliding gear, then remove gear from output shaft hub.

CAUTION: Do not attempt to remove the first-second-reverse hub from output shaft. The hub and shaft are assembled and machined as a matched set during manufacture to insure concentricity. If any output shaft gear is replaced, the countershaft gear must also be replaced to maintain proper gear mesh and avoid noisy operation.

8. Lubricate output shaft and gear bores, then install and align first-second synchronizer sleeve on output shaft hub using reference marks.
9. Install three first-second synchronizer inserts and two insert springs in first-reverse synchronizer sleeve, engaging tang end of springs in same synchronizer insert but positioning open ends of springs to face 180 degrees from each other. Align sleeve and hub using reference marks.
10. Install blocking ring and second gear on mainshaft, then install tabbed thrust washer and second gear snap ring on mainshaft, being sure washer tab is properly seated in mainshaft notch.
11. Install blocking ring and first gear on output shaft, then install first gear roll pin in output shaft.
12. Using tool J-2995 or equivalent, and arbor press, install rear bearing on output shaft.
13. Install first gear thrust washer, then install third gear, third and fourth gear synchronizer hub inserts and sleeve on shaft, making sure hub offset faces forward.
14. Install thrust bearing washer on forward end of output shaft.

Shift Control Housing, Fig. 66

Disassembly

1. Place selector arm plates and shift rail in neutral position, then turn shift rail counterclockwise until selector arm disengages from selector arm plates and selector arm roll pin is accessible.
2. Pull shift rail rearward until selector contacts first-second shift fork and remove selector arm roll pin using pin punch, then remove shift rail.
3. Remove shift forks, selector arm plates, selector arm, and roll pin and interlock plate, then remove shift rail oil seal and "O" ring using suitable tool.
4. Using punch and hammer, remove shift rail plug.
5. Remove nylon inserts and selector arm plates from shift forks, noting position of inserts and plates for assembly reference.

Assembly

1. Install nylon inserts and selector arm plates in shift forks.
2. Coat edges of shift rail plug with sealer and install, then lubricate shift rail and bores and insert shift rail in cover. Install rail until end of rail is flush with inside edge of cover.
3. Position first-second shift fork (larger of the two forks) in cover with offset facing rear of cover and push shift rail through fork.
4. Position selector arm and C-shaped interlock plate in cover and insert shift rail through arm with widest part of interlock plate facing away from cover and selector arm roll pin hole facing downward and toward rear of cover.
5. Position third-fourth shift fork in cover with offset facing rear of cover, being sure third-fourth shift fork selector arm plate is positioned under first-second shift fork selector arm plate.
6. Insert shift rail through third-fourth shift fork and into front shift rail bore in cover, then rotate shift rail until selector arm plate at forward end of rail faces away from, but is parallel to cover.
7. Align roll pin holes in selector arm and shift rail, then install roll pin, being sure it is flush with surface of selector arm to prevent contact between pin and selector arm plates during shifting.
8. Install "O" ring in groove of shift rail oil seal, then install shift rail oil seal:
   a. Install tool J-26628-2 or equivalent over threaded end of shift rail.
   b. Lubricate lip of oil seal and slide over tool and onto shift rail.
   c. Seat seal in transmission cover using tool J-26628-1 or equivalent.

Transmission, Assembly

1. Coat countershaft front bearing outer cage with suitable sealer and install countershaft front bearing flush with case using arbor press.
2. Lubricate countershaft tabbed thrust washer and install so that tab engages proper depression in case.
3. Tipping case on end, install countershaft in front bearing bore, then insert countershaft rear bearing spacer.
4. Using tool J-29895 or equivalent, and suitable mallet, lubricate and install countershaft rear bearing so that it extends .125 in. (3 mm.) beyond case surface.
5. Position reverse idler gear in case with shift lever groove facing rear of case. Install reverse idler shaft from rear of case and install retaining roll pin in shaft.
6. Install assembled output shaft in case.
7. Using tool J-2995 or equivalent, and arbor press, install front clutch shaft bearing on clutch shaft.
8. Lubricate pilot roller bearings and install in clutch shaft.
9. Install thrust bearing and race in clutch shaft.
10. Install fourth gear blocker ring on output shaft, then install rear output shaft bearing race.
11. Install clutch shaft in case and engage shaft in third-fourth synchronizer sleeve and blocking ring.
12. Using tool J-26625 or equivalent, install replacement oil seal in front bearing cap.
13. Using tool J-29184 or equivalent install new oil seal in rear adapter housing.
14. Install front bearing race in cap and install cap.
15. Install reverse lever, pivot pin, and retaining "C" clip, being sure that reverse lever fork is engaged in reverse idler gear. Coat pivot pin threads with sealer.
16. Lubricate countershaft rear bearing race and thrust bearing and install in adapter housing.
17. Install adapter housing temporarily. Do not seal housing or torque bolts at this time.
18. Turn transmission case on end and mount dial indicator on adapter housing with indicator stylus on end of output shaft.
19. Turn clutch and output shaft and zero dial indicator.
20. Pull upward on output shaft to remove end play. Read end play dimension on indicator.

NOTE: To completely eliminate output shaft and clutch shaft end play, bearings must be preloaded from .001–.005 in. (0.03–0.13 mm.)

21. Select shim package measuring .001–.005 in. thicker than measured end play.
22. Place transmission on work bench horizontally and remove front bearing cap and race, then add shims to bearing cap to obtain preload and install clutch shaft bearing race in cap.
23. Install sealant on case mating surface of front bearing cap and install cap noting alignment marks, and torque retaining bolts to 15 ft. lbs.
24. Recheck end play. There must be no end play.
25. Remove adapter housing.
26. Move shift forks on transmission cover and synchronizer rings inside transmission to the neutral position.
27. Apply suitable sealant to cover mating surface of transmission.
28. Lower cover assembly onto case while aligning shift forks and synchronizer sleeves. Center cover to engage reverse relay lever and install dowel bolts in cover. Install remaining bolts in cover and torque all cover bolts to 9 ft. lbs.

NOTE: The offset lever-to-shift rail roll pin is in a vertical position when steps 26 and 28 are performed properly.

29. Apply sealant to adapter housing-to-transmission case mating surface.
30. Install adapter housing over output shaft and shift rail so that the shift rail just enters shift cover opening.
31. Install detent spring into offset lever. Place ball in neutral guide plate detent. Apply pressure on ball with detent spring and offset lever and slide offset lever on shift rail and seat adapter housing against transmission case.

NOTE: The offset lever and shift rail roll pin holes should be aligned in a vertical position following step 31.

32. Install adapter housing retaining bolts and torque to 25 ft. lbs.
33. Install roll pin in shift rail and offset lever and install damper sleeve in offset lever, then coat backup lamp switch threads with suitable sealant and install switch in case.

MODEL T5

Transmission, Disassembly, Fig. 67

1. Remove drain plug and drain lubricant.

CAUTION: Except for fill plug and gearshift lever attaching bolts, all threaded holes and bolts used in the model T5 transmission case have metric threads. When using replacement bolts be sure they are the same size and length as the originals.

2. Using pin punch and hammer, remove offset lever to shift rail attaching roll pin.
3. Remove adapter housing-to-transmission case bolts, then remove housing and offset lever as assembly.

NOTE: Do not attempt to remove offset lever while adapter housing is still bolted in place. The lever has a positioning lug engaged in the housing detent plate which prevents moving the lever far enough forward for removal.

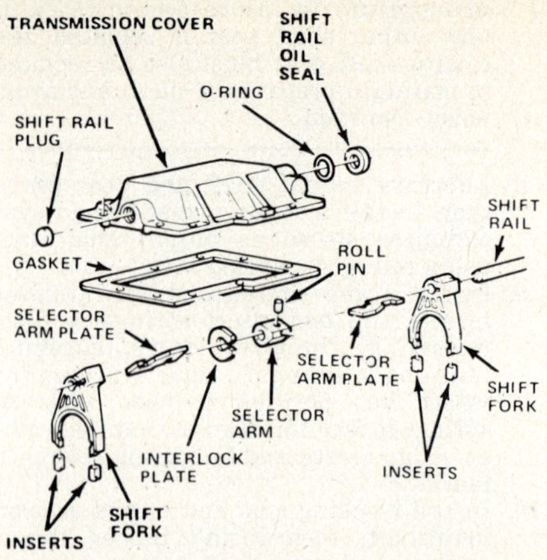

Fig. 66 Shift Control Housing Assembly, T4 & T5 Transmission

4. Remove detent ball and spring from offset lever and remove roll pin from adapter housing or offset lever, then remove plastic funnel, thrust bearing race, and thrust bearing from rear of countershaft. (The countershaft rear thrust bearing, washer, and plastic funnel may be found either on the end of countershaft or inside adapter housing.)
5. Remove transmission cover and shift fork assembly attaching bolts, noting location of the two alignment type dowel bolts for assembly reference, and remove cover.
6. Using hammer and punch, remove roll pin from fifth gear shift fork, being sure to place wood block under the fork during removal of roll pin to prevent damage to fifth gear/reverse shift rail.
7. Remove fifth gear synchronizer snap ring, shift fork, fifth gear synchronizer sleeve, blocking ring, and fifth speed drive gear from rear of countershaft.
8. Remove fifth gear insert retainer synchronizer springs and inserts from sleeve and hub, marking position of sleeve and hub for assembly reference.
9. Using tool J-25215 or equivalent, remove snap ring and fifth speed driven gear from rear of output shaft.
10. Using suitable tool, mark position of front bearing cap on front of transmission case for assembly reference.
11. Remove front bearing cap bolts and front bearing cap, then remove front race and end play shim(s) from cap and remove oil seal from cap using screwdriver.
12. Turn clutch shaft so that flat surface on main drive gear faces countershaft and remove clutch shaft from transmission case, then remove 15 clutch shaft needle bearings, thrust bearing, and race.
13. Remove output shaft rear bearing race, then tilt output shaft assembly upward and remove assembly from transmission case.
14. Using suitable tool, unhook over center link spring from rear of transmission case.
15. Remove "C" clip attaching reverse lever and fork assembly-to-reverse lever pivot pin.
16. Turn fifth gear-reverse shift rail clockwise (as viewed from top of transmission) to disengage rail from reverse lever assembly, then remove rail from rear of case.
17. Remove fork assembly and reverse lever pivot pin and detach reverse lever from reverse idler gear, then remove reverse lever and fork assembly from transmission case.
18. Remove rear countershaft snap ring and spacer, then, inserting brass drift through clutch shaft opening in front of transmission case and using arbor press, press countershaft assembly rearward to remove rear countershaft bearing. When properly installed, bearing identification numbers face outward.
19. Move countershaft assembly rearward inside case, then tilt upward and remove from case. Remove front countershaft thrust washer from case, noting washer position for assembly reference.
20. Remove countershaft rear bearing spacer, then remove roll pin from forward end of reverse idler shaft using hammer and punch.
21. Remove reverse idler shaft and gear from transmission case, noting gear position for assembly reference.
22. Using arbor press, remove countershaft front bearing from transmission case.
23. Using tool J-29721 and J-22912-01 or equivalent, remove clutch shaft front bearing.
24. Using suitable tool, remove rear adapter housing seal.

Output Shaft Service

Disassembly

1. Remove thrust bearing washer from front end of output shaft.
2. Mark third-fourth synchronizer hub and sleeve for assembly reference, then remove blocking ring, sleeve, and hub as assembly.
3. Remove third-fourth synchronizer insert springs and remove inserts, then remove sleeves from hubs.
4. Remove third gear from shaft, then remove snap ring retaining second gear on shaft and remove tabbed second gear thrust washer and second gear.
5. Using tool J-29721 and adapters 293-39 or equivalent, remove output shaft rear bearing.
6. Remove first gear thrust washer, first gear roll pin, first gear, and blocking ring.
7. Mark alignment of first-second gear synchronizer sleeve and output shaft hub for assembly reference, then remove insert spring and inserts from first-reverse sliding gear and remove gear from output shaft hub.

CAUTION: Do not try to remove the first-second-reverse hub from the output shaft as the hub and shaft are assembled and machined as a matched set during manufacture to insure concentricity.

Assembly

NOTE: If any output shaft gear is replaced, the countershaft gear must also be replaced to maintain proper gear mesh and avoid noisy operation.

1. Lubricate output shaft and gear bores, then install and align first-second synchronizer sleeve on output shaft hub using assembly reference marks.
2. Install 3 first-second synchronizer inserts and 2 insert springs in first-reverse synchronizer sleeve. Engage tang end of each spring in same synchronizer insert but

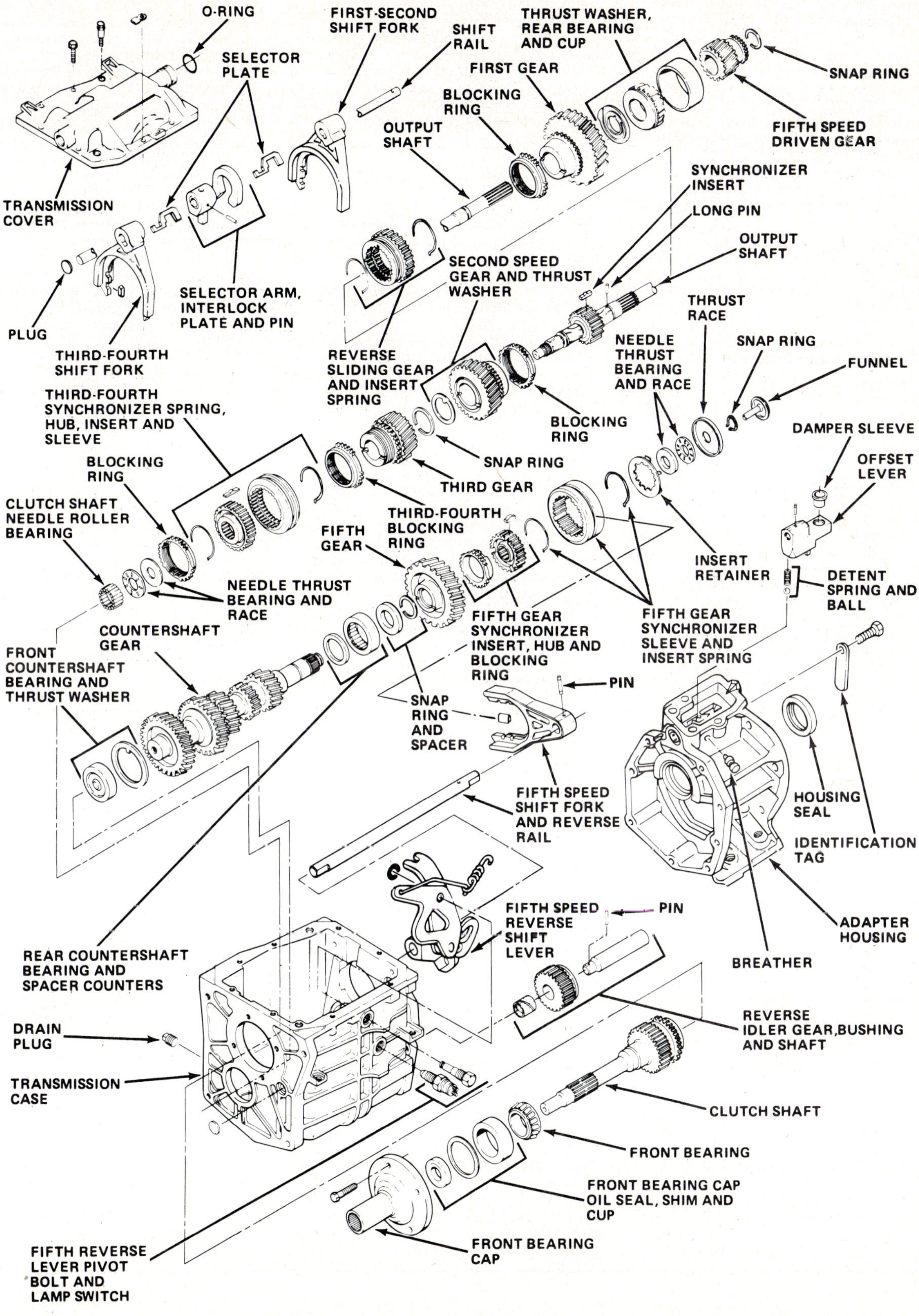

Fig. 67 T5 Five Speed Transmission, Disassembly

position open ends of insert springs to face 180 degrees from each other. Check alignment of sleeve and hub using assembly reference marks.
3. Install blocking ring and second gear on mainshaft, then install tabbed thrust washer and second gear retaining snap ring on mainshaft, making sure washer tab is properly seated in mainshaft notch.
4. Install blocking ring and first gear on output shaft and install first gear roll pin in output shaft.
5. Using tool J-2995 or equivalent, and suitable press, install rear bearing on output shaft and install first gear thrust washer.
6. Install third gear, third and fourth gear synchronizer hub inserts, and sleeve on shaft with hub offset facing forward, then install thrust bearing washer on forward end of shaft.

Shift Control Housing, Fig. 65

Disassembly

1. Place selector arm plates and shift rail in neutral position.
2. Turn shift rail counterclockwise until selector arm disengages from selector arm plates and selector arm roll pin is accessible, then pull shift rail rearward until selector arm contacts first-second shift fork.
3. Using suitable punch, remove selector arm roll pin and remove shift rail.
4. Remove shift forks, selector arm plates, selector arm, roll pin, and interlock plate, then remove shift rail oil seal and "O" ring using screwdriver.
5. Using hammer and punch, remove shift rail plug, then remove selector arm plates and nylon inserts from shift forks, noting position of plates and inserts for assembly reference.

Assembly

1. Install selector arm plates and nylon inserts in shift forks, then coat edges of shift rail plug with suitable sealer and install plug.
2. Lubricate shift rail and bores and insert shift rail in cover so that end of rail is flush with inside edge of cover.
3. Position first-second shift fork (larger of two forks) in cover with offset facing rear of cover and push shift rail through fork.
4. Position selector arm and "C" shaped interlock plate in cover and insert shift rail in cover with widest part of interlock plate facing away from cover and selector arm roll pin hole facing downward and toward rear of cover.
5. Position third-fourth shift fork in cover with offset facing rear of cover. Third-fourth shift fork selector arm plate must be positioned under first-second shift fork selector arm plate.
6. Insert shift rail through third-fourth shift fork and into front shift rail bore in cover, then turn shift rail until selector arm plate at forward end of rail faces away from, but is parallel to cover.
7. Align roll pin holes in selector arm and shift rail, then install roll pin flush with surface of selector arm to prevent contact between pin and selector arm plates during shifts.
8. Install "O" ring in groove of shift rail oil seal and install shift rail oil seal thusly:
   a. Install tool J-26628-2 over threaded end of shift rail.
   b. Lubricate lip of oil seal and slide seal over tool and onto shift rail.
   c. Using tool J-26628-1 or equivalent, seat oil seal in transmission cover.

Transmission, Assembly

1. Coat countershaft front bearing with suitable sealant and, using arbor press, install countershaft front bearing flush with case.
2. Lubricate countershaft tabbed thrust washer and install so tab engages proper depression in case.
3. Tip transmission case on end and install countershaft in front of bearing bore, then install countershaft rear bearing spacer.
4. Lubricate rear bearing and, using tool J-29895 and J-33032 or equivalent, install bearing so that it extends .125 in. (3 mm.) beyond case surface. Proper tool must be used to prevent needles from catching on countershaft shoulder.
5. Position reverse idler gear in case with shift lever groove facing rear of case and install reverse idler shaft from rear of case. Install retaining roll pin in shaft.
6. Install output shaft assembly in case.
7. Using tool J-2995 or equivalent and arbor press, install front clutch shaft bearing on clutch shaft, then lubricate pilot roller bearings and install in clutch shaft.
8. Install thrust bearing and race in clutch shaft, then install rear output shaft bearing race cap.
9. Install fourth gear blocking ring on output shaft, then install clutch shaft in case and engage in third-fourth synchronizer sleeve and blocking ring.
10. Using tool J-26625 or equivalent, install new oil seal in front bearing cap.
11. Install front bearing race in cap and temporarily install front bearing cap.
12. Coat pivot bolt threads with sealer, then install fifth speed-reverse lever, pivot bolt, and retaining "C" clip. Engage reverse lever fork in reverse idler gear.
13. Install fifth speed driven gear and retaining snap ring on rear of output shaft, then install countershaft rear bearing spacer and retaining snap ring.
14. Install fifth speed gear on countershaft, then insert fifth speed-reverse rail through opening in rear of case and install in reverse fifth speed lever, turning rail during installation to facilitate engagement with lever.
15. Install fifth speed-reverse lever over center spring, then assemble fifth gear synchronizer sleeve, insert springs, and insert retainer using assembly reference marks.
16. Install plastic inserts in notches on either side of fifth speed shift fork, then put assembled fifth gear synchronizer sleeve on fifth speed shift fork and slide onto countershaft and fifth speed-reverse rail, making sure roll pin holes in fifth-reverse rail and fifth speed shift fork are aligned.
17. Placing assembled fifth speed-reverse rail and shift fork on block of wood, install retaining roll pin, then install thrust race against fifth speed synchronizer hub and install retaining snap ring.
18. Lubricate needle type thrust bearing and thrust race and install bearing against race on countershaft.
19. Install thrust race over thrust bearing and insert plastic funnel into hole in end of countershaft gear.
20. Install adapter housing temporarily without sealing or torquing bolts, then turn transmission case on end and mount dial indicator on adapter housing with stylus on end of output shaft.
21. Turn clutch and output shaft to zero dial indicator, then pull upward on output shaft until end play is eliminated. To completely eliminate end play, bearings must be preloaded from .001–.005 in. (0.03–0.13 mm.)
22. Select shim pack measuring .001–.005 in. thicker than measured end play and, placing transmission horizontally on workbench, remove front bearing cap and race, then add shim to bearing cap and install clutch shaft bearing race cap.
23. Apply suitable sealant on case mating surface of front bearing cap, then align assembly reference marks and install front bearing cap, torquing retaining bolts to 15 ft. lbs.
24. Recheck end play to make sure there is no end play.
25. Remove adapter housing and, using tool J-29184 or equivalent, install adapter housing rear seal.
26. Move shift forks on transmission cover and synchronizer rings inside transmission to the neutral position.
27. Apply suitable sealant to cover mating surface of transmission and lower cover assembly onto case, aligning shift forks and synchronizer sleeves, then center cover on case to engage reverse relay lever and install dowel bolts in cover.
28. Install remaining bolts and torque all cover bolts to 9 ft. lbs.

NOTE: The offset lever-to-shift rail roll pin hole should be in a vertical position after completion of steps 26 through 28.

29. Apply suitable sealant to adapter housing-to-transmission case mating surface and install adapter housing over output shaft and shift rail so that shift rail just enters shift cover opening.
30. Install detent spring into offset lever and place ball in neutral guide plate detent. Applying pressure on ball with detent spring and offset lever, slide offset lever on shift rail and seat adapter housing against transmission case so that the offset lever and shift rail roll pin holes are aligned and in a vertical position.
31. Install and torque adapter housing retaining bolts to 25 ft. lbs.
32. Install roll pin in offset lever and shift rail, then install damper sleeve in offset lever.
33. Coat backup lamp switch threads with suitable sealant and install switch in case.

HEAVY DUTY CLUTCHES

Lipe DP Clutch

As shown in Figs. 1 and 2, DP type clutches are single and two plate direct pressure units. The nameplate attached to the flywheel ring (cover) identifies the cover plate as to the serial number and model number. Driven discs are stamped with an assembly number.

This type clutch can be serviced easily without the use of special tools. It is of the dry disc, push type construction, with no internal clutch adjustment to compensate for wear. The only adjustment is made by adjusting the clutch pedal free travel, which should be maintained at 1½".

Servicing the Clutch

After the clutch is disassembled, clean all parts thoroughly. Inspect all parts as indicated below.

Pressure Plate

The pressure plate must be replaced if:

1. Friction surface is severely heat checked.
2. It is warped in excess of .015".
3. Driving slots are worn.
4. Pin holes are elongated.

The pressure plate may be salvaged if the friction surface cleans up by regrinding any amount up to a maximum of 1/32" from "Dimension C", Fig. 3, and chart, Fig. 4. *Using a heat checked or warped pressure plate will cause the facings to disintegrate quickly.*

Intermediate Plate

The intermediate plate in the two plate clutches may not be used if the driving slots are worn or deep cracks are present on either friction surface. If the clearance between the contact face of the driving pins in the flywheel is over .012", it is recommended that the intermediate plate and driving pins be replaced. If the driving slots or pins are not worn and the plate is not warped over .015", it may be reground. However, it is recommended that the grinding be kept to a minimum with not over .015" being removed from either side for a total of .030" (see chart, Fig. 4 for thickness specifications).

Pressure Springs

It is of vital importance that the pressure springs have proper tension. They should be checked for compression weight in accordance with the weight given in the chart, Fig. 4.

CAUTION: Underweight springs will reduce clutch capacity and possibly cause clutch slippage; therefore, to insure best clutch performance, it is recommended that new springs be installed.

Eyebolts, Release Levers and Links

It is extremely important that the pin be a tight press fit in the eyebolt. If looseness is detected, replace with a new eyebolt assembly. If the pin is worn, replace with a new eyebolt assembly. The eyebolt and adjusting nut threads must be free fitting, but not excessively loose.

If the release lever pin holes are worn or elongated, replace levers. If release lever buttons are worn, replace with new buttons.

If link holes are worn or elongated, replace links.

Disc Assembly

The disc should be replaced if the hub splines are worn, disc is distorted, disc is cracked or broken. If the disc is acceptable for relining, use a Lipe Facing Kit to obtain correct rivets and the proper facing material of original specifications, Fig. 4.

Fig. 6 illustrates the correct riveting for types of facings to be used. The left view shows the correct method of riveting fabric facings, using a star set anvil, resulting in the split end of the rivet conforming to the tapered counterbore of the facing. The roll set type rivet should be used for installing metallic facings, Fig. 6, right view.

Lubrication

Place a small amount of high melting point grease on the driving lugs of the flywheel ring (cover) and mating slots in the pressure plate, as well as the pins.

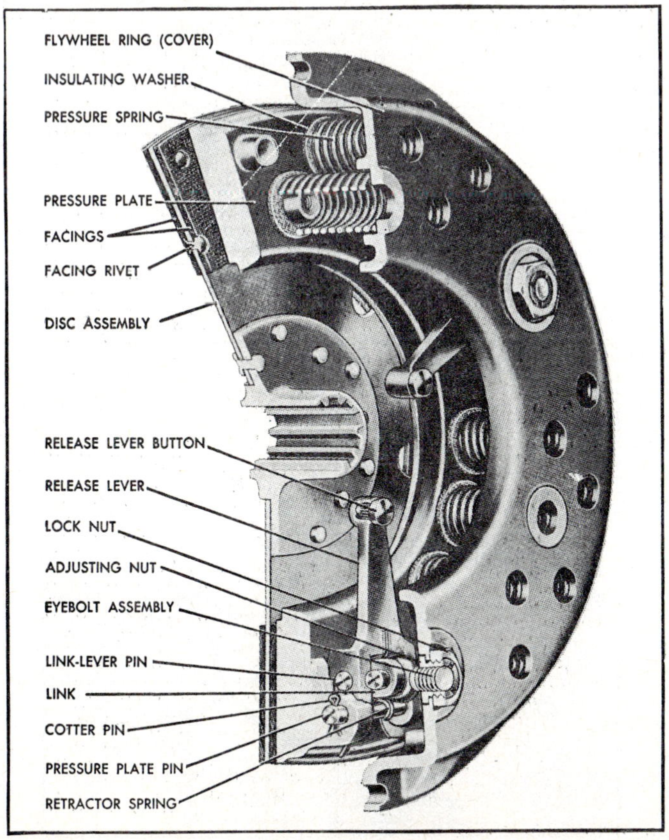

Fig. 1 Lipe DP single plate clutch

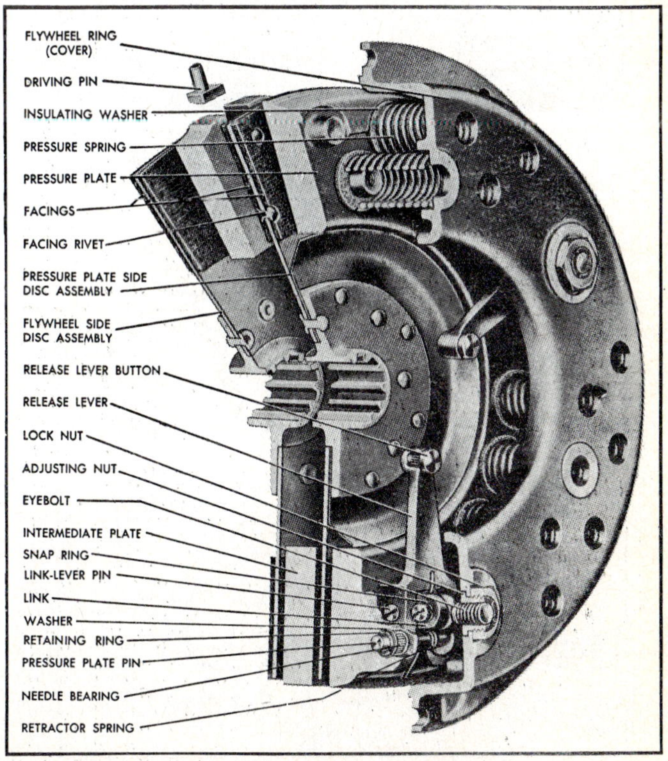

Fig. 2 Lipe DP two plate clutch

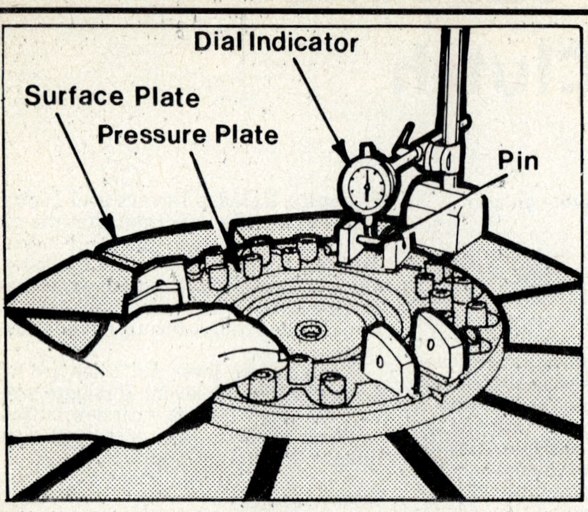

Fig. 3 Checking Dimension "C" on DP clutch

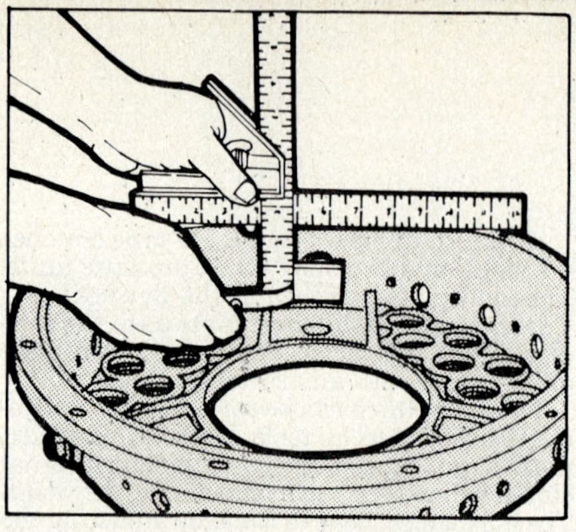

Fig. 5 Aligning drive pins in two plate clutch cover

Flywheel Ring (Cover)

The flywheel ring must be replaced if any cracks are present, driving lugs are worn, or if it is distorted.

The 14″ and 15″ single plate units and 15″ two plate clutches have milled lugs in the flywheel which engage with machined surfaces of the release lever bosses on the pressure plates.

The 14″ two plate unit has replaceable driving pins while in the 13″ single plate the pressure plate engages the cover through slots machined in the cover. A check of the clearance or wear of the driving surfaces in either cover can be made by using a new pressure plate as a gauge. There should not be less than .004″ or more than .014″ movement possible between the two mating parts. A tight fit may hamper smooth engagement and release while too loose a fit may cause noise.

In replacing the driving pins in the 14″ two plate cover, extreme care must be taken to assure accurate alignment. A straight edge and a square placed as indicated in Fig. 5 will serve to properly align the driving pins. It is recommended that a .0015″ feeler gauge be used as an aid. Pins may be turned with a 1½″ open end wrench and must be held securely while tightening the lock nuts. The driving pins must be a drive fit into the flywheel ring. Nuts should be torqued to 30 ft-lbs.

CAUTION: Burred edges on flange surface should be filed so as to assure flange being drawn tight to rim of engine flywheel at time of clutch installation. Replace retractor springs with new springs.

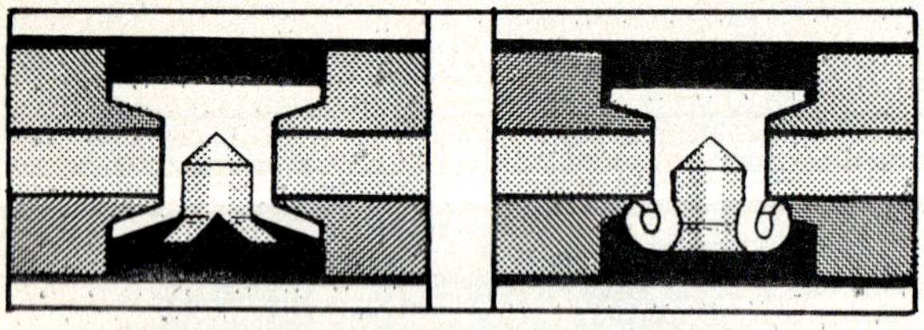

Fig. 6 Left view shows correct riveting for fabric or non-metallic facings; right view shows proper riveting for metallic facings

Clutch Assembly

1. Place pressure plate on arbor press with friction surface down.
2. Assemble adjusting nut to eyebolt assembly so that 3 or 4 threads are showing on inside of adjusting nut (use new links only).
3. Assemble the two links to eyebolt, install same to release lever assembly, using link lever pin, Fig. 7.
4. At this point, install retractor spring with the hook end seated in groove of lever, running under eyebolt assembly to groove in pressure plate.
5. Then assemble to pressure plate, inserting pressure pin through coiled loop of retractor spring. Make sure cotter pin hole of pressure plate pin is parallel to pressure plate and install cotter pin, Fig. 8.

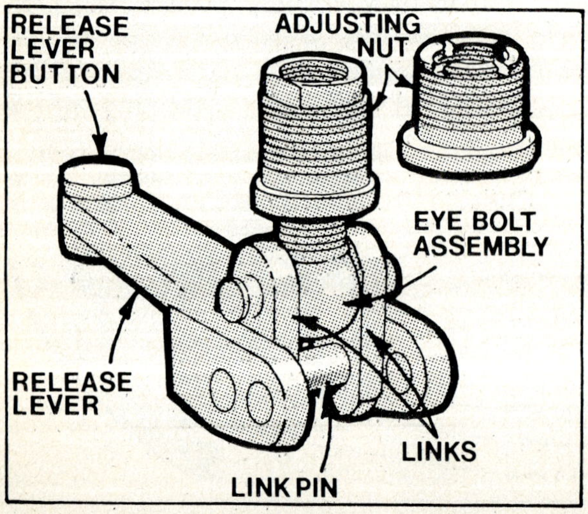

Fig. 7 Assemble eyebolt and related components as shown

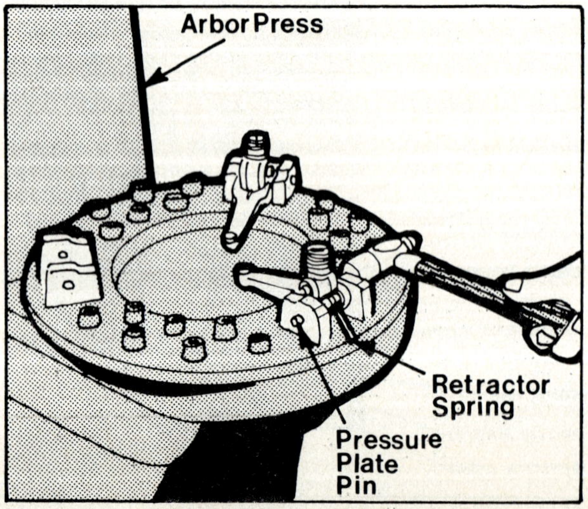

Fig. 8 Assembling release levers to pressure plate

Model Number	Dimension "A"	Flywheel Depth + .005" − .005"	Facing Thickness	PRESSURE SPRING DATA: Quantity	Color	Minimum Weight✦	PRESSURE PLATE DATA: Dimension "C" New
140-51-374	1"	1.187"	3/16"	21	Dark Maroon	130 lbs. @ 1 27/32"	1.687"
140-51-490	1"	1.187"	3/16"	21	Bronze	250 lbs. @ 1 31/32"	1.687"
140-51-496	1"	1.187"	3/16"	21	Dark Maroon	130 lbs. @ 1 27/32"	1.687"
140-51-501	1"	1.187"	3/16"	21	Aluminum	225 lbs. @ 1 31/32"	1.687"
140-51-505	1"	1.187"	3/16"	21	Aluminum	225 lbs. @ 1 31/32"	1.687"
150-51-506	1 1/8"	1.750"	3/16"	27	Purple	170 lbs. @ 2 1/16"	1.875"
140-51-519	1"	1.187"	7/32"	21	Dark Maroon	130 lbs. @ 1 27/32" ✦	1.687"
140-51-520	1"	1.187"	3/16"	21	Dark Maroon	130 lbs. @ 1 27/32"	1.687"
140-51-521	1"	1.594"	3/16"	21	Pink	120 lbs. @ 2 3/32"	1.968"
140-51-526	1"	1.187"	3/16"	21	Pink	120 lbs. @ 2 3/32"	1.687"
150-51-529	1 1/8"	1.750"	3/16"	21	Purple	170 lbs. @ 2 1/16"	1.875"
140-52-531†	1"	1.406"	5/32"	21	Dark Maroon	130 lbs. @ 1 27/32"	1.687"
130-51-532	31/32"	Flat	5/32"	20	Dark Maroon	130 lbs. @ 1 27/32"	1.406"
150-51-534	1 1/8"	1.750"	3/16"	21	Purple	170 lbs. @ 2 1/16"	1.875"
140-51-535	1"	1.187"	3/16"	21	Light Green	150 lbs. @ 1 27/32"	1.687"
140-51-536	1"	1.187"	3/16"	21	Light Green	150 lbs. @ 1 27/32"	1.687"
140-51-537	1"	1.187"	3/16"	21	Light Green	150 lbs. @ 1 27/32"	1.687"
140-51-538	1"	1.187"	3/16"	21	Dark Maroon	130 lbs. @ 1 27/32"	1.687"
140-51-539	1"	1.187"	3/16"	21	Dark Maroon	130 lbs. @ 1 27/32"	1.687"
150-51-544**	5/16"	1.750"	3/16"	27	Purple	170 lbs. @ 2 1/16"	1.875"
130-51-547	31/32"	Flat	5/32"	20	Dark Maroon	130 lbs. @ 1 27/32"	1.406"
140-51-549	1"	1.594"	3/16"	21	Pink	120 lbs. @ 2 3/32"	1.968"
150-51-551	15/16"	1.593"	3/16"	21	Purple	170 lbs. @ 2 1/16"	1.875"
130-51-552	31/32"	Flat	5/32"	12	Dark Maroon	130 lbs. @ 1 27/32"	1.406"
140-51-556	15/16"	1.187"	7/32"	18	Dark Maroon	130 lbs. @ 1 27/32" ✦	1.687"
150-52-557††	1 1/8"	2.937"	5/32"	24	Purple	170 lbs. @ 2 1/16"	1.875"
150-51-558**	1 1/8"	1.750"	3/16"	27	Purple	170 lbs. @ 2 1/16"	1.875"
150-51-562**	5/16"	1.750"	3/16"	21	Purple	170 lbs. @ 2 1/16"	1.875"
150-51-564**	1 1/8"	1.750"	3/16"	21	Purple	170 lbs. @ 2 1/16"	1.875"
150-51-565**	1 1/8"	1.750"	3/16"	21	Purple	170 lbs. @ 2 1/16"	1.875"
150-51-570**	1 1/8"	1.750"	3/16"	18	Purple	170 lbs. @ 2 1/16"	1.875"
140-51-575	1"	1.187"	3/16"	18	Dark Maroon	130 lbs. @ 1 27/32"	1.687"
150-51-579**	1 1/8"	1.750"	3/16"	27	Bronze	250 lbs. @ 1 31/32"	1.875"
150-51-580	1 1/8"	1.750"	3/16"	21	Purple	170 lbs. @ 2 1/16"	1.875"
140-52-581†	1"	1.406"	5/32"	15	Dark Maroon	130 lbs. @ 1 27/32"	1.687"
140-51-583	1"	1.594"	3/16"	21	Blue	150 lbs. @ 2 3/32"	1.968"
150-51-586**	1 1/8"	1.750"	3/16"	18	Purple	170 lbs. @ 2 1/16"	1.875"
150-52-587††	1 1/8"	2.937"	5/32"	18	Purple	170 lbs. @ 2 1/16"	1.875"
130-51-588	31/32"	Flat	5/32"	16	Dark Maroon	130 lbs. @ 1 27/32"	1.406"
140-52-589†	1"	1.406"	5/32"	12	Dark Maroon	130 lbs. @ 1 27/32"	1.687"
150-51-595	1 3/8"	1.312"	7/32"	24	Purple	170 lbs. @ 2 1/16"	1.875"
140-51-599	1"	1.187"	3/16"	21	Light Green	150 lbs. @ 1 27/32"	1.687"
150-51-608	1"	1.594"	3/16"	24	Purple	170 lbs. @ 2 1/16"	1.875"
150-51-634**	1 1/8"	1.750"	3/16"	21	Purple	170 lbs. @ 2 1/16"	1.875"
150-51-654**	5/16"®	1.750"	3/16"	27	Purple	170 lbs. @ 2 1/16"	1.875"
150-51-655	15/16"	1.594"	3/16"	21	Purple	170 lbs. @ 2 1/16"	1.875"
150-51-656	15/16"	1.594"	3/16"	24	Purple	170 lbs. @ 2 1/16"	1.875"
150-51-657**	5/16"®	1.750"	3/16"	21	Purple	170 lbs. @ 2 1/16"	1.875"
150-51-658**	1 1/8"	1.750"	3/16"	27	Purple	170 lbs. @ 2 1/16"	1.875"
150-51-659	1 1/8"	1.750"	3/16"	21	Purple	170 lbs. @ 2 1/16"	1.875"
150-51-660**	1 1/8"	1.750"	3/16"	18	Purple	170 lbs. @ 2 1/16"	1.875"
150-51-661**	1 1/8"	1.750"	3/16"	24	Purple	170 lbs. @ 2 1/16"	1.875"
150-51-662**	5/16"®	1.750"	3/16"	18	Purple	170 lbs. @ 2 1/16"	1.875"
150-51-672	1 3/8"	1.312"	7/32"	24	Purple	170 lbs. @ 2 1/16"	1.875"
150-51-685**	5/16"®	1.750	3/16"	24	Purple	170 lbs. @ 2 1/16"	1.875"
150-52-696	25/32"	2.796"	5/32"	18	Purple	170 lbs. @ 2 1/16" ✦	1.875"
150-51-714	1 1/8"	1.750"	3/16"	24	Purple	170 lbs. @ 2 1/16"	1.875"
155-51-728	1 3/8"	1.312"	7/32"	18	Purple	170 lbs. @ 2 1/16"	1.875"
150-52-739	1 1/8"	2.937"	5/32"	18	Purple	170 lbs. @ 2 1/16"	1.875"
150-52-745	1 1/8"	2.937"	5/32"	24	Purple	170 lbs. @ 2 1/16"	1.875"
150-52-747	25/32"	2.796"	5/32"	21	Purple	170 lbs. @ 2 1/16" ✦	1.875"
150-52-748	25/32"	2.796"	5/32"	24	Purple	170 lbs. @ 2 3/32" ✦	1.875"
155-51-766	1 3/8"	1.312"	7/32"	21	Purple	170 lbs. @ 2 1/16"	1.875"
155-51-767	1 3/8"	1.312"	7/32"	24	Purple	170 lbs. @ 2 1/16"	1.875"
155-51-768	1 3/8"	1.312"	7/32"	27	Purple	170 lbs. @ 2 1/16"	1.875"
150-51-771	1 1/8"	1.750"	3/16"	18	Light Green	150 lbs. @ 1 27/32"	1.875"
150-51-775	1 1/8"	1.750"	3/16"	15	Purple	170 lbs. @ 2 1/16"	1.875"
140-51-814	1 3/16"	1.594"	3/16"	21	Blue	150 lbs. @ 1 27/32"	1.986"

**Also Applicable to engine flywheel depth of 1.812" with 7/32" facing thickness.

†Intermediate Plate Thickness .754 + .000 − .003

†† Intermediate Plate Thickness .832 + .000 − .003

✦Height listed is not necessarily the working height when assembled.

Fig. 4 Lipe DP clutch chart

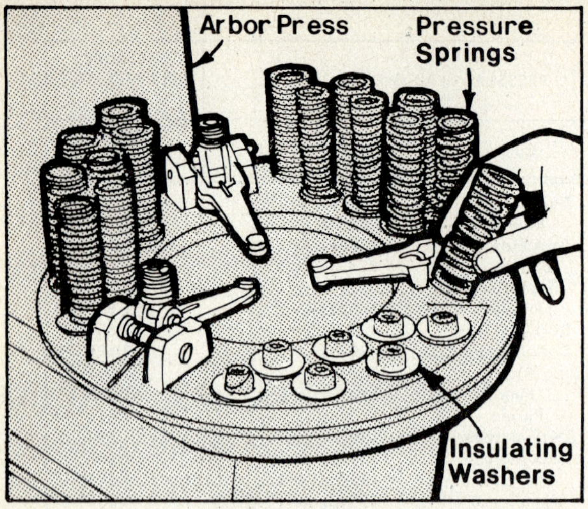

Fig. 9 Installing insulating washers and pressure springs

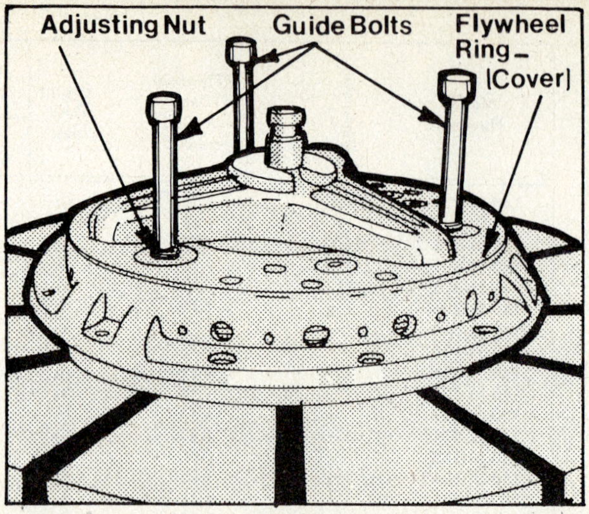

Fig. 10 Installing flywheel ring (cover) and guide bolts

6. Place insulating washers over bosses on pressure plate and install pressure springs, Fig. 9.
7. Place flywheel ring (cover) over assembly. If original parts are used, line up driving lugs marked "O" with driving slot marked "O".
8. To facilitate assembly, install guide bolts into adjusting nut until contact is made with end of eyebolt, Fig. 10.
9. Compress flywheel ring (cover) gradually, making sure that the three driving lugs enter properly into the driving slots in the pressure plate, and the pressure springs are properly seated on the pressure plate and in the seats provided in the flywheel ring (cover). Pull adjusting nut through holes in cover with guide bolts previously installed.
10. Remove guide bolts and assemble adjusting screw locknuts, making sure they are drawn tight to cover, Fig. 11.
11. Install hold-down bolts with standard washers in holes provided in flywheel ring (cover) and tapped holes in pressure plate (3 required), Fig. 11.

NOTE: An alternate procedure is to use a 7/16" wood block between release lever and underside of cover.

12. The press can now be released and the clutch cover assembly removed.

Clutch Installation

Inspect Engine Flywheel

1. Check flywheel for correct depth, Fig. 4.
2. Make sure release bearing and flywheel pilot bearing are in useable condition. Pilot bearing should be a finger press fit in flywheel and on transmission drive shaft.
3. Friction face of flywheel should be smooth and clean.

CAUTION: Friction face of flywheel must be free of heat cracks, score marks and foreign matter, as the presence of any one of these conditions will cause the clutch facings to quickly disintegrate.

Installing Single Plate Clutch

1. Try cover in flywheel before inserting driven disc to make sure it is a free fit in flywheel.
2. Install driven disc in flywheel, making sure long end of hub faces cover.
3. To facilitate assembly of transmission to engine, the use of a spare splined pilot shaft to align the driven disc while bolting the clutch in place is essential.

NOTE: The side of the driven disc that faces the clutch pressure plate has a chamfer on the end of the splined hub to facilitate the insertion of the transmission drive shaft.

4. Bolt the clutch to the flywheel and screw in each bolt until it contacts cover plate. Then gradually tighten opposite bolts until assembly is drawn up tight and completely into flywheel.
5. Remove the three hold-down bolts previously installed.

NOTE: It is very important that the correct capscrews be used when attaching the 13" cover to flywheel. The shank of the attaching capscrews are a close fit to the holes in this cover and the shank extends into the counterbore in the flywheel. If the flywheel has been excessively remachined, this counterbore may become too shallow, which will necessitate replacing the flywheel.

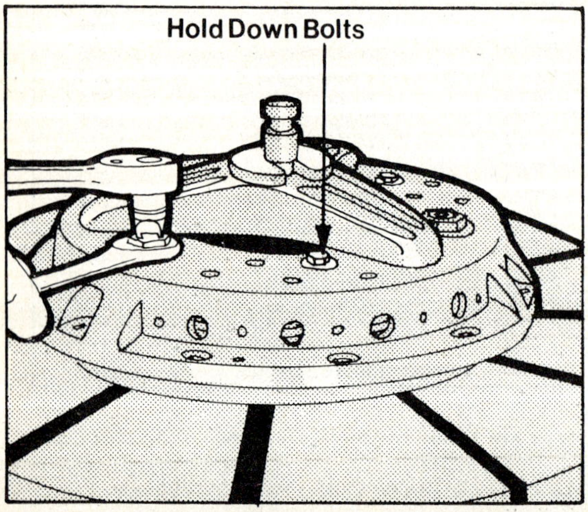

Fig. 11 Installing adjusting screw lock nuts and hold-down bolts

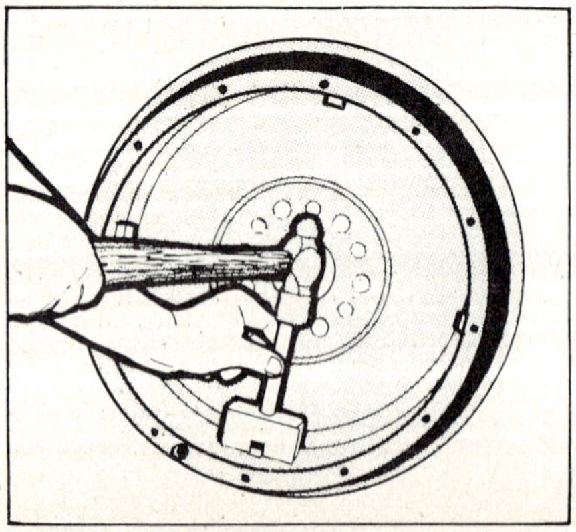

Fig. 12 Aligning drive pins in flywheel, using Lipe tool T-10670

Fig. 13 Aligning drive pins in flywheel, using machinist square and feeler gauge

Installing Two Plate Clutch

1. The four intermediate plate driving pins must be removed if the flywheel is to be machined. Pins with worn contact faces must be replaced.
2. The driving pins must be installed with the contact face square to the friction face of the flywheel.

NOTE: To insure perfect alignment, it is recommended that a tool, Fig. 12, be used. If this tool is not available, place a machinist square firmly on the flywheel rim and make sure the contact face of the driving pins have even contact with the square, Fig. 13.

3. The 14" two plate clutch driving pins are held in place by locknuts on the threaded end projecting through the flywheel which must be securely tightened. The 15" two plate driving pins may be locked in place by cup point Allen screws located in the bottom of the tapped holes used for attaching the cover to the flywheel.
4. After the driving pins are properly aligned and secured, and a new pilot bearing installed, the clutch is ready to be installed into the flywheel. It is essential that a splined shaft of the correct size for the pilot bearing and driven discs be used to align the driven discs before the cover is installed to the flywheel.
5. Install one disc in the flywheel and then place the intermediate plate on the driving pins and check for lash or free play at the driving slots. The intermediate plate should fit over the driving pins with not less than .007" clearance at the contact faces, which will permit free movement of the intermediate plate. If the clearance exceeds .012", the intermediate plate may cause some noise when the clutch is disengaged.
6. With the splined alignment shaft holding the driven discs in position, the cover assembly can now be attached to the flywheel. Attaching capscrews should be evenly tightened until cover is drawn tight to the flywheel.

NOTE: New cover assemblies have wooden blocks positioned between release levers and cover which facilitate assembly of cover to flywheel. Make sure these blocks are removed after clutch installation.

7. If the cover has been repaired or rebuilt, the lever heights should be checked after installation to flywheel. Use a straightedge and 6" scale as an "A" dimension gauge as indicated in Fig. 14. Refer to Fig. 4 for correct lever setting.
8. Always start with the levers low and keeping some tension on the lock nuts at all times, turn both the lock nut and adjusting nuts in a clockwise direction to raise the levers to the desired setting. When the correct height has been obtained, hold the adjusting nut firmly while tightening the lock nut.

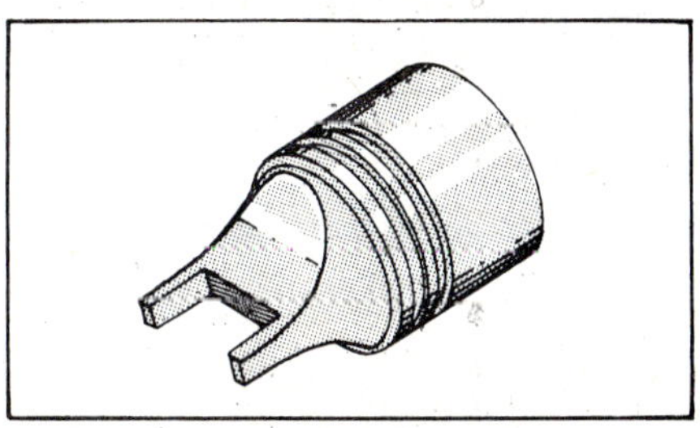

Fig. 15 Lipe tool for holding slotted adjusting nut

Adjusting to Dimension "A"

Satisfactory operation of the clutch is absolutely dependent upon the accuracy of the release lever adjustment since the pressure plate must operate parallel with the engine flywheel.

Set the release levers to the proper Dimension "A" by turning the adjusting nuts and measuring with a straightedge and scale. The release lever buttons should be the same distance from the ground surface of the bosses on the flywheel ring (cover). After Dimension "A" has been set, tighten lock nuts securely, at the same time holding the adjusting nut from turning. Use the tool shown in Fig. 15 for holding slotted adjusting nut.

On the stamped DP cover, Dimension "A" is obtained in the same manner as described for the cast cover above. Use the straightedge and scale as shown in Fig. 16 with the straightedge in the grooves adjacent to the adjusting nuts. The straightedge must also be at right angles to and over the end of the release lever. The release lever buttons must be the same distance from the grooves in the flywheel ring (cover).

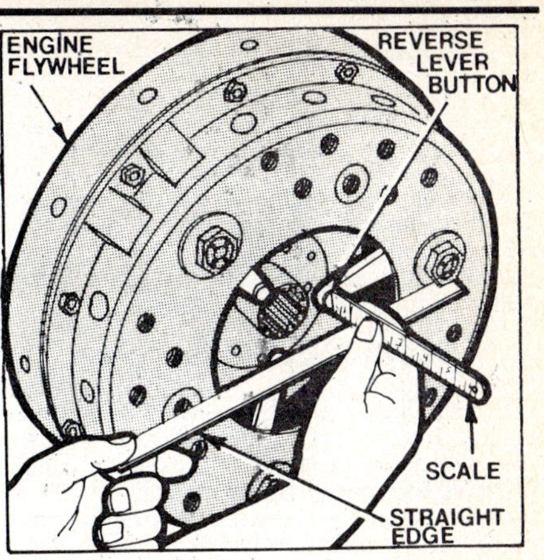

Fig. 14 Checking Dimension "A" on clutches with cast cover

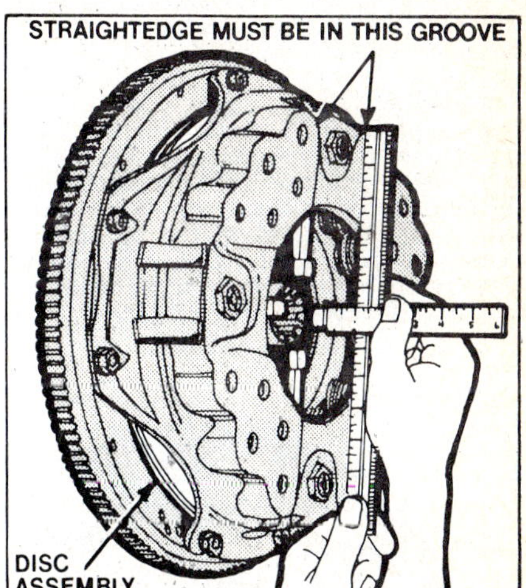

Fig. 16 Checking Dimension "A" on clutches with stamped cover

Lipe DLB Clutch

The Lipe DLB clutch is a direct pressure, dry disc design, incorporating a push type release bearing, Fig. 1 and 2. The DLB units are available in a 13 inch, two plate design and a 14 inch, single plate design.

Clutch Pedal Adjustment

Proper clutch pedal free play is approximately 1½". If inspection indicates that the free play is less than ½", immediate adjustment of the clutch pedal linkage should be made to restore 1½". This 1½" free play normally results in ⅛" clearance between clutch release levers and release bearing.

Servicing the Clutch

After the clutch is disassembled, clean all parts thoroughly. Inspect all parts as indicated in the following procedure. If in doubt about any part, replace with a new one.

Pressure Plate

The pressure plate must be replaced if friction surface is severely heat checked, it is warped in excess of .015", driving surfaces are worn, or pin holes are elongated.

It is recommended that not over 1/32" be removed from friction face by regrinding. See Fig. 3 and refer to Fig. 4 for thickness of new pressure plates, "Dimension C".

Install new bearings by pressing on the end of the bearing on which part numbers are stamped. This prevents damage to bearing case.

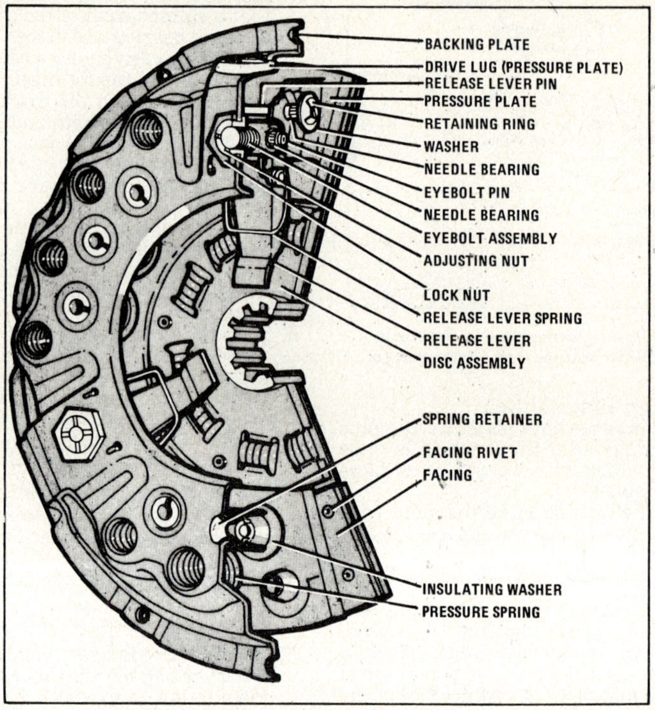

Fig. 1 Lipe DLB single plate clutch

Intermediate Plate

The intermediate plate in two plate clutches should not be re-used if there is over .014" clearance between the driving lugs and the mating slots in the flywheel. It is recommended that grinding of the intermediate plate be kept to a minimum with not over .015" being removed from either side for a total of .030". See Fig. 4 for thickness specifications.

Flywheel Ring (Cover)

Inspect flywheel ring (cover) for distortion or cracks. Bolting flanges should be flat to within .015" when checked on a surface plate.

Pressure Springs, Fig. 5

It is of vital importance that the pressure springs be checked for compression weight in accordance with the weight listed in the chart, Fig. 4. Under-weight springs will reduce clutch capacity and possibly cause clutch slippage. Therefore, to insure best clutch performance, new springs should be installed.

Eyebolt and Adjusting Nut

When installing new bearings, use a suitable tool and press on the end of the bearing having numbers stamped thereon. This will prevent damage to bearing cage.

In cover assemblies not incorporating needle bearings, the eyebolt pin hole is elongated

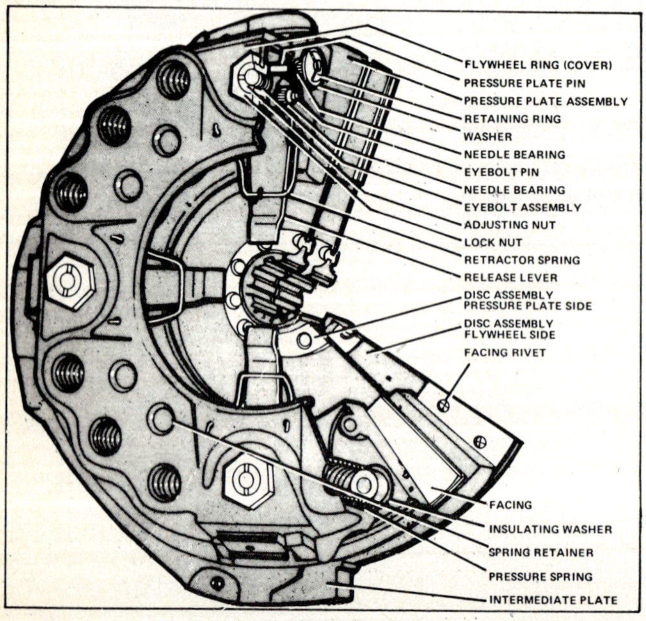

Fig. 2 Lipe DLB two plate clutch

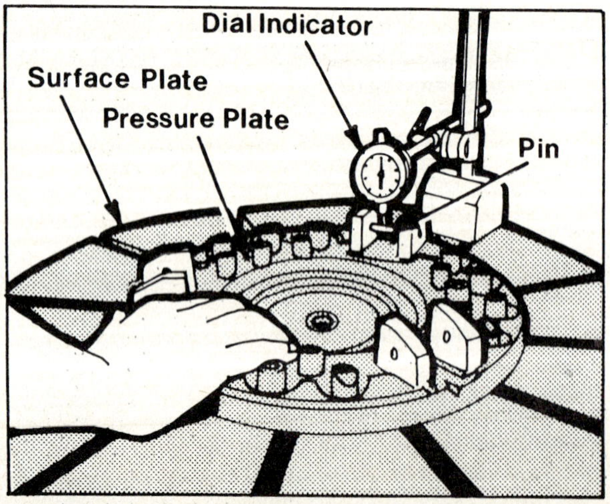

Fig. 3 Checking Dimension "C" on DLB clutches

Model	Dimension "A"	Flywheel Depth +.005" —.005"	Facing Thickness	Pressure Spring Data: Quantity	Pressure Spring Data: Color	Pressure Spring Data: Minimum Weight Lbs. @ Inch	Pressure Plate Data: Dimension "C" New
130-62-702	1 1/16	1.062	5/32	16	Dk. Maroon Stripe	133 @ 1 27/32	1.531
130-62-987	1 1/16	1.062	5/32	12	Dk. Maroon Stripe	133 @ 1 27/32	1.531
130-302-1140①	1 1/16	1.062	5/32	12	Lt. Green Stripe	153 @ 1 27/32	1.531
130-302-1508	1 1/16	1.062	5/32	16	Dk. Maroon Stripe	133 @ 1 27/32	1.531
130-302-1544	1 1/16	1.062	5/32	12	Dk. Maroon Stripe	133 @ 1 27/32	1.531
140-301-1086	29/32	Flat	3/16	18	Dk. Maroon Stripe	133 @ 1 27/32	1.687
140-301-1107	29/32	Flat	11/64	21	Lt. Green Stripe	153 @ 1 27/32	1.687

①—Some pressure plate assemblies were manufactured with 16 pressure springs and can be identified by a blue dot on the pressure plate cover. If a 12 spring unit is being reworked and 16 springs are desired, a blue dot should be added to the cover. The 16 springs used are 8 light green stripe (Tension—153 lbs. @ 1 27/32 inch) and 8 dark maroon stripe (Tension—133 lbs. @ 1 27/32 inch) positioned as shown in Fig. 5.

Fig. 4 Lipe DLB clutch chart

to allow the pin to move as the clutch is disengaged. If this hole indicates wear the eyebolt should be replaced. Eyebolt and adjusting nut threads must be free fitting but not excessively loose.

Release Lever Assembly

The release levers must be replaced if the holes for the eyebolt and pressure plate pins are worn, or if worn severely by the release bearing.

Release levers used in needle bearing clutches have elongated holes for eyebolt pins, whereas levers for standard pin clutches have round holes for both the pressure plate and eyebolt pins.

Driven Disc Assembly

The disc should be replaced if the hub splines are worn, disc is distorted, cracked or broken. If the disc is acceptable for lining, use Lipe Facing Kit to obtain correct rivets and facing material.

Make sure proper riveting method is used. See Fig. 6 in the DP clutch section for riveting method with fabric and metallic facings.

NOTE: Field relining of dampener or cushioned type disc assemblies is not recommended. Facing Kits, including original equipment facings with proper rivets are only available for non-cushioned clutches. Other component parts, such as marcels (cushioned segments) or coil springs are not available for service replacement. The rigid type disc without cushioned facings may be relined if the steel disc is not distorted or cracked and the hub splines are not worn over .010".

Clutch Assembly

1. Apply a small amount of NGLI #2 multi-purpose, lithium grease on all pins. Also, on two plate clutches, apply grease on driving lugs of flywheel ring and mating slots in pressure plate.

NOTE: Do not over lubricate.

2. Support pressure plate in an arbor press with suitable blocks so the flywheel ring (cover) can be pressed over the pressure plate without contacting base of press.
3. Assemble adjusting nuts to eyebolts, then to the release levers using new pins.
4. Assemble levers to pressure plate and install release lever pins.

NOTE: Clutches with needle bearings require flat washers installed on both ends of the pins.

5. Position insulating washers over spring bosses on pressure plate, then install pressure springs on bosses. Refer to Fig. 5 if two types (Color coded) springs are used.

NOTE: If the base of the spring boss is equipped with webs, do not use the insulating washers.

6. Assemble release lever springs on flywheel ring (cover) and position flywheel ring over pressure springs, ensuring each spring is properly positioned in the ring.

NOTE: Be sure to align the "O" marks on the pressure plate and flywheel ring.

7. Press flywheel ring (cover) into position ensuring the pressure plate driving lugs are engaged in the flywheel ring slots.
8. Pull adjusting nuts through flywheel ring

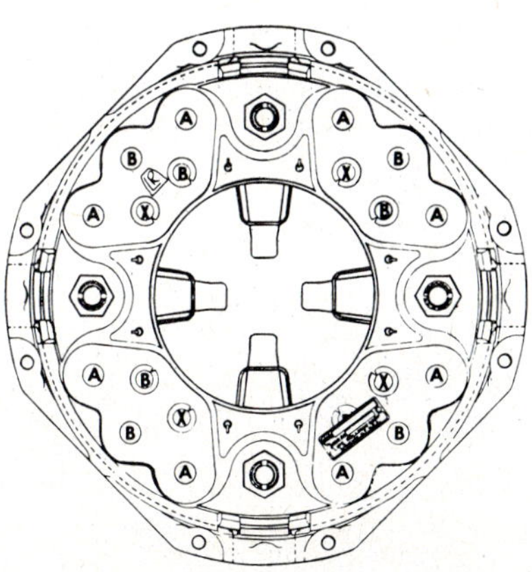

Ⓐ LOCATION OF PRESSURE SPRING w/DARK MAROON STRIPE
Ⓑ LOCATION OF PRESSURE SPRING w/LIGHT GREEN STRIPE
Ⓧ DO NOT ASSEMBLE SPRINGS ON THESE LOCATORS

Fig. 5 Pressure spring location

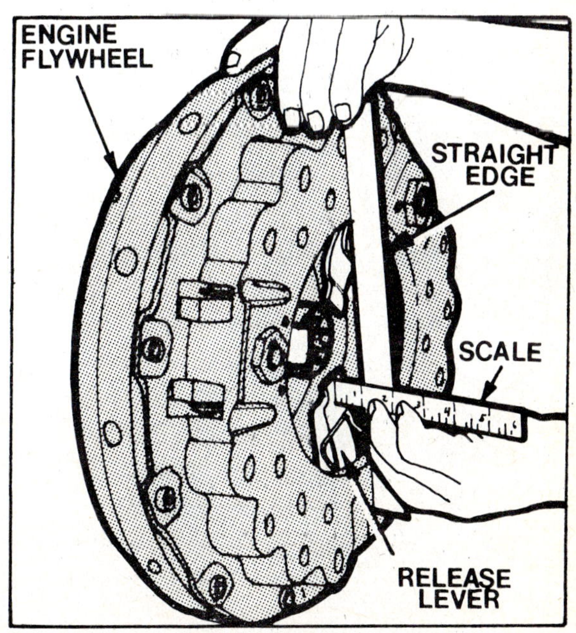

Fig. 6 Checking Dimension "A" on DLB clutches

(cover) and install lock nuts.

9. Before releasing press, install four 1/4" 20 × 2 1/4" bolts on 13 inch units or three 1/4" 20 × 2 1/4" bolts on 14 inch units through flywheel ring (cover) into the tapped holes in the pressure plate.
10. The release levers must be adjusted to the proper height and lock nuts tightened as follows:
    a. Dimension "A" is the distance from the top of the flywheel ring (cover) to ends of release levers. (Applies only to field repaired or rebuilt clutches.)
    b. See chart, Fig. 4, for correct "A" dimension for unit being serviced. Satisfactory operation of the clutch depends greatly upon the accuracy of the lever setting as this controls the parallel movement of the pressure plate.
    c. A straightedge and scale will provide a means of checking the "A" dimension. Fig. 6 shows the special tools available for slotted adjusting nuts.
    d. Adjusting the lever height is facilitated by starting with the levers low and with lock nut just snug. Turn adjusting nut and lock nut as a unit clockwise until the lever is raised the desired height. Then hold slotted nut firmly and tighten lock nut.
11. Check recessed type flywheels for proper depth as listed in chart, Fig. 4.

Lipe DPB Clutch

The Lipe DPB clutch is a direct pressure unit. The clutch requires no maintenance other than maintaining release bearing clearance which is accomplished by normal linkage adjustment, Figs. 1 and 2.

Identifying Models

The name plate attached to the flywheel ring (cover) identifies the cover plate as to the serial number and model number. For example, model number 120-61-615 is interpreted as follows: 120 indicates that it is a 12" clutch. The '61' between the dashes indicates the clutch to be a single plate type DPB unit, and 615 is the parts list number. A '61' indicates that the assembly is equipped with needle bearings; a '62' indicates a two plate assembly with needle bearings.

There are four types of disc assemblies available with the numbering system set up to identify each type. An assembly number is stamped on the disc.

Clutch Pedal Adjustment

Proper clutch pedal free play is approximately 1 1/2". If inspection indicates that the free play is less than 1/2", immediate adjustment of the clutch pedal linkage should be made to restore to 1 1/2". This 1 1/2" free play normally results in 1/8" clearance between clutch release levers and release bearing.

Some two plate DPB assemblies are equipped with intermediate plate separator springs and adjusting screws. These units may require periodic adjustment to maintain the .020" clearance recommended clearance between the end of the adjusting screws and the intermediate plate.

The adjustment must be made with the clutch in the engaged position. The adjusting screws should be turned clockwise until they just contact the intermediate plate. At this point a 1/2 turn (counterclockwise) of the screws will produce a .020" clearance. Lock the adjusting screws in position by tightening the lock nuts.

NOTE: When replacing or repairing the clutch assembly, it is recommended that the separator springs and adjusting screws be removed from the assembly before installation.

Servicing the Clutch

After the clutch is disassembled, clean all parts thoroughly. Inspect all parts as indicated in the following procedure. If in doubt about any part, replace with a new one.

Pressure Plate

The pressure plate must be replaced if friction surface is severely heat checked, it is warped in excess of .015", driving surfaces are worn, or pin holes are elongated.

It is recommended that not over 1/32" be removed from friction face by regrinding. See

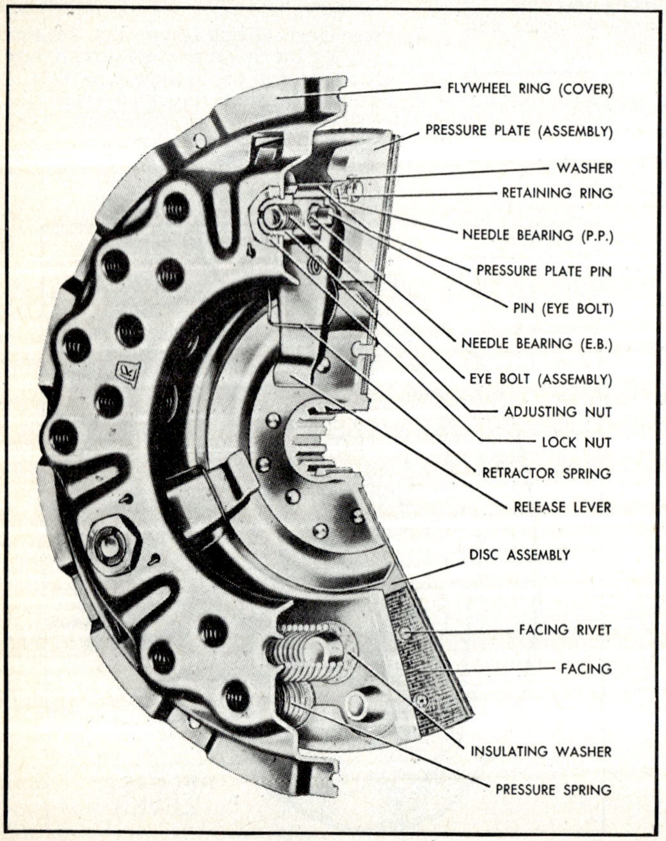

Fig. 1 Lipe DPB single plate clutch

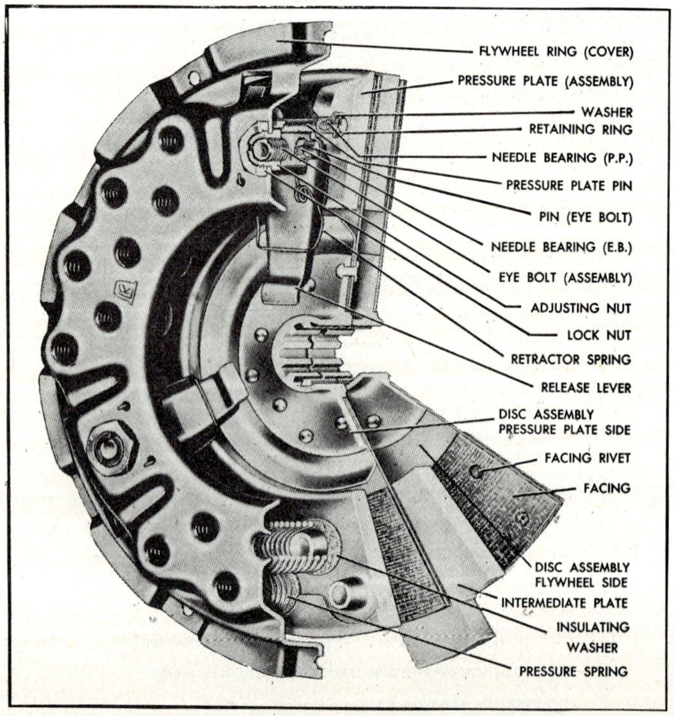

Fig. 2 Lipe DPB two plate clutch

Model Number	Dimension "A"	Flywheel Depth +.005" −.005"	Facing†† Thickness	PRESSURE SPRING DATA			PRESSURE PLATE DATA
				Quantity	Color	Minimum Weight in lbs.§	Dimension "C" New
120-61-615	1 3/16"	Flat	5/32"	12	No Color	170 @ 1 27/32"	1.468
130-61-619	1 1/16"	Flat	5/32"	16	Dark Maroon	130 @ 1 27/32"	1.531
140-61-623	29/32"	Flat	3/16"	21	Light Green	150 @ 1 27/32"	1.687
130-62-625**	1 1/16"	1.062	5/32"	20	Dark Maroon	130 @ 1 27/32"	1.531
120-61-627	1 3/16"	Flat	5/32"	12	Light Green	150 @ 1 27/32"	1.468
120-61-628	1 1/16"	Flat	5/32"	12	Red Stripe	170 @ 1 27/32"	1.468
120-62-629*	1 3/16"	1.062"	5/32"	12	Light Green	150 @ 1 27/32"	1.468
120-61-631	1 3/16"	Flat	5/32"	12	Light Green	150 @ 1 27/32"	1.468
120-61-633	1 3/16"	Flat	5/32"	12	Light Green	150 @ 1 27/32"	1.468
140-61-635	29/32"	Flat	3/16"	21	Dark Maroon	150 @ 1 27/32"	1.687
130-61-642	1 1/16"	Flat	5/32"	20	Dark Maroon	130 @ 1 27/32"	1.531
130-61-647	1 1/16"	Flat	5/32"	20	Dark Maroon	130 @ 1 27/32"	1.531
140-61-648	29/32"	1.187	3/16"	21	Light Green	150 @ 1 27/32"	1.687
140-62-650†	29/32"	1.093	5/32"	15	Dark Maroon	130 @ 1 27/32"	1.687
140-61-651	29/32"	1.187	3/16"	21	Light Green	150 @ 1 27/32"	1.687
140-61-652	29/32"	Flat	3/16"	21	Dark Maroon	130 @ 1 27/32"	1.687
140-61-653	29/32"	1.187	3/16"	21	Dark Maroon	130 @ 1 27/32"	1.687
140-62-665†	29/32"	1.093	5/32"	18	Dark Maroon	130 @ 1 27/32"	1.687
140-61-666	29/32"	1.187	3/16"	18	Dark Maroon	130 @ 1 27/32"	1.687
120-61-667	1 3/16"	Flat	5/32"	12	Red Stripe	180 @ 1 27/32"	1.468
130-61-669	1 3/16"	Flat	5/32"	16	Dark Maroon	130 @ 1 27/32"	1.531
130-61-670	1 3/16"	Flat	5/32"	20	Dark Maroon	130 @ 1 27/32"	1.531
120-62-675*	1 3/16"	1.062	5/32"	09	Light Green	150 @ 1 27/32"	1.468
140-61-684	27/32"	1.187	7/32"	18	Dark Maroon	130 @ 1 27/32"§	1.687
140-61-689	29/32"	Flat	3/16"	18	Dark Maroon	130 @ 1 27/32"	1.687
140-61-695	29/32"	Flat	3/16"	21	Dark Maroon	130 @ 1 27/32"	1.687
150-61-697	1"	1.750	3/16"	24	Purple	170 @ 2 1/16"	1.937
130-62-699**	1 1/16"	1.062	5/32"	20	Dark Maroon	130 @ 1 27/32"	1.531
130-62-700**	1 1/16"	1.062	5/32"	16	Dark Maroon	130 @ 1 27/32"	1.531
140-62-701	29/32"	1.093	5/32"	18	Dark Maroon	130 @ 1 27/32"	1.687
130-62-702**	1 1/16"	1.062	5/32"	16	Dark Maroon	130 @ 1 27/32"	1.531
140-61-703	29/32"	1.187	7/32"	21	Dark Maroon	130 @ 1 27/32"§	1.687
140-62-704†	29/32"	1.093	5/32"	21	Dark Maroon	130 @ 1 27/32"	1.687
140-61-706	29/32"	Flat	3/16"	21	Light Green	150 @ 1 27/32"	1.687
140-61-707	29/32"	1.187	3/16"	21	Dark Maroon	130 @ 1 27/32"	1.687
150-62-708	1"	1.093	5/32"	24	Purple	170 @ 2 1/16"	1.937
140-61-711	29/32"	1.187	3/16"	18	Dark Maroon	130 @ 1 27/32"	1.687
140-61-712	29/32"	1.187	3/16"	18	Dark Maroon	130 @ 1 27/32"	1.687
140-61-713	29/32"	Flat	3/16"	18	Dark Maroon	130 @ 1 27/32"	1.687
140-61-718	29/32"	1.187	3/16"	18	Dark Maroon	130 @ 1 27/32"	1.687
150-62-719	1"	1.093	5/32"	18	Purple	170 @ 2 1/16"	1.937
140-61-720	29/32"	1.187	3/16"	14	Dark Maroon	130 @ 1 27/32"	1.687
140-61-721	29/32"	Flat	3/16"	15	Dark Maroon	130 @ 1 27/32"	1.687
140-61-723	29/32"	1.187	3/16"	12	Light Green	150 @ 1 27/32"	1.687
130-61-724	1 1/16"	Flat	5/32"	12	Dark Maroon	130 @ 1 27/32"	1.531
130-61-725‡	7/8"	Flat	5/32"	20	Dark Maroon	130 @ 1 27/32"	1.531
120-61-726	1 1/16"	Flat	5/32"	12	No Color	170 @ 1 27/32"	1.468
140-62-727	29/32"	1.093	5/32"	15	Dark Maroon	130 @ 1 27/32"	1.687
150-61-731	1"	1.750	3/16"	18	Purple	170 @ 2 1/16"	1.937
130-61-734	1 1/8"	Flat	5/32"	20	Dark Maroon	130 @ 1 27/32"	1.531
140-62-735†	29/32"	1.093	5/32"	21	Light Green	150 @ 1 27/32"	1.687
130-62-736	29/32"	1.477	5/32"	12	Light Green	150 @ 1 27/32"	1.750
150-61-738	1"	1.750	3/16"	21	Purple	170 @ 2 1/16"	1.937
150-62-743	1"	1.093	5/32"	21	Purple	170 @ 2 1/16"	1.937
140-61-761	29/32"	1.187	3/16"	21	Dark Maroon	130 @ 1 27/32"	1.687
140-61-773	29/32"	1.187	3/16"	21	Dark Maroon	130 @ 1 27/32"	1.687
150-61-776	1"	1.750	3/16"	15	Purple	170 @ 2 1/16"	1.937
140-62-778	29/32"	1.093	5/32"	21	Dark Maroon	130 @ 1 27/32"	1.687
140-62-779	29/32"	1.093	5/32"	18	Dark Maroon	130 @ 1 27/32"	1.687
140-62-780	29/32"	1.093	5/32"	15	Dark Maroon	130 @ 1 27/32"	1.687
140-62-781	29/32"	1.093	5/32"	21	Light Green	150 @ 1 27/32"	1.687
140-61-785	29/32"	Flat	3/16"	15	Dark Maroon	130 @ 1 27/32"	1.687

Fig. 3 Lipe DPB clutch chart (Part 1 of 2)

Continued

Model Number	Dimension "A"	Flywheel Depth +.005" −.005"	Facing†† Thickness	PRESSURE SPRING DATA			PRESSURE PLATE DATA
				Quantity	Color	Minimum Weight in lbs.§	Dimension "C" New
170-62-789	1 7/32"	1.562	5/32"	32	Orange	355 @ 2 17/32"	1.875
120-62-791	1 3/16"	1.062	5/32"	9	Light Green	150 @ 1 27/32"	1.468
150-62-794	1"	1.093	5/32"	15	Purple	170 @ 2 1/16"	1.937
130-62-798	29/32"	1.477	5/32"	12	Light Green	150 @ 1 27/32"	1.750
130-61-802	1 3/16"	Flat	5/32"	16	Dark Maroon	130 @ 1 27/32"	1.531
140-61-805	23/32"	.976	3/32"	18	Dark Maroon	130 @ 1 27/32"	1.687
120-61-806	1 3/16"	Flat	5/32"	12	Dark Maroon	130 @ 1 27/32"	1.468
130-61-807	1 3/16"	Flat	5/32"	12	Dark Green Stripe	90 @ 1 27/32"	1.531
				4	Dark Maroon	130 @ 1 27/32"	
130-61-808	1 3/16"	Flat	5/32"	20	Dark Green Stripe	90 @ 1 27/32"	1.531
130-61-809	1 3/16"	Flat	5/32"	16	Dark Maroon	130 @ 1 27/32"	1.531
140-61-810	29/32"	1.187	3/16"	15	Dark Green Stripe	90 @ 1 27/32"	1.687
				3	Light Green	150 @ 1 27/32"	
140-61-811	29/32"	1.187	3/16"	6	Dark Maroon	130 @ 1 27/32"	1.687
				15	Dark Green Stripe	90 @ 1 27/32"	
120-61-812	1 3/16"	Flat	5/32"	6	Light Green	150 @ 1 27/32"	1.468
				6	Dark Green Stripe	90 @ 1 27/32"	
120-61-813	1 3/16"	Flat	5/32"	9	Dark Maroon	130 @ 1 27/32"	1.468
				3	No Color	170 @ 1 27/32"	
130-61-817	1 1/16"	Flat	5/32"	16	Dark Maroon	130 @ 1 27/32"	1.531
140-61-818	29/32"	Flat	3/16"	21	Light Green	150 @ 1 27/32"	1.687
140-61-819	29/32"	Flat	3/16"	18	Dark Maroon	130 @ 1 27/32"	1.687
170-61-821	1 7/32"	.396	5/32"	24	Orange	355 @ 2 17/32"	1.875

INTERMEDIATE PLATE THICKNESS — * 12" − .676 − .000 + .003 — ** 13" − .676 − .000 + .003 — † 14" − .754 − .000 + .003

†† Facings on 12" and 13" cushioned driven discs are 9/64" thick.

‡ "A" Dimension was 1 1/16". § Height listed not necessarily working height when assembled.

Fig. 3 Lipe DPB clutch chart (Part 2 of 2)

Fig. 4 and refer to Fig. 3 for thickness of new pressure plates.

Install new bearings by pressing on the end of the bearing on which part numbers are stamped. This prevents damage to bearing case.

Intermediate Plate

The intermediate plate in two plate clutches should not be re-used if there is over .012" clearance between the driving lugs and the mating slots in the flywheel. It is recommended that grinding of the intermediate plate be kept to a minimum with not over .015" being removed from either side for a total of .030". See Fig. 3 for thickness specifications.

Flywheel Ring (Cover)

1. Inspect flywheel ring (cover) for distortion or cracks. Bolting flanges should be flat to within .015" when checked on a surface plate.
2. All DPB flywheel rings or covers are of stamped construction. The only wear points are the slots which engage the driving lugs of the pressure plate. If these driving surfaces are worn to the extent that, with a new pressure plate installed and an excess of .012" clearance is present, replace the cover assembly.
3. A clearance of at least .004" is required between driving surfaces of the flywheel ring (cover) and pressure plate. An excessive amount of clearance may cause noise or allow the pressure plate to shift off center and create an unbalanced condition.
4. Some DPB cover assemblies are piloted to the flywheel by an accurately machined O.D. of the flywheel ring (cover) which fits into a recess or rim on the flywheel. In other DPB units the assembly is piloted by body fitted capscrews which hold the assembly onto the flywheel. In this case the tapped holes in the flywheel will have a reamed recess to receive the special attaching screws. It is important that the correct capscrews be used where the cover is piloted by this method.

Pressure Springs

It is of vital importance that the pressure springs be checked for compression weight in accordance with the weight listed in the chart, Fig. 3. Under-weight springs will reduce clutch capacity and possibly cause clutch slippage. Therefore, to insure best clutch performance, new springs should be installed.

Eyebolt and Adjusting Nut

When installing new bearings, use a suitable tool and press on the end of the bearing having numbers stamped thereon. This will prevent damage to bearing cage.

In cover assemblies not incorporating needle bearings, the eyebolt pin hole is elongated to allow the pin to move as the clutch is disengaged. If this hole indicates wear the eyebolt should be replaced. Eyebolt and adjusting nut threads must be free fitting but not excessively loose.

Release Lever Assembly

The release levers must be replaced if the holes for the eyebolt and pressure plate pins

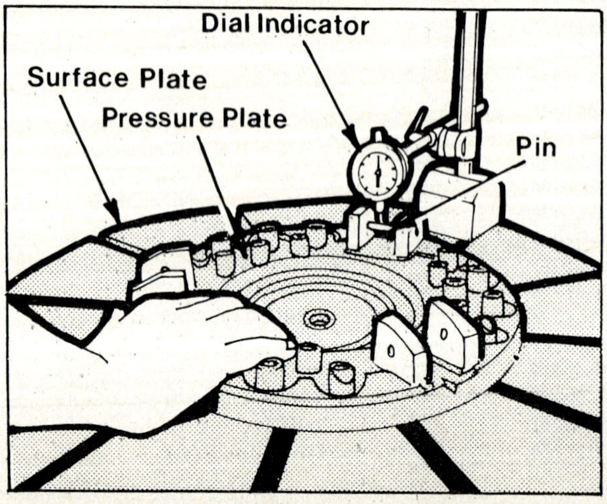

Fig. 4 Checking Dimension "C" on DPB clutches

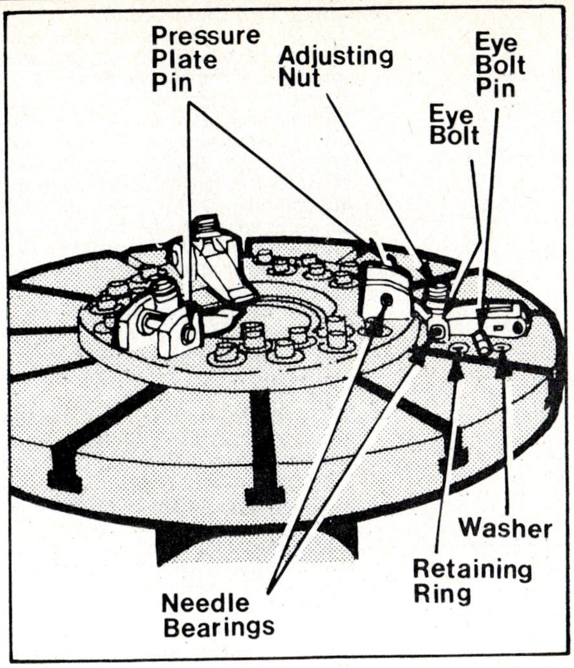

Fig. 5 Assembly of eyebolts and release levers

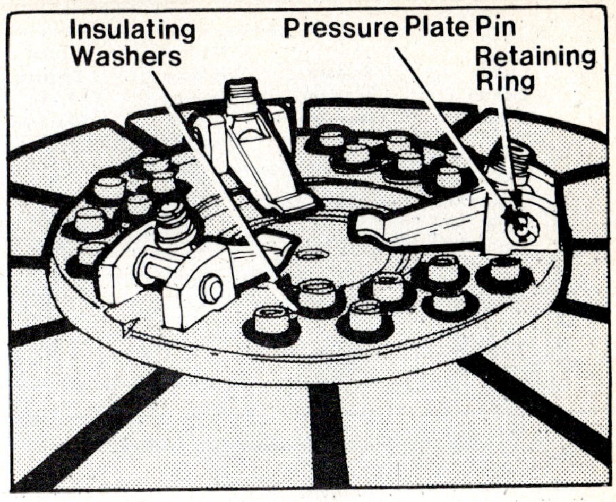

Fig. 6 Installation of release levers

are worn, or if worn severely by the release bearing.

Release levers used in needle bearing clutches have elongated holes for eyebolt pins, whereas levers for standard pin clutches have round holes for both the pressure plate and eyebolt pins.

Driven Disc Assembly

The disc should be replaced if the hub splines are worn, disc is distorted, cracked or broken. If the disc is acceptable for lining, use Lipe Facing Kit to obtain correct rivets and facing material.

Make sure proper riveting method is used. See Fig. 6 in the DP clutch section for riveting method with fabric and metallic facings.

NOTE: Field relining of dampener or cushioned type disc assemblies is not recommended. Facing Kits, including original equipment facings with proper rivets are only available for non-cushioned clutches. Other component parts, such as marcels (cushioned segments) or coil springs are not available for service replacement. The rigid type disc without cushioned facings may be relined if the steel disc is not distorted or cracked and the hub splines are not worn over .010″.

Clutch Assembly

1. Place pressure plate on arbor press with friction face down.
2. Assemble adjusting nuts to eyebolts, then assemble to release lever, using new pins, Fig. 5.
3. Assemble levers to pressure plate and install retaining rings on pressure plate pins, Fig. 6.
4. Place insulating washers over spring bosses on pressure plate and install pressure springs, Fig. 6.
5. Install new retractor springs to flywheel ring (cover), Fig. 7.
6. Place cover over pressure srpings, making sure each spring is properly positioned. If original pressure plate and cover are being used, line up the "O" marks on pressure plate and cover.
7. The cover can now be pressed into position, using care to engage driving lugs into slots of cover to avoid distortion.
8. After cover has been pressed into position, the adjusting nuts may be pulled up through the holes in the cover and lock nuts installed, Fig. 8.
9. The 12″ cover should have three 1/4″ × 20 × 2 1/4 capscrews and the 14″ cover should have three 3/8″ × 16 × 2 1/4″ capscrews installed through cover into tapped holes in pressure plate to hold it under compression when assembling to flywheel. All other DPB covers can be held in compression by placing 7/16″ wooden blocks between release lever and cover.
10. The press can now be released and cover assembly removed for attachment to flywheel where levers must be adjusted to proper height, and lock nuts tightened securely. See Dimension "A" on chart, Fig. 3.

Clutch Installation

Check Engine Flywheel

1. Check recessed type flywheels for proper

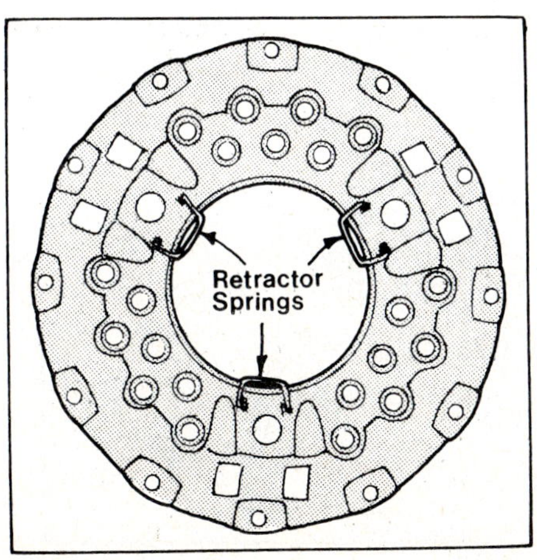

Fig. 7 Installation of retractor springs

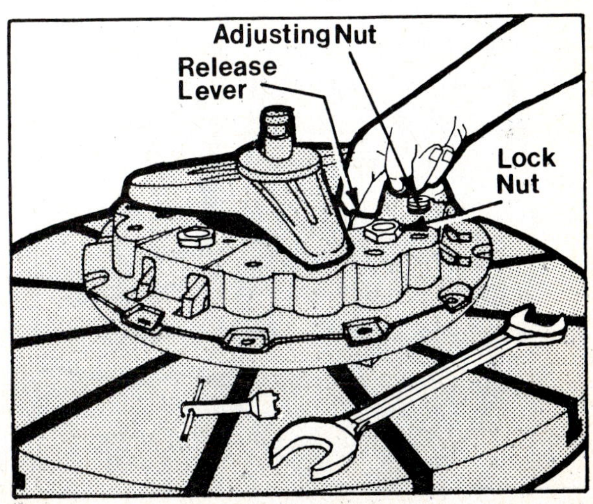

Fig. 8 Installation of lock nuts

Clutch Size & Type	Release Bearing Clearance	Normal Release Bearing Travel	Total Bearing Travel Required
12" Single Plate	1/8"	11/32"	15/32"
13" Single Plate	1/8"	7/16"	9/16"
14" Single Plate	1/8"	7/16"	9/16"
15" Single Plate	1/8"	7/16"	9/16"
17" Single Plate	1/8"	7/16"	9/16"
12" Two Plate	1/8"	1/2"	5/8"
13" Two Plate	1/8"	1/2"	5/8"
14" Two Plate	1/8"	1/2"	5/8"
15" Two Plate	1/8"	1/2"	5/8"
17" Two Plate	1/8"	1/2"	5/8"

Fig. 9 DPB clutch release travel chart

depth as listed in chart, Fig. 3.

2. Replace release bearing and carrier, and flywheel pilot bearing. Pilot bearing should be a hand press fit in the flywheel recess and transmission drive gear. The release yoke or fork should contact the release bearing carrier pads evenly to prevent a bind on the front bearing cap extension.
3. The friction face of the flywheel must be free from heat cracks, score marks or taper. The presence of any of these conditions will have an adverse effect on clutch function and life.

Installing Single Plate Clutch

1. Check cover assembly on flywheel without the disc for proper fit.
2. Install driven disc, making sure that it is properly positioned and insert an aligning shaft.

CAUTION: Normally the long end of the driven disc hub extends toward the transmission, but in a few instances the long end may face toward the engine. Make sure that the disc hub does not come within 5/32" of the pilot bearing.

3. Bolt cover to flywheel, tightening each capscrew gradually until cover is drawn up tight. Remove wooden blocks or hold-down capscrews before removing aligning shaft.
4. See Fig. 9 for release travel specifications.

NOTE: Some cover assemblies are piloted to the flywheel by capscrews with a machined shank which fits the reamed holes in the cover. The machined shank extends through the cover into a recessed or counterbored hole in the flywheel. If the flywheel has been remachined several times to remove heat cracks or scores, the counterbored holes may not be deep enough to allow complete tightening of the capscrews. This will necessitate replacing the flywheel.

Installing Two Plate Clutch

1. The DPB two plate units have an intermediate plate with driving lugs which fit into mating slots of the flywheel.
2. Before installing one of these units, the intermediate plate should be set into the driving slots and clearances checked. A minimum of .006" is recommended to allow for free movement of the intermediate plate.
3. In some vehicles, the two plate assemblies are equipped with separator springs and adjusting screws. When necessary for such an assembly to be removed for repairs or replacement, the separator springs should be removed from the flywheel and the adjusting screws from the replacement unit before the installation is made.
4. The two driven discs will be marked either flywheel side or pressure plate side and must be so installed. In no instances shall the hubs be allowed to approach the pilot bearing or come together within 5/32".
5. When installing two plate clutches, it is essential that a splined shaft or an aligning shaft be used with a key to properly align the two driven discs. Otherwise the instructions for installing these clutches are the same as for single plate clutches.

NOTE: To obtain complete disengagement of DPB clutches, the release levers must be actuated a specified distance from the engaged position. Assuming the release bearing clearance is 1/8" (between levers and bearing) when clutch is in the engaged position, the release bearing must move forward 1/8" to take up this clearance before it contacts the release levers and then move the normal amount of travel required to disengage the clutch as shown in the chart, Fig. 9.

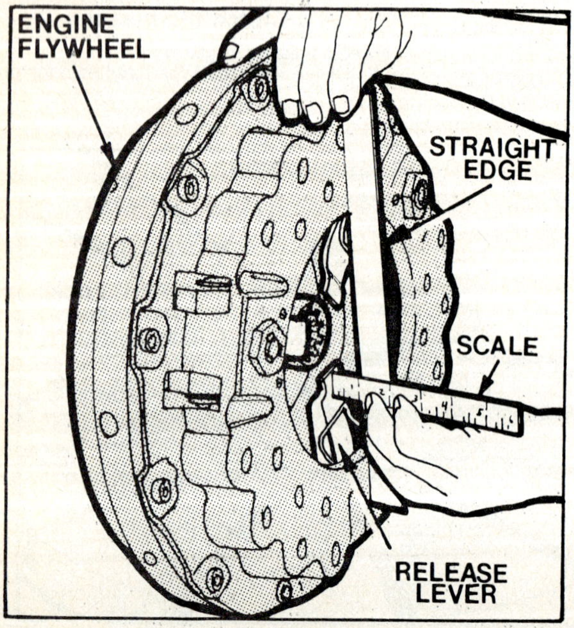

Fig. 10 Checking Dimension "A"

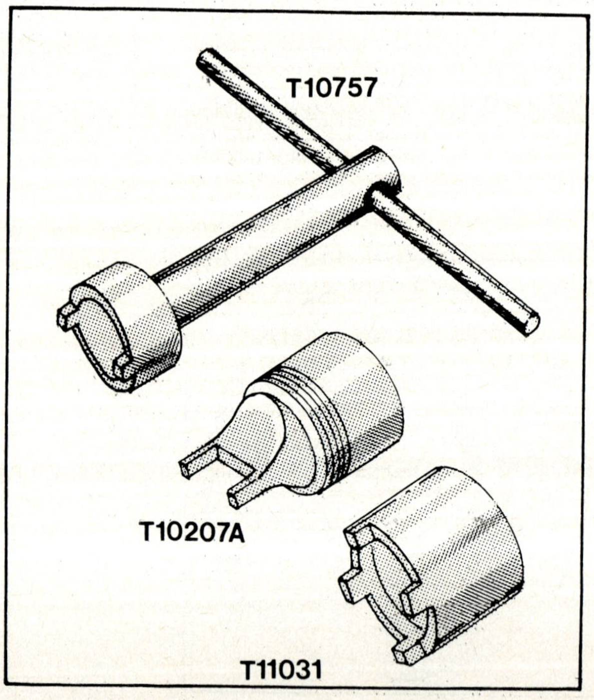

Fig. 11 Lipe tools for slotted adjusting nuts

Adjusting Lever Heights

Dimension "A"

1. This dimension is the distance from the top of the flywheel ring (cover) to ends of release levers, Fig. 10. (Applies only to field repaired or rebuilt clutches.)
2. See chart, Fig. 3, for correct "A" dimension for unit being serviced. Satisfactory operation of the clutch depends greatly upon the accuracy of the lever setting as this controls the parallel movement of the pressure plate.
3. A straightedge and scale will provide a means of checking the "A" dimension. Fig. 11 shows the special tools available for slotted adjusting nuts.
4. Adjusting the lever height is facilitated by starting with the levers low and with lock nut just snug. Turn adjusting nut and lock nut as a unit clockwise until the lever is raised the desired height. Then hold slotted nut firmly and tighten lock nut.

Lipe ML Clutch

The Lipe ML clutch is a multiple lever unit of the adjustable, dry disc, push type construction, Figs. 1 and 2. Maintenance can be accomplished easily without the use of special tools. Adjustment, to compensate for facing wear, is accomplished by removing shims. This operation also restores the original spring pressure and does not disturb the running balance of the clutch.

Identifying Models

The model and serial numbers of the clutch cover plate are stamped on the machined surface of the cover or on the name plate. Driven disc assemblies are stamped with a model number.

Periodic Inspection

It is recommended that the clutch be kept in proper adjustment by periodic inspection of the Dimension "A", which is referred to in Fig. 3 and in the clutch adjustment chart.

Clutch Adjustment

Checking Dimension "A"

1. With clutch in engaged position, place Gauge T-2187 between face of clutch release sleeve and release bearing, Fig. 4.
2. Turn thumb screw of gauge to just contact ground surface of clutch cover.
3. As shown in Fig 3, Dimension "A" is the distance from underside of gauge to end of thumb screw.
4. Block clutch pedal in released position. This prevents bending of adjusting straps and facilitates loosening of adjusting nuts.
5. Loosen adjusting nuts five full turns.
6. Remove clutch pedal block or jack to allow the adjusting plate to move back for easy removal of shims.
7. Use sharp nose pliers or cotter pin puller to remove shims, Fig. 5. Removing one shim from each stud reduces Dimension "A" by approximately 3/32". Likewise, adding one shim increases Dimension "A" by 3/32". *Make sure that an equal quantity of shims remains on each stud before proceeding.*
8. Again block clutch pedal to disengage clutch. This prevents bending of adjusting straps and facilitates tightening of adjusting nuts.
9. Tighten adjusting nuts, remove pedal block or jack and recheck Dimension "A"

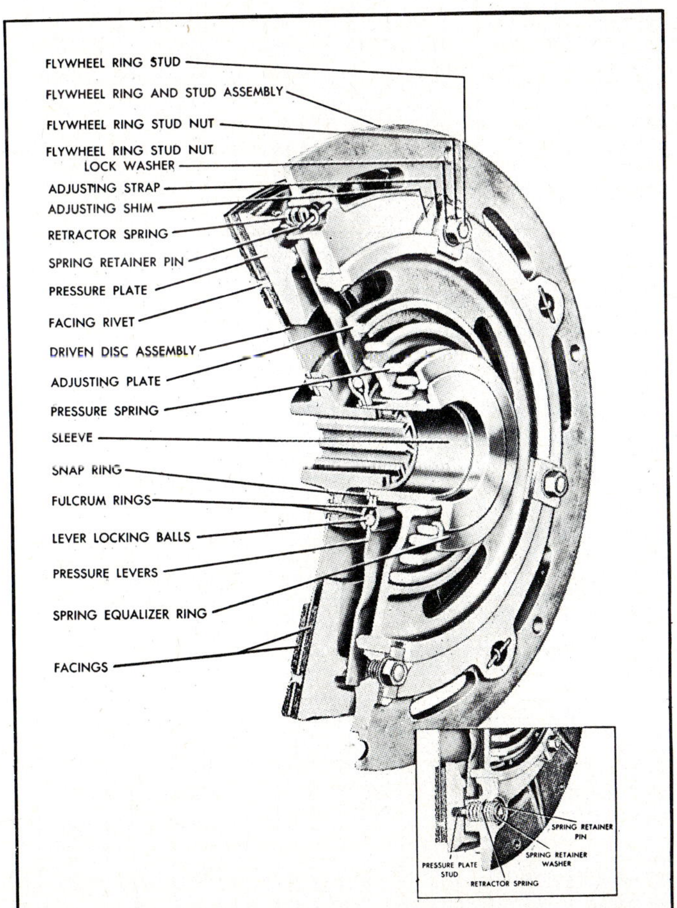

Fig. 1 Lipe ML 12", 13", 14" single plate clutch. Inset shows type of pressure plate retractor spring on 15" single plate and 14" and 15" two plate clutches

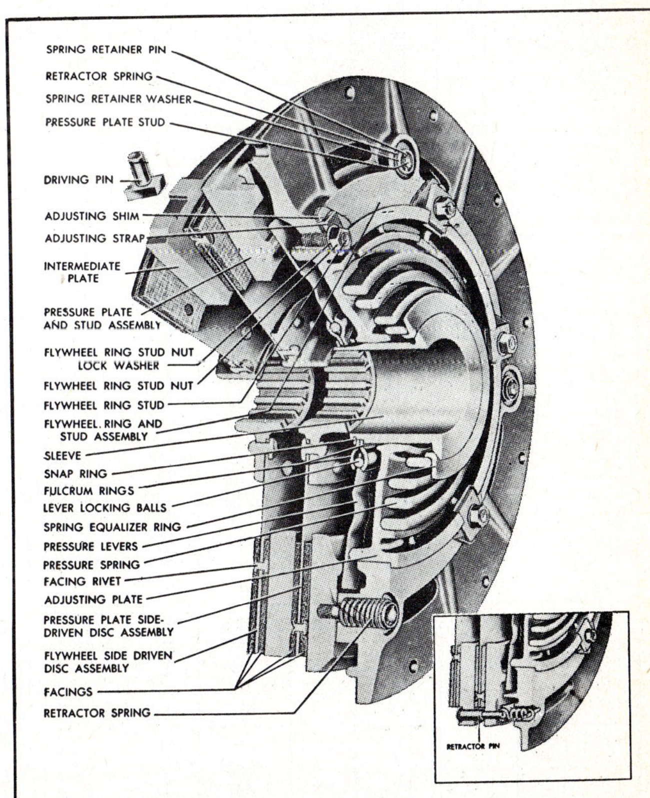

Fig. 2 Lipe ML 14" and 15" two plate clutch. Inset shows type of pressure plate retractor spring and pin on 13" two plate clutch

HEAVY DUTY CLUTCHES

Model Number	Inch Size	Dimension "A" +1/16" −0"	Quantity of Shims	Flywheel Depth	Facing Thickness	Pressure Sprg. Reference No.	Pressure Plate Reference No.
120-1-018	12	1"	6	1.920"	5/32"	9	A
130-1-021	13	1 1/16"	8	1.500"	5/32"	2	B
140-1-022	14	1 3/16"	8	1.312"	5/32"	10	D
140-1-023	14	1 3/16"	8	1.312"	5/32"	10	D
130-1-025	13	1"	6	1.843"	9/64"	2	F
140-1-026	14	1 1/4"	6	1.812"	9/64"	4	D
140-1-027	14	1 1/16"	7	1.812"	9/64"	10	D
150-1-028	15	1 3/8"	6	1.812"	5/32"	4	E
130-1-029	13	2 1/8"	6	1.843"	9/64"	3	F
120-1-030	12	1 1/16"	4	1.875"	1/8"	9	A
130-1-033	13	1"	6	1.843"	9/64"	10	F
140-1-038	14	1 1/16"	7	1.812"	9/64"	2	D
120-1-039	12	1 1/16"	4	1.875"	1/8"	10	A
140-1-041	14	1 1/4"	6	1.812"	9/64"	6	D
130-1-043	13	1"	6	1.843"	9/64"	2	B
120-1-044	12	1"	6	1.920"	5/32"	10	A
130-1-045	13	1 1/16"	7	1.500"	5/32"	10	B
130-2-054	13-2	1 3/16"	8	2.937"	1/8"	6	G
130-1-063	13	1"	6	1.812"	9/64"	10	F
140-1-086	14	1 1/16"	7	1.812"	9/64"	10	D
130-1-087	13	1"	6	1.843"	9/64"	2	B
140-1-089	14	1 5/16"	8	1.312"	5/32"	6	D
130-1-092	13	1"	6	1.843"	9/64"	10	F
130-2-096	13-2	1 5/16"	8	2.937"	1/8"	4	C
140-1-102	14	1 1/16"	7	1.812"	9/64"	10	D
130-1-103	13	1"	6	1.843"	9/64"	9	F
120-1-112	12	1"	6	1.920"	5/32"	10	H
140-1-114	14	1 1/4"	6	1.812"	9/64"	4	D
140-1-116	14	1 3/16"	8	1.312"	5/32"	10	I
130-1-117	13	1 1/16"	8	1.500"	5/32"	10	F
140-1-126	14	1 1/16"	7	1.812"	9/64"	10	K
140-1-127	14	1 1/4"	6	1.812"	9/64"	4	K
130-1-128	13	1"	6	1.843"	9/64"	2	F
130-2-132	13-2	1 1/4"	8	2.937"	1/8"	6	C
140-1-138	14	1 1/4"	6	1.812"	9/64"	15	D
150-1-145	15	1 1/8"	8	1.812"	5/32"	6	E
150-1-146	15	1 1/8"	7	1.812"	5/32"	4	E
150-1-147	15	1 1/8"	7	1.812"	5/32"	6	E
150-1-148	15	1 1/8"	8	1.812"	5/32"	4	E
150-1-149	15	1 1/8"	8	1.812"	5/32"	8	E
150-1-150	15	1 1/8"	8	1.812"	5/32"	13	E
150-1-151	15	1 1/8"	7	1.812"	5/32"	15	E
150-1-159	15	1 1/8"	7	1.812"	5/32"	8	E
150-2-162	15-2	1 1/8"	7	2.937"	5/32"	4	E
150-2-163	15-2	1 1/8"	7	2.937"	5/32"	8	E
150-1-171	15	1 1/8"	7	1.812"	5/32"	6	E
140-1-175	14	1 1/16"	7	1.812"	9/64"	10	K
130-1-176	13	2 1/8"	6	1.843"	9/64"	3	F
150-2-177	15-2	1 1/8"	7	2.937"	5/32"	4	E
150-2-178	15-2	1 1/8"	12	2.937"	5/32"	8	E
150-2-179	15-2	1 1/8"	12	2.937"	5/32"	6	E
140-2-181	14-2	1 1/16"	8	2.937"	9/64"	4	J
140-2-182	14-2	1 1/8"	7	2.937"	9/64"	8	J
150-1-187	15	1 1/8"	7	1.812"	5/32"	4	E
150-1-191	15	1 3/8"	14	1.812"	5/32"	8	E
150-1-192	15	1 3/8"	14	1.812"	5/32"	23	E
150-1-193	15	1 3/8"	14	1.812"	5/32"	8	E
140-1-195	14	1 1/16"	6	1.812"	9/64"	2	D
150-1-199	15	1 1/8"	7	1.812"	5/32"	4	E
150-1-200	15	1 1/8"	8	1.812"	5/32"	8	E
120-1-226	12	1 1/16"	8	1.875"	1/8"	2	A
130-1-227	13	1"	8	1.843"	9/64"	10	F
130-1-232	13	1"	8	1.843"	9/64"	10	F
140-1-234	14	1 3/16"	8	1.312"	5/32"	4	I
150-1-241	15	1 1/8"	7	1.812"	5/32"	4	E
150-1-247	15	1 1/8"	8	1.812"	5/32"	13	E
150-2-261	15-2	1 1/8"	12	2.937"	5/32"	8	E
140-1-266	14	1 3/16"	9	1.187"	9/64"	4	M
120-1-287	12	1 1/8"	8	1.875"	5/32"	10	A
140-1-293	14	1 1/8"	8	1.812"	9/64"	10	D
150-1-295	15	1 3/16"	7	1.812"	5/32"	4	E
120-1-300	12	1 1/8"	8	1.875"	1/8"	9	A
150-1-301	15	1 3/16"	8	1.812"	5/32"	8	E
120-1-302	12	1 1/8"	8	1.875"	1/8"	2	A
130-1-307	13	1 1/16"	6	1.812"	9/64"	10	F
140-1-308	14	1 3/8"	6	1.812"	9/64"	4	D
150-1-310	15	1 3/16"	8	1.812"	5/32"	6	E
140-1-311	14	1 1/8"	8	1.812"	9/64"	2	D
140-1-312	14	1 3/8"	9	1.812"	9/64"	6	D
120-1-314	12	1 1/16"	6	1.920"	5/32"	10	H
140-1-315	14	1 1/4"	11	1.187"	9/64"	4	M
150-1-316	15	1 3/16"	7	1.812"	5/32"	4	E
150-1-317	15	1 3/16"	7	1.812"	5/32"	13	E
130-1-318	13	1 1/16"	6	1.843"	9/64"	9	F
130-1-319	13	1 1/16"	6	1.843"	9/64"	2	F
130-1-320	13	1 1/16"	6	1.843"	9/64"	10	F
150-2-321	15-2	1 3/16"	12	2.937"	5/32"	6	E
150-2-322	15-2	1 3/16"	12	2.937"	5/32"	4	E
150-2-323	15-2	1 3/16"	12	2.937"	5/32"	8	E
150-1-326	15	1 3/16"	7	1.812"	5/32"	13	E
140-2-327	14-2	1 1/8"	8	2.937"	9/64"	4	J
140-2-328	14-2	1 1/8"	7	2.937"	9/64"	8	J
150-2-329	15-2	1 3/16"	12	2.937"	5/32"	8	E
130-2-334	13-2	1 3/8"	8	2.937"	1/8"	4	C
120-1-331	12	1 1/16"	6	1.920"	5/32"	9	A
140-1-332	14	1 3/8"	7	1.812"	9/64"	10	I
130-1-333	13	1 1/16"	6	1.843"	9/64"	10	F
130-1-334	13	1 1/16"	6	1.843"	9/64"	2	B
140-1-335	14	1 1/16"	8	1.812"	9/64"	10	K
140-1-336	14	1 3/8"	9	1.812"	9/64"	4	K
150-1-337	15	1 3/16"	7	1.812"	5/32"	8	E
150-1-338	15	1 3/16"	7	1.812"	5/32"	6	E
150-1-339	15	1 3/16"	7	1.812"	5/32"	4	E
120-1-340	12	1 1/8"	8	1.875"	1/8"	10	A
120-1-341	12	1 1/8"	8	1.875"	5/32"	9	A
120-1-342	12	1 1/8"	8	1.875"	5/32"	2	A
150-1-343	15	1 3/8"	14	1.812"	5/32"	23	L
150-1-344	15	1 3/8"	14	1.812"	5/32"	8	L
130-2-347	13-2	1 3/8"	8	2.937"	1/8"	6	C
120-1-359	12	1 1/8"	8	1.875"	5/32"	9	N
120-1-360	12	1 1/8"	8	1.875"	5/32"	2	N
120-1-361	12	1 1/8"	8	1.875"	5/32"	10	N
130-1-362	13	1 1/16"	7	1.843"	5/32"	10	B
150-2-371	15-2	1 3/16"	12	2.937"	5/32"	6	L
150-2-372	15-2	1 3/16"	12	2.937"	5/32"	4	L
150-2-373	15-2	1 3/16"	12	2.937"	5/32"	8	L
140-2-378	14-2	1 1/8"	8	2.937"	9/64"	6	J
150-1-387	15	1 3/16"	11	1.750"	5/32"	8	E
130-41-397	13	1 1/16"	8	1.500"	5/32"	74	F
140-31-400	14	1 1/4"	8	1.312"	5/32"	73	I
130-1-407	13	1 1/16"	8	1.843"	5/32"	10	F
150-2-411	15-2	1 3/16"	12	2.937"	5/32"	8	L
150-41-420	15	1 1/8"	8	1.812"	5/32"	73	E
140-1-425	14	1 1/8"	8	1.812"	9/64"	10	I
150-1-465	15	1 3/16"	7	1.812"	5/32"	8	E
150-1-466	15	1 3/16"	8	1.812"	5/32"	13	E
150-1-467	15	1 3/16"	8	1.812"	5/32"	13	E
150-1-468	15	1 3/16"	7	1.812"	5/32"	8	E
150-2-469	15-2	1 3/16"	12	2.937"	5/32"	8	L
150-2-470	15-2	1 3/16"	12	2.937"	5/32"	8	L

Continued

Model Number	Inch Size	Dimension "A" +1/16" −0"	Quantity of Shims	Flywheel Depth	Facing Thickness	Pressure Sprg. Reference No.	Pressure Plate Reference No.
140-2-471	14-2	1 1/8"	7	2.937"	9/64"	8	J
140-1-474	14	1 3/8"	9	1.812"	5/32"	8	D
140-41-475	14	1 1/4"	9	1.812"	5/32"	73	I
140-41-483	14	1 1/4"	8	1.312"	5/32"	73	I
150-41-486	15	1 1/8"	8	1.812"	5/32"	73	E
130-41-487	13	1 1/8"	8	1.500"	5/32"	74	F
140-41-495	14	1 1/4"	9	1.812"	5/32"	73	I
140-1-504	14	1 3/8"	9	1.812"	5/32"	4	I
140-1-508	14	1 1/4"	11	1.187"	9/64"	4	M
140-1-509	14	1 3/8"	9	1.812"	9/64"	4	I
140-1-510	14	1 3/8"	6	1.812"	9/64"	4	D
150-2-514	15-2	1 3/16"	12	2.937"	5/32"	4	L
130-2-515	13-2	1 3/8"	8	2.937"	1/8"	4	C
150-1-516	15	1 3/16"	8	1.812"	5/32"	4	E
140-2-517	14-2	1 1/8"	7	2.937"	9/64"	4	J
150-1-518	15	1 3/16"	11	1.750"	5/32"	13	E
150-1-528	15	1 3/16"	8	1.812"	5/32"	8	E
130-1-540	13	1 7/16"	9	1.843"	5/32"	4	F
150-2-546	15-2	1 3/16"	12	2.937"	5/32"	8	L
140-1-555	14	1 5/16"	11	1.187"	5/32"	4	M
140-1-611	14	1 1/8"	8	1.812"	5/32"	10	I
130-1-621	13	2 3/16"	9	1.843"	5/32"	3	F
150-1-693	15	1 3/16"	8	1.812"	5/32"	13	L
140-2-765	14-2	1 1/8"	8	2.937"	5/32"	8	J
140-1-822	14	1 3/8"	9	1.812"	5/32"	4	P

MODEL NO.—Is the number stamped on machined surface of the cover, on which adjusting shims are supported, or on the name plate.

INCH SIZE—Refers to the approximate diameter of the driven disc assembly or pressure plate, which determines the clutch size.

DIMENSION "A"—Is the dimension indicated as "A" on the assembly drawing (Fig. 7) and (Fig. 2), and has a limit from plus 1/16" to minus 0".

QUANTITY OF SHIMS—Is the approximate number of adjusting shims required on each stud, to give correct DIMENSION "A". This applies only when the clutch is new in a flywheel of the correct depth and when correct facings are used on the driven disc assembly. This quantity may vary one shim. Fewer shims are required in a used or worn clutch. OBTAIN DIMENSION "A" REGARDLESS OF NUMBER OF SHIMS REQUIRED.

FLYWHEEL DEPTH—Is the correct dimension from the flywheel friction face to the face of the rim. This dimension must not vary more than plus or minus .005"

FACING THICKNESS—Is the correct dimension of each facing on the disc assembly.

PRESSURE SPRING REFERENCE NO.—Is the reference number listed in the chart below to correspond with the LIPE part number.

Reference Number	Lipe Pressure Spring Number	Spring Color	Pounds Minimum Weight
2	C5-2	Yellow	360 Lbs. at 1 1/4"
3	C5-3	———	360 Lbs. at 2 1/4"
4	C5-4	Black	480 Lbs. at 1 1/4"
6	C5-6	Green	440 Lbs. at 1 1/4"
8	C5-8	Orange & White	575 Lbs. at 1 1/4"
9	C5-9	Blue	330 Lbs. at 1 1/4"
10	C5-10	Red	415 Lbs. at 1 1/4"
13	C5-13	Orange & Blue	630 Lbs. at 1 1/4"
15	C5-15	Red & Green	400 Lbs. at 1 1/4"
23	C5-23	Green & White	375 Lbs. at 1 1/4"
73	C5-73	Brown (Rd. Wire)	650 Lbs. at 1 1/8"
74	C5-74	Orange (Rd. Wire)	490 Lbs. at 1 1/8"

PRESSURE PLATE REFERENCE NO.—Is the reference number listed in the chart below to correspond with the LIPE part number.

Reference Number	Lipe Pressure Plate Number	Distance From Top of Fulcrum Edge to Friction Surface (New)
A	C1-2	.914"
B	C1-3	1.000"
C	C1-6	1.088"
D	C1-7	.975"
E	AC1-9	.977"
F	C1-16	1.013"
G	C1-19	1.026"
H	C1-23	.926"
I	C1-24	.980"
J	AC1-25	1.046"
K	C1-27	.975"
L	AC1-42	.977"
M	C1-61	.918"
N	C1-76	.916"
O	C1-118	.980"
P	C1-140	.980"

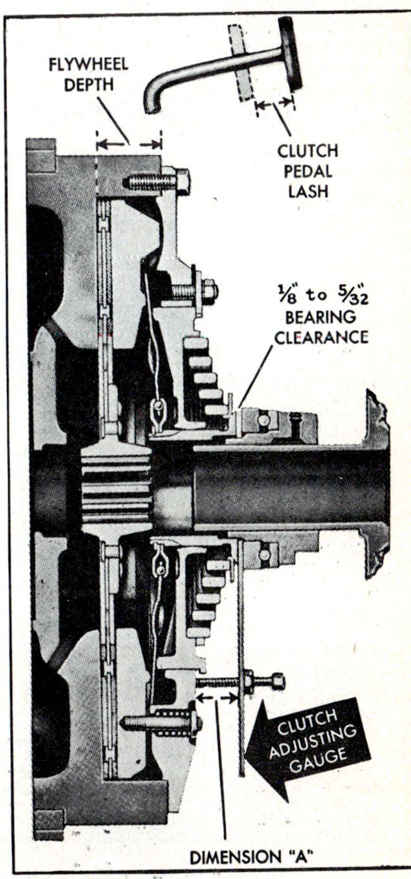

Fig. 3 Clutch specifications and adjustments for chart identification

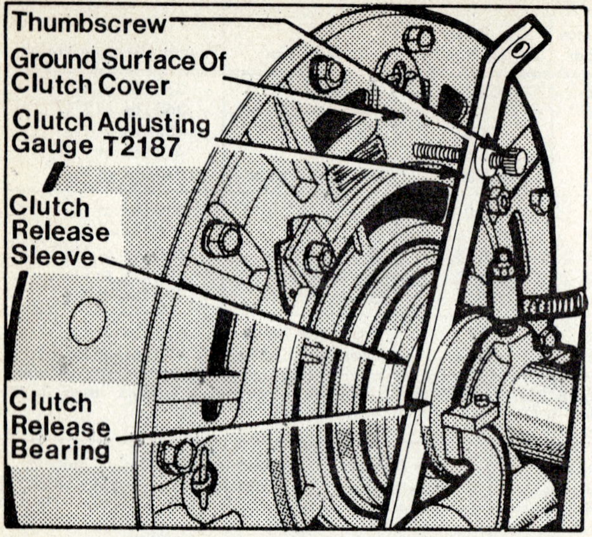

Fig. 4 Installation of clutch adjusting gauge

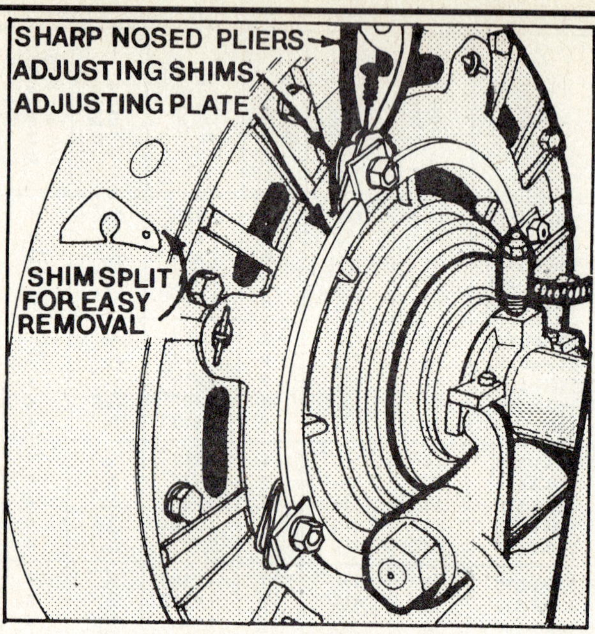

Fig. 5 Removing adjusting shims

before proceeding.

10. Check clearance between clutch release sleeve and release bearing, Fig. 6. If clearance is not 1/8", adjust vehicle clutch pedal linkage to obtain this 1/8" clearance. This 1/8" clearance normally results in 1½" of clutch pedal free travel.

Clutch Service

1. Place clutch cover assembly on table of press with assembly resting on pressure plate, Fig. 7.
2. Compress pressure spring.
3. Remove four retractor spring retaining pins, being careful not to over-stretch springs.
4. Place assembly on table of press with assembly resting on face of release sleeve, Fig. 8.
5. Using metal yoke as shown (or suitable blocks of wood) compress assembly and lock press in position.
6. Remove snap ring from groove of sleeve. The fulcrum rings, pressure levers and lever locking balls can now be removed.
7. Slowly release press to allow pressure spring, sleeve and remaining parts to be removed.

Inspection & Repair

After clutch is disassembled, clean all parts thoroughly. Inspect all parts as indicated below and replace any part that appears doubtful.

Pressure Plate

1. Inspect fulcrum edge for excessive wear. Compare fulcrum edge with cross section view of a new fulcrum edge, Fig. 9. If fulcrum edge "land" exceeds 1/16" in width, replace with a new pressure plate. *Any amount worn from fulcrum edge directly reduces total available clutch wear.*
2. Inspect friction surface of pressure plate for burning, heat cracking and distortion. If friction surface is badly scored, heat checked, warped, or dished in excess of .015", install a new pressure plate. *A scored or distorted pressure plate will cause the facings to disintegrate quickly.*

Flywheel Ring

1. Inspect flywheel ring for cracks; if present, replace with new one.
2. Inspect slots for indentation caused by pressure plate lugs. If indented, replace with a new flywheel ring.
3. Inspect flanged surface for burred edges. Burred edges on flange surface should be filed to assure flange being drawn tight to rim of engine flywheel when clutch is being installed.

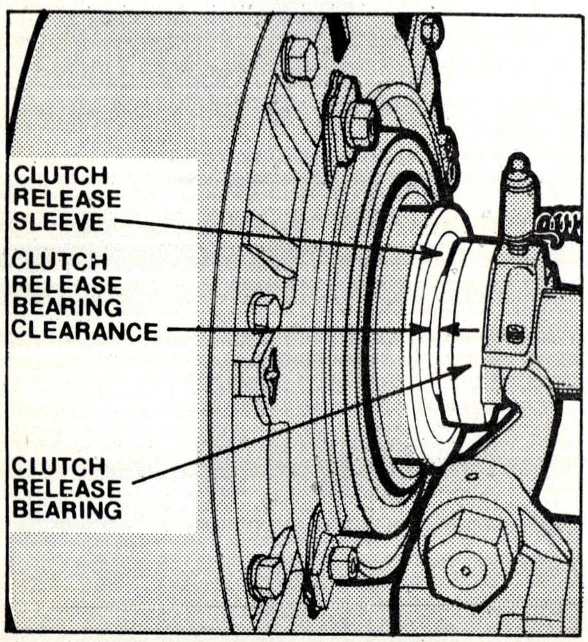

Fig. 6 Checking clearance between clutch release sleeve and release bearing

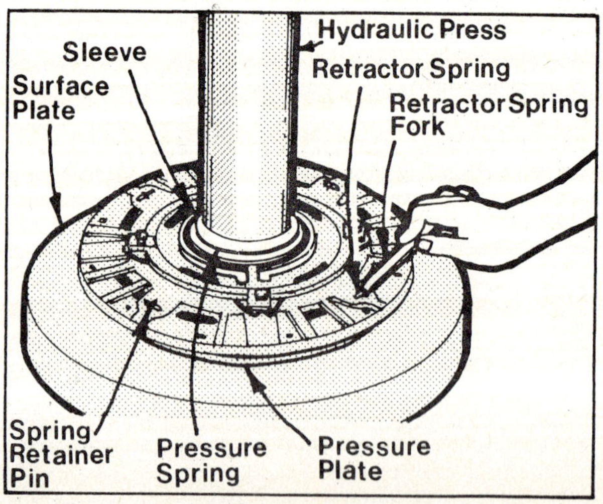

Fig. 7 Clutch cover mounted on press

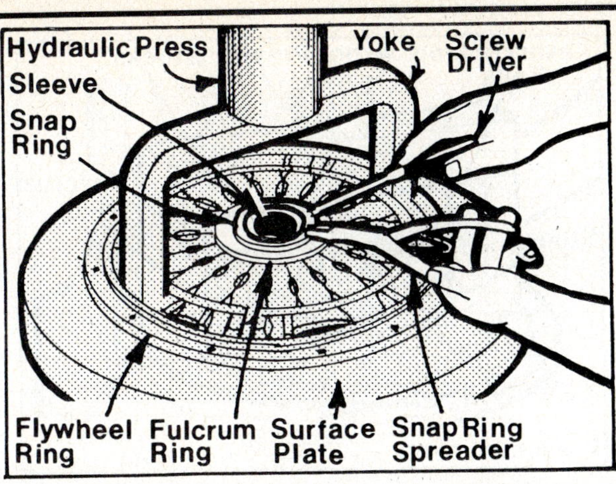

Fig. 8 Removing snap ring from sleeve groove

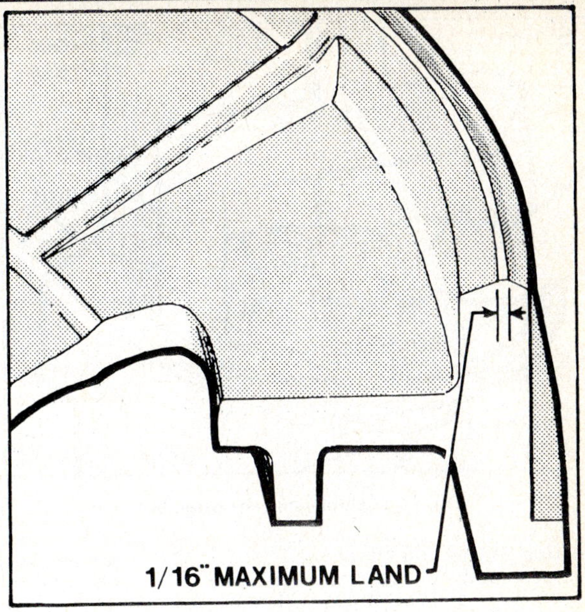

Fig. 9 Cross section view of a new fulcrum edge

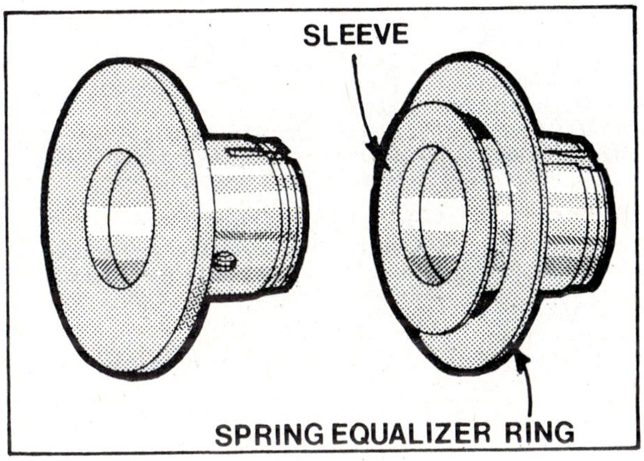

Fig. 10 Construction of release sleeves

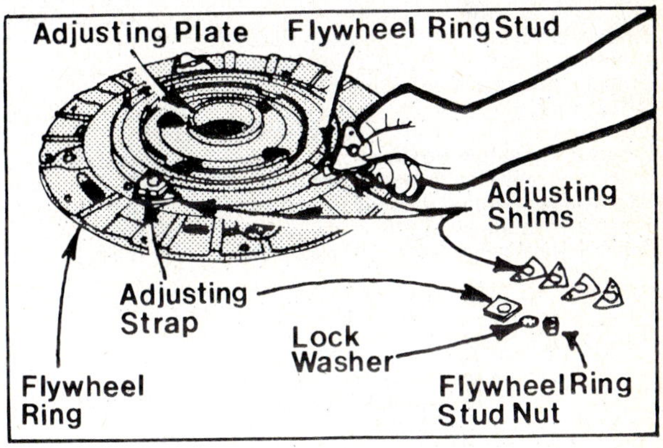

Fig. 11 Installing adjusting plate and shims

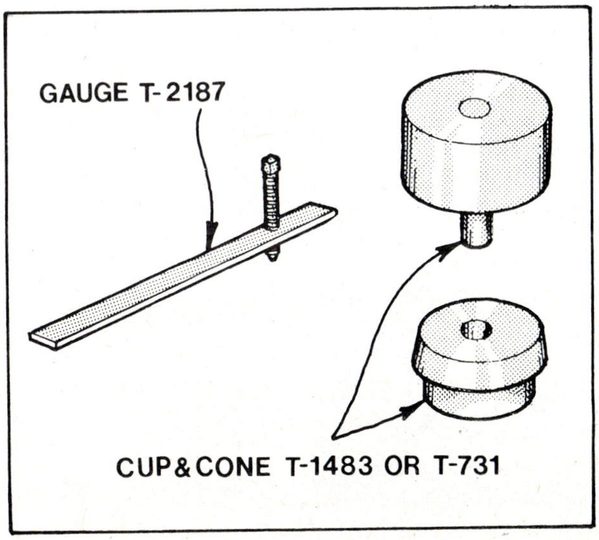

Fig. 12 Clutch overhaul tools

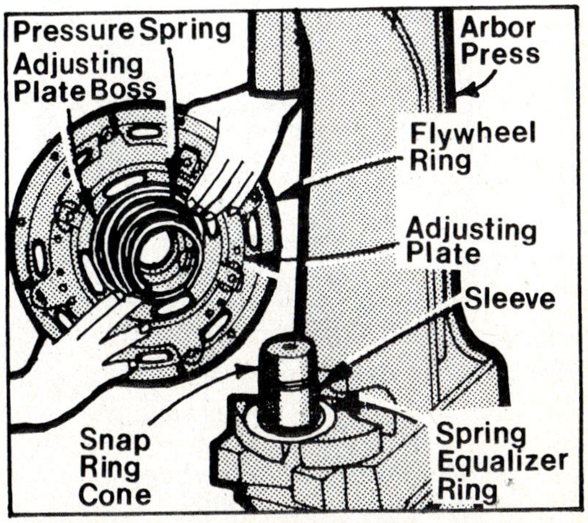

Fig. 13 Installing pressure spring in adjusting plate and assembling to sleeve

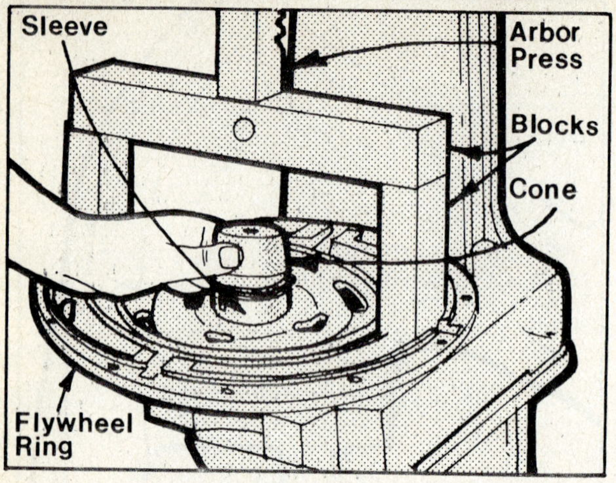

Fig. 14 Compressing the assembly

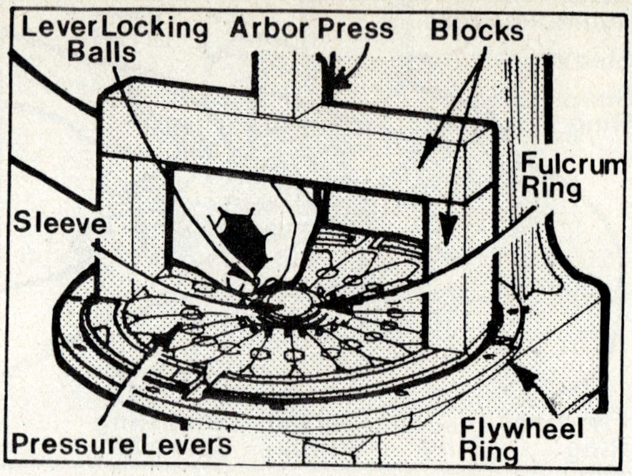

Fig. 15 Installing clutch lever locking balls

Adjusting Straps and Plate

1. Inspect straps for wear and bent condition. Worn or bent straps will prevent proper clamping of adjusting plate to flywheel ring.
2. Inspect fulcrum edge for excessive wear as shown in Fig. 9.
3. Inspect bore of adjusting plate for excessive wear. Release sleeve must be a free sliding fit in bore.
4. Inspect bottom of adjusting strap slots for excessive wear. Worn slots will prevent proper clamping of adjusting plate to flywheel ring.

Sleeve and Pressure Spring

1. Inspect sleeve for excessive wear or indentations of flange. If wear is indicated, replace with a new sleeve.
2. Inspect snap ring groove for wear, nicks or burrs. Any of these conditions will prevent proper seating of snap ring.
3. It is recommended that all one-piece type sleeves be replaced with the improved equalizer type sleeve and mating spring equalizer ring, Fig. 10.
4. Pressure springs should be checked for compression weight in accordance with the weights given in the *ML Clutch Chart.* Under-weight springs will cause excessive clutch slippage.

Fulcrum Rings and Pressure Levers

1. If fulcrum rings are worn or warped, install new rings.
2. Inspect pressure levers to see that they are not bent or excessively worn from contact with fulcrum edges of adjusting plate, pressure plate and fulcrum rings. If these conditions are present, install new levers.

Intermediate Plate and Driving Pins

1. On two plate clutches, inspect friction surfaces of intermediate plate for heat cracks, scoring or distortion. If any of these conditions exist, replace with a new plate.
2. Inspect driving slots of intermediate plate for wear; also driving pins. If either is worn, replace with new parts.

Driven Disc Assembly

Inspect disc assembly for cracks, worn splines, and warped or dished condition. If any of these conditions exist, install a new disc assembly.

Clutch Assembly

1. Assemble adjusting plate to flywheel ring. Install adjusting shims, Fig. 11, using the number of shims required on each stud to obtain the proper Dimension "A" (See ML Clutch Chart).
2. Place sleeve on arbor press with face down. Set equalizer ring over sleeve on to beveled edge of sleeve flange with flat side of ring down. Insert cone portion of snap ring assembling tool, Fig. 12, in end of sleeve.
3. Place pressure spring on adjusting plate with end of large coil abutting the small

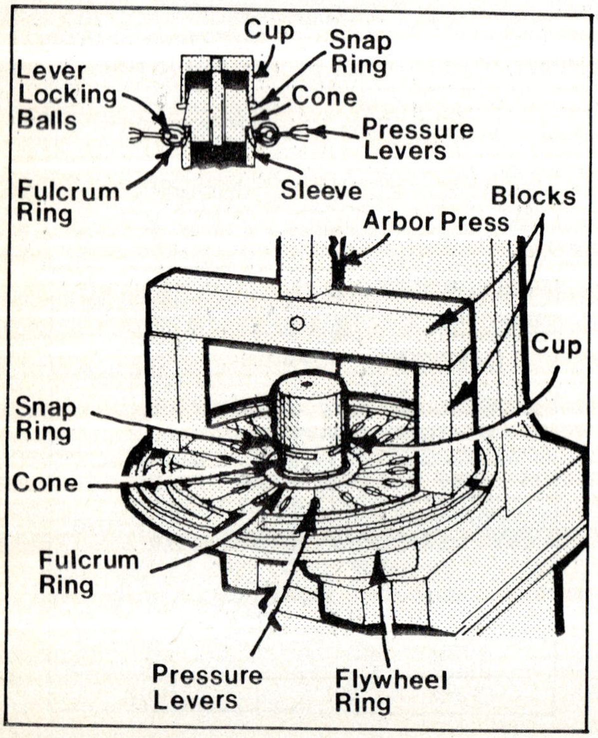

Fig. 16 Installing clutch snap ring

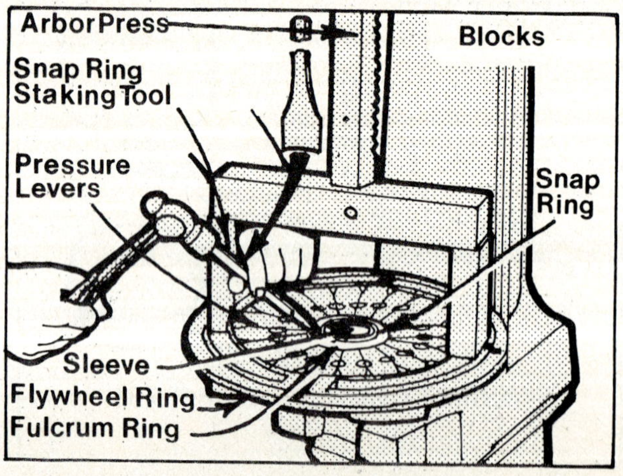

Fig. 17 Locking clutch snap ring

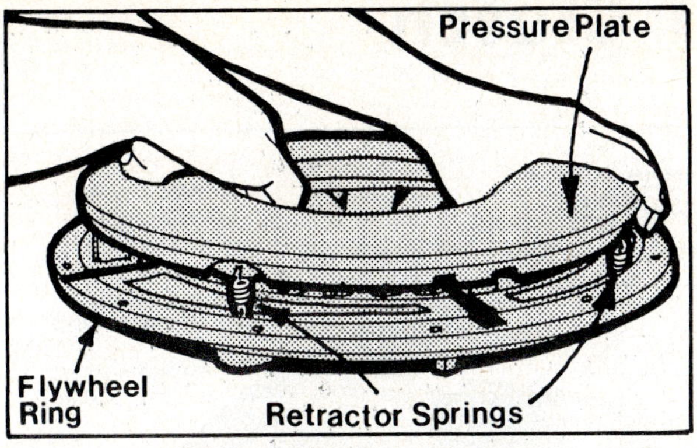

Fig. 18 Installing retractor springs

boss in the adjusting plate, Fig. 13.

4. Place flywheel ring, adjusting plate and pressure spring over the sleeve, seating small coil of spring on spring equalizer ring.
5. Using block system shown in Fig. 14, slowly compress assembly to allow sleeve to slip through bore of adjusting plate. *Be sure equalizer ring is seated properly on top coil of pressure spring.*
6. Place one fulcrum ring, cupped side up, over sleeve until it is seated firmly.
7. Arrange twenty pressure levers around sleeve. If ribbed levers are used, the raised side of lever must face pressure plate.
8. Using lever locking balls of proper size, insert one in hole at end of each lever, Fig. 15.
9. Install other fulcrum ring, cupped side down, over sleeve, making sure levers do not overlap.
10. Reinsert coned tool into sleeve and set a new snap ring over coned tool with open ends of snap ring at right angles to keyways of sleeve. Place cupped tool on snap ring and tap until snap ring is seated into sleeve groove, Fig. 16.
11. Remove cup and cone from sleeve and, with snap ring staking tool, Fig. 17, starting opposite open ends of snap ring, work each way until snap ring is seated well into groove of sleeve.

CAUTION: Test each of the twenty levers to see that they are locked securely by a ball and that they do not overlap. Release press and remove blocks.

12. Hook four retractor springs in pressure plate, Fig. 18. Place pressure plate and springs in position on the assembly, with driving lugs of pressure plate fitting into driving slots of flywheel ring.

NOTE: The driving lugs should fit free in the slots of flywheel ring with a clearance of .004″ to .006″. If original pressure plate is used, reassemble with driving lug marked "O" in driving slot marked "O". Turn assembly over and compress to facilitate installing retainer pins in retractor springs. This will prevent spring from becoming over-stretched.

Clutch Installation

Single Plate Clutch

1. Try cover plate assembly in flywheel before inserting driven disc to make sure it is a free fit in flywheel.
2. After installing driven disc, bolt clutch to flywheel and screw in each bolt until it contacts cover plate. Then gradually tighten opposite bolts until assembly is drawn tight to flywheel.
3. Check for Dimension "A" as outlined previously.
4. To facilitate the assembly of transmission to engine, the use of a spare splined pilot shaft to align the driven disc while bolting the clutch in place is essential. The pressure plate end of the driven disc hub has been chamfered to facilitate the insertion of the drive shaft, whereas the flywheel end of the hub is left plain.

Two Plate Clutch

1. Inspect the four intermediate driving pins to make certain that shank of pins are press fitted into flywheel rim, and that heads are squarely aligned with friction face. Use tool shown in Fig. 19 to facilitate the alignment of the driving pins.
2. Make sure that there is clearance between heads of driving pins and driving slots in intermediate plate by placing intermediate plate over pins.

NOTE: The intermediate plate should fit over the driving pins with perceptible circumferential lash before clutch is assembled to flywheel. A clearance of .007″ to .011″ is required.

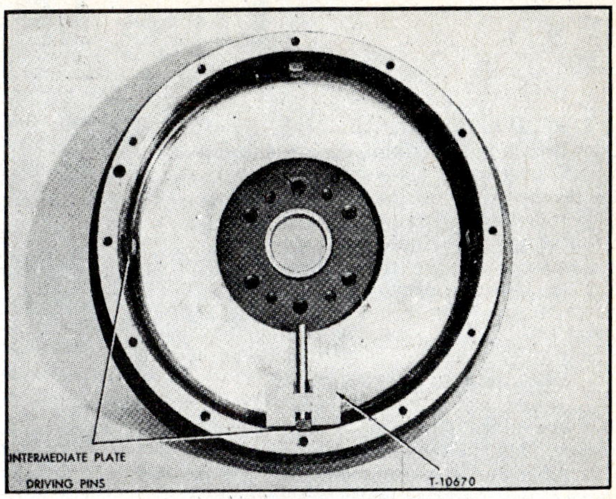

Fig. 19 Driving pin alignment gauge for two plate clutches

3. The 13″ two plate clutch incorporates retractor spring pins to retract intermediate plate when clutch is released. This necessitates the driven disc next to the pressure plate being in place before the intermediate plate is assembled to the cover plate.
4. The two discs for the two plate clutch may or may not be identical or interchangeable. In most cases they will not be interchangeable and will be marked either flywheel side or pressure plate side. They must be positioned according to the mark next to or adjacent to the respective part. In no instance shall the hubs be allowed to approach pilot bearing or come together within 5/32″.
5. Bolt cover to flywheel. Screw in each bolt until it contacts cover plate. Then gradually tighten opposite bolts until assembly is drawn tight to flywheel.
6. Check for proper Dimension "A" as listed in the *ML Clutch Chart.*
7. Use an aligning shaft when installing transmission to prevent damage to driven discs which will cause clutch drag.

NOTE: To obtain complete disengagement of ML clutches, the sleeve must be moved a specified distance from the engaged position. Assuming the release bearing clearance (distance between sleeve and bearing) to be 1/8″ when clutch is in the engaged position, the release bearing must move forward 1/8″ to take up this clearance before it contacts the sleeve and then move the normal amount of travel to disengage the clutch as shown in Fig. 20.

Clutch Size & Type	Release Bearing Clearance	Normal Sleeve Travel	Total Bearing Travel Required
12″ ML Single Plate	1/8″	3/8″	1/2″
13″ ML Single Plate	1/8″	7/16″	9/16″
14″ ML Single Plate	1/8″	7/16″	9/16″
14″ RML Single Plate	1/8″	5/16″	7/16″
15″ ML Single Plate	1/8″	1/2″	5/8″
15″ RML Single Plate	1/8″	1/2″	5/8″
13″ ML Two Plate	1/8″	1/2″	5/8″
14″ ML Two Plate	1/8″	1/2″	5/8″
15″ ML Two Plate	1/8″	1/2″	5/8″

Fig. 20 Lipe ML clutch release travel chart

Lipe LP & PT Clutch

The Lipe LP & PT clutch are of the a pull-type, dry disc, non-adjustable design, incorporating an adjustable sleeve connecting the release levers to the release bearing. Adjustment of this sleeve compensates for facing wear, maintains pedal lash and provides operating clearance for the clutch brake. Anti-friction bearings are provided in the release lever system resulting in low release load on the pedal.

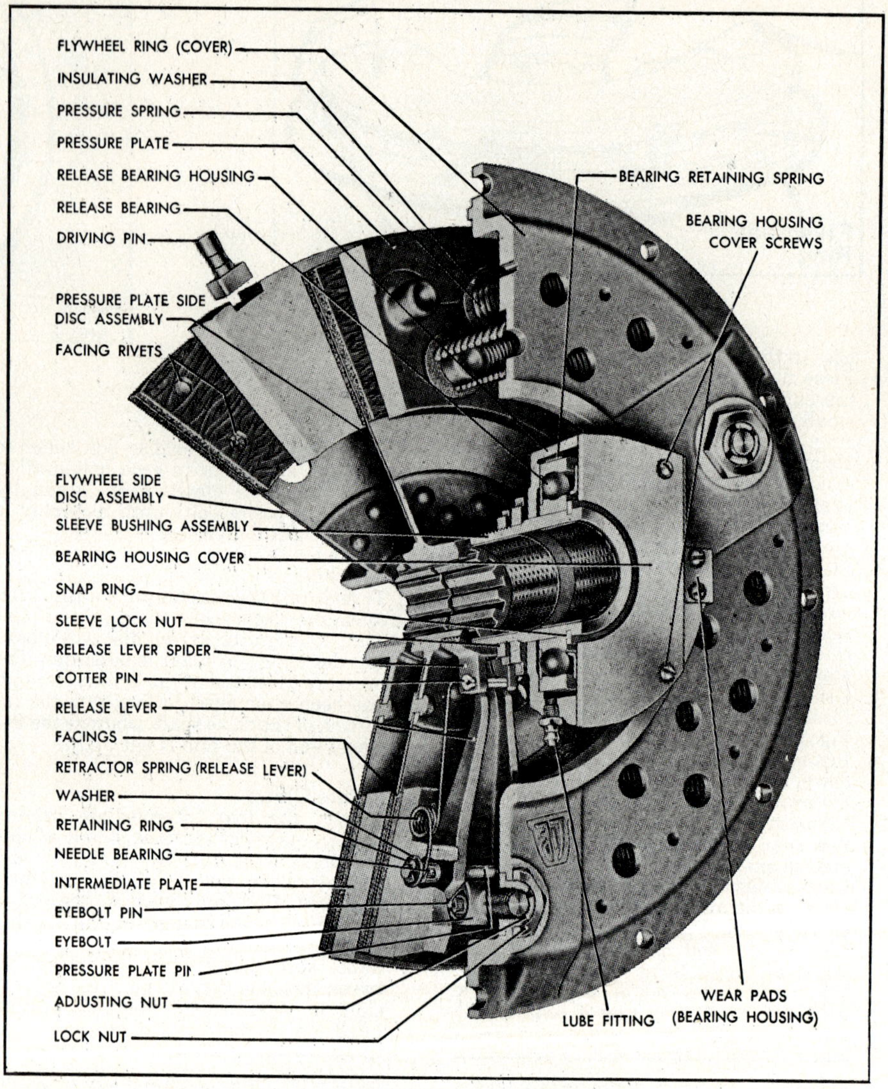

Fig. 1 Type PT two plate clutch

Identifying Models

The name plate attached to the flywheel ring (cover) identifies the cover plate as to the serial number and part number. Driven disc assemblies are stamped with an assembly number. See Chart, Fig. 2 for clutch data.

Clutch Pedal Adjustment

NOTE: Do not adjust the external linkage to compensate for clutch wear. The external linkage should only be reset to provide proper free-travel after the internal clutch adjustment has been made.

Internal Clutch Adjustment

1. Unlock the sleeve locknut.
2. Turn the slotted adjusting nut to obtain 1/2 inch clearance keeping a slight tension on the release bearing.
3. Lock the sleeve locknut.

The proper clutch pedal "free travel" is approximately 1 1/2". If inspection indicates the free travel of the clutch pedal is less than 1/2", immediate adjustment of the release sleeve should be made to restore the proper 1 1/2" free travel, Fig. 3. This 1 1/2" free travel normally results in 1/8" clearance between the release yoke and the bearing carrier wear pads or at the cross shaft adjustment, Fig. 4.

Linkage Adjustment

1. Be sure release bearing sleeve is properly adjusted and locked before attempting to adjust the external linkage to set the pedal lash.
2. If vehicle is equipped with a hydraulic release, refer to the vehicle chapter in this manual or to the vehicle manufacturer's maintenance manual for specifications relating to master and slave cylinder adjustments.
3. Master cylinder operating lever must be adjusted so that the piston can retract fully and uncover the compensating port.
4. The cross shaft lever should travel the same distance on each side of the vertical centerline of the cross shaft, Fig. 4.

Transmission Removal

When removing the transmission from the vehicle, disconnect the external linkage as it will be necessary to allow the release yoke to turn up and over the release bearing housing as the transmission is being moved away from the engine. Clutch may now be removed for inspection.

Flywheel Inspection

1. Remove worn intermediate plate drive pins.

NOTE: Most flywheels have socket head set screws holding these pins in place which must be loosened before the driving pins can be removed. If the engine flywheel has no provision for set screws, it is recommended that four holes be drilled, tapped and set screws installed. Use 5/16" drill and 3/8" × 16 tap.

2. Flywheel friction surface must be free of heat checks, scoring and taper. For proper depth, see *PT Clutch Chart,* Fig. 2.
3. Replace flywheel pilot bearing. If pilot bearing bore is worn, re-bore and sleeve to obtain proper bearing fit.
4. If flywheel is being replaced on engine crankshaft, it should be dial indicated for engine manufacturer's specifications for acceptable runout.

Clutch Service

Disassemble

1. Remove bearing housing and sleeve from spider.
2. Remove cotter pins from lever spider. Lower lever spider assembly and turn it sufficiently to clear levers and remove.
3. Place cover assembly on arbor press with pressure plate down.
4. Compress assembly to relieve spring tension and remove lock nuts.
5. Release press slowly; then remove flywheel ring (cover), pressure springs and insulating washers.
6. Remove retaining rings from pressure plate pins and remove pins to disassemble release levers and eyebolts.

Release Sleeve and Bearing

1. To disassemble, remove housing cover.
2. Remove snap ring, then press sleeve and bushing from bearing.
3. Remove wear pads from bearing housing.

Inspection

After cover is disassembled, clean all parts thoroughly. Inspect all parts as indicated below and replace any that appear doubtful.

Pressure Plate Model Number	Dimension "A" Inch	Flywheel Depth ± .005	Facing Thickness Inch	Pressure Spring Data			Pressure Plate Date Dimension "C" New Inch
				Quantity	Color	Minimum Spring Pressure Lbs. @ Inch	
140-72-760	$1\frac{9}{32}$	2.937	$\frac{3}{16}$	21	Dk. Maroon	130 @ $1\frac{27}{32}$	1.875
140-72-786	$1\frac{9}{32}$	2.937	$\frac{3}{16}$	18	Dk. Maroon	130 @ $1\frac{27}{32}$	1.875
140-72-787	$1\frac{9}{32}$	2.937	$\frac{3}{16}$	15	Dk. Maroon	130 @ $1\frac{27}{32}$	1.875
140-72-788	$1\frac{9}{32}$	2.937	$\frac{3}{16}$	21	Lt. Green	150 @ $1\frac{27}{32}$	1.875
140-72-816	$1\frac{9}{32}$	2.937	$\frac{3}{16}$	12	Dk. Maroon	130 @ $1\frac{27}{32}$	1.875
140-72-901	$1\frac{9}{32}$	2.937	$\frac{3}{16}$	21	Dk. Maroon	130 @ $1\frac{27}{32}$	1.875
140-72-917	$1\frac{9}{32}$	2.937	$\frac{3}{16}$	18	Dk. Maroon	130 @ $1\frac{27}{32}$	1.875
140-72-919	$1\frac{9}{32}$	2.937	$\frac{3}{16}$	21	Lt. Green	150 @ $1\frac{27}{32}$	1.875
140-72-920	$1\frac{9}{32}$	2.937	$\frac{3}{16}$	21	Dk. Maroon	130 @ $1\frac{27}{32}$	1.875
140-72-1041	$1\frac{9}{32}$	2.937	$\frac{3}{16}$	6 12	Lt. Green & Red	150 @ $1\frac{27}{32}$ 180 @ $1\frac{27}{32}$	1.875
140-702-1138	$1\frac{1}{8}$	2.937	$\frac{3}{16}$	9 12	Lt. Green & Dk. Maroon	150 @ $1\frac{27}{32}$ 130 @ $1\frac{27}{32}$	1.875
140-702-1139	$1\frac{1}{8}$	2.937	$\frac{3}{16}$	9 12	Lt. Green & Dk. Maroon	150 @ $1\frac{27}{32}$ 130 @ $1\frac{27}{32}$	1.875
155-702-1498	$1\frac{15}{32}$	FLAT	.184 .169	24	Purple	174 @ $2\frac{1}{16}$	1.906
155-72-970 155-702-1495	$1\frac{15}{32}$	FLAT	.184 .169	18	Purple	174 @ $2\frac{1}{16}$	1.906
155-702-1497	$1\frac{15}{32}$	FLAT	.184 .169	18	Purple	174 @ $2\frac{1}{16}$	1.906

Fig. 2 Type LP & PT clutch chart

Pressure Plate

The pressure plate must be replaced if the friction surface is severely heat checked, it is warped in excess of .015, or if there is excessive wear at the driving slots.

The pressure plate may be salvaged if the friction surface cleans up by regrinding, not to exceed $\frac{1}{32}$" maximum from Dimension "C", Fig. 6, (see *PT Clutch Chart* for specification).

Always replace needle bearings by pressing on that end of the bearing on which part numbers are stamped. This prevents damage to bearing case.

All Other Parts

Inspect and repair all other parts as outlined for the DP two plate clutch section of this chapter.

Clutch Assembly

1. Place pressure plate on arbor press face down, Fig. 7.
2. Assemble adjusting nuts to eyebolts, then use new pins to assemble to release lever.
3. Install release levers, using new pins and retaining rings. Check proper positions of retractor springs, Fig. 7.
4. Place insulating washers over spring bosses and install pressure springs.
5. Press flywheel ring (cover) into position, using care to engage driving lugs into mating slots of pressure plate as springs are being compressed.
6. Pull adjusting nuts up through holes in cover and install lock nuts, Fig. 8.
7. Install three $\frac{3}{8}$" × 16 × $2\frac{1}{4}$" hold-down capscrews through cover into tapped holes in pressure plate. This holds pressure springs in a compressed state to facilitate assembly to engine flywheel.
8. Release press and remove clutch.

Release Bearing and Housing

1. To assemble, place release bearing housing over sleeve and bushing assembly and press bearing onto sleeve and install snap ring. Be sure bearing is tight to snap ring (see Fig. 5).
2. Install bearing retaining spring in housing opposite grease fitting hole.
3. Hand pack bearing and housing with high temperature lubricant such as Esso "Alert" 275 or equivalent. *Chassis or all purpose lubricants are not satisfactory for clutch release bearings.*

Clutch Installation

1. Be sure flywheel and intermediate plates are in good condition. Also check for free fit of cover on flywheel.
2. Be sure that intermediate drive pins are a

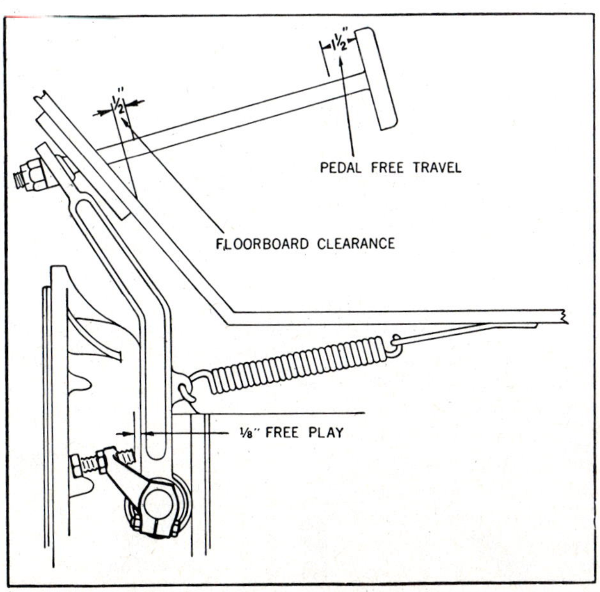

Fig. 3 Adjusting clutch pedal linkage

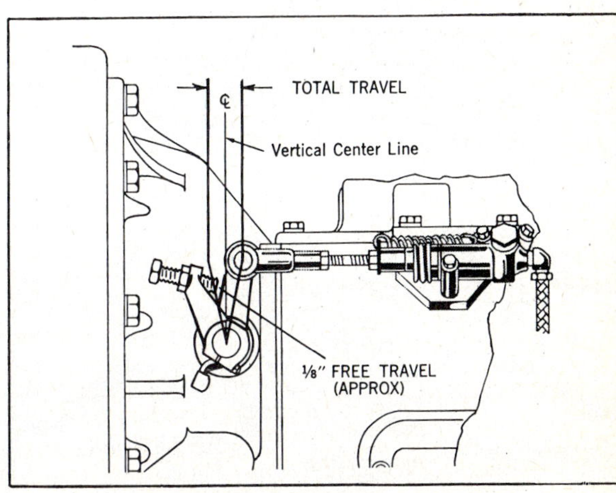

Fig. 4 Hydraulic release linkage (typical)

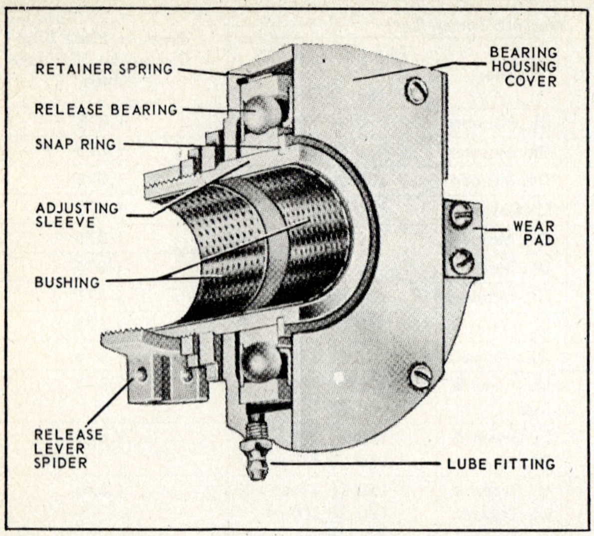

Fig. 5 Release sleeve and bearing assembly

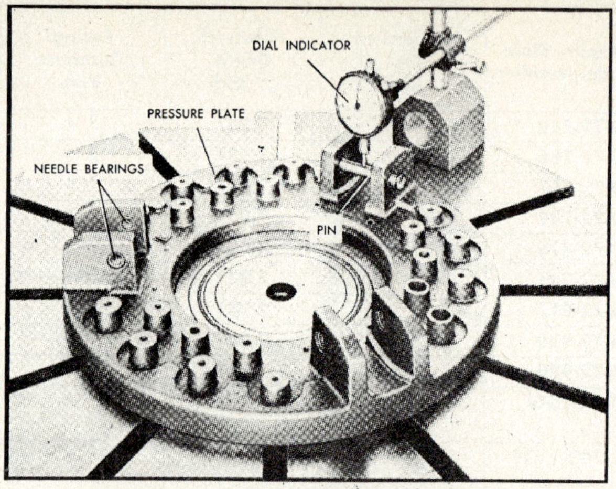

Fig. 6 Checking Dimension "C"

free fit in holes in flywheel rim and that heads are perfectly square to friction surface. Lipe Tool shown in Fig. 9 will assist in installing new pins and assure that they are properly aligned. If tool mentioned is not available, a machinist's square and feeler gauge may be used to check for squareness.

3. After pins are installed, try intermediate plate for fit over pins. With new pins and new intermediate plate, there should be .006" to .014" clearance. Try plate in different positions and select the one that provides the greatest freedom of movement. Mark the plate for location to the flywheel and then lock the drive pins in place with the socket head or Allen set screws.
4. Driven discs are marked "Pressure Plate Side" and "Flywheel Side" to indicate their proper positions.
5. The cover assembly may now be installed on the flywheel, but before tightening capscrews, install an aligning shaft through the two hubs and into the pilot bearing.
6. Cover capscrews may now be tightened progressively around the flywheel to 35 to 40 ft-lbs torque.
7. Remove the three hold-down bolts from cover before removing the aligning shaft or disc assemblies may fall out of position.

Clutch Adjustment

1. If cover assembly is new or a genuine Lipe exchange unit, the levers will not require adjustment.
2. If cover assembly has been repaired or rebuilt locally, the lever heights must now be adjusted to the correct "A" dimension as listed in the chart, Fig. 2. A hook type rule, Fig. 10, will facilitate taking the "A" dimension measurement. The tool shown in Fig. 11 is available for turning the slotted adjusting nut.
3. The best method of adjusting PT release levers is to make an initial adjustment so that the levers are all high or with an "A" dimension of about 1". Then with the lock nut just tight enough to prevent lost motion, the lock nut and adjusting nut may be turned clockwise as a unit until the exact "A" dimension is obtained. Then hold the adjusting nut still and tighten lock nut to 60 ft-lbs torque.
4. Install release sleeve and bearing assembly into spider. Be sure to include sleeve lock nut and thread sleeve well into spider.

Transmission Installation

1. Place the two brake friction discs on the transmission clutch shaft with the keyed steel disc between the friction discs. *These parts are not used with a synchronized transmission.*
2. Raise transmission and enter spline shaft into release sleeve assembly.
3. Square up transmission and move it forward to enter driven disc hubs, at the same time allow the release yoke to pass over the top of the release bearing housing and fall into its normal operating position on the bearing housing wear pads.

NOTE: An alternate method of installing the transmission to the engine is as follows: Instead of installing the release bearing and sleeve assembly into the release lever spider, the release sleeve and bearing is placed on the transmission with the release yoke in its operating position. Then after the transmission has been raised into position and started forward, the adjusting sleeve must be threaded into the release lever spider. If this method is used, care must be taken that the transmission is not moved up to the engine faster than the sleeve and spider are threaded together or cotter pins may be damaged.

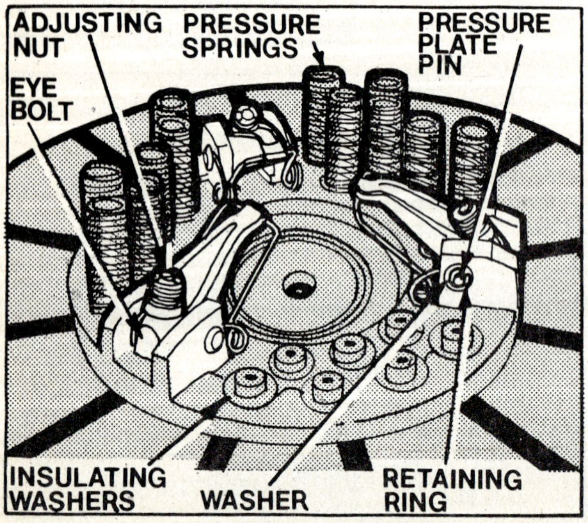

Fig. 7 Pressure plate assembly

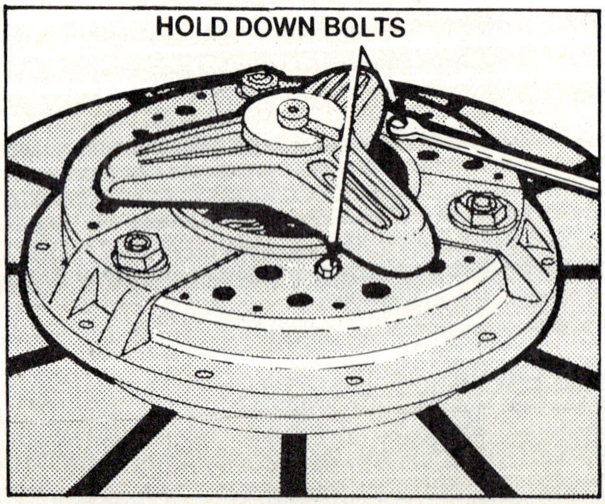

Fig. 8 Installation of hold-down bolts and lock nuts

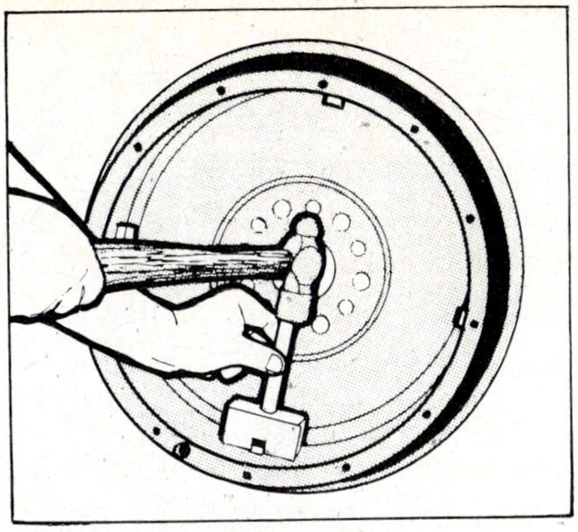

Fig. 9 Driving pin alignment gauge (T-10670)

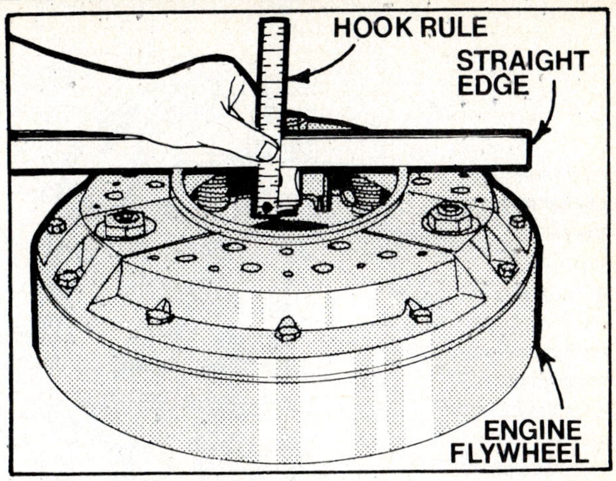

Fig. 10 Checking Dimension "A"

Release Bearing Adjustment

1. After the transmission has been securely attached to the engine, the release sleeve must be adjusted until there is 3/8" to 7/16" between the contact surface of the release bearing housing and the front brake disc.

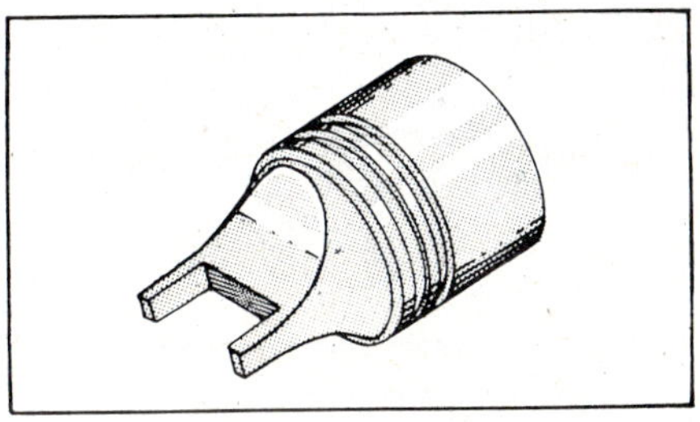

Fig. 11 Lipe Tool T-10207A for slotted adjusting nut

CAUTION: Keep tension on release yoke while turning adjusting nut. Do not attempt to secure the sleeve lock nut to the release lever spider until the pedal linkage has been connected and adjusted for "free travel".

2. Following the preliminary adjustment of the pedal linkage, check the release bearing travel by pressing down on the clutch pedal. The release bearing housing should contact the clutch brake disc just before the clutch pedal makes contact with the floor board. Depress the pedal a few times and recheck the 3/8" to 7/16" distance between the release bearing housing and brake disc.
3. Hold or block the clutch pedal down while the lock nut is turned up to release lever spider and securely tightened.

CAUTION: Always disengage the clutch while the lock nut is being locked or unlocked but have the clutch engaged while turning the release sleeve to obtain the 3/8" to 7/16" distance which must be rechecked and adjusted as facing wear occurs.

Lipe TC Clutch

The Lipe TC "Constant Capacity" clutch is of the non-adjustable, dry disc type which automatically compensates for facing wear, Figs. 1 and 2. As facings of the disc assembly wear through normal use, the toggle lever angle, which is formed by the release lever and toggle link, increases in efficiency as it straightens out and induces an increased pressure against the facings. Thus a "Constant" clutch torque capacity is maintained throughout the life of the clutch facings.

NOTE: If inspection indicates that clutch pedal free travel is less than 1/2", immediate adjustment of the clutch linkage should be made to restore the proper 1 1/2" free travel of the pedal.

Clutch Removal

1. Install two 3/8" × 16 × 3" hold-down bolts as shown in Fig. 3, through holes in spring retainer, and thread these bolts into the tapped holes in the pressure plate provided for this purpose. Tighten sufficiently to compress the springs and free the levers.
2. Remove capscrews holding clutch to flywheel and remove clutch.

Clutch Disassembly

1. Unfasten and lift off flywheel ring (cover).
2. Place clutch on table of an arbor or drill press. Place block across the spring retainer and hold springs in a compressed position so the hold-down bolts can be removed; then slowly release the spring tension.
3. Release lever and link sub-assemblies can now be removed from pressure plate by removing cotter pins at ends of pressure plate pins.
4. To remove toggle link from release lever, drive out locking pin.
5. The needle bearings in the release levers can be removed by the use of a suitable driver and arbor press; the same applies to the removal of the fulcrum guide buttons.

Inspection & Repairs

Pressure Plate

It is recommended that not more than 1/32" be machined from the pressure plate friction surface. If not more than this amount or any portion thereof is removed, the clutch should be reassembled, using hole "A" of toggle link (see Fig. 10).

IMPORTANT: When machining amounts in excess of 1/32", it is essential that resurfaced dimension "B" (shown in chart) be obtained and clutch reassembled using hole "B" of toggle link. Dimension "B" is measured from the top of the pin to the friction surface of the pressure plate. It is important that these instructions be followed, otherwise short clutch life or incomplete clutch release (drag) will result.

Flywheel Ring (Cover)

Inspect cover for wear of driving lugs and cover pin holes. Cover should fit freely in slots of pressure plate with approximately .006" clearance between lugs and mating slots in pressure plate. Cover should be checked for distortion by placing it flange face down on surface plate. If cover incorporates driving pins, make certain pins are tight and squarely aligned.

Lever and Link Assembly

Referring to Fig. 4, inspect lever for condition of toggle link hole; if elongated or worn, replace lever. Check milled slot in underside of lever; if excessively worn by fulcrum studs, replace lever. Latest design release lever incorporates a removable fulcrum guide button, as shown in Fig. 4, that can easily be

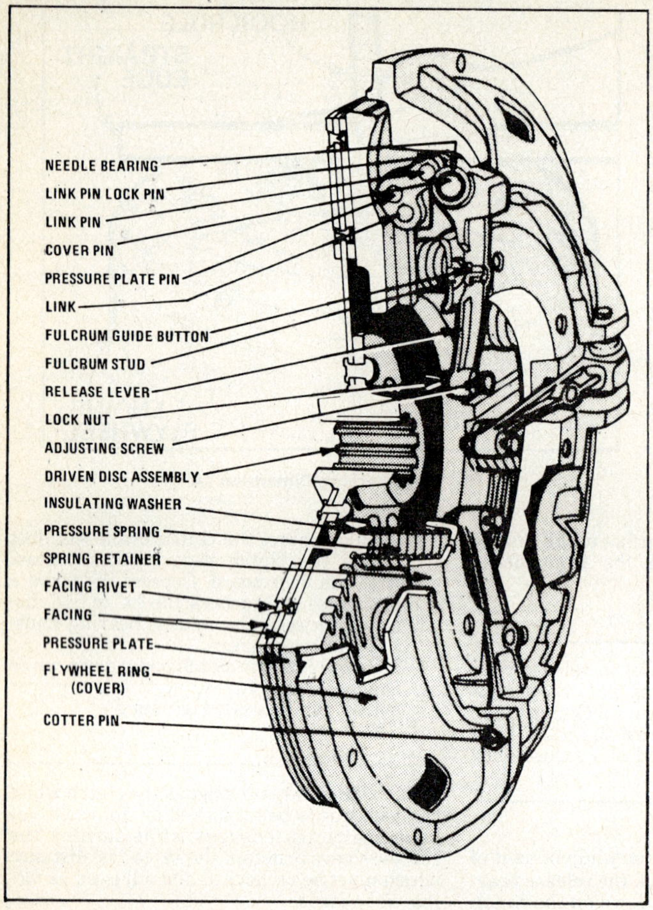

Fig. 1 Lipe TC single plate clutch

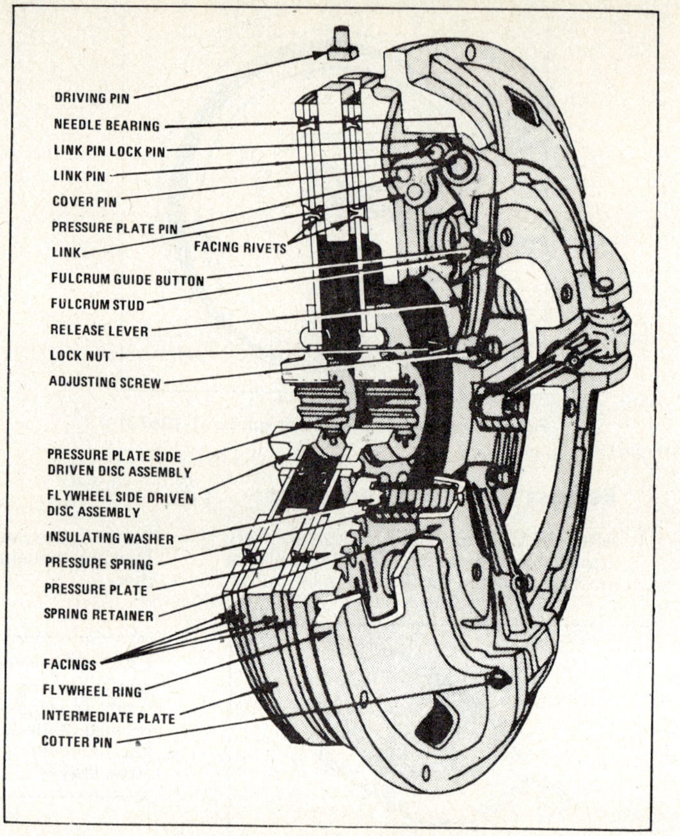

Fig. 2 Lipe TC two plate clutch

replaced if worn. If needle bearings do not rotate freely, replace.

When installing new needle bearings, press them in flush with lever. Make certain they are installed in lever with part number (stamped on end of bearing cage) facing outward, as pressing on other end of bearing cage will damage bearing. Pack needle bearing with a small amount of high melting point grease. Use correct grease and do not overlubricate.

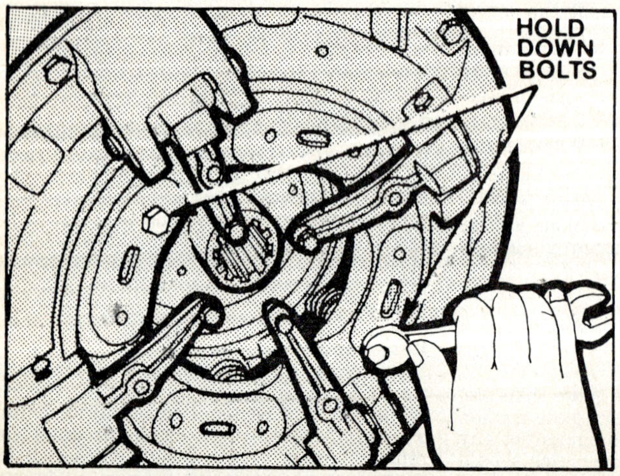

Fig. 3 Installing hold-down bolts

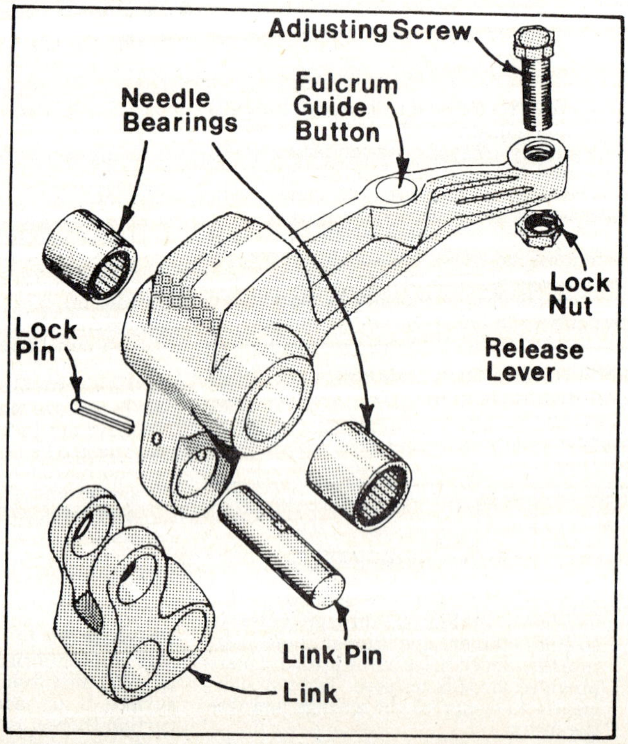

Fig. 4 Lever link and associated parts

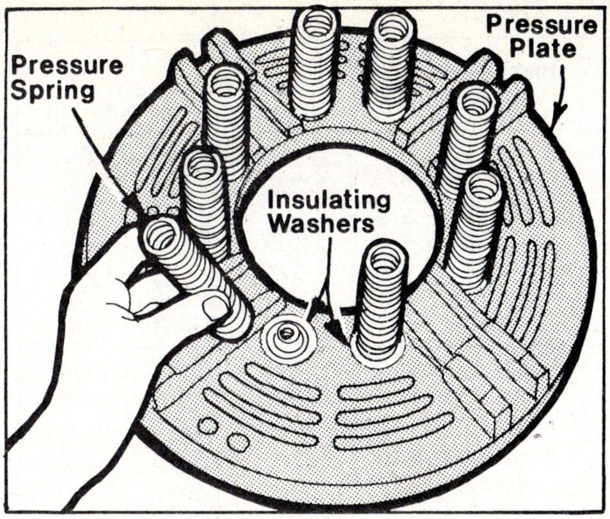

Fig. 5 Placing springs over insulating washers

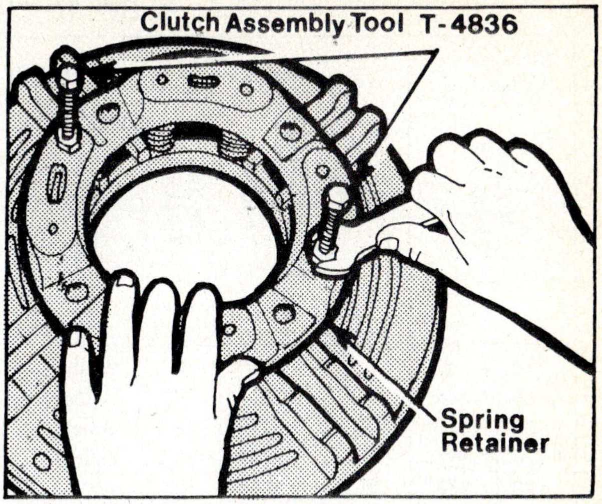

Fig. 6 Compressing pressure springs

Notice condition of lever adjusting screws. If they are worn where contacted by release bearing, install new screws in levers. Inspect fulcrum studs in spring retainer for wear. Replace if wear is indicated.

Driven Discs

Inspect and reface driven discs as outlined for other model clutches, being sure to use the proper riveting method for fabric and metallic facings.

Clutch Assembly

1. Place a small amount of high melting point grease on driving lugs of cover and mating slots in pressure plate, as well as pins and fulcrum studs. Do not over-lubricate as grease on facings will cause clutch to grab or slip.
2. Place insulating washers over spring locating bosses on pressure plate, Fig. 5. Use two washers under each spring if refaced pressure plate is installed to compensate for material ground off. Do not use two washers unless pressure plate has been refaced to dimension shown in chart under column "B Dimension Refaced".
3. Place springs over insulating washers on pressure plate, Fig. 5.
4. As shown in Fig. 6, install spring retainer over pressure springs and compress springs with the hold-down bolts used when clutch was removed. Draw nuts down against top of spring retainer to compress springs to an overall height of approximately 2″.

NOTE: An alternate method is to use an arbor or drill press to compress the springs. Place a block across the spring retainer. Install the hold-down bolts to hold the springs fully compressed. The hold-down bolts or the Lipe tool, Fig. 7, must remain in place until clutch has been bolted to flywheel.

5. As shown in Fig. 8, assemble toggle link to release lever, placing lever in slot of toggle link and insert link pin. Align milled notch of link pin with lock pin hole in lever with use of suitable tool before driving in lock pin. Lock pin should be driven in flush with lever surface.

THREE HOLE LINK: The proper link hole to use will depend on whether or not the pressure plate has been refaced. Toggle link hole "A", Fig. 10, is assembled to the pressure plate if a NEW pressure plate is used. Hole "B" is assembled only if the pressure plate has been refaced. See Chart indicating correct new and refaced pressure plate dimensions with respect to each clutch model.

6. Assemble sub-assembly, Fig. 9.
7. Place flywheel ring (cover) into position, aligning lugs in underside of cover with mating slots in pressure plate. If original pressure plate is used, be sure that cover registers with aligning "O" mark stamped on O.D. of both cover and pressure plate.
8. Referring to Fig. 11, start longest cover pins through cover, at the same time lifting or placing small blocks of proper thickness between top of work bench and underside of cover flange so as to align two opposite levers, and drive cover pins through needle bearings in levers into place.

Clutch Installation

1. Be sure hold-down bolts are used to hold springs compressed and free from release levers before clutch is assembled to flywheel.
2. Make sure flywheel is the correct depth and friction face is smooth and clean (see Chart for flywheel depth specifications).
3. Try cover plate assembly in flywheel before inserting driven disc to make certain clutch pilot diameter is a free fit in flywheel.
4. Install driven disc with pressure plate end of hub that has chamfered or pointed ends on splines to assist in the insertion of transmission drive shaft toward you.
5. Install spare splined shaft or clutch aligning arbor through disc hub into flywheel

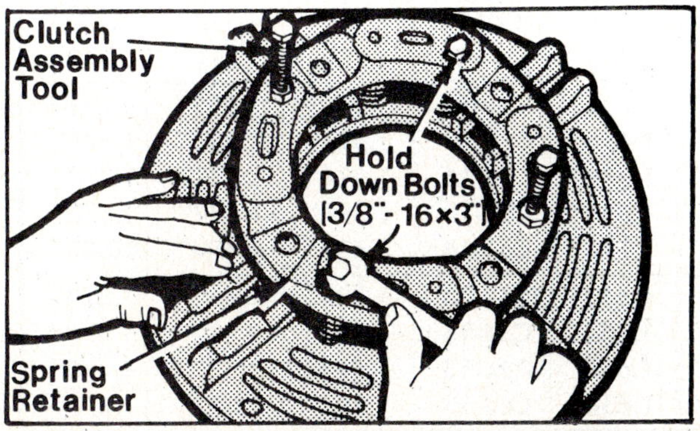

Fig. 7 Clutch assembly tool and hold-down bolts

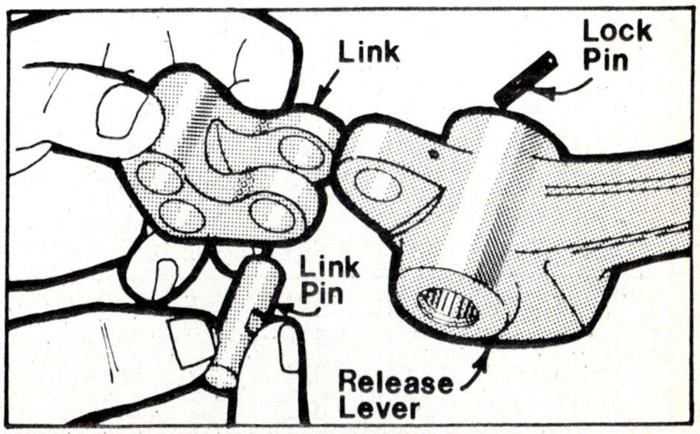

Fig. 8 Installing link to lever

LIPE TC CLUTCH DATA CHART

Model Number†	Dimension "A"	Flywheel Depth +.007″ −.000″	Facing Thickness	PRESSURE SPRING DATA		DIMENSION "B" (New)		PRESSURE PLATE REFACING DATA (Dimension "B" Refaced)		
				Color	Minimum Weight	New ±.002″	Assemble Using Hole "A" of Link	Reassemble Using Hole "A" of Link C64-20	Reassemble Using Hole "B" of Link C64-20 or C64-16	Reassemble Using Hole "B" of Link C64-13
170-11-239	11/32″	1.218″	3/16″	White	125 @ 2¼″	1.656″	C64-16	1.594″	1.532″	1.470″
155-11-244	15/32″	1.781″	5/32″	White	125 @ 2¼″	1.718″	C64-16	1.656″	1.594″	1.532″
170-12-248*	11/32″	2.321″	5/32″	Green	240 @ 2¼″	1.656″	C64-13	1.594″	1.532″	1.470″
170-11-264	11/32″	1.218″	3/16″	Yellow	170 @ 2¼″	1.656″	C64-16	1.594″	1.532″	1.470″
170-12-277*	11/32″	2.321″	5/32″	White	125 @ 2¼″	1.656″	C64-13	1.594″	1.532″	1.470″
170-12-278*	11/32″	2.321″	5/32″	Brown	115 @ 2¼″	1.656″	C64-13	1.594″	1.532″	1.470″
170-11-279	11/32″	1.218″	3/16″	Green	240 @ 2¼″	1.656″	C64-16	1.594″	1.532″	1.470″
170-11-281	3/8″	Flat	3/16″	Red	100 @ 2⅜″	1.656″	C64-16	1.594″	1.532″	1.470″
155-11-282	15/32″	1.625″	5/32″	Red	100 @ 2⅜″	1.500″	C64-16	1.438″	1.376″	1.314″
155-11-284	15/32″	1.625″	5/32″	Red	100 @ 2⅜″	1.500″	C64-16	1.438″	1.376″	1.314″
155-11-285	15/32″	1.781″	5/32″	Red	100 @ 2⅜″	1.718″	C64-16	1.656″	1.594″	1.532″
155-11-289	15/32″	1.625″	5/32″	Red	100 @ 2⅜″	1.500″	C64-16	1.438″	1.376″	1.314″
170-11-349	11/32″	1.218″	3/16″	White	125 @ 2¼″	1.656″	C64-16	1.594″	1.532″	1.470″
155-11-350	15/32″	1.781″	5/32″	White	125 @ 2¼″	1.718″	C64-16	1.656″	1.594″	1.532″
170-12-351*	11/32″	2.321″	5/32″	Brown	115 @ 2¼″	1.656″	C64-13	1.594″	1.532″	1.470″
170-12-352*	11/32″	2.321″	5/32″	White	125 @ 2¼″	1.656″	C64-13	1.594″	1.532″	1.470″
170-12-353*	11/32″	2.321″	5/32″	Yellow	170 @ 2¼″	1.656″	C64-13	1.594″	1.532″	1.470″
155-11-354	15/32″	1.781″	5/32″	Red	100 @ 2⅜″	1.718″	C64-16	1.656″	1.594″	1.532″
170-11-355	11/32″	1.218″	3/16″	Yellow	170 @ 2¼″	1.656″	C64-16	1.594″	1.532″	1.470″
170-11-367	11/32″	1.218″	3/16″	Red	100 @ 2⅜″	1.656″	C64-16	1.594″	1.532″	1.470″
170-12-368*	11/32″	2.321″	5/32″	Brown	115 @ 2¼″	1.656″	C64-13	1.594″	1.532″	1.470″
170-12-369*	11/32″	2.321″	5/32″	White	125 @ 2¼″	1.656″	C64-13	1.594″	1.532″	1.470″
170-12-370*	11/32″	2.321″	5/32″	Yellow	170 @ 2¼″	1.656″	C64-13	1.594″	1.532″	1.470″
155-11-377	15/32″	1.781″	5/32″	Yellow	170 @ 2¼″	1.718″	C64-16	1.656″	1.594″	1.532″
155-11-379	15/32″	1.781″	5/32″	Lt. Blue	150 @ 2¼″	1.718″	C64-16	1.656″	1.594″	1.532″
170-11-383	11/32″	1.156″	3/16″	Gray	280 @ 2⅝″	1.656″	C64-16	1.594″	1.532″	1.470″
170-12-388*	11/32″	2.321″	5/32″	Green	240 @ 2¼″	1.656″	C64-13	1.594″	1.532″	1.470″
170-11-389	11/32″	1.218″	3/16″	White	125 @ 2¼″	1.656″	C64-16	1.594″	1.532″	1.470″
170-11-452	11/32″	1.218″	3/16″	Lt. Blue	150 @ 2¼″	1.656″	C64-16	1.594″	1.532″	1.470″
155-11-457	15/32″	1.781″	5/32″	Black	200 @ 2¼″	1.718″	C64-16	1.656″	1.594″	1.532″
170-11-458	11/32″	1.218″	3/16″	Black	200 @ 2¼″	1.656″	C64-16	1.594″	1.532″	1.470″
170-11-460	11/32″	1.218″	3/16″	Yellow	170 @ 2¼″	1.656″	C64-16	1.594″	1.532″	1.470″
170-12-477*	3/4″	2.321″	5/32″	Green	240 @ 2¼″	1.656″	C64-13	1.594″	1.532″	1.470″
170-12-479*	11/32″	2.321″	5/32″	Lt. Blue	150 @ 2¼″	1.656″	C64-13	1.594″	1.532″	1.470″
170-12-481*	11/32″	2.321″	5/32″	Black	200 @ 2¼″	1.656″	C64-13	1.594″	1.532″	1.470″
170-12-482*	11/32″	2.321	5/32″	Green	240 @ 2¼″	1.656″	C64-13	1.594″	1.532″	1.470″
170-12-522*	11/32″	2.321	5/32″	Brown	115 @ 2¼″	1.656″	C64-13	1.594″	1.532″	1.470″
170-11-523	3/4″	1.218	3/16″	Gray	280 @ 2⅝″	1.656″	C64-16	1.594″	1.532″	1.470″
170-11-527	11/32″	1.218	3/16″	Gray	280 @ 2⅝″	1.656″	C64-16	1.594″	1.532″	1.470″
155-11-613	15/32″	1.781	5/32″	Green	240 @ 2¼″	1.718″	C64-16	1.656″	1.594″	1.532″
170-12-640*	11/32″	2.321	5/32″	Yellow	170 @ 2¼″	1.656″	C64-13	1.594″	1.532″	1.470″
170-11-664	11/32″	1.218	3/16″	Black	200 @ 2¼″	1.656″	C64-16	1.594″	1.532″	1.470″
170-12-686*	11/32″	2.321	5/32″	Brown	115 @ 2¼″	1.656″	C64-13	1.594″	1.532″	1.470″
170-12-690*	11/32″	2.312	5/32″	Lt. Blue	150 @ 2¼″	1.656″	C64-13	1.594″	1.532″	1.470″

Continued

LIPE TC CLUTCH DATA CHART—Continued

Model Number†	Dimension "A"	Flywheel Depth +.007" −.000"	Facing Thickness	PRESSURE SPRING DATA		DIMENSION "B" (New)		PRESSURE PLATE REFACING DATA (Dimension "B" Refaced)		
				Color	Minimum Weight	New ±.002"	Assemble Using Hole "A" of Link	Reassemble Using Hole "A" of Link C64-20	Reassemble Using Hole "B" of Link C64-20 or C64-16	Reassemble Using Hole "B" of Link C64-13
170-12-709*	11/32"	2.312"	5/32"	White	125 @ 2 1/4"	1.656"	C64-13	1.594"	1.532"	1.470"
170-12-710*	3/4"	2.312"	5/32"	Black	200 @ 2 1/4"	1.656"	C64-13	1.594"	1.532"	1.470"
170-12-715*	3/4"	2.312"	5/32"	Lt. Blue	150 @ 2 1/4"	1.656"	C64-13	1.594"	1.532"	1.470"
170-12-722*	11/32"	2.321"	5/32"	Gray	280 @ 2 5/8"	1.656"	C64-13	1.594"	1.532"	1.470"
170-12-733*	3/4"	2.321"	5/32"	Yellow	170 @ 2 1/4"	1.656"	C64-13	1.594"	1.532"	1.470"
170-12-737*	11/32"	2.321"	5/32"	Black	200 @ 2 1/4"	1.656"	C64-13	1.594"	1.532"	1.470"
170-12-741*	11/32"	2.321"	5/32"	Lt. Blue	150 @ 2 1/4"	1.656"	C64-13	1.594"	1.532"	1.470"
170-12-744*	3/4"	2.321"	5/32"	Black	200 @ 2 1/4"	1.656"	C64-13	1.594"	1.532"	1.470"
170-12-746*	11/32"	2.321"	5/32"	Green	240 @ 2 1/4"	1.656"	C64-13	1.594"	1.532"	1.470"
170-12-777*	3/4"	2.321"	5/32"	Gray	280 @ 2 5/8"	1.656"	C64-13	1.594"	1.532"	1.470"
170-12-799	3/4"	2.321"	5/32"	Gray	280 @ 2 5/8"	1.656"	C64-13	1.594"	1.532"	1.470"

* If the intermediate plate also requires refacing, remove the difference between dimension new and the dimension refaced as a total amount of material to be removed from the intermediate plate and pressure plate friction surfaces. Thickness of new intermediate plate is .749". Reassemble as indicated.

† Model number is the number stamped on the name plate attached to cover.
Dimension "A" refers to proper setting of release lever adjusting screw.
Dimension "B" is the distance from top of pin to ground surface of pressure plate.

pilot bearing to center disc assembly.
6. Install cover plate assembly to flywheel.

NOTE: Push inner ends of release levers in toward flywheel before starting to tighten the opposite capscrews attaching clutch to flywheel. Screw in each capscrew until it contacts clutch cover. Then gradually tighten the opposite capscrew until clutch is drawn tight to flywheel. Do not force clutch into flywheel with attaching capscrews as the clutch may be damaged. Remove hold-down bolts from clutch.

Adjusting Release Levers

Using a straight-edge and scale, set adjusting screws in ends of release levers to the proper Dimension "A" (see chart). Dimension "A" is the distance from ground surface of raised lever bosses of flywheel ring (cover) to heads of lever adjusting screws when clutch is bolted to flywheel and in the engaged position. The contact points of these adjusting screws should be set a uniform distance from underside of straight-edge. The adjusting screws must be in the same plane within 1/32". To change the position of the adjusting screws, loosen lock nuts and turn screw as required to obtain "Dimension A".

Vehicle Clutch Linkage

Check vehicle clutch linkage for excessive wear and resulting lost motion that can cause incomplete release of clutch (clutch drag) and improper release bearing clearance, which causes clutch slippage. Replace worn parts.

IMPORTANT: Set the vehicle clutch release linkage so as to provide 1/8" release bearing clearance between contact face of bearing and ends of clutch release lever adjusting screws. This usually results in 1 1/2" of clutch pedal free travel.

Two Plate Clutches

1. Check driving pins and intermediate

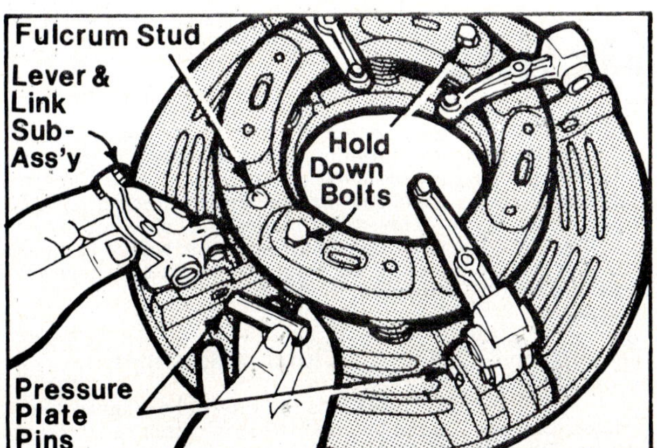

Fig. 9 Installing link to lever sub-assembly

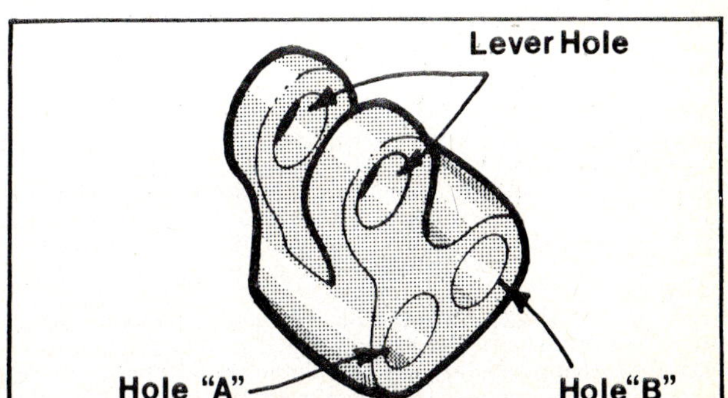

Fig. 10 Toggle link

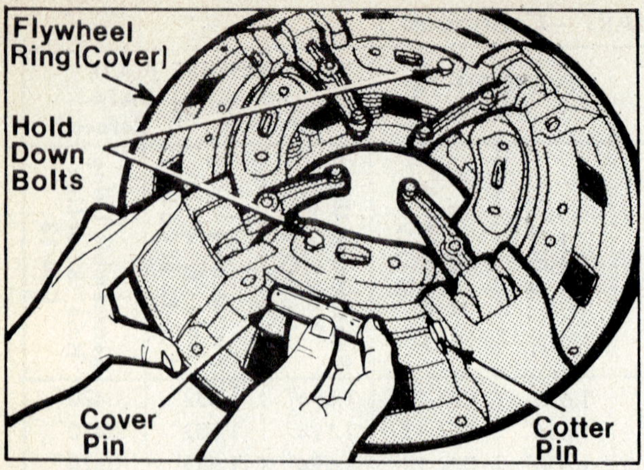

Fig. 11 Installing cover pins

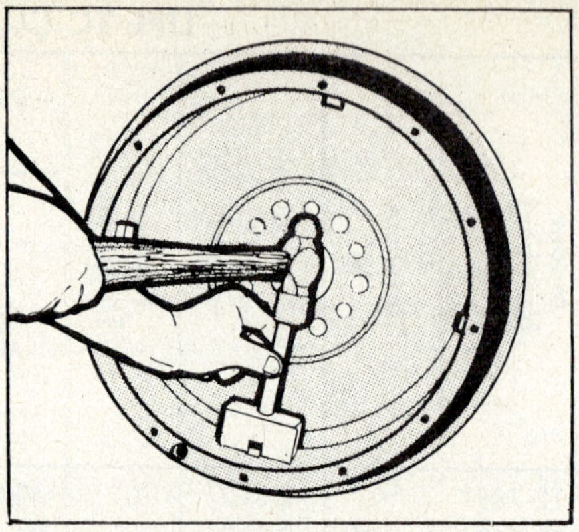

Fig. 12 Aligning drive pin in flywheel using Lipe tool T-10670

plate for wear.

2. Driving pins should be installed in flywheel with a press fit and squarely aligned with friction face of flywheel. It is recommended that the tool shown in Fig. 12 be used. If this tool is not available, place a machinist's square firmly against the rim of the flywheel and make certain the contact face of the driving pins have even contact with the square, Fig. 13.
3. Check clearance between heads of driving pins and driving slots in mating intermediate plate over the pins, making certain the plate fits freely before clutch is assembled to engine. If excessive clearance exists, new parts should be installed.
4. Try clutch in flywheel before installing driven discs and intermediate plate to be sure clutch pilot diameter is a free fit in flywheel bore.
5. When installing intermediate plate, be sure its rotation will be in the direction of the arrows appearing on one side of the plate near the bore.
6. When installing the driven discs, note that the one marked "Flywheel Side" should be installed with this mark next to the friction face of the flywheel. The disc marked "Pressure Plate Side" is installed with this mark next to the friction face of the pressure plate. This is important to prevent hub interference.
7. Use a spare spline shaft to align driven discs so as to make the installation of the transmission easier.

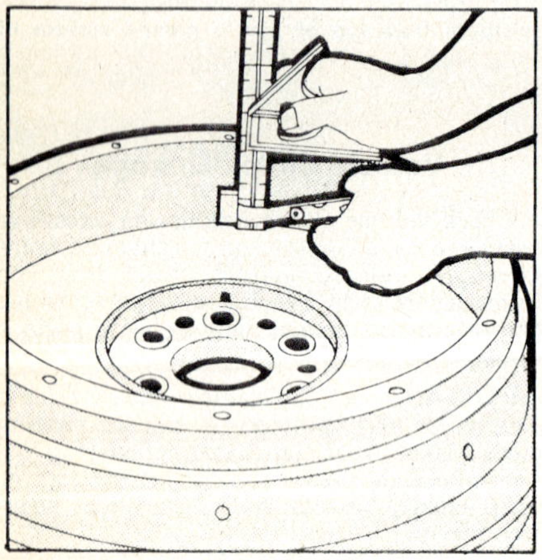

Fig. 13 Aligning drive pin in flywheel using machinist's square and feeler gauge

Release Lever Travel

To obtain complete disengagement of TC clutches, the release levers must be actuated a specified distance from the engaged position. Assuming the release bearing clearance to be 1/8" when the clutch is in the engaged position, the release bearing must move forward 1/8" to take up this clearance before it contacts the release levers, and then move the normal amount of travel required to disengage the clutch as shown in the chart, Fig. 14.

Clutch Size and Type	*Release Bearing Clearance*	*Normal Release Bearing Travel*	*Total Bearing Travel Required*
15½ TC Single Plate	⅛"	7/16"	9/16"
17 TC Single Plate	⅛"	7/16"	9/16"
17-2 TC Two Plate	⅛"	½"	⅝"

Fig. 14 TC clutch release travel chart

Spicer Clutches

Description

These clutches, Figs. 1, 2 and 2A are of the dry disc, adjustable, pull type, multiple lever units utilizing centrally located pressure springs entirely isolated from the pressure plate itself. They are available with one or two plates, with the driven disc mounted on the spline shaft of the transmission main drive gear.

In the 13", 14" single-plate and 13", 14" two-plate with angle positioned pressure springs models, the intermediate drive plate is mounted inside a "cup" type flywheel and carried on four drive pins (six pins for the 14" two-plate with angle positioned pressure springs) mounted in the flywheel, Figs 3, 3A, and 3B. However, in the 15½" models, the intermediate drive plate has external lugs which engage slots in the flywheel ring mounted on a "flat" type flywheel, Figs. 3C and 3D.

In all models the pressure plate is driven by four drive lugs or drive pins which mate with four slots in the clutch flywheel ring.

On all models, the pressure plate also carries four pull-back springs and anchor pins to retract the pressure plate when the clutch is disengaged.

The clutch release bearing rotates continuously since the inner race turns with the release sleeve and flywheel. However, the clutch release bearing carries a thrust load only when the clutch is released. Bronze, graphite-impregnated bushings are used to reduce friction when the flywheel and main drive gear are turning at different speeds.

The spring plates of all clutches are ventilated and incorporate air scoops to circulate cooling air through the clutch whenever the engine is running.

Clutch Adjustments

The need for an internal adjustment is indicated by a reduction in clutch pedal free travel. In general the clutch should be adjusted for approximately 1½" of free pedal travel. When the free pedal travel diminishes to ½", the

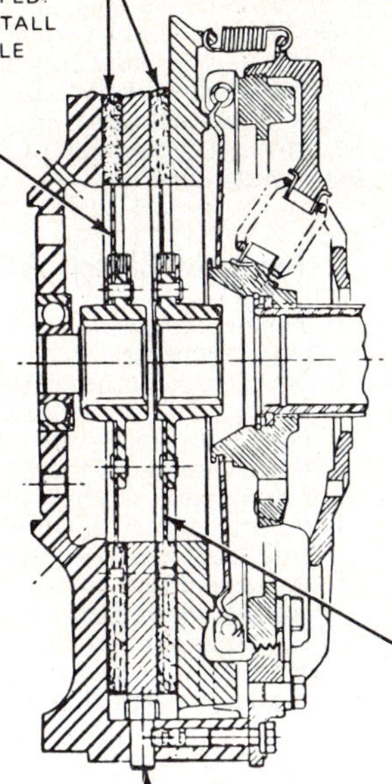

Fig. 1 Sectional view 14" clutches. 1970-80

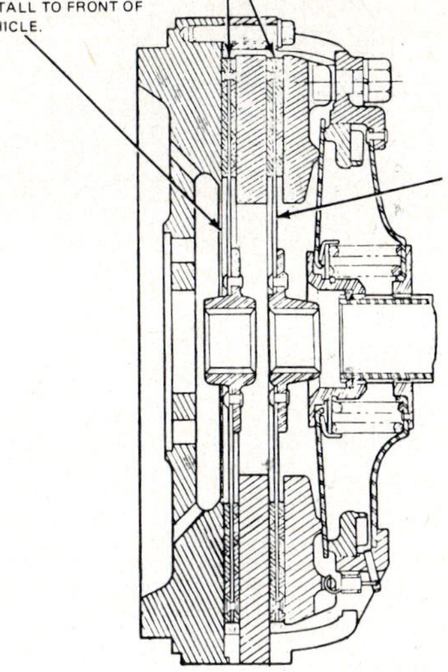

Fig. 2 Sectional view 15½" clutches (Typical) 1973–77

internal adjustment must be made.

NOTE: Adjusting the external clutch linkage to restore the original free pedal travel will not compensate for an improperly adjusted clutch. Make the internal adjustment first and then check and reset the external linkage, if required.

Adjustment with Clutch Brake

If the clutch installation is used with a clutch brake or a non-synchronized transmission, the internal clutch adjustment should be checked by measuring the clearance (gap) between the clutch release bearing housing cover and the clutch brake, Fig. 4. The average adjustment with clutch brake is given below with a tolerance of plus or minus 1/32":

13" two plate	1/2"
14" single plate	7/16"
14" two plate	1/2"
15½" single plate	9/16"
15½" two plate	1/2"

On clutch brake applications the release bearing housing cover should contact the clutch brake washer when the clutch pedal is at least one inch from the floor. Idle the engine and make sure the clutch releases fully before the clutch brake comes into action. Gears should be shifted freely without noticeable drag of clutch plates when clutch is released. If the clutch does not release before the clutch brake comes into action, make an internal clutch adjustment.

Adjustment without Clutch Brake

If the clutch installation is used with a synchronized transmission or without the clutch brake, the internal adjustment should be checked between the clutch release bearing housing and the clutch spring hub, Fig. 4. The average adjustment is given below with an allowable tolerance of plus or minus 1/32":

13" two plate	23/32"
14" single plate	3/8"
14" two plate	19/32"
15½" single plate	17/32"
15½" two plate	19/32"

With Hydraulic Release

If the vehicle is equipped with a hydraulic clutch release cylinder the truck manufacturer's instructions should be followed. However, points that should be kept in mind when setting the clutch release mechanism are:

1. That there be 1/8" free travel somewhere in the linkage.
2. That the hydraulic clutch release lever travels equally on each side of the vertical centerline.

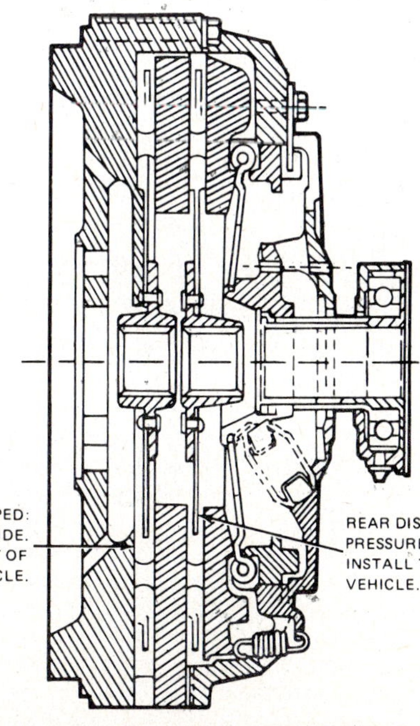

Fig. 2A Sectional view 15½" clutches (Typical) 1978–80

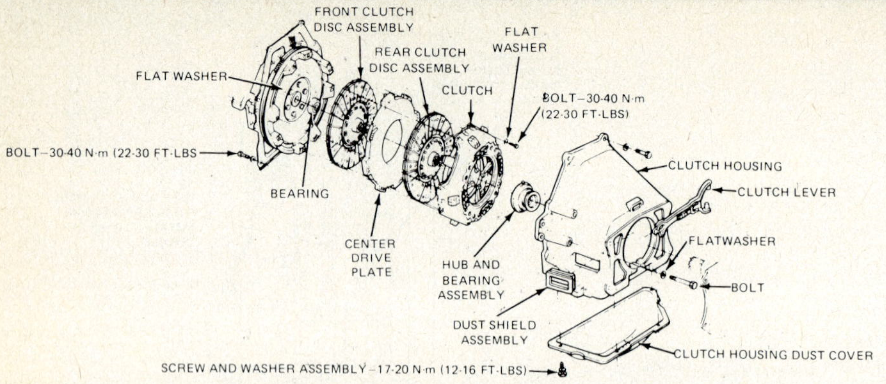

Fig. 3 Exploded view 13" double disc clutch (Typical). 1970–80

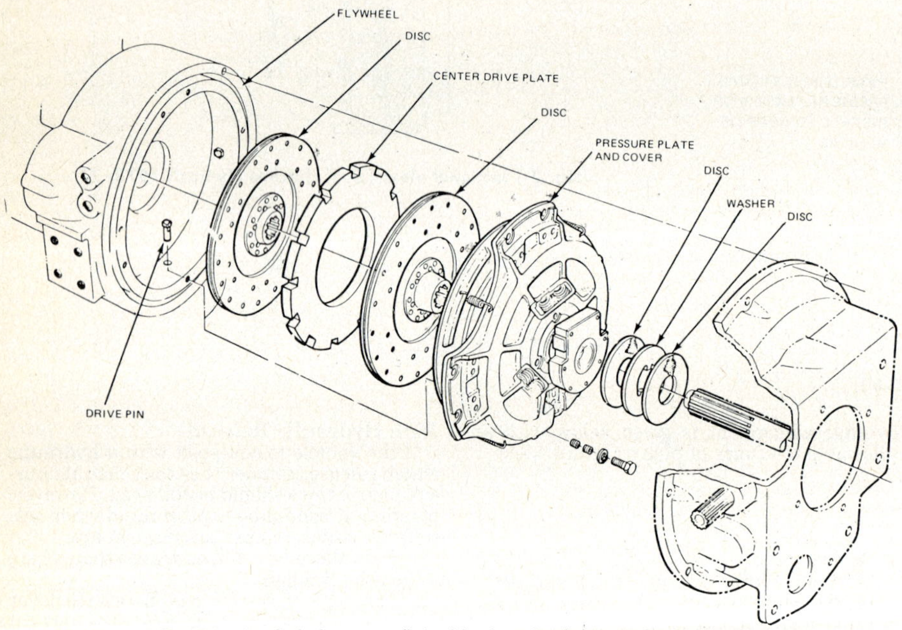

Fig. 3A Exploded view 14" double disc clutch (Typical). 1970–77

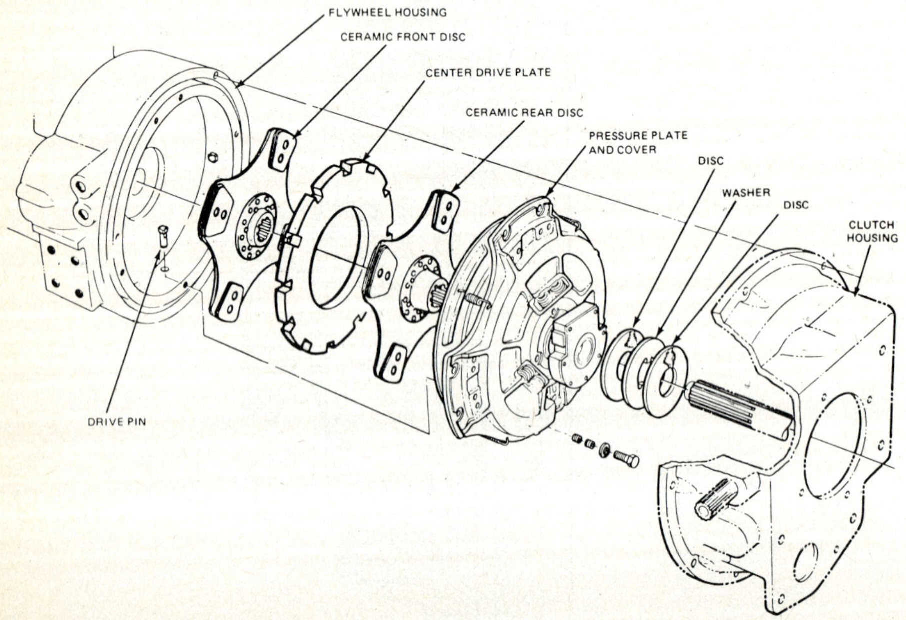

Fig. 3B Exploded view 14" double disc clutch (Typical) 1978–80

3. That the slave cylinder travel is sufficient to give a clean clutch release.

On some applications the external clutch lever is twice as long as the clutch release yoke. This gives a two to one ratio, or the slave cylinder piston must travel one inch to give 1/2" clutch release travel. In any case, an average of 1/2" clutch release bearing travel should be provided for proper release of all Spicer heavy duty clutches.

Internal Clutch Adjustments

1. Remove inspection cover at bottom of clutch housing.
2. Measure clearance between release bearing housing and clutch brake or spring plate hub, according to the installation. If the clearance is more or less than that listed above, readjust clutch as follows:
3. Rotate engine flywheel until adjusting ring lock is exposed, then remove lock.
4. Release clutch by blocking clutch pedal down towards floor.
5. Use a pry bar to turn adjusting ring. Turn counterclockwise to move release bearing housing toward the flywheel: turn clockwise to move housing away from flywheel. Rotation or movement of one lug position will move release bearing approximately .020".
6. Unblock clutch pedal to engage clutch.
7. Recheck clearances as in Step 2.
8. After clutch has been properly adjusted, install locking device in notch provided in adjusting ring.
9. Check clearance between release yoke fingers and release bearing housing pads, or clearance between clutch release lever adjusting arm and linkage if an external arm is used. This clearance should be 1/8". It is important that this clearance be provided to allow for facing wear and clutch pedal free travel. When clutch is properly adjusted, the free pedal travel should be approximately 1 1/2".

Remove & Disassemble

All Except Units with Angle Positioned Pressure Springs

A suitable sling or transmission jack should be used to properly support and maintain the engine-transmission alignment when removing or installing a transmission on the engine. Do not allow the front end of the transmission to drop and hang unsupported in the splined hubs of the clutch discs to avoid bending or distorting the driven discs. Disconnect the external linkage from the clutch release arm to permit the clutch release yoke to turn up and pull free of the release bearing thrust pads.

1. Prior to actual clutch removal, assemble a clutch disc aligning tool to the driven disc and release bearing assembly. Also remove two clutch-flywheel mounting capscrews from the upper part of the flywheel and insert two guide studs (5" long) in these holes.
2. After the transmission has been removed, insert two 3/4" spacer blocks of wood between the clutch release bearing and spring plate hub, Fig. 5. This facilitates removal of clutch-to-flywheel capscrews and protects pullback springs from excessive tension.
3. Remove clutch retaining capscrews and remove pressure plate and flywheel ring assembly.
4. Remove driven disc or discs and intermediate plate.
5. Remove drive pins from flywheel. If Allen

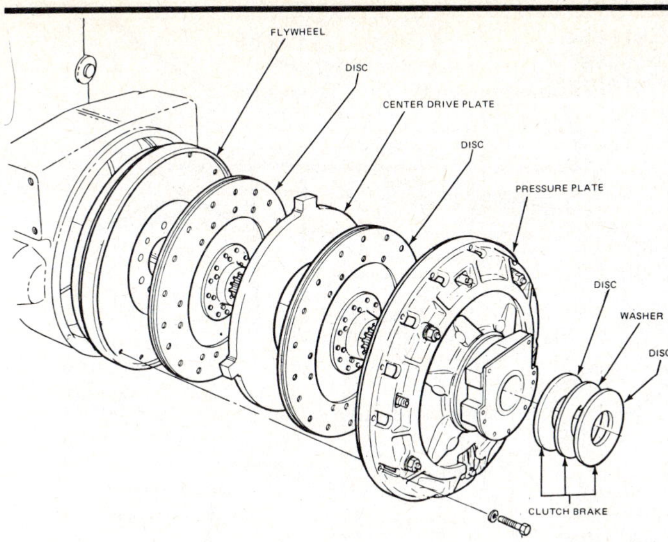

Fig. 3C Exploded view 15½" clutches (Typical). 1973–77

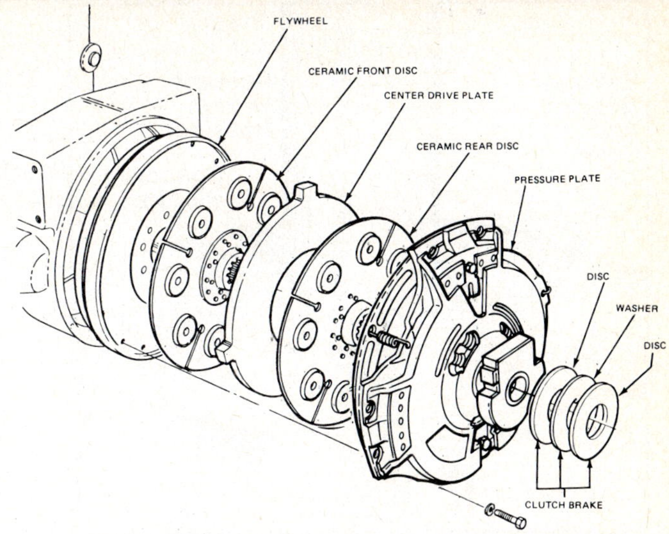

Fig. 3D Exploded view 15½" clutches (Typical). 1978–80

head set screws are used to retain drive pins, they must be removed to free drive pins.

6. To disassemble pressure plate, flywheel ring and lever assembly, first unhook the pull-back springs and remove pressure plate.
7. Press down on two places on outside of flywheel ring and remove spacer blocks.
8. Place assembly under arbor press with release bearing down. Use a spacer so pressure is exerted on release sleeve.
9. Press down on release sleeve retainer, Fig. 6, and remove split lock rings.
10. Remove spring ring retainer, Fig. 7.
11. Lift off spring ring and clutch levers, Fig. 8.
12. Clean and inspect all parts and discard any part that appears doubtful.

Units with Angle Positioned Pressure Springs

1. With transmission removed, install a clutch disc aligning tool into the driven disc and release bearing. Install two ¾ inch wood blocks between clutch release bearing housing and clutch flywheel rings, Fig. 9, as the clutch retaining bolts are loosened. These blocks will relieve the heavy spring load.
2. Remove all bolts and slide clutch assembly back with caution keeping aligning tool in place thus retaining disc and intermediate plate.
3. Carefully remove aligning tool, rear disc, intermediate plate and front disc.
4. Remove ring lock bolt, then using a screwdriver pry out ring lock.
5. With driven discs and intermediate plate removed, place clutch assembly upside down on flat surface then remove the four return springs, Fig. 25 and lift off the pressure plate.
6. Turn adjusting ring and lever assembly counterclockwise until free of flywheel, Fig. 10.
7. Remove snap ring from release sleeve retainer and install three ⅜" × 5" threaded rods through clearance holes in release sleeve retainer and into holes in flywheel ring, Fig. 11.

NOTE: Threaded rods must be flush with back of flywheel ring to prevent interference with bearing cover.

8. Install nuts onto rods. Place assembly on an arbor press with a short piece of tubing 2½ inch to 2¾ inch outside diameter used to support the release sleeve. Compress retainer until drive lugs bottom against flywheel ring and tighten nuts against retainer, Fig. 12.

NOTE: Do not support against clutch release bearing cover rivet heads.

9. Raise arbor, tilt assembly and remove wood blocks.
10. Support release sleeve as in step 8 and remove half ring locks, Fig. 13. Remove release sleeve and bearing assembly through rear of clutch.
11. Compress retainer to relieve load on nuts, then back off nuts and relieve load from springs. Remove springs and pivots, Fig. 14.
12. Remove cotter pins from pivot pins then the pins and levers.

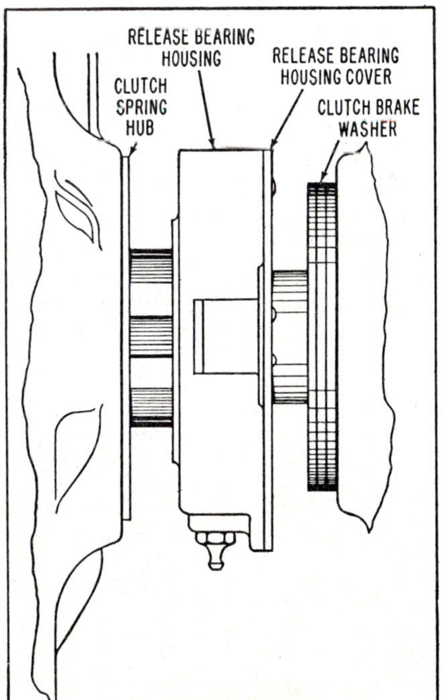

Fig. 4 Clutch adjusting points

Assemble

13" & 14" Units Without Angle Positioned Springs, Fig. 15

1. Place two flat springs in top side of bearing housing, one on top of the other. Use only one spring on the 14" single plate clutch.
2. Place release bearing in bearing housing, compressing springs.
3. Place bearing cover on housing and secure with two short rivets. The two hardened thrust plates as well as the bearing cover are assembled with four long rivets. Secure assembly with bolts, lock washers and nuts.
4. Install grease nipple if open bearings are used. Omit grease nipple and plug this opening if double-sealed bearings are used.
5. Press clutch release sleeve assembly in bearing housing, using an arbor press, Fig. 16.
6. Assemble four pull-back spring anchors in flywheel ring. Hook pull-back springs in anchors and clinch with pliers so anchors and springs do not fall out.
7. Screw adjusting ring into flywheel ring with adjusting notches down. Be sure adjusting ring turns freely. Powdered graphite mixed with oil and applied to threads will insure easy adjustment after clutch has been in service. Set adjusting ring to the dimension shown in Fig. 17 from fulcrum of ring to face of flywheel ring. Secure adjusting ring with lock, lockwasher and lock screw.
8. Continue with assembly by referring to Figs. 18 through 25.

15½" Units Without Angle Positioned Springs

NOTE: Inasmuch as an exploded view of these units and suitable detailed illustrations are not currently available, refer to Fig. 2 and earlier 15½" clutch pictures for identification of parts referred to below.

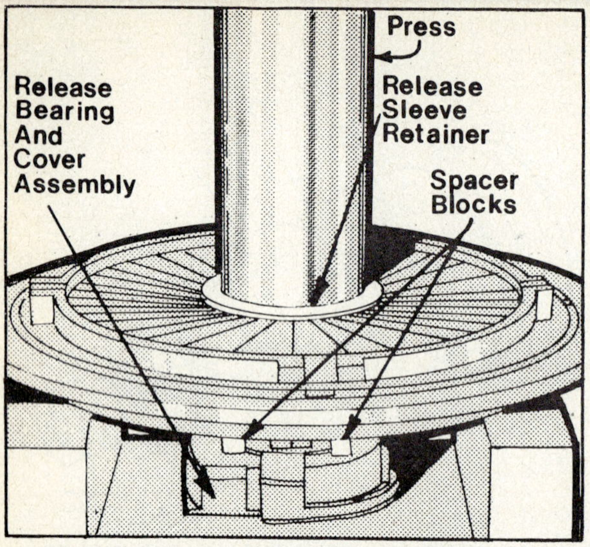

Fig. 5 Position or spacer blocks

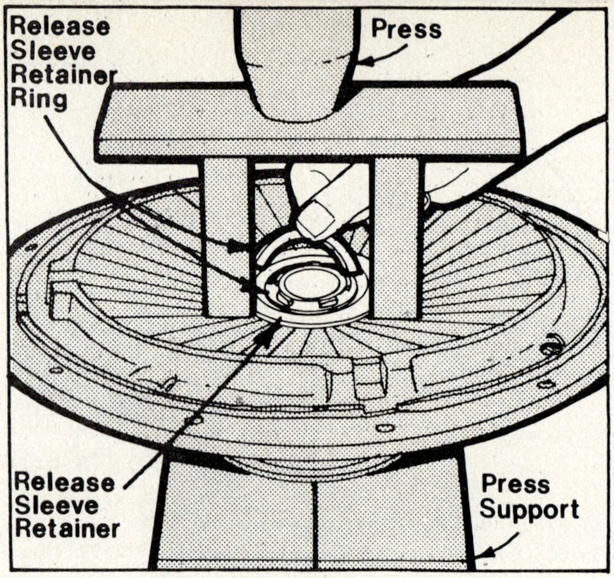

Fig. 6 Release sleeve retainer split ring

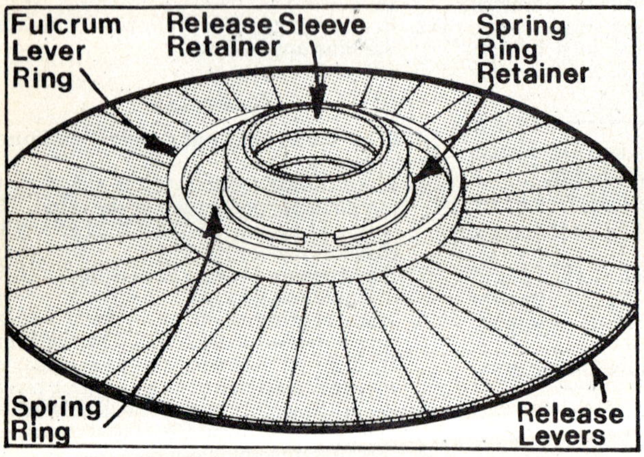

Fig. 7 Release sleeve retainer and lever assembly

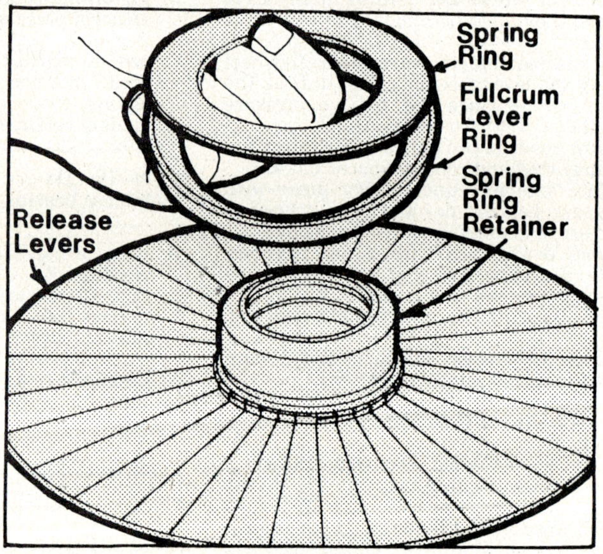

Fig. 8 Spring ring and fulcrum lever ring

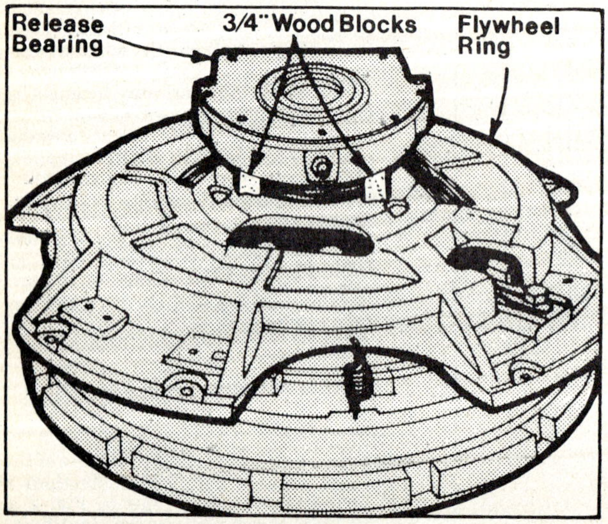

Fig. 9 Wood blocks positioned between release bearing and flywheel ring

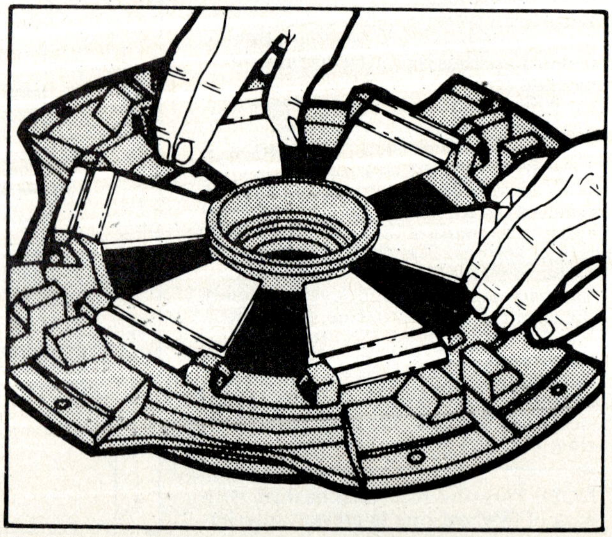

Fig. 10 Removing adjusting ring

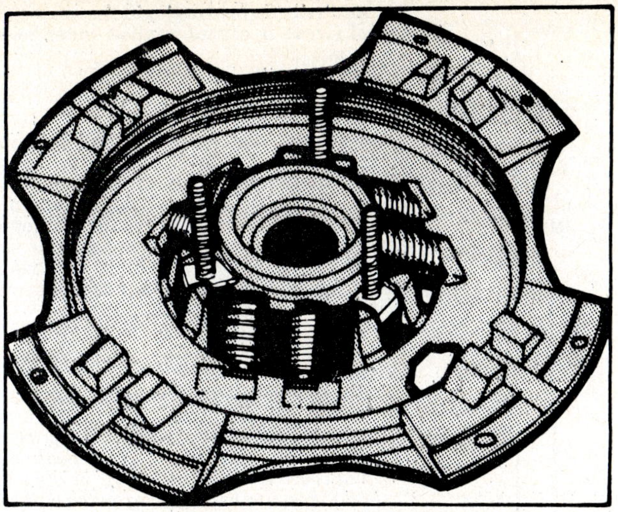

Fig. 11 Flywheel ring with threaded rods installed

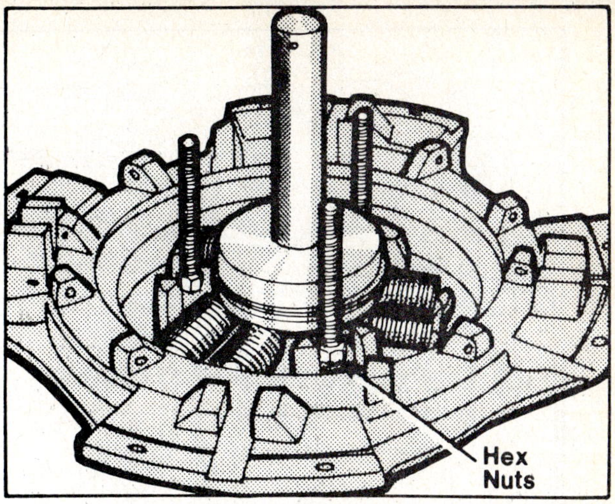

Fig. 12 Retainer ring in compressed position

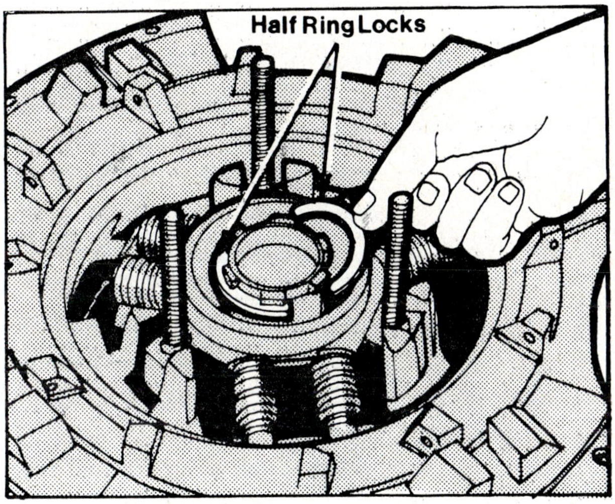

Fig. 13 Removing of installing half ring locks

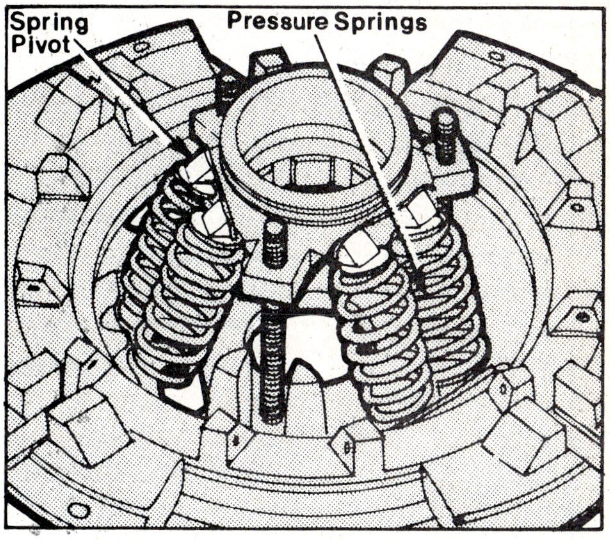

Fig. 14 Removing or installing pressure springs

Release Sleeve Retainer & Lever

1. Place release sleeve retainer on bench with flanged end down and assemble 35 levers and hooked end up around retainer. *This clutch is so designed that sufficient space for one more lever may be available; however, do not use more than 35 levers.*
2. Install release lever fulcrum ring and spring ring down over hooked end of levers.
3. Install spring ring retainer to hold sub-assembly intact during assembly.

Flywheel Ring & Spring Plate

1. Coat threads of adjusting ring with powdered graphite mixed with oil. Assemble adjusting ring to flywheel ring with "notches" down. Preset adjusting ring 2⅞" below face of flywheel ring. Be sure adjusting ring turns freely in flywheel ring. Turn flywheel ring over and lock adjusting ring in position.
2. Set pressure plate on wooden bridge on bed of arbor press. Position four drive pins in slots and wedge with .004" to .008" shim stock.
3. Place flywheel ring and plate sub-assembly down over pressure plate and drive pins, aligning threaded end of drive pins with bosses of flywheel ring. Tap ring to start all drive pins securely. Use a short piece of tubing and, moving progressively, press each drive pin into place a little at a time. Avoid bending or forcing pins into position with one pass of the press.
4. Clamp clutch sub-assembly to bed of press with block of wood on spring plate hub. Use a long bolt through a flywheel mounting bolt hole to prevent the assembly from turning. Assemble lock nuts and lock washers to drive pins. Torque all lock nuts to 145 ft-lbs.
5. Remove assembly from press and set upside down on bench. *Use care when removing clutch ring from press to prevent pressure plate from dropping free and causing personal injury or damage to pressure plate.*
6. Pry pressure plate free of driving pins (if necessary) and remove shim stock wedges. Check and make sure pressure plate slides freely on end of drive lugs. Realign drive pins if necessary to free up pressure plate.

Release Bearing & Sleeve

1. Assemble two flat springs inside and across top of release bearing housing with ends of springs inside cast relief sections. Hand pack bearing with proper lubricant and assemble bearing to housing, compressing flat springs. Seat bearing against machined flats inside housing.
2. Coat edge of release housing with a suitable sealer and assemble cover to housing with five short rivets. Assemble new hardened thrust plates to housing with long rivets. Rivet heads must be set tight to maintain grease-tight seat between bearing housing and cover. Assemble grease fitting to housing and fill cavity with proper grease.

Assemble 1¾" 10 Spline Unit

Place release bearing and cover on press with cover up. Press release sleeve sub-assem-

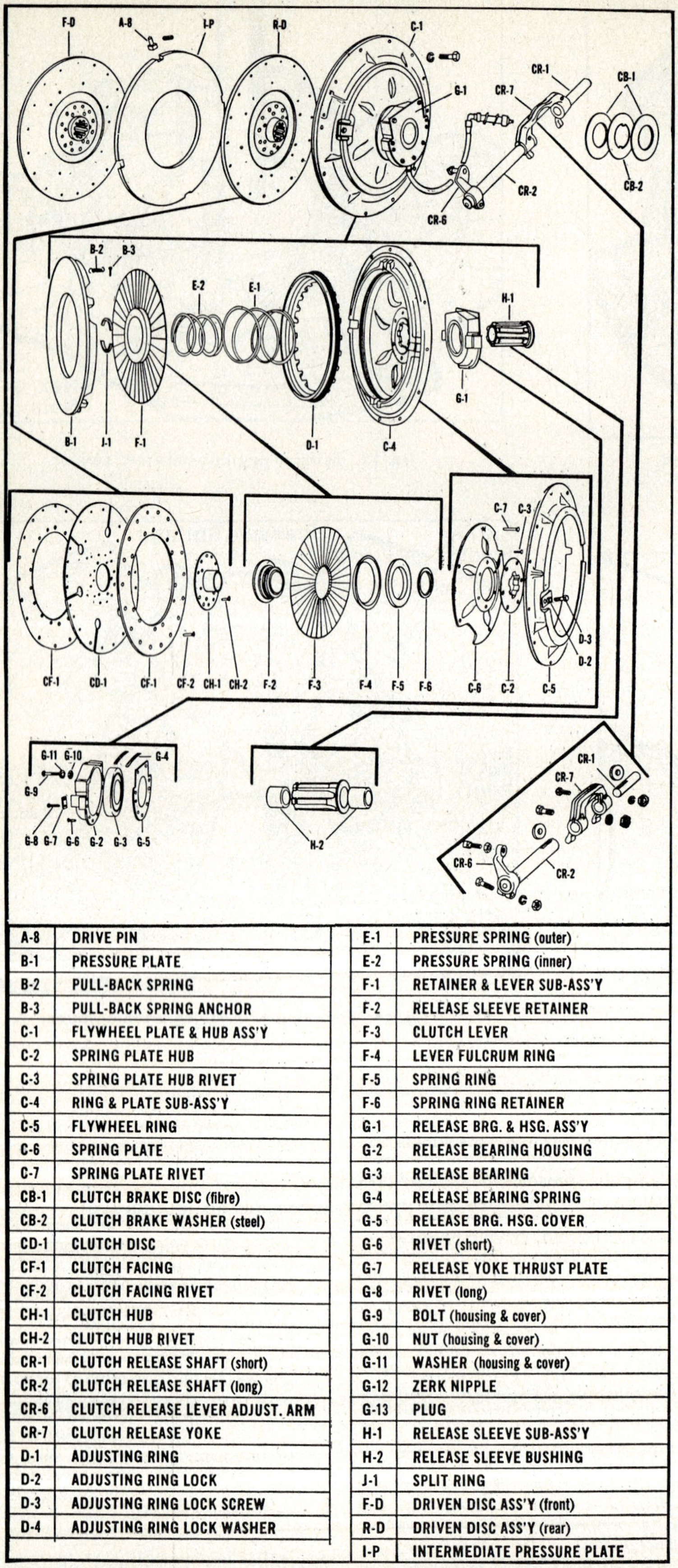

A-8	DRIVE PIN	E-1	PRESSURE SPRING (outer)
B-1	PRESSURE PLATE	E-2	PRESSURE SPRING (inner)
B-2	PULL-BACK SPRING	F-1	RETAINER & LEVER SUB-ASS'Y
B-3	PULL-BACK SPRING ANCHOR	F-2	RELEASE SLEEVE RETAINER
C-1	FLYWHEEL PLATE & HUB ASS'Y	F-3	CLUTCH LEVER
C-2	SPRING PLATE HUB	F-4	LEVER FULCRUM RING
C-3	SPRING PLATE HUB RIVET	F-5	SPRING RING
C-4	RING & PLATE SUB-ASS'Y	F-6	SPRING RING RETAINER
C-5	FLYWHEEL RING	G-1	RELEASE BRG. & HSG. ASS'Y
C-6	SPRING PLATE	G-2	RELEASE BEARING HOUSING
C-7	SPRING PLATE RIVET	G-3	RELEASE BEARING
CB-1	CLUTCH BRAKE DISC (fibre)	G-4	RELEASE BEARING SPRING
CB-2	CLUTCH BRAKE WASHER (steel)	G-5	RELEASE BRG. HSG. COVER
CD-1	CLUTCH DISC	G-6	RIVET (short)
CF-1	CLUTCH FACING	G-7	RELEASE YOKE THRUST PLATE
CF-2	CLUTCH FACING RIVET	G-8	RIVET (long)
CH-1	CLUTCH HUB	G-9	BOLT (housing & cover)
CH-2	CLUTCH HUB RIVET	G-10	NUT (housing & cover)
CR-1	CLUTCH RELEASE SHAFT (short)	G-11	WASHER (housing & cover)
CR-2	CLUTCH RELEASE SHAFT (long)	G-12	ZERK NIPPLE
CR-6	CLUTCH RELEASE LEVER ADJUST. ARM	G-13	PLUG
CR-7	CLUTCH RELEASE YOKE	H-1	RELEASE SLEEVE SUB-ASS'Y
D-1	ADJUSTING RING	H-2	RELEASE SLEEVE BUSHING
D-2	ADJUSTING RING LOCK	J-1	SPLIT RING
D-3	ADJUSTING RING LOCK SCREW	F-D	DRIVEN DISC ASS'Y (front)
D-4	ADJUSTING RING LOCK WASHER	R-D	DRIVEN DISC ASS'Y (rear)
		I-P	INTERMEDIATE PRESSURE PLATE

Fig. 15 Exploded view of 13″ and 14″ clutches

bly into release bearing, making sure shoulder of sleeve is seated against inner race of bearing.

Assemble 2″ 10 Spline Unit

Set release sleeve sub-assembly vertically on bed of press, splined end down. Position release bearing assembly on sleeve, with cover up. Use a 2 3/8″ thin wall tubing and press inner race of bearing onto ground diameter of sleeve. *Do not press on rivet heads or cover during this assembly.* Assemble snap ring to end of sleeve. Make sure inner race of bearing is seated firmly against snap ring.

Completing Assembly

1. Support release sleeve with bearing assembly on the press bed with a piece of 2 1/2″ × 2 3/4″ tubing about 3″ long. *Do not support this assembly on the release bearing cover as this will result in distortion and excessive end loading on the ball bearing.*
2. Position flywheel ring and spring plate assembly, with adjusting ring and pressure plate drive pins in place over release sleeve.
3. Assemble inner and outer pressure springs to release sleeve and resting on spring plate.
4. Assemble release sleeve retainer and lever sub-assembly, with splined hub end of retainer down and inside of pressure springs.
5. Use a small wooden bridge to compress pressure springs and permit assembly of half-ring locks.
6. Seat half-ring locks securely in groove of release sleeve and make sure retainer is not cocked or stuck, but slides back freely into its proper position when the press load is released.
7. Turn clutch over on press and rest end of release sleeve on piece of 2 1/2″ tubing. Use the 2 × 4 bridge to compress the pressure springs and install two 3/4″ blocks of wood between release bearing housing and spring plate hub.
8. Set pressure plate on the 2 × 4 bridge with pull-back springs crimped to pressure plate. Stand springs up and set flywheel ring down over pressure plate, using care to align pull-back springs with cast slots provided in flywheel ring.
9. Check to make sure pull-back spring anchor pins are screwed completely into cover. Use long-nose pliers and hook springs to anchor pins.
10. Using the rear disc as a spacer, assemble intermediate plate to clutch flywheel ring. Use a feeler gauge to check driving lug clearance in slots. Slot clearance should be .006″ to .011″.

Units with Angle Positioned Pressure Springs

1. Coat all threads of adjusting ring and flywheel ring with molykote or equivalent.
2. Assemble adjusting ring to flywheel ring with notches down. Pre-set adjusting ring by placing two pivot pins in adjusting ring bosses directly opposite each other. Using two straight edges, measure dimension from flywheel ring mounting surface to straight edge, this distance should be about 1/2 inch for 14 inch clutches, Fig. 26, or 2 11/16 for 15 1/2 inch clutches, Fig. 27.
3. Install three threaded rods into flywheel ring and install hex nut on opposite side of ring and place spring pivots into flywheel ring and retainer. Place release

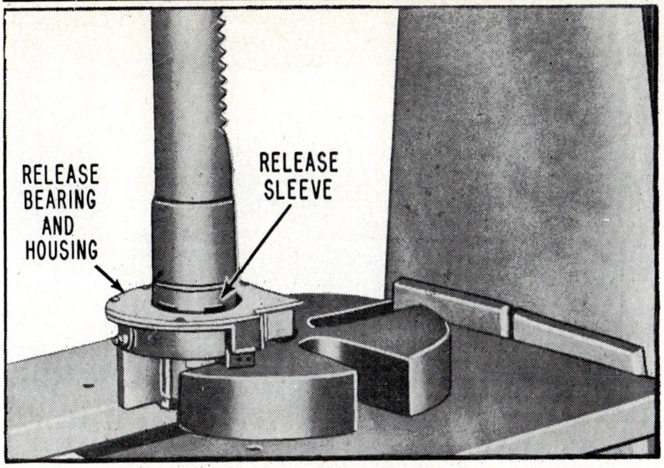

Fig. 16 Pressure clutch release sleeve into bearing housing

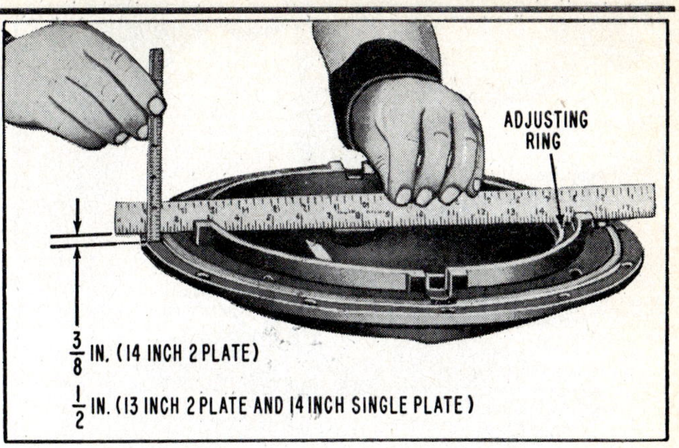

Fig. 17 Adjusting ring settings, measured from fulcrum of ring to face of flywheel ring

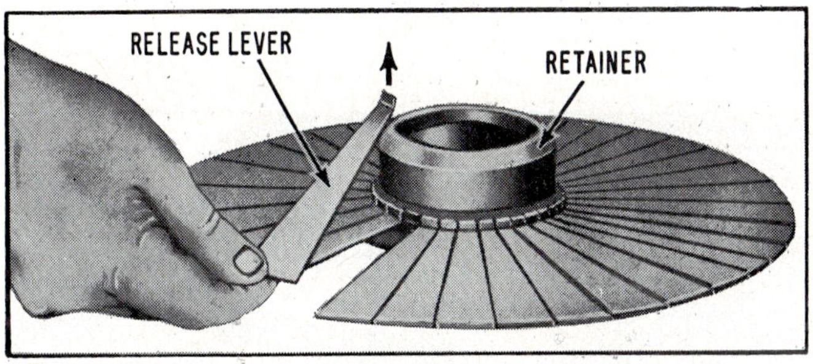

Fig. 18 Place retainer on bench with flange down and assemble levers with hook up around retainer

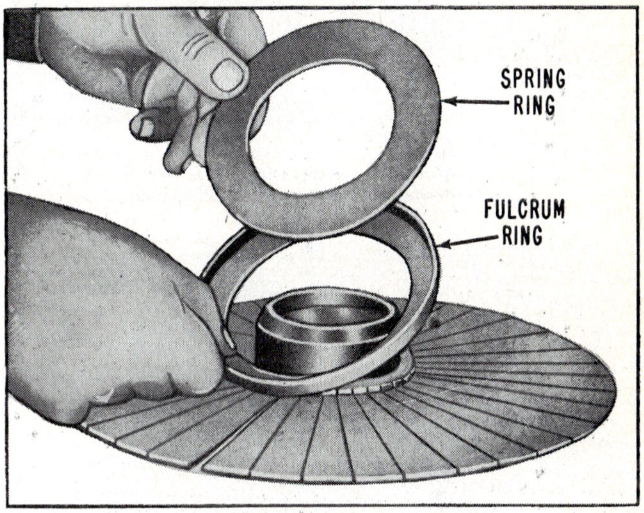

Fig. 19 Install fulcrum ring and spring ring

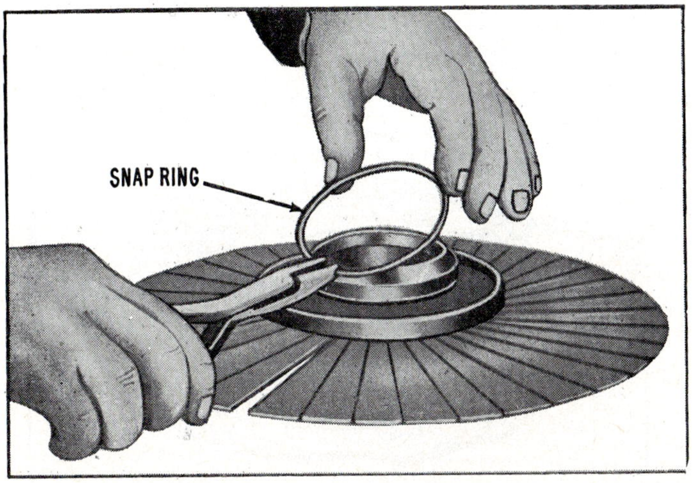

Fig. 20 Install snap ring

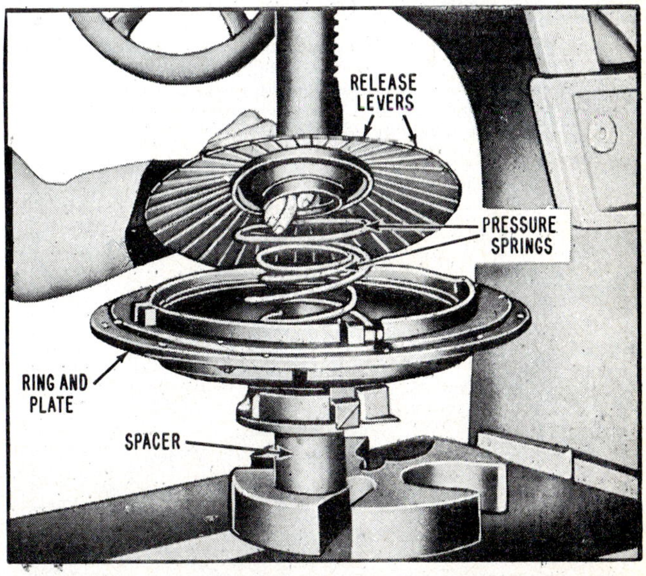

Fig. 21 Support release sleeve and bearing (sleeve up) on press with a piece of 2" tubing, 3" long. Position ring and plate down over release sleeve. Assemble inner and outer pressure springs around release sleeve. Position retainer and lever with hub of retainer down and inside springs

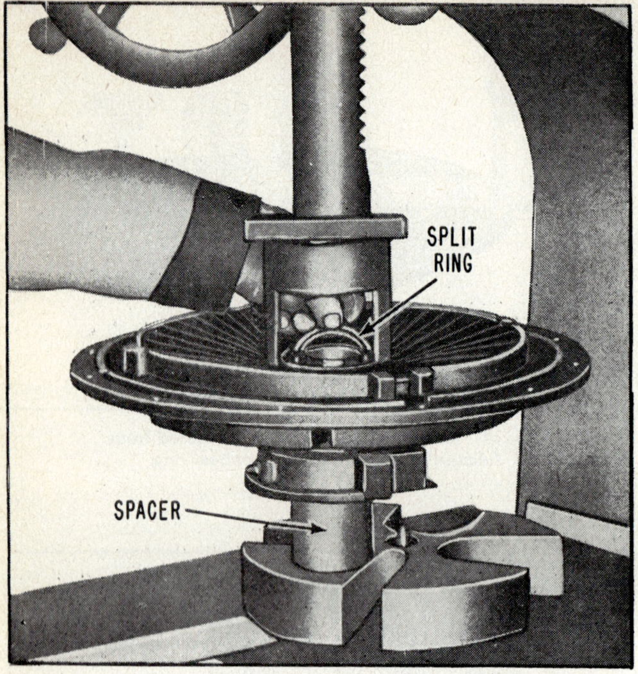

Fig. 22 Use two 3/4" × 6" long blocks of wood to compress springs. Lock lever retainer to release sleeve by installing split ring in groove of sleeve

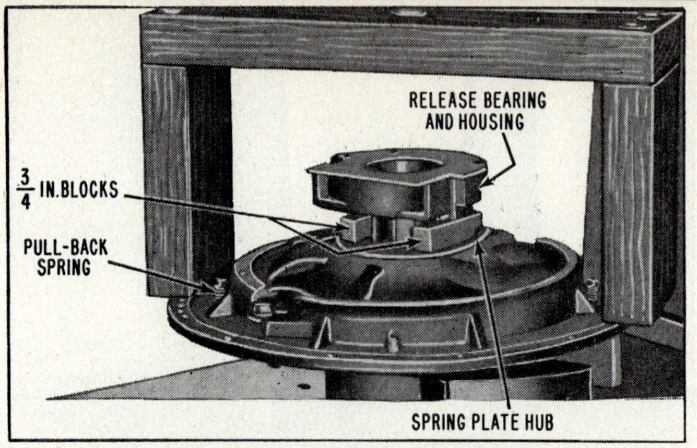

Fig. 23 Turn assembly over and rend end of sleeve on piece of 2" tubing. Use pieces of 2 × 4 placed as shown to compress clutch springs, and install two 3/4" blocks as shown

Fig. 24 Remove from press. Place clutch on bench with release bearing down. Install pressure plate to flywheel ring. Check clearance between driving lugs on pressure plate and drive slots in flywheel ring. Clearance should be .004" to .008"

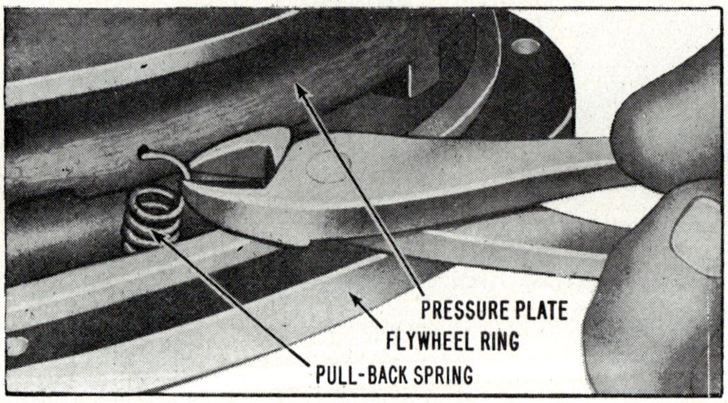

Fig. 25 Hook four pull-back springs in holes in pressure plate

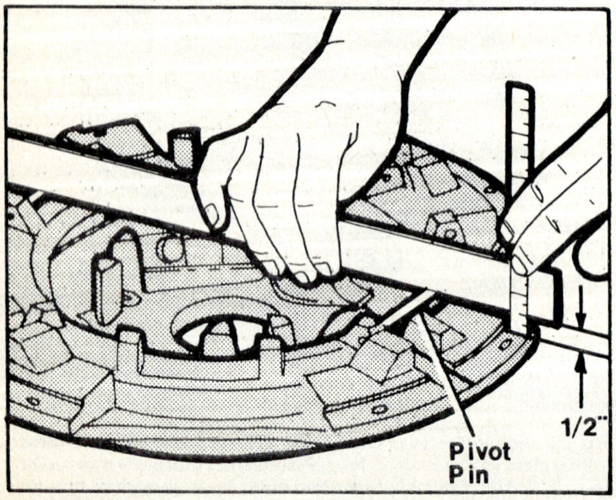

Fig. 26 Pre-setting adjusting ring. Except 15 1/2" units

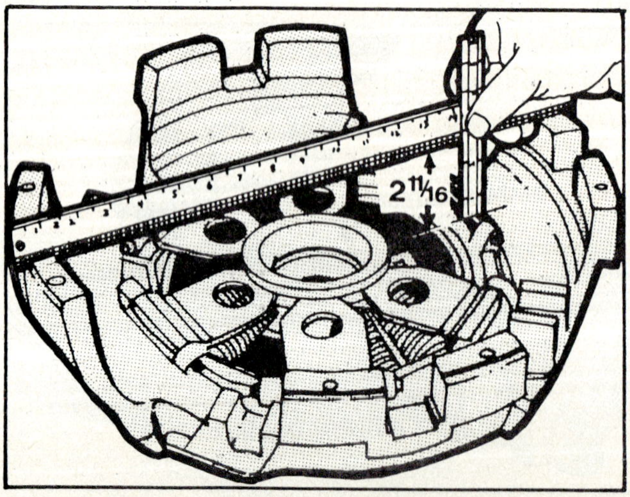

Fig. 27 Pre-setting adjusting ring. 15 1/2" units

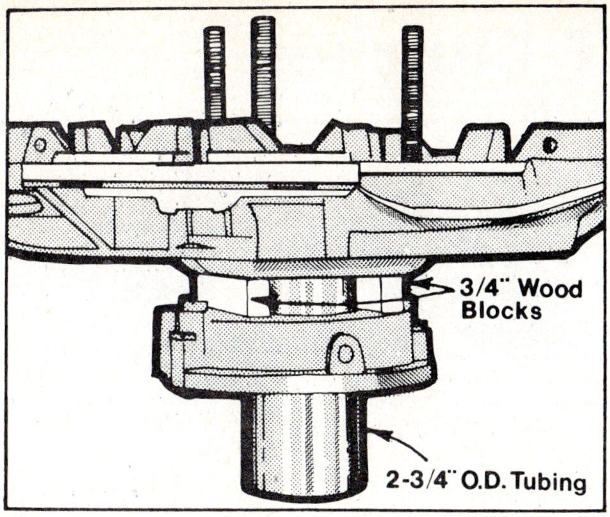

Fig. 28 Supporting sleeve with tube and wood blocks

Fig. 29 Installing and checking drive pins for squareness

sleeve retainer in position guided by rod, and install pressure springs, Fig. 14.

4. Place hex nuts on the three threaded rods and draw down enough to hold release sleeve retainer in place.
5. Place flywheel ring and release sleeve retainer assembly on arbor press. Depress retainer against flywheel ring then turn nuts on threaded rods tightly against flywheel ring, Fig. 12.

NOTE: Make certain that springs are seated in pivots.

6. Remove assembly from press then install release sleeve and bearing assembly through flywheel ring and release sleeve retainer. Install half ring locks, Fig. 13.
7. Place the 3/4 inch wood blocks between flywheel ring and release bearing housing and place assembly on arbor press supporting sleeve on tubing previously used. Compress retainer to relieve load and remove threaded rods and wood blocks, Fig. 28.
8. Install snap ring above half ring locks and place levers between adjusting ring bosses, with the narrow end in groove of release sleeve retainer. Move adjusting ring as necessary to insert pivot pins and return adjusting ring to previous position.

NOTE: Raised area of lever fulcrum must be facing pressure plate. Each pivot pin has a flat surface on the head, this pivot pin surface must be mated with milled surface on adjusting ring boss.

9. Install cotter pins in pivot pin holes from outside in, flatten cotter pin head to insure clearance between flywheel ring lugs and adjusting ring.
10. Place pressure plate in drive slots or flywheel ring. Check clearance between driving lugs on pressure plate and drive slots in flywheel ring. This clearance should be .004–.008 inch, Fig. 24.
11. Install the four return springs, Fig. 25 and the adjusting ring lock and bolt.
12. Install drive pins in flywheel making sure that shanks of drive pins are press fit in the flywheel rim and the heads are square with friction face, Fig. 29.

Clutch Installation

13", 14" and 14" Two-Plate with Angle Positioned Pressure Springs

1. On all except the 14" Two-Plate with angle positioned pressure springs, install four pins in flywheel, making sure shanks of pins are press fit in flywheel rim and heads are square with friction face.
2. Check clearance of slots in intermediate plate by positioning plate on drive pins. Measure clearance between drive pins and slots, which should be .006" to .010", and measured on same side of pins. If alignment and clearance are satisfactory, remove intermediate plate and lock drive pins in place with Allen head screws if provided in design.
3. Place front driven disc in flywheel with hub of flange towards transmission. On 14" single plate clutch, install long hub of clutch disc towards transmission.
4. Install intermediate plate in flywheel, locating drive slots on drive pins.
5. Install rear driven disc with long hub of clutch disc towards transmission.
6. Position pressure plate and flywheel ring on flywheel and start capscrews to hold in place. Insert an aligning arbor or spare transmission main drive gear through clutch and align both discs. Tool should be inserted into pilot bearing to insure central location of discs.
7. Tighten clutch to flywheel capscrews evenly, progressing around the flywheel to prevent cocking and binding of flywheel ring pilot. Tighten capscrews to 35–40 ft-lbs torque.
8. Remove wooden blocks from between release bearing housing and flywheel ring. Then remove spline aligning tool.

Installing 15½" Units

1. Insert spline aligning tool through clutch to align clutch discs and keep parts in place during assembly of flywheel.
2. Position assembly on flywheel, making sure that pilot end of aligning tool has entered pilot bearing.
3. Tighten capscrews gradually and evenly around flywheel. Final tightening to be 35–40 ft-lbs torque.
4. Remove wooden blocks from between clutch release bearing housing and flywheel housing. Then remove aligning tool and install transmission.

Installing Transmission

1. If clutch brake is used, place two fiber discs with a steel washer between them on main drive gear.
2. Rotate clutch release yoke so that top is back towards transmission.
3. Shift transmission into gear.
4. Position clutch release bearing housing so that flat section is at top.
5. Install transmission. Turn companion flange to align splines of drive gear with clutch driven discs. As transmission is entering clutch and flywheel, turn fingers of clutch release yoke down over clutch release bearing housing into their proper position.
6. After transmission is secured to engine, check internal clutch adjustment as outlined at the beginning of this discussion. After setting the internal adjustment, adjust external clutch linkage to provide 1½" free pedal travel.

Long CF Clutches

DESCRIPTION

Long CF clutches are of the single-plate, dry-disc type consisting of two basic assemblies, the driven disc and the clutch cover assembly, Fig. 1. The driven disc is splined to the transmission main drive gear. The cover assembly is bolted to the engine flywheel. The pressure plate is driven by the cover through the release lever mounting lugs which extend through openings in the cover. Coil type pressure springs between the cover and pressure plate force the pressure plate against the driven member when the clutch is engaged.

No adjustment for wear is provided in the clutch itself. An individual adjustment is provided for locating each lever. In manufacturing the adjusting nuts are locked in place and should never be disturbed unless the clutch is dismantled for replacement of parts.

The clutch release bearing and support assembly slides on the transmission main drive gear bearing retainer. The release bearing on vehicles having a release fork instead of a lever is equipped with a lubrication fitting which is accessible after removing the clutch underpan. The pilot bearing, pressed into a recess in the flywheel, carries the forward end of the transmission main drive gear. This bearing requires lubrication only at time of overhaul.

The release fork (if used) is ball stud mounted in the clutch housing, with inner end of fork engaging a groove in the release bearing collar for actuating the release levers. The clutch is disengaged by a foot pedal and, in some cases, hydraulically connected to the release fork or lever.

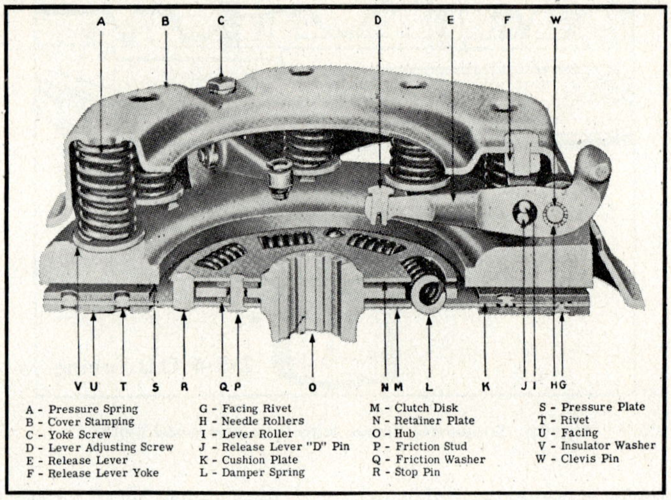

Fig. 1 Cutaway view of a typical long CF clutch

Disassembly

1. Mark pressure plate and clutch cover so they may be reassembled in correct position to maintain the original balance.
2. Place cover and pressure plate assembly on arbor press bed with pressure plate resting on spacers 3/8" thick, Fig. 2. Place a bar across the cover as shown, clearing all thrust plate bolts and yoke adjusting nuts. Turn arbor press screw down against bar and force cover down against press bed, compressing the pressure springs.
3. Remove two capscrews attaching each thrust plate to clutch cover. Remove adjusting nut from each release lever yoke. Remove thrust plates.
4. Slowly release arbor press until pressure of springs is relieved. Remove clutch cover from pressure plate. Remove pressure springs, then remove pressure plate and release lever assembly from arbor press bed and place on bench.
5. It is not necessary to remove yokes from levers or to remove levers from pressure plate unless necessary to replace parts.

Inspection & Repairs

Prior to inspection, wash all parts in cleaning solvent except the driven disc assembly. Wipe off the driven disc assembly. Inspect all clutch components and also flywheel and pilot bearing.

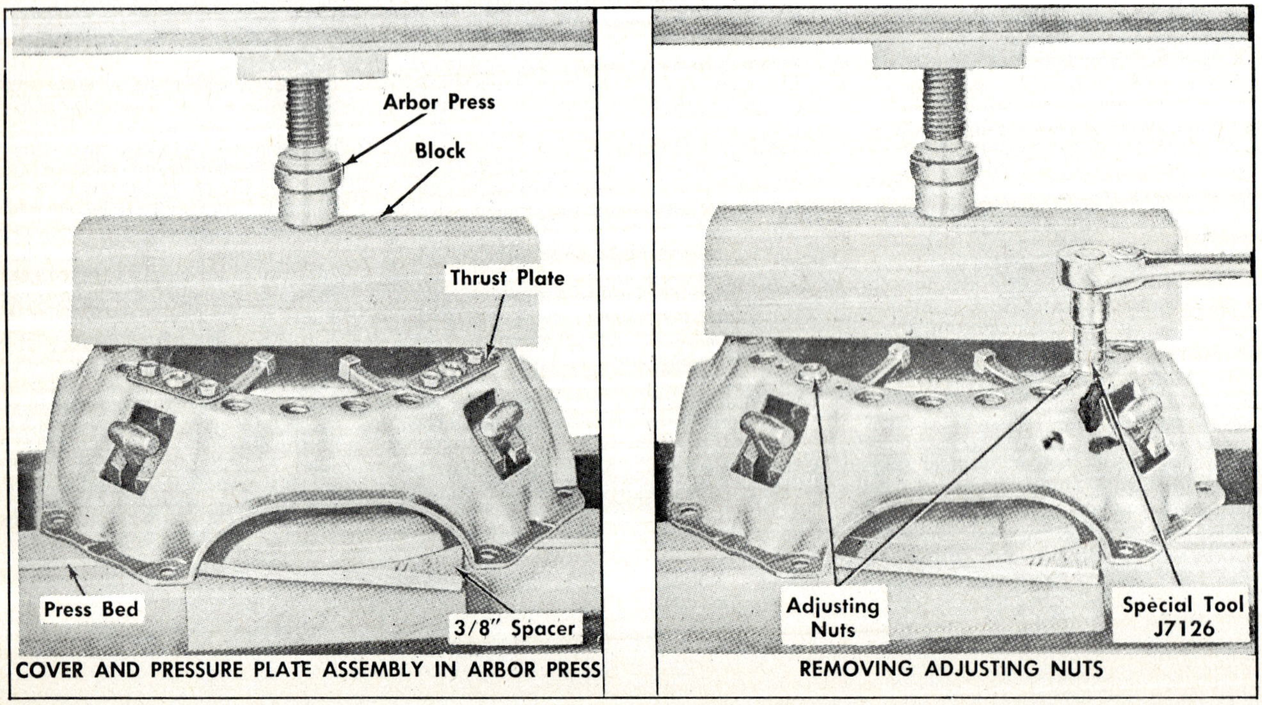

Fig. 2 Disassembling cover and pressure plate

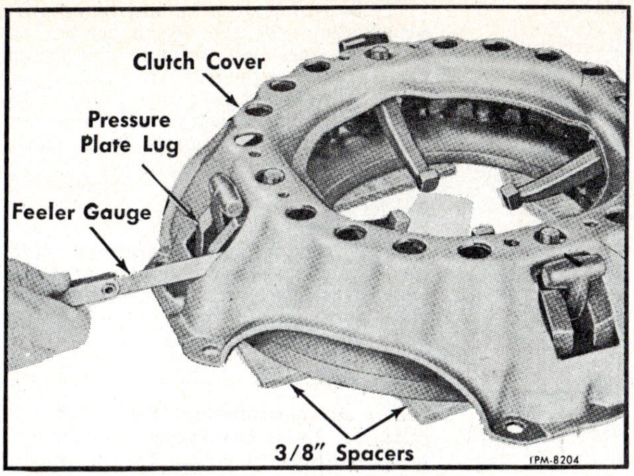

Fig. 3 Checking lug-to-cover clearance

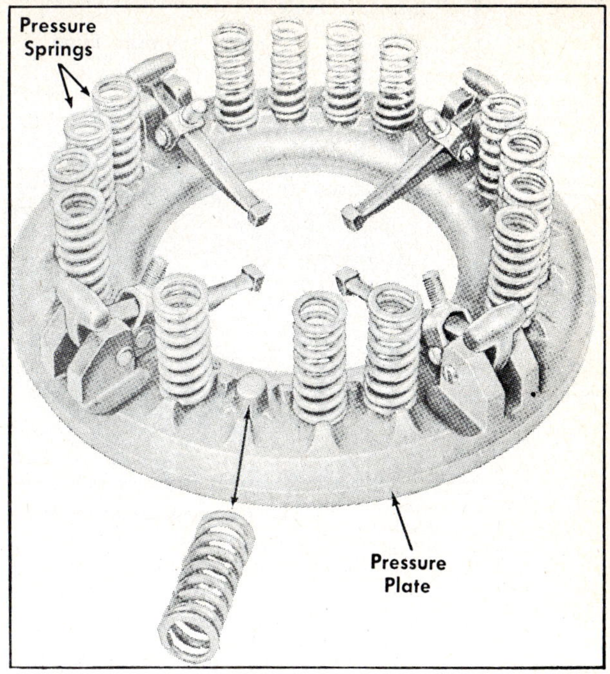

Fig. 5 Pressure springs installed

Pressure Plate

The pressure plate must be replaced if the friction surface is severely heat checked or scored; if it is warped in excess of .015" or if the driving lug slots are worn or the pin holes elongated.

The pressure plate may be salvaged if the friction surface cleans up by removing a maximum of 1/16" of metal by machining.

Clutch Cover

Inspect clutch cover for distortion or cracks. Clutch cover flange should be flat within .015" when checked on a surface plate.

Place cover over pressure plate and check clearance between driving lugs on pressure plate and edge of slots in cover, Fig. 3. Clearance should be not less than .008" or more than .015".

Release Levers

1. Check ends of release levers that contact release bearing support for wear. If excessive wear is evident, or if necessary to remove levers and yokes, remove as follows:
2. Grind off swaged end of pins securing release levers in pressure plate lugs and remove pins. Remove release levers and needle bearing rollers from pressure plate. Remove pins securing yokes on release levers in same manner.
3. Inspect needle bearing bores in release levers for wear or damage. Replace levers or bearing rollers if damage is evident.
4. A pilot pin is required for installing release levers and yokes. Chamfer both ends of pin. Lay one release lever on its side on bench and place pilot pin in release yoke hole. Coat needle bearings with Lubriplate 320 or equivalent, then insert rollers around pilot pin, Fig. 4. Position release lever yoke over release lever, then insert new pin through yoke

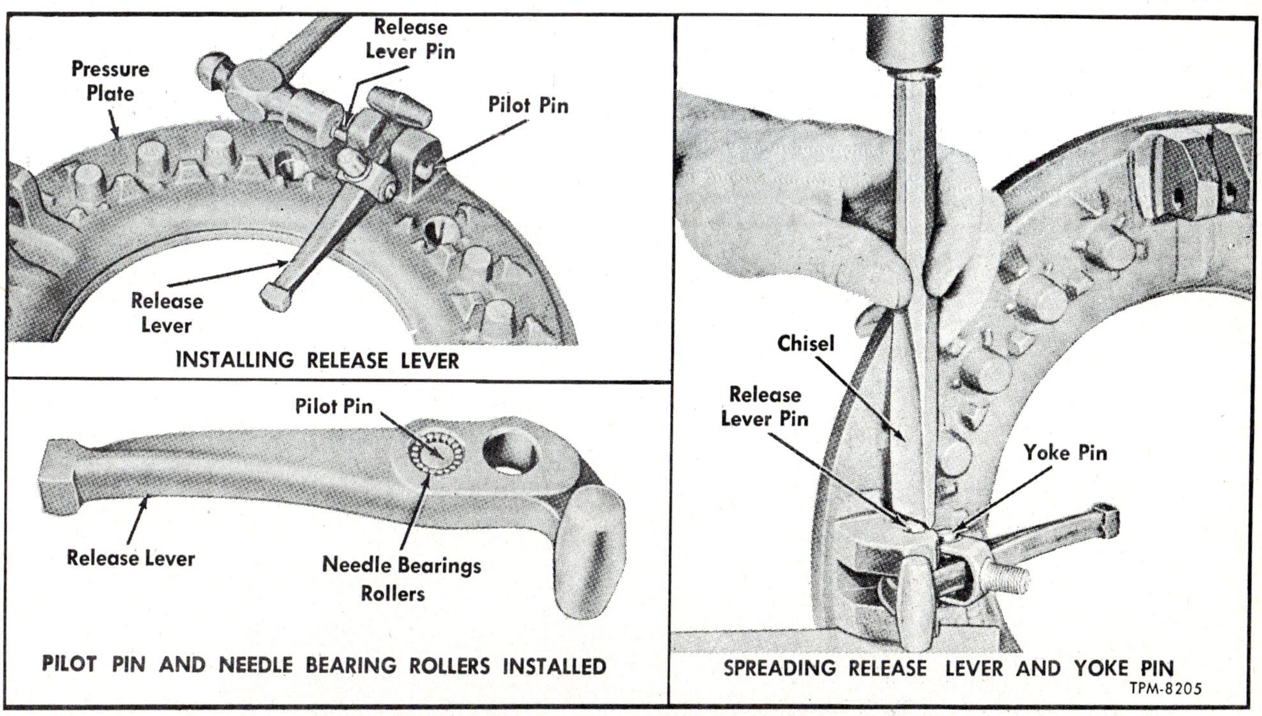

Fig. 4 Assembling and installing release lever

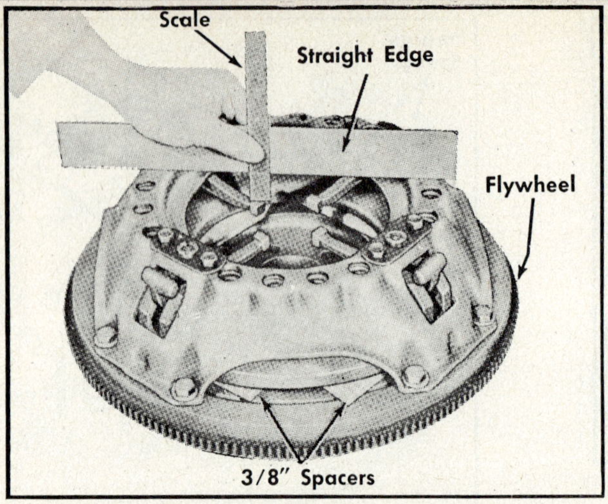

Fig. 6 Adjusting release levers

The foregoing remarks may be used if the specifications for the specific clutch being worked on are not available. For example, Fig. 6 illustrates the method of adjusting a 13" clutch. The adjustment to be made on this specific clutch model calls for a scale dimension of 1 1/32" from edge of straightedge to crown of lever tip.

Final Assembly

After completing adjustments, tighten thrust plate capscrews. Install wood blocks between release levers and cover plate to keep plate retracted until clutch is installed. Remove clutch from flywheel.

NOTE: If a new pressure plate or cover was installed, the complete assembly must be balanced within one-ounce-inch by attaching balance weights as required to side of cover with capscrews. Balance weights are available in three sizes—light, medium and heavy.

---

and bearings in lever, pushing out pilot pin. Use a chisel to spread end of pin to secure in place. Do not spread pin far enough to bend or distort yoke.

5. Install release lever in lugs on pressure plate, using pilot pin in same manner. Assemble and install the other release levers and yokes in same manner.

Assembly

1. Place pressure plate face down on arbor press bed on 3/8" thick spacers, (see Fig. 2).

NOTE: If face of pressure plate was machined, shims of a thickness equal to the amount of metal removed must be installed over each spring post on pressure plate. Shims can be made locally.

---

2. Place a pressure spring on each spring post on pressure plate, Fig. 5.
3. Place clutch cover over springs, with release lever counterweights and yokes extending through holes in cover, and with marks made prior to disassembly aligned. If a new unmarked pressure plate or cover is used, the cover and pressure plate should be balanced after completely assembled.
4. Place bar across cover, clearing all thrust plate capscrew holes and release lever yokes.
5. Turn arbor press screw down against bar and force cover down against press bed, compressing springs. Install adjusting nut on each release lever yoke until yoke end is flush with end of nut.
6. Place thrust plate over each release lever adjusting nut and attach with two capscrews, but do not tighten. Remove assembly from press.

Release Levers, Adjust

Satisfactory operation of the clutch is dependent upon the accuracy of the lever adjustment, since the pressure plate must be parallel with the flywheel. The method of checking the position of the levers depends upon the equipment available. Regardless of how the levers are adjusted, they must be in the same plane within .005".

To make the adjustment, remove the flywheel from the engine of the vehicle being worked on. Use a new driven disc or a gauge plate of the exact same thickness as the driven disc to be used in the clutch being serviced. Adjust the levers by screwing the adjusting nut on each lever until all levers are in the same plane within .005".

Installation

1. Place driven disc against engine flywheel, being sure portion of hub that goes to the front is placed to the front. While holding driven disc in place, move clutch cover assembly into position against driven disc. Insert an aligning tool or an old transmission main drive gear through the driven disc hub and into pilot bearing.
2. Install cover-to-flywheel bolts and lock washers. Tighten bolts alternately one turn at a time to compress clutch pressure springs evenly and to prevent possible distortion of cover flange. Remove aligning tool and wood blocks (if used) between levers and cover.
3. If release fork is used, install fork and fork ball stud by pushing it on until ball snaps into recess in fork.
4. Coat outside and inside grooves of release bearing support with a small quantity of high temperature grease.
5. Before installing transmission on vehicles using release fork instead of a release yoke, release bearing and support assembly must be placed in fingers of release fork. Install transmission assembly.

MINI-PAK CLUTCH AIR-HYDRAULIC UNIT

This unit, Fig. 1, is a clutch assistor consisting of a combined compressed air and hydraulic assembly used to supply power for clutch operation on vehicles equipped with a hydraulic clutch cylinder. Clutch actuation is obtained with low clutch pedal effort while retaining "drive feel" during clutch engagement. In the event of power failure, clutch actuation is made in the usual manner except that increased pedal effort is required by the driver. The assembly consists of the following basic components:

1. A compressed air cylinder containing an air pressure operated piston and push rod.
2. A hydraulic cylinder and piston.
3. An air pressure control valve containing a hydraulic pressure operated piston, an air pressure poppet, an atmospheric pressure poppet valve and an air pressure control diaphragm.

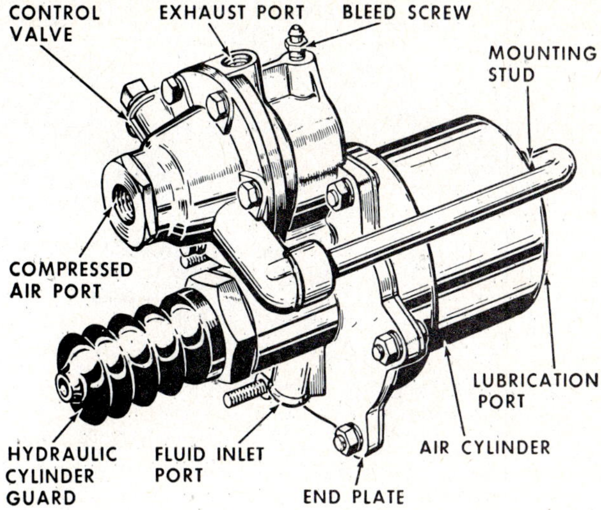

Fig. 1 Clutch air-over-hydraulic power unit

Operation

To disengage the clutch, the clutch pedal is depressed which forces fluid out of the clutch hydraulic cylinder under pressure into the clutch assistor hydraulic cylinder and control valve piston chamber, Fig. 2. When fluid pressure acting upon the control hydraulic piston reaches a predetermined value, the valve poppet closes the atmospheric port and opens the air pressure port. This action admits compressed air to the air pressure side of the air cylinder, Fig. 3. The force from the air pressure applied to the air piston is transmitted through the push rod to the hydraulic piston of the clutch assistor and through the clutch push rod to the clutch release lever to disengage the clutch.

To engage the clutch, the driver gradually releases the foot pressure on the clutch pedal. As pressure is reduced in the clutch assistor hydraulic cylinder, pressure will also be reduced in the clutch assistor hydraulic cylinder and control valve chamber. The air pressure applied to the air pressure piston will also be reduced. Fluid is then returned from the clutch assistor hydraulic cylinder in accordance with release of air pressure. When foot pressure on the clutch pedal is released sufficiently to allow the control valve to close the air pressure port and reopen the atmospheric port of the clutch assistor, the clutch is fully engaged.

Adjustment

The free pedal travel will be slight (approximately 1/4") and can be compared to the free travel that is acceptable for the pedal of hydraulic brake systems. The clutch assistor should be adjusted to provide complete engaging and disengaging of the clutch when the pedal is depressed and released. An average of 1/2" clutch release bearing travel should be provided for proper release of the clutch. The stroke of the clutch assistor hydraulic cylinder should be approximately 1 13/16".

Repairs

When overhauling the clutch assistor, refer to Fig. 4 as a guide. Be very careful when handling hydraulic system parts to prevent their coming in contact with mineral oils or grease. *Do not handle hydraulic piston cups or seals with greasy hands.* When overhauling the clutch assistor, always use a repair kit and install all the parts contained in the kit regardless of the condition of the old parts. If replacement of the air cylinder leather piston packing is necessary, use a piston packing kit in addition to the repair kit.

Lubrication

Do not lubricate the clutch assistor until it has been permanently installed on the vehicle. This is a safeguard. The clutch assistor should be lubricated every 10,000 miles or

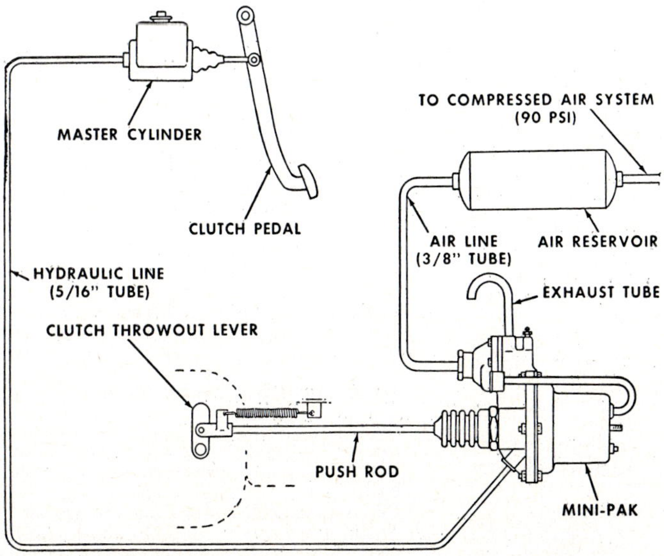

Fig. 2 Schematic diagram of typical air-hydraulic power unit

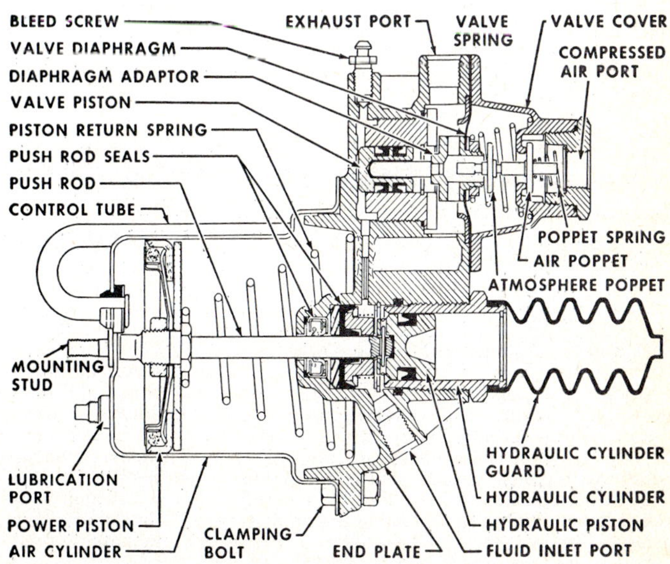

Fig. 3 Sectional view of air-hydraulic power unit

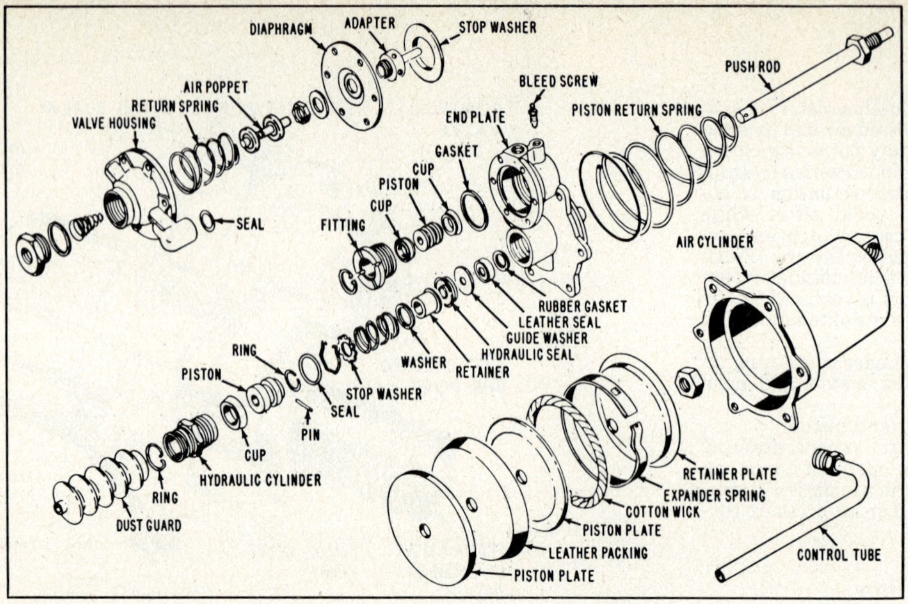

Fig. 4 Exploded view of air-hydraulic power unit

every six months, whichever occurs first, with vacuum cylinder oil.

To lubricate, remove the 1/8" pipe plug located in the end of the air cylinder shell and inject vacuum cylinder oil through this port until oil begins to run out, then replace pipe plug. *During lubrication, the engine should be shut off and the clutch assistor should be in the released position.*

TRANSFER CASES

INDEX

DESCRIPTION

A transfer case is a gear box located between the main transmission and the rear axle. Its purpose is to transfer power from the transmission to the front driving axle as well as the reae driving axle. It also provides an extra gear reduction (LO) in the power train of the vehicle.

Most units are available with a Power Take Off and a parking brake. Every unit incorporates a Front Axle Declutch which is used to drive the front axle whenever the vehicle encounters steep grades or rough terrain. These accessories are actuated by separate shift levers located in the cab of the vehicle.

Fig. 1 shows a typical hook-up of a three-shaft transfer case, while Figs. 2 and 3 illustrate a typical hook-up of a four-shaft transfer case used with a 4 × 4 and a 6 × 6 vehicle, respectively.

TIMKEN/ROCKWELL THREE SHAFT UNIT

Disassembly

1. Referring to Figs. 4 and 5, remove drain plug and drain lubricant.
2. Remove power take-off unit and wire shims together for reassembly.
3. Remove front axle declutch unit.
4. Remove input and output shaft yokes or companion flanges.
5. Remove range shift shaft and fork.
6. Remove shift cover. Cover is spring-loaded and should be held down during removal to prevent injury. Take out detent ball, plunger and spring.
7. Loosen shift fork nut and screw shift shaft out of fork and slide it out of case.
8. Loosen set screw in mainshaft driving clutch and remove clutch and oil dam.
9. Drive tapered pins from cover and remove cover.
10. Lift mainshaft assembly from case.
11. Lift out axle drive shaft assembly and then remove idler assembly.

Disassemble Power Take-Off

1. Remove yoke from power take-off shaft.
2. Remove oil filler plug from cage. Then remove detent capscrew, spring and ball.
3. Remove oil seal case, shims and gasket.
4. Through oil filler plug hole, loosen shift fork set screw. Disconnect shift lever and pull shift shaft from housing.

5. Use a bronze drift and drive power take-off shaft from housing. Remove shift fork and sliding clutch. *Do not disassemble further unless bearings need replacing.*

Disassemble Front Axle Declutch Unit

1. With suitable puller, remove declutch shaft yoke.
2. Remove bearing cage from housing.
3. Remove detent parts from housing.
4. Unscrew filler plug and loosen shift fork set screw.
5. Slide shift shaft out of housing. If necessary, remove shift and shaft oil seal.
6. Remove sliding clutch and shift fork.
7. If necessary, press apart bearing cage assembly. Remove outer bearing cup and oil seal.

Disassemble Shaft Assemblies

1. To disassemble mainshaft, press off front bearing and remove sliding gear. Press drive gear, bearing spacer and rear tapered bearing from shaft. Remove ball bearings and spacer from drive gear.
2. To disassemble axle drive shaft, remove set screw and pull off driving clutch. Press rear bearing and gear from shaft. Press off front bearing. Remove pilot bushing from shaft with a suitable puller if necessary.
3. To disassemble idler shaft, remove snap ring at low gear end. Slide off speedometer drive gear and remove Woodruff key. Remove front bearing snap ring. Press off low gear and remove spacer. Invert assembly and remove bearing retainer. Press off high gear and bearing.

Disassemble Case & Cover

1. Remove mainshaft front seal cage and idler shaft front bearing cap from transfer case housing.
2. Remove idler shaft bearing cover.
3. Pull axle drive shaft rear bearing cage, using puller screws. Wire shims together for reassembly. If necessary, press mainshaft rear bearing cup from cover.

Assemble Axle Drive Shaft

1. Referring to Fig. 6, press pilot bushing into bore at front of shaft.
2. Press front bearing into place.
3. Press driven gear and rear bearing into place, using a suitable sleeve that bears against inner race.
4. Slide driving clutch onto shaft and install set screw. Screw must enter recess in shaft.

Assemble Idle Shaft

1. Referring to Fig. 7, press low gear and tapered bearing into place on shaft. *Use a sleeve that will clear bearing journal.*
2. Install bearing snap ring, and Woodruff key in keyway.
3. Place speedometer drive gear on shaft and install outer snap ring.
4. Invert assembly and install spacer and high gear. Place bearing on shaft and press into position. Install bearing retainer, capscrews and lock wire.

Assemble Mainshaft

1. Referring to Fig. 8, tap rear ball bearing into gear with suitable sleeve against outer race.
2. Position bearing spacer in gear and tap front ball bearing into gear.
3. Press gear assembly onto shaft until it seats against shoulder on shaft.
4. Position spacer and press rear tapered bearing onto shaft.
5. Install sliding gear and front bearing. *Do not install oil dam and power take-off driving clutch until after transfer case cover has been attached.*

Assemble Front Axle Declutch Unit

1. Referring to Fig. 9, hold shift fork in place inside declutch housing and install sliding clutch. Position fork in groove.
2. Slide shift shaft through housing bore and into shift fork. Install and tighten set screw. Install oil filler plug.
3. Install detent ball, spring, lockwasher and lock screw in housing. Tighten lock screw.
4. Install inner and outer bearing cups in declutch bearing cage.
5. Press inner bearing on shaft with suitable sleeve.
6. Position bearing spacer on shaft and then install bearing cage.
7. Press outer bearing onto shaft. *Do not install oil seal at this time.*
8. Mount cage in vise and install yoke or flange and tighten nut to specified torque.
9. Check bearing adjustment by using a dial indicator at the end of the shaft as shown in Fig. 10. *Since the specified bearing adjustment is zero end play and zero preload, there should be no reading on the dial gauge.*
10. If there is no end play in the assembly, turn the shaft by hand to check for bearing preload. There should be no more than a slight drag on the shaft.
11. If there is end play in the assembly, use a thinner spacer between the bearings; if there is preload, use a thicker spacer.

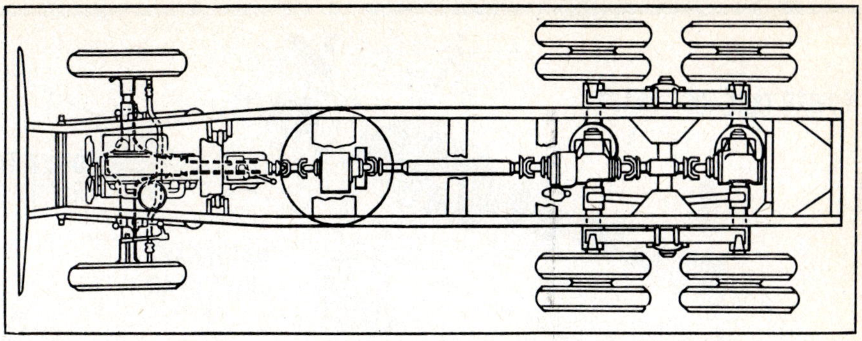

Fig. 1 Typical hook-up of a three shaft transfer case on a 6 × 6 vehicle

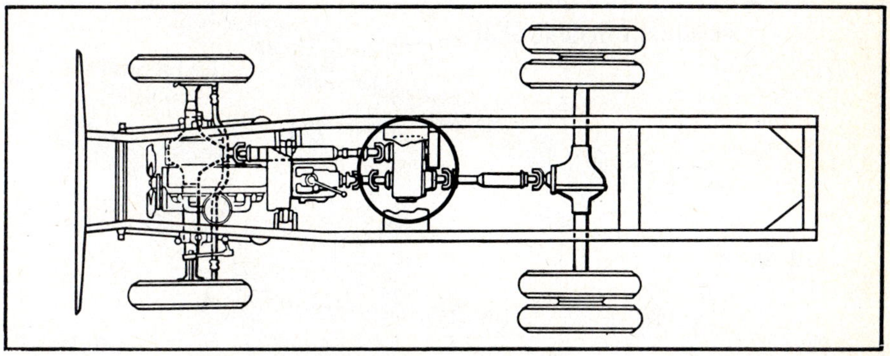

Fig. 2 Typical hook-up of a four shaft transfer case on a 4 × 4 vehicle

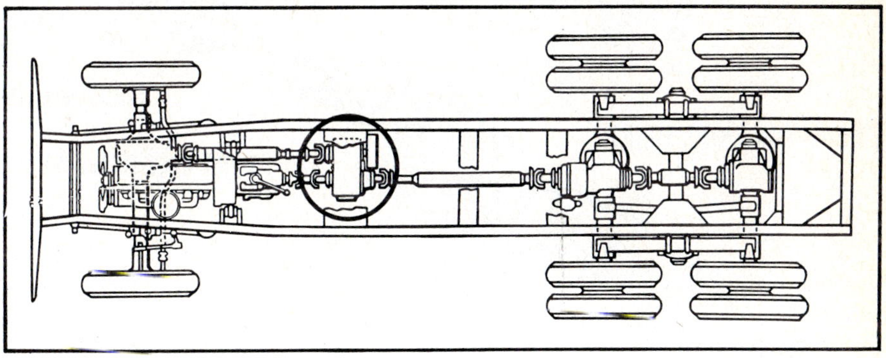

Fig. 3 Typical hook-up of a four shaft transfer case on a 6 × 6 vehicle

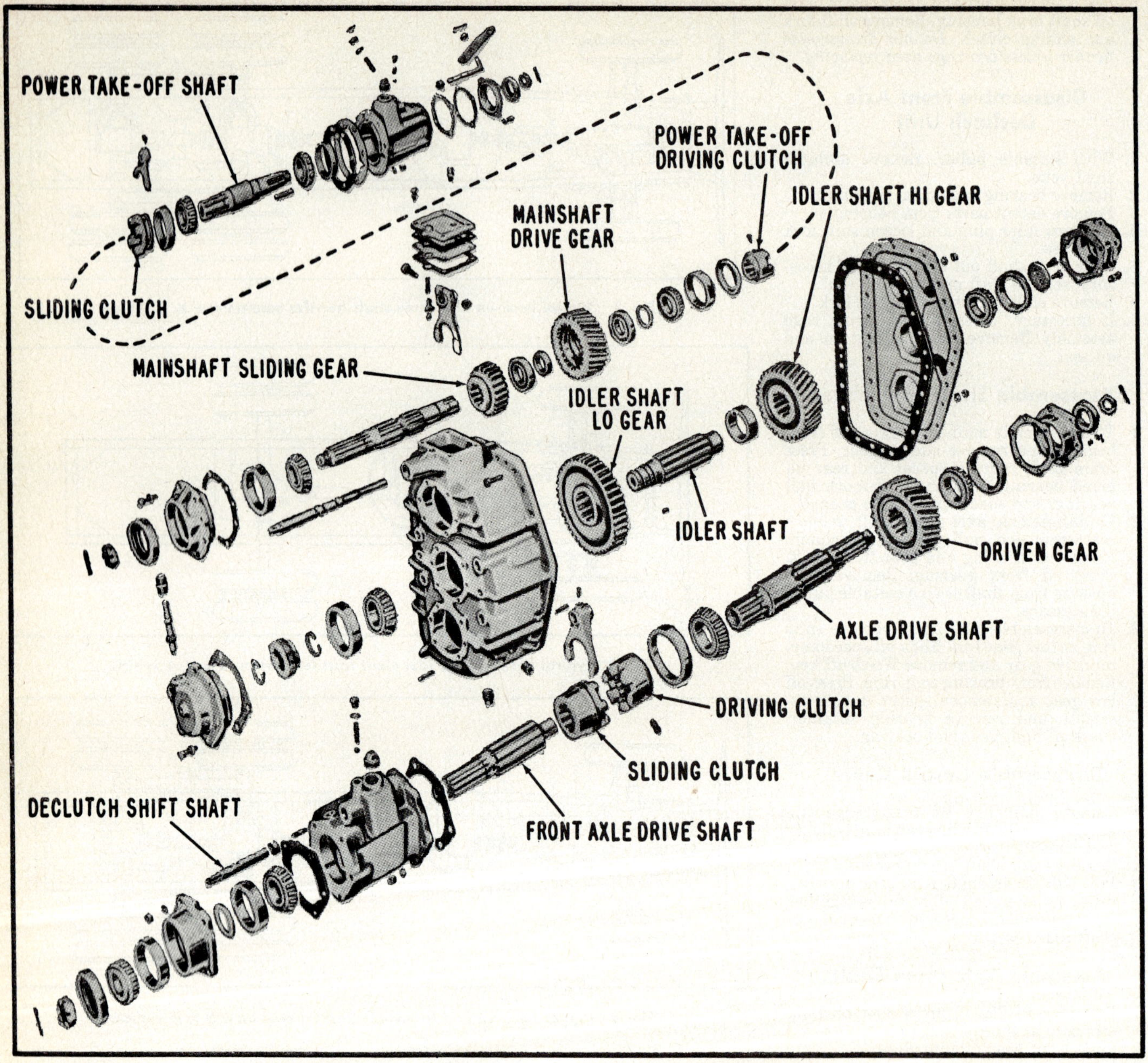

Fig. 5 Disassembled view of a Timken three shaft transfer case

12. Once the proper adjustment is obtained, remove the yoke or flange and install oil seal in bearing cage.
13. Position a new gasket over studs at front of declutch housing.
14. Line up front axle shaft with sliding clutch and slide shaft and bearing cage into housing.
15. Install lockwashers and stud nuts, tightening nuts to specified torque. Install yoke or flange and tighten nut to specified torque.

Assemble Power Take-Off

1. Referring to Fig. 11, press inner bearing cup into P.T.O. housing with a suitable sleeve.
2. Hold shift fork and sliding clutch in place inside housing.
3. Slide shaft assembly (with inner and outer bearings) into housing. Check to see that shift fork is seated in sliding clutch groove.
4. Install outer tapered bearing cup with a suitable sleeve.
5. Slide clutch shaft into housing and

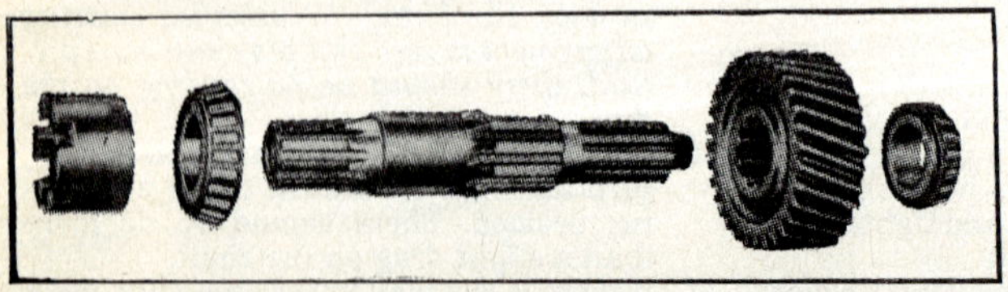

Fig. 6 Axle drive shaft components. Timken three shaft case

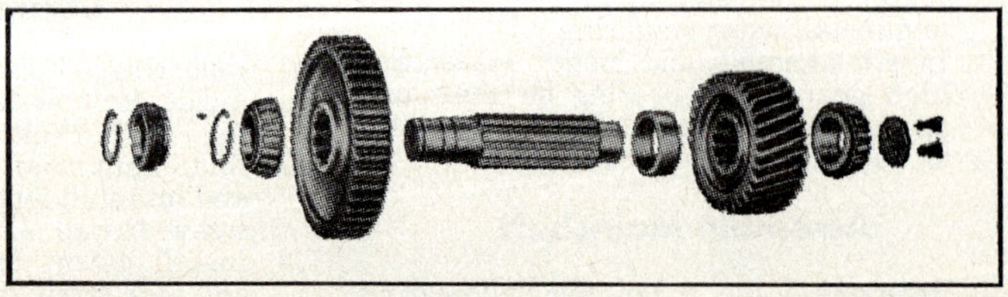

Fig. 7 Idler shaft components. Timken three shaft case

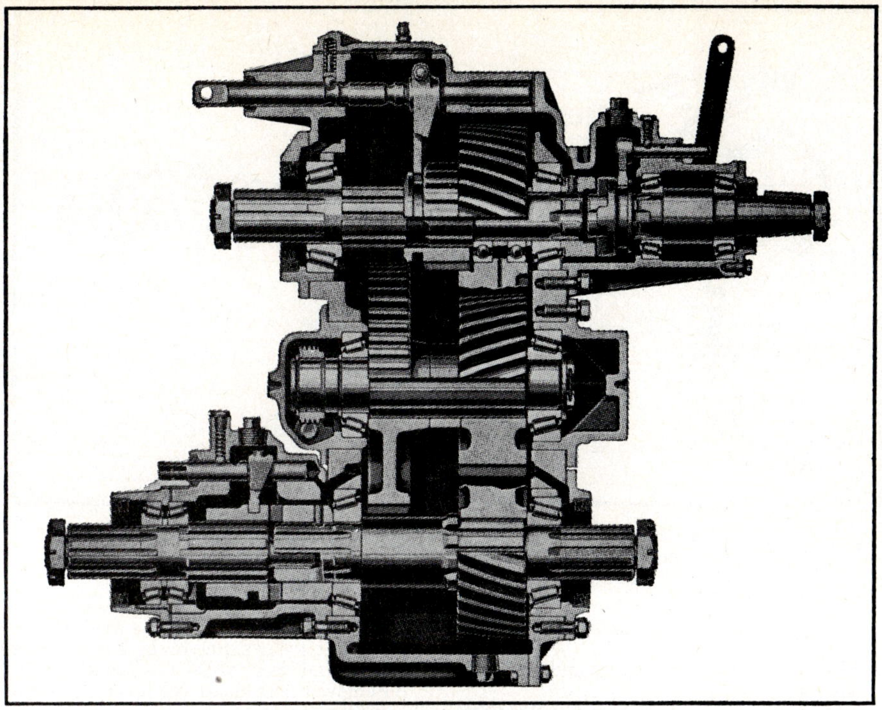

Fig. 4 Timken three shaft transfer case

Fig. 8 Mainshaft components. Timken three shaft case

Fig. 9 Front axle declutch unit. Timken three shaft case

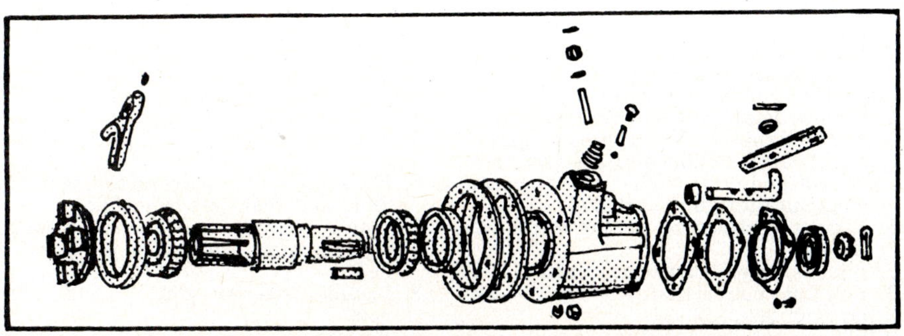

Fig. 11 Power take-off components. Timken three shaft case

through shift fork.
6. Line up recess in shift shaft and install shift fork set screw and tighten.
7. Install detent ball, spring, lockwashers and capscrew. Tighten capscrew.
8. Attach shift lever to shift shaft.
9. Position new gasket and install enough bearing adjustment shims to set up end play in the assembly.
10. Attach seal cage to housing and tighten capscrews.
11. Using a dial indicator at end of shaft, Fig. 12, check amount of end play in the assembly. End play should be .003–.005″.
12. Take off seal cage and remove enough shims to establish proper end play. For example, if dial reading is .009″, remove .005″ of shims to arrive at an adjustment of .004″.
13. Reinstall seal cage and tighten capscrews to specified torque. Install yoke or flange and tighten nut to specified torque. See Fig. 13 for optional pump type power take-off.

Installation of Shafts

1. Position axle drive shaft in case.
2. Tilt axle drive shaft to one side and install idler shaft. Then work both assemblies into position.
3. Position mainshaft in case and work into position.
4. Reassemble transfer case cover (if disassembled). Use original shim packs under idler shaft and axle drive shaft rear bearing caps.
5. Install transfer case cover but do not tighten nuts at this time. Install dowels; then tighten nuts to specified torque.
6. Install front axle declutch unit with new gasket. Tighten stud nuts to specified torque.
7. Position oil dam and P.T.O. driving clutch on mainshaft and install set screw.
8. Install new gasket and shim pack. Then mount P.T.O. assembly on cover. Tighten stud nuts to specified torque.

Checking Bearing Clearances

1. Mount a dial indicator at forward end of mainshaft and measure end play in assembly. Specified bearing adjustment is .003–.005″ end play. To adjust, alter shim pack located between P.T.O. and transfer case cover.
2. In a similar manner, measure axle drive shaft end play with a dial indicator at rear end of shaft. End play is also .003–.005″ and is adjusted by altering shims.
3. Idler shaft end play can be checked by inserting a pry bar through shift cover hole at top of case and working idler shaft low gear back and forth. No more than a

Fig. 10 Checking end play of axle drive shaft

Fig. 12 Checking end play of power take-off unit

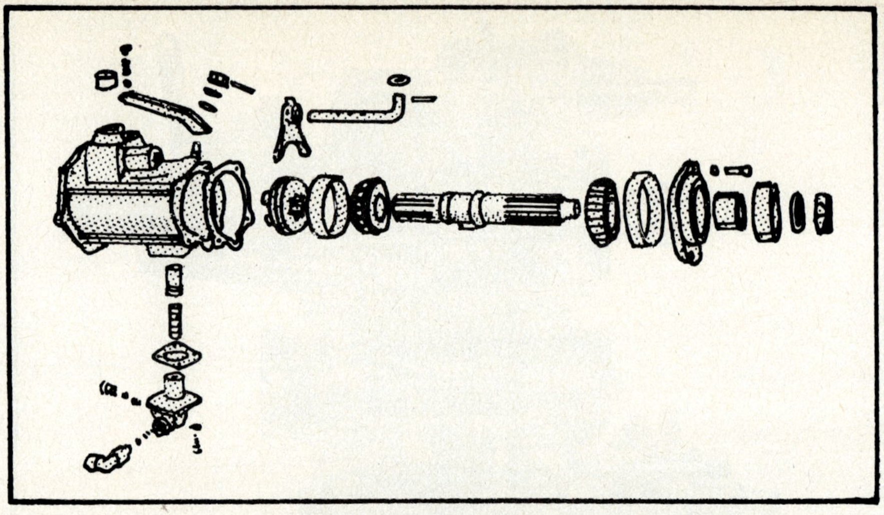

Fig. 13 Pump type P.T.O. Pump is actuated by an eccentric located on the P.T.O. shaft to assure proper lubrication during "stationary" operation of the P.T.O.

very slight bit of end play should be felt.

Final Assembly

1. Position shift fork in transfer case so that it fits in groove of mainshaft sliding gear.
2. Slide shift shaft into case and screw it into shaft fork.
3. Tighten bolt in shift fork and secure with lockwashers and nut.
4. Install detent ball in transfer case. Position plunger in detent spring and install in case.
5. Place shift fork cover gasket, breather plate and cover on case. Press cover down to compress detent plunger spring. Hold down cover and install lockwashers and capscrews.
6. Install yoke or flange on mainshaft and tighten nut to specified torque.
7. Install companion flange at rear of axle drive shaft. Tighten nut to specified torque.

BOLT AND STUD NUTS				
LOCATION ON UNIT	DIA. INCH	NO. THDS.	TORQUE—LB FT. MIN.	MAX.
Case and Cover	3/8″	16	38	42
Main Shaft and Rear Bearing Cap	7/16″	14	60	66
P.T.O. to Cover	7/16″	14	60	66
Drive Shaft Rear Bearing Cap	7/16″	14	60	66
Declutch to Case	7/16″	14	60	66
Declutch Housing Cap	3/8″	16	38	42
Main Shaft Yoke Nut	1¼″	18	300	400
Declutch Shaft Yoke Nut	1¼″	18	300	400
Drive Shaft Yoke Nut	1¼″	18	300	400
P.T.O. Shaft Yoke Nut	1″	20	300	400
CAP SCREWS				
LOCATION ON UNIT	DIA. INCH	NO. THDS.	TORQUE—LB. FT. MIN.	MAX.
Shift Cover	3/8″	16	38	42
Idler Shaft Front and Rear Bearing Caps	7/16″	14	60	66
Main Shaft Front Bearing Cap	7/16″	14	60	66
Idler Shaft Bearing Retainer	7/16″	14	60	66
P.T.O. Housing Cap	5/16″	18	22	24

Fig. 14 Torque specifications for three shaft transfer case. Torques given are for parts coated with machine oil. For dry (or "as received") parts, increase torque 10%. For parts coated with multipurpose gear oil, decrease torque 10%. Nuts on studs use same torque as for driving the stud

Lubrication

Install and tighten drain plug. Turn transfer case upright and pour ½ pint of recommended lubricant through filler plug opening. Add some oil to P.T.O. and front axle declutch unit. Do not fill unit to specified level until it is installed in the vehicle. Test operation of transfer case in all ranges.

TIMKEN/ROCKWELL T-223H & T-233 FOUR SHAFT UNITS

Disassemble

1. Referring to Fig. 15, remove all yokes or flanges.
2. Remove P.T.O. assembly or rear cover if P.T.O. is not used.
3. Remove transfer case cover.
4. Remove detent balls and springs.
5. Pull both shift shafts out through front of case if possible. Otherwise, use a soft metal drift and drive shafts and expansion plugs out through rear of case.
6. Remove oil seals from front of case if necessary.
7. Lift out range and declutch forks.
8. Pull front output shaft and declutch collar from case.
9. Slide a bar through front output shaft gear and tap out small and large expansion plugs located in brake section.
10. Remove brake drum, brake hub retaining ring and hub with a suitable puller. Remove brake shoes and lever, and brake backing plate.
11. Remove idler shaft front and rear bearing caps. Wire forward cap shims together for reassembly.
12. Remove bearing retainer plate from front end of idler shaft and press shaft out through rear of case. Lift out low gear, spacer and front bearing.
13. Remove high gear. Tap front bearing cup from case. Remove snap ring and press rear bearing from idler shaft.
14. Remove front output gear rear bearing cap and gasket.
15. Take off bearing snap ring and tap front output gear into case. Reach through cover opening and lift out gear and bearing. Press off radial bearing.
16. Take off rear output shaft front and rear bearing caps. Wire shims together for reassembly.
17. Remove input shaft front bearing cover. Wire shim pack together for reassembly.
18. Using a block of wood between sliding gear and case, press out input shaft and front bearing. Do not pound on shaft if gear becomes bound. Lift out sliding gear, drive gear and spacer. Remove front bearing from shaft.

Assemble P.T.O. Unit

1. If this unit was disassembled, refer to Fig. 16 and proceed as follows:
2. Install ball bearings on P.T.O. shaft.
3. Position wiper against inner bearing. Install snap ring in groove of outer bearing.
4. Slide shaft into housing and install sliding clutch on inner end.
5. Position new gasket and install oil seal cage.
6. Place shift fork in groove of sliding clutch. Slide shift shaft into bore of fork.
7. Line up recess in shaft with set screw and tighten screw. Lock wire set screw to fork.
8. Install expansion plug in shift shaft hole.
9. If necessary, replace shift shaft oil seal in front of housing.

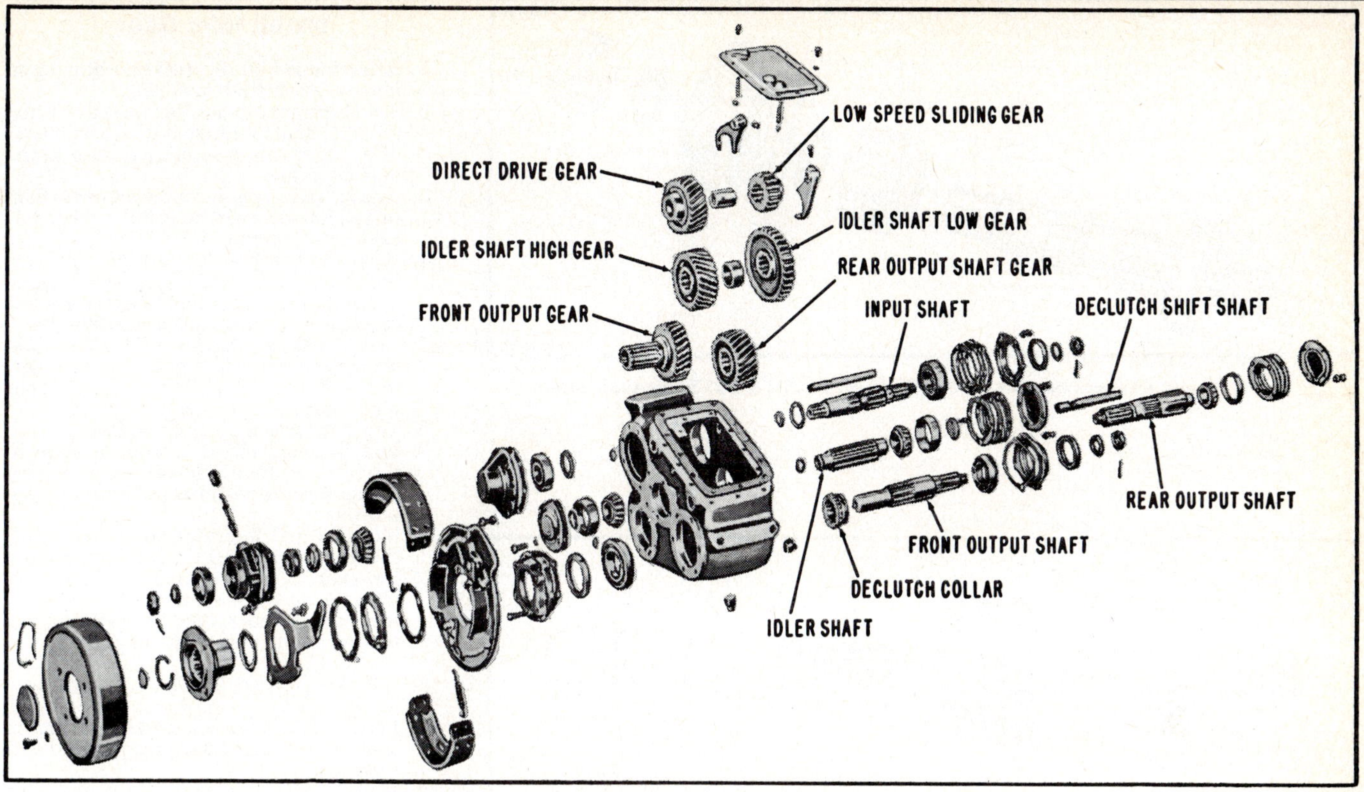

Fig. 15 Timken/Rockwell T-223H & T-233 "clover leaf" four shaft transfer case

Assemble Input Shaft

1. Referring to Fig. 17, install front bearing on shaft with shielded side against shoulder.
2. Start shaft into case. Mount sliding gear, spacer and drive gear (with bushing) on shaft.
3. Tap shaft into position using a suitable sleeve against inner race of front bearing.
4. Position original shim pack plus .010" more to make sure there is end play in the assembly. Install front cover and tighten capscrews.
5. Place thrust washer on shaft and install rear bearing with shielded side toward inside.
6. Position new gasket and attach the P.T.O. or cover to rear side of case.
7. Mount a dial indicator on the unit. Set the stem against the front end of the input shaft and check the amount of end play in the assembly. Remove enough shims from under the bearing cap to arrive at an adjustment of .003–.005" end play.
8. Reposition bearing cap, insert capscrews and torque to specifications. Remove P.T.O. for convenience in handling transfer case.

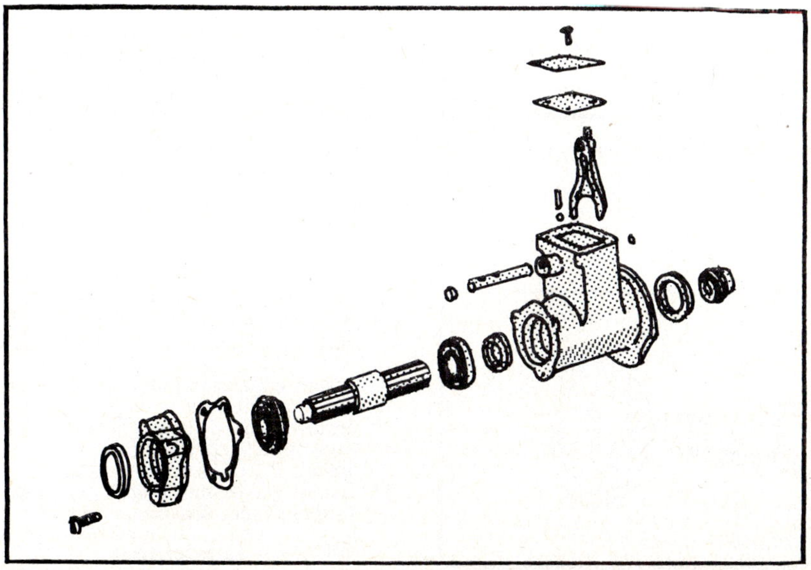

Fig. 16 Power take-off for Timken/Rockwell T-223H & T233 four shaft transfer case

Install Rear Output Shaft

1. Referring to Fig. 18, press front bearing onto shaft with suitable sleeve.
2. Hold rear output gear in position inside case and slide shaft through it.
3. Install front bearing cup and original shim pack plus .010". Install bearing cover with lockwashers and tighten capscrews.
4. Press rear bearing on shaft with a suitable sleeve. Then tap bearing cup into position.
5. Install spacer and speedometer drive gear over shaft.
6. Position new gasket, bearing cap and oil seal over shaft. Install lockwashers and capscrews and tighten to specified torque.
7. Rotate shaft to seat bearings. Mount a dial indicator against rear end of shaft to check the amount of end play in the assembly. Remove enough shims from under the front bearing cap to obtain an adjustment of zero end play and zero preload.

Install Front Output Gear

The front output gear must be installed in the case before the idler assembly is installed.

1. Install ball bearing on hub of gear.
2. Position gear and bearing in case and install snap ring on bearing.
3. Install rear bearing cap over new gasket. Insert capscrews and lockwashers and tighten to specified torque.

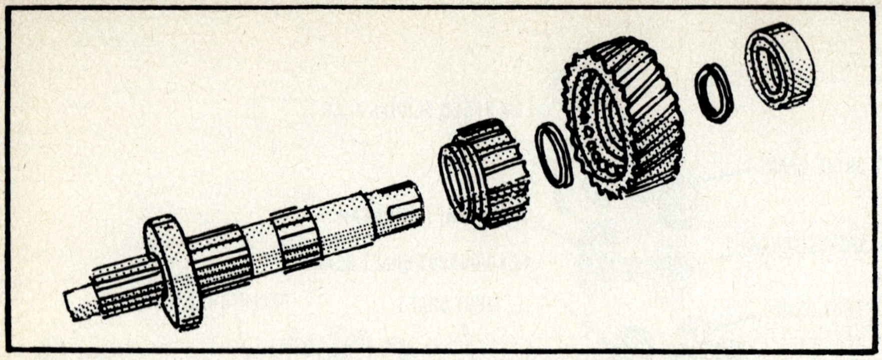

Fig. 17 Input shaft components. Timken/Rockwell T-223H & T-233 four shaft case

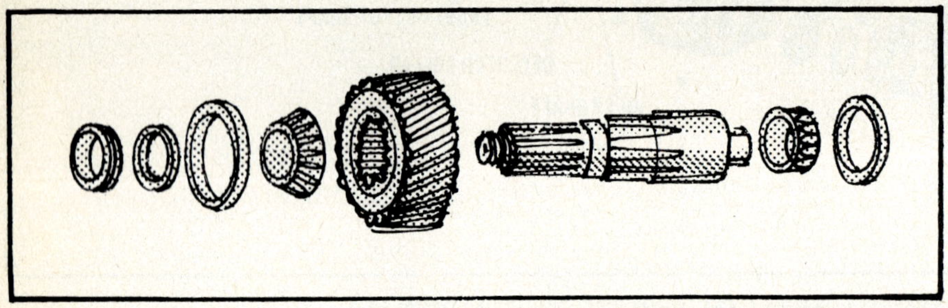

Fig. 18 Rear output shaft components. Timken/Rockwell T-223H & T-233 four shaft case

Fig. 19 Idler gear components. Timken/Rockwell T-223H & T-233 four shaft case

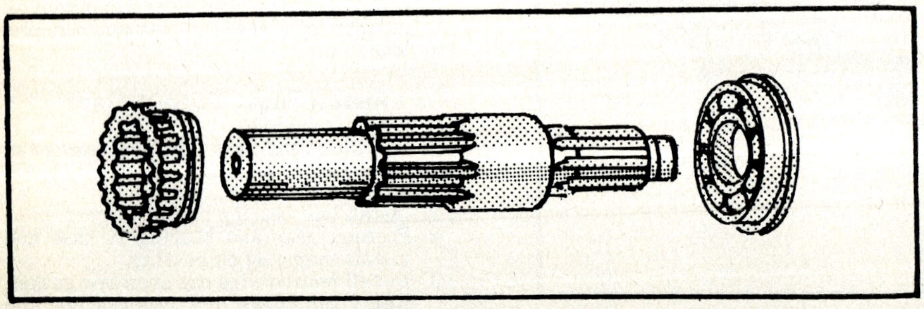

Fig. 20 Front output shaft components. Timken/Rockwell four shaft case

Install Idler Gear

1. Referring to Fig. 19, press rear bearing on idler shaft. Install snap ring.
2. Hold high gear in position inside case and tap idler shaft through it with a soft hammer. Short hub side of gear goes to the outside (rear).
3. Install rear spacer on shaft and then install low gear with long hub toward outside (front) of case.
4. With shaft in position, install rear bearing cup in case. Use a new gasket and install rear bearing cap, lockwashers and capscrews. Tighten to specified torque.
5. Drive front bearing onto idler shaft. Hold shaft rigid to avoid damaging rear bearing and cup. Install bearing retainer plate. Lock wire capscrews.
6. Tap front bearing cup into place.
7. Install enough shims to set up end play in the idler assembly. Install front bearing cap and secure with lockwashers and capscrews.
8. Mount a dial indicator on the case with the stem set against the inside face of the low gear. Check the amount of end play by working the assembly back and forth with two pry bars. Remove enough shims to obtain .003–.005" end play.

Install Front Output Shaft

1. Referring to Fig. 20, install bearing on shaft with bearing snap ring toward outside.
2. Install sliding collar on shaft and slide shaft into case.
3. With shaft in position, install new gasket, bearing cap and oil seal, lockwashers and capscrews. Tighten to specified torque.

Install Parking Brake

1. Position new gasket and mount backing plate, deflector and stamped washer. Insert capscrews and star washers and tighten to specified torque.
2. Position brake lever on backing plate.
3. Position brake shoes on backing plate with actuating pawl in web slot.
4. Hook up brake shoe return springs.
5. Slide brake hub over splines of front output gear. Install retaining ring.
6. Install brake drum. Tighten capscrews to specified torque and lock with wire.
7. Install expansion plugs. The smaller plug goes in the bore of the front output gear and the larger in the brake drum.

Install Shift Shafts

1. If necessary, install new shift shaft oil seals in case.
2. Position declutch fork in collar.
3. Grease declutch shift shaft and slide it into case and through fork.
4. With shift shaft in position, insert and tighten set screw and lock with wire.
5. Position range shift fork in sliding gear.
6. Slide range shift shaft through fork. Install and tighten set screw and lock with wire.
7. Install expansion plugs at rear of case.
8. Place detent balls and springs in case.
9. Install cover with new gasket and tighten capscrews to specified torque.
10. Install P.T.O. with new gasket.
11. Install yokes or flanges and nuts and torque to specifications.

Lubrication

Install drain plug. Turn transfer case up-

LOCATION ON UNIT	DIAMETER	NO. THREADS	TORQUE—LB. FT	
			Min.	Max.
Top cover	3/8″	16	38	42
Bearing caps (all)	3/8″	16	38	42
Brake drum	7/16″	14	60	66
Brake mounting	7/16″	14	60	66
P.T.O. to case	3/8″	16	38	42
P.T.O. bearing cap to housing	3/8″	16	38	42
P.T.O top cover	3/8″	16	38	42
YOKE OR FLANGE NUTS				
Input shaft	1″	20	300	400
Front output shaft	1″	20	300	400
Rear output shaft	1″	20	300	400
P.T.O shaft	1″	20	300	400

Fig. 21 Torque specifications for four shaft transfer case. Torques given apply to parts coated with machine oil. For dry (or "as received") parts, increase torque 10%. For parts coated with multipurpose gear oil, decrease torque 10%. Nuts on studs use same torque as for driving the stud

right and pour 1/2 pint of recommended gear lubricant through filler plug opening. Add some oil to P.T.O. Do not fill unit to specified level until it is installed in the vehicle.

COLEMAN MODELS 22 & 42

Model 22 declutcher type transfer case transmits power to the front axle only when required. It does not provide differential action between front and rear axles. The transfer case must be disengaged except when needed for traction.

Model 42 transfer case transmits power to the front axle at all times with full differential action between front and rear output shafts. The differential may be locked out for positive drive front and rear. The differential should always be unlocked under conditions of good traction.

Both models are two speed units (direct and low) and are shifted by two shift levers through the cab floor.

Disassembly

1. Position unit on bench with front housing down. Remove parking brake parts.
2. Turn assembly over and remove front output shaft end yoke.
3. Remove input shaft end yoke.
4. Remove external parts from case.
5. Remove snap ring and use a suitable puller to remove output shaft, Fig. 22.
6. Remove front output shaft sliding lock gear from front housing.
7. Remove center shaft front bearing cap and shim set from front housing. *Tag shim set for reassembly.*
8. Remove input shaft front bearing cap and shim set from front housing. *Tag shim set for reassembly.*
9. Turn case over and remove output shaft bearing case and shim set. *Tag shim set for reassembly.*

NOTE: On Model 42, rear output shaft, ball bearing assembly and speedometer drive gear will come out with bearing cap. After removing bearing cap and shaft, remove differential assembly from rear of case.

On Model 22, pull rear output shaft assembly out of case rear housing. Output shaft consists of shaft, outer bearing, speedometer drive gear, bearing spacer, rear output shaft gear, inner bearing assembly, pilot bushing, and two output shaft gear keys.

10. Remove center shaft bearing cap from rear housing.
11. Remove P.T.O. cover.
12. Using a spanner wrench, remove first input shaft high gear rear bearing nut. Then remove second shaft high gear rear bearing nut.
13. Remove input shaft high gear bearing from shaft.
14. Remove P.T.O. adapter from rear housing.
15. Separate front and rear housings.
16. Remove declutcher or differential shift bar and fork from rear housing. Remove high-low shift bar and fork.
17. Remove input shaft from high speed gear, and high speed gear from rear housing.
18. Remove center shaft assembly.
19. Remove shift bars from forks.
20. Using a plastic hammer, tap rear output shaft inner bearing cup from rear housing. Tap center shaft rear bearing cup out of rear housing.
21. Tap center shaft front bearing cup and input shaft front bearing cup out of front housing.

Reassembly

After disassembling sub-assemblies as required to install new parts, reassemble the transfer case in the reverse order of disassembly, observing the following:

1. Use all new gaskets and oil seals.
2. When installing shims and bearing caps, be sure oil drain-back holes and passages are not covered.
3. Great care should be taken in setting tapered roller bearings so they are neither too tight nor too loose. As a general guide, the front output shaft bearing should be set from .003″ to .005″ loose.
4. The input shaft high gear rear bearing should be set from .000″ to .002″ clearance.
5. Both center shaft bearings should be set from .003″ to .005″ loose.
6. The lower rear output shaft bearing should be set from .003″ to .005″ loose.
7. To adjust bearings, tighten bearing caps to a snug fit. Then measure the gap between inside of bearing cap and housing with a feeler gauge. Add the foregoing dimensions to the measured gap for total shim thickness. Shims are available in .005″, .0075″ and .020″ thicknesses.

DANA MODEL 21

This manually-shifted transfer case controls the power from the engine and transmission to the front driving axle. There are two positions for the transfer case: front axle drive engaged and front axle drive disengaged.

As shown in Fig. 25, when the transfer case front shaft clutch collar is in neutral, power from the main transmission drives the transfer case input shaft gear. The shaft gear drives the idler shaft and the gear that free

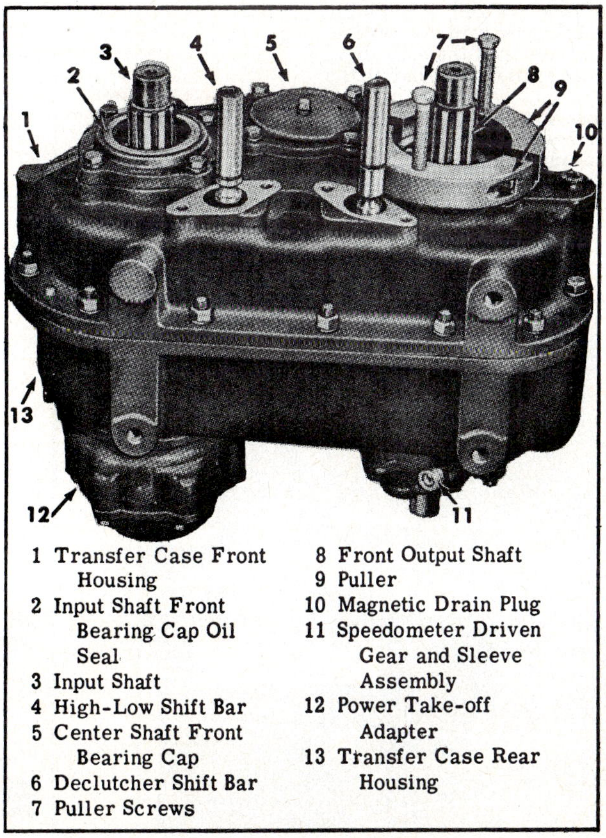

1 Transfer Case Front Housing
2 Input Shaft Front Bearing Cap Oil Seal
3 Input Shaft
4 High-Low Shift Bar
5 Center Shaft Front Bearing Cap
6 Declutcher Shift Bar
7 Puller Screws
8 Front Output Shaft
9 Puller
10 Magnetic Drain Plug
11 Speedometer Driven Gear and Sleeve Assembly
12 Power Take-off Adapter
13 Transfer Case Rear Housing

Fig. 22 Removing front output shaft assembly. Coleman units

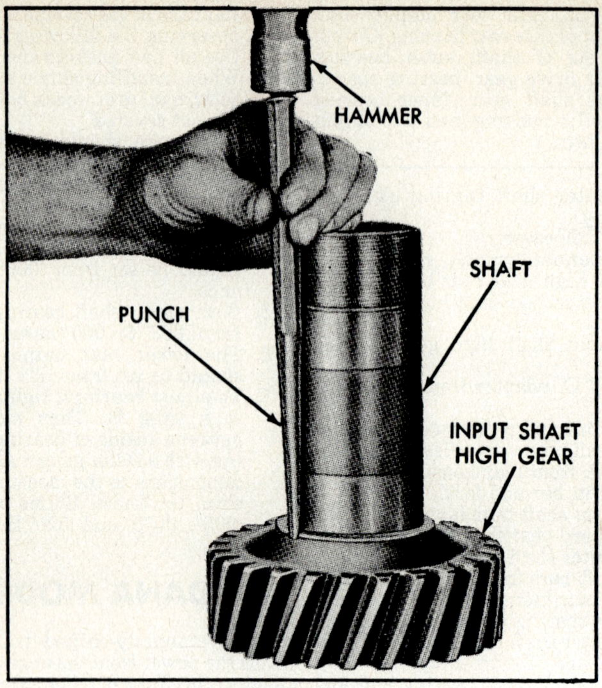

Fig. 23 Removing input shaft high gear bearing cup. Coleman units

runs on the front output shaft.

There is only power delivered to the rear axle because of the direct hook-up of the main transmission to the rear wheel output shaft.

When the transfer case shift lever is shifted into the 4-wheel drive position, it pulls the front output shaft collar forward into engagement with the clutch teeth of the main drive gears. This locks the front output shaft to the idler shaft gear.

The power from the main transmission flows from the drive gear in two directions: Direct drive to the rear axle flows through the rear output shaft. Direct drive to the front axle flows through the idler shaft drive gear through the front output shaft gear direct to the front axle.

Disassembly

1. Remove case cover, Fig. 26.
2. Remove set screw securing shift fork to rail. Tap shift rail rearward, then remove shift rail cap from rear of case.
3. Remove shift rail and fork.
4. Remove detent spring and ball that engages front drive shift rail.
5. Remove front and rear flanges.
6. Remove rear bearing retainer and output shaft as an assembly.
7. Remove front and rear idler shaft covers.
8. Drive idler shaft and rear idler bearing rearward and out of case. Then lift front bearing and idler gear from case.
9. Remove front output shaft bearing retainer and gasket. Remove retainer seal if worn or damaged.
10. Remove front output shaft rear cover and shims. Tie shims together for reassembly.
11. Tap end of front output shaft toward front of case to remove front bearing cup. Remove rear bearing cup by tapping front output shaft rearward.
12. Wedge front output shaft front bearing away from main drive gear to allow removal of snap ring from its groove in shaft. Then drive output shaft and rear bearing out of case.
13. Lift sliding gear, main drive gear, front bearing, thrust washer and snap ring from case.
14. Remove shift rail seal from case bore if it is worn or damaged.

Reassembly

After disassembling sub-assemblies as required to install new parts, Figs. 27, 28, 29, reassemble transfer case as follows:

1. While holding drive gear, sliding gear and thrust washer in case, install front output shaft through gears and washer

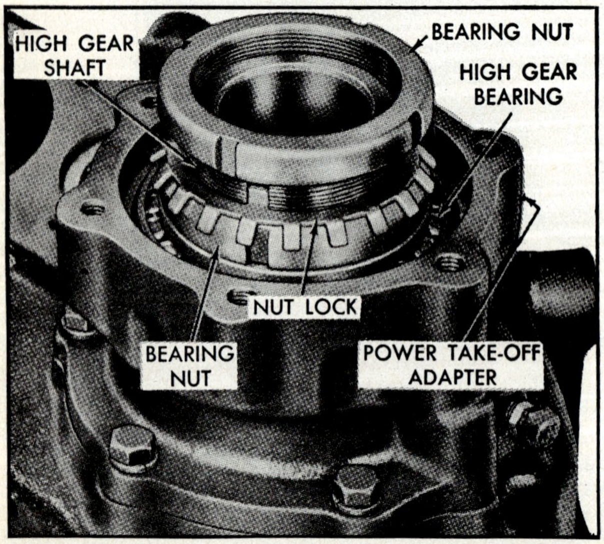

Fig. 24 Installing input shaft high gear bearing nuts. Coleman units

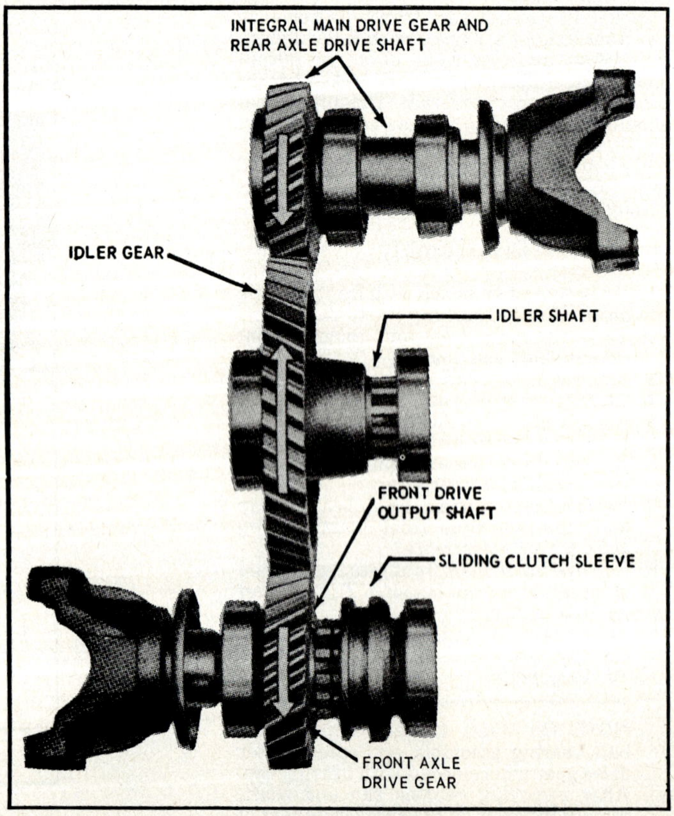

Fig. 25 Ford single speed transfer case gear train

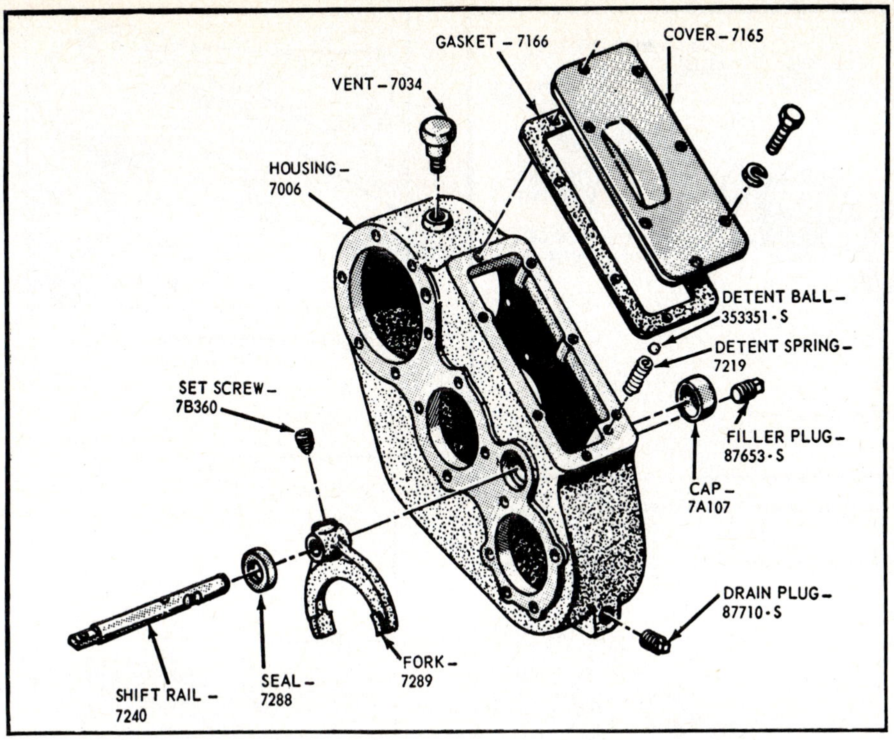

Fig. 26 Dana model 21 transfer case housing and related parts

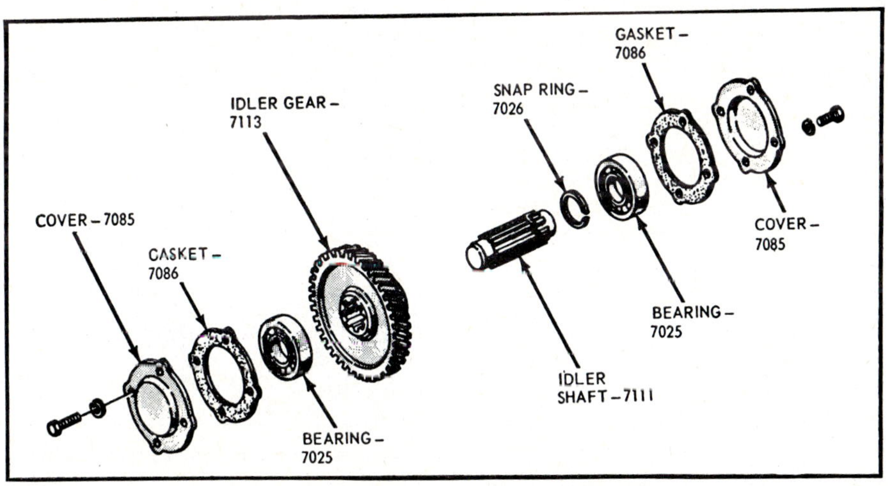

Fig. 27 Dana model 21 transfer case idler shaft assembly

from rear of case, Fig. 29. Install snap ring.

2. Install front output shaft rear bearing cup.
3. Install front output shaft rear bearing cover and shims to case. Apply gasket sealer to attaching bolts and tighten bolts. With cover installed, apply gasket sealer to outer edges of adjusting shims.
4. Install front output shaft front bearing on shaft. Install front bearing cup.
5. If front bearing retainer oil seal was removed, install a new seal. Install bearing retainer and gasket to case.
6. Install flange, O-ring, washer and nut on front output shaft.
7. Install dial indicator on front drive output shaft and check end play. End play should be .003″ to .005″. If necessary, increase or decrease shim pack thickness at front output shaft rear cover.
8. Install idler gear in case and install idler shaft through gear. Install front bearing on shaft.
9. Install front and rear idler covers and gaskets to case.
10. Install a new shaft rail seal in case if it was removed.
11. Install shift rail detent ball and spring in top of case.
12. Slide shift rail into case, and position shift fork on rail as rail enters case. Depressing ball and spring will allow rail to pass. Secure fork to rail with set screw. Install shift rail cap.
13. Install rear output shaft and bearing retainer to case.
14. Install flange, O-ring, washer and nut on rear output shaft.

NEW PROCESS MODEL 205

Disassembly, Fig. 41

1. Loosen rear output shaft yoke nut, then remove rear output shaft housing bolts, housing and retainer assembly from case, Fig. 30.
2. Remove rear output shaft retaining yoke and nut then remove shaft assembly from housing.
3. Remove and discard snap ring using tool J-23432, then remove thrust washer and washer pin.
4. Remove tanged bronze washer, gear needle bearing (32 per row), spacer and second row of needle bearings from rear output shaft.
5. Remove tanged bronze thrust washer, pilot rollers (15), retainer ring, washer, oil seal retainer, ball bearing, speedometer gear and spacer from rear shaft.
6. Discard all gaskets and press out bearing from rear output shaft.
7. Remove locknut, washer and yoke from front output shaft.
8. Remove front bearing retainer attaching bolts and retainer, then the rear bearing retainer attaching bolts from front output shaft.
9. Tap output shaft, gear assembly and rear bearing retainer from case with a rubber mallet, Fig. 31.
10. Remove synchronizer from output high gear, then the washer and bearing from case.
11. Remove and discard gear retaining ring from shaft, Fig. 32.
12. Remove thrust washer, pin, gear, needle bearings (32 per row) and spacer from shaft.

NOTE: If front output shaft rear bearing must be replaced, remove bearing by supporting the cover and pressing out the bearing. Press new bearing into position using a piece of pipe or wood until bearing is flush with opening.

13. Remove two poppet nuts, springs and, using a magnet, the poppet balls from top of transfer case.
14. Using a ¼ inch punch, drive cup plugs into case.
15. Place both shift rails in the neutral position and drive shift fork pins into case, Fig. 33.
16. Remove clevis pins, shift rail link and shift rails, Fig. 34. Remove upper (range) rail first, then the lower (4-wheel) rail.
17. Remove shift forks, sliding clutch, front output high gear, washer, bearing, shift rail cup plugs and pins from case.
18. Remove snap ring in front of bearing, then, using a rubber mallet, tap shaft and bearing out of case.
19. Tip case on power takeoff opening and remove two interlock pins from inside of case.
20. Remove idler gear shaft nut and rear cover.
21. Remove idler gear shaft using a rubber mallet and tool J-23429, Fig. 35.
22. Roll idler gear to front output shaft opening and remove gear from case, then remove bearing cups from gear.

Assembly, Fig. 41

1. Press bearing cups in idler gear using tools J-9276-2 and J-8092.

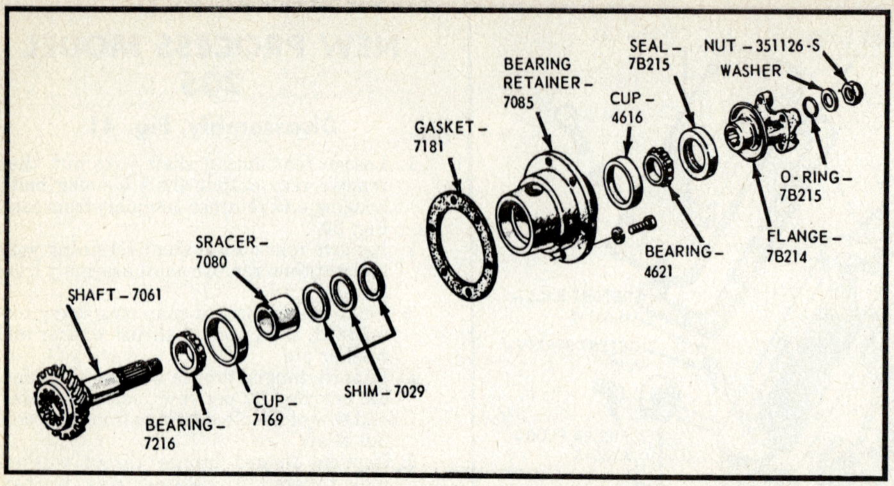

Fig. 28 Dana model 21 transfer case rear drive output shaft assembly

Fig. 32 Removing gear retaining ring. New Process

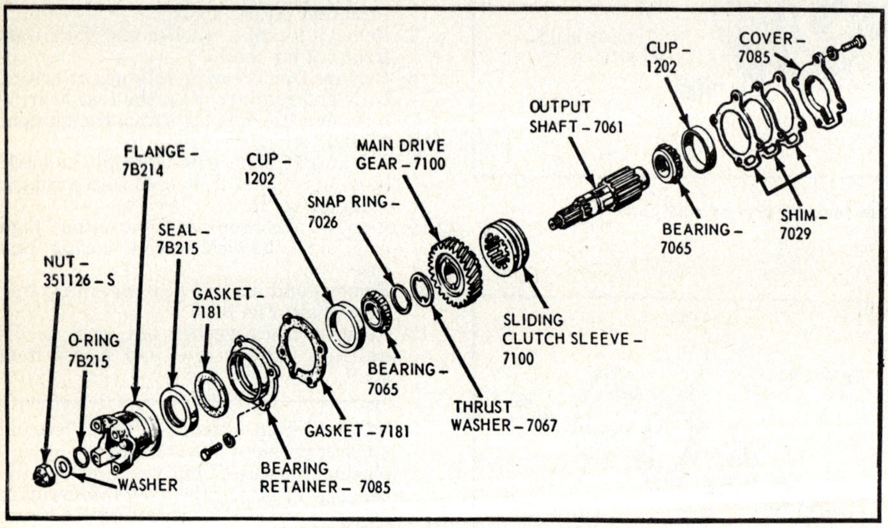

Fig. 29 Dana model 21 transfer case front output shaft assembly

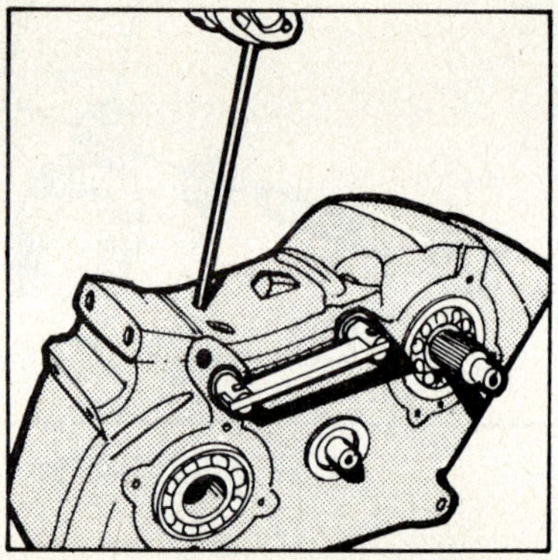

Fig. 33 Removing shift fork pins. New Process

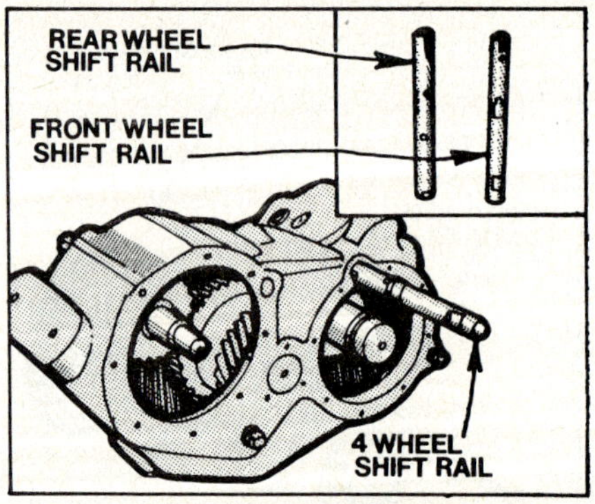

Fig. 34 Shift rail removed. New Process

2. Install the two bearing cones, spacer, shims and idler gear on dummy shaft J-23429 with bore up. Check that gear end play is .001–.002 inch.
3. Install idler gear assembly, large end first, with dummy shaft into case through front output bore.
4. Install idler shaft from large bore side and drive through using a rubber mallet.
5. Install washer and new locknut. Torque locknut to 150 ft. lbs. Check for end play and free rotation.
6. Install idler shaft cover and gasket. Torque bolts to 20 ft. lbs. Flat on cover must be adjacent to front output shaft rear cover.
7. Press the two rail seals, with metal lip outward, into case, then install interlock pins through large bore or power takeoff opening.
8. Install front output drive shift rail into case from back, slotted end first, with poppet notches up.

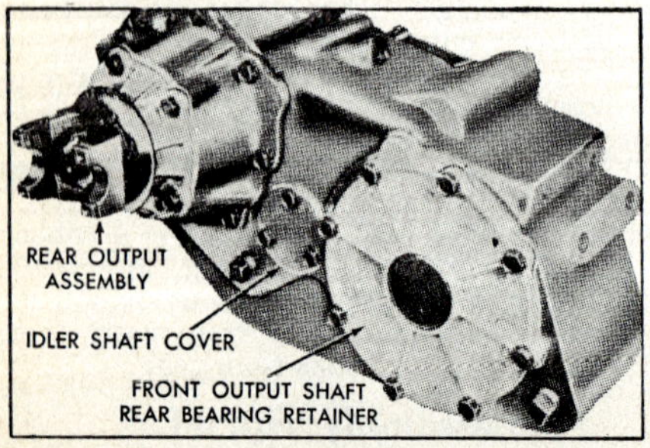

Fig. 30 Rear view of transfer case. New Process

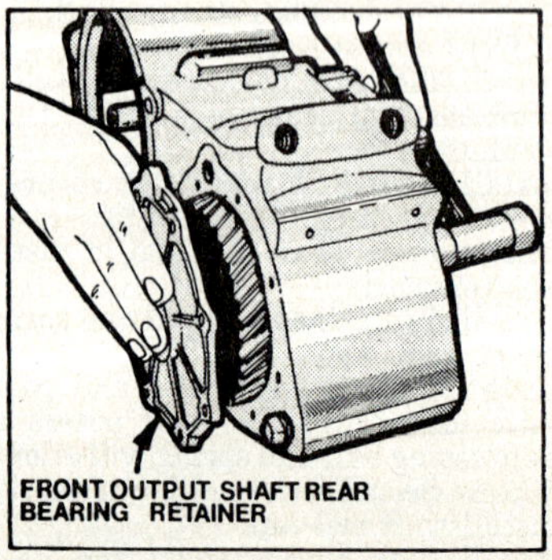

Fig. 31 Removing front output shaft rear bearing retainer. New Process

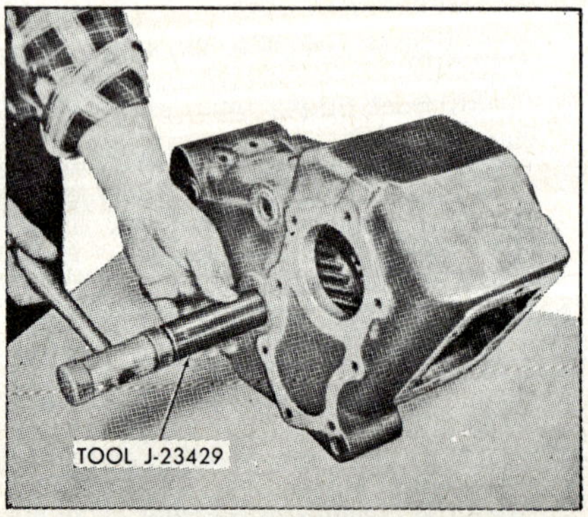

Fig. 35 Removing idler gear shaft. New Process

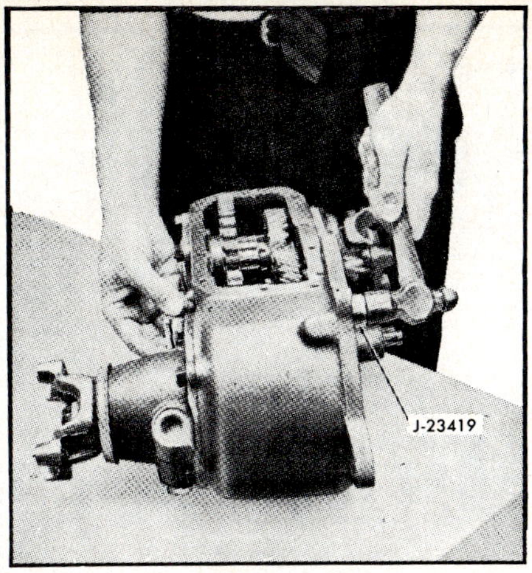

Fig. 37 Removing intermediate gear shaft. Dana Model 20

9. Install shift fork, with long end inward, into rail, then push rail through to neutral position.
10. Install input shaft bearing and shaft into case, then install range rail into case from front with poppet notches up.
11. Install sliding clutch onto fork and place over input shaft in case, then push range rail through to neutral position.
12. Install new lock pins through holes at top of case and drive them into the forks.

NOTE: When installing range rail lock pin, tip case on power takeoff opening.

13. Install two rows of needle bearings (32 each), separated by a spacer, in the front low output gear and apply an adequate amount of grease to hold in place.
14. Secure front output shaft in a soft-jawed vise, spline end down.
15. Install front low gear over shaft with clutch gear facing down, then install thrust washer pin, thrust washer and a new snap ring using tool J-23432. The snap ring opening should be opposite the pin.
16. Install front wheel high gear and washer in case.
17. Install synchronizer in shift fork, then place fork and rail in front wheel drive (4-H) position with clutch teeth meshed with teeth of front wheel high gear.
18. Line up washer, high gear and synchronizer with bearing bore, then install front output shaft and low gear assembly through high gear assembly.
19. Install new seal in bearing retainer using tool J-22836, then install front output bearing and retainer in case.
20. Thoroughly clean and grease rollers in front output rear bearing retainer and install onto case with one gasket. Apply suitable sealant to bolts, then install them and torque to 30 ft. lbs.
21. Install front output yoke, washer and locknut. Torque locknut to 1500 ft. lbs.
22. Install two rows of needle bearings (32 each), separated by a spacer, into low output gear and apply an adequate amount of grease to hold in place.
23. Install thrust washer, with tang down, onto rear output shaft in clutch gear groove.
24. Install output low gear onto shaft with clutch teeth facing down, then install thrust washer over gear with tab pointing up and away from gear.
25. Install washer pin and large thrust washer over shaft and pin. Rotate washer until tab fits into slot approximately 90° away from pin, then install snap ring using tools J-23423 and J-23423-1. Check that end play is .002–.027 inch.
26. Grease pilot bore or rear output shaft and install the 15 needle bearings, then the thrust washer and new snap ring in bore.
27. Clean, grease and install new bearing in retainer housing using tool J-23431.
28. Install housing onto output shaft assembly, then install spacer, speedometer gear and bearing.
29. Install rear bearing retainer seal using tool J-21359 or J-22834-2.
30. Install bearing retainer assembly with one or two gaskets. Torque bolts to 30 ft. lbs.
31. Install yoke, washer and locknut on output shaft, then place range rail in high and install output shaft and retainer assembly on case. Torque housing bolts to 30 ft. lbs.
32. Install power takeoff cover and gasket. Torque bolts to 15 ft. lbs.
33. Install and seal cup plugs at rail pin holes.
34. Install drain and filler plugs and torque to 30 ft. lbs.
35. Install shift rail cross link, clevis pins and lock pins.

DANA MODEL 20

Disassembly

1. Remove cover from bottom of case.
2. Remove poppet hole retainer cap screws from the outside of the front bearing retainer.
3. Remove intermediate gear shaft lock plate at rear side of transfer case.
4. Using dummy shaft, drive intermediate shaft out of rear of case, Fig. 37. Remove intermediate gear and thrust washers from case.
5. Remove complete rear output shaft sub assembly, Fig. 38.
6. Remove front output shaft yoke.
7. Loosen dog set screw from 2-wheel drive shaft fork. Rotate shift rail to pre-load poppet ball and pull rail out of the housing. Remove shift fork and clutch gear from case. Remove poppet ball and spring from retainer.
8. Remove front output shaft rear cover and shims.
9. Remove front output shaft bearing retainer and gasket.
10. Tap on front end of front output shaft removing shaft from rear of case, Fig. 39. Remove gears and washers as they clear shaft. Remove rear bearing cup. Remove front bearing cap as required, using suitable puller.
11. Press rear bearing from shaft, Fig. 40.
12. Loosen dog set screw from the 4-wheel drive shift fork. Swing fork and gear toward cover opening and lift out gear. Rotate shift rail to pre-load poppet ball and pull rail out of retainer. Remove fork as it clears shift rail and remove poppet ball and spring from retainer.
13. Remove shift rail seals with a suitable puller.
14. Remove front and rear drive bearing retainer seals.

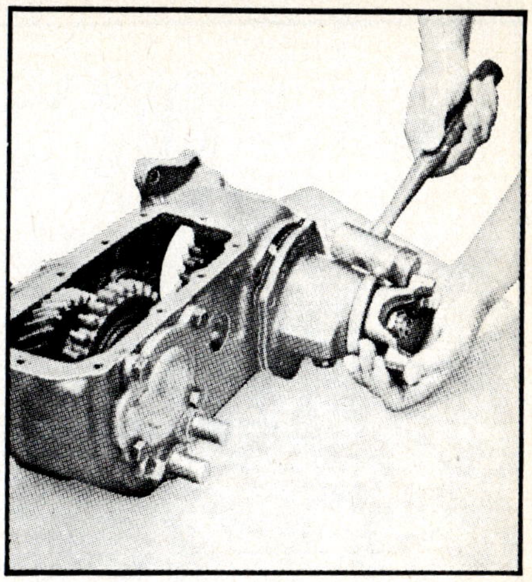

Fig. 38 Removing rear output shaft sub-assembly. Dana Model 20

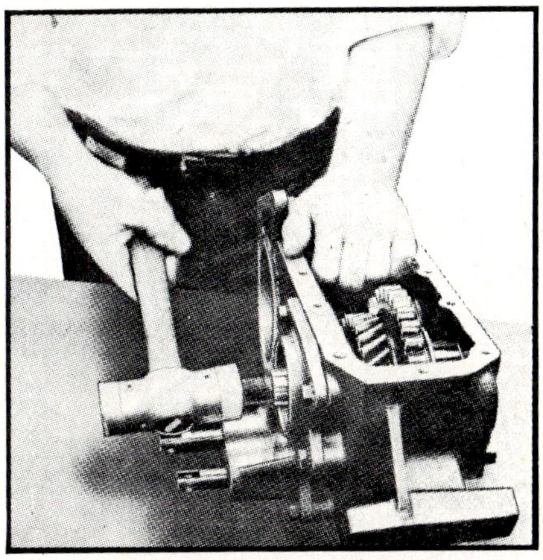

Fig. 39 Removing front output shaft. Dana Model 20

Assembly, Fig. 41

The procedure for reassembly is largely the reverse of the disassembly. However, observe the following:

1. When installing front wheel drive shift fork, set screw boss should face front of case.
2. Rear wheel drive clutch gear shift collar should face rear of case.
3. When installing shift rail poppet balls and springs, use long (yellow) spring on rear wheel drive shift rail and short (red) spring on front wheel shift rail.
4. Torque front output shaft yoke lock nut to 200–250 ft.-lbs.

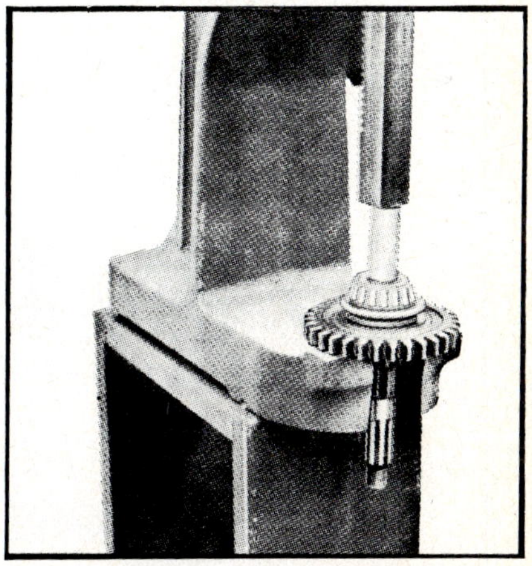

Fig. 40 Removing front output shaft rear bearing. Dana Model 20

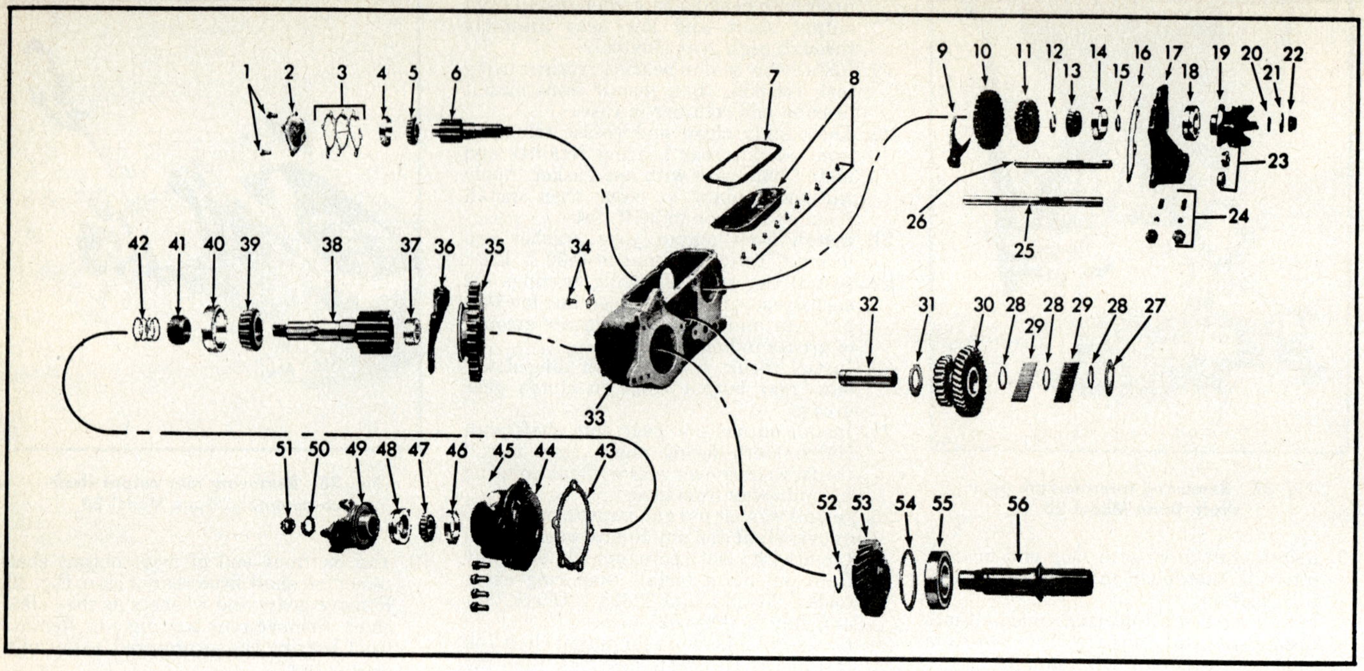
1
2
3
4
5
6
7
8
9
10
11
12
13
14
15
16
17
18
19
20
21
22
23
24
25
26
27
28
29
30
31
32
33
34
35
36
37
38
39
40
41
42
43
44
45
46
47
48
49
50
51
52
53
54
55
56

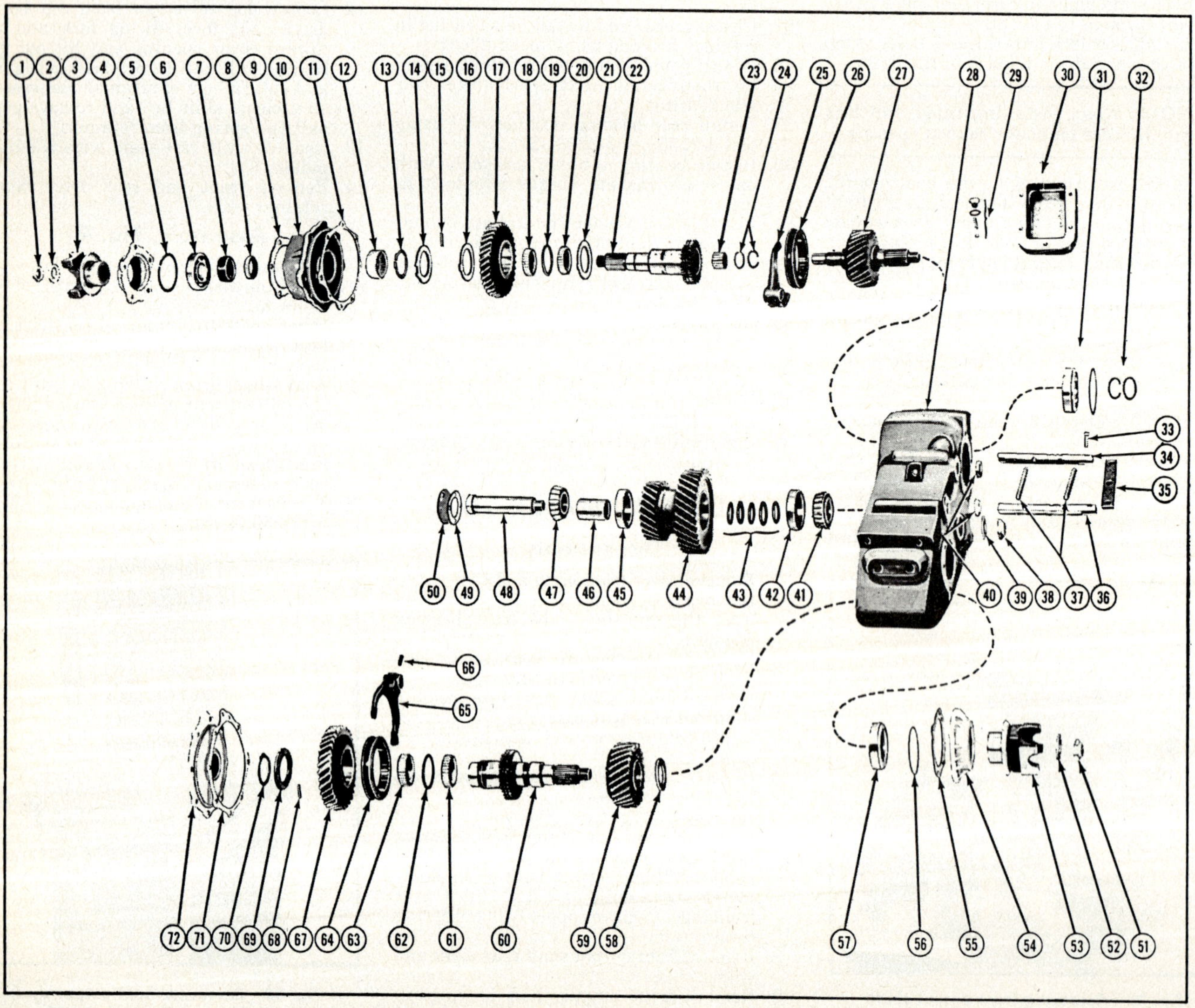
1
2
3
4
5
6
7
8
9
10
11
12
13
14
15
16
17
18
19
20
21
22
23
24
25
26
27
28
29
30
31
32
33
34
35
36
37
38
39
40
41
42
43
44
45
46
47
48
49
50
51
52
53
54
55
56
57
58
59
60
61
62
63
64
65
66
67
68
69
70
71
72

1. Cover Bolts (4)
2. Front Output Shaft Rear Cover
3. Front Output Rear Cover Shims
4. Front Output Rear Bearing Cup
5. Front Output Rear Bearing
6. Front Output Shaft
7. Bottom Cover Gasket
8. Bottom Cover and Bolts
9. Front Wheel Drive Shift Fork
10. Front Output Sliding Gear
11. Front Output Gear
12. Thrust Washer
13. Front Output Bearing
14. Front Output Bearing Cup
15. Spacer
16. Front Output Bearing Retainer Gasket
17. Front Output Bearing Retainer
18. Seal
19. Output Yoke Assembly
20. "O" Ring
21. Washer
22. Locknut
23. Shifter Rail Seals
24. Poppet Ball, Spring and Cap Screw
25. Rear Wheel Drive Shift Rail
26. Front Wheel Drive Shift Rail
27. Tanged Thrust Washer
28. Spacer
29. Needle Bearings
30. Intermediate Gears
31. Tanged Thrust Washer
32. Intermediate Shaft
33. Transfer Case
34. Intermediate Shift Lockplate and Bolt
35. Rear Wheel Sliding Gear
36. Rear Wheel Drive Shift Fork
37. Pilot Bearing
38. Rear Output Shaft
39. Rear Output Front Bearing
40. Rear Output Front Bearing Cup
41. Speedometer Drive Gear
42. Rear Output Shaft Shims
43. Rear Retainer Gasket
44. Rear Output Bearing Retainer
45. Rear Retainer Bolts (10)
46. Rear Output Rear Bearing Cup
47. Rear Output Rear Bearing
48. Rear Output Retainer Seal
49. Rear Output Yoke Assembly
50. Spacer
51. Locknut
52. Snap Ring (Input Shaft)
53. Direct Drive Gear
54. Snap Ring (Part of Adapter Assembly)
55. Bearing (Part of Adapter Assembly)
56. Input Shaft

Fig. 36 Exploded view of Dana Model 20 transfer case

1. Rear Output Shaft Locknut
2. Washer
3. Yoke
4. Bearing Retainer and Seal Assembly
5. Snap Ring
6. Bearing
7. Speedometer Gear
8. Spacer
9. Gasket
10. Housing
11. Gasket
12. Bearing
13. Snap Ring
14. Thrust Washer
15. Thrust Washer Lock Pin
16. Thrust Washer (Tanged)
17. Low Speed Gear
18. Needle Bearings
19. Spacer
20. Needle Bearings
21. Tanged Washer
22. Rear Output Shaft
23. Needle Bearings
24. Washer and Retainer
25. Shift Fork
26. Sliding Clutch
27. Input Shaft
28. Transfer Case
29. Poppet Plug, Spring and Ball
30. P.T.O. Gasket and Cover
31. Input Shaft Bearing and Snap Ring
32. Snap Ring and Rubber 'O' Ring
33. Shift Link Clevis Pin
34. Range Shift Rail
35. Shift Rail Connector Link
36. Front Wheel Drive Shift Rail
37. Interlock Pins
38. Rear Idler Lock Nut
39. Washer
40. Shift Rail Seals
41. Idler Shaft Bearing
42. Bearing Cup
43. Shims
44. Idler Gear
45. Bearing Cup
46. Spacer
47. Idler Shaft Bearing
48. Idler Shaft
49. Cover Gasket
50. Rear Cover
51. Front Output Shaft Locknut
52. Washer
53. Yoke
54. Bearing Retainer and Seal
55. Gasket
56. Snap Ring
57. Front Bearing
58. Thrust Washer
59. Front Wheel High Gear
60. Front Output Shaft
61. Needle Bearings
62. Spacer
63. Needle Bearing
64. Sliding Clutch Gear
65. Shift Fork
66. Roll Pin
67. Front Output Low Gear
68. Thrust Washer Lock Pin
69. Thrust Washer
70. Snap Ring
71. Rear Cover Gasket
72. Rear Cover and Bearing

Fig. 41 Exploded view of New Process Model 205 transfer case

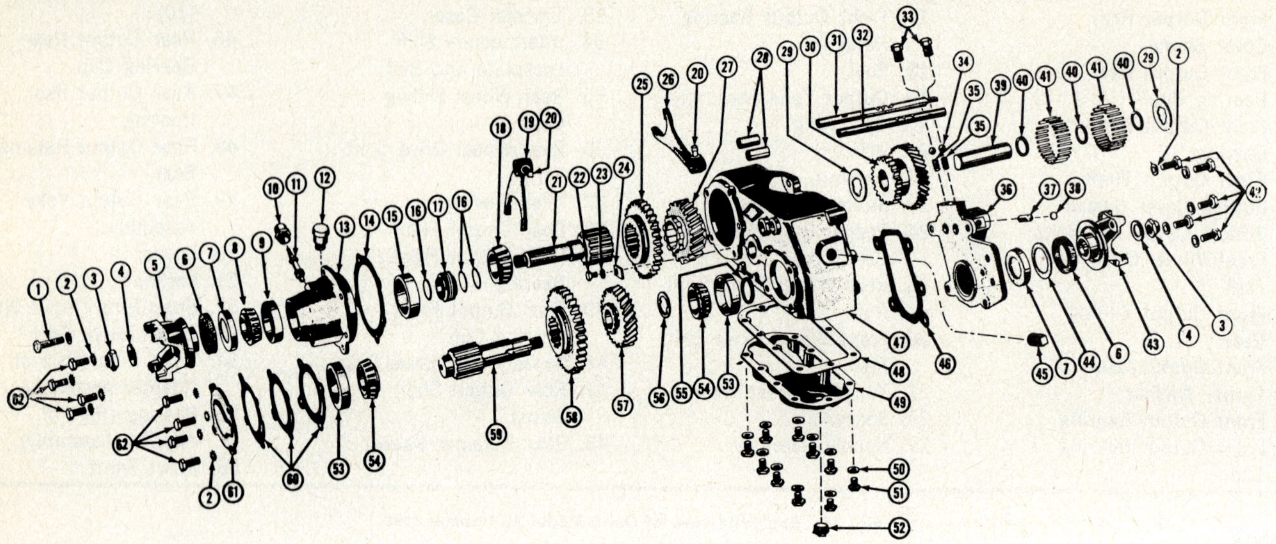

1—Long Bolt
2—Lock Washer
3—Nut
4—Flat Washer
5—Rear Propeller Shaft Yoke
6—Felt Seal
7—Yoke Oil Seal
8—Cone and Roller
9—Bearing Cup
10—Speedometer Driven Gear Sleeve
11—Speedometer Driven Gear
12—Breather
13—Rear Bearing Cap
14—Rear Bearing Cap Gasket
15—Bearing Cup
16—Shims
17—Speedometer Drive Gear
18—Cone and Roller
19—Outer Shift Fork
20—Shift Fork Bolt
21—Rear Output Shaft
22—Bolt
23—Lock Washer
24—Lock Plate
25—Mainshaft Sliding Gear
26—Inner Shift Fork
27—Mainshaft Gear
28—Shift Rod Caps
29—Thrust Washer
30—Intermediate Gear
31—Inner Shift Rail
32—Outer Shift Rail
33—Poppet Plug
34—Poppet Ball
35—Poppet Spring
36—Shift Rail Housing
37—Shift Rail Interlock
38—Plug
39—Intermediate Shaft
40—Bearing Spacer
41—Needle Bearings
42—Bolt
43—Front Propeller Shaft Yoke
44—Oil Seal Gasket
45—Oil Fill Plug
46—Shift Rail Housing Gasket
47—Case
48—Bottom Cover Gasket
49—Bottom Cover
50—Lock Washer
51—Bolt
52—Oil Drain Plug
53—Bearing Cup
54—Cone and Roller
55—Thrust Washer
56—Thrust Washer
57—Front Output Shaft Gear
58—Front Output Shaft Sliding Gear
59—Front Output Shaft
60—Shims
61—Cover Plate
62—Bolt

Fig. 43 Spicer dual range Model 20 transfer case. Exploded view

5. Intermediate shaft is a press fit into case front bore. Align lock plate slot in shaft with bolt hole before installing shaft in front bore.
6. Torque rear output shaft yoke lock nut to 200–250 ft.-lbs.

SPICER MODELS 20 & 21

Disassembly, Figs. 43 & 44

1. Remove rear bearing cap bolts and rear bearing cap assembly.
2. Remove bottom cover and lock plate.
3. Install a dummy shaft, tool W-280 and drive intermediate shaft out from rear of case, then remove intermediate gear through bottom of case.

NOTE: Tool W-280 should be centered in intermediate gear assembly to prevent interference with thrust washers.

4. Remove front output shaft nut, washer and yoke, Fig. 45. Remove oil seal with a suitable puller.
5. Remove rear cover and with a mallet, tap front output shaft rearward to drive rear bearing cup from case.
6. On Model 20, cut safety wire and remove inner shift fork bolt. Tap inner shift rod to the rear and remove shift rod cap, then drive inner shift rod out from front of case. Remove main gear and inner shift fork.
7. On all models remove shift rod housing from case.
8. Remove front bearing cone from front output shaft using a wedge forcing it between bearing and front output shaft gear. Remove snap ring, then the output shaft through rear of case. Remove output shaft gear, thrust washer and bearing roller assembly through bottom of case.
9. On Model 20, remove output shaft sliding gear from outer shift fork. On Model 21, remove output sliding clutch gear from front wheel drive shift fork. On all models, remove fork.

NOTE: Model 21 has a shift fork stop between the shift fork and rear of case.

Reassembly

1. Place outer or front wheel drive shift rod partially into case and install outer or front wheel drive shift fork on rod, aligning shift rod fork bolt hole with countersunk hole in rod. Install and torque shift fork bolt to 12 to 15 ft. lbs. and safety wire bolt.
2. On Model 20, place front output shaft sliding gear in shift fork with gear facing front of case. On Model 21, install sliding clutch gear with bevel facing front of case.
3. Install rear cone and roller onto front output shaft, position output shaft gear in case and install output shaft. Install front cone, roller and front and rear bearing cups.
4. Install shift rod housing and on Model 21, hold shift rod interlock ball down when housing is placed over shift rod.
5. Install rear cover shim set and cover. Check output shaft bearing adjustment using a dial indicator. Bearing adjustment should be .002–.005 inch. If not,

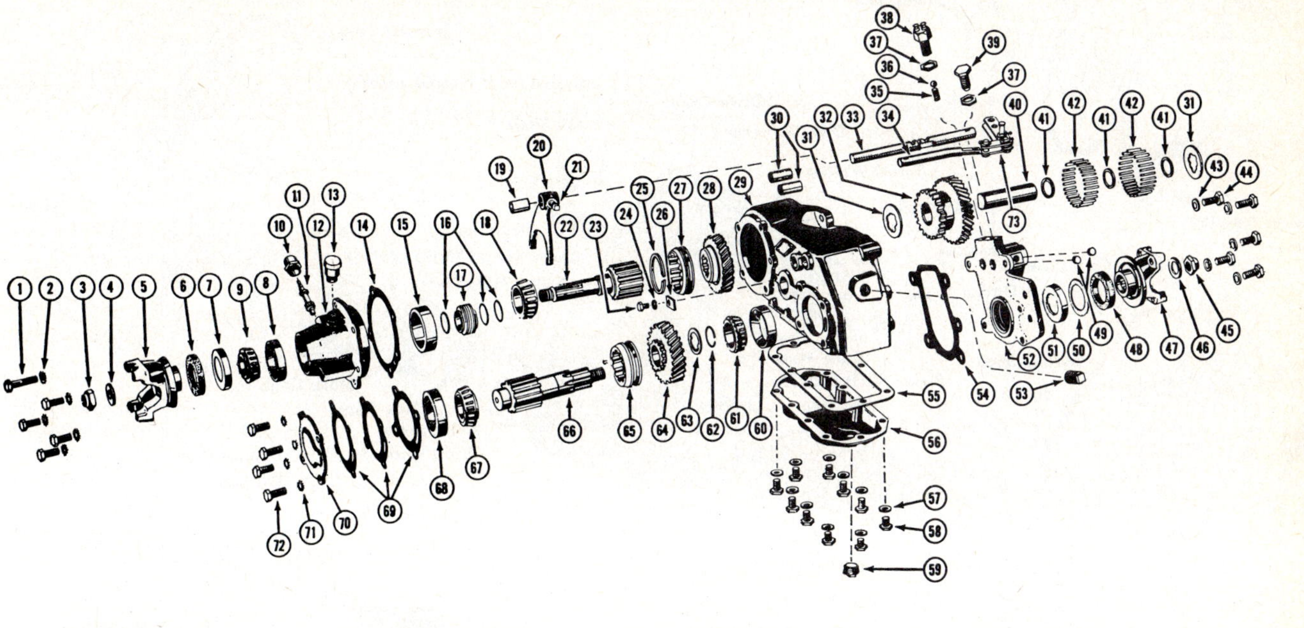

1—Bolt
2—Lockwasher
3—Nut
4—Flat Washer
5—Rear Propeller Shaft Yoke
6—Felt Seal
7—Yoke Oil Seal
8—Bearing Cup
9—Cone and Roller
10—Speedometer Driven Gear Sleeve
11—Speedometer Driven Gear
12—Rear Bearing Cap
13—Breather
14—Rear Bearing Cap Gasket
15—Bearing Cup
16—Shims
17—Speedometer Drive Gear
18—Cone and Roller
19—Shift Rod Stop
20—Front-Wheel-Drive Shift Fork
21—Shift Rod Bolt
22—Rear Output Shaft
23—Bolt
24—Lockwasher
25—Snap Ring
26—Lock Plate
27—Stationary Clutch Gear
28—Mainshaft Gear
29—Case
30—Shift Rod Caps
31—Thrust Washer
32—Intermediate Gear
33—Front-Wheel-Drive Shift Rod
34—Dummy Shift Rod
35—Spring
36—Shift Rod Interlock Ball
37—Lock Nut
38—Indicator Light Switch
39—Bolt
40—Intermediate Shaft
41—Bearing Spacer
42—Needle Bearings
43—Lockwasher
44—Bolt
45—Nut
46—Flat Washer
47—Front Propeller Shaft Yoke
48—Felt Seal
49—Plug
50—Oil Seal Gasket
51—Yoke Oil Seal
52—Front Bearing Cap
53—Oil Fill Plug
54—Shift Housing Gasket
55—Bottom Cover Gasket
56—Bottom Cover
57—Lockwasher
58—Bolt
59—Oil Drain Plug
60—Bearing Cup
61—Cone and Roller
62—Snap Ring
63—Thrust Washer
64—Front Output Shaft Gear
65—Sliding Clutch Gear
66—Front Output Shaft
67—Cone and Roller
68—Bearing Cup
69—Shims
70—Cover Plate
71—Lockwasher
72—Bolt
73—Shift Linkage

Fig. 44 Spicer single range Model 21 transfer case. Exploded view

change rear cover shims to obtain specified clearance. Shims are available in thicknesses of .003", .010" and .031".

6. On Model 20, position outer shift rod to allow shift rod interlocks to enter detents in rod when inner shift rod is installed in shift rod housing. Start inner shift rod into housing and install inner shift fork into case.
7. On all models, install mainshaft gear on fork with gear facing front of case and push shift rod into case and through fork, aligning countersunk hole in rod with shift fork bolt hole. Install and torque shift rod bolt to 12–15 ft. lbs. and safety wire bolt.
8. Install rear thrust washer and hold in place by starting intermediate shaft into case. Install front thrust washer and hold in place with grease. Align tang on thrust washers with grooves in case.
9. Install intermediate gear into case and drive intermediate shaft into gear, in turn forcing dummy shaft out of case. Install intermediate shaft lock plate and torque bolt to 12–15 ft. lbs.
10. Install rear bearing cap assembly and torque bolts to 28–32 ft. lbs.
11. Install shift rod cups approximately 3/8 inch into case.
12. Install lower cover gasket and cover, torque bolts to 12–15 ft. lbs.
13. Install front and rear yoke oil seals, oil seal gaskets and felt oil seals.
14. Install front and rear propeller shaft yokes and torque nuts to 225–250 ft. lbs.

WARNER QUADRA-TRAC

Case Disassembly

1. Remove front and rear output shaft yokes.
2. If unit is not equipped with reduction unit, remove power take-off cover and the sealing ring from transfer case rear cover, Fig. 46.
3. Remove cover from transfer case. Position cover with drive chain downward and place a 2×4×6 inch block of wood under sprocket.
4. If not equipped with reduction unit, expand snap ring securing drive hub and sleeve to drive sprocket rear splines and remove hub and sleeve.
5. If equipped with reduction unit, remove pinion cage from drive sprocket rear splines.
6. Remove cover from drive sprocket and differential, slide drive sprocket toward differential and remove chain.

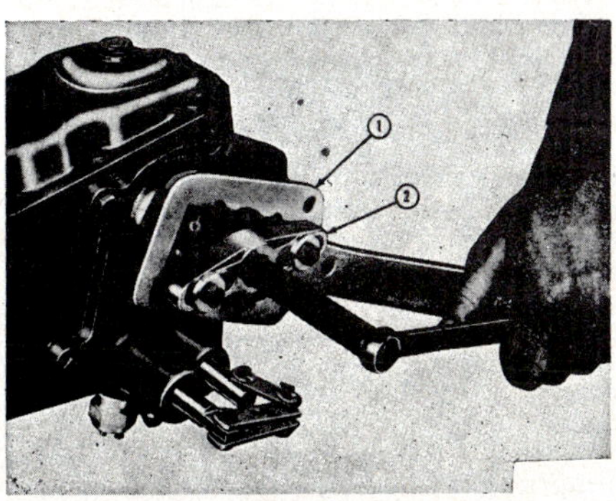

①Tool C-3281. ②Tool W-172.

Fig. 45 Removing universal joint flange. Spicer Models 20 & 21 transfer cases

Sub-Assembly Service

Differential Disassembly

1. Apply identification markings on case sprocket and end caps to ensure proper assembly.
2. Remove screws attaching front end cap to case sprocket, then remove end cap, thrust washers, preload springs, brake cone and side gear from case sprocket. Repeat for rear end cap, Fig. 46.
3. Remove pinion shaft lock pin and drive out the pinion shaft from case sprocket with a suitable drift.

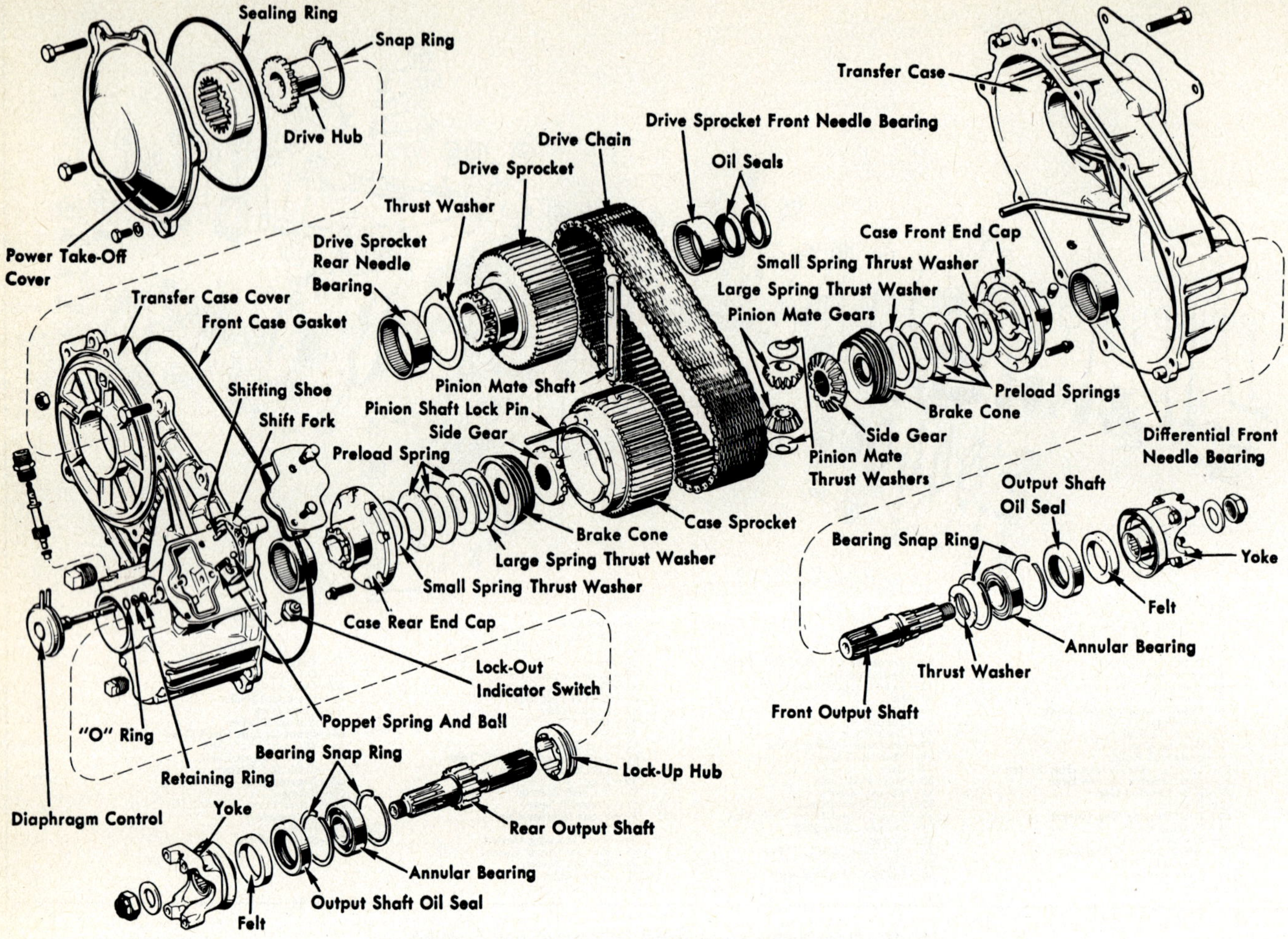

Fig. 46 Warner Quadra-Trac transfer case. Disassembled

Inspection

Clutch, thrust surfaces and shaft bores may be polished. Small smooth score marks and machine marks are acceptable. Case sprocket teeth will show a polished wear pattern, whereas pinion and side gears will have a "rough machined" appearance. The pinion shaft should be straight and fit snugly into case sprocket. The thrust washers should be flat and preload springs should be dished approximately 3/32 inch. Clean all parts of differential before assembly.

Differential Reassembly

During reassembly, bearing and thrust surfaces must be coated with lubricant, Jeep P/N 8123004.

1. Install pinion shaft sprocket approximately 3 inches into case and place pinion thrust washers and gears on shaft in proper sequence. Fully drive shaft into sprocket, aligning lock pin holes in shaft and case.
2. Slide pinion gears apart and mesh either front or rear side gear with pinion gears. Install brake cone over side gear followed by large thrust washer, preload springs, small thrust washer and end cap, Fig. 46.
3. Install pinion shaft lock pin and mesh remaining side gear with pinion gears. Install brake cone over side gear followed by large thrust washer, preload springs, small thrust washer and end cap.
4. Install front and rear output shafts into differential and rotate shafts so both shafts are aligned having entered the brake cone splines and tighten end cap screws.

Bearing Replacement

Refer to Figs. 47 thru 50 when replacing differential or drive sprocket needle bearings. Install drive sprocket rear needle bearings with removal tools, Fig. 50, with tool #W361-1 installed into the case bore.

Front and rear output shaft annular bearings are retained by selected thickness snap rings. The outer snap ring is available in the following thickness ranges: .060"–.063", .064"–.066", .067"–.069" and .070"–.072". The inner snap ring is available only in a thickness of .060–.063 inch. These bearings can be removed by hand or with a brass drift, if necessary. When removing the rear output shaft bearing, remove speedometer gear, then the outer snap ring and bearing. When installing outer snap ring, install proper snap ring to provide .001–.003 inch bearing end play.

Diaphragm Control, Shift Fork & Lock-Up Hub Disassembly

1. Remove vent cover and sealing ring.
2. Remove shift fork retaining rings from diaphragm control rod.
3. Remove dowel pin and insert a small magnet into dowel pin hole, Fig. 51. Remove diaphragm control from case cover, detent ball and spring.

NOTE: The diaphragm control is held by the detent ball and spring.

---

4. Remove shift fork, shifting shoes and lock-up hub.

Diaphragm Control, Shift Fork & Lock-Up Hub Reassembly

1. Lubricate and install shifting shoes and lock-up hub into shift fork.
2. Install fork and hub assembly into case cover with care not to separate lock-up hub from fork, Fig. 52.
3. Slide diaphragm control into case cover, past the fork but not further than the detent ball hole. Insert detent spring and ball into hole, using a 1/4 inch punch, depress detent ball and slide diaphragm control rod past ball.
4. Install shift fork retaining rings and diaphragm control retaining dowel pin.
5. Install sealing ring and vent cover.

Case Reassembly

1. Place drive sprocket on a 2×4×6 inch block of wood with differential assembly

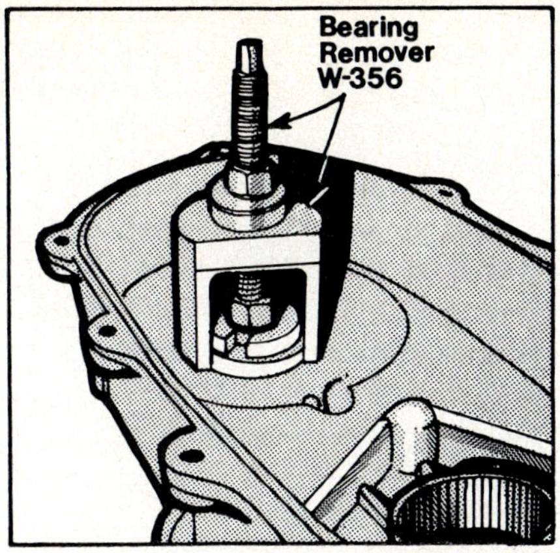

Fig. 47 Warner Quadra-Trac differential drive sprocket needle bearngs removal (typical)

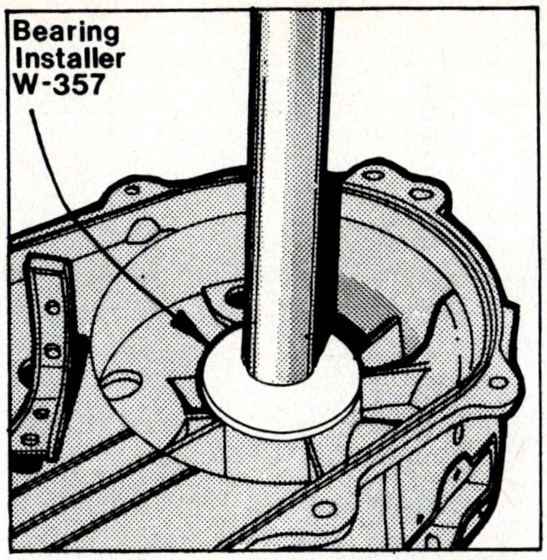

Fig. 48 Warner Quadra-Trac differential needle bearings installation (typical)

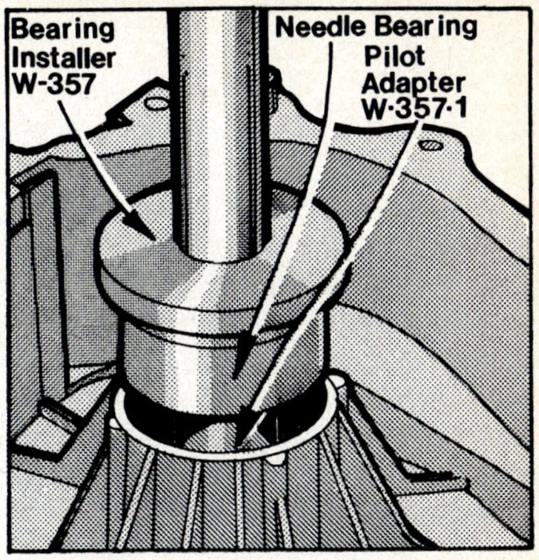

Fig. 49 Warner Quadra-Trac drive sprocket front needle bearing installation

about 2 inches from sprocket with front end down. Place drive chain around sprocket and differential, ensure chain is engaged with sprocket and differential teeth. Remove slack from chain, Fig. 53.

2. Install rear output shaft into differential and shift lock-up hub to rear of case cover.
3. With grease holding drive sprocket thrust washer in position on cover. Align and position cover on drive sprocket and differential. Rotate output shaft to align with lock-up nub.
4. Assemble drive hub, drive sleeve and snap ring. If not equipped with reduction unit, install drive hub and sleeve assembly on drive sprocket, ensuring snap ring seats properly.
5. If equipped with reduction unit, ensure oil baffle is positioned properly and install pinion cage and snap ring.
6. Install front output shaft, thrust washer and front case gasket onto case.
7. Insert oil tube into case bore located at front output shaft bearing boss, then insert a 5/16 × 6 inch rod into oil tube.
8. Position case onto drive sprocket and differential assembly. Using the 5/16 × 6 inch rod to align oil tube with case cover. Install and torque case to cover screws to 15–25 foot lbs.

NOTE: Rotate drive sleeve to ensure drive sprocket thrust washer is correctly positioned. Sleeve should turn without binding.

9. Install power take-off sealing ring and

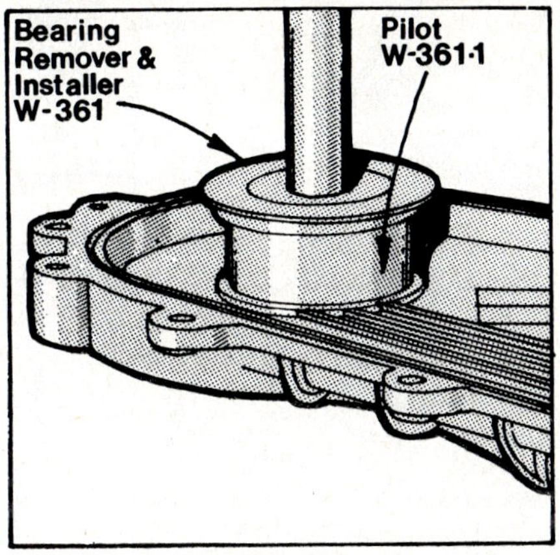

Fig. 50 Warner Quadra-Trac drive sprocket rear needle bearing removal

cover.

10. Install speedometer gear onto rear output shaft and the front and rear output shaft oil seals into case bores.
11. Install front yoke and torque nut to 90–150 ft. lbs.

WARNER QUADRA-TRAC REDUCTION UNIT

Disassembly

1. Remove power take-off cover and gasket.
2. Remove reduction mainshaft snap ring and spacer, pull reduction mainshaft and sun gear assembly forward out of housing. Remove mainshaft needle bearings, Fig. 54.
3. Remove ring gear, reduction collar plate, pinion cage lock plate, shift collar hub and the reduction collar hub assembly from housing.
4. Using a mallet, tap shift collar hub from pinion cage lock plate. Remove pinion cage lock plate and needle bearing, ring gear, reduction collar plate and shift collar hub.
5. Remove reduction collar hub and needle bearings from shift collar hub.
6. Remove snap rings retaining reduction collar plate hub and ring gear to reduction collar plate.
7. Remove needle bearing and direct drive sleeve from reduction shift collar.
8. With the control lever move reduction shift collar to center detent, move collar away from shift fork and disengage fork. Move fork rearward to direct drive detent and collar toward fork, align outer teeth on collar and inner teeth on reduction holding plate. Move fork and collar forward to reduction detent and remove reduction shift collar.

Fig. 51 Warner Quadra-Trac diaphragm control removal

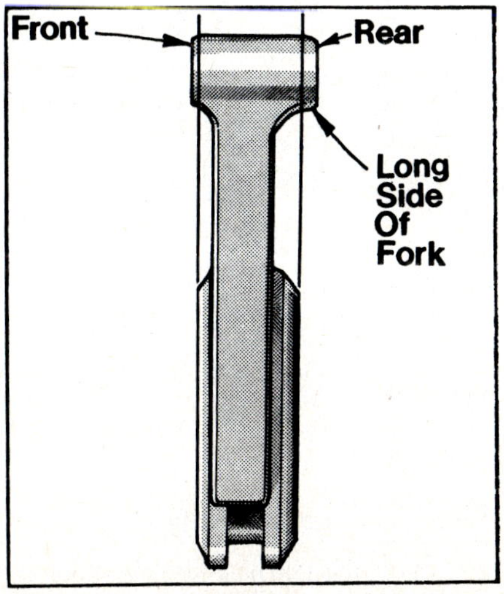

Fig. 52 Warner Quadra-Trac lock-up hub & shift fork assembly

Fig. 53 Warner Quadra-Trac drive chain installation

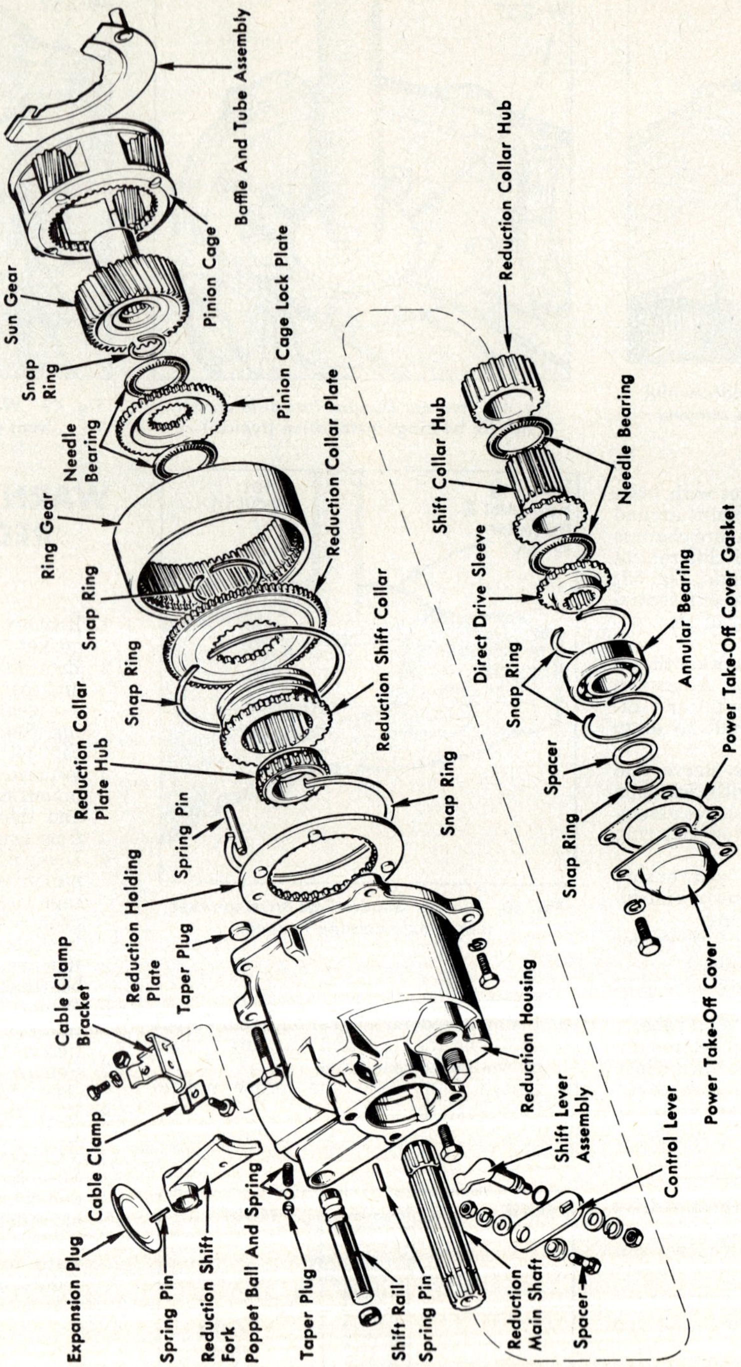

Fig. 54 Warner Quadra-Trac reduction unit. Disassembled

9. Remove rear snap ring and annular bearing.
10. Remove shift fork locating spring pin, large expansion plug and shift rail taper plugs.
11. Remove control lever from shift lever.
12. Drive spring pin out from shift fork and rail with a 3/16 inch punch, slide rail out of fork and remove fork.
13. Remove shift rail poppet ball, drive poppet taper plug into rail bore and remove plug and poppet spring.
14. Remove shift lever retaining pin and lever assembly.
15. Remove snap ring and reduction holding plate.

Reassembly

1. Install reduction holding plate with locating pin indexed in reduction housing shift fork locating spring pin holes in plate aligned with holes in the housing, Fig. 55. Install holding plate snap ring with tabs forward.
2. Install shift lever assembly into housing with lever and rearward. Place "O" ring in groove on shift lever shaft and move shift lever assembly inward to a position allowing installation of shift lever locating taper pin.
3. Install shift rail with grooved end first into rail rear bore and rotate rail so flat side is adjacent to poppet spring. Slide rail so shift fork is meshed with shift

lever assembly and rail, move rail through fork so end of rail is even with edge of poppet bore.
4. Place poppet ball on end of spring and with a spring pin, depress ball and slide shift rail over ball as far as possible. Remove pin, slide rail to first detent position and rotate rail flat side facing shift lever assembly and spring bore aligning with spring pin bore in shift fork. Align spring pin holes in shift rail and fork, install spring pin flush with fork.
5. Install shift rail taper plugs, poppet bore taper plug and rail cover expansion plug.
6. Install shift fork locating spring pin and control lever.
7. Place shift fork in center detent. Mesh reduction shift collar outer teeth with reduction holding plate inner teeth. Move fork to rear detent and shift collar away from fork, aligning groove in collar with shift fork. Move collar toward fork, engaging collar groove with fork.
8. Install direct drive sleeve into reduction shift collar with needle bearing surface and pointed ends of outer teeth forward. Adequately lubricate needle bearing and install against direct drive sleeve.
9. Install needle bearing and reduction collar hub onto shift collar hub. Install ring gear, reduction collar plate and hub assembly onto shift collar hub.

NOTE: A needle bearing is not used between reduction collar plate hub and reduction collar hub.

10. Install needle bearing onto shift collar hub and between reduction collar plate hub. Install pinion cage lock plate on shift collar hub so lock plate is snug against needle bearing. Install assembly into housing, rotating ring gear or pinion cage lock plate to align splines. Place needle bearing on shift collar hub and pinion cage lock plate.
11. Install reduction mainshaft and sun gear assembly into shift collar hub through the direct drive sleeve and annular bearing, rotating assembly to align splines. Use a drift to tap mainshaft fully rearward.

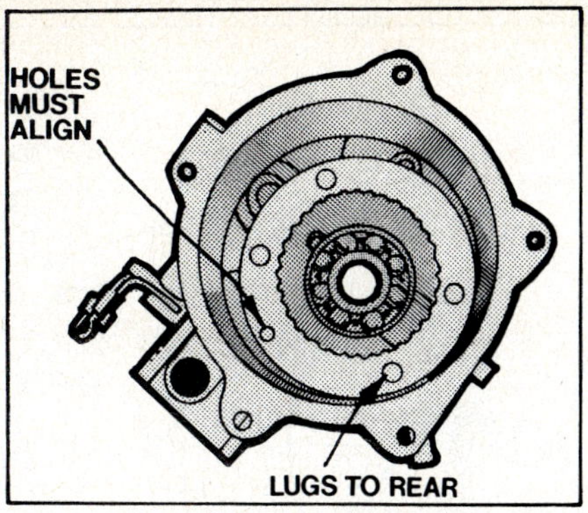

Fig. 55 Warner Quadra-Trac reaction unit holding plate installation

12. Install rear spacer and snap ring. Snap rings are available in various thickness ranges of .089″-.091″, .092″-.094″, .095″-.097″, .099″-.101″ and .103″-.105″. Install appropriate snap ring to provide .004-.009 inch spacer clearance.
13. Install power take-off gasket and cover and torque cover screws to 15-25 ft. lbs.

NEW PROCESS MODEL 203

Case Disassembly

1. Remove front output rear cover located on side of range selecter and drain lubricant, Figs. 56 and 57.
2. Remove rear output shaft retaining nut, washer and flange, Fig. 58.

NOTE: Tap dust cover shield rearward on shaft to gain clearance to remove bolts and install tool.

3. Remove front bearing retainer, Fig. 59, then using a hoist, position assembly on blocks in an upright position.
4. Remove rear output shaft retaining bolts then disengage assembly from transfer case and slide carrier unit from shaft.

NOTE: Install a suitable size band type hose clamp on input shaft to prevent losing bearings when removing input shaft assembly.

5. Raise shift rail, drive out pin retaining shift fork to rail and remove shift rail poppet ball plug, gasket spring and ball from case, Figs. 57 and 59.

NOTE: Use a small magnet to remove ball from case.

6. Push shift rail down, lift up on lockout clutch and remove shift fork from clutch assembly.
7. Remove front output shaft rear bearing retainer bolts and tap on front of shaft or carefully pry retainer from shaft and discard gasket. Recover any roller bearings which may fall from rear cover.

NOTE: If replacing bearing, press bearing out. Press new bearing in until it is flush with output bearing rear cover.

8. Pry out output shaft front bearing then disengage front output shaft from chain and remove shaft from transfer case, Fig. 59.
9. Remove intermediate chain housing bolts and remove intermediate housing from range box.
10. Remove chain from intermediate housing, then remove lockout clutch, drive gear and input shaft from range box.

NOTE: Install a suitable size band type hose clamp on end of input shaft to prevent losing roller bearings which may fall out if clutch assembly is pulled off input shaft.

11. Pull up on shift rail and disconnect rail from link then remove input shaft assembly from range box.

Inspection

With the transfer case totally disassembled into sub-assemblies. These sub-assemblies

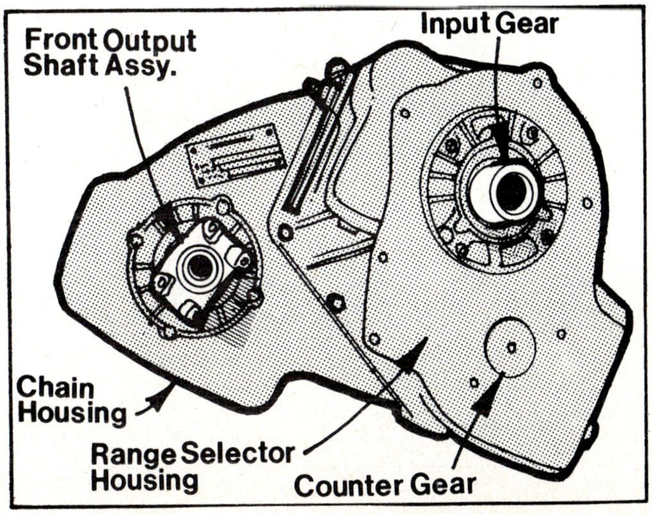

Fig. 56 New Process 203 transfer case. Front view

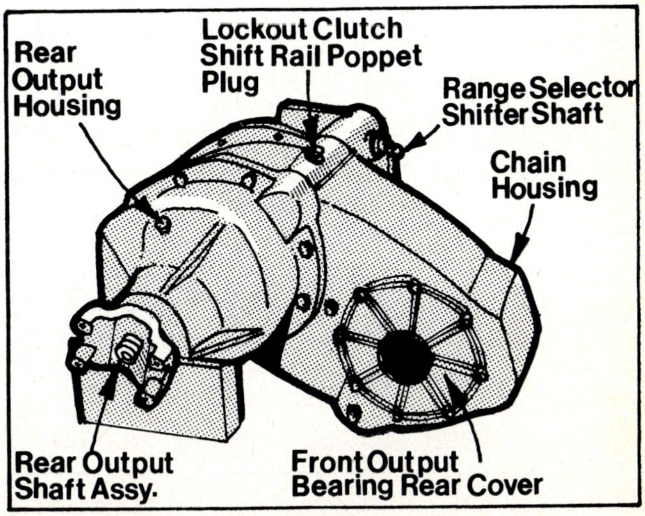

Fig. 57 New Process 203 transfer case. Rear view

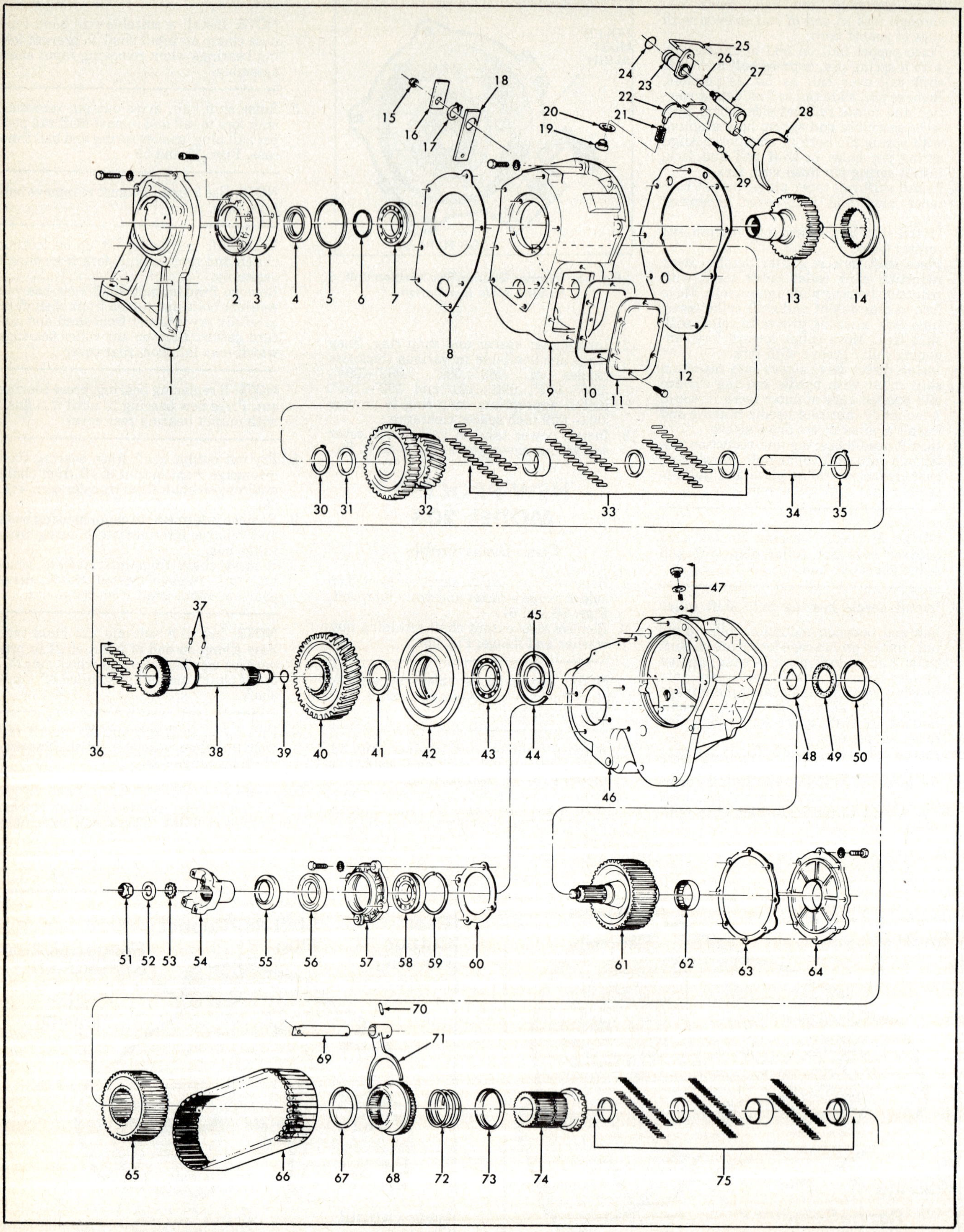
1
2
3
4
5
6
7
8
9
10
11
12
13
14
15
16
17
18
19
20
21
22
23
24
25
26
27
28
29
30
31
32
33
34
35
36
37
38
39
40
41
42
43
44
45
46
47
48
49
50
51
52
53
54
55
56
57
58
59
60
61
62
63
64
65
66
67
68
69
70
71
72
73
74
75

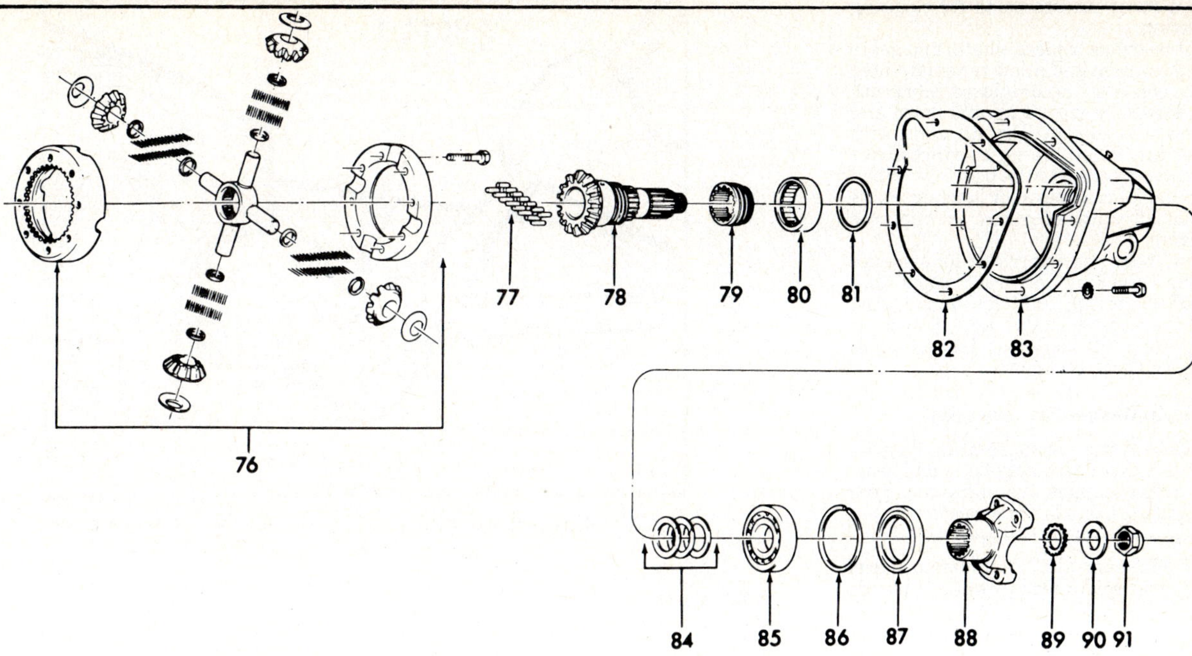

1. Adapter
2. Input Gear Bearing Retainer
3. Input Gear Bearing Retainer Gaske
4. Input Gear Bearing Retainer Seals
5. Bearing Outer Ring
6. Bearing to Shaft Retaining Ring
7. Input Gear Bearing
8. Adapter to Selector Housing Gasket
9. Range Selector Housing (Range Box)
10. P.T.O. Cover Gasket
11. P.T.O. Cover
12. Selector Housing to Chain Housing Gasket
13. Main Drive Input Gear
14. Range Selector Sliding Clutch
15. Shift Lever Lock Nut
16. Range Selector Shift Lever
17. Shift Lever Retaining Ring
18. Lockout Shift Lever
19. Detent Plate Spring Plug
20. Detent Plate Spring Plug Gasket
21. Detent Plate Spring
22. Detent Plate
23. Lockout Shifter Shaft
24. "O" Ring Seal
25. Lockout Shaft Connector Link
26. "O" Ring Seal
27. Range Selector Shifter Shaft
28. Range Selector Shift Fork
29. Detent Plate Pivot Pin
30. Thrust Washer
31. Spacer (short)
32. Range Selector Counter Gear
33. Countergear Roller Bearings and Spacers (72 Bearings Req'd.)
34. Countergear Shaft
35. Thrust Washer
36. Input Shaft Roller Bearings (15 Req'd.)
37. Thrust Washer Pins (2 Req'd.)
38. Input Shaft
39. "O" Ring Seal
40. Low Speed Gear and Bushing
41. Thrust Washer
42. Input Shaft Bearing Retainer
43. Input Shaft Bearing
44. Input Shaft Bearing Retaining Ring (Large)
45. Input Shaft Bearing Retaining Ring
46. Chain Drive Housing
47. Lockout Shift Rail Poppet Plug, Gasket, Spring and Ball.
48. Thrust Washer
49. Lubricating Thrust Washer
50. Retaining Ring
51. Flange Lock Nut
52. Washer
53. Seal
54. Front Output Yoke
55. Dust Shield
56. Front Output Shaft Bearing Retainer Seal
57. Front Output Shaft Bearing Retainer
58. Front Output Shaft Bearing
59. Bearing Outer Ring
60. Bearing Retainer Gasket
61. Front Output Shaft
62. Front Output Shaft Rear Bearing
63. Front Output Rear Bearing Retainer Cover Gasket
64. Front Output Rear Bearing Retainer
65. Drive Shaft Sprocket
66. Drive Chain
67. Retaining Ring
68. Sliding Lock Clutch
69. Lockout Shift Rail
70. Shift Fork Retaining Pin
71. Lockout Shift Fork
72. Lockout Clutch Spring
73. Spring Washer Cup
74. Front Side Gear
75. Front Side Gear Bearing and Spaces (123 Bearings Req'd.)
76. Differential Carrier Assembly (132 Bearings Req'd.)
77. Rear Output Shaft Roller Bearings (15 Req'd.)
78. Rear Output Shaft
79. Speedometer Drive Gear
80. Rear Output Shaft Front Roller Bearing
81. Oil Pump "O" Ring Seal
82. Rear Output Housing Gasket
83. Rear Output Housing
84. Shim Pack
85. Rear Output Rear Bearing
86. Bearing Retainer
87. Rear Output Shaft Seal
88. Rear Output Flange
89. Rear Output Shaft Rubber Seal
90. Washer
91. Flange Nut

Fig. 59 New process 203 transfer case legend & exploded view of differential carrier assembly & rear output housing (first design)

should be further disassembled for cleaning and inspection.

Place all bearings, rollers, shafts and gears in cleaning solution and allow to remain long enough to loosen all accumulated lubricant. Bearings should be moved up and down and turned slowly in solvent. Dry all parts with compressed air. When drying bearings direct air so as to avoid spinning the bearings. When cleaning transfer case, cover or housing, make certain that all traces of gasket are removed from surfaces.

Inspect all bearings and rollers for evidence of chipping, cracking, or worn spots that would render bearing unfit for further use. Bearings are non-adjustable and if worn or damaged, must be replaced. Inspect shaft splines, drive chain and gears.

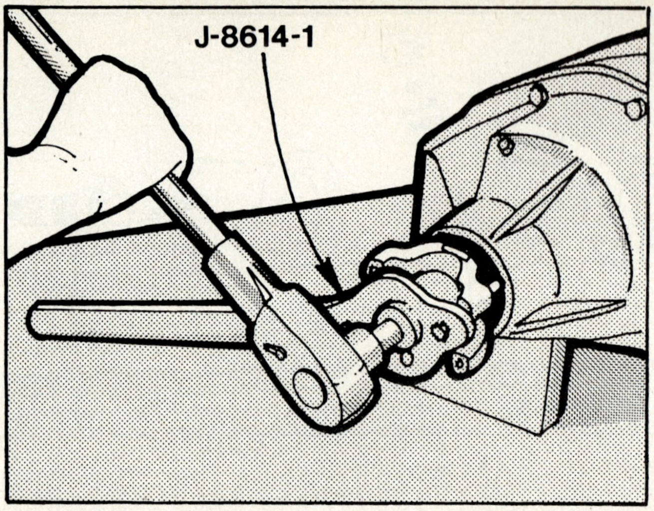

Fig. 58 Removing rear output flange. New Process 203 transfer case

Sub-Assembly Service

Differential Carrier Disassemble, Fig. 59

1. Remove carrier assembly bolts and separate carrier sections, then lift pinion gear and spider assembly from carrier.

NOTE: Make certain that undercut side of pinion gear spider faces toward front side gear.

2. Remove pinion thrust washers, pinion roller washers, pinion gears and roller bearings from spider unit.
3. Clean, inspect and replace parts as necessary.

Differential Carrier Reassembly

1. Using petroleum jelly, insert roller bearings into pinion gears.

NOTE: A total of 132 bearings are required with 33 bearings in each pinion.

2. Install pinion roller washer, pinion gear, roller washer and thrust washer on each leg of spider.
3. Install spider assembly in front half of carrier with undercut surface of spider thrust spacer facing toward gear teeth.
4. Align marks on carrier sections and position carrier halves together then install retaining bolts and torque to 45 ft. lbs.

Lockout Clutch Disassembly, Fig. 59

1. Remove front side gear from input shaft assembly and remove thrust washer, roller bearings (123 total) and spacers from front side gear bore. Note position of spacers to facilitate reassembly.
2. Remove the snap ring retaining the drive sprocket to the clutch assembly, then slide the drive sprocket out from the front side gear and remove the lower snap ring.
3. Remove sliding gear, spring and spring cup washer from front side gear.
4. Clean, inspect and replace parts as necessary.

Lockout Clutch Reassembly

1. Install spring cup washer, spring and sliding clutch gear on front side gear.
2. Install snap ring retaining sliding clutch to front side gear.
3. Using petroleum jelly, load (123) roller bearings and spacers in front side gear.
4. Install thrust washer in gear end of the front side gear, then slide sprocket onto clutch splines and install retaining ring.

Input Shaft Disassembly, Fig. 59

1. Slide thrust washer and spacer from shaft, then remove snap ring retaining input bearing retainer assembly to shaft. Remove bearing retainer assembly from shaft.
2. Support low speed gear and tap shaft out from the gear and thrust washer.

NOTE: Observe thrust washer pin locations in shaft.

3. Remove snap ring retaining input bearing in bearing retainer, then tap bearing out from the retainer.
4. Remove pilot roller bearings (15 total) and O-ring from input shaft.
5. Clean, inspect and replace parts as necessary.

Input Shaft Reassembly

1. Press bearing into retainer with ball loading slots toward concave side of retainer.
2. Install large snap ring retaining bearing in the retainer.

NOTE: Use size A, B, C or D snap ring to provide tightest fit.

3. Install low speed gear on shaft with clutch end facing toward gear end of shaft and position thrust washers onto shaft, aligning slot in washer with pin in shaft. Slide or tap washer into place.
4. Position input bearing retainer onto shaft and install snap ring, holding bearing in position on shaft.

NOTE: Use size A, B, C, or D snap ring to provide tightest fit.

5. Slide spacer and thrust washer onto shaft and align spacer with locator pin.
6. Using heavy grease, install (15) roller bearings and O-ring on end of shaft.

Removing Shifter Assembly, Fig. 59

1. Remove poppet plate spring, plug and gasket, then disengage sliding clutch gear from input gear. Remove clutch fork and sliding gear from case.
2. Remove shift lever retaining nut, upper shift lever, shift lever snap ring and lower lever.
3. Push shifter shaft assembly down and remove lockout clutch connector link.

NOTE: Long end of connector link engages poppet plate.

4. Remove shifter shaft assembly and separate the inner and outer shifter shafts. Remove and discard O-rings.
5. Inspect poppet plate for damage. If necessary drive pivot shaft from case and remove poppet plate and spring from case.

Removing Input Gear Assembly, Fig. 59

1. Remove input gear bearing and remove large snap ring from bearing.
2. Tap input gear and bearing out from the case then remove snap ring retaining input shaft bearing with the shaft and remove bearing from input gear.

Removing Countergear Assembly, Fig. 59

1. Using tool J-24745, remove countershaft from cluster gear and remove cluster gear assembly.

NOTE: Recover roller bearings (72 total) from gear case and shaft.

2. Remove countergear thrust washers.
3. Clean, inspect and replace parts as necessary.

Installing Countergear Assembly

1. Using tool J-24745 and heavy grease, install (72) roller bearings and spacers in countergear bore.
2. Using heavy grease, position countershaft thrust washers in case. Engage tab on washers with slot in case thrust surface.
3. Position countergear assembly in case and install countershaft through front face of range box and into gear assembly. Flat on countershaft face should face forward and must be aligned with case gasket.

Installing Input Gear Assembly

1. Install bearing (without large snap ring) on input gear shaft positioning snap ring groove outward and install new retaining ring on shaft. Position input gear and bearing in housing.

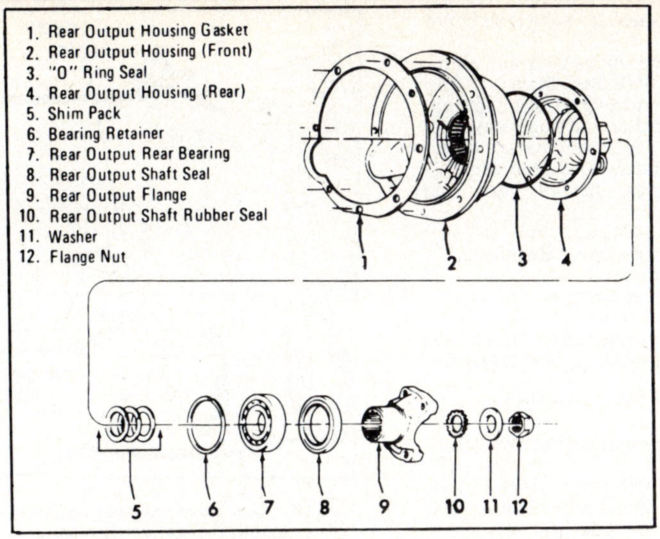

Fig. 59A New Process 203 transfer case rear output housing (second design) exploded view

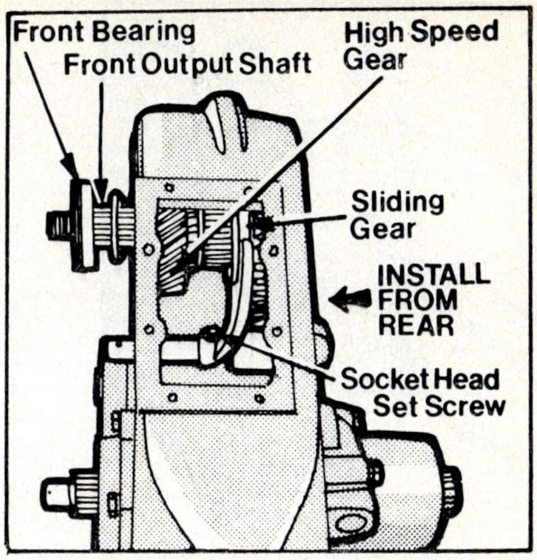

Fig. 60 Replacing front output shaft

NOTE: Use size A, B, C or D snap ring to provide tightest fit.

---

2. Install snap ring onto bearing.
3. Align oil slot in retainer with drain hole in case and install input gear bearing retainer, gasket and bolts. Torque bolts to 30 ft. lbs.

Installing Shifter Shaft Assembly

1. Install poppet plate and pivot pin assembly into housing. Use sealant on pin.
2. Install new O-rings on inner and outer shifter shafts then lubricate O-rings and assemble inner shaft into the outer shaft.
3. Push shifter shafts into housing, engaging long end of locknut clutch connector link to the outer shifter shaft before the shaft assembly bottoms out in housing.
4. Install lower shift lever, retaining ring, upper shift lever and shifter shaft retaining nut.
5. Install shift fork and sliding clutch gear. Push fork up into shifter shaft assembly to engage poppet plate, sliding gear forward onto the input shaft gear.
6. Install poppet plate spring, gasket and plug into top of housing. Check spring engagement with poppet plate.

Input Gear Bearing Replacement, Fig. 59

1. Remove bearing retainer and gasket.
2. Remove snap ring retaining bearing on the shaft. Pry bearing from case and remove it from shaft.

NOTE: Inspect input gear for burrs, scoring, heat discoloration etc. and inspect seal in retainer. Replace as necessary.

---

3. Install bearing and snap ring onto input gear shaft. Position bearing to case and tap into place using a small hammer.

NOTE: Use size A, B, C or D snap ring to provide tightest fit.

---

4. Install new snap ring on shaft and position new gasket and bearing retainer to housing. Install bolts and torque to 30 ft. lbs.

Input Gear Bearing Retainer Seal Replacement

1. Remove bearing retainer bolts, retainer and gasket.
2. Pry seal out of retainer.
3. Install seal using tool J-21359.
4. Install bearing retainer and gasket and torque bolts to 30 ft. lbs.

Rear Output Shaft Housing Disassembly, Fig. 59

1. Remove speedometer driven gear from housing and if not already removed, remove flange nut, washer and flange from shaft.
2. Using a soft hammer, tap on flange end of pinion and remove pinion from carrier. If speedometer drive gear is not on the pinion shaft, reach into carrier and remove it from housing.
3. Using a screwdriver, pry seal out and pry behind open ends of snap ring retaining rear bearing in housing and remove snap ring.
4. Tap rear bearing out, then using a long drift inserted through rear opening drive the front bearing out of housing.

Rear Output Shaft Housing Reassembly

1. Using grease, position seal in bore and press roller bearing into place until it bottoms in housing.

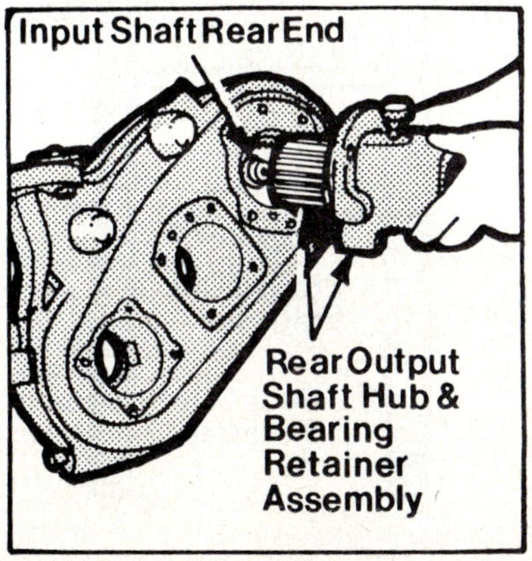

Fig. 61 Replacing rear output shaft & bearing

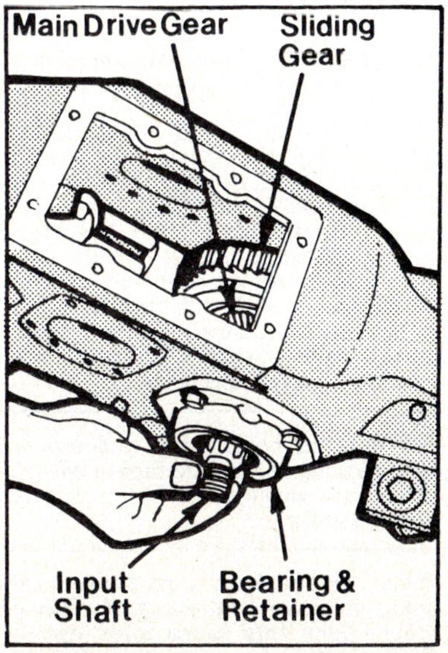

Fig. 62 Replacing input shaft

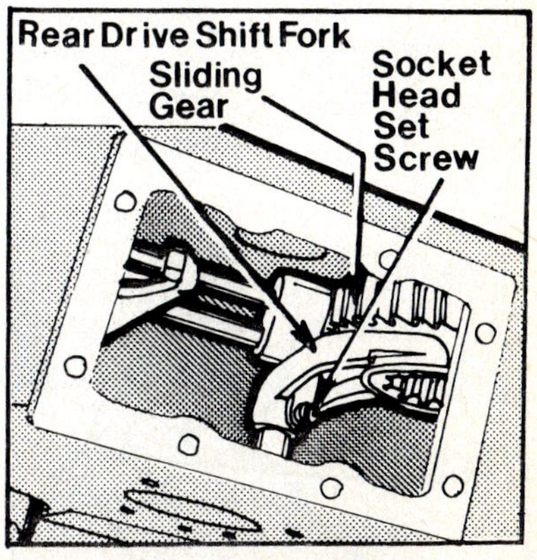

Fig. 63 Rear drive shift fork & sliding gear installation

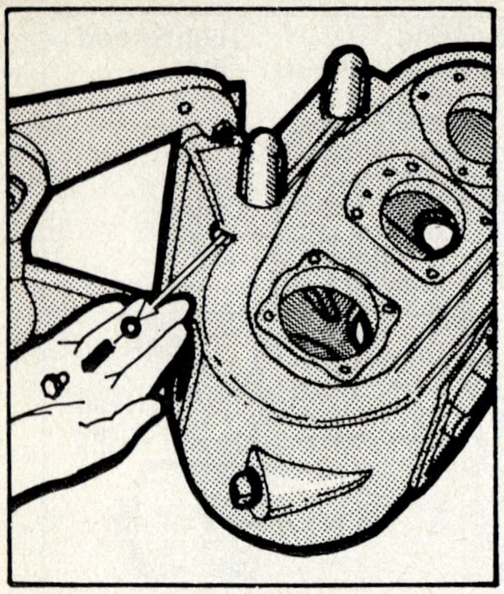

Fig. 64 Front drive shift rail detent ball, spring & plug installation

2. Tap rear bearing into place and install snap ring retaining bearing in case.

NOTE: Use size A, B, C or D snap ring to provide tightest fit.

Torque Specifications

Intermediate Case to Range Box Bolts	30 ft. lbs.
Front Output Bearing Retainer Bolts	30 ft. lbs.
Output Shaft Yoke Nuts	150 ft. lbs.
Front Output Rear Bearing Retainer Bolts	30 ft. lbs.
Differential Assembly Screws	45 ft. lbs.
Rear Output Shaft Housing	30 ft. lbs.
Poppet Ball Retainer Nut	15 ft. lbs.
Power Take Off Cover Bolts	15 ft. lbs.
Front Input Bearing Retainer Bolts	20 ft. lbs.
Filler Plug	25 ft. lbs.

3. Using tool J-22388, drive rear seal into housing until it is about 1/8–3/8 inch below housing face.

Fig. 66 Shaft rail & detent plugs installation

4. Install speedometer drive gear and shims (approximately .050 inch thick), install shaft into carrier through front opening.
5. Install flange, washer and nut. Leave nut loose until shim requirements are determined (approximately .060 inch thick).
6. Install speedometer driven gear into housing.

Front Output Shaft Bearing Retainer Seal, Replacement

1. Pry out seal from retainer.
2. Clean and inspect retainer.
3. Apply sealer to outer diameter of seal and install using tool J-22836 into retainer.

Front Output Shaft Rear Bearing, Replacement

1. Remove rear cover and gasket from transfer case.
2. Press out old bearing and press in new bearing until it is flush with opening. Use a block of wood over bearing.
3. Install cover and gasket and torque bolts to 30 ft. lbs.

Transfer Case Reassembly, Fig. 59

1. Place range box on blocks with input gear side facing down and position range box to transfer case gasket on input housing.
2. Install lockout clutch and drive sprocket assembly on input shaft assembly.

NOTE: Install a 2 inch band type clamp on end of shaft to prevent losing bearings from clutch assembly.

3. Install input shaft, lockout clutch and drive sprocket assembly into range box, aligning tab on bearing retainer with notch in gasket.
4. Connect lockout clutch shift rail to the connector link and position rail in housing bore. Rotate the shifter shaft while lowering shift rail into the housing, to prevent the link and rail from being disconnected.
5. Install chain into housing, positioning chain around outer wall of housing then install chain housing onto range box, engaging the shift rail channel of the housing to the shift rail. Position chain onto input drive sprocket.
6. Install front output sprocket into case engaging drive chain with sprocket. Rotate drive gear to aid in positioning chain onto drive sprocket.
7. Install shift fork onto clutch assembly and shift rail, then push clutch assembly into drive sprocket. Install roll pin retaining shift fork to shift rail.
8. Install front output shaft bearing, retainer, gasket and bolts.
9. Install front output shaft flange, gasket, seal, washer and nut. Install bolts in flange and tap dust shield into place.
10. Install front output shaft rear bearing retainer, gasket and bolts.

NOTE: If rear bearing was removed, place new bearing on outside face of cover and press into cover until bearing is flush with opening.

11. Install front output shaft flange, gasket, seal, washer and retaining nut. Tap dust shield back into place after installing flange bolts.
12. Install front output shaft rear bearing retainer, gasket and retaining bolts.

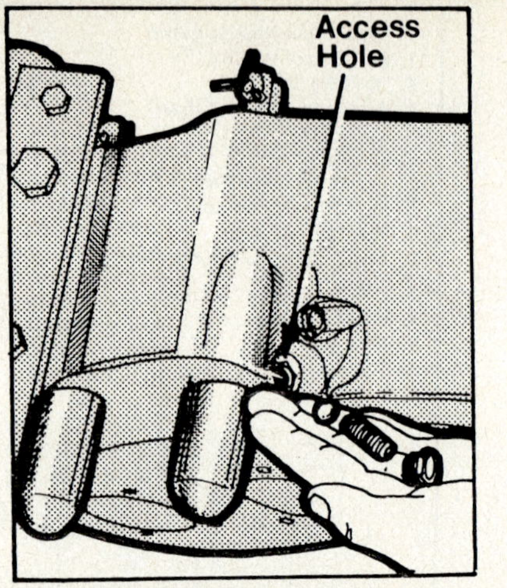

Fig. 65 Rear drive shift rail detent ball spring & plug installation

13. Install differential carrier assembly on input shaft with bolt heads facing rear of shaft.
14. Install rear output housing, gasket and bolts. Load bearings in pinion shaft.
15. Install speedometer gear and shims (about .050 inch) onto the output shaft.
16. Position the rear output of the rear assembly onto the rear output housing of the front assembly. Make sure that O-ring is properly positioned on the front section of the output housing.

NOTE: Make sure that vent is in the upward position.

17. Install flange, washer and retaining nut. Leave nut loose until shim requirements are determined.
18. Check shim requirements as follows:
    a. Make sure that output shaft retaining nut is loose about .060 inch.
    b. Install a dial indicator in such a manner as to contact the end of the output shaft.
    c. Push in rear output shaft to its full travel while rotating front output yoke and holding rear output yoke stationary. Set dial indicator to zero on the

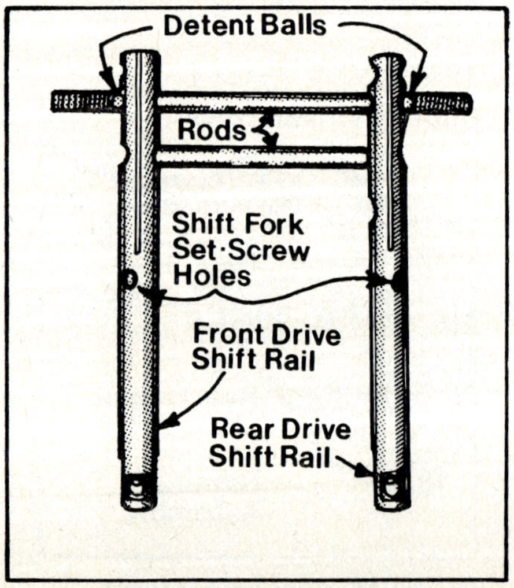

Fig. 67 Shift mechanism

highest reading obtained.

d. Pry out rear output shaft yoke until the maximum reading is obtained on the dial indicator. The allowable end play is .001-.010 inch. Add or remove shims as required to obtain the allowable end play.

e. Check to make sure that end play is not too tight after the nut has been tightened, by holding the rear yoke stationary and rotating the front yoke. If the yoke can be rotated by hand and the "bumps" are felt, the end play is acceptable.

19. Install the speedometer driven gear into the housing.
20. Install lockout clutch shift rail poppet ball, spring and plug.
21. Install poppet plate spring, gasket and plug, if not installed previously.
22. Install shift levers on shifter shafts.
23. Torque all bolts, lock nuts and plugs to specifications listed above.

DANA MODEL 24

Disassembly

1. Remove power take-off cover plate, then the idler shaft bearing retainers.
2. With a mallet, tap idler shaft and bearing rearward until rear bearing can be removed from case.
3. Remove idler shaft, two gears and spacer.
4. Remove idler shaft front bearing from case.
5. Remove flange retaining nuts from front output shaft, rear output shaft and input shaft. Then, remove the flanges and washers.
6. Remove the front and rear bearing retainers from the front output shaft.
7. Tap front output shaft and the rear bearing through the gears and case, Fig. 60, then remove high speed gear from case.
8. Remove the front output shaft front bearing and washer from case.
9. Remove set screw retaining front drive shaft fork to shift rail, then the front output shaft sliding gear from case.
10. If input shaft oil seal replacement is necessary, remove seal with a suitable puller.
11. Remove the rear output shaft bearing retainer, then the speedometer drive assembly from the retainer.
12. With a mallet, tap on the front of the input shaft to loosen the rear output shaft assembly from case, then remove the rear output shaft and bearing retainer assembly, Fig. 61.
13. Tap the input shaft through front bearing, main drive gear, sliding gear and from case, Fig. 62.
14. Remove main drive gear from case, then with a mallet, drive the input shaft front bearing from case.
15. Remove set screw retaining rear drive shift fork to shift rail, Fig. 63.
16. Remove rear output shaft sliding gear from case, Fig. 63.
17. Remove shift rail link from the two shift rails.
18. Remove retaining plug, detent spring and ball engaging the front drive shift rail detent rod, Fig. 64.
19. Remove retaining plug, detent spring and ball engaging the rear drive shift rail detent rod, Fig. 65. Then, remove the front drive detent rod access hole plug.
20. Pull front drive shaft rail to fully outward position.
21. Pull rear drive shift rail outward to permit the two detent rods to slide from case, Fig. 66.
22. Remove front and rear drive shift rails and forks from case.
23. Remove shift rail seals from case with a suitable puller.

Assembly

1. Press new oil seals into place in the case or bearing retainers.
2. If necessary, replace bearings on input shaft and the front and rear output shafts with a suitable press.
3. Slide front drive shift rail into case and position shift fork on rail.

NOTE: When installing the shift rails, the detents should be positioned as shown in Fig. 67.

4. Insert the two detent rods into case, Fig. 66, then install the rear drive shift fork, and hold the detent rods and shift fork in position as the rear drive shift rail is inserted as far as possible.
5. Pull the front drive shift rail out to the next detent and push the rear drive shift rail fully into position. Then, push the front drive shift rail back into position.
6. Install rear drive shift rail detent ball, spring and retaining plug, then the access hole plug, Fig. 65.
7. Install front drive shift rail detent ball, spring and retaining plug, Fig. 64.
8. Install shift rail link in the two shift rails.
9. Place the rear output shaft sliding gear in the shift fork and secure with set screw to rear drive shift rail, Fig. 63.
10. Install input shaft front bearing and retainer assembly. Coat retainer and attaching bolts with a suitable sealer.
11. Install main drive gear into case and slide input shaft through rear of case, main drive gear and the front bearing and retainer, Fig. 62.
12. Install roller bearing into splined hub of rear output shaft. Then, install the shaft and bearing retainer assembly, ensuring the output shaft is aligned with the input shaft, Fig. 61. Coat the case, bearing retainer and bolts with suitable sealer.
13. Place front output shaft sliding gear in the shift fork and secure with set screw to the front drive shift rail.
14. Hold sliding gear and high speed gear in position, install the front output shaft and rear bearing assembly through the two gear from the rear of case.
15. Install front output shaft rear bearing retainer and gasket. Coat retainer and bolts with a suitable sealer.
16. Install washer and front bearing, front bearing retainer and gasket. Coat retainer and bolts with a suitable sealer.
17. Install flange, washer, flange retaining nuts and cotter key on the input shaft, front output shaft and the rear output shaft.
18. Install the idler shaft gears into case, then the idler shaft and rear bearing assembly from the rear of case. Install rear bearing retainer plate, applying sealer to plate and bolts.
19. Place spacer on front of idler shaft, then the front bearing. Tap bearing into place with a mallet.
20. Install washer, retaining nut and cotter key on front of idler shaft.
21. Install idler shaft front bearing retainer plate. Apply sealer to the plate and bolts.
22. Install power take-off cover plate. Apply sealer to plate and bolts.

NEW PROCESS MODEL 208

Disassembly, Fig 68

1. Remove fill and drain plugs. Drain lubricant from transfer case.
2. Remove front and rear yokes. Discard the yoke seal washers and yoke nuts.
3. Rotate transfer case and position on wood blocks with the front case resting on the wood. If necessary cut "V" notches in the blocks so the front case rests squarely on the blocks.
4. Remove lock mode indicator switch and washer.
5. Remove detent bolt, spring and ball.
6. Mark relationship between rear retainer and case for alignment at reassembly.
7. Remove rear retainer attaching bolts, then the rear retainer attaching bolts, then the rear retainer and pump housing as an assembly. Do not pry retainer from case. If necessary, use a plastic mallet to tap retainer.
8. Remove pump housing from retainer then the pump seal from housing. Discard the seal.
9. Remove speedometer drive gear from mainshaft.
10. Remove oil pump from mainshaft noting position of pump for reassembly. The side of the pump facing toward inside of case is recessed.
11. Remove rear case to front case retaining bolts, then the rear case from front case.

NOTE: To remove rear case, insert a suitable screwdriver into th slots in the case ends and gently pry upward. Do not attempt to wedge the case halves apart at any point on the mating surfaces.

12. Remove front output shaft rear thrust bearing assembly. Note position of bearing and races for reassembly.
13. Remove driven sprocket retaining snap ring.
14. Remove drive sprocket retaining snap ring, then the thurst washer and spacer washer, if equipped.
15. Remove drive and driven sprockets and the drive chain as an assembly. Lift evenly on both sprockets to remove.
16. Remove front output shaft and front thrust bearing assembly.
17. Remove sprocket carrier stop ring and clutch spring, Fig. 69.
18. Remove sliding clutch, mode fork and mode fork spring as an assembly, Fig. 70. Then remove shift rail.
19. Remove sprocket carrier, needle bearing upper retainer, thrust washer and mainshaft needle bearings as an assembly, Fig. 71.
20. Remove mainshaft.
21. Remove annulus gear retaining ring and thrust washer, then the annulus gear and range fork as an assembly. Rotate fork counter-clockwise to disengage fork lug from range sector and lift assembly from case, Fig. 72.
23. Remove mainshaft thrust bearing from input gear, then the input gear, Fig. 74. Lift the gear straight up and out from case.
24. Remove input gear thrust bearing and race, Fig. 75. Note position of bearing and

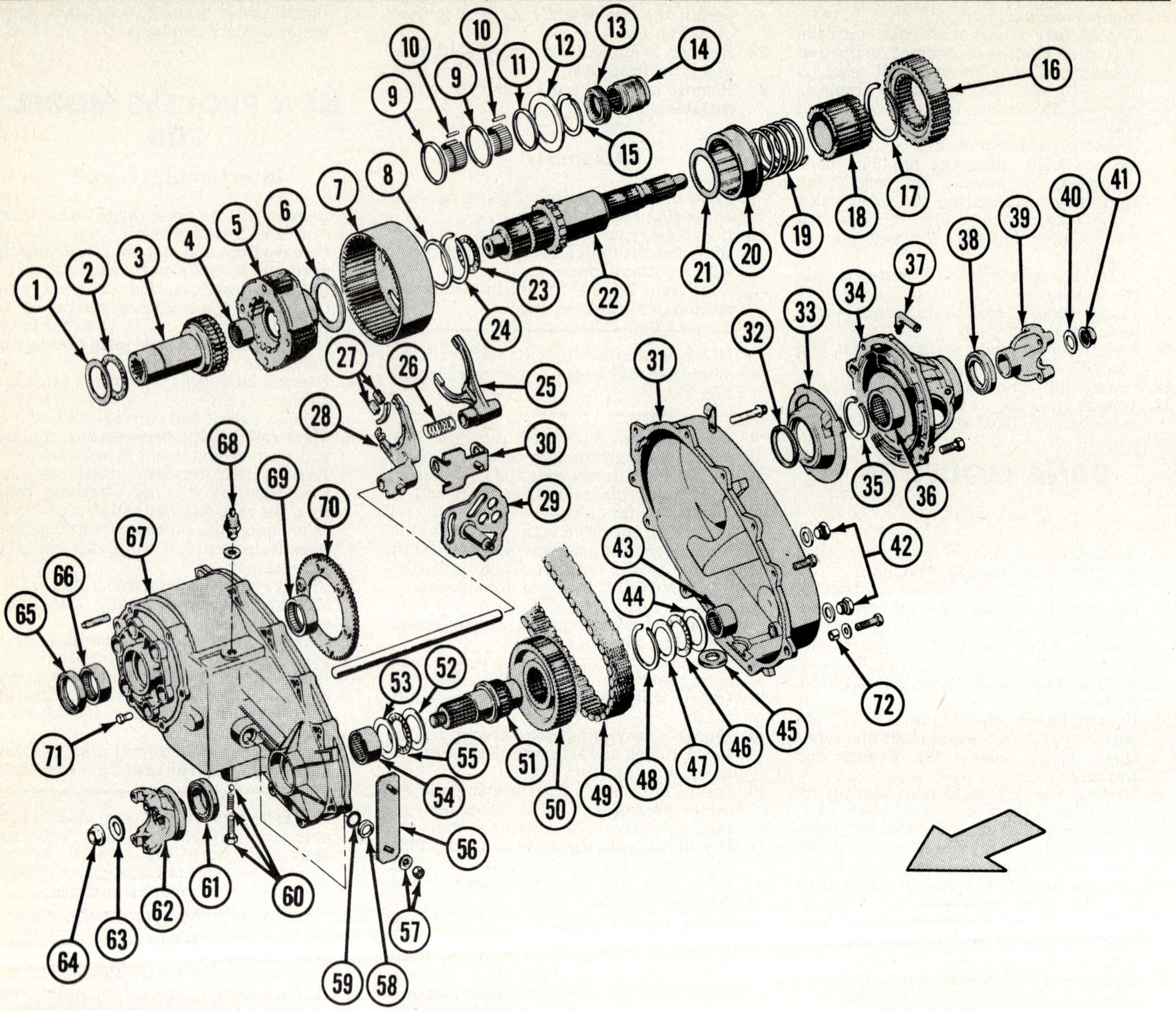

1. INPUT GEAR THRUST WASHER
2. INPUT GEAR THRUST BEARING
3. INPUT GEAR
4. MAINSHAFT PILOT BEARING
5. PLANETARY ASSEMBLY
6. PLANETARY THRUST WASHER
7. ANNULUS GEAR
8. ANNULUS GEAR THRUST WASHER
9. NEEDLE BEARING SPACERS
10. MAINSHAFT NEEDLE BEARINGS (120)
11. NEEDLE BEARING SPACER
12. THRUST WASHER
13. OIL PUMP
14. SPEEDOMETER GEAR
15. DRIVE SPROCKET RETAINING RING
16. DRIVE SPROCKET
17. SPROCKET CARRIER STOP RING
18. SPROCKET CARRIER
19. CLUTCH SPRING
20. SLIDING CLUTCH
21. THRUST WASHER
22. MAINSHAFT
23. MAINSHAFT THRUST BEARING
24. ANNULUS GEAR RETAINING RING
25. MODE FORK
26. MODE FORK SPRING
27. RANGE FORK INSERTS
28. RANGE FORK
29. RANGE SECTOR
30. MODE FORK BRACKET
31. REAR CASE
32. SEAL
33. PUMP HOUSING
34. REAR RETAINER
35. REAR OUTPUT BEARING
36. BEARING SNAP RING
37. VENT TUBE
38. REAR SEAL
39. REAR YOKE
40. YOKE SEAL WASHER
41. YOKE NUT
42. DRAIN AND FILL PLUGS
43. FRONT OUTPUT SHAFT REAR BEARING
44. FRONT OUTPUT SHAFT REAR THRUST BEARING RACE (THICK)
45. CASE MAGNET
46. FRONT OUTPUT SHAFT REAR THRUST BEARING
47. FRONT OUTPUT SHAFT REAR THRUST BEARING RACE (THIN)
48. DRIVEN SPROCKET RETAINING RING
49. DRIVE CHAIN
50. DRIVEN SPROCKET
51. FRONT OUTPUT SHAFT
52. FRONT OUTPUT SHAFT FRONT THRUST BEARING RACE (THIN)
53. FRONT OUTPUT SHAFT FRONT THRUST BEARING RACE (THICK)
54. FRONT OUTPUT SHAFT FRONT BEARING
55. FRONT OUTPUT SHAFT FRONT THRUST BEARING
56. OPERATING LEVER
57. WASHER AND LOCKNUT
58. RANGE SECTOR SHAFT SEAL RETAINER
59. RANGE SECTOR SHAFT SEAL
60. DETENT BALL, SPRING AND RETAINER BOLT
61. FRONT SEAL
62. FRONT YOKE
63. YOKE SEAL WASHER
64. YOKE NUT
65. INPUT GEAR OIL SEAL
66. INPUT GEAR FRONT BEARING
67. FRONT CASE
68. LOCK MODE INDICATOR SWITCH AND WASHER
69. INPUT GEAR REAR BEARING
70. LOCKPLATE
71. LOCKPLATE BOLTS
72. CASE ALIGNMENT DOWELS

Fig. 68 New Process model 208 transfer case, disassembled

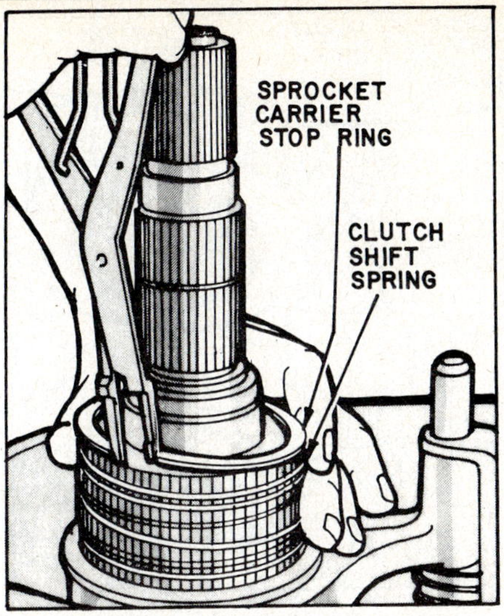

Fig. 69 Sprocket carrier stop ring & clutch spring replacement

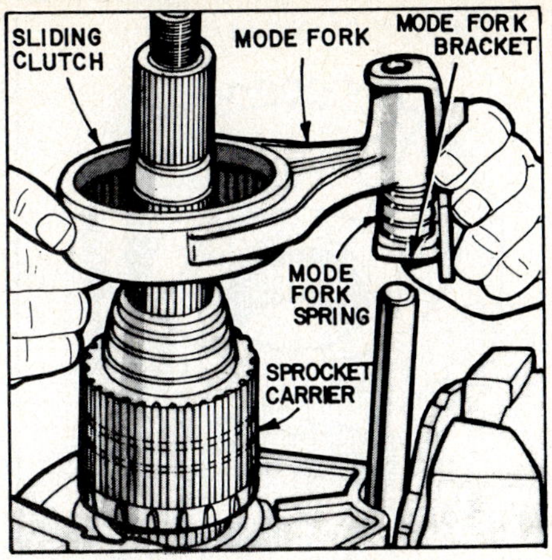

Fig. 70 Mode fork, spring, bracket & sliding clutch replacement

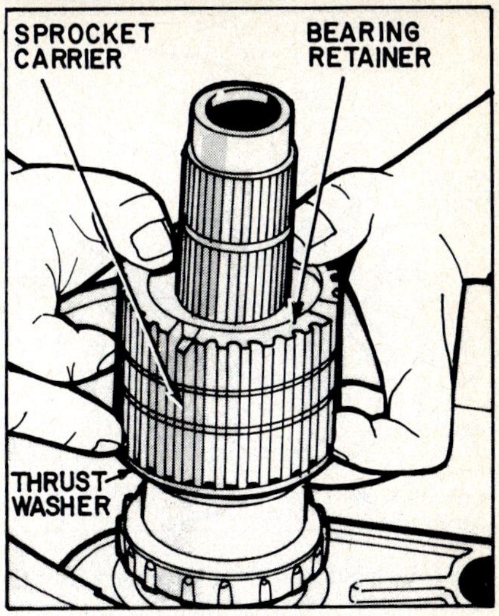

Fig. 71 Sprocket carriers, bearing retainers, thrust washer & needle bearing replacement

race for reassembly.

25. Remove range sector operating lever attaching nut and washer, then the lever. Remove sector shaft seal and seal retainer.
26. Remove range sector.
27. Inspect lock plate. If lock plate is loose, worn or damaged, remove lock plate.
28. Remove output shaft seals from front and rear case seal bores.

Sub-Assembly Service

Lock Plate, Replace

1. Remove and discard lock plate attaching bolts.
2. Remove lock plate from case, Fig. 68.
3. Coat the case and lock plate surfaces around the bolt holes with Loctite 515 sealer or equivalent.
4. Position new lock plate in the case and align bolt holes in lock plate and case.
5. Coat new lock plate attaching bolts with Loctite 271 sealer or equivalent.
6. Install and torque lock plate attaching bolts to 30 ft. lbs.

BEARING, BUSHING & SEAL REPLACEMENT

NOTE: The following bearings, bushings and seals are replaced using special service tools. However, they may also be replaced using suitable equivalent tooling and when using these tools, care should be taken as not to damage any components.

Also, all the bearings used must be correctly positioned to prevent covering the bearing oil feed holes. After replacing any bearing, check position of bearing to ensure that the oil feed hole is not obstructed or blocked by the bearing.

Rear Output Bearing & Seal, Replace

1. Remove bearing retaining snap ring and tap bearing from retainer with a mallet.
2. Remove rear seal using a suitable screwdriver or drift.
3. Install new bearing using tool J-7818. Ensure that shielded side of bearing faces toward inside of case.
4. Install bearing retaining snap ring.
5. Install new rear seal using tools J-8092 and J-29162.

Front Output Shaft Front Bearing, Replace

1. Remove bearing using tools J-8092 and J-29168.
2. Install new bearing using tools J-8092 and J-29167.
3. Remove tools and check bearing position to ensure that oil feed hole is clear.

Front Output Shaft Rear Bearing, Replace

1. Remove bearing using tools J-26941 and slide hammer J-2619-01.
2. Install new bearing using tools J-8092 and J-29163.
3. Remove tools and ensure that oil feed hole is clear and that the bearing is seated flush with edge of case bore to permit space for the thrust bearing assembly.

Input Gear Front & Rear Bearing, Replace

1. Remove both bearings simultaneously with tools J-8092 and J-29170.
2. Install new bearings one at a time. Install the rear bearing first, then the front bearing. Use tools J-8092 and J-29169.
3. Remove tools and check position of bearing to ensure that the oil feed hole is clear

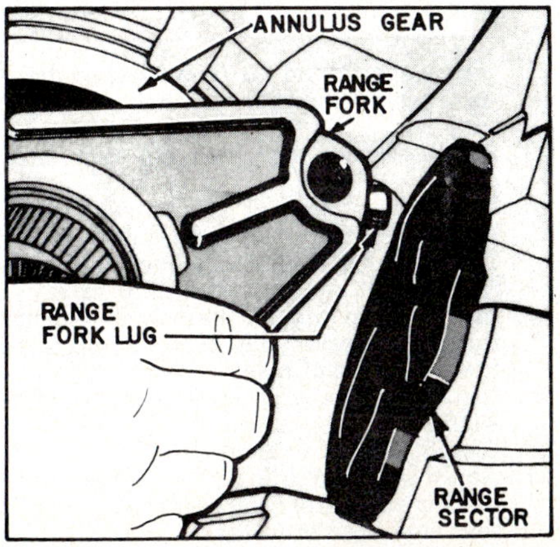

Fig. 72 Annulus gear & range fork replacement

Fig. 73 Planetary thrust washer & planetary assembly replacement

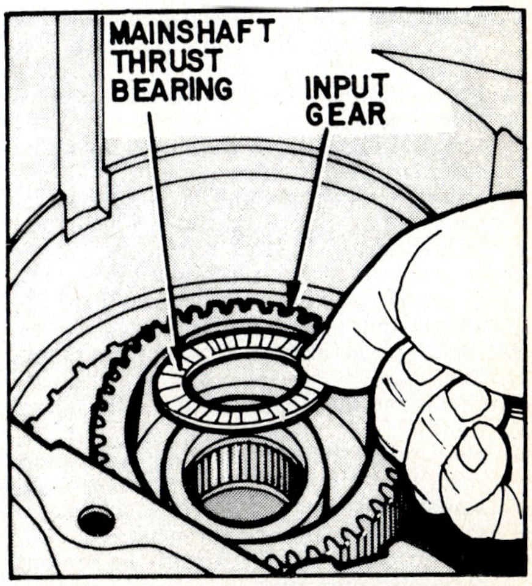

Fig. 74 Mainshaft thrust bearing & input gear replacement

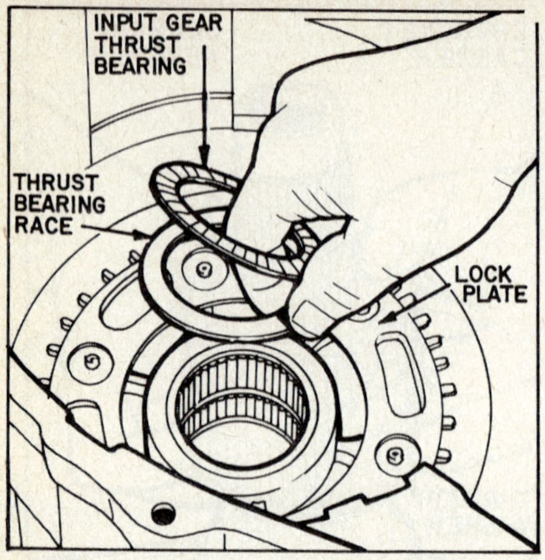

Fig. 75 Input gear thrust bearing & race replacement

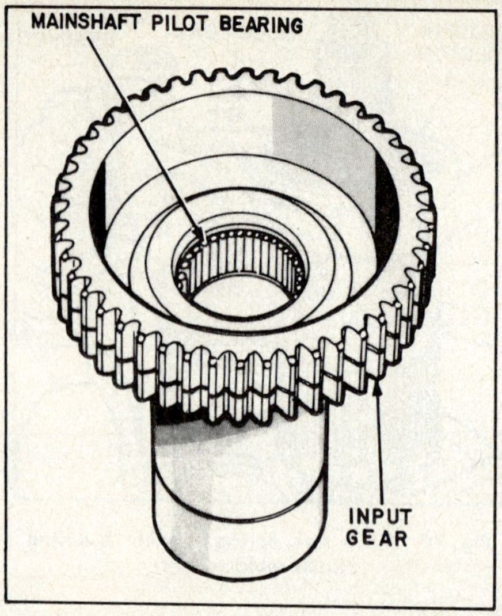

Fig. 76 Mainshaft pilot bearing installation

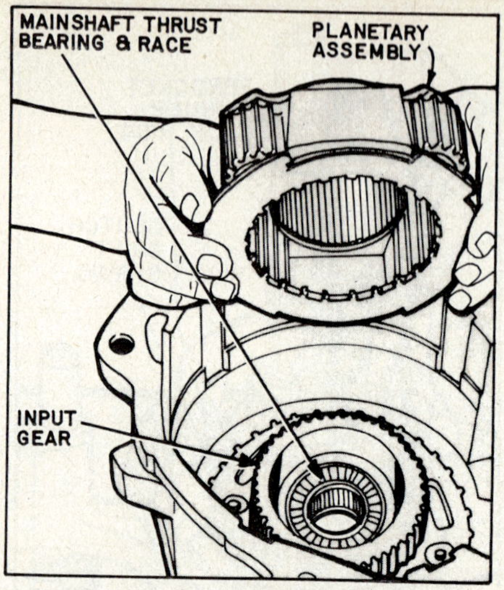

Fig. 77 Input gear, mainshaft thrust bearing & planetary installation

and that the bearings are flush with the case bore surfaces.

Mainshaft Pilot Bearing, Replace

1. If bearing cannot be removed by hand, remove by using slide hammer J-2619-01 and J-29369-1. A similar internal type blind hole bearing puller may also be used.
2. Install new bearing using tools J-8092 and J-29174.
3. Remove tools and check position of bearing to ensure that the oil feed hole is clear and that the bearing is seated flush with edge of bearing bore, Fig. 76.

Annulus Gear Bushing, Replace

1. Remove bushing with tools J-8092 and J-29185.
2. Install new bushing with tools J-8092 and J-29185-2.
3. Remove any metal chips made by bushing replacement.

Assembly, Fig. 68

NOTE: During assembly, lubricate all components with 10W-30 engine oil or petroleum jelly where indicated only. Do not use any other types of lubricant.

1. Install input gear race and thrust bearing into front case, then the input gear, Fig. 77.
2. Install mainshaft thrust bearing in input gear, Fig. 77.
3. Install range sector shaft seal and seal retainer, then the range sector.
4. Install operating lever on range sector shaft. Install and torque shaft washer and lock nut to 18 ft. lbs.
5. Install planetary assembly over input gear and ensure that the assembly is fully seated and meshed with the gear.
6. Install planetary thrust washer on planetary hub.
7. Install inserts into range fork, if removed.
8. Engage range fork into annulus gear and install annulus gear over planetary assembly. Ensure that the range fork lug is inserted fully into range sector slot.
9. Install annulus gear thrust washer and retaining snap ring.
10. Align shift rail bores in the case and range fork and install shift rail.

NOTE: The shift rail bore in the case must be dry and not contain any oil. A small amount of oil may prevent the shift rail from seating completely and also may prevent front case installation.

11. Install mainshaft. Ensure that the mainshaft thrust bearing is seated properly in the input gear before installing mainshaft.
12. Thickly coat sprocket carrier bore with petroleum jelly and position bearing retainer at center of carrier bore.
13. Coat mainshaft needle bearings with petroleum jelly and install 60 needle bearings in each end of sprocket carrier bore. 120 needle bearings are used.
14. Install bearing retainer in each end of sprocket carrier bore and position the thrust washer on the bottom of the carrier, Fig. 78.

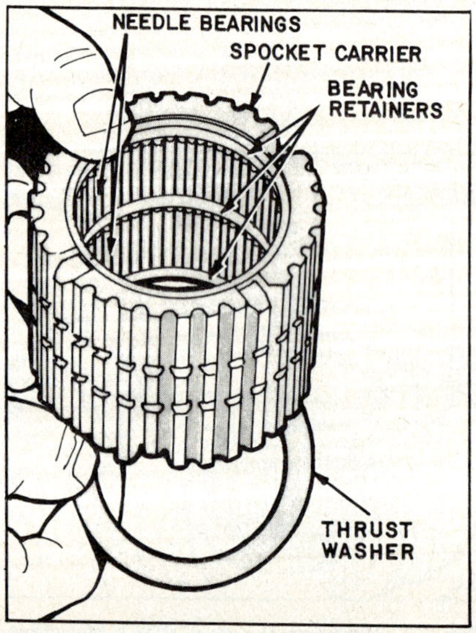

Fig. 78 Sprocket carrier components assembly

15. Align the assembled carrier and needle bearings with the mainshaft and install the assembly on the mainshaft. Use caution as not to dislodge the needle bearings.
16. Assemble the mode fork, fork spring and bracket. Engage the fork in the sliding clutch and install the assembly onto the shift rail and mainshaft.
17. Install clutch spring and stop ring on sprocket carrier.

NOTE: If the sprocket carrier has two ring grooves, install the stop ring in the upper groove only.

18. Install front output shaft front thrust bearing assembly in front case. The proper installation sequence is: thick race, thrust bearing, then thin race.
19. Install front output shaft.
20. Install sprockets and drive chain as an assembly. Position the sprockets in the chain, aligning the sprockets with the shafts and install the assembly.

NOTE: Ensure that the drive sprocket is installed with the recessed side of the sprocket facing toward the inside of the case.

21. Install spacer and thrust washer on drive sprocket, then the sprocket retaining snap ring.
22. Install driven sprocket retaining snap ring.
23. Install front output shaft rear thrust bearing assembly on front output shaft. The proper installation sequence is: thin race, thrust bearing, then thick race.
24. Install oil pump on mainshaft. Ensure that the recessed side of the pump faces toward inside of case.
25. Install speedometer drive gear on mainshaft.
26. Install magnet in front case, if removed.
27. Apply Loctite 515 sealer or equivalent to mating surface of front case. Then, install rear case onto front case.

NOTE: Ensure that the front output shaft

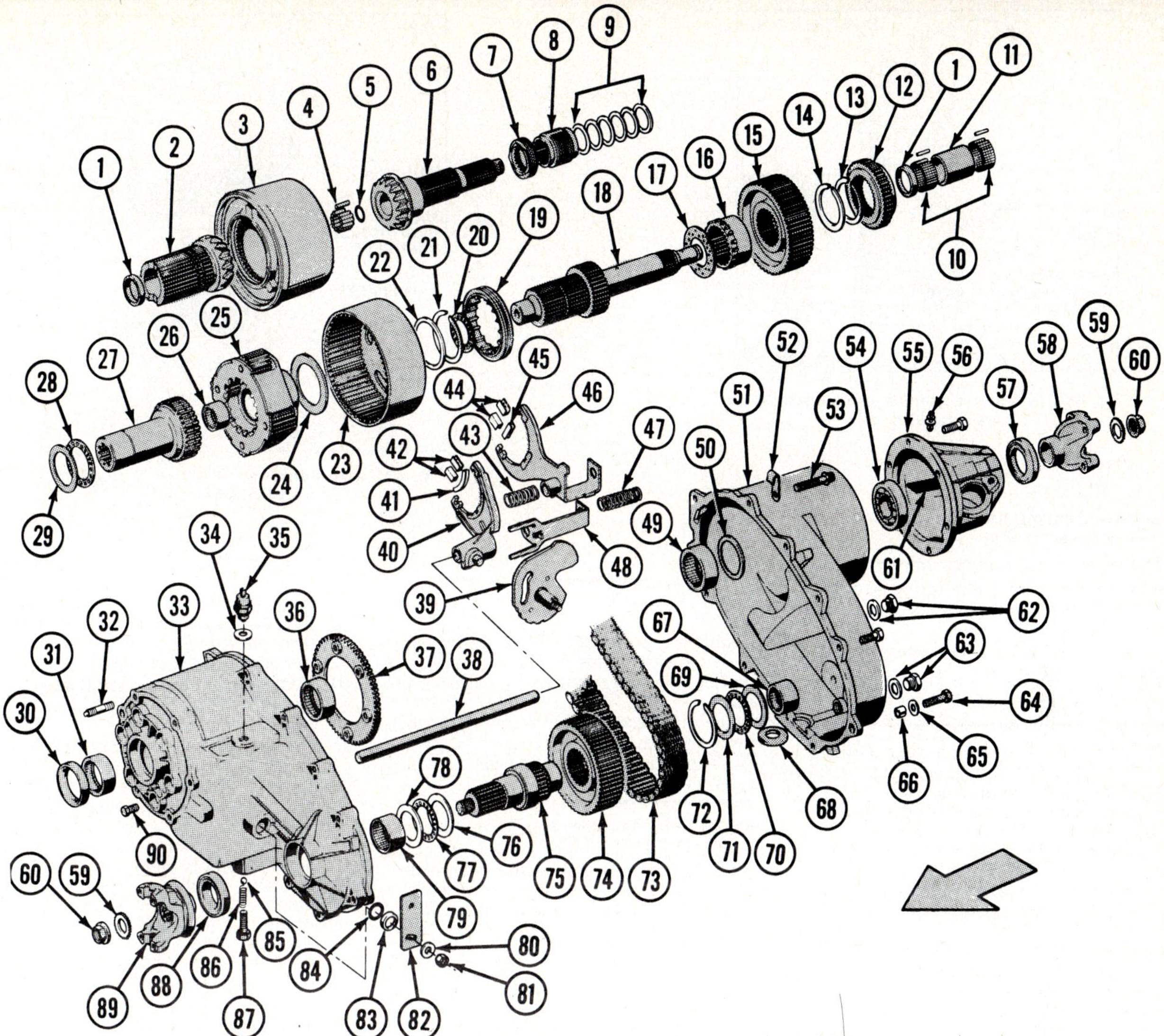

1. MAINSHAFT REAR BEARING SPACER – SHORT (2)
2. SIDE GEAR
3. VISCOUS COUPLING AND DIFFERENTIAL ASSEMBLY
4. MAINSHAFT REAR PILOT ROLLER BEARINGS (15)
5. MAINSHAFT O-RING
6. REAR OUTPUT SHAFT
7. OIL PUMP
8. SPEEDOMETER GEAR
9. DIFFERENTIAL END PLAY SHIMS (SELECTIVE)
10. MAINSHAFT NEEDLE BEARINGS (82)
11. MAINSHAFT REAR BEARING SPACER
12. CLUTCH GEAR
13. CLUTCH GEAR LOCATING RING
14. DRIVE SPROCKET LOCATING RING
15. DRIVE SPROCKET
16. SIDE GEAR CLUTCH
17. MAINSHAFT THRUST WASHER
18. MAINSHAFT
19. CLUTCH SLEEVE
20. MAINSHAFT THRUST BEARING
21. ANNULUS GEAR RETAINING RING
22. ANNULUS GEAR THRUST WASHER
23. ANNULUS GEAR
24. PLANETARY THRUST WASHER
25. PLANETARY ASSEMBLY
26. MAINSHAFT FRONT PILOT BEARING
27. INPUT GEAR
28. INPUT GEAR THRUST BEARING
29. INPUT GEAR THRUST BEARING RACE
30. INPUT GEAR OIL SEAL
31. INPUT GEAR FRONT BEARING
32. FRONT CASE MOUNTING STUD (6)
33. FRONT CASE
34. LOCK MODE INDICATOR SWITCH GASKET
35. LOCK MODE INDICATOR SWITCH
36. INPUT GEAR REAR BEARING
37. LOW RANGE LOCKPLATE
38. SHIFT RAIL
39. RANGE SECTOR
40. RANGE FORK
41. RANGE FORK INSERT
42. RANGE FORK PADS
43. MODE FORK SPRING
44. MODE FORK PADS
45. MODE FORK INSERT
46. MODE FORK
47. SHIFT RAIL SPRING
48. MODE FORK BRACKET
49. REAR OUTPUT SHAFT BEARING
50. REAR OUTPUT SHAFT BEARING SEAL
51. REAR CASE
52. WIRING CLIP
53. SPLINE BOLT
54. REAR OUTPUT BEARING
55. REAR RETAINER
56. VENT
57. OUTPUT SHAFT OIL SEAL
58. REAR YOKE
59. YOKE SEAL WASHER
60. YOKE LOCKNUT
61. VENT CHAMBER SEAL
62. FILL PLUG AND GASKET
63. DRAIN PLUG AND GASKET
64. REAR CASE BOLT
65. WASHER (2)
66. CASE ALIGNMENT DOWEL
67. FRONT OUTPUT SHAFT REAR BEARING
68. MAGNET
69. FRONT OUTPUT SHAFT REAR THRUST BEARING RACE (THICK)
70. FRONT OUTPUT SHAFT REAR THRUST BEARING
71. FRONT OUTPUT SHAFT REAR THRUST BEARING RACE (THIN)
72. DRIVEN SPROCKET RETAINING SNAP RING
73. DRIVE CHAIN
74. DRIVEN SPROCKET
75. FRONT OUTPUT SHAFT
76. FRONT OUTPUT SHAFT FRONT THRUST BEARING RACE (THIN)
77. FRONT OUTPUT SHAFT FRONT THRUST BEARING
78. FRONT OUTPUT SHAFT FRONT THRUST BEARING RACE (THICK)
79. FRONT OUTPUT SHAFT FRONT BEARING
80. WASHER
81. LOCKNUT
82. OPERATING LEVER
83. RANGE SECTOR SHAFT SEAL RETAINER
84. RANGE SECTOR SHAFT SEAL
85. DETENT BALL
86. DETENT SPRING
87. DETENT RETAINING BOLT
88. FRONT OUTPUT SHAFT SEAL
89. FRONT YOKE
90. LOCKPLATE BOLTS

Fig. 79 New Process model 219 transfer case, disassembled

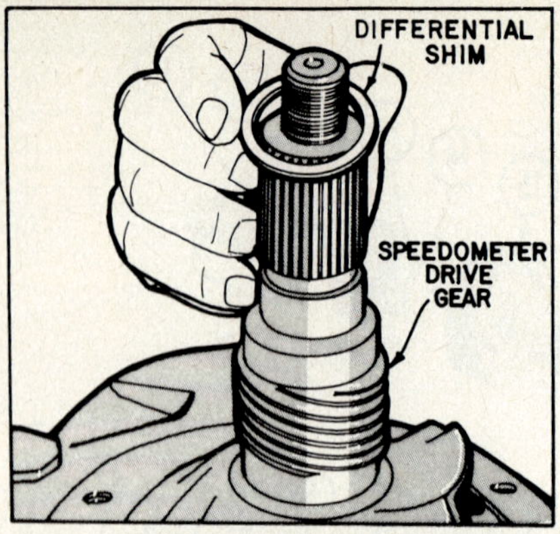

Fig. 80 Differential shim & speedometer drive gear replacement

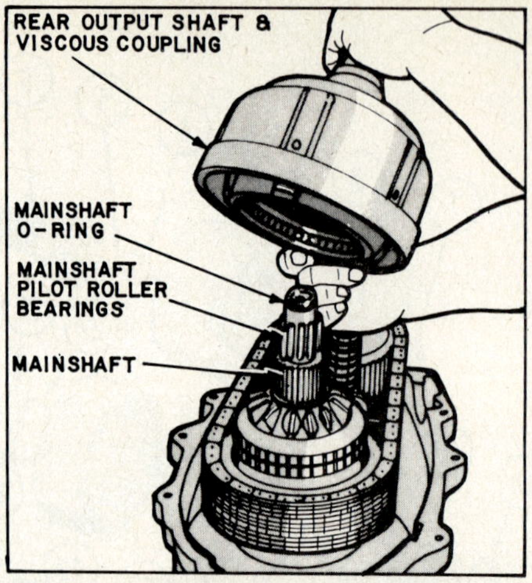

Fig. 81 Rear output shaft & viscous coupling replacement

rear thrust bearing assembly is properly seated in the rear case.

28. Align case bolt holes and alignment dowels and install retaining bolts. Alternately torque bolts to 23 ft. lbs.

NOTE: Flat washers are installed on the two bolts at opposite ends of the case.

29. Install rear output bearing in rear retainer, then install retaining snap ring.
30. Install seal into pump housing. Apply petroleum jelly to pump housing tabs. Install housing into rear retainer.
31. Apply Locktite 515 sealer or equivalent to mating surface of rear retainer. Align rear retainer and case alignment marks made during disassembly and install retainer.
32. Install and torque retainer bolts to 23 ft. lbs.
33. Install oil seal into rear retainer bore. Coat the seal lip with petroleum jelly prior to installation.
34. Install washer and indicator switch. Torque switch to 18 ft. lbs.
35. Lightly apply Locktite 515 sealer or equivalent to detent retainer bolt, then install detent ball, spring and bolt. Torque bolt to 23 ft. lbs.
36. Install drain plug and gasket. Torque drain plug to 18 ft. lbs.
37. Install oil seal in front case output shaft bore.
38. Install front and rear yokes.

NOTE: The yoke with the collar is installed on the front output shaft.

39. Install yoke seal washer and yoke nuts. Torque yoke nuts to 120 ft. lbs.
40. Install 6 pints of 10W-30 engine oil into transfer case through fill plug.
41. Install and torque fill plug to 18 ft. lbs.

NEW PROCESS MODEL 219

Disassembly, Fig. 79

1. Remove fill and drain plugs. Drain lubricant from transfer case.
2. Remove front and rear output shaft yokes. Discard yoke seal washers and yoke nuts.
3. Mark relationship between rear retainer and rear case for alignment at reassembly.
4. Remove rear retainer attaching bolts, then the rear retainer. If necessary use a mallet to loosen retainer. Do not pry from case.
5. Remove differential shim and speedometer drive gear from rear output shaft, Fig. 80. Tag the shim or shims for assembly reference.

NOTE: The speedometer drive gear is installed with the long end facing toward the case. Note the gear position for assembly.

6. Remove rear output bearing snap ring, then the bearing from retainer with a mallet.

NOTE: The rear output bearing has one shielded side. Note position for assembly.

7. Remove rear output shaft seal from rear retainer using a suitable screwdriver or a punch.
8. Position transfer case on wood blocks with the front case resting on the wood. If necessary, cut "V" notches in the blocks so the front case rests squarely on the blocks.
9. Remove rear case to front case attaching bolts, then the rear case from front case. To remove, insert suitable screwdrivers into notches at case ends to pry rear case from front case.

NOTE: The two case end bolts have flat washers and alignment dowels. Note bolt, dowel and washer locations for assembly.

10. Remove rear output shaft and viscous coupling as an assembly, Fig. 81. If necessary, tap shaft with a plastic mallet to remove.
11. Remove "O" ring seal and pilot roller bearings from mainshaft.
12. Remove rear output shaft from viscous coupling.
13. Remove shift rail spring from rail.
14. Remove plastic oil pump from shaft bore in rear case. Note pump position for reassembly. The side of the pump with the recess must face toward shaft bore when installed.
15. Remove rear output shaft bearing seal from case using a suitable screwdriver to pry seal from seal bore.
16. Remove front output shaft thrust bearing assembly, Fig. 82. Tag the assembly for installation reference.
17. Remove driven sprocket retaining snap ring.
18. Remove drive sprocket, drive chain, driven sprocket, side gear clutch and clutch gear as an assembly, Fig. 83. Place assembly on bench and mark components for reassembly reference.
19. Remove needle bearings and bearing spacers from mainshaft or side gear bore. A total of 82 needle bearings and three spacers are used.
20. Remove side gear/clutch gear assembly from drive sprocket, Fig. 83. Remove two snap rings, then the clutch gear from the side gear, Fig. 84. Note position of snap rings and gears for reassembly.
21. Remove side gear clutch, mainshaft thrust washer and the remaining (short) mainshaft needle bearing spacer.
22. Remove front output shaft and shaft thrust bearing assembly, Fig. 85. Note installation sequence of thrust bearing assembly. The proper sequence is: thin race, bearing, then thick race.
23. Remove front output shaft seal from front case using a suitable screwdriver or punch.
24. Remove shaft rail spring from shaft rail if not previously removed.
25. Remove clutch sleeve, mode fork and mode fork spring as an assembly, Fig. 86.
26. Remove mainshaft thrust washer, then the mainshaft. Pull mainshaft straight up to remove.
27. Move range operating lever downward to the last detent position, then disengage range fork lug from range sector slot.
28. Remove annulus gear retaining snap ring and thrust washer, then the annulus gear and range fork as an assembly, Fig. 87.
29. Remove planetary thrust washer from planetary assembly hub, then the planetary assembly, Fig. 88.
30. Remove mainshaft thrust bearing from input gear, then the input gear, thrust bearing and race.
31. Remove range sector detent ball and spring retaining bolt, then the ball and

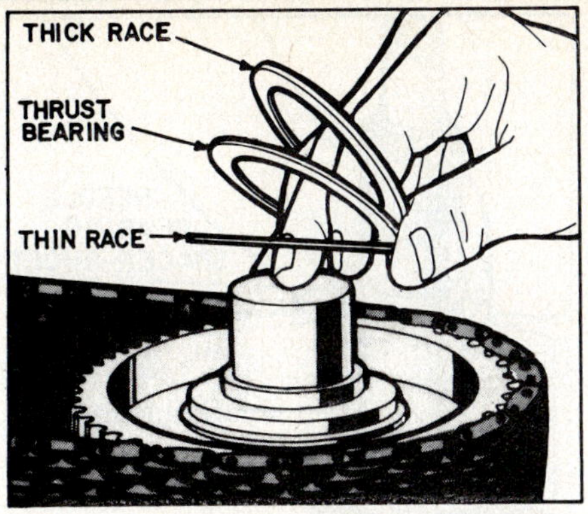

Fig. 82 Front output shaft rear thrust bearing assembly replacement

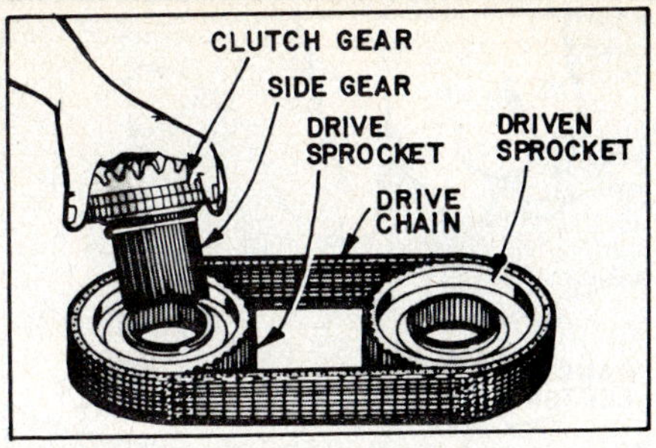

Fig. 83 Side gear, clutch gear, sprockets & chain assembly

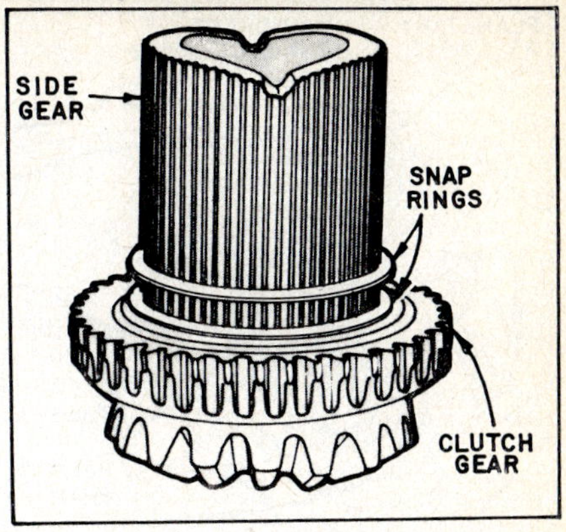

Fig. 84 Side gear components

spring.
32. Remove range sector and operating lever attaching nut and lock washer, then the lever.
33. Remove sector, then the range sector shaft "O" ring and retainer.
34. Remove input gear oil seal from front case using a suitable screwdriver or punch.

Sub-Assembly Service

Lock Plate, Replace

1. Remove and discard lock plate attaching bolts.
2. Remove lock plate from case.
3. Coat case and lock plate surfaces around bolt holes with Loctite 515 sealer or equivalent.
4. Place new lock plate in case and align bolt holes in lock plate and case.
5. Coat new lock plate attaching bolts with Loctite 271 sealer or equivalent.
6. Install and torque lock plate attaching bolts to 30 ft. lbs.

BEARING & BUSHING REPLACEMENT

NOTE: The following bearings and bushings are replaced using special service tools. However, they may also be replaced using suitable equivalent tooling and when using these tools, care should be taken as not to damage any components.

Also, all the bearings used must be correctly positioned to prevent covering the bearing oil feed holes. After replacing any bearing, check position of bearing to ensure that the oil feed hole is not obstructed or blocked by the bearing.

Rear Output Shaft Bearing, Replace

1. Remove bearing using tools J-8092 and J-29165.
2. Install new bearing with tools J-8092 and J-29166.
3. Remove tools and check position of bearing to ensure that the oil feed hole is clear.

Front Output Shaft Front Bearing, Replace

1. Remove bearing with tools J-8082 and J-29168.
2. Install new bearing with tools J-8092 and J-29167.
3. Remove tools and check position of bearing to ensure that the oil feed hole is clear.

Front Output Shaft Rear Bearing, Replace

1. Remove bearing with tools J-26941 and J-2619-01.
2. Install new bearing with tools J-8092 and J-29163.
3. Remove tools and check position of bearing to ensure that oil feed hole is clear and that the bearing is seated flush with the edge of the bore in the case to permit clearance for thrust bearing assembly.

Input Gear Front & Rear Bearings, Replace

1. Remove both bearings simultaneously with tools J-8092 and J-29170.
2. Install new bearing one at a time. First, install the rear bearing, then the front bearing. Bearings are installed with tools J-8092 and J-29169.
3. Remove tools and check position of bearings to ensure that the oil feed holes are clear and that the bearings are flush with the case bore surfaces.

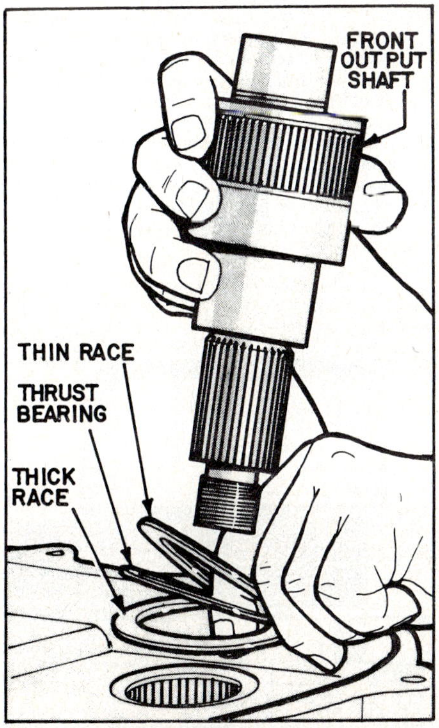

Fig. 85 Front output shaft & shaft front thrust bearing replacement

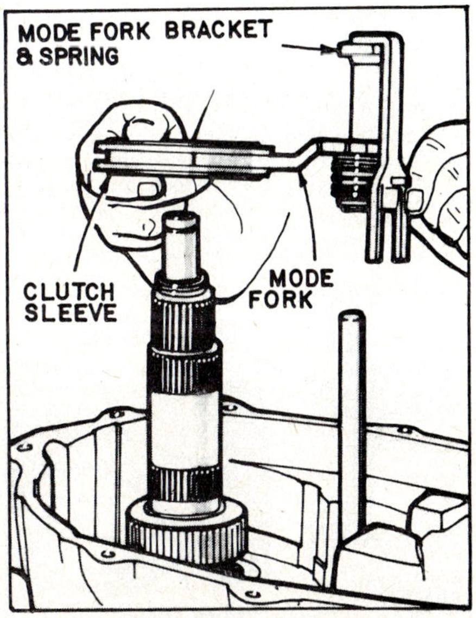

Fig. 86 Clutch sleeve & mode fork replacement

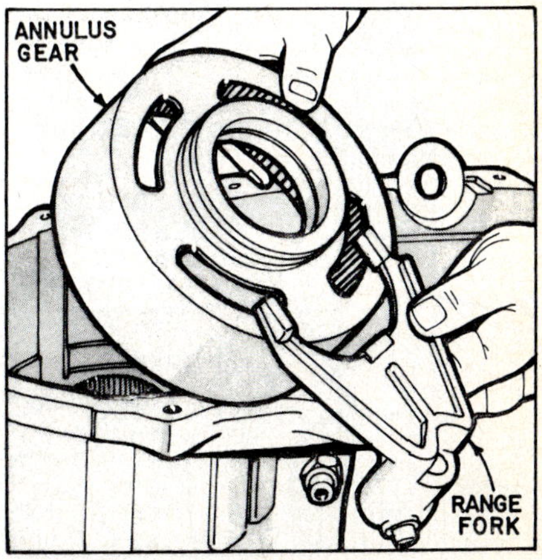

Fig. 87 Annulus gear & range fork replacement

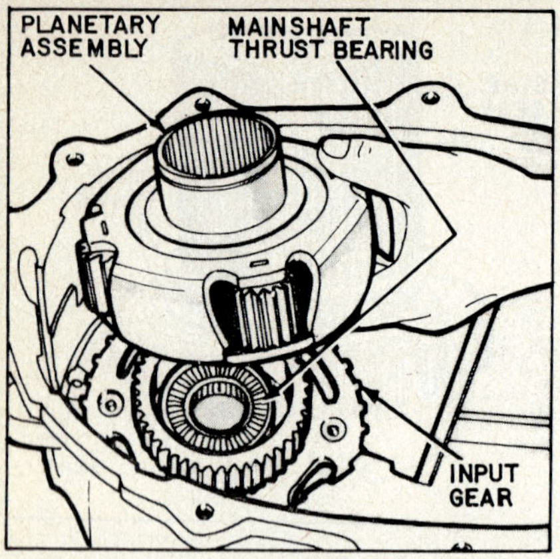

Fig. 88 Planetary assembly replacement

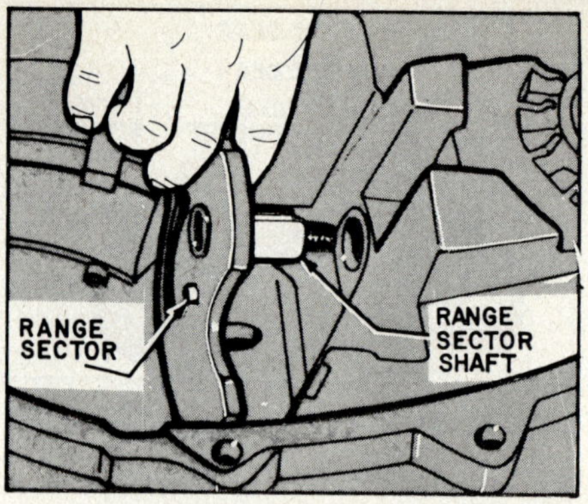

Fig. 89 Range sector installation

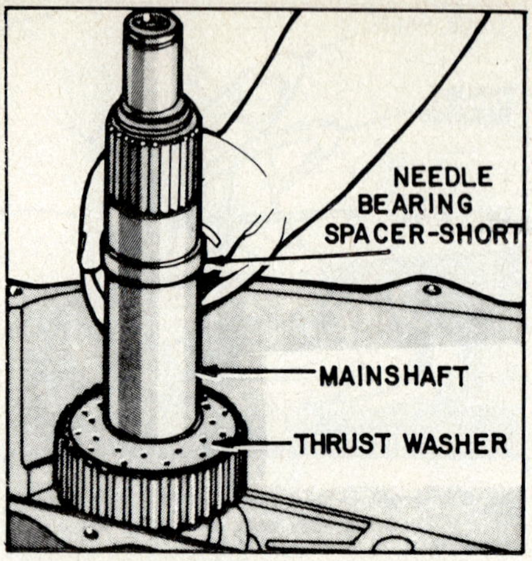

Fig. 90 Mainshaft installation

Mainshaft Front Pilot Bearing, Replace

1. If bearing cannot be removed by hand, use tools J-2619-01 and J-29369. A similar internal type blind hole bearing puller may also be used.
2. If necessary, install new bearing with tools J-8092 and J-29174.
3. Remove tools and check position of bearing to ensure that the oil feed hole is clear and that the bearing is seated flush with edge of bearing bore.

Rear Output Bearing, Replace

1. Remove snap ring, then the bearing using a mallet or brass punch.
2. Install new bearing with tools J-8092 and J-7818.

NOTE: Ensure that the shielded side of the bearing faces toward the inside of the transfer case after installation.

3. Install bearing snap ring.

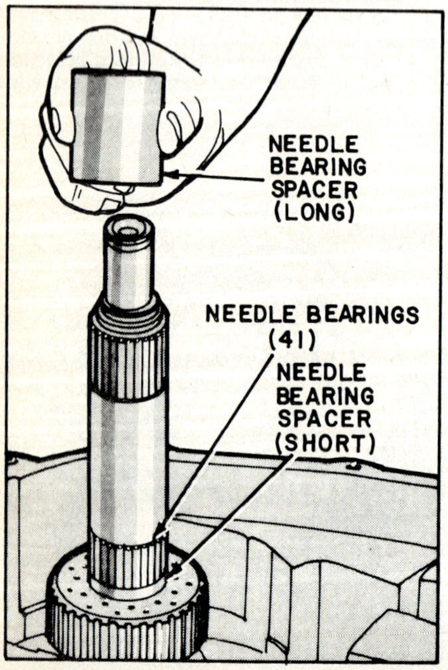

Fig. 91 Mainshaft needle bearings & spacer installation

4. Install new seal with tool J-29162.

Annulus Gear Bushing, Replace

1. Remove bushing with tools J-8092 and J-29185.
2. Install new bushing using the same tools.
3. Remove any metal chips caused by bushing replacement.

Assembly, Fig. 79

NOTE: During assembly, lubricate all components with 10W-30 engine oil or petroleum jelly where indicated only. Do not use any other types of lubricant.

1. Install new input gear and rear output shaft bearing oil seals. Seat the seals flush with the edge of seal bore or in seal groove in the case. Coat seal lips with petroleum jelly after installation.
2. Install input gear thrust bearing race in the case counterbore.
3. Install input thrust bearing on input gear, then install the gear and bearing assembly into case.
4. Install mainshaft thrust bearing into bearing recess in input gear.
5. Install planetary assembly on input gear. Ensure that the planetary pinion teeth mesh fully with the input gear.
6. Install planetary thrust washer on planetary hub.
7. Install new sector shaft "O" ring and retainer in shaft bore in the case.
8. Install range sector in front case, Fig. 89. Install operating lever on sector shaft, then the lever attaching washer and lock nut on shaft. Torque lock nut to 17 ft. lbs.
9. Install detent ball, spring and retaining bolt into front case detent bore. Torque retaining bolt to 22 ft. lbs.
10. Move range sector to last detent position.
11. Assemble annulus gear and range fork. Install the assembly on and over planetary assembly. Ensure that the annulus gear is meshed fully with planetary pinions.
12. Insert range fork lug in range sector detent slot.
13. Install annulus thrust washer and annulus retaining ring on annulus gear hub.
14. Align mainshaft thrust washer in input gear, if necessary.
15. Install mainshaft and ensure that the shaft is seated fully in the input gear, Fig. 90.
16. Install mainshaft thrust washer on mainshaft.
17. Install short mainshaft needle bearing spacer on the shaft, Fig. 91.
18. Liberally coat mainshaft needle bearing surface with petroleum jelly and to all 82 needle bearings. Install 41 needle bearings on the shaft, Fig. 91. Ensure that bearings are in the vertical position and seated on the short spacer.
19. Install the long mainshaft needle bearing spacer on the shaft, Fig. 91. Lower the spacer onto previously installed needle bearings. Avoid displacing bearings.
20. Align shift rail bore in the case with bore

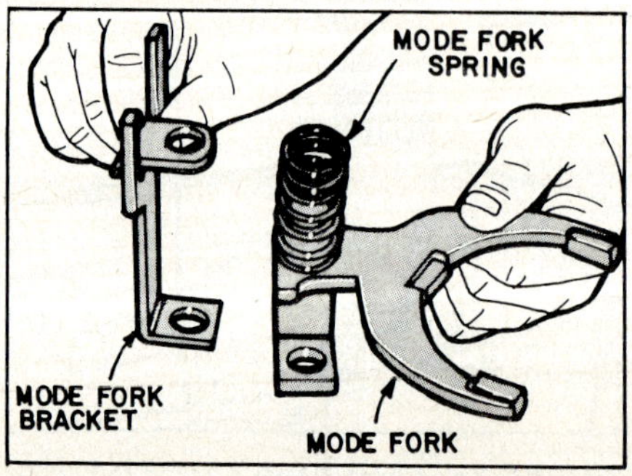

Fig. 92 Mode fork, spring & bracket assembly

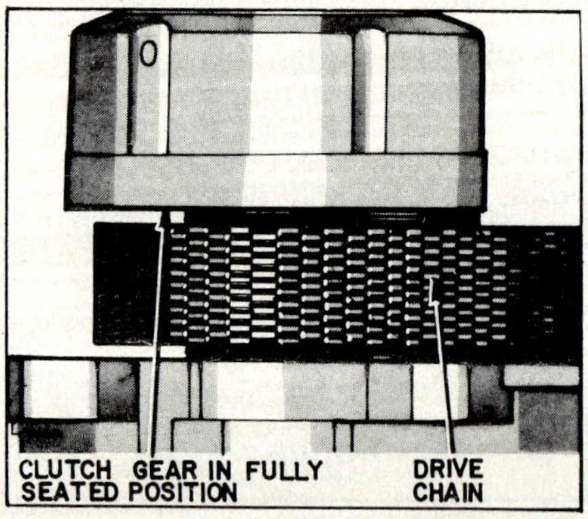

Fig. 93 Seating viscous coupling on clutch gear

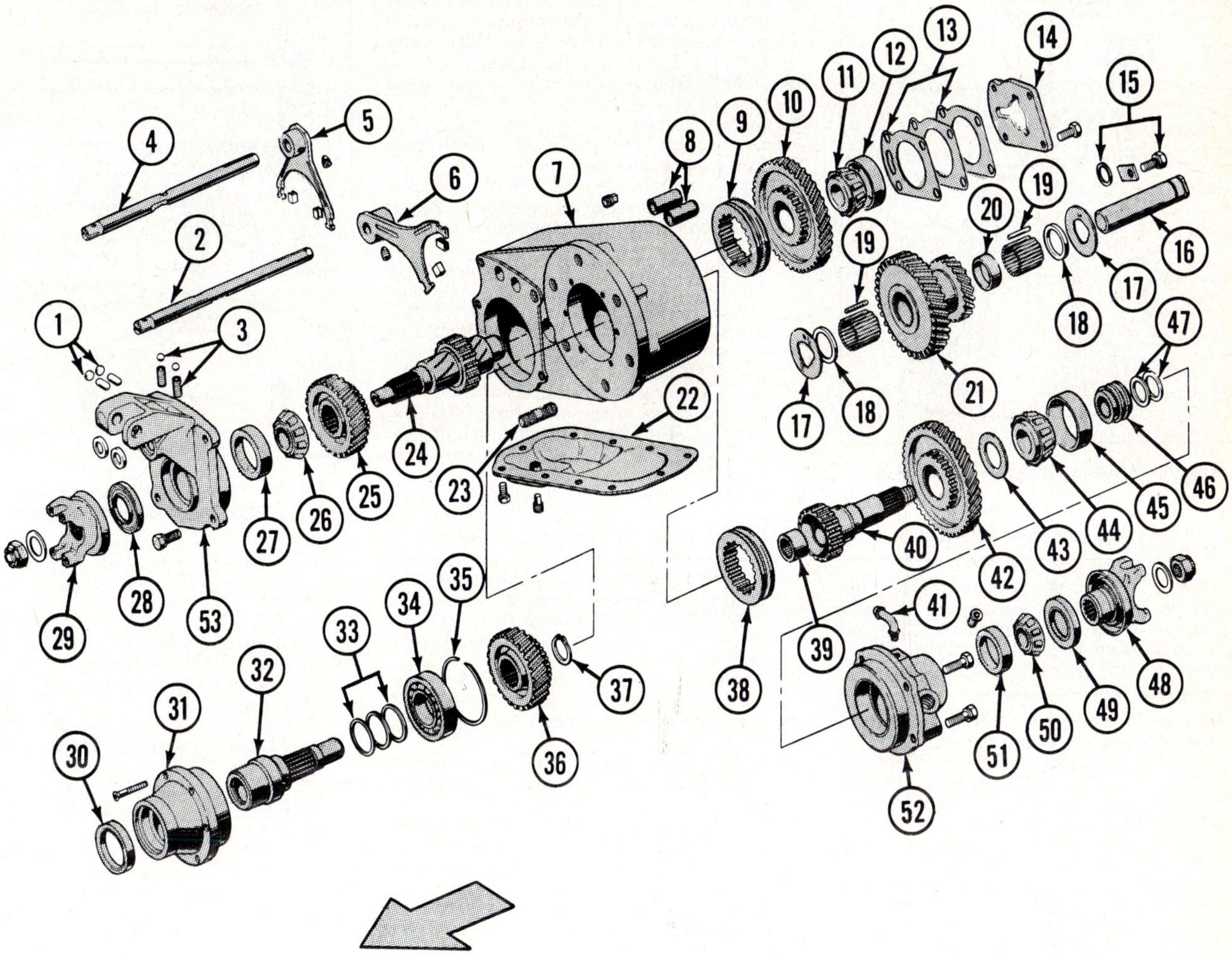

1. INTERLOCK PLUGS AND INTERLOCKS
2. SHIFT ROD – REAR OUTPUT SHAFT FORK
3. POPPET BALLS AND SPRINGS
4. SHIFT ROD – FRONT OUTPUT SHAFT FORK
5. FRONT OUTPUT SHAFT SHIFT FORK
6. REAR OUTPUT SHAFT SHIFT FORK
7. TRANSFER CASE
8. THIMBLE COVERS
9. CLUTCH SLEEVE – FRONT OUTPUT SHAFT
10. CLUTCH GEAR – FRONT OUTPUT SHAFT
11. BEARING – FRONT OUTPUT SHAFT REAR
12. RACE – FRONT OUTPUT SHAFT BEARING
13. END PLAY SHIMS – FRONT OUTPUT SHAFT
14. COVER PLATE
15. LOCK PLATE, BOLT AND WASHER
16. INTERMEDIATE GEAR SHAFT
17. THRUST WASHER
18. BEARING SPACER (THIN)
19. INTERMEDIATE GEAR SHAFT NEEDLE BEARINGS
20. BEARING SPACER (THICK)
21. INTERMEDIATE GEAR
22. BOTTOM COVER
23. STUD (CASE-TO-TRANS.)
24. FRONT OUTPUT SHAFT
25. FRONT OUTPUT SHAFT GEAR
26. FRONT OUTPUT SHAFT BEARING (FRONT)
27. FRONT OUTPUT SHAFT BEARING RACE
28. OIL SEAL
29. FRONT YOKE
30. SEAL
31. SUPPORT – INPUT SHAFT
32. INPUT SHAFT
33. SHIMS
34. INPUT SHAFT BEARING
35. INPUT SHAFT BEARING SNAP RING
36. REAR OUTPUT SHAFT GEAR
37. SNAP RING
38. CLUTCH SLEEVE – REAR OUTPUT SHAFT
39. INPUT SHAFT REAR BEARING (NEEDLE) (OR PILOT BEARING)
40. REAR OUTPUT SHAFT
41. VENT
42. CLUTCH GEAR – REAR OUTPUT SHAFT
43. THRUST WASHER
44. BEARING – REAR OUTPUT SHAFT FRONT
45. RACE – REAR OUTPUT SHAFT BEARING
46. SPEEDOMETER DRIVE GEAR
47. END PLAY SHIMS
48. REAR YOKE
49. REAR OUTPUT SHAFT OIL SEAL
50. BEARING – REAR OUTPUT SHAFT REAR
51. BEARING RACE
52. REAR BEARING CAP
53. FRONT BEARING CAP

Fig. 95 Dana model 300 transfer case, disassembled

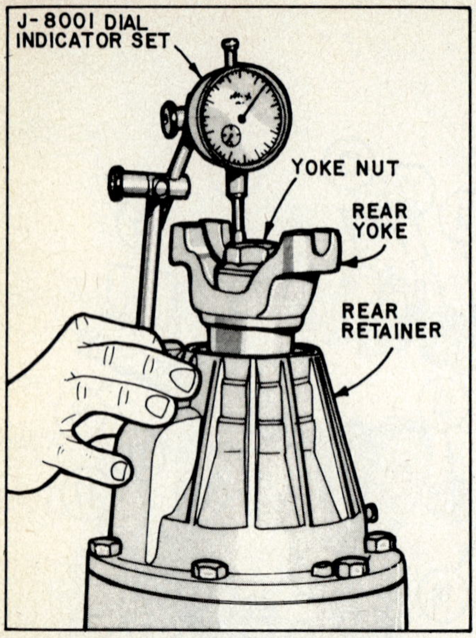

Fig. 94 Checking differential end play

in range fork and install shift rail.

NOTE: The shift rail bore in the case must be dry and not contain any oil. A small amount of oil may prevent the shift rail from seating completely and also may prevent rear case installation.

21. Assemble mode fork, mode fork spring and mode fork bracket, Fig. 92.
22. Install clutch sleeve in mode fork. Ensure that the sleeve is positioned so the I.D. numbers on the sleeve face upward after the sleeve is installed.
23. Align clutch sleeve and mode fork assembly with shift rail and install assembly onto shift rail and mainshaft. Ensure that the clutch sleeve is meshed with the mainshaft gear.
24. Lubricate the remaining 41 mainshaft needle bearings and place bearings on shaft.
25. Install side gear clutch on mainshaft with clutch gear teeth facing downward. Ensure that the gear teeth mesh with the clutch sleeve.
26. Install the remaining short mainshaft needle bearing spacer. Avoid displacing the bearings.
27. Install front output shaft front thrust bearing assembly into front case. The correct assembly sequence is: thick race, thrust bearing, then thin race.
28. Install front output shaft into front case.
29. Install clutch gear on side gear. The tapered side of the clutch gear teeth must face toward side gear teeth.
30. Install clutch gear and drive sprocket locating snap rings on side gear. Install the snap rings so the rings face each other.
31. Position drive and driven sprockets in drive chain and install the assembled side and clutch gears in the drive sprocket.
32. Install assembled drive chain, sprockets and side gear on mainshaft and front output shaft. Align sprockets with shafts, then, keeping the assembly level, lower the assembly onto both shafts simultaneously. Avoid displacing mainshaft needle bearings during chain and sprockets installation.
33. Install driven sprocket retaining snap ring.
34. Install front output shaft rear thrust bearing assembly onto front output shaft. The proper installation sequence is: thin race, thrust bearing, then thick race.
35. Install shift rail spring on shift rail.
36. Install new "O" ring on mainshaft pilot bearing hub.
37. Thickly coat mainshaft pilot roller bearing hub and pilot roller bearings with petroleum jelly. Install rollers onto shaft.
38. Install rear output shaft in viscous coupling. Ensure that the shaft is seated fully.
39. Install assembled viscous coupling and rear output shaft on mainshaft. Align mainshaft pilot hub with pilot bearing bore in rear output shaft and lower assembly onto mainshaft. Avoid displacing pilot roller bearings during installation.
40. Align clutch gear teeth with viscous coupling teeth and seat the coupling fully onto the clutch gear, Fig. 93.

NOTE: When properly installed, the clutch gear teeth will not be visible or extend past the coupling.

41. Install magnet into front case, if removed.
42. Thoroughly clean mating surfaces of front and rear cases. Apply Loctite 515 sealer or equivalent to mating surface of front case and to all case attaching bolts.
43. Install rear case onto front case. Align case dowels and install case attaching bolts. Torque attaching bolts to 22 ft. lbs.

NOTE: The two case end bolts require flat washers.

44. Install oil pump on rear output shaft and seat into case. Install the pump so the recessed side faces toward inside of case.
45. Install speedometer drive gear and differential shim on output shaft.
46. Install vent chamber seal in rear retainer, if removed.
47. Align and install rear retainer on rear case. Install retainer bolts finger tight only.
48. Install yoke onto rear output shaft. Install yoke nut finger tight only.
49. Install a dial indicator onto retainer with the plunger contacting the top of the yoke nut, Fig. 94.
50. Install yoke on front output shaft and rotate the front shaft 10 complete revolutions.
51. Rotate front output shaft again and note end play reading on dial indicator. End play should be .002 to .010 inch. If end play is not within specifications, remove rear retainer and add or remove differential shims as required. Recheck end play.
52. Remove both output shaft yokes.

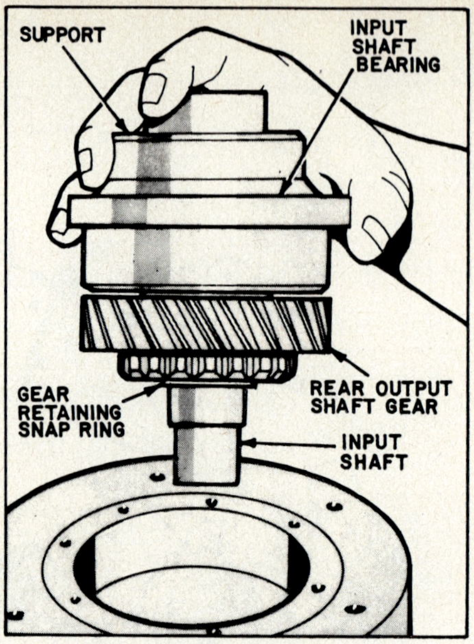

Fig. 96 Front support, input shaft & rear output shaft gear replacement

Fig. 97 Shift fork setscrew replacement

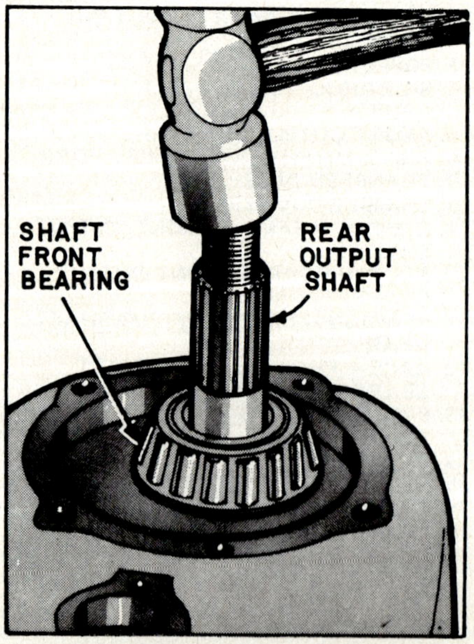

Fig. 98 Rear output shaft front bearing removal

Fig. 99 Front output shaft front bearing removal

53. Install new front and rear yoke seals if not previously installed.
54. Remove rear retainer bolts and apply Loctite 515 sealer or equivalent to mating surface of retainer and to the bolts. Install and torque retainer bolts to 22 ft. lbs.
55. Install new yoke seal washers on output shafts, then the yokes. Install and torque new yoke nuts to 110 ft. lbs.
56. Install and torque drain plug to 18 ft. lbs.
57. Install 4 pints of 10W-30 engine oil into transfer case through fill plug.
58. Install and torque fill plug to 18 ft. lbs.

DANA MODEL 300

Disassembly, Fig. 95

1. Remove shift lever assembly.
2. Remove bottom cover. The bottom cover is coated with a sealant. Use a putty knife to break the seal, then work the knife around the edge to loosen and remove cover.
3. Remove front and rear yokes. Discard yoke nuts.
4. Remove socket head screws attaching the input shaft support to the case, then the support, rear output shaft gear and input shaft as an assembly, Fig. 96. The support is coated with a sealant. Use a putty knife to break the seal, then work knife around the edge to loosen and remove support.
5. Remove rear output shaft clutch sleeve from case.
6. Remove and discard snap ring retaining the rear output shaft gear on input shaft. Remove the gear.
7. Remove and discard input shaft bearing snap ring.
8. Remove input shaft and bearing from support. Tap end of input shaft with a mallet to aid removal.
9. Remove input shaft bearing and end play shims from shaft using a suitable press.
10. Remove input shaft oil seal from support. Discard the seal.
11. Remove intermediate shaft lock plate bolt and lock plate.
12. Remove intermediate shaft. Tap shaft from case using a suitable mallet and punch.
13. Remove and discard intermediate shaft "O" ring.
14. Remove intermediate gear assembly and thrust washers.

NOTE: The thrust washers have locating tabs which must fit into notches in the case at reassembly.

15. Remove needle bearings and bearing spacers from intermediate gear.

NOTE: There are 48 needle bearings and 3 bearing spacers in the intermediate gear.

16. Remove rear bearing cap attaching bolts, then the cap. Tap on the output shaft to aid cap removal. The rear bearing cap is coated with a sealant. Use a putty knife to break the seal, then work knife around edge to loosen and remove cap.
17. Remove end play shims and speedometer drive gear from rear output shaft.
18. Remove and discard rear output shaft oil seal. Remove bearings and races from rear bearing cap.
19. Remove setscrews, retaining front and rear output shaft shift forks on shift rods, Fig. 97.
20. Remove shift rods. Insert a punch through clevis pin holes in rods and rotate rods while pulling out from case.

NOTE: When the shift rods are free of the front cap, avoid losing the shift rod poppet balls and springs.

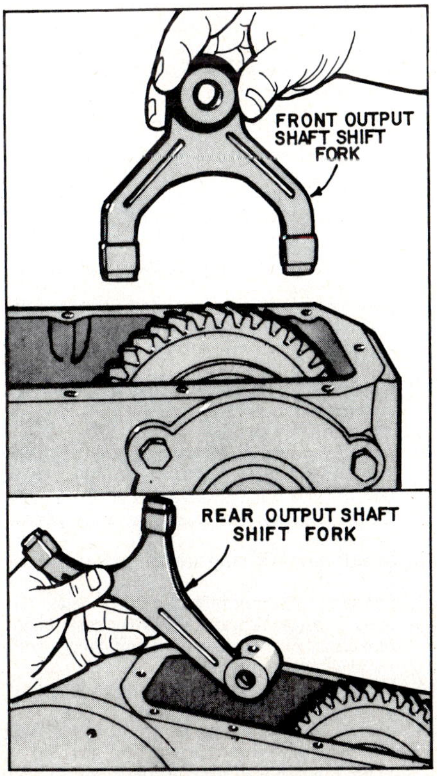

Fig. 101 Shift fork installation

Fig. 100 Front output shaft rear bearing installation

21. Remove shift forks from case.
22. Remove bolts attaching front cap to case, then the cap. The front cap is coated with a sealant. Use a putty knife to break the seal, then work knife around edge to loosen and remove cap.
23. Remove front output shaft and shift rod oil seals from front cap.
24. Remove front bearing race from front bearing cap with tool J-29168.
25. Remove cover plate bolts, then the plate and end play shims from case.
26. Move front output shaft toward front of case.
27. Remove front output shaft rear bearing race from case.
28. Remove rear output shaft front bearing, Fig. 98. Position case on wood blocks. Seat clutch gear on case interior surface and tap shaft from bearing with a suitable mallet.

NOTE: If bearing is difficult to remove, use a suitable press.

29. Remove rear output shaft front bearing, thrust washer, clutch gear and output shaft from case.
30. Remove front output shaft rear bearing using a suitable press, Fig. 99.

NOTE: Support the case with wood blocks located at either side of case core to prevent damage to case.

31. Remove transfer case from press.
32. Remove front output shaft, clutch gear and sleeve, and the shaft rear bearing from case.
33. Remove front output shaft front bearing with tool J-22192-01 and a suitable press.
34. Remove front output shaft gear from shaft.
35. Remove input shaft rear needle bearing from rear output shaft with tool J-29369-1. Support output shaft in a vise during bearing removal.
36. Remove shift rod thimbles.

Assembly, Fig. 95

1. Apply Loctite 220 sealer or equivalent to shift rod thimbles, then install the thimbles.
2. Install front output shaft gear onto front

output shaft. Ensure that the clutch teeth on the gear face face toward shaft gear teeth.
3. Install front bearing on front output shaft using a suitable press. Ensure that the bearing is seated against the gear.
4. Install front output shaft into case, then the clutch sleeve and clutch gear on shaft.
5. Install front output shaft rear bearing with a suitable press, Fig. 100.

NOTE: Install an old yoke nut on the shaft to prevent thread damage.

6. Install input shaft rear beedle bearing in rear output shaft with tool J-19179.
7. Place rear output shaft clutch gear in case and insert rear output shaft into gear.
8. Install thrust washer and front bearing on rear output shaft using a suitable press.
9. Install shims and bearing on input shaft using a suitable press.
10. Install new input shaft oil seal into input shaft support using tool J-19184.
11. Install input shaft and bearing in support. Install new bearing snap ring.
12. Install rear output shaft gear on input gear. Install new gear retaining snap ring.
13. Using a feeler gauge, measure clearance between input gear and the gear retaining snap ring. Clearance should not exceed .003 inch. If clearance exceeds specification, disassemble input shaft and add shims between input shaft and shaft bearing.
14. Install clutch sleeve on rear output shaft.
15. Apply Loctite 515 sealer or equivalent to mating surface of input shaft support and install assembled support, shaft and gear into case. Use two support bolts to align the support on case and tap support into position with a mallet.
16. Install and torque socket head screws in support to 10 ft. lbs.
17. Install rear bearing cap front bearing race using tool J-9276-3.
18. Install rear bearing cap rear bearing using tool J-29182.
19. Position rear output shaft rear bearing in rear bearing cap.
20. Install rear output shaft yoke oil seal with tool J-25160.
21. Install speedometer gear and end play shims on rear output shaft.
22. Apply Loctite 515 sealer or equivalent to mating surface of rear bearing cap, then install the cap. Use two cap bolts to align bolt holes and tap rear cap into place with a suitable mallet. Install and torque cap bolts to 35 ft. lbs.
23. Install rear output shaft yoke. Torque new lock nut to 120 ft. lbs.
24. Check rear output shaft end play as follows:
    a. Install a dial indicator onto bearing cap and position indicator so the plunger contacts end of shaft.
    b. Move output shaft back and forth to check end play. End play should be .001–.005 inch.
    c. If end play is not within specifications, remove or add shims between speedometer drive gear and output shaft rear bearing.
25. Install front output shaft rear bearing race.
26. Install front output shaft end play shims and cover plate. Install and torque cover plate bolts to 35 ft. lbs.

NOTE: Apply Loctite 220 sealer or equivalent to bolt threads prior to installation.

27. Install front output shaft front bearing race using tools J-8092 and J-29181.
28. Install front output shaft yoke oil seal using tool J-25160.
29. Install shift rod oil seals using tool J-25167.
30. Apply Loctite 515 sealer or equivalent to mating surface of front bearing cap, then install the cap. Use two bolts to align cap and case bolt holes and tap cap into position with a mallet. Torque cap bolts to 35 ft. lbs.
31. Check front output shaft end play as follows:
    a. Seat rear bearing cup against cover plate by tapping end of front output shaft with a mallet.
    b. Install a dial indicator on front bearing cap so the plunger contacts end of output shaft.
    c. Move output shaft back and forth to check end play. End play should be .001–.005 inch.
    d. If end play is not within specifications, remove or add shims between cover plate and case. If shims are added, seat rear bearing cup as outlined previously before rechecking end play.
32. Install front output shaft yoke. Install and torque new lock nut to 120 ft. lbs.
33. Insert front and rear output shaft shift forks into case, Fig. 101.
34. Install front output shaft shift rod poppet ball and spring into front bearing cap.
35. Compress poppet ball and spring and install front output shaft shift rod partially into case. Insert front output shaft shift rod through shift fork. Align setscrew hole in shift fork, then install and torque setscrew to 14 ft. lbs.
36. Install rear output shaft shift rod poppet ball and spring into front bearing cap.
37. Compress poppet ball and spring and install rear output shaft shift rail partially into case.

NOTE: Before installing shift rail, ensure that the front output shaft shift rod is in the neutral position and that the interlocks are seated in the front bearing cap bore.

38. Insert rear output shaft shift rod through shift fork. Align setscrew holes in fork and rod. Install and torque setscrew to 14 ft. lbs.
39. Insert tool J-25142 in intermediate gear and install needle bearings and spacers into gear.
40. Install intermediate gear thrust washers into case. Ensure that the washer tangs are aligned with grooves in case. Secure thrust washers in position with petroleum jelly.
41. Install new "O" ring seal on intermediate shaft.
42. Position intermediate gear into case.
43. Install intermediate shaft in case core. Tap shaft into gear until shaft forces tool J-25142 from case.
44. Install intermediate shaft lock plate and bolt. Torque bolt to 23 ft. lbs.
45. Install bottom cover. Apply Loctite 515 sealer or equivalent to mating surface of cover. Install and torque cover bolts to 15 ft. lbs.

BORG WARNER 1345

Disassembly

1. Drain fluid by removing filler plug from case.
2. Remove both output shaft yoke nuts, then the front and rear output yoke, Fig. 102.
3. Remove 4-wheel drive indicator switch.
4. Remove case bolts and separate cover from case using a screwdriver.
5. Remove magnetic chip collector from bottom of case half.
6. Remove shift collar hub from the rear output shaft, then compress shift fork spring and remove the upper and lower spring retainers from shaft, Fig. 103.
7. Lift out 4-wheel drive lockup fork and shift collar as a complete assembly, then remove thrust washer.
8. Remove snap ring and thrust washer from front output shaft, then lift the chain, drive and driven sprockets vertically from the output shafts.
9. Remove front output shaft from case.
10. Remove oil pump attaching screws, then the oil pump rear cover, pick-up tube, filter and pump body, 2 pump pins, pump spring and oil pump front cover from rear output shaft.
11. Remove snap ring securing the bearing retainer inside case, then lift the rear output shaft while tapping on the bearing retainer with a soft mallet. Remove the rear output shaft and bearing retainer from case.

NOTE: When the retainer is removed, 2 dowel pins will fall into the case.

12. Remove rear output shaft from bearing retainer and then, if necessary, press needle bearing assembly from bearing retainer.
13. Remove C-clip securing shift cam to shift actuating lever, then the shift lever retaining screw and shift lever.

NOTE: When removing the shift lever, the shift cam will disengage from shift lever shaft and may release the detent ball and spring from case.

14. Remove planetary gear set, shift rail, shift cam, input shaft and shift forks as an assembly, Fig. 104.
15. Remove spacer washer from bottom of case.
16. Drive out plug from the detent spring bore using a drift.

Sub-Assembly Service

Planetary Gear Set Disassembly

1. Slide input shaft rearward from gear set, Fig. 105.
2. Remove snap ring, shift hub and planetary gear case from annulus gear.
3. Release locking plate from hub, then remove shift hub snap ring and T-shaped lock key. Remove shift hub.
4. Remove outer fiber washer, sun gear and inner fiber washer. Rotate inner fiber washer slightly upon removal to allow positioning tabs to clear planetary gears.

Planetary Gear Set Assembly

1. Place a new inner fiber washer into planetary gear housing and then install sun gear.

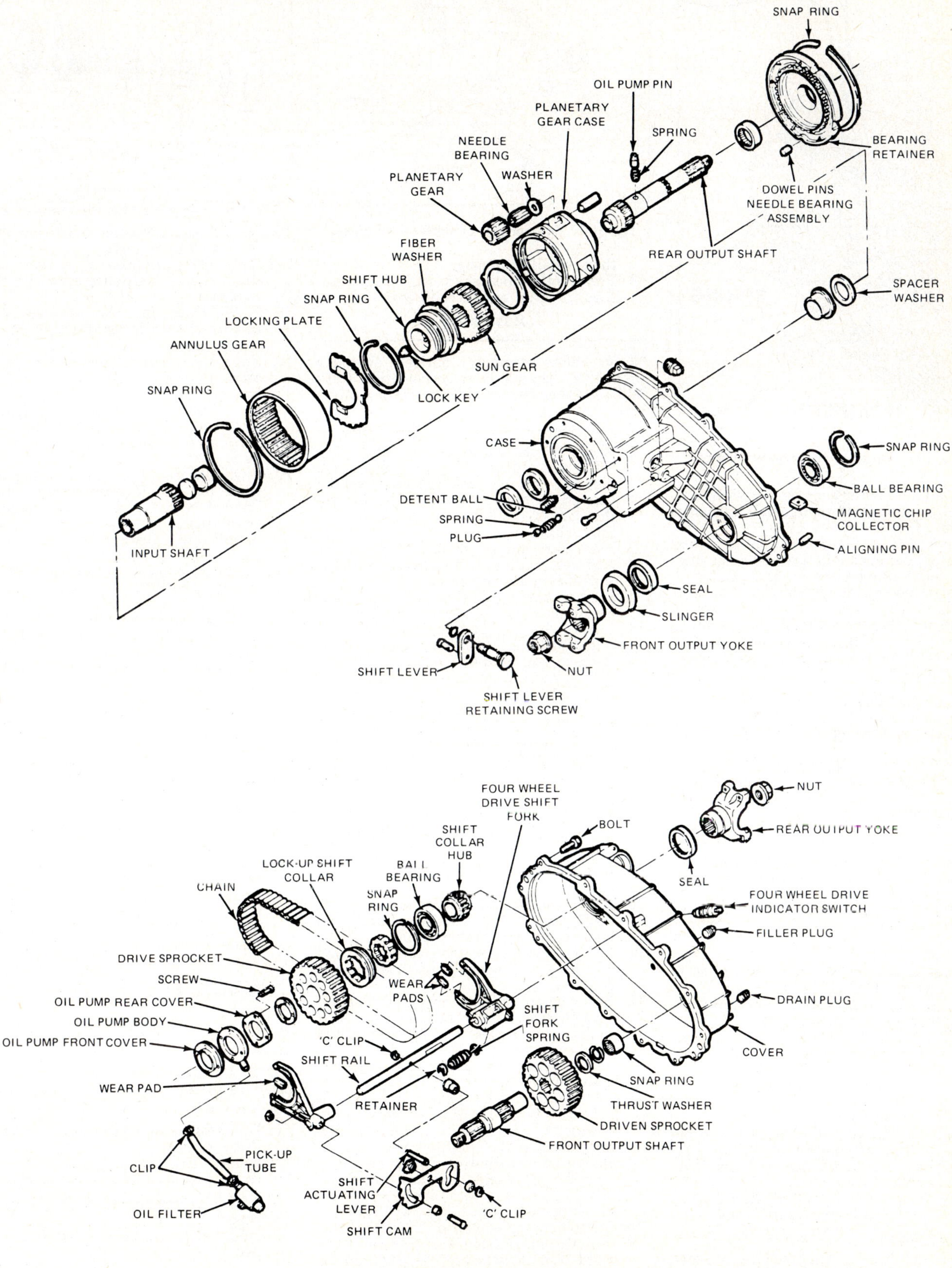

Fig. 102 Borg Warner 1345 transfer case exploded view

Fig. 103 Removal of spring retainers

2. Coat a new outer fiber washer with vaseline or equivalent, then install washer on hub.
3. Place hub in planetary gear cage and install T-shaped lock key and snap ring.
4. Install locking plate on shift hub with dished side facing toward planetary gear set.
5. Lower planetary assembly into annulus gear. Be sure the tabs on locking plate engage annulus gear teeth, then install the snap ring.

Cover, Disassembly

1. Remove snap ring retaining rear output shaft ball bearing assembly in the cover.
2. Turn cover over and remove the rear output shaft seal using seal remover 1175-AG and slide hammer T50T-100-A.
3. Remove speedometer drive gear.
4. Press rear output shaft ball bearing from cover.
5. Remove speedometer gear adapter.
6. Remove front output shaft inner needle bearing from cover using seal remover 0TC 33864 and slide hammer T50T-100-A.

Cover, Assembly

1. Press a new needle bearing into cover using tool T80T-7127-B.
2. Press a new ball bearing assembly into cover, then install snap ring.
3. Turn the cover and install the speedometer drive gear.
4. Install a new output shaft seal.
5. Install speedometer gear adapter.

Case, Disassembly

1. Remove snap ring retaining front output shaft ball bearing assembly in case.
2. Remove output shaft seal and both input shaft seals.
3. Press out the front output shaft bearing and input shaft bushing from case.

Case, Assembly

1. Press a new input shaft bushing into case, making sure that the lug is in downward position.
2. Install a new output shaft ball bearing,

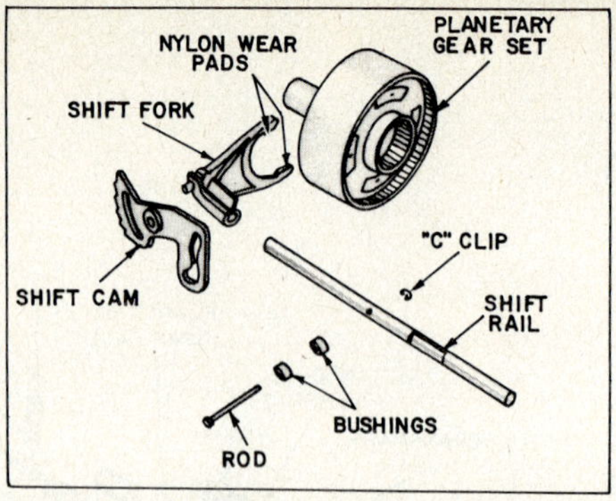

Fig. 104 Components of shifter mechanism

then install snap ring.

3. Press both input shaft seals into case, then the front output shaft seal.

Assembly

NOTE: Before assembly, lubricate all parts with ATF meeting Ford specification ESP-M2C138-CJ or Dexron II, Series D or equivalent.

1. Assemble planetary gear set, shift rail, shift cam, input shaft and shift fork together as a unit, making sure that the boss on the shift cam is installed toward the case. Install spacer washer on the input shaft.
2. Place rear output shaft into planetary gear set, making sure shift cam engages shift fork actuating pin.
3. Position case on its side and install rear output shaft and planetary gear set.
4. Install shift rail in case, then the outer roller bushing into guide in case.
5. Remove rear output shaft and position the shift fork in neutral position.
6. Place shift control lever shaft through the cam and install the clip ring. Be sure that the shift control lever is facing downward and is parallel to the front face of case.
7. Check shift fork and planetary gear engagement. The unit should operate freely.
8. Press new needle bearing into bearing retainer using tool T80T-7127-C, if the bearing was removed during disassembly.
9. Insert output shaft through bearing retainer.
10. Insert rear output shaft pilot into input shaft rear bushing. Align dowel holes and lower the bearing into position.
11. Install dowel pins, then the snap ring retaining the bearing retainer in the case.
12. Insert detent ball and spring into detent bore in case. Coat seal plug with RTV sealer or equivalent and drive plug into case until the lip of the plug is 1/32 inch below the surface of the case. Peen the case over the plug in 2 places.
13. Install oil pump front cover over the output shaft with the flanged side facing downward. The word "Top" must be facing the top of the transfer case.
14. Install oil pump spring and 2 pump pins with flat side facing outward in the output shaft. Push in both pins to install the oil pump body, pickup tube and filter.
15. Place oil pump rear cover on output shaft with the flanged side facing outward. The word "Top" must be positioned toward the top of case. Apply Loctite or equivalent to oil pump bolts, then install in pump cover and torque to 3–3.3 ft. lbs.

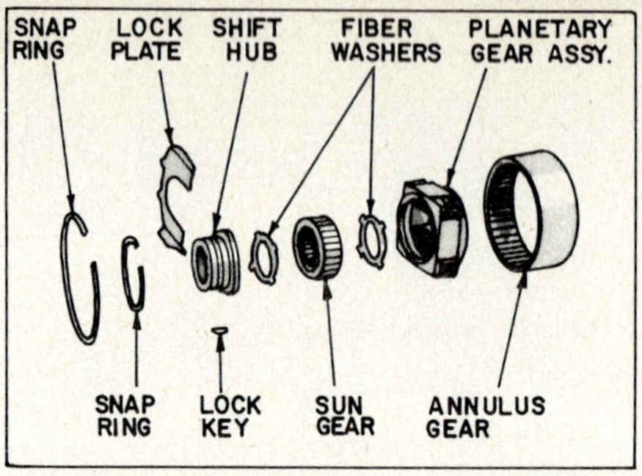

Fig. 105 Planetary gear set exploded view

16. Install thrust washer onto rear output shaft next to oil pump.
17. Place drive sprocket onto front output shaft, then install snap ring and thrust washer.
18. Install chain onto the drive and driven sprockets, then lower the unit into position in case.
19. Engage 4-wheel drive shift fork on shift collar, then slide shift fork over shift shaft and shift collar over rear output shaft. Be sure that the nylon wear pads are installed on shift fork tips and the tapered portion of the shift collar is facing rearward.
20. Push 4-wheel drive shift spring downward and install upper spring retainer. Then push spring upward and install lower retainer.
21. Install shift collar hub on rear output shaft.
22. Apply RTV sealer or equivalent to case mounting surface, then lower the cover over rear output shaft while aligning shift rail with the blind hole in cover. Also be sure the front output shaft is fully seated in its support bearing. Install and torque bolts to 40–45 ft. lbs. Allow 1 hour curing time for gasket material prior to operating vehicle.
23. Install 4-wheel drive indicator switch and torque to 8–12 ft. lbs.
24. Press oil slinger onto the front yoke, then install front and rear output shaft yokes. Coat the nuts with Loctite or equivalent and torque to 100–130 ft. lbs.
25. Refill transfer case with 6 pts. of ATF. Torque level and drain plugs to 6–14 ft. lbs. Torque fill plug to 15–25 ft. lbs.
26. Install transfer case into vehicle, then start engine and check transfer case for correct operation. Stop engine and check fluid level. Fluid should drip from level hole. If fluid flows out in a stream, the pump may not be operating properly.

MACK TC-15 & TC-150

Disassembly

NOTE: Do not disturb parts having heavy press fits unless replacement is necessary.

1. Drain fluid and remove case from vehicle, then mount in a suitable overhaul stand.

NOTE: To gain access to the shifter rail poppet ball capscrew, the unit should be mounted from the left hand mounting pad.

2. Remove oil lines and oil filter bracket with filter.

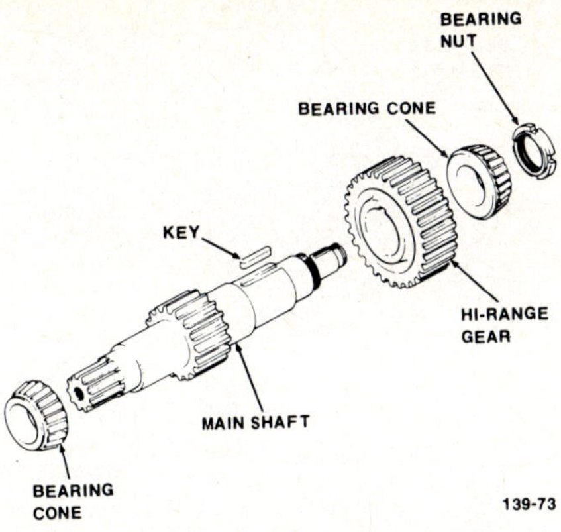

Fig. 106 Mainshaft exploded view

3. Position unit so that the front drive housing is facing upward. Remove front driving shaft yoke clamp plate capscrew and clamp plate, then the drive yoke. Remove front driving shaft bearing cover assembly and adjusting shims. Note number and thicknesses of shims for reassembly.
4. Remove front driveshaft bearing cover, then the bearing clamp plate capscrew and clamp plate.
5. Remove front drive housing end plate capscrews and dowel bolts. Then, carefully remove front drive housing end plate assembly with front drive shaft outer bearing cone. Place end plate in a press and using a suitable driver, remove front drive shaft bearing retainer.
6. Remove front driving shaft assembly.
7. Remove front driveshaft inner bearing cone and selective bearing spacer from shaft.
8. Remove front drive gear and gear snap ring from front drive shaft.
9. Remove front drive shifter assembly from front drive housing, then the front drive sliding clutch from front drive shaft.
10. Remove mainshaft front drive flange clamp plate capscrew and clamp plate, then the drive flange.
11. Remove front drive housing capscrews and dowel bolts. Then, carefully remove front drive housing.
12. Position the case so that the rear is facing upward. Remove oil pump rear cover, oil pump and adjusting shims. Note the number and thicknesses of shims for reassembly.
13. Remove rear driveshaft yoke clamp plate capscrew and clamp plate, then the drive yoke and rear driveshaft cover assembly.
14. Remove rear driveshaft bearing retainer capscrews. Install puller screws in holes provided in retainer and remove rear drive shaft and bearing retainer assembly from the case by applying pressure evenly and alternately to these screws.
15. Remove countershaft rear bearing cover and adjusting shims. Note the number and thicknesses of shims for reassembly.
16. Remove rear end plate capscrew and dowel bolts, then the rear end plate.
17. Remove 6 planet gears from planet gear cage.
18. Remove planet cage capscrews and the planet cage. Remove front driveshaft.
19. Remove mainshaft assembly from case.

NOTE: If the mainshaft is to be disassembled, the rear bearing nut must be loosened before removing the mainshaft from case. Place a soft iron wedge between the countershaft and mainshaft gears, then loosen nut using tool J24561.

20. Remove shifter rail poppet ball capscrew, located at the right hand mounting pad, then the poppet ball spring and poppet ball.
21. Remove Hi-Lo range shifter fork setscrews through rear case opening. Position the case horizontally and remove shifter rail from front of case.
22. Position the case vertically and remove Hi-Lo range shifter fork.
23. Remove the countershaft rear bearing clamp plate capscrew (L.H. thread) and end plate. Insert an eye bolt into tapped hole provided in the end of shaft and remove the countershaft assembly with the aid of a hoist.
24. Position the case so that the front is facing upward. Remove mainshaft and countershaft front bearing cover.
25. Position the case horizontally. Place a soft iron wedge between the driveshaft gear and case, remove the driveshaft bearing nut (L.H. thread) using tool J24560. Remove the driveshaft front bearing and selective bearing spacer.
26. Position the case vertically. Remove driveshaft assembly and thrust washer.
27. Position the case so that the front is facing upward. Remove the driveshaft bearing retainer capscrews. Install puller screws in the holes provided in the retainer and remove the retainer with outer race of bearing from case by applying pressure evenly and alternately to these screws.
28. Position the case so that the rear is facing upward, then remove driveshaft gear.

Sub-Assembly Service

NOTE: All moving parts should be coated with SAE 30 oil prior to assembly.

Mainshaft Disassembly

1. Remove front bearing from shaft using a suitable puller. Remove rear bearing nut which was previously loosened, Fig. 106.
2. Press mainshaft from rear bearing and Hi-range gear, then remove Hi-range gear key.

Mainshaft Assembly

1. Press front bearing cone onto shaft using a suitable driver until thick end of bearing seats firmly against shaft shoulder.
2. Turn the shaft over and insert gear key into shaft. Press Hi-range gear onto shaft with hub of gear facing toward rear of shaft. The gear should seat firmly against the shaft shoulder.
3. Press rear bearing cone onto shaft until thick end of bearing seats firmly against the gear hub, then install rear bearing nut.

Countershaft Disassembly

1. Remove countershaft rear bearing using a suitable puller. Then remove thrust washer, thrust washer lock pin, and Hi-range gear with bearings and spacer from shaft. Remove Hi-Lo range sliding clutch, Fig. 107.
2. Turn the countershaft over and remove front bearing clamp plate capscrew and clamp plate.
3. Remove countershaft front bearing using a suitable puller. Then remove the main drive gear, thrust washer ring, split thrust washers, Lo-range gear with roller bearings and spacer from shaft.

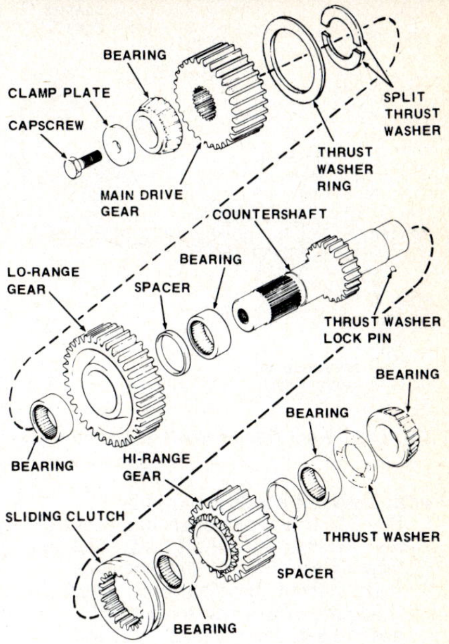

Fig. 107 Countershaft exploded view

Countershaft Assembly

1. Assemble Lo-range gear rear roller bearing, spacer and front roller bearing into the gear. Then install Lo-range gear assembly onto front of shaft with clutch teeth facing toward sliding clutch teeth of shaft.
2. Install main drive gear split thrust washers into groove of shaft, then press thrust ring over thrust washers. Install main drive gear onto shaft splines with hub of gear facing toward Lo-range gear.
3. Press countershaft front bearing onto shaft until thick end of bearing seats firmly against main drive gear. Install front bearing clamp plate and coat the capscrew with Loctite No. 262. Install and torque capscrew to 530–570 ft. lbs.
4. Turn the countershaft over and install Hi-Lo range sliding clutch onto shaft.
5. Assemble Hi-range gear front roller bearing, spacer and rear roller bearing into the gear. Then install Hi-range gear assembly onto rear of shaft with clutch teeth facing toward sliding clutch.
6. Insert rear bearing spacer lock pin into hole provided in the shaft, then slide spacer over lock pin.
7. Press countershaft rear bearing onto shaft until thick end of the bearing seats firmly against the spacer. Check the Hi-range and Lo-range gear end play. End play should be .008–.018 inch.

Main Drive Shaft Disassembly

1. Remove the driveshaft rear bearing.

Main Drive Shaft Assembly

1. Place the main driveshaft in a press. Press the rear bearing onto shaft using a suitable driver until thick end of bearing seats firmly against shaft shoulder.

Front Driving Shaft Disassembly

1. Remove front driving shaft rear bearing using a suitable puller, Fig. 108.
2. Turn the shaft over and pull the front bearing, bearing spacer and speedometer drive worm together from the shaft.

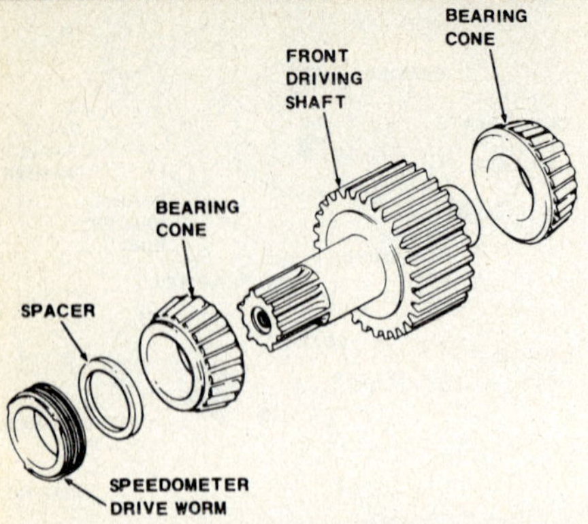

Fig. 108 Front driving shaft exploded view

Front Driving Shaft Assembly

1. Place front driving shaft in a press. Press rear bearing onto shaft until thick end of bearing seats firmly against gear.
2. Install front bearing onto shaft until thick end of bearing seats firmly against gear. Then install front bearing spacer and speedometer worm drive onto front of shaft.

Rear Drive Shaft Disassembly

1. Place rear driveshaft in a press. Press shaft from the outer bearing cone and bearing retainer, then remove the bearing selective spacer from shaft, Fig. 109.
2. Remove inner bearing cone from shaft using a suitable puller.
3. Remove bearing cups from bearing retainer using a suitable puller.

Rear Drive Shaft Assembly

1. Place rear driveshaft bearing retainer in a press. Press the inner bearing cup into retainer using a suitable driver until thick end of the cup seats firmly against internal flange.
2. Turn the retainer over and in the same manner, press outer bearing cup into retainer. Remove retainer from press.
3. Place rear driveshaft in a press. Press front driveshaft rear bushing into pilot end of shaft using a suitable driver.
4. Turn the rear driveshaft over and press inner bearing cone onto shaft until thick end of bearing seats firmly against shaft shoulder.
5. Install selective bearing spacer on shaft.
6. Place the retainer assembly over assembled parts on rear driveshaft and seat inner bearing cup onto inner bearing cone. Then, press outer bearing cone onto shaft.
7. Establish bearing end play as follows:
   a. Place rear driveshaft in a press with a tubular driver seated against the outer bearing cone, then apply a load of 5,000 lbs. Check for .001–.005 inch end play using a dial indicator, Fig. 110.
   b. If the end play is not within specifications, use a thinner or thicker spacer to adjust end play.
8. Remove the retainer assembly from press and install a new O-ring into groove provided in hub of retainer.

Front Driveshaft Bearing Retainer Disassembly

1. Remove driveshaft inner and outer bearing cups from retainer using a suitable puller, Fig. 111.

Front Driveshaft Bearing Retainer Assembly

1. Place retainer in a press. Press inner bearing cup into retainer using a suitable driver until thick end of the cup seats firmly against centered internal flange.
2. Turn the retainer over and in the same manner, press outer bearing cup into retainer.
3. Establish bearing end play as follows:
   a. Insert inner and outer bearing cones with selector bearing spacer into retainer, then place retainer in a press. Using a suitable driver seated against the bearing cone, apply a load of 5,000 lbs. Check for .001–.005 inch end play using a dial indicator, Fig. 112.
   b. If the end play is not within specifications, use a thinner or thicker spacer to adjust end play.

Oil Pump Disassembly

1. Remove end plate, then the outer and inner rotors.

Oil Pump Assembly

1. Place inner and outer rotors in pump housing recess.
2. Check the running clearance. Clearance should be .004–.006 inch, Fig. 113.
3. Check the side clearance. Clearance should be .0015–.0030 inch.
4. Install end plate with 4 setscrews. Make sure the rotors turn freely after end plate has been installed, then stake set screws to pump housing.

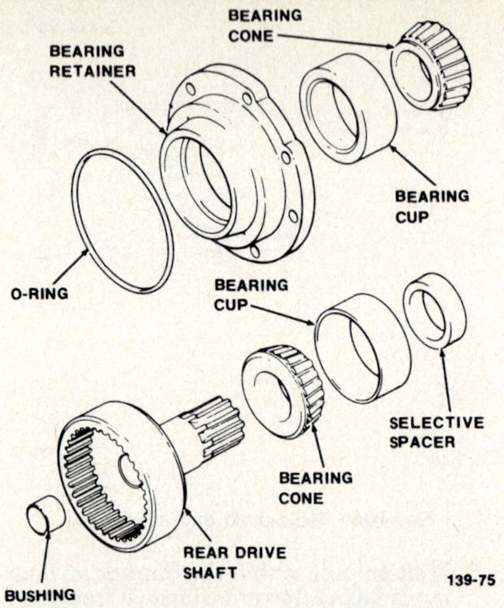

Fig. 109 Rear drive shaft exploded view

Assembly

NOTE: Replace all oil seals if a complete overhaul is performed.

1. Position the case vertically with rear facing upward, then install driveshaft gear.
2. Turn the case over with front facing upward. Install driveshaft bearing retainer using stud guides, then remove stud guides and install the retainer capscrews. Torque capscrews to 55–75 ft. lbs.
3. Turn the case over with rear facing upward. Place driveshaft gear thrust washer onto face of driveshaft gear, then begin moving the driveshaft downward into case. Insert the shaft through thrust washer and gear until rear bearing cone seats in the cup.
4. Position the case horizontally and install driveshaft selective bearing spacer and front bearing cone onto front end of shaft.
5. Place a soft iron wedge between driveshaft gear and case, then coat driveshaft front bearing nut with Loctite No. 262 or equivalent. Install and torque nut (L.H. thread) to 800–1200 ft. lbs. using tool J24560.

Fig. 110 Checking rear drive shaft end play

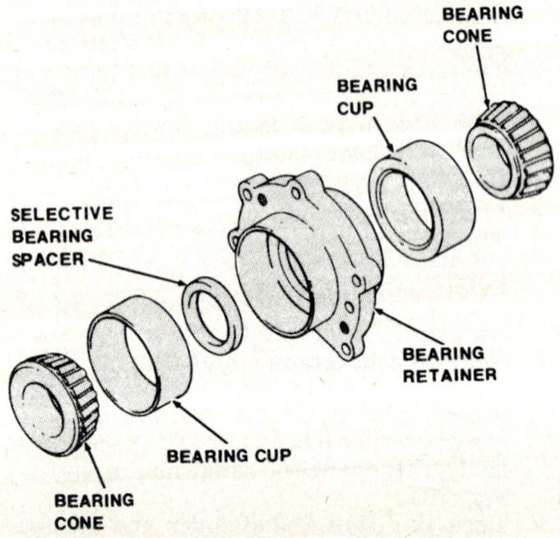

Fig. 111 Front drive shaft bearing retainer exploded view

Fig. 112 Checking front drive shaft bearing end play

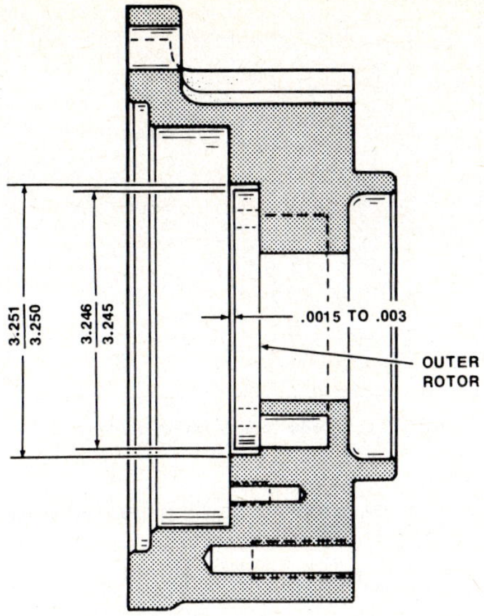

Fig. 113 Oil pump sectional view

6. Position the case vertically with rear facing upward. Check driveshaft bearing end play using a dial indicator. The end play should be .001-.005 inch. Use a thinner or thicker selective bearing spacer to adjust end play.

NOTE: When correct end play is established, place the case horizontally and stake the lip of the driveshaft front bearing nut to shaft.

7. Turn the case over with front facing upward, then install mainshaft and countershaft front bearing cover with a gasket. Install and torque cover capscrews to 45-65 ft. lbs.

NOTE: If the mainshaft and countershaft bearings were replaced, the mainshaft and countershaft front bearing cups must be installed before installing the mainshaft and countershaft assemblies. The bearing cup can be positioned to the correct depth by installing the cover with gasket onto the bearing cup, then pressing the bearing cup into the case with cover until the cover seats against the case.

8. Turn the case over with rear facing upward and install countershaft assembly into case. Install countershaft rear bearing clamp plate and coat the capscrew with Loctite No. 262 or equivalent. Install and torque capscrew (L.H. thread) to 530-570 ft. lbs.
9. Engage Hi-Lo range shifter fork into groove provided in Hi-Lo range sliding clutch.
10. Position the case horizontally, then begin to install the shifter rail into hole provided in front of case. Insert rail through hub of fork, then install and torque fork setscrews to 35-45 ft. lbs.
11. Install shifter rail poppet ball, spring and capscrew into hole provided at the right hand mounting pad.
12. Position the case vertically with rear facing upward, then install mainshaft assembly.

NOTE: If the mainshaft was disassembled, the rear bearing nut must be torqued. Place a soft iron wedge between the countershaft and mainshaft gears. Coat the nut with Loctite No. 262 and torque to 800-1200 ft. lbs. using tool J24561.

13. Install front driveshaft into inner bore of driveshaft.
14. Align planet cage capscrew holes with driveshaft capscrew holes, then mesh planet cage external splines with driveshaft internal splines. Install and torque the capscrews to 55-75 ft. lbs.
15. Install planet gears into planet cage.
16. Install a new rear end plate gasket, then carefully install rear end plate onto the case. Install and torque rear end plate to case capscrews and dowel bolts to 55-75 ft. lbs.
17. Install countershaft rear bearing cover and adjusting shims. Then install and torque capscrews to 45-65 ft. lbs.
18. Establish countershaft rear bearing end play as follows:
    a. Remove countershaft rear bearing cover oil line elbow, then insert a dial indicator spindle into bearing oil cover hole until spindle contacts edge of clamp plate screw.
    b. Use a pry bar placed through the rear driveshaft bearing retainer opening to check end play. If the end play is not .003-.008 inch, use thinner or thicker shims to adjust end play, Fig. 114.
    c. Remove dial indicator and reinstall oil line elbow.
19. Install rear driveshaft assembly. Install and torque the rear driveshaft capscrews to 45-65 ft. lbs.
20. Install rear driveshaft cover assembly with capscrews and torque screws to 18-28 ft. lbs.
21. Install rear driveshaft yoke, yoke clamp plate with capscrew coated with Loctite No. 262 or equivalent, then torque capscrew to 530-570 ft. lbs.
22. If mainshaft bearings were replaced, install and seat a new mainshaft rear bearing cup into end plate bore. Position adjusting shims over rear end of mainshaft and on the rear end plate. Align key in oil pump with slot in end of mainshaft. Then install oil pump onto mainshaft.

NOTE: Before installing oil pump, be sure rotors turn freely without binding.

23. Install and torque the oil pump capscrews to 45-65 ft. lbs. Then install oil pump rear cover with capscrews and torque screws to 18-28 ft. lbs.
24. Turn the case over with front facing upward, then install mainshaft front drive yoke, yoke clamp plate and coat capscrew with Loctite No. 262 or equivalent. Torque capscrew to 530-570 ft. lbs.
25. Check mainshaft bearing end play using a dial indicator. End play should be .003-.008 inch. If the end play is not within specifications, use thinner or thicker shims between rear end plate and oil pump to adjust.
26. Install front drive housing, then the front drive housing capscrews and dowel bolts and torque to 55-75 ft. lbs.
27. Install front drive sliding clutch onto front driveshaft.
28. Position gasket onto front drive shifter assembly, then install shifter assembly while engaging shifter fork into groove provided in front drive sliding clutch. Install and torque shifter capscrews to 10-15 ft. lbs.

Fig. 114 Checking countershaft bearing end play

29. Install front drivegear snap ring into groove provided in shaft, then the front drivegear with teeth facing rearward onto front driveshaft.
30. Install front driveshaft inner bearing cone and selective bearing spacer onto shaft.

NOTE: Correct bearing spacer thickness was obtained when establishing bearing end play described in "Sub-Assembly Service" section under "Front Driveshaft Bearing Retainer Assembly".

31. Install front driving shaft assembly.
32. Install a new front driveshaft bearing retainer to end plate gasket, then place front drive housing end plate in a press and install front driveshaft bearing retainer assembly using a screwdriver. Install end plate assembly onto front drive housing, then the front driveshaft outer bearing cone into retainer. Install and torque end plate to front drive housing capscrews and dowel bolts to 55-75 ft. lbs.
33. Install front driveshaft bearing clamp plate and coat the capscrew with Loctite No. 262 or equivalent. Torque screw to 530-570 ft. lbs.
34. Install a new front drive bearing retainer to bearing cover gasket, then install the front driveshaft bearing cover with capscrews. Torque capscrews to 30-40 ft. lbs.
35. Install front driving shaft adjusting shims and bearing cover. Then install and torque capscrews to 18-28 ft. lbs.
36. Install front driving shaft drive yoke, drive yoke clamp plate and coat the capscrew with Loctite No. 262 or equivalent, then torque capscrew to 530-570 ft. lbs.
37. Check the front driving shaft bearing end play using a dial indicator. End play should be .001-.005 inch. If the end play is not within specifications, add or remove shims to adjust end play.
38. Install transfer case oil lines and oil filter bracket with filter.
39. Install transfer case into vehicle and fill with GO-F gear oil to the level of filler plug hole (approx. 20 pts.), then check transfer case operation.

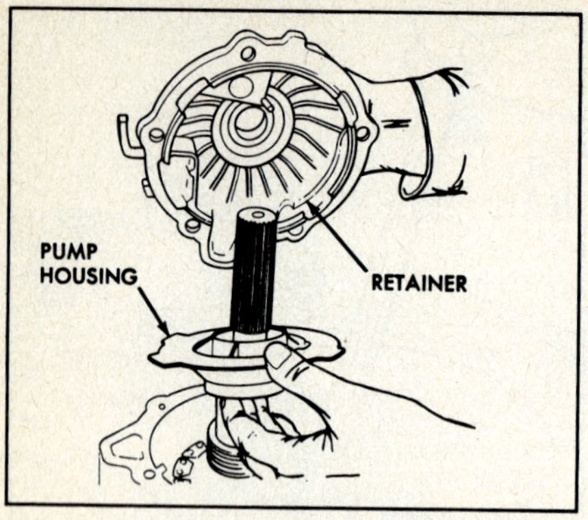

Fig. 115 Pump housing & retainer removal

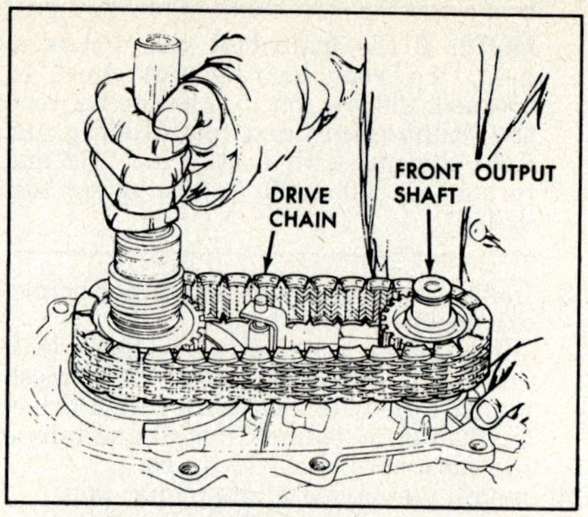

Fig. 116 Front output shaft & chain replacement

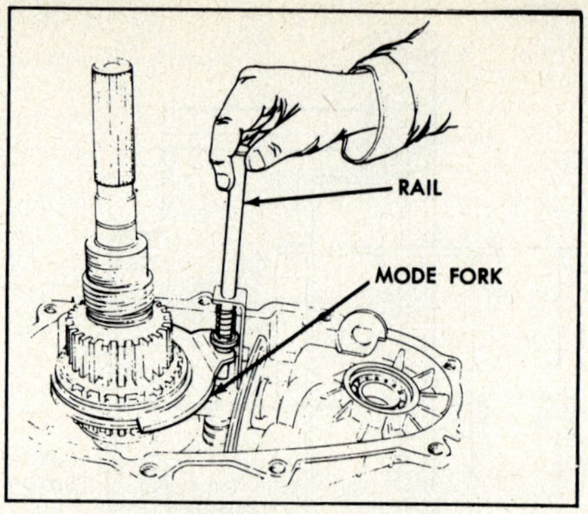

Fig. 117 Mode fork & rail replacement

NEW PROCESS MODEL 207

Disassembly

1. Remove fill and drain plugs and front yoke. Discard yoke seal washer and nut.
2. Turn transfer case on end and support front case on wooden blocks, then shift to "4 Lo" position.
3. Remove extension housing, then remove and discard mainshaft rear bearing snap ring.
4. Remove rear retainer and pump housing, Fig. 115. Discard pump housing seal.
5. Remove speedometer drive gear and the pump gear from mainshaft.
6. Remove rear case attaching bolts and separate rear case from front case by prying with screwdrivers at slots provided in case ends.
7. Remove front output shaft and chain as an assembly.

NOTE: Raise the mainshaft slightly to provide clearance for removing the output shaft, Fig. 116.

8. Lift mode fork rail until it clears range fork, then rotate mode fork and rail and remove from case, Fig. 117.
9. Remove mainshaft, then the planetary assembly with range fork from case, Fig. 118.
10. Remove planetary thrust washer, input gear, thrust bearing and front thrust washer from case, Fig. 119.
11. Remove shift sector detent spring, retaining bolt, the shift sector, shaft and spacer from case, Fig. 120.
12. Remove locking plate from case, Fig. 121.
13. Remove inner gear pilot bearing using tool J–29369–1 and a slide hammer.
14. Remove front output shaft seal, input shaft seal and rear extension seal using a brass drift or screwdriver.
15. Remove the two front input shaft caged roller bearings from case using tool J–33841 and J–8092, Fig. 122.
16. Remove front output shaft rear bearing using tool J–29369–2 with J–33367 or a slide hammer, Fig. 123.
17. Remove rear mainshaft bearing from rear retainer, then remove front output shaft bearing snap ring and the bearing.
18. Remove extension housing bushing using tools J–33839 and J–8092.

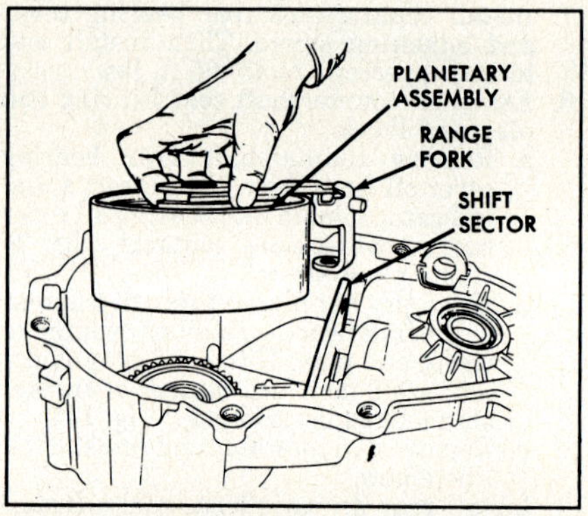

Fig. 118 Planetary gear assembly & range fork replacement

Sub-Assembly Service

Mainshaft Disassembly

1. Remove speedometer gear and pump gear from mainshaft, Fig. 124.
2. Remove synchronizer hub snap ring, then the synchronizer hub.
3. Remove drive sprocket, then, using tools J–33826 and J–8092, remove the two caged roller bearings from sprocket.
4. Remove synchronizer keys and retaining rings from synchronizer hub.

Planetary Gear Disassembly

1. Remove snap ring, then remove and discard outer thrust ring, Fig. 125.
2. Separate planetary assembly from annulus gear, then remove and discard inner thrust ring.

Planetary Gear Assembly

1. Install a new inner thrust ring on planetary assembly, then install planetary assembly into annulus gear.
2. Install a new outer thrust ring and the snap ring in the planetary gear.

Mainshaft Assembly

1. Install front drive sprocket bearing using tools J–33828 and J–8092 until tool bottoms out and bearing is flush with front surface.
2. Reverse tool on J–8092 and install rear bearing into sprocket until tool bottoms out and bearing is recessed in bore, Fig. 126.

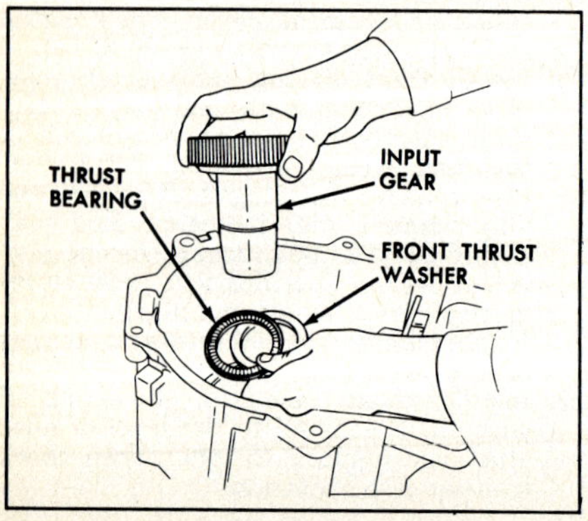

Fig. 119 Input gear replacement

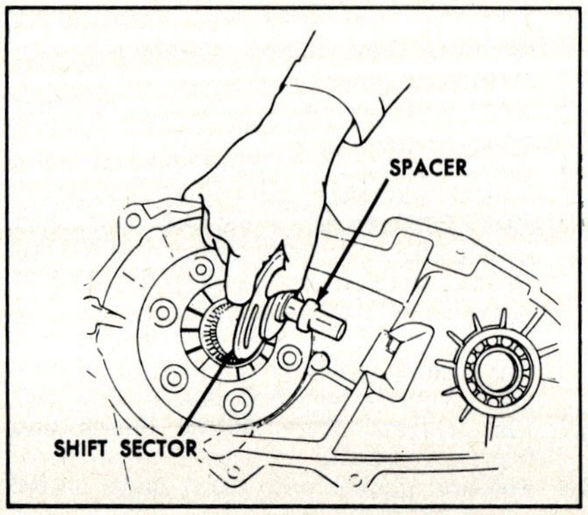

Fig. 120 Shift sector removal

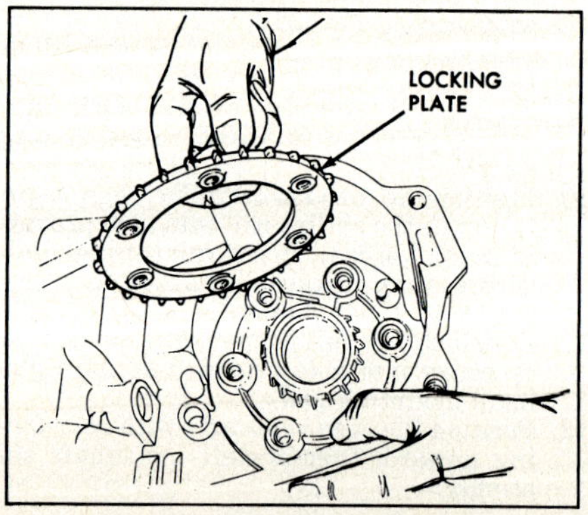

Fig. 121 Locking plate removal

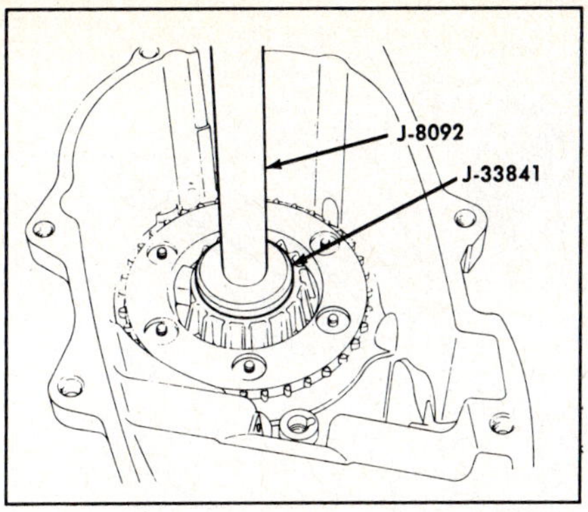

Fig. 122 Front input shaft bearing removal

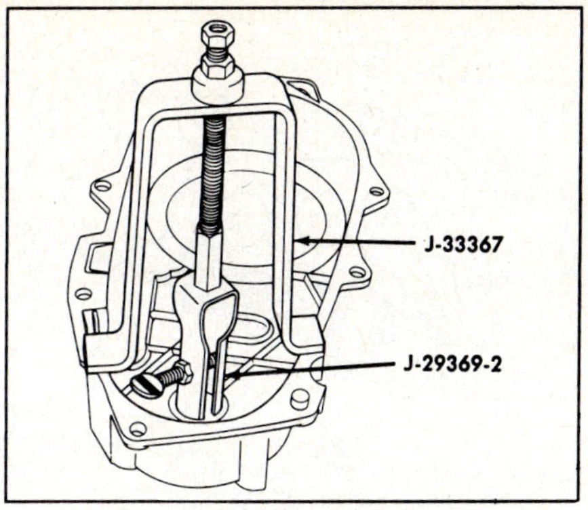

Fig. 123 Front output shaft rear bearing removal

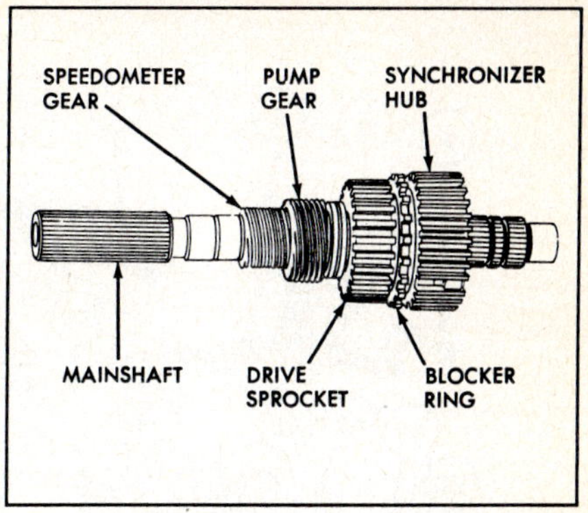

Fig. 124 Mainshaft components

3. Install thrust washer, drive sprocket, blocker ring and synchronizer hub on mainshaft.
4. Install hub with a new snap ring on mainshaft.
5. Install pump gear and speedometer gear on mainshaft.

Assembly

NOTE: All bearings used in the transfer case must be properly aligned to avoid blocking the oil passages. After installing bearings, check their position to make sure that oil passages are clear.

1. Install lock plate in case. Coat case and lock plate surfaces around bolt holes with Loctite 515, or equivalent. Install attaching bolts and torque to 20–30 ft. lbs.
2. Install input shaft roller bearings into case using tool J–33830 and J–8092, Fig. 127.
3. Install front output shaft rear bearing using tools J–33832 and J–8092.
4. Install front output shaft front bearing using tools J–33833 and J–8092.
5. Install front output shaft bearing snap ring, then install the seal using tool J–33834.
6. Install input shaft seal using tool J–33831.
7. Install spacer on shift sector shaft, then install sector and shift lever in case. Torque shift lever retaining nut to 15–20 ft. lbs.
8. Install shift sector detent spring and retaining bolt.

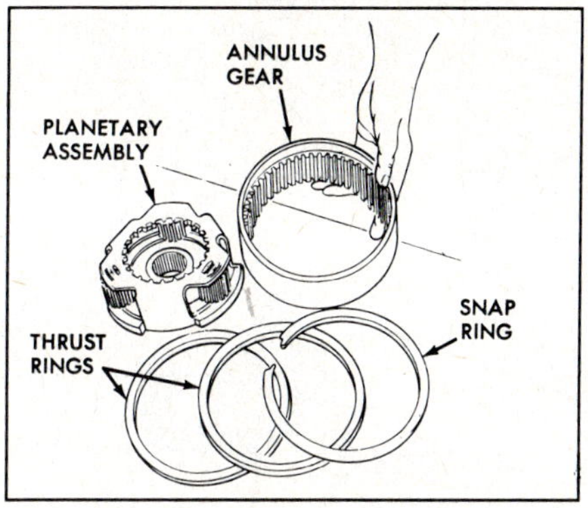

Fig. 125 Planetary gear assembly

9. Install pilot bearing into input gear using tools J–33829 and J–8029.
10. Install input gear front thrust washer, thrust bearing and input gear in case, Fig. 119.
11. Install planetary gear thrust washer on input gear and range fork on planetary assembly, then install planetary assembly into case, Fig. 118.
12. Install mainshaft into case, ensuring that thrust washer is aligned with input gear and planetary assembly, Fig. 128.
13. Install mode fork on synchronizer sleeve so that it is aligned with range fork, then slide mode fork down through range fork until rail is seated in bore of case, Fig. 117.
14. Position drive chain around front output shaft and drive sprocket, then install front output shaft in case, Fig. 116.
15. Install magnet into pocket in case.
16. Apply a 1/8 inch bead of Loctite 515 or equivalent to front case mating surface, then position rear case on front case dowel pins. Install case bolts and torque to 20–25 ft. lbs., then install two bolts and washers into the dowel pin holes.
17. Install output bearing into rear retainer using tools J–33833 and J–8092.
18. Install seal in pump housing using tool J–33835, then apply petroleum jelly to pump housing tabs and install housing in rear retainer.
19. Apply a 1/8 inch bead of Loctite 515 or equivalent to mating surface of rear retainer, then install retainer onto case. Torque bolts to 15–20 ft. lbs.
20. Install a new snap ring on mainshaft, then install bushing in extension housing using tools J–33826 and J–8092.
21. Install a new extension housing seal using tool J–33843.
22. Apply a 1/8 inch bead of Loctite 515 or equivalent to mating surface of extension housing, then install housing onto case. Torque bolts to 20–25 ft. lbs.
23. Install front yoke on output shaft using a new seal washer and nut. Torque nut to 90–130 ft. lbs.
24. Install drain and fill plugs and torque to 30–40 ft. lbs.

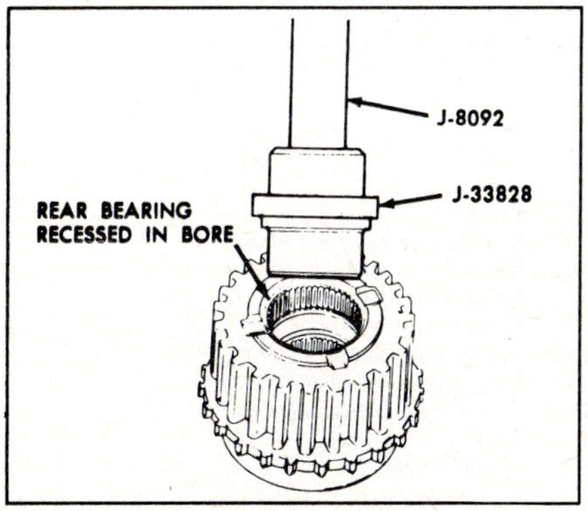

Fig. 126 Drive sprocket rear bearing installation

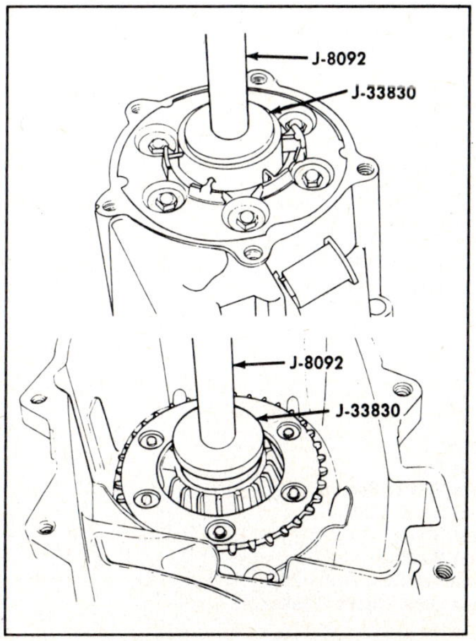

Fig. 127 Input shaft bearing installation

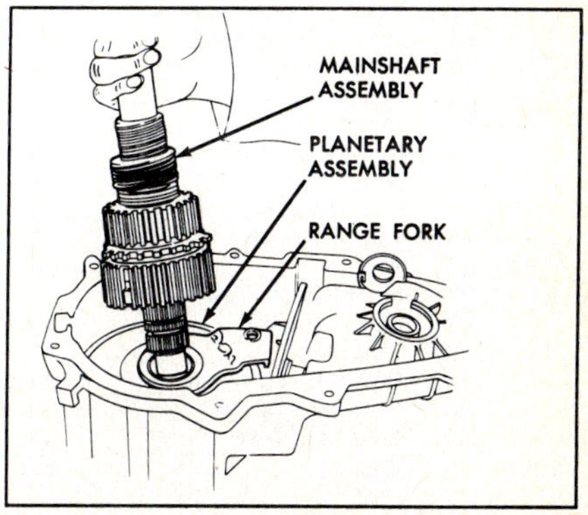

Fig. 128 Mainshaft installation

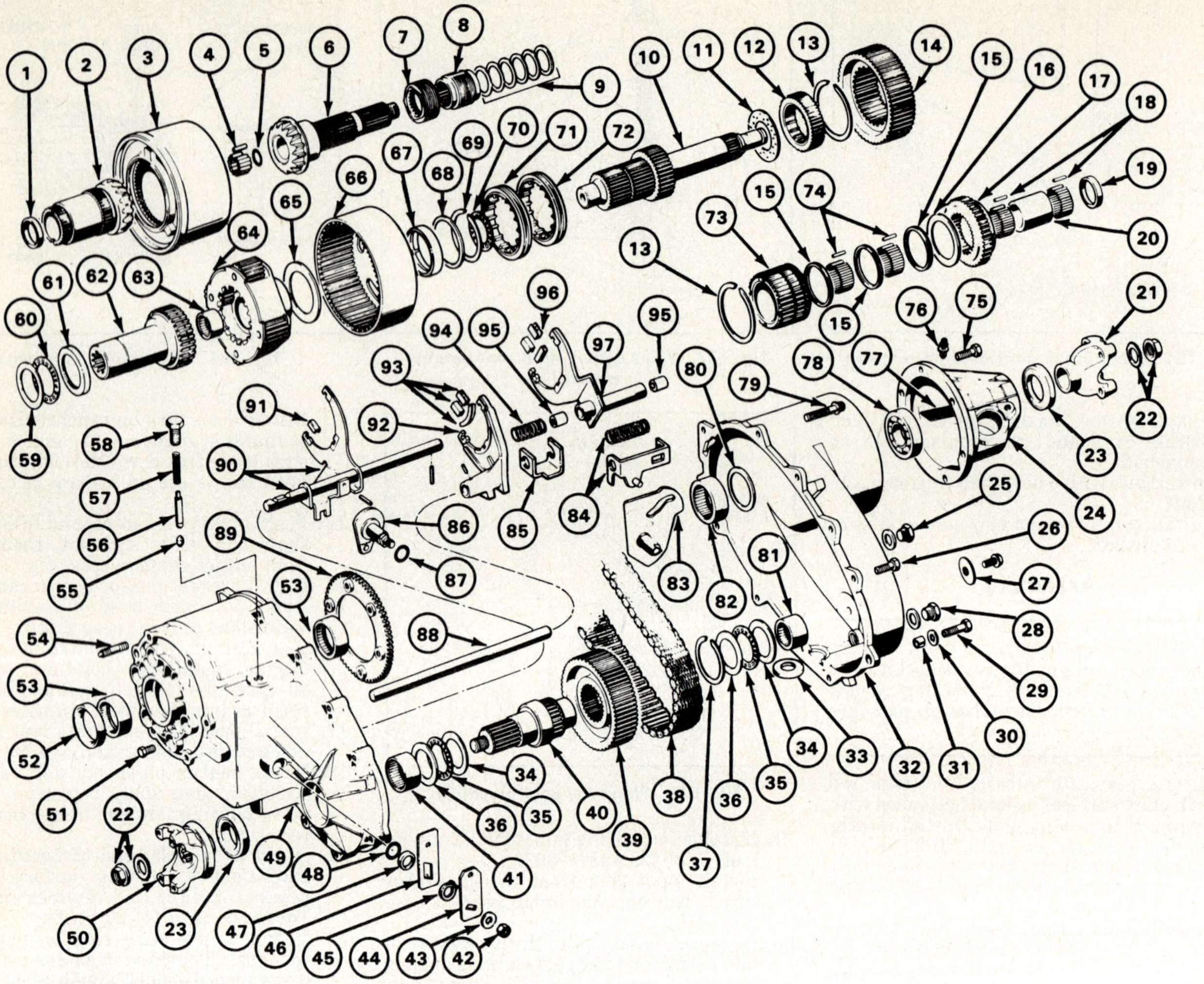

1. Spacer
2. Side Gear
3. Viscous Coupling
4. Pilot Bearing Rollers
5. O-Ring Seal
6. Rear Output Shaft
7. Oil Pump
8. Speedometer Drive Gear
9. Shim Kit
10. Mainshaft
11. Mainshaft Thrust Washer
12. Side Gear Clutch Gear
13. Retaining Ring
14. Sprocket
15. Spacer
16. Sprocket Thrust Washer
17. Viscous Clutch Gear
18. Side Gear Roller (82)
19. Spacer (Short)
20. Spacer (Long)
21. Rear Yoke
22. Nut and Seal Washer
23. Seal
24. Rear Retainer
25. Plug Assembly
26. Bolt
27. Identification Tag
28. Plug Assembly
29. Dowel Bolt
30. Dowel Bolt Washer
31. Case Half Dowel
32. Rear Half Case
33. Magnet
34. Front Output Shaft Bearing Assembly Race (Thick)
35. Front Output Shaft Bearing Assembly Thrust
36 Front Output Shaft Bearing Assembly Race (Thin)
37. Retaining Ring
38. Chain
39. Driven Sprocket
40. Front Output Shaft
41. Front Output Front Bearing
42. Nut
43. Washer
44. Mode Lever
45. Snap Ring
46. Range Lever
47. O-Ring Retainer
48. O-Ring Seal
49. Front Half Case
50. Front Output Yoke
51. Low Range Plate Bolt
52. Input Shaft Oil Seal
53. Input Shaft Bearing
54. Stud
55. Ball
56. Plunger
57. Plunger Spring
58. Screw
59. Input Race
60. Input Thrust Bearing
61. Input Race (Thick)
62. Input Shaft
63. Input Bearing
64. Planetary Gear Assembly
65. Input Gear Thrust Washer
66. Annulus Gear Assembly
67. Annulus Bushing
68. Thrust Washer
69. Retaining Ring
70. Thrust Bearing
71. High Range Sliding Clutch Sleeve
72. Mode Sliding Clutch Sleeve
73. Carrier
74. Carrier Rollers (120)
75. Rear Retainer Bolt
76. Vent
77. Vent Seal
78. Output Bearing
79. Bolt
80. Seal
81. Front Output Rear Bearing
82. Output Shaft Inner Bearing
83. Range Sector
84. Range Bracket (Outer) and Spring
85. Range Bracket (Inner)
86. Mode Sector
87. O-Ring Seal
88. Range Rail
89. Low Range Lockout Plate
90. Mode Fork, Rail and Pin
91. Mode Fork Pad
92. Range Fork
93. Range Fork Pads
94. Range Bracket Spring (Inner)
95. Locking Fork Bushing
96. Locking Fork Pads
97. Locking Fork

Fig. 129 Exploded view of New Process model 229 transfer case

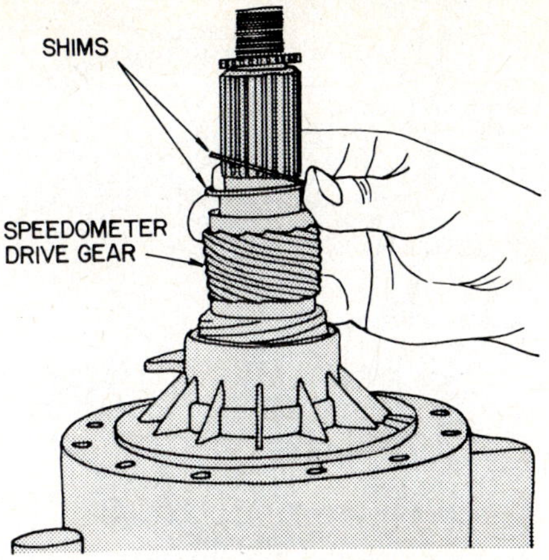

Fig. 130 Differential shims and speedometer drive gear replacement

NEW PROCESS MODEL 229

Disassembly, Fig. 129

1. Remove fill and drain plugs and drain lubricant from transfer case.
2. Remove front and rear yoke nuts and seal washers. Discard washers.
3. Mark relationship between front and rear yokes and propeller shafts for reference during reassembly, then remove yokes.
4. Support transfer case on wood blocks with V-notches cut in blocks to provide clearance for front case mounting studs.
5. Mark relationship between rear retainer and rear case for reference during reassembly, then remove rear retainer attaching bolts and pry retainer off case by inserting screwdrivers in slots provided in case.
6. Remove differential shim(s) and speedometer drive gear from rear output shaft, Fig. 130.
7. Remove rear case to front case attaching bolts, then separate the case halves by prying with screwdrivers in slots at case ends.
8. Remove thrust bearing and races from front output shaft, Fig. 131. Note position of bearing and races for reference during reassembly.

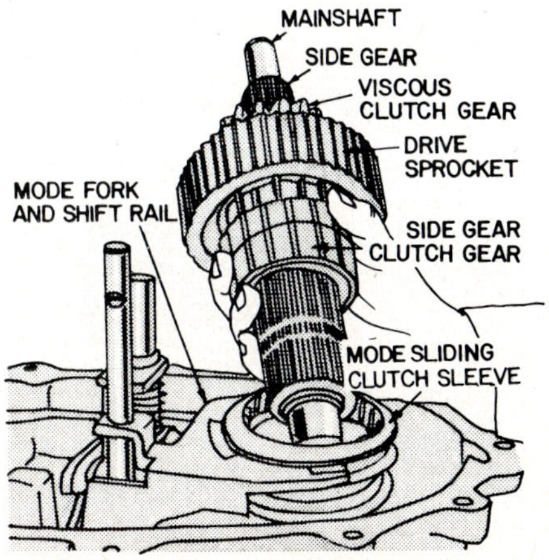

Fig. 133 Mainshaft assembly replacement

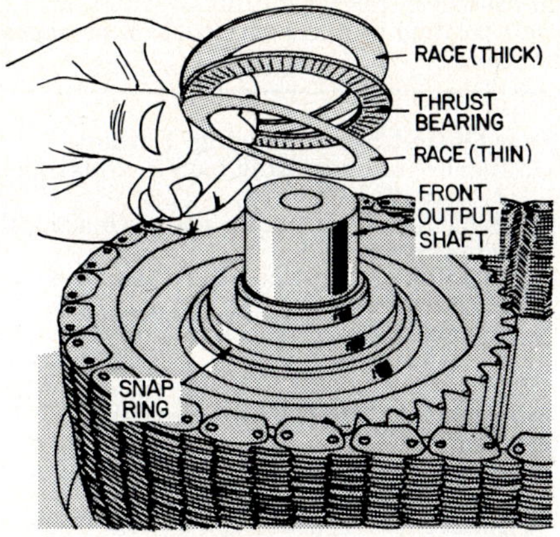

Fig. 131 Front output shaft thrust bearing replacement

9. Remove oil pump from rear output shaft for reference during reassembly.
10. Remove mainshaft pilot bearing rollers (15) from shaft, or from coupling if rollers fell off during rear output shaft removal.
11. Remove "O" ring from end of mainshaft, then the viscous coupling from mainshaft and side gear, Fig. 132.
12. Remove front output shaft, driven sprocket and drive chain assembly. To ease removal, lift the assembly upward, tilt front shaft towards mainshaft and slide chain off drive sprocket.
13. Remove front output shaft front thrust bearing from case or shaft, then the drive chain from front output shaft and sprocket.
14. Remove snap ring retaining driven sprocket to front output shaft, then remove sprocket from shaft. Mark relationship between shaft and sprocket for reference during reassembly.
15. Remove mainshaft, side gear, clutch gear, drive sprocket and spline gear as an assembly, Fig. 133.
16. Remove mode fork, shift rail and clutch sleeve as an assembly, Fig. 134. Mark relationship between sleeve and fork for reference during reassembly and remove sleeve from fork.
17. Remove locking fork, clutch sleeve, fork brackets and fork springs as an assembly, then disassemble components for cleaning and inspection.
18. Remove range sector detent screw, spring, plunger and ball.

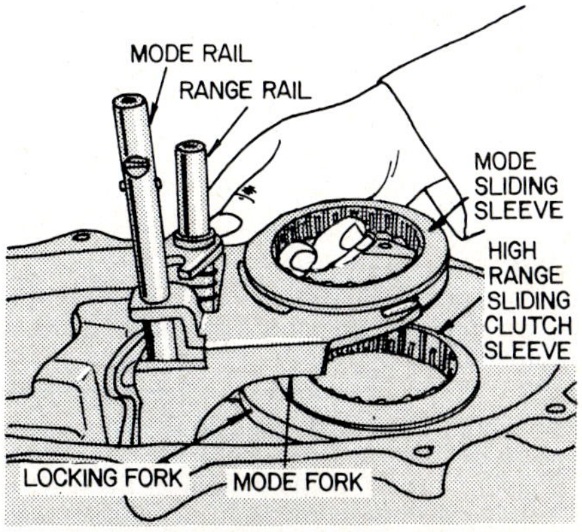

Fig. 134 Mode fork, shift fork and clutch sleeve removal

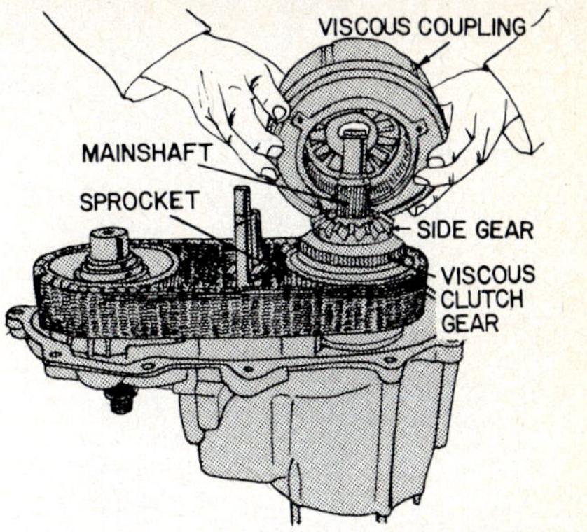

Fig. 132 Viscous coupling replacement

19. Disengage low range fork lug from range sector slot.
20. Remove annulus gear snap ring and thrust washer, then the annulus gear, range fork and rail as an assembly. Disassemble components for cleaning and inspection.
21. Remove thrust washer from planetary assembly hub, then remove planetary gear assembly.
22. Remove mainshaft thrust bearing from input gear, then the input gear and the input gear thrust bearing and race.
23. Remove range sector and operating lever attaching nut and lockwasher, then remove lever, the range sector and shaft, range sector "O" ring and retainer.
24. Remove sprocket clutch gear and side gear from mainshaft.
25. Remove needle bearings (82) and two bearing spacers from mainshaft. Note position of spacers for reference during reassembly.
26. Remove spline gear and thrust washer, side gear, and clutch gear and thrust washer from sprocket carrier and sprocket, Fig. 135.
27. Remove clutch gear and thrust washer from side gear.
28. Remove one sprocket carrier snap ring, then remove drive sprocket from carrier.

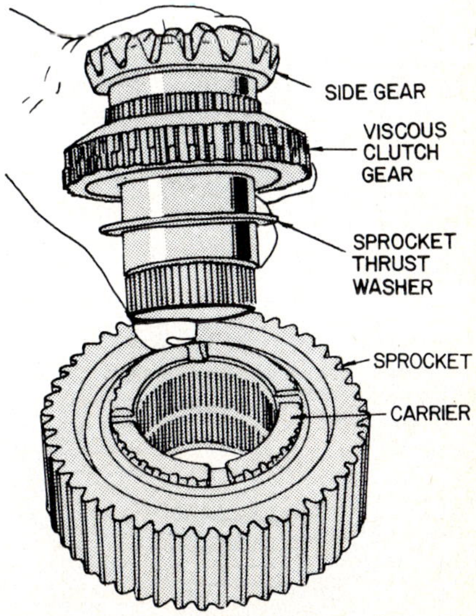

Fig. 135 Side & clutch gear replacement

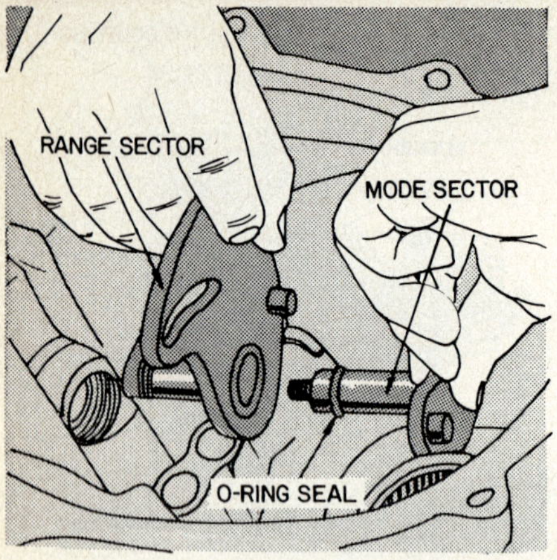

Fig. 136 Range sector installation

29. Remove three bearing spacers and all sprocket carrier needle bearings from carrier.

NOTE: The sprocket carrier and mainshaft needle bearings are different sizes and care must be taken to avoid mixing them.

30. Remove rear output bearing and rear yoke seal from retainer. Note position of bearing for reference during reassembly.
31. Remove input gear and front yoke seals from case. Pry seals from case using a screwdriver.

Sub-Assembly Service

NOTE: All bearings used in the transfer case must be properly aligned to avoid blocking the oil passages. After installing bearings, check their position to make sure that oil passages are clear.

Rear Output Shaft Bearing, Replace

1. Remove bearing using tools J-26941 and J-2619-01.
2. Remove rear output lip seal with a small screwdriver.
3. Install a new lip seal, then the new bearing using tools J-8092 and J-29166.
4. Remove tools and check that oil feed hole is not blocked.

Front Output Shaft Front Bearing, Replace

1. Remove bearing using tools J-8092 and J-29168.
2. Install new bearing using tools J-8092 and J-29167.
3. Remove tools and check that oil feed hole is not blocked.

Front Output Shaft Rear Bearing, Replace

1. Remove bearing using tools J-26941 and J-2619-01.
2. Install new bearing using tools J-8092 and J-29163.
3. Remove tools and check that oil feed hole is not blocked and that bearing is seated flush with edge of bore in case to permit clearance for thrust bearing assembly.

Input Gear Front & Rear Bearings, Replace

1. Remove both bearings simultaneously using tools J-8092 and J-29170.
2. Install bearings one at a time, rear bearing first, using tools J-8092 and J-29169.
3. Remove tools and check that oil feed hole is not blocked and that bearings are flush with case bore surfaces.
4. Install new oil seal using tool J-29162.

Mainshaft Pilot Bushing, Replace

1. Remove bushing using tools J-2619-01 and J-29369-1.
2. Install new bearing using tools J-8092 and J-29174.
3. Remove tool and check that oil feed hole is not blocked.

Rear Output Bearing & Rear Yoke Seal

1. Remove bearing and seal using a brass drift and hammer.
2. Install new bearing using tools J-8092 and J-7818.

NOTE: The shielded side of the bearing must face the case interior.

3. Install new seal in retainer using tool J-29162.

Assembly, Fig. 129

NOTE: During assembly, lubricate all internal components with Dexron II or petroleum jelly if specified. Do not use any other types of lubricant.

1. Install new input gear and rear output shaft bearing seals and apply petroleum jelly to seals. Seals must seat flush with edge of seal bore or in seal groove in case.
2. Install input gear thrust bearing race in case counterbore and thrust bearing on input gear, then install gear and bearing in case.
3. Install mainshaft thrust bearing in bearing recess in input gear.
4. Install planetary assembly on input gear. Ensure that planetary pinion teeth are fully meshed with input gear.
5. Install new sector shaft "O" ring on mode sector shaft, then install mode sector through range sector.
7. Install range sector in front case, Fig. 136.
8. Install operating lever and snap ring on range sector shaft, then the lever attaching washer and locknut on mode sector shaft. Torque locknut to 17 ft. lbs.
9. Assemble annulus gear, range fork and rail, then install the assembly on and over planetary assembly, Fig. 137. Ensure annulus gear is fully meshed with planetary pinions.
10. Install annulus thrust washer and annulus retaining ring on annulus gear hub.
11. Install detent spring, plunger, ball and screw in front case detent bore. Torque screw to 22 ft. lbs.
12. Assemble and install locking fork, fork brackets, fork springs and clutch sleeve. Ensure that lug on fork is fully seated in range sector detent.

NOTE: The splines in the locking clutch sleeve and the mode clutch sleeve are different, Fig. 138. Make sure that the correct sleeve is installed in the corresponding shift fork.

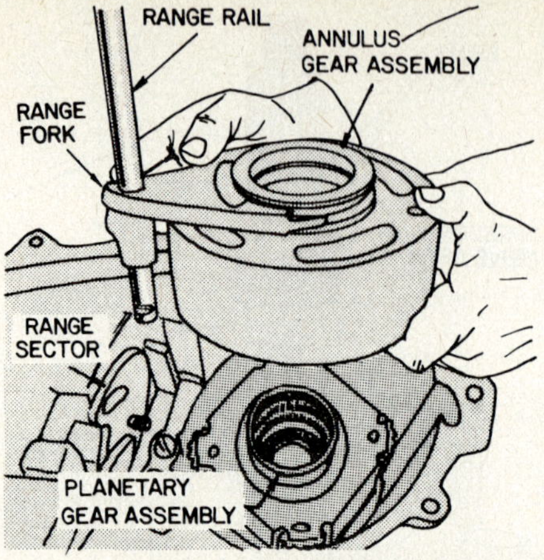

Fig. 137 Annulus gear assembly installation

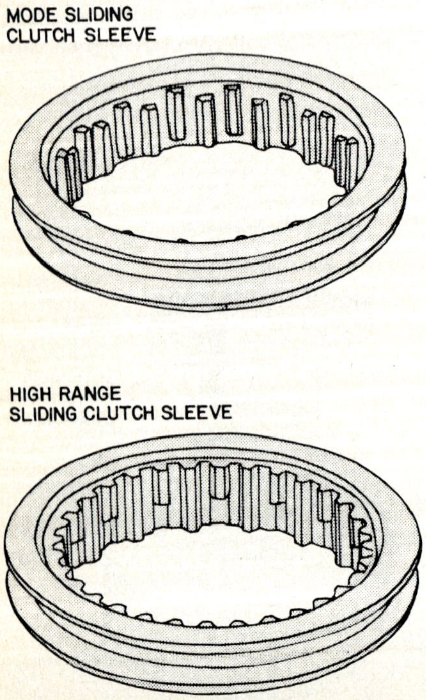

Fig. 138 Mode & locking clutch sleeve identification

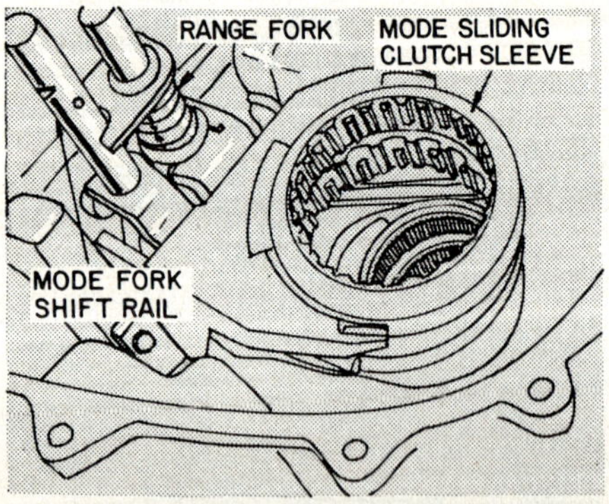

Fig. 139 Mode fork, shift rail & mode clutch sleeve installation

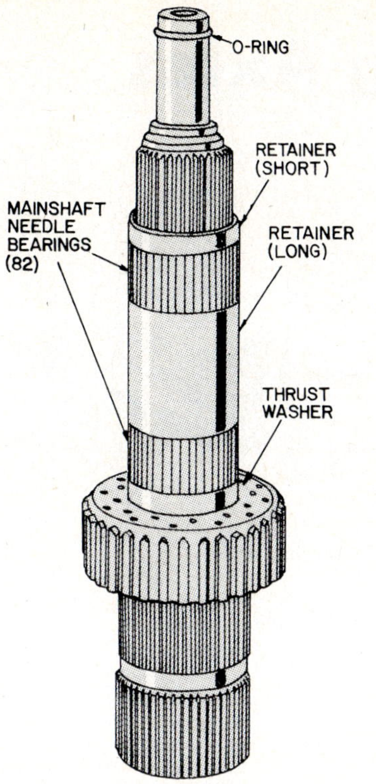

Fig. 140 Mainshaft needle bearing installation

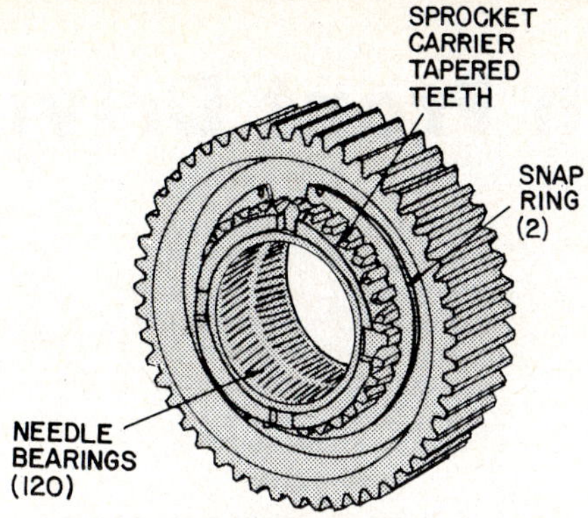

Fig. 141 Sprocket carrier & drive sprocket assembly

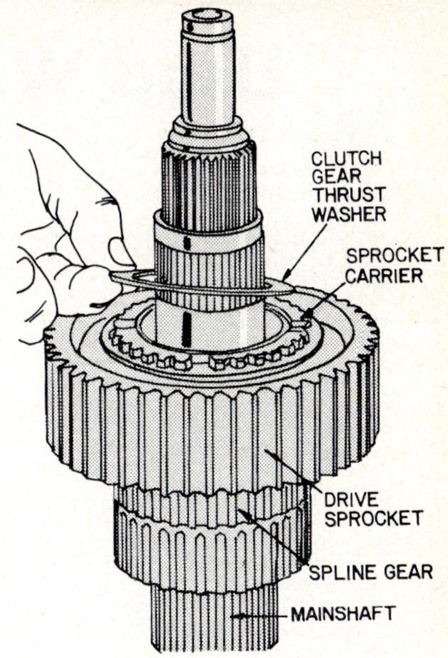

Fig. 142 Mainshaft assembly

13. Install range fork lug in range sector detent notch, then place range sector in high range position.
14. Assemble and install mode fork, shift rail and mode clutch sleeve, Fig. 139.
15. Install thrust washer, new "O" ring, needle bearings and bearing spacers on mainshaft, Fig. 140.
16. Coat shaft bearing surface and all 82 needle bearings with petroleum jelly.
17. Install 41 needle bearings and the long bearing spacer, then install remaining 41 bearings and the short spacer, Fig. 140.
18. Install spline gear on mainshaft, using care not to displace bearings.
19. Install sprocket carrier in drive sprocket using reference marks made during disassembly.

NOTE: Assemble the sprocket carrier and drive sprocket so that the tapered teeth on the carrier are on the same side as the recess in the drive sprocket, Fig. 141.

20. Install sprocket carrier bearings and spacers, Fig. 141.
21. Coat carrier bore and all 120 needle bearings with petroleum jelly.
22. Install center spacer and position 60 bearings in each end of carrier, then install the two remaining spacers, one on each side of carrier.
23. Install sprocket carrier and drive shaft on mainshaft, Fig. 142. Ensure recessed side of drive sprocket faces downward.
24. Install clutch gear thrust washer on mainshaft, Fig. 142.
25. Install clutch gear on side gear, Fig. 135. Ensure that tapered edge of clutch gear faces side gear teeth.
26. Install side and clutch gear assembly on mainshaft. Ensure side gear is fully seated in sprocket carrier.
27. Install mainshaft and gear assembly in case, Fig. 133. Ensure mainshaft is fully seated in input gear.
28. Install driven sprocket and snap ring on front output shaft using reference marks made during disassembly.
29. Install front output shaft front thrust bearing, then the thick race, bearing and thick race in case.
30. Install drive chain on driven sprocket, then lift and tilt driven sprocket and chain and install opposite end of chain on drive sprocket. Align front output shaft with shaft bore in front case and install shaft in case. Ensure shaft thrust bearing is seated in case.
31. Install front output shaft rear thrust bearing assembly on front output shaft, Fig. 131. Install thin race first, then the bearing and thick race.
32. Install viscous coupling on side gear and clutch gear, Fig. 132. Ensure coupling is fully seated on clutch gear, clutch gear is flush with coupling and gear teeth are visible.
33. Apply petroleum jelly to mainshaft pilot bearing surface and all 15 pilot roller bearings, then install bearings on shaft.
34. Install rear output shaft on mainshaft and into viscous coupling. Shaft must be completely seated in coupling. If not, tap into position with a rubber mallet. Use caution not to displace pilot bearings during shaft installation.
35. Install oil pump on rear output shaft, then a new rear output shaft bearing oil seal in rear case.
36. Apply bead of Loctite 515 sealer or equivalent to rear case mating surface and install magnet in case, if removed.
37. Install rear case on front case. Ensure alignment dowels on ends of front case are lined up with bolt holes in rear case.

NOTE: If the rear case will not seat fully in the front case, check for the following: oil in range fork rail bore; front output shaft rear thrust bearing assembly not properly aligned with rear case; mainshaft not fully seated; rear case and oil pump misaligned.

38. Install case attaching bolts and torque to 23 ft. lbs. Ensure that flat washers are used on bolts at case ends with alignment dowels.
39. Install speedometer drive gear and differential shims on rear output shaft, Fig. 130.
40. Install rear retainer on rear case. Tighten bolts securely but not to final torque.
41. Install front and rear output shaft yokes and original yoke nuts. Tighten nuts finger tight only.
42. Install a dial indicator onto rear retainer so that plunger contacts rear yoke nut, Fig. 94.
43. Rotate front output shaft 10–12 revolutions and zero the dial indicator, then rotate shaft one more revolution and note dial indicator reading.
44. End play should be .002–.010 inch. If not, remove retainer and add or remove shims as necessary to bring end play within limits.
45. Remove front and rear yokes and rear retainer. Discard original yoke nuts.
46. Apply Loctite 515 sealer or equivalent to retainer mating surface and retainer bolts, then install retainer and torque bolts to 23 ft. lbs.
47. Install yokes with new seal washers and nuts. Torque nuts to 120 ft. lbs. Hold yokes with tool J-8614-01 while tightening nuts.
48. Apply sealer to detent bolt, then install ball, spring and bolt. Torque bolt to 23 ft. lbs.
49. Install drain plug and washer and torque plug to 18 ft. lbs.
50. Fill transfer case with 6 pints of Dexron II transmission fluid and install fill plug with washer. Torque fill plug to 18 ft. lbs.

FRONT WHEEL LOCKING HUBS

Used With Front Driving Axles

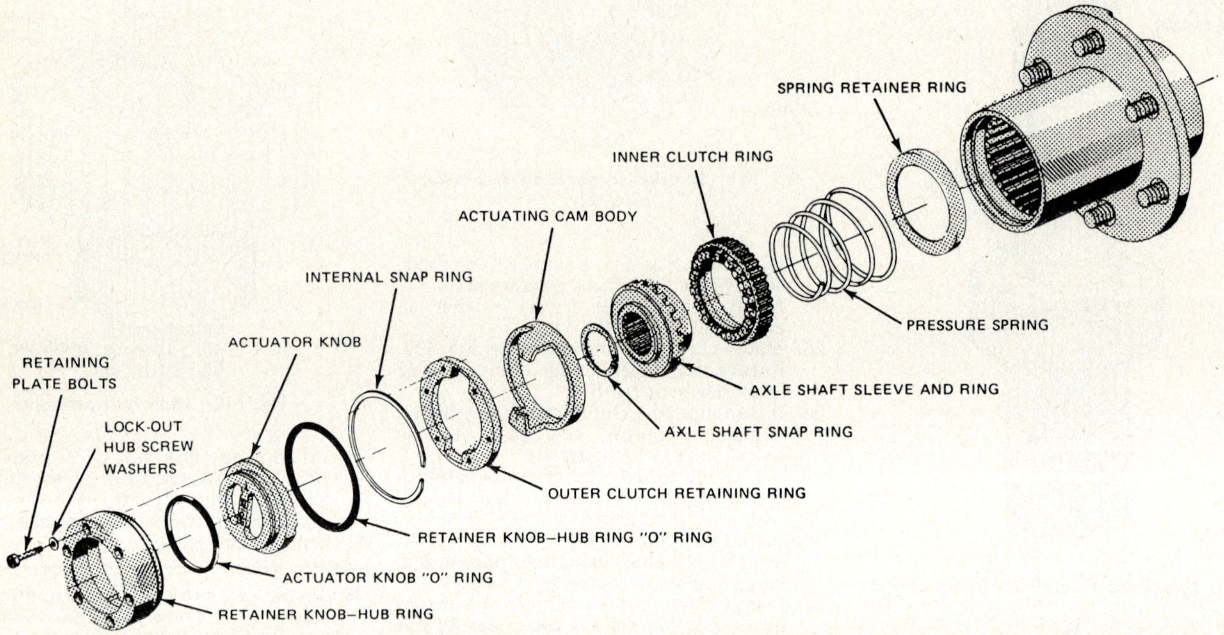

Fig. 1 Spicer external type Hub-Lok disassembled

SPICER EXTERNAL TYPE HUB-LOK

Description

As shown in Fig. 1, the splines on the inside diameter of the axle shaft sleeve and ring assembly mesh with the axle shaft splines. The assembly is retained on the axle shaft with a snap ring. The splines on the outside diameter of the inner clutch ring assembly mesh with the wheel hub splines. Therefore, when the actuator knob is turned towards the "L" position, the actuating cam body is forced outward towards the hub end, allowing the inner clutch to be forced under spring tension towards the axle shaft sleeve and ring assembly until the inner clutch assembly teeth are engaged in the axle shaft sleeve and ring assembly teeth, locking the axle and hub.

Operation

"L" Position

When the transfer case is shifted into the position for driving the front axle, turn the actuating knob so that it is aligned with the letter L. If the clutch teeth do not engage with the knob turned to this position, the clutch teeth are butted and a slight movement of the wheel in either direction complete the lock. The front axle will now drive the wheel.

"F" Position

When the transfer case is to be shifted into the position for driving the rear axle only, turn the actuating knob so that it is aligned with the letter F. This will disengage the clutch teeth and thus unlock the wheel hub from the axle shaft. The wheel will now turn free on the axle.

CAUTION: *Be certain that the transfer case is shifted into two-wheel drive position before disengaging the Hub-Lok.*

Removal

1. Remove hub screws and washers, then slide gear hub housing away from hub, Fig. 1.
2. Remove inner metal gasket, gear hub housing and the outer gasket.
3. With actuator knob in "Lock" position, remove clutch gear snap ring, clutch gear and pressure spring.
4. With actuator knob in the "Free" position, drive cam lock pin from assembly with a suitable punch.
5. Remove actuating cam from knob, then the knob retaining snap ring and the knob from knob retainer.
6. Remove axle shaft snap ring, inner clutch gear and bushing.

Installation

1. Lubricate bushing and splines of inner clutch gear then assemble the inner clutch gear to bushing and install assembly onto axle shaft. Ensure inner clutch gear splines are aligned with axle shaft splines.
2. Install axle shaft snap ring.
3. Lubricate "O" ring groove in actuator knob with "O" ring lubricant and install "O" ring onto actuator knob.
4. Install actuating knob into retainer with knob in "Free" position, then the retaining snap ring.
5. Install actuating cam onto knob, aligning the cam ears with retainer slots.
6. Install cam lock pin through cam groove and actuating knob holes. Ensure pin ends are flush with outside diameter of cam.
7. Turn actuator cam to "Lock" position and lubricate cam grooves.
8. Install pressure spring, outer clutch gear and snap ring. Ensure snap ring is properly seated in the cam groove.
9. Turn actuator knob to "Free" position and lubricate outer spline and teeth of outer clutch gear.
10. Install outer retainer gasket, gear hub housing and inner metal masket, then install two retaining screws to position the assembly.
11. Install hub assembly to axle and tighten retainer screws. Turn actuator knob to "Lock" position.
12. Install remaining retainer screws and torque all screws to 30–35 ft. lbs.

SPICER INTERNAL TYPE HUB-LOK

Referring to Fig. 2, the parts shown can be replaced individually. However, the actuator knob, actuator cam body, axle shaft sleeve and ring and the inner clutch ring are exposed to wear when the Hub-Lok is engaged. These

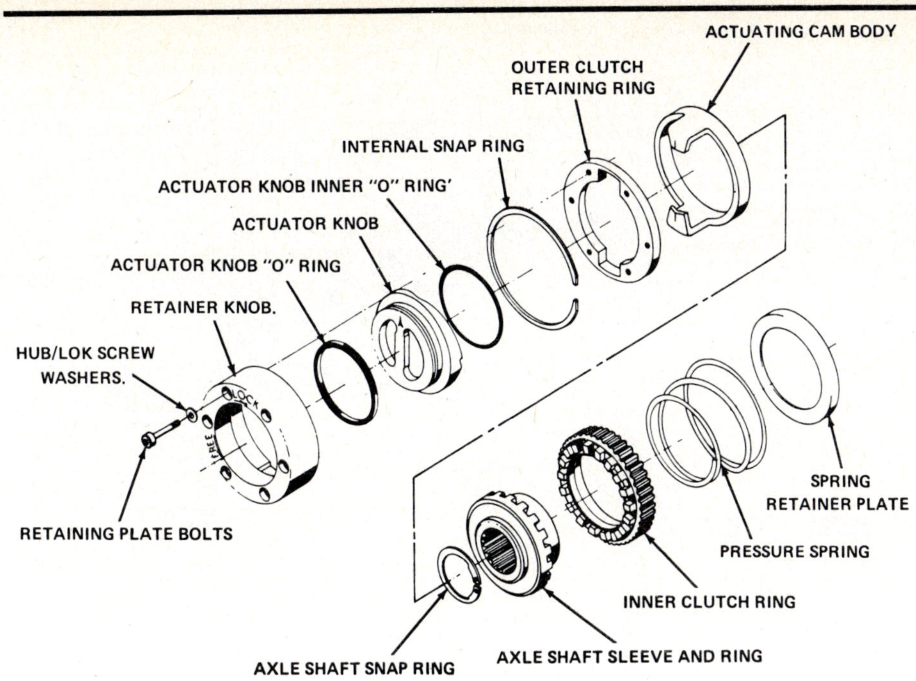

Fig. 2 Spicer internal type Hub-Lock disassembled

parts should be checked carefully during inspection.

If the actuator knob is worn or damaged but the actuating cam ring is in good condition, then replace only the knob. Replace both parts only if both show evidence of wear or damage.

The same attention should be given to the axle shaft sleeve and ring and the inner clutch ring. If wear or damage is evident on the axle shaft sleeve and ring but the inner clutch shows none, then replace only the axle shaft sleeve and ring. Replace both parts only if they both show signs of wear or damage.

Removal

1. Remove hub screws and washers, then the hub ring and knob.
2. Remove internal snap ring from hub groove.
3. Remove cam body ring and clutch retainer assembly from hub and disassemble components.
4. Remove axle shaft snap ring.
5. Remove drive gear and clutch gear.
6. Remove coil spring and spring retainer ring.

Installation

1. Lubricate hub inner splines with suitable grease.
2. Install spring retainer ring with recessed undercut side facing inboard. Ensure the ring seats against bearing.
3. Install coil spring with large end positioned against retainer ring.
4. Lubricate and install axle shaft sleeve and ring and the inner clutch ring. Ensure the teeth are meshed for ease of assembly. Also, it may be necessary to rock the hub to align the splines.
5. Install axle shaft snap ring. Ensure snap ring is fully seated in the shaft groove.
6. Install actuating cam body ring into outer clutch retaining ring, then install assembly into hub.
7. Install internal snap ring. Ensure snap ring is fully seated in the hub groove.
8. Lubricate cam ears with Lubriplate and apply a small amount of "O" ring lubricant in groove of actuating knob. Install outer "O" ring.
9. Install knob in hub ring, then the assembly on axle with the knob in locked position. Install retaining screws and ensure retaining ring is positioned properly in the hub.
10. Torque lock-out hub screws to 35–40 inch lbs.

CUTLAS SELECTIVE DRIVE HUB

The outstanding operational advantage of this type hub, Fig. 3, is its "preset" feature. Turning the control knob from two-wheel drive to four-wheel drive "presets" the hub to automatically engage the axle shaft when the transfer case is shifted into four-wheel drive. The axle shaft is not engaged until the shift into four-wheel drive is made.

Knurling on the control knob provides a firm gripping surface. An O-ring seal under the control knob keeps out all foreign matter. The control knob makes only a quarter turn, and snaps in place for positive positioning.

WARN MANUAL TYPE LOCKING HUB

Description, Figs. 4 & 5

Turning the locking hub clutch control to the "L" or "4" position connects the wheel hubs to the axle shafts. Therefore, when the transfer case is driving the front axle and the locking hubs are in the "L" or "4" position the wheels are driven by the axle shafts. When the transfer case is not driving the front axle, the locking hub clutch controls are turned to the "F" or "2" position to disconnect the wheel hubs from the axle shafts. The wheels now rotate freely but the axle shafts remain stationary.

On any vehicle equipped with manually operated locking hubs extreme care must be taken to ensure the locking hub dials are fully engaged or disengaged to prevent possible damage to the locking hubs. If the locking hubs are not properly maintained, full engagement of the control dials may become impossible due to excessive dirt and dust. Any time the vehicle is driven through water deep enough to cover the hubs the locking hubs should be removed and cleaned then lubricated thoroughly.

These hubs are serviced as an assembly or sub assembly only (hub body or hub clutch assembly only). Do not attempt to disassemble these units. If an entire hub or sub assembly is defective, replace the entire assembly or sub assembly. The hubs may be removed for cleaning and inspection purposes and for periodic lubrication only.

Model M243

Removal, Fig. 4

1. Remove bolts and lockwashers securing hub body to axle hub and remove gasket and hub body.

CAUTION: Do not turn the hub control dial after removing the hub body.

2. Remove retaining ring, hub clutch and bearing assembly.
3. Clean and inspect all components. Replace as necessary.

Installation

CAUTION: Do not turn the hub control dial until after the hub control dial has been reinstalled. The hub clutch nut and cup can be severely damaged if dial is rotated while the hub is off the vehicle.

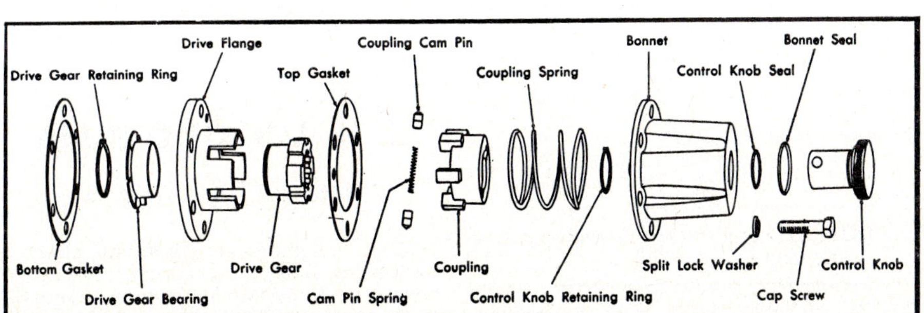

Fig. 3 Cutlas Selective Drive hub disassembled

1. Apply a light coat of lubricant to all hub components.
2. Install bearing assembly and hub clutch with retaining ring on axle shaft.
3. Install hub body with gasket. Install tabbed lockwashers and bolts in hub body and axle and torque bolts to 30 ft. lbs.
4. Raise and support front of vehicle, with dial in free position, rotate wheels. If wheels drag, check hub installation.
5. Lower vehicle.

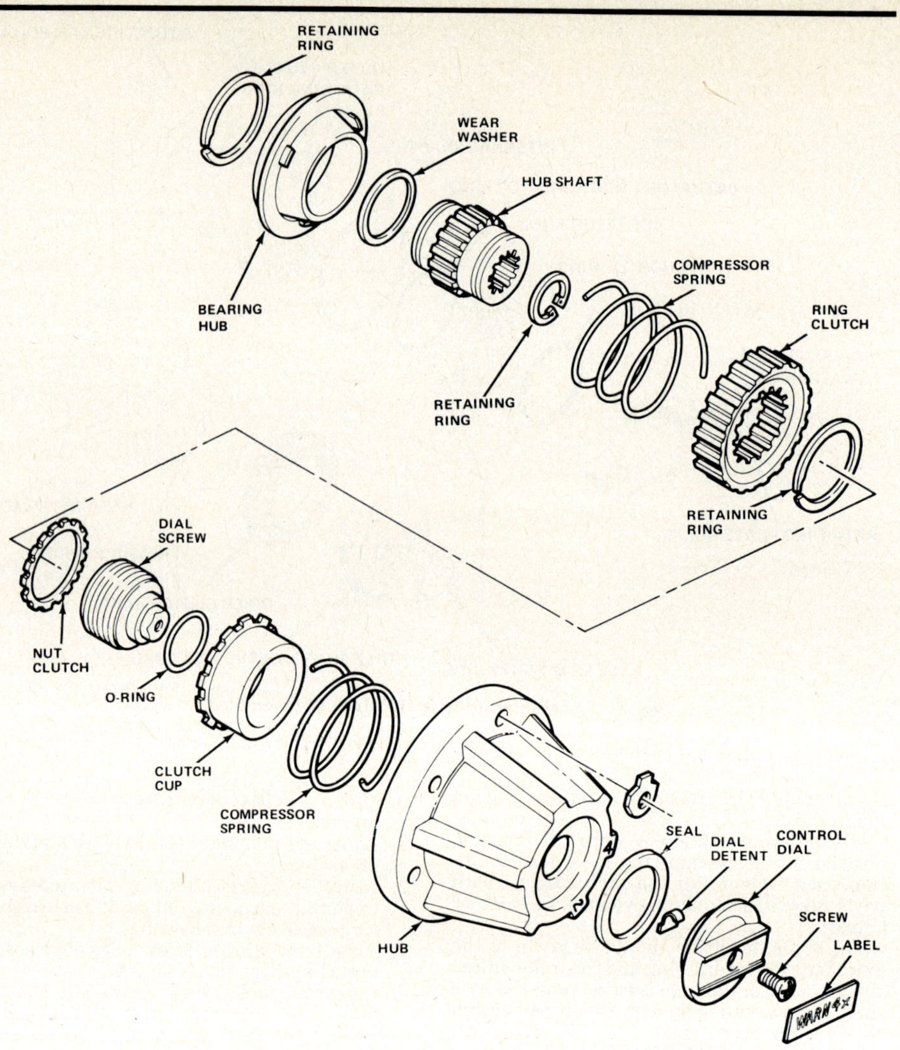

Fig. 4 Warn model M243 front locking hub disassembled

Model M247

Removal, Fig. 5

1. Remove socket head screws, axle hub retaining ring and axle shaft retaining ring. Remove hub clutch assembly.
2. Clean and inspect all components. Replace as necessary.

Installation

1. Apply a light coat of lubricant to all hub components.
2. Install hub clutch, axle hub retaining ring and axle shaft retaining ring.
3. Install new O-ring in hub body and position in clutch. Install socket head screws and torque to 30 inch lbs.
4. Raise and support front of vehicle, turn dials to free position and rotate wheels. If wheels drag, check hub installation.
5. Lower vehicle.

WARN AUTOMATIC TYPE LOCKING HUB

Description, Figs. 6 & 7

With the automatic locking hub dial control in the "Auto" position and the transfer case in two wheel drive operation, the front wheels will remain free wheeling. Whenever the vehicle is operating under power in four wheel drive the front hubs will automatically engage in either forward or reverse in case of rear wheel slippage. If the front axle is not powered as in a case of downhill descents the front wheels will automatically free wheel. Therefore, to provide engine braking where necessary in four wheel drive operation the front hub dials must be positioned in the "Lock" position.

All transfer case shifting when the automatic locking hubs are in "Auto" position must be done with the vehicle stopped and the transmission in neutral.

Removal & Installation

Early Type, Fig. 6

1. Remove six capscrews from hub cap assembly, then remove assembly from wheel hub.
2. Remove capscrew from end of axle shaft, then remove lock ring from wheel hub. Slide body assembly out from wheel hub.

NOTE: Do not pack the automatic hub lock with grease.

---

3. Unscrew all four setscrews in spindle locknut until the heads are flush with the edge of the locknut. Then using tool No. T80T-4000-V remove outer spindle locknut.
4. Reverse procedure to install. Pin on inner bearing adjusting nut should enter one of the slots on lockwasher. Torque outer spindle lock nut to 15–20 ft. lbs. with tool T80T-4000-V. Tighten down all four set screws. Two of the set screws should thread in until they bottom on the inner bearing adjusting nut. Push in on body assembly until the friction shoes are on top of the spindle outer locknut. Install lock ring.
5. Apply locking compound to capscrew and install in axle shaft. Torque to 35–50 ft. lbs.
6. Install cap on spindle and torque to 30–35 inch lbs. Raise and support front of vehicle, position dialing mechanism on both hubs to "lock" position. While spinning wheels, check to see if axle is spinning. If not, recheck installation.

Late type, Fig. 7

1. Remove five capscrews retaining hub cap to wheel hub, then remove hub cap.

NOTE: Do not allow ball bearing, race or spring to drop out of cap.

---

2. Remove rubber sealing ring, then remove seal bridge retainer from lock ring.
3. Using a suitable pair of needle nose pliers, remove lock ring while pulling hub lock from wheel hub.
4. Install hub lock into hub ensuring the large tangs are lined up with the lock washer and the outside and inside diameter splines are in line with the hub and axle shaft splines.
5. Using a suitable pair of needle nose pliers, install lock ring while pushing hub lock assembly into hub.
6. Install seal bridge retainer, narrow end first.
7. Install rubber sealing ring on hub cap, then install hub cap ensuring ball bearing, race and spring are in place.
8. Install cap screws. Torque evenly to 40–50 inch lbs.

I. H. LOCKING HUBS

Manual Type—Fig. 8

When the hub is engaged, full power is transmitted to both front wheels. Disengagement, with front axle not driving, allows the front wheels to "Free Wheel" and the axle shafts and differential gears to remain idle, saving unnecessary wear.

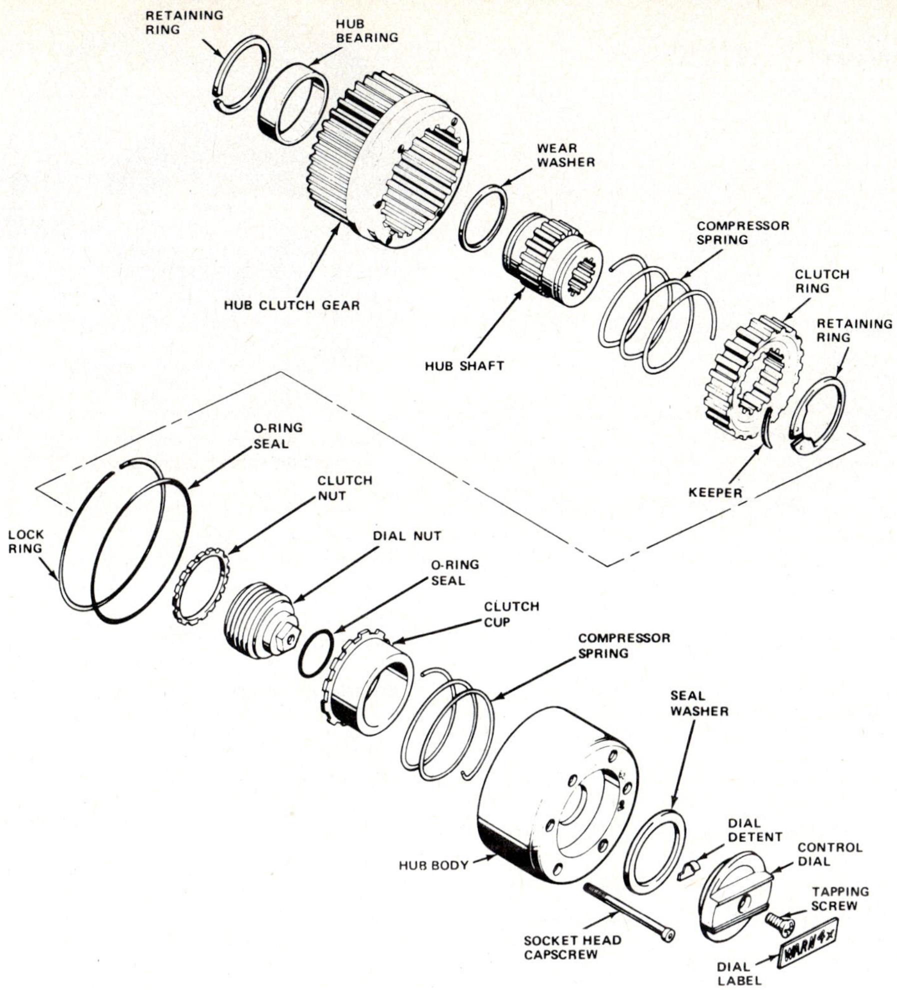

Fig. 5 Warn model M247 front locking hub disassembled

Lock-o-matic Type—Fig. 9

This type hub, when set in the "Free" position, automatically locks the front wheel and axle shaft together the moment torque is applied to the front axle. The hub controls, therefore, do not have to be changed regardless of whether the vehicle is being operated in two- or four-wheel drive. It is necessary to set the control in the "Lock" position only when engine braking is required (down steep hills, on ice, etc.).

Operation

To engage the locking hubs, turn brass controls (one on each hub) clockwise to "Lock" position. Arrow in center of controls must point directly at dot located on rim of hub. You can feel the brass control "seat" itself when it is properly positioned. If the arrow does not point directly at the dot, the control will not seat itself. Thus, the gears will not completely engage and the pressure may force off the end of the hub.

To disengage the hubs, turn brass controls counter-clockwise to "Free" position. Here again, the arrow must point directly at the dot on the rim, otherwise the gears may rake against each other.

When the controls are properly positioned, the gears are completely engaged or disengaged and units will not be damaged.

Precautions

1. Use fingers only to turn controls. If controls do not move freely with finger effort, move vehicle slightly in either direction in two-wheel drive, standard gear range. If hubs do not now turn freely, look for external damage or dirt around controls. *Do not force controls with tools.*
2. Do not drive vehicle unless controls on both hubs are properly positioned and both are set the same.
3. To avoid excessive torque loads on the rear axle, do not drive vehicle equipped with *manual locking hubs* in the low range of transfer case with hubs set in "Free" position. This does not apply to the Lock-O-Matic hub because it automatically locks the wheel with the axle shaft when torque is applied. Therefore, no increased load is placed on the rear axle.
4. During vehicle operation, arrow in center of control must always point directly to one of the dots on the rim of the hub. Also, both hubs must be set the same.

Removal

1. Remove bolts holding clutch body to hub body.

CAUTION: *Use only a thin wall socket to remove these bolts. Heavier sockets may force in recessed wall.*

2. Lift off clutch body, holding it erect so as not to let the drive pins slide out of body.
3. Remove lock ring holding hub body to axle shaft and pull off hub body.
4. If necessary to remove drag shoe from axle spindle, loosen hex-head set screw and unscrew drag shoe from spindle.

Manual Type, Disassemble, Fig. 10

NOTE: The spring-loaded manual type locking hub disassembly procedure is the same as the nonspring-loaded type hub except that the clutch ring assembly is one assembly and the ring and drive pin cannot be removed from clutch ring nor can the clutch screw be separated from the control assembly.

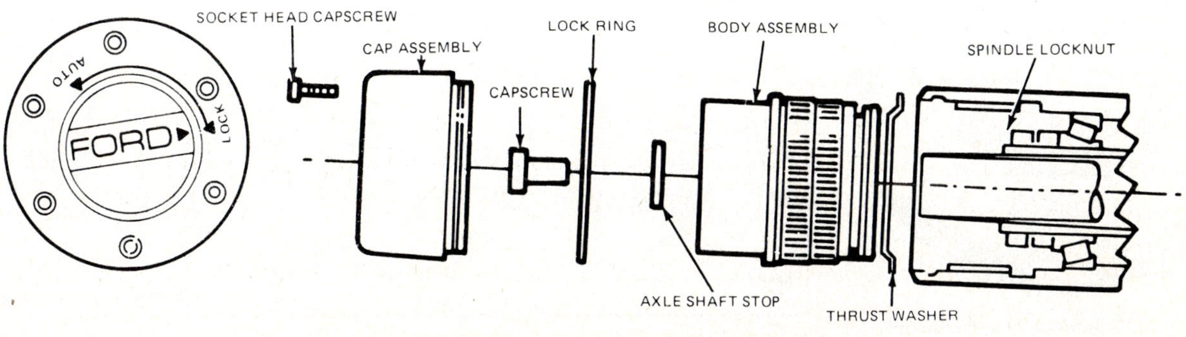

Fig. 6 Warn automatic locking hub exploded view. Early type

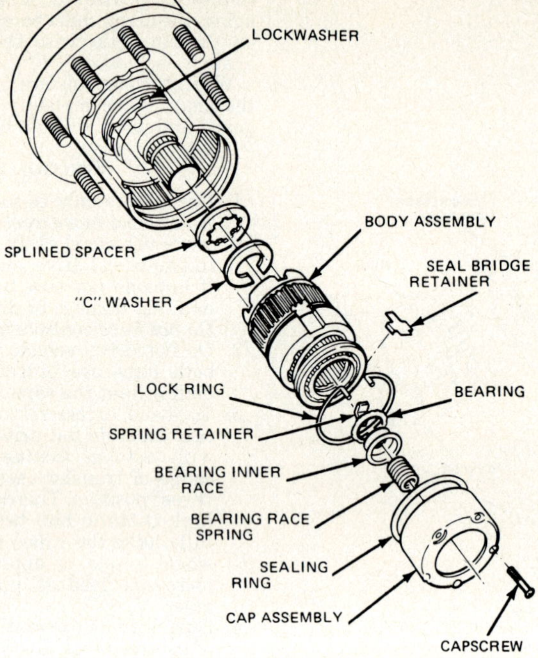

Fig. 7 Warn automatic locking hub exploded view. Late type

Fig. 8 I.H. (International Harvester) Locking Hub (manual type)

1. Remove snap ring and pull axle shaft hub out of hub body, noting from which side of hub body the axle shaft hub is extracted. If necessary needle bearing assembly may be removed from hub body.
2. Using a small chisel, remove staked-over metal retaining clutch hex-head screw, being careful not to damage screw.
3. Remove hex-head screw with Allen wrench. Lift out clutch screw and clutch ring. Clutch ring may be separated from clutch screw by unscrewing it.
4. Turn clutch body face down to remove the twelve drive pins.
5. Remove disc from bottom center of clutch body.
6. Place thumb inside clutch body on center of control assembly and push outward.
7. Remove O-ring seal and outer oil seal from control dial. Dowel pin may be lifted out of control dial.

NOTE: *The poppet ball and spring located in the control dial cannot be serviced; therefore, if damaged, the control assembly must be replaced.*

Manual Type, Assemble, Fig. 10

1. Place axle shaft hub into hub body and secure with snap ring in groove in end of axle shaft hub.
2. Apply a small amount of chassis lube on bearing side and in grooves of control assembly.
3. Install new O-ring seals on control assembly.
4. Place disc on inside (clutch side) of clutch body.
5. Assemble clutch screw into clutch ring from back side. Be certain clutch screw works freely. If it is sticky in any position, tap lightly on back side.
6. Drop in clutch ring and screw assembly.

NOTE: *Clutch screw must be flush with back edge of clutch ring.*

7. Insert twelve drive pins. *Try clutch ring for free siding fit on drive pins. If it does not move freely from top to bottom, lift out clutch ring and screw assembly and turn it to another position. If it still does not move freely, it should be removed and the clutch ring, body and drive pins examined for damage.*
8. Apply a light grade of chassis lube to inside face of clutch body and disc (from front side of clutch body). Hold hand over drive pins to prevent their falling out.
9. Position control assembly with dowel pin into face of clutch body so the arrow stops on the dot marked "Free". Install and tighten the 1/4" flat-head screw into the control assembly.
10. To check clutch screw setting, turn control from "Lock" to "Free" position and back several times. Control should snap into both positions. In "Free" position, clutch ring should just clear bottom of clutch body.

NOTE: *If clutch ring is set too far above bottom of clutch body when set in the "Free" position, axle shaft hub will rub face of clutch ring.*

11. Stake dowel pin and flat-head screw.
12. Turn control to "Free" position and apply

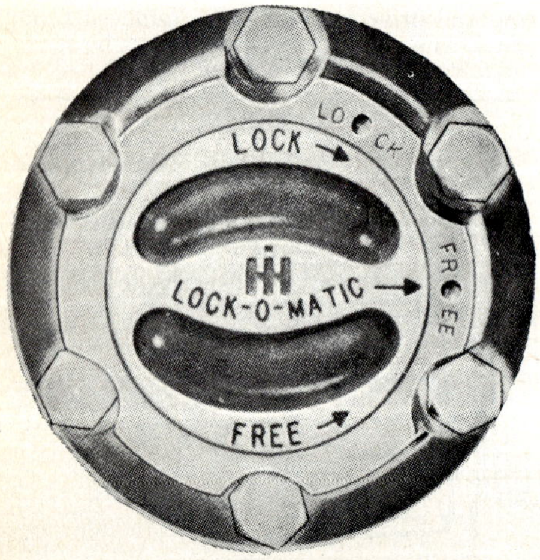

Fig. 9 I.H. (International Harvester) (Lock-O-Matic)

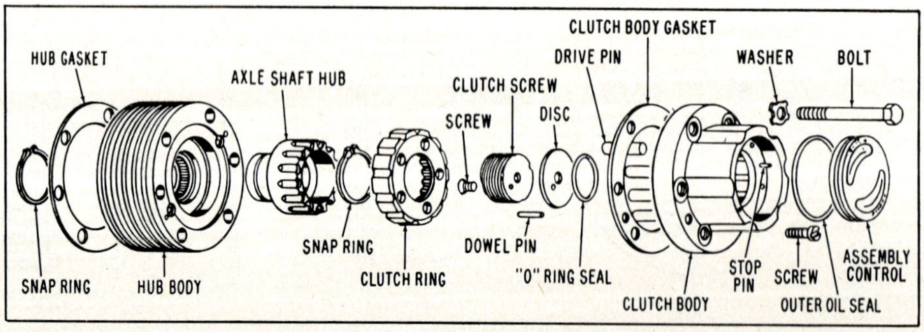

Fig. 10 I.H. (International Harvester) manual type locking hub disassembled

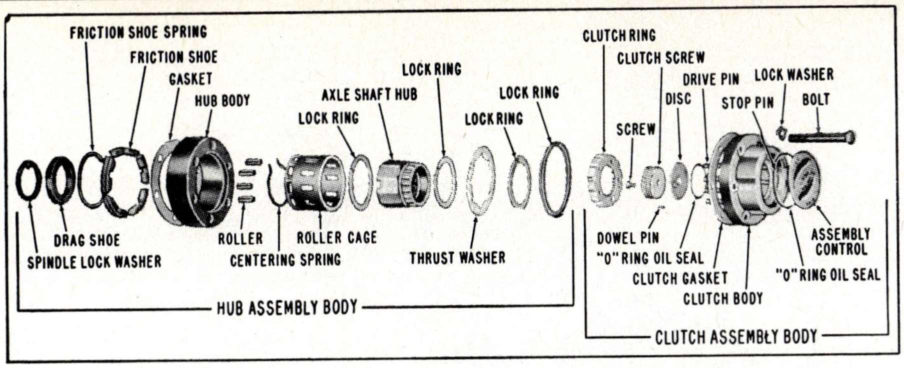

Fig. 11 I.H. (International Harvester) Lock-O-Matic disassembled

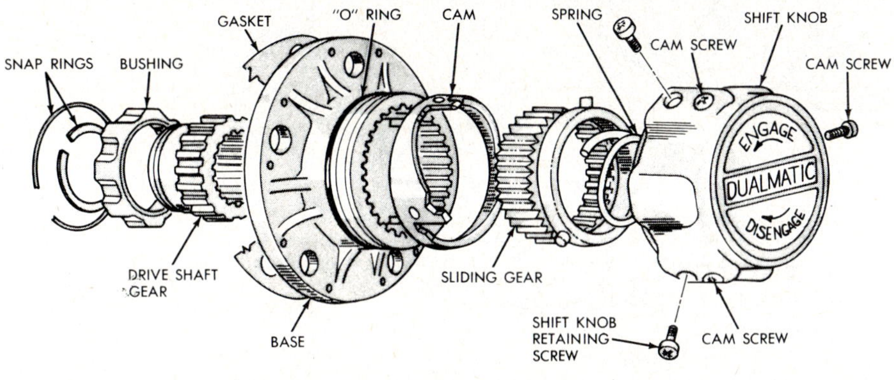

Fig. 13 Dualmatic locking hub exploded view

a thin coating of light grade chassis lube around clutch screw and drive pins.

Lock-O-Matic Type, Disassemble

1. Referring to Fig. 11, remove lock ring retaining axle shaft hub into hub body. Pull axle shaft hub and roller cage assembly out of hub body, noting side of hub from which gear teeth extend.
2. Remove ten rollers from roller cage. Take out centering spring.
3. Remove lock ring located in second groove on axle shaft hub. This permits axle shaft hub to be separated from roller cage.
4. Pull axle shaft hub out of roller cage.

NOTE: *The corner points of axle shaft hub must be aligned with grooves in friction shoes to permit removal of hub.*

5. Remove locking ring located in roller cage in end opposite friction shoes.
6. Remove friction shoe spring and lift out friction shoes from roller cage.
7. Remove lock ring and thrust washer from hub body.
8. Disassembly of the Lock-O-Matic clutch body is identical to that of the manual type.

Lock-O-Matic Type, Assemble

1. Referring to Fig. 11, install thrust washer in hub body. Place lock ring in its groove in hub body to hold thrust washer in position.
2. Position friction shoes into roller cage. Install spring.
3. Install lock ring into end opposite friction shoes of roller cage.
4. Insert axle shaft hub into roller cage against lock ring.

NOTE: *Points on axle shaft hub must be aligned with grooves in friction shoes to permit entry.*

5. Install lock ring into second groove on axle shaft hub.
6. Slip centering spring into back side of axle shaft hub.

NOTE: *The two ends of spring go into the slot in cage which is longer than the rest.*

7. Place a small amount of light grade chassis lube into all slots of cage and into bore of hub body.
8. Hold hub and cage assembly with clutch end up and insert ten rollers into cage slots.
9. Slip axle shaft hub and cage assembly into hub body from back side. Install lock ring into remaining groove in axle shaft hub.

NOTE: *To hold axle shaft hub while lock ring is being installed, place it on a piece of 2¼" diameter bar stock.*

10. Assemble clutch body in the same manner described for the manual type clutch body.

BORG WARNER AUTOMATIC LOCKING HUB

SERVICE NOTE:

Complaints of automatic locking hubs not engaging or disengaging on 1981 Chrysler vehicles can be caused by the installation of the incorrect drag sleeve retainer washer, Fig. 12. To determine if the incorrect washer was installed, remove the automatic locking hub, then inspect the drag sleeve retainer washer located behind the wheel bearing lock nut. There are four tabs on the drag sleeve retainer washer that should be visible, Fig. 12. If the four tabs are not visible, the incorrect drag sleeve retainer washer is installed and should be replaced with the correct part.

Description

The Automatic Locking Hub, shown in Fig. 12, engages or disengages to lock the front axle shaft to the hub of the front wheel. Engagement occurs whenever the vehicle is operated in 4WD. Disengagement occurs whenever 2WD has been selected and the vehicle is moving rearward. Disengagement will not occur when the vehicle is moved rearward if 4WD is selected and the hub has already been engaged.

Before disassembling a unit for a complaint of abnormal noise, note the following:

a. To obtain all-wheel drive, the transfer case lever must be placed in 4L or 4H, at which time the hub locks will automatically engage.
b. To unlock the hubs, shift the transfer case lever to 2H, then slowly reverse the vehicle direction approximately 10 feet.
c. An incomplete shift from 2WD to 4WD, or disengagement of only one hub lock may cause an abnormal sound from the front axle. Shift to 4WD to stop the noise, then unlock the hubs as previously stated.

Removal

1. Remove 5 screws retaining the cover to outer clutch housing.
2. Remove cover, seal, seal bridge and bearing components.
3. Compress the wire retaining ring and remove the remaining components from the hub.

Disassembly

1. Remove snap ring from groove in hub sleeve, then turn clutch gear until it falls into disengagement with the outer clutch housing. Lift and tilt the drag sleeve to unlock the tangs of the brake band from window of inner cage, then remove the drag sleeve and brake assembly.

NOTE: The brake band must never be removed from the drag sleeve. The spring tension of the brake band can be changed if the coils are over-expanded and could affect hub operation.

2. Remove snap ring from groove in outer clutch housing, then pry plastic outer cage free from inner cage while inner cage is being removed.

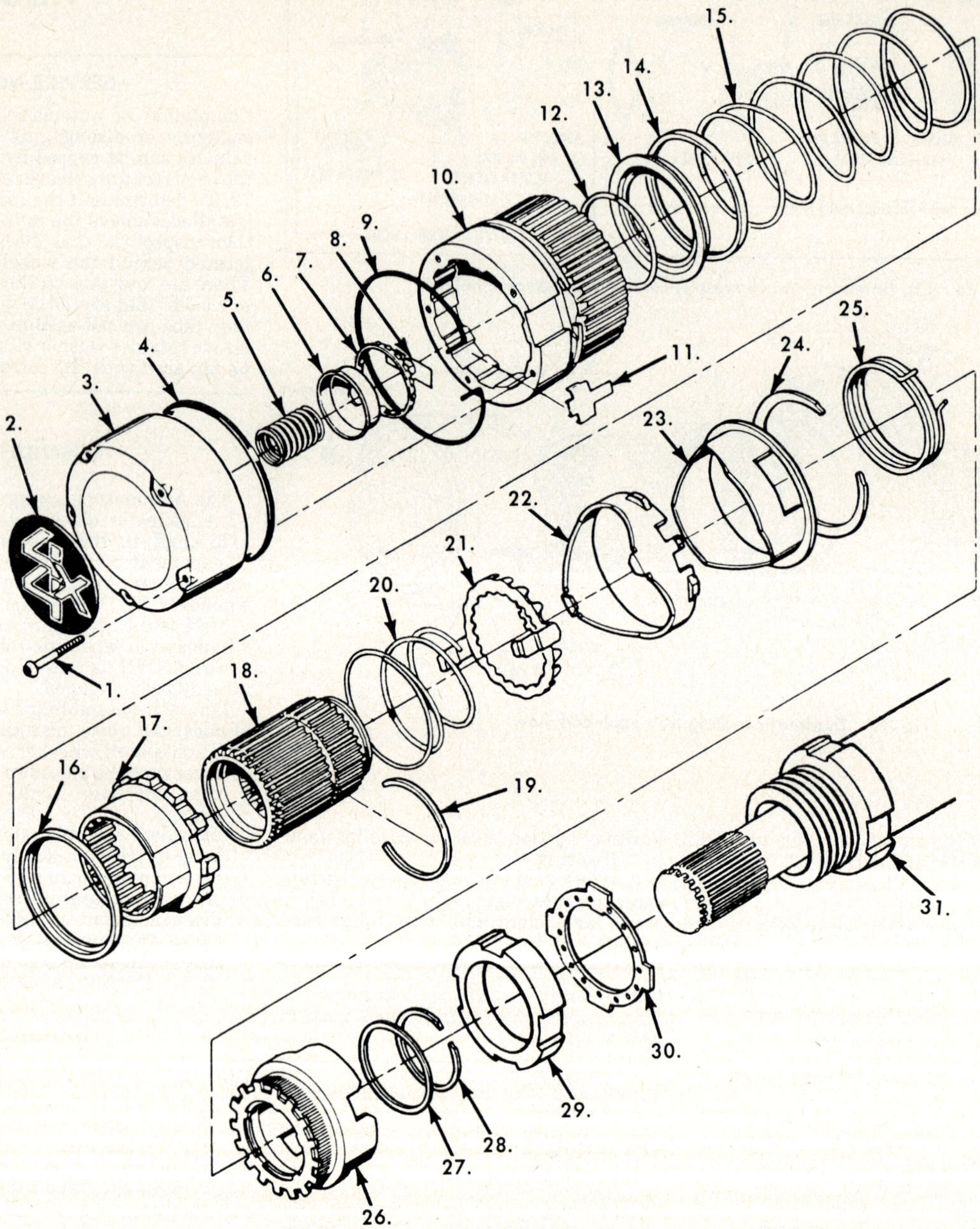

1. Machine Screw
2. Cover Plate
3. Cover
4. Sealing Ring
5. Bearing Race Spring
6. Bearing Inner Race
7. Bearing
8. Bearing Retainer Clip
9. Wire Retaining Ring
10. Outer Clutch Housing
11. Seal Bridge — Retainer
12. Retaining Ring
13. Spring Support Washer
14. Spring Retainer
15. Return Spring
16. Spring Retainer
17. Clutch Gear
18. Hub Sleeve
19. "C" Type Retaining Ring
20. Conical Spring
21. Cam Follower
22. Outer Cage
23. Inner Cage
24. Snap Ring
25. Brake Band
26. Drag Sleeve and Detent
27. Small Spacer
28. Retaining Ring
29. Lock Nut
30. Drag Sleeve Retainer Washer
31. Adjusting Nut, Wheel Bearing

Fig. 12 Borg Warner automatic locking hub exploded view

3. Pry plastic outer cage tabs free from groove in outer clutch housing, then remove outer cage.
4. Remove clutch sleeve and attached components from outer clutch housing.
5. Compress return spring and hold the spring with clamps, then position the assembly in a vise so that the vise holds both ends of the clutch sleeve. Remove retaining ring.
6. Remove clamps holding the return spring, then slowly open the vise to permit the release of the return spring in a safe manner. Remove retainer seat, spring and spring support washers from the hub sleeve.
7. Remove C-ring from clutch sleeve. It is necessary to position sleeve assembly so that the C-ring ends are aligned with legs of cam follower, allowing removal between the legs.
8. Remove conical spring from between cam follower and clutch gear, then separate cam follower from clutch gear. Do not pry the legs of the cam follower apart.

Assembly & Installation

1. Install tangs of cam follower over flats of clutch gear, then compress conical spring and position with large diameter located against clutch gear.
2. Position clutch gear assembly over splines of hub sleeve. The teeth of the cam follower should be located at the end of the hub sleeve with no splines, and the clutch gear and spring should slide freely over splines of hub sleeve.
3. Install C-ring into groove of hub sleeve, then install a spring retainer over each end of return spring.
4. Position one end of return spring with retainer against shoulder of clutch gear, then place support washer on end of the return spring. Compress return spring and install retainer ring into groove of hub sleeve.
5. Place assembled components into outer housing. The cam follower should be positioned with legs facing outward.
6. Install 3 of the cover screws into 3 holes of the outer clutch housing. These screws will support the component to permit the clutch hub to drop down so that the tangs of the brake band can be assembled.
7. Carefully install the plastic outer cage into outer clutch housing with ramps facing toward cam follower. The small external tabs of the plastic cage should be located in wide groove of outer clutch housing.
8. Install steel inner cage into the outer cage, aligning tab of outer cage with window of the inner cage, then install retaining ring into groove of outer clutch housing above outer cage.
9. The brake band and drag are serviced as a complete assembly. Install one of the 2 tangs of the brake band on each side of the lug of the outer cage, located in the window of the steel inner cage. It will be necessary to tilt these parts to engage the tangs in this position as the drag sleeve is positioned against the face of the cam follower.
10. Remove the 3 screws and rest the end of the hub sleeve on a suitable support, then install washer and snap ring above drag sleeve.

NOTE: The following steps may be completed as the hub is installed into vehicle.

11. Install wire retaining ring into groove of unsplined end of clutch housing. The tangs of retainer ring should point away from splined end of clutch housing.
12. Hold the tangs together and install the 2 bent down tabs of seal bridge over tangs. The seal bridge holds the wire retainer ring in a clamped position in groove of outer clutch housing. Install O-ring into groove of outer clutch housing and over the seal bridge.
13. Lubricate and install the bearing over the inner race. The steel ball bearing should be visible when the bearing is properly installed.
14. Install bearing retainer clip into hole in outer race, then install bearing and retainer assembly in end of hub sleeve. Install seal ring over outer clutch housing.
15. Install bearing race spring into bore of cover, then install cover and spring assembly. Align holes in cover to holes in outer clutch housing and install 5 screws.
16. Install O-ring over seal bridge to prevent it from dislodging during handling prior to the hub bearing being installed into the vehicle.
17. The hub and attached parts should turn freely after installation. The 5 cover screws must be loosened to install the hub into the vehicle. After installation, torque the cover screws to 40–50 inch lbs.

DUALMATIC LOCKING HUBS

Removal & Assembly

NOTE: Do not remove cam (outer) screws from cover. If cam screws are removed, misalignment could occur causing damage to cam.

1. Turn to engage position.
2. Apply pressure to the face of the shift knob, remove three shift knob retaining screws, Fig. 13.
3. Pull shift knob from mounting base and remove snap ring from axle shaft.
4. Remove capscrews and lockwashers from mounting flange.
5. Remove locking hub assembly from rotor hub and discard gasket.
6. Wash parts in mineral spirits and blow dry with compressed air. Inspect components for damage.

Assembly & Installation

1. Lubricate parts with multi-purpose lubricant part no. 2932524 or equivalent.
2. Install new gasket and locking hub onto rotor hub.
3. Install capscrews and lockwashers and torque to 30–40 ft. lbs.
4. Install axle shaft snap ring and shift knob on mounting base.
5. Align the splines by pushing inward on shift knob and turning it clockwise to lock it in position.
6. Install and tighten three shift knob retaining screws.

AUTOMATIC TRANSMISSIONS

INDEX

Allison AT-540, 543 & 545

DESCRIPTION

This transmission, Fig. 1, has four forward speeds and reverse. Shifting within the four ranges selected by the operator is fully automatic. A simple 3 element torque converter transmits power from the engine to the transmission gearing. The torque converter serves as both a fluid coupling and a torque multiplier. Ratios for the four forward speeds and reverse are established by planetary gearing which is controlled by hydraulic clutches. All gearing is in constant mesh.

TROUBLE SHOOTING GUIDE

Automatic Shifts at Too High Speed

1. Governor valve stuck.
2. Vacuum modulator hose to engine kinked or leaking (light throttle shifting delayed).
3. Vacuum modulator failed.
4. Mechanical actuator cable kinked, broken or improperly adjusted.
5. Mechanical actuator malfunctioning.

Automatic Shifts at Too Low Speed at Full Throttle

1. Governor valve stuck.
2. Governor spring weak.
3. Mechanical actuator cable kinked, broken or improperly adjusted.
4. Mechanical actuator malfunctioning.

Low Main Pressure in All Ranges

1. Low oil level.
2. Oil filter element clogged.
3. Seal ring on oil pickup tube leaking or missing.
4. Main pressure regulator valve spring weak.
5. Control valve body leakage.
6. Valves sticking.
7. Oil pump worn or damaged.

Low Main Pressure in First Gear, Normal Pressure in Other Ranges

1. First gear circuit of control valve body leakage.
2. Excessive leakage at first-and-reverse piston seals.

Buzzing Noise Occurring Intermittently

1. Low oil level.
2. Air leak at oil intake pipe.
3. Clogged filter.
4. Aerated oil.

Excessive Creep in First and Reverse Gears

1. Engine idle speed too high.

Low Lubrication Pressure

1. Oil level low.
2. Excessive internal oil leakage.
3. Cooler lines restricted or leaking.

Oil Leaking into Converter Housing

1. Engine crankshaft rear oil seal leakage.
2. Charging oil pump, lip type seal at converter leaking.
3. Seal ring around O.D. of oil pump leaking.
4. Cracked weld in converter assembly, leaking.

Transmission Heating Up in All Ranges

1. Oil level low.
2. Oil level high.
3. Engine cooling system restricted.
4. Oil cooler lines restricted.
5. Broken parts in converter.

No Response to Movement of Shift Lever

1. Range selector linkage unhooked.
2. Range selector linkage defective or broken.
3. Main pressure low.
4. Range selector not engaged at control valve.

High Stall Speed

1. Oil level low.
2. Clutch pressure low.
3. Forward clutch slipping (forward).
4. First-and-reverse clutch slipping.

Low Stall Speed

1. Engine not performing efficiently (may be due to high altitude).
2. Broken converter parts.

Rough Shifting

1. Manual selector linkage out of adjustment.
2. Control valves sticking.
3. Vacuum modulator valve sticking.
4. Vacuum modulator hose kinked or leaking.
5. Engine idle speed too fast.
6. Governor valve malfunctioning.
7. Mechanical actuator cable kinked, broken or improperly adjusted.
8. Mechanical actuator malfunctioning.

Engine Overspeeds on Full Throttle Upshift

1. Piston seals leaking or clutch plates slipping in range involved.
2. Forward clutch piston seals or clutch plates slipping (all upshifts).
3. Broken seal rings on front support hub.
4. Sticking governor valve.

Excessive Slippage or Chatter in One Range

1. Clutch slippage in that range clutch.
2. Excessive oil leakage in range piston seals.
3. Oil leakage in valve components for that range.

Dirty Oil

1. Failure to change oil at specified intervals.
2. Excessive heat.
3. Clutch failure.
4. Damaged oil filter.

Oil Leak at Output Shaft

1. Faulty or missing seal at output flange.

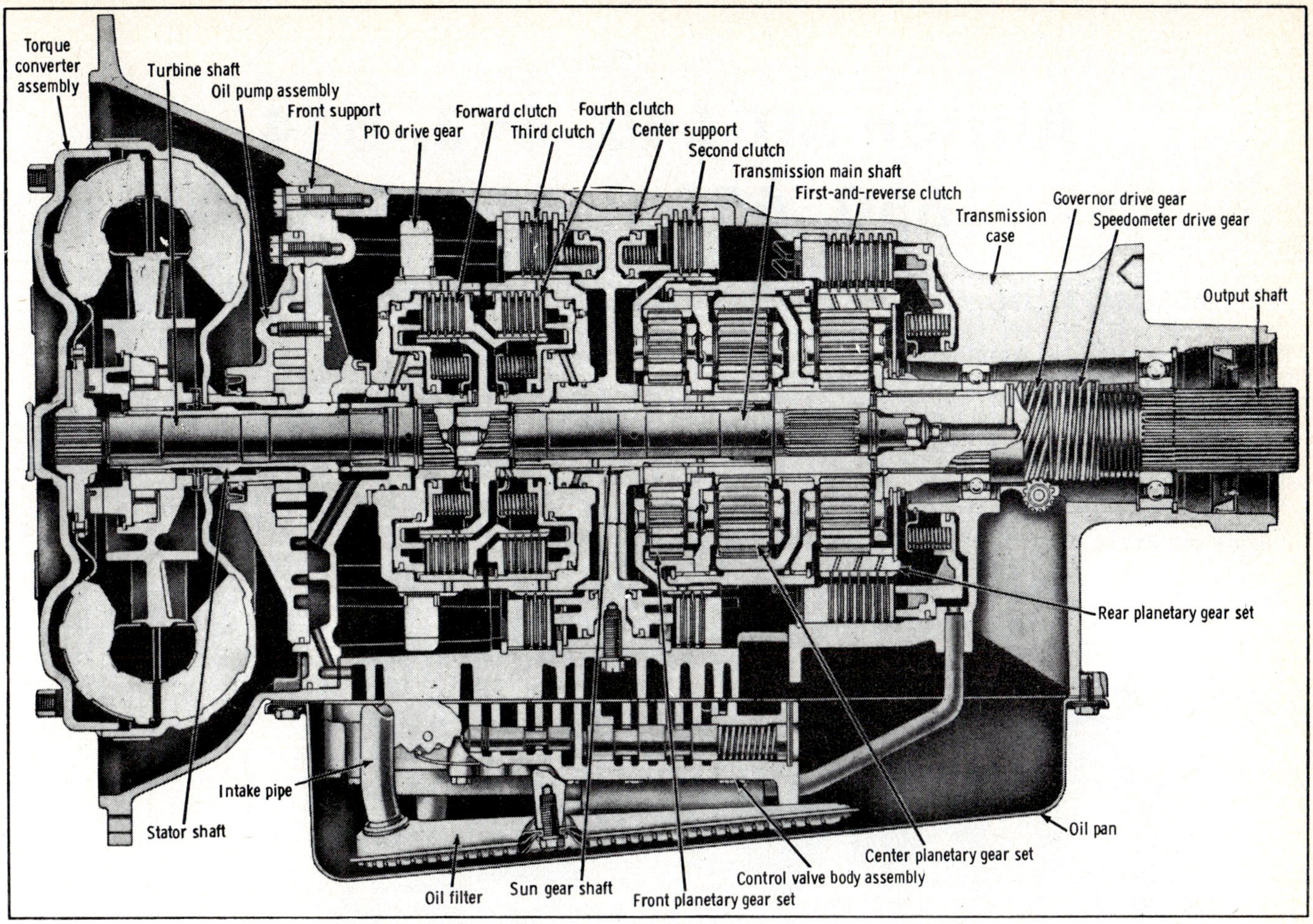

Fig. 1 Sectional view of AT 540 & 543 four speed automatic transmission

Slippage in All Forward Gears

1. Oil level low.
2. Clutch pressure low.
3. Forward clutch slipping.

Slippage in Fourth and Reverse Gear Only

1. Fourth clutch slipping.
2. Broken seal rings on support assembly hub.

Slippage in Reverse and First Gear; Proper Function in Other Forward Gears

1. First-and-reverse clutch slipping.

Vehicle Moves Forward in Neutral

1. Range selector linkage out of adjustment.
2. Forward clutch failed and dragging.

Vehicle Moves Backward in Neutral

1. Range selector linkage out of adjustment.
2. Fourth clutch failed and dragging.

Throws Oil Out of Transmission Filler Tube

1. Dipstick loose.
2. Oil level too high.
3. Breather stopped up.

MAINTENANCE

NOTE: Some 1983 models will have a spin-on type automatic transmission oil filter mounted on the truck chassis frame rails as standard equipment. The transmission oil and filter should be changed after the first 6000 miles of operation of a new vehicle or after transmission overhaul, and thereafter every 24,000 miles.

Change the spin-on filter using the following procedure:

1. Unscrew and discard the filter from the bracket mounted cover assembly.
2. Apply a light coating of transmission fluid to the new filter base gasket.
3. Screw new filter onto bracket mounted cover assembly. Torque to 13 ft. lbs.
4. Top off transmission with one quart of transmission fluid. Check transmission level, refer to "Oil level".

Oil Level

The transmission oil level should be checked every 1000 miles. Oil should be added only when the level is below the ADD mark. The safe operating level is from the ADD mark to the FULL mark. One quart of oil changes the level about ½" on the dipstick.

In order to check the oil level, the transmission must be at operating temperature (160° to 220°F). Shift the transmission through all ranges to fill all cavities and passages. Vehicle must be level and transmission in Neutral when checking oil.

Changing Oil

The oil and oil filter should be changed every 25,000 miles or 12 months, whichever occurs first. The transmission should be at operating temperature when oil is drained. This will insure quicker and better drainage.

Remove the oil fill tube from the oil pan and allow oil to drain. Unfasten the oil pan and replace the internal oil filter. Replace the pan gasket with a new one when installing the pan. Only Dexron fluid is recommended for use in this transmission.

NOTE: Governor oil filter screen should be cleaned or replaced at each oil change.

Allison MT-640, 643, 644, 650, 653 & 654CR

DESCRIPTION

The MT-640, 643 and 644, 653 & 654CR Fig. 1, has four forward speeds and one reverse. The MT-650, 643 and 644, 653 and 654CR Fig. 2, has five forward speeds and one reverse. Shifting within any of the forward ranges selected by the operator is fully automatic.

A simple three element torque converter transmits power to the transmission gearing. The torque converter which serves as both a fluid coupling and a torque multiplier, has a stall ratio of 2.3:1.

A lockup clutch automatically locks the turbine to the flywheel. When the vehicle reaches a certain speed, hydraulic pressure automatically applies the clutch. When the lockup clutch is applied, the engine output is directed to the transmission gearing at a ratio of 1:1. A decrease in speed automatically releases the lockup clutch.

The planetary gear sets are controlled by six hydraulic-applied clutches. All gearing is in constant mesh.

TROUBLE SHOOTING GUIDE

Automatic Shifts at Too High Speed

1. Governor valve stuck.
2. Vacuum signal valve spring adjustment too tight.

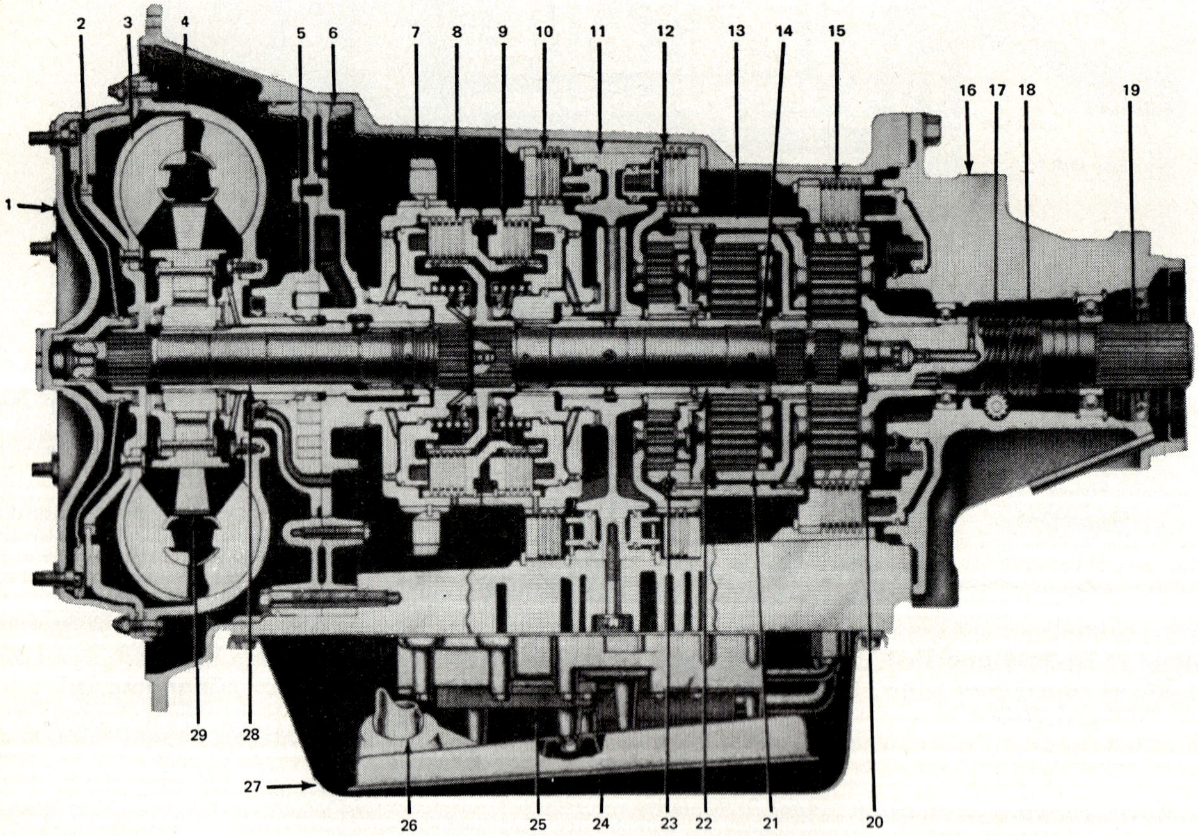

1 - Converter cover
2 - Lockup clutch
3 - Converter turbine
4 - Converter pump
5 - Oil pump
6 - Front support and valve assembly
7 - Power takeoff gear
8 - Forward clutch
9 - Fourth clutch
10 - Third clutch
11 - Center support
12 - Second clutch
13 - Planetary connecting drum
14 - Transmission main shaft
15 - First clutch
16 - Transmission rear cover
17 - Governor drive gear
18 - Speedometer drive gear
19 - Output shaft
20 - Rear planetary gear set
21 - Center planetary gear set
22 - Center sun gear shaft
23 - Front planetary gear set
24 - Oil filter
25 - Control valve assembly
26 - Intake pipe
27 - Oil pan
28 - Turbine shaft
29 - Converter stator

Fig. 1 Sectional view of MT-640, 643, 644 four speed automatic transmission

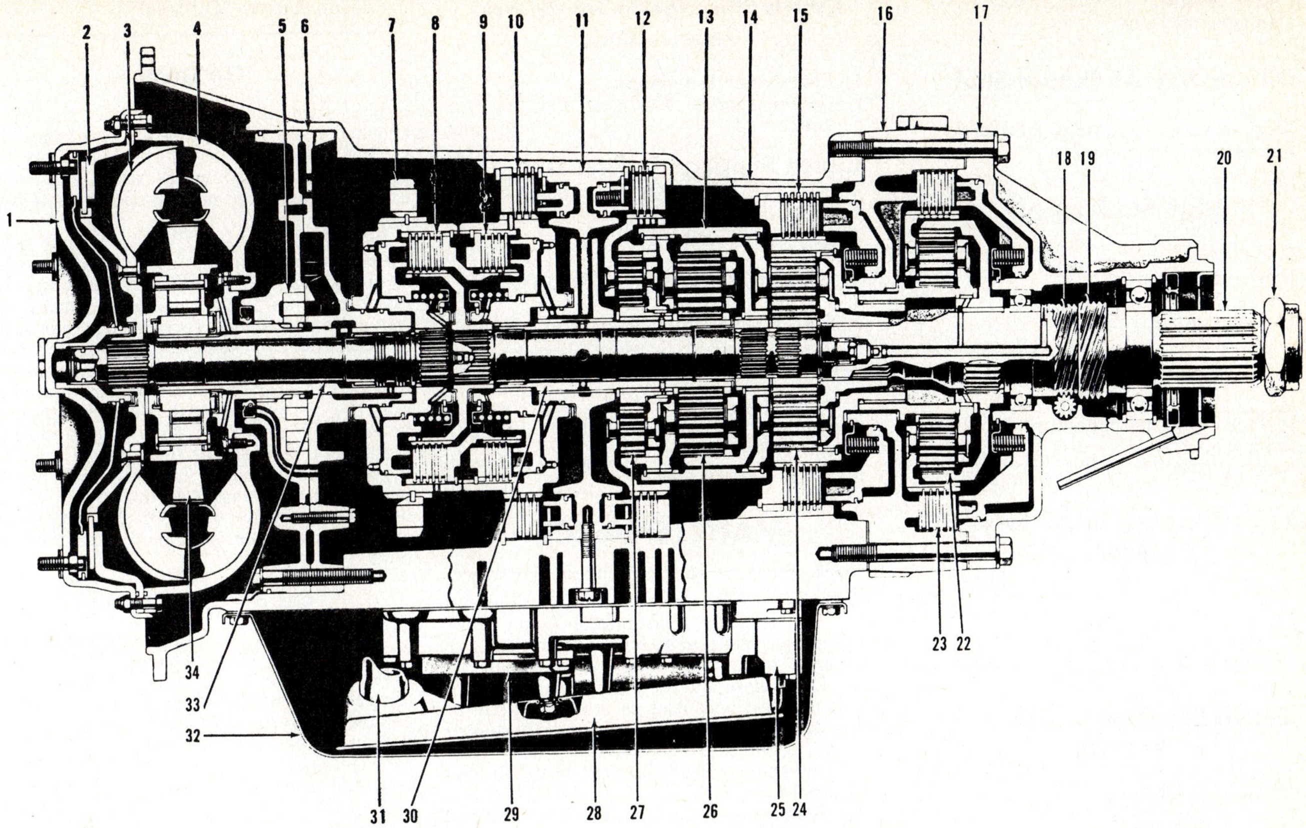

1 - Converter cover
2 - Lockup clutch
3 - Converter turbine
4 - Converter pump
5 - Oil pump
6 - Front support and valve assembly
7 - Power takeoff gear
8 - Forward clutch
9 - Fourth clutch
10 - Third clutch
11 - Center support
12 - Second clutch
13 - Planetary connecting drum
14 - Transmission housing
15 - First clutch
16 - Adapter housing
17 - Rear cover
18 - Governor drive
19 - Speedometer drive gear
20 - Output shaft
21 - Output flange nut
22 - Low planetary gear set
23 - Low clutch
24 - Rear planetary gear set
25 - Low shift valve
26 - Center planetary gear set
27 - Front planetary gear set
28 - Oil filter
29 - Control valve assembly
30 - Center sun gear shaft
31 - Intake pipe
32 - Oil pan
33 - Turbine shaft
34 - Converter stator

Fig. 2 Sectional view of MT-640, 653, 654CR five speed automatic transmission

3. Valves sticking.

Automatic Shifts at Too Low Speed

1. Low oil level.
2. Oil filter clogged.
3. Intake pipe seal leaking or missing.
4. Main pressure regulator valve spring weak.
5. Control valve body leakage.
6. Trimmer valves, relay valves or main pressure regulator valves stuck.
7. Worn or damaged oil pump.

Low Main Pressure in One Operating Range, Normal in Other Ranges

1. Governor valve stuck.
2. Governor spring weak.
3. Vacuum signal valve spring adjustment.
4. Stuck modulator valve.
5. Modulator valve spring adjustment too loose.

Low Main Pressure in All Ranges

1. Leakage in clutch apply circuit for specific range.
2. Excessive leakage at clutch seals for specific range.

Excessive Creep in First and Reverse

1. Idle setting too high.

Low Lubrication Pressure

1. Oil level too low.
2. Excessive internal leakage.
3. Cooler lines restricted or leaking.
4. Lubrication valve spring weak.

Oil Leaking Into Converter Housing

1. Worn converter pump hub seal.
2. Converter pump hub worn at seal area.
3. Worn rear engine seal.

Transmission Overheating in All Ranges

1. Oil level too low.
2. Oil level too high.
3. Restricted oil cooler.

No Response To Shift Lever Movement

1. Shift selector linkage disconnected.
2. Shift selector linkage broken or defective.
3. Low main pressure.
4. Shift selector not engaged at control valve.

Rough Shifting

1. Shift selector linkage incorrectly adjusted.
2. Control valves sticking.
3. Sticking modulator valve or spring incorrectly adjusted.
4. Modulator actuator cable kinked or incorrectly adjusted.

Dirty Oil

1. Oil not changed at proper intervals.
2. Overheating.

3. Clutch failure.
4. Damaged oil filter.

Oil Leakage At Output Shaft

1. Worn or damaged output shaft seal.
2. Flange worn at seal area.

High Stall Speed

1. Low oil level.
2. Low clutch pressure.
3. Forward clutch slipping.
4. First clutch slipping.

Low Stall Speed

1. Inefficient engine operation (may be due to high altitude).
2. Broken converter parts.

Clutch Slippage in All Forward Gears

1. Low oil level.
2. Clutch (main) pressure low.
3. Forward clutch slipping.
4. Front support hub seals worn or broken.

Clutch Slippage in First and Reverse

1. First clutch slipping.

Clutch Slippage in Second or Third (MT 640); or Third or Fourth (MT 650) Only

1. Clutch for specific range slipping.

Clutch Slippage in Fourth and Reverse Only

1. Fourth clutch slipping.
2. Center support hub seal worn or broken.

Clutch Slippage in First Only (MT 650)

1. Low clutch slipping.

Vehicle Moves in Neutral

1. Shift selector linkage out of adjustment.

Oil Overflowing From Filler Tube

1. Loose dipstick.
2. Oil level too high.
3. Clogged breather.
4. Worn dipstick gasket.

MAINTENANCE

NOTE: Some 1983 models will have a spin-on type automatic transmission oil filter mounted on the truck chassis frame rails as standard equipment. The transmission oil and filter should be changed after the first 6000 miles of operation of a new vehicle or after transmission overhaul, and thereafter every 24,000 miles.

Change the spin-on filter using the following procedure:

1. Unscrew and discard the filter from the bracket mounted cover assembly.
2. Apply a light coating of transmission fluid to the new filter base gasket.
3. Screw new filter onto bracket mounted cover assembly. Torque to 13 ft. lbs.
4. Top off transmission with one quart of transmission fluid. Check transmission level, refer to "Oil level".

Oil Level

The transmission oil level should be checked every 6000 miles. Oil level should be at the "Full" mark or slightly below, but never above the "Full" mark when engine and transmission are at normal operating temperatures.

In order to check oil level on early model units, the transmission must be at operating temperature (160°-220°F) and the vehicle in a level position. Apply parking brake and momentarily select each gear, then place selector lever in Neutral and check oil level. Oil level should be between Add and Full marks. On later model units, a cold oil check may be performed with fluid temperature between 60° and 120°. With engine idling in neutral, oil level should be between "cold run" marks on dipstick. A hot oil check may be performed with engine idling in neutral and fluid temperature at 160°–200°. Oil level should be between "hot run" marks on dipstick.

Oil & Filter Change

Oil and filter should be changed every 25,000 miles or 12 months, which ever occurs first. The transmission should be at operating temperature when oil is changed. This will ensure quicker and better drainage.

Remove oil pan drain plug and gasket or fill tube from pan and allow oil to drain thoroughly. Remove oil pan attaching bolts and pan and gasket. Replace internal oil filter, then install oil pan, new gasket and drain plug or fill tube. Only Dexron fluid is recommended for use in this transmission.

NOTE: Both governor oil filter screens should be cleaned or replaced at each oil change.

Allison HT-740, 750

DESCRIPTION

The 740 units consist of four forward speeds and the 750 units consist of five forward speeds. Shifting is fully automatic on the HT-740D, Fig. 1, 750CRD, Fig. 2, models, while the HT-750DRD, Fig. 3, models must be manually shifted out of first gear, but the remaining ranges are fully automatic.

A simple three element torque converter transmits power from the engine to the transmission gearing. The dual purpose torque converter is a torque multiplier and a fluid coupling. Depending on application, the three combinations of converters which are used provide a stall ratio of 2.8:1, 2.2:1 or 2.1:1.

The lockup clutch which locks the turbine element of the torque converter to the flywheel is automatically applied by hydraulic pressure when the vehicle attains sufficient speed. A decrease in speed automatically releases the clutch. When the lockup clutch is applied, the engine output is directed to the transmission gearing at a ratio of 1:1.

Some HT-700D units incorporate a built in hydraulic retarder which is primarily a vaned rotor attached to the torque converter turbine output shaft. When the manually operated control valve admits oil into the vaned chamber surrounding the rotor, the retarder assists in slowing the vehicle.

These transmissions are provided with provisions for mounting a power take-off unit either on the transmission housing or converter housing. The transmission mounted PTO units are converter driven, while the converter housing mounted PTO units are engine driven.

On HT-740 units, ratios for the forward speeds are established by three planetary gear sets and are controlled by five hydraulically applied clutches. On HT-750 units, ratios for the forward speeds are established by four planetary gear sets and are controlled by six hydraulically applied clutches. All gearing on both the HT-740, 750 are in constant mesh.

TROUBLE SHOOTING GUIDE

Automatic Shifts Occur at Too High Speed

1. Shift signal valve spring too tight.
2. Governor valve stuck.
3. Valves sticking.

Automatic Shifts Occur at Too Low Speed

1. Governor valve stuck.
2. Weak governor spring.
3. Shift signal valve spring adjustment too loose.
4. Stuck modulator valve.
5. Modulator valve spring adjustment too loose.

Low Main Pressure In All Ranges

1. Low oil level.
2. Clogged oil filter element.
3. Seal at oil intake pipe (filter output) leaking or missing.
4. Weak main pressure regulator valve spring.
5. Control valve body leakage.
6. Trimmers, relays, and/or main pressure regulator valves stuck.
7. Worn or damaged oil pump.

Low Main Pressure In One Operating Range, Normal In All Other Ranges

1. Leakage in clutch apply circuits for specific range.
2. Excessive leakage at clutch piston seals for specific range.

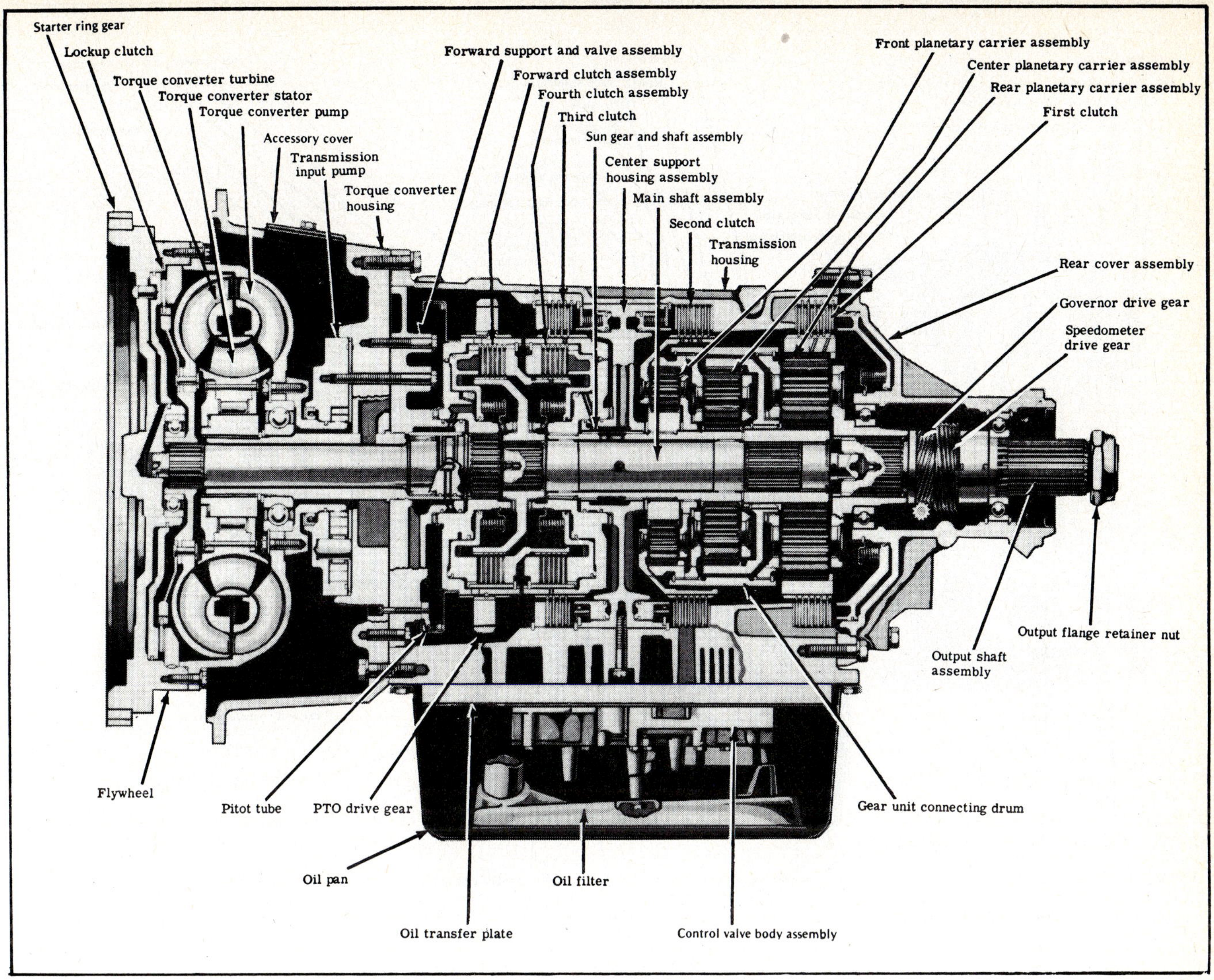

Fig. 1 HT-740 transmission

Excessive Creep in First and Reverse Gears

1. Idle throttle setting too high.

Low Lubrication Pressure

1. Oil level too low.
2. Excessive internal oil leakage.
3. Restricted or leaking cooler lines.
4. Weak lubrication valve spring.

Oil Leaking Converter Housing

1. Converter hub seal worn.
2. Converter hub worn at seal area.
3. Engine rear seal worn.

Transmission Overheating In All Ranges

1. Low oil level.
2. High oil level.
3. Restricted cooler at oil or coolant side.

No Response to Shift Lever Movement

1. Range selector linkage disconnected.
2. Worn or defective range selector linkage.
3. Low Main Pressure.
4. Range selector disengaged at control valve.

Rough Shifting

1. Range selector linkage maladjusted.
2. Sticking control valves.
3. Sticking modulator valve or spring adjustment too tight.
4. Modulator actuator cable kinked or out of adjustment.

Contaminated Oil

1. Failure to change oil at proper intervals.
2. Overheating.
3. Damaged oil filter.
4. Clutch failure.

Oil Leaking At Output Shaft

1. Output flange seal worn or damaged.
2. Output flange worn at seal area.

High Stall Speed

1. Low oil level.
2. Low clutch pressure.
3. Slipping forward clutch (forward).
4. First clutch slipping.
5. Fourth clutch slipping (reverse).
6. Low clutch slipping (5 speed).

Low Stall Speed

1. Engine not performing efficiently. (May be due to high altitude).
2. Damaged converter.

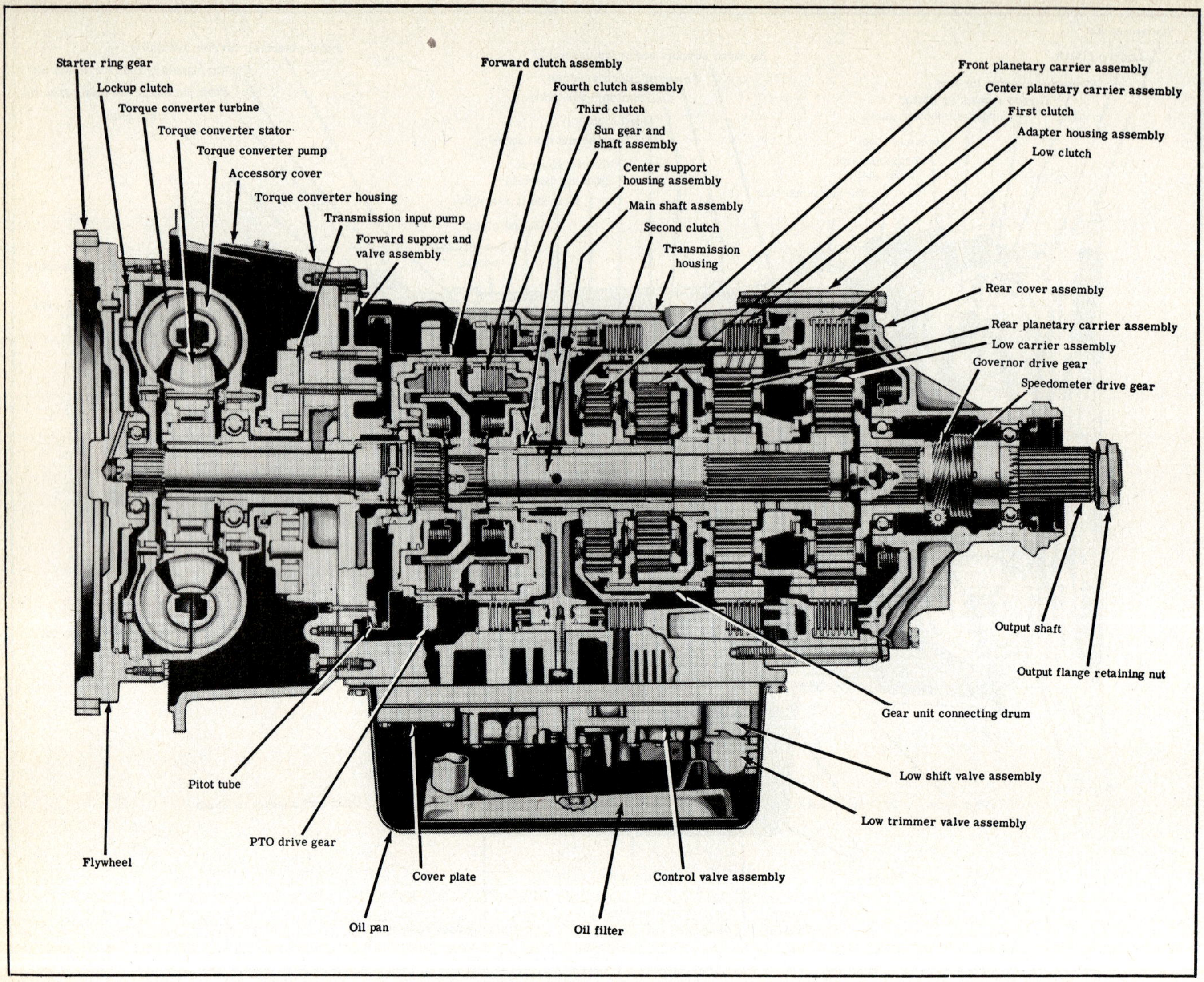

Fig. 2 HT-750 CRD transmission

Clutch Slippage In All Forward Gears

1. Low oil level.
2. Low clutch (main) pressure.
3. Foward clutch slipping.
4. Seal rings on front support hub worn or broken.

Clutch Slippage In First and Reverse Only (740 D units)

1. First clutch slippage.

Clutch Slippage In All Forward Gears But No Slippage In Reverse

1. Forward clutch slipping.

Clutch Slippage In Fourth and Reverse Gears Only (740D units)

1. Fourth clutch slipping.

Clutch Slippage In Third Gear Only (740D units)

1. Third clutch slipping.

Clutch Slippage In Second Gear Only (740D units)

1. Second clutch slipping.

Clutch Slippage In Third Gear Only (750 CRD, DRD units)

1. Third clutch slipping.

Clutch Slippage In Third Gear Only (750 CRD, DRD units)

1. Second clutch slipping.

Clutch Slippage In Second Gear Only (750 CRD units)

1. First clutch slipping.

Clutch Slippage In Second and Reverse Gears Only (750 CRD units)

1. First clutch slipping.

Clutch Slippage In Fifth and Reverse Gears Only (750 CRD, DRD units)

1. Fourth clutch slipping.

Clutch Slippage In First Gear Only (750 DRD units)

1. Low clutch slipping.

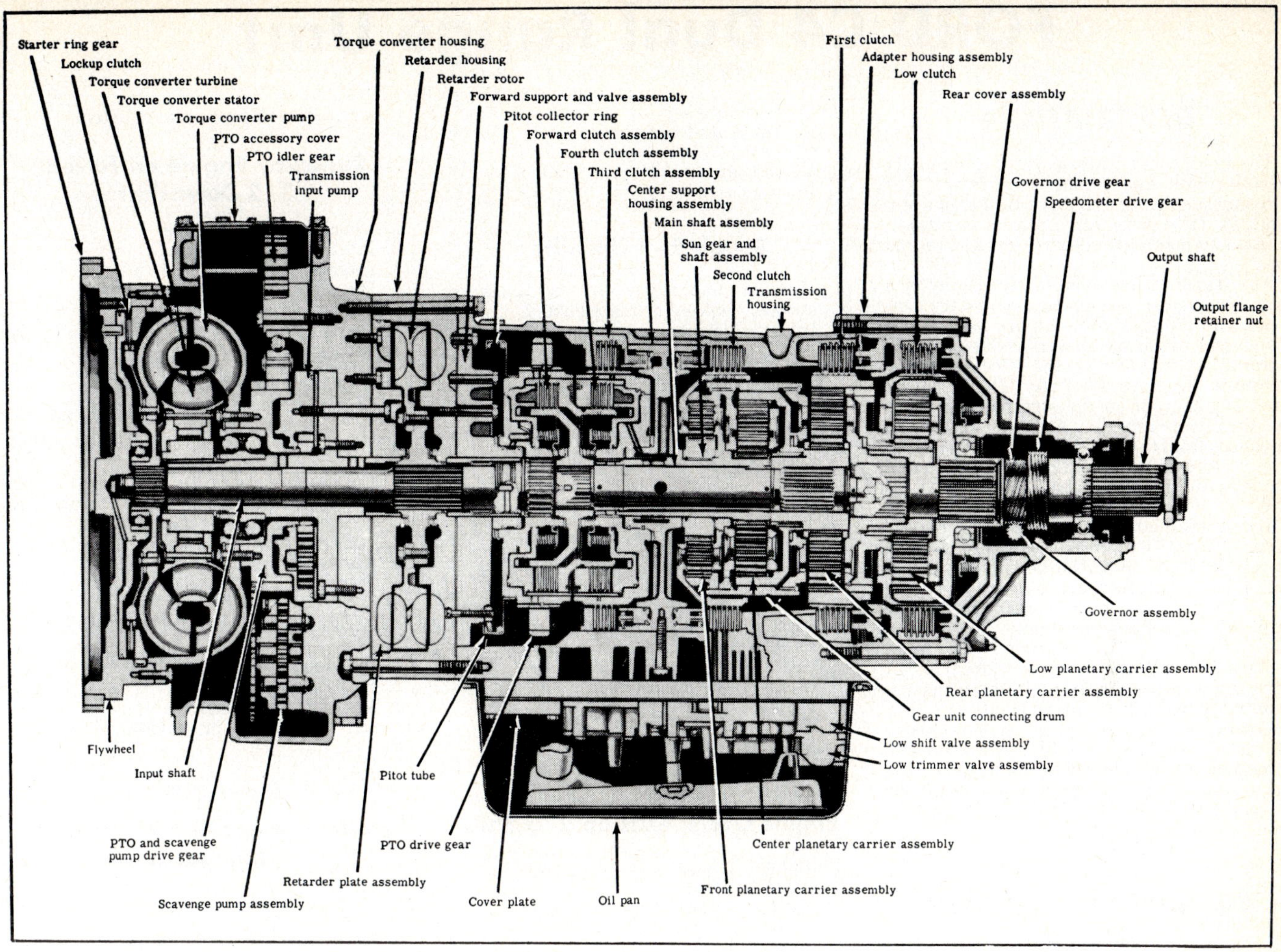

Fig. 3 HT-750 DRD transmission

Vehicle Moves in Neutral

1. Range selector linkage out of adjustment.
2. Forward or fourth clutch will not release.

Oil Thrown From Filler Tube

1. Loose dipstick.
2. Oil level too high.
3. Clogged breather.
4. Worn dipstick gasket.
5. Improper dipstick marking.

MAINTENANCE

NOTE: Some 1983 models will have a spin-on type automatic transmission oil filter mounted on the truck chassis frame rails as standard equipment. The transmission oil and filter should be changed after the first 6000 miles of operation of a new vehicle or after transmission overhaul, and thereafter every 24,000 miles.

Change the spin-on filter using the following procedure:

1. Unscrew and discard the filter from the bracket mounted cover assembly.
2. Apply a light coating of transmission fluid to the new filter base gasket.
3. Screw new filter onto bracket mounted cover assembly. Torque to 13 ft. lbs.
4. Top off transmission with one quart of transmission fluid. Check transmission level, refer to "Oil level".

Oil Level

The transmission oil level should be checked every 1000 miles. Oil should be added only when the level is below the ADD mark. The safe operating level is from the ADD mark to the FULL mark. One quart of oil changes the level about ½" on the dipstick.

In order to check the oil level, the transmission must be at operating temperature (160° to 200° F). Shift the transmission through all ranges to fill all cavities and passages. Vehicle must be level and transmission in Neutral when checking oil.

Changing Oil

Oil should be changed each 50,000 miles or 12 months for on highway vehicles and 25,000 miles, 800 hours or 6 months for off highway vehicles. Internal filter should be changed at first oil change and each 500,000 miles thereafter if equipped with external oil filter. On models less external oil filter, change internal filter at each oil change. External oil filter should be changed after first 5000 miles and each 50,000 miles for on highway vehicles or 25,000 for off highway vehicles thereafter. The transmission should be at operating temperature when oil is drained. This will insure quicker and better drainage.

Remove fill tube from oil pan on early models or drain plug on late models and allow oil to drain. On all models, unfasten the oil pan and replace the internal oil filter. Replace the pan gasket with a new one when installing the pan. Only Dexron fluid is recommended for us in this transmission.

FORD C4 Dual Range Unit

DESCRIPTION

The main control incorporates a manually selective first and second gear range. The transmission features a drive range that provides for fully automatic upshifts and downshifts, and manually selected low and second gears.

The transmission consists essentially of a torque converter, a compound planetary gear train, two multiple disc clutches, a one-way clutch and a hydraulic control system, Fig. 1.

For all normal driving the selector lever is moved to the green dot under "Drive" on the selector quadrant on the steering column or on the floor console. As the throttle is advanced from the idle position, the transmission will upshift automatically to intermediate gear and then to high.

With the throttle closed the transmission will downshift automatically as the car speed drops to about 10 mph. With the throttle open at any position up to the detent, the downshifts will come in automatically at speeds above 10 mph and in proportion to throttle opening. This prevents engine lugging on steep hill climbing, for example.

When the selector lever is moved to "L" with the transmission in high, the transmission will downshift to intermediate or to low depending on the road speed. At speed above 25 mph, the downshift will be from high to intermediate. At speeds below 25 mph, the downshift will be from high to low. With the selector lever in the "L" position the transmission cannot upshift.

TROUBLE SHOOTING GUIDE

The items to check for each trouble symptom are arranged in a logical sequence that should be followed for quickest results.

Rough Initial Engagement in D1 or D2

1. Engine idle speed.
2. Vacuum diaphragm unit or tubes restricted, leaking or maladjusted.
3. Check control pressure.
4. Pressure regulator.
5. Valve body.
6. Forward clutch.

1–2 or 2–3 Shift Points Erratic

1. Check fluid level.
2. Vacuum diaphragm unit or tubes restricted, leaking or maladjusted.
3. Intermediate servo.
4. Manual linkage adjustment.
5. Governor.
6. Check control pressure.
7. Valve body.
8. Make air pressure check.

Rough 1–2 Upshifts

1. Vacuum diaphragm unit or tubes restricted, leaking or maladjusted.
2. Intermediate servo.
3. Intermediate band.
4. Check control pressure.
5. Valve body.
6. Pressure regulator.

Rough 2–3 Upshifts

1. Vacuum diaphragm unit or tubes restricted, leaking or maladjusted.
2. Intermediate servo.
3. Check control pressure.
4. Pressure regulator.
5. Intermediate band.
6. Valve body.
7. Make air pressure check.
8. Reverse-high clutch.
9. Reverse-high clutch piston air bleed valve.

Dragged Out 1–2 Shift

1. Check fluid level.
2. Vacuum diaphragm unit or tubes restricted, leaking or maladjusted.
3. Intermediate servo.
4. Check control pressure.
5. Intermediate band.
6. Valve body.
7. Pressure regulator.
8. Make air pressure check.
9. Leakage in hydraulic system.

Engine Overspeeds on 2–3 Shift

1. Manual linkage.
2. Check fluid level.
3. Vacuum diaphragm unit or tubes restricted, leaking or maladjusted.
4. Reverse servo.
5. Check control pressure.
6. Valve body.
7. Pressure regulator.
8. Intermediate band.
9. Reverse-high clutch.
10. Reverse-high clutch piston air bleed valve.

No 1–2 or 2–3 Shift

1. Manual linkage.
2. Downshift linkage, including inner lever position.
3. Vacuum diaphragm unit or tubes restricted, leaking or maladjusted.
4. Governor.
5. Check control pressure.
6. Valve body.
7. Intermediate band.
8. Intermediate servo.
9. Reverse-high clutch.
10. Reverse-high clutch piston air bleed valve.

No 3–1 Shift in D1 or 3–2 Shift in D2

1. Governor.
2. Valve body.

No Forced Downshifts

1. Downshift linkage, including inner lever position.
2. Valve body.
3. Vacuum diaphragm unit or tubes restricted, leaking or maladjusted.

Runaway Engine on Forced 3–2 Downshift

1. Check control pressure.
2. Intermediate servo.
3. Intermediate band.
4. Pressure regulator.
5. Valve body.
6. Vacuum diaphragm unit or tubes restricted, leaking or maladjusted.
7. Leakage in hydraulic system.

Rough 3–2 or 3–1 Shift at Closed Throttle

1. Engine idle speed.
2. Vacuum diaphragm unit or tubes restricted, leaking or maladjusted.
3. Intermediate servo.
4. Valve body.
5. Pressure regulator.

Shifts 1–3 in D1 and D2

1. Intermediate band.
2. Intermediate servo.
3. Vacuum diaphragm unit or tubes restricted, leaking or maladjusted.
4. Valve body.
5. Governor.
6. Make air pressure check.

No Engine Braking in 1st Gear —Manual Low

1. Manual linkage.
2. Reverse band.
3. Reverse servo.
4. Valve body.
5. Governor.
6. Make air pressure check.

Slips or Chatters in 1st Gear—D1

1. Check fluid level.
2. Vacuum diaphragm unit or tubes restricted, leaking or maladjusted.
3. Check control pressure.
4. Press regulator.
5. Valve body.
6. Forward clutch.
7. Leakage in hydraulic system.
8. Planetary one-way clutch.

Slips or Chatters in 2nd Gear

1. Check fluid level.
2. Vacuum diaphragm unit or tubes restricted, leaking or maladjusted.
3. Intermediate servo.
4. Intermediate band.
5. Check control pressure.
6. Pressure regulator.
7. Valve body.
8. Make air pressure check.
9. Forward clutch.
10. Leakage in hydraulic system.

Slips or Chatters in R

1. Check fluid level.
2. Vacuum diaphragm unit or tubes restricted, leaking or maladjusted.
3. Reverse band.
4. Check control pressure.
5. Reverse servo.

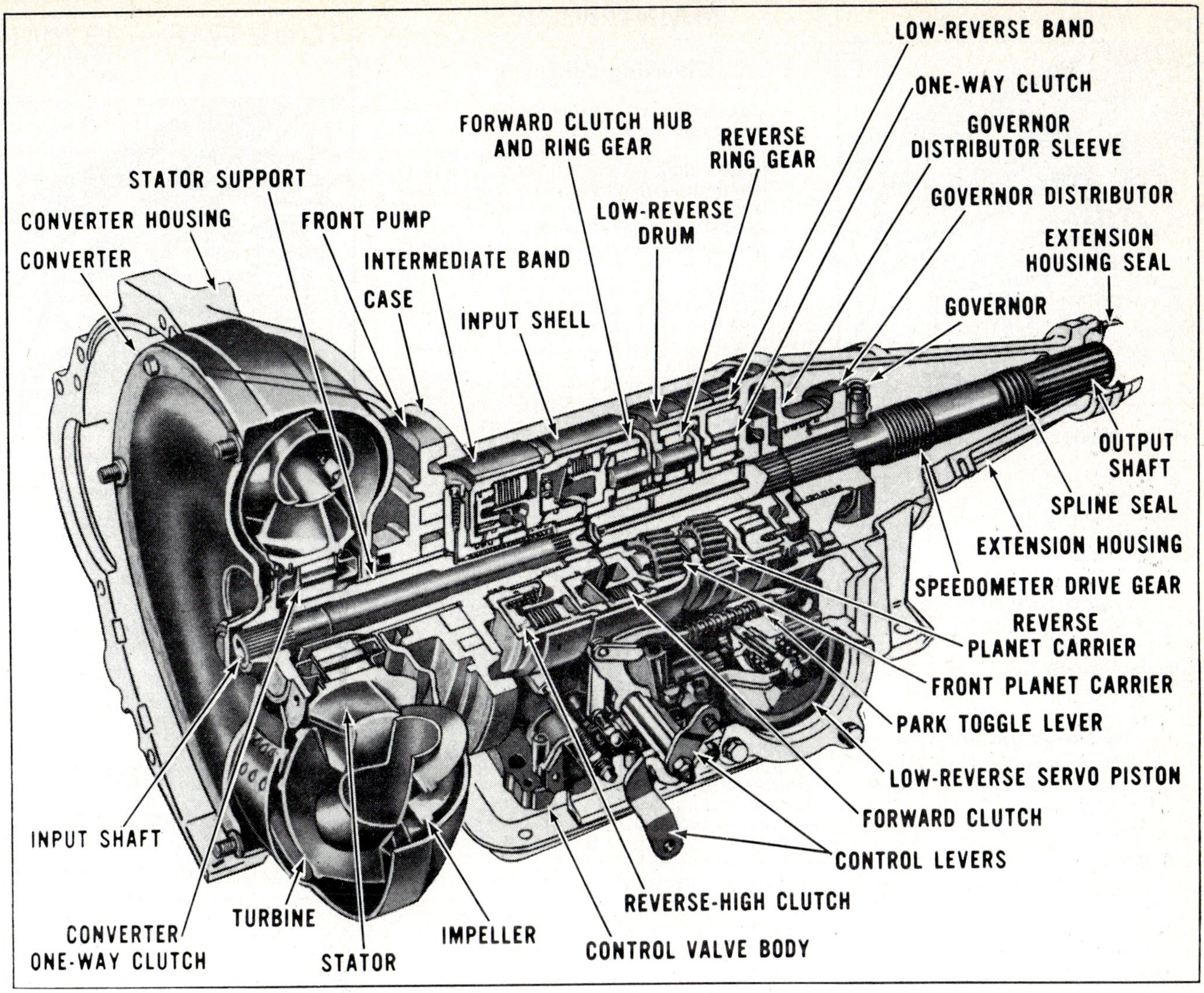

Fig. 1 C4 dual range automatic transmission

6. Pressure regulator.
7. Valve body.
8. Make air pressure check.
9. Reverse-high clutch.
10. Leakage in hydraulic system.
11. Reverse-high piston air bleed valve.

No Drive in D1 Only

1. Check fluid level.
2. Manual linkage.
3. Check control pressure.
4. Valve body.
5. Make air pressure check.
6. Planetary one-way clutch.

No Drive in D2 Only

1. Check fluid level.
2. Manual linkage.
3. Check control pressure.
4. Intermediate servo.
5. Valve body.
6. Make air pressure check.
7. Leakage in hydraulic system.
8. Planetary one-way clutch.

No Drive in L Only

1. Check fluid level.
2. Manual linkage.
3. Check control pressure.
4. Valve body.
5. Reverse servo.
6. Make air pressure check.
7. Leakage in hydraulic system.
8. Planetary one-way clutch.

No Drive in R Only

1. Check fluid level.
2. Manual linkage.
3. Reverse band.
4. Check control pressure.
5. Reverse servo.
6. Valve body.
7. Make air pressure check.
8. Reverse-high clutch.
9. Leakage in hydraulic system.
10. Reverse-high clutch piston air bleed valve.

No Drive in Any Selector Position

1. Check fluid level.
2. Manual linkage.
3. Check control pressure.
4. Pressure regulator.
5. Valve body.
6. Make air pressure check.
7. Leakage in hydraulic system.
8. Front pump.

Lockup in D1 Only

1. Reverse-high clutch.
2. Parking linkage.
3. Leakage in hydraulic system.

Lockup in D2 Only

1. Reverse band.
2. Reverse servo.
3. Reverse-high clutch.
4. Parking linkage.
5. Leakage in hydraulic system.
6. Planetary one-way clutch.

Lockup in L Only

1. Intermediate band.
2. Intermediate servo.
3. Reverse-high clutch.
4. Parking linkage.
5. Leakage in hydraulic system.

Lockup in R Only

1. Intermediate band.

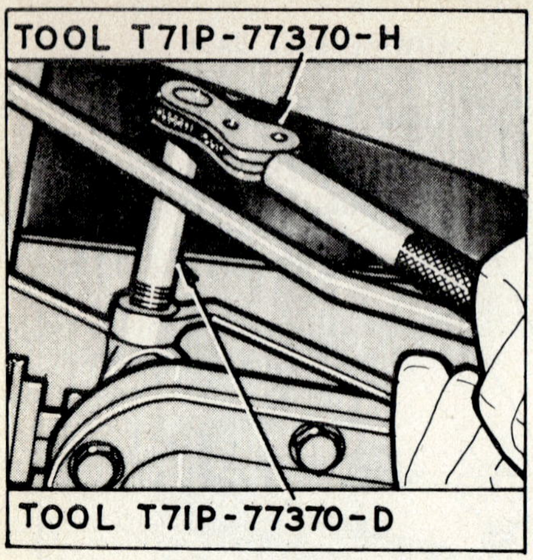

Fig. 2 Intermediate band adjustment

2. Intermediate servo.
3. Forward clutch.
4. Parking linkage.
5. Leakage in hydraulic system.

Parking Lock Binds or Does Not Hold

1. Manual linkage.
2. Parking linkage.

Maximum Speed Too Low, Poor Acceleration

1. Engine performance.
2. Brakes bind.
3. Converter one-way clutch.

Noisy in N or P

1. Check fluid level.
2. Pressure regulator.
3. Front pump.
4. Plantetary assembly.

Noisy in All Gears

1. Check fluid level.
2. Pressure regulator.
3. Planetary assembly.
4. Forward clutch.
5. Front pump.
6. Planetary one-way clutch.

Car Moves Forward in N

1. Manual linkage.
2. Forward clutch.

MAINTENANCE

Checking Oil Level

1. Make sure car is on a level floor.
2. Apply parking brake firmly.
3. Run engine at normal idle speed. If transmission fluid is cold, run engine at a fast idle until fluid reaches normal operating temperature. When fluid is warm, slow engine to normal idle speed.
4. Shift selector lever through all positions, then place lever at "P". Do not shut down engine during fluid level checks.
5. Clean all dirt from dipstick cap before removing dipstick from filler tube.
6. Pull dipstick out of tube, wipe it clean and push it all the way back in tube.
7. Pull dipstick out of tube again and check fluid level. If necessary, add enough fluid to raise the level to the "F" mark on dipstick. Do not overfill.

Drain & Refill

NOTE: The Ford Motor Company recommends the use of an automatic transmission fluid with Qualification No. M2C-33F, type F, for 1974–79 models and M2C-138-CJ or Dexron II for 1980–81 models (on container). The recommended fluid is said to have a greater coefficient of friction and greater ability to handle maximum engine torques without band or clutch slippage.

Normal maintenance and lubrication requirements do not necessitate periodic fluid changes. If a major failure has occurred in the transmission, it will have to be removed for service. At this time the converter and transmission cooler must be thoroughly flushed to remove any foreign matter.

When filling a dry transmission and converter, install five quarts of fluid. Start engine, shift the selector lever through all ranges and place it a R position. Check fluid level and add enough to raise the level in the transmission to the "F" (full) mark on the dipstick.

When a partial drain and refill is required due to front band adjustment or minor repair, proceed as follows:

1. Loosen and remove all but two oil pan bolts and drop one edge of the pan to drain the oil.
2. Remove and clean pan and screen.
3. Place a new gasket on pan and install pan and screen.
4. Add three quarts of fluid to transmission.
5. Run engine at idle speed for about two minutes.

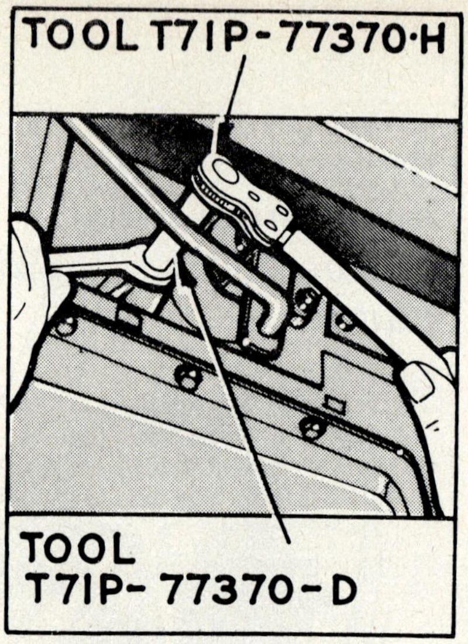

Fig. 3 Low-reverse band adjustment

6. Check oil level and add oil as necessary.
7. Run engine at a fast idle until it reaches normal operating temperature.
8. Shift selector lever through all ranges and then place it in R position.
9. Add fluid as required to bring the level to the full mark.

BANDS, ADJUST

Intermediate Band

1. Loosen lock nut several turns.
2. With tool shown in Fig. 2 tighten adjusting screw until tool handle clicks. *This tool is a pre-set torque wrench which clicks and overruns when the torque on the adjusting screw reaches 10 ft-lbs.*
3. Back off adjusting screw exactly 1¾ turns.
4. Hold adjusting screw from turning and tighten lock nut.

Rear Band

1. Loosen lock nut several turns.
2. Tighten adjusting screw until tool handle clicks, Fig. 3. *Tool shown is a pre-set torque wrench which clicks and overruns when the torque on the* adjusting screw reaches 10 ft-lbs.
3. Back off adjusting screw exactly 3 full turns.
4. Hold adjusting screw from turning and tighten lock nut.

Ford FMX Cast Iron Case

DESCRIPTION

This transmission, Fig. 1, is a three speed unit which provides automatic upshifts and downshifts. First and second gears may be selected manually. This unit consists of a torque converter, planetary gear train, multiple disc clutches and hydraulic control system.

TROUBLE SHOOTING GUIDE

Rough Initial Engagement

1. Idle speed.
2. Vacuum unit or tubes.
3. Front band.
4. Check control pressure.
5. Pressure regulator.
6. Valve body.

Shift Points High, Low or Erratic

1. Fluid level.
2. Vacuum unit or tubes.
3. Manual linkage.
4. Governor.
5. Check control pressure.

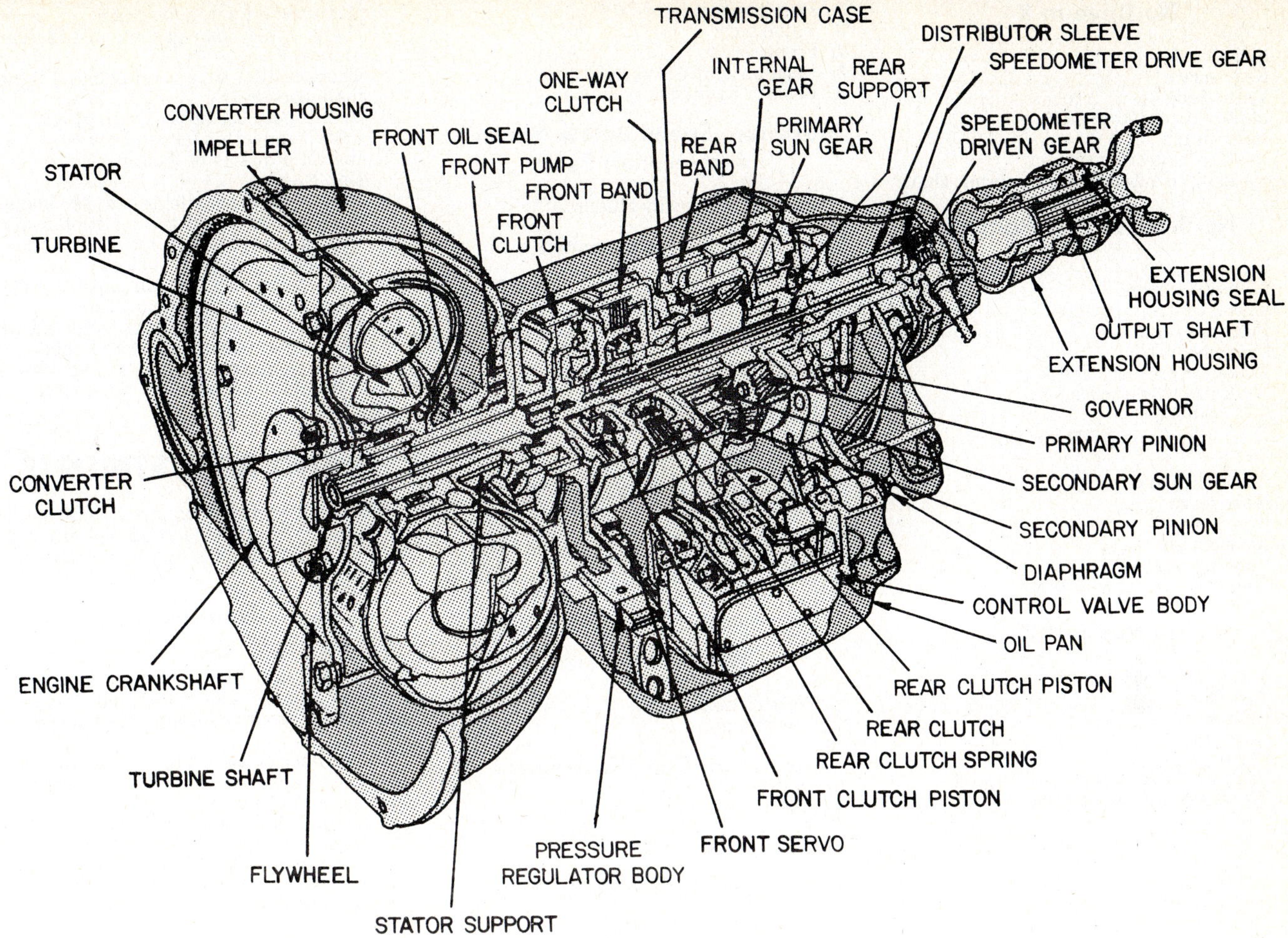

Fig. 1 Ford FMX three speed transmission with cast iron case.

6. Valve body.
7. Downshift linkage.

Rough 2–3 Shift

1. Manual linkage.
2. Front band.
3. Vacuum unit or tubes.
4. Pressure regulator.
5. Valve body.
6. Front servo.

Engine Overspeeds, 2–3 Shift

1. Vacuum unit or tubes.
2. Front band.
3. Valve body.
4. Pressure regulator.

No 1–2 or 2–3 Shifts

1. Governor.
2. Valve body.
3. Manual linkage.
4. Rear clutch.
5. Front band.
6. Front servo.
7. Leakage in hydraulic system.
8. Pressure regulator.

No Forced Downshifts

1. Downshift linkage.
2. Check control pressure.
3. Valve body.

Rough 3–2 or 3–1 Shifts

1. Engine idle speed.
2. Vacuum unit or tubes.
3. Valve body.

Slips or Chatters in 2nd

1. Fluid level.
2. Vacuum unit or tubes.
3. Front band.
4. Check control pressure.
5. Pressure regulator.
6. Valve body.
7. Front servo.
8. Front clutch.
9. Leakage in hydraulic system.

Slips or Chatters in 1st

1. Fluid level.
2. Vacuum unit or tubes.
3. Check control pressure.
4. Pressure regulator.
5. Valve body.
6. Front clutch.
7. Leakage in hydraulic system.
8. Fluid distributor sleeve in output shaft.
9. Planetary one-way clutch.

Slips or Chatters in Reverse

1. Fluid level.
2. Rear band.
3. Check control pressure.
4. Pressure regulator.
5. Valve body.
6. Rear servo.
7. Rear clutch.
8. Vacuum unit or tubes.
9. Leakage in hydraulic system.
10. Fluid distributor sleeve in output shaft.

No Drive in D or D2

1. Valve body.
2. Make air pressure check.
3. Manual linkage.
4. Front clutch.
5. Leak in hydraulic system.
6. Fluid distributor sleeve in output shaft.

No Drive in D1

1. Manual linkage.
2. Valve body.
3. Planetary one-way clutch.

No Drive in L

1. Manual linkage.
2. Front clutch.
3. Valve body.
4. Make air pressure check.
5. Leak in hydraulic system.
6. Fluid distributor sleeve in output shaft.

No Drive in R

1. Rear band.
2. Rear servo.
3. Valve body.
4. Make air pressure check.
5. Rear clutch.
6. Leak in hydraulic system.
7. Fluid distributor sleeve in output shaft.

No Drive in Any Range

1. Fluid level.
2. Manual linkage.
3. Check control pressure.
4. Pressure regulator.
5. Valve body.
6. Make air pressure check.
7. Leak in hydraulic system.

Lockup in D or D1

1. Manual linkage.
2. Rear servo.
3. Front servo.
4. Rear clutch.
5. Parking linkage.
6. Leak in hydraulic system.

Lockup in D2

1. Manual linkage.
2. Rear band.
3. Rear servo.
4. Rear clutch.
5. Parking linkage.
6. Leak in hydraulic system.
7. Planetary one-way clutch.

Lockup in R

1. Front band.
2. Front servo.
3. Front clutch.
4. Parking linkage.
5. Leak in hydraulic system.

Lockup in L

1. Front band.
2. Pressure regulator.
3. Valve body.
4. Rear clutch.
5. Parking linkage.
6. Leak in hydraulic system.

Parking Lock Binds or Won't Hold

1. Manual linkage.
2. Parking linkage.

Unable to Push Start

1. Fluid level.
2. Manual linkage.
3. Pressure regulator.
4. Valve body.
5. Rear pump.
6. Leak in hydraulic system.

Transmission Overheats

1. Oil cooler and connections.
2. Pressure regulator.
3. Converter one-way clutch.

Engine Runaway on Forced Downshift

1. Front band.
2. Pressure regulator.
3. Valve body.
4. Front servo.
5. Vacuum unit or tubes.
6. Leak in hydraulic system.

Maximum Speed Below Normal, Acceleration Poor

1. Converter one-way clutch.

No 3–1 Downshift

1. Engine idle speed.
2. Vacuum unit or tubes.
3. Valve body.

Noise in Neutral

1. Pressure regulator.
2. Front clutch.
3. Front pump.

Noise in 1–2–3 or R

1. Pressure regulator.
2. Planetary assembly.
3. Front clutch.
4. Rear clutch.
5. Front pump.

Noise in Reverse

1. Pressure regulator.
2. Front pump.

Noise on Coast in Neutral

1. Rear pump.

MAINTENANCE

Adding Fluid

The fluid level in the transmission should be checked at 1000-mile intervals. Make sure that the car is standing level, and firmly apply the parking brake.

Run the engine at normal idle speed. If the transmission fluid is cold, run the engine at fast idle speed until the fluid reaches normal operating temperature. When the fluid is warm, slow the engine down to normal idle speed, shift the transmission through all ranges and then place the lever or button at P.

Clean all dirt from the transmission fluid dipstick cap before removing the dipstick from the filler tube. Pull the dipstick out of the tube, wipe it clean and push it all the way back into the tube.

Pull the dipstick out again and check the fluid level. If necessary, add enough Type F Automatic Transmission Fluid to the transmission to raise the fluid level to the F (full mark) on the dipstick.

Changing Fluid

The transmission fluid should be changed at 24,000-mile intervals. The procedure for changing fluid is as follows:

1. Remove cover from lower front side of converter housing.
2. Remove one of the converter drain plugs. Then rotate the converter 180 deg. and remove the other plug. *Do not attempt to turn the converter with a wrench on the converter stud nuts as there is danger of stripping threads as well as skinning your knuckles on the bell housing.*
3. When all fluid has drained, remove and clean the oil pan and screen.
4. Using a new pan gasket, install screen and pan.
5. Connect filler tube to oil pan and tighten fitting securely.
6. Install both converter drain plugs.
7. Install converter housing cover.
8. Install 5 quarts of Automatic Transmission Fluid.
9. Run engine at idle speed for about 2 minutes; then add the additional quantity of oil required for the particular transmission being serviced.
10. Run engine at a fast idle until it reaches normal operating temperature.
11. Shift the transmission through all positions; then place it at P and check fluid level. If necessary, add enough fluid to bring the level to the F mark on the dipstick.

BAND ADJUSTMENTS

The front and rear bands of the transmission should be adjusted at 15,000 mile intervals or as operation of the transmission dictates.

Front Band

1974–77

1. Drain fluid from transmission, remove and clean oil pan and screen.
2. Loosen front servo adjusting screw locknut.
3. Pull back on actuating rod and insert a 1/4 inch spacer between adjusting screw and servo piston stem.
4. Tighten adjusting screw to 10 inch-lbs. torque. Remove spacer and tighten adjusting screw an additional 3/4 turn. Hold adjusting screw stationary and tighten locknut securely.
5. Install oil pan with new gasket and add fluid to transmission.

Rear Band

NOTE: On all 1974–77 models, there is no access hole in the floor pan to adjust the rear band. With the use of special tools this band can be adjusted externally as follows:

1. Loosen rear bank adjusting screw locknut. A special tool is required to gain access in limited space.
2. Tighten adjusting screw until special tool clicks. It is preset to overrun when torque reaches 10 ft. lbs.
3. Back off adjusting screw 1½ turns. **NOTE:** Severe damage may result if the adjusting screw is not backed off exactly 1½ turns.
4. Hold adjusting screw stationary and tighten locknut securely.

OIL PRESSURE REGULATOR

Remove oil pan and screen. Maintain constant pressure on spring retainer to prevent damage to springs and remove retainer from bosses on oil pressure regulator body. Remove springs and pilots. Remove the three pipes. Unfasten and remove the oil pressure regulator from the transmission case.

CONTROL VALVE

To remove the assembly, loosen the adjustment on the front and rear bands 5 or 6 turns.

Loosen front servo attaching screws. Remove cap screws and washers which attach control valve to case. Align throttle and manual levers to permit removal of control valve. Disengage front servo tubes from control valve and lift valve assembly from case.

FRONT & REAR SERVOS

To remove the front servo, remove the cap screw which holds it to the case. Hold the actuating lever strut with one hand and lift the servo from the case.

To remove the rear servo, take out the attaching cap screws. Then hold the anchor strut and lift the servo from the case.

EXTENSION HOUSING SEAL

After removing the drive shaft and telescopic shield, the seal may be pulled out of the extension housing.

Before installing the new seal, inspect the sealing surface of the universal joint yoke for scores. If scores are evident, replace the yoke. Inspect the counterbore in the housing for burrs. Polish all burrs with crocus cloth.

To install the new seal, position it in the bore of the extension housing with the felt side of the seal to the rear. The seal may be driven into the housing with a special tool designed for the purpose.

OIL DISTRIBUTOR

With Bolted Distributor & Sleeve

After removing the extension case remove the spacer from the transmission output shaft and slide the distributor toward the rear of the transmission. Note that the tube spacer is located in the center tube.

Remove the three tubes and spacer from the distributor. Remove the screws which attach the distributor to the sleeve and separate these parts.

Inspect the distributor and sleeve for burrs on the mating surfaces and obstructed fluid passages. Check the fit of the tubes in the distributor. Inspect the distributor sleeve for wear and scores in the sleeve bore.

To assemble, align the distributor and sleeve and install the cap screws. Install the tubes in the distributor with the spacer installed on the center tube.

With One Piece Distributor & Sleeve

After removing the extension housing, remove the distributor drive gear snap ring. Remove distributor gear, taking care not to lose the gear drive ball. Remove distributor sleeve and pipes from the transmission. Inspect the 4 seal rings on the output shaft for wear or breakage, and replace if necessary. Inspect the distributor sleeve for wear and the tubes for proper alignment and fit into the distributor sleeve.

With tubes installed in the distributor sleeve, install distributor on output shaft (chamfer forward) sliding the distributor over the seal rings and at the same time guiding tubes into the case. Install speedometer drive ball and gear and install snap ring.

GOVERNOR

Remove the governor inspection cover from the extension housing. Rotate the drive shaft to bring the governor body in line with the inspection hole. Remove the two screws which attach the governor body to the counterweight, and remove the body.

Remove the valve from the new governor body. Lubricate the valve with automatic transmission fluid. Install the valve in the body, making sure the valve moves freely in the bore. Install the body in the counterweight making sure the fluid passages in the counterweight and body are aligned.

Ford C6 Dual Range Unit

DESCRIPTION

As shown in Fig. 1, the transmission consists essentially of a torque converter, a compound planetary gear train controlled by one band, three disc clutches and a one-way clutch, and a hydraulic control system.

TROUBLE SHOOTING GUIDE

No Drive In Forward Speeds

1. Manual linkage adjustment.
2. Check control pressure.
3. Valve body.
4. Make air pressure check.
5. Forward clutch.
6. Leakage in hydraulic system.

Rough Initial Engagement in D, D1, D2 or 2

1. Engine idle speed too high.
2. Vacuum diaphragm unit or tubes restricted, leaking or maladjusted.
3. Check control pressure.
4. Valve body.
5. Forward clutch.

1–2 or 2–3 Shift Points Incorrect or Erratic

1. Check fluid level.
2. Vacuum diaphragm unit or tubes restricted, leaking or maladjusted.
3. Downshift linkage, including inner lever position.
4. Manual linkage adjustment.
5. Governor defective.
6. Check control pressure.
7. Valve body.
8. Make air pressure check.

Rough 1–2 Upshifts

1. Vacuum diaphragm unit or tubes restricted, leaking or maladjusted.
2. Intermediate servo.
3. Intermediate band.
4. Check control pressure.
5. Valve body.

Rough 2–3 Shifts

1. Vacuum diaphragm or tubes restricted leaking or maladjusted.
2. Intermediate servo.
3. Check control pressure.
4. Intermediate band.
5. Valve body.
6. Make air pressure check.
7. Reverse-high clutch.
8. Reverse-high clutch piston air bleed valve.

Dragged Out 1–2 Shift

1. Check fluid level.
2. Vacuum diaphragm unit or tubes restricted, leaking or maladjusted.
3. Intermediate servo.
4. Check control pressure.
5. Intermediate band.
6. Valve body.
7. Make air pressure check.
8. Leakage in hydraulic system.

Engine Overspeeds on 2–3 Shift

1. Manual linkage adjustment.
2. Check fluid level.
3. Vacuum diaphragm unit or tubes restricted, leaking or maladjusted.
4. Intermediate servo.
5. Check control pressure.
6. Valve body.
7. Intermediate band.
8. Reverse-high clutch.
9. Reverse-high clutch piston air bleed valve.

No 1–2 or 2–3 Shift

1. Manual linkage adjustment.
2. Downshift linkage including inner lever position.
3. Vacuum diaphragm unit or tubes restricted, leaking or maladjusted.
4. Governor.
5. Check control pressure.
6. Valve body.
7. Intermediate band.
8. Intermediate servo.
9. Reverse-high clutch.
10. Leakage in hydraulic system.

No 3–1 Shift In D1, 2 or 3–2 Shift In D2 or D

1. Governor.
2. Valve body.

No Forced Downshifts

1. Downshift linkage, including inner lever position.
2. Check control pressure.
3. Valve body.

Runaway Engine on Forced 3–2 Shift

1. Check control pressure.

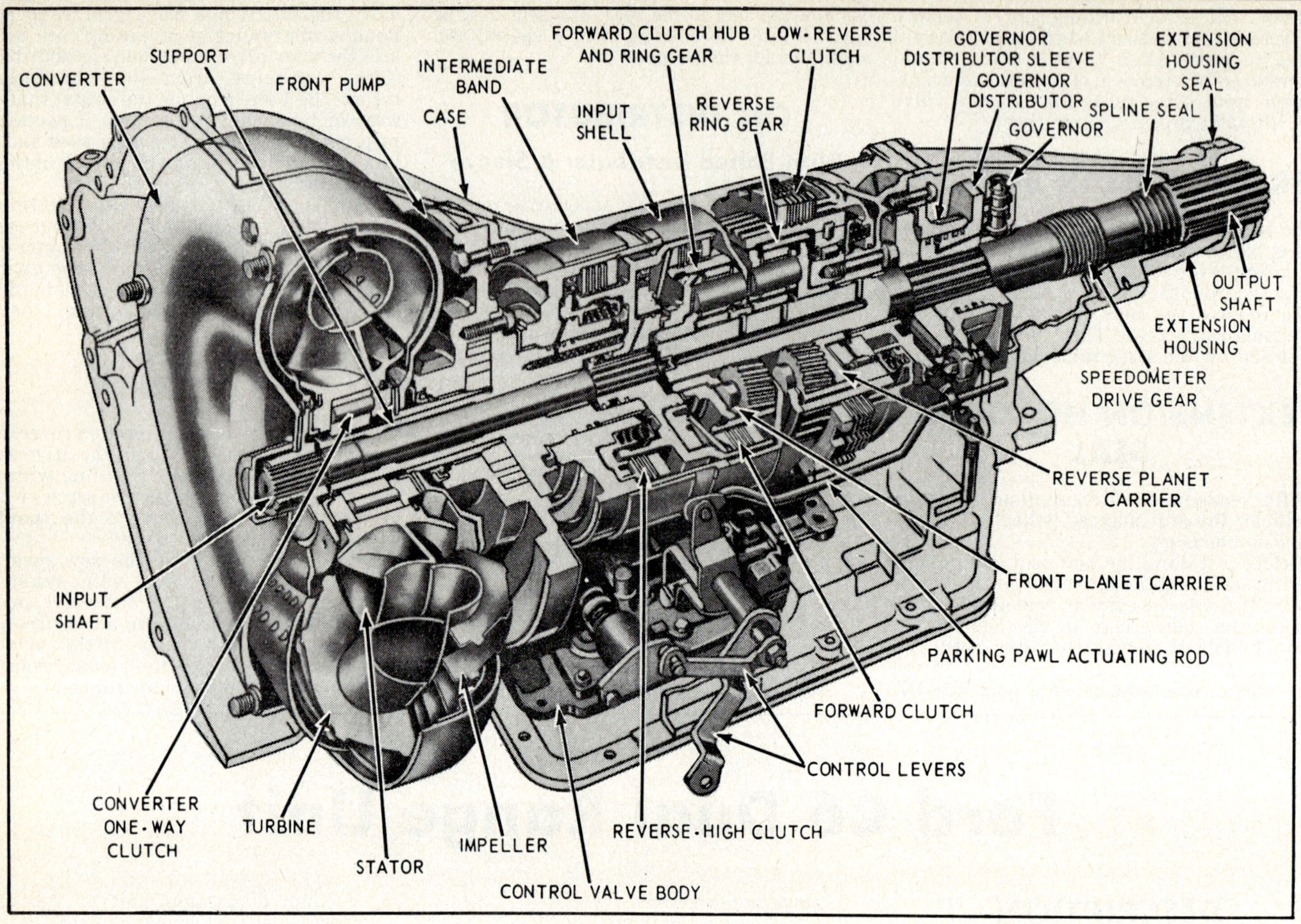

Fig. 1 C6 dual range automatic transmission

2. Intermediate servo.
3. Intermediate band.
4. Valve body.
5. Vacuum diaphragm unit or tubes restricted, leaking or maladjusted.
6. Leakage in hydraulic system.

Rough 3–2 Shift or 3–1 Shift at Closed Throttle

1. Engine idle speed.
2. Vacuum diaphragm unit or tubes restricted, leaking or maladjusted.
3. Intermediate servo.
4. Check control pressure.
5. Valve body.

Shifts 1–3 in D, D1, 2, D2

1. Intermediate band.
2. Intermediate servo.
3. Valve body.
4. Governor.
5. Make air pressure check.

No Engine Braking in 1st Gear—Manual Low Range

1. Manual linkage adjustment.
2. Low-reverse clutch.
3. Valve body.
4. Governor.
5. Make air pressure check.
6. Leakage in hydraulic system.

Creeps Excessively

1. Engine idle speed too high.

Slips or Chatters In 1st Gear, D1

1. Check fluid level.
2. Vacuum diaphragm unit or tubes restricted, leaking or maladjusted.
3. Check control pressure.
4. Valve body.
5. Forward clutch.
6. Leakage in hydraulic system.
7. Planetary one-way clutch.

Slips or Chatters In 2nd Gear

1. Check fluid level.
2. Vacuum diaphragm unit or tubes restricted, leaking or maladjusted.
3. Intermediate servo.
4. Intermediate band.
5. Check control pressure.
6. Valve body.
7. Make air pressure check.
8. Forward clutch.
9. Leakage in hydraulic system.

Slips or Chatters In Reverse

1. Check fluid level.
2. Vacuum diaphragm unit or tubes restricted, leaking or maladjusted.
3. Manual linkage adjustment.
4. Low-reverse clutch.
5. Check control pressure.
6. Valve body.
7. Make air pressure check.
8. Reverse-high clutch.
9. Leakage in hydraulic system.
10. Reverse-high clutch piston air bleed valve.

No Drive In D1 or 2

1. Manual linkage adjustment.
2. Check control pressure.
3. Valve body.
4. Planetary one-way clutch.

No Drive In D, D2

1. Check fluid level.
2. Manual linkage adjustment.
3. Check control pressure.
4. Intermediate servo.
5. Valve body.
6. Make air pressure check.
7. Leakage in hydraulic system.

No Drive In L or 1

1. Check fluid level.
2. Check control pressure.
3. Valve body.
4. Make air pressure check.
5. Leakage in hydraulic system.

No Drive In R Only

1. Check fluid level.

2. Manual linkage adjustment.
3. Low-reverse clutch.
4. Check control pressure.
5. Valve body.
6. Make air pressure check.
7. Reverse-high clutch.
8. Leakage in hydraulic system.
9. Reverse-high clutch piston air bleed valve.

No Drive In Any Selector Position

1. Check fluid level.
2. Manual linkage adjustment.
3. Check control pressure.
4. Valve body.
5. Make air pressure check.
6. Leakage in hydraulic system.
7. Front pump.

Lockup In D1 or 2

1. Valve body.
2. Parking linkage.
3. Leakage in hydraulic system.

Lockup In D2 or D

1. Low-reverse clutch.
2. Valve body.
3. Reverse-high clutch.
4. Parking linkage.
5. Leakage in hydraulic system.
6. Planetary one-way clutch.

Lockup In L or 1

1. Valve body.
2. Parking linkage.
3. Leakage in hydraulic system.

Lockup In R Only

1. Valve body.
2. Forward clutch.
3. Parking linkage.
4. Leakage in hydraulic system.

Parking Lock Binds or Does Not Hold

1. Manual linkage adjustment.
2. Parking linkage.

Transmission Overheats

1. Oil cooler and connections.
2. Valve body.
3. Vacuum diaphragm unit or tubes restricted, leaking or maladjusted.
4. Check control pressure.
5. Converter one-way clutch.
6. Converter pressure check valves.

Maximum Speed Too Low, Poor Acceleration

1. Engine performance.
2. Car brakes.
3. Forward clutch.

Transmission Noisy In N and P

1. Check fluid level.
2. Valve body.
3. Front pump.

Noisy In 1st, 2nd, 3rd or Reverse

1. Check fluid level.
2. Valve body.
3. Planetary assembly.
4. Forward clutch.
5. Reverse-high clutch.
6. Planetary one-way clutch.

Car Moves Forward In N

1. Manual linkage adjustment.
2. Forward clutch.

Fluid Leak

1. Check fluid level.
2. Converter drain plugs.
3. Oil pan gasket, filler tube or seal.
4. Oil cooler and connections.
5. Manual or downshift lever shaft seal.
6. 1/8" pipe plugs in case.
7. Extension housing-to-case gasket.
8. Extension housing rear oil seal.
9. Speedometer driven gear adapter seal.
10. Vacuum diaphragm unit or tubes.
11. Intermediate servo.
12. Engine rear oil seal.

MAINTENANCE

Checking Oil Level

1. Make sure vehicle is on a level floor.
2. Apply parking brake firmly.
3. Run engine at normal idle speed. If transmission fluid is cold, run engine at a fast idle until fluid reaches normal operating temperature. When fluid is warm, slow engine to normal idle speed.
4. Shift selector lever through all positions, then place lever at "P." Do not shut down engine during fluid level checks.
5. Clean all dirt from dipstick cap before removing dipstick from filler tube.
6. Pull dipstick out of tube, wipe it clean and push it all the way back in tube.
7. Pull dipstick out of tube again and check fluid level. If necessary, add enough fluid to raise the level to the "F" mark on dipstick. Do not overfill.

Drain & Refill

NOTE: The Ford Motor Company recommends the use of an automatic transmission fluid with Qualification No. M2C-33F for 1974–76 models and M2C-138-CJ or Dexron II for 1977–83 models (on container). The recommended fluid is said to have a greater coefficient of friction and greater ability to handle maximum engine torques without band or clutch slippage.

Normal maintenance and lubrication requirements do not necessitate periodic fluid changes. If a major failure has occurred in the transmission, it will have to be removed for service. At this time the converter and transmission cooler must be thoroughly flushed to remove any foreign matter.

1. To drain the fluid, loosen pan attaching bolts and allow fluid to drain.
2. After fluid has drained to the level of the pan flange, remove pan bolts working from rear and both sides of pan to allow it to drop and drain slowly.
3. When fluid has stopped draining, remove and clean pan and screen. Discard pan gasket.
4. Using a new gasket, install pan.
5. Add 3 quarts of recommended fluid to transmission through filler tube.
6. Run engine at idle speed for 2 minutes, and then run it at a fast idle until it reaches normal operating temperature.
7. Shift selector lever through all positions, place it at "P" and check fluid level.
8. If necessary, add enough fluid to transmission to bring it to the "F" mark on the dipstick.

BAND ADJUSTMENT

NOTE: When making the intermediate band adjustment, the lock nut must be discarded and a new one installed each time the band is adjusted.

1. Raise truck on hoist and support on suitable jack stands.
2. Loosen the locknut on the adjusting screw several turns.
3. Torque the screw to 10 ft.-lbs., or until the adjuster wrench overruns.
4. Back the screw off exactly 1½ turns.
5. Hold the adjustment and torque the locknut to the 35–45 ft.-lbs. on all except 1980–82 vehicles, 22–29 ft. lbs. on 1980 vehicles, 40 ft. lbs. on 1981 vehicles, or 35–40 ft. lbs. on 1982 vehicles.

OIL PAN & CONTROL VALVE

Removal

1. Raise truck on hoist or jack stands.
2. Loosen and remove all but two oil pan bolts from front of case and drop rear edge of pan to drain fluid. Remove and clean pan and screen.
3. Unfasten and remove valve body.

Installation

1. Position valve body to case, making sure that selector and downshift levers are engaged, then install and torque attaching bolts to 95–125 inch lbs.
2. Using a new pan gasket, secure pan to case and torque bolts to 12–16 ft. lbs.
3. Lower truck and fill transmission to correct level with specified fluid.

INTERMEDIATE SERVO

Removal

1. Raise truck and remove engine rear support-to-extension housing bolts.
2. Raise transmission high enough to relieve weight from support.
3. Remove support (1 bolt).
4. Lower transmission.
5. Disconnect muffler inlet pipe from exhaust manifolds and let pipe hang free.
6. Place drain pan beneath servo.
7. Remove servo cover-to-case bolts.
8. Loosen band adjusting screw locknut.
9. Remove servo cover, piston, spring and gasket from case, *screwing band adjusting screw inward as piston is removed. This insures that there will be enough tension on the band to keep the struts properly engaged in the band end notches while the piston is removed.*

Replacing Seal

1. Apply air pressure to port in servo cover to remove piston and stem.
2. Remove seals from piston.
3. Remove seal from cover.
4. Dip new seals in transmission fluid.
5. Install seals in piston and cover.
6. Dip piston in transmission fluid and install in cover.

Installation

1. Position new gasket on servo cover and spring on piston stem.
2. Insert piston stem in case. Secure cover with bolts, taking care to back off band adjusting screw while tightening cover bolts. Make sure that vent tube retaining clip is in place.
3. Connect muffler inlet pipe to exhaust manifolds.
4. Raise transmission high enough to install engine rear support. Secure support to extension housing. Lower transmission as required to install support-to-crossmember bolt.
5. Remove jack and adjust band.
6. Lower truck and replenish fluid as required.

EXTENSION HOUSING & GOVERNOR

Removal

1. Raise truck and drain transmission.
2. Disconnect drive shaft from rear axle flange and remove from transmission.

NOTE: On 4 × 4 models, the transfer case must be removed.

3. Disconnect speedometer cable from extension housing.
4. Remove two nuts that secure engine rear mount to crossmember.
5. Raise transmission with a jack just high enough to relieve weight from crossmember. Remove crossmember.
6. Remove engine rear support.
7. Lower transmission to permit access to extension housing bolts. Remove bolts and slide housing off output shaft.
8. Disconnect governor from distributor (4 bolts) and slide governor off output shaft.

Installation

1. Secure governor to distributor flange.
2. Position new gasket on transmission.
3. Secure extension housing to case.
4. Secure engine rear support to case.
5. Install crossmember.
6. Lower transmission and remove jack. Then install and torque engine rear support-to-extension housing bolts.
7. Install speedometer cable and install drive shaft.
8. Replenish transmission fluid.

Ford C5 Unit

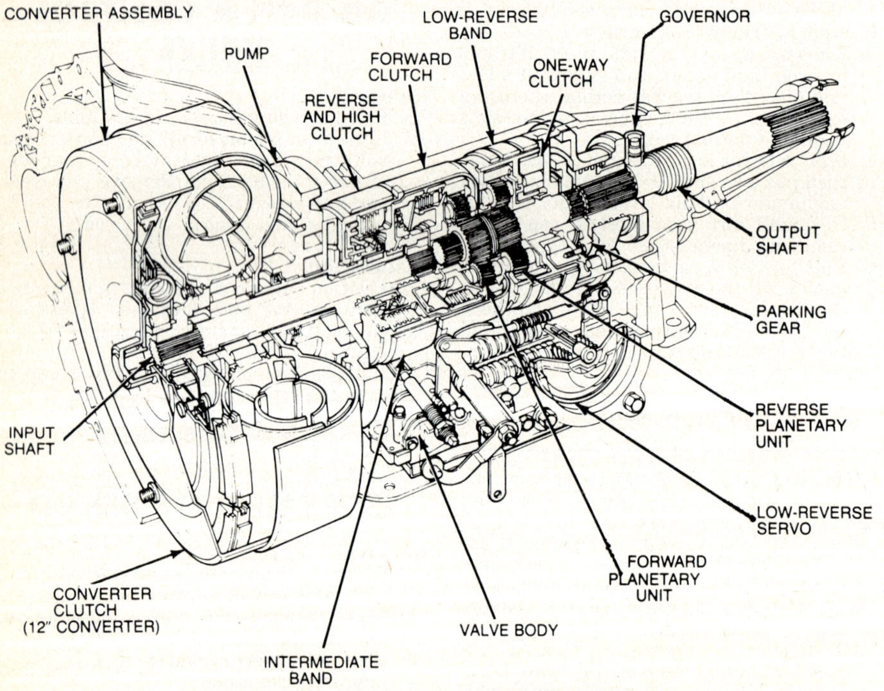

Fig. 1 Ford C5 automatic transmission

DESCRIPTION

The C5 transmission, Fig. 1, is a fully automatic transmission with three speeds and one reverse. It consists of a welded torque converter assembly, a two unit planetary gear train, and a hydraulic system to control gear selection and automatic shift. The 12-inch torque converter, used with larger displacement engine, has a converter clutch. The planetary gear train is a Simpson design with two gear sets in series and a common sun gear. It is the same gear train used in model C4 with minor changes. Two friction clutches and two bands are used to control the gear operation.

The torque converter is coupled to the engine crankshaft, and transmits engine power into the gear train. The output shaft drives the rear wheels through a conventional driveshaft and rear axle. Gear reductions needed to match the engine to the axle take place in the planetary gear train and in the torque converter.

The C5 transmission is closest in resemblance to the C4 which it replaced in production.

Some major differences from the C4 transmission are in the hydraulic system. Several new valves are incorporated along with a new timing valve body. Also the converter relief valve is moved from the reactor support (in the pump assembly) to the timing valve body. Thus, C4 oil pump assemblies must not be used to service the C5. The C5 12 inch converter with the converter clutch is also different than the C4 conventional converter.

In general, C4 special tools will service the C5. One new seal protector tool is needed to install the reverse and high clutch piston. The tool number is T82L-77404-A.

TROUBLE SHOOTING GUIDE

Slow Initial Engagement

1. Improper fluid level.
2. Damaged or improperly adjusted linkage.
3. Contaminated fluid.
4. Improper clutch and band application or low main control pressure.

Rough Initial Engagement in Either Forward or Reverse

1. Improper fluid level.
2. High engine idle.
3. Automatic choke on (warm temp.).
4. Looseness in the driveshaft U-joint or engine mount.
5. Incorrect linkage adjustment.
6. Improper clutch or band application, or oil control pressure.

No or Delayed Forward Engagement

1. Improper fluid level.
2. Manual linkage, misadjusted or damaged.
3. Low main control pressure.
4. Valve body bolts, loose or too tight.
5. Valve body, dirty or sticking valve.
6. Forward clutch assembly burnt or damaged.
7. Forward clutch assembly piston seals worn or cut.
8. Forward clutch assembly cylinder ball check not seating.
9. Forward clutch assembly stator support seal ring grooves, damaged or worn.

No or Delayed Reverse Engagement

1. Improper fluid level.
2. Low main control pressure in reverse.
3. Manual linkage misadjusted or damaged.
4. Valve body, dirty or sticking valve.

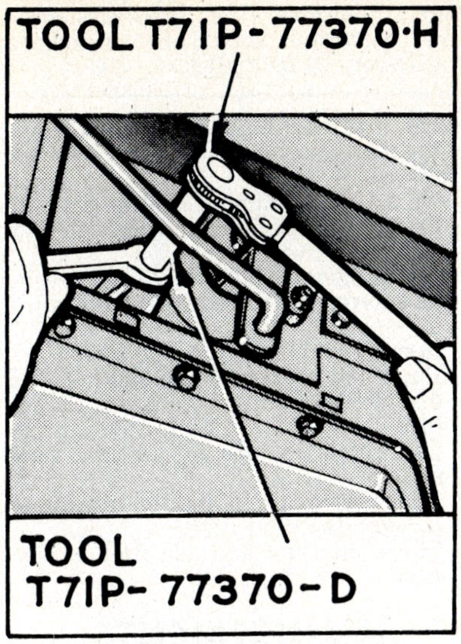

Fig. 2 Intermediate Band Adjustment

5. Valve body bolts loose or too tight.
6. Reverse clutch assembly, burnt or worn.
7. Reverse clutch assembly piston seals, worn or cut.
8. Reverse clutch assembly piston ball not seating.
9. Reverse clutch assembly stator support seal rings or ring grooves, worn or damaged.

No or Delayed Reverse Engagement and/or No Engine Braking in Manual Low (1)

1. Low reverse band or servo piston burnt or worn.
2. Low reverse servo seal worn or cut.
3. Low reverse servo bore damaged.
4. Low reverse servo piston sticking in bore.
5. Low reverse band, line pressure low.
6. Low reverse bands out of adjustment.
7. Polished or glazed band or drum.

No Engine Braking in Manual Second Gear

1. Improper fluid level.
2. Linkage out of adjustment.
3. Intermediate band out of adjustment.
4. Improper band or clutch application, or oil pressure control system.
5. Intermediated servo leaking.
6. Polished or glazed band or drum.

No or Delayed Engagement Forward and Reverse

1. Pump gear damaged (no engagement).
2. Output shaft broken (no engagement only).
3. Turbine shaft or input shaft broken (no engagement only).

Forward Engagement Slip, Shudders or Chatters

1. Improper fluid level.
2. Manual linkage misadjusted or damaged.
3. Low main control pressure.
4. Valve body bolts, loose or too tight.
5. Valve body dirty or sticking valve.
6. Forward clutch piston ball check not sealing.
7. Forward clutch piston seal cut or worn.
8. Contamination blocking forward clutch feed hole.
9. Low (planetary) one way clutch damaged.

Reverse Engagement Slip, Shudders or Chatters

1. Improper fluid level.
2. Low main control pressure in reverse.
3. Reverse servo or servo bore damaged.
4. Low (planetary) one-way clutch damaged.
5. Reverse clutch drum bushing damaged.
6. Reverse clutch stator support seal rings or ring grooves worn or damaged.
7. Reverse clutch piston seal cut or worn.
8. Reverse band out of adjustment or damaged.
9. Looseness in the driveshaft U-joints or engine mounts.

No Drive, Slips or Chatters in First Gear "D"

1. Damaged or worn one-way clutch.

No Drive, Slips or Chatters in Second

1. Improper fluid level.
2. Damaged or improperly adjusted linkage.
3. Intermediated band out of adjustment.
4. Improper band or clutch application, or oil pressure control.
5. Damaged or worn servo and/or internal leaks.
6. Dirty or sticking valve body.
7. Polished or glazed intermediate band or drum.

Start Up in Second or Third

1. Improper fluid level.
2. Damaged or improperly adjusted linkage.
3. Improper band and/or clutch application, or oil pressure control system.
4. Damaged or worn governor, governor sticking.
5. Valve body loose.
6. Dirty or sticking valve body.
7. Cross leaks between valve body and case mating surface.

Shift Points Incorrect

1. Improper fluid level.
2. Improper vacuum hose routing or leaks.
3. Improper operation of EGR system.
4. Throttle out of adjustment.
5. Improper clutch or band application, or oil pressure control system.
6. Damaged or worn governor.
7. Dirty or sticking valve body.

All Upshifts Harsh, Delayed or No Upshifts

1. Improper fluid level.
2. Manual linkage misadjusted or damaged.
3. Governor sticking.
4. Main control pressure too high.
5. Valve body bolts loose or too tight.
6. Valve body dirty or valves sticking.
7. Vacuum leak to diaphragm unit.

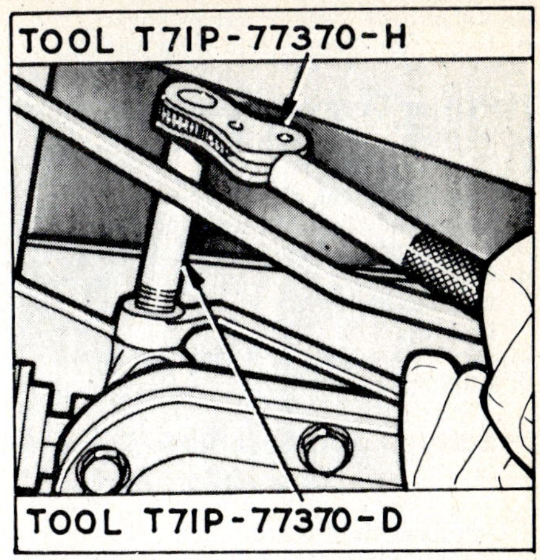

Fig. 3 Low-Reverse Band Adjustment

All Upshifts Early or Sluggish

1. Improper fluid level.
2. Low main control pressure.
3. Valve body loose or too tight.
4. Valve body valve sticking.
5. Governor valve sticking.

No Low to Second Upshift

1. Improper fluid level.
2. Manual linkage misadjusted or damaged.
3. Governor valve sticking.
4. Valve body bolts loose or too tight.
5. Valve body dirty or sticking valves.
6. Intermediate clutch or band and or servo assembly burnt.
7. Intermediate piston seals worn or cut.
8. Intermediate piston not positioned properly.
9. Intermediate clutch improper stack up.
10. Low line pressure in intermediate clutch or band.

Rough, Harsh, or Delayed Upshift Low to Second

1. Governor valve sticking.
2. Improper fluid level.
3. Poor engine performance.
4. Main control pressure too high.
5. Valve body bolts loose or too tight.
6. Valve body dirty or valves sticking.
7. Intermediate band out of adjustment.
8. Damaged intermediate servo.
9. Engine vacuum leak.

Early, Soft or Slipping Low to Second Upshift

1. Improper fluid level.
2. Low main control.
3. Valve body bolts loose or too tight.
4. Valve body dirty or valves sticking.
5. Governor valve sticking.
6. Incorrect engine performance.
7. Intermediate band out of adjustment.
8. Damaged intermediate servo or band.
9. Polished or glazed band or drum.

No Second to Third Upshift

1. Low fluid level.
2. Low main control pressure to direct clutch.
3. Valve body bolts too loose or too tight.

4. Valve body dirty or valves sticking.
5. Converter damper hub weld broken.

Harsh or Delayed Second to Third Upshift

1. Low fluid level.
2. Valve body bolts loose or too tight.
3. Valve body dirty or valves sticking.
4. Damaged or worn intermediate servo release and high clutch piston check ball.
5. Incorrect engine performance.
6. Engine vacuum leak.

Early or Soft Second to Third Upshift

1. Improper fluid level.
2. Valve body bolts loose or too tight.
3. Valve body dirty or valves sticking.

Erratic Shifts

1. Improper fluid level.
2. Throttle linkage binding or sticking.
3. Valve body bolts loose or too tight.
4. Valve body dirty or valves sticking.
5. Governor valve sticking.
6. Output shaft collector body seal rings (large cast iron) worn or cut.

Shifts From Low to Third in "D"

1. Improper fluid level.
2. Intermediate band out of adjustment.
3. Damaged intermediate servo and/or internal leaks.
4. Polished or glazed band or drum.
5. Improper band or clutch application, or oil pressure control system.
6. Valve body dirty or valves sticking.

Engine Over Speeds on Second to Third Upshift

1. Improper fluid level.
2. Linkage out of adjustment.
3. Improper band or clutch application, or oil pressure control system.
4. Damaged or worn high clutch and/or intermediate servo.
5. Valve body dirty or valves sticking.

Rough or Shudder Third to Low Shift at Closed Throttle

1. Improper fluid level.
2. Incorrect engine idle or performance.
3. Improper linkage adjustment.
4. Improper clutch or band application, or oil pressure control system.
5. Improper governor operation.
6. Valve body dirty or valves sticking.

No Forced Downshift

1. Improper fluid level.
2. Kickdown linkage out of adjustment.
3. Damaged internal kickdown linkage.
4. Damaged or misadjusted (short) throttle linkage.
5. Valve body dirty or valves sticking.
6. Dirty or sticking governor.

Engine Runway on Third to Second Shift

1. Improper fluid level.
2. Linkage out of adjustment.
3. Intermediate band out of adjustment.
4. Improper band or clutch application, or oil pressure control system.
5. Damaged or worn intermediate servo.
6. Polished or glazed band or drum.
7. Valve body dirty or valves sticking.

Shift Efforts High

1. Manual shift linkage damaged or misadjusted.
2. Inner manual lever nut loose.
3. Manual lever retainer pin damaged.

No Start in "P"

1. Manual linkage misadjusted.
2. Plug connector for the neutral start switch does not fit properly.
3. Neutral start switch plunger travel, inadequate.

No Start in "P" and "N"

1. Plug connector for the neutral start switch does not fit properly.

Transmission Overheats

1. Improper fluid level.
2. Incorrect engine idle or performance.
3. Improper band or clutch application, or oil pressure control system.
4. Restriction in cooler or lines.
5. Seized converter one-way clutch.
6. Valve body dirty or valves sticking.

MAINTENANCE

Checking Oil Level

1. With engine idling, foot brake applied and vehicle on level surface, move selector lever through each range, pausing in each position.
2. Place selector in Park and apply parking brake. Leave engine running during fluid level check.
3. Clean dirt from transmission fluid dipstick cap and remove dipstick. Wipe dipstick and push back into tube making sure it is fully seated.
4. Pull dipstick out and check fluid level. With transmission at operating temperature, fluid level should be between arrows. With transmission cool, fluid level should read between inner holes. Use only Type H fluid, Ford spec. ESP M2C166-H. Do not overfill.
5. Insert dipstick, making sure it is fully seated.

NOTE: If transmission fluid is at operating temperature, the fluid level on the dipstick should be between the arrows. If fluid level is checked at room temperature, the fluid level on the dipstick should be between the middle and top holes.

Drain & Refill

NOTE: Normal maintenance and lubrication requirements do not necessitate periodic fluid changes. If a major failure has occurred in the transmission, it will have to be removed for service. At this time, the converter, transmission cooler and cooler lines must be thoroughly flushed to remove any foreign matter.

If vehicle is operated under continuous or severe conditions, the transmission and torque converter should be drained and refilled every 20,000 miles.

When a partial drain and refill is required due to an in-vehicle repair operation, proceed as follows:

1. On all except Ranger models, disconnect fluid filler tube from transmission oil pan and drain fluid.
2. On Ranger models, loosen oil pan bolts and lower one edge of pan to drain fluid.
3. Remove and clean pan and screen.
4. Place new gasket on pan and install pan and screen. Torque bolts to 12–16 ft. lbs.
5. Connect filler tube to pan and tighten fitting.
6. Add three quarts of fluid to transmission through the filler tube.
7. Check oil level and add fluid as necessary.

ADJUSTMENTS

Intermediate Band

1. Clean all dirt from the band adjusting screw area. Remove and discard the locknut.
2. Install a new locknut on the adjusting screw.
3. Tighten adjusting screw using Band Adjusting Ratchet T71P-77370-H and Socket T71P-77370-D or equivalent, Fig. 2. Tool T71P-77370-H or equivalent, is a pre-set torque wrench which will click when the torque on the adjusting screw reaches 10 ft. lbs.
4. Back off adjusting screw exactly 4 1/4 turns.
5. Hold the adjusting screw from turning and tighten locknut to 40 ft. lbs.

Low Reverse Band

1. Clean all dirt from the band adjusting screw area. Remove and discard the locknut.
2. Install a new locknut on the adjusting screw.
3. Tighten adjusting screw using Band Adjusting Ratchet T71P-77370-H and Socket T71P-77370-D or equivalent, Fig. 3. Tool T71P-77370-H or equivalent, is a pre-set torque wrench which will click when the torque on the adjusting screw reaches 10 ft. lbs.
4. Back off adjusting screw exactly three full turns. Hold adjusting screw from turning and tighten locknut to 40 ft. lbs.

Neutral Start Switch

1. Loosen neutral start switch attaching bolts.
2. Set the manual lever in park position. Insert a 3/32" drill bit through the switch hole. Move the switch as necessary to allow the drill bit to seat fully.
3. Tighten switch attaching bolts to 8 ft. lbs. and remove drill bit.

VALVE BODY, REPLACE

1. Raise vehicle.
2. Loosen pan bolts and drain transmission fluid.
3. Remove oil pan bolts, pan, and gasket.
4. Shift transmission lever to Park, and remove two bolts attaching detent spring to valve body and case.

5. Remove oil pan screen.
6. Remove remaining valve body bolts. Hold manual valve in valve body and remove valve body from case. Manual valve must be hold to prevent it from being bent or damaged.
7. Clean and remove all gasket material from pan and pan mounting face. Remove and discard nylon shipping plug, if found in pan.
8. Position valve body in case. Make sure inner downshift lever is between downshift lever stop and downshift valve. The two lands on end of manual valve must engage actuating pin on manual detent lever. Install seven valve body bolts.
9. Position detent spring on lower valve body and install spring-to-case bolt.
10. Hold detent spring roller in center of manual detent lever and install detent spring-to-lower valve body bolt. Torque bolt to 80–120 inch lbs.
11. Torque valve body bolts to 80–120 inch lbs.
12. Position screen and torque bolt to 30 inch lbs.
13. Install pan using new gasket and torque bolts to 12–16 ft. lbs.
14. Lower vehicle and fill transmission with fluid. Check pan area for leakage.

EXTENSION HOUSING SEAL, REPLACE

1. Raise and support vehicle.
2. Remove the driveshaft.
3. Remove extension housing seal using Seal Remover T74P-77248-A and Slide Hammer T50T-100-A or equivalent.
4. Reverse procedure to install using Seal Installer T61L-7657-A or equivalent.

EXTENSION HOUSING, REPLACE

1. Raise vehicle and remove driveshaft.
2. Using suitable jack, support transmission.
3. Remove speedometer cable from extension housing.
4. Remove engine rear support to crossmember nuts.
5. Raise transmission and remove rear support bolts. Remove crossmember.
6. Loosen extension housing bolts and let transmission drain.
7. Remove the six extension housing bolts and vacuum tube clip and remove extension housing.
8. Reverse procedure to install, noting the following torques: extension housing bolts, 28–40 ft. lbs.; crossmember nuts, 35–50 ft. lbs.; rear support bolts, 25–35 ft. lbs.

GOVERNOR, REPLACE

1. Refer to "Extension Housing, Replace" and remove extension housing.
2. Remove bolts holding governor housing to governor distributor.
3. Slide governor away from distributor body and off output shaft.
4. Reverse procedure to install. Torque governor bolts to 80–120 inch lbs.

LOW-REVERSE SERVO, REPLACE

1. Raise and support vehicle.
2. Loosen low-reverse band adjusting screw lock nut. Torque band adjusting screw to 10 ft. lbs. to prevent band strut from falling when reverse servo piston assembly is removed.
3. Disengage neutral switch harness from clips on servo cover.
4. Remove servo cover bolts, servo cover, and seal from case.
5. Remove servo piston from case. If seal is bad, piston must be replaced.
6. Install piston in case. Install cover with new seal. Use two 5/16-18 × 1 1/4 bolts to position cover against case. Install two cover bolts, remove two locating bolts and install remaining bolts. Torque to 12–20 ft. lbs.
7. Position neutral switch harness in clips.
8. Adjust low-reverse band. Refer to "Bands, Adjust" procedure.

NOTE: If band cannot be adjusted properly, low-reverse band struts are not in position. Remove oil pan and valve body. Position struts and install valve body and pan. Adjust band.

9. Lower vehicle and check transmission fluid level.

Chrysler Corp. Loadflite & A-345

DESCRIPTION

Loadflite

These transmissions, Figs. 1 and 2, combine a torque converter with a fully automatic three speed gear system. The converter housing and transmission case are an integral aluminum casting. The transmission consists of two multiple disc clutches, an overrunning (one-way) clutch, two servos and bands and two planetary gear sets to provide three forward speeds and reverse.

The common sun gear of the planetary gear sets is connected to the front clutch by a driving shell that is splined to the sun gear and to the front clutch retainer.

The hydraulic system consists of a single oil pump and a valve body that contains all the valves except the governor valve.

Venting of the transmission is accomplished by a drilled passage through the upper part of the front pump housing.

The torque converter is attached to the engine crankshaft through a flexible driving plate. The converter is cooled by circulating the transmission fluid through an oil-to-water type cooler located in the radiator lower tank. The converter is a sealed assembly that cannot be disassembled.

A-345

This transmission, Figs. 3 and 4 except for the compounder, lower valve body and lack of an internal parking mechanism, is basically the same as the Loadflite A-727, and may be serviced as such except for the following procedures described in this chapter.

This transmission consists of a sealed torque converter, a fully automatic four speed transmission, four multiple disc clutches, two overrunning clutches, two servos and bands, three planetary gear sets and a hydraulic system.

TROUBLE SHOOTING GUIDE

Harsh Engagement In D-1-2-R

1. Engine idle speed too high.
2. Hydraulic pressures too high or too low.
3. Low-reverse band out of adjustment.
4. Accumulator sticking, broken rings or spring.
5. Low-reverse servo, band or linkage malfunction.
6. Worn or faulty front and/or rear clutch.

Delayed Engagement In D-1-2-R

1. Lower fluid level.
2. Incorrect manual linkage adjustment.
3. Oil filter clogged.
4. Hydraulic pressures too high or low.
5. Valve body malfunction or leakage.
6. Accumulator sticking, broken rings or spring.
7. Clutches or servos sticking or not operating.
8. Faulty front oil pump.
9. Worn or faulty front and/or rear clutch.
10. Worn or broken input shaft and/or reaction shaft support seal rings.
11. Aerated fluid.

Runaway or Harsh Upshift and 3-2 Kickdown

1. Low fluid level.
2. Incorrect throttle linkage adjustment.
3. Hydraulic pressures too high or low.
4. Kickdown band out of adjustment.
5. Valve body malfunction or leakage.
6. Governor malfunction.
7. Accumulator sticking, broken rings or spring.
8. Clutches or servos sticking or not operating.
9. Kickdown servo, band or linkage malfunction.
10. Worn or faulty front clutch.
11. Worn or broken input shaft and/or reaction shaft support seal rings.

No Upshift

1. Low fluid level.
2. Incorrect throttle linkage adjustment.
3. Kickdown band out of adjustment.
4. Hydraulic pressures too high or low.
5. Governor sticking.
6. Valve body malfunction or leakage.
7. Accumulator sticking, broken rings or spring.
8. Clutches or servos sticking or not operating.

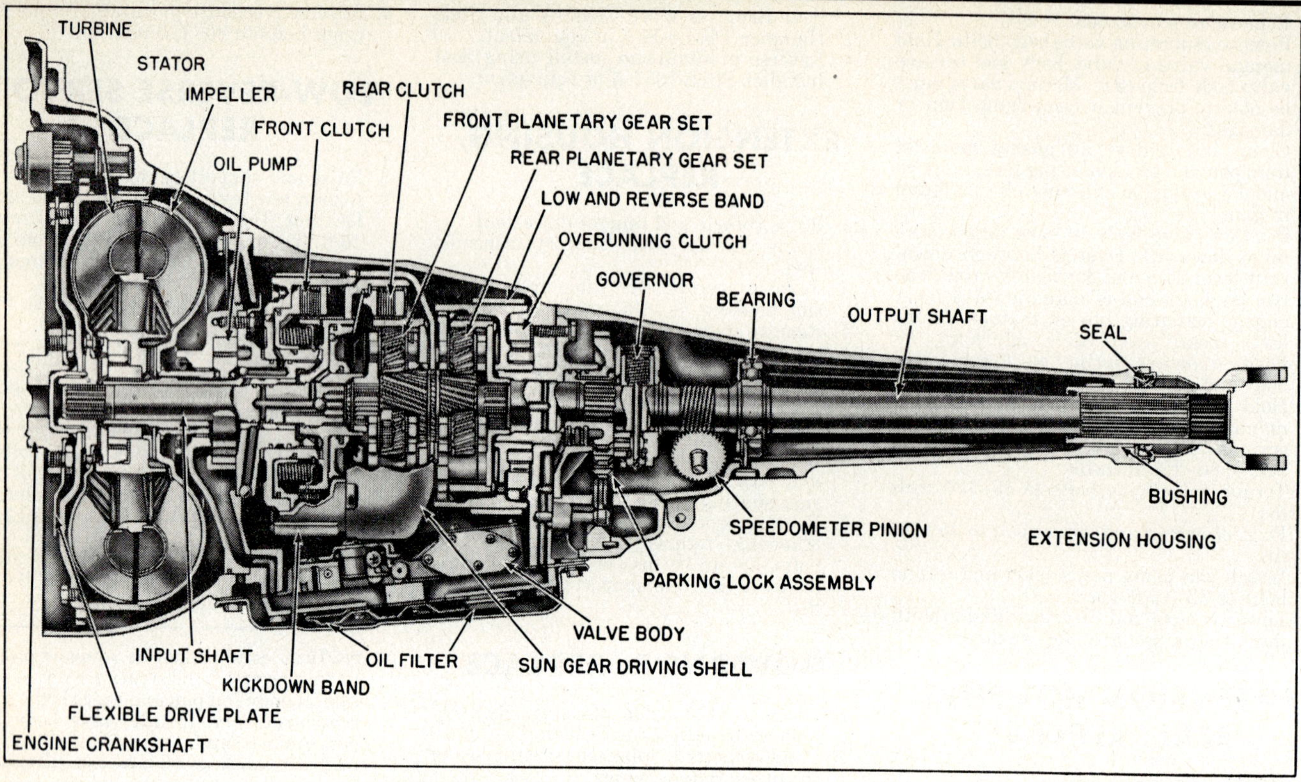

Fig. 1 Model A-727 Loadflite transmission. (Typical)

9. Faulty oil pump.
10. Kickdown servo, band or linkage malfunction.
11. Worn or faulty front clutch.
12. Worn or broken input shaft and/or reaction shaft support seal rings.

No Kickdown or Normal Downshift

1. Incorrect throttle linkage adjustment.
2. Incorrect gearshift linkage adjustment.
3. Kickdown band out of adjustment.
4. Hydraulic pressure too high or low.
5. Governor sticking.
6. Valve body malfunction or leakage.
7. Accumulator sticking, broken rings or spring.
8. Clutches or servos sticking or not operating.
9. Kickdown servo, band or linkage malfunction.
10. Overrunning clutch not holding.

Erratic Shifts

1. Low fluid level.
2. Aerated fluid.
3. Incorrect throttle linkage adjustment.
4. Incorrect gearshift control linkage adjustment.
5. Hydraulic pressures too high or low.
6. Governor sticking.
7. Oil filter clogged.
8. Valve body malfunction or leakage.
9. Clutches or servos sticking or not operating.
10. Faulty oil pump.
11. Worn or broken input shaft and/or reaction shaft support seal rings.

Slips In Forward Drive Positions

1. Low oil level.
2. Aerated fluid.
3. Incorrect throttle linkage adjustment.
4. Incorrect gearshift control linkage adjustment.
5. Hydraulic pressures too low.
6. Valve body malfunction or leakage.
7. Accumulator sticking, broken rings or spring.
8. Clutches or servos sticking or not operating.
9. Worn or faulty front and/or rear clutch.
10. Overrunning clutch not holding.
11. Worn or broken input shaft and/or reaction shaft support seal rings.

Slips In Reverse Only

1. Low fluid level.
2. Aerated fluid.
3. Incorrect gearshift control linkage adjustment.
4. Hydraulic pressures too high or low.
5. Low-reverse band out of adjustment.
6. Valve body malfunction or leakage.
7. Front clutch or rear servo sticking or not operating.
8. Low-reverse servo, band or linkage malfunction.
9. Faulty oil pump.

Slips In All Positions

1. Low fluid level.
2. Hydraulic pressures too low.
3. Valve body malfunction or leakage.
4. Faulty oil pump.
5. Clutches or servos sticking or not operating.
6. Worn or broken input shaft and/or reaction shaft support seal rings.

No Drive In Any Position

1. Low fluid level.
2. Hydraulic pressures too low.
3. Oil filter clogged.
4. Valve body malfunction or leakage.
5. Faulty oil pump.
6. Clutches or servos sticking or not operating.

No Drive In Forward Drive Positions

1. Hydraulic pressures too low.
2. Valve body malfunction or leakage.
3. Accumulator sticking, broken rings or spring.
4. Clutches or servos, sticking or not operating.
5. Worn or faulty rear clutch.
6. Overrunning clutch not holding.
7. Worn or broken input shaft and/or reaction shaft support seal rings.

No Drive In Reverse

1. Incorrect gearshift control linkage adjustment.
2. Hydraulic pressures too low.
3. Low-reverse band out of adjustment.
4. Valve body malfunction or leakage.
5. Front clutch or rear servo sticking or not operating.
6. Low-reverse servo, band or linkage malfunction.
7. Worn or faulty front clutch.

Drives In Neutral

1. Incorrect gearshift control linkage adjustment.
2. Incorrect control cable adjustment.
3. Valve body malfunction or leakage.
4. Rear clutch inoperative.

Drags or Locks

1. Kickdown band out of adjustment.

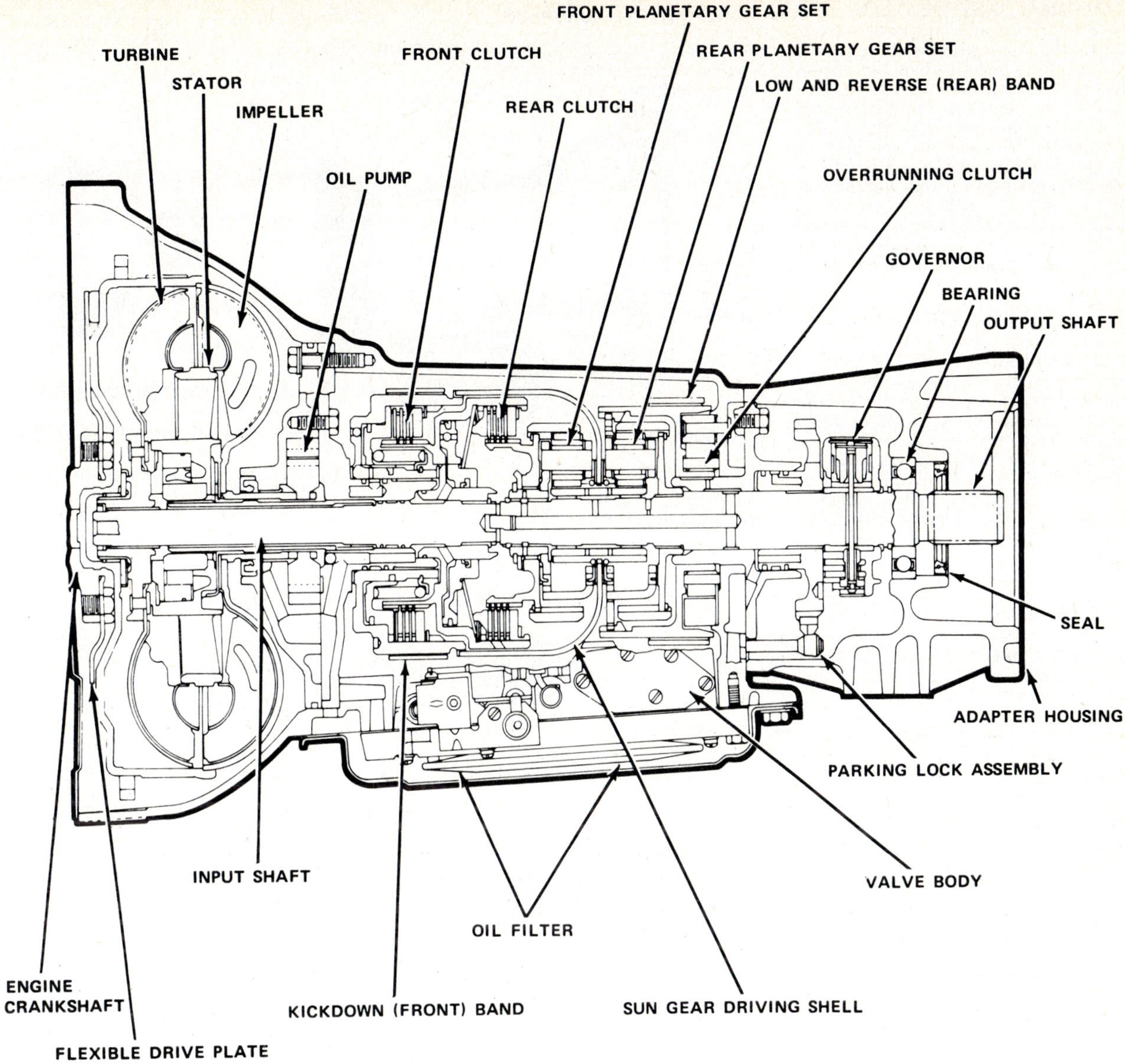

Fig. 2 Model 904 & 999 Loadflite transmission. (Typical)

2. Low-reverse band out of adjustment.
3. Kickdown and/or low-reverse servo, band or linkage malfunction.
4. Front and/or rear clutch faulty.
5. Planetary gear sets broken or seized.
6. Overrunning clutch worn, broken or seized.

Grating, Scraping or Growling Noise

1. Kickdown band out of adjustment.
2. Low-reverse band out of adjustment.
3. Output shaft bearing and/or bushing damaged.
4. Governor support binding or broken seal rings.
5. Oil pump scored or binding.
6. Front and/or rear clutch faulty.
7. Planetary gear sets broken or seized.
8. Overrunning clutch worn, broken or seized.

Buzzing Noise

1. Low fluid level.
2. Pump sucking air.
3. Valve body malfunction.
4. Overrunning clutch inner race damaged.

Hard to Fill, Oil Flows Out Filler Tube

1. High fluid level.
2. Breather clogged.
3. Oil filter clogged.
4. Aerated fluid.

Transmission Overheats

1. Low fluid level.
2. Kickdown band adjustment tight.
3. Low-reverse band adjustment too tight.
4. Faulty cooling system.
5. Cracked or restricted oil cooler line or fitting.
6. Faulty oil pump.
7. Insufficient clutch plate clearance in front and/or rear clutches.

Starter Will Not Energize In Neutral or Park

1. Incorrect gearshift control linkage adjustment.
2. Faulty or incorrectly adjusted neutral starting switch.
3. Broken lead to neutral switch.

MAINTENANCE

Service Note

It has been found that an occasional "no-drive" condition, generally occurring after making the first stop when a vehicle is cold can be caused by incorrect transmission oil level. In cases where this condition is encountered, it is essential that the transmission oil level be checked and corrected as outlined below.

If the no-drive condition still exists with the correct oil level, the selector lever adjustment should be checked.

After the above corrections have been made, if the no-drive condition still exists, it is suggested that the transmission be removed and the front pump disassembled for inspection before attempting any further repairs. Inspect the pump inner rotor and front support for wear, especially where the pinion rubs the support. The clearance specified for the front pump rotors and the face of the housing is .001" to .0025". If this clearance is increased in any way, such as wear, out-of-flatness, foreign objects between housing and support, the clearance will increase and the front pump will not be able to maintain capacity.

Adding Oil

To check the oil level, apply the parking brake and operate the engine at idle speed. Move selector lever into each position ending in the neutral position. A properly filled transmission should read near the "add one pint" mark when the fluid temperature is 70° F., and near but not over the "full" mark at 180° F. which is normal operating temperature. Then add oil as necessary to bring the oil to the prescribed level.

Changing Oil

Oil should be changed every 32,000 miles. Vehicles that operate continuously with abnormal loads should have more frequent periodic maintenance. Transmission should not be idled in gear for long periods. When refilling, use only fluids labeled Dexron II.

1. Remove drain plug (if equipped) from transmission oil pan and allow oil to drain.

NOTE: *If the oil pan does not have a drain plug, loosen pan bolts and tap pan with a soft mallet to break it loose, permitting fluid to drain.*

2. On 1974–77 models, remove flywheel access plate, remove torque converter drain plug and allow to drain. Replace plug.
3. Remove transmission oil pan, clean intake screen and pan, and reinstall.
4. Add 6 quarts on 1974–77 or 4 quarts on 1978–83 units of automatic transmission fluid through filler tube.
5. Start engine and add approximately one quart while engine is idling.
6. Allow engine to idle for about two minutes. Then with parking brake applied, select each range momentarily, ending with the "N" position.
7. Add oil as necessary to bring to proper level.

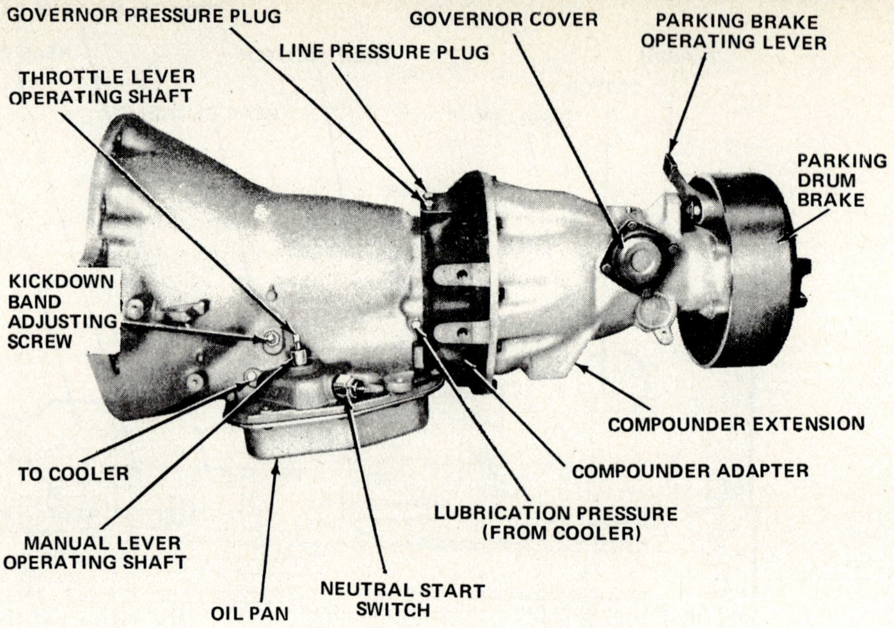

Fig. 3 A-345 transmission assembly. Left side

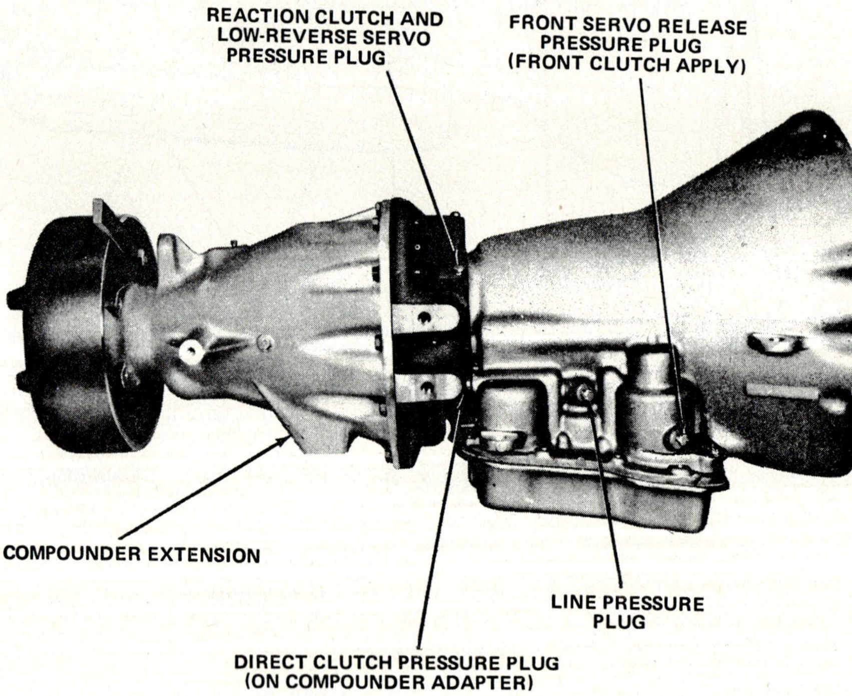

Fig. 4 A-345 transmission assembly. Right side

BANDS, ADJUST

Kickdown Band

The kickdown band adjusting screw is located on the left side of the transmission case near the throttle lever shaft.

1. Loosen lock nut and back off approximately five turns. Check adjusting screw for free turning in transmission case.
2. Using an inch-pound torque wrench, tighten the band adjusting screw to a reading of 72 inch lbs.
3. On A-345 units, backoff adjusting screw 2 turns. On Loadflite units, backoff adjusting screw the number of turns listed in Fig. 5.
4. While holding adjusting screw in position, tighten lock nut.

Low and Reverse Band

1. Raise vehicle, drain transmission and remove oil pan.
2. Loosen adjusting screw lock nut and back off nut approximately five turns. Check adjusting screw for free turning in the lever.
3. Using an inch-pound torque wrench, tighten band adjusting screw to a reading of 72 inch lbs.
4. On Dodge and Plymouth models, backoff adjusting screw 2 turns, except on 1981–82 A-904T and A-999 model transmissions, back off screw 4 turns. On 1974–82 International models, backoff adjusting screw 2¼ turns. On 1981 Jeep CJ-7 models with 4-151 engine (model 904 transmission) backoff adjusting screw 7 turns from 41 inch lbs. On 1980–82 CJ-7 models (model 999 transmission) backoff adjusting screw 4 turns from 72 inch lbs. On 1980–82 Jeep Cherokee, Wagoneer, and Truck models (model 727 transmission) backoff adjusting screw 2 turns from 72 inch lbs. On all models, tighten locknut while holding adjusting screw in position.
5. Install oil pan and fill transmission with fluid.

Model	Year	Engine	Number of Turns
DODGE & PLYMOUTH			
B100-300, CB, MB300-400 & Voyager	1974–80	6 Cyl.	2½
		V8	2
	1981–83	All	2½
Ramcharger & Trail Duster	1974–79	Except V8-440	2½
		V8-440	2
	1980	6 Cyl.	2½
		V8	2
	1981–83	All	2½
D, P, S & W 100-800	1974–77	Except V8-440	2½
		V8-440	2
	1978–79	Except 6-243 Diesel & V8-440	2½
		6-243 Diesel & V8-440	2
	1980	6 Cyl.	2½
		V8	2
	1981–83	All	2½
M300-600 & RM300-400	1973–79	V8-318 & 360	2½
		V8-413 & 440	2
	1980	6 Cyl.	2½
		V8	2
INTERNATIONAL			
All	1974–82	All	2¼
JEEP			
CJ-7	1981	4-151	2½
CJ-7	1980–83	6-258 & V8-304	2
Cherokee, Wagoneer & Truck	1980–83	6-258 & V8-360	2½

Fig. 5 Kickdown band adjustment chart

Turbo Hydra-Matic 400 & 475

DESCRIPTION

This transmission, Fig. 1, is a fully automatic unit consisting primarily of a three-element hydraulic torque converter and a compound planetary gear set. Three multiple-disc clutches, two one-way clutches, and two bands provide the friction elements required to obtain the desired functions of the planetary gear set.

NOTE: The two one-way clutches mentioned above consist two roller clutches.

The torque converter, the multiple-disc clutches and the one-way clutches couple the engine to the planetary gears through oil pressure, providing three forward speeds and reverse. The torque converter, when required, supplements the gears by multiplying engine torque.

Torque Converter

The torque converter is of welded construction and is serviced as an assembly. The unit is made up of two vaned sections, or halves, that face each other in an oil-filled housing. The pump half of the converter is connected to the engine and the turbine half is connected to the transmission.

When the engine makes the converter pump revolve, it sends oil against the turbine, making it revolve also. The oil then returns in a circular flow back to the converter pump, continuing this flow as long as the engine is running.

Stator

The convertor also has a smaller vaned section, called a stator, that funnels the oil back to the converter pump through smaller openings, at increased speed. The speeded up oil directs additional force to the engine-driven converter pump, thereby multiplying engine torque. In other words, without the stator, the unit is nothing more than a fluid coupling.

The stator assembly in some transmissions is a variable pitch unit. The stator blades are operated at either of two positions: maximum or high angle, and minimum or low angle.

Maximum or high angle means greater redirection of the oil and increased engine speed, and torque multiplication for maximum performance. At engine idle, it reduces the converter's efficiency, reducing "creep". Minimum or low angle results in a more efficient converter for cruising operation.

TROUBLE SHOOTING GUIDE

Oil Pressure High

1. Vacuum line or fittings clogged or leaking.
2. Improper engine vacuum.
3. Vacuum leak in vacuum operated accessory.
4. Vacuum modulator.
5. Modulator valve.
6. Water in modulator.
7. Pressure regulator.
8. Oil pump.
9. Governor.
10. Malfunction in detent downshift system.

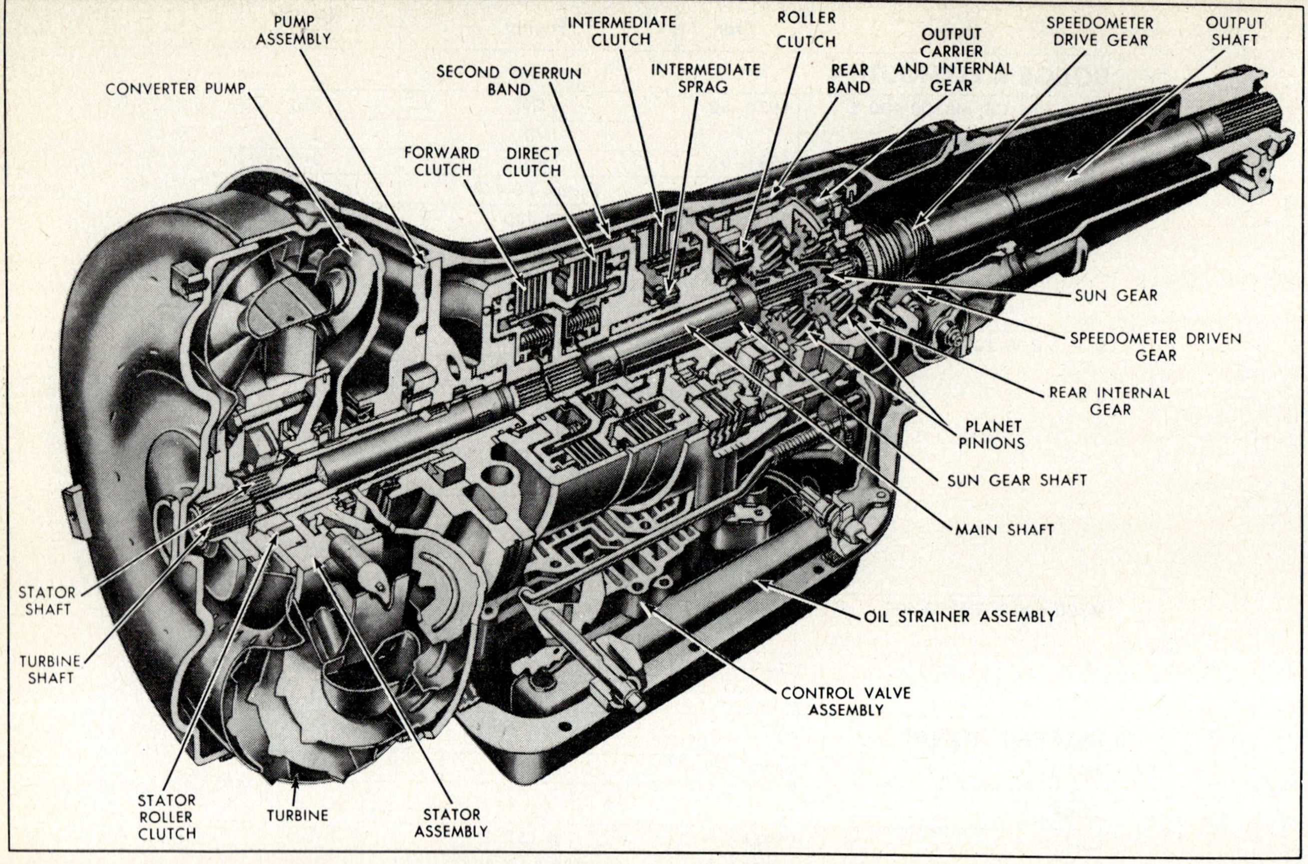

Fig. 1 Turbo-Hydra-Matic "400 & 475" transmission

Check for shorted detent wiring, detent solenoid stuck open, detent feed orifice in spacer plate blocked or restricted, loose detent solenoid, damaged detent valve bore plug or detent regulator valve pin too short.

Oil Pressure Low

1. Low oil level.
2. Defective vacuum modulator.
3. Filter blocked or restricted, or incorrect filter assembly.
4. O-ring seal on intake pipe and/or grommet omitted or damaged.
5. Split or leaking oil intake pipe.
6. Malfunction in oil pump. Check for stuck pressure regulator and/or boost valve, weak pressure regulator valve spring, insufficient spacers in pressure regulator, excessive gear clearance, gears damaged, worn or incorrectly installed, pump cover-to-body gasket mispositioned, defective or mismatched pump body-to-pump cover.
7. Internal circuit leakage.
8. Case porosity.
9. Intermediate clutch cup plug leaking or mispositioned.
10. Low-reverse check ball mispositioned or missing. This will cause no reverse and no overrun braking in low range.

No Drive In Drive Range

1. Low oil level (check for leaks).
2. Manual control linkage not adjusted properly.
3. Low oil pressure. Check for blocked strainer, defective pressure regulator, pump assembly or pump drive gear. See that tangs have not been damaged by converter.
4. Check control valve assembly to see if manual valve has been disconnected from manual lever pin.
5. Forward clutch may be stuck or damaged. Check pump feed circuits to forward clutch, including clutch drum ball check.
6. Roller clutch assembled incorrectly.

1–2 Shift At Full Throttle Only

1. Detent switch may be sticking or defective.
2. Detent solenoid may be stuck open, loose or have leaking gasket.
3. Control valve assembly may be leaking, damaged or incorrectly installed.
4. Third-to-second shift valve stuck.
5. Case porosity.

1st Speed Only—No 1–2 Shift

1. Governor valve may be sticking.
2. Driven gear in governor assembly loose, worn or damaged.
3. The 1-2 shift valve in control valve assembly stuck closed. Check governor feed channels for blocks, leaks, and position. Also check control valve body gaskets for leaks and damage.
4. Intermediate clutch plug in case may be leaking or blown out.
5. Check for porosity between channels and for blocked governor feed channels in case.
6. Center support oil rings missing, or damaged. Orifice plug missing.
7. Intermediate clutch piston seals missing, incorrectly installed or damaged.

No 2–3 Shift—1st & 2nd Only

1. Detent solenoid may be stuck open.
2. Detent switch may not be properly adjusted.
3. Control valve assembly may be stuck, leaking, damaged, or incorrectly installed.
4. Control valve body gaskets leaking, damaged or incorrectly installed.
5. Check direct clutch case center support for broken, leaking or missing oil rings.
6. Check clutch piston seals and piston ball check in clutch assembly.

Moves Forward In Neutral

1. Manual control linkage improperly adjusted.
2. Manual valve disconnected or broken.
3. Inside detent lever pin broken.
4. Transmission oil pressure leaking into forward clutch apply passage.
5. Burned forward clutch plates.
6. Forward clutch does not release.

No Drive In Reverse or Slips In Reverse

1. Check oil level.
2. Manual control linkage improperly adjusted.
3. Vacuum modulator assembly may be defective.

4. Vacuum modulator valve sticking.
5. Strainer may be restricted or leaking at intake.
6. Regulator or boost valve in pump assembly may be sticking.
7. Low oil pressure.
8. Rear servo and accumulator may have damaged or missing servo piston seal ring.
9. Reverse band burned out or damaged. Determine that apply pin or anchor pins engage properly.
10. Direct clutch may be damaged or may have stuck ball check in piston.
11. Forward clutch does not release.
12. Low-reverse ball check missing from case.
13. Control valve body malfunctioning. Check for leaking, damaged or incorrectly installed gaskets, 2-3 shift valve stuck open, or restricted reverse feed passage.

Slips In All Ranges & On Starts

1. Check oil level.
2. Vacuum modulator defective.
3. Modulator valve sticking.
4. Strainer assembly plugged or leaking at neck.
5. Pump assembly regulator or boost valve sticking.
6. Leaks from damaged gaskets or cross leaks from porosity of case.
7. Forward and direct clutches burned.
8. Low oil pressure.

Slips 1–2 Shift

1. Incorrect oil level.
2. Vacuum modulator valve sticking.
3. Vacuum modulator defective.
4. Pump pressure regulator valve defective.
5. Porosity between channels in case.
6. Control valve assembly.
7. Pump-to-case gasket may be mispositioned.
8. Intermediate clutch plug in case may be missing or leaking excessively.
9. Intermediate clutch piston seal missing or damaged.
10. Intermediate clutch plates burned.
11. Front or rear accumulator oil ring may be damaged.
12. Leak in center support feed circuit, excessive leak between center support tower and bushing, blocked center support orifice bleed plug hole, center support bolt not properly seated.
13. Raised ridge around case center support bolt not allowing control valve assembly to seat properly.
14. Low oil pressure.

Slips 2–3 Shift

1. Items 1 through 6 under Slips 1-2 Shift will also cause 2-3 shift slips.
2. Direct clutch plates burned.
3. Oil seal rings on direct clutch may be damaged permitting excessive leaking between tower and bushing.

Rough 1–2 Shift

1. Modulator valve sticking.
2. Modulator assembly defective.
3. Pump pressure regulator or boost valve stuck or inoperative.
4. Control valve assembly loosened from case, damaged or mounted with wrong gaskets.
5. Intermediate clutch ball missing or not sealing.
6. Porosity between channels in case.
7. Rear servo accumulator assembly may have oil rings damaged, stuck piston, broken or missing spring or damaged bore.

Rough 2–3 Shift

1. Items 1, 2 and 3 under Rough 1-2 Shift will also cause rough 2-3 shift.
2. Front servo accumulator spring broken or missing. Accumulator piston may be sticking.

No Engine Braking In Second Speed

1. Front servo or accumulator oil rings may be leaking.
2. Front band may be broken or burned out.
3. Front band not engaged on anchor pin and/or servo pin.

No Engine Braking In Low Range

1. Low-reverse check ball may be missing from control valve assembly.
2. Rear servo may have damaged oil seal ring, bore or piston; leaking, apply pressure.
3. Rear band broken, burned out or not engaged on anchor pins or servo pin.

No Part Throttle Downshifts

1. Vacuum modulator assembly.
2. Modulator valve.
3. Regulator valve train.
4. Control valve assembly has stuck 3-2 valve or broken spring.

No Detent Downshifts

1. Detent switch needs fuse, connections tightened or adjustment.
2. Detent solenoid may be inoperative.
3. Detent valve train in control valve assembly malfunctioning.

Low or High Shift Points

1. Oil pressure. Check vacuum modulator assembly, vacuum line connections, modulator valve, and pressure regulator valve train.
2. Governor may have sticking valve or feed holes that are leaking, plugged or damaged.
3. Detent solenoid may be stuck open or loose.
4. Control valve assembly. Check detent, 3-2, and 1-2 shift valve trains, and check spacer plate gaskets for positioning.
5. Check case for porosity, missing or leaking intermediate plug.

Won't Hold In Park

1. Manual control linkage improperly adjusted.
2. Internal linkage defective; check for chamfer on actuator rod sleeve.
3. Parking pawl broken or inoperative.
4. Parking pawl return spring missing, broken or incorrectly installed.

Noisy Transmission

1. Pump noises caused by high or low oil level.
2. Cavitation due to plugged strainer, porosity in intake circuit or water in oil.
3. Pump gears may be damaged.
4. Gear noise in low gear of Drive Range - transmission grounded to body.
5. Defective planetary gear set.
6. Clutch noises during application can be worn or burned clutch plates.

Forward Clutch Plates Burned

1. Check ball in clutch housing damaged, stuck or missing.
2. Clutch piston cracked, seals damaged or missing.
3. Low line pressure.
4. Manual valve mispositioned.
5. Restricted oil feed to forward clutch.
6. Pump cover oil seal rings missing, broken or undersize; ring groove oversize.
7. Case valve body face not flat or porosity between channels.
8. Manual valve bent and center land not properly ground.

Intermediate Clutch Plates Burned

1. Constant bleed orifice in center support missing.
2. Rear accumulator piston oil ring damaged or missing.
3. 1-2 accumulator valve stuck in control valve assembly.
4. Intermediate clutch piston seal damaged or missing.
5. Center support bolt loose.
6. Low line pressure.
7. Intermediate clutch plug in case missing.
8. Case valve body face not flat or porosity between channels.
9. Manual valve bent and center land not ground properly.

Direct Clutch Plates Burned

1. Restricted orifice in vacuum line to modulator.
2. Check ball in direct clutch piston damaged, stuck or missing.
3. Defective modulator bellows.
4. Center support bolt loose.
5. Center support oil rings or grooves damaged or missing.
6. Clutch piston seals damaged or missing.
7. Front and rear servo pistons and seals damaged.
8. Manual valve bent and center land not cleaned up.
9. Case valve body face not flat or porosity between channels.
10. Intermediate sprag clutch installed backwards.
11. 3-2 valve, 3-2 spring or 3-2 spacer pin installed in wrong location in 3-2 valve bore.

MAINTENANCE

Checking & Adding Fluid

Fluid level should be checked at every engine oil change. The full ("F") and "ADD" marks on the transmission dipstick are one pint apart and determine the correct fluid level at normal operating temperature (180°F.). *Careful attention to transmission oil temperature is necessary as proper fluid level at low operating temperatures will be below the "ADD" mark on the dipstick. Proper fluid level at higher operating temperatures will rise above the "F" mark.*

Fluid level must always be checked with the car on a level surface, and with the engine running to make certain the converter is full.

To determine proper fluid level, proceed as follows:

1. Operate engine at a fast idle for about 1½ minutes with selector lever in park ("P") position.
2. Reduce engine speed to slow idle and check fluid level.
3. With engine running, add fluid as required to bring it to the proper level.

NOTE: An early change to a darker color from the usual red color and or a strong odor that is usually associated with overheated fluid is normal, and should not be treated as a positive sign of needed maintenance or unit failure.

The normal maintenance schedule for drain and refill of this type fluid remains unchanged at 24,000 miles under normal service and 12,000 miles under severe operating conditions. On 1976–78 vehicles, the drain interval is extended to 60,000 under normal service on Light Duty Emissions vehicles and 24,000 miles on Heavy Duty Emissions vehicles and 15,000 miles under severe operating conditions on Light Duty Emissions vehicles, or 12,000 miles on Heavy Duty Emissions vehicles. On 1979–83 vehicles, the drain interval is 100,000 miles under normal service and 15,000 miles under severe operating conditions on Light Duty Emissions vehicles, or 12,000 miles on diesel and/or Heavy Duty Emissions vehicles.

CAUTION: *Do not overfill as foaming might occur when the fluid heats up. If fluid level is too low, especially when cold, complete loss of drive may result after quick stops. Extremely low fluid level will result in damage to transmission.*

Changing Fluid

1. Raise vehicle and support with suitable jack stands.
2. Place drain pan under transmission oil pan, then remove oil pan attaching bolts from front and side of pan.
3. Loosen rear pan attaching bolts approximately four turns, then pry oil pan loose with a screwdriver and allow fluid to drain.
4. Remove remaining bolts and separate oil pan from transmission.
5. Drain remaining fluid from pan, then clean pan with a suitable solvent and thoroughly dry with compressed air.
6. On 1974–81 vehicles, remove oil filter assembly and the O-ring from intake pipe. Install new O-ring and filter and torque filter retaining bolt to 10 ft. lbs., then install a new strainer-to-valve body gasket, strainer and two screws.
7. On 1982–83 vehicles, remove screen/filter-to-body bolts, then the screen/filter and gasket. Clean screen in solvent and dry with compressed air. If applicable, paper or felt filter should be replaced. Install screen/filter assembly using a new O-ring and torque attaching bolt to 10 ft. lbs.
8. Install oil pan with a new gasket. Torque attaching bolts to 12 ft. lbs.

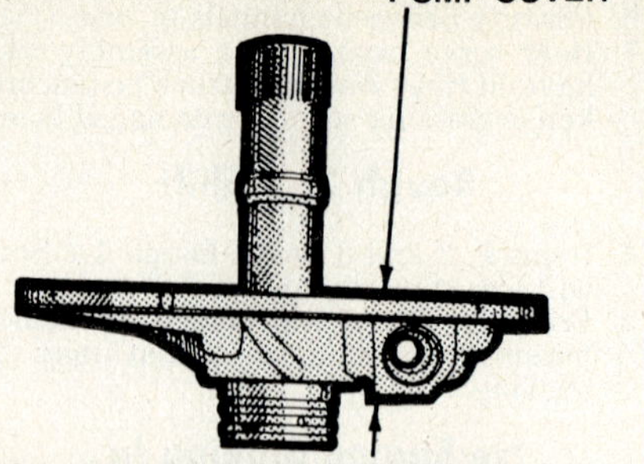

Fig. 2 Pressure regulator identification

9. Lower vehicle and add 5 pints of Dexron II or equivalent transmission fluid through filler tube.
10. With transmission in Park, start engine and run at idle.
11. Move transmission through each gear range and return to Park position, then recheck fluid level. Add additional fluid to bring level to ¼ inch below add mark on dipstick on 1974–81 vehicles, or between dimples on dipstick on 1982–83 vehicles.

Adding Fluid to Fill Dry Transmission & Converter

1. Add 9 pints of transmission fluid through filler tube.
2. Operate engine at high idle and move transmission lever through each gear range.
3. After 1–3 minutes of idle operation with transmission in Park, recheck fluid level. Add additional fluid to bring level to ⅜ inch below add mark on dipstick on 1974–82 vehicles or between dimples on dipstick on 1982–83 vehicles.

IN VEHICLE REPAIRS

Pressure Regulator Valve

NOTE: A solid type pressure regulator valve must be used only in a pump cover with a "Squared Off" (machined) pressure regulator boss, Fig. 2. A pressure regulator valve with oil holes and an orifice cup plug may be used with either type pump.

1. Remove bottom pan and strainer.
2. Using a screwdriver or steel rod, compress regulator boost valve bushing against pressure regulator spring.

CAUTION: Pressure regulator spring is under extreme pressure and will force valve bushing out of bore when snap ring is removed if valve bushing is not held securely.

3. Continue to exert pressure on valve bushing and remove snap ring. Gradually release pressure on valve bushing until spring force is exhausted.
4. Carefully remove regulator boost valve bushing and valve, and pressure regulator spring. Be careful not to drop parts as they will fall out if they are not held.
5. Remove pressure regulator valve and spring retainer. Remove spacers if present.
6. Reverse procedure to install.

Control Valve Body

1. Remove bottom pan and strainer.
2. Disconnect pressure switch lead wire.
3. Remove control valve body attaching screws and detent roller spring assembly. *Do not remove solenoid attaching screws.*
4. Remove control valve body and governor pipes. If care is used in removing control valve body, the six check balls will stay in place above spacer plate.
5. Remove governor pipes and manual valve from control valve body.
6. Reverse procedure to install.

Governor

1. Remove governor cover and discard gasket.
2. Withdraw governor from case.
3. Reverse procedure to install, using a new gasket.

Modulator & Modulator Valve

1. Remove modulator attaching screw and retainer.
2. Remove modulator assembly from case and discard O-ring seal.
3. Remove modulator valve from case.
4. Reverse procedure to install, using a new O-ring seal.

Parking Linkage

1. Remove bottom pan and oil strainer.
2. Unthread jam nut holding detent lever to manual shaft.
3. Remove manual shaft retaining pin from case.
4. Remove manual shaft and jam nut from case.
5. Remove O-ring seal from manual shaft.
6. Remove parking actuator rod and detent lever assembly.
7. Remove parking pawl bracket, pawl return spring and pawl shaft retainer.
8. Remove parking pawl shaft, O-ring seal and parking pawl.
9. Reverse procedure to install, using new seals and gasket.

Rear Seal

1. Remove propeller shaft.
2. Pry out seal with screwdriver.
3. Install new seal with a suitable seal driver.
4. Install propeller shaft.

Turbo Hydra-Matic 200, 200C

Description

The Turbo Hydra-Matic 200 transmission, Fig. 1, is fully automatic and consists of a three element torque converter and a compound planetary gear set. Three multiple disc clutches, a roller clutch and a band provide the required friction elements to obtain the desired function of the planetary gear set.

The Turbo Hydra-Matic 200C, incorporates a converter clutch assembly consisting of a three element torque converter, and a converter clutch, Fig. 2. The converter clutch is splined to the turbine assembly and when operated, applies against the converter cover providing a mechanical direct drive coupling of the engine to the planetary gears. When the converter clutch is released, the assembly operates as a normal torque converter. The converter clutch is applied only when transmission is in third gear, vehicle speed is above 30 mph, engine coolant temperature is above 130° F, engine vacuum is above 3 in. Hg. and brake pedal is released.

TROUBLE SHOOTING GUIDE

No Drive in Drive Range

1. Low oil level.
2. Manual linkage maladjusted.
3. Low oil pressure due to:
   a. Restricted or plugged oil screen.
   b. Oil screen gasket improperly installed.
   c. Oil pump pressure regulator.
   d. Pump drive gear tangs damaged by converter.
   e. Case porosity in intake bore.
4. Forward clutch malfunctioning due to:
   a. Forward clutch not applying due to cracked piston, damaged or missing seals, burned clutch plates, snap ring not in groove.
   b. Forward clutch seal rings damaged or missing on turbine shaft, leaking feed circuits due to damaged or mispositioned gasket.
   c. Clutch housing check ball stuck or missing.
   d. Cup plug leaking or missing from rear of turbine shaft in clutch apply passage.
   e. Incorrect forward clutch piston assembly or incorrect number of clutch plates.
5. Roller clutch malfunctioning due to missing rollers or springs or possibly galled rollers.

Oil Pressure High Or Low

1. Throttle valve cable maladjusted, binding, disconnected or broken.
2. Throttle lever and bracket improperly installed, disconnected or binding.
3. Throttle valve shift valve, throttle valve or plunger binding.
4. Pressure regulator valve and spring malfunctioning due to:
   a. Binding valve.
   b. Incorrect spring.
   c. Oil pressure control orifice in pump

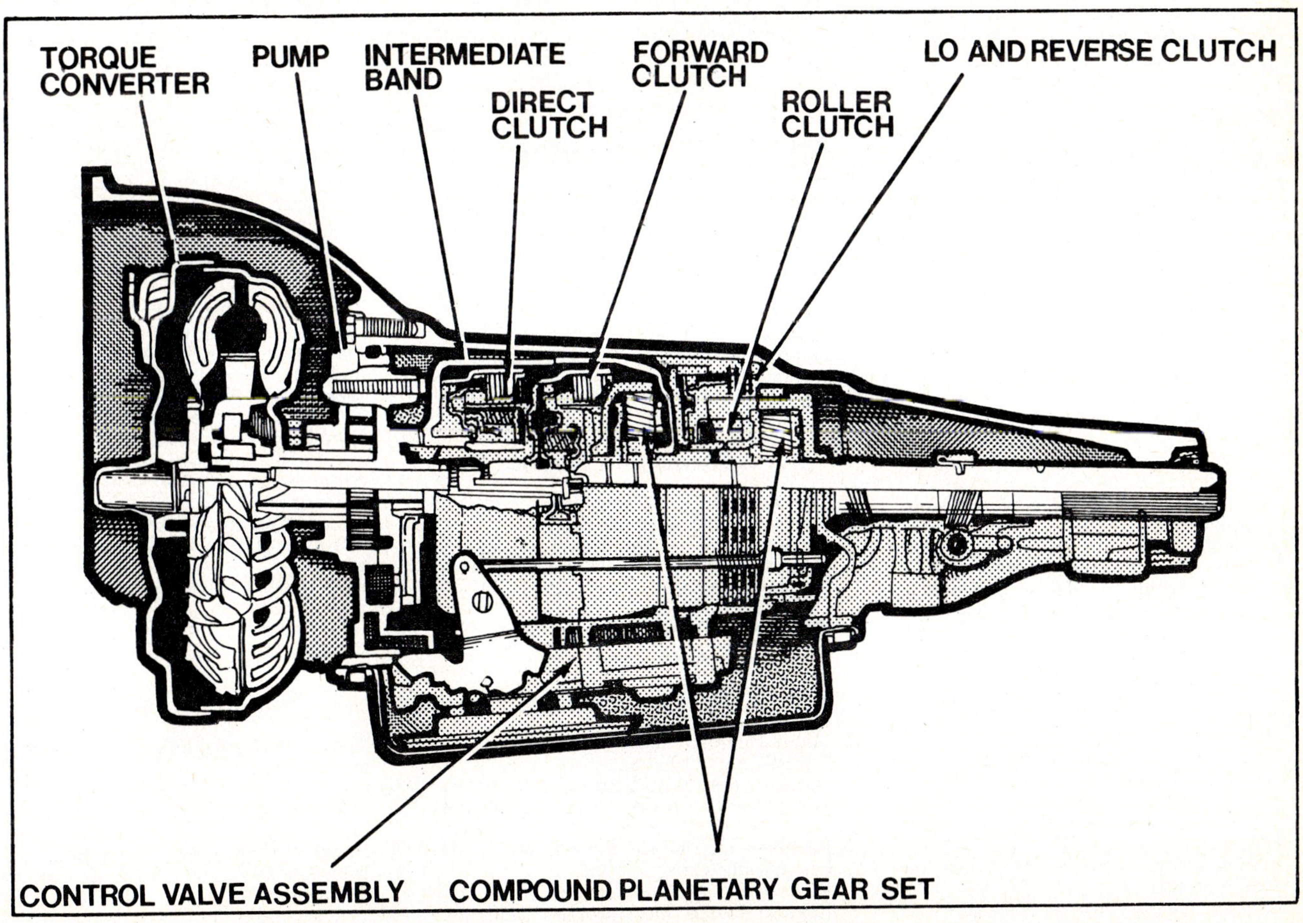

Fig. 1 Turbo-Hydramatic 200 transmission

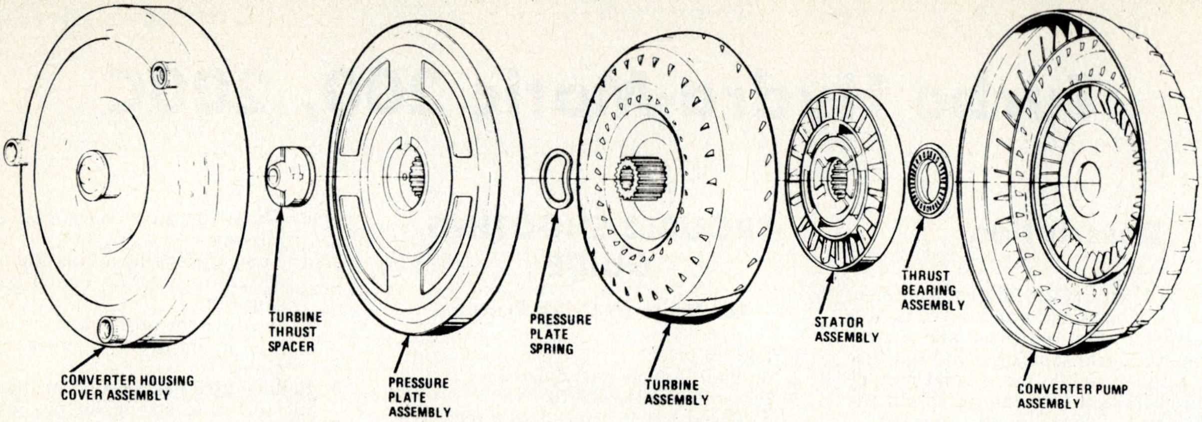

Fig. 2 Torque converter clutch. T.H.M. 200C

cover plugged, causing high oil pressure.
   d. Pressure regulator bore plug leaking.
5. Manual valve disconnected.
6. Intermediate boost valve binding, causing oil pressures to be incorrect in 2nd and low ranges.
7. Orifice in spacer plate at end of intermediate boost valve plugged.
8. Reserve boost valve binding, causing pressure to be incorrect in reverse only.
9. Orifice in spacer plate at end of reverse boost valve plugged.

1—2 Shift At Full Throttle Only

1. Throttle valve cable maladjusted, binding, disconnected or broken.
2. Throttle lever and bracket assembly binding or disconnected.
3. Throttle valve exhaust ball lifter or number 5 check ball binding, mispositioned or disconnected.

NOTE: If number 5 ball is fully seated, it will cause full throttle valve pressure regardless of throttle valve position.

4. Throttle valve and plunger binding.
5. Valve body gaskets leaking, damaged or incorrectly installed.
6. Porous control valve assembly.

First Speed Only, No 1—2 Shift

1. Due to governor and governor feed passages:
   a. Plugged governor oil feed orifice in spacer plate.
   b. Plugged orifice in spacer plate that feeds governor oil to the shift valves.
   c. Balls missing in governor assembly.
   d. Governor cover O-ring missing or leaking. If governor cover O-ring leaks, an external oil leak will be present and there will be no upshift.
   e. Governor shaft seal missing or damaged.
   f. Governor driven gear stripped.
   g. Governor weights binding.
   h. Governor assembly missing.
2. Control valve assembly 1–2 shift valve or 1–2 throttle valve stuck in downshift position.
3. Porosity in case channels or undrilled 2nd speed feed holes.
4. Excessive leakage between case bore and intermediate band apply ring.
5. Intermediate band anchor pin missing or disconnected from band.
6. Missing or broken intermediate band.
7. Due to intermediate servo assembly:
   a. Servo to cover oil seal ring damage or missing.
   b. Porous servo cover or piston.
   c. Incorrect intermediate band apply pin.
   d. Incorrect cover and piston.

1st & 2nd Only, No 2—3, Shift

1. 2–3 shift valve or 2–3 throttle valve stuck in downshift position.
2. Direct clutch feed orifice in spacer plate plugged.
3. Valve body gaskets leaking, damaged or incorrectly installed.
4. Porosity between case passages.
5. Pump passages plugged or leaking.
6. Pump gasket incorrectly installed.
7. Rear seal on pump cover leaking or missing.
8. Direct clutch oil seals missing or damaged.
9. Direct clutch piston or housing cracked.
10. Direct clutch plates damaged or missing.
11. Direct clutch backing plate snap ring out of groove.
12. Intermediate servo to case oil seal broken or missing on intermediate servo piston.
13. Intermediate servo exhaust hole in case between servo piston seals plugged or undrilled.

Moves Forward In Neutral

1. Manual linkage maladjusted.
2. Forward clutch does not release.
3. Cross leakage between pump passages.
4. Cross leakage to forward clutch through clutch passages.

No Drive in Reverse or Slips in Reverse

1. Throttle valve cable binding or maladjusted.
2. Manual linkage maladjusted.
3. Throttle valve binding.
4. Reverse boost valve binding in bore.
5. Low overrun clutch valve binding in bore.
6. Reverse clutch piston cracked, broken or has missing seals.
7. Reverse clutch plates burned.
8. Reverse clutch has incorrect selective spacer ring.
9. Porosity in passages to direct clutch.
10. Pump to case gasket improperly installed or missing.
11. Pump passages cross leaking or restricted.
12. Pump cover seals damaged or missing.
13. Direct clutch piston or housing cracked.
14. Direct clutch piston seals cut or missing.
15. Direct clutch housing ball check, stuck, leaking or missing.
16. Direct clutch plates burned.
17. Incorrect direct clutch piston.
18. Direct clutch orifices plugged in spacer plate.
19. Intermediate servo to case seal cut or missing.

Slips 1—2 Shift

1. Aerated oil due to low level.
2. 2nd speed feed orifice in spacer plate partially blocked.
3. Improperly installed or missing spacer plate gasket.
4. 1–2 accumulator valve stuck, causing low 1–2 accumulator pressure.
5. Weak or missing 1–2 accumulator valve spring.
6. 1-2 accumulator piston seal leaking or spring missing or broken.
7. Leakage between 1-2 accumulator piston and pin.
8. Incorrect intermediate band apply pin.
9. Excessive leakage between intermediate band apply pin and case.
10. Porous intermediate servo piston.
11. Servo cover to servo seal damaged or missing.
12. Incorrect servo and cover.
13. Throttle valve cable improperly adjusted.
14. Shift throttle valve or throttle valve binding.
15. Intermediate band worn or burned.
16. Case porosity in 2nd clutch passages.

Rough 1—2 Shift

1. Throttle valve cable improperly adjusted or binding.
2. Throttle valve or plunger binding.
3. Shift throttle or 1–2 accumulator valve binding.
4. Incorrect intermediate servo pin.
5. Intermediate servo piston to case seal damaged or missing.
6. 1–2 accumulator oil ring damaged piston stuck, bore damaged or spring broken or missing.

Slips 2–3 Shift

1. Low oil level.
2. Throttle valve cable improperly adjusted.
3. Throttle valve binding.
4. Direct clutch orifice in spacer plate partially blocked.
5. Spacer plate gaskets improperly installed or missing.
6. Intermediate servo to case seal damaged.
7. Porous direct clutch feed passages in case.
8. Pump to case gasket improperly installed or missing.
9. Pump passages cross feeding, leaking or restricted.
10. Pump cover oil seal rings damaged or missing.
11. Direct clutch piston or housing cracked.
12. Direct clutch piston seals cut or missing.
13. Direct clutch plates burned.

Rough 2-3 Shift

1. Throttle valve cable improperly installed or missing.
2. Throttle valve or throttle valve plunger binding.
3. Shift throttle valve binding.
4. Intermediate servo exhaust hole undrilled or plugged between intermediate servo piston seals.
5. Direct clutch exhaust valve number 4 check ball missing or improperly installed.

No Engine Braking In 2nd Speed

1. Intermediate boost valve binding in valve body.
2. Intermediate-Reverse number 3 check ball improperly installed or missing.
3. Shift throttle valve number 3 check ball improperly installed or missing.
4. Intermediate servo to cover seal missing or damaged.
5. Intermediate band off anchor pin, broken or burned.

No Engine Braking In 1st Speed

1. Low overrun clutch valve binding in valve body.

NOTE: The following conditions will also cause no reverse.

2. Low-reverse clutch piston seals broken or missing.
3. Porosity in low-reverse piston or housing.
4. Low-reverse clutch housing snap ring out of case.
5. Cup plug or rubber seal missing or damaged between case and low-reverse clutch housing.

No Part Throttle Downshift

1. Throttle plunger bushing passages obstructed.
2. 2–3 throttle valve bushing passages obstructed.
3. Valve body gaskets improperly installed or damaged.
4. Spacer plate hole obstructed or undrilled.
5. Throttle valve cable maladjusted.
6. Throttle valve or shift throttle valve binding.

Low or High Shift Points

1. Throttle valve cable binding or disconnected.
2. Throttle valve or shift throttle valve binding.
3. Number 1 throttle shift check ball improperly installed or missing.
4. Throttle valve plunger, 1–2 or 2–3 throttle valves binding.
5. Valve body gaskets improperly installed or missing.
6. Pressure regulator valve binding.
7. Throttle valve exhaust number 5 check ball and lifter, improperly installed, disconnected or missing.
8. Throttle lever binding, disconnected or loose at valve body mounting bolt or not positioned at the throttle valve plunger bushing pin locator.
9. Governor shaft to cover seal broken or missing.
10. Governor cover O-rings broken or missing.

NOTE: Outer ring will leak externally and the inner ring will leak internally.

11. Case porosity.

Will Not Hold In Park

1. Manual linkage maladjusted.
2. Parking pawl binding in case.
3. Actuator rod or plunger damaged.
4. Parking pawl damaged.
5. Parking bracket loose or damaged.
6. Detent lever nut loose.
7. Detent lever hole worn or damaged.
8. Detent roller to valve body bolt loose.
9. Detent roller or pin damaged, incorrectly installed or missing.

Converter Clutch Applied In All Ranges, Engine Stalls When Transmission Is Put In Gear (T.H.M. 200C)

1. Converter clutch valve in pump sticking in apply position.

Converter Clutch Applies Erratically (T.H.M. 200C)

1. Vacuum switch malfunction.
2. Release orifice at pump blocked or restricted.
3. Damaged turbine shaft O-ring.
4. Converter malfunctioning, clutch pressure plate warped.
5. O-ring damaged at solenoid.
6. Solenoid bolts loose.

Converter Clutch Applies At A Very Low or High 3rd Speed Gear (T.H.M. 200C)

1. Governor switch malfunction.
2. Governor malfunction.
3. High line pressure.
4. Converter clutch valve sticking or binding.
5. Solenoid inoperative or shorted to case.

MAINTENANCE

To check fluid, drive vehicle for at least 15 minutes to bring fluid to operating temperature (200° F). With vehicle on a level surface and engine idling in Park and parking brake applied, the level on the dipstick should be at the "F" mark. To bring the fluid level from the ADD mark to the FULL mark requires 1 pint of fluid. If vehicle cannot be driven sufficiently to bring fluid to operating temperature, the level on the dipstick should be between the two dimples on the dipstick with fluid temperature at 70° F.

If additional fluid is required, use only Dexron or Dexron II automatic transmission fluid.

NOTE: An early change to a darker color from the usual red color and or a strong odor that is usually associated with over-heated fluid is normal and should not be considered as a positive sign of required maintenance or unit failure.

CAUTION: When adding fluid, do not overfill, as foaming and loss of fluid through the vent may occur as the fluid heats up. Also, if fluid level is too low, complete loss of drive may occur especially when cold, which can cause transmission failure.

Every 60,000 miles, on 1977–78 vehicles or every 100,000 miles on 1979–83 vehicles, the oil should be drained, the oil pan removed, the screen cleaned and fresh fluid added. For vehicles subjected to more severe use such as heavy city traffic especially in hot weather, prolonged periods of idling or towing, this maintenance should be performed every 15,000 miles.

Draining Bottom Pan

1. Remove front and side oil pan attaching bolts, then loosen the rear oil pan attaching bolts.
2. Carefully pry oil pan loose and allow fluid to drain into a suitable container.
3. Remove the oil pan and gasket, then remove the screen attaching bolts and remove screen.
4. Thoroughly clean oil screen and oil pan with solvent.
5. Install oil screen using a new gasket and torque attaching bolts to 6–10 ft. lbs., then install oil pan using a new gasket and torque attaching bolts to 10–13 bolts.
6. Add 3 quarts of fluid on 1977–78 vehicles or 3½ quarts of fluid on 1979–83 vehicles then with engine idling and parking brake applied, move selector lever through each range and return selector lever to PARK.
7. Check fluid level and add fluid as required to bring level between the two dimples on the dipstick.

Adding Fluid To Dry Transmission and Converter

1. Add 4½ quarts of lfuid on 1977–78 vehicles or 5 quarts of fluid on 1979–83 vehicles.
2. With transmission in PARK and parking brake applied, start the engine and place carburetor on fast idle cam.
3. Move shifter lever through each range, then with transmission in PARK, add additional fluid as required to bring the level between the two dimples on the dipstick.

IN VEHICLE REPAIRS

Valve Body Assembly

1. Drain transmission fluid, then remove oil

pan and screen.
2. Remove detent cable retaining bolt and disconnect cable.
3. Remove throttle lever and bracket assembly. Use care to avoid bending throttle lever link.
4. Remove detent roller and spring assembly.
5. Support valve body and remove retaining bolts, then while holding manual valve, remove valve assembly, spacer plate and gaskets as an assembly to prevent dropping the five check balls.

NOTE: After removing valve body assembly, the intermediate band anchor band pin, and reverse cup plug may be removed.

6. To install control valve reverse removal procedure and torque all valve body bolts to 8 ft. lbs.

CAUTION: Assure that intermediate band anchor pin is located on intermediate band prior to installation of valve body, as damage to transmission may result.

Governor

1. Disconnect battery ground cable and remove cleaner.
2. If necessary, on vehicles with air conditioning, remove the five heater core cover screws, then disconnect the electrical connectors and position heater core aside.
3. Disconnect exhaust pipe and allow to hang down.
4. Support transmission, then remove transmission rear support bolts and propeller shaft and lower transmission until enough clearance is obtained to remove governor.
5. Remove governor retainer ring and cover, then remove governor and washer.

Fig. 3 Removing or installing pressure regulator

NOTE: If governor to case washer falls into transmission, use a small magnet to remove it. If it cannot be easily removed, replace the washer with a new one.

6. To install governor, reverse removal procedure.

CAUTION: Do not attempt to hammer governor assembly into case, as damage to governor, case or cover may result.

Pressure Regulator Valve

1. Drain transmission fluid, then remove oil pan and screen.
2. Using a small screwdriver or tool J-24684, Fig. 3, compress regulator spring.
3. Remove retaining ring and slowly release spring tension.
4. Remove pressure regulator bore plug, valve, spring and guide.
5. To assemble, install pressure regulator spring, guide and valve with stem end first and bore plug with hole side out.
6. Using a small screwdriver or tool J-24684, Fig. 3, compress regulator spring and install retaining ring.

Turbo Hydra-Matic 250, 250C, 350, 350C

DESCRIPTION

The Turbo Hydra-Matic 250, 350, Figs. 1 and 2, are fully automatic three speed transmissions consisting of a three element torque converter and a compound planetary gear set. The Turbo Hydra-Matic 250C and 350C, also incorporate a torque converter clutch, Fig. 3. The Turbo Hydra-Matic 350 transmission has four multiple-disc clutches, two roller clutches and a band to provide the required friction elements to obtain the desired function of the planetary gear set. The Turbo Hydra-Matic 250 transmission uses an adjustable intermediate band in place of the intermediate clutch found in the Turbo Hydra-Matic 350. Also, the Turbo Hydra-Matic 250 has three multiple-disc clutches and one roller clutch.

The friction elements couple the engine to the planetary gears through oil pressure, providing three forward speeds and one reverse.

The three element torque converter is of welded construction and is serviced as an assembly. The unit consists of a pump or driving member, a turbine or driven member and a stator assembly. When required, the torque converter supplements the gears by multiplying engine torque.

The Turbo Hydra-Matic 250C and 350, the converter clutch assembly consist of a three element torque converter, with the addition a converter clutch, Fig. 3. The converter clutch is splined to the turbine assembly and when operated, applies against the converter cover providing a mechanical direct drive coupling of the engine to the planetary gears. When the converter clutch is released, the assembly operates as a normal torque converter. The converter clutch is applied only when transmission is in third gear, vehicle speed is above 30 mph, engine coolant temperature is above 130°F, engine vacuum is above 3 in. Hg. and brake pedal is released.

TROUBLE SHOOTING GUIDE

No Drive In Drive Range

1. Low oil level (check for leaks).
2. Manual control linkage improperly adjusted.
3. Low oil pressure due to blocked strainer, defective pressure regulator, pump assembly or pump drive gear. See that tangs have not been damaged by converter. Check case for porosity in intake bore.
4. Check control valve assembly to be sure manual valve has not been disconnected from inner lever.
5. Forward clutch may be stuck or damaged. Check pump feed circuits to forward clutch, including clutch drum ball check.
6. Roller clutch assembly broken or damaged.

Oil Pressure High or Low

High Pressure:
1. Vacuum line or fittings leaking.
2. Vacuum modulator.
3. Modulator valve.
4. Pressure regulator.
5. Oil pump.

Low Pressure:
1. Vacuum line or fittings obstructed.
2. Vacuum modulator.
3. Modulator valve.
4. Pressure regulator.
5. Governor.
6. Oil pump.

1–2 Shift At Full Throttle Only

1. Detent valve may be sticking or linkage may be misadjusted.
2. Vacuum line or fittings leaking.
3. Control valve body gaskets leaking, damaged or incorrectly installed. Detent valve train or 1–2 valve stuck.
4. Check case for porosity.

First Speed Only, No 1–2 Shift

T.H.M. 250 & 350
1. Governor valve may be sticking.
2. Driven gear in governor assembly loose, worn or damaged. If driven gear shows damage, check output shaft drive gear for nicks or rough finish.
3. Control valve governor feed channel blocked or gaskets leaking. 1–2 shift valve train stuck closed.
4. Check case for blocked governor feed channels or for scored governor bore which will allow cross pressure leak. Check case for porosity.
5. Intermediate clutch or seals damaged.
6. Intermediate roller clutch damaged.

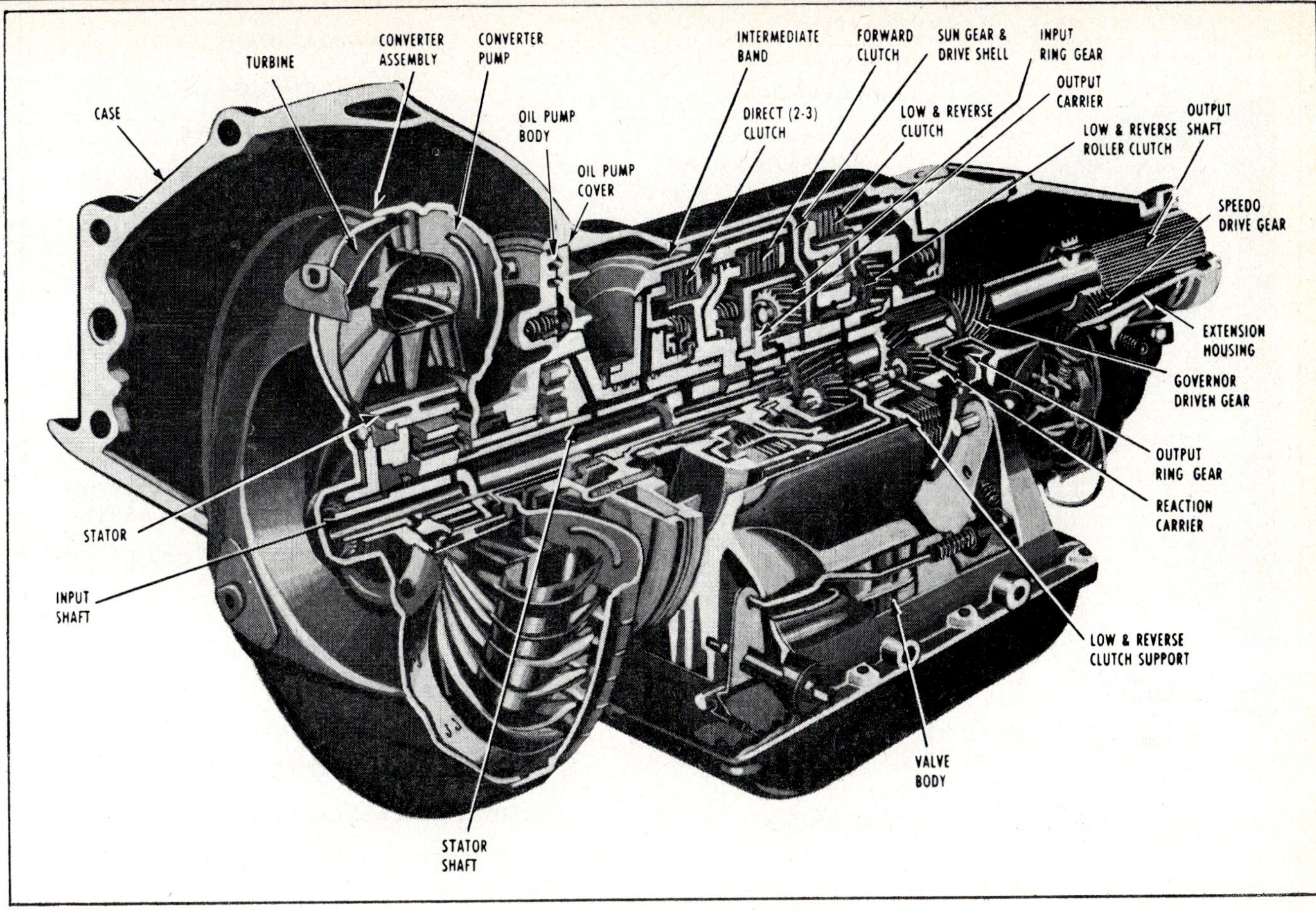

Fig. 1 Turbo Hydra-Matic "250" transmission

T.H.M. 250

1. Intermediate servo piston seals damaged, missing or installed improperly.
2. Intermediate band improperly adjusted.
3. Intermediate servo apply rod broken.

1st & 2nd Only, No 2–3 Shift

1. Control valve 2–3 shift train stuck. Valve body gaskets leaking, damaged or improperly installed.
2. Pump hub-to-direct clutch oil seal rings broken or missing.
3. Direct clutch piston seals damaged. Piston ball check stuck or missing.

No First Speed

T.H.M. 250

1. Intermediate band adjusted too tightly.
2. 1–2 shift valve stuck in upshift position.

T.H.M. 350

1. Excessive number of clutch plates in intermediate clutch pack.
2. Incorrect intermediate clutch piston.

Moves Forward In Neutral

1. Manual linkage misadjusted.
2. Forward clutch not releasing.

No Drive In Reverse or Slips In Reverse

1. Low oil level.
2. Manual linkage misadjusted.
3. Modulator valve stuck.
4. Modulator and reverse boost valve stuck.
5. Pump hub-to-direct clutch oil seal rings broken or missing.
6. Direct clutch piston seal cut or missing.
7. Low and reverse clutch piston seal cut or missing.
8. Number 1 check ball missing.
9. Control valve body gaskets leaking or damaged.
10. 2–3 valve train stuck in upshifted position.
11. 1–2 valve train stuck in upshifted position.
12. Intermediate servo piston or pin stuck so intermediate overrun band is applied.
13. Low and reverse clutch piston out or seal damaged.
14. Direct clutch plates burned—may be caused by stuck ball check in piston.
15. Forward clutch not releasing.

Slips In All Ranges

1. Low oil level.
2. Vacuum modulator valve defective or sticking.
3. Filter assembly plugged or leaking.
4. Pressure regulator valve stuck.
5. Pump to case gasket damaged.
6. Check case for cross leaks or porosity.
7. Forward clutch slipping.

Slips 1–2 Shift

T.H.M. 250 & 350

1. Low oil level.
2. Vacuum modulator assembly defective.
3. Modulator valve sticking.
4. Pump pressure regulator valve defective.
5. 2–3 accumulator oil ring damaged or missing. 1–2 accumulator oil ring damaged or missing. Case bore damaged.
6. Pump to case gasket mispositioned or damaged.
7. Check for case porosity.
8. Intermediate clutch piston seals damaged. Clutch plates burned.

T.H.M. 250

1. Intermediate servo piston seals damaged or missing.
2. Burned intermediate band.

T.H.M. 350

1. 2–3 accumulator oil ring damaged or missing.

Rough 1–2 Shift

T.H.M. 250 & 350

1. Vacuum modulator, check for loose fittings, restrictions in line or defective modulator assembly.
2. Modulator valve stuck.
3. Valve body regulator or boost valve stuck.
4. Pump to case gasket mispositioned or damaged.
5. Check case for porosity.
6. Check 1–2 accumulator assembly for damaged oil rings, stuck piston, broken or missing spring, or damaged case bore.

T.H.M. 250

1. Intermediate band improperly adjusted.

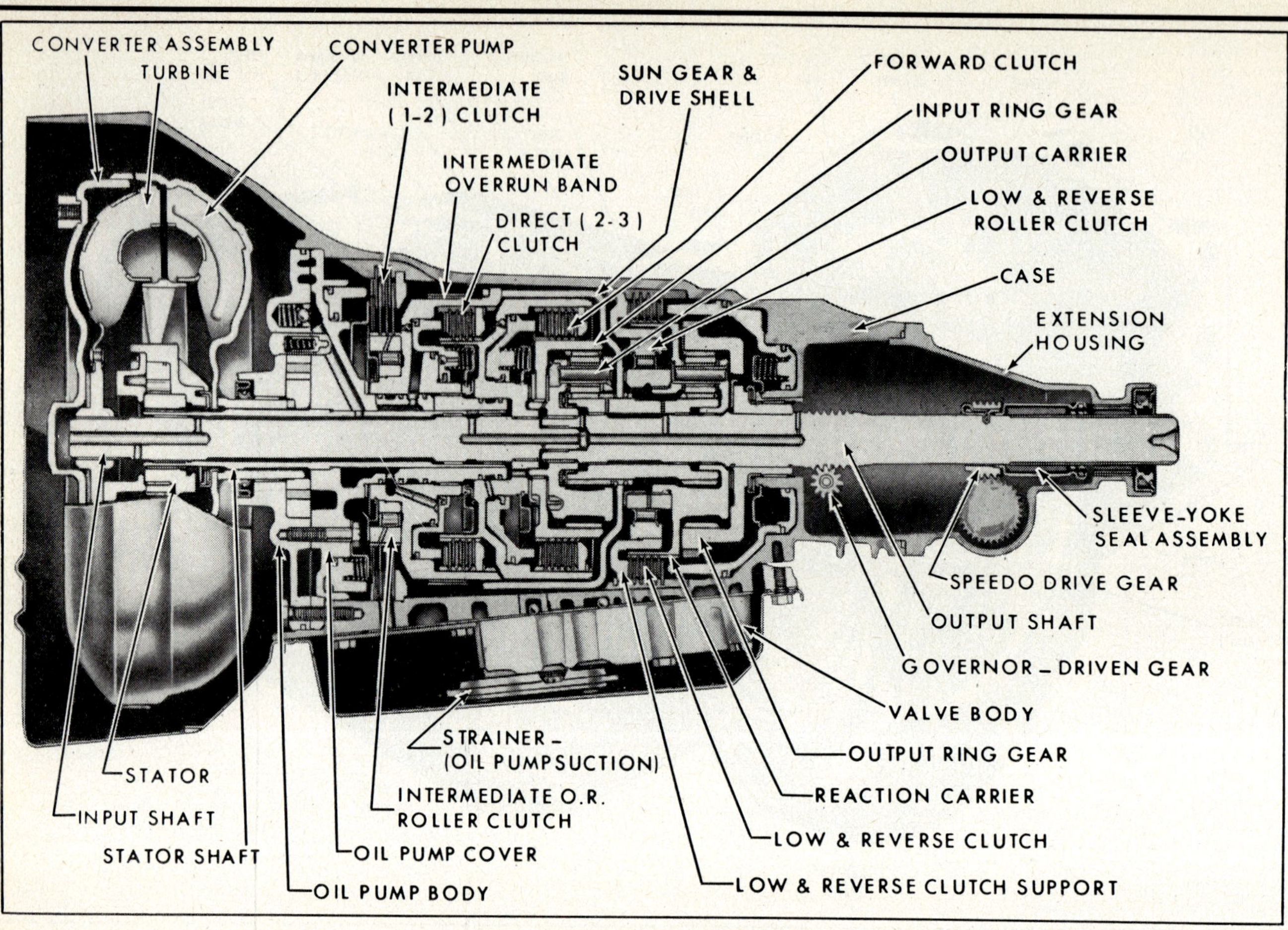

Fig. 2 Turbo Hydra-Matic "350" transmission

2. Improper or broken servo spring.

T.H.M. 350

1. Burned intermediate clutch plates.
2. Improper number of intermediate clutch plates.

Slips 2–3 Shift

1. Low oil level.
2. Modulator valve or vacuum modulator assembly defective.
3. Pump pressure regulator valve or boost valve; pump to case gasket mispositioned.
4. Check case for porosity.
5. Direct clutch piston seals or ball check leaking.

Rough 2–3 Shift

1. High oil pressure. Vacuum leak, modulator valve sticking or pressure regulator or boost valve inoperative.
2. 2–3 accumulator piston stuck, spring broken or missing.

No Engine Braking In Second Speed

1. Intermediate servo or 2–3 accumulator oil rings or bores leaking or accumulator piston stuck.
2. Intermediate overrun band burned or broken.
3. Low oil pressure: Pressure regulator and/or boost valve stuck.

No Engine Braking In 1st Speed

1. Manual low control valve assembly stuck.
2. Low oil pressure: Pressure regulator and/or boost valve stuck.
3. Low and reverse clutch piston inner seal damaged.

No Part Throttle Downshift

1. Oil pressure: Vacuum modulator assembly, modulator valve or pressure regulator valve train malfunctioning.
2. Detent valve and linkage sticking, disconnected or broken.
3. 2–3 shift valve stuck.

No Detent Downshifts

1. 2–3 valve stuck.
2. Detent valve and linkage sticking, disconnected or broken.

Low or High Shift Points

1. Oil pressure: Check engine vacuum at transmission end of modulator pipe.
2. Vacuum modulator assembly vacuum line connections at engine and transmission, modulator valve, pressure regulator valve train.
3. Check governor for sticking valve, restricted or leaking feed holes, damaged pipes or plugged feed line.
4. Detent valve stuck open.
5. 1–2 or 2–3 valve train sticking.
6. Check case for porosity.

Won't Hold In Park

1. Manual linkage misadjusted.
2. Parking brake lever and actuator assembly defective.
3. Parking pawl broken or inoperative.

Burned Forward Clutch Plates

1. Check ball in clutch drum damaged, stuck or missing.
2. Clutch piston cracked, seals damaged or missing.
3. Low line pressure.
4. Pump cover oil seal rings missing, broken or undersize; ring groove oversize.
5. Transmission case valve body face not flat or porosity between channels.

Burned Intermediate Clutch Plates

T.H.M. 350

1. Intermediate clutch piston seals damaged or missing.
2. Low line pressure.
3. Transmission case valve body face not flat or porosity between channels.

Burned Intermediate Band

T.H.M. 250

1. Intermediate servo piston seals damaged or missing.
2. Low line pressure.

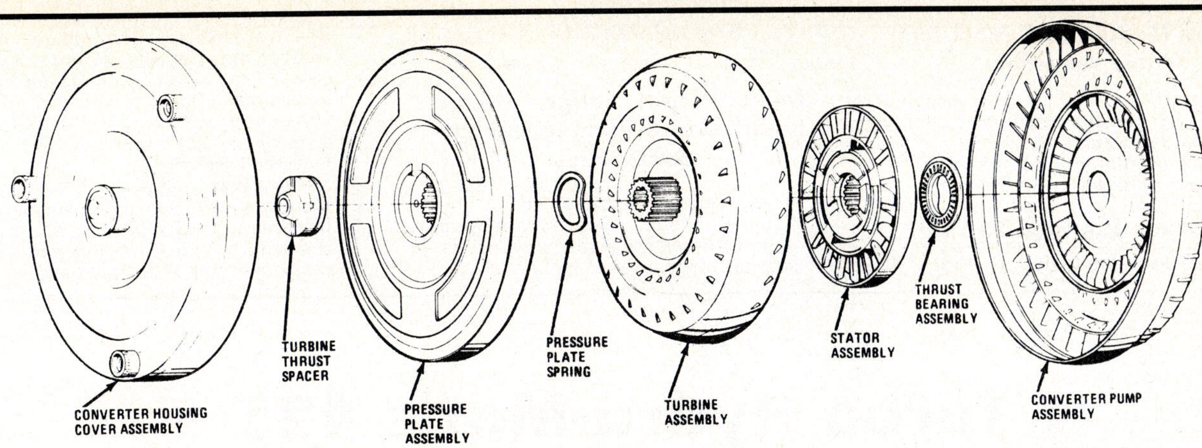

Fig. 3 Torque converter clutch. T.H.M. 250C & 350C

3. Transmission case valve body face not flat or porosity between channels.

Burned Direct Clutch Plates

1. Restricted orifice in vacuum line to modulator.
2. Check ball in clutch drum damaged, stuck or missing.
3. Defective modulator.
4. Clutch piston cracked, seals damaged or missing.
5. Transmission case valve body face not flat or porosity between channels.

Noisy Transmission

NOTE: Before checking transmission for noise, ensure noise is not coming from water pump, alternator or any belt driven accessory.

Park, Neutral & All Driving Ranges

1. Low fluid level.
2. Plugged or restricted screen.
3. Damaged screen to valve body gasket.
4. Porosity in valve body intake area.
5. Transmission fluid contaminated with water.
6. Porosity at transmission case intake port.
7. Improperly installed case to pump gasket.
8. Pump gears are damaged.
9. Driving gear assembled backwards.
10. Crescent interference in pump.
11. Damaged or worn oil pump seals.
12. Loose converter to flywheel bolts.
13. Damaged converter.

1st, 2nd And/Or Reverse Gear

1. Planetary gears or thrust bearings damaged.
2. Damaged input or output ring gear.

Acceleration in any Gear

1. Transmission case or transmission oil cooler lines contacting underbody.
2. Broken or loose engine mounts.

Squeal At Low Vehicle Speed

1. Speedometer driven gear shaft seal requires lubrication or replacement.

Converter Clutch Applied In All Ranges; Engine Stalls When Transmission Is Put In Gear

T.H.M. 250C & 350C

1. Converter clutch valve in pump sticking in apply position.

Converter Clutch Applies Erratically

T.H.M. 250C & 350C

1. Vacuum switch malfunction.
2. Release orifice at pump blocked or restricted.
3. Damaged turbine shaft O-ring.
4. Converter malfunctioning, clutch pressure plate warped.
5. O-ring damaged at solenoid.
6. Solenoid bolts loose.

Converter Clutch Applies At A Very Low or High 3rd Speed Gear

T.H.M. 250C & 350C

1. Governor switch malfunction.
2. Governor malfunction.
3. High line pressure
4. Converter clutch valve sticking or binding.
5. Solenoid inoperative or shorted to case.

MAINTENANCE

Fluid should be checked every 6,000 miles with engine idling, selector lever in neutral position, parking brake set and transmission at operating temperature. Use only General Motors Dexron transmission fluid when adding oil. Do not overfill.

Every 24,000 miles on 1974 vehicles, 30,000 miles on 1975–76 vehicles, 60,000 miles on 1977–78 vehicles or 100,000 miles on 1979–83 vehicles, drain transmission oil sump. After replacing sump, add approximately 5½ pints fluid. Check fluid and add as necessary to bring level to full mark on dipstick with engine at operating temperature.

NOTE: Chevrolet and GMC models are using a revised type Dexron fluid. An early change to a darker color from the usual red color and or a strong odor that is usually associated with overheated fluid is normal, and should not be treated as a positive sign of needed maintenance or unit failure.

The normal maintenance schedule for drain and refill of this type fluid remains unchanged under normal service at 24,000 miles for 1974 vehicles, 30,000 miles on 1975–76 vehicles, 36,000 miles on 1977–78 vehicles and 100,000 miles on 1979–83 vehicles. If vehicle is operated under severe conditions, such as frequent trailer towing, fluid should be changed at the following intervals: all 1974 and 1975–83 with Heavy Duty Emissions and/or diesel engine, 12,000 miles; 1975–83 with Light Duty Emissions, 15,000 miles.

Valve Body Assembly

1. Remove oil pan and strainer.
2. Remove retaining pin to disconnect downshift actuating lever bracket, remove valve body attaching bolts and detent roller and spring assembly.
3. Remove valve body assembly while disconnecting manual control valve link from range selector inner lever.

CAUTION: Do not drop valve.

4. Remove manual valve and link from valve body assembly.
5. Reverse procedure to install.

Governor

1. Where necessary, remove shift linkage and transmission to crossmember bolts.
2. Raise transmission with jack and remove crossmember. Lower transmission enough to remove governor.
3. Remove governor cover retainer and cover.
4. Remove governor.

Intermediate Clutch Accumulator Piston Assembly

1. Remove two oil pan bolts adjacent to accumulator piston cover, install compressor on oil pan lip and retain with these two bolts.
2. Compress intermediate clutch accumulator piston cover and remove retaining ring piston cover and O ring from case.
3. Remove spring and intermediate clutch accumulator piston.

Vacuum Modulator & Modulator Valve Assembly

1. Disconnect vacuum hose from modulator stem and remove vacuum modulator screw and retainer.
2. Remove modulator and its O ring.
3. Remove modulator valve from case.

Extension Housing Oil Seal

1. Remove propeller shaft.
2. Pry out lip seal with screwdriver or small chisel.

Manual Shaft, Range Selector Inner Lever & Parking Linkage Assemblies

1. Remove oil pan and strainer.
2. Remove manual shaft to case retainer and unthread jam nut holding range selector inner lever to manual shaft.
3. Remove jam nut and remove manual shaft from range selector inner lever and case. *Do not remove manual shaft lip seal unless replacement is required.*
4. Disconnect parking pawl actuating rod from range selector inner lever and remove bolt from case.
5. Remove bolts and parking lock bracket.
6. Remove pawl disengaging spring.
7. If necessary to replace pawl or shaft, clean up bore in case and remove shaft retaining plug, shaft and pawl.

Turbo Hydra-Matic 425

DESCRIPTION

This transmission is a fully automatic unit used for front wheel drive applications, Fig. 1. It consists primarily of a three-element hydraulic torque converter, dual sprocket and chain link assembly, compound planetary gear set, three multiple disc clutches, a sprag clutch, a roller clutch, two band assemblies, and a hydraulic control system.

Torque Converter

The torque converter consists of a pump or driving member, a turbine or driven member and a stator or reaction member.

The stator is mounted on a one-way roller clutch which allows it to overrun when not used as a reaction member.

The torque converter couples the engine to the planetary gear set through the use of a drive sprocket, a chain link assembly, and a driven sprocket. Clockwise engine torque turns the drive sprocket clockwise. This, in turn, drives the driven sprocket in a clockwise direction. This in effect is a reverse in the direction of engine torque due to the side mounting of the gear unit.

Planetary Gear Set

The gear set provides three forward ratios and reverse.

TROUBLE SHOOTING GUIDE

NOTE: In many of the following diagnosis procedures, it is recommended that air pressure be applied to help in determining if the seal, rings or pistons are stuck, missing or damaged. Therefore, when air is applied, listen carefully for escaping air and piston action as air is applied to a particular area.

No Drive In "D" Range

1. Low oil level. Check for external leaks or vacuum modulator diaphragm leaking.
2. Manual linkage maladjusted. Correct alignment in manual lever shift quadrant.
3. Low oil pressure.
4. Oil strainer O-ring seal missing or damaged, neck weld leaking, strainer blocked.
5. Oil pump pressure regulator stuck or inoperative. Pump drive gear tangs damaged by converter.
6. Case porosity in intake bore.
7. Control valve. Manual valve disconnected from manual lever pin. (Other shift lever positions would also be affected.)
8. Forward clutch does not apply. Piston cracked; seals missing or damaged. These defects can be checked by removing the valve body and applying air pressure to the drive cavity in the case valve body face. Missing, damaged or worn oil rings on driven support housing can also be checked in this manner at the same time because they can also cause the forward clutch not to apply. Clutch plates burned.
9. Roller clutch inoperative. Rollers worn, damaged springs, or damaged races. May be checked by placing selector lever in "L" range.

No Drive In "R" or Slips In Reverse

1. Low oil level.
2. Manual linkage.
3. Oil pressure. Vacuum modulator defective, modulator valve sticking.
4. Restricted strainer, leak at intake pipe or O-ring seal. Pressure regulator or boost valve sticking.
5. Control valve body gaskets leaking or damaged (other malfunctions may also be indicated). Low-reverse check ball missing from case (this will cause no overrun braking in low range). The 2-3 valve train stuck open (this will also cause 1-3 upshifts in drive range). Reverse feed passage not drilled; also check case passages. Apply air to reverse passage in case valve body face.
6. Rear servo and accumulator. Servo piston seal ring broken or missing. Apply air pressure to drilled hole in intermediate clutch passage of case valve body face to check for piston operation and excessive leakage. Band apply pin too short (this may also cause no overrun braking or slip in overrrun braking in low range).
7. Rear band burned, loose lining, apply pin or anchor pin not engaged; band broken.
8. Direct clutch outer seal damaged or missing. Clutch plates burned (may be caused by stuck ball check in piston).
9. Forward clutch does not release (will also cause drive in neutral range).

Drive In Neutral

1. Manual linkage maladjusted.
2. Forward clutch does not release (this condition will also cause no reverse).

1st Speed Only—No 1–2 Upshift

1. Governor valve sticking; driven gear loose, damaged or worn. If driven gear shows signs of water or damage, check output flange drive gear for nicks or rough finish.
2. Control valve. The 1–2 shift valve train stuck closed. Dirt, chips or damaged valve in 1–2 shift valve train. Governor feeds channels blocked or leaking; pipes out of position. Valve body gaskets leaking or damaged. Case porosity between oil channels. Governor feed passage blocked.
3. Intermediate clutch. Case center support oil rings missing, broken or defective. Clutch piston seals missing, improperly assembled, cut or damaged. Apply air to intermediate clutch passage located in case valve body face to check for these defects.

1-2 Shift Obtained Only At Full Throttle

1. Detent switch sticking or defective.
2. Detent solenoid loose, gasket leaking, sticks open, electrical wire pinched between cover and casting.
3. Control valve body gasket leaking or damaged. Detent valve train stuck.

1st & 2nd Speeds Only No 2–3 Shift

1. Detent solenoid stuck open (the 2–3 shift would occur at very high speeds) may be diagnosed as no 2–3 shift.
2. Detent switch sticking or defective.
3. Control valve body. The 2–3 valve train stuck with dirt or foreign material. Valve body gaskets leaking or damaged.
4. Direct clutch. Case center support oil rings missing or broken. Clutch piston seals missing, improperly assembled, cut or damaged; piston ball check stuck or missing. Apply air to direct clutch passage in case valve body face to check these conditions.

Slips In All Ranges

1. Oil level incorrect.
2. Low oil pressure. Vacuum modulator defective or valve sticking. Oil strainer plugged or leaks at neck; O-ring (case to strainer) missing or damaged. Pressure regulator or boost valve sticking.
3. Case cross channel leaks; porosity.
4. Forward, intermediate and direct clutches slipping. Clutch plates burned.

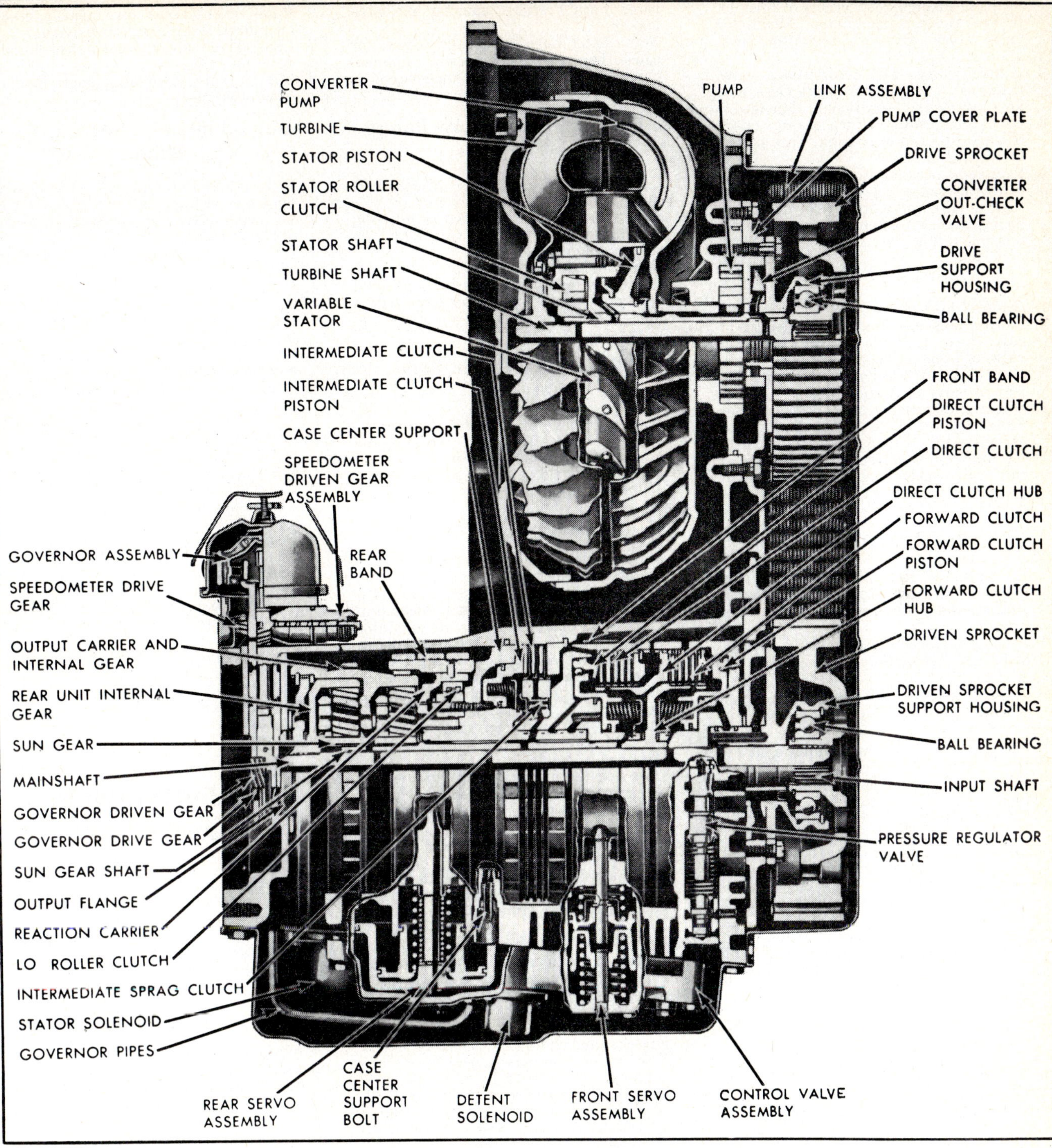

Fig. 1 General Motors front wheel drive Turbo Hydra-Matic transmission

Always look for a primary defect that would cause clutch plates to burn. (Missing feed holes, seals and oil rings, etc., are primary defects).

5. Roller clutch rollers worn; springs or cage damaged, and worn or damaged races (operates normally in low and reverse ranges).

Slips 1–2 Shift

1. Oil level incorrect.
2. Low oil pressure. Look for defective vacuum modulator or valve sticking. Pump Pressure regulator valve stuck.
3. Front servo accumulator piston cracked or porous, oil rings damaged or missing.
4. Control valve. The 1–2 accumulator valve train (may cause a slip-bump shift). Porous valve body or case valve body face.
5. Rear servo accumulator oil ring missing or damaged; case bore damaged; piston cracked or damaged.
6. Case porous between oil passages.
7. Intermediate clutch lip seals missing, cut or damaged. Apply air pressure to intermediate clutch passage in case valve body face to check. Clutch plates burned. Case center support leaks in feed circuits (oil rings damaged or grooves damaged) or excessive leak between tower and bushing.

Rough 1–2 Shift

1. Oil pressure. Check vacuum modulator

for loose fittings, restrictions in line; defective vacuum modulator. Modulator valve stuck. Pressure regulator boost valve stuck.
2. Control valve. 1–2 accumulator valve train; valve body-to-case bolts loose; gaskets inverted, off location, or damaged.
3. Case. Intermediate clutch passage check ball missing or not seating. Case porous between channels.
4. Rear servo accumulator piston stuck. Apply air pressure to 1–2 accumulator passage in case valve body face(you should hear the servo piston move). Broken or missing spring; bore scored or damaged.

Slips 2–3 Shift

1. Oil level high or low.
2. Low oil pressure. Modulator defective or valve sticking. Pump pressure regulator valve or boost valve sticking.
3. Control valve. Accumulator piston pin leak at valve body end.
4. Direct clutch piston seals leaking. Case center support oil seal rings damaged or excessive leak between tower and bushing. Apply air to direct clutch passage in case valve body face. If air comes out intermediate passage, center support is defective.

Rough 2–3 Shift

1. Oil pressure high. Vacuum modulator defective or valve sticking. Pump pressure regulator valve or boost valve stuck or inoperative.
2. Front servo accumulator spring missing or broken; accumulator piston stuck.

Shifts Occur at Too High or Too Low Car Speed

1. Oil pressure. Vacuum modulator defective or valve sticking. Leak in vacuum line (engine to transmission). Vacuum modulator line fitting on carburetor blocked. Pump pressure regulator valve or boost valve train stuck.
2. Governor valve stuck or sticking. Feed holes restricted or leaking; pipes damaged or mispositioned.
3. Detent solenoid stuck open or loose on valve body (will cause late shifts).
4. Control valve. Detent valve train sticking; 3–2 valve train sticking; 1–2 shift valve stuck; 1–2 detent valve sticking open (will probably cause early 2–3 shift).
5. Spacer plate gaskets inverted or mispositioned; orifice holes missing or blocked; check balls missing or mislocated.
6. Case porous in channels or foreign material blocking channels.

No Detent Downshift

1. Detent switch mispositioned or electrical connections loose.
2. Solenoid defective or electrical connections loose.
3. Control valve detent valve train stuck.

No Engine Braking—Super Range 2nd Speed

1. Front servo or accumulator piston rings broken or missing. Case or valve body bores worn oversize, causing excessive leakage.
2. Front band worn or burned (check for cause); band end lugs broken or damaged; band lugs not engaged on anchor pins or servo apply pin (check for cause).

No Engine Braking—Low Range 1st Speed

1. Control valve low-reverse check ball missing from case.
2. Rear servo oil ring damaged or missing; piston damaged or porous, causing a leak in apply pressure.
3. Rear band lining worn or burned (check for cause); band end lugs broken; band ends not engaged on anchor pin or servo apply pin. These items will also cause slip in reverse or no reverse.

Will Not Hold Car In Park Position

1. Manual linkage maladjusted (external).
2. Parking brake lever and actuator rod assembly defective (check for proper actuator spring action). Parking pawl broken or inoperative.

Poor Performance or Rough Idle

1. Stator switch defective or maladjusted.
2. Stator solenoid defective or wire ground to solenoid housing; electrical connection loose; stator valve train stuck (located in valve body); oil feed circuit to stator restricted or blocked (check feed hole in stator shaft); converter-out check valve broken or missing (reed valve located in cover plate under drive support housing).
3. Turbine shaft converter return passage not drilled; oil seal rings broken, worn or missing.
4. Case porous in feed circuit channels or foreign material blocking feed circuit.
5. Converter assembly defective.

Transmission Noise

1. Pump noise. Oil level high or low; water in oil, driving gear assembled upside down; driving or driven gear teeth damaged.
2. Gear noise (1st gear drive range). Check planetary pinions for tooth damage. Check sun gear and front and rear internal gears for tooth finish or damage.
3. Clutch noise during application. Check clutch plates.
4. Sprocket and chain link assembly. Chain link too long (sounds similar to popcorn popping). There will be a rough burr along teeth or drive sprocket if chain link is too long; replace chain link and drive sprocket. Drive or driven sprocket teeth damaged. Engine mounts worn or damaged.

Burned Forward Clutch Plates

1. Check ball in clutch housing damaged, stuck or missing.
2. Clutch piston cracked, seals damaged or missing.
3. Low line pressure.
4. Manual valve mispositioned.
5. Restricted oil feed to forward clutch.
6. Pump cover oil seal rings missing, broken or undersize or ring groove oversize.
7. Case valve body face not flat or porosity between channels.
8. Manual valve bent and center land not ground properly.

Burned Intermediate Clutch Plates

1. Rear accumulator piston oil ring damaged or missing.
2. 1-2 accumulator valve stuck in control valve assembly.
3. Intermediate clutch piston seals damaged or missing.
4. Center support bolt loose.
5. Low line pressure.
6. Intermediate clutch plug in case missing.
7. Case valve body face not flat or porosity between channels.
8. Manual valve bent and center land not ground properly.

Burned Direct Clutch Plates

1. Restricted orifice in vacuum line to modulator.
2. Check ball in direct clutch piston damaged, stuck or missing.
3. Defective modulator bellows.
4. Center support bolt loose.
5. Center support oil rings or grooves damaged or missing.
6. Clutch piston seals damaged or missing.
7. Front and rear servo pistons and seals damaged.
8. Manual valve bent and center land not cleaned up.
9. Case valve body face not flat or porosity between channels.
10. Intermediate sprag clutch installed backwards.

MAINTENANCE

Adding Oil

The fluid level should be checked at every engine oil change interval, and should be changed at 12,000 mile intervals. The fluid level should be checked with the selector lever in PARK position, engine running at the idle speed and vehicle on a level surface. *The filler tubes comes out from the final drive housing but it is for the transmission.*

Changing Oil

When changing transmission oil, first add 4 quarts, start the engine, and add oil to bring the fluid level to the FULL mark on the dipstick. Total capacity after overhaul is 10 quarts. Use Dexron automatic transmission oil.

NOTE: An early change to a darker color from the usual red color and or a strong odor that is usually associated with over-heated fluid is normal, and should not be treated as a positive sign of needed maintenance or unit failure.

The normal maintenance schedule for drain and refill of this transmission is 12,000 miles.

Ford Automatic Overdrive Transmission

TRANSMISSION IDENTIFICATION

Year	Truck Model	Trans. Code	Engine Model
1981–82	F-100	T	V8-302
	F-150	T	V8-302
	F-250	T	V8-302

DESCRIPTION

This unit is a 4 speed automatic transmission incorporating an integral overdrive feature. With selector lever in 1 position, the transmission will start and remain in first gear until the selector lever is moved to another position. In 3 position, the transmission will automatically shift through 1-2-3 range, but will not engage overdrive. In D position, the transmission will automatically select the appropriate time to shift into overdrive (4th gear). The design of the transmission features a split torque path in third gear, where 40% of the engine torque is transmitted hydraulically through the torque converter and 60% is transmitted mechanically through solid connections (direct drive input shaft) to the driveshaft. When transmission is in overdrive (4th gear), 100% of engine torque is transmitted through the direct drive input shaft.

The transmission consists essentially of a torque converter assembly, compound planetary gear train and a hydraulic control system, Fig. 1. For gear control the transmission has four friction clutches, two one-way roller clutches and two bands. Overdrive is accomplished by the addition of a band to lock the reverse sun gear while driving the planet carrier. The torque converter operation is similar to other types of automatic transmission, but has an added damper assembly and input shaft for 3rd gear and overdrive. The direct drive input shaft couples the engine directly to the direct clutch. This shaft is driven by the torque converter cover through the damper assembly which cushions engine shock to the transmission.

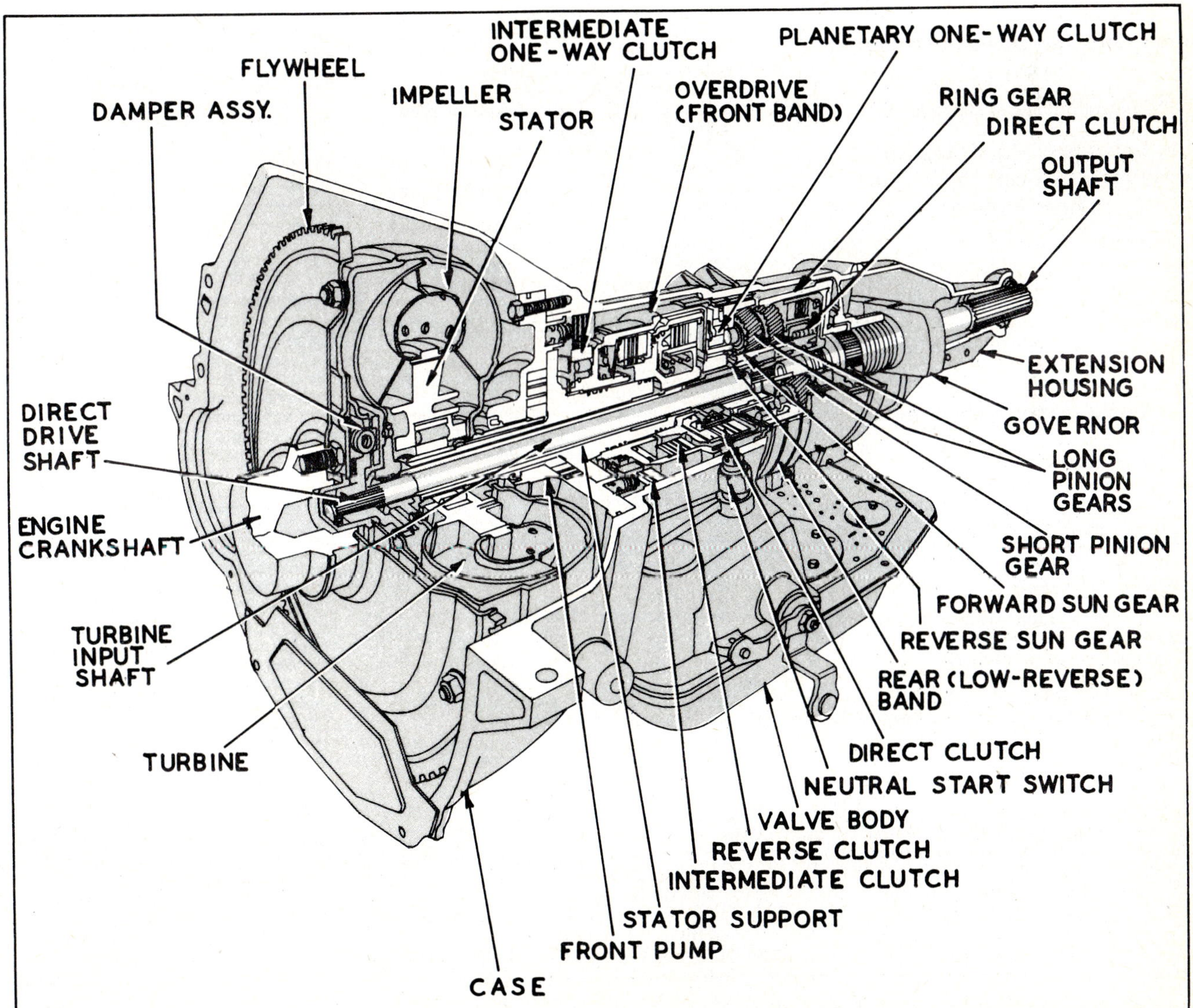

Fig. 1 Sectional view of Ford Automatic Overdrive Transmission

TROUBLE SHOOTING GUIDE

Early &/or Soft Upshifts

1. TV control linkage set too short.

Slip on Light Throttle Into and Out of O/D

1. TV control linkage set too short.

No Kickdown Function at Appropriate Speed

1. TV control linkage set too short.

Harsh Idle Engagement After Engine Warm Up

1. TV control linkage set too long.

Shift Clunk When Throttle is Backed Off After Full or Heavy Throttle

1. TV control linkage set too long.

Harsh Coasting Downshifts Out of O/D

1. TV control linkage set too long.

Delayed and Harsh Upshifts, Especially at Light to Moderate Acceleration

1. Interference prevents return of TV linkage.
2. TV control rod disconnected.
3. Clamping bolt at lower end of control rod loose.
4. Linkage lever control spring broken or disconnected.

MAINTENANCE

Checking Oil Level

1. With transmission at operation temperature, park vehicle on level surface.
2. Operate engine at idle speed with parking brake applied and move selector lever through each detent position. Return selector lever to Park.
3. With engine idling, remove dipstick and check fluid level. Fluid level should be between arrows on dipstick.
4. Add fluid as necessary to bring fluid to proper level. Use only fluid meeting Ford Qualification No. M2C-138-CJ of Dexron II.

Drain & Refill

1. Raise the vehicle and place a drain pan under the transmission.
2. Loosen the pan attaching bolts and drain the fluid.
3. Remove and clean the pan discarding the filter, filter gasket, oil pan gasket, and valve body gasket.
4. Install new gaskets and the pan.
5. Add three quarts of fluid to the transmission through the filler tube.
6. Run the engine and move the selector lever through the detents checking the fluid level as described above.

IN-VEHICLE REPAIRS

Control Valve Body

1. Raise and support vehicle, drain transmission fluid, then remove transmission pan, gasket and filter.
2. Remove detent spring attaching bolt, then the spring.
3. Remove valve body to case attaching bolts, then the valve body.
4. Reverse procedure to install. Use suitable guide pins to align valve body to case.

Overdrive Servo Assembly

1. Remove valve body as previously described.
2. Compress overdrive servo piston cover with a suitable tool, then remove snap ring retainer.
3. Apply compressed air to servo piston release passage and remove the overdrive servo piston cover and spring. Remove piston from cover, then the rubber seal from piston and cover.
4. Install new servo piston and cover seals on the servo piston and cover.
5. Lubricate all seals, piston and piston bore with transmission fluid.
6. Install servo piston into cover, then the return spring into servo piston.
7. Install overdrive piston assembly into overdrive servo bore.
8. Compress overdrive piston using suitable tool, then install snap ring retainer.
9. Install valve body, filter, pan and gasket. Refill transmission to proper fluid level.

Reverse Servo Assembly

1. Refer to "Overdrive Servo Assembly" procedure for replacement. Apply compressed air to the servo piston release passage to remove servo piston from case.

NOTE: Reverse servo piston is under spring pressure. Use caution when removing servo piston cover.

3-4 Accumulator Piston

1. Remove valve body as previously described.
2. Compress 3-4 accumulator piston cover, then remove snap ring retainer.
3. Release cover slowly, then remove piston cover, return spring and piston. Some models do not use a spring.
4. Remove seal from 3-4 accumulator cover and piston and inspect for damage and wear.
5. Install new seals on 3-4 accumulator cover, if necessary. Lubricate cover pocket of case with transmission fluid.
6. Install 3-4 accumulator piston and return spring into case, then the cover.
7. Compress cover using suitable tool, then install snap ring. Ensure cover is reseated snugly against snap ring.
8. Install valve body, filter, pan and gasket. Refill transmission pan to proper fluid level.

2-3 Accumulator Piston

1. Refer to "3-4 Accumulator Piston" procedure for replacement.

Extension Housing

1. Raise and support vehicle.
2. Disconnect parking brake cable from equalizer, if necessary.
3. Disconnect drive shaft from rear axle flange, then remove drive shaft from transmission.
4. Disconnect speedometer cable from extension housing.
5. Remove engine rear support to extension housing attaching bolts. On Lincoln Continental, remove reinforcement plate.
6. Support transmission with suitable jack and raise transmission enough to remove weight from rear engine support.
7. Remove engine rear support from crossmember, then lower transmission and remove extension housing attaching bolts. Slide extension housing from output shaft and allow fluid to drain.
8. Reverse procedure to install.

Governor

1. Remove extension housing as described above.

NOTE: If governor body only is being removed, proceed to step 4.

2. Remove governor to output shaft retaining snap ring.
3. Remove governor assembly from output shaft using suitable tool. Remove governor driveball.
4. Remove governor to counterweight attaching screws. Remove governor from counterweight.
5. Reverse procedure to install.

Internal & External Shift Linkage

1. Raise and support vehicle.
2. Drain transmission fluid from pan, then remove pan and gasket.
3. Disconnect shift rod at transmission manual lever, then the throttle valve linkage at transmission.
4. Disconnect inner throttle lever spring. Remove detent spring.
5. Hold outer throttle lever, then loosen outer throttle lever attaching nut. Remove attaching nut and lock washer.
6. Remove outer throttle lever seal, then the manual lever roll pin.
7. Remove outer manual lever attaching bolt, then the outer manual lever.
8. Remove inner throttle lever and spring.
9. Remove inner manual lever and park pawl actuating rod.
10. Remove manual lever oil seal.
11. Reverse procedure to install. Adjust transmission manual linkage and throttle linkage as outlined previously.

Turbo Hydra-Matic 700-R4

DESCRIPTION, FIG. 1

The model 700-R4 is fully automatic transmission consisting of a 3-element hydraulic torque converter with the addition of a converter clutch.

Also two planetary gear sets, five multiple-disc type clutches, two roller or one-way clutches and a band are used which provide the friction elements to produce four forward speeds, the last of which is overdrive.

The torque converter, through oil, couples the engine power to the gear sets and hydraulically provides additional torque multiplication when required. Also, through the converter clutch, the converter drive and driven members operate as one unit when applied providing mechanical drive from the engine through the transmission.

The gear ratio changes are fully automatic in relation to the vehicle speed and engine torque. Vehicle speed and engine torque are directed to the transmission providing the proper gear ratio for maximum efficiency and performance at all throttle openings.

A hydraulic system pressurized by a variable capacity vane type pump provides the operating pressure required for the operation of the friction elements and automatic controls.

TROUBLE SHOOTING GUIDE

Oil Pressure High or Low

1. Pump assembly pressure regulator valve binding, dirty or damaged spring.
2. T.V. and reverse boost plug and bushing are dirty, sticking, damaged or incorrectly assembled.
3. Pump assembly pressure relief ball not seated or damaged.
4. Pump assembly slide sticking.
5. Pump assembly not regulating.
6. Excess rotor clearance in pump assembly.
7. Manual valve not engaged or damaged.
8. T.V. exhaust valve binding or damaged.
9. Throttle lever and bracket assembly, misassembled, binding, damaged or check valve missing.
10. Valve body throttle valve or plunger sticking.
11. Valve body T.V. limit valve sticking.
12. Throttle link, not engaged, damaged, incorrect, burr on upper end or hanging on T.V. sleeve.
13. Filter, restricted, has missing "O" ring or hole in intake pipe.

High or Low Shift Points

1. T.V. cable binding or not set.
2. Improper external linkage travel.
3. Binding throttle valve or plunger.
4. T.V. modulator up or down valve sticking.
5. Valve body gaskets or spacer plate mispositioned or damaged.
6. T.V. limit valve sticking.
7. Pump assembly, sticking pressure regulator valve, T.V. boost valve.
8. Pump slide sticking.

First Speed Only—No Upshift

1. Sticking governor valve.
2. Governor driven gear is damaged.
3. Governor driven gear retainer pin missing.
4. Nicks or burrs on output shaft.
5. Correct governor retainer pin in case (longer or shorter).
6. Burrs on governor sleeve.
7. Burrs on governor case.
8. Governor weights and springs damaged.
9. 1-2 Shaft valve sticking.
10. Valve body gaskets or spacer plate are mispositioned.

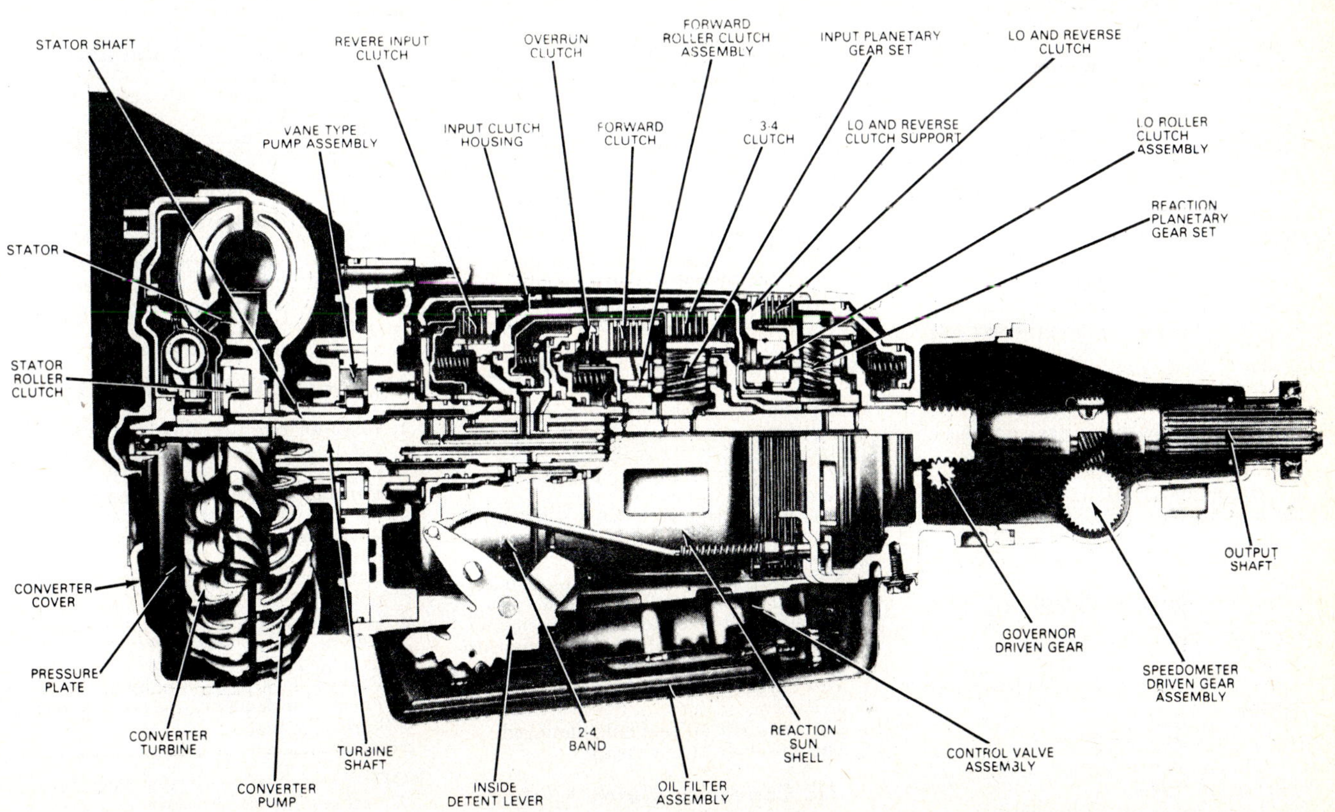

Fig. 1 700-R4 Automatic transmission

11. Valve body pad—porosity and or damaged lands.
12. Restricted or damaged governor screen.
13. 2-4 Servo apply passages, servo apply pin and pin hole in case, restricted or damaged.
14. Damaged or missing servo piston seals.
15. 2-4 Band assembly burned, band anchor pin not engaged.
16. 2-4 Band assembly apply end broken.

Slips in First Gear

1. Forward clutch plates burned.
2. Porosity—forward clutch piston.
3. Forward clutch seals, cut or damaged.
4. Damaged forward clutch housing.
5. Forward clutch internal leak.
6. Forward clutch housing check ball damage.
7. Low oil or oil pressure.
8. Accumulator valve sticking.
9. Valve body lands or interconnected passages damaged.
10. Valve body gasket, spacer plate damaged or mispositioned.
11. Binding internal T.V. linkage.
12. 1-2 Accumulator piston assembly, piston or bore porous.
13. 1-2 Accumulator piston assembly seals, cut or damaged.
14. Leak between piston and pin.
15. Missing or broken accumulator spring.

1-2 Full Throttle Shifts Only

1. T.V. Cable not connected.
2. T.V. Cable too long or short.
3. Throttle lever bracket assembly misassembled or binding.
4. Missing exhaust check valve.
5. Throttle link not connected, burr on upper end or hanging on T.V. sleeve.
6. Throttle valve or plunger hanging or sticking in full open position.
7. Inter connected passages—pump, case or valve body restricted or damaged.

1-2 Slip or Rough

1. Throttle lever and bracket assembly, damaged or incorrectly installed.
2. Throttle valve or bushing, sticking.
3. Sticking 1-2 shift valve train.
4. Valve body gasket or spacer plate, mispositioned.
5. Sticking line bias valve.
6. Sticking accumulator valve.
7. Sticking T.V. limit valve.
8. Incorrect 2-4 servo apply pin.
9. 2-4 Servo oil seal rings or seals, damaged.
10. 2-4 Servo bores, damaged.
11. 2-4 Servo oil passages, restricted or missing.
12. 2nd Accumulator piston seal, damaged.
13. Accumulator spring, missing.
14. 2nd Accumulator bores, damaged.
15. 2nd Accumulator piston, porous.
16. 2nd Accumulator oil passages, restricted or missing.
17. Burned 2-4 band.

2-3 Slip or Rough

1. 2-3 Shift valve train, sticking.
2. Accumulator valve sticking.
3. Valve body gasket or spacer plate mispositioned.
4. Throttle valve sticking.
5. T.V. Limit valve sticking.
6. 3-4 Clutch plates burned or excessive clutch plate travel.
7. 3-4 Piston seals, cut or damaged.
8. 3-4 Piston porosity.
9. 3-4 Piston exhaust ball open.
10. Apply passages restricted.
11. 3-4 Clutch check ball capsule, damaged or misassembled.

3-4 Slip or Rough

1. 3-4 Accumulator spring, missing.
2. 3-4 Piston porosity.
3. Accumulator feed passages, restricted.
4. 3-4 Accumulator piston oil seal ring, broken.
5. Accumulator case bore damaged.
6. Servo band apply incorrect.
7. Servo piston seals damaged or missing.
8. Servo piston bores damaged.
9. Servo piston porosity.
10. 3-4 Clutch burned. (Refer to 2-3 slip for other clutch diagnosis).
11. 2-4 Band burned.
12. Valve body 2-3 shift valve train, sticking.
13. Accumulator valve sticking.
14. Valve body gaskets or spacer plate mispositioned.
15. Valve body throttle valve sticking.
16. T.V. limit valve sticking.

No Reverse or Slips in Reverse

1. Forward clutch will not release.
2. Manual linkage improperly adjusted.
3. Pump assembly reverse boost plug, sticking.
4. Valve body gaskets or saucer plate mispositioned.
5. Lo reverse clutch piston seals cut or damaged.
6. Lo reverse clutch apply passages restricted or missing.
7. Lo reverse clutch plates burned.
8. Lo reverse clutch cover plate loose or cover plate gasket damaged.
9. Reverse input clutch plates burned.
10. Reverse input clutch piston seals cut or damaged.
11. Reverse input clutch apply passage restricted or missing.
12. Reverse input clutch housing exhaust ball and capsule, damaged.

No Part Throttle Downshifts

1. Binding external or internal linkage.
2. Valve body T.V. modulator downshift valve binding.
3. Valve body throttle valve binding.
4. Valve body throttle valve bushing, feed hole restricted or missing.
5. Valve body check ball #3 mispositioned.

No Overrun Braking Manual 3-2-1

1. External manual linkage not properly adjusted.
2. Overrun clutch plates burned.
3. Overrun clutch inner or outer piston seals damaged.
4. Overrun clutch piston exhaust ball sticking or missing.
5. Overrun clutch piston porosity.
6. Valve body gaskets or spacer plate mispositioned or orifice holes plugged.
7. Valve body 4-3 sequence valve sticking.
8. Valve body check balls 3, 9 or 10 mispositioned.
9. Turbine shaft oil feed passages restricted or missing.
10. Turbine shaft oil seal ring damaged.
11. Turbine shaft plug missing.

No Converter Clutch Apply

1. 12 Volts not being supplied to the transmission.
2. Defective transmission outside electrical connector.
3. Defective inside electrical connectors, wiring harness, solenoid.
4. Defective electrical ground inside transmission.
5. Defective pressure switch or improper connection.
6. Solenoid not grounded.
7. Valve body converter clutch shift or throttle valve sticking.
8. Valve body casting or spacer plate in converter clutch valve area are mispositioned or damaged.
9. Converter clutch apply valve, stuck or installed backwards.
10. Pump assembly signal oil orifice, restricted or missing.
11. Pump assembly "O" ring on solenoid damaged or missing.
12. Pump to case gasket, damaged or mispositioned.
13. Pump assembly cup plug missing from apply passage.
14. Pump assembly orifice plug missing from the cooler input passage.
15. High or uneven bolt torque on cover to body.
16. Converter clutch stop valve or retainer ring not installed properly.

Converter Shudder

1. Converter clutch pressure plate damaged.
2. Check ball on end of turbine shaft damaged.
3. Sticking converter clutch shift valve in valve body.
4. Sticking converter clutch apply valve in valve body.
5. Restricted converter clutch apply passage.
6. Low oil or oil pressure.
7. Engine not tuned properly.

No Converter Release

1. Converter clutch apply valve stuck in the open position.
2. "O" ring or check ball in the end of the turbine shaft damaged.
3. Internal converter damaged.

Drives in Neutral

1. Forward clutch burned or not releasing.
2. Manual linkage or manual valve incorrectly set, or disconnected internal linkage.
3. Case interconnected passage.

No Park or Will Not Hold in Park

1. Actuator rod assembly bent or damaged.
2. Actuator rod spring, binding or improper crimp.
3. Parking lock pawl return spring, damaged or not assembled properly.
4. Actuator rod is not attached to inside detent lever.
5. Parking brake bracket damaged or bolts not torqued.
6. Inside detent lever nut, not torqued.
7. Detent roller improperly installed or damaged.
8. Parking lock pawl, binding or damaged.
9. Parking lock pawl interference with the lo reverse piston.

NOTE: Be careful to always check oil level, T.V. cable and oil pressure following each trouble shooting procedure.

MAINTENANCE

Adding Oil

To check fluid, drive vehicle for at least 15 minutes to bring fluid to operating temperature (190°–200° F.). With vehicle on a level surface and engine idling in Park and parking brake applied, the level on the dipstick should be at the Full Hot mark. To bring the fluid level from the ADD mark to the FULL mark requires one pint of fluid. If vehicle cannot be driven sufficiently to bring fluid to operating temperature, the level on the dipstick should be between the two dimples on the dipstick with fluid temperature at 65°–85° F.

If additional fluid is required, use only Dexron II automatic transmission fluid.

NOTE: An early change to a darker color from the usual red color and or a strong odor that is usually associated with overheated fluid is normal and should not be considered as a positive sign of required maintenance or unit failure.

CAUTION: When adding fluid, do not over fill, as foaming and loss of fluid through the vent may occur as the fluid heats up. Also, if fluid level is too low, complete loss of drive may occur especially when cold, which can cause transmission failure.

Every 100,000 miles, the oil should be drained, the oil pan removed, the screen cleaned and fresh fluid added. For vehicles subjected to more severe use such as heavy city traffic especially in hot weather, prolonged periods of idling or towing, this maintenance should be performed every 15,000 miles.

Changing Oil

1. Raise vehicle and position drain pan under transmission pan.
2. Loosen rear pan attaching bolts approximately four turns.
3. Carefully pry transmission pan loose with a screwdriver and allow fluid to drain.
4. Remove pan attaching bolts, pan and pan gasket.
5. Drain remaining fluid from pan, then clean pan with solvent and dry with compressed air.
6. Remove transmission screen. Remove O-ring seal from intake pipe or case bore.
7. Thoroughly clean screen assembly with solvent and dry with compressed air.
8. Install O-ring on intake pipe, then install screen retainer.
9. Install gasket on pan, then install pan and torque attaching bolts to 12 ft. lbs.
10. Lower vehicle and add approximately 5 qts. of Dexron II type transmission fluid through filler tube.
11. Start engine and operate at idle speed, then move selector lever through each range.
12. Place transmission in Park position and check fluid level.

TRANSAXLES

CHRYSLER A-460 & A-465 FOUR & FIVE SPEED MANUAL SHIFT TRANSAXLES

Description

Gear reduction, ratio selection and differential functions are combined into one unit, Figs. 1 and 2. The A-460 four speed transaxle case and the A-465 five speed transaxle case are both a one piece aluminum die casting utilizing fully synchronized forward gears.

The synchronizers on both transaxles use a "winged strut" design to prevent the struts from popping out of position.

When working on these transaxles, it is advisable to wear safety goggles.

Disassembly

1. Remove differential cover bolts and differential cover.
2. Remove differential bearing retainer bolts, then rotate differential bearing retainer using tool L-4435, or equivalent. Remove retainer.
3. Remove extension housing bolts, then remove differential assembly and extension housing.
4. Remove selector shaft housing and, on model A-460, rear end cover.
5. On 5-speed model A-465, proceed as follows:
   a. Remove 5th speed shifter pin, detent ball and spring.
   b. Remove fill plug and rear end cover.
   c. Remove 5th speed synchronizer strut retainer plate snap ring, then the retainer plate.
   d. Remove 5th speed synchronizer assembly, shift fork with shift rail, and the intermediate shaft 5th speed gear, Fig. 3.
   e. Remove input shaft 5th speed gear snap ring and the gear.
   f. Remove bearing support plate bolts and the support plate, Fig. 4.
6. On model A-460, remove large snap ring from intermediate shaft rear bearing.
7. Remove bearing retainer plate by tapping with plastic hammer, Fig. 5.
8. On model A-460, remove 3–4 shift fork rail. On all models, remove reverse idler gear shaft and reverse idler gear, Fig. 6.
9. Remove intermediate gear shaft and input gear shaft assemblies.
10. Remove shift forks and pads from intermediate gear shaft assembly.
11. Remove clutch release bearing shaft E clip, then remove clutch release shaft, Fig. 7.
12. Remove input shaft retainer and seal.
13. Remove reverse shift lever "E" clip and reverse shift lever.

Sub Assembly Service

Transaxle Case

1. Remove input shaft front bearing cup using tools C-4171, C-4656 and suitable press.
2. Remove intermediate shaft bearing strap, then remove intermediate shaft bearing using tool C-4660, or equivalent.
3. Install intermediate shaft bearing using tools C-4657, C-4171 and suitable press.
4. Install input shaft front bearing cup using tools C-4171, C-4655, or equivalents and suitable press.

Intermediate Shaft Assembly

NOTE: The 1–2 and 3–4 shift forks and synchronizer stop rings are interchangeable. If original parts are to be reused, install in original positions.

---

1. Remove intermediate shaft rear bearing snap ring, Fig. 8.
2. Remove intermediate shaft rear bearing, using suitable puller.
3. Remove 3–4 synchronizer hub snap ring, then, 3–4 synchronizer hub and third speed gear, using suitable puller.
4. Remove retaining ring and split thrust washer.
5. Remove second speed gear and stop ring, then, 1–2 synchronizer hub snap ring.
6. Remove first speed gear, stop ring and 1–2 synchronizer assembly.
7. Remove first speed gear thrust washer, Fig. 9.
8. Install first speed gear thrust washer, first speed gear, stop ring and 1–2 synchronizer assembly.
9. Install 1–2 synchronizer hub snap ring, second speed gear stop ring and second speed gear.
10. Install split thrust washer and retaining ring.
11. Install 3–4 synchronizer hub, using suitable press, and install third speed gear.
12. Install 3–4 synchronizer hub snap ring and intermediate shaft rear bearing, using tool C-4672, or equivalent and suitable press.
13. Install intermediate shaft rear bearing snap ring.

NOTE: Check to ensure that all gears rotate freely and have a minimum of .003 inch (.076 mm) end play.

Selector Shaft Housing

1. Remove selector shaft oil seal.
2. Remove selector shaft E clips, Figs. 10 and 11, then remove selector shaft. Disassemble remaining components of housing shown in Figs. 10 and 11.
3. Assemble selector shaft housing components in reverse order of removal, then install selector shaft.
4. Install selector shaft E clips, then install selector shaft oil seal, using tool C-4662 or equivalent.

Synchronizer Assemblies

Refer to Figs. 12, 13 and 14 for servicing of synchronizer assemblies.

Differential Bearing Retainer Fig. 15

1. Remove differential bearing retainer oil seal.
2. Remove differential bearing retainer cup

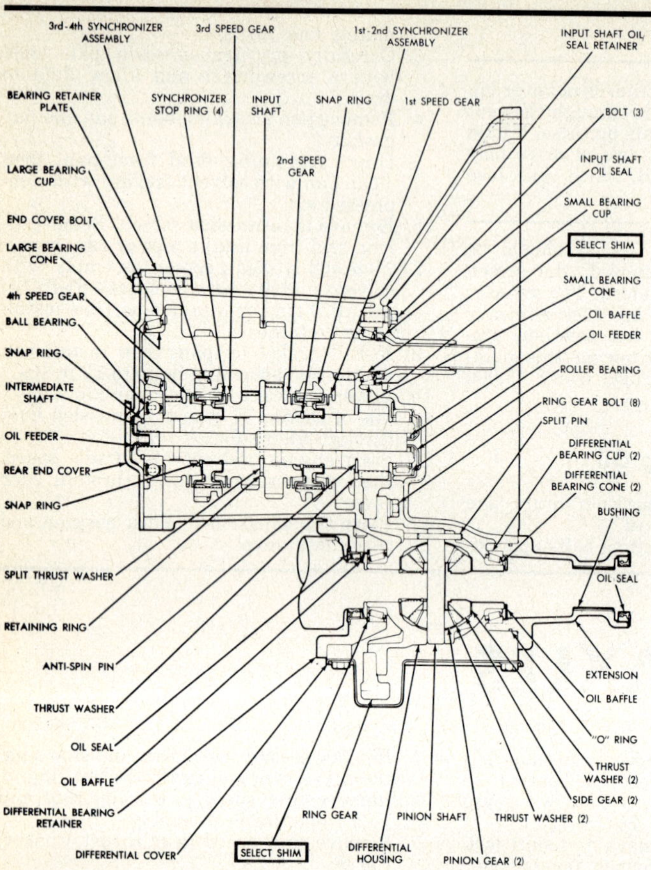

Fig. 1 Sectional view of Chrysler A-460 4-speed manual transaxle

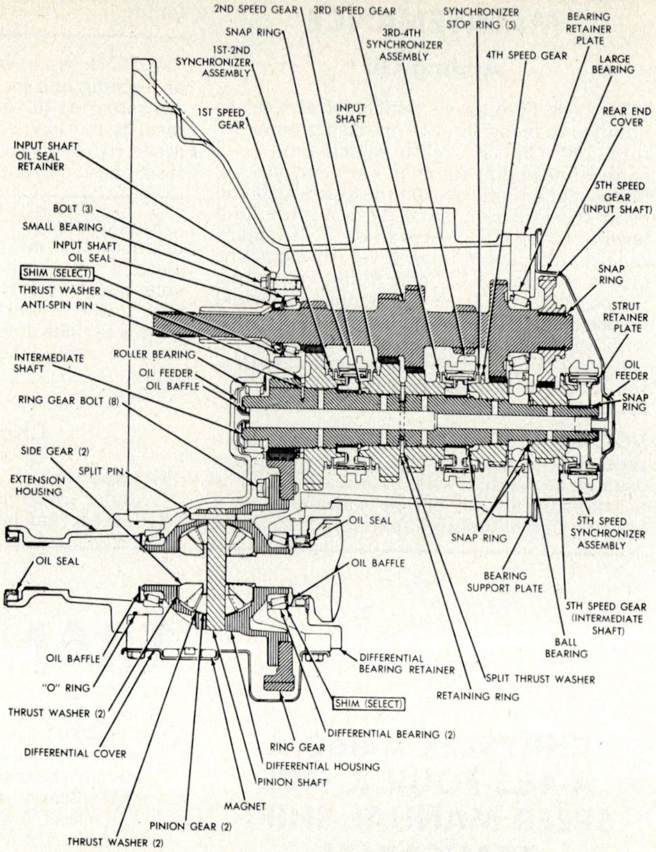

Fig. 2 Sectional view of Chrysler A-465 5-speed manual transaxle

and oil baffle using tool L-4518 or equivalent.

3. Install oil baffle, then install shim and bearing retainer cup, using tools C-4171 and L-4520, or equivalents. Shim is a select fit. To determine required shim thickness, refer to "Bearing Adjustments."
4. Install bearing retainer oil seal.

Extension Housing, Fig. 16

1. Remove extension housing oil seal, then remove extension housing and oil baffle, using tool L-4518, or equivalent.
2. Install oil baffle, using tools C-4171 and L-4520, or equivalents and suitable press.
3. Install bearing cup using same tools and suitable press.

Input Shaft

NOTE: Input shaft bearing shim thickness must be checked if any of the following parts are replaced: transaxle case, input shaft seal retainer, bearing retainer plate, rear end cover, input shaft or input shaft bearings. Refer to "Bearing Adjustments" for procedure.

1. Remove input shaft rear bearing cone, using tool C-293 and adapter C-293-45, or equivalents.
2. Remove input shaft front bearing cone, using tool C-293, or equivalent.
3. Remove input shaft bearing rear bearing cup from bearing retainer plate, using tool C-4171 and L-4520 and suitable press.
4. Install input shaft rear bearing cone, using tool C-4652, or equivalent on A-460 model or L-4507 on A-465 model.
5. Install input shaft front bearing cone, using tool C-4662, or equivalent.
6. Install rear end cover on model A-460, or bearing support plate on model A-465 onto bearing retainer plate, then support plate on blocks of wood. Install input shaft rear bearing cup into bearing retainer plate using tools, L-4520, C-4171 and suitable press. Check to ensure that bearing cone bottoms on rear end cover.

Differential

NOTE: Differential bearing shim thickness must be checked if any of the following parts are replaced: transaxle case, differential bear-

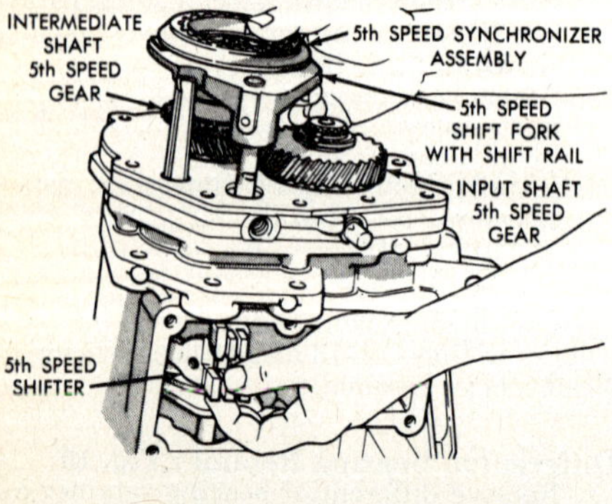

Fig. 3 5th speed synchronizer and shift fork removal. A-465 transaxle

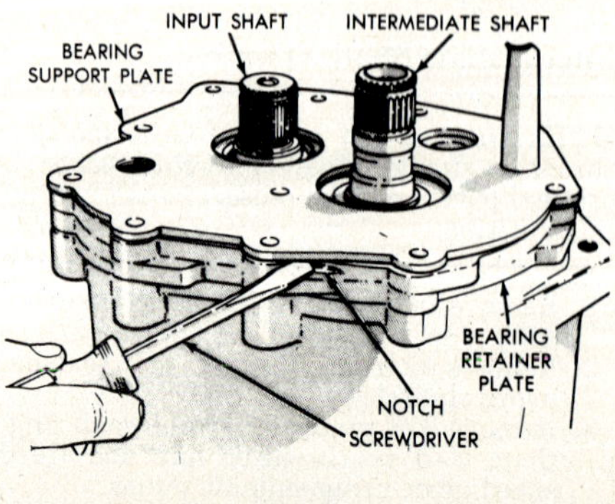

Fig. 4 Bearing support plate removal. A-465 transaxle

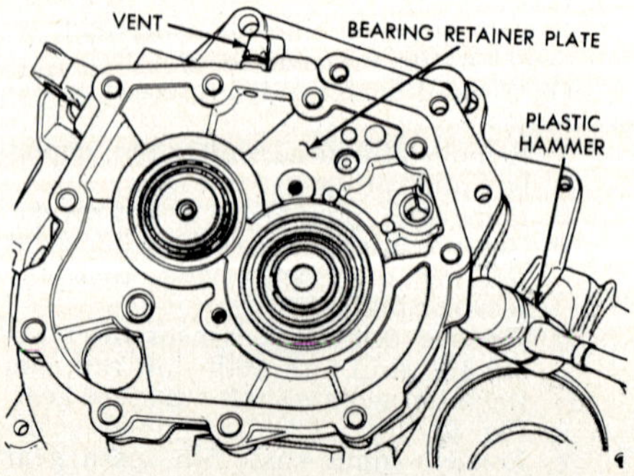

Fig. 5 Bearing retainer plate removal

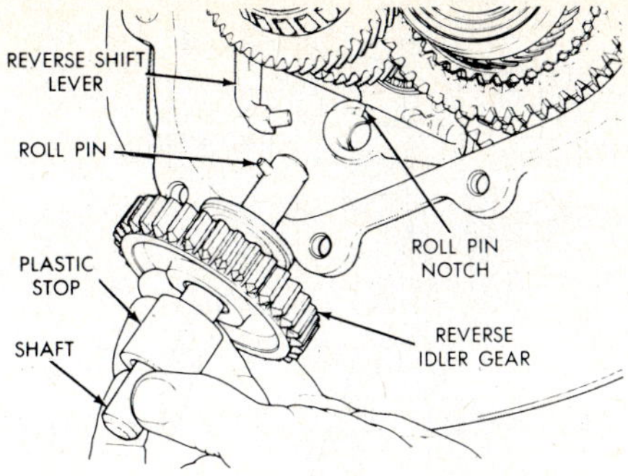

Fig. 6 Reverse idler gear, shaft and stop removal

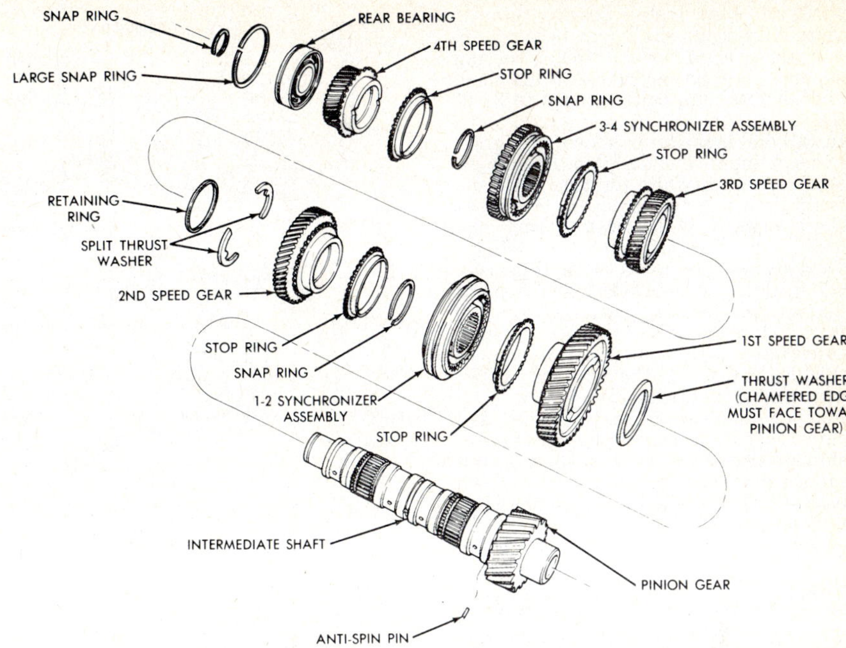

Fig. 8 Exploded view of intermediate shaft assembly

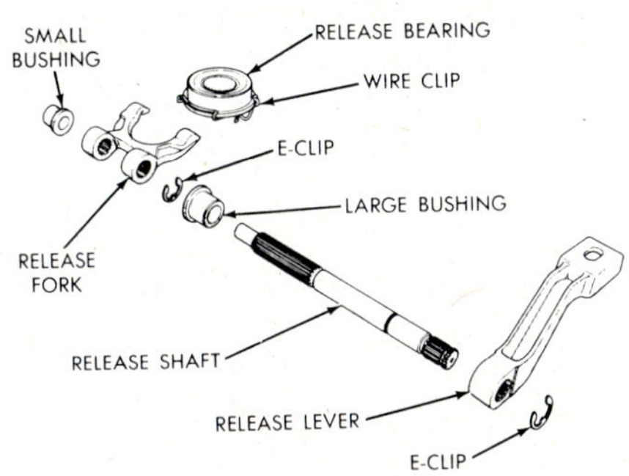

Fig. 7 Clutch release shaft components, disassembled

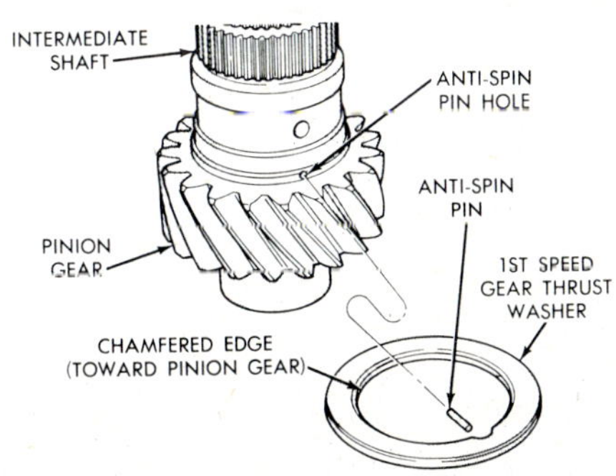

Fig. 9 First speed gear thrust washer removal

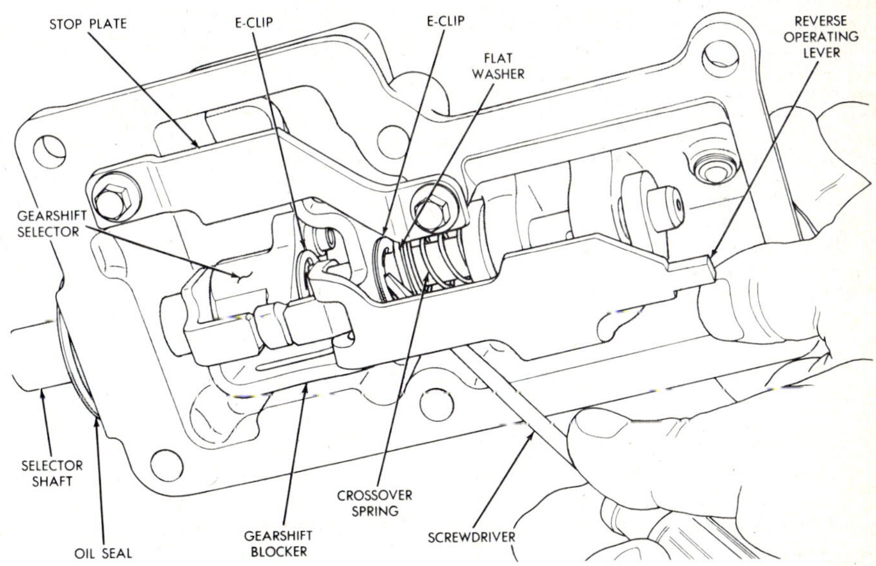

Fig. 10 Selector shaft disassembly. A-460 transaxle

ing retainer, differential case or differential bearings. Refer to "Bearing Adjustment" for procedure.

1. Remove differential bearing cones, using tools L-4406-1 and L-4406-3, or equivalents, Fig. 17.
2. Remove ring gear from differential case.
3. Using suitable punch, remove pinion shaft split pin.
4. Remove pinion shaft, then remove pinion gears, side gears and thrust washers from differential case.
5. Reverse procedure to assemble. Use tools C-4171 and L-4410 and suitable press to install differential bearing cones.

Bearing Adjustments

Input Shaft Bearing Endplay

NOTE: If input shaft bearing shim was reinstalled, press input shaft front bearing cup forward, using tools L-4656 and C-4171 or equivalents, then press bearing cup back into case using tools L-4655 and C-4171. This is necessary to properly position bearing cup and must be performed prior to endplay adjustment.

1. Select a shim that will allow .001–.010 inch endplay. This shim size may be determined by measuring original shim in input shaft seal retainer and selecting a shim .010 inch thinner than original.
2. Install shim selected onto bearing cap, then install input shaft seal retainer.

NOTE: Tighten input shaft seal retainer bolts alternately and evenly until input shaft seal retainer is bottomed against case. Torque bolts to 21 ft. lbs.

3. Lubricate input shaft bearings with

transmission fluid, then install input shaft into case. Install bearing retainer plate with input shaft rear bearing cup and end cover on model A-460, or bearing support plate on model A-465 installed, and all retaining bolts torqued to 21 ft. lbs.
4. Install dial indicator as shown in Fig. 18, to check input shaft endplay. Push input shaft rearward while rotating back and forth, then zero dial indicator. Pull input shaft forward while rotating back and forth, and record dial indicator reading.
5. To determine required shim thickness, add thickness of shim selected in Step 1 to dial indicator reading, then subtract .002 inch. Select shims as required to obtain shim that is within .0016 inch of required thickness. Refer to chart in Fig. 19 for available shim thickness.
6. Remove input shaft seal retainer and shim selected in step 1. Install shim or shims selected in step 5, then install input shaft seal retainer with a 1/16 inch bead of R.T.V. sealant and torque retaining bolts to 21 ft. lbs.
7. Using tool L-4508, or equivalent and an inch pound torque wrench, check input shaft rotating torque. Torque should be one to five inch lbs. for new bearings or one inch lb. for used bearings. If rotating torque is greater than specified, install .0016 inch thinner shim. If rotating torque is lower than specified, install .0016 inch thicker shim.

NOTE: Refer to note at beginning of this procedure and perform steps described whenever a thinner shim is installed. This will insure that the input shaft front bearing cup is installed the proper distance into the case.

Differential Bearing Preload

1. Remove bearing cup and shim from differential bearing retainer.
2. Select a shim that will allow .001–.010 inch endplay. This shim size may be determined by measuring thickness of original shim in differential bearing retainer and selecting a shim .015 inch thinner than original.
3. Install shim into differential bearing retainer, then install bearing cup.

NOTE: It is not necessary to install oil baffle when performing this adjustment.

4. Lubricate differential bearings with transmission fluid and install differential assembly into transaxle case. Install extension housing and differential bearing retainer and torque all bolts to 21 ft. lbs.
5. Position transaxle onto suitable work bench with bellhousing facing downward. Install dial indicator as shown in Fig. 20.
6. Press downward on ring gear while rotating ring gear back and forth to seat bearings. Zero dial indicator, then push upward on ring gear while rotating ring gear back and forth. Note dial indicator reading.
7. To determine required shim thickness, add thickness of shim selected in step 2 to dial indicator reading, then add .010 inch. Select shims as required to obtain shim thickness that is within .002 inch of required thickness. Refer to chart in Fig. 21 for available shim thicknesses.
8. Remove differential bearing retainer, bearing cup and shim. Install oil baffle and selected shim or shims, then install bearing cup into differential bearing retainer.
9. Install differential bearing retainer with a 1/16 inch bead of R.T.V. sealant, then the extension housing and torque all bolts to 21 ft. lbs. Apply suitable sealer to housing prior to installing.
10. Using tool L-4436, or equivalent and an inch pound torque wrench, check rotating torque of differential assembly. Rotating torque should be 9–14 inch lbs. for new bearings, or 6 inch lbs. for used bearings. If rotating torque is greater than specified, install a shim .002 inch thinner. If rotating torque is lower than specified, install a shim .002 inch thicker.

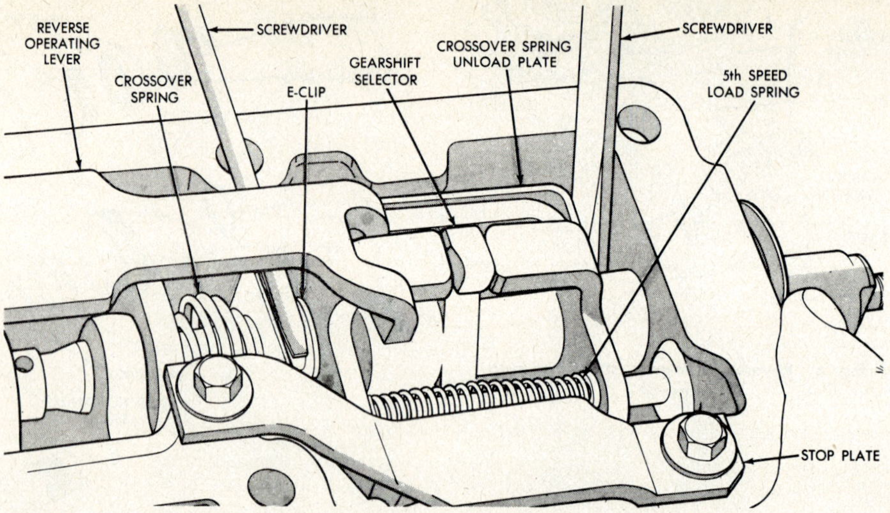

Fig. 11 Selector shaft disassembly. A-465 transaxle

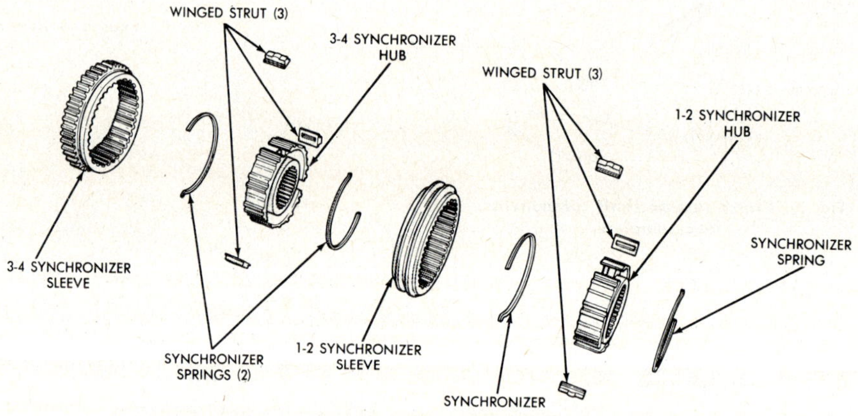

Fig. 12 Synchronizer assembly. Exploded view

Assembly

1. Install reverse shift lever and "E" clip.
2. Install input shaft oil seal using tool C-4674, or equivalent.
3. Install input shaft retainer and clutch release shaft. Install clutch release shaft "E" clip.
4. Install shift forks onto intermediate shaft assembly, then install intermediate shaft assembly and input shaft assembly into transaxle case.
5. Install reverse idler gear and shaft.
6. On A-460 models, install 3–4 shift fork rail. On all models, install bearing retainer plate.
7. On A-465 models, proceed as follows:
   a. Install bearing support plate, then the input shaft 5th speed gear with snap ring.
   b. Install intermediate shaft 5th speed gear, 5th speed synchronizer assembly and shift fork with shift rail.
   c. Install 5th speed synchronizer strut retainer plate with snap ring, then install rear end cover and fill plug.
   d. Install 5th speed detent ball and spring and the 5th speed shifter pin.
8. On A-460 models, install intermediate shaft rear bearing large snap ring.

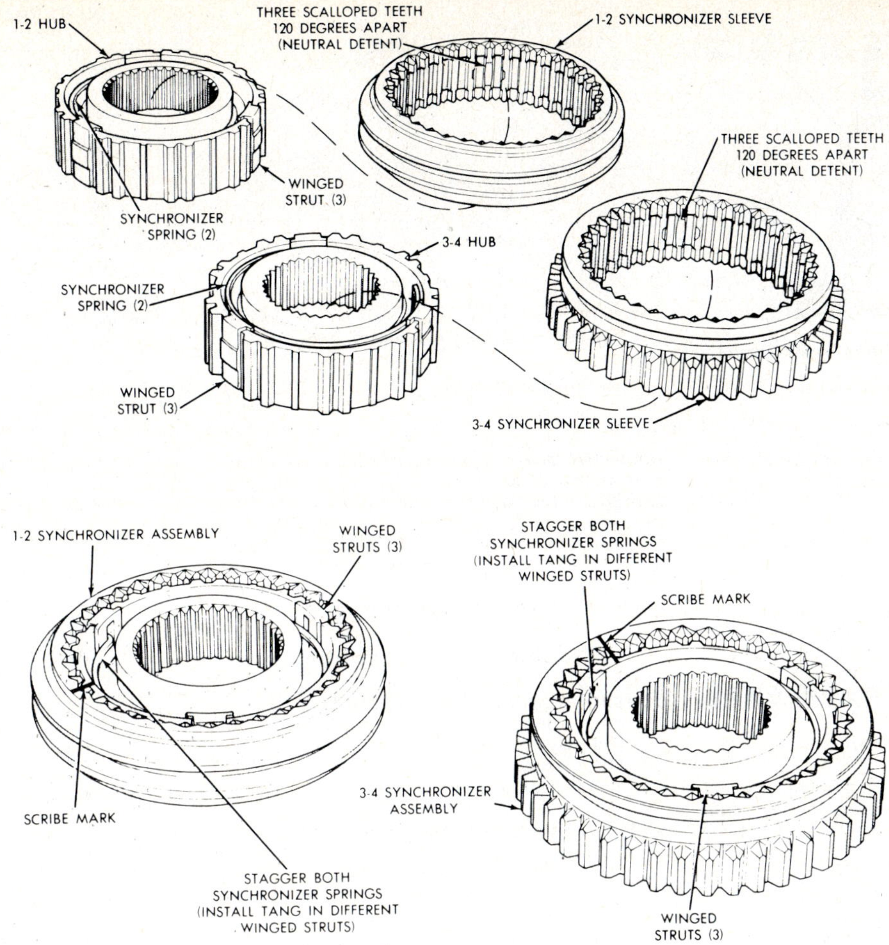

Fig. 13 Synchronizer assembly. Disassembled

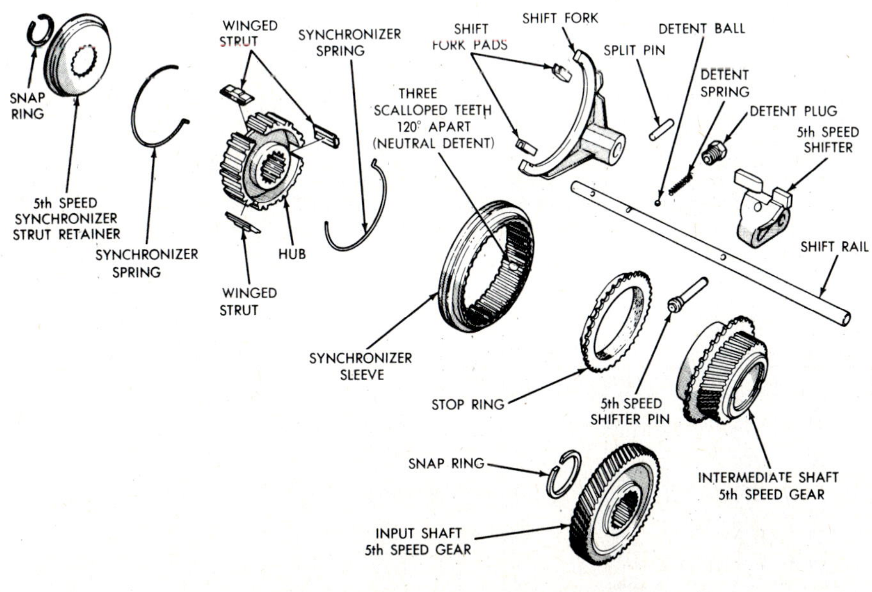

Fig. 14 5th speed synchronizer assembly. A-465 transaxle

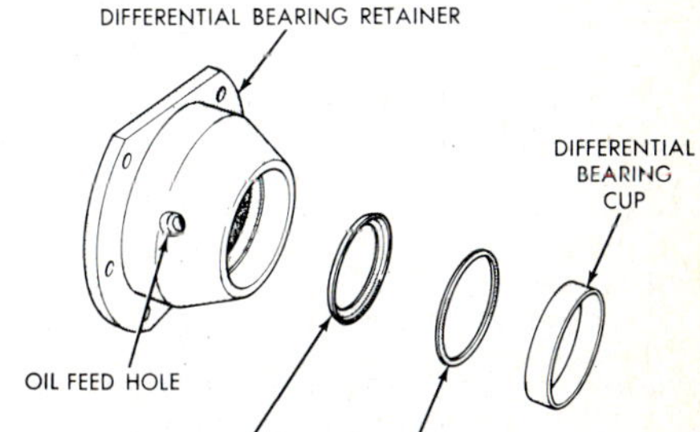

Fig. 15 Differential bearing retainer. Disassembled

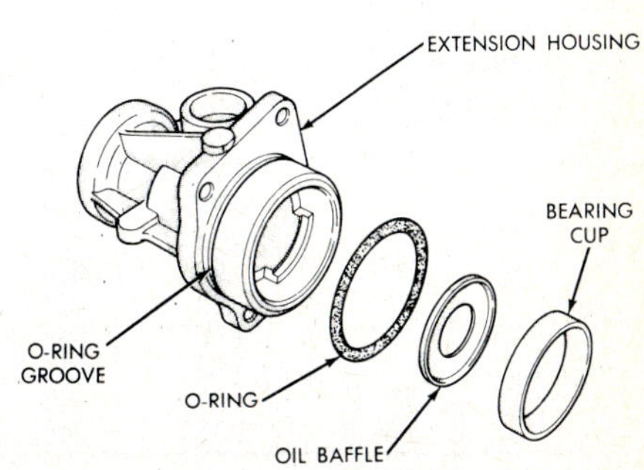

Fig. 16 Extension housing. Disassembled

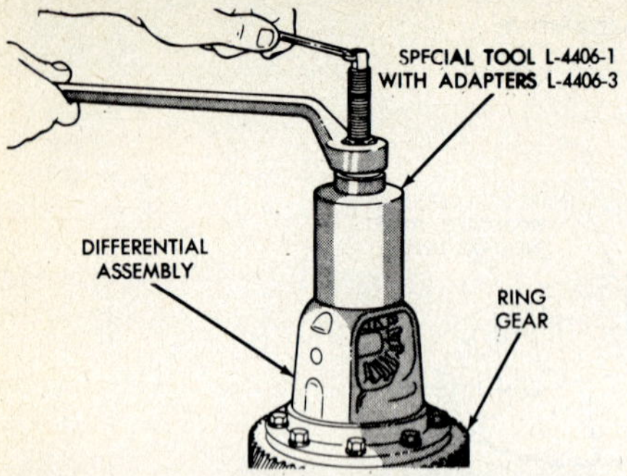

Fig. 17 Differential bearing cone removal

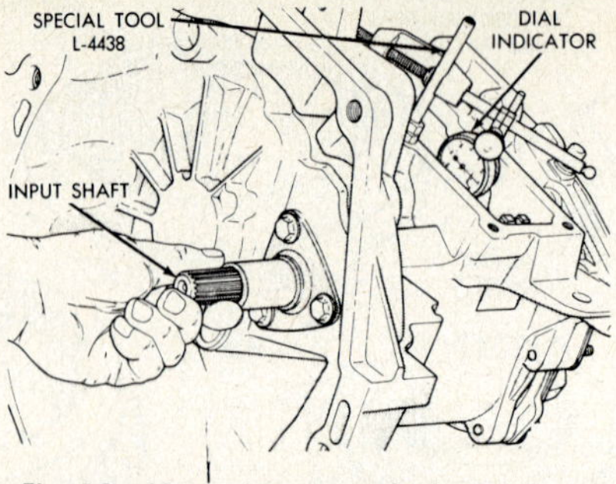

Fig. 18 Measuring input shaft endplay

INPUT SHAFT SHIM CHART

mm	mm	inch
.62		.024
.66		.026
.70		.028
.74		.029
.78		.031
.82		.032
.86		.034
.90		.035
.94		.037
.98		.039
1.02		.040
1.06		.042
1.10		.043
1.14		.045
1.18		.046
1.22		.048
1.26		.050
1.30		.051
1.34		.053
1.36	(.66 + .70)	.054
1.40	(.66 + .74)	.055
1.44	(.70 + .74)	.057
1.48	(.70 + .78)	.059
1.52	(.74 + .78)	.060
1.56	(.74 + .82)	.061
1.60	(.78 + .82)	.063
1.64	(.78 + .86)	.065
1.68	(.82 + .86)	.066
1.72	(.82 + .90)	.068
1.76	(.86 + .90)	.069

Fig. 19 Input shaft bearing selective shim chart

9. On A-460 models, install rear end cover. On all models, install selector shaft housing, differential assembly and extension housing.
10. Install differential bearing retainer and cover.

CHRYSLER A-413 TORQUEFLITE AUTOMATIC TRANSAXLE

Description

These transaxles combine a torque converter, automatic 3 speed transmission, final drive gearing and differential combined into one unit. The torque converter, transaxle and differential assemblies are housed in an integral aluminum die cast housing, Fig. 22.

NOTE: On 1982 models, the differential oil sump is separate from the transaxle pump. Be sure differential oil lever is 1/8 to 3/8 inch below the oil filler hole on the differential cover. On 1983 models, the differential oil sump is integral with the transaxle sump, and separate filling of the differential is not necessary.

The torque converter is connected to the crankshaft through a flexible drive plate. Converter cooling is accomplished by an oil-to-water type cooler, located in the radiator side tank. The torque converter cannot be disassembled.

The transaxle consists of two multiple disc clutches, an overrunning clutch, two servos, a hydraulic accumulator, two bands and two planetary gear assemblies to provide three forward and one reverse gear. The sun gear is connected to the front clutch retainer. The hydraulic system consists of an oil pump, and a single valve body which contains all of the valves except the governor valves. Output torque from the main drive gears is transferred through helical gears to the transfer shaft. An integral ring gear on the transfer shaft drives the differential ring gear.

Trouble Shooting Guide

Harsh Engagement From N to D or R

1. High idle speed.
2. Defective or leaking valve body.
3. High hydraulic pressure.
4. Worn or damaged rear clutch.

Delayed Engagement From N to D or R

1. Low hydraulic pressure.
2. Defective or leaking valve body.
3. Low-reverse servo, band or linkage malfunction.
4. Low fluid level.
5. Incorrect gearshift linkage adjustment.
6. Clogged transmission oil filter.
7. Faulty oil pump.
8. Worn or damaged input shaft seal rings.
9. Aerated fluid.
10. Low idle speed.
11. Worn or damaged reaction shaft support seal rings.
12. Worn or defective front clutch.
13. Worn or defective rear clutch.

Runaway Upshifts

1. Low hydraulic pressure.
2. Defective or leaking valve body.
3. Low fluid level.
4. Clogged transmission oil filter.
5. Aerated fluid.
6. Incorrect throttle linkage adjustment.
7. Worn or damaged reaction shaft support seal rings.
8. Kickdown servo, band or linkage malfunction.
9. Worn or faulty front clutch.

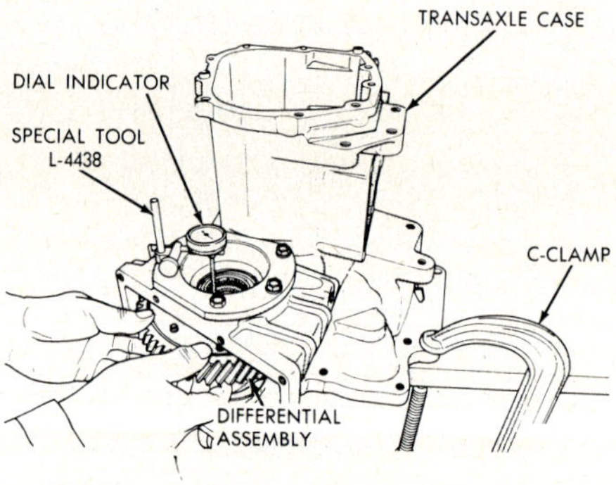

Fig. 20 Measuring differential bearing endplay

No Upshift

1. Low hydraulic pressure.
2. Defective or leaking valve body.
3. Low fluid level.

Required Shim Combination	Total Thickness		Required Shim Combination	Total Thickness		Required Shim Combination	Total Thickness	
mm	mm	Inch	mm	mm	Inch	mm	mm	Inch
.50	.50	.020	.50 + .70	1.20	.047	1.00 + .70	1.70	.067
.75	.75	.030	.50 + .75	1.25	.049	1.00 + .75	1.75	.069
.80	.80	.032	.50 + .80	1.30	.051	1.00 + .80	1.80	.071
.85	.85	.034	.50 + .85	1.35	.053	1.00 + .85	1.85	.073
.90	.90	.035	.50 + .90	1.40	.055	1.00 + .90	1.90	.075
.95	.95	.037	.50 + .95	1.45	.057	1.00 + .95	1.95	.077
1.00	1.00	.039	.50 + 1.00	1.50	.059	1.00 + 1.00	2.00	.079
1.05	1.05	.041	.50 + 1.05	1.55	.061	1.00 + 1.05	2.05	.081
.50 + .60	1.10	.043	1.00 + .60	1.60	.063	1.05 + 1.05	2.10	.083
.50 + .65	1.15	.045	1.00 + .65	1.65	.065			

Fig. 21 Differential bearing endplay selective shim chart

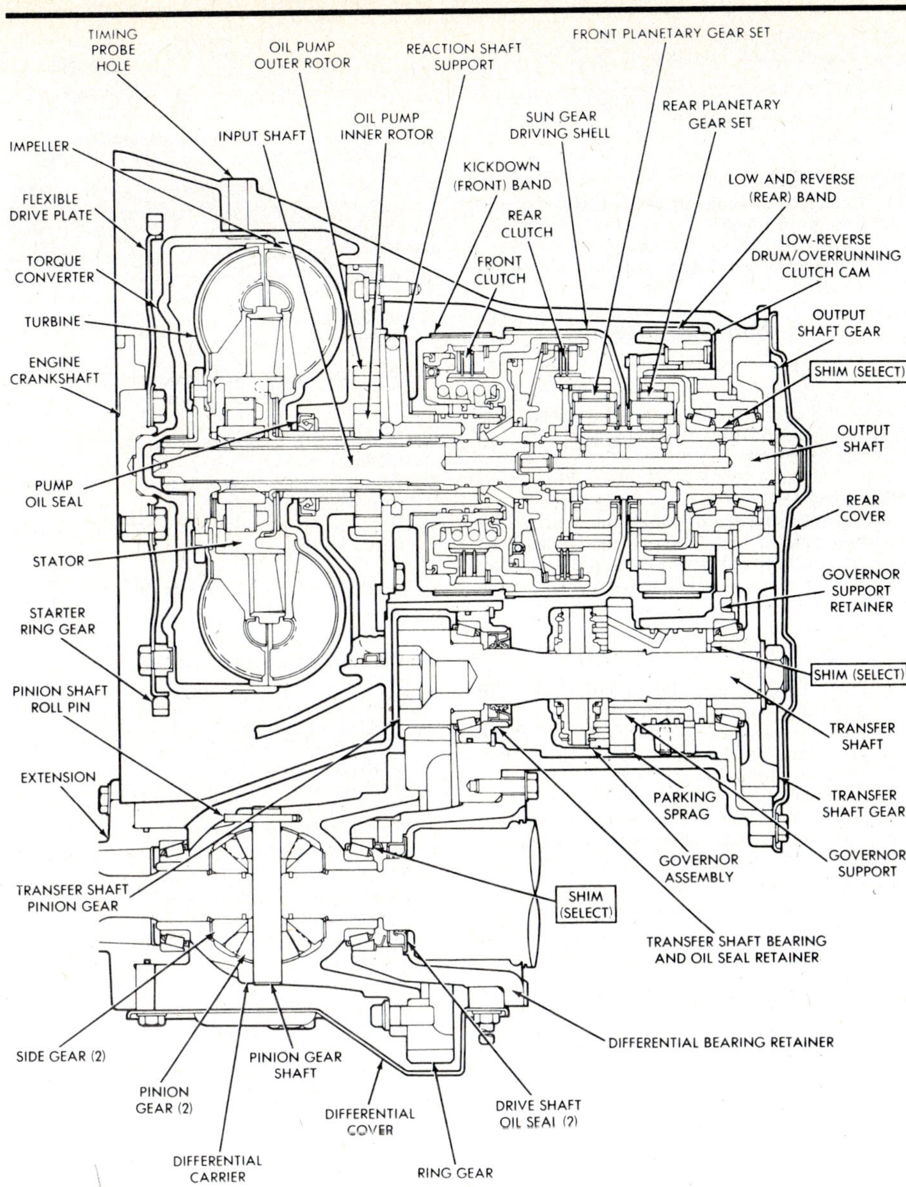

Fig. 22 Sectional view of A-413 automatic transaxle. Typical

4. Incorrect gearshift linkage adjustment.
5. Incorrect throttle linkage adjustment.
6. Worn or damaged governor support seal rings.
7. Faulty governor.
8. Kickdown servo, band or linkage malfunction.
9. Worn or faulty front clutch.

3-2 Kickdown Runaway

1. Low hydraulic pressure.
2. Defective or leaking valve body.
3. Low fluid level.
4. Aerated fluid.
5. Incorrect throttle linkage adjustment.
6. Kickdown band adjustment.
7. Worn or damaged reaction shaft support seal rings.
8. Kickdown servo, band or linkage malfunction.
9. Worn or faulty front clutch.

No Kickdown or Normal Downshift

1. Defective or leaking valve body.
2. Incorrect throttle linkage adjustment.
3. Faulty governor.
4. Kickdown servo, band or linkage malfunction.

Erratic Shifts

1. Low hydraulic pressure.
2. Defective or leaking valve body.
3. Low fluid level.
4. Incorrect gearshift linkage adjustment.
5. Clogged transmission oil filter.
6. Faulty oil pump.
7. Aerated fluid.
8. Incorrect throttle linkage adjustment.
9. Worn or damaged governor support seal rings.
10. Worn or damaged reaction shaft support seal rings.
11. Faulty governor.
12. Kickdown servo, band or linkage malfunction.
13. Worn or faulty front clutch.

Slips in 1, 2 or D

1. Low hydraulic pressure.
2. Defective or leaking valve body.
3. Low fluid level.
4. Incorrect gearshift linkage adjustment.
5. Clogged transmission oil filter.
6. Faulty oil pump.
7. Worn or damaged input shaft seal rings.
8. Aerated fluid.
9. Incorrect throttle linkage adjustment.
10. Overrunning clutch not holding.
11. Worn or faulty rear clutch.
12. Overrunning clutch worn damaged or seized.

Slips in R Only

1. Low hydraulic pressure.
2. Low-reverse band adjustment.
3. Defective or leaking valve body.
4. Low-reverse servo, band or linkage malfunction.
5. Low fluid level.
6. Incorrect gearshift linkage adjustment.
7. Faulty oil pump.
8. Aerated fluid.
9. Worn or damaged reaction shaft seal rings.
10. Worn or faulty front clutch.

Slips in All Ranges

1. Low hydraulic pressure.
2. Defective or leaking valve body.
3. Low fluid level.
4. Clogged transmission oil filter.
5. Faulty oil pump.
6. Worn or damaged input shaft seal rings.
7. Aerated fluid.

No Drive in Any Range

1. Low hydraulic pressure.
2. Defective or leaking valve body.
3. Low fluid level.
4. Clogged transmission oil filter.
5. Faulty oil pump.
6. Planetary gear sets damaged or seized.

No Drive 1, 2 or D

1. Low hydraulic pressure.
2. Defective or leaking valve body.
3. Low fluid level.
4. Worn or damaged input shaft seal rings.
5. Overrunning clutch not holding.
6. Worn or faulty rear clutch.
7. Planetary gear sets damaged or seized.
8. Overrunning clutch worn, damaged or seized.

No Drive in R

1. Low hydraulic pressure.
2. Low-reverse band adjustment.
3. Defective or leaking valve body.
4. Low-reverse servo, band or linkage malfunction.
5. Incorrect gearshift linkage adjustment.
6. Worn or damaged reaction shaft support seal rings.
7. Worn or faulty front clutch.
8. Worn or faulty rear clutch.
9. Planetary gear sets damaged or seized.

Drive in Neutral

1. Defective or leaking valve body.
2. Incorrect gearshift linkage adjustment.
3. Insufficient clutch plate clearance.
4. Worn or faulty rear clutch.
5. Rear clutch dragging.

Drags or Locks

1. Low-reverse band adjustment.

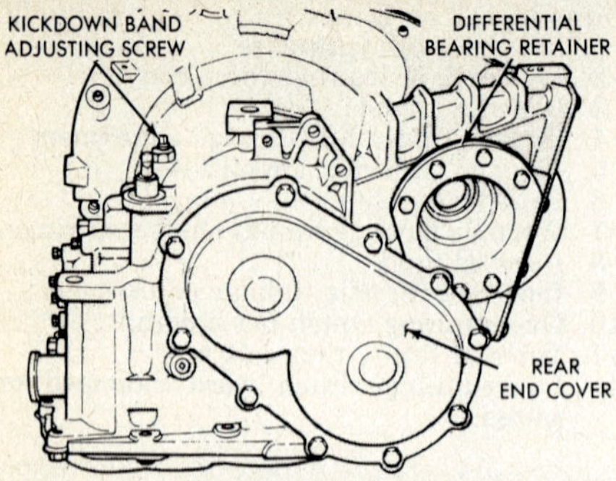

Fig. 23 Kickdown band adjusting screw location

2. Kickdown band adjustment.
3. Planetary gear sets damaged or seized.
4. Overrunning clutch worn, damaged or seized.

Hard to Fill (Oil Blows Out Filler Tube)
1. Clogged transmission oil filter.
2. Aerated fluid.
3. High fluid level.
4. Breather clogged

Transmission Overheats
1. Stuck switch valve.
2. High idle speed.
3. Low hydraulic pressure.
4. Low fluid level.
5. Incorrect gearshift adjustment.
6. Faulty oil pump.
7. Kickdown band adjustment too tight.
8. Faulty cooling system.
9. Insufficient clutch plate clearance.

Harsh Upshifts
1. Low hydraulic pressure.
2. Incorrect throttle linkage adjustment.
3. Kickdown band adjustment.
4. High hydraulic pressure.

Delayed Upshift
1. Incorrect throttle linkage adjustment.
2. Kickdown band adjustment.
3. Worn or damaged governor support seal rings.
4. Worn or damaged reaction shaft support seal rings.
5. Faulty governor.
6. Kickdown servo, band or linkage malfunction.
7. Worn or faulty front clutch.

Grating, Scraping or Growling Noise
1. Low-reverse band out of adjustment.

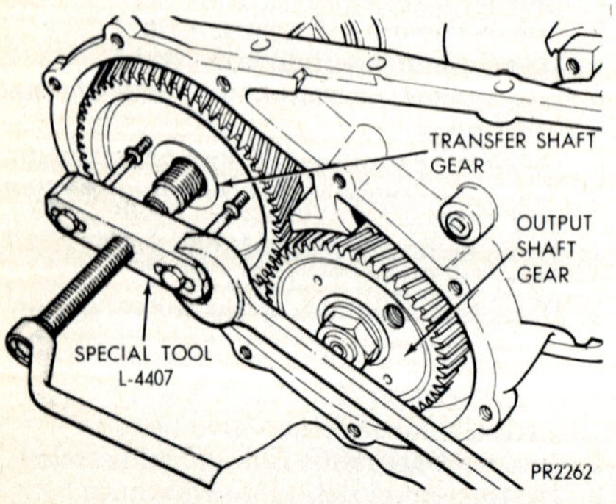

Fig. 25 Removing transfer shaft gear

2. Kickdown band adjustment.
3. Output shaft bearing or bushing damaged.
4. Planetary gear sets damaged or seized.
5. Overrunning clutch worn, damaged or seized.

Buzzing Noise
1. Defective or leaking valve body.
2. Low fluid level.
3. Aerated fluid.
4. Overrunning clutch inner race damaged.

Maintenance

Adding Oil

To check fluid level, apply the parking brake and operate engine at idle speed with transmission in Neutral or Park position. Add fluid as necessary.

Changing Oil

Fluid and filter changes are not required for average vehicle use. Severe usage such as commerical type usage or prolonged operation in city traffic, requires that fluid be changed and bands adjusted every 15,000 miles.

Whenever factory fill fluid is changed, only fluid of the type labeled Dexron II should be used.

1. Raise vehicle and place a suitable drain pan under transmission oil pan.
2. Loosen transmission oil pan attaching bolts and allow fluid to drain, then remove oil pan.
3. Replace oil filter and adjust bands, if necessary, and install oil pan and gasket.
4. Add four quarts of approved automatic transmission fluid through the filler tube.
5. Start engine and allow to idle for at least two minutes on 1982 models, or at least six minutes on 1983 models, then with parking brake applied, move selector lever momentarily to each position. Place selector lever in Neutral or Park and check fluid level. Add fluid to bring level to Add mark.
6. Recheck fluid level after transmission has reached operating temperature. The level should be between Add and Full marks.

Bands, Adjust

Kickdown Band
1. Loosen lock nut and back off nut approximately five turns, Fig. 23.
2. Using tool No. C-3380-A and adapter C-3705, tighten band adjusting screw to 47 to 50 inch lbs. If adapter C-3705 is not used, tighten adjusting screw to 72 inch lbs.
3. Back off adjusting screw 2¾ turns. While holding adjusting screw in this position, torque lock nut to 35 ft. lbs.

Low Reverse Band, Adjust
1. Loosen locknut and back off nut approximately five turns.
2. Torque adjusting nut to 41 inch lbs.
3. Back off adjusting nut 3½ turns, and torque locknut to 20 ft. lbs. on 1982 models, or 10 ft. lbs. on 1983 models.

Valve Body, Replace

1. Loosen transmission oil pan attaching bolts and allow transmission to drain, then remove oil pan.

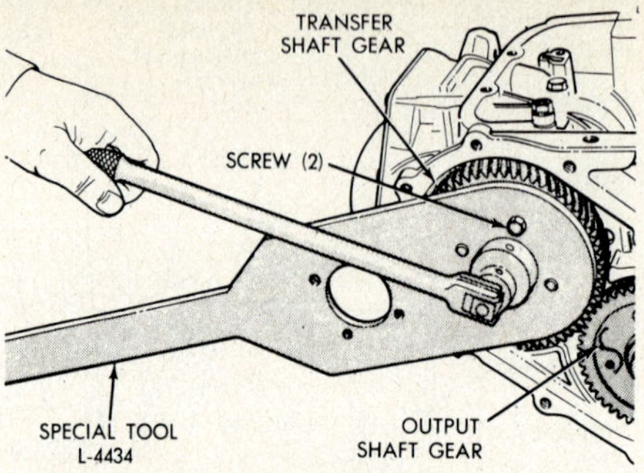

Fig. 24 Loosening transfer shaft gear retaining nut

2. Remove oil filter attaching screws and oil filter.
3. Using a screwdriver, remove E-clip, then remove parking rod.
4. Remove seven valve body attaching bolts, valve body and governor oil tubes.

Governor & Transfer Shaft Oil Seal, Replace

1. Remove rear cover attaching bolts and rear cover.
2. Using tool No. L-4434, remove transfer shaft gear retaining nut, Fig. 24.
3. Using tool No. L-4407, remove transfer shaft gear and shim, Fig. 25.
4. Remove governor support retainer, then low-reverse band anchor pin.
5. Remove governor assembly.
6. Remove transfer shaft retainer snap ring. Using tool No. L-4512 and a suitable puller, remove transfer shaft and retainer assembly.
7. Remove transfer shaft retainer from shaft.
8. Using a screwdriver, remove oil seal from transfer shaft retainer.
9. Using tools Nos. L-4520 and C-4171, press oil seal into shaft retainer, Fig. 26.
10. Reverse procedure to install. Torque transfer shaft gear retaining nut to 200 ft. lbs.

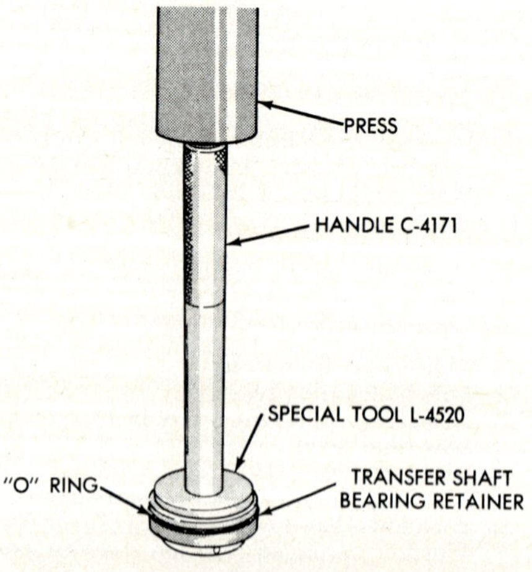

Fig. 26 Installing oil seal into shaft retainer

INDEX

INTERNATIONAL ADAPTIVE BRAKE SYSTEM

This system is designed to prevent any wheel from locking up during brake applications above a speed of about 5 mph. The system reduces skid potential of a locked wheel and still maintains brake pressure for maximum stopping effort. The end result is to improve directional control and steerability of the vehicle, and in many cases to reduce the distance required to bring the vehicle to a stop.

The major components of this system are shown in Fig. 1. These include a mechanically driven speed sensor at each rear wheel, an electronic control unit located on the dash panel left side in the passenger compartment and a pressure modulator in the left side of the engine compartment.

The speed sensors incorporate a stationary permanent magnet and coil in one case and a tone wheel, Fig. 2, which are attached to a mounting bracket. Part of the mounting bracket forms a spring to hold the tone wheel knurled shaft in contact with the rubber drive band pressed into the drive ring. The drive ring, located between the axle flange and drum, rotates with the rear wheel.

The pressure modulator, Fig. 3, incorporates a vacuum chamber, bypass tube, end plate, air valve, bypass valve and pressure modulator switch. Engine vacuum is applied to the modulator. The hydraulic circuit for the rear brakes is routed from the master cylinder, through the brake warning light switch to the pressure modulator and on to the rear brakes.

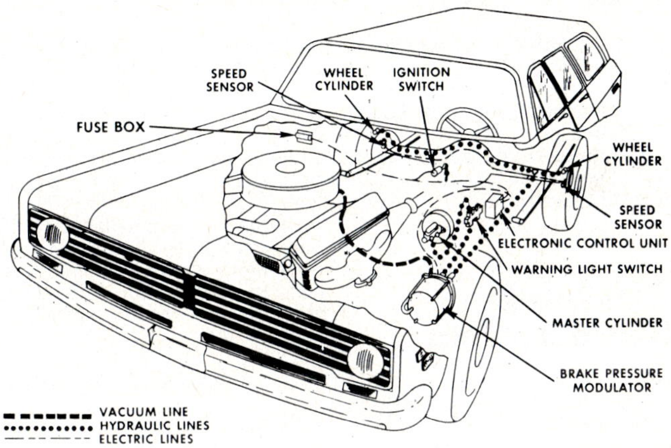

Fig. 1 International Harvester anti-skid brake system components

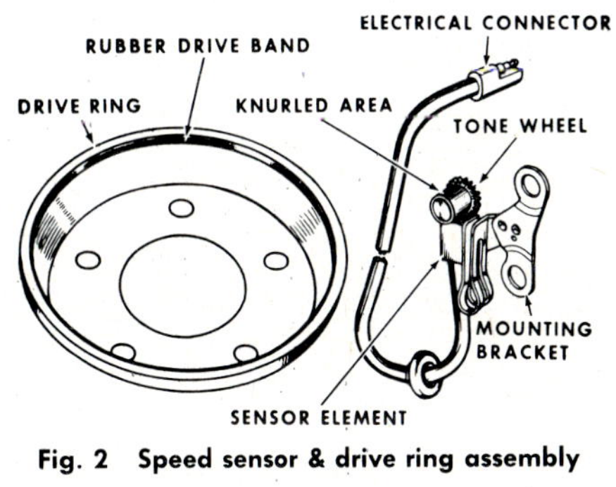

Fig. 2 Speed sensor & drive ring assembly

Operation

Engine Running—Vehicle Not In Motion

The wheel sensors do not generate any signals for transmission to the electronic control unit when the vehicle is not in motion. Hence, the electronic control unit sends no commands to the pressure modulator.

Engine Running—Vehicle In Motion

With the vehicle in motion, alternating current voltage is generated at each rear wheel sensor and sent to the electronic control unit. The electronic control unit processes the signals received from the rear wheel sensors to sample the speed of each rear wheel. If the brakes are not applied or of they are applied lightly, the electronic control unit does not send any commands to the pressure modulator.

When the brakes are applied with greater force, the electronic control unit, based on wheel sensor signals, determines the rate at which each rear wheel is decelerating. If the rate is great and might produce wheel slippage or lockup, the electronic control unit sends a command to the modulator that controls the braking for the rear wheel or wheels concerned.

Exercise Cycle

If the engine is started with the brake pedal depressed, the pressure modulator goes through an exercise cycle. When the ignition switch is turned from OFF to START the modulator cycles once. This cycle insures that the system is working properly. The cycle can be heard under some conditions, but should not be a cause for concern.

Warning System

This anti-skid system includes a secondary system to warn the driver of certain types of failures in the system. The warning system uses the brake warning light. The warning light will be on during the following conditions:

1. If the pressure modulator is activated in the absence of a brake light signal.
2. If the electronic control unit sends a sig-

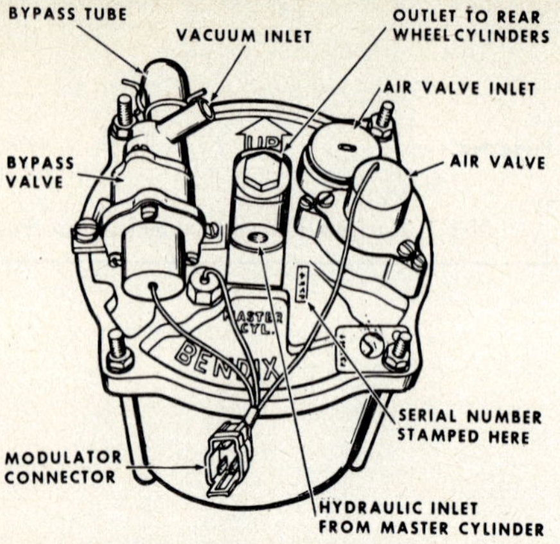

Fig. 3 Pressure modulator

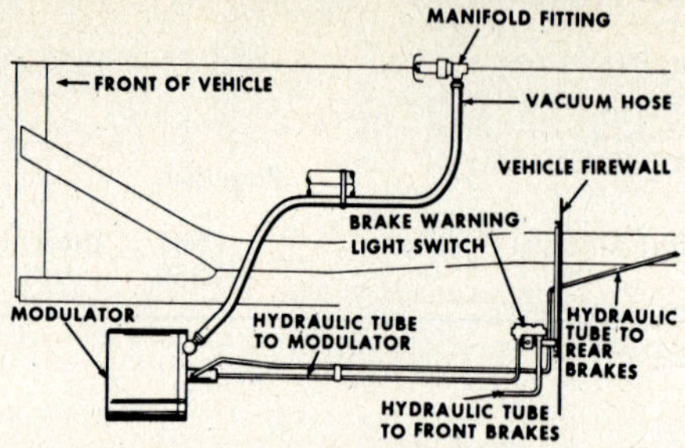

Fig. 5 Vacuum & hydraulic system routing

nal to open the air valve on a modulator in the absence of a signal to open the bypass valve.

3. If the electrical continuity of the air valve lead wire is broken.
4. If a blown fuse, loose connection or open circuit occurs between the ignition switch and the electronic control unit.

Disabling Anti-Skid System

If necessary parts or service is not available, the system should be disabled. The brakes will then function the same as those on trucks without this system. Proceed as follows:

Disconnect the electronic control unit (all three connectors), Fig. 4, and the vacuum supply hose at the manifold fitting, Fig. 5. The fitting and the hose to the modulator should be plugged to prevent entry of foreign material.

An additional step may be necessary to disable the system for one type of failure that can occur. If the modulator diaphragm plate sticks in the retracted position, hydraulic pressure is cut off from the rear brakes. To determine if this malfunction is present, have a helper depress the brake pedal and open a rear brake bleeder screw. If the fluid runs freely, the shut-off valve is open. If not, the valve is closed and the modulator must be bypassed. This can be accomplished as follows:

Disconnect both hydraulic lines from modulator, bend lines slightly and connect the two lines using an adapter and a union. This is done in conjunction with disconnecting the electrical and vacuum systems.

Trouble-Shooting

Brake Warning Lamp (Red) Test

1. Turn ignition switch to "Start" position, if warning lamp does not light when engine cranks, replace lamp bulb or repair warning circuit as necessary.
2. If warning lamp remains lit after bulb is replaced or circuit repaired, proceed to next step to determine if malfunction is in anti-skid system or basic brake system. If warning lamp is lit while engine cranks and goes out when engine starts, proceed to "Electrical Circuit Tests."
3. To determine if malfunction is in anti-skid system or basic brake system when warning lamp remains "On", disconnect electrical lead from brake warning switch. If warning lamp goes out, malfunction is in basic brake system and must be repaired. If warning lamp remains lit, proceed to "Electrical Circuit Tests".

Anti-Skid Brake Cycling Indicator Lamp (Green) Test

1. Disconnect cycling lamp connector, Fig. 4, and connect a jumper wire from the battery positive terminal to lamp side of connector. If lamp does not light, replace bulb and socket assembly.
2. If lamp lights in step 1, remove jumper wire and connect cycling lamp connector. Check anti-skid system by road testing vehicle on gravel or similar surface at a reasonable speed but at a minimum of 20 MPH. Apply sufficient brake pedal pressure to normally lock wheels and observe cycling lamp. If cycling lamp flickers and rear wheels appear to roll at low speed then lock, the anti-skid brake system is satisfactory. If not, proceed to "Electrical Circuit Tests".

Electrical Circuit Tests

Step 1: Checks at Electronic Control Unit

Make measurements listed in Fig. 6 at the Electronic Control Unit connectors, Fig. 7. The connectors should be disconnected except those measurements indicated with a double

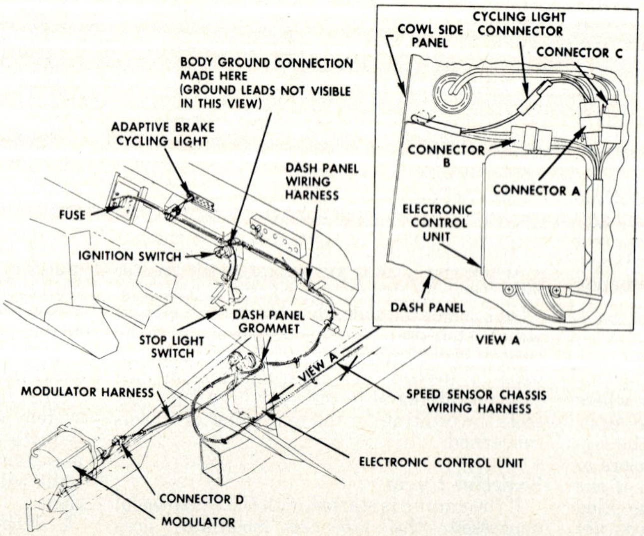

Fig. 4 Anti-skid brake system body & cab wiring schematic

asterisk (**). All readings must be obtained from the terminals on the side of the connector attached to the brake harness modulator cable assembly and speed sensor assembly. If any reading is not within specifications, refer to indicated step for additional measurements or repairs.

Step 2: Modulator Switch Circuit

NOTE: Make measurements in sequence of steps listed below to identify faulty components. When faulty component has been identified, reconnect all connectors except A, B and C at ECU, Fig. 4. Repeat measurement in Step 1 which produced incorrect reading. This repeat measurement is necessary to insure malfunction still exists prior to replacing any component.

a. Disconnect connector D, Fig. 4, and on modulator lead side of connector, measure resistance between Pin 3 and the ground. If zero ohms is indicated, replace modulator cable. If any reading above zero ohms is obtained, modulator or modulator switch is faulty and modulator must be replaced.

Step 3: Air Valve & Bypass Valve Circuits

Refer to NOTE under "Step 2; Modulator Switch Circuit" and proceed as follows:

a. With connector D, Fig. 4, disconnected, measure resistance between pin which had incorrect reading and the ground. If reading obtained is 3 to 5 ohms, replace modulator cable assembly. If reading obtained is not within specifications, replace modulator.

Step 4: Electronic Control Unit Ground Circuit

a. Disconnect body ground wire from under instrument panel, clean wire end eye and ground point on body where eye is secured and reconnect the ground wire.
b. Measure resistance between Pin 2 on brake harness side of connector B and the ground. If zero ohms is indicated, the ground connection was unsatisfactory and the malfunction has been corrected. If any reading above zero ohms is obtained, replace brake harness.

Step 5: Power Supply Circuit

Refer to NOTE under "Step 2: Modulator Switch Circuit" and proceed as follows:

a. Check for blown 10 amp fuse in-line between ignition switch and connector and replace if necessary. If fuse is satisfactory, ensure circuit is properly connected to ignition switch and battery voltage is present at ignition switch terminal.
b. If fuse and circuit connections are satisfactory and battery voltage is present at switch terminal but not at Pin 4, connector B with ignition switch "On", replace brake harness.

Step 6: Stop Light to Electronic Control Unit Circuit

Refer to NOTE under "Step 2: Modulator Switch Circuit" and proceed as follows:

a. Measure DC voltage between output side of stop light switch and the ground with the brakes applied. If battery voltage is obtained, replace brake harness. If reading obtained is below battery voltage, measure DC voltage between battery side of stop light switch and the ground. If battery voltage is obtained, adjust stop light switch position or replace switch as necessary. If reading is below battery voltage, check wiring to stop light switch and repair as necessary.

Step 7: Voltage Feedback

Refer to NOTE under "Step 2: Modulator Switch Circuit" and proceed as follows:

a. Check stop lamp operation and correct if necessary. Repeat measurement which produced incorrect reading. If reading is still above zero volts, check tail light-stop light assemblies for proper ground connections and repair as necessary.

Step 8: Electronic Control Unit Circuits

If reading obtained is not within specifications, the electronic control unit is faulty, requiring replacement.

Step 9: Warning Light Circuit

Refer to NOTE under "Step 2: Modulator Switch Circuit" and proceed as follows:

a. Separate connectors E and F, Fig. 8, then on the speed sensor side of the connectors, repeat measurement which produced incorrect reading.
b. If reading is as specified in Fig. 6, the speed sensor cable is defective and must be replaced. If reading is not within specifications, disconnect speed sensor connector in line with connector which had faulty reading. Measure resistance between the two terminals and between each terminal and the ground.
c. If all readings are correct, replace rear axle sensor harness. If any reading is not within specifications, the corresponding speed sensor is faulty, requiring replacement.

Modulator Test

a. With engine operating, disconnect connector D, Fig. 4, and on modulator side of connector, measure resistance between Pin 3 and the ground.
b. If reading obtained is above zero ohms, replace modulator.

Vacuum & Hydraulic Tests

Step 1: Vacuum Supply Check

a. Ensure all vacuum hose connections are proper, start engine and check for vacuum leaks. Repair as necessary.
b. Connect a vacuum gauge between modulator vacuum hose and modulator inlet. Gauge should indicate normal engine vacuum. If not, check for vacuum leak or tune engine as necessary.

Step 2: Air Valve & Bypass Valve Operation

a. With connector D, Fig. 4, disconnected, start engine and observe vacuum reading.

NOTE: Air valve and bypass valve should not be energized for more than two minutes with a five minute cooling-off period between since valve solenoids may be damaged by excessive heat.

b. Connect a jumper wire between battery positive terminal and Pin 1 on modulator side of connector. Vacuum reading should drop. If not, replace modulator.
c. Connect a second jumper wire from the battery positive terminal and Pin 2 on the connector, leaving the first jumper wire in place. The vacuum reading should increase to reading previously noted in step a. If not, replace modulator. If vacuum reading does increase, proceed to Step 3.

Step 3: Hydraulic Shut-Off Valve

a. Install a suitable hydraulic pressure gauge in modulator output and bleed air from system and gauge.
b. Start engine and apply brake pedal to obtain a 500-600 PSI reading.

NOTE: The air and bypass valves should not be energized for more than two minutes with a five minute cooling-off period between since the valve solenoids may be damaged by excessive heat.

c. With brake pedal still depressed, connect jumper wires between battery positive terminal and Pins 1 and 2 on modulator side of connector D. The hydraulic pressure should drop to zero and if not, replace modulator.
d. Hold brake pedal depressed and keep jumper wires connected for a two minute period. Hydraulic pressure should not rise during this period. If the pressure did rise, replace modulator.

Step 4:

If results of all three steps are normal, the electronic control unit is faulty, requiring replacement.

Hydraulic Output Test

a. Connect a suitable hydraulic pressure gauge in modulator output and bleed air from system and gauge.
b. Disconnect connector D, Fig. 4, and connect a jumper wire between battery positive terminal and Pin 2 on modulator side of connector D.

NOTE: Air and bypass valves should not be energized for more than two minutes with a five minute cooling-off period between since the valve solenoids may be damaged by excessive heat.

c. Start engine and depress brake pedal to obtain a pressure of 600 PSI. If hydraulic pressure does not rise when brake pedal is depressed, replace modulator.

False Cycling Test

Step 1: Speed Sensor Inspection

a. Remove rear wheel and drum and ensure sensor mounting bracket holds sensor knurled wheel against drive ring rubber drive band. Remove drive ring.
b. Check tone wheel teeth, sensor mounting bracket and sensor lead for damage. Ensure sensor wheel can be easily turned.
c. Inspect rubber drive band pressed into drive ring for slippage inside ring or for damage. Replace drive ring if necessary.
d. Inspect sensor knurled wheel for deposits of foreign material. If deposits are excessive, clean knurled area with a knife.
e. Install drive ring, drum and wheel.

Step 2: Functional Check

a. Drive over a rough road (gravel road, brick road or similar surface) with light pressure on brake pedal until the false cycling occurs. Keep speed under 15 MPH.
b. Connect a jumper wire between Pins 1 and 3 of connector C with the connector joined.
c. Disconnect one speed sensor at a time and drive over same area and under same conditions which produced false cycling.
d. When false cycling stops, replace disconnected speed sensor. Remove jumper wire from connector C.

Between	and	Measure	Ignition Switch	Brake Pedal	Reading Should Be	If not, refer to
Connector A-3 Pin						
Pin 1	Ground	Ohms	Off	Released	3.0-5.0	Step 3
Pin 2	Ground	Ohms	Off	Released	3.0-5.0	Step 3
Pin 3	Ground	Ohms	Off	Released	0	Step 2
Pin 2**	Ground	DC Volts	On	Applied	0	Step 8
Pin 1**	Ground	DC Volts	On	Applied	0	Step 8
Connector B-4 Pin						
Pin 1	Ground	DC Volts	Off	Applied	12*	Step 6
Pin 1***	Ground	DC Volts	Off	Released	0 w/Headlights ON	Step 7
Pin 2**	Pin 2 Connector C	Ohms	Off	Released	0	Step 8
Pin 3	Ground	DC Volts	On	Released	12*	Step 9
Pin 2	Ground	Ohms	Off	Released	0	Step 4
Pin 4	Ground	DC Volts	On	Released	12*	Step 5
Connector C-3 Pin						
Pin 1	Ground	Ohms	Off	Released	Above 50,000	Step 10
Pin 1	Pin 2	Ohms	Off	Released	250-350	Step 10
Pin 2	Ground	Ohms	Off	Released	Above 50,000	Step 10
Pin 3	Ground	Ohms	Off	Released	Above 50,000	Step 10
Pin 3	Pin 2	Ohms	Off	Released	250-350	Step 10

*Should be same as battery voltage.
**Make measurements with all connectors connected.
***Make measurements with headlights on.

Fig. 6 Circuit checks at electronic control unit

e. If both speed sensors function properly, the electronic control unit is defective, requiring replacement.

ROCKWELL-STANDARD SKID-TROL

Computer/Control Valve

The computer/control valve assembly, Figs. 9 and 10, functions as a normal brake relay valve and as a solenoid operated modulative valve. During normal operation, the valve supplies air to the brake chambers. This air pressure is equal to the pressure supplied from the treadle (brake) valve. Air pressure from the treadle valve is transmitted through the solenoid to the top of the diaphragm, in turn, forcing the diaphragm downward, opening the poppet valve and permits supply air to flow through the delivery ports to the brake chamber.

When Skid-Trol action is initiated by the computer which has sensed imminent wheel lock, the air pressure transmitted to the brake chambers is modulated through solenoid action. When the solenoid is energized by the computer, air pressure from the treadle valve is blocked off and the existing air is directed to exhaust. The diaphragm then moves upward, in turn, closing the poppet supply seat, opening the exhaust seat and directs brake chamber air to exhaust.

The computer is a fully solid state unit encased in the control valve body and is totally sealed. The computer circuitry operates on either positive or negative ground and includes provisions for two wheel speed sensor inputs, a solenoid input a 12 volt D.C. power input and a failure warning indication circuit. The wheel speed sensors send velocity information to the computer. The computer, in turn, processes this data and directs the control valve to provide the necessary corrective action to adjust brake pressure. The internal failure sensing circuitry monitors the system for proper functioning. When a failure is detected, this failure sensing circuitry causes an internal fuse to open and returns the brakes on the affected axle to standard manual control.

Wheel Speed Sensors & Rotors

The wheel speed sensors used with this system are self-generating electromagnetic devices which generate a signal frequency directly proportional to wheel speed. A pulse is generated each time a tooth of the slotted rotor passes the sensor. The sensor consists of a permanent magnet, two coils and an output cable placed in the adjusting spring and into the housing Fig. 11. The sensor is automatically adjusted to the proper position when the rotor pushes against the sensor as the hub and drum assembly is installed.

The rotor is mounted against the inner surface of the brake drum in place of the grease slinger. The assembly rotates with the wheel and drum with the minimum allowable radial and axial runout. The radial runout of the rotor depends on the concentricity between the wheel bearings and axle spindle. Also, if the total axial runout of the rotor exceeds .020 inch, the output signal generated by the sensor will be affected, in turn, feeding an incorrect pulse to the computer.

Diagnosis & Testing

To properly diagnose and test the Skid-Trol System, the use of Service Aid Tester, tool C-4368, is required. However, the following tests may be performed without the tester:

System Failure Warning Lamp Test

With the engine cranking, the system failure warning lamp should light. If not, check for burned out lamp bulb and wiring for proper connections. If lamp bulb and wiring are satisfactory, perform relay test.

Axle System Failure Test

This test is performed to determine which axle system has failed when the system failure warning lamp is lit.

1. With an ohmeter, check resistance across the relay terminals with the following color coded wires:
   Relay No. 1—Light green wire & brown wire
   Relay No. 2—Dark blue wire & green wire
   Relay No. 3—Dark green wire & tan wire
2. Any relay producing an infinite reading indicates the axle system that has failed. If no infinite reading is obtained at any relay, one relay may be defective. Perform relay test to determine defective relay.

Relay Test

1. With ignition "Off", note location of wiring connectors, then disconnect the connectors.
2. With an ohmmeter, check resistance between the terminals which had the red wire and purple wire connected. Resistance reading should be zero. If not, replace relay.
3. Check resistance between the two remaining terminals. Resistance should be 50 ohms or less. If not, replace relay.

Computer Voltage Test

1. Check battery voltage and note reading.
2. Disconnect computer power cable and turn ignition switch to "Run".
3. Check voltage between black wire and white wire of the disconnected power cable. If voltage obtained is not within one volt of battery voltage, check the 10 amp circuit breaker and the red and white wires for opens, shorts and proper connections.

Relay Voltage Tests

Relay Coil Wire Test

1. Note battery voltage and with computer power cable connected, disconnect the following wires from the relays:
   Relay No. 1—Light green wire & brown wire
   Relay No. 2—Dark blue wire & green wire
   Relay No. 3—Dark green wire & tan wire
2. Turn ignition switch to "Run".
3. Check voltage across the disconnected wires of each relay. Voltage reading should be within one volt of battery voltage. If not, check wiring for opens, shorts and proper connections.

NOTE: If the following wires are shorted to ground, the computer fuse will blow:

Relay No. 1—Light green wire

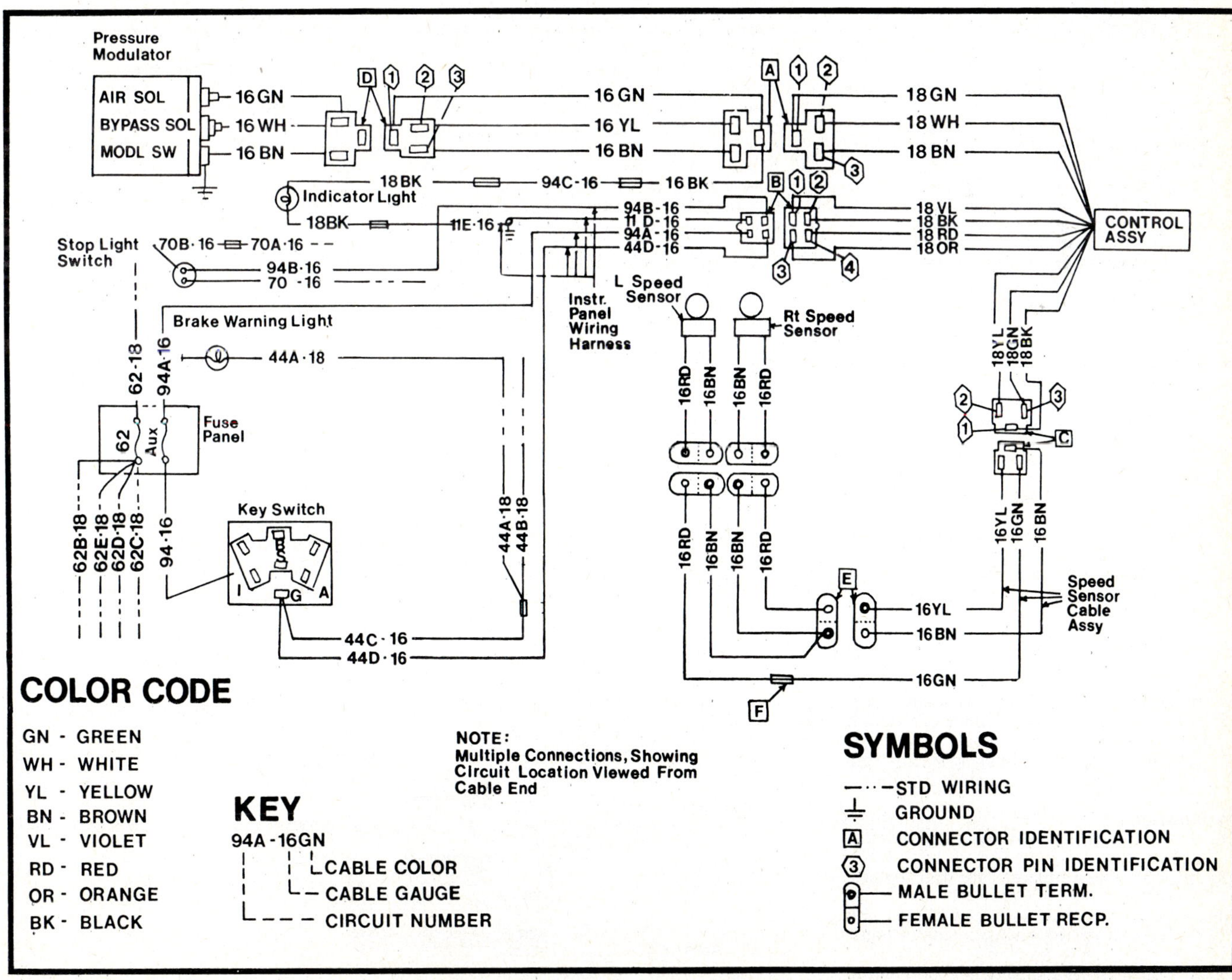

Fig. 7 Anti-skid brake system wiring diagram

Relay No. 2—Green wire
Relay No. 3—Tan wire

Relay Contact Wire Test
1. Note battery voltage and disconnect red and purple wires from relays.
2. Turn ignition switch to "Run".
3. Check voltage between disconnected red and purple wires of each relay. Voltage readings should be within one volt of battery voltage. If not, check red and purple wires for opens, shorts and proper connections.

Service

Computer Or Control Valve, Replace

Removal
1. Disconnect vehicle power source and relieve air pressure from reservoir supplying axle system.
2. Remove computer cover to upper housing screws and pull computer and cover assembly from housing. If replacing computer, disconnect power and sensor connectors from cover, Fig. 10.
3. Disconnect solenoid leads from computer, Fig. 12, and position computer aside or remove from vehicle.
4. Disconnect all air lines from control valve, Fig. 10, then remove screws securing control valve and the control valve from vehicle.

Installation
1. Install control valve and attaching screws.
2. Connect air lines to valve. Recharge sully reservoir and check air lines for leaks.
3. Inspect computer cover gasket for damage and replace, if necessary. Use a suitable adhesive to bond gasket to cover.
4. Connect computer solenoid leads.
5. Install computer and cover assembly into upper housing and loosely install attaching screws. Ensure solenoid leads are not pinched between cover and upper housing.
6. Check valve for proper operation by applying brake and observing brake action. No audible leakage should be noted during application and air pressure should rapidly relieve when the brakes are released.
7. Connect power cable connector, if necessary.
8. Torque computer cover screws to 100 inch lbs.
9. Connect sensor connectors, if necessary.

Wheel Sensor, Replace

Removal
1. Cut nylon tie straps securing sensor cable to air delivery hose, however, only up to the inline connector.
2. Disconnect sensor to computer cable at inline connector. Do not twist connector.
3. Slide sensor cable and connector through hole in spider or dust shield.
4. Remove sensor and adjusting spring from housing by prying out spring, Fig. 13.

Installation
1. Slide new spring over back and sides of sensor.
2. Inspect rubber support caps and replace, if necessary.
3. Install sensor and spring over pins and press until spring engages side of housing.
4. The initial sensor setting is approximately 7/16 (0.44 inch) inch from housing, Fig. 14. The sensor may be inserted to this setting or a plastic stop can be installed over the pins. If plastic stop is used, remove stop prior to hub and drum installation.
5. Insert connector through hole in brake spider or dust shield and install grommet.
6. Route sensor cable along air delivery hose. Ensure cable does not interfere with any brake component or the hub. Secure cable to air delivery hose at 8 inch intervals.
7. When the wheel is installed, the sensor should be positioned to the maximum setting in the housing. As the wheel nuts are torqued, the rotor acting on the sensor face, forces the sensor back to the correct setting. Ribs, located on the sensor face, will wear slightly during the first few wheel revolutions until clearance is obtained between sensor and rotor.

Rotor Service

1. Raise vehicle and remove hub, drum and tire.
2. Assemble tools on hub and drum, Figs. 15 and 16.
3. Zero dial indicator and using the cone clamp handle, pivot dial indicator around rotor surface, Fig. 17.
4. If runout exceeds .020 inch, replace rotor as follows:
   a. Remove screws and washers securing rotor to brake drum, then the rotor.
   b. Apply a suitable seal to rotor and brake drum contact surfaces.
   c. Install rotor on brake drum and tighten screws.

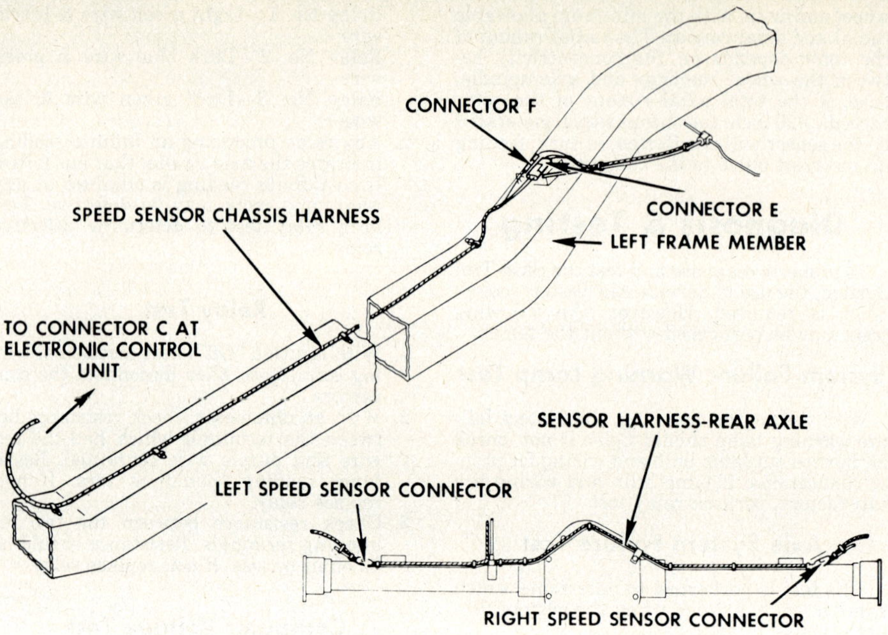

Fig. 8 Anti-skid brake system axle & chassis wiring schematic

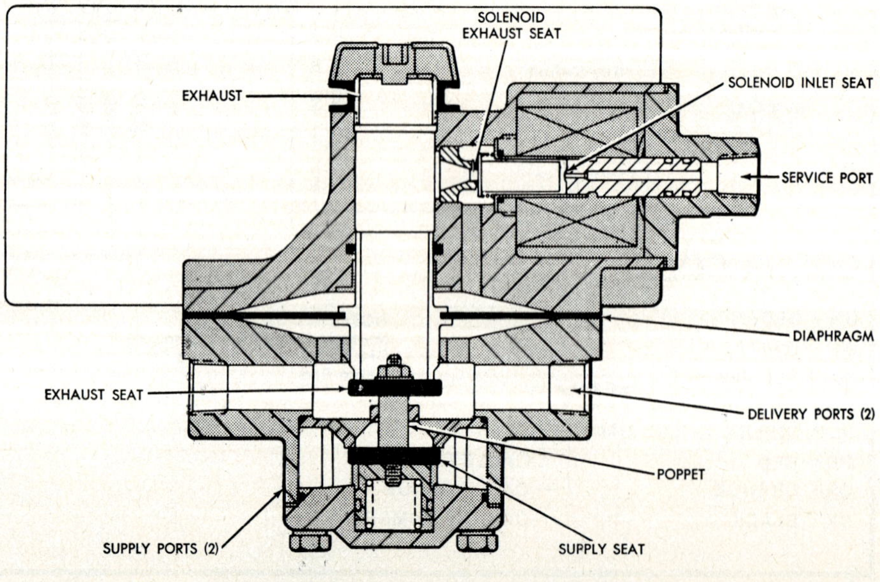

Fig. 9 Cross-sectional view of computer/control valve

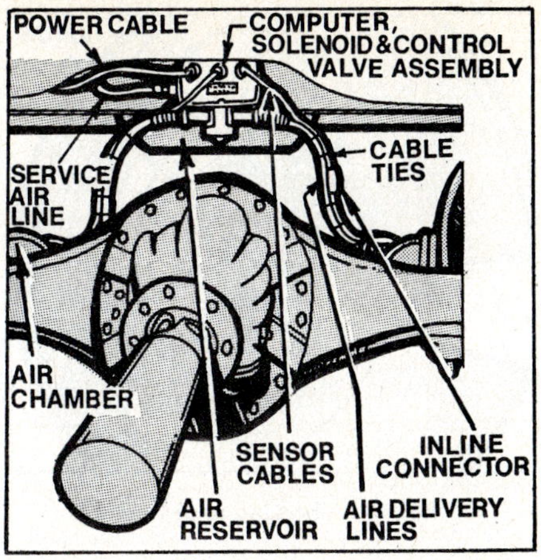

Fig. 10 Computer/Control Valve location

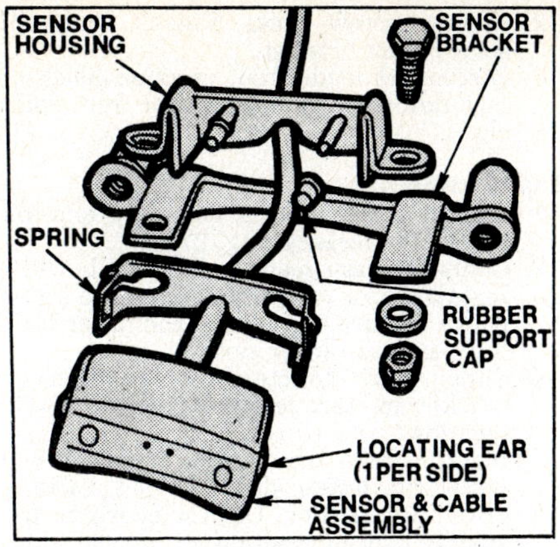

Fig. 11 Sensor components

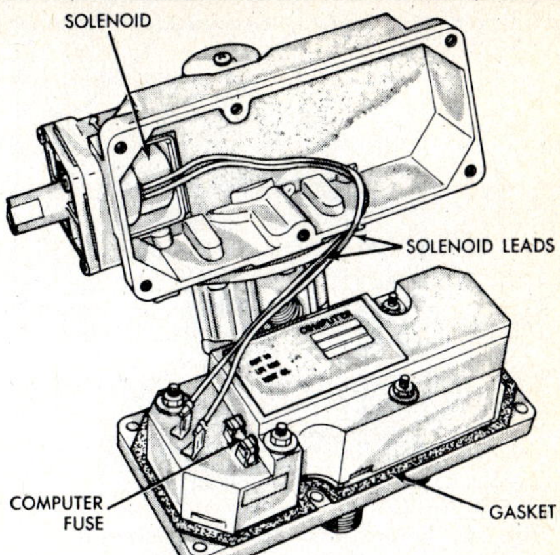

Fig. 12 Solenoid lead connections

EATON & KELSEY-HAYES ANTI-SKID

These two anti-skid systems are basically similar in design and operation, however, the individual system components are not interchangeable, except for the disc assemblies. If a specific system is used on one axle of a truck, that system must be used on all axles. The computer module on the Eaton system is color coded black and gold. The Kelsey-Hayes computer module is color coded blue. Both systems incorporate the following components:

Sensor

The sensor assembly is comprised of the sensor and mounting bracket. On wheels equipped with disc brakes, the sensor is mounted inside one of the bolts securing the brake adapter to the wheel spindle, Fig. 18. On wheels equipped with drum brakes, the sensor is mounted on the brake backing plate, Fig. 19.

The Eaton sensor is mounted on the bracket with spring loaded screw packs and contains a magnet and two coils. The Kelsey-Hayes sensor is friction-fitted in the bracket.

The sensor transmits a voltage to the computer as the notches in the disc assembly "Excite" the sensor. The voltage frequency is directly proportional to wheel speed.

Disc Assembly

On drum brakes, the disc assembly is mounted inside the drum with attaching bolts. On disc brakes, the disc assembly is pressed inside the brake rotor. A ring of notches on the disc are used to "Excite" the sensor as the wheel rotates, Fig. 20.

Monitor Assembly & Warning Lamp

The monitor assembly is an electronic unit that receives signals from the axle modulators, in turn activating the warning lamp in event of a system failure. The warning lamp, mounted on the dash panel, is used to indicate a malfunction in the skid control system.

The warning lamp should light momentarily when the ignition switch is turned "ON". If the lamp stays lit for more than a few seconds, a malfunction is indicated. The bulb and lamp circuit should be checked if the lamp fails to light when the ignition switch is turned "On".

Warning Lamp & Power Relays

The warning lamp relay is used to activate the warning lamp if no power is available to the system. The power relay provides power to the modulators and the monitor assembly. The power relay may be installed specifically for the skid-control system or may be an accessory relay already mounted in the truck.

Modulator Assembly

Anti-Lock Valve Assembly

The anti-lock valve assembly is a combined relay and solenoid valve. The relay valve applies and releases the brakes, responding to foot valve application. The solenoid valve exhausts the air pressure to release the brakes, when the computer determines from the sensor signals that a wheel is locking up.

Computer Assembly

The Solid-state computer determines if a wheel is locking up by receiving sensor signals which are directly the wheel deceleration rate. If a wheel lock-up is sensed, the computer energizes the solenoid valve momentarily, relieving brake pressure, in turn releasing the brakes. The system can cycle up to eight.

Testing & Diagnosis

Modulator Operation & Leak Test

1. Chock wheels and fully charge air brake system.
2. Apply brakes several times and check brake operation at all wheels.
3. With a soap solution, leak test exhaust port under the valve. Slight leakage is permitted as follows:
   a. Brakes applied—a one inch bubble in 3 seconds.
   b. Brakes released—a one inch bubble in 5 seconds.
4. On Eaton systems, separate computer from valve, then leak test solenoid mounting with soap solution. With the brakes fully applied, no leakage is permitted.
5. Replace valve if excessive leakage is noted or if it functions improperly.

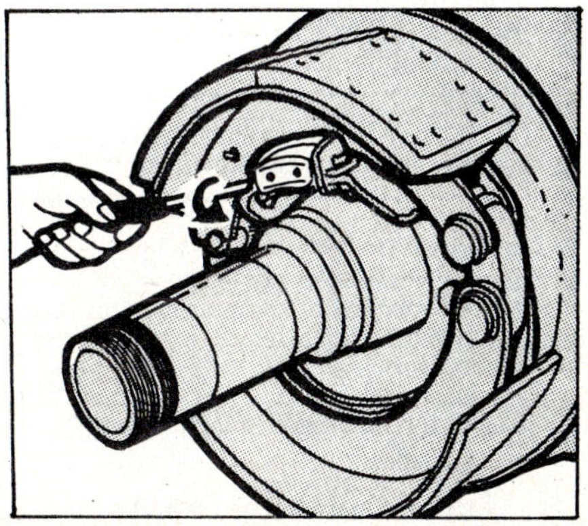

Fig. 13 Sensor removal

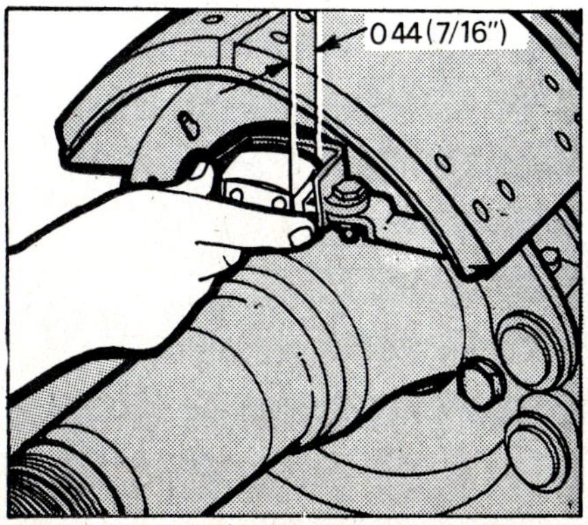

Fig. 14 Sensor installation

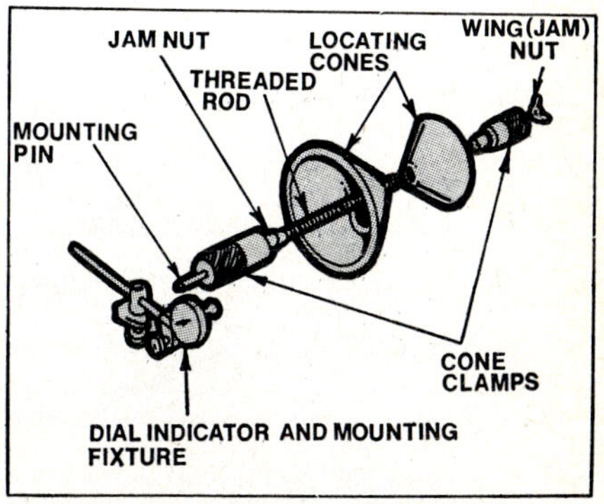

Fig. 15 Rotor runout tools

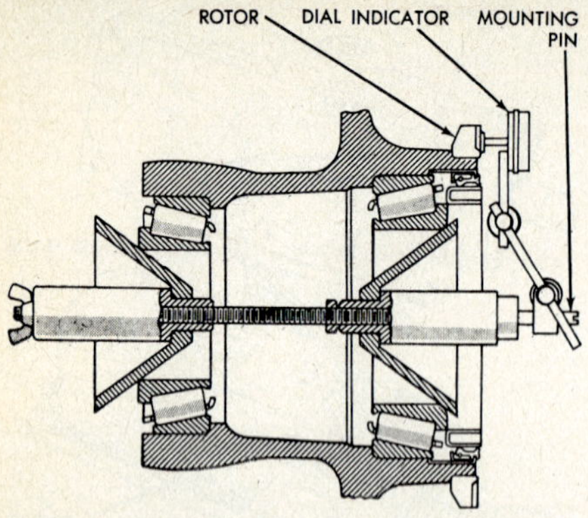

Fig. 16 Rotor runout tools positioned in hub

Warning Lamp Operation Test

1. Turn ignition switch to "On" and observe warning lamp. Lamp should light for 2–4 seconds and go out. If lamp remains on, the cause may be the warning relay powering the warning lamp circuit, axle system malfunction or a defective warning monitor assembly.
2. If lamp flashes on and off in step 1, a loose connection is indicated.
3. If lamp fails to light in step 1, replace bulb and recheck. If lamp still does not light, check for proper ground and power to lamp. If voltage is present at the monitor "P" terminal and no voltage at the "W" terminal for 2–4 seconds after turning the ignition switch "On", replace monitor assembly.

Electrical Diagnosis

Refer to the following Figures for electrical diagnostic procedures:
Fig. 21—Kelsey-Hayes monitor assembly diagnosis.
Fig. 22—Kelsey-Hayes modulator and sensor diagnosis.
Fig. 23—Eaton or Kelsey-Hayes power supply diagnosis.
Fig. 24—Eaton monitor assembly diagnosis.
Fig. 25—Eaton modulator and sensor diagnosis.

Service

Eaton Sensor, Replace

Removal
1. Remove wheel and brake drum.
2. Remove screw packs, Fig. 26 and pull sensor from bracket.
3. Disconnect leads from junction block or modulator and remove sensor from vehicle.

Installation
1. Inspect gauge blocks for wear and replace, if necessary, Fig. 27.
2. Connect sensor leads to junction block.
3. Install sensor on bracket, then the screw packs. Ensure screw pack and lower body fully engage sensor slot.
4. Slide sensor to outermost position on bracket and torque screw packs to 35–50 inch lbs.
5. Install brake drum and wheel. When installing brake drum, avoid rocking drum into position. Use the bearings and spindle nut to position drum. The disc assembly will position and wear the sensor until the proper clearance is obtained.

Kelsey-Hayes Sensor W/Drum Brake, Replace

Removal
1. Remove wheel and brake drum.
2. Remove bracket attaching bolts and separate bracket from brake backing plate, Fig. 19.
3. Push sensor from bracket and disconnect leads from junction block.
4. Remove sensor from vehicle.

Installation
1. Connect sensor leads to junction block.
2. Insert sensor in bracket and place sensor in outermost position in bracket.
3. Install bracket and mounting screws.
4. Install brake drum and wheel. When installing brake drum, avoid rocking drum into position. Use the bearings and spindle nut to position drum. The disc assembly will properly position sensor.

Kelsey-Hayes Sensor W/Disc Brakes, Replace

Removal
1. Disconnect sensor leads at junction block.
2. Remove sensor lead bracket from spindle, Fig. 18.
3. Unscrew sensor from adapter and spindle.
4. Push sensor from bolt.

Installation
1. Insert sensor into bolt and push sensor out to contact disc when bolt and sensor assembly is installed. Then, install the assembly.
2. Install sensor lead bracket on spindle.
3. Connect sensor leads to junction block.

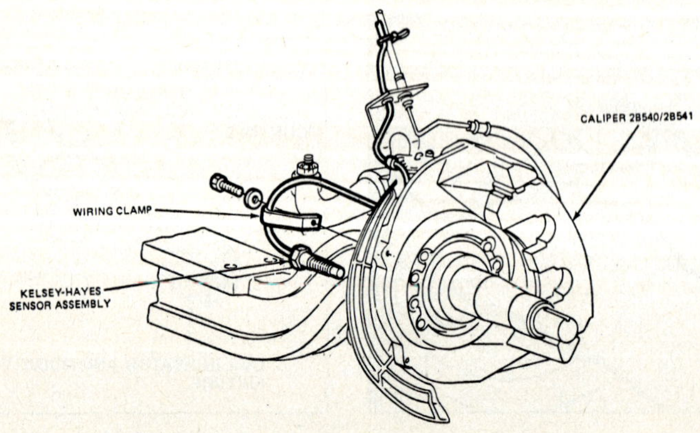

Fig. 18 Sensor installation with disc brakes.

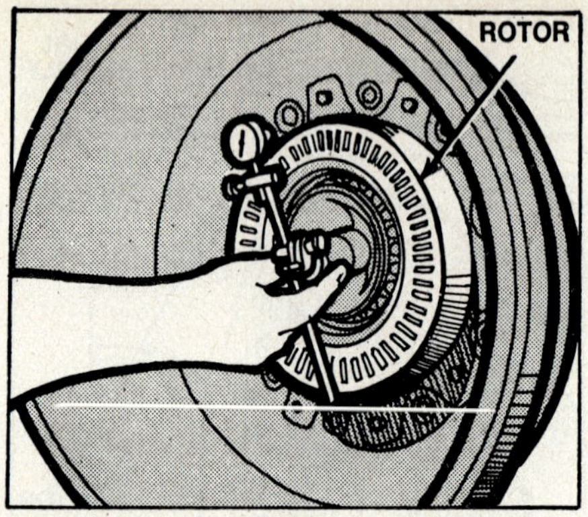

Fig. 17 Checking rotor runout

Disc Assembly, Replace

The disc assembly is serviceable only on wheel with drum brakes.

Removal
1. Remove wheel and brake drum.
2. Remove screws securing disc assembly to brake drum, Fig. 19, then the disc assembly.

Installation
1. Attach disc assembly to brake drum.
2. Reposition sensor as outlined under the "Sensor, Replace" procedures.
3. Install brake drum and wheel.

Monitor Assembly, Replace

1. Disconnect wiring connectors from monitor assembly.
2. Remove monitor assembly mounting screws and the monitor assembly.
3. Reverse procedure to install.

Modulator Assembly, Replace

Removal
1. Chock wheels and ensure vehicle will not roll when brakes are released.
2. Drain air system reservoir.
3. Remove cover from computer module or terminal housing.
4. Note location of all electrical connections to computer module, then disconnect the electrical connections.
5. Note location of all air lines to modulator, then disconnect the air lines.
6. Remove modulator mounting bolts and the modulator.

NOTE: It may be necessary to remove the modulator mounting bracket from frame.

Installation
1. Replace damaged air lines.
2. Install modulator on mounting bracket.
3. Connect electrical leads and air lines to modulator.
4. Perform "Modulator Operation & Leak Test" as outlined under "Testing and Diagnosis".

Eaton Computer, Replace

1. Remove through bolts from front of com-

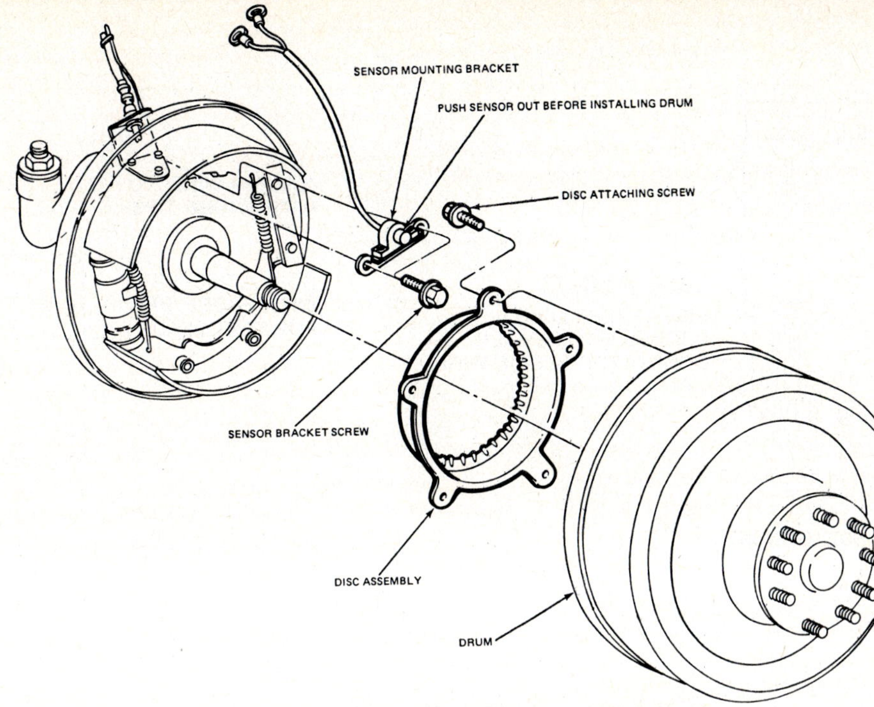

Fig. 19 Sensor installation with drum brakes (Typical)

puter module.
2. Separate computer from valve and disconnect solenoid leads, Fig. 28.
3. Note wiring connections inside computer housing, then disconnect the wiring.
4. Remove spring clip and separate harness from computer housing.
5. Reverse procedure to install.

Kelsey-Hayes Computer, Replace

r1. Remove bolts securing computer to valve, Fig. 29.
2. Separate computer from valve, using caution not to stretch solenoid wires.
3. Disconnect solenoid leads and pull dust cover over solenoid leads, Fig. 30.
4. Remove screw securing computer mounting plate to valve.
5. Reverse procedure to install. When installing mounting bolts, the longer bolts are installed at the top of the valve. Torque mounting bolts to 50 inch lbs.

AC (GENERAL MOTORS) ANTI-SKID

This anti-skid control system is an axle by axle system in which both wheel brakes on a single axle are modulated simultaneously. The main components incorporated in the system are the wheel speed sensors, modulator, electrical cables and connectors, and a cab warning unit. Each axle has a modulator and each wheel assembly has a wheel speed sensor. The modulator is the logic center for each axle for the operation of the axle wheel lock control components. The modulator analyzes signals from the wheel speed sensors, in turn signaling the solenoid to release air brake pressure is wheel lock-up is imminent.

If a malfunction occurs in an axle system, the malfunction has no effect on other axle systems. The affected axle system will then function as a normal air brake. The cab warning lamp will be lit for the specific malfunction.

This system incorporates an external wheel speed sensor, mounted on a bracket bolted to the axle housing or to the brake spider. The brake drum has notched teeth on the outer edge. When the wheel assembly rotates, AC voltage is generated in the wheel sensor which is then transmitted to the modulator. This AC voltage is dependent upon the rate of change of the magnetic field. The magnetic field is altered three factors: clearance between the sensor and drum, wheel speed and the width of the teeth on the brake drum.

Logic Component Operation, Fig. 31

The wheel speed sensors (A and B) transmit AC electrical signals to the logic center. The logic center incorporates a self check circuit (E) which checks both wheel speed sensors, the battery power source (C) and the condition of the brake release solenoid (N). If a malfunction is present, a signal is transmitted to the cab warning lamp (D) and the release solenoid (N) so brake release cannot occur (G).

The wheel speed signals are then combined (F) after the self check and this combined signal is continuously analyzed to determine if wheel lock control is necessary. When the speed indicates that the wheels are decelerating (II), this deceleration rate is transmitted to a companion circuit (J). If the deceleration rate is too great, the logic "OR" (L) signals the solenoid (N) to energize and release air brake pressure until the deceleration rate is acceptable. Simultaneously, the combined wheel

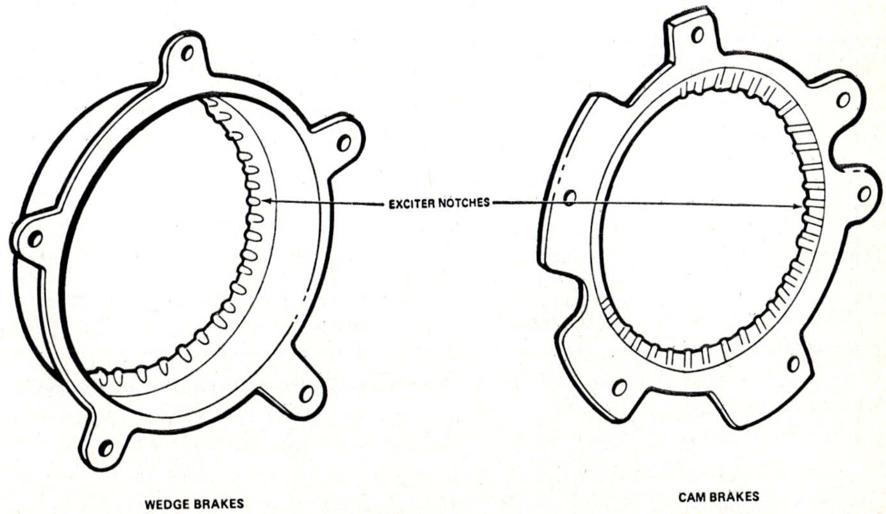

Fig. 20 Disc assembly

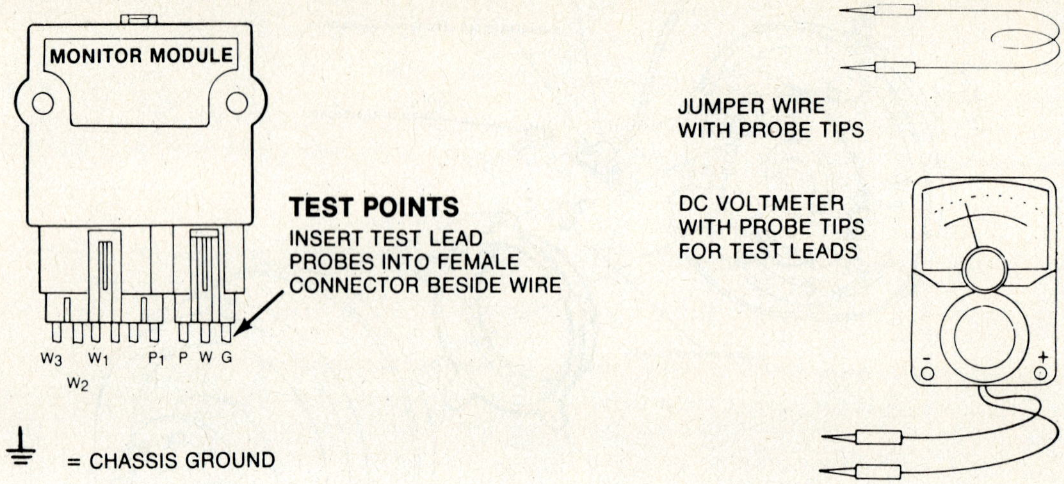

TEST	PROCEDURE	RESULTS
A. Power to Monitor	1. Voltmeter: + to P – to ⏚ 2. Ignition on 3. 11-14 Volts required	Not Okay — check power supply circit (Fig. 20) and blue wire to relay. Okay — check monitor ground (B).
B. Monitor Ground	1. Voltmeter: + to P – to G 2. Ignition on 3. 11-14 volts required	Not Okay — repair ground and check warning light. Okay — check power out (C).
C. Power Out to Warning Relay	1. Voltmeter: + to P1 – to ⏚ 2. Ignition on 3. 11-14 volts required	Not Okay — replace monitor assembly. Okay — check for warning signal in (D).
D. Warning Signal In	1. Ingition on 2. Jumper from P (or any hot terminal) to W1, W2, W3 in turn. 3. Drive vehicle 15 mph or faster if vehicle speed is involved. 4. Light should go out for terminal receiving a warning signal	Warning Light out — test modulator sending warning signal. Light stays on — check warning signal out(D1).
D1. Warning Signal Out	1. Ignition 2. Voltmeter: + to W – to ⏚ 3. Zero volts required	Voltage out — replace monitor Okay (no voltage) — check warming relay (E).
E. Warning Relay	1. Ignition on 2. Volmeter: + to green wire terminal of relay – to ⏚	Okay — check relay ground. Repair ground or replace relay. Not Okay — repair green wire from monitor PI to relay.

NOTE: Replace monitor assembly if:
(1) Warning relay and W1, W2, W3 signals are okay, but light stays on.
(2) Light doesn't come on momentarily with ignition on (see "Warning Lamp Operation Test").

Fig. 21 Kelsey-Hayes monitor assembly diagnosis

TEST POINTS

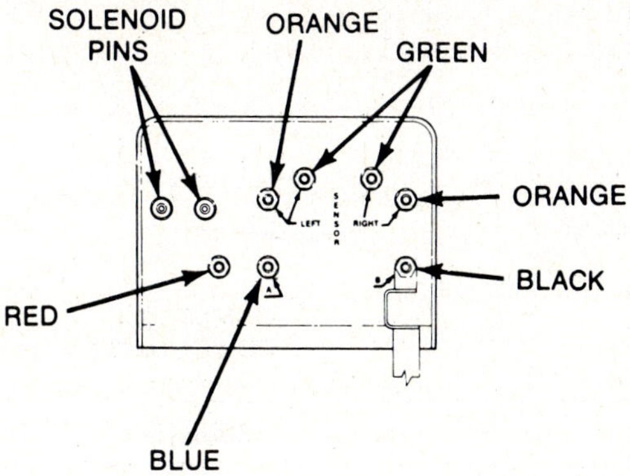

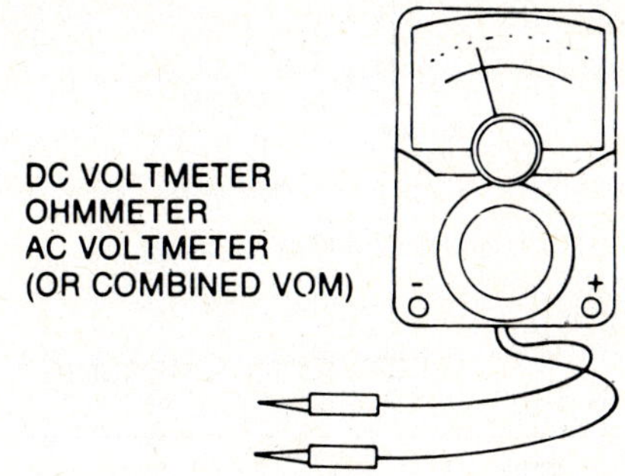

⏚ = CHASSIS GROUND

NOTE: VOLTAGES ARE DC UNLESS SPECIFIED AC

TEST	PROCEDURE NOTE: Voltages are DC unless specified AC	RESULTS
F. Warning Signal Out	1. Remove red wire 2. Voltmeter: + to FAIL − to CHASSIS 3. Ignition on 4. Okay signal is 6-14 volts	Less than 6 volts — check power in to modulator (G). Okay — repair red wire from modulator to monitor and check warning light.
G. Power in to Modulator	1. Ignition on 2. Voltmeter: + to HOT − to ⏚ 3. 11-14 volts required	Okay — check ground (H). Not Okay — repair blue wire to splice in harness (if other modulators OK) or check out power supply (Fig. 20).
H. Modulator Ground	1. Ignition on 2. Voltmeter: + to HOT − to CHASSIS 3. 11-14 volts required	Okay — check solenoid resistance (I). Not Okay — repair ground.
I. Solenoid Resistance	1. Ohmmeter leads to solenoid pins 2. 4-6 ohms required	Not Okay — replace valve assembly. Okay — check sensor output voltage (J).
J. Sensor Output Voltage (Test both Sensors)	1. Jack up wheels and release brakes. Ignition off. 2. AC Voltmeter: + to ORANGE − to GREEN 3. Turn wheel 30 rpm (one turn every 2 seconds) 4. 0.2 volts AC required	Not Okay — check sensor resistance (K). Okay — check for sensor short (L).
K. Sensor Resistance (Test Low-Output Sensor)	1. Remove orange and green wires 2. Ohmmeter leads to wire eyelets. 3. 4000Ω to 6000Ω required	Okay — remove wheel and reposition sensor. Not Okay — repair open sensor wire or replace sensor.
L. Short Sensor or Wire	1. Ohmmeter leads: sensor Wire and ⏚ 2. High resistance (10,000Ω) required	Okay — replace computer module if all other tests are okay. Not Okay — repair shorted sensor wire. If not shorted, replace sensor.

Fig. 22 Kelsey-Hayes modulator & sensor diagnosis

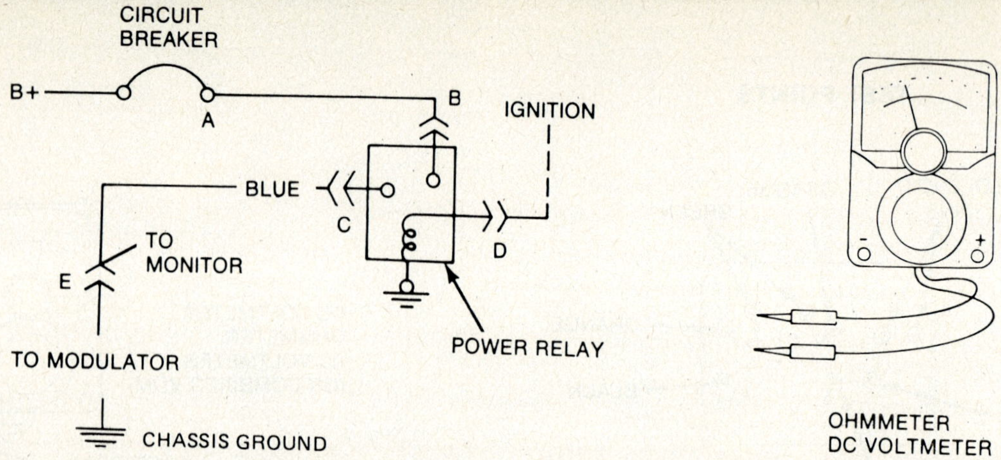

TEST	PROCEDURE	RESULTS
M. Power out of Breaker	1. Voltmeter: + to A − to ⏚ 2. 12-14 volts required.	Okay — check for power from relay (O). Not Okay — check for short to ground.
N. Short to Ground	1. Circuit breaker open 2. Ohmeter leads: A and ⏚ 3. Should be infinite resistance	Okay — replace breaker. Zero Resistance — trace and repair short circuit.
O. Power out from Relay	1. Ignition on 2. Voltmeter: + to C − to ⏚ 3. 11-14 volts required	Okay — check blue wiring to monitor and modulators; tests A and G. Not Okay — check battery power in (P).
P. B+ Power in to Relay	1. Voltmeter: + to B − to ⏚ 2. 11-14 Volts required	Okay — check ignition power in (Q). Not Okay — repair circuit between relay and breaker.
Q. Ignition Power in to Relay	1. Ignitionon 2. Voltmeter: + to D − to ⏚ 3. 12-14 volts required	Okay — check relay ground. If okay, replace relay and recheck warning lamp. Not Okay — locate and repair open ignition power circuit.
NOTE: Refer to wiring schematics for connector locations.		

Fig. 23 Eaton or Kelsey-Hayes power supply diagnosis

speed signal is analyzed by the slip level detector (I). If the wheel slip is too great, a decision in the companion circuit (K) is made the logic "OR" (L) again signals the solenoid (N) to release the air brake pressure.

The logic "OR" (L) will signal for a brake release if either the wheel deceleration level or the wheel slip level is too great since either condition will cause wheels to lock up without the wheel lock control. When wheel speed recovers, the brake release signal is removed and the brakes are re-applied until a release condition is detected.

Diagnosis & Testing

Refer to Figs. 32 through 54 for diagnosis procedures. Procedures for both volt-ohmmeter and AC test plug are outlined.

Service

Wheel Sensor, Replace

1. Loosen sensor jam nut, then remove sensor from mounting bracket, Fig. 55.
2. Remove clips and ties from wiring harness, then disconnect harness connector and remove sensor.
3. Connect harness connector to sensor and install harness clips and ties, then position sensor in mounting bracket.
4. If original is to be installed, assemble sensor to bracket so that sensor is in contact with drum edge, then back sensor out one turn. This should provide a sensor to drum clearance of .040 to .060 inch. Torque jam nut to 30 to 40 ft. lbs.
5. If a new sensor is to be installed, assemble sensor to mounting bracket so that sensor is contact with drum edge, then torque jam nut to 30 to 40 ft. lbs. New sensors incorporate a fiber spacer which is removed within the first few revolutions of the drum, leaving the proper sensor to drum clearance.

Modulator, Replace

1. Block wheels, then drain air system.
2. Disconnect air lines and electrical connections from modulator.
3. Remove mounting nuts and bolts and modulator.
4. Reverse procedure to install, torque modulator attaching bolts to 25 to 30 ft. lbs.

Modulator Disassemble

1. Remove four outer cover attaching screws, Fig. 56, using tools J-25360-5, 9.
2. Using a rubber mallet, separate controller and cover assembly from lower casting.

NOTE: Use care not to damage sealing surfaces of casting.

3. Disconnect two solenoid leads from controller terminals, then remove controller and cover assembly.
4. Remove four lower cover to valve assembly attaching screws using tools J-25360-5, 9.

NOTE: It is not necessary to remove lower

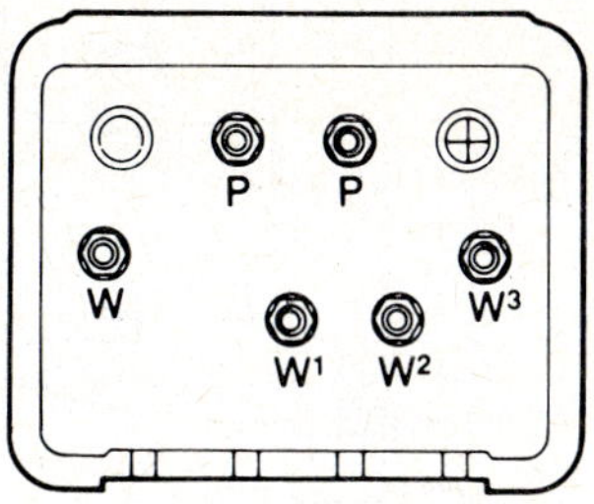

FRONT COVER REMOVED

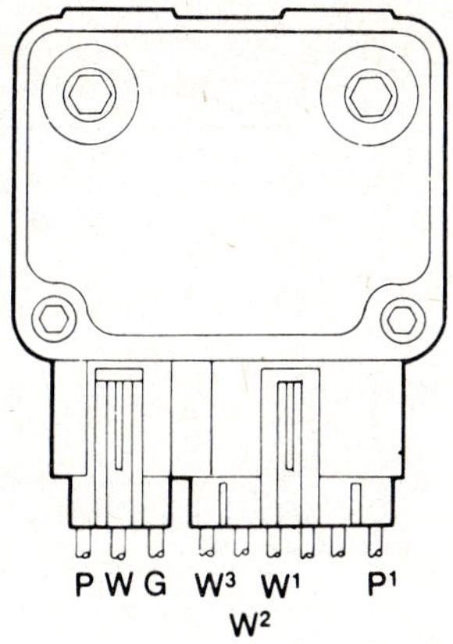

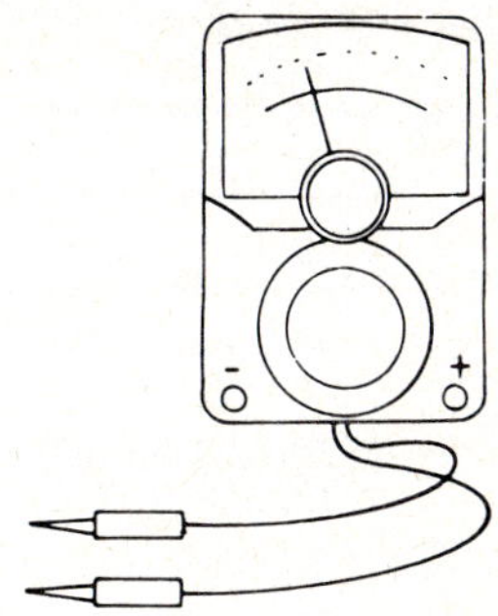

DC VOLTMETER
WITH PROBE TIPS
FOR TEST LEADS

TEST POINTS

INSERT TEST LEAD PROBES INTO FEMALE CONNECTOR BESIDE PLUG OR REMOVE COVER TO EXPOSE TEST POINTS.

⏚ = CHASSIS GROUND

TEST	PROCEDURE	RESULTS
A. Power to Monitor	1. Voltmeter: + to P – to ⏚ 2. Ignition on 3. 11-14 volts required	Not Okay — check power supply circuit (Fig. 20) and blue wire to power relay. Okay — check ground (B).
B. Monitor Ground	1. Voltmeter: + to P – to G 2. Ingition on 3. 11-14 volts required	Not Okay — ground monitor and recheck warning lamp. Okay — check power out (C).
C. Power out to Warning Relay	1. Voltmeter: + to PI – to ⏚ 2. Ignition on 3. 11-14 volts required	Not Okay — replace monitor assembly. Okay — check for warning signal in.
D. Warning Signal In	1. Ignition on 2. Voltmeter: + to W1, W2, W3 – to ⏚ 3. 3-8 volts required for OK signal	Not Okay — test modulator sending not-okay signal Okay — check warning signal out (D1).
D1. Warning Signal Out	1. Ignition on 2. Voltmeter: + to W – to ⏚ 3. Zero volts required	Voltage Out — replace monitor. Okay (no voltage) — check warning relay (E).
E. Warning Relay	1. Ignition on 2. Voltmeter: + to green wire terminal of relay – to ⏚ 3. 11-14 volts required	Okay — check relay ground. Repair ground or replace relay. Not Okay - repair green wire from monitor PI to relay.
NOTE: Replace monitor assembly if: (1) Warning relay and W1, W2, W3 signals are okay, but light stays on. (2) Light doesn't come on momentarily with ignition on (see "Warning Lamp Operation Test").		

Fig. 24 Eaton monitor assembly diagnosis

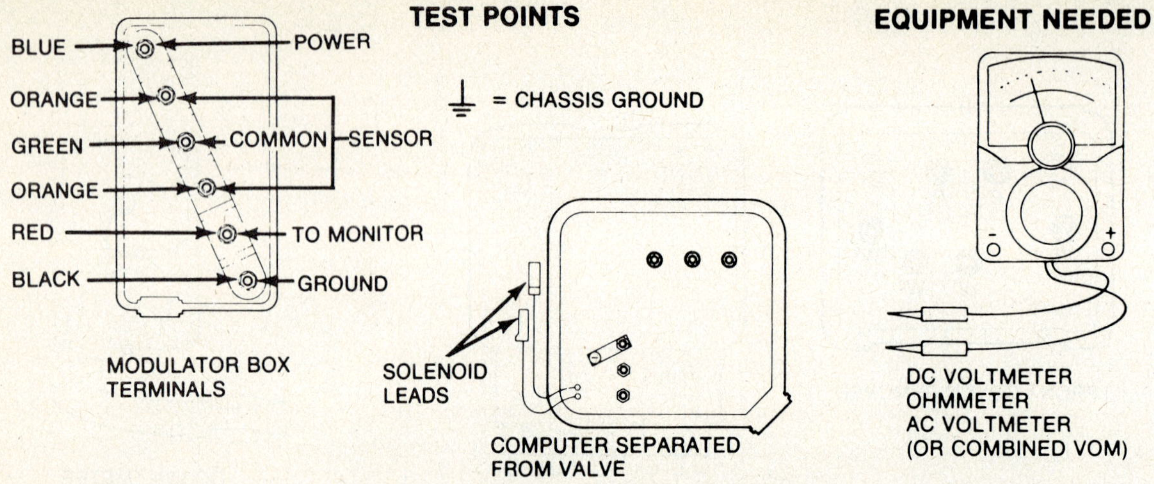

TEST	PROCEDURE NOTE: Voltages are DC unless specified AC	RESULTS
F. Warning Signal Out	1. Ignition on 2. Voltmeter: + to RED – to ⏚ 3. 3-8 volts required for okay	Okay — repair red wire to monitor and check warning light. Not Okay — check power in (G).
G. Power in to Modulator	1. Ignition on 2. Voltmeter: + to BLUE – to ⏚ 3. 11-14 volts required	Okay — check modulator ground (H). Not Okay — repair blue wire to splice in harness (if other modulators okay) or check out power supply (Fig. 20).
H. Modulator Ground	1. Ignition off 2. Ohmmeter leads to computer and ⏚ 3. Zero ohms required.	Okay — check solenoid resistance (I) and solenoid terminal resistance (I1). Not Okay — Repair ground.
I. Solenoid Resistance	1. Separate computer from valve 2. Ohmmeter to solenoid terminals 3. 4-6 ohms required	Okay — check lead resistance (I1). Not Okay — replace valve assembly.
I1. Solenoid Lead Resistance (In Computer)	1. Solenoid leads off 2. Ohmmeter: to solenoid leads on computer 3. 10,000 ohms or higher required	Okay — check sensor output voltage (J). Not Okay — reverse ohmmeter leads and recheck Still Not Okay — replace computer module.
J. Sensor Output Voltage (Test both Sensors)	1. Jack up wheels and release brakes. Ignition off. 2. AC voltmeter: + to ORANGE – to GREEN 3. Turn wheel (43 rpm for 20-inch; 56 rpm for 15 inch) 4. 0.3 volts AC required	Okay — check for shorted sensor (L). Not Okay — check sensor resistance (K).
K. Sensor Resistance (Test Low-Output Sensor)	1. Remove orange and green leads 2. Ohmmeter: to wire eyelets 3. 500-1500 ohms required	Okay — remove wheel and re-position sensor. Not Okay — repair open sensor wire or replace sensor.
K1. Computer Sensor Terminal Resistance	1. Green wire off terminal. 2. Ohmmeter: open terminal and ⏚ 3. 10,000 ohms required	Okay — check for shorted sensor or wire (L). Not Okay — replace computer module.
L. Shorted Sensor or Wire	1. Ohmmeter: GREEN wire and ⏚ 2. Both sensors connected 3. 3000 ohms required	Okay — replace computer module if all tests above are okay. Not Okay — repair shorted sensor wire. If not shorted, replace sensor.

Fig. 25 Eaton modulator & sensor diagnosis

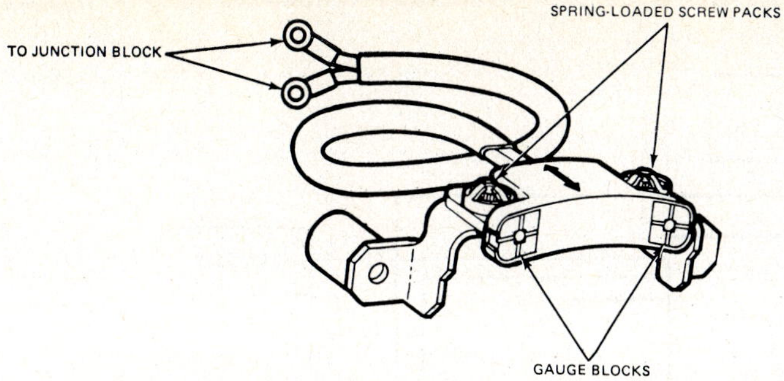

Fig. 26 Eaton sensor assembly

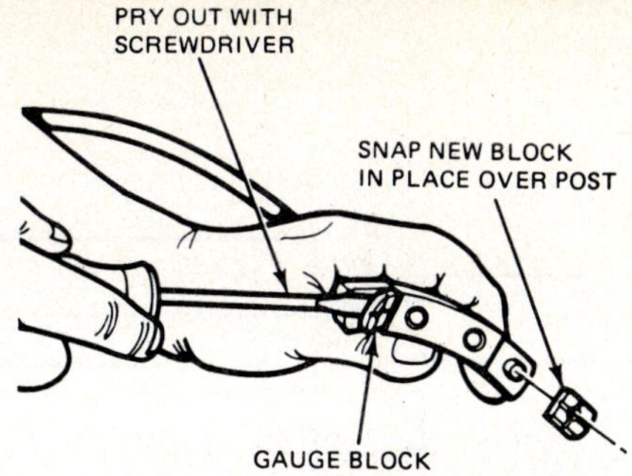

Fig. 27 Replacing Eaton sensor gauge blocks

cover assembly unless entire valve assembly is to be replaced.

5. Using a rubber mallet, separate lower cover from valve assembly.

NOTE: Use care not to damage sealing surfaces of casting.

6. Remove two long and two short bolts attaching solenoid valve body and relay valve body.
7. Separate relay valve body from solenoid valve body, then remove O ring seal from between valve bodies and discard.
8. Using suitable pliers remove flow restrictor from relay valve body.
9. Remove O ring from flow restrictor and discard.
10. Pull relay valve piston from valve body.
11. Remove conical spring and O ring from piston and discard.
12. Remove screw retaining seal to piston, then separate components and discard molded rubber valve assembly.
13. Remove snap ring and washer, then pull out exhaust valve body, spring and modulation tube.
14. Remove and discard exhaust valve body and modulation tube O rings and spring.
15. Remove seal retainer from modulation tube and discard seal.

NOTE: Use care not to damage distort seal retainer.

16. Remove screw from exhaust valve body then separate components and discard exhaust diaphragm.
17. Remove allen head screw from solenoid valve body.
18. Remove solenoid using tool No. J-25360-1.
19. Remove three check valve cover to solenoid valve body attaching screws, then separate cover from valve body.
20. Remove spring and button check valve, then using a small screw driver carefully pry out metal valve seat retainer and remove spring and check valve.
21. Using a screw driver through check valve

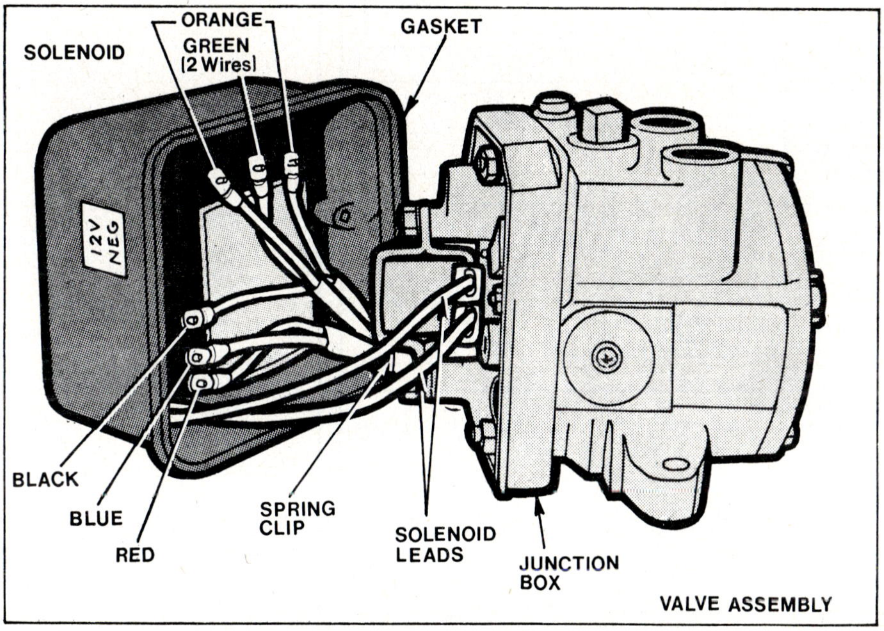

Fig. 28 Eaton computer replacement

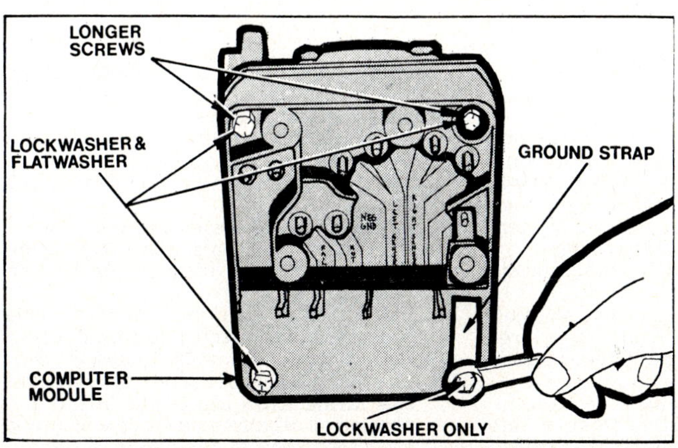

Fig. 29 Kelsey-Hayes computer replacement

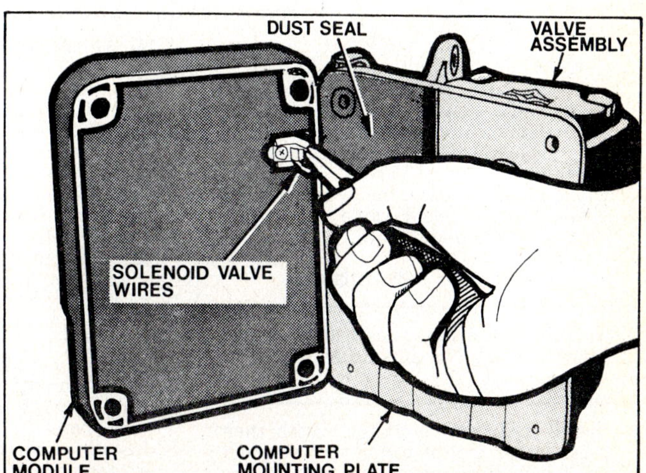

Fig. 30 Kelsey-Hayes computer solenoid lead connections

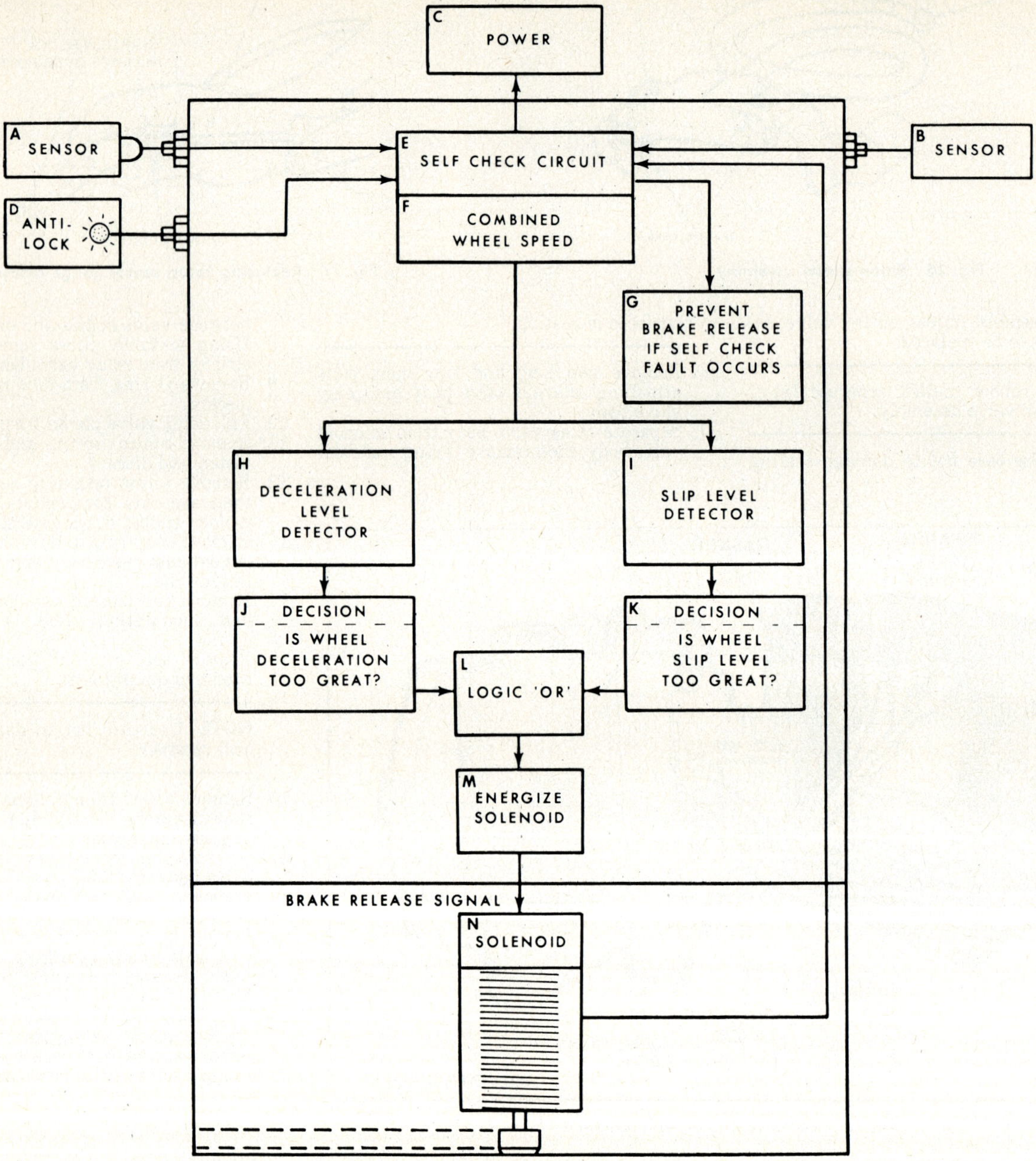

Fig. 31 AC logic component operation

opening, press on flat surface of seat and remove solenoid valve components.
22. Pull out metal retainer and remove rubber solenoid exhaust valve.
23. Using suitable pliers, remove restriction piston from solenoid valve body, then remove O ring and discard.

Modulator Assemble

NOTE: Lubricate all new seals and O rings with Dow Corning No. 33 silicone grease or equivalent.

1. Install O rings on restriction piston, then press either end of piston into solenoid valve body until piston bottoms, Fig. 56.
2. Position rubber solenoid exhaust in metal retainer, then press retainer and exhaust in until retainer snaps into place.
3. Install O rings on valve body and press small end into solenoid valve body through solenoid bore until seated.
4. Install O rings on solenoid valve stem and cage, then install spring on valve stem.
5. Install valve stem end without spring through open end of valve cage, then install diaphragm seal on end of stem protruding from other side of cage.
6. Install cage and valve stem into solenoid bore, Fig. 57.
7. Place rubber button check valve on small end of conical spring, then insert large end of spring into large diameter check valve bore.
8. Install O ring on metal valve seat retainer, then press open end of retainer into large diameter check valve bore until fully seated.
9. Insert button check valve into long spring, then install rounded end of check valve into small diameter check valve bore in solenoid valve body.
10. Install O ring on check valve cover, then position spring on cover boss and install cover. Torque attaching screws to 55 to 65 in. lbs.
11. Install O ring on solenoid, then thread solenoid into solenoid valve body using tool No. J-25360-1. Torque solenoid to 66 to 96 in. lbs.
12. Place tension ball into solenoid valve body hole and install allen head screw. Torque screw to 20 to 30 in. lbs.
13. Position washer over screw with concave side facing screw head, then install diaphragm on exhaust valve body and torque attaching screw to 20 to 25 in. lbs.
14. Install seal and retainer on modulation

PROBLEM

WARNING LIGHT COMES ON AND STAYS ON

WARNING LIGHT LOCATED IN THE INSTRUMENT PANEL, OPERATES WHEN THE IGNITION SWITCH IS "ON" OR POWER IS APPLIED TO THE WHEEL LOCK CONTROL SYSTEM.

WARNING LIGHT COMES ON FOR UP TO 5 SECONDS WHEN IGNITION SWITCH IS TURNED ON TO CHECK BULB AND WARNING CIRCUIT.

WARNING LIGHT IS NORMALLY "OFF" WHEN SYSTEMS ARE FUNCTIONAL.

VOLT-OHMMETER METHOD

WARNING LIGHT COMES ON AND STAYS ON WHEN A PROBLEM IS DETECTED IN ONE OR MORE AXLE SYSTEMS.

TROUBLE SHOOTING PROCEDURE LOCATES FAULTY AXLE SYSTEM/SYSTEMS AND ISOLATES PROBLEM TO WARNING LIGHT, POWER CABLE, SENSOR CABLE, SENSOR, MODULATOR, HARNESS ADAPTER, OR VEHICLE WIRING.

	STEP	PROCEDURE	WARN LIGHT	RESULTS	NEXT STEP
LOCATE FAULTY AXLE	1	TURN IGNITION SWITCH "ON". DISCONNECT POWER CABLE FROM EACH MODULATOR ASSEMBLY, ONE AT A TIME, UNTIL WARNING LIGHT GOES "OFF". (DO NOT RECONNECT CABLES UNLESS INSTRUCTED.) 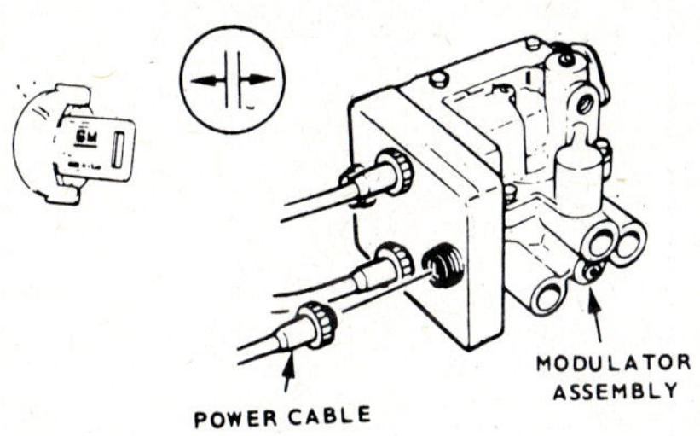	OFF	WARNING LIGHT WILL GO "OFF" WHEN A FAULTY SYSTEM IS DISCONNECTED. CONTINUE TROUBLESHOOTING THAT AXLE SYSTEM.	2
			ON	IF WARNING LIGHT STAYS "ON" WHEN ALL SYSTEMS ARE DISCONNECTED, PROBLEM IS IN WARNING LIGHT CIRCUIT.	1A
	1A	DISCONNECT POWER CABLES AT WIRING HARNESS ADAPTER, ONE AT A TIME, UNTIL THE WARNING LIGHT GOES "OFF". 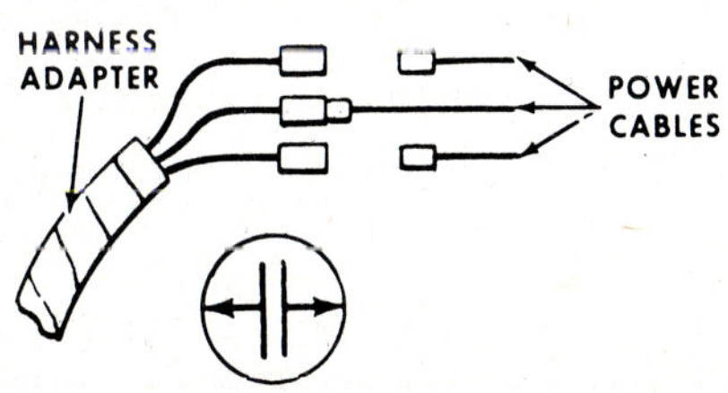	OFF	WARNING LIGHT GOES "OFF" WHEN POWER CABLE IS DISCONNECTED, PROBLEM IS IN THAT POWER CABLE.	1B
			ON	IF ALL POWER CABLES HAVE BEEN DISCONNECTED AND WARNING LIGHT IS STILL "ON", PROBLEM IS SHORT TO NEGATIVE IN WIRING HARNESS ADAPTER, CAB WIRING, OR WARNING LIGHT.	1C
	1B	REPLACE POWER CABLE. RECONNECT POWER CABLES AT WIRING HARNESS ADAPTER, ONE AT A TIME, MAKING SURE THE WARNING LIGHT REMAINS "OFF".	OFF	WARNING LIGHT CIRCUIT IS "OK."	5
			ON	IF WARNING LIGHT COMES "ON" WHEN A POWER CABLE IS RECONNECTED, PROBLEM IS IN THAT POWER CABLE.	(1B)
	1C	DISCONNECT INSTRUMENT PANEL (I/P) CONNECTOR FROM WIRING HARNESS ADAPTER. 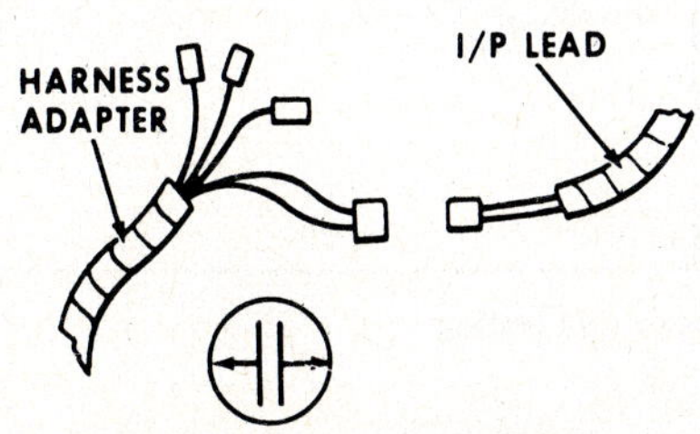	OFF	WARNING LIGHT GOES "OFF" WHEN I/P LEAD IS DISCONNECTED, PROBLEM IS IN WIRING HARNESS ADAPTER.	1D
			ON	WARNING LIGHT REMAINS "ON", PROBLEM IS IN WARNING LIGHT OR CAB WIRING BETWEEN INDICATOR LIGHT AND HARNESS ADAPTER.	1E

Fig. 32 AC anti-skid diagnosis and testing

	STEP	PROCEDURE	WARN LIGHT	RESULTS	NEXT STEP
LOCATE FAULTY AXLE	1D	REPLACE WIRING HARNESS ADAPTER. RECONNECT I/P LEAD. RECONNECT POWER CABLES AT ADAPTER ONE AT A TIME, MAKING SURE THE WARNING LIGHT REMAINS "OFF".	OFF	WARNING LIGHT CIRCUIT IS "OK."	5
			ON	IF WARNING LIGHT COMES "ON" WHEN ADAPTER IS CONNECTED, PROBLEM IS IN ADAPTER.	(1D)
	1E	DISCONNECT WARNING LIGHT FROM CAB WIRING.	OFF	PROBLEM IS IN CAB WIRING BETWEEN INDICATOR LIGHT AND ADAPTER HARNESS.	1F
			ON	WARNING INDICATOR LIGHT IS PROBLEM.	1G
	1F	REPLACE OR REPAIR WIRING. RECONNECT I/P LEAD AND POWER CABLES AT ADAPTER ONE AT A TIME, MAKING SURE THE WARNING LIGHT REMAINS "OFF".	OFF	WARNING LIGHT CIRCUIT IS "OK".	5
			ON	IF WARNING LIGHT COMES "ON" WHEN LAMP IS CONNECTED TO CAB WIRING, PROBLEM REMAINS IN WIRING.	1F
	1G	REPLACE WARNING INDICATOR LIGHT. RECONNECT I/P LEAD AND POWER CABLES AT ADAPTER ONE AT A TIME, MAKING SURE THE WARNING LIGHT REMAINS "OFF." 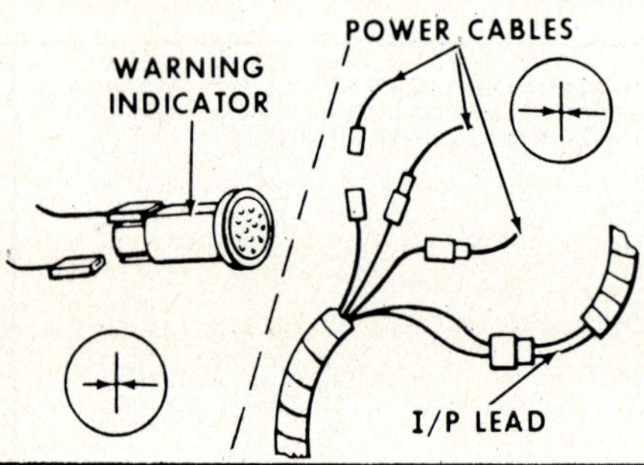	OFF	WARNING LIGHT CIRCUIT IS "OK".	5
			ON	IF WARNING LIGHT COMES "ON" WHEN I/P LEAD IS CONNECTED, PROBLEM IS IN WARNING LIGHT.	(1G)
POWER CABLE	2	MEASURE D.C. VOLTAGE AT POWER CABLE TERMINALS AS SHOWN. 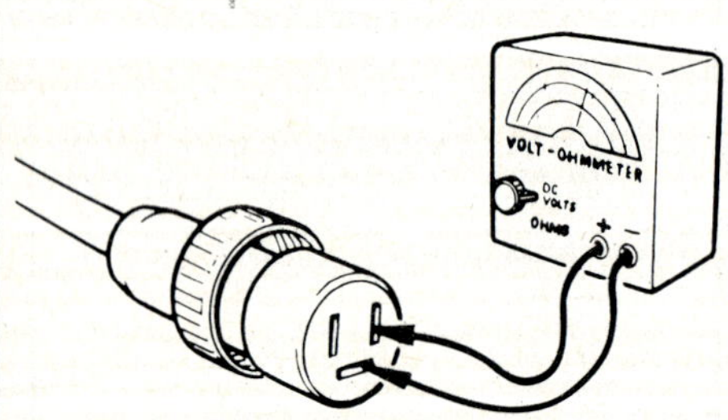	—	VOLTAGE GREATER THAN 10 V.D.C. MODULATOR IS GETTING POWER. PROBLEM IN SENSOR, SENSOR CABLES, OR MODULATOR.	3
			—	VOLTAGE LESS THAN 10 V.D.C. MODULATOR IS NOT GETTING POWER. PROBLEM IS IN POWER CABLE, POWER CABLE CONNECTIONS TO CAB WIRING, OR CAB WIRING.	2A
	2A	TO INVESTIGATE FURTHER, MEASURE D.C. VOLTAGE AT CAB END OF POWER CABLE AS SHOWN. (DO NOT DISCONNECT POWER CABLE FROM WIRING HARNESS ADAPTER). 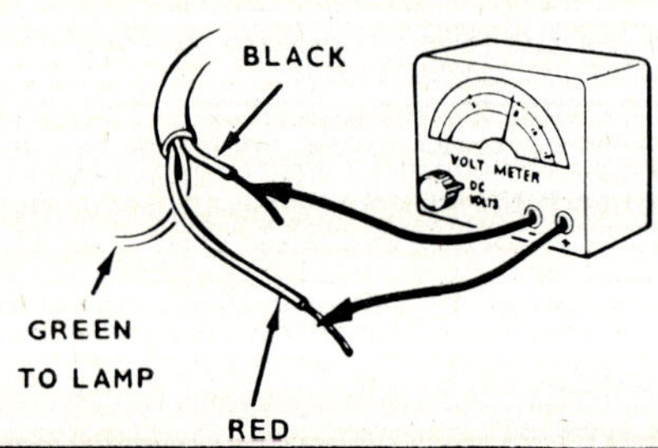	—	VOLTAGE GREATER THAN 10 V.D.C. PROBLEM IS IN POWER CABLE	2B
			—	VOLTAGE LESS THAN 10 D.C.V. PROBLEM IS IN CAB OR VEHICLE WIRING.	2C

Fig. 33 AC anti-skid diagnosis and testing

	STEP	PROCEDURE	WARN LIGHT	RESULTS	NEXT STEP
POWER CABLE	2B	REPLACE POWER CABLE AND CHECK VOLTAGE PER STEP 2 PROCEDURE. IF VOLTAGE IS LESS THAN 10 V.D.C. THERE IS STILL A POWER CABLE PROBLEM REPAIR OR REPLACE. IF VOLTAGE IS GREATER THAN 10 V.D.C. RECONNECT POWER CABLE TO MODULATOR. 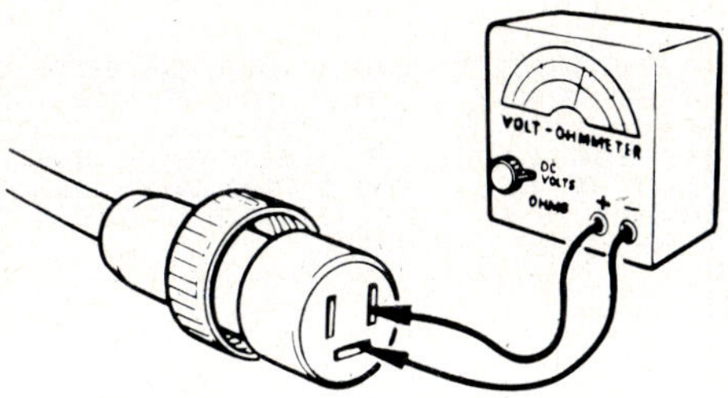	OFF	SYSTEM IS OK.	5
			ON	POWER CABLE PROBLEM HAS BEEN CORRECTED. - - - - - - - - - - - OTHER PROBLEM IS IN SENSORS, SENSOR CABLES OR MODULATOR.	3
LOCATE FAULTY AXLE	2C	CHECK AND REPAIR TRUCK ELECTRICAL SYSTEM INCLUDING FUSE OR CIRCUIT BREAKER, BATTERY, AND BATTERY TERMINALS AND ALTERNATOR OR GENERATOR. AFTER REPAIRING PROBLEM IN VEHICLE ELECTRICAL SYSTEM, RECONNECT POWER CABLE TO MODULATOR. 	OFF	SYSTEM IS "OK".	5
			ON	THERE IS STILL A PROBLEM WITH THE AXLE BEING EVALUATED.	2
SENSOR & SENSOR CABLE	3	DISCONNECT SENSOR CABLES FROM BOTH SENSORS. MEASURE RESISTANCE OF EACH SENSOR AS SHOWN: 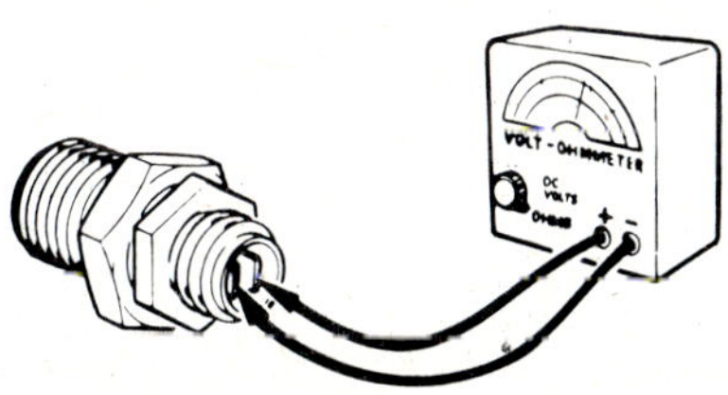	—	RESISTANCE BETWEEN 1000 & 2000 OHMS.	3A
			—	RESISTANCE **NOT** BETWEEN 1000 & 2000 OHMS. SENSOR IS FAULTY – REPLACE & RECHECK PER THIS STEP.	(3)
	3A	MEASURE RESISTANCE **OF EACH SENSOR** FROM SENSOR TERMINAL TO SENSOR BODY AS SHOWN. 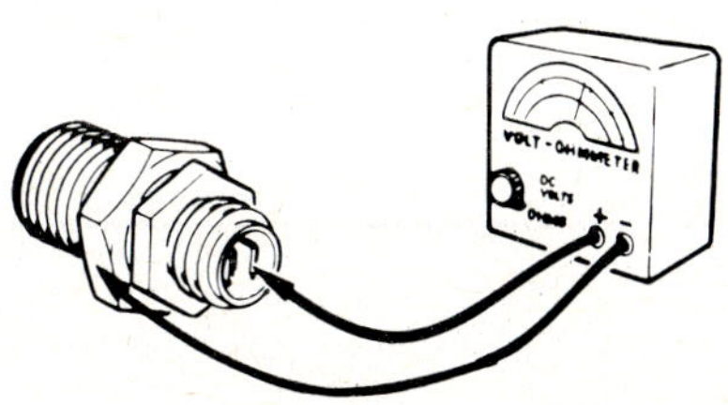	—	RESISTANCE GREATER THAN 100,000 OHMS. SENSOR IS GOOD.	3B
			—	RESISTANCE LESS THAN 100,000 OHMS. SENSOR IS FAULTY – REPLACE & RECHECK PER STEP 3.	3

Fig. 34 AC anti-skid diagnosis and testing

	STEP	PROCEDURE	WARN LIGHT	RESULTS	NEXT STEP
SENSOR & SENSOR CABLE	3B	RECONNECT SENSOR CABLES TO BOTH SENSORS. DISCONNECT BOTH SENSOR CABLES FROM MODULATOR & MEASURE RESISTANCE AT TERMINALS OF **BOTH** SENSOR CABLES AS SHOWN:	—	BOTH CABLES HAVE A RESISTANCE BETWEEN 1000 & 2000 OHMS.	3C
		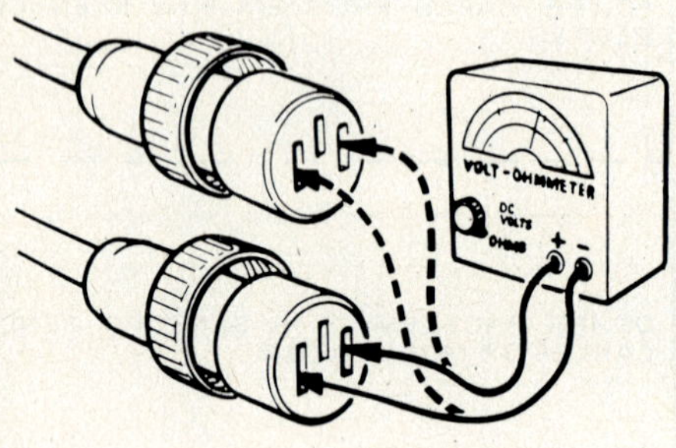	—	RESISTANCE OF ONE CABLE (OR BOTH CABLES) IS NOT BETWEEN 1000 & 2000 OHMS. RECHECK CONNECTION BETWEEN SENSOR & SENSOR CABLE. IF RESISTANCE IS STILL NOT BETWEEN 1000 & 2000 OHMS, SENSOR CABLE IS FAULTY REPLACE CABLE & RECHECK PER THIS STEP.	(3B)
	3C	MEASURE RESISTANCE AT TERMINALS OF **BOTH** SENSOR CABLES AS SHOWN.	—	BOTH CABLES HAVE A RESISTANCE GREATER THAN 100,000 OHMS. SENSOR & SENSOR CABLES ARE GOOD.	3D
		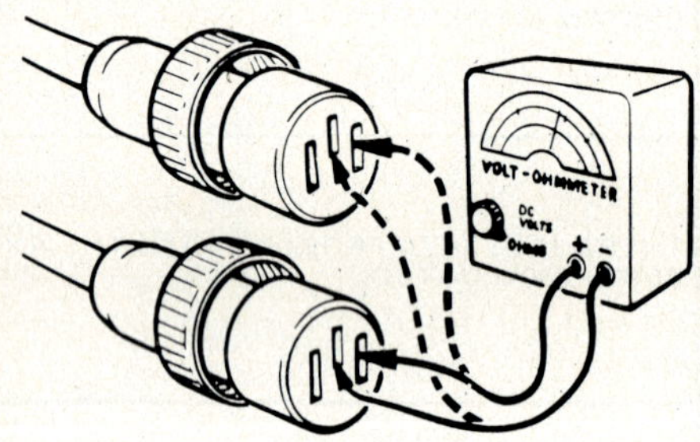	—	RESISTANCE OF ONE CABLE (OR BOTH CABLES) IS LESS THAN 100,000 OHMS. SENSOR CABLE IS FAULTY – REPLACE CABLE AND RECHECK PER STEP 3B.	3B
	3D	RECONNECT BOTH SENSOR CABLES TO THE MODULATOR. ALSO RECONNECT THE POWER CABLE. (CONNECT SENSOR CABLES BEFORE CONNECTING POWER CABLE).	OFF	SYSTEM IS "OK"	5
		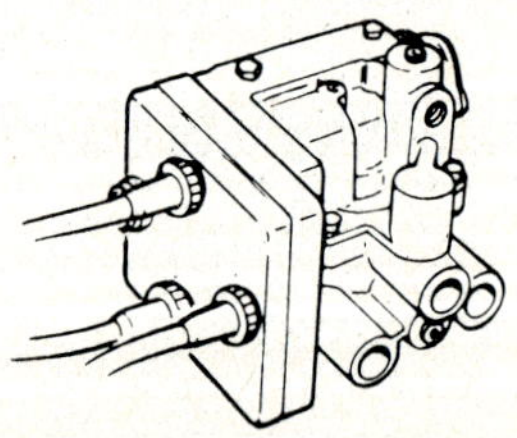	ON	SENSOR & SENSOR CABLES HAVE BEEN CHECKED AND FOUND GOOD. PROBLEM MUST BE IN MODULATOR.	4
REPLACE MODULATOR	4	REPLACE MODULATOR ASSEMBLY, DISCONNECT AND RECONNECT ALL HOSES AND CABLES PROPERLY.	OFF	SYSTEM IS "OK"	5
		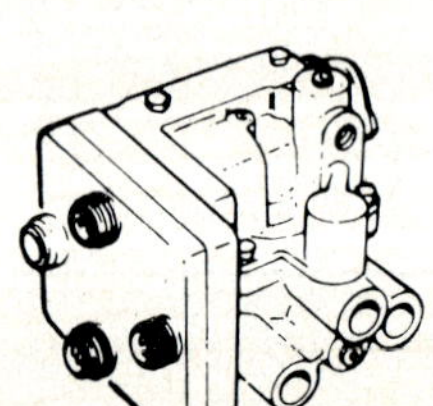	ON	THERE IS STILL A PROBLEM WITH THE SYSTEM ON THIS AXLE. CONTINUE TROUBLE-SHOOTING.	3

Fig. 35 AC anti-skid diagnosis and testing

	STEP	PROCEDURE	WARN LIGHT	RESULTS	NEXT STEP
LOCATE OTHER FAULTY AXLES	5	PLUG IN REMAINING POWER CABLES TO OTHER AXLE/S, ONE AT A TIME, AND CHECK THE WARNING LIGHT AFTER EACH CABLE IS PLUGGED IN.	OFF	SYSTEM IS "OK".	—
		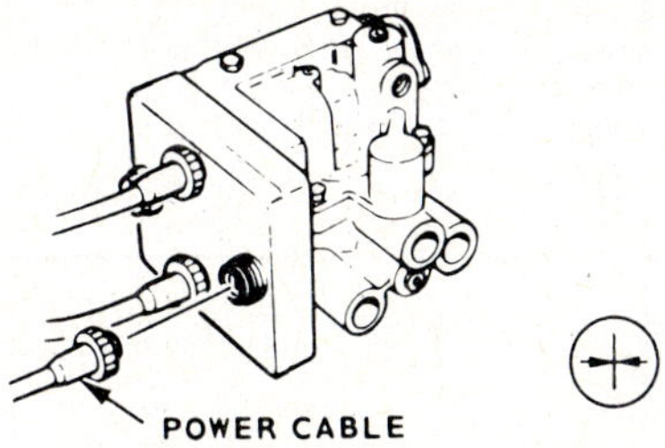	ON	IF WARNING LIGHT COMES ON AND STAYS ON AFTER PLUGGING IN POWER CABLE. THER IS SOMETHING WRONG WITH THAT AXLE SYSTEM ALSO.	2

PROBLEM

"WARNING LIGHT OFF" & "NO BULB-CHECK"

WARNING LIGHT COMES ON FOR UP TO 5 SECONDS WHEN IGNITION SWITCH IS TURNED ON TO CHECK BULB AND WARNING CIRCUIT. THIS IS NORMAL AND IS CALLED THE "BULB–CHECK" WHEN THERE IS NO "BULB–CHECK", TROUBLESHOOTING IS REQUIRED.

VOLT-OHMMETER METHOD

WHEN THERE IS NO "BULB–CHECK", THE MOST PROBABLE CAUSE IS A LOOSE OR BURNED-OUT WARNING LAMP. IF THE LAMP CIRCUIT IS FUNCTIONAL, EVERY AXLE SYSTEM HAS A PROBLEM IN THE POWER CABLE AND/OR MODULATOR.

	STEP	PROCEDURE	WARN LIGHT	BULB CHECK	RESULT	NEXT STEP
IN-LINE FUSE	1	CHECK IN-LINE 3 AMP FUSE. REPLACE IF NECESSARY. TURN ON IGNITION AND OBSERVE FOR "BULB-CHECK."	—	OFF	"BULB-CHECK" IS FUNCTIONAL. RETURN VEHICLE TO SERVICE.	—
					NO "BULB-CHECK", CONTINUE TROUBLESHOOTING.	2
			—	ON	SEE PROCEDURE "WARNING LIGHT COMES ON AND STAYS ON."	—

Fig. 36 AC anti-skid diagnosis and testing

tube.
15. Install O rings on modulation tube and valve body.
16. Install modulation tube, seal and first, into relay valve body, then position spring over tube.
17. Press exhaust valve body into relay valve body until fully seated, then install washer and snap ring, Fig. 58.
18. Position molded rubber valve assembly inside of relay piston with rib against piston, then install and torque retaining screw to 70 to 80 in. lbs.
19. Install O ring on relay piston.
20. Position small end of conical spring over lugs in piston cavity in relay valve body, then press relay piston into bore.
21. Install O ring on flow restrictor, then press flow restrictor into bore on relay valve body.
22. Install O ring on solenoid valve body, position relay and solenoid valve bodies to align rate control valve restriction bores and attaching bolt holes. Press valve bodies together until seated and install four attaching screws. Torque attaching screws to 110 to 130 in. lbs.
23. Insert solenoid leads through lower cover gasket and large oval hole in cover with concave side of cover facing away from valve assembly.
24. Install cover on valve body and install and torque attaching screws to 140 to 160 in. lbs. using tools 25360-5, 9.
25. Position gasket on cover and controller assembly, then connect two stator leads to controller terminals.
26. Carefully position excess solenoid leads inside lower cover and place controller and cover assembly on valve assembly. Assemble with Aux stamp on cover closest to modulator mounting surface.
27. Install attaching screws and torque to 16 to 20 ft. lbs. using tools J-25360-5, 9.

STEP	PROCEDURE	WARN LIGHT	BULB CHECK	RESULT	NEXT STEP
2 WARNING LAMP & CAB WIRING TEST	DISCONNECT I/P CONNECTOR FROM ADAPTER: JUMPER FROM GROUND TO GREEN WIRE IN I/P CONNECTOR. 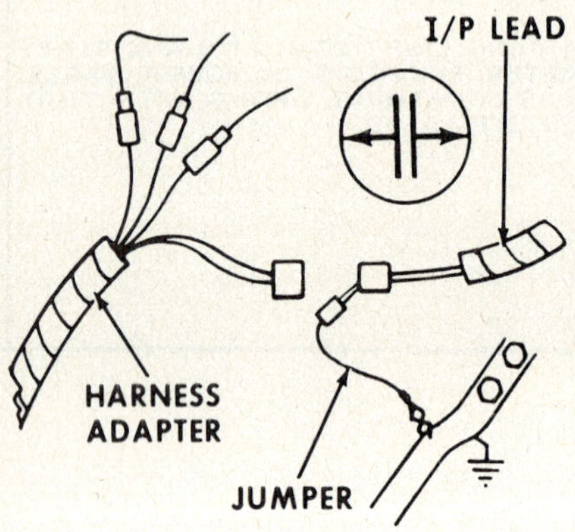	OFF	—	PROBLEM IS IN LAMP, LAMP SOCKET OR CAB WIRING.	2A
		ON	—	LAMP AND LAMP WIRING OK. EVERY AXLE SYSTEM HAS A PROBLEM IN THE POWER CABLE AND/OR MODULATOR.	3
3 AXLE SYSTEM TEST	REPEAT THIS STEP FOR EVERY AXLE SYSTEM.	ON	—	VOLTAGE GREATER THAN 10 V.D.C. POWER CABLE IS OK. PROBLEM IS IN MODULATOR.	3A
	DISCONNECT POWER CABLES FROM ALL MODULATORS. MEASURE D.C. VOLTAGE AT POWER CABLE AS SHOWN – OBSERVE POLARITY.	OFF	—	PROBLEM IS IN POWER CABLE REGARDLESS OF VOLTAGE MEASURED.	3B
	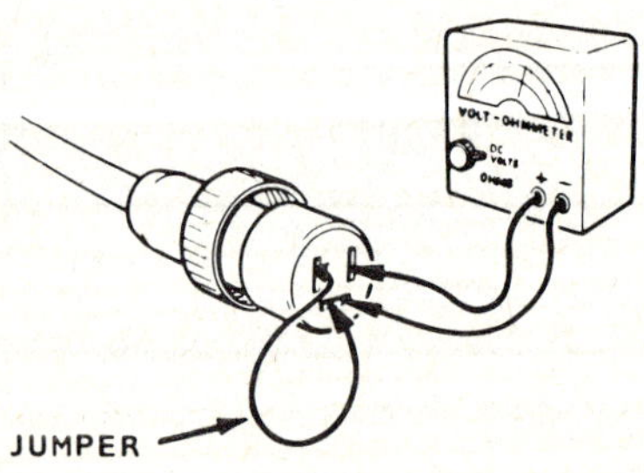	ON OR OFF	—	METER POINTER DEFLECTS DOWNSCALE (INDICATING REVERSE POLARITY) PROBLEM IS **REVERSE POLARITY** IN POWER CABLE OR IN CAB WIRING.	3C
2A WARNING LAMP & CAB WIRING	CORRECT WARNING LAMP AND/OR CAB WIRING PROBLEM BY REPAIRING OR REPLACING FAULTY COMPONENTS. TURN IGNITION "ON" AND LOOK FOR "BULB-CHECK".	—	YES	"BULB-CHECK" IS FUNCTIONAL. RETURN VEHICLE TO SERVICE.	—
		—	NO	CONTINUE TROUBLESHOOTING.	2

Fig. 37 AC anti-skid diagnosis and testing

STEP		PROCEDURE	WARN LIGHT	BULB CHECK	RESULT	NEXT STEP
3A	REPLACE MODULATOR	REPLACE MODULATOR. DISCONNECT AND RE-CONNECT ALL HOSES AND CABLES PROPERLY. TURN ON IGNITION AND LOOK FOR "BULB-CHECK".	—	YES	THIS AXLE SYSTEM OK. CONTINUE TROUBLESHOOTING NEXT AXLE.	3
					IF THIS IS FINAL AXLE SYSTEM, RECONNECT ALL POWER CABLES AND RETURN VEHICLE TO SERVICE.	—
			—	NO	PROBLEM IS STILL IN MODULATOR.	(3A)
3B	POWER CABLE	REPLACE POWER CABLE AND RECONNECT TO MODULATOR. TURN IGNITION "ON" AND LOOK FOR "BULB-CHECK".	—	YES	THIS AXLE SYSTEM IS OK. CONTINUE TROUBLESHOOTING NEXT AXLE.	3
					IF THIS IS FINAL AXLE SYSTEM, RECONNECT ALL POWER CABLES AND RETURN VEHICLE TO SERVICE.	—
			—	NO	CONTINUE TROUBLESHOOTING THIS AXLE.	3
3C	CAB WIRING	CORRECT REVERSE POLARITY PROBLEM IN POWER CABLE OR IN CAB WIRING, AND RE-CONNECT POWER CABLE TO MODULATOR. TURN IGNITION "ON" AND LOOK FOR "BULB-CHECK".	—	YES	THIS AXLE SYSTEM IS OK. CONTINUE TROUBLESHOOTING NEXT AXLE.	3
					IF THIS IS FINAL AXLE SYSTEM, RE-CONNECT ALL POWER CABLES AND RETURN VEHICLE TO SERVICE.	—
			—	NO	CONTINUE TROUBLESHOOTING THIS AXLE.	3

Fig. 38 AC anti-skid diagnosis and testing

AIRBRAKE – WLC MODULATOR ELECTRICAL SERVICE PROCEDURE ("SOLENOID" AND/OR "CONTROLLER & COVER ASSEMBLY" ONLY)

VOLT-OHMMETER METHOD

	STEP	INSTRUCTIONS		NEXT STEP
DISASSEMBLY	1	REMOVE CONTROLLER & COVER ASSEMBLY & DISCONNECT SOLENOID LEADS.	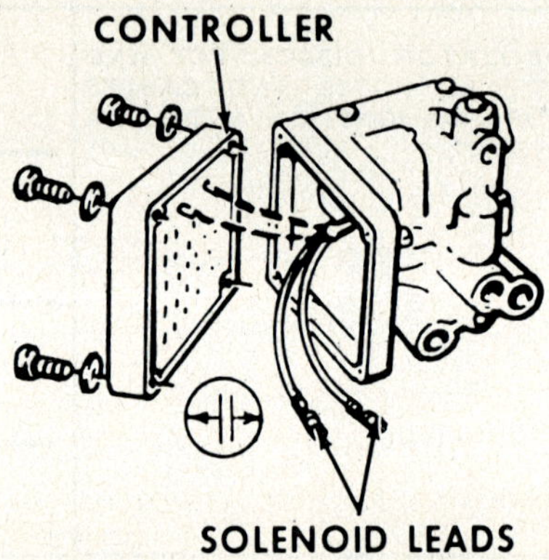	2
SOLENOID	2	TAKE SOLENOID LEADS AND MOMENTARILY CONNECT THEM TO A 12 VOLT BATTERY. 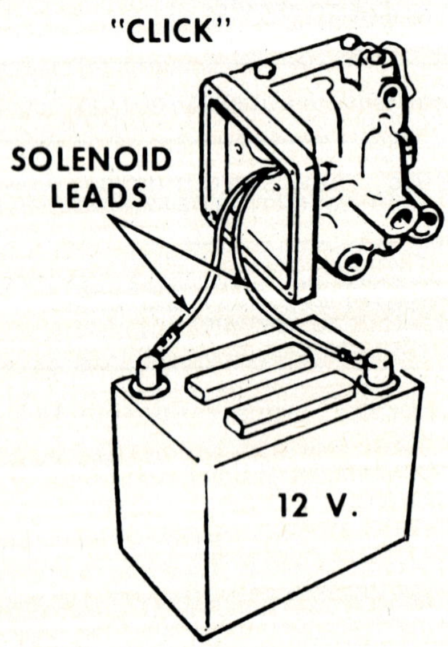	SOLENOID IS OK IF YOU HEAR A "CLICK" WHEN LEADS ARE CONNECTED AND ANOTHER "CLICK" WHEN THEY ARE DISCONNECTED. PROBLEM IS IN THE CONTROLLER & COVER ASSEMBLY.	5
			IF "CLICKS" ARE NOT HEARD, PROBLEM IS IN SOLENOID.	3

Fig. 39 AC anti-skid diagnosis and testing

<table>
<tr><th></th><th>STEP</th><th colspan="2">INSTRUCTION</th><th>NEXT STEP</th></tr>
<tr><td rowspan="3">SOLENOID</td><td>3</td><td>REPLACE THE SOLENOID.</td><td>CONTINUE WITH NEXT STEP –</td><td>4</td></tr>
<tr><td rowspan="2">4</td><td rowspan="2">TEST NEW SOLENOID IN SAME MANNER AS SHOWN IN STEP NO. 2.</td><td>SOLENOID IS OK IF A "CLICK" IS HEARD WHEN LEADS ARE CONNECTED AND ANOTHER "CLICK" WHEN THEY ARE DISCONNECTED. REASSEMBLE THE CONTROLLER & COVER ASSEMBLY AND RETEST MODULATOR.</td><td>6</td></tr>
<tr><td>IF NO "CLICKS" ARE HEARD, PROBLEM IS STILL IN SOLENOID.</td><td>3</td></tr>
<tr><td>CONTROLLER</td><td>5</td><td colspan="2">INSTALL NEW CONTROLLER & COVER ASSEMBLY.</td><td>6</td></tr>
<tr><td rowspan="4">TESTING REPAIRED MODULATOR ASSEMBLY</td><td rowspan="2">6</td><td rowspan="2">TEST THE REPAIRED MODULATOR BY SIMULATING A VEHICLE INSTALLATION ON THE BENCH AS FOLLOWS: CONNECT TWO SPEED SENSORS, TWO SPEED SENSOR CABLES, A BATTERY, AND A 12 VOLT TROUBLE LIGHT AS SHOWN.
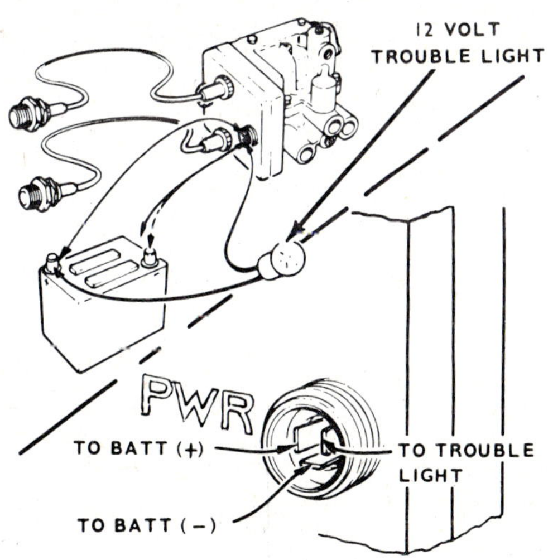

WHEN THE LEAD IS CONNECTED TO THE NEGATIVE TERMINAL OF THE BATTERY, A NORMAL "BULB-CHECK" SHOULD OCCUR.</td><td>NORMAL BULB – CHECK IS OBSERVED. (THE TROUBLE LIGHT COMES ON FOR UP TO 5 SECONDS AND THEN GOES OFF.)</td><td>7</td></tr>
<tr><td>NO "BULB-CHECK" IS OBSERVED OR TROUBLE LIGHT REMAINS ON. DOUBLE CHECK ALL ELECTRICAL CONNECTIONS. IF THE LIGHT REMAINS ON, THERE IS STILL A PROBLEM IN THE MODULATOR ASSEMBLY.</td><td>1</td></tr>
<tr><td rowspan="2">7</td><td rowspan="2">DISCONNECT ONE SENSOR CABLE FROM MODULATOR TO SIMULATE A FAULT.</td><td>IF THE TROUBLE LIGHT COMES "ON" AND STAYS ON, THE WARNING SYSTEM HAS CORRECTLY DETECTED THE FAULT.</td><td>8</td></tr>
<tr><td>IF THE TROUBLE LIGHT DOES NOT COME "ON", THE WARNING CIRCUIT IS NOT OPERATING.</td><td>5</td></tr>
</table>

Fig. 40 AC anti-skid diagnosis and testing

	STEP	INSTRUCTION		NEXT STEP
TESTING REPAIRED MODULATOR ASSEMBLY	8	RECONNECT THE SENSOR CABLE TO THE MODULATOR.	IF THE TROUBLE LIGHT GOES "OFF", THE MODULATOR IS OPERATING PROPERLY AND MAY BE RETURNED TO SERVICE.	—
			IF THE TROUBLE LIGHT REMAINS "ON", DOUBLE CHECK ALL TEST CONNECTIONS. IF LIGHT REMAINS "ON", THERE IS A PROBLEM IN THE CONTROLLER & COVER ASSEMBLY.	5

PROBLEM

WARNING LIGHT COMES ON AND STAYS ON

TRACTOR WARNING LIGHT LOCATED IN THE INSTRUMENT PANEL, OPERATE WHEN THE IGNITION SWITCH IS "ON" OR POWER IS APPLIED TO THE WHEEL LOCK CONTROL SYSTEM.

WARNING LIGHT COMES ON FOR UP TO 5 SECONDS WHEN IGNITION SWITCH IS TURNED ON TO CHECK BULB AND WARNING CIRCUIT.

WARNING LIGHT IS NORMALLY "OFF" WHEN SYSTEMS ARE FUNCTIONAL.

TEST PLUG METHOD

WARNING LIGHT COMES ON AND STAYS ON WHEN A PROBLEM IS DETECTED IN ONE OR MORE AXLE SYSTEMS.

TROUBLE SHOOTING PROCEDURE LOCATES FAULTY AXLE SYSTEM/SYSTEMS AND ISOLATES PROBLEM TO WARNING LIGHT, POWER CABLE, SENSOR CABLE, SENSOR, MODULATOR, HARNESS ADAPTER OR VEHICLE WIRING.

	STEP	PROCEDURE	WARN LIGHT	RESULTS	NEXT STEP
LOCATE FAULTY AXLE	1	TURN IGNITION SWITCH "ON", DISCONNECT POWER CABLE FROM EACH MODULATOR ASSEMBLY, ONE AT A TIME, UNTIL WARNING LIGHT GOES "OFF". (DO NOT RECONNECT CABLES UNLESS INSTRUCTED.)	OFF	WARNING LIGHT WILL GO "OFF" WHEN A FAULTY SYSTEM IS DISCONNECTED. CONTINUE TROUBLESHOOTING THAT AXLE SYSTEM.	2
			ON	IF WARNING LIGHT STAYS "ON" WHEN ALL SYSTEMS ARE DISCONNECTED, PROBLEM IS IN WARNING LIGHT CIRCUIT.	1A

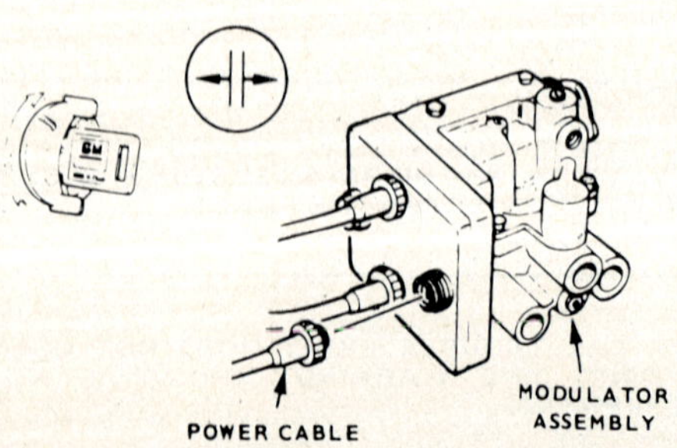

Fig. 41 AC anti-skid diagnosis and testing

	STEP	PROCEDURE	WARN LIGHT	RESULTS	NEXT STEP
LOCATE FAULTY AXLE	1A	DISCONNECT POWER CABLES AT WIRING HARNESS ADAPTER ONE AT A TIME, UNTIL THE WARNING LIGHT GOES "OFF". 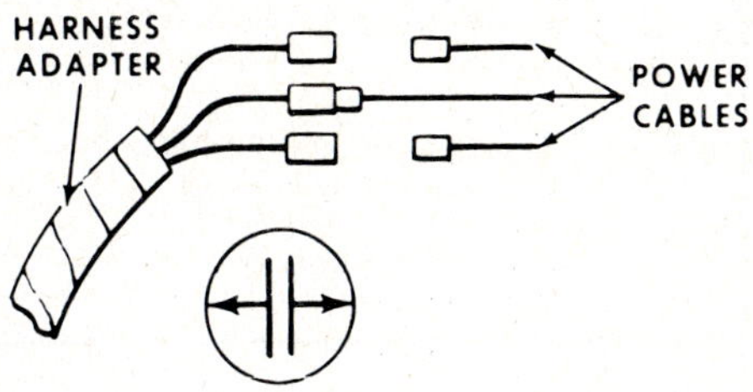 	OFF	IF WARNING LIGHT GOES "OFF" WHEN POWER CABLE IS DISCONNECTED, PROBLEM IS IN THAT POWER CABLE.	1B
			ON	IF ALL POWER CABLES HAVE BEEN DISCONNECTED AND WARNING LIGHT IS STILL "ON", PROBLEM IS SHORT TO NEGATIVE IN WIRING HARNESS ADAPTER, CAB WIRING, OR WARNING LIGHT.	1C
	1B	REPLACE POWER CABLE. RECONNECT POWER CABLES AT WIRING HARNESS ADAPTER, ONE AT A TIME, MAKING SURE THE WARNING LIGHT REMAINS "OFF".	OFF	WARNING LIGHT CIRCUIT IS "OK".	5
			ON	IF WARNING LIGHT COMES "ON" WHEN A POWER CABLE IS RECONNECTED, PROBLEM IS IN THAT POWER CABLE.	(1B)
	1C	DISCONNECT INSTRUMENT PANEL (I/P) CONNECTOR FROM WIRING HARNESS ADAPTER. 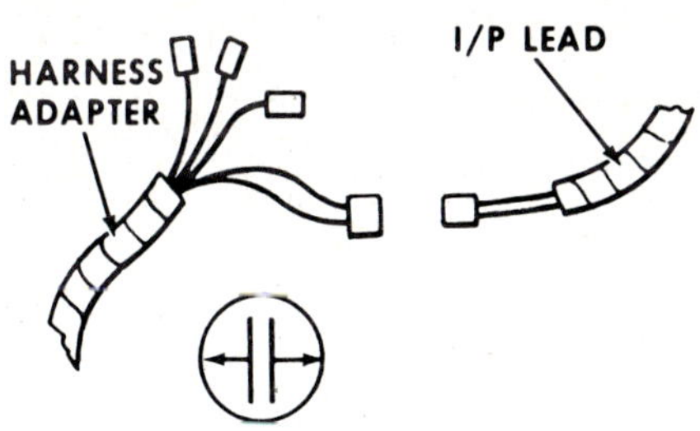 	OFF	WARNING LIGHT GOES "OFF" WHEN I/P LEAD IS DISCONNECTED, PROBLEM IS IN WIRING HARNESS ADAPTER.	1D
			ON	WARNING LIGHT REMAINS "ON", PROBLEM IS IN WARNING LIGHT OR CAB WIRING BETWEEN INDICATOR LIGHT AND HARNESS ADAPTER.	1E
	1D	REPLACE WIRING HARNESS ADAPTER. RECONNECT I/P LEAD. RECONNECT POWER CABLES AT ADAPTER ONE AT A TIME, MAKING SURE THE WARNING LIGHT REMAINS "OFF".	OFF	WARNING LIGHT CIRCUIT IS "OK".	5
			ON	IF WARNING LIGHT COMES "ON" WHEN ADAPTER IS CONNECTED. PROBLEM IS IN ADAPTER.	(1D)
	1E	DISCONNECT WARNING LIGHT FROM CAB WIRING.	OFF	PROBLEM IS IN CAB WIRING BETWEEN INDICATOR LIGHT AND ADAPTER HARNESS.	1F
			ON	WARNING INDICATOR LIGHT IS PROBLEM.	1G
	1F	REPLACE OR REPAIR WIRING. RECONNECT I/P LEAD AND POWER CABLES AT ADAPTER ONE AT A TIME, MAKING SURE THE WARNING LIGHT REMAINS "OFF".	OFF	WARNING LIGHT CIRCUIT IS "OK".	5
			ON	IF WARNING LIGHT COMES "ON" WHEN LAMP IS CONNECTED TO CAB WIRING, PROBLEM REMAINS IN WIRING.	(1F)

Fig. 42 AC anti-skid diagnosis and testing

	STEP	PROCEDURE	WARN LIGHT	RESULTS	NEXT STEP
LOCATE FAULTY AXLE	1G	REPLACE WARNING INDICATOR LIGHT. RECONNECT I/P LEAD AND POWER CABLES AT ADAPTER ONE AT A TIME, MAKING SURE THE WARNING LIGHT REMAINS "OFF". 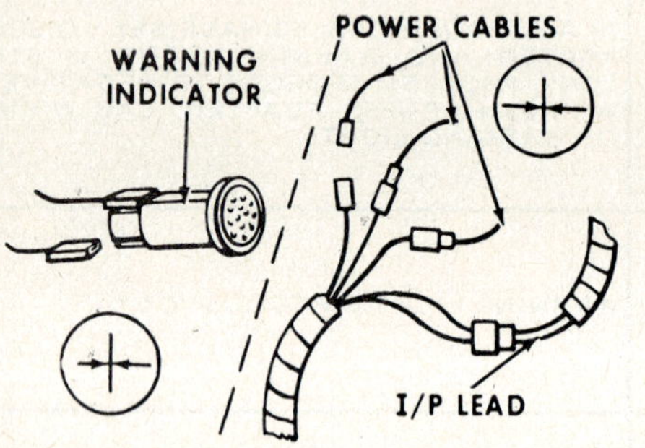 	OFF	WARNING LIGHT CIRCUIT IS "OK".	5
			ON	IF WARNING LIGHT COMES "ON" WHEN I/P LEAD IS CONNECTED, PROBLEM IS IN WARNING LIGHT.	(1G)
POWER CABLE TEST	2	TAKE POWER CABLE AND CONNECT IT TO POWER PLUG FROM SERVICE TOOL KIT. 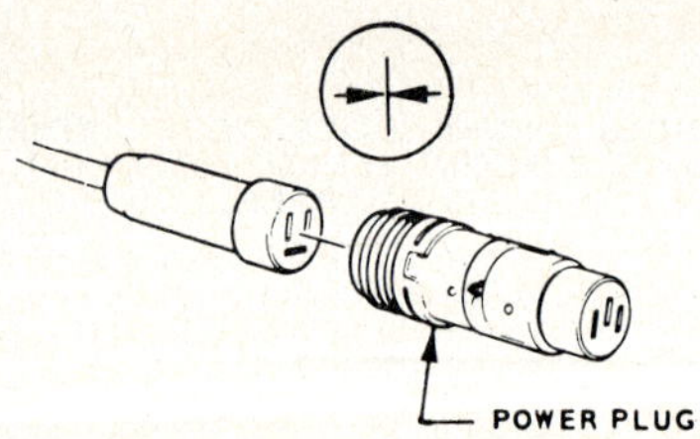 	—	IF GREEN LIGHT IN POWER PLUG COMES "ON", MODULATOR IS GETTING POWER. PROBLEM IS IN THE SENSOR, SENSOR CABLES, OR MODULATOR.	3
			—	IF GREEN LIGHT IN POWER PLUG IS "OFF", MODULATOR IS NOT GETTING POWER. PROBLEM IS IN POWER CABLE, POWER CABLE CONNECTIONS TO CAB WIRING, OR CAB WIRING.	2A
POWER CABLE	2A	TO INVESTIGATE FURTHER, TAKE POWER PLUG AND PLUG IT TOGETHER WITH POWER ADAPTER PLUG. TOUCH RED PIN TO RED WIRE AND BLACK PIN TO BLACK WIRE AT OTHER END OF POWER CABLE (IN CAB). 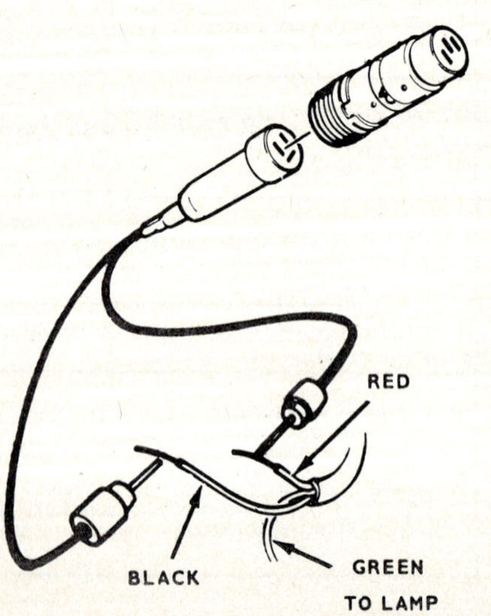 	—	IF GREEN LIGHT IN POWER PLUG COMES "ON", PROBLEM IS IN POWER CABLE.	2B
			—	IF GREEN LIGHT IN POWER PLUG IS "OFF", PROBLEM IS IN CAB OR VEHICLE WIRING.	2C

Fig. 43 AC anti-skid diagnosis and testing

	STEP	PROCEDURE	WARN LIGHT	RESULTS	NEXT STEP
POWER CABLE	2B	REPLACE POWER CABLE. INSTALL POWER PLUG IN MODULATOR END OF NEW POWER CABLE TO MAKE SURE POWER WILL NOW BE AVAILABLE TO MODULATOR. (GREEN LIGHT IN POWER PLUG SHOULD BE "ON". IF IT ISN'T, THERE IS STILL A POWER CABLE PROBLEM. REPAIR OR REPLACE.) REMOVE POWER PLUG AND CONNECT POWER CABLE TO THE MODULATOR. 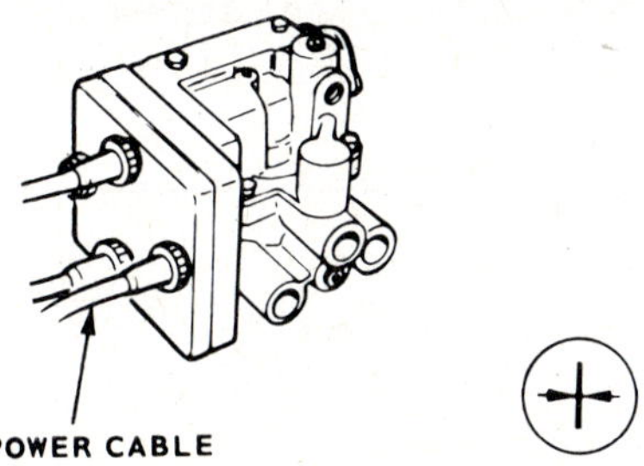	OFF	SYSTEM IS "OK".	5
			ON	PROBLEM IS IN SENSORS, SENSORS CABLES, OR MODULATOR.	3
	2C	CHECK AND REPAIR TRUCK ELECTRICAL SYSTEM INCLUDING FUSE OR CIRCUIT BREAKER, BATTERY, AND BATTERY TERMINALS AND ALTERNATOR OR GENERATOR. AFTER REPAIRING PROBLEM IN VEHICLE ELECTRICAL SYSTEM, RECONNECT POWER CABLE TO MODULATOR. 	OFF	SYSTEM IS "OK".	5
			ON	THERE IS STILL A PROBLEM WITH THE AXLE BEING EVALUATED.	2
SENSOR TEST	3	RECONNECT POWER CABLE AND DISCONNECT BOTH SENSOR CABLES FROM MODULATOR. TAKE BOTH BLUE SENSOR PLUGS FROM SERVICE TOOL KIT AND PLUG INTO SENSOR TERMINALS ON MODULATOR. (CYCLE IGNITION KEY "OFF-ON" TO PREVENT LATCHING OF WARNING LIGHT.) 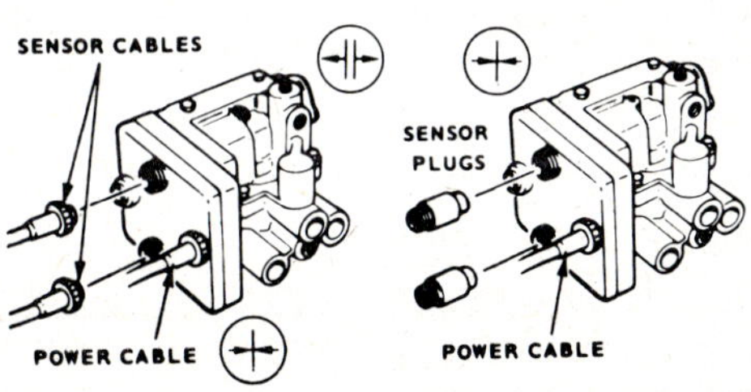	OFF	PROBLEM IS IN SENSORS OR SENSOR CABLES.	3A
			ON	CONTINUE WITH NEXT STEP –	4

Fig. 44 AC anti-skid diagnosis and testing

	STEP	PROCEDURE	WARN LIGHT	RESULTS	NEXT STEP
SENSOR & SENSOR CABLES	3A	REMOVE ONE BLUE SENSOR PLUG AND RECONNECT LEFT SENSOR CABLE. 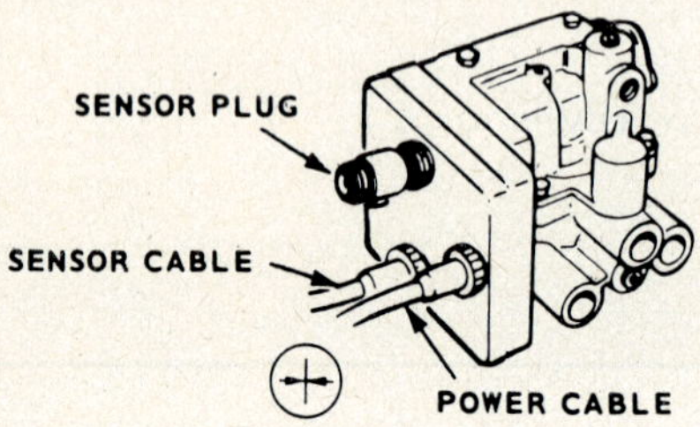	OFF	LEFT SENSOR AND SENSOR CABLE ARE OK. PROBLEM IS IN RIGHT SENSOR OR SENSOR CABLE.	3B
			ON	PROBLEM IS IN LEFT SENSOR OR SENSOR CABLE.	3F
	3B	REMOVE REMAINING BLUE SENSOR PLUG AND RECONNECT RIGHT SENSOR CABLE TO MODULATOR. (NOT NECESSARY IF PREVIOUS STEP WAS 3J). DISCONNECT RIGHT SENSOR CABLE FROM SENSOR AND CONNECT IT TO BLUE SENSOR PLUG. 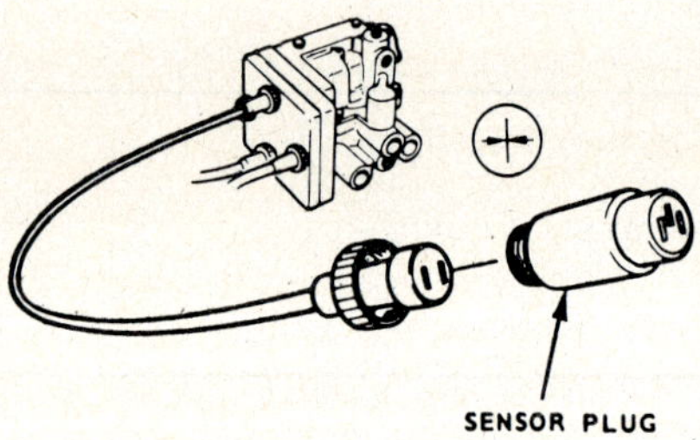	OFF	PROBLEM IS IN RIGHT SENSOR.	3C
			ON	PROBLEM IS IN RIGHT SENSOR CABLE.	3D
	3C	REPLACE RIGHT SENSOR. REMOVE BLUE SENSOR PLUG (NOT NECESSARY IF PREVIOUS STEP WAS 3E). AND REINSTALL SENSOR CABLE TO SENSOR.	OFF	SYSTEM IS "OK".	5
			ON	PROBLEM IS STILL IN RIGHT SENSOR OR CABLE CONNECTIONS. REPAIR OR REPLACE.	(3C)
	3D	REPLACE RIGHT SENSOR CABLE. REINSTALL BLUE SENSOR PLUG IN END OF CABLE. 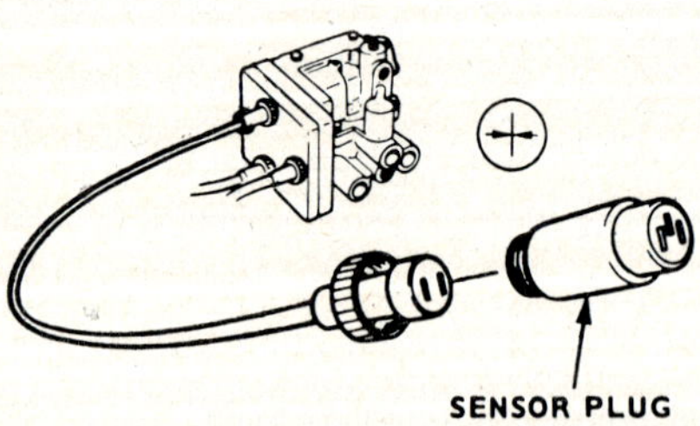	OFF	NEW RIGHT SENSOR CABLE IS "OK".	3E
			ON	PROBLEM IS STILL IN RIGHT SENSOR CABLE OR CABLE CONNECTIONS. REPAIR OR REPLACE.	(3D)
	3E	REMOVE BLUE SENSOR PLUG AND RECONNECT NEW RIGHT SENSOR CABLE TO RIGHT SENSOR.	OFF	SYSTEM IS "OK".	5
			ON	PROBLEM IS IN RIGHT SENSOR.	3C

Fig. 45 AC anti-skid diagnosis and testing

SENSOR & SENSOR CABLES (CONTINUED)

STEP	PROCEDURE	WARN LIGHT	RESULTS	NEXT STEP
3F	DISCONNECT LEFT SENSOR CABLE FROM SENSOR AND CONNECT IT TO BLUE SENSOR PLUG. 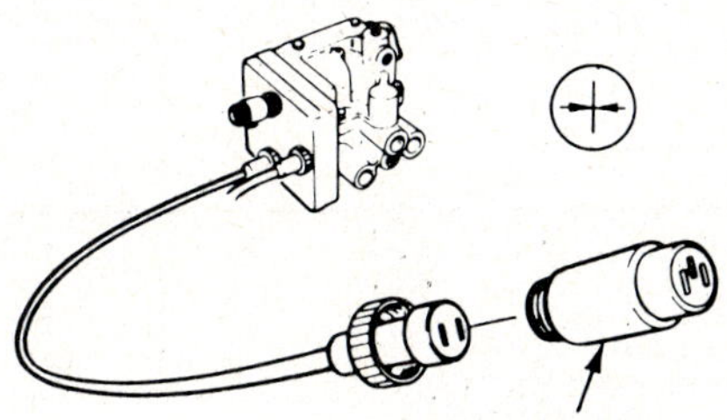 SENSOR PLUG	OFF	PROBLEM IS IN LEFT SENSOR.	3G
		ON	PROBLEM IS IN LEFT SENSOR CABLE.	3H
3G	REPLACE LEFT SENSOR. REMOVE BLUE SENSOR PLUG (NOT NECESSARY IF PREVIOUS STEP WAS 3D and REINSTALL SENSOR CABLE TO SENSOR.	OFF	LEFT SENSOR AND SENSOR CABLE "OK".	3J
		ON	PROBLEM IS STILL IN LEFT SENSOR OR CABLE CONNECTIONS. REPAIR OR REPLACE.	(3G)
3H	REPLACE LEFT SENSOR CABLE. REINSTALL BLUE SENSOR PLUG IN END OF CABLE. 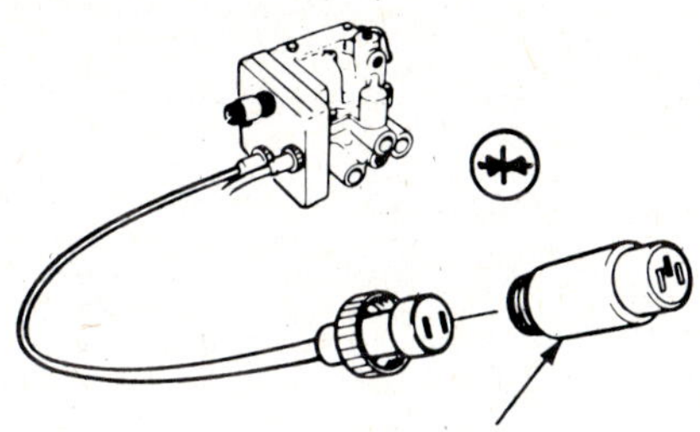 SENSOR PLUG	OFF	NEW LEFT SENSOR CABLE IS "OK".	3I
		ON	PROBLEM IS STILL IN LEFT SENSOR CABLE OR CABLE CONNECTIONS. REPAIR OR REPLACE.	(3H)
3I	REMOVE BLUE SENSOR PLUG AND CONNECT NEW LEFT SENSOR CABLE TO LEFT SENSOR.	OFF	LEFT SENSOR AND SENSOR CABLE ARE "OK".	3J
		ON	PROBLEM IS IN LEFT SENSOR.	3G
3J	REMOVE BLUE SENSOR PLUG FROM MODULATOR AND REINSTALL RIGHT SENSOR CABLE. 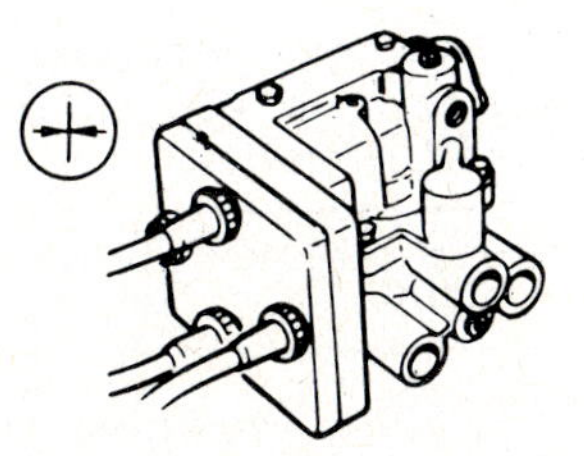	OFF	SYSTEM IS "OK".	5
		ON	PROBLEM IS IN RIGHT SENSOR OR SENSOR CABLE.	3B

Fig. 46 AC anti-skid diagnosis and testing

	STEP	PROCEDURE	WARN LIGHT	RESULTS	NEXT STEP
REPLACE MODULATOR	4	REPLACE MODULATOR ASSEMBLY, DISCONNECT AND RECONNECT ALL HOSES AND CABLES PROPERLY. 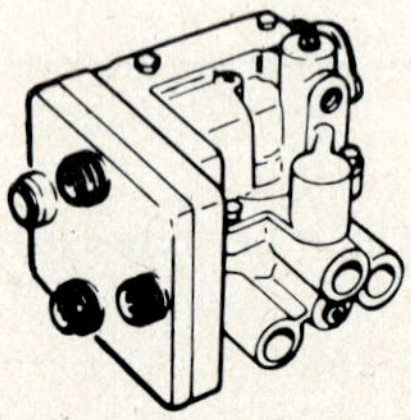	OFF	SYSTEM IS "OK".	5
			ON	THERE IS STILL A PROBLEM WITH THE SYSTEM ON THIS AXLE. CONTINUE TROUBLE-SHOOTING.	3
LOCATE OTHER FAULTY AXLES	5	PLUG IN REMAINING POWER CABLES TO OTHER AXLE/S, ONE AT A TIME, AND CHECK THE WARNING LIGHT AFTER EACH CABLE IS PLUGGED IN. 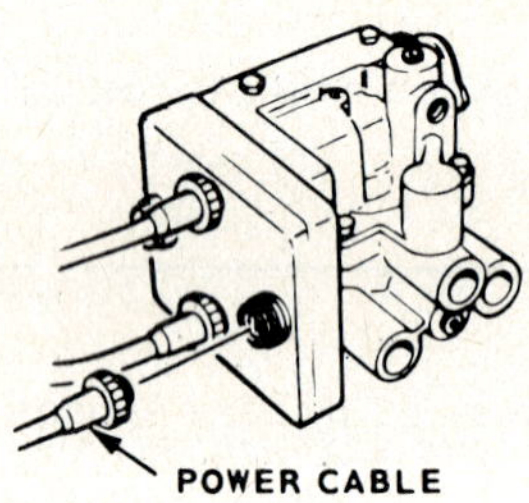	OFF	SYSTEM IS "OK".	—
			ON	IF WARNING LIGHT COMES ON AND STAYS ON AFTER PLUGGING IN POWER CABLE, THERE IS SOMETHING WRONG WITH THAT AXLE SYSTEM ALSO.	2

PROBLEM

"WARNING LAMP OFF" & "NO BULB-CHECK"

WARNING LIGHT COMES ON FOR UP TO 5 SECONDS WHEN IGNITION SWITCH IS TURNED ON TO CHECK BULB AND WARNING CIRCUIT. THIS IS NORMAL AND IS CALLED THE "BULB-CHECK". WHEN THERE IS NO "BULB-CHECK", TROUBLE-SHOOTING IS REQUIRED.

TEST PLUG METHOD

WHEN THERE IS NO "BULB-CHECK", THE MOST PROBABLE CAUSE IS A LOOSE OR BURNED-OUT WARNING LAMP, IF THE LAMP CIRCUIT IS FUNCTIONAL, EVERY AXLE SYSTEM HAS A PROBLEM IN THE POWER CABLE AND/OR MODULATOR.

	STEP	PROCEDURE	WARN LIGHT	BULB CHECK	RESULT	NEXT STEP
IN-LINE FUSE	1	CHECK IN-LINE 3 AMP FUSE. REPLACE IF NECESSARY. TURN ON IGNITION AND OBSERVE FOR "BULB-CHECK".	—	OFF	"BULB-CHECK" IS FUNCTIONAL. RETURN VEHICLE TO SERVICE.	
					NO "BULB-CHECK", CONTINUE TROUBLE-SHOOTING.	2
			—	ON	SEE PROCEDURE "WARNING LIGHT COMES ON AND STAYS ON".	—

Fig. 47 AC anti-skid diagnosis and testing

STEP	PROCEDURE	WARN LIGHT	BULB CHECK	RESULT	NEXT STEP
2 WARNING LAMP & CAB WIRING TEST	DISCONNECT I/P CONNECTOR FROM ADAPTER: JUMPER FROM GROUND TO GREEN WIRE IN I/P CONNECTOR. 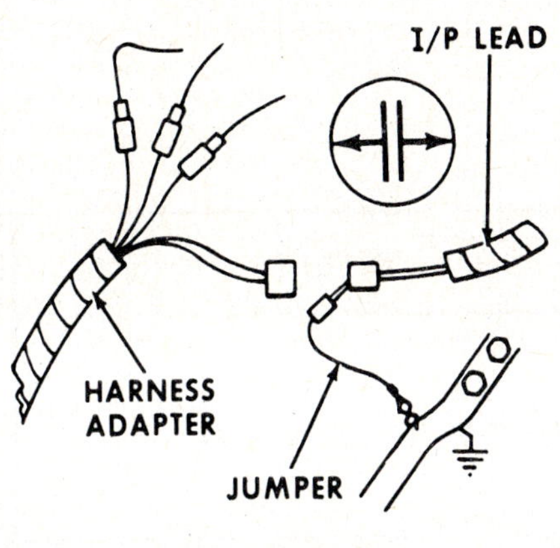	OFF	—	PROBLEM IS IN LAMP, LAMP SOCKET, OR CAB WIRING.	2A
		ON	—	LAMP AND LAMP WIRING OK, EVERY AXLE SYSTEM HAS A PROBLEM IN THE POWER CABLE AND/OR MODULATOR.	3
3 AXLE SYSTEM TEST	REPEAT THIS STEP FOR EVERY AXLE SYSTEM. DISCONNECT POWER CABLES FROM ALL MODULATORS. CONNECT THE POWER PLUG (FROM THE SERVICE TOOL KIT) TO THE POWER CABLE OF THE AXLE SYSTEM YOU ARE TROUBLESHOOTING. 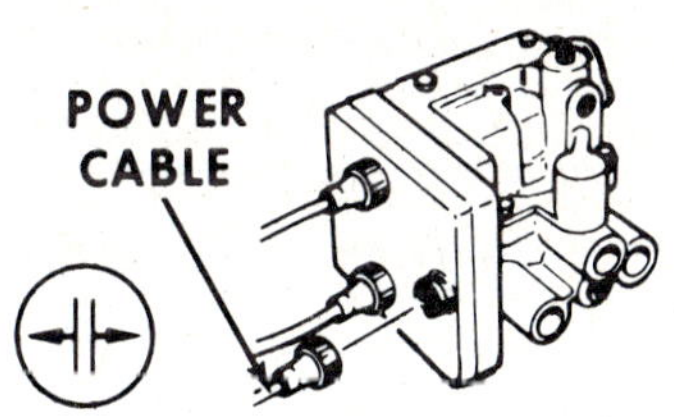 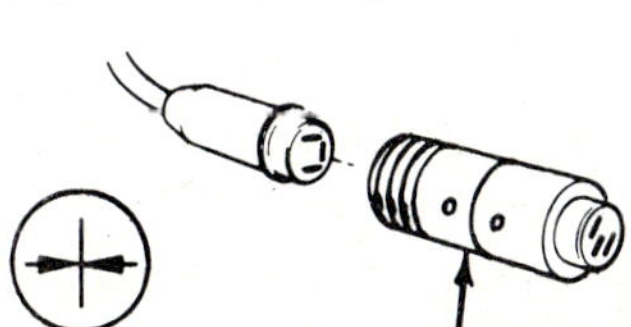	ON	—	IF GREEN LIGHT IN POWER PLUG IS "ON", POWER CABLE IS OK. PROBLEM IS IN MODULATOR.	3A
		OFF	—	IF GREEN LIGHT IN POWER PLUG IS "ON", OR BOTH POWER PLUG LIGHTS ARE "OFF", PROBLEM IS IN POWER CABLE.	3B
		ON OR OFF	—	IF RED LIGHT IN POWER PLUG IS "ON" PROBLEM IS REVERSE POLARITY IN CABLE OR IN CAB WIRING.	3C
2A WARNING LAMP & CAB WIRING	CORRECT WARNING LAMP AND/OR CAB WIRING PROBLEM BY REPAIRING OR REPLACING FAULTY COMPONENTS. TURN IGNITION "ON" AND LOOK FOR "BULB-CHECK".	—	YES	"BULB-CHECK" IS FUNCTIONAL. RETURN VEHICLE TO SERVICE.	—
		—	NO	CONTINUE TROUBLESHOOTING.	2

Fig. 48 AC anti-skid diagnosis and testing

STEP	PROCEDURE	WARN LIGHT	BULB CHECK	RESULT	NEXT STEP
3A REPLACE MODULATOR	REPLACE MODULATOR. DISCONNECT AND RE-CONNECT ALL HOSES AND CABLES PROPERLY. TURN ON IGNITION AND LOOK FOR "BULB-CHECK".	—	YES	THIS AXLE SYSTEM OK. CONTINUE TROUBLESHOOTING NEXT AXLE.	3
				IF THIS IS FINAL AXLE SYSTEM, RECONNECT ALL POWER CABLES AND RETURN VEHICLE TO SERVICE.	—
		—	NO	PROBLEM IS STILL IN MODULATOR.	(3A)
3B POWER CABLE	REPLACE POWER CABLE AND RECONNECT TO MODULATOR. TURN IGNITION "ON" AND LOOK FOR "BULB-CHECK".	—	YES	THIS AXLE SYSTEM IS OK. CONTINUE TROUBLESHOOTING NEXT AXLE.	3
				IF THIS IS FINAL AXLE SYSTEM, RECONNECT ALL POWER CABLES AND RETURN VEHICLE TO SERVICE.	—
		—	NO	CONTINUE TROUBLESHOOTING THIS AXLE.	3
3C CAB WIRING	CORRECT REVERSE POLARITY PROBLEM IN POWER CABLE OR IN CAB WIRING, AND RECONNECT POWER CABLE TO MODULATOR. TURN IGNITION "ON" AND LOOK FOR "BULB-CHECK".	—	YES	THIS AXLE SYSTEM IS OK. CONTINUE TROUBLESHOOTING NEXT AXLE.	3
				IF THIS IS FINAL AXLE SYSTEM, RECONNECT ALL POWER CABLES AND PUT VEHICLE BACK IN SERVICE.	—
		—	NO	CONTINUE TROUBLESHOOTING THIS AXLE.	3

Fig. 49 AC anti-skid diagnosis and testing

AIR BRAKE – WHEEL LOCK CONTROL MODULATOR – ELECTRICAL SERVICE PROCEDURE

("Solenoid" and/or "Controller and Cover Assembly" only)

TEST PLUG METHOD

	STEP	PROCEDURE		NEXT STEP
DISASSEMBLY	1	REMOVE CONTROLLER AND COVER ASSEMBLY AND DISCONNECT SOLENOID LEADS.	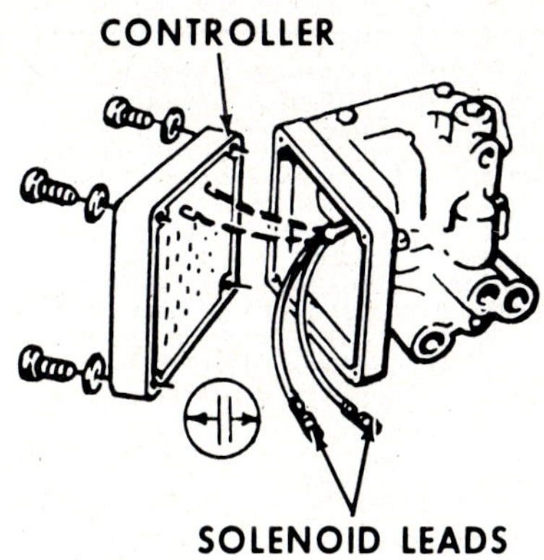	2
SOLENOID	2	TAKE SOLENOID LEADS AND MOMENTARILY CONNECT THEM TO A 12 VOLT BATTERY. 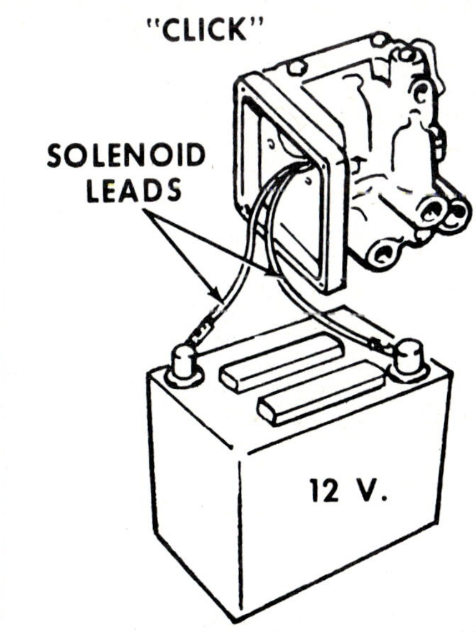	SOLENOID IS OK IF YOU HEAR A "CLICK" WHEN LEADS ARE CONNECTED AND ANOTHER "CLICK" WHEN THEY ARE DISCONNECTED. PROBLEM IS IN CONTROLLER AND COVER ASSEMBLY.	5
			IF "CLICKS" ARE NOT HEARD – PROBLEM IS IN SOLENOID.	3
	3	REPLACE THE SOLENOID.	CONTINUE WITH NEXT STEP –	4
	4	TEST NEW SOLENOID IN SAME MANNER AS DESCRIBED IN STEP 2.	SOLENOID IS OK IF A "CLICK" IS HEARD WHEN LEADS ARE CONNECTED AND ANOTHER "CLICK" WHEN THEY ARE DISCONNECTED. REASSEMBLE THE CONTROLLER AND COVER ASSEMBLY AND RETEST MODULATOR.	6
			IF NO "CLICKS" ARE HEARD, PROBLEM IS STILL IN SOLENOID.	3

Fig. 50 AC anti-skid diagnosis and testing

	STEP	PROCEDURE		NEXT STEP
CONTROLLER	5	REPLACE THE CONTROLLER AND COVER ASSEMBLY.		6
TESTING REPAIRED MODULATOR ASSEMBLY	6	TEST THE REPAIRED ASSEMBLY BY INSTALLING TWO "SENSOR" PLUGS (FROM TROUBLESHOOTING KIT) INTO SENSOR TERMINALS AND "MODULATOR TEST" PLUG INTO POWER TERMINAL. USE "POWER" AND "POWER ADAPTER" PLUGS FROM THE KIT TO SIMULATE TRACTOR WARNING LIGHT. CONNECT LEADS AS SHOWN BELOW – A NORMAL "BULB-CHECK" SHOULD BE OBSERVED WHEN THE CIRCUIT IS COMPLETED BY CONNECTING BLACK LEAD TO NEGATIVE TERMINAL OF BATTERY. 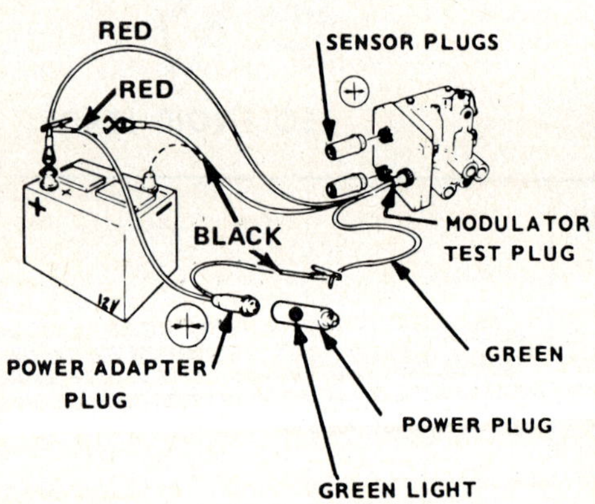	NORMAL "BULB-CHECK" IS OBSERVED (THE GREEN LIGHT IN THE POWER PLUG COMES ON FOR 2-5 SECONDS AND THEN GOES OFF.	7A
			NO "BULB-CHECK", OR – LAMP IN POWER PLUG REMAINS "ON". CHECK ALL TEST PLUGS AND OTHER ELECTRICAL CONNECTIONS. IF LAMP REMAINS "ON", THERE IS STILL A PROBLEM IN MODULATOR ASSEMBLY.	1
	7	REMOVE ONE SENSOR PLUG TO SIMULATE A FAULT.	GREEN LAMP SHOULD GO "ON" INDICATING A FAULT. THIS IS NORMAL.	8
			GREEN LAMP DOES NOT COME "ON" INDICATING SELF-CHECK CIRCUIT IS NOT OPERATING. REMOVE CONTROLLER AND COVER ASSEMBLY PER STEP 1 AND REPLACE PER STEP 5.	5
	8	REINSTALL SENSOR PLUG BACK ON SENSOR TERMINAL.	GREEN LAMP SHOULD GO "OFF". THIS IS NORMAL AND INDICATES MODULATOR ASSEMBLY IS OPERATIONAL – RETURN TO SERVICE.	—
			GREEN LAMP REMAINS "ON". DOUBLE CHECK ALL TEST PLUGS AND OTHER ELECTRICAL CONNECTIONS. IF LAMP IS STILL "ON", REPEAT MODULATOR SERVICE PROCEDURE.	1

Fig. 51 AC anti-skid diagnosis and testing

AC WHEEL LOCK CONTROL
SERVICE TOOL KIT CHECKOUT

SERVICE TOOL KIT COMPONENTS ARE FUNCTIONAL IF THE FOLLOWING REQUIREMENTS ARE MET. IF THE REQUIREMENTS ARE NOT MET, REPLACE THE COMPONENT.

TEST	PROCEDURE	REQUIREMENTS
SENSOR PLUGS	MEASURE RESISTANCE AT BOTH ENDS OF EACH SENSOR PLUG AS SHOWN 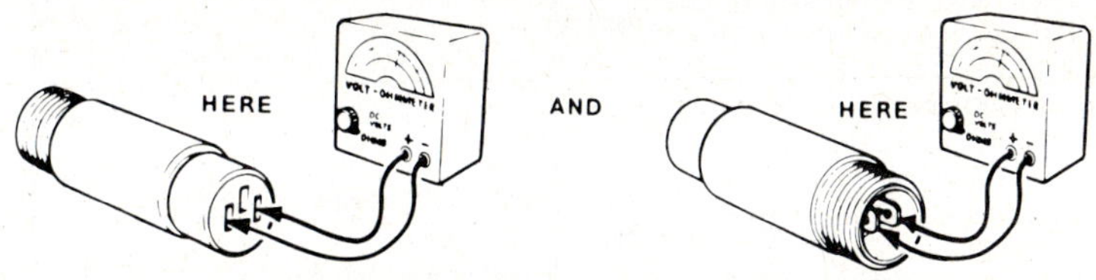	RESISTANCE SHOULD BE BETWEEN 1000 AND 2000 OHMS
POWER ADAPTOR PLUG	MEASURE RESISTANCE OF POWER ADAPTOR PLUG IN TWO PLACES AS SHOWN 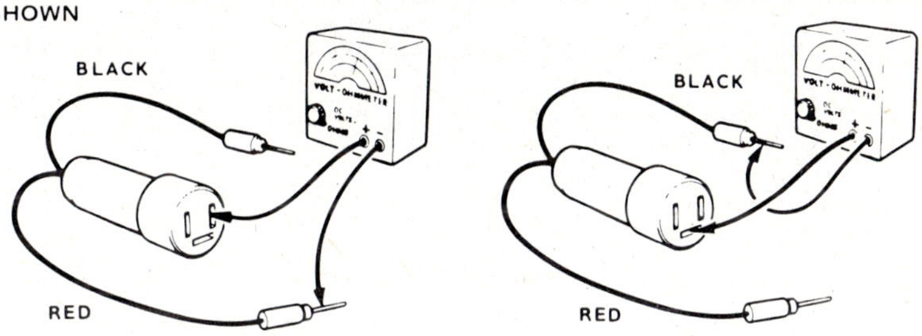	RESISTANCE SHOULD BE LESS THAN 1.0 OHM
	CONNECT 12 VOLTS TWO PLACES ONE AT A TIME AS SHOWN 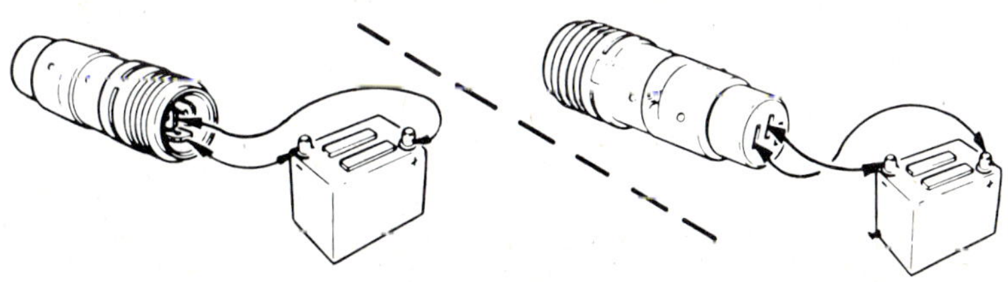	GREEN LIGHT SHOULD BE "ON" RED LIGHT SHOULD BE "OFF"
POWER PLUG	CONNECT 12 VOLTS TWO PLACES ONE AT A TIME AS SHOWN 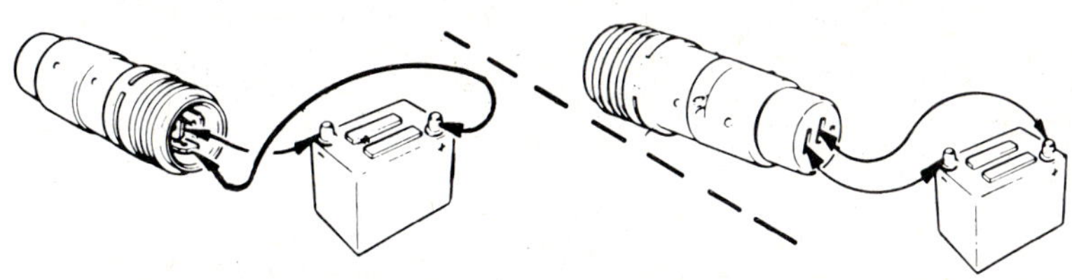	RED LIGHT SHOULD BE "ON" GREEN LIGHT SHOULD BE "OFF"
	MEASURE RESISTANCE AS SHOWN 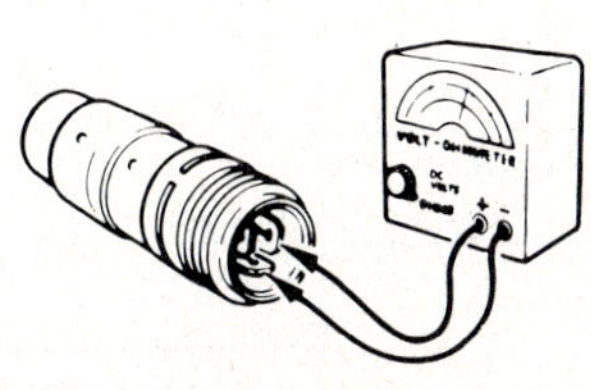	RESISTANCE SHOULD BE LESS THAN 1.0 OHM

Fig. 52 AC anti-skid diagnosis and testing

<table>
<tr><th>TEST</th><th>PROCEDURE</th><th>REQUIREMENTS</th></tr>
<tr><td>SENSOR ADAPTOR PIGTAIL</td><td>MEASURE RESISTANCE OF EACH WIRE IN PIGTAIL AS SHOWN
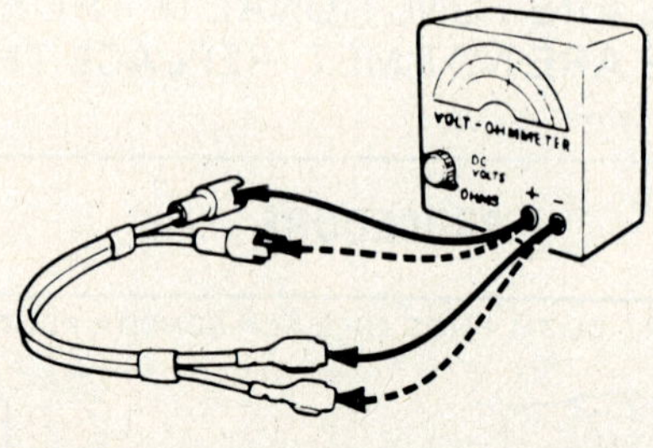</td><td>RESISTANCE SHOULD BE LESS THAN 1.0 OHM</td></tr>
<tr><td>MODULATOR TEST PLUG</td><td>MEASURE RESISTANCE OF MODULATOR TEST PLUG THREE PLACES AS SHOWN
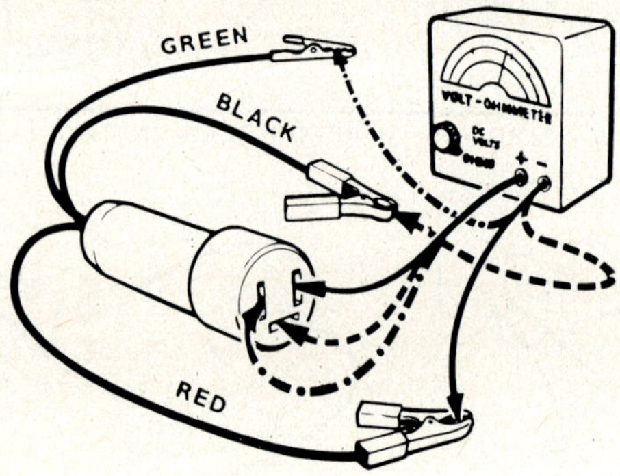</td><td>RESISTANCE SHOULD BE LESS THAN 1.0 OHM</td></tr>
</table>

MODULATOR PNEUMATIC CHECKOUT PROCEDURE

<table>
<tr><th></th><th>STEP</th><th>PROCEDURE</th><th>RESULT</th><th>NEXT STEP</th></tr>
<tr><td>CHARGE AIR</td><td>1</td><td>BLOCK VEHICLE AND FULLY CHARGE AIR BRAKE SYSTEM</td><td>PRESSURE SHOULD BE SUFFICIENT TO REMOVE SPRING BRAKES.</td><td>2</td></tr>
<tr><td rowspan="2">BRAKE APPLICATION</td><td rowspan="2">2</td><td rowspan="2">MAKE SEVERAL BRAKE APPLICATIONS
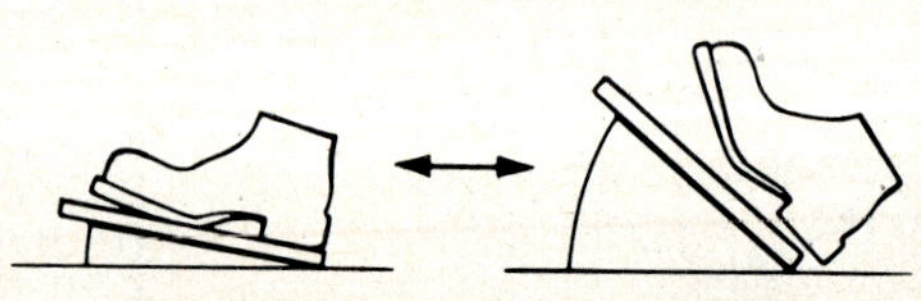</td><td>BRAKE CHAMBERS SHOULD APPLY AND RELEASE AT ALL WHEELS.
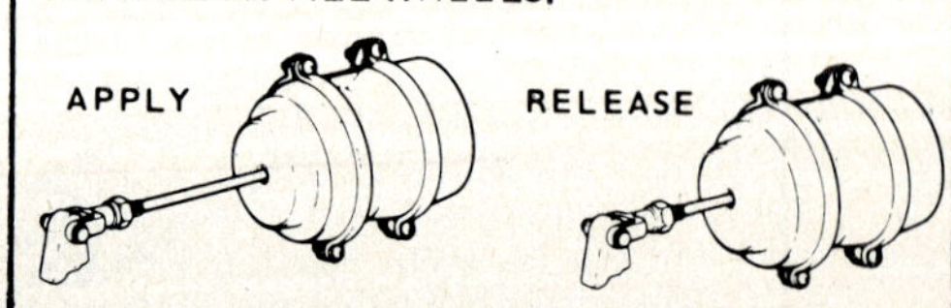</td><td>3</td></tr>
<tr><td>VALVE DOES NOT FUNCTION</td><td>6</td></tr>
</table>

Fig. 53 AC anti-skid diagnosis and testing

	STEP	PROCEDURE	RESULT	NEXT STEP
RELEASE LEAKAGE	3	RELEASE THE BRAKES. COAT EXHAUST PORT OF VALVE WITH SOAP SOLUTION. 	A 1 INCH DIAMETER BUBBLE IN 5 SECONDS IS ACCEPTABLE.	4
			EXCESSIVE LEAKAGE	6
APPLICATION LEAKAGE	4	MAKE AND HOLD A SERVICE BRAKE APPLICATION. COAT THE EXHAUST PORT OF VALVE WITH SOAP SOLUTION. 	A 2 INCH DIAMETER BUBBLE IN 3 SECONDS IS ACCEPTABLE.	5
			EXCESSIVE LEAKAGE	6
VALVE LEAKAGE	5	MAKE AND HOLD A SERVICE BRAKE APPLICATION. COAT THE RELAY-SOLENOID VALVE MATING SURFACE WITH A SOAP SOLUTION. 	NO LEAKAGE. PNEUMATIC SYSTEM OK.	—
			LEAKAGE	6
OVERHAUL VALVE	6	REMOVE MODULATOR FROM VEHICLE AND OVERHAUL PER VALVE OVERHAUL PROCEDURE. REINSTALL ON VEHICLE.	REPEAT MODULATOR PNEUMATIC CHECKOUT PROCEDURE	1

Fig. 54 AC anti-skid diagnosis and testing

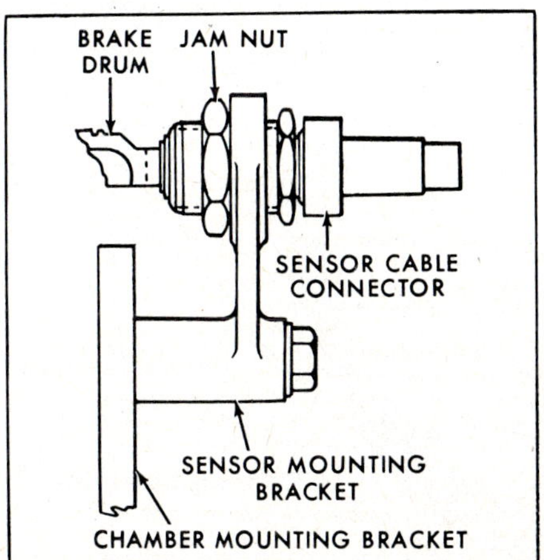

Fig. 55 Wheel speed sensor

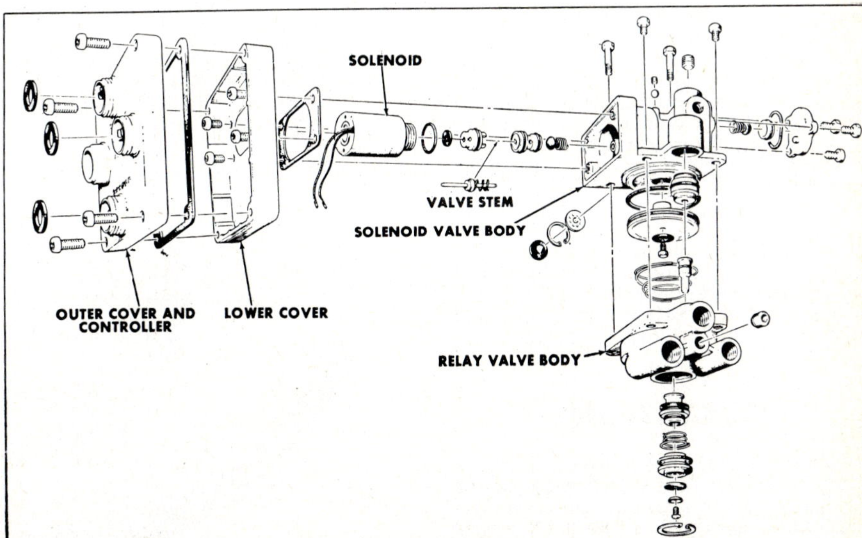

Fig. 56 AC modulator disassembled

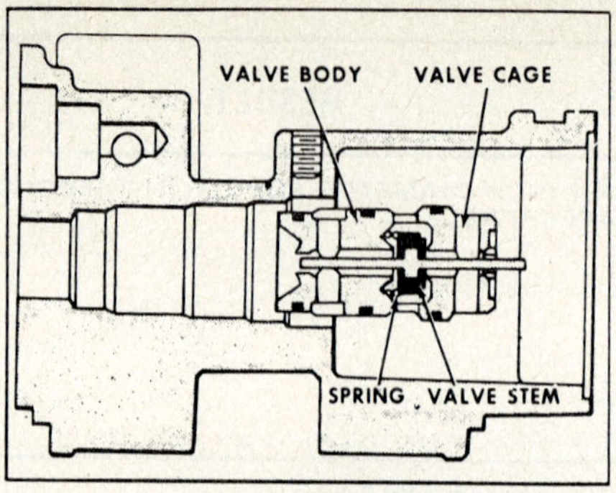

Fig. 57 Solenoid valve, stem and cage installation

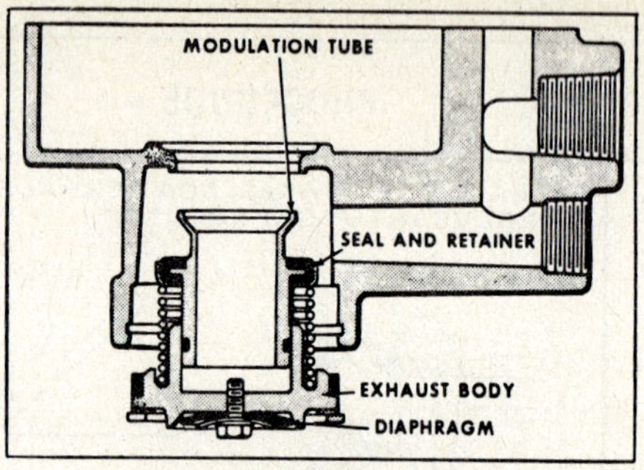

Fig. 58 Modulator tube and exhaust valve installation

BENDIX TWINPLEX BRAKES

DESCRIPTION

The Twinplex brake, Fig. 1, is a wedge actuated mechanical brake. The torque spider, Fig. 2, is the basic foundation of the unit and includes guide pads, guide bosses and the two actuators which are integral with the spider. A service brake air chamber or spring brake chamber is mounted on each actuator. The actuator, Fig. 3, is comprised of an adjusting piston assembly, anchor piston and a wedge and roller assembly.

The adjusting piston incorporates the piston, adjusting nut, adjusting screw, adjusting screw link and automatic adjuster parts. The adjusting screw link has a slot for the shoe web and on the opposite end, a machined ramp to accept the wedge roller. A boot, boot protector and protector retainer are installed in the adjusting piston end of the actuator.

The anchor piston, Fig. 4, has a machined ramp for the wedge roller, a groove for the dust boot and a slot for the shoe web. An indexing pin is pressed into each piston and the pistons are identified right hand and left hand and are not interchangeable.

The wedge and roller cannot be disassembled and must be replaced as an assembly. The wedge, made from hardened steel and chrome plated, is available in angles of 10°, 12°, 14° or 18°, depending upon application. The wedge angle is stamped on the side of the wedge for identification.

The brake shoes have an anchor end and an adjusting end which is marked "ADJ END" and is installed in the adjusting screw link slot. Both shoes are self-energizing during forward or reverse braking and are held in contact against the adjusting screw links and anchor pistons by shoe-to-shoe springs.

OPERATION

When the brakes are applied, air pressure enters the service brake chamber, Fig. 3, and actuates the diaphragm and push rod. The wedge shaft, bottomed in the push rod, is pushed between the adjusting and anchor pistons, causing the pistons to ride outward on the rollers and forcing the brake shoes against the brake drum. If air pressure is lost, the spring brake chamber mechanically applies the wedge brake.

The adjusting piston incorporates an adjusting nut, adjusting screw and adjusting nut cap. An overload spring secures the adjusting nut cap to the adjusting nut. The adjusting screw is coupled with the adjusting nut and a slotted link is mated with the brake shoe web. A detent spring prevents the adjusting screw from turning during automatic adjuster operation.

When the brakes require an adjustment, the pistons will move far enough outward for the adjuster levers to pick up the next tooth on the adjuster cap, Fig. 5. When the brakes are released and the shoes retract, the pistons are pressed back into the actuator, causing the adjuster lever to pivot downward on the adjuster lever pin. This pivoting action rotates the adjuster nut cap and the adjuster nut one tooth. The adjuster screw does not rotate with the adjuster nut, therefore, moves outward when the adjuster nut rotates. The

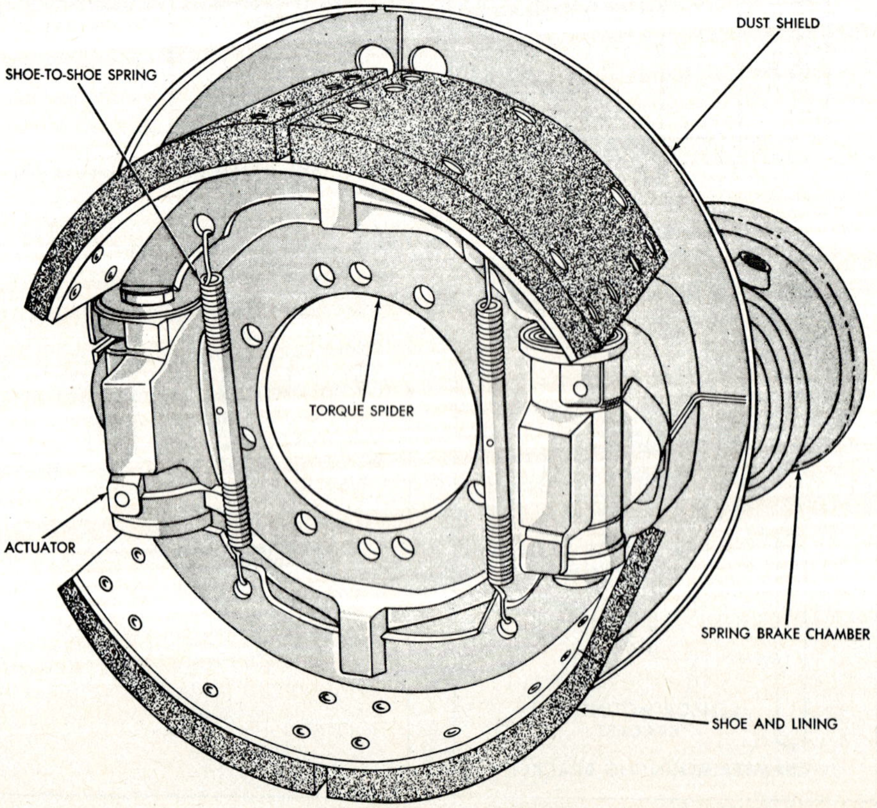

Fig. 1 Twinplex brake assembly

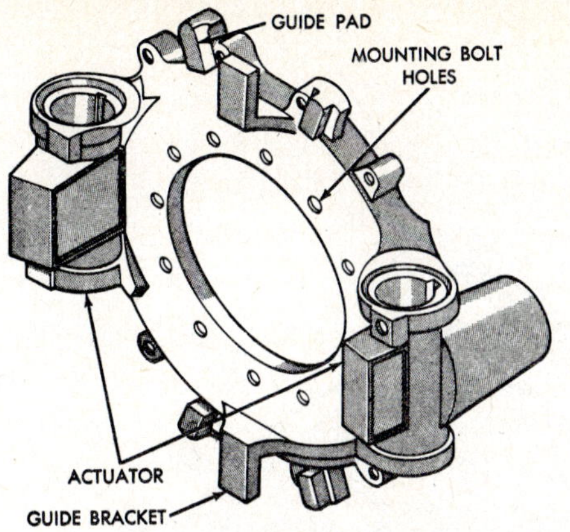

Fig. 2 Torque spider

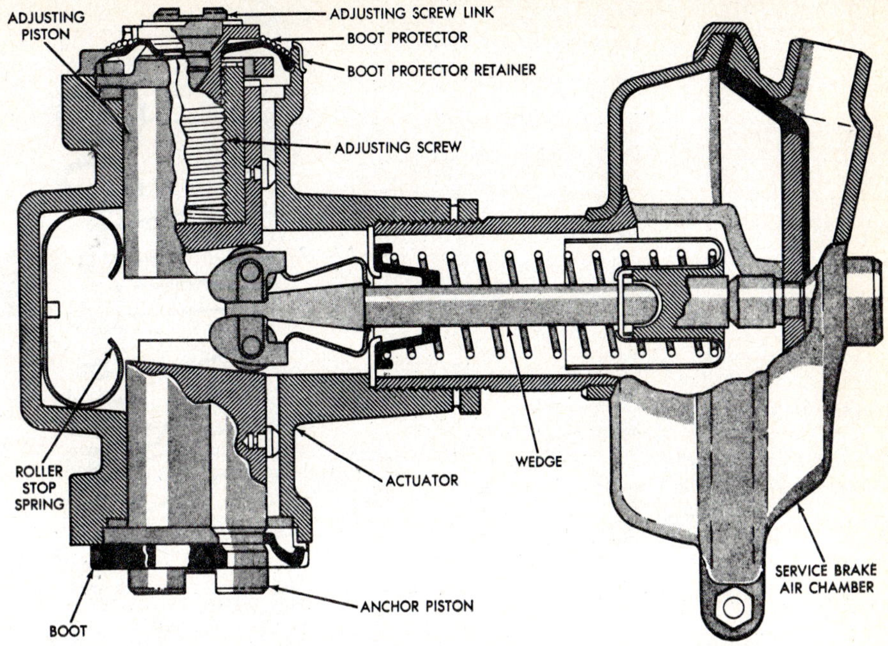

Fig. 3 Brake actuator & air chamber

brake shoes are expanded .001 inch when the adjuster nut rotates one tooth position.

BRAKE SHOES

Removal

If air pressure is present in the system, ensure that service brakes are not applied and air pressure is not applied to the service brake air chamber. If air pressure is removed from the system and the brakes are equipped with spring brake chambers, release spring brakes, permitting the shoes to retract, thereby providing clearance from drum removal.

1. Remove wheel and tire assemblies, then the hub and drum. To retract brakes, if necessary, remove dust shield and back off star wheel at each adjusting screw.
2. Remove brake shoe return springs with a suitable tool, then the brake shoes from the torque spider.

Installation

1. Lubricate adjusting screw link shoe slots with a suitable lubricant. Also, lubricate anchor piston slots and the torque spider guide pads and brackets with lubricant.
2. Install shoes on torque spider, then the shoe return springs. Install a spring damper over each return spring.

NOTE: If brake drum is resurfaced .080 inch over original drum diameter, the standard adjusting screws must be replaced with service adjusting screws (P/N 323128).

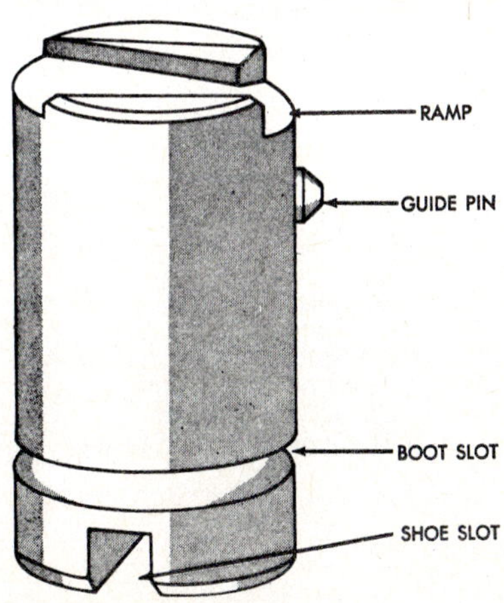

Fig. 4 Anchor piston

3. Install drum and hub, then the wheel and tire assemblies.

NOTE: If oversized linings are installed and the linings are too thick for drum installation, the linings may be ground on the vehicle using an axle attached grinding tool. Grind linings .020 to .030 inch under drum diameter.

---

4. Rotate wheel in a forward direction and apply the brakes 25 to 30 times, thereby adjusting brakes.

PISTONS, WEDGE & ROLLER ASSEMBLIES

Disassembly

1. Disconnect air hose from service brake air chamber. It is permitted to separate air chamber halves in place of disconnecting air hose.
2. Straighten edges of air chamber tube lockwasher, then loosen check nut on air chamber tube.
3. Unscrew air chamber from actuator.
4. Remove wedge and roller assembly from actuator housing, Fig. 6.

NOTE: The wedge and roller is not disassembled for service. If any part is worn or damaged, replace complete assembly.

---

5. Remove boot protector retainer and boot protector from actuator.
6. Unseat adjusting piston boot from actuator and pull adjusting piston assembly from actuator bore.
7. Unseat anchor piston boot from actuator and pull anchor piston from actuator bore, then remove dust boot from piston.
8. Remove adjuster lever and spring from actuator, Fig. 7.
9. Remove adjusting screw and nut assembly from adjusting piston, then unscrew nut from screw.
10. Remove automatic adjuster overload spring from top of adjusting nut and cap assembly and separate nut from cap.
11. Pry adjusting screw link from adjusting screw.

Assembly

1. If removed, press a new adjuster lever pin in place, ensuring new pin fits snugly in actuator hole.
2. If removed, install a new roller stop spring in actuator, Fig. 3.
3. Assemble adjuster nut cap and adjuster nut and align slots. Install a small drill bit or rod through alignment holes in cap and nut for proper alignment when installing overload spring.
4. Install overload load spring in slots of adjuster nut and cap with open end of

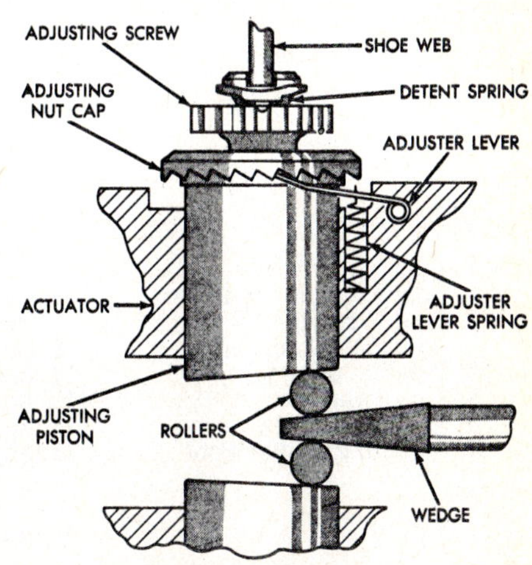

Fig. 5 Automatic adjuster

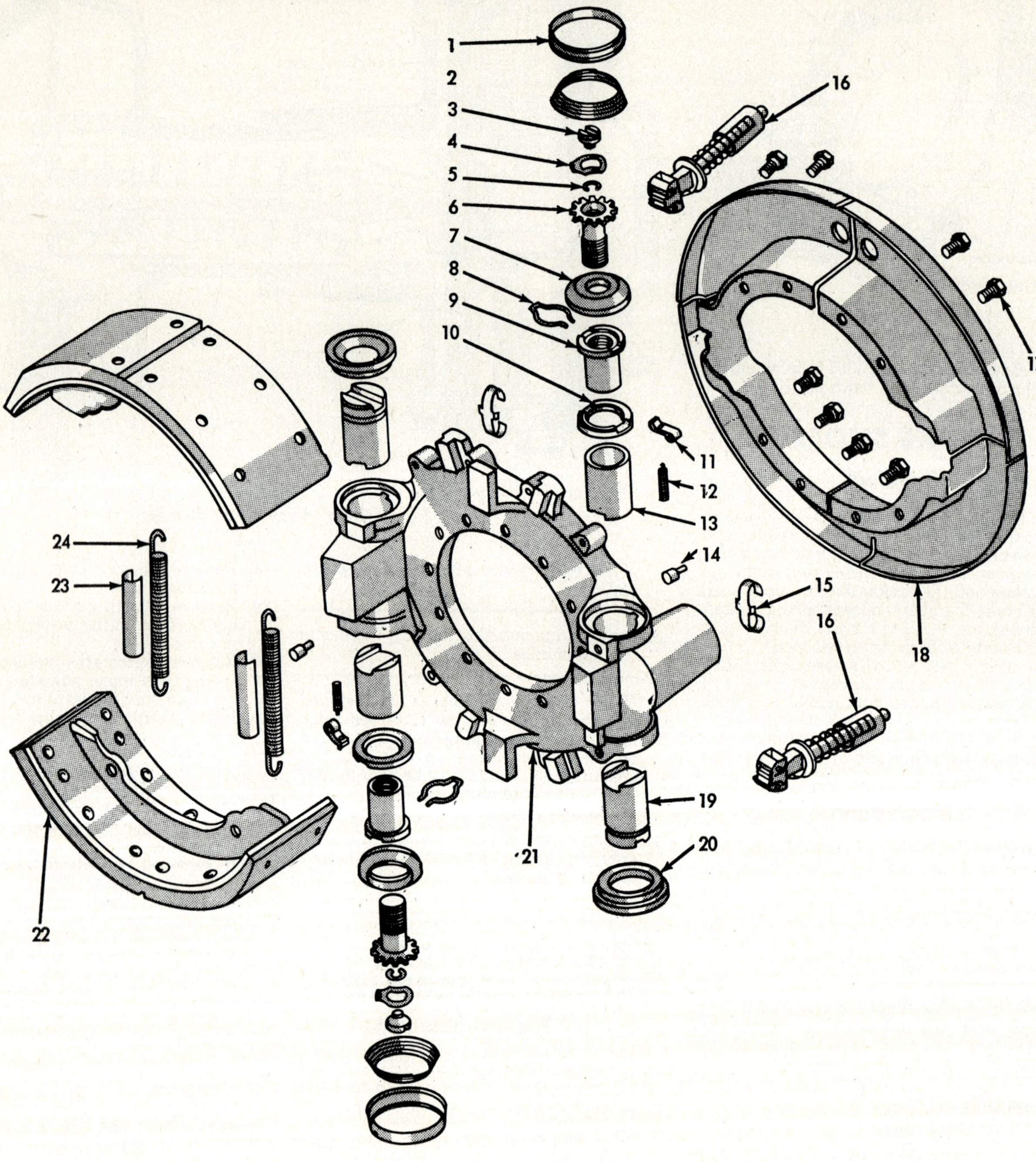

1. Boot Protector Retainer
2. Boot Protector
3. Adjusting Screw Link
4. Detent Spring
5. Link Retainer
6. Adjusting Screw
7. Adjusting Piston Boot
8. Automatic Adjuster Overload Spring
9. Adjusting Nut
10. Adjusting Nut Cap
11. Automatic Adjuster Lever
12. Adjuster Lever Spring
13. Adjusting Piston
14. Adjusting Lever Pin
15. Roller Stop Spring
16. Wedge and Roller Assembly
17. Dust Shield Screw
18. Dust Shield
19. Anchor Piston
20. Anchor Piston Boot
21. Spider
22. Shoe and Lining
23. Spring Damper
24. Shoe-to-Shoe Spring

Fig. 6 Twinplex brake, disassembled

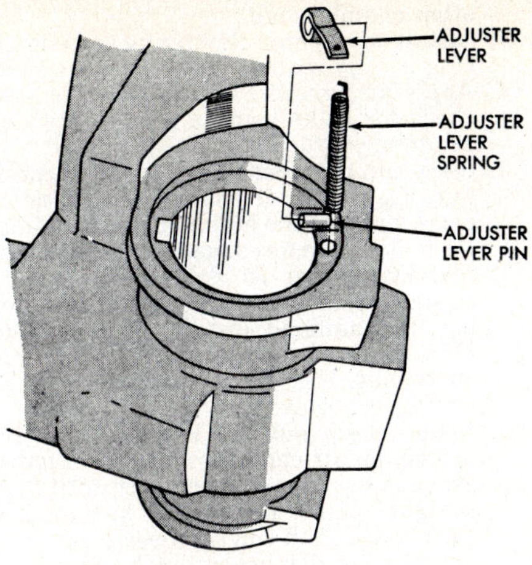

Fig. 7 Adjuster lever & spring

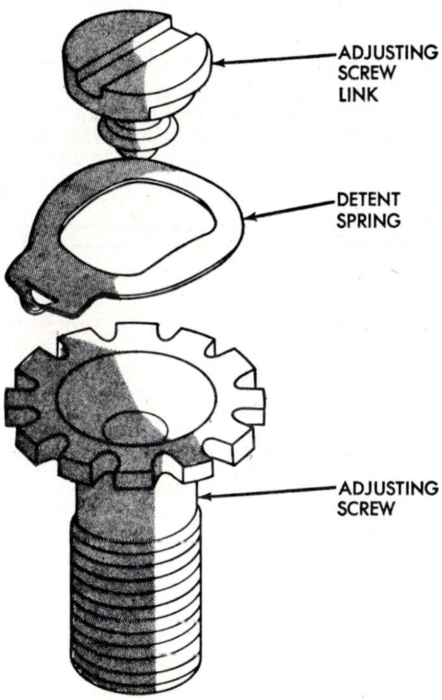

Fig. 9 Adjuster screw link & detent spring

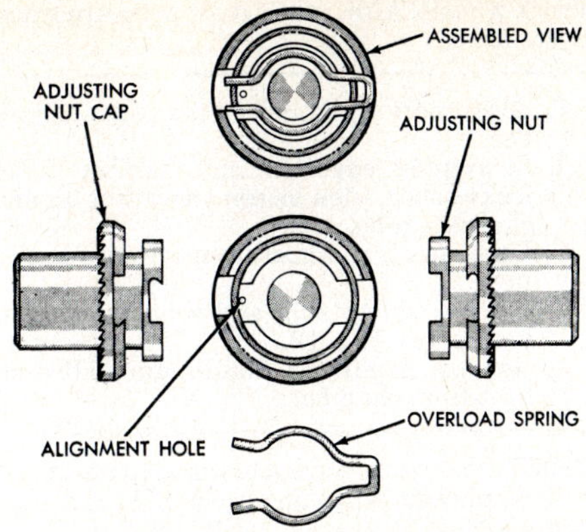

Fig. 8 Adjuster nut, cap & overload spring

spring toward alignment hole, Fig. 8. To check overload spring action, hold adjuster nut with overload spring upward and turn adjuster nut cap clockwise. The cap should slightly rotate clockwise, against the overload spring, but not counter-clockwise.

5. Install retaining ring in adjuster screw link groove.
6. Install detent spring on adjuster screw link so detent will engage star wheel teeth when link is pressed into adjuster screw, Fig. 9.
7. Lubricate adjuster screw hollow end with suitable lubricant and install screw link into adjuster screw, ensuring link is fully pressed into the screw.
8. Install boot on adjuster screw and adequately lubricate adjuster screw and adjuster nut threads with suitable lubricant.
9. Rotate adjuster screw into nut until bottomed, then back screw out one full turn.
10. Lubricate outside surface of adjuster nut and inside surface of adjuster piston, then install adjuster nut assembly into adjuster piston.
11. Install dust boot in anchor piston groove, using caution not to damage boot.
12. Insert adjuster lever spring hook into hole of adjuster lever, Fig. 7. Coat spring and lever with suitable lubricant. Install adjuster spring into actuator well, then the adjuster lever on the lever pin.
13. Adequately lubricate adjuster piston outside diameter, ramp on piston bottom, and actuator bore with suitable lubricant, then install adjuster piston into actuator bore, aligning piston guide pin with slot in bore.

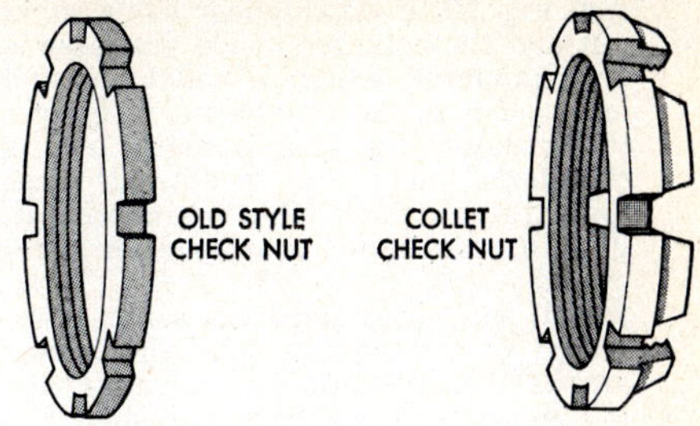

Fig. 10 Air chamber tube check nut identification

14. Adequately lubricate anchor piston outside diameter, ramp on piston bottom, and anchor piston end of actuator bore with suitable lubricant, then install anchor piston into actuator bore, aligning piston guide pin with slot in bore.
15. Place boot on actuator and, using a 1⅝ inch socket or similar tool and a mallet, seat metal ring of boot in actuator. Ensure boot is not damaged and that it is fully seated.
16. Install boot protector on adjuster piston boot and press boot retainer over boot protector.
17. Lubricate wedge and roller assembly with suitable lubricant and install assembly into actuator.
18. Some air chambers have a collet check

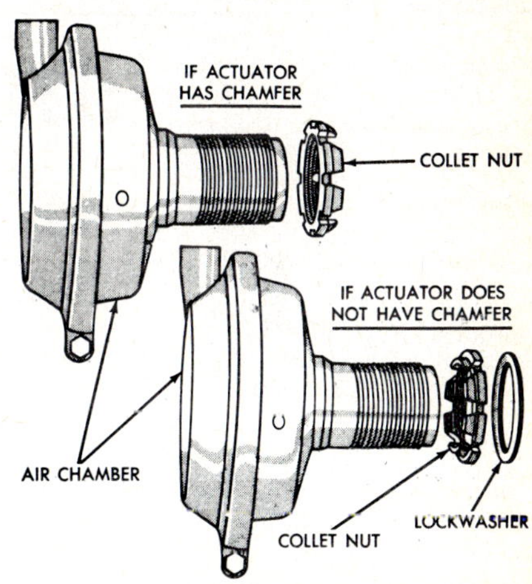

Fig. 11 Air chamber tube collet check nut installation

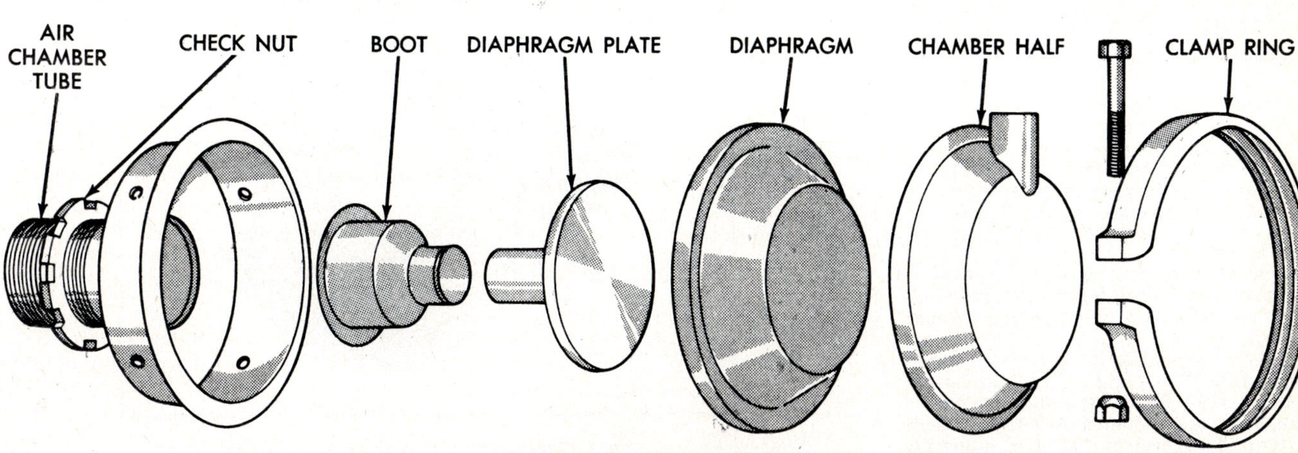

Fig. 12 Brake air chamber

nut, Fig. 10. If air chamber has a collet nut and the actuator inside diameter is not chamfered, assemble collet nut and lockwasher so flat side of nut is against the actuator, Fig. 11. If actuator bore is chamfered, install nut so flat side faces toward air chamber. If old type check nut, Fig. 10, is installed a lockwasher is required.

19. Install check nut and lockwasher, if required, on air chamber tube, then screw air chamber tube into actuator until tube is tight against wedge stop washer. Ensure wedge shaft end is properly seated in air chamber diaphragm push rod.
20. Torque air chamber tube check nut to 150 ft. lbs. against lockwasher or actuator. Bend lockwasher outer edge, if used, into check nut notches.
21. Connect air hose to air chamber.

NOTE: Do not back air chamber tube out of actuator to align hose connection. Loosen clamp ring securing chamber halves and rotate inner chamber to align hose connection. Use care not to crimp or turn air chamber diaphragm.

BRAKE AIR CHAMBERS

Removal & Disassembly

CAUTION: Park vehicle on level surface and chock wheels. Also, exhaust air pressure from system.

1. Disconnect air hose from brake air chamber.
2. Straighten edges of air chamber tube lockwasher, then loosen check nut on air chamber tube.
3. Unscrew air chamber from actuator housing.
4. Remove clamp ring and separate chamber halves, Fig. 12.
5. Remove diaphragm, diaphragm plate and boot from chamber.

NOTE: It is recommended to install a new boot and diaphragm when an air chamber is disassembled.

Assembly & Installation

NOTE: The diaphragm must be glued to the diaphragm plate and the boot be glued to that portion of the air chamber with a suitable adhesive.

1. Lubricate hollow end of diaphragm plate push rod with suitable lubricant and insert push rod through boot.
2. Ensure diaphragm lip is properly positioned around chamber edge and install other chamber half.
3. Install and tighten clamp ring and ensure ring is fully seated.
4. Apply air pressure to the chamber and check for leaks using a soap and water solution.
5. Some air chambers have a collet check nut, Fig. 10. If air chamber has a collet check nut and the actuator inside diameter is not chamfered, assemble collet nut and lockwasher so flat side of nut is against the actuator, Fig. 11. If actuator bore is chamfered, install nut so flat side faces toward air chamber. If old type check nut, Fig. 10, is installed, a lockwasher is required.
6. Install check nut and lockwasher, if required, on air chamber tube, then screw air chamber tube into actuator until tube is tight against wedge stop washer. Ensure wedge shaft end is properly seated in air chamber diaphragm push rod.
7. Torque air chamber tube check nut to 150 ft. lbs. against lockwasher or actuator. Bend lockwasher outer edge, if used, into check nut notches.

NOTE: Do not back air chamber tube out of actuator to align hose connections. Loosen clamp ring securing chamber halves and rotate inner chamber to align hose connections. Use care not to crimp or turn air chamber diaphragm.

STOPMASTER BRAKES

INDEX

DESCRIPTION

The Stopmaster Brake is a wedge actuated brake. The brake power units can be either air chambers or hydraulic clyinders. They can also be equipped with Fail-Safe units to provide parking and emergency braking. The brake power unit forces a wedge between two rollers and two plungers. This causes the plungers to spread apart and push the brake shoes against the brake drum.

There are several variations of Stopmaster Brakes in use. Some have two power units per brake and others only one. Where two power units are used, there may be a Fail-Safe unit on one or both. The brake support may be either a cast spider or a stamped backing plate. On the cast spiders, the plunger housings may be either integral or bolted on. Brake shoe adjustment may be either manual or automatic.

The basic Stopmaster Brakes are identified as model RDA, RDH, RSA or RSH with letters denoting the following: R—Stopmaster Brake; S—Single Actuated; A—Air operated; H—Hydraulic operated. Figs. 1, 2, 3 and 3A illustrate four types of brakes.

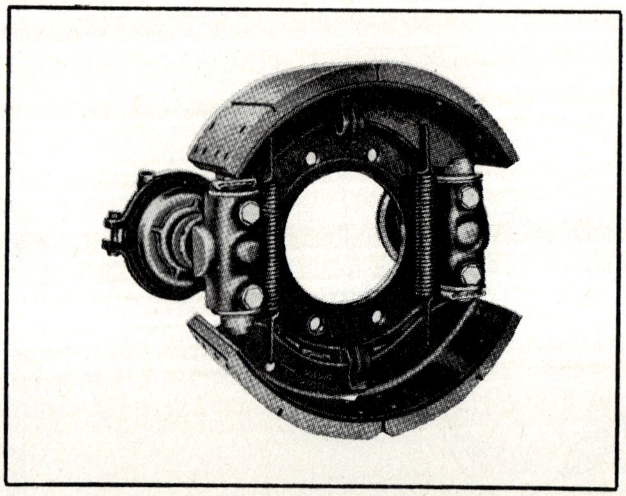

Fig. 1 Stopmaster FDA brake with bolted on spider and integral plunger housings

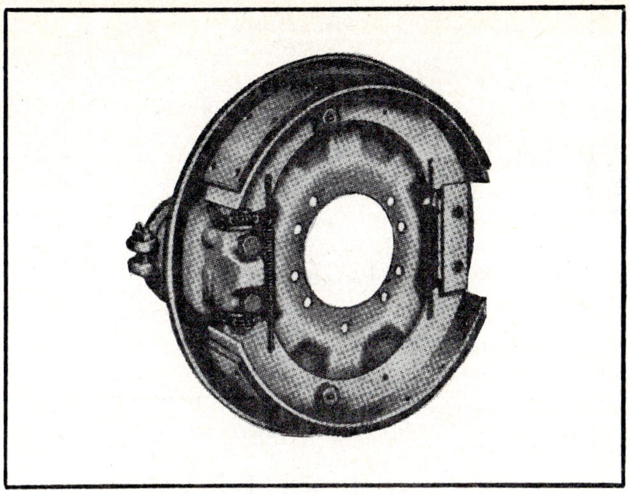

Fig. 2 Stopmaster RSA brake, backing plate mounted and bolted on plunger housing

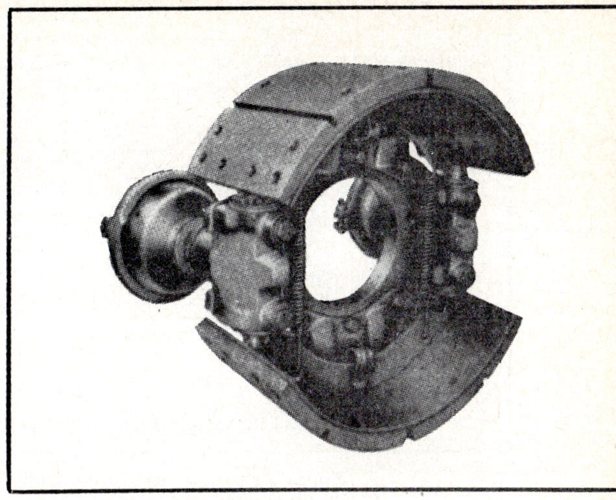

Fig. 3 Stopmaster RSA brake with welded on spider and bolted on plunger housings

Stopmaster Actuating System

Fig. 4 shows one of the actuating systems of an air-operated, double-actuated Stopmaster RDA brake. All parts are shown in the off position.

This system has an air chamber power unit threaded into the wedge bore of the plunger housing. The socket in the end of the diaphragm push rod connects the air chamber to the wedge rod. The wedge retracting spring acts as the return spring for both the wedge and the diaphragm. A pair of rollers are held in place on the wedge head by a retaining cage. The rollers are also engaged in corresponding slots in the inner ends of the plungers. The unslotted portions of the inner ends of the plungers are resting on abutments in the plunger housing. The outer ends of the plungers are engaged with and supporting the brake shoes.

On a double-actuated brake, each of the two actuating systems has one anchor (solid) plunger and one adjustable plunger (as illustrated). On a single actuated brake the one actuating system would have two adjustable plungers. All of the plungers are retained in the housings and the roller slots are kept in proper alignment by means of guide screws that engage slots in the side of the plungers. A hydraulic brake would have a hydraulic cylinder threaded into the plunger housing (in place of air chamber). The hydraulic piston would connect with the wedge rod.

When the brake is actuated, the air chamber pushes the wedge head deeper in between the rollers. This spreads the rollers and plungers apart and pushes the brake shoes outward. Initially all the plungers are lifted off of the plunger abutments and momentarily suspended. As the brake shoes contact the drums, the drum drags the shoes and the suspended plungers around with it. This causes the plunger at the trailing end of each shoe to reseat on its abutment and thus absorb and transfer the brake torque to the brake support. When the brake is released, the wedge spring returns the wedge and diaphragm to the off position. At the same time, the shoe return springs push the raised plungers back to the abutments.

Adjustable Plungers

Two types of adjustable plungers are in use—manual and automatic. The manually adjusted plunger has an adjusting bolt threaded into the plunger itself. The bolt head is scalloped to facilitate adjustment and carries a detent arrangement that engages the brake shoe web and prevents accidental rotation of the bolt.

On the automatically adjusted plunger, Fig. 5, the adjusting bolt is threaded into an adjusting sleeve which in turn is free-fitted inside the plunger itself. The plunger guide screw is replaced by a hollow cap screw, a spring and an adjusting pawl which also serves as the plunger guide. The end of the adjusting pawl has saw-tooth type teeth which engage corresponding helical teeth on the outside of the adjusting screws.

As the brake is actuated, the plunger, sleeve and bolt move outward and the sloping face of the teeth on the adjusting sleeve lifts the adjusting pawl against the spring. When the brake is released, all the parts return to their starting points. As the lining wears, the plunger stroke and resulting pawl lift gradually increases until the pawl climbs over and drops into the next tooth space. This time, when the brake is released and the plunger is pushed back into its bore, the upright face of

Fig. 3A Stopmaster RDA brake with welded on spider and bolted on plunger housings

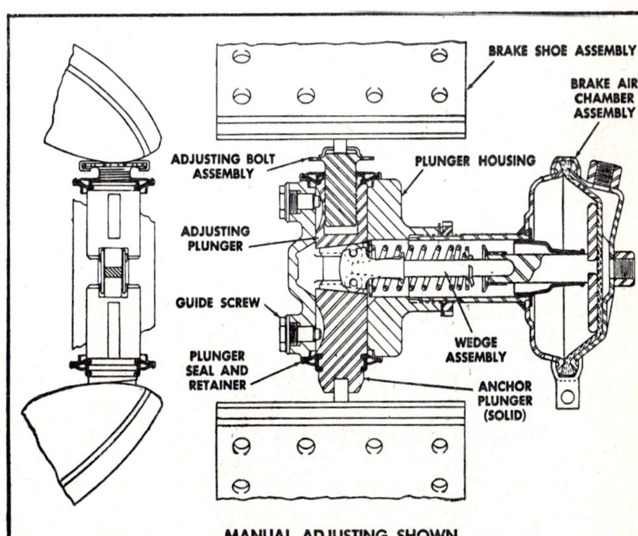

Fig. 4 Actuating system of air operated RDA brake (3/4" stroke wedge assembly shown)

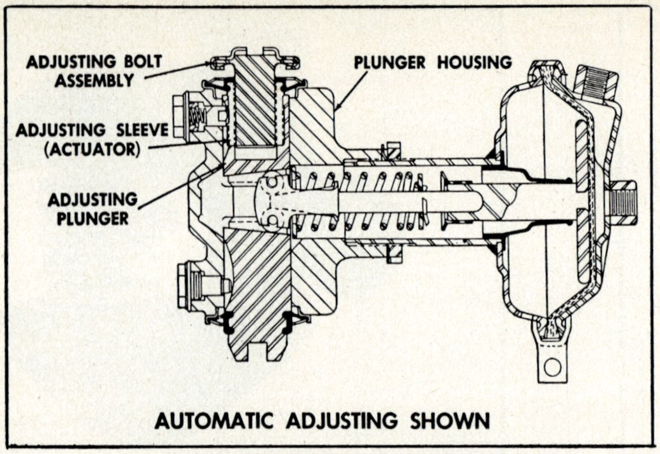

Fig. 5 Actuating mechanisms with automatically adjusted plunger (1 3/4" stroke wedge assembly shown)

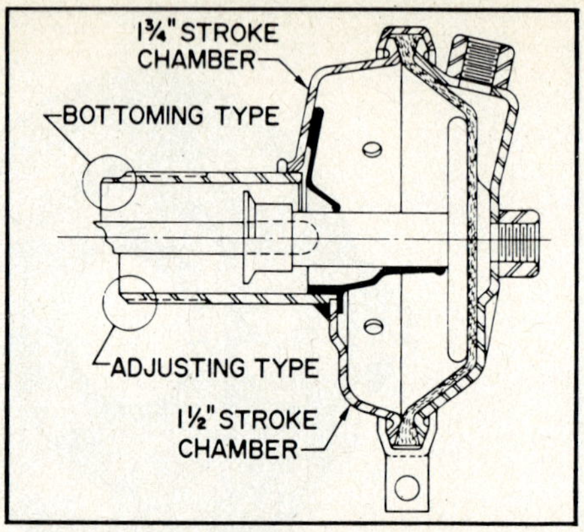

Fig. 6 Power unit adjustment

the pawl teeth causes the adjusting sleeves to rotate and advance the adjusting bolt. This reduces the lining clearance and the cycle starts over again. The automatic adjuster operates only in forward vehicle direction.

Power Unit Adjustment

The air chamber (or hydraulic cylinder) should be screwed into the wedge bore of the plunger housing to such a depth that the wedge is ready to lift the plungers off of the abutment seats at the first movement of the diaphragm (or piston). This provides the least lost motion and maximum useful chamber (or piston) stroke.

Current power units are designed to "bottom out" in the wedge bore and provide this optimum adjustment automatically. (The bottoming type units have a short unthreaded portion on the leading end, Fig. 6.) Earlier power units must be adjusted manually by screwing the power unit into the wedge bore deep enough for the wedge to spread the plungers so they can be pushed back and forth in the plunger housing. Then the power unit is backed out until the back and forth plunger movement disappears. *The newer bottoming type units can be used as replacements for the earlier units to obtain the bottoming feature.*

Wedge Alignment

When the power unit is removed from the brake, the wedge assembly may also become dislodged. Before reinstalling the power unit, reposition the wedge assembly so that the rollers and roller cage are engaged in the plunger slots, Fig. 7. On newer brake assemblies this is accomplished automatically by simply aligning the two ears on the wedge spring retainer with corresponding grooves in the wedge bore of the plunger housing. On older assemblies, the wedge head must be aligned manually so that the rollers engage the plunger slots properly. In either case, proper alignment can be checked by pushing on the wedge rod while visually checking for shoe and plunger lift.

Fail-Safe Units

The Fail-Safe unit is a spring powered brake actuator that assembles piggy-back on the air chamber. As shown in Fig. 8, Fail-Safe units are produced in three types, Custom, Standard and Super. When 70 psi or more air pressure is applied against the piston, the spring will be held in a compressed position. When the air pressure is removed, the spring will push the piston against the diaphragm plate and apply the brake.

On the Super Fail-Safe units an internal venting system working in conjunction with a one-way breathing arrangement on the cap allows system air to fill the vacuum behind

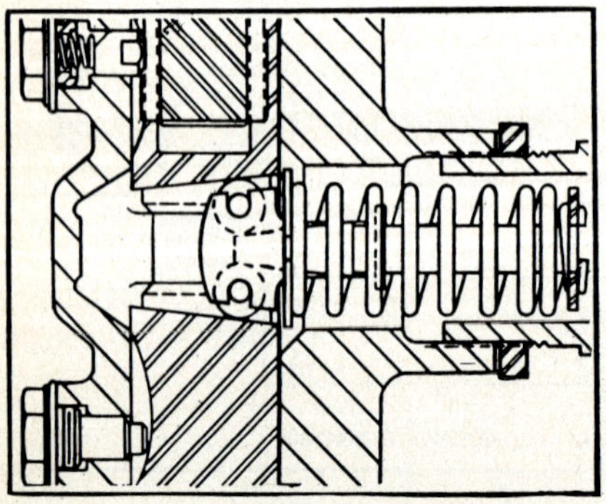

Fig. 7 Wedge alignment

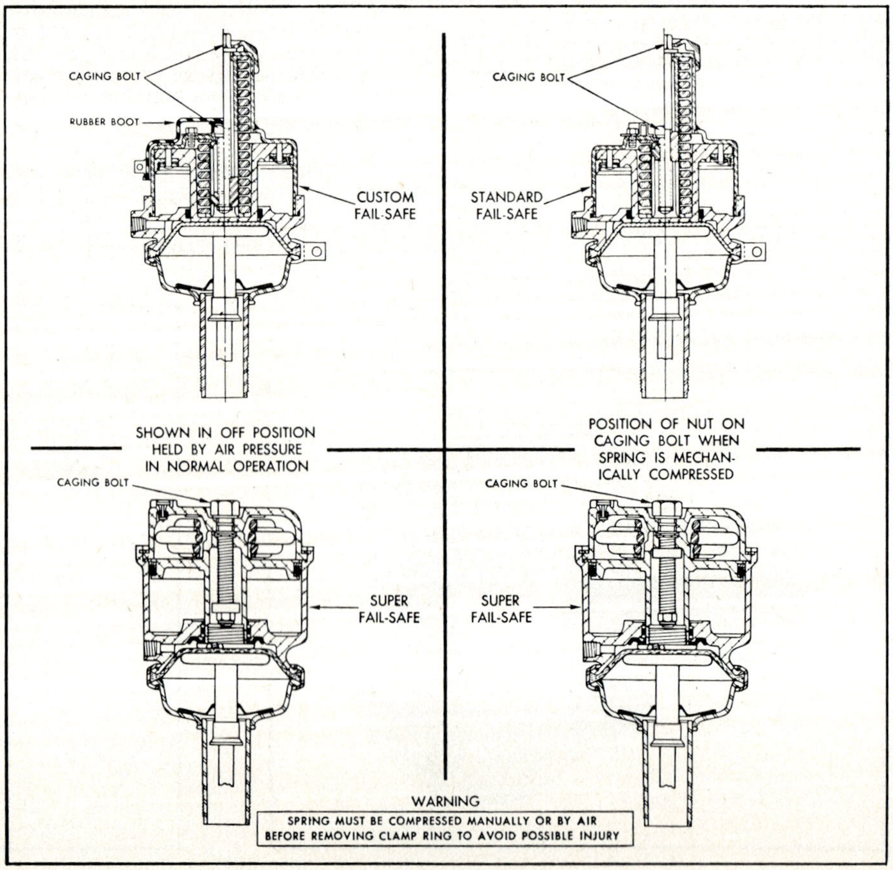

Fig. 8 Fail-Safe units

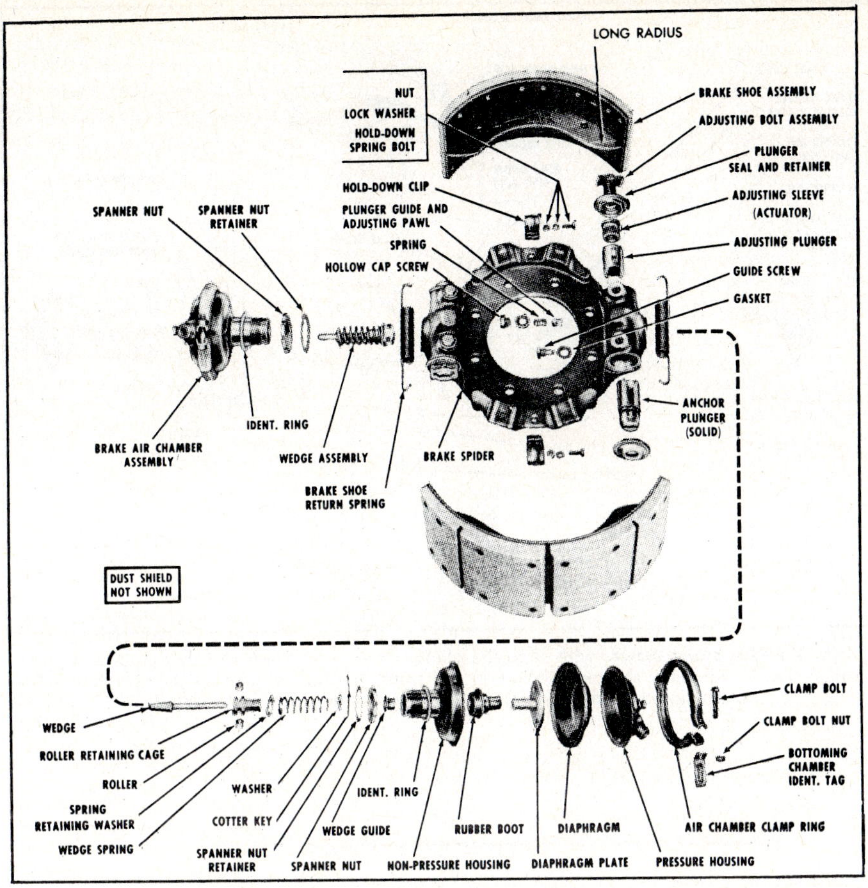

The RDA brake shown, with integral plunger housings, employs a cast spider mounting that bolts to a flange on the axle housing. Adjusters may be manual or automatic (automatic shown)

the system when the Fail-Safe is actuated and keeps the seal from direct atmospheric contamination.

All types of Fail-Safe units are equipped with manual caging bolts to permit safe handling for service work. These bolts are sealed and prevented from rotating due to vibration.

Safety Precaution

When the brakes are equipped with Fail-Safe units (or other auxiliary spring power units), cage the power springs before starting any disassembly or removal of wheels and drums. After parts are all reassembled and in place, uncage the power springs before returning the vehicle to service. When a vehicle is disabled due to low or lost air pressure, block wheels and cage the power springs before moving the vehicle.

Caging and Uncaging Fail-Safe Units

On Custom units, the head of the caging bolt is exposed at all times beyond the plastic cap. However, the cap should be backed off three turns before turning caging bolt, otherwise the bolt may damage the cap.

On Standard units, first loosen the boot clamp screw and remove the rubber boot (if one is used). Then loosen the caging bolt-lock screws and swing the lock out of the way.

The power spring on both types is caged (compressed) by turning the caging bolt clockwise approximately 18 full turns. Do not force the bolt beyond its normal stop. Uncaging (releasing) the spring is accomplished by reversing the procedure—again do not force the bolt beyond its normal stop. If desired, both the caging and uncaging operation can be made easier by applying air pressure to the Fail-Safe chamber to take the spring load off the caging bolt.

On Custom units, after uncaging the spring, retighten the plastic cap by hand. On

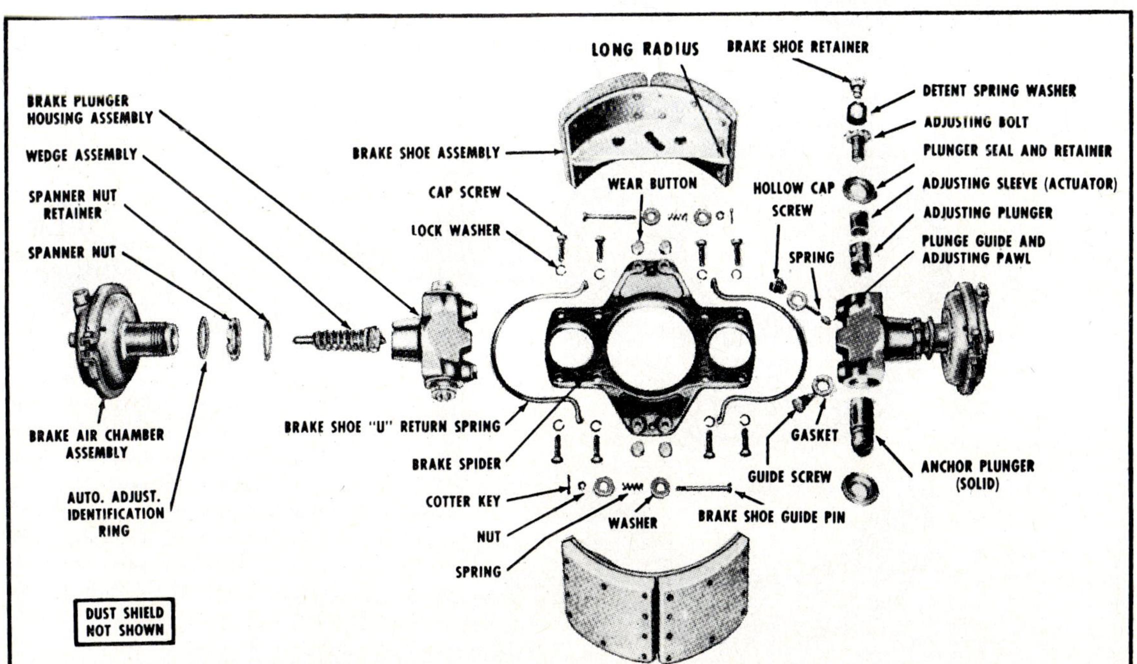

The RDA brake shown, with separate plunger housings, employs a cast closed end spider that is welded to the axle. The actuation system is removed by removing four capscrews and brake air chamber assembly. This is 12¼" diameter trailer brake and employs an automatic adjuster

Standard units, after uncaging the spring, swing bolt-lock back and secure in place. Clean cap and boot and install boot with vent slot at bottom.

TROUBLE SHOOTING

If Brakes are Poor or Do Not Apply
Check system pressure at source and at brakes for possible restriction or leak in brake lines, valves, etc. Brakes out of adjustment. Leaking diaphragm or wheel cylinder cup.

Uneven Braking or Lining Wear
Check for ruptured diaphragm. Wedge rod out of push rod socket. Rollers and cage out of plunger socket. Corroded or frozen plungers. Brakes out of adjustment. Grease on lining. Glazed lining. Shoes installed backward.

Automatic Adjusters Not Working
Adjusting pawl installed backward. Pawl spring collapsed or missing. Bolt frozen in adjusting sleeve. Detent damaged, allowing bolt to rotate with sleeve. Adjusting plunger in wrong position (should be at leading end of shoe).

Fail-Safe Unit Not Holding
Power spring not fully released (uncaged). Brakes out of adjustment. Hold-off air not releasing fully. Power spring broken.

Brake Dragging
Low Fail-Safe holding air pressure (need 70 psi minimum). Improper connection of emergency line at Fail-Safe. Leaking lines or Fail-Safe seals.

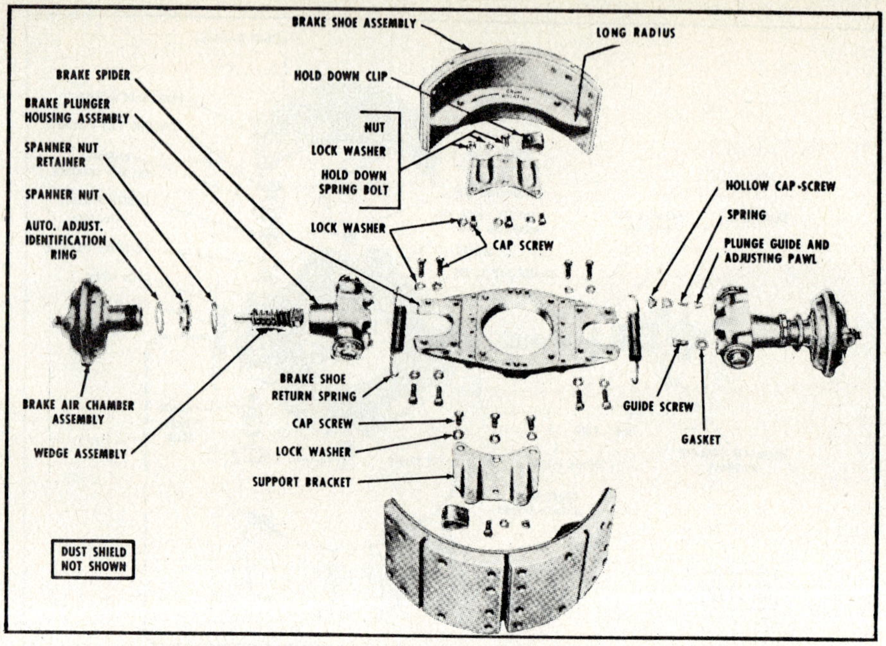

The RDA brake shown, with separate plunger housings, employs a cast open end spider that is welded to the axle. The open end spider permits removal of the complete actu- action system at disassembly. This brake employs automatic adjusters

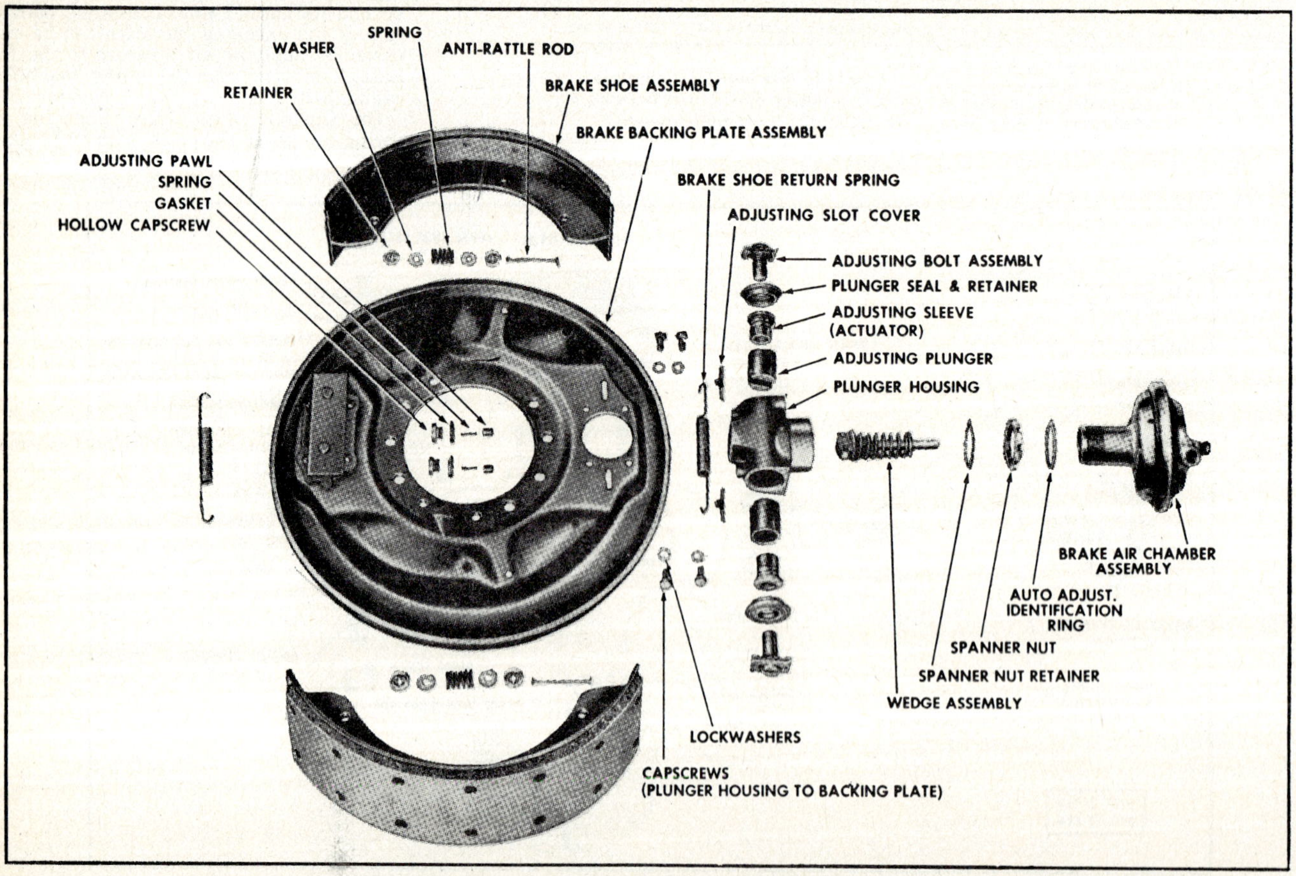

The RSA brake shown, with separate plunger housing, has a stamped backing plate that bolts to a flange on the axle housing. Adjuster may be manual or automatic (automatic shown)

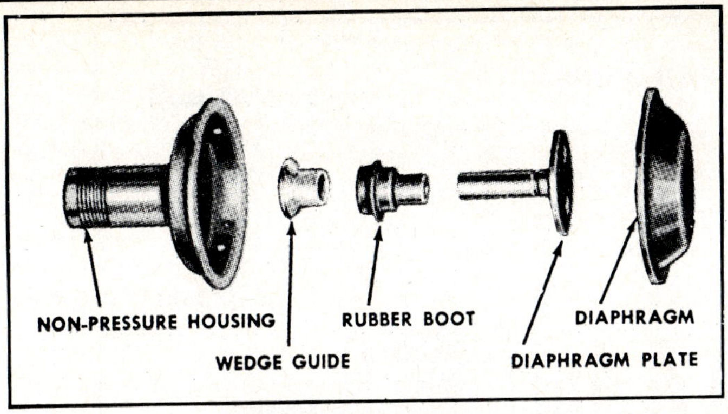

Fig. 9 Air chamber diaphragm parts layout

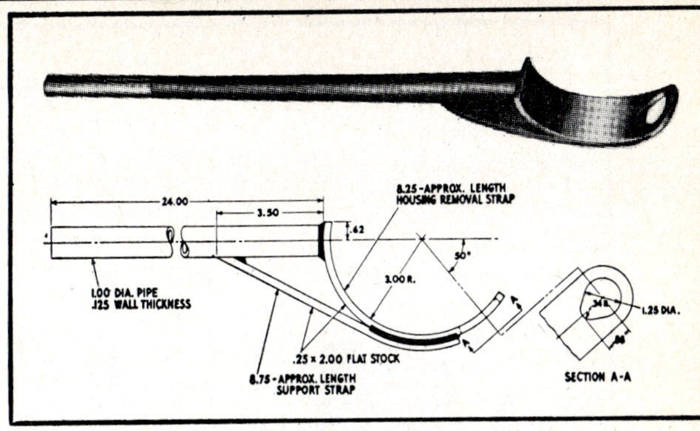

Fig. 10 Special wrench for removal of Fail-Safe unit housing

BRAKE ADJUSTMENTS

Refer to "Brake Adjustments" section located elsewhere in this manual.

POWER UNIT SERVICE

Air Chamber Diaphragm

The following procedures are used to replace the diaphragm and boot with the non-pressure housing on the brake assembly.

1. Cage power spring if Fail-Safe unit is used.
2. Remove air lines and tag emergency line.
3. Remove clamp ring nuts, bolts and Fail-Safe warning tag, and bottoming chamber tags, if used.
4. Spread clamp ring. Hold diaphragm on non-pressure half and remove pressure half or Fail-Safe head.
5. Carefully remove diaphragm while holding diaphragm plate against wedge rod. This will prevent wedge assembly from coming out of engagement with the plungers.

NOTE: If wedge assembly backs out of plunger anytime during the entire procedure, remove non-pressure housing from brake to replace diaphragm.

---

6. Continue to hold plate and inspect boot. If boot is torn or not attached to housing, strip old boot from housing and carefully remove plate assembly off wedge rod and out of non-pressure housing tube. (Boot and wedge guide will remain on plate push rod.)
7. Remove wedge guide and old boot from push rod. Inspect guide for wear and replace if necessary.

Reassemble, Fig. 9

1. Install new boot on diaphragm plate push rod and press wedge guide all the way on to end of push rod.
2. Clean non-pressure housing with cement thinner or similar material in area where boot is to be cemented.
3. Apply cement to housing around tube end. Position plate push rod assembly into tube. Carefully engage wedge rod so as not to pull it out of plungers.
4. While holding plate against wedge, press boot into position for cementing.
5. Install new diaphragm over plate and onto non-pressure housing while pushing plate against wedge. Assemble pressure half and clamp ring in reverse manner of disassembly.
6. Connect air lines to proper chamber and Fail-Safe ports. Make a full pressure brake application and check for air leaks. Road test for brake performance.

FAIL-SAFE UNIT SERVICE

Disassemble—Standard & Custom

1. Remove air chamber diaphragm.
2. Secure Fail-Safe unit by hexagon part of cap in a vise and remove housing by unscrewing it from cap with a spanner or strap wrench, Fig. 10.
3. Pull piston assembly out of cap. Piston can be pushed out by turning caging bolt and release spring.
4. Remove rubber washer and O-ring from Fail-Safe unit (use small screwdriver or other pointed tool.)

NOTE: Do not disassemble the cap and spring assembly or the piston assembly. If parts are damaged, the complete assembly must be replaced.

Assemble, Fig. 11

1. Inspect and check for worn parts that may need replacement. Inspect threads, bore, and end of cap and spring assembly carefully for nicks or scratches that would affect sealing. Clean all parts. DO NOT use solvent to clean leather piston seal, rubber boot or plastic cap (Custom units). Use hot soapy water.
2. Wipe internal parts clean with a soft cloth and re-grease thoroughly with recommended grease. The power spring and all mating parts should be packed with grease.
3. Install rubber washer and O-ring seal in proper grooves in Fail-Safe housing. The rubber washer must fit completely and squarely in its groove.
4. Install piston and seal assembly halfway into the cap.
5. Clamp hexagon part of cap securely in a vise and place Fail-Safe housing onto piston and seal assembly. Push housing down to engage threads on cap.
6. To start threads, turn housing counterclockwise until thread engagement is felt. Then turn housing at least three turns clockwise to insure proper thread engagement.
7. Tighten housing onto cap until it bottoms, using wrench used for disassembly.

Disassemble—Super Fail-Safe Unit

Follow steps 1 to 4 under diaphragm replacement covered previously and then proceed as follows:

1. Referring to Fig. 12, release foot set screw and unscrew foot from piston. Remove piston seal and sealing compound from joint at cap and housing.
2. Take out lock ring and separate cap, spring and piston assembly from housing. Remove rubber seal from housing, if used.
3. Remove hex nut and snap ring from release bolt. Secure cap, spring and piston assembly in arbor press or clamping fixture.
4. Turn release bolt until square nut falls free, release pressure of press or fixture until spring is unloaded. Assembly can now be removed from press or fixture. Separate cap from spring and piston.
5. Remove back-up ring and "O" ring from piston groove.
6. Remove two small back-up rings and "O"

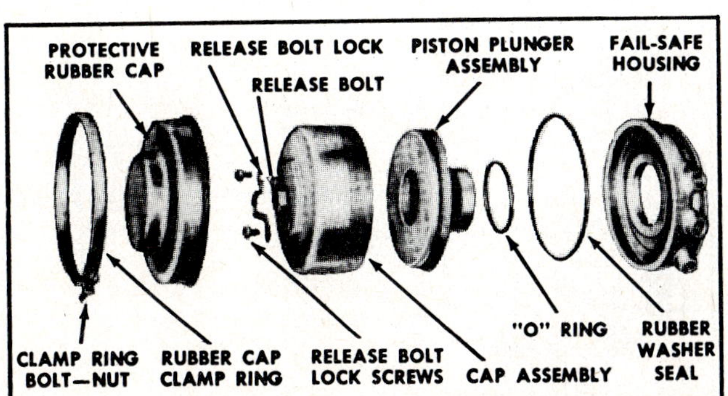

Fig. 11 Standard and Custom Fail-Safe unit disassembled

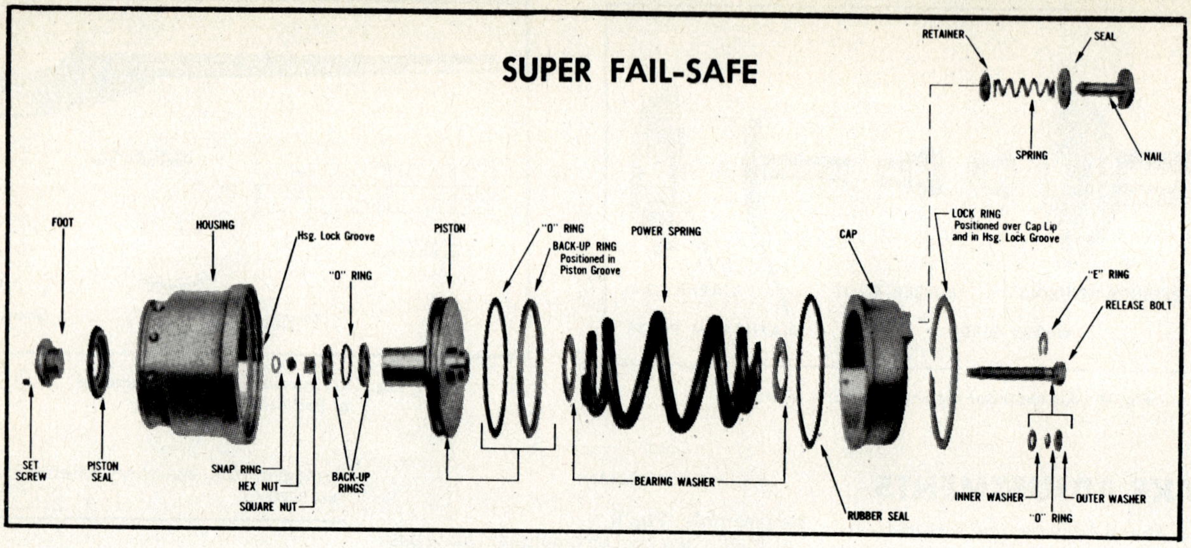

Fig. 12 Super Fail-Safe unit disassembled

ring from I.D. in bottom of housing and the two bearing spring washers, located at top of piston and inside top of cap.

7. Take out cotter pin or snap ring from release bolt. This permits removal of inner washer from release bolt.
8. Take release bolt out of cap. Remove "O" ring from groove of bolt and take off outer flat washer.
9. Reverse removal procedure to assemble.

Bench Test for Leaks

Plug one Fail-Safe port and apply air pressure (150 psi maximum) to other Fail-Safe port. Apply soapy water to check for leaks in three places. Bubbles at release bolt indicate leakage at leather piston seal, and bubbles at piston plunger indicate leakage at O-ring. These leaks can generally be corrected by cycling the piston (with spring uncaged) by applying and releasing air pressure. Bubbles at the cap threads indicate leakage past the rubber washer. This leakage is corrected by reassembling the cap and housing and properly installing the rubber washer.

When leak test is completed, install Fail-Safe unit onto non-pressure half of power unit in reverse order of disassembly.

REMOVING POWER UNIT FROM BRAKE

To remove either an air or hydraulic power unit from the brake assembly, proceed as follows:

1. If Fail-Safe unit is used, cage the power spring.
2. Remove brake lines and tag emergency line.
3. Determine type of power unit used, bottoming or adjusting type. Bottoming air chambers have an identification tag fastened to the clamp ring bolt and bottoming hydraulic cylinders have the word "bottoming" cast on housing.
4. If adjusting type unit is used, carefully mark the position of the assembly on the first exposed thread on the housing tube with a scribe or punch and hammer to aid reassembly.
5. Using a drift and hammer, straighten retainer and then loosen spanner nut.

NOTE: On welded-on spider-mounted brakes with open ends, the entire actuation system can be taken off at this time as one assembly, if desired, by removing four spider-to-plunger housing capscrews. However, the hub and drum and dust shield will have to be taken off before removing the actuation system.

---

6. Unscrew power unit from plunger housing. At this time, remove and inspect wedge assembly, and repair as necessary.

AIR CHAMBERS

To service an air chamber, follow the procedure covered under *Air Chamber Diaphragm,* disregarding wedge assembly precautions.

Note that adjusting type non-pressure housing is interchangeable with bottoming type housing and may be converted at this time.

HYDRAULIC CYLINDERS

Disassemble, Fig. 13

1. After removing cylinder from brake, remove bleeder screw and drain all fluid.
2. Place threaded end of cylinder on a cloth or wood surface and push internal parts out with air pressure. If parts are frozen, hydraulic pressure may be necessary.
3. Remove piston, cup, seal, and ball spacer from cylinder. Early piston design use a post to retain the cup seal and act as a spacer. To service, remove the piston post extension.

Assemble, Fig. 13.

1. Clean all parts in hydraulic fluid. If cylinder bore is corroded, scored or scratched, it must be replaced. Inspect piston for nicks, scoring or corrosion and polish with crocus cloth. Inspect end of bleeder screw for marks that would prevent it from sealing.
2. Lubricate cylinder bore with clean hydraulic fluid that is to be used with this system.
3. Install ball spacer in bottom of cylinder.
4. Install cup seal into cylinder, carefully entering lip of cup seal into cylinder bore, with flat end of cup out.
5. Install piston (flat end first) into cylinder and push piston and cup down to bottom of bore. Then install bleeder screw.

WEDGE ASSEMBLY

Disassemble, Fig. 14

1. Remove wedge from plunger housing by pulling it straight out.
2. Remove cotter key or "E" washer (earlier design) from wedge shaft while holding

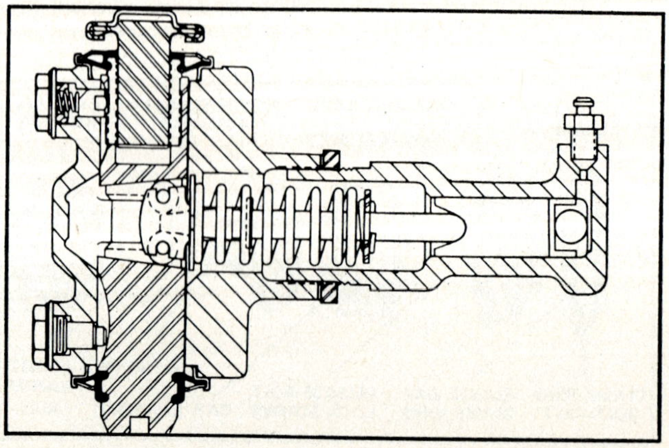

Fig. 13 Hydraulic cylinder

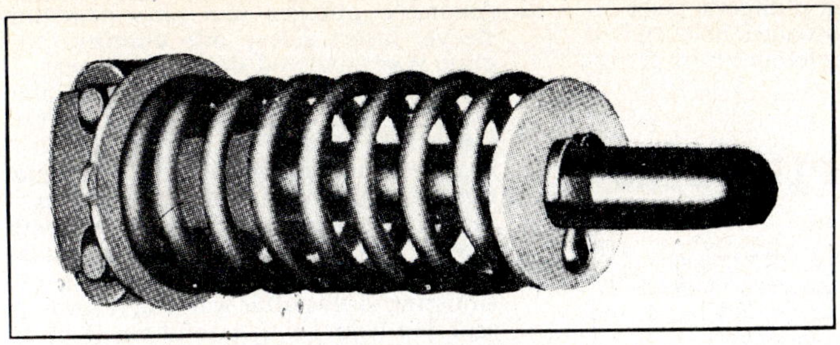

Fig. 14 Wedge assembly

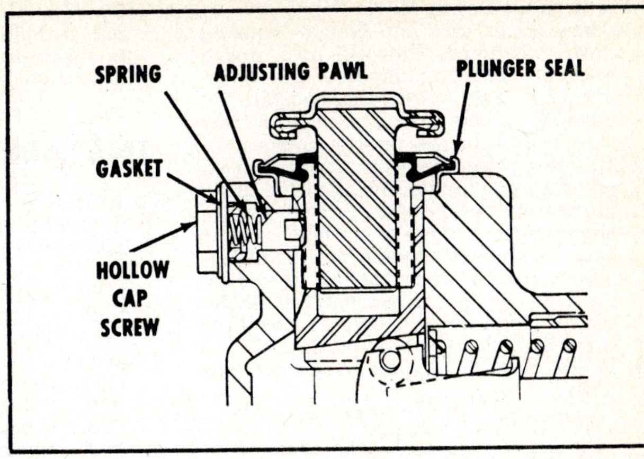

Fig. 15 Plunger Housing

spring compressed by hand.

3. Remove wedge retainer washer, wedge spring and retainer.
4. Insert a thin-bladed screwdriver between one flat of the wedge head and roller retainer cage. Spread cage open just far enough to remove rollers.

IMPORTANT: Do not attempt to drive wedge through the rollers and cage or force the rollers through the slightly closed slots of the cage. This will permanently damage the cage.

---

5. Remove roller retainer cage by sliding it off wedge shaft.

Assemble, Fig. 14

1. Clean all parts thoroughly and inspect. Check angled faces of wedge to see that they are free from pits or marks. If "E" washer type lock is used, inspect wedge shaft lock groove to see that it is sharp and clean. Inspect roller retainer cage to make sure it will contain rollers properly, and inspect the spring for marks that would cause breakage.
2. Insert wedge shaft into roller retainer cage so angled faces of wedge head are exposed.
3. Insert a thin-bladed screwdriver between flat of wedge head and roller retainer cage. Spread cage open just far enough to insert roller (journal hub into cage slot). Tip roller into cage and position other journal hub into slot of retainer. Install other roller in same manner and remove screwdriver. *Do not force rollers through ends of cage slots.*
4. Install spring retainer over wedge shaft and position centrally over cage and roller assembly.

NOTE: Current production and service replacement use a spring retainer that has two tabs or protrusions on the O.D. These tabs serve to align the wedge assembly as it is installed into the plunger housing by engaging grooves in the wedge bore. If the plunger housing is not equipped with such grooves, remove the tabs at the break-off marks and file the O.D. of the retainer smooth.

---

5. Install wedge spring over wedge shaft (large coil first). Add spring washer and compress spring by hand far enough to expose cotter key hole or "E" lock groove and install cotter or "E" lock.
6. Install wedge assembly into plunger housing. Check for correct roller-plunger engagement by first pushing on wedge rod by hand while checking for plunger and shoe lift; second, measure standout of wedge rod from end of threaded housing bore. When properly assembled, wedge standout is 2¼".

INSTALLING POWER UNIT TO BRAKE

Check position of plunger housing to make certain wedge assembly is properly seated. Be sure to replace automatic adjusting identification ring (if used) on power unit tube. Thread spanner nut onto power unit tube and install spanner nut retainer.

Bottoming Type Power Unit

1. Screw power unit into plunger housing until it bottoms (spanner nut loose).
2. Align connection ports with brake lines, if necessary, unscrewing power unit not more than one full turn.
3. Connect brake lines.
4. Make and hold a full pressure brake application. At this time position spanner nut retainer so it will engage plunger housing slot and drive spanner nut with a drift and hammer until it is tight against the retainer and plunger housing. Release brake pressure.
5. Using a drift and hammer, peen section of retainer into one slot of spanner nut.
6. Check for leaks at all connections.
7. Bleed hydraulic brakes (if used).
8. After installing drums, uncage Fail-Safe units.

Power Unit with Depth Mark

1. Screw power unit into plunger housing several turns. Turn spanner nut toward plunger housing so depth mark on threads is just exposed. Continue turning power unit into plunger housing until it bottoms on spanner nut and retainer.
2. Continue procedures 3 to 6 for bottoming type power units.

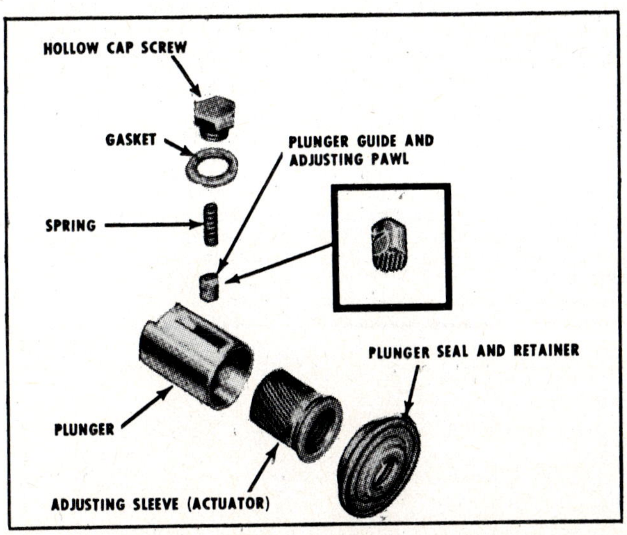

Fig. 16 Automatic adjusting plunger

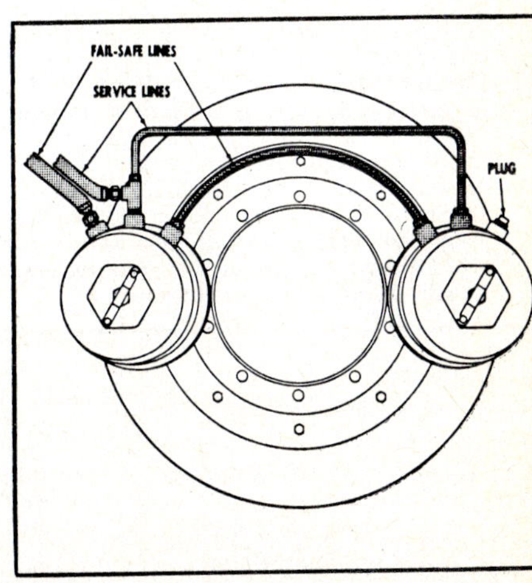

Fig. 17 Manifolding of Fail-Safe chambers

Units Without Depth Mark

1. Screw power unit into plunger housing until it bottoms. This will push wedge assembly between plungers and lift them off their seats inside the housing. By pushing on one shoe or plunger, second shoe or plunger will be seen to move.
2. Unscrew power unit one turn. Push on one shoe or plunger and then the other alternately, observing movement of opposite plunger. If there is movement of the opposite plunger, unscrew power unit another turn and continue this procedure until no plunger movement can be detected. This point is usually two or three turns from the bottomed position.
3. Follow Steps 3 to 6 for bottoming type units.

PLUNGER HOUSING

Disassemble

1. With brake shows removed, remove guide screws and gaskets from plunger housing. If brake is automatic adjusting, remove hollow capscrew, gasket, spring and adjusting pawl, Fig. 15.
2. Pry plunger seals loose and remove anchor (solid) plungers, adjusting plungers, adjusting sleeves (actuator), and adjusting bolt. If a four-piece adjusting bolt is used it should also be disassembled.

Assemble

1. Inspect and clean all parts, including housing plunger and seal bores and shoe rest pads. DO NOT solvent clean any rubber parts. Wire brush plunger parts and adjusting bolt threads to remove caked-on dirt or corrosion.
2. Carefully inspect plunger seals and gaskets for tears, cuts or deterioration and replace as required. Also check the angled plunger roller faces for pits, grooves or nicks and replace if necessary.

NOTE: RD spider mounted brakes have one adjusting plunger and one anchor (solid) plunger per plunger housing. The anchor plungers are marked on the shoe slot end—"R" for right-hand brakes and "L" for left-hand brakes. Do not mix.

Assemble Seals On Plungers

1. Apply film of grease to inside surfaces of seals.
2. On anchor plungers, inspect nose for burrs. Mask brake show web slot in plunger nose with masking tape to protect seals.
3. Carefully push double lip seal onto plunger, stretching outer seal lip over plunger nose end until inner seal lip is completely in the second plunger groove and the outer seal lip is in the first plunger groove. Remove masking tape. *Brakes employing single grooved plungers and single lip seals are assembled in the same manner except masking tape is not used.*
4. On manual adjusting plungers, push inner seal lip over threaded hole end of plunger until lip completely enters plunger seal groove.

INSTALLING PLUNGERS

RD Brakes

1. Coat all plunger bores with grease.
2. Make sure anchor plungers marked "L" and "R" are installed in left-hand and right-hand brakes, respectively.
3. Coat entire plunger with grease, packing cavity behind seal, and insert plunger and seal into housing with key-way slot aligned with guide screw hole. Make certain plunger goes all the way into plunger bore and seats on bosses on bottom.
4. Make certain guide screw end enters plunger key-way slot so plunger can slide freely in housing bore. Tighten guide screws to 15-20 ft-lbs torque.

RS Brake With Manual Adjust Plungers

1. Apply grease to outside of plunger and install plunger and seal into housing in same manner described for anchor plungers. Also follow procedures for seating plunger seal and assembling gasket and guide screw.
2. Coat adjusting bolt threads with grease and turn bolt into adjusting plunger by working it through hole in outer seal flap, being careful not to pinch seal on threads. Turn adjusting bolt in just short of seal. DO NOT bottom bolt on seal.

NOTE: If four-piece adjusting bolt is used, assemble parts in the following manner after adjusting bolt has been installed into plunger. A. Install detent washer on retainer, making sure flats are aligned, then install spring ring in groove in retainer. B. Turn open end of spring ring so that it faces dimple in detent washer. C. Install detent and retainer so that open end of spring ring and detent dimple engages the bolt last. Push assembly into bolt with thumb pressure only. DO NOT FORCE as it will damage spring ring.

RS Brake with Automatic Adjust Plungers

1. Grease inside and outside surfaces of both adjusting plunger and adjusting sleeve, Fig. 16. Place plunger into housing, aligning plunger key-way slot with guide hole and assemble adjusting pawl, gasket, spring, and hollow capscrew into plunger housing. Turn capscrew in one or two turns.

NOTE: The adjusting pawl has teeth and flats on one end and a chamfered edge on the other end. Coat pawl with grease and insert it into guide hole, teeth first, and with chamfer toward brake shoe. This aligns pawl and sleeve teeth and flats with keyway slot.

---

2. Assemble plunger seal onto adjusting sleeve. Insert sleeve into plunger, wiggling it so pawl and sleeve teeth mesh. If necessary, back off capscrew to enable meshing of pawl and sleeve teeth. Push sleeve into plunger until it bottoms.
3. Seat seal into plunger housing and hand-tighten capscrew into plunger housing.
4. Proper meshing of pawl and sleeve teeth can be checked at this point. Grease adjusting bolt threads and thread it into adjusting sleeve after working it through seal flap hole, being careful not to pinch seal in threads. Thread adjuster bolt into sleeve until it bottoms. A clicking sound and a ratcheting feel will indicate meshing of teeth. Turn bolt out three turns; if there is no clicking sound or ratcheting feel, this indicates proper meshing.
5. Tighten hollow capscrew to a torque of 15-20 ft-lbs.
6. Coat adjusting bolt with grease and thread it into the sleeve after working it through the seal flap hole, being careful not to pinch the seal on the threads. Turn bolt in just short of the seal. DO NOT bottom it on the seal. Pull the bolt outward slightly against the seal to insure entry of seal lip in sleeve groove. Push bolt and sleeve back into plunger.

NOTE: On earlier models of front brakes (backing plate mounted) a finger spring type of adjusting bolt lock was used. To assemble, first install plunger parts, press in seal, turn bolt into plunger or adjusting sleeve and then install finger spring and guide bolt. Lock guide bolt by peening lock plate to bolt boss and head.

The installation of plungers with single lip seals is the same as for double lip seals except with automatic adjusters. The single lip is installed on the adjusting sleeve first. Grease the sleeve, seal and adjusting plunger and install in the plunger housing. Then assemble the adjusting pawl, spring and hollow capscrew with gasket. Drive the seal into the housing. Screw the adjusting bolt into the sleeve until it bottoms and perform the check for adjusting pawl tooth mesh as outlined previously. Then back the bolt out about 1/2 turn.

MANIFOLDING

For all tandem axle applications, four Standard Fail-Safe chambers on one axle of the tandem or four Custom Fail-Safe chambers located one (per brake) at each corner of the tandem is recommended, Fig. 17.

Any single Fail-Safe chamber configuration, either Standard, Custom or Super, requires automatic show adjustment to work within the range of the lift provided by single wedge.

Before operating the vehicle, the caging bolt must be turned counter-clockwise as far as it will go (about 18 turns) to release the Fail-Safe spring. This allows the Fail-Safe unit to function in the event of air pressure failure.

BRAKE ADJUSTMENTS

INDEX

BENDIX DUO-SERVO BRAKE

This type brake, Fig. 1, is self-energizing in both forward and reverse. Each brake has one wheel cylinder mounted at the top of the backing plate below the single anchor pin. The anchor pin holds the wheel cylinder in place, being inserted through a flange on the wheel cylinder casting. Shoe return springs hold the upper end of the shoes against the pin with the brakes released. A guide plate, installed on the anchor pin outside of the shoe webs, holds the upper end of the shoes in alignment. Brake shoe lower guides, riveted to the backing plate, are also used on some models.

Brake adjustment is made at the adjusting screw in the adjusting link assembly. The link spring holds the shoe ends in place in the link ends. The link spring also serves as a lock for the notched adjusting screw.

Each brake shoe is held against three bearing surfaces on the backing plate by the hold-down pins and springs. Lining on the secondary (rear) shoe is longer than the lining on the primary (front) shoe.

Unless the vehicle has a transmission parking brake, the rear brake assemblies have a parking brake lever attached by a cable to the secondary shoes. A strut between the primary and secondary shoes assures braking action by both shoes. Parking brake action is entirely mechanical and operates independently of the hydraulic brake system.

Adjustment, Fig. 1

Before attempting to make any adjustments, be sure the wheel bearings are properly adjusted and that the brake backing plates are tight.

1. Jack up wheels clear of floor and release parking brake.
2. Remove adjusting hole covers.
3. Insert screwdriver or special adjusting tool in slot of backing plate to engage star wheel adjusting screw.
4. Move outer end of tool toward axle, expanding brake shoes to a point where the wheel can just be turned by hand.
5. Parking brake cable (if used) adjustment should be made at this time. With brake shoes still expanded, disconnect parking brake cable at intermediate lever end. With parking brake lever applied approximately one inch from fully released position, pull cables by hand to remove all slack. Then adjust cables as required at yoke ends so that clevis pins can be inserted.
6. Back off star wheel adjusting screw until brake drums turn freely. It may be necessary to tap the backing plate lightly to permit brake shoes to assume a central position.
7. Pull parking brake lever back until rear wheel can just be turned by hand. Check rear wheels for equal brake drag. If drag is unequal, loosen the tight brake until drag is equal on both rear wheels.

Anchor Pin Adjustment

On some models, an adjustable anchor pin is provided for when brakes have been relined or when brake action is unequal or severe. Some vehicles have the adjustable anchor on all four brake assemblies while others have it on the front wheels only.

1. Loosen nut on anchor pin at inner side of backing plate, then retighten nut 1/4 turn beyond finger-tight.
2. Remove adjusting hole cover from backing plate and turn star wheel adjuster to expand brake shoes until a heavy drag is felt when turning the brake drum by hand.
3. While continuing to expand shoes, periodically tap backing plate adjacent to the anchor pin to cause the anchor pin to center itself. When brake drag can no longer be relieved by tapping on the backing plate, hold anchor pin and tighten lock nut securely.

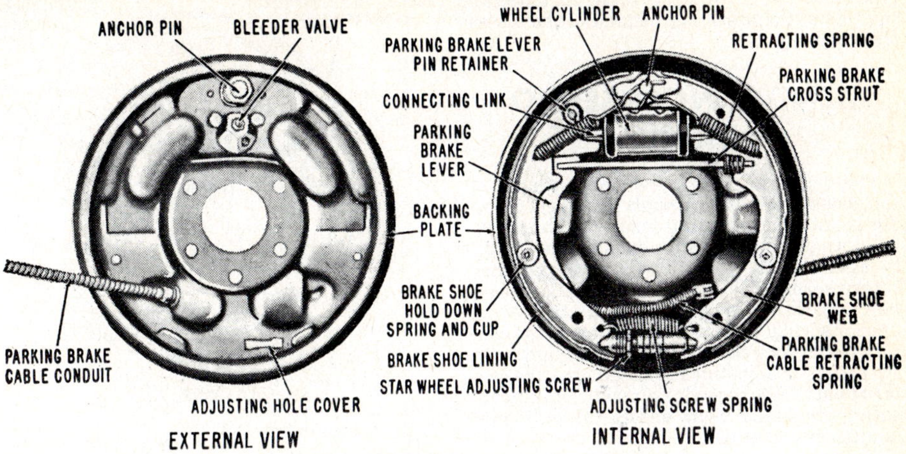

Fig. 1 Bendix duo-servo single anchor brake with star wheel adjustor

BENDIX UNI-SERVO BRAKE

This type brake, Fig. 2, is essentially the same as the duo-servo type except that a single end wheel cylinder is used instead of the double end type shown in Fig. 1. Adjustments are also the same as outlined for the duo-servo type.

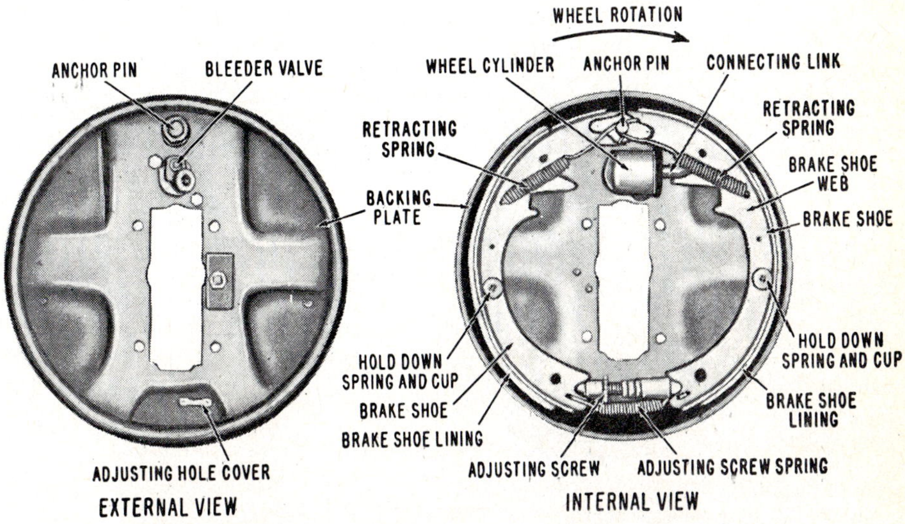

Fig. 2 Bendix uni-servo single anchor brake with star wheel adjuster

It should be noted that some models of this type brake had the wheel cylinder mounted on the side of the assembly instead of the top as shown in Fig. 2.

BENDIX NON-SERVO BRAKE

This type brake, Fig. 3, has the upper end of the shoes extending through the wheel cylinder boots and contacting inserts in the wheel cylinder pistons. The shoe ends are held firmly against the pistons by the brake shoe return springs. The lower end of the shoes are held against a fixed anchor plate by the anchor spring. A hold-down spring at the center of each shoe holds the shoes in alignment. An eccentric adjustment located near the center of each shoe provides a means of compensating for lining wear and is the only shoe adjustment required.

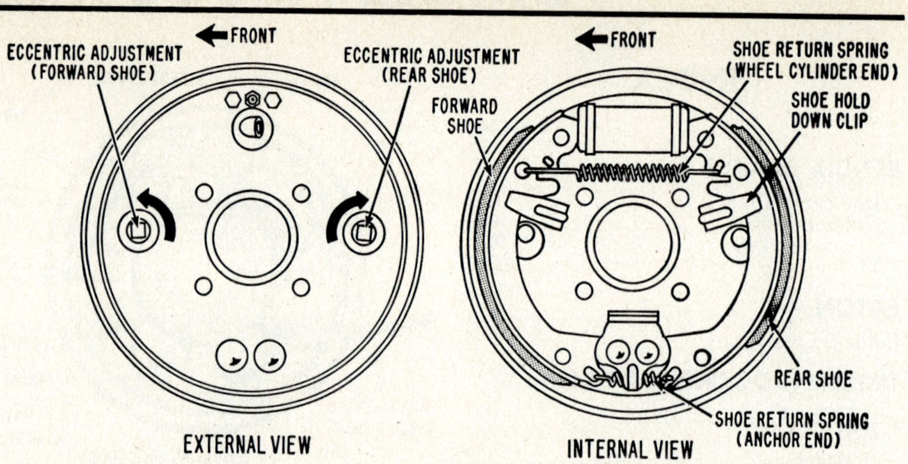

Fig. 3 Bendix non-servo brake of the floating shoe type with an eccentric cam adjustment for each shoe

Adjustment

1. With wheels raised clear of floor and hand brake released, turn forward shoe adjusting cam until a heavy drag occurs. Then turn the cam in the opposite direction until the drum is just free of drag.
2. Repeat with the rear shoe.

BENDIX WEDGE TYPE

This unit, Fig. 3A is an air actuated, non-servo, heavy duty brake assembly with an automatic adjuster assembly, Fig. 3B. A slotted clip is assembled to the end of the adjusting screw and mates with the shoe web. The adjusting screw which is threaded into the adjusting nut is secured by a clip so it does not rotate. When the adjusting nut turns, the adjusting screw moves outward and expands the brake shoes .001 inch per each adjusting nut tooth rotation.

Adjustment

After brakes have been relined, an initial adjustment is obtained by driving the vehicle forward and applying the brakes. If unit is equipped with a spring brake, continue applying brakes until spring brake locks the wheel. On units without a spring brake, continue applying brakes until a .015–.030 inch brake shoe to drum clearance is obtained. Access to linings for measurement or inspection can be obtained by removing dust shield.

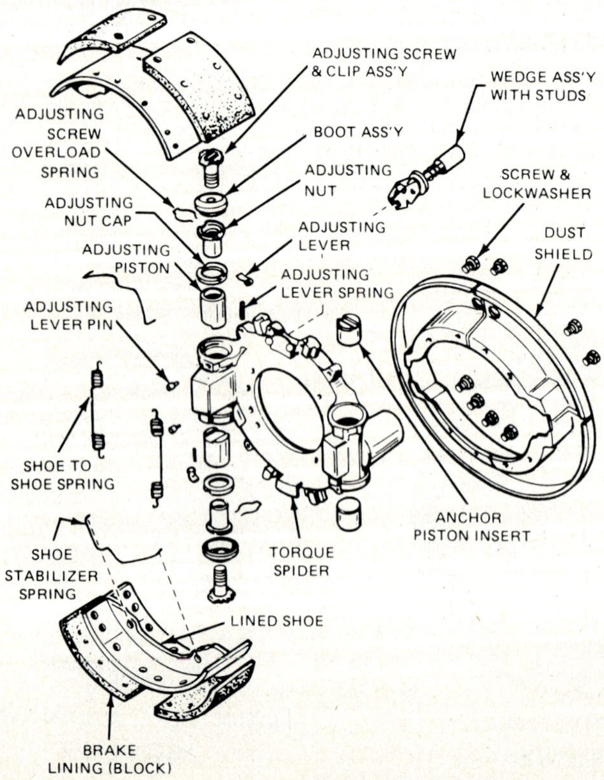

Fig. 3A Bendix non-servo wedge brake exploded view

EATON S-CAM BRAKE

The Eaton S-Cam brake, Fig. 3C, is an air actuated brake equipped with rollers on the cam end of the shoes. The shoe anchor ends rest on a single pin secured in position by two shoe retaining springs. The actuating cam is integral with the camshaft and the camshaft is supported on two plastic bushings in the bracket.

Adjustment

Adjustment of the brakes is made at the slack adjuster. With brakes released, check that the angle formed by the slack adjuster arm and brake pedal push rod is greater than 90°, Fig. 3D. Set all slack adjusters at the same angle, then apply brakes and check that the angle is still greater than 90°. If angle is less than 90°, adjust brakes as follows:

1. Raise and support vehicle at wheel to be adjusted.
2. Rotate slack adjuster worm shaft until brake drags, then back off adjuster until wheel rotates freely.

NOTE: Lock sleeve must be pushed in before worm shaft can be rotated. Check that sleeve is pushed in sufficiently to clear hex end of worm shaft before rotating shaft.

---

3. Check that wheel rotates freely with brakes released. Ensure that lock sleeve engages hex end of worm shaft. If necessary, pry sleeve out with a screwdriver. Lubricate lock sleeve and end of worm shaft with wheel bearing grease.

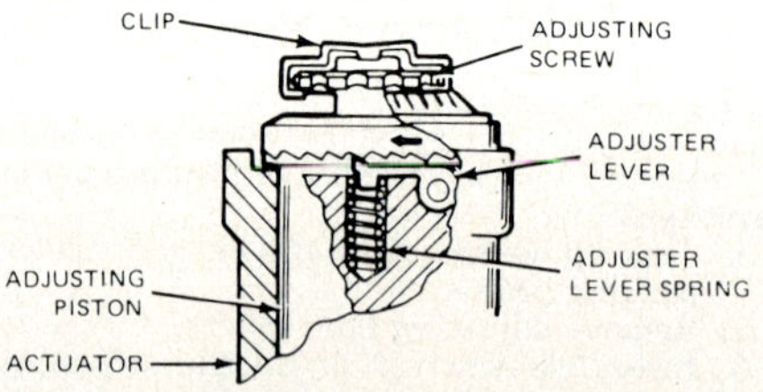

Fig. 3B Automatic adjuster assembly. Bendix non-servo wedge brake

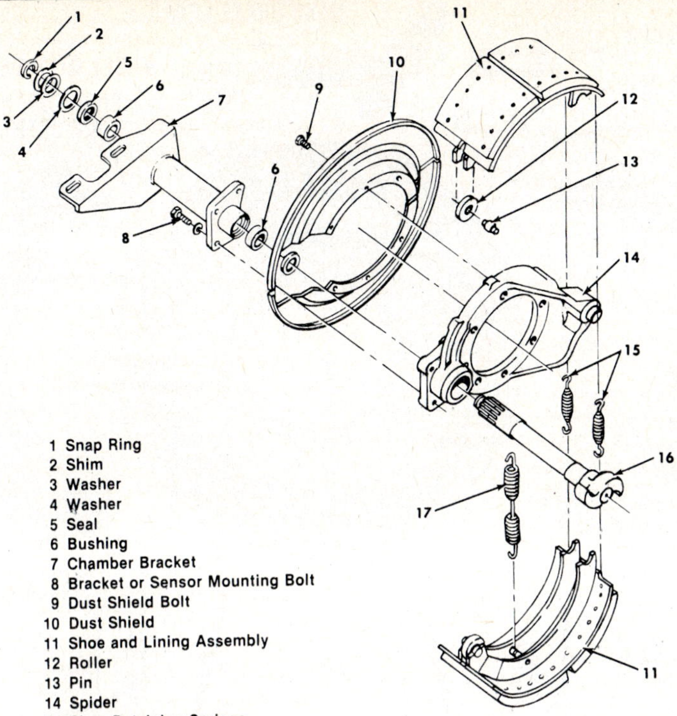

1 Snap Ring
2 Shim
3 Washer
4 Washer
5 Seal
6 Bushing
7 Chamber Bracket
8 Bracket or Sensor Mounting Bolt
9 Dust Shield Bolt
10 Dust Shield
11 Shoe and Lining Assembly
12 Roller
13 Pin
14 Spider
15 Shoe Retaining Springs
16 Camshaft
17 Shoe Release Spring

Fig. 3C Eaton S-cam brake assembly exploded view

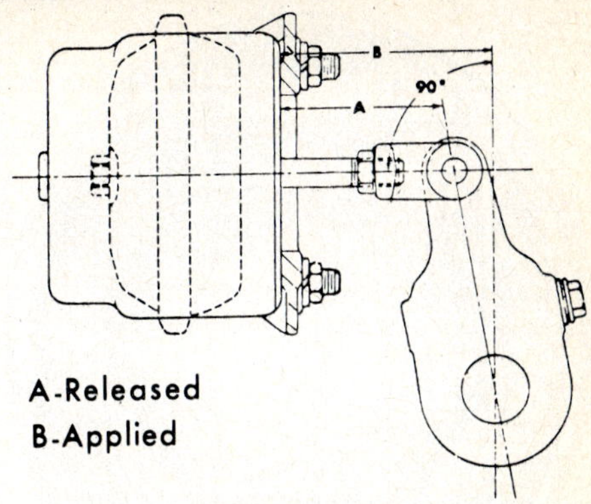

Fig. 3D Adjusting slack adjuster and push rod angle. Eaton S-cam brake

TIMKEN (ROCKWELL) "DH" BRAKE

This brake, Fig. 4, commonly referred to as the Duplex Brake, is an hydraulically-actuated dual primary assembly. It features identical shoe and liner assemblies and four identical return springs which simplify assembly and disassembly, Fig. 5.

Adjustment

Brake is adjusted from the back face of the backing plate at the slotted holes. Insert screwdriver through slotted hole against ratchet lugs of adjusting bolts and move end of handle up or down as a lever, depending on rotation desired.

Shoe liners should be brought out against the brake drum until a slight drag can be felt while the drum is in rotation. Adjusting bolt should then be backed off until drum can rotate freely. Brake shoes should be adjusted individually.

TIMKEN (ROCKWELL) "FSH" BRAKE

This brake, Fig. 6, is a floating shoe hydraulic assembly. Actuation permits the shoes to center themselves in the drum with equal effectiveness in either direction. This brake is supplied either with or without a built-in mechanical parking brake.

This brake is also built with automatic adjustment for use on special applications. On the automatic type, one actuation of the foot pedal sets the automatic adjustment. No further adjustment of the brake is required during the life of the brake lining. Fig. 7 shows the components that make up the assembly.

Adjustment

The adjustable type brake has two eccentric cams that may be adjusted with a wrench on the adjusting bolts at the back face of the backing plate.

Upon reassembly, first actuate the brake to center the shoes in the drum. Then adjust the liners out until a slight drag can be felt while the drum is in rotation. Back off the adjusting bolt until the drum can rotate freely.

Subsequent adjustments to compensate for

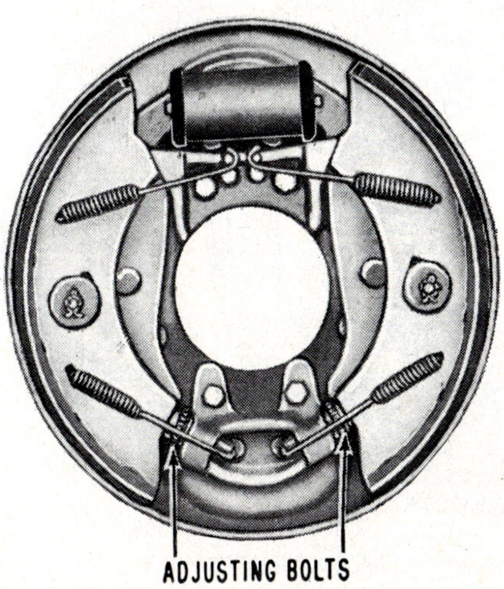

Fig. 4 Timken (Rockwell) "DH" series brake

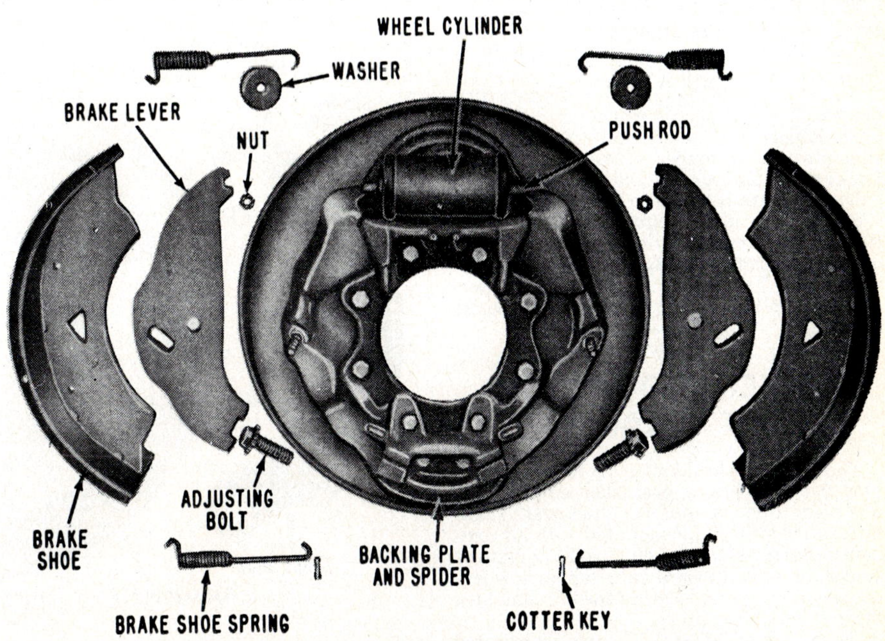

Fig. 5 Timken (Rockwell) "DH" series brake components

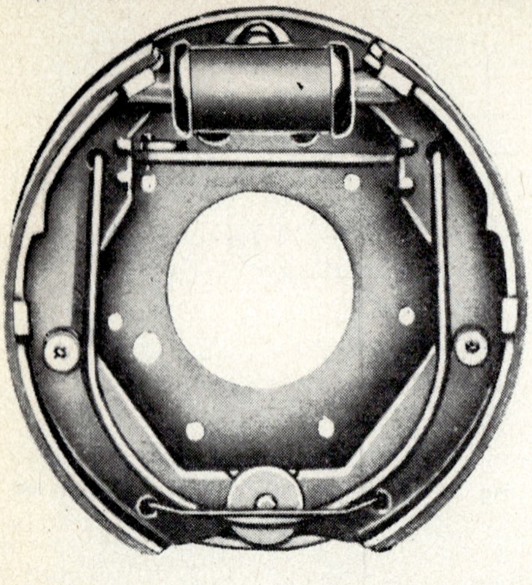

Fig. 6 Timken (Rockwell) "FSH" series brake

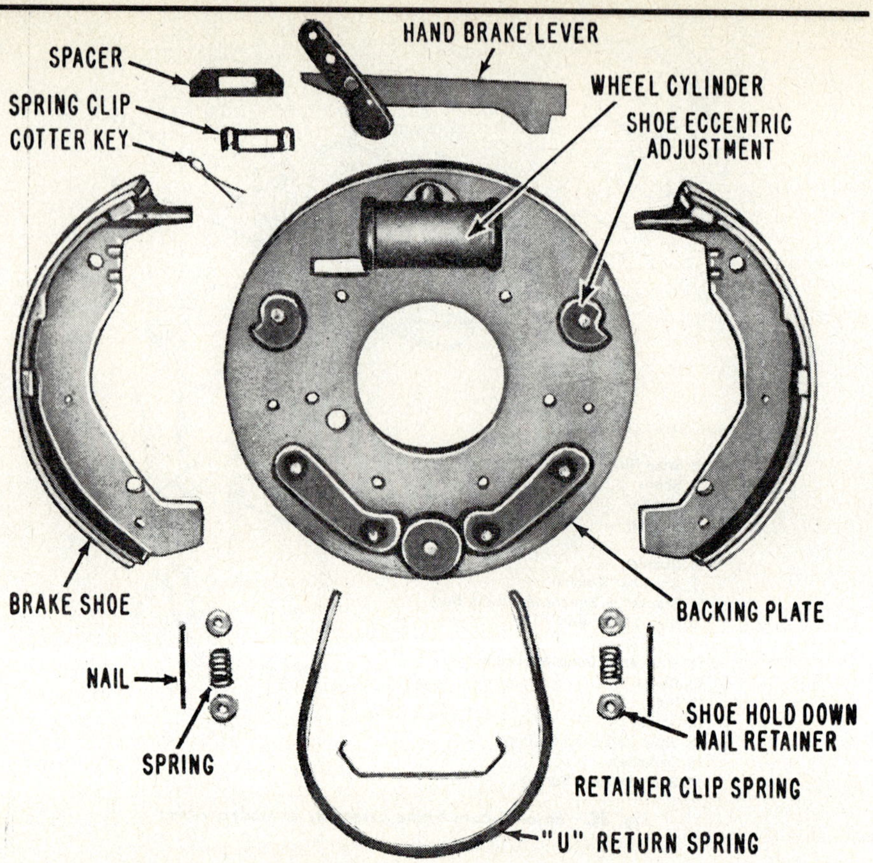

Fig. 7 Timken (Rockwell) "FSH" series brake components. When equipped with the parking lever feature, an upper coil spring is generally used rather than the "U" spring

lining wear may be made by moving the shoe in or out by turning the eccentric cam in the required direction.

TIMKEN (ROCKWELL) "H" BRAKE

The brake, Fig. 8, is a light duty, two shoe type mounted on the backing plate which also serves as a dust shield. Adjustable anchor pins provide a means of centering the brake shoe arc in relation to the drum. Secondary or minor adjustments are made by rotating the eccentric cam which bears on the brake shoe flange or pin in the shoe web.

Adjustment

Following overhaul or when new linings are installed, the initial adjustment should be carefully made to locate properly the curvature of the lining to the drum and to obtain the proper clearance.

Each shoe must be adjusted to center the brake shoe arc in relation to the drum. Back off on both anchor pins and cams and adjust each shoe individually.

Adjust cam to bring lining into contact with drum and rotate anchor in enough to relieve drag. Repeat until additional rotation of anchor pin will no longer relieve drag, Fig. 9. Lock anchor pin lock nut and back off cam just enough to permit wheel to turn freely.

Subsequent adjustments to compensate for lining wear are made with the eccentric cam only. Turn cam to bring lining into contact with drum. Then back off just enough to permit free rolling drum. Repeat on opposite shoe.

TIMKEN (ROCKWELL) "T" BRAKE

The "T" series brake, Fig. 10, is an air actuated two shoe brake, each shoe employing a one piece liner of uniform thickness. These brakes are either unit mounted with all component parts assembled on a backing plate or spider, or have inboard cam supports where operation warrants locating the air chambers toward the axle centers.

Some sizes of "T" brakes employ fabricated shoes with hardened steel cam roller followers Fig. 10, while other sizes employ either heat-treated malleable cast shoes with hardened cam follower wear pads, Fig. 11, or cam roller followers. Anchor pins of two different designs are used in various sizes of "T" series brakes. Some sizes use eccentric pins while others use fixed anchor pins.

The "T" series brakes are equipped with

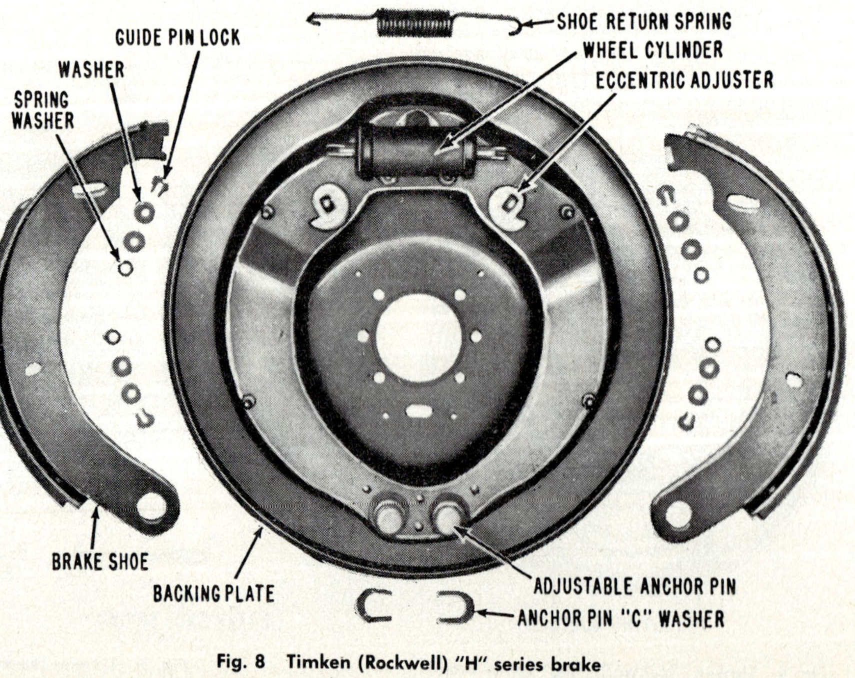

Fig. 8 Timken (Rockwell) "H" series brake

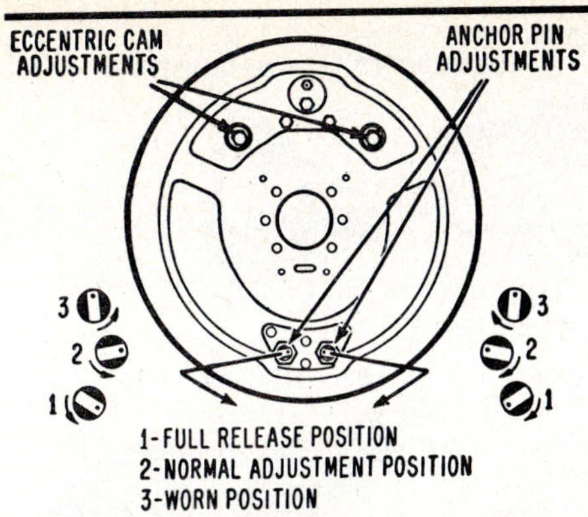

Fig. 9 (Rockwell) Timken "H" series brake adjustments

enclosed adjustable levers when shoe liner thickness does not exceed 3/8". The enclosed adjustable lever on these will permit the maximum liner wear. "T" series brakes with thicker liners, up to and including 1/2", are equipped with slack adjusters. Fig. 12 illustrates the component parts of the assembly.

Adjustment

Less Slack Adjuster

On the "T" Fabricated Steel Brake, each shoe must be adjusted to center the brake shoe arc with the drum.

1. Rotate both anchor pins to the full release position.
2. Adjust cam to bring the liners in with drum and rotate anchor pins just enough to relieve drag. Repeat until additional rotation of anchor pins will no longer free drag.
3. Tighten anchor pin lock nuts and back off cam to minimum running clearance. To compensate for liner wear on both the fabricated steel and the cast assemblies, adjust cam at cam lever adjusting bolt to bring liners in contact with drum and back off to minimum running clearance.

With Slack Adjuster

On vehicles equipped with slack adjusters, refer to "Adjustments" under "Eaton S-Cam Brake".

TIMKEN (ROCKWELL) "P" BRAKE

The automotive "P" series brake is an air actuated two shoe brake, each shoe employing

Fig. 11 Timken (Rockwell) "T" series air actuated brake with cam follower wear pads

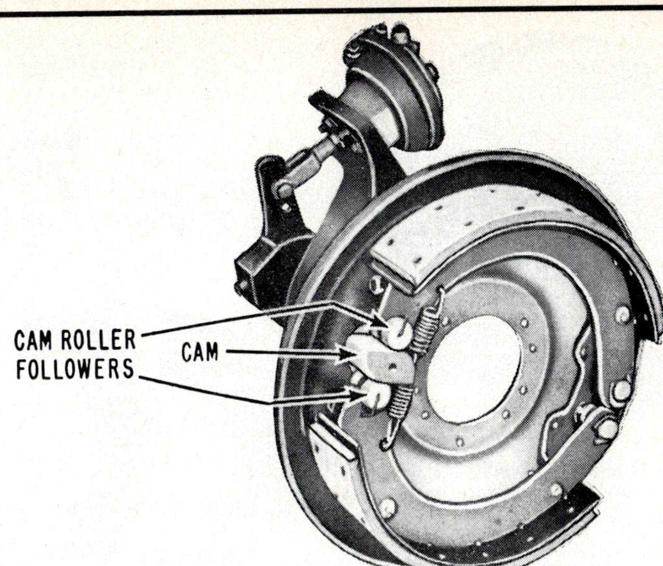

Fig. 10 Timken (Rockwell) "T" series air actuated brake with cam roller followers

two 3/4" tapered block liners, Fig. 13. These shoes are fabricated steel, mounted on individual anchor pins and supported by open type spiders. The air chamber mounts on a bracket that bolts directly to the spider, thus making the brake completely unit mounted.

Automotive "P" series brakes may also be equipped with inboard cam supports where operation or axle design requires mounting the air chambers toward the axle centers. Dust shields are available for use when protection becomes desirable.

The "P" series brakes are actuated by "S" type constant lift cams which are forged integrally with shaft and supported in nylon bushings. Cam pressure is applied by cam roller followers. Fig. 14 shows the brake components.

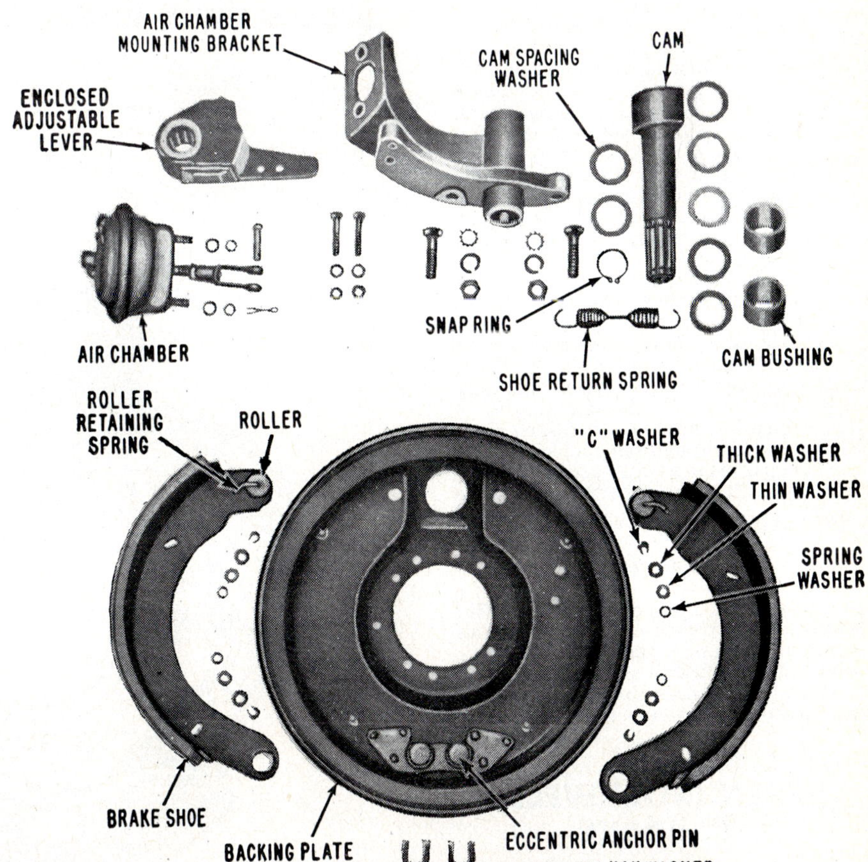

Fig. 12 Timken (Rockwell) "T" series brake components

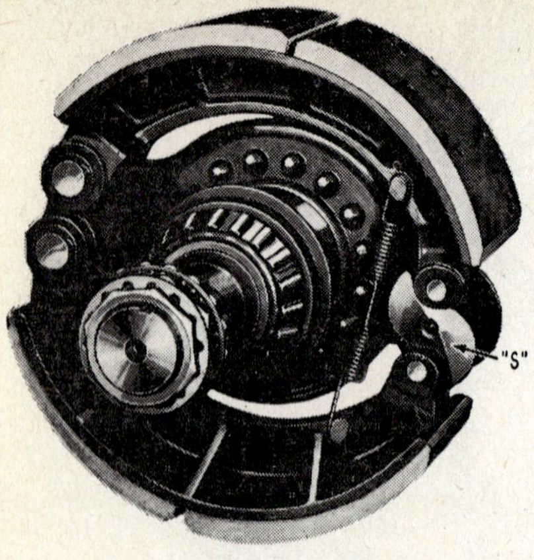

Fig. 13 Timken (Rockwell) "H" series brake adjustments

Adjustment

New liners should be circle ground to .070" less than drum diameter. Adjust cam as required to obtain 80% contact.

Adjust slack adjusters or levers to obtain 3/4" travel at 60 lbs. pressure. When travel increases to 1½" due to wear, readjust to 3/4".

TIMKEN (ROCKWELL) HEAVY DUTY "P" BRAKE

The Heavy Duty "P" series brake, Fig. 15, is an air actuated two shoe brake designed for heavy duty and special equipment where greater shoe and drum areas are necessary. This heavy duty series ranges from 18" to 22" and in width from 4" to 7".

The shoes are of heat treated malleable castings mounted on open spiders by individual anchor pins. Each shoe has hardened cam pressure pads and employs two 3/4" tapered block liners.

The camshaft and air chamber support bracket is mounted directly on the brake spider, making the assembly complete as a unit. However, the heavy duty brake may also be equipped with inboard cam supports where operation or axle design requires mounting the air chambers toward the axle centers.

Cam pressure is applied to the shoe hardened wear pads through the double "S" cam which is supported in needle bearings. Fig. 16 illustrates the component parts of the brake assembly.

Adjustment

New liners should be circle ground to .070" less than the drum diameter. Adjust cam as required to obtain 80% contact.

Adjust slack adjusters or levers to obtain 3/4" travel at 60 lbs. pressure. When travel increases to 1½" due to wear, readjust to 3/4".

TIMKEN (ROCKWELL) "Q" BRAKE

For adjustment of this air actuated two shoe brake, Fig. 15A, refer to "Adjustments", under "Eaton S-Cam Brake".

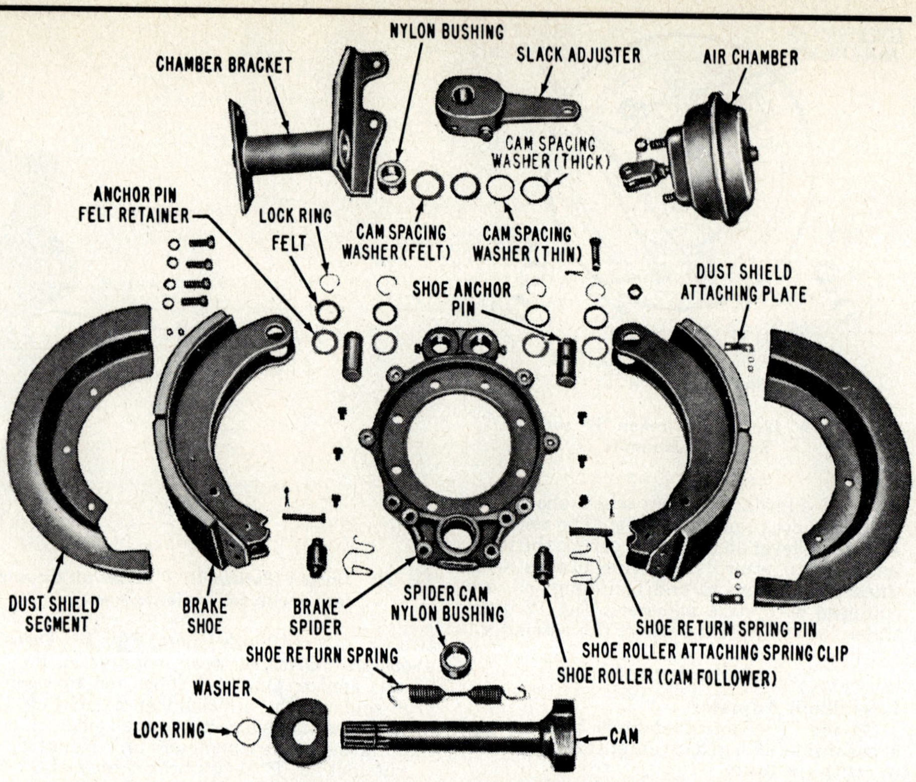

Fig. 14 Timken (Rockwell) "P" series brake components

TIMKEN (ROCKWELL) STOPMASTER BRAKE

The Stopmaster Brake is a wedge actuated brake. The brake power units can be either air chambers or hydraulic cylinders. They can also be equipped with Fail-Safe units to provide parking and emergency braking. The brake power unit forces a wedge between two rollers and two plungers. This causes the plungers to spread apart and push the brake shoes against the brake drum.

There are several variations of Stopmaster Brakes in use. Some have two power units per brake and others only one. Where two power units are used, there may be a Fail-Safe unit on one or both. The brake support may be either a cast spider or a stamped backing plate. On the cast spiders, the plunger housings may be either integral or bolted on. Brake shoe adjustment may be either manual or automatic.

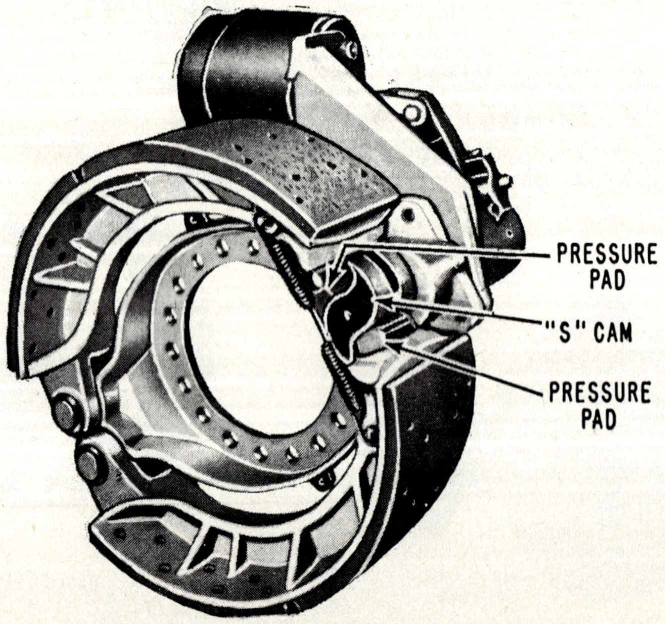

Fig. 15 Timken (Rockwell) "P" series heavy duty brake

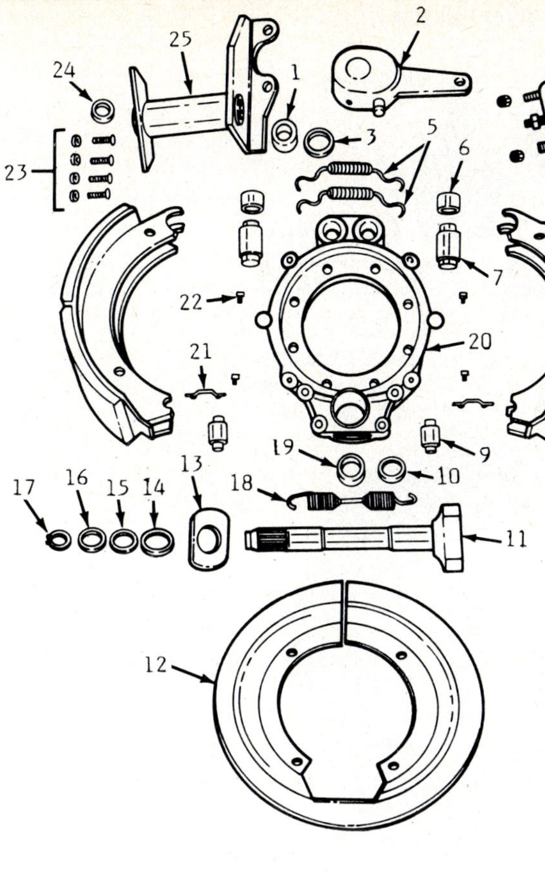

1. CAMSHAFT BUSHING
2. SLACK ADJUSTER
3. GREASE SEAL
4. BRAKE CHAMBER
5. SHOE RETAINING SPRINGS
6. ANCHOR PIN BUSHING
7. ANCHOR PIN
8. SHOE AND LINING
9. ROLLER
10. GREASE SEAL
11. CAMSHAFT
12. DUST SHIELD
13. CAMHEAD WASHER
14. CAMSHAFT WASHER
15. THIN SPACING WASHER
16. THICK SPACING WASHER
17. LOCK RING
18. RETURN SPRING
19. CAMSHAFT BUSHING
20. SPIDER
21. RETURN SPRING PINS
22. DUST SHIELD CAPSCREW
23. BRACKET CAPSCREWS & WASHERS
24. BRACKET GASKET
25. CHAMBER BRACKET

Fig. 15A Timken (Rockwell) "Q" series brake assembly exploded view

Adjustment

Manual Adjusters

1. Hoist vehicle free of ground.
2. Remove cover from adjusting slot (two places on each brake).

NOTE: On RS type brakes the adjusting slots are above and below the single power unit. On RD type brakes the adjusting slots are below the forward and above the rear power unit. If star-wheel adjusting bolts are not found at these positions, the brake has been assembled on the wrong side of the vehicle.

---

3. Adjusting bolts have right-hand threads. With an adjusting spoon, turn star-wheel until heavy drum drag is developed. Then back off the bolt to a very light drag of the drum. Repeat for other brake shoe.
4. Replace dust covers in adjusting slots. Repeat for other brakes.

Automatic Adjusters

Check drum to lining clearance with feeler gauge. If clearance is more than .060", adjust brake manually (same as above) and schedule the vehicle for brake service.

WAGNER FLAT CAM BRAKES

This air actuated brake, Fig. 16A, consists of two shoes secured on anchor pins by lock rings. Contact between cam ends of the shoes and the cam is made through rollers. Two guide pins on each shoe maintain shoes in alignment with backing plate. Refer to "Adjustments" under "Eaton S-Cam Brake" for adjustment procedure.

WAGNER NON-SERVO BRAKE

This type brake, Fig. 17, has an adjustable anchor pin and an eccentric cam for each shoe. Note that the brake shown has a stepped type hydraulic cylinder with one end smaller than the other. However, other models have both ends of the cylinder the same diameter.

Minor Adjustment

1. With wheels raised clear of floor and hand brake in released position, rotate cam "A" outward until front shoe contacts drum and a definite drag is felt when wheel is turned by hand. Then back off adjustment until wheel is free.
2. Repeat procedure with cam "B".

Major Adjustment

1. With wheels and drums installed, rotate cam "A" until there is a brake drag felt when wheel is turned by hand.
2. Turn shoe anchor "C" in the direction of forward rotation of the wheel until a definite brake drag is felt. Note position of the anchor. Turn anchor "C" back in the opposite direction until the same amount of brake drag is felt and again note the position of the anchor. The turn anchor "C" back until it is halfway between the two extremes, producing a free wheel. This will indicate that the shoe is roughly centralized.
3. Check both toe and heel of the shoe with a feeler gauge to determine the next steps to be taken in moving the adjustments.
4. Adjust cam "A" and anchor "C" until a clearance of .005" is obtained at the heel and a clearance of .010" is obtained at the toe end of the shoe as measured about 1½" from ends of lining.
5. Tighten lock nut on anchor "C" securely. Recheck clearances at shoe ends to make sure adjustments have not altered.
6. Repeat the foregoing operations on the rear shoe by adjusting cam "B" and anchor "D", substituting the word reverse for the word forward.

WAGNER "F" BRAKE

This type brake, Fig. 18, is self-centering and has two identical brake shoes arranged on the backing plate so that their toes are diagonally opposite. Two single end hydraulic cylinders are arranged so that one cylinder is mounted between each shoe toe and the opposite shoe heel.

The two hydraulic cylinder pistons apply equal amounts of hydraulic force to each shoe toe. Each cylinder body is shaped to provide an anchor block for the opposite shoe heel. Each cylinder anchor block serves as a shoe stop centering point and provides the fulcrum around which the shoe heel pivots when the brakes are applied.

Adjustment

1. With wheels jacked clear of floor, place wrench on one adjusting cam stud to adjust one shoe. Rotate wrench in the direction of forward wheel rotation until lining drags on drum.
2. Move wrench slightly in opposite direction until wheel is free to turn; then move wrench an additional 7 to 10 degrees to provide running clearance (7 to 10 degrees is equal to 1" to 1½" of travel at end of an 8-inch wrench).
3. Place wrench on opposite adjusting cam stud and adjust second shoe in the same manner.

WAGNER "FA" & "F2A" BRAKE

This type brake, Fig. 19, has two single end wheel cylinders per wheel. The closed ends have slotted anchor surfaces which supports the opposite brake shoe. Brake shoes are identical and are all energized by forward rotation.

Manual Adjustment

Brake shoes may be manually expanded by two 5/8" hex-head, friction spring locked studs, which are exposed on the backcam located under the shoe table. Torque limit for manual adjustment should not exceed 120 in. lbs.

Automatic Adjustment

On each shoe an automatic adjuster lever is pivot-pinned to the inner side of the shoe web and rests against the manual adjuster cam. A drum contact plug in the center of the shoe

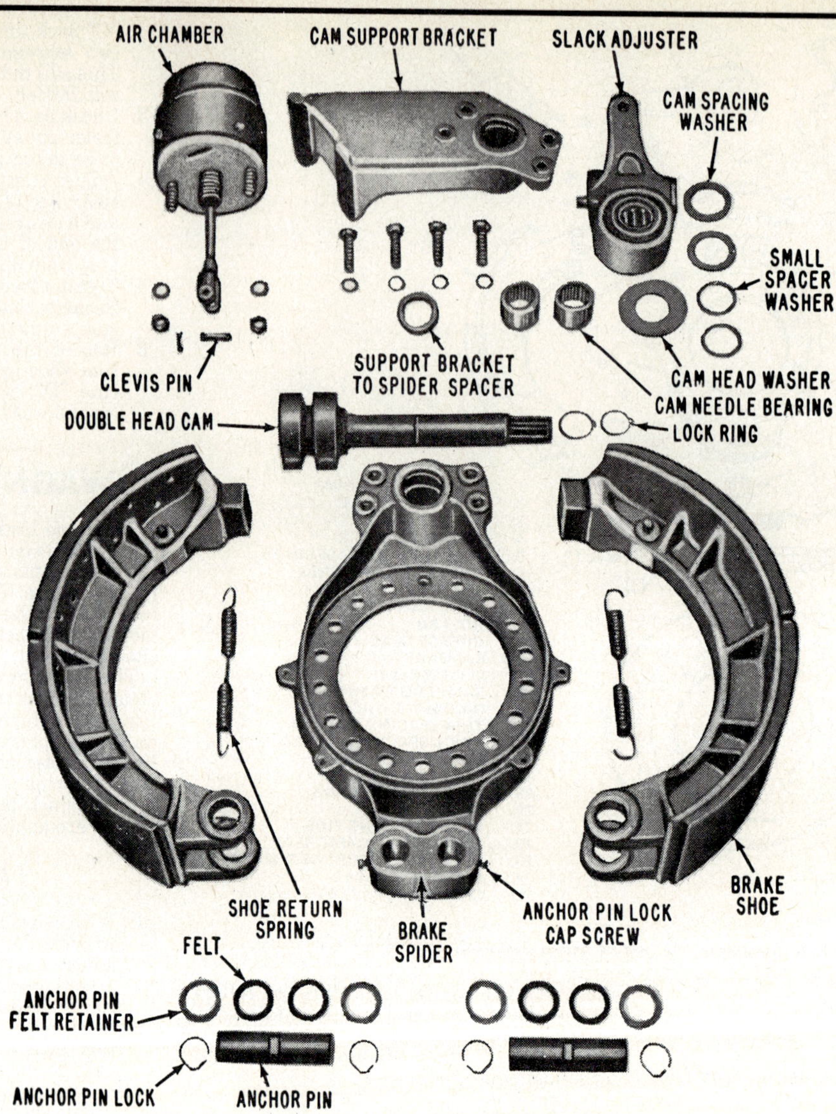

Fig. 16 Timken (Rockwell) heavy duty "P" series brake components

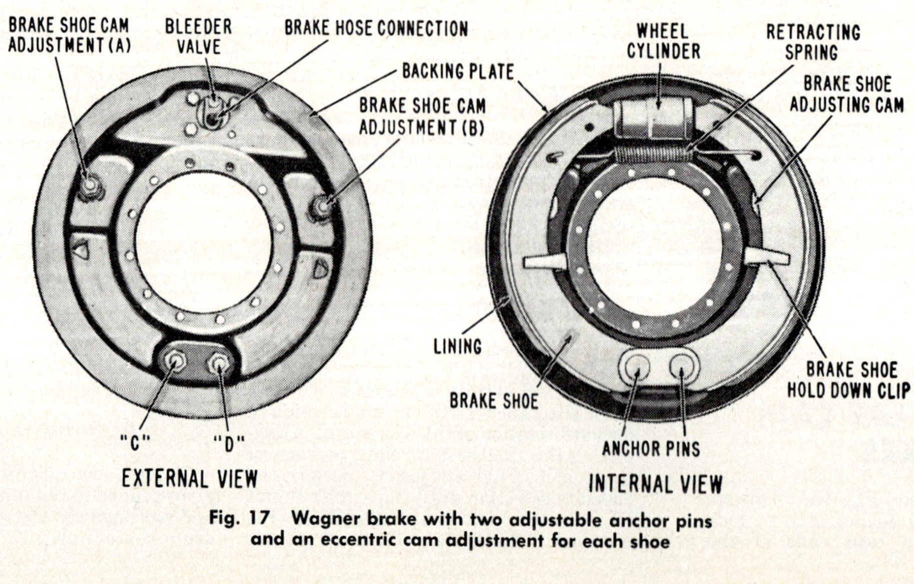

Fig. 17 Wagner brake with two adjustable anchor pins and an eccentric cam adjustment for each shoe

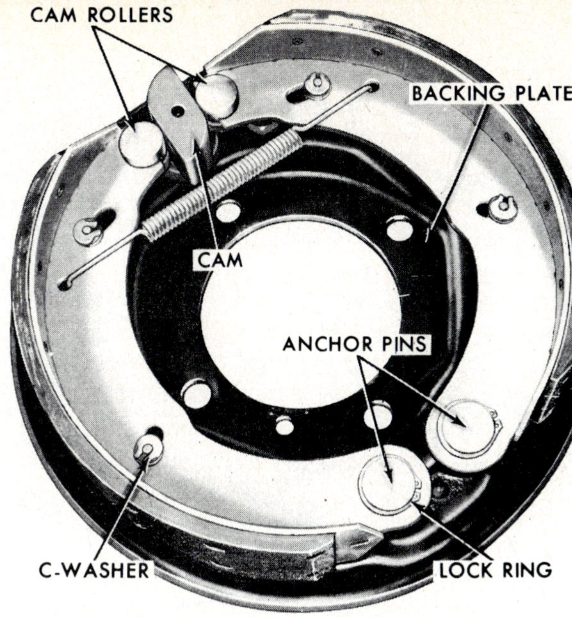

Fig. 16A Wagner flat cam brake assembly

tables has a shank pinned to the automatic adjuster lever on the outside of the web.

A spring loaded serrated wedge slides on the lever pivot pin, between a washer and the shoe web, under the shank of the drum contact plug entering the space between the plug pin and the wedge guide which is also pinned to the shoe. An adjuster torsion spring is hooked over the plug pin, lever pivot pin and the edge of the shoe table.

As brake lining wears, the drum depresses the drum contact plug which swings the adjuster lever away from the shoe table. Thereupon, the spring loaded wedge moves to keep the gap between the plug pin and the wedge guide closed, holding the lever and shoe in adjusted position. At maximum lining wear, the drum contact plug pin bottoms in the slot in the shoe web, stopping the automatic adjusting process.

WAGNER "FR-3 & FR3A BRAKE

The FR-3 brake is equipped with two double end wheel cylinders which apply hydraulic pressure to both the toe and heel of the two self centering brake shoes, Fig. 19A. The shoes anchor at either the toe or the heel depending on direction of drum rotation. The brake anchor supports and backing plates are bolted to the axle housing flange. The supports are equipped with removable slotted anchor pins at the shoe heels. Adjusting screws act as anchors in the reverse direction of rotation. Each adjusting screw is threaded into or out of its support by means of an adjusting wheel, which are accessible through slots located in backing plates.

The FR-3A brake is identical to the FR-3 brake except that it is equipped with an automatic adjusting mechanism, Fig. 19B. The shoes maintain adjustment by identical link crank adjuster systems. During a reverse brake application, the shoe heel moves away from its forward acting anchor causing a pivoting motion in the link crank system. An adjustable eccentric on the shoe web acts as a hinge for a short link which transfers this motion to a crank lever secured to the forward acting anchor. From this crank, motion is transferred through a longer link to an adjusting lever mounted on the opposite anchor bracket, allowing a pawl on the lever to mesh with the star wheel. This motion pivots the adjusting lever backwards and overcomes the force of the adjusting spring connected between lever finger and the spring pin. As the lining wears and clearance increases, the adjuster lever pawl engages the next tooth on the star wheel.

Adjustment

1. Raise and support vehicle, then remove adjusting hole covers from the backing plate.
2. Using suitable adjusting tool, rotate adjusting wheel until lining drags on drum.
3. Rotate adjusting wheel in opposite direc-

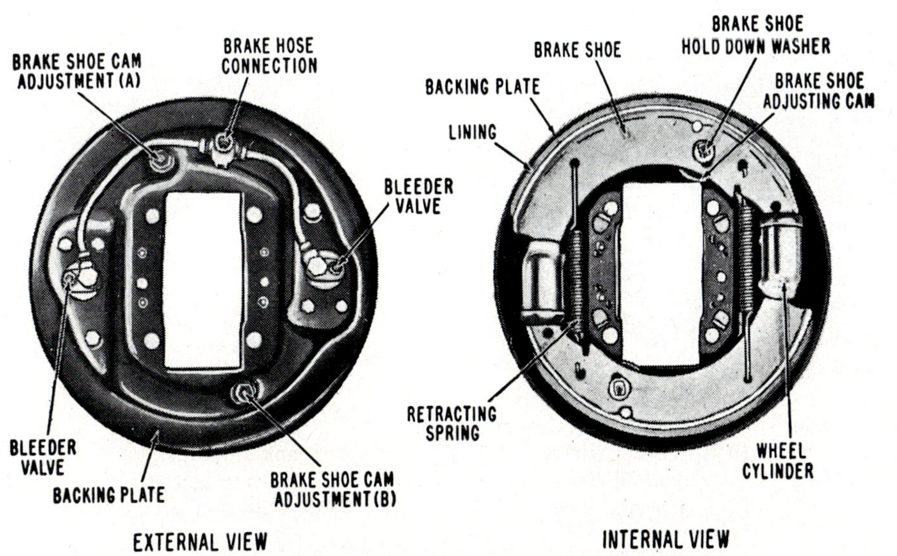

Fig. 18 Wagner "F" type brake. It should be noted that some models have the wheel cylinders mounted near the top and bottom of the assembly with the adjusting cams located at the sides

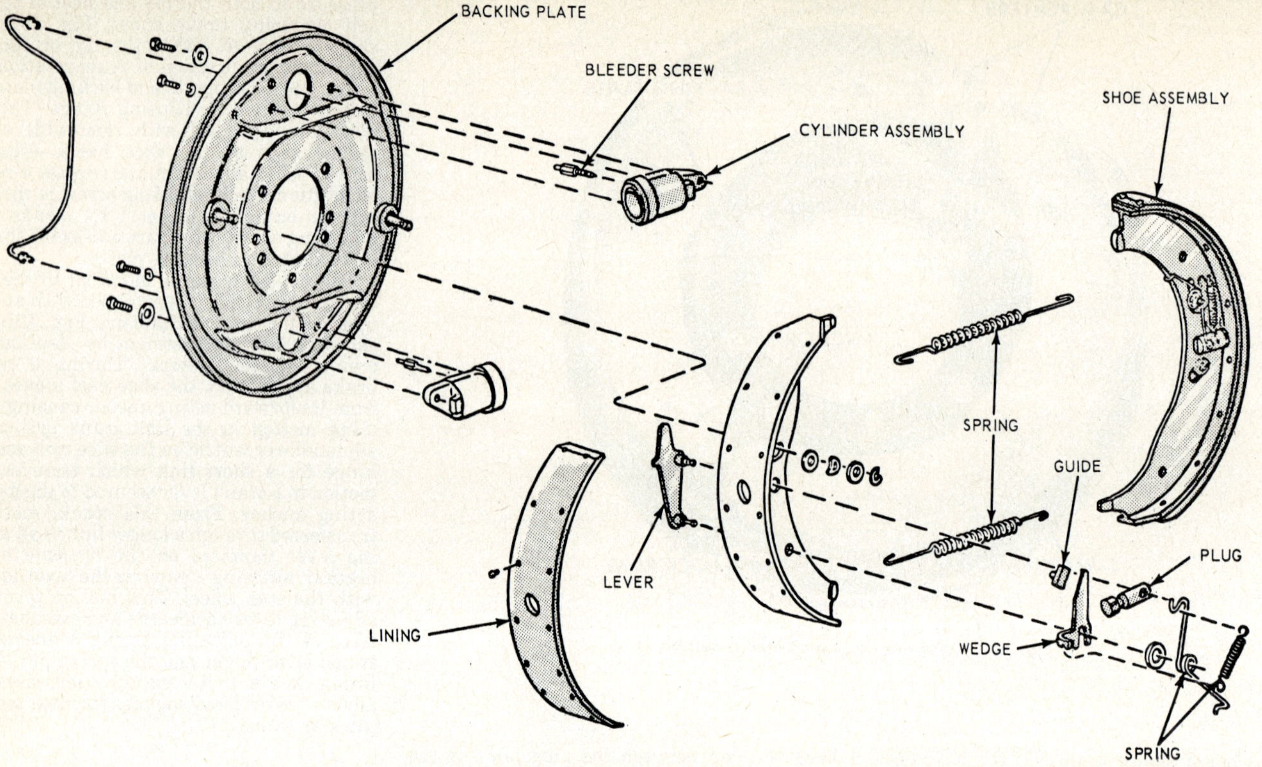

Fig. 19 Wagner "FA" & "F2A" self-adjuster front wheel brake

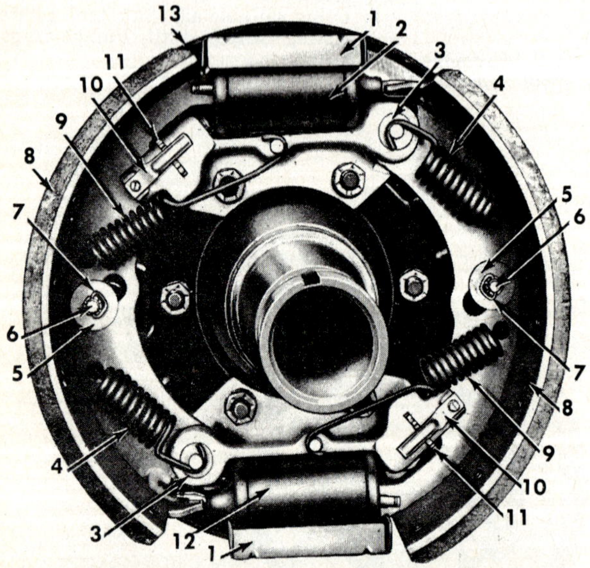

1 Wheel Cylinder Heat Shield
2 Upper Wheel Cylinder
3 Anchor Pin
4 Return Spring (Short)
5 Guide Washer
6 Guide Bolt
7 Guide Bolt Nut Lock Wire
8 Brake Shoe and Lining Assembly
9 Return Spring (Long)
10 Adjusting Wheel Lock Spring
11 Adjusting Wheel
12 Lower Wheel Cylinder
13 Backing Plate

Fig. 19A Wagner FR-3 type brake assembly

1 Brake Shoe and Lining Assembly
2 Hold-down Bolt, Washer, Nut, and Lock Wire
3 Crank Link (Long)
4 Eccentric and Screw
5 Shoe Return Spring (Short)
6 Adjuster Crank
7 Crank Link (Short)
8 Wheel Cylinder Cover
9 Wheel Cylinder
10 Backing Plate
11 Adjuster Spring
12 Starwheel
13 Starwheel Crank
14 Shoe Return Spring (Long)
15 Anchor Pin
16 "C" Washer

Fig. 19B Wagner FR-3A type brake assembly

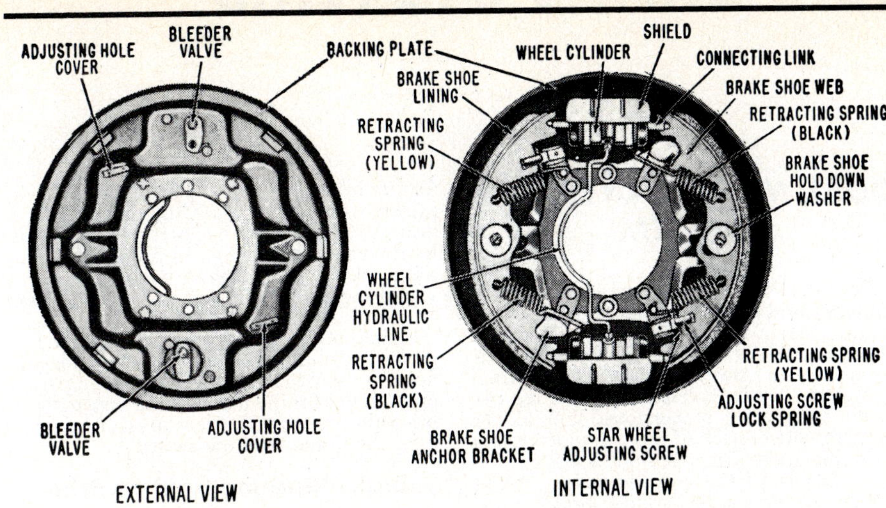

Fig. 20 Wagner twin action brake. Some models have the wheel cylinders mounted at approximately 5 and 11 o'clock positions. Also, some have brake shoe retracting springs of unequal lengths

tion until wheel rotates freely.

4. Perform final adjustment as follows: On "FR-3 brakes, back off adjusting wheel 3 notches for worn linings or 5 notches for new linings; On FR-3A brakes rotate adjuster 9–10 notches to position adjuster running clearance.
5. Install adjusting hole covers, lower vehicle and check brakes for proper operation.

WAGNER TWIN ACTION BRAKE

This type brake, Fig. 20, is equipped with two double end hydraulic cylinders which apply hydraulic pressure to both the toe and heel of two identical, self-centering shoes. The shoes anchor at either toe or heel, depending upon the direction of drum rotation. Brake anchor supports and backing plates are riveted to the axle housing flange. The supports have slotted anchor pins at the shoe heels, and adjusting screws at the shoe toes. The adjusting screws act as anchors in the reverse direction of rotation. Each adjusting screw is threaded into or out of its support by means of an adjusting wheel. The adjusting wheels are accessible through slots in the backing plate.

Adjustment

1. With wheels raised clear of floor, remove adjusting holes covers from backing plate.
2. Use a screwdriver or special adjusting tool through opening in backing plate and turn rear adjusting screw until a light drag is felt as wheel is turned. Then back off until drum turns freely.
3. Repeat adjustment at front adjusting screw. Install adjusting hole covers. Fig. 21 shows the twin action brake that employs a self-adjuster.

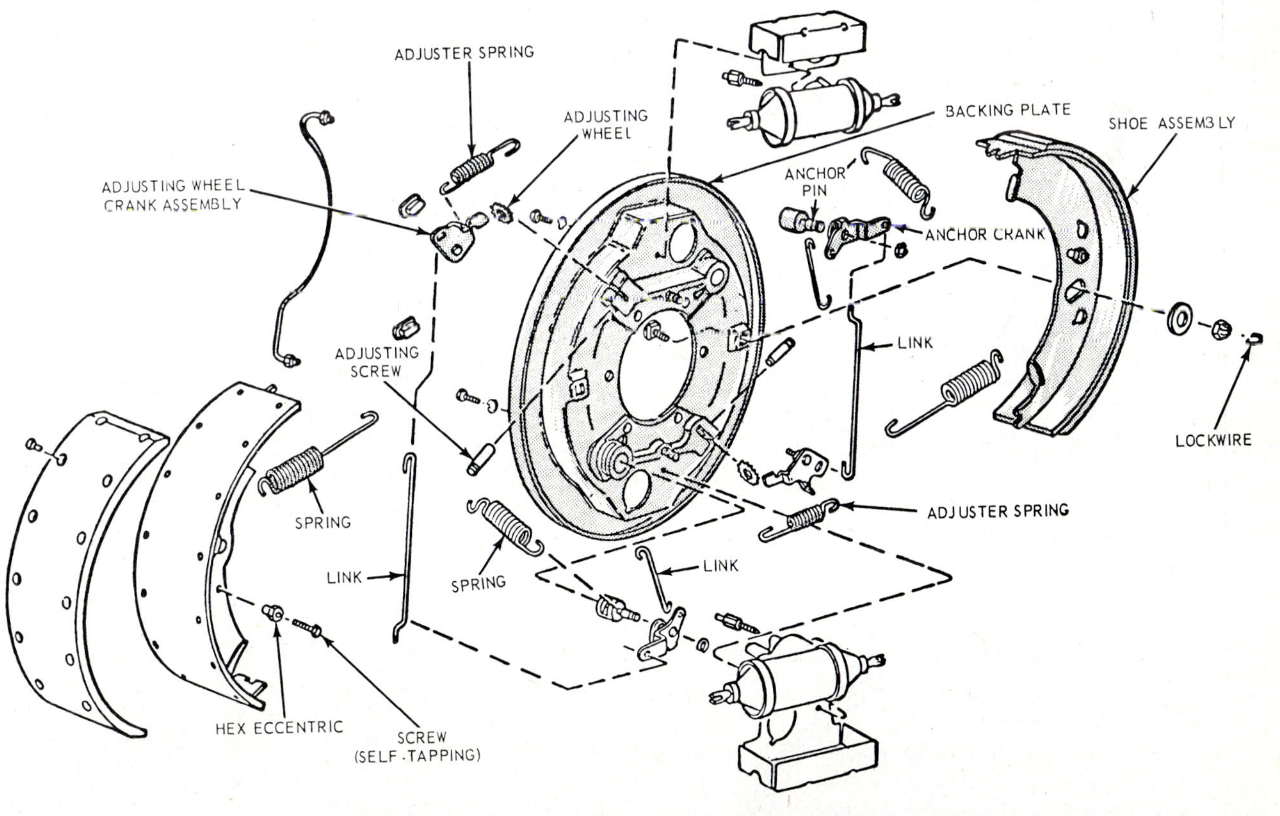

Fig. 21 Wagner twin action self-adjuster brake

HYDRAULIC BRAKE SYSTEMS

SINGLE MASTER CYLINDER SYSTEM

Depressing the brake pedal moves the master cylinder push rod and piston, forcing hydraulic fluid out through a check valve, Fig. 1. This fluid flows through the hydraulic lines into the wheel cylinders, forcing the wheel cylinder pistons outward from the center of the cylinder and expanding the brake shoes and linings against the brake drums.

When the brake pedal is quickly released, the master cylinder piston returns to the released position faster than fluid returns from the lines. Holes in the piston head allow fluid to pass from the rear to the front of the piston head, past the primary cup to fill the space.

At the same time (when the pedal is released) the brake shoe return springs force the wheel cylinder pistons to return toward the center of the wheel cylinder (released position). Fluid forced out of the wheel cylinders by this action returns to the master cylinder by overcoming the pressure of the master cylinder piston spring which holds the check valve closed. As this fluid returns, the excess portion will return to the reservoir through the compensating port which is uncovered when the master cylinder piston is in the released position. The piston spring will close the check valve when the pressure in the lines is reduced to 8 to 12 lbs, maintaining a slight pressure in the lines at all times. The purpose of this pressure is to keep the wheel cylinder cups from leaking fluid and to reduce the possibility of air entering the system.

DUAL MASTER CYLINDER SYSTEM

When the brake pedal is depressed, both the primary (front brake) and the secondary (rear brake) master cylinder pistons are moved simultaneously to exert hydraulic fluid pressure on their respective independent hydraulic system. The fluid displacement of the two master cylinders is proportioned to fulfill the requirements of each of the two independent hydraulic brake systems, Figs. 2 and 3.

If a failure of the rear (secondary) brake system should occur, initial brake pedal movement causes the unrestricted secondary piston to bottom in the master cylinder bore. Primary piston movement displaces hydraulic fluid in the primary section of the dual master cylinder to actuate the front brake system.

Should the primary (front) system fail, initial brake pedal movement causes the unrestricted primary piston to bottom out against the secondary piston. Continued downward movement of the brake pedal moves the secondary piston to displace hydraulic fluid in the rear brake system to actuate the rear brakes.

The increased pedal travel and the increased pedal effort required to compensate for the loss of the failed portion of the brake system provides a warning that a partial brake system failure has occurred. When the ignition switch is turned on, a brake warning light on the instrument panel provides a visual indication that one of the dual brake systems has become inoperative.

Should a failure of either the front or rear brake hydraulic system occur, the hydraulic fluid pressure differential resulting from pressure loss of the failed brake system forces the valve toward the low pressure area to light the brake warning switch.

Brake Warning Light Switches

There are three basic types of brake warning light switches as shown in Figs. 4, 5 and 6, and usually they form a common electrical circuit with the parking brake light.

When a pressure differential occurs between the front and rear brake systems, the valves will shuttle toward the side with the low pressure.

As shown in Fig. 4, movement of the differential valve forces the switch plunger upward over the tapered shoulder of the valve to close the switch contacts and light the dual brake warning lamp, signaling a brake system failure.

In Fig. 5, the valve assembly consists of two

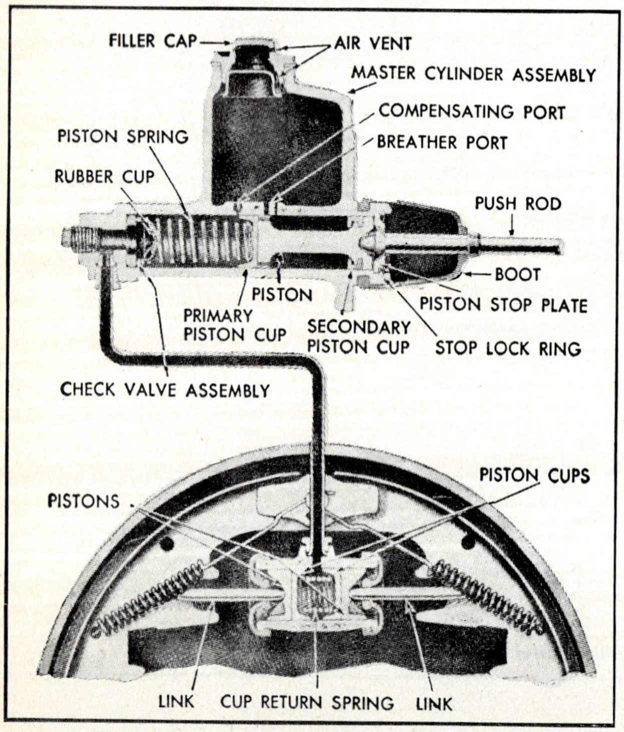

Fig. 1 Diagram of a typical hydraulic brake system

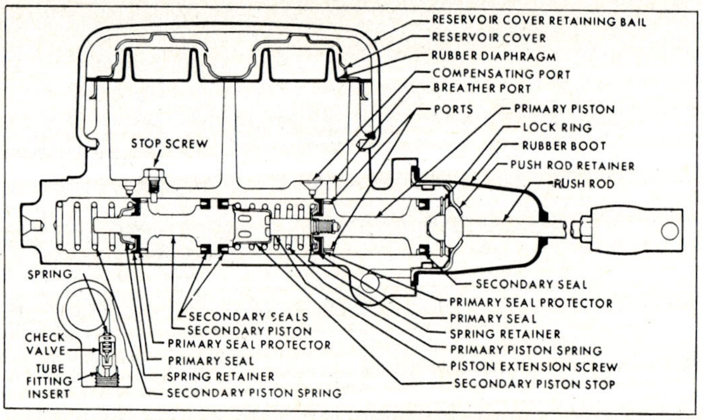

Fig. 2 Delco-Moraine dual master cylinder (typical)

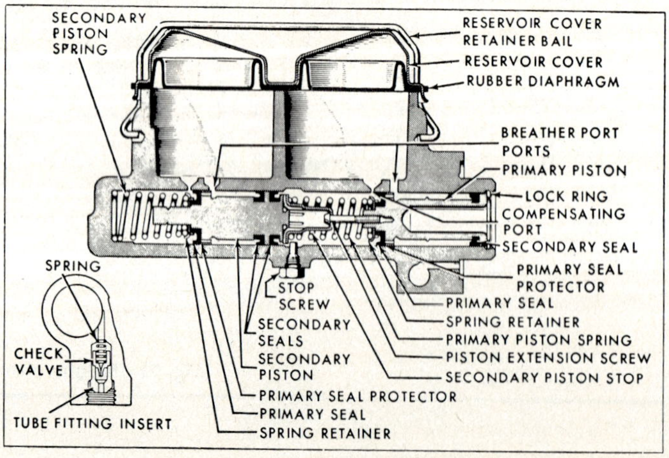

Fig. 3 Bendix dual master cylinder (typical)

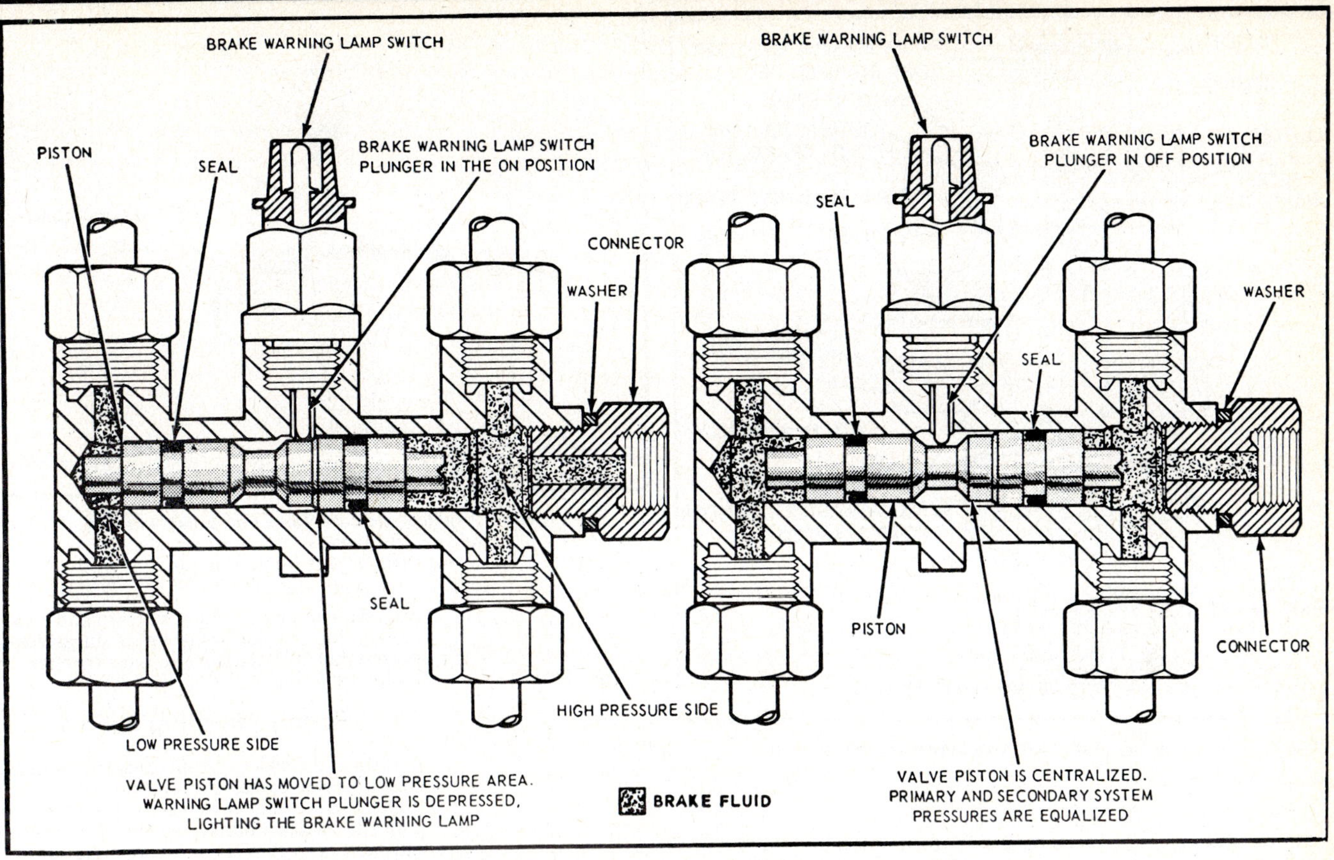

Fig. 4 Pressure differential valve and brake warning light switch

valves in a common bore that are spring-loaded toward the centered position. The spring-loaded switch contact plunger rests on top of the valves in the centered position (right view). When a pressure differential occurs between the front and rear brake systems, the valves will shuttle toward the side with the low pressure. The spring-loaded switch plunger is "triggered", and the ground circuit for the warning light is completed, lighting the lamp (left view).

In Fig. 6, as pressure fails in one system the other system's normal pressure forces the piston to the inoperative side, contacting the switch terminal, causing the warning light on the instrument panel to glow.

Testing Warning Light System

If the parking brake light is connected into the service brake warning light system, the brake warning light will flash only when the parking brake is applied with the ignition switch ON. The same light will also glow should one of the two service brake systems fail when the brake pedal is applied.

To test the system, turn the ignition ON and apply the parking brake. If the lamp fails to light, look for a burned out bulb, disconnected socket, a broken or disconnected wire at the switch.

Fig. 7 is an exterior view of one of these switches. They are usually mounted on the left frame side rail or on the brake pedal bracket.

To test the brake warning system, raise the vehicle and open a wheel bleeder valve while a helper depresses the brake pedal and observes the warning light on the instrument panel. If the bulb fails to light, look for a burned out bulb, disconnected socket, or a broken or disconnected wire at the switch. If the bulb is not burned out, and the wire continuity is proven, replace the brake warning switch.

COMBINATION VALVE

The combination valve, Fig. 8, is a metering valve, failure warning switch, and a propor-

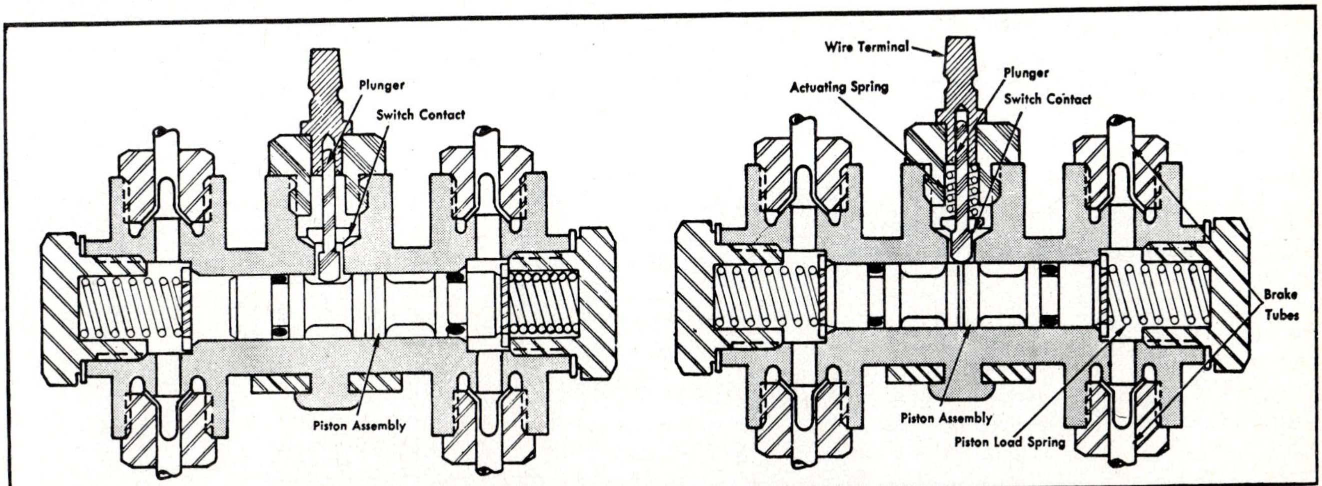

Fig. 5 Pressure differential valve and brake warning light switch

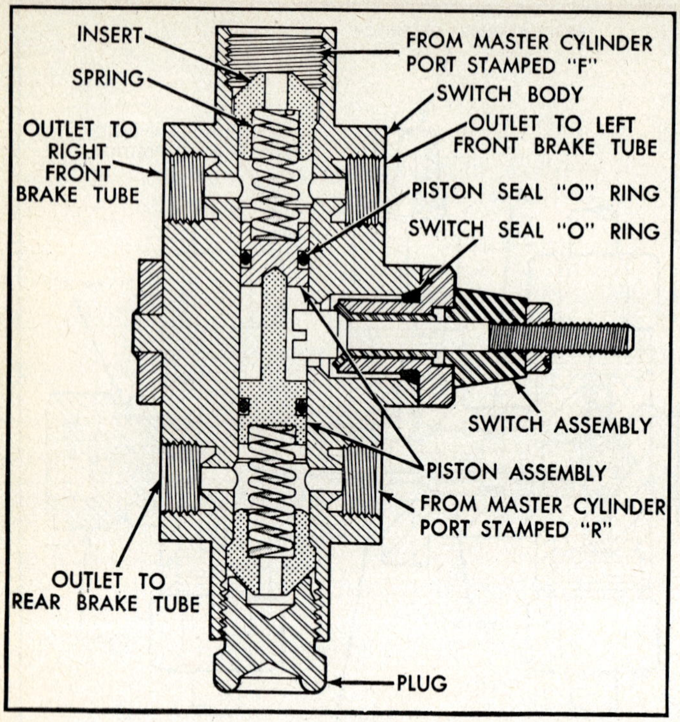

Fig. 6 Pressure differential valve and brake warning light switch

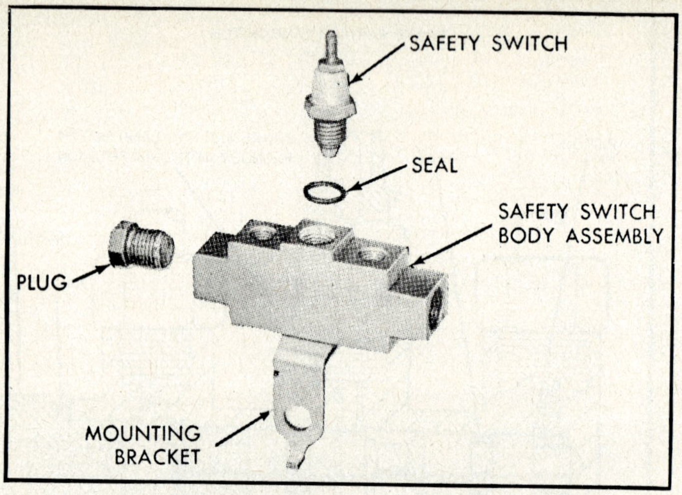

Fig. 7 Typical pressure valve and brake warning light switch. These switches are usually mounted on left frame side rail

tioner in one assembly and is used on disc brake applications. The metering valve delays front disc braking until the rear drum brake shoes contact the drum. The failure warning switch is actuated in event of front or rear brake system failure, in turn activating a dash warning lamp. The proportioner balances front to rear braking action during rapid deceleration.

Metering Valve

When the brakes are not applied, the metering valve permits the brake fluid to flow through the valve, thus allowing the fluid to expand and contract with temperature changes.

When the brakes are initially applied, the metering valve stem moves to the left, preventing fluid to flow through the valve to the front disc brakes. This is accomplished by the smooth end of the metering valve stem contacting the metering valve seal lip at 4 to 30 PSI, Fig. 9. The metering valve spring holds the retainer against the seal until a predetermined pressure is produced at the valve inlet port which overcomes the spring pressure and permits hydraulic pressure to actuate the front disc brakes, Fig. 10. The increased pressure into the valve is metered through the valve seal, to the front disc brakes, producing an increased force on the diaphragm. The diaphragm then pulls the pin, in turn pulling the retainer and reduces the spring pressure on the metering valve seal. Eventually, the pressure reaches a point at which the spring is pulled away by the diaphragm pin and retainer, leaving the metering valve unrestricted, permitting full pressure to pass through the metering valve.

Failure Warning Switch

If the rear brake system fails, the front system pressure forces the switch piston to the right, Fig. 11. The switch pin is then forced up into the switch, completing the electrical circuit and activates the dash warning lamp.

When repairs are made and pressure returns to the system, the piston moves to the left, resetting the switch. The detent on the piston requires approximately 100 to 450 PSI to permit full reset of the piston. In event of front brake system failure, the piston moves

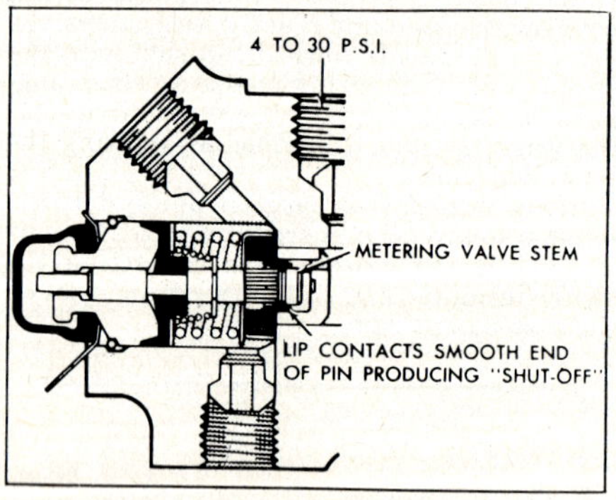

Fig. 9 Metering valve, initial braking

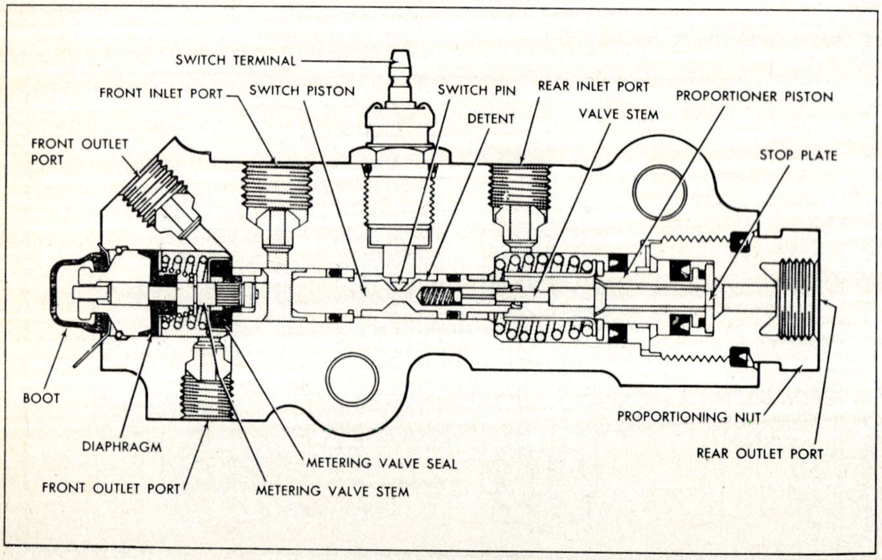

Fig. 8 Combination valve

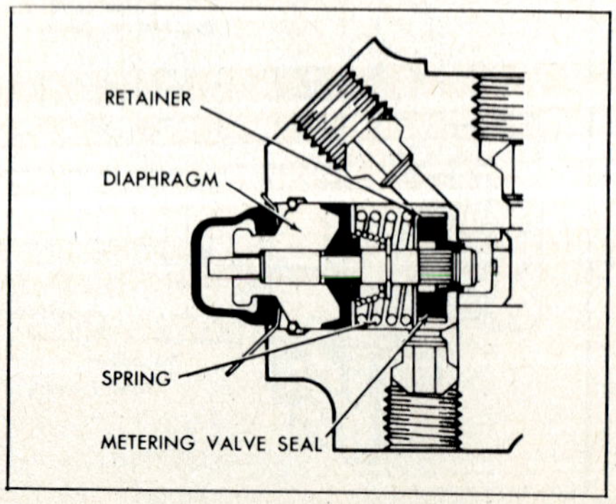

Fig. 10 Metering valve, continued braking

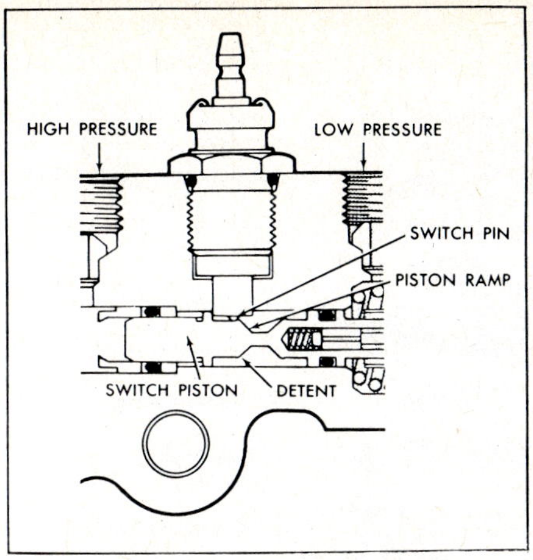

Fig. 11 Failure warning switch, rear system failure

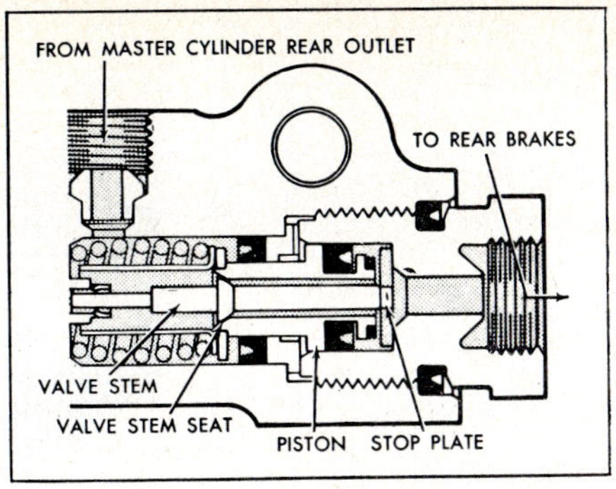

Fig. 12 Proportioner, rapid deceleration

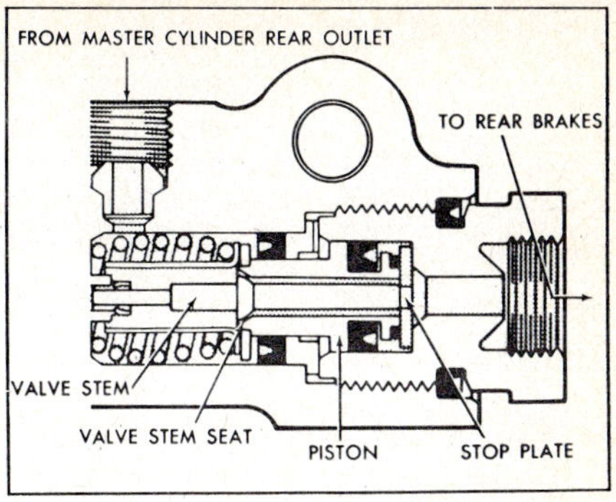

Fig. 13 Proportioner, normal braking

to the left and the same sequence of events are followed as for rear system failure except the piston resets to the right.

Proportioner

During rapid deceleration, a portion of the vehicle's weight is transferred to the front wheels. This resultant loss of weight at the rear wheels must be compensated for to avoid early rear wheel skid. The proportioner reduces the rear brake system pressure, delaying rear wheel skid. The pressure developed within the valve acts against the large end of the piston, overcoming the spring pressure and moves the piston to the left, Fig. 12. The piston then contacts the stem seat and proportions (restricts) line pressure through the valve.

During normal braking action, the proportioner is not functional. Brake fluid flows into the proportioner between the piston center hole and the valve stem, through the stop plate and to the rear brakes. Spring pressure loads the piston, during normal braking, causing it to rest against the stop plate, Fig. 13.

Master Cylinder Service

Figs. 14 to 17 show an array of dual and single master cylinders. With cylinder removed from vehicle, and from brake booster if so equipped, remove the covers and disassemble the unit as suggested by the illustration of the unit being serviced.

When disassembled, wash all parts in alcohol ONLY. Use an air hose to blow out all passages, orifices and valve holes. Air dry and place parts on clean paper or lint-free cloth. Inspect master cylinder bore for scoring, rust, pitting or etching. Any of these conditions will require replacement of the housing. Inspect master cylinder pistons for scoring, pitting or distortion. Replace piston if any of these conditions exist.

If either master cylinder housing or piston is replaced, clean new parts with alcohol and blow out all passages with air hose.

Examine reservoirs for foreign matter and check all passages for restrictions. If there is any suspicion of contamination or evidence of corrosion, completely flush the hydraulic system as outlined below.

When overhauling a master cylinder, use all parts contained in the repair kit. Before starting reassembly, dip all cups, seals, pistons, springs, check valves and retainers in alcohol and place in a clean pan or on clean paper. *Wash hands with soap and water to prevent contamination of rubber parts from oil, kerosene or gasoline.* During assembly, dip all parts in clean, heavy duty brake fluid.

Inspect through side outlet of dual master cylinder housing to make certain cup lips do not hang up on edge of hole or turn back, which would result in faulty operation. A piece of 3/16" rod with an end rounded off will be helpful in guiding cups past hole.

BLEEDING BRAKES

NOTE: Pressure bleeding is recommended for all hydraulic brake systems.

The bleeding operation itself is fairly well standardized. First step in all cases is cleaning the dirt from the filler cap before removing it from the master cylinder. This should be done thoroughly.

Pressure bleeding is fastest because the master cylinder doesn't have to be refilled several times, and the job can be done by one man. Pressure bleeding equipment of the diaphragm type is recommended. If using equipment which is not of the diaphragm type, make sure to prevent air from the pressure tank from entering the lines, do not shake the tank while air is being added to the

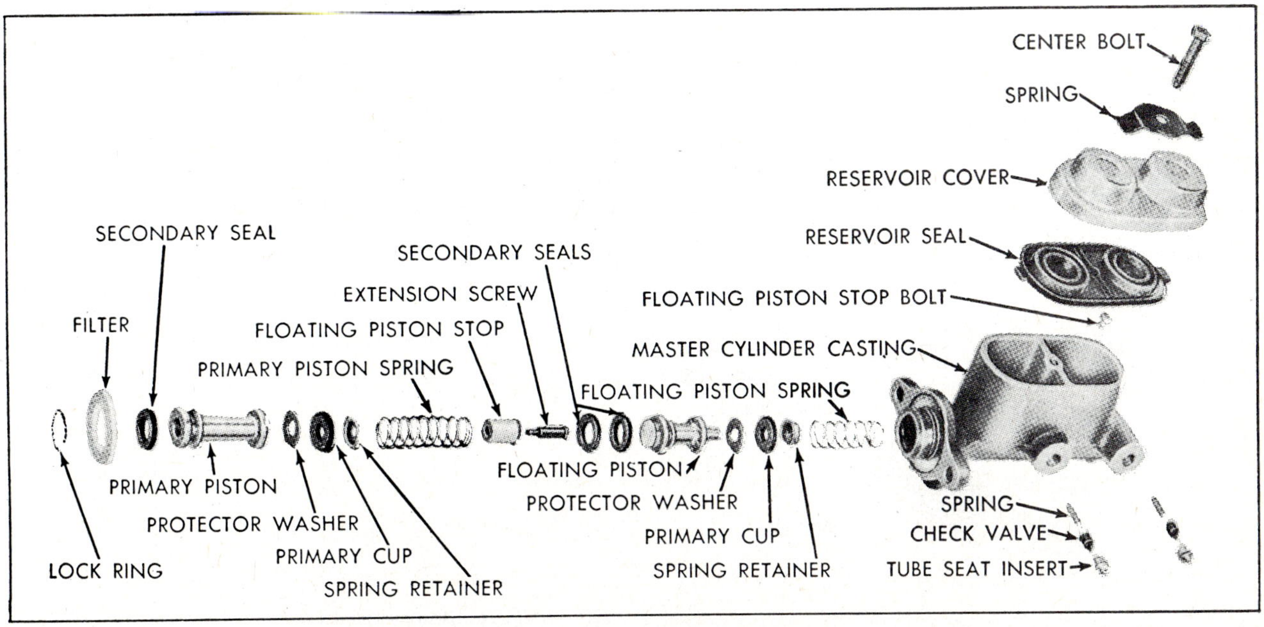

Fig. 14 Disassembled view of a typical Delco-Moraine dual master cylinder

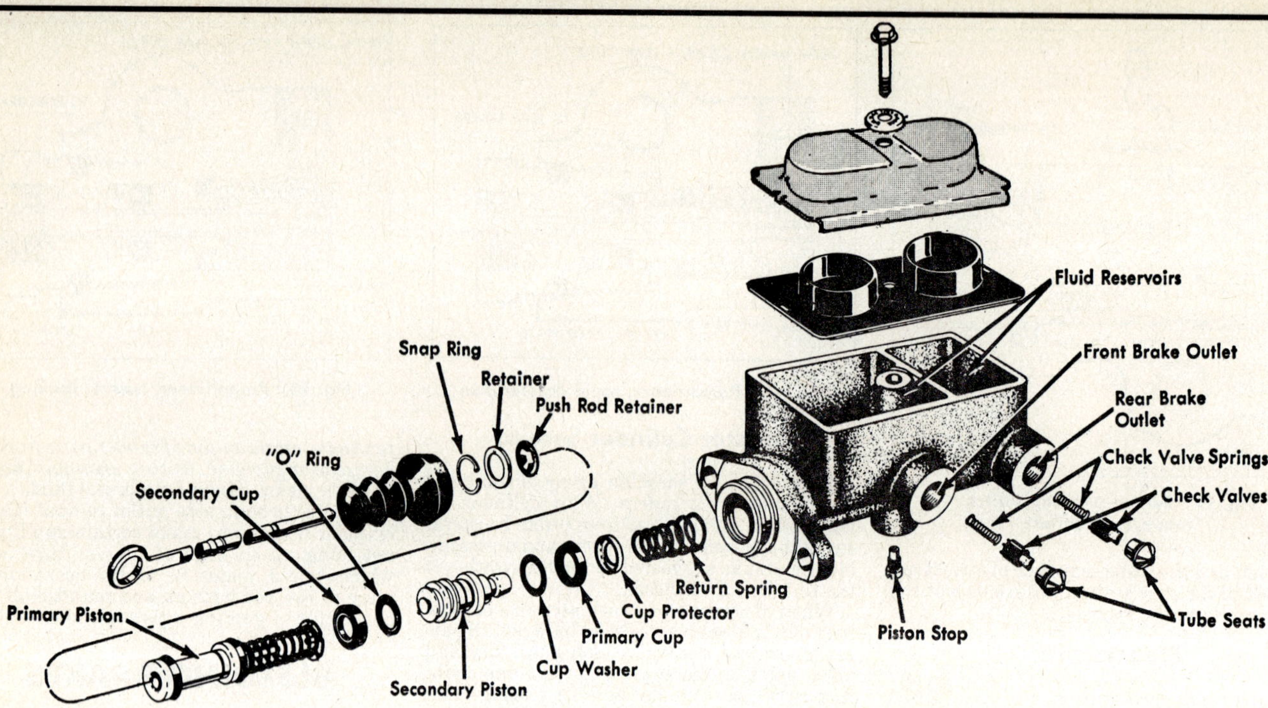

Fig. 15 Disassembled view of a typical Bendix dual master cylinder

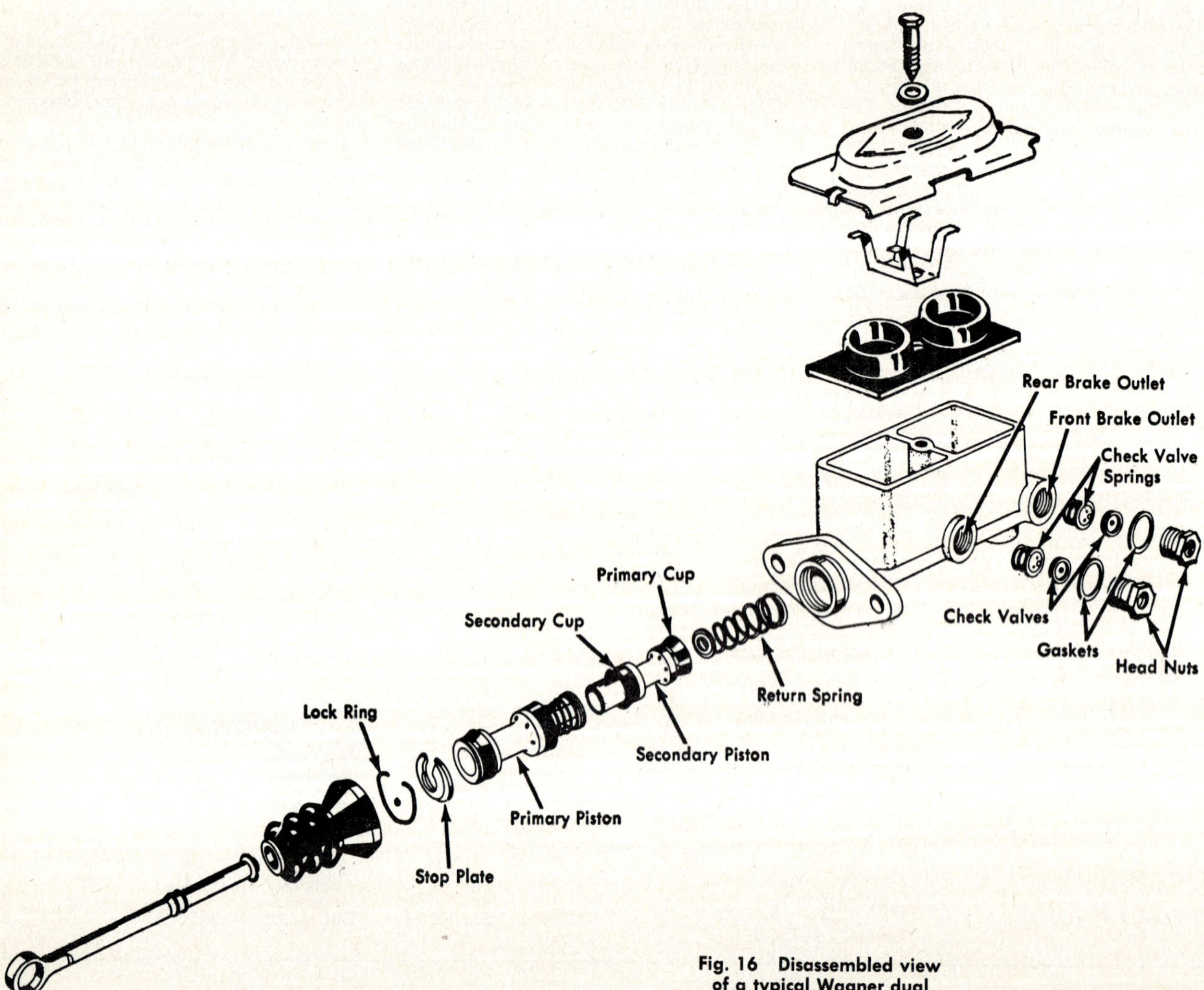

Fig. 16 Disassembled view of a typical Wagner dual master cylinder

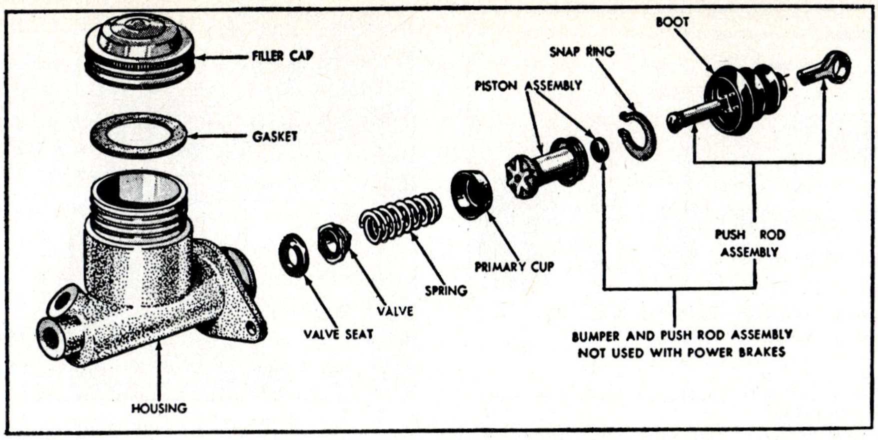

Fig. 17 Typical single brake master cylinder

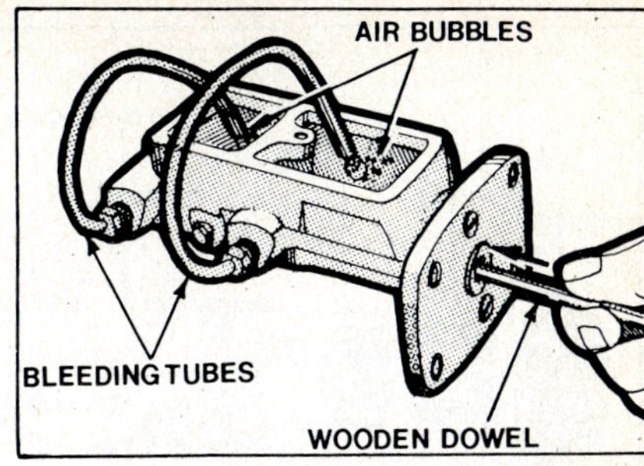

Fig. 17A Bleeding master cylinder used in conjunction with disc brakes

tank or after it has been pressurized. Set the tank in the required location, bring the air hose to the tank, and do not move it during the bleeding operation. The tank should be kept at least one-third full.

NOTE: On vehicles equipped with disc brakes, the brake metering valve or combination valve must be held in its position, using the recommended tool, (C-4121 for Chrysler Corp., J-22742 for Ford Motor Co. and J-23709 or J-23770 for General Motors).

If air does get into the fluid, releasing the pressure will cause the bubbles to increase in size, rise to the top of the fluid, and escape. Pressure should not be greater than about 35 lb. per sq. in.

NOTE: On vehicles equipped with plastic reservoirs, do not exceed 25 psi. bleeding pressure.

When bleeding without pressure, open the bleed valve three-quarters of a turn, depress the pedal a full stroke, then close bleeder screw and allow pedal to return slowly to its released position. Repeat this operation several times until expelled brake fluid flows in a solid stream without air bubbles. On vehicles with power brakes, first reduce the vacuum in the power unit to zero by pumping the brake pedal several times with the engine off before starting to bleed the system.

Pressure bleeding, of course, eliminates the need for pedal pumping. Chrysler Corp. suggests that, when pressure is used, the bleeder valve should be opened and closed intermittently at about four-second intervals. This gives a whirling action to the fluid in the wheel cylinder, and helps expel the air.

At one time, some makers recommended that a clean container be used for the drained fluid, so that the fluid could be reused. All now agree that drained fluid should be discarded. Care should be taken not to spill brake fluid, since this can damage the finish of the vehicle.

Flushing is essential if there is water, mineral oil or other contaminants in the lines, and whenever new parts are installed in the hydraulic system. Fluid contamination is usually indicated by swollen and deteriorated cups and other rubber parts.

Bleeding is necessary on all four wheels if air has entered the system because of low fluid level, or the line or lines have been disconnected. If a line is disconnected at any one wheel cylinder, that cylinder only need be bled. Of course, on brake reline jobs, bleeding is advisable to remove any air or contaminants.

Master cylinders equipped with bleeder valves should be bled first before the wheel cylinders are bled. In all cases where a master cylinder has been overhauled, it must be bled. Where there is no bleeder valve, this can be done by leaving the line (or lines) loose, actuating the brake pedal to expel the air and then tightening the line (or lines).

NOTE: After overhauling a dual master cylinder used in conjunction with disc brakes, it is advisable to bleed the cylinder before installing it on the vehicles. The reason for this recommendation is that air may be trapped between the master cylinder pistons because there is only one residual pressure valve (check valve) used in these units.

The recommended precedure for Chrysler Line vehicles is as follows:

1. Clamp master cylinder in a vise and attach the special Bleeding Tubes (Tool No. C4029) Fig. 17A. *Be sure that the residual pressure valve is on the end of the tube in the large capacity reservoir as shown. This keeps the brake fluid from being syphoned out of the reservoir while bleeding.*
2. Fill both reservoirs with approved brake fluid.
3. Using a wooden stick or dowel (vehicles with power brakes) depress push rod slowly and allow the pistons to return under pressure of the springs. Do this several times until all air bubbles are expelled.
4. Remove bleeding tubes from cylinder and install cover and gasket.
5. Install master cylinder on vehicle and bleed wheel cylinders, preferably with a pressure bleeder.

Alternate Method

1. Support assembly in a vise and fill both reservoirs with brake fluid.
2. Loosely install a plug in each outlet port of the cylinder. Depress push rod several times until air bubbles cease to appear in the brake fluid.
3. Tighten plugs and attempt to depress the piston. Piston travel should be restricted after all air is expelled.
4. Install master cylinder on vehicle and bleed wheel cylinders, preferably with a pressure bleeder.

Testing Dual Master Cylinders

Be sure that master cylinder compensates in both ports. This can be done by applying the brake pedal lightly (engine running with power brakes), and observing for brake fluid squirting up in the reservoirs. This may only occur in the front chamber. To determine if the rear compensating port is open, pump up the brakes rapidly and then hold the pedal down. Have an observer watch the fluid in the rear reservoir while the pedal is raised. A disturbance in the fluid indicates that the compensating port is open.

Wheel Bleeding Sequence

Difference of opinion as to whether the longest or shortest line should be bled first still exists. To be safe, use the sequence given below, recommended by the vehicle manufacturers:

Chrysler Corp.	RR-LR-RF-LF
Ford Company	RR-LR-RF-LF
General Motors	LF-RF-LR-RR
American Motors	RR-LR-RF-LF

Dual Master Cylinder Bleeding Notes

Ford Motor Co.

NOTE: Some vehicles use a self-centering valve. After any bleeding operation, turn ignition switch to ACC or ON position and depress brake pedal. Valve will center itself.

General Motors

On vehicles equipped with combined drum and disc brakes, the pressure differential valve must be held in its open position while bleeding the brakes using a pressure type bleeder. This is done by depressing the plunger and holding it in place using a special tool. When using pressure bleeding equipment follow the recommendations of the manufacturer.

American Motors

Before bleeding brakes, disconnect the switch terminal wire and remove nylon switch terminal, contact plunger actuating spring, and nylon plunger with contact.

In the event the valve has "triggered", the valve centering spring pressure may hold the switch plunger. If this happens, apply a slight amount of brake pedal pressure while releasing the plunger from the valve body.

After the bleeding operation, assemble the plunger spring and install valve with contact down. Install the nylon terminal and connect warning light wire to valve terminal. In the event brake fluid leaks from the center termi-

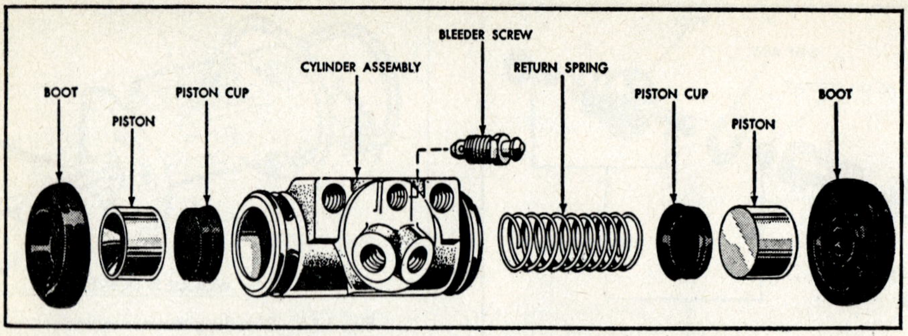

Fig. 18 Disassembled view of a typical wheel cylinder

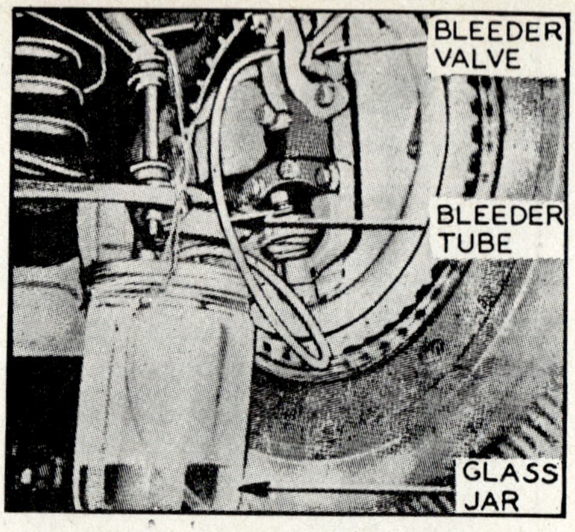

Fig. 19 Bleeding wheel cylinder

nal body opening when the terminal is removed, replace the valve assembly.

Chrysler Corp.

Some Chrysler built vehicles with disc brakes are equipped with front disc brake pressure metering valves.

The purpose of the metering valve is to provide a better match of the front disc brakes with the rear drum brakes, resulting in improved braking balance *in light pedal applications*.

Gravity bleed and pedal methods are not affected by the presence of the metering valve. However, pressure bleeding is influenced by the metering valve.

Bleed pressure, which is normally about 35 psi, is high enough to cause the metering valve to close, which stops the flow of fluid to the front brakes. However, the valve can be held open manually by depressing the pressure release plunger (located at the bottom of the valve) in its uppermost position by hand or secured with masking tape while bleeding the brakes.

CAUTION: Under no conditions should a rigid clamp, wedge or block be used to secure the plunger as this can cause an internal failure in the valve. It should be noted that the pressure release plunger of the valve is already in its uppermost position when there is no pressure present.

WHEEL CYLINDERS

1. Remove wheel, drum and brake shoes.
2. Disconnect hydraulic line at wheel cylinder. *Do not pull metal line away from cylinder as the cylinder connection will bend metal line and make installation difficult. Line will separate from cylinder when cylinder is moved away from brake backing plate.*
3. Remove screws holding cylinder to brake plate and remove cylinder.

Overhaul

1. Referring to Fig. 18 as a guide, remove boots, pistons, springs and cups from cylinder.
2. Place all parts, except cylinder casting, in alcohol. Wipe cylinder walls with alcohol.
3. Examine cylinder bore. A scored bore may be honed providing the diameter is not increased more than .005". Replace worn or damaged parts from the repair kit.
4. Before assembling, wash hands with soap and water only as oil, kerosene or gasoline will contaminate rubber parts.
5. Lubricate cylinder wall and rubber cups with brake fluid.
6. Install springs, cups, pistons and boots in housing.
7. Wipe end of hydraulic line to remove any foreign matter.
8. Place hydraulic cylinder in position. Enter tubing into cylinder and start connecting fitting.
9. Secure cylinder to backing plate and then complete tightening of tubing fitting.
10. Install brake shoes, drum and wheel.
11. Bleed system as outlined previously, and adjust brakes.

FLUSHING SYSTEM

It may sometimes become necessary to flush out the system due to the presence of mineral oil, kerosene, gasoline, etc., which will cause swelling of rubber piston cups and valves so they become inoperative. The procedure is as follows:

1. Attach bleeder tube and open bleeder valve at left front wheel, Fig. 19.
2. Flush out system thoroughly with clean alcohol, pumping the fluid from the master cylinder and out of the wheel cylinder bleeder valve.
3. Repeat Steps 1 and 2 at remaining wheel cylinders. To insure thorough flushing, about a 1/2 pint of alcohol should be bled through each wheel cylinder.
4. Replace all rubber parts in master and wheel cylinders. Thoroughly clean cylinders and pistons in alcohol before installing new parts.
5. After installing parts, fill system with recommended brake fluid and flush system of cleaning solution and then bleed brakes. In doing this, pump brake fluid from wheel chylinder bleeder valves until clear brake fluid flows from bleeder tube and then, if necessary, continue until no air bubbles emerge from bleeder tube.

HYDRAULIC TUBING

Steel tubing is used to conduct hydraulic pressure to the brakes. All fittings, tubing and hose should be inspected for rusted, damaged or defective flared seats. The tubing is equipped with a double flare or inverted seat to insure more positive seating in the fitting. To repair or reflare tubing, proceed as follows:

1. Using the tool shown in Fig. 20 or its equivalent, cut off the damaged seat or damaged tubing.
2. Ream out any burrs or rough edges showing on inside edges of tubing. This will

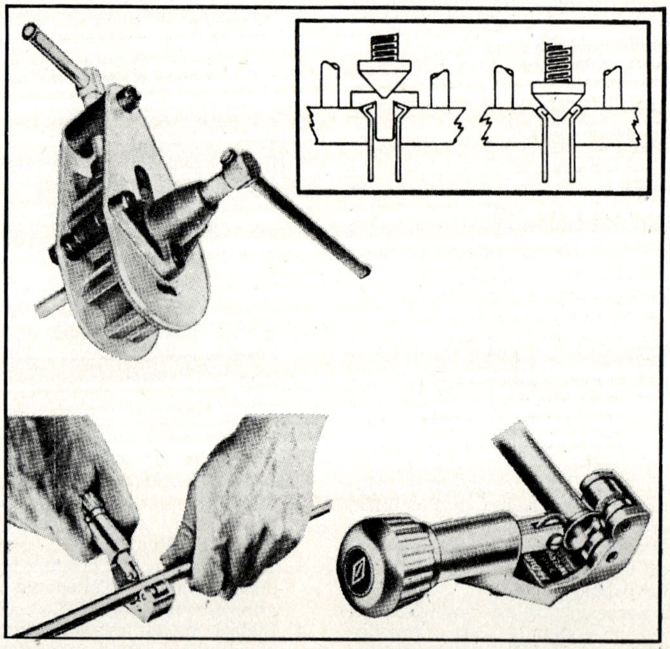

Fig. 20 Cutting and flaring hydraulic brake tubing

make the ends of the tubing square and insure better seating of the flared end. *Before flaring tubing, place a compression nut on tubing.*
3. Open handles of flaring tool and rotate jaws of tool until mating jaws of tubing size are centered in the area between vertical posts.
4. Slowly close handles with tubing inserted in jaws but do not apply heavy pressure to handle as this will lock tubing in place.
5. Referring to Fig. 20, place gauge on edge over end of tubing and push tubing through jaws until end of tubing contacts recessed notch of gauge matching size of tubing.
6. Squeeze handles of flaring tool and lock tubing in place.
7. Place proper size plug of gauge down in end of tubing. Swing compression disc over gauge and center tapered flaring screw in recess in disc.
8. Lubricate taper of flaring or screw and screw in until plug gauge has seated in jaws of flaring tool. This action has started to invert the extended end of tubing.
9. Remove gauge and apply lubricant to tapered end of flaring screw and continue to screw down until tool is firmly seated in tubing.
10. Remove tubing from flaring tool and inspect the seat. If seat is cracked, cut off cracked end and repeat flaring operation.

PARKING BRAKES

INDEX

EXTERNAL CONTRACTING BAND TYPE

This type brake, Fig. 1, operates on a drum mounted on the companion flange at the rear of the transmission mainshaft. The brake band is anchored to a bracket that is attached to the transmission case. The linkage is so arranged that both ends of the brake band are contracted equally. The braking effort is equalized to each wheel through the differential.

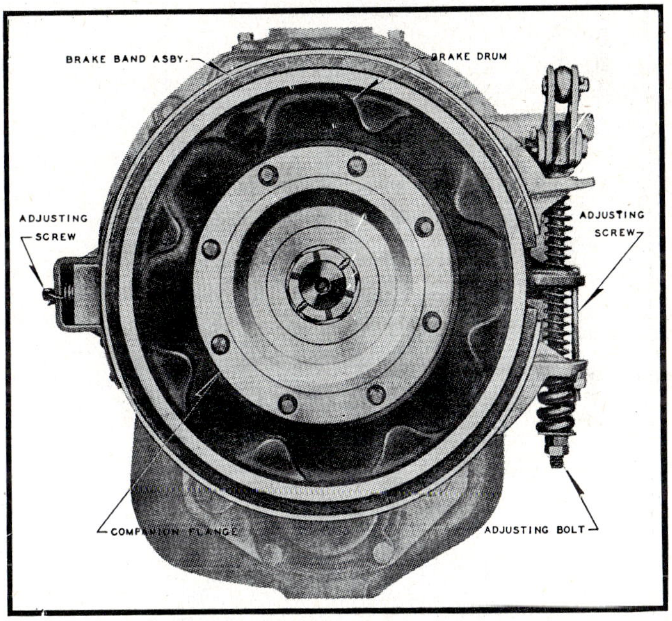

Fig. 1 External contracting band type parking brake

Disassembly

1. Remove hand brake lever.
2. Remove brake adjusting screw after removing nuts from screw.
3. Remove jam nuts, adjusting bolt washer and spring from bottom of adjusting bolt.
4. Remove cam lever pin from spacer links, cam levers and adjusting bolt.
5. Remove adjusting bolt, release springs and cam operating shoe.
6. Remove anchor adjusting screw.
7. Remove brake band assembly, also anchor clip spring.

Reassembly

1. Reline brake band with proper lining.
2. Insert anchor clip spring into anchor support and install brake band assembly.
3. Install anchor adjusting screw.
4. Insert adjusting bolt through cam operating shoe, upper brake band bracket, upper release spring, locating bracket hole, lower release spring and lower brake band bracket.
5. Install cam lever pin through spacer links, cam levers, and adjusting bolt hook end. Secure with new cotter pin.
6. Install adjusting spring, washer and jam nuts on lower end of adjusting bolt.
7. Insert adjusting screw into locating bracket and through lower brake band bracket and install nuts.
8. Install draw rod end and secure with clevis pin and cotter pin.
9. Adjust hand brake as directed below and secure anchor adjusting screw with lock wire.

Adjustment

1. Three places are provided for adjustment. Where the band is supported by a bracket is a screw to adjust the band clearance at this point. Turning the screw to the right moves band and lining toward drum. Proper clearance is 1/32". After adjustment is completed, lock with wire.
2. Pull up lower half of band with bolt holding end of band to brake support. Allow 1/32" clearance and lock this adjustment with the locknut on the bolt.
3. Adjust clearance for upper half of band by means of the jam nuts at the bottom of the brake adjusting bolt and lock the two jam nuts together.

INTERNAL EXPANDING TYPE BRAKE

This type brake, Fig. 2, is mounted at the rear of the transmission. The brake support plate is attached to the transmission rear bearing cap or retainer. The brake drum is mounted between universal joint flanges. At the start of braking action, brake shoes are forced out against the drum by a lever-operated cam. The self-energizing action of both shoes then completes brake application.

Adjustment

A brake adjustment should be made before it becomes necessary to pull the hand brake lever to the limit of its travel to obtain full brake application.

1. Jack up at least one wheel. Block wheels and release hand brake.
2. Remove clevis pin connecting pull rod and relay lever. This will assure freedom for full shoe release.
3. Rotate drum to bring an access hole into line with adjusting screw at bottom of shoes.
4. Expand shoes by rotating adjusting screw with screwdriver inserted through hole in drum, Fig. 3. Move outer end of screwdriver away from drive shaft. Continue adjustment until a snug fit is obtained on a .010" feeler gauge placed between the drum and adjusting end of each shoe.
5. Place parking lever in fully release position. Take up slack in brake linkage by

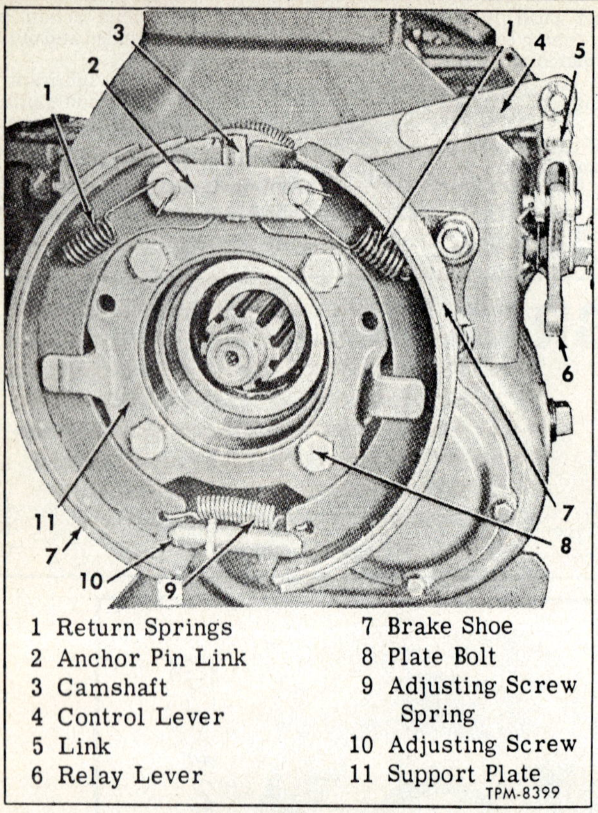

Fig. 2 Internal expanding type brake installed

Fig. 3 Internal expanding type brake adjustment

pulling back on control lever just enough to overcome spring tension. Adjust clevis of pull rod to line up with hole in relay lever. Insert clevis pin and cotter pin, then tighten clevis lock nut.

Brake Shoe Service

NOTE: Fig. 3 illustrates the brake with the output shaft flange removed. It is not necessary to remove the flange to remove brake shoes. The flange is removed for clarity of illustration.

1. Jack up at least one wheel. Block wheels and release parking brake.
2. Disconnect U-joint at brake drum.
3. Unfasten and remove yoke flange.
4. Lift off brake drum.
5. Remove shoe return springs and anchor pin link from cam ends of brake shoes far enough to permit removal of support plate. Remove shoes, then separate by removing adjusting screw and spring.
6. After shoes have been relined, position brake shoes on lever arms and install brake shoe pins, retainers and capscrews.
7. Place spring and washers on tie rod. Swing rear lever arm forward and insert tie rod through hole in bottom of arm.
8. Move lever to released position and install tension spring.
9. Install adjusting nut and lock nut and adjust brakes as outlined above.

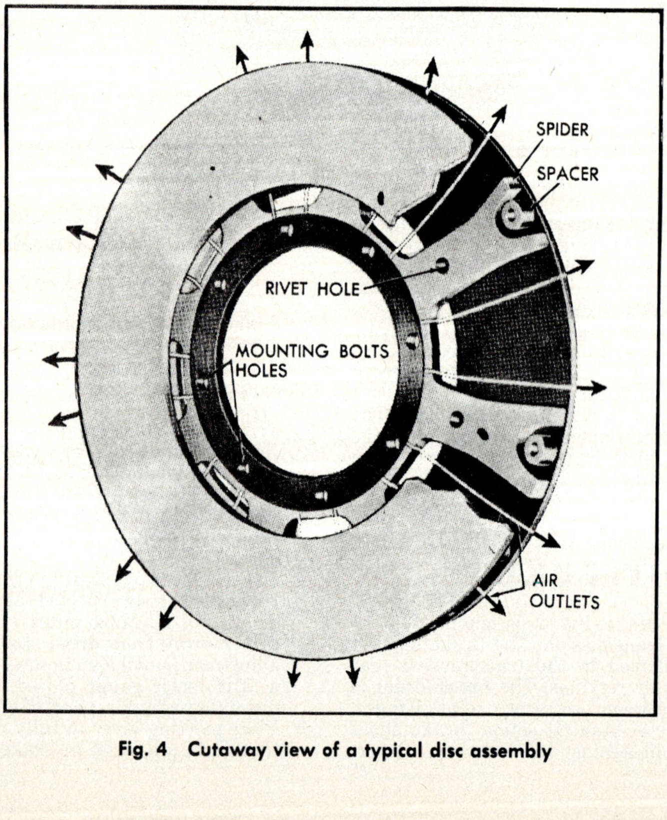

Fig. 4 Cutaway view of a typical disc assembly

DISC TYPE BRAKE

All Exc. Tru-Stop

The disc type parking brake, Figs. 4 and 5, consists of a spider and spacers located between two disc plates and brake shoe assemblies. The disc assembly is mounted to the drive line between the companion flange and universal joint. The spider and spacers separate the disc plates and insure adequate cooling and prevent warping during continuous heavy duty operation. When the brake is applied, a squeezing action of the brake shoes against the discs provides the braking effort needed.

Disassembly

1. Disconnect brake cable at brake lever.
2. Remove locking and adjusting nuts from adjusting rod.
3. Remove clevis pin from lower end of operating arm.
4. Remove operating lever and adjusting rod from brake shoe levers.
5. Remove brake lever spring and washers from between brake shoe levers.
6. Remove small brake shoe retracting

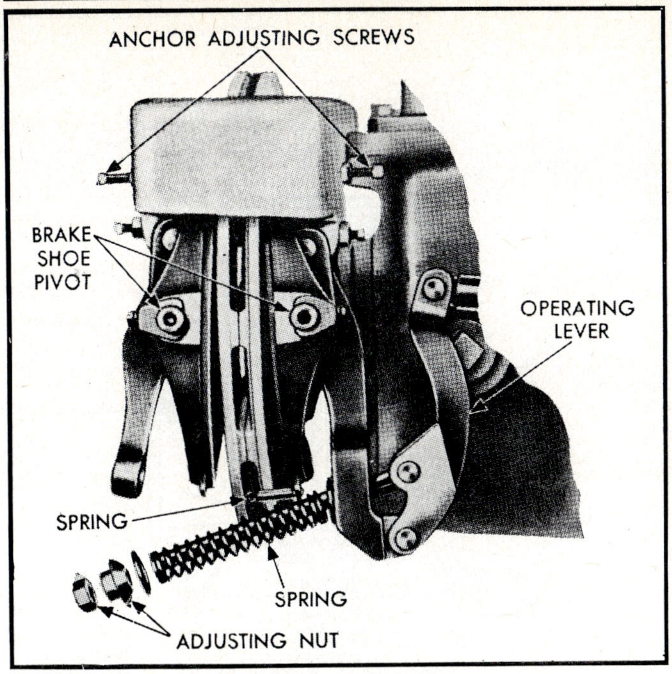

Fig. 5 Disc type parking brake, exc. Tru-Stop

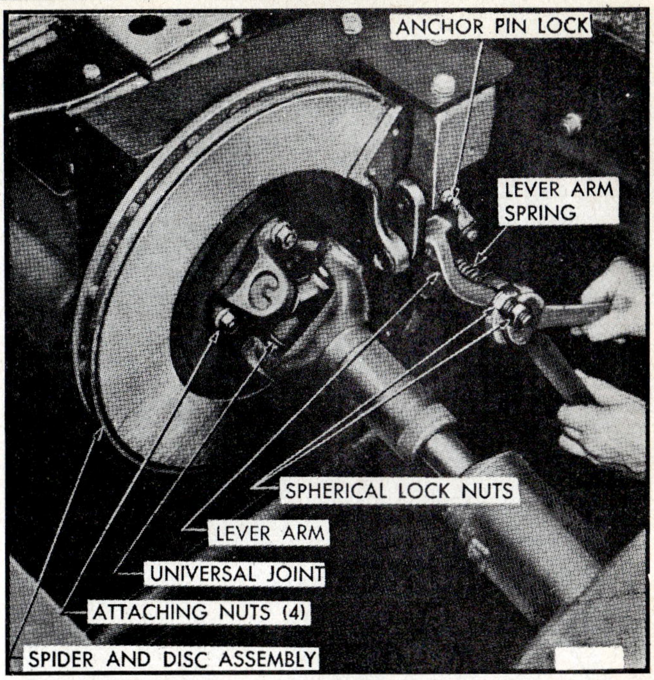

Fig. 6 Adjusting disc type parking brake, exc. Tru-Stop

spring.

7. Remove rear brake shoe pivot locking bolt and clamp.
8. Remove rear brake shoe pivot pin and brake shoe from lever assembly.
9. Remove front brake shoe pivot locking bolt and clamp.
10. Remove front brake shoe pivot pin and brake shoe.

Reassembly

Clean all moving parts with solvent and air dry for inspection. Inspect all linkage for worn parts. If inspection reveals worn pins or bushings, replace both pin and bushing.

1. Install front brake shoe, pivot pin, clamp and lock screw.
2. Install rear brake shoe, pivot pin, clamp and lock screw.
3. Install brake shoe retracting spring.
4. Install spring and washers between brake shoe operating levers.
5. Install adjusting rod through front lever spring and rear brake shoe lever.
6. Install spherical nut and lock nut on adjusting rod.
7. Install clevis pin and cotter pin in operating linkage and lever.
8. Connect cable and operating lever.

Adjustment

1. Disconnect cable from operating lever.
2. Disconnect U-joint from flange so disc is free to turn.
3. Adjust both brake anchor screws, Fig. 6, so that a slight drag may be felt when passing a .030" feeler gauge between top end of brake shoe and disc.
4. Leaving the feeler gauge in place, tighten spherical adjusting nut until a slight drag may be felt on a .030" feeler gauge inserted at lower end of both brake shoes.
5. While holding adjusting nut, tighten lock nut.
6. Remove all feeler stock and revolve brake disc and check for dragging.
7. Install U-joint.

Tru-Stop

This disc type parking brake, Fig. 7, is mounted between the propeller shaft flange and the auxiliary transmission shaft companion flange. The ventilated brake disc is located between the brake shoes and when the brake is applied, the shoes are forced against the disc.

Brake Shoe Removal

1. Remove tie rod locknut and adjusting nut, then the tension spring.
2. Place in-cab parking brake lever to the applied position so rear lever arm can be swung clear of tie rod, then remove tie rod washers and spring.
3. Remove brake shoe pin retainer screws, retainers, brake shoe pins and brake shoes.

Brake Shoe Installation

1. Place brake shoes on lever arms and install brake shoe pins, pin retainers and screws.
2. Install spring and washers on tie rod, then swing rear lever arm forward and insert tie rod through hole in arm.
3. Place in-cab parking brake lever in released position and install tension spring, adjusting nut and locknut.

Adjustment

1. Disconnect brake cable or rod clevis from brake lever.
2. Tighten adjusting nut until spring pressure forces lever against front lever arm.
3. Install a 1/32 inch shim between rear brake shoe and disc, then tighten adjusting nut until front brake shoe firmly contacts disc, however, still allowing for shim removal.
4. Ensure tension spring is properly installed and adjust parallel adjusting screws until front and rear shoes are parallel with disc surfaces. This adjustment provides 1/64 inch clearance between front and rear brake shoes and disc at all points.
5. Ensure in-cab parking brake lever is in fully released position and install clevis on brake lever. Adjust clevis on cable or rod so clevis can be installed without changing position of lever.

MGM SHORTSTOP BRAKE

This type brake, Fig. 8, is used on tandem axle trucks equipped with Bendix Twinplex wedge-type air brakes. Shortstop brake units provide mechanical spring actuation of the Twinplex wedge brakes whenever air pressure fails. During normal operation, the Shortstop power spring is held compressed by air pressure under the piston in the unit. When air pressure fails, the power spring forces the piston inward. This motion is transmitted by the release bolt to the push rod and pressure plate to move the piston assembly in the non-pressure air chamber, applying the wedge in the service brakes. The release bolt permits releasing the service brakes if normal air pressure cannot be restored.

Two of the three inlets on the Shortstop unit transmit air pressure to the service brake air chamber for normal service brake application. The third inlet in the center transmits air pressure to the piston chamber to compress the power spring. If air pressure fails in the Shortstop piston chamber, the power spring applies the brakes and, unless pressure can be reestablished, the service brakes must be released as follows in order to move the truck:

1. Referring to Fig. 8, remove bolt and nut holding cap and filter to the housing and remove cap and filter.
2. Using a deep socket wrench, unscrew release bolt until it turns freely. The brake shoe pressure is now released. Follow this procedure until all Shortstop units are released.
3. When air pressure has been restored,

1 Brake Support Bracket
2 Parallel Adjusting Screws
3 Front Lever Arm Pin
4 Pin Retaining Screw
5 Brake Cable Clevis
6 Brake Lever
7 Brake Shoe Pin Retainer
8 Brake Shoe Pin
9 Front Brake Shoe
10 Rear Brake Shoe
11 Front Lever Arm
12 Brake Disc
13 Tension Spring
14 Spring
15 Rear Lever Arm
16 Adjusting Nut
17 Tie Rod

Fig. 7 Tru-Stop disc type parking brake

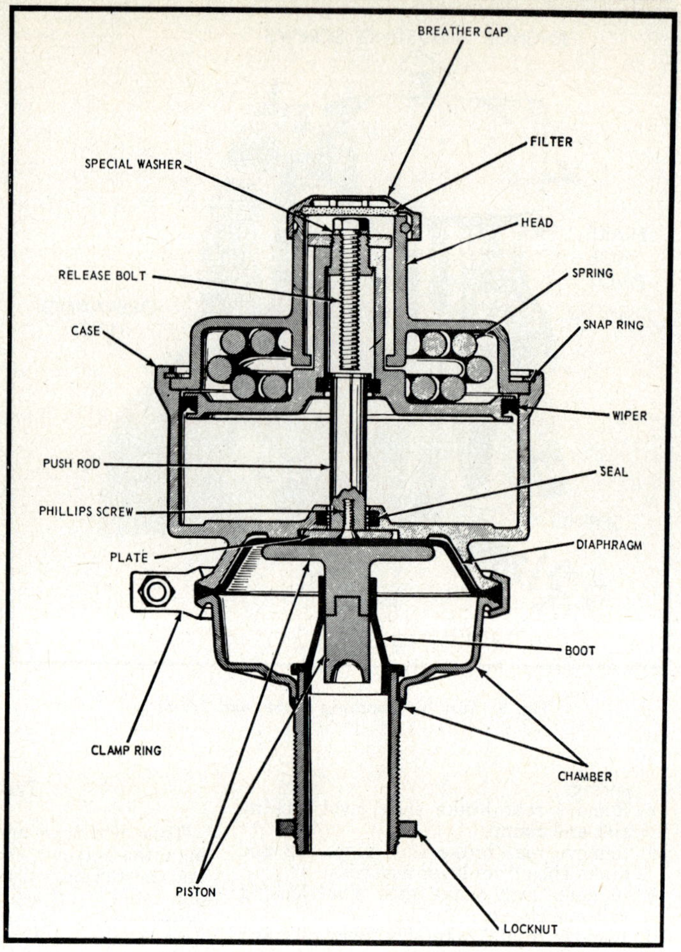

Fig. 8 MGM Shortstop emergency and parking brake

screw the release bolt in until the lockwasher holds it securely. This MUST be done to regain Shortstop protection.

Parking Operation

For parking operation, a control valve is provided in the cab. Pulling the valve knob out will release the air from the Shortstop pressure chamber, allowing the spring pressure to apply the brakes. The spring will be compressed again and the brakes released as soon as the parking valve is pushed in and/or system air pressure is restored by the compressor.

Shortstop Unit, Replace

1. Disconnect all air lines to unit.
2. Remove breather cap and filter, then unscrew release bolt completely, using a deep socket wrench.
3. Remove clamp ring fastening Shortstop unit to service brake chamber and remove Shortstop unit.
4. To install, first make sure release bolt is completely unscrewed.
5. Position Shortstop unit to service brake chamber and install clamp ring.
6. Connect air lines, charge Shortstop chamber with air and screw release bolt in tight. Install cap and filter.

Shortstop Unit, Overhaul

1. Referring to Fig. 9, use a sharp thin-bladed penknife or screwdriver to remove the solder from the head of the Phillips screw.
2. Position Phillips head screwdriver and seat it in screw slot with aid of a light hammer.
3. Remove Phillips head screw and pressure plate.
4. Remove release bolt and washer and insert a screwdriver through release bolt hole. Hold screwdriver firmly against top of push rod and charge the spring brake piston chamber with air (70 psi minimum).
5. Remove screwdriver and unscrew guide washer, using a drift punch and hammer.
6. Exhaust the air pressure. Remove sealant covering and snap ring between head and case.
7. Place spring brake assembly in an arbor press prior to final disassembly. Press must have a 6" stroke remaining after unit is placed in press.
8. Take up slight tension on press. Do not attempt to press head into case.
9. Using a screwdriver, remove snap ring from groove in case.
10. Relax press slowly. The spring will push head out of case. When spring is free, remove entire assembly from press.
11. Remove head and spring from case.
12. Grasp piston by its hub and pull it straight out of case.
13. Wash all parts in suitable non-oily solvent.
14. Inspect all parts for wear or damage. Replace with new parts or refinish with emery or crocus cloth, if necessary.

Reassembly

1. With all parts perfectly clean, apply a low temperature moly-type lubricant to all inside surfaces.
2. Lightly lubricate seal groove in piston, also collar seal and push rod seal.
3. If removed from piston, the push rod is installed by working the flanged end through the seal in the center of the piston.
4. Install piston assembly in the case, starting at a 45° angle. Slowly bring piston hub to 90° and install push rod in seal.
5. Install spring on piston and head on spring.
6. Place assembly in an arbor press and slowly compress spring. Use care, making

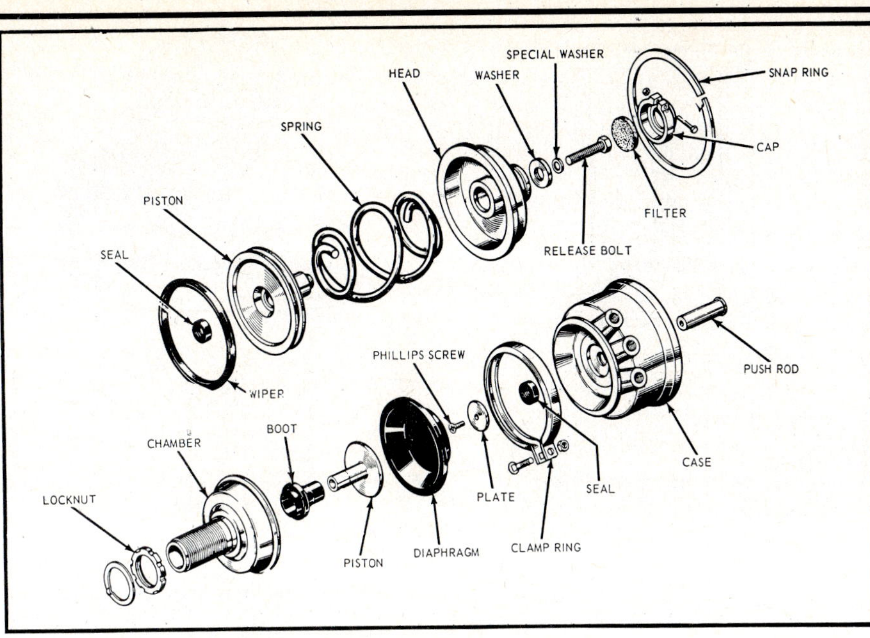

Fig. 9 Shortstop brake unit disassembled

sure piston hub enters head opening correctly. Continue to compress assembly, making sure lip of head seats against shoulder of case correctly.

7. Install snap ring, tapping both ends to insure complete seating.
8. Relax press pressure and remove assembly from press.
9. Using a screwdriver through release bolt hole, press down firmly on push rod while charging the cylinder with air (70 psi minimum).
10. Remove screwdriver and install guide washer on insert in piston hub. Tighten washer to 30 ft-lbs minimum, and check for air leaks as follows:

Checking for Air Leaks

1. Charge unit with air. Apply oil around guide washer. If bubbles appear, replace the main seal.
2. With unit charged with air, apply oil around push rod in diaphragm end of case. If bubbles appear, replace push rod seal.
3. Install pressure plate on push rod. Use emery cloth to smooth burrs off head of Phillips head screw, if present.
4. Clean snap ring groove area with solvent and apply a sealant to cover the snap ring. If possible, allow sealant to dry overnight before installing unit on truck.

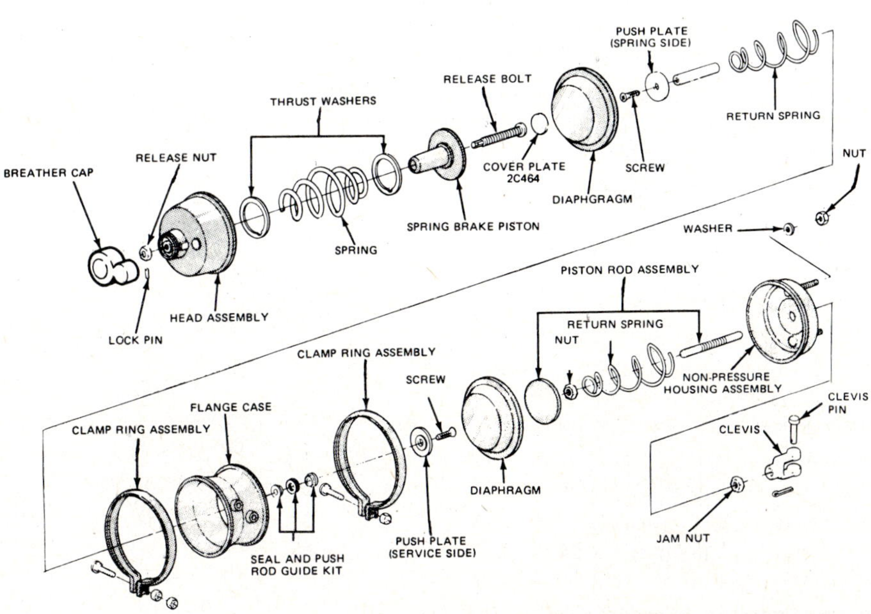

Fig. 10 Disassembled view of MGM Stopgard brake

MGM STOPGARD BRAKE

Description

This dual diaphragm spring brake chamber, Fig. 10, is used on S-cam type rear wheel service brakes only.

The Stopgard brake chamber assembly consists of two separate brake chambers and diaphragms, each having its own application and release mechanisms. The major difference between the two chambers is the service chamber applies the brakes by air pressure and releases the brakes by spring pressure, while the spring brake applies the brakes by spring pressure and releases the brakes by air pressure.

During normal brake applications, air pressure from the brake valve works against the service brake diaphragm, applying the brakes. Air reservoir pressure against the spring brake chamber diaphragm and piston restrains the emergency parking brake spring.

During parking or in an emergency brake application, air is exhausted from the spring brake chamber, permitting the emergency parking brake spring to expand, applying the brakes.

Manual Release

Loss of air pressure will result in automatic application of the brakes. If it is necessary to move the vehicle before air pressure can be restored or to disassemble the unit, the emergency parking brake spring must be manually compressed as follows:

1. Block the wheels to prevent vehicle from rolling.
2. Open breather cap, Fig. 10.
3. Using a suitable 15/16 inch wrench, loosen the release bolt approximately 30 turns, or until the brake shoes release from the drum. If the release bolt binds, apply a suitable penetrating oil to the threads and allow it to soak.

NOTE: If the vehicle has 85 psi of air or more, manually releasing the spring brakes can be made easier if the emergency parking brakes are released, thereby using air pressure to compress the spring. After releasing the emergency parking brake, follow the same procedure as above. After the release bolt has been loosened 30 turns, or when the brake shoes release from the drum, the air pressure can be removed from the spring brake chamber by placing the cab brake control in the "apply" position.

4. To manually reset the emergency parking brake spring, coat the release bolt with chassis lubricant, then tighten the release bolt until the nut bottoms against the head assembly. Torque the nut to 50 ft. lbs.
5. Reinstall the breather cap.

NOTE: Do not operate vehicle without breather cap installed. Replace the breather cap every 50,000 miles or more frequently in severe service applications.

Stopgard Unit, Replace

1. Block the wheels to prevent vehicle from rolling. Expel any air from system.

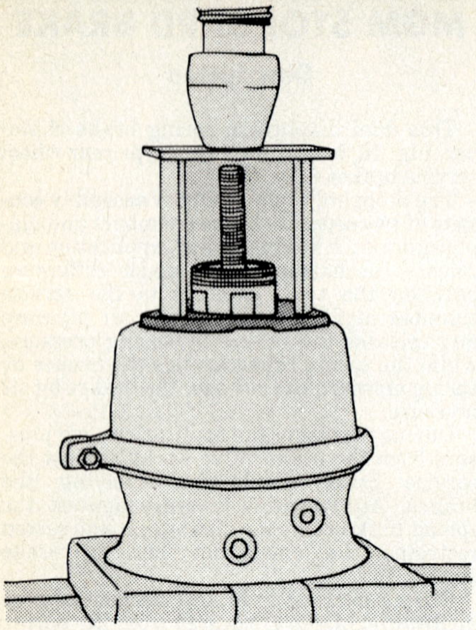

Fig. 11 Positioning Stopgard unit in press

2. Manually release the spring brake. Refer to "Manual Release".
3. Remove air hoses from service and spring brake chambers. Tag hoses for proper installation.
4. Disconnect service brake rod yoke from slack adjuster.
5. Using suitable tool, remove brake chamber mounting nuts from studs, then remove brake chamber.
6. To reinstall, position brake chamber on mounting bracket.
7. Install mounting nuts and lockwashers. Torque to 150 ft. lbs.
8. Connect service brake rod yoke to slack adjuster.

NOTE: The distance from the brake rod yoke centerline to the mounting face of the chamber should be maintained at 2.60 ± 1/8 inch.

---

9. Manually reset the emergency parking brake spring. Refer to "Manual Release".
10. Reconnect air hoses to their respective inlets.
11. Charge system and check for leaks.

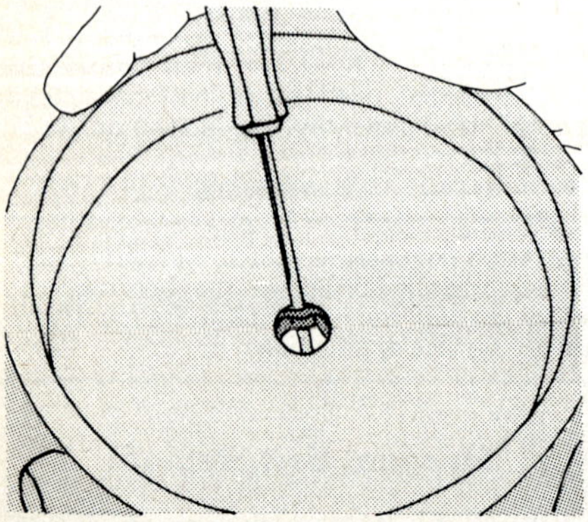

Fig. 14 Removing push rod seal

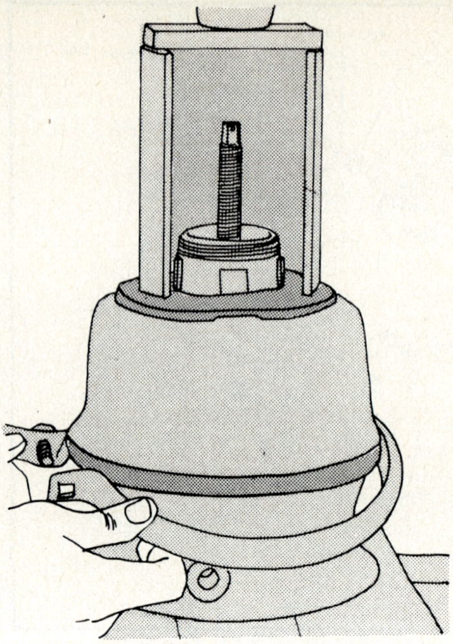

Fig. 12 Removal & installation of spring brake chamber clamp ring

Stopgard Unit, Disassemble

CAUTION: The following procedures must be followed exactly as stated to prevent the possibility of personal injury.

---

1. Ensure the spring brake is manually released. Refer to "Manual Release".
2. Position a suitable pair of locking pliers on service brake pushrod, Fig. 10, to retain piston rod assembly in housing.
3. Loosen service brake chamber clamp nut(s), then separate service brake chamber from flange case by removing clamp ring.
4. Using a suitable punch, remove lock pin from spring brake chamber release bolt, then remove nut.
5. Position spring brake unit in a suitable hydraulic press, using a slit tube and flat plate as shown in Fig. 11.

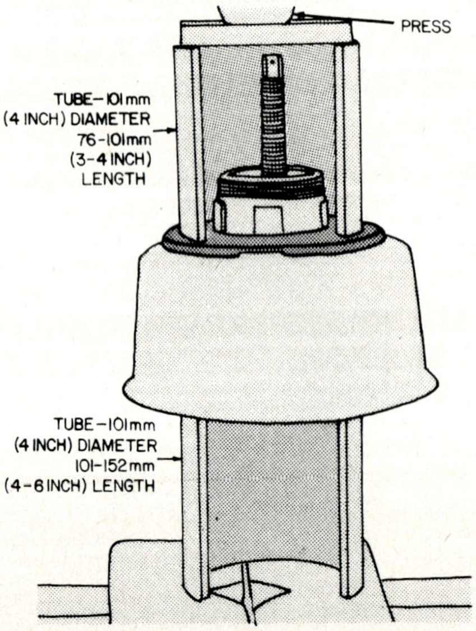

Fig. 15 Positioning spring brake sub-assembly in press

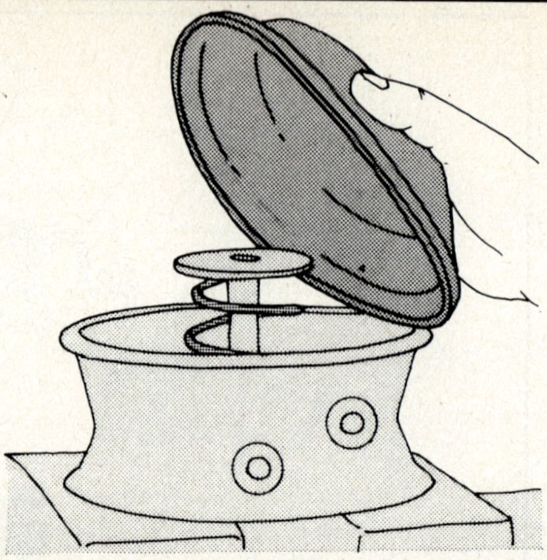

Fig. 13 Removal & installation of spring brake diaphragm

NOTE: The hydraulic press ram must have a minimum of 8 inches of upward travel after it has contacted the flat plate.

---

6. Apply very light pressure with the ram on the plate, then remove the spring brake chamber clamp ring, Fig. 12.
7. Very slowly release press ram pressure. At this point, there should be no outward pressure from the spring brake chamber and the ram and plate should separate almost at once.

NOTE: If tension remains between the press ram and plate, either the spring brake chamber is not fully manually released, or damage has occurred in the release mechanism. If the spring brake chamber is released, but pressure is present, continue to slowly release ram pressure until the emergency parking brake spring is fully extended and parts are free.

---

8. Lift off the spring chamber head, spring and piston sub-assembly and set aside.

NOTE: Handle this sub-assembly carefully because the powerful emergency parking brake spring is compressed between

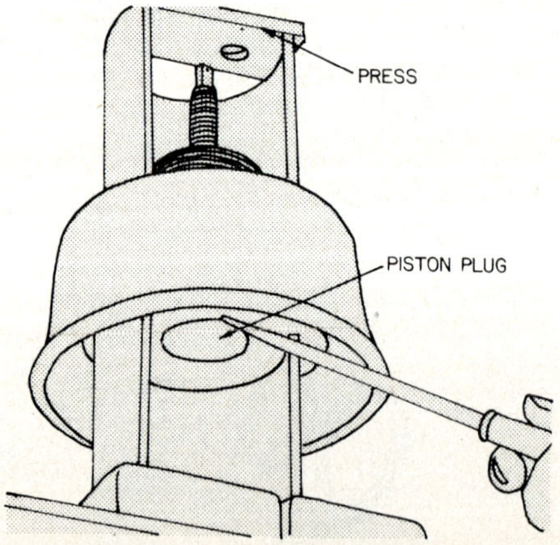

Fig. 16 Removing piston plug

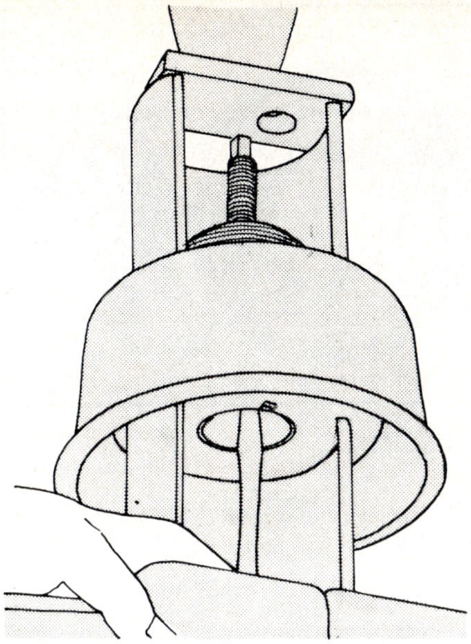

Fig. 17 Removal & installation of release bolt

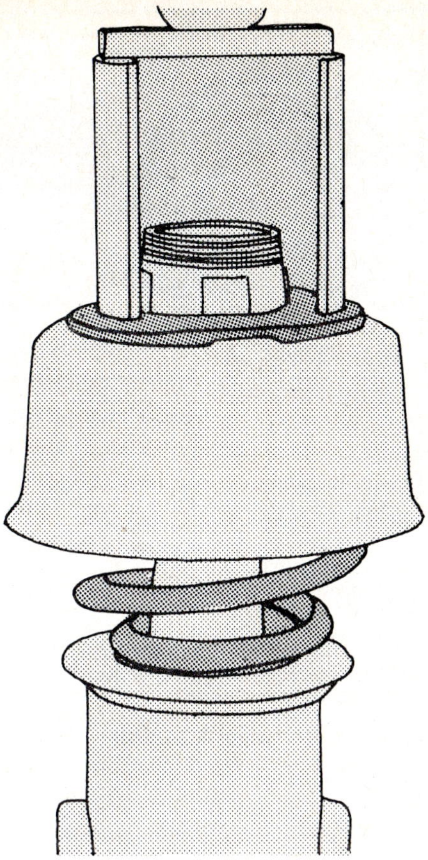

Fig. 18 Removing or applying ram pressure

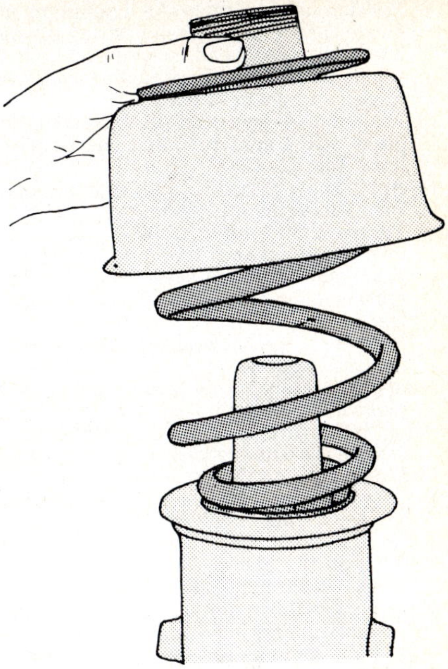

Fig. 19 Assembly & disassembly of spring brake sub-assembly

the piston and head.

9. Remove diaphragm from flange case, Fig. 13. Inspect flange case, return spring, push plate and screw and push rod seal and guide rings, Fig. 10. If any of these parts require service, proceed to step 10. Otherwise, for further disassembly, proceed to step 13.
10. With one hand holding the push plate, remove screws securing the push plates to the push rod. If the screw is too tight to remove in this manner, use a pair of padded pliers to grip the push rod while removing the screws.

NOTE: Care must be taken when removing screws. The push plates hold the return spring under light tension.

11. Remove all parts from flange case and wash thoroughly.
12. Remove guide rings from flange case, then using a small screwdriver in the groove behind the push rod seal, pry the seal out of the groove, Fig. 14.
13. Position the spring chamber, spring and piston sub-assembly in a press as shown in Fig. 15. Position a 4 inch diameter tube, approximately 3½ inches long with a slit as shown in Fig. 15, under the sub-assembly supporting the piston. Position another 4 inch diameter tube, approximately 5 inches long with a slit, between the spring chamber head and the flat plate under the press ram, Fig. 15. Do not attempt to disassemble this unit without the proper tubes.

NOTE: The hydraulic press ram must have a minimum of 8 inches of upward travel after it has contacted the flat plate.

14. Check the sub-assembly for proper alignment in the press, then apply very light pressure on the press ram.
15. Using a suitable screwdriver, pop off the piston plug in the center bottom of the piston, Fig. 16.
16. Using a suitable broad bit screwdriver, remove the release bolt, Fig. 17.

NOTE: If the release bolt is too tight to remove, apply slightly more pressure with the press ram.

17. With release bolt removed, gradually release ram pressure, Fig. 18, until ram is free of contact with spring brake chamber head, Fig. 19.
18. Remove spring brake chamber head, spring and piston assemblies from press.

Cleaning & Inspection

Wash all parts in suitable cleaning solvent. Inspect all parts for wear or damage. Any questionable parts should be replaced. Service brake and spring brake diaphragms should be replaced and push rod seal assembly should be replaced.

Stopgard Unit, Assemble

1. Apply a suitable chassis lubricant generously to the inside of the spring brake chamber head, outside and inside piston shank and to the release bolt.
2. Position the slit tube on the press base, then position parts removed on tube in the following order; piston thrust washer, emergency parking brake spring, thrust washer, spring brake chamber, slit tube and flat plate, Fig. 19.
3. Check up and down alignment of all parts, then standing to side of press, take up tension very slowly on press ram, Fig. 18.

NOTE: While compressing spring, check up and down alignment continually. If misalignment occurs, ease press ram and repeat step 3.

4. When spring is fully compressed between the piston and head, and the piston is bottomed against the head, replace the release bolt, Fig. 17.
5. Tighten the bolt until approximately 3 inches of the bolt extends through the head insert.
6. Ease tension on the press ram slowly until the ram is separated from the sub-assembly.
7. Remove sub-assembly from press, then install release nut and lock pin on release bolt.
8. Reinstall the piston plug in the center hole of the piston. Firmly tap plug with a hammer until the plug moves over center to a flat position.
9. Apply suitable chassis lubricant to pushrod and flange case seal inner groove, then carefully install seal. Seat seal in groove.
10. Install guide rings on both sides of seal.

NOTE: The wide flange of guide rings faces away from seal.

11. Lightly lubricate push rod and install into seal. Install smaller push plate on push rod using suitable locking compound to lock screw.
12. Place push rod and case sub-assembly on a 3 inch diameter support. Position return spring in case, then compress spring with large push plate enough to install screw through push plate and into push rod. Apply suitable locking compound to screw and install. Place bottom push plate in a suitable soft jaw vise and tighten top push plate screw. This will tighten both screws securely.
13. Ensure that return spring and large push plate are in proper alignment with flange case.

14. Position new spring brake diaphragm on flange case, Fig. 13, then position spring brake chamber head, spring and piston sub-assembly over diaphragm. Check for proper alignment of parts.
15. Position case and head sub-assemblies in press with a split tube and flat plate on head, Fig. 11.
16. Apply light pressure to plate with press ram and reinstall clamp ring, Fig. 12. Torque clamp nut(s) to 30 ft. lbs., then using a suitable soft hammer, tap around the clamp circumference to ensure proper seating and retorque clamp nut(s) to 30 ft. lbs.
17. With the spring brake assembly still mounted in press, apply a shop air line of approximately 115 psi pressure to spring plate inlet. Using a soap solution, check for leaks. If a leak is discovered, disassemble the unit, reposition the parts and reassemble.
18. After passing leak test, retighten clamps to 30 ft. lbs. and remove unit from press.
19. Position the piston rod assembly in the non-pressure housing assembly, Fig. 10.
20. Compress the piston enough to compress the spring and install a suitable pair of locking pliers on service brake push rod.
21. Position new service brake diaphragm on flange case, then position service brake chamber over diaphragm.
22. Install clamp ring. Torque clamp nut(s) to 30 ft. lbs., then using a suitable soft hammer, tap around the clamp circumference to ensure proper seating and retorque clamp nut(s) to 30 ft. lbs.
23. Apply a shop air line of approximately 115 psi pressure to service brake inlet. Using a soap solution, check for leaks. If a leak is discovered, disassemble service brake unit, reposition the parts and reassemble.
24. Remove air supply. Reinstall unit in vehicle, refer to "Replace".

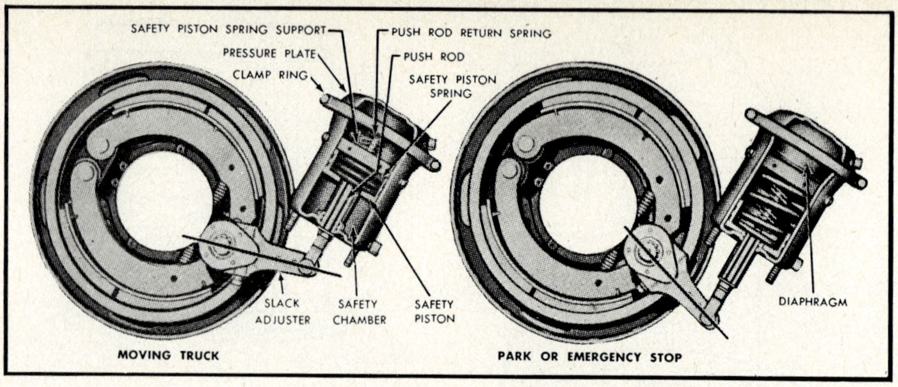

Fig. 20 Maxibrake operation

MAXIBRAKE

This device, Fig. 20, is available on vehicles with full air brakes for emergency stops or parking. The air brake chamber (used on rear axle brakes only) is much like the conventional brake chamber; it differs only in that it contains an independently spring-loaded piston which, when released due to low air pressure, will apply the service brakes.

Normal air pressure applied to the safety chamber and piston keeps the piston spring compressed, while the service brakes continue to operate normally.

A manual control valve, located in the cab, exhausts the air from the safety chamber so that the safety piston spring pressure can be used for emergency stops.

Remove & Install

Release all air from the system and then disconnect air lines from brake chamber. Disconnect push rod yoke from slack adjuster. Remove mounting nuts and take off brake chamber.

To install, position brake chamber on mounting bracket and install retaining nuts. Connect push rod yoke to slack adjuster. Connect air lines to brake chamber and build up the air pressure.

Disassembly

1. Referring to Fig. 21, first mark both the pressure plate and cylinder with relation to clamping ring and air line fittings so that unit can be reassembled in the same position.
2. Remove clevis yoke and flanged nut from push rod and install a 3/4" OD × 5/8" ID flat washer and a 5/8" nut on push rod. Tighten nut until plate end of push rod is seated on safety piston spring support. *It is necessary to compress push rod return spring and safety piston spring so the unit may be disassembled safely without being hampered by spring pressure.*
3. Remove clamp ring, pressure plate and diaphragm from cylinder.
4. Unfasten safety piston spring support from cylinder body and remove complete inner assembly (8 screws).
5. Place push rod and piston assembly on work bench and remove nut and washer used to compress springs. With spring tension removed, disassemble push rod, return spring, safety piston, spring and seat.
6. Remove push rod seal from bore of piston and O-ring seals from piston and cylinder.

Reassembly, Fig. 21

NOTE: Always be sure that the correct return spring is used in any brake chamber. Also be sure that the brake chamber on the opposite side of the axle has the same return spring, otherwise uneven braking will result.

If a new diaphragm is installed in the brake chamber on one side of the axle, a new one should be installed in the corresponding brake chamber on the opposite side also, otherwise uneven braking will result.

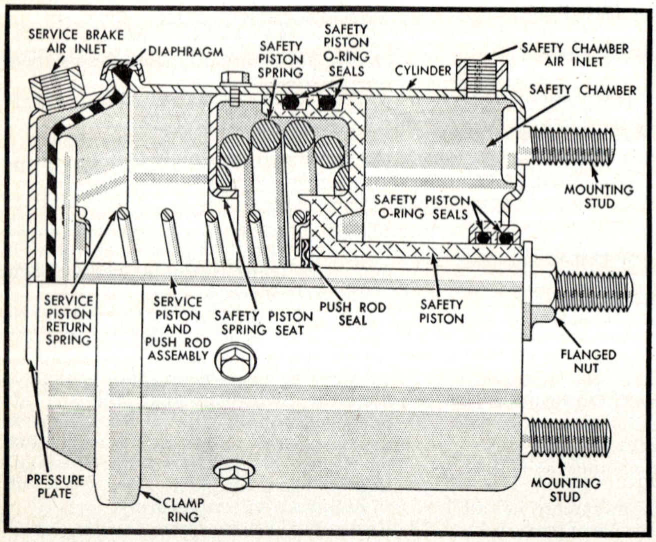

Fig. 21 Maxibrake air brake chamber (sectional view)

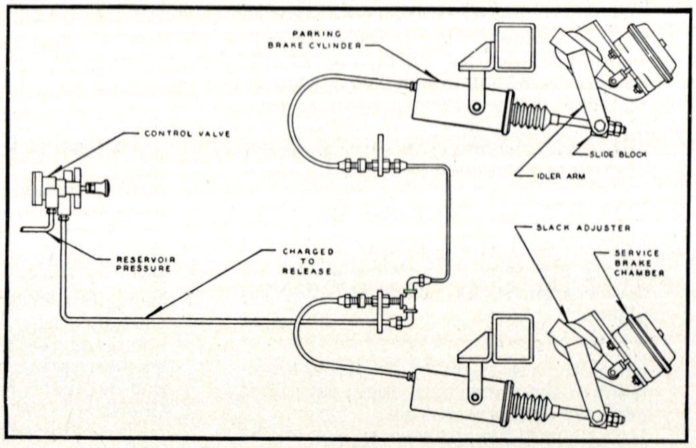

Fig. 22 MGM spring-type parking brake system

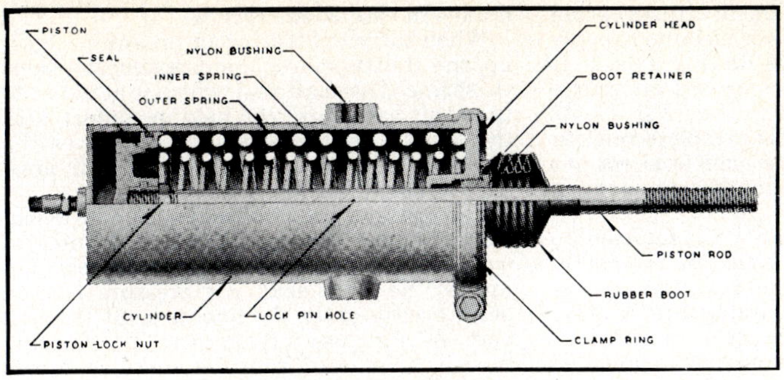

Fig. 23 Parking brake cylinder

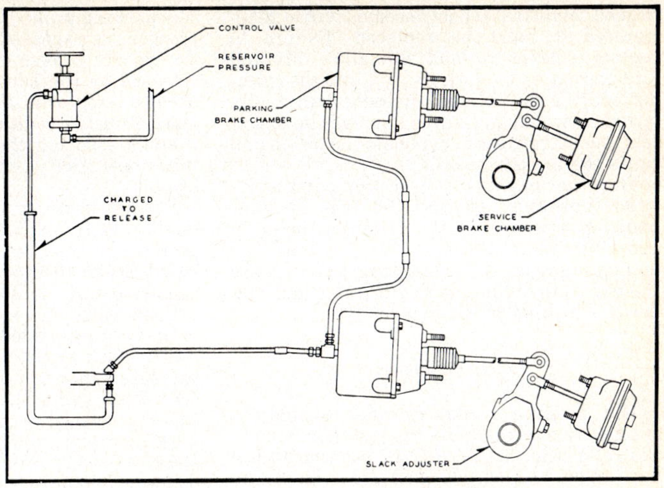

Fig. 24 Berg "Airo-Mech" spring-type parking brake system

1. Coat safety piston and inside of cylinder with Lubriplate or other suitable lubricant.
2. Rest push rod upright on a flat surface and assemble in order: push rod return spring, safety piston spring seat and spring, and safety piston. Compress push rod and springs by installing nut and flat washer used in disassembly procedure.
3. Install push rod seal in piston bore and install O-ring seals on piston and in cylinder.
4. Position piston and spring assembly in cylinder. Align threaded holes in piston spring support with holes in side of cylinder, and install retaining capscrews.
5. Position diaphragm and pressure plate with clamp ring to cylinder; align marks made prior to disassembly and tighten clamp ring.
6. Remove nut and washer used to compress springs. Connect air lines to brake chamber and build up air pressure.
7. With safety chamber under pressure, and safety piston in retracted position, run flange nut up snug on push rod and install clevis yoke.

NOTE: Maintain distance of 2$^{13}/_{16}$" between centerline of clevis pin holes in yoke and mounting flats of cylinder. Tighten flange nut against yoke to lock it in place.

TROUBLE SHOOTING AIR SYSTEM

Slow Pressure Build-Up In Reservoirs

1. Leaking application valves, check valve or relief valve.
2. Leaking compression discharge valves.
3. Leaking lines or connections.
4. Weak unloader piston return springs.
5. Clogged compressor air cleaner.
6. Worn compressor pistons and rings. Carbon in discharge line.
7. Bent or kinked line.
8. Insufficient discharge valve travel.

Quick Loss of Reservoir Pressure when Engine is Stopped

1. Leaking relief valve.
2. Leaking compression discharge valves.
3. Leaking tubing or connections.
4. Leaking application valves or check valve.
5. Leaking governor.

Compressor Not Unloading

1. Unloader pistons leaking or sticking.
2. Restriction in passages from governor to unloader.
3. Governor not operating properly.

Slow Brake Application, Slow Release or Inefficient Brakes

1. Low system pressure.
2. Brake chamber push rod travel excessive.
3. Restriction in line.
4. Leaking brake chamber diaphragm.
5. Poor brake lining or drum condition.
6. Leaking brake valve or weak rubber spring in valve.
7. Binding cam or camshafts.
8. Quick release valve leaking or sticking.

MGM SPRING BRAKE

This parking brake mechanism, Fig. 22, utilizes spring pressure as a separate power medium to apply the service brakes on the driving rear axle. The system is completely self-contained, with an individual cylinder applied to two or more rear wheel brakes in the service brake system. It complements the basic brake system but does not replace it. The cylinders are held inactive by the same compressed air source used to apply the service brakes. Approximately 90 psi is required to hold this brake in the fully released position, thereby assuring that the vehicle is not put into operation with insufficient service brake air pressure.

Operation

Two or four wheel brakes on the rear driving axle are provided with a springloaded air cylinder, connected to the slack adjuster through an idler arm with slide block, Fig. 22. Air pressure applied to the cylinders by a driver controlled valve, compresses the springs in the cylinders holding the piston rods in the extended position, thereby allowing the service brake system to function independently. When a normal foot brake application is made and the service brake chambers actuate the slack adjuster, the idler arm and slide block ride freely on the extended piston rod.

Control Valve

The parking brake control valve, located in the cab, is both manual and automatic. The control valve knob should be pushed "in" for park and pulled "out" for drive. In the case of an air loss, detected by the driver, the control valve knob can be pushed in, exhausting the air pressure in the cylinder, allowing the springs to retract the piston rod and apply the brakes. Even through this brake is primarily installed a parking brake, an emergency brake application will automatically occur any time the system pressure drops below a safe level, whether or not the control valve knob is pushed in. If the system pressure is exhausted while the vehicle is parked, it will be necessary to build pressure up to 90 psi before pulling out the control valve knob to release the brakes.

Brake Cylinder

Each parking brake cylinder is attached to a gimbal hanger bracket with two trunnion bolts. The trunnion bolts pass through the gimbal bracket and pilot into nylon bushings that are pressed into the bosses on each side of the cylinder housing. This allows the cylinder to oscillate, as required, when piston and rod are in operation. A piston, fitted with a neoprene seal is installed in the cylinder bore. Coiled springs, compressed between the piston rear face and cylinder head, provide the necessary energy for applying the parking brake. A rubber boot, covering the piston rod opening in the cylinder head, prevents the entrance of dirt and water into the cylinder.

Two springs are in the position shown in Fig. 23 when there is no air pressure in the parking brake system due to either a manual or an automatic brake application. When air pressure in the cylinder is reduced to approximately 80 psi the piston rod moves one-half inch. Air in the cylinder must be reduced to approximately 70 psi in order for the rod to move 1½ inches. With normal service brake adjustment it is at this point that the brake shoes contact the drum. Brakes are fully applied when cylinder pressure drops to approximately 40 psi.

Maintenance

This brake system requires very little maintenance other than periodic visual inspection. It is important to see that the rubber boot, covering the piston rod opening in the cylinder head, is in place on the retainer and in good condition. If boot becomes torn or deteriorated it must be replaced. Failure to replace a defective boot will allow dirt and foreign matter to work its way into the cylinder, scoring the polished cylinder wall. Badly pitted or scored walls will damage the piston seal, permitting air to leak past the piston and cause the brakes to drag.

The other important inspection point is the nylon bushings, pressed into the cylinder housing bosses, and the trunnion bolts. If excessive wear develops, replace bushings and trunnion bolts. Failure to do so will cause the bosses to break due to vibration, and render the parking brake inoperative.

Adjustment

The parking brake cylinder should be charged with air to move the piston and rod to the fully released or "drive" position before adjustment is made. Loosen and back off the two 3/4" nuts on end of piston rod. Adjust the service brakes through the slack adjusters. Then tighten the two 3/4" nuts with fingers until the large nut rests firmly against the slide block in the idler arm, then lock the jam nut.

NOTE: Adjustment to compensate for brake lining wear should be made at the slack adjusters—never by tightening the nuts on the end of the piston rod.

Lubrication

Every six months fill cylinders with air and insert lock pin in piston rod, Fig. 23. Remove air hose and pour one ounce of S.A.E. 10 engine oil through the tapped hole. Replace the air hose, fill cylinder with air and remove lock pin. Turn piston rod 1/4 turn by using a wrench on the 3/4" nuts at end of rod. Do not use pipe wrench to turn rod.

Keep piston rod clean and lightly lubricate it with Lubriplate or equivalent to prevent slide block from binding when service brakes are in operation.

BERG "AIRO-MECH" SPRING BRAKE

The parking brake control valve handle is pulled up to park and pushed down to release. The parking brake mechanism utilizes spring pressure as a separate power medium to apply the service brakes on the driving axle. It is released by the same compressed air source used to apply the service brakes, Fig. 24. At least 40 psi is required for complete release, thereby assuring that the vehicle is not put into operation with insufficient service brakes.

Parking brake spring tension is sufficient to hold a fully loaded vehicle on maximum grades permitted on modern highways, provided brake drums and lining are in good condition and shoes are properly adjusted.

Each brake on the driving axle is provided with a spring-loaded air chamber, connected to the slack adjuster by a flexible cable. Air pressure applied to the chambers by a driver-controlled valve, compresses the springs, thereby releasing the brakes. Although the parking brake chambers are connected to the slack adjusters, the parking brake system operates independently of the service brakes. The control valve is so designed that it is both manual and automatic. In case of an air loss below a safe level, the air in the chambers is exhausted and the brakes are automatically applied.

Brake Chamber, Fig. 25

Each brake chamber is mounted on a bracket which is clamped to a special rear spring shackle bolt. A piston, fitted with an O-ring to form an air-tight seal, is installed in the chamber bore. Three coil springs, compressed between the piston rear face and chamber cover, provide the necessary energy for applying the parking brake. A flexible cable, connecting the piston rod and slack adjuster, transmits the spring pressure to the brake cam.

The springs are in the position shown in Fig. 25 (720 lbs. pressure) when there is no air pressure in the parking brake system due to either a manual or an automatic brake application. When air pressure in the system reaches approximately 25 psi, the piston begins to move and compress the springs. At 50 psi the piston has reached the end of its travel (contacting shoulder on cover); springs are fully compressed and brakes are fully released.

Control Valve, Fig. 26

The parking brake control valve is both manual and automatic. It serves as a manual control for releasing and applying the parking brake and also automatically applies the brakes in case of emergency.

When sufficient tractor air pressure is built up, the shuttle moves, compressing the shuttle spring. The shaft seal, which is attached to the shaft assembly, and held in contact with the shuttle by the shaft spring, moves the shuttle valve, thus containing the air pressure in chamber "A".

The operator then pushes the handle downward, causing the shuttle seal seat (integral with shaft) to contact the shuttle seal and unseat the shaft seal. Air pressure is then permitted to enter the hollow shaft through port "A" and pass through port "B", entering the air lines leading to the brake chambers and releasing the parking brakes.

The air pressure in chambers "A" and "B" is now equal as they are directly connected by the ports in the hollow shaft. Air pressure in the entire brake system is now the same. If pressure anywhere in the parking or service brake system drops to approximately 35 psi, the shuttle spring in the control valve expands and moves the shuttle, unseating the shuttle seal seat, thereby permitting the air to escape through the exhaust port and apply the brakes.

In a manual brake application the operator lifts the handle which draws the shuttle seal seat away from the shuttle seal in order to permit the air pressure to exhaust. In an automatic brake application the shuttle spring forces the shuttle away from the seal seat.

Adjustment

Each cable should have a small amount of slack in it to eliminate the possibility of a partial brake application due to the rise and fall of the axle on rough surfaces.

Service brakes should be properly adjusted and the parking brake chambers charged with air to the "released" position before adjustment can be made. Adjustment to compensate for lining wear should be made at the slack adjusters—not on the cables. No general lubrication of the parking brake system is required.

BERG "SHUR-BRAKE" SPRING BRAKE

This parking brake is the spring-applied air-release type which acts in conjunction with the rear wheel brake shoes, Fig. 27. The parking brake control valve handle is pulled out to park and pushed in to relase.

This system operates in a similar manner to the Berg Airo-Mech type described above. At least 45 psi is required for complete release, thereby assuring that the vehicle is not put into operation with insufficient service brake air supply.

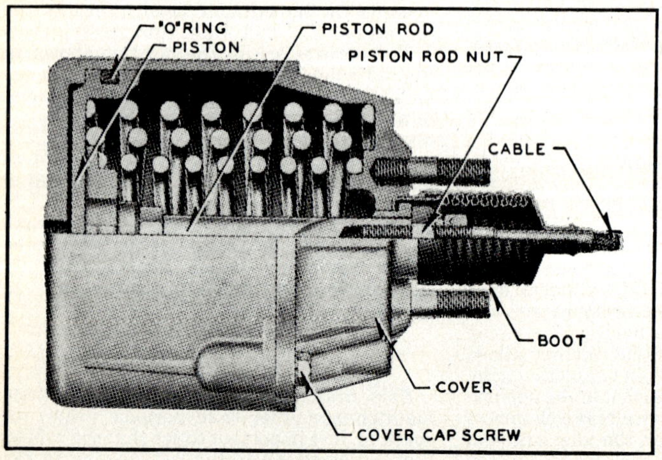

Fig. 25 Parking brake chamber

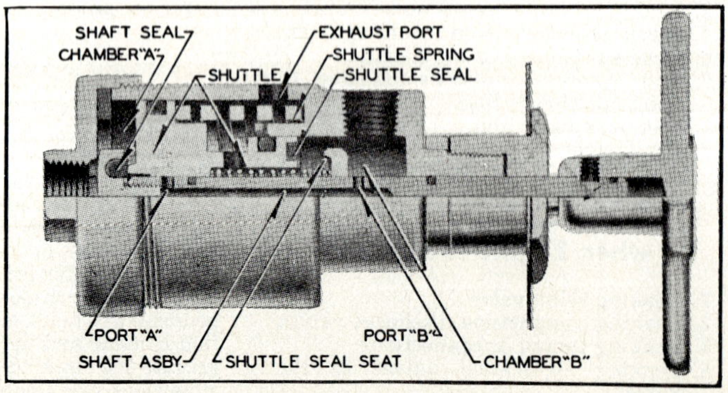

Fig. 26 Hand control valve

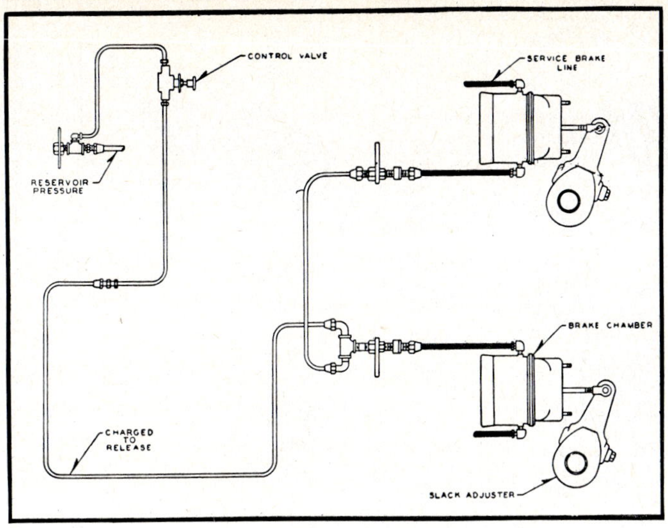

Fig. 27 Berg "Shur-Brake" spring-type parking brake system

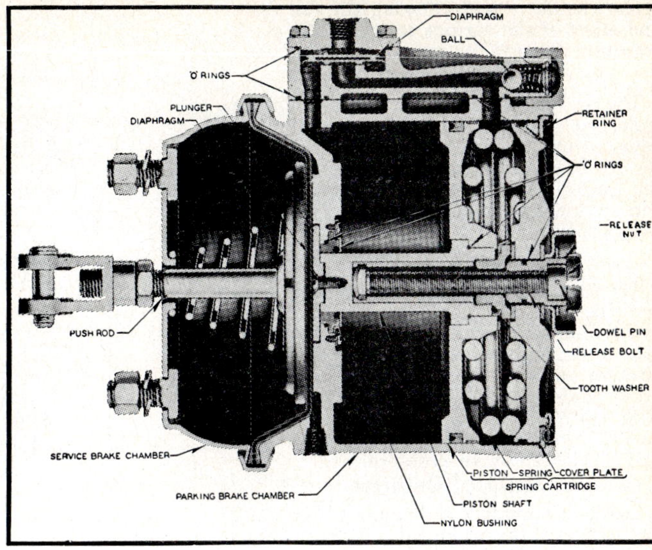

Fig. 28 Brake chamber with quick release valve

Brake Chambers, Fig. 28

Each unit contains two separate air tight chambers, the parking brake chamber and service brake chamber, each supplied by its own air pressure source. A barrel-shaped coil spring, compressed between the piston's rear face and the spring plate, provides the necessary energy for applying the spring brake. A piston shaft, passing through an air-tight nylon bushing in the chamber separation wall and bearing on the pressure face of the service brake diaphragm, transmits the spring pressure to the slack adjuster through the service brake rod and clevis assembly. A rubber boot (not used when the brake is equipped with a quick release valve) is fitted over the rear lip of the brake chamber cylinder to prevent dirt and water entering the unit.

Quick Release Valve, Fig. 29

Mounted to the parking brake chamber, an integral quick release valve is used on some brake chamber assemblies to control the flow of air pressure within the chamber.

A function of the valve is to equalize air pressure within the unit to that of the vehicle's system pressure. In doing so, the excess chamber pressure is exhausted to the atmosphere through the valve's exhaust port. Being a normal function of the unit, it should not be confused with spring brake malfunction.

At least once a year, the quick release valve should be removed from the parking brake chamber and disassembled. Check valve diaphragm and rubber-coated ball check. If tears or nicks are present, replace. Pay particular attention to the feather edge around the outer circumference of the diaphragm. A repair kit is available to recondition the valve.

Hand Control Valve, Fig. 30

The parking brake control valve is both manual and automatic. It serves as a manual control for releasing and applying the parking brakes and also automatically applies the brakes in case of emergency (vehicle air pressure below 38 psi).

When sufficient pressure is built up, the operator pushes the control valve knob in, causing the piston seat to move in, sealing the shuttle and closing the air passage from the outlet port to exhaust. Further inward movement of the knob unseats the shuttle from the valve body seat, permitting the air pressure on the inlet side to pass through the valve to the outlet port and hence to the parking brake chambers. A minimum pressure of 40 psi must be available at the inlet port to counteract the spring pressure and thus hold the piston in against the shuttle.

Should the inlet pressure fall below 40 psi, the spring force acting on the piston seat will move the piston assembly and knob outward, causing the shuttle spring to follow the piston seat until the shuttle contacts the valve body seat. Further movement of the piston results in losing contact with the shuttle and opens the air passage from the outlet port to exhaust, thus evacuating the pressure from the

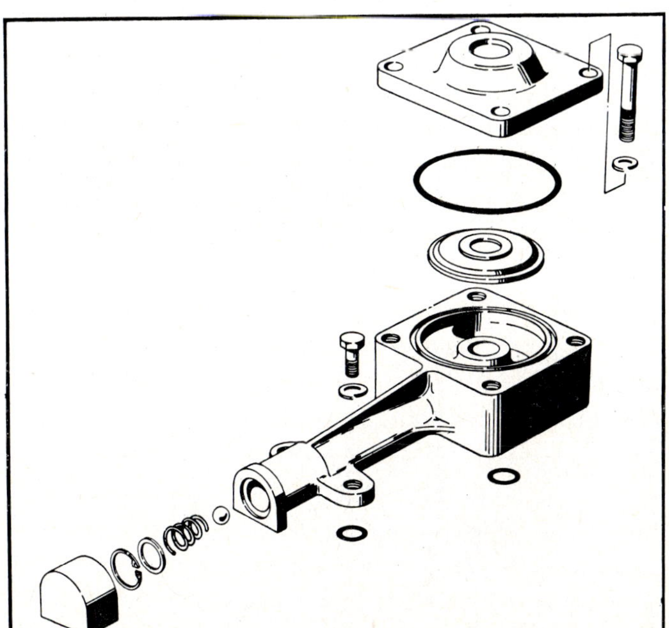

Fig. 29 Exploded view of quick release valve

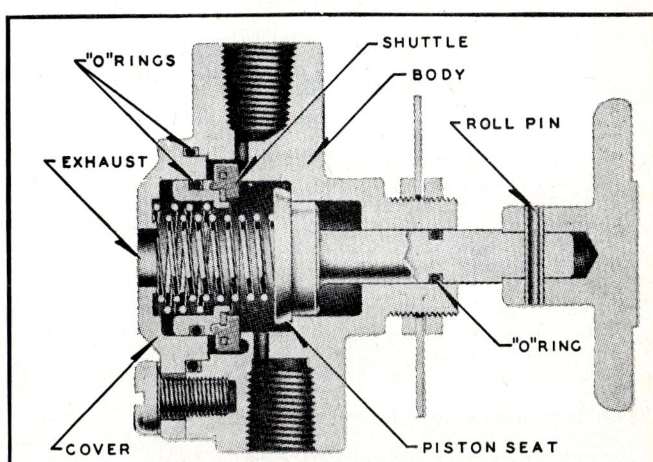

Fig. 30 Hand control valve

brake chambers and allowing the full force and effect of the springs to act on the slack adjusters through the service brake linkage.

Adjustment

The parking brakes must be placed in the "released" position and the service brakes adjusted in the normal manner. Other than service brake adjustment, no adjustment of the parking brake system is required. No general lubrication of the parking brake system is required.

ANCHOR LOK

The Anchor Lok brake system is used on all vehicles with S cam brake assemblies. The brake chamber assembly consists of two separate air chambers, each with its own diaphragm and push rod, Fig. 31. The service brake chamber applies the brake by air pressure and releases it by spring pressure. The parking brake is applied by spring pressure and is released by air pressure.

Manual Release

Loss of air pressure will result in automatic application of the brakes. If it is necessary to move the vehicle before air pressure can be restored, or to disassemble the unit, the emergency parking brake spring must be manually compressed as follows:

1. Block the wheels to prevent vehicle from rolling. Expel any air from system.
2. Remove the release stud tool and nut from the carrying pocket on the chamber assembly.
3. Remove the parking brake spring chamber access plug, Fig. 31.
4. Install release stud tool through opening from which access plug was removed in spring chamber and into the spring pressure plate, Fig. 32. Turn the release stud 1/4 turn to engage the tangs on the release stud into the slot in the pressure plate.
5. Assemble release stud washer and nut onto release stud.

NOTE: While installing nut on release stud, ensure release stud tangs stay engaged with spring pressure plate.

6. Tighten release stud nut with a suitable wrench until emergency parking spring is fully compressed. Do not overtighten release stud nut, maximum torque is 50 ft. lbs.
7. To reset charge system with air, then back off release stud nut and remove release stud.

NOTE: Do not loosen or remove release stud until the parking chamber is completely assembled and securely clamped.

Anchor Lok Unit, Replace

1. Block wheels to prevent vehicle from rolling. Expel any air from system.
2. Manually release spring brake. Refer to "Manual Release".
3. Disconnect service brake rod yoke from slack adjuster.
4. Remove air hoses from service and spring brake chambers. Tag hoses for proper installation.
5. Remove mounting nuts, then remove unit from vehicle.
6. To reinstall, position unit on mounting bracket and install lockwashers and nuts. Torque nuts to 90 ft. lbs.
7. Connect service brake rod yoke to slack adjuster.

NOTE: The distance from the brake rod yoke centerline to the mounting face of the chamber should be maintained at 2.60 ± 1/8 inch.

8. Reconnect air lines to their respective inlets.
9. Build up system pressure and check for leaks.
10. Manually reset emergency parking brake spring. Refer to "Manual Release".

Anchor Lok Unit, Disassemble

NOTE: The following procedures can be performed with the unit installed on vehicle.

1. Block wheels to prevent vehicle from rolling. Expel any air from system.
2. Manually release spring brake. Refer to "Manual Release.
3. Remove the parking brake housing to adapter clamps.
4. Remove the parking brake housing, spring and pressure plate as an assembly.

NOTE: The spring housing, spring and pressure plate are serviced as an assembled unit.

5. Remove parking brake diaphragm.
6. Remove adapter to service chamber clamp, then remove service chamber.
7. Remove service brake diaphragm.

Anchor Lok Unit, Assemble

NOTE: The following procedures can be performed with the unit installed on vehicle.

1. Position service brake diaphragm squarely on service brake chamber.
2. Position adapter on service brake chamber then install chamber to adapter clamp. Torque bolts evenly to 25 ft. lbs.
3. Position parking brake diaphragm squarely on adapter.
4. Position parking brake housing, spring and pressure plate assembly on adapter and install chamber to adapter clamp. Torque bolts evenly to 25 ft. lbs.
5. Charge system with air, then using a suitable soap solution, check for leaks.
6. Manually reset emergency parking brake spring. Refer to "Manual Release".

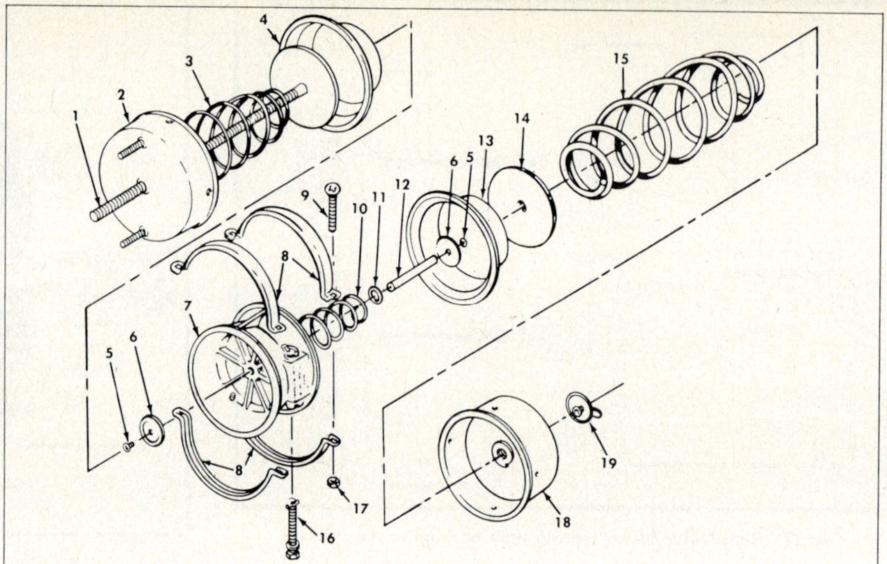

1 Push Rod Assembly
2 Service Brake Spring Housing
3 Service Diaphragm Return Spring
4 Service Brake Diaphragm
5 Screw
6 Adapter Plate
7 Adapter Assembly
8 Clamp Ring
9 Clamp Bolt
10 Parking Diaphragm Return Spring
11 O-Ring
12 Adapter Push Rod
13 Parking Brake Diaphragm
14 Parking Brake Spring Plate
15 Parking Brake Spring
16 Spring Caging Stud
17 Nut
18 Parking Brake Spring Housing
19 Rubber Cap

Fig. 31 Exploded view of Anchor Lok brake

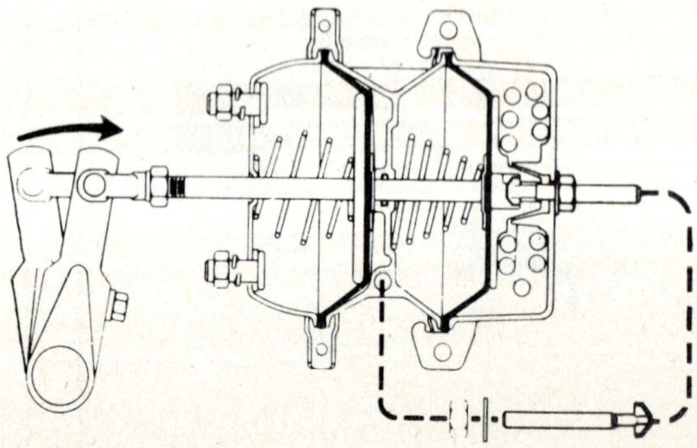

Fig. 32 Positioning release stud for compressing parking brake spring

HYDRAULIC DISC BRAKES

INDEX

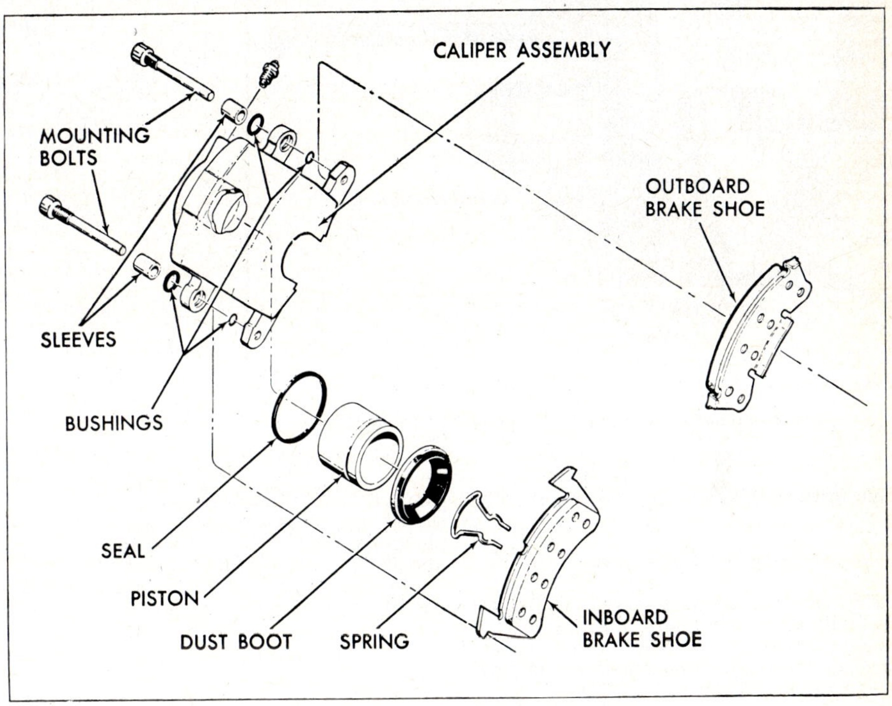

Fig. 1 Single piston caliper. Exploded

TROUBLE SHOOTING

Excessive Pedal Travel

1. Shoe and lining knock back after violent cornering or rough road travel.
2. Piston and shoe and lining assembly not properly seated or positioned.
3. Air leak or insufficient fluid in system or caliper.
4. Loose wheel bearing adjustment.
5. Damaged or worn caliper piston seal.
6. Improper booster push rod adjustment.
7. Shoe out of flat more than .005″.
8. Rear brake automatic adjusters inoperative.
9. Improperly ground rear brake shoe and lining assemblies.

Brake Roughness or Chatter; Pedal Pumping

1. Excessive lateral run-out of rotor.
2. Rotor excessively out of parallel.

Excessive Pedal Effort

1. Frozen or seized pistons.
2. Brake fluid, oil or grease on linings.
3. Shoe and lining worn below specifications.
4. Proportioning valve malfunction.
5. Booster inoperative.
6. Leaking booster vacuum check valve.

Pull, Uneven or Grabbing Brakes

1. Frozen or seized pistons.
2. Brake fluid, oil or grease on linings.
3. Caliper out of alignment with rotor.
4. Loose caliper attachment.
5. Unequalized front tire pressure.
6. Incorrect front end alignment.
7. Lining protruding beyond end of shoe.

Brake Rattle

1. Excessive clearance between shoe and caliper or between shoe and splash shield.
2. Shoe hold-down clips missing or improperly positioned.

Heavy Brake Drag

1. Frozen or seized pistons.
2. Operator riding brake pedal.
3. Incomplete brake pedal return due to linkage interference.
4. Faulty booster check valve holding pressure in hydraulic system.
5. Residual pressure in front brake hydraulic system.

Caliper Brake Fluid Leak

1. Damaged or worn caliper piston seal.
2. Scores in cylinder bore.
3. Corrosion build-up in cylinder bore or on piston surface.
4. Metal clip in seal groove.

No Braking Effect When Pedal is Depressed

1. Piston and shoe and lining assembly not properly seated or positioned.
2. Air leak or insufficient fluid in system or caliper.
3. Damaged or worn caliper piston seal.
4. Bleeder screw open.
5. Air in hydraulic system or improper bleeding.

Rear Brakes Locking On Application

On brake system equipped with a proportioning or rear pressure regulator valve, should the valve malfunction rear brakes may receive excess pressure, resulting in wheel lock-up.

SERVICE PRECAUTIONS

Brake Lines & Linings

Remove one of the front wheels and inspect the brake disc, caliper and linings. (The wheel bearings should be inspected at this time and repacked if necessary).

Do not get any oil or grease on the linings. If the linings are worn to within .030″ of the surface of the shoe, replace both sets of shoe and lining assemblies. It is recommended that both front wheel sets be replaced whenever a respective shoe and lining is worn or damaged. Inspect and, if necessary, replace rear brake linings also.

If the caliper is cracked or fluid leakage through the casting is evident, it must be replaced as a unit.

Shoe & Lining Wear

If visual inspection does not adequately determine the condition of the lining, a physical check will be necessary.

To check the amount of lining wear, remove a wheel from the car, the caliper from the steering knuckle, and the shoe and lining assemblies. Three thickness measurements should be taken (with a micrometer) across the middle section of the shoe and lining; one reading at each side and one reading in the center.

When a shoe and lining assembly has worn to a thickness of .180″, it should be replaced. If shoes do not require replacement, reinstall

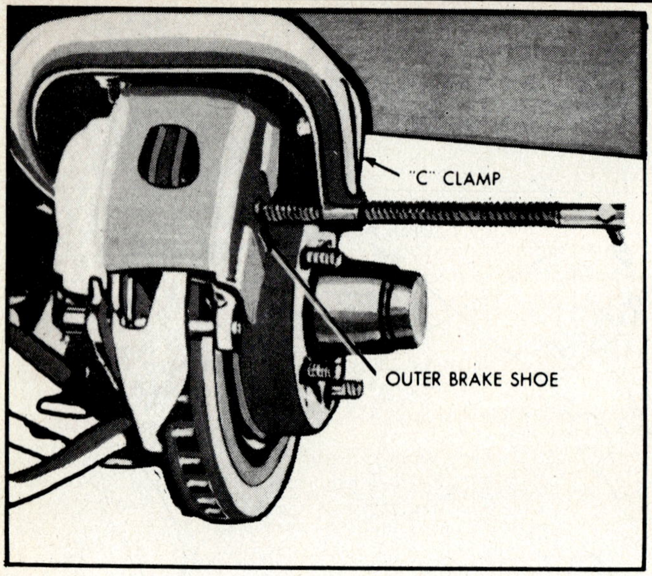

Fig. 2 Compressing piston and shoes with "C" clamp

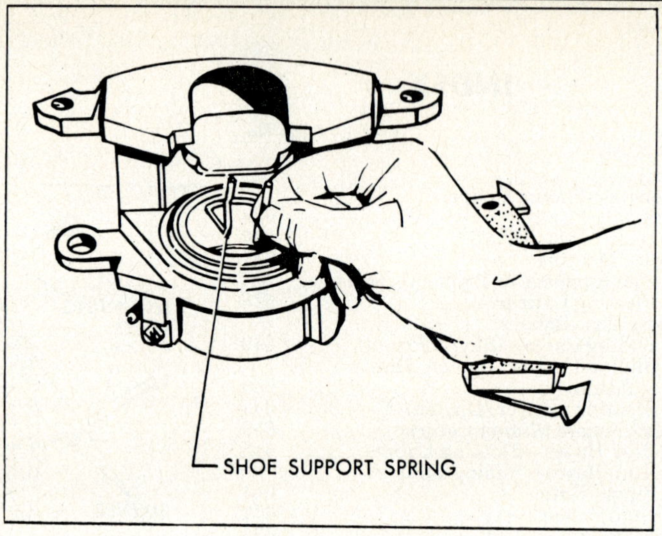

Fig. 3 Installing support spring

them in their original inner and outer positions.

Brake Roughness

The most common cause of brake chatter on disc brakes is a variation in thickness of the disc. If roughness or vibration is encountered during highway operation or if pedal pumping is experienced at low speeds, the disc may have excessive thickness variation. To check for this condition, measure the disc at 12 points with a micrometer at a radius approximately one inch from edge of disc. If thickness measurements vary by more than .0005", the disc should be replaced with a new one.

Excessive lateral runout of braking disc may cause a "knocking back" of the pistons, possibly creating increased pedal travel and vibration when brakes are applied.

Before checking the runout, wheel bearings should be adjusted. The readjustment is very important and will be required at the completion of the test to prevent bearing failure. Be sure to make the adjustment according to the recommendations given.

Brake Disc Service

Servicing of disc brakes is extremely critical due to the close tolerances required in machining the brake disc to insure proper brake operation. In manufacturing brake discs, tolerances of the rubbing surfaces for flatness is .001" and usually for parallelism .0005". Lateral runout of the faces should not exceed .004 to .005" in most cases although the limit on Ford Company products is only .002".

The maintenance of these close controls of the shape of the rubbing surfaces is necessary to prevent brake roughness. In addition, the surface finish must be non-directional and maintained at a micro inch finish. This close control of the rubbing surface finish is necessary to avoid pulls and erratic performance and promote long lining life and equal lining wear of both left and right brakes.

In light of the foregoing remarks, refinishing of the rubbing surfaces should not be attempted unless precision equipment, capable of measuring in micro inches (millionths of an inch) is available.

To check runout of a disc, mount a dial indicator on a convenient part (steering knuckle, tie rod, disc brake caliper housing) so that the plunger of the dial indicator contacts the disc at a point one inch from the outer edge. If the total indicated runout exceeds specifications, install a new disc.

General Precautions

1. Grease or any other foreign material must be kept off the caliper, surfaces of the disc and external surfaces of the hub, during service procedures. Handling the brake disc and caliper should be done in a way to avoid deformation of the disc and nicking or scratching brake linings.
2. If inspection reveals rubber piston seals are worn or damaged, they should be replaced immediately.
3. During removal and installation of a wheel assembly, exercise care so as not to interfere with or damage the caliper splash shield, the bleeder screw or the transfer tube.
4. Front wheel bearings should be adjusted to specifications.
5. Be sure vehicle is centered on hoist before servicing any of the front end components to avoid bending or damaging the disc splash shield on full right or left wheel turns.
6. Before the vehicle is moved after any brake service work, be sure to obtain a firm brake pedal.
7. The assembly bolts of the two caliper housings should not be disturbed unless the caliper requires service.

Inspection of Caliper

Should it become necessary to remove the caliper for installation of new parts, clean all parts in alcohol, wipe dry using lint-free cloths. Using an air hose, blow out drilled passages and bores. Check dust boots for punctures or tears. If punctures or tears are evident, new boots should be installed upon reassembly.

Inspect piston bores in both housings for scoring or pitting. Bores that show light scratches or corrosion can usually be cleaned with crocus cloth. However, bores that have deep scratches or scoring may be honed, provided the diameter of the bore is not increased more than .002". If the bore does not clean up within this specification, a new caliper housing should be installed (black stains on the bore walls are caused by piston seals and will do no harm).

When using a hone, be sure to install the hone baffle before honing bore. The baffle is used to protect the hone stones from damage. Use extreme care in cleaning the caliper after honing. Remove all dust and grit by flushing the caliper with alcohol. Wipe dry with clean lint-less cloth and then clean a second time in the same manner.

Bleeding Disc Brakes

NOTE: Chrysler Corp. recommends only pressure bleeding for all models.

The disc brake hydraulic system can be bled manually or with pressure bleeding equipment (except as noted above). On vehicles with disc brakes the brake pedal will require more pumping and frequent checking of fluid level in master cylinder during bleeding operation.

Never use brake fluid that has been drained from hydraulic system when bleeding the brakes. Be sure the disc brake pistons are returned to their normal positions and that the shoe and lining assemblies are properly seated. Before driving the vehicle, check brake operation to be sure that a firm pedal has been obtained.

Testing Proportional Valve

When a premature rear wheel slide is obtained on a brake application, it usually is an indication that the fluid pressure to the rear wheels is above the 50% reduction ratio for the rear line pressure and that a malfunction has occurred within the proportioning valve.

To test the valve, install gauge set in brake line between master cylinder and proportioning valve, and at output end of proportioning valve and brake line. Be sure all joints are fluid tight.

Have a helper exert pressure on brake pedal (holding pressure). Obtain a reading on master cylinder output of approximately 800 psi. While pressure is being held as above, reading on valve outlet should be 530–570 psi. If the pressure readings do not meet these specifications, the valve should be removed and a new valve installed.

DELCO-MORAINE SINGLE PISTON

The caliper assembly, Fig. 1, slides on its mounting bolts. Upon brake application, fluid pressure against the piston forces the inboard shoe and lining assembly against the inboard side of the disc. This action causes the caliper assembly to slide until the outboard lining comes into contact with the disc. As pressure builds up, the linings are pressed against the disc with increased force.

Caliper Removal

1. Siphon enough brake fluid out of the master cylinder to bring fluid level to 1/3 full to avoid fluid overflow when the caliper piston is pushed back into its bore.
2. Raise vehicle and remove front wheels.
3. Using a "C" clamp, as illustrated in Fig. 2, push piston back into its bore.
4. Remove two mounting bolts and lift caliper away from disc.

Brake Shoe Removal

1. Remove caliper assembly as outlined above.
2. Remove inboard shoe. Dislodge outboard shoe and position caliper on the front suspension so the brake hose will not support the weight of the caliper.
3. Remove shoe support spring from piston.
4. Remove two sleeves from inboard ears of the caliper.
5. Remove four rubber bushings from the grooves in each of the caliper ears.

Brake Shoe Installation

1. Lubricate new sleeves, rubber bushings, bushing grooves and mounting bolt ends with Delco Silicone Lube or its equivalent.
2. Install new bushings and sleeves in caliper ears.

NOTE: Position the sleeve so that the end toward the shoe is flush with the machined surface of the ear.

3. Install shoe support spring in piston cavity, Fig. 3.
4. Position inboard shoe in caliper so spring ends centrally contact shoe edge. Initially, this will place the shoe on an angle. Push upper edge of shoe down until shoe is flat against caliper. When properly seated, spring ends will not extend past shoe more than .100".
5. Position outboard shoe in caliper with shoe ears over caliper ears and tab at bottom of shoe engaged in caliper cutout.
6. With shoes installed, lift caliper and rest bottom edge of outboard lining on outer edge of brake disc to be sure there is no clearance between outboard shoe tab and caliper abutment.
7. Using a 1/4" × 1" × 2 1/2" metal bar to bridge caliper cutout, clamp outboard shoe to caliper with a "C" clamp.
8. Bend both ears of outboard shoe over caliper until clearance between shoe ear and caliper (measured at both the edge and side of the caliper) is .005" or less, Fig. 4.
9. Remove "C" clamp and install caliper.

Disassembling Caliper

1. Remove caliper as outlined above.
2. Disconnect hose from steel line, remove U shaped retainer and withdraw hose from frame support bracket.
3. After cleaning outside of caliper, remove brake hose and discard copper gasket.
4. Drain brake fluid from caliper.
5. Pad caliper interior with clean shop towels and use compressed air to remove piston, Fig. 5.

NOTE: Use just enough air pressure to ease piston out of bore. Do not blow piston out of bore.

CAUTION: Do not place fingers in front of piston in an attempt to catch or protect it when applying compressed air. This could result in serious injury.

6. Carefully pry dust boot out of bore.
7. Using a small piece of wood or plastic, remove piston seal from bore.

NOTE: Do not use a metal tool of any kind to remove seal as it may damage bore.

8. Remove bleeder valve.

Assembling Caliper

1. Lubricate caliper piston bore and new piston seal with clean brake fluid. Position seal in bore groove.
2. Lubricate piston with clean brake fluid and assemble a new boot into the groove in the piston so the fold faces the open end of the piston, Fig. 6.
3. Using care not to unseat the seal, insert piston into bore and force the piston to the bottom of the bore.
4. Position dust boot in caliper counterbore and install, Fig. 7.

NOTE: Check the boot installation to be sure the retaining ring moulded into the boot is not bent and that the boot is installed below the caliper face and evenly all around. If the boot is not fully installed, dirt and moisture may enter the bore and cause corrosion.

5. Install the brake hose in the caliper using a new copper gasket.
6. Install shoes and re-install caliper assembly.

Caliper Installation

1. Position caliper over disc, lining up holes in caliper with holes in mounting bracket. If brake hose was not disconnected during removal, be sure not to kink it during installation.
2. Start mounting bolts through sleeves in inboard caliper ears and the mounting bracket, making sure ends of bolts pass under ears on inboard shoe.

NOTE: Right and left calipers must not be interchanged.

3. Push mounting bolts through to engage holes in the outboard ears. Then thread mounting bolts into bracket.
4. Torque mounting bolts to 30–40 ft. lbs.
5. If brake hose was removed, reconnect it and bleed the calipers.

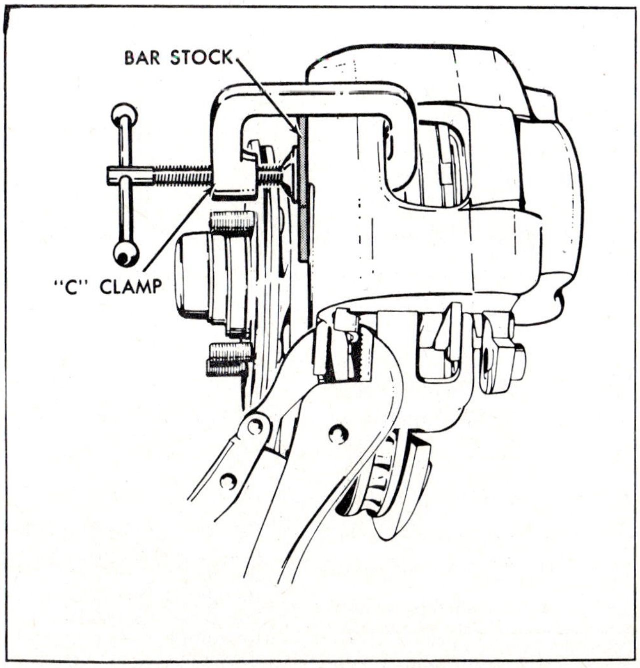

Fig. 4 Fitting shoe to caliper

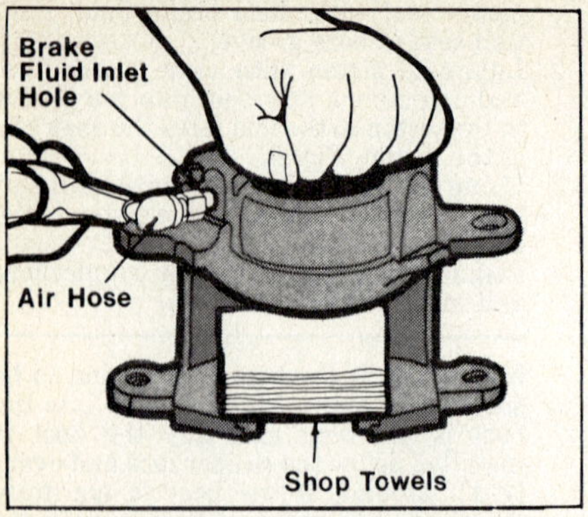

Fig. 5 Removing piston from caliper

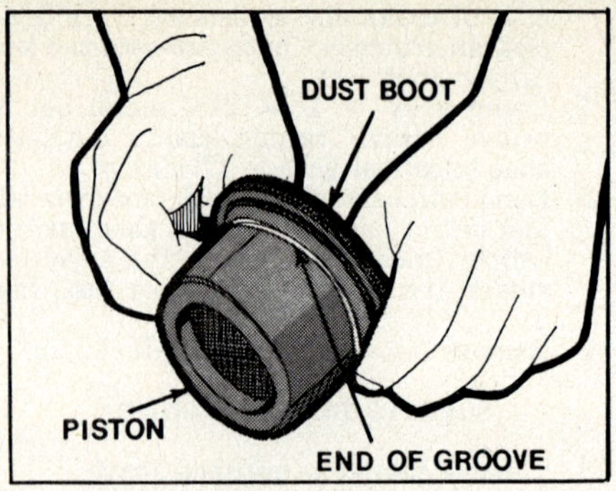

Fig. 6 Installing boot to piston

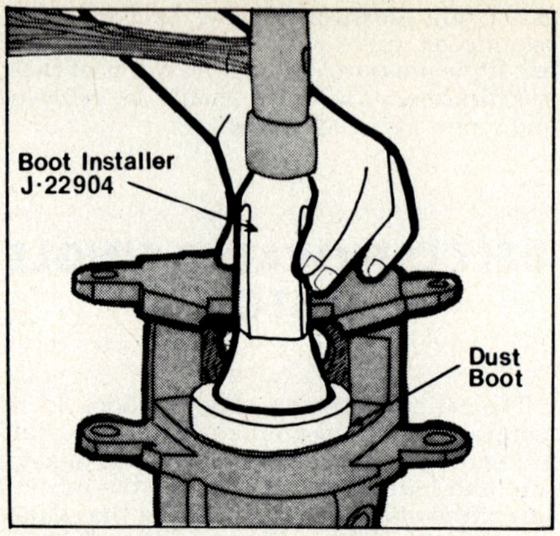

Fig. 7 Installing boot to caliper

6. Replace front wheels, lower vehicle and add brake fluid to master cylinder to bring level to 1/4″ from top.

NOTE: Before moving vehicle, pump brake pedal several times to be sure it is firm. Do not move vehicle until a firm pedal is obtained.

DODGE RAMPAGE & PLYMOUTH SCAMP FLOATING CALIPER

The single piston floating caliper disc brake assembly consists of the hub and disc brake rotor assembly, caliper, shoes and linings, splash shield and adapter, Fig. 8.

The caliper assembly floats on two rubber bushings riding on two steel guide pins threaded into the adapter. The bushings are inserted on the inboard portion of the caliper. Two machined abutments on the adapter position and align the caliper fore and aft. Guide pins and bushings control caliper and piston seal movement to assist in maintaining proper shoe clearance.

All braking force is taken directly by the adapter.

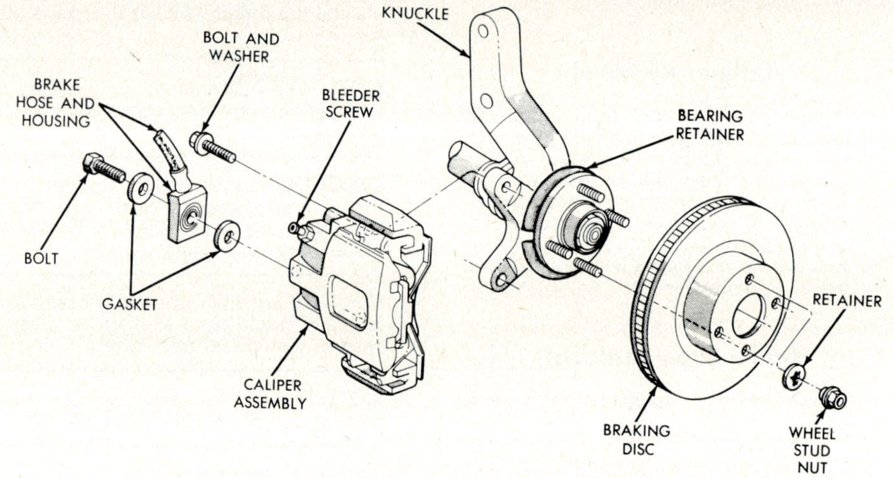

Fig. 8 Floating caliper disc brake. Dodge Rampage & Plymouth Scamp

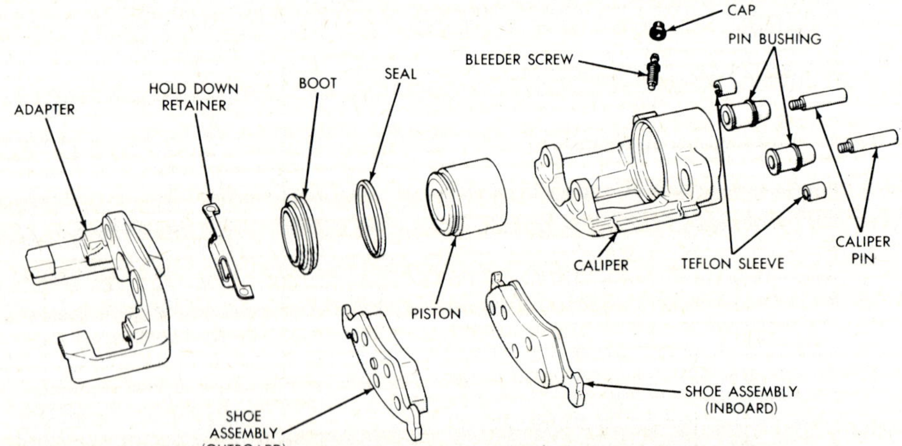

Fig. 9 Disassembled view of disc brake caliper assembly. Dodge Rampage & Plymouth Scamp

Brake Shoe & Lining, Replace

1. Raise and support vehicle, then remove wheel and tire assemblies.
2. Remove hold down spring from caliper assembly by pressing spring outward.
3. Loosen, but do not remove, caliper guide pins, then remove caliper from disc. Inboard shoe will remain inside caliper. Support caliper assembly to prevent damage to hydraulic brake hose.

NOTE: Remove caliper guide pins only if bushings or sleeves are to be replaced.

4. Remove inboard shoe from caliper and outboard shoe from adapter.
5. Push caliper piston into bore.

NOTE: Remove some brake fluid from master cylinder reservoir to prevent overflowing when piston is pushed into bore.

6. Install new inboard shoe into caliper with retainer positioned in piston bore.
7. Install outboard shoe onto adapter.
8. Position caliper over brake disc and adapter, then torque guide pins to 18–22 ft. lbs.
9. Install hold down spring, then install tire and wheel assemblies and lower vehicle to ground.
10. Check master cylinder reservoir for proper level of brake fluid and add as necessary.

Caliper Overhaul

Disassemble, Fig. 9

1. Remove caliper assembly as described under Brake Shoe & Lining, Replace.
2. With brake hose attached to caliper, carefully depress brake pedal to push piston out of caliper bore. Prop brake pedal to

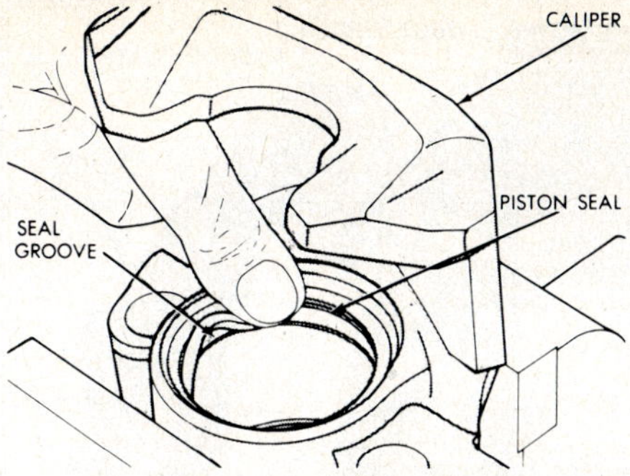

Fig. 10 Installing piston seal. Dodge Rampage & Plymouth Scamp

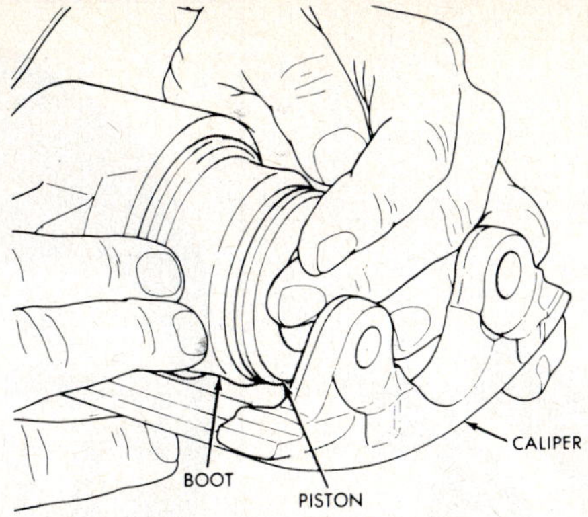

Fig. 11 Installing caliper piston. Dodge Rampage & Plymouth Scamp

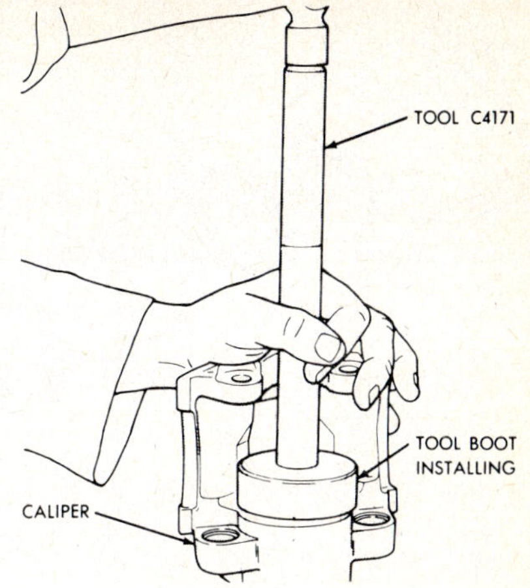

Fig. 12 Installing piston dust boot. Dodge Rampage & Plymouth Scamp

any position below first inch of brake pedal travel to prevent brake fluid loss.

3. If pistons are to be removed from both calipers, disconnect brake hose at frame bracket after removing piston, then cap brake line and repeat procedure to remove piston from other caliper.
4. Disconnect brake hose from caliper.
5. Mount caliper in a soft jawed vise.
6. Support caliper and remove dust boot and discard.
7. Using a small wooden or plastic stick, remove seal from groove in piston bore and discard.
8. Using a suitable tool, remove bushings from caliper.

Inspection

1. Clean all components using alcohol or other suitable cleaning solvent, then blow dry using compressed air. With compressed air blow out drilled passages and bores.
2. Inspect piston bore for pitting or scoring. Light scratches or corrosion can usually be cleared with crocus cloth. Bores that have deep scratches or scoring should be honed with tool No. C-4095, providing bore diameter is not increased by more than .001 in. If scratches or scoring cannot be cleared up, or if caliper bore is increased by more than .001 in., replace caliper housing.

NOTE: When using hone C-4095, coat hone and caliper bore with clean brake fluid. After honing carefully clean boot and seal grooves with a stiff non-metallic brush. Flush caliper with clean brake fluid and wipe dry with a clean lintless cloth, then flush and wipe caliper dry again.

---

3. Replace piston if found to be scored, pitted or if plating is severely worn or if caliper bore was honed. Black stains on steel piston are caused by piston seal and are not cause for replacing piston.

Assemble, Fig. 9

1. Mount caliper in a soft jawed vise.
2. Lubricate piston seal with clean brake fluid and install seal in caliper bore groove, Fig. 10. Ensure seal is properly seated.
3. Position dust boot over piston.
4. Install piston into bore by pushing it evenly past piston seal until it bottoms in bore, Fig. 11.
5. Position dust boot in counterbore, then using a hammer and tools C-4682 and C-4171 or equivalent, drive boot into counterbore, Fig. 12.

8. Compress flanges of guide pin bushings and install bushings on caliper housing. Ensure that bushing flanges extend evenly over caliper housing on both sides.

NOTE: Remove teflon sleeves from guide pin bushings prior to installing bushings into caliper. After bushings are installed into caliper, reinstall teflon sleeves into bushings.

---

9. Connect brake hose to brake line at frame bracket.
10. Install caliper on vehicle as described under Brake Shoe & Lining, Replace.
11. Check brake fluid level of master cylinder reservoir, then open caliper bleed screw and bleed brake system. Continue bleeding procedure until firm pedal is obtained.

FORD RANGER SLIDING CALIPER

Front Caliper & Lining Removal

Fig. 13

1. Remove portion of brake fluid from larger master cylinder reservoir to avoid fluid overflow when caliper piston is depressed. Discard removed fluid.
2. Raise and support vehicle and remove front wheel and tire assemblies.
3. Working on either side of vehicle, place an eight inch "C" clamp or equivalent on caliper and tighten clamp to bottom caliper piston in cylinder bore, then remove clamp.

NOTE: Do not use screwdriver or similar tool to pry piston away from rotor.

---

4. Two types of caliper pins are used: a single tang type and a double tang type. The pin removal process depends upon the direction of the pin, bolt head. Remove upper caliper pin first.
5. If bolt head is on outside of caliper use following procedure.
   a. From inner side of caliper, tap bolt within caliper pin until bolt head on outer side of caliper shows a separation between bolt head and caliper pin.

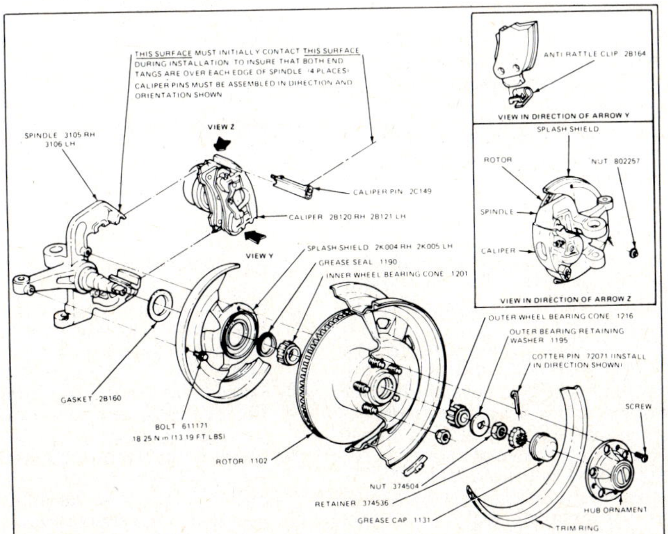

Fig. 13 Disc Brake Assembly. Ford Ranger

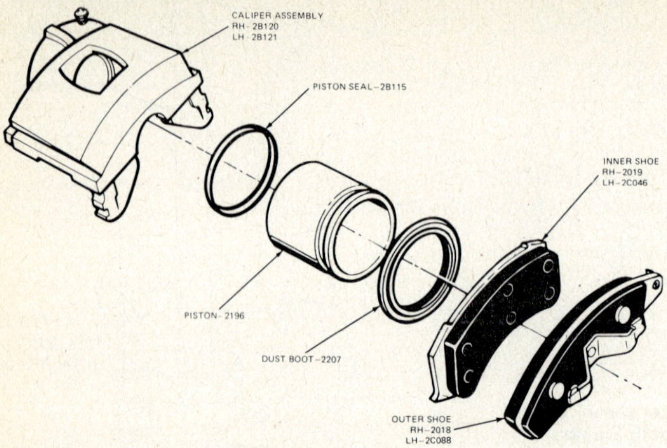

Fig. 14 Disc Brake Caliper Assembly. Ford Ranger

b. Using suitable cutter, remove bolt head from bolt.
c. Depress tab on bolt head end of upper caliper pin with a screwdriver while tapping on pin with hammer until tab is depressed by v-slot.
d. Place one end of suitable punch against end of caliper pin and drive pin out of caliper toward inside of vehicle. Do not use edged tool to drive out pin.

6. If the nut end of the bolt is on the outside of the caliper, use the following procedure.
   a. Remove nut from bolt.
   b. Depress lead tang on end of upper caliper pin with screwdriver while tapping on pin with hammer until lead tang is depressed by v-slot.
   c. Place one end of punch against end of caliper pin and drive caliper pin out of caliper toward inside of vehicle. Do not use edged tool to drive caliper pin.
7. Remove lower caliper pin using procedure in steps 5 and 6.
8. Remove caliper from rotor and remove brake hose from caliper, then remove outer lining, anti-rattle clips and inner lining.

Front Caliper & Lining Installation

1. Place new anti-rattle clip on lower end of inner shoe, being sure tabs on clip are properly positioned and clip is fully seated.
2. Place inner shoe and anti-rattle clip in shoe abutment with anti-rattle clip tab against shoe abutment and loop type spring away from rotor. Compress anti-rattle clip and slide upper end of shoe into position.
3. Install outer shoe, being sure torque buttons on shoe spring clip are firmly seated in matching holes in caliper.
4. Clean mounting surfaces of caliper and, using suitable lubricant, lubricate caliper grooves, then install new caliper pins. Pin must be installed with lead tang in first, bolt head facing outward. Place lead tang in V-slot mounting surface and drive into caliper until drive tang is flush with caliper assembly, then install nut and torque to 32–47 ft. lbs.
5. Install brake hose on caliper.
6. Repeat steps 3–8 of removal procedure and steps 1–5 of installation procedure on opposite side of vehicle.
7. Bleed brakes, then install wheel and tire assemblies and torque nuts to 85–115 ft. lbs.
8. Lower vehicle, check brake fluid level and replenish as necessary, then check brakes for proper operation.

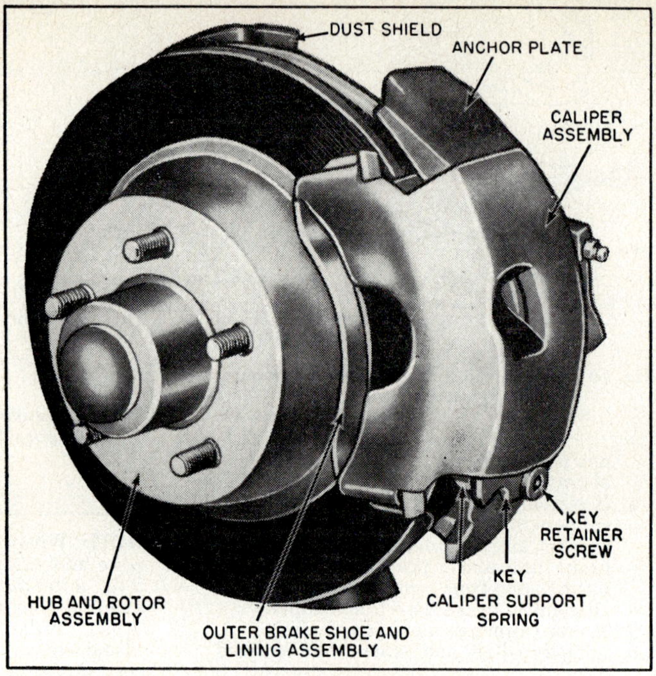

Fig. 15 Ford single piston sliding caliper disc brake

Front Caliper, Overhaul

Fig. 14

1. Remove caliper assembly as described in caliper and lining removal.
2. Placing cloth over piston to prevent damage to piston, apply air pressure to fluid port in caliper with suitable tool to remove piston. If piston is seized and cannot be forced from caliper, tap lightly around piston while applying air pressure.

CAUTION: Piston may develop considerable force from pressure build-up.

---

3. Remove dust boot from caliper assembly, then remove and discard rubber piston seal of cylinder.
4. Using suitable solvent, clean all metal parts, then clean out and dry grooves and passageways with compressed air.
5. Check cylinder bore and piston for damage or excessive wear and replace piston if it is pitted, scored, or corroded, or the plating is worn away.
6. Apply film of clean brake fluid on new caliper piston seal and install it in cylinder bore, being sure seal does not become twisted but is firmly seated in groove.
7. Install new dust boot by setting flange squarely in outer groove of caliper bore.
8. Coat piston with brake fluid and install in cylinder bore, spreading dust boot over piston as it is installed, seating dust boot in piston groove.
9. Install caliper as described in front caliper and lining installation.

FORD SINGLE PISTON SLIDING CALIPER

The caliper assembly is made up of a sliding caliper housing assembly and an anchor plate, Fig. 15.

The anchor plate is bolted to the wheel spindle arm. Two angular machined surfaces on the upper end of the caliper housing contact mating machined surfaces of the anchor plate. A steel, plated key and a caliper support spring is fitted between the angular machined surfaces of the lower end of the caliper and the machined surface of the anchor plate. The key is held in position with a retaining screw. The caliper is held in position against the mating surfaces of the anchor plate by means of the caliper support spring. A brake shoe anti-rattle spring clip is provided on the anchor plate at the lower end of the inner brake shoe and lining assembly. The inner and outer brake shoe assemblies are not interchangeable.

The sliding caliper contains a single cylinder and a piston with a molded dust boot to seal the cylinder bore from contamination. A square section rubber piston seal is positioned in a groove in cylinder bore to provide sealing between cylinder and piston.

Caliper Removal

1. Raise car and support with safety stands. Block both rear wheels if a jack is used.
2. Remove wheel and tire assembly from hub.
3. Disconnect brake hose from caliper.
4. Remove retaining screw from caliper retaining key, Fig. 15.
5. Slide caliper retaining key and support spring either inward or outward from anchor plate. Use hammer and drift, if necessary, to remove the key and caliper support spring. Use care to avoid damaging the key.
6. Lift caliper assembly away from anchor

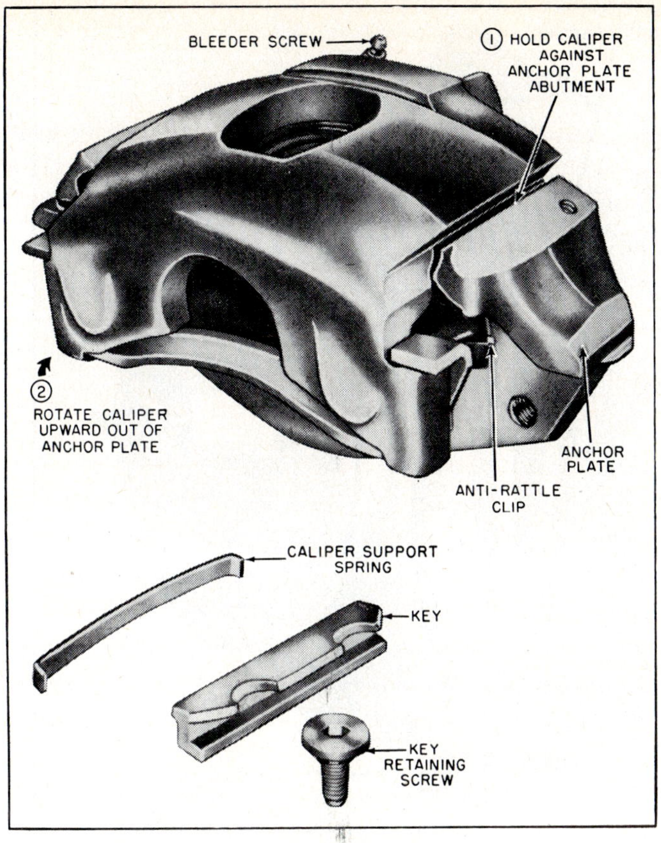

Fig. 16 Removing caliper assembly

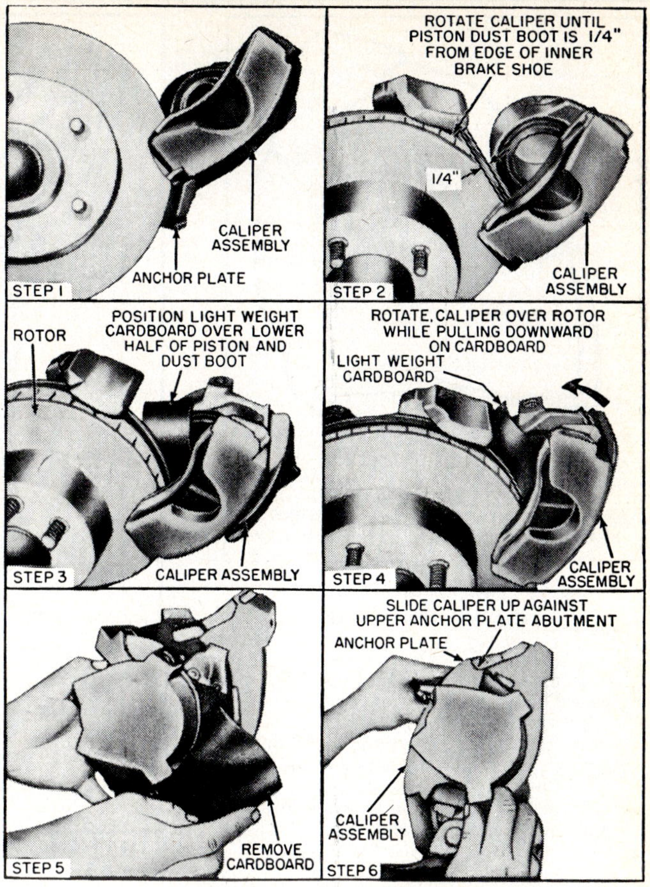

Fig. 18 Installing caliper assembly

plate by pushing caliper down against anchor plate and rotate upper end upward out of anchor plate, Fig. 16.

7. Remove inner shoe and lining from anchor plate. The brake shoe anti-rattle clip (inner shoe only) may become displaced at this time and if so, reposition it on anchor plate. Fig. 17. Tap lightly on outer shoe and lining to free it from caliper.
8. Clean caliper, anchor plate and rotor assemblies and inspect them for signs of fluid leakage, wear or damage. If either lining is worn to within 1/32" of any rivet head, both shoe and lining assemblies must be replaced. Also, if necessary to replace shoes and lining on one wheel, they must be replaced on both wheels to maintain equal brake action.

Caliper Installation

1. If new shoe and lining assemblies are to be installed, use a 4" C-clamp and a block of wood 1¾" × 1" and about ¾" thick to seat the caliper piston in its bore. This must be done to provide clearance for the caliper to fit over new shoes when installed.
2. Be sure brake shoe anti-rattle clip is in place on lower inner brake shoe support on anchor plate with pigtail of clip toward inside of anchor plate. Position inner shoe and lining on anchor plate with lining toward rotor, Fig. 17.
3. Install outer shoe and lining with lower flange ends against the caliper leg abutments and the brake shoe upper flanges over the shoulders on caliper legs. The shoe upper flanges fit tightly against the shoulder machined surfaces. If the same brake shoes and linings are to be used, be sure they are installed in their original positions.
4. Remove C-clamp if used, from the caliper (the piston will remain seated in its bore).
5. Position caliper housing lower V groove on anchor plate lower abutment surface, Fig. 18. Refer to Figs. 18 and 19 to complete assembly following steps shown. Connect brake hose, bleed brakes and replace wheel.

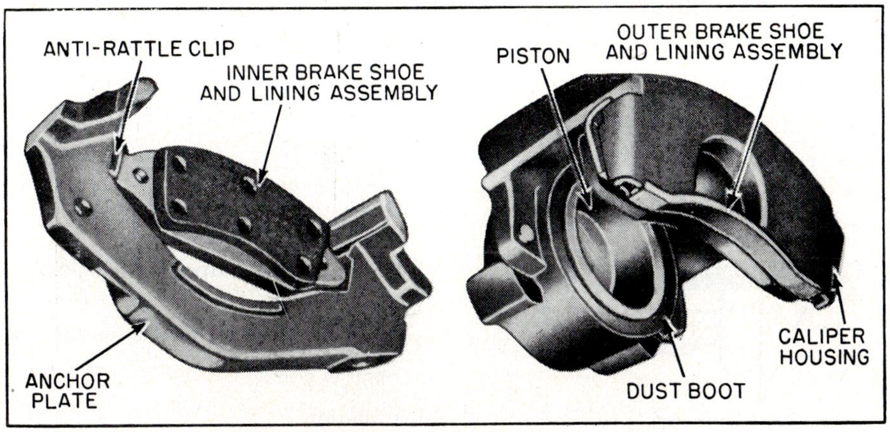

Fig. 17 Caliper and outer shoe removed from anchor plate

Brake Shoe & Lining, Replace

The procedure to replace the shoe and lining assemblies is the same as the caliper removal discussed previously with the exception that it is not necessary to disconnect the brake hose. Use care to avoid twisting or stretching the brake hose.

Hub & Rotor, Remove

1. Remove caliper and shoes as previously described. If no repairs are necessary on the caliper it is not necessary to disconnect the brake hose. The caliper can be temporarily secured to the upper suspension arm. Do not remove the anchor plate and be careful not to stretch or twist the brake hose.
2. Remove grease cap from wheel spindle

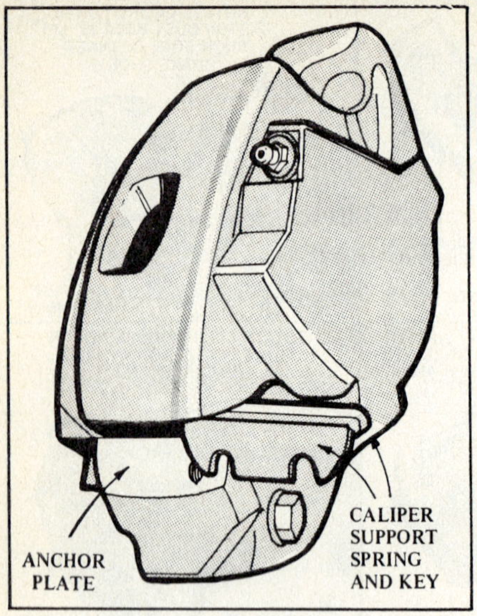

Fig. 19 Installing caliper support spring and retaining key

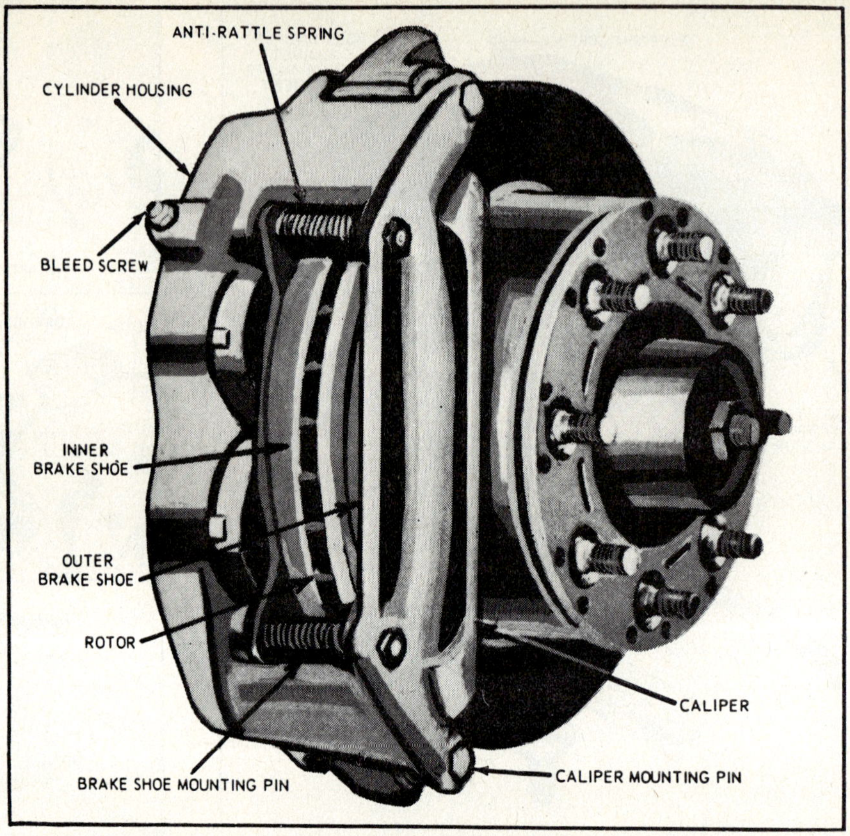

Fig. 20 Floating caliper disc brake

and remove cotter pin and nut lock from wheel bearing adjustment nut.

3. Remove wheel bearing adjusting nut and grasp the hub and rotor and pull it out far enough to loosen the washer and outer wheel bearing. Then push it back in and remove the washer, outer wheel bearing and remove the hub and rotor.

FORD (DAYTON) FLOATING CALIPER

The disc brake, Fig. 20, is the floating caliper design with two pistons on one side of the rotor. It is a two piece unit consisting of the caliper and cylinder housing. The caliper is mounted to the anchor plate on two mounting pins and bushings in the anchor plate. The bushings and pins are protected by boot type seals.

Two brake shoe and lining assemblies are used in each caliper, one on each side of the rotor. The shoes are identical and are attached to the caliper with two mounting pins.

The cylinder housing contains the two pistons. The pistons and cylinder bores are protected by boot seals which are fitted to a groove in the piston and attached to the cylinder housing with retainers. The cylinder assembly is attached to the caliper with two cap screws and washers.

The anchor plate is bolted directly to the spindle. It positions the caliper assembly over the rotor forward of the spindle.

Removal, Shoe & Lining, Replace

Replace shoe and lining assemblies when worn to a minimum of 1/16" thickness. Combined thickness of shoe and lining 1/4" minimum.

1. Remove the shoe and lining mounting pins, anti-rattle springs and old shoe and lining assemblies.

Installation

1. Remove master cylinder cover.
2. Loosen piston housing to caliper mounting bolts sufficiently to allow installation of new shoes. Do not move pistons.

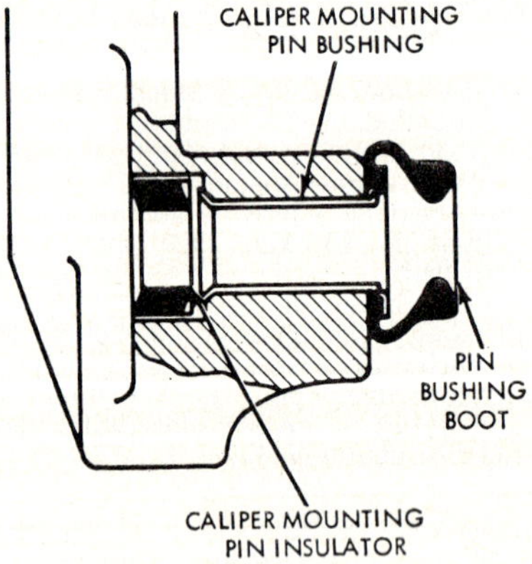

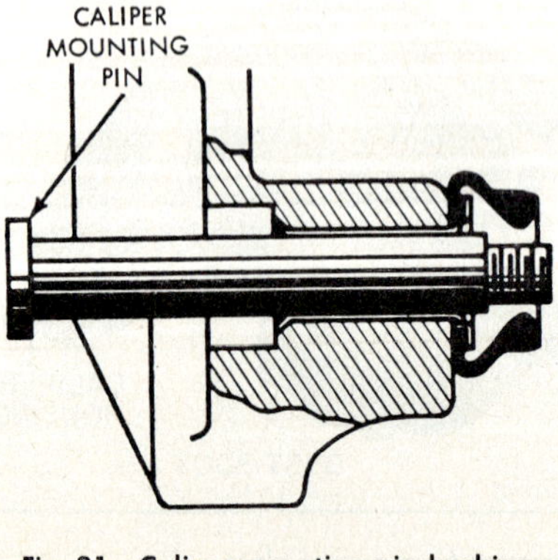

Fig. 21 Caliper mounting pin bushing

3. Install new shoes, shoe mounting pins and anti-rattle springs. Be sure spring tangs are located in holes provided in shoe plates.
4. Tighten shoe mounting pins to 17–23 ft. lbs.
5. Reset pistons to correct location in cylinders by placing shims or feeler gauge of .023–.035" between shoe plate of outboard shoe and the caliper. Then retighten piston housing to caliper mounting bolts. Keep cylinder housing square with caliper. Check master cylinder for fluid overflow.
6. Loosen housing to caliper mounting bolts and remove shims. Tighten mounting bolts to 155–185 ft. lbs.
7. Check master cylinder fluid level and replace cover.

Caliper Removal

1. Raise front of vehicle and remove wheel cover.
2. Remove wheel and tire assembly.
3. Remove pins and nuts retaining caliper to anchor plate.
4. Disconnect brake hose from caliper and remove caliper.
5. Reverse procedure to install and bleed brakes.

Caliper Mounting Pin Bushing

Removal

1. Crimp chamfered end of bushing and remove it from anchor plate boss with a pair of pliers.

Installation

1. Assemble pin bushing boot onto caliper mounting pin bushing.

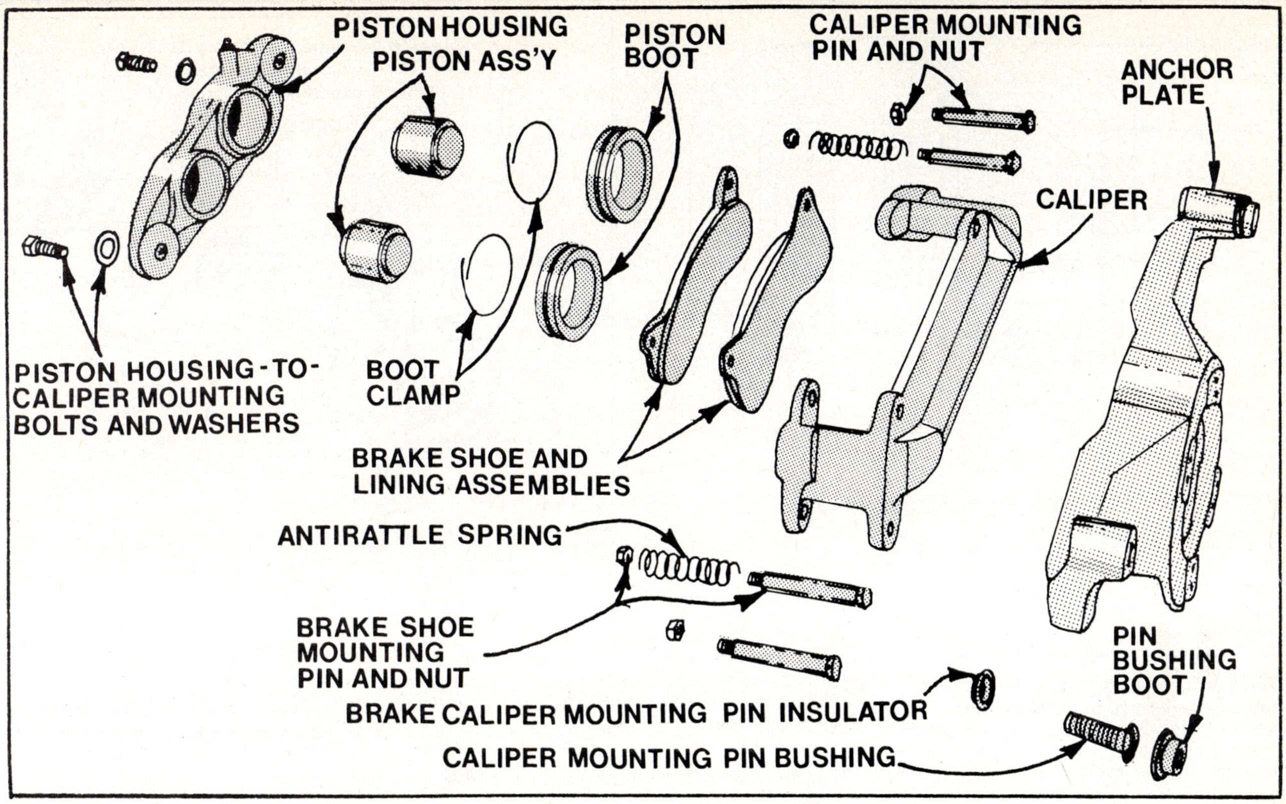

Fig. 22 Ford (Dayton) floating caliper disc brake disassembled (Typical)

2. Insert pin bushing into anchor plate boss, Fig. 21.
3. Peen end of bushing to conform to chamfered end of anchor plate boss. Use a large ball bearing or similar object.
4. Insert caliper mounting pin into bushing from chamfered end and tap lightly around circumference of pin with a rubber mallet to insure that bushing conforms to anchor plate boss and pin is free in bushing.
5. Insert caliper mounting pin insulator into large end of anchor plate boss.
6. Lubricate bushing with chassis lubricant.

Caliper Service

If caliper is leaking, the piston assemblies must be removed from the housing and replaced, Fig. 22. If cylinder bores are scored, corroded or excessive wear is evident, the piston housing must be replaced. Do not hone the cylinder bores as piston assemblies are not available for oversize bores. The piston housing must be removed from the caliper for replacement.

Disc Hub & Rotor Removal

1. Remove caliper as outlined previously.
2. Remove the dust cap, cotter pin, nut, washer and outer bearing and remove rotor from spindle.
3. Remove inner bearing cone and seal.
4. Reverse procedure to install.

KELSEY-HAYES FLOATING CALIPER

This type brake is a floating caliper, single piston, ventilated unit, actuated by the hydraulic system, Fig. 23. The caliper assembly, Fig. 24, is made up of a floating caliper assembly and an anchor plate. The anchor plate is bolted to the wheel spindle arm by two bolts. The caliper is attached to the anchor plate through two spring steel stabilizers. The caliper slides on two guide pins which also attach to the stabilizers. A single piston is used. The cylinder bore contains a piston with a molded rubber dust boot to seal the cylinder bore from contamination and also to return the piston to the released position when hydraulic pressure is released. Also a rubber piston seal is used to provide sealing between cylinder and piston.

Fig. 23 Single piston disc brake (typical)

Service Precautions

In addition to the precautions described at the beginning of this chapter, the following must be observed.

1. If the piston is removed for any reason the piston seal must be replaced.
2. During removal and installation of a wheel assembly, use care not to interfere with and damage the caliper splash shield or the bleeder screw fitting.
3. Be sure the vehicle is centered on the hoist before servicing any front end components to avoid bending or damaging the rotor splash shield on full right or left wheel turns.
4. The proportioning valve should not be disassembled or adjustments attempted on it.
5. The wheel and tire must be removed separately from the brake rotor.
6. The caliper assembly must be removed from the spindle prior to removal of shoe and lining assembly.
7. Do not attempt to clean or restore oil or grease soaked brake linings. When contaminated linings are found, linings must be replaced in complete axle sets.

Remove & Disassemble Caliper

1. Raise vehicle and remove front wheels.
2. Remove inboard shoe hold down clips and slide outer retaining clips from shoe retaining pins, Fig. 25.
3. Disconnect brake line.
4. Remove guide pins and bolts, if necessary, and remove caliper assembly from stabilizer.
5. Lift caliper away from disc.
6. Slide shoes and lining assemblies out of caliper.

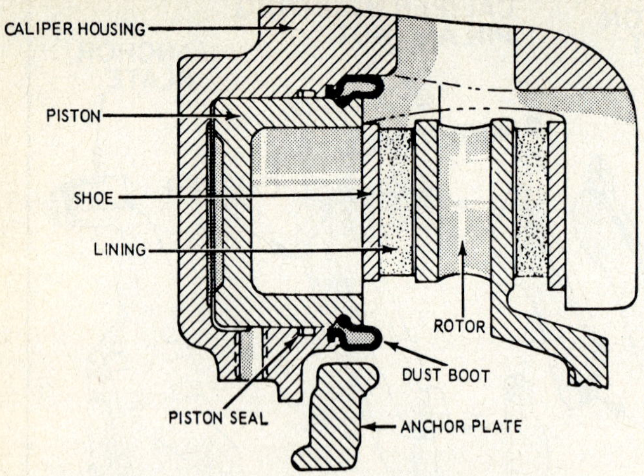

Fig. 24 Single piston disc brake caliper (typical)

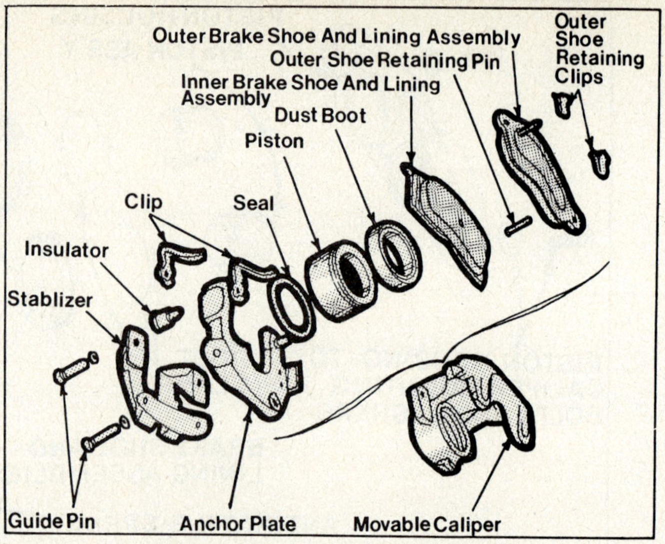

Fig. 25 Single piston disc brake caliper disassembled (typical)

7. Remove guide pin insulators from anchor plate.

NOTE: If necessary to remove the piston, apply air pressure to the fluid port in the caliper, Fig. 26, to remove the piston. Place a cloth over the piston to prevent damage to the piston. If the piston is seized and cannot be forced from the caliper, tap lightly around the piston before applying air pressure. Care should be taken because the piston can develop considerable force due to pressure build-up.

---

8. Remove dust boot from caliper.
9. Remove rubber piston seal from cylinder and discard it.

Assemble & Install Caliper

1. Apply a film of clean brake fluid to the new caliper piston seal and install it in cylinder bore. Be sure seal does not become twisted and that it is seated fully in the groove.
2. Install a new dust boot by setting the flange squarely in the outer groove of the caliper bore.
3. Coat piston with brake fluid and install in cylinder bore. Spread dust boot over piston as it is installed. Seat dust boot in piston groove.
4. Position inner brake shoe so that ears of shoe rests on top of anchor plate bosses and beneath hold-down springs.
5. Install new caliper guide pin insulators in anchor plate.
6. Position caliper on anchor plate.
7. Install guide pins loosely in anchor plate, being sure guide pins are free of oil, grease or dirt.
8. Install caliper on spindle.

Brake Shoes & Linings, Install

NOTE: When new shoe and lining assemblies are being installed to replace worn linings it will be encessary to push the piston all the way into the caliper bore. This will displace fluid from the caliper into the master cylinder reservoir. Check the primary (front) brake system reservoir level and remove fluid to approximately half full before replacing brake shoes. This will prevent overflow. Do not re-use the removed fluid.

---

1. Install new caliper guide pin insulators in anchor plate.
2. Position caliper in anchor plate.
3. Install caliper guide pins loosely in anchor plate, being sure they are free of oil, grease or dirt.
4. Position outer brake shoe on caliper and install two retaining pins and clips.
5. Install inner brake shoe so that ears of shoe are on top of anchor plate bosses and under shoe hold-down springs.
6. Position shoe and lining assemblies so that caliper can be placed over rotor. Rotate hammer handle between linings to provide proper clearance.
7. Install caliper over rotor and on spindle. Install and tighten the two caliper bolts, tightening the upper bolt first. Install safety wire and twist ends at least five turns. With moderate pressure applied to brake pedal, tighten stabilizer attaching screws and caliper guide pins.

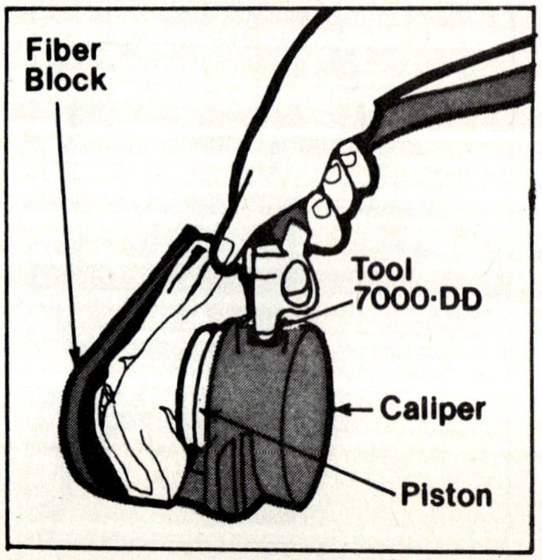

Fig. 26 Removing piston with air pressure

KELSEY-HAYES SINGLE PISTON SLIDING CALIPER

This sliding caliper single piston system uses a one piece hub and is actuated by the hydraulic system and disc assembly, Fig. 27. Alignment and positioning of the caliper is achieved by two machined guides or "ways" on the adaptor, while caliper retaining clips allow lateral movement of the caliper, Fig. 28. Outboard shoe flanges are used to position and locate the shoe on the caliper fingers, Fig. 29, while the inboard shoe is retained by the adaptor, Fig. 30. Braking force applied onto the outboard shoe is transferred to the caliper, while braking force applied onto the inboard shoe is transferred directly to the adaptor.

A square cut piston seal provides a hydraulic seal between the piston and the cylinder bore, Fig. 27. A dust boot with a wiping lip installed in a groove in the cylinder bore and piston, prevents contamination in the piston and cylinder bore area. Adjustment between the disc and the shoe is obtained automatically by the outward relocation of the piston as the inboard lining wears and inward movement of the caliper as the outboard lining wears.

Caliper Removal

1. Raise the vehicle and remove front wheel.
2. Remove caliper retaining clips and anti-rattle springs, Fig. 28.
3. Remove caliper from disc by slowly sliding caliper assembly out and away from disc.

NOTE: Use some means to support caliper. Do not let caliper hang from hydraulic line.

Brake Shoe Removal

1. Remove caliper assembly as outlined above.
2. Remove outboard shoe by prying between the shoe and the caliper fingers, Fig. 31,

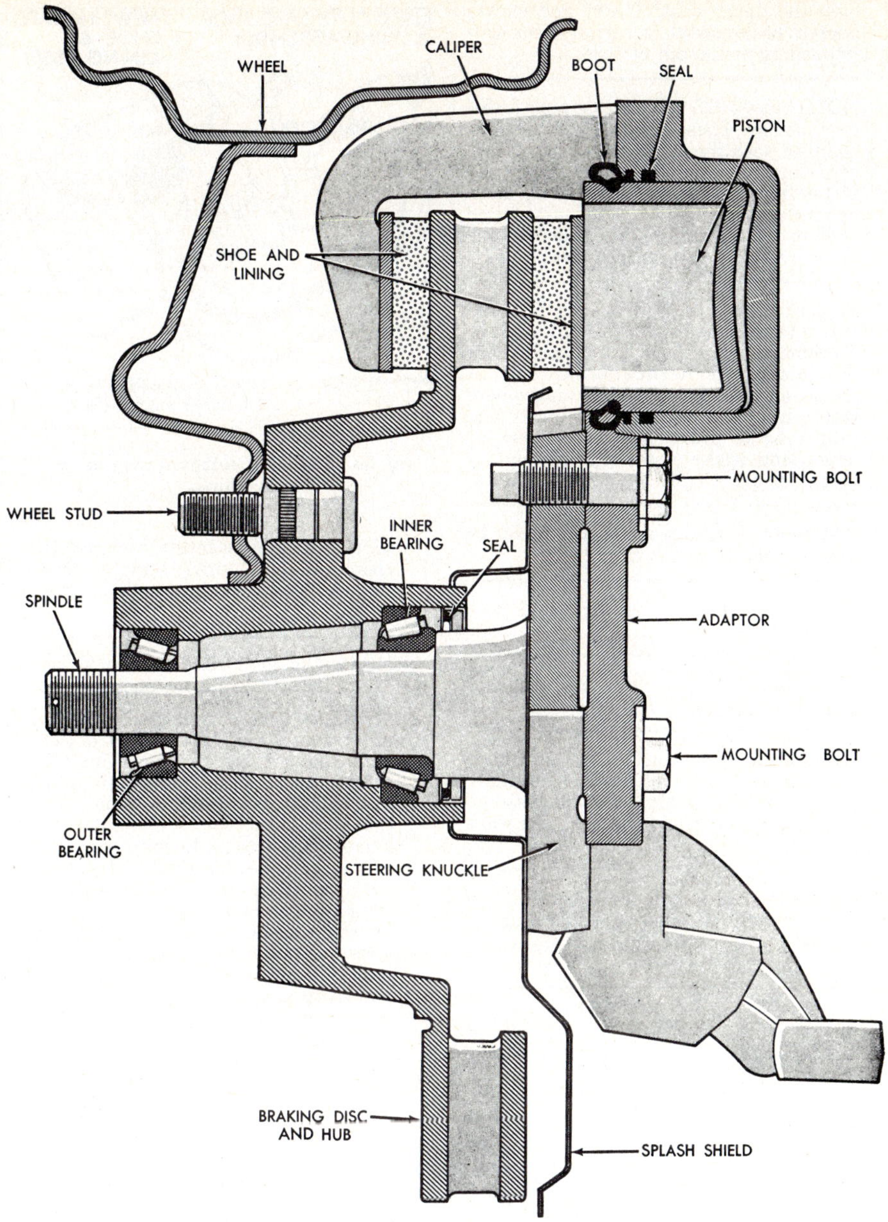

Fig. 27 Sectional view or Kelsey-Hayes single piston sliding caliper disc brake

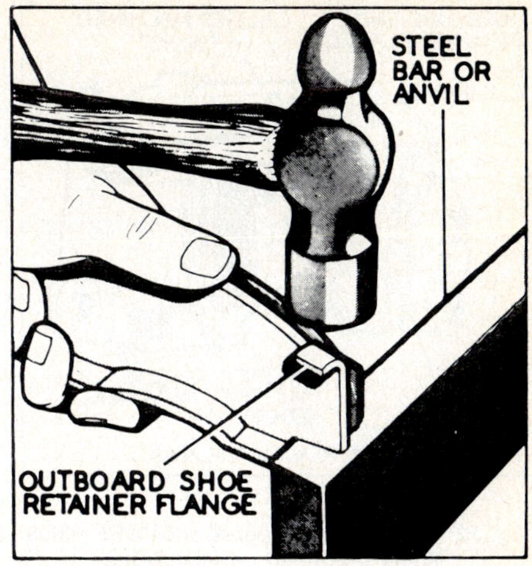

Fig. 29 Fitting outboard shoe retaining flange

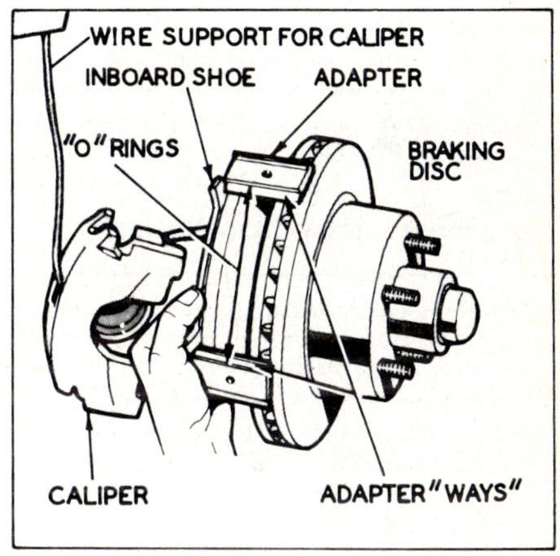

Fig. 30 Replacing inboard shoe

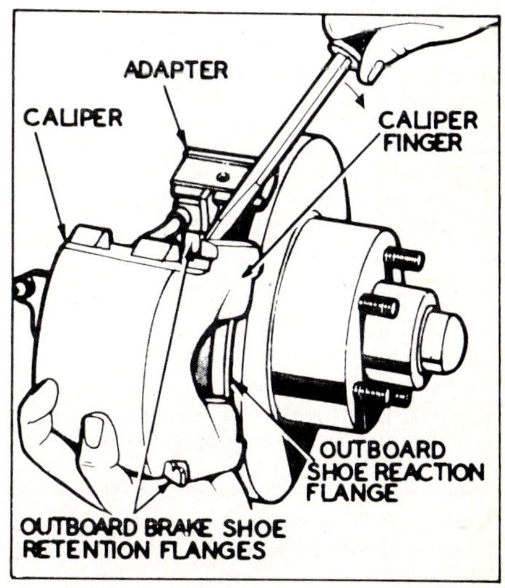

Fig. 31 Removing outboard shoe

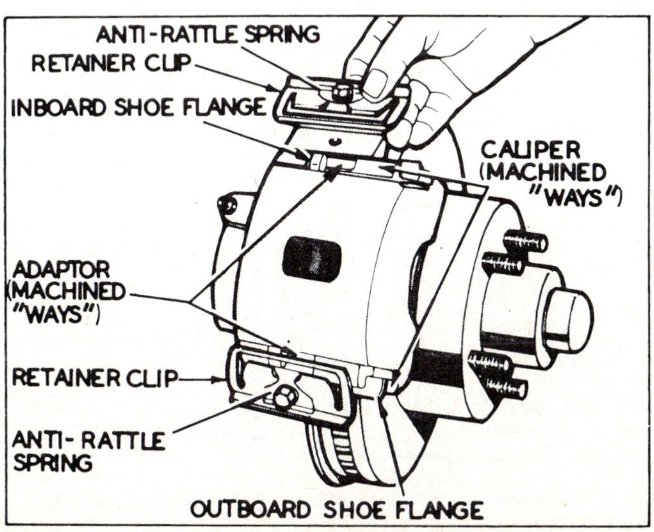

← Fig. 28 Caliper mounted "ways" & assembly retention

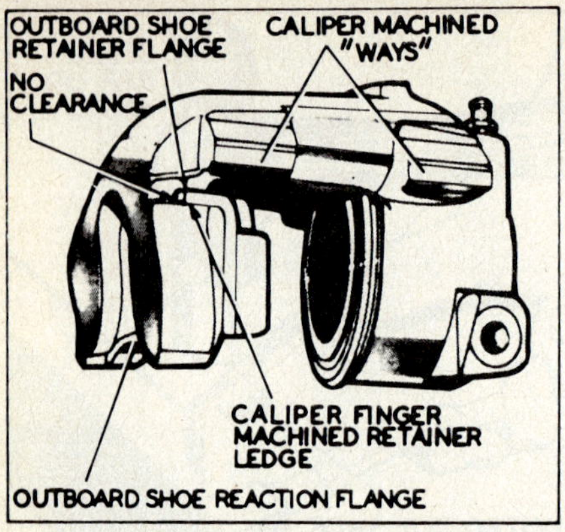

Fig. 32 Positioning outboard shoe onto caliper finger machined retainer ledge

since flanges on outboard shoe retain caliper firmly.

NOTE: Caliper should be supported to avoid damage to the flexible brake hose.

3. Remove inboard brake shoe from the adaptor, Fig. 30.

Brake Shoe Installation

NOTE: Remove approximately 1/3 of the brake fluid out of the reservoir to prevent overflow when pistons are pushed back into the bore.

1. With care, push piston back into bore until bottomed.
2. Install new outboard shoe in recess of caliper.

NOTE: No free play should exist between brake shoe flanges and caliper fingers, Fig. 32.

If up and down movement of the shoe shows free play, shoe must be removed and flanges bent to provide a slight interference fit, Fig. 29. Reinstall shoe after modification, if shoe can not be finger snapped into place, use light "C" clamp pressure, Fig. 33.

3. Position inboard shoe with flanges inserted in adaptor "ways," Fig. 30.
4. Carefully slide caliper assembly into adaptor and over the disc while aligning caliper on machined "ways" of adaptor.

NOTE: Make sure dust boot is not pulled out from groove when piston and boot slide over the inboard shoe.

5. Install anti-rattle springs and retaining clips and torque retaining screws to 180 inch-pounds.

NOTE: The inboard shoe anti-rattle spring is to be installed on top of the retainer spring plate, Fig. 28.

Caliper Disassembly

1. With caliper and shoes removed as described previously, place the caliper onto the upper control arm and slowly depress brake pedal, in turn hydraulically pushing piston out of bore.

NOTE: Some 1975 vehicles use plastic caliper pistons. These pistons can be interchanged with the 3.1 bore caliper assembly. However, do not attempt to pry the piston from the caliper, as this may damage the dust boot groove in the piston, and lead to subsequent brake fluid leakage and corrosion of the caliper bore.

2. Support pedal below first inch of pedal travel to prevent excessive fluid loss.
3. To remove piston from the opposite caliper, disconnect flexible brake line at frame bracket, from vehicle side where piston has been removed previously and plug tube to prevent pressure loss. By depressing brake pedal this piston can also be hydraulically pushed out.

CAUTION: Air pressure should never be used to remove piston from bore.

4. Mount caliper in a vise equipped with protector jaws.

CAUTION: Excessive vise pressure will distort caliper bore.

5. Remove the dust boot, Fig. 34.
6. Insert a suitable tool such as a small, pointed wooden or plastic object between the cylinder bore and the seal and work seal out of the groove in the piston bore.

NOTE: A metal tool such as a screwdriver should not be used since it can cause damage to the piston bore or burr the edges of the seal groove.

Caliper Assembly

1. Before installing the new piston seal in groove of bore, dip seal in Ucon LB1145Y24 lubricant or equivalent.

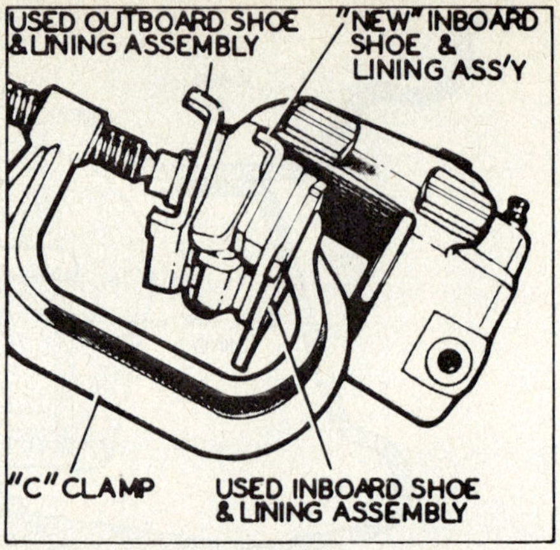

Fig. 33 Installing outboard shoe using "C" clamp

Work seal gently into the groove (using clean fingers) until seal is properly seated, make sure that seal is not twisted or rolled.

NOTE: Old seals should never be reused.

2. With new piston boot lubricated generously with Ucon LB1145Y24 or equivalent. Using finger pressure, install into caliper by pushing into outer groove of the caliper bore. When properly positioned in groove boot will snap into place. Double check to make sure boot is properly installed and seated by running finger around the inside of the boot.
3. Plug high pressure inlet to caliper and bleeder screw hole and coat piston with a generous amount of lubricant. Spread boot with finger and work piston into boot while pressing down on piston. As piston is depressed, entrapped air below piston will force boot around piston and into its groove.
4. Remove the plug and apply uniform force to the piston (avoid cocking piston) until piston bottoms in bore.
5. Install caliper and shoes as described under "Brake Shoe Installation."

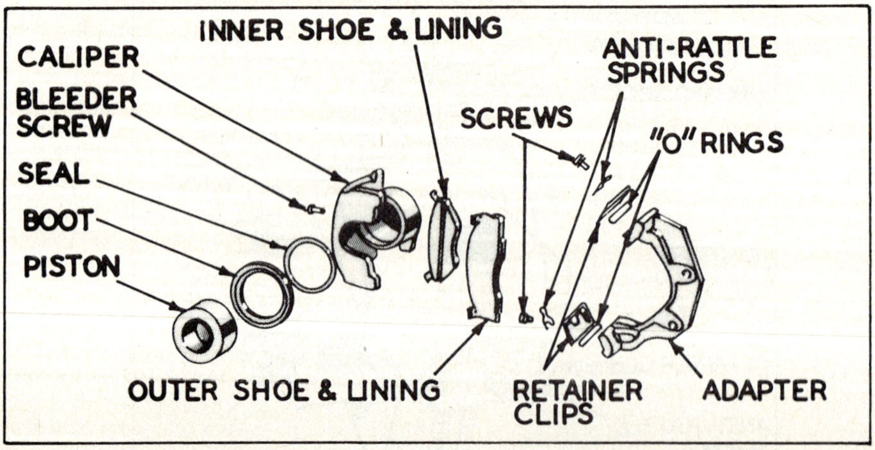

Fig. 34 Exploded view of a Kelsey-Hayes single piston sliding caliper disc brake

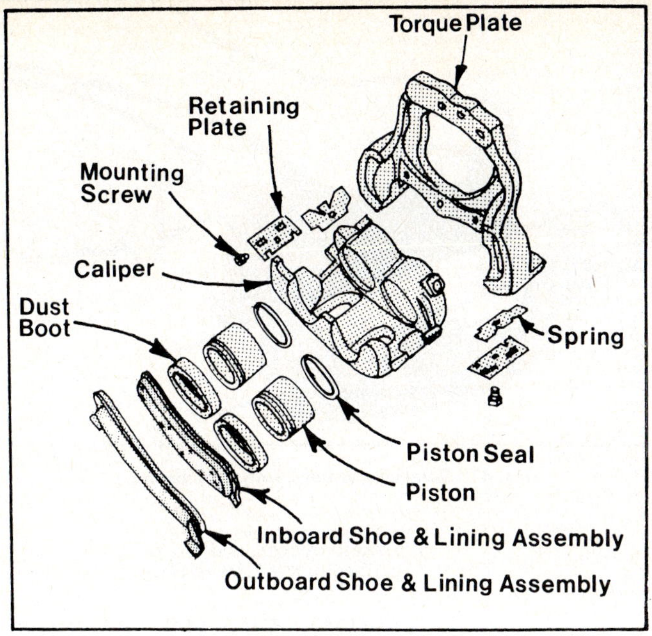

Fig. 35 Exploded view of a Kelsey-Hayes dual piston sliding caliper disc brake

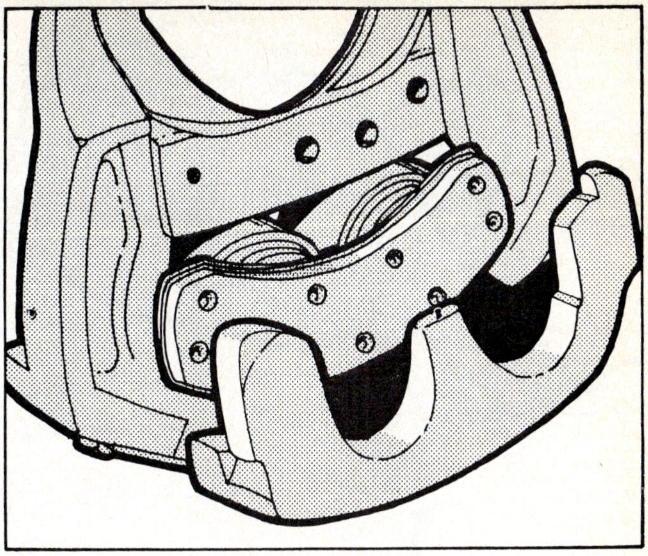

Fig. 36 Inboard brake shoe

KELSEY-HAYES DUAL PISTON SLIDING CALIPER

This sliding caliper dual piston system uses a one piece hub and is actuated by the hydraulic system and disc assembly Fig. 35. Alignment and positioning of the caliper is achieved by two machined guides or "ways" on the torque plate, while the caliper retaining clips allow lateral movement of the caliper. Outboard shoe flanges are used to position and locate the shoe on the caliper fingers while the inboard shoe is retained by the torque plate, Fig. 36. However, on some applications, the outboard shoe and lining assembly is retained by a hold-down spring, pin and cup. Braking force applied onto the outboard shoe is transferred to the caliper, while braking force applied onto the inboard shoe is transferred directly to the torque plate.

A square cut piston seal provides a hydraulic seal between the piston and the cylinder bore. A dust boot with a wiping lip installed in a groove in the cylinder bore and piston, prevents contamination in the piston and cylinder bore area. Adjustment between the disc and the shoe is obtained automatically by the outward relocation of the piston as the inboard lining wears and inward movement of the caliper as the outboard lining wears.

Caliper Removal

1. Remove brake fluid from master cylinder to bring fluid level to 1/3 full.
2. Raise vehicle and remove front wheel.
3. Remove retaining plate mounting screws, then the retaining plates and springs, Fig. 37.
4. Remove caliper from disc by slowly sliding caliper assembly out and away from disc.

NOTE: Use some means to support caliper. Do not let caliper hang from hydraulic line.

Brake Shoe Removal

1. Remove caliper assembly outlined above.
2. Remove inboard brake shoe from anchor plate, Fig. 38.
3. Tap lightly on outboard brake shoe to free from caliper. The brake shoe may be pried from the caliper using a suitable tool.

NOTE: On some applications, remove the outboard shoe and lining assembly hold-down spring, pin and cup.

Brake Shoe Installation

1. Install new inboard brake shoe into torque plate.
2. Push pistons back into bore until bottomed using a "C" clamp and block of wood, Fig. 39.
3. Install outboard brake shoe with lower flange ends against caliper leg abutments and upper flanges over shoulder over caliper legs. Then secure brake shoe by crimping in position using vise grips, Fig. 40.

NOTE: On some applications, install the

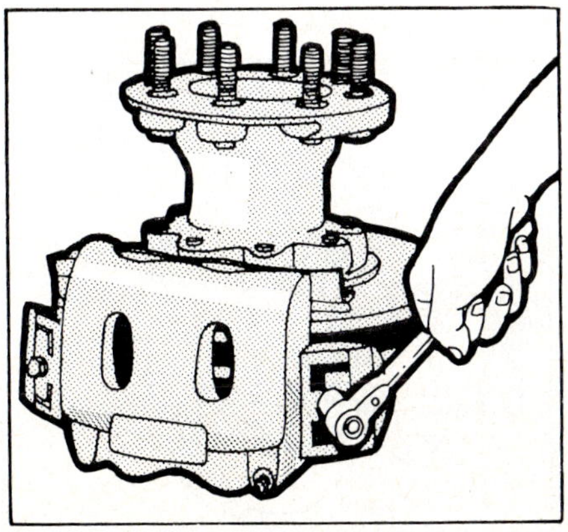

Fig. 37 Removing retaining plate screws

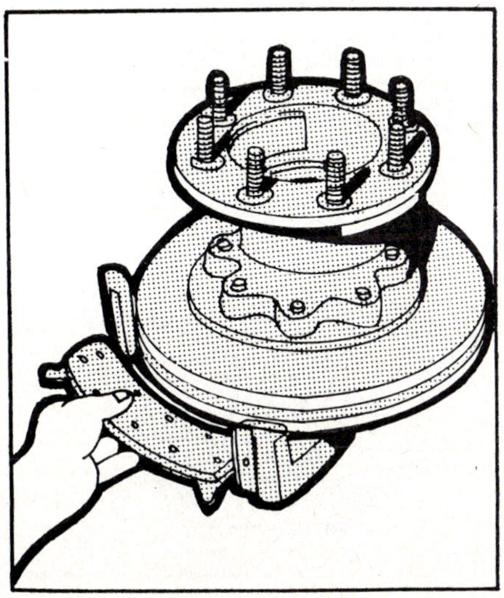

Fig. 38 Removing inboard brake shoe

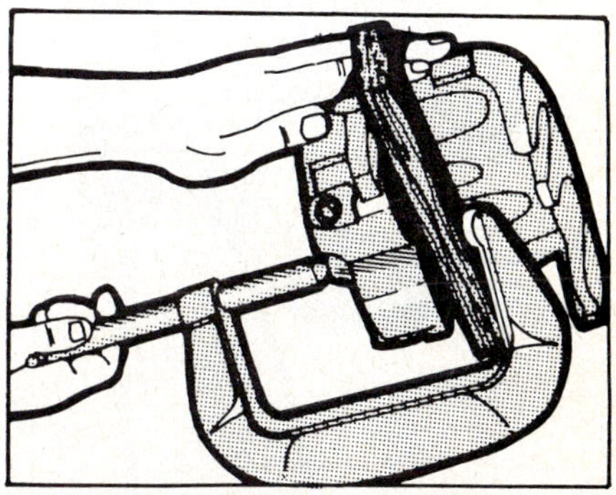

Fig. 39 Installing pistons with "C" clamp

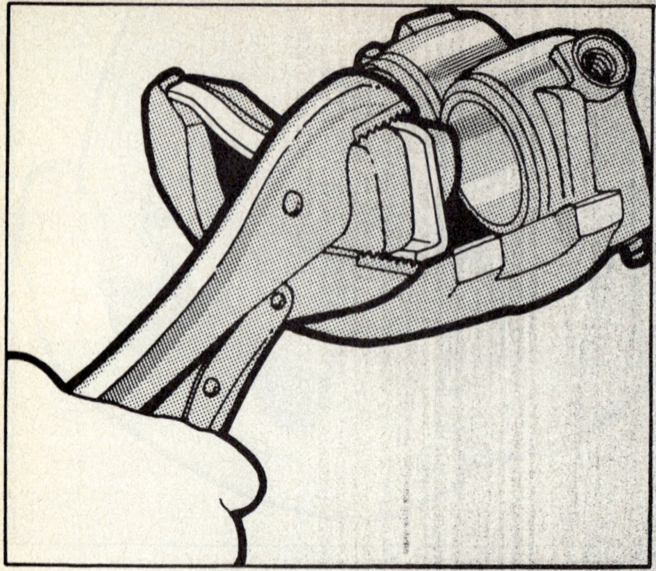

Fig. 40 Securing outboard brake shoe

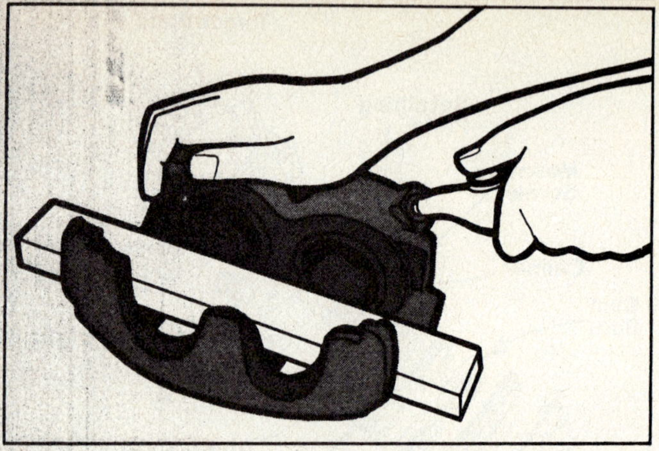

Fig. 41 Removing pistons with air pressure

outboard shoe and lining assembly hold-down spring, pin and cup.

4. Install caliper onto torque plate.
5. Install retaining plate springs and retaining plate and torque mounting screws to 17–22 ft. lbs.

Caliper Disassembly

1. Remove caliper and brake shoes as previously outlined.
2. To remove pistons, apply air pressure to the fluid port of the caliper to ease pistons from bore, Fig. 41. Use a block of wood and a shop towel to protect pistons from damage.

NOTE: Use just enough air pressure to ease pistons from bores since they may be damaged if violently blown out. Also, do not attempt to catch pistons with fingers since this could result in personal injury.

3. Remove piston boots, Fig. 35.
4. Insert a suitable tool such as a small, pointed wooden or plastic object between the cylinder bore and the seal and work seal out of groove in piston bore.

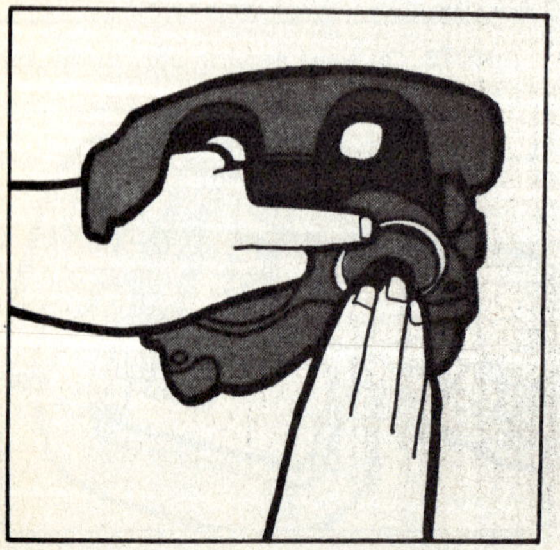

Fig. 42 Piston seal installation

NOTE: A metal tool such as a screwdriver should not be used since it can cause damage to the piston bore or burr the edges of the seal groove.

Caliper Assembly

1. Before installing new piston seal in groove of bore, dip in clean brake fluid. Work seal gently into groove (using clean fingers) until seal is properly seated and ensure seal is not twisted or rolled, Fig. 42.
2. Install dust boot in cylinder groove in similar manner as piston seal described previously.
3. Coat pistons with clean brake fluid and install pistons into bore. Work dust boot around piston using a small flat plastic tool.
4. Press piston straight into caliper bore until bottomed. The dust boot should slide up the piston and enter its groove in the piston.

NOTE: Apply uniform force to piston to avoid cocking piston in bore.

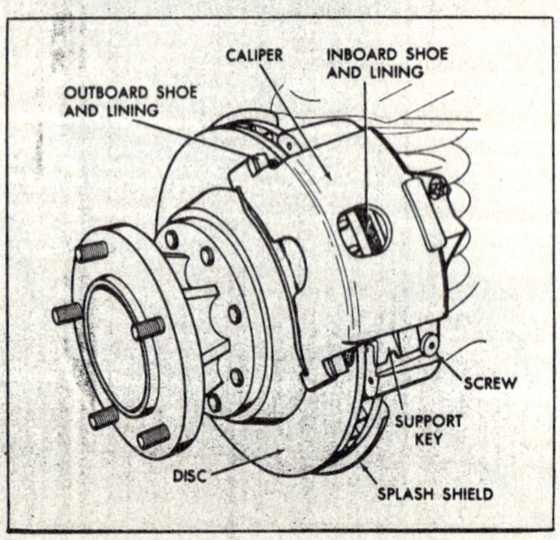

Fig. 43 Bendix single piston sliding caliper disc brake

5. Install caliper and brake shoes as described under "Brake Shoe Installation".

BENDIX SINGLE PISTON SLIDING CALIPER

The Bendix single piston sliding caliper disc brake is standard equipment on some truck applications, while on other trucks it is used only in Hydro-Boost power brake applications with both the front disc brakes and 4-wheel disc brakes.

NOTE: On the 4-wheel disc brakes the front & rear systems are identical.

The sliding caliper attaches to and slides on the steering knuckle. The caliper assembly is held in place with a support key and spring Fig. 43. A screw prevents the key from sliding on the steering knuckle. The linings are riveted to the brake shoes. The inboard and outboard brake shoes are not interchangeable.

Caliper Removal

1. Siphon two-thirds of brake fluid from master cylinder reservoir serving front disc brakes.
2. Raise vehicle, support on jackstands and remove front wheels.
3. Bottom the caliper piston in bore. Insert a screwdriver between inboard shoe and piston, then pry piston back into bore. The piston can also be bottomed in the bore with a large "C" clamp, Fig. 44.

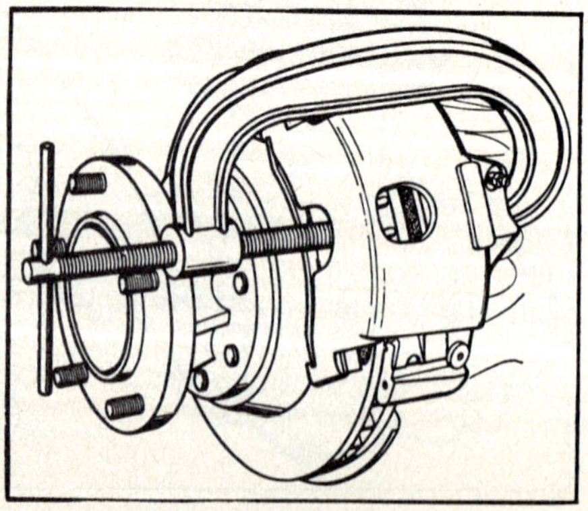

Fig. 44 Bottoming piston in bore

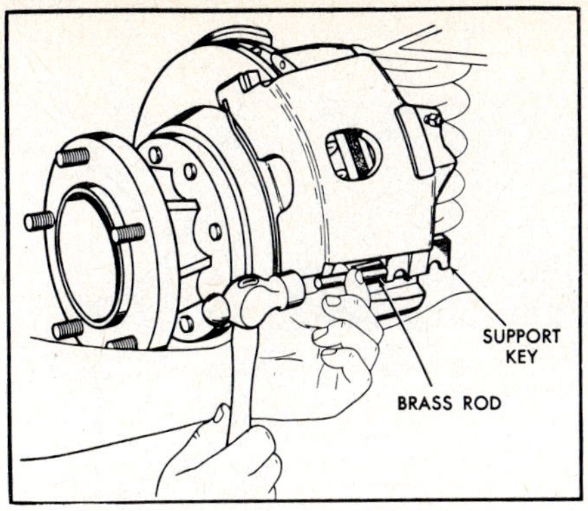

Fig. 45 Removing caliper support key

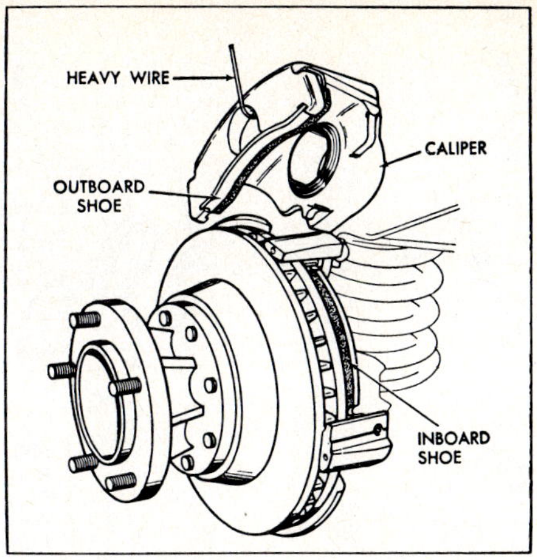

Fig. 46 Removing caliper from disc

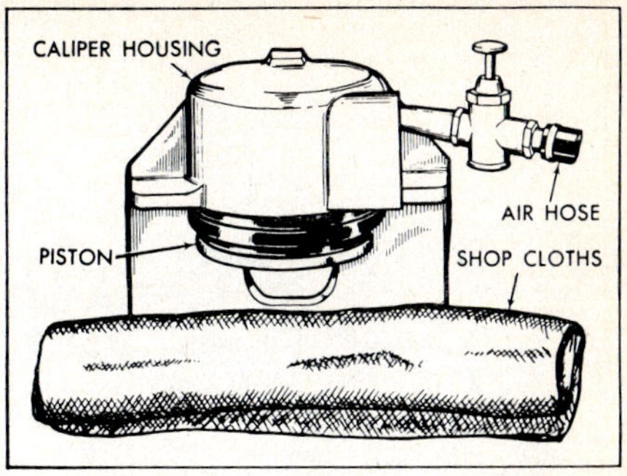

Fig. 47 Removing caliper piston

4. Remove support key retaining screw.
5. Drive caliper support key and spring from steering knuckle with a suitable drift and hammer, Fig. 45.
6. Lift caliper from anchor plate and off rotor, Fig. 46. Hang caliper from coil spring with wire. Do not allow caliper to hang from brake hose.
7. Remove inboard brake shoe from steering knuckle, then the anti-rattle spring from the brake shoe.
8. Remove outboard brake shoe from caliper, it may be necessary to loosen the brake shoe with a hammer to permit shoe removal.

Caliper Disassembly

1. Drain brake fluid from caliper.
2. Position caliper with shop cloths, Fig. 47, and apply compressed air to fluid inlet port to ease piston from bore.

NOTE: Do not attempt to catch piston or to protect it when applying compressed air since personal injury is possible.

3. Remove boot from piston, then the piston seal from bore, Fig. 48. Use wooden or plastic tool to remove piston seal since metal tools may damage piston.
4. Remove bleeder screw.

Caliper Assembly

1. Coat square cut piston seal with clean brake fluid, then install seal into piston bore. Work seal into groove with clean fingers.
2. Install and torque bleeder screw to 100 inch lbs.
3. Lubricate boot and tool J-24548 with clean brake fluid, then place dust seal on tool, allowing 1/4 inch of tool to extend past small lip of boot Fig. 49.
4. Place dust seal and tool over piston bore, then work large lip of boot into seal groove, Fig. 50. Ensure dust seal is fully seated.
5. Lubricate caliper piston and insert through tool. Center piston in bore and use a hammer handle to apply pressure to install piston halfway into bore, Fig. 50.
6. Remove tool J-24548 and seat small lip of boot in caliper piston groove, then bottom piston in bore.

Brake Shoe & Lining, Replace

The procedures to remove & install the brake shoe and lining assemblies are outlined under "Caliper Removal" and "Caliper Installation". It is not necessary to disconnect the brake hose, however, use caution not to twist or kink hose.

Caliper Installation

1. Clean and lubricate sliding surfaces of caliper and the anchor plate with Delco Silicone Lube, P/N 5459912, or equivalent.
2. Install inboard brake shoe anti-rattle spring on brake shoe rear flange, ensure looped section of clip is facing away from rotor.
3. Install inboard brake shoe in steering knuckle.
4. Install outboard brake shoe in caliper. Ensure the shoe flange is seated fully into outboard arms of caliper. It may be necessary to use a "C" clamp to seat the shoe.
5. Place caliper assembly over rotor and position in steering knuckle. Ensure dust boot is not torn or mispositioned by inboard brake shoe during caliper installation.
6. Align caliper with steering knuckle abutment surfaces, then insert support key and spring between abutment surfaces at the trailing end of caliper and steering knuckle. With a hammer and brass drift, drive caliper support key and spring into position, then install and torque support key retaining screw to 12–18 ft. lbs.
7. Refill master cylinder to within one inch of rim. Press brake pedal several times to seat shoes.
8. Install front wheels and lower vehicle.

BENDIX DUAL PISTON SLIDING CALIPER

On this sliding caliper dual piston system, the caliper is mounted on an anchor plate and the rotor is attached to the hub, Fig. 51. Two machined grooves in the anchor plate position and align the caliper. Retainer keys are used to hold caliper in the grooves and allow for lateral movement of the caliper. The outboard shoe uses flanges to position and locate the shoe on the caliper while the inboard shoe is held in position by the anchor plate. Braking force applied to the outer shoe is transferred directly to the caliper while braking force applied to the inner shoe is transferred directly to the anchor plate.

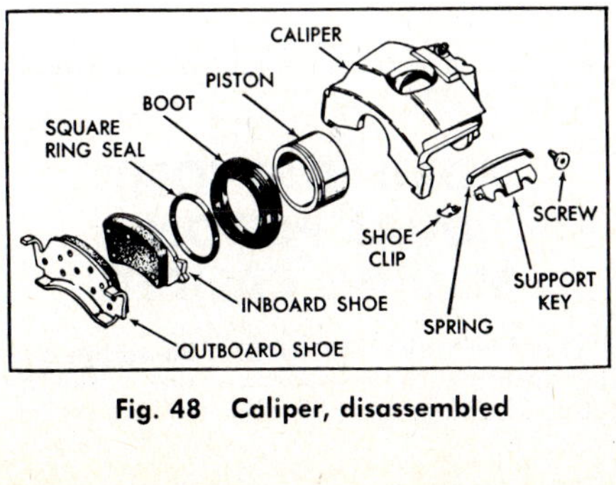

Fig. 48 Caliper, disassembled

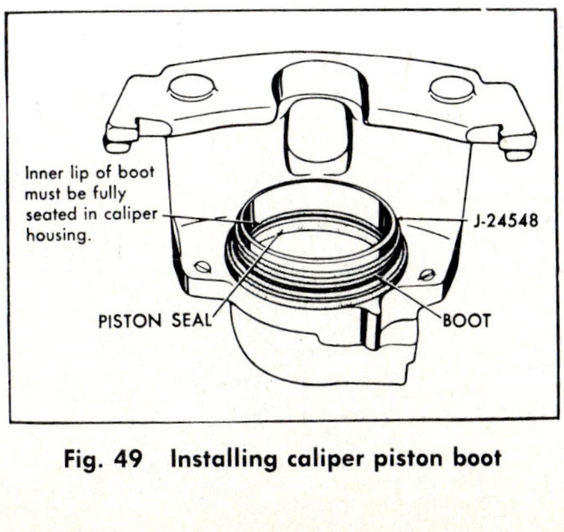

Fig. 49 Installing caliper piston boot

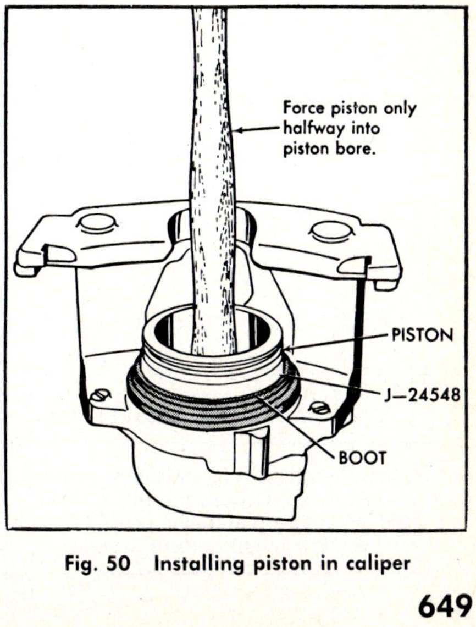

Fig. 50 Installing piston in caliper

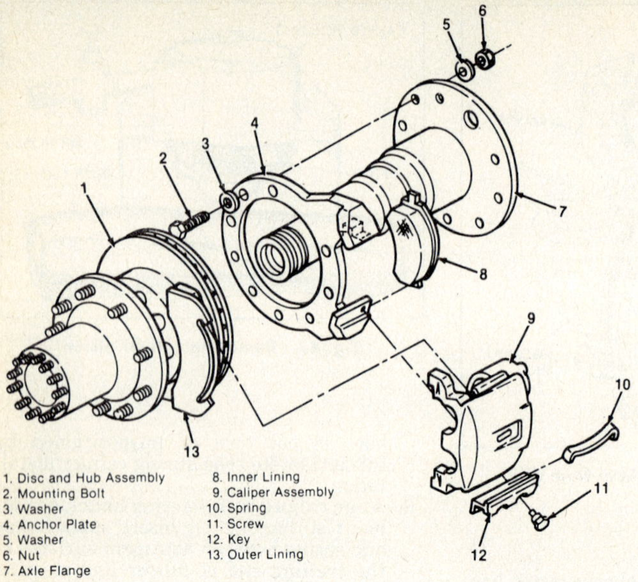

Fig. 51 Exploded view of a Bendix dual piston sliding caliper brake assembly

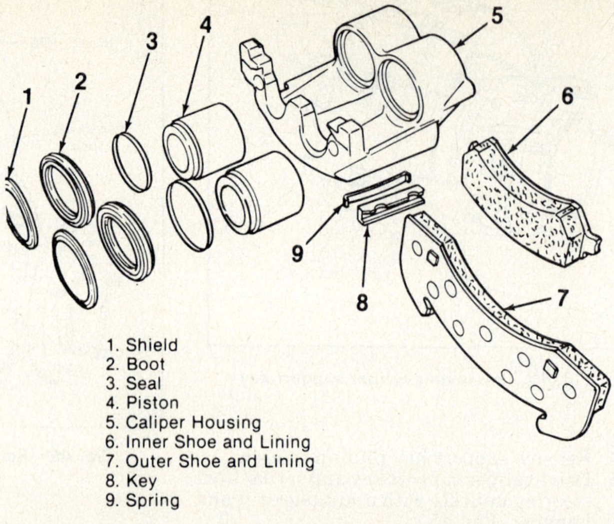

Fig. 52 Exploded view of a Bendix dual piston sliding caliper

The caliper is a one piece casting with piston cylinder bores on the inboard side, Fig. 52. A square rubber seal is located in a machined groove in the cylinder bore to provide a hydraulic seal between phenolic pistons and cylinder wall. A dust boot installed in the cylinder bore keeps contamination from cylinder wall and piston. Adjustment between disc and shoe is obtained automatically by the outward relocation of piston as the inboard lining wears and the inward movement of the caliper as the outboard lining wears.

Caliper Removal

1. Remove brake fluid from master cylinder until about 1/3 full.
2. Raise vehicle and remove wheel.
3. Remove key retaining screw, key and spring. Lift caliper assembly off hub and disc assembly.

NOTE: It may be necessary to bottom caliper pistons in bore in order to provide removal clearance. Place large C-clamp on caliper piston housing and outer shoe, Fig. 44. Watch for possible overflow at reservoir when bottoming pistons.

Brake Shoe Removal

1. Remove caliper assembly as outlined above.
2. Remove inner shoe from anchor plate.
3. Remove outer shoe and lining by sliding on groove to notch in caliper housing and lifting lining out, Fig. 53.

Brake Shoe Installation

1. Install new inner brake shoe into anchor plate, Fig. 51.
2. Push pistons into bores until bottomed.

NOTE: Pistons can be bottomed either by hand or with a C-clamp and a block of wood, Fig. 44. Watch for possible reservoir overflow when bottoming pistons.

3. Install new outer brake shoe into caliper, Fig. 53.
4. Install caliper over disc and into anchor plate grooves. Install caliper key, spring and retaining screw. Torque retaining screw to 15 ft. lbs.

Caliper Disassembly

1. Remove caliper and brake shoes as previously outlined.
2. To remove pistons, apply air pressure to the fluid port of the caliper to ease pistons from bore, Fig. 41. Use a block of wood and a shop towel to protect pistons from damage.

NOTE: Use just enough air pressure to ease pistons from bores since they may be damaged if violently blown out. Also, do not attempt to catch pistons with fingers since this could result in personal injury.

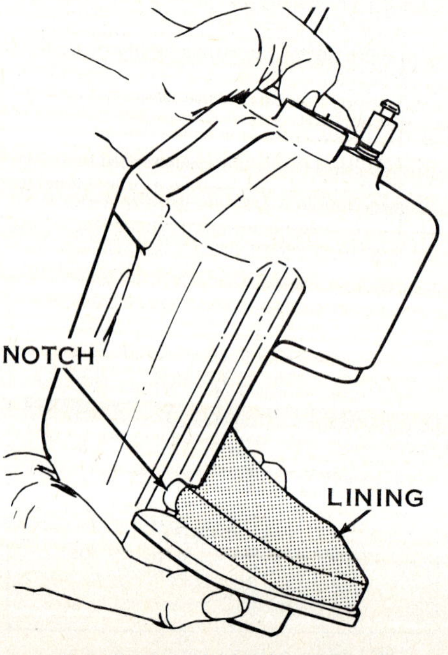

Fig. 53 Outer brake shoe removal

3. Remove heat shield and piston boot from piston. Discard boots.
4. Insert a suitable tool such as a small, pointed wooden or plastic object between the cylinder bore and the seal and work seal out of groove in piston bore.

NOTE: A metal tool such as a screwdriver should not be used since it can cause damage to the piston bore or burr the edges of the seal groove.

Caliper Assembly

1. Coat new piston seal in clean brake fluid. Carefully work seal into cylinder groove until properly seated.
2. Lubricate piston and bore with clean brake fluid. Lubricate new dust boot with brake fluid and install over base of piston with lip facing down. Boot should be extended past base of piston.
3. Place piston and boot on caliper bore. Working from rear around to front, insert boot lip into recess of piston bore using small flat wooden or plastic tool. When boot is seated, push piston into bore. Boot will slide on piston. Check to see if boot is evenly seated.

NOTE: Use care when installing piston into bore to avoid dislodging piston seal. Push in evenly and rotate slightly until bottomed.

4. Install heat shield around piston with lip facing out.

FORD (DAYTON) DUAL PISTON SLIDING CALIPER

This disc brake, Fig. 54, is of the sliding caliper design with two pistons on one side of the rotor. The caliper which slides on the anchor

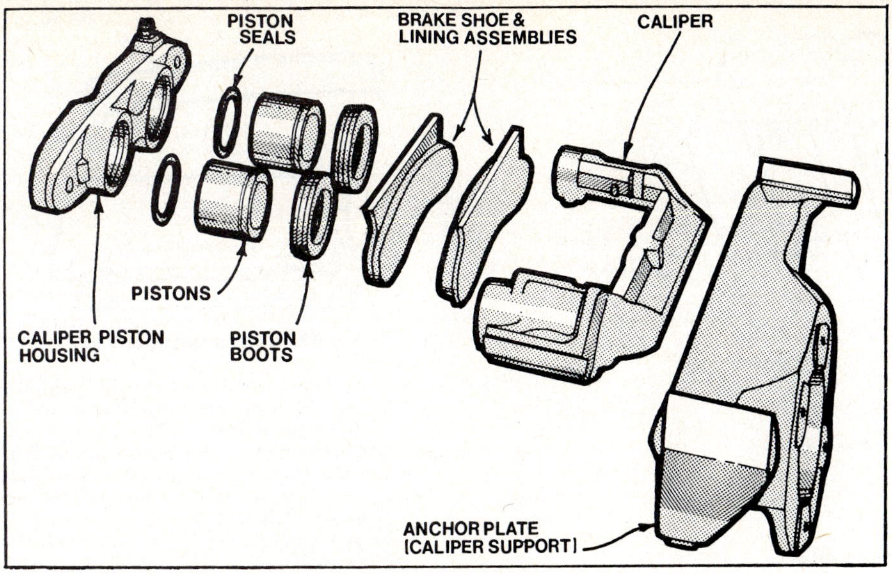

Fig. 54 Ford (Dayton) Dual Piston Sliding Caliper disc brake

plate is retained by a key and spring. A key retaining screws maintains the key and spring in proper position. Two brake shoes and lining assemblies, one on each side of the rotor, are used and are not identical. The brake shoes slide on the caliper bridge and one anti-rattle clip is used on both shoes. The cylinder housing contains the two pistons. The pistons and caliper bores are protected by boot seals fitted to a piston groove and attached to the cylinder housing. The cylinder assembly is attached to the caliper with cap screws and washers. The anchor plate and shield are bolted to the spindle. The anchor plate positions the caliper over the rotor forward of the spindle.

Brake Shoe & Lining, Replace

Removal

1. To avoid fluid overflow when pistons are pushed into caliper, remove some brake fluid from master cylinder.
2. Raise vehicle and remove wheel assembly.
3. Remove key retaining screw, Fig. 55, then using a brass drift and hammer, drive out the key and spring, Fig. 56.
4. Remove caliper and its support by rotating the key and spring end out and away from the rotor. Slide opposite end of caliper clear of the slide in the support and off the rotor. Place caliper on tie rod or axle.

NOTE: Do not allow the brake hose to support weight of caliper, as this can damage the hose.

5. Remove caliper brake shoe anti-rattle spring and inner and outer shoe and lining assemblies.
6. Clean and inspect caliper assembly. Thoroughly clean areas of caliper and support that contact during the sliding action of the caliper.

Installation

1. Place used inner lining and shoe assembly over the pistons, then place a C-clamp on the caliper housing midway between the two pistons over the lining and shoe assembly, Fig. 57. Tighten the clamp until pistons are bottomed in caliper, then remove C-clamp and lining and shoe assembly.
2. Install the new inner and outer shoe and lining assemblies, and the anti-rattle spring.
3. Position caliper rail into the slide on the support and rotate the caliper onto rotor.
4. Position the key and spring, Fig. 58, then install the sub-assembly between the caliper and support. Note that the spring is between the key and caliper and that spring tangs overlap the ends of the key.
5. If necessary use a screwdriver to hold caliper against support assembly, then using a hammer, drive the key and spring into position aligning the correct notch with the existing hole in the support.
6. Install the key to support retaining screw and torque to 12–20 ft. lbs.
7. Install wheel assembly and lower vehicle. Check brake fluid level and add as necessary with heavy duty brake fluid.
8. Depress brake pedal several times to seat linings on rotor. Do not move vehicle until a firm brake pedal has been obtained.

Caliper, Replace

Removal

1. Perform steps 1 to 5 outlined under "Brake Shoe & Lining, Replace" Removal procedure.
2. Disconnect flexible hose from caliper.
3. Remove caliper assembly.

Installation

1. Perform steps 1 and 2 outlined under "Brake Shoe & Lining, Replace" installation procedure.
2. Connect flexible hose to caliper.
3. Perform steps 3 to 6 outlined under "Brake Shoe & Lining, Replace" installation procedure, then after bleeding brakes, proceed with the remainder of the procedure.

Caliper Overhaul

Disassembly

1. Drain brake fluid from caliper.
2. Remove brake shoe and lining assemblies.
3. To remove pistons, apply air pressure to the caliper fluid port to ease the pistons from the bores. Use a block of wood and a

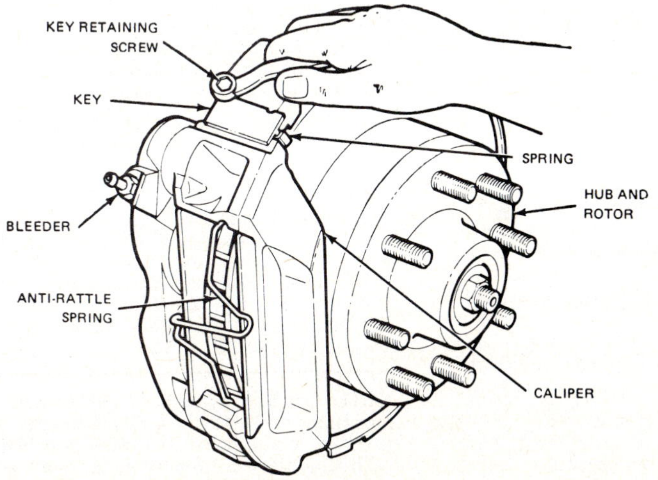

Fig. 55 Removing key retaining screw

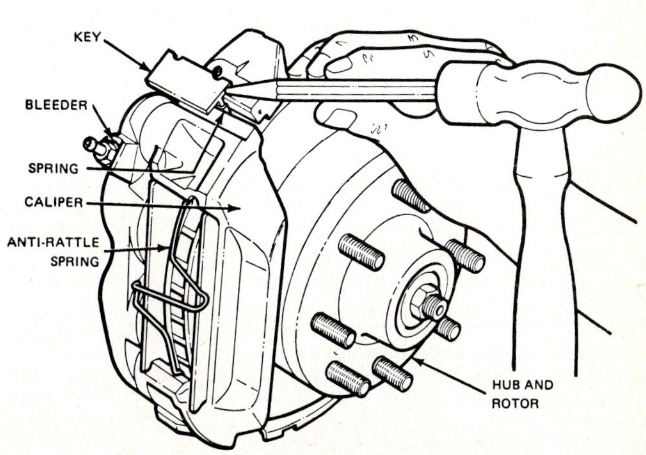

Fig. 56 Removing spring and key

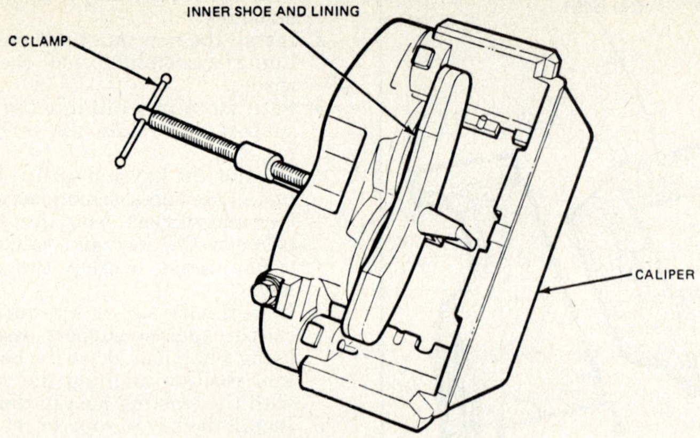

Fig. 57 Bottoming caliper pistons

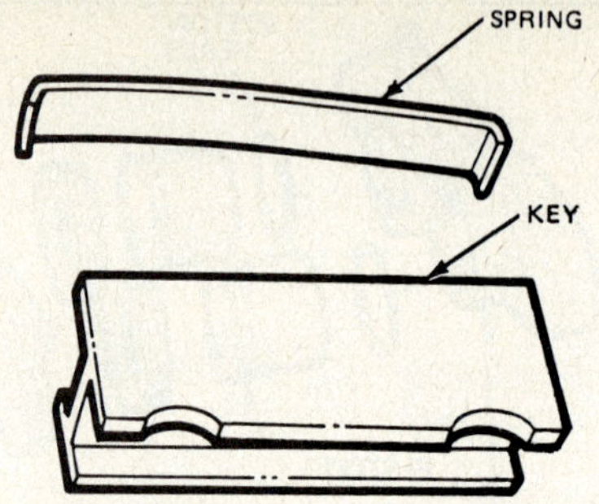

Fig. 58 Caliper spring and key

shop towel to protect pistons from damage.
4. Remove cylinder housing to caliper attaching bolts and separate housing from caliper.
5. Remove piston seals.

Assembly

1. Lubricate cylinder bores and new piston seals with clean brake fluid.
2. Install piston seals in cylinder bore grooves.
3. Lubricate dust boot retaining lips with clean brake fluid and install retaining lips in the boot retaining grooves in the cylinder bores.
4. Lubricate piston with clean brake fluid and insert pistons into dust boots. Start pistons into cylinders until located beyond the piston seals.
5. Place a block of wood over one piston and press piston into cylinder using caution not to damage the piston seals or piston. Press the second piston into the bore.
6. Place piston housing on caliper and install and torque mounting bolts to 155–185 ft. lbs.

AIR DISC BRAKES

LINING & SUPPORT BRACKET, REPLACE, FIG. 1

Removal

1. Exhaust air from system and drain all reservoirs.

CAUTION: Do not proceed until spring brake chamber is mechanically caged.

2. Back off manual or automatic slack adjuster by rotating slack adjustment screw counterclockwise one turn.
3. Disconnect air hoses at actuator and remove four nuts from studs through support bracket.
4. Remove spring clips and lift support bracket and caliper clear of disc and adaptor assembly.

Disassembly

1. Remove support bracket from caliper and remove carrier and liner assemblies.

NOTE: Failure to replace linings with 1/8 inch or less remaining can result in brake failure.

2. Turn adjustment screw clockwise until piston is exposed. Clean exposed area of piston. Retract piston until it bottoms out.
3. Inspect support bracket for wear. Maximum allowable wear on any surface is .015 inch.

Assembly

1. Install carrier and liner assemblies in caliper.
2. Place support bracket into carrier, liner assemblies and caliper. Position slots in support bracket to face away from caliper.
3. Position caliper assembly over disc until support bracket engages torque plate studs.
4. Install spring clips and nuts. Torque nuts to 118 ft. lbs.
5. Measure clearances between support bracket and caliper, Fig. 2. If clearances are exceeded parts must be replaced.
6. Connect air hoses to actuator and service grease fitting on slack adjuster and caliper.
7. Perform brake adjustment procedure as described in this chapter.
8. Uncage spring chamber.

CALIPER OVERHAUL & SEAL REPLACEMENT

Caliper Removal & Disassembly

1. Remove caliper as described in "Lining & Support Bracket Replace".
2. Remove slack adjuster cotter and clevis pins and remove actuator clevis.
3. Remove nuts from actuator studs and actuator from mounting bracket.
4. Remove retaining ring, adjuster, washers, and seal from power screw shaft.
5. Remove cap assembly, power screw shaft and piston assembly from caliper.
6. Remove power screw shaft and piston assembly from cap assembly. Unscrew piston assembly from power screw shaft. Remove thrust bearing and piston seal.
7. Inspect journal bearing in cap assembly for wear. Maximum inside diameter cannot exceed 1.510 inches.

Caliper Assembly & Installation

1. Install piston seal into caliper.
2. Apply suitable lubricant to face of thrust bearing and install over splined end of power screw shaft with inside diameter lip facing away from thrust collar.
3. Fill piston assembly thread to bottom of threads with grease and screw power screw shaft into piston assembly.

NOTE: Grease should squeeze out past power screw shaft. If not, add more grease.

4. Coat sides of piston with grease and slide assembly into cap assembly, shaft end first.
5. Install gasket between cap assembly and caliper. Bolt cap assembly to caliper. Refer to Figs. 3 and 4 for washer size. Torque bolts to 140 ft. lbs.

NOTE: Grade 8 nyloc bolts can be used four times before the locking feature deteriorates. These bolts should be punched each time they are removed, and may be used with up to three punch marks.

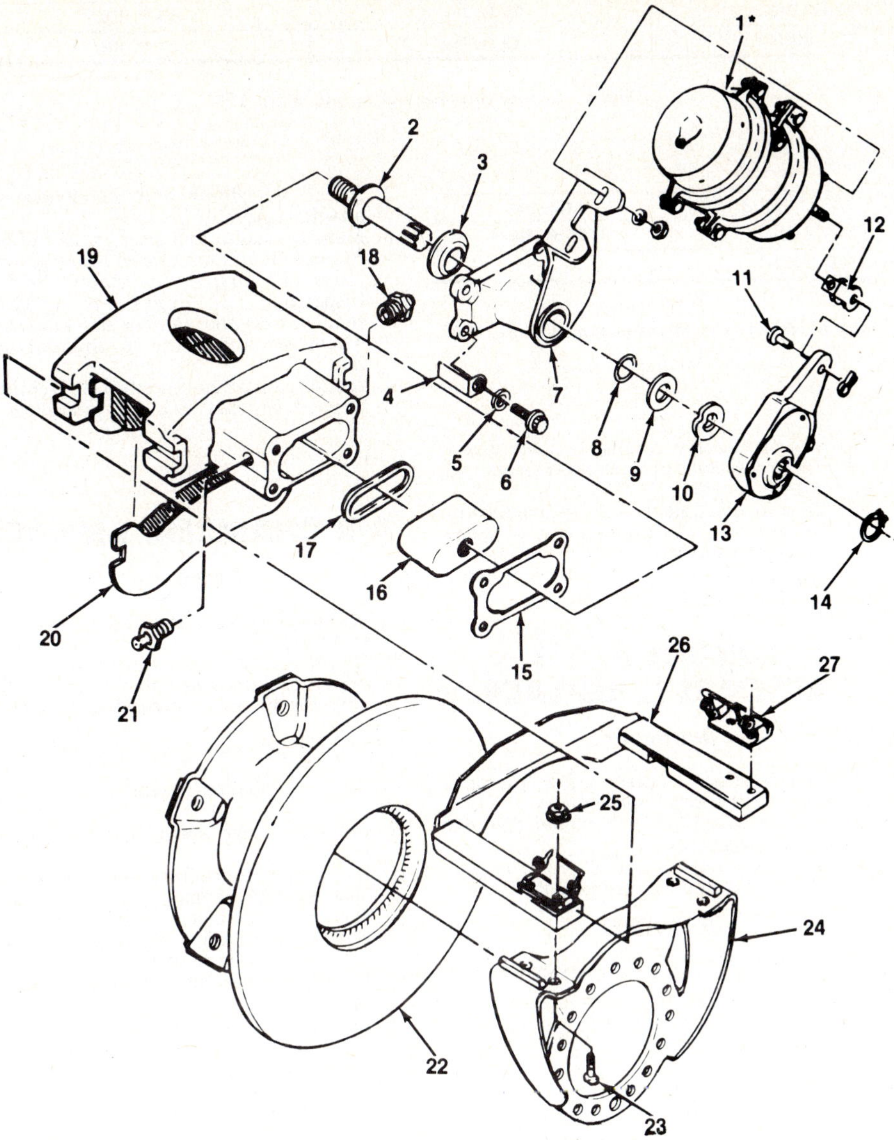

1.* Actuator (Brake Chamber)
2. Power Screw Shaft
3. Thrust Bearing
4. Instruction Plate
5. Cap Screw Washer
6. Cap Screw, 12 point
7. Mounting Bracket and Cap Assembly
8. Preformed Packing
9. Flat Washer
10. Wave Washer
11. Clevis Pin
12. Clevis
13. Slack Adjuster
14. Retaining Ring
15. Cap Gasket
16. Piston
17. Piston Seal
18. Pressure Relief Fitting
19. Caliper
20. Shoe and Lining Assembly
21. Lube Fitting
22. Disk and Adapter Assembly
23. Plate to Bracket Stud
24. Torque Plate Assembly
25. Stud Flange Nut
26. Support Bracket Assembly
27. Spring Clip

* Actuators (Brake Chambers) used with disc brakes have heavy duty return springs and cannot be replaced with standard actuators.

Fig. 1 BF Goodrich Air Disc Brake

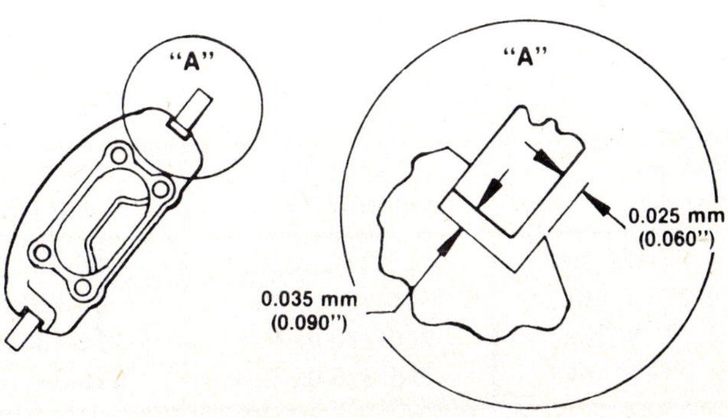

Fig. 2 Caliper to support bracket clearances

6. Install preformed seal, flat washer and one spring washer over power screw shaft.
7. Place actuator on cap assembly in slot closest to power screw shaft opening. Do not torque bolts at this time. Apply coat of grease to mounting spline of slack adjuster and install on power screw shaft aligning slack adjusting arm with actuator clevis. Slack adjuster adjusting screw must be on opposite side of applied actuating force.
8. Torque service chamber nuts using standard flat washers. Refer to Figs. 5 and 6 and "Automatic Slack Adjuster Installation" procedure.
9. Service lube fittings on slack adjuster and caliper.
10. Install lining and bracket assembly according to "Lining & Support Bracket Replace" procedure.

Bolt Size	Hardened Washer Size
5/8–11	1.31 × .66 × .07–.12 thk.
3/4–10	1.38 × .81 × .09–.11 thk.

Fig. 4 Torque plate bolt & washer chart. 1982

Stud	Phosphate/Oil	Cadmium Plate/Wax
5/8–11UNC	98 ft. lbs.	60 ft. lbs.
7/16–20UNF	33 ft. lbs.	23 ft. lbs.

Fig. 5 Service chamber nut torque chart. 1981

DISC & ADAPTER ASSEMBLY, REPLACE

Disassembly

1. Remove support bracket and caliper as outlined in "Lining & Support Bracket, Replace".
2. Remove support bracket from caliper and remove carrier and liner assemblies.
3. Remove wheel/hub and mated disc and adapter assembly from axle.
4. Remove disc and adapter assembly from wheel/hub.

NOTE: When refinishing disc surface, cut should be made only as far as a diameter of 9.650–9.750 inches. This will ensure full lining contact with resurfaced area and not interfere with weld area.

Assembly

1. Place disc and adapter assembly on wheel/hub. Refer to Figs. 7 thru 10 for proper order of assembly and torque values.
2. Install wheel/hub and mated disc and adapter assembly on axle.

Inspection, Fig. 11

NOTE: Before checking disc for runout, eliminate all wheel bearing end play by tightening adjusting nut. Nut must be just free enough to allow disc to rotate. Readjust after inspection.

1. Measure radial runout of disc. Rework disc if runout exceeds .035 inch TIR.
2. Measure axial runout of disc. Rework disc if runout exceeds .015 inch TIR new or .020 inch TIR used.
3. Measure exciter axial runout. Runout should not exceed .035 inch TIR.
4. Measure disc face parallelism. Rework disc if parallelism exceeds .010 inch TIR.

Stud	Torque
5/8–11UNC	90 ft. lbs.
7/16–20UNF	33 ft. lbs.

Fig. 6 Service chamber nut torque chart. 1982

5. Inspect disc for cracks. Replace disc if cracks exceed 1/2 inch across O.D. or crack extends to weld.
6. Measure disc thickness at thinnest point. Replace disc if measurement is below dimension stamped on O.D. of disc.
7. Inspect disc for dishing. Resurface disc if dishing exceeds .060 inches.

TORQUE PLATE ASSEMBLY, REPLACE

Removal

1. Remove caliper as described in "Lining & Support Bracket, Replace".
2. Remove wheel/hub and mated disc and adapter assembly.
3. Mark position of torque plate on axle flange then remove torque plate.

Assembly

1. Mount torque plate assembly to axle flange, Fig. 3.
2. Install wheel/hub and mated disc and adapter assembly on axle.
3. Measure radial and axial runout of disc. Runout should not exceed limits given in "Disc & Adapter Assembly Replace, Inspection". If limits are exceeded, remove disc and adapter assembly from wheel/hub and locate in different position.
4. Install caliper as described in "Lining & Support Bracket, Replace" procedure.

AIR CHAMBER, REPLACE

Removal

1. Manually cage spring brake chamber, if equipped.
2. Disconnect air hoses at actuator.
3. Separate actuator clevis from slack adjuster by removing cotter and clevis pins.
4. Remove actuator from cap assembly.

Installation

1. Manually cage spring brake, if equipped.
2. Measure distance from center of clevis to surface of air chamber, Fig. 12. Distance should be as listed in Fig. 13. If not, remove clevis pin, loosen locking pin and adjust clevis as required. Tighten clevis locking nut and connect slack adjuster to clevis.
3. Mount brake actuator on cap assembly with washers and nuts and position in slot closest to power screw shaft opening. If automatic slack adjuster is used, position control arm strap under mounting nut and washer. Do not torque nuts at this time.
4. Reconnect clevis to slack adjuster. Torque chamber nuts to value indicated in Figs. 5 and 6 and "Automatic Slack Adjuster Installation" procedure.
5. Connect air hoses to actuator.
6. Adjust brakes as outlined further on.

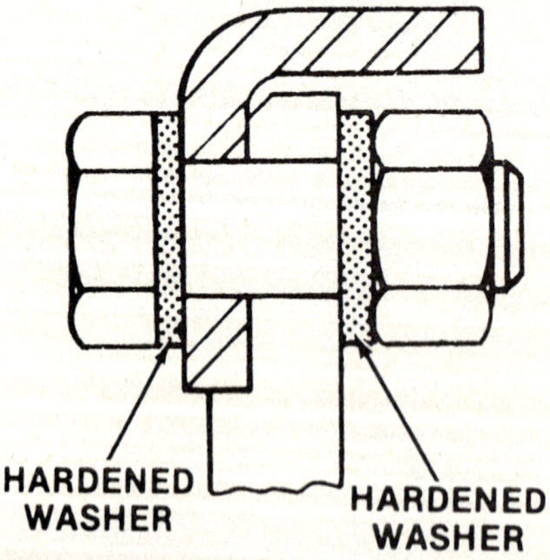

Bolt Size	SAE Grade	Hardened Washer Size	Phosphate/ Oil Hardware	Cad Plate/ Wax Hardware
1/2-13	8	1.06 x .53 x .07-.12 thk.	95-110 lb. ft.	60-70 lb. ft.
1/2-20	8	1.06 x .53 x .07-.12 thk.	110-125 lb. ft.	65-75 lb. ft.
5/8-18	8	1.31 x .66 x .07-.12 thk.	200-230 lb. ft.	125-145 lb. ft.
3/4-16	8	1.38 x .81 x .09-.11 thk.	325-375 lb. ft.	225-260 lb. ft.

Fig. 3 Torque plate bolt & washer chart. 1981

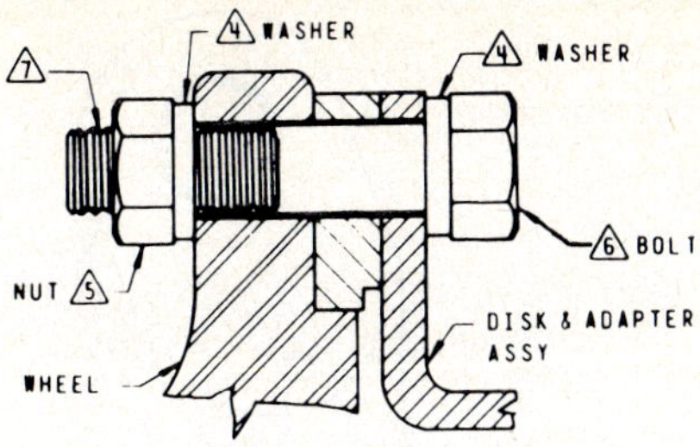

3/4 NOMINAL SIZE

4 WASHERS: 3/4 NOM. SIZE CONFORMING TO ANSI B27.2 TYPE B NARROW SERIES HARDENED R_C 38-53. (O.D. 1.38 I.D. .81 AND THICKNESS .112/.090)

5 NUT:

3/4-16 SAE GRADE 8 ASSEMBLY TORQUE	
PHOSPHATE/OIL	CAD PLATED/WAX
325-375 LBS-FT	225-260 LBS-FT

6 BOLT: 3/4-16 SAE GRADE 8 PHOSPHATE/OIL

7 MIN 2 THREADS EXPOSED AFTER ASSEMBLY

Fig. 7 3/4 Inch through bolt application & torque charts

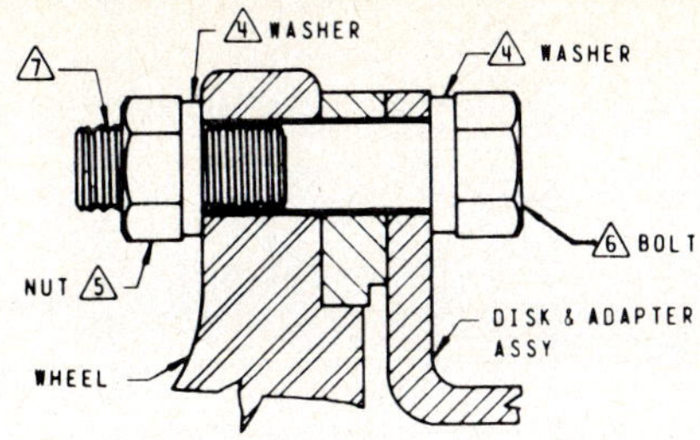

5/8 NOMINAL SIZE

4 WASHERS: 5/8 NOM. SIZE CONFORMING TO ANSI B27.2 TYPE A SAE SERIES HARDENED R_C 38-53 (O.D. 1.31; I.D. .66 AND THICKNESS .121/.074)

5 NUT:

5/8-18 SAE GRADE 8 ASSEMBLY TORQUE	
PHOSPHATE/OIL	CAD. PLATE/WAX
200-230 LBS-FT	125-145 LBS-FT

6 BOLT: 5/8-18 SAE GRADE 8 PHOSPHATE/OIL

7 MIN 2 THREADS EXPOSED AFTER ASSEMBLY

Fig. 8 5/8 Inch through bolt application & torque chart

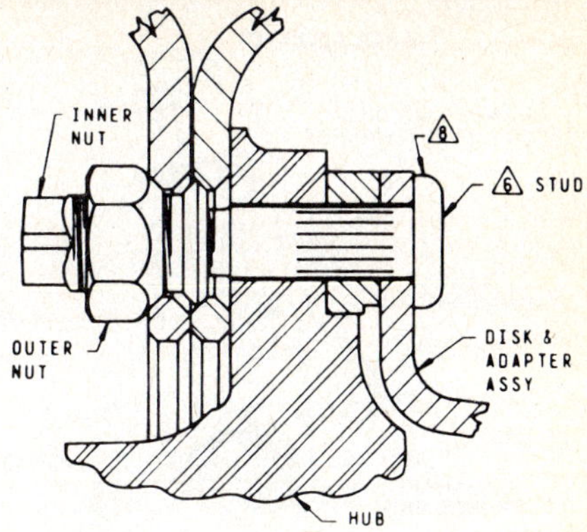

8 DURING INSTALLATION ASSURE THAT STUD HEAD FACE IS FULLY SEATED WITH DISK & ADAPTER FLANGE FACE

6 HEADED STUD: SAE GRADE 8 PHOSPHATE/OIL

Fig. 9 Single ended stud application chart

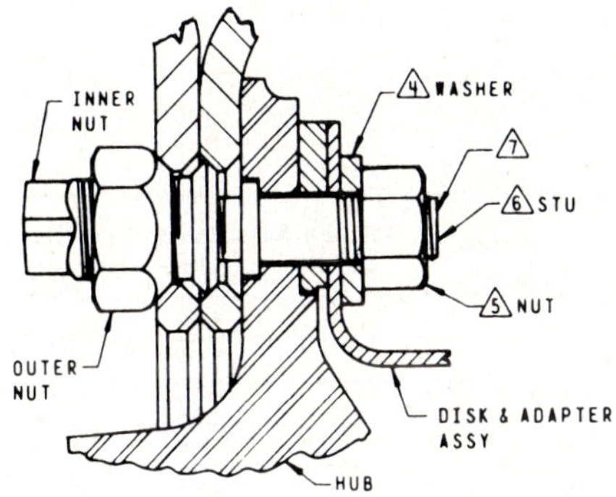

3/4 NOMINAL SIZE (HUB END)

4 WASHER: 3/4 NOM. SIZE CONFORMING TO ANSI B27.2 TYPE B NARROW SERIES HARDENED R_C 38-53 (O.D. 1.38 I.D. .81 AND THICKNESS .112/.090)

5 NUT:

3/4-16 SAE GRADE 8	ASSEMBLY TORQUE
PHOSPHATE/OIL	CAD PLATE/WAX
325-375 LBS-FT.	225-260 LBS-FT.

6 STUD: 3/4-16 SAE GRADE 8 (HUB END) PHOSPHATE OIL

7 MIN. 2 THREADS EXPOSED AFTER ASSEMBLY

Fig. 10 Double ended stud application & torque chart

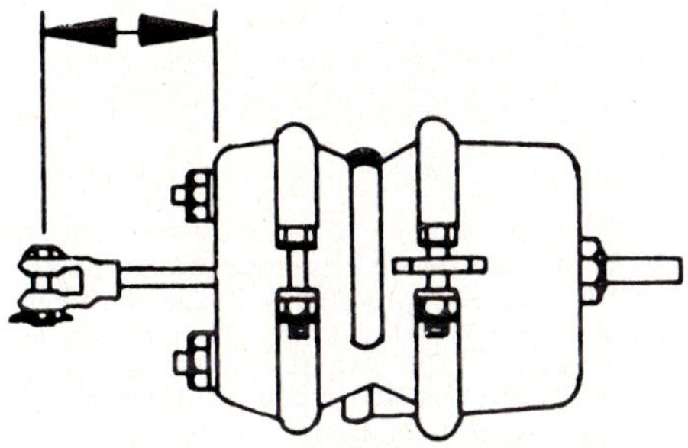

Fig. 12 Measuring manual slack adjuster setting

Brake Chamber Type	Clevis Setting In Inches (± 1/16")
9, 12, 16	2 3/4
20, 24, 30	2 9/16

Fig. 13 Manual slack adjusting clevis pin setting chart

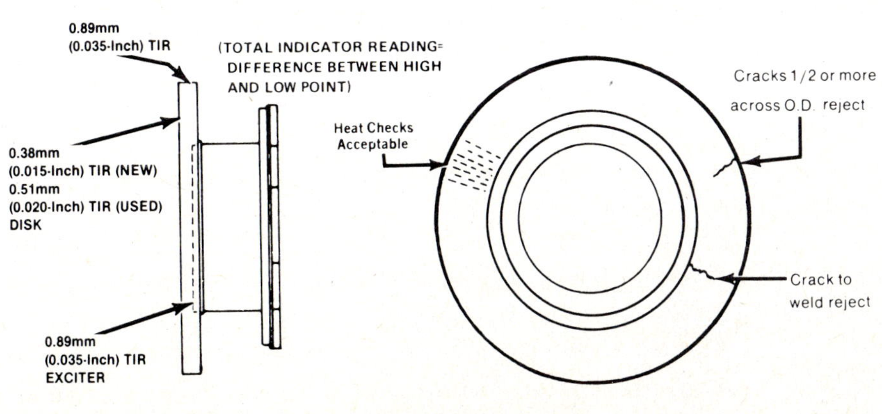

Fig. 11 Brake rotor inspection

AUTOMATIC SLACK ADJUSTER INSTALLATION, FIG. 14

1. Position automatic slack adjuster on camshaft splined end with manual adjusting nut pointing away from brake chamber. Install spline washer and snap ring.
2. Place anchor strap on control arm stud. Guide anchor strap slot onto chamber mounting stud. Torque flange nut on control arm to 11 ft. lbs.
3. Move slack adjuster toward clevis by turning manual adjusting nut clockwise. When holes in clevis line up with holes in adjusting arm, install clevis pin and cotter pin.
4. Place control arm in full release position by forcing it in direction of arrow (A), Fig. 14. Hold control arm in this position and install chamber nut and washer. Torque to values indicated in Figs. 5 and 6.

NOTE: A larger flat washer may be required under existing washer on a 7/16 stud to avoid washer deformation due to control arm strap slot.

BRAKE ADJUSTMENTS

Automatic Slack Adjuster

Cycle service brake pedal and observe rotation of adjusting hex during brake release. This indicates that an excessive clearance exists and the slack adjuster is making an adjustment to compensate for this clearance. On each subsequent brake release, the amount of adjustment and push rod travel will be decreased until the desired clearance is obtained. No further adjustment is required.

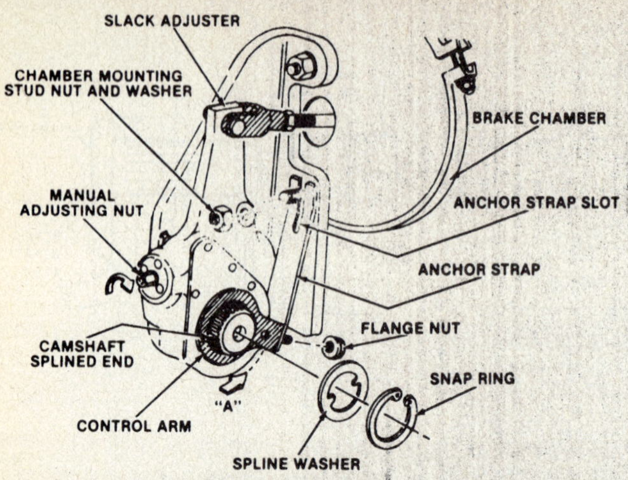

Fig. 14 Automatic slack adjuster installation

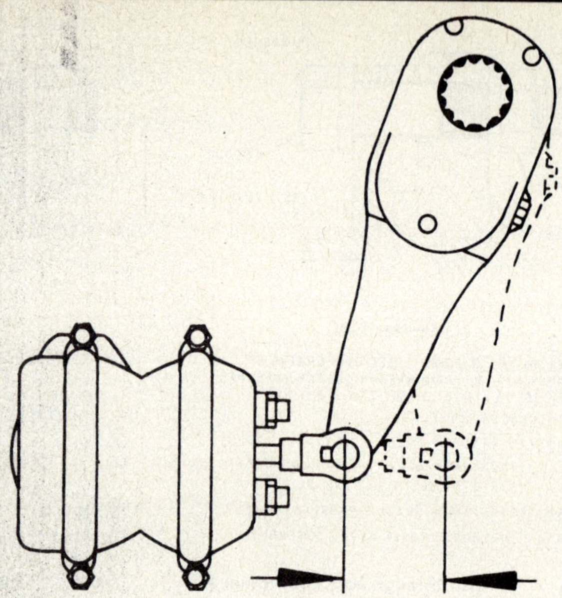

Fig. 15 Measuring brake chamber stroke

Manual Slack Adjuster

Clevis Setting

1. Fully cage spring brake manually or with 100 psi. minimum of air pressure.
2. Measure distance from center of clevis to surface of air chamber, Fig. 12. The distance between center of clevis pin and air chamber surface should be as listed in Fig. 13.
3. If distance is not within limits, remove clevis pin attaching clevis to slack adjuster, then loosen clevis lock nut and adjust clevis to obtain specified distance.
4. Tighten clevis lock nut and reconnect clevis to slack adjuster.

Air Chamber Stroke

1. Fully cage spring brake manually or with 100 psi. minimum of air pressure.
2. Rotate slack adjuster adjusting screw until brake pads are firmly seated against brake rotor, then back off adjusting screw 3/4 to 1 turn.
3. To determine air chamber stroke, apply 100 psi. of air pressure to brake chamber and release. The air chamber stroke is determined by the difference between clevis pin position when air pressure is applied and released, when measured from the air chamber surface, Fig. 15. Rotate slack adjuster adjusting screw until specified air chamber stroke is obtained, Fig. 16.
4. Uncage spring chamber and check brakes for proper operation.

Brake Chamber Type	Stroke Adjustment In Inches (5" Manual Slack Adjuster)
9	1 1/4
12	1 3/8
16	1 1/2
20 2024	1 5/8
24 2424	1 3/4
30 3030	1 7/8

Fig. 16 Brake chamber stroke setting chart

BRAKE BOOSTERS

INDEX

BOOSTER BLEEDING

Bleeding hydraulic brake systems which include a brake booster requires that the booster be bled first before proceeding with the operation on the wheel cylinders. (Bleeding instructions are given in the *Hydraulic Brake System* chapter.)

As shown in the illustrations, Figs. 1 to 6, all boosters have a bleed screw on the control valve (screw No. 1) and some have another bleed screw on the hydraulic cylinder (screw No. 2). Start the bleeding operation with the control valve bleed screw, then the hydraulic cylinder screw and, finally, bleed the wheel cylinders.

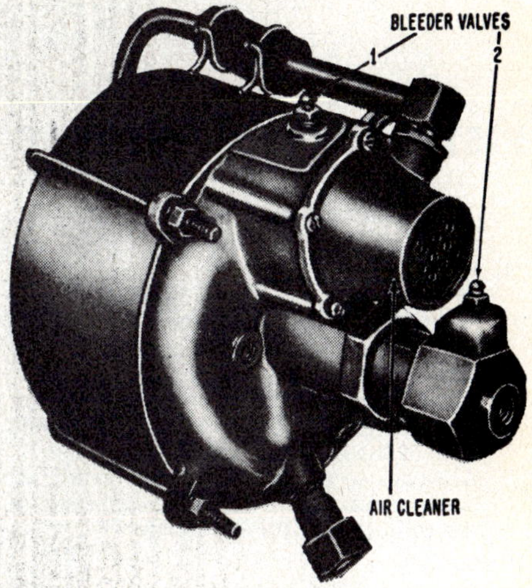

Fig. 1 Single 6¾" diameter Hydrovac with integral air cleaner

Fig. 2 Single 6¾" Hydrovac

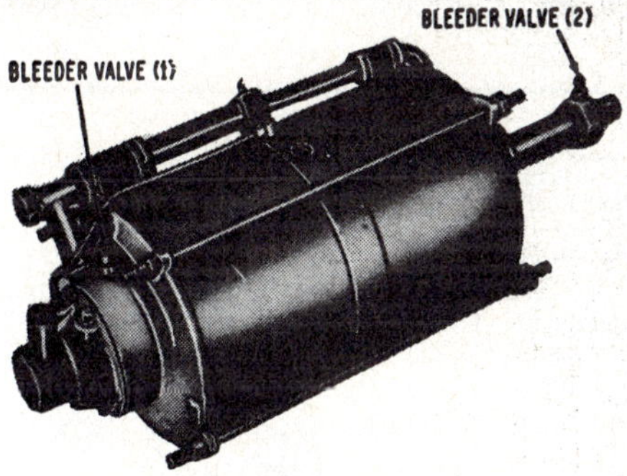

Fig. 4 Tandem 9½" Hydrovac

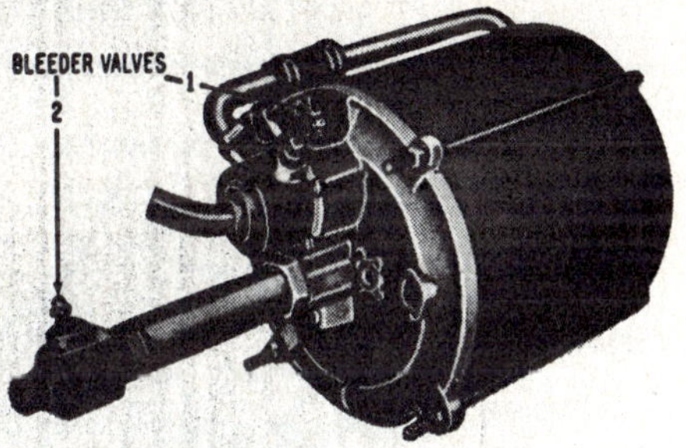

Fig. 3 Single 9½" Hydrovac

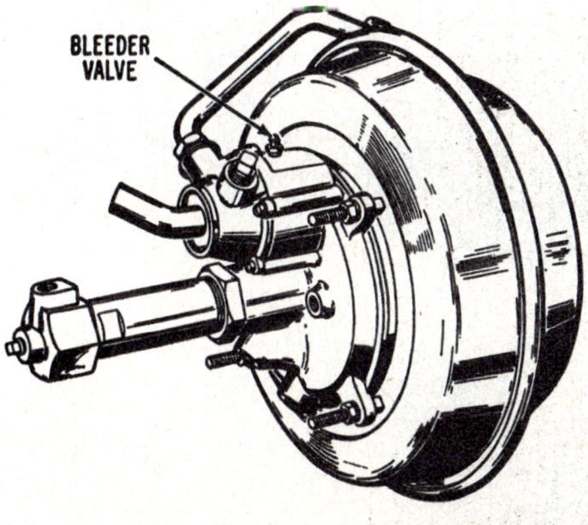

Fig. 5 Diaphragm type power cylinder. Note that only one bleeder valve is used

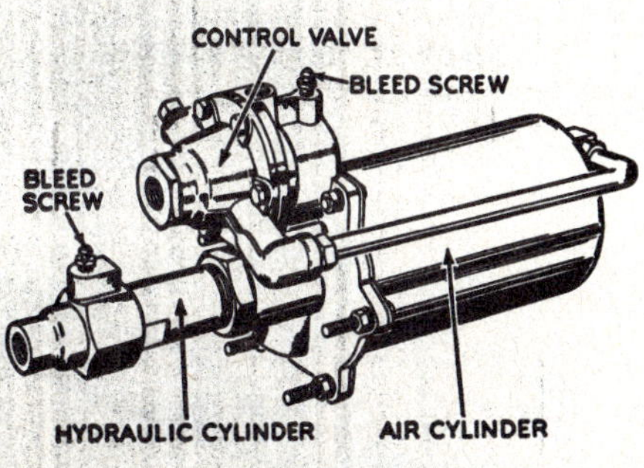

Fig. 6 Air-Pak power cylinder

Bendix Brake Boosters

PISTON TYPE HYDROVAC

Disassembly Notes

1. Before separating the main components of the booster, scribe a line across the cylinder shell and end plate so these parts can be reassembled in their original position.
2. Before separating the vacuum piston from the end plate, use hook-type clamps to hold the spring compressed.
3. Always use a repair kit and install all parts contained in the kit regardless of the condition of the old parts.

Reassembly

1. Press push rod leather seal into end plate with lip of seal toward hydraulic cylinder side of end plate, Fig. B1.
2. Assemble push rod seal parts into end plate as shown. Install stop washer with chamfered side toward end plate, the seal cup with lip of cup away from stop washer, retainer washer with flat side next to cup and small end of spring as shown. Place washer against spring and assemble snap ring into groove. Install new cylinder end seal.
3. Assemble hydraulic control valve piston retainer ring and stop washer into the fitting, Fig. B1.
4. Dip piston cups in brake fluid and assemble them on piston. On 6¾" diameter boosters, assemble one piston cup, spacer and the other cup on the piston and secure with the C-washer. Pinch ends of C-washer together. Lip of both piston cups must point *toward* C-washer end of piston. On 9½" units, assemble piston cups into grooves of piston. Cups must be assembled with their lips pointing *away* from each other. Insert piston and cups into the fitting, with the hole end of piston next to stop washer.
5. Place new rubber gasket over threaded end of fitting, thread fitting into end plate, and tighten fitting securely.
6. Thread nut on guide tube to limit of threads, Fig. B2.
7. Assemble reinforcing plate and seal over threads of guide tube. *Use care not to damage seal.*
8. Position guide tube and reinforcing plate assembly against cylinder shell. From inside shell, install and thread nut onto guide tube until it is flush with end of tube. Stake nut securely in two places, and tighten tube nut against reinforcing plate to insure a good seal between guide tube and cylinder shell.
9. Fabricate a vacuum piston ring by cutting a one-inch section from an old cylinder shell of correct size. Place assembly ring, Fig. B3, on bench and assemble vacuum piston parts in the assembly ring. Install larger diameter piston plate with chamfered side of hole up. Lip of packing should be *up* on 6¾" units and *down* on 9½" units. The seal ring and small diameter piston plate is installed with chamfered side of hole down.
10. Cut wicking to required length and assemble it against inner face of piston packing lip.
11. Position expander ring inside wicking with gripper points up and notch at loop end of expander ring under clip near opposite end of expander ring.
12. Assemble retainer plate with cut-out portion over loop of expander ring.
13. Hold guide in vertical position and assemble flat washer over threaded end.
14. With assembly ring still in position over vacuum piston, turn piston and ring upside down and assemble them into guide.

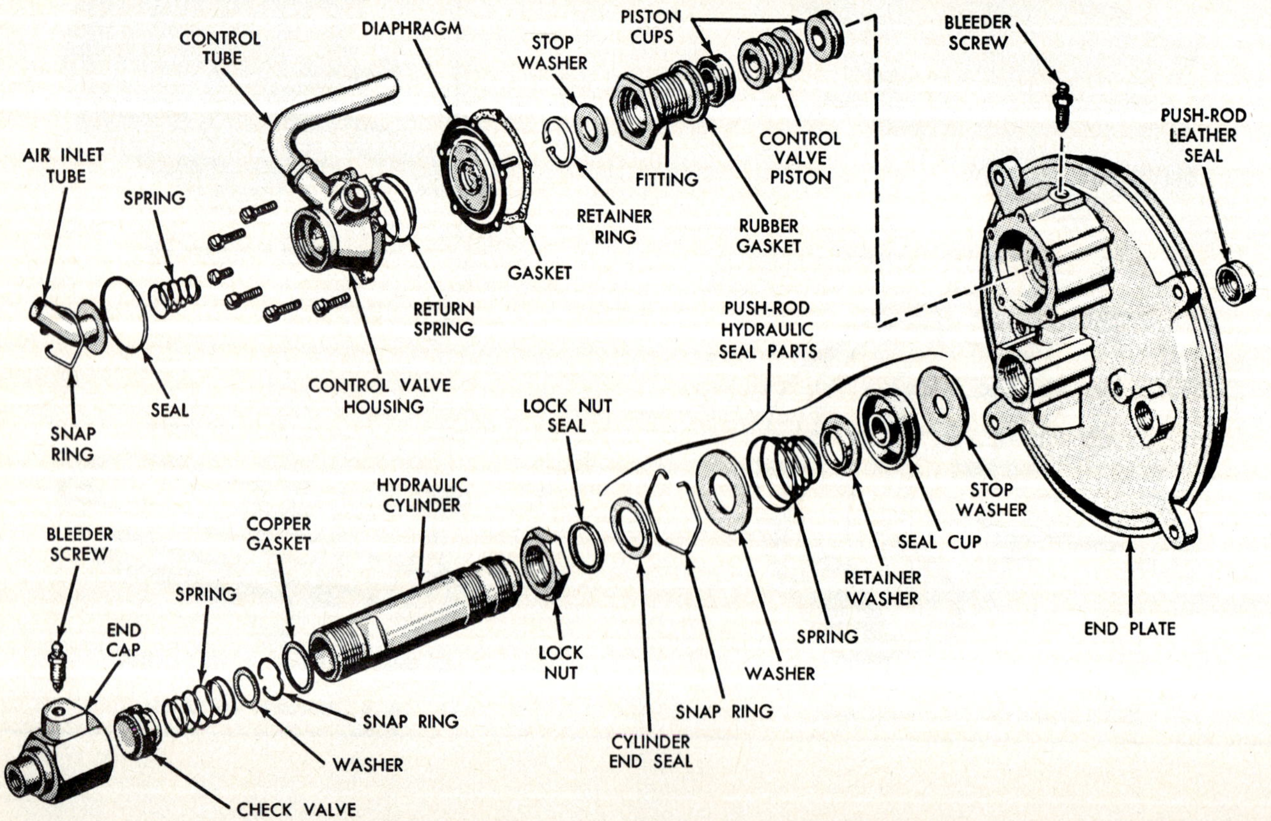

Fig. B1 End plate disassembled. Piston type Hydrovac

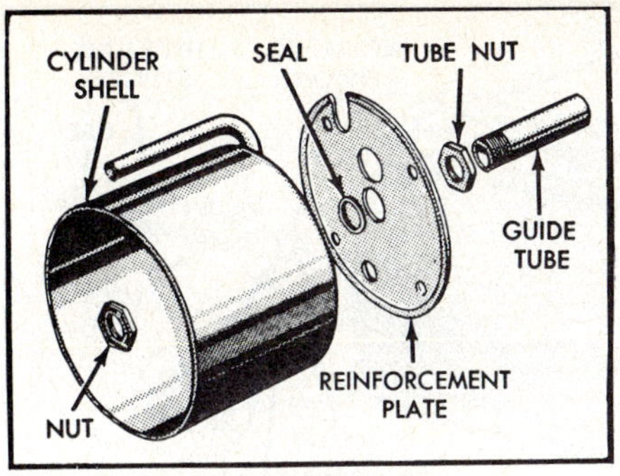

Fig. B2 Cylinder shell and guide tube disassembled

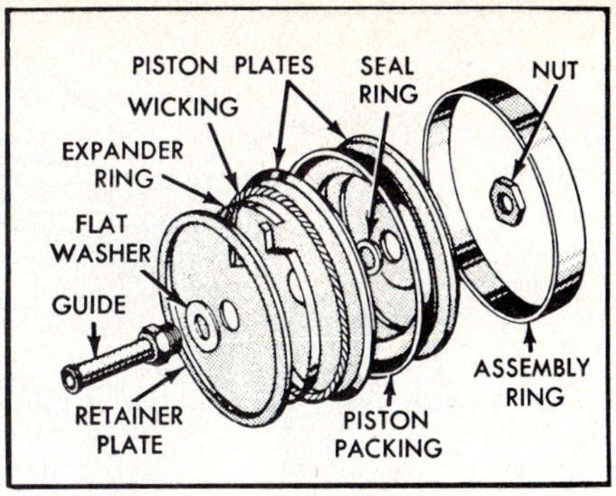

Fig. B3 Vacuum piston disassembled

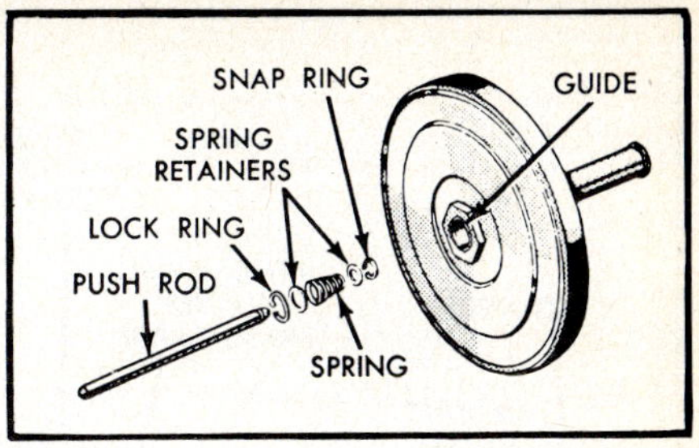

Fig. B4 Push rod parts

To prevent damage to seal ring, remove larger diameter piston plate while guiding seal ring over threads.

15. Replace piston plate and assemble nut onto guide. Tighten nut securely and stake in two places. *Do not remove assembly ring until piston is installed in cylinder shell.*
16. On 9½" units, clamp hexagonal section of guide firmly in a vise. Assemble push rod attaching parts over beveled end of push rod, Fig. B4, starting with lock ring. Use care in assembling snap ring in groove at lower end of push rod to avoid distorting lock ring.
17. On 9½" units, insert end of push rod in recess at end of guide. Attach push rod to guide by compressing spring and inserting lock ring in groove of guide. Be sure lock ring is firmly seated in groove.
18. Assemble check valve parts in end cap, Fig. B1, and secure with lock ring. Be sure lock ring is firmly seated in groove.
19. Place a new copper gasket in end cap. Thread hydraulic cylinder into end cap with milled flats next to end cap. Tighten cylinder securely.
20. Install bleeder screw into end cap. Thread lock nut on cylinder to limit of thread. Position a new lock nut seal over two sections of thread.
21. Mount vacuum piston and guide in vise at hex portion of guide. Place piston return spring over push rod with small end of spring next to piston, Fig. B5.
22. Carefully guide push rod through leather seal of end plate. Compress spring and install hook clamps to hold spring compressed.
23. Position hydraulic piston on push rod and secure with retainer pin. Slide retainer spring over hole in piston to hold retainer pin in place.
24. Carefully guide hydraulic cylinder over lip of piston cup, and thread cylinder into end plate (hand tight). Remove hook clamps and unit from vise.
25. Mount cylinder shell in vise (at guide tube nut) with cylinder shell in vertical position.
26. Insert rubber stopper into control line tube, Fig. B6.
27. With top face of vacuum piston flush with top edge of assembly ring, line up assembly ring with cylinder shell and guide vacuum piston into shell. *On 9½" units, if vacuum piston is permitted to tip slightly, vacuum piston guide will not enter guide tube.*
28. As soon as guide has entered tube, lift assembly ring to permit inspection of piston packing. If leather packing has started to enter cylinder shell without any part of packing folded back, remove rubber stopper, press vacuum piston into cylinder shell about 2" and replace rubber stopper. If inspection reveals that any part of packing has been folded back, repeat above operation.
29. With rubber stopper still in control line tube, slide assembly ring over end plate, Fig. B6, and remove rubber stopper.
30. Assemble a new rubber gasket in groove of end plate and align end plate with cylinder shell at scribe marks. Install hook bolts and tighten uniformly.
31. The use of guide pins threaded into end plate will simplify assembly of spring, diaphragm, gasket and control valve housing, and will reduce possibility of damaging diaphragm during assembly. Guide pins may be made by cutting heads from 8–32 × 2½" machine screws.
32. Install guide pins. Position a new gasket over pins and install diaphragm, return spring and control valve housing, Fig. B1.
33. Hold control valve housing and parts against end plate and remove one guide pin at a time, replacing with a screw and lock washer. Tighten screws securely.
34. Assemble vacuum hose on control tube and tighten hose clamp.
35. Thread hydraulic cylinder into end plate (hand tight). Align bleeder screw in end cap with bleeder screw in end plate. Tighten lock nut securely. *Do not use a wrench when tightening hydraulic cylinder in end plate.*
36. Make sure that all bolts, nuts, washers and screws are in place and securely tightened.

TANDEM TYPE PISTON HYDROVAC

NOTE: When disassembling, refer to the illustrations and observe the following: After removing the hydraulic cylinder and seal, place booster in an arbor press and insert a brass drift through atmospheric port in rear end plate. Press vacuum cylinder piston until hydraulic piston is exposed at hydraulic cylinder opening in end plate. Compress push rod pin retaining spring (on hydraulic piston), remove retaining pin and hydraulic piston from push rod. Disassemble hydraulic piston and remove booster from vise.

After disassembling the end plates, cylinder shells and center plate, force the center plate and rear vacuum piston together. Then insert a rod through the hole in the piston rod to hold the piston return spring in the compressed position. Fabricate two assembly rings from an old cylinder. Place assembly ring over front piston and remove nut securing front piston to rod. Then, remove piston

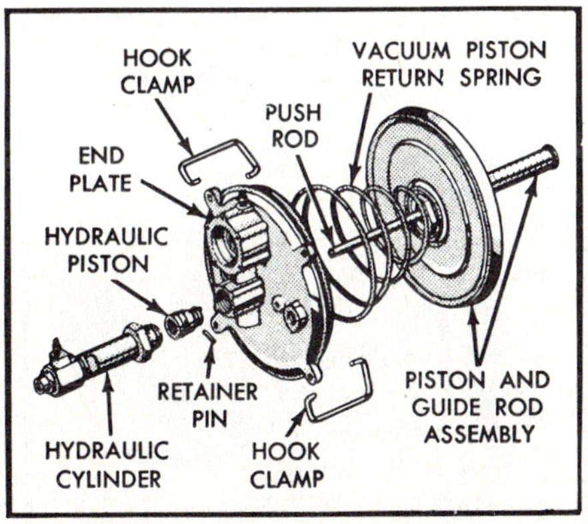

Fig. B5 Vacuum piston, end plate and hydraulic cylinder

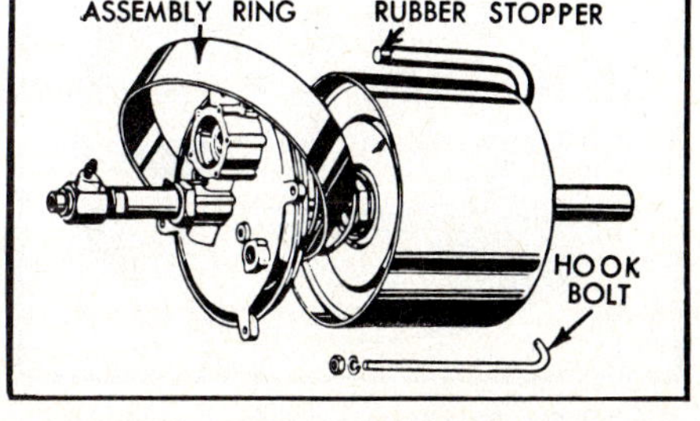

Fig. B6 Assembly ring removal

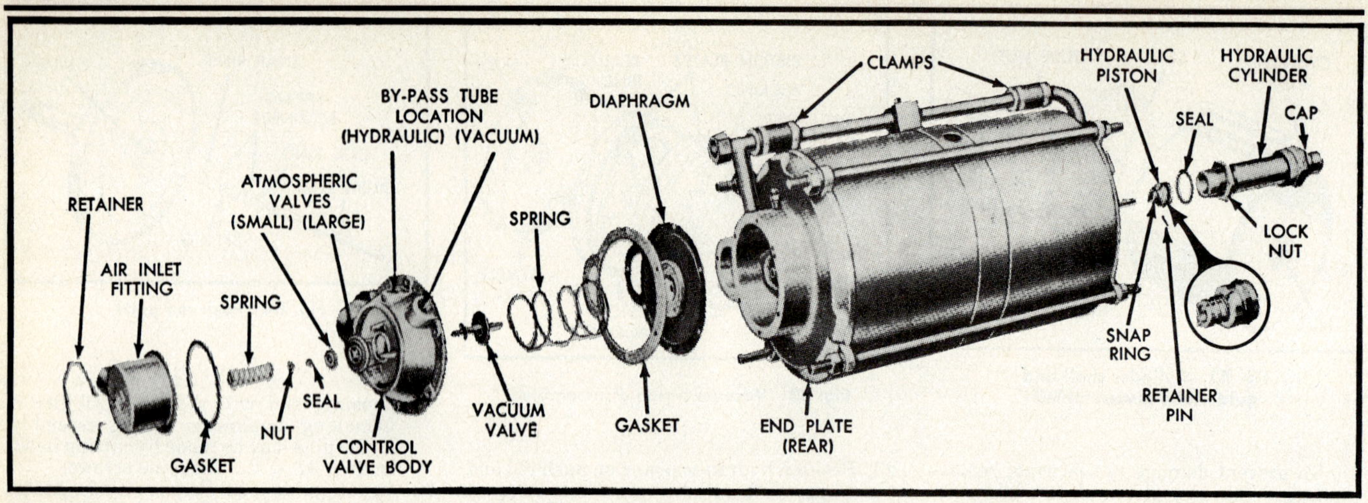

Fig. B7 Tandem piston type Hydrovac

rod nut, push rod retainer pin and push rod. Remove front piston, but keep parts assembled in assembly ring.

Reassembly, Fig. B7

1. Assemble rear vacuum piston by installing nut on rear vacuum piston end of piston rod with flat side of nut upward, Fig. B8.
2. Position larger diameter piston plate on piston rod with chamfered side of hole at the top. Guide rubber seal ring over threads of piston rod.
3. Place assembly ring on flat surface and install leather packing with lip side upward. Then position smaller diameter piston plate, with chamfered side of hole downward, in the ring.
4. Cut a new piece of wick to the required length. Then place it against inner face of leather packing lip. Assemble wick retainer (with gripper points upward) against wick and hook notched end of retainer under clip near opposite end of ring.
5. Position cut-out of retainer plate over loop of wick retainer.
6. Hold piston parts in assembly ring, assemble them on end of piston rod, then install nut on top of piston assembly. Tighten nut until it is flush with end of piston rod. Stake nut securely in two places. Clamp staked nut firmly in vise. Then tighten nut on opposite side of piston plate solidly against piston plate.
7. Place fast application valve shaft and rod seals, Fig. B9, into center plate. Piston valve shaft seal must be flush with bottom of hole. Piston rod seal should rest against shoulder of center plate.
8. Position center plate as shown in Fig. B10. Then place return spring on top of seal with small end at top. Install a bullet-nosed tool on threaded end of valve shaft and insert valve shaft through seal. Position gasket on center plate.
9. Place valve seat plate (with seat side downward) on gasket and install screws and lockwashers.

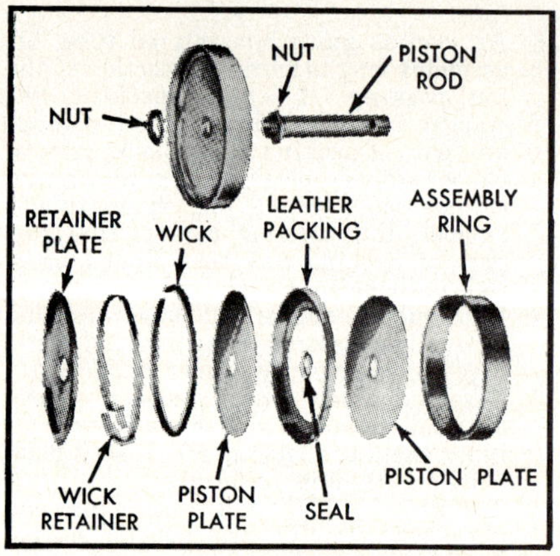

Fig. B8 Rear vacuum piston and rod assembly

10. Turn center plate over. Place lower diaphragm plate (rounded edge up) on valve shaft. Then place diaphragm gasket on top of plate, Fig. B11. Position diaphragm on top of gasket so that screw holes and by-pass hole index with identical holes in center plate. Install other diaphragm plate with rounded edge facing diaphragm.
11. Install nut on valve shaft. Use a screwdriver to prevent shaft from turning and tighten nut. Stake nut securely at two opposite points.
12. Position gasket and valve cover on center plate then install screws and lockwashers.
13. Place return spring over piston rod with small end of spring at bottom, Fig. B12.
14. Carefully guide piston rod through seal in center plate with piston stop flanges of center plate facing upward. Press center plate down against spring and insert a rod in the piston rod.
15. Assemble large end of push rod in end of piston rod and install pin.
16. Thread nut on piston rod with flat side of nut upward to limit of threads.
17. If front piston was disassembled to replace leather piston packing, cotton wicking or other parts, assemble piston parts in assembly ring in same order as rear piston, Fig. B8. Hold parts in ring and turn assembly ring over. Remove large diameter piston plate and seal. With assembly ring still in place, guide remaining piston parts over end of push rod and against piston rod nut, Fig. B13. Careful-

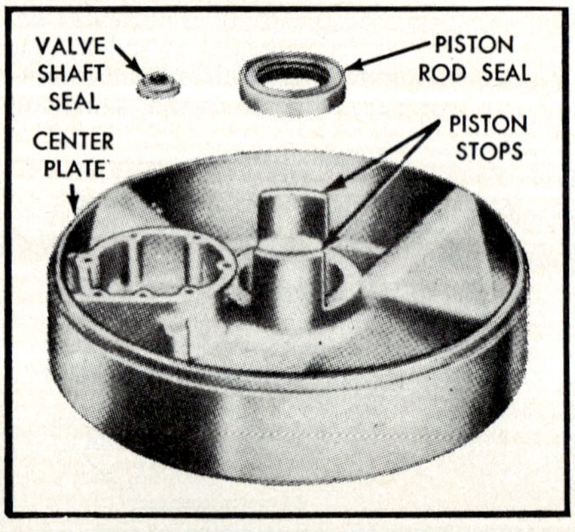

Fig. B9 Fast application valve and piston rod seals

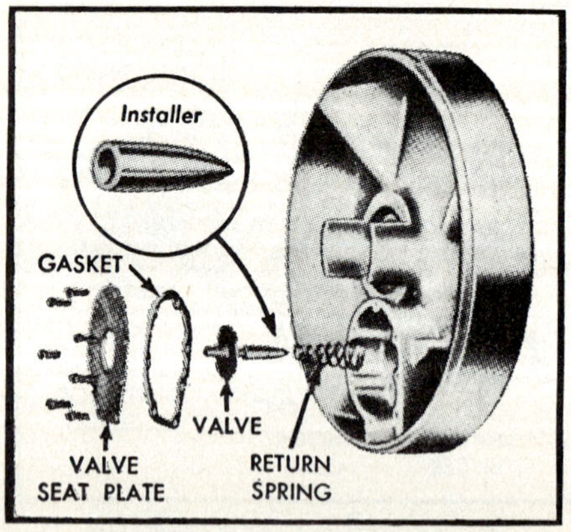

Fig. B10 Fast application valve and seal

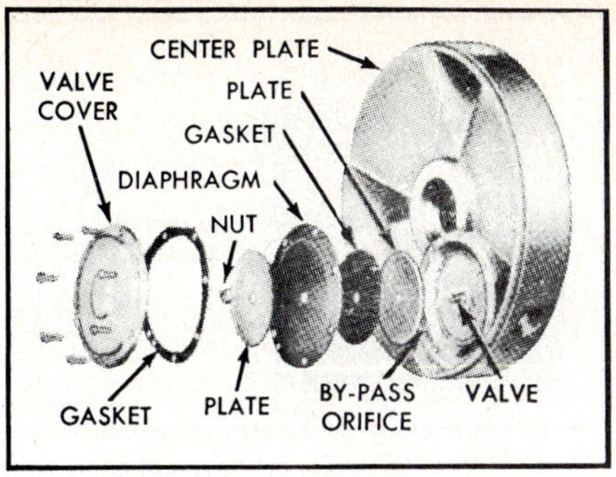

Fig. B11 Fast application valve and diaphragm

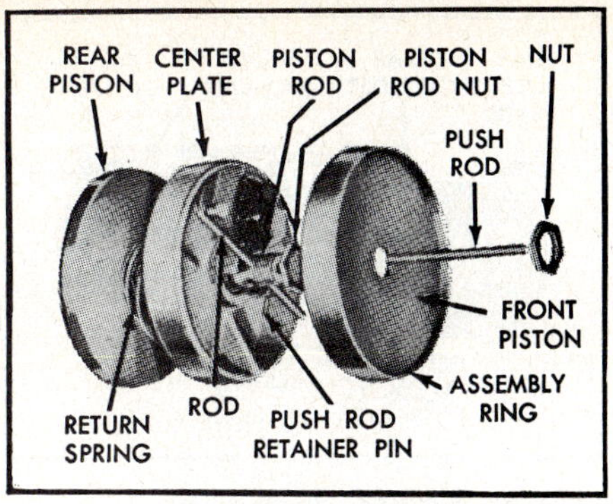

Fig. B12 Push rod piston and center plate

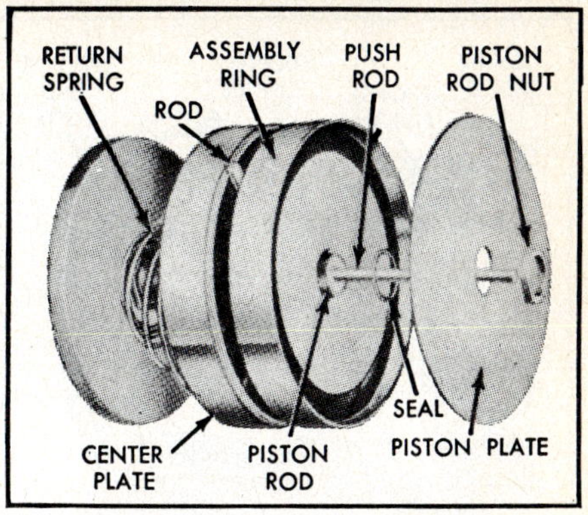

Fig. B13 Rear vacuum piston, return spring and outer plate assembly

ly install seal over threads of piston rod. Place larger diameter piston plate on piston rod with chamfered side of hole toward seal.
18. Install piston rod nut (flat side downward) on end of piston rod. Tighten nut until it is flush with face of piston rod. Stake nut securely at opposite points.
19. Hold piston rod nut in a vise or with a wrench and tighten inner nut securely against piston. *Use care when tightening inner nut to prevent retainer plate from shifting.*
20. Remove assembly ring. Then remove rod holding return spring compressed.
21. Hold end cap in a vise and assemble parts as shown in Fig. B14. Small end of spring must be placed inside check valve clips.
22. Install new copper gasket in end cap. Hydraulic cylinder must be assembled with milled flats next to end cap. Tighten hydraulic cylinder solidly in end cap and thread lock nut on hydraulic cylinder up to limit of threads.
23. Install lock nut seal (if used) in groove of cylinder tube. Install bleeder screw in cap.
24. Press leather seal into hydraulic cylinder bore of front end of plate (from inner side of plate) with lip of seal toward outer end of plate, Fig. B15.
25. Install push rod seal and related parts in order shown in Fig. B15. Chamfered side of washer is down, lip of push rod seal is up, flat side of flange washer is next to seal. Install retaining washer.
26. Install snap ring in inner groove of end plate.
27. Place rear end plate on holding fixture, Fig. B16, and install stop washer with flat side in valve fitting. Install retainer.
28. Dip hydraulic piston cups in brake fluid. Assemble cups on pistons with lips of cups positioned away from each other.

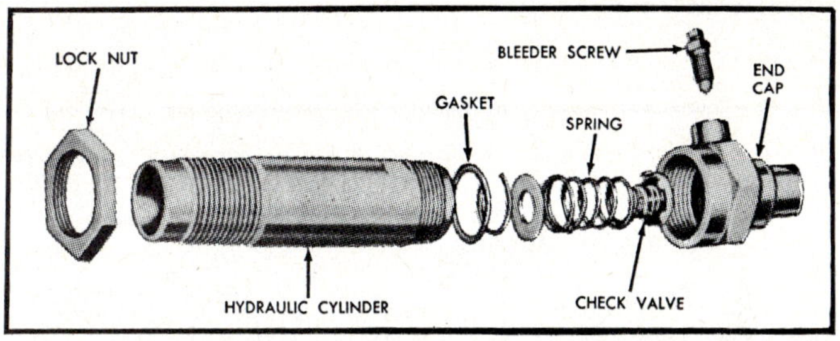

Fig. B14 Hydraulic cylinder, check valve and end cap

29. Insert piston into valve fitting with open end of piston toward stop washer.
30. Install new gasket on stop fitting. If fitting has no groove install copper gasket; if fitting has a groove use rubber seal gasket.
31. Install valve fitting in end plate. Tighten fitting equipped with rubber gasket firmly. Torque fitting equipped with copper gasket to 325–330 ft-lbs.
32. Assemble vacuum control valve parts in control body in sequence shown in Fig. B7. Install a new lead washer.
33. Install tee fitting and tubes on center plate. Small diameter pipe on vacuum tee should be toward front end plate.
34. Dip front piston leather packing and wicking in vacuum cylinder oil and allow excess oil to drain off.
35. Position a new gasket on front edge of center plate. Install front cylinder to center plate, Fig. B17. Coat interior of cylinder with vacuum oil. Carefully guide push rod through seal in front end plate. At same time, align vacuum tube in end plate with vacuum tube on center plate.
36. Position new gasket on front end plate and place on front cylinder.
37. Slide hose in place to connect two vacuum tubes.
38. Position a new gasket on outer edge of center plate, Fig. B18. Coat interior of rear cylinder with vacuum cylinder oil. Then tip front cylinder and end plate assembly at a 45-degree angle to prevent damage to rear piston leather packing. Carefully push rear cylinder over rear piston and onto center plate.
39. Place a new gasket on rear end plate. Then install end plate on rear cylinder, aligning end plate vacuum tube and center plate vacuum tube.
40. Install cylinder studs and tighten nuts securely.
41. Position vacuum hoses on tubes and tighten clamps.
42. Connect hydraulic by-pass tube to front and rear end plates. Install tube clip and screw to center plate.
43. To assembly hydraulic piston parts, place large end of spring in retainer cup and install check ball in piston body behind spring.
44. Dip new piston cup in brake fluid. Install cup on piston with lip of cup toward check ball.
45. Place booster assembly in arbor press. Insert a brass drift through atmospheric port in rear end plate. Press hydraulic cylinder piston until piston rod is exposed at front end plate. Assemble hydraulic piston on push rod and install retaining pin. Remove booster from press. Install seal on hydraulic cylinder.
46. Carefully guide hydraulic cylinder over piston cup and thread cylinder into end

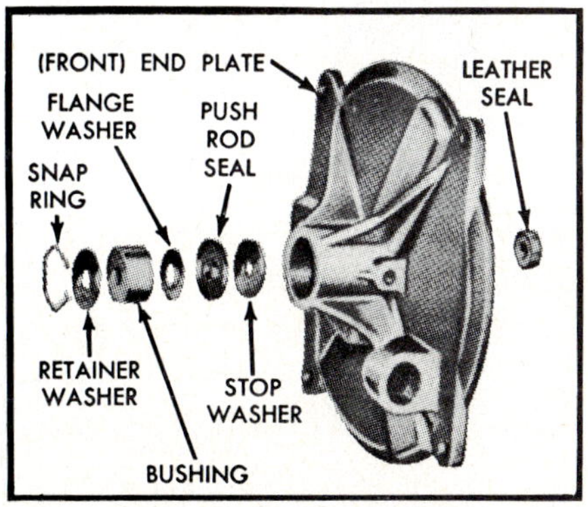

Fig. B15 Hydraulic cylinder push rod seals

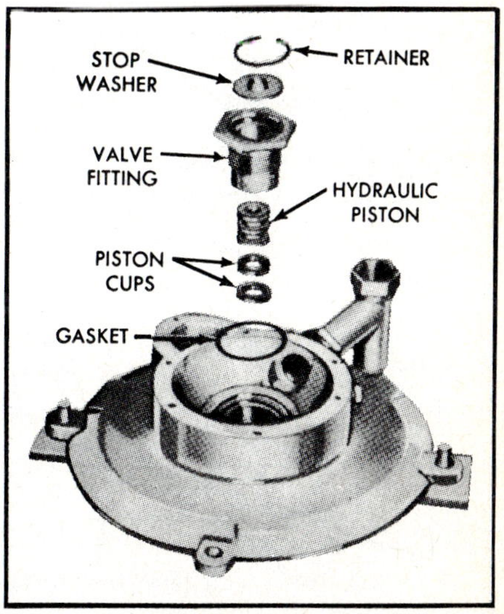

Fig. B16 Hydraulic control valve piston and fitting

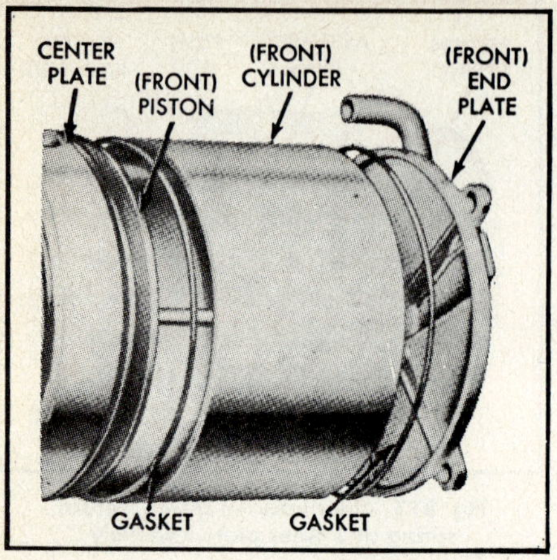

Fig. B17 Front cylinder and end plate

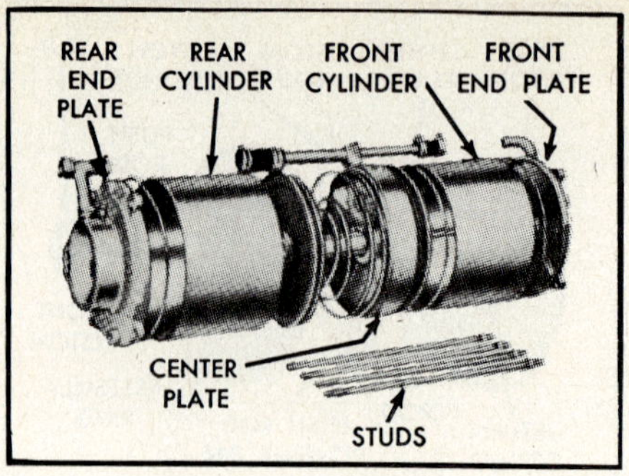

Fig. B18 Power cylinder

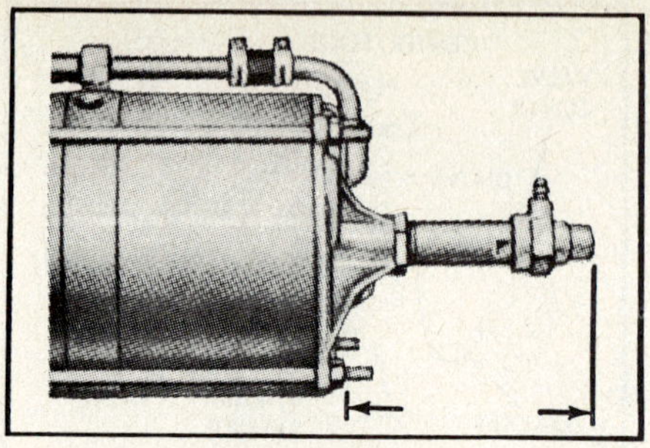

Fig. B19 Hydraulic cylinder adjustment

Atlast models	7 3/16″
Modul models	8 13/16″
Dreadnaught models	10 5/8″

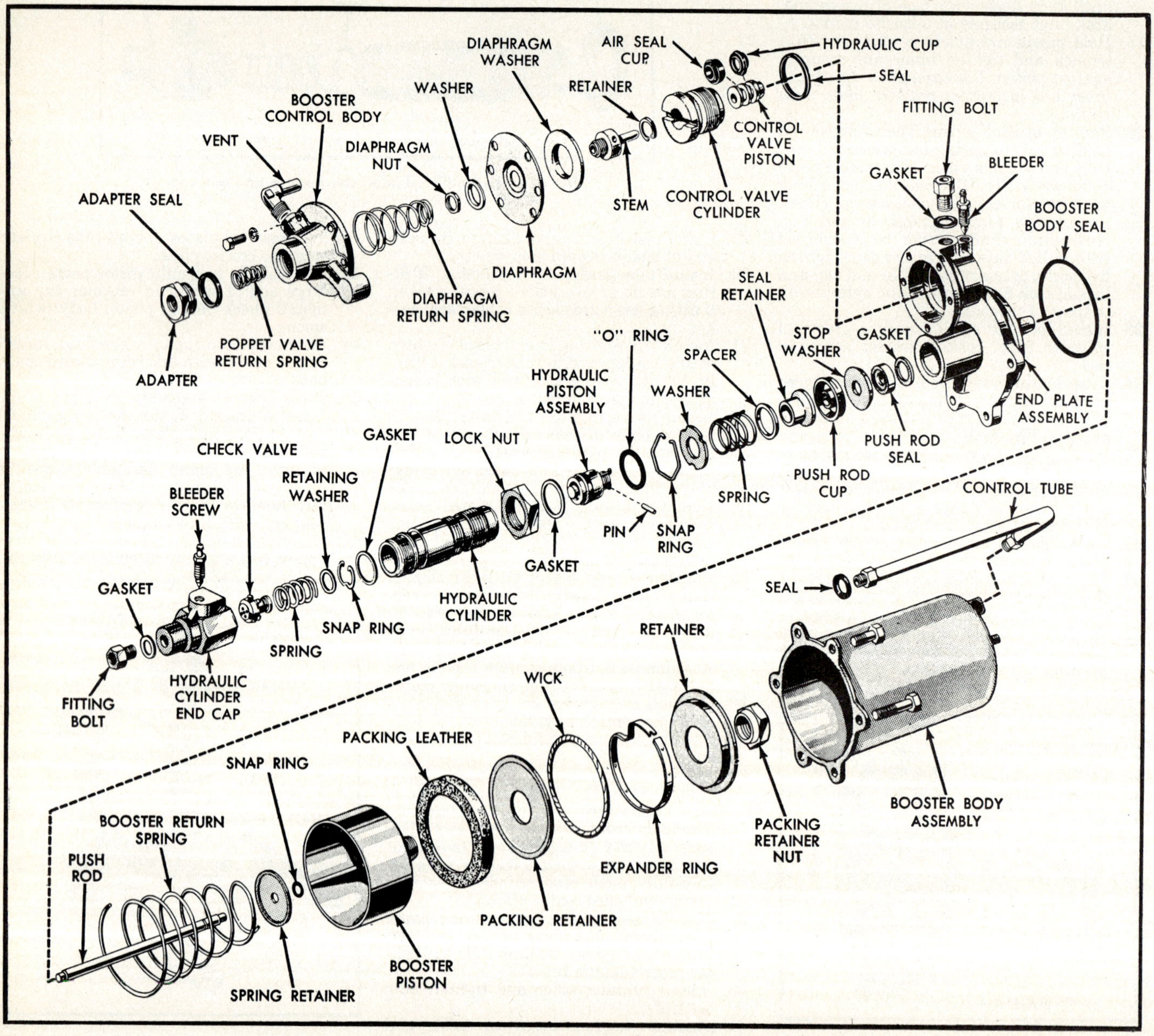

Fig. B20 Air-Pak power unit

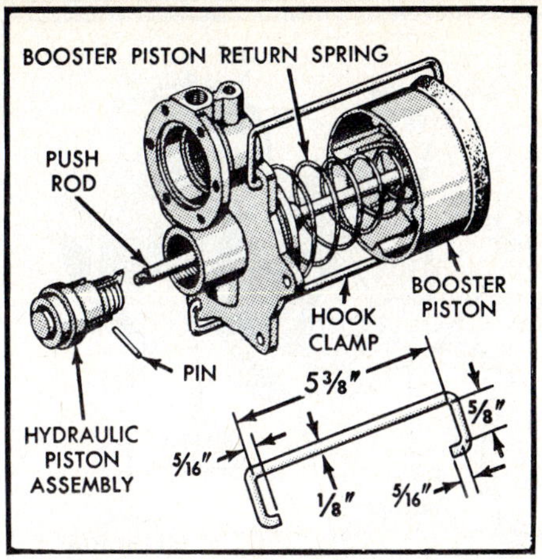

Fig. B21 Push rod pin removal

plate. Align bleeder screw in end cap with bleeder screw in control valve. Adjust hydraulic cylinder to the dimension given for the unit being serviced, Fig. B19, plus or minus 1/2 turn.

47. Install 5 guide pins (made from 8–32 × 2 1/2″ machine screws with heads cut off) in rear end plate.
48. Install air inlet fitting gasket and spring in control valve body, then install retainer.
49. Install diaphragm, inserting its stem into piston hole. Place diaphragm spring, gasket and control valve body on top of diaphragm.
50. Remove guide pins one at a time, and replace each with an attaching screw and new lock washer. Tighten screws progressively and firmly.
51. Install hydraulic and vacuum by-pass tubes.
52. See that all bolts, nuts, screws and plugs are in place, and that all tubes, clamps and fittings are firmly tightened.
53. Remove lube plugs from end and center plates. Add vacuum cylinder oil to level of filler plug holes. Install and tighten plugs.

BENDIX AIR-PAK POWER UNIT

Disassembly Notes

CAUTION: The length of the hydraulic cylinder, measured from the outer end of the cylinder end cap to the facing surface of the end plate varies according to the size of the Air-Pak. Therefore, before disassembling the unit, make a note of this dimension (see Fig. B24) so it can be adjusted properly when the unit is reassembled.

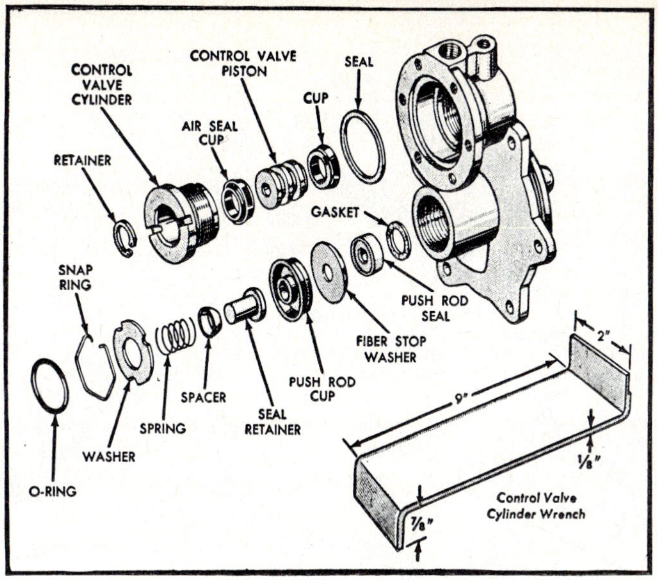

Fig. B22 End plate disassembled

1. Before separating the main components of the unit, Fig. B20, scribe a line on end plate and control body and also on the end plate and booster control body to assure proper alignment of these parts during reassembly.
2. Disassemble the unit as suggested by the illustrations, observing the following:
3. To disengage push rod from hydraulic piston, place piston over booster return spring, Fig. B21. Compress return spring by pressing toward end plate to move push rod forward. Remove pin retaining hydraulic piston to push rod. Remove hydraulic piston.
4. When removing booster piston, return spring and push rod from end plate, a hook clamp may be used to hold return spring in compressed position while removing pin, Fig. B21. Clamp should hook through one of the end plate holes and over end of booster piston. Clamp may be fabricated to dimensions shown in Fig. B21.
5. To remove push rod seal and gasket from end plate, thread a 1/4″ pipe tap into the seal. Drive seal out from booster body end of end plate, using a short piece of rod against pipe tap.
6. With special tool, Fig. B22, remove control valve cylinder from end plate.
7. To disassemble booster piston, use a holding fixture to prevent damaging piston. Dimensions for fabricating fixture are shown in Fig. B23.

Reassembly

NOTE: Always use a repair kit when servicing an Air-Pak unit, and use all parts contained in the kit regardless of the condition of the old parts. Do not allow any of the parts to come in contact with mineral oil or grease.

1. To aid in reassembling the booster piston, make an assembly ring from an old booster body of the same diameter as the one being serviced. Fig. B25 shows a sample assembly ring.
2. With assembly ring flat on work bench, insert a new packing leather inside ring with lip up, Fig. B23. Place packing retainer over packing leather with cupped edge pointing toward packing

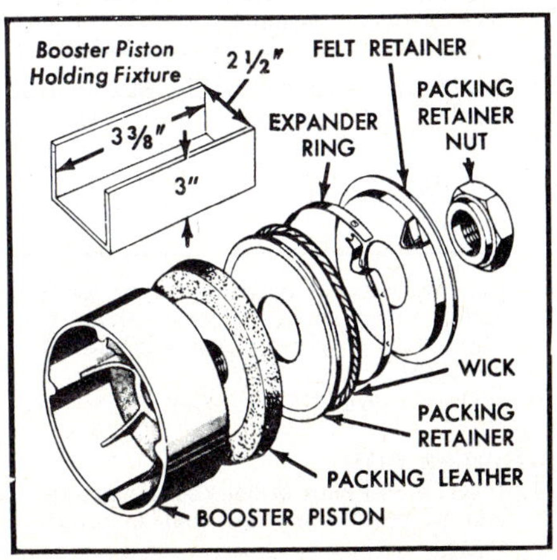

Fig. B23 Booster piston disassembled

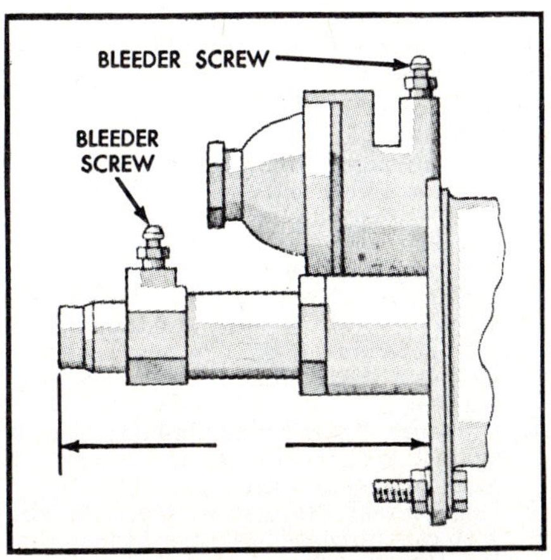

Fig. B24 Hydraulic cylinder length setting

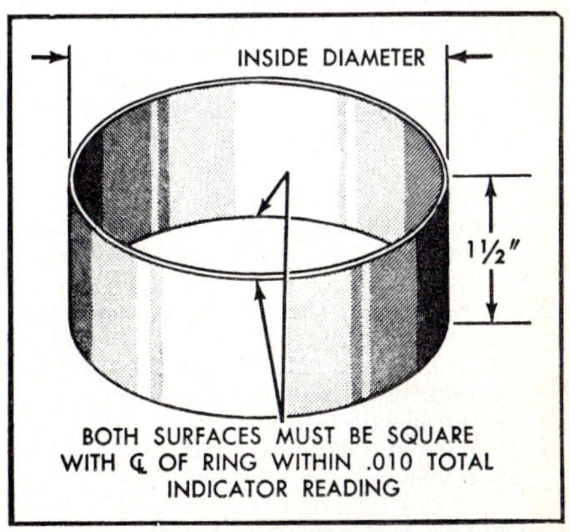

Fig. B25 Assembly ring made from old booster body

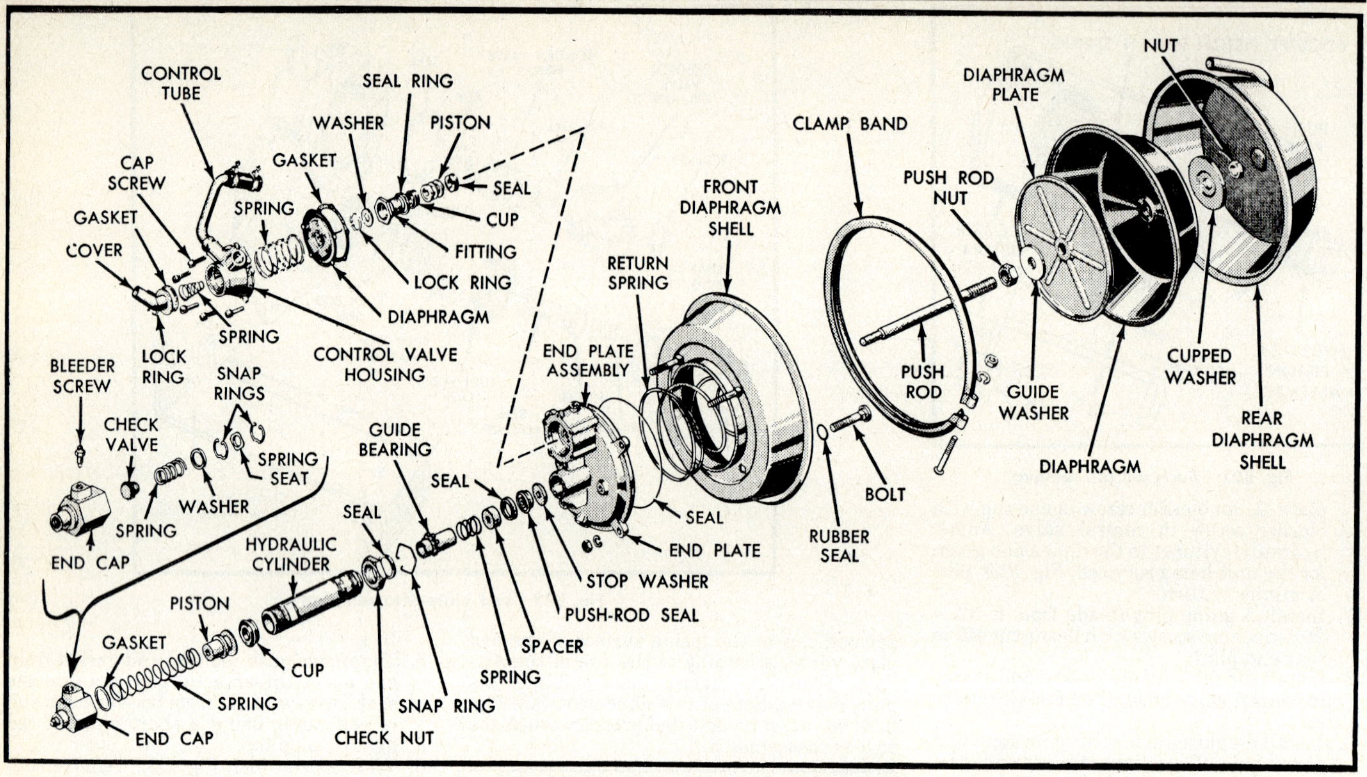

Fig. B26 Diaphragm type Hydrovac disassembled

leather.
3. Cut a new wick to proper length and dip it in vacuum cylinder oil. Assemble wick against inner lip of packing leather.
4. Position a new expander ring on inner side of wick (gripping points up). Engage notch at loop end of ring with hook near opposite end of ring.
5. Position retainer over expander ring with cut-out in retainer lining up with loop in expander ring.
6. Place assembly ring with assembled parts over threaded boss of booster piston. Install packing retainer nut finger tight.
7. Clamp piston holding fixture in vise. Place booster piston and assembly ring on holding fixture. Tighten packing retainer nut securely. *Do not remove assembly ring from piston until assembly is to be inserted into booster body.*
8. Remove piston assembly and holding fixture from vise.

HYDRAULIC CYLINDER END CAP

1. Clamp end cap in vise.
2. Insert check valve, spring (with small end inside tabs of check valve) and retaining washer in cap bore. Compress spring and insert a new snap ring. Be sure snap ring is engaged in its recess.

HYDRAULIC CYLINDER

1. Place new copper gasket in end cap. Screw hydraulic cylinder into end cap and tighten.
2. Screw lock nut onto inner threaded portion of cylinder. Dip a new gasket in brake fluid and place it between threaded portions of cylinder.
3. Install bleeder screw.
4. Remove end cap and hydraulic cylinder from vise.

CONTROL VALVE CYLINDER

1. Dip a new control valve piston, hydraulic cup and air seal cup in brake fluid. Assemble hydraulic cup in recess of valve piston (at chamfered end of piston). Lip of hydraulic cup must be pointing toward chamfered end of piston.
2. Assemble air seal (identified by red dot) in other recess of valve piston with lip pointing opposite to that of hydraulicc cup.
3. Insert retainer into recess in notched end of control valve cylinder.
4. Place piston assembly into valve cylinder from opposite side of retainer ring groove, with hole end of piston next to retainer.

DIAPHRAGM

1. Place a new diaphragm over threaded end of stem with concave side of diaphragm toward threaded end.
2. Place flat side of washer facing nut. Screw diaphragm nut over threaded end of stem finger tight.
3. Clamp diaphragm nut in a vise. Tighten assembly by using a drift in one of the adapter holes.
4. Remove diaphragm assembly from vise and stake nut.

END PLATE

1. Clamp flange of end plate in vise.
2. Position a new seal gasket and a leather push rod seal into recess in end of base of end plate with push rod seal lip pointing toward seal gasket.
3. Seat push rod seal in recess by using a length of 7/8" or 15/16" rod and tapping rod with fiber mallet. Insert a new fiber stop washer with chamfered side toward push rod seal.
4. Dip a new cup in brake fluid and place it against stop washer with lip of cup pointing away from stop washer.
5. Position seal retainer and spacer in bore with counterbore of retainer next to cup, Fig. B22. Install spring and washer.
6. Hold assembled parts in place by installing stop ring into its recess in end plate bore. Dip new O-ring in brake fluid and insert it in recess in bore of end plate.
7. Dip a new control valve cylinder seal in brake fluid and install it in recess of control valve cylinder.
8. Screw control valve cylinder into end plate. Tighten cylinder using tool shown in Fig. B22.
9. Install bleeder screw, fitting bolts and gasket.
10. Dip hydraulic piston in brake fluid. Lip of cup must point toward hydraulic cylinder end cap.
11. Slide push rod with booster return spring through end plate seals and cup.
12. Place booster piston with assembly ring over end of push rod and spring retainer.
13. Compress booster return spring and slide hydraulic piston assembly onto other end of push rod, Fig. B21. Line up hydraulic piston, pull back piston spring, and install pin in push rod hole. Release pressure on booster return spring and remove booster piston assembly.

NOTE: The hook clamp, Fig. B21, may be used to hold booster return spring in compressed position while installing pin. Clamps should hook through one side of holes in end plate and over end of booster piston.

14. Carefully guide hydraulic cylinder and end cap over lip of hydraulic piston cup. Then thread cylinder into end plate finger tight.
15. Remove end plate from vise.

BOOSTER BODY

1. Insert a new booster body seal in groove of end plate. Make sure seal is properly seated.
2. Apply a light coat of vacuum cylinder oil to inner walls of booster body. *Do not use*

engine oil.

3. Insert booster piston into booster body (wick end first) being careful not to damage or roll over leather lip. Remove assembly ring.
4. Assemble booster body to end plate by placing body over return spring. Line up scribed marks on body and end plate. Secure body to end plate, using two short bolts at each upper end of body.
5. Install remaining bolts, tightening each evenly and uniformly.
6. Adjust length of hydraulic cylinder to its original position as measured before unit was disassembled, Fig. B25.
7. Align bleeder screw on hydraulic cylinder end fitting with bleeder screw on end plate. Tighten lock nut securely.
8. Install fitting bolt and gasket into end of hydraulic cylinder end cap.

BOOSTER CONTROL BODY

1. Place a new seal on adapter. Position small end of spring in body bore over poppet valve stem and install adapter.
2. Remove body from vise.
3. Place diaphragm washer in recess in end plate, with flat side facing outward.
4. Position diaphragm return spring and diaphragm in booster control body.
5. Position booster control body on end plate. Align diaphragm holes and control body-to-end plate alignment marks. Install bolts and lockwashers and tighten bolts securely.
6. Install a new seal on control body end of air control tube. Connect air control tube to booster control body and to booster body.

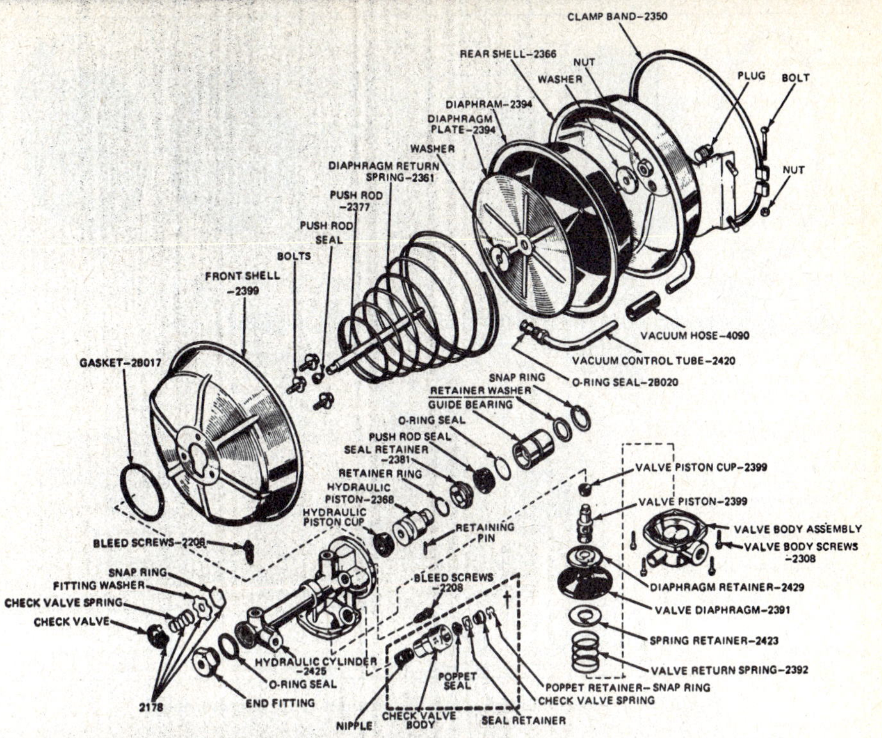

† USED ON SPLIT SYSTEM INSTALLATION ONLY.

Fig. B27 Frame mounted single diaphragm type, disassembled

SINGLE DIAPHRAGM TYPE HYDROVAC, FIG. B26

NOTE: Before disassembling the booster, scribe a line across the two halves of the diaphragm chamber and also across the end plate and control valve housing so that these parts can be reassembled in their original position. When reassembling use all the parts contained in the repair kit regardless of the condition of the old parts.

Reassembly

1. If the push rod and diaphragm were disassembled, thread the push rod nut on the rod to the limit of the thread. Flat side of nut should face diaphragm. Position guide washer and diaphragm plate on push rod. Position diaphragm and cupped washer with concave side of washer next to diaphragm. Install and stake nut in place.
2. If front diaphragm shell bolts were removed, installed bolts, using rubber seals. Stake bolts to shell. Place rear shell on bench. Position push rod and diaphragm on shell. Install front shell and align front and rear shells to scribed mark and to bead on diaphragm. Install clamp band over rim of two shells and tighten securely.
3. Dip control valve piston cup and seal in brake fluid and position cup with lip away from hole end of piston. Position seal on piston with lip flare opposite that of cup flare.
4. Insert piston into fitting. Place seal on fitting, then install fitting on end plate. Install washer and lock ring.
5. Position gasket, diaphragm and spring on end plate.
6. Position control valve housing over spring. Align scribed marks on control valve housing and end plate. Install capscrews and tighten securely.
7. Position spring in valve housing with small end toward valve housing. Install gasket cover and lock ring.
8. Install a new ring seal in groove of end plate.
9. Place small end of return spring over guide washer of diaphragm.
10. Position end plate over return spring and align scribed marks located on end plate and diaphragm chamber. Install nuts and lockwashers and tighten nuts uniformly.
11. Slide hose over control tube and tighten clamp.
12. Install check valve, spring, washer, snap ring, spring retainer and second snap ring in end cap.
13. Install lock nut and new seal on hydraulic cylinder.
14. Dip piston cup in brake fluid. Position cup on piston with lip of cup toward small diameter end of piston.
15. Position return spring over small end of piston. Insert spring and piston into hydraulic cylinder.
16. Install stop washer with chamfered edge toward diaphragm. Install push rod seal with flared edge toward hydraulic cylinder, then the spacer with chamfered edge toward seal, the spring, and guide bearing with serrated end outward. Install snap ring and a new seal in end plate.
17. Screw cylinder into end plate. Align bleeder screw on end cap with bleeder screw on end plate, then tighten lock nut.
18. Position a new copper washer in end cap on check nut and screw hydraulic cylinder into end cap. Tighten cylinder securely.

SINGLE DIAPHRAGM TYPE HYDROVAC, FIG. B27

NOTE: Before disassembling unit, scribe marks across front and rear shells, across flange of hydraulic cylinder and across flanges of valve body and housing. Diaphragm assembly should be removed only if damaged parts are to be replaced.

Reassembly

1. Install check valve, spring, washer and snap ring on end fitting, then assemble O-ring and end fitting on hydraulic cylinder.
2. Install cups, back to back on control valve piston, then assemble piston, diaphragm retainer and valve diaphragm.

NOTE: Make certain that inner bead of diaphragm is seated in piston groove.

3. If removed, install small coil end of poppet spring over valve poppet and install parts in valve body. Assemble valve seal (flat side facing up) and poppet retainer on stem of valve poppet. Invert valve body and rest valve poppet on a small wood block. Using a soft hammer, tap small retainer lightly until retainer snap locks on stem. Assemble plastic cover in valve body.
4. Install spring retainer (flange facing down) on spring in valve body then place piston and diaphragm on spring retainer and press outer bead of diaphragm into groove of valve body.

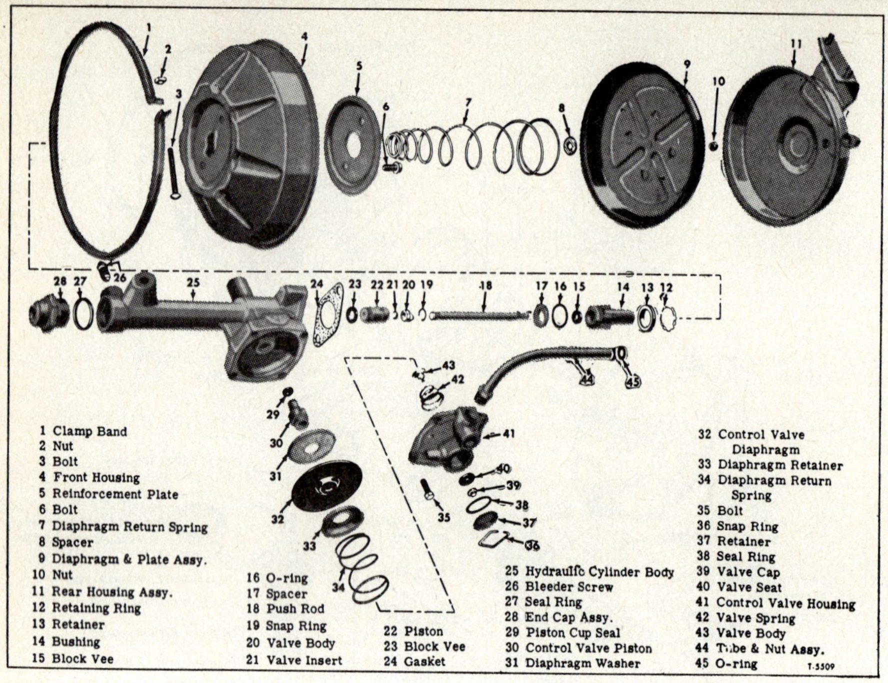

Fig. B28 Frame mounted single diaphragm type, disassembled

5. Coat control valve piston with brake fluid and assemble piston in control valve cylinder. Align scribe marks on valve body, install and tighten screws.
6. Place new seal on clean wood block with rubber side facing down. Rest push rod vertically on seal stem and strike threaded end of push rod with soft hammer to seat stem in push rod.

NOTE: Make certain that shoulders of push rod and seal are in contact.

7. Install onto push rod the snap ring, retainer, washer, guide bearing with O-ring in groove, seal cup and seal retainer on push rod.
8. Install piston onto push rod and secure with retaining pin and ring. Coat cup with brake fluid and install it on piston.
9. Assemble nut, washer, diaphragm plate, diaphragm, washer and nut on push rod.
10. Install the gasket in groove in flange of hydraulic cylinder, aligning the cut-out in front shell with port in hydraulic cylinder, install and tighten screws.
11. Install large coil end first of return spring against diaphragm plate. Coat hydraulic cylinder bore with brake fluid and work piston, cups and seal into cylinder bore.
12. Roll back outer edge of diaphragm and press against diaphragm to compress return spring. Guide parts on push rod into cylinder bore and secure with snap ring.

NOTE: Make certain that snap ring is securely seated before relieving pressure on spring.

13. Lightly coat both sides of diaphragm with silicone lubricant or talcum powder then align scribe marks on front and rear shells and press rear shell flange and diaphragm bead into place against front shell flange.
14. Install band clamp nut and bolt. Tap band lightly with a rawhide mallet while torquing nut to 30–40 in. lbs.
15. Install a new seal on vacuum tube and install tube and hose on real shell tube and torque tube nut to 80–120 in. lbs.

SINGLE DIAPHRAGM TYPE HYDROVAC, FIG. B28

NOTE: Before disassembling unit, scribe marks on rear diaphragm housing, clamp band and front housing, on front diaphragm housing and hydraulic cylinder and on control valve housing and hydraulic cylinder body.

CAUTION: When removing diaphragm and plate assembly, be very careful as outer edge of diaphragm plate is very sharp and can cause serious cuts if not handled properly.

Reassembly

1. Place control valve spring on body and install in housing, then install valve seat from other end and secure it with valve cap (press fit).
2. Install seal ring and retainer in housing and secure with snap ring, then assemble diaphragm retainer, diaphragm, diaphragm washer, control piston and cup seal.
3. Install control valve diaphragm assembly in cylinder body and place control valve diaphragm return spring into housing.
4. Install housing and spring on cylinder body and tighten the four bolts alternately.
5. Install block vee on piston and valve insert in valve body, then assemble valve body, piston and push rod and secure with snap ring.
6. Install spacer and O-ring on push rod and vee block in bushing, then install bushing and washer (where required) on push rod.
7. Install push rod, piston and bushing assembly in hydraulic cylinder body. Secure with snap ring or retainer (torque retainer to 15–20 ft. lbs.) and retaining ring.
8. Place gasket, front housing and reinforcement plate on hydraulic cylinder body and tighten the two bolts.
9. Install diaphragm return spring, spacer, diaphragm, plate assembly and lock nut on push rod.
10. Install tube and nut assembly on control valve and install O-ring.
11. Position rear housing against front housing and secure with clamp band, then install mounting bracket on hydraulic cylinder body (or check valve end cap, if used).

DASH MOUNTED, TANDEM DIAPHRAGM TYPE

NOTE: Before disassembling unit, scribe marks across front and rear shells of power section and across mounting bracket.

Reassembly, Fig. B29

1. Place rear shell on wood blocks, with studs down, and press new seal, with plastic bearing face first, into rear shell.

NOTE: The top outside flange of seal should be pressed in .305 inch (about 5/16 inch) below flat shell surface next to seal cavity.

2. Place front diaphragm on front plate. Apply a light film of silicone lubricant on outside surface of front plate hub and liberally on center plate bore seal, then install large O-ring seal on the center plate and press O-ring against the four bosses around outside diameter of center plate.
3. Carefully guide center plate and seal assembly (seal side first) onto front plate hub. Apply a light coat of silicone lubricant on front and rear bearing surfaces of valve plunger, avoiding the rubber grommet inside plunger. Assemble valve plunger around rubber bumper and set spring and plunger assembly in recess of front plate hub, with grommet side facing up.
4. With vacuum seal firmly placed against shoulder on outside of front plate hub, set rear plate with threaded bore facing down, over the valve plunger and screw rear plate into front plate hub. Place a 1 1/16 inch hex stock bar in vise and place plate assembly with front plate down, on hex bar. Using air channel slot or rear plate edges, torque plates to 120–180 in. lbs.
5. Remove plate assembly from vise and install rear diaphragm onto rear plate and over lip of center plate. Assemble rear diaphragm spring retainer over rear diaphragm and lip of center plate, then press retainer onto center plate until it seats against shoulder of center plate.
6. Coat inside wall of rear shell with talcum powder and liberally apply silicone lubricant to bearing seal in rear shell and to the scalloped cut-outs on edge of front

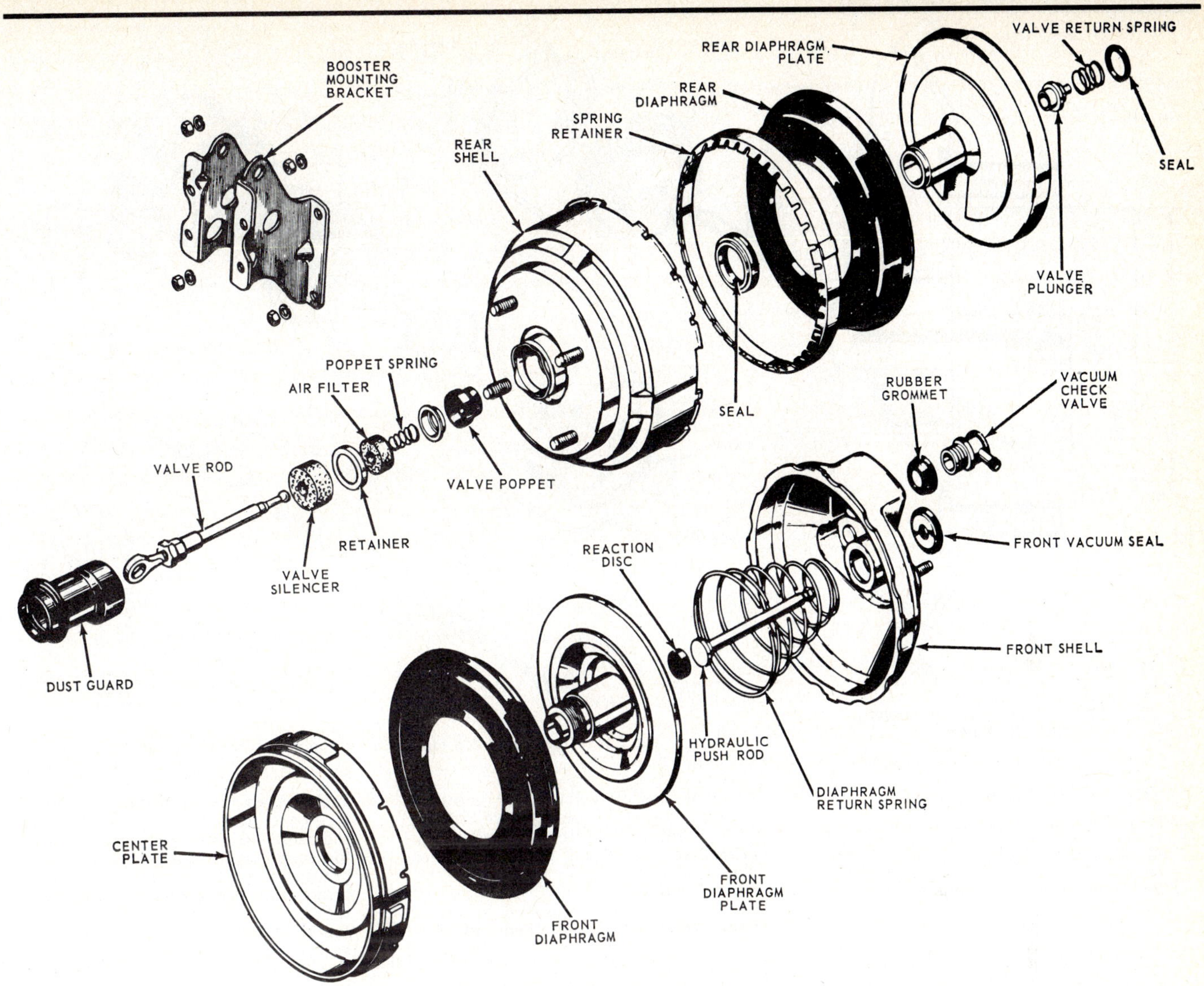

Fig. B29 Dash mounted tandem diaphragm type, disassembled

shell.

NOTE: When assembling diaphragm and plate assembly in rear shell, the rear diaphragm and center plate lugs must be aligned between lances on rear shell.

7. With valve rod eyelet clamped in a vise, carefully guide valve housing sleeve through bearing seal inside of the rear shell, maintaining diaphragm and plates in correct alignment. Ease outer rim of front diaphragm into rear shell until outer rim of front diaphragm is under each of the retaining lances on rear shell.
8. Place large end of a diaphragm spring over front plate hub and position rear shell onto spring so that scribe marks will be aligned when the shells are twist-locked together (use a flat bar on rear shell to turn shells).

NOTE: Make certain that cut-outs on front shell are aligned with retaining lances on rear shell.

9. Guide rim of front shell into rear shell and press firmly downward on rear shell until edge of front shell clears the lances on rear shell, then twist the rear shell clockwise in relation to front shell until contact's with stops. Remove unit from vise.
10. Liberally apply silicone lubricant to entire surface of rubber reaction disc and to piston end of hydraulic push rod, then place reaction disc on piston end of push rod. Apply a light coat of silicone lubricant on rod of push rod, keeping lubricant away from threads.

NOTE: Do not allow lubricant on the adjustment screw or threads.

11. Install push rod with reaction disc on piston end into front plate hub cavity, then twist push rod to make sure that reaction disc is seated in front plate hub and air bubbles are eliminated between hub, disc and push rod piston.
12. Assemble seal (support plate side first) over adjustment screw end of push rod, then press seal into front shell recess until seal bottoms against shell. If vacuum check valve was removed, coat a new grommet with alcohol and press it into front shell, making certain that grommet is seated in shell.
13. Coat shoulder of check valve and inside of grommet with alcohol and press check valve into grommet until entire circumference of check valve flange bears against grommet.
14. Coat poppet valve with alcohol and assemble it into valve housing with small end of poppet first, then coat the poppet retainer with alcohol and assemble it into the housing with flange facing out. Press in against retainer to make certain that shoulder on the retainer is positioned inside poppet.
15. Assemble retainer, valve silencers and filters and valve return spring over ball end of valve rod, then coat rubber grommet and ball end of valve rod with alcohol. Guide the spring, filters and silencers into valve housing and assemble the ball end of valve rod in valve plunger. Using a soft mallet, tap end of valve rod to lock ball end of rod in valve plunger grommet, then press filters into place and assemble

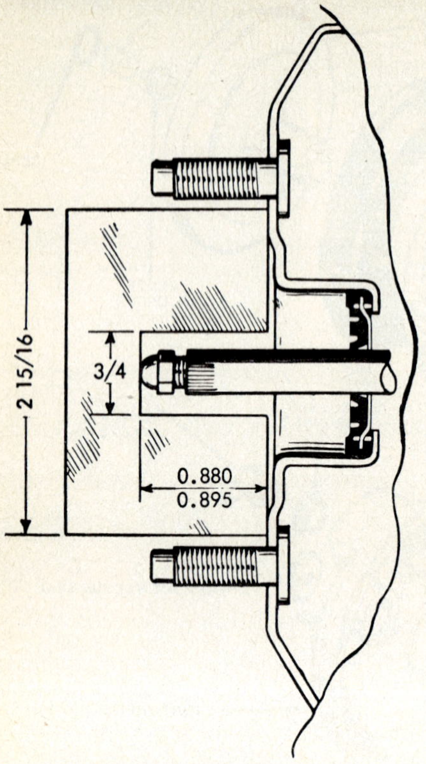

Fig. B30 Push rod gauge dimension and adjustment

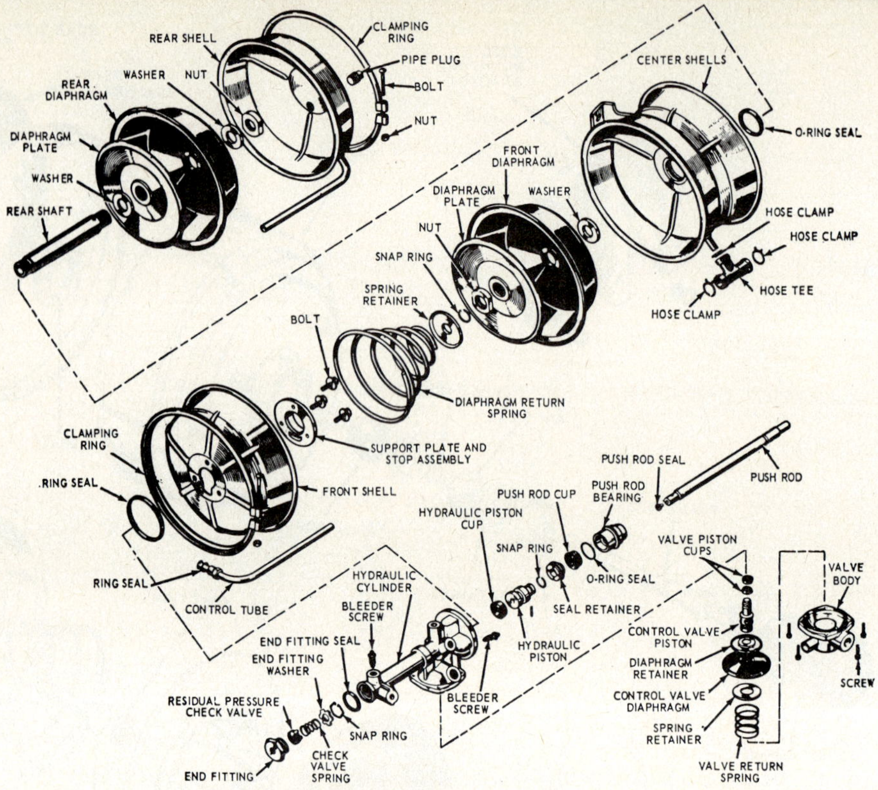

Fig. B31 Frame mounted tandem diaphragm type, disassembled

retainer on end of valve housing, being careful not to chip the plastic.

16. Coat small diameter of dust guard with alcohol and assemble over valve rod eye, using care not to damage the guard. Press guard against valve housing and seat large diameter end of guard over scalloped flange of rear flange.
17. Using gauge shown in Fig. B30, check distance from end of hydraulic push rod to master cylinder mounting face of front shell. If this distance is not within .880–.895 inch, hold the serrated end of the push rod with pliers and turn the adjusting screw to obtain the correct dimension.

FRAME MOUNTED TANDEM DIAPHRAGM TYPE

NOTE: Before disassembling unit, scribe marks across both clamping rings, on shells, across top surface of hydraulic cylinder flange and onto front shell and across flanges of control valve housing below hydraulic cylinder.

Reassembly, Fig. B31

1. Assemble piston cups back to back in grooves of control valve piston.
2. Slide diaphragm retainer (flange side first) onto other end of control valve piston. Coat inside diameter of diaphragm with alcohol then slide it over end of piston and seat it against retainer. Assemble spring retainer with flange side facing away from diaphragm.
3. Assemble control valve piston and diaphragm assembly onto valve return spring, then place spring into valve body around vacuum poppet guides and press diaphragm bead firmly into groove on flange of body.
4. Coat control valve piston and cups with brake fluid and insert them into the control valve bore of the hydraulic cylinder. Align scribe marks on flanges of valve body and housing. Install the four retaining screws and torque to 40–60 in. lbs.
5. Assemble washer, rear diaphragm plate, rear diaphragm and washer on end of shaft with hole. Install nut on shaft. Grip nut with a vise then insert a drift through holes in shaft and torque nut to 10–15 ft. lbs.
6. Install O-ring seal into groove in hub of center shells, then coat seal, bearing surface and outside of rear shaft with silicone lubricant.
7. Install rear shaft through bearing hub and seal of center shell, then assemble washer, front diaphragm and diaphragm plate on end of shaft, install and torque nut to 10–15 ft. lbs.
8. To install a new seal on push rod, place seal on wood block with rubber side facing down. Rest end of push rod vertically over seal stem and using a rawhide mallet, strike end of push rod to seat the seal.

NOTE: Make certain that shoulders of seal and rod are in contact.

9. Coat push rod, push rod bearing and all hydraulic parts with brake fluid, then install support plate and piston stop assembly, push rod bearing with O-ring installed in its groove, push rod cup and seal retainer onto push rod.
10. Install snap ring on piston, but not into its groove, then connect piston to the push rod using valve plunger pin and slide piston snap ring into its groove to cover the plunger pin hole. Coat piston cup with brake fluid and install it onto the piston.
11. Install residual check valve, check valve spring and washer into the end fitting then install end fitting and washer on hydraulic cylinder and torque to 50–85 ft. lbs.
12. Place ring seal in groove around flange of hydraulic cylinder and slide small diameter end of diaphragm spring over piston end of push rod and coat hydraulic cylinder bore and piston with brake fluid.
13. Bottom small coil end of return spring against spring retainer on push rod and place large coil end of spring inside front shell with hydraulic piston inserted through hole in shell.
14. Slightly compress return spring, then guide hydraulic piston, seal retainer, push rod cup and push rod bearing into cylinder bore, then seat hydraulic cylinder flange against the front shell with ring seal in place.
15. Place support plate and stop assembly on other side of shell and install the three bolts that retain stop assembly and the front shell to the hydraulic cylinder. Relieve pressure on spring.
16. Using talcum powder or silicone lubricant, lightly coat both sides of front and rear diaphragm beads, then guide rear shaft onto push rod and align scribe marks on shell flanges next to front diaphragm. Press shells together to seat bead of front diaphragm in shell flanges around the circumference.
17. Install clamping ring, bolt and nut on shell flanges, then using a rawhide mallet, lightly tap ring while torquing nut to

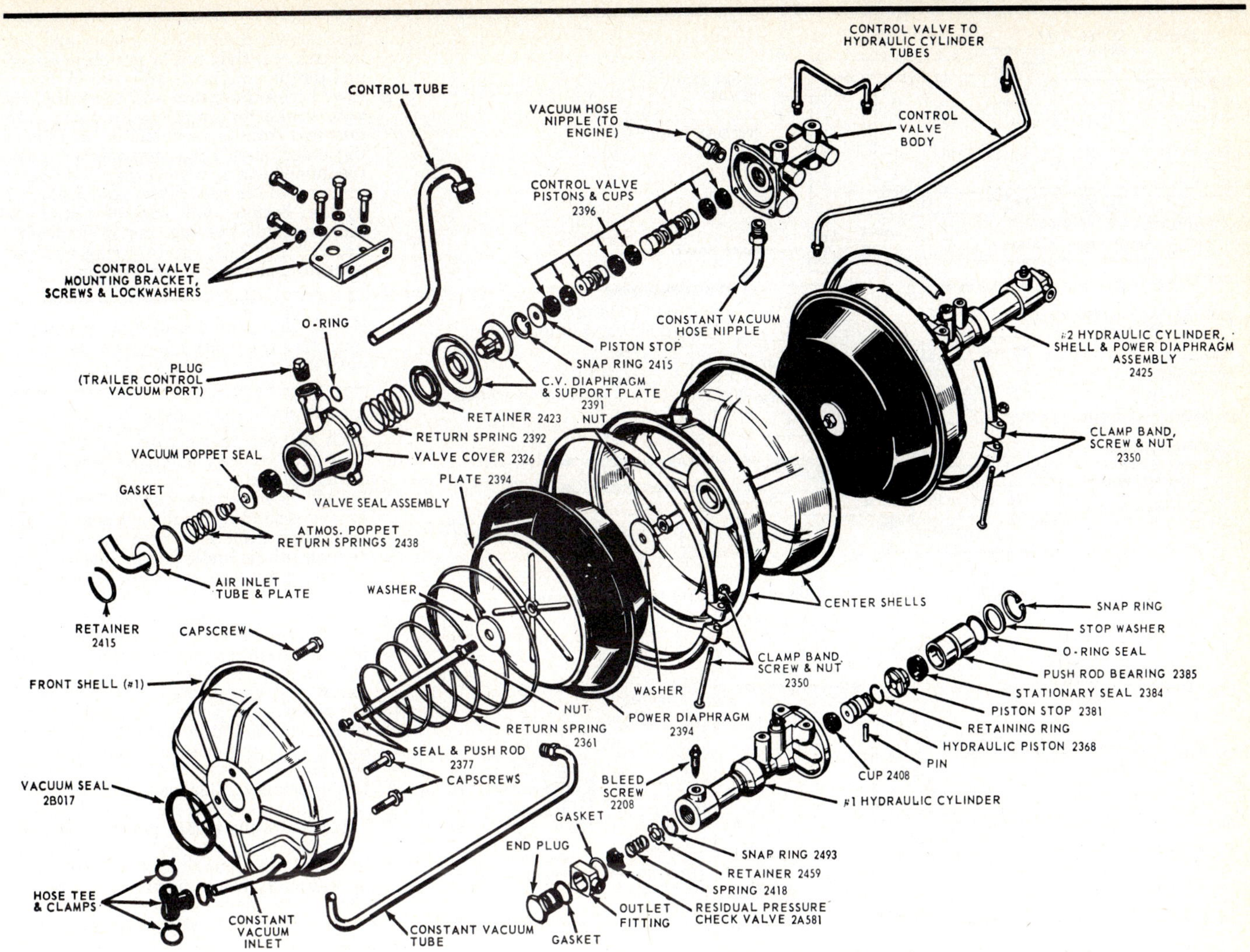

Fig. B32 Frame mounted Duo-Vac dual diaphragm type, disassembled

30–40 in. lbs.

18. With scribe marks aligned on rear shells and rear diaphragm bead in shell flanges around the circumference, press rear shells together and install the clamping ring onto the shell flanges. Torque nut in same manner as in previous step.
19. Connect hose tee to control tube on rear shell and onto hose nipple on center shell using hose clamps, then assemble ring seal on end of control tube to hose tee with a hose clamp. Connect tube nut into control valve body and torque tube nut to 80–120 in. lbs.
20. If pipe plug was removed, install it in rear shell and torque to 12–20 ft. lbs. If bleeder screws were removed, install them into hydraulic cylinder and torque to 10–15 ft. lbs.

NOTE: Tube seat inserts in input and output hydraulic ports should not be removed, except if replacing a damaged tube seat.

DUO-VAC DUAL DIAPHRAGM TYPE

Disassembly Notes Fig. B32

1. Scribe one mark across left and fitting outlet adapter and hydraulic cylinder in front of bleed screw. Scribe another mark on top of hydraulic cylinder flange and on front shell. Finally, scribe one mark on top of front shell across clamp band onto center shell.
2. Mark opposite half of power unit in same places as above, using two scribe marks instead of one at each location.
3. If control valve is to be disassembled, scribe a mark across air inlet tube base plate and flange of valve cover, across flanges of valve cover, body and across flange and bosses of mounting bracket and valve body.

NOTE: Do not disassemble power diaphragm, or push rod assembly unless required for replacement of damaged parts.

Reassembly, Fig. B32

Control Valve

1. Install mounting bracket onto the control valve with scribe marks aligned, then place control valve mounting bracket in vise with bore facing up.
2. Coat cups with brake fluid, assemble them in pairs and back to back onto pistons. Coat control valve bore with silicone lubricant then coat piston and cups with brake fluid and install them into bore, (either end first). Insert cups into bore and press piston to the bottom.
3. Coat other piston with brake fluid and install it flat end first into bore until it bottoms against first piston. Install stop washer and secure with a snap ring.
4. Install poppet assembly in valve cover through the air inlet side. Install small coil end of conical spring around top of metal stem of small atmospheric poppet. Install large return spring (light blue) on large atmospheric poppet.
5. Coat gasket with powdered graphite and install it against shoulder of air inlet opening. Seat air inlet tube and plate assembly on plate and valve cover. Secure tube and plate with snap ring.

NOTE: Make certain that all corners of snap ring are seated in groove.

6. Coat inside diameter of control valve with brake fluid and slide diaphragm onto hub of plastic plate. Both heads of diaphragm should be facing up with stem of plate facing down. Install metal diaphragm and spring retainer onto hub with retainer flanges facing up and bottom of retainer holding inner bead of diaphragm in groove around plate hub.

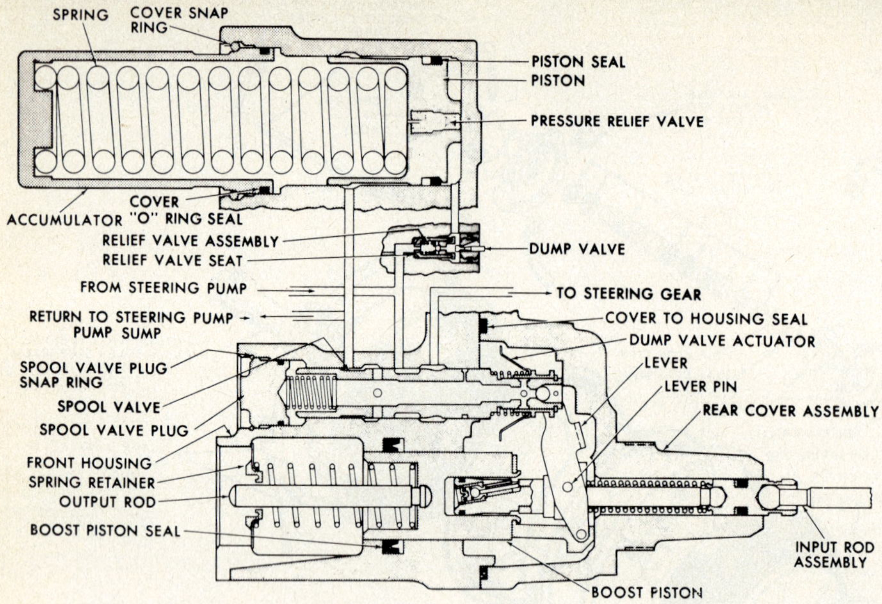

Fig. B33 Sectional view of Bendix Hydro-Boost with spring type accumulator

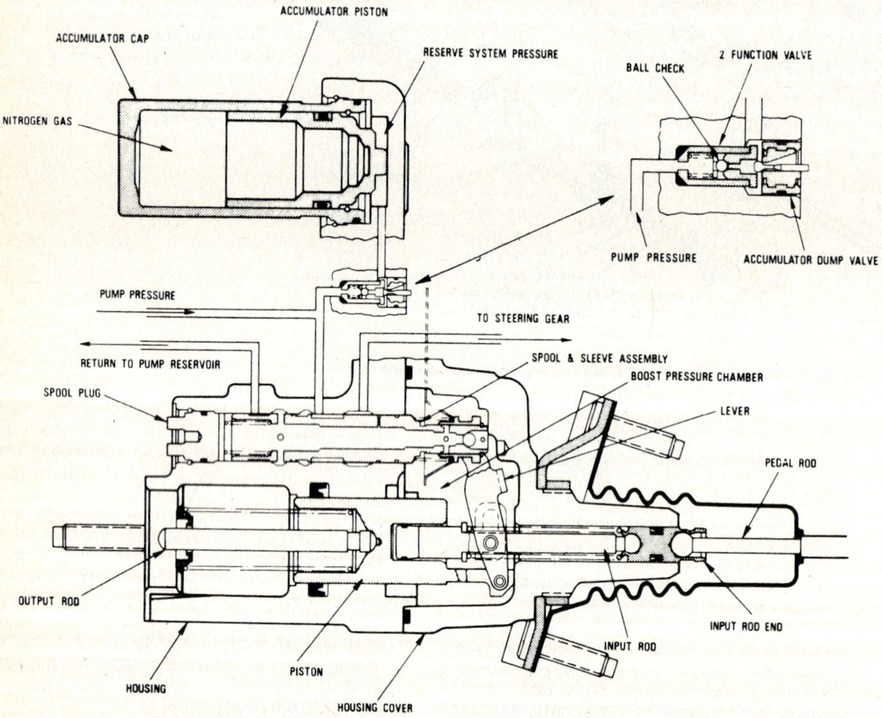

Fig. B34 Sectional view of Bendix Hydro-Boost with pneumatic type accumulator

7. Coat outer bead of diaphragm with brake fluid and slide diaphragm return spring (orange) onto hub of plate. Bottom spring in metal retainer and install diaphragm, plate, retainer and spring assembly into control valve cover spring end first. Seat bead of diaphragm in groove around cover flange, install four cover to valve body screws and torque to 60–90 in. lbs.

NOTE: Make certain that stem of plastic diaphragm plate bottoms in cavity in front control valve piston and that scribe marks on flanges of cover and body are aligned.

8. Remove assembly from vise and install O-ring seal in groove of control tube port.

Hydraulic Cylinder

1. Install new push rod seal if removed. Place seal on wood block with rubber side of seal facing down. Hold push rod above seal with hole in push rod resting on stem of seal. Tap threaded end of push rod with soft mallet until end of push rod bottoms against shoulder of seal.
2. If removed, assemble power diaphragm parts on push rod, screw the thick nut (if used) with undercut side first onto push rod, until it bottoms on threads. Install washer, support plate (curled flange side first), power diaphragm, washer and nut on push rod. Grip nut or flat on push rod and torque end nut to 160–200 in. lbs.
3. Coat hydraulic piston and bearing parts with brake fluid and install them onto push rod from the seal end in the following order; snap ring, stop washer, push rod guide bearing with O-ring in its outer groove and seal cavity end last. Followed by the, bottom (flat side) of stationary seal in guide bearing, piston stop with grooved side last, retainer ring and the hydraulic piston and cup assembly.

NOTE: The lips of seal and piston cup must face the hydraulic piston end of the push rod.

4. Install pin through holes in piston and push rod, secure pin with retainer ring. Seat ring securely in groove around hydraulic piston.
5. Install new square vacuum seal into groove around hydraulic cylinder flange, then align front shell with scribe marks on hydraulic cylinder flange. Install three retaining screws and torque to 130–230 in. lbs.
6. Coat bore of hydraulic cylinder, piston, seal and bearing parts with brake fluid. Grip hydraulic cylinder collar in vise with bleed screw facing up, cylinder horizontal and hub facing out. Install large coil end first, of power diaphragm return spring over hydraulic piston and push rod and seat spring against diaphragm support plate.
7. Lift assembly, then seat small end of spring against front shell and carefully guide hydraulic piston assembly, piston stop, stationary seal and push rod guide bearing with O-ring into bore of hydraulic cylinder. Press firmly against diaphragm to compress diaphragm return spring, then seat stop washer against push rod guide bearing and secure it with snap ring inside groove in hub of hydraulic cylinder.

NOTE: Before releasing pressure on diaphragm, make certain that snap ring is properly seated to avoid personal injury and damage to parts.

8. Install residual pressure check valve, spring and retainer into end fitting and secure with snap ring, then install end of spring inside check valve against metal disc and bottom rubber face of valve in the fitting. Place retainer washer with indented center flange inside outer spring coil and seat snap ring securely into groove.
9. Slide new copper gasket on end fitting followed by correctly marked (1 or 2 scribe marks) side outlet adapter and a second new copper gasket. Install end fitting assembly into output end of hydraulic cylinder, aligning the scribe marks on outlet and cylinder, torque the fitting to 50–85 ft. lbs.
10. Repeat above steps, to reassemble the other hydraulic cylinder.

Complete Unit

1. With diaphragm facing up, grip hydraulic cylinder collar and front shell assembly in vise. Lightly coat bead of diaphragm, flanges of front shell and center shell with

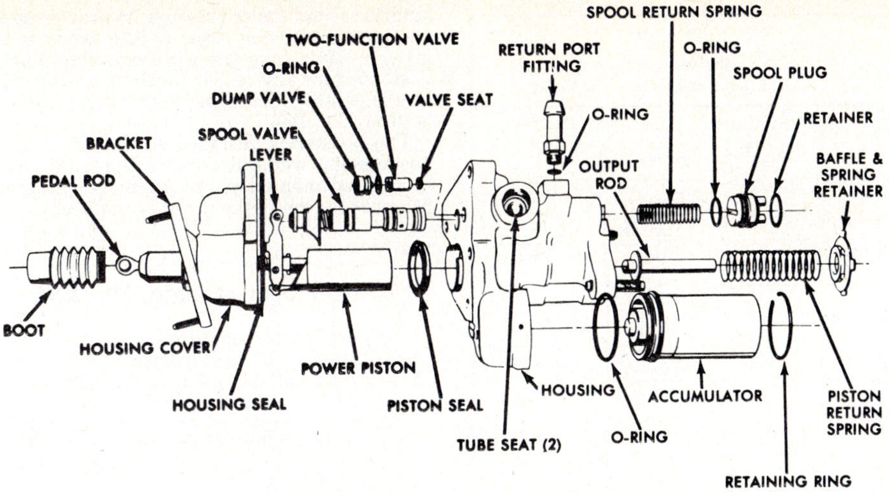

Fig. B35 Hydro-Boost unit disassembled

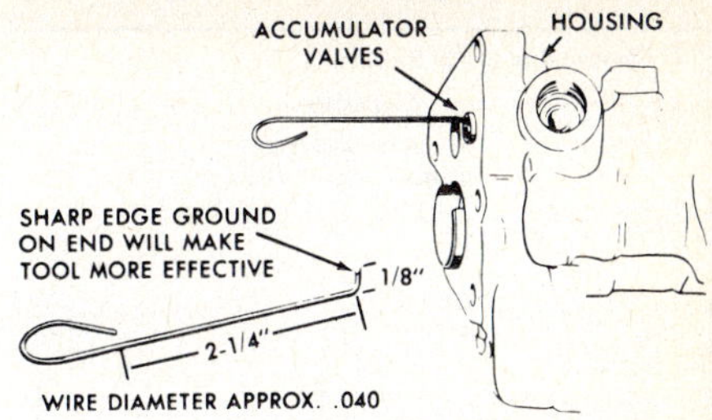

Fig. B36 Removing accumulator valves

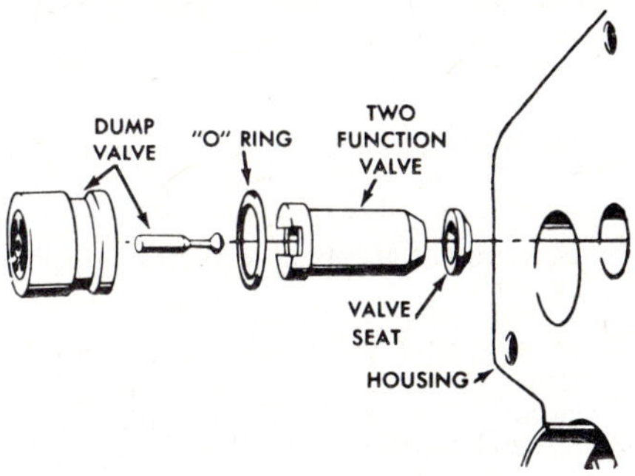

Fig. B37 Accumulator valves

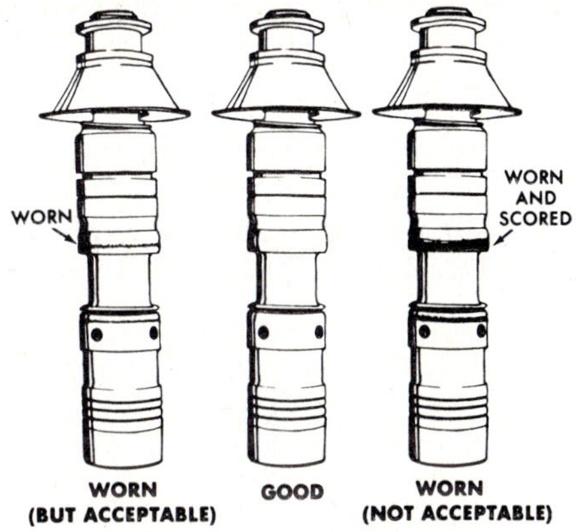

Fig. B38 Spool valve inspection

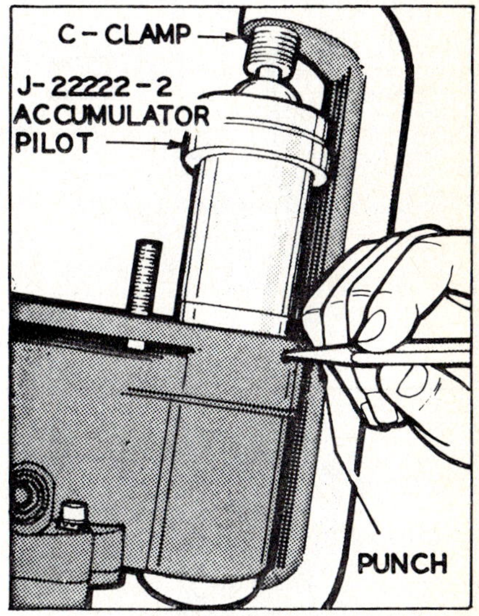

Fig. B39 Unseating accumulator retaining snap ring. Spring type units

talcum powder or silicone lubricant.

2. Press diaphragm bead into front shell flange and seat flange of center shell against it.

NOTE: Make certain that bead is fully seated around circumference against both shell flanges.

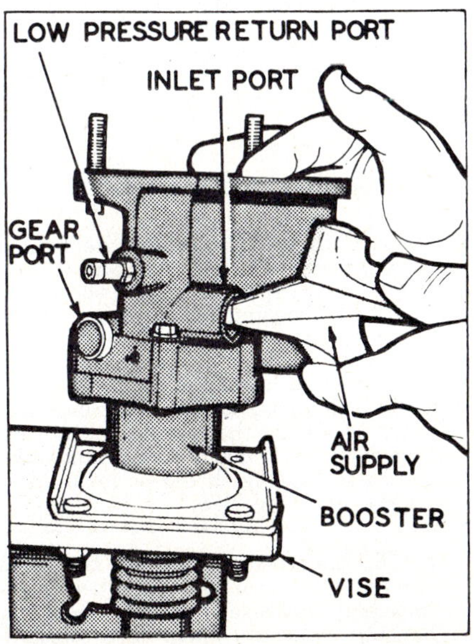

Fig. B40 Removing accumulator piston with compressed air. Spring type units

3. Align matching scribe marks on shell flanges, clamp band, screw, part number tag and nut, As viewed from end fitting of hydraulic cylinder with bleed screw at 12 o'clock. The opening in clamp band must be at 7 o'clock, the vacuum inlet tube on front shell must be at 4:30 o'clock and the control vacuum inlet port (atmospheric inlet) on center shell must be next to the clamp band. Torque clamp nut to 30–40 in. lbs., then using a rawhide mallet, tap on band at three points 90 degrees apart and retorque clamp nut to 30–40 in. lbs.

NOTE: If vacuum ports and tubes are not in locations described above, disassemble shells and remove hydraulic cylinder. Rotate shell to correct position and repeat steps 1, 2 and 3. If a vertical position of unit makes assembly of #2 hydraulic cylinder difficult, reposition unit, by positioning collar of #1 cylinder in a horizontal position or by positioning collar of #2 cylinder in a vertical position.

4. Coat bead of #2 diaphragm and shell flanges as described above and assemble #2 hydraulic cylinder to center shell with bead of diaphragm fully seated against both shell flanges around circumferences and with matching scribe marks aligned on shell flanges.
5. Install clamp band with opening aligned with the other clamp band opening and torque clamp nut to 30–40 in. lbs. tap on band as in step 3 and retorque clamp nut to 30–40 in. lbs.
6. Install control valve assembly, bolt and washer onto the #1 hydraulic cylinder.

NOTE: The valve must be on same side as the 12 inch vacuum tube extending from #1 front shell.

7. Connect short piece of 14 inch hydraulic tubing from top rear port (next to front shell) of control valve body with the input port on top of #1 hydraulic cylinder. Torque both tube nuts to 135–150 in. lbs.
8. Connect long piece of 14 inch hydraulic tubing from top front port of control valve

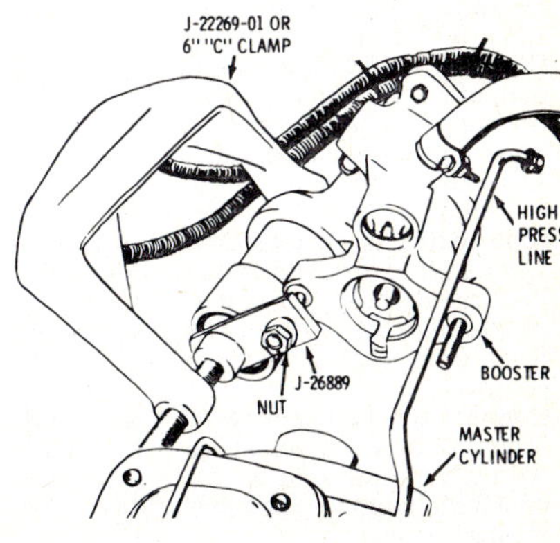

Fig. B41 Removing pneumatic type accumulator

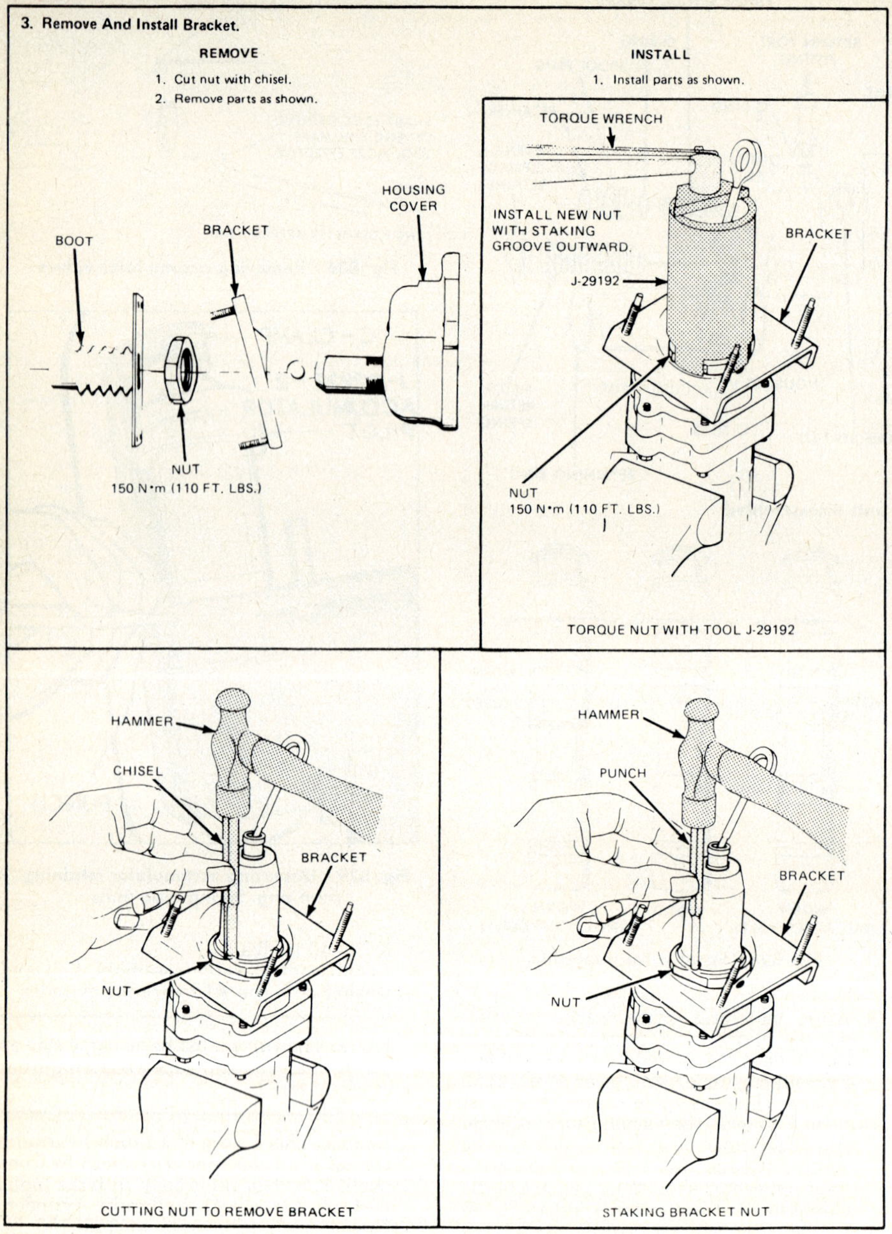

Fig. B42 Bracket, removal and installation

body with the input port on top of #2 hydraulic cylinder. Torque tube nuts to 135–150 in. lbs.

9. Connect the 58 inch tube to control tube port of valve cover, sliding end of tube through O-ring. Connect tube end with control vacuum port on top of center shell and torque tube nut to 100–140 in. lbs.
10. Install short part of 12 inch tube with tube seated in port on bottom of control valve body. Torque tube nut to 80–120 in. lbs. Install hose tee and hose clamp on this tube and the 12 inch tube extending from front shell. Secure hose tee with clamps.
11. Connect the longer piece of the 12 inch tube into hose tee and secure with clamp. Attach tube nut end to the vacuum port on #2 shell and torque tube nut to 80–120 in. lbs.

NOTE: If any tubing does not reach from control valve ports to power unit ports, check steps 1 through 6 to determine whether shell or control valve parts must be repositioned.

HYDRO-BOOST

Description

The Hydro-Boost system, Figs. B33 and B34, provides an additional cylinder in the brake system. This cylinder contains no brake fluid. The Hydro-Boost cylinder is hydraulically operated, by pressurized fluid from the power steering pump, providing power assist to operate a dual master cylinder brake system.

The booster is composed of two sections; the linkage section and the power section, Fig. B34.

On early units, a spring accumulator incorporated in the Hydro-Boost, Fig. B33, stores sufficient fluid under pressure, to provide several power assisted stops in the event that pressure fluid from the power steering pump is stopped for any reason (such as a ruptured line). On late units, a pneumatic accumulator is used, Fig. B34.

The booster is designed so that if a total absence of power assist occurs, the brakes can be applied manually, but somewhat greater pressure on the brake pedal is required.

Booster Disassemble, Fig. B35

1. Secure booster mounting bracket in a vise with pedal rod facing downward.
2. With a small screwdriver, disengage tabs of baffle and spring retainer from ledge inside opening near master cylinder mounting flange of the booster. Remove retainer piston return springs and output rod from opening.
3. Position drain pan under booster housing, then remove five housing to housing cover attaching bolts.
4. Carefully lift housing from cover, leaving spool valve and power piston assembly attached to housing cover.

NOTE: If accumulator is to be serviced, refer to Accumulator Service.

---

5. Remove power piston seal from housing bore.
6. Remove seal from housing cover.
7. If accumulator valve is to be removed, fabricate tool from .040 in. diameter wire as shown in Fig. B36. Remove dump valve by catching tool under pin guide near center of valve, then remove two function valve and seat, Fig. B37.
8. Remove hose fitting O-ring, if necessary.
9. Remove spool valve and spring from power piston assembly by rotating valve out of lever arm. Refer to Fig. B38, and inspect spool valve. If spool valve is found to be defective, the valve assembly must be replaced.
10. Inspect power piston. If scratches can be felt with finger nail, replace power piston.
11. Using diagonal cutters, cut off end of connecting pin, then push pin out using a suitable punch and remove piston.
12. If necessary, remove spool valve plug retaining ring and spool valve plug.

Cleaning and Inspection

1. Clean all parts in a suitable solvent being careful to avoid losing small parts.
2. Inspect valve spool and valve spool bore in booster housing for corrosion, nicks, scoring or other damage. Discoloration of the spool or bore, particularly in the grooves is not harmful.
3. If the valve spool or the spool bore has nicks or scoring that can be felt with a fingernail, particularly on the lands, the spool and housing should be replaced as an assembly, Fig. B38.

NOTE: The clearance between the valve spool and the spool bore of the housing is important. Because of this, the spool and housing make are made as a selective assembly and therefore can only be replaced as an assembly.

---

4. Inspect the input rod and piston assembly for corrosion, nicks, scoring or excessive wear. If the piston is damaged, the input rod and piston assembly should be replaced.
5. Inspect piston bore in booster housing for corrosion, nicks, scoring or other damage. If the bore is damaged, the valve spool and housing should be replaced as an assembly.

Booster Assemble, Fig. B35

CAUTION: Be sure to keep all parts clean until reassembly. Re-wash at reassembly if there is any occasion to doubt cleanliness. Lubricate all seals and metal friction parts with power steering fluid. Whenever the booster is disassembled, all seals, spiral snap rings, check valve and ball should be replaced. All of these parts are included in a seal kit.

1. If removed, install spool valve and O-ring, then push spool valve in and install snap ring.
2. Slide connecting bracket on piston into small yoke in lever and push pin through hole.
3. Using tool No. J-26410, mushroom end of pin. Ensure lever moves freely with no binding.
4. Secure booster mounting bracket in a vise with pedal rod facing downward.
5. Install housing seal in groove on housing cover.
6. Install power piston seal in housing with flat side facing toward master cylinder mounting flange.
7. If removed, install O-ring and return hose fitting.
8. If removed, install accumulator valve seat with cup side facing upward, Fig. B37. The seat can be forced to the bottom of the bore by installing the two function valve. Insert dump valve over two function valve making certain that dump valve plunger is held in position until installation is completed.
9. Insert spool valve spring and spool valve assembly into housing bore.
10. Pull upward on power piston and extend lever to accept sleeve on spool valve, then push seal protector onto power piston until it seats.
11. While holding lever extended, position housing with spool valve assembly directly over rear cover and slide lever pins into slot on sleeve.
12. Center power piston in bore and lower housing down onto rear cover. Push housing and cover together, then remove tool No. J-25083.
13. Install housing to cover attaching bolts. Torque bolts to 20 ft. lbs.
14. Install output rod, spring and spring retainer.
15. Secure baffle and spring retainer using a 7/8 in. deep socket.
16. If removed, install accumulator as described under Accumulator Service.

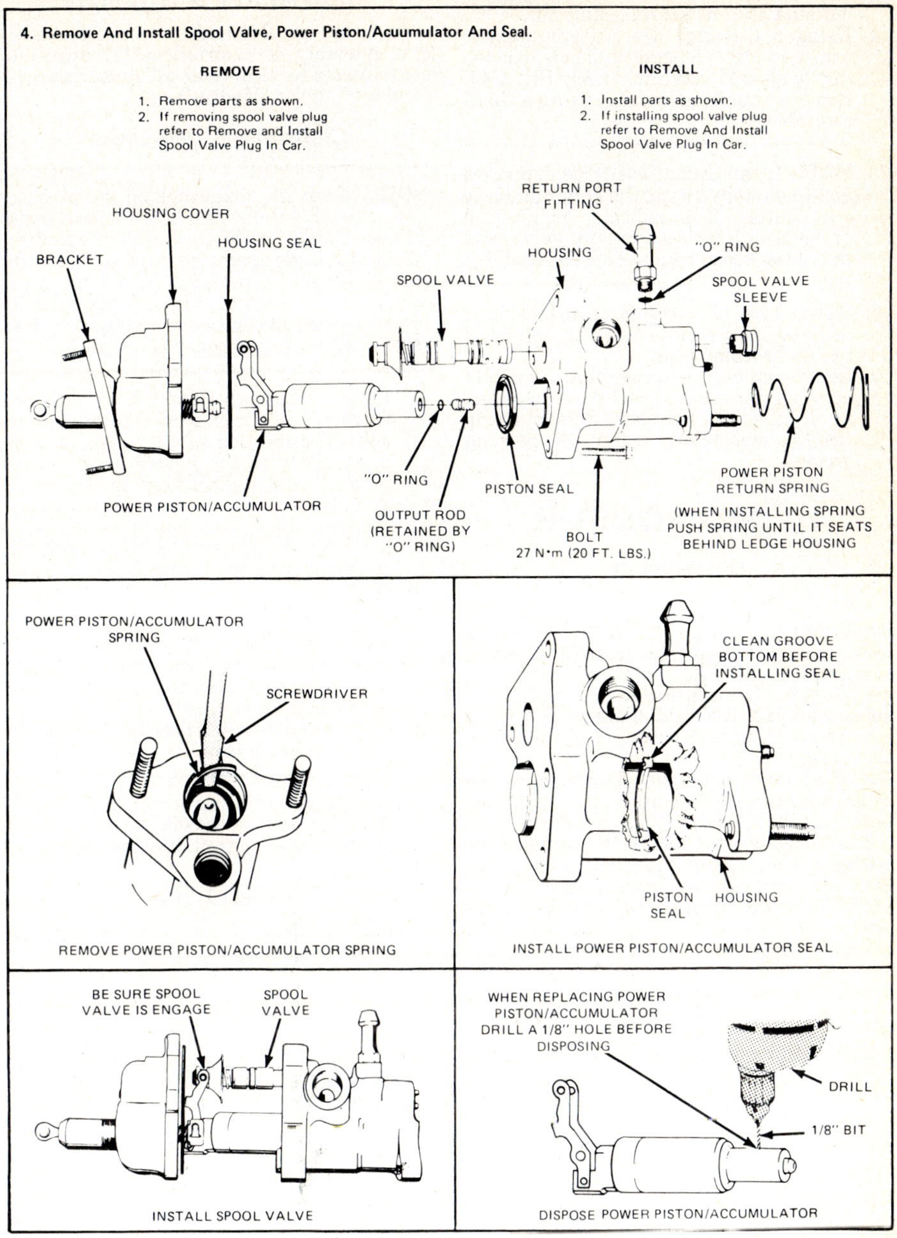

Fig. B43 Spool valve, power piston/accumulator and seal, assembly & disassembly

Accumulator Service

Spring Type Accumulator

1. Secure booster mounting bracket to a vise with pedal rod facing downward.
2. Pump pedal rod several times to ensure accumulator is depleted.
3. Using a suitable C-clamp and tool J-22222-2, depress accumulator spring cap and insert a punch into small hole in front housing to unseat retainer ring, Fig. B39. Remove retainer ring.
4. Slowly and carefully back off C-clamp until spring tension on cap is relieved, then remove clamp, spring cap and accumulator spring.
5. Plug gear and return ports, then while holding hand over accumulator piston, pressurize booster with compressed air through the inlet port, Fig. B40. Piston will back out of housing bore. If compressed air is not available, use a hook fabricated from a stiff piece of wire. Engage hook in piston inlet hole, then using a suitable tool for leverage, wrap wire around tool and pry against housing to remove piston from bore.
6. Disassemble booster as described under Booster Disassemble.
7. Using a small diameter wire hook, remove charging valve and seat.
8. Reverse procedure to assemble. Lubricate components with clean power steering fluid. If piston cannot be seated in bore, pump pedal rod several times to relieve trapped air pressure below piston. Use care not to disturb relief valve when installing accumulator spring in piston.

Pneumatic Accumulator

NOTE: Do not attempt to repair or apply heat to the accumulator. Before discarding an inoperative accumulator, drill a 1/16 in. in diameter hole through the end of the accumulator pan. Do not drill through the piston end. Always wear safety glasses when performing service on the accumulator.

1. Pump pedal rod several times to ensure accumulator is depleted.
2. Secure brake booster mounting bracket in a vise with pedal rod facing downward.

3. Place tool No. J-26889 over master cylinder stud and install retaining nut.
4. Using a C-clamp, depress accumulator can, then insert a punch into hole in housing to unseat retaining ring, Fig. B41. Remove ring from housing using a small screwdriver.

NOTE: If can cannot easily be depressed approximately .10 in., the accumulator is still charged. This indicates an internal problem with the accumulator valves and the brake booster must be disassembled.

5. Slowly back off C-clamp until tension is relieved, then remove C-clamp, retaining cap and accumulator.
6. Reverse procedure to assemble. Lubricate accumulator can seal with power steering fluid. Ensure accumulator can retaining ring is completely seated in housing groove.

HYDRO-BOOST II

Description

Hydro-Boost II is a hydraulically operated power brake booster. The hydraulic brake booster consists of an open center spool valve and a hydraulic cylinder combined into a single housing. The power steering pump provides the hydraulic fluid pressure to operate both the power brake booster and the power steering gear.

A dual master cylinder is bolted to the brake booster and is operated by a push rod projecting from the booster cylinder bore.

The Hydro-Boost II also has an accumulator system which stores sufficient fluid under pressure to provide one power assisted brake application in the event of steering pump failure.

Disassembly & Assembly

Refer to Figs. B42 and B43 for disassembly and assembly of Hydro-Boost II unit. For accumulator service, refer to "Pneumatic Accumulator" under "Hydro-Boost."

Cleaning & Inspection

NOTE: Keep all disassembled parts clean until assembly. Lubricate all seals and metal friction points with power steering fluid. When unit is disassembled, all seals and tube inserts should be replaced. These parts are available in replacement kits. If any accumulator valve components are damaged or lost, replace all valve components.

1. Inspect spool valve and spool valve bore for corrosion, nicks, scoring or other damage. Discoloration of the spool or bore, particularly in the groove areas, is not harmful and replacement is not necessary.
2. If spool valve or bore has nicks or scoring that can be felt with a fingernail, the entire booster should be replaced as an assembly.

NOTE: The clearance between the spool valve and the spool valve bore is important. Because of this, the spool valve and the housing make up a selective assembly. The spool valve is selected to match the spool valve bore.

3. Inspect piston for scratches and nicks. If scratches on the outside surface can be felt with a fingernail, replace piston.

HYDRO MAX

The Hydro Max power brake system consists of a hydraulic powered booster, a reserve system electric motor pump and a warning system. Fluid from the power steering system enters the inlet port of the booster, flows through the pressure valve, power piston and the flow switch, then exits through the return port. When the brakes are applied, a pressure valve restricting flow through the power piston is activated. This pressure acting on the power piston applies force to the master cylinder. The pressure regulator limits maximum internal booster pressure developed during application of the brakes, and retains pressure for steering functions. Fluid flow through the flow switch opens the reserve motor pump electrical circuit during normal operation. A check valve in the motor pump prevents backflow through the motor pump during normal operations. If normal flow from the steering pump is interrupted, the electric motor pump provides power for reserve stops. The flow switch closes energizing a power relay, providing electrical power to the pump. During reserve operation, fluid is retained in the booster by the inlet port check valve. The motor pump circulates fluid within the booster assembly with the number of brake applications limited by the electrical capacity of the vehicle. Manual braking is possible if both the power and reserve systems malfunction.

Hydro Max Booster, Replace

1. Disconnect battery ground cable and brake pedal push rod from brake pedal.
2. Disconnect flow switch and electric motor pump electrical connectors.
3. Disconnect supply and return lines from booster inlet port and outlet port and plug all open lines and fittings.
4. Remove master cylinder from booster, then remove nuts securing booster to dash panel.
5. Remove booster from dash panel.
6. Reverse procedure to install.

Delco Moraine Brake Boosters

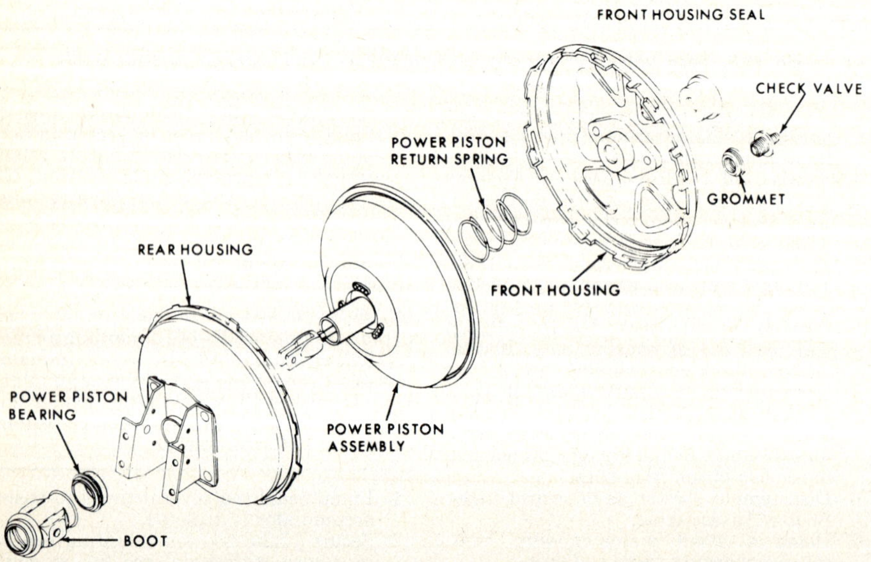

Fig. D1 Delco Moraine single diaphragm power brake unit exploded view

SINGLE DIAPHRAGM TYPE—FIGS. D1 & D2

This power brake unit is a combination vacuum-hydraulic brake booster of the vacuum suspended type which utilizes engine intake manifold vacuum and atmospheric pressure for its power.

The unit consists of a vacuum power section and a hydraulic master cylinder section. The vacuum power section contains a power piston with rolling diaphragm mechanism and power piston return spring.

The control valve is made up of an air valve and floating vacuum control valve assembly. The reaction mechanism consists of a hydraulic piston, reaction plate and a series of levers. The valve operating rod, which operates the air valve, projects from the power section and is connected to the brake pedal linkage.

The hydraulic push rod operates against the master cylinder piston. A split system (tandem piston) type master cylinder is incorporated into some units. The front half of the master cylinder in the split system operates the rear brakes while the rear half of the master cylinder operates the front brakes (see HYDRAULIC BRAKE SYSTEM chapter).

A vacuum check valve, attached to the front vacuum chamber and connected to the intake manifold, traps vacuum in the power unit at the highest level of vacuum.

Operation

As the brakes are applied by the driver, the valve operating rod and control piston move forward in the power piston assembly to compress the valve return spring and bring the poppet valve into contact with the vacuum valve seat in the valve housing to *close* the vacuum port. Any additional movement of the valve operating rod in the applied direction moves the control valve away from the poppet valve to *open* the atmospheric port and admit air through the air filter and passages to the chamber at the right of the vacuum power piston assembly. With vacuum on the left side of the diaphragm and atmospheric pressure on the right side of the diaphragm, a force is developed to move the vacuum power piston assembly, hydraulic push rod, and hydraulic piston to the left to close the compensating port and force hydraulic fluid under pressure through the residual check valve and brake tubes into the brake wheel cylinders.

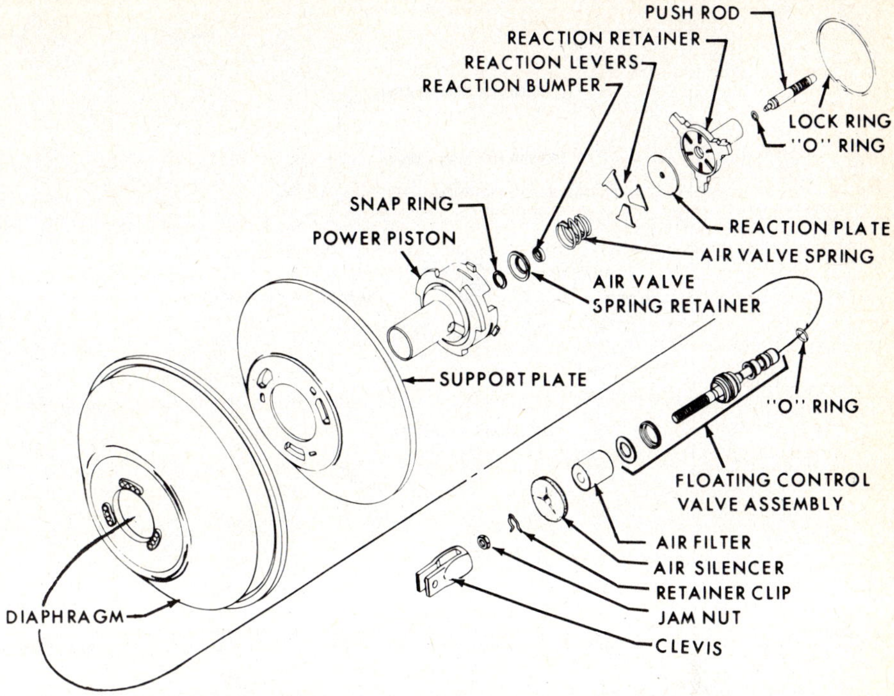

Fig. D2 Delco Moraine single diaphragm power piston assembly exploded view

Trouble Shooting

Hard Pedal
1. Internal vacuum leak.
2. Faulty control valve.

Brakes Grab
1. Faulty control valve.

Slow or No Release
1. Faulty push rod adjustment.
2. Bind in linkage.

TANDEM DIAPHRAM TYPE—FIGS. D3 & D4

This unit has a vacuum power chamber that consists of a front and rear shell, a housing divider, front and rear diaphragm and plate assemblies, a hydraulic push rod and a diaphragm return spring.

The unit operates in much the same manner as the Single Diaphragm unit described above. The diaphragm and plate assemblies utilize the pressure differential created by engine intake manifold vacuum and atmospheric pressure to assist the hydraulic push rod.

Trouble Shooting

Hard Pedal
1. Internal Vacuum Leak.
2. Faulty Control Valve.

Brakes Grab
1. Faulty Control Valve.

Slow or No Release
1. Faulty push rod adjustment.
2. Bind in linkage.

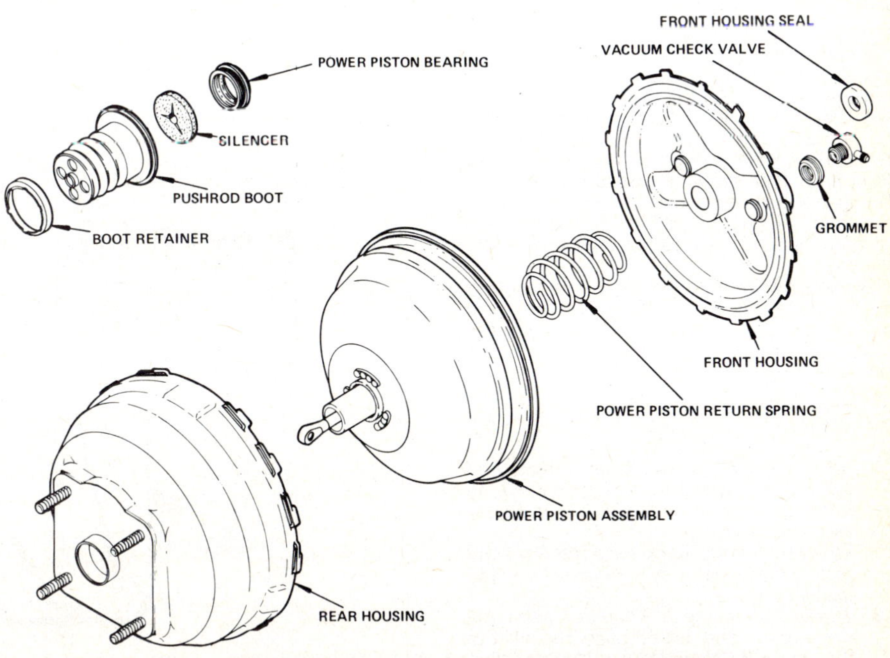

Fig. D3 Delco Moraine tandem power brake unit exploded view

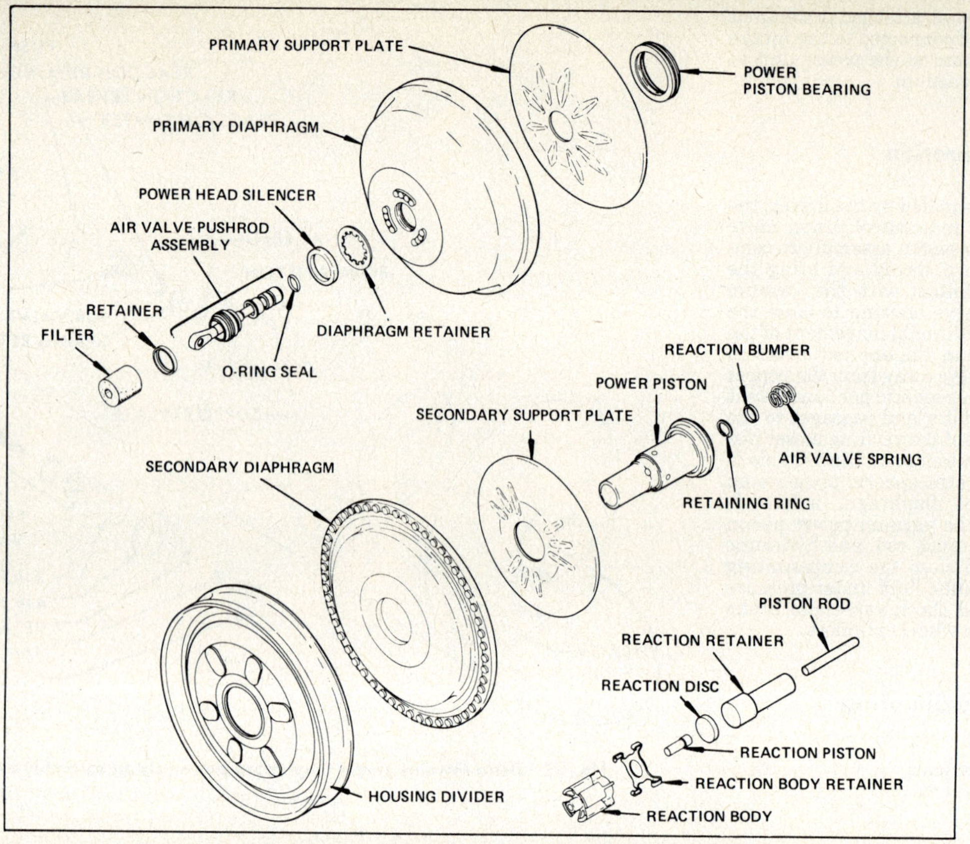

Fig. D4 Delco Moraine tandem power piston assembly exploded view

DUAL POWER BOOSTER—FIG. D5

Description

The Dual Power booster is a combination vacuum and hydraulic booster mounted in tandem. The vacuum booster operates off engine vacuum, or in some cases off a vacuum pump. The hydraulic booster utilizes power steering pump pressure.

Application of the brake pedal forces the vacuum booster to function as a conventional vacuum booster, the hydraulic booster is activated by overtravel of the vacuum valving in the vacuum booster. This overtravel affects a hydraulic control valve, restricting hydraulic fluid flow. This causes a pressure increase in the hydraulic booster pressure chamber, forcing the power piston and piston rod outward to exert boosted force to the master cylinder pistons.

An integral relief valve in the power head allows hydraulic pressure to bypass the power piston when total vacuum and hydraulically boosted pressure approaches 2000 psi. brake line pressure.

Vacuum Booster Disassembly Notes

1. After removal of unit from vehicle and prior to disassembly, scribe a mark on front and rear housings to facilitate assembly, Fig. D5.
2. When removing clamp securing front and rear sections, use care as booster is spring loaded.
3. Always use a repair kit when servicing these units, and use all parts included in the repair kit regardless of the condition of old parts.

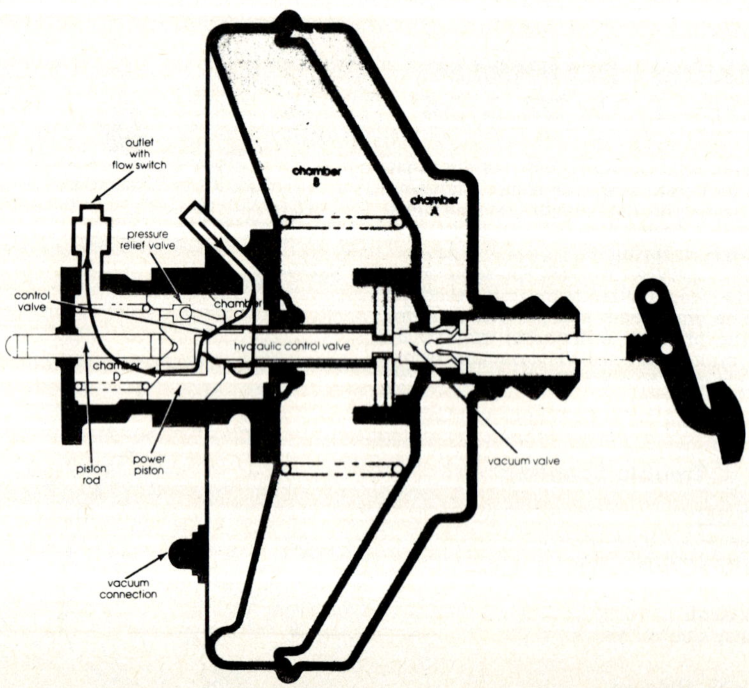

Fig. D5 Delco Moraine Dual Power Booster

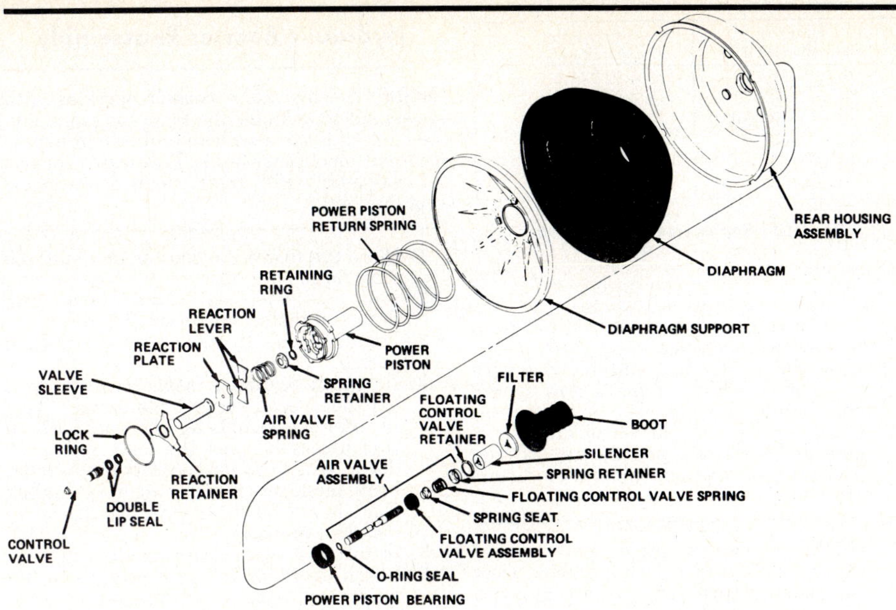

Fig. D6 Delco Moraine Dual Power vacuum booster disassembled

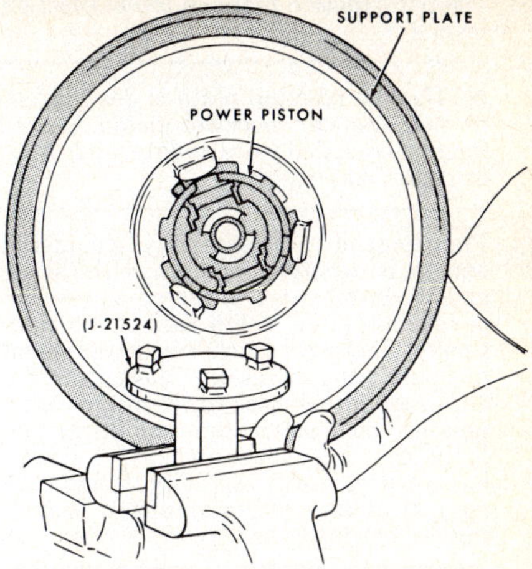

Fig. D7 Positioning power piston on holding tool

4. Clean parts in denatured alcohol only, use clean, dry, filtered compressed air to dry parts. Blow out all passages.
5. Disassemble the vacuum booster as suggested by Fig. D6.

Vacuum Booster Reassembly, Fig. D6

1. Lightly lubricate air valve O-ring and air valve surface with lubricant supplied in repair kit. Install O-ring in second groove of air valve.
2. Place power piston wrench, No. J-21524 in a vise. Position power piston on the wrench with the three tool lugs fitting into the three notches in power piston, Fig. D7.
3. Install air valve push rod assembly, valve end first, into power piston tube until it is seated.
4. Lightly lubricate the outside diameter of a new floating valve assembly with lubricant supplied in repair kit, then assemble spring seat, spring (small end first), and spring retainer to the floating control valve assembly. Install floating control valve assembly into power piston tube.
5. Using a suitable tool, press floating control valve assembly to seat in power piston tube and install snap ring on air valve.
6. Install silencer onto power piston tube.
7. Lightly lubricate inside diameter of diaphragm with silicone lubricant and assemble to support plate. Ensure the edge of center hole fits into groove of diaphragm flange.
8. Pull diaphragm away from support plate so support plate can be gripped with hands. Lightly lubricate with silicone the part of diaphragm that contacts power piston.
9. Position support plate and diaphragm over power piston tube. Press down and turn support plate counterclockwise until the locking lugs of the power piston contact the stops of support plate.
10. Turn piston and diaphragm assembly over and place in a soft jawed vise, but do not tighten. Position air valve spring retainer over snap ring.

NOTE: Inspect position of reaction plate in power piston before placing reaction levers and spring in position.

11. Place air valve return spring, small end down, on spring retainer, then place the reaction levers into position seating them in slots provided in power piston. The narrow ends will rest on the top of air valve return spring.
12. Lightly lubricate control valve and control valve seals and install seals with lip towards chamfered end of valve, then assemble control valve, valve sleeve, and reaction retainer.

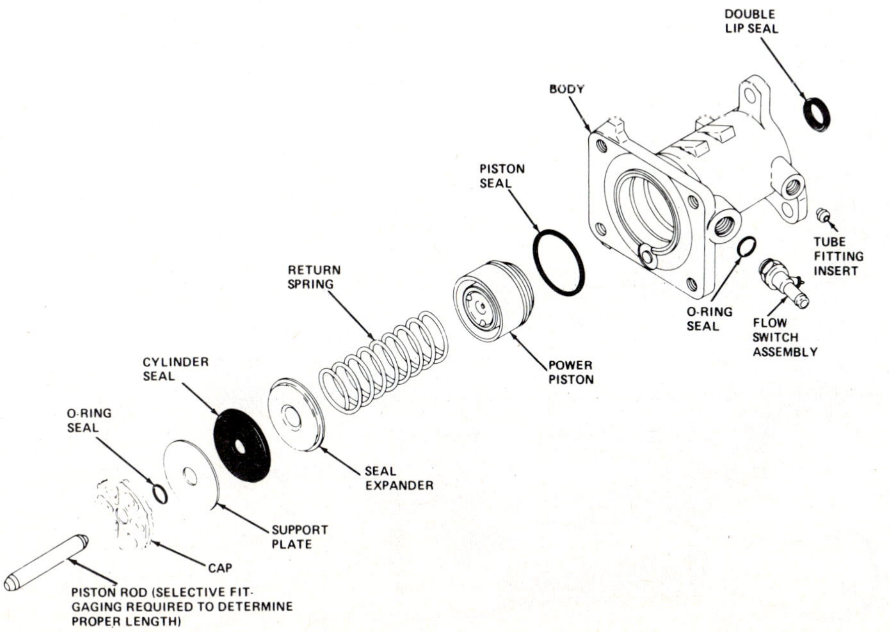

Fig. D8 Delco Moraine Dual Power hydraulic booster disassembled

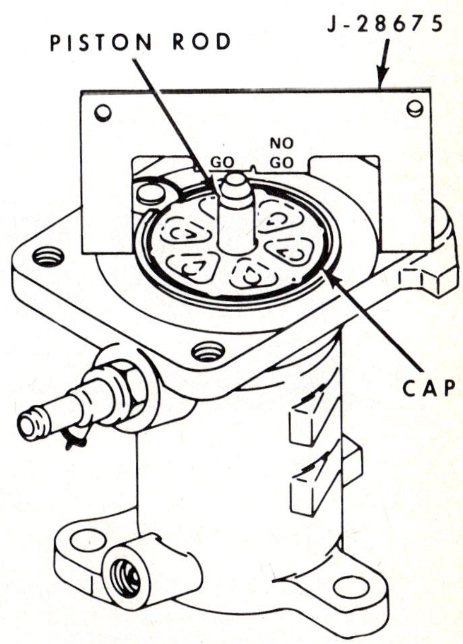

Fig. D9 Checking piston rod length

13. Position reaction plate on top of reaction levers.

NOTE: The reaction plate is designed to fit only one way in power piston. Do not force the reaction plate as damage to power piston may result.

14. Place control valve assembly, chamfered end up, on reaction plate. Place the small end of control valve in the hole in center of reaction plate. Align ears of the reaction retainer with notches in power piston and push reaction retainer down until the ears seat in notches, while maintaining pressure on reaction retainer install the large lock ring over control valve assembly, position lock ring so both ends are against divider of power piston locking lug.

NOTE: Reaction plate should be flat on reaction levers and ears on the reaction retainer seated fully in piston notches.

15. Lubricate and install two new grommets in front housing. Install new check valve and elbow.
16. Install a new power piston bearing in the rear housing so flange and thin lip project out stud side of the housing. Lubricate the inside diameter of the bearing.
17. Place front housing on a flat clean surface and install piston return spring in center.
18. Lightly lubricate power piston tube, then assemble to rear housing by pressing the power piston tube gently through rear bearing.
19. Align rear housing and power piston assembly with front housing so scribe marks line up, carefully lower rear housing towards front housing while ensuring the return spring seats around the power piston and the outside diameter of the diaphragm is properly seated.
20. Reinstall clamp. Torque clamp to 60 in. lbs.

Hydraulic Booster Disassembly, Fig. D8

1. Using a suitable container, drain fluid from unit by pulling out pushrod with a twisting motion. Turn the end of the unit down and allow any fluid to drain from the pushrod opening.
2. Place the booster in a suitable vise with the flow switch up, remove flow switch and O ring.
3. Using a small screwdriver or other suitable tool, carefully pry the cap off the booster. Use care not to bend or distort cap.

NOTE: The support plate, seal and seal expander are under spring tension from the piston return spring. Care must be taken to ensure these parts are restrained or personal injury may occur.

4. Carefully remove support plate, seal and seal expander (if not previously removed).
5. Remove piston return spring and piston.
6. Remove seal from small diameter end of body.
7. Remove power piston seal.

Cleaning & Inspection

1. Clean all metal parts except flow switch in denatured alcohol or power steering fluid only. Use only dry, filtered air to dry parts.
2. Inspect all parts for stripped threads, cracks or distortion. Replace as necessary.
3. Inspect cylinder bore for excessive scoring, nicks, burrs or corrosion. If any of these conditions exist, replace unit. Do not attempt to use any abrasives on any internal parts.
4. Use only new rubber parts from repair kit. Be sure to replace all parts included repair kit.
5. Replace flow switch if it leaks or is faulty.

Length, Millimeters	Length, Inches
102.05–102.21	4.018–4.024
102.63–102.79	4.041–4.047
103.21–103.37	4.063–4.070
103.79–103.95	4.086–4.093
104.37–104.53	4.109–4.115
104.96–105.12	4.132–4.139

Fig. D10 Replacement Piston Rods

Hydraulic Booster Reassembly

NOTE: The hydraulic booster operates with power steering fluid only. Prior to reassembly be sure all parts are clean and free from contamination of other fluids. Lubricate all parts at reassembly with fresh, clean power steering fluid only.

1. Place the hydraulic booster in a suitable vise, large bore up.
2. Lubricate piston seal with power steering fluid and install on power piston.
3. Install piston in bore and piston return spring.
4. Lubricate cylinder seal. Position seal expander, seal, support plate and O ring in order of assembly against cap. Position this assembly over the body and push down. When the cap contacts the cylinder body, tap it into place using a soft mallet. Cap should be flush with casting when installed properly.
5. Install flow switch with new O ring.
6. Lubricate piston rod with power steering fluid and install in booster. Secure hydraulic booster to vacuum booster.
7. Press down piston rod firmly ensuring the rod and all internal parts are seated in both boosters, using tool No. J-28675. Check that piston rod length is within specifications, Fig. D9.
8. If piston rod is not within specifications, select a new piston rod to bring the measurement within limits. Six piston rods are available with lengths as shown in the "Replacement Piston Rods" chart, Fig. D10.

HY-POWER HYDRAULIC BOOSTER

Description, Fig. D11

The Hy-Power power brake system consists of a hydraulic powered booster, a reserve system electric motor pump and a warning system. Fluid from the power steering system enters the inlet port of the booster, flows through the pressure valve, power piston and the flow switch, then exits through the return port. When the brakes are applied, a pressure valve restricting flow through the power piston is activated. This pressure acting on the power piston applies force to the master cylinder. The pressure regulator limits maximum

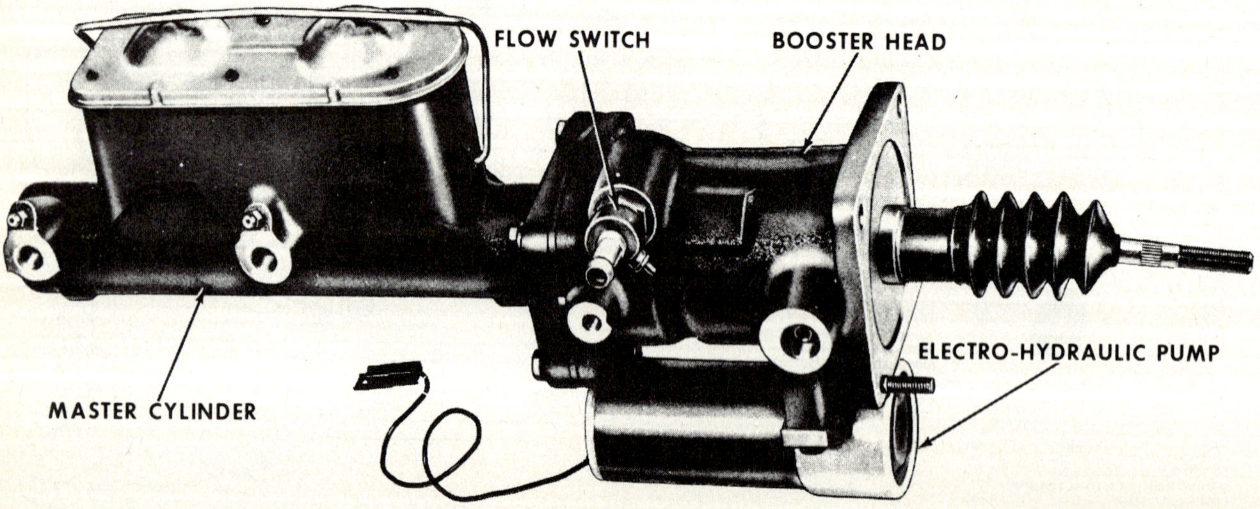

Fig. D11 Hy-Power hydraulic booster, typical

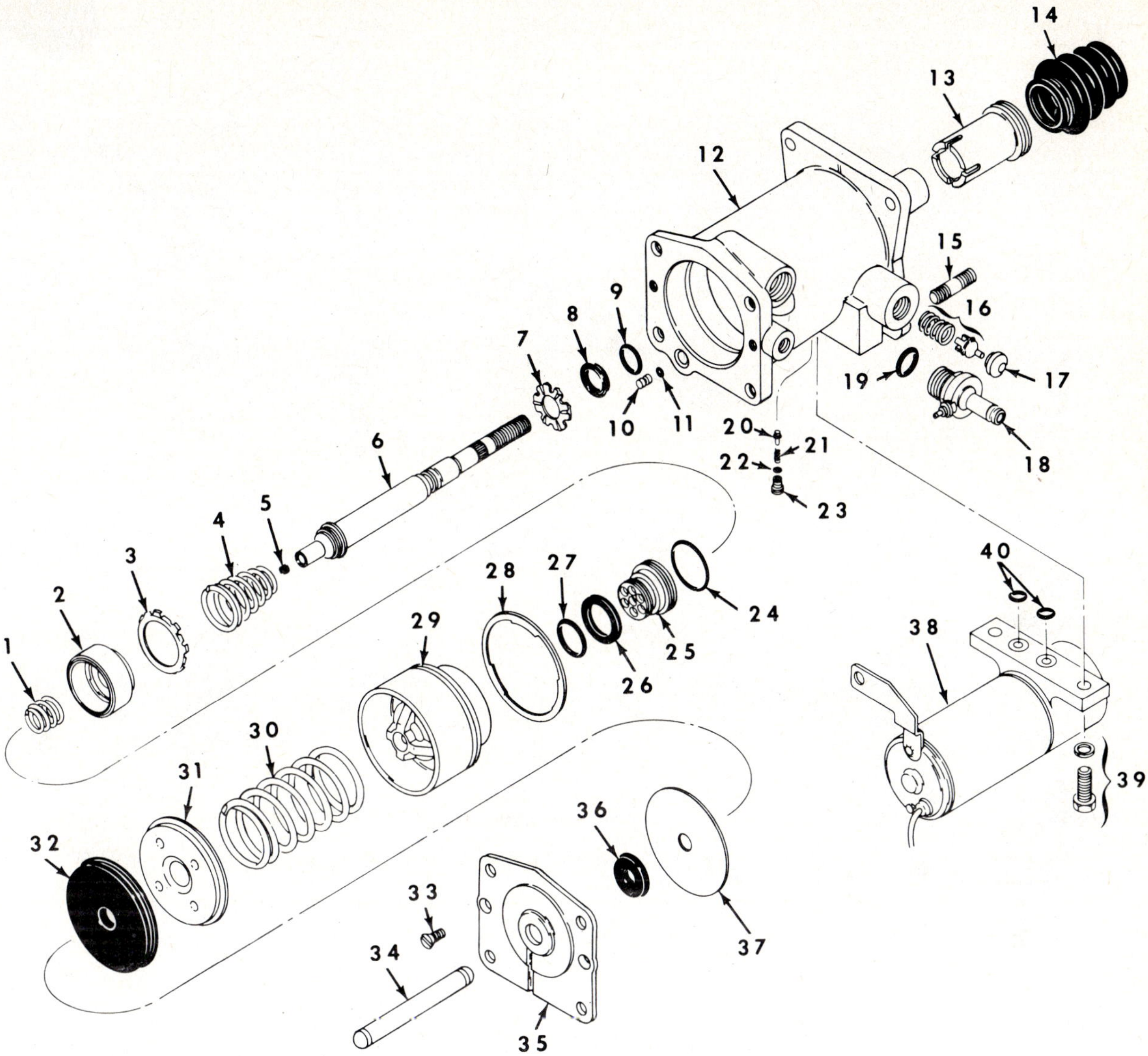

1. Reaction Piston Spring
2. Valve Seat
3. Valve Seat Retainer
4. Valve Return Spring
5. Valve to Reaction Piston Snubber
6. Control Valve/Push Rod Assembly
7. Valve Snubber (Bumper)
8. V-Lip Seal
9. O-ring Seal
10. By-Pass Plug
11. By-Pass Plug O-ring
12. Booster Head Housing
13. Sleeve
14. Boot
15. Mounting Stud
16. Check Valve
17. Tube Seat
18. Flow Switch
19. Flow Switch O-ring
20. Relief Valve Stem
21. Relief Valve Spring
22. Valve O-ring
23. Relief Valve Screw
24. Reaction Piston O-ring
25. Reaction Piston
26. Reaction Piston to Power Piston Snubber
27. Reaction Piston O-ring
28. Power Piston Seal
29. Power Piston
30. Power Piston Return Spring
31. Expander
32. Cylinder Seal
33. Cover Retaining Screw
34. Output Push Rod (Piston Rod)
35. End Cover
36. Push Rod V-Lip Seal
37. Seal Support Plate
38. Electro-Hydraulic Pump
39. Electro-Hydraulic Pump Mounting Bolt & Washer
40. O-ring

Fig. D12 1974 to mid 1976 Hy-Power hydraulic booster disassembled

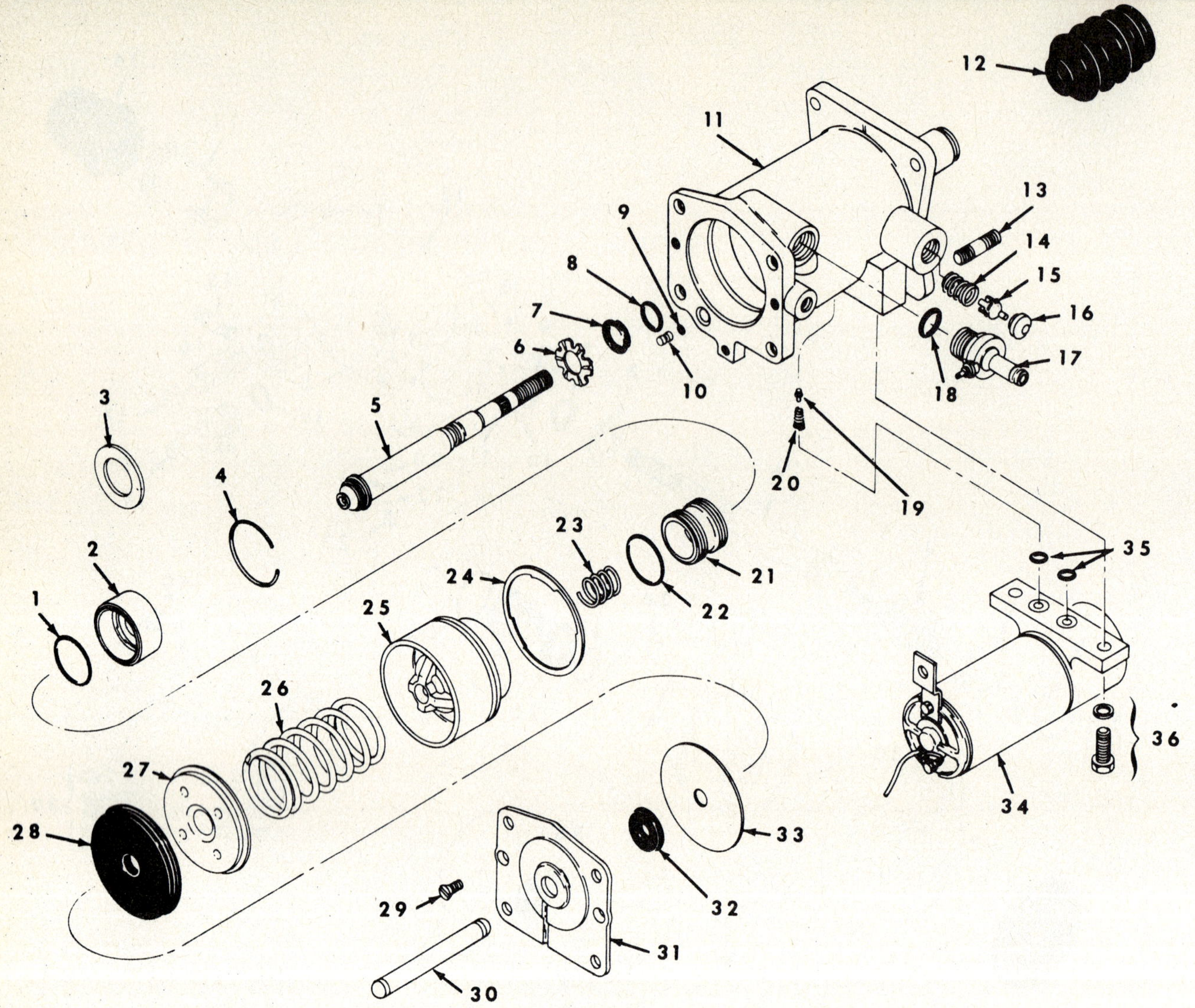

1. O-ring	13. Mounting Stud	25. Power Piston
2. Valve Seat	14. Check Valve Spring	26. Return Spring
3. Flow-Thru Valve (Washer), Early Models Only	15. Check Valve Stem	27. Expander
4. Retainer	16. Tube Seat	28. Cylinder Seal
5. Control Valve/Push Rod Assembly	17. Flow Switch	29. Cover Screw
6. Valve Snubber (Bumper)	18. Flow Switch O-ring	30. Output Push Rod (Piston Rod)
7. V-Lip Seal	19. Relief Valve Stem	31. End Cover
8. O-ring	20. Relief Valve Spring	32. V-Lip Seal
9. By-Pass Plug O-ring	21. Reaction Piston	33. Support Plate
10. By-Pass Plug	22. Reaction Piston O-ring	34. Electro-Hydrulic Pump
11. Booster Head Housing	23. Reaction Spring	35. O-rings
12. Boot	24. Power Piston Seal	36. EH Pump Mounting Bolt & Washer

Fig. D13 Mid 1976 to 1977 Hy-Power hydraulic booster disassembled

internal booster pressure developed during application of the brakes, and retains pressure for steering functions. Fluid flow through the flow switch opens the reserve motor pump electrical circuit during normal operation. A check valve in the motor pump prevents backflow through the motor pump during normal operations. If normal flow from the steering pump is interrupted, the electric motor pump provides power for reserve stops. The flow switch closes energizing a power relay, providing electrical power to the pump. During reserve operation, fluid is retained in the booster by the inlet port check valve. The motor pump circulates fluid within the booster assembly with the number of brake applications limited by the electrical capacity of the vehicle. Manual braking is possible if both the power and reserve systems malfunction.

Hy-Power Booster Disassembly Notes

1. Refer to Figs. D12, D13 and D14 when disassembling these units.
2. Remove the flow switch using a thin blade open end wrench only, otherwise switch damage may result.
3. Do not attempt to disassemble the Electro-Hydraulic Pump.
4. To remove booster end cover, tool No. J-24862 must be used. If tool No. J-24862 is not used, serious injury may result as the piston return spring is under a 60 pound load.

Cleaning & Inspection

1. Clean all metal parts except flow switch and Electro-Hydraulic pump in dena-

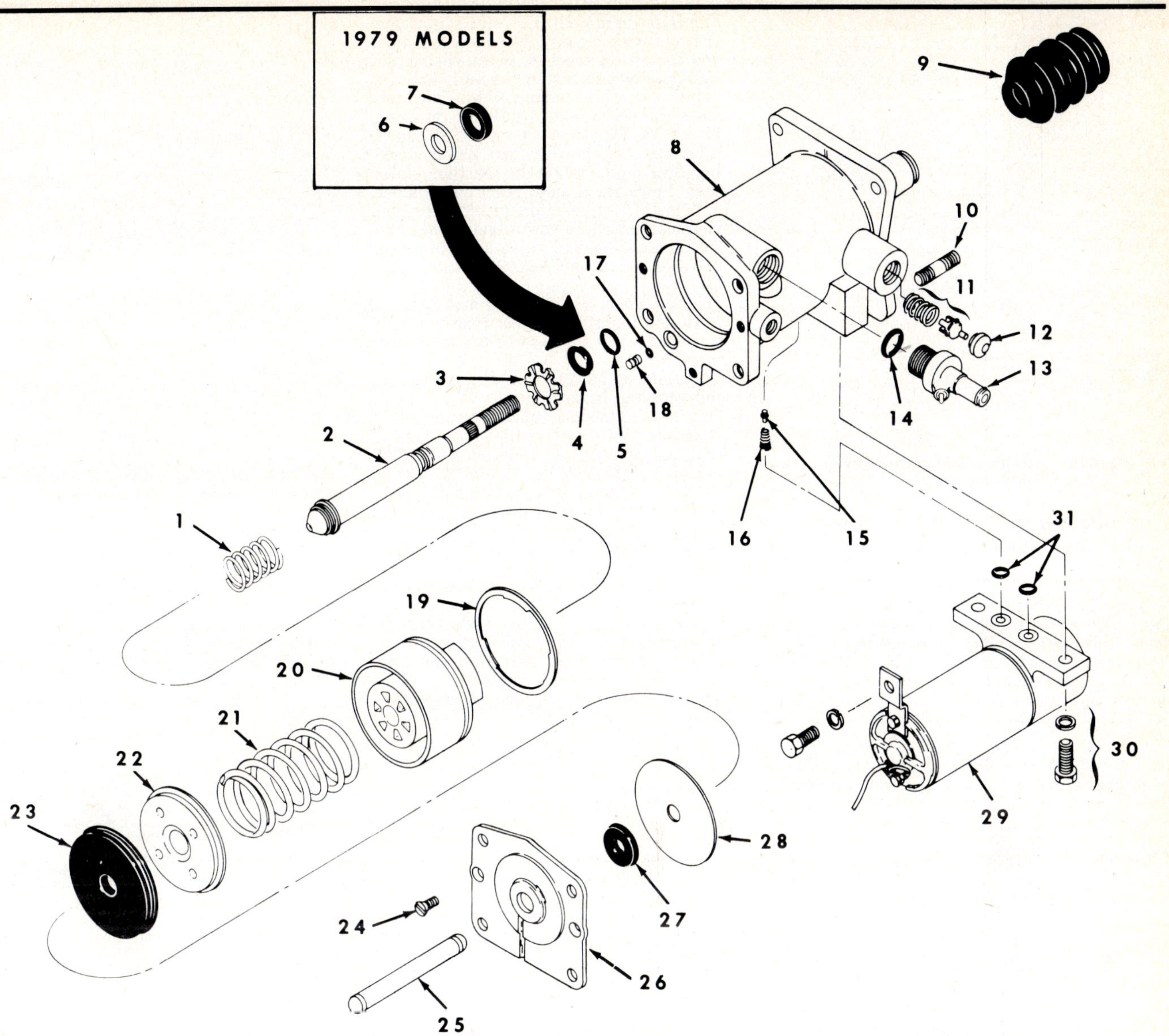

1. Return Spring
2. Control Valve/Push Rod Assembly
3. Snubber (Bumper)
4. V-Lip Seal
5. O-ring
6. Washer
7. V-Lip Seal
8. Booster Head
9. Boot
10. Mounting Stud
11. Check Valve & Spring
12. Tube Seat
13. Flow Switch
14. O-ring
15. Relief Valve Stem
16. Relief Valve Spring
17. By-Pass Plug O-ring
18. By-Pass Plug
19. Power Piston Seal
20. Power Piston
21. Return Spring
22. Seal Expander
23. Cylinder Seal
24. Cover Screw
25. Output Push Rod (Piston Rod)
26. Cover
27. Piston Rod Seal
28. Support Plate
29. Electro-Hydraulic Pump
30. Pump Mounting Bolt & Washer
31. O-rings

Fig. D14 1978—79 Hy-Power hydraulic booster disassembled

tured alcohol or power steering fluid only. Use only dry, filtered air to dry parts and to blow out passages.

2. Inspect all parts for stripped threads, cracks or distortion. Replace as necessary.
3. Inspect cylinder bore for excessive scoring, nicks, burrs or corrosion, if any of these conditions exist, replace unit. Do not attempt to use any abrasives on any internal parts.
4. Use only new rubber parts from repair kit, be sure to replace all parts included in repair kit.
5. Replace flow switch if it leaks or is faulty.
6. Inspect the conical seating surface on the pushrod and valve for nicks, burrs or corrosion. Inspect the mating surface in the valve seat in the same manner.
7. Inspect the pushrod and valve ball joint is free to swivel within its limits and there is no perceptible evidence of wear. Also ensure the stop staked to the valve near the conical end is not loose.
8. Inspect the contact surfaces between the body and pump for nicks and burrs or any damage that would cause leakage between the two surfaces.
9. If the valve seat in the piston assembly for mid-1976 to 1977 model units is replaced and the unit was originally equipped with a flow through valve, discard the flow through valve. It is not required and is not used with the replacement valve seat.

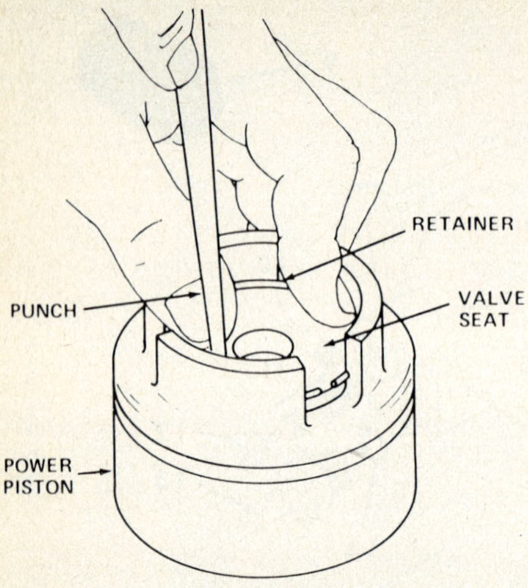

Fig. D15 Installing valve seat retainer, 1978–79 models

Hy-Power Booster Reassembly, Figs. D12, D13 & D14

NOTE: The Hy-Power hydraulic booster operates with power steering fluid only. Prior to reassembly be sure all parts are clean and free from contamination of other fluids. Lubricate all parts at reassembly with fresh, clean power steering fluid only.

1. On 1978–79 models position valve seat in piston. Ensure the face with the valve seat is facing up, then install retainer, Fig. D15. On mid 1976–77 units lubricate O rings with clean power steering fluid and install on reaction plate, then position the reaction piston assembly and the valve seat in the piston. While pushing down the valve seat, install the retainer, Fig. D16. On 1974 to mid 1976 units lubricate O rings with clean power steering fluid and install on reaction plate, then lubricate a new reaction snubber and install on reaction piston. Position piston assembly in power piston then the reaction spring and valve seat, pressing down on valve seat install retainer, Fig. D17.
2. Position check valve spring and stem in body and install tube seat insert. Use a spare tube nut to press in the tube seat insert.
3. Lubricate by-pass plug O ring and position on plug. Install plug.
4. On 1974–78 models, lubricate and install O ring and V lip seal in valve bore of body, sealing lips must face large bore. On 1979 models, lubricate V lip seal and install in valve bore using seal installer J-28681, seal must be installed flush with sealing lips facing large bore.
5. Install washer in counterbore above seal.
6. On 1974 to mid 1976 models, install control snubber in end of control valve/pushrod assembly. On all models install snubber on control valve/pushrod by starting bumper on threaded end and sliding up against the retainer.
7. Install control valve/pushrod assembly into body, carefully working the threaded end through the valve bore and seals.
8. Install valve return spring.
9. Lubricate piston seal and install on piston, install piston in bore.
10. Place the hydraulic booster in a suitable vise, large bore up.
11. Lubricate and install piston rod seal in end cover. The seal lips must face the inside of the cover when installed.

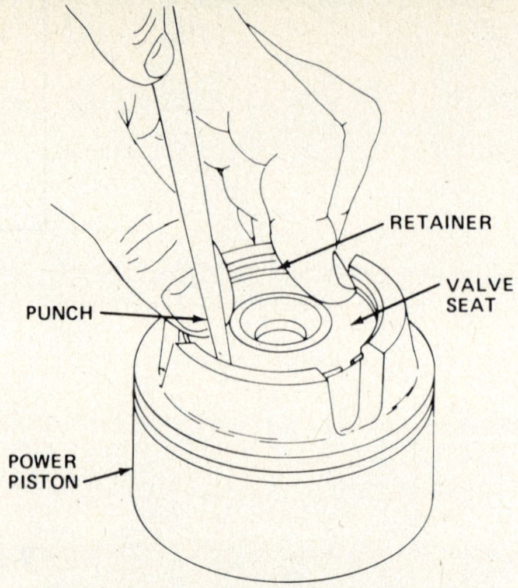

Fig. D16 Installing valve seat retainer, mid 1976 to 1977 models

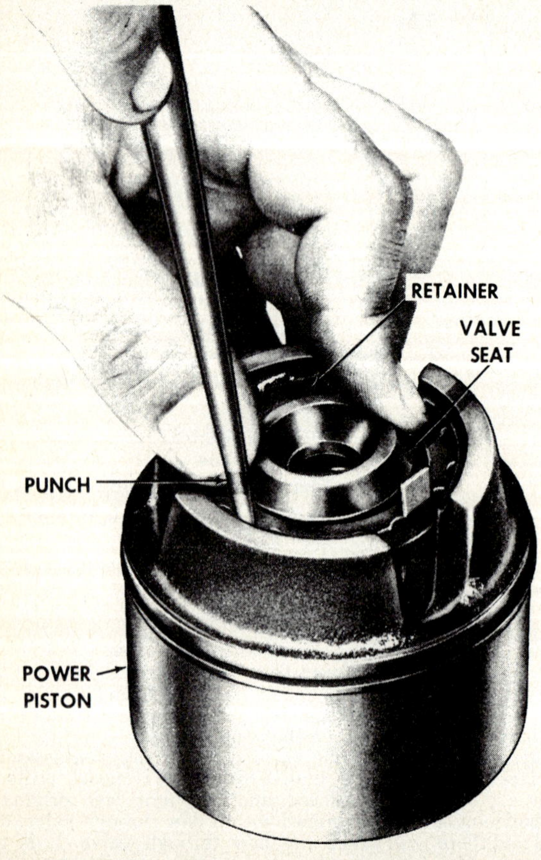

Fig. D17 Installing valve seat retainer, 1974 to mid 1976 models

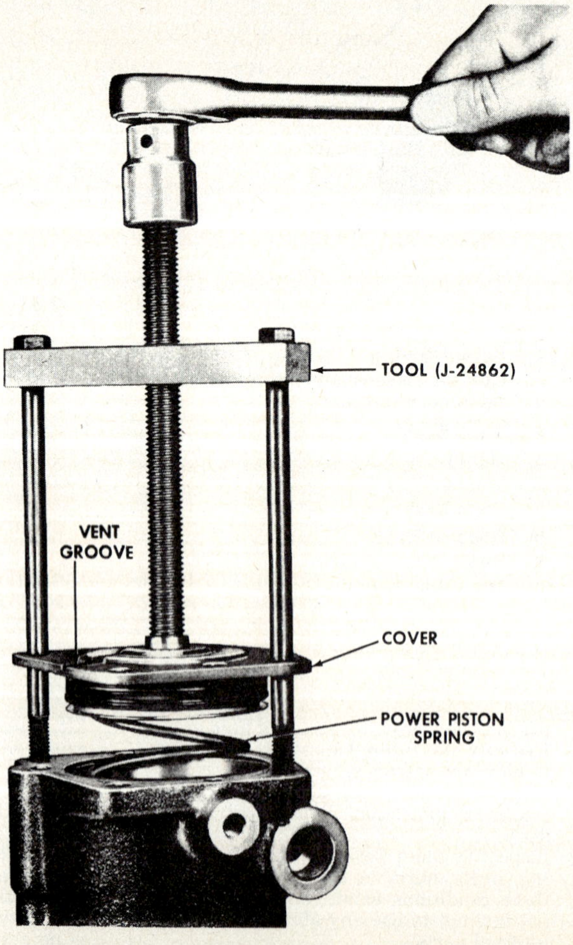

Fig. D18 Installing end cover assembly

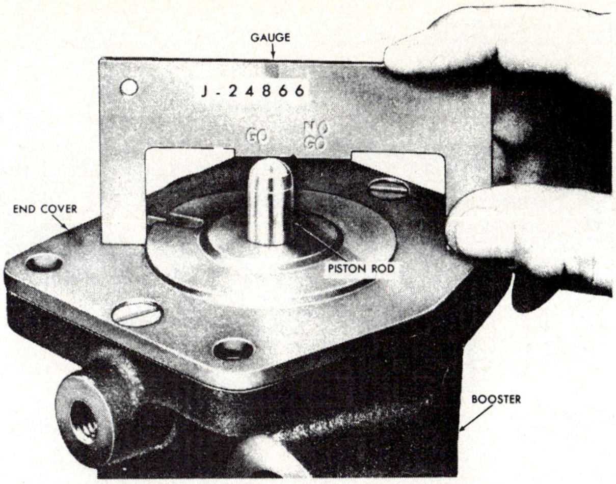

Fig. D19 Checking piston rod length

Length, Millimeters	Length, Inches
111.88–112.05	4.405–4.411
112.06–112.23	4.412–4.418
112.24–112.41	4.419–4.426
112.42–112.59	4.426–4.443
112.60–112.77	4.443–4.440
112.78–112.94	4.440–4.446
112.95–113.12	4.447–4.454
113.13–113.30	4.454–4.461

Fig. D20 Replacement Piston Rods

12. Lubricate cylinder seal then position support plate, cylinder seal and seal expander against cover in order of assembly. Install tool J-24862 to align parts during installation. Position assembly over body on the return spring, ensure the cover is positioned so the vent groove will be on the same side as the Electro-Hydraulic pump mounting surface.
13. Using tool No. J-24862, thread the legs into body and press the parts together, Fig. D18.
14. Install end cover retaining screws, and remove tool No. J-24862.
15. Reposition unit in vise so Electro-Hydraulic mounting flange is facing up. Install relief valve stem into body, notched end first. Position relief valve spring into body on top of relief valve. On 1974 to mid 1976 models, lubricate relief valve screw O ring and position on relief valve screw, install screw.
16. Install Electro-Hydraulic pump on flange. Torque two 9/16 inch bolts to 23 ft. lbs.
17. On mid 1976 and later models torque pump bracket attaching bolt to 117 in. lbs.
18. Install boot or sleeve (if equipped).
19. Install flow switch. Torque to 25 ft. lbs.
20. Lubricate piston rod and install using a twisting motion until it is seated.
21. Press down piston rod firmly, ensuring all internal parts are seated. Using tool No. J-24866 check that piston rod length is within specifications, Fig. D19.
22. If piston rod is not within specifications, select a new piston rod to bring the measurement within limits. Eight piston rods are available with lengths as shown in the "Replacement Piston Rods" chart, Fig. D20.

Midland-Ross Brake Boosters

DASH-MOUNTED SINGLE DIAPHRAGM TYPE

Disassembly Notes

CAUTION: The booster push rod is provided with an adjustment screw to maintain the correct relationship between booster control valve plunger and master cylinder piston. Failure to maintain this relationship will prevent the master cylinder piston from completely releasing hydraulic pressure and can cause the brakes to drag. Therefore, inasmuch as the adjustment varies between one vehicle and another, measure and note for future reference the distance between master cylinder mounting surface and end of screw. Master cylinder must be removed to gain access to push rod (see Fig. M7).

1. Disassemble the booster as suggested by Fig. M1.
2. Scribe a line across booster body and end plate before separating these parts to ensure proper alignment on reassembly.
3. Do not remove valve operating rod from control valve plunger unless plunger or rod is to be replaced. To remove, hold rod firmly and force plunger off rod, breaking plastic retainer. Remove all pieces of broken plastic retainer from groove in plunger if plunger is to be used again with replacement rod.
4. Always use a repair kit when reassembling, and use all parts contained in the kit regardless of condition of old parts.

Reassembly, Fig. M1

1. If valve operating rod was removed from plunger, assemble new plastic retainer to end of rod, Fig. M1. Insert rod into plunger so that retainer engages groove in plunger.
2. Install "Block Vee" type seal and O-rings on valve plunger, Fig. M2.
3. Insert control valve plunger into control valve hub from rear of hub, Fig. M3.
4. Assemble atmospheric valve, return spring and reaction load ring to valve plunger and hub.
5. Push control valve plunger forward and reaction load ring backward against return spring in order to install retainer in groove of plunger.
6. Install O-ring in groove at front side of diaphragm. Assemble valve plunger and hub to diaphragm so that operating rod and small diameter end of hub enter front side of diaphragm and protrude from rear side.
7. Install "Block Vee' type seal in hub rear seal adapter, with sealing lip toward rear. Slide seal and adapter over rear end of valve hub so that large diameter side of adapter bears against diaphragm.
8. Install two plastic plunger guides in their grooves on valve plunger, Fig. M4.
9. Install rubber reaction ring in valve hub so that ring locating knob indexes in notch in hub with ring tips toward front, Fig. M4.
10. Assemble reaction lever and ring, then install assembly in valve hub, Fig. M4.
11. Assemble reaction cone and cushion ring to push rod and secure to rod with retainer, Fig. M1. Install push rod to valve hub so that valve plunger indexes in push rod, Fig. M4.
12. Assemble bellows retainer to rear fold of bellows. Install bellows stay and two support rings. Plastic stay is located in center fold of bellows with support ring in fold at each side of stay, Fig. M1. Assemble bellows protector to front fold.
13. Position bellows over push rod against front side of diaphragm. Secure bellows to diaphragm by installing retaining ring, Fig. M5. Make sure retainer ring is fully seated.
14. Install O-ring in groove at front side of end plate. Assemble bellows, control valve and diaphragm to end plate by inserting valve hub through front side of end plate with small diameter side of seal adapter protruding from rear side of end plate, Fig. M1.
15. Install large C-ring to rear seal adapter at rear side of end plate.
16. Install diaphragm return spring in booster body, and assembly end plate to booster

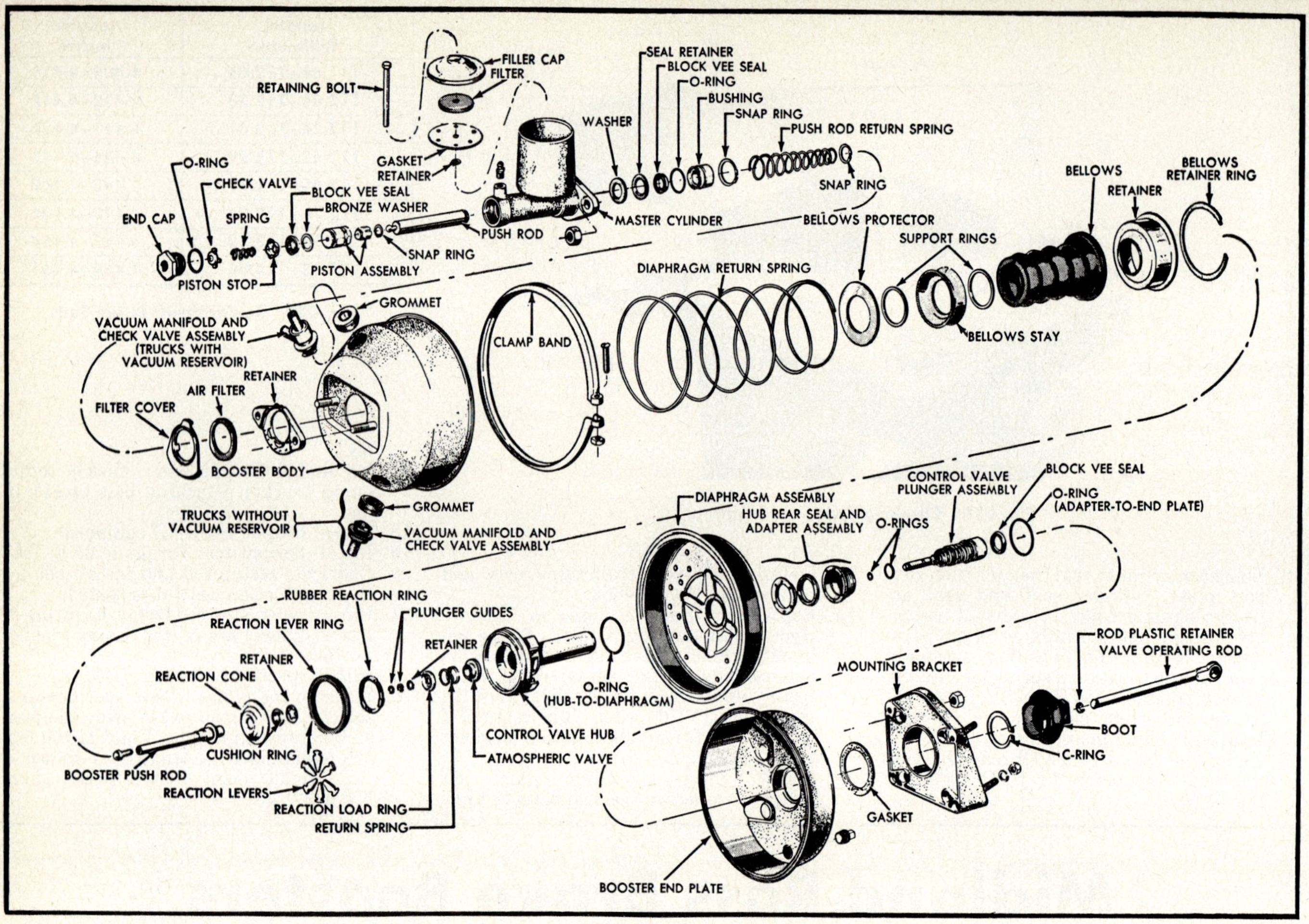

Fig. M1 Midland dash-mounted diaphragm type booster

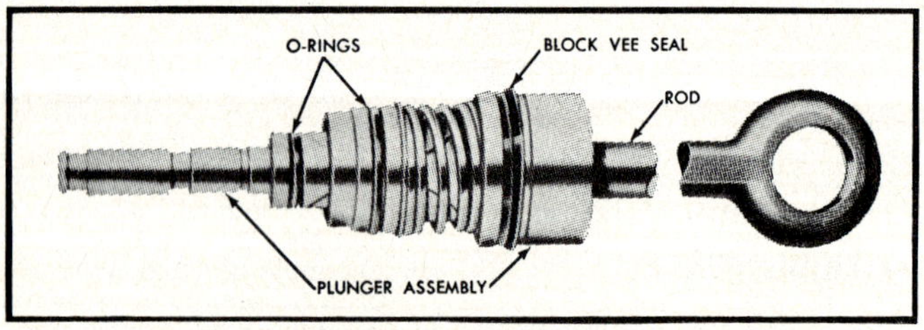

Fig. M2 Valve operating rod end plunger assembly

body so that scribed marks are aligned. *Be sure that lip of diaphragm is evenly positioned between retaining flanges of booster body and end plate.* Secure booster body, diaphragm and end plate together with clamp band and tighten clamp bolt and nut.

17. Pull front lip of bellows through booster body and position it around outer face of booster body, Fig. M6.
18. Place gasket on end plate, slide mounting bracket over end plate studs and install retaining nuts, Fig. M1.
19. Slide rubber boot over valve operating rod and engage boot with groove on mounting bracket.

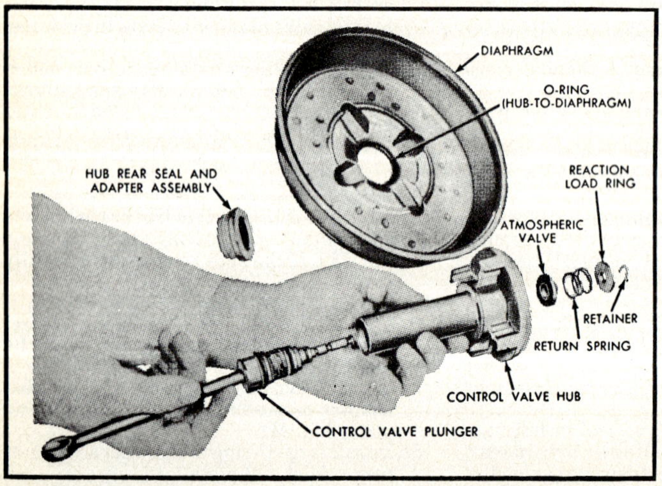

Fig. M3 Control valve components to hub and diaphragm

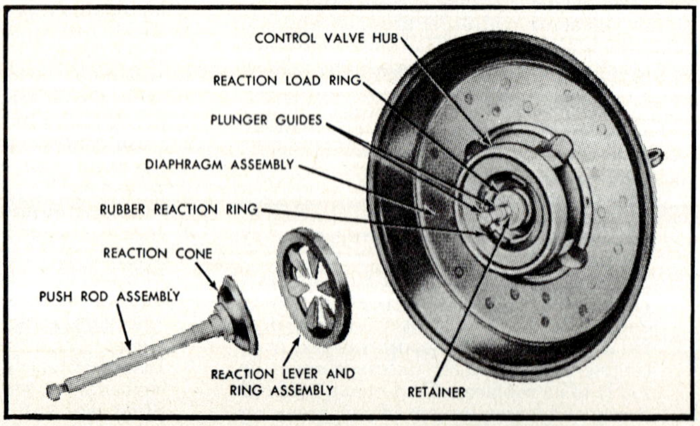

Fig. M4 Reaction components and push rod to valve hub

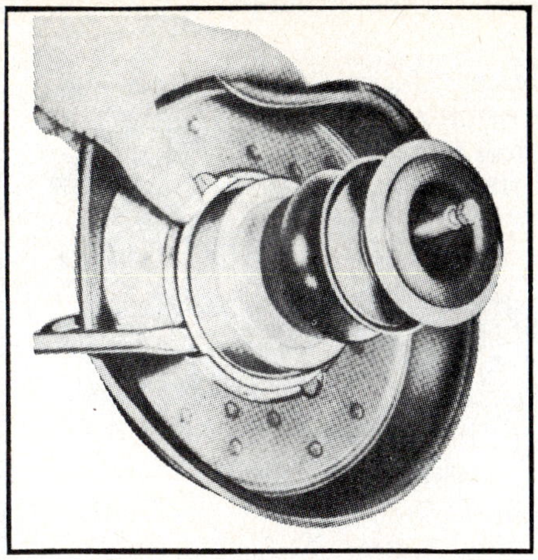

Fig. M5 Bellows installation

20. Install rubber grommet in vacuum port in booster body. Large diameter side of grommet should be to outside of booster. Force vacuum manifold and check valve through grommet. Do not push grommet into vacuum chamber.
21. Install air filter cover, filter and retainer on booster body.
22. Check booster push rod adjustment and, if necessary, adjust to same dimension measured before unit was disassembled, Fig. M7.

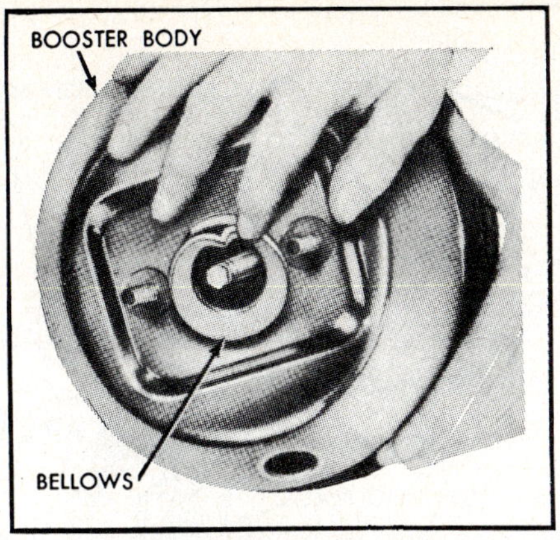

Fig. M6 Bellows-to-booster body engagement

FRAME MOUNTED, SINGLE DIAPHRAGM TYPE

Disassembly Notes

1. Before separating the major components

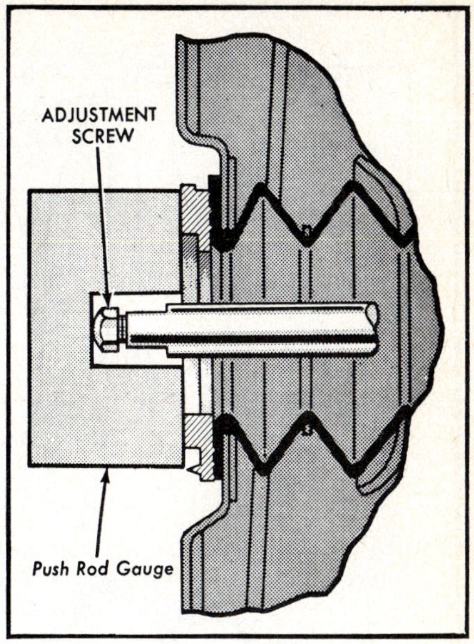

Fig. M7 Push rod adjustment. Measure adjustment before disassembly of booster to be sure it is correct when booster is installed

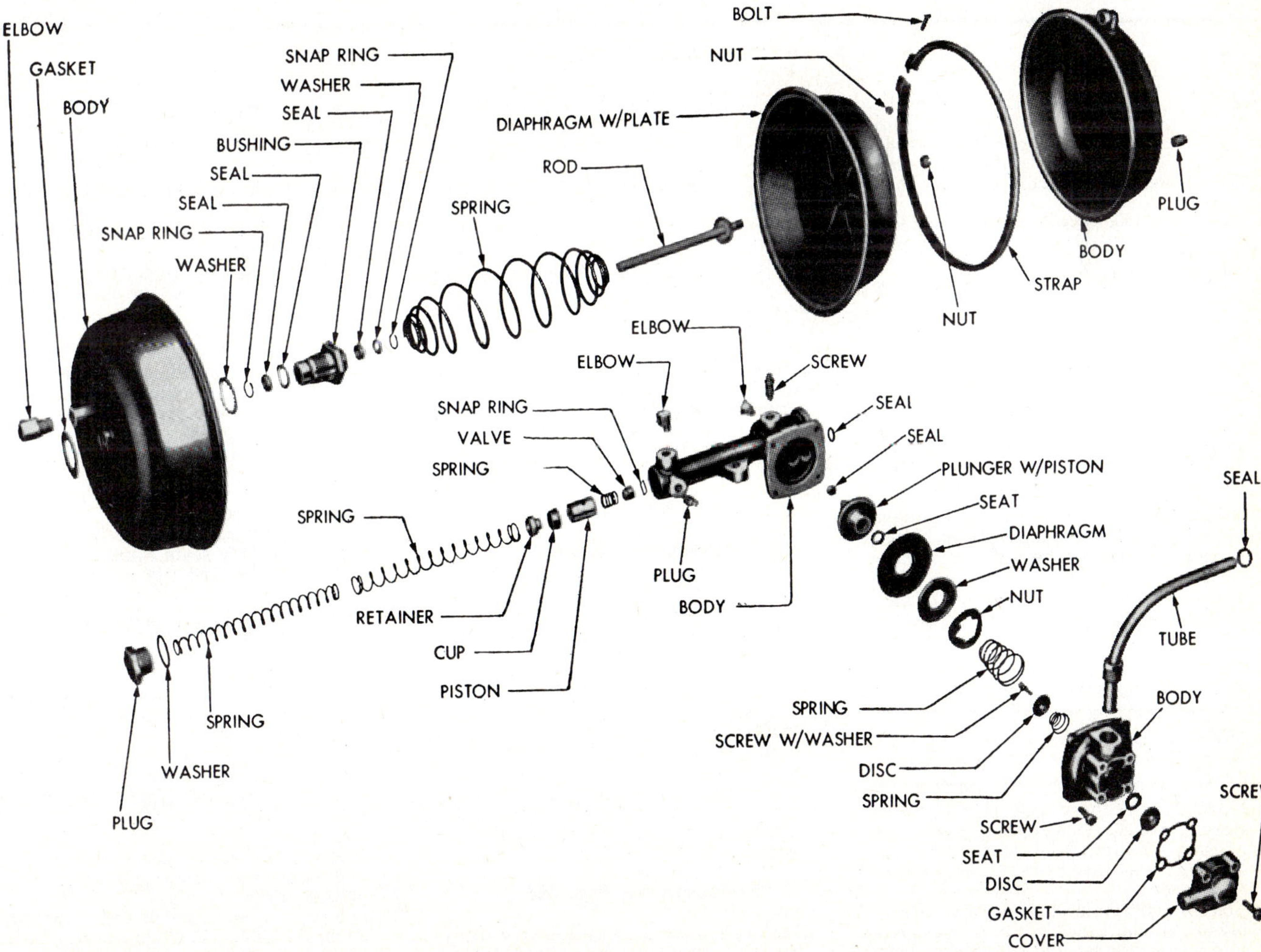

Fig. M8 Frame mounted single diaphragm type, disassembled

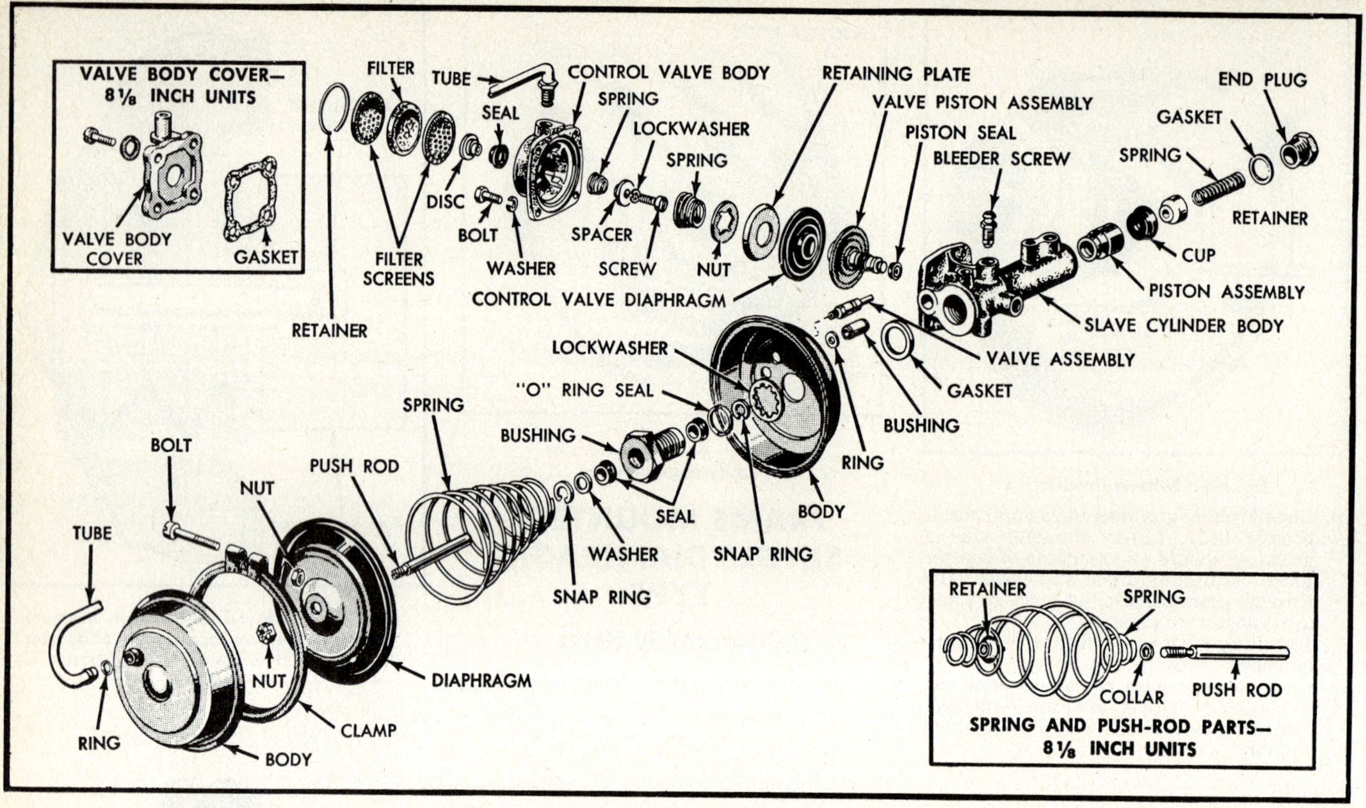

Fig. M9 Hy-Power vacuum booster disassembled

of the unit, mark both halves of the diaphragm body with a scriber. Also mark the flanges of the control valve body and slave cylinder body so these parts can be reassembled in their original positions.

2. Scribe a line on the side of valve body cover and valve body before removing cover.
3. Always use a repair kit when servicing these units, and use all parts in the kit regardless of the condition of the old parts.

Reassembly, Fig. M8

1. Install control rod diaphragm, by-pass valve, atmospheric and vacuum springs and diaphragm pressure plate assembly.

NOTE: When replacing rod in diaphragm assembly, hold push rod in radius blocks as close to ring groove as possible. Use sealant on threads making sure that proper length rod and proper size diaphragm is used.

2. Install seals on push rod bushing with lip of seals facing slave cylinder body.

NOTE: If rear chamber is dented, it must be replaced as brakes will not release.

3. Place slave cylinder body in a vise and install gasket, then insert piston rod housing through front half of booster body and into slave cylinder body. Tighten securely.
4. Place diaphragm and press plate assembly in rear chamber, then insert push rod into slave cylinder and clamp chambers together with clamp strap making certain to align scribe marks.
5. Insert gauge C-13456 firmly against rod bushing and make certain that push rod length is within indications on gauge.
6. Install slave cylinder piston with open end facing chamber body and install piston cup with lip of cup toward end plug.
7. Install spring retainer and return springs using a new copper gasket and tighten plug.
8. Lubricate piston cups with brake fluid and install control valve piston.
9. Install diaphragm return spring with small coil toward diaphragm, then install control valve body making certain that scribe marks are aligned.

HY-POWER DIAPHRAGM TYPE

Disassembly Notes

1. Before separating the major components of the unit, mark both halves of the diaphragm body with a scriber. Also mark the flanges of the control valve body and slave cylinder body so these parts can be reassembled in their original positions, Fig. M9.
2. Scribe a line on the side of valve body cover and valve body before removing cover.
3. Always use a repair kit when servicing these units, and use all parts in the kit regardless of the condition of the old parts.

Reassembly, Fig. M9

1. Position a new seal in control valve body. Assemble spring and spacer in valve body and secure with screw and lock washer. Tighten screw securely.
2. Position control valve diaphragm and plate and secure with retainer nut.
3. Install seal on control valve piston.
4. Insert check valve spring, valve and retainer in hydraulic piston. Be sure check valve floats freely and does not bind in bore.
5. Install transfer bushing, rubber seal and gasket on front body of slave cylinder body.
6. Install push rod seals, washer and snap rings in push rod bushing. Both push rod seals should be installed with open end of seal facing slave cylinder body. Install lock washer over end of bushing, and install bushing seal.
7. Mount slave cylinder body in a vise. Position front body over end of slave cylinder, inserting transfer bushing with proper hole in front body.
8. Thread push rod bushings in place. Be sure front body seats squarely on slave cylinder body, and tighten bushing securely.
9. Coat piston bores in slave cylinder body with brake fluid. Dip hydraulic cylinder piston, seals, spring retainer and spring in brake fluid.
10. Install hydraulic piston in slave cylinder bore with recessed end (large bore of piston) toward push rod bushing. Carefully position piston cup with recessed end toward large spring, spring retainer and spring on top of piston. *On frame-mounted booster, install spring seat in spring coils.*
11. Position new copper gasket on end plug and screw plug in slave cylinder. Tighten plug securely.
12. Dip control valve piston and diaphragm in brake fluid. Position control valve spring on diaphragm with small end of spring over piston boss.

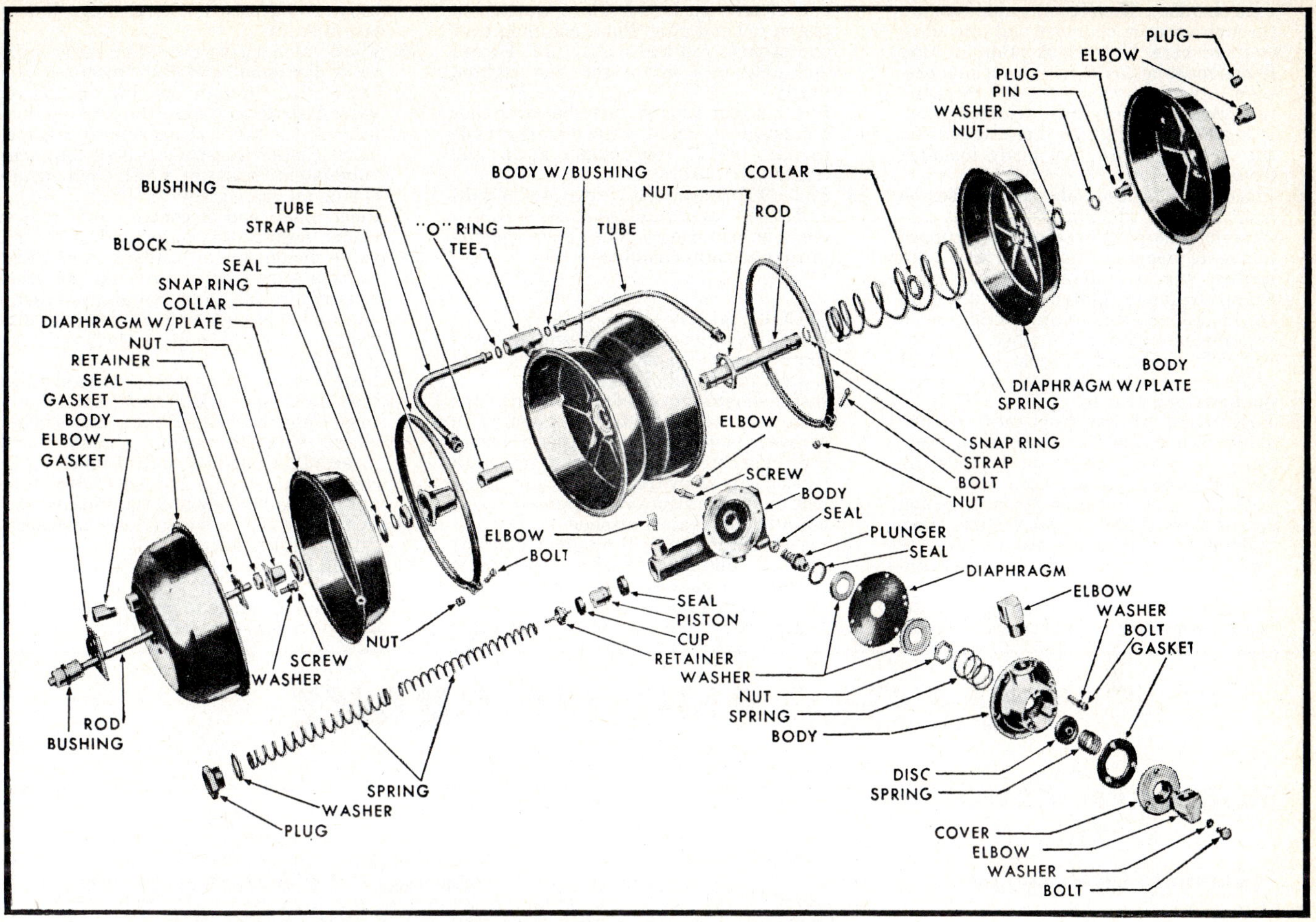

Fig. M10 Double diaphragm booster disassembled

13. Position control valve body over spring align scribe marks, install valve body on slave cylinder body and secure with attaching bolts and lock washers. Tighten bolts securely.
14. Position new gasket on control valve body. Place valve body cover over gasket and secure cover with attaching bolts and lock washers.
15. Position collar over threaded end of push rod. Place spring retainer in spring as shown in insert of Fig. M9. Insert push rod and collar in coils of spring and through retainer. Place diaphragm over threaded end of push rod and secure with push rod nut. After nut is tightened, coat threads of push rod with shellac to prevent leakage. Coat push rod with brake fluid.
16. Place return spring over push rod bushing.
17. Place rear body on top of diaphragm with scribed marks on rear body aligned with mark on front body. Compress return spring and install and tighten clamp, making certain diaphragm bead is properly positioned between two halves of body.
18. Install by-pass tube.

DOUBLE DIAPHRAGM TYPE

Disassembly Notes

1. Before separating main components, scribe marks on front and rear vacuum chambers to assist in aligning parts on reassembly.
2. Always use a repair kit, and use all the parts contained in the kit regardless of the condition of the old parts.

Reassembly, Fig. M10

1. Install outer seal in groove in slave cylinder piston with lip of seal toward small end of piston.
2. Install spring and valve in end of piston and hook piston onto end of front push rod.
3. Insert round end of rear push rod into rear end of front push rod. Place two halves of collar in front push rod and install snap ring.
4. Install seal in flange end of push rod guide bushing, with lip of seal pointing inward. Install washer over seal and secure with snap ring.
5. Install seal in bore in other end of bushing, with lip of seal pointing outward.
6. Install O-ring in groove in guide bushing.

VACUUM CHAMBER ASSEMBLY

1. Install piston rod guide bushing in intermediate vacuum chamber. Insert bushing through vacuum chamber from external control line tee-fitting port side. Clamp a socket in vise and set hex head of guide bushing in socket. Install nut on guide bushing and tighten firmly, using another socket. Stake nut at bushing threads at three points after tightening.
2. Install snap ring in groove in rear end of piston rod, then install collar on rod with chambered side of hole over snap ring. Install rear diaphragm over piston rod, then thread nut onto end of rod. Clamp nut in vise and use 1/4" bar stock through holes in piston rod to tighten rod into nut. After tightening, stake edge of nut into threads on piston rod at three points.
3. Install large end of diaphragm return spring under clips on rear diaphragm plate.
4. With piston rod near nut still clamped in vise, install intermediate vacuum chamber over piston rod with piston rod guide bushing inside small end of diaphragm spring. Press down on intermediate vacuum chamber, compressing spring to expose snap ring groove in piston rod. Install snap ring, collar, front diaphragm and nut. Tighten nut firmly and stake in place at three points.
5. Mount slave cylinder body in vise with outer end down. Outer end must be exposed so that piston and push rods can be inserted later.
6. Place slave cylinder piston stop washer in slave cylinder bore. Make sure washer seats against shoulder in cylinder bore.
7. Place front vacuum chamber to slave cylinder gasket on slave cylinder body, with all holes aligned. Position front vacuum chamber on slave cylinder and install cap screw with lock washer. *Do not tighten cap screw.*
8. Place guide bushing gasket on front vac-

uum chamber. Then insert guide bushing through vacuum chamber and into slave cylinder bore. Cut-out of guide bushing flange must be positioned at vacuum passage in vacuum chamber. Attach piston rod guide to slave cylinder with four socket head screws. Tighten these screws and the one cap screw previously installed loosely.

9. Place intermediate vacuum chamber on front vacuum chamber with front diaphragm bead positioned between cupped flange of vacuum chambers, and with scribed marks aligned. Install front clamp strap and tighten clamp nut firmly. Install external control tube tee in intermediate vacuum chamber. When tee is firmly tightened, it must be at right angle to vacuum chamber with small diameter end to rear.
10. With rear push rod, front push rod and slave cylinder piston assembled as previously directed, insert push rods up through slave cylinder bore. Rear push rod must enter vacuum chamber piston rod as slave cylinder piston enters slave cylinder bore. Use a wood screwdriver handle to push piston up into slave cylinder bore.
11. Press down on rear diaphragm plate, compressing diaphragm spring, to expose end of rear push rod. Place washer on piston rod end plug. Place end plug over end of push rod and install pin. Thread end plug into piston rod and tighten firmly.
12. Position rear vacuum chamber over rear diaphragm, aligning scribed marks made earlier. Install rear clamp strap and tighten nut securely.
13. Place O-ring over end of rear control tube and insert tube into tee. Install rubber elbow on rear control tube and in opening in rear vacuum chamber.

CONTROL VALVE

1. Install new piston seal on small end of control valve piston with lip of seal toward small end of piston.
2. Install diaphragm on control valve piston with bead at inner diameter seated in groove in piston. Place diaphragm washer over piston against diaphragm with cupped side away from diaphragm. Place new diaphragm nut over piston and press down firmly against washer.
3. Insert metering wire through hole in side of piston. Both ends of wire must be bent over so wire cannot fall out, but must be free to move in hole.
4. Insert piston into bore in slave cylinder body, using care not to damage lips of piston seals.
5. Place valve spring over stem on vacuum valve disc (small end first). Insert stem of valve disc through opening in control valve body from inner side, making sure large end of spring seats squarely against body. Place atmosphere valve disc over valve stem (seal side first), then install O-ring on valve stem.
6. Place small end of control valve piston return spring over outer end of control valve piston. Install control valve body, making sure control valve spring seats squarely in body, and that groove around outer edge of body engages bead on diaphragm. Attach body to slave cylinder with capscrews.
7. Install cover and gasket on control valve body and attach with screws.
8. Insert end of control tube into rubber elbow or tee in vacuum chamber, and insert other end into control tube port in control valve body. Thread control tube nut into valve body and tighten firmly.
9. Install end cap with gasket in end of slave cylinder and tighten firmly.
10. Install bleeder screws in top of slave cylinder body.
11. Cover all openings to prevent entrance of dirt during installation.

JACOBS ENGINE BRAKE

MODEL APPLICATION CHART

Cummins Diesel Engines

Model 20

NHC-4	NH-220	NTE-235
NH-180	NHS-6	NTO-6
NHE-180	NHRS-6	NRT-6
NH-195	NT-4	NRTO-6
NHE-195	NT-6	NTF-6

Models 25 & 25A

NHE-220	NHE-225	NH-250
NT-280	NT-335	NT-380
NTC-260, 335	NHC-225, 250	NHCT-240, 270

Model 25B

N-927	NH-250	NHE-220, 225
NHCT-270	NHD-230	NT-280, 300
NHF-240, 265	NHTF-295	NT-335, 380
NTA-370, 420	NTC-290	NTC-335-350
NTF-295, 365	NHC-290	NHC-225, 250

Models 30 & 30E

NTC-250, 290, 350, 400
Power Torque-270
NTCC-250, 290, 350, 400

Model 59

V6-140	V6-155	V8-185
V8-210	V8-225	

Model K200

Model 903

V-903	VT-903

Model K1150

Detroit Diesel Engines

Model 53A

3-53	4-53	6V-53	8V-53

Model 71

4-71	6V-71	12V-71
6-71	8V-71	16V-71

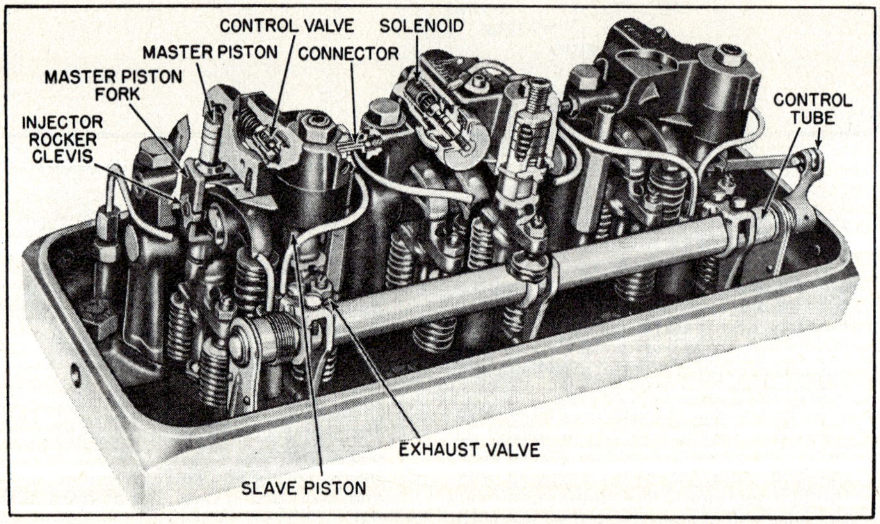

Fig. 1 Jacobs engine brake. Typical

Mack Diesel Engines

Models 67 & 675

673	675	711

INTRODUCTION

The Jacobs Engine Brake, Fig. 1, is a device which, when energized, effectively converts a power producing diesel engine into a power absorbing air pump. This transfer of motion is accomplished when desired through a master-slave piston arrangement which opens the exhaust valves near the top of the compression stroke, releasing the compressed cylinder charge to exhaust.

The blowdown of compressed air to atmospheric pressure prevents the return of energy to the piston on the expansion stroke, the effect being a net energy loss since the work done in compressing the charge is not returned during the expansion process.

ADJUSTMENTS

Slave Piston, Model 20

1. Loosen and back off locknut.
2. Using Allen wrench, back off slave piston adjusting screw until piston spring "lets go" of adjusting screw (valve piston will move up into housing under influence of its spring until it seats in its bore, remov-

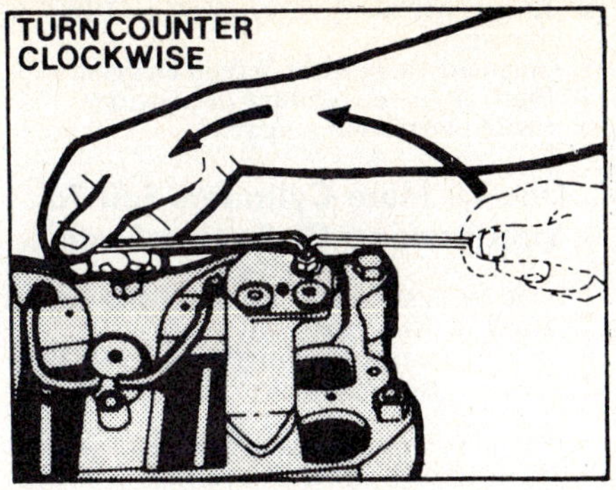

Fig. 2 Slave piston adjustment. Model 20

Fig. 3 Slave piston adjustment Models 25, 25A, 25B, 59 & 71 (typical)

Fig. 4 Slave piston adjustment. Models 25, 25A & 25B

ing spring load from adjusting screw.)

3. With engine at operating temperature and idling at 550 RPM, turn adjusting screw down slowly until slave piston contacts crosshead as indicated by the twitch in the handle of the Allen wrench.

NOTE: Do not attempt to make this adjustment by holding the locknut and turning the adjusting screw through both the locknut and the brake housing.

4. Back adjusting screw out exactly 1/2 turn and tighten locknut, Fig. 2.

Slave Piston, Models 25, 25A, 25B, 30 & 30E

NOTE: Adjustment is made statically with engine cold or warm.

1. Loosen locknuts and with an Allen wrench, back slave piston adjusting screws out of housing until the slave pistons seat in the bores.
2. Rotate engine until "1-6 V.S." mark on accessory drive pulley aligns with the timing mark on gear case cover.
3. The exhaust valves on No. 1 and 6 cylinders are now closed. Adjust clearance between slave piston and crosshead to .018 inch on these cylinders, Fig. 3.
4. Rotate engine until the exhaust valves of No. 2 and 5 cylinders are closed and adjust clearance between slave piston and crosshead to .018 inch.
5. Continue to rotate engine until the exhaust valves of No. 3 and 4 cylinders are closed and adjust clearance between slave piston and crosshead to .018 inch.

Slave Piston, Model 53A

With exhaust valves fully closed and the injector spring fully depressed, adjust clearance between the exhaust valve bridge and the exhaust valve stems to .002 inch, Fig. 4. Then, back off adjusting screw 1 1/2 turns. This provides approximately .064 inch operating clearance.

Slave Piston, Model 59

With exhaust valves closed, adjust clearance between crosshead and slave piston to .038 inch, Fig. 3.

Slave Piston, Model 71

NOTE: Before making adjustment, ensure exhaust valves are closed and injector is in the delivery position.

On standard timed engines, the slave piston clearance is .066 inch. On advanced timed engines, the slave piston clearance is .071 inch. Determine type of engine timing before attempting adjustment.

2 valve engines: Insert correct size feeler gauge between slave piston foot and valve stem cap and adjust to specified clearance, Fig. 3.

4 valve engines: Insert correct size feeler gauge between slave piston foot and exhaust valve bridge and adjust to specified clearance, Fig. 3.

NOTE: A new setting as of August 1980 for slave piston has been released. The setting of .059 inches is to provide an increase in performance. However, two criteria must be met before the setting can be used:

1. Rocker brackets must be of the latest style with a larger pad area for better engine brake housing support. The new bracket was introduced in November 1977. The part number is 5103903.
2. The model 71A/92A engine brake housings (P/N 007077 and 00708) only can be used at the .059 inch setting.

Slave Piston, Models 67 & 675

With exhaust valves fully closed, turn slave piston adjusting screw in until zero clearance is obtained between exhaust valve stem cap and slave piston foot. Then back off adjusting screw 3/4 turn.

Slave Piston, Model 903

Install a dial indicator so plunger rests on back side of slave piston, Fig. 5. With exhaust valves closed and the crosshead loose, drive slave piston down with adjusting screw until fingers of piston contact the lug on bottom of crosshead. Zero the dial indicator and back off the adjusting screw until dial indicator measures .023 ± .001 inch.

Slave Piston, Model K200 & K1150

Loosen exhaust valve rocker lever adjusting screw and back off adjustment. Place feeler gauge between slave piston feet and top of Jacobs crosshead. Turn both adjusting screws in alternately and evenly using 1/8 turn increments until crosshead and feeler gauge make contact with slave piston.

Fig. 5 Slave piston adjustment. Model 903

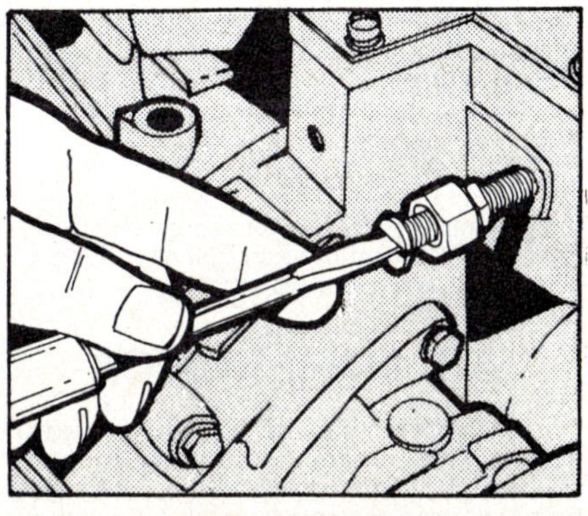

Fig. 6 Adjusting buffer screw. Models 53A & 71 (typical)

Fig. 7 Installing buffer switch. Model 71

Make sure that adjusting screw sockets are bottomed on the valve stems. This keeps crosshead level to obtain the proper slave cylinder clearance. Adjust to .018 inch. Hold the adjusting screws and lock nuts. Torque nuts to 34–36 ft. lbs. Readjust exhaust valve rocker lash.

Buffer Screw Adjustment, Models 53A & 71

With engine running, adjust buffer screw to stop governor "hunt" at idle speed, Fig. 6. Do not increase engine speed above normal idle by forcing buffer screw past point where governor hunt has been eliminated. Tighten locknut. On Model 71, attach buffer switch to buffer screw and position switch to clear other engine components, Fig. 7.

Buffer Switch Adjustment, Model 71

This adjustment is to be made only with tool #2859. Switches are factory set and should not be disturbed except for switch or diode replacement.

1. Insert threaded portion of buffer screw into open end of tool so that buffer spring is close to, but does not touch the plunger.
2. Tighten lock nut so tool is tight on buffer screw.
3. Push plunger in so it rests on buffer spring and tighten wing screw on plunger.
4. Move micro switch in until it clicks on then back out until it clicks off. Hold this position and tighten switch mounting screws.
6. With switch in this position, a .010" feeler gauge inserted between switch plunger and switch actuator should actuate the switch.

If, after the above adjustment the engine brake does not operate correctly (in the no fuel range only) check fuel rack, idle speed and buffer screw adjustments.

TROUBLE SHOOTING

Engine Fails To Start

1. Solenoid valves stuck in on position.

Sudden Drop In Engine Lube Oil Pressure

1. Oil inlet supply seal missing or damaged.
2. Upper solenoid valve seal missing or damaged.
3. Fuel pipe leakage.

One Or Two Cylinders Fail To Brake

1. Slave piston control valve stuck in "OFF" position.
2. Slave piston control valve failure.
3. Slave piston adjustment incorrect.
4. Engine brake housing oil connectors or seals leaking.

Solenoids Won't Control Brake Operation

1. Center solenoid valve seal missing or damaged.

Solenoids Will Not Energize

1. Blown fuse.
2. Automatic switches fail to close.
3. Incorrect electrical power source.

Engine Brake Slow To Operate

1. Lube oil cold and/or too thick.
2. Lower solenoid valve seal missing or damaged.
3. Solenoid valve filter screen clogged.
4. Control valves binding in housing.
5. Switch operation sluggish.

One Or More Cylinders Fail To Stop Braking Or Engine Stalls

1. One or more slave piston control valves stuck in "ON" position.
2. Solenoid valve sticking in on position.
3. Center solenoid valve seal missing or damaged.
4. Solenoid valve exhaust plugged.
5. Switch stuck in "ON" position or misadjusted.
6. Back or beefer switch set too tight.

No Engine Braking

1. Blown fuse.
2. Low engine oil pressure.
3. Switches misadjusted or defective.

Engine Misses Or Loses Power

1. Slave piston adjustment too tight.

Control Valve Covers Broken Or Bent—Models 20 & 25A

1. Slave piston control valve ball check in adjoining cylinder (same cylinder) leaking or dirty.
2. Control valve under bent or broken cover malfunctioning.

Excessive Lube Oil Consumption, Model 20

1. Solenoid exhaust oil entering intake manifold from vent hole in rocker box.

AIR BRAKES

INDEX

Air brake equipment on trucks and truck tractors provides a means of controlling the brakes through the medium of compressed air. The equipment consists of a group of devices as shown in Fig. 1. Some of these devices maintain a supply of compressed air, some direct and control the flow of compressed air, and others transfer the energy of compressed air into the mechanical force and motion necessary to apply the brakes.

Different sizes and types of devices are in use on different types of vehicles to meet operating requirements. Following are the devices comprising a typical air brake system with a brief description of the function of each device along with service requirements.

AIR COMPRESSOR

The compressor supplies the compressed air to operate the brakes. Two types are in general use, the piston type and the rotary type. Both types run continuously while the engine is operating.

In the piston type, Figs. 2 and 3, the actual compression of air is controlled by the governor which, acting in conjunction with the unloading mechanism in the compressor cylinder block, starts and stops the compression of air by loading or unloading the compressor when the pressure in the air brake system reaches the desired minimum of 95–105 pounds or maximum of 110–120 pounds.

In the rotary type, Figs. 4 and 5, the actual compression of air is regulated by the control valve (governor) which starts or stops air compression by "loading" or "unloading" the compressor as required. This action estab-

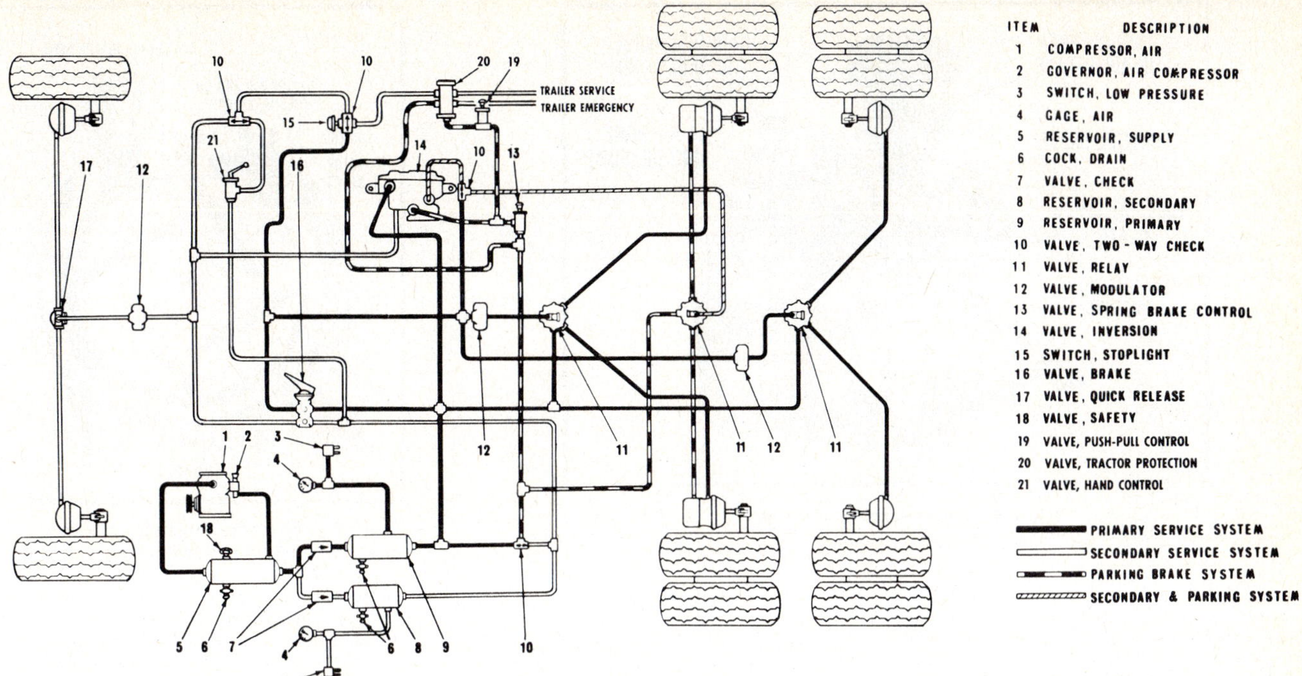

Fig. 1 Typical air brake system

lishes an intermittent pumping cycle which keeps reservoir tank pressure within the desired range.

GOVERNORS
B-W Governors

The compressor controls the air pressure in the reservoir by actuating the unloading mechanism in the compressor when the compressor reaches the maximum limit of 120 psi. It permits compression of air to continue when reservoir pressure falls below the 95 psi minimum.

The governor is mounted on the compressor by a bracket. A passage in the bracket carries the air pressure from the governor to the compressor unloading mechanism. *On some applications the governor is mounted directly on the compressor.*

B-W Type O-1 Governor

Operation

The O-1 type governor, Fig. 5A, operates on the principle of a bourdon tube, a curved, flattened metal tube which straightens when under internal pressure.

Air pressure enters the governor through the reservoir and is always present below the lower valve and in the spring (bourdon) tube. As air pressure increases, the spring tube straightens and the load on the lower valve decreases.

When reservoir air pressure reaches the cut-out pressure setting of the governor (100–105 PSI), the force of the spring tube on the upper and lower valves is low enough to permit the air pressure to raise the lower

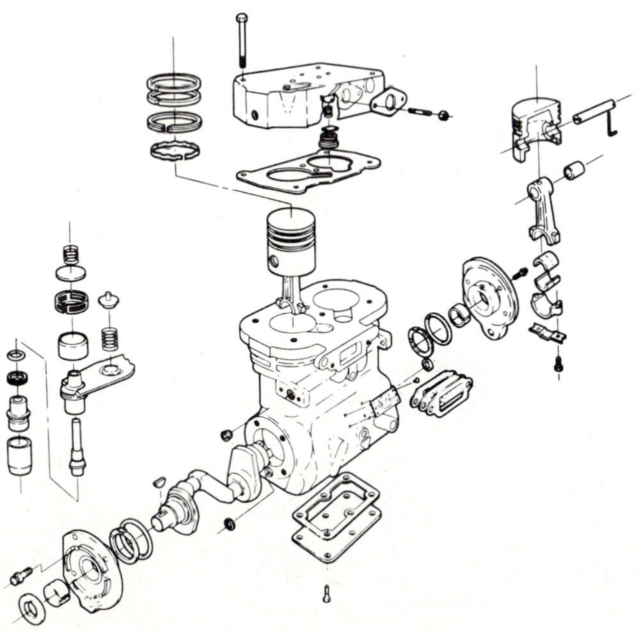

Fig. 2 Piston type air compressor. Tu-F10501

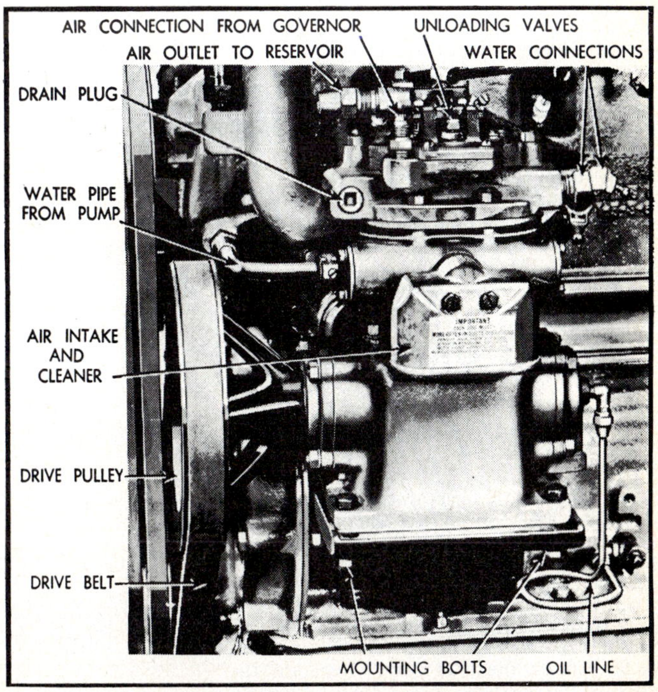

Fig. 3 Side-mounted piston type air compressor

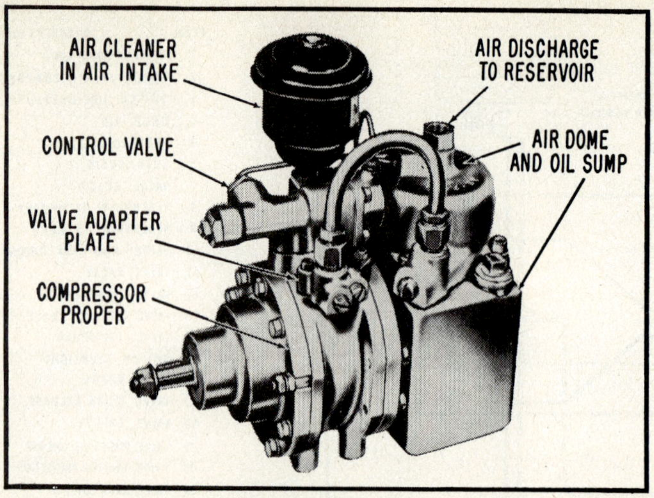

Fig. 4 Rotary type air compressor

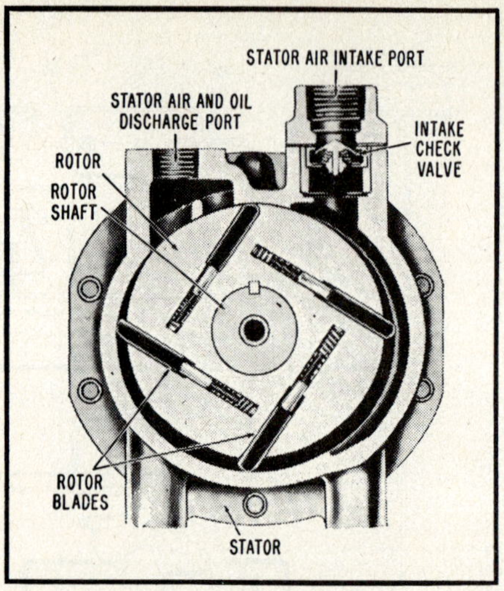

Fig. 5 Rotor and stator of rotary type air compressor

valve off the seat. When the lower valve is unseated, the upper valve is seated and the exhaust port closes, permitting air pressure to flow to the air compressor unloading mechanism. When reservoir air pressure falls to the cut-in pressure setting of the governor (80–85 PSI), the force of the spring tube on the upper valve increases and unseats the upper valve. Reservoir air pressure is prevented from passing through the governor, and air pressure in the unloading diaphragm cavity in the compressor cylinder head escapes through the exhaust port in the governor. The unloading valves in the compressor cylinder head close and compression is resumed.

Adjustment

Normal cut-out pressure of the governor is 100–105 PSI and normal cut-in pressure is 80–85 PSI. Cut-in and cut-out pressures may be adjusted by removing the cover and turning the adjusting screw. To raise a pressure setting, loosen the locknut and turn the adjusting screw clockwise as viewed from the top. To lower a pressure setting, turn the adjusting screw counterclockwise. If the governor cannot be adjusted to meet specifications, it must be replaced.

The range between cut-in and cut-out pressures may be adjusted by removing shims beneath the upper valve guide to increase the range or by adding shims to decrease the range.

B-W Type D Governor

Operation

As air pressure increases, the diaphragm and stem assembly moves against the pressure setting spring, Fig. 6. When reservoir pressure reaches the governor cutout point (120 pounds) the exhaust valve is seated and the inlet valve is opened, permitting air pressure to flow to the air compressor unloading mechanism. As reservoir pressure is reduced to 95 pounds, the inlet valve closes and the exhaust valve opens, thus allowing the compression of air to be resumed.

Adjustment

Normal cut-out pressure of the governor is 110 to 120 pounds and normal cut-in pressure is 95 to 105 pounds. Cut-in and cut-out pressures may be adjusted by removing the cap nut, cotter pin and then turning the adjusting nut. Turn nut clockwise to raise the setting, and counterclockwise to lower the setting. *The range between cut-out and cut-in pressures if fixed at approximately 20 pounds and cannot be adjusted.*

Service

Prior to reassembly lightly coat the valve parts, valve bores and O-rings with recommended grease (B-W 240176).

1. Install new O-ring seals on cap nuts, inlet valve seat, diaphragm stem, and in groove of exhaust stem bore of body. Grease lightly before installing.
2. Install conical spring and exhaust stem in body.
3. Install diaphragm on stem, making sure that bevelled side of each follower face diaphragm. Fasten in place with nut and cotter pin.
4. Install diaphragm and stem assembly. Fasten body and spring cage together with four screws.
5. Install cylindrical filters.
6. Install and tighten filter cap nuts. Make certain O-rings and rubber washers are in place.

NOTE: Attach a dial indicator to flanges of body and spring cage with indicator stem contacting end of exhaust stem. Pull diaphragm stem outward as far as possible and set indicator at zero. Then push diaphragm stem inward to limit of its travel and note reading on indicator. Total travel should be within .060" to .098". Install shims, inlet valve seat and valve and repeat procedure with dial indicator zeroed on end of inlet valve. Add or remove shims under inlet valve seat until travel is .030" to .040".

7. Install valve and seat retaining springs. Then install body cap-nut with O-ring.
8. Install spring and spring seat and assemble nut on stem.

NOTE: Mount governor on suitable test bench or on vehicle and build up reservoir pressure from zero. Note pressure at which air starts to escape from spring

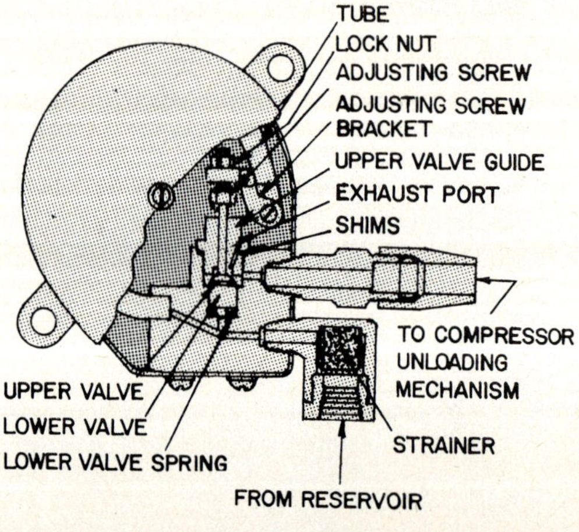

Fig. 5A Type O-1 compressor governor

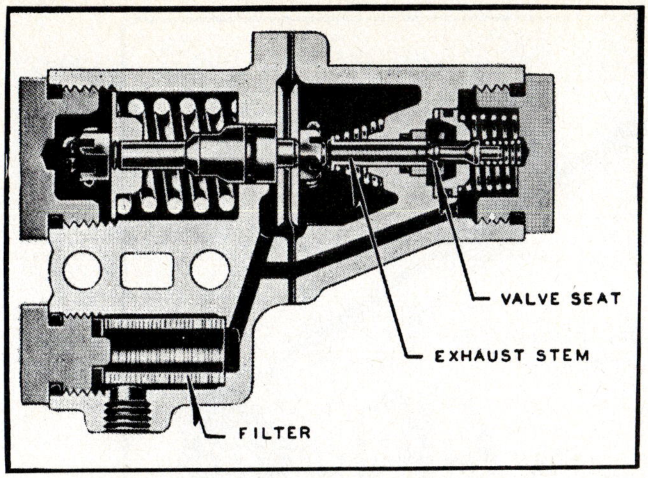

Fig. 6 Type D compressor governor

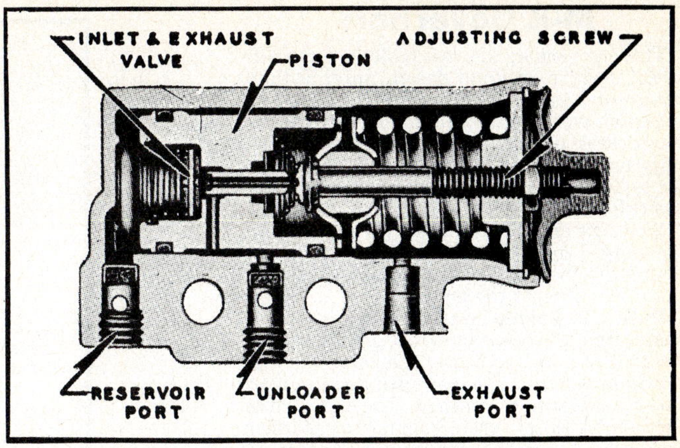

Fig. 7 Type D-2 compressor governor

cage. If pressure is below 110 pounds, turn adjusting nut clockwise; if pressure is above 120 pounds, turn counterclockwise. After final adjustment, leakage should start at 110–120 pounds. Lock with new cotter pin.

9. Install spring cage cap nut with O-ring in place.

B-W Type D-2 Governor

Operation

Referring to Fig. 7, reservoir air pressure exerts force on bottom of piston and inlet-exhaust valve. Piston movement is prevented by the adjusting screw and spring assembly as long as reservoir pressure does not exceed the spring pressure above piston.

When reservoir pressure does exceed spring pressure, the piston lifts. The exhaust stem does not move since it butts against the head of the stationary adjusting screw. The inlet-exhaust valve moves with the piston until it contacts exhaust stem (closing exhaust passage in stem). Continued piston movement unseats inlet-exhaust valve from its seat in piston.

Air pressure is then permitted to pass around inlet-exhaust valve, through passage in piston, and out unloader port to unloading mechanism in compressor. Spring pressure above piston forces piston downward when reservoir pressure becomes less than spring pressure. As piston moves downward, it contacts inlet-exhaust valve, closing air inlet. Continued downward movement unseats the inlet-exhaust valve from the exhaust stem, permitting air in unloader to escape and compression of air to resume.

Service

Coat parts with B-W 240176 grease and reassemble governor as follows:

1. Install O-ring in counterbore in piston. Install retaining washer and stake securely.
2. Install exhaust stem spring in piston.
3. Install exhaust stem in piston bore by passing it through spring and O-ring.
4. Install inlet-exhaust valve in piston.
5. Install inlet-exhaust valve spring against valve and snap outer end in groove provided.
6. Install O-rings in piston grooves. Be sure they are not twisted.
7. Install piston in housing bore with exhaust stem up.
8. Install adjusting screw and spring assembly. Then install snap ring.

NOTE: If pressure adjustment is necessary, loosen lock nut and turn adjusting screw clockwise to decrease and counterclockwise to increase reservoir air pressure. When satisfactory, tighten lock nut and install dust cover.

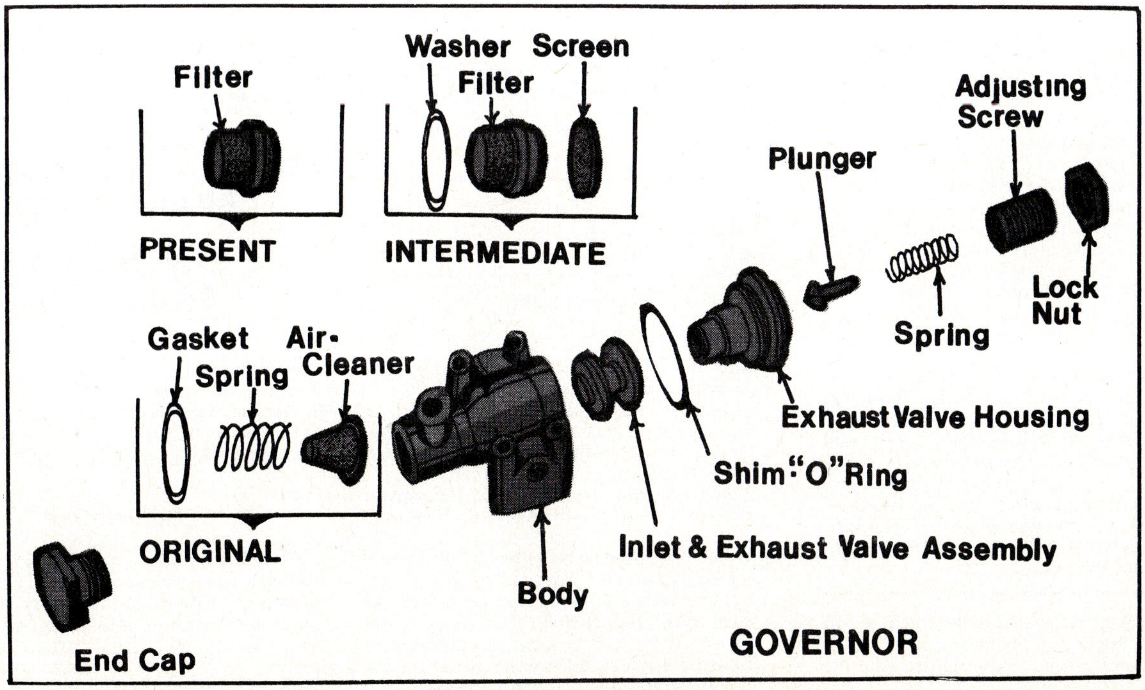

Fig. 7A M-R compressor governor

M-R Governors

These governors are available in three different designs: original design, intermediate design and the present design, Fig. 7A. The governor, working in conjunction with the air compressor unloader mechanism, automatically maintains a pre-set air pressure in the air system. The operation of all governors is basically the same.

With the engine off and no air pressure in the system, the governor spring maintains the plunger in the down position, thereby seating the inlet valve on the body.

With the engine operating, air pressure passes from the reservoir to the governor port. As air pressure increases, the inlet valve is slowly raised from the seat until maximum or cut-out pressure is reached. The inlet valve then snaps to full open, raising the plunger and compressing the spring, in turn seating the exhaust valve. The reservoir pressure is passed through the inlet valve orifice and around the exhaust valve, filling the center cavity of the governor and passing through the connecting tube to the upper cavity of the unloader assembly.

When reservoir air pressure decreases to minimum or cut-in pressure, the inlet valve is seated by governor spring pressure acting against the plunger and also unseats the exhaust valve. This action permits air pressure in the unloader and governor to exhaust through a port in the adjusting screw.

Service

Original & Intermediate Design

1. Install inlet and exhaust valves or the assembly (intermediate design) into body, Fig. 7A.
2. Install plunger and governor spring.
3. Place three shims on exhaust valve housing and install housing by turning housing into body until tight.
4. Rotate adjusting screw inward nine complete turns while holding valve housing secure and loosely install adjusting screw lock nut.

NOTE: Adjusting screw should protrude approximately 3/8 inch from exhaust valve housing.

5. Install screen or air cleaner into body. Assemble filter (intermediate design), washer and end cap, then install the assembly into body.
6. Mount governor on a suitable test bench and build reservoir pressure to 115 PSI. If governor cuts-out before 115 PSI is reached, turn adjusting screw clockwise one complete turn and repeat this step.
7. Slowly rotate adjusting screw counterclockwise until governor cuts out.
8. Bleed air from reservoir until governor cuts in. Cut-in pressure should be 93–98 PSI.
9. If cut-in pressure is not within specifications, remove exhaust valve housing, adjusting screw, lock nut and shims. If cut-in pressure was below 93 PSI, add shims to the assembly. If cut-in pressure was above 98 PSI, remove shims. The addition or removal of one shim is equal to a range change of approximately 4 PSI. Install exhaust valve housing, adjusting screw and lock nut.

NOTE: Pressure differential between cut-in and cut-out must not be less than 15 PSI. Recheck cut-out pressure and adjust if necessary.

Present Design

1. Install valve assembly into valve body, Fig. 7A, then the plunger and spring.
2. Lubricate exhaust valve housing "O" ring and install on housing. Turn exhaust valve housing into body until bottomed on valve assembly, then turn back 3/4 turn and loosely install locknut on exhaust valve housing.
3. Turn adjusting screw in nine complete turns.

NOTE: Adjusting screw should protrude approximately 3/8 inch from exhaust valve housing.

4. Assemble filter and end cap, then install in body.
5. Mount governor on a suitable test bench and build reservoir pressure to 115 PSI. If governor cuts-out before 115 PSI is reached, turn adjusting screw clockwise one complete turn and repeat this step.
6. Slowly rotate adjusting screw counterclockwise until governor cuts out.
7. Bleed air from reservoir until governor cuts in. Cut-in pressure should be 93–98 PSI.
8. If cut-in pressure is below 93 PSI, hold adjusting screw stationary and rotate exhaust valve housing approximately 1/6 turn counter-clockwise. If cut-in pressure is above 98 PSI, rotate exhaust valve housing 1/6 turn clockwise. Rotating exhaust valve housing 1/6 turn results in a range change of approximately 5 PSI.

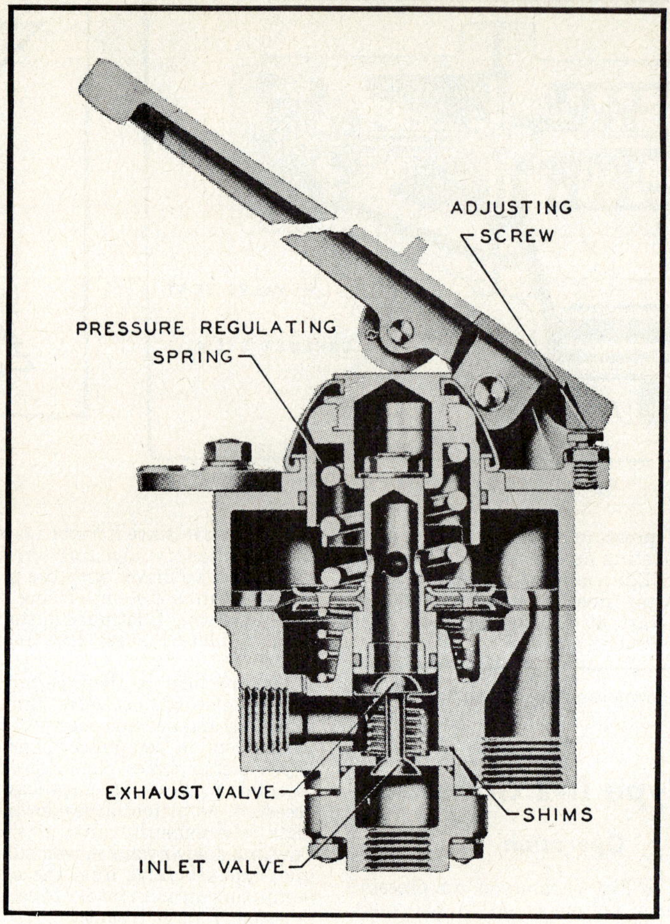

Fig. 8 Type D-1 brake application valve

NOTE: Pressure differential between cut-in and cut-out must not be less than 15 PSI. Recheck cut-out pressure and adjust if necessary.

9. Tighten exhaust valve housing and adjusting screw lock nuts.

BRAKE APPLICATION & BRAKE VALVES

B-W Type D-1 Brake Valve

This brake valve, Fig. 8, is actuated by a brake treadle. Operation of the brake treadle regulates the movement of an inlet valve and an exhaust valve which, in turn, control the air pressure being delivered to or released from the brake chambers.

Adjustment

If brake valve does not release promptly or does not fully release, it indicates exhaust valve is not opening sufficiently. This can be caused by improper adjustment of the adjusting screw (screw being turned out too far, causing lever roller to exert too much force against spring cage when lever is in released position) or by insufficient shims being used

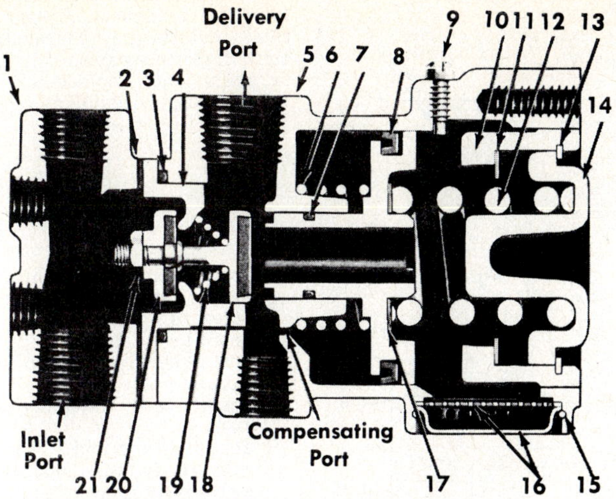

1 Inlet Cover
2 Inlet Cover Gasket
3 Valve Cage O-ring
4 Valve Cage
5 Valve Body
6 Piston Spring
7 Piston O-ring
8 Piston Seal
9 Piston Retaining Screw
10 Valve Piston
11 Piston Cup Spacers
12 Graduating Spring
13 Snap Ring
14 Piston Cup
15 Snap Ring
16 Exhaust Filter Screens
17 Graduating Spring Spacers
18 Exhaust Valve
19 Valve Spring
20 Inlet Valve
21 Valve Retaining Nut

Fig. 8A Midland-Ross N-20804 & N-20826 application valve

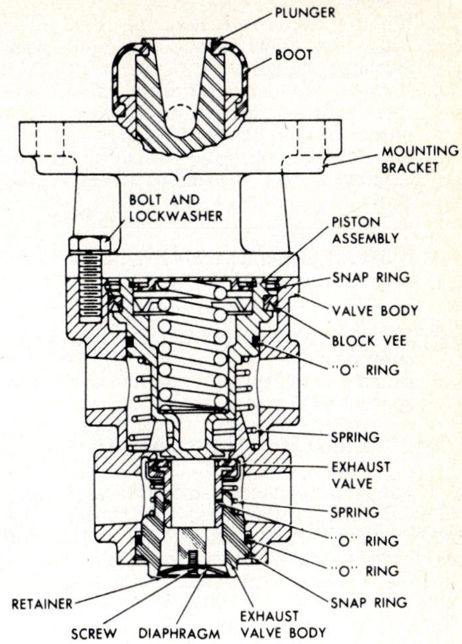

Fig. 8B Midland-Ross N-30028G application valve

between intake valve seat and body of valve.

If brake valve does not apply promptly, it indicates intake valve is not opening sufficiently. This can be caused by too many shims being used between intake valve seat and brake valve body. To adjust, proceed as follows:

1. Make sure intake valve is seated.
2. Place a depth gauge against intake valve by inserting gauge through reservoir supply port in intake valve body.
3. Apply brake valve and measure maximum intake valve travel which should be .148" to .156". Add shims between intake valve seat and body of brake valve to decrease travel. Remove shims to increase travel.
4. Turn adjusting screw until roller on lever just barely contacts top of spring cage. *Make sure spring cage is up as far as it will go.* Then lock adjusting screw.
5. Test adjustment by placing a .109" thickness gauge between roller and spring cage with brake released. The air pressure delivered by the brake valve should be from two to seven pounds.

NOTE: If brake valve does not graduate the delivered pressure properly, check to make sure bleed hole to cavity immediately below diaphragm is not restricted.

Midland-Ross Models N-20804, N-3974-A & N-20826

Description

The Midland-Ross application valve models N-20826 and N-20804 are identical except the N-28026 models incorporate a breather tube installation at the exhaust opening in place of a filter screen which is used on N-20804 models. The N-3974-A application valve is the N-20804 base valve which incorporates a treadle and plate assembly.

Disassembly, Fig. 8A

1. Remove piston retaining screw, and piston assembly from valve body.
2. Mark inlet cover and valve body to assure correct installation of cover during assembly, then remove inlet cover bolts, cover, gasket and valve cage from valve body.
3. Secure exhaust valve to avoid turning then remove nut, inlet valve, exhaust valve and valve spring from cage.
4. Remove snap ring and filter screens or exhaust tube from valve body.
5. Remove "O" ring and piston seal from piston. Do not disassemble piston unless necessary to replace parts.

Cleaning & Inspection

Clean all parts using a suitable solvent, and blow dry using compressed air. Inspect valve body, internal bores and external surfaces for evidence of wear and corrosion. Slight defects may be repaired with crocus cloth. Lubricate parts using lube No. 5459912 or equivalent. Replace all components not considered serviceable as required. Replace all "O" rings and seals.

Assembly

1. Install piston seal and "O" ring into grooves in piston. Seal lip must point toward small end of piston.
2. Install piston spring and piston assembly into valve body. Push piston into valve body until piston cup is flush with valve body. Install and tighten piston retaining bolt. Check piston for free play in valve body.
3. Install exhaust valve, valve spring, inlet valve and nut onto valve cage.
4. Position "O" ring over valve cage, then install cage and cage bolts onto valve body.
5. Install gasket, inlet cover and bolts onto valve body.
6. Install filter screens or exhaust tube into exhaust opening in valve body and secure with snap ring.

Midland-Ross Models N-20837C & N-30028G

Disassembly, Fig. 8B

1. On models N-30028G, remove boot and plunger.
2. On models N-30028G, remove bolts and washers, then separate mounting bracket from control valve.

CAUTION: Piston assembly is under spring pressure, slowly release snap ring when removing piston spring.

---

3. Depress piston assembly, then remove snap ring.
4. Remove block V and "O" ring from piston

assembly.
5. Remove piston return spring from valve body. Then remove snap ring from lower section of valve.
6. Remove exhaust valve from valve body, then remove screw and separate diaphram from exhaust body.
7. Remove "O" ring, valve return spring and exhaust valve assembly from valve body.

NOTE: If piston assembly requires disassembly, proceed as follows.

8. Using an arbor press, remove piston cap snap ring.
9. Remove cap, spring, reaction ring and spacers from piston assembly.

Cleaning & Inspection

Clean all parts using a suitable solvent, and blow dry using compressed air. Inspect valve body, internal bores and external surfaces for evidence of wear and corrosion. Slight defects may be repaired with crocus cloth. Lubricate parts using lube No. 5459912 or equivalent. Replace all components not considered serviceable as required. Replace all "O" rings and seals.

Assembly

NOTE: Steps 1 and 2 apply only if piston was disassembled.

1. Install large diameter shims and reaction ring into piston body. Adjust shim thickness to obtain 0.170–0.180 inch gap between piston cap and snap ring.
2. Remove cap and install small diameter shims. Install spring, cap and snap ring. Adjust small diameter shim thickness to obtain 67–73 lbs. preload on piston cap.
3. Install "O" ring and block V into piston grooves. Lip of block V must be near small end of piston.
4. Install large end of piston return spring into valve body.
5. Install piston assembly and snap ring into valve body.
6. Install "O" ring into lower valve body cavity.
7. Install exhaust valve assembly into lower body bore, then install exhaust valve return spring into exhaust valve.
8. Install exhaust valve body assembly and secure with snap ring.
9. Install exhaust valve diaphragm, retainer and screw.
10. Install mounting bracket, bolts and washers onto valve body.
11. Install plunger and plunger boot.

Midland-Ross Models N-30028L & N-30099AE

Disassembly, Fig. 8C

1. For models N-30028L, remove boot and plunger.
2. For models N-30028L, remove bolts and washers from mounting bracket, then separate mounting bracket from control valve.
3. Mark upper and lower valve bodies for valve assembly.
4. Remove bolts, then separate valve bodies. Remove piston, spring and "O" ring from lower valve body.

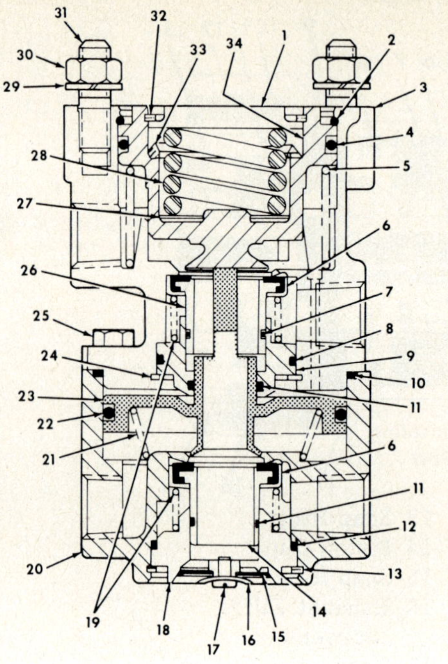

Key	Description	Key	Description
1	Spring Retainer	18	Exhaust Valve Body
2	Retainer ring	19	Spring
3	Valve Body	20	Valve Body
4	O-ring	21	Spring
5	Spring	22	O-ring
6	Seal Retainer	23	Piston
7	O-ring	24	Retaining Ring
8	O-ring	25	Screw
9	Exhaust Valve Body	26	Valve
10	O-ring	27	Shim(s)
11	O-ring	28	Spring
12	O-ring	29	Washer
13	Retaining Ring	30	Hex Nut
14	Inlet valve	31	Stud
15	Diaphragm	32	Retainer ring
16	Washer	33	Reaction ring
17	Screw	34	Piston

Fig. 8C Midland-Ross N-30099AE application valve

5. Remove snap ring from lower valve body, then remove spring, inlet valve and "O" ring.
6. Remove "O" ring from piston.
7. Remove screw, washer and diaphragm from exhaust valve body.
8. Remove "O" ring and seals from inlet valve body.
9. Remove snap ring, exhaust valve body, spring and valve from upper valve body.
10. Remove "O" rings from the exhaust valve body, then remove "O" ring and seals from exhaust valve.

CAUTION: Piston spring is under a spring pressure, carefully remove piston.

11. Depress piston and remove retaining ring.
12. Remove piston and spring from upper valve body.
13. Remove "O" ring from piston, then position piston in a soft-jaw vise, and remove spring retainer, spring, reaction ring and shims from piston.

Cleaning & Inspection

Clean all parts using a suitable solvent, and blow dry using compressed air. Inspect valve body, internal bores and external surfaces for evidence of wear and corrosion. Slight defects may be repaired with crocus cloth. Lubricate parts using lube No. 5459912 or equivalent. Replace all components not considered serviceable as required.

Assembly

1. Install exhaust diaphragm, washer and screw onto exhaust valve body. Torque screw to 7 inch lbs.
2. Install seal and "O" ring onto valve.
3. Install seal retainer on tube and position around valve seal.
4. Install valve assembly into lower valve body.
5. Install spring and "O" ring onto exhaust valve body, then install snap ring into valve body groove.
6. Install rubber seal onto valve, then install seal retainer onto tube and position around valve seal.
7. Position valve and seal assembly into upper valve body.
8. Install spring over tube and onto seal retainer.
9. Install inner and outer "O" rings onto exhaust valve body, then install exhaust valve body into valve body and install snap ring into valve body groove.
10. Install "O" ring onto piston, then position conical spring into lower valve body.
11. Install piston assembly into lower valve body.
12. Position "O" ring onto upper body, then align marks on the valve bodies and push valves together. Install valve body bolts and torque to 55 inch lbs.
13. Install shims into piston, then install reaction ring onto ridge provided on piston.
14. Install spring into piston, then install spring retainer onto piston and spring.
15. Using an arbor press, install spring retainer and snap ring onto piston.

NOTE: Use care not to damage exhaust seat of piston.

16. Install "O" ring onto piston.
17. Install spring, piston assembly and retaining ring into upper body.
18. Install mounting bracket, plunger and boot onto valve body.

B-W Type E Brake Valves

E-2 and E-3 brake valves have an insert type combination inlet-exhaust valve that can easily be removed for service without disconnecting any air lines or removing the valve. A diaphragm type exhaust check valve in the bottom of the inlet-exhaust valve prevents entry of dirt at the exhaust port, Fig. 9.

The supply ports in the lower portion of the valve are charged with reservoir pressure whereas the delivery ports in the upper portion of the valve are charged with air pressure at application only.

The E-5 brake valve, Fig. 9A, is a suspended pedal operated type.

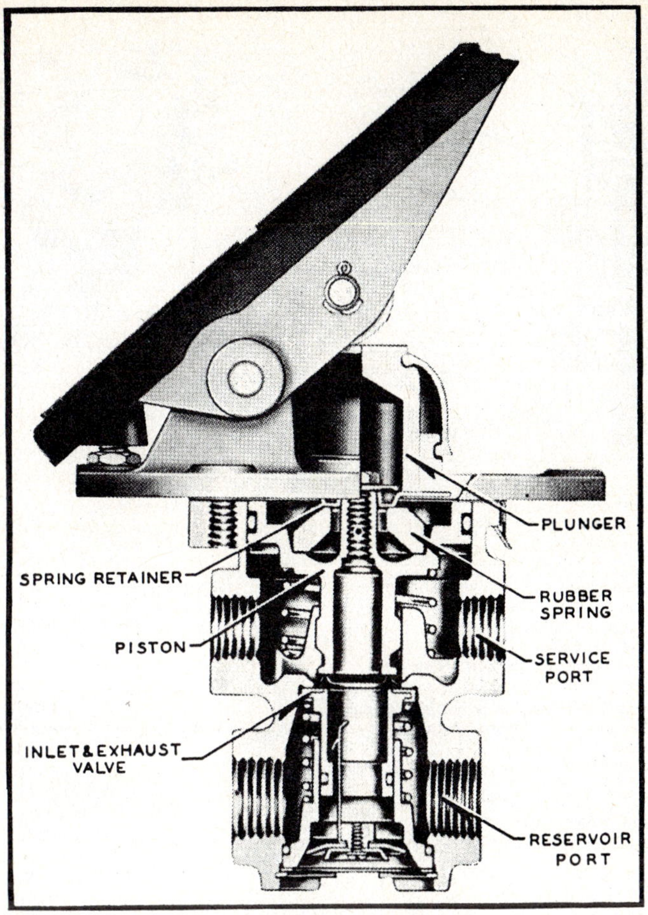

Fig. 9 Type E-3 brake valve

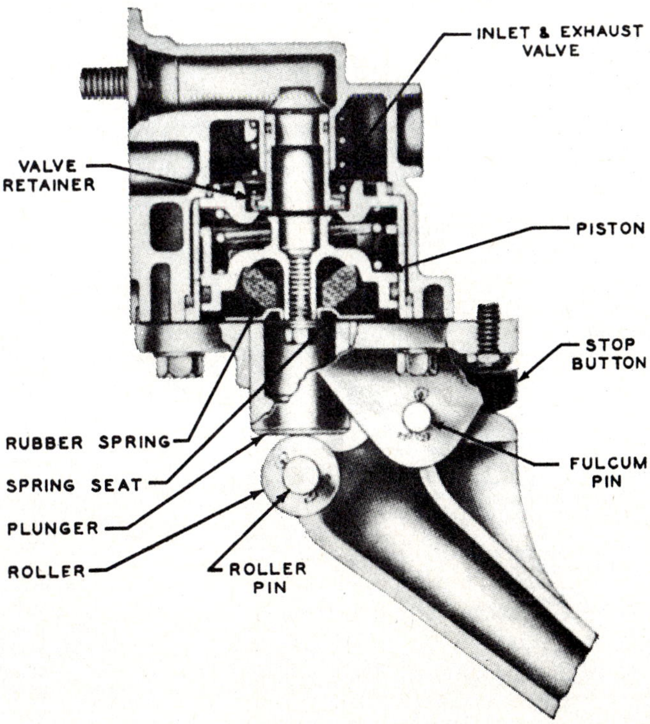

Fig. 9A Type E-5 brake valve

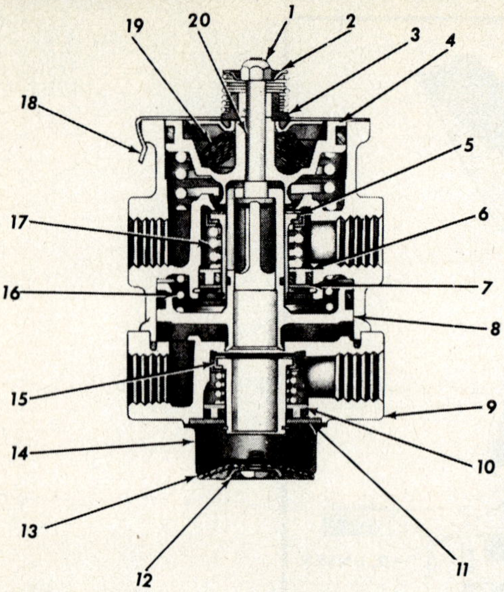

1. Nylon insert Nut
2. Spring Guide
3. Spring Seat Nut
4. Primary (No. 1) Piston
5. No. 1 Inlet and Exhaust Valve
6. O-Ring Retainer (No. 1)
7. Washer (some models)
8. Relay Piston
9. Lower Body
10. O-Ring Retainer (No. 2)
11. Washer
12. Exhaust Diaphragm Washer
13. Exhaust Diaphragm
14. Exhaust Cover
15. No. 2 Inlet and Exhaust Valve
16. Relay Piston Spring (some models)
17. Primary (No. 1) Piston Return Spring
18. Primary (No. 1) Piston Retainer
19. Rubber Spring
20. Stem

Fig. 9B Treadle type brake valve

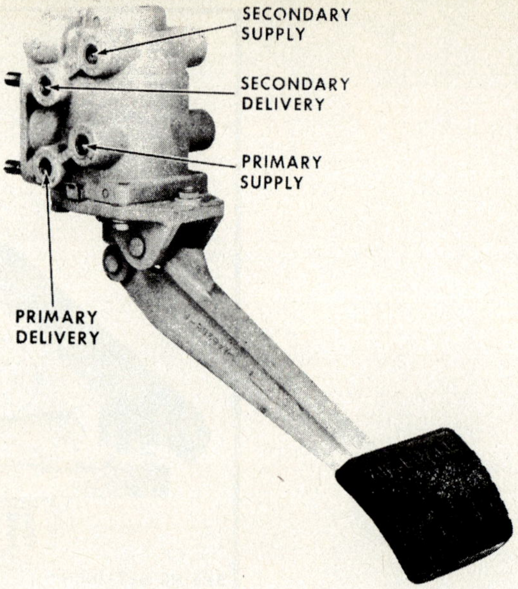

Fig. 9C Suspended type brake valve

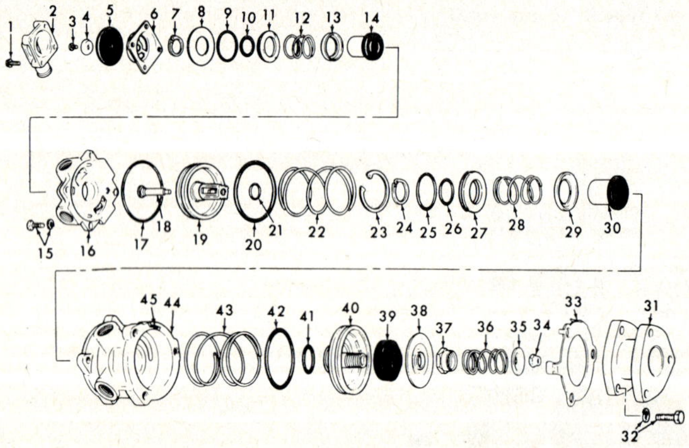

1 Exhaust Cover Screw
2 Exhaust Cover (Some Models)
3 Diaphragm Washer Screw
4 Diaphragm Washer
5 Diaphragm
6 Exhaust Cover (Some Models)
7 Retaining Ring
8 Washer
9 Outer O-ring
10 Inner O-ring
11 O-ring Retainer
12 No. 2 Inlet & Exhaust Valve Spring
13 No. 2 Inlet & Exhaust Valve Retainer
14 No. 2 Inlet & Exhaust Valve Assembly
15 Cap Screw & Washer
16 Lower Body
17 Lower Body O-ring
18 Stem
19 Relay Piston
20 Piston O-ring
21 Small Diameter Piston O-ring
22 Relay Piston Spring
23 No. 1 Inlet & Exhaust Valve Assembly Retaining Ring
24 No. 1 Inlet & Exhaust Valve Retaining Ring
25 Outer O-ring
26 Inner O-ring
27 O-ring Retainer
28 No. 1 Inlet & Exhaust Valve Spring
29 No. 1 Inlet & Exhaust Valve Retainer
30 No. 1 (Primary) Inlet & Exhaust Valve Assembly
31 Adapter (Spacer) (Some Models)
32 Adapter to Upper Body Bolt & Washer
33 No. 1 (Primary) Piston Retainer
34 Nylon - Insert Nut
35 Spring Guide
36 Stem Spring
37 Spring Seat Nut
38 Spring Seat
39 Rubber Spring
40 No. 1 (Primary) Piston
41 Rubber Washer
42 O-ring
43 Piston Return Spring
44 Upper Body
45 Vent Filter

Fig. 9D Type E-6 brake application valve exploded view

The treadle type E-6, Fig. 9B and the suspended type E-7, Fig. 9C brake valves are equipped with two separate supply and delivery circuits for service and emergency braking. The primary circuit is that portion of the valve between spring seat which contacts the plunger and relay piston. The secondary circuit is that portion between relay piston and exhaust cavity. Both primary and secondary circuits of brake valve use a common exhaust check valve.

Service

Type E-2, E-3 & E-5

Prior to reassembly, lightly coat valve parts, valves bores and O-rings with B-W 240176 grease. Reassemble as follows, referring to Fig. 10.

1. Position rubber spring in piston with concave side down.
2. Place retainer on rubber spring.
3. Center flat washer on rubber spring retainer.
4. Pass capscrew through flat washer and install in piston. Tighten to 50 ft-lbs torque.
5. Install O-ring in piston groove.
6. Place piston return spring in valve body.
7. Install piston assembly with care so that O-ring is not damaged or twisted in the process.
8. Install retainer and snap ring prongs in place while depressing piston.
9. Attach adapter plate to valve body with three screws.
10. Position preload spring in exhaust check valve seat.
11. Install exhaust diaphragm so that it is seated on hub against bottom of preload spring. Install diaphragm washer, lockwasher and screw.
12. Install O-ring on body of exhaust check valve seat.
13. Install O-ring on inlet-exhaust valve.
14. Install inlet-exhaust valve spring on shoulder of exhaust valve seat. Position spring seat washer on inlet-exhaust valve.
15. Slide inlet-exhaust valve into exhaust check valve seat so that prong of preload

spring snaps over shoulder in inlet-exhaust valve.
16. Install insert in inlet-exhaust valve.
17. Install identification ring.
18. Install snap ring.
19. Place boot on plunger and install plunger in adapter bore after coating outside of plunger with B-W 239377 grease. Slip boot over hub of adapter plate.

Type E-5
1. Install exhaust check diaphragm in valve body.
2. Install inlet valve stem "O" ring and retainer, then install "O" rings on inlet valve seat.
3. Assemble piston by installing rubber spring, spring seat, washer cap screw and "O" ring.
4. Install inlet and exhaust valve, inlet valve seat with tab on seat aligning with valve body slot, and piston assembly, then install valve retainer, ensuring tabs snap into position.

Installation

After attaching valve to mounting bracket, check to see that roller contacts plunger. Loosen locknut behind rubber stop button and turn adjusting screw, if necessary.

Connect wires to stop light switch, and air lines to valve fittings. After reservoir pressure is built up, check air line connections for leakage with soap suds.

NOTE: Rebuilt valves should be tested before putting vehicle into service. Connect an air gauge into service line and note pressure at all treadle positions during both application and release. Pressure should remain steady. Approximately full reservoir pressure should register at full application. Coat exhaust port with soap suds and check for leakage. In both fully applied and fully released positions, leakage should not exceed a one-inch soap bubble in one second.

Type E-6 Service

Removal
1. Chock wheels.
2. Drain air completely from all reservoirs.
3. Disconnect and mark to assist in installation, all supply and delivery lines at brake valve.
4. Mark and remove connections from valve.
5. On E-6 brake valves, remove bolts on inner bolt circle, then remove valve. On E-7 brake valves, remove nuts on exterior side of dash board, and remove valve.
6. Reverse procedure to install valve.

Disassembly, Fig. 9D
1. Remove bolts securing treadle assembly to brake valve.
2. Remove bolt washer, exhaust diaphragm and exhaust cover.
3. Remove bolts that secure exhaust cover to lower body, then remove secondary inlet and exhaust valve assembly from lower body.
4. Remove bolts and disengage lower body from upper body.

CAUTION: Before proceeding to the following steps, nut and stem are used to contain primary piston return spring. Combined forces of these springs are approximately 50 ft. lbs. Use care when removing nut.

5. Remove rubber seal ring from lower body, locking tabs from primary piston retainer, stem, stem spring, spring guide and nut from primary piston.
6. Remove relay piston, relay piston spring, primary piston and primary piston return spring from upper body.
7. Disassemble primary piston by turning spring seat nut counterclockwise. Separate spring seat nut, spring seat, rubber spring and remove piston "O" ring.
8. Remove large and small "O" rings from relay piston.
9. Remove retaining ring securing primary inlet and exhaust valve in upper body and remove valve assembly.

Cleaning & Inspection

Clean all parts using a suitable solvent, and blow dry using compressed air. Inspect valve body, internal bores and external surfaces for evidence of wear and corrosion. Slight defects may be repaired with crocus cloth. Lubricate parts using lube No. 5459912 or equivalent. Replace all components not considered serviceable as required. Replace all "O" rings and seals.

Assembly
1. Lubricate all "O" rings, seals, piston bores and contact surfaces.
2. Install primary inlet, exhaust assembly and snap ring into upper body.
3. Install "O" rings onto relay piston.
4. Install rubber spring and spring seat onto primary piston.
5. Install primary piston spring seat nut to spring seat and turn clockwise until top surface of spring seat is even with top surface of piston.
6. Install primary piston and return spring into upper body bore.
7. Install primary piston, then install spring seat onto return spring and press piston into body bore.
8. Install relay piston return spring into upper body.
9. Place stem through exhaust passage of lower body so threaded portion extends through the primary piston. Use a screwdriver to engage and secure slotted head of stem. Position stem spring and spring guide onto primary piston.
10. Install nut on stem, and torque to 20–30 inch lbs.
11. Install primary piston retainer onto piston, ensuring locking tabs have engaged outer lip of body.
12. Install secondary inlet and exhaust valve assembly into lower body.
13. Install exhaust cover onto lower body and torque screws to 20–30 inch lbs.
14. Install exhaust diaphragm onto exhaust cover using bolt and lock washer.
15. Install seal ring into lower body and secure lower to upper body using bolts and lockwashers, torque bolts to 70–100 inch lbs.
16. Install treadle assembly to brake valve using bolts and lockwashers; torque bolts to 80–100 inch lbs.
17. Test brake valve, refer to the service procedure to ensure correct operation.

Type E-7 Service

Disassembly
1. Remove bolts and pedal assembly.
2. Disengage locking tabs from retainer and remove retainer.
3. Remove lock nut from static piston and pull piston assembly from body.
4. Remove upper static piston, secondary inlet and exhaust valve from body.
5. Using screwdriver, remove locking groove from primary piston spring.
6. Remove locking nut, spring stem guide and stem spring.
7. Remove spring seat nut, spring seat and rubber spring.

CAUTION: Before proceeding to the next step it should be noted that spring tension of the primary piston return spring and relay piston return spring combined is approximately 50 ft. lbs. Care should be taken when removing them.

8. Remove primary piston and relay piston return springs.
9. Remove stems and "O" rings from relay and primary pistons.
10. Remove snap ring from lower piston and remove primary inlet and exhaust valve assembly.
11. Remove snap ring from upper static piston assembly and remove secondary inlet and exhaust valve assembly.

Cleaning & Inspection

Clean all metal parts in a suitable cleaning solvent and blow dry. Inspect parts for excessive wear and corrosion. Inspect valve seats for nicks or burrs and springs for wear and corrosion. Replace all rubber seals and any part not considered serviceable.

Assembly
1. Lubricate all "O" rings, "O" ring grooves, piston bores and contact surfaces.
2. Install secondary inlet, exhaust valve assembly and snap ring into upper static piston. Ensure snap ring is engaged onto groove of upper static piston bore.
3. Install "O" ring and upper static piston assembly into valve body.
4. Install primary inlet and exhaust valve assembly into lower static piston and secure with snap ring. Ensure snap ring is engaged onto groove on lower static piston.
5. Install "O" rings into grooves of lower static piston assembly.
6. Install "O" rings on relay and primary pistons.
7. Install rubber spring, spring seat and nut onto primary piston. Turn nut clockwise until top surface of spring seat is parallel with top surface of piston.
8. Install relay and primary piston return springs into the lower static piston. Compress both primary and secondary pistons in lower static piston and insert stem through exhaust passage of relay and primary pistons. Using a screwdriver install stem spring and spring guide onto spring seat nut.
9. Compress guide spring assembly and install lock nut onto stem, torque nut to 20–30 inch lbs.
10. Install lower static piston assembly and snap ring onto valve body. Ensure locking tabs on snap ring engage on valve body bosses.
11. Install bolts and pedal assembly. Ensure plunger is in contact with spring seat.
12. Test brake valve, refer to the service procedure to ensure correct operation.

Service Note

E-6 & E-7 Brake Valves

Check delivery pressure of both primary

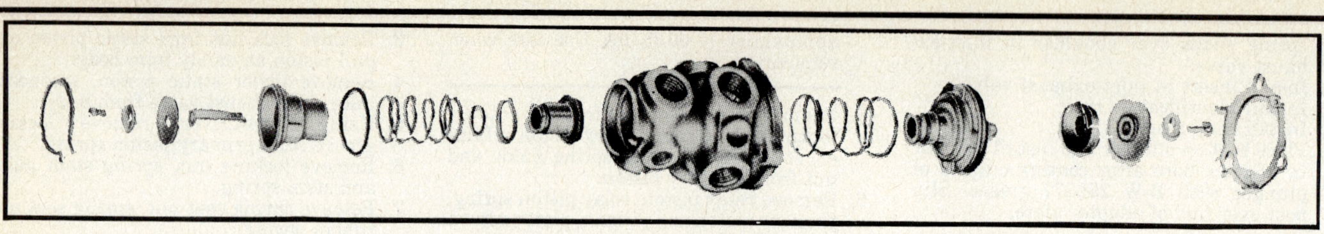

Fig. 10 Exploded view of E-3 brake valve

and secondary systems using a suitable test gauge. Depress pedal or treadle to several positions between fully released and applied positions. Check delivered pressure to ensure that pressure varies proportionately with corresponding movements of the brake pedal. After a full application is released, reading should fall off to zero. Primary system delivery pressure will be about 2 psi greater than secondary pressure with both supply reservoirs at same pressure. This is a normal condition for these valves.

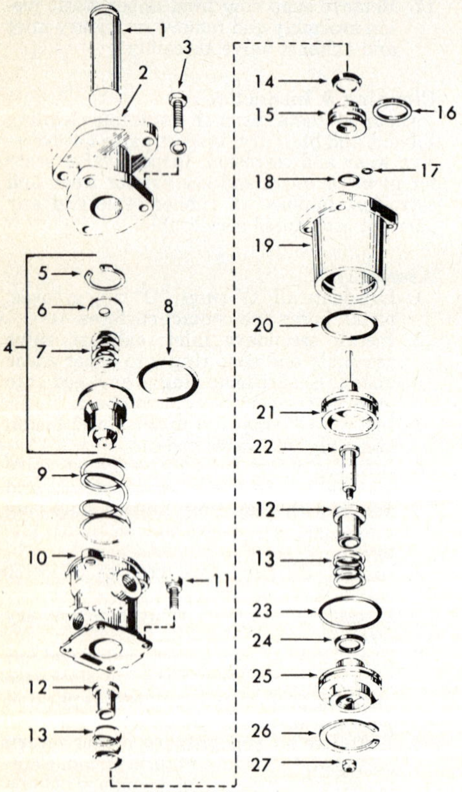

1. Plunger
2. Mounting Flange
3. Bolt—Lockwasher
4. Piston Assembly
5. Snap Ring
6. Washer
7. Metering Spring
8. "O" Ring
9. Piston Return Spring
10. Primary Valve Body
11. Bolt—Lockwasher
12. Valve Poppet
13. Poppet Spring
14. "V" Packing
15. Primary Valve Guide
16. Guide Seal
17. "O" Ring
18. "O" Ring
19. Secondary Valve Body
20. "O" Ring
21. Secondary Piston
22. Valve Stem
23. "O" Ring
24. "O" Ring
25. Secondary Valve Guide
26. Elastic Stop Nut
27. Snap Ring

Fig. 10A Wagner Electric FGB application valve

CAUTION: A change in vehicle braking characteristics or a low pressure reading may indicate a malfunction in either pneumatic and mechanical braking systems. Even though vehicle air brake system may continue to function, vehicle should not be operated until the necessary alterations and/or repairs have been made. Always check vehicle braking systems after performing work and before returning vehicle to service.

Wagner Electric Model FGB Brake Valve

Disassembly, Fig. 10A

1. Remove mounting flange bolts and separate valve mounting flange from primary valve body. Remove plunger from valve mounting flange.
2. Remove metering spring retaining ring, washer, spring, primary piston and piston return spring. Remove "O" ring from primary piston.
3. Remove primary to secondary valve body bolts, and separate assemblies.
4. Remove primary guide, poppet spring and inlet-exhaust valve poppet, then remove V packing and seal from the valve guide.
5. Remove "O" ring seal and "O" ring from secondary housing stem, then remove "O" ring from primary and secondary valve housings.
6. Remove secondary valve snap ring from lower portion of secondary valve housing, then push secondary piston stem downward to remove piston and valve guide assemblies from valve body.
7. Remove elastic stop nut, secondary valve guide, valve stem, inlet-exhaust valve poppet and poppet spring from valve body. Remove "O" rings from piston and valve guide.

Cleaning & Inspection

Clean all parts using a suitable solvent, and blow dry using compressed air. Inspect valve body, internal bores and external surfaces for evidence of wear and corrosion. Slight defects may be touched up with crocus cloth. Lubricate parts using lube No. 5459912 or equivalent. Replace all components not considered serviceable as required. Replace all "O" rings and seals.

Assembly

1. Install "O" rings into secondary piston, secondary valve guide and inside upper bore of secondary valve guide.
2. Install inlet-exhaust valve poppet, spring, secondary valve guide and elastic stop nut onto valve stem.
3. Install piston assembly, secondary valve assembly and snap ring onto valve body. Ensure snap ring is properly seated in its groove in valve housing.
4. Install "O" rings onto secondary valve housing piston stem and air passage.
5. Install seal and V packing onto primary valve guide.
6. Install inlet-exhaust valve poppet, spring and primary valve guide assembly onto lower portion in the primary valve body.
7. Install primary and secondary valve bodies and secure with bolts.
8. Install "O" ring, piston return spring, piston assembly, metering spring, washer and snap ring onto valve body. Ensure snap ring has properly seated in its groove in the valve housing.
9. Install plunger and valve mounting flange onto valve body, then secure assembly with bolts and washers.

B-W DOUBLE CHECK VALVE & STOP LIGHT SWITCH

The shuttle type double check valve portion of this combination valve provides a means for making a separate trailer brake application.

The left end, Fig. 11, is attached to a service port of the brake valve. A service line from the hand control valve is connected to the opposite end. A service line leading to the trailer brakes is connected to the outlet port. The stop light switch is mounted on top of the valve opposite the outlet port.

Entry of air from the treadle valve causes the shuttle to seal the opposite inlet port and direct air to the trailer brakes. Simultaneously, another service line leading from the brake valve actuates the tractor brakes. When the hand control valve is applied, the shuttle seals the inlet port leading from the treadle valve and directs air to the trailer only. In either case, air pressure actuates the stop light switch.

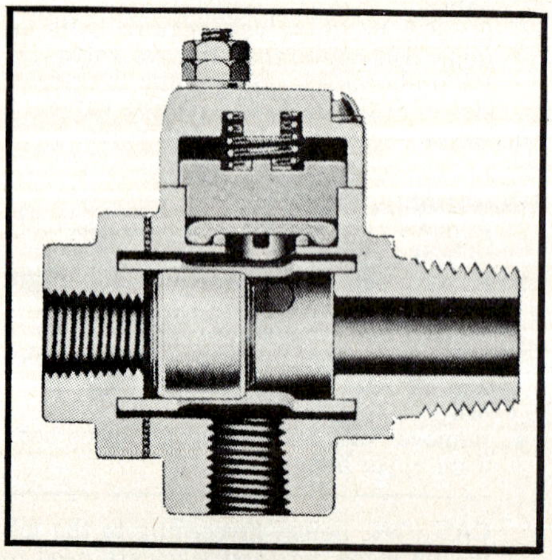

Fig. 11 Double check valve and stop light switch

NOTE: A leaking double check valve would cause an air loss during brake applications due to air escaping through the open exhaust port of the valve connected to the opposite inlet port. Service air from the treadle valve would escape through the exhaust port of the hand control valve or vice versa.

Testing

Periodically test the serviceability of the valve by applying and releasing the brake valve and noting that the brakes apply and release on both the tractor and trailer. Apply and release the hand control valve, noting that the brakes apply and release on the trailer only. When the treadle valve is applied, there should be no leakage at the exhaust port of the hand control valve, or at the exhaust port of the treadle valve when the hand control valve is applied.

Service

The double check valve and stop light switch should be serviced when the brake valve is removed from the vehicle.

Remove end cap, cap gasket and shuttle. If the shuttle guide or the sealing ends of shuttle show excessive wear, service by replacement. If valve seats in cap or housing or corroded or pitted excessively, replace valve. *Do not lubricate double check valve.*

B-W STOP LIGHT SWITCH

The stop light switch shown in Fig. 12 is used mainly in truck applications and is similar to the switch that is combined with the double check valve. Both type switches are actuated by air pressure and complete the stop light circuit each time a brake application is made.

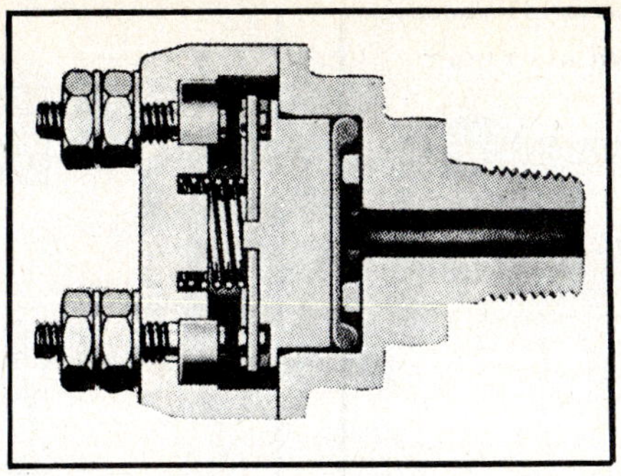

Fig. 12 Stop light switch

As the switch should make contact at three to seven pounds air pressure, it is important that the contacts are clean and that the diaphragm is not worn and permitting leakage. The diaphragm can be removed from the switch body (without removing switch from vehicle) by making a slight brake application.

When reassembling the switch, be sure the diaphragm faces as shown in Fig. 12, and that the contact strip is installed on the piston so that the round side of contacts face the terminal cover. Lubricate diaphragm, piston and body bore with B-W 239377 grease.

B-W SR-1 INVERSION VALVE

The inversion valve shown in Fig. 12A is used exclusively on straight truck models with dual air brake systems. The valve's purpose is to allow a modulated spring brake application if air loss should occur in the primary portion of system. The valve senses loss of air pressure in primary system allowing spring brakes on the rear axle to be applied or released corresponding with front service brake application or release by the operator. The inversion valve also permits air pressure to enter the parking brake chamber, releasing the spring brakes.

Valve Service

At least once a year remove, disassemble, clean and inspect all components. If wear and deterioration are found, install new components. It is recommended that all rubber seals be replaced.

Valve Testing

1. Position parking control valve in the apply position, ensure spring brake actuators apply. Install test guage in the delivery port of the inversion valve and check to ensure that spring brake actuators release fully, when control valve is placed in release position.
2. With parking control valve in released position, note gauge pressure reading. The reading should be the same as reading on instrument panel. If reading is incorrect, valve must be replaced.
3. Place parking control valve in apply position, gauge reading should drop to zero. A slow release of pressure could indicate a faulty single check valve within the modulating valve. At approximately 28–35 psi spring parking brakes should be applied.
4. Place parking control valve in release position, then drain number one service reservoir completely.
5. Apply foot brake valve several times and note pressure reading on gauge as pressure decreases each time foot brake valve is applied. After foot brake valve has been applied several times, pressure on gauge will drop to the point where release of the spring brake actuators will no longer occur.
6. With air system fully charged and parking brake control valve in release posi-

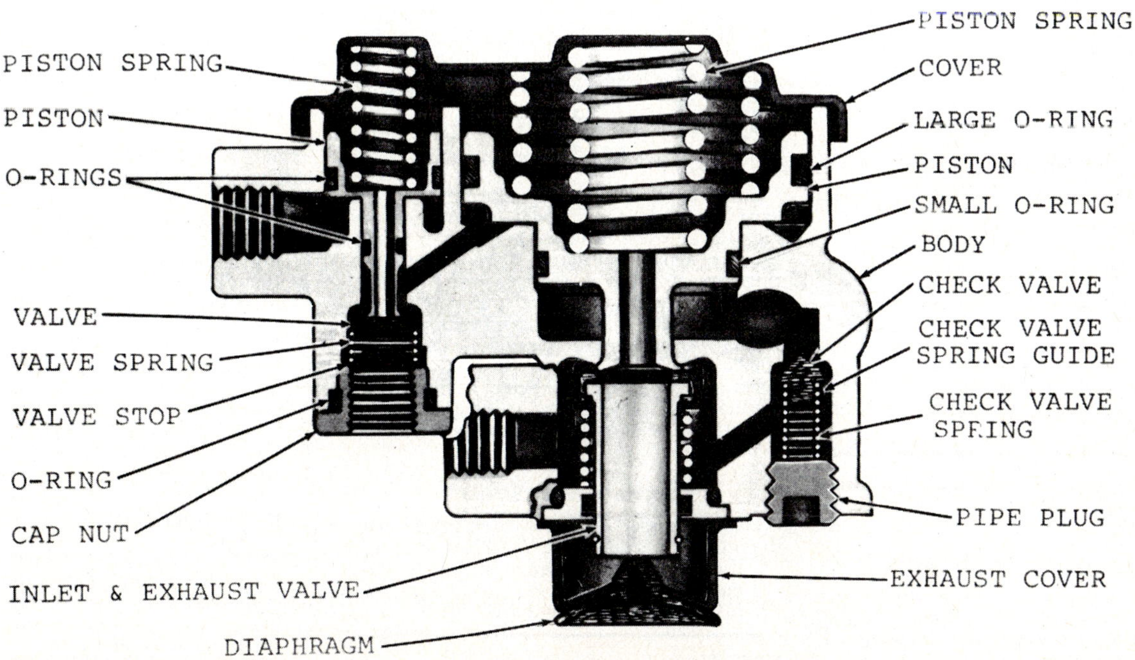

Fig. 12A Sectional view of B-W SR-1 inversion valve

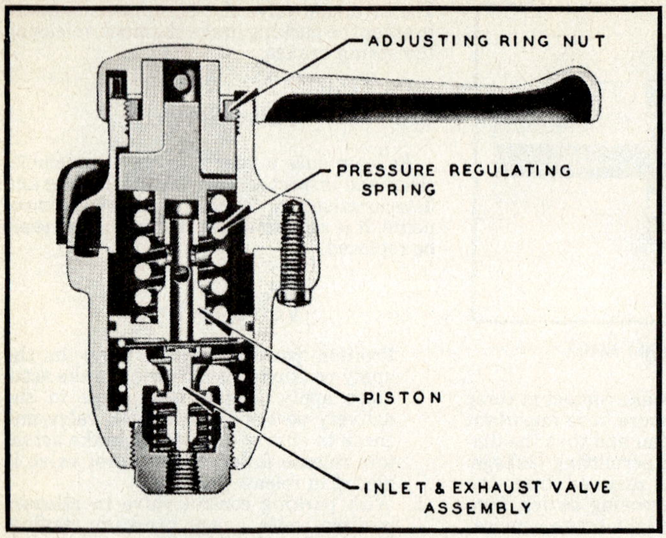

Fig. 13 TC hand control valve

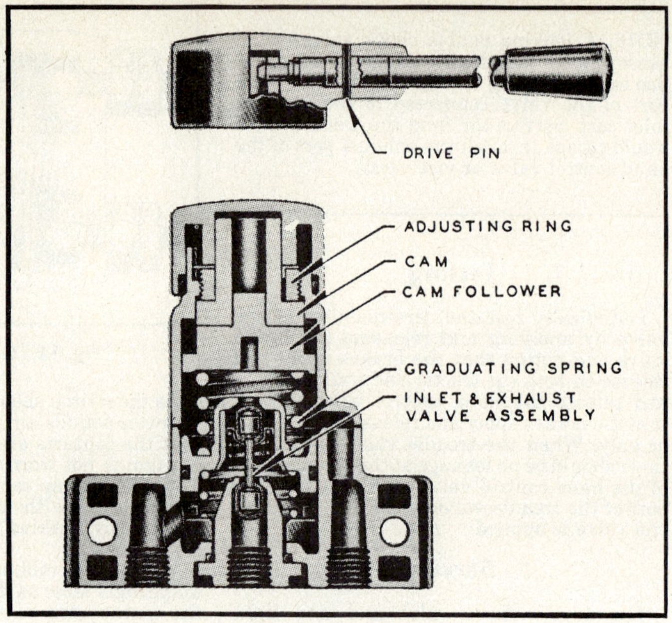

Fig. 14 TC-2 (type 1) hand control valve

tion, use a suitable leak detector gauge and place gauge over valve cover opening and exhaust port to detect air leakage. Slight leakage is permitted. If there is excessive leakage replace inversion valve.

Valve, Replace

1. Apply parking brakes and drain all air reservoirs.
2. Mark to assist in installation all air lines, then disconnect lines from inversion valve.
3. Remove bolts and inversion valve.
4. Reverse procedure to install. Before releasing vehicle for service, refer to testing procedure and perform steps as outlined.

Valve Disassembly

1. Refer to Fig. 12A, and remove socket head pipe plug, check valve spring, spring guide and check valve.
2. Remove screws and exhaust cover.
3. Remove exhaust diaphragm, then inlet and exhaust valve assemblies from cover.
4. Remove inlet and exhaust valve cap nut, and remove cap nut "O" ring.
5. Remove valve stop, valve spring, inlet and exhaust valves.
6. Remove bolts, lockwashers and cover from body.

CAUTION: Cover is under a spring tension and should be secured by clamps while removing bolts. Loosen bolts and clamps slowly and evenly until spring tension decreases, and cover can be removed from body safely.

7. Remove three piston springs from cover.
8. Remove large and small pistons, then the "O" rings.

Cleaning & Inspection

Clean all parts using a suitable cleaning solvent and blow dry using compressed air. Any components that are found to be unserviceable should be replaced.

Valve Assembly

1. Lubricate all contact surfaces, "O" rings and seals.
2. Install check valve, valve spring guide and valve spring into body.
3. Apply a suitable sealant to the socket head pipe plug and install it into body. Torque plug to 130–170 inch lbs.
4. Install inlet and exhaust valve assemblies into valve body and install exhaust diaphragm into exhaust cover.
5. Install cover onto body, torque cover screws to 20–30 inch lbs.
6. Install exhaust valve, valve spring and valve stop into body.
7. Install "O" rings and cap nut onto body, torque nut to 100–125 inch lbs.
8. Install "O" rings onto the large and small pistons and install into body.
9. Install springs into pistons.
10. Secure cover to body using bolts and lockwashers, torque bolts to 50–80 inch lbs.

HAND CONTROL VALVES

B-W Type Hand Control Valves

Either of three types of hand control valves have been used. One, model TC, is shown in Fig. 13, and the others, model TC-2 (types 1&2) are shown in Figs. 14 & 14A. Adjusting rings are provided in all types for adjusting delivery pressure.

Every 50,000 miles the hand control valve should be checked for graduation. Every 100,000 disassemble the control valve, clean and inspect all parts and replace as necessary. Whenever the hand control valve is disassembled, the inlet and exhaust valves should be replaced.

Service—TC Valve

Prior to reassembling the valve, lightly coat the internal parts with B-W 240176 grease.

1. If adjusting ring is removed, install with marks in alignment.
2. Install adjusting ring lock.

NOTE: If pressure adjustment is necessary, do not install adjusting ring lock and handle until adjustment has been made. Valve should deliver 75 to 85 pounds. Turn adjusting ring clockwise to increase pressure, and counterclockwise to decrease it.

3. Align either groove with lock plunger and install cam follower.
4. Install cam. Rotate cam follower so that cam will be at its lowest point ("OFF" position).
5. Position piston return spring in valve body.
6. Place new O-ring in groove and install piston.
7. Position graduating spring on cam.
8. Attach valve cover to body (4 screws).
9. Position handle on cam follower with pin holes in alignment. Install Allen head pin.
10. Place O-ring in position in valve body and install inlet valve screw cap together with inlet valve spring and inlet-exhaust valve.

Service—TC-2 (Type 1) Valve

Prior to reassembling valve, lightly coat internal parts with B-W 240176 grease.

1. Position piston return spring in valve body.
2. Slide piston into valve body after installing O-ring on it.
3. Install adjusting ring in cover and align marks that should have been made before removing.
4. Install adjusting ring lock.
5. Install cam follower.
6. Install cam. Rotate follower to permit cam to seat on it.
7. Position graduating spring on piston.
8. Position gasket on valve body.
9. Attach cover to valve body with two special capscrews. Make sure gasket is positioned properly.
10. Position O-ring on valve cover.
11. Install valve head on follower (with handle holes in alignment) and over valve cover. Carefully push O-ring under valve head.
12. Slide valve handle into valve head and follower after installing O-ring on handle. Install drive pin.

NOTE: If pressure adjustment is necessary, do not install drive pin until adjustment has been made. Remove adjusting ring lock and turn ring clockwise to increase delivery pressure, and counterclockwise to decrease. Valve should deliver 75 to 85 pounds pressure.

13. Slide inlet-exhaust valve into valve body after installing O-ring. Install screws and lockwashers.
14. Test valve for leakage in same manner as testing foot brake valve.

Service-TC-2 (Type 2) Valve

Prior to reassembling valve, lightly coat internal parts with lubricant supplied in repair kit.

1. Install inlet valve on stem, place valve and stem assembly in supply port, then insert a suitable tool in supply port to hold inlet valve on its seat.
2. Install spring and valve guide over stem and compress spring, then assemble exhaust valve on stem.
3. Position piston return spring in body and fit a new O ring to piston.
4. Install piston in body assembly.
5. Position adjusting ring in cover and tighten it down until flush with top of cover.
6. Install cam and cam follower in cover.
7. Install piston graduating spring in body.
8. Position new cover to body gasket and assemble cover, install cover retaining screws, torque to 85 in. lbs.
9. Install adjusting ring washer rotating adjusting ring as necessary to align washer notch with valve body notch.
10. Install head seal O ring and head, then install handle O ring and install handle into valve head. Secure with spiral pin.

NOTE: If pressure adjustment is necessary, it can be adjusted by removing the handle, head and adjusting ring washer. By rotating the adjusting ring clockwise, pressure will increase. To lower pressure, turn the adjusting ring counter-clockwise, then reassemble unit.

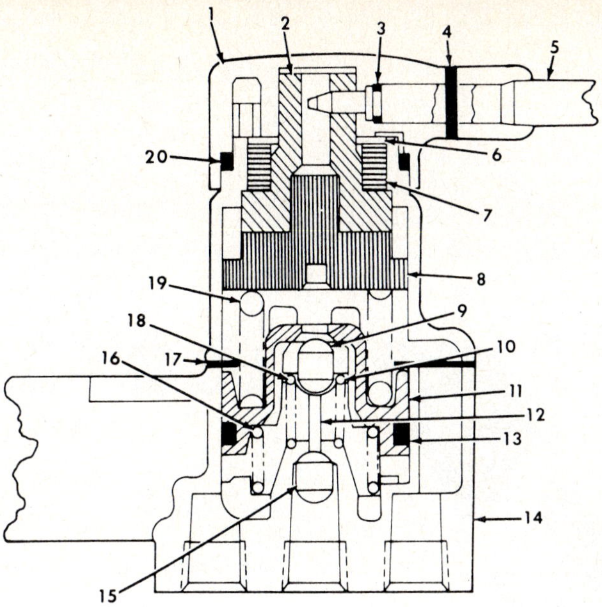

1 Head	11 Piston
2 Cam Follower	12 Valve Stem
3 O-ring	13 O-ring
4 Spiral (Roll) Pin	14 Body
5 Handle	15 Inlet Valve
6 Adjusting Ring Lock Washer	16 Piston Return Spring
7 Adjusting Ring	17 Gasket
8 Cam	18 Valve Spring
9 Exhaust Valve	19 Graduating Spring
10 Valve Guide	20 O-ring

Fig. 14A TC-2 (type 2) hand control valve

M-R Type Hand Control Valves

Two types of M-R hand control valves are used. Type N-30060 is shown in Fig. 14B and type N-30107 as shown in Fig. 14C. Both type valves use an adjusting bearing to adjust delivery pressure.

Service—Midland-Ross N-30060 Valve

NOTE: A repair kit should always be used when overhauling these units. Old parts should always be replaced even if they appear to be in good condition. Parts contained in repair kits include: valve disc, valve retainer, piston and valve return springs, graduating spring, block vee and all O rings.

1. Place discharge valve return spring in body, then install discharge valve assembly.
2. Compress discharge valve return spring then install inlet valve disc assembly, secure with elastic stop nut.
3. Install piston return spring. Install piston block vee, then install piston assembly in valve body.
4. Install graduating spring in valve body.
5. Position adjusting bearing in cam housing, then install cam and camshaft. Place assembly in position on valve body. Secure by installing collar.
6. To adjust exhaust valve gap, turn adjusting bearing in while checking for inlet valve motion through the end cap opening in valve body. Note the position of adjusting bearing at which the inlet valve starts to move, back adjusting bearing out ½ to ¾ turn. Lock adjustment with set screw in cam housing.
7. Install end cap O ring and install on body.
8. Lubricate friction rings with molybdenum disulfide powder and assemble over cam housing, then install lever assembly and secure with nut.

NOTE: The handle position is fully adjustable. To adjust loosen the collar, then rotate the cover and cam housing to the desired position, then retighten collar.

9. Install cover and name plate, secure with cover screw.

Service—Midland Ross N-30107 Valve

Prior to disassembly, mark wide spline on camshaft in relation to valve body. Prior to reassembly, lubricate valve body bore and piston O ring with Delco Silicone Lube No. 5459912 or equivalent.

1. Position spring on discharge valve assembly with small diameter of spring facing valve assembly. Insert assembly through valve body and place valve disc over discharge valve stem, secure with nut and torque to 6 in. lbs.
2. Position O ring on ledge in valve body, then place disc over O ring and secure with snap ring.
3. Install spring in large bore of valve body, then install O ring on piston and insert in valve body.
4. Place spring in piston cavity.
5. Install cam in valve body. Ensure tabs on cam enter slots in body.
6. Position camshaft in cam. Ensure the wide spline on camshaft lines up with scribe marks which should have been made previous to disassembly. If not, remove cam and rotate 180° and reinstall.
7. Place friction O ring in bearing groove.

NOTE: Do not lubricate friction O ring or bearing with any lubricants. These surfaces must provide friction for the valve to work properly.

8. Thread bearing into valve body and tighten. Connect 100–125 psi air supply to inlet port and a pressure gauge to delivery port.
9. Move handle slowly to applied position, pressure should start to build up on pressure gauge after 10–15° of handle motion. Full supply pressure should be obtained after 60° of handle rotation.

NOTE: If handle travel is excessive before

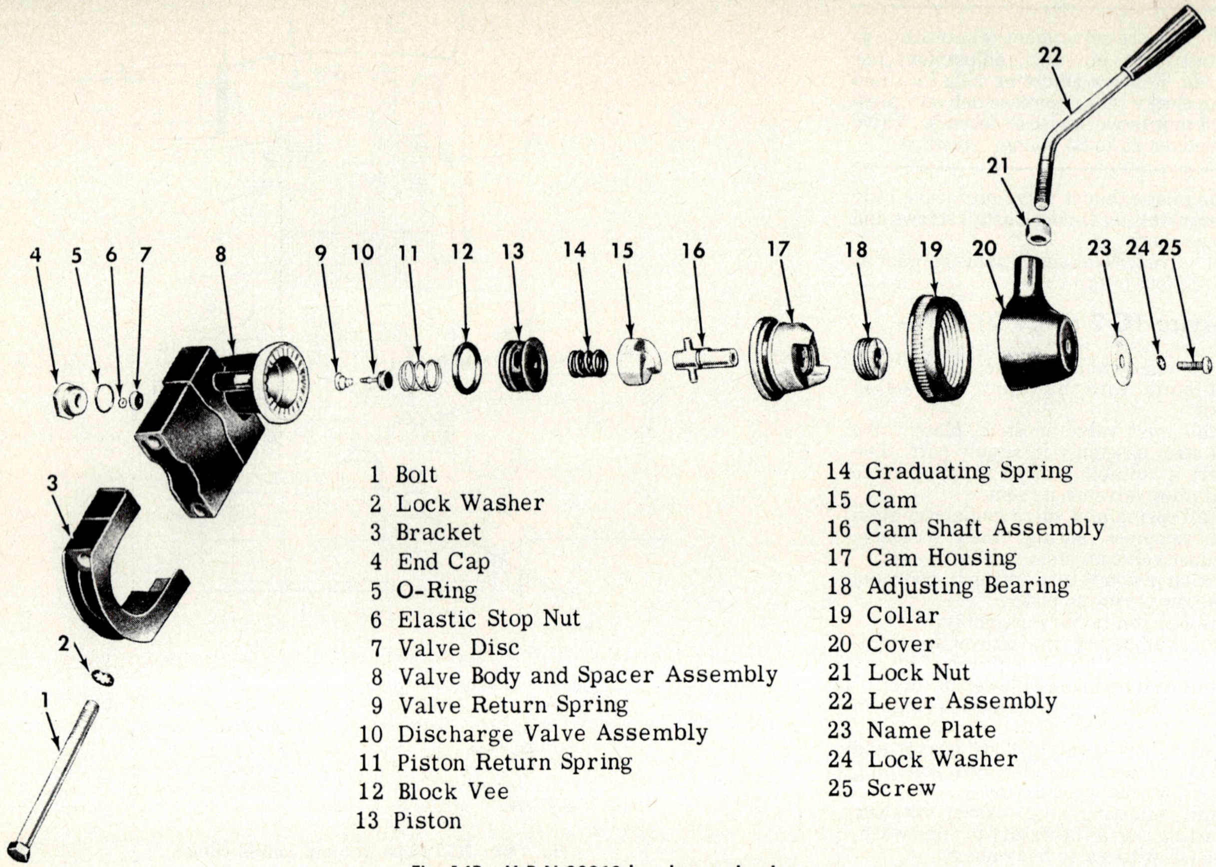

1	Bolt	14	Graduating Spring
2	Lock Washer	15	Cam
3	Bracket	16	Cam Shaft Assembly
4	End Cap	17	Cam Housing
5	O-Ring	18	Adjusting Bearing
6	Elastic Stop Nut	19	Collar
7	Valve Disc	20	Cover
8	Valve Body and Spacer Assembly	21	Lock Nut
9	Valve Return Spring	22	Lever Assembly
10	Discharge Valve Assembly	23	Name Plate
11	Piston Return Spring	24	Lock Washer
12	Block Vee	25	Screw
13	Piston		

Fig. 14B M-R N-30060 hand control valve

delivery begins, or full supply pressure is not reached, bearing must be tightened further into valve body. If air pressure is delivered in the fully released position, the bearing should be loosened.

10. For proper exhaust valve gap, with handle in the fully released position and air supply connected to inlet port, slowly tighten bearing into valve housing until air just starts to flow from the delivery port. Back the bearing off ½ turn, bearing can now be slightly rotated in either direction to allow lock assembly to be installed properly.
11. Place lock assembly on bearing so lock pin is aligned next to notch in valve body. Secure lock assembly with screw in first screw hole clockwise of lockpin. Torque screw to 6 in. lbs.
12. Torque control handle screw to 13 in. lbs.

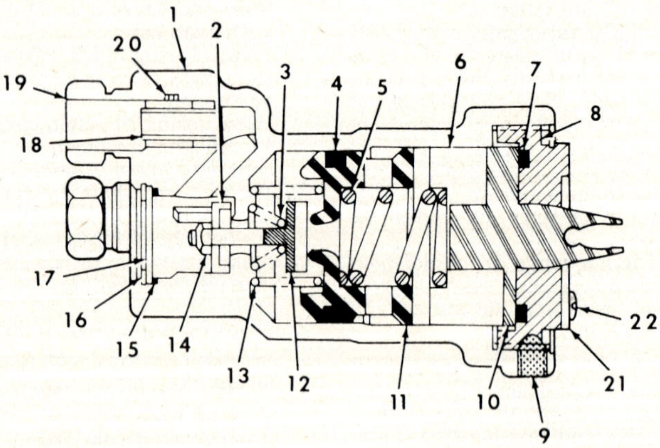

1	Valve Body	12	Discharge Valve Assembly
2	Valve Disc Assembly	13	Spring
3	Spring	14	Nut
4	O-ring	15	O-ring
5	Spring	16	Snap Ring
6	Cam	17	Disc
7	O-ring	18	Tube Insert
8	Bearing	19	Tube Nut
9	Set Screw	20	Sleeve
10	Cam Shaft	21	Lock Assembly
11	Piston	22	Screw

Fig. 14C M-R N-30107 hand control valve

BRAKE RELAY VALVES

Relay valves are used to operate the brakes on the rear axle of six wheel tractors, trucks or long wheel base trucks.

It is operated by air pressure from the foot brake valve. The air pressure delivered to the brake chambers is the same as the pressure delivered to the relay valve.

With a direct air supply from the reservoir, the relay acts as a remote brake valve to speed up the application and release of brakes.

Every 100,000 miles the relay valve should be disassembled, cleaned and inspected for wear. Replace parts as necessary.

B-W Brake Relay Valve Type R-1

Disassembly, Fig. 15.

1. Remove 4 bolts and lock washers securing adapter to valve body. Remove adapter and gasket; discard gasket.
2. Remove 6 bolts securing valve cover to valve body. Remove cover, diaphragm, diaphragm ring, diaphragm "O" ring (at passage "D"), and diaphragm guide from valve body. Remove filter from by-pass port in cover.

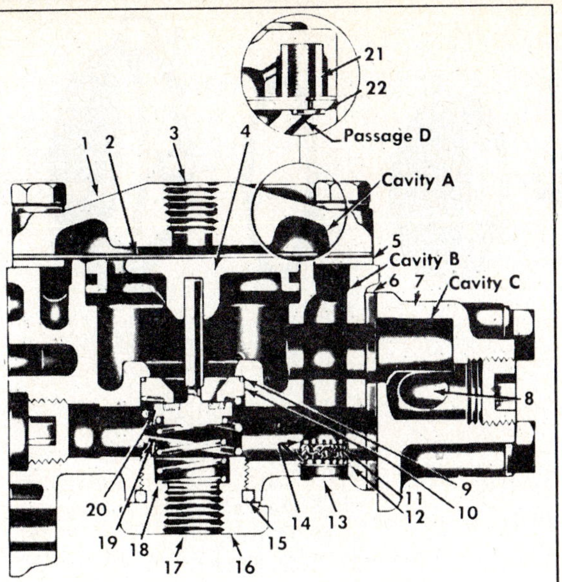

1 Cover
2 Diaphragm
3 Inlet from Application Valve
4 Diaphragm Guide
5 Diaphragm Ring
6 Gasket
7 Adapter
8 Outlet to Brake Chambers
9 O-ring
10 Supply Valve Seat
11 Curled Hair
12 Snap Ring
13 Exhaust Port
14 Strainer Plates
15 Cap Nut O-ring
16 Supply Valve Cap Nut
17 Supply from Air Tank
18 Supply Valve Spring
19 Supply Valve Seat Spring
20 Supply Valve
21 Filter
22 O-ring

Fig. 15 B-W brake relay valve (R-1)

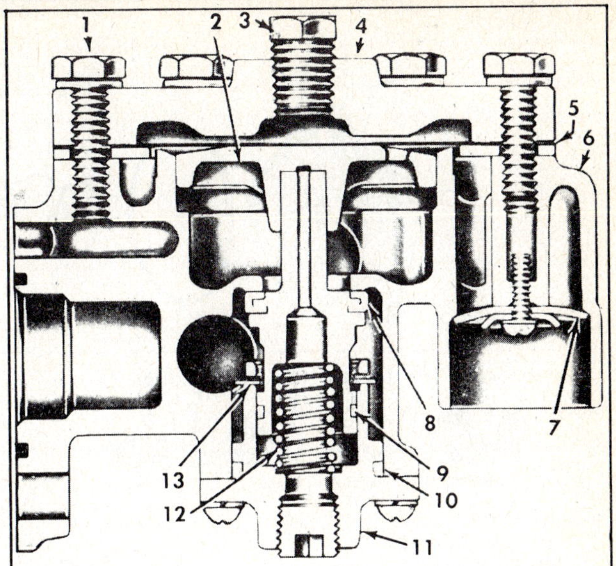

1 Cover Bolts
2 Diaphragm Guide
3 Plug
4 Cover
5 Diaphragm
6 Valve Body
7 Exhaust Diaphragm
8 Inlet Valve Assy.
9 O-Ring
10 O-Ring
11 Inlet Valve Guide Cap
12 Valve Spring
13 Retaining Ring

Fig. 15A B-W brake relay valve (R-5)

3. Remove supply valve cap nut from bottom of valve body. Remove supply valve spring, supply valve, supply valve seat spring and supply valve seat from valve body.
4. Remove snap ring, strainer plates, and strainer from exhaust port in valve body.

Inspection & Repair

1. Thoroughly clean all parts in a suitable cleaning fluid. Make sure exhaust port strainer plates and strainer are thoroughly cleaned.
2. Examine diaphragm seat at top of valve body. Seat must be smooth and free from scratches or corrosion. If only slightly scratched or corroded, seat may be repaired by lapping on flat surface with crocus cloth. After lapping, edge of seat must be rounded off with crocus cloth. The original shape of the seat should be maintained as closely as possible, and not more than 0.005" should be removed from the seat.
3. Examine supply valve and supply valve seat. If damaged or badly worn, replace with new parts.

Assembly

NOTE: A repair kit is available containing parts which should always be replaced when overhauling a relay valve. Parts contained in repair kit are: cover gasket, strainer, "O" ring, diaphragm, adapter gasket, filter, supply valve and lubricant. Discard these parts and replace with new parts during assembly.

Prior to assembly, apply a light coat of lubricant supplied in the kit, lube #5459912 or an equivalent lube to the valve bores and "O" rings.

1. Lay valve body on bench, bottom side up.
2. Place supply valve seat and grommet in body bore.
3. Insert long end of supply valve through valve seat and position valve on seat.
4. Place supply valve spring on valve.
5. Place supply valve seat retainer spring on valve seat.
6. Install supply valve cap nut and tighten cap nut firmly.
7. Turn assembly over. Install diaphragm guide in body and over supply valve stem.
8. Place new diaphragm "O" ring in valve body around passage D.
9. Place diaphragm ring on valve body, with small bleed hole in ring aligned with passage D.
10. Place new diaphragm on ring and guide, with large hole in diaphragm aligned with bleed hole in diaphragm ring.
11. Install new by-pass port filter in cover. Install valve cover on valve body with both D passages aligned.
12. Install bolts and lock washer, tighten evenly and firmly.
13. Install adapter on side of valve body using a new gasket. Firmly tighten adapter attaching bolts.

B-W Brake Relay Valve Type R-5

Disassembly, Fig. 15A.

1. Remove 6 cover bolts and remove cover.
2. Remove diaphragm and diaphragm ring, discard diaphragm.
3. Remove diaphragm follower guide.
4. Remove 4 screws from inlet valve guide cap and assembly.
5. Depress inlet valve stem in cap assembly and remove valve stem retaining ring.

NOTE: Avoid damage to guide surfaces when removing ring.

6. Remove inlet valve assembly; discard inlet valve.
7. Remove inlet valve spring from valve guide cap; discard spring.
8. Remove screw and washer from exhaust diaphragm and diaphragm from valve body; discard diaphragm.
9. Remove "O" ring from guide cap.

Inspection & Repair

1. Thoroughly clean all metal parts in suitable cleaning fluid; dry thoroughly.
2. Examine interior bores of valve body and inlet valve guide cap, polish with crocus cloth if slightly scratched or corroded. Replace with new part if damage is excessive.
3. Examine inlet valve seat for evidence of wear, deterioration, cracks and scratches.

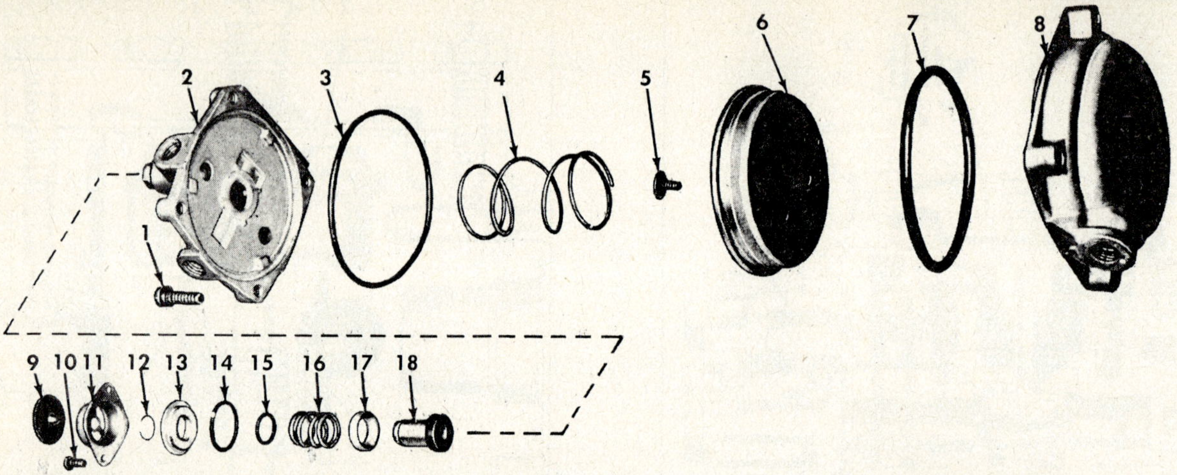

1 Bolt	7 O-Ring	13 Valve Cap
2 Valve Body	8 Valve Cover	14 O-Ring
3 O-Ring	9 Exhaust Check	15 O-Ring
4 Piston Return Spring	10 Bolt	16 Valve Return Spring
5 Inlet and Exhaust Valve Seat	11 Exhaust Cover	17 Spring Seat
6 Piston	12 Snap Ring	18 Inlet and Exhaust Valve

Fig. 15B B-W brake relay valve (R-6)

Assembly

NOTE: A repair kit is available containing parts which should always be replaced when overhauling a relay valve. Parts contained in repair kit are: relay diaphragm, exhaust diaphragm, inlet valve, valve spring, "O" rings and lubricant. Discard these parts and replace with new parts during assembly.

Prior to assembly, apply a light coat of lubricant that is supplied in repair kit, lube #5459912 or equivalent to the valve bore and "O" rings.

1. Install new "O" rings on inlet valve and valve guide cap.
2. Install new exhaust diaphragm in valve body and secure with washer and screw.
3. Place new inlet valve spring in valve guide cap.
4. Place inlet valve spring in valve body, depress stem and install retaining ring.
5. Install inlet valve spring and guide cap in valve body and secure with 4 screws.
6. Install diaphragm follower guide on valve stem.
7. Position diaphragm ring and diaphragm.
8. Install cover and secure with 6 bolts.

B-W Brake Relay Valve Type R-6

Disassembly, Fig. 15B.

1. Scribe reference marks on cover and valve body to aid in reassembly. Remove four bolts securing cover and remove, then remove piston return spring.
2. Remove piston and O ring from cover (use air pressure if necessary), then separate O ring from piston.
3. Remove inlet-exhaust valve seat.
4. Remove O ring from valve body.
5. Remove bolts securing exhaust cover to valve body.
6. Remove exhaust cover then remove exhaust check valve from cover.
7. Press inlet-exhaust valve from valve body.
8. Remove inlet-exhaust valve snap ring, then separate valve cap, return spring and spring seat from valve. Remove O ring from valve cap.

Inspection & Repair

1. Inspect interior bores of valve body and inside of valve cover for scratches and pits. If damage is excessive, replace with new parts, otherwise a piece of crocus cloth can be used to polish the surfaces.
2. Inspect valve and seat for signs of wear, deterioration, cracks and scratches. Replace as necessary.
3. Inspect valve and piston springs. Compare tension and height with known good springs. Replace if not similar.
4. Inspect exhaust check valve. If worn, cracked or deteriorated, replace.

Reassembly, Fig. 15B.

NOTE: A repair kit should always be used when overhauling these units. Old parts should always be replaced even if they appear to be in good condition. Repair kits contain: O rings, inlet-exhaust valve, valve retainer, valve spring, snap ring, exhaust check valve, exhaust valve seat and lubricant.

1. Install O ring in valve cap then assemble inlet-exhaust valve; install spring seat, return spring and valve cap on valve, secure with snap ring.
2. Press inlet-exhaust valve into valve body.
3. Install exhaust check valve in exhaust cover and install cover.
4. Install valve body O ring then piston O ring.
5. Install valve seat in piston then piston assembly in cover.
6. Position piston return spring in valve body.
7. Position piston and cover assembly over valve body, align reference marks.
8. Press piston and cover towards valve body while compressing spring and secure cover to body with four bolts.

B-W Brake Relay Valve Type R-8

Disassembly, Fig. 15C.

1. Scribe reference marks on cover and body to aid in reassembly. Remove four bolts securing cover and remove, then separate the two halves and remove the piston spring.
2. Remove sealing ring from body groove.
3. Remove relay piston and O ring from cover.
4. Remove service port cap nut and O ring from cover and remove the quick release diaphragm from service port.
5. Remove screw securing exhaust diaphragm and washer to the exhaust cover.
6. Remove four screws securing the exhaust cover to the body.
7. Remove inlet and exhaust valve from body.
8. Disassemble the inlet and exhaust valve insert by removing the retaining ring and separating the washer, O ring retainer and O rings, valve spring, valve retainer and inlet and exhaust valve, Fig. 15D.

Inspection & Repair

1. Inspect all metal parts for scratches, scoring and corrosion. If damage is excessive, replace parts as necessary.
2. Inspect all valve seats in the unit for scratches, scoring and corrosion that could cause excessive leakage during operation. If damage is excessive, replace parts as necessary.

NOTE: A repair kit should always be used when overhauling these units. Old parts should always be replaced even if they appear to be in good condition. All rubber parts and springs should be replaced with new parts from the repair kit.

Prior to assembly, lubricate all O rings, O

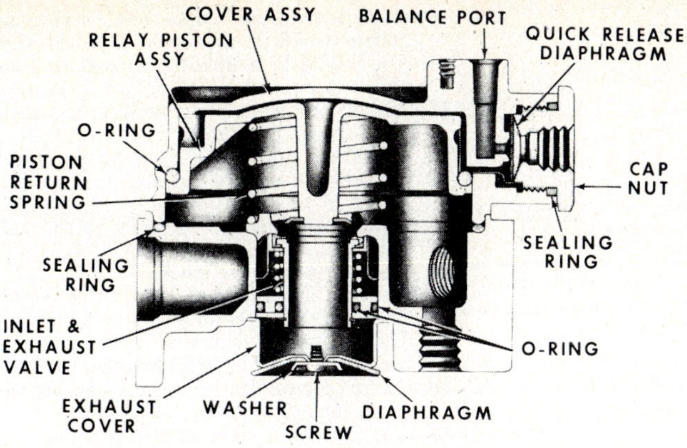

Fig. 15C brake relay valve (R-8)

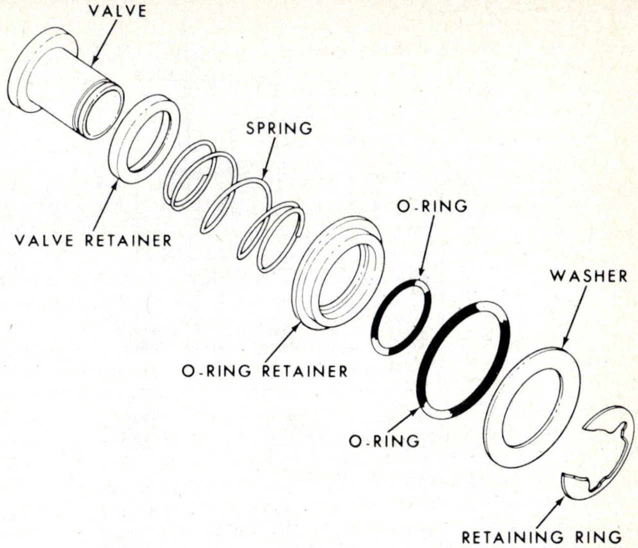

Fig. 15D B-W Brake relay valve inlet & exhaust valve insert components

ring grooves, piston bores and metal to metal moving surfaces with a barium base lubricant.

Reassembly, Fig. 15C.

1. Install O ring on service port cap nut.
2. Position quick release diaphragm in valve, ensuring it is seated correctly between guide rings in cover.
3. Install and torque the service port cap nut to 150 in. lbs.
4. Position the O ring on relay piston then install piston in valve cover.
5. Position inlet exhaust valve, rubber side down on a clean flat surface. Install valve retainer and the spring on the valve, Fig. 15D. Continue assembling inlet exhaust valve as shown in Fig. 15D. Install completed assembly in the body bore.
6. Position exhaust cover on body. Install four cover securing screws and torque to 24 in. lbs.
7. Reinstall exhaust diaphragm on exhaust cover. Secure with screw and washer, torque to 20 in. lbs.
8. Install piston return spring on piston in cover, then install the valve body seal ring.
9. Align cover and body reference marks made previous to disassembly and install cover to body. Install securing screws, torque to 150 in. lbs.

M-R Brake Relay Valve Type N-30106A & N-30100AH

Disassembly, Figs. 15E & 15F.

After valve is disconnected and removed from the vehicle, inspect it visually for damage. If casting is broken or shows fractures, replace the valve.

1. Mark the valve outside cover, mounting bracket, and body to show part alignment for assembly.
2. Remove the 4 cover cap screws.
3. Remove valve cover and piston assembly from valve body.
4. Remove exhaust valve body snap ring from valve body.
5. Remove washer, exhaust valve body, spring and modulation tube assembly from valve body.
6. Remove "O" ring seal from exhaust valve body.
7. Remove exhaust valve body screw, washer and diaphragm from exhaust valve body.
8. Remove "O"-ring seal, seal retainer and valve seal from modulation tube.
9. Remove exhaust valve screw, valve seal and square cut seal (on M-R N-30100AH relay valves remove "O"-ring) from piston.
10. Remove "O"-ring seal from valve cover, except on M-R N-30106A relay valve.
11. Remove valve, valve seat and "O"-ring seal from valve cover, except on M-R N-30106A relay valve.

Inspection & Repair

Wash all metal parts in mineral spirits or other suitable cleaning solvent and then dry. Wipe all rubber parts clean for inspection.

Examine rubber valve seals, "O"-ring, and exhaust port diaphragm. "O"-ring seals must not be cut or show surface nicks or gouges.

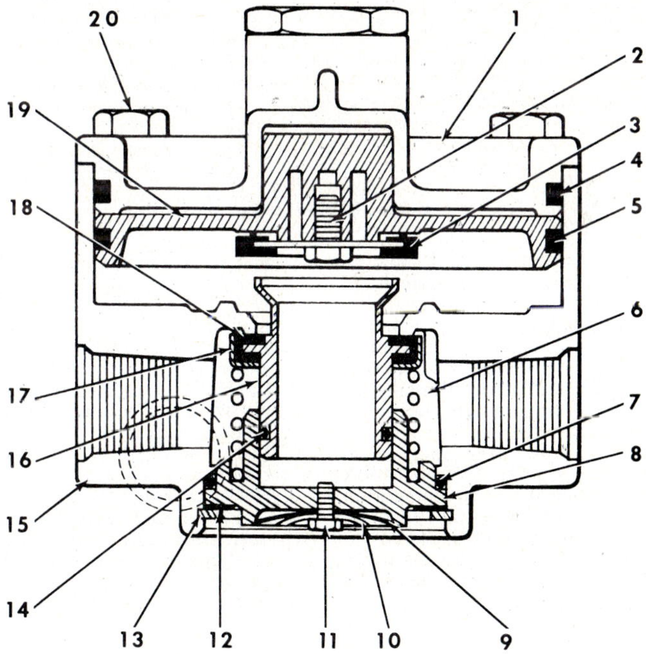

1 Cover	11 Exhaust Valve Body Screw
2 Valve Screw	12 Retaining Washer
3 Exhaust Valve Seal	13 Retaining Ring
4 Cover Sq. Cut Seal	14 O-Ring Seal
5 Piston Sq. Cut Seal	15 Valve Body
6 Modulation Tube Spring	16 Modulation Tube
7 O-Ring Seal	17 Seal Retainer
8 Exhaust Valve Body	18 Inlet Valve Seal
9 Exhaust Port Diaphragm	19 Piston
10 Washer	20 Cover Cap Screw

Fig. 15E M-R N-30106A brake relay valve

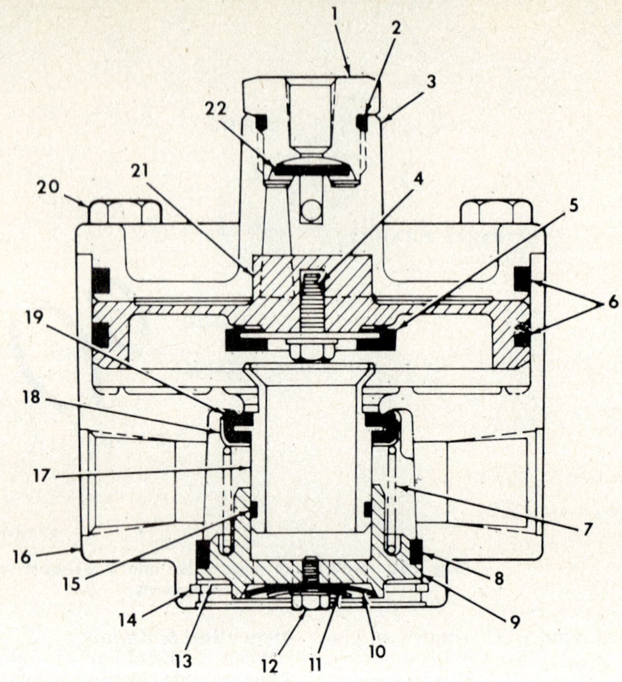

Key	Description	Key	Description
1	Valve Seat	12	Screw
2	O-Ring Seal	13	Washer
3	Valve Cover	14	Retaining Ring
4	Screw	15	O-Ring Seal
5	Valve Seal	16	Valve Body
6	O-Ring Seal	17	Modulation Tube
7	Spring	18	Seal Retainer
8	O-Ring Seal	19	Valve Seal
9	Exhaust Valve Body	20	Screw
10	Diaphragm	21	Piston
11	Washer	22	Valve Disc

Fig. 15F M-R N-30100AH brake relay valve

Diaphragm should be flexible and show no cuts, tears or distortion. Inlet valve seal and valve must be smooth on valve face area. There must not be any nicks, cuts, dents or imperfections visually evident in the rubber. Exhaust valve seal face must be flat and smooth in area contacting modulation tube seat. Rubber should be firm and not show separation from metal disc.

Examine exhaust valve body, relay valve cover and body, and piston. On M-R N-30106A relay valve measure seal groove diameter of piston. Diameter must be 3.375/3.379 inches. If not within specification, replace piston.

Castings should be free of cracks and voids. All surfaces for "O"-ring seals must be smooth, round and free of deep scratches, dents or gouges. There should be no evidence of metal to metal galling of aluminum surfaces where piston slides into valve body. Valve body flanges, threads and snap ring groove should not be distorted. Valve body seat must be flat and smooth, showing no cracks or distortion.

Examine modulation tube. Flange for valve seal must not be dented or bent. Seat at end must be flat and smooth, showing no dents or distortion.

Examine springs, snap ring, retainer, washer and screws. Parts should be free of distortion and corrosion. Replace parts that show any of the above defects.

Assembly

1. Install exhaust port diaphragm, washer and screw to exhaust valve body. Torque screw to 23 in. lbs.
2. Install inlet valve seal and "O"-ring on modulation tube.
3. Place seal retainer over tube and around valve seal.
4. Carefully place modulation tube assembly into valve body so that inlet valve seal rests on body seat.
5. Install spring over tube to rest on seal retainer.
6. Place "O"-ring over exhaust valve body.
7. Align exhaust valve body to spring and push body into relay valve body firmly and hold to prevent spring from pushing exhaust body out.
8. Install washer and install snap ring into valve body.
9. Place exhaust valve seal on piston.

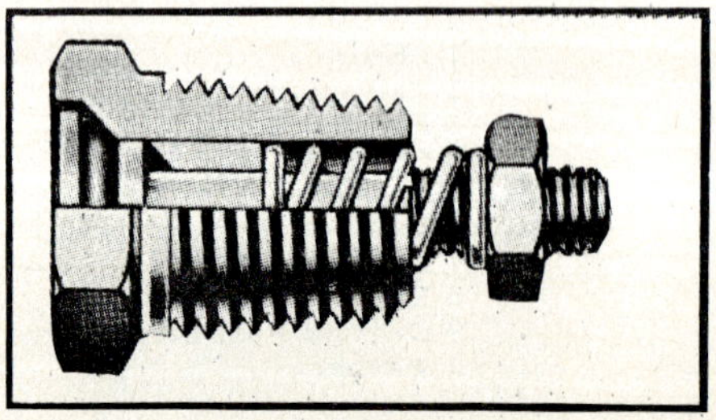

Fig. 16 Relief valve

10. Install exhaust valve seal screw and torque to 79 in. lbs. on M-R N-30100AH. On the M-R N-30106A torque to 17 in. lbs.
11. Install the square cut seal on the piston M-R N-30106A only. On the M-R N-30100AH install "O"-ring on the piston.
12. Install the square cut seal on the valve cover M-R N-30106A only. On the M-R N-30100AH install "O"-ring on the valve cover.
13. Install piston into cover.
14. Install "O"-ring seal onto valve seat on M-R N-30100AH only.
15. Install valve seat and valve disc into valve cover on M-R N-30100AH only.
16. Align piston and cover assembly and push into valve body.
17. Position valve mounting bracket on cover and install four screws. Torque screws to 135 in. lbs. on M-R N-30100AH only. On M-R N-30106A torque to 75 in. lbs.

RELIEF VALVE

The pressure relief or safety valve, Fig. 16, protects the air brake system against excessive pressure. The relief valve is installed in the reservoir and resembles a pipe plug in appearance. It has a hex head, and consists of a brass body, a poppet type valve, an O-ring, a pressure spring and an adjusting nut. The valve is pre-set to open at 140–150 psi.

Should the brake system pressure rise to about 140 pounds (usually resulting from a malfunctioning or improperly adjusted governor, or an inoperative compressor unloading mechanism) the pressure within the reservoir overcomes spring pressure. The valve is then unseated and air pressure in excess of the spring setting is exhausted.

Service

At least once a year, remove, disassemble and clean the relief valve. Replace unserviceable parts or the entire valve, if necessary.

Reset the valve to release at 145 psi. Pressure setting is adjusted by turning the adjusting screw. Turning in a clockwise direction with raise setting, and turning counterclockwise will lower setting. After installing valve, check for leakage.

RELEASE VALVES

Quick Release Valve

These valves, Figs. 17 and 17A, are used to hasten the release of air pressure from the front and rear wheel brake chambers.

When the brakes are applied, air entering at the top deflects the diaphragm and passes to the brake chambers, closing the exhaust port seat. In release, the pressure on the top of the diaphragm is released by the brake valve, whereupon the brake chamber pressure deflects the diaphragm upward, opening the exhaust port so that the chamber pressure passes to the atmosphere. This speeds up the release and saves time which, without the release valve, would be expended in passing the brake chamber pressure up to the brake valve for release.

Service

Every 100,000 miles, or yearly, the valve should be removed, disassembled, cleaned and checked for proper operation.

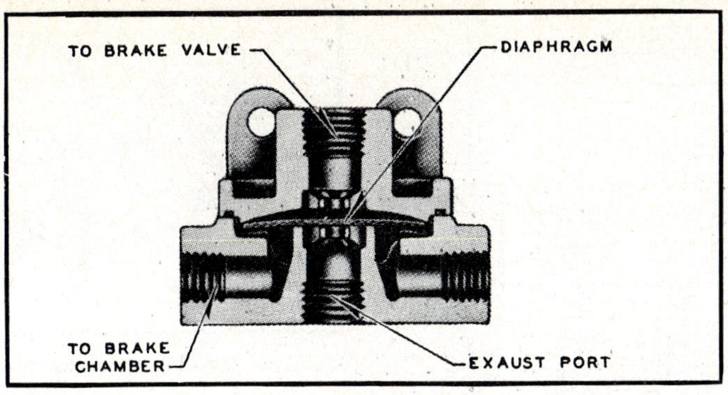

Fig. 17 Bendix QR-1 quick release valve (not required on rear axle when relay valve is used)

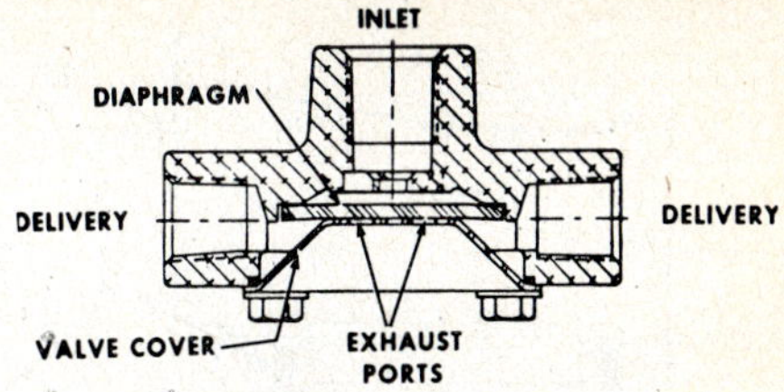

Fig. 17A Midland-Ross quick release valve

Testing Valve

1. Apply, then release brakes. When brakes are released, check that air pressure is quickly exhausted through exhaust port of valve. If not, check exhaust port for restrictions.
2. Test valve for leakage using Leak Detector Tester SE-2326, or equivalent connected to the exhaust port.
3. Apply, then release brakes. Check that valve releases immediately when brakes are released.
4. Leakage may not exceed a one inch soap bubble in one second on Bendix QR-1 valve, or a one inch soap bubble in six seconds on Midland-Ross valves.
5. If leakage exceeds allowable limits, the valve must be repaired or replaced.

NOTE: Excessive leakage may be caused by dirt in the valve or a defective diaphragm.

Disassembly

1. Remove four screws and cover.
2. On Bendix QR-1 Type valve, remove spring and spring seat, if so equipped.
3. Remove diaphragm and cover "O" ring.

Cleaning & Inspection

1. Clean all parts in a suitable solvent.
2. Inspect diaphragm for wear or deterioration, paying particular attention to lower contact area.
3. Inspect cover exhaust seat for pitting or nicks. A fine piece of emery cloth may be used to dress defects.
4. On Bendix QR-1 Type valve equipped with valve spring, check spring and seat for wear or corrosion.
5. Repair or replace parts as needed to return valve to a serviceable condition.

Assembly

1. On Bendix QR-1 Type valve equipped with valve spring and seat, install spring in valve body.
2. Position diaphragm and "O" ring in valve body.
3. Assembly valve cover on body and install the four retaining screws.

Relay-Quick Release Valve

The air relay-quick release valve, Fig. 17B, is a piston type air operated valve used to deliver and exhaust air being supplied to brake chambers. It is also designed to provide a quick means for applying and releasing the service brakes. The valve is remotely controlled by air pressure from the foot valve.

When the brake valve is applied the relay valve permits direct air pressure, supplied from the air reservoir, to pass and apply the air chambers. When the brakes are released the relay valve also permits the air chamber pressure to exhaust out the valve and into the atmosphere.

The air pressure from the brake valve is also influenced by the action of modulator valve in the anti-lock control.

The relay valve is mounted on the frame close to the brake chambers it serves. When the brake valve is applied, air pressure enters the relay valve at the control port and into cavity "A". Air pressure above piston moves it down and seats the exhaust valve against the modulation tube closing off exhaust passages. Further downward movement of the piston forces modulation tube down and opens inlet valve by moving away from valve seat of body. Reservoir air pressure enters cavity "B" through reservoir port and then passes through delivery ports and to service brake chambers.

Air is blocked from exhausting through modulation tube center. As air pressure builds in cavity "B" it tends to move the piston upward. When the air pressure above and below piston approaches a balanced condition, piston moves upward until inlet valve closes against valve body seat.

If air pressure coming into the control port is held momentarily at braking condition, the relay valve attains a "lap" position. Lap position implies that both inlet and exhaust valves are seated simultaneously. When the valve is in this "lap" position it is instantly responsive to any change in pressure from the brake control valve to increase or decrease

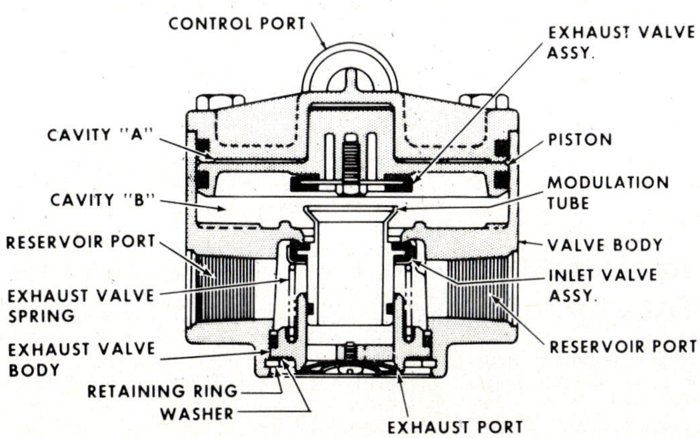

Fig. 17B Relay-quick release valve

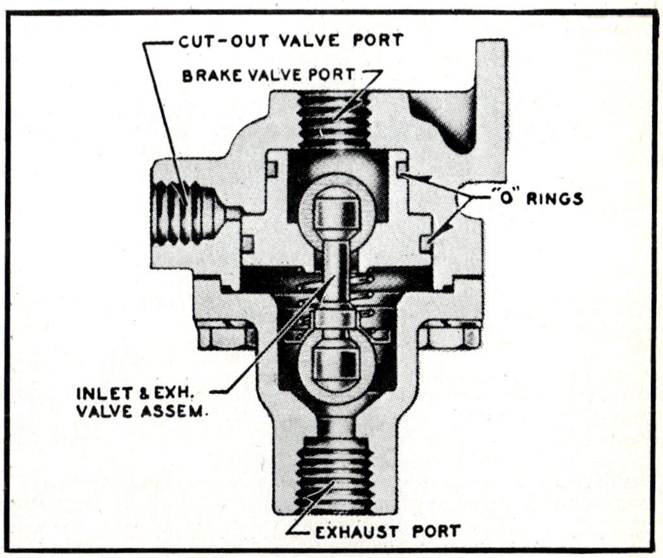

Fig. 18 Limiting quick release valve

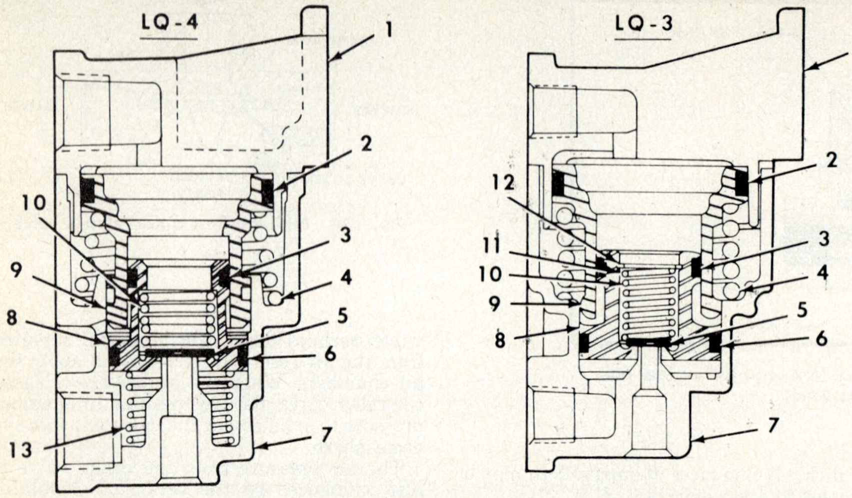

1. Body
2. O-Ring
3. O-Ring, S.D.
4. Outer Piston Spring
5. Valve
6. O-Ring, L.D.
7. Cover
8. Inner Piston
9. Outer Piston
10. Inner Piston Spring
11. Washer
12. Retaining Ring
13. Hold Off Spring (When Used)

Fig. 18A Automatic front brake limiting valves (ratio valves)

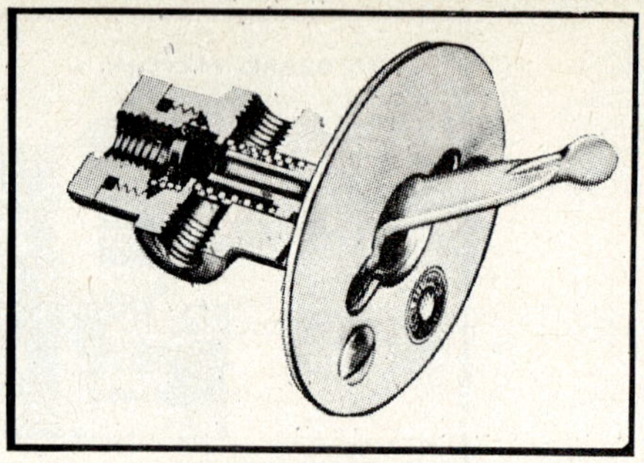

Fig. 19 Two-way valve

vehicle braking. Air pressure outputs of relay valve are directly proportional to the signal air pressure at control port.

When the brake valve is released, air pressure acting above piston is removed. Air pressure in cavity "B" continues to move the piston upward and opens the exhaust valve seat. Air pressure from service brake chambers returns through the delivery ports into cavity "B" and exhausts through modulation tube passage and out exhaust port.

Service

Once a year or every 100,000 miles remove the relay-quick release valve, disassemble, inspect and clean all parts. It is recommended that all rubber parts be replaced and any other parts that show signs of wear. Clean all metal parts in mineral spirits or other suitable solvent.

Limiting Quick Release Valve

A limiting quick release valve, Fig. 18, is used in conjunction with the two-way valve shown in Fig. 19. It is used on air brake systems where driver-controlled front wheels brakes are desired.

The design of the limiting quick release valve is such that it reacts according to the setting of the two-way valve by admitting one-half treadle valve delivery pressure to the front wheel brake chambers when the two-way valve is set for slippery road. Front wheel brakes receive full delivery pressure when the two-way valve is set for dry road.

The limiting quick release valve is usually mounted on a frame front crossmember. A service line from the treadle valve is connected to the inlet port of the two-way valve. Another service line is connected to the inlet port (top) of the limiting quick release valve. The service line from the output port of the two-way valve is connected to the limiting quick release valve at the port facing the rear. The two side ports of the limiting quick release valve are connected to the front brake chambers.

The limiting quick release valve is usually mounted on a frame front crossmember. A service line from the treadle valve is connected to the inlet port of the two-way valve. Another service line is connected to the inlet port (top) of the limiting quick release valve. The service line from the output port of the two-way valve is connected to the limiting quick release valve at the port facing the rear. The two side ports of the limiting quick release valve are connected to the front brake chambers.

The limiting quick release valve, beside providing for one-half reduction of front wheel braking pressures, also serves as the front axle quick release valve.

Service

Once a year, the valve should be disassembled, cleaned and inspected. The gasket, piston O-rings, and inlet and exhaust valves should be replaced. The inlet and exhaust valves are identical and are a snap fit over the ends of the stem.

Slightly pitted bores can be reconditioned by polishing. If bores or valve seats are deeply pitted, service is by replacement of body, cover and piston, as required. Prior to reassembly, lubricate valve with B-W 240176 grease.

AUTOMATIC FRONT BRAKE LIMITING VALVE (RATIO VALVE)

Limiting valve, Fig. 18A, is mounted to a bracket, front crossmember, or side rail near front axle. The valve is used on front axles to limit the application pressure to brake chambers during normal service braking.

The supply and delivery ports on the valve are identified with embossed lettering. The single exhaust port is also identified but contains no threads. Two 5/16 mounting holes are provided for installation.

NOTE: When mounted on the vehicle, the exhaust port of the LQ-3 should be down and pointing to the road surface.

The supply port of the valve is connected to the delivery port of the front service portion of the brake application valve and the delivery port of the valve is connected to the quick release valve.

Pressure delivered to the valve supply port is reduced by 50% as it passes through the valve and out the delivery port. The 50% reduction occurs only when brake application pressures to the supply port of the valve is between 0 psi and 40 psi. Brake applications between 40 psi and 60 psi to the valve supply port are reduced by less than 50% and applications above 60 psi are not reduced at all. An example would be: 65 psi delivered to the supply port results in 65 psi out the valve delivery port.

NOTE: This valve requires the use of a limiting quick release valve when used with type 20 or 24 brake chambers.

Service

Every 3600 operating hours, 100,000 miles, or yearly, disassemble, clean and inspect all parts. If signs of wear are evident, install new parts. Wash all metal parts with mineral spirits or suitable cleaning solvent and dry. Replace all rubber seals.

Prior to assembling the valve lube all the "O"-rings, "O"-ring grooves, piston bores and any metal moving surfaces with a light film of silicone lubricant.

TWO-WAY VALVE

The two-way valve, Fig. 19, is an "OFF" (released) and "ON" (applied) valve used to manually control other air devices as the limiting quick release valve and tractor protection valve.

Either of two types of two-way valves are used to control the limiting quick release valve. The lever type, Fig. 19, is mounted on the dash, whereas the cable-operated type is mounted on the left frame rail and remotely controlled from the cab.

The valve shown in Fig. 19 is also used with the tractor protection valve. On some applications, it is used in place of the control valve shown in Fig. 21, and provides manual control only.

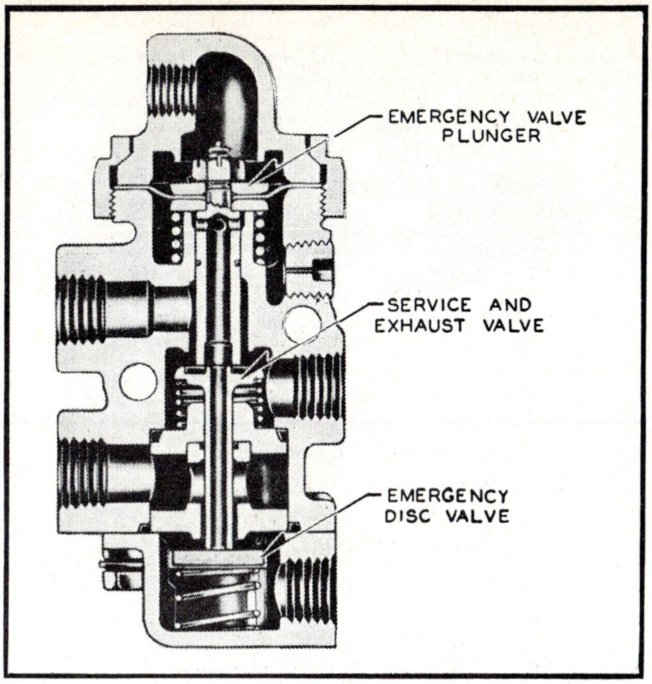

Fig. 20 Tractor protection valve

The operation of the two-way valve in conjunction with the limiting quick release valve is explained above. When the two-way valve is used with the tractor protection valve, the released ("OFF") position is for "emergency," and the applied ("ON") position is for "normal."

Service

At least once each year the valve should be disassembled, cleaned and inspected. Replace O-rings and inlet valve. Lubricate valve with B-W 240176 grease.

After reassembling valve, test for leakage in both applied and released positions by coating the exhaust opening with soap suds. Leakage should not exceed one-inch soap bubble in five seconds.

TRAILER AIR SUPPLY VALVE

The trailer supply valve is basically two non-modulating valves in a single knob control assembly. This valve provides for an in cab control of the tractor protection valve, and functions in conjunction with the tractor protection valve. The trailer air supply valve is panel mounted, and has a push-pull type operation. The button to operate the valve must be manually depressed, but will automatically come out and bleed off air pressure if air supply is below 40 psi. A 50 psi or greater supply air pressure is required to keep button in operating position, and it will remain in operating position until supply air pressure falls below 40 psi. In order to disconnect tractor from trailer, button is pulled out manually. Pulling button closes the plunger inlet and unseats exhaust valve. Air which was flowing out delivery port is routed back to the valve and is bled off through vent in mid-section of valve. In the event of trailer breakaway or a malfunction of the trailer supply line, button on valve will come out and retain approximately 40–65 psi. air pressure in tractor air system.

Actual air pressure retained is dependent upon severity of the malfunction, and the vehicle's air system.

Service

The valve should be checked every 300 operating hours or 10,000 miles whichever occurs first. Check valve for proper operation and air leakage. Check supply lines for evidence of wear and corrosion. The valve should be overhauled every 7,200 operating hours or 200,000 miles whichever occurs first. Disassemble, clean and inspect internal components. Install new parts where they are determined to be worn or damaged. Replace all "O" rings and seals.

Cleaning & Inspection

Clean all metal parts using a suitable solvent, and blow dry using compressed air. Inspect valve body, internal bores and external surfaces of plunger for evidence of wear and corrosion. Slight defects may be touched up with crocus cloth. Lubricate parts using lube No. 5459912 or equivalent. Replace all components not considered serviceable as required. Replace all "O" rings and seals.

Testing Valve

To perform the following test, an accurate vehicle test gauge installed in the spare hose coupling is required. The vehicle's dash gauge should be checked for accuracy prior to performing the testing procedure described below.

1. Chock the vehicle's wheels.

NOTE: Do not use the vehicle's service air brakes.

---

2. With trailer supply control valve button out and system parking control valve in bleed off position, or the brakes in applied position, increase air system pressure to governor cut-out pressure.
3. Apply a soap soultion to exhaust vent or bleed off vent of valve, and check for bubbles which will represent a leak.
4. Depress control button on valve and repeat step 3.
5. Install test gauge and coupling on trailer supply coupling at rear of tractor. Position system parking control valve into brakes released position. An immediate rise in pressure should be noticed on test gauge. The pressure registered on test gauge should equal dash gauge or system pressure.
6. Apply soap solution to exhaust port of valve and check for leakage.
7. With the tractor air system at governor cut-out pressure, shut off engine and disconnect test gauge and hose coupling from trailer supply line coupling. The button on valve will come out and leakage from trailer supply line will stop at approximately 40–65 psi.
8. Reconnect test gauge to trailer supply line and depress button on valve. Open drain cocks on service reservoirs slightly, then note what pressure at which valve comes out and exhausts tractor-trailer supply line. This should occur at approximately 20–45 psi air system pressure. If for any reason trailer supply valve fails to function as described in steps 2 through 8, replace valve.

Valve, Replace

Conventional Cab Models

1. Chock vehicle wheels.
2. Drain all air from reservoirs.
3. Drive and remove roll pin from valve knob.
4. For aluminum conventional cab models, remove six plate bolts and plate.
5. For steel conventional cab models, remove bolts retaining panel to instrument panel, then remove panel.
6. Disconnect air pressure lines at valve.
7. Thread mounting nut from valve, and remove valve.
8. Reverse procedure to install valve. Check parking brake application and release.

Tilt Cab Models

1. Chock vehicle wheels.
2. Drain all air from reservoirs.
3. Remove center access panel from behind the instrument panel, or remove valve and gauge panel.
4. Disconnect air pressure line from valve.
5. Drive and remove roll pin from valve knob.
6. Thjread mounting nut from valve, and remove valve.
7. Reverse procedure to install valve.

TRACTOR PROTECTION VALVE

This valve provides a means of preserving air pressure in sufficient amount to stop the tractor in the event of a trailer breakaway. The system also provides for an automatic trailer brake application if an air loss in the tractor or trailer system is undetected by the operator. (The operator is normally forewarned by the low pressure indicator before this happens).

The tractor protection valve, Fig. 20, in effect, is a set of remotely controlled cut-out cocks with both manual and automatic controls. The valve is manually controlled by a push-pull control valve, Fig. 21, which is connected to the tractor reservoir. When the control valve is set in the "charge" position, air pressure is admitted to the control port of the tractor protection valve, thereby opening the service and emergency lines to the trailer.

When the control valve is set in the "emergency" position, air pressure in the emergency lines between the control valve and tractor protection valve, and also between the tractor protection valve and the trailer relay emergency valve is exhausted. The emergency features of the trailer relay emergency valve then cause a trailer brake application.

The control valve should be set in the "emergency" position when disconnecting a trailer and in the "charge" position immediately after connecting one.

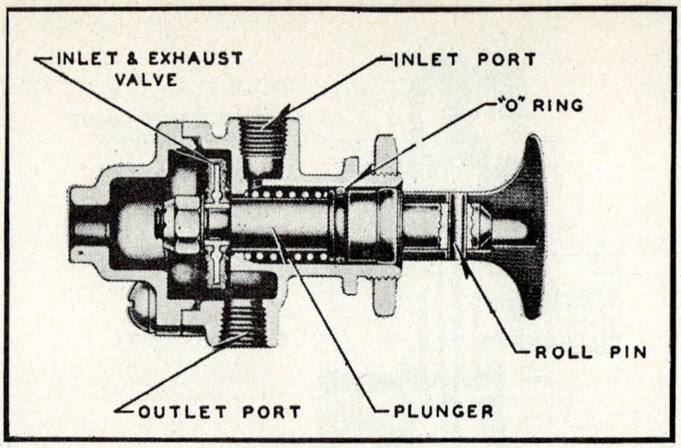

Fig. 21 Control valve

Service

At least once a year the tractor protection valve should be disassembled and cleaned. Rubber parts and others subject to wear should be replaced as required. Prior to reassembly coat all internal parts (except springs and diaphragm) with B-W 240176 grease.

1. Position O-ring in valve body bore.
2. Install disc valve seat as an assembly with service and exhaust valve guide, service and exhaust valve spring, and the small O-ring.
3. Install O-ring in counterbore at bottom of valve body.
4. Install emergency disc valve spring in cover.
5. Position disc valve on its spring with rubber side away from spring.
6. Install cover.
7. Position exhaust plunger spring in valve body.
8. Install exhaust valve plunger assembly in body after installing followers, diaphragm, nut, and plunger stem O-ring.
9. Position control cap on body with control port facing same as exhaust port. Install and tighten cap nut securely to prevent leakage.

Testing Valve—With Two-Way Valve

1. Block vehicle wheels to prevent vehicle from moving.
2. Drain brake system.
3. Close drain cocks.
4. Set two-way valve in "normal" position.
5. Connect a test gauge in tractor protection valve service port or trailer service line coupling.
6. Start engine to build up air pressure in system.
7. When system pressure reaches 30–40 psi, make and hold a brake application. Note pressure on test gauge. A full application should show full or nearly full reservoir pressure. A partial application should be proportional to it.
8. Release brakes. No air should be passing through emergency line until system pressure reaches 45–55 psi.
9. Set two-way valve in "emergency" position and remove test gauge from service line and connect it into emergency line.
10. Set two-way valve in "normal" position. Continue to build up reservoir pressure until it reaches maximum. Stop engine.
11. Set two-way valve in "emergency" position. Test gauge should drop to zero.
12. Remove test gauge. There should be no leakage at the emergency line. Make and hold a brake application. There should be no leakage at the service line. Release brake.
13. Reconnect test gauge to emergency line and set control valve in "charge" position. Test gauge should register reservoir pressure.
14. Turn on ignition key and open reservoir drain cock. Low pressure buzzer should be activated at about 60 psi. Tractor protection valve should vent the emergency line at 30–40 psi which would actuate trailer relay emergency valve if trailer was connected.
15. Disconnect test gauge. Make and hold a brake application. There should be no noticeable leakage at either the service or emergency ports of the tractor protection valve below 25 psi system pressure.

Bendix Model TP-2

Disassembly, Fig. 21A

NOTE: Mark position of control port and tractor emergency port location relative to the valve body in such a way that mark will not be removed during cleaning of parts.

---

1. Remove screws and washers from cover, then remove cover from body.
2. Remove emergency valve and spring from cover.
3. Remove service and exhaust valve, valve seat, retainer and spring.
4. Remove "O" rings, then the control cap nut and control cap.
5. Remove exhaust valve plunger, diaphragm and spring from body.

Cleaning & Inspection

Clean all metal parts using a suitable solvent and blow dry with compressed air. Inspect all moving parts for wear or damage and bores for an out of round condition. Inspect diaphragm and rubber valves for wear or deterioration and springs for damage or corrosion. Check that valve seats are smooth with no nicks or burrs. Replace all rubber components and any other parts which are not serviceable.

Assembly, Fig. 21A

1. Lubricate all "O" rings, bores, mating surfaces and cap nut threads with a suitable silicone lubricant.
2. Secure valve in a soft jawed vise with control cap end up.
3. Install "O" ring in groove of stem of exhaust valve plunger and diaphragm assembly.

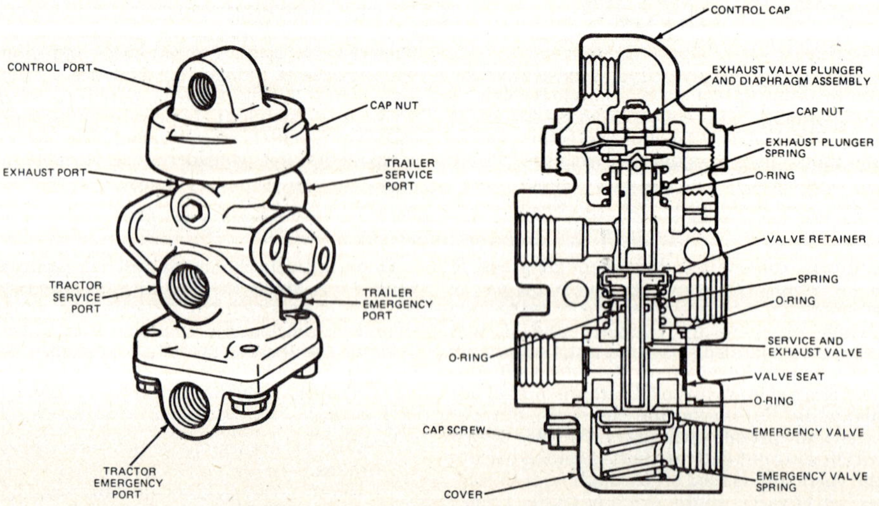

Fig. 21A Bendix TP-2 tractor protection valve

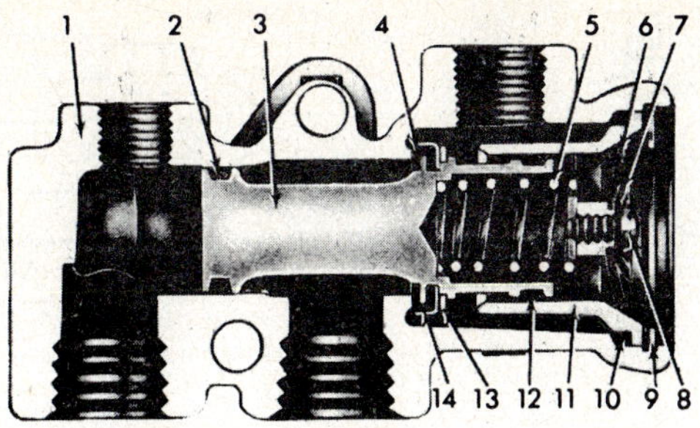

1. Valve Body
2. Plunger "O" Ring
3. Valve Plunger
4. Valve
5. Inlet Valve Preload Spring
6. Diaphragm
7. Diaphragm Washer
8. Screw
9. Retainer Ring
10. Check Valve "O" Ring
11. Exhaust Check Valve Seat
12. Plunger "O" Ring
13. Retainer Ring
14. Valve Retainer

Fig. 21B Bendix TP-3 tractor protection valve

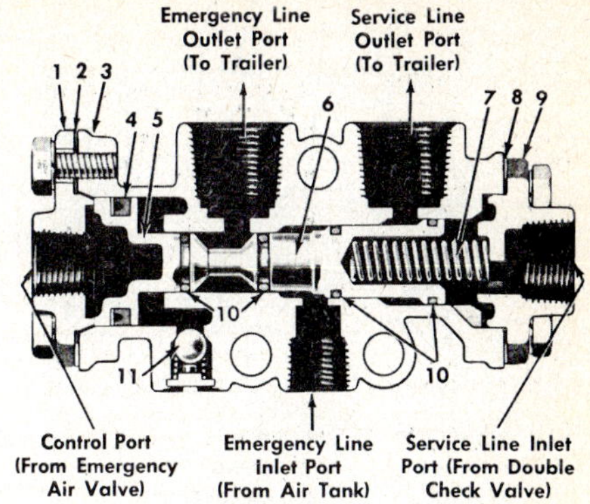

Fig. 21C Midland-Ross N-20592 tractor protection valve

4. Install exhaust plunger spring, exhaust plunger and diaphragm assembly. Plunger should have a snug sliding fit in bore.

NOTE: Both surfaces of diaphragm and mating surfaces of body and control cap must be free of lubricant.

---

5. Install control cap by aligning marks made during disassembly, then insert a 1/4 inch pipe 4–6 inches long in control port of a cap. Firmly apply pressure to cap with pipe and tighten cap nut to 50 ft. lbs., then remove pipe from cap. The cap must not turn while tightening the nut.
6. Secure valve in a soft jawed vise with control cap end down.
7. Install small "O" ring in valve body.
8. Install "O" ring in small bore of valve seat, then position heavier of two remaining springs over protrusion on small bore end of valve seat.
9. Install valve retainer on stem of service and exhaust valve, then slide valve stem through center of spring and into small bore of valve seat.
10. Install valve, seat and spring assembly in valve body and hold in place.
11. Install "O" ring in body, then position emergency valve spring on cover and place emergency valve on top of spring. Metal side of valve must be in contact with spring.
12. Install cover, aligning reference marks made during disassembly, and check that "O" ring and emergency valve are properly positioned. Torque screws to 100 inch lbs.

Bendix Model TP-3

Disassembly, Fig. 21B

1. Remove snap ring, washer and all valve components from plunger body.
2. Remove screw, washer and diaphragm from valve housing.
3. Remove exhaust check valve seat, inlet valve spring and valve plunger from valve housing.
4. Remove "O" ring from check valve seat and "O" rings from plunger.
5. Remove retainer ring, retainer and valve from plunger.

Cleaning & Inspection

Clean all metal parts using a suitable solvent, and blow dry using compressed air. Inspect valve body, internal bores and external surfaces of plunger for evidence of wear, scoring or corrosion. Slight defects may be touched up with crocus cloth. Lubricate parts using lube No. 5459912 or equivalent. Replace all components not considered serviceable as required. Replace all "O" rings and seals.

Assembly, Fig. 21B

1. Install valve, valve retainer and retainer ring on plunger.
2. Install "O" rings on plunger and "O" ring onto check valve seat.
3. Assemble inlet valve preload spring, exhaust check valve seat and valve plunger into valve housing.
4. Install diaphragm, washer and screw into valve housing.
5. Insert valve parts into valve body.
6. Position washer on exhaust check valve seat and install snap ring.

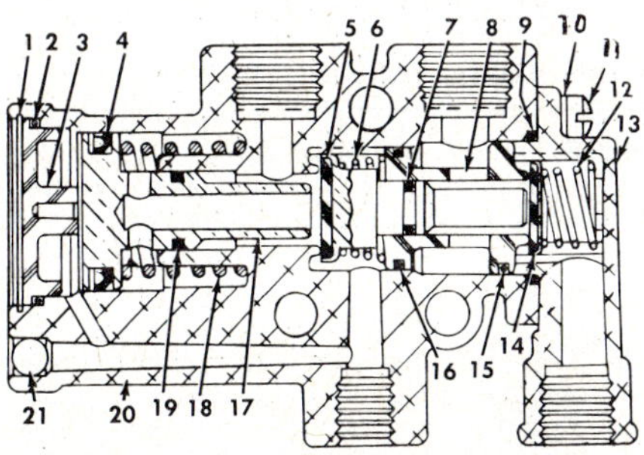

1 Snap Ring	8 Spool	15 O-ring
2 O-ring	9 O-ring	16 O-ring
3 Piston Stop	10 Lock Washer	17 Piston
4 Block Vee	11 Screw	18 Spring
5 Exhaust Valve Assy.	12 Spring	19 O-ring
6 Spring	13 End Cap	20 Body
7 O-ring	14 Valve	21 Ball

Fig. 21D Midland-Ross N-30073 tractor protection valve

Midland-Ross Model N-20592

Disassembly, Fig. 21C

1. Remove screws and washers from control end cover, then separate end cover from body.
2. Remove control plunger from bore, then remove and discard all seal cups and "O" rings.
3. Remove screws and washers from service line end cover, then separate end cover from body.
4. Remove plunger spring and valve plunger from bore, and discard "O" rings.
5. Remove exhaust check end cap, spring and ball.

Cleaning & Inspection

Clean all metal parts using a suitable solvent, and blow dry using compressed air. Inspect valve body, plunger surfaces and bores for cracks or other signs of damage. Slight defects may be touched up with crocus cloth. Lubricate parts using lube No. 5459912 or equivalent. Replace all components not considered serviceable as required.

Assembly, Fig. 21C

1. Carefully install new seal cup and "O" ring seals into grooves of control plunger. Lip of seal cup should point in a direction away form the "O" rings. Insert plunger small end first into the control end of valve.
2. Position new gasket over pilot and onto control end cover. Install cover pilot into control end of valve, then install screws and washers onto end cover and body.
3. Install new "O" ring seals into grooves of the valve plunger. Insert plunger, small end first into the service end of the valve, and install plunger spring.
4. Position new gasket over pilot onto service line end cover, and install cover pilot into service end of valve. Install screws and washers into end cover and body.
5. Insert exhaust check valve and spring into exhaust port. Install and positive stake end caps.

Midland-Ross Model N-30073

Disassembly, Fig. 21D

1. Remove snap ring, plastic piston stop and "O" ring from valve body.
2. Remove piston assembly from valve body, then block V and remove "O" ring from piston.
3. Remove piston return spring, then re-

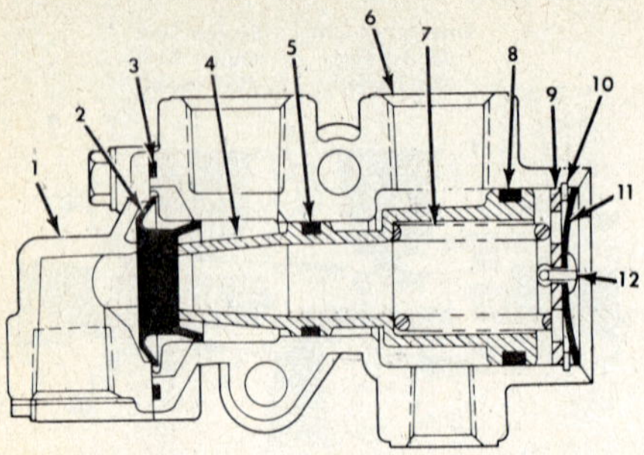

1. END CAP
2. VALVE
3. O-RING
4. PISTON
5. O-RING
6. BODY
7. SPRING
8. O-RING
9. EXHAUST DISC
10. SNAP RING
11. EXHAUST DIAPHRAGM
12. RIVET

Fig. 21E Midland-Ross N-20949 tractor protection valve

move screws and washers from front of valve.

4. Remove end cap from the valve, then remove "O" ring from end cap.
5. Remove inlet valve disc and inlet valve spring from end cap.
6. Using a drift and hammer, tap on exhaust valve, and remove plastic valve spool from exhaust valve.
7. Remove "O" rings from the plastic spool.
8. Remove exhaust valve spring and exhaust valve from housing.

Cleaning & Inspection

Clean all metal parts using a suitable solvent, and blow dry using compressed air. Inspect valve body, end cap, internal bores and external surfaces for evidence of wear and deterioration. Slight defects may be touched up with crocus cloth. Lubricate parts using lube No. 5459912 or equivalent. Replace all components not considered serviceable as required. Replace all "O" rings and seals.

Assembly, Fig. 21D

1. Install exhaust valve assembly and exhaust valve spring into valve body.
2. Install "O" rings onto plastic valve spool, and install spool into valve body.
3. Install inlet valve spring, disc and "O" rings onto end cap.
4. Place end cap assembly into position on valve body, then secure with screws and washers.
5. Install piston return spring onto valve body.
6. Install a new block V and "O" ring onto piston. Lip on V should point in a direction away from the "O" ring, then install piston assembly into valve body.
7. Install "O" ring, plastic piston stop and snap ring into valve body.

Midland-Ross Model N-20949AA

Disassembly, Fig. 21E

1. Remove snap ring, spring and disc from valve body.
2. Remove end cap screws, end cap, "O" ring and valve diaphragm from body.
3. Remove piston from valve body, then remove "O" rings from piston.

Cleaning & Inspection

Clean all metal parts using a suitable solvent, and blow dry using compressed air. Inspect valve body, internal bores and external surfaces for evidence of wear and deterioration. Slight defects may be touched up with crocus cloth. Lubricate parts using lube No. 5459912 or equivalent. Replace all components not considered serviceable as required. Replace all "O" rings and seals.

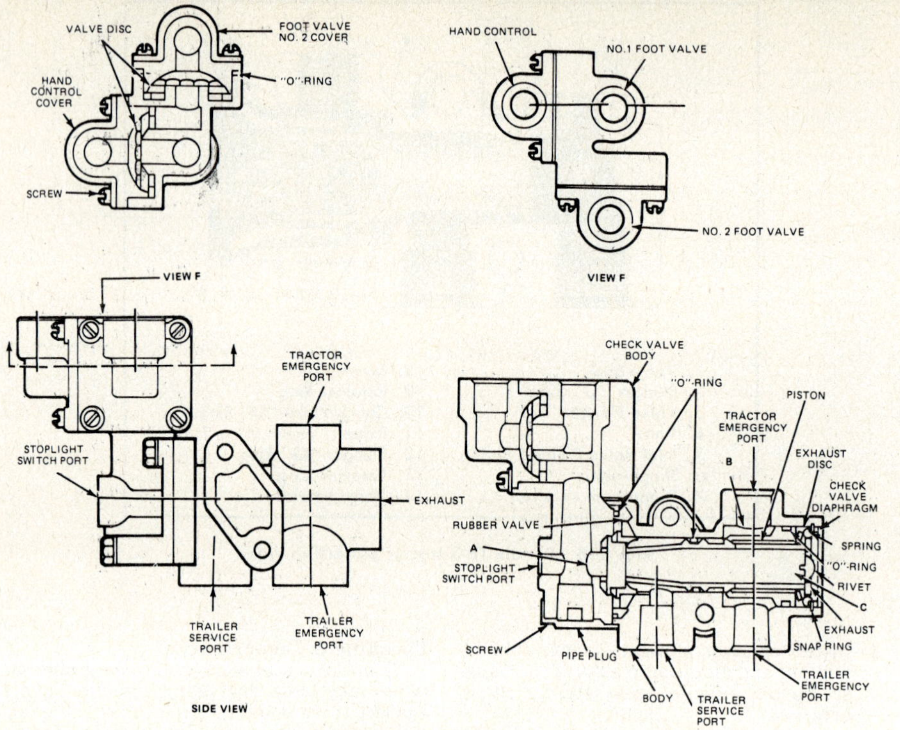

Fig. 21F Midland-Ross N-30162 tractor protection valve

Assembly

1. Install valve diaphragm so diaphragm lays flat on flutes of valve body, then install "O" ring into groove on flange body.
2. Install screws and end cap, torque bolts to 90 ft. lbs.
3. Install "O" rings onto piston, then install piston into valve body.
4. Install spring, exhaust disc, and snap ring onto valve body.

Midland-Ross Model N-30162

Disassembly, Fig. 21F

NOTE: Before beginning disassembly, check valve for damage. If casting is broken or fractured, the valve must be replaced.

1. Remove snap ring, exhaust disc and spring.
2. Remove eight retaining screws, covers and "O" rings.
3. Remove four remaining screws, check valve body, "O" ring and valve from body.
4. Remove piston from exhaust side of body, being careful not to cut "O" ring on snap ring.
5. Remove "O" rings from piston.

Cleaning & Inspection

Clean all parts using a suitable solvent, and blow dry with compressed air. Inspect covers, check valve body, piston, body, valve seat and snap ring groove for any wear or damage. Any damaged components must be replaced. The rubber valve, valve discs and "O" rings may be reused if not damaged.

Assembly, Fig. 21F

1. Lubricate all seals and sliding surfaces with Dow Corning molykote U-paste lubricant, or equivalent.
2. Install valve in body so valve lays flat on flutes of body.
3. Install "O" ring in body, then assemble check valve body to valve body using four screws.
4. Install "O" rings in valve and valve discs onto valve seats in check valve body.
5. Align valve cover and check valve body cover and assemble with eight screws.
6. Install "O" rings on piston and slide piston into body.
7. Install piston and exhaust disc, then install snap ring while pressing exhaust disc into valve body.

TRACTOR PROTECTION VALVE MANIFOLD

The tractor protection valve manifold is mounted on a bracket which is secured to the cab cowl. The valve incorporates integral double check valves and a quick release valve in one body. This valve works in conjunction with trailer supply valve, and receives operational signals from hand control valve, front brake system and rear brake system.

Valve, Replace

1. Check vehicles wheels.
2. Release all air pressure from system.
3. Remove power steering reservoir if so equipped, and other components, as nec-

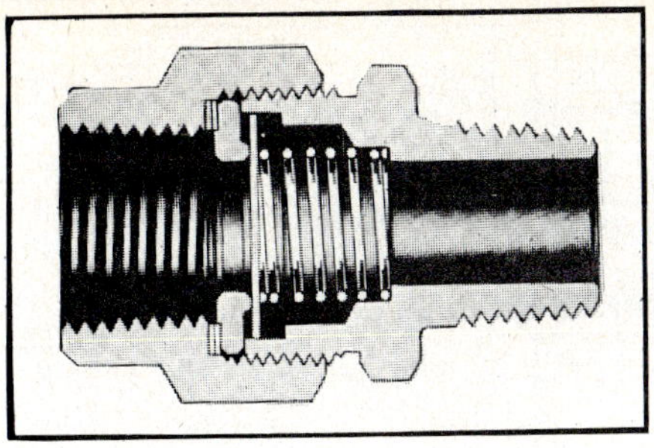

Fig. 22 Single check valve (B-W)

essary, to gain access to valve and air line connections.
4. Remove air lines at valve.
5. Remove bolts retaining valve mounting bracket to cowl.
6. Remove bolts and valve from bracket.
7. Reverse procedure to install valve.

CONTROL VALVE

The push-pull type tractor protection control, Fig. 21, is basically an "on" and "off" valve, but differs from the two-way valve in that it also has an automatic pressure release.

This particular control valve releases automatically at approximately 20 psi. The valve has three pipe ports—one supply and two delivery (only one delivery port is used). The unthreaded port at the bottom is the exhaust.

The shank of the control button has a red warning ring that is visible when the valve is set in the "emergency" position.

Operation

A line from the reservoir is connected to the supply port. When the system pressure reaches about 50 pounds, the operator pushes the control button in (charge position). Reservoir air pressure then flows through the valve and out the delivery port. The button will automatically stay in after 50 pounds of pressure has been built up.

To apply the valve the operator pulls the button out. The inlet seat seals off the supply of air and the exhaust opens, exhausting the air in the line between the push-pull control valve and tractor protection valve.

If a drop in system pressure takes place that is unnoticed by the operator, and the tractor

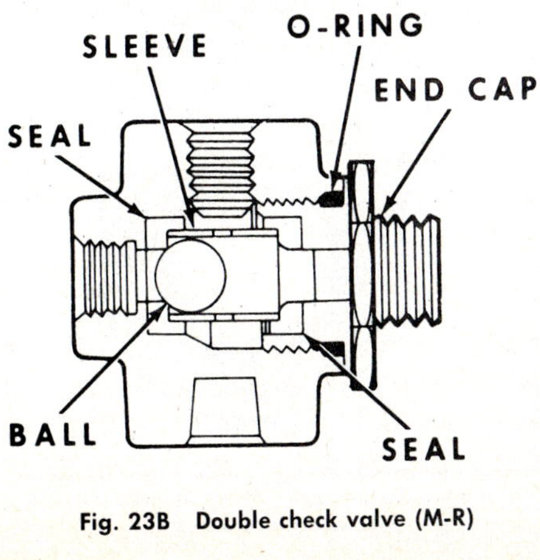

Fig. 23B Double check valve (M-R)

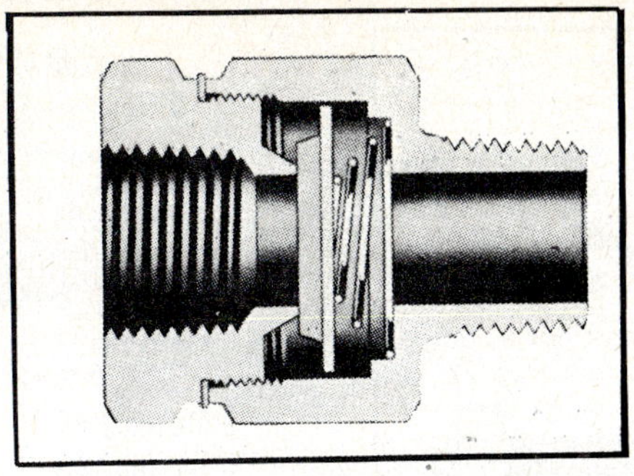

Fig. 23 Single check valve (Midland)

protection valve fails to vent, the control valve will automatically exhaust when system pressure drops to approximately 20 psi. The tractor protection valve will then seal off the service and emergency lines to the trailer.

Service

Every six months the control valve should be removed, disassembled, cleaned and inspected. Worn parts should be replaced as required.

To remove the valve from the cab, first block vehicle wheels and bleed off brake system. Drive the roll pin out of the valve button and plunger. Remove button and mounting nut. Remove valve from cab.

Disassemble valve by removing two screws and separate cover and body. Remove plunger nut, washer and inlet and exhaust valve. Slide plunger out of valve body. Remove O-ring from plunger and spring from valve body.

When reassembling the valve, tighten plunger nut to 30–40 inch-lbs torque. Lubricate O-ring and bearing surfaces with Barium base grease (B-W 240176).

Testing Valve

With system pressure at zero, build up the pressure and hold the control valve plunger in. The control valve plunger should remain in when system pressure is above 50 psi. With the system pressure built up, pull the control valve plunger out and note that the delivery air is exhausted through the exhaust port of the control valve. Then push the button in and note that air is again delivered through the valve.

With the control valve plunger out and the red indicator on the button showing, check the exhaust port for leakage past the inlet valve. With system pressure above approximately 70 psi and the control button in (colored indicator not visible) check the exhaust port for leakage by the exhaust valve. Check around the plunger for leakage by the plunger grommet. Also check for gasket leakage between cover and body.

CHECK VALVES

Single Check Valve

The single check, or one-way valve, Figs. 22 and 23, is installed in the line between the two compartments of the reservoir tank and permits flow of air in one direction only.

The check valve is opened by air passing from the wet tank compartment to the dry tank compartment. If pressure in the wet tank falls below that of the dry tank, the check valve remains closed, thereby preserving air pressure in the dry tank.

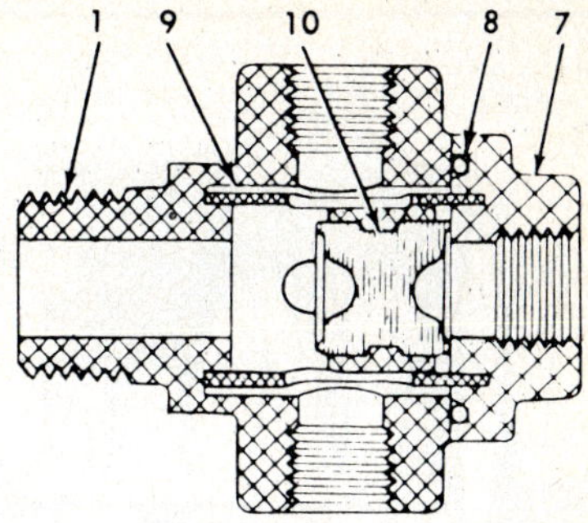

7 Cap
8 O-ring
9 Guide
10 Shuttle Valve
11 Gasket

Fig. 23A Double check valve (B-W)

Periodically the valve should be removed, disassembled, cleaned and inspected. Worn parts should be replaced as necessary. Do not lubricate valve.

Test the valve by applying air pressure in the direction opposite normal flow and check for leakage. Valve should hold a pressure of 200 psi.

Double Check Valve

The double check valve, Figs. 23A & 23B, is used to direct air from two sources into a common line. A movable shuttle on the inside, seals off the inlet of one of the supply lines, depending on air pressure against it from the opposite end. Air pressure is never cut off from the outlet port.

A double check valve is used to tie the front and rear air systems together, the spring brakes and hand control valves. The valve is

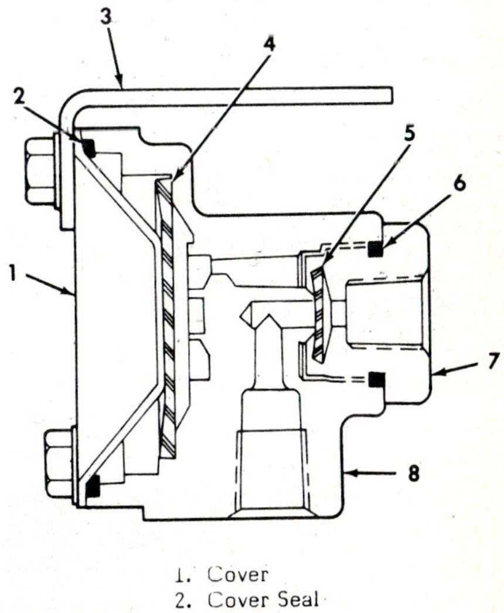

1. Cover
2. Cover Seal
3. Bracket
4. Diaphragm
5. Valve Disc
6. O-Ring
7. Valve Seat
8. Body

Fig. 23C M-R N-20952 double check valve and quick release combination valve

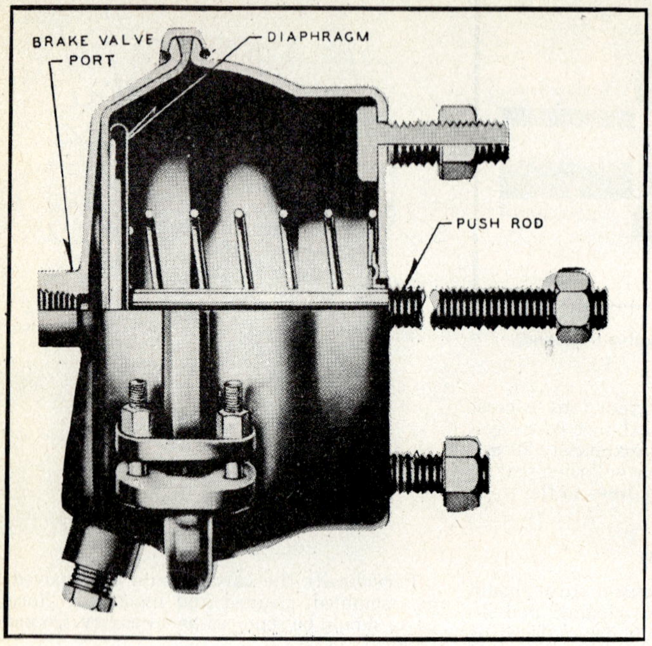

Fig. 24 Diaphragm type brake chamber

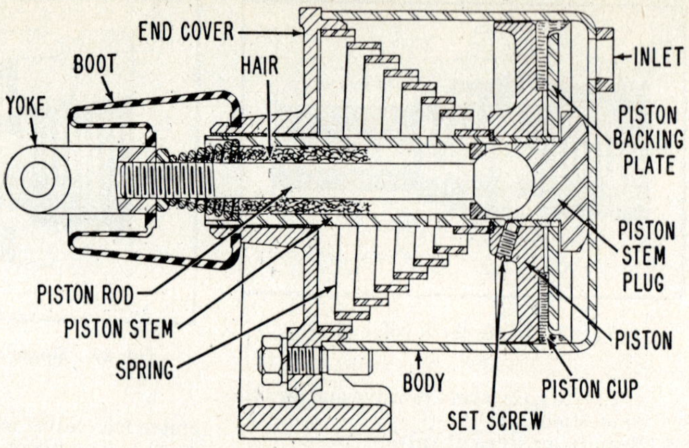

Fig. 25 Piston type brake chamber

also used in the trailer service line on some models.

Double check valve should be removed, disassembled, and cleaned at regular intervals. Inspect for cracks, wear, or deterioration.

Clean all metal parts in mineral spirits or other suitable solvent, and replace all seals. Prior to assembly, lubricate guide sleeve and "O"-ring lightly.

M-R N-20952, Double Check Valve and Quick Release Combination Valve

This valve, Fig. 23C, is used on a vehicle with trailer brake controls. It is plumbed into the parking brake circuit and is mounted near the rear axle. The purpose of the valve is to prevent brake pressure compounding and be a rapid air release location for the spring brake chambers.

With park brake valve applied, air pressure enters the supply (control) port and seals the check valve diaphragm against its seat and air flows to the spring brake chambers and cages the springs. If a service brake application is made with the park brakes set (springs released) air pressure enters the balance port, unseats the check valve diaphragm, closing the control port inlet, and flows to the spring brake chambers, partially caging the power spring. This function prevents brake compounding.

Valve should be removed, disassembled, and cleaned at regular intervals. Inspect parts for wear or deterioration.

Clean all metal parts in mineral spirits or other suitable solvent, replace all seals.

LOW PRESSURE INDICATOR

The low pressure indicator consists of a pressure switch and an audible buzzer. It is a safety device designed to give the operator an automatic warning when the reservoir air pressure is (or when it drops) below a safe limit for continuous braking operation.

The switch is electrically connected in series with the buzzer and ignition switch. The electrical circuit is closed until there is at least 60 psi air pressure in the reservoirs. The device is inoperative when the ignition switch is off.

Service

If the indicator fails to operate, by-pass the switch with a jumper wire to determine whether the switch or buzzer is faulty. Service is by replacement only as adjustments to either are not recommended.

BRAKE CHAMBERS

A typical diaphragm type brake chamber is shown in Fig. 24. A piston type, Fig. 25, and a Rotochamber, Fig. 26, are also used. They all convert the energy of compressed air into the mechanical force necessary to expand the brake shoes against the brake drums.

Air pressure delivered to the brake chambers by the brake valve moves the diaphragm (or piston) and push rod outward, applying the

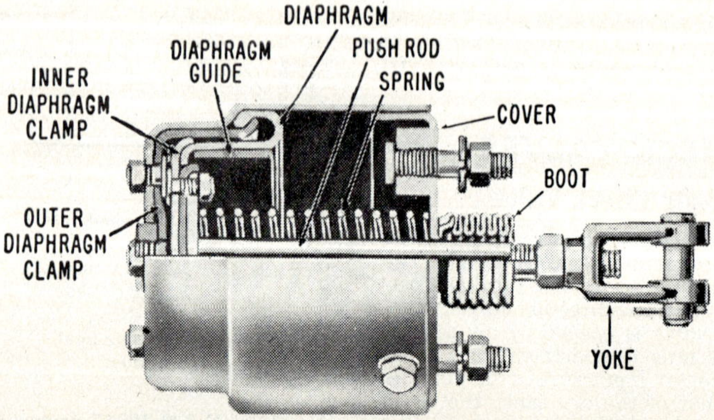

Fig. 26 Roto brake chamber

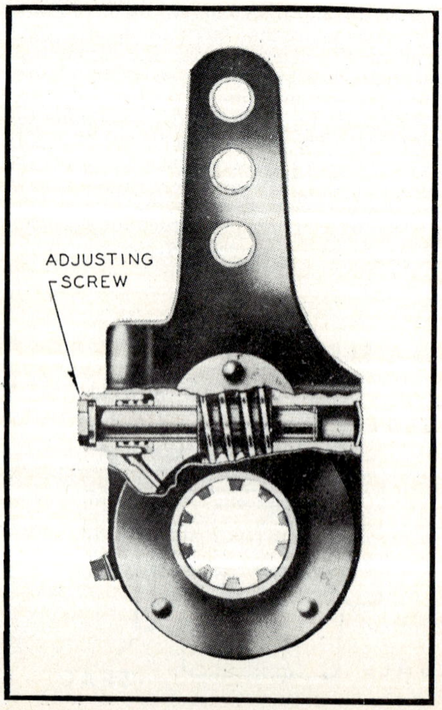

Fig. 27 Slack adjuster

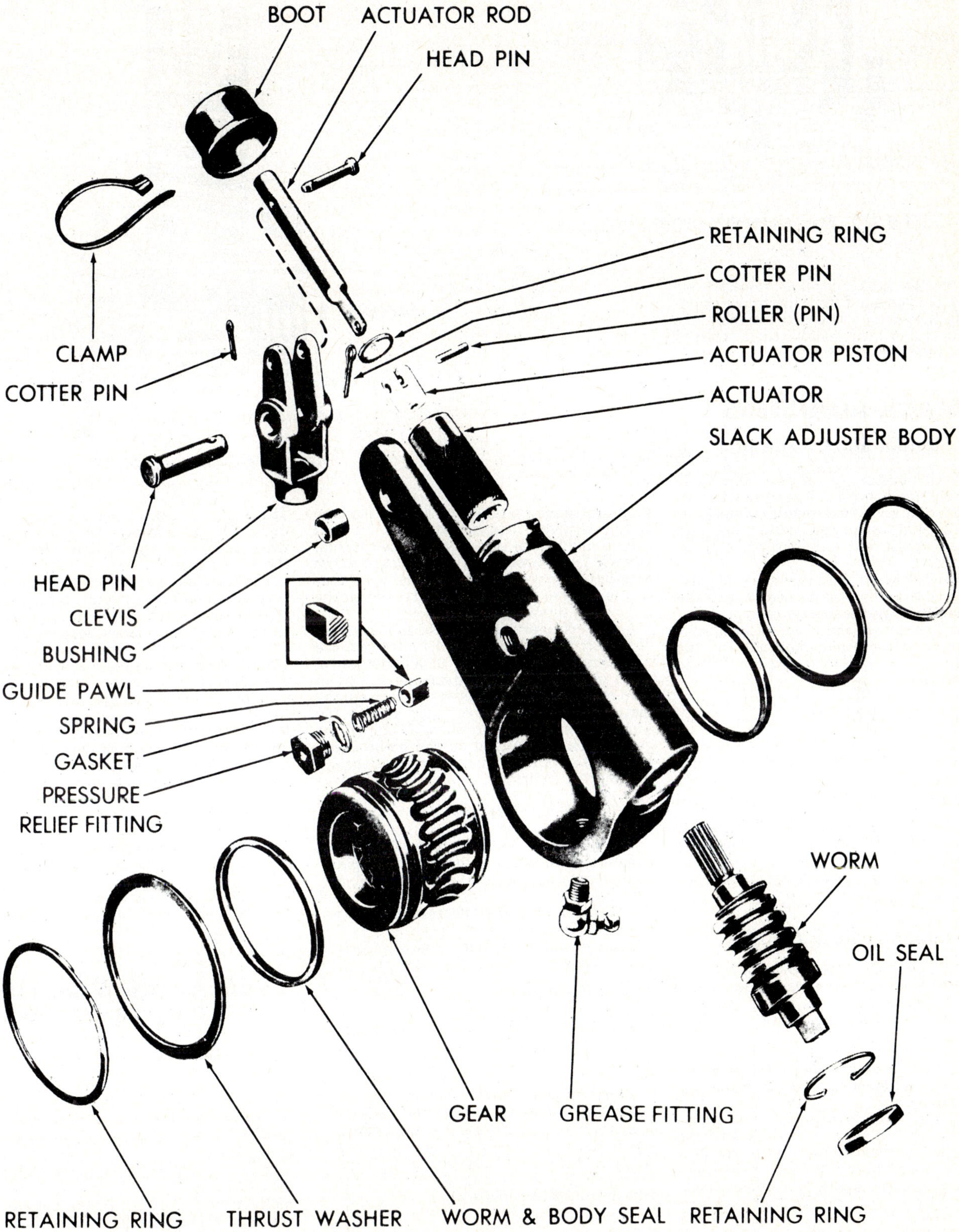

Fig. 27A Automatic slack adjuster

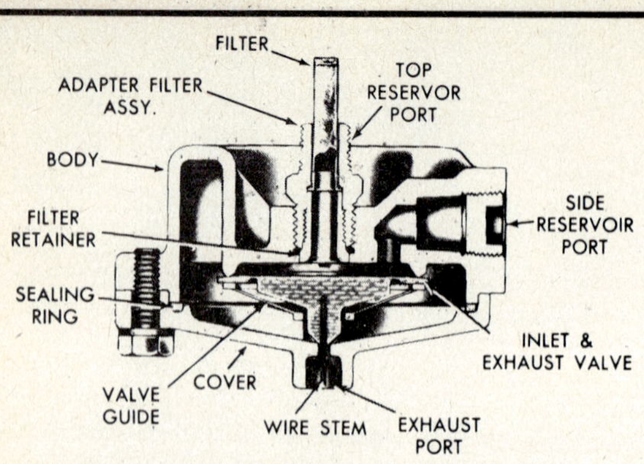

Fig. 27B Reservoir moisture ejection valve

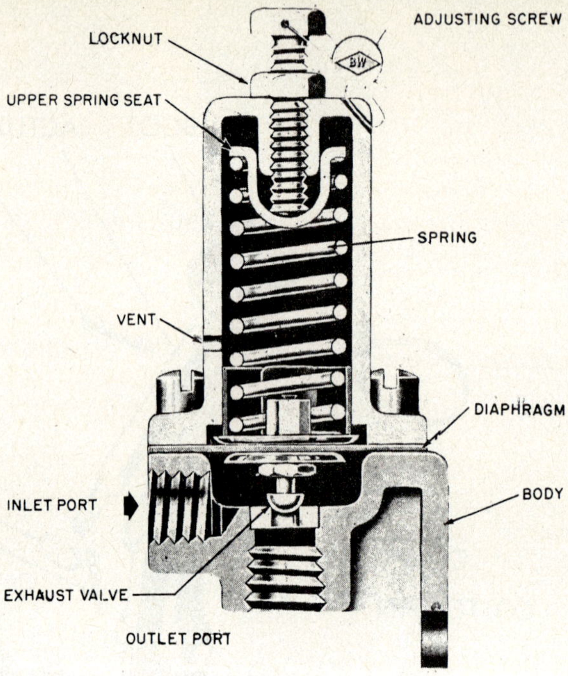

Fig. 27C Pressure regulating valve

brake. As the brake pedal is released, air pressure is exhausted from the brake chambers and the springs return the diaphragms (or pistons) and push rods to their released position, releasing the brakes.

SLACK ADJUSTERS

Manual Slack Adjuster

The slack adjuster, Fig. 27, is an adjustable lever, including a worm and gear, so arranged that it not only serves as a regular brake lever during normal brake application but also provides a quick and easy method of brake adjustment.

In normal braking, the entire slack adjuster operates as a unit, rotating bodily with the brake camshaft as the brakes are applied or released. The splined end of the slack adjuster is mounted on the brake camshaft and the other end is connected to the push rod of the brake chamber.

Service

The slack adjuster requires no attention other than periodic inspection of the bronze bushing fitted in the hole in the slack adjuster arm. If worn excessively, replace the bushing.

Sufficient lubrication is provided when initially assembled and, therefore, lubrication of the slack adjuster should be confined only to such a time as is necessary to service or overhaul the entire unit.

Automatic Slack Adjuster

The automatic slack adjuster, Fig. 27A, is designed to maintain proper cam brake chamber stroke adjustment and lining to drum clearance during the normal service life of the brake lining.

The brake chamber push rod is equipped with clevis which has two sets of holes permitting the connection of the slack adjuster arm and actuator.

When the brakes are applied the brake chamber push rod moves the slack adjuster arm, which in turn rotates the brake cam shaft thus applying the brakes. At the same time the brakes are applied the brake chamber push rod clevis lifts the actuator rod which is connected to the actuator piston by a roll pin. The actuator piston moves up and down each time the brakes are applied and released.

As the brake lining wears more travel of brake chamber push rod stroke is required and the actuator piston will also travel further, contracting a retaining ring in the bore of the actuator. The actuator has internal splines and helical type teeth on the external surface. The adjusting pawl is indexed to the external teeth on the actuator and tooth contact is assured through spring tension of the hollow cap screw and pawl spring.

As the actuator is lifted through movement of the actuator piston and actuator rod, the adjusting pawl will index in another set of teeth on the actuator, advancing it and in turn rotating the worm assembly through the mating splines in the actuator and on the worm assembly. As the worm advances, the gear which is mounted on the brake group cam shaft advances, thus maintaining the lining to drum clearance.

Service

Every 24,000 miles check brake lining wear to determine if brake lining requires replacement. Check for correct chamber stroke, also the condition of boot seal at actuator rod for cuts and tears, replace if necessary. Lubricate slack adjusters.

Every 48,000 miles remove pawl capscrew, pawl spring and pawl assembly. Examine pawl for grease retention. If grease is in good condition, perform normal lubrication operation. If grease is hardened or pawl is dry and shows extreme wear, remove slack adjuster, disassemble, clean and inspect all internal parts, relube and install new minor repair kit. Be sure slack adjuster has been lubricated.

Every 100,000 miles remove the slack adjusters, disassemble, clean and inspect all internal parts. It is suggested that both the minor and major repair kits be used to repair the slack adjusters. If any damage is found on the worm, gear or housing, the complete slack adjuster must be replaced.

RESERVOIR & DRAIN COCKS

Either of two types of air tanks may be used. The dual type tank consists of one tank having two separate compartments of equal size, whereas two individual tanks are used with the second type.

The front section (or wet tank) of the dual type tank and the rear tank of the individual type are connected to the air compressor and serve to provide a place where the air, heated during compression, may cool, and oil and water vapors condense. The rear section (or dry tank) of the dual type tank and the front tank of the individual type are charged from the wet tank by external piping and supply the air brake system—and any other air devices used. A one-way check valve in the line between the two tanks prevents air loss in case of leakage between the wet tank and compressor.

Each air tank is equipped with a drain cock. On the dual type tank, the upper one is for the front (wet) tank and the other is for the rear (dry) tank. Both type tanks should be drained regularly—the frequency depending on local humidity and the condition of the compressor. In many cases, daily draining will be necessary.

RESERVOIR MOISTURE EJECTOR VALVE

Description, Fig. 27B

The automatic drain valve ejects moisture and contaminants from the reservoir in which it is connected. It operates 100% automatically with no manual assistance or control lines of any kind.

The reservoir moisture ejector valve has a die cast aluminum body and cover and is normally mounted either in the bottom of the reservoir, using the top port of the drain valve or in the end of an end drain reservoir, using the side port of the valve.

For vehicles operating in subfreezing temperatures, the valve is also available with a heater and thermostat integral with the cover.

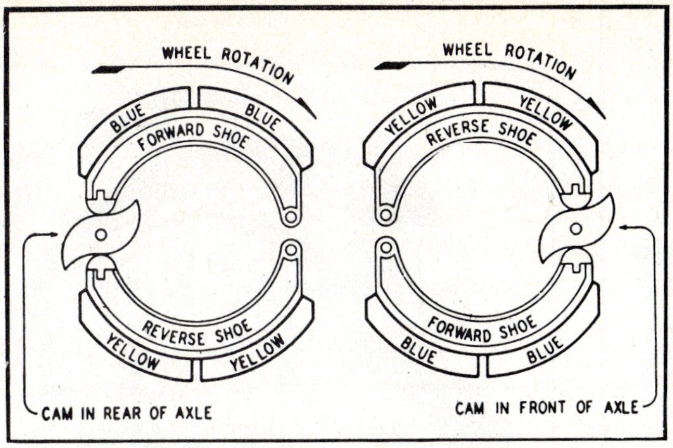

Fig. 28 Combination lining installation

Operation

The inlet valve remains open while pressure builds in the system until maximum governor cut-out pressure is reached. The spring action of valve guide in sump cavity closes the inlet. The inlet and exhaust valves are now closed.

When reservoir pressure drops approximately 2 psi, air pressure in the sump cavity opens exhaust valve allowing moisture and contaminants to be ejected from sump cavity until pressure in sump cavity drops sufficiently to close the exhaust valve. The length of time the exhaust valve remains open and the amount of moisture and contaminants ejected depends upon the sump pressure and reservoir pressure drop that occurs each time air is used from the system.

The valve can be manually drained by inserting a tool in the exhaust port and pushing up against the wire and holding that position until moisture is drained.

Service

Every six months the reservoir moisture ejector valve should be removed, disassembled, cleaned, inspected and lubricated. Worn parts should be replaced as necessary.

NOTE: Early production reservoir moisture ejector valves were equipped with a top reservoir port filter. When servicing these units this filter may be removed and discarded.

Prior to reassembly, lightly lubricate inlet valve seat with grease. Do not apply oil to the inlet and exhaust valve.

PRESSURE REGULATING VALVE

A pressure regulating valve, Fig. 27C, controls the pressure of a volume of compressed air. It remains closed until its setting is reached, at which it opens and passes all excess air through the outlet port to maintain a constant pressure. Formerly, the most common use was a simple function of controlling reservoir air pressure. Presently it is chiefly used in a protected Air Brake System to control the amount of air pressure taken from the air brake system to operate auxiliary devices. In a protected air brake system the pressure regulating valve is connected between the air brake reservoir and the auxiliary device reservoir. The valve is usually at 60 pounds which provides at least 60 psi in the air brake system before air is supplied to an auxiliary device.

Check valve may be connected across the pressure regulating valve to permit air to return from the auxiliary device reservoir to the air brake reservoir if pressure in the brake reservoir falls below that in the auxiliary device reservoir. This makes any air pressure in either reservoir available to operate the air brake system before the operation of an auxiliary device.

In a rapid buildup system a quick buildup of air pressure for brake operation is required. A pressure regulating valve is connected between a small reservoir and the large, main reservoir. Pressure in the small reservoir will rapidly build up to setting of the pressure regulating valve and is available for use before any air is passed to the main reservoir for continuing operation of the brake system.

Service

Disassemble the pressure regulating valve, clean and inspect all parts. Prior to assembly replace any damaged or worn parts. Whenever pressure regulating valve is disassembled always replace the diaphragm.

TUBING, HOSE & FITTINGS

Every 20,000 miles cover the tubing, hose and fittings with soap suds. If leakage is evident, the tubing should be repaired or replaced. Hose must be replaced. Leakage at fittings should be stopped by tightening the connections.

BRAKE ASSEMBLIES

IMPORTANT: The procedure which follows outlines the general procedure for disassembling and reassembling of brakes and for brake adjustments. For details on the various types of brakes in use, refer to the *Brake Adjustment Chapter*.

Disassembly

1. To remove shoes and replace lining, first raise vehicle until wheels are free to rotate.
2. Release hand brake.
3. Loosen slack adjusters to free shoes in drum.
4. Remove wheel or hub.
5. Remove pull-back spring or springs.
6. Remove brake shoe assemblies from anchor pins after locks, anchor plate, lock screws or whatever method is used to hold them in place, is removed. After removal the shoes may be relined or replaced.

Reassembly

1. Install brake shoe assemblies and lock in place on anchor pins.
2. Install pull-back spring or springs.
3. Install wheel or hub.
4. Adjust wheel bearings.
5. Adjust brake and lower vehicle.

Combination Linings

Combination type linings, having a different coefficient of friction for forward (primary) and reverse (secondary), shoes, are frequently used. For ready identification primary linings are usually painted blue on the edge and secondary linings are painted yellow. When applying new brake linings it is important to establish correctly the forward and reverse shoes for each brake assembly, and then apply the linings as indicated in Fig. 28.

To identify the forward or reverse shoe, each brake must be considered as a separate unit. If the camshaft is behind the axle, the top shoe is the forward acting shoe. If the camshaft is ahead of the axle, the bottom shoe is the forward acting shoe.

Brake Adjustments

NOTE: As covered in the *Brake Adjustment Chapter* several different types of service brakes are used and the methods of adjustment vary slightly as outlined in the following.

Before making any brake adjustments, always check wheel bearing adjustment. Raise the vehicle until wheels are free to rotate. Release hand brake and proceed as follows:

Front Brakes

1. Referring to Fig. 29, tighten slack adjusters until brake shoes are tight against drum.
2. Back off slack adjuster until wheel turns freely without brake drag.
3. Loosen the two anchor pin lock nuts and turn eccentric anchor pin until there is .005″ clearance between brake drum and heel of each shoe, measured with a feeler gauge.
4. Clearance between drum and toe of each shoe should measure .010″. Adjust slack adjuster to obtain this clearance.

NOTE: Where the clearance is greater on one shoe than the other, loosen the camshaft bracket bolts and tap the camshaft up or down as required to equalize this clearance.

5. Following the adjustment outlined in Step 4, it may be necessary to reset the drum-to-heel adjustment.
6. When adjustment is completed, tighten anchor pin lock nuts, bring sure to hold the anchor pin stud with a wrench to prevent altering the adjustment.

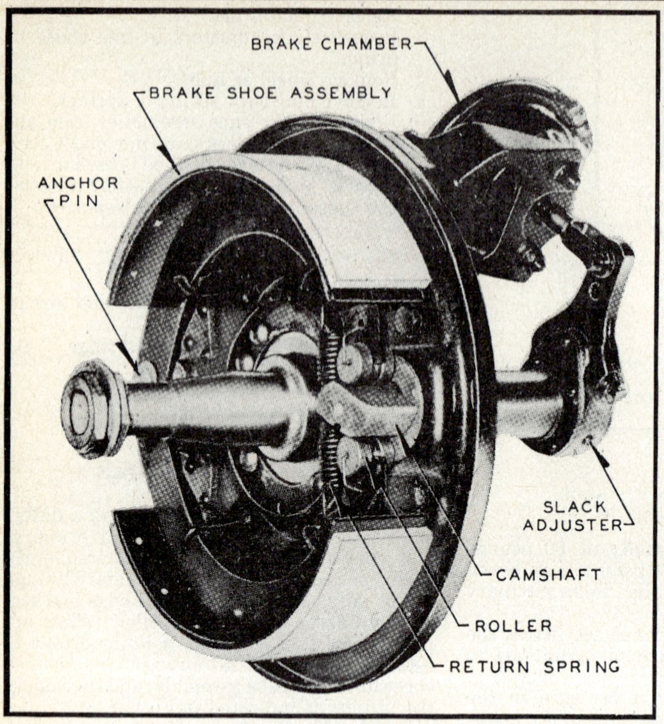

Fig. 29 Typical front brake

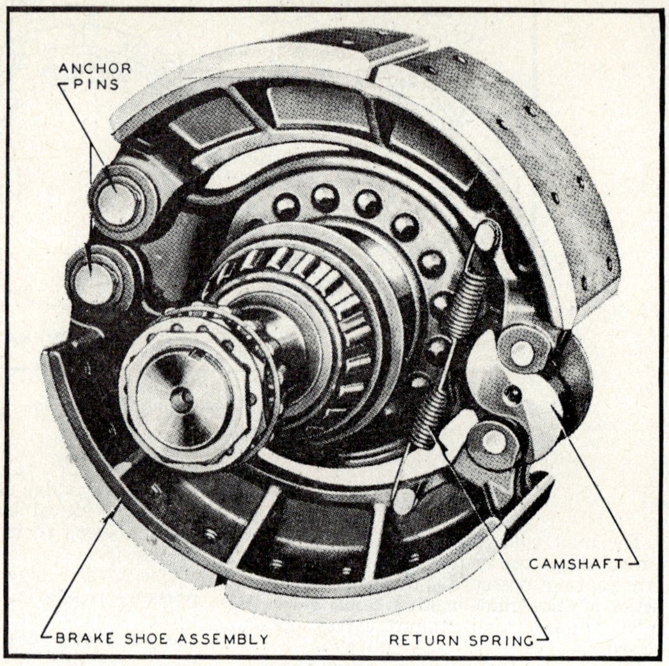

Fig. 30 Typical rear brake

Rear Brakes

1. Referring to Fig. 30, tighten brakes with slack adjusters until shoes are tight against drum.
2. Back off slack adjuster until wheel turns freely without brake drag.

NOTE: Adjustment of the brakes before operating the vehicle should be made as follows: With the system charged to normal operating pressure, connect a test gauge to any convenient point in the brake chamber line and test the delivered pressure by depressing the brake treadle to the floor. This pressure should not be appreciably less than the reservoir pressure.

Check brake chamber push rods to be certain that each is fully released. A positive method is to remove the clevis pin from the push rod yoke and check alignment of holes in yoke and slack adjuster arm. If necessary, make adjustment so that slack adjuster arm and brake chamber push rod form an angle of 90 degrees at the mid-point of slack adjuster travel. In released position, angle formed should be greater than 90 degrees. All slack adjusters on the vehicle should be mounted at the same angle and have the same amount of travel. All brake chambers and slack adjusters should be of the same size and type.

AIR SUPPLY VALVE

Bendix Models PP-1 & PP-8, Figs. 31 & 32

Disassembly

NOTE: The PP-1 and PP-8 model valves are basically the same, except the PP-1 has an internal spring. Also the mounting nut thread is different on each valve.

1. Remove bolts, then separate valve body from cover.
2. Remove "O" ring from valve cover.
3. Remove nut, washer and valve from plunger.
4. Refer to the "Cleaning & Inspection" procedure for servicing valve.

Assembly

1. Position plunger spring into valve body, then install "O" ring onto plunger and the plunger into valve body.
2. Position new "O" ring into valve body.
3. Install valve, washer and nut onto plunger, then install cover and bolts.

Bendix Model PP-2, Fig. 33

Disassembly

1. Secure plunger assembly from turning, while removing nut at bottom of valve.
2. Remove plunger, spring and grommet from valve body.
3. Remove cover bolts, and separate cover from valve body.
4. Remove cover seal grommet, and inlet-exhaust valve from body.
5. Remove piston, lower piston grommet and upper piston grommet from piston.
6. Refer to the "Cleaning & Inspection" procedure for servicing valve.

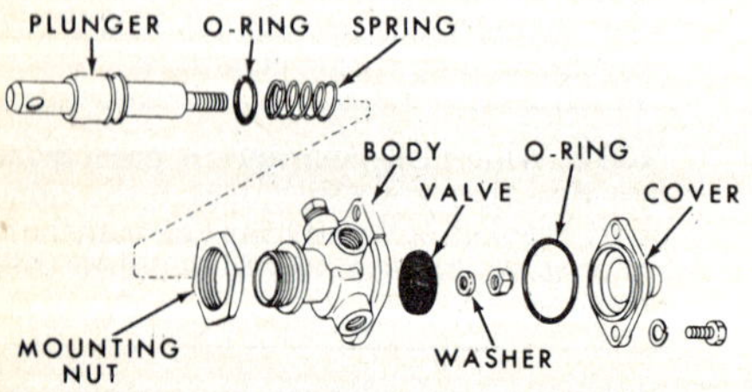

Fig. 31 Bendix PP-1 air supply valve

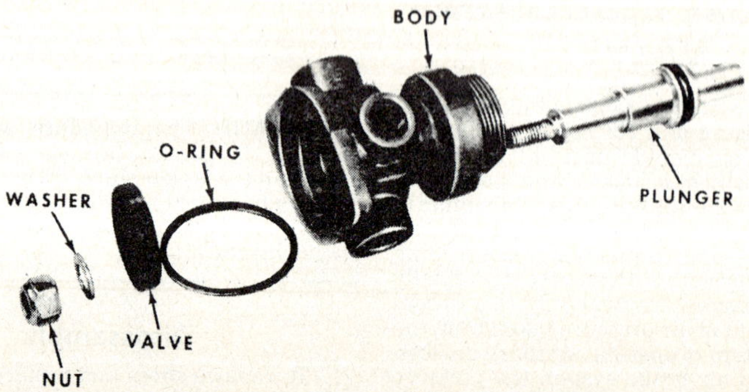

Fig. 32 Bendix PP-8 air supply valve

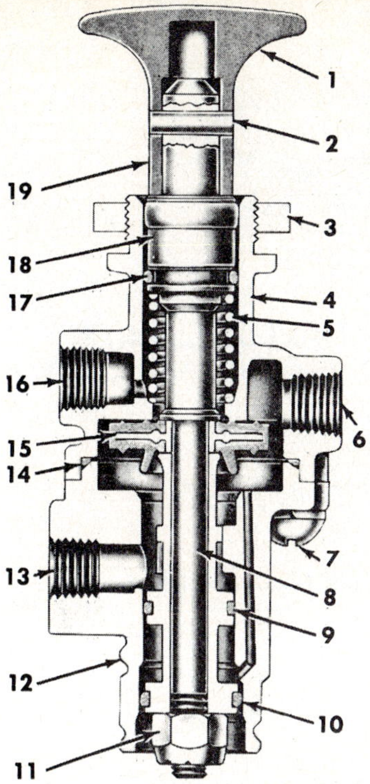

1 Button
2 Spirol Pin
3 Mounting Nut
4 Valve Cover
5 Spring
6 Delivery Port
7 Cover Screw
8 Piston
9 Piston Grommet (Small)
10 Piston Grommet (Large)
11 Nut
12 Valve Body
13 Exhaust Port
14 Seal Grommet
15 Inlet and Exhaust Valve
16 Supply Port
17 Plunger Grommet
18 Plunger
19 Warning Ring (Red)

Fig. 33 Bendix PP-2 air supply valve

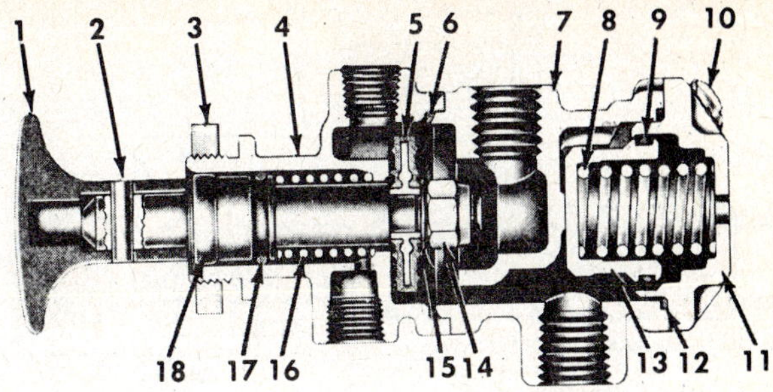

1 Knob
2 Pin
3 Nut
4 Valve Body
5 Inlet and Exhaust Valve
6 Valve Body O-Ring
7 Valve Lower Body
8 Spring
9 Piston O-Ring
10 Cover Screw
11 Lower Body Cover
12 Lower Body O-Ring
13 Piston
14 Plunger Nut
15 Plunger Washer
16 Valve Spring
17 Plunger O-Ring
18 Plunger

Fig. 34 Bendix PP-3 air supply valve

Assembly

1. Install grommet onto plunger.
2. Position spring onto plunger, then install assembly into valve body.
3. Install inlet-exhaust valve over protruding end of plunger, then position cover grommet into valve body.
4. Install cover bolts and cover onto valve body.
5. Install pistons grommets, large grommet in bottom piston groove and small grommet in upper piston groove.
6. Install piston assembly into valve body.
7. Depress and secure plunger from turning, while installing plunger stem nut. Torque stem nut to 35 ft.lbs.

Bendix Model PP-3, Fig. 34

Disassembly

1. Remove bolts, then separate valve body, lower body and cover.
2. Remove piston spring and piston.
3. Remove "O" rings from piston, lower body and valve body.
4. Remove plunger nut, washer and valve.
5. Pull plunger from upper body, then remove plunger spring and "O" ring.
6. Refer to the "Cleaning & Inspection" procedure for servicing the valve.

Assembly

1. Position plunger spring into the valve body, then install "O" ring onto plunger and plunger into the valve body.
2. Position valve and washer onto plunger, then install nut.
3. Install "O" rings onto the piston, lower body and the valve body.
4. Install piston and spring onto the lower body.
5. Assemble valve body, lower body, cover and bolts.

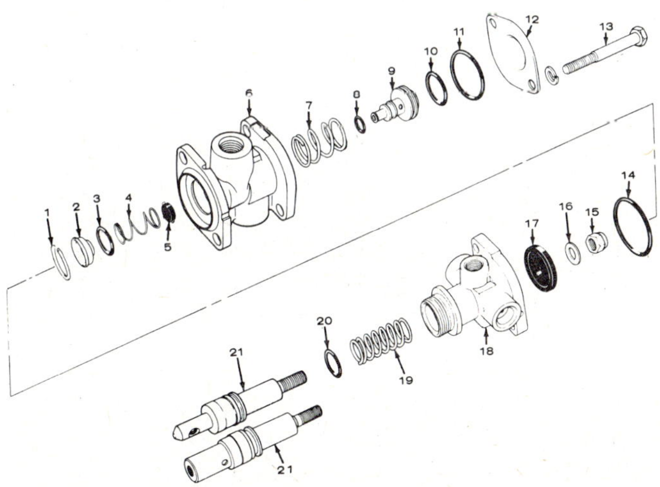

1 Snap Ring
2 O-ring Retainer
3 O-ring
4 Return Spring
5 Control Inlet Exhaust Valve
6 Valve Body
7 Piston Return Spring
8 Piston O-ring
9 Control Piston
10 Piston O-ring
11 O-ring
12 End Plate
13 Bolt
14 Seal Ring
15 Plunger Nut
16 Washer
17 Plunger Valve
18 Upper Body
19 Plunger Spring
20 Plunger O-ring
21 Plunger

Fig. 35 Bendix PP-7 air supply valve

Bendix Model PP-7, Fig. 35

Disassembly

1. Mark the upper and lower body halves, then remove bolts and separate the two body halves.
2. Remove the square cut sealing ring between upper and lower body halves.
3. Remove end plate and "O" from lower body.
4. Remove control piston and piston return spring from valve body.
5. Remove large and small diameter "O" rings from piston.
6. Remove snap ring from lower valve body, then remove "O" ring retainer and the "O" ring.
7. Remove control inlet-exhaust valve return spring from valve housing.
8. Remove nut, inlet-exhaust valve and washer from plunger.
9. Remove plunger, plunger return spring and "O" ring.
10. Refer to the "Cleaning & Inspection" procedure for servicing the valve.

Assembly

1. Install control inlet-exhaust valve and return spring into lower valve body.
2. Install "O" ring, "O" ring retainer and snap ring into valve body.
3. Install end plate seal ring into lower

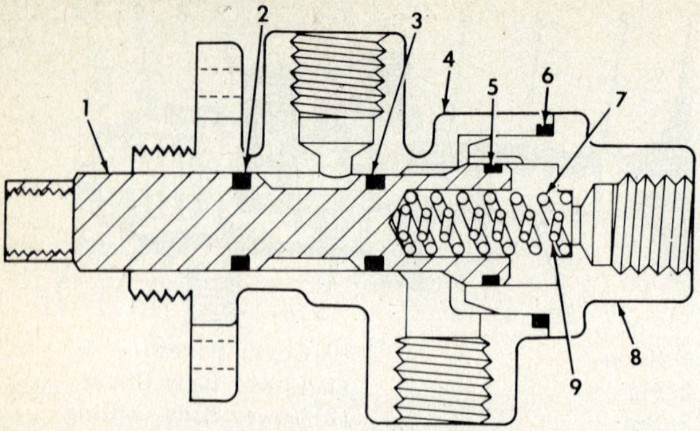

1. Piston (Plunger)
2. Piston O-Ring
3. Piston O-Ring
4. Body
5. Piston O-Ring
6. Cover O-Ring
7. Outer Spring
8. Cover
9. Inner Spring

Fig. 36 Midland-Ross N-14488 air supply valve

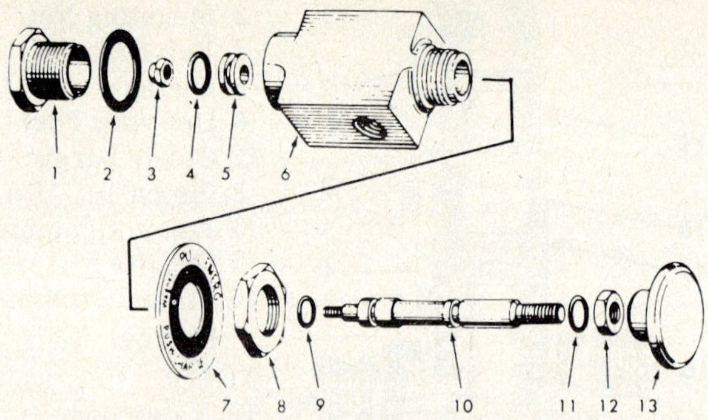

1. Valve End Cap
2. Gasket
3. Nut
4. "O" Ring
5. Piston
6. Valve Body
7. Name Plate
8. Nut
9. "O" Ring
10. Valve Stem
11. "O" Ring
12. Locknut
13. Knob

Fig. 37 Wagner Electric Model PB

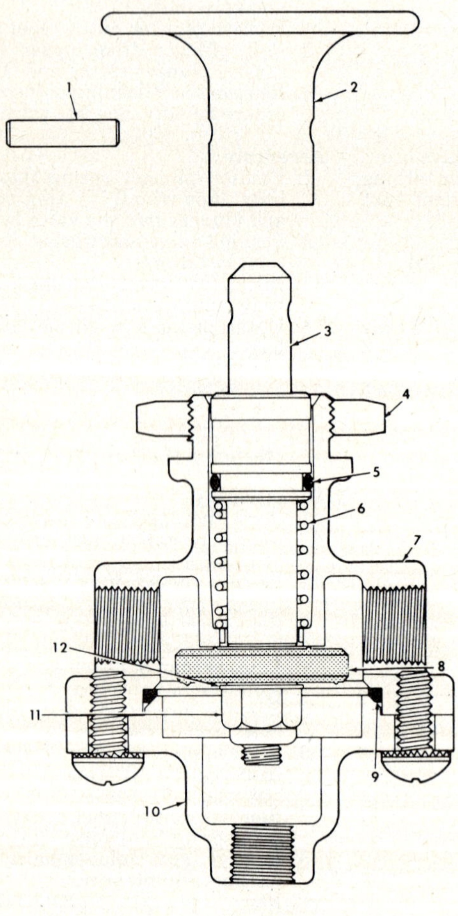

1. Roll Pin
2. Knob
3. Plunger
4. Mounting Nut
5. O-Ring
6. Plunger Spring
7. Body
8. Valve (Poppet)
9. Cover Seal Ring
10. Cover
11. Plunger Nut
12. Washer

Fig. 38 Williams air controls WM762A2A air supply valve

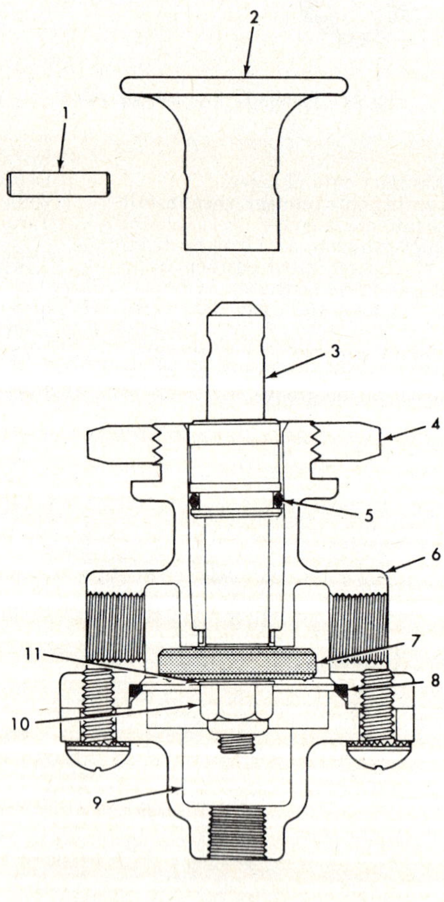

1. Roll Pin
2. Knob
3. Plunger
4. Mounting Nut
5. O-Ring
6. Body
7. Valve (Poppet)
8. Cover Seal Ring
9. Cover
10. Plunger Nut
11. Washer

Fig. 39 Williams air controls WM763A4A air supply valve

valve body.
4. Install plunger "O" ring, return spring and plunger into upper valve body.
5. Depress plunger until threaded portion extends enough to install inlet-exhaust valve, washer and nut. Torque nut to 35 ft.lbs.
6. Install square cut seal ring into the upper body.
7. Install large and small diameter "O" rings on control piston, then position upper and lower bodies together. Ensure marks align properly.
8. Install control piston return spring and control piston into lower body.
9. Install end plate and bolts onto lower body. Torque bolts to 35 inch lbs.

Midland-Ross Model N-14488, Fig. 36

Disassembly
1. Remove bolts and cap from valve body.
2. Remove "O" ring from cap and two springs from body.
3. Remove piston and "O" rings from valve body.
4. Refer to the "Cleaning & Inspection" procedure for servicing the valve.

Assembly
1. Install "O" rings and piston into the valve body.
2. Install springs and "O" ring into valve body.
3. Install cap and bolts onto valve body.

Wagner Electric Model PB, Fig. 37

Disassembly
1. Remove end cap and gasket from valve body.
2. Remove and discard elastic stem nut, then pull out valve stem knob, valve stem and "O" ring.
3. Remove knob and nut from the valve stem.
4. Remove valve stem piston from valve cap end of valve body, then remove "O" ring.
5. Remove valve mounting nut and name plate.
6. Refer to "Cleaning & Inspection" procedure for servicing valve.

Assembly
1. Install valve nameplate and mounting nut.
2. Install "O" rings onto valve stem.
3. Install nut and valve knob onto valve stem.
4. Install valve stem and knob assembly into valve body.
5. Position "O" ring onto valve stem piston, then install piston assembly into valve body.
6. Install valve stem nut, then install gasket and end cap onto valve body.

Williams Model WM762A2A & WM763A4A, Figs. 38 & Fig. 39

Disassembly

NOTE: The WM762A2A and WM763A4A model valves are basically the same valve, except the WM762A2A valve incorporates an internal spring and the mounting thread nut on each one is different.

1. Remove bolts and separate cover from valve body.

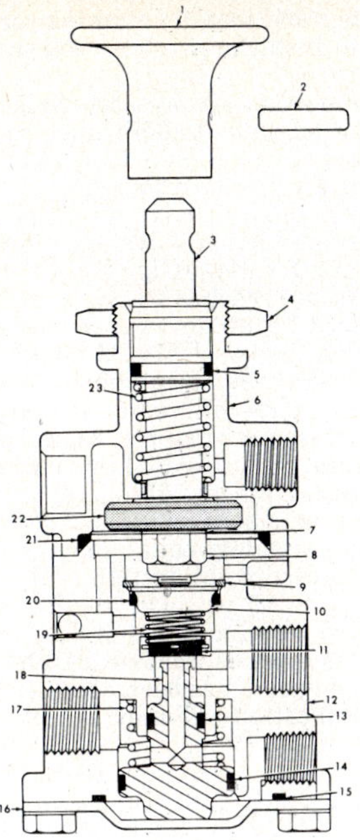

1. Knob
2. Roll Pin
3. Plunger
4. Mounting Nut
5. O-Ring
6. Upper Body
7. Washer
8. Plunger Nut
9. Snap Ring
10. Spring Retainer
11. Inlet Exhaust Valve (Poppet)
12. Lower Body
13. O-Ring
14. O-Ring
15. Cover O-Ring Seal
16. Cover
17. Spool Spring
18. Control Piston (Spool)
19. Valve Spring
20. O-Ring
21. Upper to Lower Body Seal Ring
22. Valve (Poppet)
23. Plunger Spring

Fig. 40 Williams air controls WM764A6A air supply valve

2. Remove cover seal "O" ring.
3. Remove nut, washer and valve poppet from plunger.
4. Remove plunger, "O" ring and spring if so equipped.
5. Refer to "Cleaning & Inspection" procedure for servicing the valve.

Assembly
1. Position plunger spring if so equipped into the valve body.
2. Install "O" ring onto plunger, then install plunger into valve body.
3. Position cover seal "O" ring into valve body.
4. Install valve poppet and washer onto plunger, then install nut.
5. Install bolts and cover onto the valve body.

Williams WM764A6A, Fig. 40

Disassembly
1. Mark for reference for installation the upper and lower valve bodies.
2. Remove bolts and cover from lower valve body.
3. Remove "O" ring from lower valve body cover.
4. Separate valve bodies, and remove seal "O" ring.
5. Remove control piston spool and return spring from valve body.
6. Remove large and small diameter "O" rings from the piston.
7. Remove snap ring from the lower valve body, then remove spring retainer and "O" ring.
8. Remove inlet-exhaust valve and return spring from the lower body.
9. Remove nut from plunger, then remove washer, valve poppet, plunger, return spring and "O" ring.
10. Refer to "Cleaning & Inspection" procedure for servicing valve.

Assembly
1. Position "O" ring onto the plunger, then install plunger and return spring into upper valve body.
2. Install valve poppet, washer and nut onto the valve body.
3. Install inlet-exhaust valve poppet and spring.
4. Position "O" ring into the body, then install spring retainer and snap ring.
5. Position "O" rings onto control piston spool, then install return spring and piston into valve body.
6. Place seal ring into upper valve body, then install upper and lower bodies together aligning marks made at disassembly.
7. Place cover "O" ring seal into lower body groove, then install cover bolts and cover onto the valve body assembly.

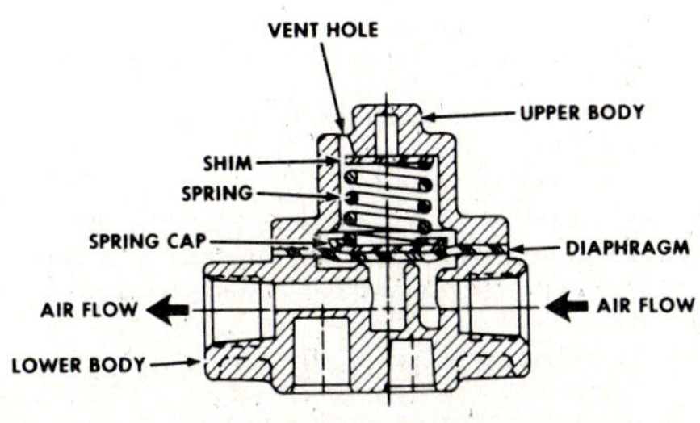

Fig. 41 Pressure protection valve

Cleaning & Inspection

Clean all metal parts using a suitable solvent, and blow dry them using compressed air. Inspect valve body, end cap, internal bores and external surfaces for evidence of wear and deterioration. Slight defects may be touched up with crocus cloth. Lubricate parts using lube No. 5459912 or equivalent. Replace all components not considered serviceable as required. Replace all "O" rings and seals.

PRESSURE PROTECTION VALVE

The pressure protection valve, Fig. 41, functions to cut off the air supply to any air operated device as long as the air supply is below a specified rating.

When air supply pressure is below 65–75 psi, the spring holds the diaphragm on its seat, restricting air flow through the valve. When air pressure overcomes spring force, the diaphragm moves off its seat and air flows through the valve.

If air pressure falls approximately 20 psi below opening pressure, spring tension will overcome the air pressure and move the diaphragm onto its seat. The air supply will be cut off in order to protect the air system from a complete loss of air through the use of auxiliary equipment.

Service

Every 100,000 miles, or yearly, the valve should be removed, disassembled and cleaned. If the diaphragm is worn or deteriorated, it should be replaced.

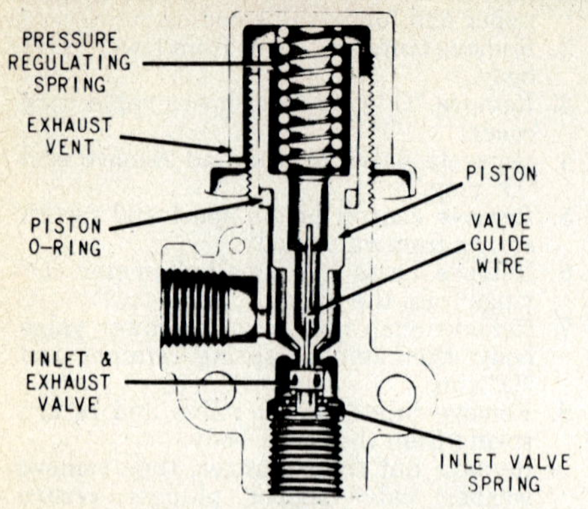

Fig. 42 Pressure reducing (limiting) valve

Testing Valve

1. Inspect vent hole for obstructions.
2. Bleed off main air supply, then disconnect air outlet side of valve and connect an air pressure gauge with a shut-off valve.
3. Start engine and, as air pressure builds up, observe gauge on instrument panel as well as gauge at valve outlet.
4. Air pressure should not be indicated until main air reservoir pressure reaches 65–75 psi, at which time both gauges should indicate approximately equal pressures.
5. Stop engine and open shut-off valve on test gauge. Air should exhaust until vehicle air pressure drops to a value approximately 20 psi below valve opening pressure, and then stop.
6. If valve performs as described, it is operating properly and can be returned to service.

Disassembly

1. Remove four screws securing upper and lower bodies together.

NOTE: Exercise extreme caution when separating body sections due to spring tension acting on the case.

---

2. Remove diaphragm spring cap or diaphragm protector, spring and shim from upper body.

Cleaning & Inspection

1. Clean all parts in a suitable solvent.
2. Inspect vent hole in upper valve body for obstructions.
3. Inspect diaphragm for wear or damage, paying particular attention to lower contact area.
4. Inspect spring for distortion.
5. Inspect diaphragm seat for nicks or pitting. Slight defects can be dressed with emery cloth. Excessive pitting necessitates replacement of complete valve assembly.
6. If any parts other than diaphragm seat are excessively worn, they may be replaced individually.

Assembly

1. Install same number of shims in upper body as were removed during disassembly.
2. Install spring, spring cap and diaphragm on upper body.
3. Force upper and lower body halves together and install retaining screws.
4. Before installing valve, retest as follows:
   a. Connect an air pressure gauge with a shut-off valve to both inlet and outlet sides of valve, then connect to an outside air pressure source.
   b. Open shut-off valve at outlet of pressure protection valve and slowly open shut-off valve at air source until air pressure reaches 55 psi.
   c. Check for air leakage at valve outlet using a soap solution. A slight amount of leakage is acceptable.
   d. Close shut-off valve at outlet and increase air pressure. No air pressure should be indicated at outlet until pressure reaches 65–75 psi at inlet. At this time, both gauges should indicate approximately equal air pressure.
   e. If pressure protection valve opens prior to 65 psi, add another shim to valve. If valve remains closed after 75 psi is reached, remove a shim.
   f. When valve has been adjusted to meet specifications, it may be returned to service.

PRESSURE REDUCING (LIMITING) VALVE

The pressure reducing (limiting) valve, Fig. 42, reduces air pressure and maintains it at a constant preset pressure below that of the supply pressure. Some valves are lockwired and sealed to prevent tampering.

The pressure setting of the valve is determined by the setting of the adjusting cap which applies pressure to the spring on top of the piston. When incoming air pressure on the bottom of the piston overcomes spring tension on top of the piston, the piston moves upward and the inlet valve is seated, closing off the supply of air. As pressure in the delivery line drops, the inlet valve moves off its seat and air flows out the delivery port.

If pressure in the delivery line exceeds the pressure setting of the pressure limiting valve, air pressure below the piston will overcome spring tension above the piston. The piston will move up from the exhaust valve and air will pass by the exhaust valve, through the hollow piston and valve guide and be vented through the slot in the adjusting cap. When spring tension overcomes air pressure below the piston, the exhaust valve is seated and pressure in the delivery line is equal to the setting of the pressure limiting valve.

Service

Every 6 months or 50,000 miles, perform the "Valve Closing Check." Every 100,000 miles or yearly, perform the "Valve Opening Check."

Valve Closing Check

1. Connect a test gauge to delivery line and observe at what pressure inlet valve closes.
2. If pressure delivered by pressure limiting valve varies more than 5 psi ± from pressure setting of valve, the valve should be adjusted by turning the adjusting cap.
3. After adjusting valve, tighten locknut and replace seal and lockwire if used.

Valve Opening Check

1. Connect a test gauge to delivery line and observe at what pressure inlet valve opens.
2. If pressure delivered by pressure limiting valve varies more than 5 psi ± from pressure setting of valve, the valve should be adjusted by turning the adjusting cap. Turn cap clockwise to raise delivery pressure.
3. Check exhaust vent in adjusting cap for leakage. A slight amount of leakage is acceptable. However, a large amount of leakage necessitates replacement of the valve.
4. If leakage is not excessive, tighten locknut, replace seal and lockwire, if used, and return valve to service.

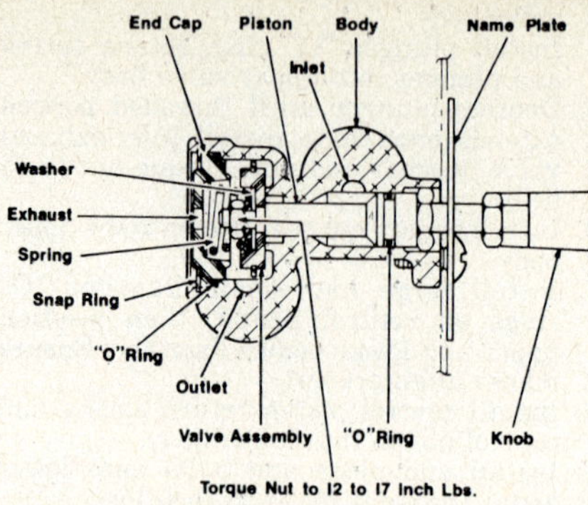

Fig. 43 Midland-Ross parking brake control valve

PARKING BRAKE CONTROL VALVE

The Midland-Ross and Bendix PP-1 parking brake control valves, Figs. 43 and 44, are manually operated, push-pull type valves. When the valve is pushed in, the valve is open and the brakes are released. When the valve is pulled out, supply air is cut off, delivered air is exhausted to the atmosphere and the parking (spring) brakes are applied. These valves also incorporate a plunger pressure sensing assembly which will automatically apply the parking brakes if brake system air pressure falls to a predetermined level.

The Bendix TW-3 valve, Fig. 45, functions similarly to the M-R and PP-1 valves, except that it is operated by a lever, rather than a push-pull knob. This valve does not have the capability of automatically applying the parking brakes in the event of low air system pressure.

Testing Valve

Midland-Ross & Bendix PP-1

1. Connect an outside air pressure source with a manual shut-off valve to inlet port and air pressure gauge with shut-off valve to outlet port of valve.
2. With parking brake control valve pulled out, apply approximately 120 psi to inlet port.
3. Shut off manual valve and check for air leakage using a soap solution.
4. No leakage is allowed between body castings on both type valves and none is allowed around piston on M-R valve.
5. Permissible leakage at exhaust port of M-R valve is a one inch soap bubble in six seconds. On PP-1 valve, leakage at exhaust port must not exceed a one inch soap bubble in five seconds with valve pulled out, or a one inch soap bubble in three seconds with valve pushed in.

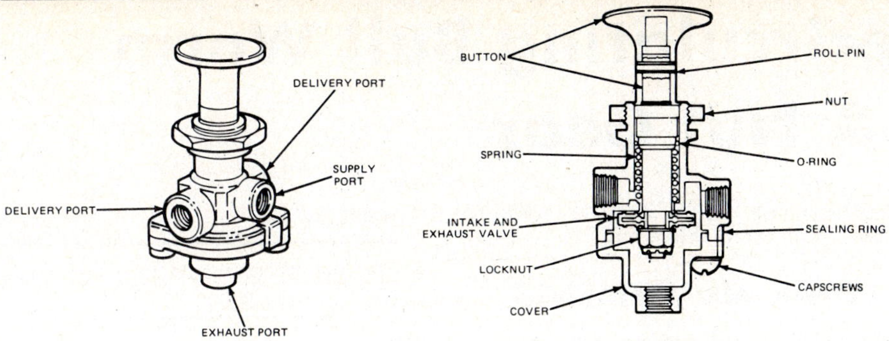

Fig. 44 Bendix PP-1 parking brake control valve

6. Apply approximately 120 psi air pressure to valve with valve pushed in.
7. Slowly reduce pressure. The valve should automatically pop out when air pressure drops to 7–35 psi on M-R valve, or 35–45 psi on PP-1 valve.
8. If valve does not perform as specified, it must be repaired or replaced.

Bendix TW-3

1. Connect an outside air pressure source to supply port and an air pressure gauge to delivery port of valve. If there are two delivery ports, block one.
2. Apply 100 psi air pressure to valve.
3. With valve released, check for leakage at exhaust holes with a soap solution. No leakage is permitted.
4. With valve in the applied position, supply air pressure should be indicated on gauge and there must be no leakage at exhaust holes.
5. If valve does not perform as specified, it must be repaired or replaced.

Disassembly

Midland-Ross

NOTE: Before beginning disassembly, check valve for damage. If the casting is broken or fractured, the valve must be replaced.

1. Remove end cap snap ring, end cap, "O" ring and spring.
2. Remove nut, washer and valve assembly.
3. Remove piston and knob from valve body and remove "O" ring from piston.

Bendix PP-1

1. Remove retaining screws from lower cover and remove cover.
2. Remove sealing ring, then insert a small punch in roll pin hole in stem.
3. Remove locknut, inlet-exhaust valve and spring and the "O" ring.

Bendix TW-3

1. Drive pin out of body, then remove operating lever, plunger and plunger spring.
2. Remove supply cap nut, inlet valve and spring.
3. Remove "O" rings from plunger and supply cap nut.

Assembly

Midland-Ross

1. Lubricate "O" rings, "O" ring sealing surfaces and piston-to-body mating surfaces with a suitable silicone lubricant.
2. Install "O" ring on piston and insert piston into body.
3. Install valve assembly over piston with washer and nut.
4. Install spring with small end against valve disc.
5. Install "O" ring in end cap and push end cap into body.
6. While holding end cap in valve body, install snap ring in valve body groove.

Bendix PP-1

1. Install new "O" ring, then the valve and spring.
2. Install and tighten locknut, then assemble lower cover using a new sealing ring.

Bendix TW-3

1. Lubricate body bore, plunger, "O" rings and cap nut threads with a suitable silicone lubricant.
2. Install valve and spring in body.
3. Place "O" ring on capnut and install capnut.
4. Install plunger spring from top of body, then install plunger with "O" ring.

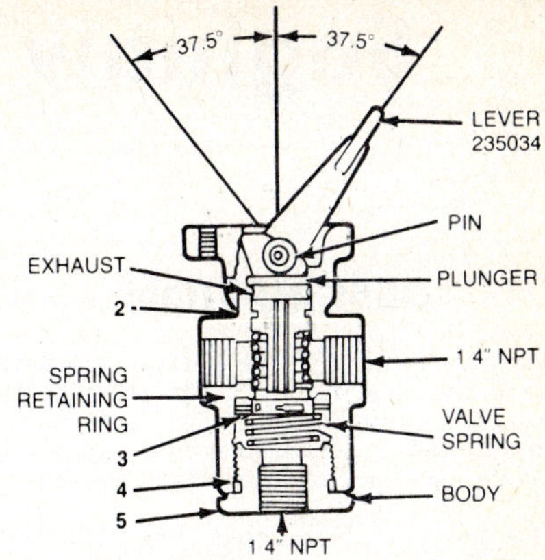

Fig. 45 Bendix TW-3 parking brake control valve

5. While depressing plunger, install lever cam in slot in body, line up holes in body and insert pin.

ALCOHOL EVAPORATORS

The alcohol evaporator permits vaporized alcohol to be drawn into the air brake system to prevent freezing of any moisture in the system. The use of an alcohol evaporator does not eliminate the need for daily draining of all air reservoirs.

Operation

When the air compressor is in the compressing cycle, there is a partial vacuum at the compressor intake. A line connects the evaporator and the compressor intake and a vacuum is created above the alcohol in the alcohol reservoir. Depending on the type of installation, air at or above atmospheric pressure enters the evaporator and passes through the tube immersed in the alcohol. The air passing through the alcohol causes the alcohol to bubble. The vapor formed by this bubbling is induced into the compressor intake and goes into the air system.

Service

Every 25,000 miles inspect all fittings for tightness and check that bubbles are present in alcohol when compressor is in the compressing cycle. Every 50,000 miles check alcohol strainer, if equipped, and clean or replace as necessary. Every 100,000 miles disassemble and clean evaporator and check valve, if equipped, and replace all gaskets and rubber parts.

MANUAL STEERING GEARS
Saginaw Recirculating Ball Worm & Nut Gear

DESCRIPTION

As shown in Fig. 1, the worm on the lower end of the steering shaft and the ball nut which is mounted on the worm have mating spiral grooves in which steel balls circulate to provide a low friction drive between worm and nut.

Two sets of balls are used, ranging in number from approximately 20 to 30 to a set, depending upon the size of the gear unit. Each set of balls operate independently of the other. The circuit through which each set of balls circulates includes the grooves in the worm and ball nut and a ball return guide attached to the outer surface of the nut.

When the wheel and steering shaft turn to the left, the ball nut is moved downward by the balls which roll between the worm and nut. As the balls reach the outer surface of the nut, they enter the return guides which direct them across and down into the ball nut where they enter the circuit again.

When a right turn is made, the ball nut moves upward and the balls circulate in the reverse direction.

The teeth of the ball nut engage teeth on the sector which is forged integral with the pitman shaft, Fig. 2. The teeth on the ball nut are made so that a "high point" or tighter fit exists between the ball nut and pitman shaft sector teeth when the front wheels are in the straight-ahead position. The teeth on the sector are tapered slightly so that a proper lash may be obtained by moving the pitman shaft endwise by means of a lash adjuster screw which extends through the gear housing side cover.

The head of the last adjuster and the selectively fitted shim fit snugly into a T-slot in the end of the pitman shaft so that the screw also controls end play of the shaft. The screw is locked by an external lock nut.

GEAR ADJUSTMENTS

There are two adjustments on the steering gear: worn bearing preload and pitman shaft overcenter preload, Fig. 3.

IMPORTANT

Never attempt to adjust the steering gear while it is connected to the steering linkage. The gear must be free of all outside load in order properly to make any steering gear adjustment.

Preliminary

1. Tighten steering gear mounting bolts.
2. Disconnect steering linkage from steering arm or gear.
3. Turn wheel slowly from one extreme to the other.

CAUTION: Never turn the wheel hard against the stopping point in the gear as damage to the ball nut assembly may result.

---

4. Steering wheel should turn freely and smoothly throughout its entire range.

NOTE: Roughness indicates faulty internal parts requiring disassembly of the gear unit. Hard pull or binding indicates an excessively tight adjustment of the worm bearings, or excessive misalignment of the steering shaft. Any excessive misalignment must be corrected before the gear can be properly adjusted.

Checking Worm Bearing Preload

1. Turn steering wheel gently in one direction until it stops, to position gear away from "high point" load.
2. Attach inch pound torque wrench to steering wheel nut and check torque required to turn wheel steadily in the range where lash exists between ball nut and pitman shaft sector.
3. If preload is not 16 inch lbs., refer to "Adjust Worm Bearing Preload" and adjust

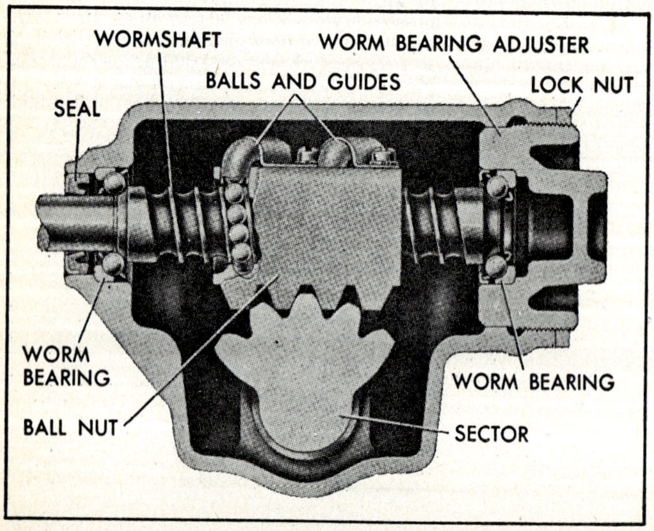

Fig. 1 Recirculating ball-worm and nut steering gear (typical)

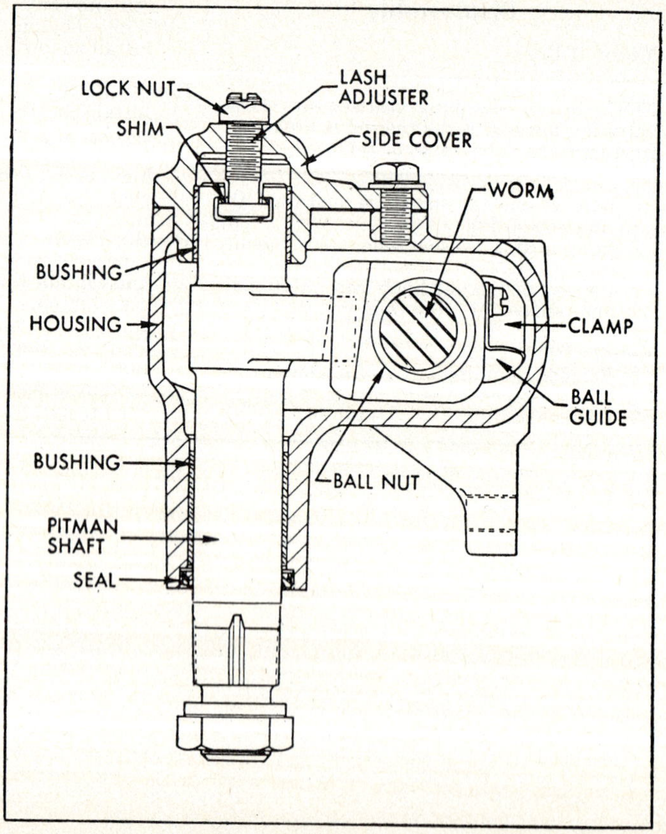

Fig. 2 Recirculating ball-worm and pitman (cross) shaft

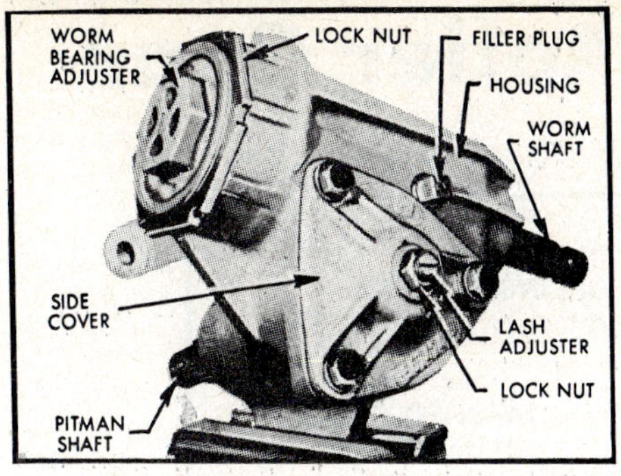

Fig. 3 Steering gear adjustments

Fig. 4 Checking adjustment with spring scale

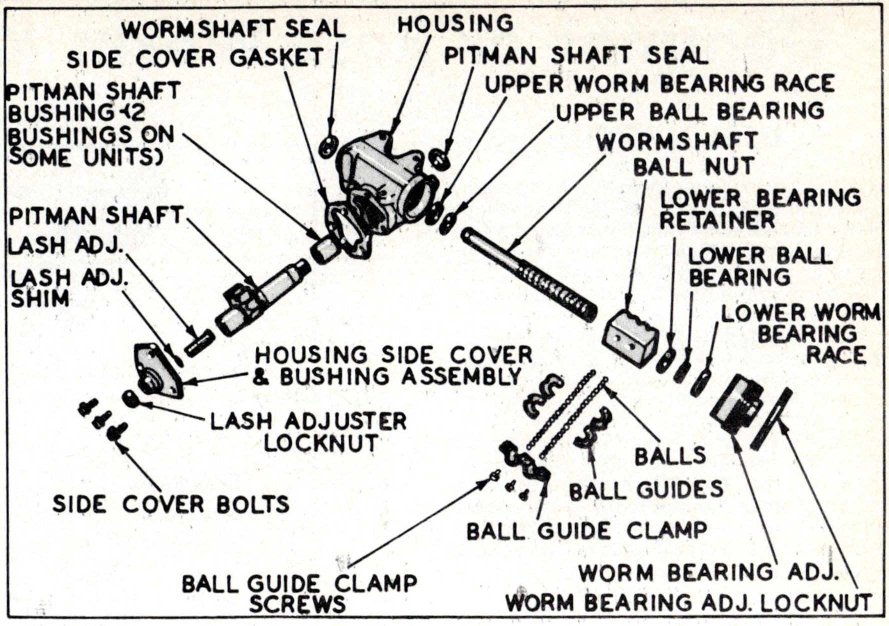

Fig. 5 Steering gear disassembled

preload.

Adjust Worm Bearing Preload

1. Loosen worm bearing adjust lock nut, using a drift, Fig. 3.
2. Turn bearing adjuster as required to bring adjustment within specified limits.
3. Tighten lock nut and recheck preload.

Checking Pitman Shaft Overcenter Preload

1. Turn steering wheel from one extreme to the other while counting the total turns, then turn wheel back 1/2 the number of turns. This positions steering gear on "high point" where a preload should exist between ball nut and pitman shaft teeth.
2. Check pull at wheel rim with spring scale, Fig. 4. Take highest reading on scale as wheel is turned through center position. This should be between 7/8 and 1 1/2 pounds on light duty models and 2 3/4 to 3 1/4 pounds on heavy duty gears.
3. If adjustment is not satisfactory, adjust as follows:

Adjust Pitman Shaft Overcenter Preload

1. Loosen lock nut and turn pitman shaft lash adjuster screw as required to bring adjustment within proper limits.
2. After tightening lock nut, rotate steering wheel back and forth through the "high point" and through the entire range to check for tight spots.

NOTE: If lash cannot be removed at the "high point" or if gear load varies greatly and feels rough, the gear should be removed for inspection of internal parts.

3. Attach linkage to steering gear when adjustments have been completed.

STEERING GEAR REPAIRS

Disassemble

1. Referring to Fig. 5, loosen adjusting screw lock nut and remove housing side cover by removing adjusting screw.
2. Loosen lock nut and back off worm bearing adjuster several turns, then remove housing end cover and gasket.
3. Remove lower thrust bearing, steering shaft and upper bearing from housing.
4. Remove ball return guide clamps and guides from ball nut. Turn ball nut over and remove ball nut from steering shaft worm.

Inspection of Parts

1. Clean and inspect all ball and roller bearings and races, including race in housing.
2. Inspect pitman shaft bushings in gear housing and end cover. If bushings are worn excessively, replace bushings.
3. It is advisable to replace pitman shaft grease seal in housing to avoid possible leakage of lubricant. Seal must be installed with feather edge toward inside of housing.
4. Inspect pitman shaft for wear or pits in bearing races, which would require replacement of shaft.
5. Check shaft for straightness.
6. Inspect teeth of ball nut and pitman shaft. If scored or excessively worn it is advisable to replace both parts to insure proper mating of teeth.
7. Check serrations of pitman shaft; if twisted replace shaft.
8. Check fit of pitman shaft adjusting screw and shim in slot in end of pitman shaft. With shim in place, screw head must be free to turn in slot with zero to .002" end play. If end play is excessive, selectively fit a new shim, which are available in four different thicknesses.

Reassemble

NOTE: Lubricate all seals, bushings, bearings and gears with multi-purpose gear lube prior to installation.

1. Position ball nut over worm shaft so that deep side of teeth will be toward side cover when installed in gear housing.
2. Install exactly 1/2 the number of balls in each circuit, rocking worm shaft slightly to aid in installing balls.
3. Place about six balls in each return guide, using grease to hold balls in place.
4. Install return guides, clamp and screw.
5. Rotate worm through its complete travel several times to be sure balls are installed correctly and rotate freely.
6. Place upper bearing on worm shaft and slide worm shaft assembly into housing.
7. Place lower bearing in worm bearing adjuster and install bearing retainer.
8. Install adjuster assembly and lock nut in housing. Tighten adjuster only enough to hold worm bearings in place. Final adjustment will be made later.
9. Turn worm shaft until center groove in ball nut lines up with center of pitman shaft bushing.
10. Install pitman shaft and lash adjuster with shim so that center tooth meshes with center groove in ball nut.
11. Install side cover with gasket on lash adjuster by turning adjuster counterclockwise.
12. Install side cover bolts and washers.
13. Turn lash adjuster so that teeth on shaft and ball nut engage but do not bind.
14. Install lash adjuster lock nut loosely.
15. To protect pitman shaft seal from damage, cover shaft splines with masking tape. Slide new seal into place and seat it against shoulder in housing.
16. Install new worm shaft seal flush with surface of housing.
17. Fill gear housing with multi-purpose lubricant and adjust gear assembly as outlined previously.

Gemmer or Ross Worm & Roller Gear

DESCRIPTION

In this type steering gear, Fig. 6, the worm is integral with the steering shaft and is supported on each end by opposed tapered roller bearings. The triple tooth roller is attached to the roller shaft by means of a steel shaft. Two needle bearing assemblies are installed between this shaft and the roller.

The roller shaft is mounted in the steering gear housing on two needle bearing assemblies which are pressed into the housing. The housing cover is attached to the housing by four cap screws. An adjustment screw, mounted in the cover, controls roller shaft end play and worm and roller mesh adjustment.

The steering wheel and roller shaft arm (pitman arm) are splined to the steering shaft and roller shaft respectively. Both the pitman arm and steering wheel have master splines to insure correct installation.

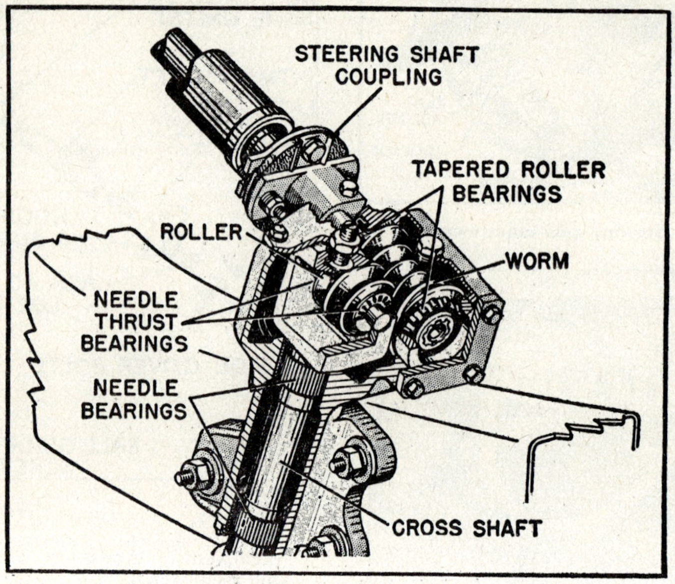

Fig. 6 Worm and roller steering gear (typical)

ADJUSTMENTS

Worm End Play, Adjust

1. Free the steering gear of all load by disconnecting the linkage and loosening the steering column braces.
2. Loosen the four cover screws about 1/8".
3. Use a knife to separate the top shim, passing the blade all the way around between the shims, being careful not to damage the remaining shims.
4. Remove one shim at a time between inspections to remove the end play.
5. The adjustment is correct when there is not end play and no stiffness in the steering gear throughout the complete range of its travel.

Roller Shaft End Play, Adjust

1. Turn the steering gear to either extreme and back off 1/8 of a turn.
2. Gripping the pitman arm at the hub, the roller shaft should rotate freely without end play.
3. If end play exists, adjust as required by means of the roller shaft adjusting screw in the side cover.
4. Be sure to tighten the lock nut securely and inspect for end play and free rotation throughout the entire range of steering gear travel.

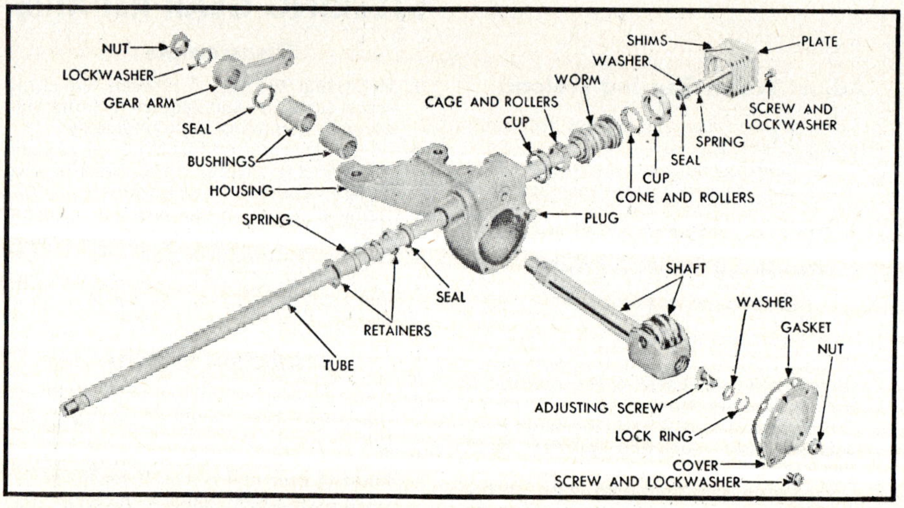

Fig. 7 Worm and roller gear unit disassembled

Worm & Roller Mesh, Adjust

1. Loosen the roller shaft adjusting screw lock nut.
2. With the steering gear in its central position (linkage disconnected) tighten the roller shaft adjusting screw just enough to remove play between the roller shaft roller tooth and worm.
3. Check this by the amount of play felt at the pitman arm. It is better to leave a slight amount of play at this point than to tighten too much.
4. If tightened beyond the point where the lash is removed, serious results will occur which will cause poor steering operation.
5. Tighten the adjusting screw lock nut.

GEAR REPAIRS

1. Referring to Fig. 7, use a suitable puller to remove gear oil seal from housing. If shaft is corroded or dirty, clean the portion between oil seal and serrations to avoid binding in bearings.
2. Place a suitable arbor over cross shaft threads while withdrawing cross shaft, following with the arbor. The arbor will keep bearing rollers from dropping out of their cages.
3. Remove cross shaft adjusting screw lock nut. Remove cover and shims from bottom cover gasket and cross shaft.
4. Remove shaft-worm, bearings and cups.
5. If necessary, drive needle bearings from housing.
6. Clean all parts and inspect for wear.
7. Assemble parts without lubrication. Lubrication should be done after adjustments are completed.
8. If either of the worm thrust bearings is damaged, replace both bearings. Use new oil seals.

Ross Cam & Lever Gear

DESCRIPTION

These steering gears, Figs. 8 to 11, are of the cam and lever type in which the cam is mounted on ball bearings and the lever shaft turns on steel backed bronze bushings.

As the cam is turned by the wheel tube, the follower stud or studs are pulled along the cam groove causing the lever arm to rotate the lever shaft. As an aid to easy steering the lever shaft stud or studs which engage the cam are mounted on tapered roller bearings, Figs. 10 and 11. The groove in the cam is cut shallower in the straight-ahead driving position to provide closer adjustment between the studs and cam where most of the steering action occurs.

Cam End Play, Adjust

Free the steering gear of all load by disconnecting the linkage and loosening the steering column braces. Then adjust as follows:

Shim Adjustment Type—Loosen the lock nut and adjusting screw in the side cover to free the studs in the cam groove. Remove the upper cover stud nuts and raise the housing upper cover to permit removal of adjusting shims, which are .002", .003" and .010" thickness. Clip and remove one thin shim, tighten down the cover and check the adjustment. There should be a slight drag but the steering wheel should turn freely with the thumb and forefinger lightly gripping the rim. If necessary, remove or replace shims until the adjustment is correct.

Plug Adjustment Type—Loosen the housing side cover adjusting screw to free the stud in the cam groove. Turn down the cam adjusting plug, located at the top of the gear housing, until there is a barely perceptible drag so that the steering wheel can be turned freely with the thumb and forefinger lightly gripping the rim.

Lever Shaft Backlash, Adjust

1. Centralize the steering gear by turning the wheel all the way to the right. Then starting from this point count the number of turns required to reach the end of travel to the left. Turn the wheel back half this number of turns to the mid-position.
2. Tighten the side cover adjusting screw until a very slight drag is felt through the mid-position when turning the steering wheel slowly from one side to the other. The gear should not bind in any position but a slight drag should be felt in the mid-position only.
3. After proper adjustment is secured, tighten the lock nut and give the gear a final check for binding.

Stud Roller Bearing, Adjust

In most cases the foregoing adjustments will be adequate but it is sometimes necessary to adjust the stud roller bearing units as they should be preloaded at all times. A heavy drag should be felt when revolving the studs by hand. If a torque wrench is available the bearings should be adjusted so that the torque required to revolve the stud is as follows:

Lever Shaft Dia.	Torque, In. Lb.
1 3/8"	3 to 4
1 1/2"	3 to 4
1 5/8"	4 to 8
1 3/4"	5 to 11

1. To adjust, remove lever shaft from housing, wash the bearings in solvent and blow dry with compressed air, using care not to spin the bearings. Lubricate the bearings with light oil.
2. Straighten out the prong of the locking washer to release the stud nut. Remove the nut and install a new locking

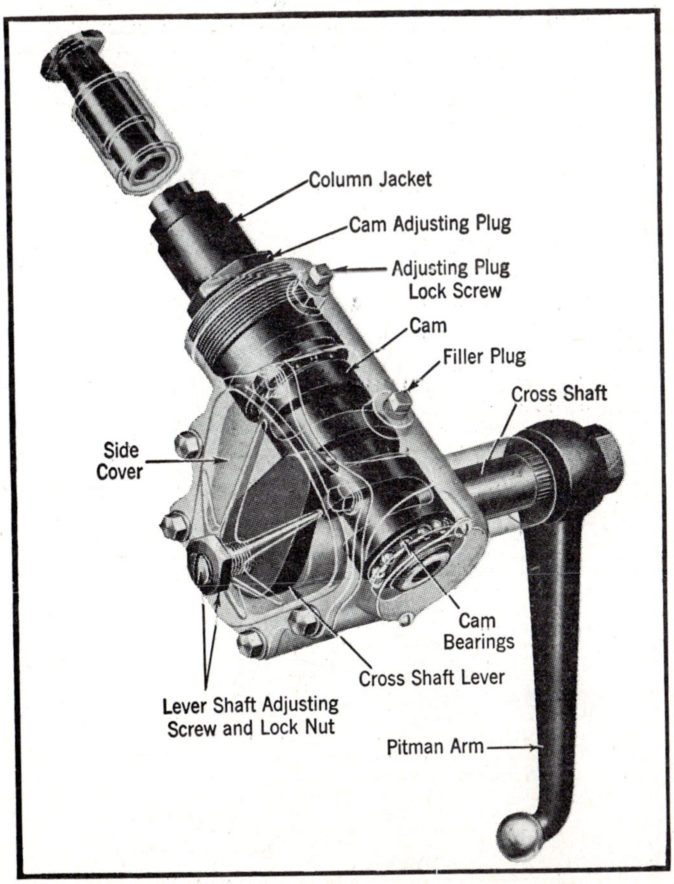

Fig. 8 Cam and single lever steering gear

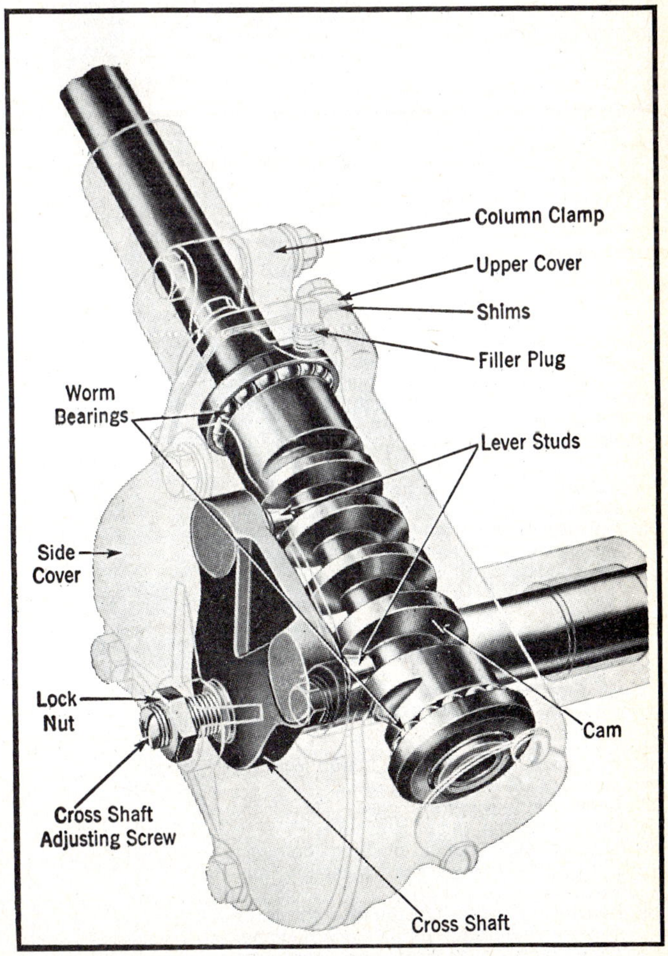

Fig. 9 Cam and twin lever steering gear

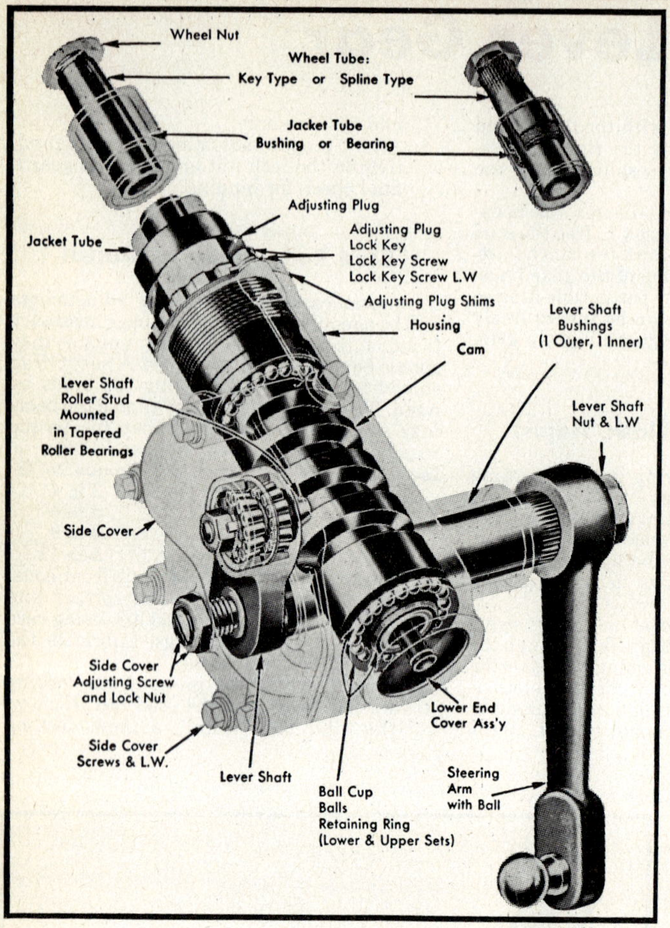

Fig. 10 Single lever gear unit with roller mounted stud

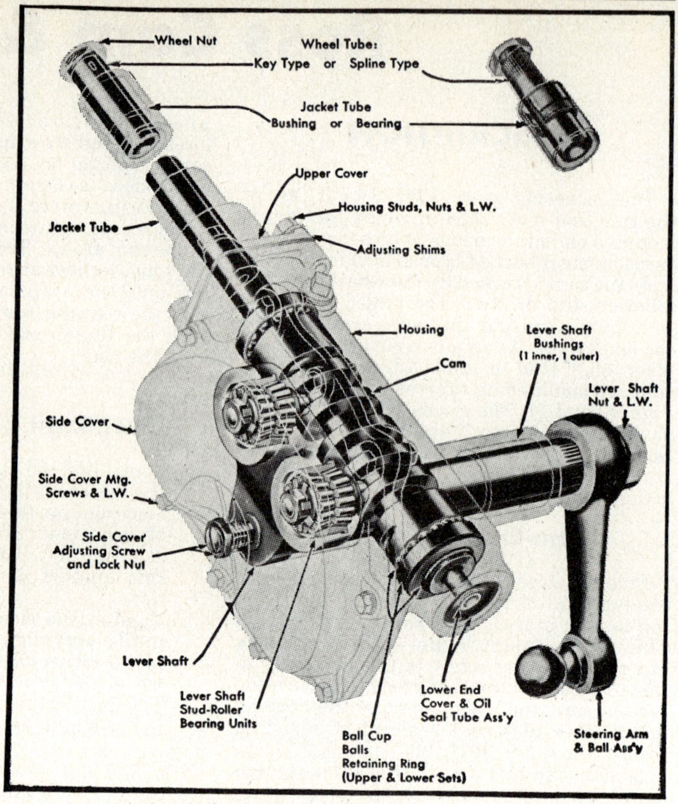

Fig. 11 Twin lever gear unit with roller mounted studs

washer.

3. While holding the stud nut from turning with a spanner wrench on the locking washer, tighten the nut as required to revolve the stud in accordance with the table above.
4. When proper adjustment has been secured, bend the prong of the washer against a side of the nut to keep it from turning.
5. Lubricate the bearings with gear oil and assemble the lever shaft in the steering gear housing.

GEAR REPAIRS

Disassemble

1. Using Fig. 12 as a guide, loosen lock nut and back off adjusting screw.
2. Remove side cover and gasket.
3. Clean serrated end of cam lever shaft to prevent seal and/or bushing damage. Then pull cam lever shaft assembly from housing.
4. Remove top cover and adjusting shims (or unscrew adjusting plug) from housing. Then remove steering post and cam assembly.
5. Remove lower cam bearing retainer ring

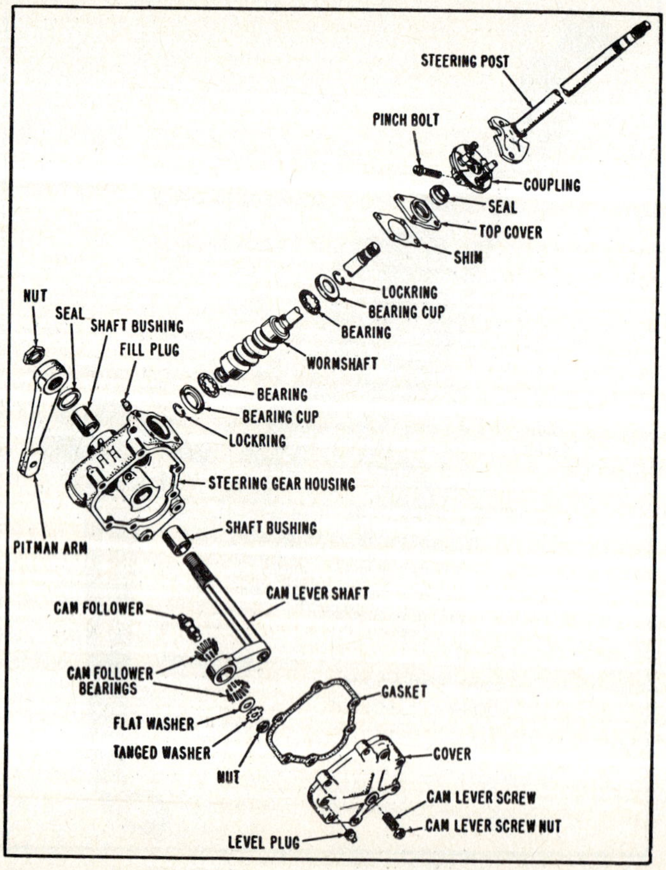

Fig. 12 Single lever gear unit disassembled (typical)

and slip lower bearing cup and ball retainer assembly off shaft. Remove upper bearing in same manner.

6. Remove cam lever shaft oil seal from gear housing. If cam lever bushings are to be replaced, use a suitable arbor and press inner and outer bushings from housing.
7. Remove cam follower from lever shaft by bending down locking tang. Then remove tanged washer and retainer washer and slip bearings and stud out of lever.

Reassemble

1. If cam lever bushings were removed from housing, install new bushings, using an arbor press and a suitable arbor. The bushings should then be line-reamed to fit the cam lever shaft. Use a suitable arbor and install a new cam lever shaft oil seal.
2. Place the inner set of bearing rollers around cam follower stud. Hold rollers in place with heavy grease.
3. Carefully insert stud and assembled rollers into lever and assemble outer set of rollers, using heavy grease. Then install retainer washer, a new tanged lock washer and nut.
4. Adjust gear as outlined above and fill gear housing with specified lubricant.

Koyo Recirculating Ball Worm & Nut Gear

DESCRIPTION

In this steering gear, a bushing in the sector cover and a needle bearing in the housing support the sector shaft, Fig. 13. A continuous row of ball bearings are used to mate the threads of the ball nut to the threads of the worm shaft, Fig. 14. The ball bearings recirculate through ball return guides. The ball nut has gear teeth cut on one face which mesh with gear teeth on sector shaft. When the steering gear is turned, the worm shaft rotates causing the ball nut to rotate up or down the worm shaft. This movement of the ball nut causes the sector shaft to rotate and swing the Pitman arm.

GEAR ADJUSTMENTS

Worm Bearing Preload, Check

NOTE: Before checking worm gear preload, check to ensure intermediate shaft flex coupling is not distorted and steering column is aligned.

1. Remove pitman arm from ball stud.
2. Lubricate worm shaft seal with drop of automatic transmission fluid.
3. Remove horn pad assembly from steering wheel and turn wheel slowly to either stop.
4. Using inch pound torque wrench on steering wheel nut, measure torque required to turn steering wheel 1½ turns at a constant speed. Torque should be 2–6 inch lbs. Readjust worm gear preload if not within specifications.

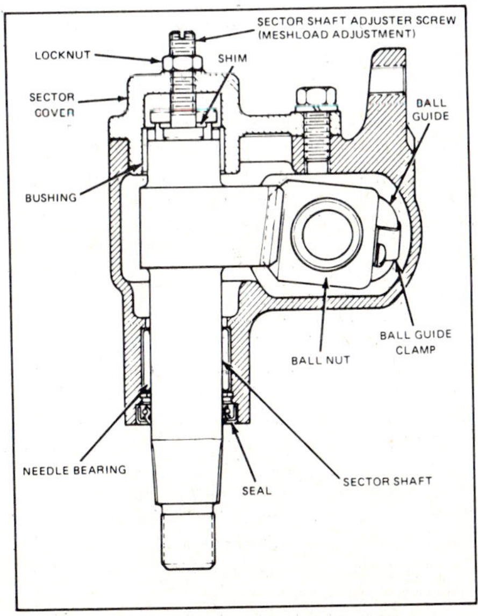

Fig. 13 Side view of Koyo manual steering gear

Worm Bearing Preload, Adjust

1. Remove steering gear from vehicle.
2. Torque sector shaft cover bolts to 40 ft. lbs.
3. Loosen worm bearing adjuster plug lock nut and tighten worm bearing adjuster nut until all end play has been removed. Lubricate worm shaft seal with drop of automatic transmission fluid.
4. Turn worm shaft to full right position. Measure left turn torque required to rotate worm shaft 1½ turns. Worm bearing preload reading should be 5–6 inch lbs. Tighten or loosen adjusting nut until this preload is obtained.

Pitman Shaft Overcenter Preload, Check

1. Disconnect pitman arm at ball stud and remove horn pad from steering wheel.
2. Turn steering wheel from one extreme to the other while counting total number of turns. Turn wheel back one half the number of turns so gear is centered.
3. Place inch lbs. torque wrench on steering wheel nut. Rotate steering shaft 90° either way across center. Observe highest reading. Overcenter preload should be 4–10 inch lbs. and at least 2 inch lbs. over worm bearing preload check torque.

Pitman Shaft Overcenter Preload, Adjust

1. Perform steps 1 through 3 of "Worm Bearing Preload, Adjust."
2. Rotate worm shaft from one extreme to the other while counting total number of turns. Turn worm shaft back halfway to center gear.
3. Using inch lbs. torque wrench, turn worm shaft approximately 90° across center. Highest reading should be within 9–10 inch lbs. and at least 4 inch lbs. over preload. Turn sector shaft adjusting screw until this torque is achieved.
4. Hold sector shaft adjusting screw and tighten lock nut to 25 ft. lbs.

STEERING GEAR REPAIR

Disassembly

1. Rotate worm shaft gear through entire length of travel, counting number of turns. Turn back exactly halfway, centering gear.
2. Remove sector cover bolts, sector cover and shaft. Remove shaft from cover by turning screw clockwise, Fig. 15. Do not loose screw shim.
3. Remove worm bearing adjuster locknut, worm bearing adjuster and worm shaft

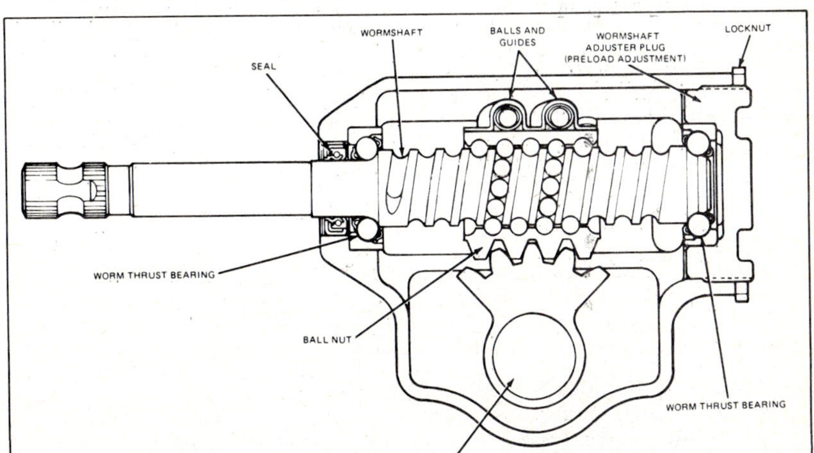

Fig. 14 Top view of Koyo manual steering gear

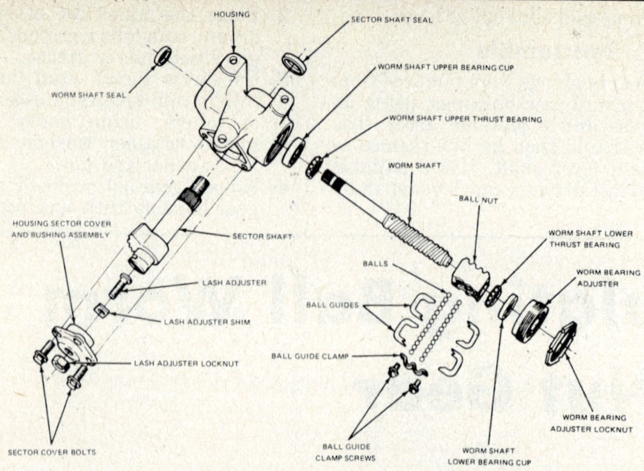

Fig. 15 Exploded view of Koyo manual steering gear

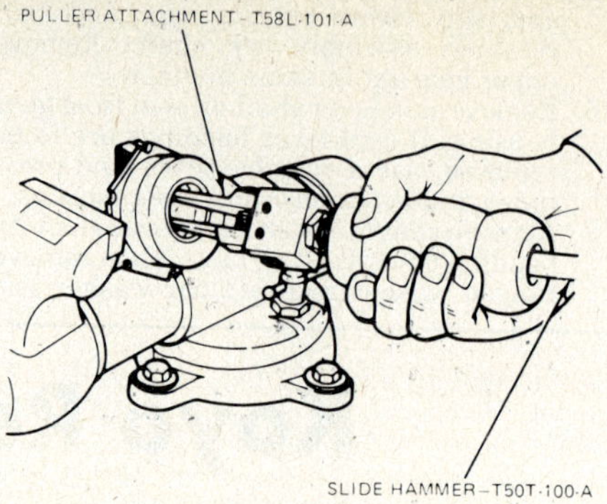

Fig. 16 Worm bearing adjuster nut cup removal

thrust bearing.

4. Remove worm shaft and ball assembly from housing, ensuring ball does not travel to either end of shaft.
5. Pull out sector shaft and worm shaft seals and discard seals, Fig. 15.

NOTE: Individual parts of steering gear assembly are not serviceable. If worm gear cannot rotate freely in ball nut, replace entire assembly. Do not disassemble for repair.

6. Remove adjuster nut bearing cup from worm bearing adjuster using puller tool T58L-101-A and slide hammer, Fig. 16.
7. Remove housing bearing cup using bearing driver and hammer, Fig. 17.

NOTE: Bushing in sector cover is not serviceable. Sector cover is serviced as a unit only. Sector shaft needle bearing is not serviceable, except as a unit with housing. Do not attempt to remove needle bearing.

Inspection of Parts

NOTE: Clean all parts in cleaning solvent prior to inspection.

1. Inspect all bearings and races. Replace as necessary.
2. Inspect bushing in sector cover. If bushing is worn, replace cover.
3. Inspect teeth on ball nut for chipping. If teeth are defective, replace ball nut and worm shaft as an assembly.
4. Inspect ball nut for free rotation on worm shaft. If sticking or binding is found, replace entire assembly.
5. Inspect housing for cracks and damage. Inspect needle bearings inside housing for missing or damaged needles. Replace housing if any defects are found.
6. Inspect sector shaft gear teeth for chipping, wear or metal fatigue. Replace if defective.
7. Check clearance between sector adjusting screw head and bottom of sector shaft T-slot, Fig. 18. If clearance is more than .004 in., add shims until clearance is within .004 in.
8. Hold sector shaft adjusting screw and rotate sector shaft, Fig. 18. Sector shaft should rotate freely. If sector shaft does not rotate freely, increase T-slot clearance by adding an additional shim. Do not exceed .004 in. clearance.

Assembly

1. Install new worm shaft bearing cup, Fig. 19. Install new adjuster plug bearing cup, Fig. 20. Install sector shaft seal in housing until it bottoms out, Fig. 21. Tap worm shaft seal into housing until it is flush with housing, Fig. 22.
2. Place housing assembly in vise with sector cover opening up.
3. Apply steering gear lubricant to sector cover bushing, sector shaft needle bearings and worm shaft bearings. Use lubricant C3AZ-19578-A or equivalent.
4. Place worm shaft thrust bearing over splined end of worm shaft. Insert worm shaft and ball nut assembly into housing, Fig. 23. Push splined end of worm shaft through bearing cup and seal.
5. Place worm shaft thrust bearing in adjuster plug bearing cup. Install adjuster plug and locknut into housing so that

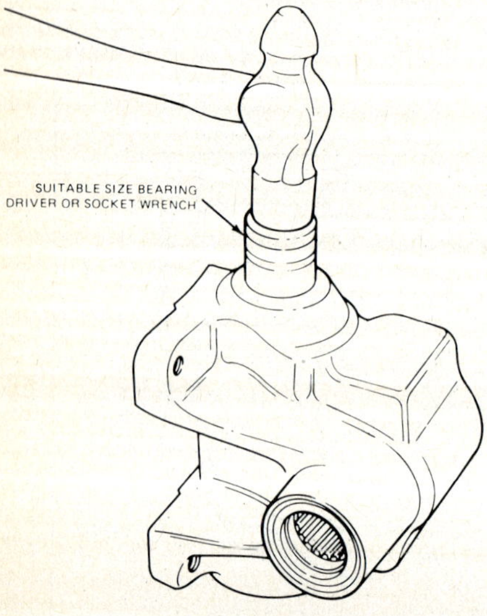

Fig. 17 Housing bearing cup removal

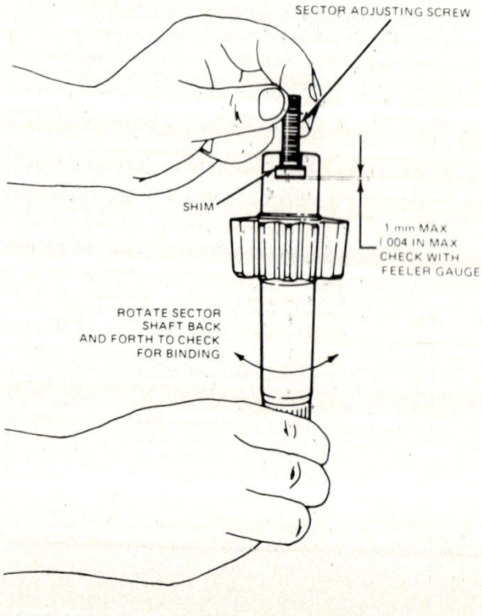

Fig. 18 Inspection of sector shaft

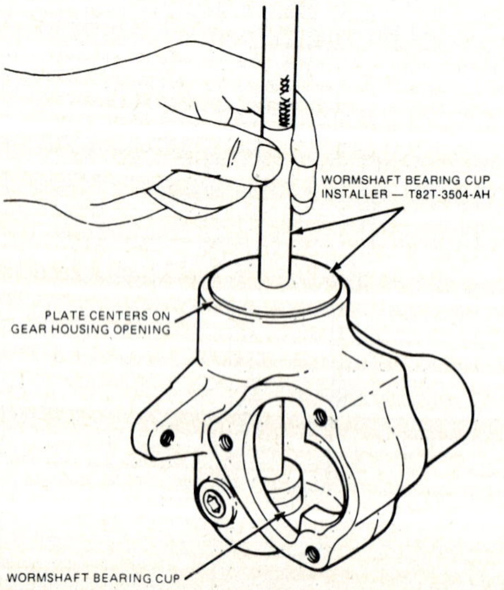

Fig. 19 Installation of worm shaft housing bearing cup

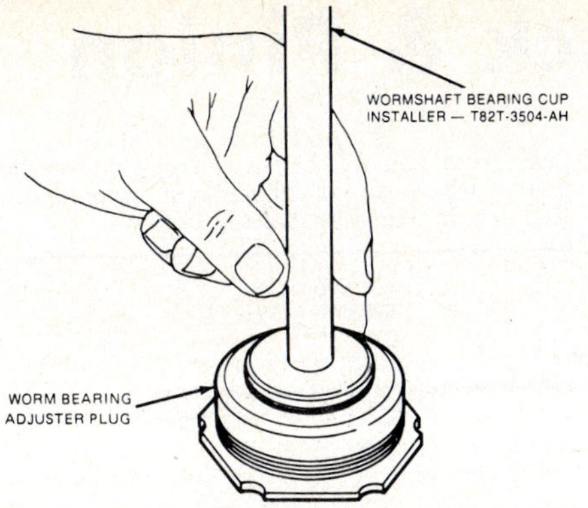

Fig. 20 Installation of worm bearing adjuster nut cup

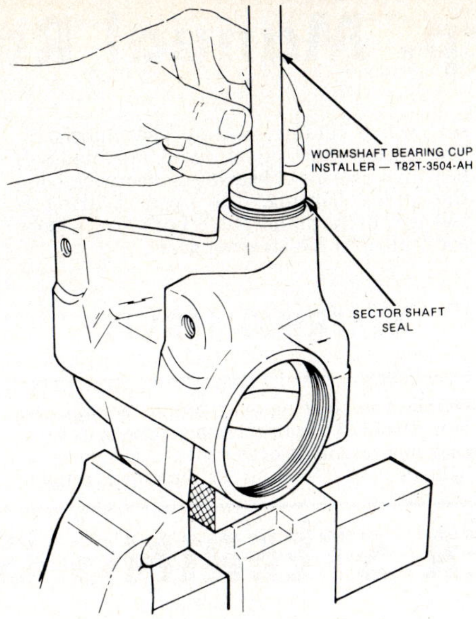

Fig. 21 Installation of sector shaft seal

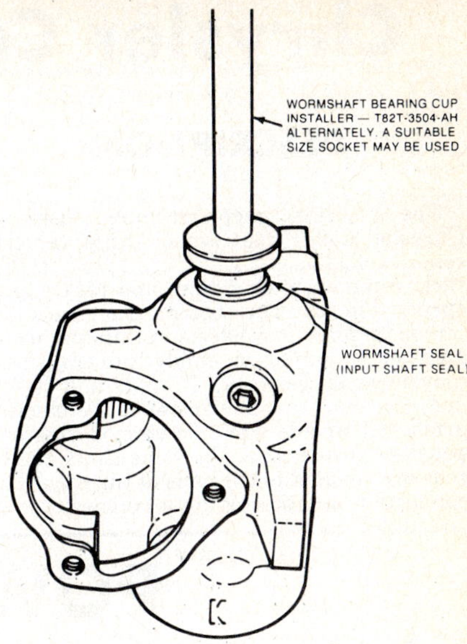

Fig. 22 Installation of worm shaft seal

nearly all end play has been removed from worm shaft.

6. Insert sector shaft adjusting screw and shim(s) into T-slot.
7. Pack steering gear with 10.2 oz. of C3AZ-19578-A grease or equivalent. Rotate ball nut until it is near end of its travel. Pack housing with as much grease as possible without losing it out sector shaft opening. Rotate worm shaft to other end of its travel and pack more grease into housing. Center ball nut on worm shaft.
8. Insert sector shaft assembly into housing so that center tooth of sector gear enters center rack tooth space in ball nut. Rotate ball nut teeth slightly upward to aid in alignment of gear teeth.
9. Pack remaining grease into housing.
10. Apply bead of sealant D6AZ-19562-A or equivalent to mating surfaces of sector cover and housing.
11. Place sector cover on sector adjusting screw and turn screw counterclockwise until sector cover is flush with housing, Fig. 24.
12. Install sector cover to housing attaching bolts. Check for proper lash between sector shaft and worm shaft. After lash is set, torque bolts to 32–40 ft. lbs.

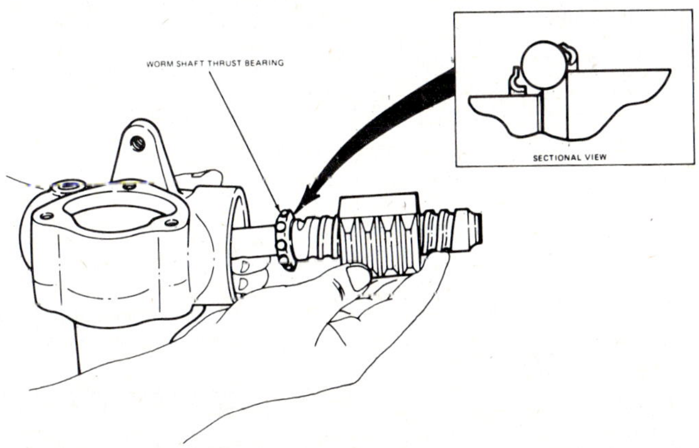

Fig. 23 Installation of worm shaft and ball nut assembly into gear housing

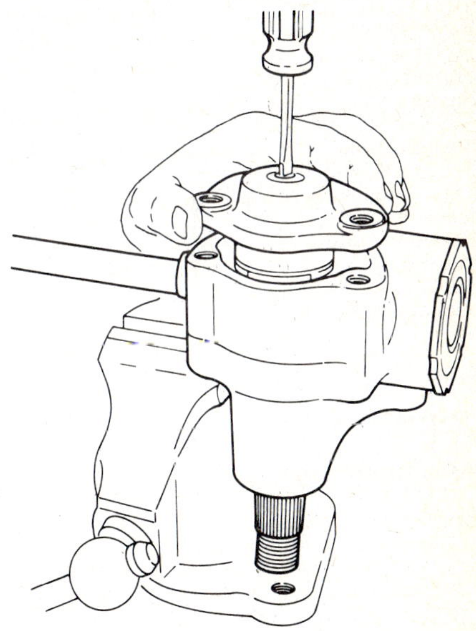

Fig. 24 Installation of sector cover

Chrysler Corp. Manual Rack & Pinion

DESCRIPTION

The steering gear assembly, used on Chrysler Corp. front wheel drive vehicles, consists of a housing which contains a toothed rack, a pinion, rack slipper and rack slipper spring, Fig. 25. The steering gear rack and pinion assembly converts rotational movement of the pinion assembly into transverse movement of the rack. Tie rods and tie rod ends transmit this movement to the steering arms and wheels while accommodating suspension movement at the same time. The tie rods are coupled to the ends of the rack. This connection is protected by a bellows type oil seal which retains steering gear lubricant. The pinion runs on straddle mounted-ball bearings. The lower bearing is incorporated in the pinion housing. The upper bearing is swagged to the pinion shaft. Lock to lock stops are built into the steering gear.

STEERING SERVICE

NOTE: The steering gear cannot be adjusted or repaired. If a malfunction occurs, the entire rack and pinion assembly must be replaced. If bellows seal is leaking or damaged, remove gear from vehicle. Drain, clean and relube gear with 1/4 pt. of MS-564490 hypoid oil. Replace damaged or leaking bellows.

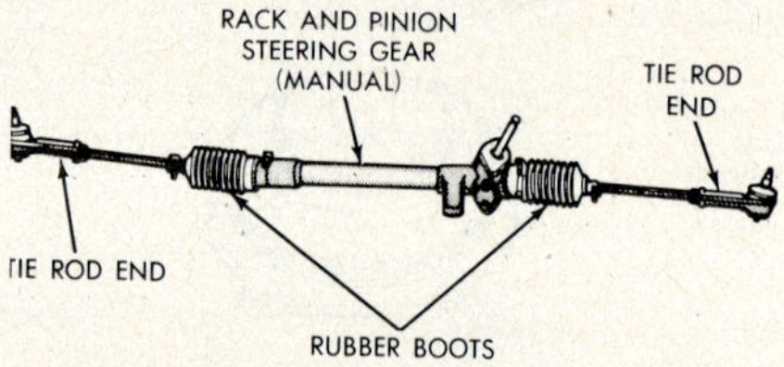

Fig. 25 Chrysler Corp. rack & pinion steering gear. 1982—83 Rampage & Scamp

POWER STEERING GEARS

CONTENTS

TROUBLE SHOOTING

Hard Steering

HYDRAULIC FLUID OVERFLOW

1. Reservoir too full.
2. Filter out of place.

STANDING OR PARKING

1. Fluid level too low.
2. Loose or glazed pump belt.
3. Check tires: Pressure, wear, alignment, balance.
4. External leakage.
5. Linkage loose, tight, binding, friction.
6. Check entire hydraulic system.

AT ALL TIMES

1. Fluid level low.
2. Pump belt loose or glazed.
3. Check tires: Pressure, wear, alignment, balance.
4. External leakage.
5. Mechanical linkage loose, tight, binding, friction.
6. Check entire hydraulic system.

PART OF THE TIME

1. Fluid level low.
2. Pump belt loose or glazed.
3. Check tires: Pressure, wear, alignment, balance.
4. External leakage.
5. Mechanical linkage loose, tight, binding, friction.
6. Steering gear needs adjustment.
7. Interference.
8. Check entire hydraulic system.

"LUMPY" EFFECT OR SHIMMY

1. Fluid level too low.
2. External leakage.
3. Mechanical linkage loose, tight, binding, friction.
4. Defective pump.

JERKS OR SURGES

1. Pump belt loose or glazed.

WHEN TRYING TO STEER "FAST"

1. Fluid level too low.
2. Pump belt loose or glazed.
3. Check tires: Pressure, wear, alignment, balance.
4. External leakage.
5. Check entire hydraulic system.

IN ONE DIRECTION

1. Gear out of adjustment.
2. Interference.
3. Defective control valve.

AT CERTAIN HAND WHEEL POSITIONS

1. Steering gear out of adjustment.
2. Interference.

WITH ABNORMAL NOISES

1. Fluid level too low.
2. Pump belt loose or glazed.
3. Defective pump.

WANDER, DRIFT, "DIVE", "DART"

1. Check tires: Pressure, wear, alignment, balance.
2. Mechanical linkage loose, tight, binding, friction.

LACK OF RECOVERY

1. Mechanical linkage loose, tight, binding, friction.
2. Gear out of adjustment.
3. Interference.
4. Check entire hydraulic system.
5. Defective control valve.

MAINTENANCE

Filling Hydraulic System

1. Fill reservoir with recommended fluid to indicated level and replace cap.
2. Start engine and idle, turning wheels from one extreme to the other three times. Shut off engine and refill reservoir.
3. Continue with Step 2 until system maintains its proper indicated oil level.
4. Run engine at faster speed and continue turning wheels slowly from one extreme to the other and back again for about five minutes to check for leaks. This will bleed the system of air. Again refill the reservoir to the indicated level.

Belt Tension

Pump belt tension is usually obtained by

loosening an adjusting screw and moving the pump upward and outward. The proper tension is obtained when the belt can be depressed about 3/8" halfway between the two pulleys. Tighten pump mounting bolts securely after the adjustment is made.

Power Steering Pump Section

ROTOR TYPE PUMP

This unit, Fig. 1, consists of the reservoir, pressure relief valve, filter element, and the rotating pump gearing. The pump reservoir illustrated has a filler cap with a fluid level checking gauge.

Disassemble

1. Remove reservoir filler cap.
2. Loosen reservoir mounting bolt and remove cover.
3. Remove filter element.
4. Drain oil from reservoir.
5. Remove return stud from body, Fig. 2.
6. Remove reservoir-to-body mounting screws.
7. Remove reinforcement plate and reservoir.
8. Remove four O-ring seals, Fig. 3.
9. Remove intake O-ring seal retainer.
10. Remove cover-to-body mounting screws and separate body from cover, Figs. 4 and 5.
11. Remove gasket, and O-ring seal from body, Fig. 5.
12. Scribe a line on inner and outer rotors for reassembly purposes, Fig. 6.
13. Remove outer and inner rotors.
14. Remove inner rotor drive pin from shaft.
15. Remove bearing retaining ring with snap ring pliers.
16. Remove shaft and bearing from body.
17. Press bearing from shaft.
18. Remove seal from body, using a suitable puller.
19. To remove the combination flow control and relief valve in the pump cover, remove the spring retainer car fitting and circular section rubber O-ring seal and discard seal, Fig. 7.
20. Lift out flow control valve spring.
21. Tap cover on wooden block to remove combination flow control and relief valve.
22. To remove pressure relief valve and spring from flow control valve body, place valve in vise. Place punch against relief valve and remove spring pressure. Remove retaining ring, Fig. 8, using internal snap ring pliers. Then remove relief valve and spring, Fig. 9.

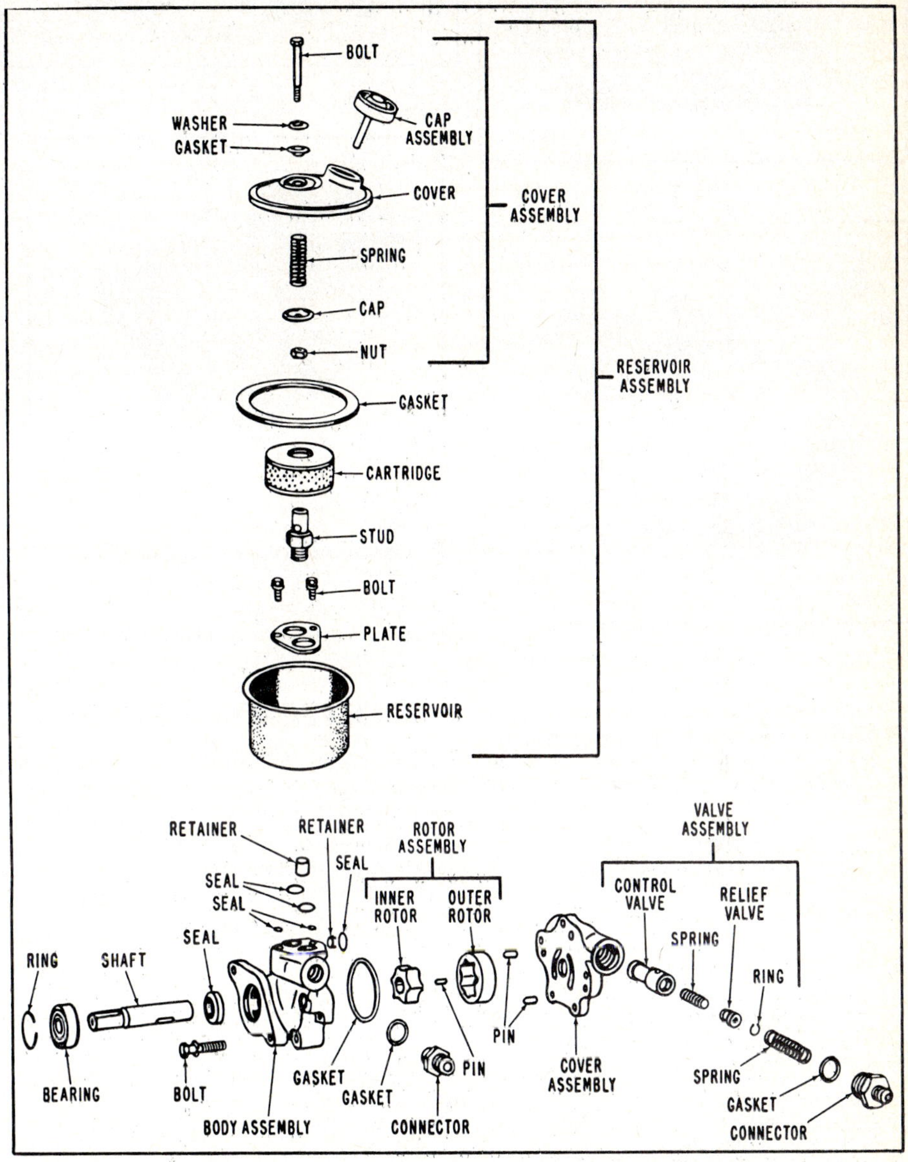

Fig. 1 Rotor type power steering pump disassembled

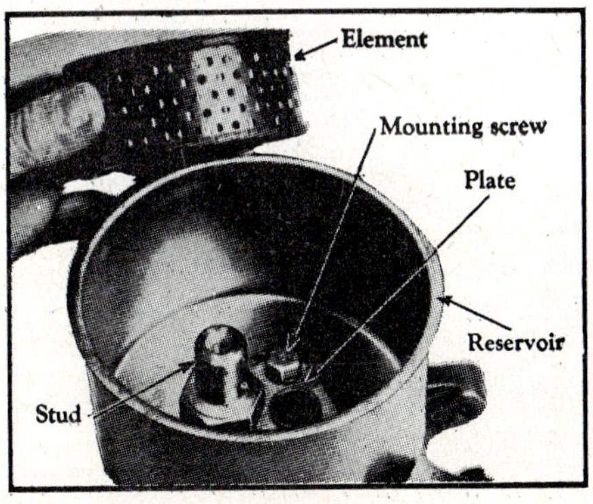

Fig. 2

Reassembly

Clean all parts and air dry only. Coat all parts with a light film of oil as assembly progresses. This will assure initial lubrication of parts.

1. The relief valve spring weight should be checked with a spring tester before preassembling the flow control valve. The spring load is usually between 14 and 16 pounds when compressed to a length of 1 3/16".
2. Preassemble the flow control valve as shown in Figs. 8 and 9.
3. Install flow control valve in pump cover.
4. Install flow control valve spring.
5. Install new O-ring seal on flow control valve cap fitting and install into pump cover.
6. Install new seal with lip toward rotor into pump body, using a suitable seal driver.
7. Press bearing onto shaft until bearing seats against shoulder on shaft.
8. Install shaft and bearing into pump body.
9. Install bearing retaining ring.
10. Install inner rotor drive pin in shaft groove.
11. Install inner rotor, making sure scribe line is facing upward.
12. Install outer rotor, making sure scribe

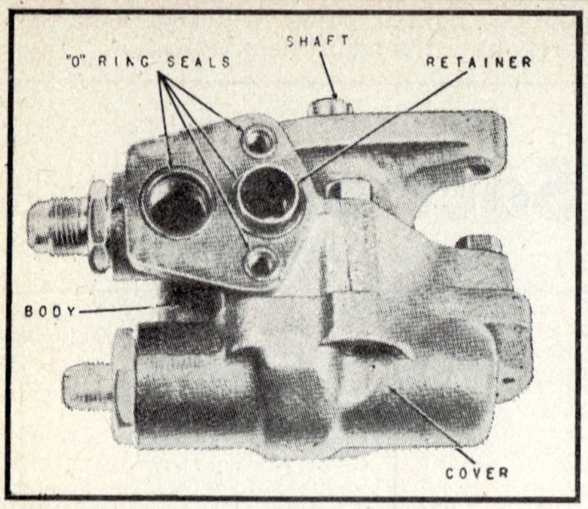

Fig. 3

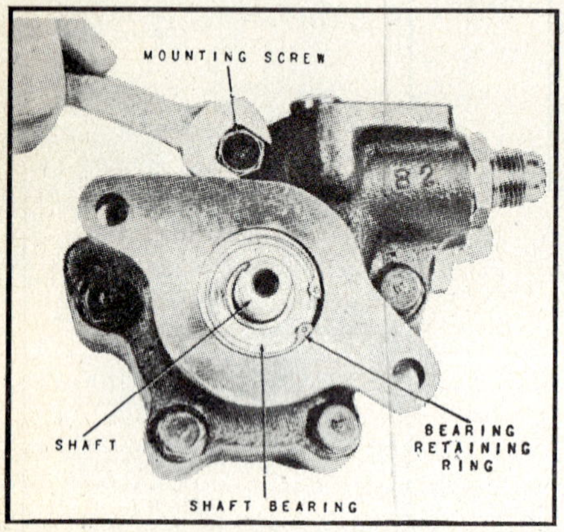

Fig. 4

line is facing upward.

13. After rotors have been nested in pump body, the tooth nose clearance should be checked with a feeler gauge as shown in Fig. 10. If the clearance exceeds .008", replace rotors with a new set. Also the end clearance should be checked as shown in Fig. 11, and in the event that it exceeds .0025", the rotors must be replaced with a new set.
14. Install new O-ring seal in body.
15. Install new body-to-cover gasket.
16. Position cover on body and torque cover screws to 20–25 ft-lbs.
17. Rotate pump shaft and check for free rotation.
18. Install O-ring seal retainer into body.
19. Position four new O-ring seals on pump body to mount reservoir.
20. Position reservoir over the four O-ring seals and pump body.
21. Place reservoir reinforcing plate into position.

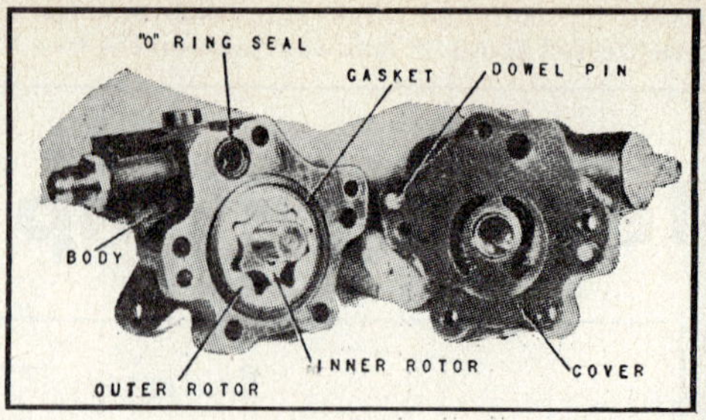

Fig. 5

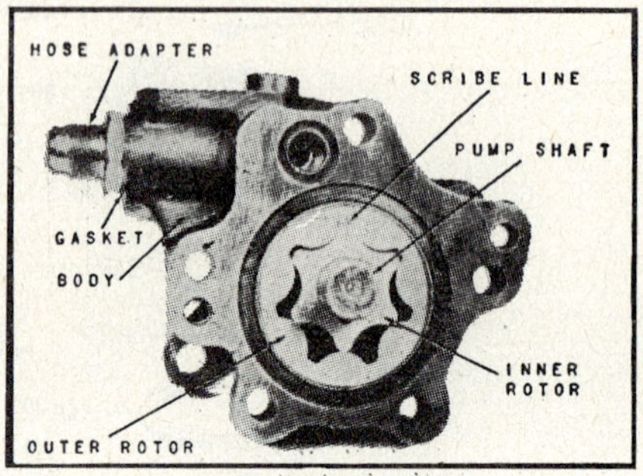

Fig. 6

22. Secure reservoir and plate to body with two screws and lockwashers.
23. Install pump return stud in body.
24. Install new filter element.
25. Install new cover gasket with cover and tighten securely.

ROLLER TYPE PUMP

This type pump, Fig. 12, consists of reservoir, pressure relief valve, filter element, and the rotating pump cam.

Thoroughly clean exterior of pump and reservoir, using care so that dirt does not enter pump or reservoir upon disassembly. During all assembly operations, extreme caution must be used to prevent any dirt from entering pump. All parts should be lightly oiled during reassembly to provide initial lubrication.

Disassembly

1. Remove reservoir cover, gasket, filter element spring, spring seat and filter element.
2. Remove $^5/_{16}$" stud with stud puller or two nuts, Fig. 13.

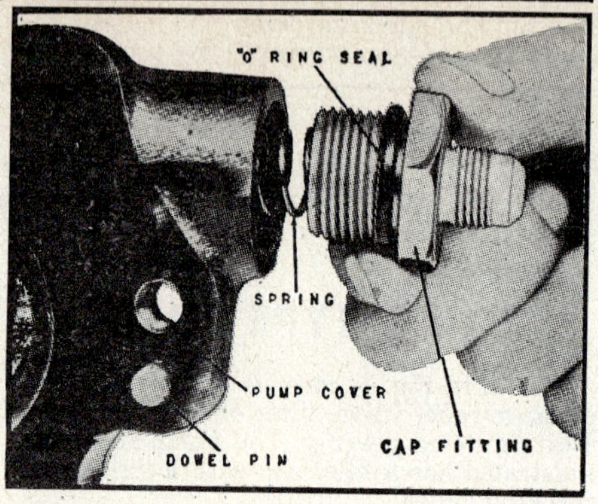

Fig. 7

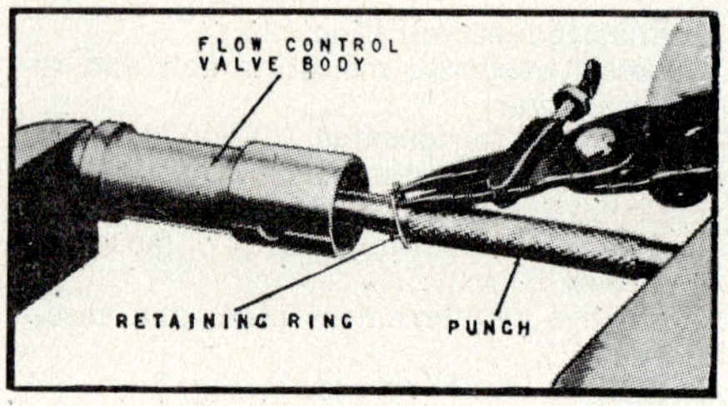

Fig. 8

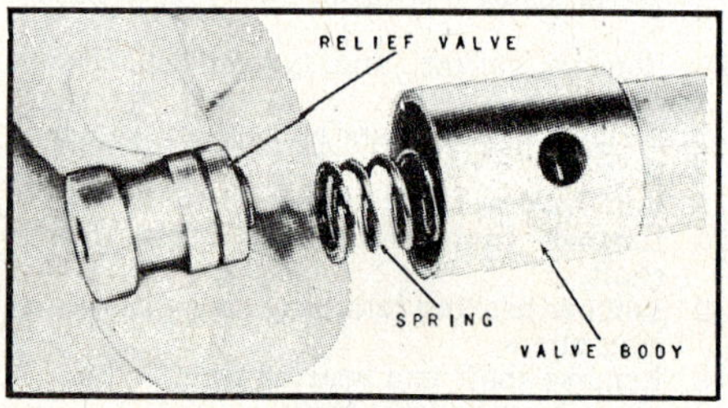

Fig. 9

3. Remove stud that retains filter element support, Fig. 14.
4. Remove reinforcing plate and reservoir.
5. Remove four O-ring seals between reservoir and pump body and discard seals, Fig. 15.
6. Remove cover-to-body mounting screws and studs, Fig. 16. Separate cover from body, Fig. 17, lifting cover vertically from body to prevent internal parts from falling out.
7. Remove O-ring seals from grooves in pump housing, Fig. 17, and discard

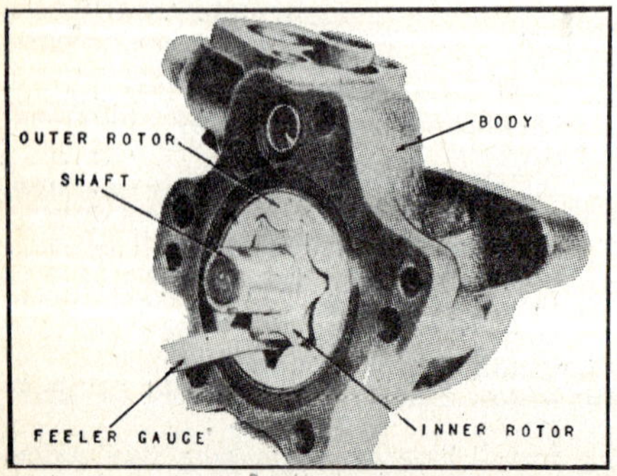

Fig. 10

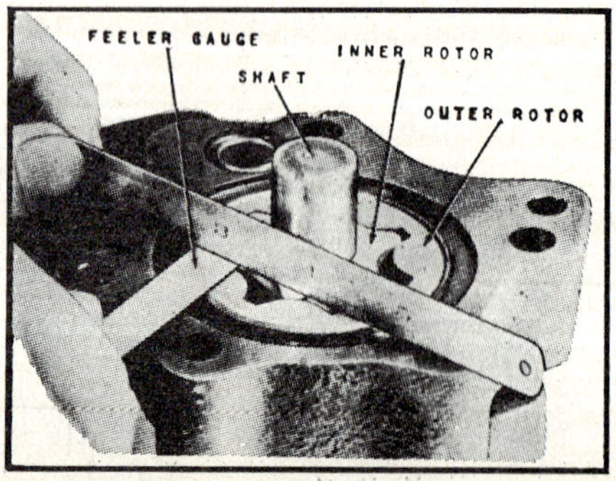

Fig. 11

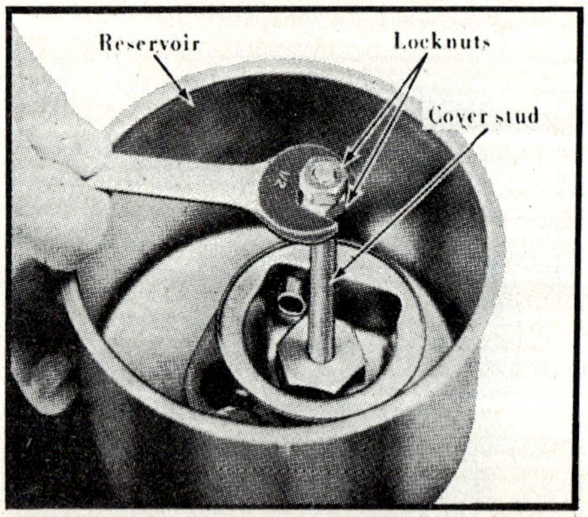

Fig. 13

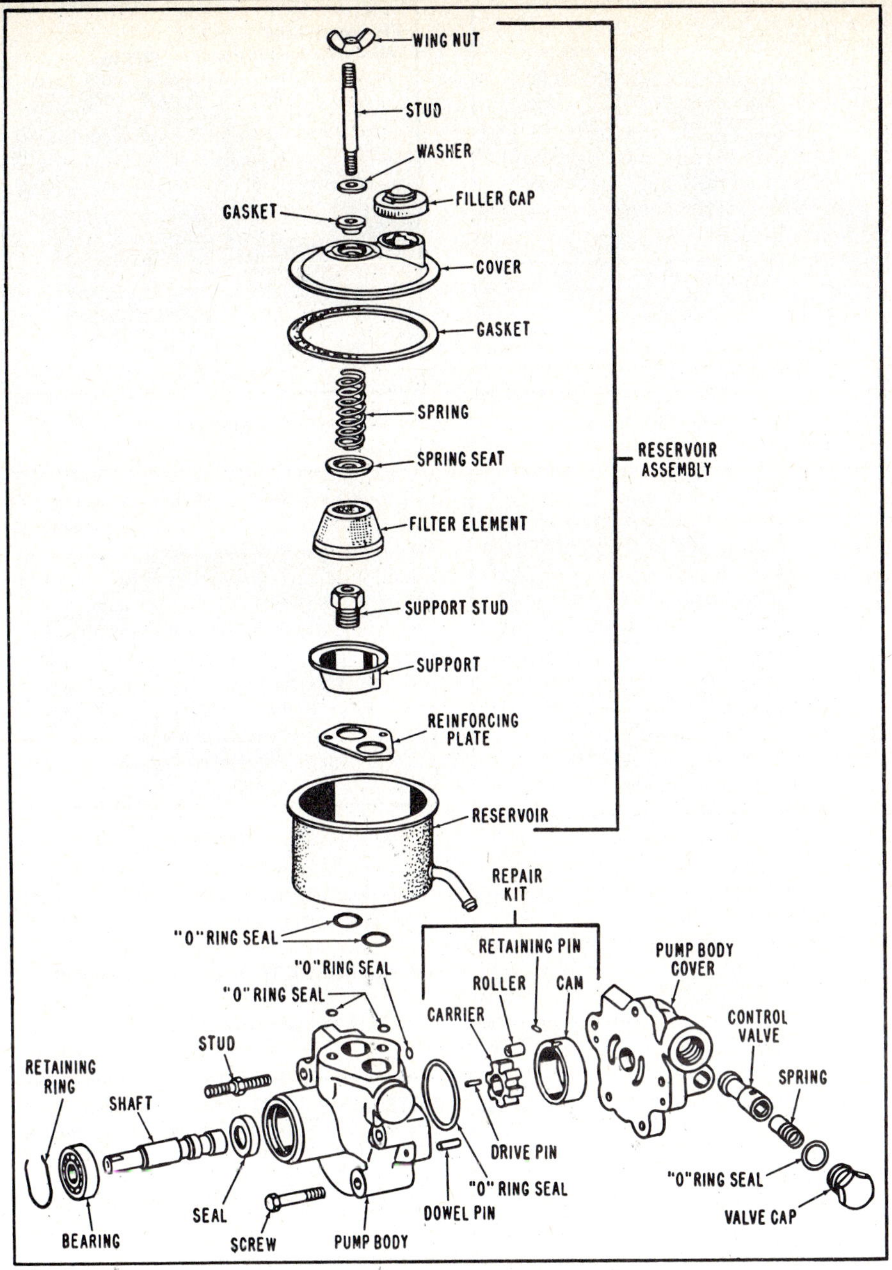

Fig. 12 Roller type power steering pump disassembled

seals.

8. Before removing carrier, mark a line across the carrier and cam, Fig. 18. This will permit correct reassembly as the carrier has a drive side and must be installed as shown. Do not scratch surfaces when making marks.
9. Remove rollers, carrier and drive pin from shaft. Then lift out cam and cam retaining pin, Fig. 19.
10. Remove bearing retaining ring. Remove shaft and bearing from body, and press bearing from shaft.
11. Remove seal from body, using a suitable seal puller.
12. To remove combination flow control and relief valve in pump cover, Fig. 20, remove spring retainer cap and O-ring seal; discard seal.
13. Lift out flow control valve spring.
14. Tap cover on wooden block to remove combination flow control and relief valve, Fig. 21. *The relief valve is an internal part of the flow control valve and is not to be disassembled.*

Reassembly

1. Install flow control valve into pump cover, making sure exposed ball enters bore first.
2. Install flow control valve spring.
3. Install new O-ring seal on flow control valve retainer cap and install into pump cover. Torque retainer cap to 30–35 ft-lbs.
4. Install new seal with lip toward carrier into pump body, using a suitable seal driver.
5. Press bearing onto pump shaft until bearing seats against shoulder on shaft.
6. Install shaft and bearing into body.
7. Install bearing retainer ring in groove in body, making sure ring is seated properly.
8. Install cam retaining pin into body.
9. Install cam into body, making sure notch in cam indexes with retaining pin and cam is seated firmly in pocket.
10. Install carrier drive pin into groove in pump shaft.
11. Install carrier into pump body, making sure line previously marked on carrier indexes with line marked on cam.
12. Install rollers into pockets of carrier.
13. After the cam, carrier and rollers have been placed in pump body, the end clearance of carrier and rollers in body should be checked, Fig. 22. If end clearance exceeds .002", replace carrier or rollers.
14. Install new O-ring seals in body.

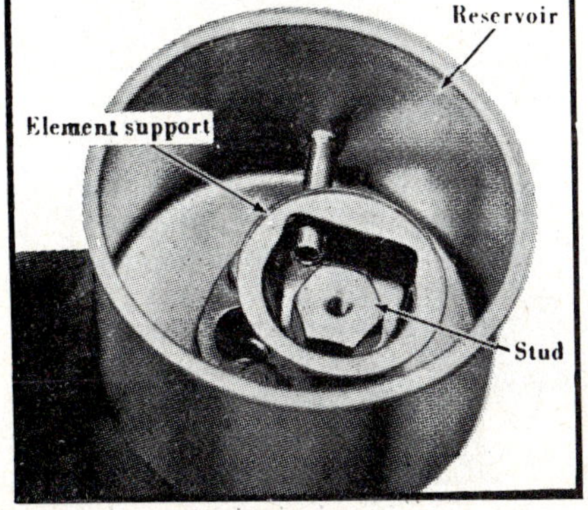

Fig. 14

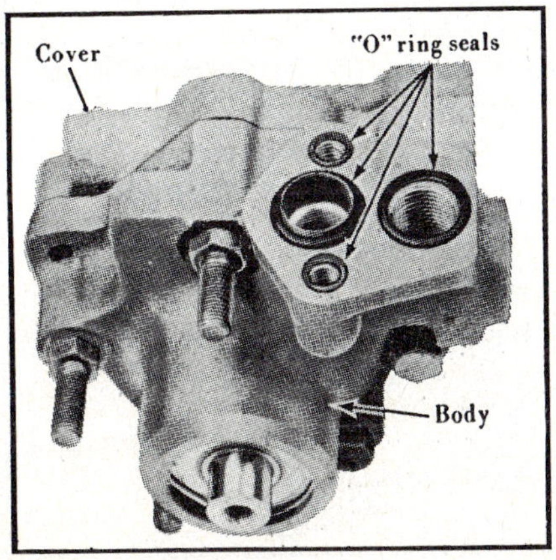

Fig. 15

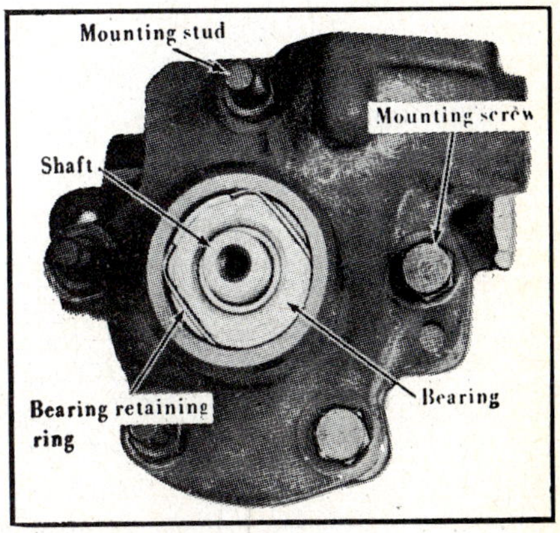

Fig. 16

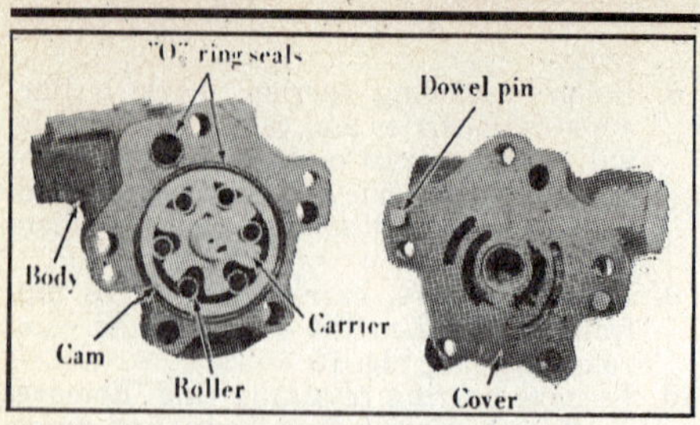

Fig. 17

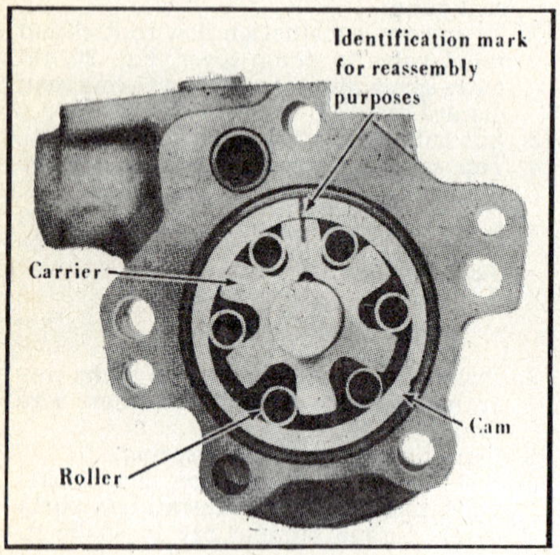

Fig. 18

15. Position cover on body and torque screws and studs to 20–25 ft-lbs. Rotate pump shaft and check for free rotation.
16. Position four new O-ring seals on pump body to mount reservoir.
17. Position reservoir over four O-ring seals and body.
18. Secure reservoir and reinforcing plate and torque screws to 30–35 ft-lbs.

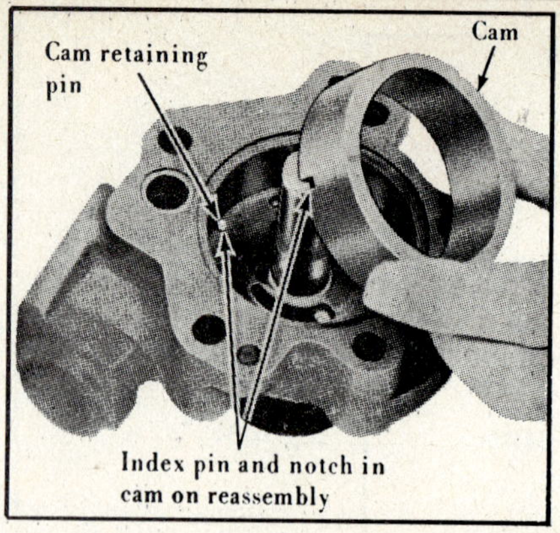

Fig. 19

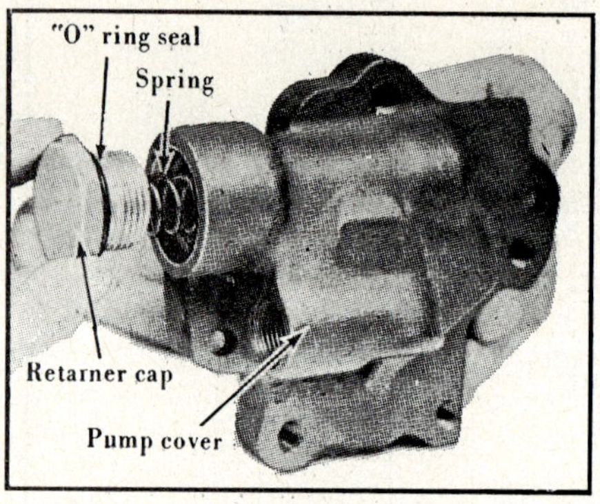

Fig. 20

19. Install filter element support and torque stud to 30–35 ft-lbs.
20. Install 5/16" stud into support stud and torque to 40–60 inch-lbs.
21. Place new filter element on support. Position spring seat on top of filter. Install spring over stud and onto spring seat.

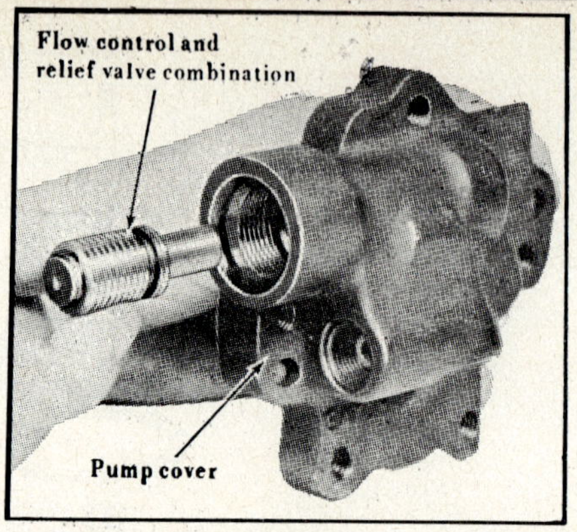

Fig. 21

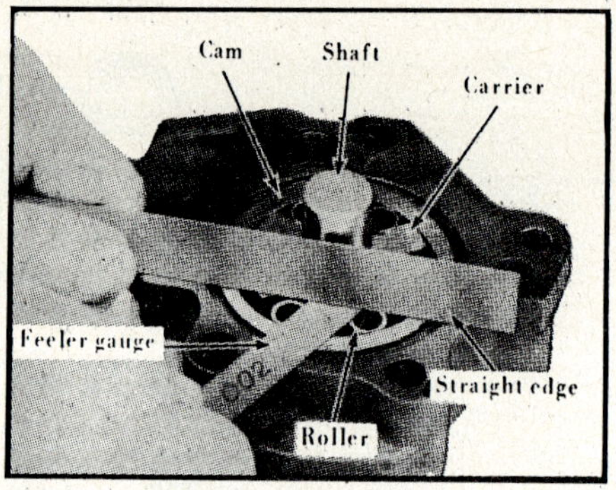

Fig. 22

Install reservoir cover with new gasket. Install reservoir cap.

SLIPPER TYPE PUMP

Except Ford Model CII

NOTE: *Disassembly of the slipper type pump, Fig. 23, is not recommended as the internal parts of the pump are not serviced separately. The only parts that are serviced are the pump assembly, filler cap, reservoir, O-ring and gasket, pump shaft oil seal, flow control plug and O-ring, snap ring and relief valve.*

Disassembly

1. Clean exterior of pump.
2. Remove filler cap and drain reservoir.
3. Remove brackets, reservoir screws, gas-

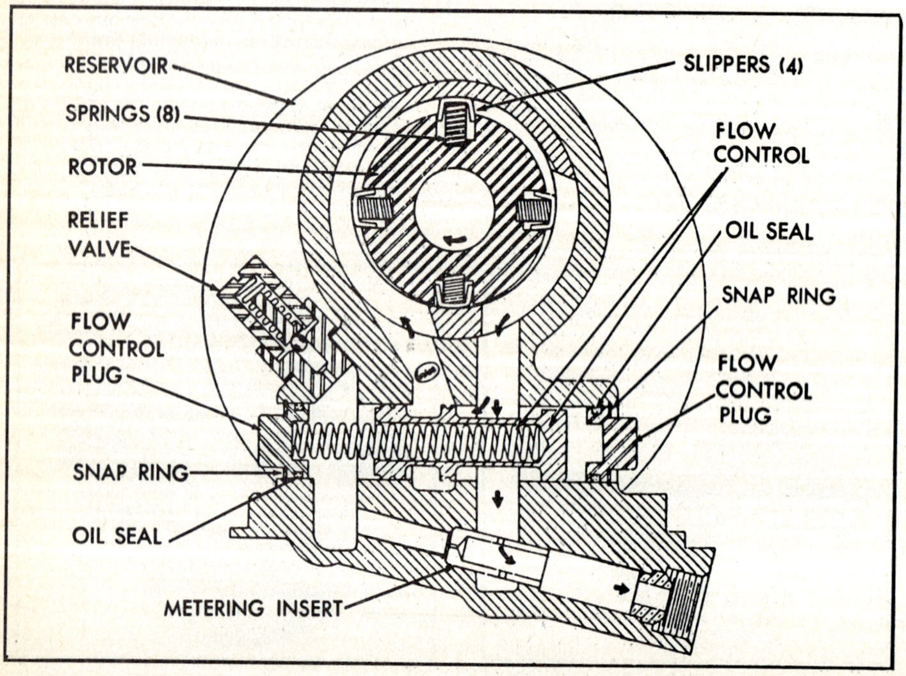

Fig. 23 Slipper type power steering pump

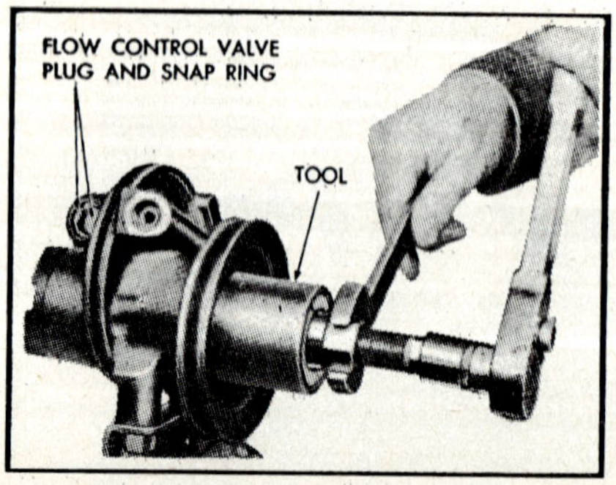

Fig. 24 Removing pump pulley

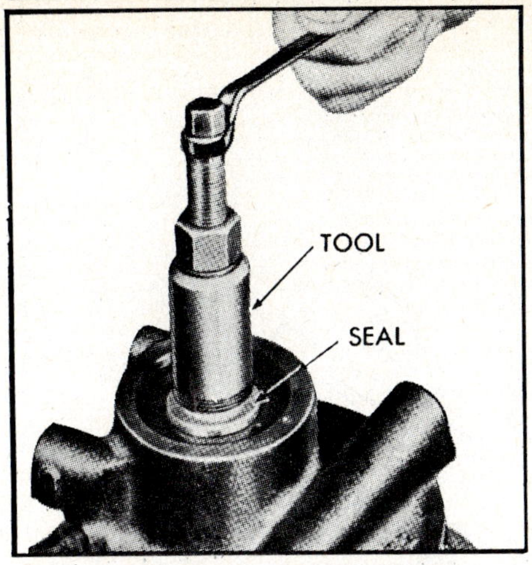

Fig. 25 Removing pump oil seal

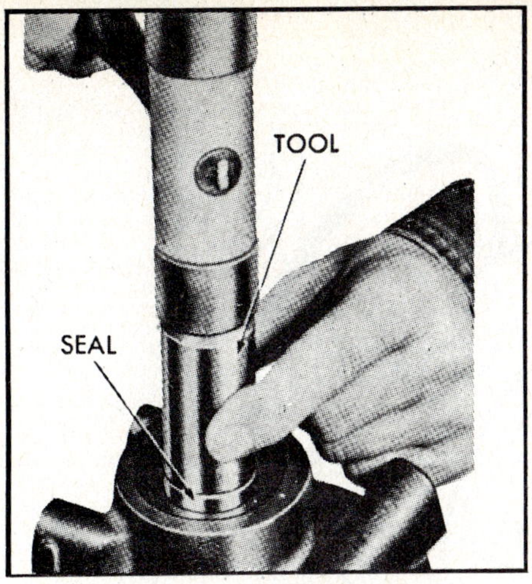

Fig. 26 Installing oil seal

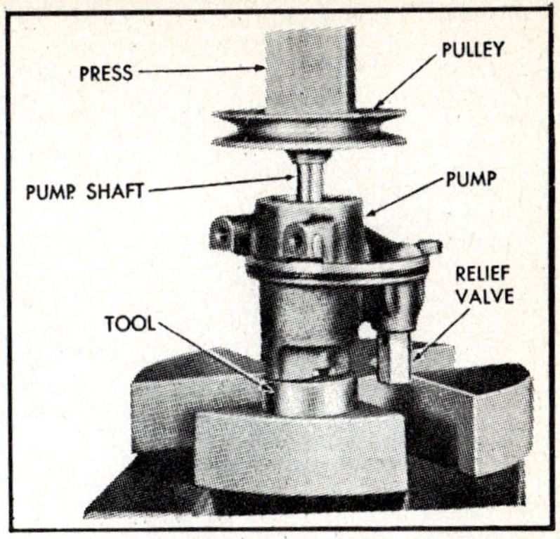

Fig. 27 Installing pump pulley

kets and O-ring.

4. Using spacer washers between front bracket and pump, reinstall front bracket for use as a holding fixture. Clamp bracket in a vise.
5. Use a suitable puller to remove pulley, Fig. 24.
6. Remove relief valve and gasket.
7. Remove flow control valve plug snap ring.
8. Remove flow control valve plug and O-ring by gripping end of plug with needle nose pliers.

CAUTION: *The valve is spring loaded.*

9. Remove oil seal by threading a suitable tool far enough to engage metal portion of seal. Then by turning the center puller screw while holding the tool body will force the seal from the pump, Fig. 25.

Inspection

Clean parts in mineral spirits or other suitable solvent. Discard body-to-reservoir O-ring and flow control valve plug O-ring. Pump shaft should turn freely and should be smooth at seal contact area. Flow control valve bore and valve should be smooth, free of scores or scratches. Valve must operate freely in bore.

NOTE: *Small scratches can be removed with crocus cloth. Do not round off the square edges on the valve as they are vitally important to this type of valve. The housing bore should not be honed. If the bore is scratched or worn, the pump should be replaced.*

Reassembly

1. Install flow control valve spring and valve. Lubricate and assemble a new O-ring on the flow control valve plug and install plug.
2. Install snap ring in bore.
3. Install new oil seal with lip of seal toward pump.
4. Install pressure relief valve and gasket. Tighten plug to 30 ft-lbs torque.
5. Remove front bracket. Support pump body on a holding fixture, Fig. 26. Pump must be supported in the manner shown when installing the seal so that all pressing force will be applied to the shaft only; otherwise the pump body and rotor will be damaged.
6. Install pulley with a heavy duty arbor press, Fig. 27. Press on pulley hub only

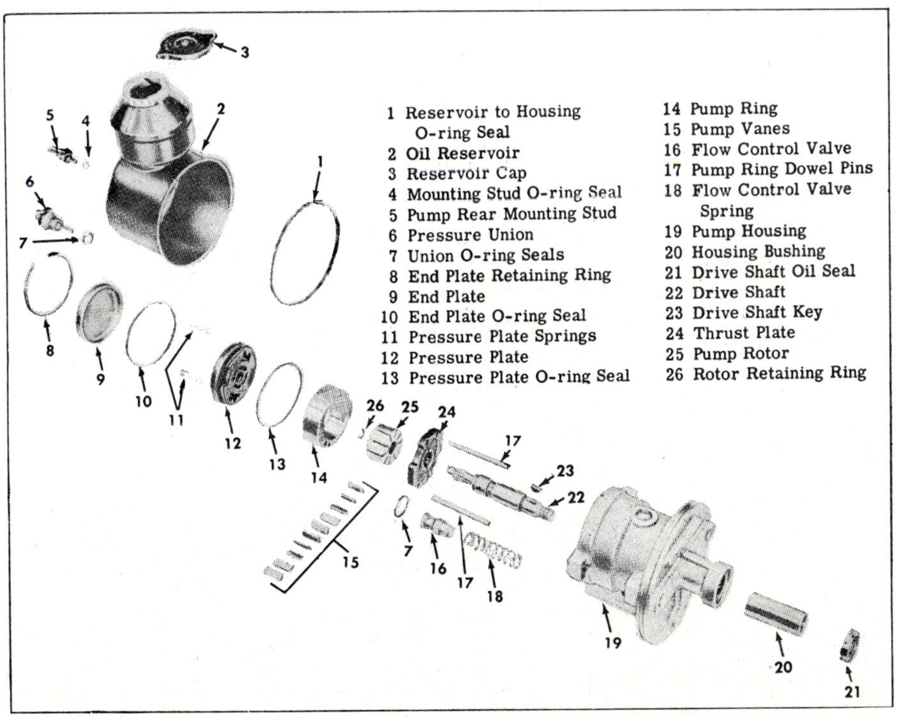

Fig. 28 Vane type power steering pump disassembled

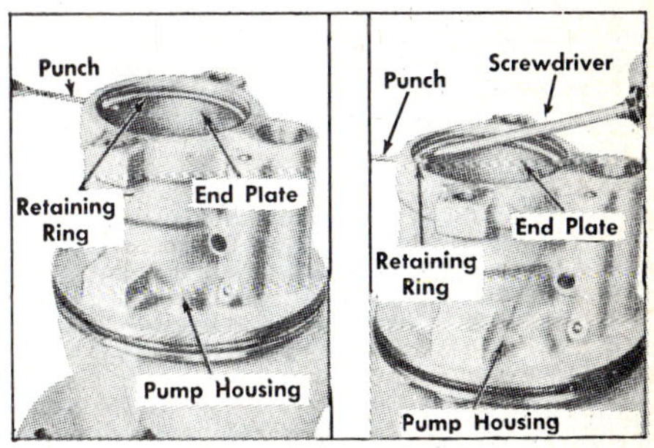

Fig. 29 Removing end plate retaining ring

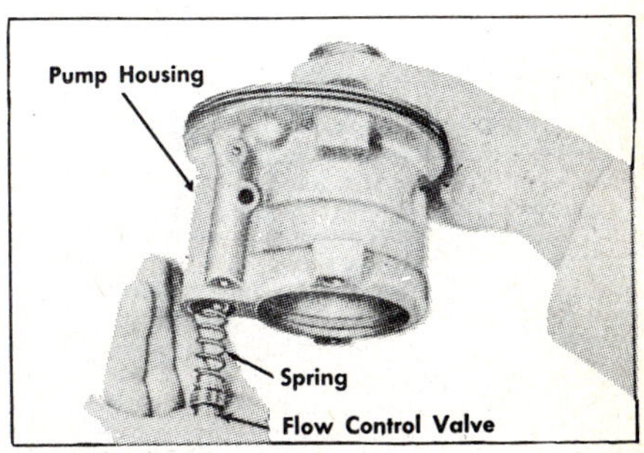

Fig. 30 Removing flow control valve and spring

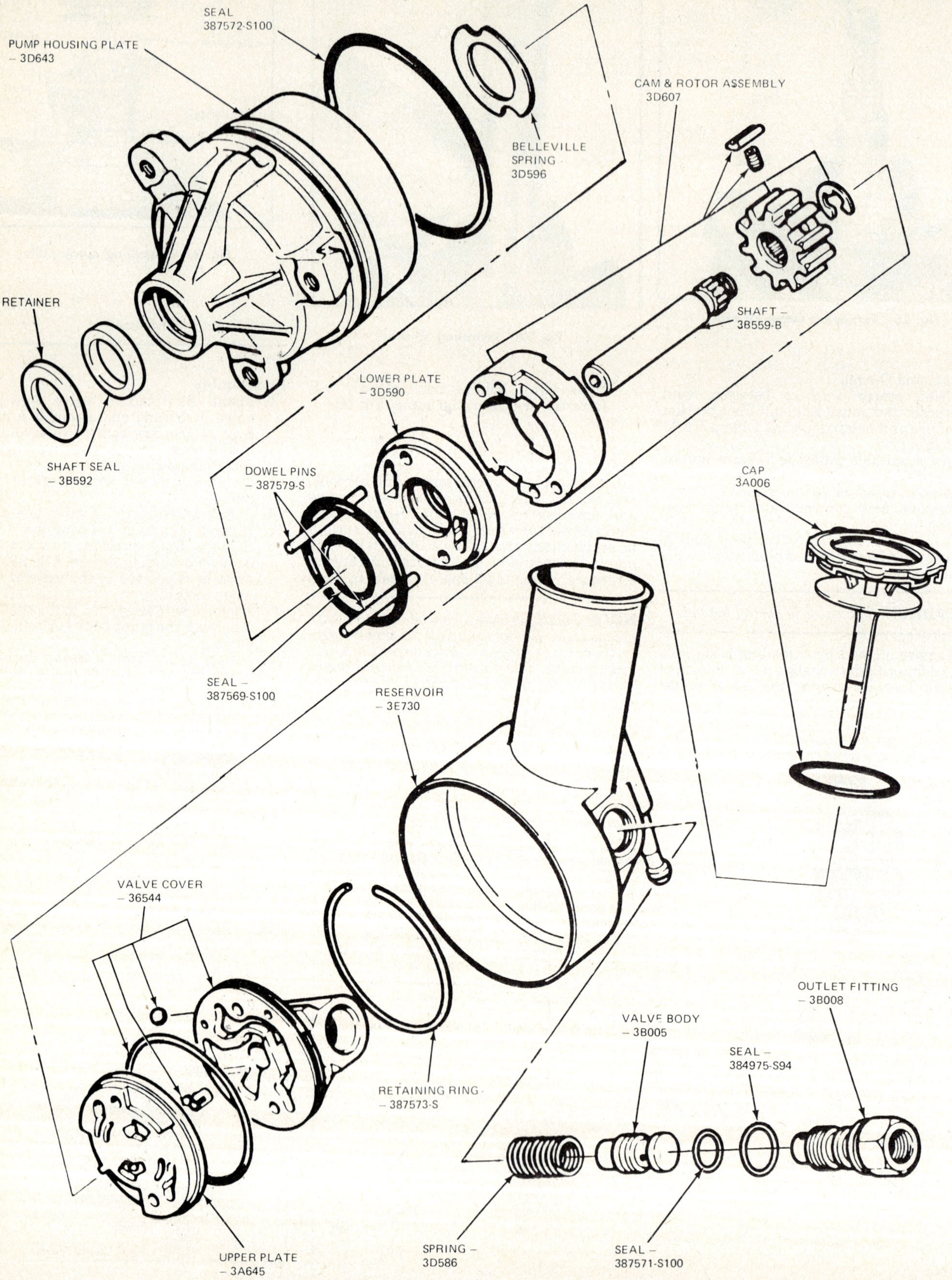

Fig. 27A Disassembled view of Ford Model CII power steering pump

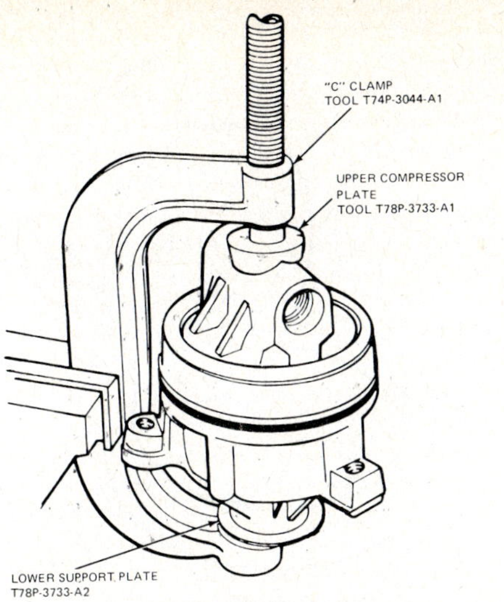

Fig. 27B Positioning pump in C clamp. Ford Model CII

until hub is flush with end of pump shaft.

7. Lubricate large O-ring and reservoir cap screw gasket and install both on pump body.
8. Install reservoir and pump brackets. Torque screws to 10 ft-lbs.

Ford Model CII

Disassembly, Fig. 27A

1. Remove pulley from pump.
2. Remove outlet fitting, flow control valve and flow control valve spring from pump then remove reservoir.
3. Place a suitable C-clamp in a vise.
4. Position lower support plate T78P-3733-A2 over pump rotor shaft.
5. Install upper compressor plate T78P-3733-Al into upper portion of C-clamp.
6. While holding compressor tool, place pump assembly into C-clamp with rotor shaft facing downward, Fig. 27B.

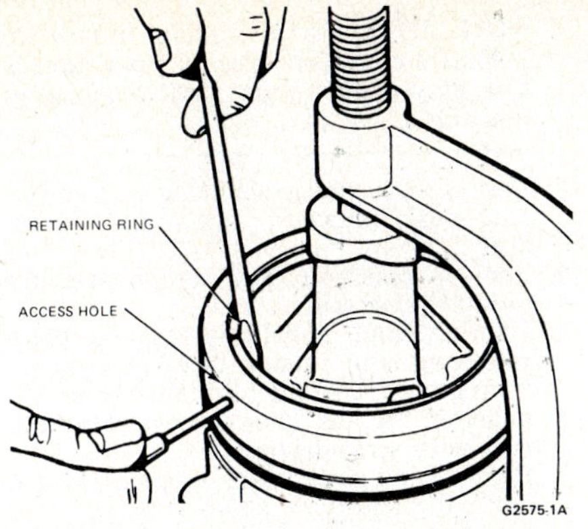

Fig. 27C Removing valve cover retaining ring. Ford Model CII

7. Tighten C-clamp until a slight bottoming of valve cover is observed.
8. Through small hole located on side of pump housing, insert a suitable drift and push inward on valve cover snap ring. While pushing inward on snap ring, place a screwdriver under snap ring edge and remove ring from housing, Fig. 27C.
9. Loosen C-clamp and remove upper compressor plate T78P-3733-A2, then remove pump assembly.
10. Remove pump valve cover and O-ring.
11. Remove rotor shaft, upper plate, cam and rotor assembly and two dowel pins.
12. Remove lower plate and spring, by tapping housing on a flat surface.
13. Using a suitable screwdriver, remove rotor shaft seal.

Assembly, Fig. 27A

1. Position rotor on rotor shaft splines with triangle detent on rotor counter bore facing upward.
2. Install snap ring into groove on end of rotor shaft.
3. Position insert cam over rotor. Ensure recessed notch on insert cam is facing upward.
4. With rotor extended upward apporoximatcly half out of cam, insert spring into rotor picket, Fig. 27D.
5. Use a slipper to compress spring, then install slipper with groove facing cam, Fig. 27E.
6. Perform steps 4 and 5 on slipper cavity beneath opposite inlet recess.
7. While holding cam stationary, index rotor left or right one space and install another spring and slipper until all ten rotor cavities have been filled. Use car when turning rotor springs and slippers remain in position.
8. Apply Loctite No. 242 or 271 adhesive or equivalent to outside diameter of seal and Locquic NF or T primer or equivalent to seal bore in housing. Install rotor shaft seal using tool No. T78P-3733-A3. Using a plastic mallet, drive seal into bore until properly seated.
9. Position pump plate on flat surface with pulley side facing downward.
10. Install two dowel pins and spring into housing.

NOTE: Spring must be inserted with dished surface facing upward.

---

11. Lubricate inner and outer O-ring seals with power steering fluid, then install seals on lower pressure plate.

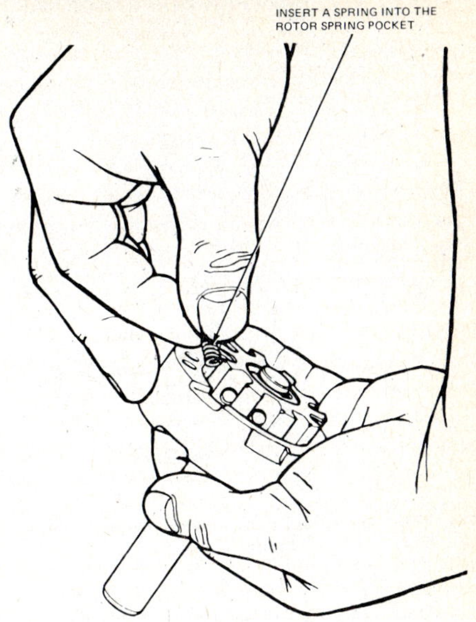

Fig. 27D Installing slipper rings. Ford Model CII

12. Install lower pressure plate into housing and over dowel pin with O-ring seals facing toward front of pump. Position assembly on C-clamp. Place tool No. T78P-3733-A3 into rotor shaft hole and press on lower plate lightly until it bottoms in pump housing. This will seat the outer O-ring seal.
13. Install cam, rotor, rotor and slippers and rotor shaft assembly into pump housing over dowel pins. When installing assembly into pump housing, stepped holed must be used for dowel pins and notch in cam insert must be toward reservoir and approximately 180 degrees opposite square mounting lug on housing. Fig. 27F.
14. Position upper pressure plate over dowel pins with recess directly over recessed notch on cam insert and approximately 180 degrees opposite square mounting lug, Fig. 27G.
15. Lubricate O-ring seal with power steering fluid. then position O-ring on valve

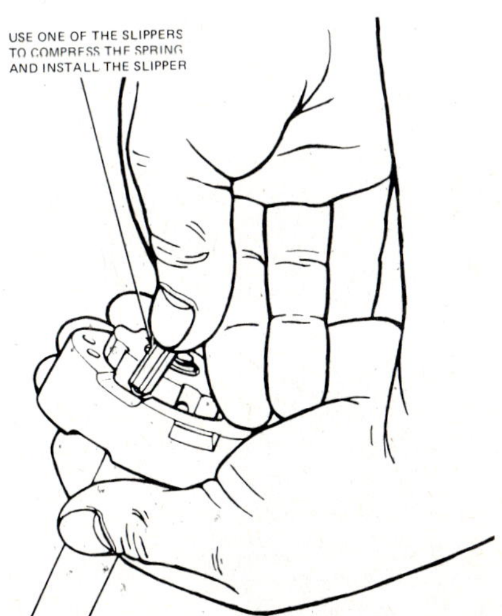

Fig. 27E Installing slippers. Ford Model CII

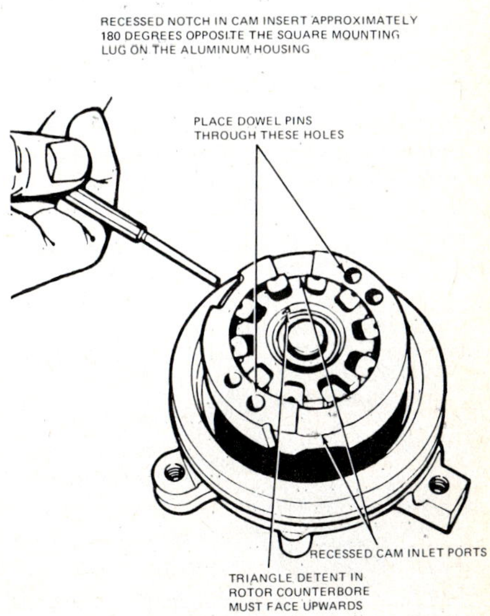

Fig. 27F Assembling cam, slippers and rotor. Ford Model CII

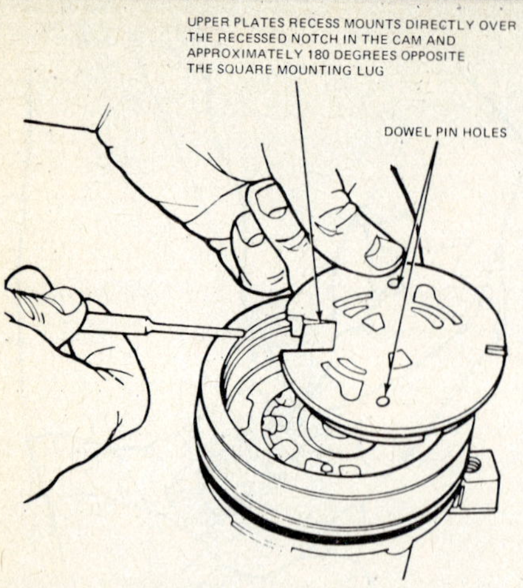

Fig. 27G Installing upper pressure plate. Ford Model CII

cover. Ensure plastic baffle is securely in position on valve cover. A coat of vaseline may be used to hold baffle in position.

16. Insert valve cover over dowel pins. Ensure outlet fitting hole in valve cover is aligned with square mounting lug on housing, Fig. 27H.
17. Place assembly in C-clamp and compress valve cover into pump housing until snap ring groove on housing is exposed.
18. Install valve cover snap ring in pump housing. Ensure snap ring ends are near access hole in pump housing.
19. Remove pump assembly from C-clamp.
20. Lubricate O-ring seal with power steering fluid, then place seal on pump housing.
21. Install reservoir on pump housing.
22. Install flow control valve and spring into valve cover.
23. Lubricate O-ring seals with power steering fluid, then place seals on outlet fitting.
24. Install outlet fitting on valve cover. Torque outlet fitting to 25 to 34 ft. lbs. Use care not to cock flow control valve when installing. Do not force valve forward otherwise damage to housing may result.

VANE TYPE PUMP

Disassembly

1. Referring to Fig. 28, carefully mount pump in vise with soft jaws and remove reservoir cap.

NOTE: *When clamping pump in vise, do not exert excessive force on front hub as this may distort drive shaft bushing in housing.*

2. Remove rear pump mounting stud and O-ring seal, pressure union and O-ring seal.
3. Remove reservoir, and second pressure union O-ring seal.
4. Using a small punch, depress end plate retaining ring enough to allow removal from groove. Use the 1/8" diameter hole in housing, Fig. 29. Then remove retaining ring with screwdriver.
5. Remove end plate which is spring loaded and will usually sit above housing level. If end plate sticks, a slight rocking action should free it. If rocking action fails to free plate, use a magnet.
6. Remove two pressure springs from pump housing.
7. Remove flow control valve and spring by inverting housing. Do not disassemble flow control valve as it is serviced as a unit and pre-set at the factory, Fig. 30.
8. Remove drive shaft key from slot in shaft. Then with end of shaft pointed downward, press down until shaft is free.
9. Turn assembly over and remove drive shaft and rotary group.
10. Remove rotor retaining ring from groove in drive shaft. Remove rotor and thrust plate from shaft, Fig. 31.
11. Remove and discard O-ring seals from pump housing. Remove drive shaft oil seal only if inspection shows necessity for replacement.

Inspection

1. Clean all parts except the drive shaft oil seal in cleaning fluid. The seal will be damaged if immersed in cleaning fluid.
2. Check fit of vanes in slots of rotor for tightness or excessive looseness. Vanes must fit snugly but slide freely in slots in rotor. Tight fit of vanes in rotor can usually be corrected by thorough cleaning. Replace rotor if excessive looseness exists between rotor and vanes. Replace vanes if worn or scored.
3. Examine machined surfaces of pump ring for roughness or wear. Replace ring if roughness cannot be corrected with crocus cloth.
4. Inspect thrust plate, pressure plate and

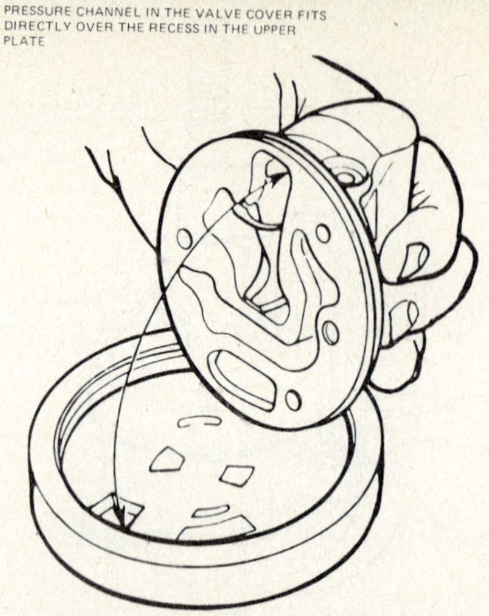

Fig. 27H Installing valve cover. Ford Model CII

end plate for wear, scores or other damage.

5. Inspect pump housing for cracks or damage. Check housing for evidence of wear or scoring.
6. Check all springs for free length, distortion or collapsed coils.
7. Inspect located dowel pins for distortion.
8. Examine outer diameter of flow control valve for scoring or roughness. Slight damage may be cleaned up with crocus cloth. Check valve assembly for freedom of movement in bore of pump housing.
9. Check all oil passages in pump parts for obstruction. Use a piece of tag wire to clean out holes.
10. Check bushing in pump housing for wear or damage.

Reassembly

1. Lubricate new O-ring seals with vaseline and place in pump housing.
2. If drive shaft oil seal was removed during disassembly, lubricate new seal with vaseline. Then install seal in pump body front hub, using a suitable seal driver.
3. Position thrust plate and rotor on drive shaft. Then install rotor retaining ring in groove of drive shaft, Fig. 31.

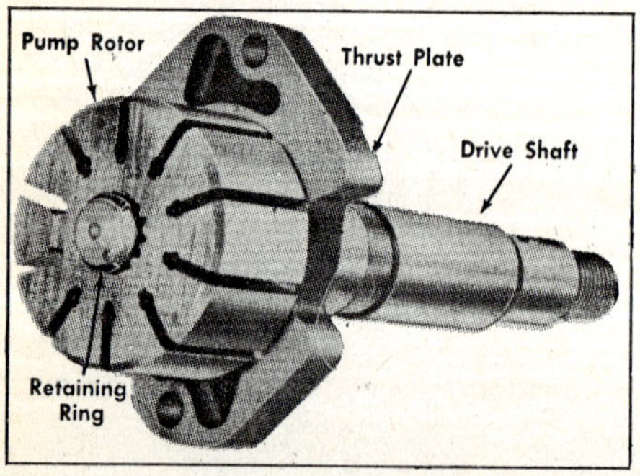

Fig. 31 Thrust plate and rotor installed on drive shaft

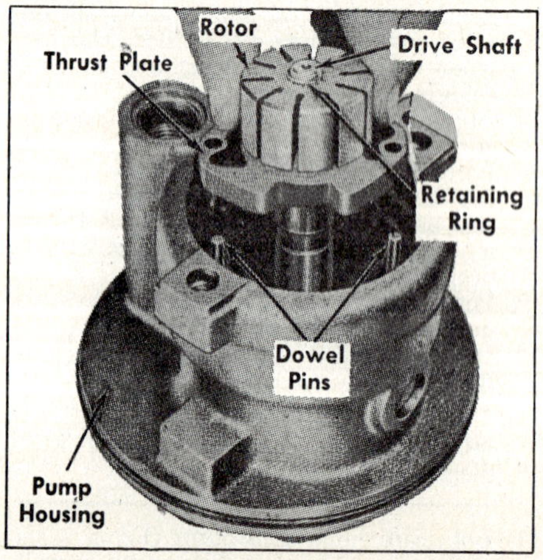

Fig. 32 Installing drive shaft assembly

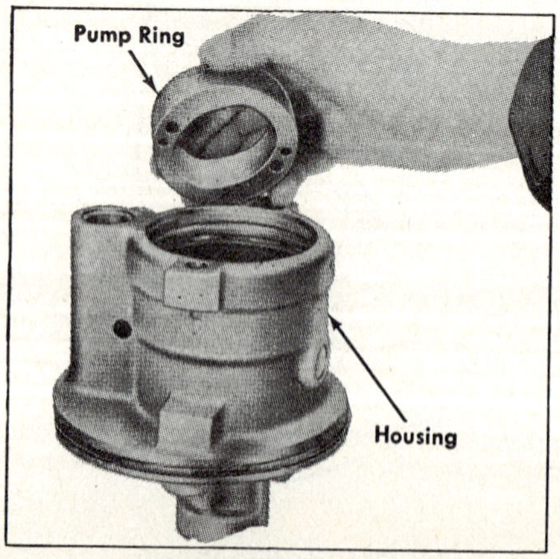

Fig. 33 Installing pump ring

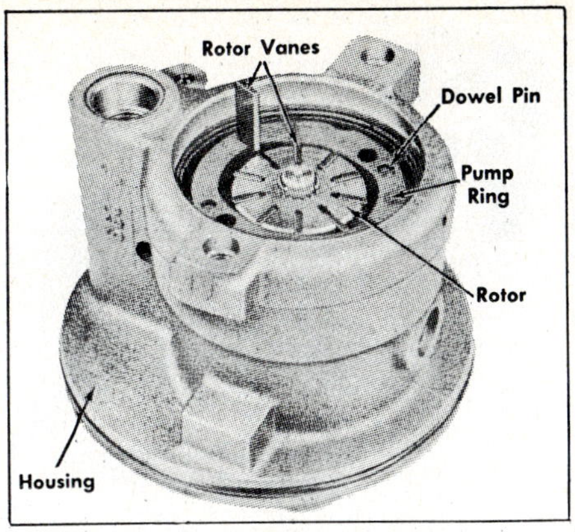

Fig. 34 Installing rotor vanes

4. Place pump housing on work bench as shown in Fig. 32 and install two pump ring dowel pins in bore of housing.
5. Install drive shaft in housing, making sure shaft seats properly.
6. Install pump ring in housing over two dowel pins with rotation arrow on ring to rear of pump housing, Fig. 33.

NOTE: *Arrow on outer edge of pump ring points in direction of pump rotation (counterclockwise when viewed from rear of pump).*

7. Install rotor vanes in slots of pump rotor, with radius edge of vanes toward outside of rotor, Fig. 34.
8. Position pump housing on two wood blocks and push parts down into place as shown in Fig. 35.
9. Lubricate outside diameter and chamfer of pressure plate with vaseline to insure against damage to O-ring seals in housing. Then install pressure plate in housing over dowel pins.

NOTE: *Ported face of pressure plate goes toward pump ring, Fig. 35.*

10. Using a 2¾" diameter sleeve to apply pressure to outer edge of pressure plate only, press pressure plate into seat.

NOTE: *Pressure plate will travel about 1/16" to seat. Do not press or hammer on center of pressure plate as to do so will cause permanent distortion resulting in pump failure.*

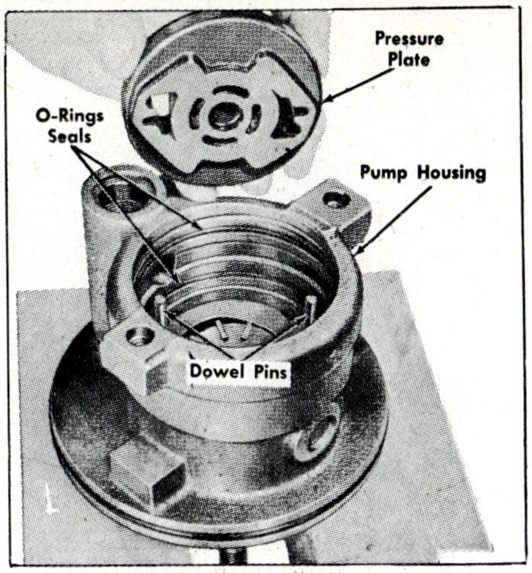

Fig. 35 Installing pressure plate

11. Install two pressure plate springs, one over each dowel pin in pump housing, Fig. 36.
12. Lubricate outside diameter and chamfer of end plate with vaseline to insure against damage to O-ring seal in housing. Then press end plate into housing. While holding end plate down in housing, install end plate retaining ring. Be sure end plate is completely and solidly seated in groove of housing, Fig. 37.
13. Remove assembly from press and place on work bench. Install flow control spring and valve in bore of housing. Make sure hex head screw goes into housing bore, Fig. 38.
14. Position pressure union O-ring seal in bore of housing over flow control valve, Fig. 39.
15. Mount pump housing in vise with soft jaws and install reservoir.
16. Install pressure union and new O-ring seal in reservoir. Tighten union.
17. Install mounting stud and new O-ring seal in reservoir. Tighten stud.
18. Install drive shaft key in slot in shaft. Support shaft on opposite side while installing key. Install reservoir cap.

SLEEVE TYPE PUMP

Disassembly, Fig. 40

1. Drain oil from pump and reservoir. Cap discharge and return line fittings to exclude dirt from pump.
2. Clean pump and mount it in a vise.

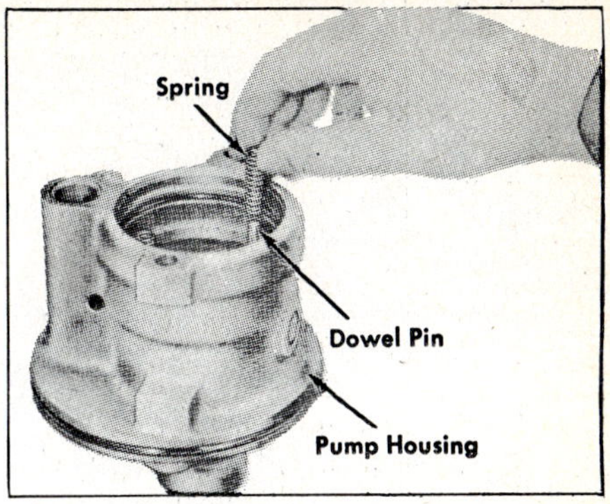

Fig. 36 Installing pressure plate springs

3. Remove reservoir.
4. Loosen capscrews at inlet end cap.
5. Remove flow valve spring retainer fitting.

CAUTION: *Flow and plunger springs are under pressure. Use care when removing cap and retainer to prevent spring and fitting from escaping due to spring force.*

6. Remove flow valve and spring.
7. Remove and disassemble both end caps from inlet cap body and pump housing. Plunger spring will tend to push cylinder blocks out of body. Use care when disassembling.
8. The cylinder blocks will push out of body slightly due to pressure of plunger spring between blocks. Remove cylinder drive blocks and sleeves from pump body as well as cylinder plunger spring.
9. Remove ball bearing retainer ring from housing, using snap ring pliers.
10. Remove shaft and bearing.

Inspection

1. Wipe bearing and shaft assembly with clean, lint-free cloths. Do not soak in cleaning solvent as the lubricants sealed in the bearing may become diluted by solvent.
2. Inspect shaft for wear and bearing for roughness or noisy operation.
3. If bearing must be replaced, remove shaft key and press bearing from shaft away from splined or serrated end of shaft.
4. Examine retaining ring groove in housing.

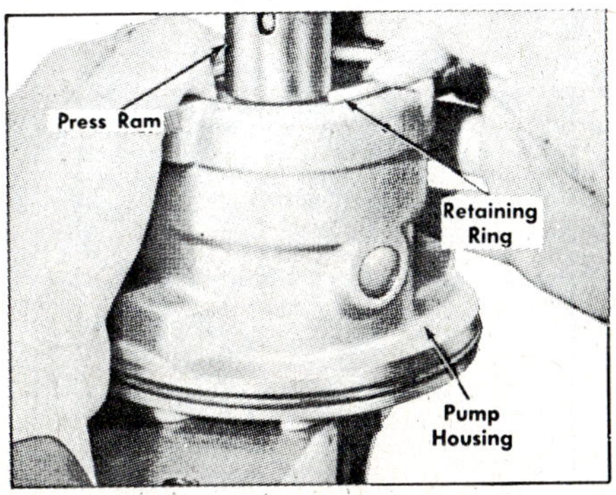

Fig. 37 Installing end plate and retaining ring

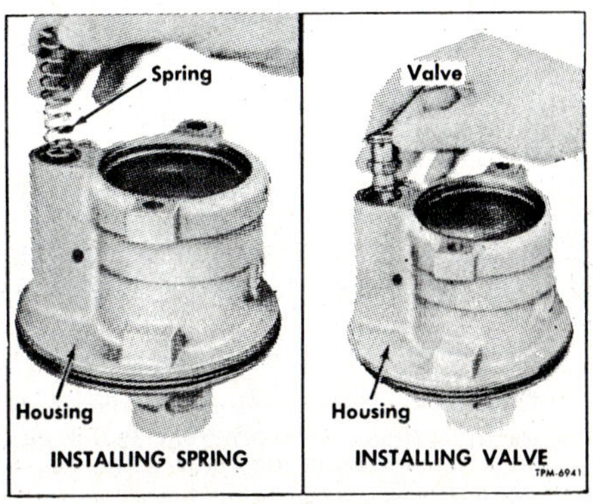

Fig. 38 Installing flow control valve and spring

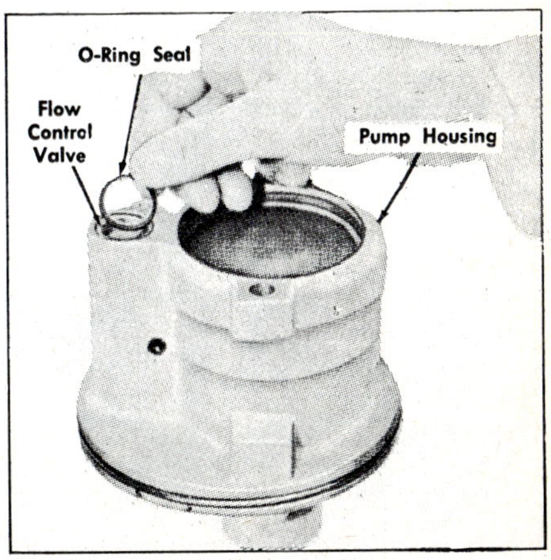

Fig. 39 Installing pressure union O-ring seal

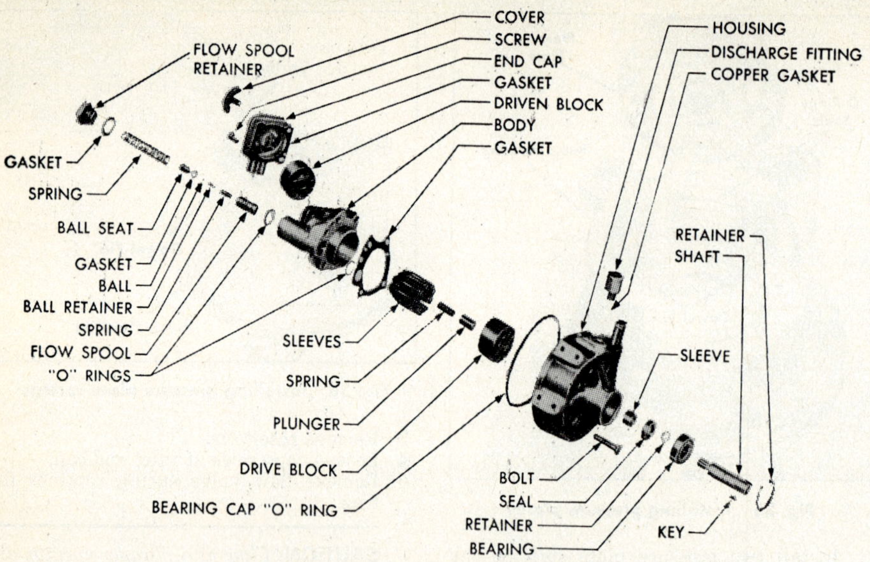

Fig. 40 Sleeve type power steering pump

5. If ball bearing is to be replaced, support bearing on its inner race and press shaft through bearing until retaining ring stops against inner race of bearing. Retaining ring must always be placed between bearing and splined or serrated end of shaft.
6. Check fit of sleeves in cylinder block bores. Sleeves must slide freely.
7. Examine mating surfaces of sleeves and bores. Heavy scoring, if present, can impair pump efficiency. Cylinder blocks showing such scoring must be replaced. Hairline markings seen on sleeves are normal.
8. Examine flow valve spool and valve liner. Slight hairline scratches are permissible. Burrs from heavy gouges or scratches which may cause flow spool to stick must be removed. This may be done by polishing with a smooth stone.

CAUTION: *Do not break sharp edges of spool or liner.*

9. Insert flow valve body. By means of a pencil or other such rod which will not mar internal diameter of valve liner, move flow spool back and forth to inspect. On each pass, rotate spool slightly. Spool must slide freely. If spool sticks or drags, remove it from liner. Remove dirt, nicks or burns, using above caution and check by repeating procedure.
10. Examine shaft seal in bearing cap; if worn or damaged, replace seal.

NOTE: *Lips of seal must point toward casting. Examine running surfaces of bearing cap and inlet end cap. If heavy scratches or gouges are present or if port edges are damaged, the part should be replaced.*

11. Press seal in until shoulder stops against bearing cap casting. Care must be used to see that seal is properly aligned. This operation should be performed in an arbor press.
12. Check pressure relief valve which is located in flow valve spool. The valve must not leak at any pressure below 850 psi. Evidence of leakage will be seen at small holes in side of flow spool. If relief valve leaks, disassemble by removing valve seat and checking for dirt or nicks.
13. Clean all valve parts and reassemble, replacing valve seat if nicked or damaged. Check pressure again. If pressure still falls below minimum requirement, disassemble valve again and replace pressure relief spring or place a 1/32" washer between spring and ball retainer. To correct pressures exceeding the maximum requirement, remove valve seat and place a 1/32" washer between valve seat and flow spool body.
14. Examine pump body for signs of unusual wear or damage.

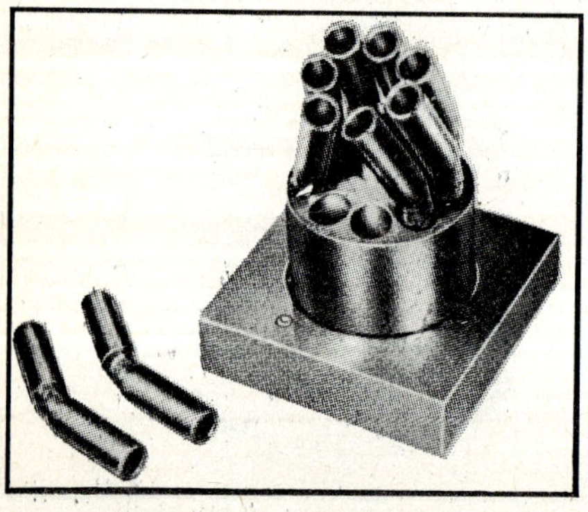

Fig. 42 Assembling sleeves in drive cylinder

Reassembly, Fig. 40

1. Press drive shaft and bearing into bearing cap. Check rotation of shaft to make certain it turns freely. Use a sleeve which bears against outer race of ball bearing and press sub-assembly in place.

CAUTION: *Make certain that the shaft bearing sub-assembly are properly aligned, otherwise the shaft may push sleeve bearing out of bearing cap or may scratch or mar inside diameter of sleeve bearing.*

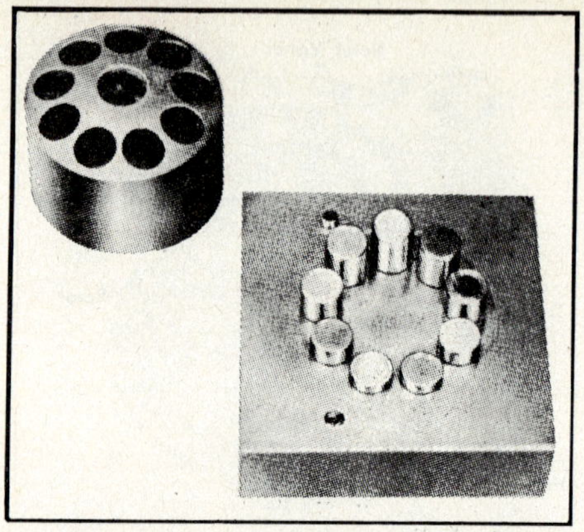

Fig. 41 Drive block assembling fixture

2. Using assembly fixture (C-3602) shown in Fig. 41, place driven cylinder block (without spline or serration) on fixture.
3. Lubricate (light oil) and insert plunger spring, plunger and seven sleeves in drive block, Fig. 42.
4. Place pump body, square end down, over cylinder drive block and fixture locating pins. Use a pointed probe to align sleeves to a uniform spacing and install remaining sleeves.
5. Position drive splined block with serration over sleeves. Sighting through bores in drive block for alignment, lower cylinder block until it engages the two sleeves in the forward position.
6. Again use a pointed probe slightly smaller in diameter than the sleeve bores, Fig. 43, and correct the alignment of sleeves in the 5 and 7 o'clock positions, at the same time guiding cylinder block downward. Continue this procedure to the 4 and 8 o'clock positions until all sleeves are aligned and engaged. The block is then pushed in all the way.

CAUTION: *Do not force the cylinder block in place. Proper alignment of the block and sleeves will allow the block to be pushed easily into place without excessive pressure. Tighten body to housing.*

7. Remove body and cylinder block from fixture, using care to see that blocks are not forced out of body by plunger spring.
8. Assemble a new end cap gasket on pump body. Install end cap on body and tighten capscrews finger tight.

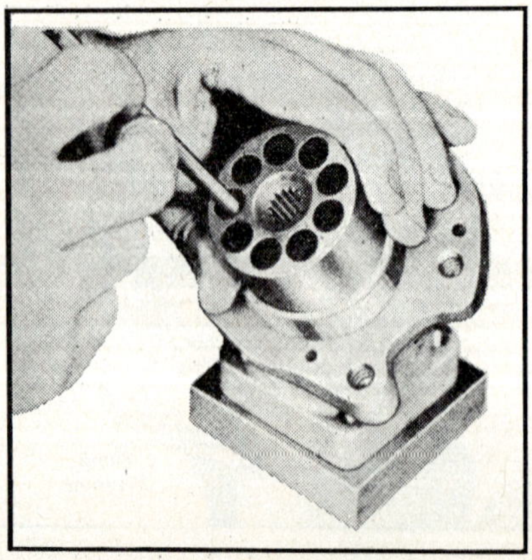

Fig. 43 Installing drive block

9. Insert a new O-ring in counterbore at flow valve liner in body. Install a new bearing cap gasket, and assemble bearing cap to body.

NOTE: *It may be necessary to exert hand pressure on bearing cap to get drive block down into body after bearing cap is seated.*

10. Tighten capscrews finger tight. Grip end of pump shaft in vise and rotate pump. Pump should rotate freely without binding. After making sure pump rotates freely, tighten capscrews uniformly to 25 ft-lbs. Check pump again for rotation.
11. Mount pump in vise in vertical position, gripping on bearing hub.
12. Install flow valve spool with the 3/16" land down.
13. Place flow valve spring on top of spool. Replace O-ring on flow spring retainer fitting. Compress flow spring with fitting and screw fitting into place. Tighten to 20 ft-lbs.
14. Install and align a new reservoir O-ring on bearing cap. Lubricate O-ring for ease in assembling reservoir. Align reservoir and push it into place, applying force around its outside diameter.

CAUTION: *Do not attempt to align or pull reservoir into place on the angular boss with the 1/4" capscrews.*

15. Install shaft key and pulley. Tighten pulley attaching bolt to 20 ft-lbs.
16. Refill pump with proper fluid.

BORG WARNER PUMP

Disassembly, Fig. 44

1. Clamp pump in a vise with soft jaws and remove Woodruff key.
2. Remove front and rear covers from pump body by removing cap screws.
3. Slide bearings, drive and driven gears from pump body and remove "O" rings from both covers.
4. Remove pressure loading seals and with a brass drift, drive seal from front cover.
5. Remove control valve and orifice plug from rear cover.

Control Valve, Fig. 45

1. Remove snap ring and slide spool and spring from control valve.
2. Unscrew relief valve seat and remove ball, spring guide and spring.
3. Tap shims from spool bore.
4. Reverse procedure to assemble control valve.

Reassembly, Fig. 44

1. Apply light coat of Permatex No. 3 to front cover bore and press in new seal.
2. Clamp front cover into vise with outside face downward and install new "O" ring and pressure loading seals.
3. Place body on front cover and install front bearings, drive and driven gears, and rear bearings.
4. Install control valve and orifice plug in rear cover.
5. Place new "O" ring on rear cover, assemble rear cover to pump body and install cap screws.
6. Install Woodruff key.

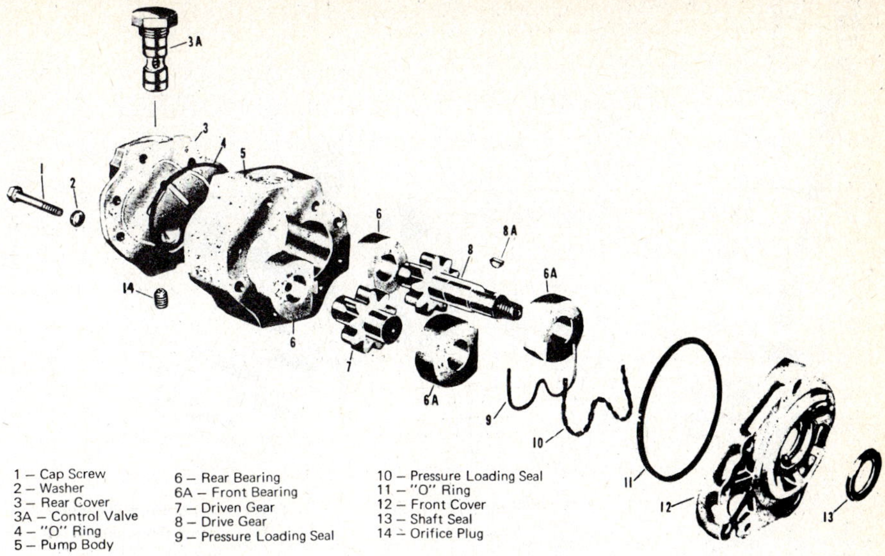

Fig. 44 Borg Warner pump, disassembled

EATON PUMP, EXC. MODELS B-165 & 209

Disassembly, Fig. 46

1. Remove coupling assembly from pump shaft.
2. Clamp pump in a vise with soft jaws and remove cover, "O" ring and "O" ring retainer.
3. Remove pump shaft, key, snap ring and inner rotor from pump body.
4. Turn body over and tap outer rotor out of body.
5. Remove oil seal from body and flow control valve from pump cover.

Reassembly, Fig. 46

1. Install new oil seal and inner rotor and shaft assembly into pump body.
2. Install outer rotor and "O" rings in body and thrust washer in cover.
3. Install cover on pump body and flow control valve into cover.
4. Install coupling assembly onto shaft.

EATON PUMP MODELS B-165 & 209

Disassembly, Fig. 46A

1. Disconnect high pressure hose from pump and return hose from reservoir. Remove pump from vehicle.
2. Drain balance of oil from pump and using pump mounting bracket place pump in vise.
3. On pulley type pumps, remove pulley with appropriate puller. On direct drive pumps with a threaded shaft end, remove the nut used on the shaft, washer and gear.

NOTE: Do not drive shaft out of pulley unless pump is suspended freely and the pulley held securely.

4. Remove two hex head nuts and spacers and two flat sealing washers from back of reservoir assembly and the two 3/8-16 attaching studs, if equipped. If assembly uses bolts, remove bolts. If assembly uses

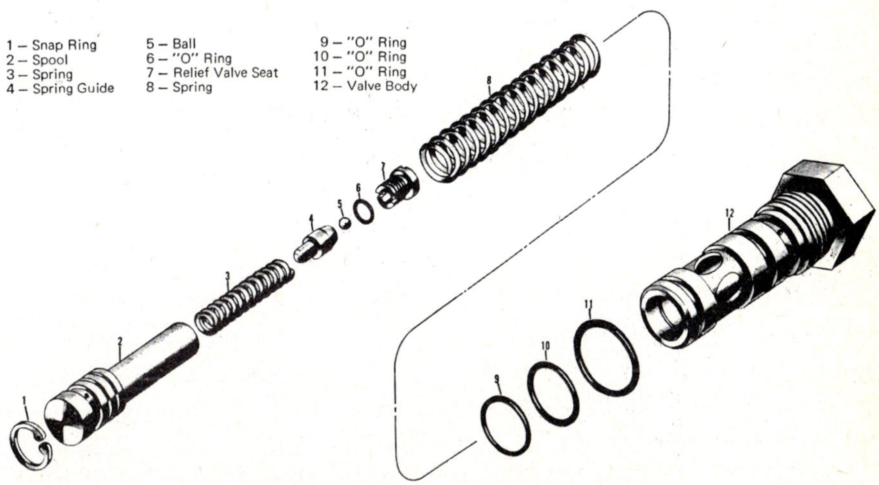

Fig. 45 Borg Warner pump flow control valve

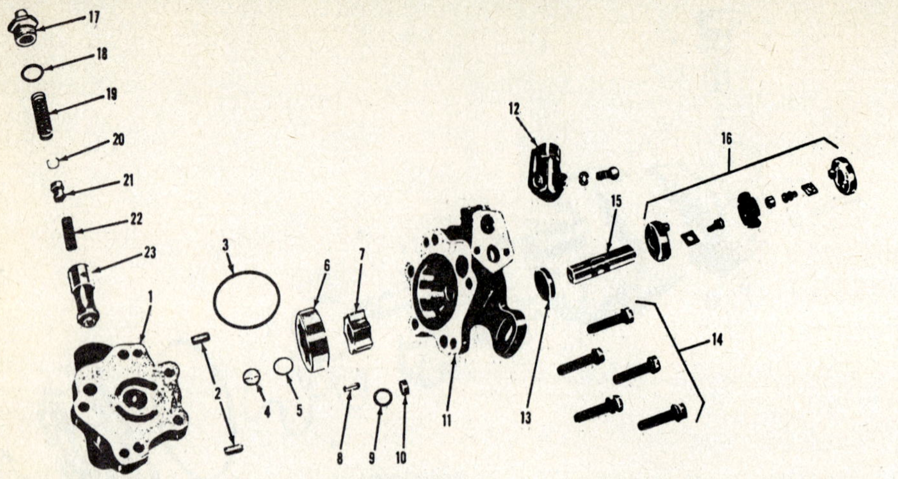

1. Cover
2. Cover Dowel
3. Body O-Ring
4. Thrust Washer
5. Snap Ring
6. Outer Rotor
7. Inner Rotor
8. Drive Pin
9. Bypass O-Ring
10. Bypass O-Ring Retainer
11. Body
12. Outlet Adapter
13. Oil Seal
14. Cover Screws
15. Pump Shaft
16. Coupling Assembly
17. Hose Connector
18. Connector O-Ring
19. Flow Control Valve Spring
20. Valve Retainer Snap Ring
21. Relief Valve
22. Relief Valve Spring
23. Flow Control Valve

Fig. 46 Eaton pump, disassembled

rear strap brackets, remove nuts, brackets, spacers and studs.

5. Mark position of reservoir to pump base and remove by slipping side of reservoir until it rotates slightly. Remove two flat washers from inside of reservoir.

NOTE: Do not pry on edge with any tool. Once reservoir has been loosened it must be removed and new gaskets installed.

6. Loosen, but do not remove, valve cap, separate relief valve cap (if equipped), large threaded retainer ring from center of pump body, jam nut under discharge fitting noting position of elbow and bolts that attach mounting bracket. Remove pump from vise.
7. Remove mounting bracket and discharge fitting from pump. Remove woodruff key from pump shaft, if so equipped.
8. Remove valve cap, flow control valve spring and valve assembly. If equipped also remove separate relief valve cap, shims, spring and valve.
9. Remove large threaded retainer from center of pump body.
10. Remove cover and, by lightly tapping on front end of pump shaft, bushing subassembly.
11. Remove rear end plate.

NOTE: The two end plates are different from each other and should be identified for proper reassembly.

12. Remove shaft, carrier and twelve loose rolls.

NOTE: Do not remove carrier from shaft at this time.

13. Remove cam ring by lifting straight up. Mark up side of cam ring for proper reassembly.
14. Remove long locating pin (two pins on some models of which one will come out with cam), front end plate and two preformed O-rings. If body has two pin holes record which hole pin is removed from.
15. Remove large reservoir seal from pump body and shaft oil seal.
16. Remove O-ring from discharge fitting, and on pumps equipped with separate relief valves, remove O-ring from cap.

Inspection

Clean all parts in suitable solvent and air dry. If ball plug in valve cap has blown out, the ball can be driven back into the drilled angle hole to a depth of 1/4 inch below cast surface. Replace carrier if ends are worn or scored and wear pattern on any of roll drive faces is greater than .09 inch in width. On models with relief valve assembly, inspect valve for wear. Valve wear should not exceed .06 inch depth.

Assembly

1. Install new shaft oil seal into pump body bushing sub-assembly and pre-formed gaskets into front end plate.

NOTE: On some models pre-formed gaskets can be replaced by O-rings that can be formed into grooves. Assemble with O-ring side down.

2. Assemble front end plate into pump body by inserting long locating dowel pin through slot in plate and into dowel hole in pocket of pump body.

NOTE: If two pin holes exist, be sure to install pin in original hole as noted previously.

3. Install cam ring into pump body pocket over locating pin.

NOTE: Be sure cam ring is in original position. On models equipped with two cam pins, install second pin in cam.

4. Lubricate face of end plate, shaft bushing, and I.D. wall of cam ring.
5. If replacing only the carrier, remove snap ring nearest shaft end. Note tooth form orientation before removing carrier. Carrier drive pin must be in position when installing new carrier. Install new carrier in same orientation and snap ring in groove closest to carrier.

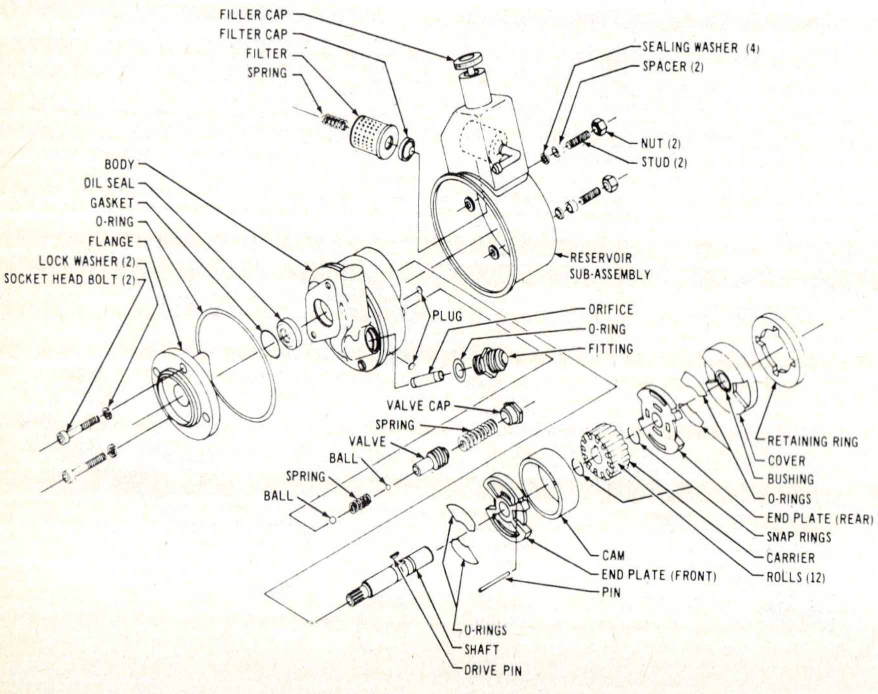

Fig. 46A Eaton models B-165 & 209 pump exploded view

6. If replacing only the shaft, remove snap rings holding carrier from shaft. Note tooth form orientation before removing carrier. Install one snap ring closest to center of carrier, then install carrier and drive pin on shaft. Install second snap ring in groove closest to carrier.
7. If replacing both carrier and drive shaft, use old sub-assembly as a guide. Install snap rings in grooves closest to carrier ends.
8. Lubricate oil seal and install shaft and carrier into pump body. Insert twelve rolls into carrier and lubricate carrier face and shaft with steer-system oil.
9. Assemble gaskets or O-rings into rear end plate. Using a small amount of grease to hold gaskets or O-ring into end plate, insert end plate into pump body with gasket or O-ring side up.
10. Insert cover & bushing sub-assembly over shaft and onto dowel pin for proper location.

NOTE: End plates, open slots of body and cover sub-assembly should align if properly assembled.

11. Thread retainer ring into pump body finger tight. Install up side of ring up. Lubricate valve bore and insert valve assembly into bore with long small diameter end first. Place flow control spring in valve bore and thread valve cap in finger tight.
12. Install O-ring onto discharge fitting and thread into pump body finger tight. Assemble pump mounting bracket or flange to front of pump. Install O-rings.

NOTE: Face gaskets have been replaced by O-rings in almost all cases. Do not install rear only or front and rear fastened brackets at this time.

13. Complete assembly of pump and reservoir by placing in vise. Tighten mounting bolts to 30–50 ft. lbs. Tighten direct drive type bolts to 30–50 ft. lbs.
14. Tighten valve cap and separate relief valve cap to 40–55 ft. lbs. Tighten large retainer ring into pump body to 90–120 ft. lbs. torque.

NOTE: It is important that this retainer ring be tight.

15. Locate discharge fitting elbow to its original position and torque jam nuts to 30–45 ft. lbs. Check shaft to determine if it rotates freely and smoothly.
16. Position gasket into groove around body.

NOTE: Gasket must be assembled with yellow side out.

17. Remove oil filter from reservoir by inserting screwdriver behind spring and prying out. Remove loose filter guide from end of return tube on inside of reservoir.
18. Replace filter by first sliding filter guide over end of return tube. Position open end of filter over filter guide. Insert spring into closed end of filter and place in original position in reservoir.

NOTE: On some units a spacer should be placed over guide tube first, followed by filter guide.

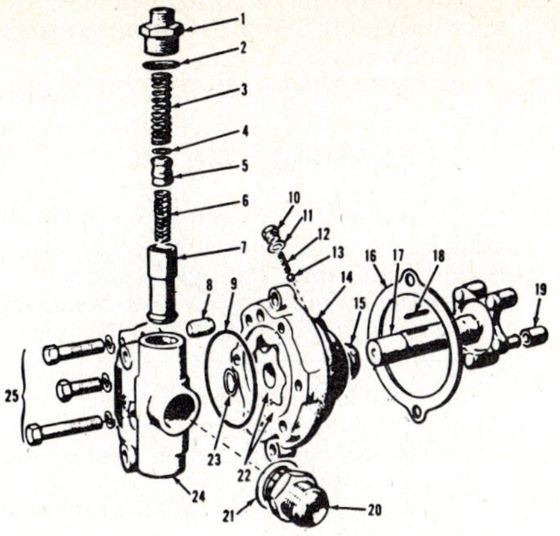

1. Connector
2. O-Ring
3. Flow Control Valve Spring
4. Lock Ring
5. Pressure Regulator Piston
6. Pressure Regulator Spring
7. Flow Control Valve
8. Dowel
9. Housing O-Ring
10. Valve Plug
11. Gasket
12. Spring
13. Valve Ball
14. Housing
15. Oil Seal
16. Gasket
17. Pump Shaft
18. Key
19. Rubber Bushing
20. Connector
21. Connector Gasket
22. Rotor Assembly
23. Pump Shaft Lock Ring
24. Cover
25. Cover Screws

Fig. 47 Mack-Scania pump, disassembled

19. Install studs in back of pump body and assemble two flat sealing washers over studs. Torque studs to 1–10 ft. lbs.
20. Place reservoir assembly over pump assembly. On units without rear mounting brackets thread two lock nuts on reservoir studs and tighten to 25–35 ft. lbs. On units with rear mounting brackets, install spacer, bracket and lock nut and torque lock nuts to 25–35 ft. lbs. Install front bracket bolts and torque to 30–50 ft. lbs. Install woodruff drive key in pump shaft, if so equipped.
21. Install pulley if so equipped. Some pulleys are a slip fit except for last 1/4 inch which is a press fit. Tighten this type pulley nut to 15–20 ft. lbs. On pulleys using a nut and washer torque nuts to 25–30 ft. lbs. For Detroit Diesel engine units with a shaft drive gear, install coupling, nut and washer and torque to 25–40 ft. lbs.

MACK-SCANIA PUMP

Disassembly, Fig. 47

1. Remove cover from housing.
2. Remove housing "O" ring, lock pin from shaft and rotor assembly from housing.
3. Remove key and shaft, then with a suitable puller remove oil seal from housing.
4. Remove connector and flow control spring from valve body.

NOTE: Use care when removing connector from valve body as it is spring loaded by the flow control spring.

5. Turn cover over and tap valve assembly out.
6. Remove lock ring, piston and spring from valve body.

Reassembly, Fig. 47

1. Install spring, piston and lock ring into valve body.
2. Install valve body, spring, "O" ring and connector into cover.
3. Install new oil seal, rotor and shaft into housing, aligning keyway on shaft with slot in rotor, then install key and lock ring.
4. Install new "O" ring on housing, then install cover.

PESCO PUMP

Disassembly, Fig. 48

1. Remove hex head cap screws and cover from pump body.
2. Press shaft seal from cover and remove cover bearings, body and bearing seal

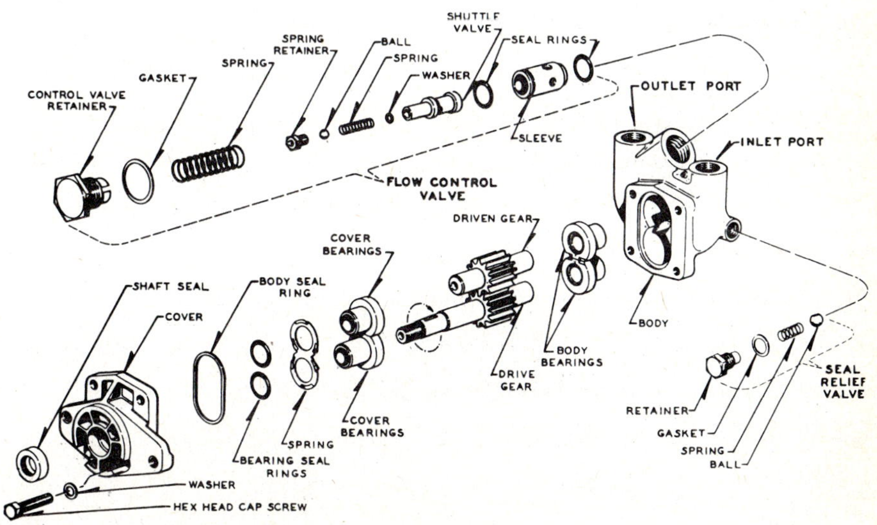

Fig. 48 Pesco pump, disassembled

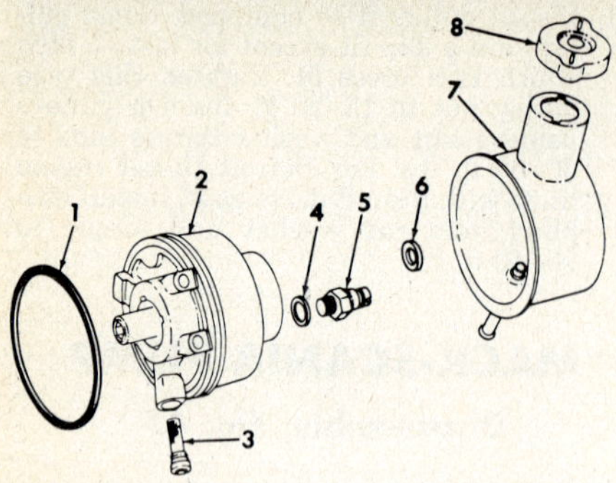

1 SEAL, O-RING
2 PUMP, ASM.
3 SCREEN
4 GASKET
5 VALVE, RELIEF
6 GASKET
7 RESERVOIR
8 CAP, W/GASKET

Fig. 49 TRW pump disassembled

rings and spring.
3. Remove drive and driven gears and body bearings.
4. Remove flow control valve retainer, gasket and spring, then shuttle valve assembly from body.
5. Remove sleeve from body, then "O" rings from sleeve.
6. Remove seal relief valve retainer, gasket, spring and ball from body.

Reassembly, Fig. 48

1. Install seal relief valve ball, spring, gasket and retainer into body.
2. Install new "O" rings on sleeve, then sleeve into body.
3. Insert shuttle valve, spring, gasket and control valve retainer into sleeve and body.
4. Install body bearings, drive and driven gears and cover bearings.
5. Place spring on cover bearings, ensuring lips of springs face bearing flanges, and install bearing seal and body seal rings.
6. Apply light coat of Permatex No. 3 to cover bore, then press new shaft seal into cover.
7. Install cover and secure with hex head cap screws.

TRW PUMP

Disassembly, Fig. 49

1. Using suitable puller, remove pulley from shaft.
2. Remove reservoir mounting bolt and gasket from rear of pump. Pry reservoir from pump and discard O-ring.
3. If necessary, remove relief valve and gasket.

Reassembly, Fig. 49

1. Using new gasket, install relief valve in pump body and torque to 20–25 ft. lbs.
2. Install pulley onto shaft as follows:
   a. Position pump on press with rear end supported on a 5/8 inch socket.

CAUTION: If rear end of pump is not supported, damage to internal pump parts will result when pulley is pressed onto shaft.

   b. For 6-258 engines, press pulley until flush with shaft.
   c. For DV-550B engines, press pulley until shaft is recessed 5/32 inch from pulley. To obtain this dimension, place two 3/4 inch outside diameter by 5/32 inch thick flat washers on shaft end. Washers will position pulley so that shaft is recessed the specified 5/32 inch.
3. Lightly coat new O-ring with petroleum jelly or power steering fluid and install O-ring into pump body groove.
4. Install reservoir onto pump body and secure with mounting bolt and copper washer.

VICKERS V-200 PUMP

Disassembly, Fig. 50

1. Remove mounting flange and clamp pump body in a vise with cover end up.
2. Note location of cover port, then remove cover and "O" ring.
3. Remove spring, pressure plate and elliptical ring.
4. Remove vanes from rotor, then rotor from shaft.
5. Remove shaft key and outer bearing snap ring and tap shaft from body.
6. Press outer bearing from shaft, remove shaft seal and press inner bearing from body.

Reassembly, Fig. 50

1. Press outer bearing onto shaft and inner bearing into body.

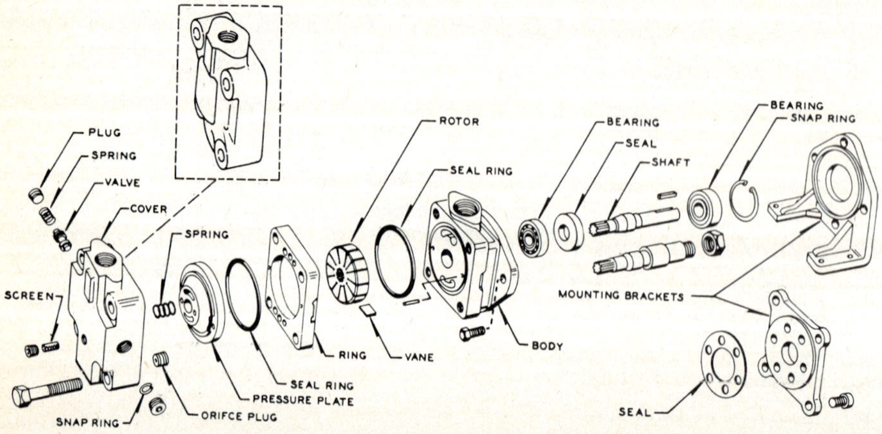

Fig. 50 Vickers V-200 pump, disassembled

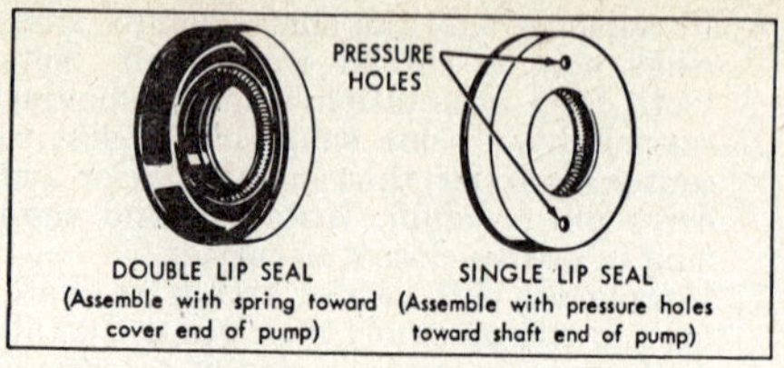

Fig. 51 Vickers V-200 pump, shaft seal installation

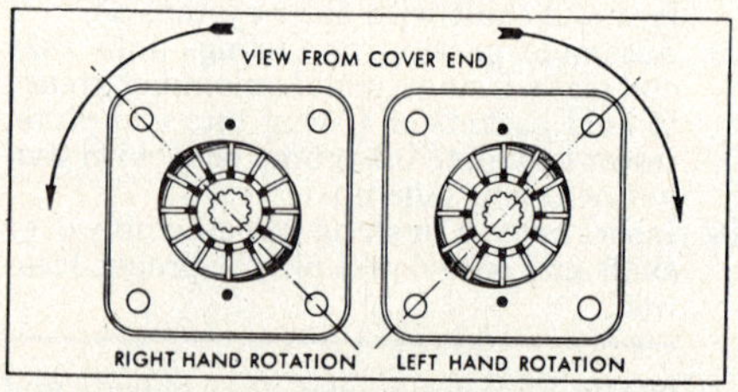

Fig. 52 Vickers, V-200 pump, elliptical ring positioning

2. Correctly install shaft seal into pump body, Fig. 51.
3. Slide shaft into body until bearing is fully seated and install snap ring.
4. Install new "O" rings in pump body and cover.
5. Install elliptical ring in proper position, Fig. 52.
6. Install rotor on shaft and insert vanes in slots with radius edges facing cam ring.
7. Install pressure plate, spring, cover and cover screws.
8. Turn shaft to check for internal binding and install key and mounting flange.

VICKERS EXC. V-200 PUMP

Disassemble, Fig. 53

1. Clamp pump mounting flange in a vise with soft jaws.
2. If pump is equipped with a reservoir proceed as follows: Remove wing nut, washer, reservoir cover and gasket. Remove the stud and nut from the reservoir. Remove two screws, lock washers, and baffle from the reservoir. Separate reservoir and gasket from pump.
3. If pump is not equipped with a reservoir proceed as follows: Remove three manifold attaching screws and the copper washer and separate the manifold and gasket from the pump.
4. With pump clamped in a soft jaw vise, remove four cover attaching screws and separate the cover and pressure plate spring from pump body.
5. Hold the pump ring and remove the pressure plate from pump body.
6. Remove the pump ring, locating pins, rotor and vanes, and the two "O" rings from the pump body.
7. Remove the snap ring from the shaft end of pump housing.
8. Press the shaft and needle bearing out of housing.
9. Remove shaft seal from the housing.
10. If the bearing is worn or damaged, remove the snap ring from the shaft and press the shaft from the bearing.
11. Place the pump cover in a soft jaw vise and drive the retaining pin from the cover with a pin punch, Fig. 54. Do not allow

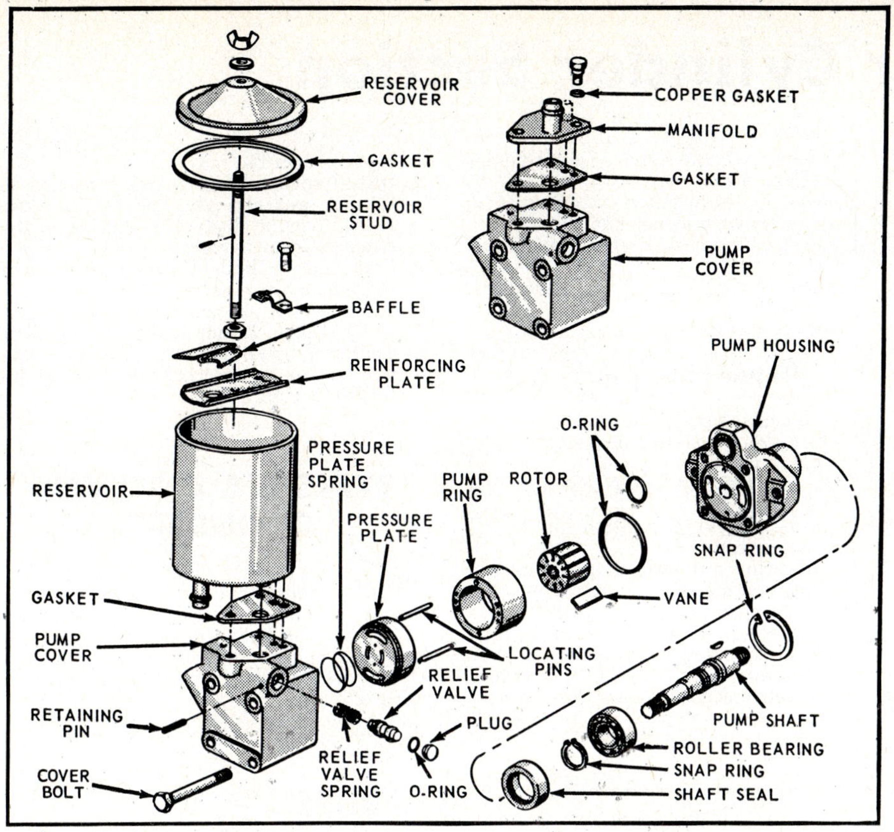

Fig. 53 Vickers power steering pump exc. V-200 pump disassembled

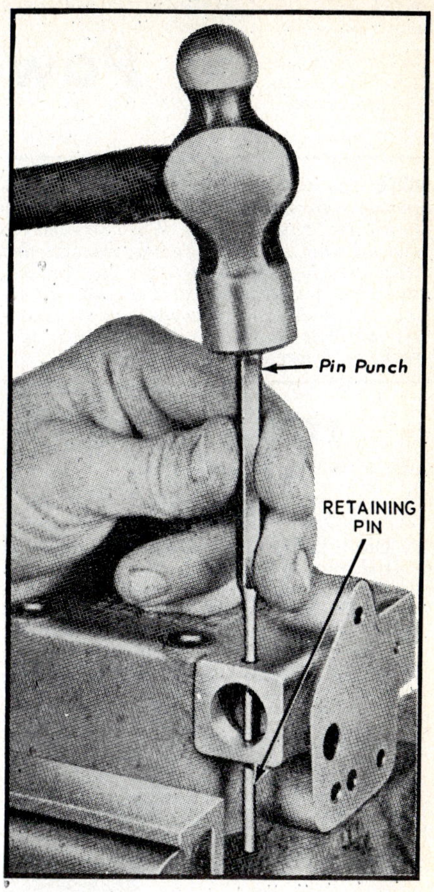

Fig. 54 Removing relief valve retaining

the relief valve and plug to fall out of the cover and become damaged.

12. Remove the plug, relief valve and spring from the cover. The relief valve can be pushed from the cover through a hole inside the cover. Remove the pressure fitting from the cover.

Reassembly, Fig. 53

During assembly, immerse all parts in clean hydraulic oil and replace all seals, gaskets and "O" rings.

1. Position the shaft seal to the pump body, being careful not to damage the seal. Press the seal in until it engages the shoulder in the body.
2. Install the small snap ring on the pump shaft and press the bearing on the shaft.
3. Install the shaft and bearing in the pump body and install the snap ring.
4. Install the two pump ring locating pins in the valve body and install the pump ring.

NOTE: The pump ring must be installed with the arrow on the outside diameter of the ring pointing in the direction of pump rotation. This is important because the direction of pump rotation varies with engine usage.

5. Install the rotor with the chamfered edge of the splined hole toward the pump body.
6. Install the vanes with their radius edge toward the center of the rotor, Fig. 55.
7. Install the pressure plate and "O" rings.
8. Install the pressure plate spring and cover and tighten screws to 25–30 ft. lbs.
9. Position the relief valve spring in the bore. Insert the relief valve in the bore with the hex toward the spring.
10. Install the plug with the "O" ring in the bore and install a new retaining pin.
11. If equipped with a reservoir, install same in the reverse order of removal.
12. If not equipped with a reservoir install the manifold in the reverse order of removal.
13. Rotate the pump shaft to check for free operation and install the pressure fitting in the pump cover.

Fig. 55 Vane installation

Power Cylinder Section

NOTE: *The following material covers a miscellaneous collection of power cylinders to be found on all types of trucks. When necessary to service a unit, first identify it by the exploded views furnished with the test. During the overhaul procedure, remove and discard all O-rings and seals, using new ones upon reassembly. Lubricate all O-rings and seals with light engine oil before installation.*

TYPE A POWER CYLINDER

Disassembly, Fig. 1

1. Drain hydraulic fluid by running piston rod from end to end.
2. Use a suitable spanner wrench to loosen cover about 1/8 of a turn. This will bend out the staked portion of outer cylinder. The cylinder should then be tapped lightly with a mallet in the staked area to straighten out the threads.
3. Unscrew cover and remove from cylinder barrel and piston rod.
4. Remove seal from piston rod.
5. Remove spring retainer and spring.
6. Loosen packing from around bearing and remove.
7. Remove bearing from cylinder barrel and piston rod.
8. Remove rings from piston.

Reassembly, Fig. 1

1. Install rings to piston and rod.
2. Using a piston installer sleeve, insert piston rod in barrel.
3. Slide bearing onto piston rod and into barrel.
4. Using new packing, position into barrel and around bearing.
5. Install spring with large diameter next to bearing, making sure it rests in its seat.
6. Insert spring retainer into spring.
7. With a suitable thimble positioned on threaded end of rod and over wrench flats, install a new piston rod seal, making sure the side on which is molded "this side out" is started on the rod.
8. Install cover, using spanner wrench to tighten securely. The face of the cover should now be flush with the end of the outer cylinder barrel.
9. Stake outer cylinder barrel into a wrench slot in the cover, using a blunt nosed tool. Select a point on the cylinder not previously staked. Minor leakage may occur around cylinder threads. However, this will stop as the packing swells after contacting the oil.

TYPES B & C POWER CYLINDERS

These cylinders have the bearing welded to the cylinder barrel. Also the cylinder shown in Fig. 2 has a socket welded to the head end of the cylinder. The only components that are serviced are the piston rod seals, socket seats, spring and plug.

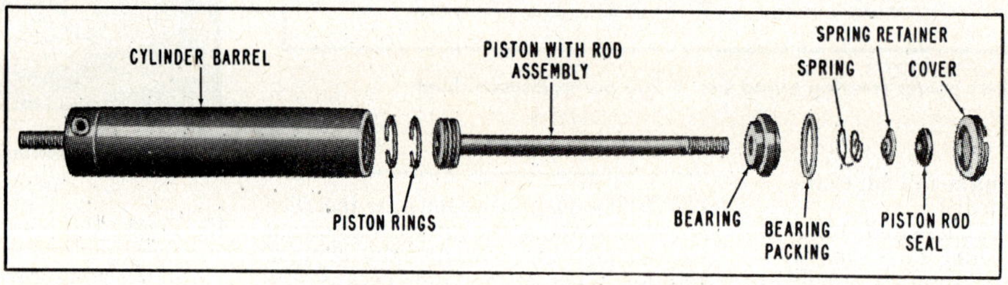

Fig. 1 Type A power cylinder mounted on tie rod

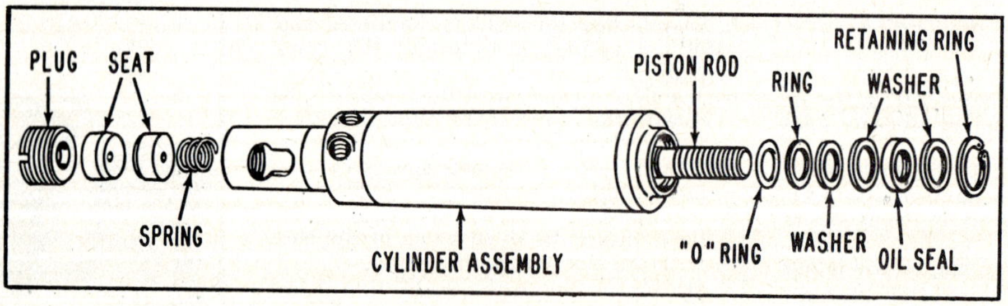

Fig. 2 Type C power cylinder mounted on tie rod

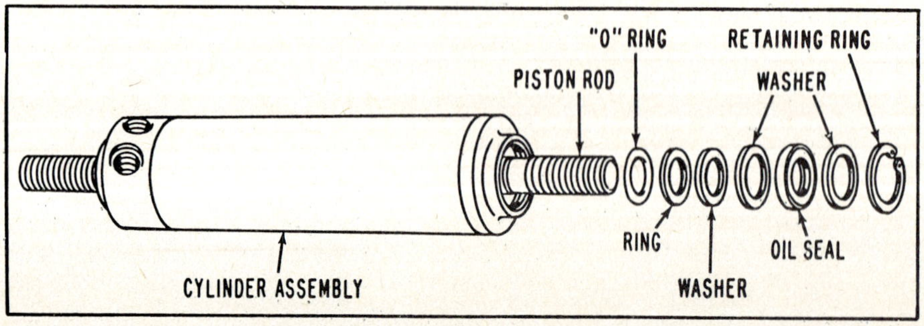

Fig. 3 Type B power cylinder mounted on tie rod

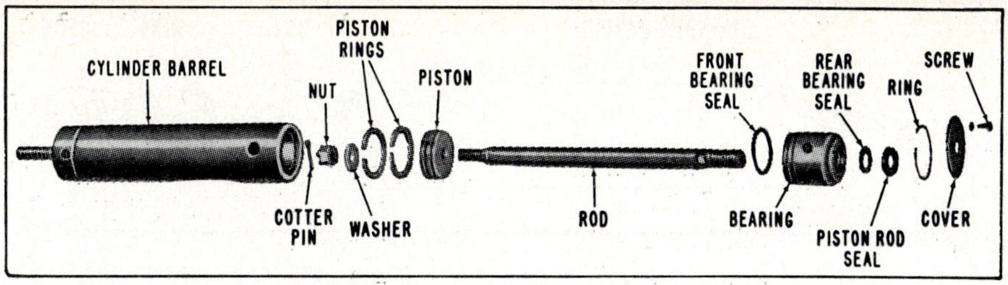

Fig. 4 Type D power cylinder mounted on tie rod

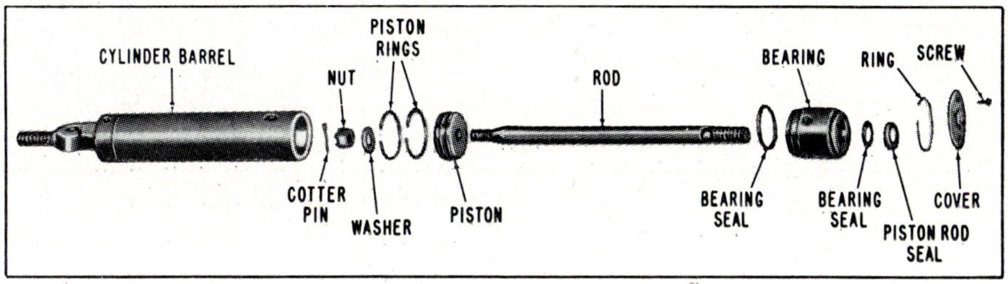

Fig. 5 Type E power cylinder mounted on frame side rail

Disassembly, Figs. 2 & 3

1. Drain fluid by running piston rod from end to end.
2. Remove internal snap ring from bearing.
3. Pull piston rod outward to disengage seals from cylinder bearing. In the event the seals remain in the bearing bore, an awl may be used to pry the seals out. However, when using this method, care must be taken not to scratch the piston rod.
4. Remove seals, back-up ring, and retaining rings from piston rod.
5. Remove socket plug seats and spring from opposite end of cylinder.
6. Wash socket seats and spring in cleaning solvent and inspect for damage or wear.

Reassembly, Figs. 2 & 3

1. Install spring seat in socket end of cylinder.
2. Install inner seat into socket with concave side facing outward.
3. Position outer seat with concave side facing inward into socket.
4. Loosely assembly plug into socket.
5. Cover threaded end of piston rod with cellophane or plastic tape to prevent damage to seals during installation.
6. Install O-ring seal over piston rod. Then slip on the back-up ring, small retaining ring, large retaining ring, seal, large retaining ring. Secure assembled parts with snap ring.

TYPES D, E, F POWER CYLINDERS

Disassembly, Figs. 4, 5, 6

1. Drain hydraulic fluid by running piston rod from end to end.
2. Remove cover.
3. Using a drift punch and hammer, work around the bearing, tapping it lightly into the cylinder barrel past inner edge of snap ring. (This item does not apply to Figs. 4 and 5.)
4. Insert punch in hole along side cylinder barrel to disengage lock ring from its groove, and remove lock ring.
5. Pull on piston rod end to remove rod and internal parts from cylinder barrel. When removing the parts from the piston rod, slide them off the piston end of the rod.
6. Remove piston rod nut, piston and rings.
7. Remove bearing.
8. Remove O-ring seal and back-up ring from bearing.
9. Remove piston seal and inner O-ring seal from bearing.
10. Wash all parts in cleaning solvent and inspect for damage or wear.

Reassembly, Figs. 4, 5, 6

NOTE: *As all parts are machined to very close limits, use extreme care in handling. Damaged sealing edges or burrs will not permit the unit to function properly. The by-pass valve is an internal part of the piston assembly and in the event it does not function properly, it will be necessary to replace the piston assembly as components of the by-pass valve are not supplied separately. To prevent damage to any of the parts, it is recommended that assembly be made from the piston end of the rod.*

1. Assemble inner O-ring seal and piston rod seal in bearing.
2. Assemble back-up ring and O-ring seal to bearing and slide over piston rod down toward threaded end of rod.
3. Assemble rings on piston, then install piston to rod.
4. Install piston nut and torque to 225–230 ft-lbs. *If tightening nut without use of torque wrench, be careful of over-tightening as this may cause piston to swell and prevent assembly in barrel.*
5. Using a piece of shim stock .010″ thick wrapped around piston and rings, slide complete piston rod assembly into cylinder barrel. Push assembly into cylinder as far as possible.
6. Insert bearing into cylinder barrel, taking precautions not to damage bearing O-ring seal against hydraulic line fitting hole in cylinder. To assist in pushing O-ring past this hole, use a piece of wood having a round edge to work the O-ring past the edge.
7. Working around the bearing with a drift punch and hammer, move the bearing past the inner edge of the snap ring groove in the cylinder barrel. This applies only to cylinder shown in Fig. 6.
8. Install lock ring in its groove.
9. Pull back on piston rod until bearing contacts edge of snap ring.
10. Install end plate and secure with lockwashers and screws.

TYPE G POWER CYLINDER

Disassembly, Fig. 7

1. Drain hydraulic fluid by running piston rod from end to end.
2. Remove cover.
3. Using a punch and hammer, work around bearing, tapping it lightly into cylinder barrel to permit snap ring removal.
4. Insert punch in hole along side of cylinder barrel to disengage snap ring from its groove; remove snap ring.
5. Pull on piston rod end to remove rod and internal parts from cylinder barrel. When removing parts from piston rod, slide them off piston end of rod.
6. Remove piston rod nut, piston and rings.
7. Remove bearing.
8. Remove O-ring seal and back-up ring from bearing.
9. Remove snap ring from front of bearing.
10. Remove retaining ring, seal, washers, back-up ring and O-ring from bearing.
11. Remove socket plug seats and spring from opposite end of cylinder.
12. Wash all parts and inspect for damage or

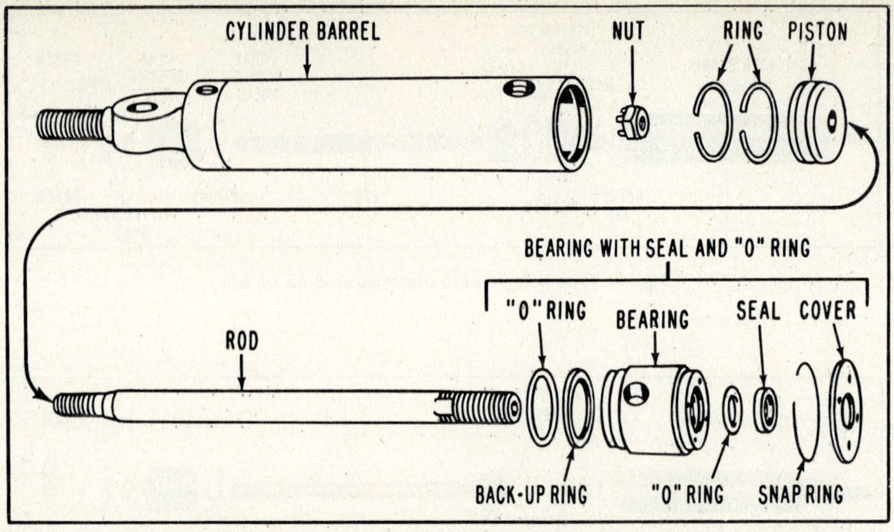

Fig. 6 Type F power cylinder mounted on frame side rail

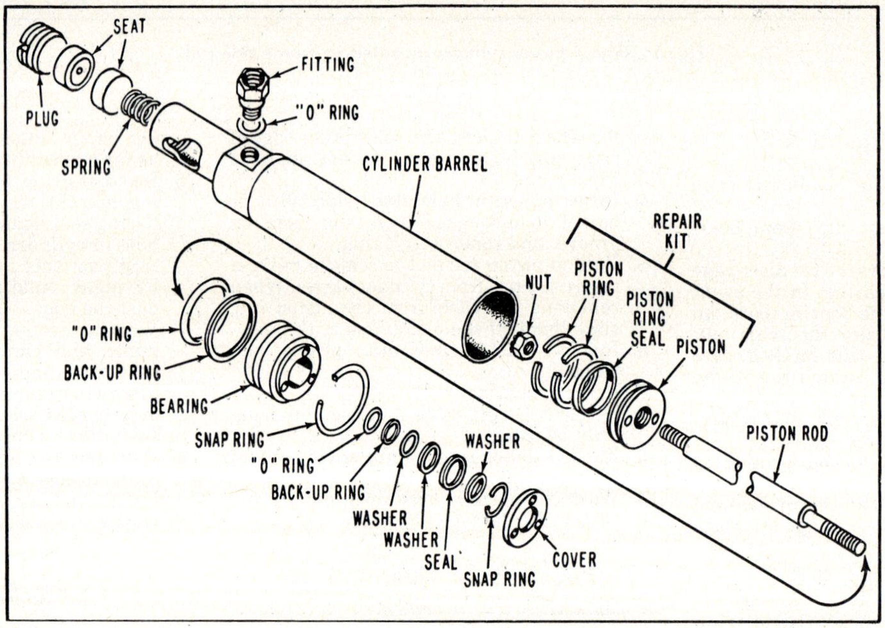

Fig. 7 Type G power cylinder mounted on frame side rail and connected to pitman arm or to a steering arm

wear.

Reassembly, Fig. 7

NOTE: *As all parts are machined to very close limits, use extreme care in handling them. Damaged sealing edges or burrs will not permit the unit to function properly. To prevent damage to any of the parts, assembly should be made from the piston end of the rod.*

1. Assemble O-ring, back-up ring, washers, seal (lip toward front of bearing) and retaining ring.
2. Install snap ring in bearing groove.
3. Install back-up ring and O-ring to rear of bearing.
4. Slide bearing over piston rod down threaded end of rod.
5. Assemble seal and piston rings to piston, then install piston to rod.
6. Install piston nut and tighten securely.
7. Slide complete piston rod assembly into cylinder barrel. Push assembly into cylinder as far as possible.
8. Insert bearing into cylinder barrel, taking precautions not to damage bearing O-ring against edge of cylinder barrel.
9. Working around bearing with a drift punch and hammer, move bearing past inner edge of snap ring groove in cylinder barrel.
10. Install lock ring in its groove.
11. Pull back on piston rod until bearing contacts edge of snap ring.
12. Install end plate and secure with lockwashers and screws.
13. Install seat spring in socket end of cylinder.
14. Install inner seat into socket with concave side facing outward.
15. Position outer seat with concave side facing inward into socket.
16. Loosely assemble socket plug into socket.

TYPE H CHEVROLET & GMC POWER CYLINDER

NOTE: *This power cylinder, Fig. 8, is used on all models of the above truck makes when equipped with Saginaw Rotary Valve and Spool Valve Power Steering Gears. The piston rod end of the cylinder is attached to the steering relay tie rod, idler lever and tie rod, or drag link, depending on the truck application. The opposite end of the cylinder is attached to a bracket which is bolted to the frame or front axle.*

Disassembly, Fig. 8

NOTE: *The only parts that can be replaced are the piston rod seals and the socket end ball stud seats and adjuster screws.*

1. Remove adjuster screw and ball stud seats from both ends of power cylinder.
2. Loosen clamp bolt on outside of socket end assembly.
3. Using a wide blade screwdriver to keep the piston from turning, unthread socket end from piston rod.
4. Force piston rod in and out of power cylinder to drain remaining fluid.
5. Remove scraper retainer snap ring from groove in piston rod guide assembly.
6. Apply air pressure to retraction port in guide assembly while at the same time holding a finger over the extension port. This will dislodge wiper ring assembly, scraper retainer, O-ring seal, and piston rod seal from guide.
7. Clean all parts with cleaning solvent and dry with compressed air. Replace all parts that are not in first class condition.

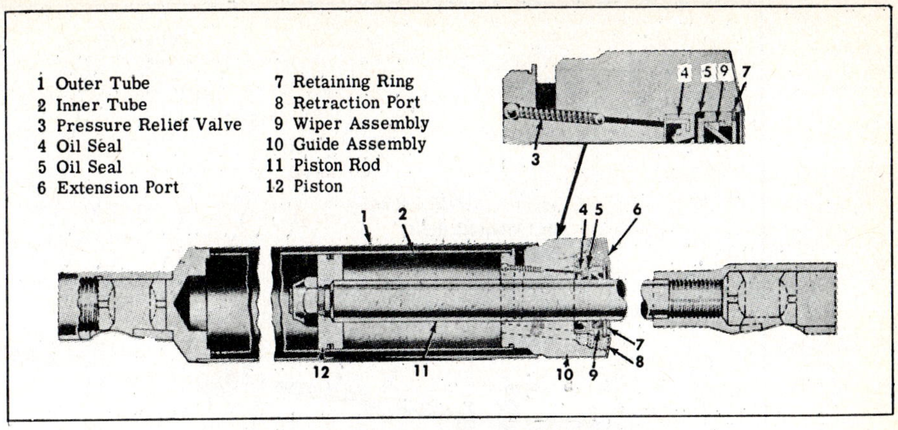

Fig. 8 Type H power cylinder used on Chevrolet and GMC trucks equipped with Saginaw Rotary and Spool Valve power steering gears

Reassembly, Fig. 8

1. Lubricate lip of piston rod seal with a thin layer of grease containing zinc oxide #3, and insert it into guide assembly with the "U" of the cup toward bottom of opening in guide.
2. Install scraper retainer O-ring seal.
3. Install scraper retainer in guide with "U" side pointing out.
4. Insert wiper ring assembly. Then install retaining snap ring. Make sure ring is well seated in groove of guide.
5. Thread socket end assembly on piston rod until it shoulders against rod.
6. Tighten clamp bolt securely.
7. At both ends of power cylinder, install ball seats. Then thread adjuster screws into the sockets loosely. Adjust after installed on vehicle.

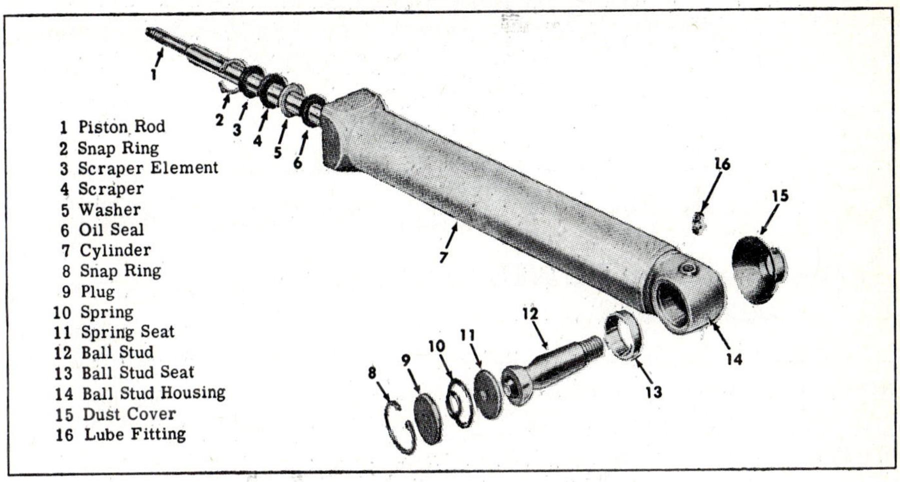

Fig. 9 Type I power cylinder used on Chevrolet and GMC trucks equipped with linkage type power steering

TYPE 1 CHEVROLET & GMC POWER CYLINDER

NOTE: *This type power cylinder, Fig. 9, is used on the above truck makes with linkage type power steering. There are only two units that can be disassembled: They are the piston rod seal assembly and the ball stud assembly.*

Disassembly, Fig. 9

1. Remove snap ring at piston rod end of power cylinder.
2. Pull piston rod out of cylinder, being careful not to spray oil out of cylinder ports.
3. Remove piston rod seal, scraper, washer, and scraper element from piston rod.
4. Depress plug. Then remove snap ring from groove in ball stud housing.
5. Remove plug, spring, spring seat, and ball stud. Remove grease fitting.
6. If necessary to remove ball stud seat from housing, seat will have to be pressed out.

Inspection

1. Clean all parts except seals in solvent and dry with compressed air.
2. Check seat spring for free length, distortion or collapsed coils.
3. Check all parts for wear or damage and replace any that are not in first class condition.

Reassembly, Fig. 9

1. If previously removed, press new ball stud seat in housing.
2. Position ball stud, spring seat, spring, and plug in housing. Depress plug and install snap ring in housing groove.
3. At opposite end of power cylinder, position seal, scraper, washer and scraper element on piston rod.
4. Move parts into position in cylinder. Then install retaining snap ring.

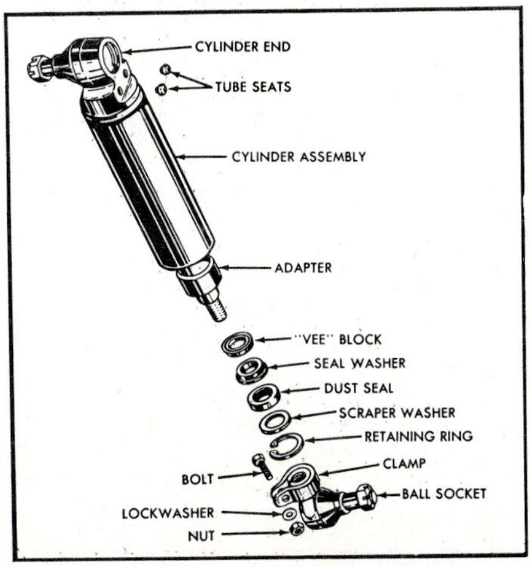

Fig. 10 Type J power cylinder used on Dodge trucks with linkage type power steering

TYPE J DODGE POWER CYLINDER

NOTE: *This type power cylinder, Fig. 10, is used with trucks having linkage type power steering. Service procedure is limited to replacement of the seals only.*

1. To disassemble refer to Fig. 10 and loosen clamp on vertical ball stud.
2. Place wrench on flats, located on end of piston rod, and remove vertical ball stud.
3. Remove retaining ring with snap ring pliers.
4. Remove scraper washer, dust seal, seal back-up washer and "V" block seal.

NOTE: *If dust seal or "V" block seal have surface breaks or cuts they should be replaced. The scraper washer should be a close fit (.001–.0035") on the piston rod.*

5. Reverse foregoing procedure to assembly cylinder.

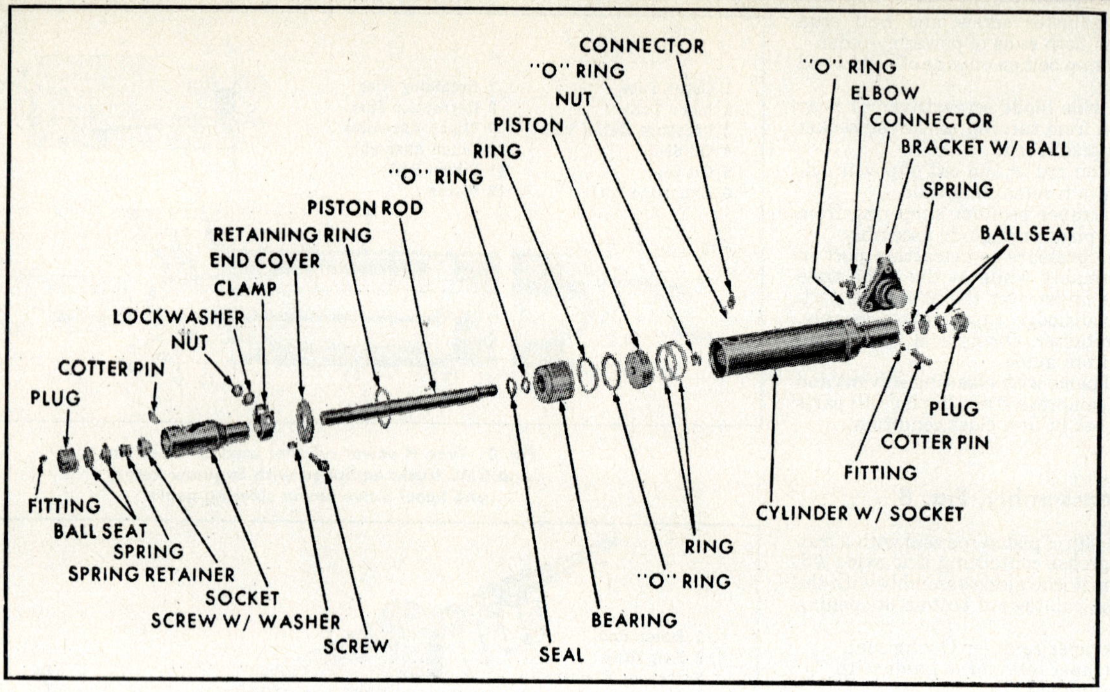

Fig. 11 Type K power cylinder used on Dodge trucks in conjunction with Ross power steering gear with built-in concentric control valve

TYPE K DODGE POWER CYLINDER

NOTE: *This type power cylinder, Fig. 11 is used in conjunction with the Ross power steering gear with built-in concentric control valve. Service on the latter unit is given further on.*

Disassembly, Fig. 11

1. Loosen socket clamp at steering gear arm end and remove socket from piston rod.

NOTE: *Socket is threaded on piston rod. Scribe a mark on the piston rod to insure that socket will be reassembled on piston rod to maintain proper power piston rod adjustment.*

2. Remove four screws attaching cylinder end cover to cylinder bearing.
3. Remove bushing retaining snap ring.
4. Remove hose connector.
5. Lubricate area around retaining ring groove with a liberal amount of light engine oil. Then carefully remove bearing with leather back-up ring and O-ring. Rotate bearing slightly while slowly pulling outward on bearing.
6. Carefully withdraw piston rod and piston with rings. Rotate piston rod slowly when approaching bearing retainer groove in cylinder so that piston rings will drop into retainer groove.
7. Remove piston retaining lock nut from piston rod and remove piston assembly.

NOTE: *If ball socket or springs require replacement, scribe a mark on plugs and sockets to maintain proper ball socket adjustment upon reassembly.*

Inspection

1. Clean all parts and air dry.
2. Inspect cylinder and piston for excessive wear.
3. Cylinder bore size for the 2¾" bore cylinder is 2.747–2.750". For the 2¼" bore cylinder the bore size is 2.247–2.250". Piston ring gap should be .003–.005".
4. Make sure piston poppet works freely in piston.
5. Inspect piston rod for wear or damage.
6. Inspect bushing for wear in bearing bore. Replace all seals.

Reassembly, Fig. 11

1. Install piston on piston rod and install lock nut. Torque nut to 100–125 ft-lbs for the 2¾" bore cylinder and 60–55 ft-lbs for the 2¼" bore cylinder.
2. Install piston rings and space gaps 15 degrees apart.
3. Lubricate piston and rings liberally with automatic transmission fluid.
4. Install piston and piston rod assembly into cylinder bore.

NOTE: *Enter piston carefully to clear bushing retainer ring groove. Use four pieces of .0015" feeler stock to assist in installation of piston assembly.*

5. Position piston until it bottoms in cylinder.
6. Install back-up washer and O-ring on bearing.
7. For the 2¾" bore cylinder install the O-ring and seal in bushing with lip of seal toward bushing. For the 2¼" bore cylinder, install O-ring seal, back-up leather washer, flat washer, steel retaining ring, oil seal, seal retaining ring and snap ring in that order.
8. Lubricate bushing and seals and carefully install bushing over piston rod and insert it in cylinder.
9. Install bushing retaining ring, making

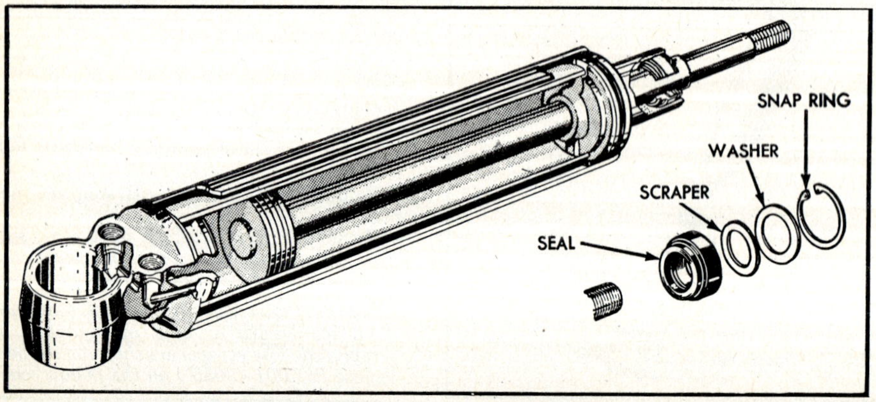

Fig. 12 Type L power cylinder used on Ford trucks with linkage type power steering

sure it is seated properly in cylinder groove.
10. Install cylinder cover and attaching screws.
11. Install socket clamp on steering gear arm socket.
12. Thread socket to scribe marks made at disassembly.
13. Install clamp screw, lockwasher and nut and tighten nut to 40 ft-lbs.
14. Install spring retainer, ball seats and plugs.
15. Install spring, ball seat, support bracket ball, ball seat and plug.
16. Lubricate ball and seats and tighten plug until ball moves freely and without any bind.

TYPE L FORD POWER CYLINDER

NOTE: *This type power cylinder is used on trucks having a separately mounted hydraulic control valve. The only service required is to replace the seal.*

1. Referring to Fig. 12, remove snap ring from end of cylinder.
2. Pull piston out all the way to remove retainer washer, bronze scraper washer and oil seal.
3. After lubricating seal with automatic transmission fluid, position bronze scraper washer in seal.
4. Position seal and retainer washer on piston, and install seal and washer in cylinder, using suitable deep socket.
5. Install retaining snap ring.

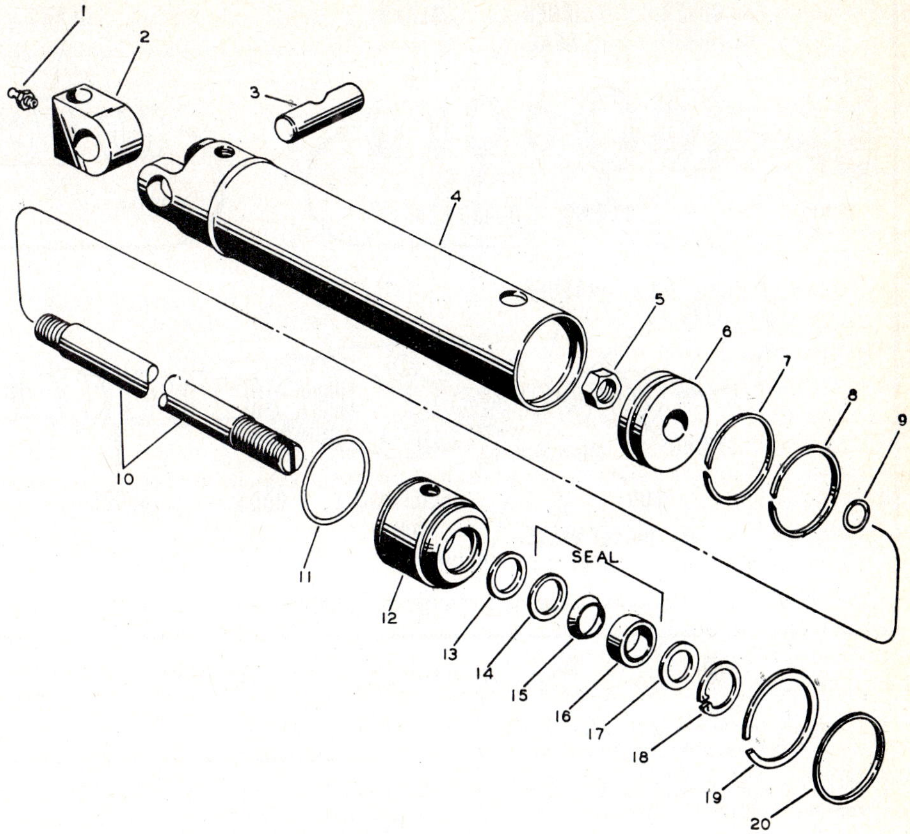

1 – Fitting
2 – Universal Block
3 – Pin
4 – Body
5 – Piston Nut
6 – Piston
7 – Piston Ring
8 – Piston Ring
9 – Copper Seal
10 – Rod
11 – "O" Ring
12 – Cylinder Head
13 – Snap Ring
14 – Neoprene Seal
15 – Nylon Wedge Seal
16 – Seal
17 – Washer
18 – Snap Ring
19 – Cylinder Tube Snap Ring
20 – Cylinder Head Snap Ring

Fig. 13 Type M power cylinder used on White trucks

TYPE M POWER CYLINDER

Disassembly, Fig. 13

1. Remove cylinder head snap ring, push cylinder head in and remove cylinder tube snap ring.
2. Pull rod, head and piston assembly from housing.
3. Remove snap ring, washer, seal assembly, washer and cylinder head.
4. Remove piston nut, piston and copper seal.

Inspection

1. Clean all parts and air dry.
2. Inspect cylinder and piston for wear. Replace piston rings as necessary.
3. Inspect seal assembly for broken or notched edges and replace as necessary.
4. Inspect pin and universal block for wear. Pin should fit snugly in cylinder base and universal block.

Reassembly

1. Install copper seal, piston and piston nut on rod.
2. Compress piston rings and install piston and rod into body.
3. Install washers and seal assembly into cylinder head and secure with snap ring.
4. Lubricate and install "O" ring on cylinder head and secure with snap ring.
5. Install snap ring in body and pull rod outward, forcing the head against snap ring, then install the cylinder head snap ring.

Control Valve Section

This section covers service procedures on the three general types of control valves which are:

1. Type used with linkage type power steering but mounted separately from the power cylinder.
2. Type used with linkage type power steering but mounted in combination with the power cylinder. This design is commonly known as an "In Line Booster."
3. Type built into the steering gear itself.

NOTE: *Since the major parts of the valve are machined to very close limits, they should be handled carefully to prevent damage to the sealing edges. Damaged or burred edges will prevent the valve from functioning properly. Cleanliness throughout the entire servicing procedure cannot be over-emphasized. Parts should be cleaned with a suitable solvent, washed and dried. Always use new O-ring seals when reassembling. Lubricate all seals and O-rings with light engine oil during reassembly.*

TYPES A, B, C, D CONTROL VALVES

Disassembly, Figs. 1, 2, 3, 4

1. Remove dust seal.

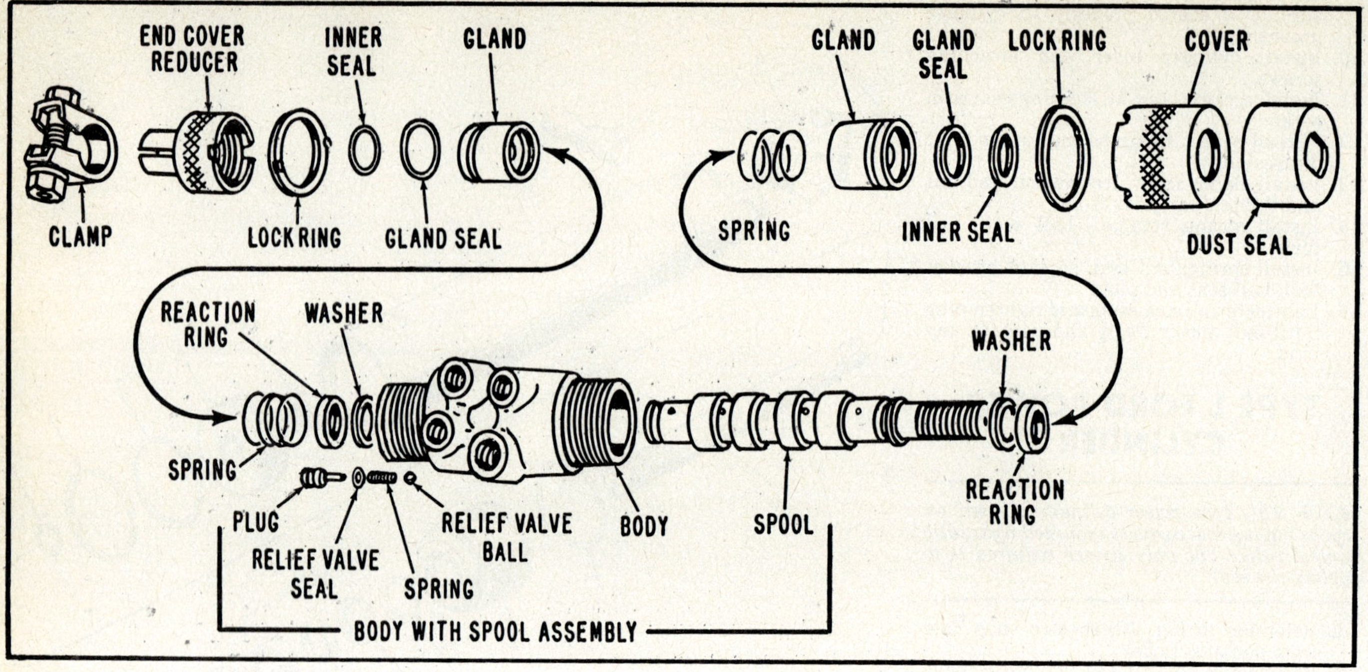

Fig. 1 Type A control valve mounted between pitman arm and steering gear

2. Using a screwdriver or chisel, unlock end cover caps by straightening lock ring tabs.
3. Unscrew and remove both end covers from valve body.

NOTE: *Use extreme care so as not to damage mating surfaces on valve body when removing caps. Also the use of a strap wrench will prevent damage to the exterior finish of the caps.*

---

4. Remove lock rings from each end of valve body assembly.
5. Remove gland and spring from valve body by slightly pushing on threaded end of valve spool.

NOTE: *On the valve shown in Fig. 4, it will be necessary to remove the "C" washer and plain washer before the O-ring, gland and spring can be removed.*

---

6. On valve shown in Fig. 2, remove ring.
7. Remove O-ring from end of valve spool (end from which gland was removed).
8. Remove gland and spring from opposite end of valve by pulling on threaded end of valve spool. When removing spool, one

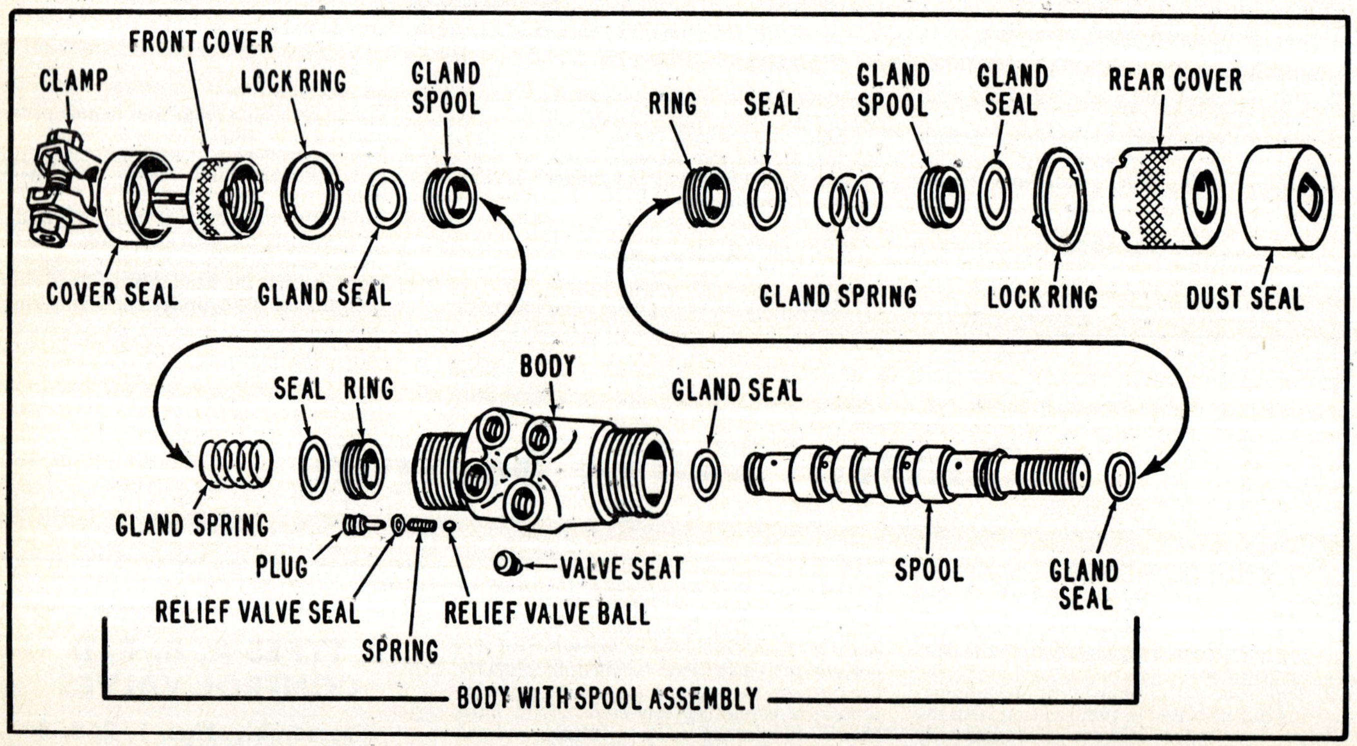

Fig. 2 Type B control valve mounted between pitman arm and steering gear

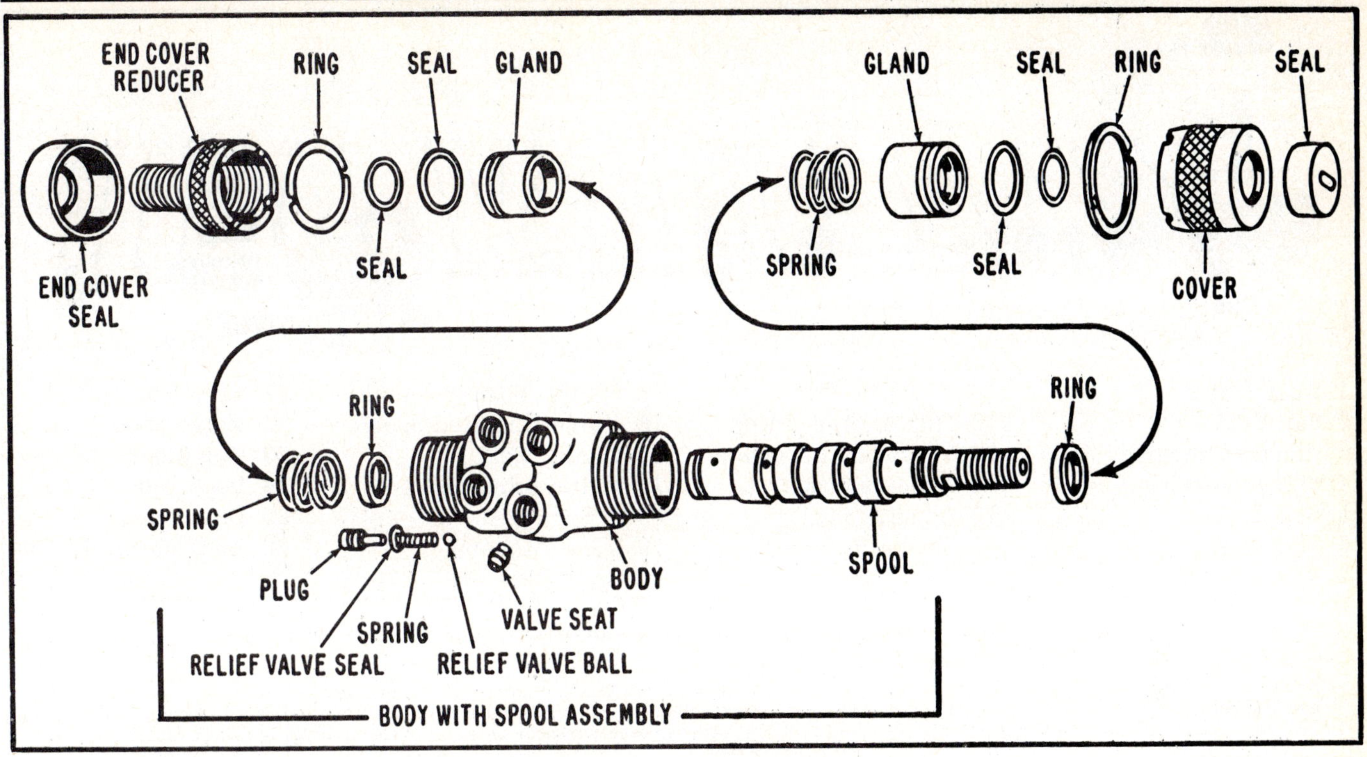

Fig. 3 Type C control valve mounted on frame side rail

reaction ring and spacing washer (if used) will probably remain in valve body which can easily be removed. The other ring and spacing washer (if used) will remain on the spool.

NOTE: *Check the relation of the threaded end of valve spool to ports in valve body before removing spool. This is important for reassembly.*

---

9. Remove remaining O-ring and reaction ring from valve spool. Match reaction rings with their respective glands.
10. Remove O-rings from glands.
11. Components of the relief valve unit should not require servicing except for possible replacement of the small O-ring should leakage be evident. Remove relief valve plug. Remove O-ring seal from plug. Remove relief valve spring and steel ball.

Reassembly

NOTE: *Since very little wear takes place in these valves, the only parts that need replacing in addition to the O-rings are the centering springs, lock rings, and occasionally an end cover or clamp assembly. Inspect the valve spool bore for nicks, burrs or scoring. Slight score marks may be removed with crocus cloth provided it is used carefully. When using on valve spool, use extreme care not to round off the sharp edge portion as this is vitally important to this type of valve.*

---

1. Assemble steel ball into relief valve bore in valve body, then slide in spring.
2. Assemble new O-ring seal to plug assembly and insert into valve body. Press plug in until it is flush with outside surface of valve body. After the valve has been completely assembled, the relief valve is held in place by the end cover lock ring.
3. Slide spool in valve body. Be sure spool is inserted in same position it was in before disassembly. Use care to prevent damage to sealing edges. After spool has been pushed through valve body, assemble one reaction ring and one O-ring on spool.
4. Assemble remaining reaction ring and O-ring seal on spool.
5. Assemble springs on each end of spool.
6. Assemble O-ring seals on glands.
7. Install glands (as matched to the respective reaction rings) to each end of spool.
8. Install lock rings over threaded end of valve body, inserting lugs of ring into milled slots of valve body.
9. Assemble end caps until a tight fit has been obtained. Lock end caps by driving lock ring tabs into one of milled slots of caps and valve body.
10. Assemble dust seal.

TYPE E CONTROL VALVE

Disassembly, Fig. 5

1. Pry dust cover out of valve housing.
2. Remove lock nut from end of valve shaft.

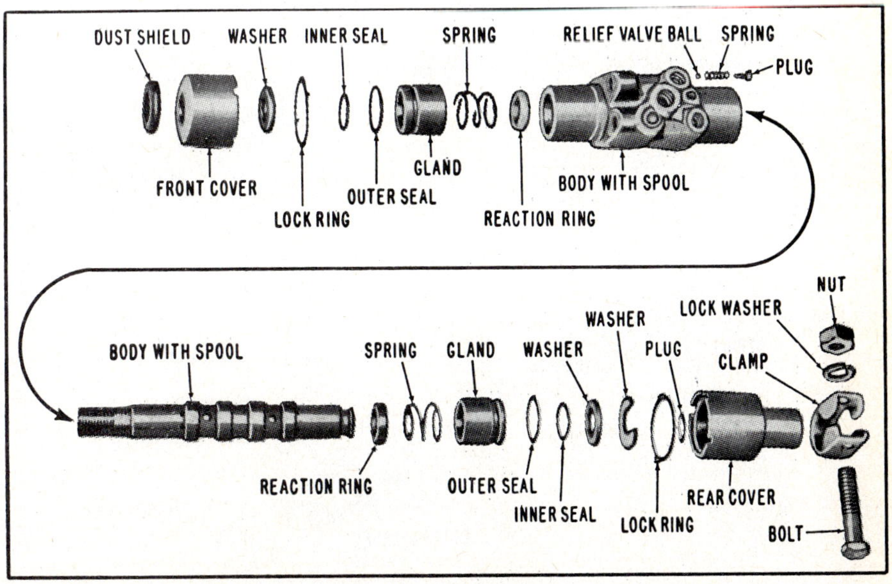

Fig. 4 Type D control valve mounted on frame side rail

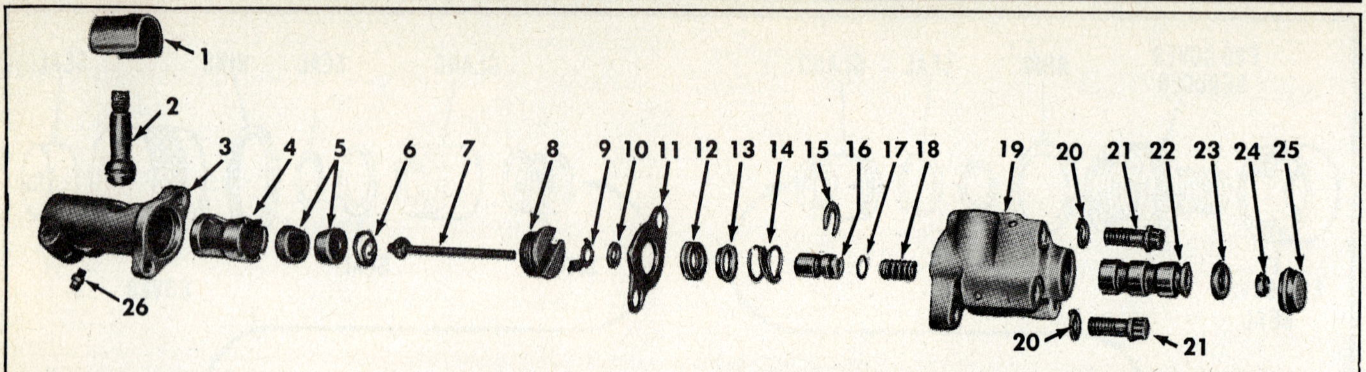

1 Dust Shield
2 Ball Stud
3 Adapter Housing
4 Sleeve Bearing
5 Ball Seats
6 Seat Spring
7 Valve Shaft
8 Adjuster Plug
9 Plug to Sleeve Key
10 Valve Shaft Washer
11 Annulus Spacer
12 Annulus Seal
13 Spring Retainer
14 Valve Spring
15 Spring Thrust Washer
16 Valve Reaction Spool
17 O-ring Seal
18 Valve Adjustment Spring
19 Valve Housing
20 Lock Washers
21 Attaching Bolts
22 Valve Spool
23 Vee Block Seal
24 Lock Nut
25 Dust Cover
26 Lubrication Fitting

Fig. 5 Type E control valve used on Chevrolet and GMC trucks with linkage type power steering

3. Remove two valve housing-to-adapter housing bolts.
4. Lift valve housing and valve spool from adapter housing.
5. Lift valve spool and "V" block seal out of valve housing. Remove "V" block seal from groove in valve spool and discard.
6. Remove valve adjustment spring, reaction spool, spring thrust washer, reaction spring, spring retainer and annulus seal from valve shaft. Discard annulus seal.
7. Remove O-ring seal from groove in valve reaction spool. Remove spring thrust washer from groove in reaction spool. Discard O-ring seal.
8. Remove annulus spacer, shaft washer and plug-to-sleeve key from valve shaft.
9. Using a screwdriver, turn ball stud adjuster plug out of adapter housing. Remove plug and valve shaft. Use care not to nick top surface of adjuster plug.
10. Remove spring and one ball seat from adapter housing.
11. Remove ball stud, second ball seat and sleeve bearing from adapter housing.

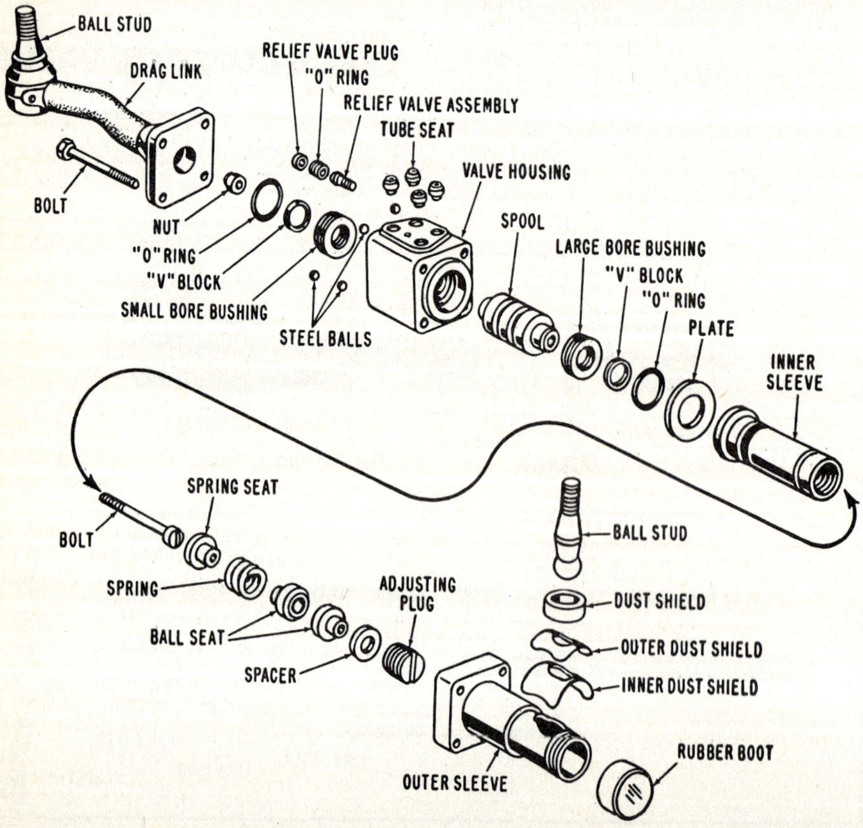

Fig. 6 Type F control valve used on Dodge trucks with linkage type power steering

Inspection

1. With all parts clean and dry, check valve spring, adjustment spring and seat spring for free length, distortion or collapsed coils.
2. Inspect adapter housing and valve housing for cracks or other damage.
3. Inspect all threads for crossed or stripped condition.
4. Discard all seals and replace all parts not in first class condition.

Reassembly, Fig. 5

1. Install sleeve bearing and first ball seat in adapter housing.
2. Install ball stud, second ball seat and spring seat in adapter housing. Note that small coil of spring goes down.
3. With adapter housing mounted in a vise, insert valve shaft through slot in adjuster plug. Then install shaft and plug in adapter housing.
4. Turn adjuster plug in tight. Then back it off until slot lines up with notches in sleeve bearing.
5. Install plug-to-sleeve key on valve shaft. Be sure small tangs on end of key engage notches in sleeve bearing.
6. Position annulus spacer over valve shaft and on adapter housing.
7. Position new O-ring seal in groove of valve reaction spool.
8. Install new annulus seal (lip up) on valve shaft.
9. Position spring retainer, valve spring, valve reaction spool and valve adjustment spring on valve shaft in that order.
10. Install spring thrust washer in groove of valve reaction spool. Side of washer with chamfer should be up.
11. Position new "V" block seal in groove of valve spool with lip down. Then install spool and seal in valve housing, being careful not to jam spool in housing.
12. Position valve housing and spool on

adapter housing. Be sure side ports of housing are on same side as ball stud in adapter. Attach housing to adapter and tighten securely.

13. Depress valve spool and turn lock nut on valve shaft about four turns.
14. Using a plastic hammer, tap new dust cover into bore of valve housing.

TYPE F CONTROL VALVE

Disassembly, Fig. 6

1. Remove bolts attaching drag link to valve assembly. If the drag link ball stud is exceptionally loose, the entire drag link assembly must be replaced.
2. Remove sliding sleeve boot and adjusting plug from inner sleeve.
3. Remove ball seal dust shields, ball stud, ball seal, spacer, spring and spring seat.
4. Hold head of spool bolt with screwdriver in slotted head and remove lock nut from spool bolt. Then remove bolt, inner sleeve and plate.
5. Remove spool bushings by pressing on alternate ends of spool projections by hand enough to expose seals which are assembled in large and small bore bushings. If external seals and spool projection seals show wear or surface breaks, they should be replaced.
6. Remove spool, using care to protect spool edges and bore of valve housing from damage.

CAUTION: *Do not clean spool or valve housing with cloth. Remove foreign matter with dry compressed air.*

Reassembly, Fig. 6

1. Carefully assemble valve spool into valve housing bore with the large projection end of spool facing sliding sleeve and small projection end facing drag link.
2. Assemble large bore bushing and small bore bushing with their respective "V" block and O-ring seals inserted in their respective ends of valve housing.
3. Install inner sleeve into outer sleeve and install spool bolt in hole at flange end of sliding sleeve.
4. Place stop plate over spool bolt and insert bolt in spool and install lock nut. Make sure stop plate is positioned between sliding sleeve and valve housing.
5. Tighten lock nut and back it off about 1/4 turn until spool can be rotated without any bind.
6. Position drag link with ball stud on same side as valve ports and install four through bolts. Tighten bolts securely and install lock wire.
7. Install sliding sleeve spring seat, spring ball seat, ball stud, ball seat, spacer and adjusting plug in that order.
8. Install tube seats.
9. Tighten adjusting plug and back off just enough to align holes in outer sleeve with seat in adjusting plug and secure with cotter pin.
10. Install rubber boot and lock wire.
11. Install sliding sleeve inner dust shield, outer dust shield and stud outer dust shield.

TYPE G CONTROL VALVE

Disassemble, Fig. 7

1. Remove four housing bolts and drag link assembly.
2. After unthreading ball stud plug, remove outer seat and spacer, ball stud, and inner seat, spring and spring seat.
3. After removing and discarding nut from spool bolt, remove spacer washer, centering spring seats, spring and sleeve, bushing seal plate and shim. *Note that spacer washer and shim are not identical and note their relative positions.*
4. Separate valve housing from outer sleeve and remove plate bushing.
5. Remove spool bolt and inner sleeve from outer sleeve.
6. Press spool to free seals from housing and remove seal and spool. Remove spool carefully to avoid nicks or scratches.
7. If necessary to replace check valve or its O-ring, remove plug by means of a piece of brass rod inserted through return port.

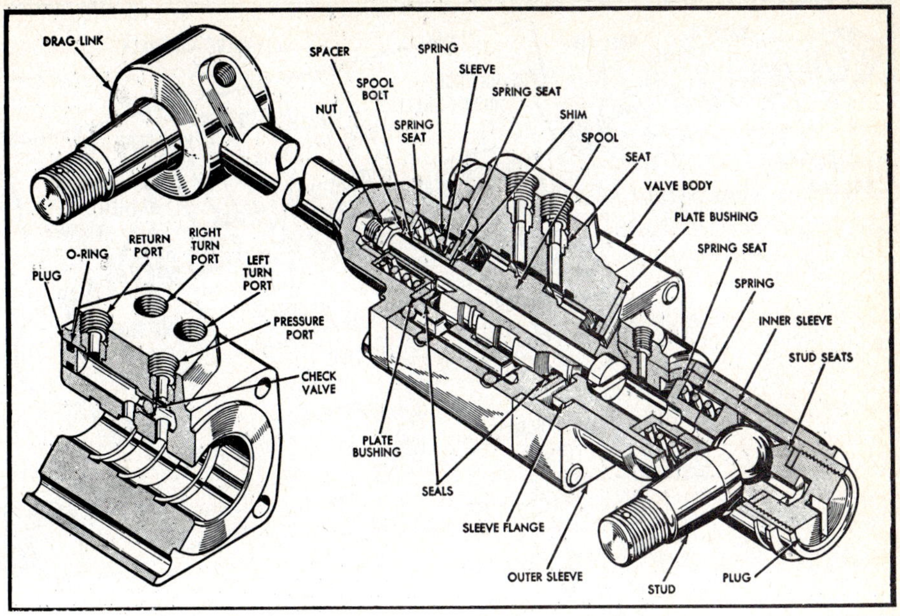

Fig. 7 Type G control valve used on Ford C and H Series trucks

Reassembly, Fig. 7

1. After lubricating spool seals and spool with automatic transmission fluid, carefully position spool in valve body. *Grooved valley between lands must be toward drag link.*
2. Position seals. Inner lip of each seal should be toward valve body.
3. Position spool bolt and inner sleeve in outer sleeve.
4. Position plate bushing in outer sleeve and place valve body next to outer sleeve.

NOTE: *Note that the valve body face has a shoulder that mates with outer sleeve face.*

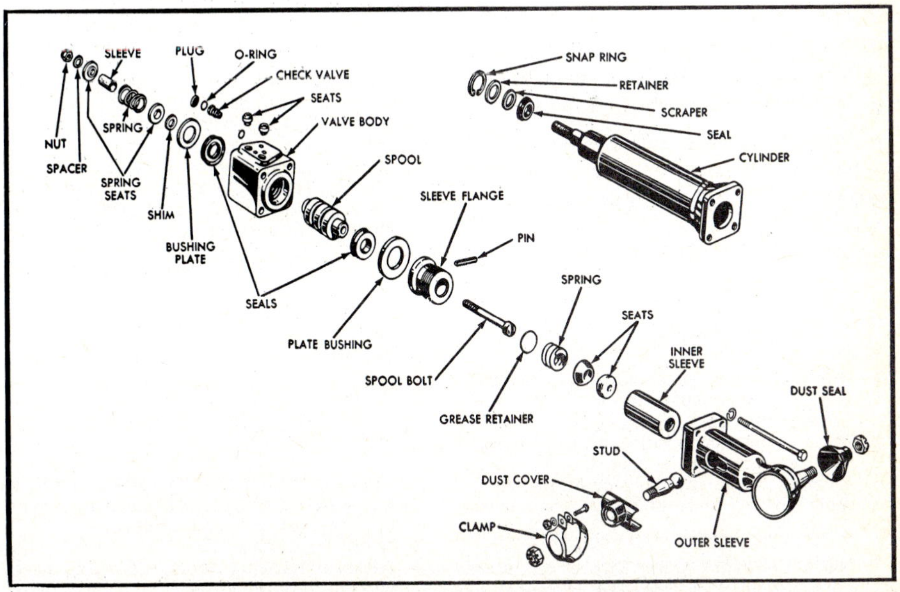

Fig. 8 Type H control valve used on Ford E and T Series trucks. This is an "In Line Booster" consisting of a combined power cylinder and control valve

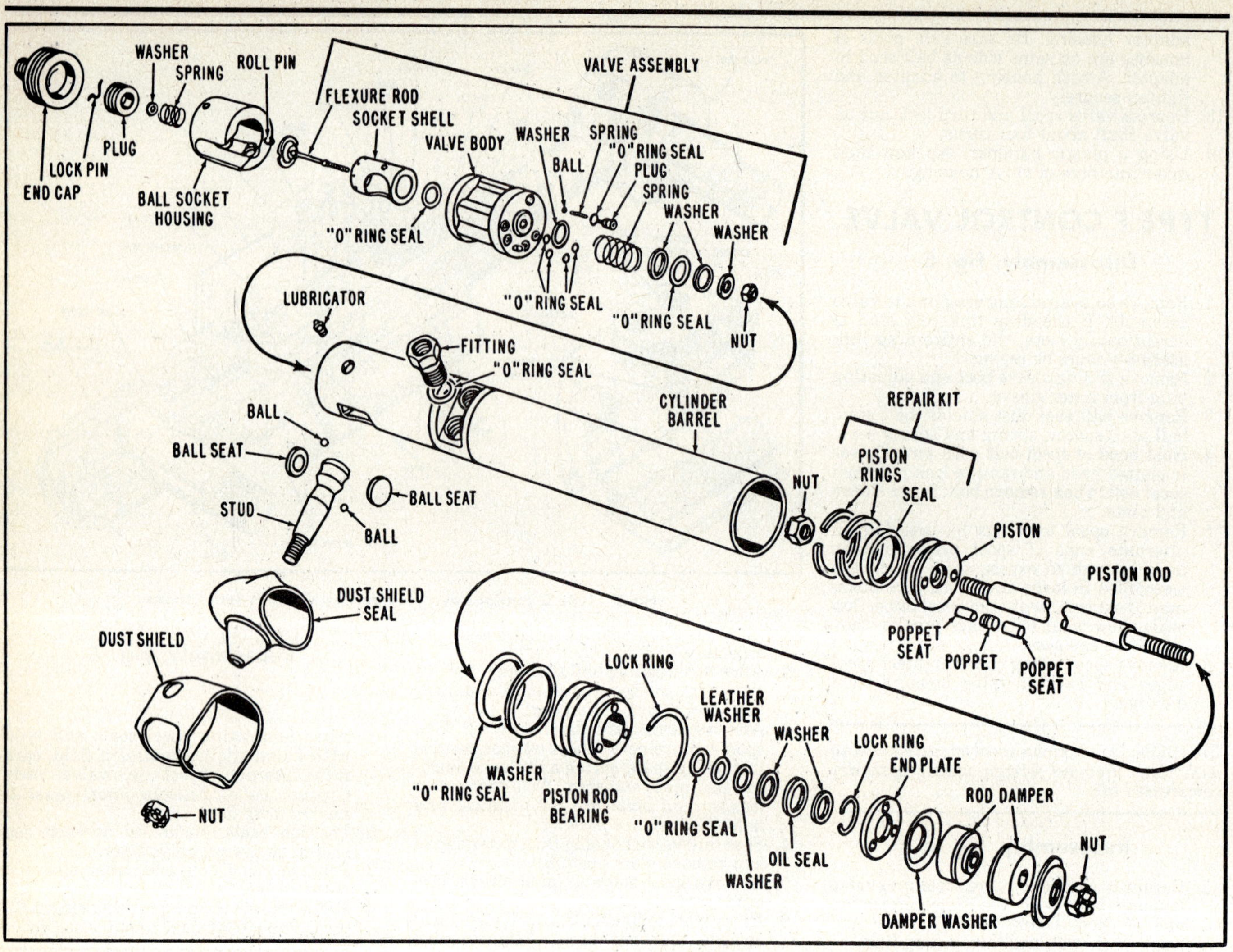

Fig. 9 Type I "In Line Booster" used on International and other trucks

5. Position shim (ground surface toward spool), bushing seal plate, sleeve, spring, spring seats and spacer washer on spool bolt. Install a new nut.
6. Position ball stud and its seat components in inner sleeve.
7. Tighten ball stud plug until spring is fully compressed, then back it off to the nearest cotter pin hole and insert cotter pin.
8. Position drag link against valve body and install housing bolts.

TYPE H CONTROL VALVE

"In Line Booster" Combining Control Valve with Power Cylinder Disassembly, Fig. 8

1. After removing dust cover clamp, remove four valve body bolts and separate cylinder assembly from valve body.
2. After removing and discarding nut from spool bolt, remove spacer washer, outer spring seat, spring, sleeve, inner spring seat, bushing plate and shim.

NOTE: *Note that spacer washer and shim are identical, and note their relative position.*

3. Separate valve body from outer sleeve and remove plate bushing.
4. Pull out sleeve bolt as far as possible, and remove lock pin from sleeve plate.
5. Unthread sleeve plate from inner sleeve, and remove stud, its seat and spring assembly, and the inner sleeve.
6. Press on spool to free spool seals from housing, and remove seals and spool. Remove spool carefully to avoid nicks and scratches.
7. If necessary to replace check valve or plug O-ring, remove plug by means of an appropriately bent brass rod inserted through return port.
8. Before removing any part of cylinder piston seal assembly, carefully clean area and remove snap ring.
9. Use air pressure applied at one of the ports in the mounting face of the cylinder to dislodge seal, scraper and retainer.

Reassembly, Fig. 8

1. Lubricate new seal with automatic transmission fluid and install seal, scraper and retainer, using a deep socket.
2. Install snap ring.
3. After lubricating spool seals and spool with automatic transmission fluid, carefully position spool in valve body. *The larger diameter bearing surface of spool must be next to cylinder.*
4. Position seals. Inner lip of each seal must be toward valve body.
5. Position inner sleeve, stud and its seat, and spring in outer sleeve.
6. Completely thread sleeve plate (with spool bolt) into inner sleeve, and then back off plate to nearest aligning holes for lock pin. Insert pin.
7. Position plate bushing in housing. Then position valve body next to outer sleeve. *Note that the valve body face has a shoulder that mates with outer sleeve face.*
8. Position shim (ground surface toward spool), bushing plate, inner spring seat, sleeve, spring, outer seat and spacer washer on spool bolt in that order. Install and tighten a new nut.
9. Position cylinder against valve body, install and tighten bolts.

TYPE I CONTROL VALVE

"In Line Booster" Combining Control Valve with Power Cylinder Disassembly of Power Cylinder, Fig. 9

1. Drain oil from cylinder and valve by

moving piston rod in and out from one extreme end of its travel to the other.

2. Disassemble cylinder parts on threaded end of piston rod by removing slotted nut, cushion retainers and cushions.
3. Remove end plate.
4. Using a drift and hammer, work around bearing, tapping it lightly into cylinder barrel about 1/4".
5. Insert punch in hole along side of cylinder barrel to disengage lock ring from its groove. Remove lock ring.
6. Pull on piston rod end to remove internal parts from cylinder. When removing parts from piston rod, slide them off piston end of rod.
7. Remove piston rod nut. Hold rod by the two wrench flats just above outer threaded end to prevent damage to finish on rod O.D.
8. Slide off piston and remove ring and seal.
9. Check action of poppet valve for leakage. If defective, replace piston assembly.
10. With bearing removed from rod, remove O-ring and back-up washer from O.D. of bearing.
11. Remove sealing parts in bearing by first removing lock ring, then retainer washer, oil seal, spacing washers, leather back-up washer and O-ring.

Disassembly of Control Valve, Fig. 9

1. With slotted nut removed from ball stud, remove dust seal cover and dust seal.
2. Straighten out the one staked place and remove end cap.
3. Remove lock pin and unscrew adjusting plug.
4. From inside plug remove spring and washer.
5. Remove outer ball seat.
6. Remove ball stud and inner ball stud seat. Do not lose steel balls.
7. Unscrew grease fitting.
8. Pull out valve sub-assembly, using care not to damage threads in outer end of valve housing when pulling this assembly. As this sub-assembly is removed, the four small O-rings may or may not remain in counterbores on end of valve body. If not they will be found on face of cylinder head.
9. To simplify disassembling of flexure rod, socket shell, valve body and spool sub-assembly, rassemble ball stud against inner ball seat, install outer ball seat in socket shell and screw adjusting plug in. This will tend to keep flexure rod from turning when unscrewing elastic nut.
10. After above disassembly has been completed, remove ball stud, seats and adjusting plug from socket shell.
11. After elastic nut has been removed, washer, centering washers, seal, spring, washer, socket shell and flexure rod can be disassembled.
12. Remove spool from valve body by pushing out from end opposite counterbore O-ring seats in valve body.
13. Remove plug assembly.

Reassembling Power Cylinder, Fig. 9

1. Lubricate all seals and O-rings with light engine oil before installing.
2. Assemble O-ring to inside of bearing, then leather back-up ring and steel spacing washers. Assemble oil seal with its lip toward outside. Assemble retainer washer and secure oil seal parts with lock ring.
3. Assemble back-up ring on O.D. of bearing, then O-ring on pressure side of bearing.
4. To prevent damage to rod seal, assemble bearing to rod from piston end of rod. Slide bearing over piston rod toward outer end of rod.
5. Assemble seal in piston ring groove of piston and rings over seal.
6. Assemble piston to end of rod and secure with nut.
7. Slide piston into cylinder barrel. Push assembly into cylinder halfway, and assemble bearing into cylinder, taking care not to damage O-ring as it slides past retaining ring groove in cylinder.
8. Lock bearing in place with retaining ring in cylinder groove.
9. Check for freedom of binding by moving piston rod back and forth.
10. Assemble end plate and tighten screws securely.
11. Install inner cushion retainer with cup side out. Assemble inner and outer rubber cushions, then outer cushion retainer with cup side in. Secure assembly with slotted nut.

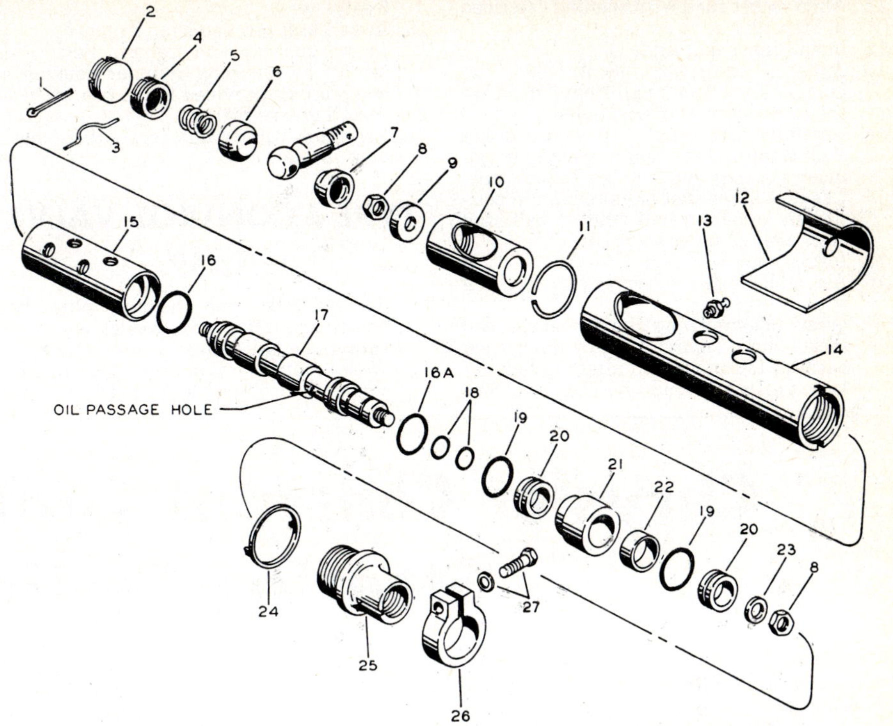

1 – Cotter Pin
2 – Housing Plug
3 – Ball Socket End Plug Lock
4 – Socket End Plug
5 – Spring
6 – Ball Seat
7 – Ball Seat
8 – Nut
9 – Bearing
10 – Ball Socket
11 – Lock Ring
12 – Cover
13 – Fitting
14 – Housing
15 – Valve Body
16 – "O" Ring
16A – "O" Ring
17 – Spool
18 – "O" Ring
19 – "O" Ring
20 – Reaction Rings
21 – Gland
22 – Spacer
23 – Washer
24 – Lock Ring
25 – Drag Link Adapter
26 – Drag Link Clamp
27 – Lockwasher & Bolt

Fig. 10 Type I control valve disassembled

Reassembly of Valve, Fig. 9

1. Assemble O-ring to groove in spool and insert spool in valve body from end opposite the four small O-ring seats. This will prevent O-ring from being cut by sharp inner edges of body.
2. If by-pass unit has been removed, place new O-ring on plug. Stand body on end and drop ball into hole. Place spring over plug stem and insert in hole until flush with end face of valve body.
3. To assist easier assembly of centering washers, spring, seal, washer and elastic nut, assemble flexure rod to socket shell, and assemble in socket shell the inner ball seat, ball stud, outer ball seat and adjusting plug. Tighten securely. Place this assembly in a vise, holding at threaded end of socket shell.
4. Slide valve body and spool assembly over flexure rod, being sure the four counterbored holes are on top.
5. Assemble centering washer, O-ring centering washer and heavy washer. Press into valve body carefully until elastic nut can be started on threaded end of flexure rod. Using a box wrench push down on top of washer, while at the same time rotating valve body back and forth until O-ring and outer centering washer enter valve body completely.
6. As this is being done, tighten elastic nut until parts seat solidly. To assure all parts are correctly aligned, hold valve body in hand with threaded end of flexure rod against a solid object and push to check freedom of valve spool.
7. Disassemble ball stud, seats, and adjusting plug from socket shell.
8. Assemble four O-rings to valve body.
9. Slide valve body with flexure rod into cylinder housing. *Roll pin assembled in valve body must fit in dowel hole in face of cylinder head.*
10. Assemble socket housing over socket shell, aligning slotted opening in socket housing with rectangular opening in cylinder housing.

11. Align socket shell with opening described in Step 10.
12. Install inner ball seat.
13. Apply a small amount of heavy cup grease to the two small holes in socket shell and insert the steel balls.
14. Carefully insert ball stud into socket shell through rectangular opening in cylinder housing and ball socket housing. Use care to see that balls remain in position in groove in ball stud as ball stud rests against inner ball seat.
15. Install outer ball seat.
16. Assemble washer and spring into opening of adjusting plug.
17. Screw adjusting plug into threaded end of socket shell. Tighten plug by hand, then back off to the nearest aligning hole for lock. Ball stud should be firm but move freely.
18. Install lock pin and snap in place.
19. Install end cap and tighten. Secure assembly by crimping cylinder housing at one place as provided for in cap.
20. Install grease fitting.
21. Assemble rubber dust seal, then metal dust seal cover.

TYPE J CONTROL VALVE

Fig. 10

1. Remove cotter pin, housing plug, ball socket end plug lock, socket end plug, spring and ball seats.
2. With control valve held in a vise, straighten crimped lock ring in notch of drag link adapter, then remove adapter.
3. Insert a drift into spool oil passage, preventing spool rotation, remove spool nut, ball socket and bearing.
4. Remove nut from pool gland end, then washer, gland, reaction rings and spacer.

NOTE: Some units incorporate a spring instead of the spacer.

5. Slide spool toward gland end of body and remove "O" ring seal, then move spool toward ball stud side and remove the other "O" ring.
6. Remove spool from gland end of body, then remaining "O" rings.
7. Reverse procedure to reassemble.

Chrysler Constant Control Power Steering Section

The Constant Control Full Time Power Steering consists of a hydraulic pressure pump, a power steering gear and connecting hoses. The power steering gear consists of a gear housing, containing a gear shaft and sector shaft; a power piston with gear teeth milled into the side of the piston which is in constant mesh with the gear shaft sector teeth; and a worm shaft which connects the steering wheel to the power unit piston through a coupling.

Steering Gear Repairs

Prior to disassembly, clean the gear assembly thoroughly in a suitable solvent. Crocus cloth may be used to remove small nicks and burrs provided it is used carefully. When used on the steering gear valve, use extreme care not to round off the sharp edge portions of the two lands located between the valve drilled holes. Remove and discard all "O" ring seals, using new ones lubricated with fluid when reassembling.

Disassemble

1. Drain steering gear by rotating wormshaft from stop to stop.
2. Remove valve body and three "O" rings, Fig. CC1.
3. Remove valve lever by prying under spherical head, Fig. CC2. *Use care not to collapse slotted end of valve lever as this will destroy bearing tolerances of head.*
4. Remove gear shaft cover nut, Fig. CC3.
5. Rotate wormshaft to position sector shaft teeth at center of piston travel, then loosen power train nut.
6. Install tool C-3786 on sector shaft threaded end and slide tool into housing until tool and shaft engages bearings.
7. Rotate wormshaft to left stop, compressing power train components, then remove power train nut and housing head tang washer. Using a screwdriver, pry on piston teeth to remove complete power train, Fig. CC6.

NOTE: The cylinder head, center race and spacer assembly and the housing head must be kept in close contact since the reaction rings may become disengaged from grooves in cylinder head and housing. Also, the center spacer may separate from center race and jam in the housing, in turn damaging the spacer and/or housing when removing the power train.

8. Remove worm shaft upper oil seal, Fig. CC7, and disassemble column jacket support. Reaction seal may be removed from groove in face of jacket support by blowing air pressure into ferrule chamber. Make sure passage from ferrule chamber to upper reaction chamber is unobstructed.

Column Jacket Support, Assemble

Install worm shaft upper oil seal with sealing lip toward bearing, Fig. CC8. Tool C-3650 should be used to drive seal until tool bottoms on casting to obtain proper compression of rubber seal. Lubricate reaction seal and install in groove in face of column jacket support with flat side of seal out.

Cylinder Head

Disassemble—Remove two "O" rings in outer grooves in head. Remove "O" ring in groove in face of cylinder head with air pressure into oil hole located in groove between two "O" ring grooves, Fig. CC9. Replace cylinder head seal if necessary, Fig. CC10. Check oil passage in ferrule for obstruction. Check lands of cylinder head for burrs.

Assemble—Lubricate and install two large "O" rings in grooves on head. Install lower reaction seal in groove in face of head. The small "O" ring for ferrule groove should be installed after worm shaft bearing preload has been established, otherwise "O" ring will be damaged by reaction springs.

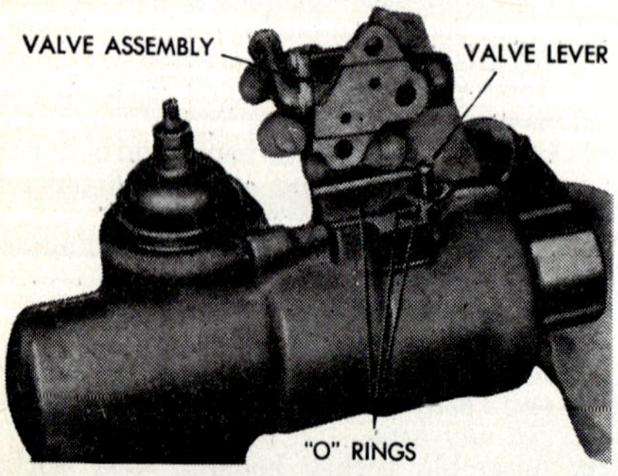

Fig. CC1 Removing or installing valve body assembly

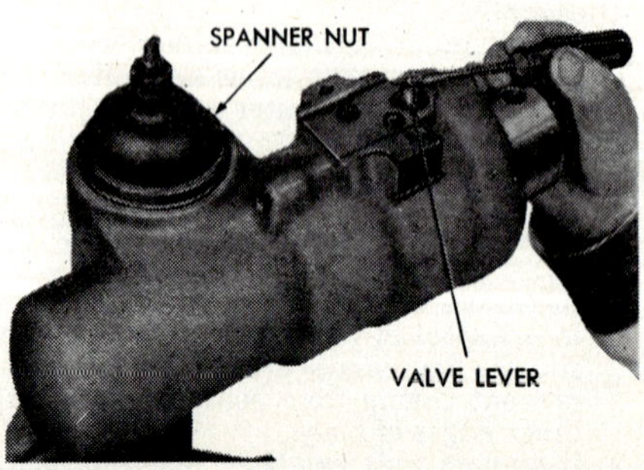

Fig. CC2 Removing valve lever

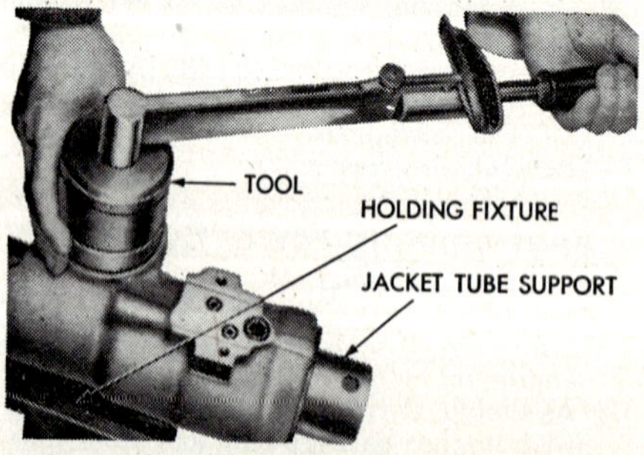

Fig. CC3 Removing or installing gear shaft cover nut

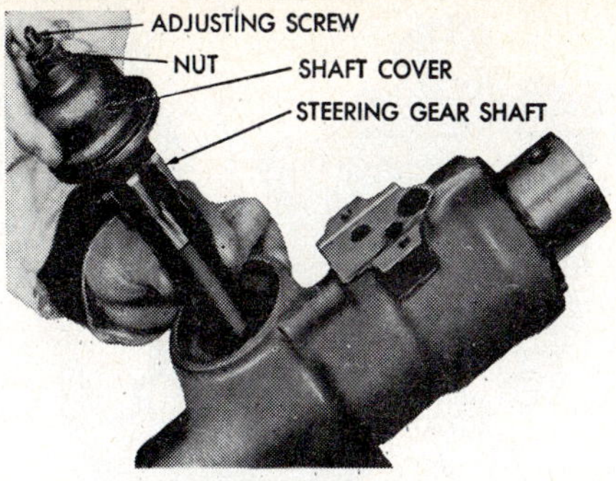

Fig. CC4 Removing or installing gear shaft & cover

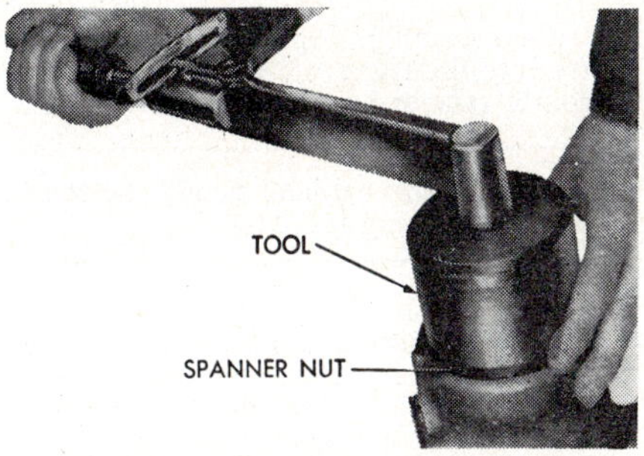

Fig. CC5 Removing steering column support nut

Steering Valve

Disassemble—Remove outlet fitting, spring, valve piston and cushion spring. Remove valve body (2 screws) from steering valve. Shake out valve piston. If spool valve or valve housing is damaged, replace complete valve assembly. Do not remove valve end plug unless inspection indicates a leak at seal.

Assemble—Fig. CC11. Install steering valve in housing. Valve lever hole should align with steering gear valve lever opening in bottom of valve housing. Valve must fit smoothly in housing without striking or binding. If valve end plug was removed installed new seal and tighten plug to 25 lb. ft. torque. Install cushion spring, then lubricate and install valve piston, nose end first, into valve body. Install spring on top of piston and outlet plug and torque to 20 ft. lbs. Assemble control valve body to steering valve body and torque attaching screws to 95 inch lbs. Install new copper sealing gasket and fitting in threaded hole on top of valve body and torque to 30 lb. ft.

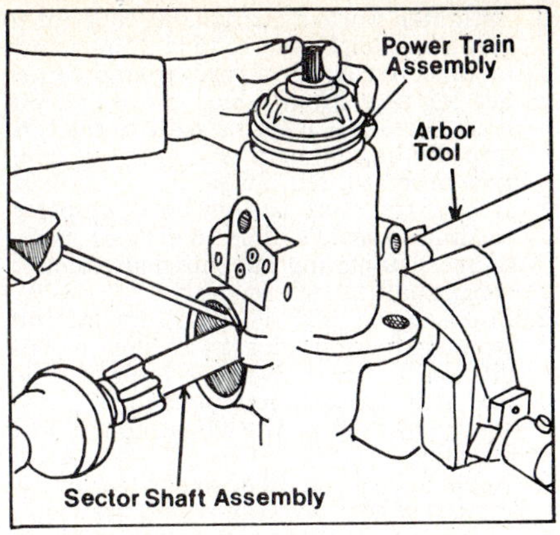

Fig. CC6 Removing or installing power train

Gear Shaft

Disassemble—Remove adjusting screw lock nut and unscrew cover from adjusting screw. Remove screw and washer from, "T" slot in end of gear shaft. Remove date tag gasket from top of cover and large "O" ring from base of cover.

Assemble—Place adjusting screw washer over screw and slide both into "T" slot of shaft. Screw cover into adjusting screw until gear shaft bottoms in cover. Install date tag gasket over adjusting screw into position at top of cover. Install lock nut on adjusting screw but do not tighten. Install large "O" ring in groove on lower face of gear cover.

Steering Gear Housing

Disassemble—The equipment shown in Fig. CC12 should be used for this operation. After removing the oil seal snap ring and back-up ring, remove the oil seal by sliding the threaded portion of the tool's adapter over end of gear shaft and install nut section of tool (C-3350) on shaft. Maintain pressure on adapter with nut of tool while turning adapter, forcing it into seal until it has bottomed in seal. Apply two half rings and retainer over both portions of tool. As hex nut is removed from shaft seal will be pulled from housing.

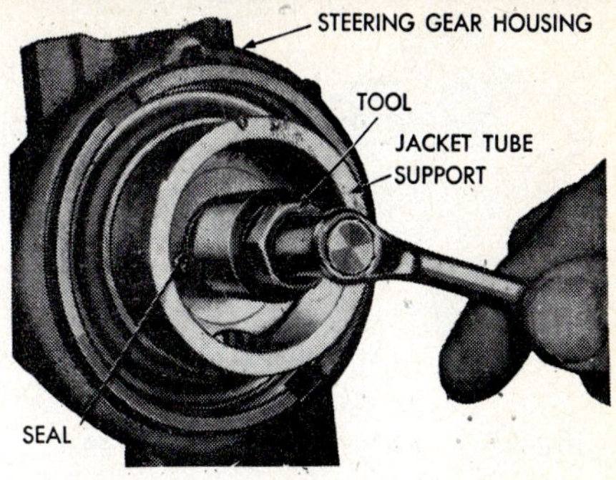

Fig. CC7 Removing worm shaft upper oil seal

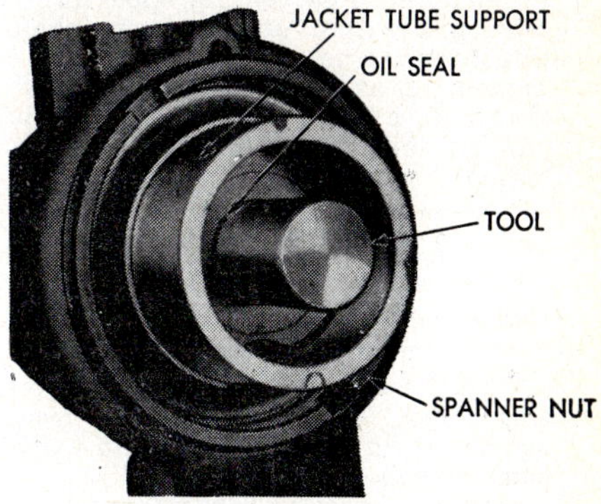

Fig. CC8 Installing worm shaft upper oil seal

Assemble—Install gear shaft oil seal in housing (lip of seal toward needle bearing) using Tool C-3350. Place adapter against seal and tool nut on threaded end of gear shaft. Tighten tool nut until adapter shoulder contacts housing, Fig. CC13. Install oil seal back-up ring, oil seal and snap ring.

Power Train, Assemble

Fig. CC14. Position piston assembly on bench with worm shaft up. Slide cylinder head (ferrule up) on worm shaft and against piston flange, making sure gap on worm shaft ring is closed to avoid breaking ring. Lubricate and install in the following order:

1. Lower thrust bearing race (thick).
2. Lower thrust bearing.

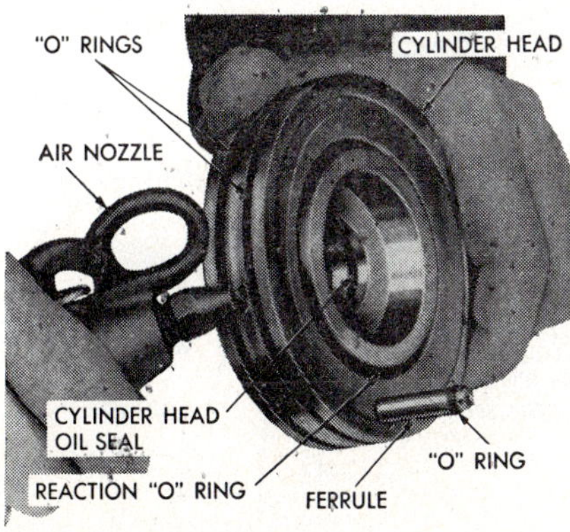

Fig. CC9 Removing reaction ring from cylinder head

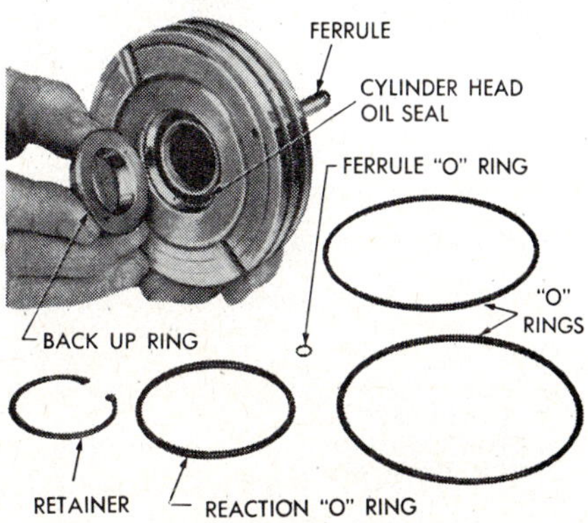

Fig. CC10 Removing cylinder head set

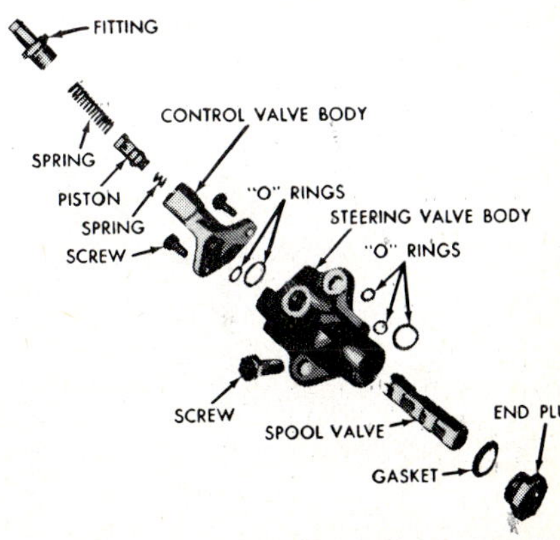

Fig. CC11 Control valve disassembled

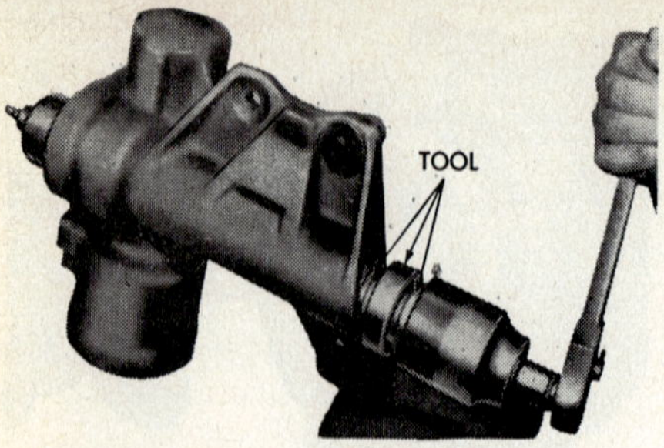

Fig. CC12 Removing gear shaft oil seal

5. Center bearing race, indexing control hole with hole in center bearing race.
6. Install upper thrust bearing, upper thrust bearing race (thin) and new worm shaft thrust bearing nut.
7. Tighten nut as follows: Turn worm shaft clockwise 1/2 turn and hold shaft in this position while tightening nut to 50 lb. ft. torque.

NOTE: *If worm shaft is turned more than 1/2 turn cylinder head seal will clear oil ring on worm shaft.*

8. Always check position of worm shaft oil ring before bottoming cylinder head against worm piston shoulder to avoid damaging oil ring.
9. Loosen worm bearing adjusting nut.
10. Retighten worm bearing adjusting nut to give a bearing torque of 16 to 24 oz. Check by placing rounds of cord around center bearing race. Make a loop in one end and hook the loop of a spring scale in the cord loop. Pulling on cord will cause bearing race to rotate, Fig. CC15.
11. If adjusting nut is tightened properly, reading on scale will be 16 to 24 oz. Stake flange of adjusting nut into depression in worm shaft to lock securely.
12. Stake upper part of adjusting nut into wormshaft knurled area in four places, 90° apart. Check bearing preload after first staking operation and adjust, if necessary, by tapping the adjusting nut with a mallet.
13. Apply a torque of 20 ft. lbs. to the adjusting nut in both directions. The nut should not move, ensuring nut was properly staked.
14. Install center bearing spacer over bearing race to engage dowel pin with slot in center bearing race.

NOTE: *Make sure valve lever hole in center bearing race and center spacer are properly aligned.*

Steering Gear, Assemble

1. With steering gear housing fastened in holding fixture, lubricate power train housing bore with fluid and install power train assembly, Fig. CC6. Face piston teeth to the right and valve lever hole in center race and spacer in "Up" position.

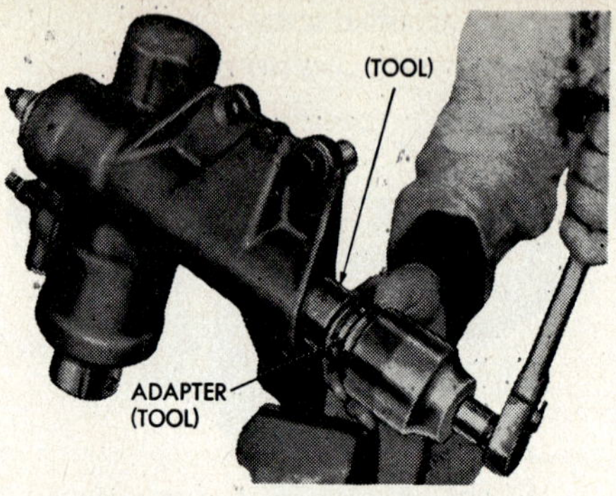

Fig. CC13 Installing gear shaft oil seal

NOTE: Ensure cylinder head is bottomed against housing shoulder.

2. With valve lever hole in center bearing race and spacer aligned with valve level hole in gear housing, install valve lever (double bearing end first) into center bearing race and spacer, ensuring valve lever slots are parallel to wormshaft, Fig. CC16. It may be necessary to gently tap end of valve lever to seat lower pivot point

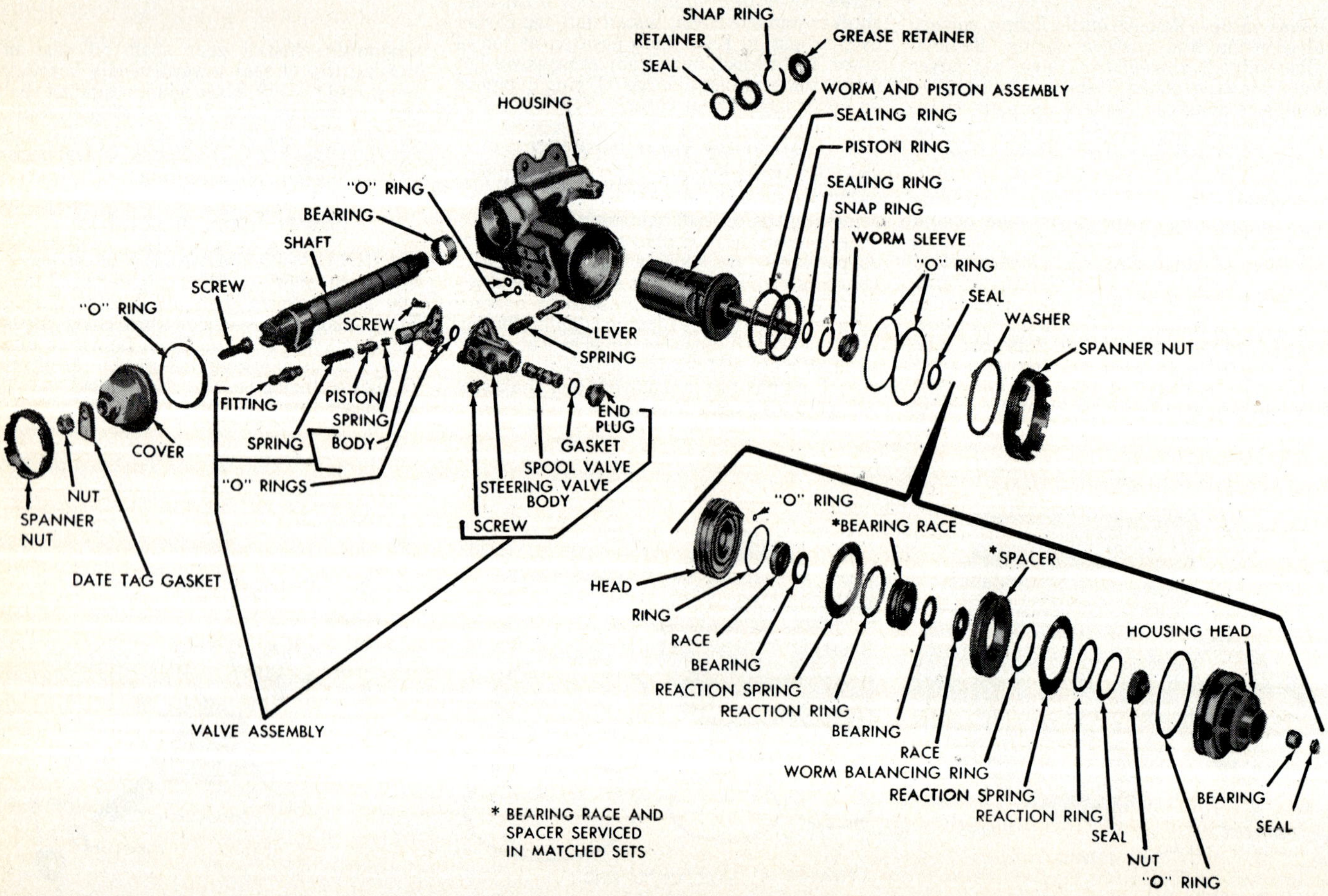

Fig. CC14 Steering gear disassembled

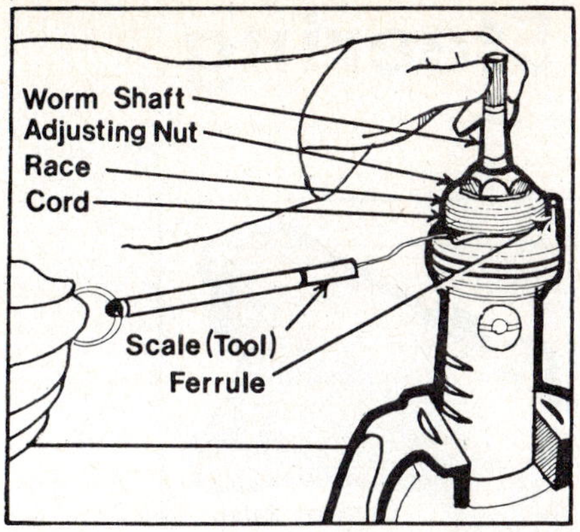

Fig. CC15 Checking worm shaft bearing preload

in center race.

3. Install housing head tang washer, indexing with groove in housing, and steering column support nut and torque nut to 150–200 ft. lbs. Ensure valve lever remains centered in housing hole by rotating wormshaft until piston bottoms in both directions and observe valve lever action. The valve lever must be in center of hole and return to the center position when wormshaft torque is relieved.
4. Install valve lever spring (small end first), position power piston at center of travel and install sector shaft and cover assembly, indexing sector teeth with piston rack teeth. Ensure sector shaft cover "O" ring is installed properly.
5. Install and torque cover nut to 110–200 ft. lbs.
6. Install valve body onto housing with valve lever entering hole in valve spool. Ensure "O" rings are installed properly, then install and torque valve retaining screws to 7 ft. lbs.

Tests and Adjustments

1. Fill reservoir to level mark.
2. Connect test hoses to hydraulic pump on car with pressure gauge installed between pump and steering gear to register pressures.
3. Start engine and operate at idle to bring steering gear to normal operating temperature.
4. Center the valve until unit is not self-steering. Tap valve body retaining screw heads to move valve body up on steering housing or end plug to move valve body down.
5. Rotate wormshaft through range of travel to bleed air from system, then refill reservoir.
6. With steering gear on center, tighten sector shaft adjusting screw until backlash is removed from steering arm, Fig. CC17. If the power train was removed, tighten adjusting screw an additional $1\frac{1}{4}$ turns, then tighten locknut.
7. Operate unit through several cycles, aligning piston rack and sector teeth.
8. With gear on center, readjust sector shaft backlash. Loosen adjusting screw until backlash is present, then tighten adjusting screw until backlash is removed. Tighten adjusting screw an additional $\frac{3}{8}$ to $\frac{1}{2}$ turn and torque locknut to 28 ft. lbs.
9. Recenter valve body.

TROUBLE SHOOTING

Hard Steering

1. Tires not properly inflated.
2. Low oil level in reservoir.
3. Loose pump belt.
4. Oil on pump belts.
5. Steering linkage needs lubrication.
6. Power steering pump output low.
7. Cross shaft adjustment too tight.
8. Pressure control valve stuck in closed position.
9. External oil leaks.
10. Defective or damaged valve lever.
11. Dirt or chips in steering gear.
12. Damaged column support worm shaft bearings.
13. Damaged thrust bearings or excessive preload adjustment.
14. Rough or hard to turn worm and piston assembly.
15. Excessive internal leakage.

Poor Recovery From Turns

1. Tires not properly inflated.
2. Steering linkage binding.
3. Improper wheel alignment.
4. Damaged or defective steering tube bearings.
5. Steering column jacket and steering gear not properly aligned.
6. Improper cross shaft mesh adjustment.
7. Pressure control valve piston stuck open.
8. Column support spanner nut loose.
9. Defective or damaged valve lever.
10. Improper worm thrust bearing adjustment.
11. Burrs or nicks in reaction ring grooves in cylinder head or column support.
12. Defective or damaged cylinder head worm shaft seal ring.
13. Dirt or chips in steering gear unit.
14. Rough or catchy worm and piston assembly.

Self-Steering or Leads to Either Side

1. Tires not properly inflated.
2. Improper wheel alignment.
3. Steering wheel off center when car is traveling straight ahead.
4. Valve body out of adjustment.
5. Valve lever damaged.
6. Column support spanner nut loose.
7. Return turn reaction seal damaged.
8. Column coupling bottomed.

Temporary Increase In Effort When Turning Steering Wheel

1. Low oil level.

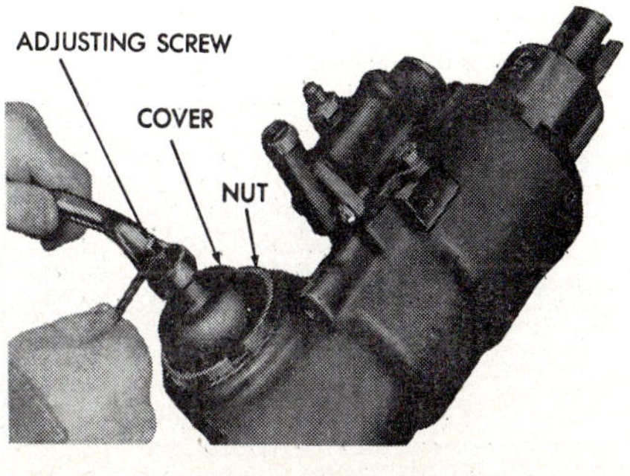

Fig. CC17 Adjusting sector shaft backlash

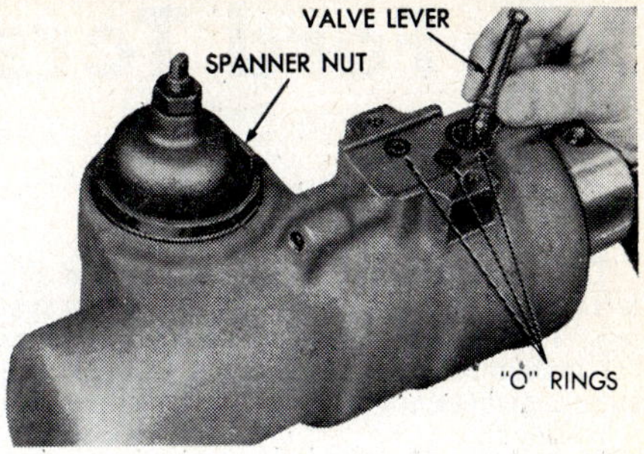

Fig. CC16 Installing valve lever

2. Loose pump belt.
3. Oil on pump belts.
4. Binding steering linkage.
5. Engine idle too slow.
6. Defective power steering pump.
7. Air in system.
8. External adjustment.
9. Improper cross shaft adjustment.
10. Excessive internal leakage.

Excessive Steering Wheel Free Play

1. Improper cross shaft adjustment.
2. Column support spanner nut loose.
3. Improper worm thrust bearing adjustment.
4. Improper front wheel bearing adjustment.

Lack of Assistance in One Direction

1. Oil leaking past worm shaft cast iron seal ring or ferrule "O" ring.

Lack of Assistance in Both Directions

1. Broken "O" ring on worm piston.
2. Piston end plug loose.
3. Pump belt slipping.
4. Pump output low.

Noises

1. Buzzing noise in neutral only is caused by sticking pressure control valve.
2. Noisy power pump.
3. Damaged hydraulic lines.
4. Pressure control valve sticking.
5. Improper sector shaft adjustment.
6. Air in system.

Lack of Effort, Left Turn

1. Left turn reaction seal "O" ring missing, damaged or worn.
2. Left turn reaction oil passage not drilled in housing or cylinder head.
3. Left turn reaction ring sticking in cylinder head.

Lack of Effort, Right Turn

1. Right turn U-shaped reaction seal missing, damaged or worn.
2. Right turn reaction oil passage not drilled in housing head or ferrule pin.
3. Right turn reaction ring sticking in housing head.

Lack of Effort in Both Directions

1. Improper sector shaft adjustment.
2. Pressure plates installed on wrong side of reaction rings.

Ford Torsion Bar Power Steering

DESCRIPTION

The power steering unit, Figs. F1, F2 & F3, is a torsion bar type of hydraulic-assisted system. This system furnishes power to reduce the amount of turning effort required at the steering wheel. It also reduces road shock and vibrations.

The unit includes a worm and one piece rack-piston which is meshed to the gear teeth on the steering sector shaft. The unit also includes a hydraulic valve, valve actuator, input shaft and torsion bar assembly which are mounted on the end of the worm shaft and operated by a twisting action of the torsion bar.

The spool valve type gear, Fig. F2, is designed with the one piece rack-piston, worm and sector shaft in the one housing and the valve spool in an attaching housing. The Rotary Valve type gear, Fig. F3, uses a one piece rotary valve and sleeve assembly. Therefore no centering shims are used and no adjustment is possible. This makes possible internal fluid passages between valve and cylinder, thus eliminating all external lines and hoses except the pressure and return hoses between pump and gear.

The power cylinder is an integral part of the gear housing. The piston is double acting in that fluid pressure may be applied to either side of the piston.

A selective metal shim, used only on the spool valve type gear, Fig. F2, is located in the valve housing of the gear is for the purpose of tailoring the steering gear efforts. If efforts are not satisfactory they can be changed by increasing or decreasing shim thickness as follows:

Efforts heavy to the left—increase shim thickness. Efforts light to the left—decrease shim thickness. A change of one shim size will increase or decrease steering efforts approximately 2 inch lbs. on 1974 models. Shims are available in the following selective thicknesses:

.0057 to .0063″
.0077 to .0083″
.0097 to .0103″
.0117 to .0123″
.0137 to .0143″

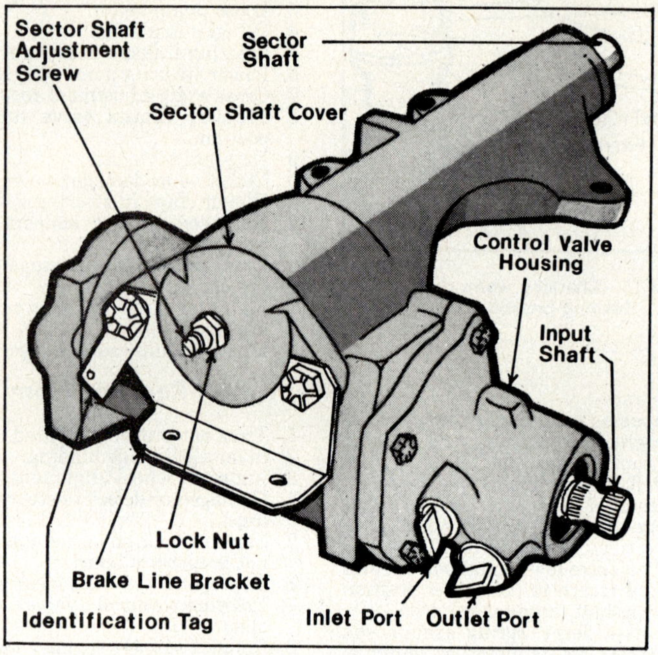

Fig. F1 Ford torsion bar power steering gear

Operation

The operation of the hydraulic control valve spool is governed by the twisting of a torsion bar. All effort applied to the steering wheel is transmitted directly through the input shaft and torsion bar to the worm and piston. Any resistance to the turning of the front wheels results in twisting of the bar. The twisting of the bar increases as the front wheel turning effort increases. The control valve spool, or the rotary valve sleeve actuated by the twisting of the torsion bar, directs fluid to the side of the piston where hydraulic assistance is required.

As the torsion bar twists, its radial motion is transferred into axial motion by helical threads. Thus, the valve is moved off center, and fluid is directed to one side of the piston or the other.

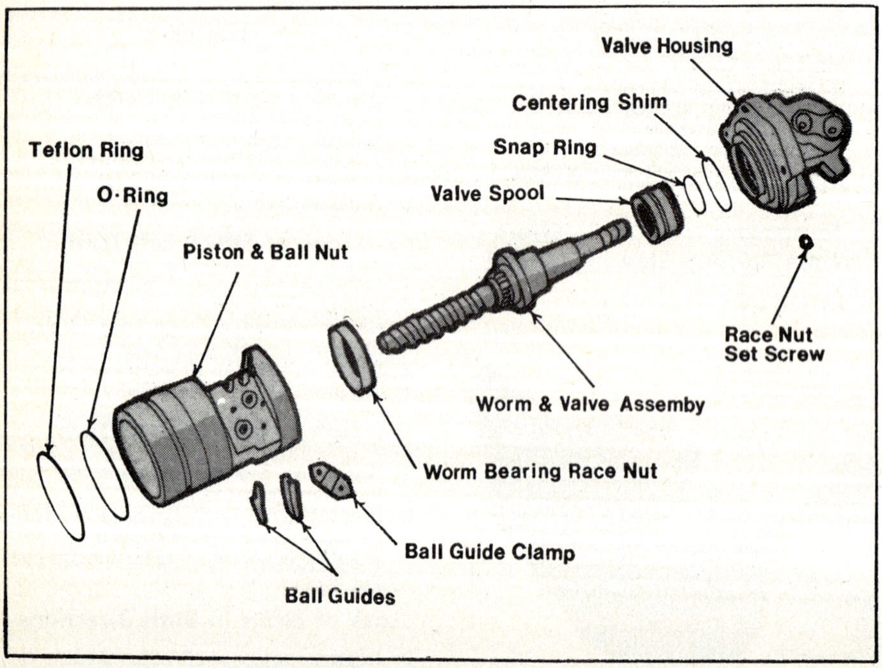

Fig. F2 Ball nut & valve housing disassembled. 1974 Spool valve type

IN-VEHICLE ADJUSTMENTS & REPAIRS

Valve Spool Or Rotary Valve Centering Check

1. Install a 2000 psi pressure gauge in pressure line between pump outlet port and steering gear inlet port. *Make sure that valve on gauge is in fully open position.*
2. Check fluid level in reservoir and replen-

ish as required.

3. Start engine and cycle steering wheel from stop-to-stop to bring steering lubricant up to normal operating temperature. Stop engine and recheck reservoir. Add fluid as necessary.
4. With engine running at a fast idle speed (1000 rpm) and steering wheel centered, attach an inch-pound torque wrench to steering wheel retaining nut. Apply sufficient torque to wrench in each direction (either side of center) to get a gauge reading of 250 psi.
5. *Valve Spool type gear*
   The torque reading should be the same in both directions. If the difference between readings exceeds 4 inch lbs, remove steering gear from vehicle and change thickness of valve centering shim. On 1974 units, use as many shims as necessary, not exceeding .030 inch total thickness. If steering effort is heavy to the left, increase shim thickness and if steering effort is light to the left, decrease shim thickness.
   Rotary Valve type gear
   The torque reading should be the same in both directions. If the difference between readings exceeds 4 inch lbs., remove steering gear from vehicle and replace the shaft and control assembly.

NOTE: The "out-of-vehicle" procedure for centering check is the same as for the "in-vehicle" except the torque and simultaneous pressure reading must be made at the right and left stops instead of either side of center.

Steering Gear Adjustments

Preload (thrust bearing adjustment) and worm-to-rack preload cannot be changed in service. The only adjustment that can be performed is the total overcenter position load to eliminate excessive lash between sector and rack teeth.

1. Disconnect pitman arm from sector shaft.
2. Disconnect fluid return line at reservoir and cap reservoir return line pipe.
3. Place end of return line in a clean container and cycle steering wheel in both directions as required to discharge fluid from gear.
4. Remove ornamental cover from wheel hub and turn steering wheel 45° from left stop.
5. Using an inch-lb torque wrench on steering wheel nut, determine torque required to rotate shaft slowly through an approximately 1/8 turn from the 45° position.
6. Turn steering gear back to center, then determine torque required to rotate shaft back and forth across center position.
7. Loosen adjuster nut and turn adjusting screw until reading is 11 to 12 inch lbs. greater than torque measured 45° from stop. Hold screw in place and tighten locknut.
8. Recheck readings and replace pitman arm and steering wheel.
9. Connect fluid return line and replenish reservoir.

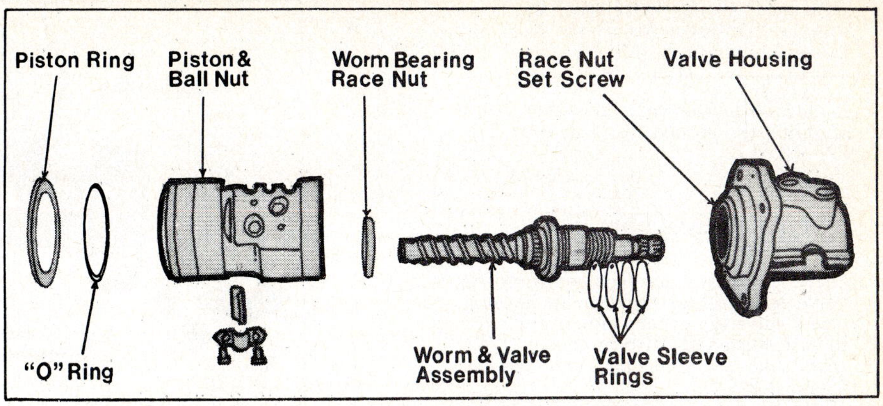

Fig. F3 Ball nut & valve housing disassembled. 1974–83 Rotary valve type

STEERING GEAR REPAIRS

Disassembly

1. Hold steering gear over drain pan in an inverted position and cycle input shaft six times to drain remaining fluid from gear.
2. Remove lock nut and brass thrust washer from adjusting screw.
3. Turn input shaft to either stop, then turn it back approximately 1 3/4 turns to center the gear.
4. Remove two sector shaft cover studs, brake line bracket and identification tag.
5. Tap lower end of sector shaft with a soft-faced hammer to loosen it, then lift cover and shaft from housing as a unit. Discard O-ring.
6. Turn sector shaft cover counterclockwise off adjuster screw.
7. Remove four valve housing attaching bolts. Lift valve housing from gear housing while holding piston to prevent it from rotating off worm shaft. Remove valve housing and lube passage O-rings and discard.
8. Stand valve body and piston on end with piston end down. Rotate input shaft counterclockwise out of piston, allowing balls to drop into piston.
9. Place a cloth over open end of piston and turn it upside down to remove balls.
10. Remove ball guide clamp screws from ball nut and remove clamp and guides, Figs. F2 & F3.
11. Hold valve body in a fixture and loosen the Allen head race nut screw from the valve housing and remove the worm bearing race nut.
12. Carefully slide input shaft, worm and valve out of valve housing.

CAUTION: Do to the close diametrical

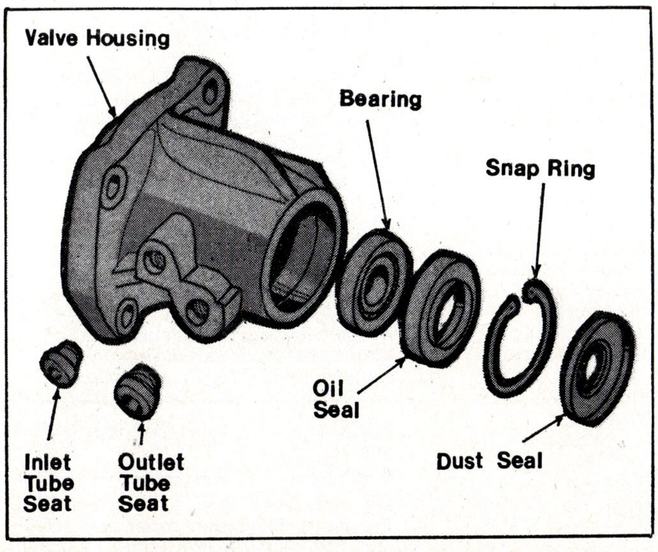

Fig. F4 Valve housing disassembled

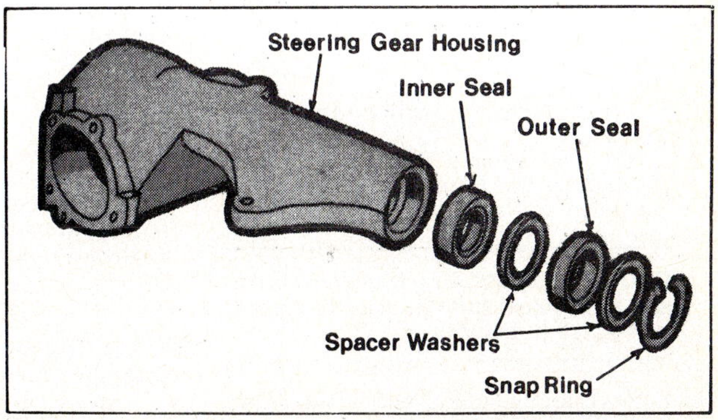

Fig. F5 Steering gear housing disassembled

clearance between valve spool and housing, the slightest cocking of the spool may cause it to jam in housing.

13. Remove shim from valve housing if disassembling the spool valve type gear, Fig. F2.

Reassembly

1. On 1974 units, place required thickness valve spool centering shim in housing, (spool valve type gear only), Fig. F2.
2. Install worm and valve in housing.
3. Install race nut in housing and torque to 55–65 ft-lbs.
4. Install race nut set screw and torque to 15–25 in. lbs.
5. Place piston on bench with ball guide holes facing up. Insert worm shaft into piston so that first groove is in alignment with hole nearest to center of piston, Fig. F6.
6. Place ball guide into piston. Place 27 to 29 balls in guide, (depending on piston design), turning worm clockwise (viewed from input end of shaft). If all balls have not been fed into guide upon reaching right stop, rotate input shaft in one direction and then in the other while installing remaining balls. After balls have been installed, do not rotate input shaft or piston more than 3½ turns off the right stop to prevent balls from falling out of circuit.
7. Secure guides to ball nut with clamp.
8. Position new lube passage O-ring in counterbore of gear housing.
9. Apply vaseline to teflon seal.
10. Place a new O-ring on valve housing.
11. Slide piston and valve into gear housing, being careful not to damage teflon seal.
12. Align lube passage in valve housing with one in gear housing, and install but do not tighten attaching bolts at this time.
13. Rotate ball nut so that teeth are in same plane as sector teeth. Torque four valve housing attaching bolts to 35 to 45 ft-lbs.
14. Position sector shaft cover O-ring in gear housing. Turn input shaft as required to center piston.
15. Apply vaseline to sector shaft journal, then position sector shaft and cover into gear housing. Install the brake line bracket, identification tag and air conditioner line mounting bracket (if so equipped) and sector shaft cover bolts, torquing them to 55–70 ft-lbs.
16. Attach an inch lb. torque wrench to input shaft and adjust mesh load as outlined previously.

Valve Housing, Replace

1. Referring to Fig. F4, remove dust seal and snap ring.
2. Invert housing and from opposite seal end, gently tap bearing and seal out of housing.

CAUTION: Exercise care when inserting and removing tool to prevent damage to valve bore in housing.

3. Remove oil inlet and outlet tube seats with "EZ-out" if damaged.
4. Coat tube seats with vaseline and position them in housing. Install and tighten tube nuts to press seats to proper location.
5. Coat bearing and seal surface in housing with a film of vaseline.
6. Position bearing in valve housing. Seat bearing in housing with tool T65P-3524-A1 making sure that bearing rotates freely.
7. Dip new oil seal in gear lube. Then seal facing outward. Drive seal into housing until outer edge of seal does not quite clear snap ring.
8. Place snap ring in housing. Then drive on ring with tool T65P-3524-A1 until snap ring seats in its groove to locate seal properly.
9. Place dust seal in housing with dished side (rubber side) facing out. Drive dust seal in place so that it is located behind undercut in input shaft when it is installed.

Fig. F6 Assembling piston on worm shaft

Worm & Valve (Valve Sleeve), Replace

1974 Spool Valve Type Gear

1. Remove snap ring from actuator.
2. Slide control valve spool off worm shaft, Fig. F2.
3. Install valve spool evenly and slowly with a slight oscillating motion into flanged end of valve housing with valve identification groove between spool lands outward. Check for freedom of valve movement within housing working area. Spool should enter housing bore freely and fall by its own weight.
4. If valve spool is not free, check for burrs at outward edges of working lands in housing and remove with a hard stone.
5. Check valve for burrs and, if necessary, stone valve in a radial direction only.
6. Remove valve spool from housing.
7. Slide spool onto worm shaft, making sure that the beveled ID of the snap ring is away from the spool.
8. Install snap ring to retain spool.

1975–83 Rotary Valve Type Gear

1. Remove valve sleeve rings, Fig. F3, from sleeve by inserting the blade of a small pocket knife carefully under them and cutting them off, without scratching the valve sleeve.
2. With the worm end of the worm and valve sleeve assembly mounted into a soft-jawed vise, install mandrel tool T75L-3517-A1 over the sleeve and slide one valve sleeve ring over the tool.
3. Slide the pusher tool T75L-3517-A2 over the mandrel then rapidly push down on the pusher tool, forcing the ring down the ramp and into the fourth groove of the valve sleeve. Repeat this step three more times, each time adding one of the spacers, tool T75L-3517-A3, under the mandrel tool.

NOTE: Adding the spacer each time will align the mandrel tool with the next groove of the valve sleeve.

4. After installing all four valve sleeve rings, apply a light coat of gear lubricant to the sleeve and rings.
5. Slowly install the sizing tube tool T75L-3517-A4, over the sleeve valve end of the worm shaft and onto the valve sleeve rings.

NOTE: Make sure that the rings turn freely in the grooves.

Piston & Ball Nut, Replace

1. Remove teflon ring and O-ring from piston and ball nut, Figs. F2 & F3.
2. Dip a new O-ring in gear lube and install on piston and ball nut.
3. Install new teflon ring on piston and ball nut, being careful not to stretch it any more than necessary.

Gear Housing, Replace

1. Remove snap ring and spacer washer, Fig. F5, from lower end of gear housing.
2. Remove lower seal from housing and like spacer washer from housing.
3. Remove upper seal in same manner.

NOTE: Some housings have only one seal and one spacer.

4. Dip both sector shaft seals in gear lube.
5. Position sector shaft inner seal (widest one) into housing with lip facing inward. Press into place with tool T65P-3576-B.

Place a 0.090 inch spacer washer on top of seal.
6. Place outer seal in housing with lip facing inward and press into place. Place a 0.090 inch spacer washer on top of seal and apply more lubricant.
7. Position snap ring in housing and press it in place to properly locate seals and engage snap ring in the groove.

TROUBLE SHOOTING

Hard Steering

1. Low or uneven tire pressure.
2. Improper gear adjustment.
3. Improper wheel alignment.
4. Low fluid level.
5. Twisted or bent suspension parts, frame and linkage components.
6. Tight wheel bearings.
7. Steering spindle bent.
8. Pump belt out of adjustment.
9. Pump output low.
10. Air in system.
11. Valve spool or rotary valve out of adjustment.
12. Valve spool or rotary valve sticking.
13. Steering linkage binding.

Hard Steering, Straight Ahead

1. Steering adjustment too tight.
2. Steering gear shaft binding.

Hard Steering While Turning or Parking

1. Oil level low.
2. Pump pressure low.
3. Pressure loss in steering gear due to leakage past "O" rings.
4. Pressure loss between valve spool and sleeve (valve spool type gear only.)
5. Pressure loss past piston ring or scored housing bore.

Loose Steering

1. Loose wheel bearings.
2. Loose tie rod ends or linkage.
3. Worn ball joints.
4. Worn suspension parts.
5. Insufficient mesh load.
6. Insufficient worm bearing preload.
7. Valve spool or rotary valve out of adjustment.

Erratic Steering

1. Oil or brake fluid on brake lining.
2. Out of round brake drums.
3. Improperly adjusted brakes.
4. Under-inflated tires.
5. Broken spring or other details in suspension system.
6. Improper caster adjustment.
7. Fluid level low.

Binding or Poor Recovery

1. Steering gear shaft binding.
2. Steering gear out of adjustment.
3. Steering linkage binding.
4. Valve spool or rotary valve binding due to dirt or burred edges.
5. Valve spool or rotary valve out of adjustment.
6. Interference at sector shaft and ball stud.

Loss of Power Assist

1. Pump inoperative.
2. Hydraulic lines damaged.
3. Power cylinder damaged.
4. Valve spool or rotary valve out of adjustment.

Loss of Power Assist In One Direction

1. Valve spool or rotary valve out of adjustment.

Noisy Pump

1. Air being drawn into pump.
2. Lines touching other parts of car.
3. Oil level low.
4. Excessive back pressure caused by obstructions in lines.
5. Excessive wear of internal parts.

Poor Return of Steering Gear to Center

1. Valve spool or rotary valve sticking.
2. Valve spool or rotary valve out of adjustment.
3. All items given under "Binding or Poor Recovery."

Steering Wheel Surge While Turning

1. Valve spool or rotary valve sticking.
2. Excessive internal leakage.
3. Belt slippage.

Ross Power Steering Section

SEMI-INTEGRAL GEAR

These power steering gears, Figs. R1, R2 and R3, are of the semi-integral type with either a built-in or externally mounted hydraulic control valve on or within the steering gear. The steering is then teamed with a fluid reservoir, power steering pump, and either an axle or frame mounted cylinder to provide a power assist for the steering system. Effort applied to the steering wheel actuates the control valve which directs fluid from the pump to the power cylinder. Fig. R4 illustrates the operation of built-in concentric control valve.

NOTE: *There is no adjustment to make on the control valve. The steering gear itself, however, is adjusted in the same manner as the comparable mechanical steering gear.*

Removal of Gear

Because of the different kinds of steering columns used on various model trucks, some variation in the removal procedure can be expected. However, before removing the gear, regardless of type, always note the hook-up of the hydraulic lines. Identify them by tagging connecting lines and marking the valve ports into which each connects. The general procedure which follows can be adapted to both the solid and jointed type steering columns. However, where a jointed column is being used, it is not necessary to remove the steering wheel and column.

1. On solid type columns, remove steering wheel. On jointed column models, loosen nut on steering shaft collar and remove steering column shaft from gear.
2. Use a puller to remove pitman arm.
3. Disconnect hydraulic lines at control valve. Plug all openings to keep out dirt.
4. Unfasten gear from its mounting and remove from chassis.

EXC. HPS-70

Disassembly

1. Loosen lock nut and unscrew lever shaft adjusting screw a few turns.
2. Remove gear housing side cover.
3. Slide lever shaft from housing after making sure that there are no burrs on outer end of shaft to damage bushings and seal in housing.
4. To remove control valve, identify position and face of valve next to housing with a point file or punch.
5. Remove lower end cover.
6. Remove adjusting nuts by straightening bent prong of lock washer. Remove tongued spacer washer. Remove bearings and O-rings, noting especially the position of each part of bearing.
7. Remove control valve. Do not damage any polished surfaces or the edges of the lands on spool and body while handling. Protect all internal surfaces and parts by covering valve. Any small dirt particle will affect proper operation of unit.
8. Remove upper bearings and O-rings in same manner as lower ones, noting their position.
9. Remove steering gear upper cover and gasket and pull cam and shaft from housing.

Inspection

Inspect the steering gear in the usual manner as one would the manual steering gear.

NOTE: *The control valve spool and valve are selectively fitted. The spool has identifying grooves in one end. Mark the valve body end where the identifying groove in the spool was positioned. If this is not done, maximum efficiency of the valve cannot be expected.*

1. Handle valve parts carefully to avoid damage. Note the condition of the edges of lands on spool and body. If edges are damaged, leakage will result.
2. When removing spool, plungers and springs, place parts on lint-free cloth.

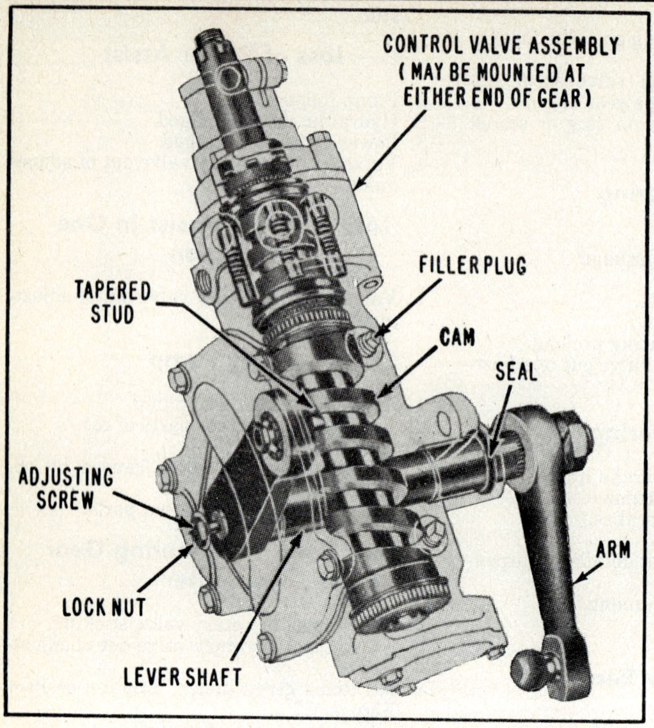

Fig. R1 Ross power steering gear with a built-in concentric type control valve

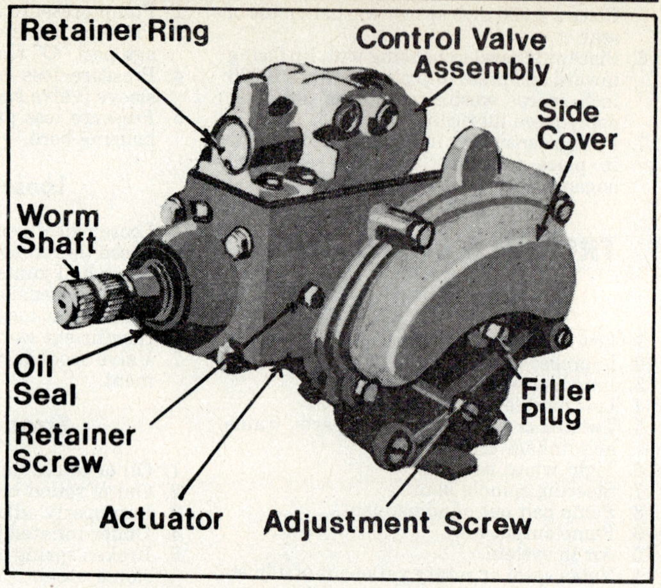

Fig. R2 Ross model HPS-70 semi-integral power steering gear with externally mounted control valve (Typical of Dodge)

Cover to protect from dust and dirt.
3. Inspect all other parts carefully, giving special attention to spline in steering arm and steering arm ball. Replace all defective parts.

Reassembly, Fig. R5

1. Slide cam with bearings assembled into housing. Be sure unit turns freely without binding.
2. Install new gasket and upper cover.
3. Assemble shim pack to lever shaft and install shaft in housing. Do not damage oil seal in end of housing trunnion. A cut seal lip will permit lubricant to leak out.
4. Assemble gasket and side cover, but do not adjust at this time.
5. Install thrust washer, thrust bearing and bearing race over lower end of camshaft. Be certain that bearing race is installed so that O-ring side faces toward control valve. Position O-ring seals.
6. Assemble control valve unit on end of gear, making sure plungers and springs do not fall out. Note that valve is positioned properly relative to port openings and as previously marked before disassembly. Plug valve port openings to prevent entrance of dirt.
7. Assemble lower O-rings and bearings.
8. Slide on washer with internal lug and new lock washer with eight external lugs. Assemble adjusting nut and adjust as covered below under "Adjustments".
9. Install lower end cover and new gasket and secure with bolts and lock washers.

MODEL HPS-70

Steering Gear, Fig. R6

1. Loosen cross shaft adjusting screw locknut and back off adjusting screw a few turns, then remove housing side cover and gasket.
2. Slide levershaft from steering gear housing. First make certain that there are no burrs on outer end of shaft which may damage levershaft bushings. Keep shim pack intact.
3. Remove control valve and spool assembly, attaching screws and valve as an assembly, then remove actuator housing gasket and actuator lever.
4. Remove upper cover and actuator housing attaching screws and slide cover from cam and tube assembly.
5. Remove actuator retaining screw and slide actuator housing from actuator. Remove actuator and cam tube as an assembly, then remove steering gear end cover and gasket.
6. Remove valve actuator and cam tube as an assembly, then remove steering gear end cover and gasket.
7. Remove cam and tube bearing and bearing retaining ring.

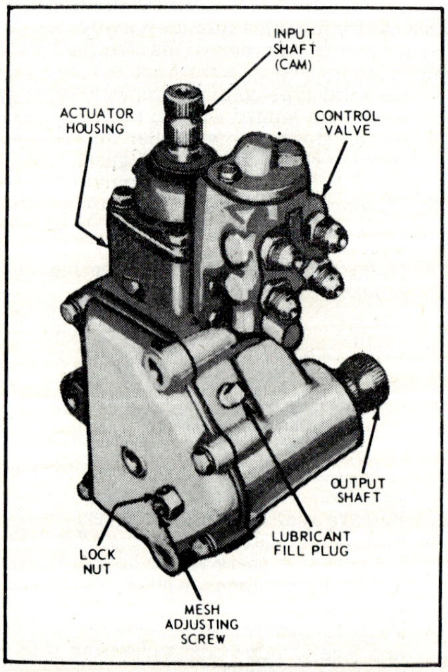

Fig. R3 Ross model HPS-70 semi-integral power steering gear with externally mounted control valve (Typical of Ford)

Valve Actuator and Cam Disassembly

1. Place lower end of cam and tube assembly in a soft jawed vise.
2. Straighten adjusting lockwasher tangs and remove adjusting nut and lockwasher.
3. Remove tongued spacer washer, upper thrust washers and thrust bearing.
4. Remove upper centering washer from end of actuator.
5. Remove actuator, using care not to lose centering screws that are in actuator.
6. Remove lower centering washer, thrust washer and thrust bearing.

Control Valve, Disassembly

CAUTION: The major parts of the control valve are precision machined to close tolerences. The body and spool are selectively fitted and are not separately replaceable. If either part needs replacing, the control valve must be replaced as a unit. Sealed edges of the valve sleeve and spool should not be broken as excessive leakage and reduced hydraulic power will result. Use care when handling parts to prevent damage.

1. Remove retainer ring, cover plate and O-ring seal.
2. Remove valve body housing end cover and O-ring seals from valve body.
3. Remove elastic stop nut and washer from end of clevis rod assembly and pull rod out of spool.
4. Push spool out in same direction to permit removal of centering washers, O-ring and centering spring.
5. Using care not to damage seal, remove O-ring from spool using a pointed tool.
6. If necessary, remove by-pass valve parts, plug assembly, spring and ball.
7. Inspect spool and valve body for scoring by dirt. Wear should be negligible because spool operates in circulating.

Control Valve, Reassembly Fig. R7

NOTE: Refer to caution under "Control Valve, Disassembly."

1. Install O-ring on spool, then lubricate spool using SAE 10 oil and assemble spool into valve body being sure end with O-ring is towards clevis end of valve.

NOTE: When assembling spool into valve body, a twisting motion applied to the spool will aid in assembly. Before assembly, make certain that all parts are thoroughly cleaned.

2. Install centering washer, spring, centering washer, O-ring, and centering washer into control valve body.
3. Install clevis rod into spool and assemble washer and lock nut onto rod. Torque nut to 125–150 inch lbs.
4. If by-pass valve parts were removed, replace O-ring on by-pass plug before installing plug in body, then install steel ball, spring, and by-pass plug into control valve body.

Steering Gear and Valve Actuator, Reassembly Fig. R6

NOTE: Replace all gaskets and seals.

1. If removed, install needle bearings and retainers into end of steering gear housing, being careful not to press bearing too hard against bearing retaining ring.
2. Install tube and cam in a soft jawed vise and assemble lower thrust bearing over tube, seating bearing against upper end of cam.
3. In order, install the following parts; thrust washer, centering washer, actuator, actuator centering springs, centering washer, thrust washer, thrust bearing, thrust washer, tongued washer, new lockwasher (pronged with 13 external lugs) and adjusting nut.
4. To adjust thrust bearing, torque adjusting nut to 75 inch lbs. and rotate cam in relation to actuator. Back off adjusting nut to next tang of pronged lockwasher. Torque required to rotate cam in relation to actuator should be 1/2 to 3 inch lbs.

NOTE: This adjustment is similar to a wheel bearing adjustment and should provide a light preload of needle thrust bearings without lash or heavy drag.

5. Assemble cam and tube with actuator assembly in housing, making certain that cam rotates and oscillates freely in housing.
6. Position actuator gasket and install actuator housing over actuator.
7. Position actuator housing and assemble actuator retainer screw and seal washer. Make certain that screw engages horizontal slot in actuator.
8. Position actuator housing upper cover gasket, then fasten upper cover and actuator housing, using the four long mounting screws.
9. Assemble shim pack onto levershaft and install levershaft into steering housing, using care not to damage oil seal in end of housing trunnion.
10. Position side cover gasket and install side cover, then install lever shaft adjusting screw and lock nut.
11. Adjust lever stud in cam groove as follows:
   a. Tighten side cover adjusting screw until a slight drag (high spot) is felt when turning gear through mid-position. If high spot cannot be felt, remove shims from shim pack until it can be felt.

NOTE: This is a delicate and important adjustment, therefore, remove only enough shims to permit feel of high spot.

   b. When adjusting to a high spot, back off adjusting screw 1/16 turn and lock adjustment with lock nut.
   c. If high spot can be felt without removal of shims, additional shims may be needed in shim pack. A positive way of checking is to add shims until high spot cannot be felt, then remove shims until high spot can be felt.
   d. After lock nut has been tightened, check adjustment by turning gear through its full travel (from extreme left to extreme right).
   e. Install actuator lever into actuator housing, making sure stud end of lever seats in groove of actuator and position slot in other end of lever so that

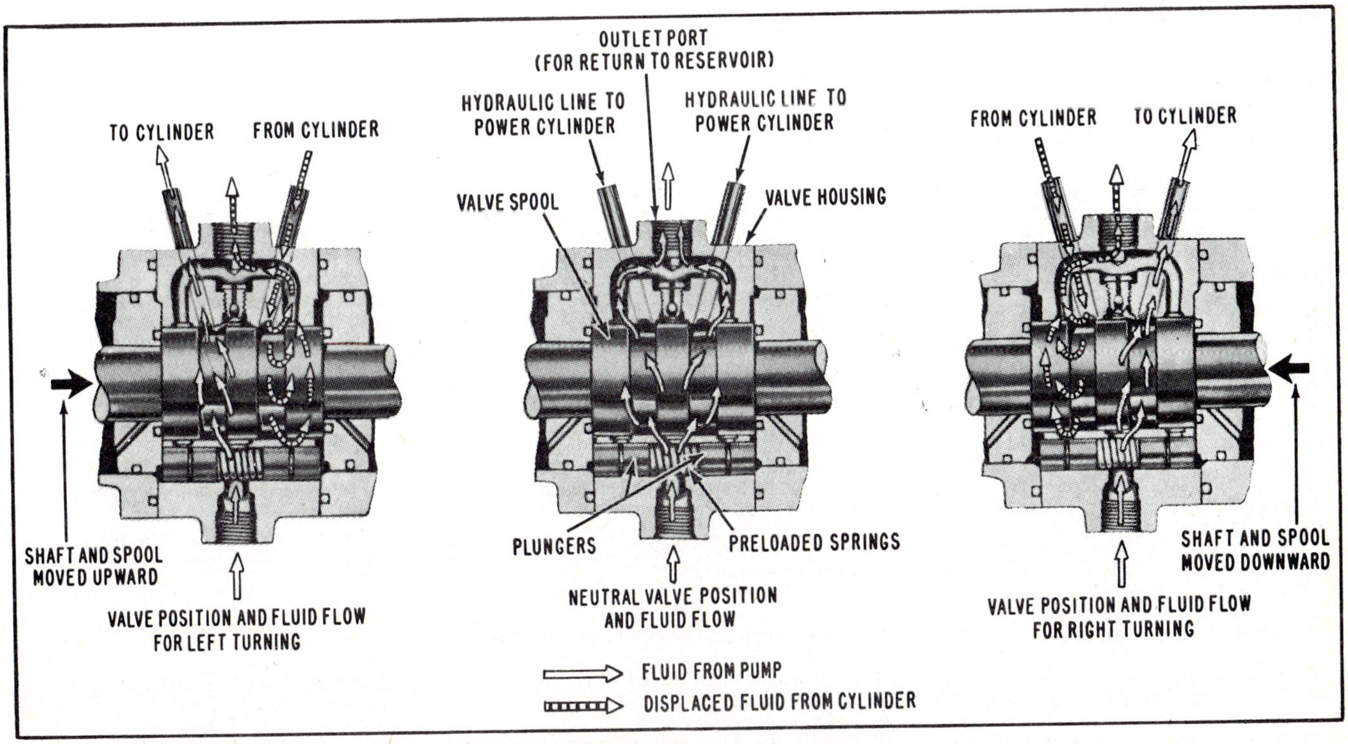

Fig. R4 Ross built-in concentric control valve operation

pin in clevis of flexure rod will fit freely into lever when mounting valve on actuator housing.

f. Mount gasket and control valve on actuator housing making certain clevis pin fits freely into slot of actuator lever. Start all four mounting screws and tighten lightly in rotation before finally torquing to 10–15 ft. lbs.

NOTE: Overtorquing may cause valve spool to be pulled off center by actuator level interference with clevis pin.

g. Be sure spool actuates (moves axially in both directions) before installing end covers. This can be done by placing steering gear arm on lever shaft, then turning worm shaft to move steering arm against a stop and applying sufficient force to actuate valve spool. Reverse arm against an opposite stop to actuate valve spool in either direction.

h. Assemble O-ring on end of control valve body and install valve housing end cover and tighten screws securely.

i. Assemble O-ring, cover plate and retaining ring into clevis rod end of control valve housing.

12. Install end cover gasket and end cover.

Adjustments, All Models

Thrust Bearing Adjustment

1. Turn gear off its center position to free stud in cam groove.
2. Remove upper cover.
3. Remove adjusting nut and spacer washer.
4. Test freedom of nut on camshaft by running it on camshaft without the eight-lug washer in place until the nut seats firmly against thrust bearing using only the fingers to run the nut. If this cannot be done, the threads are fouled and must be cleared with a thread file or other means.
5. Reassemble parts, using the tongued spacer washer (with internal lug) and eight-pronged lock washer.
6. Turn on adjusting nut and tighten to 10 ft-lbs. Back off nut 10 to 20 degrees, which can be done by moving the nut relative to the washer approximately the width of one of the lugs. Observe the lugs nearest in alignment with a pair of flats and bend lugs tightly against flats.
7. Install upper cover.

Stud in Cam Groove Adjustment

1. Tighten side cover. Adjust screw until a slight drag is felt when slowly turning gear through mid-position. No drag is actually necessary. It is suggested as a positive means of knowing that the backlash is removed. The drag should be very slight.
2. A shim pack is used between lever and housing to prevent taper pin from being pulled into cam groove by external forces on lever shaft. The shim pack consists of .003″, .010″ and .020″ shims between two washers. To determine proper thickness of shim pack, proceed as follows:
   a. Adjust high spot and then place gear on high spot.
   b. Remove side cover.
   c. Hold shaft in position with hand with tapered pin pressed in cam groove.
   d. If shim pack can be turned with fingers, add shims until pack cannot be turned and then remove one .003″ shim.
   e. Reassemble side cover and adjust as per Step "a" above. If high spot cannot be felt, remove one shim at a time until desired adjustment is obtained.
3. Lock the adjustment. While holding adjusting screw, tighten lock nut. Recheck adjustment through full travel of steering gear.

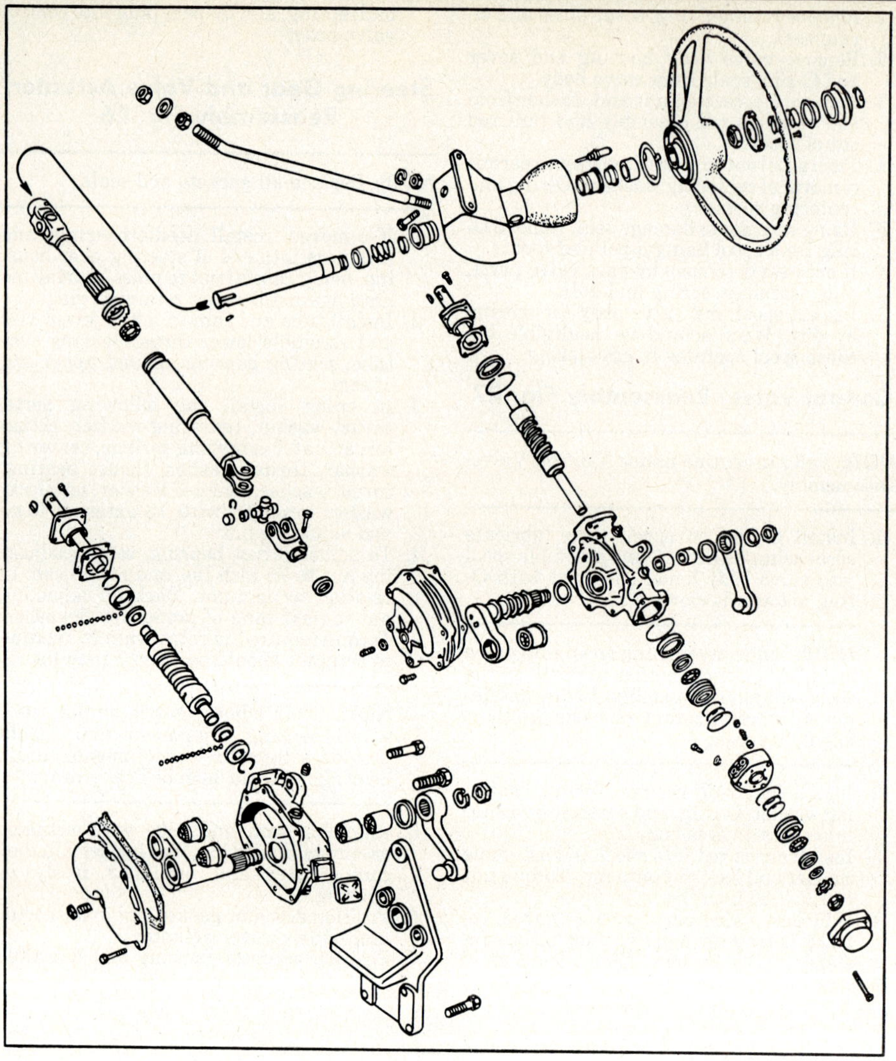

Fig. R5 Disassembled view of Ross steering gear with concentric control valve

Cam & Actuator Assembly Thrust Bearings Adjustment, Model HPS-70

1. Turn gear off its center position to free stud in cam groove, then remove upper cover retaining screws and remove cover. Place 3/8 inch spacers on screws and reinstall screws without cover.
2. Straighten lockwasher prong, remove nut, lockwasher, upper thrust washers and thrust bearing.
3. Finger tighten nut on camshaft making sure there is no thread interference. If necessary clean threads to insure a free turning nut.
4. Reassemble thrust bearing, thrust washers, tongued lockwasher and nut.
5. Torque adjusting nut to 75 inch lbs.
6. Rotate cam with respect to actuator and back off adjusting nut to first locking tang.
7. Measure torque required to rotate cam with respect to actuator. If torque exceeds 3 inch lbs., back off adjusting nut to next locking tang and recheck torque. Final torque should be 1/2 to 3 inch lbs.
8. Bend lockwasher tang and reinstall cover and retaining screws.
9. Connect steering column joint and drag link.

Stud Roller Bearing Adjustment, Model HPS-70

The preceding adjustments should be sufficient in nearly every instance, but in some cases it may be necessary to adjust stud roller bearing lever in unit shaft. In order to make this adjustment, steering gear must be removed and shaft removed from gear.

1. Break off bent prong from locking washer.
2. Lubricate all parts with SAE 90 Multi-Purpose Gear Lubricant.
3. Tighten stud adjusting nut to 4–6 inch lbs.
4. Rotate stud several times in both directions and back off adjusting nut, then tighten adjusting nut to 1 1/2 inch lbs.
5. Lock adjusting nut by bending unused prong of locking washer over nut.

POWER STEERING WITH TOGGLE TYPE CONTROL VALVE

This steering gear, Fig, R8, a standard Ross cam and lever type incorporating a hydraulic toggle type valve. Steering effort applied to the steering wheel actuates the valve which, in turn, directs hydraulic fluid from a pump to a power cylinder located in the steering linkage. Fig. R9 illustrates the operation of the control valve.

NOTE: *Steering gear adjustments are similar to that outlined for the Ross steering gear with the concentric type control valve. There is no adjustment on the control valve. However, when clamped to the actuator housing the valve spool must not be pulled off center.*

Control Valve Removal

1. Loosen lock nut and unscrew lever shaft adjusting screw a few turns.
2. Remove housing side cover.
3. Slide lever shaft from housing, having first made sure that there are no burrs on the outer end of shaft to damage bushing and seal in housing. Keep shim pack intact.
4. Remove screws holding valve to actuator housing and remove valve.

Control Valve Disassembly

The major parts of the valve, which are the body and spool, are machined to very close limits and with precision machined edges. The spool and valve body are selectively fitted according to size of O.D. of spool and I.D. of body. Therefore, these two parts are not supplied separately. If either needs replacing, the entire valve assembly must be replaced.

1. Remove retainer ring, cover plate and O-ring seal.
2. Remove end cover and O-ring seals from valve body.
3. Remove elastic stop nut and washer from end of flexure rod and pull flexure rod out of spool.
4. Push spool out in same direction to permit removal of centering washers, O-ring and centering spring.
5. Remove O-ring from spool. Use pointed instrument but be careful not to damage seal.
6. By-pass valve parts, plug assembly, spring and ball may be removed if desired.
7. Inspect spool. Inspect body internally. Check for scoring due to dirt in system.

Reassemble Steering Gear, Fig. R10

1. If needle bearings in ends of housing have been removed, replace them. Take care not to press bearing too hard against retaining ring.
2. Pre-assemble actuator assembly on cam and stub shaft as follows:
   a. First be sure that threads of nut and camshaft are free of interference by running nut onto camshaft with finger pressure only. If nut cannot be turned all the way with the fingers, the threads are fouled and must be cleaned with a thread file or other means.

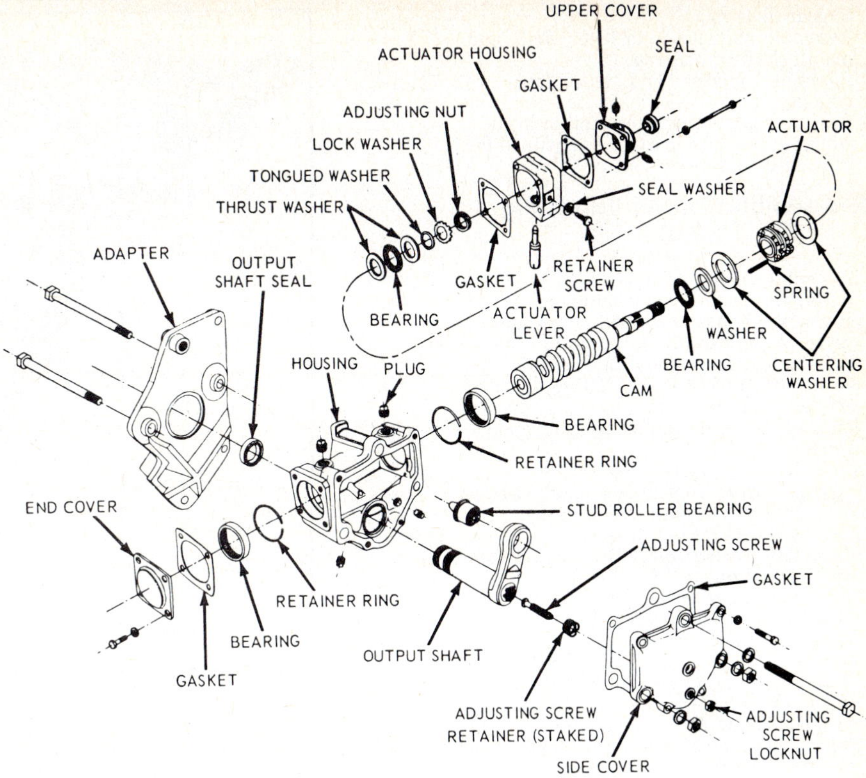

Fig. R6 Disassembled view of Ross model HPS-70 power steering gear

   b. Assemble needle bearing over stub shaft and seat against upper end of cam.
   c. Assemble actuator in sequence shown in Fig. R10.
   d. Adjust thrust bearing as outlined for the gear with concentric control valve.
3. Assemble cam in housing, being sure cam rotates and oscillates freely in housing.
4. Assemble gasket to top of housing.
5. Assemble actuator housing over actuator.
6. Position actuator housing for location of valve mounting. Assemble retainer screw and washer. Be sure screw engages horizontal slot in actuator.
7. Assemble gasket to actuator housing.
8. Assemble upper cover.
9. Fasten upper cover and actuator housing

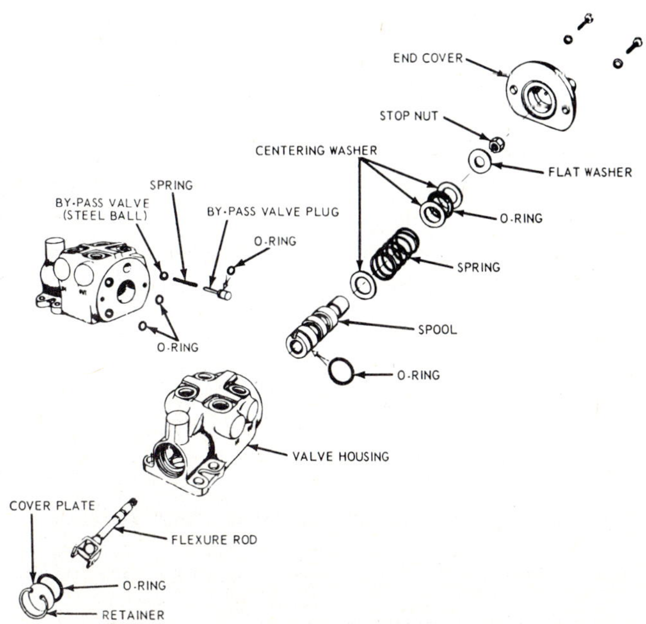

Fig. R7 Control valve housing

to gear housing.
10. Assemble shim pack to lever shaft and install shaft in housing.
11. Assemble gasket and side cover to housing.
12. Assemble adjusting screw and lock nut.
13. Adjust screw in side cover as outlined for gear with concentric control valve.

Reassemble Control Valve, Fig. R10

1. Assemble O-ring on spool.
2. Assemble spool in valve body, being sure end with O-ring is toward clevis end of valve. When assembling spool in body, a twisting motion applied to the spool will be helpful. Before reassembling, be sure all parts are thoroughly clean. Light engine oil should be applied to spool and O-rings.
3. Assemble into valve body the centering washer, spring, centering washer, O-ring and centering washer in that order.
4. Assemble flexure rod in spool.
5. Assemble washer and nut to flexure rod and tighten nut to 125–150 inch-lbs.
6. If by-pass valve parts were removed, replace O-ring on plug before assembling in body. Assemble ball, spring and plug in body.
7. Assemble actuator lever in actuator housing, making sure stud end of lever seats in groove of actuator and position slot in other end of lever so that pin in clevis of flexure rod will fit freely into it when mounting valve.
8. Place gasket on actuator housing and mount valve, making sure clevis pin fits freely into slot of actuator lever.
9. Start mounting screws and tighten lightly in rotation before applying the final tightening torque of 10–15 ft-lbs. *Careless tightening may cause valve spool to be pulled off center by actuator lever interference with clevis pin.*
10. Be sure spool moves freely in both directions before assembling end covers as follows: Place steering gear arm on lever shaft. Move steering arm against stop. Apply sufficient effort to actuate spool. Reverse arm against opposite stop to actuate spool in other direction.
11. Assemble O-rings on end of body and install end cover and tighten to valve body.
12. Assemble O-ring, cover plate and retaining ring.

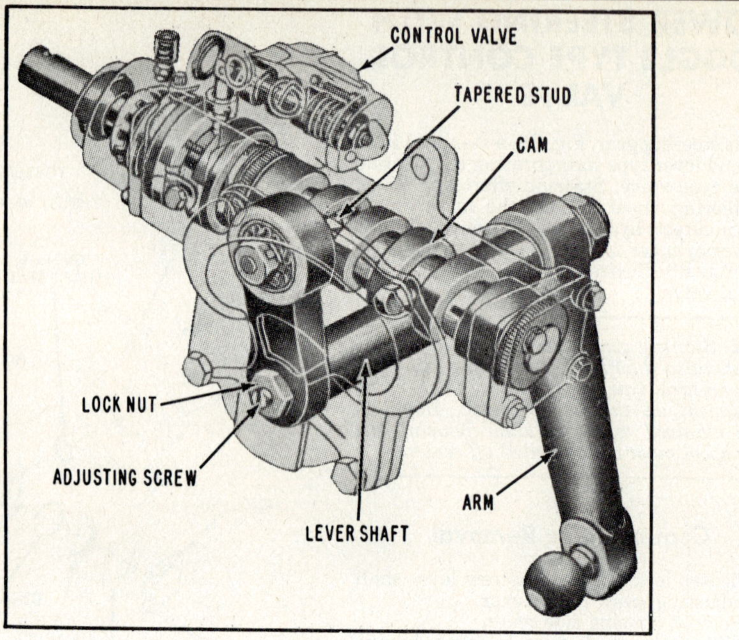

Fig. R8 Ross power steering gear with an integral toggle type control valve

HF54 & HF64 INTEGRAL TYPE GEARS

These fully integral steering gears incorporate a hydraulic control valve and power cylinder, Figs. R11 and R12. Power is transmitted to the output shaft by gear teeth on the rack piston mating with the sector gear. Fluid flow from the pump is directed to the power cylinder by the control valve which is concentric with the input shaft. Fig. R8 illustrates the operation of the concentric control valve.

Disassembly

1. Rotate input shaft to position sector shaft index mark perpendicular to centerline of gear (straight ahead position).
2. Remove side cover attaching screws, tap end of sector shaft with a mallet to disengage side cover seal and allow gear housing to drain. Remove side cover and sector shaft as an assembly.
3. Remove sector shaft seal adapter screws and adapter from housing.
4. Remove control valve adapter screws and the control valve and rack piston assembly from housing.
5. Remove sector shaft adjusting screw lock

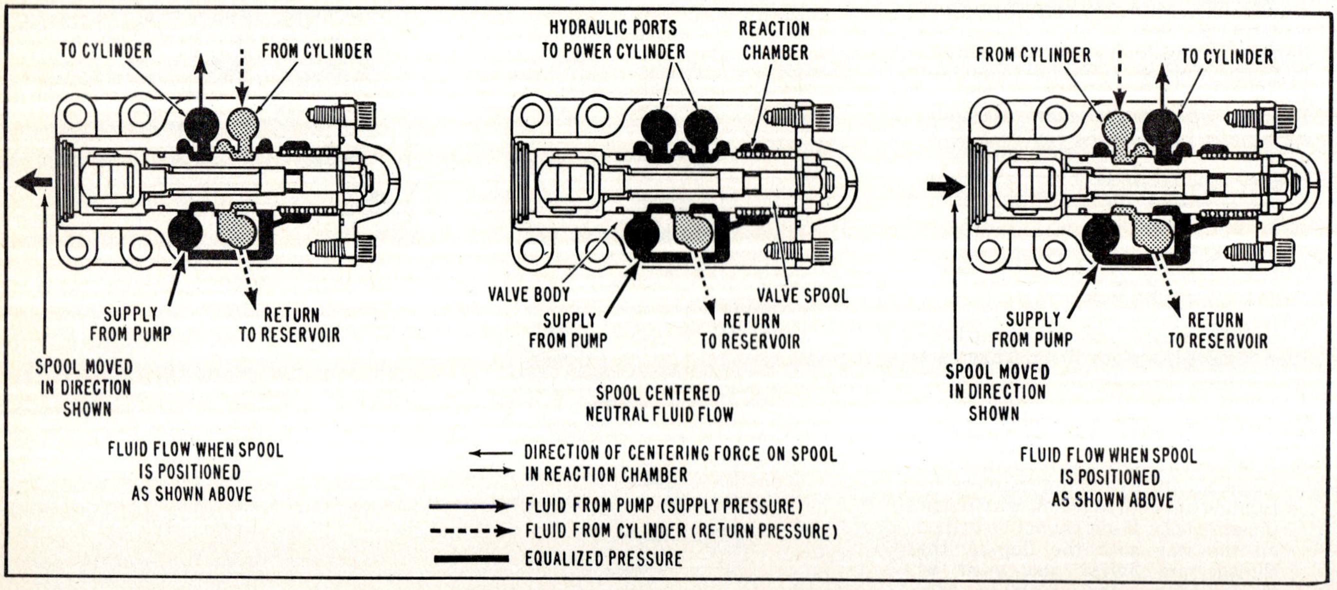

Fig. R9 Toggle type control valve operation

nut and rotate adjusting screw clockwise to remove from side cover.
6. Remove unloader valve retainer, unloader valves, rod and spring from rack piston.
7. Inspect rack piston assembly for damage and, if damage is not evident, do not disassemble rack piston assembly. If rack piston assembly is damaged, disassemble as follows:
   a. Remove ball guide retainer, guide and balls. Tilt rack piston over a clean pan and rock worm shaft to empty all the balls.
   b. Remove worm shaft from rack piston.
8. Clamp input shaft in a soft-jawed vise and remove snap ring, washer, bronze washer, cup, seal and washer. Cut teflon cup from shaft, if necessary.
9. Remove valve cover dirt and water seal, valve cover screws and valve cover from control valve.
10. Unstake adjuster nut lock washer and remove adjuster nut.
11. Remove lock washer, internal tang washer, small bearing race, thrust bearing and large bearing race. Remove control valve and adapter from input shaft.
12. Remove spiral lock ring, seal and washer from valve cover adapter counterbore.
13. The valve should not be disassembled since the body and spool are machined to close tolerances with precision edges. If the valve body or spool is worn or damaged, replace the complete valve assembly. However, if during gear disassembly, the valve parts drop out, reassemble the valve as follows:
   a. Clean all parts with a petroleum-base solvent and blow dry. Do not wipe dry with a cloth since lint may cause binding and sticking of valve components.
   b. Insert spool into valve with machined identification groove in ID of spool facing toward gear housing, Fig. R13.
   c. Insert the six solid centering plunger sets with reaction springs into body with small hole in each plunger facing outward.

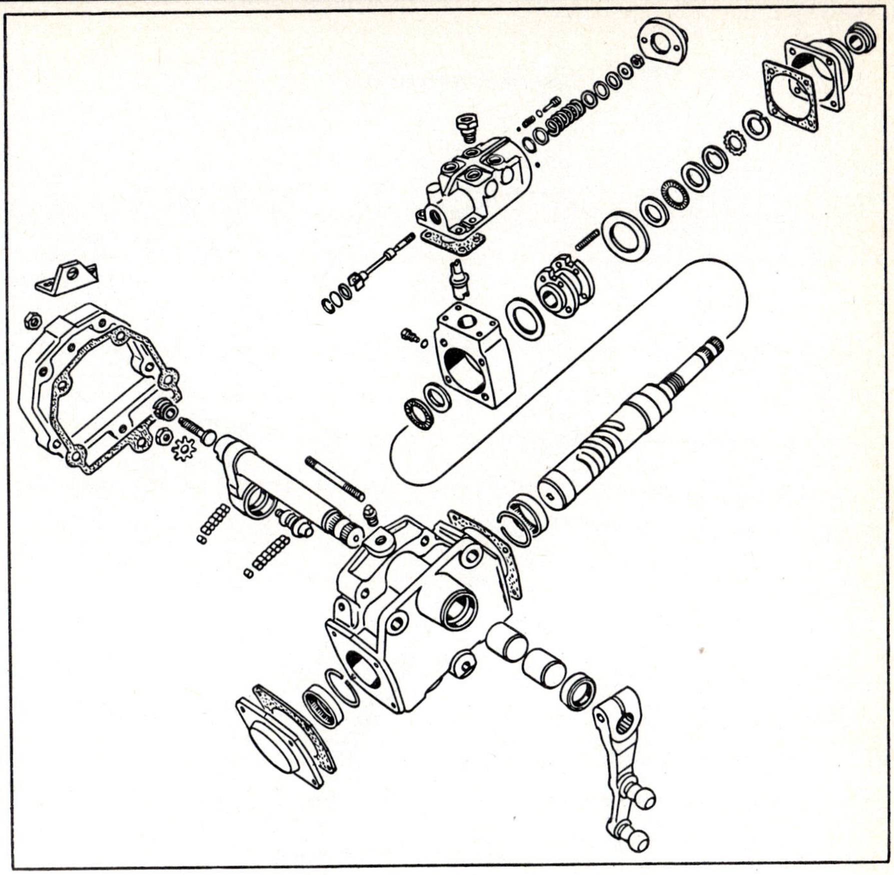

Fig. R10 Disassembled view of a Ross power steering gear with an integral toggle type control valve

Assembly

1. On HF-54 gears, if sector shaft bearing was removed, install snap ring in gear housing outboard side.
2. On all gears, press sector shaft bearing, if removed, into housing until outer surface is flush with housing.
3. Lubricate unloader valve pressure adjusting screw "O" ring and slide into groove on non-threaded end of adjusting screw.
4. Thread adjusting screw into housing lower end until approximately 7/8 inch of the screw protrudes from housing. Install and tighten adjusting screw lock nut.
5. Secure input shaft in a soft-jawed vise, permitting access to either end of shaft.
6. On HF-54 gears, install large thrust race, thrust bearing, control valve with cylinder ports facing toward the shoulder, small bearing race, internal tang washer, lock washer and thrust bearing adjustment nut.
7. On HF-64 gears, install thrust bearing race washer, spacer, control valve with cylinder ports facing toward shoulder, needle thrust bearing, thrust bearing race washer, tang washer, lock washer and thrust bearing adjustment nut.
8. On all gears, torque adjustment nut to 20 ft. lbs., then loosen one-half to one lock washer prongs. Bend one prong into adjustment nut slot. Control valve should rotate freely on input shaft with a torque of 2–3 1/2 in. lbs. and without any end play.
9. Install new washer, seal and spiral lock ring in valve cover adapter counterbore with lip of seal facing toward spiral lock ring.
10. Lubricate valve cover seal and cylinder port seals, then place seals into control valve cover adapter recesses on the surface adjacent to control valve.
11. Reposition input shaft in vise to secure serrated end.
12. Slide adapter over input shaft, align cylinder port seals with control valve ports and install one attaching bolt finger tight.
13. Assemble steel washer, rubber seal, teflon cup with lip facing toward seal, bronze washer and retaining washer. Compress washer and seal and install snap ring on input shaft end, ensuring snap ring is fully seated in groove and recessed area of retaining washer.
14. Clamp rack piston in a soft-jawed vise with ball guide holes facing upward.
15. Expand piston ring and install in piston groove.
16. On HF-54 gears, clean unloader valve threads and install valve into rack piston. Clean seat threads and apply a drop of Lock-tite. Assemble seat with lock washer and torque to 10–15 ft. lbs., then bend lock washer against one flat of seat.
18. On all gears, lubricate input shaft seal and install input shaft into rack piston bore.
19. Install sixteen balls into ball guides while rotating input shaft counterclockwise. Alternate black spacer balls and polished steel balls. Coat ball return guides with grease and install six remaining balls. Secure guide retaining clip to rack piston and torque screws to 8–10 ft. lbs. Bend prong of lock washer against flat of screw.
20. Lubricate sector shaft adjusting screw and place screw head in end slot of sector shaft, then install screw retainer. Tighten retainer to permit rotation of screw with noticeable end play. Stake retainer and check rotational effort.
21. Press new pressure relief plug, if removed or damaged, into side cover until flush with surface.
22. On HF-64 gear, install snap ring into side cover, then on all gears install steel washer with taper facing toward snap ring, leather washer and two piece seal. Ensure the words "Oil Side" are visible on seal after installation.
23. Thread sector shaft adjusting screw and sector shaft assembly into side cover. On HF-54 gears, ensure that the shaft seal has not been mispositioned.
24. On all gears, install outer seal on seal adapter, then the leather washer and inner seal, ensuring the words "Oil Side" are visible on seal after installation.
25. Press input shaft needle bearing, if removed, into control valve cover. Press bearing from part number end to a depth of 1 1/8 inch from valve cover face. Ensure rollers are free to rotate.
26. Install control valve cover seal with lip

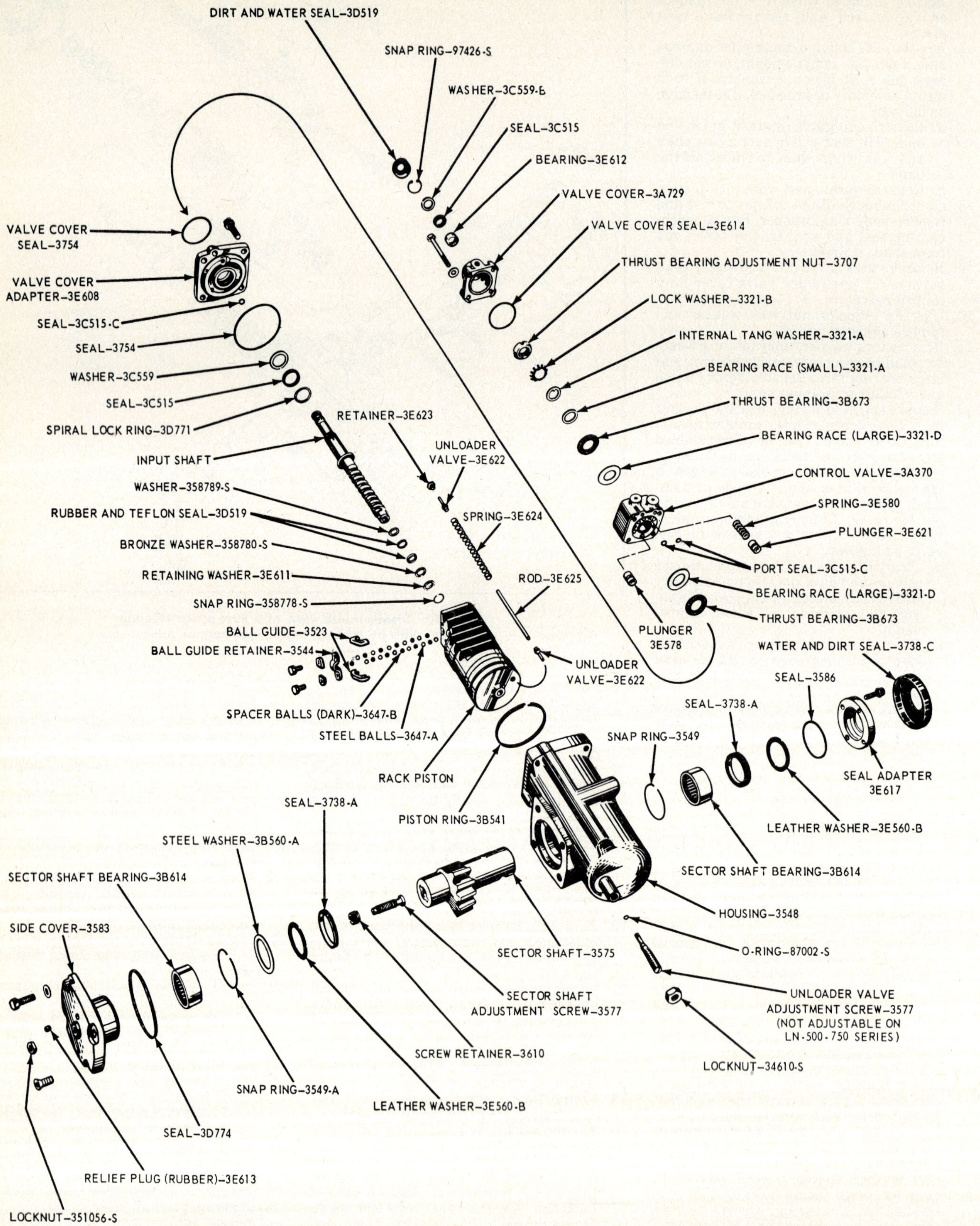

Fig. R11 Disassembled view of Ross model HF-54 integral power steering gear

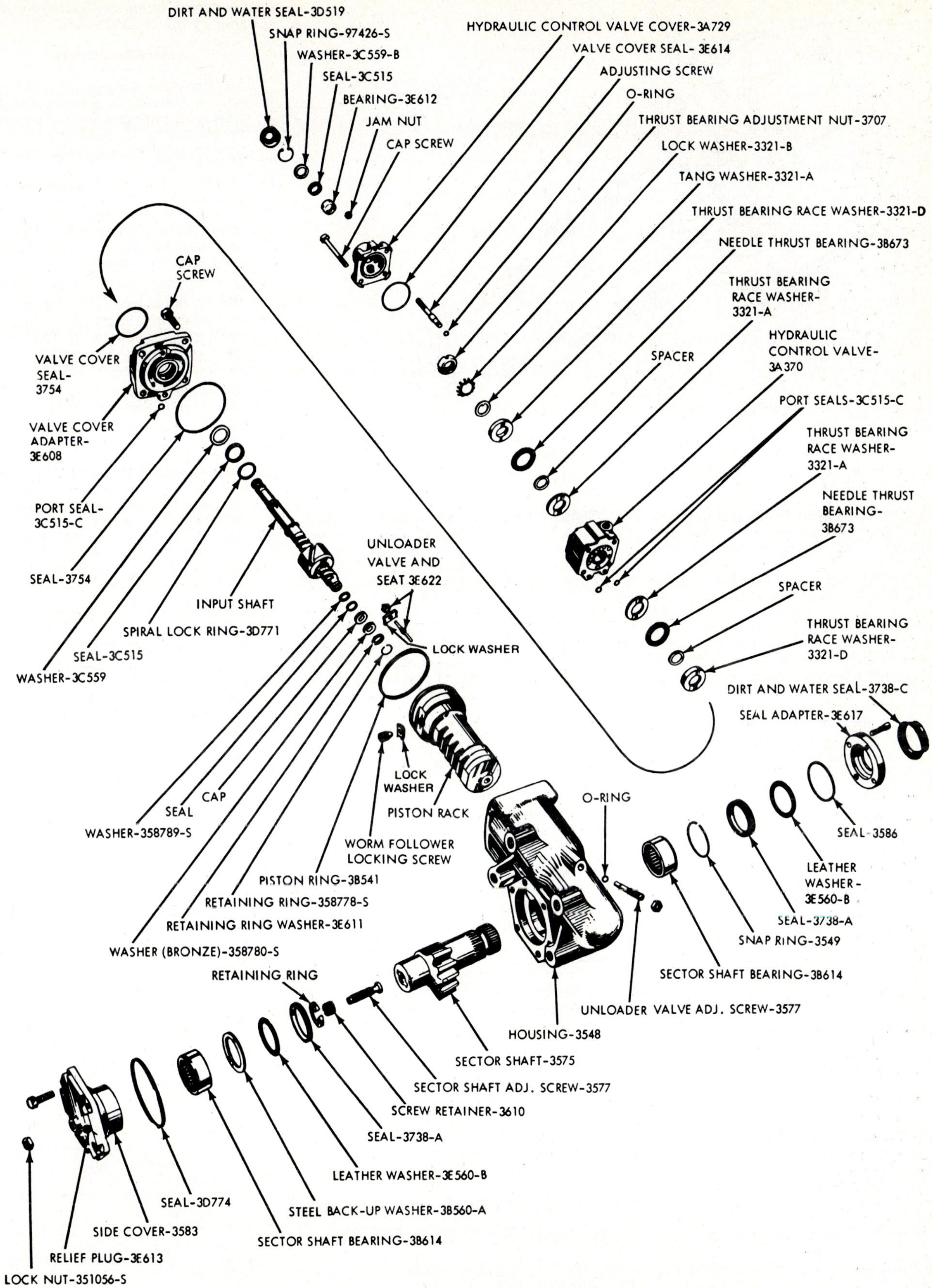

Fig. R12 Disassembled view of Ross model HF-64 integral power steering gear

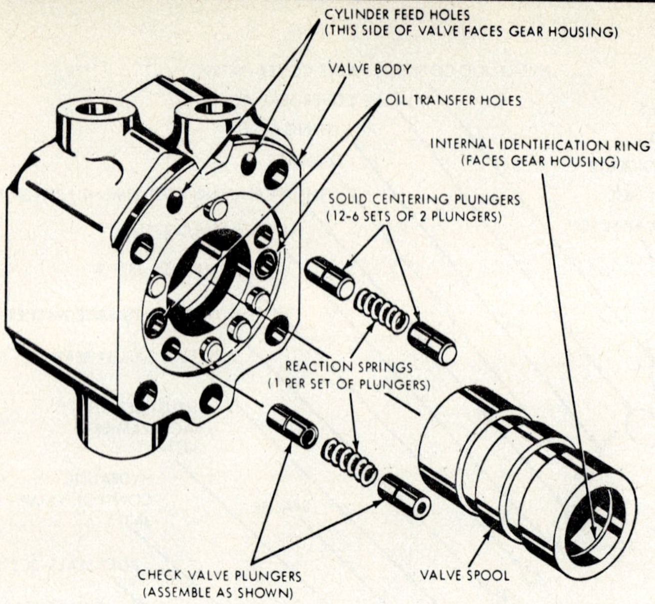

Fig. R13 Hydraulic control valve assembly

facing toward needle bearing. Lubricate washer and install on cover, then the snap ring, thereby securing seal and washer.

27. Fill new dirt and water seal with a suitable lubricant and install on control valve cover.
28. Lubricate gear housing bore and start rack piston into bore. Compress ring and move piston into position so piston teeth are visible through side cover opening. Install adapter to housing bolts and remove bolt installed in step 12.
29. Lubricate and install new valve cover seal into valve cover recess, then slide valve cover onto input shaft and install attaching bolts.
30. Rotate input shaft to align rack piston center tooth (marked tooth) with side cover opening.
31. Lubricate and install new side cover "O" ring on cover, then install side cover and sector shaft assembly onto gear housing, ensuring center tooth meshes with center space (marked space). Install and torque side cover screws to 45–55 ft. lbs. on HF-54 gears or 150–170 ft. lbs. on HF-64 gears.
32. Cover sector shaft serrations with tape to protect adapter seal and install adapter over sector shaft and onto gear housing.
33. Pack adapter outer seal with suitable lubricant and install seal on adapter.

Adjustments

Sector Shaft

1. Disconnect drag link from pitman arm.
2. Center steering wheel and check for free movement between sector shaft and rack piston. If the pitman arm moves freely, remove gear from vehicle to adjust.
3. Loosen sector shaft adjusting screw lock nut.
4. Rotate input shaft through full travel at least five times and torque sector shaft adjusting screw to 15–20 in. lbs. while rotating input 90 degrees each side of center.
5. Loosen adjusting screw one turn and measure the torque required to rotate input shaft 90 degrees each side of center. Tighten adjusting screw lock nut until snug, then tighten adjusting screw to provide an increased rotating torque of 2–4 in. lbs. 45 degrees each side of center. Torque adjusting screw lock nut to 20–25 ft. lbs.

NOTE: The input torque of the assembled gear with hydraulic oil drained should not exceed 15 in. lbs. for full travel of the output shaft.

Unloader Valve

This adjustment is made for right turn only on HF-54 gears and both turns on HF-64 gears.

1. Install a suitable pressure gauge between pump and steering gear pressure port.
2. Idle engine, turn steering to right stop and note pressure reading.
3. Loosen unloader valve adjusting screw lock nut and turn adjusting screw clockwise to obtain a pressure reading 400 psi below reading noted above.

NOTE: Do not hold steering in extreme position longer than 15 seconds since heat damage may occur to the pump.

4. On HF-64 gears, repeat procedure for left turn.
5. On HF-54 gears, torque adjusting screw lock nut to 17–22 ft. lbs. On HF-64 gears, torque upper adjusting screw lock nut to 15–20 ft. lbs. and the lower adjusting screw lock nut to 20–25 ft. lbs.
6. Remove pressure gauge.

HFB52, HFB64 & HFB70 INTEGRAL TYPE GEAR

The HFB52, HFB64 & HFB70 power steering units, Fig. R14 are fully integral power steering units incorporating a hydraulic control valve, a hydraulic power cylinder and a manual steering mechanism. The control valve is of rotary design. Oil flow from an engine driven pump is directed to the power cylinder by the control valve.

Disassembly

1. Position steering gear in a soft-jaw vise with wormshaft in a horizontal position. Check timing mark located on end of sector shaft, Fig. 15, position mark in a vertical position with steering gear in center of gear travel.
2. Using suitable screwdriver, remove seal from housing.
3. Loosen jam nut on sector shaft adjusting screw, then place a pan under steering gear unit for fluid to drain, and remove bolts from side cover.

NOTE: Bolts incorporate a special ring located on bolt head for sealing. If bolts need to be replaced, replacement bolts must be same as the original.

4. Remove side cover. Then using a wooden drift and hammer, remove sector shaft from gear housing. Ensure all bearing rollers are in place after sector shaft is removed.
5. Remove sector shaft adjusting screw nut, then using a suitable screwdriver, remove sector shaft adjusting screw and sector shaft from side cover.
6. Remove snap ring, seal and back-up washers from side cover.
7. Remove bearing rolls from side cover.

NOTE: Bearing, bearing rollers and side cover must be replaced as one unit.

8. Remove snap ring, back-up washers and seal from gear housing.
9. Using a suitable screwdriver and a 3/4 inch socket, loosen adjusting screw sealing nut and adjusting screw approximately two turns. Using a screwdriver and wrench, loosen adjusting screw sealing nut and adjusting screw approximately two turns.
10. Remove relief valve, if equipped, from valve housing. Remove "O" and seal rings from relief valve.
11. Remove bolts and valve housing from gear housing.

NOTE: Proper valve function depends on the valve sleeve and wormshaft. Timing marks have been incorporated on these parts so they can be reassembled correctly.

12. Remove valve sleeve, "O" ring, thrust washer and thrust bearing from valve housing.
13. Remove seal, snap ring, back-up washer, seal cup and "O" ring from valve housing.
14. Remove seal and "O" rings from valve sleeve.
15. Remove rack piston and wormshaft assembly from valve housing.
16. Position rack piston and wormshaft assembly on a clean surface, and remove ball bearings.
17. Remove screws or seal rings from valve housing, if equipped.
18. Remove clip or cap from valve housing, if equipped.
19. Remove seal from cap, if equipped.
20. Remove ball return guides and balls from rack piston. Use screwdriver, if necessary to remove balls from guide.

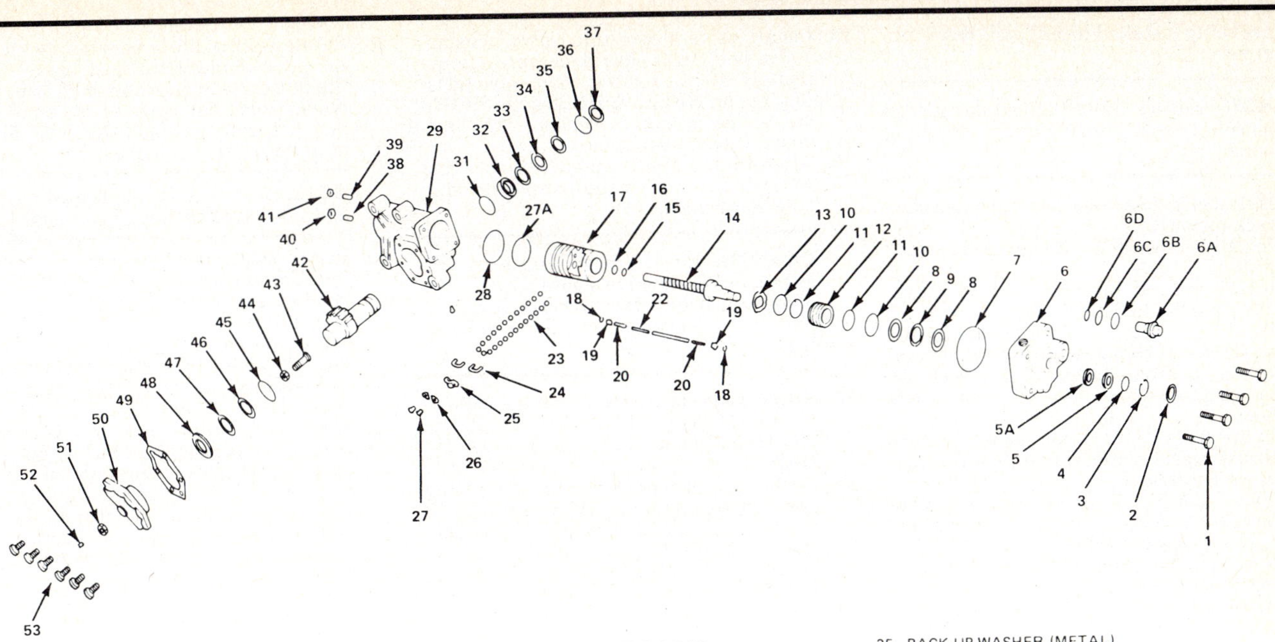

1. HEXAGON HEAD BOLT (4) (7/16-14 x 2.00)
2. SEAL
3. RETAINING RING
4. BACK-UP WASHER (STEEL)
5. SEAL (TWO-PIECE)
6. VALVE HOUSING
6A. RELIEF VALVE
7. "O" RING
8. THRUST WASHER (2)
9. THRUST BEARING
10. SEAL RING (2) (PLASTIC)
11. "O" RING (2)
12. VALVE SLEEVE
13. DRIVE RING
14. WORM SHAFT
15. "O" RING
16. SEAL RING (PLASTIC)
17. RACK PISTON
18. RETAINING RING (2)
19. POPPET SEAT (2)
20. POPPET (2)
21. SPRING
22. ROD
23. BALL (25)
24. BALL RETURN GUIDE (2)
25. CLIP
26. LOCK WASHER (2)
27. SCREW (2)
27A. "O" RING
28. SEAL RING (PLASTIC)
29. HOUSING
30. BLEED SCREW
31. RETAINING RING
32. BEARING
33. SEAL
34. BACK-UP WASHER (PLASTIC)
35. BACK-UP WASHER (METAL)
36. RETAINING RING
37. SEAL
38. ADJUSTING SCREW (1.250 LONG)
39. SEALING NUT (3/8-24)
40. ADJUSTING SCREW (1.500 LONG)
41. SEALING NUT (1/2-20)
42. SECTOR SHAFT
43. ADJUSTING SCREW
44. RETAINER
45. RETAINING RING
46. SEAL
47. BACK-UP WASHER (PLASTIC)
48. WASHER
49. SIDE COVER GASKET
50. SIDE COVER
51. NUT (1/2-20)
52. VENT PLUG
53. SPECIAL BOLT (6) (9/16-18 x 1")

Fig. R14 Disassembled view of Ross HFB52, HFB64 & HFB70 power steering gear

CAUTION: The HFB52 assembly contains 25 matched balls, the HFB64 assembly contains 27 matched balls and the HFB70 assembly contains 34 matched balls. Care must be taken not to lose them. A complete new set of balls will be required if any are lost.

21. Remove wormshaft from rack piston.
22. Place rack piston in a soft-jaw vise, then remove snap rings, poppet seats, poppet rods and spring from rack piston.

NOTE: Remove these parts only if replacement is necessary.

23. Remove seal ring and "O" ring from rack piston.
24. Remove seal ring and "O" ring from wormshaft.
25. If bearing requires replacement, remove snap ring. Using an arbor press remove bearing from valve housing.

Cleaning & Inspection

Clean all parts using a suitable solvent, and blow drying them using compressed air. Inspect all components for evidence of wear or corrosion. Replace all components not considered serviceable. Replace all "O" rings and seals. Lubricate all "O" rings, seals and thrust washers.

Assembly

1. If bearing was removed during disassembly, install bearing into valve housing.
2. Install "O" ring, seal ring, poppet seat and snap ring into rack piston. Torque poppet seats to 20–25 ft. lbs.
3. Install "O" ring and seal ring onto wormshaft.
4. Install wormshaft into rack piston.
5. Install ball return guides and balls into rack piston. Rotate wormshaft as balls

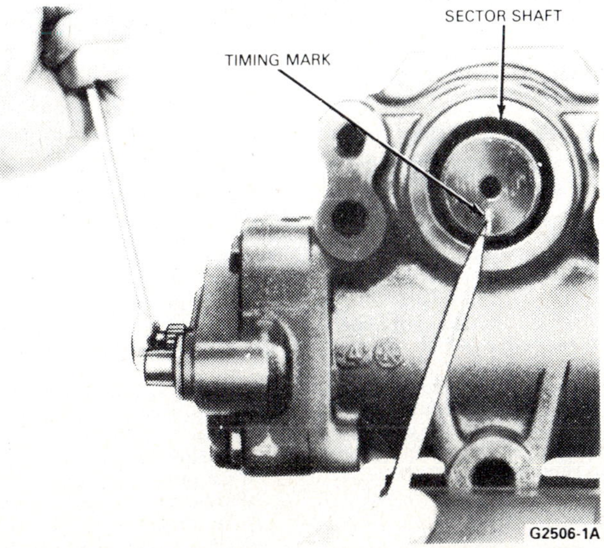

Fig. R15 Sector shaft timing mark

are being installed to pull balls down into groove.

NOTE: Ensure ball return guides stay in place in rack piston while installing balls.

6. Install clip or seal and cap, if equipped on rack piston.
7. Install lock washers, if equipped on rack piston.
8. Install bolts, if equipped on rack piston. Torque bolts to 8–12 ft. lbs.
9. Bend tabs on lock washers up against flat on screws, then apply lubricant onto seal ring area of rack piston. Install rack piston and wormshaft assembly onto valve housing.
10. Lubricate "O" ring, then install "O" ring, thrust washer and thrust bearing into valve housing.
11. Using tool No. T75L-3517-A, or equivalent, install seal onto valve sleeve and coat with gear lubricant.
12. Repeat steps 10 and 11 to install remaining seal and "O" rings.
13. After seals are positioned into valve grooves, they must be compressed or sized before inserting worm and valve sleeve assembly into gear housing.
14. Apply a light coat of gear lubricant to valve sleeve and seals. Slowly push tool No. A4 or equivalent over valve sleeve until tool bottoms out.

NOTE: Ensure seals are not bent over when sizer tool slides over them.

15. Keep sizer tool positioned over seals for a minimum of one minute.
16. Remove sizer, then apply lubricant to seal ring areas of valve sleeve.
17. Apply lubricant to end of valve sleeve, then install thrust washer onto valve sleeve.
18. Install valve sleeve into valve housing. Valve sleeve face should measure approximately .400 inch lower than valve housing face on HFB52 & HFB64. On HFB70 valve sleeve should measure .375 inch above valve housing face.
19. Align timing marks on valve sleeve and wormshaft, then install valve housing onto wormshaft.

CAUTION: Ensure drive ring teeth engage notches in valve sleeve. Use a suitable wrench to rotate wormshaft and to pull valve housing down against housing.

20. Install bolts into housing, torque bolts to 45–55 ft. lbs. for HFB52, 105–115 ft. lbs. for HFB64 and 145–155 ft. lbs. for HFB70.
21. With rack piston at center of steering gear travel, torque adjusting screw to 10 ft. lbs. Back adjusting screw out one turn, then measure torque required to rotate wormshaft through 90° each side of center of steering gear travel. Torque should not exceed 15 inch lbs. Torque adjusting screw and nut to 40–50 ft. lbs. After torquing nut, torque required to rotate wormshaft must increase 4–7 inch. lbs. above torque measure previously.
22. Apply lubricant to the bearing race and install ball bearings.
23. Install back-up washers, seal and snap ring into side cover. Ensure seal seats properly in cover, with flat side toward bearing.
24. Apply lubricant to short bearing area on sector shaft and insert into side cover. Screw adjusting screw into side cover. Side cover should rotate on sector shaft freely with no binding.
25. Install vent plug flush into side cover.
26. Loosely install nut onto adjusting screw.
27. Install gasket into side cover and sector shaft assembly.
28. Rotate wormshaft to position rack piston in center of its steering gear travel. Align the center tooth on sector shaft with third notch from seal ring, then install rack piston, sector shaft and side cover assemblies onto valve housing.

NOTE: Ensure sector shaft does not damage side cover gasket when installing.

29. Install side cover bolts onto valve housing. Torque bolts to 150–170 ft. lbs. for HFB52, 220–240 ft. lbs. for HFB64 and 150–170 ft. lbs. for HFB70.
30. Turn side cover adjusting screw to provide 23–28 inch lbs. torque at wormshaft as steering gear is rotated through 90° each side of center. Back off adjusting screw one turn and note torque required to move steering gear 90° of travel on each side of center. Turn adjusting screw in to provide an increase of 2–4 in. lbs. torque at a location 45° from each side of center. Torque jam nut to 40–45 ft. lbs. for HFB52 & HFB64 and 20–25 ft. lbs. for HFB70. After torquing nut, torque required to rotate wormshaft must not exceed 26 inch lbs. at any point of steering gear travel.
31. Position tape onto sector shaft serrations, then install back-up washers and snap ring onto sector shaft.
32. Install "O" ring, seal and "O" ring onto relief valve, then install assembly into valve housing if equipped. Torque nut to 25–35 ft. lbs.
33. Install "O" ring onto seal cup with flat side of seal cup out into valve housing. Install back-up washer into valve housing. Then using tool No. J26653 or equivalent, install seal into valve housing.
34. Install snap ring into valve housing.
35. Using brass drift and hammer, install seal into valve housing.
36. Install bleed screw into valve housing, if equipped, and torque screw to 25–35 ft. lbs.

Saginaw Power Steering Gear Section

NON-INTEGRAL ROTARY VALVE TYPE

This power steering system, Fig. S1, consists primarily of three units used in conjunction with a conventional steering gear. They are the control valve, power cylinder and hydraulic pump.

The power cylinder is actuated by the operation of the control valve which supplies hydraulic pressure on either side of the power cylinder piston as required, depending on the position of the valve. The pressure used to operate this system is supplied by a vane-type pump. (The power cylinder and pump are covered elsewhere in this chapter).

Disassembly

In most cases, complete disassembly of the steering gear will not be necessary. It is suggested that only those assemblies which are faulty be disassembled. As in repairing any hydraulically operated unit, cleanliness is of the utmost importance. Therefore, the bench, tools and parts must be kept clean at all times. Thoroughly clean the exterior of the steering gear with a suitable solvent, and when necessary, drain as much of the hydraulic fluid as possible from the unit.

Clamp Yoke Removal

1. With the gear assembly removed from the vehicle, drain as much fluid from the unit as possible. This is not necessary if only the yoke is being replaced.
2. If not previously removed, mark the stub shaft and clamp yoke before removing the yoke so they can be assembled in the same position when the gear is reassembled.
3. Remove bolt and clamp yoke from stub shaft.

Adjuster Plug, Fig. S2

Removal

1. Turn stub shaft to turn pitman arm counterclockwise until ball nut reaches end of its travel at the top.
2. Loosen and remove adjuster plug lock nut.
3. Remove adjuster plug with spanner wrench.
4. Remove upper thrust bearing from steering gear top cover.

Disassembly

1. Remove O-ring seal.
2. Remove stub shaft oil seal retaining ring and back-up washer.
3. Remove stub shaft oil seal by prying out with screwdriver.
4. Do not remove stub shaft bearing unless inspection shows necessity.

5. Replace all seals, and other parts not in first class condition.

Reassembly

1. Position adjuster plug on wood block.
2. If previously removed, install needle bearing through thrust bearing end of plug. Press against identification end of bearing until end of bearing is flush with bottom surface of stub shaft seal bore.
3. Lubricate new stub shaft oil seal with automatic transmission fluid. Then install seal with lip toward inside of plug and back-up washer far enough into bore of plug to provide clearance for retainer ring.
4. Install oil seal retaining ring, being sure it is seated properly in groove.
5. Lubricate new O-ring seal with vaseline and install on plug.

Rotary Valve Assembly, Fig. S3

The complete valve assembly is a precision unit with selectively fitted parts and is hydraulically balanced at the factory. Only those parts which are marked as service items are replaceable. If replacement of any non-serviceable parts is necessary, the complete rotary valve assembly must be replaced.

The valve assembly rarely requires service with the possible exception of replacing the valve spool dampener O-ring. Do not disassemble the valve unless absolutely necessary since this may result in damage to the assembly. If the valve spool dampener O-ring requires replacement, remove the valve spool only. Replace the O-ring and reinstall spool immediately. Do not disassemble further.

Removal

1. Remove clamp yoke and adjuster plug as outlined above.
2. Remove valve assembly by grasping stub shaft and pulling straight out. It may be necessary to use pliers to aid in removal but use care not to damage stub shaft splines.

Disassemble

1. Remove spool spring by prying on small coil with screwdriver. Do not pry against valve body as this may result in a sticky valve. Work spring onto bearing diameter of stub shaft, then slide spring off shaft.
2. The diametral clearance between valve and spool may be as little as .0004". The slightest cocking of spool may jam it in valve body. To remove the spool, hold the valve assembly in both hands with stub shaft pointing down and tap stub shaft gently against wood block. It may be necessary to push lightly on valve spool with a pencil or small brass rod by inserting pencil or rod through openings in valve cap until spool is far enough out of valve so it can be grasped by hand, Fig. S4. Withdraw spool with a steady oscillating pull to prevent jamming, Fig. S5.

NOTE: *If a slight sticking occurs, make a gentle attempt to reverse withdrawal procedure. If this does not free spool, it has become cocked in the valve body bore. Do not attempt to force spool in or out of valve body. If such is the case, continue to disassemble valve assembly and return to the spool as outlined later.*

---

3. Remove stub shaft, torsion bar and valve cap by holding valve in both hands as before, except with thumbs on valve body. Tap torsion bar slightly against work bench. This will dislodge cap from valve body-to-cap pin. The stub shaft, torsion bar and valve cap can then be removed from valve body.
4. If valve spool has become cocked as mentioned above, it can now be freed. By visual inspection on a flat surface, it can be determined in which direction the spool is cocked. A very few light taps with a small soft-faced hammer should align the spool in the bore and free it. *Do not tap the spool with anything metallic. If spool can be rotated it can be removed.*
5. Remove dampener O-ring seal from spool.
6. Do not remove valve rings or ring back-up seals unless inspection shows necessity.

Inspection

1. If valve leaks around torsion bar, the entire valve must be replaced.
2. If valve rings and back-up seals show evidence of excessive wear, carefully cut rings and seals and remove them.
3. Check pin in valve body which engages cap. If badly worn, cracked or broken, entire valve should be replaced.
4. Check worm pin groove in valve body

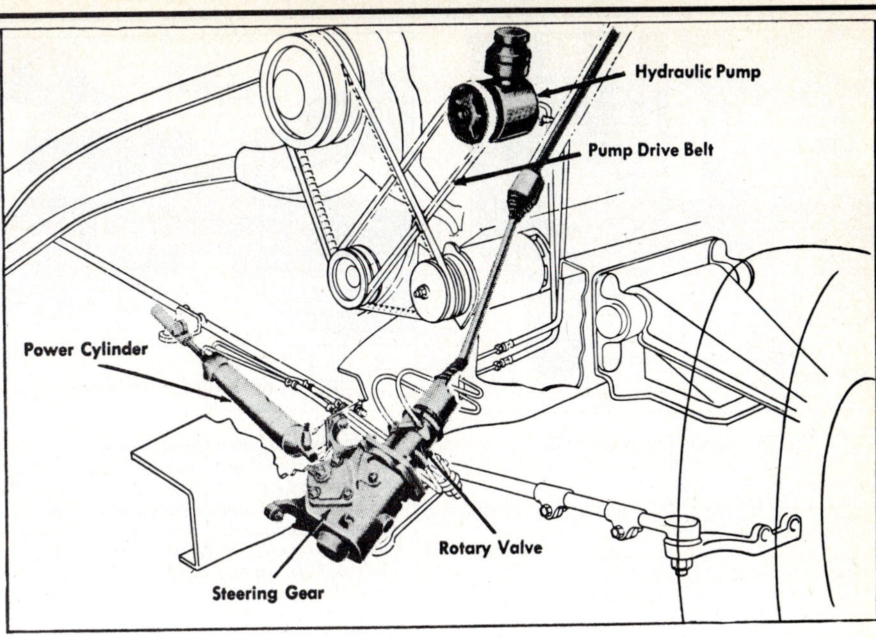

Fig. S1 Saginaw rotary valve power steering gear used on Chevrolet and GMC trucks

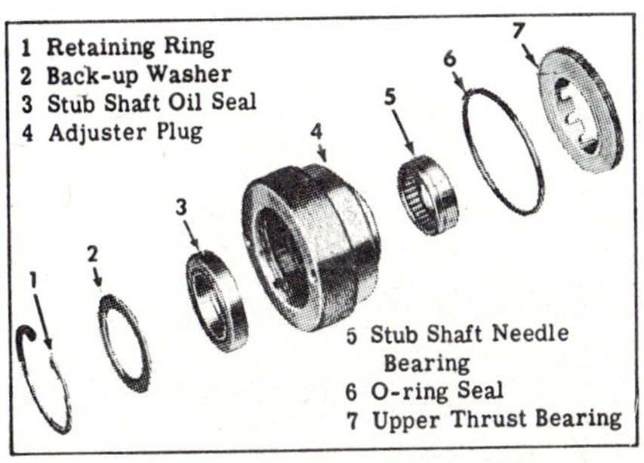

Fig. S2 Adjuster plug components and upper thrust bearing

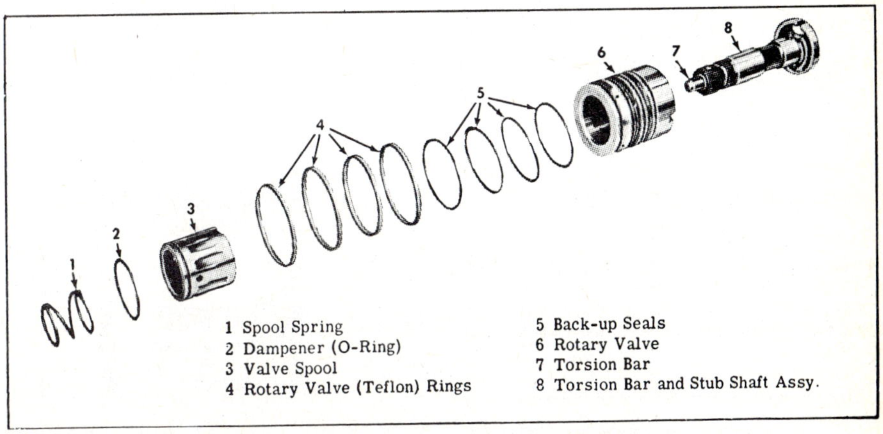

Fig. S3 Rotary valve components disassembled

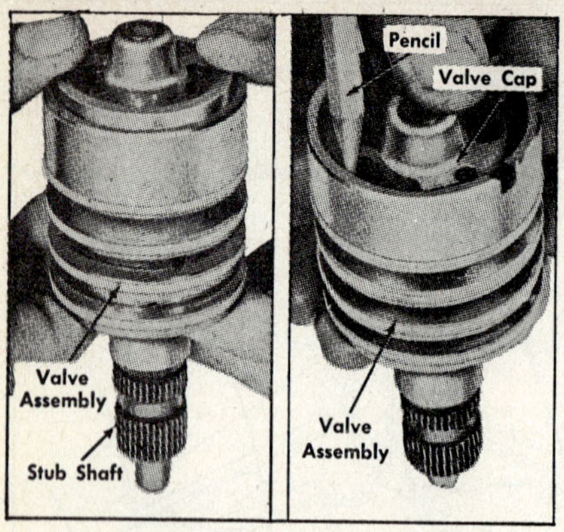

Fig. S4 Loosening valve spool

(smaller of two). If worn badly, entire valve should be replaced.

5. Check spool drive pin stub shaft if badly worn replace entire valve.
6. Examine outside diameter of spool for nicks, burrs or bad wear spots. If found, replace entire valve. A slight polish is normal on valve surfaces.
7. If nicks, burrs or bad wear spots are evident on valve, replace entire valve assembly. As on the spool a slight polish is normal on valving surfaces.
8. Check fit of spool in valve body bore without spool dampener O-ring seal and lubricant. Spool should rotate smoothly without binding or catching. If either occurs, entire valve must be replaced. A small burr on the spool or in the valve body can usually be removed with a very fine hone.
9. Check valve spool spring for free length, distortion or collapsed coils. Free length should be .82"; if it measures .79" or less, spring should be replaced.

Reassembly

1. Lubricate four new valve body (Teflon) rings and four new back-up O-ring seals with automatic transmission fluid. Assemble rings and seals on valve body, Fig. S6.

NOTE: *Do not allow seals to become twisted.*

2. Install new spool dampener O-ring seal in spool groove. Lubricate seal with automatic transmission fluid.
3. Assemble stub shaft in valve body, aligning groove in valve cap with pin in valve body. Tap lightly on cap with plastic hammer until cap is against shoulder in valve body with pin in cap groove.

NOTE: *Make sure groove and pin are in line before tapping on cap. Hold these parts together during balance of assembly procedure.*

4. Lubricate valve spool with automatic transmission fluid. Slide spool over stub shaft with notch toward valve body. Align notch with spool drive pin in stub shaft and carefully engage spool in valve body bore.

NOTE: *Because the clearance between spool and valve body is very small, extreme care must be taken when assembling these*

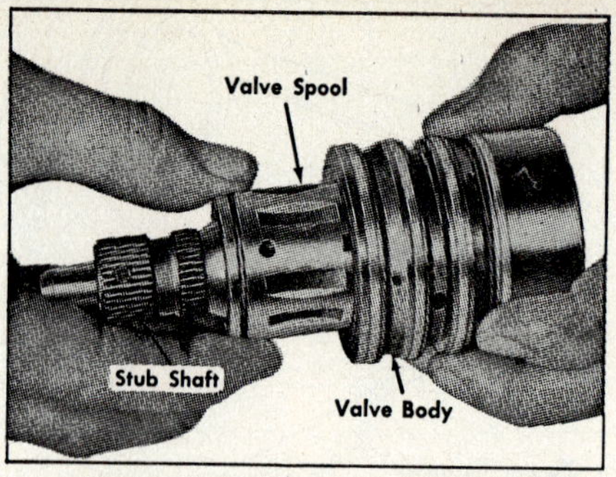

Fig. S5 Removing valve spool

parts. Push the tool evenly and slowly with a slight oscillating motion until spool reaches drive pin. Rotate spool slowly with pressure until notch engages pin. Before pushing spool completely in, make sure spool dampener O-ring seal is evenly distributed in spool groove. Slowly push spool completely in, using extreme care not to cut or pinch O-ring seal.

5. Position seal protector, Fig. S7, or equivalent over stub shaft. Slide valve spool spring over seal protector and work spring down until it is seated in undercut part of stub shaft. Take care not to scratch sealing surface of stub shaft.

NOTE: *If during assembly, stub shaft and cap assembly is allowed to slip out of engagement with valve body pin, spool will be permitted to enter valve body too far. Dampener O-ring seal will expand into valve body oil grooves, preventing withdrawal of spool. Attempt to withdraw spool with slight pull and much rotary motion. If this does not free spool after several tries, make sure spool is free to rotate. Then place valve body on a flat surface with notched end up and tap spool with wooden or plastic rod until O-ring seal is cut and spool can be removed. Replace dampener O-ring seal with new part and proceed with assembly as described above.*

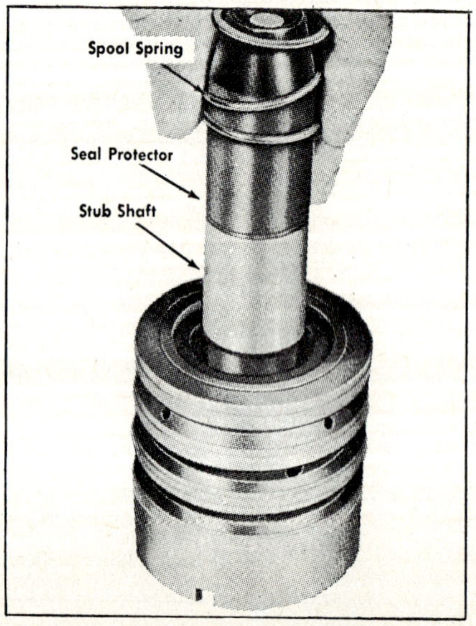

Fig. S7 Installing valve spool spring

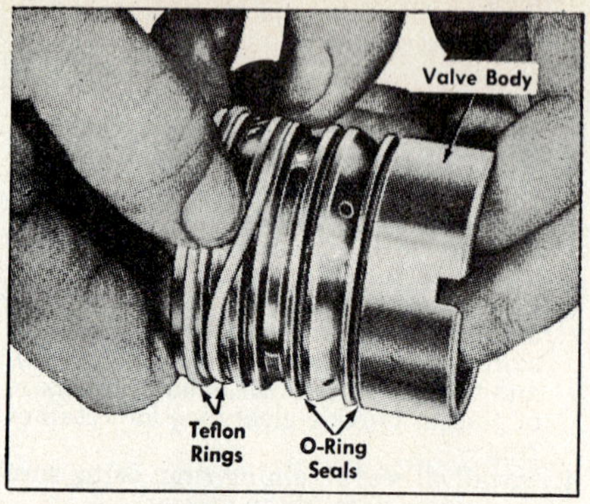

Fig. S6 Installing valve rings

Steering Gear

Removal

1. Mark clamp bolt on intermediate or lower shaft and steering gear stub shaft so gear can be installed in the same position.
2. Disconnect linkage from pitman arm.
3. Unfasten steering gear shaft from stub shaft.
4. Raise shaft and clamp yoke off stub shaft.
5. On some vehicles it may be necessary to remove lower or intermediate shaft to provide clearance for steering gear removal.
6. Drain as much fluid from gear as possible.
7. Disconnect control valve-to-pump tubes.
8. Unfasten gear from frame and remove.

Disassemble, Fig. S8

1. Remove top cover with worm and ball nut as a unit. Remove lower worm thrust bearing. Hold assembly as shown in Fig. S9 with hand holding ball nut from turning and top cover from dropping.
2. Try action of ball nut on worm shaft. Nut must rotate freely with no evidence of binding or roughness. Tape shaft worm at both ends of ball nut to keep nut from running up or down on shaft.
3. Remove top cover from shaft worm.

CAUTION: *Do not hold worm shaft and ball nut in a vertical position as ball nut will travel by its own weight to end of shaft. If ball nut sharply strikes either end of shaft worm, ball guides will be damaged.*

4. Using a screwdriver and hammer, remove worm seal and back-up washer from top cover.
5. Further disassembly should be confined to that found necessary as a result of inspection.

Reassembly, Fig. S8

1. Position seal back-up washer in top cover over bearing.
2. Lubricate worm seal and press it in to seat against back-up washer. Seal should be positioned so its lip faces away from back-up washer.
3. Insert worm through top cover.
4. Mount gear housing in vise so that top

cover opening is up.

5. Lubricate lower worm thrust bearing and position it in cup in lower end of gear housing. Install bearing retainer.
6. Position new top cover gasket on gear housing.
7. Remove tape from worm shaft and rotate ball nut so it reaches end of shaft.
8. Lift top cover, worm and ball nut by grasping top cover and worm. Turn shaft into a vertical position so that ball nut is at bottom. Guide assembly into gear housing until end of shaft contacts lower thrust bearing. In so doing, rotate worm so ball nut return guide clamp faces back-up adjuster opening in housing.
9. Bolt top cover to gear housing and torque to 35–45 ft-lbs.
10. Install pitman shaft and side cover.
11. Align valve body drive pin on worm shaft with narrow pin slot in valve body.

CAUTION: *Do not push against stub shaft as this may cause stub shaft and cap to push out of valve body, allowing valve spool dampener seal to slip in valve body oil grooves. Valve assembly should be pushed in by pressing against body with fingers. Be sure valve body is properly seated before installing and adjusting the adjuster plug. Return port in valve housing should be fully visible when valve body is properly seated.*

Adjustments

Thrust Bearing Preload

1. Install seal protector over end of stub shaft, Fig. S10.
2. Install adjuster plug.
3. Tighten adjuster plug up snugly with spanner wrench, Fig. S11, then back off 1/4 turn.
4. Attach inch-pound torque wrench to stub shaft and measure rotary valve drag, Fig. S11.
5. Adjust bearing so that preload is one to three inch-pounds torque in excess of rotary valve drag while swinging torque wrench back and forth through a 60-degree arc.

NOTE: *Due to the low torque adjustment and because it would be difficult to get an accurate reading on wrench, do not use a torque wrench having a maximum torque reading over 100 inch-pounds. When taking torque readings, take the reading while pulling torque wrench to the right and a reading pulling to the left. Total both readings and take one-half of this total as the average torque.*

6. Using a suitable wrench, install adjuster plug lock nut and tighten to 50–65 ft-lbs. torque.
7. Recheck thrust bearing preload. Total adjustment plus seal drag must not exceed 8 inch-pounds.
8. Install pitman shaft seal so that lips of seal face inside of gear housing.

Adjusting Back-up Adjuster

On gears having a back-up adjuster, install adjuster in gear housing. Then tighten adjuster until it bottoms against ball nut return guide clamp. Back off 1/8 to 1/4 turn and secure in place with adjuster lock nut.

Adjust Pitman Shaft Preload

After steering gear is completely assembled, find center or straight-ahead position of worm by rotating worm through full travel, counting the number of turns and reversing one-half the number of turns counted.

When taking the following torque readings, take a reading pulling the torque wrench to the right and a reading pulling to the left. Total both readings and take one-half of the total as the average torque.

Using the equipment shown in Fig. S12, adjust lash adjuster so that torque is between four and eight inch-pounds in excess of the total bearing preload and valve drag readings previously recorded. The pitman shaft preload reading should be taken with the gear on center and while rotating the steering shaft through an arc of not more than 10 degrees.

Tighten lash adjuster nut to 25–35 ft-lbs. The final over-center reading, which represents the total of the valve and seal drag, worm bearing preload, and lash adjuster preload, should not exceed 16 inch-pounds.

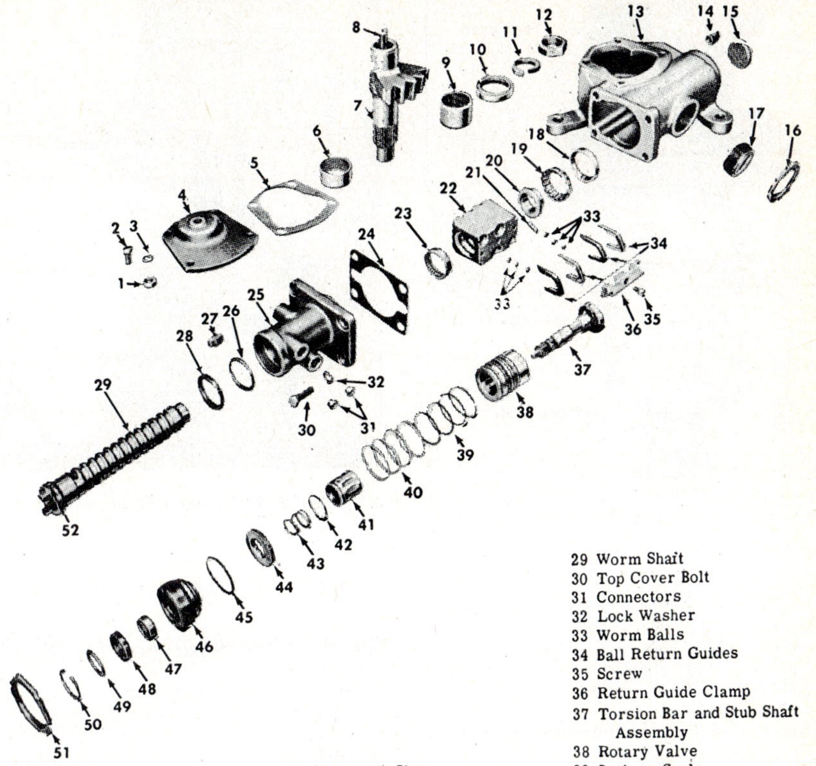

1 Lock Nut
2 Bolt
3 Lock Washer
4 Side Cover
5 Side Cover Gasket
6 Needle Bearing
7 Pitman Shaft and Sector Gear
8 Lash Adjuster
9 Needle Bearing
10 Oil Seal
11 Lock Washer
12 Nut
13 Steering Gear Housing
14 Drain Plug
15 Expansion Plug
16 Lock Nut
17 Back-up Adjuster
18 Bearing Cup
19 Lower Thrust Bearing
20 Thrust Bearing Cone
21 Stop
22 Ball Nut
23 Needle Bearing
24 Top Cover Gasket
25 Top Cover
26 Back-up Washer
27 Filler Plug
28 Worm Oil Seal
29 Worm Shaft
30 Top Cover Bolt
31 Connectors
32 Lock Washer
33 Worm Balls
34 Ball Return Guides
35 Screw
36 Return Guide Clamp
37 Torsion Bar and Stub Shaft Assembly
38 Rotary Valve
39 Back-up Seals
40 Rotary Valve (Teflon) Rings
41 Valve Spool
42 Dampener O-ring
43 Spool Spring
44 Upper Thrust Bearing
45 O-ring Seal
46 Adjuster Plug
47 Stub Shaft Needle Bearing
48 Stub Shaft Oil Seal
49 Back-up Washer
50 Retaining Ring
51 Lock Nut
52 Valve Body Drive Pin

Fig. S8 Saginaw rotary valve power steering disassembled

SPOOL VALVE POWER STEERING GEAR

This steering gear is a recirculating ball bearing, worm and sector nut type. A spool type control valve, Fig. S13, is mounted concentrically on the worm shaft and bolted to the lower end of the steering gear housing.

The spool valve is held in neutral position by ten centering plungers and five springs in the valve housing which bear against the thrust bearing and at the same time against the adapter and valve covers.

Removal

1. Remove steering wheel.
2. Disconnect all control linkage, horn parts and brackets attached to steering column.
3. Remove hydraulic connections to control valve. Cover or plug exposed tubes and ports to prevent dirt entering into hydraulic system.
4. Remove pitman arm with a suitable puller.
5. Remove steering gear from vehicle.

Spool Valve Removal, Fig. S13

1. Scribe a mark on valve cover, valve body,

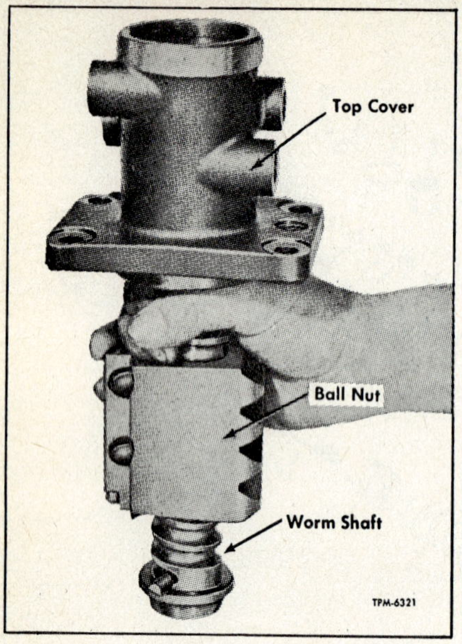

Fig. S9 Top cover, worm and ball nut

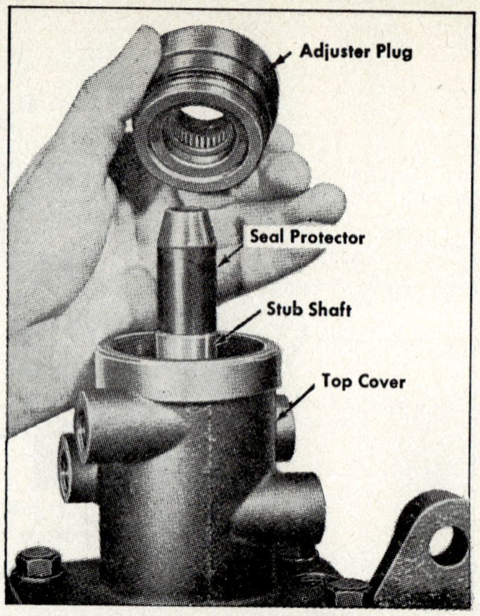

Fig. S10 Installing adjuster plug

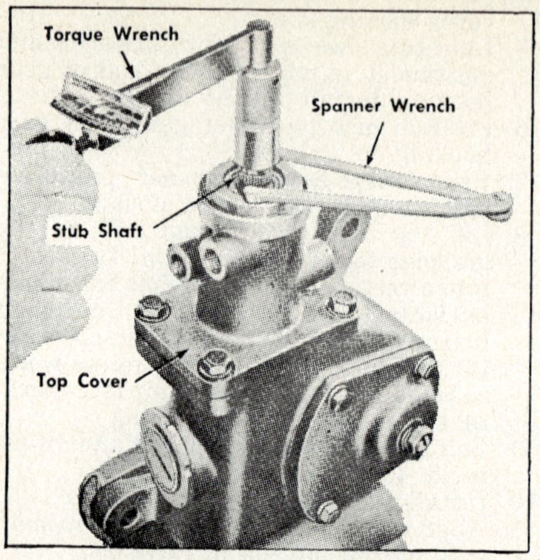

Fig. S11 Adjusting thrust bearing preload

adapter, and steering gear housing to insure proper positioning of parts upon reassembly.

2. Remove valve cover.
3. Remove control valve worm bearing adjuster nut.
4. Remove valve spool preload spring from worm shaft.
5. Remove lower small bearing race, bearing and large bearing race. Keep these parts together so they can be reinstalled together upon reassembly.
6. Remove valve housing, spool, plungers and centering springs as a unit.
7. Remove upper large bearing race, bearing and small bearing race, keeping these parts together so they can be reinstalled together upon reassembly.
8. Remove adapter from gear housing and from worm shaft.

Spool Valve, Disassembly, Fig. S13

1. Remove valve centering plungers and springs from valve body.
2. Remove valve spool from valve housing, noting which end of spool has a groove or counterbore on the I.D. so parts can be reassembled properly.

Spool Valve, Reassemble, Fig. S13

1. Coat valve centering plungers and valve spool with grease containing zinc oxide #1.
2. Install valve spool in valve body with groove or counterbore at same end in relation to valve body as noted upon disassembly.
3. Do not force spool into valve body. When spool is properly aligned it will drop into place. Forcing spool will damage both spool and valve bore.
4. Install valve centering springs into their bores in valve body.
5. Install a plunger at each end of centering springs.

Steering Gear

The procedure for disassembling and reassembling the steering gear is similar to that for the mechanical gear of the same design. However, adjustments are somewhat different, the procedure for which follows:

Adjustments

Before adjusting the steering gear, be sure to check the following important items:

1. The thrust bearings have small races away from valve faces.
2. Scribed marks on valve body, adapter and steering gear housing are aligned.
3. Install valve spool preload spring on worm shaft.
4. Install a new adjuster plug nut finger tight.
5. Install valve clamping ring and valve end cover-to-adapter bolts. Tighten bolts to 15–20 ft-lbs.

NOTE: *A valve clamping ring, Fig. S14,*

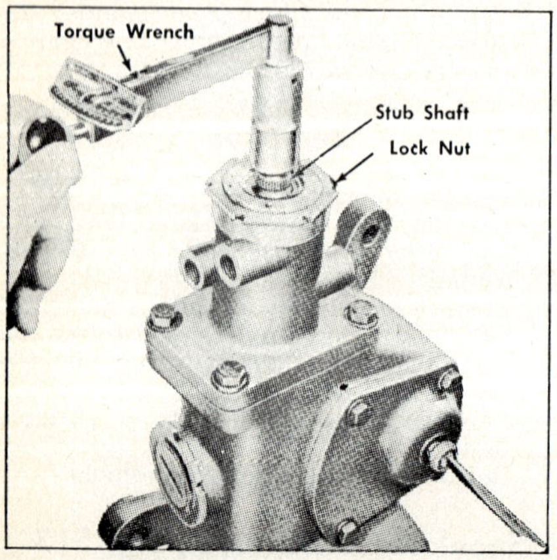

Fig. S12 Adjusting pitman shaft preload

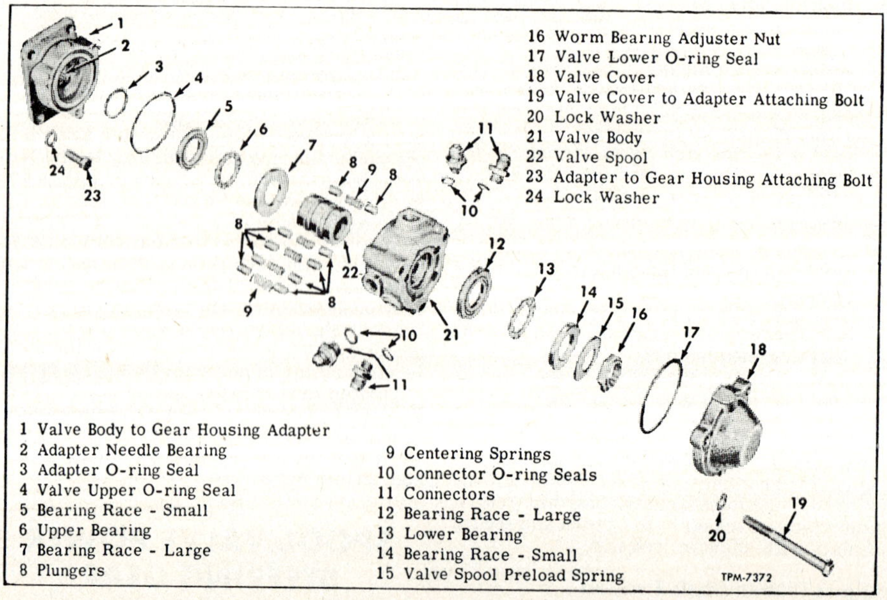

Fig. S13 Spool valve components disassembled

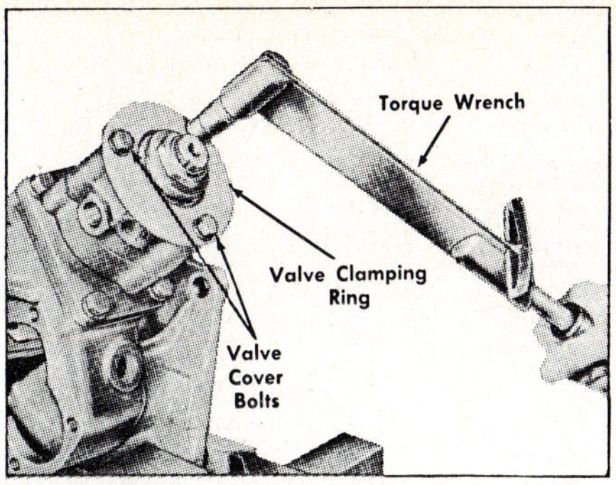

Fig. S14 Adjusting worm bearing preload

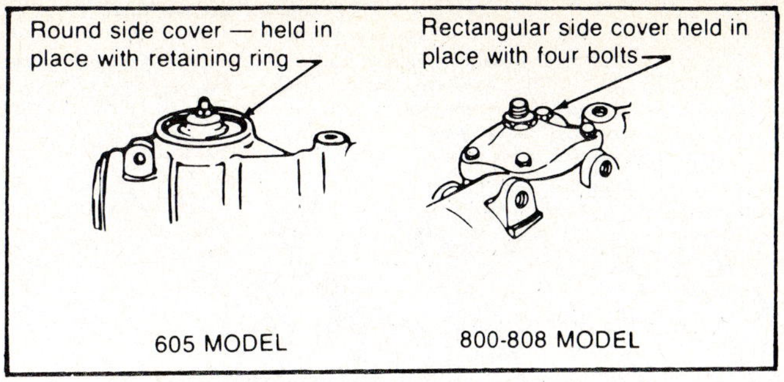

Fig. S15 Steering gear identification

can be made from a valve cover by cutting out the middle portion and leaving outer ring only.

6. Install steering wheel. Hold wheel on center and tighten adjuster nut firmly to remove all end play of valve spool.
7. Back off nut and retighten lightly. Release steering wheel, making sure nut exerts light pressure against spring loaded plungers, Fig. S14.
8. Tighten nut until a load of 1/2 to 1 lb. is measured with a spring scale on the rim of the 18″ steering wheel. Stake nut in place.
9. Remove valve clamping ring and position valve cover with marks scribed previously with related parts.
10. Install attaching bolts and torque to 15–20 ft-lbs.
11. With lash adjuster screw and shim in place in slotted end of pitman shaft, start side cover over end of shaft. Insert screwdriver into hole in side cover to engage slot in screw.
12. Turn steering shaft until ball nut is in approximate center of shaft worm. Center tooth of pitman shaft sector gear must enter center tooth space of ball nut rack.
13. Insert pitman shaft into housing, meshing teeth as described.
14. Turn lash adjuster to pull cover over end of shaft. Back off adjuster screw to permit lash between sector and ball nut.
15. Install and tighten side cover bolts.
16. With gear on center, adjust pitman shaft adjuster screw so that pull on the 18″ diameter steering wheel is 1 1/4 to 2 pounds through a 20-degree arc over center. Tighten lock nut.
17. Total pull on 18″ wheel for all adjustments should be 1 1/8 to 2 pounds over center.
18. Install pitman shaft oil seal.

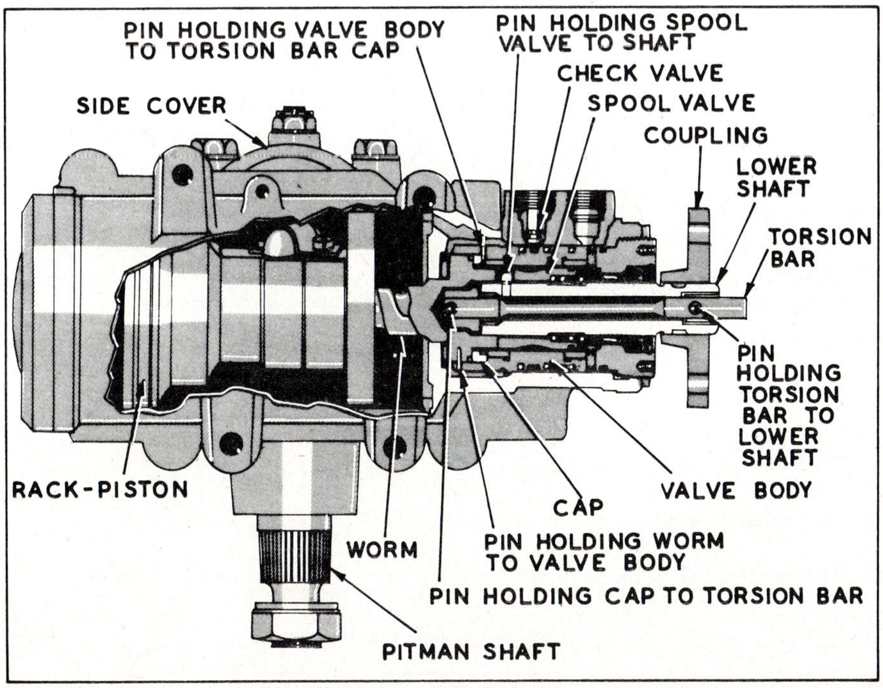

Fig. S16 Model 800-808 Saginaw steering gear

INTEGRAL ROTARY VALVE TYPE, SINGLE PISTON

Description

There are two types of Saginaw integral rotary valve type single piston steering gears. The two different types of gears can be identified by the configuration of the side covers. The Model 605 gear has a round side cover, secured with a retaining ring, Fig. S15. The Model 800-808 gear has a rectangular side cover, secured with four bolts, Fig. S15. Also, note that the Model 800-808 gear incorporates balls in the rack piston, Figs. S16 and S17, while the Model 605 gear does not, Fig. S18.

The power steering gear operates entirely by displacing oil to provide hydraulic pressure assists only when turning. As the entire gear assembly is always filled with oil, all internal components of the gear are immersed in oil, making periodic lubrication unnecessary. In addition this oil acts as a cushion to absorb road shocks that may be transmitted to the driver.

The steering shaft, hydraulic valve, worm and rack piston nut are all in line, making a compact and space saving steering gear. All oil passages are internal except the pressure and return hoses between gear and pump.

Maintenance

Four factors affect power operation of the steering system: fluid level and condition, drive belt tension, pump pressure and steering gear adjustment. These should always be checked before any major service operations are performed.

Conditions such as, hard or loose steering, road shock or vibrations are not always due to steering gear or pump, but are often related instead to such factors as low tire pressure and front end alignment. These factors should be checked and corrected before any adjustment of the steering gear is made.

Checking Fluid Level

1. Run engine to normal operating temperature, then shut it off. Remove reservoir filler cap and check oil level on dipstick. Level should be up to "Full" mark on dipstick.
2. If level is low, add power steering fluid to proper level on dipstick and replace filler cap.

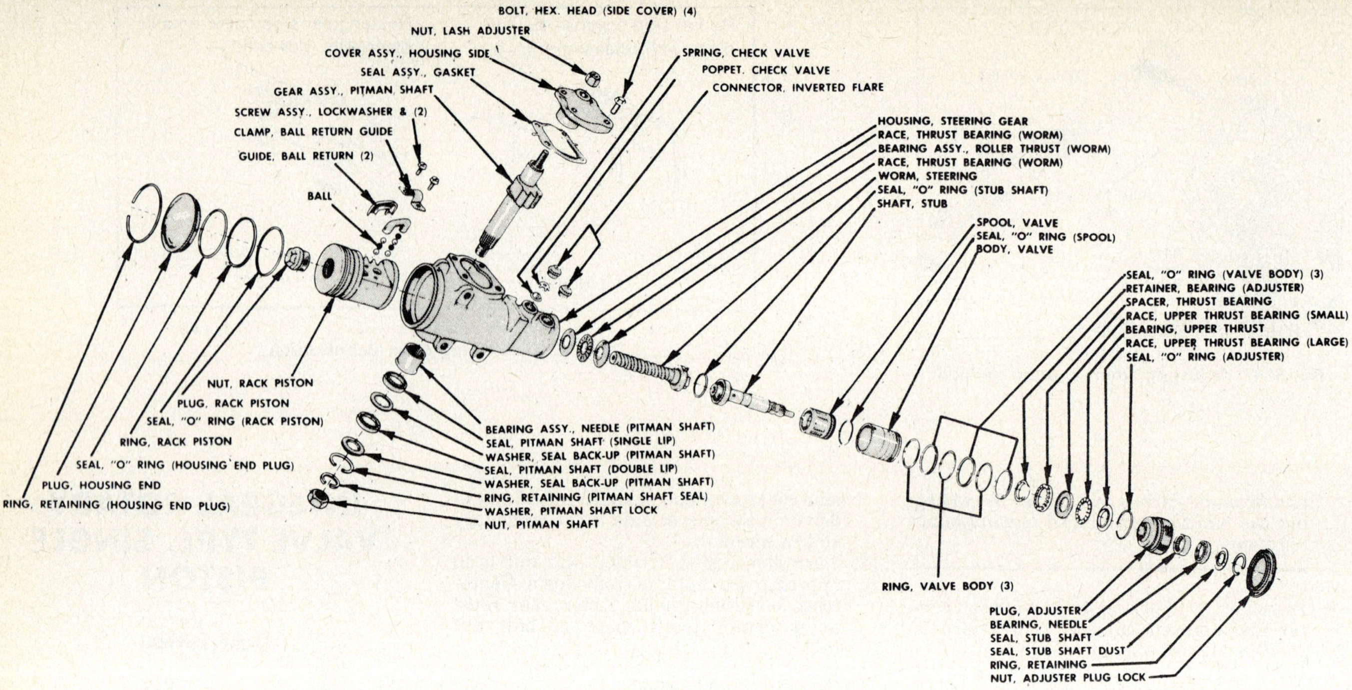

Fig. S17 Model 800-808 power steering gear disassembled

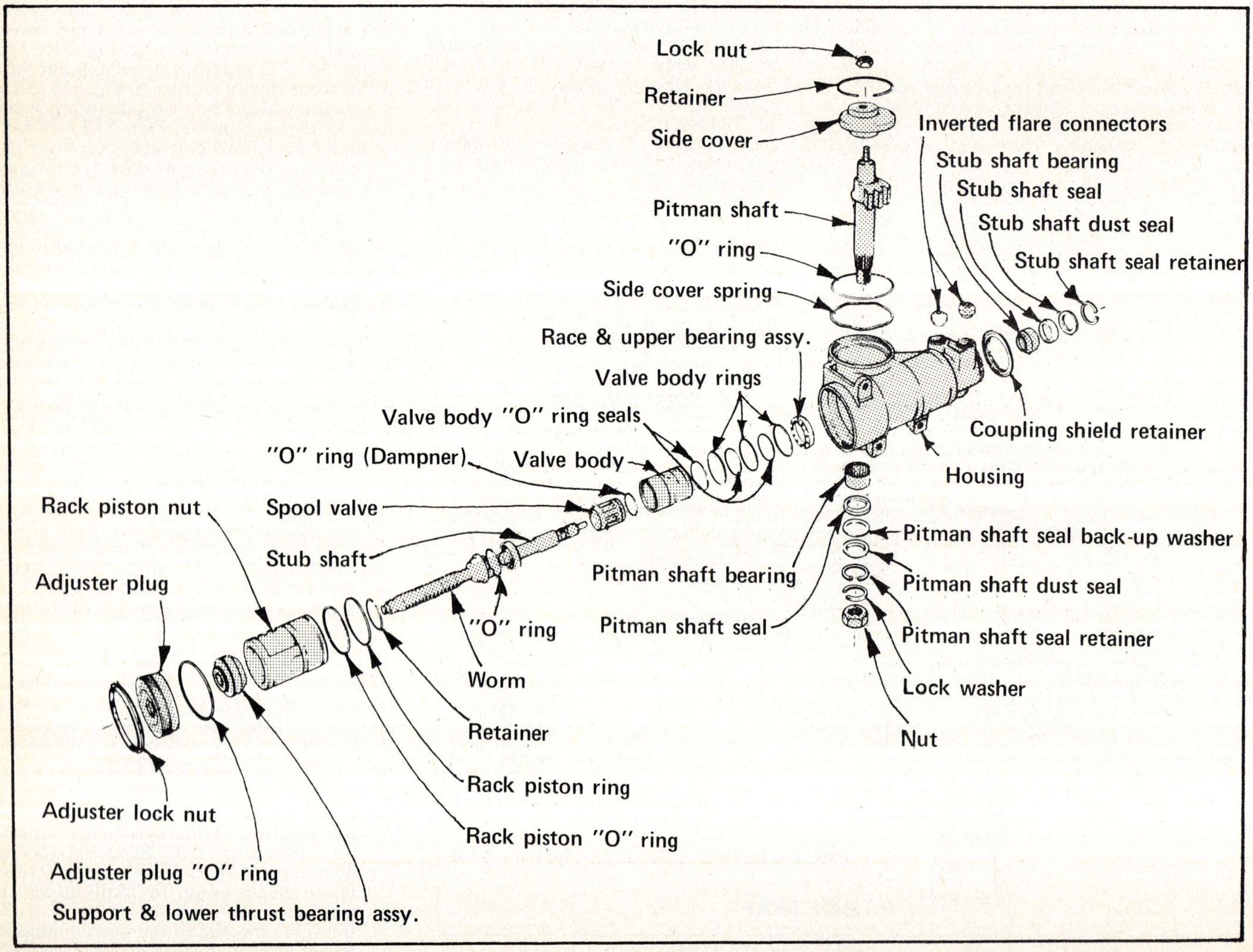

Fig. S18 Model 605 power steering gear disassembled

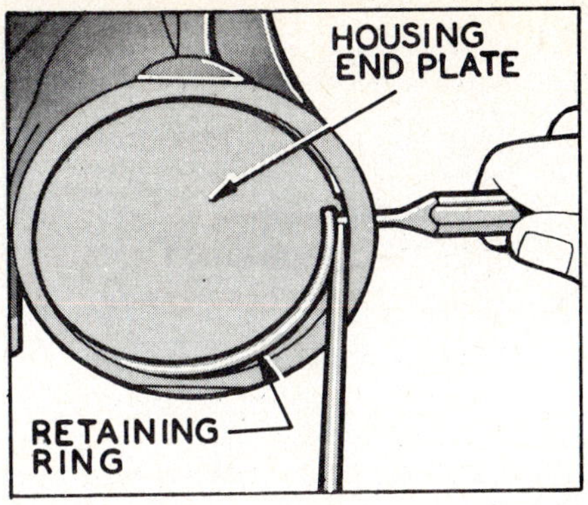

Fig. S19 Removing end plug retaining ring

NOTE: *When adding less than one pint of fluid, it is permissible to use automatic transmission fluid. When adding larger quantities or when making a complete fluid change, it is recommended that special power steering fluid be used.*

3. When checking fluid level after system has been serviced, air must be bled from the system as follows:

Bleeding System

1. With wheels turned all the way to the left, add the recommended fluid to the "Add" mark on dipstick.
2. Start engine. While running at fast idle, recheck fluid level and, if necessary, add fluid to the "Add" mark on dipstick.
3. Bleed system by turning wheels from side to side without hitting stops. Maintain fluid level just above internal pump casting. Fluid with air in it will have a light tan appearance. This air must be eliminated from fluid before normal steering action can be obtained.
4. Return wheels to center position and continue to run engine for two or three minutes, then shut it off.
5. Road test vehicle to make sure steering functions normally and is free from noise.

Fig. S20 Installing rack-piston arbor tool in end of worm

6. Recheck fluid level, making sure level is at the "Full" mark.

Model 800-808 Steering Gear

Disassemble

1. Referring to Fig. S17, position gear in vise with gear housing end plug facing up.
2. Rotate gear housing end plug retaining ring so that one end of ring is over hole in housing. Spring one end of ring with a punch to allow screwdriver to be inserted to lift ring out, Fig. S19.
3. Rotate coupling flange counterclockwise until rack-piston just forces end plug out of housing. Remove plug from housing.

CAUTION: *Do not rotate any further than necessary or balls with fall out of their circuit and pitman shaft teeth and rack-piston will become disengaged.*

4. Remove rack-piston end plug, using a 1/2" square drive.

NOTE: *To make removal easier, tap rack-piston end plug with a plastic mallet to unseat threads. This is important as end plug is tightened to 75–80 ft-lbs during assembly and could break during removal if not handled carefully.*

5. Remove lock nut from pitman shaft adjuster screw and discard.
6. Remove four side cover retaining screws and washers from cover.
7. Rotate pitman shaft adjuster screw with an Allen wrench until side cover is lifted free of housing.
8. Separate side cover from pitman shaft; discard side cover O-ring seal.
9. Turn coupling flange until pitman shaft teeth are centered in housing.
10. Tap end of pitman shaft with a soft mallet and slide pitman shaft out of housing.
11. Remove housing end plug O-ring seal and discard.
12. Insert Rack-Piston Arbor into rack-piston against end of worm, Fig. S20. Turn coupling flange counterclockwise, while holding tool tightly against worm, to force rack-piston onto arbor, and remove rack-piston from gear housing.
13. Remove stub shaft-to-coupling flange retaining screw and remove flange.
14. Remove adjuster plug lock nut by breaking it loose with hammer and punch, and remove lock nut from housing.
15. Loosen adjuster plug and remove from housing, Fig. S21.
16. Remove valve assembly by grasping stub shaft and pulling out.
17. Remove worm, lower thrust bearing and races from upper end of housing.

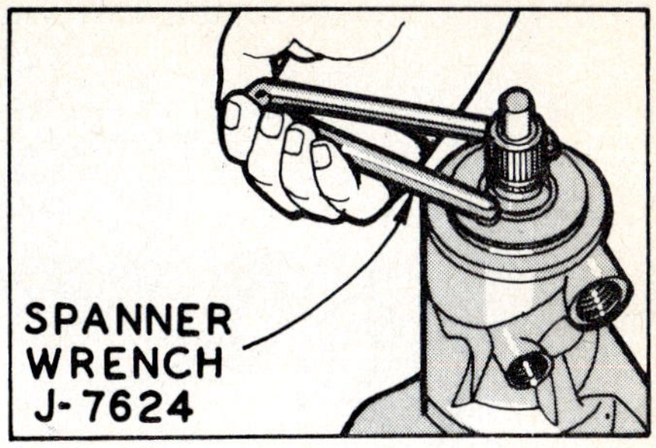

Fig. S21 Removing adjuster plug

Replacing Valve Spool Dampener "O" Ring (Only If Gear Squawks)

NOTE: *The rotary valve assembly includes the valve body, valve spool and stub shaft assembly. All these parts are precison units and are hydraulically balanced at the factory. Under no circumstances are parts in this unit to be replaced or interchanged with other parts or units. If unit parts are scored or damaged the entire rotary valve assembly is to be replaced.*

1. To replace the valve spool dampener "O"

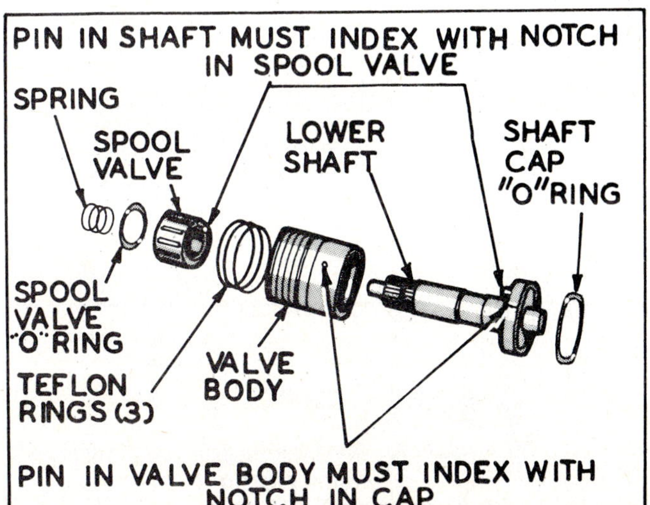

Fig. S22 Rotary valve parts

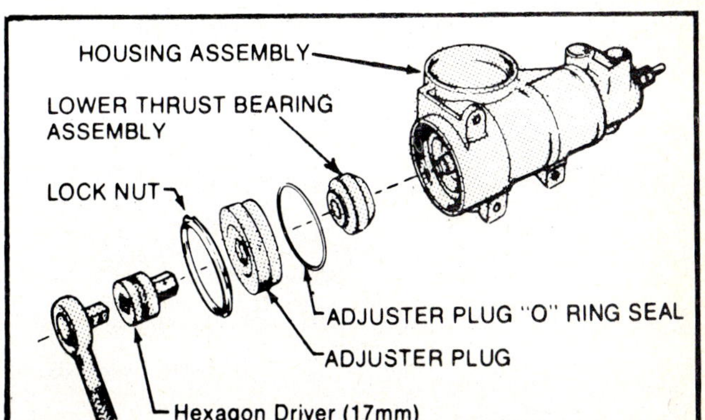

Fig. S23 Lower thrust bearing & adjuster assembly. Model 605

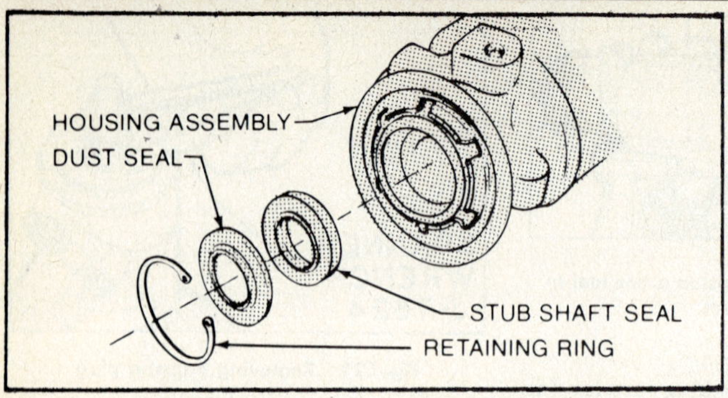

Fig. S24 Stub shaft seals & snap ring. Model 605

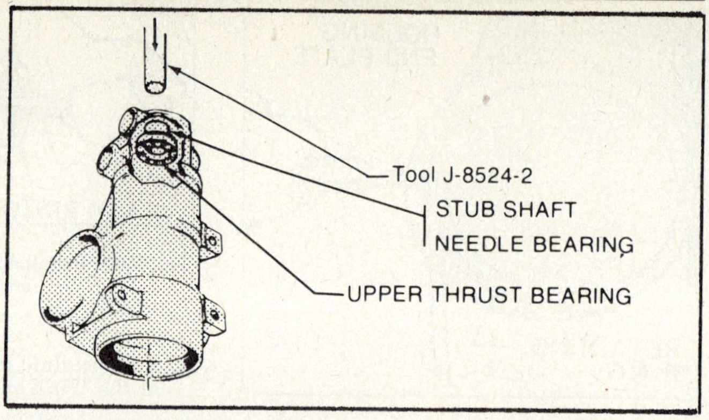

Fig. S25 Stub shaft bearing removal. Model 605

ring, work spool spring into bearing diameter of stub shaft and remove spool spring.
2. Tap end of stub shaft gently against workbench to remove valve spool. *The diametrical clearance between the valve body and spool may be as low as .0004". The slightest cocking of the spool may jam it in the valve body.*
3. Remove valve spool dampener "O" ring.
4. Install new "O" ring in valve spool groove, then lubricate seal in automatic transmission fluid. Do not allow seal to twist in groove.
5. With notch end of spool towards valve body, install spool, aligning spool notch with pin in stub shaft, Fig. S22.
6. *Because of the small clearance between spool and valve body, extreme care must be taken when assembling these parts. Push spool evenly and slowly with a slight oscillating motion until spool reaches drive pin. Before pushing spool completely in, make sure dampener "O" ring seal is evenly distributed in spool groove. Slowly push spool completely in, with extreme care taken not to cut or pinch "O" ring seal.*
7. Slide spool spring over stub shaft and work spring into position.

Disassemble Rotary Valve

1. If used, work spool spring onto bearing diameter of stub shaft and remove spool spring.
2. Tap end of stub shaft gently against workbench to remove valve spool.

NOTE: *Because of the slight clearance between valve body and spool, the slightest cocking of spool may jam it in valve body. If slight cocking occurs, make a gentle attempt to reverse removal procedure. If this does not free spool, it has become cocked in valve body bore and may be removed later.*

---

3. Remove and discard valve spool dampener "O" ring.
4. Remove stub shaft, torsion bar (small diameter bar extending through stub shaft) and valve cap by tapping end of torsion bar lightly with a plastic hammer. This will dislodge cap from valve body cap pin, Fig. S22. Do not disassemble stub shaft as these parts are pinned together and serviced only as an assembly.
5. If valve spool has become cocked as mentioned above, first inspect parts to determine in which direction the spool is cocked. A few very light taps with a plastic hammer should align and free the spool in the bore. Remove and discard "O" ring dampener seal from spool.
6. Carefully remove valve body Teflon rings and ring back-up "O" ring seals.

Disassemble Housing

1. Remove pitman, shaft outer dust seal retaining ring.
2. Remove outer dust seal.
3. Remove seal (double lip) by inserting offset screwdriver between seal and back-up washer and prying out of housing.
4. Remove back-up washer.
5. Remove seal (single lip) by cutting and collapsing seal.
6. Remove pitman shaft needle bearings (if necessary) with a suitable driver.
7. If connectors are to be removed, tap threads in holes of connectors using 5/16-18 NF tap. Remove connectors by using threaded bolt into tapped holes with washer and nut as extractor.

Model 605 Steering Gear

Disassemble

1. Remove preload adjuster nut.

NOTE: Preload adjusting nut has a left hand thread.

---

2. Center gear by rotating stub shaft, then remove side cover snap ring, pitman shaft, side cover, "O" ring seal and spring.
3. Remove locknut, adjuster plug and "O" ring seal, then the lower thrust bearing assembly, Fig. S23.
4. Push stub shaft to remove rack piston and valve assembly.
5. Remove stub shaft snap ring, dust seal and shaft seal, then the needle bearing and upper thrust bearing, if necessary, Figs. S24 and S25.
6. Remove pitman shaft snap ring, dust seal, washer and oil seal, then the needle bearing, if necessary, Fig. S26.

Disassemble Rack Piston, Valve & Worm Assembly

1. Remove valve and worm from rack, then the teflon ring and "O" ring seal from rack piston, Fig. S27.
2. Remove snap ring and worm shaft from valve, Fig. S28.
3. To disassemble spool valve, Fig. S29, proceed as follows:
   a. Hold valve and tap stub shaft on wooden block to loosen shaft cap.
   b. To remove stub shaft, pull cap out 1/4 inch, then disengage pin on shaft from hole in spool, Fig. S30.

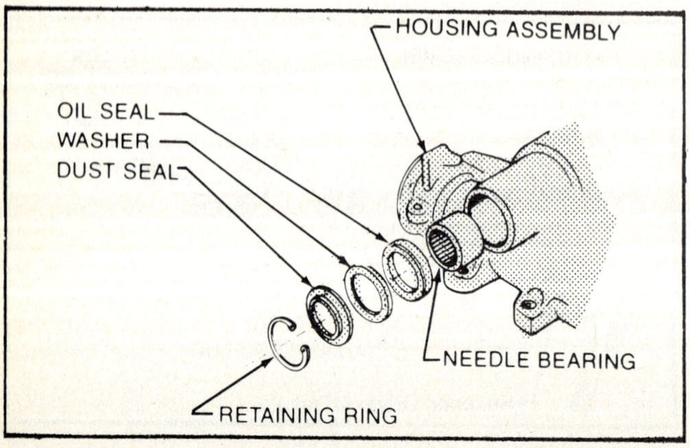

Fig. S26 Pitman shaft seals & bearings. Model 605

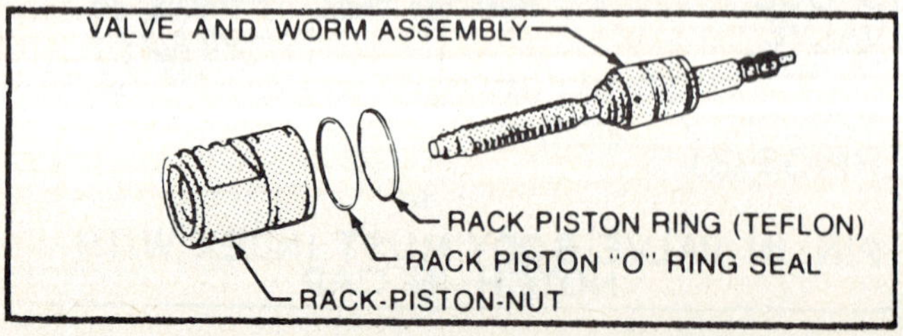

Fig. S27 Valve & worm removed from rack. Model 605

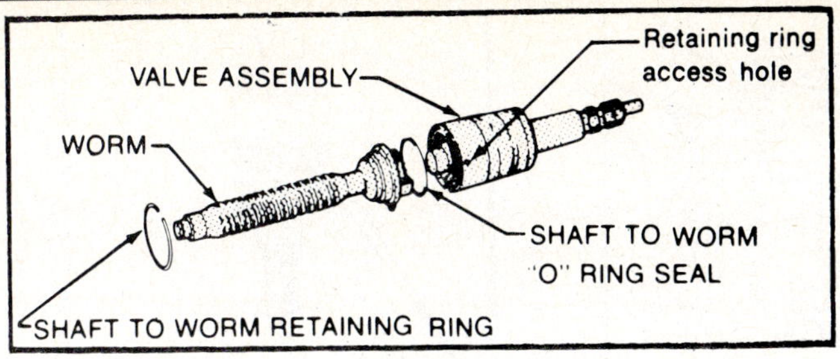

Fig. S28 Valve & worm disassembled. Model 605

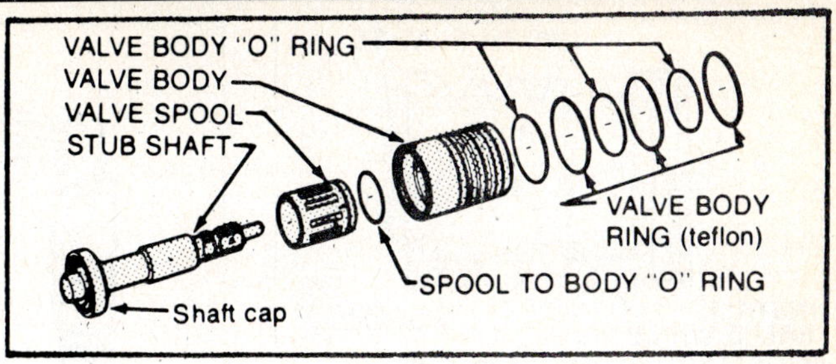

Fig. S29 Spool valve assembly. Model 605

c. Rotate and lift spool from valve body.
d. Remove teflon rings and "O" rings from valve body and spool.

Model 800-808 Assemble Sub-Assemblies

Refer to Fig. S17 and lubricate all parts as they are assembled.

1. Screw lash adjuster through side cover until cover bottoms on pitman shaft gear. Install lash adjuster lock nut while holding lash adjuster with 7/32" allen wrench.
2. Mount adjuster plug in vise with soft jaws.
3. If it has been removed, assemble needle bearing by pressing towards thrust bearing end of adjuster plug against identification end of bearing. End of bearing to be flush with bottom surface of stub shaft seal bore.
4. Install stub shaft far enough to provide clearance for dust seal and retaining ring. Install new dust seal with rubber surface outward. Install new retaining ring.
5. Assemble large thrust bearing race, thrust bearing, small race and spacer (with grooves up) on adjuster plug, Fig. S31, and secure with retainer.

Assemble Rotary Valve

1. Assemble one valve body Teflon ring back-up "O" ring seal in each groove in valve body, being sure seals do not become twisted.
2. Assemble valve Teflon rings in ring grooves over "O" ring seals by carefully slipping rings over valve body. The rings may appear loose or twisted in the grooves but the heat of the oil during subsequent operation will cause them to straighten.

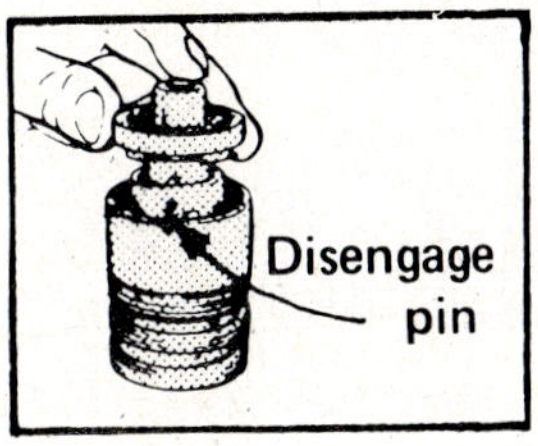

Fig. S30 Stub shaft removal. Model 605

3. Install valve spool dampener "O" ring seal in valve spool groove, being sure it is not twisted.
4. Insert stub shaft through valve spool, engaging spool locking pin.
5. Pull valve spool and stub shaft assembly into valve body, aligning stub shaft notch and valve body pin.
6. If used, slide spool spring over stub shaft and place into position.
7. Lubricate cap-to-worm "O" ring and install in valve body. *During assembly of the valve, if the stub shaft and cap is allowed to slip out of engagement with the valve body pin, the spool will be permitted to enter the valve body too far. The dampener "O" ring seal may expand into valve body oil grooves, preventing removal of spool.* If this happens, remove spool spring and disassemble rotary valve. Press on spool until "O" ring seal is cut and can be removed. Install new "O" ring and reassemble.

Assemble Housing

1. With stamped end of needle bearing facing outward, drive bearing into bore from outside of housing until flush. Make sure bearings rotate freely.
2. Lubricate cavity between lips of pitman shaft (double lip) seal with power steering fluid.
3. Lubricate and install pitman shaft seals as shown in Fig. S32. Make sure seal lips are properly positioned, retaining ring is seated.
4. If connectors were removed, install new ones by driving them into place.

Assemble Rack-Piston & Worm

1. Lubricate and install new ring back-up seal and Teflon piston ring on rack-piston nut, being careful ring and seal do not twist during installation.
2. Insert worm into rack-piston nut to bearing shoulder.
3. Align ball return guide holes with worm groove. Load balls into guide hole nearest the teflon piston ring while slowly rotating worm to left, feeding balls through circuit, Fig. S33. If balls are installed properly, the worm should turn out of the rack-piston nut.
4. Fill one-half of ball return guide with remaining balls. Place other guide over balls and plug each end with heavy grease to prevent balls from falling out when installing guides into rack-piston nut.
5. Insert guides into guide holes of rack-piston nut. Make sure black ball in guide is installed next to white ball in rack piston. Guides should fit loosely.

Check Worm Preload

The worm groove is ground with a high point in the center. When the rack-piston nut passes over this high point, a preload of a 1 to 4 inch pounds torque should be obtained.

1. With worm pointing up, clamp rack-piston nut in a vise with soft jaws (do not hold rack-piston nut in area of Teflon ring).
2. Place valve on worm, engaging worm drive pin.
3. Rotate worm until it extends 1¼" from rack-piston nut to thrust bearing face,

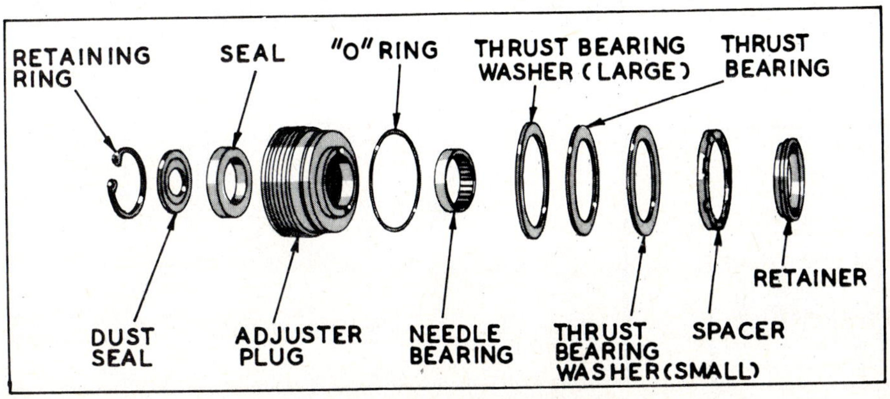

Fig. S31 Adjuster plug parts

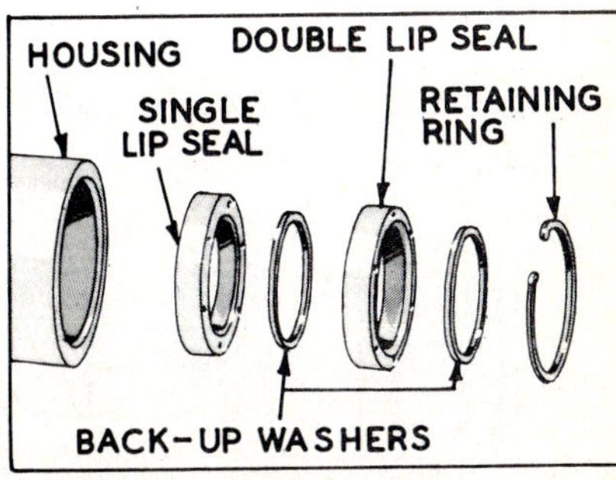

Fig. S32 Pitman shaft seals & washers

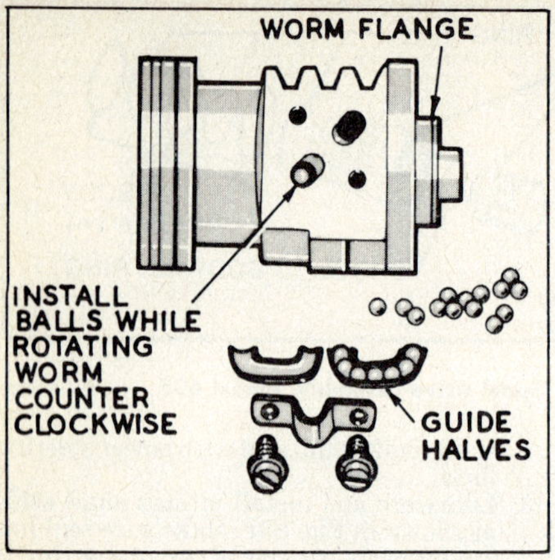

Fig. S33 Loading rack-piston nut

Fig. S34. This is the center position.

4. Attach an inch pound torque wrench with socket on stub shaft, Fig. S34. Oscillate wrench through a total arc of approximately 60° in both directions several times and take a reading. The highest reading obtained with worm rotating should be between 1 and 4 inch pounds. Take a torque reading pulling the torque wrench to the right and a reading pulling the wrench to the left. Total both readings and take one-half of this total as the average torque. *Do not use a torque wrench having maximum torque reading of more than 100 inch pounds.*

Refitting Rack-Piston Balls

Do not refit balls unless the steering is loose. If such is the case, a thrust adjustment and over-center adjustment should correct the problem if it lies in the steering gear. If balls are pitted or rough, then select the proper ball size for proper adjustment.

If the torque reading obtained above is too high or too low (on new balls only), disassemble and reassemble, using the next size smaller (or larger) balls and recheck worm preload.

A rack-piston nut with a ball size of 7 does not have a number stamped on the flat surface. For ball sizes other than 7, the ball size is stamped on the flat surface of the rack-piston nut. Ball sizes are numbered from 6 to 11, with 6 being the smallest and 11 the largest. Ball sizes are graduated in increments of .00008" from .28117" to .28157".

Model 605 Rack Piston, Valve & Worm Assembly

Assemble

1. To assemble spool valve, Fig. S29, proceed as follows:
   a. Install three teflon & three "O" ring seals on spool.
   b. Lubricate spool with power steering fluid, then rotate spool white installing.
   c. Install stub shaft into valve body.

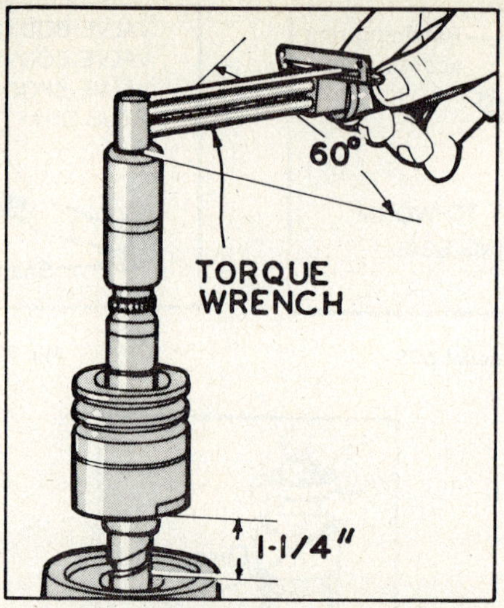

Fig. S34 Checking worm preload

NOTE: Notch must fully engage pin and cap must seat against shoulder, Fig. S35.

---

2. Assemble worm shaft on valve, Fig. S28.

NOTE: Install snap ring with "curved in end" on same side as access hole.

---

3. To install valve and worm in rack:
   a. Install teflon ring and "O" ring seal, Fig. S27.
   b. Hold rack teeth and worm drive pin in position, Fig. S36.
   c. With light pressure to hold worm thread against piston, turn worm and valve counter-clockwise slowly until one click is felt, then turn clockwise to complete assembly.
   d. Check relation of stub shaft and rack piston by threading worm and valve into rack piston until valve face and rack piston face are flush. Referring to Fig. S37, flats on stub shaft and piston must be parallel. If not parallel, disassemble and repeat steps b and c.

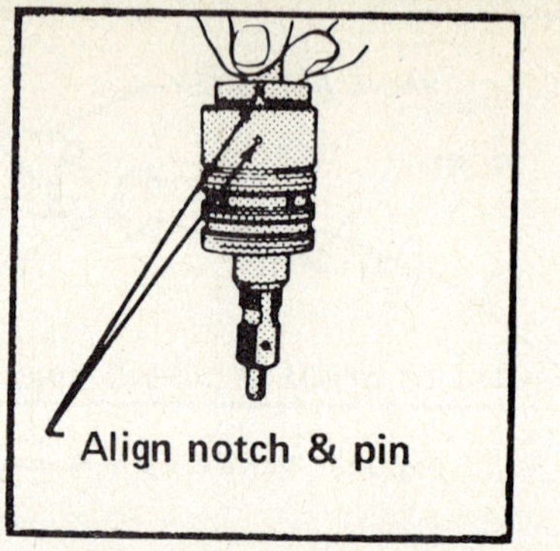

Fig. S35 Stub shaft installation. Model 605

Model 800-808 Steering Gear, Assemble

Install Rack-Piston Worm and Valve

1. Install valve and worm in housing as a unit, Fig. S38.
2. Install new "O" rings on adjuster plug.
3. Install adjuster plug on stub shaft in gear housing finger tight, Fig. S39.
4. Holding a suitable Teflon ring compressor sleeve tightly against shoulder of gear housing, insert rack-piston nut and arbor into housing, holding arbor unit it contacts worm end, Fig. S40.
5. Holding arbor tight against worm, turn stub shaft flange and worm to draw ball nut onto worm and into housing until arbor is free. Be certain that no balls drop out.
6. Remove arbor and sleeve.

Install Rack-Piston Nut End Plug

1. Install new "O" ring seal on end plug, being careful not to allow seal to twist in groove.
2. Install end plug into rack-piston nut by pressing into place.
3. Install end plug retaining ring, being sure ring is bottomed in its groove.

Install Housing Lower End Plug

1. Install new housing end plug "O" ring seal.

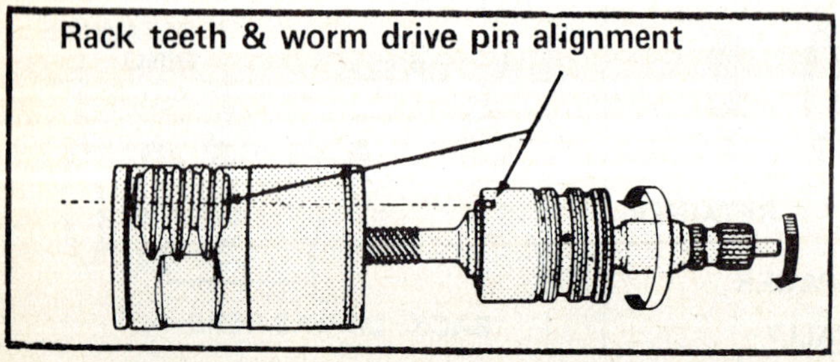

Fig. S36 Installing valve & worm into rack. Model 605

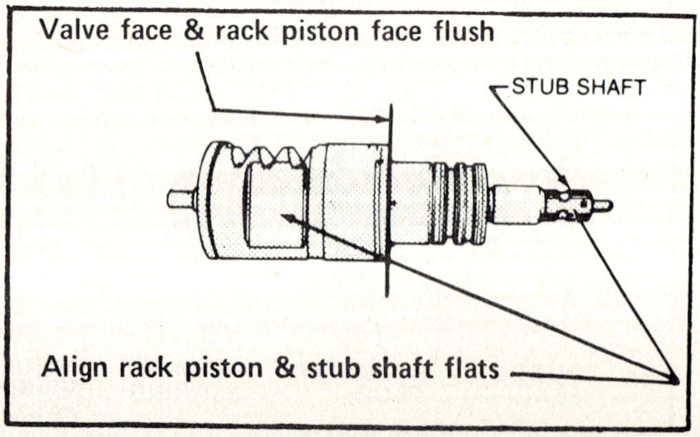

Fig. S37 Stub shaft & rack positioning. Model 605

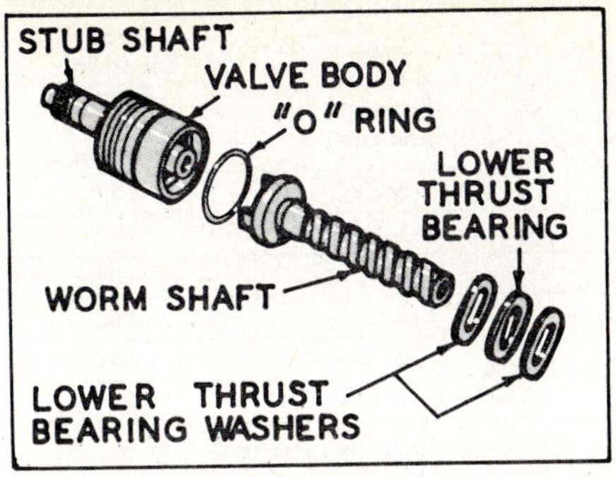

Fig. S38 Worm shaft & rotary valve parts

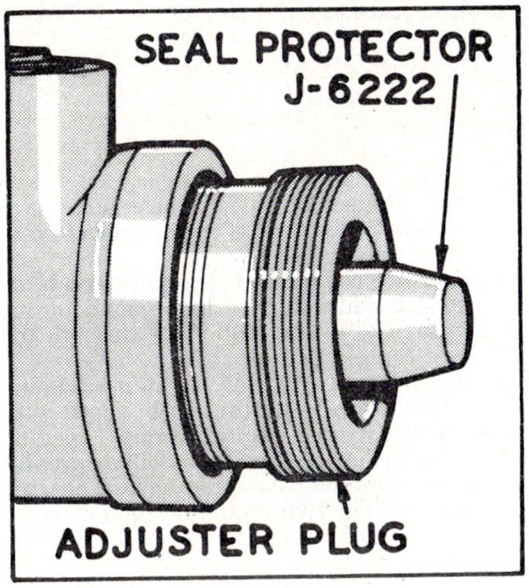

Fig. S39 Installing adjuster plug

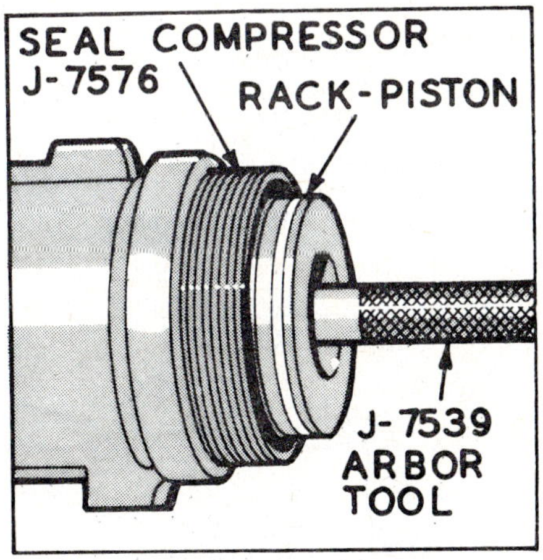

Fig. S40 Installing rack-piston seal compressor

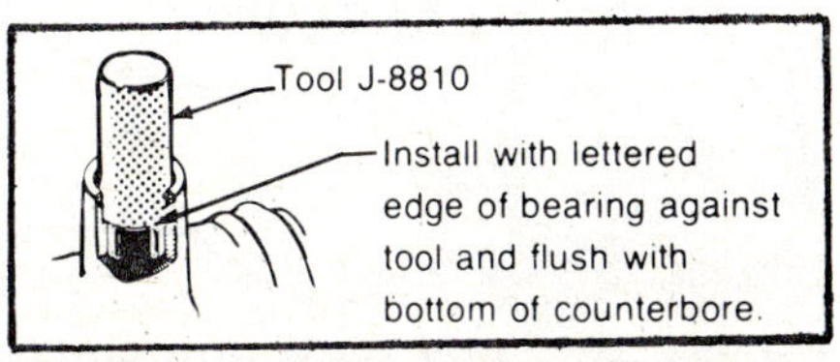

Fig. S41 Pitman shaft bearing installation. Model 605

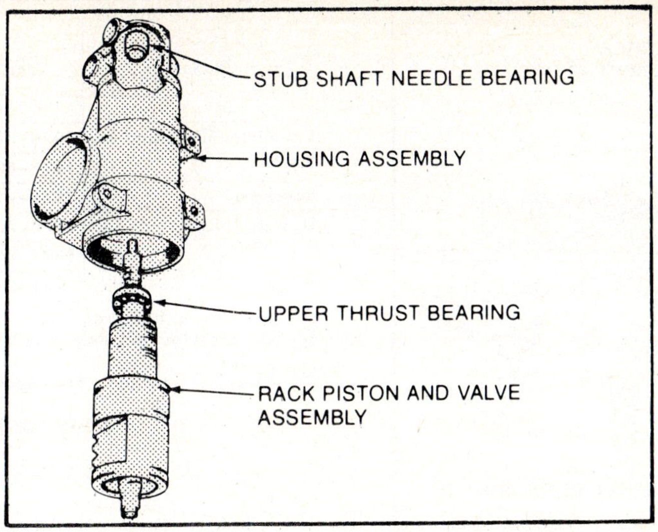

Fig. S42 Stub shaft bearing installation. Model 605

2. Insert end plug into gear housing and seat against "O" ring. Slight pressure may be necessary to seat end plug properly.
3. Install end plug retainer ring so end of ring extends over and at least 1/2" beyond ring removal assist hole.

Install Pitman Shaft Gear and Side Cover

1. Install stub shaft flange and turn steering worm until center groove of rack-piston is aligned with center of pitman shaft needle bearings.
2. Install new side cover "O" ring.
3. Install pitman shaft gear so that center tooth of gear meshes with center groove of rack-piston. Make sure side cover "O" ring is in place before pushing cover against housing.
4. Install and tighten side cover screws.

Install Stub Shaft Flange

Rotate stub shaft slightly over two complete turns from either extreme right or left turn. This will place gear in center position, which will place torsion bar pin through stub shaft in a straight up and down position.

Rotate flange so the bolt with the large head is in top position. Visually align the block tooth in the serrations of the flange with the torsion bar pin and install flange on stub shaft. Tighten flange clamping bolt.

Sub-Assemblies, Install

1. Position gear housing in vise with adjuster plug end facing up.
2. Lubricate worm shaft, lower thrust bearing and races with power steering fluid, then position thrust bearing and race on worm.
3. Align valve body drive pin on worm with narrow pin slot on valve body. Be sure "O" ring seal between valve body and worm head is installed.
4. Position valve body and worm shaft in housing as a unit.

CAUTION: *Do not push against stub shaft as this might cause stub shaft and cap to pull out of valve body, allowing spool seal to slip into valve body oil grooves. Valve assembly can be installed by pushing on outer diameter of valve body housing with the fingers of both hands. Make certain that Teflon rings are not binding on inside of housing. Valve is properly seated when oil return hole in gear housing is fully visible.*

---

5. Place a suitable seal protector over end of stub shaft, Fig. S39.
6. Lubricate new adjuster plug "O" ring seal with power steering fluid and install in groove on adjuster plug.
7. Install adjuster plug over end of stub shaft and tighten just enough to make certain that all parts are properly seated in gear housing. Remove seal protector.
8. Install adjuster plug lock nut loosely on adjuster plug.
9. Install seal compressor in gear housing, holding it tightly against shoulder in housing, Fig. S40.
10. Insert rack-piston into housing until arbor engages worm. Turn stub shaft clockwise, using a 3/4" wrench on 1974–77 units, and 11/16 in. wrench on 1978–83 units, to draw rack-piston into housing. When piston ring is in housing bore, arbor and seal compressor can be removed.
11. Turn stub shaft as necessary until middle rack groove in rack-piston is aligned with center of pitman shaft needle bearing.
12. Lubricate new side cover "O" ring seal and install in groove in face of side cover.
13. Assemble side cover on pitman shaft by screwing cover onto adjusting screw until cover bottoms on pitman shaft.
14. Install pitman shaft so that center tooth in sector meshes with center groove of rack-piston. Make sure side cover "O" ring is in place before pushing side cover down on gear housing.
15. Install side cover screws and lockwashers.
16. Hold adjuster screw with Allen wrench and install new adjuster lock nut halfway on adjuster screw.
17. Install rack-piston end plug in rack-piston and tighten plug to 75–80 ft. lbs.
18. Lubricate new housing end plug "O" ring seal with power steering fluid and install in gear housing.
19. Insert end plug into gear housing and seat against "O" seal. Slight tapping with

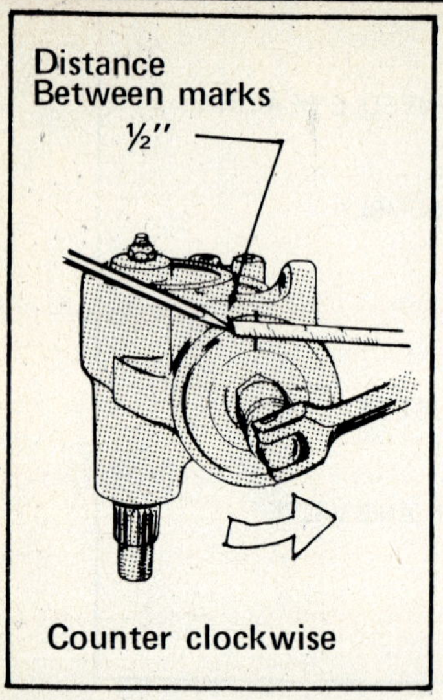

Fig. S43 Thrust bearing preload adjustment

Models	Year	Thrust Bearing Preload Inch Lbs.	Over Center Preload Inch Lbs.	Total Steering Gear Preload Inch Lbs.
Caballero, El Camino & Sprint	1974–80	3–4④	5②③	14
	1981–82	3–4④	4–5②③	—
Ranchero	1974–79	4–8	9	18
Lt. Duty Trucks	1974–76	½–2①	3–6	14
	1977–80	½–2①	5	14
	1981–82	3–4④	4–5③⑤	—

①—In excess of ball drag.
②—In excess of ball drag and thrust bearing preload.
③—Except for new gears. New gears are torqued to 6–10 inch lbs., see note 2 or 5.
④—In excess of valve assembly and seal drag.
⑤—In excess of combined thrust bearing preload.

Fig. S45 On-bench over-center preload specification chart

a mallet may be necessary to seat end plug properly.
20. Snap end plug retainer ring into place with fingers. Slight tapping may be required to bottom ring in housing securely. Install coupling flange.

Model 605 Steering Gear Assembly

1. If pitman shaft needle bearing was removed, install with lettered edge facing outward and flush with bottom of counter bore with a suitable tool, Fig. S41.
2. Install pitman shaft oil seal, washer and snap ring, Fig. S26.
3. If removed, install stub shaft needle bearing and upper thrust bearing, Fig. S42.
4. Apply anhydrous calcium grease to stub shaft seal, then install shaft seal, dust seal and snap ring, Fig. S24.
5. Install rack piston and valve assembly, then lower thrust bearing, "O" ring seal, adjuster plug and lock nut, Fig. S23.

NOTE: See procedures for thrust bearing preload adjustment before tightening lock nut.

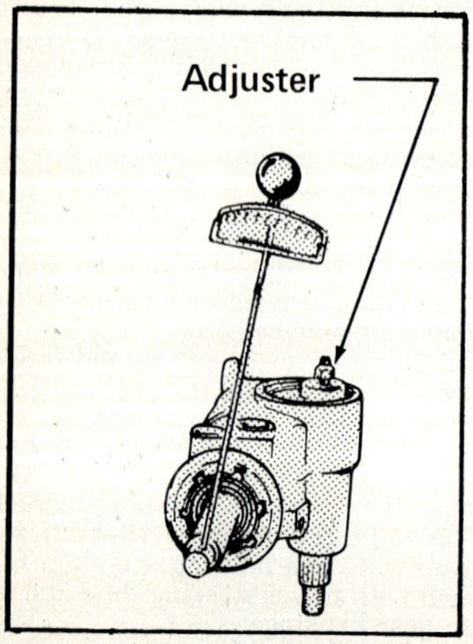

Fig. S44 Over-center preload adjustment

6. Install side cover spring and "O" ring seal.
7. Install pitman shaft gear so center tooth of gear meshes with center groove of rack piston, then depress side cover and install snap ring.

NOTE: Open end of snap ring should be ½ inch from access hole.

Model 605 Bench Adjustments

Thrust Bearing Preload
1. Tighten adjuster plug until bottomed and torque to 30 ft. lbs.
2. Draw a line on gear housing and plug, and a second line ½ inch counter clockwise on housing only, Fig. S43.
3. Turn adjuster back until mark on adjuster aligns with second mark on housing, then tighten lock nut.

Over-Center Preload
1. Back off preload adjuster until it stops, then turn in one turn.

NOTE: Preload adjuster has left hand thread.

2. Rotate stub shaft from stop to stop, counting number of turns, then back exactly ½ number of turns to locate center of travel.
3. Using an inch pound torque wrench, measure torque required to rotate stub shaft. Then turn adjuster until rotating torque is increased 6 to 10 inch lbs. Torque adjuster lock nut to 20 ft. lbs., Fig. S44.

Model 800-808 Bench Adjustments

Thrust Bearing Preload
1. Tighten adjuster plug until bottomed and torque to 20 ft. lbs.
2. Draw a line on gear housing and adjuster plug, and a second line ½ inch counter clockwise on housing only, Fig. S43.
3. Turn adjuster plug counterclockwise until mark on adjuster aligns with second mark on housing. Tighten locknut while holding plug securely to prevent rotation.

Over-Center Preload
1. Rotate stub shaft from stop to stop, counting number of turns, then turn back exactly ½ number of turns to locate center of travel.
2. Measure combined ball and thrust bearing preload using an inch pound torque wrench to rotate stub shaft through the over center position.
3. Referring to Fig. S45, tighten adjusting screw while rotating stub shaft with torque wrench until torque reading obtained is specified inch pounds greater than combined ball and thrust bearing preload, Fig. S46. Note that total steering gear preload must not exceed specifications.

Trouble Shooting

Hard Steering
1. Frozen steering shaft bearings.
2. Lower coupling flange rubbing against adjuster.
3. Steering adjustment tight.
4. Low oil level in pump.
5. Loose pump belt.
6. Insufficient oil pressure.
7. Improper tire inflation.
8. Flow control valve sticking.
9. Pump leaking internally.
10. Steering gear leaking internally.

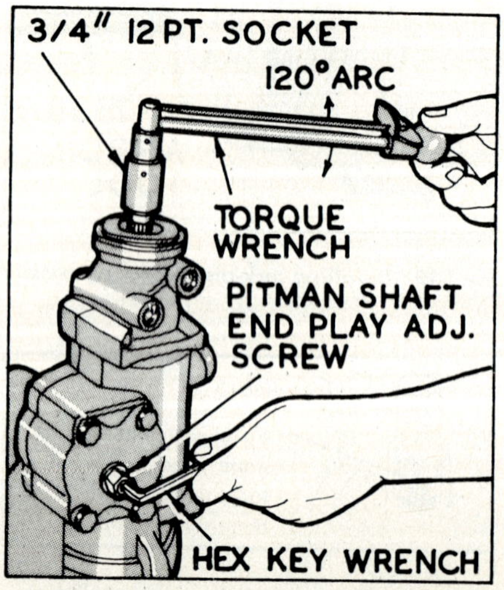

Fig. S46 Adjusting over-center preload

Poor Return of Steering
1. Frozen steering shaft bearings.
2. Lower coupling flange rubbing against adjuster.
3. Tires not inflated properly.
4. Incorrect caster and toe-in.
5. Tight steering linkage.
6. Steering gear or column misalignment.
7. Tight suspension ball joints.
8. Steering adjustment tight.
9. Thrust bearing adjustment tight.
10. Tight sector-to-rack piston adjustment.
11. Rack piston nut and worm preload too tight.
12. Sticky valve spool.
13. Lack of lubricant in suspension joints and steering linkage.
14. Steering wheel rubbing against directional signal housing.
15. Rubber spacer in shift tube is rubbing steering shaft.

Car Leads to One Side
1. Front end misalignment.
2. Imbalanced or badly worn valve.
3. Improper tire inflation.
4. Radial tire pull.

Momentary Increase In Effort When Turning Wheel Fast
1. Low oil level in pump.
2. Pump belt slipping.
3. High internal leakage.
4. Insufficient pump pressure.

External Oil Leaks
1. Loose hose connections.
2. Damaged hose.
3. Side cover O-ring seal.
4. Pitman shaft seals.
5. Housing end plug seal.
6. Adjuster plug seals.
7. Torsion bar seal.
8. Defective housing.

Steering Gear Noise
1. A rattle or chuckle noise caused by loose over-center adjustment.
2. A hissing sound caused by gear being loose on frame.

Excessive Wheel Kickback or Loose Steering
1. Lash in steering linkage.
2. Air in system.
3. Excessive lash between pitman shaft sector and rack piston.
4. Loose thrust bearing adjustment.
5. Ball nut and worm preload.
6. Steering gear loose on frame.
7. Steering linkage worn.
8. Improper front wheel bearing adjustment.
9. Poppet valve worn.

Wheel Surges or Jerks
1. Loose pump belt.
2. Fluid level low.
3. Steering linkage hitting engine oil pan at full turn.
4. Sticky flow control valve.

Hard Steering When Parking
1. Loose pump belt.
2. Low oil level in reservoir.
3. Lack of lubrication in linkage or front suspension.
4. Tires not properly inflated.
5. Insufficient oil pressure.
6. Low oil pressure due to restriction in hoses.
7. Low oil pressure due to worn piston ring or scored housing bore.
8. Pressure loss due to leakage at valve rings, valve body-to-worm seal or rack piston end plug seal.
9. Pressure loss due to loose fit of spool in valve body or leaky valve body.

Valve Squawk
1. Cut or worn dampener ring on valve spool.
2. Loose or worn rotary valve parts.
3. Loose belt.
4. Glazed belt.
5. Flow control valve sticking or binding.

No Effort Required to Turn
1. Broken torsion bar.

INTEGRAL DUAL PISTON GEAR

Description

This power steering gear is a recirculating ball type, Fig. S47, and incorporates a dual cylinder system in which a double geared sector shaft is actuated by two rack-pistons moving in opposite directions. The primary rack-piston contains the worm gear assembly.

The open center rotary type four-way control valve directs oil to either side of the rack-pistons. The valve is operated by a torsion bar, attached to the valve body at one end and to the valve spool through the stub shaft at the other. When the torsion bar twists, the valve spool is displaced in relation to the valve body, thereby operating the valve.

Adjustments

In-vehicle adjustments of this power steering gear is not recommended since the hydraulic fluid may cause a variable in the adjustment. For proper adjustment, remove steering gear from vehicle, drain fluid and mount in a suitable vise.

Thrust Bearing Adjustment
1. Remove adjuster plug locknut and rotate adjuster plug clockwise until plug and thrust bearing are firmly bottomed.
2. Mark gear housing in-line with a hole in adjuster plug, Fig. S48. Measure 3/16 to 1/4 inch counter-clockwise and place a second mark on gear housing.
3. Rotate adjuster plug counter-clockwise until hole in plug aligns with second mark.
4. Install and tighten adjuster plug lock-nut while maintaining adjuster plug position.
5. With an inch pound torque wrench, rotate stub shaft to right stop, then back 1/4 turn, noting rotating torque. The reading should be between 4 and 10 in. lbs., if not, repeat procedure.

Over-Center Adjustment
1. With steering gear centered, loosen sector shaft adjuster screw locknut, tighten adjuster screw and re-tighten locknut.
2. Measure over-center torque at stub shaft by rotating shaft with an inch pound torque wrench through a 180 degree arc either side of center and note torque reading when passing over-center, Fig. S49.
3. Adjust over-center preload until torque reading is 4–8 in. lbs. greater than reading obtained in above step on new gear assemblies or 4–5 in. lbs. greater on used gear assemblies.
4. Total over-center preload on new gears should not exceed 18 in. lbs. or 14 in. lbs. on used gears.

NOTE: Total over-center preload includes thrust bearing, over-center and drag torques.

Steering Gear Disassembly

1. Position end plug retaining ring so gap is approximately 1/4 inch from hole in housing on both primary and secondary cylinders. Then, using a small punch inserted through housing hole to dislodge retaining ring, pry ring from housing groove.
2. Rotate stub shaft counter-clockwise until end cover is forced from primary cylinder.

NOTE: Do not rotate stub shaft any farther than necessary for cover removal since the rack-piston balls may drop into the piston chamber.

3. Remove end cover "O" ring from housing groove.
4. Remove rack-piston end plug by first loosening plug using a brass drift and a hammer, then rotate end plug out of rack-piston with a 1/2 inch square drive.
5. Remove sector shaft adjuster screw locknut, side cover bolts and, using a 1/4 inch Allen wrench, rotate adjuster screw clockwise until side cover separates from sector shaft.
6. Rotate stub shaft counter-clockwise until sector shaft teeth are disengaged from rack-piston.

NOTE: If the secondary piston end cover is free, proceed to step 8.

7. Index sector shaft so it clears rack-piston completely, Fig. S50. Then, rotate stub shaft clockwise until rack-piston bottoms in gear housing and slowly turn stub shaft counter-clockwise to engage sector shaft end tooth with center tooth space of primary rack, Fig. S51. Continue turning stub shaft until the leading tooth of sector shaft forces the secondary rack against end cover, in turn pushing the end cover from housing.
8. Remove secondary cylinder rack-piston. Do not remove end plug unless plug is to be replaced.
9. Rotate stub shaft clockwise until sector shaft teeth are disengaged from rack-piston, then index sector shaft so it clears rack completely, allowing free movement of rack.
10. Insert ball retaining tool into rack-piston bore with tool pilot seated into end of

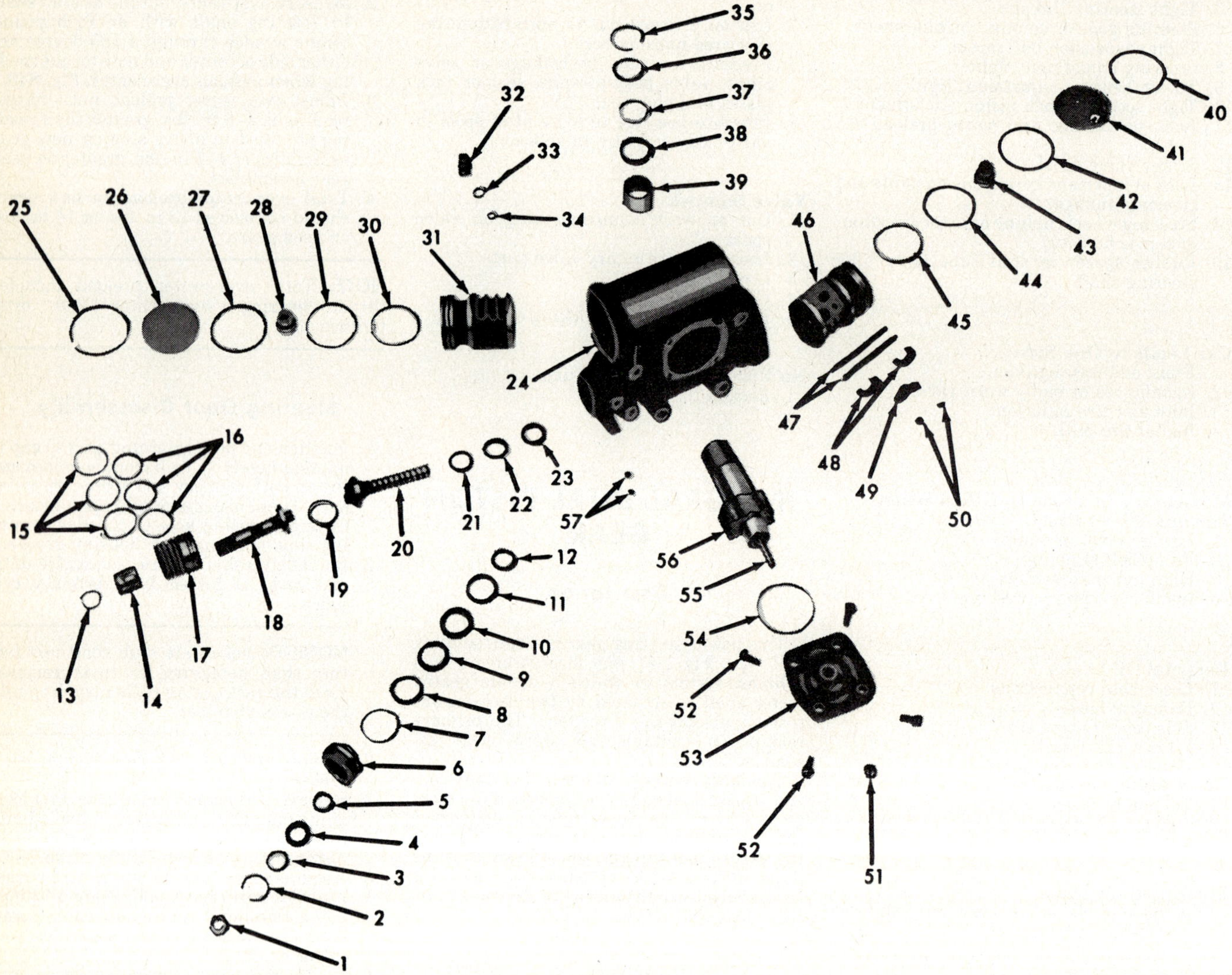

1. Locknut
2. Retaining Ring
3. Back-Up Washer
4. Stub Shaft Seal
5. Needle Bearing
6. Adjuster Plug
7. "O" Ring
8. Thrust Race (Upper)
9. Thrust Bearing
10. Thrust Race
11. Spacer
12. Retainer
13. Dampener "O" Ring
14. Valve Spool
15. Teflon "O" Rings
16. Back-Up "O" Rings
17. Valve Body
18. Stub Shaft
19. Cap to Body "O" Ring
20. Steering Worm
21. Thrust Bearing Race
22. Thrust Bearing
23. Thrust Bearing Race
24. Housing
25. Retaining Ring
26. Housing End Plug
27. End Plug "O" Ring
28. Rack Piston End Plug
29. Teflon "O" Ring
30. Back-Up "O" Ring
31. Rack Piston
32. Relief Valve
33. "O" Ring
34. "O" Ring
35. Retaining Ring
36. Dust Seal
37. Back-Up Washer
38. Oil Seal
39. Needle Bearing
40. Retaining Ring
41. Housing End Plug
42. End Plug "O" Ring
43. Rack Piston End Plug
44. Teflon "O" Ring
45. Back-Up "O" Ring
46. Rack Piston
47. Balls
48. Ball Return Guides
49. Clamp
50. Lockwasher & Screw Assemblies
51. Lock-Nut
52. Side Cover Bolts
53. Side Cover
54. Side Cover "O" Ring
55. Preload Adjuster Screw
56. Sector Shaft
57. Connectors

Fig. S47 Integral dual piston power steering gear disassembled

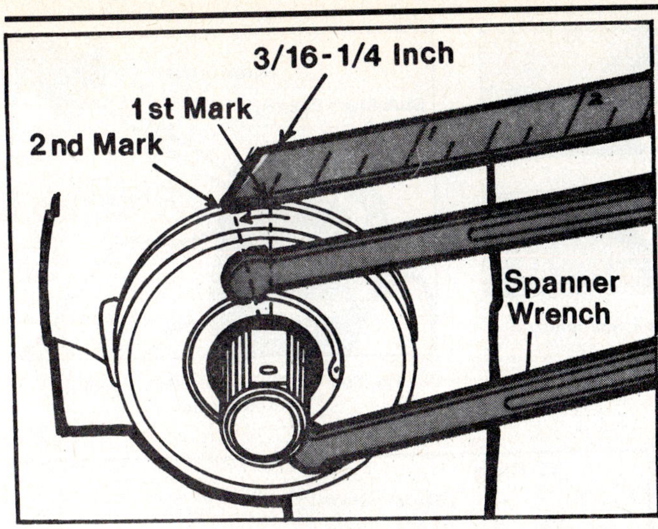

Fig. S48 Thrust bearing adjustment

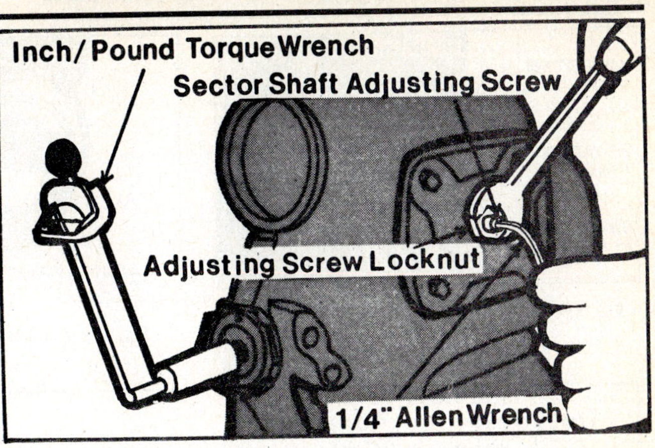

Fig. S49 Over-center adjustment

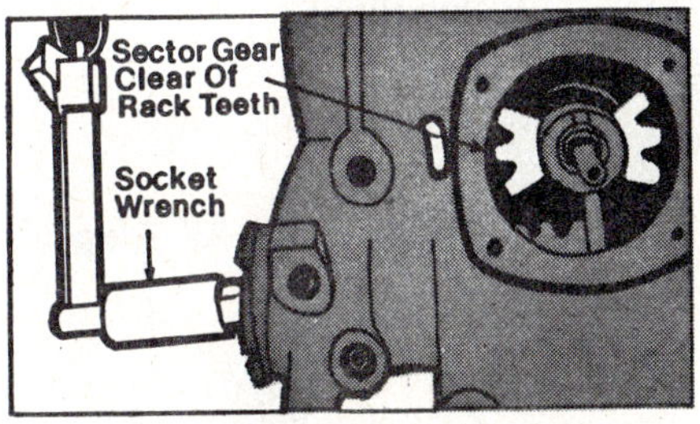

Fig. S50 Indexing sector shaft to clear rack teeth

Fig. S51 Sector shaft end tooth aligned with rack center space

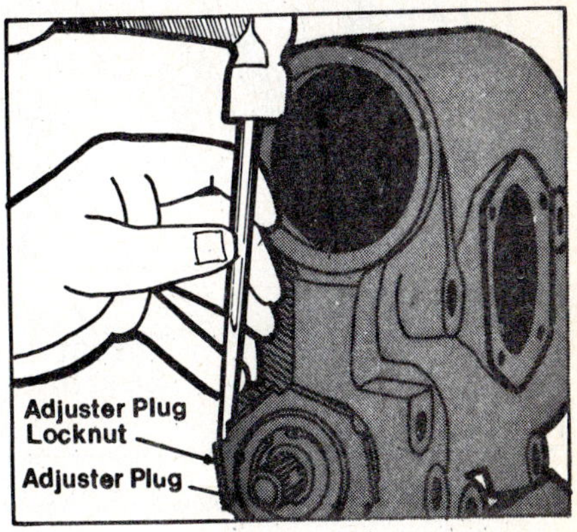

Fig. S52 Adjuster plug locknut removal

worm. Rotate stub shaft counter-clockwise, holding tool firmly against worm, to force rack-piston onto tool. Remove rack-piston and the ball retaining tool from gear housing.

11. Index sector shaft teeth to clear housing, then remove sector shaft.
12. Remove adjuster plug locknut, Fig. S52, and adjuster plug, Fig. S53.
13. Remove valve and worm assembly, ensuring both races and thrust bearing are removed from gear housing, Fig. S54, then separate worm from valve, Fig. S55.

Sub-Assembly Service
Adjuster Plug Disassembly

1. If oil seal only is to be replaced, install adjuster plug into housing finger tight. Remove retaining snap ring, Fig. S56, back-up washer and pry oil seal from adjuster plug bore, Fig. S41.
2. If thrust bearing only is to be placed, pry thrust bearing retainer from plug bore, Fig. S58, and remove spacer, washer, bearing and second washer.
3. With thrust bearing assembly remove from plug, drive needle bearing, oil seal, and dust seal from adjuster plug, Fig. S59.

Valve & Stub Shaft Disassembly

1. Remove cap to wormshaft "O" ring, Fig. S60.
2. Hold valve assembly with stub shaft facing downward, then tap stub shaft lightly against work bench until cap is free from valve body approximately 1/4 inch, Fig. S61.

NOTE: If shaft assembly is pulled out too far, the valve spool may become cocked and bind in the valve body.

---

3. Disengage shaft pin, Fig. S62, and remove spool by rotating it while removing, Fig. S63. If binding occurs, realign valve spool, then remove.
4. Remove damper "O" ring.

NOTE: The teflon rings generally do not require replacement unless severely worn or damaged. However, when replacing the teflon rings, replace the "O" rings also.

Primary Rack-Piston Disassembly

1. Remove ball return guide clamp screws, then the guide, arbor tool and recirculating balls.

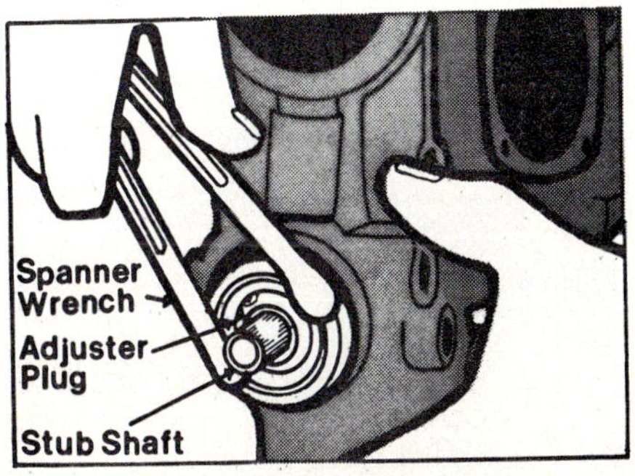

Fig. S53 Adjuster plug removal

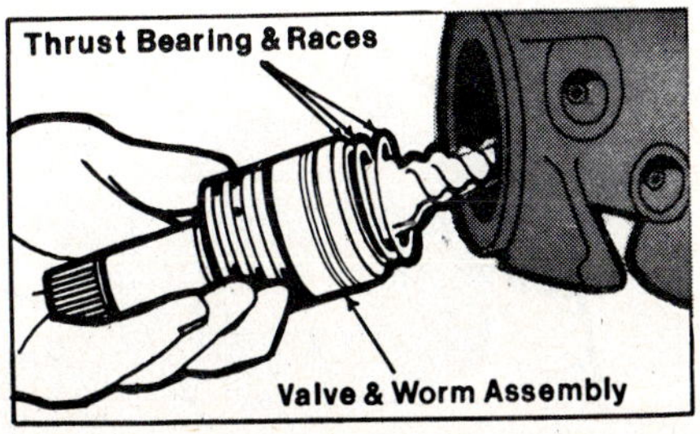

Fig. S54 Worm & valve assembly removal

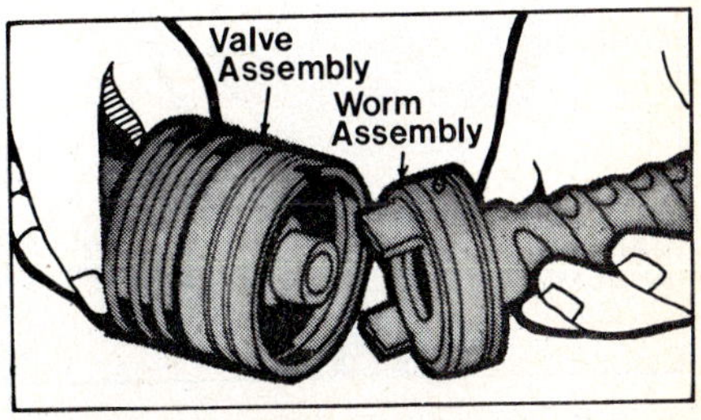

Fig. S55 Separating worm from valve

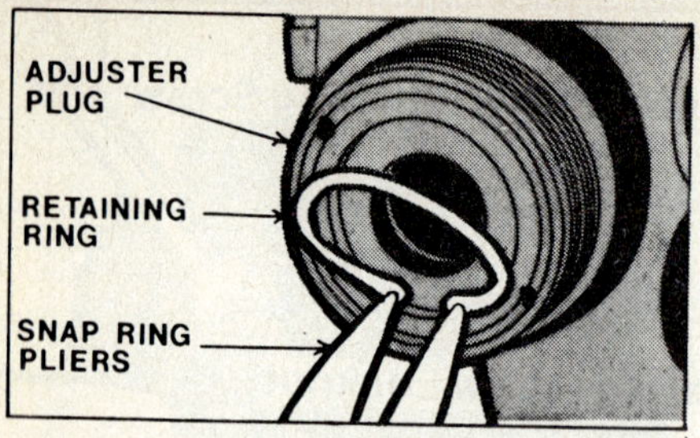

Fig. S56 Adjuster plug snap ring removal

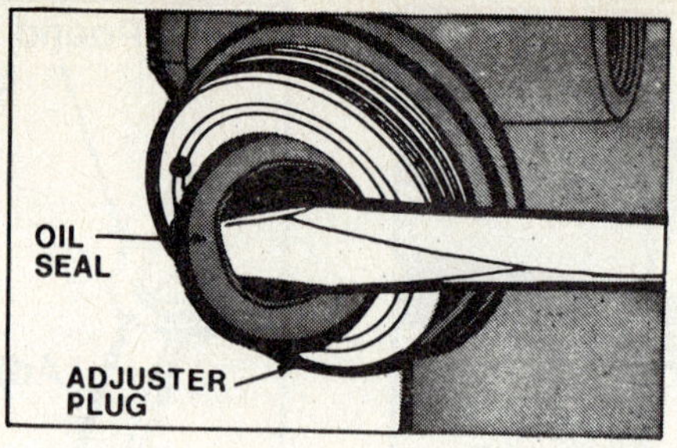

Fig. S57 Adjuster plug oil seal removal

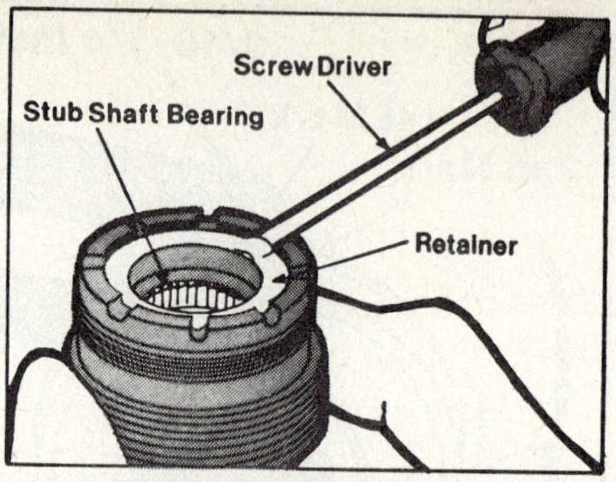

Fig. S58 Adjuster plug thrust bearing retainer removal

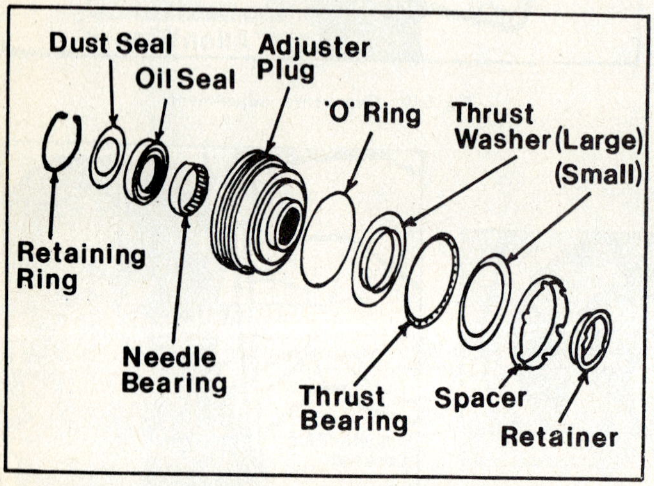

Fig. S59 Adjuster plug disassembled

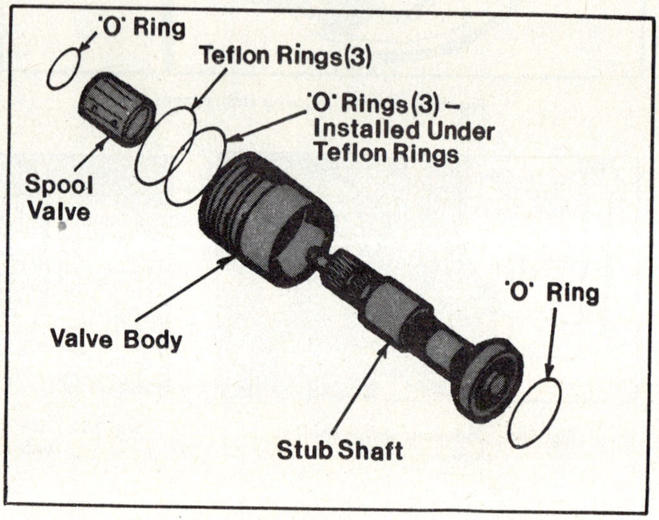

Fig. S60 Valve & stub shaft assembly

Hose Connector Replacement

NOTE: This procedure may be performed on or off the vehicle.

1. Using a 5/16 × 18 tap, cut threads into connector seat. Only cut two or three threads to prevent check valve damage.
2. Using a 5/15 × 18 bolt, nut and flat washer, thread the bolt into the threaded connector. Hold the bolt stationary and turn the nut to bear against the flat washer. The washer will lodge against the gear housing, causing the bolt to move out, pulling the connector seat from the housing.
3. Clean gear housing and ensure no metal chips or dirt remain.
4. Remove and discard poppet check valve and spring from pressure port.
5. Install new check valve spring in pressure port with large end facing downward. Ensure spring is properly seated in pressure port counterbore.
6. Install new check valve over spring with tangs facing downward and valve centered on small end of spring.
7. Using grease to hold connector seat on check valve, drive the seat into place.

Sector Shaft Needle Bearing & Seals Replacement

1. If seals only are to be replaced, remove snap ring retainer and outer steel washer. Pry dust seal from housing, remove inner steel washer, then pry out inner seal.
2. To remove needle bearing and seals, press bearing and seals from housing, using a suitable tool to apply pressure from cover side of housing.
3. Lubricate needle bearing, then using tool J-24654, press bearing into housing until tool bottoms on housing hub.
4. Lubricate inner double lipped seal and place seal on tool J-24656 with seal identification facing toward tool handle. Drive seal into bore until tool bottoms.

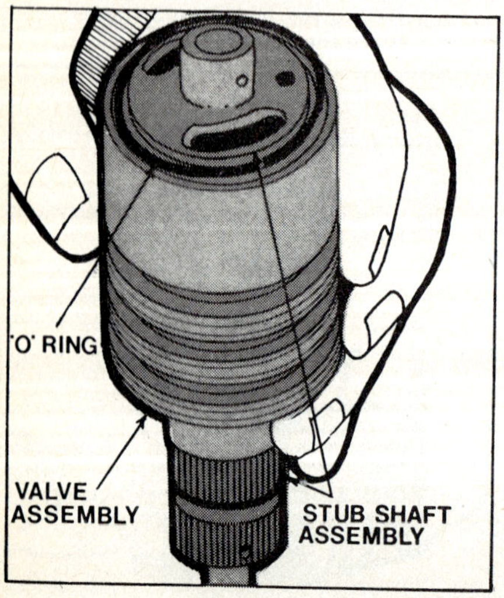

Fig. S61 Freeing valve body from stub shaft

Fig. S62 Shaft pin disengaged

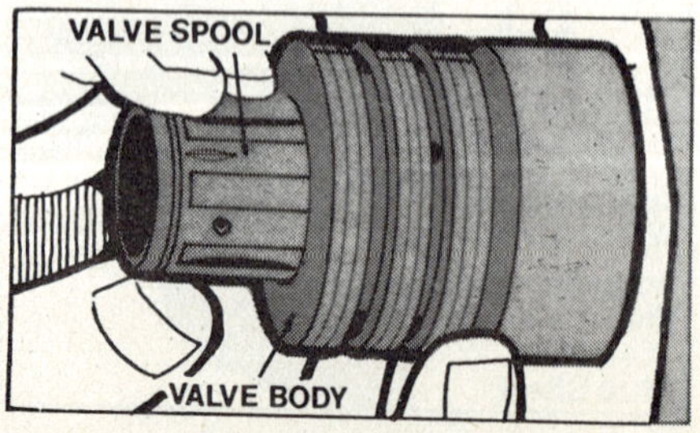

Fig. S63 Valve spool removal

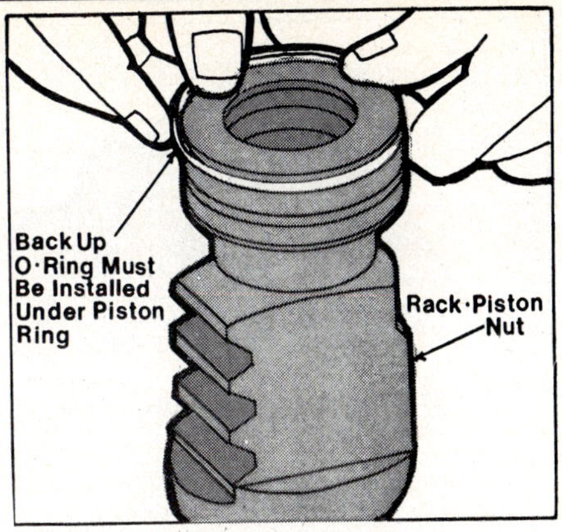

Fig. S64 Rack-piston teflon seal & "O" ring installation

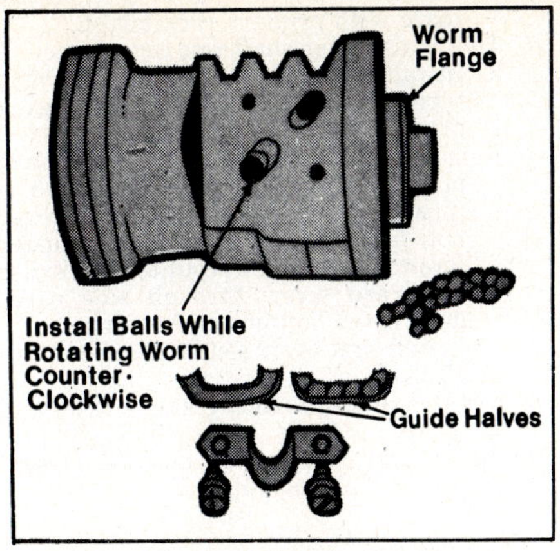

Fig. S65 Installing balls into rack-piston

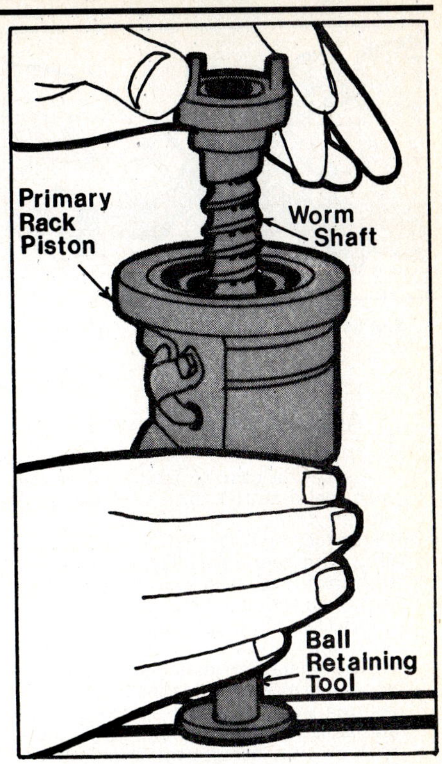

Fig. S66 Ball retaining tool installation

5. Install back-up washer into bore and place dust seal onto tool J-24656 with sealing lip facing toward tool handle, then press seal into bore until seal just clears snap ring groove.
6. Install snap ring retainer into bore groove, ensuring snap ring is properly seated in groove.

Primary Rack Piston Assembly

1. Lubricate and install new teflon seal and "O" ring into rack-piston groove, Fig. S64, if removed.
2. Slide wormshaft into rack-piston and rotate worm to align worm groove with ball return guide hole nearest piston ring.
3. Lubricate balls and while rotating wormshaft clockwise, load 28 balls into guide hole nearest piston ring, Fig. S65. Alternately install the black and silver balls since the black balls are .0005 inch smaller than the silver balls.
4. Place the remaining six balls into the ball return guide, using the same alternate sequence as in the previous step. Retain balls with a suitable grease and install the ball guide into rack-piston holes, then the ball guide clamp and torque retaining screws to 6 ft. lbs.
5. Remove wormshaft while installing the ball retaining tool, Fig. S66 as described under "Steering Gear Disassembly".

Secondary Rack-Piston Assembly

1. Lubricate and install new teflon ring and "O" ring into rack-piston groove, Fig. S64, if removed.
2. Slide wormshat into rack-piston and rotate worm to align worm groove with ball return guide hole nearest piston ring.
3. Lubricate balls and while rotating wormshaft clockwise, load 28 balls into guide hole nearest piston ring, Fig. S65. Alternately install the black and silver balls since the black balls are .0005 inch smaller than the silver balls.
4. Place the remaining six galls into the ball return guide, using the same alternate sequence as in the previous step. Retain balls with a suitable grease and install the ball guide into rack-piston holes, then the ball guide clamp and torque retaining screws to 6 ft. lbs.
5. Remove wormshaft while installing the ball retaining tool, Fig. S66 as described under "Steering Gear Disassembly."

Valve & Stub Shaft Assembly

1. Lubricate and install new valve body teflon rings and "O" rings into valve body grooves, if removed.
2. Lubricate and install valve spool damper "O" ring over valve spool.
3. Lubricate valve spool and body and rotate spool slightly while pushing spool into valve body, until shaft pin hole is visible from opposite end.
4. Slide stub shaft into spool until shaft pin can be inserted into valve spool hole.
5. Align notch in shaft cap with pin in valve body, Fig. S67, then press valve spool and shaft assembly into valve body.

NOTE: The valve body pin must mate with the cap notch prior to valve body installation.

---

6. Install new "O" ring in valve body shaft cap end.

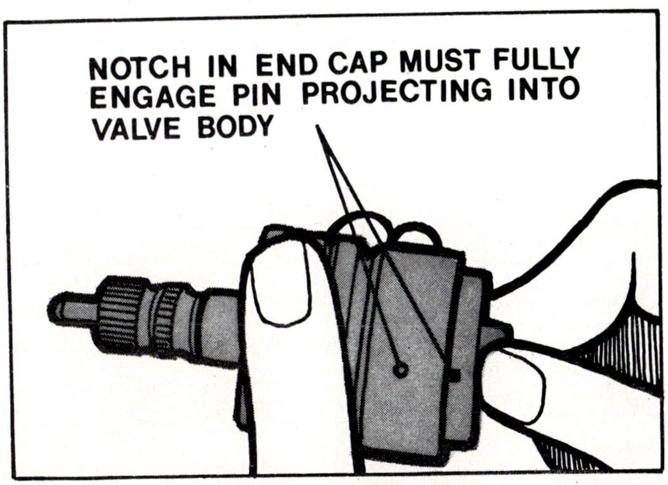

Fig. S67 Notch & pin alignment

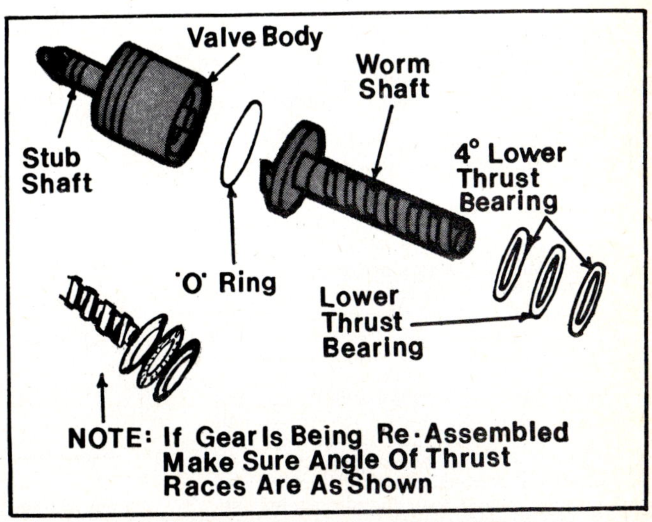

Fig. S68 Worm & valve body with thrust washers

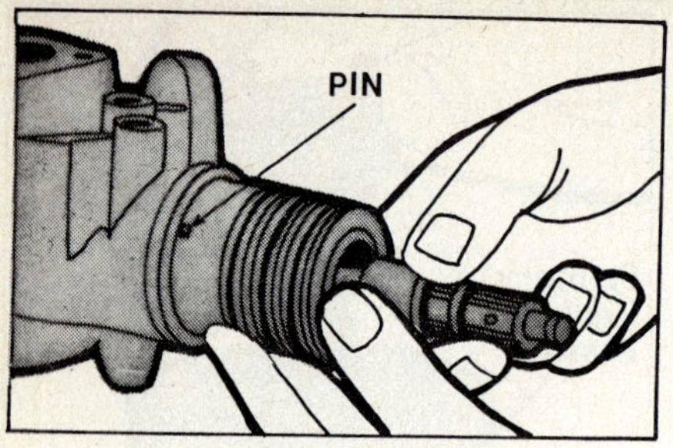

Fig. S69 Valve to wormshaft alignment

Adjuster Plug Assembly

1. Press needle bearing into adjuster plug with tool J-8524-1 until tool bottoms in plug.
2. Using tool J-85241, seat new oil seal into plug with sealing lip facing away from tool. Install the dust seal using the same tool, then the snap ring retainer.
3. Lubricate and install new adjuster plug "O" ring in plug groove. Place large thrust washer with flat side facing toward plug on plug hub, then install thrust bearing, small thrust washer with flange side facing away from plug and the spacer with grooves facing away from bearing.
4. Install bearing retainer by tapping lightly on flat surface of retainer.

NOTE: Projections must not extend beyond spacer when seating retainer. Also, the spacer must be free to rotate.

Steering Gear Reassembly

1. Lubricate worm, thrust bearing and washers with power steering fluid.
2. Install thrust washer, bearing and second thrust washer over end of worm, Fig. S68. Note position of conical thrust races, if equipped.
3. Lubricate valve body, teflon rings and new cap to body "O" ring, then install "O" ring in valve body, seated against lower shaft cap.
4. Align narrow notch in valve body with pin in worm and install valve and shaft assembly into gear housing, Fig. S69. The valve body is installed properly when housing oil return hole is completely uncovered, Fig. S70.
5. Seat adjuster plug against valve body in housing and adjust thrust bearing preload as outlined under "Adjustments".
6. To install rack-pistons:
   a. Lubricate rack-piston teflon ring with power steering fluid and with ball retaining tool in place, push rack-piston into housing with rack teeth facing toward sector shaft opening until the tool bottoms against worm.
   b. Using a 3/4 inch 12 point socket, rotate stub shaft clockwise to thread rack-piston onto worm with ball retaining tool held against worm. Remove ball retaining tool and center the rack teeth in sector shaft opening.
   c. Install secondary rack-piston in opposite cylinder with end plug facing outward and rack teeth facing toward sector shaft opening. Align center tooth space with primary rack-piston.
7. To install sector shaft and side cover:
   a. Install sector shaft into housing with tapered teeth of sector shaft engaging primary rack-piston teeth.
   b. Install new side cover "O" ring, then place side cover on housing. Insert a 1/4 inch allen wrench through side cover hole and into sector shaft adjuster screw. Turn screw counter-clockwise to thread screw through side cover until cover bottoms on housing.
   c. Install and torque side cover bolts to 35 ft. lbs.
8. Install and torque rack-piston end plug to 75 ft. lbs.
9. Install end covers with new "O" rings, then the retaining rings.

SAGINAW RACK & PINION

Description

This power steering gear assembly, Fig. S71, incorporates a rack piston which is permanently attached to the rack and sealed in the gear housing by a piston ring. One oil line is attached to the gear housing on each side of the rack piston and ring assembly. The rack housing at the pinion end is sealed by an inner rack seal, and at the piston end by an inner and outer cylinder bulkhead, a bulkhead seal and an "O" ring, all held in place by a bulkhead retaining ring.

Operation

A rotary-type valve body is used to control the hydraulic steering assist. Fluid under pressure is directed to the gear housing and into the valve body. The valve body then directs fluid to the power cylinder.

A spool valve, connected to the stub shaft rotates within the valve body. Fluid directional passages, machined into the spool valve, are aligned with fluid passages in the valve body as the spool valve rotates. Fluid is directed through these passages, into either side of the power cylinder through the externally mounted oil lines.

Steering Gear, Disassemble, Fig. S71

1. Mark position of tie rod end on tie rod, then loosen jam nut and remove tie rod end.
2. Remove jam nut, then using pliers to expand outer bellows clamp, remove clamp from bellows.

NOTE: An outer clamp is not used on 1983 vehicles. To prevent damage to the bellows when removing it, note the following. Slide the small end of the bellows towards the rack until the groove is exposed. Place a small rubber band in the groove, then slide the small end of the boot over the rubber band and off the threaded end of the tie rod.

3. Cut inner bellows clamp, then remove clamp from gear.
4. Mark breather tube location, then remove bellows from gear housing.

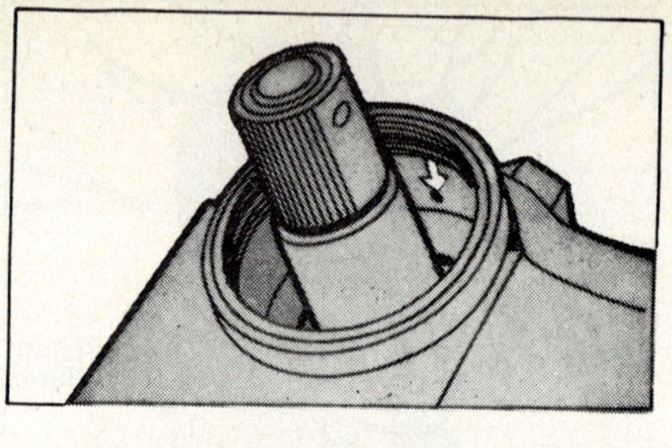

Fig. S70 Valve body properly seated

5. Remove shock dampener ring from tie rod housing and slide ring back on rack.
6. Using a wrench to prevent rack from turning, place a wrench on tie rod housing flats and turn housing counterclockwise until tie rod assembly separated from rack, Fig. S72.
7. Loosen adjuster plug lock nut, then turn adjuster plug counterclockwise to remove plug from housing. Remove spring and rack bearing from housing bore.
8. Remove retaining ring from stub shaft.
9. Remove pinion shaft lock nut dust cover.
10. While holding stub shaft in place, remove lock nut from pinion shaft.

NOTE: If stub shaft is not held in place while removing lock nut, damage to pinion teeth may result.

11. If only stub shaft seals are to be replaced, using a press, press on threaded end of stub shaft until shaft is flush with ball bearing assembly. Complete removal of valve and pinion assembly is not necessary to replace stub shaft seals.
12. To remove valve and pinion assembly, rotate stub shaft until rack is centered in housing, with pinion fully engaged. Mark position of stub shaft flat on housing for reassembly. Using a press, press on threaded end on pinion shaft until it is possible to remove valve and pinion shaft assembly.
13. If necessary, remove valve body seal rings.
14. Insert a suitable punch into access hole and remove bulkhead retaining ring.
15. If only bulkhead, bulkhead O-ring seal or bulkhead rack seal are to be replaced, remove cylinder lines and plug both cylinder line holes in cylinder using two M12 × 1.25 bolts, tighten bolts finger tight. Position drain pan under gear, then rotate stub shaft to move rack toward right stop to force bulkhead out of housing. If inner rack seal or piston ring are to be replaced the rack must be removed from the housing.
16. Remove rack from housing, then remove piston ring from rack.
17. Remove inner rack seal from housing using tool L-4627, Fig. S73.
18. Remove lower pinion bearing retaining ring, then using a suitable drift, tap lower pinion bearing from housing, Fig. S74.
19. Using a suitable drift, tap upper pinion bushing and seal from housing, Fig. S75.
20. Using a suitable flare wrench, loosen fittings and remove cylinder lines from housing. Also remove line fitting O-rings.
21. Mark breather tube location on housing for reassembly, then remove breather tube and grommet.

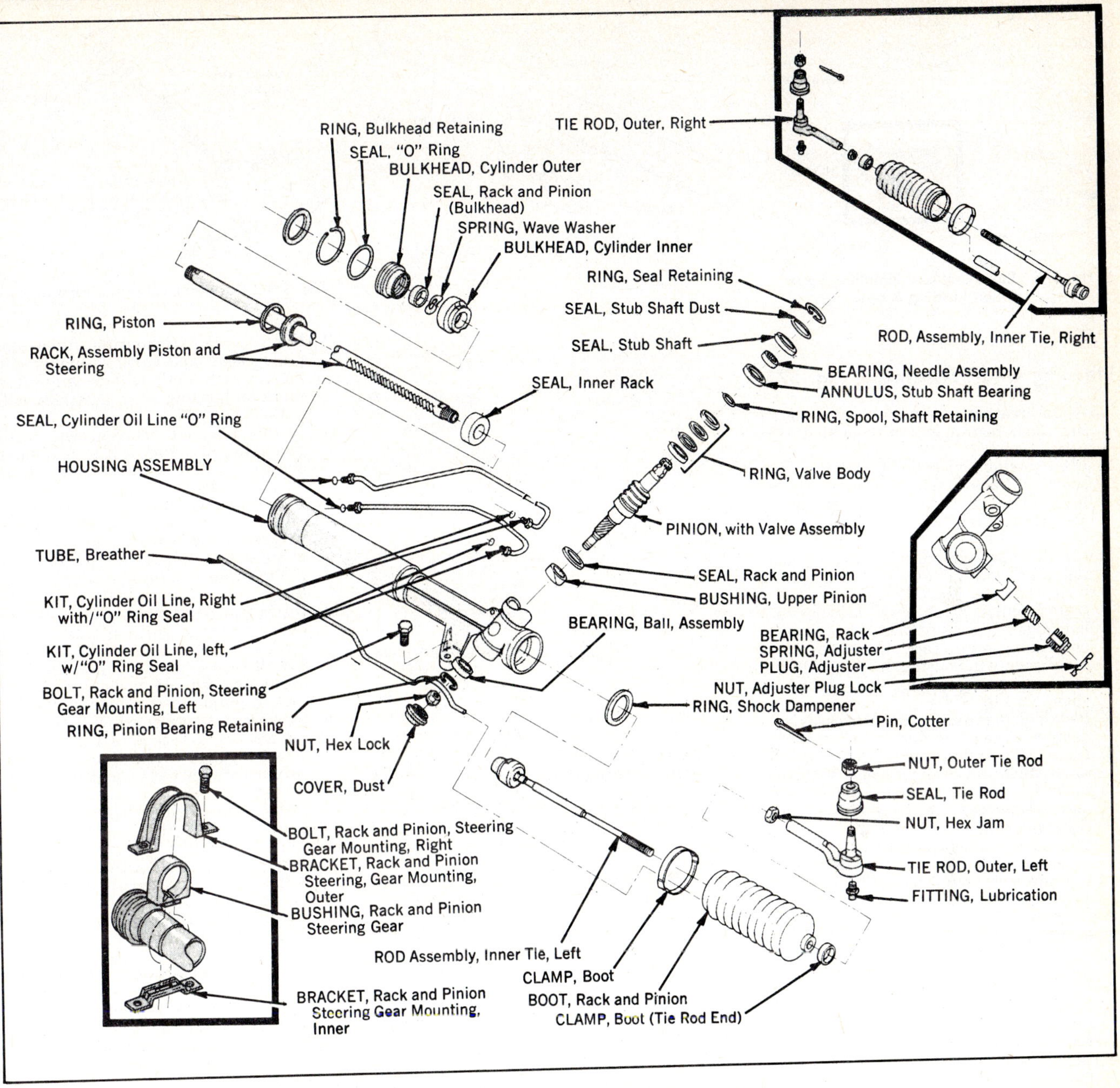

Fig. S71 Exploded view of Saginaw rack & pinion steering gear

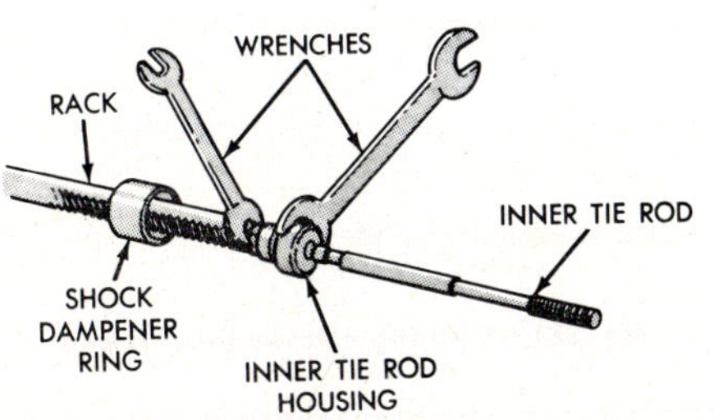

Fig. S72 Removing tie rod from rack assembly

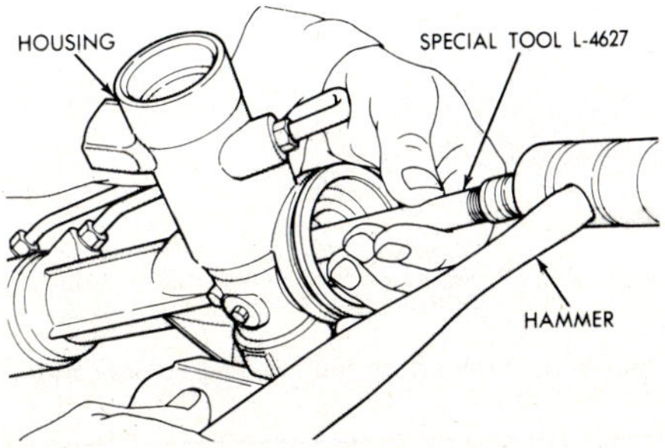

Fig. S73 Removing inner rack seal

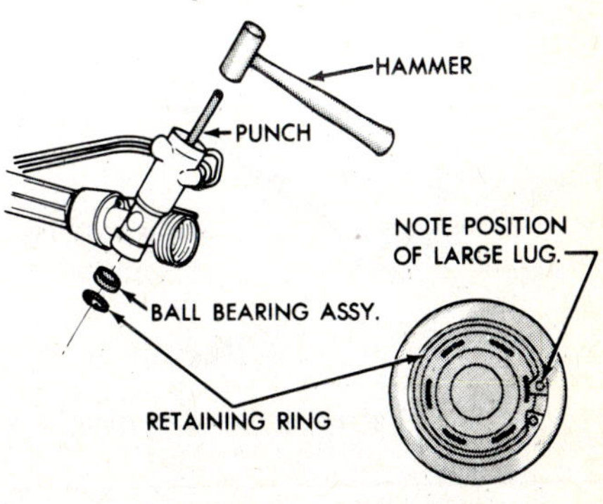

Fig. S74 Removing pinion lower bearing

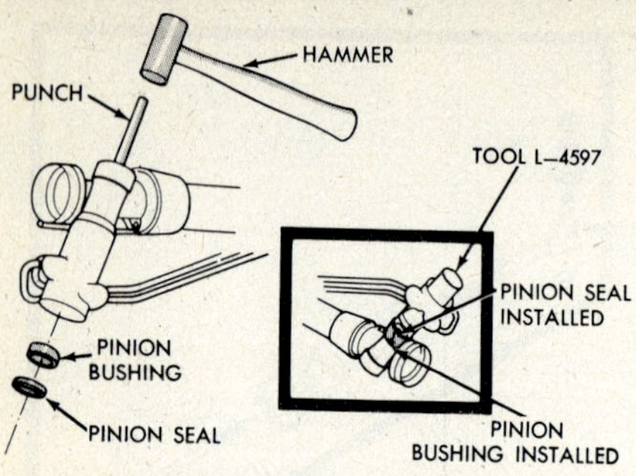

Fig. S75 Removing and installing upper pinion bushing & seal

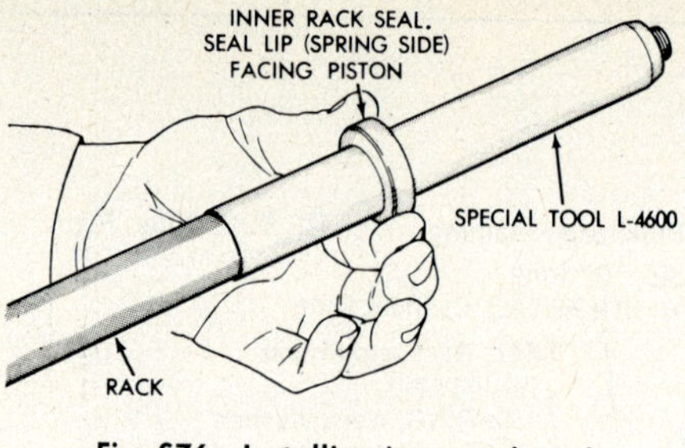

Fig. S76 Installing inner rack seal

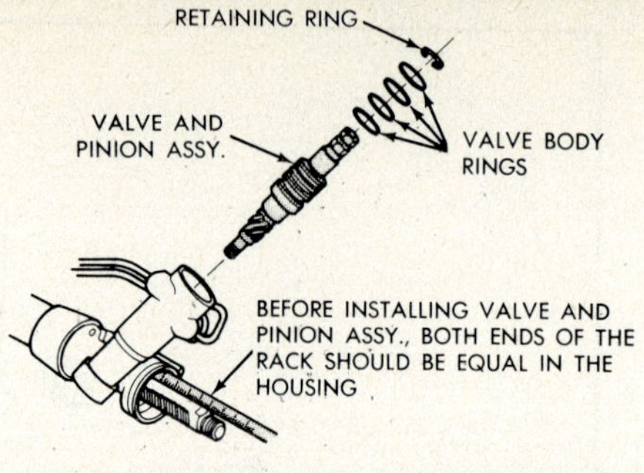

Fig. S77 Installing valve & pinion assembly

Steering Gear, Assemble, Fig. S71

1. Align breather tube with mark on housing, then install grommet so that tabs lock.
2. Place O-rings on cylinder lines and assemble cylinder lines to housing. Tighten fittings finger tight to ensure proper alignment, then torque fittings to 18 ft. lbs.
3. Install upper pinion bushing into housing, then install pinion seal with seal lip facing upward, Fig. S75.
4. Install pinion shaft lower bearing into housing, using a suitable socket to press on bearing outer race. Use care not to cock bearing in housing. Install retaining ring and position large lug of ring as shown in Fig. S74.
5. Install piston ring on rack. Use care not to damage piston ring during installation.
6. Place tool L-4600 around rack teeth. Slide inner rack seal over protector with seal lip facing piston, Fig. S76. Remove tool and bottom seal on rack piston. Slide retaining ring on rack and into seal, then slide rack and seal in housing.
7. With rack and seal positioned in housing, tap rack with a soft faced mallet to seat seal.
8. Coat seal with power steering fluid, then position seal protector over end of rack and install bulkhead. Use care not to damage seals when installing bulkhead.
9. Install bulkhead retaining ring. Ensure that open end of ring is approximately 1/2 in. from access hole. Also ensure that retaining ring is properly seated.
10. Install seal rings on valve body, if removed, Fig. S77.

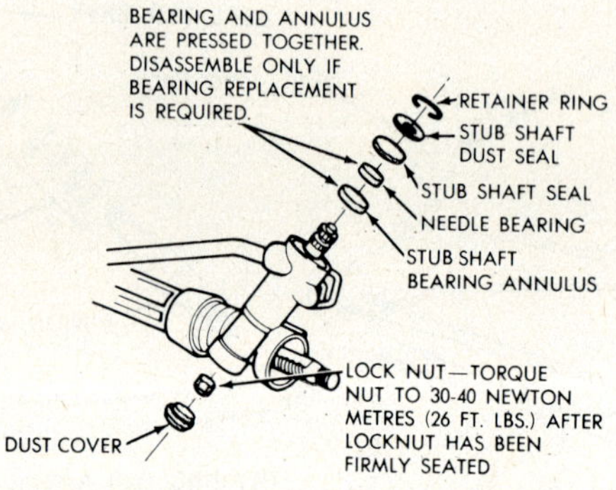

Fig. S78 Installing stub shaft bearing & seals

NOTE: If a condition of reduced power assist is encountered when the engine is started cold, it can be caused by improper rack and pinion valve rings. The proper rings are colored green. Any other color ring should be replaced by using Repair Package No. 5205297.

Wear marks in the housing valve bore are considered to be normal and do not necessitate replacement of the housing.

11. Center rack in housing, then install pinion shaft and valve assembly so that flat on stub shaft will be aligned with mark on housing when assembly is fully seated, Fig. S77.
12. Install stub shaft bearing, seal, dust seal and retaining ring. While holding stub shaft in position, torque lock nut to 26 ft. lbs., Fig. S78.
13. Install rack bearing, spring and adjuster plug. Turn adjuster plug clockwise until it bottoms, then back plug off 40 to 60 degrees. Install lock nut and torque to 50 ft. lbs. while holding adjuster plug in position.
14. Install tie rod assembly on rack. While holding rack in position with a wrench, torque tie rod housing to 70 ft. lbs., Fig. S79. If rack is not held in position when tightening tie rod housing, damage to rack or pinion teeth may result.
15. Support rack and housing, then stake tie rod housing to rack flats, Fig. S80. Using a feeler gauge check clearance between stakes and rack flats. Clearance must not exceed .010 in.
16. Install shock damper over tie rod housing.
17. Align breather tube with marks made during disassembly, then install bellows with over housing lip with hole in bellows aligned with breather tube. Position inner clamp on bellows, then secure clamp using tool No. C-4124, Fig. S81.
18. Position bellows outer clamp (if equipped), and jam nut on tie rod. Install tie rod end in the same position as removed. Torque jam nut to 50 ft. lbs. on 1982 vehicles, or 55 ft. lbs. on 1983 vehicles, then, on 1982 vehicles, install outer bellows clamp. Ensure bellows is not twisted.

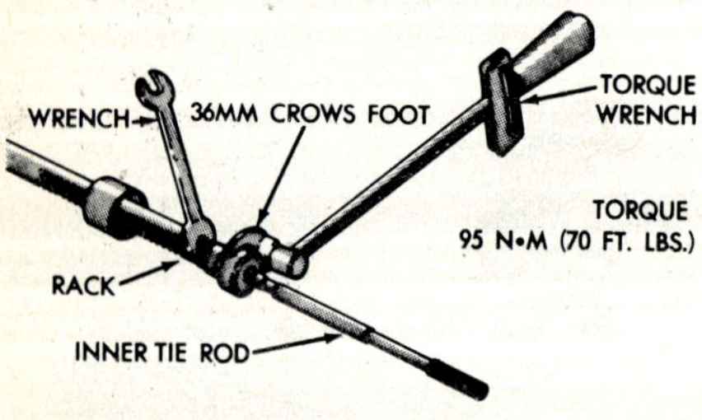

Fig. S79 Tightening tie rod housing

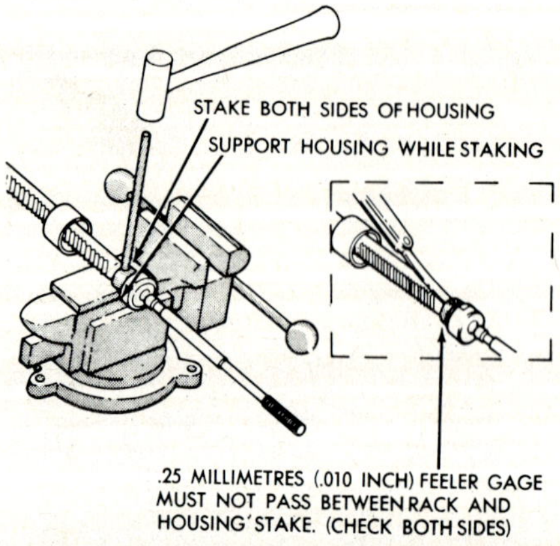

Fig. S80 Staking tie rod housing to rack flats

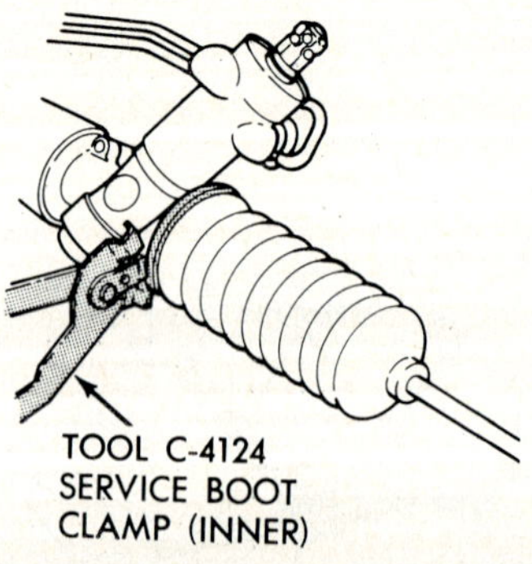

Fig. S81 Securing bellows inner clamp

Sheppard Integral Power Steering Section

DESCRIPTION

This system is of the integral type and incorporates the steering shaft, power cylinder and control valve into a single assembly, Fig. 1. The actuating shaft is geared to the control valve, housed within the piston. When the shaft is turned, the control valve moves axially, directing oil flow to either piston end, in turn forcing the piston through the cylinder. The piston rack rotates the pinion, output shaft and the Pitman arm.

DISASSEMBLY, FIG. 1

1. Remove pitman arm using a suitable puller.

NOTE: The pitman arm may be retained by two socket head retaining set screws, a self-locking nut or by a split nut and cap screw.

2. Scribe reference marks between bearing cap, cylinder head and housing for proper reassembly.

NOTE: The cylinder and bearing cap will fit either end of housing.

3. Remove pinion gear cover to housing attaching bolts.
4. Clean the exposed part of the output shaft to prevent bearing damage, then using a mallet, tap end of output shaft to loosen cover.
5. Slide output shaft pinion gear and cover assembly from housing, Fig. 2.
6. Pull output shaft from cover. Remove and discard cover "O" ring.
7. Loosen relief valve plunger jamb nuts and remove plungers from cylinder head and bearing cap. Remove and discard square ring seals or gaskets.

NOTE: Late production units may have slotted and recessed relief plungers. These plungers are removed or adjusted using a straight bladed screwdriver.

Fig. 2 Output shaft pinion gear removal

8. Remove bolts from bearing cap. Then rotate actuating shaft to free bearing cap from piston and housing.
9. While supporting bearing cap, rotate actuating shaft until shaft is free from piston, Fig. 3. Remove and discard gasket or seal ring. Some units use square ring seals in place of gaskets on cylinder head and bearing cap.
10. Remove cylinder head and discard gasket or seal ring.
11. Slide piston assembly from housing bore, Fig. 4.

NOTE: Early production units may utilize piston rings, Fig. 5. If so, remove and discard piston rings since the rings are no longer required. However, the 372/382 series steering gears with the step bore housing and piston flange will use the piston ring.

12. Pull locking pin from housing, then using a suitable spanner wrench, remove bearing retaining nut from cap.
13. Drive actuating shaft and bearing from bearing cap using a mallet or press.

NOTE: The actuating shaft and bearing are serviced as an assembly. Also, the actuating shaft may be of several types. The thread may be a machine cut thread or a rolled thread. The rolled thread may appear rough and flaky or have a split in the thread. This is a normal condition for a rolled thread.

14. The standard and short series bearing caps utilize three seals for high pressure sealing and actuating shaft protection. The high pressure seal is protected by a dirt and a salt seal. The grease cavity between the dirt and salt seals are filled with chassis lube to prevent the entry of dirt and salt. Flush the cavity by forcing grease through the cavity with a low pressure grease gun.
15. Press the three seals from the bearing cap by pressing the seals toward the inside of the bearing cap.

NOTE: The back-up washers will also be removed and should be retained for reinstallation with the new seals. Also, early production units use two seals in the bearing cap and the back-up washer is machined into the bearing cap. Use a screwdriver to pry the dirt seal from the bearing cap on these units, then using a suitable punch placed through the access holes, drive the high pressure seal from the bearing cap, Fig. 6.

16. On standard high pressure, short series and 372/382 series units, remove the pinion gear roll pin, then drill out the retaining pin. Press pinion gear from shaft.

NOTE: If the retaining pin cannot be drilled out, it can be sheared with approximately 10 tons of pressure. Drive half of the pin out of the pinion gear and drill out the remaining half from the shaft.

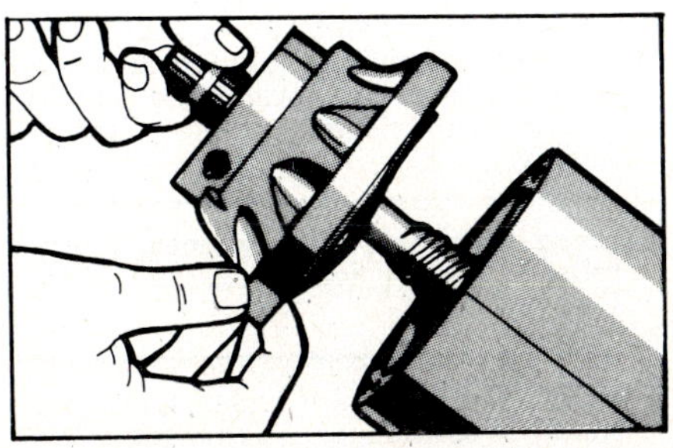

Fig. 3 Bearing cap & actuating shaft removal

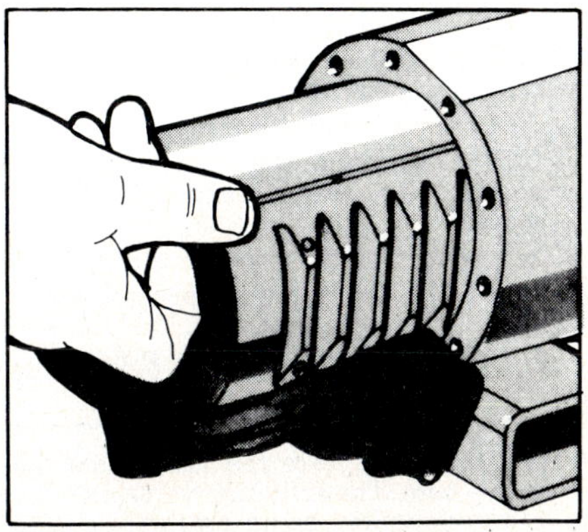

Fig. 4 Piston assembly removal

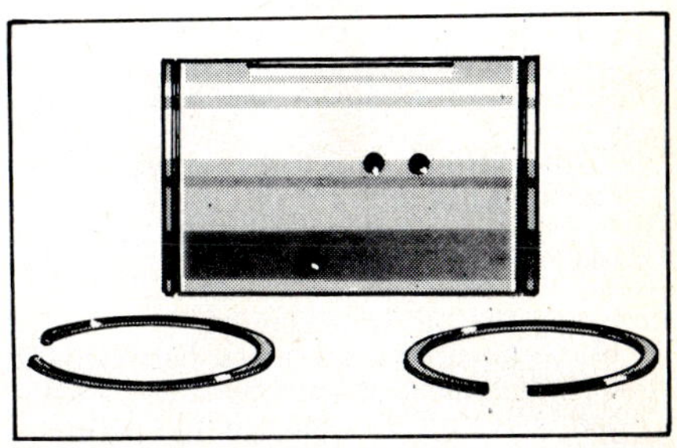

Fig. 5 Piston assembly, early production units

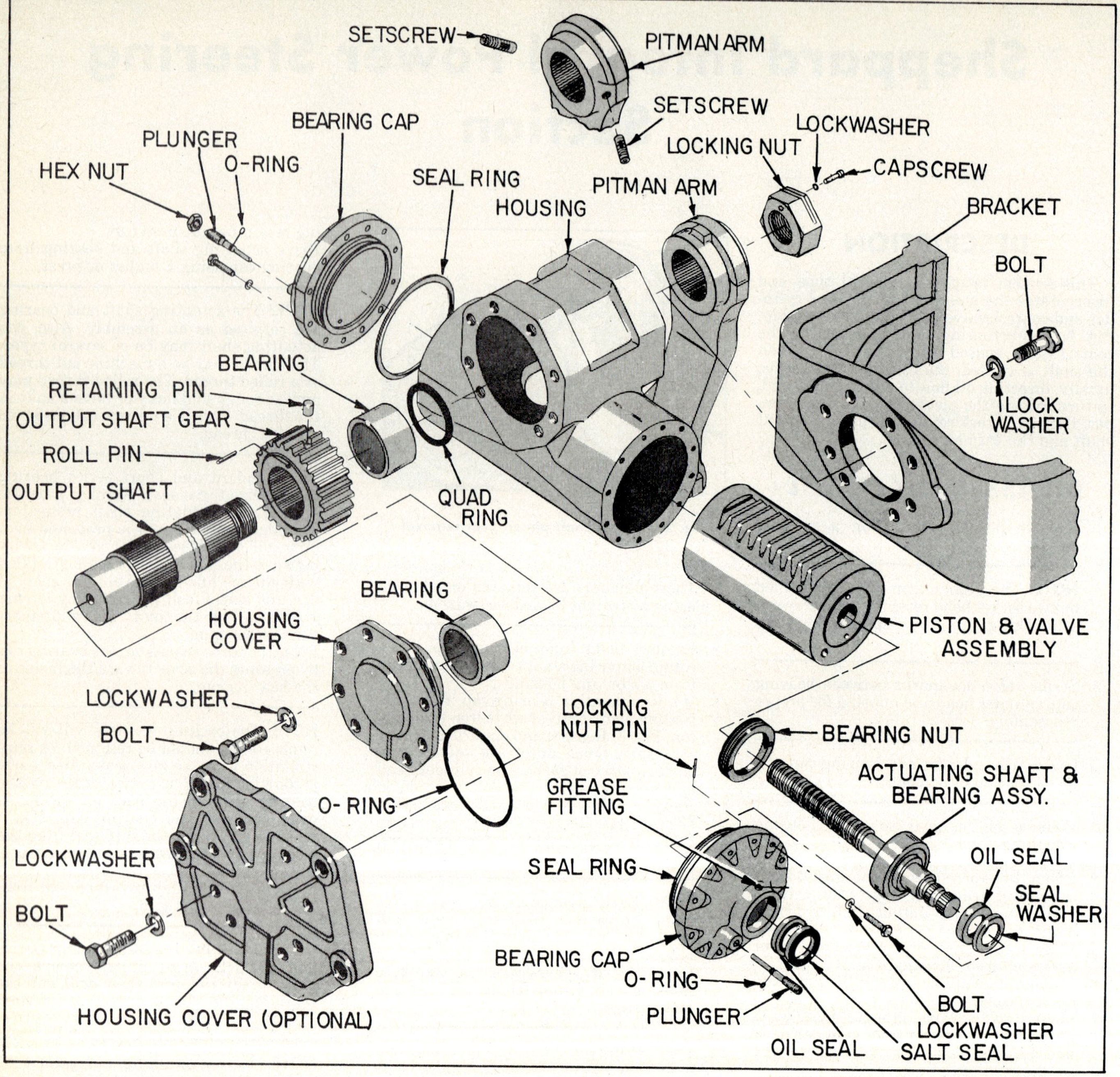

Fig. 1 Sheppard power steering gear, disassembled

17. On low pressure units, remove the two pinion gear cap screws with an allen wrench and rotate pinion gear retaining nut counter-clockwise to remove. Then, press pinion gear from shaft.
18. On all units, remove and discard Quad ring oil seal from the gear housing, Fig. 7.
19. With a suitable punch, drive bronze bushing from housing, Fig. 8.
20. Remove bronze bearing with a suitable tool.

Standard Piston, Disassembly

1. Make an alignment mark between the top adjustment nut and piston so the nut can be installed in the original position.
2. Remove pin securing adjustment nut to piston.
3. With a suitable spanner wrench, remove top adjustment nut from piston, Fig. 9.

NOTE: The bottom adjustment nut, located inside the piston, should not be removed.

4. Remove the top reversing spring from the nose of the actuating valve, Fig. 10, then pull the actuating valve from piston, Fig. 11.

NOTE: Do not force the valve from the piston bore. The actuating valve and piston are serviced as an assembly due to critical tolerances.

5. Remove actuating valve positioning pin from piston.

NOTE: The positioning pin on late production units incorporate an "O" ring oil seal. Also, the reversing springs are the only servicable components of the piston assembly.

6. Using a suitable allen wrench, remove relief ball seats, relief balls and spring, Fig. 12.

NOTE: The balls are under spring tension. Use caution when removing to prevent loss or personal injury.

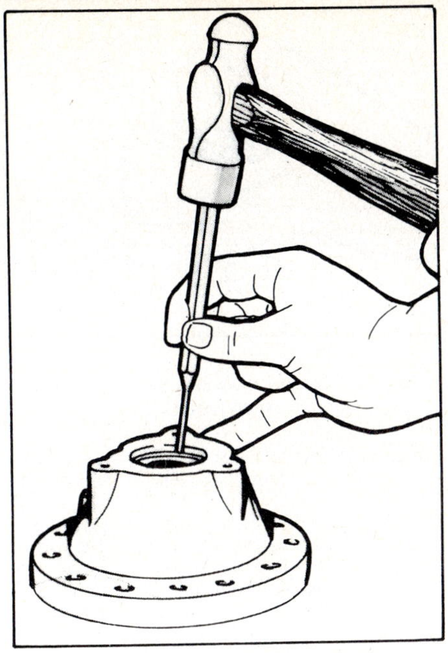

Fig. 6 High pressure seal removal

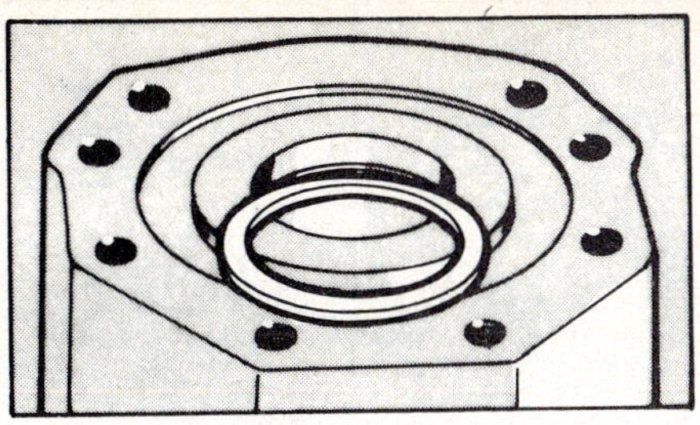

Fig. 7 Quad ring removal

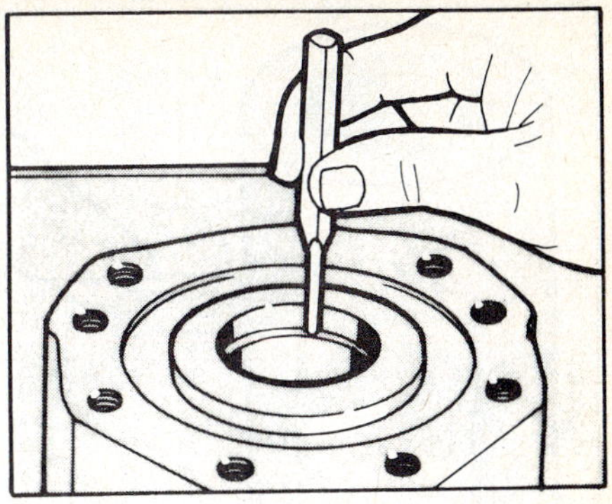

Fig. 8 Bronze bearing removal

Fig. 9 Piston top adjusting nut removal

Fig. 10 Piston top reversing spring removal

7. In event of severe oil contamination, remove and clean piston bottom plug. On reinstallation, the plug must be pinned.

Short Series Piston, Disassembly

NOTE: The short series units use an inverted bearing cap and the piston is machined to permit the bearing to travel inside the top of the piston, Fig. 13. Also, the top adjusting nut is factory adjusted and must not be removed.

1. Remove bottom plug locking pin. The pin may be drilled out or may be sheared when removing the plug. Using a suitable spanner wrench, remove bottom plug from piston, Fig. 14.
2. Pry adjusting nut locking pin through piston and pull out with a suitable pair of pliers, Fig. 15.
3. Mark position of adjusting nut and piston so the nut can be installed in original position.
4. Using a suitable spanner wrench, remove valve adjusting nut, Fig. 16.
5. Remove the double reversing springs, then pull actuating valve from piston bore.
6. Remove valve positioning pin with a suitable screwdriver. Remove and discard "O" ring.
7. Remove the two bottom reversing springs, Fig. 17.
8. Using a suitable allen wrench, remove relief ball seats, relief balls and spring from piston.

Fig. 11 Actuating valve removal

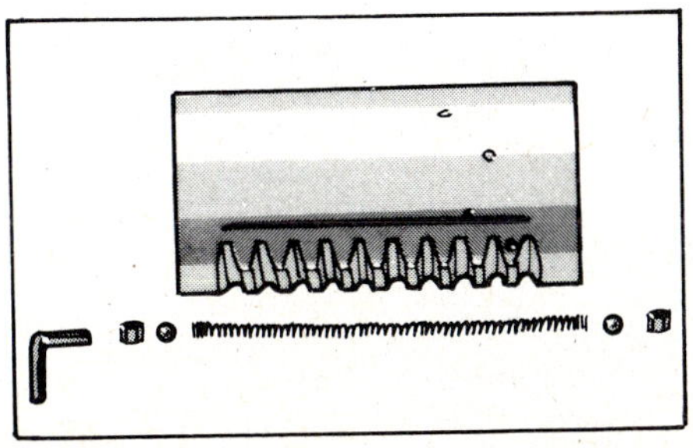

Fig. 12 Piston relief ball assembly removal

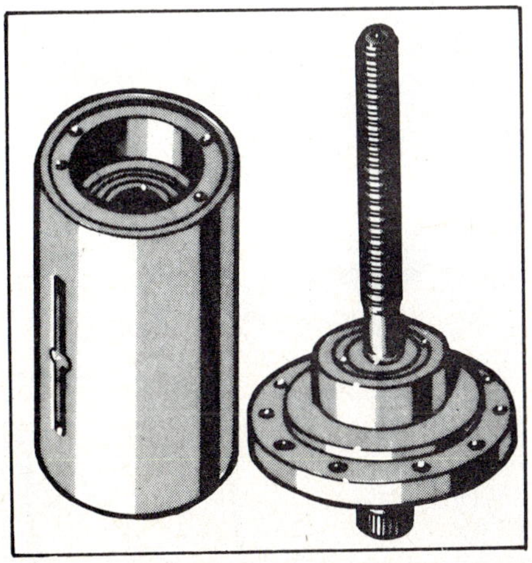

Fig. 13 Bearing cap & piston assembly, short series

372-382 Series Piston (Step Bore), Disassembly, Fig. 18

These pistons are disassembled in the same manner as the standard piston except the piston ring flange is factory located and installed and must not be removed. Also, the piston is replaceable.

ASSEMBLY

1. Press new bushings into gear housing and cover. The bushing should be installed so the inside face of the bushing is flush with the inside face of the gear housing. Also, the bushing face must be flush with the cover except on model 592 units and 392/492 units. On model 592 units, the bushing should be recessed 5/16 inch from the cover and face on 392/492 units, the bushing should be recessed 1/2 inch from the cover face.

Fig. 14 Bottom piston plug removal, short series

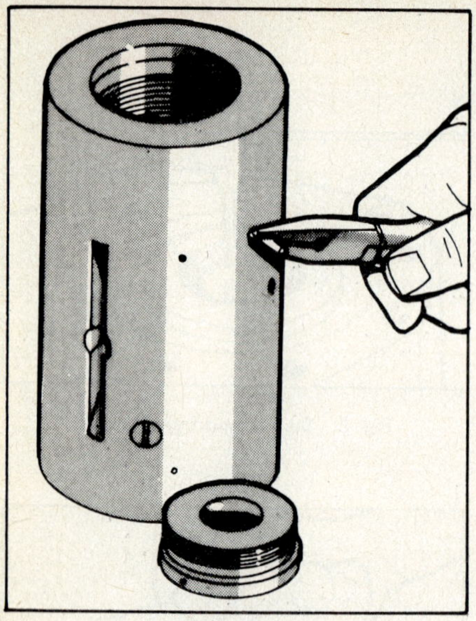

Fig. 15 Valve adjusting nut locking pin removal, short series

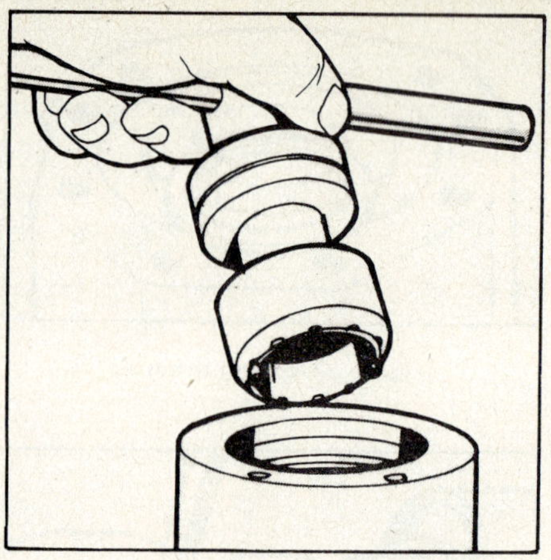

Fig. 16 Valve adjusting nut removal, short series

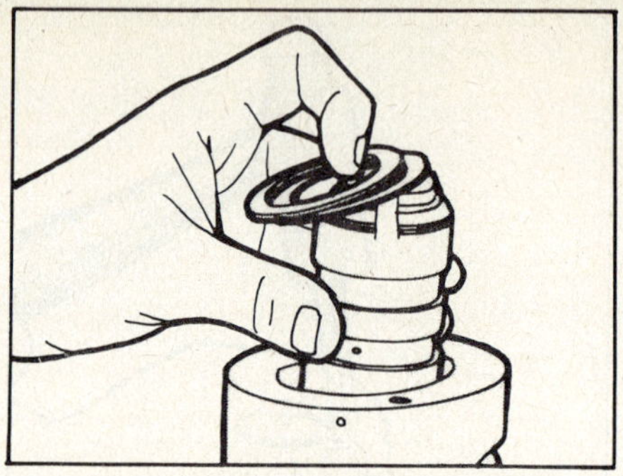

Fig. 17 Bottom reversing spring removal, short series

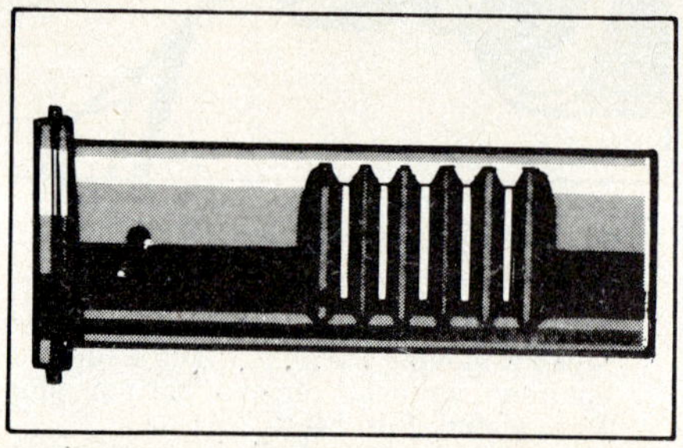

Fig. 18 372-382 step bore piston

NOTE: When installing the bushing in the output shaft bore of the housing and the bushing is easily installed, encountering little press resistance, remove the bushing. Then, clean the bushing of all oil and grease and the housing bore of foreign material by scraping with a brass sheet. Coat the housing bore and the outside diameter of the bushing with Loctite High Strength Retainer Compound and reinstall the bushing. Allow a minimum of 6 hours curing time, then remove excess compound with naphtha.

The bronze bushings are presized and boring or honing is not required.

2. Install a new Quad ring into gear housing groove, Fig. 19.
3. Install the relief valve spring and one relief ball into valve bore, then install one of the valve seats. The valve seat must be tight and flush with or slightly below the end surface of the piston. Install the second relief ball and seat.
4. Place one of the reversing springs in bottom of valve bore. The spring should be centered since the valve must enter the inside diameter of the spring.
5. Install the valve positioning pin into the piston, Fig. 20. Rotate the pin inward with a screwdriver until the pin is below the surface of the piston. The flats must enter into the piston 1/4 inch to engage the mating slot in the valve.
6. Slide actuating valve into piston so the slot on end of valve is positioned over the pin.
7. Place reversing spring onto shoulder of valve end.
8. Install valve adjustment nut. Rotate nut clockwise into the piston until the nut is against the spring. Align the marks made during disassembly on nut and piston. Install the locking pin and ensure that the pin is below the outside edge of the piston.
9. On short series pistons, install piston plug and the locking pin. If the piston plug enters farther into the unit than the original locking pin hole, drill a new hole 180° opposite original pin position and install pin.
10. Install bearing cap seals. Press the seals into position as shown in Fig. 21.

NOTE: On early production steering gears, the high pressure seal is installed with the lip facing inside and the dirt seal with the lip toward the outside.

11. Adequately lubricate the seals.
12. Press actuating shaft assembly into the bearing cap.
13. Install bearing retaining nut. Insert the locking pin through the hole in the bearing cap and into hole in nut. If a new nut is used, drill a 3/32 inch diameter hole in nut. The nut must be in place to drill the hole. Drill through the locking pin hole in the bearing cap and 3/16 inch into the nut.
14. Align the timing mark on the pinion gear with marked spline on output shaft, Fig. 22. Press gear onto shaft, if necessary. Install the gear retaining nut. The nut must be tight against the output shaft

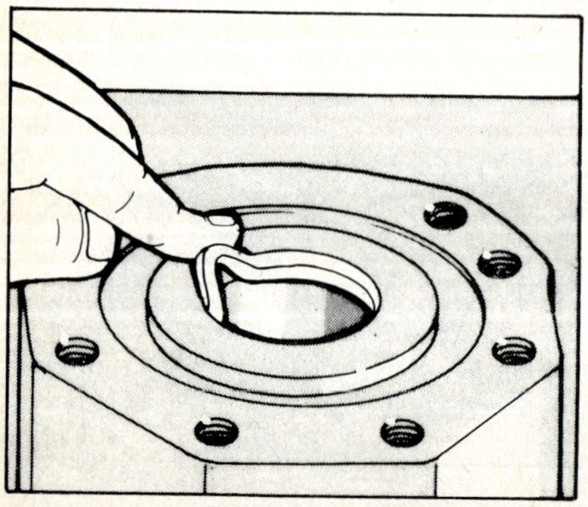

Fig. 19 Quad ring installation

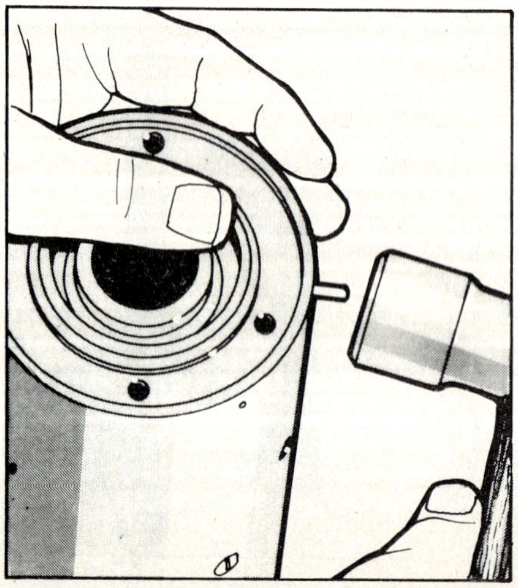

Fig. 20 Valve positioning pin installation

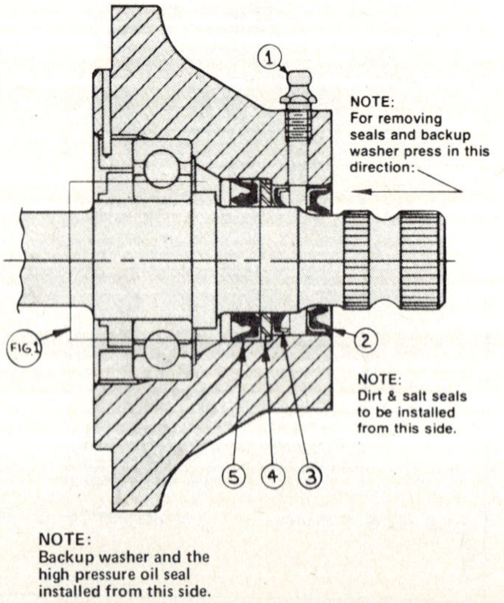

Fig. 21 Bearing cap seal installation

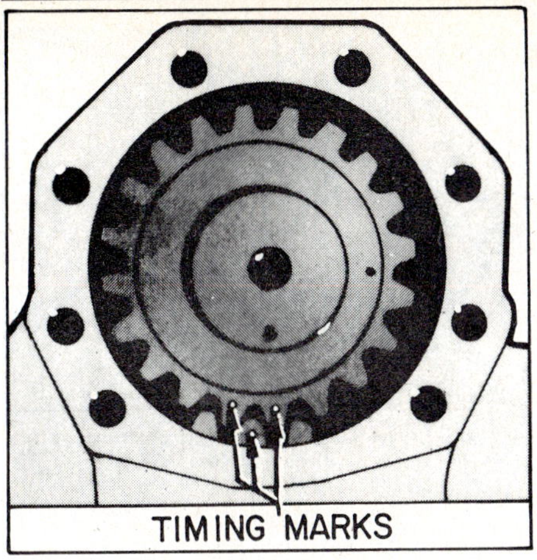

Fig. 22 Pinion gear timing marks aligned

Fig. 23 Miter gear timing marks, miter gearbox

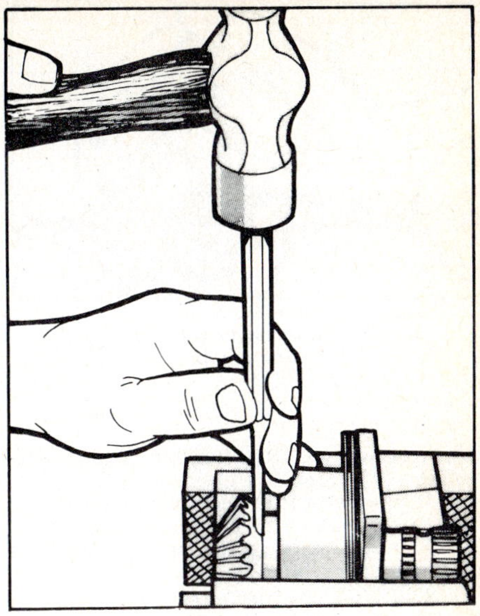

Fig. 24 Miter gear roll pin removal, miter gearbox

gear. Install capscrews into retaining nut. If holes do not align, back off retaining nut until the holes align.

NOTE: On high pressure units, align shaft and gear retaining pin holes and press gear onto shaft.

15. Install retaining pin through gear and seat into shaft, then install roll pin.
16. Lightly coat cylinder bore with engine oil.
17. Install piston into cylinder bore with the opening for the actuating shaft facing toward the bearing cap end of housing.
18. Align timing mark on piston rack with center of pinion gear housing.
19. Lubricate the housing bearing and Quad ring, then slide output shaft through housing. Align timing marks on pinion gear with timing mark on rack.
20. Install cylinder head with new gasket or seal. Align marks made during disassembly and tighten bolts.
21. Using a new gasket or seal on bearing cap, thread actuating shaft into valve. Align punch marks on gear housing and bearing cap. Rotate shaft until the cap is positioned on end of cylinder. Check the plunger hole alignment with valve seat in the piston. Install and tighten bearing cap bolts.
22. Place a new "O" ring on housing cover, then install cover assembly. It may be necessary to use a mallet to tap cover into place. Install and torque cover bolts.
23. Install new "O" rings on relief valve plungers. Install plunger and turn plungers inward approximately six turns.
24. Install pitman arm onto output shaft.

Fig. 25 Miter gear removal, miter gearbox

MITER GEARBOX SERVICE

Disassembly

1. Remove cover bolts and lock washers, then the cover from housing.
2. Note timing marks on shaft and gear, then drive roll pin from miter gear, Figs. 23 & 24.
3. Pull miter gear from actuating shaft, Fig. 25
4. Remove locking pin from cover, Fig. 26.
5. Rotate the bearing retaining nut counterclockwise from cover, Fig. 27, then remove bearing and actuating shaft assembly from cover, Fig. 28.
6. Drive actuating shaft from bearing with a mallet.
7. If replacement is necessary, remove seal from cover.
8. Remove grease fitting from housing.
9. Remove bolt from inside housing, Fig. 29.
10. Remove bolts and lock washers securing housing to bearing cap.

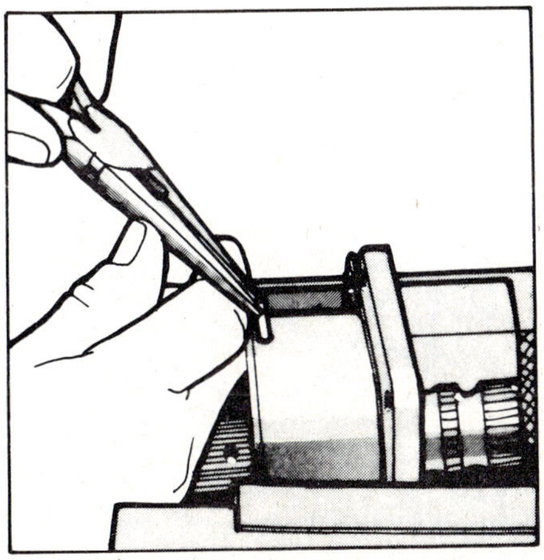

Fig. 26 Cover locking pin removal, miter gearbox

Fig. 27 Bearing retaining nut removal, miter gearbox

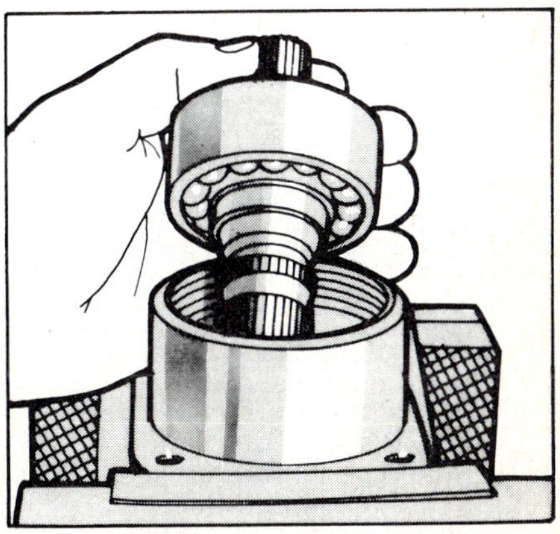

Fig. 28 Bearing & actuating shaft removal, miter gearbox

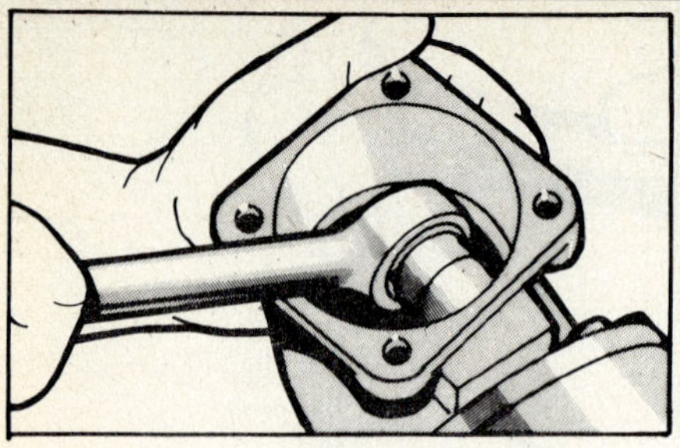

Fig. 29 Inside housing bolt removal, miter gearbox

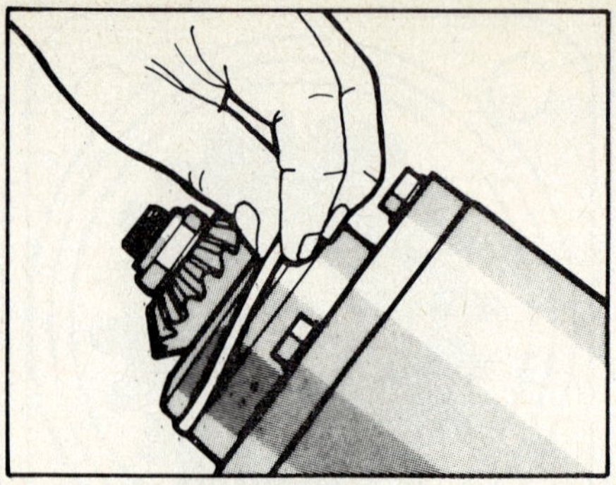

Fig. 30 Bearing cap "O" ring removal, miter gearbox

Fig. 31 Bearing cap locking pin removal, miter gearbox

11. Tap housing from bearing with a mallet.
12. If replacement is necessary, remove "O" ring from bearing cap groove, Fig. 30.
13. Remove retaining nut and washer from actuating shaft.
14. Pull miter gear from actuating shaft.
15. Remove locking pin from bearing cap, Fig. 31. Place a mark on the lock nut to indicate proper pin hole location.
16. Place an alignment mark between lock nut and housing, then remove lock nut from bearing cap.
17. Remove shims from top of bearing, Fig. 32.
18. Remove housing bolts and lock washers.
19. Remove bearing cap and shaft assembly, turning the shaft from the valve. Note the timing marks on shaft and valve for reassembly.
20. Remove actuating shaft and bearing assembly from bearing cap, using either a mallet or a press, if necessary.
21. If damaged, remove locking pin, then the fixed plunger from bearing cap.
22. If replacement is required, remove high pressure oil seal from bearing cap.

NOTE: On early production units, the seal must be pried from bearing cap. On late production units, two holes are drilled in the bearing cap to facilitate the use of a pin punch for seal removal, Fig. 33.

Assembly

1. If removed, press new oil seal into bearing cap.
2. If removed, install fixed plunger into bearing cap. Install locking pin so the pin is positioned slightly below bearing cap surface.
3. Press bearing onto actuating shaft.
4. Install actuating shaft and bearing assembly into bearing cap. It may be necessary to use a press. Avoid damaging oil seal when installing the assembly.
5. Rotate actuating shaft into valve and align timing marks between shaft and valve.
6. Install bearing cap to housing lock washers and bolts.
7. Place shims on upper bearing surface, Fig. 32.
8. Install bearing retaining nut into bearing cap, Fig. 34.
9. Align mark on lock nut with pin hole and install locking pin through bearing cap into bearing retaining nut.
10. Install miter gear on actuating shaft, with marks aligned, Fig. 23, then the washer and retaining nut.
11. Install new "O" ring in bearing cap groove, if removed.
12. Install housing into bearing cap. Use a mallet to tap into position.
13. Install housing to bearing cap lockwashers and bolts.
14. Install the bolt inside housing.
15. Install grease fitting into housing.
16. If removed, install seal into cover.
17. With a mallet, drive actuating shaft into bearing. Use a press, if necessary.
18. Install bearing and actuating assembly into cover. Use a press, if necessary.
19. Rotate bearing retaining nut clockwise into cover.
20. Install locking pin through cover into bearing retaining nut.
21. Position miter gear on actuating shaft, aligning marks on gear and shaft.
22. Install miter gear roll pin.
23. Install cover assembly onto housing.
24. Install cover to housing lock washers and bolts.
25. Fill miter box housing with suitable lubricant.

ADJUSTMENTS

Relief Valve Plungers, Adjust

1. Install pitman arm onto output shaft with marks aligned. Install drag link, if removed.
2. Rotate bearing cap plunger and cylinder head plunger into the gear until bottomed.
3. Start engine and operate at fast idle speed.
4. With full weight of vehicle on the wheels, turn steering wheel until the bearing cap plunger opens the relief valve. Do not force the wheel.
5. Adjust bearing cap plunger outward and continue to turn steering wheel until 1/8 inch clearance is maintained at the wheel stops. Then, lock the plunger with jam nut.
6. Repeat steps 4 and 5 to adjust cylinder head plunger.

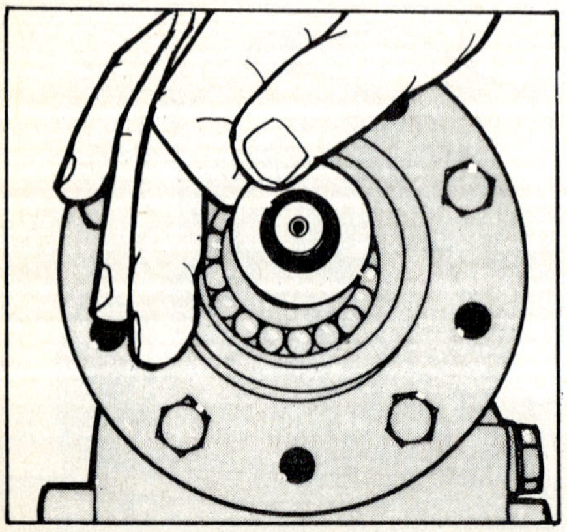

Fig. 32 Shim replacement, miter gearbox

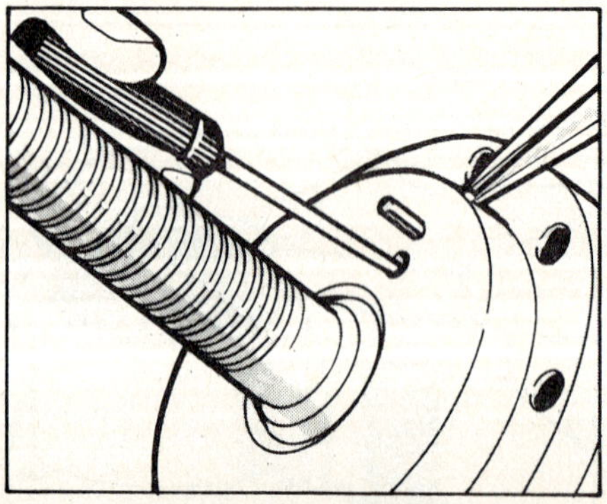

Fig. 33 High pressure seal removal, late production miter gearbox

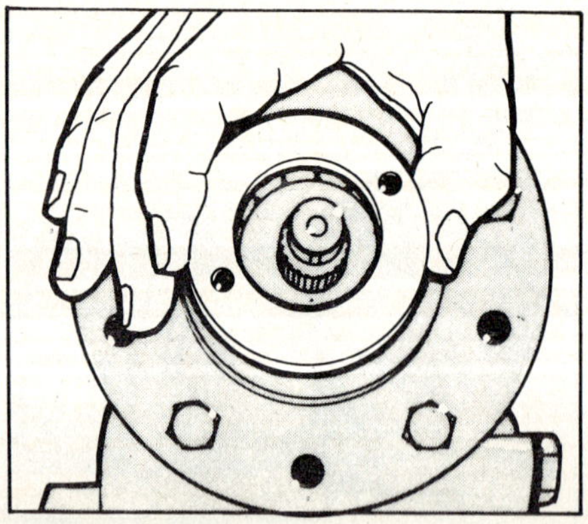

Fig. 34 Bearing retaining nut installation, miter gearbox

Garrison Power Steering Section

DESCRIPTION

The main components of this system are a hydraulic pump, fluid reservoir, steering valve and booster cylinder. The pump supplies hydraulic fluid to the steering valve which directs fluid to the booster cylinder to provide the degree and direction of power assistance required. A typical arrangement is shown in Fig. 1.

Adjusting Steering Valve Ball Socket Clearance

NOTE: Check fluid in reservoir before starting adjustment procedures.

1. Raise front wheels off ground. Start engine so pump is operating and turn wheels to full right and left several times.
2. Raise engine speed to 750–800 RPM to insure centering valve is at neutral position. Turn end plug until .060 inch gap is obtained between end plug and ball socket.

NOTE: Measure gap with round thickness gage using inspection hole in valve housing.

3. Adjust end plug inward until first set of holes lines up with slot in end plug. Install cotter pin.

Centering Booster Cylinder

1. Loosen cylinder cross-steering clamp and turn wheels completely to the right.

NOTE: There must be at least 1/3 turn reserve in steering gear and at least 1/2 inch clearance between inside of tire and pitman arm.

2. Push cylinder rod all the way in, then out 1/8 inch. Tighten cylinder clamp and set right hand knuckle wheel stop.
3. Turn wheels completely to the left so cylinder rod is fully extended. Turn wheels back to the right so cylinder rod is retracted 1/8 inch. Set left hand knuckle wheel stop.

Steering Valve

Disassembly

1. Disconnect hydraulic lines from valve noting position of each line. Remove cotter pin and end plug, ball socket end plug lock and socket end plug. Remove valve from steering arm.
2. Disconnect drag link from steering arm.
3. Place valve in vise in center of housing and remove snap ring from drag link adapter, then remove drag link adapter with suitable wrench.
4. Remove ball socket and ball socket bearing.

NOTE: Place a punch into hole of piston to keep it from turning.

5. Remove valve body from housing. Place valve in vise and remove nut from gland end of piston. Remove washer and, while holding piston in body, remove gland, two reaction rings, and spacer.

NOTE: O-ring seals on piston will be damaged if piston and gland are removed together.

6. Move piston toward gland end of body and remove O-ring seal. Move piston in opposite direction and remove O-ring seal from opposite end. Slide piston from body.

Cleaning & Inspection

Clean all parts using a suitable solvent and blow dry all parts. Make visual inspection of all parts for evidence of wear or corrosion. Light scratches can be removed with crocus cloth. O-ring seals should be examined carefully and replaced if damaged.

NOTE: Do not round off edges of ports on piston or body as this could result in excessive leakage and reduced hydraulic power.

Assembly

1. Place valve body into vise and insert piston into gland end of body.
2. Place O-ring on piston opposite gland end first, then move piston just far enough to install O-ring on gland end of piston.
3. Position shoulder on gland end of piston flush with bottom of counterbore in body. Install two O-ring seals on gland end of piston and two O-ring seals on two reaction rings.
4. Install one reaction ring on piston, insert gland into body and place second reaction ring on end of piston.

NOTE: If valve gland has compression spring, insert in gland between the two reaction rings and assemble over piston. If valve gland uses a spacer it should be installed between the two reaction rings and assembled in body.

5. Install washer and nut and torque to 30–35 ft. lbs.

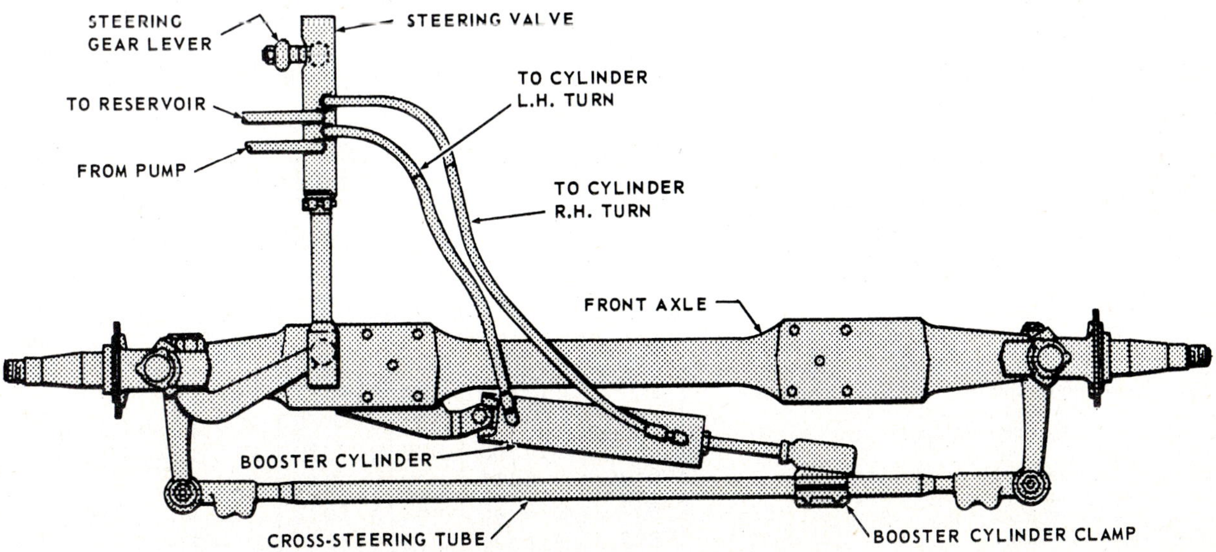

Fig. 1 Typical Garrison Power Steering Arrangement

Parker-Hannifin Power Steering Pump Section

D, H, & M SERIES

Disassembly, Fig. 1

1. Scribe line across pump for reassembly. Remove hex screws and cap screws and key from drive shaft. Separate front and rear covers from mid-section. Remove wear plate and thrust plate.

NOTE: Mark front cover island next to pressure vent hole in heat shield, gasket, and V-seal to act as guide in reassembly.

4. Remove phenolic heat shield, paper compound gasket, rubber V-seal, and O-rings.
5. Remove shaft seal, flow control valve, and relief valve.

Inspection

1. Clean and inspect all parts for excessive wear or damage.
2. Measure shafts of drive and driven gears. Minimum acceptable is .4998 inch in D series pump, .7492 inch in H series and .9365 inch in M series.
3. Measure gear O.D. Minimum acceptable in D series is 1.2395 inch, 1.7410 inch in H series and 2.1047 inch in M series. For minimum gear widths refer to Fig. 2.

NOTE: Gear assemblies may be replaced separately if only one is damaged.

4. Check seal pressurizing valve in driven gear. There should be slight resistance but valve should be free to move.
5. Measure bearing bore diameter in front and rear covers. Bore diameter should not exceed .5015 in D series, .7518 in H series and .9394 in M series. If diameter exceeds limits, replace cover. Bearings are not supplied separately.
6. Check shaft seal. Replace only if evidence of leaking or excessive wear is present.

NOTE: Bearings must be below cover faces and show no signs of contact with snap rings on gear shaft. If bearings are scored, rough or show signs of heat discoloration, replace cover assemblies.

7. Check center section gear bores for wear on inlet side of pump. Also check for wear ridge at thrust plate end of gear bore. Ridge should not exceed 1/32 inch. Lightly lap faces to remove nicks or burrs. Do not break inside edges.
8. Inspect bronze wear surfaces of wear plate and thrust plate for roughness or heat discoloration. Replace plates if wear ridges exceed .0005 inch.

Assembly

1. Install shaft seal into front cover with

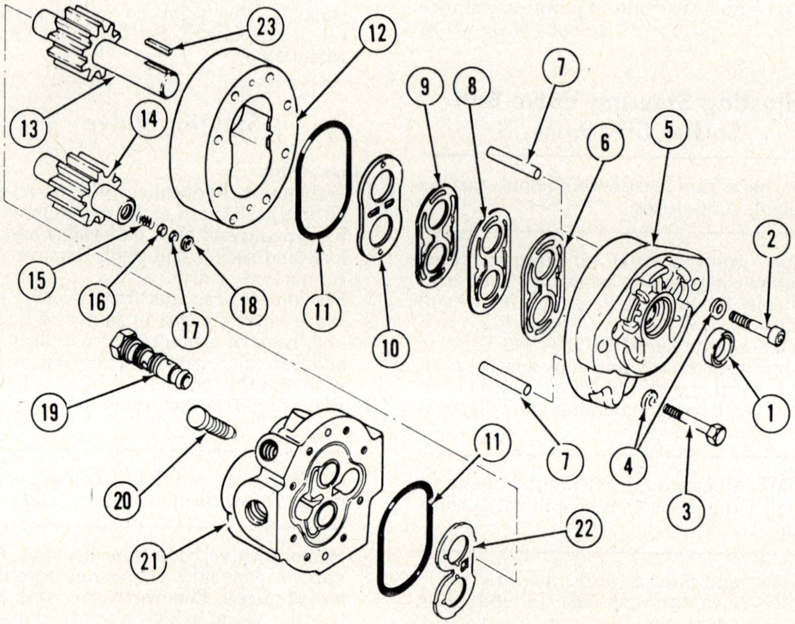

1. SHAFT SEAL
2. CAP SCREWS
3. HEX SCREWS
4. WASHERS
5. FRONT COVER
6. V-SEAL
7. DOWEL PINS
8. GASKET
9. HEAT SHIELD
10. WEAR PLATE
11. "O" RINGS
12. CENTER SECTION
13. DRIVE GEAR ASSEMBLY
14. DRIVEN GEAR ASSEMBLY (INCLUDES 15, 16, 17, 18)
15. SPRING (H&M ONLY)
16. PLATE (H&M ONLY)
17. RING (H&M ONLY)
18. SNAP RING (H&M ONLY)
19. FLOW CONTROL ("H" FLOW DIVIDER PUMP ONLY)
20. RELIEF VALVE ("H" FLOW DIVIDER PUMP ONLY)
21. BACK COVER
22. THRUST PLATE (H&M ONLY)
23. KEY (WHERE REQUIRED)

Fig. 1 Parker-Hannifin D, H and M Series Pumps

Size	"D" Series Min. Gear Width (Inch)	Size	"H" Series Min. Gear Width (Inch)	Size	"M" Series Min. Gear Width (Inch)
D05	0.1875	H20	0.3624	M06	0.6883
D07	0.2770	H25	0.4531	M07	0.8605
D09	0.3463	H31	0.5663	M09	1.0756
D11	0.4309	H39	0.7079	M11	1.3446
D14	0.5412	H49	0.8849	M14	1.6807
D17	0.6655	H62	1.1072		
D22	0.8597	H77	1.3840		
D27	1.0562				

Fig. 2 Minimum Gear Width Chart

spring loaded lip facing inward.
2. Install V seal, gasket and heat shield into front cover.

NOTE: All these parts contain a vent hole which must be lined up with scribe mark on island which was made previously in step 1 of "Disassembly". Lips of V seal shall face toward cavity.

3. Install front cover O-ring. Install wear plate into oval cavity of front cover positioned so that bronze surface will be against gears and small vent hole will be in line with hole in heat shield.
4. Install drive gear and driven gear into front cover.

NOTE: Apply oil to drive end of shaft to prevent damage to shaft seal. Check to see if shaft seal lip and spring have been pushed out by drive shaft.

5. Install center section over gears until it engages wear plate. Position center section so previously scribed lines on housing exterior are in line with those on front cover. Small slot in mid-section should line up with small vent hole in wear plate.
6. Install dowel pins and add oil to gear cavities. Rotate gears to distribute oil.
7. Place thrust plate over gears in mid-section with bronze face toward gears and open side toward inlet.
8. Install O-ring into back cover and position back cover onto mid-section so that previously scribed lines on all sections line up.
9. Torque housing retaining screws to 190–210 in. lbs. on D & H series and to 355–375 in. lbs. for M series.
10. Add oil to both ports of pump and check to see if drive shaft is free to rotate. Maximum torque required to turn shaft is 100 in. lbs.

DISTRIBUTORS, STANDARD

The ignition system can be divided as follows, Fig. 1.

1. Battery to supply current.
2. Ignition wiring to carry current to the units in the system.
3. Ignition switch to control the circuit.
4. Ignition coil to increase the voltage delivered to the spark plugs.
5. A distributor to distribute current to each cylinder.
6. Spark plugs to ignite the fuel in each cylinder.

But inasmuch as the "Tune Up" chapter deals with such service as comes within the province of tuning up an engine—such as batteries, spark plugs, etc.—this chapter will discuss the functions and service requirements of the distributor itself, together with any additional data not included in the "Tune Up" chapter.

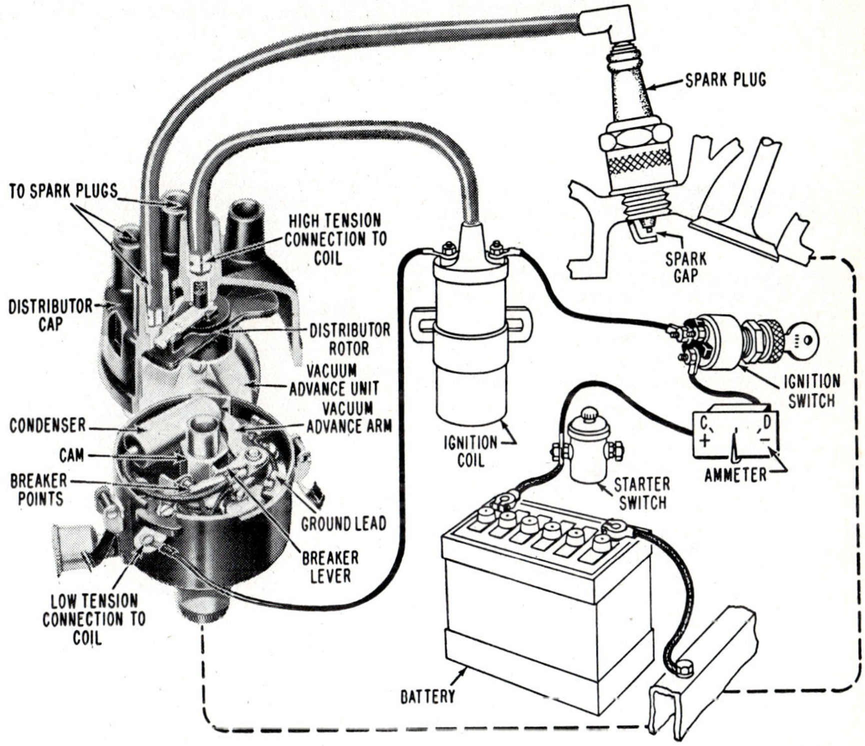

Fig. 1 Typical ignition system

BREAKER CONTACT POINTS

Contact Analysis

The normal color of points should be a light gray. If the contact surfaces are black it is usually caused by oil vapor or grease from the cam. If they are blue, the cause is usually excessive heating due to improper alignment, high resistance or open condenser circuit.

If the contacts develop a crater or depression on one point and a high spot of metal on the other, the cause is an electrolytic action transferring metal from one contact to the

Fig. 2 Showing how metal from one contact transfer to the other

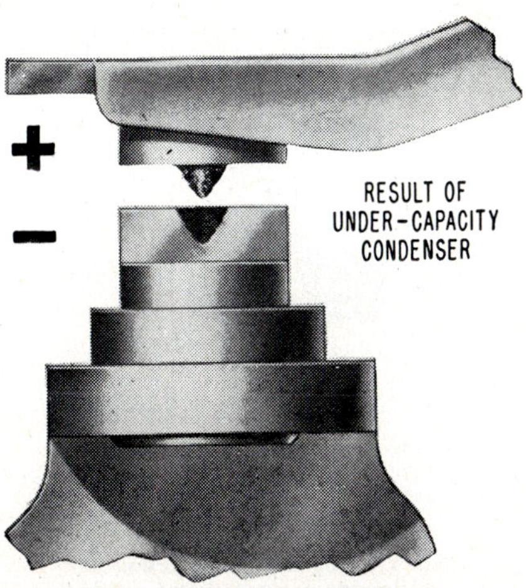

Fig. 3 Mound on positive point

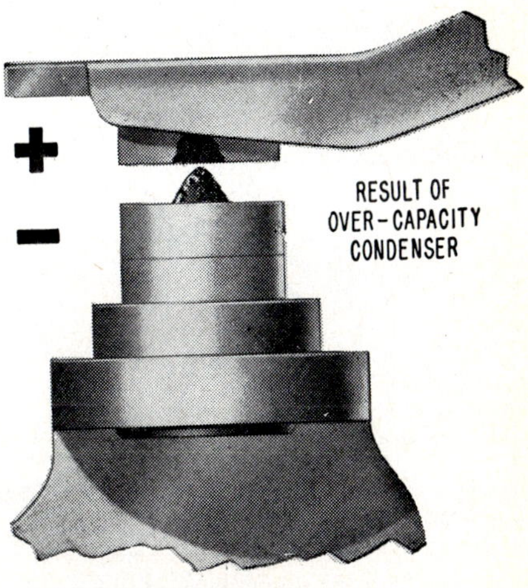

Fig. 4 Mound on negative point

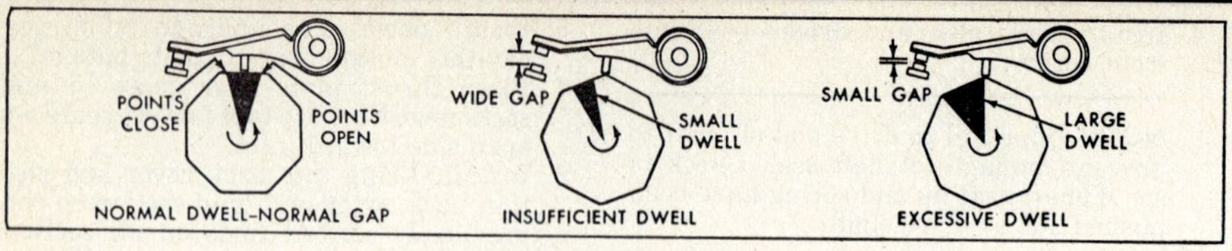

Fig. 5 Cam angle or dwell

other, Fig. 2. This can be the result of some unusual operation of the vehicle. A slow-speed driver in city traffic or door-to-door delivery vehicles will be one extreme, and high-speed, long distance driving would be the other extreme. It may also be due to an unbalanced ignition system, which can sometimes be improved by a slight change in condenser capacity. If the mound is on the positive point, Fig. 3, install a condenser of greater capacity; if on the negative point, Fig. 4, use a condenser of lesser capacity.

One of the most common causes of point failure is the presence of oil or grease on the contact surfaces, usually from over-lubrication of the wick at the top of the cam or too much grease on the rubbing block of the breaker arm.

Breaker Point Gap

Correct contact point opening is important, especially during starting and low speed operation. If points are set too close, arcing and burning will occur, causing hard starting and poor low speed performance. If points are set too wide, the cam angle or dwell will be too small to allow saturation of the coil at high engine speeds, resulting in weak spark.

Contact point opening has a direct bearing on cam angle or dwell which is the number of degrees that the breaker cam rotates from the time the points close until they open again, Fig. 5. The cam angle or dwell increases as point opening is decreased and vice versa. In view of the importance of point opening to low speed engine performance and cam dwell to high speed engine performance, the cam angle or dwell should be checked after adjusting and aligning points. This is done on a distributor testing machine, following the manufacturers instructions.

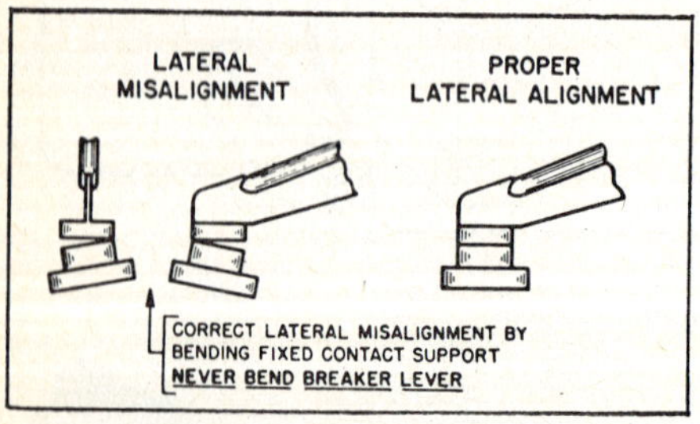

Fig. 7 Breaker point alignment

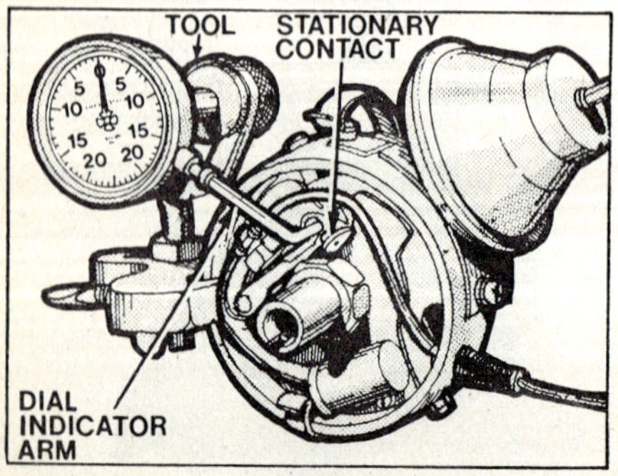

Fig. 8 Dial indicator for measuring breaker gap

Breaker Arm Spring Tension

Breaker arm spring tension is important. If the tension is too great the arm will bounce, causing an interruption of the current in the coil and misfiring. If the spring tension is too little, the rubbing block will not follow the cam, causing a variation in cam dwell. The spring tension should always be set at the high limit as given in the *Distributor Specifications* chart in truck chapter, as it will be reduced as the rubbing block wears.

Hook a spring scale on the breaker arm and pull in a straight line as shown in Fig. 6. Take a reading as the points start to separate under the slow and steady pull of the scale. If the tension is not within specification, loosen the screw that holds the end of the point spring and slide the end of the spring in or out as necessary. Tighten the screw and recheck the spring tension.

Breaker Point Alignment

Check alignment of points with points closed, Fig. 7. Align new points where necessary but do not attempt to align used points. Instead, replace used points where serious misalignment is observed. After aligning points, adjust point gap.

Adjusting Breaker Gap

Specifications for breaker gap, *as measured with a wire gauge*, is listed in the *Tune Up Specifications* in truck chapters. However, if at all possible, this opening should be set on a distributor testing machine or with a dial indicator, Fig. 8. This not only eliminates the possibility of an incorrect gap setting but if the points are slightly rough but otherwise in alignment, there is the danger of obtaining an incorrect gap, as shown in Fig. 9.

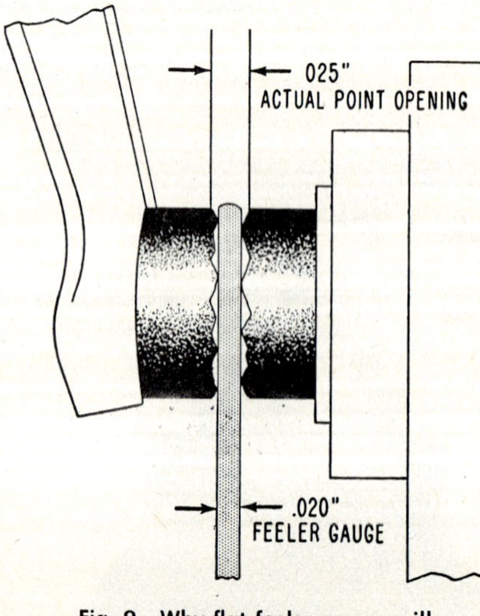

Fig. 9 Why flat feeler gauge will not provide accurate point spacing if points are rough

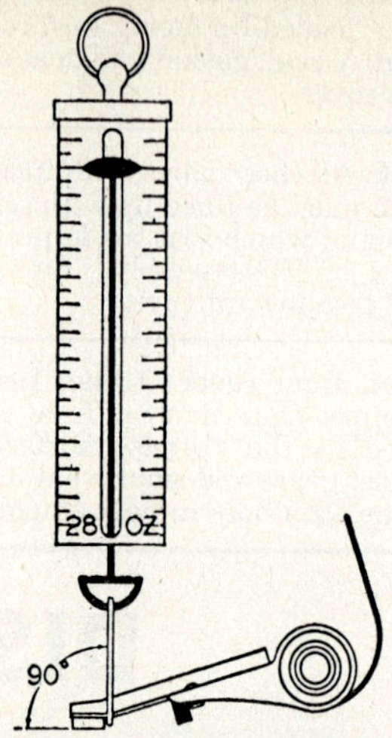

Fig. 6 Measuring breaker arm spring tension

The advantage of a distributor testing machine is that it not only measures cam angle or dwell but it also uncovers irregularities between cam lobes, point bounce, alignment of rubbing block with cam, alignment of contacts and breaker arm spring tension.

Auto-Lite & Chrysler Single Point Set Adjustment

Adjustment of breaker gap is accomplished by loosening the lock screw in the stationary point and moving the adjusting screw as required to obtain the correct breaker gap.

Delco-Remy Internal Adjustment Breaker Points

An opening in the point set assembly allows for easy dwell angle or breaker point adjustment, Fig. 10.

Delco-Remy External Adjustment Breaker Points

With engine running at idle speed, the breaker gap is adjusted by first raising the window provided in the cap and inserting a "hex" wrench into the adjusting screw, Fig. 11. Turn the adjusting screw clockwise until

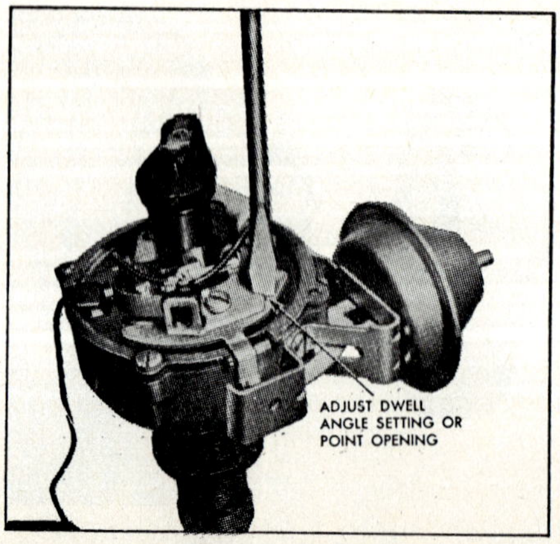

Fig. 10 Breaker point setting

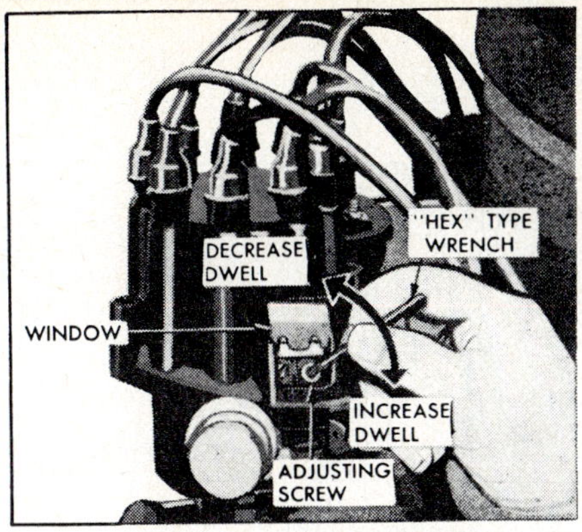

Fig. 11 Adjusting breaker point gap through window in distributor cap

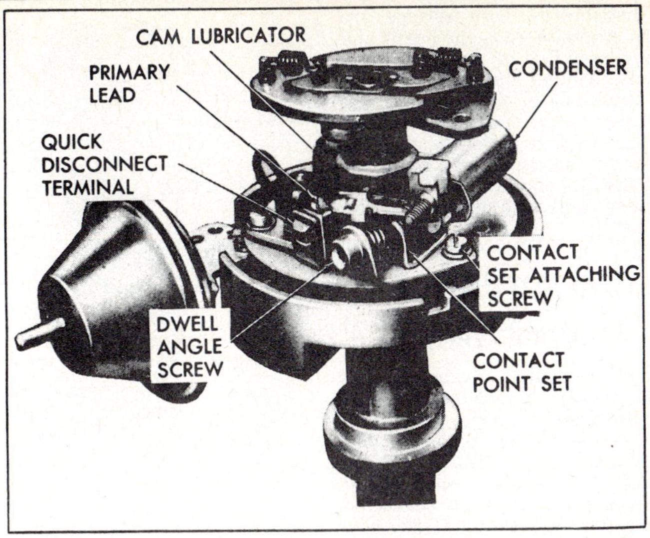

Fig. 12 Delco-Remy one piece breaker point & condenser (Uni-Set)

the engine begins to misfire. Then give the wrench one-half turn in the opposite direction which will provide the proper breaker gap. If a dwell meter is to be used, turn the adjusting screw until the correct angle is obtained.

NOTE: Some General Motors V-8 trucks may be equipped with the new Uni-Set breaker points (one piece points and condenser). If the Uni-Set points and condenser is not available for service the following alternatives can be applied.

1. A conversion kit is available (Part No. 1876065) which contains a breaker plate, contact set, radio shields, condenser and wick.
2. If the existing breaker plate has tapped holes for the condenser clamp and radio shields, it need not be changed and the conversion kit is not necessary. You can install the previous two piece points and condenser but you must obtain and install the radio shields.

Holley Breaker Points

The breaker point set is attached to the movable breaker plate. A slot in the stationary point bracket allows for easy breaker point adjustment, Fig. 13.

CONDENSER

A condenser should not be condemned because the points are burned or oxidized. Oil vapor, or grease from the cam, or high resistance may be the cause of such a condition.

Condensers should be tested with a good condenser tester for leakage, breakdown, capacity, and resistance in series in the condenser circuit. Manufacturers of condenser testers furnish complete instructions as to their use.

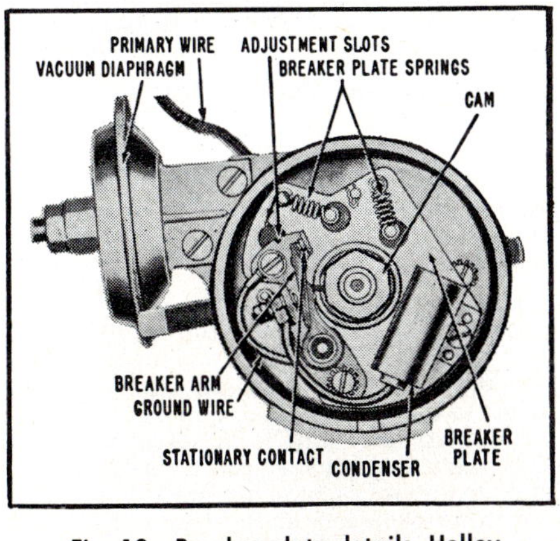

Fig. 13 Breaker plate details. Holley Loadamatic distributor. Note slot for adjusting breaker gap

CENTRIFUGAL ADVANCE

Except for Holley "full vacuum" distributors, all other distributors utilize an automatic advance mechanism which functions by virtue of centrifugal weights. Some distributors employ both centrifugal and vacuum advance mechanisms, while others make use of only the centrifugal mechanism.

When engine speed increases, the spark must be introduced in the cylinder earlier in the cycle in order that the fuel charge can be ignited and will have time to burn and deliver its power to the piston. To provide this spark advance based on engine speed, the centrifugal governor mechanism is used.

This mechanism, Fig. 14, consists of centrifugal advance weights which throw out against spring tension as the engine speed increases. This movement imparts, through a toggle arrangement, rotational motion to the breaker cam, causing it to rotate a number of degrees with respect to the distributor drive shaft. This causes the lobes on the cam to close and open the contact earlier in the cycle so that the spark is induced and is delivered to the cylinder earlier with respect to the position of the upward moving piston.

Fig. 14 Top view of Delco-Remy distributor with breaker plate removed to show centrifugal advance mechanism

In servicing the distributor, all weights should be removed from the hinge pins, cleaned and checked for excessive wear, either in the weights or pins, or the plate which is slotted for the movement of the pins on top of the governor weights. Replacement should be made if there is any appreciable wear in the slots, as any wear at this point would change the characteristic of the spark advance.

If these parts are in good condition, the hinge pins should be lubricated before being reassembled, by greasing the hinge pins and filling the pockets in the governor weights with grease. Do not use vaseline for this purpose as its melting point is comparatively low.

When installing new centrifugal governor assemblies, it is important that the spacer washers between the housing and shaft be installed correctly. If incorrectly installed, the governor assembly will be too high, causing it to rub against the bottom of the breaker plate.

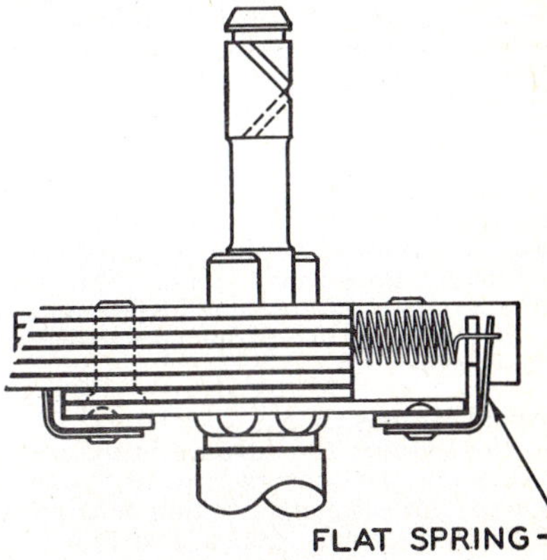

Fig. 15 Flat spring used on some governors to provide a rapid spark advance

On some distributors, both springs are alike, while on others there is one heavy and one light spring, as in Fig. 14. Another combination that may be found is an additional flat spring on the outside of the outer spring posts, Fig. 15. As the governor speed is increased, the flat springs are first pulled against the posts by the eyes of the coil springs to provide a rapid spark advance of a few degrees before the coil springs pull against the spring posts.

VACUUM ADVANCE

Conventional Type

The two types of vacuum advance mechanisms used on Auto-Lite and Delco-Remy distributors are illustrated in Figs. 17 and 18. Both types make use of a spring-loaded diaphragm which is connected through linkage to the distributor. The spring loaded side of the diaphragm is air tight and is connected through a vacuum line to the carburetor above the throttle plate so that idling performance will not be affected.

When the throttle is open vacuum from the intake manifold is introduced into the vacuum advance mechanism and the diaphragm is pulled against the spring causing the distributor to advance.

In Fig. 16 the mechanism is attached to the distributor breaker plate so that the breaker plate rotates. In Fig. 17, the mechanism is connected to the distributor body so that the entire distributor moves. In both cases, the rotational movement carries the contact points around to an advanced position so that the breaker cam closes and opens the points earlier in the cycle.

See Holley Distributor Service for vacuum advance data for these units.

Emission Control Types

Chrysler C.A.P.
The Chrysler vacuum advance units incorporate a vacuum advance control valve in the advance vacuum circuit to provide the necessary retard during closed throttle operation.

Chrysler C.A.S.
Some Chrysler C.A.S. engines have a solenoid incorporated in the distributor vacuum advance mechanism to retard the ignition timing when the throttle is closed. At closed throttle, electrical contacts on the carburetor throttle stop, with idle adjusting screw in the closed position, cause the distributor solenoid to energize. This retards the ignition timing to provide reduced exhaust emissions under hot idle conditions. Cold or part throttle starting is not penalized because the distributor solenoid is not energized unless the hot idle adjusting screw is against the throttle stop contact. Timing must be set at closed throttle to give accurate setting.

Ford Dual-Diaphragm
This unit consists of two independent diaphragms. The outer diaphragm uses carburetor vacuum to advance timing. The inner diaphragm uses intake manifold vacuum to provide additional retard during closed throttle operation.

General Motors CCS
In this system, the advance is the ported type, that is the vacuum take-off is located above the throttle plate(s) so that during periods of closed throttle operation there is little or no vacuum reaching the advance unit and timing is retarded. As soon as the throttle is cracked, vacuum reaches the advance unit and timing is advanced.

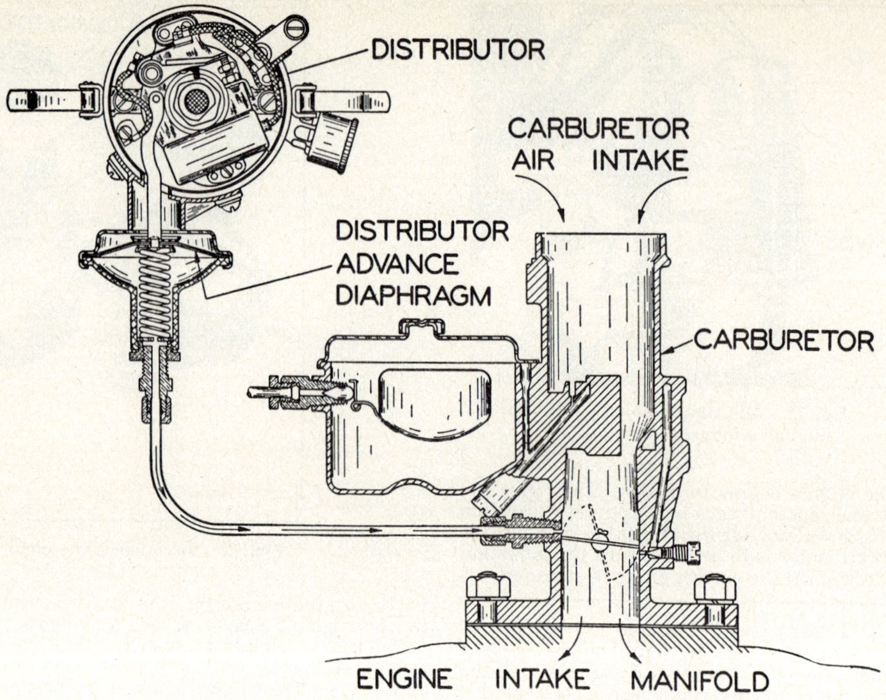

Fig. 16 Auto-Lite vacuum advance mechanism of type mounted of distributor. Breaker plate is supported on a ball bearing, and breaker plate alone rotates in the housing as vacuum conditions

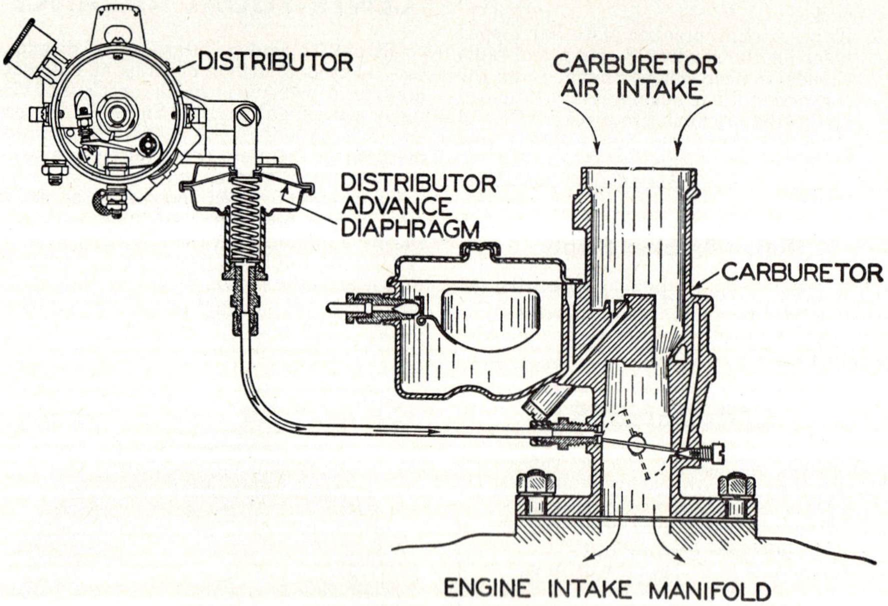

Fig. 17 Auto-Lite vacuum advance mechanism of type clamped around distributor so that entire distributor is rotated as vacuum conditions change

NOTE: Some models of the Chevrolet V8-307 engine use the conventional type advance unit.

Prestolite IBP Distributors

Used on some Dodge trucks, this distributor uses a very different type vacuum spark advance. Unlike other systems which either rotate the breaker plate or the entire distributor when manifold vacuum is high, the IBP breaker plate is pivoted in such a way that the points swing in an arc about the cam when the vacuum advance unit is in operation. Thus, cam angle and breaker point gap change as high manifold vacuum advances the spark. For this reason cam angle should be checked with the vacuum line disconnected and the point gap checked or adjusted when the vacuum advance unit is in full retard position.

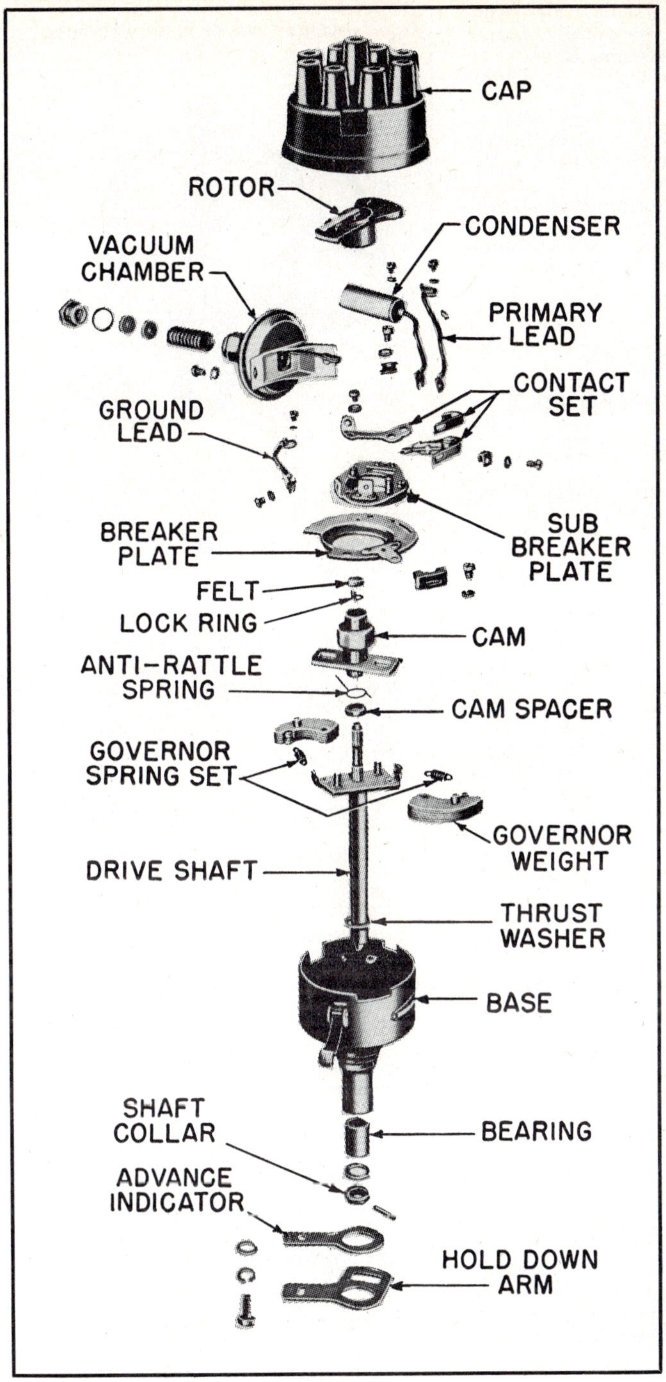

Fig. 18 Disassembled view of a typical Prestolite distributor

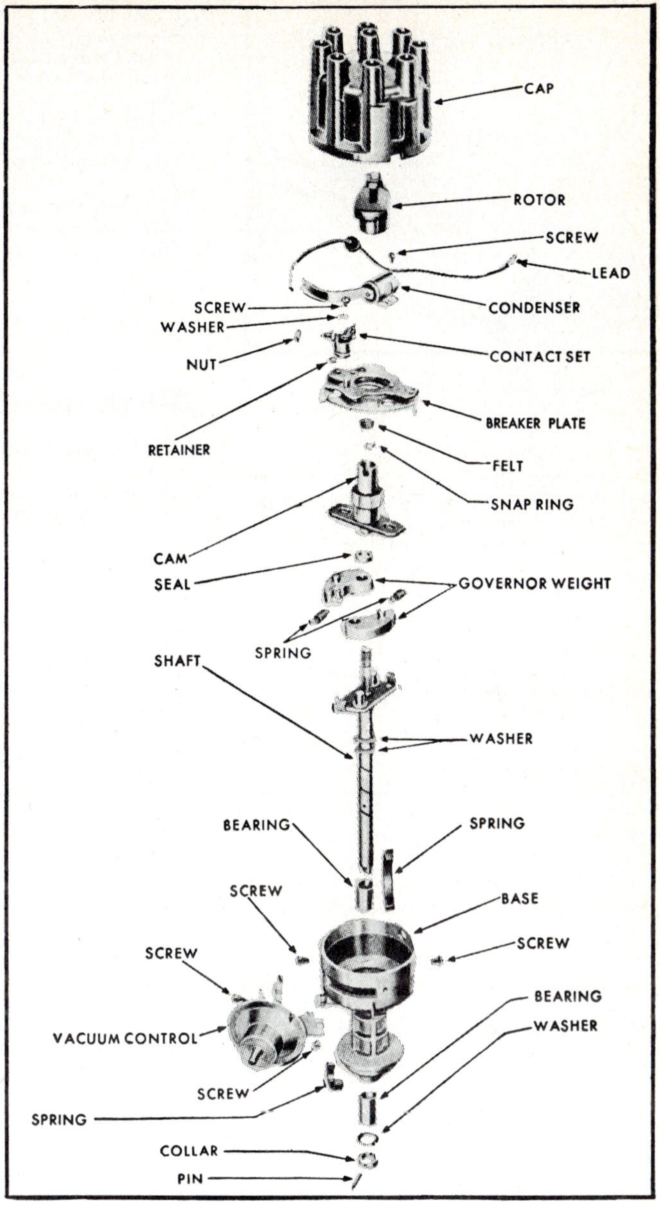

Fig. 19 Disassembled view of a typical Chrysler Corp. (V-8 distributor)

PRESTOLITE & CHRYSLER DISTRIBUTOR SERVICE

Shaft & Bushing Wear Test

1. Remove distributor from vehicle and clamp distributor in a vise. Use extreme caution not to damage distributor.
2. Install a dial indicator on housing so plunger rests against moveable contact arm when rubbing block is on the highest point of a cam.
3. Place a wire loop around distributor shaft and hook a spring scale on the other end of the loop. Apply a one pound pull in line with indicator plunger and read movement on indicator. Movement must not exceed .006 inch. If movement exceeds limit, replace either housing or shaft on Chrysler distributors or bushing or shaft on Prestolite distributors to bring movement back within tolerance.

If the distributor has been disassembled, reassemble as follows, referring to Figs. 18 and 19 for guidance.

1. Check operation of centrifugal weights and weight springs for distortion. Lubricate governor weights.
2. Inspect all bearing surfaces and pivot pins for roughness, binding, or excessive looseness.
3. Install cam spacer (chamfered end down) on distributor shaft.
4. Slide cam and yoke on distributor shaft. Engage weight lugs with slots in yoke as shown in Fig. 20. Install cam retaining clip, being sure it is properly seated in distributor shaft groove.
5. Lubricate and install two concave washers for Prestolite distributors, or a single flat thrust washer for Chrysler distributors. Position washers on shaft and slide shaft into distributor body. On Chrysler six cylinder distributors, if drive gear is worn or damaged, replace as follows:
   a. Install thrust washer and old gear on shaft and install pin. Scribe a line centered between two gear teeth from center to edge of shaft and in line with center of rotor electrode.
   b. Remove pin and gear. Clean burrs from around pin hole and install new gear.

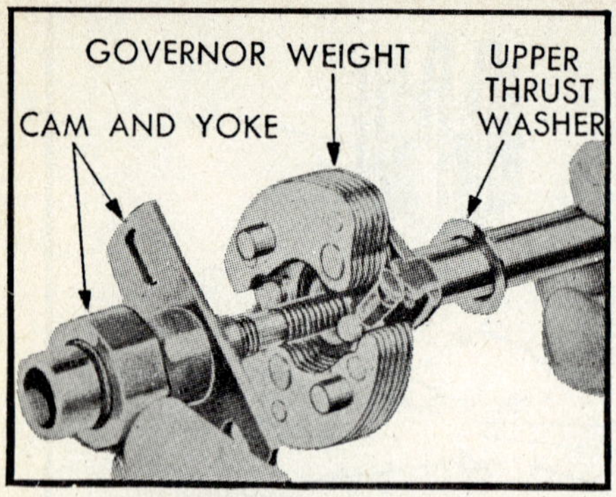

Fig. 20 Distributor shaft details

c. Place a .007 inch feeler gauge between gear and thrust washer and drill a .124–.129 inch hole in gear and shaft about 90 degrees from old hole in shaft and with scribe line centered between gear teeth and in line with center of rotor electrode as shown in Fig. 21.

NOTE: If new pin hole location appears to interfere with shaft oil groove, rotate gear to centerline of next pair of gear teeth and align with scribe line on shaft.

d. Install pin.
6. Position lower thrust washer and drive collar on lower end of shaft and install retainer pin.
7. Install oiler wick and oiler.
8. Install breaker plate assembly, align condenser lead, breaker point spring, primary lead and install attaching screw.
9. Install felt wick in top of distributor cam.
10. Attach vacuum advance unit arm to breaker plate and install retainer. Install vacuum unit attaching screws and washers.
11. Test breaker arm spring tension and adjust breaker gap.
12. Lubricate felt pad in top of distributor cam with 3 to 5 drops of light engine oil and install rotor.

DELCO-REMY DISTRIBUTOR SERVICE

External Adjustment Type

If the distributor has been disassembled, refer to Fig. 22 for guidance when reassembling.

1. Place gasket on shaft housing.
2. Place felt washer around bushing in housing.
3. Install vacuum advance unit.
4. Install breaker plate in housing and spring retainer on upper bushing.
5. Install condenser.
6. Install breaker point set.
7. Install cam and weight base assembly on shaft. If lubrication in grooves at top of shaft was removed during disassembly, replace with Plastilube #2 or its equivalent.
8. Install shaft and cam weight assembly in housing.

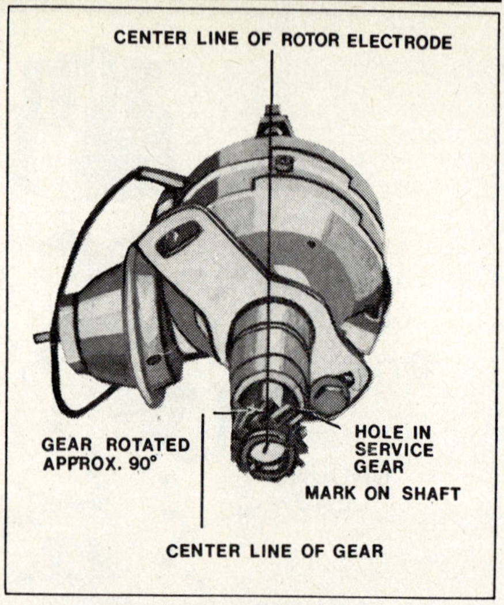

Fig. 21 Gear tooth alignment, Chrysler 6 cyl. distributor

9. Using a pin, install driven gear to shaft.
10. Install advance weights and springs.
11. Install cam lubricator.
12. Install rotor.
13. Check breaker arm spring tension and adjust breaker gap.

Internal Adjustment Type

If the distributor has been disassembled,

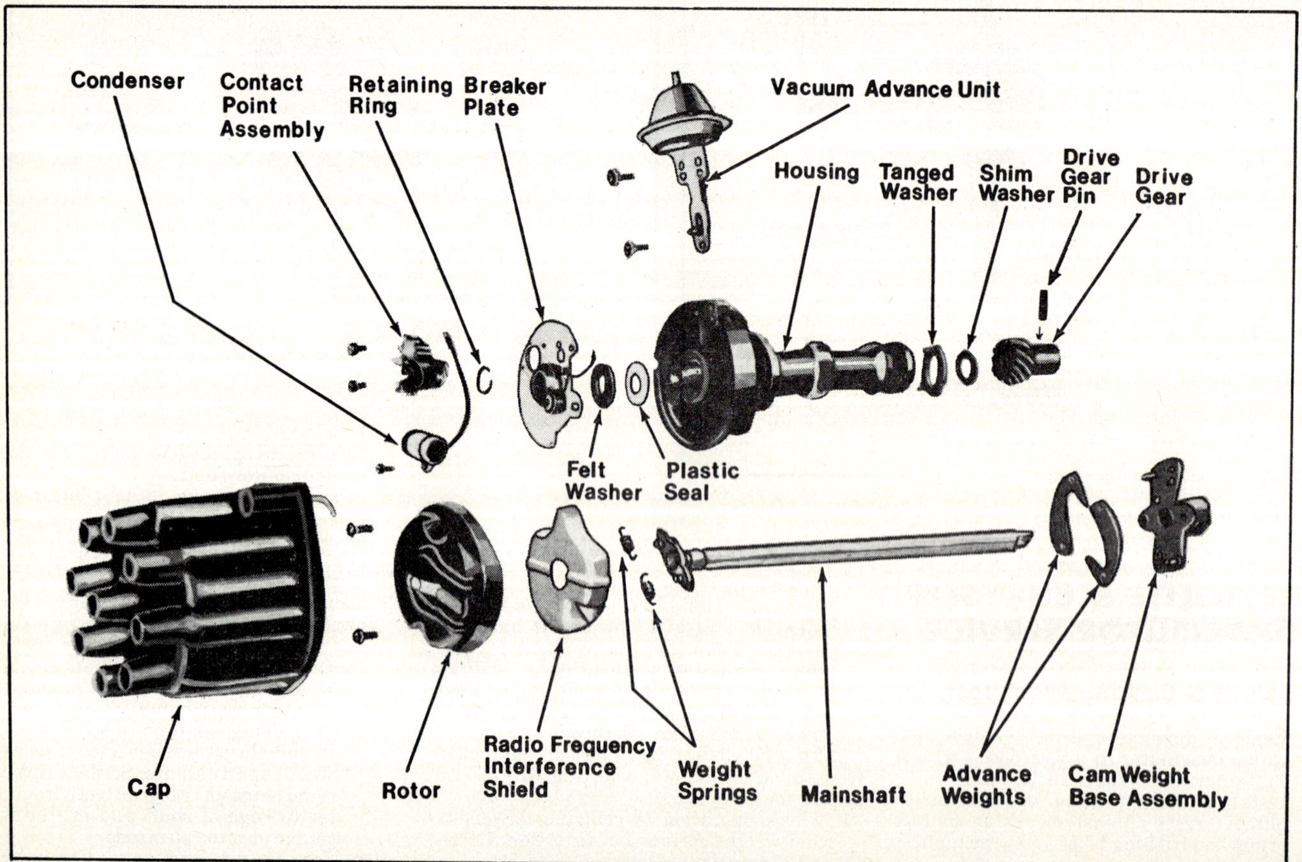

Fig. 22 Delco-Remy external adjustment distributor

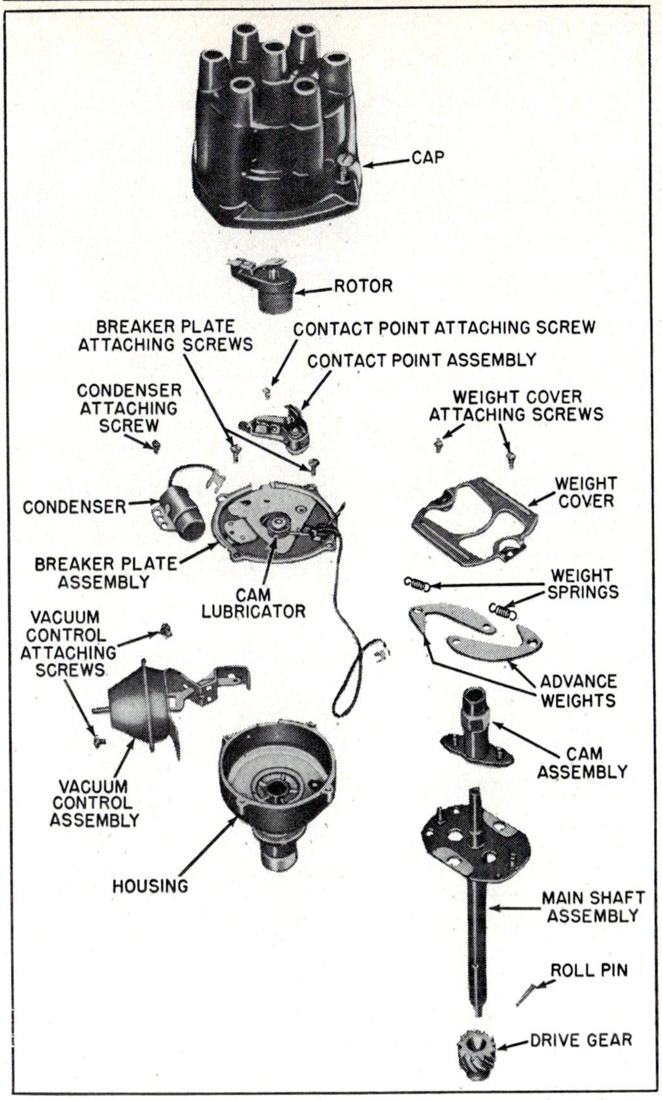

Fig. 23 Delco-Remy internal adjustment distributor

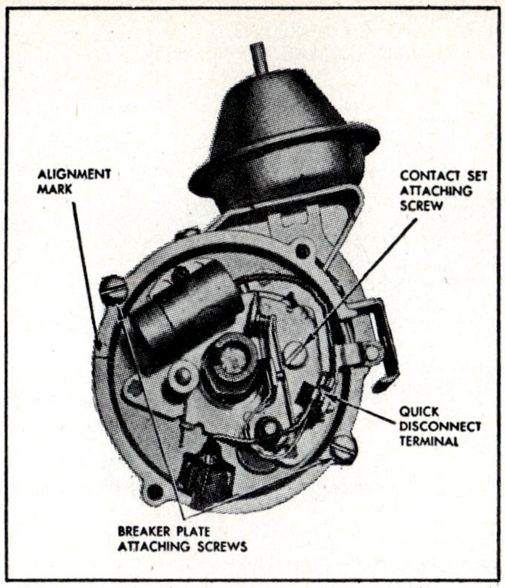

Fig. 24 Breaker plate installation. Delco-Remy internal adjustment distributor

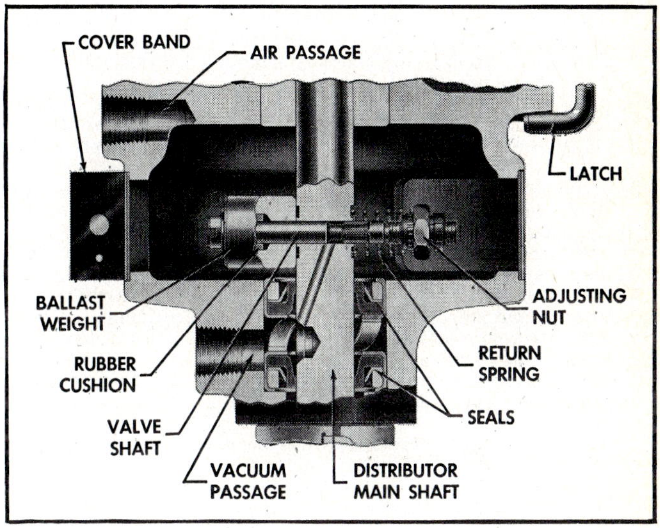

Fig. 25 Delco-Remy distributor speed governor mechanism

refer to Fig. 23. When assembling, Fig. 24 shows the details of the breaker plate and attaching parts.

1. Replace cam assembly to shaft. Lubricate top end of shaft with light engine oil prior to replacing.
2. Install weights on their pivot pins. Install weights, weight cover and stop plate.
3. Lubricate shaft and install in housing.
4. Install thrust washers and driven gear to shaft and secure with roll pins. Check to see that shaft turns freely. Install driven gear with mark on hub in line with rotor segment.
5. Install breaker plate.
6. Attach condenser and breaker point set in proper location with appropriate attaching screws, Fig. 24. Connect primary and condenser leads to breaker point set quick disconnect terminal. *Contact point set pilot must engage matching hole in breaker plate.*
7. Attach vacuum control assembly to distributor housing, using upper mounting holes.
8. Adjust breaker arm spring tension and breaker gap.
9. Install rotor.

D-R EXTERNAL ADJUSTMENT TYPE DISTRIBUTOR WITH BUILT-IN SPEED GOVERNOR

Except for the built-in maximum speed governor, Fig. 25, this distributor is similar to the standard external adjustment type.

The governor mechanism, encased in a chamber in the distributor housing, is assembled on and revolves with the distributor shaft. Disassembly, assembly and adjustment of the governor mechanism is made through external openings in the casting that are covered with a removable cover band.

The governor assembly consists of a valve shaft extending through the distributor main shaft. One end of the shaft has a slotted head, whereas the opposite end is threaded to take the adjusting nut. Assembled to the slotted head of the valve shaft is a plain washer, a ballast weight, and a rubber shock absorbing sleeve inside the weight. Assembled on the opposite end is a return spring and a self-locking adjusting nut.

Fig. 26 shows a simplified diagram of the air flow and vacuum connections between the carburetor and distributor. An air line connects the carburetor air cleaner to the governor chamber in the distributor housing. A vacuum line connects the lower opening in the distributor housing to the carburetor actuator.

When the distributor shaft revolves, the centrifugal force of the weight pulls against the spring, causing the valve shaft to move transversely in the main shaft. The slots on the valve shaft move into the main shaft of the distributor, gradually restricting the flow of air. When this restriction occurs, vacuum is created in the carburetor actuator, causing it to move the butterfly valve in the carburetor, thereby governing the engine speed. When the valve shaft in the distributor does not restrict air flow, air is brought from the air cleaner through the distributor. An overriding feature on the carburetor butterfly valve allows the foot pedal to control the butterfly valve position except when the vacuum diaphragm pull is sufficient to overcome the opposing spring force.

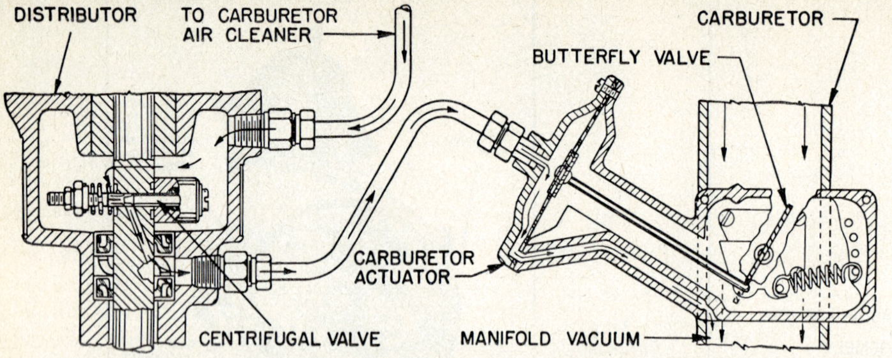

Fig. 26 Air flow and vacuum connections between carburetor and Delco-Remy governed distributor (typical)

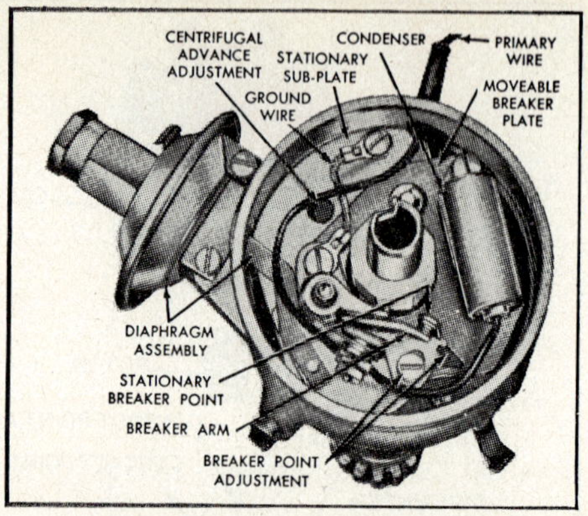

Fig. 27 Breaker plate installation. Autolite/Motorcraft dual advance distributor

Governor Adjustment

The distributor governor mechanism can be adjusted off the vehicle to an approximate setting by using an approved tester and the following procedure:

1. With the "spinner" valve (centrifugal valve) adjusted so that at least two threads on the valve are showing behind the nut, mount the distributor in the tester.
2. Screw vacuum adapter into lower vacuum opening on distributor housing.
3. With vacuum hose pinched or bent double to prevent leakage through the hose, adjust vacuum so that gauge reads 5.0" Hg. Release hose, then repeat adjustment until reading returns to 5.0" Hg each time the hose is pinched.
4. Attach vacuum hose to adapter on distributor and start tester. Increase speed until vacuum reading reaches a maximum value. This maximum value will vary, depending on the type "spinner" parts used, but will probably be above 3.8" Hg.
5. After this maximum reading has been reached, slowly decrease distributor speed until vacuum gauge pointer falls back .1" Hg. The speed at which this .1" Hg drop occurs will be the approximate no-lead governed speed. (Bear in mind that distributor rpm is half engine rpm when making this adjustment.)
6. Final adjustment must be made with the distributor on the vehicle. Road testing is necessary for accurate full-throttle governing; however, it can be assumed that full-throttle governed speed will be about 200 rpm below no-lead governed speed. Therefore, the governed speed can be checked by operating the engine at high speed in neutral.
7. To adjust distributor governing mechanism, stop the engine, remove the cover band from the distributor and turn the centrifugal valve adjusting nut, Fig. 26, holding opposite end of shaft with a screwdriver as adjusting nut is turned.
8. To increase speed limit setting, turn adjusting nut clockwise, and vice-versa.
9. When the cover band is replaced, be sure it covers the openings and is tight so as to prevent dirt from entering the chamber.

AUTOLITE/MOTORCRAFT DUAL ADVANCE DISTRIBUTOR

This distributor, Figs. 27 and 28, is similar to conventional design in that both a centrifugal advance mechanism is provided to regulate ignition timing according to speed and a vacuum advance unit to regulate ignition timing according to load. However, unlike other make distributors, the centrifugal advance mechanism can be adjusted through a slot in the breaker plate.

Adjust centrifugal advance before adjusting vacuum advance. If the correct advance is not indicated when tested on a distributor machine, bend one spring bracket with a screwdriver through the hole in the breaker plate, Fig. 29. Bend bracket away from distributor shaft to decrease advance and toward shaft to increase advance. Identify bracket after adjustment is made. After an adjustment has been made to one spring, check the minimum advance point again. Then operate distributor at the specified rpm to give an advance just

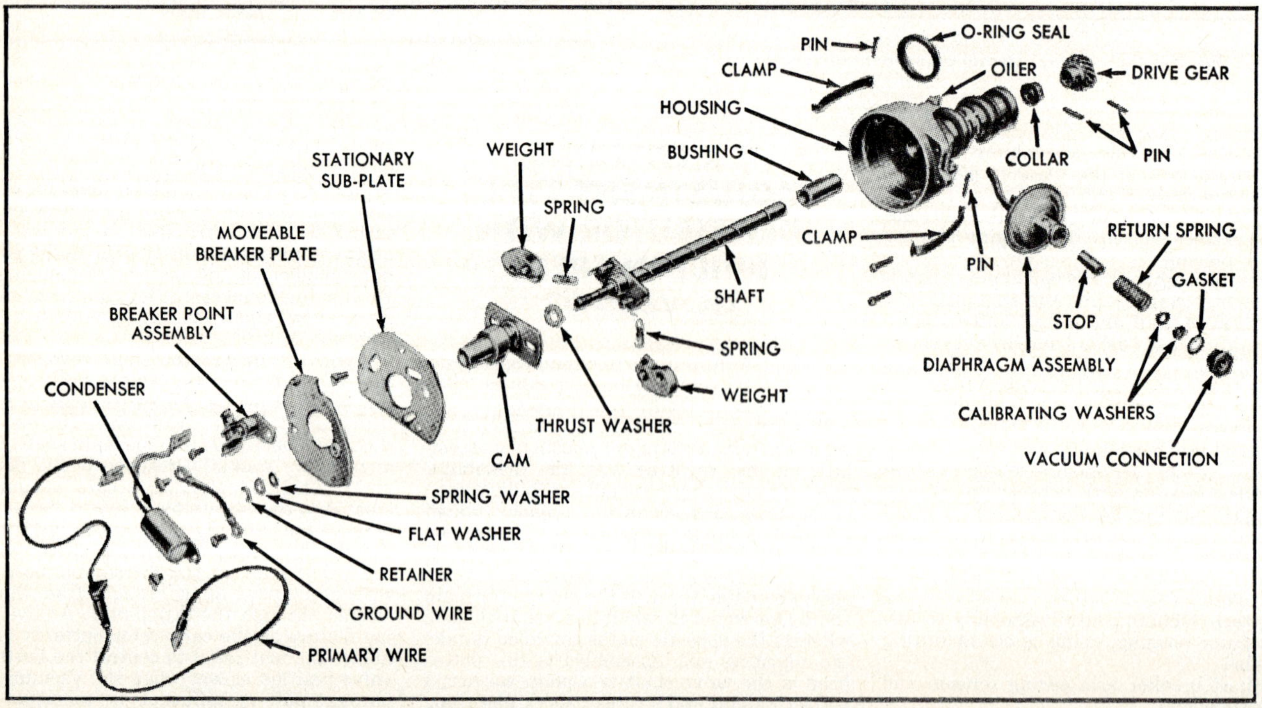

Fig. 28 Disassembled view of Autolite/Motorcraft dual advance distributor

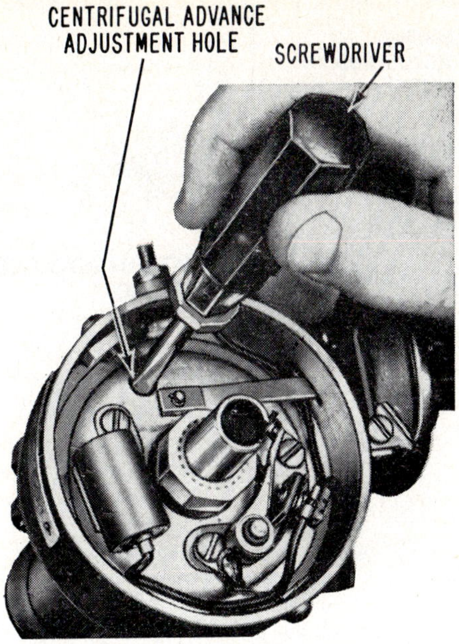

Fig. 29 Adjustment of centrifugal advance

below maximum. If this advance is not up to specifications, bend the other spring bracket to give the correct advance.

Vacuum advance on all 1970–71 distributors and 1972 units used on Super Duty engines can be adjusted by changing the calibrated washers between the vacuum chamber spring and nut, Fig. 30. The addition of one washer will decrease advance and the removal of a washer will increase advance.

Vacuum advance on 1972 distributors except with Super Duty engines and on all 1973–74 units can be adjusted by turning an allen head screw inside the vacuum unit, Fig. 31. Turning the screw clockwise increases advance and counter-clockwise decreases advance.

Distributor Service

If the distributor has been disassembled, refer to Fig. 28 for guidance upon reassembly.

1. Oil shaft and slide it into distributor body.
2. Place collar in position on shaft and align holes in collar and shaft, then install a new pin.
3. Install distributor cap clamps.
4. Check shaft end play with feeler gauge placed between collar and base of distributor. If shaft end play is not within .024–.035", replace shaft and gear.
5. Fill grooves in weight pivot pin with ball bearing grease.
6. Position weights in distributor.
7. Install weight springs, being sure proper weight, spring and adjustment bracket are assembled together.
8. Install upper thrust washer.
9. Fill grooves in upper portion of distributor shaft with ball bearing grease.
10. Install cam assembly, being sure that slots in cam engage pins in weights.
11. Install cam retainer. Apply a light film of cam lubricant to cam lobes. Saturate wick with 10W engine oil. Install wick in cam.
12. Position stationary sub plate in distributor. Install one end of ground wire under plate retaining screw close to diaphragm mounting flange.
13. Position movable breaker plate in distributor. Install spring washer on pivot pin. Place flat washer on spring washer. Be sure protruding edges of spring washer are facing upward. Install retainer.
14. Install new breaker point assembly. Install ground wire on breaker point attaching screw furthest from point adjustment slot.
15. Install condenser.
16. Working from inside to outside of distributor housing, pass primary wire through opening in distributor. Pull wire through opening until locating stop is flush with inside of distributor.
17. Connect condenser wire and primary wire to breaker points.
18. Position diaphragm and hook its link over pin on breaker plate. Install diaphragm attaching screws. Secure link with retainer. Install oil seal.
19. Adjust breaker arm spring tension, align and adjust breaker points and check and adjust cam dwell, centrifugal and vacuum advance.

Fig. 31 Vacuum advance adjustment. Autolite/Motorcraft 1972 exc. Super Duty & all 1973–74 distributors

HOLLEY "ROTO-VANCE" DISTRIBUTOR

This distributor, Fig. 32, functions as a standard ignition unit and in addition acts as a control unit for the engine speed governor valves located below the carburetor. Fig. 33.

The distributor-governor consists of an adjusting nut, a calibrating spring and governor weight. Below governing speed the weight cannot overcome the spring tension and the air inlet port is open. Thus air is permitted to flow through the distributor shaft, the governor air line and through the engine governor. This air is at atmospheric pressure.

With an increase of engine speed and distributor shaft rotational speed, centrifugal force is exerted on the weight and it moves out against spring tension. Continued increases of engine speed moves the weight far enough to close off the air inlet port. Closing the air inlet port permits vacuum pressure to be impressed on one side of the engine governor diaphragm. The opposite side of the diaphragm is vented to atmospheric pressure through a connecting line to the fresh air inlet line. The resulting pressure differential on the diaphragm causes the throttle valve to be moved towards a partially closed position against the throttle valve plate governor spring tension.

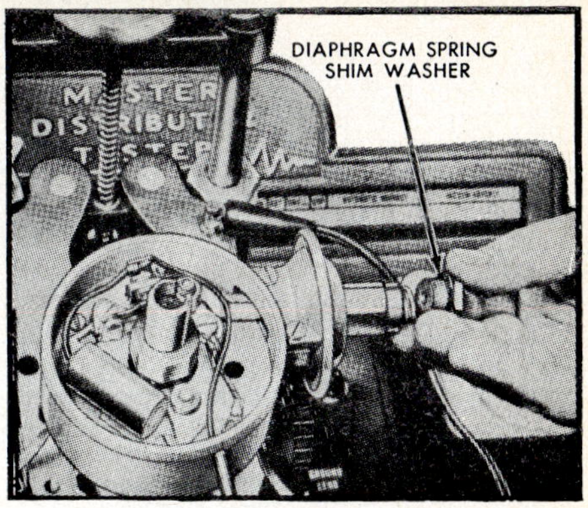

Fig. 30 Vacuum advance adjustment. Autolite/Motorcraft all 1970–71 & 1972 exc. Super Duty distributors

A decrease in engine speed allows the distributor governor valve weight spring to push the weight towards the distributor shaft, thus opening the inlet port and neutralizing the pressure differential on the diaphragm. The governor throttle plate is returned to its fully open position by the engine governor spring.

Automatic Advance Adjustment

Install distributor on a test bench and remove the adjusting hole cover plug. Check the operation of the automatic advance against specifications given in Truck Chapters. Adjust the weight springs by screwing in or out on the adjusting nuts, Fig. 34, turning both nuts the same amount. Turning in on the nuts will increase the tension and the rpm required to move the weights and will give a retarded advance. Loosening the nuts will allow the weights to move at a slower rpm and will cause the timing to advance earlier. *Be*

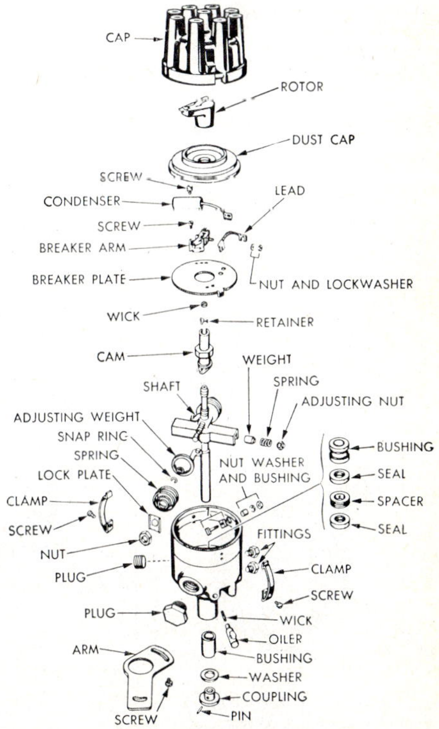

Fig. 32 Holley "Roto-Vance" (carburetor governor control) distributor

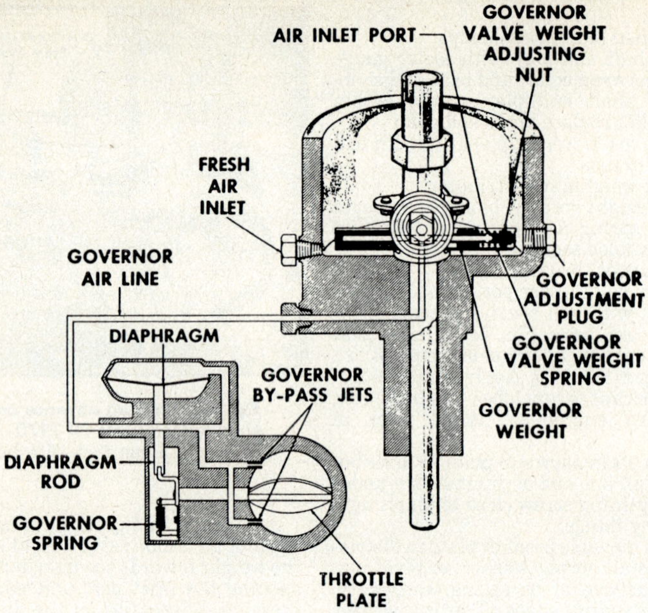

Fig. 33 Diagrammatic view of governor system. Holley "Roto-Vance" distributor

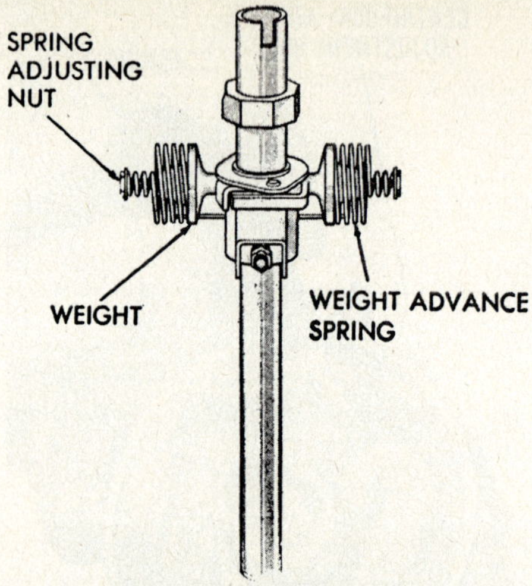

Fig. 34 Adjusting advance weight springs. Holley "Roto-Vance" distributor

sure nut flats index lock plate. Where the initial setting is correct but the advance does not follow specifications it is an indication that the advance springs are faulty or the distributor cam is binding on the shaft.

CAUTION: Care should be taken when adjusting the advance mechanism to approximate the specified curve as closely as possible. The spark timing of the engine in operating range is determined directly by the accuracy of the distributor advance mechanism. *It is important to remember that a one-degree error in the distributor advance settings will produce a two-degree error in spark timing when operating in the engine.*

Testing Engine Speed Governor Valve In Test Stand

Operate the test stand vacuum pump. Pinch the end of the vacuum hose and record the highest vacuum reading obtained. Connect the vacuum line to the lowest fitting on the distributor. Make certain that no leaks are present.

Without rotating the distributor shaft, note the vacuum reading on the gauge. If the passages in the distributor are free of obstructions, a desirable very low reading should be obtained. An undesirable high reading will indicate that obstructions in the passages are present, caused by dirt or other causes. The distributor should then be thoroughly checked to determine cause and corrections made.

NOTE: Some test machines have a pipe cleaner type of restrictor. For these tests this restrictor should be removed or an inaccurate reading will be obtained. The vacuum line in the machine must be open.

Rotate the distributor shaft and gradually increase the speed while observing the vacuum gauge. When the distributor shaft speed reaches the rpm necessary to move the governor weight out on its shaft and close the air inlet port, the vacuum reading will "jump" to a high reading almost instantly. The vacuum reading at this rpm and at all higher speeds should be within one or two inches (Hg) of the reading recorded at the start of the test. If the vacuum reading is within the specified one to two inches of the previous reading, it is an indication that the weight valve is sealing the inlet port and that no noticeable leaks are apparent.

Decrease the distributor shaft speed gradually while observing the vacuum gauge. When the speed drops slightly below the governor closing speed, the air inlet port should open, causing a rapid drop in vacuum.

HOLLEY 1500 & 1510 DISTRIBUTORS

This distributor is designed so that it is available with centrifugal advance, vacuum advance, a governor control valve and a tachometer drive, or with any combination of these features. The governor valve is identical in function to that used with the "Roto-Vance" distributor. The distributor also features a fixed breaker plate and a movable plate on which the breaker points are mounted, Fig. 35.

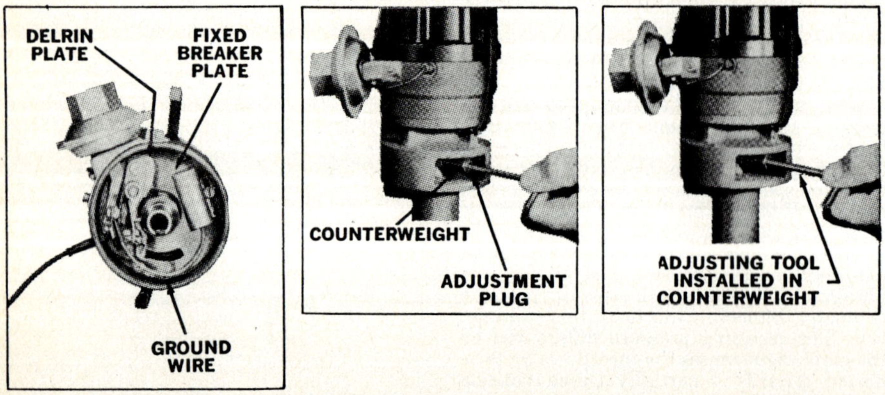

Fig. 35 Holley 1500 distributor

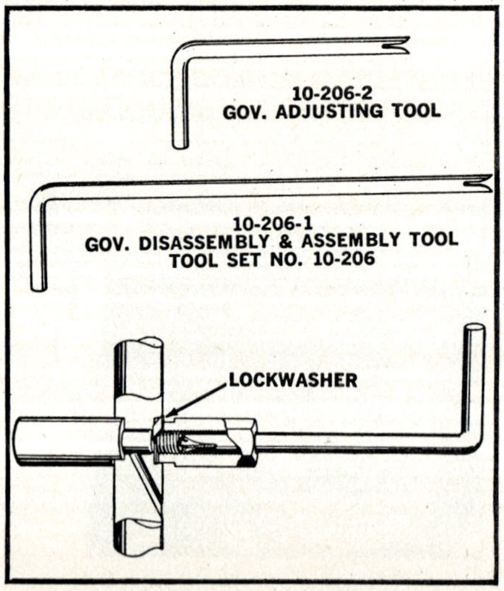

Fig. 36 Governor service tools for Holley 1500 distributor

Distributor Service

The movable breaker plate can easily be removed as a unit with the old breaker points. New breaker points can then be installed and aligned with ease and reinstalled as an assembly in the distributor. Spring tension and breaker gap are tested and adjusted in the conventional manner.

Governor, Adjust

For adjustment on the engine, the shorter tool shown in Fig. 36 is used because other engine components are in the way. The longer tool is necessary for disassembly and assembly of the governor valve. These tools are available from any Holley distributor.

1. Remove governor service band and gasket assembly.
2. Remove hex head plug at adjusting end of valve, using a 1/8" Allen wrench.
3. Install slotted end of tool over spring adjusting nut.

NOTE: The spring type adjusting screw cannot be adjusted with a screwdriver or by any other means except by this special tool.

Disassemble Governor Valve

1. Remove service band and gasket.
2. Bend ears of lockwasher from governor counterweight.
3. Remove hex head plug.
4. Slip a 7/16" deep socket over the long disassembly tool and loosen counterweight.
5. Hold tool in fixed position and unscrew counterweight by rotating deep socket with the fingers.
6. When spring adjusting screw is completely out of governor counterweight, let the socket and counterweight slide up the shaft of the tool, and push the valve spring and adjusting nut out of the opening in the opposite side of the distributor.

ELECTRONIC IGNITION SYSTEMS

INDEX

American Motors Breakerless Inductive Discharge (BID) Ignition System

DESCRIPTION

The BID ignition system incorporates four major units; an electronic control unit, ignition coil, distributor and high tension wires, Fig. 1. The electronic control unit is a solid-state, moisture resistant module with the components sealed in a potting compound to resist vibration and environmental conditions. Since the control unit has an internal current regulator, a resistance wire or ballast resistor is not necessary in the primary circuit. Battery voltage is applied to the ignition coil positive terminal when the ignition switch is in the "On" or "Start" position, therefore, an ignition system bypass is not required in this system. The primary coil circuit is electronically regulated by this unit.

The ignition coil is of standard construction and requires no special service. The function of the ignition coil in the BID ignition system is the same as for conventional ignition systems.

The distributor is conventional except the contact points, condenser and cam are replaced by a sensor and trigger wheel, and since no wearing occurs between the trigger wheel and sensor, dwell angle remains constant and requires no adjustment. The sensor is a small coil of fine wire and receives an alternating current signal from the electronic control unit. The sensor develops an electromagnetic field used to detect the presence of

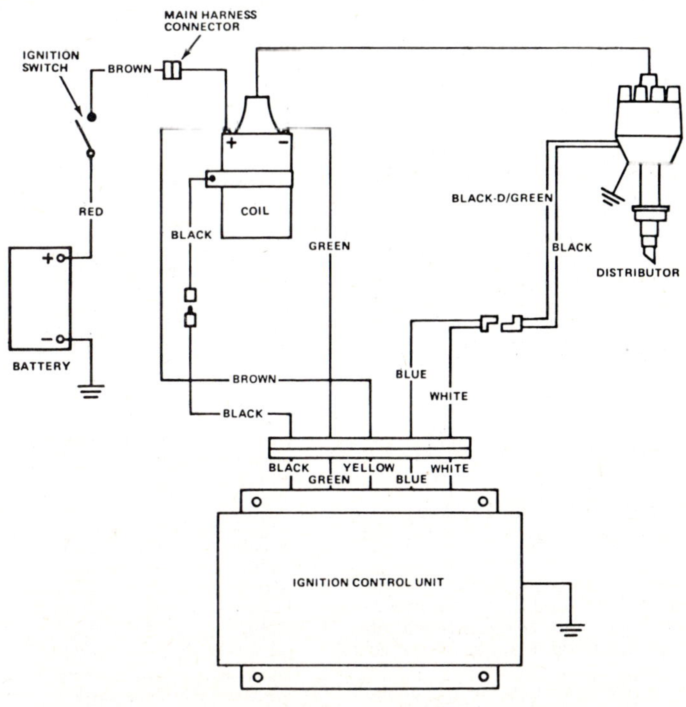

Fig. 1 American Motors BID ignition system wiring

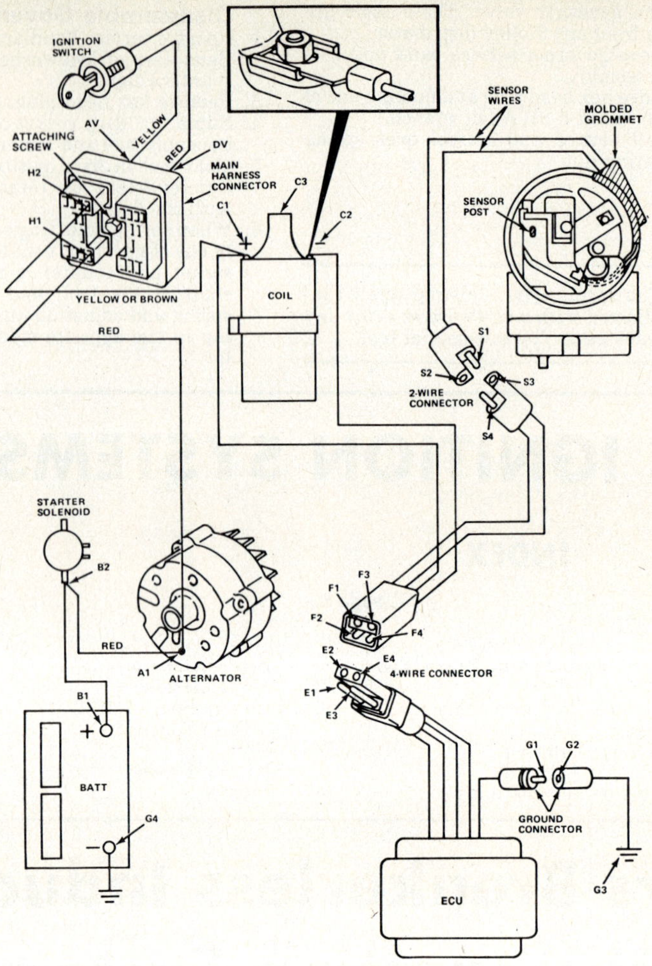

Fig. 2 BID ignition system checkpoints

metal which are the leading edges of the trigger wheel teeth.

OPERATION

When the ignition switch is placed in the "Start" or "Run" position, the control unit is activated. An oscillator within the control unit excites the sensor coil, in turn developing the electromagnetic field. When a leading edge of a trigger wheel tooth enters the electromagnetic field, the tooth reduces the sensor oscillation strength to a predetermined level, in turn activating the demodulator circuit. The demodulator circuit controls a power transistor located in series with the coil primary circuit. The power transistor switches the coil primary circuit off, thereby inducing a high voltage in the coil secondary winding. The high voltage is then delivered to the spark plugs through the distributor rotor, cap and high tension wires.

TROUBLESHOOTING, FIG. 2

1. Disconnect Electronic Control Unit (ECU) ground wire and the four wire connector. Using a small wire brush and solvent, clean the terminals. Leave connectors disconnected.
2. Disconnect battery cables, then momentarily move ignition switch to START and allow it to move to ON.
3. Measure resistance of entire ignition feed circuit by connecting an ohmmeter to battery positive B1 and F3 terminal in the 4-wire connector.
4. If resistance is less than 1 ohm, tighten main harness connector attaching screw to fully seat connector. If resistance is 1 ohm or more, isolate trouble area by connecting the ohmmeter and measuring the resistance of each portion of the ignition feed circuit between the following terminals: B1 and B2, B2 and A1, A1 and H1, F3 and C1, C1 and H2, H1 and H2, AV and DV. Clean, tighten or reposition connectors as needed.
5. Inspect coil primary connections for looseness and proper assembly. Wire terminals must be between channel washer and nut, channel washer tabs must be facing up. Reposition and tighten as needed.
6. Connect an ohmmeter between F3 and F4 terminals to measure coil primary circuit resistance. If resistance is 1–2 ohms, proceed to step 7. If resistance is less than 1 ohm, replace coil and proceed to step 8. If resistance is more than 2 ohms, isolate trouble and insure that the resistance between the following terminals is as follows: C1 and C2; 1–2 ohms, F3 and C1; 0 ohms, F4 and C2; 0 ohms. Replace coil or repair trouble area and proceed to step 7. If coil was replaced, proceed to step 8.
7. Measure coil secondary resistance by removing coil secondary wire from coil and connecting ohmmeter between C1 and C3.

NOTE: Set ohmmeter to the 1,000 ohm scale before testing secondary resistance.

If resistance is 9,000–15,000 ohms, reconnect coil secondary wire and proceed to step 8. If resistance is less than 9,000 ohms or more than 15,000 ohms, replace coil and reconnect coil secondary wire and proceed to Step 8.

8. Remove distributor cap, rotor and dust cover and check for 1.6–2.4 ohms resistance between F1 and F2 terminals.

NOTE: Pull and flex sensor wires, firmly squeeze molded sensor grommet at distributor, and apply firm side to side pressure on sensor post while checking resistance.

If resistance is as specified and steady, proceed to step 10. If resistance is too high or low, or if needle fluctuates or wavers, proceed to step 9.

9. Disconnect 2 wire connector and check for 1.6–2.4 ohms resistance between S1 and S2 terminals.

NOTE: Pull and flex sensor wires, firmly squeeze molded sensor wire grommet at distributor, and apply firm side to side pressure on sensor post.

If resistance is as specified and steady, proceed to step 10. If resistance is too high or two low, or if needle fluctuates or wavers, replace sensor and proceed to step 10.

10. Using a small wire brush and solvent, clean terminals S1, S2, S3 and S4.
11. Measure resistance of ECU ground circuit by connecting an ohmmeter between terminal G2 and battery negative cable G4. Resistance should be 0 ohms. Clean and tighten connections as needed.
12. Using pliers, squeeze terminals, E1, E2, E3, E4 and G1 until terminals have a distinct oval shape, thereby assuring a tight fit when terminals are connected.
13. Using petroleum jelly, coat the male terminals in the 4 wire connector and ground connectors and around the outer edge of the terminal end of the ECU 4 wire connector.
14. Connect the 4 wire and ground connectors.
15. Connect Pulse Simulator J-25331 to the S3 and S4 terminals and reconnect battery cables. Remove coil secondary wire from distributor and place end of wire ½ inch from ground, then with ignition switch ON, operate simulator and observe for spark across the ½ inch gap. If spark jumps gap, proceed to step 17. If spark does not jump gap, proceed to step 16.
16. Disconnect wire from coil negative terminal, then connect one pulse simulator clip to the coil negative terminal and connect the remaining clip to the ground. With ignition switch ON, operate pulse simulator and observe for spark across the ½ inch gap when button is released. If spark jumps the gap, replace ECU unit and proceed to step 17. If spark does not jump gap, replace coil before proceeding to step 17.

NOTE: If ECU unit is replaced, squeeze and lubricate the connectors as described in steps 12 and 13.

17. Disconnect pulse simulator. Before connecting 2-wire connector, squeeze and lubricate terminals as described in steps 12 and 13. Connect coil negative wire.
18. Inspect distributor cap for cracks and carbon tracks. Replace as necessary. Install dust cover, rotor, distributor cap and coil secondary wire.

NOTE: If the malfunction still exists after performing the above procedure, connect an engine oscilloscope and measure the ignition dwell. A scope must be used because a dwell meter will not accurately measure the dwell on this system. Start the engine and observe the dwell readings, then lower the hood being careful not to squeeze the oscilloscope leads between the fender and hood and allow the engine and wiring harness to warm up for 10 minutes. Prior to, during and after the engine warm up period, the dwell must be 23° on V8 engines and 32° for 6 cylinder engines and the dwell must not vary more than 3° at any time. If the dwell readings are as specified at all times, the malfunction may be caused by the fuel system. If the fuel system checks out OK and the malfunction still exists, replace the ECU unit and recheck the dwell. If the dwell readings are still not as specified above, replace the ignition wiring harness and recheck the dwell above. If the dwell is below specifications, disconnect the 4-wire connector and measure the sensor circuit resistance and integrity as described in step 8. If sensor circuit resistance and integrity are within specifications, replace ECU unit. If sensor circuit resistance is too high or too low, or if needle fluctuates, replace the sensor. Be sure to squeeze and lubricate the terminals as described in steps 12 and 13.

COMPONENT REPLACEMENT

1. Place distributor in a suitable holding fixture and remove cap, rotor and dust shield, Fig. 3.
2. Using a small gear puller, remove trigger wheel. Ensure puller jaws are gripping trigger wheel inner shoulder to prevent trigger wheel damage. Also, use a thick flat washer or nut as a spacer and do not press against small center shaft.
3. Loosen sensor locking screw approximately three turns, lift sensor lead grommet from distributor bowl and pull sensor leads from slot around sensor spring pivot pin. Release sensor spring, ensure spring clears sensor leads and slide sensor from bracket.

NOTE: The sensor locking screw utilizes a tamper proof head design and requires tool J-25097 for removal. However, if special tool is not available, use a small needlenose plier to remove screw. The service (replacement) sensor has a standard slotted head screw.

4. If vacuum control unit is to be replaced, remove retaining screw and vacuum unit.
5. Install new vacuum control unit and assemble sensor, sensor guide, flat washer and retaining screw.

NOTE: Install retaining screw far enough to hold assembly together and ensure it does not protrude past bottom of sensor.

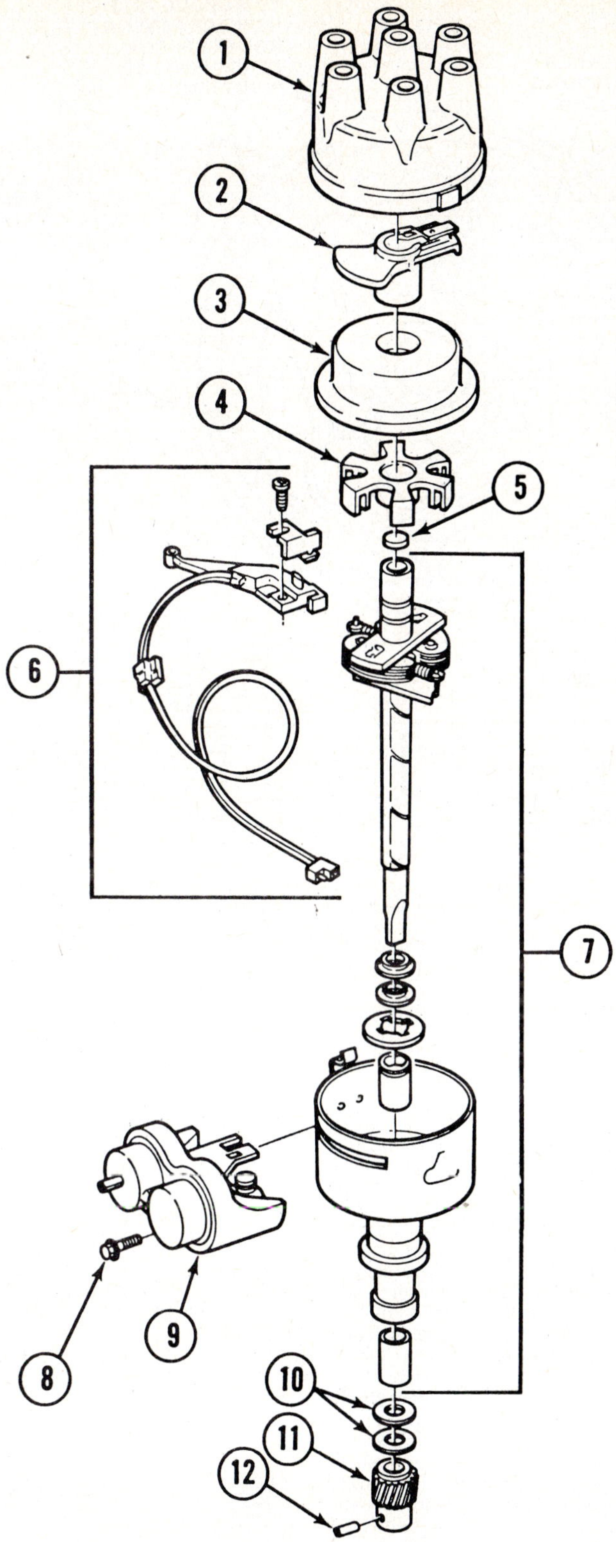

1. DISTRIBUTOR CAP
2. ROTOR
3. DUST SHIELD
4. TRIGGER WHEEL
5. FELT
6. SENSOR ASSEMBLY
7. HOUSING
8. VACUUM CONTROL SCREW
9. VACUUM CONTROL
10. SHIM
11. DRIVE GEAR
12. PIN

Fig. 3 Exploded view of BID distributor

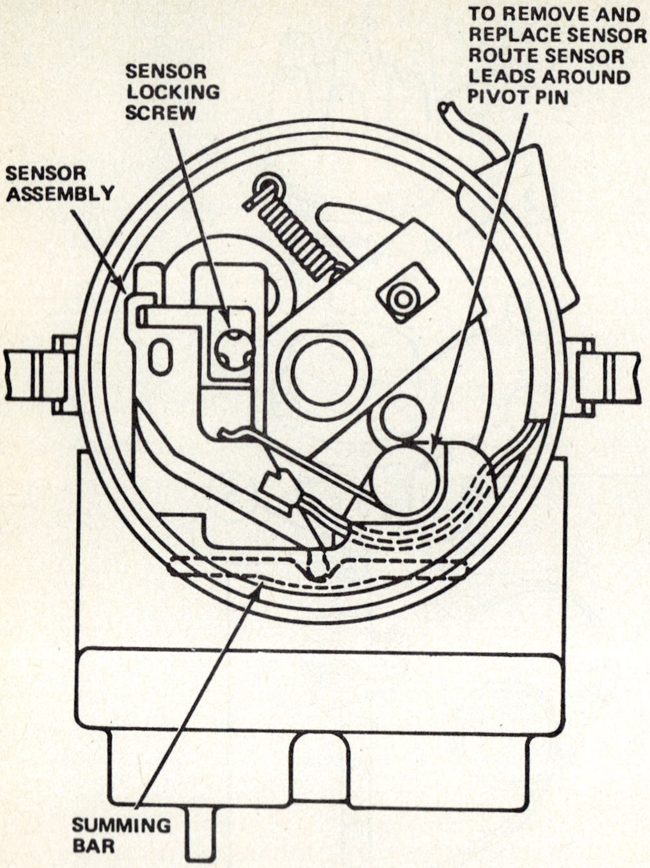

Fig. 4 Sensor installation

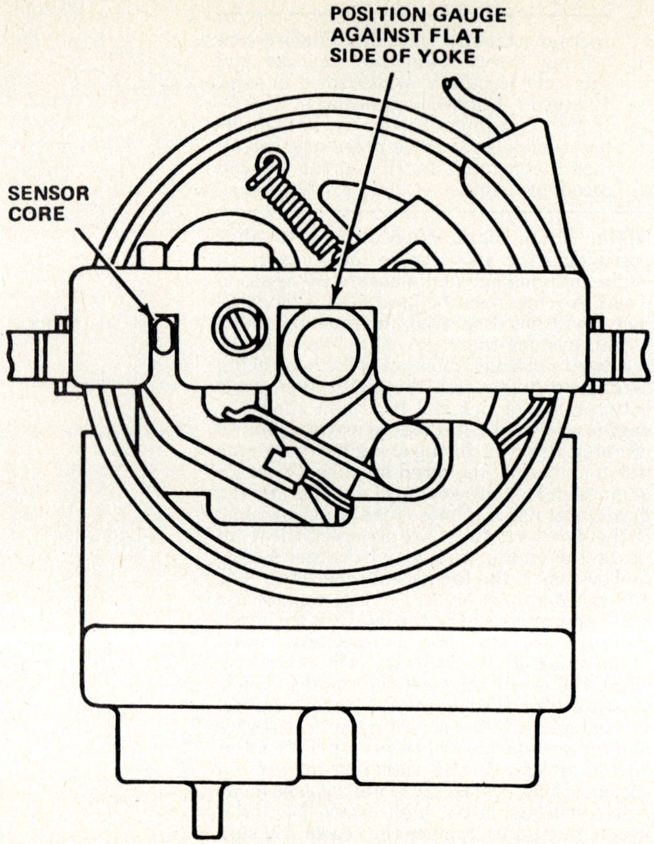

Fig. 5 Positioning sensor

6. If vacuum control has been replaced and original sensor is being used, replace special head screw with standard slotted head screw.
7. Install sensor assembly on vacuum chamber bracket, ensuring tip of sensor located properly in summing bar. Place sensor spring on sensor and route sensor leads around spring pivot pin, Fig. 4. Install sensor lead grommet and position leads away from trigger wheel.
8. Install sensor positioning gauge over yoke, ensure gauge is against flat of shaft, and move sensor sideways until gauge can be positioned. Snug retaining screw and check sensor position by removing and installing gauge, Fig. 5. When gauge can be removed and replaced without sensor side movement, sensor is positioned properly. Tighten retaining screw and check sensor position.
9. Place trigger wheel on yoke and check if sensor core is positioned approximately in center of trigger wheel legs. Bend a .050 inch gauge wire to dimension specified in Fig. 6, and place between trigger wheel legs and sensor base. Press trigger wheel onto yoke until legs contact gauge wire.
10. Apply 3 to 5 drops of light engine oil to felt wick in top of yoke, then install dust shield, rotor and cap.

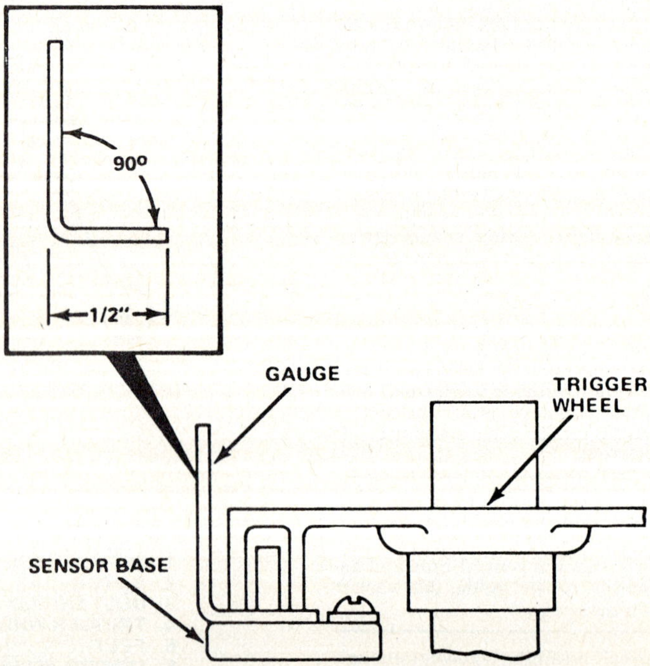

Fig. 6 Trigger wheel installation

American Motors Solid State Ignition System

DESCRIPTION

This ignition system, Figs. 1 & 2, is used on 1978–82 vehicles. The solid state ignition system consists of the ignition switch, electronic ignition control unit, ignition coil, primary resistance wire and bypass, distributor, spark plugs and on 1982–83 six cylinder models with electronic feedback system, an electronic spark retard.

The electronic ignition control unit is a solid state, moisture resistant module. The component parts are sealed in a potting material to resist vibration and environmental conditions. The control unit is incorporated with reverse polarity protection and transient voltage protection.

The distributor incorporates a sensor and trigger wheel. Current flowing through the ignition coil creates a magnetic field in the primary windings. When the circuit is opened, the magnetic field collapses and induces a high voltage in the coil secondary windings. This circuit is electronically controlled by the electronic ignition control unit. The distributor sensor and trigger wheel provide the signal to operate the control unit. The trigger wheel is mounted on the distributor shaft and has one tooth for each cylinder. The sensor, a coil of fine wire mounted to a permanent magnet, develops an electromagnetic force that is sensitive to the presence of ferrous metal. The sensor detects the trigger wheel teeth as the teeth pass the sensor. When a trigger wheel tooth approaches the pole piece of the sensor, it reduces the reluctance of the magnetic field, increasing field strength. Field strength decreases as the tooth moves away from the pole piece. This increase and decrease of field strength generates an alternating current which is interpreted by the electronic ignition control unit. The control unit then opens and closes the ignition coil primary circuit.

Since there are no contacting surfaces and no wear occurs, dwell angle requires no adjustment. The dwell angle is electronically controlled by the electronic ignition control unit. When the coil circuit is switched open, an electronic timer in the control unit keeps the circuit open only long enough for the spark to discharge. Then, it automatically closes the ignition coil primary circuit.

TROUBLE SHOOTING

Secondary Circuit Test

1. Disconnect coil wire from distributor cap and, using insulated pliers, hold wire approximately 1/2 inch from a good engine ground.
2. Crank engine and observe wire for spark. If no spark occurs, proceed to Step 5. If spark occurs, proceed to Step 3.
3. Reconnect coil wire to distributor cap. Remove wire from one spark plug.

CAUTION: Do not remove wires from spark plugs on cylinders 3 and 5 on 1978–79 six cylinder models or cylinders 1 and 5 on 1980–83 models. On eight cylinder models, do not remove spark plug wires from cylinders 3 or 4. Removing these wires during testing

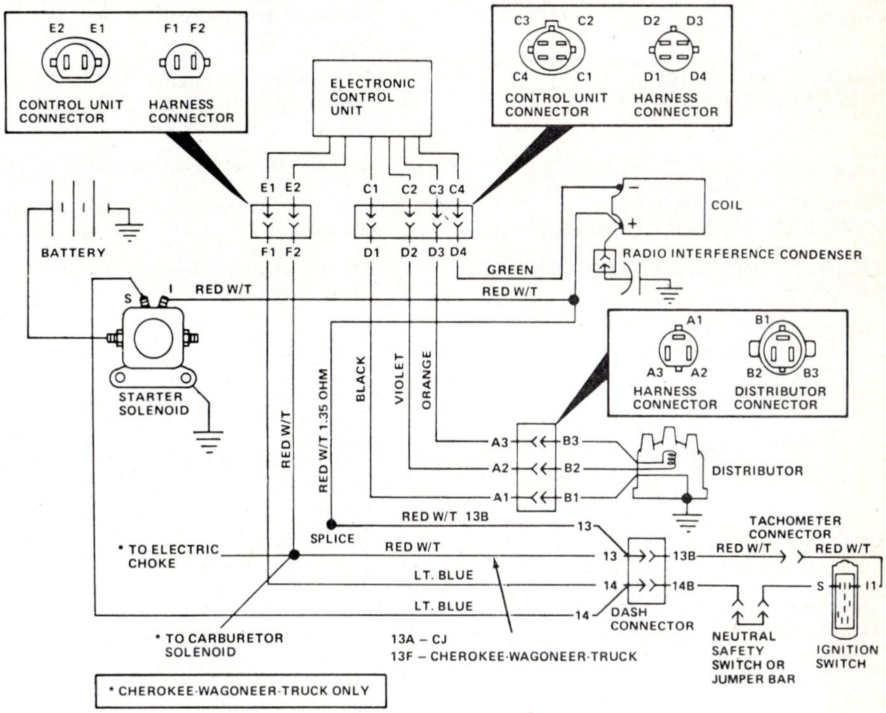

Fig. 1 Schematic view of American Motors solid state ignition system. All except 1982–83 models with electronic feedback system

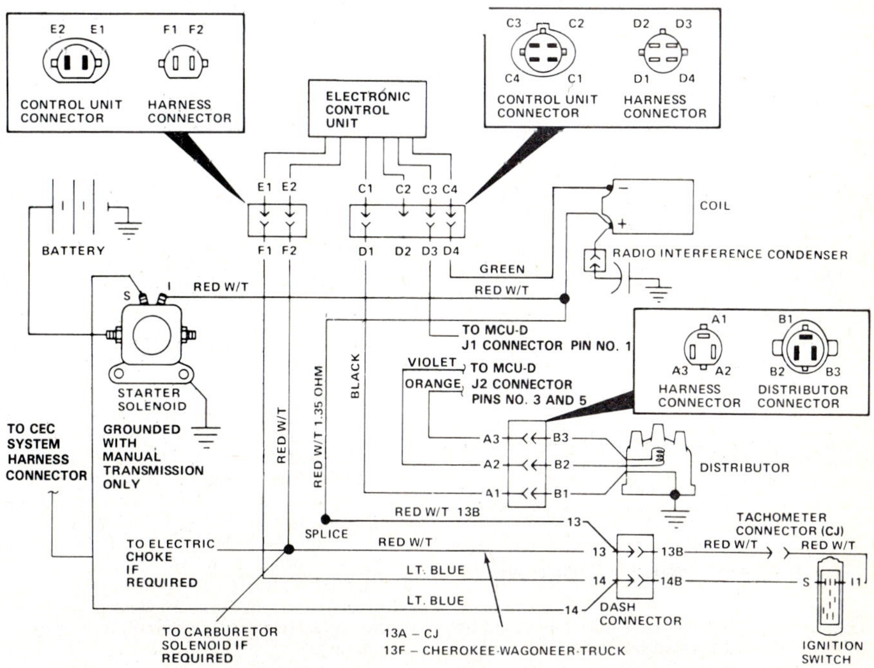

Fig. 2 Schematic view of American Motors solid state ignition system. 1982–83 6 cyl. models with electronic feedback system

Condition	Possible Cause	Correction
ENGINE FAILS TO START (NO SPARK AT PLUGS)	1. No voltage to ignition system.	1. Check battery, ignition switch and wiring. Repair as required.
	2. Electronic Control Unit ground lead inside distributor open, loose or corroded.	2. Clean, tighten or repair as required.
	3. Primary wiring connectors and fully engaged.	3. Clean and fully engage connectors.
	4. Coil open or shorted.	4. Test coil. Replace if faulty.
	5. Electronic Control Unit defective.	5. Replace Electronic Control Unit.
	6. Cracked distributor cap.	6. Replace cap.
	7. Defective rotor.	7. Replace rotor.
ENGINE BACKFIRES BUT FAILS TO START	1. Incorrect ignition timing.	1. Check timing. Adjust as required.
	2. Moisture in distributor.	2. Dry cap and rotor.
	3. Distributor cap faulty.	3. Check cap for loose terminals, cracks and dirt. Clean or replace as required.
	4. Ignition wire not in correct firing order.	4. Install in correct order.
ENGINE RUNS ONLY WITH KEY IN START POSITION	1. Open in resistance wire or excessive resistance.	1. Repair resistance wire.
ENGINE CONTINUES TO RUN WITH KEY OFF	1. Defective starter solenoid.	1. Replace solenoid.
	2. Defective ignition switch.	2. Replace switch.
ENGINE DOES NOT OPERATE SMOOTHLY AND/OR ENGINE MISFIRES AT HIGH SPEED	1. Spark plugs fouled or faulty.	1. Clean and gap plugs. Replace as required.
	2. Ignition cables faulty.	2. Check cables. Replace as required.
	3. Spark advance system(s) faulty.	3. Check operation. Repair as required.
	4. "I" terminal shorted to starter terminal in solenoid.	4. Replace solenoid.
	5. Trigger wheel pin missing.	5. Install pin.
	6. Distributor wires installed in wrong firing order.	6. Install wires correctly.
EXCESSIVE FUEL CONSUMPTION	1. Incorrect ignition timing.	1. Check timing. Adjust as required.
	2. Spark advance system(s) faulty.	2. Check operation. Repair as required.
ERRATIC TIMING ADVANCE	1. Faulty vacuum advance assembly.	1. Check operation. Replace if required.
	2. Centrifugal weights sticking.	2. Remove dirt, corrosion.
TIMING NOT AFFECTED BY VACUUM	1. Defective vacuum advance unit.	1. Replace vacuum advance unit.
	2. Advance unit adjusting screw too far counterclockwise.	2. Turn screw clockwise to bring advance curve within specifications.
	3. Sensor pivot corroded.	3. Clean pivot.
INTERMITTENT OPERATION	1. Loose or corroded terminals.	1. Clean and tighten terminals. Apply electrical grease.
	2. Defective sensor.	2. Perform sensor tests.
	3. Defective control unit.	3. Perform control unit tests.
	4. Loose ground connector in distributor.	4. Clean and tighten connection.
	5. Wires to distributor shorted together or to ground.	5. Check for frayed, pinched or burned wires.

Fig. 3 American Motors solid state ignition system service diagnosis chart

could result in sensor damage.

4. Using insulated pliers, hold wire approximately ½ inch from a good engine ground. Crank engine and observe wire for spark. If spark occurs, check for fuel system problems or incorrect ignition timing. If no spark occurs, check for defective distributor cap, rotor or spark plug wires.
5. If no spark occurs at coil wire, measure coil wire resistance. If resistance is greater than 10,000 ohms, replace wire.
6. If malfunction still exists, proceed to the following tests or diagnosis procedures.

Intermittent Failure Diagnosis

Since intermittent failure may be caused by loose or corroded terminals, defective components, poor ground connections or defective wiring, it is necessary to check all wiring connections in the ignition system. Also, refer to Fig. 3 for further diagnosis.

Ignition Coil Primary Circuit Test

1. Turn ignition switch "On" and connect a voltmeter between ignition coil positive terminal and the ground. If voltage is 5.5–6.5 volts, proceed to Step 2. If battery voltage is noted, proceed to Step 4. If voltage is below 5.5 volts, disconnect condenser lead. If voltage is not 5.5–6.5 volts, replace condenser. If voltage is still not within specifications, proceed to Step 6.
2. Turn ignition switch to "Start" and measure voltage at coil positive terminal while cranking engine. If battery voltage is present while cranking engine, the ignition coil primary circuit is satisfactory. If voltage present is less than battery voltage, proceed to Step 3.
3. Check for shorted or open circuit in wire attached to starter "I" terminal. Check for defective starter solenoid. Repair as necessary.
4. Place ignition switch in "On" position, disconnect wire from starter solenoid "I" terminal and measure voltage at ignition coil positive terminal. If voltage drops to 5.5–6.5 volts, replace starter solenoid. If voltage remains constant at battery voltage, connect a jumper wire between ignition coil negative terminal and the ground. If voltage drops to 5.5–6.5 volts, proceed to Step 5. If not, repair defective resistance wire and repeat Step 2.
5. Check continuity between ignition coil negative terminal and terminal "D4", Figs. 1 and 2. Also, check continuity between terminal "D1" and the ground. If continuity is present, replace electronic ignition control unit. If continuity is not present, locate and repair open circuit.
6. Turn ignition switch "Off" and measure resistance between ignition coil positive terminal and the dash connector "13", Figs. 1 and 2. If resistance is greater than 1.40 ohms, repair or replace resistance wire. If resistance is 1.30–1.40 ohms, proceed to Step 7.
7. With ignition "Off", measure resistance between dash connector "13" and ignition switch terminal "I1", Figs. 1 and 2. If resistance is greater than .1 ohm, check and repair terminal connections at dash

connector or defective wiring.

Coil Test

1. Inspect ignition coil for oil leaks, exterior damage and carbon tracks. If satisfactory, proceed to Step 2. If not, replace ignition coil.
2. Disconnect ignition coil connector and connect ohmmeter between coil terminals. If resistance is 1.13–1.23 ohms at 75° F. or 1.5 ohms at 200° F., proceed to Step 3. If not, replace ignition coil.
3. Connect ohmmeter between ignition coil center tower and the plus or minus terminal. Resistance should be 7700–9300 ohms. If not replace ignition coil.

Sensor & Control Unit Test

1. Disconnect the four wire connector at the control unit, Figs. 1 and 2. Disconnect coil wire from center tower of distributor and hold wire approximately 1/2 inch from a good engine ground with insulated pliers, then turn ignition "On". If spark is observed at coil wire, proceed to next step. If not, proceed to Step 5.
2. Measure resistance between terminals "D2" and "D3" of the harness connector, Figs. 1 and 2. If resistance is 400–800 ohms, proceed to Step 6. If not, proceed to next step.
3. Disconnect and connect the three wire connector at distributor and measure resistance between terminals "D2" and "D3" of the harness connector, Figs. 1 and 2. If resistance is now between 400–800 ohms, proceed to Step 6. If not, disconnect three wire connector at distributor and proceed to next step.
4. Measure resistance between terminals "B2" and "B3" of the distributor connector, Figs. 1 and 2. If resistance is not 400–800 ohms, repair or replace harness between three wire and four wire connector.
5. Connect an ohmmeter between terminal "D1" of the harness connector, Figs. 1 and 2, and the battery negative terminal. If reading is below .002 ohm, repeat Step 2. If not, check for an improper ground. Check ground cable resistance, distributor to engine block resistance and distributor ground screw to terminal "D1" resistance, Figs. 1 and 2.
6. Using a voltmeter connected between terminals "D2" and "D3" of the harness connector, Figs. 1 and 2, observed reading while cranking engine. If voltmeter reading fluctuates, it indicates proper sensor and trigger wheel operation. If not, the trigger wheel is defective or the distributor is not rotating.

Ignition Feed To Control Unit Test

NOTE: Perform the "Ignition Coil Primary Circuit Test" before performing this test.

1. Disconnect two wire connector from control unit and connect a voltmeter between terminal "F2" and the ground, Figs. 1 and 2. Turn ignition "On". If voltmeter reading is within .2 volts of battery voltage, replace control unit and proceed to Step 3. If voltmeter reading is not within .2 volt of battery voltage, proceed to next step.
2. Locate and repair cause of voltage reduction as noted in Step 1. Check for a corroded dash connector or defective ignition switch. Then, check for spark at coil wire. If not, replace control unit.
3. Connect the two wire connector at control unit and disconnect the four wire connector from control unit. Connect an ammeter between terminal "C1" and the ground, Figs. 1 and 2. If ammeter reading is 1 ± .1 amp., the system is satisfactory. If ammeter reading is higher or lower, replace module.

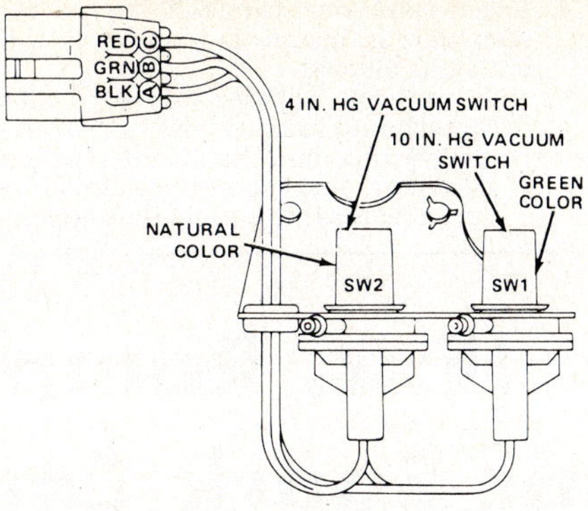

Fig. 4 Vacuum switch assembly

Current Flow Test

1. Remove connector from coil.
2. Depress plastic barb and remove positive wire from connector. Remove negative wire in same manner.
3. Connect an ammeter between coil positive terminal and the disconnected positive wire.
4. Connect a jumper wire between coil negative terminal and the engine ground.
5. Turn ignition "On" and note ammeter reading. Reading should be approximately 7 amps. and not exceeding 7.6 amps. If reading exceeds 7.6 amps., replace ignition coil.
6. Remove jumper wire from coil negative terminal and connect the coil green wire to negative terminal. Ammeter reading should be approximately 4 amps. If reading is less than 3.5 amps., check for poor connections at the three wire and four wire connectors or a poor ground at distributor ground screw. If reading is greater than 5 amps., the control unit is defective, requiring replacement.
7. Start and run engine. Ammeter reading should be 2–2.4 amps. If not, replace control unit.

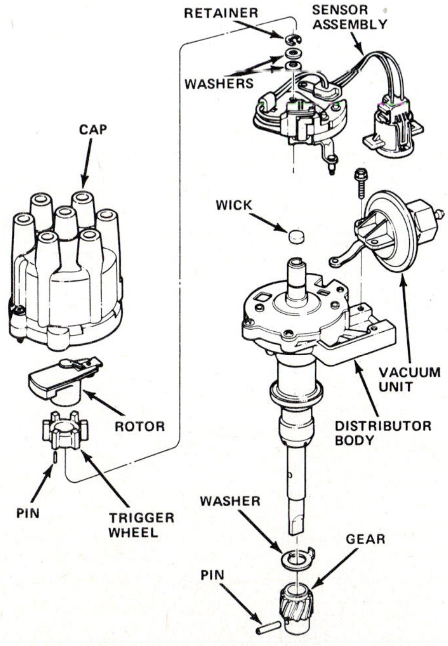

Fig. 5 American Motors distributor used with solid state ignition system (Typical)

Electronic Spark Retard Test

NOTE: This test only applies to 1982–83 models with electronic feedback system.

1. Allow engine to reach normal operating temperature.
2. Disconnect vacuum switch assembly wire connector, Fig. 4. Disconnect and plug distributor vacuum advance hose.
3. Disconnect knock sensor connector (located on intake manifold), then using a suitable jumper wire ground wire to engine block.
4. Increase engine speed to 1600 RPM, then using a suitable timing light, check timing and compare to specifications (refer to individual truck chapter). Adjust as necessary.
5. Allow engine to return to idle speed, then connect vacuum switch connector wire connector and increase engine speed to 1600 RPM while observing timing marks. Timing should advance 6° over results found in step 4. If timing did not advance, check for vacuum switch failure. If vacuum switch is satisfactory, check continuity of wiring with a suitable ohmmeter from knock sensor to Micro Computer Unit (MCU). If wiring is satisfactory, replace MCU unit and retest. If timing did advance, proceed to step 6.
6. Allow engine to return to idle speed, then connect knock sensor connector to knock sensor. Increase engine speed to 1600 RPM, then while observing timing marks with a suitable timing light, knock on intake manifold near the knock sensor. Timing should retard. If timing did not retard, check for continuity of knock sensor harness connector to J2 pin 16 of MCU connector. If satisfactory, replace knock sensor and retest. If timing did retard, the electronic spark retard is operating satisfactory. Allow engine to return to idle speed and connect vacuum hose to distributor.

DISTRIBUTOR SERVICE

Trigger Wheel & Sensor, Replace

1. Remove distributor cap and rotor, Fig. 5.
2. Remove trigger wheel with a suitable gear puller. Use a flat washer to prevent gear puller from contacting inner shaft. The trigger wheel may also be removed by using two screwdrivers to pry trigger wheel upward. Remove pin.
3. On six cylinder distributors, remove sensor retainer and washers from pivot pin on base plate.
4. On V8 distributors, remove sensor snap ring from shaft, then the retainer from vacuum unit to sensor drive pin and position vacuum unit lever aside.
5. On all distributors, remove ground screw from harness tab.
6. Remove sensor assembly from distributor housing.
7. Reverse procedure to assemble.

Vacuum Unit, Replace

1. Disconnect vacuum hose.
2. On six cylinder distributors, remove vacuum unit attaching screws and the vacu-

um unit, Fig. 5. It is necessary to tilt the vacuum unit to disengage the link from the sensor pin and also loosen the base plate screws for clearance.
3. On V8 distributors, remove distributor cap and the retainer from the sensor pin. Remove vacuum unit attaching screws and the vacuum unit, Fig. 5.
4. Reverse procedure to install. If a new vacuum unit is installed, it must be calibrated as follows:
   a. Insert an appropriate size allen wrench into vacuum hose tube of original vacuum unit. Rotate allen wrench clockwise and note the number of turns required to bottom the adjusting screw.
   b. Insert allen wrench into vacuum hose tube of replacement vacuum unit. Turn the allen wrench clockwise until the adjusting screw is bottomed, then rotate allen wrench counterclockwise the number of turns noted in the previous step.

Chrysler Corp. (Exc. Rampage & Scamp) Less Electronic Spark Control System

DESCRIPTION

This system, Figs. 1 thru 3, is composed of a magnetic distributor, an electronic control unit, a wiring harness, a production coil and a single or dual ballast resistor.

The distributor is essentially the same as the conventional type except the contacts have been replaced by a pickup coil and the cam by a reluctor. With a conventional contact type system, the voltage necessary to fire the spark plugs is developed by interrupting the current flowing through the primary of the ignition coil by opening a set of contacts. With the Electronic System, the voltage is produced the same way except that the current is interrupted by a transistor in the electronic control unit. This happens each time the control unit receives a "timing" pulse from the distributor magnetic pickup.

Since the magnetic pickup, reluctor and the control unit, which replace the contact points and cam, do not normally change or wear out with service, engine timing and dwell do not require periodic adjusting. This minimizes regular ignition maintenance of cleaning and replacing the spark plugs.

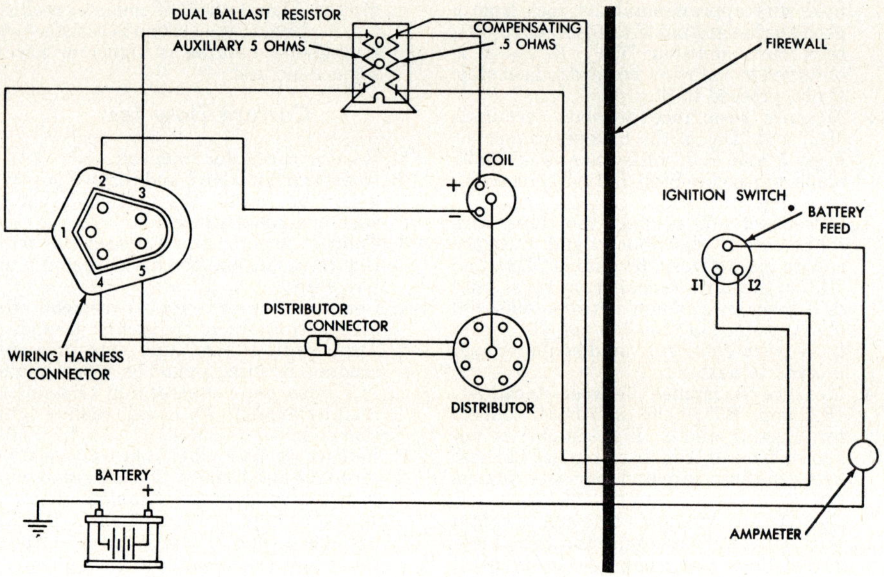

Fig. 1 Chrysler electronic ignition system wiring. 1974–79

TROUBLE SHOOTING

Engine Will Not Start—Fuel System OK

1. Wiring harness electrical terminals covered with grease.
2. Faulty ballast resistor.
3. Faulty ignition coil.
4. Faulty pickup or improper pickup air gap.
5. Faulty wiring.
6. Faulty control unit.

Engine Surges Severely—Not Lean Carburetor

1. Wiring.
2. Faulty pickup leads.
3. Ignition coil.

Engine Misses—Carburetion Good

1. Spark plugs.
2. Secondary cables.
3. Ignition coil.
4. Wiring.
5. Control unit.

SYSTEM TESTING

NOTE: To completely test components and circuits of the electronic ignition system, special

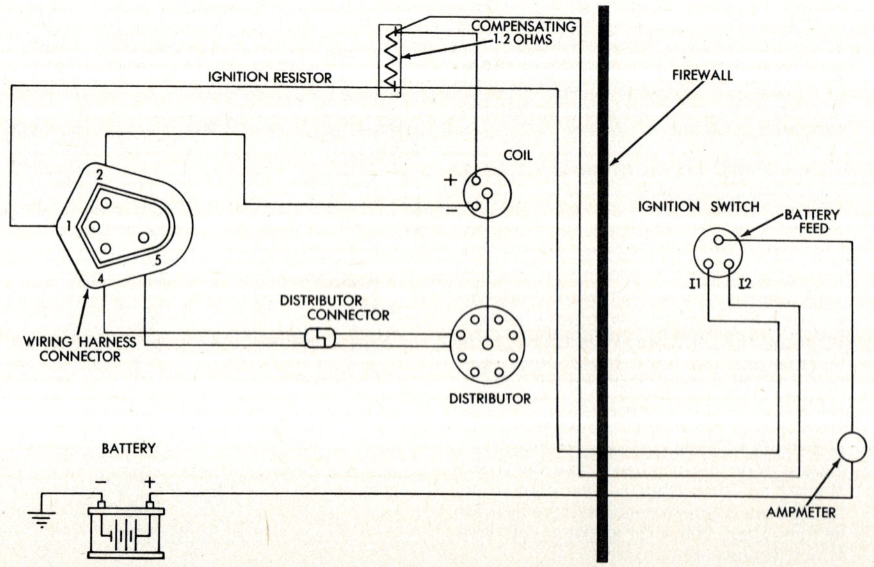

Fig. 2 Chrysler electronic ignition system wiring. 1980

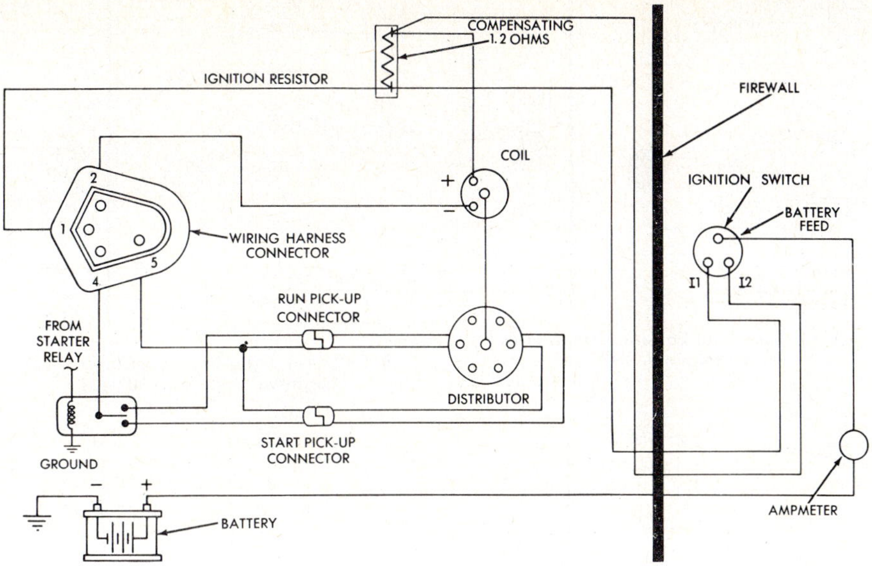

Fig. 3 Chrysler electronic ignition system wiring. 1981–83

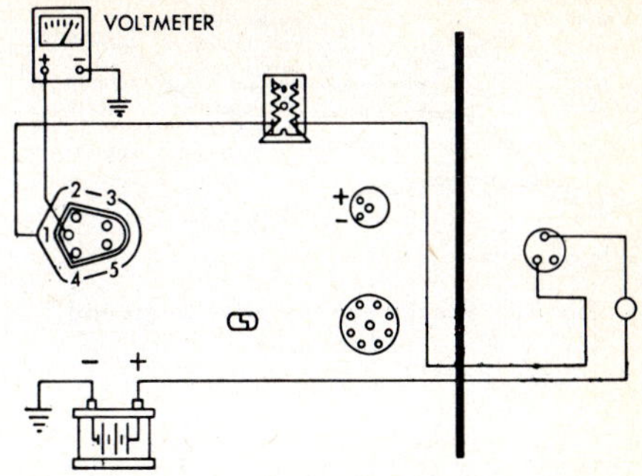

Fig. 4 Harness wiring test, No. 1 cavity. 1974–80 (typical)

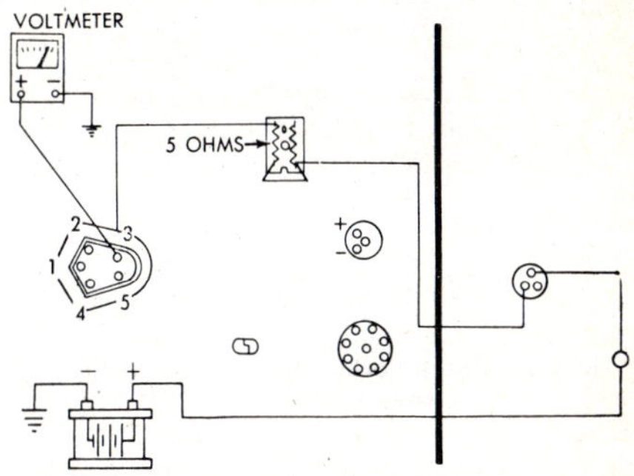

Fig. 6 Harness wiring test, No. 3 cavity. 1974–80 (typical)

testers should be used. However, in event the testers are not available, the following procedures may be utilized. A voltmeter with a 20,000 volt/ohm rating and an ohmmeter using a 1.5 volt battery for power should be used for testing. Before performing any electrical tests, ensure all wiring is properly connected.

1974–80 Models

Harness Wiring Test

1. Check battery voltage and note reading.
2. Disconnect harness connector from control unit.

CAUTION: Before disconnecting or connecting harness connector, ensure ignition switch is in the "Off" position.

3. Turn ignition switch to "On" position.
4. Connect the voltmeter between harness connector cavity No. 1 and the ground. Voltage reading should be within 1 volt of battery voltage earlier noted. If not, check circuit between cavity No. 1 and the battery, Fig. 4.
5. Connect voltmeter between harness connector cavity No. 2 and ground. Voltage reading should be within 1 volt of battery voltage noted in step 1. If not, check circuit between cavity No. 2 and battery, Fig. 5.

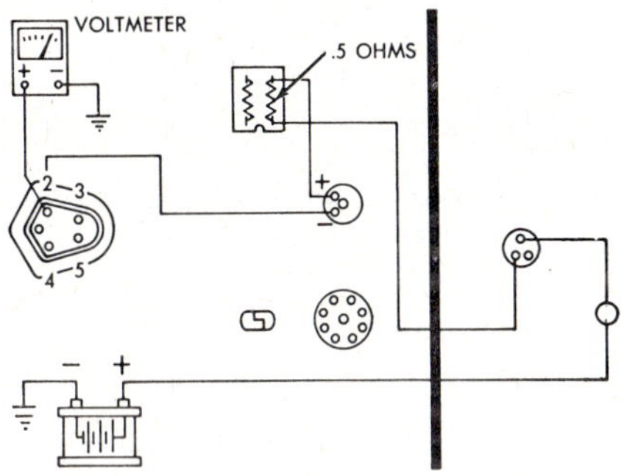

Fig. 5 Harness wiring test, No. 2 cavity. 1974–80 (typical)

6. Connect voltmeter between harness connector cavity No. 3 and ground. Voltage reading should be within 1 volt of battery voltage noted in step 1. If not, check circuit between cavity No. 3 and battery, Fig. 6.
7. Turn ignition switch to off position.

Distributor Pick-up Coil Test

1. Connect an ohmmeter between harness connector cavities numbers 4 and 5, Fig. 7. Resistance reading should be 150 to 900 ohms.
2. If reading is not as specified in above step, disconnect distributor dual lead connector and connect ohmmeter between the two leads on distributor side of connector, Fig. 8. If resistance is not between 150 and 900 ohms, replace pick-up coil.
3. Connect one ohmmeter lead to a good ground and the other lead to either connector of the distributor. If ohmmeter shows a reading, the pickup coil must be replaced.

Control Unit Ground Circuit Test

Connect an ohmmeter between control unit connector pin No. 5 and the ground, Fig. 9. If ohmmeter indicates infinite resistance, tighten bolts securing control unit to firewall and recheck resistance. If reading is still infinite, replace control unit.

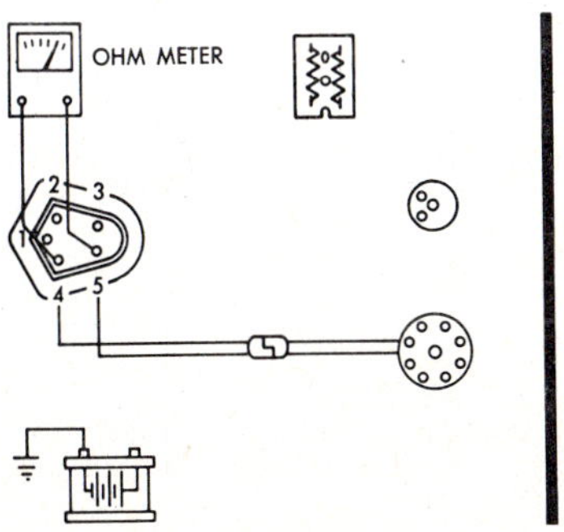

Fig. 7 Pick-up coil test, cavity Nos. 4 & 5. 1974–80 (typical)

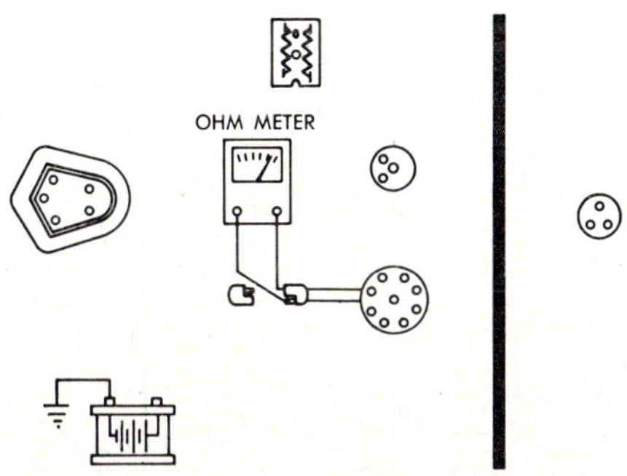

Fig. 8 Pick-up coil test, distributor lead connector. 1974–80 (typical)

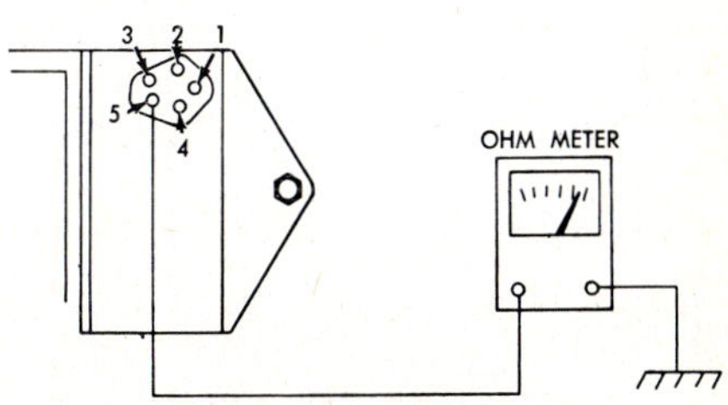

Fig. 9 Control unit ground circuit test. 1974–80 (typical)

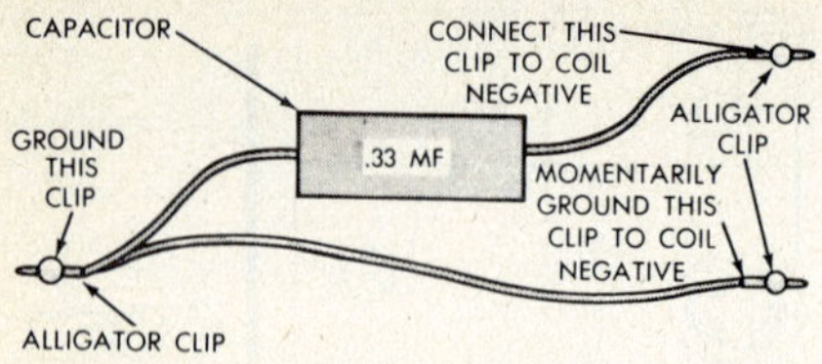

Fig. 10 Special coil negative to ground jumper wire

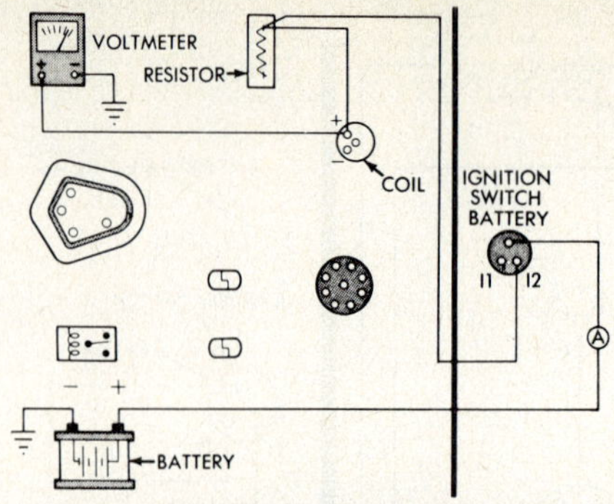

Fig. 11 Checking battery voltage at coil positive terminal. 1981–83

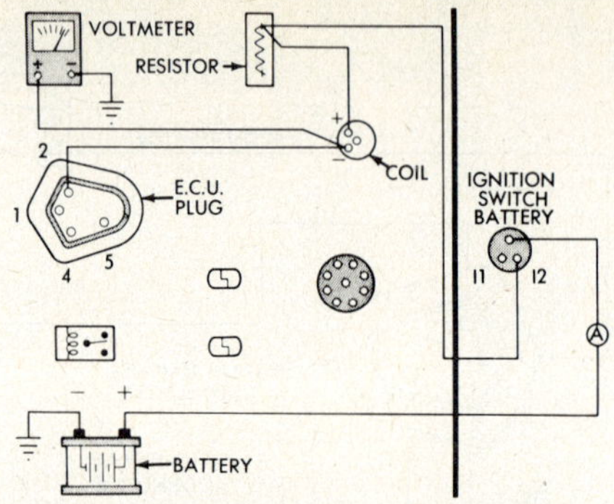

Fig. 12 Checking battery voltage at coil negative terminal. 1981–83

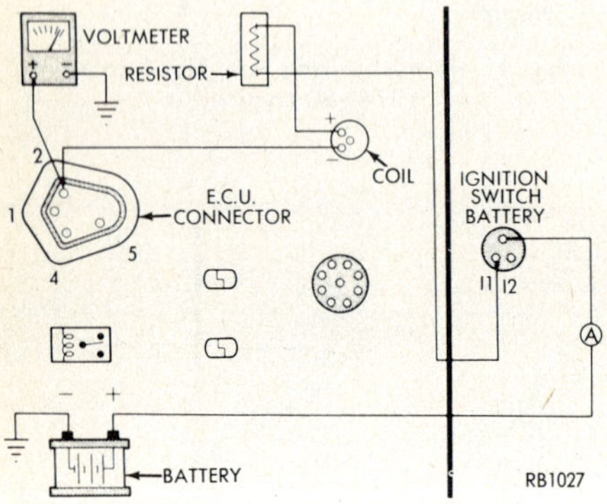

Fig. 13 Checking battery voltage at No. 2 cavity. 1981–83

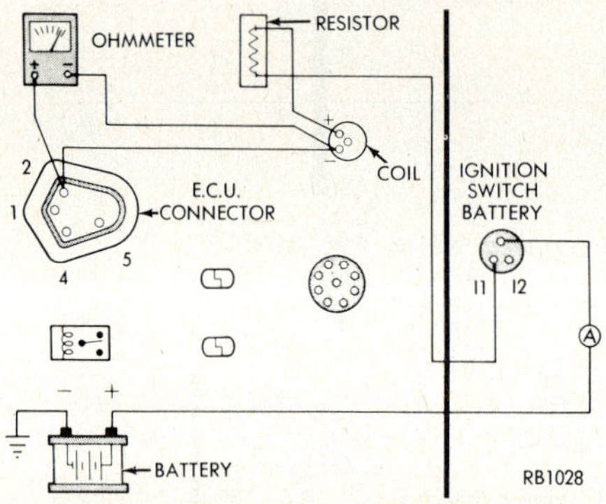

Fig. 14 Checking continuity between coil negative terminal and No. 2 cavity. 1981–83

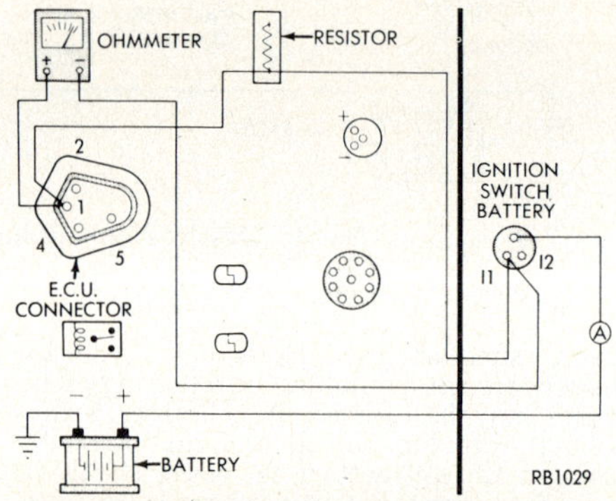

Fig. 15 Checking continuity between ignition switch and No. 1 cavity. 1981–83

1981–83 Models

1. Visually check all spark plug cables and spark plugs for proper installation or damage.
2. Check for proper installation of ignition coil primary wire at coil terminals and at ballast resistor.
3. If above checks are satisfactory, measure and record battery voltage.
4. Disconnect coil secondary lead from distributor cap. Turn ignition "On", then using a special jumper wire as shown in Fig. 10, momentarily ground distributor negative terminal while holding coil secondary lead 1/4 inch from good ground. If no spark is observed, turn ignition "Off", then disconnect four wire electrical connector at electronic control unit and proceed to next step. If spark is observed, proceed to step 10.

NOTE: Before connecting or disconnecting the harness connector from the control unit, check to ensure that the ignition switch is in the Off position.

5. Repeat step 4. If spark is observed, replace electronic control unit.
6. If no spark is observed, measure voltage at coil positive terminal. Reading obtained should be within one volt of previously recorded battery voltage.
7. If voltage reading is zero, replace starter relay and check wiring between battery positive terminal and coil, Fig. 11.
8. If voltage reading obtained in step 6 is not continuous, replace ignition resistor and repeat step 6.
9. Check battery voltage at coil negative terminal, Fig. 12. Reading obtained should be within one volt of battery voltage. If voltage reading is satisfactory, but no spark is observed when coil negative terminal is grounded, replace coil.
10. If spark was obtained in step 4 or step 9, but engine will not start, check to ensure that four-wire connector is disconnected from electronic control unit, then with ignition On check for battery voltage at cavity 2 of ECU harness connector. Reading obtained should be within one volt of battery voltage, Fig. 13.
11. If voltage reading is not satisfactory, turn ignition Off, and check for continuity between cavity 2 and coil negative terminal using suitable ohmmeter, Fig. 14. If continuity is not present, check for damaged wiring or loose connections and repair as necessary.
12. If voltage reading in step 10 is satisfactory, check for battery voltage at cavity 2 of ECU connector. If battery voltage is not

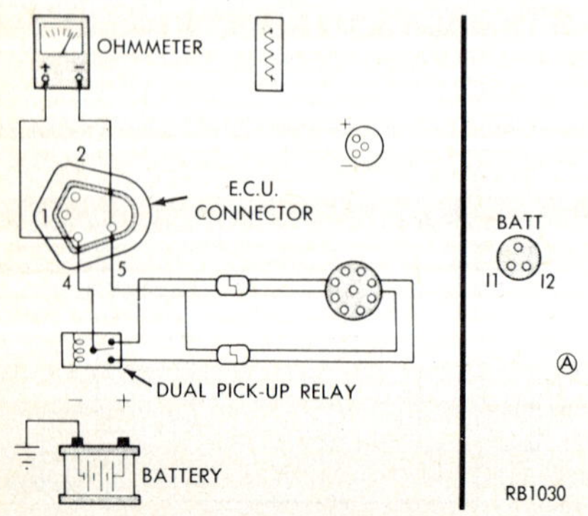

Fig. 16 Checking resistance between cavities Nos. 4 & 5. 1981–83

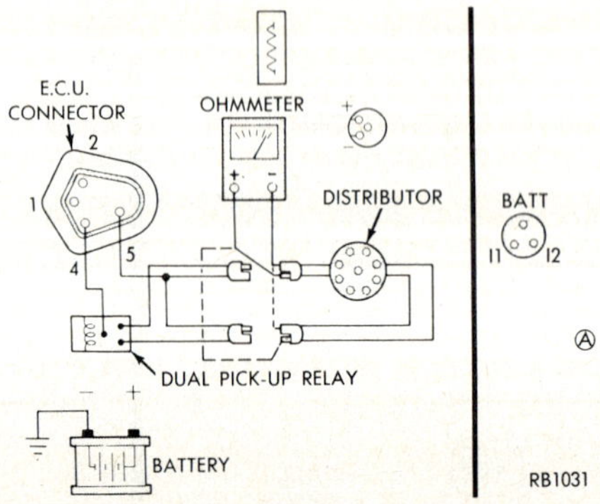

Fig. 17 Testing pick up coil resistance. 1981–83

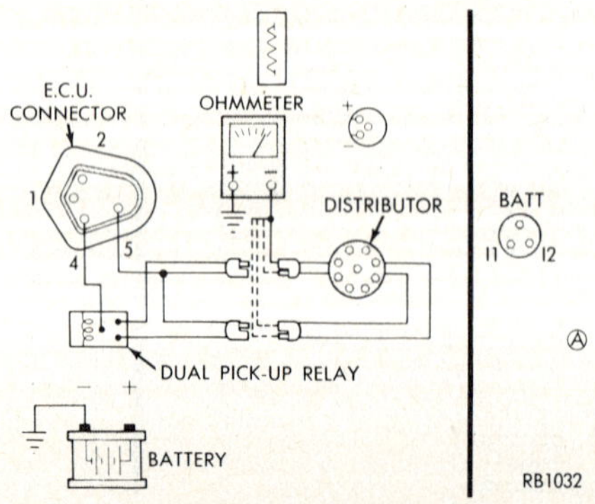

Fig. 18 Checking for short circuit at pick up coil terminal. 1981–83

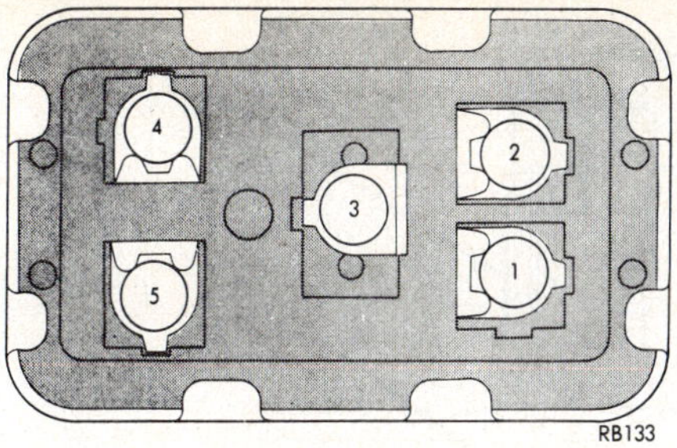

Fig. 19 Dual pick-up start run delay. 1981–82

obtained, turn ignition Off and check for continuity between cavity 1 and ignition switch using suitable ohmmeter, Fig. 15. If continuity is not present, check for damaged wiring or loose connections and repair as necessary.

13. If battery voltage is obtained at cavity 2 of ECU connector, turn ignition Off and check resistance between cavities 4 and 5 of ECU harness connector using suitable ohmmeter, Fig. 16. Reading of 150–900 ohms should be obtained.
14. If reading is satisfactory, check wiring between cavities 4 and 5 for open or short circuit and repair as necessary. If wiring is satisfactory, check for proper operation of dual pick up start run relay as described under "Dual Pick Up Start-Run Relay Test".
15. If ohmmeter reading obtained in step 13 is not between 150–900 ohms, disconnect pick up leads and measure resistance at distributor side of pick up lead, Fig. 17. If reading obtained is not between 150–900 ohms, replace pick up coils as necessary.
16. Check for short circuit at each distributor side of pick up lead by connecting ohmmeter as indicated in Fig. 18. If ohmmeter indicates pick up is shorted, replace pick up coil and adjust air gap to .006 inch.
17. If pick up coil is satisfactory, check for proper ECU ground contact by connecting ohmmeter between pin 5 of ECU and ground. If ohmmeter indicates poor ECU ground contact, check for proper ECU installation or poor ECU electrical connections.
18. If ECU is properly grounded, reconnect all electrical connections and check again for spark. If no spark is available, replace ECU.

Dual Pick up Start-Run Relay Test, 1981–83 Models

1. Disconnect two-way connector from pins 5 and 4 of dual pick up start-run relay, Fig. 19.
2. Connect suitable ohmmeter between pins 4 and 5 of relay.
3. If ohmmeter reading of 20–30 ohms cannot be obtained, replace relay.

DISTRIBUTOR SERVICE

Distributor, Replace

Removal

1. Disconnect vacuum line at distributor.
2. Disconnect distributor pickup lead(s) at wiring harness connector, then remove distributor cap.

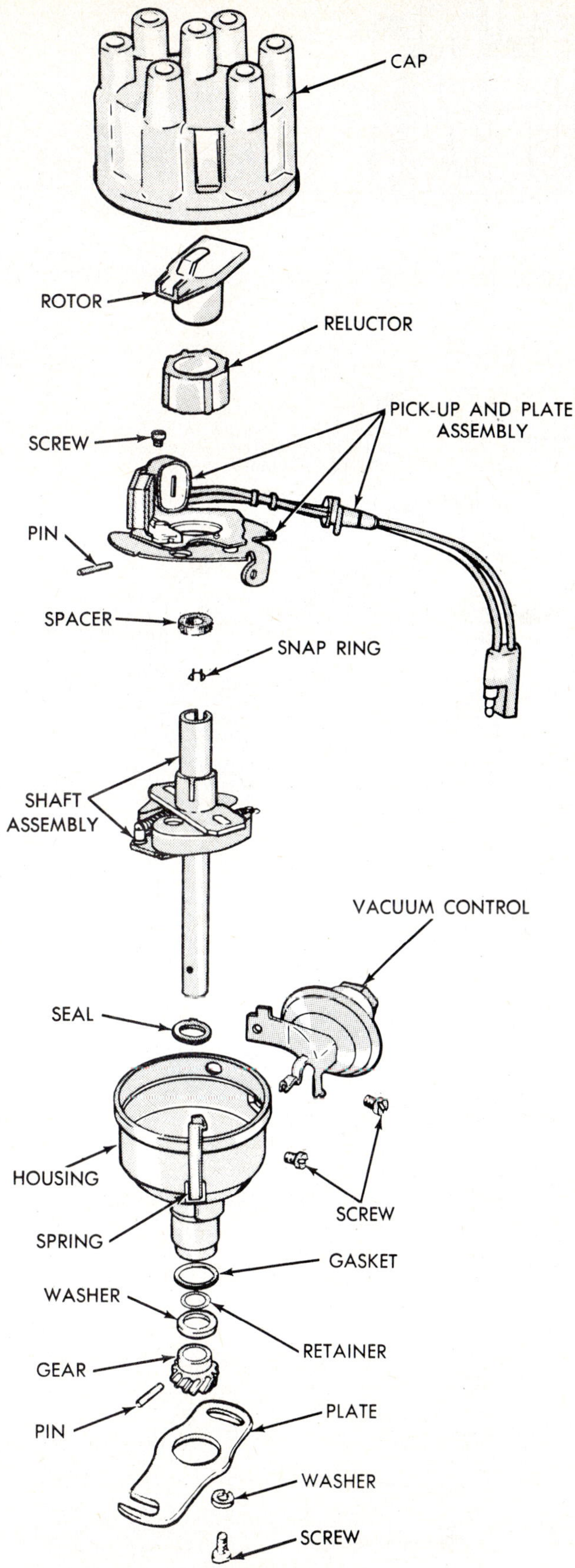

Fig. 20 Disassembled view of Chrysler electronic 6 cylinder distributor with single pick up

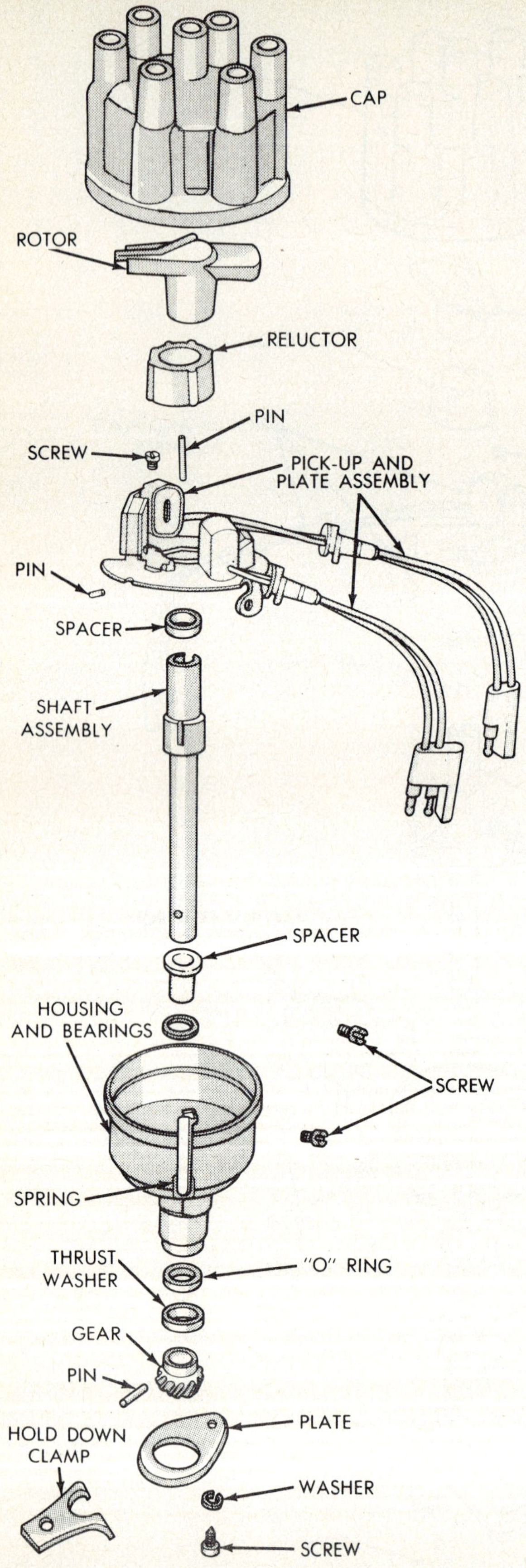

Fig. 21 Disassembled view of Chrysler electronic 6 cylinder distributor with dual pick up

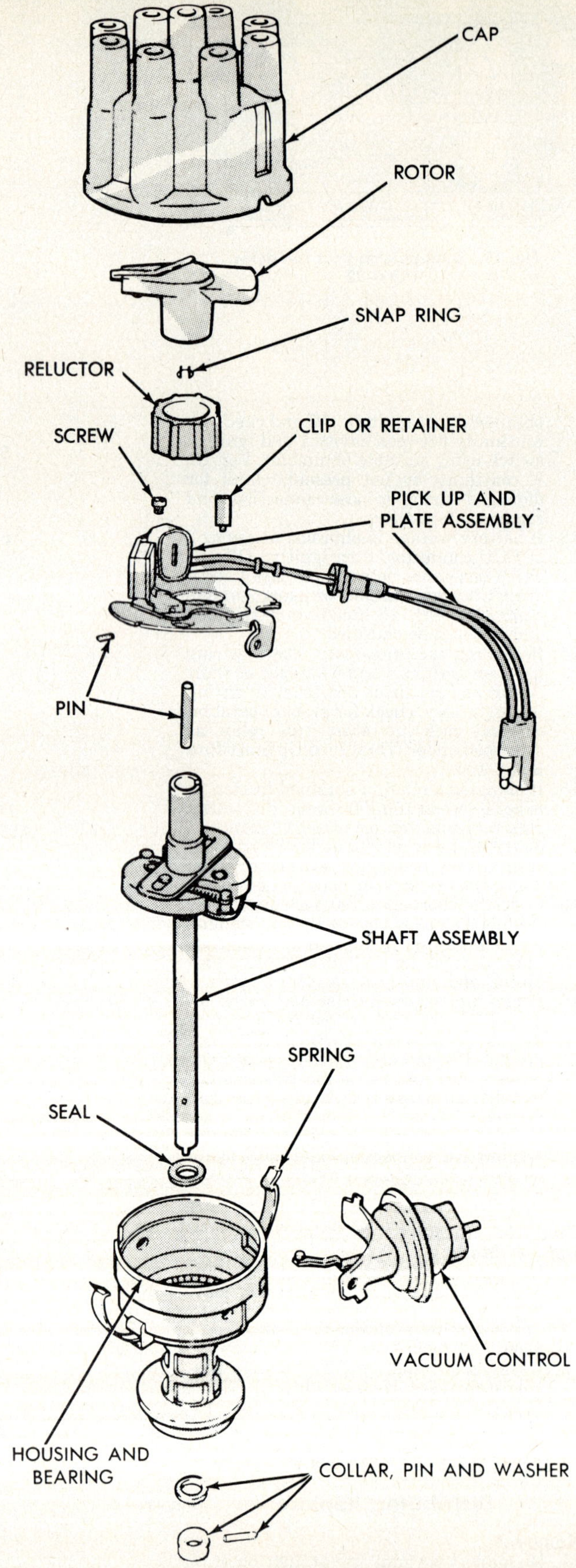

Fig. 22 Disassembled view of Chrysler electronic 8 cylinder distributor

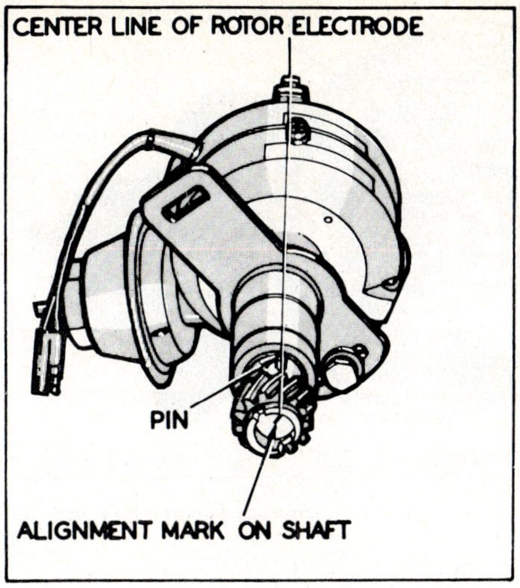

Fig. 23 Scribe line on distributor shaft. 6 cylinder units

3. Mark position of rotor on distributor body and engine block surface so that distributor can be installed in the same position.
4. Remove holddown and/or bolt and lift distributor from engine.

Installation

1. If engine was cranked after distributor was removed from engine, rotate crankshaft to bring No. 1 piston up on its compression stroke and align timing mark on crankshaft pulley with "O" (TDC) mark on timing cover.
2. With distributor gasket or O-ring in place, hold distributor over mounting pad.
3. Turn rotor to a position just ahead of the No. 1 distributor cap terminal.
4. Install distributor, engaging distributor gear with camshaft drive gear on 6 cylinder engines. On V-8 engines, engage tang of distributor shaft with slot in oil pump drive gear. With distributor fully seated on engine, rotor should be under No. 1 cap terminal.
5. Install distributor holddown and/or bolt, distributor cap, pick-up lead(s) and vacuum line.
6. Adjust ignition timing to specifications found in the individual car chapters.

Distributor Shaft & Bushing Wear Test

1. Remove distributor from vehicle and clamp distributor in a vise. Use extreme caution not to damage distributor.
2. Attach a dial indicator to housing so plunger rests against reluctor sleeve.
3. Place a wire loop around reluctor sleeve and hook a spring scale on the other end of the loop. Apply a one pound pull in line with indicator plunger and read movement on indicator. Movement must not exceed .006 inch. If movement exceeds limit, replace either housing or shaft to bring movement back within tolerance.

Distributor Disassemble

1. Remove rotor and vacuum advance unit, Figs. 20 thru 22.
2. Remove reluctor by prying up from bottom of reluctor using two screwdrivers with a maximum blade width of 7/16 in. Use care not to damage or distort reluctor teeth.
3. Remove two screws and lockwashers attaching lower plate to distributor housing, then lift out lower plate, upper plate and pick-up coil as an assembly. Do not remove distributor cap clamp springs.
4. On six cylinder units, if distributor housing, or shaft and governor assembly are to be replaced, proceed as follows:
   a. If gear is worn or damaged, scribe a line on end of shaft from center to edge, so that line is centered between two gear teeth, Fig. 23. Do not scribe line completely across shaft. Remove distributor drive gear retaining pin and slide gear off end of shaft.

   NOTE: Support hub of gear so that pin can be driven out without damaging shaft.

   b. If necessary use a file to clean burrs from around pin hole area on shaft, then remove lower thrust washer.
   c. Push shaft upward and remove from distributor body.
5. On eight cylinder units, if distributor housing, shaft, reluctor sleeve or governor weights are to replace, proceed as follows:
   a. Remove distributor shaft retaining pin and slide retainer off end of shaft.
   b. If necessary, use a file to clean burrs from around pin hole area on shaft, then remove lower thrust washer.
   c. Push shaft upward and remove from distributor housing.

Distributor Assemble

1. Lubricate and test operation of governor weights. Inspect weight springs for distortion and bearing surfaces and pins for damage.
2. Lubricate upper thrust washer and install onto shaft. Install shaft into housing, Figs. 20 thru 22.
3. On 6 cylinder units, install lower thrust washer and distributor gear and roll pin. If a replacement distributor gear is to be installed, proceed as follows:
   a. Install thrust washer and replacement gear on rotor shaft. Position pin hole in replacement gear approximately 90 degrees from hole in distributor shaft, with scribed line made during disassembly between gear teeth, Fig. 23.

   NOTE: On replacement distributor gears, the roll pin hole is located higher than the original gear roll pin hole, so that distributor shaft will not be weakened when the shaft is rotated 90 degrees and drilled to accommodate the replacement gear.

   b. Before drilling through shaft and gear, place a .007 in. feeler gauge between gear and thrust washer and observe that centerline between two gear teeth is in line with centerline of rotor electrode, Fig. 24. Drill a .124 to .129 in. hole and install roll pin.

   NOTE: Support gear hub when installing roll pin so gear teeth will not be damaged.

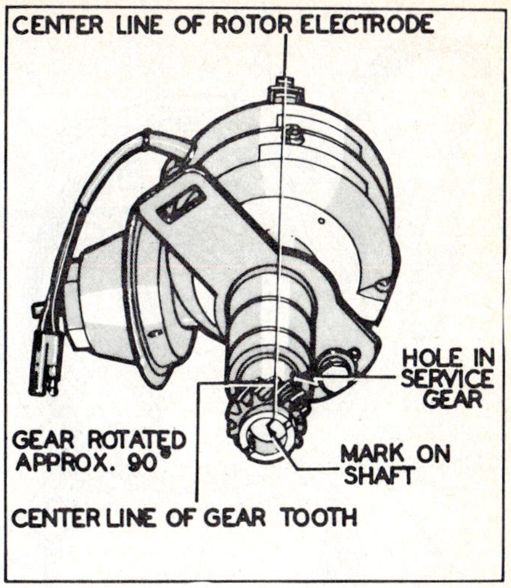

Fig. 24 Aligning gear teeth with center line of rotor electrode. 6 cylinder units

4. On eight cylinder units, install distributor shaft retainer and pin, Fig. 22.
5. On all units, install lower plate, upper plate and pick-up coil assembly, Figs. 20 thru 22.
6. Attach vacuum advance unit to pick-up plate, then install vacuum advance unit attaching screws and washers.
7. Position reluctor keeper pin into place on reluctor sleeve, then slide reluctor down reluctor sleeve and press firmly into position. Install keeper pin.
8. Lubricate felt pad located in top of reluctor sleeve with one drop of light engine oil, then install rotor.

Pick-up Replacement & Air Gap Adjustment

1. With distributor removed from vehicle, perform Steps 1 to 3 as outlined in Distributor Disassemble.
2. Remove pick-up coil and upper plate by depressing retainer clip and moving it away from mounting stud. Pick-up coil cannot be removed from upper plate.
3. Lightly lubricate upper plate pivot pin and lower plate support pins with distributor lubricant. Install upper plate pivot pin through smallest hole in lower plate and install retainer clip.

NOTE: The upper plate must ride on the support pins on the lower plate.

4. Install lower and upper plates and pick-up coil as an assembly and install distributor into vehicle.

NOTE: On dual pick up distributors, the start pick up may be identified by a two prong male connector and the run pick up may be identified by a male-female plug, Fig. 25.

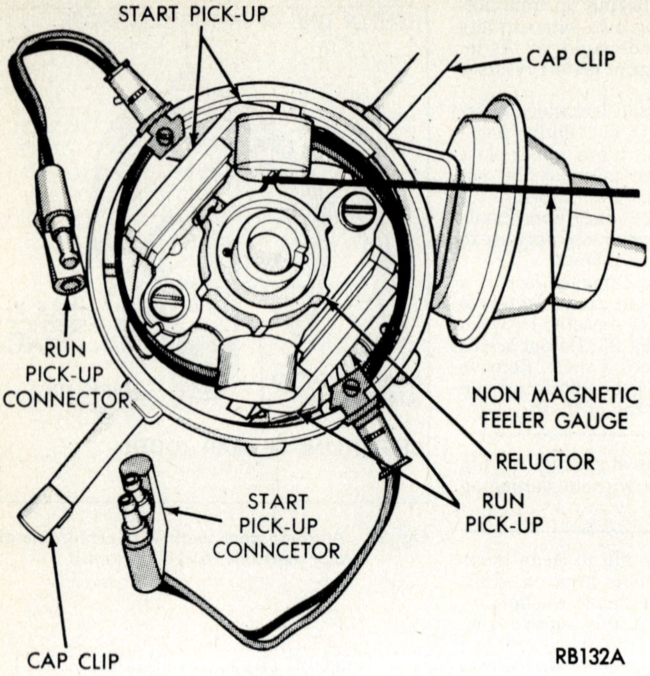

Fig. 25 Air gap adjustment. Distributors equipped with dual pick up

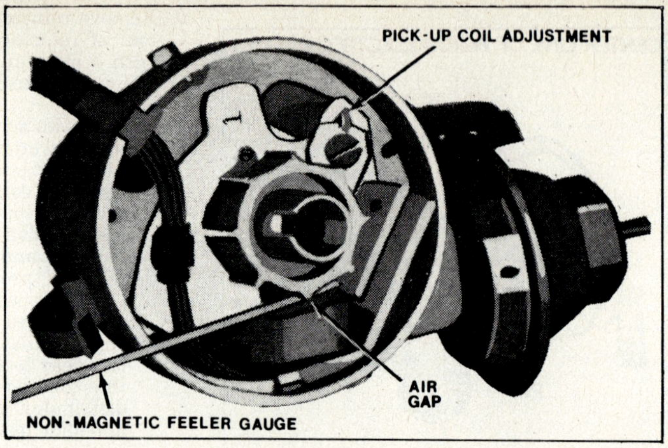

Fig. 26 Air gap adjustment. Distributors equipped with single pick up

5. To set air gap on all single pick up distributors and on start pick up of dual pick up distributors, align one reluctor tooth with pick up pole and install a .006 inch non-magnetic feeler gauge between reluctor tooth and pick up pole, Figs. 25 and 26. Rotate pick up coil until contact is made between reluctor tooth, feeler gauge and pick up pole. Tighten pick up coil hold down screw and remove feeler gauge. The feeler gauge should be removed without force. If it cannot, readjust gap.
6. To set air gap on run pick up of dual pick up distributors, first adjust start pick up as described in step 5, then adjust run pick up as described in step 5. Use a .012 inch non-magnetic feeler gauge to adjust run pick up air gap.
7. Perform a second gap check using a .008 inch feeler gauge on all single pick up distributors and on start pick up of dual pick up distributor or a .014 inch feeler gauge on run pick up of dual pick up distributors. Do not force feeler gauge between reluctor tooth and pick up pole as it is possible to do so. The feeler gauge should not be able to fit between the reluctor tooth and pick up pole if air gap is correctly set. Apply vacuum to vacuum control unit. Pick up should not contact reluctor tooth. Readjust air gap if contact occurs.

NOTE: If pick up contacts reluctor teeth on one side of shaft only, the distributor shaft is most likely bent and shaft replacement required.

Chrysler Corp. (Exc. Rampage & Scamp) With Electronic Spark Control System

SERVICE NOTE

Ignition Timing, Adjust

1981

1. Connect a tachometer to engine and a suitable timing light to No. 1 cylinder.

NOTE: Do not puncture ignition cables or boots with test probes.

2. Start engine, then apply parking brake and place transmission in Neutral. Allow engine to reach operating temperature.
3. Disconnect and plug EGR vacuum hose at EGR valve and vacuum hose at distributor.
4. Disconnect PCV valve and vapor canister purge hose from carburetor. Do not plug hoses.
5. Check idle speed and adjust as necessary, then reconnect PCV valve vacuum hose and vapor canister purge hose.
6. Loosen distributor hold down bracket screw just enough so that distributor housing can be rotated.
7. Check and adjust ignition timing as necessary.
8. Tighten distributor hold down screw, then recheck idle speed and ignition timing.
9. Remove test equipment and connect vacuum hoses.

1982

1. Connect a suitable tachometer to engine and a suitable adjustable timing light to No. 1 cylinder secondary cable.

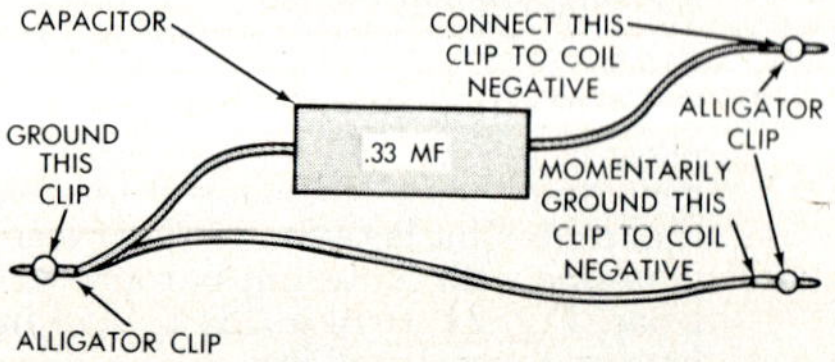

Fig. 1 Special coil negative to ground jumper wire

NOTE: Do not puncture ignition secondary cables or boots with test probes.

2. Set parking brake, start engine and place gear selector in neutral. Run engine until warm.
3. On vehicles equipped with carburetor switch, connect a jumper wire between carburetor switch and a known good ground.

NOTE: On vehicles equipped with 225 c.i. 6 cylinder with 2 bbl. carburetor and miser option, disconnect and plug vacuum line at spark control computer.

4. Ensure curb idle speed is set at or below specifications. Adjust as necessary.
5. Check ignition timing. If within ± 2° of specifications, shut engine "Off" and remove timing light, tachometer and car-

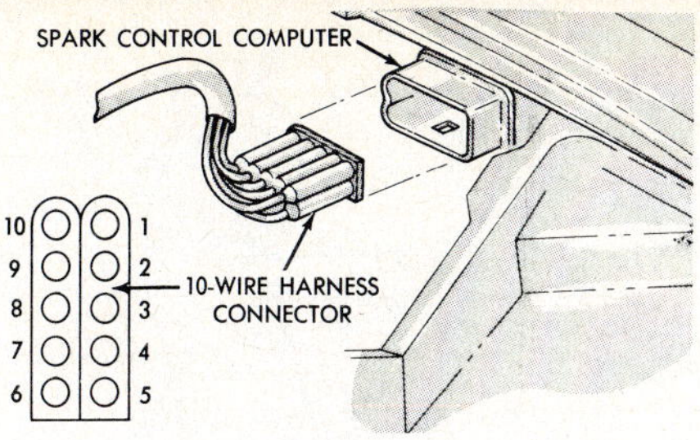

Fig. 2 10 way connector location and cavity identification

buretor ground wire. If outside of tolerance, proceed to step 6.

6. Loosen distributor hold down arm screw, rotate distributor housing until correct timing marks are lined up, ensuring idle speed is still at or below specified curb idle speed.
7. Tighten distributor hold down screw, then recheck idle speed and ignition timing.
8. Turn engine "Off" and remove timing light, tachometer and carburetor ground wire.

SYSTEM TESTING

Ignition System Starting Test

1. Using a suitable voltmeter, measure and record battery voltage.

NOTE: Battery specific gravity must be at least 1.220 temperature corrected to operate the cranking and ignition system properly.

2. Turn ignition "On", then remove coil secondary lead from distributor cap. Using a special jumper wire as shown in Fig. 1, momentarily ground distributor negative terminal while holding coil secondary lead 1/4 inch from good ground.
3. If a spark occurs, it must be constant and bright blue in color. If spark is proper color and constant, continue to momentarily ground coil negative with special jumper wire, Fig. 1, while slowly moving coil wire away from ground. If arcing occurs at the coil tower, replace coil. If the spark is weak or inconsistent or no spark at all, proceed to "Failure To Start Test". If spark is satisfactory, the ignition system is producing the proper secondary voltage. Inspect the cap, rotor, wires and plugs. If satisfactory, it will be necessary to check the fuel system and engine mechanical components.

Failure To Start Test

NOTE: Before proceeding with this test, perform "Ignition System Starting Test". Failure to do so may lead to unnecessary diagnostic time and incorrect test results.

1. Turn ignition "Off". Remove 10 way connector from base of Spark Control Computer, Fig. 2. Turn ignition "On", then holding end of coil wire about 1/4" away from a ground, momentarily short coil negative wire to ground with special jumper wire, Fig. 1. If a spark is obtained,

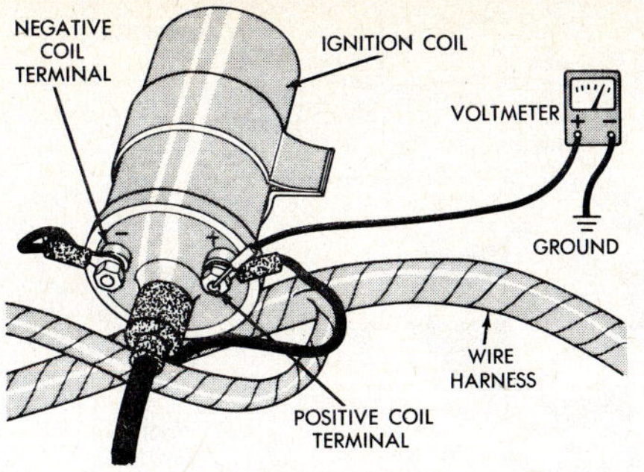

Fig. 3 Checking ignition coil positive terminal voltage

replace Spark Control Computer. If no spark is obtained, proceed to step 2.

2. Using a voltmeter, check for battery voltage at coil positive terminal with the ignition "On", Fig. 3. If voltage is within 1 volt of battery voltage, proceed to step 4. If not, proceed to step 3.
3. Using an ohmmeter, check continuity of wiring between battery and coil positive, Fig. 4, repair as necessary and repeat step 2.
4. Using a voltmeter, check for battery voltage at coil negative. If within 1 volt of battery voltage, proceed to step 5. If not, replace coil.
5. If voltage is correct, but no spark is obtained when grounding coil negative, replace coil.
6. If spark is obtained, but engine will not start, place ignition switch in "On" position. Using a suitable voltmeter, measure voltage of disconnected 10 way connector cavity 1, Fig. 2, to ground. Voltage should be within 1 volt of battery voltage. If voltage is within 1 volt of battery voltage, proceed to step 8. If not, proceed to step 7.
7. Using an ohmmeter, check continuity of wire for opens and repair as necessary (no power between coil negative and 10 way connector). Repeat step 6.
8. Place a piece of paper between curb idle adjusting screw and carburetor switch, Fig. 5, or ensure curb idle adjusting screw is not contacting carburetor switch.
9. Connect negative lead of voltmeter to a good engine ground.
10. Rotate ignition switch to "On" position and measure voltage at carburetor switch terminal, Fig. 5. If voltage is approximately 5 volts, proceed to step 12. If voltage is not at least 5 volts, place ignition switch in "Off" position and disconnect wire connector from bottom of Spark Control Computer, Fig. 2. Rotate ignition switch back to the "On" position and measure voltage at wire connector terminal 2,

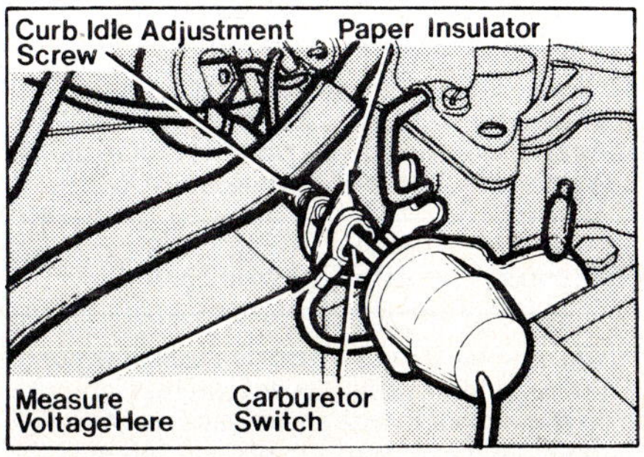

Fig. 5 Positioning paper insulator and carburetor switch terminal location. (Typical)

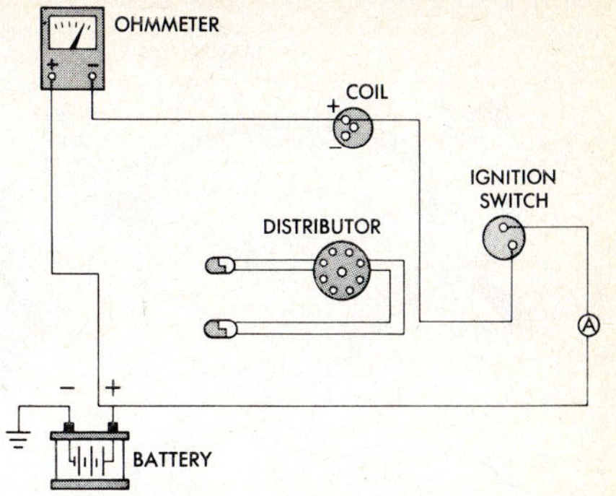

Fig. 4 Checking battery to ignition coil positive terminal for continuity

Fig. 2. Voltage should be within one volt of battery voltage noted previously. If voltage is correct proceed to step 11. If voltage is incorrect check wiring between wire terminal 2 and ignition switch for open or short circuits or improper connections.

11. With ignition switch in the "Off" position, disconnect wire connector from Spark Control Computer. Using an ohmmeter, check for continuity between connector terminal 7, Fig. 2, and carburetor switch terminal. Continuity should exist between these two points. If not, check wiring for open or short circuit or improper connections. If continuity is noted, check for continuity between connector terminal 10 and engine ground. If continuity exists, replace Spark Control Computer. If not, check wiring for open circuit or improper connections and proceed to step 12 only if engine still fails to start.
12. Place ignition switch in the "Off" position, then using an ohmmeter, measure resistance between terminals No. 5 and 9 for run pickup coil and terminals 3 and 9 for start pickup coil Fig. 2. Resistance should be between 150 and 900 ohms. If not, disconnect pickup coil leads from distributor and measure resistance at lead going into distributor. If resistance is now between 150 and 900 ohms, this would indicate an open or short circuit or improper connection between distributor connector and terminals 5 and 9 or terminals 3 and 9 of dual connector. If resistance is not within specifications, pickup coil is defective.
13. Connect one ohmmeter lead to engine ground and with the other lead check for continuity at each terminal going to the distributor. There should be no continuity. Reconnect distributor lead and proceed to step 14. If there is continuity, replace pickup coil.
14. Remove distributor cap and check air gap of pickup coil. Refer to Pickup Coil Replacement and Air Gap Adjustment and adjust gap as necessary.
15. Install distributor cap and reconnect all wiring, then start engine. If engine fails to start, replace spark control computer.
16. After installing new computer and engine still fails to start, reinstall original computer and repeat test procedures since one of the test procedures may have been performed incorrectly.

Poor Fuel Economy, Performance & Unusually High Idle Speed Tests

Engine Temperature Switch Tests, Figs. 6 & 7

Turn ignition "Off". To test temperature

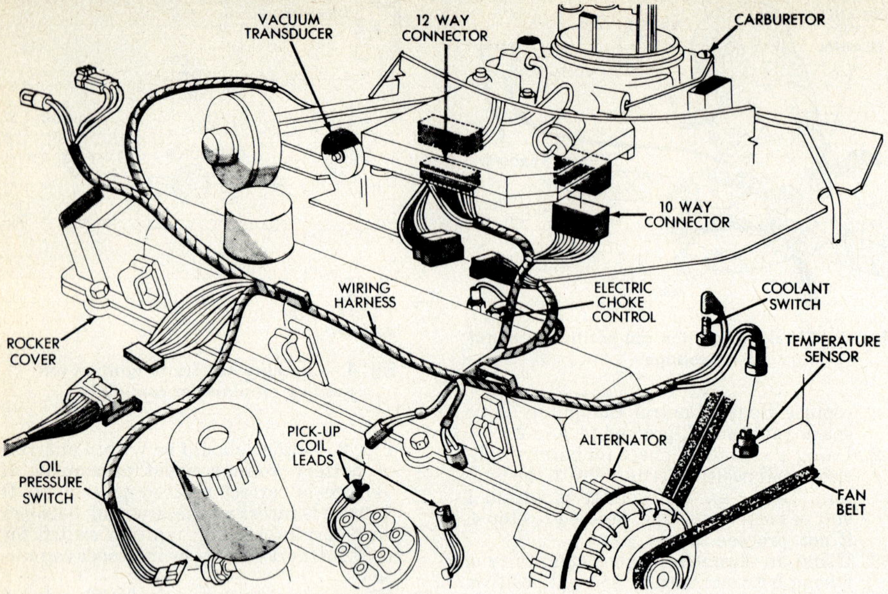

Fig. 6 Electronic Spark Advance switches and sensors. Front view

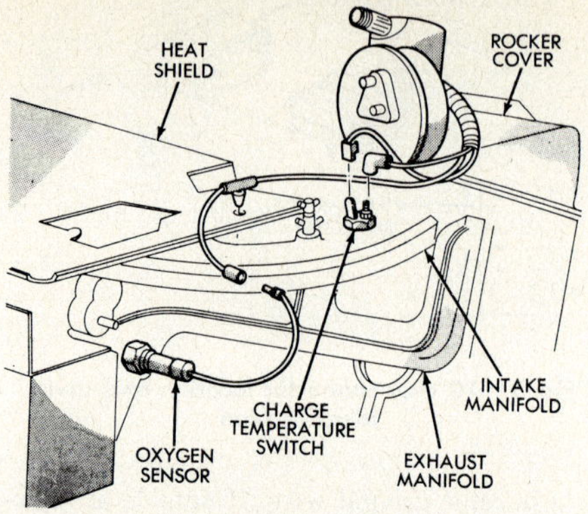

Fig. 7 Electronic Spark Advance switches and sensors. Rear view

switch, connect a suitable ohmmeter between center terminal of temperature switch and ground. To test charge temperature switch, connect ohmmeter to center terminal of charge temperature switch and switch ground terminal.

With engine cold, continuity should be present with resistance less than 100 ohms. If not, replace switch. Charge temperature switch must be cooler than 60°F. in order to perform test properly.

With engine hot, continuity should not be present. If continuity is present, replace temperature switch.

Spark Advance Of Spark Control Computer Test

1. Ensure engine is fully warmed up and the temperature sensor is connected and operating properly.
2. Set ignition timing to specifications.
3. Place a piece of paper between curb idle adjusting screw and carburetor switch, Fig. 5, or ensure curb idle adjusting screw is not contacting carburetor switch.
4. Remove and plug vacuum transducer line.
5. Using a suitable hand operated vacuum pump, apply 16" of vacuum to Spark Control Computer transducer.
6. Increase engine speed to 2000 RPM, wait 1 minute (or specified clock up time, see specifications at end of chapter). Advance specifications are in addition to basic timing advance. On systems equipped with an accumulator the specified time must be reached with the carburetor switch ungrounded before checking the specified spark advance schedule (see specifications at end of chapter).
7. If the computer fails to obtain specified settings, replace computer.

Carburetor Switch Test

NOTE: On most systems, grounding the carburetor switch eliminates all spark advance.

1. Turn ignition Off, then disconnect 10-way connector from computer.
2. Check to ensure throttle is fully closed, then check continuity between pin 7 of harness connector and ground using suitable ohmmeter, Fig. 2.
3. If continuity is not present, check for open circuit in carburetor switch wiring circuit and repair as necessary.
4. Open throttle and check continuity between pin 7 of harness connector and ground. Continuity should not be present.

EGR System Test

NOTE: The electronic EGR control is incorporated into the Spark Control Computer. Check to ensure that engine temperature switches are operating properly before proceeding with test.

1. With engine cold and ignition Off, connect one lead of suitable voltmeter to gray wire terminal on EGR solenoid. Connect second voltmeter lead to ground.
2. Start engine and check voltmeter reading. Reading should be less than one volt. Reading will remain at this value until engine reaches normal operating temperature and electronic EGR schedule has timed out (see specifications at end of chapter). EGR Solenoid will then de-energize and voltmeter will indicate charging system voltage.
3. If system does not de-energize in specified amount of time, replace solenoid.
4. If voltmeter indicates charging system voltage before EGR schedule is complete, replace Spark Control Computer.

NOTE: If vehicle is started with engine at operating temperature, EGR solenoid will be energized for length of time of time delay schedule only and will then de-energize.

Electronic Throttle Control Test

NOTE: The electronic throttle control system is incorporated in the Spark Control Computer on some models (see specifications at end of chapter). A carburetor mounted solenoid is energized whenever the air conditioning, heater, electronic heated backlite or electronic timers are activated. The timers operate with a two second time delay after the throttle is closed or with an EGR time delay after an engine start condition.

1. Connect tachometer to engine, then start engine and run until operating temperature is reached.
2. Depress accelerator, then release. Curb idle speed should increase for length of time of EGR schedule (see specifications at end of chapter).
3. Curb idle speed should increase when air conditioner, heater and/or electronic heated backlite are operated and return to normal when components are turned Off.

NOTE: Air conditioner clutch, if equipped, will cycle on and off as air conditioner operates and should not be mistaken as part of electronic control system operation.

4. If curb idle speed does not increase and decrease as described above, disconnect three-way electrical connector from carburetor.
5. Connect ohmmeter between carburetor solenoid black wire terminal and ground. Resistance should be 15–35 ohms. If not, replace solenoid.
6. Start engine and measure voltage between black wire of three-way connector and ground before EGR time delay has timed out. Voltmeter reading should indicate charging system voltage. If not, replace Spark Control Computer.
7. With engine operating and voltmeter connected as described in step 6, turn on air conditioner, heater and/or electronic heated backlite. Voltmeter should read charging system voltage after time delay has timed out. If not, check wiring between three-way harness connector and instrument panel for damage and repair as necessary.

DISTRIBUTOR, REPLACE

Refer to "Chrysler Corp. (Exc. Rampage & Scamp) Less Electronic Spark Control System" for distributor replacement procedures.

DISTRIBUTOR SERVICE

Refer to "Chrysler Corp. (Exc. Rampage & Scamp) Less Electronic Spark Control System" for distributor service procedures.

Chrysler Corp. Rampage & Scamp

SYSTEM TESTING

Ignition System Starting Test

1. Remove coil wire from distributor cap and hold end of wire approximately 1/4 inch from a good engine ground. Crank engine and observe spark at coil wire.
2. The spark at the coil wire must be constant and bright blue in color. If so, continue to crank engine and slowly move coil wire away from the ground. If arcing occurs at the coil tower, replace coil. If spark is weak, not constant or there is no spark, proceed to the "Failure To Start Test".
3. If spark is satisfactory and no arcing occurs at the coil tower, the ignition system is producing the necessary high secondary voltage. However, this voltage is transmitted to the spark plugs by the distributor rotor, cap, spark plug wires and spark plugs and must also be checked. If satisfactory, the ignition system is not at fault. It will be necessary to check the fuel system and engine mechanical components.

Failure To Start Test

NOTE: Before performing this test, perform the "Ignition System Starting Test". Failure to do so may lead to unnecessary diagnostic time and incorrect test results.

1. With a voltmeter, measure and note battery voltage. Battery specific gravity must be at least 1.220, temperature corrected, to deliver the necessary voltage to operate the cranking and ignition systems properly.
2. Remove coil wire from distributor cap.
3. Place ignition switch in "On" position, then using special jumper wire, Fig. 1, momentarily ground the ignition coil negative terminal while holding the coil wire 1/4 inch from a good engine ground. A spark should be obtained.
4. If spark was obtained, proceed to step 7. If not, turn ignition Off, then disconnect 10-way, electrical connector at Spark Control Computer, Fig. 2.

NOTE: Do not remove grease from harness connector or connector cavity since grease is used to prevent moisture from corroding the terminals. There must be at least 1/8 inch of grease on bottom of computer connector cavity. If not, apply a liberal amount of Mopar Multi-purpose grease, part number 2932524 or equivalent over end of connector plug before installation.

5. Turn ignition On, hold wire 1/4 inch from ground and momentarily ground coil negative terminal using jumper wire, Fig. 1. If spark is present, replace Spark Control Computer. If spark is not present, measure voltage at coil positive terminal using suitable voltmeter. Voltage reading should be within one volt of battery voltage.
6. If no voltage was obtained at coil positive terminal, check for open circuit between battery and coil and repair as necessary. If voltage was obtained, measure voltage at coil negative terminal. Reading should be within one volt of battery voltage. If no voltage is present or voltage is present but spark is not, replace coil.
7. If spark was obtained in step 2, but engine will not start, hold open carburetor switch with thin piece of paper and measure voltage at carburetor switch. Reading obtained should be at least five volts. If voltage reading is satisfactory, proceed to step 11. If no voltage is present, turn ignition Off, then disconnect 10-way connector from Spark Control Computer.
8. Turn ignition On and measure voltage between connector cavity 2 and ground, Fig. 2. Reading should be within one volt of battery voltage. If not, check for continuity between battery and cavity 2 using ohmmeter. If continuity is not present, check for open circuit between battery and cavity 2 and repair as necessary.
9. If battery voltage is present in step 8, turn ignition Off and check continuity between carburetor switch and cavity 7 on 10 way connector. If continuity is not present, check for open circuit between cavity 7 and carburetor switch and repair as necessary.
10. If continuity is present in step 9, check continuity between cavity 10 and ground. If continuity is present, replace Spark Control Computer. If continuity is not present between cavity 10 and ground, check for damaged wiring or an open circuit and repair as necessary. If wiring is satisfactory, but engine will not start, proceed to step 11.
11. Reconnect 10-way connector into Spark Control Computer, turn ignition On and hold coil wire 1/4 inch from ground. Disconnect distributor electrical connector and momentarily connect jumper wire between terminals A and C of connector, Fig. 3. Spark should be observed at coil wire.
12. If spark is present at coil wire, but engine will not start, replace pick up assembly and perform rotor test as described in step 18.

NOTE: When replacing Hall effect pick up assembly, check to ensure rotor shutter blades are grounded, Fig. 4. Connect one lead of suitable ohmmeter to shutter blade and other lead to ground. If continuity is not present, press downward on rotor shaft and recheck continuity. If continuity is still not present, replace rotor. Check to ensure that new rotor is marked ESA on top. Do not start engine until continuity has been obtained.

13. If no spark is present at coil wire check voltage between cavity B of distributor connector and ground, Fig. 3. Reading obtained should be within one volt of battery voltage. If battery voltage is not present, proceed to step 15. If battery voltage is present, turn ignition off and disconnect 10-way connector from Spark Advance Computer. Check for continuity between cavity C of distributor connector, Fig. 3, and cavity 9 of 10-way connector, Fig. 2. Repeat procedure between cavity A of distributor connector and cavity 5 of 10-way connector.
14. If continuity is not present, open circuit exists in harness wiring. Repair as necessary. If continuity is present, replace Spark Advance Computer.
15. If no battery voltage is present at cavity A of distributor connector, turn ignition Off, then disconnect 10-way connector from Spark Control Computer.
16. Check for continuity between cavity B of distributor connector and cavity 3 of 10-way connector. If continuity is not present, repair open circuit in wire between cavities as necessary.
17. If continuity is present turn ignition On and check for voltage between cavity 2 and 10 of 10-way connector. If battery voltage is present, replace Spark Control Computer. If battery voltage is not present, check for proper computer ground circuit connection and repair as necessary.
18. To perform rotor test, check to ensure that rotor is stamped ESA on top. Turn

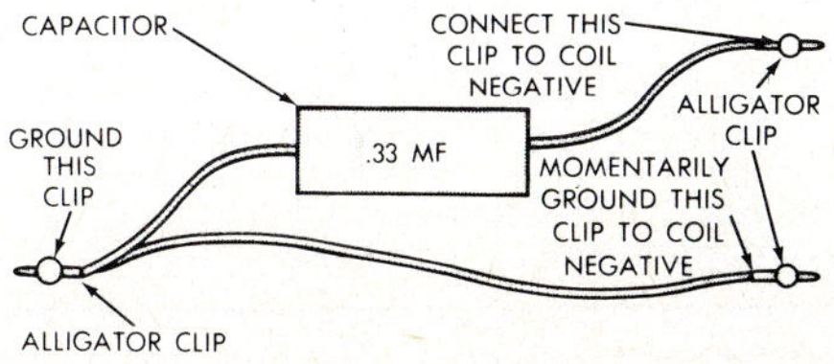

Fig. 1 Coil negative terminal jumper wire

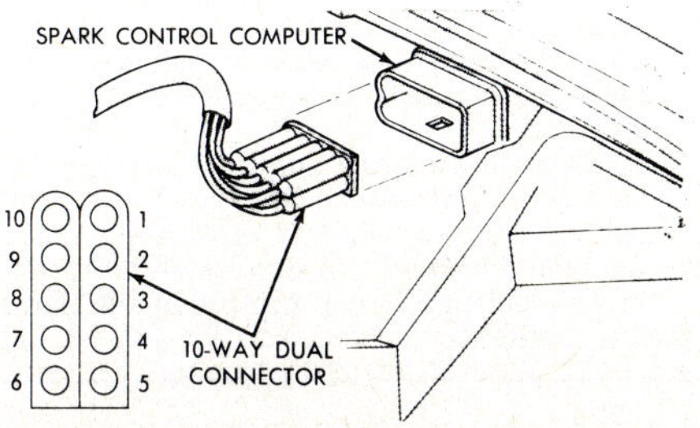

Fig. 2 Spark control computer 10-way connector

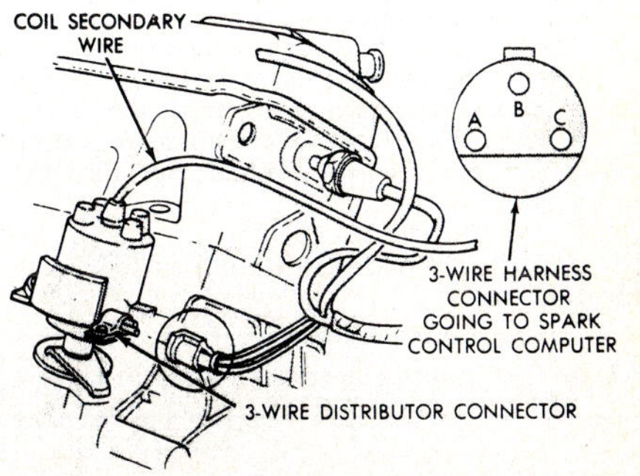

Fig. 3 Distributor 3-wire harness connector

ignition Off, then using suitable ohmmeter, check for good rotor ground contact at distributor shaft. If continuity is indicated, rotor is satisfactory. If continuity is not indicated, check to ensure that rotor is properly seated on shaft.

Poor Performance Tests

Carburetor Switch Test

1. With ignition switch in "Off" position, disconnect 10-way harness connector from "Spark Control Computer," Fig. 2.
2. With throttle completely closed, check continuity between terminal 7 and the engine ground, Fig. 2. Continuity should exist. If not, check wiring and carburetor switch.
3. With throttle opened, check continuity between terminal 7 and the engine ground, Fig. 2. Continuity should not exist.

Coolant Switch Test

1. With ignition switch in "Off" position, disconnect wire from coolant switch.
2. Check continuity between coolant switch terminal and the engine ground. The ohmmeter readings should be as follows: Engine cold, below 150°—Continuity should exist. If not, replace coolant switch. Engine hot, above 150°—No continuity should exist. If continuity exists, replace coolant switch.

Spark Advance Test

1. Run engine until normal operating temperature is reached, then disconnect carburetor switch electrical connector. Check to ensure that coolant sensor is operating as described under "Poor Performance Tests" under "Coolant Sensor Test".
2. Disconnect and plug vacuum transducer vacuum hose.
3. Apply 16 inches Hg. vacuum from outside vacuum source to vacuum transducer.
4. Increase engine speed to 2000 RPM, wait one minute, then check amount of advance timing. Refer to "Rampage & Scamp Electronic Spark Control Specification Chart" for specification.

NOTE: On certain models equipped with an accumulator, the accumulator must be allowed to time out with the carburetor switch disconnected before checking spark advance.

5. If spark advance specifications cannot be obtained, replace Spark Control Computer.

Start Advance Timing Test, 1982 Models

1. Connect an adjustable timing light to engine so total timing advance can be checked.
2. Connect a jumper wire between carburetor switch and the engine ground.
3. Start engine and adjust timing light to align specified timing marks. The timing light meter should indicate the amount of advance listed in the "Rampage & Scamp Electronic Spark Control Specification Chart."

Electronic Throttle Control Test, 1983 Models

NOTE: The electronic throttle control system

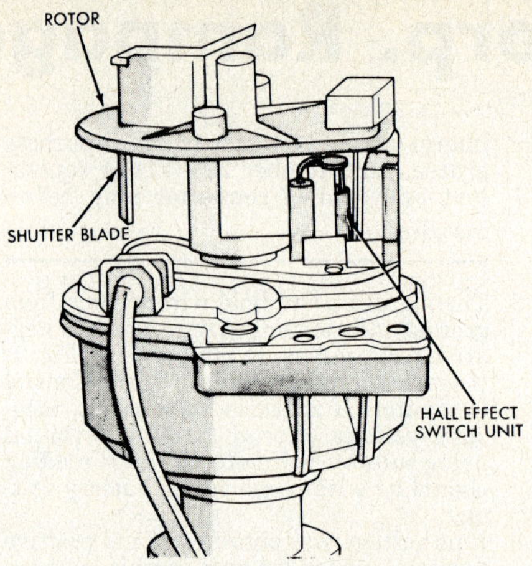

Fig. 4 Rotor shutter blades

is incorporated in the Spark Control Computer system on some models. A carburetor mounted solenoid is energized whenever the air conditioning, heater, electronic heated backlite or electronic timers are activated. The timers operate with a two second time delay after the throttle is closed or after an engine start condition.

1. Connect tachometer to engine, then start engine and run until operating temperature is reached.
2. Depress accelerator, then release. Curb idle should increase for a specified time. Refer to "1983 Rampage & Scamp Electronic Spark Control Specification Chart".
3. Curb idle speed should increase when air conditioner, heater and/or electronic heated backlite are operated and return to normal when components are turned off.

NOTE: Air conditioner clutch, if equipped, will cycle on and off as air conditioner operates and should not be mistaken as part of electronic control system operation.

4. As the air conditioner clutch cycles on and off, the solekicker plunger should extend and retract. Also, when any of the above components are on, or during a specified time after the engine is started (refer to Rampage & Scamp Electronic Spark Control Specification Chart), the solekicker should be extended. If not, check the kicker system for vacuum leaks.
5. If no vacuum leaks are found, disconnect the carburetor 3-way connector.
6. Connect a suitable ohmmeter between black wire of carburetor 3-way connector and ground. Resistance should be between 20 and 100 ohms. If not, replace solenoid.
7. Have assistant start engine, then using a suitable voltmeter, measure voltage across vacuum solenoid terminals. Voltage should be within 2 volts of charging system voltage. If not, replace Spark Control Computer.
8. Turning on the air conditioning or electronic backlite should also produce charging system voltage after the time delay has timed out. Refer to "Rampage & Scamp Electronic Spark Control Specification". If not, check wiring harness back to the instrument panel for an open circuit.

DISTRIBUTOR SERVICE

Disassembly

1. Remove rotor from shaft and disconnect pick up electrical connector.
2. Remove Hall Effect Pick Up Assembly retaining clips.
3. Remove splash shield screws, then the splash shield.
4. Remove Hall Effect Pick Up Assembly.
5. Remove distributor drive gear roll pin using punch, then remove drive gear.
6. Remove drive gear thrust washer and distributor shaft.
7. Remove distributor to block seal.

Assembly

1. Lubricate distributor housing bushings with engine oil, then install engine seal onto distributor housing.
2. Install distributor shaft thrust washer onto distributor shaft, then install shaft into distributor housing.
3. Install distributor drive gear thrust washer, then install distributor drive gear onto distributor shaft.
4. Install drive gear roll pin using punch.
5. Install Hall Effect Pick Up Assembly, then install splash shield.
6. Install Hall Effect Pick Up Assembly retaining clips, then install rotor.

SERVICE NOTE

Ignition Timing, Adjust

1. Connect a tachometer to engine and a suitable timing light to No. 1 cylinder.

NOTE: Do not puncture ignition cables or boots with test probes.

2. Start engine, then apply parking brake and place transmission in Neutral. Allow engine to reach operating temperature.
3. Disconnect and plug EGR vacuum hose at EGR valve and vacuum hose to Electronic Spark Control computer, then connect a jumper wire from carburetor switch to ground.
4. Disconnect PCV valve and vapor canister purge hose from carburetor. Do not plug hoses.
5. Check idle speed and adjust as necessary, then reconnect PCV valve vacuum hose and vapor canister purge hose.
6. Loosen distributor hold down bracket screw just enough so that distributor housing can be rotated.
7. Check and adjust ignition timing as necessary.
8. Tighten distributor hold down screw, then recheck idle speed and ignition timing.
9. Remove test equipment and connect vacuum hoses.

1981–82 CHRYSLER ELECTRONIC SPARK CONTROL SPECIFICATIONS CHART

Spark Control Computer	4145472	4145478	4145747	4145755
Basic Timing	12°	12°	16°	16°
Vacuum Advance (Range)	3–10	3–10	3–10	3–10
Accumulator Time In Minutes	0	0	0	0
Spark Advance Test @ 2000 RPM	25°±4°	25°±4°	25°±4°	25°±4°
Electronic EGR Time Delay	60 sec.	60 sec.	60 sec.	60 sec.
Electronic Throttle Control	Yes①	Yes①	Yes	No

Spark Control Computer	4145760	4145762	4145764	4289024
Basic Timing	16°	16°	12°	12°
Vacuum Advance (Range)	3–10	3–10	3–12	1–7
Accumulator Time In Minutes	0	0	0	0
Spark Advance Test @ 2000 RPM	25°±4°	25°±4°	21°±4°	32°±4°
Electronic EGR Time Delay	60 sec.	60 sec.	60 sec.	0 sec.
Electronic Throttle Control	Yes	No	Yes	No

①—Federal models only

1983 CHRYSLER ELECTRONIC SPARK CONTROL SPECIFICATIONS CHART

Spark Control Computer	4289024	4289067	4289069	4289089	4289091
Basic Timing	12°	12°	16°	12°	16°
Vacuum Advance (Range)	1–7	3–10	3–10	1–7	3–10
Accumulator Time In Minutes	0	1	1	0	0
Spark Advance Test @ 1500 RPM	25°±4°	16°±4°	4°±4°①	28°±4°	22°±4°
Electronic EGR Time Delay	0	65 sec.	65 sec.	—	60 sec.
Electronic Throttle Control	No	Yes	Yes	No	Yes

Spark Control Computer	4289097	4289099	4289075	4289077
Basic Timing	12°	12°	16°	16°
Vacuum Advance (Range)	3–10	3–10	2–14	2–14
Accumulator Time In Minutes	0	0	0	0
Spark Advance Test @ 1500 RPM	25°±4°	21°±4°	13°±4°	8°±4°
Electronic EGR Time Delay	60 sec.	60 sec.	20 sec.	20 sec.
Electronic Throttle Control	Yes	Yes	yes	Yes

①—At 2000 RPM

1982 RAMPAGE ELECTRONIC SPARK CONTROL SYSTEM SPECIFICATIONS CHART

Spark Control Computer	5213480	5213481	5213486
Basic Timing	12°	12°	12°
Vacuum Advance (Range)	0–12	0–12	1–11
Accumulator Time (In Minutes)	0	0	0
Spark Advance Test @ 2000 RPM w/16" Vacuum	45°±4°	47°±4°	44°±4°

1983 RAMPAGE & SCAMP ELECTRONIC SPARK CONTROL SYSTEM SPECIFICATIONS CHART

Spark Control Computer	5213691	5213832	5213834	5213840
Basic Timing	10°	10°	10°	6°
Vacuum Advance (Range)	2–12	2–12	2–12	2–21
Accumulator Time (In Minutes)	0	0	0	0
Spark Advance Test @ 2000 RPM w/16" Vacuum	28°±4°	28°±4°	28°±4°	34°±4°
Electronic Throttle Control	30 seconds after start 2 seconds after throttle closes	15 seconds after start 2 seconds after throttle closes	15 seconds after start 2 seconds after throttle closes	0 seconds after start 2 seconds after throttle closes

1974–83 Delco-Remy High Energy Ingition System (H.E.I.)

The H.E.I. system utilizes an all electronic module, pickup coil and timer core in place of the conventional ignition points and condenser (the condenser is used for noise suppression only). Point pitting and rubbing block wear resulting in retarded ignition timing, is eliminated.

Since the coil is part of the H.E.I. distributor there is no need for distributor-to-coil primary (breaker points to coil negative lead) or secondary lead (high voltage lead).

NOTE: On late 1975 and all 1976–83 coils, the primary and secondary windings have been separated and a secondary coil ground lead is incorporated into the assembly, Fig. 1.

The main features of H.E.I. system differentiating this system from the standard ignition system are shown in Figs. 2, 3 and 4.

The magnetic pickup consists of a rotating timer core attached to the distributor shaft, a stationary pole piece, permanent magnet and pickup coil.

When the distributor shaft rotates, the teeth of the timer core line up and pass the teeth of the pole piece inducing voltage in the pickup coil which signals the all-electronic module to open the ignition coil primary circuit. Maximum inductance occurs at the moment the timer core teeth are lined up with the teeth on the pole piece. At the instant the timer core teeth start to pass the pole teeth, the primary current decreases and a high voltage is induced in the ignition coil secondary winding and is directed through the rotor and high voltage leads to fire the spark plugs.

NOTE: Since this is a full 12 volt system it does not require a resistance wire.

The vacuum diaphragm is connected by linkage to the pole piece. When the diaphragm moves against spring pressure it rotates the pole piece allowing the poles to advance relative to the timer core. The timer core is rotated about the shaft by conventional advance weights, thus providing centrifugal advance.

CAUTION: Never connect to ground the "tach" terminals, Fig. 2, of the distributor connector as this will damage the electronic circuitry of the module.

A convenient tachometer connection is incorporated in the wiring connector on the side of the distributor, Figs. 2 and 4. However due to its transistorized design, the high energy

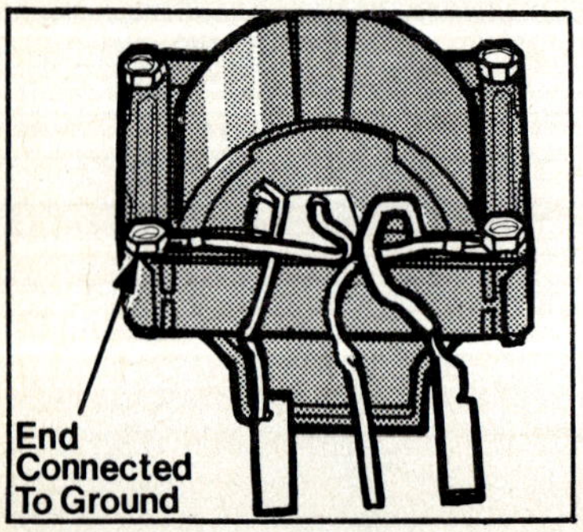

Fig. 1 H.E.I. coil, late 1975 & 1976–83 units with internal coil

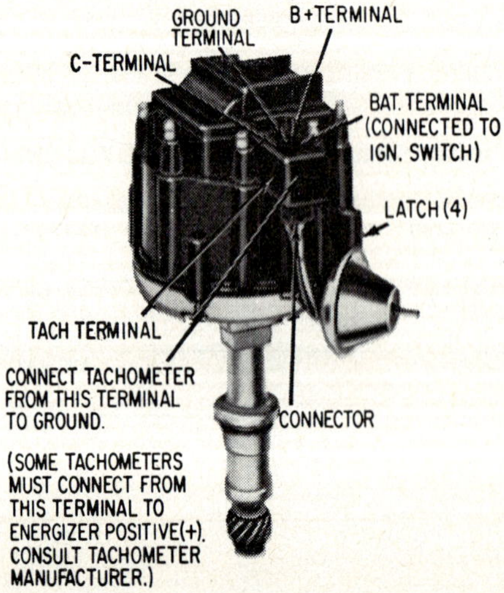

Fig. 2 H.E.I. distributor external components. Internal coil units

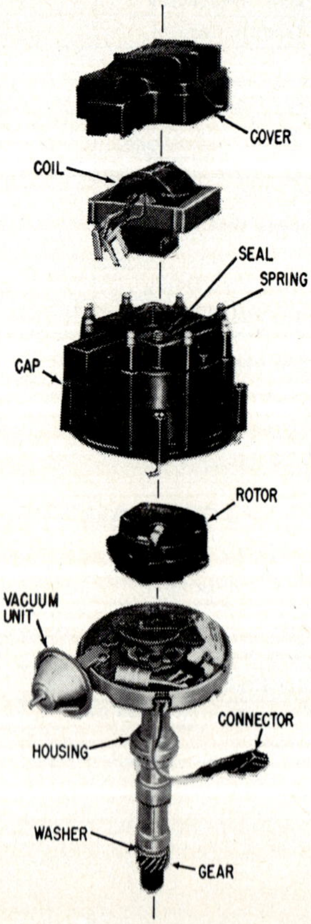

Fig. 3 H.E.I. distributor internal components. Internal coil units

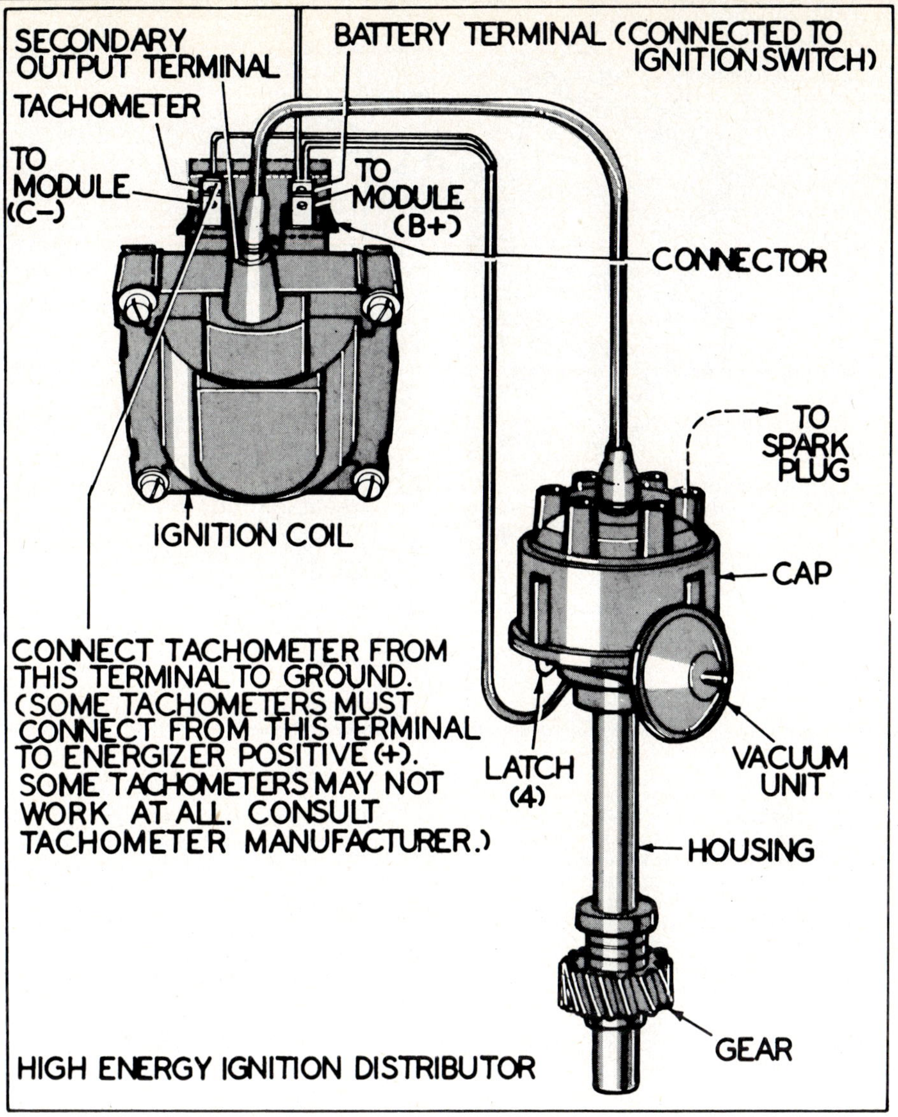

Fig. 4 HEI distributor components. External coil units

energy ignition system will not trigger some models of engine tachometers.

NOTE: When using a timing light to adjust ignition timing, the connection should be made at the No. 1 spark plug. Forcing foreign objects through the boot at the No. 1 terminal of the distributor cap will damage the boot and could cause engine misfiring.

The spark plug boot has been designed to form a tight seal around the spark plug and should be twisted 1/2 turn before removal.

1980–83 light duty models with V8-305 engines and 4 Bbl. carburetor except California models, are equipped with an Electronic Spark Control (ESC) system. The ESC system is a closed loop system that controls detonation by constantly adjusting spark timing. The degree of engine detonation controls the amount of spark retard. The ESC system is made up of a sensor, Fig. 5, that detects the presence or absence of engine detonation and the intensity. The sensor output consists of an electrical signal that is wired to the controller. The controller, Fig. 5, processes the sensor signal and applies it to the H.E.I. distributor to adjust spark timing. The controller is a hard wired signal processor/amplifier that operates in a range of 6 to 16 volts and has no memory storage provisions.

On 1982 engines equipped with automatic transmissions, a ESC vacuum switch is used, Fig. 5. Its function is to provide a momentary signal to the ESC controller during throttle tip-in conditions which then briefly retards spark timing to prevent engine detonation.

SYSTEM DIAGNOSIS

1980–83 Units With ESC

To diagnose ESC system only, use the following procedures. To diagnose H.E.I. system, refer to "1976–83 Units With Internal Coil".

Engine Cranks But Does Not Start, Figs. 6 & 7

1. Check ESC harness for shorts or opens. Inspect ESC controller connector, distributor 4 pin connector, distributor 2 blade male connector and ignition switch 2 blade female connector for proper contact. If any defects are found, repair as necessary. If all wiring is satisfactory, proceed to step 2.
2. Disconnect distributor 4 pin connector, then using a suitable jumper wire, connect pins A and C in distributor connector and attempt to start engine. If engine does not start, refer to diagnose procedures for Engine Cranks, But Does Not Start in "1976–83 Units With Internal Coil". If engine starts, proceed to step 3.
3. Remove jumper wire and reconnect distributor 4 pin connector. Turn ignition "On", then measure voltage from pin F to pin K at ESC controller 10 pin connector. If voltage is less than 7.0 volts, repair circuit between ignition switch and pin F. If voltage exceeds 7.0 volts, proceed to step 4.
4. Inspect wires in ESC harness from pins G, H, J and K of controller 10 pin harness for opens and short circuits. If any are found, repair harness as necessary. If wires are satisfactory, replace ESC controller.

Poor Engine Performance, Figs. 6 & 7

1. Disconnect distributor 4 pin connector, then using a suitable jumper wire, connect pins A and C in distributor connector. If trouble still remains check other sources of poor engine performance. If trouble has cleared up, remove jumper wire from distributor 4 pin connector and reconnect. Using a suitable jumper wire, connect pins A and B of ESC controller connector without disconnecting connector. If trouble clears up, proceed to step 2. If trouble remains, proceed to step 3.
2. On all models, remove jumper wire from ESC controller connector. On 1980–82 models, disconnect ESC controller 10 pin connector. Using a suitable ohmmeter, measure resistance from pin B to pin K in connector. Reading should be 175–375 ohms. If resistance is satisfactory, check engine for noises other than detonation causing input to sensor and repair as necessary. If resistance is out of specifications, disconnect detonation sensor wire from sensor. Using an ohmmeter, measure resistance from sensor terminal to ground. Reading should be 175–375 ohms. If resistance is satisfactory, check sensor wire, shield and sensor connection for an open circuit and repair as necessary. If resistance is out of specifications, replace sensor. On 1983 models, with engine operating at approximately 2000 RPM, disconnect detonation sensor wire from sensor. Using a suitable voltmeter, measure voltage from sensor terminal to ground. Voltage should be .08 volts or more. If voltage is out of specifications, replace sensor. If voltage is satisfactory, inspect wires from pins A, B and K of ESC controller connector for opens or short circuits. If any defects are found, repair as necessary. If wiring is satisfactory, repair sensor connector.
3. Remove jumper wire from ESC connector, then with engine running, measure voltage from pin F to pin K of ESC 10 pin connector. Voltage should be 11.6 volts. If voltage is under 11.6 volts, repair alternator charging system. If voltage is over 11.6 volts, check wires in ESC harness from pins H and K for proper connections and open circuits. If wiring is satisfactory, replace ESC controller. If wiring is defective, repair as necessary.

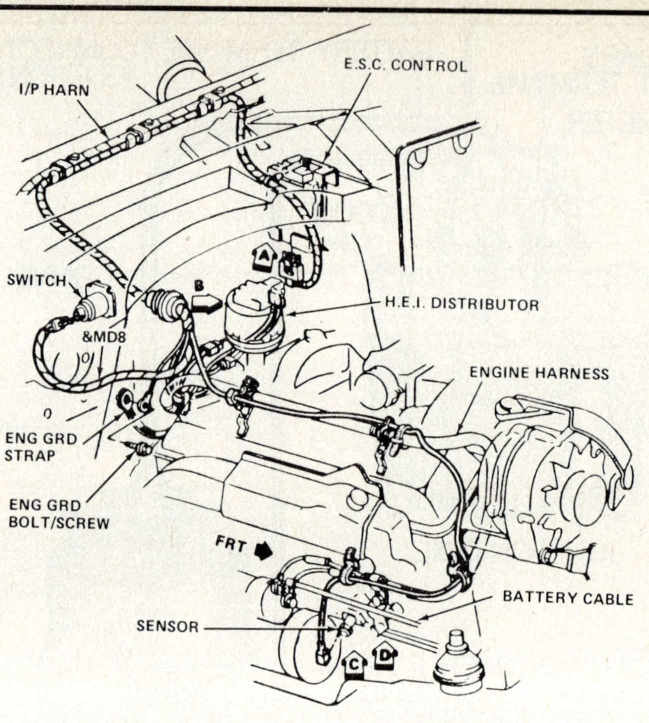

Fig. 5 ECS system component location. (Typical)

Engine Detonation, Figs. 6 & 7

1. With engine speed above 1000 RPM and transmission in "Neutral" or "Park", tap exhaust manifold lightly and repeatedly while checking for spark retard with a suitable timing light. If spark retards, check for other causes of engine detonation. If spark does not retard, proceed to step 2.
2. On all models, disconnect ESC controller 10 pin connector. On 1980–82 models, measure resistance from pin B to pin K in connector with a suitable ohmmeter. Resistance should be 175–375 ohms. If resistance is satisfactory, proceed to step 3. If resistance is outside specifications, disconnect detonation sensor wire from sensor. Using an ohmmeter, measure resistance from sensor terminal to ground. Resistance should be 175–375 ohms. If resistance is out of specifications, replace sensor. If resistance is within specifications, check wires from pins A, B and K of ESC controller connector for opens or short circuits. If wiring is satisfactory, repair sensor connector. If wiring is defective, repair as necessary.

   On 1983 models, using a suitable voltmeter, measure voltage from pin B to pin K of ESC controller 10 pin connector with engine operating at approximately 2000 RPM. Voltage should be .08 volts or more. If voltage is within specifications, proceed to step 3. If voltage is outside specifications, disconnect detonation sensor wire from sensor. Using a voltmeter, measure voltage from sensor terminal to ground. Voltage should be .08 volts or more with engine operating at approximately 2000 RPM. If voltage is outside specifications, replace sensor. If voltage is satisfactory, inspect wires from pins A, B and K of ESC controller connector for opens or short circuits. If any defects are found, repair as necessary. If wiring is satisfactory, repair sensor connector.
3. With ESC controller 10 pin connector disconnected attempt to start engine. If engine starts, replace distributor electronic ignition module. If engine does not start, reconnect ESC controller 10 pin connector. With engine running at fast idle, insert a suitable jumper wire into detonation sensor wire connector, then lay wire on top of distributor over ignition coil and check for spark retard. If spark retards, replace sensor. If spark does not retard, proceed to step 4.
4. With ignition "On", measure voltage from pin H to pin K of ESC controller connector. Voltage should be more than 0.2 volts. If voltage is satisfactory, replace ESC controller. If voltage is not satisfactory repair open circuit in ESC wiring harness from pin H.

Units with Internal Coil

Refer to diagnostic charts, Figs. 8 and 9 for ignition system diagnostic procedures.

Units with External Coil

With the wiring connector properly attached to connector at side of distributor cap and all the spark plug leads properly connected at plugs and at distributor terminals, proceed as follows:

Engine Will Not Start

1. Connect voltmeter between "Bat" terminal lead on distributor connector and ground and turn on ignition switch.
2. If voltage is zero, there is an open circuit between the distributor and the bulkhead connector; or between the bulkhead connector and the ignition switch; or between the ignition switch and the starter solenoid. Repair as required.
3. If reading is battery voltage, hold one spark plug lead with insulated pliers approximately 1/4 inch away from a dry area of engine block and crank engine. If a spark is visible, the distributor has been eliminated as source of trouble. Check spark plugs and fuel system.
4. If there is no visible spark, perform the "Component Checkout" and proceed as described further on.

Engine Starts But Runs Rough

1. Check for proper fuel delivery to carburetor.
2. Check all vacuum hoses for leakage.
3. Visually inspect and listen for sparks jumping to ground.
4. Check ignition timing
5. Check centrifugal advance mechanism for proper operation.
6. Remove spark plugs and check for unusual defects, such as very wide gap, abnormal fouling, cracked insulators (inside and out), etc.
7. If no defects are found, perform the "Component Checkout" procedure as described below.

Component Checkout

1. Remove cap and coil assembly.
2. Inspect cap, coil and rotor for spark arcover.
3. On internal coil units:
   a. Connect ohmmeter, Fig. 10, step 1. If ohmmeter reading is other than zero or very near to zero, the ignition coil must be replaced.
   b. If no ohmmeter reading was observed in step 1, reconnect ohmmeter both ways, Fig. 10, step 2. If both ohmmeter readings are infinite on high scale, replace ignition coil.
4. On external coil units:
   a. Connect ohmmeter, Fig. 11, step 1. If reading is not infinite, replace coil.
   b. Connect ohmmeter, Fig. 11, step 2. If reading is not zero or near zero, replace coil.
   c. Connect ohmmeter, Fig. 11, step 3. If reading is infinite, replace coil.
5. Connect an external vacuum source to the vacuum advance unit. Replace vacuum unit if inoperative.
6. If vacuum unit is operating properly, connect ohmmeter, Fig. 12, step 1. If ohmmeter reading on middle scale in not infinite at all times, pick-up coil must be replaced.
7. With ohmmeter connected, Fig. 12, step 2, ohmmeter reading must be within 500 to 1500 ohms at all times. If not replace pick-up coil.

NOTE: Tester J-24624 is required to test the module. If this tester is not available, and malfunction still exists after performing the above checks, replace module.

Service Note

Some 1974–75 G. M. vehicles equipped with High Energy Ignition Systems (H.E.I.) may encounter a no-start condition or engine miss. These conditions may be caused by the battery ignition wire installed in such a way that the connector in distributor cap is bent and jammed on wrong side of blade in ignition wire connector, Fig. 13. To correct this condition proceed as follows:

1. Pry latch on black ignition wire connector and remove it from distributor, then unlock tang on blade and remove connector from blade.
2. Straighten blade, if bent. Bend tang on blade outboard to insure a positive lock when blade is reinstalled.
3. Slide connector over blade aligning tang

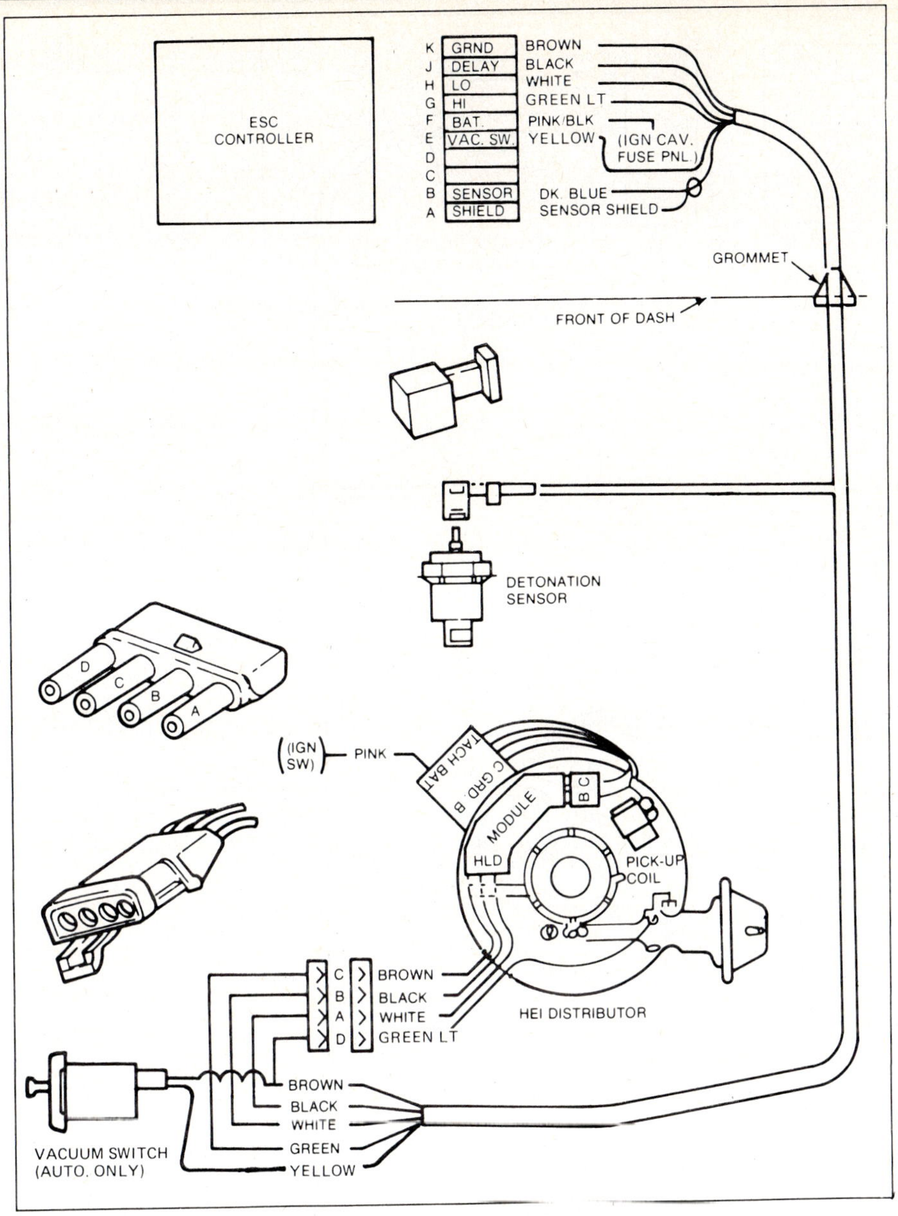

Fig. 6 ESC system wiring schematic. 1980–82 models (typical)

with groove in connector until it bottoms, then pull lightly on wire to insure a positive lock.

4. Inspect blade in distributor cap connector. If bent, straighten and center.
5. Reinstall ignition wire connector by pushing straight up until latch is locked. The blade on distributor connector should be toward lock tab on ignition wire connector. When correctly installed, the connector should move freely while in the locked position.
6. Start engine and move connector. If engine does not start or misses, recheck above procedure.

COMPONENTS REPLACE

Ignition Coil Replacement, Internal Coil Units, Fig. 3

1. Remove screws holding distributor cover to distributor cap and remove distributor cover.
2. Remove screws holding coil to cap.
3. Remove harness connector and battery wire from side of distributor cap.
4. Push coil leads out of position in cap and remove coil.
5. Reverse procedure to install.

External Coil Units, Fig. 4

1. Disconnect ignition switch to coil lead from coil.
2. Disconnect coil to distributor leads from coil.
3. Remove coil to engine retaining screws and remove coil.
4. Reverse procedure to install.

Module Replacement, Fig. 14

1. Disconnect wiring harness connector at side of distributor cap and remove distributor cap.
2. Remove rotor and disconnect wire from module terminals.
3. Remove two mounting screws and remove module.

NOTE: Two types of H.E.I. wiring harness are used. The second type is a wiring harness, connector and capacitor which is serviced as an assembly.

4. Reverse procedure to install.

CAUTION: At installation, coat bottom of new module with dielectric lubricant (furnished with new module) to aid in heat transfer into distributor housing. Failure to appy lubricant will cause excessive heat at module and premature module failure.

Pole Piece, Magnet or Pick-up Coil Replacement, Fig. 14

Removal

1. With distributor removed, disconnect wires at module terminals.
2. Remove roll pin from drive gear by driveing out with 1/8 inch diameter drift punch.
3. Remove gear, shim and the tanged washer from distributor shaft. Remove any burrs that may have been caused by removal of pin.
4. Remove distributor shaft from housing.
5. Remove washer from upper end of distributor housing.

NOTE: Bushings in the housing are not serviceable.

6. Remove three screws securing pole piece to housing and remove pole piece, magnet and pick-up coil.

Installation

1. Install pick-up coil, magnet and pole piece and loosely install three screws holding pole piece.
2. With washer installed at top of housing, install distributor shaft and rotate to check for proper clearance between pole piece teeth and timer core teeth.
3. If necessary, realign pole piece to provide adequate clearance and secure properly.
4. Install tanged washer, shim and drive gear (teeth up) to bottom of shaft. Align drive gear and install new roll pin.
5. Connect all wires to module terminals.

DISTRIBUTOR SERVICE

NOTE: If distributor has been disassembled, refer to Fig. 15 during reassembly.

1. Liberally apply silicone to bottom of module and install module and retaining screws.

CAUTION: Failure to apply silicone lubricant will cause excessive heat build-up at module and premature module failure.

2. Install capacitor and harness making certain that ground lead is under capacitor and retaining screw.
3. Install vacuum advance unit and retaining screws.
4. Place felt wick under pick-up coil plate and lubricant with engine oil, then install pick-up coil and retain with waved C-

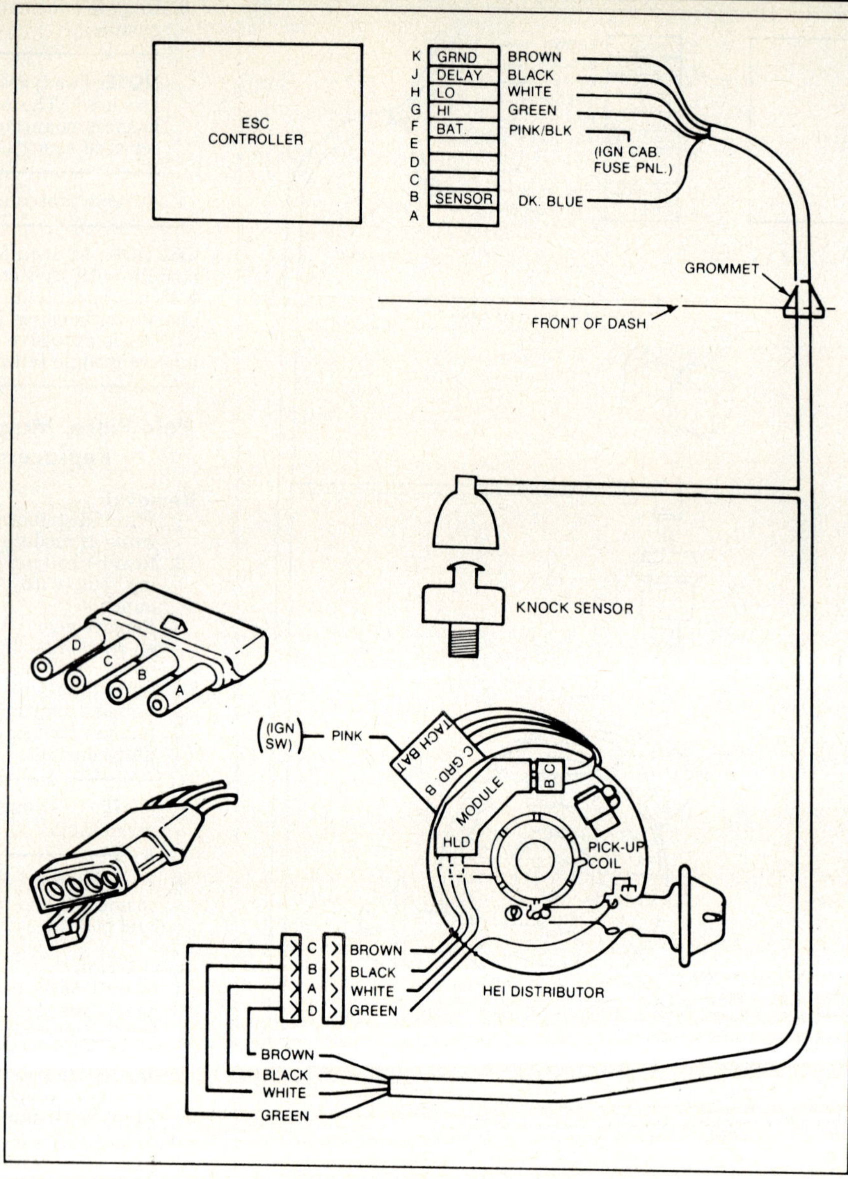

Fig. 7 ESC system wiring schematic. 1983 models.

washer.

5. Install distributor shaft into housing, then install drive gear onto shaft and install new roll pin.
6. On internal coil units, install arc seal, coil and coil cover.

GOVERNOR ELECTRONIC CONTROLLER & SOLENOID CONTROL VALVE

The electronic governor system used on medium and heavy duty vehicles equipped with HEI consists of the governor electronic controller and solenoid control valve, Figs. 16 and 17. The function of the electronic controller is to regulate current to the solenoid control from the HEI system. The function of the solenoid valve is to create governor vacuum. The electronic controller is not adjustable. In normal operation, the controller compares the HEI tachometer signal to an internal reference. The reference is proportional to the desired governor setting. When the tachometer signal is greater than the controller reference, current flows to the solenoid control valve, creating a governor vacuum signal which reduces engine RPM to the preset governor speed.

Diagnosis

Improper governing of engine RPM may be caused by a defective solenoid control valve or defective electronic controller. Using a suitable volt-ohmmeter, proceed as follows to determine the source of trouble:

1. Disconnect battery ground cable and solenoid control valve electrical connector from solenoid control valve.
2. Measure resistance across solenoid terminals. Resistance should be 25–35 ohms for vehicles with 2 bbl. carburetors or 47–53 ohms for vehicles with 4 bbl. carburetors. If resistance is satisfactory, perform step 3. If resistance is less than the values indicated, perform steps 4 through 8.
3. Check solenoid valve for internal short circuit by measuring resistance between either terminal and metal valve case. Resistance should be 800 ohms or higher. If resistance is satisfactory, connect solenoid valve electrical connector and check for proper governor operation.
4. If resistance is less than values indicated in step 2, refer to Fig. 18 and disconnect electrical harness connector from electronic controller.
5. Measure resistance between ground terminal in harness connector and chassis ground. Resistance should be three ohms or less. If high resistance or an open circuit is indicated, check for defective ground wire or terminal and repair as necessary.
6. Turn ignition switch on with engine static. Using voltmeter check for battery voltage at wiring harness connector ter-

ENGINE CRANKS, BUT WILL NOT START

IF A TACHOMETER IS CONNECTED TO THE TACHOMETER TERMINAL, DISCONNECT IT BEFORE PROCEEDING WITH THE TEST.

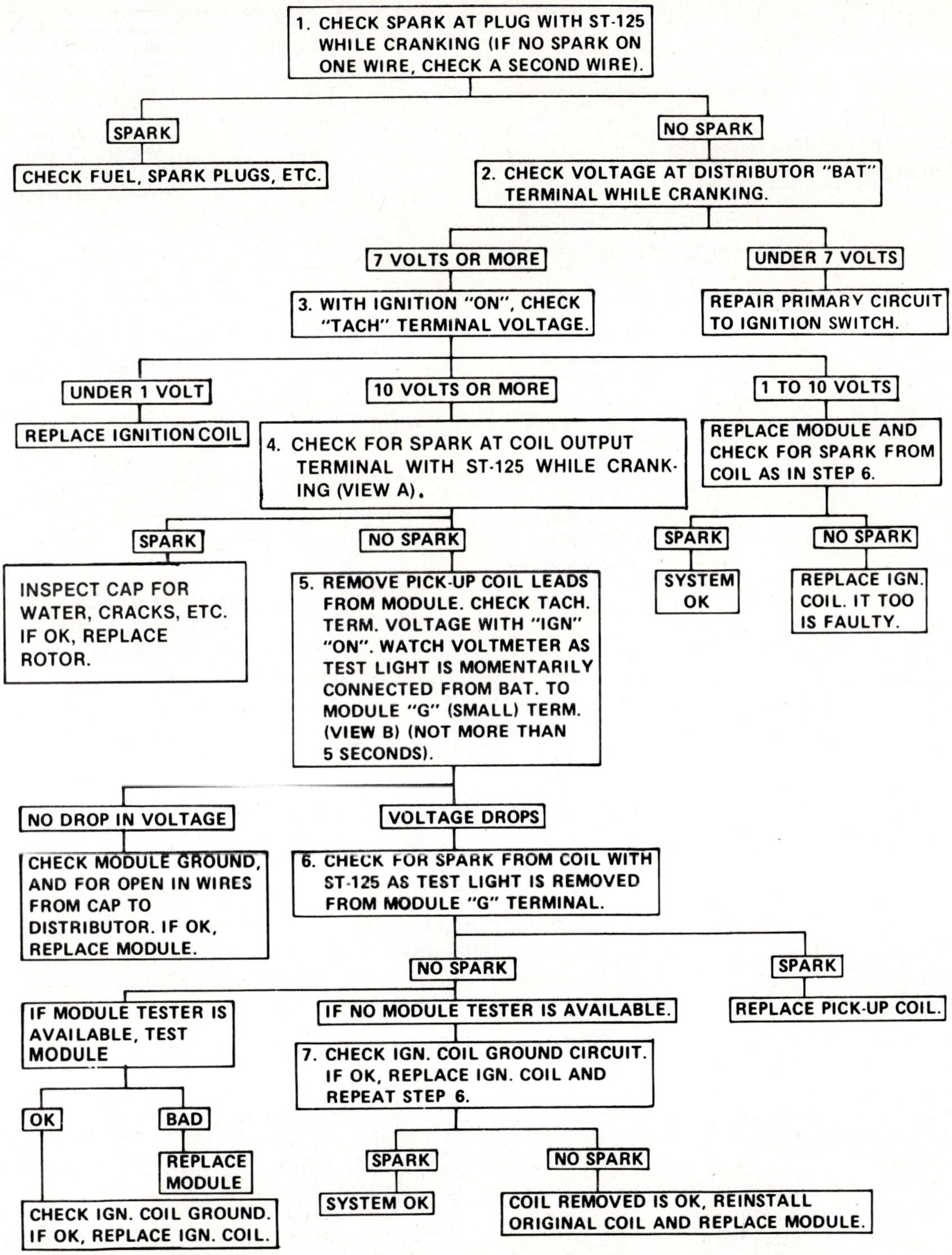

Fig. 8 H.E.I. diagnosis chart (part 1 of 2). 1976–83 units

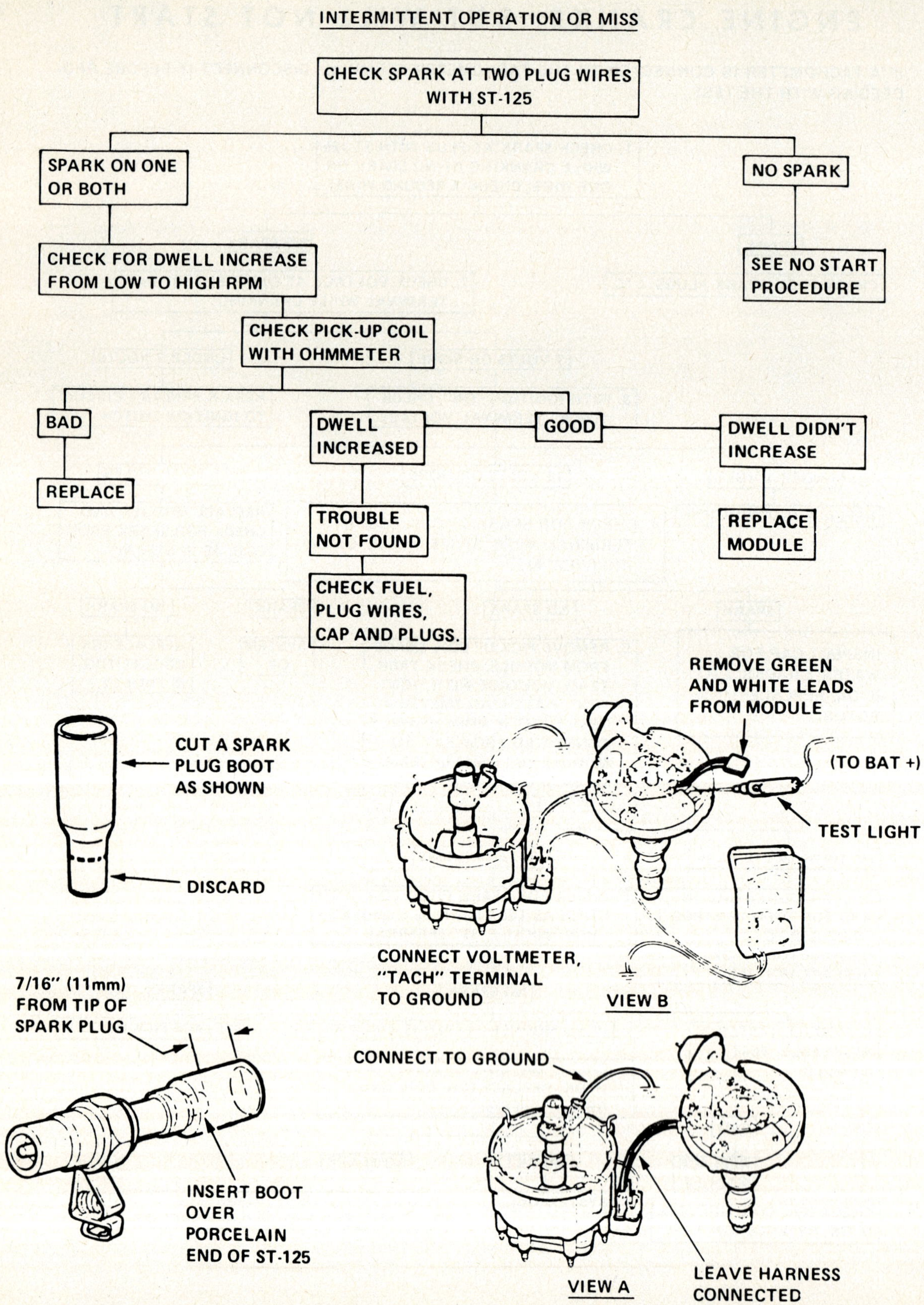

Fig. 9 H.E.I. diagnosis chart (part 2 of 2). 1976–83 units

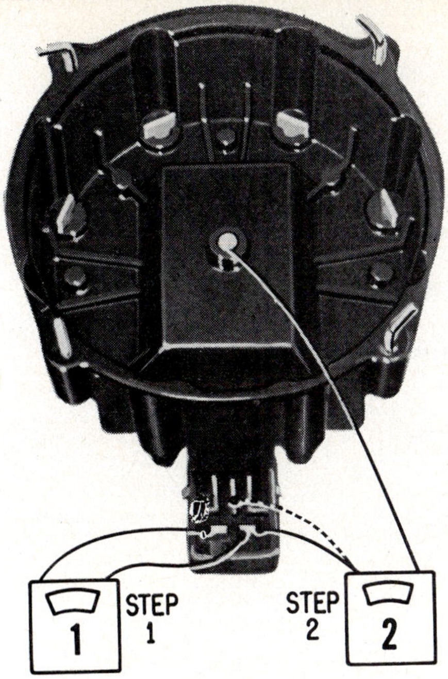

Fig. 10 H.E.I. distributor ignition coil ohmmeter tests. Internal coil units

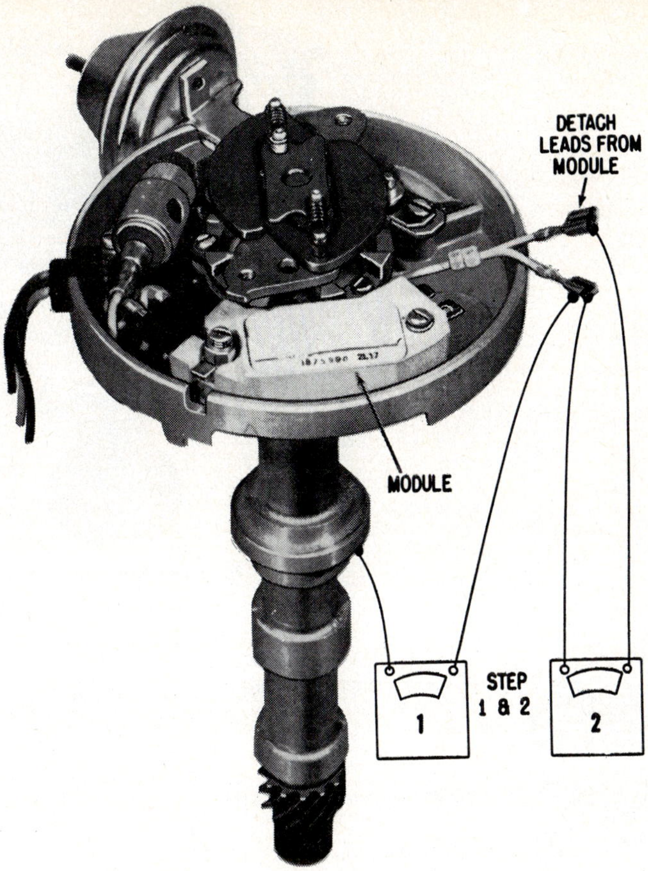

Fig. 12 H.E.I. distributor pickup coil ohmmeter test

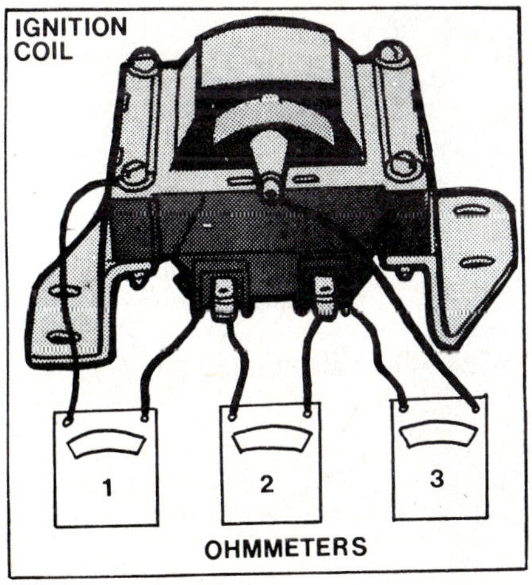

Fig. 11 HEI distributor ignition coil ohmmeter test. External coil units

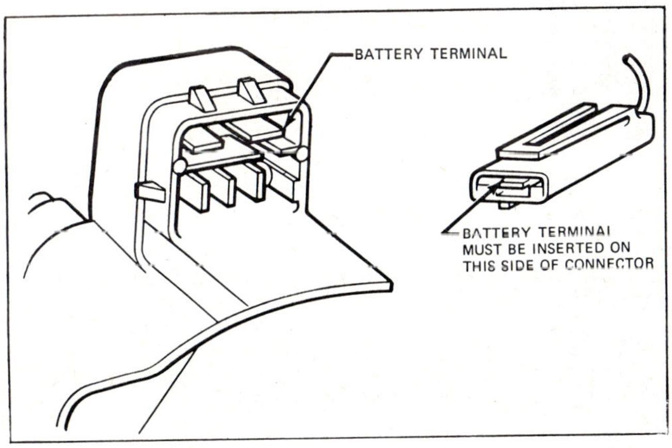

Fig. 13 H.E.I. distributor cap connector

minals 3, 4, 5, and 6 Fig. 18. If battery voltage is not present at any of these terminals, check for broken wires or loose or damaged connectors and repair as necessary.

7. Turn ignition off, then using an ohmmeter check signal input wire, 1, for continuity between distributor tachometer connector and electronic controller harness connector. If continuity cannot be obtained, check for broken wires or loose or damaged connectors.
8. If wiring harness, solenoid control valve, tubing, fuse and governor all operate satisfactorily, but governor does not govern properly, replace electronic controller.

Overspeed Warning System

The overspeed warning system warns the driver of excessive engine speed, such as during down hill operation when the wheels may begin to drive the engine. The system consists of a dash panel warning light actuated by a signal from the governor electronic controller, Fig. 17. The warning light will remain off until the engine reaches a predetermined governed speed. The light should be on with the ignition switch in the start position. If not, check for a defective light bulb. If the light remains on after the ignition switch is returned to the on position with the engine operating at idle speed, check for a defective ignition switch or governor electronic controller and replace as necessary.

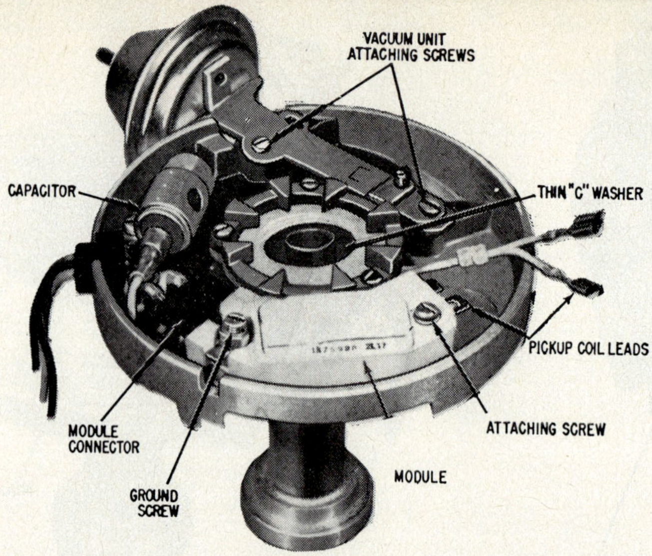

Fig. 14 H.E.I. distributor component replacement

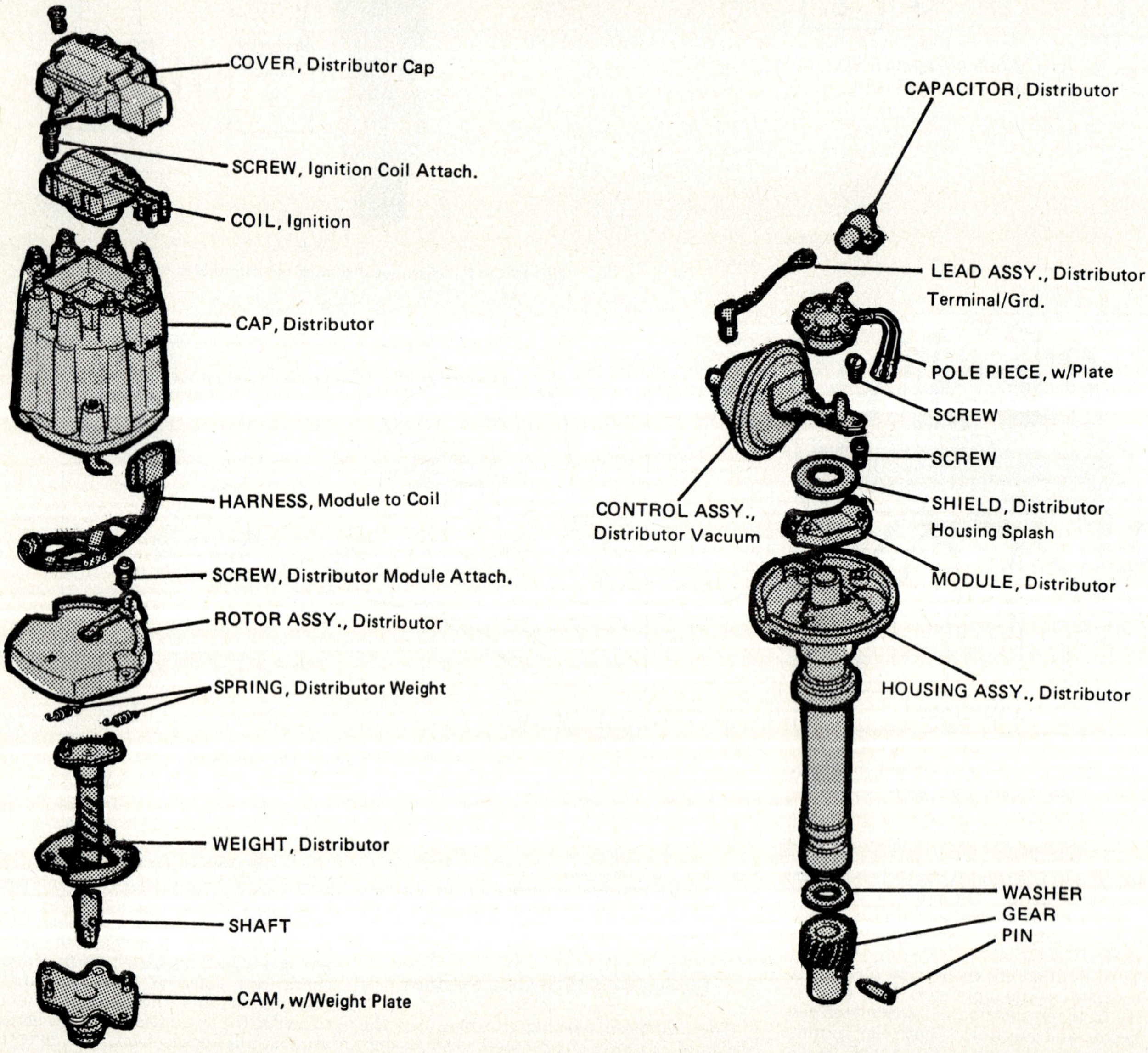

Fig. 15 Exploded view of HEI distributor (typical)

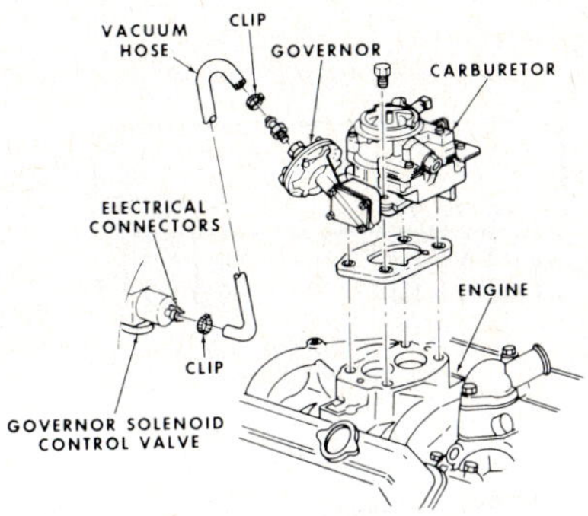

Fig. 16 Governor solenoid control valve

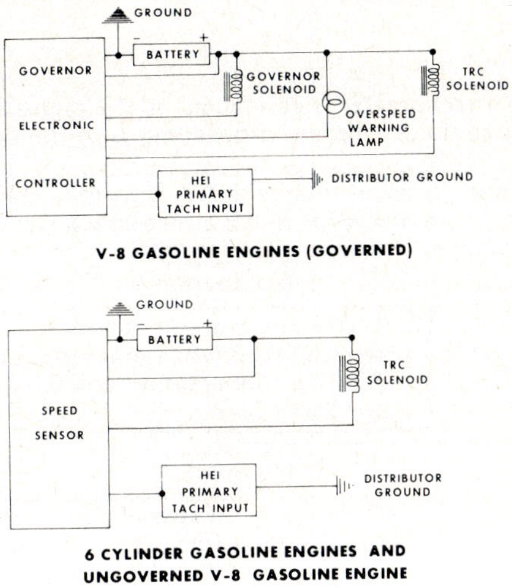

Fig. 17 Governor electronic controller wiring schematic

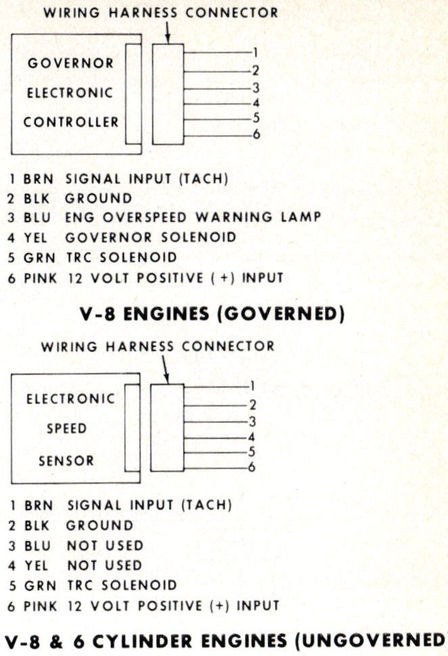

Fig. 18 Governor electronic controller terminal connector wiring identification

1981—83 Delco-Remy High Energy Ignition/Electronic Spark Timing (HEI/EST) System

DESCRIPTION

This system consists of an electronic distributor similar to the distributor used in the conventional HEI system. The ignition coil is either internally mounted in the distributor cap or mounted externally, depending on engine model, Figs. 1 and 2.

The HEI/EST system uses an Electronic Spark Timing (EST) feature to change ignition timing in response to engine operating requirements. All spark timing changes in the HEI/EST system are performed electronically by the Electronic Control Module (ECM). The ECM monitors information from various engine sensors, determines the correct spark timing and signals the distributor to change the timing as necessary. A secondary spark advance system is incorporated into the system to signal the ignition module in case of ECM failure. The HEI/EST distributor does not use a vacuum or mechanical advance device.

Some 1983 HEI/EST units incorporate a Hall Effect Switch mounted above the pick-up coil in the distributor. Also, some HEI/EST units are equipped with Electronic Spark Control (ESC), which is used to retard spark timing when detonation occurs. The spark is retarded for approximately 20 seconds, then the spark control returns to EST.

The ESC system consists of three basic components: sensor, distributor and controller. The ESC sensor is an accelerometer or magneto-strictive device, mounted at rear of engine block. It detects presence, or absence, and intensity of detonation by vibration characteristics of the engine. The sensor's output is an electrical signal which is sent to the controller. A failure of the sensor would allow no retard.

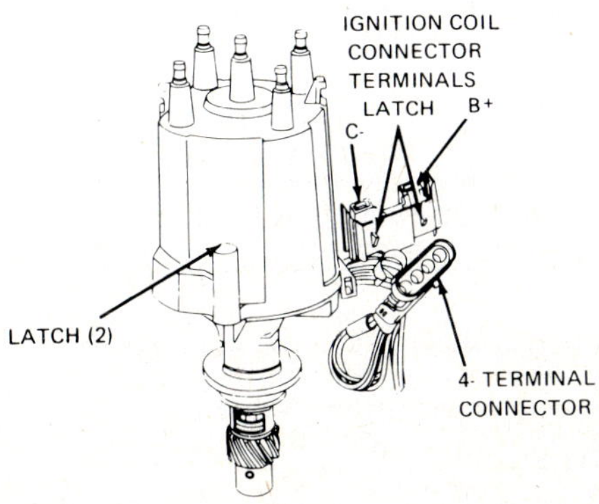

Fig. 1 HEI/EST distributor with external coil. (Typical)

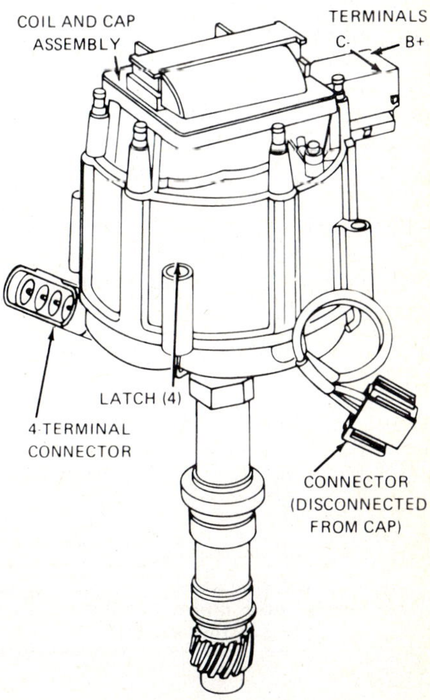

Fig. 2 HEI/EST distributor with internal coil. (Typical)

ENGINE CRANKS, BUT WILL NOT RUN

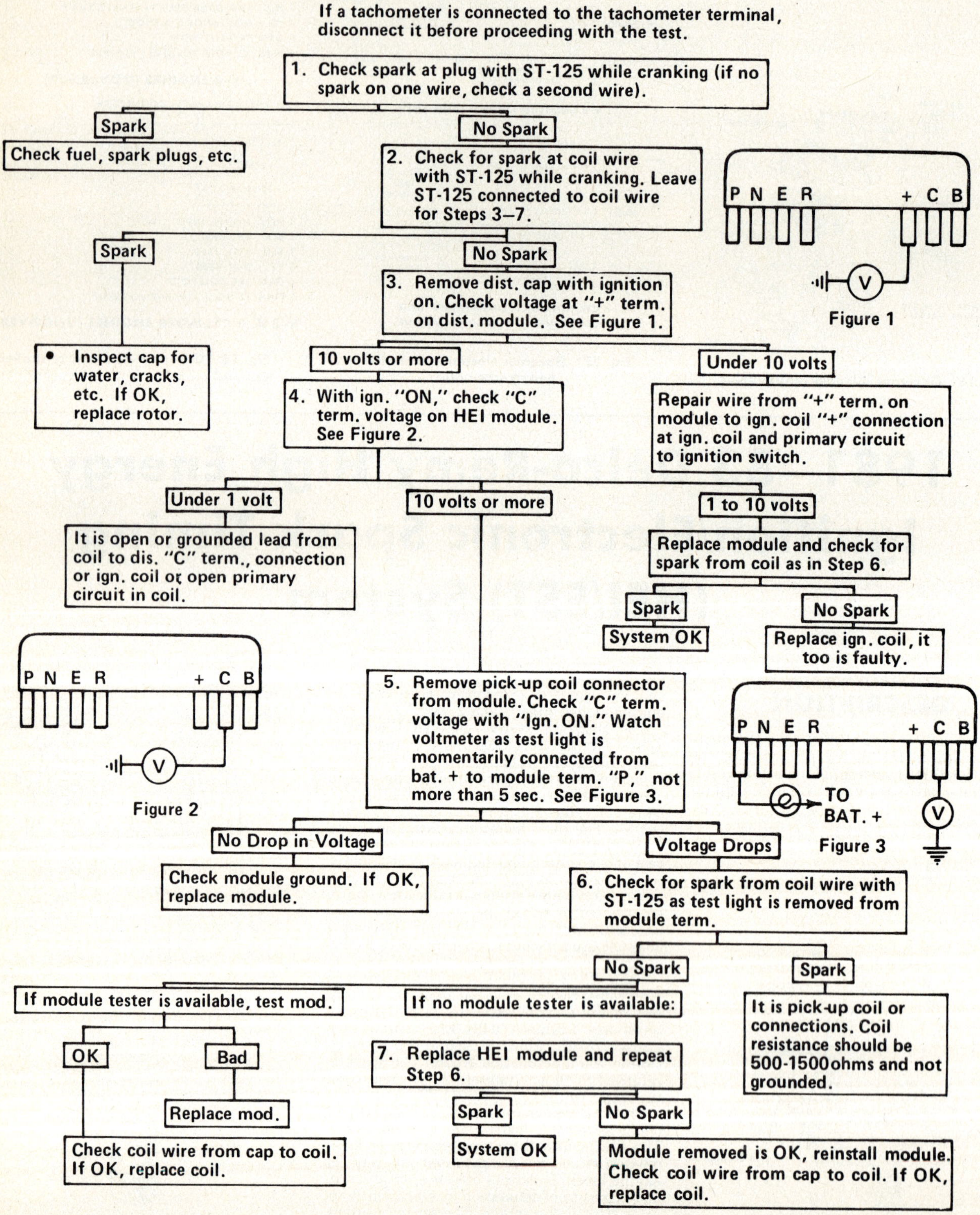

Fig. 3 HEI/EST ignition system diagnostic chart (Part 1 of 5). All models with remote coil

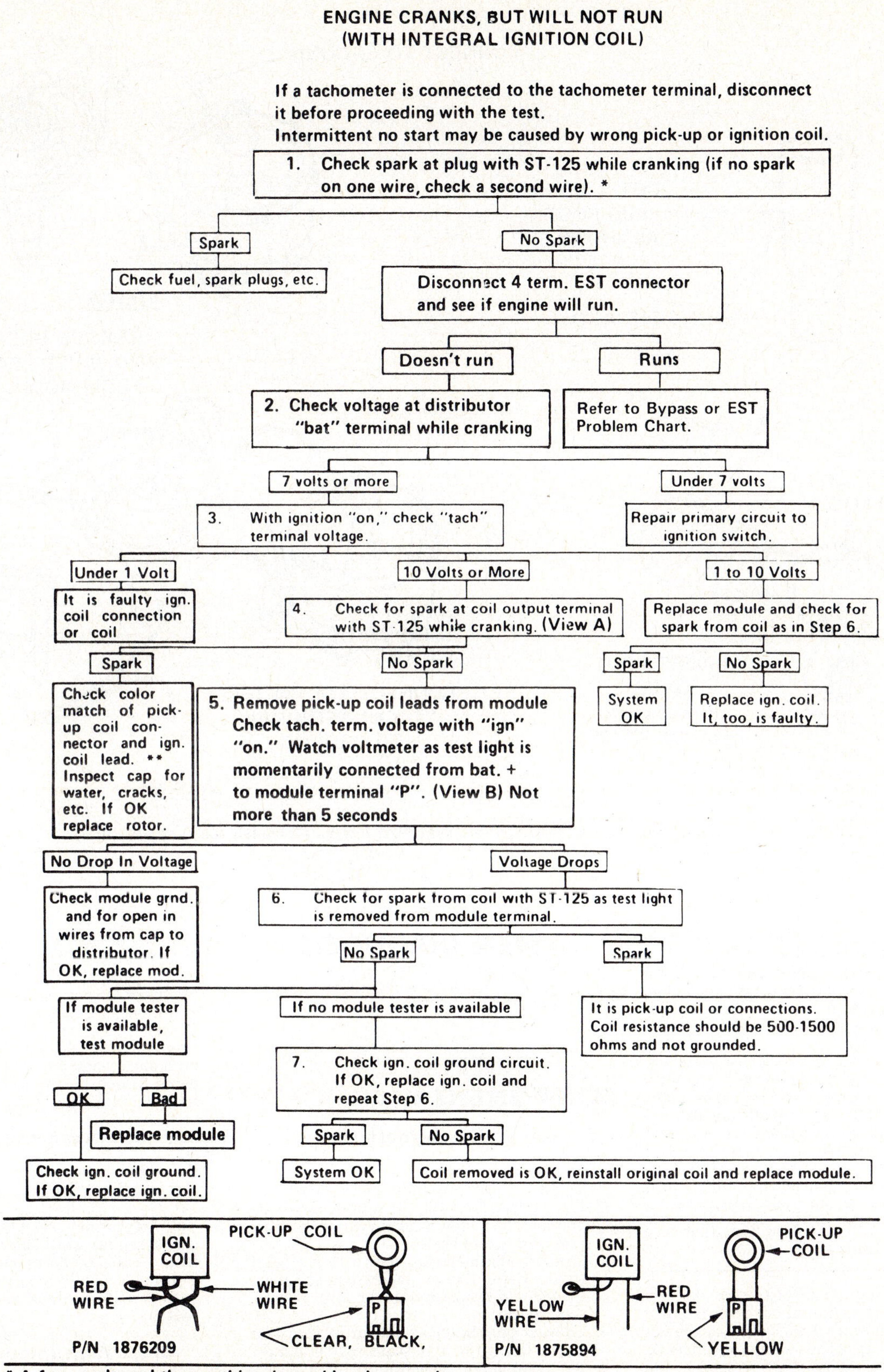

* A few sparks and then nothing, is considered no spark.

Fig. 4 HEI/EST ignition system diagnostic chart (Part 1 of 5). All models with integral coil

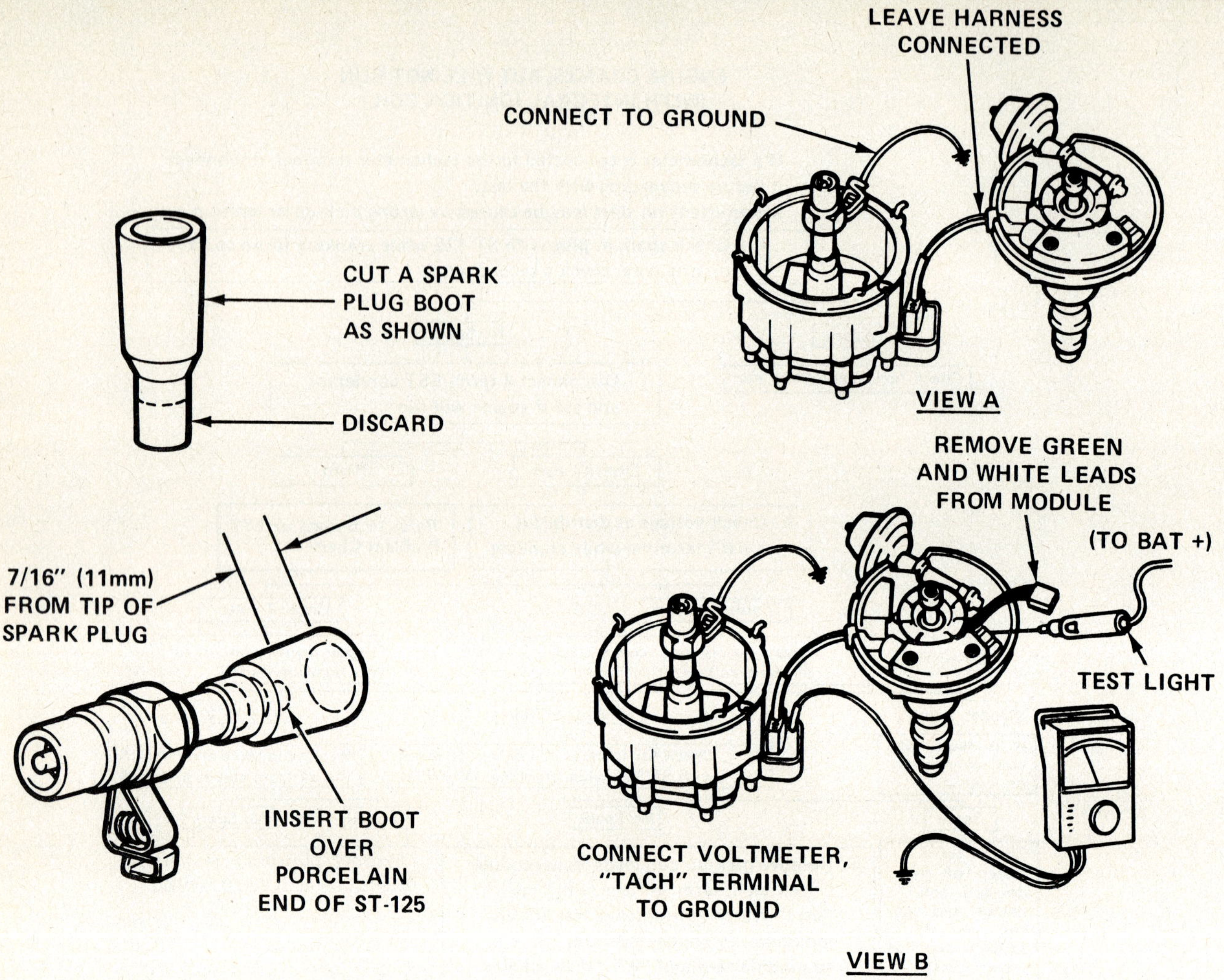

Fig. 5 HEI/EST ignition system diagnostic chart (Part 1 of 5). All models with integral coil

The distributor is an HEI/EST unit with an electronic module modified so that it can respond to ESC controller signal. The command is delayed when detonation is detected, providing the level of retard required. The amount of retard is determined by the severity of detonation.

The ESC controller processes the sensor signal into a command signal to the distributor to adjust spark timing. This is a continuous process monitoring and controlling detonation. The controller is a hard-wired signal processor and amplifier which operates from 6 to 16 volts. Controller failure would be indicated by no ignition, no retard of full retard.

NOTE: This is a full 12 volt system, therefore no resistance wire is used.

CAUTION: Never connect a wire directly between the Tach terminal of the distributor connector and ground as this will damage the electronic circuitry of the module. When using a timing light to adjust ignition timing, the connection should be made at the No. 1 spark plug. Forcing foreign objects through the boot at the No. 1 spark plug terminal will damage the boot and cause engine misfire.

SYSTEM DIAGNOSIS

Refer to HEI diagnosis charts, Figs. 3 thru 10 for diagnosis procedures.

DISTRIBUTOR COMPONENT DIAGNOSIS

Units With Remote Coil

Testing Ignition Coil

1. Connect a suitable ohmmeter as shown in Fig. 11, step 1. Using the high scale, read ohmmeter. Should read very high or infinite. If not, replace coil.
2. Connect ohmmeter as shown in Fig. 11, step 2. Using the low scale, read ohmmeter. Should read very low or zero. If not, replace coil.
3. Connect ohmmeter as shown in Fig. 11, step 3. Using the high scale, read ohmmeter. Should not read infinite. If it does, replace coil.

Testing Pickup Coil

1. Remove distributor rotor, then disconnect pickup coil leads from module.
2. Connect ohmmeter as shown in Fig. 12, step 1. Flex leads by hand to check for opens. Ohmmeter should read infinite at all times. If not, pickup coil is defective and must be replaced.
3. Connect ohmmeter as shown in Fig. 12, step 2. Repeat step 2. Ohmmeter should read one steady value between 500–1500 ohms. If not, pickup coil is defective and must be replaced.

Testing Hall Effect Switch (If Equipped), Fig. 13

1. Carefully noting polarity, connect battery and suitable voltmeter as shown in Fig. 14.
2. Voltmeter should read less than 0.5 volts. If not, replace Hall Effect switch.
3. Insert knife blade as shown in Fig. 14. Voltmeter should read within 0.5 volts of battery voltage. If not, replace Hall Effect switch.

Units With Integral Coil

Testing Ignition Coil

1. Connect a suitable ohmmeter as shown in Fig. 15, step 1. Using the low scale, read ohmmeter. Should be near zero or zero. If not, replace coil.

EST PERFORMANCE CHECK

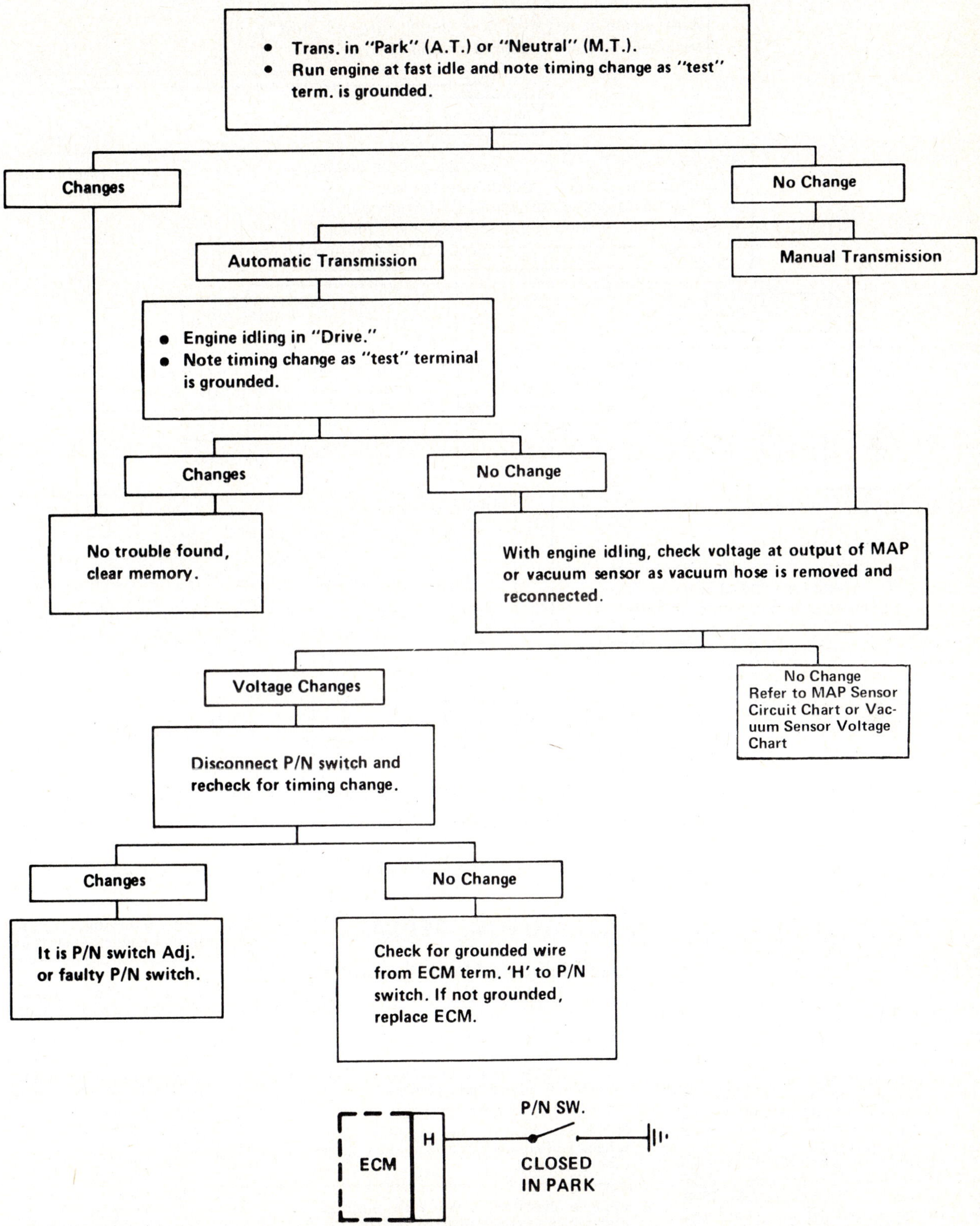

Fig. 6 EST system performance diagnosis (Part 2 of 5). All models

BYPASS OR EST PROBLEM

If car will not start and run, check for grounded EST wire to ECM terminal "12."

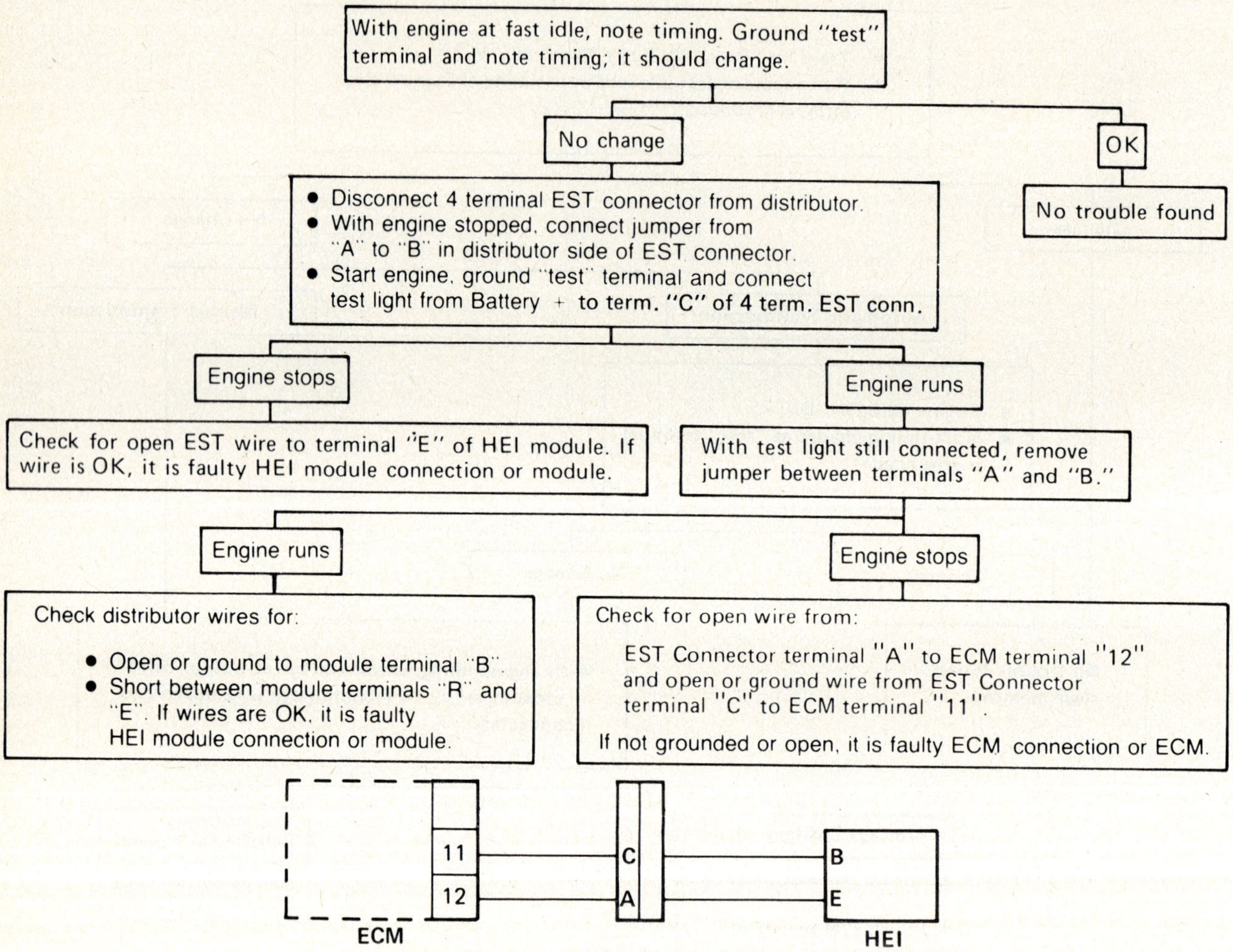

Fig. 7 Bypass or EST problem chart (Part 2 of 5). All models

2. Connect ohmmeter both ways as shown in Fig. 15, step 2. Using the high scale, read ohmmeter. If both readings are infinite, replace coil.

Testing Pickup Coil

1. Remove distributor rotor, then disconnect pickup coil leads from module.
2. Connect ohmmeter as shown in Fig. 16, step 1. Flex leads by hand to check for opens. Ohmmeter should read infinite at all times. If not, pickup coil is defective and must be replaced.
3. Connect ohmmeter as shown in Fig. 16, step 2. Repeat step 2. Ohmmeter should read one steady value between 500–1500 ohms. If not, pickup coil is defective and must be replaced.

Testing Hall Effect Switch (If Equipped)

Refer to "Units With Remote Coil".

DISTRIBUTOR, REPLACE

Models Equipped With Integral Coil Distributor

1. Disconnect battery and tachometer lead from distributor cap.
2. Disconnect coil connectors from distributor cap and 4 terminal connector.

NOTE: Do not use a tool of any kind to release coil connectors or damage to lock tabs will result.

---

3. Remove distributor cap with ignition high tension wires connected and position aside.
4. Remove distributor clamp screw and hold down clamp.
5. Mark relationship of rotor to distributor housing, then lift distributor from engine slightly until rotor stops turning. Mark relationship of new rotor position on distributor housing, then remove distributor from engine.
6. If engine was not cranked after distributor was removed from engine, proceed as follows:
   a. Position rotor to align with second mark made on distributor housing prior to removing distributor.
   b. Install distributor into engine and align rotor with first mark made on distributor housing prior to removing distributor.
   c. Install distributor cap and connect all electrical connectors.
   d. Adjust ignition timing as specified in individual car chapters.
7. If engine was cranked after distributor was removed from engine, proceed as follows:
   a. Remove No. 1 spark plug and crank engine until compression pressure is felt in No. 1 cylinder. Slowly rotate

ELECTRONIC SPARK CONTROL (ESC)
SYSTEM CHECK
ENGINE KNOCK, POOR PERFORMANCE OR POOR ECONOMY

This chart should only be used after all other causes of Spark Knock have been checked, i.e., Timing, EGR, Engine Temperature or Excessive Engine Noise.

- "Test" terminal ungrounded.
- Connect tachometer.
- Engine running at about 1500 RPM.
- Transmission in park or neutral and at normal operating temperature.
- Tap Engine Block in area of knock sensor and check for RPM drop.

RPM Drop
System OK

No RPM Drop
- Disconnect ESC.
- Check for RPM change.

RPM Drops
- Run engine at 2000 RPM.
- Check voltage between ESC harness connector pins "E" and "D" with digital voltmeter and controller disconnected. Should be over .08 volts on A.C. scale at 2000 RPM ± 100 RPM.

Not OK
Check for open in circuit from ESC connector Pin "D" to ground.

Open
Repair and recheck.

Not Open
Check for faulty connection, open or ground in wire from sensor to ESC term. "E". If circuit is OK, it is faulty sensor connection or sensor.

OK
Faulty ESC connection or ESC controller.

No RPM Change
Note "check engine" light.

Light OFF
"Ignition ON," engine stopped. Check voltage from ESC Term. "C" to ground.

Under 2 Volts
Replace ECM.

Over 2 Volts
Correct short to Bat. + in wire from ESC Term. "C" to ECM Term. "L".

Light ON
It is most likely a faulty ESC controller, but could be the knock sensor.

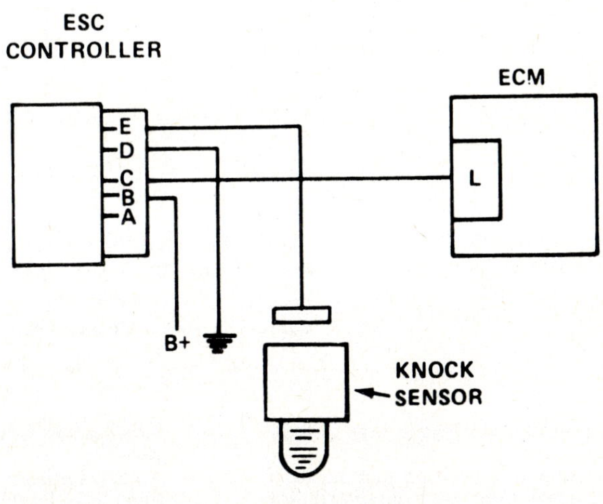

Fig. 8 ESC system diagnosis (Part 3 of 5). All models with ESC

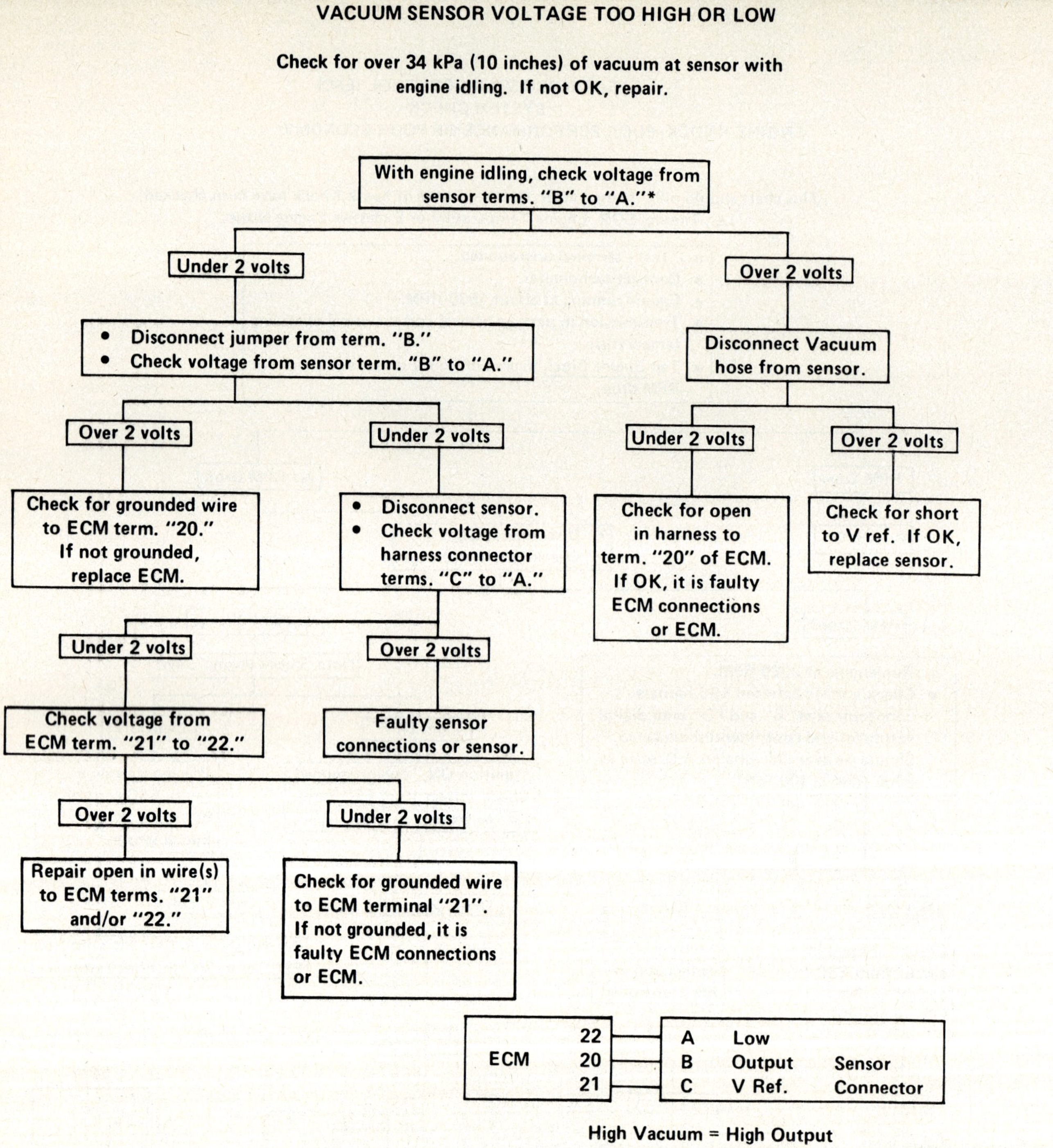

*This requires use of three jumpers between the sensor and the connector. They can be made using terminals 12014836 and 12014837.

Fig. 9 Vacuum sensor circuit chart (Part 4 of 5). All models with vacuum sensor

engine until top dead center is indicated.

b. Turn rotor to a position just ahead of No. 1 spark plug tower on distributor cap.

c. Install distributor. Connect 4 terminal connector.

d. Install distributor cap, coil connectors, tachometer lead and battery feed wire.

e. Adjust ignition timing to specifications in individual car chapters.

Models Equipped With Remote Coil Distributor

1. Disconnect distributor wire connector from coil.
2. Remove distributor cap and position aside.
3. Remove distributor hold down bolt and clamp.
4. Mark relationship of rotor to distributor housing and distributor housing to engine block, then pull distributor up

MAP SENSOR CIRCUIT

Check for over 10" vacuum at MAP sensor with engine idling. If not OK, repair hoses or connections

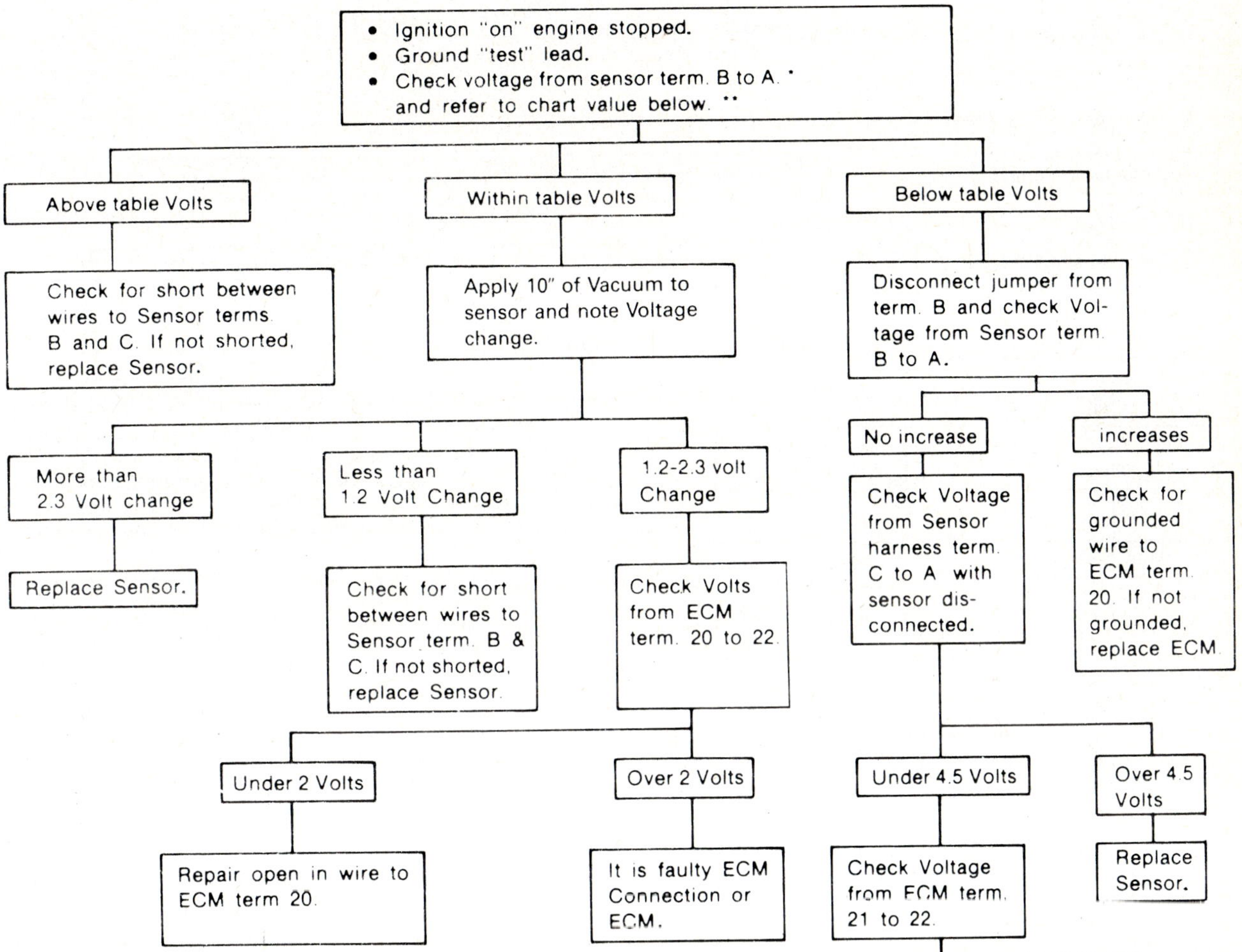

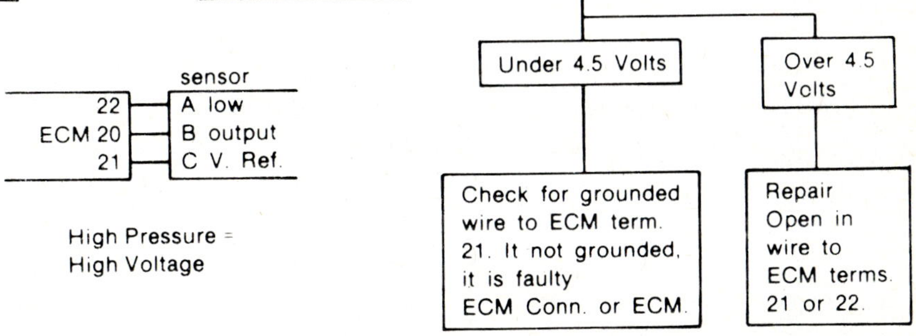

**Altitude	Voltage range
Below 1,000'	3.8 - 5.5 V
1,000- 2,000'	3.6 - 5.3 V
2,000- 3,000'	3.5 - 5.1 V
3,000- 4,000'	3.3 - 5.0 V
4,000- 5,000'	3.2 - 4.8 V
5,000- 6,000'	3.0 - 4.6 V
6,000- 7,000'	2.9 - 4.5 V
7,000- 8,000'	2.8 - 4.3 V
8,000- 9,000'	2.6 - 4.2 V
9,000-10,000'	2.5 - 4.0 V

*This requires use of three jumpers which can be made using terms. 12014836 and 12014837.

Fig. 10 MAP sensor voltage chart (Part 5 of 5). 1981 models with MAP sensor

MAP SENSOR CIRCUIT

Check for over 10" vacuum at MAP sensor with engine idling. If not OK, repair hoses or connections

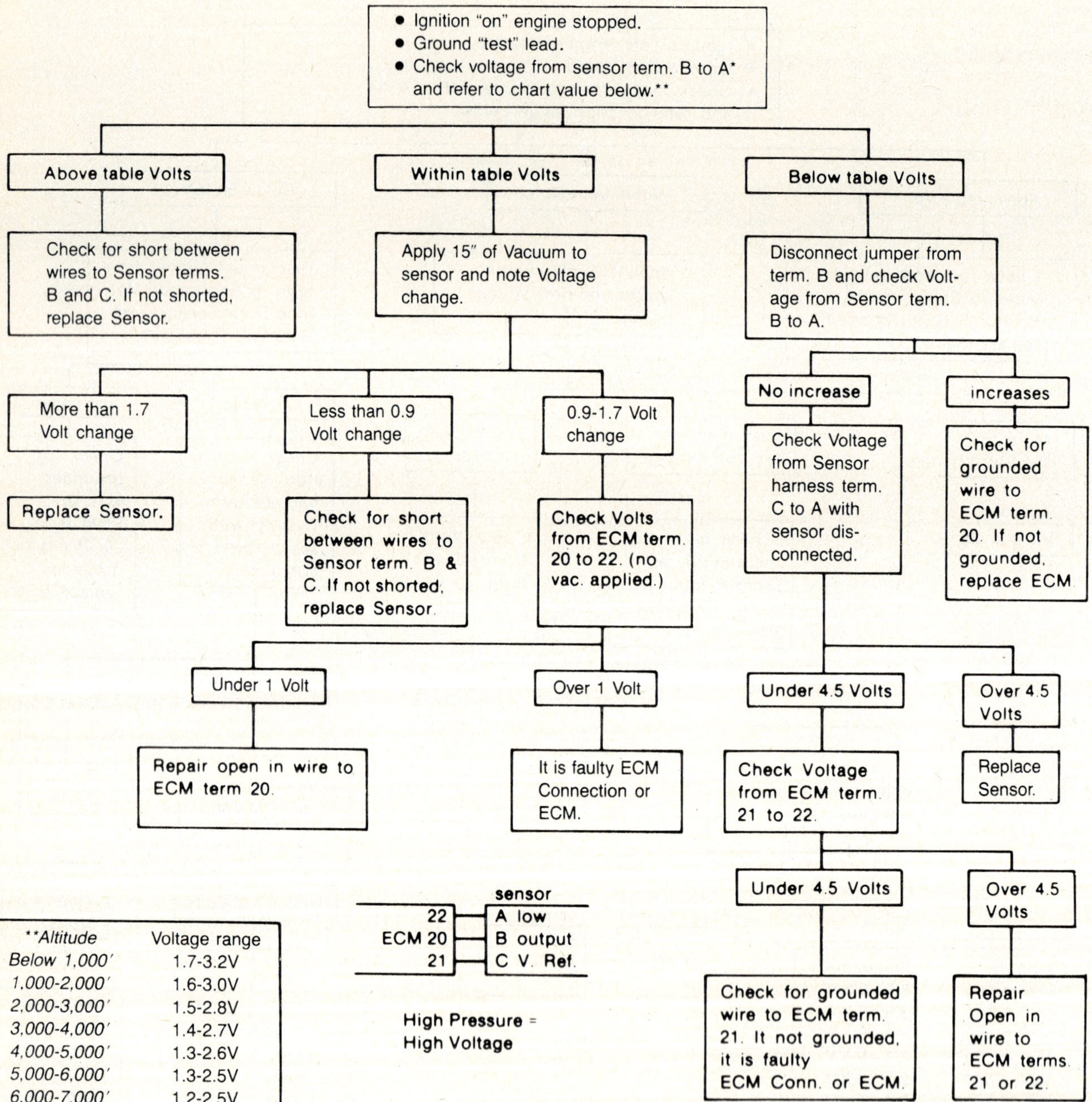

**Altitude	Voltage range
Below 1,000'	1.7-3.2V
1,000-2,000'	1.6-3.0V
2,000-3,000'	1.5-2.8V
3,000-4,000'	1.4-2.7V
4,000-5,000'	1.3-2.6V
5,000-6,000'	1.3-2.5V
6,000-7,000'	1.2-2.5V
7,000-8,000'	1.1-2.4V
8,000-9,000'	1.1-2.3V
9,000-10,000'	1.0-2.2V

*This requires use of three jumpers which can be made using terms. 12014836 and 12014837.

Fig. 10A MAP sensor voltage chart (Part 5 of 5). 1982–83 models with MAP sensor

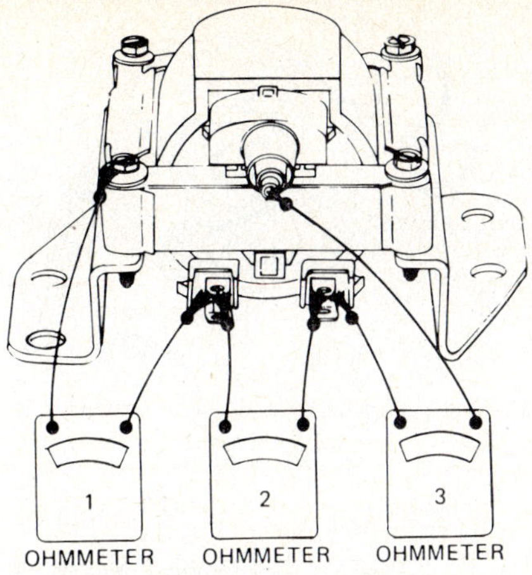

Fig. 11 Testing remote ignition coil

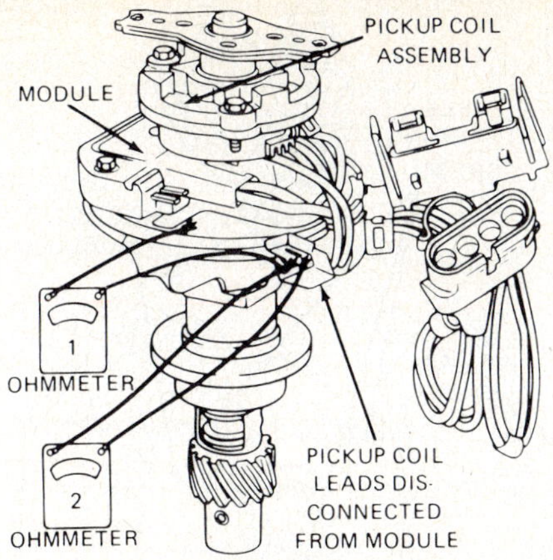

Fig. 12 Testing remote coil distributor pickup coil

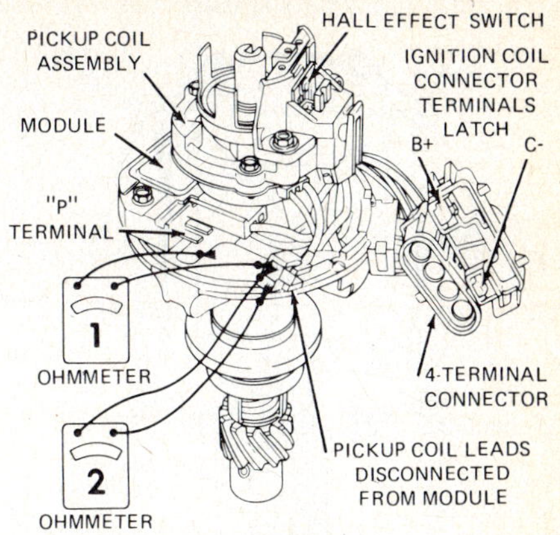

Fig. 13 Hall Effect switch equipped distributor

until rotor stops turning and mark relationship of rotor to distributor housing again.

5. Remove distributor.
6. If engine was not cranked after distributor was removed from engine, proceed as follows:
   a. Position rotor to align with second mark made on distributor housing prior to removing distributor.
   b. Install distributor into engine and align rotor with first mark made on distributor housing prior to removing distributor.
   c. Install distributor cap and connect all electrical connections.
   d. Adjust ignition timing as specified in individual car chapter.
7. If engine was cranked after distributor was removed from engine, proceed as follows:
   a. Remove No. 1 spark plug and crank engine until compression pressure is felt in No. 1 cylinder. Slowly rotate engine until top dead center is indicated.
   b. Turn rotor to a position just ahead of No. 1 spark plug tower on distributor cap.
   c. Install distributor.
   d. Install distributor cap and connect all electrical connectors.
   e. Adjust ignition timing as specified in individual car chapter.

COMPONENTS, REPLACE

Ignition Coil Replacement

Units With Internal Coil

1. Remove electrical connector from distributor cap by lifting retaining tabs.
2. Remove three coil cover attaching screws, then remove cover, Fig. 17.
3. Remove coil attaching screws, then remove ignition coil with leads from distributor cap.
4. Remove coil arc seal.
5. Reverse procedure to install.

Units With External Coil

1. Disconnect distributor wire connector from coil.
2. Disconnect coil high tension lead.

NOTE: When removing coil high tension lead, twist boot one half turn and pull on boot only to prevent damage to coil high tension lead.

3. Remove coil attaching screws and coil.
4. Reverse procedure to install.

Module Replacement

1. Remove distributor cap and rotor.
2. Remove module retaining screws, then lift module upward.
3. Disconnect module electrical connector. Note position of electrical connector prior to disconnecting. Remove module from distributor.
4. Reverse procedure to install. Check to ensure that module electrical connector is installed in proper position. If module is not being replaced, do not wipe grease from module or distributor base. If new module is to be installed, coat module and distributor base with pack of grease supplied with replacement module.

Pick Up Coil, Replace

1. Mark relationship of distributor shaft and driven gear for reference during assembly.
2. Remove roll pin and driven gear from shaft.
3. Remove distributor cap, magnetic shield (if equipped), shaft and rotor.

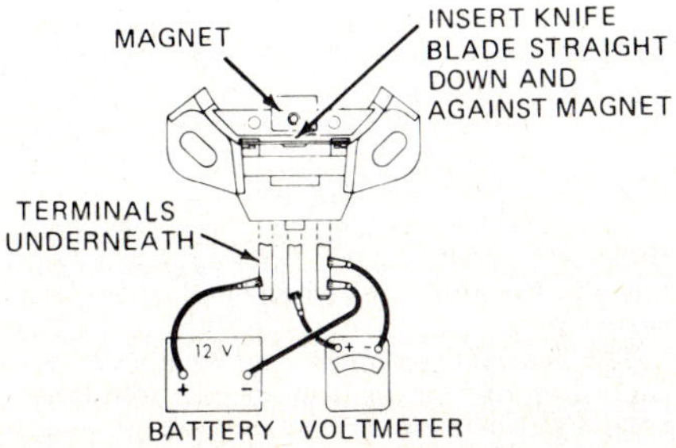

Fig. 14 Testing Hall Effect switch

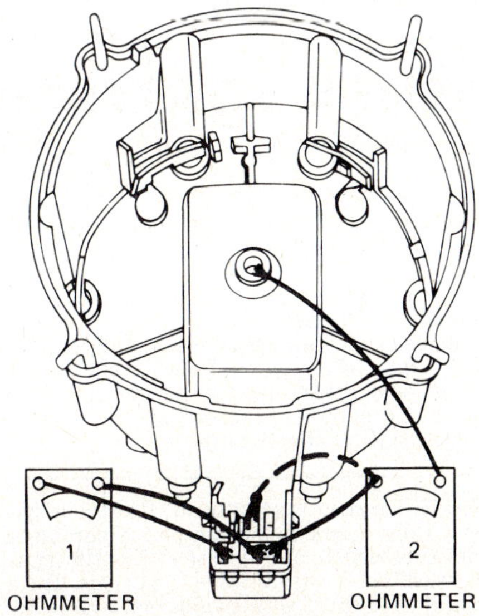

Fig. 15 Testing integral ignition coil

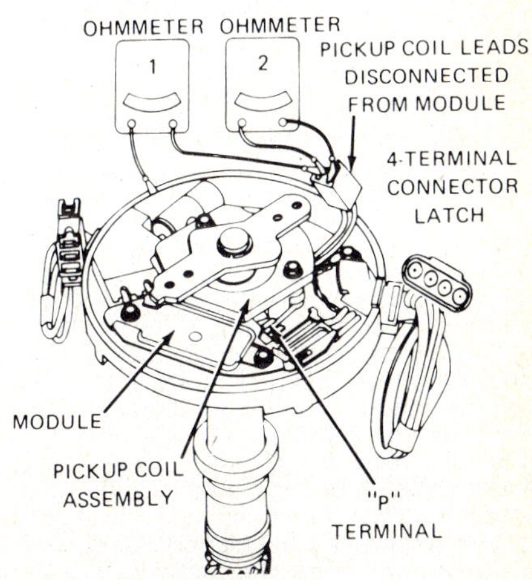

Fig. 16 Testing integral coil distributor pickup coil

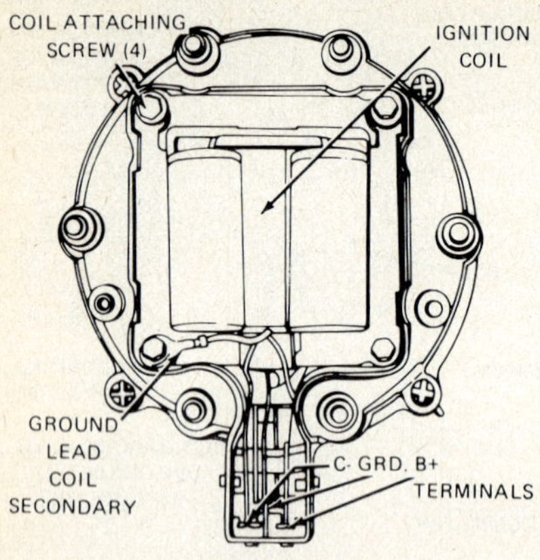

Fig. 17 Ignition coil removal. Models with integral coil

4. Remove C washer on top of pick up coil assembly, then disconnect pick up coil leads from module and remove pick up coil assembly.
5. Reverse procedure to install.

Hall Effect Switch (If Equipped), Replace

1. Remove distributor cap and rotor.
2. Remove switch retaining screws, then while pulling switch away, remove wiring connector. Remove switch.
3. Reverse procedure to install

NOTE: After Hall Effect switch is installed, spin distributor shaft to ensure teeth do not touch. If necessary loosen, then re-tighten pick up coil teeth and Hall Effect switch to eliminate contact.

DISTRIBUTOR SERVICE

Units With Internal Coil

1. Remove cap and coil as an assembly.

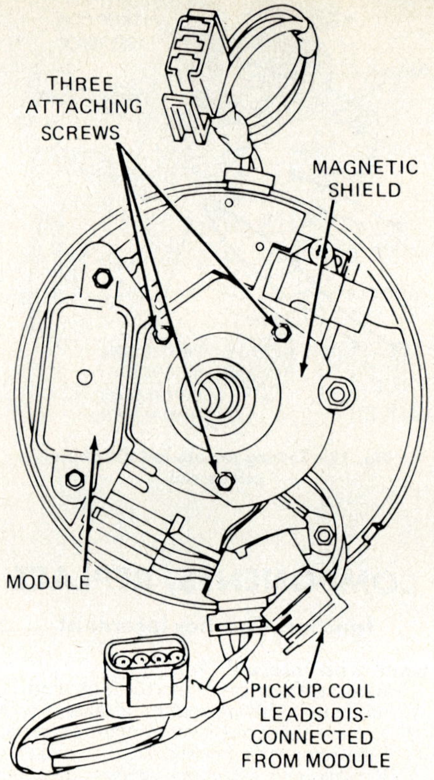

Fig. 18 Removing distributor shield. Models with integral coil

2. Remove rotor and pick up coil leads from module.
3. Mark relationship of distributor shaft and driven gear for reference during assembly.
4. Remove roll pin and driven gear.
5. Remove distributor shaft assembly.
6. Remove shield, Fig. 18.
7. Remove retaining ring, pick up coil, magnet and pole piece, Fig. 19.

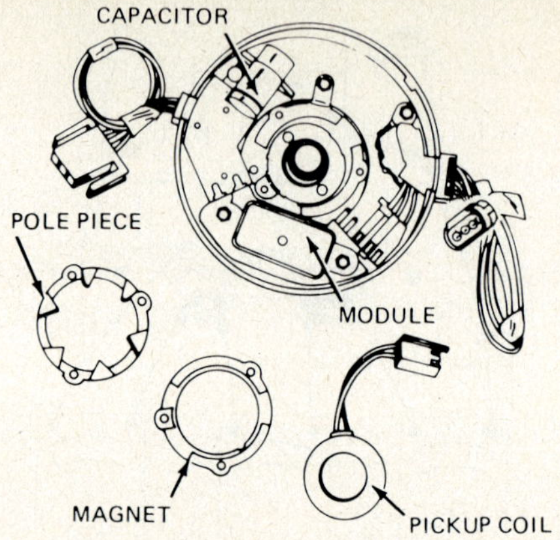

Fig. 19 Pickup coil disassembled. Models with integral coil

8. Remove module retaining screws and capacitor screw, then remove module, capacitor and harness assembly from distributor.
9. Disconnect wiring harness from module. Reverse procedure to assemble.

Units With External Coil

1. Remove rotor, then disconnect pick up coil leads from module.
2. Mark relationship of distributor shaft and driven gear for reference during assembly.
3. Remove roll pin and driven gear from distributor shaft.
4. Remove C washer retaining ring, then remove pick up coil assembly.
5. Disconnect module electrical connectors, then remove module retaining screws and the module.
6. Clean distributor base, then apply suitable silicone lubricant between module and base.
7. Reverse procedure to assemble.

Ford Breakerless (B/L) & Dura Spark I & II Solid State Ignition System

The B/L and Dura Spark I & II ignition systems do not use points and are controlled by an electronic module. The Dura Spark system uses higher spark plug voltages during the starting and running modes, permitting the use of wider spark plug gaps required to ignite the leaner air/fuel mixtures.

Two versions of the Dura Spark system are used. The Dura Spark I incorporates a new module ignition coil distributor cap and adaptor, rotor and ignition wires. The module has a built-in current regulator which reduces overheating of the coil and module during operation. The ignition coil has revised primary and secondary windings necessary to supply the higher spark plug voltages. The Dura Spark II system, Fig. 1, uses a new distributor cap and adaptor, rotor and ignition wires. The higher spark plug voltage in this system is obtained by reducing the value of the ballast resistor in the primary side of the ignition system.

The electronic module, Figs. 1, and 2, is the brain of this system and is well protected from outside elements such as heat and shock. The heat sink containing all the electronic devices is sealed in a mixture of epoxy and sand. This module cannot be disassembled and must be replaced if malfunctioning.

NOTE: On some high altitude models, a special Dura Spark II module plus a barometric pressure switch, allows base engine timing to be modified to suit altitude conditions. These units have three wire connectors, Fig. 3. All other elements and performance characteristics of this module are identical in both modes of operation to the basic Dura Spark II system.

Oil filled ignition coils are used, therefore conventional ignition coils are not to be used with these systems. The proper coil is easily identified as it is all blue and terminals are labeled differently from conventional ignition coils "BAT" (battery) and "DEC" (Distributor Electronic Control), Fig. 4.

The ignition switch energizes the module through the white wire while engine is cranking and through the red wire when engine is running.

The distributor shaft and armature rotation, Fig. 5, causes the armature poles to pass by the core of the magnetic pick-up assembly, in turn cutting the magnetic field and signaling the electronic module, Figs. 6

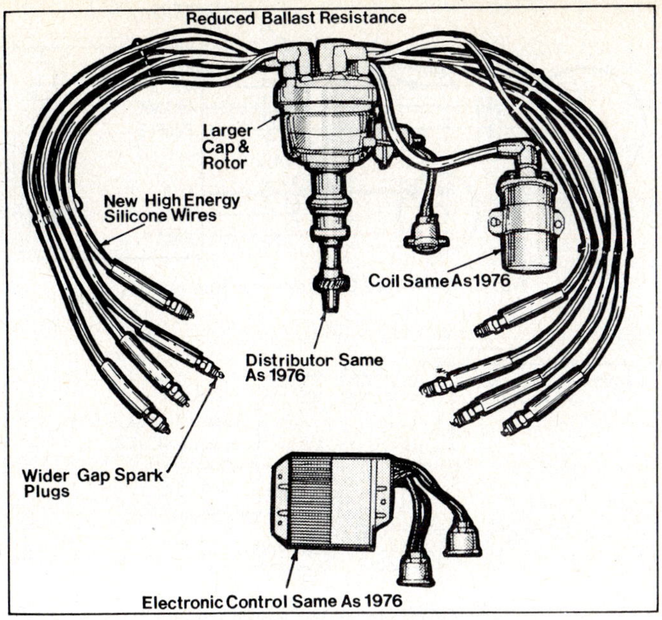

Fig. 1 1977–83 Ford Dura Spark ignition system exc. California (Typical)

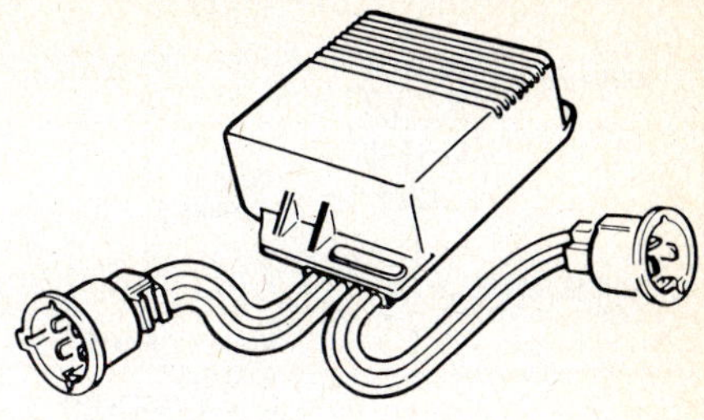

Fig. 2 Ford breakerless electronic module

thru 10, through the orange and purple wires to break the primary ignition current, thus inducing secondary voltage in the coil to fire the spark plugs. The coil is then energized again by the primary circuit and ready for the next spark cycle. This primary circuit is controlled by a timing circuit in the module.

The ignition system is protected against electrical current produced during normal vehicle operation and against reverse polarity or high voltage accidentally applied if vehicle is jump started.

Total diagnosis of the system requires only a volt-ohmmeter tester.

CAUTION: The ignition system will be damaged if other than volt-ohm test procedures are used to check alternator output. This alternator test procedure is outlined in the "Ford Autolite/ Motorcraft Alternator" section, under "Voltmeter Test."

Do not use the volt-amp test procedure or any other test that utilizes a knife switch on the battery terminal.

Since the interval between the time that the module activates the primary ignition circuit and the time the distributor signal turns it off varies with engine speed, a dwell measurement is insignificant.

SYSTEM DIAGNOSIS

If the ignition system is suspected of a malfunction inspect for loose connections and perform the "Secondary Ignition Checkout" procedure in the Trouble Shooting Chapter elsewhere in this manual. Since the secondary circuit is identical to that of a conventional ignition system.

If no spark is observed during the above test, check the ignition coil high tension wire, replace if damaged.

SERVICE NOTE

If a no start condition exists during wet weather/high humidity conditions, the cause may be high voltage leakage to ground at the coil to distributor wire boots due to excessive moisture and contamination (such as salt). To correct this condition proceed as follows:

1. Clean all moisture and dirt accumulation from exposed surfaces on the distributor cap and coil tower without removing wires from the cap or coil tower. If there is evidence of wire damage replace as necessary. Attempt to start the engine.
2. If the engine still does not start, remove the distributor to coil high tension wire and clean the coil tower, the center tower of the distributor cap and the distributor to coil wire. Apply Dow 4X or GE G-624 Silicone grease to the outer surfaces of the coil tower and the center tower of the cap and re-install the coil wire insuring that both ends are fully seated.
3. If condition persists, replace the coil to distributor high tension wire to provide new sealing qualities at the boots seals.

Secondary Circuit Test

Rotor Air Gap

1. Connect an oscilloscope with voltage pick-up on coil to distributor cable. Set oscilloscope to 30 KV parade pattern.
2. Disconnect any spark plug wire (except no. 1) and ground it firmly to engine block.

CAUTION: While engine is running, do not disconnect spark plug wires 1 or 3 on 4 cylinder engine, 1, 3, or 5 on inline 6 cylinder engine, 1 or 4 on V6 engine or 1 or 8 on V8 engine. These contacts in the distributor cap are above the stator and arcing may result when the circuit is open.

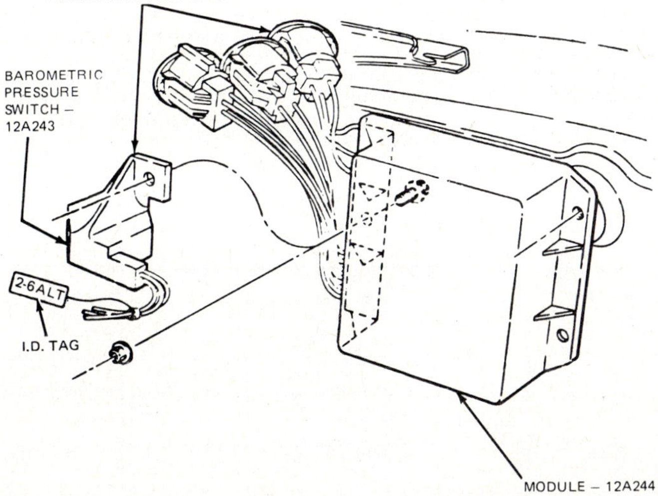

Fig. 3 Ford Dura Spark II electronic module. High altitude models with dual mode timing

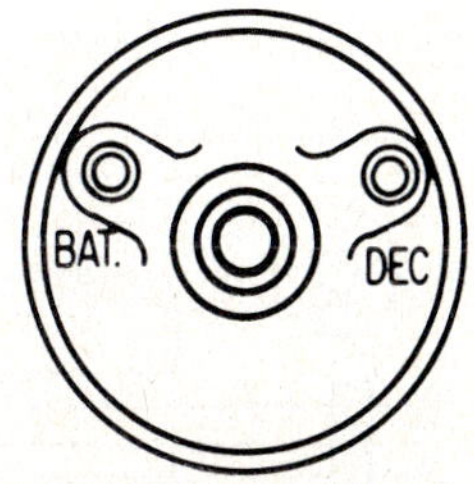

Fig. 4 Ford electronic ignition coil identification

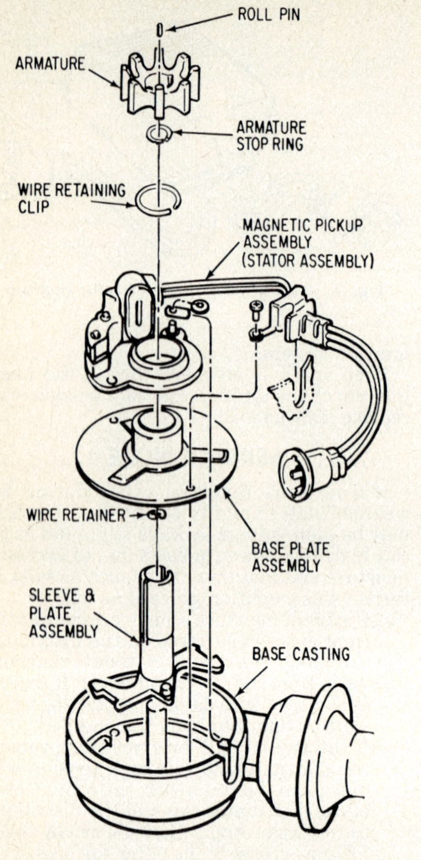

Fig. 5 Ford breakerless (B/L) distributor.

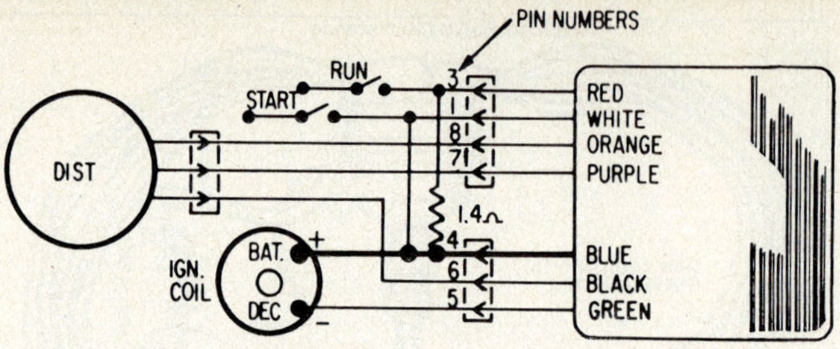

Fig. 6 1974 Ford breakerless primary circuit

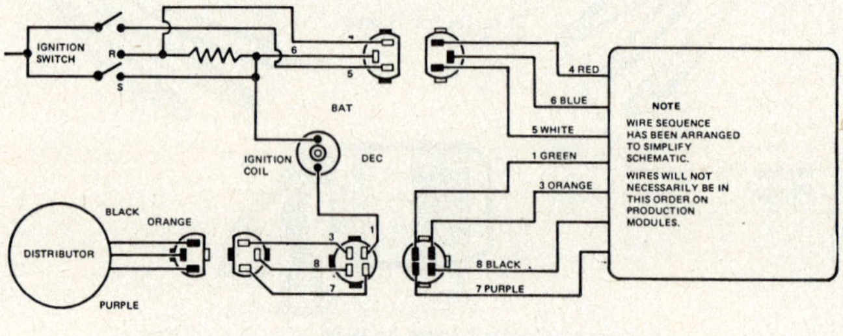

Fig. 7 1975 Ford breakerless ignition primary circuit

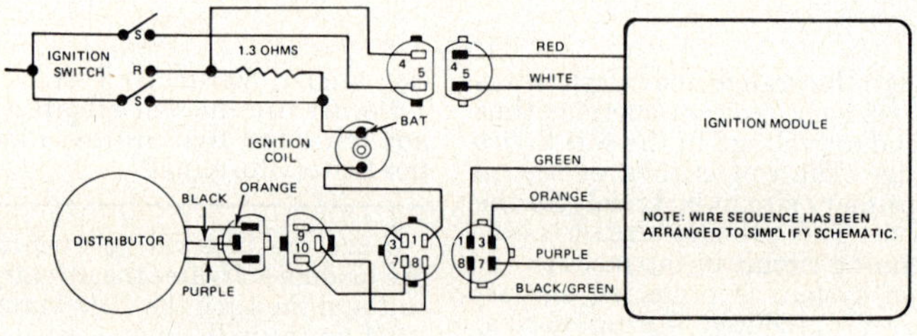

Fig. 8 1976 Ford breakerless ignition primary circuit

3. Run engine at 1000 rpm and observe height of firing spike of disconnected cylinder. Firing spike should be 8 KV or less.
4. If firing spike is 8000 volts or less, air gap resistance is within acceptable limits. If firing spike is greater than 8000 volts, replace rotor and repeat test. If still too high, replace distributor cap.

NOTE: White deposits on rotor are normal and should not be removed. If rotor is replaced, coat new rotor with silicone grease.

Spark Plug Firing Voltage Test

1. Reconnect spark plug lead, then run engine at 2000 rpm and set oscilloscope to 30 KV parade pattern.
2. Check for firing spikes which are appreciably lower or higher (50 percent) than the other cylinders.
3. Normal firing spikes should be 6–20 KV.
4. Expand pattern to inspect individual firing spikes. If all spikes are below or above specifications, check rotor, distributor cap or coil to distributor cable.
5. To check low firing voltage, disconnect spark wire from spark plug and hold it so that spark cannot arc to ground:
   a. If spike does not increase to 28,000 volts or more (coil output), check for a cracked distributor cap with a grounded terminal.
   b. If spike does not rise, replace affected spark plug.
6. To check high firing voltage, disconnect spark plug wire from spark plug and ground the wire:
   a. If high spike drops, replace spark plug.
   b. If high spike remains, check spark plug wire and its connection at distributor cap.

Spark Plug Wires Continuity Test

1. Remove distributor cap and disconnect suspected wire from spark plug or coil.
2. Using an ohmmeter, check wire resistance through distributor cap. Resistance should be 100,000 ohms or less.
3. If resistance of wire is greater than 100,000 ohms, replace wire.

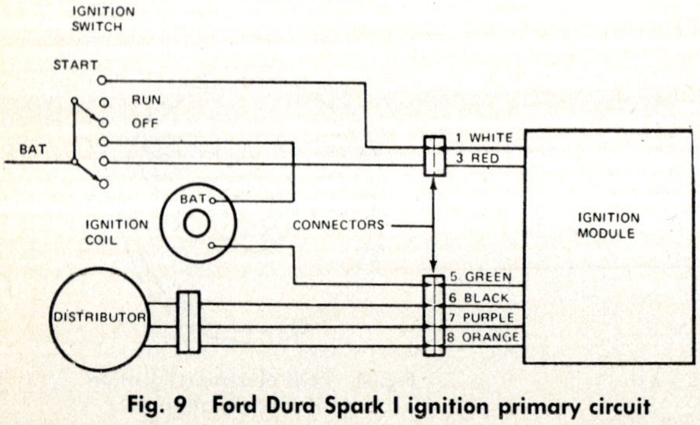

Fig. 9 Ford Dura Spark I ignition primary circuit

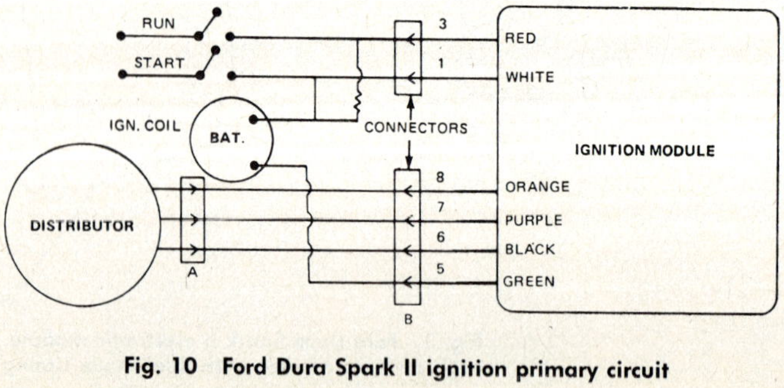

Fig. 10 Ford Dura Spark II ignition primary circuit

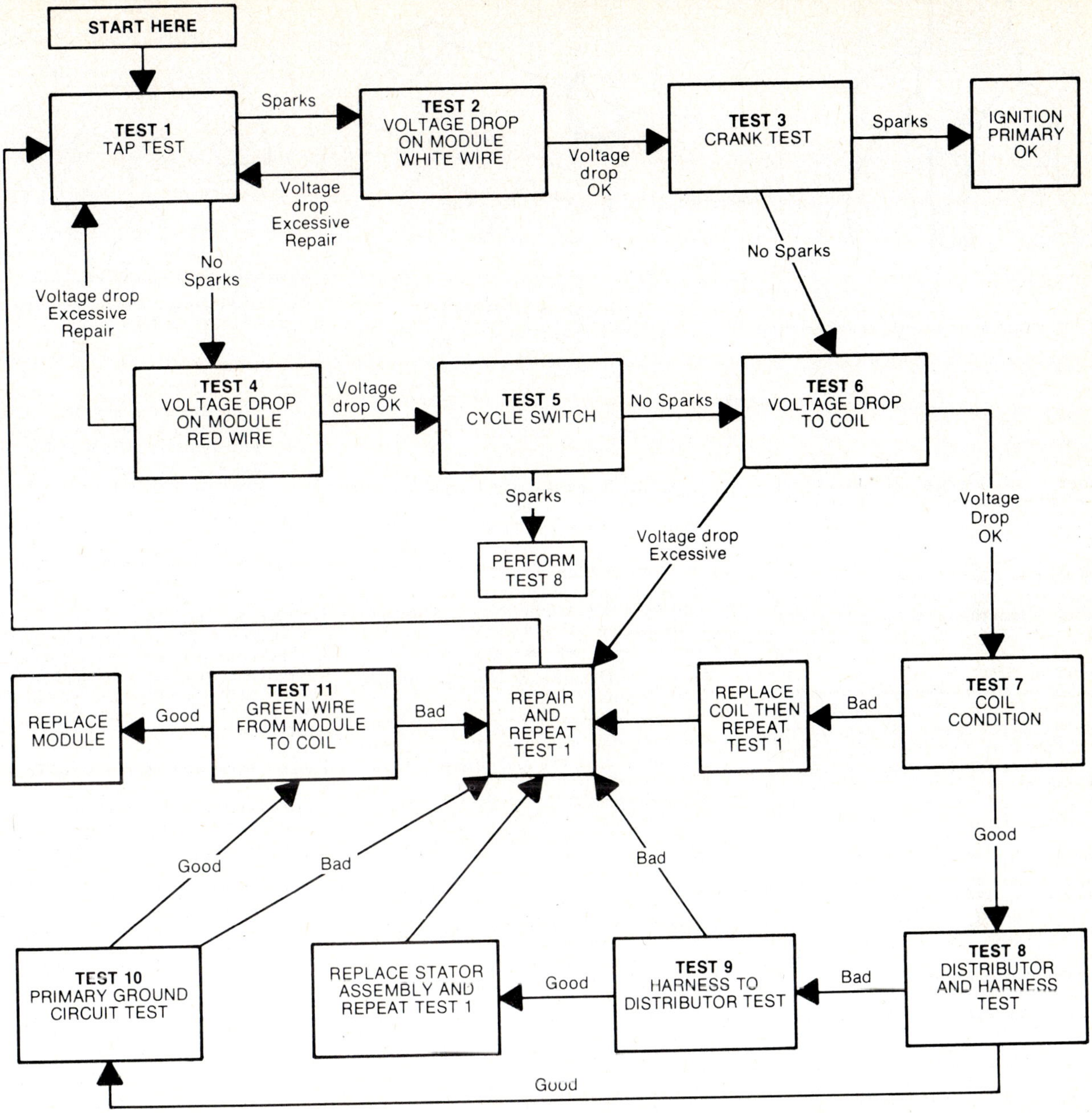

Fig. 11 System diagnosis flow chart

Spark Plug Wire Inspection

1. Clean off any deposits of road salt, dirt etc. from wires, boots, distributor cap and coil using water and a mild soap solution.
2. Inspect wires and boots for cuts, punctures, scrapes and burns. Replace as required.
3. Inspect wire terminals for corrosion. Remove corrosion with fine sandpaper.
4. Coat inside of all boots with silicone grease before reinstalling.

Primary Circuit Test

NOTE: The flow chart in Fig. 11, and test procedures provide a sequence for testing which should locate most problems in the primary ignition circuit. It is important to perform these tests in the exact order listed to check all portions of the system. A volt-ohmmeter, 16 inch jumper wire, straight pin, modified spark plug and ballast resistor, Fig. 12, are required to check the system.

Test 1, Tap Test

1. Remove distributor cap and rotor. Crank engine to align one tooth of armature with magnet in pick-up coil, Fig. 13.
2. On Dura Spark I system, install ballast resistor Motorcraft DY-36 in series with primary circuit at BAT terminal of ignition coil, Fig. 14. Using a straight pin, pierce both red and white leads of module to short these two leads together, Fig. 15.

CAUTION: Never install bypass until after the ballast resistor is in place or permanent damage to module or coil may result.

3. Remove coil wire from distributor cap and connect to a well grounded modified spark plug, Fig. 12, or hold wire 1/4" from engine. Turn ignition switch to Run and tap distributor body with a screwdriver handle.
4. If there are sparks, proceed to test 2. If there are no sparks, proceed to Test 4.

Test 2, Voltage Drop On Module White Wire

1. On Dura Spark I, remove pin from red wire only, leaving pin to make contact with white wire only, Fig. 15.

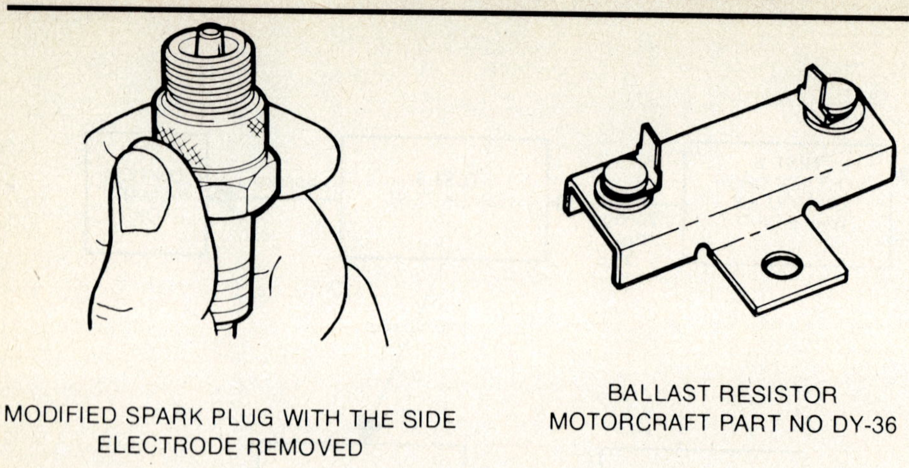

Fig. 12 Test equipment

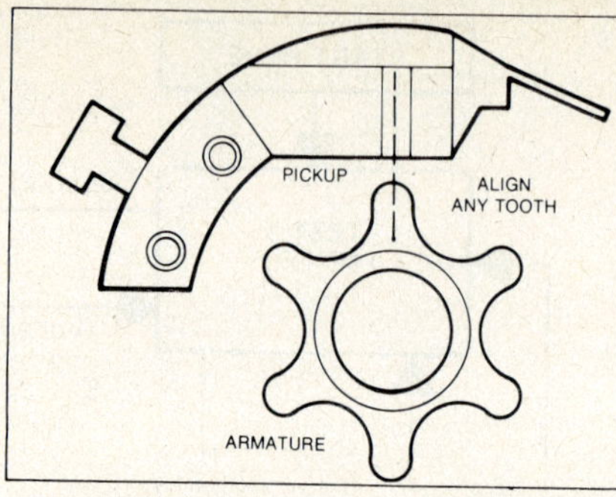

Fig. 13 Aligning armature with magnet in pick-up coil

2. Connect positive lead of voltmeter to positive battery terminal and negative lead of voltmeter to pin in white wire. Crank engine and observe voltmeter reading. Reading should not exceed 1 volt.
3. If reading is above 1 volt, repair circuit feeding white wire, then repeat Test 1. If reading is less than 1 volt, perform Test 3.

Test 3, Crank Test

1. Crank engine while checking for sparks.
2. If there are no sparks, perform Test 6. If there are sparks, ignition system is functioning properly. Remove pin and ballast resistor. Check fuel system and refer to "Testing for Intermittent Conditions".

Test 4, Voltage Drop On Module Red Wire

1. On Dura Spark I, remove pin from white wire only, leaving pin to make contact with red wire, Fig. 16.
2. On Dura Spark II, insert pin into red wire to make contact, Fig. 16.
3. Measure voltage drop on red wire of ignition module with key in On position. Connect positive lead of voltmeter to positive terminal and negative lead to red wire.
4. If reading is above 1 volt, repair circuit feeding red wire, then repeat Test 1. If reading is below 1 volt, perform Test 5.

Test 5, Cycle Switch

1. On Dura Spark I, pierce both red and white wires of module to short these two leads using a straight pin, Fig. 16.

NOTE: On Dura Spark II, do not install straight pin.

2. Cycle ignition switch On and Off or turn ignition switch On and disconnect and reconnect distributor connector. A spark should be seen at the modified spark plug each time the ignition switch is cycled, or each time the connector is disconnected.
3. If there are no sparks, perform Test 6. If there are sparks, perform Test 8.

Test 6, Voltage Drop to Coil, Fig. 17

1. Connect jumper wire from "DEC" terminal to ground. Turn ignition switch to Start and then Run positions.
2. Measure voltage drop to "BAT" terminal of coil:
   a. While cranking engine, voltage drop should be no more than 1 volt.
   b. With ignition switch at Run, voltage drop should be less than 1 volt on Dura Spark I, and less than 6.5 volts on Dura Spark II.
3. If voltage drop is within specifications, perform Test 7. If voltage drop is not within specifications, perform repairs as necessary and repeat test.

Test 7, Coil Condition

Refer to Fig. 18, for ohmmeter connections to check coil primary and secondary resistance. Resistance values should be as specified. If resistance values are within specifications, perform Test 8. If not, replace coil and repeat Test 1.

Test 8, Distributor & Harness Test, Fig. 19

1. Disconnect module connector with purple and orange wires. Test distributor by connecting a voltmeter to orange and purple wires at harness side of male connector.

NOTE: Do not use a voltmeter combined with a dwell-meter, as slight needle oscillations of 1/2 volt may not be detectable on this type of test equipment. Also, a digital volt-ohmmeter will not function on this test.

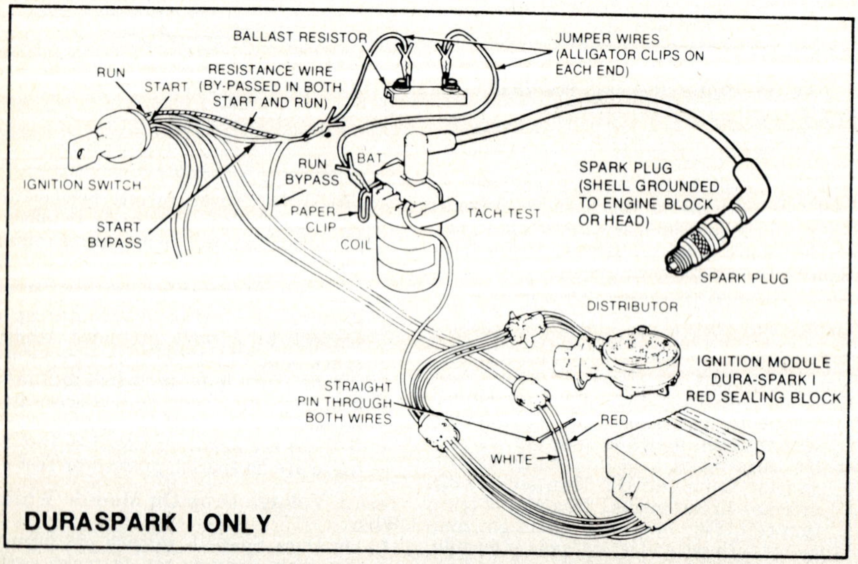

Fig. 14 Ballast resistor installed in primary circuit. Dura Spark I system

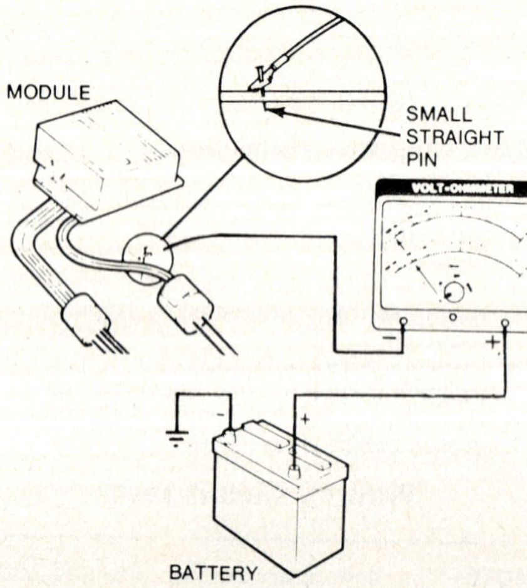

Fig. 15 Checking voltage drop on module white wire

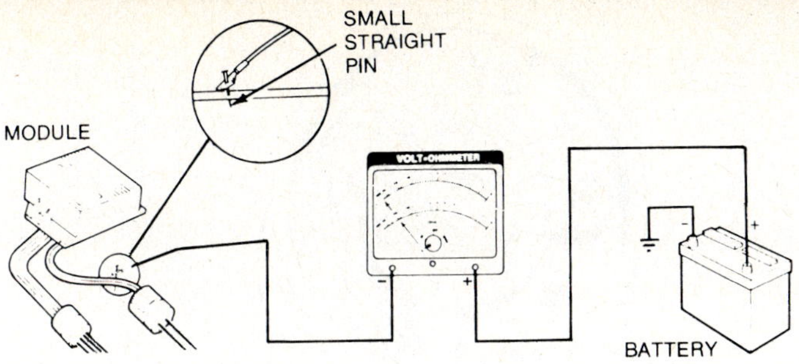

Fig. 16 Checking voltage drop on module red wire

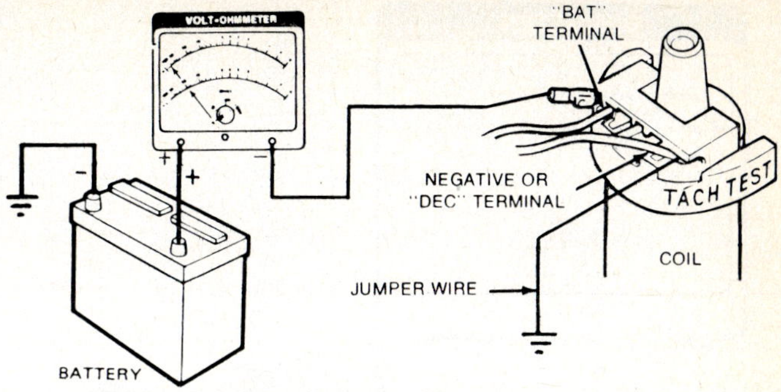

Fig. 17 Checking voltage drop to coil

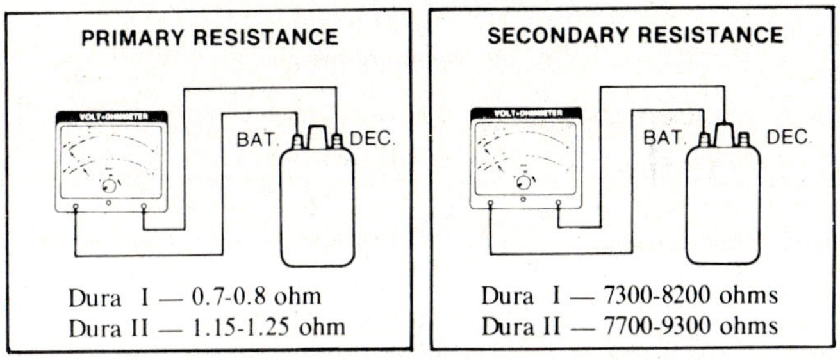

Fig. 18 Checking coil condition

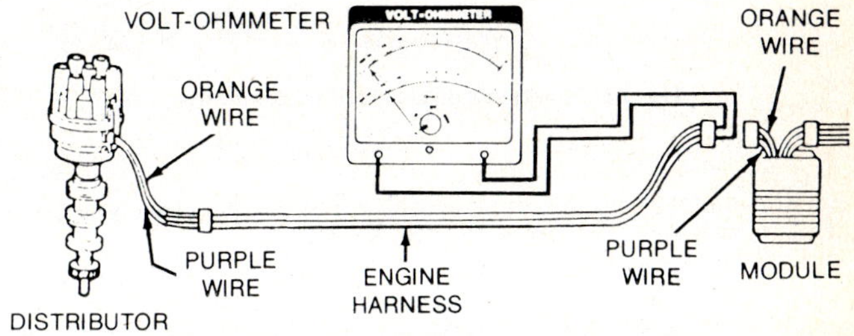

Fig. 19 Distributor and harness test

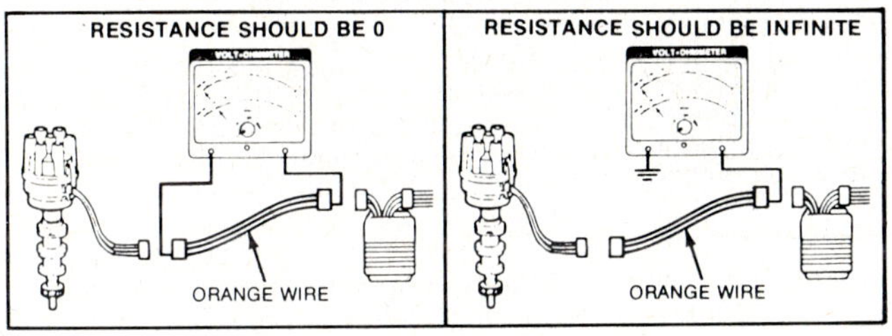

Fig. 20 Harness to distributor test

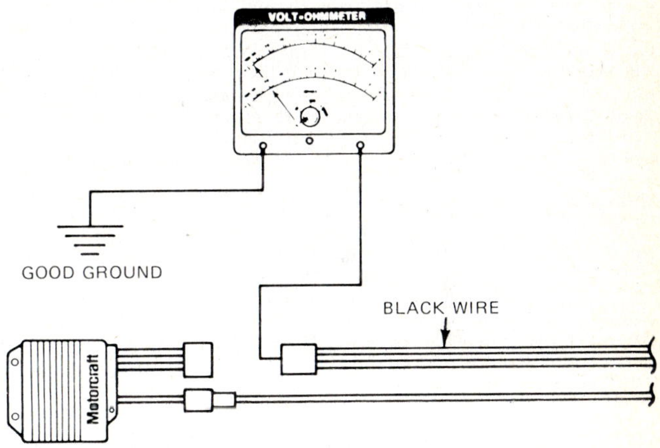

Fig. 21 Primary ground circuit test

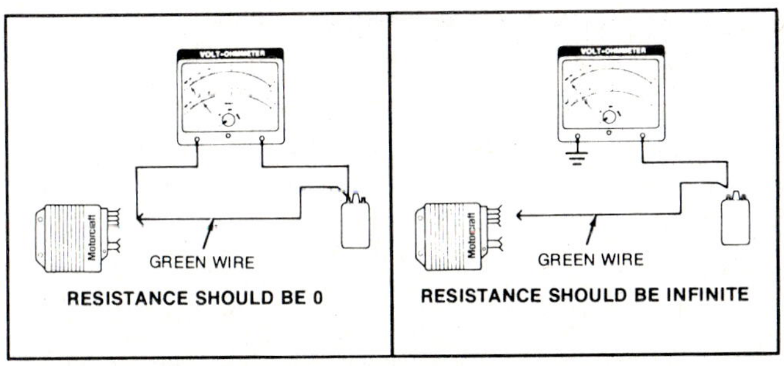

Fig. 22 Checking green wire from module to coil

CAUTION: If vehicle is equipped with a catalytic converter, disconnect air supply line between by-pass valve and manifold before cranking engine with ignition off. This will prevent damage to the catalytic converter. Also, after testing, run engine for at least three minutes before reconnecting air supply line, to clear excess fuel from exhaust system.

---

2. Set voltmeter to its lowest scale and crank engine. Meter needle should oscillate slightly (about 1/2 volt). Refer to "Distributor Testing" for additional information.
3. If voltmeter indications are as specified, perform Test 10. If not, perform Test 9.

Test 9, Harness To Distributor Test, Fig. 20

1. Check for open or shorted condition in orange and purple wires from module connector (harness side) to distributor connector (harness side).
2. If wires are not open or shorted, replace stator assembly and repeat test. If a shorted or open condition exists, make necessary repairs and repeat test.

Test 10, Primary Ground Circuit Test, Fig. 21

1. Connect an ohmmeter to black lead at harness side of module connector and a good ground.
2. With ohmmeter at its lowest scale, there should be no measurable resistance.
3. If there is no measurable resistance, perform Test 11. If resistance can be measured, check ground in distributor and wire from module to distributor. Repair or replace as necessary and repeat test.

Test 11, Green Wire From Module To Coil, Fig. 22

1. Check green wire from module to coil for open or shorted condition.
2. If wires are open or shorted, repair or replace as necessary and repeat test. If shorted or open condition does not exist, refer to "Testing For Intermittent Conditions". If problem cannot be found, replace module.

Coil Testing

If an oscilloscope is available, connect oscilloscope and disconnect one spark plug wire

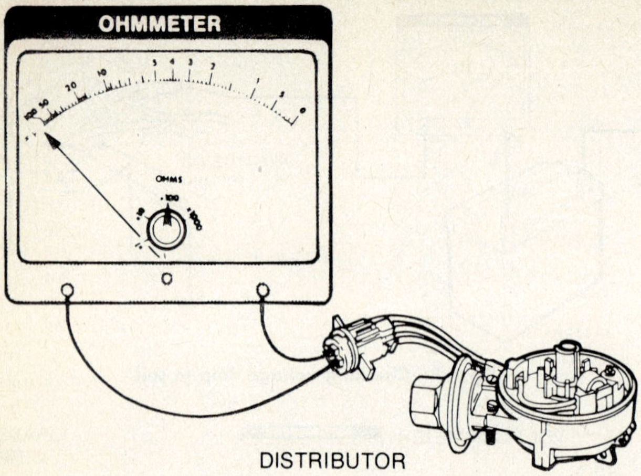

Fig. 23 Checking pick-up coil in distributor

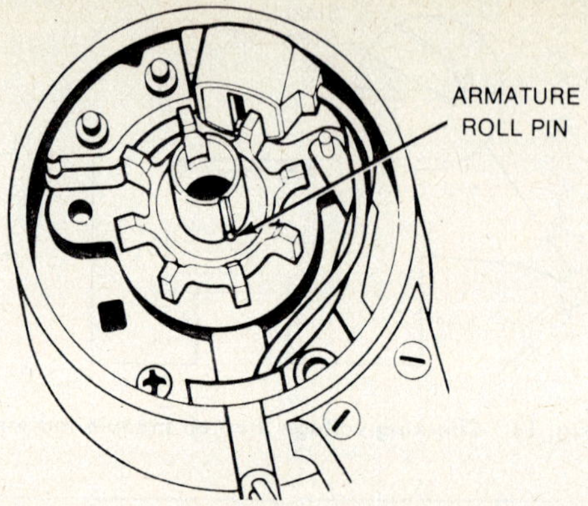

Fig. 24 Armature roll pin

not allowing it to arc to ground. Crank engine while observing coil reserve voltage. Voltage spike should be 28 KV or more.

Distributor Testing

Test 1, Fig. 23

Connect an ohmmeter to parallel blade of distributor connector. Resistance should be 400–1000 ohms. A reading of less than 400 ohms indicates a short while an infinity reading indicates an open. With ignition switch off, check resistance between orange wire and ground and purple wire and ground. Reading at each wire should be greater than 70,000 ohms.

Test 2, Fig. 24

Remove distributor cap and check that roll pin retaining armature on distributor shaft is in place. If roll pin is missing, armature can rotate in relation to distributor shaft causing timing to be out of specifications. Check for correct connection of orange and purple wires between distributor and module. If wires are reversed, distributor timing will be 22½ degrees out of phase.

Test 3

If a known good distributor is available, connect it to harness, turn ignition switch "On" and spin distributor by hand while checking for sparks.

Testing For Intermittent Conditions

If the ignition system becomes operative while performing diagnosis and a repair has not been made, it indicates that a malfunctioning component has become operative. The following procedures may locate the malfunctioning components:

With engine running, try to recreate the problem by moving wires at the coil, module, distributor and other harness connectors. Start with connectors disturbed originally. Also check ground connection in distributor. Also, disconnecting and reconnecting connectors may locate an intermittent condition.

CAUTION: Do not clean compound from connectors as it is required to prevent terminal corrosion.

Pick-Up Coil

With engine off, remove distributor cap, rotor and adapter, if used. Heat stator pick-up coil using a 250 watt heat lamp placed about 1 to 2 inches from its top surface. Apply heat for 5 to 10 minutes while observing pick-up coil continuity between the parallel blades of the disconnected distributor connector. Resistance should be 400–1000 ohms. Less than 400 ohms indicates a short while infinity indicates an open. Lightly tapping distributor with a screwdriver may reveal an intermittent condition.

Ignition Module

With engine running, heat module using a 250 watt heat lamp placed about 1 to 2 inches from module. Tapping with a screwdriver may reveal an intermittent condition.

CAUTION: Do not allow module temperature to exceed 212 degrees. This can be checked after the first 10 minutes by applying drops of water to module housing. Repeat every 2 minutes thereafter until water droplets start to boil.

If the above procedure results in an ignition malfunction, substitute a known good module. If substitution corrects the malfunction, recheck old module by reconnecting it into the circuit. Refer to "Test 1" under "Primary Circuit Test" to functionally check the original and the substitute modules.

Trouble Shooting Intermittent Conditions

Engine Starts & Runs OK But Quits When Normal Operating Temperature Is Reached

1. Run engine until normal operating temperature is reached or engine quits, whichever occurs first.
2. Crank engine while checking voltage between orange and purple wires at ignition module. With voltmeter set at lowest scale, only a slight needle movement

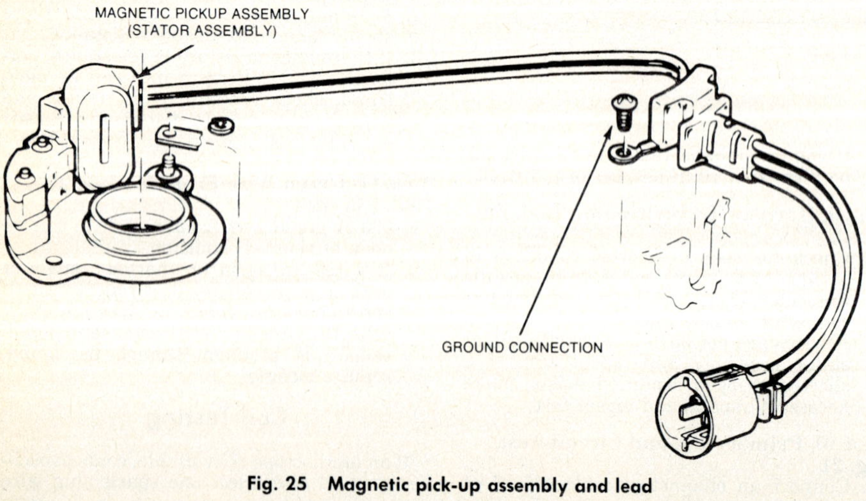

Fig. 25 Magnetic pick-up assembly and lead

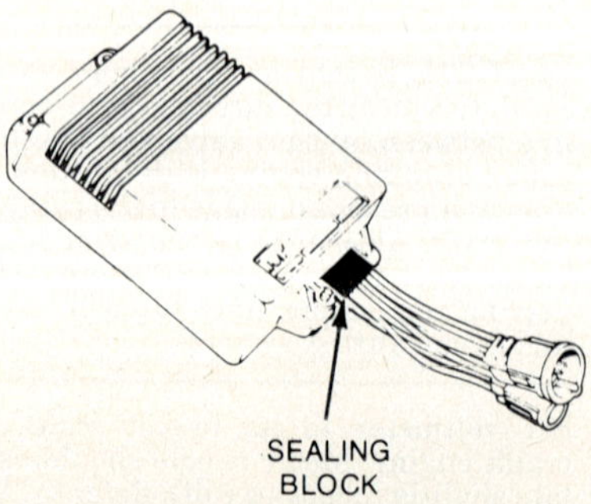

Fig. 26 Module identification

(about 1/2 volt) should be observed.
3. With ignition switch "Off", check resistance between purple and orange wires at distributor. Resistance should be 400–1000 ohms.
4. With ignition switch "Off" check resistance between purple wire at distributor and ground and between orange wire and ground. Resistance should be greater than 70,000 ohms at each wire.
5. If resistance values in step 3 and 4 are not within specifications, replace pick-up assembly.

Engine Starts Normally At Moderate Temperature But Starts Hard Or Will Not Start In Extremely Cold Temperature

1. Check for open in feed to white lead to module.

Engine Quits Intermittently With Complete Loss Of Ignition

1. Check primary circuit ground resistance at ignition module connector black wire. Resistance should be zero ohms.
2. If resistance is not zero ohms, remove distributor cap and inspect retaining screw at rubber plug where wires enter distributor, Fig. 25. A loose or cross threaded screw or a dirty or corroded connection at this screw can cause an intermittent high resistance or a complete loss of ground.

MODULE IDENTIFICATION

The ignition module can be identified as to type of ignition system by the color of the sealing block, Fig. 26.

Sealing Block Color	Type of Ignition System
Red	Dura Spark I
White	Dura Spark II (With Cranking Retard)
Blue	Dura Spark II
Yellow	Dura Spark II (Dual Mode)
Green Or Black	Early Solid State Ignition
Brown	EEC System

COMPONENT REPLACEMENT

Magnetic Pickup Assembly

Removal

1. Remove distributor cap and rotor, then disconnect distributor wiring harness plug, Fig. 5.
2. Using two screw drivers, pry armature from advance plate sleeve and remove roll pin.
3. Remove snap ring securing pickup assembly to base plate. On 6 cylinder models, remove washer and wave washer.
4. On all models, remove snap ring securing vacuum advance link to pickup assembly.
5. Remove pickup assembly ground screw and lift assembly from distributor.
6. Disconnect vacuum advance link from pickup assembly post.

Installation

1. Position pickup assembly over base plate and slide wiring harness into slot on side of distributor housing, Fig. 5.
2. On 6 cylinder models, install washers. On all models, install snap ring securing pickup assembly to base plate.
3. Position vacuum advance link on pickup assembly post and install snap ring.
4. Insert ground screw through wiring harness tab and install on base plate.
5. Install armature on advance plate sleeve, ensure roll pin is engaged in slot.
6. Install distributor rotor and cap, then connect distributor wiring harness plug to vehicle wiring harness.

Vacuum Advance Unit, Replace

1. Remove distributor cap and rotor.
2. Disconnect vacuum lines, then remove snap ring that secures vacuum advance link to pickup assembly.
3. Remove vacuum advance attaching screws, then tilt unit downward to disconnect link.
4. Carefully remove unit from distributor.
5. Reverse procedure to install.

Fixed Base Plate, Replace

1. Remove distributor cap and rotor.
2. Remove vacuum advance unit and magnetic pickup assembly.
3. Remove attaching screws and lift base plate from distributor.
4. Reverse procedure to install.

Ford Dura Spark III Solid State Ignition System

DESCRIPTION

The Dura Spark III system, Fig. 1, is used on vehicles equipped with Electronic Engine Control (EEC). The system consists of a primary side consisting of the battery, ignition switch, primary circuit, EEC system input and ignition module and a secondary side consisting of distributor cap, rotor, spark plug wires and spark plugs. When the ignition switch is in the On position, the primary circuit and the ignition coil are energized. The EEC system provides a signal which tells the ignition module to turn off the coil primary circuit. The off and on times of the primary circuit are controlled by the EEC computer. When the circuit is turned Off, the magnetic field built up in the ignition coil collapses, inducing a high voltage into the coil secondary windings. This high voltage is then delivered to the spark plugs by the secondary ignition wires.

SYSTEM DIAGNOSIS

NOTE: Before proceeding with system diagnosis, ensure battery is fully charged and all accessories are "Off". Inspect all vacuum hoses and spark plug wires for proper routing and secure connections. Inspect engine compartment ignition wiring harness and connectors for signs of insulation damage, burning, overheating and loose or broken connections. Repair as necessary.

Secondary Circuit Tests

Spark Plug Firing Voltage Test, 1981 Models

1. Connect an oscilloscope with voltage pick up on coil to distributor cable. Set oscilloscope to 30 KV parade pattern.
2. Connect timing light to No. 1 spark plug, then start engine and operate at 2000 RPM.
3. Apply 25 inches Hg. of vacuum from outside source to BP port of MAP sensor.
4. Check oscilloscope for highest firing spikes. If highest spikes are between 6 and 20 KV, system is operating properly.
5. If reading is greater than 20 KV or highest spike is 50% greater than lowest, refer to "Primary Circuit Test" under "Test 3, Ignition Coil Primary Voltage."
6. If voltage reading is less than 6 KV, refer to "Spark Plug Wires Continuity Test."
7. Check ignition timing at 2000 RPM. If reading is not between 27 and 30° BTDC, EEC system is malfunctioning and must be repaired.

Spark Plug Firing Voltage Test, 1982–83 Models

1. Connect an oscilloscope with voltage pick-up on coil to distributor wire. Set oscilloscope to parade pattern.
2. Start engine, then while slowly increasing engine RPM from idle to 2000 RPM, observe scope pattern. If the average spark plug firing voltage is 15 KV and the spark plug firing voltages do not vary more than 5 KV, the system is operating properly.
3. If the spark plug firing voltages do not vary more than 5 KV and the average spark plug firing voltage is more than 15 KV, check:
   a. Ignition coil wire for proper installation and resistance. Resistance should be less than 5000 ohms per inch. If more, replace.
   b. Spark plugs for wide gaps.
   c. Distributor cap and rotor for excessive clearance.
   d. Distributor cap and rotor for lack of silicone compound on rotor.
   e. Distributor rotor alignment. See Distributor Rotor, Replace.
4. If the spark plug firing voltages vary more than 5 KV, check:
   a. Spark plug gap(s) or worn electrodes.
   b. Correct distributor cap, adapter and rotor installation.
   c. Distributor rotor alignment. See Distributor Rotor, Replace.
5. If one or more spark plug firing voltages are unusually high, check:
   a. Disconnected spark plug wire(s).

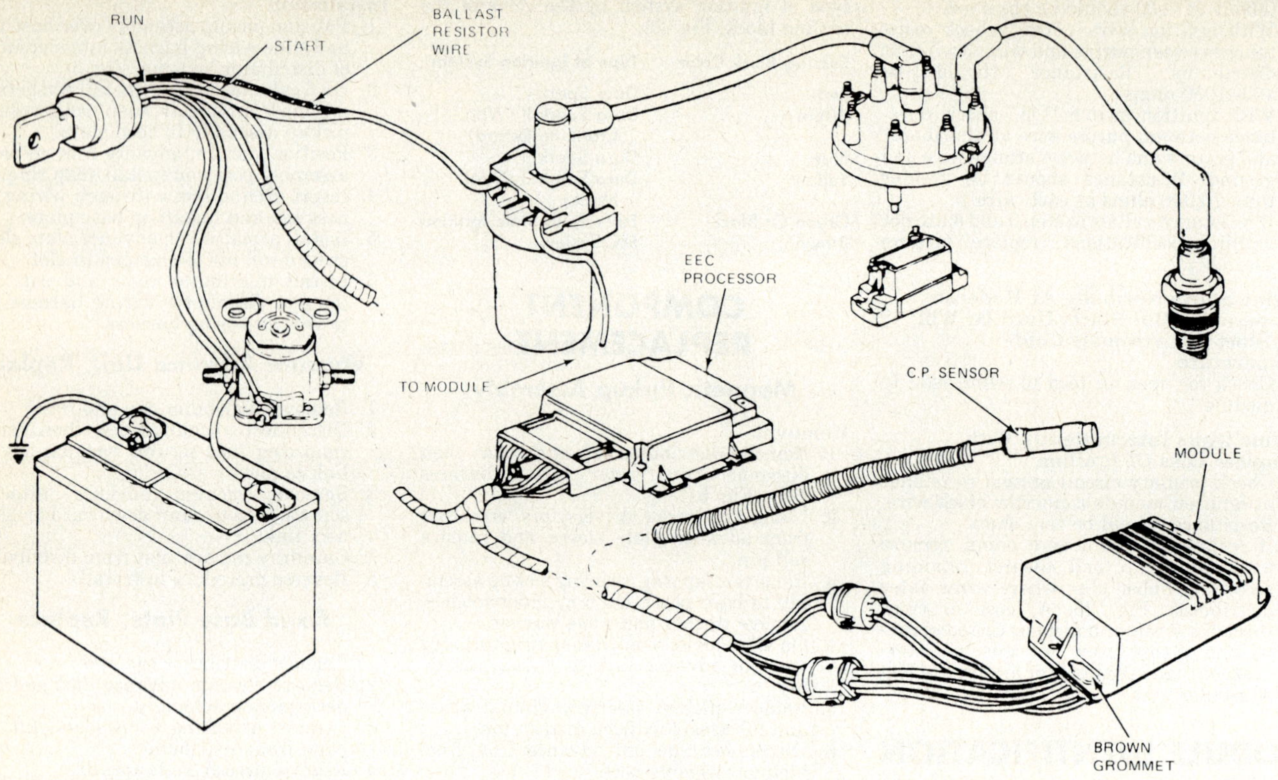

Fig. 1 Dura Spark III ignition system, typical. 1981–83

   b. Spark plug gap(s) or open plug wire(s).
6. If one or more spark plug firing voltages are unusually low, check:
   a. Spark plug(s) for fouling or narrow gap(s).
   b. Spark plug wires for grounding.
   c. Distributor cap and adapter for tracking.
7. If spark plug firing voltages are inverted, check:
   a. Ignition coil primary connector for improper installation. If installation is satisfactory, replace coil.

Spark Plug Wire Continuity Test, All Models
1. Remove distributor cap and disconnect suspected wire from spark plug or coil.
2. Using an ohmmeter, check wire resistance through distributor cap.
3. Resistance should be 5000 ohms per inch. If reading is greater than specified, replace wire.

Spark Plug Wire Inspection, All Models
1. Clean off any deposits of dirt from wires, boots, distributor cap and coil using mild soap and water solution.
2. Inspect wires, boots for cuts, punctures or other damage.
3. Inspect wire terminals for corrosion and clean with fine sandpaper.
4. Coat all boots with silicone grease before installing.

Secondary Circuit Voltage Drop, 1981 Models
1. Connect an oscilloscope with voltage pick up on coil to distributor cable.
2. Disconnect a spark plug wire and ground firmly to engine block.
3. Start engine and operate at idle speed.
4. If oscilloscope indicates reading of 8 KV, distributor cap voltage drop is satisfactory.
5. If reading is greater than 8 KV, check for damaged cap or rotor and replace.
6. If an abnormal scope pattern is obtained, check for worn or damaged distributor parts.

Primary Circuit Tests, 1981 Models

Test 1, Ignition Coil Output
1. Disconnect spark plug wire and ground firmly to engine block.
2. Crank engine while checking for spark.
3. If spark is observed, system is operating properly.
4. If no spark is observed, remove cap and crank engine to verify rotor movement.
5. If rotor does not move when engine is cranked, repair distributor as necessary.

Test 2, Trigger Test
1. Disconnect ignition module connector and install ignition diagnostic test adapter into wiring circuit, Fig. 2.
2. Disconnect ignition coil secondary wire at distributor and ground firmly to engine block. Turn ignition to Run position.
3. Touch diagnostic adapter lead to battery positive terminal. Spark should occur every time lead touches battery.
4. If spark occurs, system is operating properly. If not, refer to "Test 3, Ignition Coil Primary Voltage".

Test 3, Ignition Coil Primary Voltage
1. Turn ignition to Run position. Connect voltmeter between BATT terminal of ignition coil and ground.
2. If reading is between 5 and 8 volts, ignition system is satisfactory. If reading is greater than 8 volts, refer to "Test 7, Ground Circuit Check". If reading is less than 6 volts, refer to "Test 6, Wiring Harness Short Circuit" and to "Test 8, Ignition Coil Circuit".
3. In addition, check for open circuit in power supply circuit such as battery lead to ignition coil.

Test 4, Module Run Circuit
1. Turn ignition key to Run position.
2. Measure voltage between red module wire and engine ground by connecting straight pin to voltmeter positive lead and inserting into red wire.

NOTE: Do not allow straight pin to contact ground.

---

3. If reading is within 90% of battery voltage, system is operating properly.
4. If not, check for open circuit in run circuit wiring harness and repair as necessary.

Test 5, Start Circuit Voltage
1. On starter relays equipped with 1 terminal, disconnect cable between starter relay and starter motor.
2. On starter relays without I terminal, disconnect electrical connector at relay S terminal.
3. Hold ignition switch in Start position and measure voltage between coil BATT terminal and ground and between white module wire and ground. Connect straight pin to voltmeter lead and insert

into white wire to perform test.
4. If reading is greater than 90% of battery voltage, system is satisfactory.
5. If reading is less than 90% of battery voltage, check for open circuit in ballast resistor bypass wire or module start wire and repair as necessary. Also check for defective ignition switch.

Test 6, Wiring Harness Short Circuit
1. Disconnect ignition module, ignition coil and EEC computer connectors and inspect for damage.
2. Check for damaged primary circuit wiring and repair as necessary.
3. Place transmission in Park and set parking brake. Check resistance between module connector orange wire terminal and ground, Fig. 3.
4. Measure resistance between the following connector wire terminals:
   a. Red to white wire terminals.
   b. Red to green wire terminals.
   c. Red to orange wire terminals.
   d. White to green wire terminals.
   e. White to orange wire terminals.
5. If any resistance reading obtained is less than 70,000 ohms, repair damaged wire in that circuit as necessary.

Test 7, Ground Circuit Check
1. Disconnect ignition module connector containing black wire and check for damage.
2. Check resistance between black wire terminal in module connector and ground.
3. If resistance is less than one ohm, system is satisfactory.
4. If resistance is greater than one ohm, check for open circuit at battery ground strap connection and repair as necessary.

Test 8, Ignition Coil Circuit
1. Disconnect ignition coil and module connectors and inspect for damage.
2. Measure resistance between green wire terminal of module connector and TACH terminal on ignition coil.
3. If resistance is greater than 2 ohms, check for open circuit in TACH wire and repair as necessary.
4. If resistance is less than 2 ohms, check resistance of ballast resistor by connecting ohmmeter between BATT terminal on coil and red wire terminal in ignition module connector.
5. If resistance is less than 2 ohms, system is operating satisfactorily. If not, check for an open circuit or replace ballast resistor as necessary.

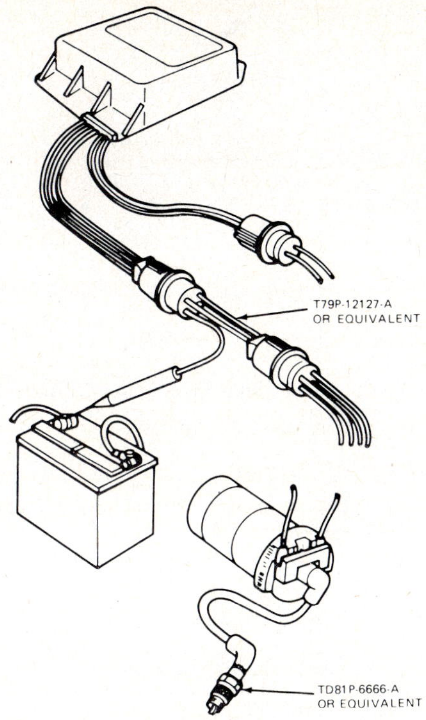

Fig. 2 Installing ignition diagnostic test adapter. 1981–83 Dura Spark III ignition system

Test 9, Module Output Check
1. Connect any connectors that were previously disconnected.
2. Connect voltmeter between coil TACH terminal and ground.
3. Set voltmeter on highest scale, then firmly ground ignition coil high tension lead to engine block.
4. Crank engine, then switch voltmeter to lowest scale.
5. If meter needle fluctuates, module is operating satisfactorily. If not, refer to "Test 10, Module Check".

NOTE: If difficulty is encountered in performing this test using a voltmeter, a test light may be substituted in its place. The light flashing would correspond to the voltmeter needle fluctuating.

Test 10, Module Check
1. Disconnect module connector containing green wire.
2. Connect voltmeter between green wire terminal of module connector and ground.
3. Turn ignition to Run position and check voltage.
4. If reading is more than 90% of battery voltage, replace ignition module.
5. If reading is less than 90% of battery voltage, refer to "Test 11, Continuity Check".

Test 11, Continuity Check
1. Check resistance between green wire terminal in harness side module connector and coil TACH terminal.
2. If ohmmeter indicates less than one ohm, system is operating satisfactorily.
3. If reading is greater than one ohm, check for open circuit in wire between coil TACH terminal and ignition module and repair as necessary.

Test 12, Coil Primary Circuit Check
1. Disconnect ignition coil electrical connector.
2. Measure resistance between coil BATT and TACH terminals.
3. If reading is between 1 and 2 ohms, primary circuit is satisfactory.
4. If not, replace coil.

Test 13, Coil Secondary Circuit Check
1. Check resistance between coil TACH terminal and high tension lead terminal.
2. If reading is between 7700 and 9600 ohms, secondary circuit is satisfactory.
3. If not, replace ignition coil.

Primary Circuit Tests, 1982 Models

Test 1, Run Circuit
1. Disconnect ignition module three wire-

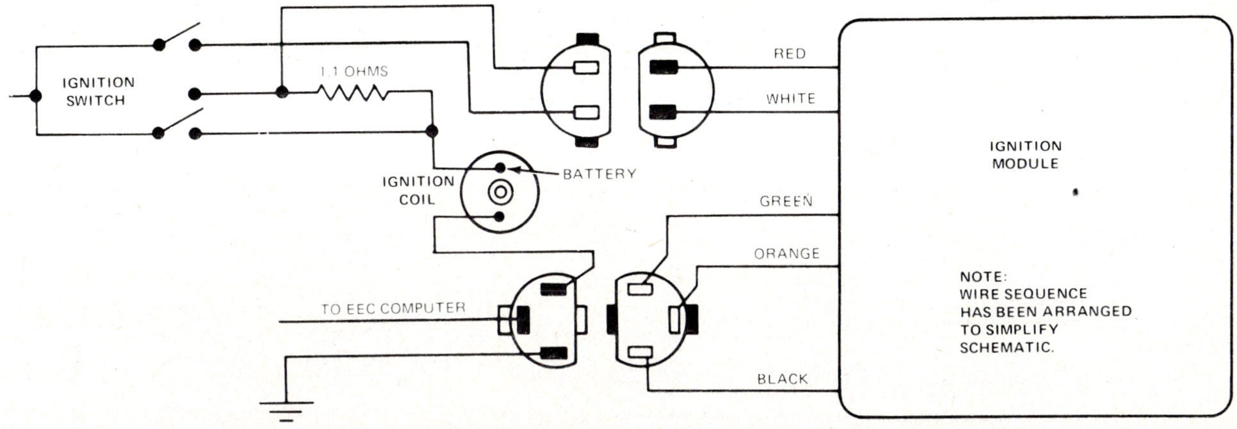

Fig. 3 Electrical schematic. 1981–83 Dura Spark III ignition system

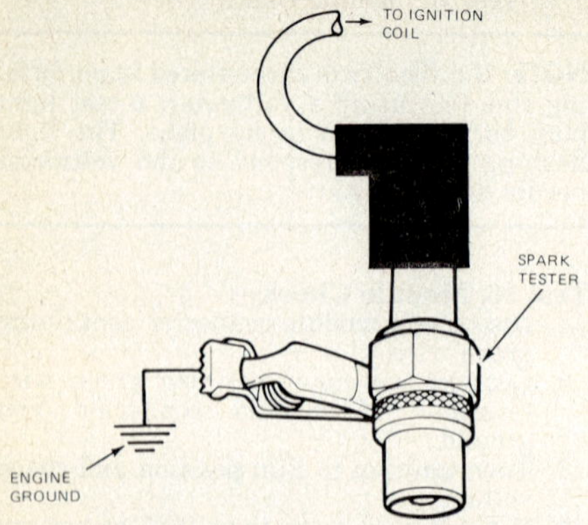

Fig. 4 Spark tester

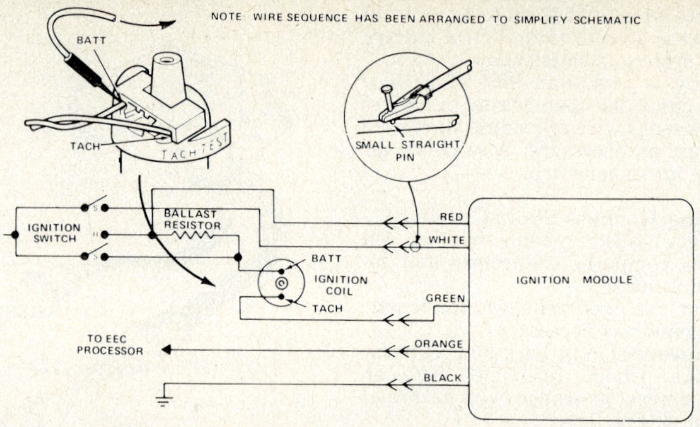

Fig. 5 Dura Spark III ignition system start voltage test connections. 1982–83

connector and install ignition diagnostic test adapter into wiring circuit, Fig. 2.
2. Connect spark tester, Fig. 4, between ignition coil and ground.
3. Turn ignition "On", then while observing spark tester, touch diagnostic test lead to battery positive terminal. Spark should occur every time lead touches battery.
4. If spark occurs, circuit is satisfactory. If not, proceed to test 2.

Test 2, Start Circuit

1. Remove diagnostic test adapter installed in test 1 but, leave spark tester attached.
2. While cranking over engine with ignition switch, observe spark tester for sparks.
3. If sparks occur, circuit is satisfactory. If not, proceed to test 3.

Test 3, Start Voltage

1. Reconnect coil wire to distributor cap.
   a. If starter relay is equipped with a I terminal, disconnect starter cable from starter relay.
   b. If starter relay is equipped with S terminal, disconnect S wire from relay.
2. With ignition "Off", install straight pin into module white wire, Figure 5.
3. Connect a suitable voltmeter between ground and straight pin.
4. Turn ignition switch to "Start" position and note voltage.
5. Move voltmeter positive lead to coil Batt. connector and note voltage with ignition switch in start "position".
6. If voltage obtained is 90% of battery voltage, circuit is satisfactory. If voltage obtained is less than 90% of battery voltage, a defective ignition switch or wiring harness is indicated.

Test 4, Ignition Coil Primary Circuit Switching

1. Disconnect ignition module three wire connector and install ignition diagnostic test adapter into wiring circuit, Fig. 2.
2. Connect a suitable test light between coil Tach terminal and engine ground.
3. Turn ignition "On", then while observing test light, touch diagnostic test lead to battery positive terminal. Test light should flash every time lead touches battery.
4. If test light flashes, circuit is satisfactory. If not, Proceed to test 6.

Test 5, Ignition Coil Secondary Resistance

1. Disconnect and inspect ignition coil electrical lead.
2. Using a suitable ohmmeter, measure resistance between coil Batt. terminal and high tension lead terminal.
3. If resistance is between 7700–10,500 ohms, coil is satisfactory. If not between limits, replace coil.

Test 6, Supply Voltage Circuits

1. Remove diagnostic test adapter installed in test 4 and test light, then:
   a. If starter relay is equipped with a I terminal, disconnect starter cable from starter relay.
   b. If starter relay is equipped with S terminal, disconnect S wire from relay.
2. With ignition "Off", install straight pins into module red and white wires, Fig. 6.

NOTE: Do not allow straight pins to contact engine ground.

3. Using a suitable voltmeter, connect negative lead to ground.
   a. With ignition in "On" position, measure voltage at red wire pin.
   b. Place ignition in "Start" position, measure voltage at white wire pin and at coil Batt. terminal.

NOTE: When measuring voltages, wiggle wires to simulate any open circuits that might exist.

4. If readings obtained are 90% of battery voltage, supply circuits are satisfactory. Proceed to test 7.
5. If readings obtained are less than 90% of battery voltage:
   a. Defective wiring harness or connectors in supply voltage circuits.
   b. Defective ignition switch.
   c. Defective ignition coil or radio interference capacitor.

Test 7, Ignition Coil Supply Voltage

1. Connect positive lead of voltmeter to coil

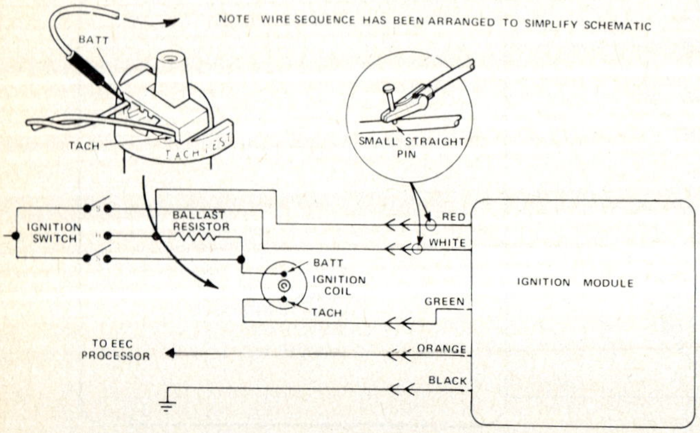

Fig. 6 Dura Spark III ignition system supply voltage test connections. 1982–83

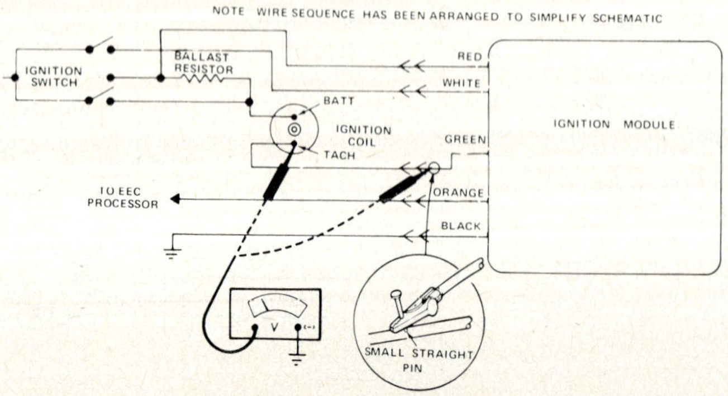

Fig. 7 Dura Spark III ignition system module to coil wire test connections. 1982–83

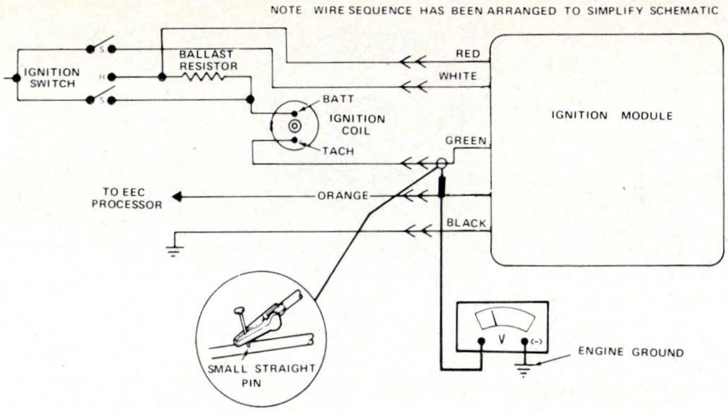

Fig. 8 Dura Spark III ignition system primary circuit continuity test connections. 1982–83

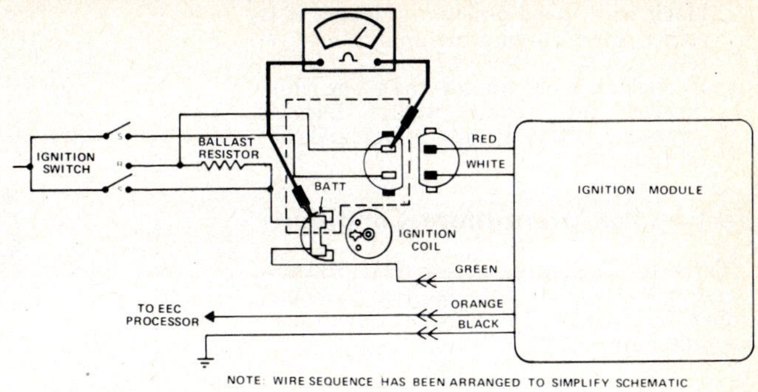

Fig. 9 Dura Spark III ignition system ballast resistor test connections. 1982–83

Batt. terminal and negative lead to ground.
2. Turn ignition switch to "On" position.
3. Observe voltmeter reading. If 6–8 volts, circuit is satisfactory. If less than 6 volts or more than 8 volts, proceed to test 9.

Test 8, Testing Tach Wire
1. Disconnect and inspect module three wire connector and coil connector.
2. Using a suitable ohmmeter, connect one lead to ground and the other to the coil connector Tach. terminal. Measure resistance.
3. If readings obtained are one ohm or less, circuit is satisfactory. If readings are more than one ohm, repair short in tach wire.

Test 9, Ignition Coil Primary Resistance
1. Disconnect and inspect ignition coil connector.
2. Using a suitable tachometer, measure resistance between coil Batt. and Tach. terminals.
3. If readings are 0.8–1.6 ohms, circuit is satisfactory. If readings are out of specifications, replace coil.

Test 10, Module to Coil Wire
1. With ignition "Off", insert straight pin in module green wire, Fig. 7.

NOTE: Do not allow straight pin to contact engine ground.

2. Using a suitable voltmeter, connect negative lead to ground.
3. Turn ignition "On", then measure voltage at green wire pin and Tach. terminal of ignition coil, Fig. 7.
4. If voltmeter shows less than ½ volt difference, circuit is satisfactory. If more than ½ volt difference, insepct and repair wiring harness between coil and module.

Test 11, Primary Circuit Continuity
1. Leave straight pin in green wire as outlined in test 10.
2. Turn ignition "On," then using a suitable voltmeter connected to ground measure voltage at green wire pin, Fig. 8.
3. If voltage exceeds 1½ volts, proceed to test 13. If voltage obtained is 1½ volts or less, proceed to test 12.

Test 12, Ballast Resistor
1. Disconnect and inspect module two wire connector and ignition coil connector, Fig. 9.
2. Using a suitable ohmmeter, measure resistance between coil connector Batt. terminal and wiring harness connector which mates with module red wire, Fig. 9.
3. If reading is 0.8–1.6 ohms, circuit is satisfactory. If reading is less than 0.8 ohms or more than 1.6 ohms, replace ballast resistor.

Test 13, Ground Circuit
1. With ignition "Off", install straight pin in module black wire, Fig. 10.

NOTE: Do not allow straight pin to touch engine ground.

2. Using a suitable ohmmeter, correct negative lead to engine ground and positive lead to black wire pin.
3. Turn ignition "On", measure voltage.
4. If voltage obtained was more than 1½ volts, proceed to test 14. If voltage obtained was less than 1½ volts, replace ignition module.

Test 14, Wiring Harness Ground Circuit
1. Disconnect module three wire connector, inspect connector for dirt and corrosion.
2. Using a suitable ohmmeter, connect one lead to engine ground. Connect other lead to wiring harness terminal which mates with black module wire terminal. Measure resistance.

NOTE: When measuring resistances, wiggle wires to simulate any open circuits that might exist.

3. If resistance is less than one ohm, inspect wiring harness connector and module

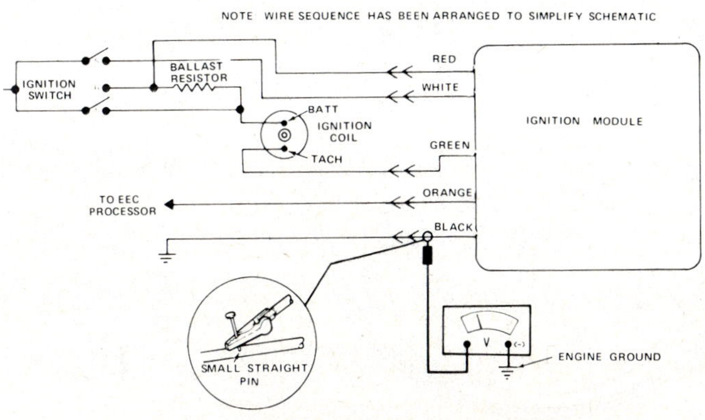

Fig. 10 Dura Spark III ignition system ground circuit test connections. 1982–83

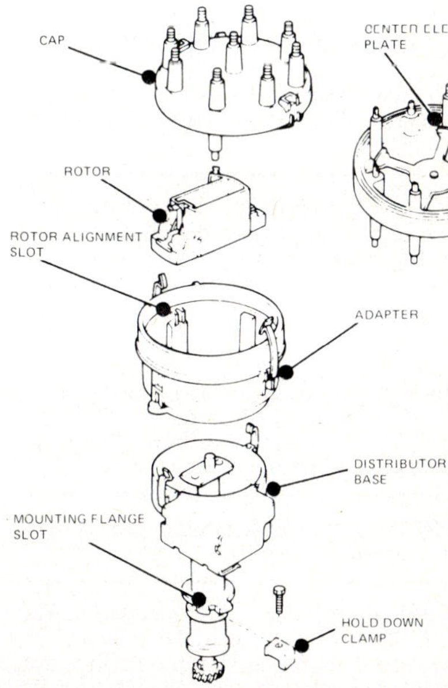

Fig. 11 Early model Dura Spark III distributor

black wire. If no problem is found, it is either intermittent or not in ignition system.

4. If resistance is greater than one ohm, inspect and repair wiring harness between ignition module and ground connection.

Testing For Intermittent Conditions

Refer to "Ford Breakerless (B/L) & Dura Spark I & II Solid State Ignition Systems" under "Testing For Intermittent Conditions" for procedure.

Troubleshooting Intermittent Conditions

Refer to "Ford Breakerless (B/L) & Dura Spark I & II Solid State Ignition Systems" under "Troubleshooting Intermittent Conditions" for procedures.

DISTRIBUTOR, REPLACE

Early Models, Fig. 11

1. Remove distributor cap. Install rotor alignment tool T78P-12200-A and remove rotor. Refer to Distributor Rotor, Replace.
2. With rotor alignment tool in position, remove distributor hold down bolt and clamp. Remove alignment tool.
3. Carefully remove distributor from engine block, noting position of rotor blades in relation to distributor housing as distributor drive gear is felt to disengage with cam gear.

NOTE: Do not rotate engine after distributor has been removed.

4. Install distributor in engine block. Ensure the distributor hold down flange slot is aligned with the clamp bolt hole and rotor upper blade slot is aligned with the slot in the distributor cap adapter.
5. Install distributor hold down bolt and clamp. Torque to 17 ft. lbs.
6. Align rotor and install distributor cap. Refer to Distributor Rotor, Replace.

Late Models, Fig. 12

1. Remove distributor cap and rotor. Position crankshaft as outlined in Distributor Rotor, Replace.

NOTE: The alignment slot in the adapter should be aligned with the alignment slot in the sleeve assembly.

2. Remove distributor hold down bolt and clamp.
3. Carefully remove distributor from engine block, noting position of the large slot in the sleeve in relation to the adapter alignment slot as distributor drive gear is felt to disengage with cam gear.

NOTE: Do not rotate engine after distributor has been removed.

4. Install distributor in engine block. Ensure the distributor hold down flange slot is aligned with the clamp bolt hole and adapter alignment slot is aligned with the large sleeve alignment slot.

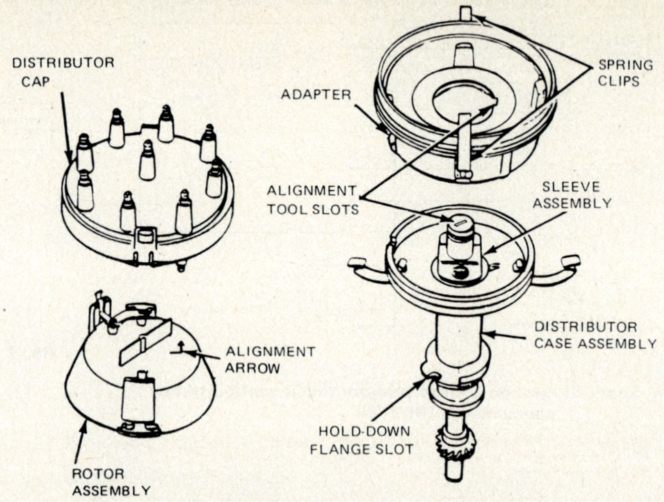

Fig. 12 Late model Dura Spark III distributor

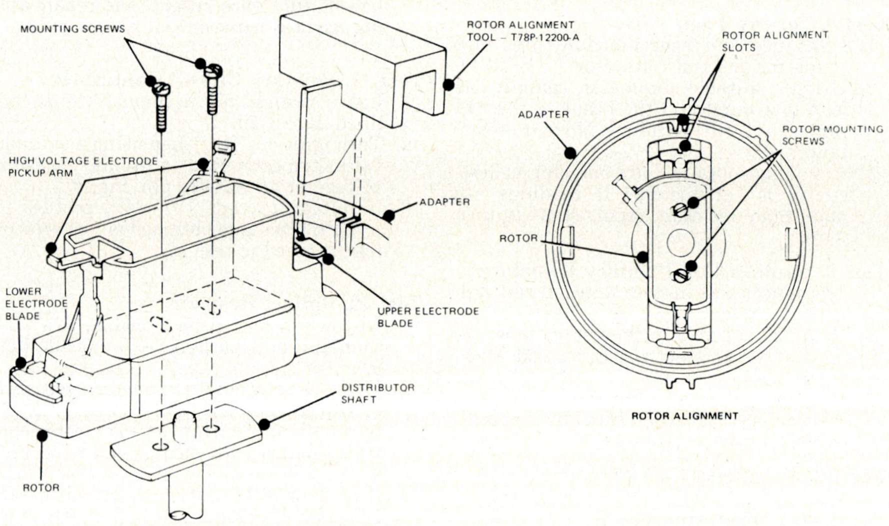

Fig. 13 Early model rotor alignment

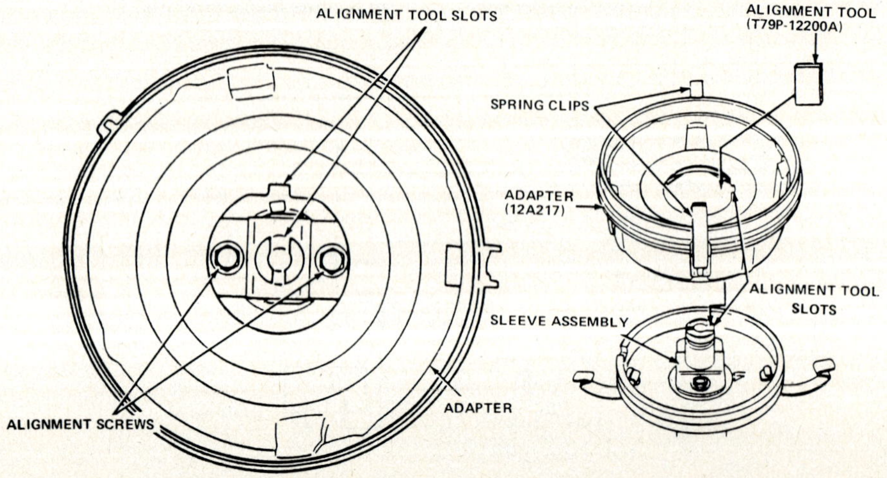

Fig. 14 Late model rotor alignment

5. Install distributor hold down bolt and clamp. Torque to 17 ft. lbs.

DISTRIBUTOR ROTOR, REPLACE

Early Models, Fig. 11

1. Remove distributor cap.
2. Position the crankshaft to align the rotor upper blade slot with the distributor cap adapter slot to enable the rotor alignment tool T78P-12200-A to fall into place. Fig. 13.

NOTE: If alignment of rotor and adapter is not possible due to damage, position No. 1 cylinder at TDC on compression stroke with timing marks aligned at O.

3. Remove rotor alignment tool, then remove rotor retaining screws and rotor.
4. Position rotor on distributor shaft with upper rotor blade slot facing adapter slot. Install, but do not tighten rotor retaining screws.
5. Position rotor alignment tool as shown in Fig. 13, then tighten rotor retaining screws to 15–20 in. lbs.
6. Remove rotor alignment tool and install distributor cap.

Late Models, Fig. 12

1. Remove distributor cap and rotor.
2. Position No. 1 piston on compression stroke, then rotate crankshaft until rotor alignment tool, T79P-12200-A or equivalent, can be inserted into alignment slots, Fig. 14.
3. Check vibration damper and timing pointer alignment marks. If timing pointer is within ±4° of TDC, alignment is satisfactory. If alignment is out of specifications, position vibration damper at TDC with cylinder No. 1 at TDC. Loosen sleeve assembly adjustment screws and rotate sleeve until rotor alignment tool can be inserted into alignment slots. Fig. 14. Torque adjustment screws to 25–35 in. lbs. and remove tool.
4. Replace rotor and distributor cap.

Holley (IH) Breakerless Ignition System (External Electronic Control Unit)

DESCRIPTION

This ignition system incorporates four major units; an electronic control unit, ignition coil, distributor and high tension wires, Fig. 1. The electronic control unit is a solid-state, moisture resistant module with the components sealed in a potting compound to resist vibration and environmental conditions. Since the control unit has an internal current regulator, a resistance wire or ballast resistor is not necessary in the primary circuit. Battery voltage is applied to the ignition coil positive terminal when the ignition switch is in the "On" or "Start" position, therefore, an ignition system bypass is not required in this system. The primary coil circuit is electronically regulated by this unit.

The ignition coil is of standard construction and requires no special service. The function of the ignition coil in this ignition system is the same as for conventional ignition systems.

The distributor is conventional except the contact points, condenser and cam are replaced by a sensor and trigger wheel, and since no wearing occurs between the trigger wheel and sensor, dwell angle remains constant and requires no adjustment. The sensor is a small coil of fine wire and receives an alternating current signal from the electronic control unit. The sensor develops an electromagnetic field used to detect the presence of metal which are the leading edges of the trigger wheel teeth.

TROUBLESHOOTING

Ensure all electrical connections are correct and battery is fully charged before proceeding with the following checks.

1. Disconnect a high tension cable from a spark plug, hold cable approximately 1/2 inch from a suitable ground and crank engine. If spark jumps between the cable and the ground, system is satisfactory.
2. If no spark occurs in step 1, disconnect coil wire from distributor cap and connect a jumper wire to the coil wire approximately 1/4–1/2 inch from distributor end and the ground, Fig. 2. Crank engine and observe for spark between the coil wire distributor end terminal and the jumper wire. If spark occurs, the distributor cap or rotor is faulty and must be replaced.
3. If no spark occurs in step 2, disconnect distributor primary wiring connector and connect tester switch SE-2503 into wiring harness, Fig. 3. Turn ignition switch "On" and press tab on tester switch. If spark occurs between coil wire distributor end and jumper wire, the distributor sensor unit is defective, requiring replacement.
4. If no spark occurs in step 3, connect a voltmeter between the coil positive terminal and the ground, Fig. 4, then turn ignition switch "On". If voltage indicated is less than battery voltage, a high resistance exists between the battery and the coil positive terminal. Check wiring and repair as necessary.
5. If indicated voltage was equal to battery voltage in step 4, connect a voltmeter between the coil negative terminal and the ground, Fig. 5, then turn ignition

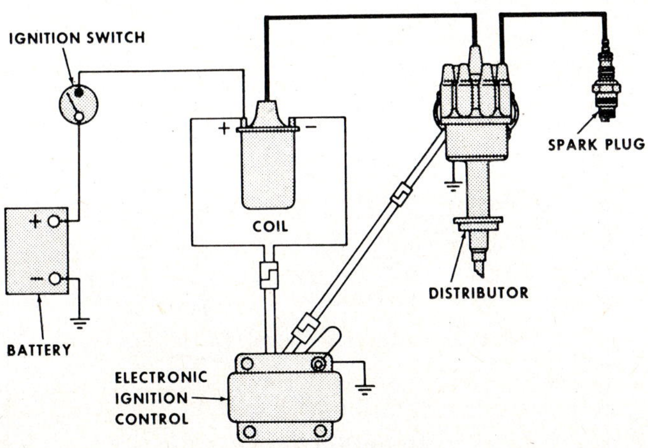

Fig. 1 Holley (International Harvester) breakerless ignition system (External electronic control unit type)

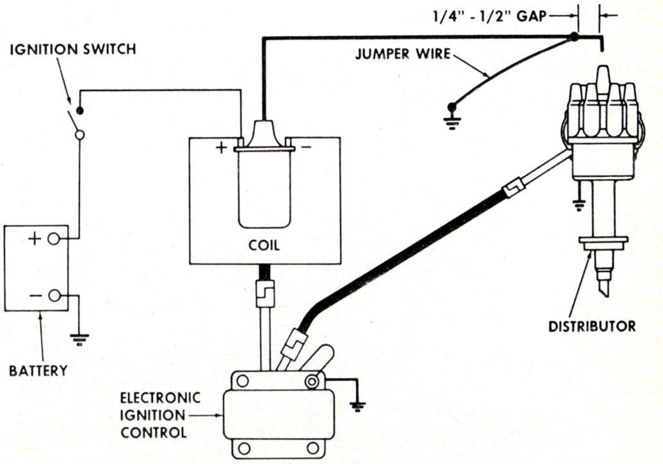

Fig. 2 Checking spark at coil wire

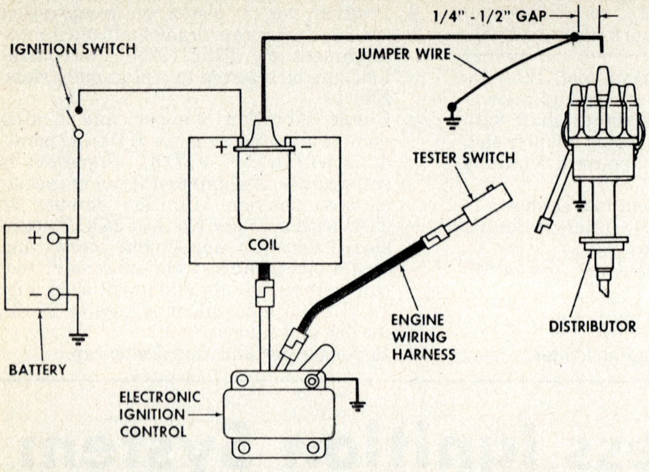

Fig. 3 Testing spark at coil wire with tester switch & jumper wire

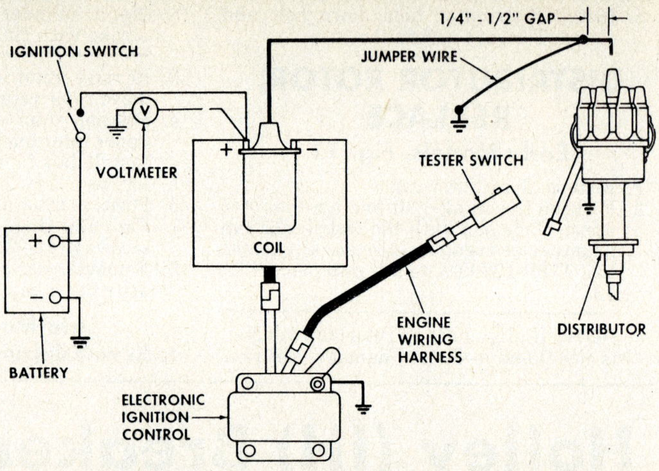

Fig. 4 Checking voltage at ignition coil positive terminal

switch "On". A voltage under 5 volts or over 8 volts indicates a defective coil and must be replaced. If indicated voltage is between 5 and 8 volts, press tab on tester switch. Voltage should increase to battery voltage. Release tester switch tab and voltage should decrease to 5 to 8 volts. If voltage does not increase or decrease, the electronic ignition control unit is defective and must be replaced. If voltage increases and decreases, but no spark occurs between coil wire distributor end and jumper wire, the ignition coil is defective, requiring replacement.

SHAFT & BUSHING WEAR TEST

1. Clamp distributor in a soft jawed vise.
2. Remove distributor rotor.
3. Install a dial indicator so the plunger contacts trigger wheel.
4. Place one end of a wire loop around rotor shaft and hook a spring scale around the other end.
5. Apply 5 pounds of pull on spring scale and observe dial indicator reading. If reading exceeds .006 inch, replace distributor shaft bushings.

DISTRIBUTOR DISASSEMBLE

1. Remove rotor and dust cover.
2. Remove vacuum advance diaphragm rod "E" ring retainer, advance unit mounting screws, then the vacuum advance unit.
3. Pull sensor wiring seal/retainer from distributor housing slot, remove sensor plate mounting screws and lift plate assembly from housing, Fig. 6.
4. Remove epoxy sealer from sensor mounting screw, then remove mounting screw, disengage wiring clip from plate and remove sensor.
5. Remove retaining spring from underside of lower plate, Fig. 7, separate upper and lower plates, then remove three nylon "Buttons" from upper plate, Fig. 8.
6. Remove felt lubrication wick and trigger wheel retainer from shaft, Fig. 9, then slide trigger wheel from shaft and remove slider blocks, Fig. 10.
7. Remove advance weight springs, advance weights, bushings and thrust washers.
8. Measure distributor shaft end play which should be .035–.040 inch. If not, replace distributor shaft and gear.
9. Drive roll pin from gear and shaft and press gear from shaft.
10. Remove distributor shaft from housing.

DISTRIBUTOR ASSEMBLE

1. Install new bushings in advance weight bores, new thrust washers on advance weight pins, then place advance weights on pins and install new advance springs.
2. Lubricate trigger wheel pilot surface with distributor cam lubricant. Grooves in pilot diameter should be filled and lands between grooves should have a thin film of lubricant. The reservoir in the center should be filled to depth of lands with lubricant.
3. Install new slider blocks on trigger wheel, then place trigger wheel on shaft pilot surface with slider block flats aligned with slots in advance weights and push trigger wheel into position.
4. Install trigger wheel retainer and felt lubrication wick.
5. Lubricate distributor shaft with engine oil, then slide shaft into distributor housing.
6. Press gear onto distributor shaft, aligning roll pin hole in gear with hole in shaft. If a new distributor shaft is used, a new gear must be installed. Press new gear onto shaft until specified shaft end play is obtained. Using roll pin hole in the gear as a guide, drill roll pin hole in shaft. Install roll pin.
7. Install three nylon "Buttons" in upper plate and assemble upper plate to lower plate and install retaining spring.
8. Place sensor on upper plate and loosely

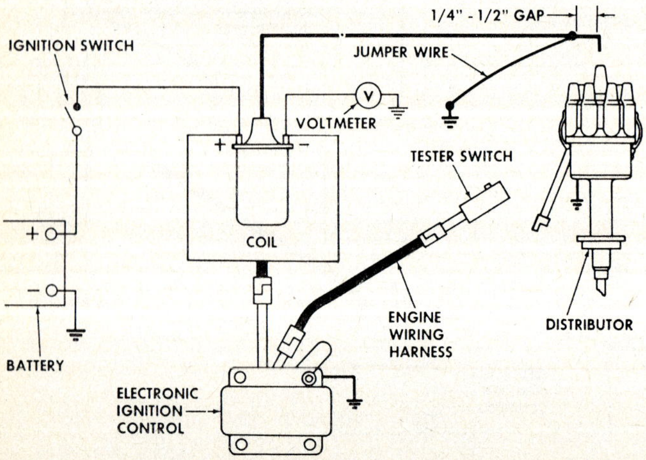

Fig. 5 Checking voltage at ignition coil negative terminal

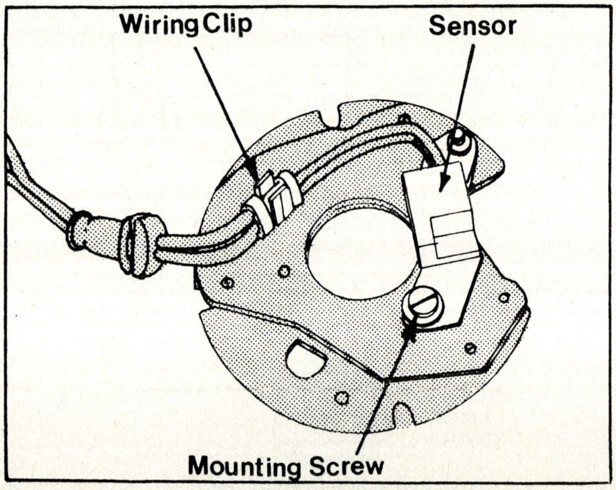

Fig. 6 Removing sensor

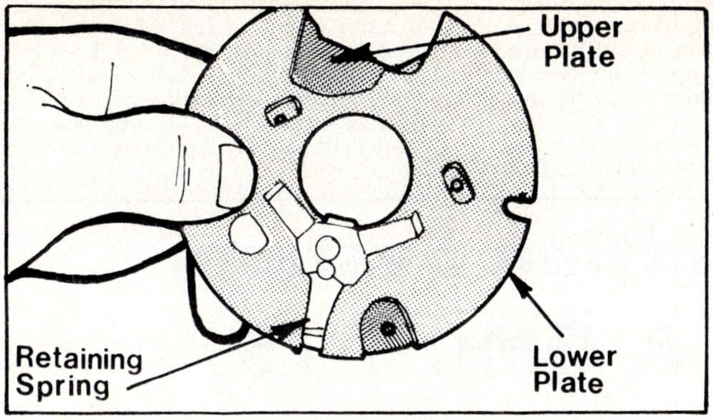

Fig. 7 Removing plate retaining spring

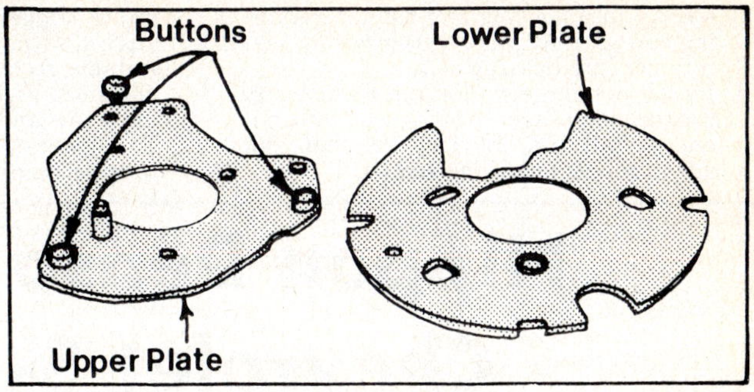

Fig. 8 Upper & Lower plates

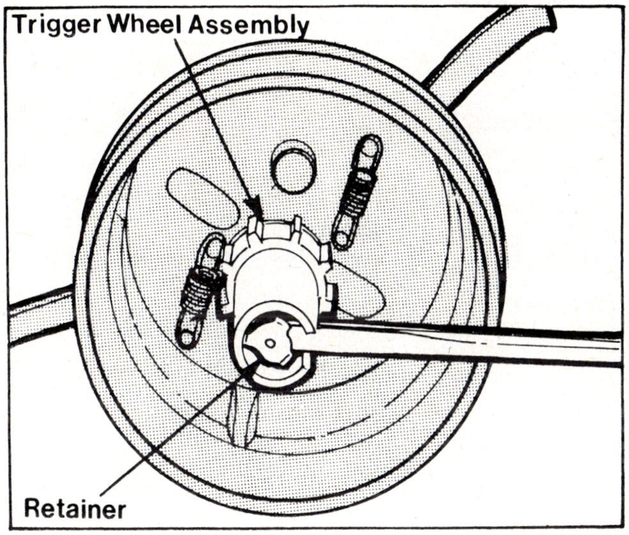

Fig. 9 Removing trigger wheel retainer

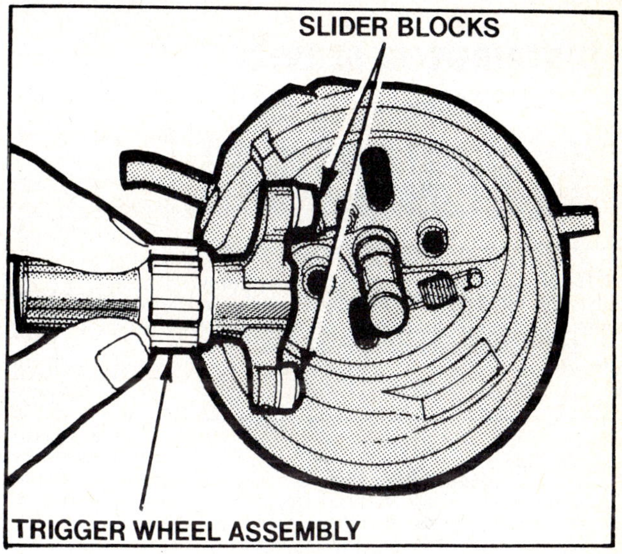

Fig. 10 Removing trigger wheel

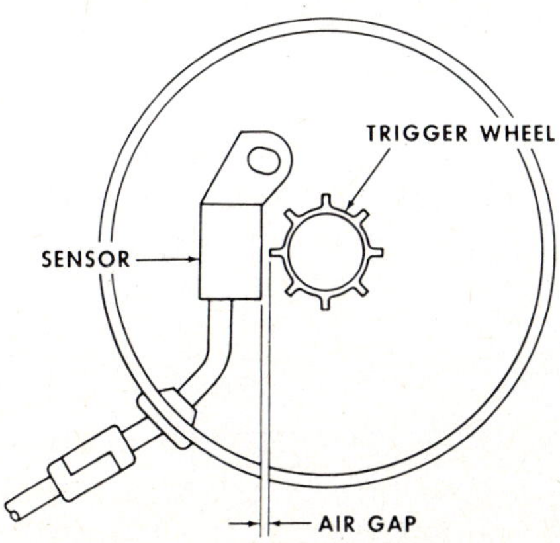

Fig. 11 Adjusting air gap

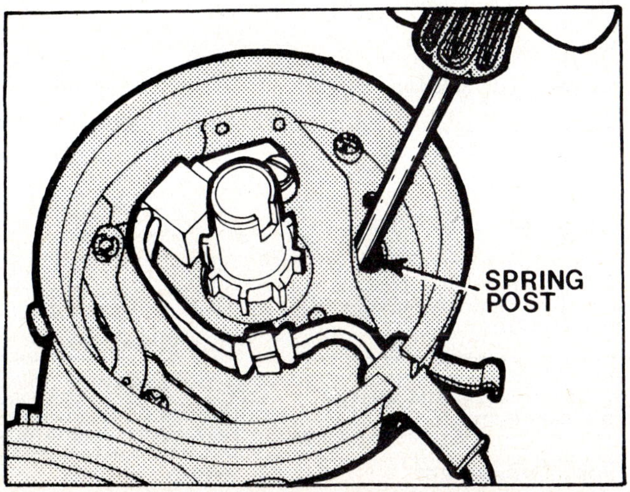

Fig. 12 Adjusting centrifugal advance

install mounting screw, install plate assembly into housing and tighten retaining screws, then place sensor wiring seal retainer into housing slot.
9. Install vacuum advance unit on housing, connect diaphragm rod to sensor plate pin and install the "E" ring retainer and advance unit mounting screws.
10. To adjust trigger wheel to sensor air gap, rotate trigger wheel until one tooth is aligned with center of sensor and, using a suitable feeler gauge, adjust air gap to .008 inch, Fig. 11. Tighten sensor mounting screw and recheck air gap.
11. To adjust centrifugal advance, install distributor on distributor test stand. Check advance at each specified test speed. If advance is not within limits, adjust by bending spring posts, Fig. 12. To increase advance, bend post toward distributor shaft thereby decreasing spring tension. Decrease advance by bending post away from shaft thereby increasing spring tension.

Nippon Denso Ignition System, Chevrolet S-10 & GMC S-15

SYSTEM DIAGNOSIS

Refer to diagnosis charts, Figs. 1 thru 7 for diagnosis procedures.

DISTRIBUTOR SERVICE

Removal

1. Disconnect battery ground cable, spark plug & ignition coil wires and remove bolt attaching distributor bracket. Remove distributor assembly.

NOTE: Mark position of rotor to distributor body and position of distributor body to engine, then pull distributor up until rotor just stops turning and again mark position of rotor. Mark position of cap tower for No. 1 lead.

Disassembly, Fig. 8

1. Remove cap, rotor and packing, and cover.
2. Remove screws attaching vacuum controller and vacuum controller from distributor housing.
3. Remove screws from harness assembly. Disconnect connectors of harness assembly from igniter unit, then remove harness assembly from distributor housing.
4. Remove reluctor from rotor shaft. Remove screws from breaker plate and remove breaker plate from distributor.
5. Remove module from breaker plate assembly. Module is an integral construction and cannot be disassembled.
6. Drive out roll pin, Fig. 9.
7. Remove governor shaft assembly from distributor housing.
8. Remove packing from governor shaft assembly and screw attaching rotor shaft assembly. Separate rotor shaft assembly from governor shaft.

NOTE: The end of the governor shaft is offset and to maintain original relationship of shaft to shaft, mark rotor to governor shaft before removal.

9. Remove governor weights and springs from governor shaft assembly.

Air Gap, Adjustment

1. Measure air gap between pole piece and stator. Air gap should be .12–.20 inch.
2. Adjust air gap by loosening the two screws and moving pole piece. Tighten screws and recheck air gap.

Governor Advance

Turn rotor shaft in counterclockwise direction by hand. If rotor shaft returns to normal position when released, governor advance is operating normally.

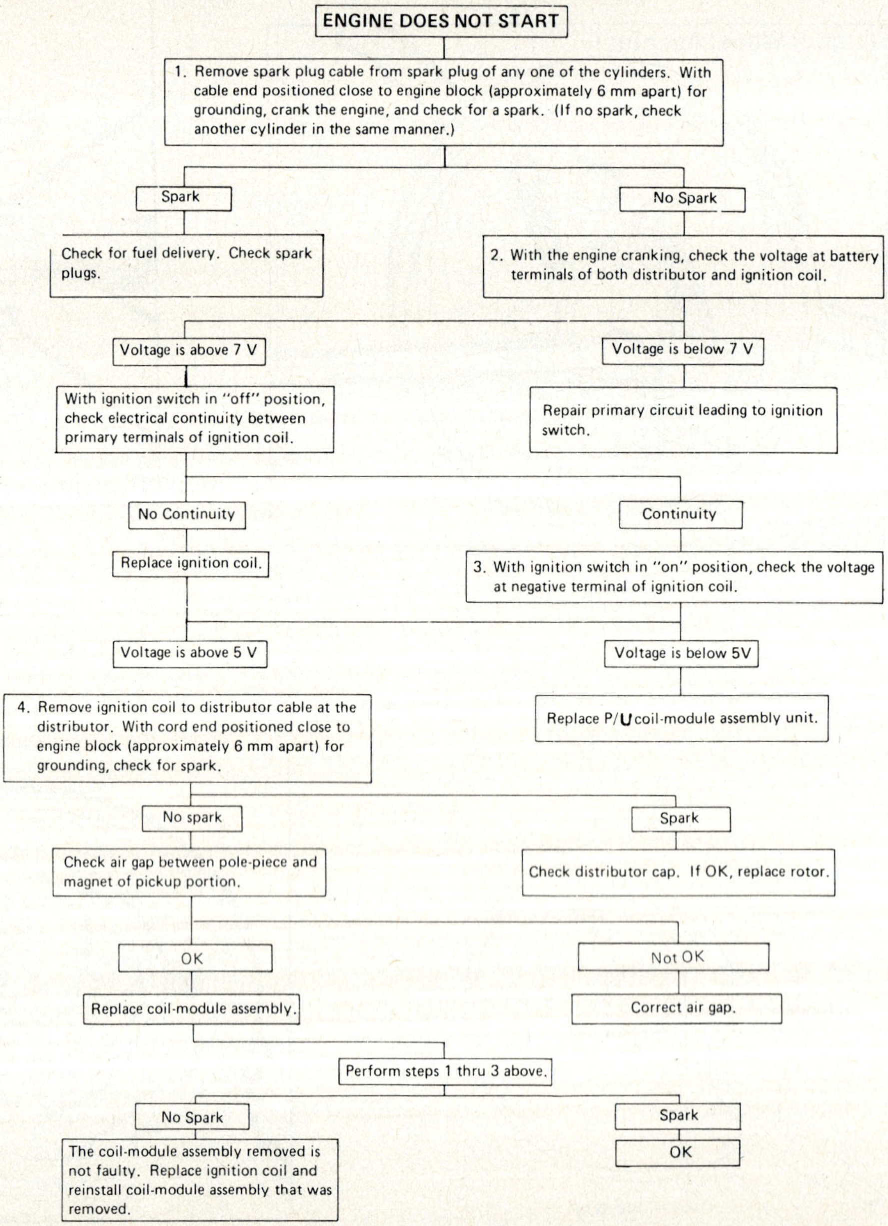

Fig. 1 Nippon Denso diagnosis chart (part 1 of 7)

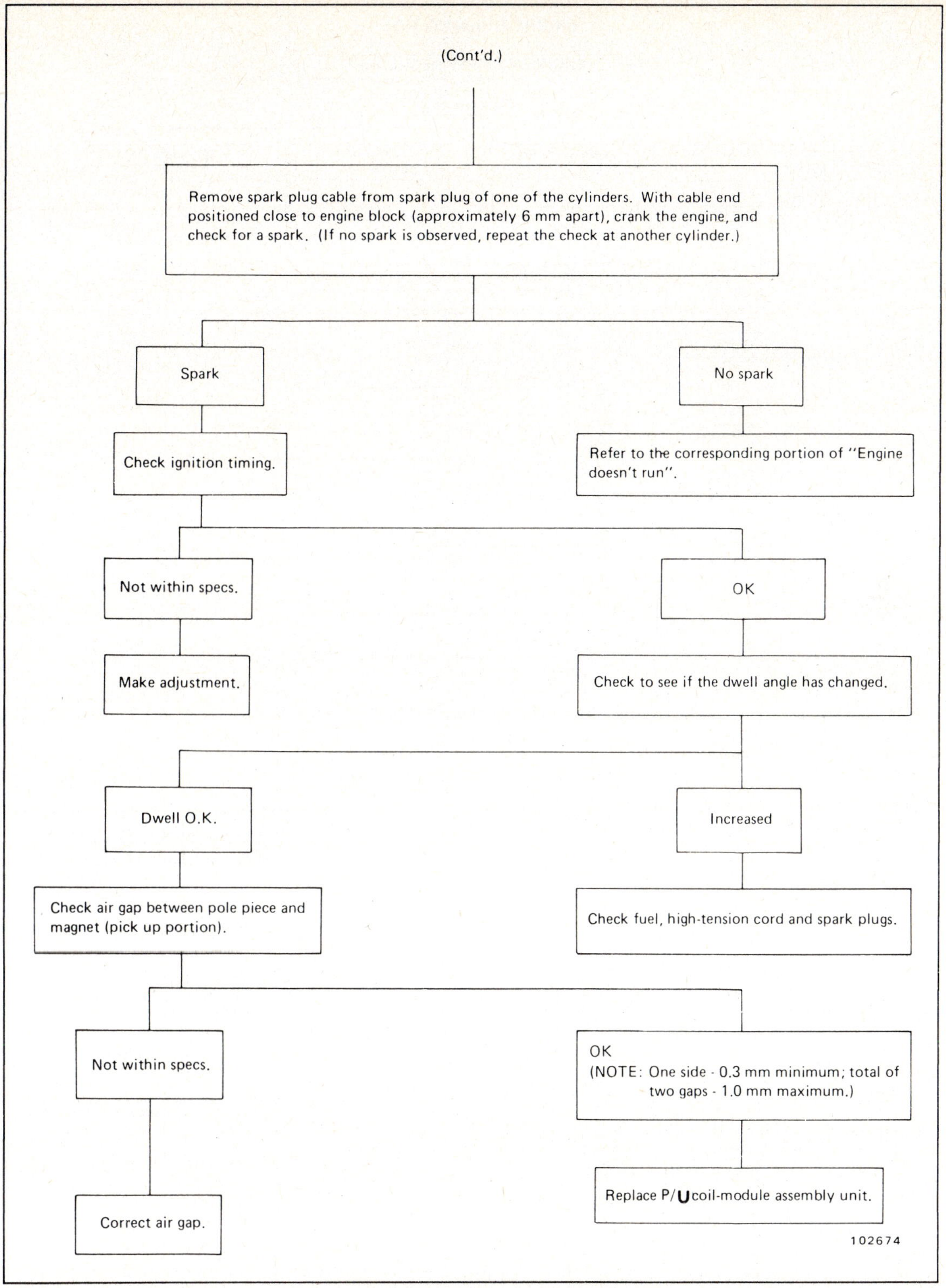

Fig. 2 Nippon Denso diagnosis chart (part 2 of 7)

Vacuum Advance

Check vacuum controller for leaks by applying vacuum with a pump.

Assembly

Reverse procedure to install. The governor springs should be fitted to spring hanger pin on governor shaft assembly with the smaller hook end turned downward. The relation between cutaway portion of rotor shaft and offset of projection at end of governor shaft should be carefully noted when installing the rotor shaft to the governor shaft assembly. The edge at cutaway portion of the base must be flush with the edge of base fixing screw slot when installing the breaker plate assembly. The roll pin should be installed so that slot is in parallel with cutaway portion of reluctor as viewed from above. Connect the harness terminal to the P/U coil-module assembly. Refer to Fig. 10. Install gasket after setting cover to housing.

Installation

Reverse procedure to install. Install distributor assembly by fitting projected end of shaft into offset slot at upper face of the oil pump drive shaft. Move rotor slightly to facilitate smooth engagement. Connect high tension cable securely. Start engine and check ignition timing.

ENGINE RUNNING ROUGH

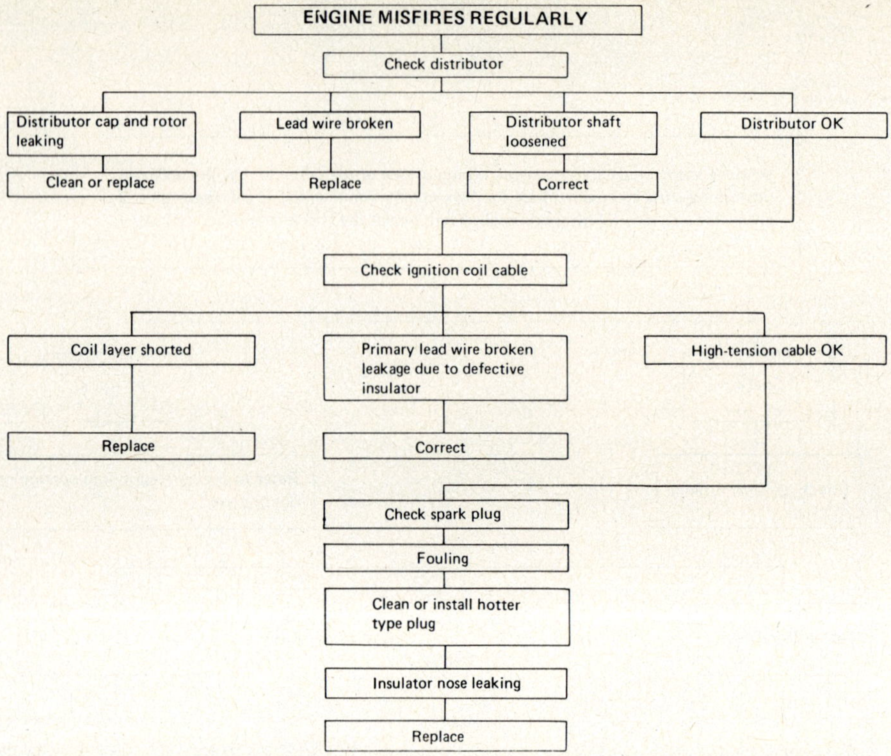

Fig. 3 Nippon Denso diagnosis chart (part 3 of 7)

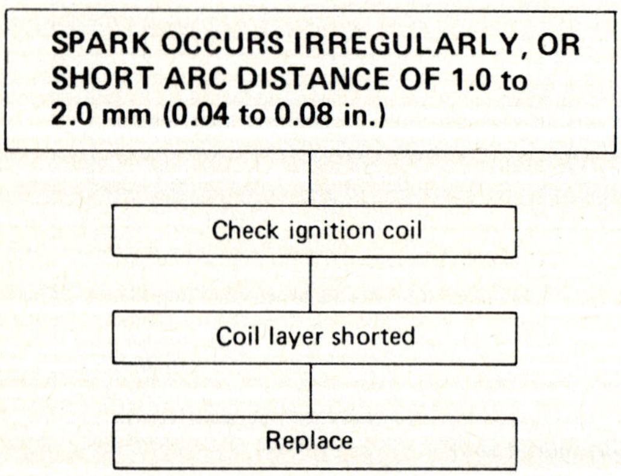

Fig. 4 Nippon Denso diagnosis chart (part 4 of 7)

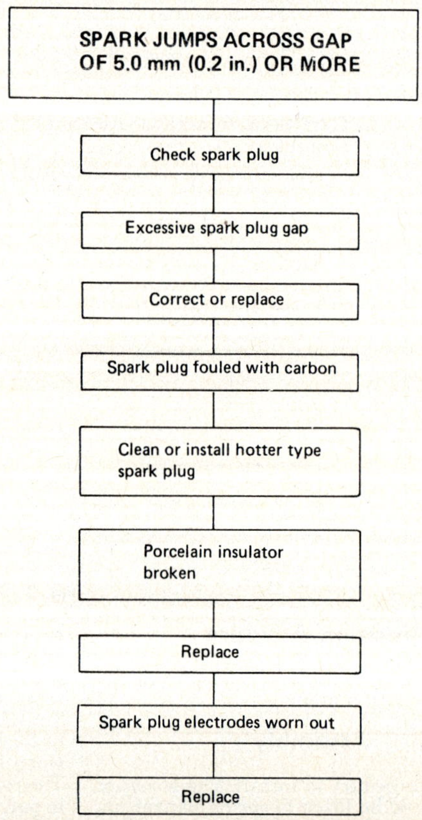

Fig 5 Nippon Denso diagnosis chart (part 5 of 7)

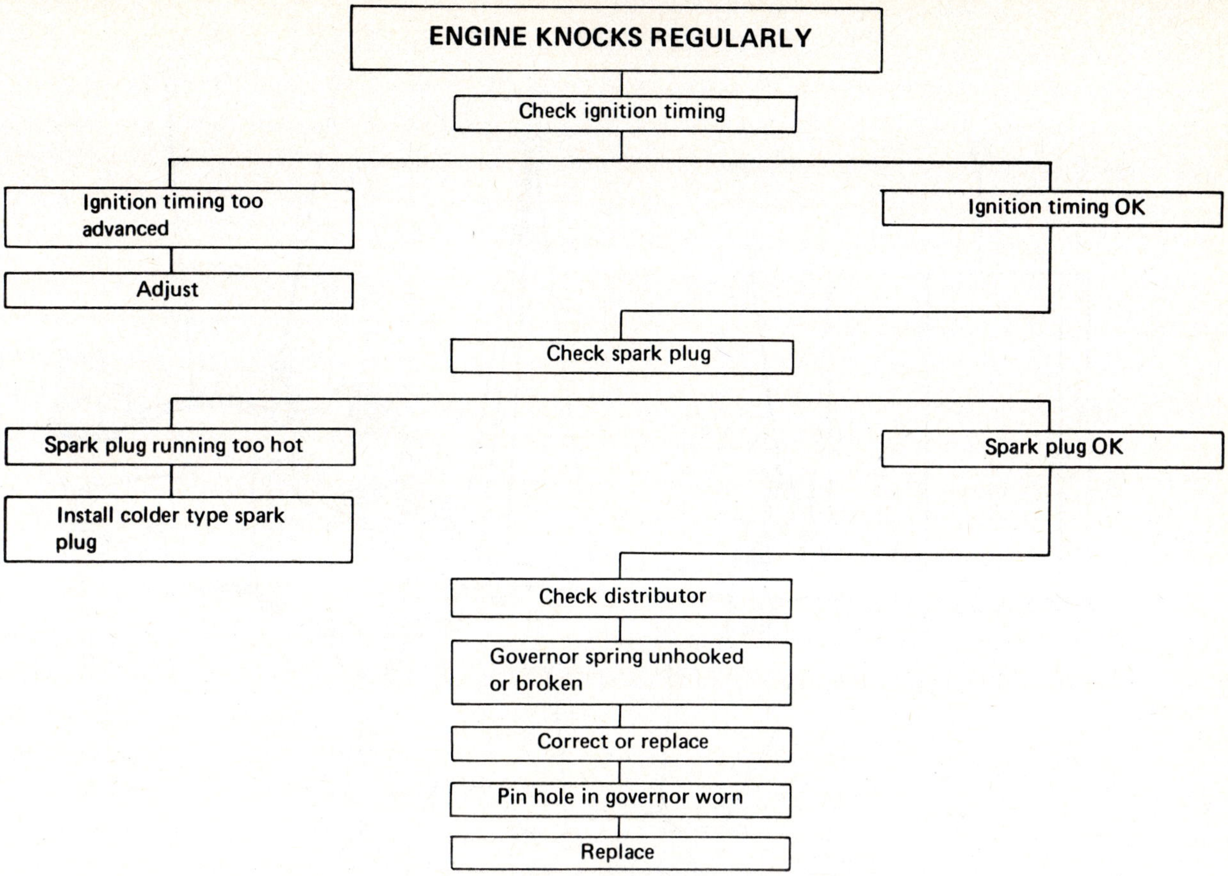

Fig. 6 Nippon Denso diagnosis chart (part 6 of 7)

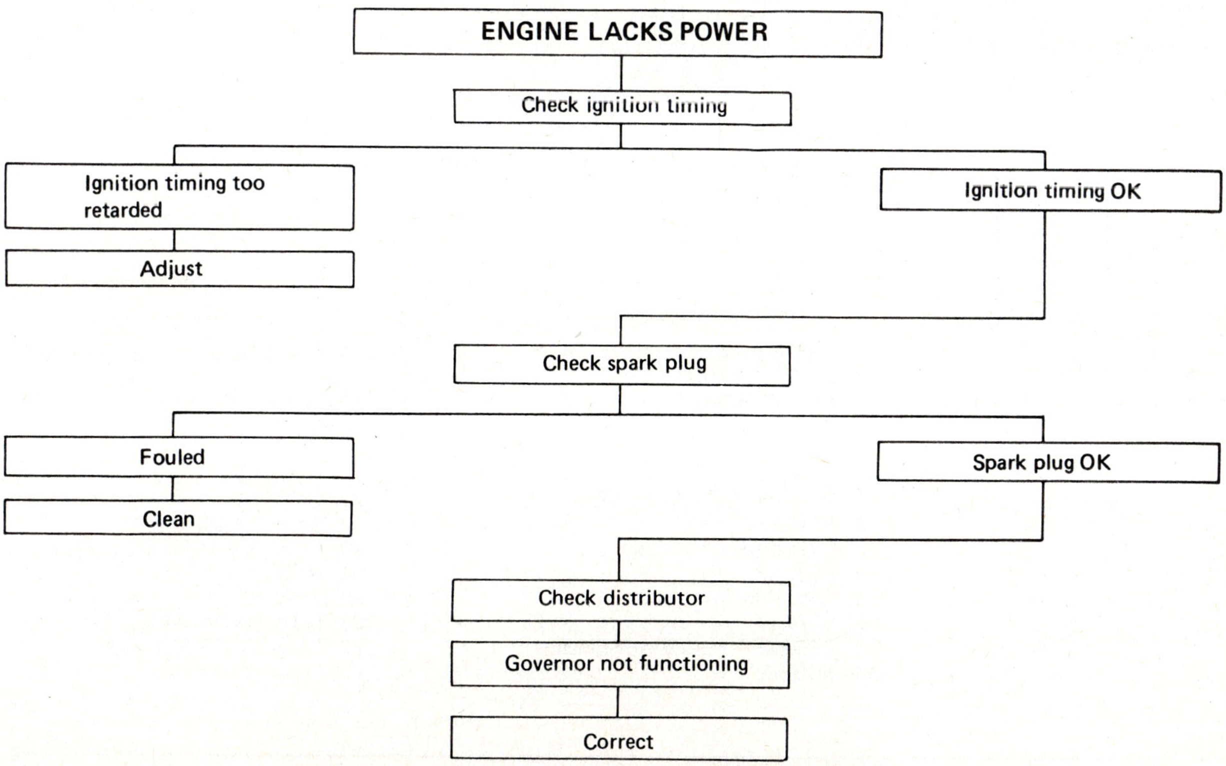

Fig. 7 Nippon Denso diagnosis chart (part 7 of 7)

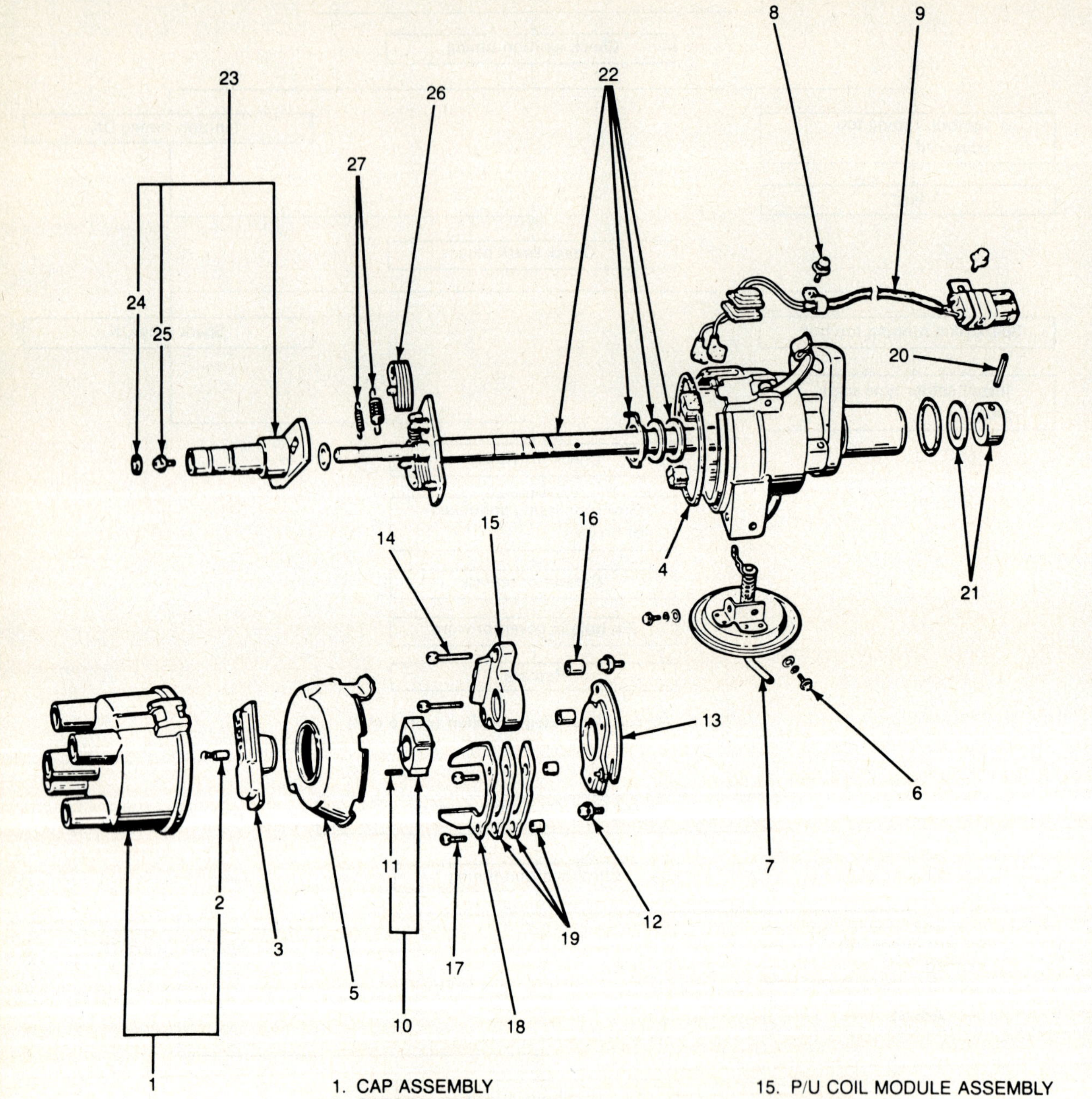

1. CAP ASSEMBLY
2. CARBON POINT
3. ROTOR HEAD
4. PACKING
5. COVER
6. SCREW
7. VACUUM CONTROL ASSEMBLY
8. SCREW
9. HARNESS ASSEMBLY
10. POLE PIECE
11. ROLL PIN
12. SCREW
13. BREAKER PLATE ASSEMBLY
14. SCREW
15. P/U COIL MODULE ASSEMBLY
16. SPACER
17. SCREW
18. STATOR
19. MAGNET SET
20. ROLL PIN
21. COLLAR
22. SHAFT ASSEMBLY
23. ROTOR SHAFT ASSEMBLY
24. PACKING
25. SCREW
26. GOVERNOR WEIGHT
27. GOVERNOR SPRING

Fig. 8 Exploded view of Nippon Denso Distributor

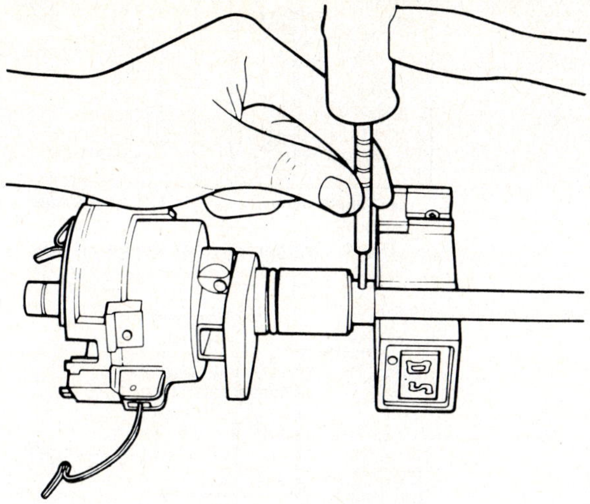

Fig. 9 Roll pin removal

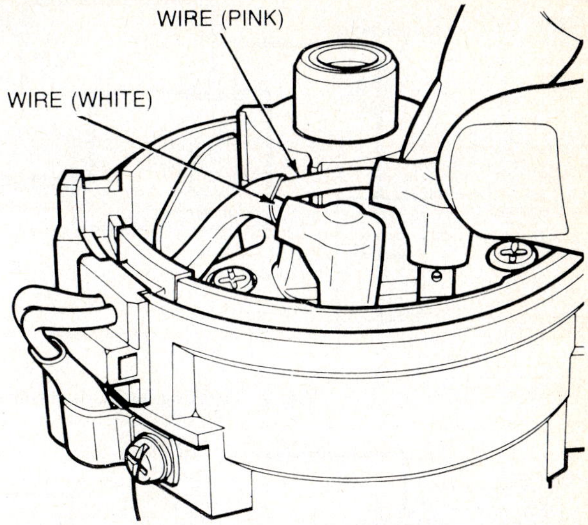

Fig. 10 Connecting harness terminal

Prestolite (IH) Breakerless Ignition System (Integral Electronic Control Unit)

DESCRIPTION

This ignition system incorporates two major units: a distributor and an ignition coil, Fig. 1. The distributor is conventional except that a trigger wheel and an electronic control unit (Circuit board and sensor) replace the usual distributor cam, contact points and condenser.

The electronic control unit is associated with the primary side of the ignition system and electronically makes and breaks the primary circuit in response to rotation of the trigger wheel. The control unit circuit board is of solid state design and its components are sealed in a waterproof and vibration resistant compound. The sensor is a small coil, wound of fine wire, which is a metal detector. The metal that the sensor detects is the teeth of the trigger wheel.

The electronic circuit board and sensor are mounted on the distributor plate assembly. The sensor leads are soldered to the circuit board. The electronic control unit (Circuit board, sensor and plates) is serviced as an assembly.

TROUBLESHOOTING

Ensure all electrical connections are correct and the battery is fully charged before proceeding with the following checks.

1. Disconnect a high tension cable from a spark plug, hold cable approximately 1/2 inch from a suitable ground and crank engine, Fig. 2. If a spark jumps between the cable and the ground, the system is satisfactory.
2. If no spark occurs in Step 1, disconnect coil wire from distributor cap. Insert an extension adapter into boot and engage into cable terminal. Hold cable approximately 1/2 inch from a suitable ground and crank engine, Fig. 3. If spark occurs, the problem may be the distributor cap, rotor or high tension leads.
3. If no spark occurs in Step 2, check trigger wheel to sensor air gap. Remove distributor cap, rotor and shield. Crank engine to align one of the trigger wheel teeth with the sensor coil, Fig. 4. Check air gap between trigger wheel tooth and sensor. The air gap should be .008 inch. Adjust air gap, if necessary, and repeat Step 2. If no spark occurs, proceed to next step.

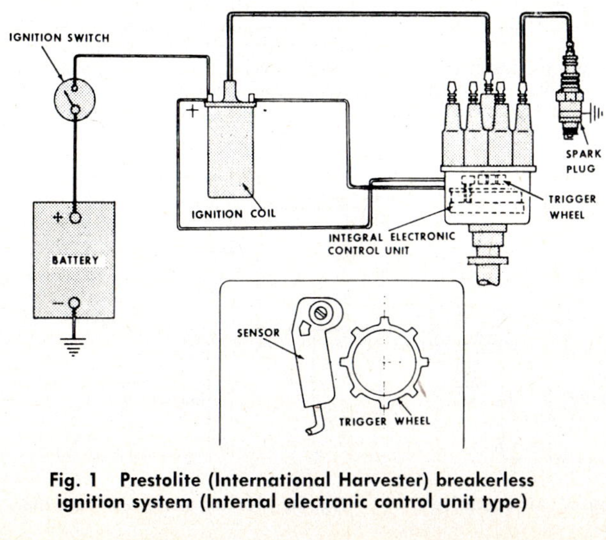

Fig. 1 Prestolite (International Harvester) breakerless ignition system (Internal electronic control unit type)

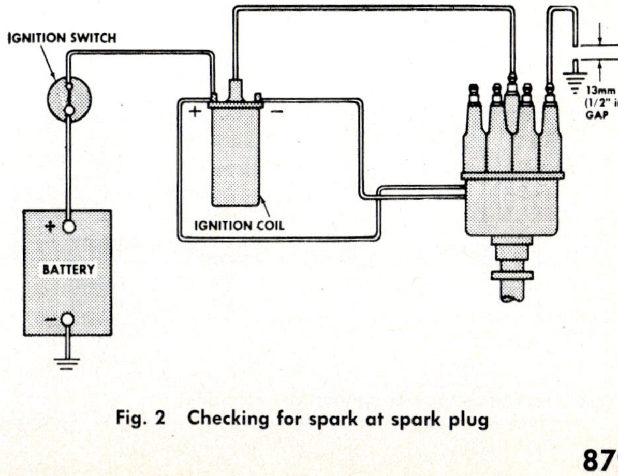

Fig. 2 Checking for spark at spark plug

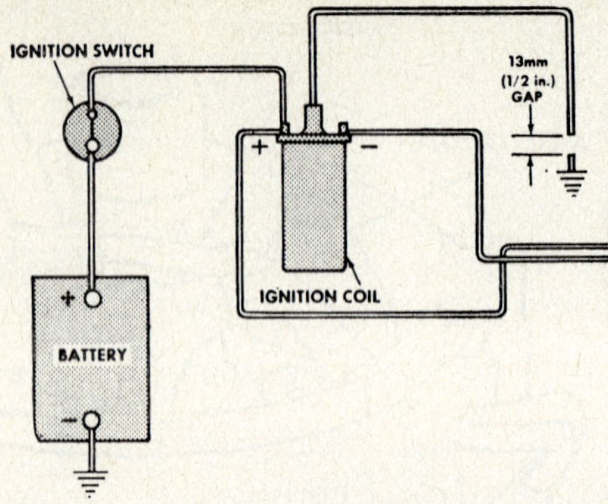

Fig. 3 Checking for spark at coil wire

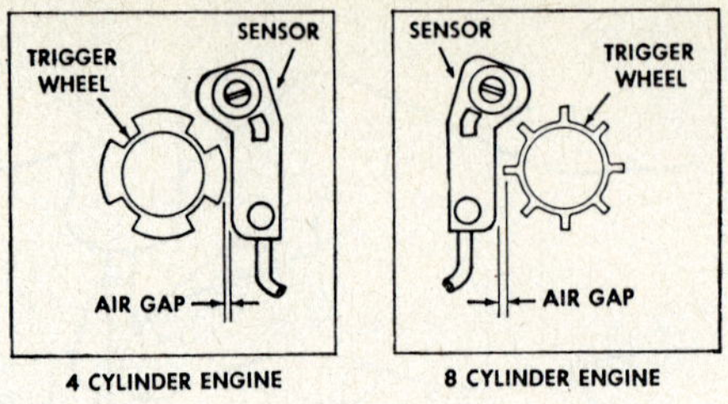

Fig. 4 Trigger wheel to sensor air gap

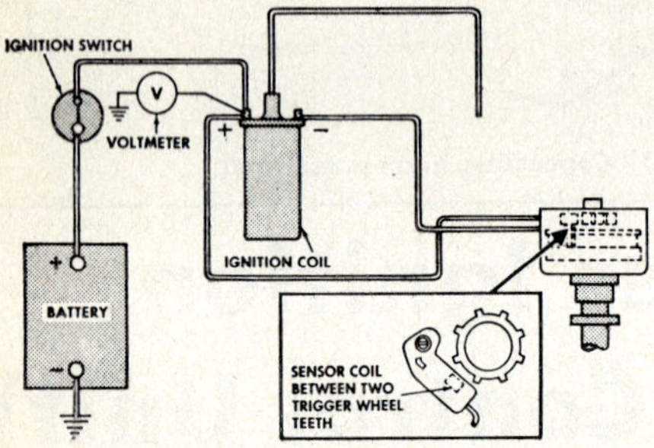

Fig. 5 Testing primary voltage

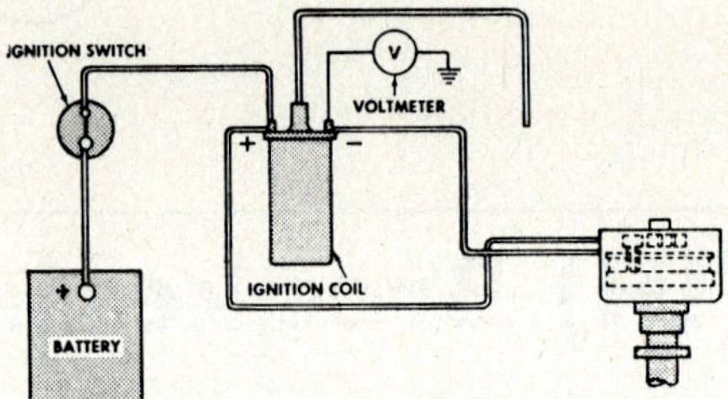

Fig. 6 Testing voltage at ignition coil negative terminal

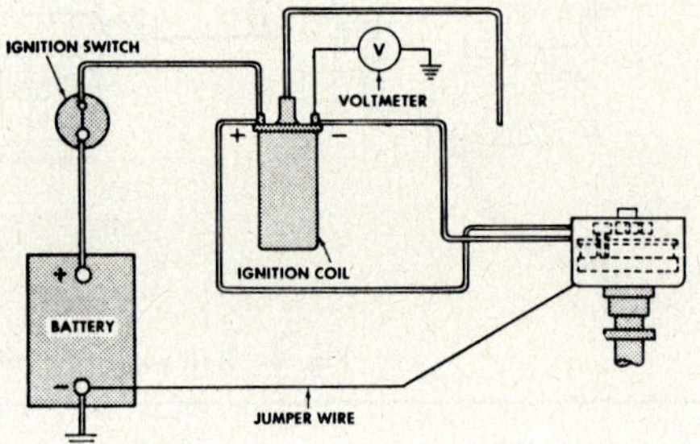

Fig. 7 Testing voltage at ignition coil negative terminal with jumper wire installed

4. Crank engine to position sensor coil between two trigger wheel teeth. Connect a voltmeter between the ignition coil positive terminal and the ground, Fig. 5. Turn ignition switch "On" and note voltmeter reading. If voltage is lower than battery voltage, a high resistance exists between the battery and the ignition coil. If battery voltage is present, proceed to next step.
5. Connect voltmeter between the ignition coil negative terminal and the ground with the sensor coil position between two trigger wheel teeth, Fig. 6. Observe voltmeter reading with ignition "On." If reading is 5 to 8 volts, proceed to next step. If reading is 12 to 13 volts, connect a jumper wire between distributor housing and battery negative terminal and observe voltmeter, Fig. 7. If the voltage remains at 12–13 volts, the electronic control unit in the distributor is defective, requiring replacement. If the voltage changes to 5 to 8 volts with the jumper wire connected, a problem exists in the ground circuit between the distributor and the battery. If the voltage is 0 to 5 volts, disconnect the voltmeter, remove the brown wire from the ignition coil negative terminal and reconnect the voltmeter between the ignition coil negative terminal and the ground, Fig. 8. If voltage remains at 0 to 5 volts, the ignition coil is defective, requiring replacement. If voltage increases to 12 to 13 volts, the electronic control unit is faulty. Reconnect the brown wire to the ignition coil negative terminal.
6. Connect voltmeter between the ignition coil negative terminal and the ground, Fig. 9. With ignition "On", place a screwdriver blade against face of sensor, Fig. 9. Voltage should increase to 12 to 13 volts. Remove the screwdriver blade and voltage should drop to 5 to 8 volts. If so, proceed to next step. If not, replace electronic control unit.
7. Disconnect coil wire from distributor cap and insert an extension adapter into boot and engage into cable terminal. Hold cable approximately 1/2 inch from a suitable ground. With ignition "On", observe for spark each time a screwdriver blade is placed across the sensor face, Fig. 10. If no spark occurs, the sensor coil is defective, requiring replacement.
8. After replacing components, remove test equipment and repeat Step 1.
9. Connect a dwell meter to engine. Operate engine and observe dwell reading. It may be necessary to readjust the trigger wheel to sensor air gap to obtain specified dwell.

Primary Voltage Drop Test, Fig. 11

1. Remove distributor cap, rotor and shield. Crank engine to position sensor coil between two trigger wheel teeth.
2. Connect voltmeter positive lead to battery positive terminal and the voltmeter negative lead to the ignition coil positive terminal.
3. Turn ignition "On" and observe voltmeter reading. A reading of less than 1 volt should be obtained.
4. Check circuit conditions by observing voltmeter while moving the connectors at the following locations: battery cables, starter solenoid battery terminal, dash panel bulkhead connector, ammeter terminals and ignition switch connections.
5. If a reading fluctuation or an upswing of

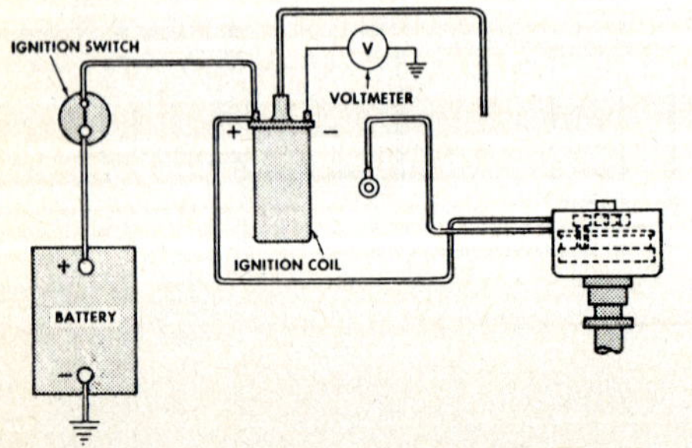

Fig. 8 Testing voltage at ignition coil negative terminal with brown wire disconnected

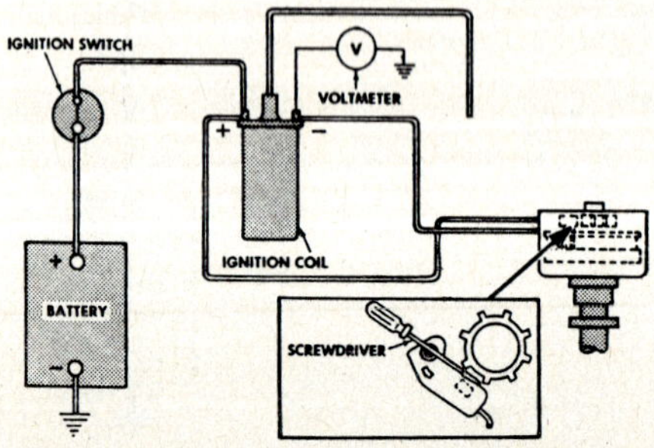

Fig. 9 Testing electronic control unit operation

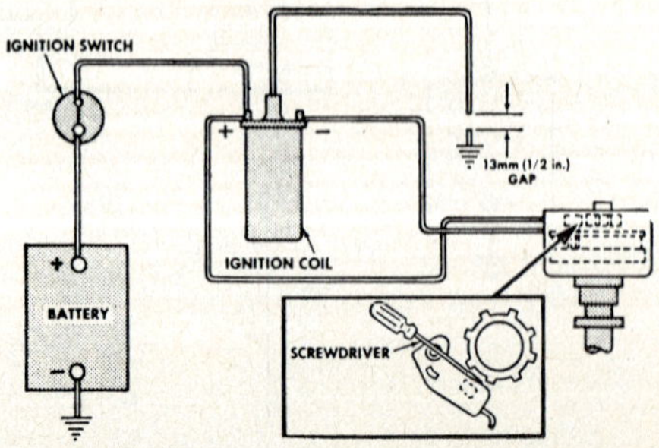

Fig. 10 Testing sensor coil operation

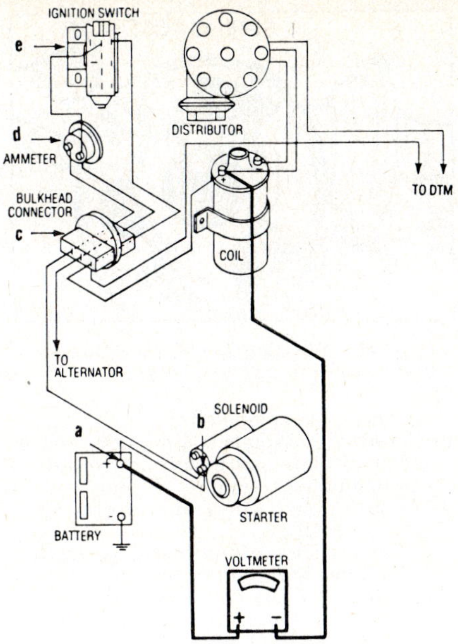

Fig. 11 Primary voltage drop test

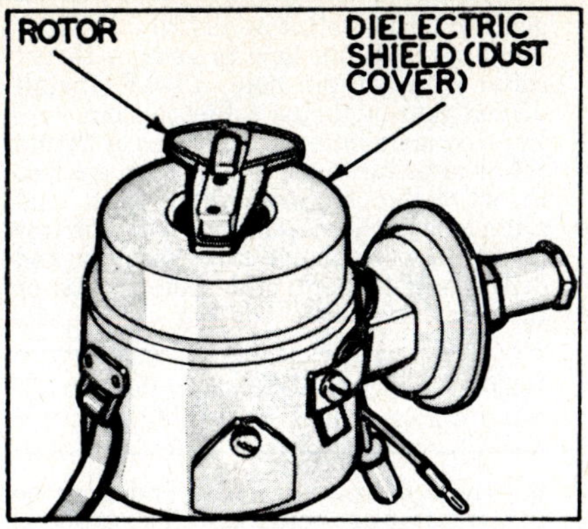

Fig. 12 Distributor rotor & shield

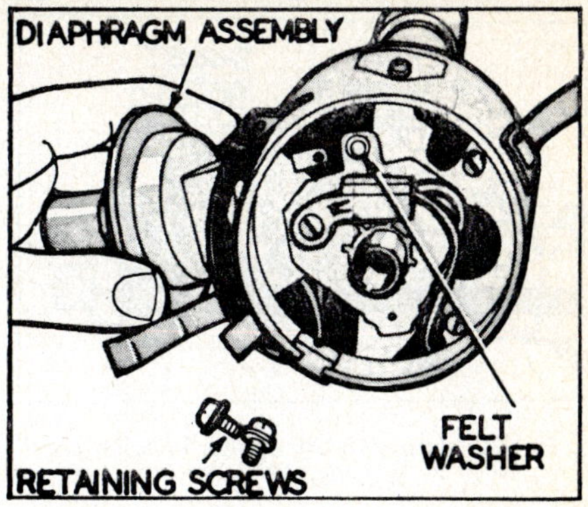

Fig. 13 Vacuum advance assembly

the voltmeter needle is observed, a poor connection exists and must be corrected.

Shaft & Bushing Wear Test

1. Clamp distributor is a soft jawed vise.
2. Remove distributor rotor and shield.
3. Install a dial indicator so the pluger contacts the trigger wheel.
4. Place one end of a wire loop around the rotor shaft and hook a spring scale around the other end.
5. Apply 1/2 pound of pull on spring scale and observe dial indicator reading. Apply 1/2 pound of pull in opposite direction and observe dial indicator reading.
6. Peak to peak side play is the total of the two readings in Step 5. If the side play exceeds .006 inch, replace distributor shaft bushings.

DISTRIBUTOR DISASSEMBLE

1. Remove rotor and shield, Fig. 12
2. Remove vacuum advance diaphragm rod "E" ring retainer, advance unit mounting screws and the vacuum advance unit, Fig. 13. Remove felt washer from diaphragm rod pin on sensor advance plate, Fig. 13.
3. Remove electronic control unit (Circuit board, sensor and plate assembly), Fig. 14, as follows:
   a. Remove distributor plate (Control unit) mounting screws.
   b. Pull wiring grommet from slot in distributor housing.
   c. Lift control unit assembly from housing.
4. Measure distributor shaft end play between thrust washer and gear, Fig. 15, or thrust washer and thrust collar, Fig. 16. If end play exceeds .035–.040 inch on 4-196 and V8-345, 392 units, .004–.018 inch on V8-404, 446 units or .002–.010 inch on V8-537 units, replace upper and lower thrust washers.
5. Drive roll pin from gear and shaft. Press gear or thrust collar and lower thrust washers from shaft.
6. Remove distributor shaft from housing. Use care not to lose the upper thrust washers.

DISTRIBUTOR ASSEMBLE

NOTE: If a new distributor shaft is being installed, a new distributor driven gear or thrust collar and new thrust washers must be installed.

1. Install distributor shaft into housing as follows:
   a. Place new upper thrust washers on shaft. Units with a tachometer drive use one thrust washer, Fig. 17, while

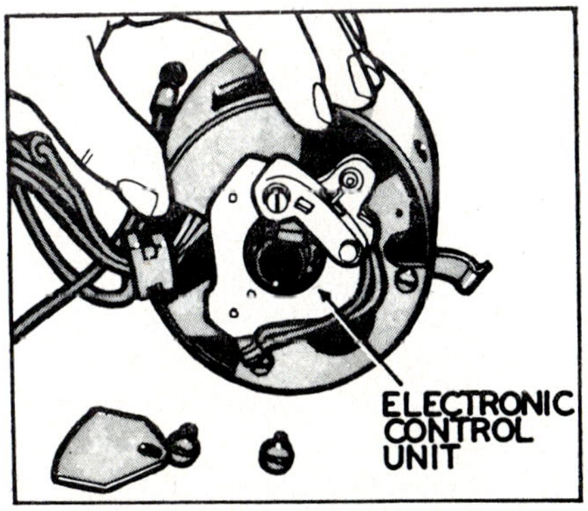

Fig. 14 Electronic control unit installation

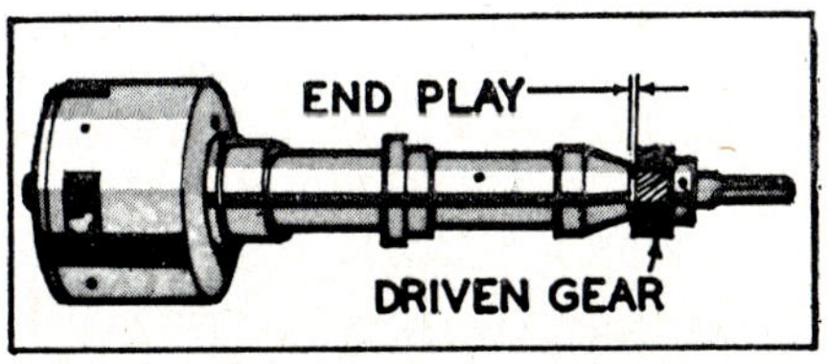

Fig. 15 Checking distributor shaft end play. Units with driven gear

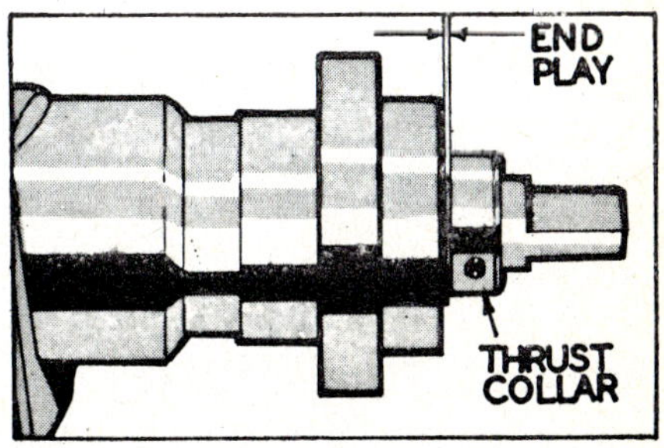

Fig. 16 Checking distributor shaft end play. Units with thrust collar

Fig. 17 Upper thrust washer installation. Units with tachometer drive

Fig. 18 Upper thrust washer installation. Units less tachometer drive

Fig. 19 Lower thrust washer installation. V8-537 units

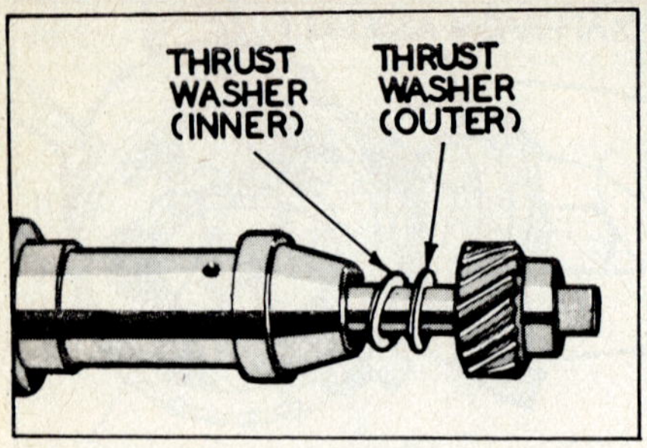

Fig. 20 Lower thrust washer installation. Exc. V8-537 units

all others use two. On units using two thrust washers, Fig. 18, install the outer washer first with the lip facing toward lower end of shaft. Install the inner washer with lip facing toward lower end of shaft and seated on the lip of the outer washer.

b. Lubricate distributor shaft with engine oil.

c. Slide distributor shaft into housing carefully to avoid damaging lower oil seal, if installed.

2. Install new lower thrust washers on shaft. Units used on V8-537 engines use one thrust washer, Fig. 19, while all others use two. On units using two thrust washers, Fig. 20, install inner washer first with the lip facing toward the distributor housing. Install outer washer with the lip facing toward the distributor housing and seated on the lip of the inner washer.

3. If original distributor driven gear or thrust collar is being installed, position gear or thrust collar on distributor shaft and align roll pin hole in gear or thrust collar with roll pin hole in shaft. Install driven gear or thrust collar roll pin.

4. If a new distributor driven gear or thrust collar is being installed, position gear or thrust collar on distributor shaft and align hole with roll pin hole in distributor shaft. Support distributor assembly and drill a hole through other side of gear or thrust collar. Install roll pin.

NOTE: A new gear or thrust collar has the roll pin hole drilled in one side only.

---

5. If a new distributor shaft and gear or thrust collar are being installed, position gear or thrust collar on distributor shaft and adjust distributor shaft end play. End play should be .035–.040 inch on 4-196 and V8-345, 392 units, .004–.018 inch on V8-404, 446 units and .002–.010 inch on V8-537 units. Then, support distributor assembly and, using the existing hole in gear or thrust collar as a guide, drill roll pin hole in distributor shaft and other side of gear or thrust collar. Install roll pin.

6. Install electronic control unit (Circuit board, sensor and plate assembly) as follows:

a. Apply a small amount of grease under the three thrust buttons.

b. Place control unit assembly and wiring grommet in distributor housing.

c. Install distributor plate (Control unit) mounting screws.

7. Place vacuum advance diaphragm assembly on distributor housing and install mounting screws.

8. Install felt washer on diaphragm rod pin of sensor advance plate. Lubricate felt

Fig. 21 Adjusting centrifugal advance

washer with engine oil.

9. Connect diaphragm rod to pin on sensor advance plate and install flat washer, if used, and rod retaining "E" clip.

10. Adjust trigger wheel to sensor air gap, Fig. 4, as follows:

a. Rotate trigger wheel until one tooth is aligned with centerline of sensor.

b. Using a feeler gauge, measure air gap between sensor and end of tooth. Air gap should be .008 inch. If not, move sensor to obtain proper air gap. Tighten sensor mounting screw and recheck air gap.

11. To adjust centrifugal advance, install distributor on distributor feet stand. Check advance at each specified speed. If advance is not within limits, adjust by bending spring posts, Fig. 21. To increase advance, bend post toward distributor shaft thereby decreasing spring tension. To decrease advance, bend spring posts away from distributor shaft thereby increasing spring tension.

12. Install shield, rotor and distributor cap.

AUTOCAR

INDEX OF SERVICE OPERATIONS

MODEL INDEX & ENGINE APPLICATION

ENGINE IDENTIFICATION—Autocar Service Parts Identification Plate is located inside of truck cab. Stamped on the plate is the model of the engine, transmission, auxiliary transmission, rear axle and front axle. This data represents only the equipment on the vehicle when it was shipped from the factory. Prior to Autocar's assimilation by The White Motor Company, identification plates were attached to engines and transmissions.

Year	Model	Engine Make	Base Engine Model②	Crankcase Refill Capacity, Qts.	Cooling System Capacity, Qts.
Diesel Engines					
1974–78	A42,B,TB	Cummins	NH-230	—	①
	A64,B,TB	Cummins	NH-230	—	①
	CK64B	Cummins	NH-230	—	①
	DC7366	Cummins	NH-230	—	①
	DC87T	Cummins	NH-230	—	①
	DC9364B	Cummins	NH-230	—	①
	DC9964	Cummins	NH-230	—	①
	DC9966	Cummins	NH-230	—	①
	DC10364B	Cummins	NH-230	—	①
	DC10366	Cummins	NH-230	—	①
	KK9364B	Cummins	NH-230	—	①
	S42,B,TB	Cummins	NH-230	—	①
	S64,B,TB	Cummins	NH-230	—	①
1978	C9364U	Cummins	NTC-230	—	①
	DC87TU	Cummins	NTC-230	—	①
	DC9364U	Cummins	NTC-230	—	①
	DC9964U	Cummins	NTC-230	—	①
	DC10364U	Cummins	NTC-230	—	①
	KS42	Cummins	NTC-230	—	①
1979–81	A42TB	Cummins	NTC-230	—	①
	A64TB	Cummins	NTC-230	—	①
	C9364B	Cummins	NTC-230	—	①
	C9364U	Cummins	NTC-230	—	①
	DC7366	Cummins	NTC-230	—	①
	DC87T	Cummins	NTC-230	—	①
	DC87TU	Cummins	NTC-230	—	①
	DC9364B	Cummins	NTC-230	—	①
	DC9364U	Cummins	NTC-230	—	①
	DC9964	Cummins	NTC-230	—	①
	DC9964U	Cummins	NTC-230	—	①

Continued

MODEL INDEX & ENGINE APPLICATION—Continued

ENGINE IDENTIFICATION—Autocar Service Parts Identification Plate is located inside of truck cab. Stamped on the plate is the model of the engine, transmission, auxiliary transmission, rear axle and front axle. This data represents only the equipment on the vehicle when it was shipped from the factory. Prior to Autocar's assimilation by The White Motor Company, identification plates were attached to engines and transmissions.

Year	Model	Engine Make	Base Engine Model ②	Crankcase Refill Capacity, Qts.	Cooling System Capacity, Qts.
	DC9966	Cummins	NTC-230	—	①
	DC10364B	Cummins	NTC-230	—	①
	DC10364U	Cummins	NTC-230	—	①
	DC10366	Cummins	NTC-230	—	①
	KM42,64③	Cummins	NTC-230	—	①
	KM42,64④	Cummins	Formula 290	—	①
	KS42,64③	Cummins	NTC-230	—	①
	KS42,64④	Cummins	Formula 290	—	①
	KL64③	Cummins	NTC-230	—	①
	KL64④	Cummins	Formula 290	—	①
	S42TB	Cummins	NTC-230	—	①
	S64TB	Cummins	NTC-230	—	①
1982	DC42B	Cummins	NTC-300	—	①
	DC44	Cummins	NTC-300	—	①
	DC64B	Cummins	NTC-300	—	①
	DC66	Cummins	NTC-300	—	①
	KS42,64	Cummins	Formula 300	—	①
	KM42,64	Cummins	Formula 300	—	①
	S42TB	Cummins	NTC-300	—	①
	S64TB	Cummins	NTC-300	—	①

①—Cooling system capacity varies depending on engine and radiator installed.

②—Optional engines available on some models.

③—1979 models.

④—Except 1979 models.

WHEEL ALIGNMENT SPECIFICATIONS

Front Axle Model	Caster, Deg.	Camber, Deg.	Toe-In, Inch	Kingpin Inclination, Deg.
DCAS	①	1	3/32	0
FD-900, FE-900	①	1	3/32	5 1/2
FE-18	①	1	3/32	0
FF921	①	1	3/32	5 1/2
FG-900	①	0	3/32	8
FU-900	①	1	3/32	0
FH-901, FL-901	①	1	3/32	5 1/2
FCA	①	1	3/32	0

Front Axle Model	Caster, Deg.	Camber, Deg.	Toe-In, Inch	Kingpin Inclination, Deg.
F3-140	①	②	3/32	③
F3W-120, 140	①	②	3/32	③
F233, F235	5	0	1/4	0
FDS-1600	①	0	—	0
F-7900	①	—	—	—
FDS-1800	①	0	3/32	0
27642	①	0	3/32	8

①—Caster angle varies with each chassis due to model, springs and suspension, steering gear, etc.

②—Right side 1/4°. Left side 1/2°.

③—Right side 6 3/4°. Left side 6 1/2°.

STARTING MOTOR SPECIFICATIONS

Unit Part No.	Rotation ①	Brush Spring Tension, Oz.	No Load Test			Torque Test		
			Amperes	Volts	R.P.M.	Amperes	Volts	Torque Ft. Lb.
1107671	C	35	112	10.6	3240	320–385	3.5	—
1108605	C	24–28	75	11.25	6000	615	5.85	29
1108607	C	24–28	75	11.25	6000	615	5.85	29
1108840	C	36–40	115	11.60	7000	570	2.30	20
1108863	C	36–40	100	23.00	8000	525	3.00	32
1108872	C	36–40	100	23.00	8000	525	3.00	32

Continued

STARTING MOTOR SPECIFICATIONS—Continued

Unit Part No.	Rotation ①	Brush Spring Tension, Oz.	No Load Test			Torque Test		
			Amperes	Volts	R.P.M.	Amperes	Volts	Torque Ft. Lb.
1108956	C	36–40	65	11.30	5500	675	4.00	30
1109510	C	36–40	65	11.40	6000	725	5.00	44
1109173	C	36–40	65	11.40	6000	725	5.00	44
1113641	C	—	75–105①	9	6000	—	—	—
1113658	C	—	130–160①	9	6000	—	—	—
1113664	C	—	130–160①	9	6000	—	—	—
1113670	C	—	130–160①	9	6000	—	—	—
1113678	C	—	130–160①	9	6000	—	—	—
1113708	C	35–40	100	23.00	8000	500	3.00	28
1113800	C	35–40	100	23.00	8000	500	3.00	28
1113818	C	80	60–90②	23	7000–10000	500	3.5	26
1113842	C	80	60–90②	23	7000–10000	500	3.5	26
1113851	C	80	60–90②	23	7000–10000	500	3.5	26
1113872	C	80	60–90②	23	7000–10000	500	3.5	26
1113876	C	80	60–90②	23	7000–10000	500	3.5	26
1113882	C	80	60–90②	23	7000–10000	500	3.5	26
1114052	C	—	120–150①	9	3750	—	—	—
1114055	C	—	140–190①	9	5500	—	—	—
1114058	C	80	140–190②	9	5500	700	1.5	15
1114070	C	—	140–190①	9	5500	—	—	—
1114072	C	80	140–190②	9	5500	700	1.5	15
1114077	C	—	120–150①	9	3750	—	—	—
1114078	C	—	120–150①	9	3750	—	—	—
1114082	C	80	90–130②	11.5	7600	600	2.5	16.5
1114086	C	80	140–190②	9	5500	700	1.5	15
1114087	C	80	115–170②	11.0	9500	700	1.5	15
1114088	C	80	140–190②	9	5500	700	1.5	15
1114096	C	—	120–150①	9	3750	—	—	—
1114098	C	80	115–170②	11.0	9500	700	1.5	15
1114110	C	—	140–190①	9	5500	—	—	—
1114128	C	—	140–190①	9	5500	—	—	—
1114129	C	—	140–190①	9	5500	—	—	—
1114132	C	80	115–170②	11.0	9500	700	1.5	15
1114135	C	—	120–150①	9	3750	—	—	—
1114137	C	—	120–150①	9	3750	—	—	—
1114143	C	—	140–190①	9	5500	—	—	—
1114144	C	—	140–190①	9	5500	—	—	—
1114145	C	—	140–190①	9	5500	—	—	—
1114154	C	—	120–150①	9	3750	—	—	—
1114157	C	—	140–190①	9	5500	—	—	—
1114158	C	—	140–190①	9	5500	—	—	—
1114159	C	—	140–190①	9	5500	—	—	—
1114161	C	—	140–190①	9	5500	—	—	—
1114163	C	—	140–190①	9	5500	—	—	—
1114193	C	—	140–190①	9	5500	—	—	—
1114720	C	—	140–190①	9	5500	—	—	—

①—As viewed from drive end: C—Clockwise. CC—Counterclockwise. ②—Includes solenoid.

BROCKWAY

INDEX OF SERVICE OPERATIONS

MODEL INDEX & ENGINE APPLICATION

ENGINE IDENTIFICATION—Engine model is stamped on plate attached to engine. Transmissions also have identification plates attached to them which gives the transmission manufacturers name and model of unit.

Truck Model	Standard Engine			Crank-case Refill Capacity Qts.	Cooling System Capacity Qts.
	Make	Mfg's. Model	Brockway Model		
TRACTOR MODELS					
147T	Continental	M6363	41BD	7	31
155T	Continental	B6427	42BXD	8	32
158T	Continental	L478	44BD	—	36
C158T	Cummins	C-160	C-160	—	30
257T	Continental	R6572	48BD	14	44
N257T	Cummins	NH-220	NH-220	—	40
258T	Continental	R6572	48FD	14	40
N258T	Cummins	NH-220	NH-220	—	42
260T	Continental	R6572	48BD	14	40
N260T	Cummins	NH-220	NH-220	—	42
C358T	Cummins	C-180	—	—	—
J358T	Cummins	V555	—	—	—
T358T	Detroit	V8185	—	—	—
N358T	Cummins	NH-230	—	—	—
N359T	Cummins	NH-230	—	—	—
E359T	Detroit	671N	—	—	—
K359T	Detroit	8V71N	—	—	—
V359T	Cummins	V903	—	—	—
CHASSIS & CAB MODELS					
128L	Continental	M6330	40BD	7	31
147L	Continental	M6363	41BD	7	31
148L	Continental	B6427	42BXD	8	32
158L	Continental	L478	44BD		36
260L	Continental	R6572	48BD	14	44
776LL	CAT	3208	—	—	—

Truck Model	Standard Engine			Crank-case Refill Capacity Qts.	Cooling System Capacity Qts.
	Make	Mfg's. Model	Brockway Model		
C158L	Cummins	C-160	C-160	—	30
J358L	Cummins	V555	—	—	—
N260L	Cummins	NH-220	NH-220	—	42
C358L	Cummins	C-180	—	—	—
T358L	Detroit	V8185	—	—	—
N358L	Cummins	NH-230	—	—	—
N359L	Cummins	NH-230	—	—	—
E359L	Detroit	671N	—	—	—
K359L	Detroit	8V71N	—	—	—
V359L	Cummins	V903	—	—	—
N360L	Detroit	NH-230	—	—	—
E360L	Detroit	NH-130	—	—	—
F360L	CAT	1674	—	—	—
K360L	Detroit	8V71N	—	—	—
V360L	Cummins	V903	—	—	—
N361L	Detroit	NH-230	—	—	—
E361L	Detroit	671N	—	—	—
F361L	CAT	1674	—	—	—
K361L	Detroit	8V71N	—	—	—
V361L	Cummins	V903	—	—	—
E527L	Detroit	671N	—	—	—
J527L	Cummins	V555	—	—	—
N527L	Cummins	NH230	—	—	—
T527L	Cummins	V8-210	—	—	—
N360T	Detroit	NH-230	—	—	—
E360T	Detroit	671N	—	—	—

Continued

MODEL INDEX & ENGINE APPLICATION—Continued

Truck Model	Standard Engine Make	Standard Engine Mfg's. Model	Standard Engine Brockway Model	Crank-case Refill Capacity Qts.	Cooling System Capacity Qts.
F360T	CAT	1674	—	—	—
K360T	Detroit	8V71N	—	—	—
V360T	Cummins	V903	—	—	—
N361T	Cummins	NH-230	—	—	—
E361T	Detroit	671N	—	—	—
F361T	CAT	1674	—	—	—
K361T	Detroit	8V71N	—	—	—
V361T	Cummins	V903	—	—	—
N457T	Cummins	NH-230	—	—	—
E457T	Detroit	671N	—	—	—
F457T	CAT	1674	—	—	—
K459T	Detroit	8V71N	—	—	—
V459T	Cummins	V903	—	—	—
E527T	Detroit	671N	—	—	—
J527T	Cummins	V555	—	—	—
N527T	Cummins	NH230	—	—	—
T527T	Cummins	V8-210	—	—	—
158LQ	Continental	L478	44BD	—	36
C158LL	Cummins	C-160	C-160	—	30
C158LQ	Cummins	C-160	C-160	—	30
260LF	Continental	R6572	48BD	14	44
N260LF	Cummins	NH-220	NH-220	—	42
260LQ	Continental	R6572	48BD	14	44
N260LQ	Cummins	NH-220	NH-220	—	42
C358LL	Cummins	C-180	—	—	—
J358LL	Cummins	V555	—	—	—
N358TL	Cummins	NH-230	—	—	—
N358LL	Cummins	NH-230	—	—	—
T358LL	Detroit	V8185	—	—	—
E359LL	Cummins	NH-230	—	—	—
E359TL	Detroit	671N	—	—	—
K359LL	Detroit	8V71N	—	—	—
K359TL	Detroit	8V71N	—	—	—
N359LL	Cummins	NH-230	—	—	—
N359TL	Cummins	NH-230	—	—	—
V359LL	Cummins	V903	—	—	—
V359TL	Cummins	V903	—	—	—
E360LL	Detroit	671N	—	—	—
E360TL	Detroit	671N	—	—	—
F360LL	CAT	1674	—	—	—
F360TL	CAT	1674	—	—	—
H360TL	Detroit	12V71N	—	—	—
K360LL	Detroit	8V71N	—	—	—
K360TL	Detroit	8V71N	—	—	—
N360LL	Cummins	NH-230	—	—	—
N360TL	Cummins	NH-230	—	—	—
V360LL	Cummins	V903	—	—	—
V360TL	Cummins	V903	—	—	—
A550L, T	Detroit	6V-53N	—	—	—
B550L, T	CAT	3306	—	—	—
E550L, T	Detroit	671N	—	—	—
J550L	Cummins	V555	—	—	—
N550L, T	Cummins	NTC230	—	—	—
A758L, T	Detroit	6V53N	—	—	—
N758L, T	Cummins	NTC230	—	—	—
E759L, T	Detroit	671N	—	—	—
N759L, T	Cummins	NTC230	—	—	—
E761L, T	Detroit	671N	—	—	—
F761L, T	CAT	1674	—	—	—
K761T	Detroit	8V71N	—	—	—
N761L, T	Cummins	NTC230	—	—	—
V761L, T	Cummins	VT903	—	—	—
A776L, T	Detroit	6V53N	—	—	—
C776L, T	CAT	3208	—	—	—
E776L, T	Detroit	671N	—	—	—
L776L, T	Detroit	6V92T	—	—	—
N776L, T	Cummins	NTC230	—	—	—

TANDEM AXLE MODELS

Truck Model	Standard Engine Make	Standard Engine Mfg's. Model	Standard Engine Brockway Model	Crank-case Refill Capacity Qts.	Cooling System Capacity Qts.
148LL	Continental	B6427	42BXD	8	32
158LL	Continental	L478	44BD	—	36
T527LL	Cummins	V8-210	—	—	—
A550LL	Detroit	6V-53N	—	—	—
B550LL, TL	CAT	3306	—	—	—
E550LL, TL	Detroit	671N	—	—	—
J550LL	Cummins	V555	—	—	—
N550LL	Cummins	NTC230	—	—	—
N550TL	Cummins	NTC250	—	—	—
A758LL	Detroit	6V53N	—	—	—
N758LL	Cummins	NTC230	—	—	—
N758TL	Cummins	NTC250	—	—	—
E759LL, TL	Detroit	671N	—	—	—
K759LL, TL	Detroit	8V71N	—	—	—
N759LL	Detroit	NTC230	—	—	—
N759TL	Cummins	NTC250	—	—	—
V759LL, TL	Cummins	VT903	—	—	—
F760LL, TL	CAT	1674	—	—	—
K760TL	Detroit	8V71N	—	—	—
N760LL	Cummins	NTC230	—	—	—
N760TL	Cummins	NTC250	—	—	—
V760TL	Cummins	VT903	—	—	—
U360LL	Detroit	12V71N	—	—	—
U360TL	Detroit	12V71N	—	—	—
E361LL	Detroit	671N	—	—	—
E361TL	Detroit	671N	—	—	—
F361LL	CAT	1674	—	—	—
F361TL	CAT	1674	—	—	—
K361LL	Detroit	8V71N	—	—	—
K361TL	Detroit	8V71N	—	—	—
N361LL	Cummins	NH-230	—	—	—
N361TL	Cummins	NH-230	—	—	—
V361LL	Cummins	V903	—	—	—
V361TL	Cummins	V903	—	—	—
E457TL	Detroit	671N	—	—	—
F457TL	CAT	1674	—	—	—
H457TL	CAT	1693TA	—	—	—
N457TL	Cummins	NH-230	—	—	—
K459TL	Detroit	8V71N	—	—	—
U459TL	Detroit	12V71N	—	—	—
V459TL	Cummins	V903	—	—	—

Continued

BROCKWAY

MODEL INDEX & ENGINE APPLICATION—Continued

Truck Model	Standard Engine			Crank-case Refill Capacity Qts.	Cooling System Capacity Qts.
	Make	Mfg's. Model	Brockway Model		
E527TL	Detroit	671N	—	—	—
J527LL	Cummins	V555	—	—	—
N527LL	Cummins	NH230	—	—	—
N527TL	Cummins	Sup. 250	—	—	—
E761LL, TL	Detroit	671N	—	—	—
F761LL, TL	CAT	1674	—	—	—
K761LL, TL	Detroit	8V71N	—	—	—
N761LL	Cummins	NTC230	—	—	—
N761TL	Cummins	NTC250	—	—	—
V761LL, TL	Cummins	VT903	—	—	—
H762LL, TL	CAT	1693TA	—	—	—
U762LL, TL	Detroit	12V71N	—	—	—
A776, LL	Detroit	6V53N	—	—	—
E776LL, TL	Detroit	671N	—	—	—
L776LL, TL	Detroit	6V92T	—	—	—
N776LL	Cummins	NTC230	—	—	—
N776TL	Cummins	NTC250	—	—	—
OFF-HIGHWAY MODELS					
260LFOH	Continental	R6572	48BD	14	44
260LROH	Continental	R6572	48BD	14	44
N260LROH	Cummins	NH-220	NH-220	H	42
N260LFOH	Cummins	NH-220	NH-220	—	42

GENERAL ENGINE SPECIFICATIONS

Engine Model	No. Cyls. & Valve Location	Bore & Stroke	Piston Displacement, Cu. In.	Compression Ratio①	Maximum Brake H. P. @ R. P. M.①	Maximum Torque, Ft. Lb. @ R. P. M.①	Normal Oil Pressure, Lbs.
38B	6 In Block	$3\frac{3}{4} \times 4\frac{3}{8}$	290	6.3	104 @ 3200	212 @ 1250	60
41BD	6 In Block	$4 \times 4\frac{13}{16}$	363	6.7	150 @ 3200	293 @ 1600	60
42BD	6 In Block	$4\frac{5}{16} \times 4\frac{7}{8}$	427	6.4	165 @ 3000	352 @ 1600	60
42BX	6 In Block	$4\frac{5}{16} \times 4\frac{7}{8}$	427	6.7	142 @ 3000	332 @ 1200	60
46B	6 In Head	$4\frac{1}{2} \times 5\frac{3}{8}$	513	5.9	180 @ 2800	400 @ 1300	60
46BD	6 In Head	$4\frac{1}{2} \times 5\frac{3}{8}$	513	6.2	220 @ 2800	412 @ 1400	60
46FD	6 In Head	$4\frac{1}{2} \times 5\frac{3}{8}$	513	6.2	220 @ 2800	412 @ 1400	60
48BD	6 In Head	$4\frac{3}{4} \times 5\frac{3}{8}$	572	6.3	230 @ 2800	484 @ 1400	60
48FD	6 In Head	$4\frac{3}{4} \times 5\frac{3}{8}$	572	6.3	230 @ 2800	484 @ 1400	60
49BD	6 In Head	$4\frac{7}{8} \times 5\frac{3}{8}$	602	6.3	232 @ 2800	482 @ 1200	60

①—Current production.

TUNE UP & VALVE SPECIFICATIONS

Engine Model	Firing Order	Spark Plugs		Ignition Timing		Cylinder Head Torque, Ft. Lb.	Valve Seat Angle, Degrees	Valve Clearance H—Hot C—Cold		Valve Spring Pressure Lb. @ In. Length
		Type ①	Gap	Timing Mark	Location			Intake	Exhaust	
38B	153624	D14	.025	1/2″ BTDC	Flywheel	70–75	②	.022C	.024C	119 @ 1½
40BD	153624	D10	.025	1/2″ BTDC	Flywheel	70–75	②	.020H	.022H	119 @ 1½
41BD	153624	D14	.025	5/16″ BTDC	Flywheel	70–75	②	.016H	.024H	119 @ 1½
42BX	153624	D10	.030	5/8″ BTDC	Flywheel	70–75	②	.016H	.022H	144 @ 1.316
42BD	153624	D10	.025	3/16″ BTDC	Flywheel	70–75	②	.016H	.024H	144 @ 1.316
46B	153624	D10	.025	1″ BTDC	Flywheel	85–95	②	.020H	.024H	①
46BD	153624	D10	.025	5/8″ BTDC	Flywheel	85–95	②	.023C	.033C	①
46FD	153624	D10	.025	5/8″ BTDC	Flywheel	85–95	②	.023C	.030H	①
48BD③	153624	D10	.025	1/2″ BTDC	Flywheel	85–95	②	.020H	.030H	①
48FD③	153624	D10	.025	1/2″ BTDC	Flywheel	85–95	②	.020H	.030H	①
48BD④	153624	D10	.020	TDC	Flywheel	85–95	②	.023C	.033C	①
48FD④	153624	D10	.020	TDC	Flywheel	85–95	②	.023C	.033C	①
49BD	153624	D10	.025	1/2″ BTDC	Flywheel	85–95	②	.020H	.030H	①

①—Champion.
②—Intake 30°, exhaust 45°.
③—Gasoline.
④—LPG.

DISTRIBUTOR SPECIFICATIONS

Unit Part No. ①	Rotation ②	Cam Angle, Degrees	Breaker Gap, In.	Condenser Capacity, Mfds.	Breaker Arm Spring Tension, Oz.	Centrifugal Advance Deg. @ R.P.M. of Dist.		Vacuum Advance		
						Advance Starts	Full Advance	Inches of Vacuum to Start Plunger Movement	Inches of Vacuum for Full Plunger Travel	Maximum Vacuum Advance, Dist. Degrees
1GC-5001-B	CC	39	.020	.25–.28	17–20	1 @ 275	15 @ 1200	None	None	None
1GC-6002-1AE	CC	39	.020	.25–.28	17–20	1 @ 400	9 @ 1250	None	None	None
1GC-6002-1AF	CC	39	.020	.25–.28	17–20	1 @ 475	6 @ 1425	None	None	None
1GC-6002-1AG	CC	39	.020	.25–.28	17–20	1 @ 425	8 @ 1425	None	None	None
1GC-6002-1E	CC	39	.020	.20–.25	17–20	1 @ 390	9 @ 1250	None	None	None
1GC-6002-1R	CC	39	.020	.20–.25	17–20	1 @ 400	8 @ 1420	None	None	None
1GC-6002-1S	CC	39	.020	.20–.25	17–20	1 @ 475	6 @ 1420	None	None	None
1GC-6010-1B	CC	39	.020	.20–.25	17–20	1 @ 465	6 @ 1415	None	None	None
1GC-6010-1C	CC	39	.020	.20–.25	17–20	1 @ 285	14 @ 1600	None	None	None
1GC-6010-1D	CC	39	.020	.25–.28	17–20	1 @ 300	14 @ 1600	None	None	None
1GC-6010-1E	CC	39	.020	.25–.28	17–20	1 @ 325	12 @ 1400	None	None	None

①—Distributor number stamped on plate riveted to housing.

②—As viewed from top, CC—Counterclockwise.

PISTON, PIN, RING, CRANKSHAFT & BEARING SPECIFICATIONS

Engine Model	Wristpin Diameter	Piston Clearance, Inch②	Ring End Gap, In. (Minimum)		Crank Pin Diameter, Inch	Rod Bearing Clearance, In.	Rod Bolt Torque, Lb. Ft.	Main Bearing Journal Diameter, In.	Main Bearing Clearance, Inch	Main Bolt Torque, Lb. Ft.
			Comp.	Oil						
38B	1.109	.005	.008	.008	2.248–2.249	.001–.0028	70–75	2.623–2.624	.0015–.0037	①
40BD	1.109	.005	.008	.008	2.248–2.249	.001–.0028	70–75	2.624–2.625	.0015–.0037	①
41BD	1.109	.002	.008	.008	2.248–2.249	.001–.0028	70–75	2.624–2.625	.0015–.0037	①
42BD	1.250	.003	.008	.008	2.498–2.499	.001–.003	100–110	2.873–2.874	.0015–.0037	100–110
42BX	1.250	.005	.008	.008	2.498–2.499	.001–.003	100–110	2.873–2.874	.0015–.0037	100–110
46B	1.500	.005	.013	.013	2.999–3.000	.001–.003	85–95	3.249–3.250	.0022–.004	85–95
46BD	1.500	.005	.013	.013	2.999–3.000	.001–.003	85–95	3.249–3.250	.0022–.004	85–95
46FD	1.500	.005	.013	.013	2.999–3.000	.001–.003	85–95	3.249–3.250	.0022–.004	85–95
48BD	1.500	.005	.013	.013	2.999–3.000	.001–.003	85–95	3.249–3.250	.0022–.004	85–95
48FD	1.500	.005	.013	.013	2.999–3.000	.001–.003	85–95	3.249–3.250	.0022–.004	85–95
49BD	1.500	.005	.013	.013	2.999–3.000	.001–.003	85–95	3.249–3.250	.0022–.004	85–95

①—1/2″ 100–110, 9/16″ 130–140.

②—Pistons removed from above.

STARTING MOTOR SPECIFICATIONS

Unit Part No. ①	Rotation ②	Brush Spring Tension, Oz.	No Load Test			Torque Test		
			Amperes	Volts	R.P.M.	Amperes	Volts	Torque Lb. Ft.
MAB-4071	C	42–53	60	5	3600	390	2	9.2
MAS-4102	C	42–53	40	10	3900	340	4	13.2
MAW-4042	C	42–53	65	5	4300	335	2	6.0
ML-4301	C	38–49	65	5	2250	370	2	10.3
ML-4303	C	42–53	65	5	2250	370	2	10.3
ML-4318	C	42–53	65	5	2250	370	2	10.3

①—Starter number stamped on plate riveted to housing.

②—As viewed from drive end; C—Clockwise.

WHEEL ALIGNMENT

Axle Model	Caster	Camber	Toe-In	Axle Model	Caster	Camber	Toe-In
32350①	1	1	1/16–1/8	FE-10A-43②	2	1	1/16–1/8
FD-901①	1	1	1/16–1/8	FG-900③	2	1	1/16–1/8

①—Kingpin angle 5½°.
②—Kingpin angle 0°.

3
Ⓚingpin angle 8°.

CHEVROLET & GMC
Exc. S/T 10 & 15

INDEX OF SERVICE OPERATIONS

CHEVROLET TRUCK MODELS & ENGINE APPLICATION

IDENTIFICATION PLATE: Located inside driver's compartment. Engine identification and other unit data on plate represents only the equipment in the vehicle when shipped from the factory.

Models	Year	Engine Make	Standard Engine Model	Crankcase Refill Capacity, Qts.	Cooling System Capacity, Qts.
BLAZER					
6 Cyl.	1974–82	Own	6-250	4	15
V8-305	1977–79	Own	V8-305	4	18
V8-350	1974–79	Own	V8-350	4	18
V8-305	1983	Own	V8-305	4	—
V8-350	1983	Own	V8-350	4	—
EL CAMINO					
1AC80, AD80	1974–77	Own	6-250	4	12½
1AC80, AD80	1976	Own	V8-305	4	18
1AC80, AD80	1974–77	Own	V8-350	4	18
1AW80, Z15	1978–79	Own	V6-200	4	15
1AW80	1980–82	Own	V6-229	4	18.5
1AW80	1980–81	Own	V8-305	4	19
1AW80, Z15	1978–79	Own	V8-305, 350	4	18
1AW80	1979–82	Buick	V6-231	4	15.5
1AW80	1979–81	Own	V8-267	4	16.7
1AW80	1983	Own	V6-229	4	15
1AW80	1983	Buick	V6-231	4	12
CONVENTIONAL GASOLINE					
C10	1974–83	Own	6-250	4	15
C10	1980–83	Own	V8-305	4	17.5
C10	1974–79	Own	V8-350	4	18
C10	1980–83	Own	V8-350	4	17.5
C20	1974–83	Own	6-250	4	15
C20	1975–83	Own	6-292	5	15
C20	1977–79	Own	V8-305	4	18
C20	1974–81	Own	V8-350	4	18
C30	1974–83	Own	6-250	4	14.6
C30	1974–79	Own	6-250	4	15
C30	1975–83	Own	6-292	5	15
C30	1974–81	Own	V8-350	4	18
C5D042	1979–83	Own	V8-350	5	21
C6D042	1979–83	Own	V8-350	5	21
C6D062①	1979–83	Own	V8-366	6	33
C7D042	1979–83	Own	V8-366	6	33
C7D064②	1979–83	Own	V8-366	6	33
CS50	1974	Own	6-250	4	16
CS50	1975–78	Own	6-292	5	17
CE50	1974–78	Own	V8-350	5	21
CE60	1974–78	Own	V8-350	5	21
CM60	1974	Own	V6-305	8	36
CS60	1974–78	Own	6-292	5	17
HM80	1974	Own	V6-432	10	37
HE80	1975–77	Own	V8-427	6	37
J8C042	1979–80	Own	V8-427	8	36.5
J8C064②	1979–80	Own	V8-427	8	36.5
CE65	1974–77	Own	V8-366	6	36
CM65	1974	Own	V6-379	10	34.5

①—6 × 2 Tandem Series
②—6 × 4 Tandem Series

Models	Year	Engine Make	Standard Engine Model	Crankcase Refill Capacity, Qts.	Cooling System Capacity, Qts.
FOUR-WHEEL DRIVE					
K10	1974–83	Own	6-250	4	15
K10	1974–81	Own	V8-350	4	18
K10	1978	Own	V8-305	4	18
K20	1974	Own	6-250	4	15
K20	1975–83	Own	6-292	5	15
K20	1974–76	Own	V8-350	4	18
K20	1980–83	Own	V8-350	4	17.5
K20	1978–79	Own	V8-305	4	18
K30	1977–83	Own	6-292	5	15
TANDEM SERIES					
ME65	1974–79	Own	V8-366	6	33
MM65	1974	Own	V6-432	10	36
JM80	1974	Own	V6-432	10	37
JE80	1975–79	Own	V8-427	6	37
STEP-VAN & FORWARD CONTROL SERIES					
G10	1974–83	Own	6-250	4	17
G10	1974–76	Own	V8-350	4	18
P10	1974	Own	6-250	4	13
P10	1975–80	Own	6-292	5	13.6
G20	1974	Own	6-250	4	13
G20	1975–78	Own	6-292	5	14.8
G20	1979–83	Own	6-250	4	17
G20	1974–76	Own	V8-350	4	18
P20	1974	Own	6-250	4	13
P20	1975–83	Own	6-292	5	13.6
P20	1974–76	Own	V8-350	4	17
G30	1974	Own	6-250	4	13
G30	1975–78	Own	6-292	5	14.8
G30	1974–76	Own	V8-350	4	18
G30	1978–81	Own	V8-350	4	18
G30	1979	Own	6-250	4	17
P30	1974	Own	6-250	4	13
P30①	1975–83	Own	6-292	5	13.6
P30	1974–76	Own	V8-350	4	17
P30②③	1977–82	Own	V8-350	4	19
P30②④	1977–82	Own	V8-454	6	25
P30②	1983	Own	V8-454	6	24.5
P4T042	1979–83	Own	6-292	4	17
P6T042	1983	Own	V8-350	5	—

①—Except Motor Home Chassis
②—Motor Home Chassis
③—125″, 137″ and 158″ wheelbase
④—178″ wheelbase

Models	Year	Engine Make	Standard Engine Model	Crankcase Refill Capacity, Qts.	Cooling System Capacity, Qts.
TILT CAB SERIES					
TE60	1974–79	Own	V8-350	5	28.1
TE65	1974–79	Own	V8-366	6	35.2
TM60	1974	Own	V6-305	8	46.6
TM65	1974	Own	V6-379	8	45
TM80	1974	Own	V6-432	10	47

Continued

CHEVROLET TRUCK MODELS & ENGINE APPLICATION—Continued

Models	Year	Engine Make	Standard Engine Model	Crankcase Refill Capacity, Qts.	Cooling System Capacity, Qts.
TILT CAB SERIES—Cont'd					
W6N042	1979–80	Own	V8-350	5	28
W7N042	1979–80	Own	V8-366	6	35
WM80	1974	Own	V6-432	10	47
SCHOOL BUS SERIES					
B6P042	1979–83	Own	V8-350	5	21
SS60	1974	Own	6-292	5	17
SM60	1974	Own	V6-305	8	36
SE60	1974–79	Own	V8-350	5	20
SG60	1974	GMC	DH478	10	29
CONVENTIONAL DIESELS					
CG60	1974	GMC	DH478	10	29
HV70	1974–78	Detroit	6V-53N	14	36.5
HY70	1975–77	Catpl'r	3208	12	50½
HY70	1978	Catpl'r	3208	10	44
JV70	1974–78	Detroit	6V-53N	14	36.5
JY70	1975–77	Catpl'r	3208	12	50½
JY70	1978	Catpl'r	3208	10	44
HC90	1974–78	Cum'ins	NTC290	28	61.5
HH90	1974	Detroit	8V-71N	24	80
HH90	1975–78	Detroit	8V-71N	27	80
HI90	1974	Detroit	6-71N	19	51.2
HI90	1975–78	Detroit	6-71N	22	51¼
HN90	1974	Cum'ins	NHC250	24	52.9
JC90	1975–78	Cum'ins	NTC-290	28	61½
JH90	1973–74	Detroit	8V-71N	24	55.7
JH90	1975–78	Detroit	8V-71N	27	80
JI90	1974–78	Detroit	6-71N	22	51¼
JJ90	1977	Detroit	6V-92TT	20	57
J9C042	1979–80	Detroit	8V-71N	22	60
J9C064①	1979–80	Detroit	8V-71N	22	60
MH90	1975–76	Detroit	8V-71N	27	73
NI-9005	1977–78	Detroit	6-71N	22	50
NH-9005	1977–78	Detroit	8V-71N	23	92
NJ-9005	1977–78	Detroit	6V-92N	18	80
NC-9005	1977–78	Cum'ins	NTC290	28	72
N9E042	1979–80	Detroit	8V-71N	23	92
N9E064①	1979–80	Detroit	8V-71N	23	92

Models	Year	Engine Make	Standard Engine Model	Crankcase Refill Capacity, Qts.	Cooling System Capacity, Qts.
N9F042	1979–80	Detroit	8V-71N	23	92
N9F064①	1979–80	Detroit	8V-71N	23	92
MH-9005	1977–78	Detroit	8V-71N	23	92
MC-9005	1977–78	Cum'ins	NTC290	28	72
MK-9005	1977–78	Detroit	8V-92N	27	92

①—Tandem Series

Models	Year	Engine Make	Standard Engine Model	Crankcase Refill Capacity, Qts.	Cooling System Capacity, Qts.
TILT DIESELS					
TG60	1974	GMC	DH478	11	38.3
TV70	1974	Detroit	6V-53N	14	31.5
WV70	1974	Detroit	6V-53N	16	31.5
FB90	1974	Cum'ins	V8-903	26	61
FI90	1974	Detroit	6-71N	19	51.2
FI90	1975–78	Detroit	6-71N	22	51¼
FJ90	1975–78	Detroit	6V-92	18	57
FC90	1974–78	Cum'ins	NTC-290	28	55
FH90	1974	Detroit	8V-71N	24	73
FH90	1975–78	Detroit	8V-71N	27	73
FN90	1974	Cum'ins	NHC-250	20	53
DB90	1974	Cum'ins	V8-903	26	61
DB90	1975–76	Cum'ins	VT-903	18	62
DC90	1974–78	Cum'ins	NTC-290	28	55
DH90	1974	Detroit	8V-71NE	24	73
DH90	1975–77	Detroit	8V-71N	27	73
DI90	1974	Detroit	6-71N	19	51.2
DI90	1975–78	Detroit	6-71N	22	51
DJ90	1975–78	Detroit	6V-92	18	57
DK90	1975–78	Detroit	8V-92	27	74
DN90	1974	Cum'ins	NHC-250	20	53
DP90	1974	Detroit	12V-71	38	98
DP90	1975–78	Detroit	12V-71N	34	98
DL90	1977–78	Cum'ins	KT450	52	71
DR90	1977–78	Catpl'r	3406	34	69
D9K042	1979–80	Detroit	8V-71N	23	72
D9K064①	1979–80	Detroit	8V-71N	23	72
D9L042	1979–80	Detroit	8V-71N	23	72
D9L064①	1979–80	Detroit	8V-71N	23	72

①—Tandem Series

GMC MODEL INDEX & ENGINE APPLICATION

ENGINE IDENTIFICATION—The GMC Service Parts Identification Plate or decal is located inside the cab or on the truck firewall. Printed on the plate or decal is the model of the engine, transmission, auxiliary transmission, service brakes, front and rear axles. This information represents only the equipment on the vehicle when it was shipped from the factory.

Model	Year	Engine Make	Standard Engine Model	Crankcase Refill Capacity, Qts.	Cooling System Capacity, Qts.
SPRINT					
Sprint	1974–77	Chevrolet	6-250	4	12
Sprint	1974–75	Chevrolet	V8-350	4	15
Sprint	1976	Chevrolet	V8-305	4	17.2
CABALLERO					
Caballero	1978–79	Chevrolet	V6-200	4	18.5
Caballero	1978–82	Buick	V6-231	4	15.5
Caballero	1980–82	Chevrolet	V6-229	4	18.5
Caballero	1983	Chevrolet	V6-229	4	15
Caballero	1983	Buick	V6-231	4	12
2 AXLE CONVENTIONAL					
C-1500	1974–83	Chevrolet	6-250	4	15
C-1500	1979–83	Chevrolet	V8-305	4	17.5
C-1500	1974–79	Chevrolet	V8-350	4	18
C-1500	1980–83	Chevrolet	V8-350	4	17.5
K-1500	1974	Chevrolet	6-250	4	13
K-1500	1975–83	Chevrolet	6-250	4	15
K-1500	1974–76	Chevrolet	V8-350	4	18
K-1500	1977–78	Chevrolet	V8-305	4	18
K-1500	1979–83	Chevrolet	V8-305	4	17.5
K-1500	1979–83	Chevrolet	V8-350	4	17.5
C-2500	1979–83	Chevrolet	6-250	4	14.8
C-2500	1979–82	Chevrolet	V8-350	4	17.5
C-2500	1977–78	Chevrolet	V8-305	4	18
C-2500	1975–83	Chevrolet	6-292	5	15
C-2500	1974–76	Chevrolet	V8-350	4	18
C-2500	1974	Chevrolet	6-250	4	15
C-2500	1974	Chevrolet	V8-454	4	26
K-2500	1974	Chevrolet	6-250	4	15
K-2500	1975–83	Chevrolet	6-292	5	15
K-2500	1974–76	Chevrolet	V8-350	4	18
K-2500	1977–78	Chevrolet	V8-305	4	18
K-2500	1979	Chevrolet	V8-350	4	18.0
K-2500	1980–83	Chevrolet	V8-350	4	17.5
C-3500	1974	Chevrolet	6-250	4	15
C-3500	1975–83	Chevrolet	6-292	5	15
C-3500	1974–76	Chevrolet	V8-350	4	18
K-3500	1977–83	Chevrolet	6-292	5	15
PS-4500	1977–78	Chevrolet	6-292	5	17
C5DO42	1979–83	Chevrolet	V8-350	5	21
CE-5000	1974–75	Chevrolet	V8-350	4	21
CS-5000	1974	Chevrolet	6-250	4	16
CS-5000	1975	Chevrolet	6-292	5	15
CS-5000	1976–78	Chevrolet	6-292	5	17
CE-5000	1975	Chevrolet	V8-350	5	18
CE-5000	1976–78	Chevrolet	V8-350	5	21
C6D042	1979–83	Chevrolet	V8-350	5	21
CM-6000	1974	Own	V6-379	8	36
CS-6000	1974–78	Chevrolet	6-292	5	17
CE-6000	1974–78	Chevrolet	V8-350	6	21
CD-6000	1977–78	Own	4-53N	14	23
CG-6000	1974	Own	V6-DH478	8	29
CD-6500	1978	Detroit	4-53T	14	23
CM-6500	1974	Own	V6-379	8	34½
CE-6500	1974–78	Chevrolet	V8-366	6	33
CY-6500	1975	Caterpillar	V8-3208	10	44
C7D042	1979–83	Chevrolet	V8-366	6	33
HE-7500	1975–77	Chevrolet	V8-427	8	31.6
HE-7500	1978	Chevrolet	V8-427	8	36½
HM-7500	1974	Own	V6-432	8	37
HV-7500	1974–78	Detroit	6V-53N	14	36½
HY-7500	1975–77	Cat.	3208	10	50
HY-7500	1978	Cat.	3208	10	44
J8C042	1979–81	Chevrolet	V8-427	8	37
J8C042	1982–83	Cat.	3208N	20	40
HC-9500	1974–76	Cummins	NTC-290	28	61½
HC-9500	1977–78	Cummins	NTC-290	36	61½
J9C042	1979–80	Detroit	8V-71N	23	60
J9C042	1981–83	Detroit	6V-92T	20	58
N9E042	1979–80	Detroit	8V-71N	23	92
N9E042	1981	Detroit	6V-92T	20	80
N9E042	1982–83	Detroit	6V-92T	20	60
N9F042	1979–80	Detroit	8V-71N	23	92
N9F042	1981–82	Detroit	8V-71N	23	60
N9F042	1983	Detroit	6V-92T	20	60
3-AXLE CONVENTIONAL					
C6D062	1979–83	Chevrolet	V8-366	6	33
C7D064	1979–83	Chevrolet	V8-366	6	33
C7D064	1979–82	Detroit	V8-500	11	32
HJ-9500	1977–78	Detroit	6V-92TT	18	57
HH-9500	1974–78	Detroit	8V-71NE	27	80
HI-9500	1974	Detroit	6-71N	19	51.2
HI-9500	1975–78	Detroit	6-71N	22	51.2
HN-9500	1974	Cummins	NHC-250	24	53
JE-7500	1975–77	Chevrolet	V8-427	8	31.6
JE-7500	1978	Chevrolet	V8-427	8	31½
JM-7500	1974	Own	V6-432	8	37
JV-7500	1974–78	Detroit	6V-53N	14	36½
JY-7500	1975–78	Cat.	3208	10	7.5
J8C064	1979–81	Chevrolet	V8-427	8	37
J8C064	1982–83	Cat.	V8-3208N	20	40
J9C064	1979–80	Detroit	8V-71N	23	60
J9C064	1981–83	Detroit	6V-92T	20	58
JJ-9500	1977–78	Detroit	6V-92TT	18	57
JH-9500	1974–78	Detroit	8V-71NE	27	80
JI-9500	1974	Detroit	6-71N	19	51
JI-9500	1975–78	Detroit	6-71N	22	51

Continued

GMC MODEL INDEX & ENGINE APPLICATION—Continued

Model	Year	Engine Make	Standard Engine Model	Crankcase Refill Capacity, Qts.	Cooling System Capacity, Qts.
JN-9500	1974	Cummins	NHC-250	24	53
JC-9500	1974–76	Cummins	NTC-290	28	61½
JC-9500	1977–78	Cummins	NTC-290	36	61½
ME-6500	1974–78	Chevrolet	V8-366	8	33.1
MM-6500	1974	Own	V6-432	8	36
MC-9500	1974	Cummins	NTC-350	28	63½
MC-9502	1978	Cummins	NTC-290	36	72
MH-9005	1977	Detroit	8V-71N	27	92
MH-9500	1974–76	Detroit	8V-71NE	27	73
MH-9502	1977–78	Detroit	8V-71N	27	92
MK-9005	1977	Detroit	8V-92	27	92
MK-9502	1977–78	Detroit	8V-92	27	92
N9E064	1977–80	Detroit	8V-71N	23	92
N9E064	1981–83	Detroit	6V-92T	20	60
N9F064	1979–80	Detroit	8V-71N	23	92
N9F064	1981–83	Detroit	6V-92T	20	60
MI-9501	1978	Detroit	6-71N	22	72
NI-9502	1977–78	Detroit	6-71N	22	72
NH-9502	1977–78	Detroit	8V-71N	27	92
NJ-9502	1977–78	Detroit	6V-92N	18	80
NC-9502	1977–78	Cummins	NTC-290	36	72
2-AXLE TILT CAB					
TE-6000	1974–75	Chevrolet	V8-350	6	29
TE-6000	1976–78	Chevrolet	V8-350	6	28.1
TE-6500	1977–78	Chevrolet	V8-366	6	35
TG-6000	1974	Own	V6-DH478	11	38½
TM-6000	1974	Own	V6-305C	8	46½
TM-6500	1974	Own	V6-379	8	45
TE-6500	1974–78	Chevrolet	V8-366	6	35
W6N042	1979–80	Chevrolet	V8-350	5	28
W7N042	1979–82	Chevrolet	V8-366	6	35
TM-7500	1974	Own	V6-432	8	47
TV-7500	1974–75	Detroit	6V-53N	16	31½
3-AXLE TILT CAB					
WM-7500	1974	Own	V6-432	8	47
WV-7500	1974–75	Detroit	6V-53N	16	31½
2-AXLE SET FORWARD TILT CAB					
D9K042	1979–80	Detroit	8V-71N	23	72
D9K042	1981–83	Detroit	6V-92T	20	51
D9L042	1979–80	Detroit	8V-71N	23	72
D9L042	1981–83	Detroit	6V-92T	20	51
FI-9502	1974	Detroit	6-71N	19	51
FI-9502	1975–78	Detroit	6-71N	22	51
FJ-9502	1975–76	Detroit	6V-92	18	73
FJ-9502	1977–78	Cummins	NTC-290	36	55
FH-9502	1974–78	Detroit	8V-71NE	27	73
FN-9502	1974	Cummins	NHC-250	24	53
FC-9502	1974–76	Cummins	NTC-290	28	55
FC-9502	1977–78	Cummins	NTC-290	36	55
FB-9502	1974	Cummins	V-903	18	61
3-AXLE SET FORWARD TILT CAB					
D9K064	1979–80	Detroit	8V-71N	23	72
D9K064	1981–83	Detroit	6V-92T	20	51
D9L064	1979–80	Detroit	8V-71N	23	72
D9L064	1981–83	Detroit	6V-92T	20	51
DI-9502	1974	Detroit	6-71N	19	51
DI-9502	1975–78	Detroit	6-71N	22	51
DH-9502	1974–78	Detroit	8V-71NE	27	73
DN-9502	1974	Cummins	NHC-250	24	53
DC-9502	1974–76	Cummins	NTC-290	28	55
DC-9502	1977–78	Cummins	NTC-290	36	55
DB-9502	1974–78	Cummins	V-903	18	62
DJ-9502	1975–78	Detroit	6V-92	18	57
DK-9502	1975–76	Detroit	8V-92	27	74
DK-9502	1977–78	Detroit	8V-92	27	60
DP-9502	1974–78	Detroit	12V-71	34	98
DL-9502	1977–78	Cummins	KT450	40	211/2
DR-9502	1977–78	Cat.	3406	32	69
UNITIZED FORWARD CONTROL					
G-1500	1979–83	Chevrolet	6-250	4	17
G-1500	1974–78	Chevrolet	6-250	4	15
G-1500	1974–75	Chevrolet	V8-350	4	16
G-1500	1976	Chevrolet	V8-350	4	18
G-2500	1979–83	Chevrolet	6-250	4	17
G-2500	1979	Chevrolet	V8-350	4	20
G-2500	1974–78	Chevrolet	6-250	4	13
G-2500	1975–78	Chevrolet	6-292	5	15
G-2500	1974–75	Chevrolet	V8-350	4	16
G-2500	1976	Chevrolet	V8-350	4	18
G-3500	1979–83	Chevrolet	6-250	4	17
G-3500	1982	Chevrolet	8-350	—	—
G-3500	1979–83	Chevrolet	V8-350	4	20
G-3500	1974–78	Chevrolet	6-250	4	13
G-3500	1975–78	Chevrolet	6-292	5	15
G-3500	1974–75	Chevrolet	V8-350	4	16
G-3500	1976	Chevrolet	V8-350	4	18
TG-31303	1975–83	Chevrolet	V8-350	4	20
TG-31303	1976–78	Chevrolet	6-292	5	15
TG-31332	1979–83	Chevrolet	V8-350	4	20
TG-31603	1975–83	Chevrolet	V8-350	4	20
TG-31632	1979–83	Chevrolet	V8-350	4	20
TG-31603	1976–78	Chevrolet	6-292	5	15
CHASSIS FORWARD CONTROL					
P-1500	1974	Chevrolet	6-250	4	13
P-1500	1975–80	Chevrolet	6-292	5	14
P-2500	1974	Chevrolet	6-250	4	13
P-2500	1979	Chevrolet	V8-350	4	16.8
P-2500	1975–83	Chevrolet	6-292	5	14
P-2500	1974–76	Chevrolet	V8-350	4	17
P-3500	1974	Chevrolet	6-250	4	13
P-3500	1975–83	Chevrolet	6-292	5	14
P-3500	1979–82	Chevrolet	V8-350	4	16.8
P-3500	1974–76	Chevrolet	V8-350	4	17
P-3500	1974–75	Chevrolet	V8-454	4	27
PS-4500	1977	Chevrolet	6-292	5	17
P4T042	1979–83	Chevrolet	6-292	5	17

Continued

GMC MODEL INDEX & ENGINE APPLICATION—Continued

Model	Year	Engine Make	Standard Engine Model	Crankcase Refill Capacity, Qts.	Cooling System Capacity, Qts.
TP30832	1983	Chevrolet	V8-454	6	—
TP-30832	1979–82	Chevrolet	V8-350	4	19
TP-30832	1975–78	Chevrolet	V8-350	4	20
TP31132	1983	Chevrolet	V8-454	6	—
TP-31132	1979–82	Chevrolet	V8-350	4	19
TP-31132	1975–78	Chevrolet	V8-350	4	20
TP-31432	1975–82	Chevrolet	V8-350	4	20
TP-31832	1979–83	Chevrolet	V8-454	6	24.5
TP-31832	1975–78	Chevrolet	V8-454	4	25
SCHOOL BUS CHASSIS					
B6P042	1979–82	Chevrolet	V8-350	5	21
SE-6000	1974–77	Chevrolet	V8-350	6	21

Model	Year	Engine Make	Standard Engine Model	Crankcase Refill Capacity, Qts.	Cooling System Capacity, Qts.
SE-6000	1978	Chevrolet	V8-350	5	25
SG-6000	1974	Own	V6-DH478	11	29
SM-6000	1974	Own	V6-305C	8	36
SS-6000	1974	Chevrolet	6-292	5	16½
MOTORHOME & TRANSMODES					
ZE06083	1974–77	Olds.	V8-455	5	21
ZE06083	1977–78	Olds.	V8-403	5	23½
ZE06581	1974–77	Olds.	V8-455	5	21
ZE06581	1977–78	Olds.	V8-403	5	23½
ZE06583	1974–77	Olds.	V8-455	5	21
ZE06583	1977–78	Olds.	V8-403	5	23½

GENERAL ENGINE SPECIFICATIONS

Year	Engine Model	Carb. Type	Bore & Stroke	Comp. Ratio	Horsepower @ R.P.M. ①	Torque Ft. Lbs. @ R.P.M. ②	Comp. Pressure Cranking Speed	Governed Speed R.P.M. Full Load	Normal Oil Pressure Lbs.
1974	6-250	1 Bore	3.875 × 3.53	8.5	100 @ 3600	175 @ 1600⑬	⑦	3600	40–60
	6-292	1 Bore	3.875 × 4.12	8.0	120 @ 3600	215 @ 2000	⑦	3600	40–60
	6-292	1 Bore	3.875 × 4.12	8.0	130 @ 3600	225 @ 2000	⑦	3600	40–60
	V6-305C	2 Bore	4.25 × 3.58	7.75	148 @ 4000	238 @ 1600	⑦	3400	57
	V8-350	2 Bore	4.00 × 3.48	8.5	145 @ 3600	205 @ 2000	⑦	—	40
	V8-350	4 Bore	4.00 × 3.48	8.5	155 @ 4000	255 @ 2400	⑦	—	40
	V8-350	2 Bore	4.00 × 3.48	8.0	160 @ 4000	265 @ 2400	⑦	4000⑫	40
	V8-366	4 Bore	3.937 × 3.76	8.0	200 @ 4000	310 @ 2800	⑦	4000⑫	40–55
	V6-379	2 Bore	4.56 × 3.86	7.5	170 @ 3600	280 @ 1600	⑦	3400	57
	V8-400	2 Bore	4.125 × 3.75	8.5	150 @ 3200	295 @ 2000	⑦	—	40
	V8-400	4 Bore	4.125 × 3.75	8.5	180 @ 3800	290 @ 2400	⑦	—	40
	V8-427	4 Bore	4.25 × 3.76	8.0	230 @ 4000	360 @ 2800	⑦	4000	40–55
	V6-432	2 Bore	4.87 × 3.86	7.5	190 @ 3200	331 @ 1600	⑦	3200	60
	V8-454	4 Bore	4.25 × 4.00	8.25	240 @ 4000	355 @ 2800	⑦	—	40
	V8-455	4 Bore	4.125 × 4.250	8.5	265 @ 4300	375 @ 2800	⑦	—	—
	V6-478M	2 Bore	5.125 × 3.86	7.0	192 @ 3200	371 @ 1400	⑦	3200	60
1975	6-250	1 Bore	3.875 × 3.53	8.25	105 @ 3800	185 @ 1200	130	—	40–60
	6-292	1 Bore	3.87 × 4.12	8.0	120 @ 3600	215 @ 2000	130	—	40–60
	6-292	1 Bore	3.87 × 4.12	8.0	130 @ 3600	225 @ 2000	130	4000	40–60
	V8-350	2 Bore	4.00 × 3.48	8.5	145 @ 3500	250 @ 2200	150	—	40
	V8-350	2 Bore	4.00 × 3.48	8.0	160 @ 4000	265 @ 2400	150	4000	30
	V8-350	4 Bore	4.00 × 3.48	8.5	155 @ 3800	250 @ 2800	150	—	40
	V8-350	4 Bore	4.00 × 3.48	8.5	160 @ 3800	250 @ 2400	150	—	40
	V8-366	4 Bore	3.937 × 3.76	8.0	200 @ 4000	305 @ 2800	⑦	4000	40–55
	V8-400	4 Bore	4.126 × 3.76	8.5	175 @ 3600	305 @ 2000	150	—	40
	V8-427	4 Bore	4.25 × 3.76	8.0	220 @ 4000	360 @ 2400	⑦	4000	40–55
	V8-454	4 Bore	4.25 × 4.00	8.15	215 @ 4000	350 @ 2400	150	—	40
	V8-454	4 Bore	4.25 × 4.00	8.15	245 @ 4000	355 @ 3000	150	—	40
	V8-454	4 Bore	4.25 × 4.00	8.15	245 @ 4000	375 @ 2800	150	—	40
	V8-455	4 Bore	4.126 × 4.250	8.5	212 @ 3400	344 @ 2400	⑦	—	30–45
1976	6-250⑭	1 Bore	3.87 × 3.53	8.25	105 @ 3800	185 @ 1200	130	3800	40–60
	6-250⑮	1 Bore	3.87 × 3.53	8.0	100 @ 3600	175 @ 1800	130	3600	40–60
	6-292⑭	1 Bore	3.87 × 4.12	8.0	120 @ 3600	215 @ 2000	130	3600	40–60
	6-292⑮	1 Bore	3.87 × 4.12	8.0	130 @ 3600	225 @ 2000	130	3600	40–60

Continued

GENERAL ENGINE SPECIFICATIONS—Continued

Year	Engine Model	Carb. Type	Bore & Stroke	Comp. Ratio	Horsepower @ R.P.M. ①	Torque Ft. Lbs. @ R.P.M. ②	Comp. Pressure Cranking Speed	Governed Speed R.P.M. Full Load	Normal Oil Pressure Lbs.
	V8-305	2 Bore	3.736 × 3.48	8.5	140 @ 3800	245 @ 2000	155	—	32–40
	V8-350⑭	2 Bore	4.00 × 3.48	8.5	145 @ 3800	250 @ 2200	150	—	40
	V8-350⑮	2 Bore	4.00 × 3.48	8.0	160 @ 4000	265 @ 2400	150	—	30
	V8-350	4 Bore	4.00 × 3.48	8.5	165 @ 3800	260 @ 2200	150	—	32–40
	V8-350⑭	4 Bore	4.00 × 3.48	8.5	165 @ 3800	260 @ 2400	160	—	40
	V8-350⑮	4 Bore	4.00 × 3.48	8.5	165 @ 3800	255 @ 2800	160	—	40
	V8-366⑯	4 Bore	3.937 × 3.76	8.0	195 @ 4000	290 @ 2800	150	4000	40–55
	V8-366⑰	4 Bore	3.937 × 3.76	8.0	200 @ 4000	305 @ 2800	150	4000	40–55
	V8-400	4 Bore	4.125 × 3.75	8.5	175 @ 3600	290 @ 2800	160	—	40
	V8-400	4 Bore	4.125 × 3.75	8.5	175 @ 3600	305 @ 2000	160	—	40
	V8-427	4 Bore	4.25 × 3.76	7.8	220 @ 4000	360 @ 2400	150	4000	40–55
	V8-454⑭	4 Bore	4.25 × 4.00	8.25	245 @ 3800	365 @ 2800	160	—	40
	V8-454⑱	4 Bore	4.25 × 4.00	8.15⑳	240 @ 3800	370 @ 2800	160	—	40
	V8-454⑲	4 Bore	4.25 × 4.00	8.15⑳	250 @ 3800	385 @ 2800	160	—	40
	V8-455	4 Bore	4.126 × 4.250	8.5	212 @ 3400	344 @ 2400	⑦	—	30–45
1977	6-250③	1 Bore	3.87 × 3.53	8.25	110 @ 3800	195 @ 1600	⑦	—	40–60
	6-250④	1 Bore	3.87 × 3.53	8.25	90 @ 3600	175 @ 1800	⑦	—	40–60
	6-292⑭	1 Bore	3.87 × 4.12	8	120 @ 3600	215 @ 2000	130	—	40–60
	6-292⑮	1 Bore	3.87 × 4.12	8	130 @ 3600	220 @ 2000	130	—	40–60
	V8-305④	2 Bore	3.74 × 3.48	8.5	140 @ 3800	235 @ 2000	⑦	—	32–40
	V8-305③	2 Bore	3.74 × 3.48	8.5	145 @ 3800	245 @ 2400	⑦	—	32–40
	V8-350㉑㉒	4 Bore	4.00 × 3.48	8.5	170 @ 3800	270 @ 2400	160	—	40
	V8-350㉑㉓	4 Bore	4.00 × 3.48	8.5	160 @ 3800	260 @ 2400	160	—	40
	V8-350③	4 Bore	4.00 × 3.48	8.5	165 @ 3800	260 @ 2400	160	—	40
	V8-350④	4 Bore	4.00 × 3.48	8.5	165 @ 3800	255 @ 2800	160	—	40
	V8-366	4 Bore	3.937 × 3.76	8	200 @ 4000	305 @ 2800	150	—	40–55
	V8-400	4 Bore	4.125 × 3.75	8.5	175 @ 3600	290 @ 2800	⑦	—	40
	V8-403	4 Bore	4.351 × 3.385	7.9	200 @ 3600	330 @ 2400	⑦	—	30–45
	V8-427	4 Bore	4.25 × 3.76	—	220 @ 4000	360 @ 2400	150	—	40–55
	V8-454③	4 Bore	4.25 × 4.00	8.5	245 @ 3800	365 @ 2800	⑦	—	40
	V8-454④㉒	4 Bore	4.25 × 4.00	8.5	240 @ 3800	370 @ 2800	⑦	—	40
	V8-454④㉓	4 Bore	4.25 × 4.00	8.5	250 @ 3800	385 @ 2800	⑦	—	40
	V8-455	4 Bore	4.126 × 4.250	8.5	212 @ 3400	344 @ 2400	⑦	—	—
1978	V6-200㉛	2 Bore	3.50 × 3.48	8.2	95 @ 3800	160 @ 2000	⑦	—	34–39
	V6-231㉛	2 Bore	3.80 × 3.40	8.0	105 @ 3800	180 @ 2000	⑦	—	37
	6-250	1 Bore	3.87 × 3.53	8.0	100 @ 3600	175 @ 1800	⑦	—	40–60
	6-292	1 Bore	3.87 × 4.12	8.0	130 @ 3600	220 @ 2000	⑦	—	40–60
	V8-305	2 Bore	3.74 × 3.48	8.5	140 @ 3800	245 @ 2400	⑦	—	40
	V8-350	4 Bore	4.00 × 3.48	8.5	165 @ 4000	270 @ 2400	⑦	—	40
	V8-366	4 Bore	3.937 × 3.76	8.0	200 @ 4000	305 @ 2800	⑦	—	40–55
	V8-400	4 Bore	4.125 × 3.75	8.5	170 @ 3600	290 @ 2800	⑦	—	40
	V8-403	4 Bore	4.35 × 3.38	7.9	200 @ 3600	330 @ 2400	⑦	—	30–45
	V8-427	4 Bore	4.25 × 3.76	8.1	220 @ 4000	360 @ 2400	⑦	4000	40–55
	V8-454	4 Bore	4.25 × 4.00	8.5	240 @ 3800	370 @ 2800	⑦	—	40
1979	V6-200㉛	2 Bore	3.50 × 3.48	8.2	94 @ 4000	154 @ 2000	⑦	—	34–39
	V6-231㉛	2 Bore	3.80 × 3.40	8.0	115 @ 3800	190 @ 2000	⑦	—	37
	6-250㉕	2 Bore	3.87 × 3.53	8.3	130 @ 4000	210 @ 2000	⑦	—	40–60
	6-250㉖	2 Bore	3.87 × 3.53	8.3	125 @ 4000	205 @ 2000	⑦	—	40–60
	6-250㉗	2 Bore	3.87 × 3.53	8.3	130 @ 4000	205 @ 2000	⑦	—	40–60
	6-292㉘㉙	1 Bore	3.87 × 4.12	7.8	115 @ 3400	215 @ 1600	⑦	—	40–60
	6-292㉚	1 Bore	3.87 × 4.12	7.8	125 @ 3600	225 @ 2000	⑦	4000	40–60
	V8-267	2 Bore	3.50 × 3.48	8.5	125 @ 3800	215 @ 2400	⑦	—	40
	V8-305㉜	2 Bore	3.74 × 3.48	8.4	140 @ 4000	240 @ 2000	⑦	—	45
	V8-305㉒㉛	4 Bore	3.74 × 3.48	8.4	160 @ 4000	235 @ 2400	⑦	—	45
	V8-305㉓㉛	4 Bore	3.74 × 3.48	8.4	155 @ 4000	225 @ 2400	⑦	—	45

Continued

GENERAL ENGINE SPECIFICATIONS—Continued

Year	Engine Model	Carb. Type	Bore & Stroke	Comp. Ratio	Horsepower @ R.P.M. ①	Torque Ft. Lbs. @ R.P.M. ②	Comp. Pressure Cranking Speed	Governed Speed R.P.M. Full Load	Normal Oil Pressure Lbs.
	V8-350⑮	2 Bore	4.00 × 3.48	8.0	161 @ 3800	275 @ 2400	⑦	4000	30
	V8-350⑭㉛	4 Bore	4.00 × 3.48	8.2	165 @ 3800	260 @ 2400	⑦	—	45
	V8-350⑤⑭㉜	4 Bore	4.00 × 3.48	8.2	165 @ 3600	270 @ 2000	⑦	—	45
	V8-350⑥⑭㉜	4 Bore	4.00 × 3.48	8.2	155 @ 3600	260 @ 2000	⑦	—	45
	V8-350⑧⑭㉜	4 Bore	4.00 × 3.48	8.3	165 @ 3800	255 @ 2800	⑦	—	45
	V8-366⑨⑯	4 Bore	3.937 × 3.76	7.6	180 @ 4000	290 @ 2400	⑦	4000	40–55
	V8-366⑩⑰	4 Bore	3.937 × 3.76	7.6	190 @ 4000	305 @ 2400	⑦	4000	40–55
	V8-400㉕	4 Bore	4.125 × 3.75	8.2	185 @ 3600	300 @ 2400	⑦	—	40
	V8-400⑪	4 Bore	4.125 × 3.75	8.2	170 @ 3600	305 @ 1600	⑦	—	40
	V8-400⑧	4 Bore	4.125 × 3.75	8.3	180 @ 3600	310 @ 2400	⑦	—	40
	V8-427⑯	4 Bore	4.25 × 3.76	7.5	195 @ 3800	340 @ 2400	⑦	4000	40–55
	V8-427⑰	4 Bore	4.25 × 3.76	7.5	210 @ 3800	350 @ 2400	⑦	4000	40–55
	V8-454⑬	4 Bore	4.25 × 4.00	7.6	205 @ 3600	335 @ 2800	⑦	—	40
	V8-454⑨	4 Bore	4.25 × 4.00	7.6	210 @ 3800	340 @ 2800	⑦	—	40
	V8-454⑮⑰	4 Bore	4.25 × 4.00	7.6	225 @ 4000	365 @ 2400	⑦	4000	40–55
1980	V6-229	2 Bore	3.736 × 3.48	8.6	115 @ 4000	175 @ 2000	⑦	—	45
	V6-231	2 Bore	3.80 × 3.4	8.0	110 @ 3800	190 @ 1600	⑦	—	37
	6-250	2 Bore	3.876 × 3.53	8.25	130 @ 4000	210 @ 2000	⑦	—	40–60
	V8-267	2 Bore	3.50 × 3.48	8.3	120 @ 3600	215 @ 2000	⑦	—	45
	6-292㉘㉙	1 Bore	3.876 × 4.12	7.8	115 @ 3400	215 @ 1600	⑦	—	50
	6-292㉚	1 Bore	3.876 × 4.12	7.8	125 @ 3600	225 @ 1600	⑦	—	50
	V8-305㉓	4 Bore	3.74 × 3.48	8.6	155 @ 4000	230 @ 2400	⑦	—	45
	V8-305㉒	4 Bore	3.74 × 3.48	8.6	155 @ 4000	240 @ 1600	⑦	—	45
	V8-350	2 Bore	4.0 × 3.48	8.0	161 @ 3800	275 @ 2400	⑦	4000	30
	V8-366⑨⑯	4 Bore	3.937 × 3.76	7.6	180 @ 4000	290 @ 2400	⑦	4000	40–55
	V8-366⑩⑰	4 Bore	3.937 × 3.76	7.6	190 @ 4000	305 @ 2400	⑦	4000	40–55
	V8-427⑯	4 Bore	4.25 × 3.76	7.5	195 @ 3800	340 @ 2400	⑦	4000	40–55
	V8-427⑰	4 Bore	4.25 × 3.76	7.5	210 @ 3800	350 @ 2400	⑦	4000	40–55
	V8-454⑰	4 Bore	4.25 × 4.00	7.6	225 @ 4000	365 @ 2400	⑦	4000	40–55
1981	V6-229㉒	2 Bore	3.736 × 3.48	8.6	110 @ 4200	170 @ 2000	⑦	—	45
	V6-231㉓	2 Bore	3.80 × 3.40	8.0	110 @ 3800	190 @ 1600	⑦	—	37
	6-250㉕	2 Bore	3.88 × 3.53	8.3	115 @ 3600	200 @ 2000	⑦	—	40–60
	6-250⑪	2 Bore	3.88 × 3.53	8.3	105 @ 3600	195 @ 1600	⑦	—	40–60
	6-292㉘㉙	1 Bore	3.88 × 4.12	7.8	115 @ 3400	215 @ 1600	⑦	—	40–60
	6-292㉚	1 Bore	3.88 × 4.12	7.8	125 @ 3600	225 @ 1600	⑦	—	40–60
	V8-267㉒	2 Bore	3.50 × 3.48	8.3	115 @ 4000	200 @ 2400	⑦	—	45
	V8-305㉒㉛	4 Bore	3.74 × 3.48	8.6	145 @ 3800	240 @ 2400	⑦	—	45
	V8-305㉕	4 Bore	3.74 × 3.48	9.2	165 @ 4400	240 @ 2000	⑦	—	45
	V8-350	2 Bore	4.00 × 3.48	8.0	161 @ 3800	275 @ 2400	⑦	—	30
	V8-350㉕	4 Bore	4.00 × 3.48	8.2	165 @ 3800	275 @ 1600	⑦	—	45
	V8-350⑪	4 Bore	4.00 × 3.48	8.2	150 @ 3600	255 @ 1600	⑦	—	45
	V8-350⑧㉒	4 Bore	4.00 × 3.48	8.3	165 @ 3800	255 @ 2800	⑦	—	45
	V8-350⑧㉓	4 Bore	4.00 × 3.48	8.3	155 @ 4000	240 @ 2800	⑦	—	45
	V8-366⑨⑯	4 Bore	3.937 × 3.76	7.6	179 @ 3800	290 @ 2400	⑦	—	40–45
	V8-366⑩⑰	4 Bore	3.937 × 3.76	7.6	190 @ 4000	305 @ 2400	⑦	—	40–45
	V8-427⑯	4 Bore	4.25 × 3.76	7.5	195 @ 3800	340 @ 2400	⑦	—	40–45
	V8-427⑰	4 Bore	4.25 × 3.76	7.5	210 @ 3800	350 @ 2400	⑦	—	40–45
	V8-454⑧	4 Bore	4.25 × 4.00	7.9	210 @ 3800	340 @ 2800	⑦	—	40
1982	V6-229㉒	2 Bore	3.74 × 3.48	8.6	110 @ 4200	170 @ 2000	⑦	—	50
	V6-231㉓	2 Bore	3.8 × 3.4	8.0	110 @ 3800	190 @ 1600	⑦	—	37
	6-250㉕	2 Bore	3.88 × 3.53	8.3	120 @ 3600	200 @ 2000	⑦	—	40–60
	6-250⑪	2 Bore	3.88 × 3.53	8.3	110 @ 3600	195 @ 2000	⑦	—	40–60
	6-292㉘㉙	1 Bore	3.88 × 4.12	7.8	115 @ 3400	215 @ 1600	⑦	—	40–60
	6-292㉚	1 Bore	3.88 × 4.12	7.8	125 @ 3600	225 @ 1600	⑦	—	40–60

Continued

GENERAL ENGINE SPECIFICATIONS—Continued

Year	Engine Model	Carb. Type	Bore & Stroke	Comp. Ratio	Horsepower @ R.P.M. ①	Torque Ft. Lbs. @ R.P.M. ②	Comp. Pressure Cranking Speed	Governed Speed R.P.M. Full Load	Normal Oil Pressure Lbs.
	V8-267㉒	2 Bore	3.50 × 3.48	8.3	115 @ 4000	205 @ 2400	⑦	—	45
	V8-305㉛	4 Bore	3.74 × 3.48	8.6	150 @ 3800	240 @ 2400	⑦	—	45
	V8-305㉒㉜㉞	4 Bore	3.74 × 3.48	9.2	160 @ 4400	235 @ 2000	⑦	—	45
	V8-305㉓㉜㉟	4 Bore	3.74 × 3.48	9.2	165 @ 4400	240 @ 2000	⑦	—	45
	V8-305㉓㉜	4 Bore	3.74 × 3.48	8.6	155 @ 4000	245 @ 1600	⑦	—	45
	V8-350	2 Bore	4.0 × 3.48	8.0	161 @ 3800	275 @ 2400	⑦	—	30
	V8-350㉔	4 Bore	4.0 × 3.48	8.2	165 @ 3800	275 @ 1600	⑦	—	45
	V8-350㉒㉝	4 Bore	4.0 × 3.48	8.3	160 @ 3800	250 @ 2800	⑦	—	45
	V8-350	4 Bore	4.0 × 3.48	8.3	155 @ 4000	240 @ 2800	⑦	—	45
	V8-366⑯	4 Bore	3.937 × 3.76	7.6	179 @ 3800	290 @ 2400	⑦	—	40–45
	V8-366⑰	4 Bore	3.937 × 3.76	7.6	190 @ 3800	305 @ 2400	⑦	—	40–45
	V8-427⑯	4 Bore	4.25 × 3.76	7.5	195 @ 3800	340 @ 2400	⑦	—	40–45
	V8-427⑰	4 Bore	4.25 × 3.76	7.5	210 @ 3800	350 @ 2400	⑦	—	40–45
	V8-454	4 Bore	4.25 × 4.00	7.9	210 @ 3800	340 @ 2800	⑦	—	40
1983	V6-229	2 Bore	3.74 × 3.48	8.6	110 @ 4200	170 @ 2000	⑦	—	50
	V6-231	2 Bore	3.80 × 3.40	8.0	110 @ 3800	190 @ 1600	⑦	—	37
	6-250㉕	2 Bore	3.88 × 3.53	8.3	120 @ 3600	200 @ 2000	⑦	—	40–60
	6-250⑪	2 Bore	3.88 × 3.53	8.3	110 @ 3600	195 @ 2000	⑦	—	40–60
	6-292	1 Bore	3.88 × 4.12	7.8	115 @ 3400	215 @ 1600	⑦	—	40–60
	V8-305㉛	4 Bore	3.74 × 3.48	8.6	150 @ 3800	240 @ 2400	⑦	—	45
	V8-305㉒㉟	4 Bore	3.74 × 3.48	9.2	165 @ 4400	240 @ 2000	⑦	—	45
	V8-305㉒㉞	4 Bore	3.74 × 3.48	9.2	160 @ 4400	235 @ 2000	⑦	—	45
	V8-305⑪	4 Bore	3.74 × 3.48	8.6	155 @ 4000	240 @ 2800	⑦	—	45
	V8-350	2 Bore	4.00 × 3.48	8.0	161 @ 3800	275 @ 2400	⑦	—	30
	V8-350③	4 Bore	4.00 × 3.48	8.2	165 @ 3800	275 @ 1600	⑦	—	45
	V8-350⑧㉒	4 Bore	4.00 × 3.48	8.3	160 @ 3800	250 @ 2800	⑦	—	45
	V8-350⑧㉓	4 Bore	4.00 × 3.48	8.3	155 @ 4000	240 @ 2800	⑦	—	45
	V8-366⑯	4 Bore	3.937 × 3.76	7.6	179 @ 3800	290 @ 2400	⑦	—	40–45
	V8-366⑰	4 Bore	3.937 × 3.76	7.6	188 @ 3800	305 @ 2400	⑦	—	40–45
	V8-427⑯	4 Bore	4.25 × 3.76	7.5	195 @ 3800	340 @ 2400	⑦	—	40–45
	V8-427⑰	4 Bore	4.25 × 3.76	7.5	210 @ 3800	350 @ 2400	⑦	—	40–45
	V8-454	4 Bore	4.25 × 4.00	8.5	210 @ 3800	340 @ 2800	⑦	—	40

①—Horsepower ratings may vary slightly depending on model application.
②—Torque ratings may vary slightly depending on model application.
③—Except heavy duty emissions.
④—Heavy duty emissions.
⑤—Light & medium duty emissions (G.V.W.R. 8500 lbs. or under) except Calif. & high altitude.
⑥—California light & medium duty emissions (G.V.W.R. 8500 lbs. or under) & high altitude emissions (G.V.W.R. 6000 lbs. or under).
⑦—Lowest reading must be within 70% of highest reading.
⑧—Heavy duty emissions (G.V.W.R. 8501 lbs. or above).
⑨—School bus chassis.
⑩—Except school bus chassis.
⑪—California light & medium duty emissions (G.V.W.R. 8500 lbs. or under).
⑫—4100 with AT 475 transmission.
⑬—Light & medium duty emissions (G.V.W.R. 8500 lbs. or under).
⑭—Light Duty.
⑮—Heavy Duty.
⑯—Single Exhaust.
⑰—Dual Exhaust.
⑱—Heavy Duty except California.
⑲—California Heavy Duty.
⑳—C-10/1500 F44 & C-10/1500, 20/2500 models, 8.25:1.
㉑—El Camino & Sprint.
㉒—Except Calif.
㉓—California.
㉔—Except C6P heavy duty chassis.
㉕—Light & heavy duty emissions (G.V.W.R. 8500 lbs. or under) except Calif.
㉖—California Series 10/1500 light duty emissions (G.V.W.R. 6000 lbs. or under).
㉗—California Series 20/2500 & 30/3500 light & medium duty emissions (G.V.W.R. 8500 lbs. or under).
㉘—P-10/1500–30/3500 & C, K-30/3500.
㉙—P4T042.
㉚—C5D042 & CD6042.
㉛—Caballero & El Camino.
㉜—Except Caballero & El Camino.
㉝—With C6P heavy duty chassis.
㉞—All models except C-10/1500.
㉟—C-10/1500.

TUNE UP SPECIFICATIONS

The following specifications are published from the latest information available. This data should be used only in the absence of a decal affixed in the engine compartment.

★ When using a timing light, disconnect vacuum hose or tube at distributor and plug opening in hose or tube so idle speed will not be affected.

● When checking compression, lowest cylinder must be within 70 percent of highest.

▲ Before removing wires from distributor cap, determine location of the No. 1 wire in cap, as distributor position may have been altered from that shown at the end of this chart.

Year & Engine	Spark Plug		Ignition Timing BTDC①★				Curb Idle Speed②		Fuel Pump Pressure
	Type	Gap	Firing Orders Fig.▲	Man. Trans.	Auto. Trans.	Location	Man. Trans.	Auto. Trans.	
1983									
V6-229	R45TS	.045	A	—	—	Damper	—	—	4½–6
V6-231	R45TS8	.080	B	—	15°	Damper	—	㉟	3 Min.
6-250	R45TS	—	C	—	—	Damper	—	—	4½–6
6-292	R44T	.035	C	8°	8°	Damper	700	700N	4½–6
V8-305⑤	R45TS	.045	D	—	—	Damper	—	—	7½–9
V8-305	R44T	.045	D	—	—	Damper	—	—	7½–9
V8-350 2 Barrel Carb.	R44T	.045	D	4°㉝	4°㉝	Damper	700	700N	7–9
V8-350 4 Barrel Carb.㊵	R45TS	.045	D	—	—	Damper	—	—	④
V8-350 4 Barrel Carb.㊷	R44T	.045	D	—	—	Damper	—	—	④
V8-366	R43T⑲	.045	D	8°	8°	Damper	700	700D	7–9
V8-427	R42T⑲	.045	D	8°	8°	Damper	700	700D	7–9
V8-454	R44T	.045	D	—	—	Damper	—	—	④
1982									
V6-229	R45TS	.045	A	—	TDC	Damper	㉟	㉟	4½–6
V6-231	R45TS8	.080	B	—	15°	Damper	㉟	㉟	3 Min.
6-250	R45TS	—	C	—	—	Damper	—	—	4½–6
6-292	R44T	.035	C	8°	8°	Damper	700	700N	4–5
V8-267	R45TS	.045	D	—	2°	Damper	—	500/600D	7½–9
V8-305⑤	R45TS	.045	D	—	6°	Damper	—	500/600D	7½–9
V8-305 Series 10-30	R45TS㊶	.045	D	—	—	Damper	—	—	④
V8-350 2 Barrel Carb.	R44T	.045	D	4°㉝	4°㉝	Damper	700	700N	7–9
V8-350 4 Barrel Carb.㊵	R45TS	.045	D	—	—	Damper	—	—	④
V8-350 4 Barrel Carb.㉚	R44T	.045	D	—	—	Damper	—	—	④
V8-366	R43T	.045	D	8°	8°	Damper	700	700N	7–9
V8-427	R42T	.045	D	8°	8°	Damper	700	700N	7–9
V8-454	R44T	.045	D	—	—	Damper	—	—	④

Continued

TUNE UP SPECIFICATIONS—Continued

The following specifications are published from the latest information available. This data should be used only in the absence of a decal affixed in the engine compartment.

★ When using a timing light, disconnect vacuum hose or tube at distributor and plug opening in hose or tube so idle speed will not be affected.

● When checking compression, lowest cylinder must be within 70 percent of highest.

▲ Before removing wires from distributor cap, determine location of the No. 1 wire in cap, as distributor position may have been altered from that shown at the end of this chart.

Year & Engine	Spark Plug		Ignition Timing BTDC①★				Curb Idle Speed②		Fuel Pump Pressure
	Type	Gap	Firing Orders Fig.▲	Man. Trans.	Auto. Trans.	Location	Man. Trans.	Auto. Trans.	
1981									
V6-229	R45TS	.045	A	6°	6°	Damper	㉟	㉟	4½–6
V6-231	R45TS8	.080	B	—	15°	Damper	—	㉟	3 Min.
6-250	R46TS	.035	C	10°	10°	Damper	450/750㊱	450/650D㊲	4½–6
6-292	R44T	.035	C	8°	8°	Damper	450/700	450/700N	4½–6
V8-267	R45TS	.045	D	—	6°	Damper	—	500/600D	7½–9
V8-305⑤	R45TS	.045	D	6°	6°	Damper	700/800	500/600D	7½–9
V8-305 2 Barrel Carb. Series 10-30	R45TS	.045	D	8°	8°	Damper	600/700	500/600D	④
V8-305 4 Barrel Carb. C10-20, K10 Exc. Calif.	R45TS㊶	.045	D	4°	㊳	Damper	㊴	500D	④
V8-305 4 Barrel Carb. C10-20 Calif.	R45TS	.045	D	—	8°	Damper	—	550/650D	④
V8-305 4 Barrel Carb. G10-20	R45TS㊶	.045	D	6°	4°	Damper	700	500D	④
V8-350 Exc. Calif.㊵	R45TS	.045	D	8°	8°	Damper	700	500/600D	④
V8-350 CK10 Calif.㊵	R45TS	.045	D	—	6°	Damper	—	550/650D	④
V8-350 CK20 Calif.㊵	R45TS	.045	D	—	㉕	Damper	—	550/650D	④
V8-350 G10-20 Calif.㊵	R45TS	.045	D	—	8°	Damper	—	550/650D	④
V8-350 CK20-30 Exc. Calif.㉚	R44T	.045	D	4°	6°	Damper	700/1600㉛	700/1600N㉛	④
V8-350 G30, P20-30 Exc. Calif.㉚	R44T	.045	D	4°	4°	Damper	700/1600㉛	700/1600N㉛	④
V8-350 Calif.㉚	R44T	.045	D	6°	6°	Damper	700/1500㉛	700/1500N㉛	④
V8-454	R44T	.045	D	4°	4°	Damper	700/1500㉛	700/1500N㉛	7½–9
1980									
V6-229	R45TS	.045	A	8°	12°	Damper	700/800	600/675D	4½–6
V6-231 Exc. Calif.	R45TSX	.060	B	—	15°	Damper	—	560/670D	3 Min.
V6-231 Calif.	R45TSX	.060	B	—	15°	Damper	—	600D	3 Min.
6-250 Exc. Calif.	R46TS	.035	C	10°	10°	Damper	450/750	450/650D	4½–6
6-250 C&G-10 Calif.	R46TS	.035	C	10°	10°	Damper	425/750	425/600D	4½–6
6-250 C&G-20 Calif.	R46TS	.035	C	10°	8°	Damper	425/750	425/600D	4½–6
6-292	R44T	.035	C	8°	8°	Damper	700	700N	4–5
V8-267	R45TS	.045	D	—	4°	Damper	—	500/600D	7½–9
V8-305 El Camino Exc. Calif.	R45TS	.045	D	4°	4°	Damper	700	500/600D	7½–9
V8-305 El Camino Calif.	R45TS	.045	D	—	4°	Damper	—	550/650D	7½–9
V8-305 Series 10-30	R45TS	.045	D	㉜	8°	Damper	600/700	500/600D	④
V8-350 2 Barrel Carb. B, C, P & W-4, 5, 6	R44T	.045	D	4°㉝	4°㉝	Damper	700	700N	7–9
V8-350 4 Barrel Carb.⑦	R45TS	.045	D	8°㉞	8°㉞	Damper	700	500/600D	④
V8-350 4 Barrel Carb. Exc. Calif.⑧	R44T	.045	D	4°	4°	Damper	700	700N	④
V8-350 4 Barrel Carb. Calif.⑧	R44T	.045	D	6°	6°	Damper	700	700N	④
V8-366	R43T	.045	D	8°	8°	Damper	700	700N	7–9
V8-400 G-20	R45TS	.045	D	—	4°	Damper	—	500/600D	④
V8-400 K-20 & G, K-30 Exc. Calif.	R44T	.045	D	—	4°	Damper	—	700N	④

Continued

TUNE UP SPECIFICATIONS—Continued

The following specifications are published from the latest information available. This data should be used only in the absence of a decal affixed in the engine compartment.

★ When using a timing light, disconnect vacuum hose or tube at distributor and plug opening in hose or tube so idle speed will not be affected.

● When checking compression, lowest cylinder must be within 70 percent of highest.

▲ Before removing wires from distributor cap, determine location of the No. 1 wire in cap, as distributor position may have been altered from that shown at the end of this chart.

Year & Engine	Spark Plug		Ignition Timing BTDC①★				Curb Idle Speed②		Fuel Pump Pressure
	Type	Gap	Firing Orders Fig.▲	Man. Trans.	Auto. Trans.	Location	Man. Trans.	Auto. Trans.	
1980—Continued									
V8-400 K-20 & G, K-30 Calif.	R44T	.045	D	—	6°	Damper	—	700N	④
V8-427	R42T	.045	D	8°	8°	Damper	700	700N	7–9
V8-454 C-20 & C, P-30	R44T	.045	D	4°	4°	Damper	700	700N	7½–9
V8-454 C7D	R42T	.045	D	8°	8°	Damper	700	700N	7–9
1979									
V6-200	R45TS	.045	A	8°	12°	Damper	700/800	600/700D	4¼–5¾
V6-231 Exc. Calif. & High Alt.	R45TSX	.060	B	—	15°	Damper	—	③	3 Min.
V6-231 Calif. & High Alt.	R45TSX	.060	B	—	15°	Damper	—	600D	3 Min.
6-250 C, G & K-10	R46TS	.035	C	10°	10°	Damper	425/750	425/600D	4½–6
6-250 C-20 & G-20, 30 Exc. Calif.	R46TS	.035	C	10°	10°	Damper	425/750	425/600D	4½–6
6-250 C-20 & G-20, 30 Calif.	R46TS	.035	C	6°	8°	Damper	425/750	425/600D	4½–6
6-292	R44T	.035	C	8°	8°	Damper	700	700N	4½–6
V8-267	R45TS	.045	D	4°	8°	Damper	600/700	500/600D	7½–9
V8-305 Exc. High Alt.⑤	R45TS	.045	D	4°	4°	Damper	700	500/600D	7½–9
V8-305 High Alt.	R45TS	.045	D	—	8°	Damper	—	600/650D	7½–9
V8-305 C, G, K-10 & C-20	R45TS	.045	D	6°	6°	Damper	600/700	500/600D	④
V8-350 2 Barrel Carb. B, C, P & W-4, 5, 6	R44T	.045	D	4°	4°	Damper	700	700N	—
V8-350 4 Barrel Carb.⑤	R45TS	.045	D	—	8°	Damper	—	600/650D	7½–9
V8-350 4 Barrel Carb.⑦	R45TS	.045	D	8°	8°	Damper	700	500/600D	④
V8-350 4 Barrel Carb.⑧	R44T	.045	D	4°	4°	Damper	700	700N	④
V8-366	R43T	.045	D	—	4°	Damper	700	700N	6
V8-400 K-10, 20 & C-20, 30	R45TS	.045	D	—	4°	Damper	—	500/600D	④
V8-400 G, K-30	R44T	.045	D	4°	4°	Damper	700	700N	④
V8-427	R42T	.045	D	8°	8°	Damper	700	700N	6
V8-454⑦	R45TS	.045	D	8°	8°	Damper	700	550/600D	7½–9
V8-454⑧	R44T	.045	D	4°	4°	Damper	700	700N	7½–9
V8-454 C-7D	R42T	.045	D	8°	8°	Damper	700	700N	6
1978									
V6-200	R45TS	.045	A	8°	8°	Damper	700	600D	4¼–5¾
V6-231	R46TSX	.060	B	15°	15°	Damper	600/800	600/670D	3 Min.
6-250 Exc. Calif. & High Alt.⑦	R46TS	.035	C	8°	8°	Damper	425/750	⑨	4–5
6-250 Calif.⑦	R46TS	.035	C	8°	10°	Damper	425/750	425/600D	4–5
6-250 High Alt.⑦	R46TS	.035	C	8°	12°	Damper	425/750	425/600D	4–5
6-250⑧	R46T	.035	C	6°	6°	Damper	450/600	450/600N	4–5
6-292⑧	R44T	.035	C	8°	8°	Damper	450/600	450/600N	4–5
6-292⑥	R44T	.035	C	8°	8°	Damper	600	600N	4–5
V8-305 Exc. Calif.⑤⑦	R45TS	.035	D	4°	4°	Damper	600	500D	7½–9
V8-305 Calif.⑤	R45TS	.035	D	—	6°	Damper	—	500/600D	7½–9
V8-305⑧	R44T	.045	D	6°	6°	Damper	700	700N	7½–9
V8-350 2 Barrel Carb.⑥	R44T	⑩	D	4°	4°	Damper	600	700N	7½–9

Continued

TUNE UP SPECIFICATIONS—Continued

The following specifications are published from the latest information available. This data should be used only in the absence of a decal affixed in the engine compartment.

★ When using a timing light, disconnect vacuum hose or tube at distributor and plug opening in hose or tube so idle speed will not be affected.

● When checking compression, lowest cylinder must be within 70 percent of highest.

▲ Before removing wires from distributor cap, determine location of the No. 1 wire in cap, as distributor position may have been altered from that shown at the end of this chart.

Year & Engine	Spark Plug		Ignition Timing BTDC①★				Curb Idle Speed②		Fuel Pump Pressure
	Type	Gap	Firing Orders Fig.▲	Man. Trans.	Auto. Trans.	Location	Man. Trans.	Auto. Trans.	
1978—Continued									
V8-350 4 Barrel Carb.⑤⑪	R45TS	.045	D	6°	6°	Damper	700	500/600D	7½–9
V8-350 4 Barrel Carb. Calif.⑤	R45TS	.045	D	—	8°	Damper	—	500/650D	7½–9
V8-350 4 Barrel Carb. High Alt.⑤	R45TS	.045	D	—	8°	Damper	—	500/600D	7½–9
V8-350 4 Barrel Carb.⑦	R45TS	.045	D	8°	8°	Damper	⑫	500/600D	7½–9
V8-350 4 Barrel Carb. Exc. Calif.⑧	R44T	.045	D	8°	8°	Damper	700	700N	7½–9
V8-350 4 Barrel Carb. Calif.⑧	R44TX	.060	D	2°	2°	Damper	700	700N	7½–9
V8-366 Exc. Calif.	R43T	.045	D	8°	8°	Damper	700	700N	8¼
V8-366 Calif.	R43TX	.060	D	8°	8°	Damper	700	700N	8¼
V8-400 Exc. Calif.⑧	R44T	.045	D	4°	4°	Damper	700	700N	7½–9
V8-400 Calif.⑧	R44T	.045	D	2°	2°	Damper	700	700N	7½–9
V8-400 Calif.⑦	R45TS	.045	D	—	4°	Damper	—	500/600D	7½–9
V8-403	R46SZ	.060	H	—	—	Damper	—	—	5½–6½
V8-427 Exc. Calif.	R42T	.045	D	8°	8°	Damper	700	700N	8¼
V8-427 Calif.	R42TX	.060	D	8°	8°	Damper	700	700N	8¼
V8-454⑦	R45TS	.045	D	8°	8°	Damper	700	500/600D	7½–9
V8-454⑧	R44T	.045	D	8°	8°	Damper	700	700N	7½–9
1977									
6-250 Exc. Calif. & High Alt.⑤	R46TS	.035	C	6°	8°	Damper	⑬	⑨	4–5
6-250 Calif.⑤	R46TS	.035	C	—	6°	Damper	—	425/550D	4–5
6-250 High Alt.⑤	R46TS	.035	C	—	10°	Damper	—	425/600D	4–5
6-250 Exc. Calif. & High Alt.⑦	R46TS	.035	C	8°	12°	Damper	425/750	⑨	4–5
6-250 Calif.⑦	R46TS	.035	C	6°	10°	Damper	425/850	425/600D	4–5
6-250 High Alt.⑦	R46TS	.035	C	—	12°	Damper	—	425/600D	4–5
6-250⑧	R46T	.035	C	6°	6°	Damper	450/600	450/600N	4–5
6-292 Series 10-30	R44T	.035	C	8°	8°	Damper	450/600	450/600N	4–5
6-292 Series 40-60	R44T	.035	C	8°	8°	Damper	600	600N	4½–5½
V8-305 Exc. Calif.⑤⑦	R45TS	.045	D	8°	8°	Damper	600/700	500/650D	7½–9
V8-305 Exc. Calif.⑧	R44T	.045	D	6°	6°	Damper	700	700N	7½–9
V8-350 2 Barrel Carb.⑥	R44T	.035	D	4°	4°	Damper	600	600N	7½–9
V8-350 4 Barrel Carb.⑤	R45TS	.045	D	—	8°	Damper	—	⑭	7½–9
V8-350 4 Barrel Carb.⑦⑪	R45TS	.045	D	8°	8°	Damper	700	500/650D	7½–9
V8-350 4 Barrel Carb. Calif.⑦	R45TS	.045	D	6°	6°	Damper	700	500/650D	7½–9

Continued

TUNE UP SPECIFICATIONS—Continued

The following specifications are published from the latest information available. This data should be used only in the absence of a decal affixed in the engine compartment.

★ When using a timing light, disconnect vacuum hose or tube at distributor and plug opening in hose or tube so idle speed will not be affected.

● When checking compression, lowest cylinder must be within 70 percent of highest.

▲ Before removing wires from distributor cap, determine location of the No. 1 wire in cap, as distributor position may have been altered from that shown at the end of this chart.

Year & Engine	Spark Plug		Ignition Timing BTDC①★				Curb Idle Speed②		Fuel Pump Pressure
	Type	Gap	Firing Order Fig.▲	Man. Trans.	Auto Trans.	Location	Man. Trans.	Auto. Trans.	
1977—Continued									
V8-350 4 Barrel Carb. High Alt.⑦	R45TS	.045	D	—	6°	Damper	—	600/650D	7½–9
V8-350 4 Barrel Carb. Exc. Calif.⑧	R44T	.045	D	8°	8°	Damper	700	700N	7½–9
V8-350 4 Barrel Carb. Calif.⑧	R44TX	.060	D	2°	2°	Damper	700	700N	7½–9
V8-366 Exc. Calif.	R43T	.045	D	8°	8°	Damper	700	700N	7½–9
V8-366 Calif.	R43TX	.060	D	8°	8°	Damper	700	700N	7½–9
V8-400 Exc. Calif.⑧	R44T	.045	D	—	4°	Damper	700	700N	7½–9
V8-400 Calif.⑧	R44T	.045	D	—	2°	Damper	700	700N	7½–9
V8-403 Exc. Calif.	R46SZ	.060	H	—	12°㉔	Damper	—	600D	5½–6½
V8-403 Calif.	R46SZ	.060	H	—	2°㉔	Damper	—	600D	5½–6½
V8-427 Exc. Calif.	R42T	.035	D	8°	8°	Damper	700	700N	7½–9
V8-427 Calif.	R42TX	.060	D	8°	8°	Damper	700	700N	7½–9
V8-454⑦	R45TS	.045	D	—	4°	Damper	—	600D	7½–9
V8-454⑧	R44T	.045	D	8°	8°	Damper	700	700N	7½–9
V8-455	R46SX	.060	H	—	8°㉔	Damper	—	600D	5½–6½
1976									
6-250⑤	R46TS	.035	C	6°	6°	Damper	425/850	⑮	3½–4½
6-250 Exc. Calif.⑦	R46TS	.035	C	⑯	10°	Damper	425/900	425/550D	3½–4½
6-250 Calif.⑦	R46TS	.035	C	6°	10°	Damper	425/1000	425/600D	3½–4½
6-292 Series 10-30	R44T	.035	C	8°	8°	Damper	450/600	450/600N	4–5
6-292 Series 40-60 Less H.E.I.⑰	R44T	.035	E	8°	8°	Damper	800	600D	4–5
6-292 Series 40-60 w/H.E.I.	R44T	.035	C	8°	8°	Damper	800	600D	4–5
V8-305 El Camino	R45TS	.045	D	6°	8°	Damper	800	600D	7–8½
V8-305 Series 10-20	R44T	.045	D	6°	6°	Damper	800	600D	7–8½
V8-350 2 Barrel Carb.⑤⑦	R45TS	.045	D	2°	6°	Damper	800	600D	7–8½
V8-350 2 Barrel Carb. Less H.E.I.⑥⑱	R44T	.035	F	4°	4°	Damper	600	600N	7½–9
V8-350 2 Barrel Carb. w/H.E.I.⑥	R44T	.035	D	4°	4°	Damper	600	600N	7½–9
V8-350 4 Barrel Carb. Exc. Calif.⑤⑦	R45TS	.045	D	8°	8°	Damper	800	600D	7–8½
V8-350 4 Barrel Carb. Calif.⑤⑦	R45TS	.045	D	6°	6°	Damper	800	600D	7–8½
V8-350 4 Barrel Carb. Exc. Calif.⑧	R44TX	.060	D	8°	8°	Damper	700	700N	7–8½
V8-350 4 Barrel Carb. Calif.⑧	R44TX	.060	D	2°	2°	Damper	700	700N	7–8½
V8-366 Exc. Calif. Less H.E.I.⑱	R43T⑲	.035	F	8°	8°	Damper	700	700N	7½–9
V8-366 Exc. Calif. w/H.E.I.	R43T⑲	.035	D	8°	8°	Damper	700	700N	7½–9
V8-366 Calif.	R43TX⑳	.060	D	8°	8°	Damper	700	700N	7½–9
V8-400⑤	R45TS	.045	D	—	8°	Damper	—	600D	7–8½
V8-400⑧	R44TX	.060	D	—	4°	Damper	—	700N	7–8½

Continued

TUNE UP SPECIFICATIONS—Continued

The following specifications are published from the latest information available. This data should be used only in the absence of a decal affixed in the engine compartment.

★ When using a timing light, disconnect vacuum hose or tube at distributor and plug opening in hose or tube so idle speed will not be affected.

● When checking compression, lowest cylinder must be within 70 percent of highest.

▲ Before removing wires from distributor cap, determine location of the No. 1 wire in cap, as distributor position may have been altered from that shown at the end of this chart.

Year & Engine	Spark Plug		Ignition Timing BTDC[1]★				Curb Idle Speed[2]		Fuel Pump Pressure
	Type	Gap	Firing Order Fig.▲	Man. Trans.	Auto Trans.	Location	Man. Trans.	Auto. Trans.	
1976—Continued									
V8-427 Exc. Calif. Less H.E.I.[18]	R42T[21]	.035	F	8°	8°	Damper	700	700N	7½–9
V8-427 Exc. Calif. w/H.E.I.	R42T[21]	.035	D	8°	8°	Damper	700	700N	7½–9
V8-427 Calif.	R42TX[22]	.060	D	8°	8°	Damper	700	700N	7½–9
V8-454[7]	R45TS	.045	D	—	[23]	Damper	—	600D	7–8½
V8-454[8]	R44T	.045	D	8°	8°	Damper	700	700N	7–8½
V8-455	R46SX	.080	H	—	8°[24]	Damper	—	600D	5½–6½
1975									
6-250 Exc. Calif.[5]	R46TX	.060	C	10°	10°	Damper	425/850	425/550D	3½–4½
6-250 Calif.[5]	R46TX	.060	C	—	10°	Damper	—	425/600D	3½–4½
6-250 Series 10-30	R46TX	.060	C	10°	10°	Damper	425/900	425/550D	3½–4½
6-292 Series 10-30	R44TX	.060	C	8°	8°	Damper	450/600	450/600N	3½–4½
6-292 Series 40-60[17]	R44T	.035	E	8°	8°	Damper	600	600N	4–5
V8-350 2 Barrel Carb.[5][7]	R44TX	.060	D	6°	6°	Damper	800	600D	7–8½
V8-350 2 Barrel Carb.[6][18]	R44T	.035	F	4°	4°	Damper	[26]	550N	7½–9
V8-350 4 Barrel Carb.[5][7]	R44TX	.060	D	6°	[27]	Damper	800	600D	7–8½
V8-350 4 Barrel Carb. Exc. Calif.[8]	R44TX	.060	D	8°	8°	Damper	600	600N	7–8½
V8-350 4 Barrel Carb. Calif.[8]	R44TX	.060	D	2°	2°	Damper	700	700N	7–8½
V8-366 Exc. Calif.[18]	R43T[19]	.035	F	8°	8°	Damper	700	700N	7½–9
V8-366 Calif.	R43TX	.060	D	8°	8°	Damper	700	700N	7½–9
V8-400 El Camino	R44TX	.060	D	—	8°	Damper	—	600D	7–8½
V8-400 Exc. Calif.[8]	R44TX	.060	D	4°	4°	Damper	700	700N	7–8½
V8-400 Calif.[8]	R44TX	.060	D	2°	2°	Damper	700	700N	7–8½
V8-427 Exc. Calif.[18]	R42T[19]	.035	F	8°	8°	Damper	700	700N	7–9½
V8-427 Calif.	R42TX	.060	D	8°	8°	Damper	700	700N	7–9½
V8-454 Exc. Calif.[5][7]	R44TX	.060	D	—	16°	Damper	—	[28]	7–8½
V8-454 Exc. Calif.[8]	R44TX	.060	D	8°	8°	Damper	500/700	500/700N	7–8½
V8-454 Calif.[8]	R44TX	.060	D	8°	8°	Damper	600	600N	7–8½
V8-455 Less H.E.I.[18]	R45S	.040	H	—	8°[24]	Damper	—	600D	5½–6½
V8-455 W/H.E.I.	R46SX	.080	J	—	8°[24]	Damper	—	600D	5½–6½
1974									
4-140[17]	R42TS	.035	H	10°	12°	Damper	700	750D	3–4½
6-250 Light Duty[17]	R46T	.035	E	8°	8°	Damper	850	600D	4–5
6-250 Heavy Duty[17]	R46T	.035	E	6°	6°	Damper	600	600	4–5
6-292[17]	R44T	.035	E	8°	8°	Damper	600	600	4–5
V6-305C[17]	CR44NS	.035	I	7½°	7½°	Damper	600	600	5–7
V8-350, 145 H.P.[18]	R44T	.035	F	TDC	8°	Damper	900	600D	7–8½
V8-350, Light Duty[18]	R44T	.035	F	8°	8°	Damper	900	600D	7–8½
V8-350, Light Duty, Exc. El Camino[18]	R44T	.035	F	[29]	12°	Damper	900	600D	7–8⅛
V8-350, Light Duty, Calif.[18]	R44T	.035	F	4°	[29]	Damper	900	600D	7–8½
V8-350 Heavy Duty[18]	R43T	.035	F	4°	4°	Damper	600	600	7–8½
V8-366[18]	R43T	.035	F	8°	8°	Damper	600	600	7–9

Continued

TUNE UP SPECIFICATIONS—Continued

The following specifications are published from the latest information available. This data should be used only in the absence of a decal affixed in the engine compartment.

★ When using a timing light, disconnect vacuum hose or tube at distributor and plug opening in hose or tube so idle speed will not be affected.

● When checking compression, lowest cylinder must be within 70 percent of highest.

▲ Before removing wires from distributor cap, determine location of the No. 1 wire in cap, as distributor position may have been altered from that shown at the end of this chart.

Year & Engine	Spark Plug		Ignition Timing BTDC(1)★				Curb Idle Speed(2)		Fuel Pump Pressure
	Type	Gap	Firing Orders Fig.▲	Man. Trans.	Auto. Trans.	Location	Man. Trans.	Auto. Trans.	
1974—Continued									
V6-379(17)	R43T	.035	I	6°	6°	Damper	575	575	5–7
V8-427(18)	R42T	.035	F	8°	8°	Damper	600	600	7–9
V6-432(17)	R43T	.035	I	6°	6°	Damper	525	525	5–7
V8-454 Light Duty(18)	R44T	.035	F	10°	10°	Damper	800	600D	7–8½
V8-454 Heavy Duty(18)	R44T	.035	F	8°	8°	Damper	700	700	7–8½
V6-478(17)	R43T	.035	I	7½°	7½°	Damper	525	525	5–6½

(1)—B.T.D.C.—Before top dead center.
(2)—Idle speed on Man. Trans. vehicles is adjusted in Neutral & on Auto. Trans. vehicles is adjusted in Drive (D) or Neutral (N) as specified. Where two idle speeds are listed, the higher speed is with the A/C or idle solenoid energized.
(3)—Less A/C, 550D RPM; with A/C, 560/670D RPM.
(4)—With vapor return line, 5½–7 psi.; less vapor return line, 7½–9 psi.
(5)—Caballero, El Camino & Sprint.
(6)—Series 40-60.
(7)—Series 10-30 light duty emissions, GVWR 6000 lbs. & under.
(8)—Series 10-30 heavy duty emissions, GVWR 6001 lbs. & above.
(9)—Less A/C, 425/550D RPM; with A/C, 425/600D RPM.
(10)—Except Calif., .035"; Calif., .045".
(11)—Except Calif. & high altitude.
(12)—Except Calif., 600 RPM; Calif., 700 RPM.
(13)—Less A/C, 425/750 RPM; with A/C, 425/800 RPM.
(14)—Except high altitude, 500/650D RPM; high altitude, 600/650D RPM.
(15)—Except Calif., 425/550D RPM; California, 425/600D RPM.
(16)—Models w/ 3 spd. man. trans., 6° BTDC; models w/4 spd. man. trans., 10° BTDC.
(17)—Breaker point gap, new .019", used .016"; dwell angle, 31–34°.
(18)—Breaker point gap, new .019", used .016"; dwell angle, 30°.
(19)—For predominately urban driving where spark plug fouling may occur, use R44T.
(20)—For predominately urban driving where spark plug fouling may occur, use R44TX.
(21)—For predominately urban driving where spark plug fouling may occur, use R43T or R44T.
(22)—For predominately urban driving where spark plug fouling may occur, use R43TX or R44TX.
(23)—Models less catalytic converter, 8° BTDC; Models w/catalytic converter, 12° BTDC.
(24)—At 1100 RPM.
(25)—Except emission control label code AAZ, 6° BTDC; emission control label code AAZ, 6° BTDC.
(26)—Models less governor, 450 RPM; models w/ governor, 550 RPM.
(27)—El Camino except Calif., 8° BTDC; El Camino California models & all Series 10-30 models, 6° BTDC.
(28)—Caballero, El Camino & Sprint, 500/600D RPM; Series 10-30, 500/650D RPM.
(29)—Exc. Suburban, 8°; Suburban, 6°.
(30)—Series 10-30 heavy duty emissions, GVWR 8501 lbs. & above.
(31)—Higher speed is throttle return control speed.
(32)—Distributor model No. 1103381, set at 8° BTDC; distributor model No. 1103369, set at 6° BTDC.
(33)—Non-governed California models, set at 2° BTDC.
(34)—C-10 & 20 series with distributor models No. 1103339, set at 6° BTDC.
(35)—Equipped w/idle speed motor.
(36)—G-10-20 w/emission control label code ADA, set at 450/800 RPM.
(37)—G-10-20 w/emission control label code AAC, set at 450/700D RPM.
(38)—Emission control label code AAH, 4° BTDC; AAN, 6° BTDC; AAS, 2° BTDC.
(39)—Except emission control label code AUS, 700 RPM; emission control label code AUS, 600 RPM.
(40)—Series 10-30 light duty emissions, GVWR 8500 lbs. & under.
(41)—On all models except California, if preignition or poor engine performance is encountered under moderately severe operating conditions, use spark plug R44TS gapped at .045 inch.

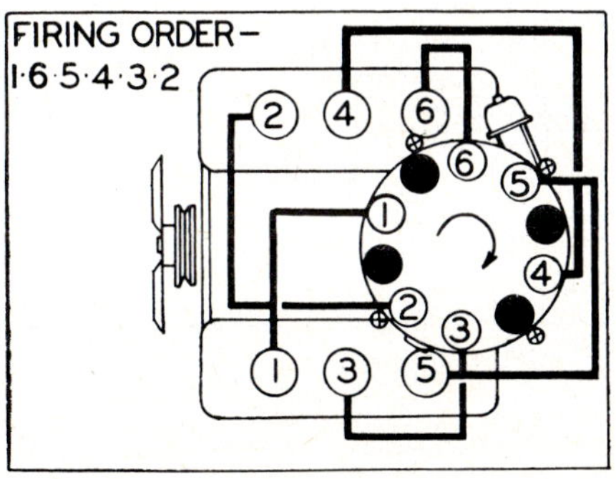

Fig. A

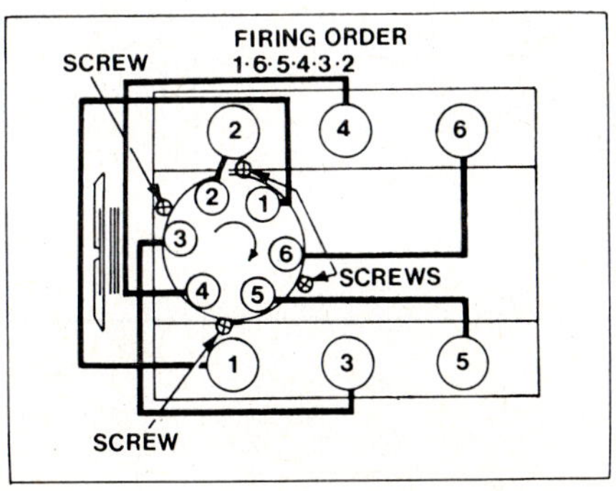

Fig. B

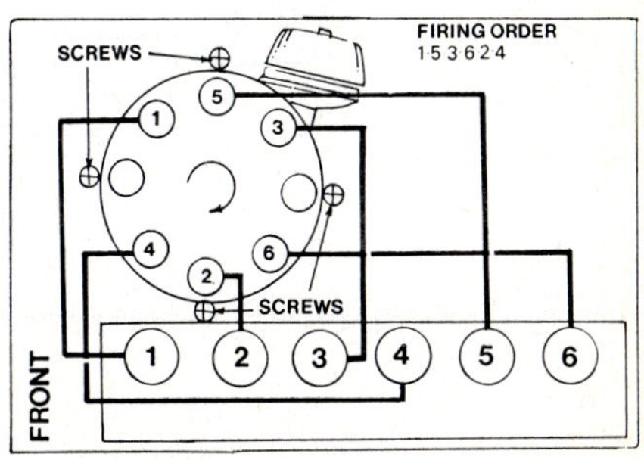

Fig. C

Continued

TUNE UP NOTES—Continued

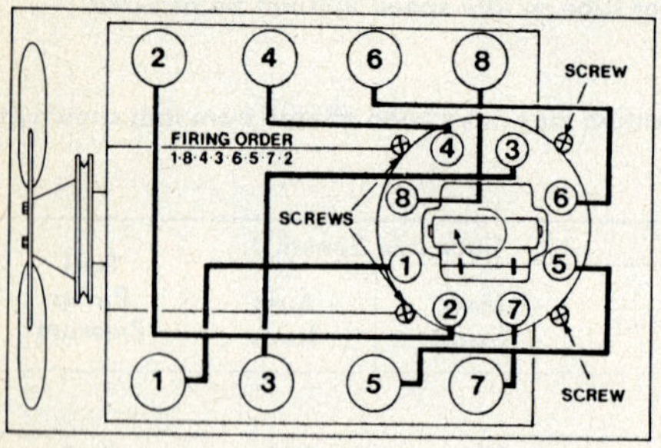

Fig. D

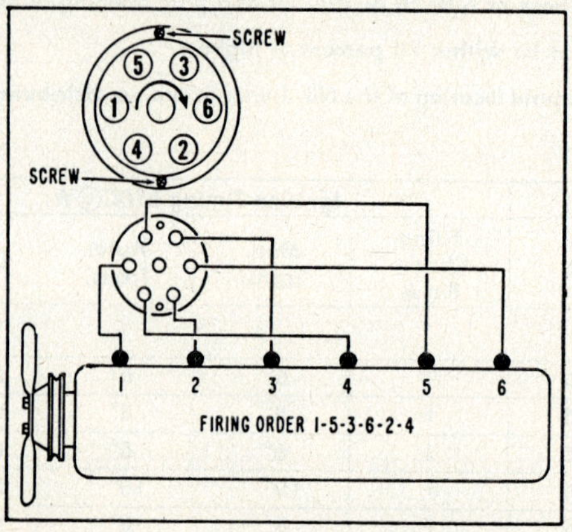

Fig. E

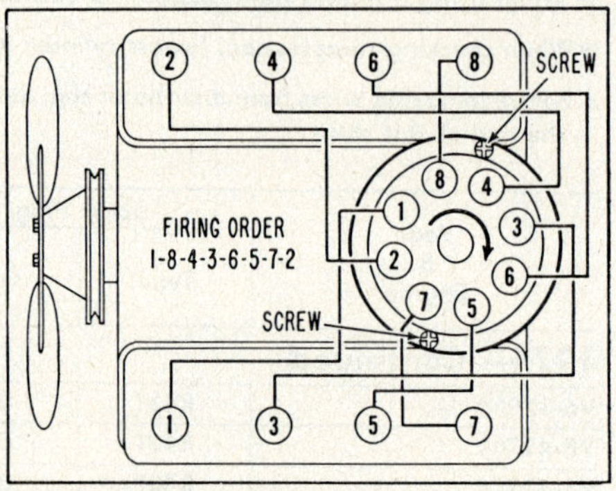

Fig. F

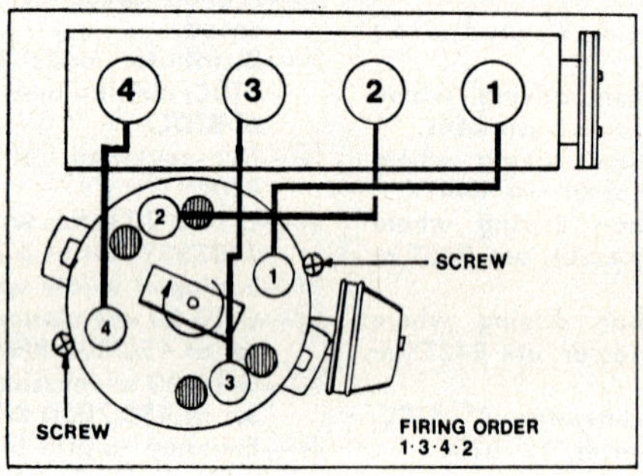

Fig. G

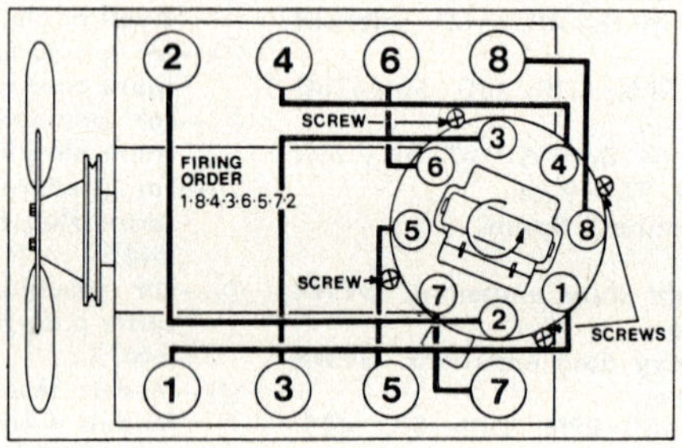

Fig. H

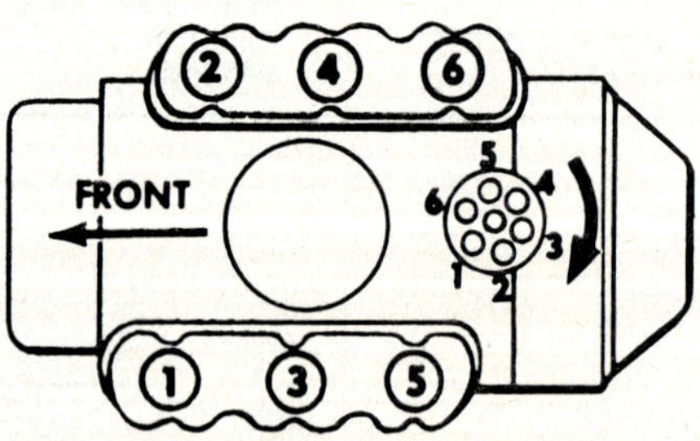

V-6 FIRING ORDER 1-6-5-4-3-2

Fig. I

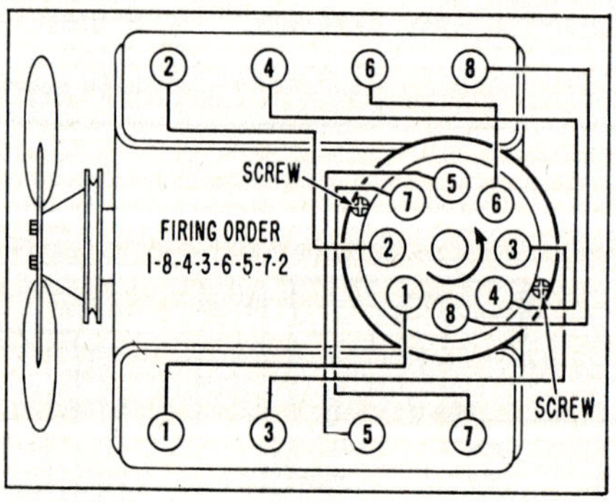

Fig. J

DISTRIBUTOR SPECIFICATIONS

Year	Model	Distributor Number	Rotation ①	Breaker Gap (New)	Dwell Angle Deg.	Breaker Arm Spring Tension	Centrifugal Advance Degrees @ R.P.M. of Distributor		Vacuum Advance	
							Advance Starts	Full Advance	Inches of Vacuum To Start Plunger	Max. Adv. Dist. Deg. @ Vacuum
1974	6-250, 10 Ser. Exc. G	1110499	C	.019	31–34	19–23	0 @ 550	12 @ 2050	7	12 @ 15½
	6-250, 10 Ser. G	1110500	C	.019	31–34	19–23	0 @ 500	12 @ 2050	7	12 @ 15½
	6-250, 20–30 Ser. Exc. G	1110520	C	.019	31–34	19–23	0 @ 550	12 @ 2050	13	5 @ 13½
	6-250, 20–30 Ser. G	1110525	C	.019	31–34	19–23	0 @ 550	12 @ 2050	13	5 @ 13½
	6-292, 20–30 Ser.	1110518	C	.019	31–34	19–23	0 @ 500	12 @ 2100	10	5 @ 13½
	V6-305C, 379, 432, 478	1112674	C	—	31–34	19–23	0–2 @ 510	10–12 @ 1750	—	—
	V8-350 40–60 Ser.⑨	1111364	C	—	28–32	19–23	0–2 @ 550	9–11 @ 1900	—	—
	V8-350 40–60 Ser.	1112149	C	—	28–32	19–23	0–2 @ 450	9–11 @ 1900	—	—
	V8-350 40–60 Ser.	1111373	C	—	29–31	19–23	0–2 @ 550	9–11 @ 1900	—	—
	8-350, C, K10, Suburban	1112093	C	.019	29–31	19–23	0 @ 550	9 @ 2100	6	7½ @ 14
	8-350, G10-30	1112847	C	.019	29–31	19–23	0 @ 550	9 @ 2100	3	7 @ 8½
	8-350 Auto. Trans. (Calif.)④	1112847	C	.019	29–31	19–23	0 @ 550	9 @ 2100	3	7 @ 8
	8-350, C, K, G	1112543	C	.019	29–31	19–23	0 @ 500	11 @ 2100	6	7½ @ 14
	8-350 Auto. Trans.③	1112848	C	.019	29–31	19–23	0 @ 450	7 @ 2100	3	7 @ 8½
	8-350 Std. Trans.③	1112844	C	.019	29–31	19–23	0 @ 500	10 @ 2100	3	7 @ 8½
	8-350 Std. Trans.⑤	1112844	C	.019	29–31	19–23	0 @ 500	10 @ 2100	3	7 @ 8½
	8-350 Std. Trans.⑥	1112849	C	.019	29–31	19–23	0 @ 500	11 @ 2100	3	7 @ 8½
	8-350 G, P–30	1112097	C	.019	29–31	19–23	0 @ 600	11 @ 2100	8	5 @ 13½
	8-366, 427	1111365	C	—	28–32	19–23	0–2 @ 625	12–14 @ 2200	—	—
	V8-366 Auto. Trans.	1111374	C	—	29–31	19–23	0–2 @ 625	12–14 @ 2200	—	—
	8-454, C, K–10	1112113	C	.019	29–31	19–23	0 @ 550	9 @ 2100	6	10 @ 15.7
	8-454, C, K, Suburban	1112504	C	.019	29–31	19–23	0 @ 550	9 @ 2100	8	8 @ 16
	8-454 C, K20–30	1112105	C	.019	29–31	19–23	0 @ 550	10 @ 2100	6	12 @ 14.3
	V8-455	1112172	CC	.016	30	—	0–2 @ 575	9 @ 1700	8–10	12½ @ 20.5
1975	6-250	1112863	C	—	—	—	0 @ 550	8 @ 2100	4	9 @ 12
	6-250 Calif.	1110650	C	—	—	—	0 @ 550	8 @ 2100	4	7 @ 12
	6-292	1112887	C	—	—	—	0 @ 550	12 @ 2100	10	5 @ 13
	6-292⑩	1110516	C	.019	31–34	—	0 @ 650	13 @ 2050	9	5 @ 13.25
	8-350	1112880	C	—	—	—	0 @ 600	11 @ 2100	4	9 @ 12
	8 350	1112888	C	—	—	—	0 @ 550	8 @ 2100	4	9 @ 12
	8-350	1112940	C	—	—	—	0 @ 600	10 @ 2100	8	7½ @ 15½
	8-350, 400 Calif.	1112884	C	—	—	—	0 @ 575	11 @ 2100	8	5 @ 13
	8-350⑨⑩	1111364	C	—	28–32	—	0 @ 550	11 @ 1900	—	—
	8-350⑩⑪	1112149	C	—	28–32	—	0 @ 450	11 @ 1900	—	—
	8-350⑩⑫	1111373	C	—	29–31	—	0 @ 550	11 @ 1900	—	—
	8-366⑬	1111378	C	—	28–32	—	0 @ 625	14 @ 2200	—	—
	8-366⑫⑬	1111379	C	—	29–31	—	0 @ 625	14 @ 2200	—	—
	8-366⑭	1112890	C	—	—	—	0 @ 600	14 @ 2200	—	—
	8-400	1112941	C	—	—	—	0 @ 500	9½ @ 1725	8	5 @ 13
	8-427⑬	1111378	C	—	28–32	—	0 @ 625	14 @ 2200	—	—
	8-427⑭	1112890	C	—	—	—	0 @ 600	14 @ 2200	—	—
	8-454 L.D.⑦	1112886	C	—	—	—	0 @ 900	6 @ 2100	4	9 @ 7
	8-454 L.D.⑧	1112943	C	—	—	—	0 @ 550	9 @ 2100	6	10 @ 15
	8-454 H.D.	1112869	C	—	—	—	0 @ 550	10 @ 2100	6	10 @ 15
	8-454 H.D.	1112494	C	—	—	—	0 @ 550	10 @ 2100	10	7 @ 17
	V8-455	1112172	CC	.016	30	—	0–2 @ 575	9 @ 1700	8–10	12½ @ 20.5

Continued

DISTRIBUTOR SPECIFICATIONS—Continued

Year	Model	Distributor Number	Rotation ①	Breaker Gap (New)	Dwell Angle Deg.	Breaker Arm Spring Tension	Centrifugal Advance Degrees @ R.P.M. of Distributor		Vacuum Advance	
							Advance Starts	Full Advance	Inches of Vacuum To Start Plunger	Max. Adv. Dist. Deg. @ Vacuum
	V8-455⑬	1112893	CC	—	—	—	0 @ 450	8 @ 1700	8–10	12 @ 20
	V8-455⑭	1112945	CC	—	—	—	0 @ 450	8 @ 1700	14–15	5 @ 19
1976	6-250 Std. Trans.	1110666	C	—	—	—	0 @ 500	10 @ 2100	4	12 @ 15
	6-250	1112863	C	—	—	—	0 @ 550	8 @ 2100	4	9 @ 12
	6-250 Calif. Std. Trans.	1110667	C	—	—	—	0 @ 500	10 @ 2100	4	7½ @ 12
	6-250 Calif. Auto. Trans.	1110650	C	—	—	—	0 @ 550	8 @ 2100	4	7½ @ 12
	6-250⑬⑮	1110666	C	—	—	—	0 @ 500	10 @ 2100	4	12 @ 15
	6-250, 292⑬	1112887	C	—	—	—	0 @ 550	12 @ 2050	10	5 @ 13
	6-292⑩	1110516	C	.019	31–34	—	0 @ 650	13 @ 2050	9	5 @ 13.25
	6-292 Calif.	1110664	C	—	—	—	0 @ 550	12 @ 2050	10	5 @ 13
	8-350 2 BBl.	1112880	C	—	—	—	0 @ 600	11 @ 2100	4	9 @ 12
	8-350 4 BBl.	1112888	C	—	—	—	0 @ 550	8 @ 1200	4	9 @ 12
	8-350 4 BBl.	1112940	C	—	—	—	0 @ 600	10 @ 2100	8	7½ @ 15½
	8-350 Calif. 4 BBl.	1112905	C	—	—	—	0 @ 600	11 @ 2100	6	7½ @ 12
	8-350, 400 Calif.	1112884	C	—	—	—	0 @ 575	11 @ 2100	8	5 @ 13
	V8-350⑨⑩	1111364	C	—	28–32	—	0 @ 550	11 @ 1900	—	—
	V8-350⑪⑩	1112149	C	—	28–32	—	0 @ 450	11 @ 1900	—	—
	V8-350⑫⑩	1111373	C	—	28–32	—	0 @ 550	11 @ 1900	—	—
	V8-366⑬⑩	1111378	C	—	29–31	—	0 @ 625	14 @ 2200	—	—
	V8-366⑫⑬⑩	1111379	C	—	28–32	—	0 @ 625	14 @ 2200	—	—
	V8-366⑭⑩	1112890	C	—	—	—	0 @ 600	14 @ 2200	—	—
	8-400	1112941	C	—	—	—	0 @ 500	9½ @ 1725	8	5 @ 13
	V8-427⑬⑩	1111378	C	—	18–32	—	0 @ 625	14 @ 2200	—	—
	V8-427⑭⑩	1112890	C	—	—	—	0 @ 600	14 @ 2200	—	—
	8-454⑦	1112886	C	—	—	—	0 @ 650	6 @ 2100	4	9 @ 7
	8-454⑧	1112494	C	—	—	—	0 @ 550	10 @ 2100	10	7½ @ 17
	8-454 Calif.	1112869	C	—	—	—	0 @ 550	10 @ 2100	6	10 @ 15
	V8-455⑬	1112893	CC	—	—	—	0 @ 450	8 @ 1700	8–10	12 @ 20
	V8-455⑭	1112945	CC	—	—	—	0 @ 450	8 @ 1700	14–15	5 @ 19
1977	6-250⑬⑮⑰	1110678	C	—	—	—	0 @ 500	10 @ 2100	4	12 @ 15
	6-250⑫⑬⑰	1110674	C	—	—	—	0 @ 550	8 @ 2100	4	9 @ 12
	6-250⑭⑮⑰	1110692	C	—	—	—	0 @ 550	10 @ 2100	4	7½ @ 12
	6-250⑫⑭⑰	1110682	C	—	—	—	0 @ 550	8 @ 2100	4	9 @ 12
	6-250, 292⑬⑯	1110679	C	—	—	—	0 @ 550	12 @ 2050	10	5 @ 13
	6-292⑭⑯	1110680	C	—	—	—	0 @ 550	12 @ 2050	10	5 @ 13
	V8-305⑰	1103252	C	—	—	—	0 @ 500	10 @ 1900	4	9 @ 12
	V8-305⑯	1103237	C	—	—	—	0 @ 500	10 @ 1900	8	5 @ 13
	V8-350⑬⑰⑱	1103253	C	—	—	—	0 @ 550	11 @ 2300	4	9 @ 12
	V8-350⑰⑲	1103246	C	—	—	—	0 @ 600	11 @ 2100	4	9 @ 12
	V8-350⑭⑰	1103254	C	—	—	—	0 @ 600	11 @ 2100	6	7½ @ 12
	V8-350⑬⑯	1103274	C	—	—	—	0 @ 600	10 @ 2100	6	7½ @ 12
	V8-350⑭⑯	1103250	C	—	—	—	0 @ 575	11 @ 2100	8	5 @ 13
	V8-350⑩	1103251	C	—	—	—	0 @ 500	10 @ 1900	—	—
	V8-366	1103241	C	—	—	—	0 @ 550	13 @ 2200	—	—
	V8-400⑬	1103249	C	—	—	—	0 @ 500	9½ @ 1725	6	7½ @ 12
	V8-400⑭	1103250	C	—	—	—	0 @ 575	11 @ 2100	8	5 @ 13
	V8-403	—	CC	—	—	—	—	—	—	—
	V8-427	1103241	C	—	—	—	0 @ 550	13 @ 2200	—	—
	V8-454⑬	1103238	C	—	—	—	0 @ 550	10 @ 2100	10	7½ @ 17
	V8-454⑭	1103240	C	—	—	—	0 @ 550	10 @ 2100	6	10 @ 15
	V8-455	—	CC	—	—	—	—	—	—	—

Continued

DISTRIBUTOR SPECIFICATIONS—Continued

Year	Model	Distributor Number	Rotation (1)	Breaker Gap (New)	Dwell Angle Deg.	Breaker Arm Spring Tension	Centrifugal Advance Degrees @ R.P.M. of Distributor		Vacuum Advance	
							Advance Starts	Full Advance	Inches of Vacuum To Start Plunger	Max. Adv. Dist. Deg. @ Vacuum
1978	V6-200(22)(13)(17)	1110696	C	—	—	—	0 @ 500	10 @ 1900	3	8 @ 6.5
	V6-231(22)(13)(17)	1110695	C	—	—	—	0 @ 1000	9 @ 1800	7	12 @ 13
	V6-231(22)(14)(17)	1110731	C	—	—	—	0 @ 1000	9 @ 1800	4	8 @ 9
	6-250(3)(13)(17)	1110717	C	—	—	—	0 @ 550	12 @ 2050	4	9 @ 12
	6-250(3)(14)(19)(15)(17)	1110715	C	—	—	—	0 @ 500	10 @ 2100	4	9 @ 12
	6-250(3)(14)(19)(12)(17)	1110719	C	—	—	—	0 @ 550	8 @ 2100	4	9 @ 12
	6-250(16)(13)(6)	1110720	C	—	—	—	0 @ 550	12 @ 2050	10	5 @ 13
	6-292(16)(13)(3)	1110720	C	—	—	—	0 @ 550	12 @ 2050	10	5 @ 13
	6-292(14)(16)(3)	1110721	C	—	—	—	0 @ 550	12 @ 2050	10	5 @ 13
	V8-305(16)(13)(4)	1103237	C	—	—	—	0 @ 500	10 @ 1900	8	5 @ 13
	V8-305(22)(2)(13)(17)	11103281	C	—	—	—	0 @ 500	10 @ 1900	4	9 @ 12
	V8-305(22)(17)	11103282	C	—	—	—	0 @ 500	10 @ 1900	4	10 @ 10
	V8-305(3)(13)(17)	1110338	C	—	—	—	0 @ 500	10 @ 1900	5	12 @ 12
	V8-350(22)(13)(17)	1103337	C	—	—	—	0 @ 550	8 @ 1200	4	12 @ 10
	V8-350(22)(14)(17)	1103285	C	—	—	—	0 @ 600	11 @ 2100	4	5 @ 8
	V8-350(3)(13)(19)(17)	1103286	C	—	—	—	0 @ 550	11 @ 2300	4	9 @ 12
	V8-350(14)(3)(17)	1103302	C	—	—	—	0 @ 550	11 @ 2300	6	7.5 @ 12
	V8-350(14)(6)(17)	1103339	C	—	—	—	0 @ 550	11 @ 2300	4	5 @ 8
	V8-350(14)(16)(3)	1103250	C	—	—	—	0 @ 575	11 @ 2100	8	5 @ 13
	V8-350(9)(13)(14)(16)	1103251	C	—	—	—	0 @ 500	10 @ 1900	—	—
	V8-350(13)(16)(3)	1103274	C	—	—	—	0 @ 600	10 @ 2100	6	7.5 @ 13
	V8-366(9)(13)(14)(16)	1103241	C	—	—	—	0 @ 550	13 @ 2200	—	—
	V8-400(16)(3)(13)	1103249	C	—	—	—	0 @ 500	9.5 @ 1725	6	7.5 @ 12
	V8-400(16)(3)(14)	1103250	C	—	—	—	0 @ 575	11 @ 2100	8	5 @ 13
	V8-400(14)(17)(3)	1103301	C	—	—	—	0 @ 500	9.5 @ 1725	4	5 @ 8
	V8-403	—	CC	—	—	—	—	—	—	—
	V8-427(9)(13)(14)(16)	1103241	C	—	—	—	0 @ 550	13 @ 2200	—	—
	V8-454(16)(13)(3)	1103238	C	—	—	—	0 @ 550	10 @ 2100	10	7.5 @ 17
	V8-454(16)(14)(3)	1103240	C	—	—	—	0 @ 550	10 @ 2100	6	10 @ 15
	V8-454(13)(14)(17)(4)	1103240	C	—	—	—	0 @ 550	10 @ 2100	6	10 @ 15
	V8-454(4)(14)(17)	1103358	C	—	—	—	0 @ 550	10 @ 2100	4	7.5 @ 10
1979	V6-200(22)(13)(15)	1110696	C	—	—	—	0 @ 500	10 @ 1900	2	8 @ 7.5
	V6-200(22)(13)(12)	1110756	C	—	—	—	0 @ 700	7 @ 1900	2	12 @ 10
	V6-231(22)(13)(20)	1110766	C	—	—	—	0 @ 1000	7.5 @ 1800	4	12 @ 12
	V6-231(22)(14)	1110767	C	—	—	—	0 @ 1000	7.5 @ 1800	3	10 @ 12
	6-250(17)(3)(13)	1110717	C	—	—	—	0 @ 550	12 @ 2050	4	9 @ 12
	6-250(17)(3)(14)	1110749	C	—	—	—	0 @ 550	8 @ 2100	4	5 @ 8
	V8-267(22)(13)(15)	1103371	C	—	—	—	0 @ 500	11 @ 2200	3	12 @ 10
	V8-267(22)(13)(12)	1103370	C	—	—	—	0 @ 650	8 @ 2100	3	12 @ 10
	6-292	1110753	C	—	—	—	0 @ 550	12 @ 2050	4	5 @ 8
	V8-305(22)(13)(19)(12)(15)	1103282	C	—	—	—	0 @ 500	10 @ 1900	4	10 @ 11
	V8-305(22)(12)(13)	1103379	C	—	—	—	0 @ 500	10 @ 1900	3	10 @ 8.5
	V8-305(22)(12)(14)	1103368	C	—	—	—	0 @ 500	10 @ 1900	4	5 @ 8
	V8-305(17)(13)(15)	1103381	C	—	—	—	0 @ 600	10 @ 2100	3	10 @ 7.5
	V8-305(17)(12)(13)(15)	1103374	C	—	—	—	0 @ 600	10 @ 2100	4	8 @ 7.5
	V8-305(17)(13)(15)	1103369	C	—	—	—	0 @ 600	10 @ 2100	3	8 @ 6.5
	V8-350(22)(19)(12)(3)	1103353	C	—	—	—	0 @ 500	11 @ 2300	4	10 @ 11
	V8-350(16)(13)(14)	1103375	C	—	—	—	0 @ 575	11 @ 2100	4	5 @ 8
	V8-350(3)(13)(15)(17)	1103372	C	—	—	—	0 @ 550	11 @ 2300	4	7 @ 8
	V8-350(14)(15)(3)(17)	1103302	C	—	—	—	0 @ 550	11 @ 2300	6	7.5 @ 12
	V8-350(12)(13)(14)(15)(17)	1103339	C	—	—	—	0 @ 550	11 @ 2300	4	5 @ 8
	V8-350(13)(12)(17)	1103353	C	—	—	—	0 @ 550	11 @ 2300	4	10 @ 10

Continued

DISTRIBUTOR SPECIFICATIONS—Continued

Year	Model	Distributor Number	Rotation ①	Breaker Gap (New)	Dwell Angle Deg.	Breaker Arm Spring Tension	Centrifugal Advance Degrees @ R.P.M. of Distributor		Vacuum Advance	
							Advance Starts	Full Advance	Inches of Vacuum To Start Plunger	Max. Adv. Dist. Deg. @ Vacuum
	V8-350⑫③⑰	1103302	C	—	—	—	0 @ 550	11 @ 2300	6	7.5 @ 12
	V8-350⑫⑭⑥⑰	1103286	C	—	—	—	0 @ 550	11 @ 2300	4	9 @ 12
	V8-350②	1103378	C	—	—	—	—	—	—	—
	V8-366②	1103377	C	—	—	—	—	—	—	—
	V8-400⑫⑬⑭⑰	1103375	C	—	—	—	0 @ 575	11 @ 2100	4	5 @ 8
	V8-400⑫⑬⑭⑰	1103301	C	—	—	—	0 @ 500	9.5 @ 1725	4	5 @ 8
	V8-427②	1103377	C	—	—	—	—	—	—	—
	V8-454②	1103377	C	—	—	—	—	—	—	—
	V8-454⑫⑬⑭⑮⑯	1103376	C	—	—	—	0 @ 550	10 @ 2100	8	5 @ 13
	V8-454⑮⑬⑭⑰	1103240	C	—	—	—	—	—	—	—
	V8-454⑫⑬⑭⑰	1103358	C	—	—	—	—	—	—	—
1980	V6-229㉒⑬⑮	1110696	C	—	—	—	0 @ 500	5 @ 1900	2	8 @ 7.5
	V6-229㉒⑫⑬	1110756	C	—	—	—	0 @ 700	7 @ 1900	2	12 @ 10
	V6-231㉒⑫⑬	1110766	C	—	—	—	0–2 @ 1000	7.5–8.5 @ 1800	4	12 @ 12
	V6-231㉒⑫⑭	1110767	C	—	—	—	0–2 @ 1000	7.5–8.5 @ 1800	3	10 @ 12
	6-250⑫⑬	1110717	C	—	—	—	0 @ 550	12 @ 2050	4	9 @ 12
	6-250⑭⑮	1110747	C	—	—	—	0 @ 550	12 @ 2050	4	7.5 @ 12
	6-250⑫⑭	1110749	C	—	—	—	0 @ 550	8 @ 2100	4	5 @ 8
	6-250⑬⑮	1110755	C	—	—	—	0 @ 550	12 @ 2050	5	8 @ 11.5
	6-292	1110753	C		—	—	0 @ 550	12 @ 2050	4	5 @ 8
	V8-267⑫⑬㉒	1103371	C	—	—	—	0 @ 500	11 @ 2200	3	12 @ 10
	V8-305⑬⑮㉒	1103282	C	—	—	—	0 @ 500	10 @ 1900	4	10 @ 11
	V8-305⑫⑭㉒	1103368	C	—	—	—	0 @ 500	10 @ 1900	4	5 @ 8
	V8-305⑬	1103369	C	—	—	—	0 @ 600	10 @ 2100	3	8 @ 6.5
	V8-305⑫⑬㉒	1103379	C	—	—	—	0 @ 500	10 @ 1900	3	10 @ 8.5
	V8-305⑬⑮	1103381	C	—	—	—	0 @ 600	10 @ 2100	3	5 @ 6.5
	V8-350⑫	1103339	C	—	—	—	0 @ 550	11 @ 2300	4	5 @ 8
	V8-350㉓	1103372	C	—	—	—	0 @ 550	11 @ 2300	4	7 @ 8
	V8-350⑬⑮	1103375	C	—	—	—	0 @ 575	11 @ 2100	4	5 @ 8
	V8-350⑭	1103420	C	—	—	—	0 @ 900	12 @ 2000	10	5 @ 13
	V8-350⑫⑭	1103423	C	—	—	—	0 @ 500	9.5 @ 1725	4	7.5 @ 10
	V8-350	1103435	C	—	—	—	0 @ 550	11 @ 2300	3	8 @ 6.5
	V8-350	1103436	C	—	—	—	0 @ 550	11 @ 2300	3	10 @ 7.5
	V8-350⑫⑬	1103439	C	—	—	—	0 @ 500	9.5 @ 1725	4	7 @ 8
	V8-400⑫⑬	1103375	C	—	—	—	0 @ 575	11 @ 2100	4	5 @ 8
	V8-400⑫⑭	1103420	C	—	—	—	0 @ 900	12 @ 2000	10	5 @ 13
	V8-400⑫⑭	1103423	C	—	—	—	0 @ 500	9.5 @ 1725	4	7.5 @ 10
	V8-400⑫⑬	1103439	C	—	—	—	0 @ 500	9.5 @ 1725	4	7 @ 8
	V8-454	1103376	C	—	—	—	0 @ 550	10 @ 2100	8	5 @ 12
	V6-229⑫⑬㉒㉓	1110754	C	—	—	—	—	—	—	—
	V6-229⑬⑮㉒㉓	1103434	C	—	—	—	—	—	—	—
	V6-231⑫⑭㉒㉓	1111386	C	—	—	—	—	—	—	—
	6-250⑬	1110589	C	—	—	—	0 @ 550	12 @ 2050	3	8 @ 6.5
	6-250⑬⑮	1110590	C	—	—	—	0 @ 550	12 @ 2050	4	7.5 @ 10
	6-250⑬⑮	1111388	C	—	—	—	0 @ 550	12 @ 2050	3	10 @ 7.5
	6-250⑬⑮	1110753	C	—	—	—	0 @ 550	12 @ 2050	4	5 @ 8
	6-250	1110749	C	—	—	—	0 @ 550	8 @ 2100	4	5 @ 8
	6-292	1110753	C	—	—	—	0 @ 550	12 @ 2050	4	5 @ 8
	V8-267⑫⑬㉒㉓	1103443	C	—	—	—	—	—	—	—
	V8-305㉒㉓	1103443	C	—	—	—	—	—	—	—

Continued

DISTRIBUTOR SPECIFICATIONS—Continued

Year	Model	Distributor Number	Rotation ①	Breaker Gap (New)	Dwell Angle Deg.	Breaker Arm Spring Tension	Centrifugal Advance Degrees @ R.P.M. of Distributor		Vacuum Advance	
							Advance Starts	Full Advance	Inches of Vacuum To Start Plunger	Max. Adv. Dist. Deg. @ Vacuum
	V8-305⑫⑮㉔	1103369	C	—	—	—	0 @ 600	10 @ 2100	3	8 @ 6.5
	V8-305⑬㉔	1103381	C	—	—	—	0 @ 600	10 @ 2100	3	5 @ 7.5
	V8-305⑬㉓㉕	1103464	C	—	—	—	—	—	—	—
	V8-305⑬⑮㉓㉕	1103465	C	—	—	—	—	—	—	—
	V8-305⑫⑭㉕	1103432	C	—	—	—	0 @ 500	7 @ 2000	4	7.5 @ 12
	V8-350⑬㉖	1103353	C	—	—	—	0 @ 550	11 @ 2300	4	10 @ 10
	V8-350⑭㉖	1103339	C	—	—	—	0 @ 550	11 @ 2300	4	5 @ 8
	V8-350⑭㉖	1103433	C	—	—	—	0 @ 550	11 @ 2300	4	7.5 @ 12
	V8-350⑬㉗	1103375	C	—	—	—	0 @ 500	9.5 @ 1725	4	7 @ 8
	V8-350⑭㉗	1103420	C	—	—	—	0 @ 500	11 @ 2100	4	7.5 @ 10
	V8-454	1103376	C	—	—	—	0 @ 550	10 @ 2100	8	5 @ 13
1982	6-250⑮	1103481	C	—	—	—	—	—	—	—
	6-250⑮	1103482	C	—	—	—	—	—	—	—
	6-250⑫	1103483	C	—	—	—	—	—	—	—
	6-250⑫	1103484	C	—	—	—	—	—	—	—
	6-250⑮	1103488	C	—	—	—	—	—	—	—
	6-250⑮	1111388	C	—	—	—	—	—	—	—
	6-292	1110753	C	—	—	—	0 @ 550	12 @ 2050	4	5 @ 8
	V6-229㉒㉓	1110584	C	—	—	—	—	—	—	—
	V6-231㉒㉓	1103470	C	—	—	—	—	—	—	—
	V8-267㉒㉓	1103460	C	—	—	—	—	—	—	—
	V8-305㉒㉓	1103460	C	—	—	—	—	—	—	—
	V8-305	1103465	C	—	—	—	—	—	—	—
	V8-305	1103487	C	—	—	—	—	—	—	—
	V8-350②	1103378	C	—	—	—	0 @ 500	10 @ 1900	—	—
	V8-350⑬㉖	1110436	C	—	—	—	—	—	—	—
	V8-350⑭㉖	1103433	C	—	—	—	—	—	—	—
	V8-350⑭㉖	1103435	C	—	—	—	—	—	—	—
	V8-350⑬㉗	1103375	C	—	—	—	—	—	—	—
	V8-350⑭㉗	1103420	C	—	—	—	—	—	—	—
	V8-366, 427	1103377	C	—	—	—	0 @ 550	13 @ 2200	—	—
	V8-454	1103376	C	—	—	—	—	—	—	—

①—As viewed from top; C, clockwise; CC, counterclockwise.
②—Medium & Heavy Duty Truck.
③—C, K-10/1500, C, K-20/2500, G-10/1500, 20/2500, G-30/3500 Sport Van.
④—C, K-10/1500, 20/2500 except Suburban.
⑤—C-10/1500 Suburban.
⑥—C, K-10/1500 except Suburban.
⑦—With catalytic converter.
⑧—Without catalytic converter.
⑨—With governor.
⑩—Heavy duty.
⑪—Without governor.
⑫—With auto. trans.
⑬—Exc. Calif.
⑭—Calif.
⑮—Std. trans.
⑯—Heavy duty emissions, GVWR 6001 & above.
⑰—Light duty emissions, GVWR 6000 & below.
⑱—Exc. high altitude engines.
⑲—High altitude engines.
⑳—High Altitude, auto. trans. & manual trans. Exc. Calif.
㉑—Auto. trans. Exc. High Altitude.
㉒—Caballero, El Camino & Sprint.
㉓—Electronic spark control.
㉔—2 barrel carb., series 10-30.
㉕—4 barrel carb., series 10-30.
㉖—Light duty emissions, GVWR 8500 lbs. & under.
㉗—Heavy duty emissions, GVWR 8501 lbs. & above.

STARTING MOTOR SPECIFICATIONS

Starter Model	Year	Brush Spring Tension ①	Free Speed Test			Resistance Test③	
			Amperes	Volts	R.P.M.	Amperes	Volts
1108195, 196	1974	—	50–75②	9	6500–10000	—	—
1108360	1974–76	35	55–80②	9	3500–6000	—	—
1108365, 367	1974	35	50–80②	9	5500–10500	—	—
1113184	1974	80	50–70②	9	—	—	—
1113686	1974	80	130–160②	9	5000–7000	—	—
1108427	1974	—	55–80②	9	3500–6000	—	—
1108430	1974	—	65–95②	9	7500–10500	—	—
1108480	1974	—	50–80②	9	5500–10500	—	—
1108485, 486	1973–76	—	35–75②	9	6000–9000	—	—
1108487	1973–76	—	35–75②	9	6000–9000	—	—
1108502	1973–74	—	65–95②	9	7500–10500	—	—
1108522, 761	1974–78	—	65–95②	9	7500–10500	—	—
1114101	1973–80	—	120–150②	9	3000–4500	—	—
1114143	1973–79	—	140–190②	9	4000–7000	—	—
1114161	1973–79	—	140–190②	9	4000–7000	—	—
1114178	1973–80	—	120–150②	9	3000–4500	—	—
1108771, 772	1975	—	50–75②	9	6500–10000	—	—
1108744, 778	1975–78	—	50–80②	9	5500–10500	—	—
1108746, 779	1975–76	—	50–80②	9	5500–10500	—	—
1108747, 780	1975–76	—	50–80②	9	3500–6000	—	—
1108748, 781	1975–76	—	65–90②	9	7500–10500	—	—
1108502, 782	1975–76	—	65–90②	9	7500–10500	—	—
1108752	1975–76	—	65–95②	9	7500–10500	—	—
1108753	1975–76	—	65–92②	9	7500–10500	—	—
1108754	1975–76	—	65–95②	9	7500–10500	—	—
1108783	1975–76	—	55–80②	9	3500–6000	—	—
1108784	1975–76	—	55–80②	9	3500–6000	—	—
1108785	1975–76	—	55–80②	9	3500–6000	—	—
1108787	1975–76	—	35–75②	9	6000–9000	—	—
1108788	1975–76	—	65–95②	9	7500–10500	—	—
1108792	1975–76	—	35–75②	9	6000–9000	—	—
1109631	1975–80	—	95–120②	20	5500–7500	—	—
1109772	1975–79	—	95–120②	20	5500–7500	—	—
1113184	1975–76	—	50–70②	9	3500–5500	—	—
1114086	1975–80	—	140–190②	9	4000–7000	—	—
1114101	1975–80	—	120–150②	9	3000–4500	—	—
1114733	1975–79	—	140–190②	9	4000–7000	—	—
1108793	1977	—	55–80②	9	3500–6000	—	—
1109030	1977	—	55–80②	9	3500–6000	—	—
1109032	1977	—	35–75②	9	6000–9000	—	—
1109033	1977	—	65–95②	9	7500–10500	—	—
1109034	1977	—	65–95②	9	7500–10500	—	—
1109035	1977	—	65–95②	9	7500–10500	—	—
1113243	1977	—	50–70②	9	3500–5500	—	—
1109056	1977–82	—	50–80②	9	7500–10500	—	—
1109798	1977–82	—	50–80②	9	5500–10500	—	—
1109052	1977–82	—	65–95②	9	7500–10500	—	—
1108776	1977–82	—	65–95②	9	7500–10500	—	—
1108780	1977–82	—	50–80②	9	3500–6000	—	—
1108779, 778	1977–82	—	50–80②	9	5500–10500	—	—
1108799	1978	—	50–80②	9	5500–10500	—	—
1109075④	1978–83	—	35–75②	9	6000–9000	—	—
1109075⑤	1978–83	—	65–95②	9	7500–10500	—	—
1109076	1978–83	—	65–95②	9	7500–10500	—	—
1109077	1978–83	—	65–95②	9	7500–10500	—	—

Continued

STARTING MOTOR SPECIFICATIONS—Continued

Starter Model	Year	Brush Spring Tension ①	Free Speed Test			Resistance Test③	
			Amperes	Volts	R.P.M.	Amperes	Volts
1109078	1978–83	—	55–80②	9	3500–6000	—	—
1109079	1978–83	—	55–80②	9	3500–6000	—	—
1109086	1978	—	95–120②	9	4000–7000	—	—
1114141	1978–83	—	140–215②	9	4000–7000	—	—
1114161	1978–80	—	140–215②	9	4000–7000	—	—
1114733	1978–83	—	140–215②	9	4000–7000	—	—
1109059	1978–79	—	65–95②	9	7500–10500	—	—
1108774	1979	—	60–88②	9	6500–10100	—	—
1108778	1979–80	—	50–80②	9	5500–10500	—	—
1109061	1979–80	—	60–85②	9	6800–10300	—	—
1109524	1980–81	—	50–80②	9	7500–11400	—	—
1114773	1980	—	140–215②	9	4000–7000	—	—
1114771	1981–83	—	140–215②	9	4000–7000	—	—
1114810	1981–83	—	140–215②	9	4000–7000	—	—
1114817	1981–83	—	120–150②	9	3000–4500	—	—
1114826	1981–83	—	140–215②	9	4000–7000	—	—
1114895	1981–83	—	140–215②	9	4000–7000	—	—
1114897	1981–83	—	140–215②	9	4000–7000	—	—
1114936	1981–83	—	120–150②	9	3000–4500	—	—
1114984	1981–83	—	140–215②	9	4000–7000	—	—
1998229	1981–83	—	65–95②	9	7500–10500	—	—
1109295	1981–83	—	210–270②	11	5000–7400	—	—
1109534	1982–83	—	50–80②	10.6	7500–11400	—	—
1998236	1982–83	—	60–85②	9	6800–10300	—	—
1109535	1983	—	45–70②	9	7000–11900	—	—
1109563	1983	—	—	10	—	—	—
1998211	1983	—	65–95②	10	7500–10500	—	—
1998241	1983	—	65–95②	10	7500–10500	—	—
1998243	1983	—	65–95②	10	7500–10500	—	—
1998244	1983	—	60–85②	10	6800–10300	—	—
1998396	1983	—	70–110②	10	6500–10700	—	—
1998397	1983	—	70–110②	10	6500–10700	—	—
1998554	1983	—	160–140②	10	4400–6300	—	—

①—Minimum
②—Includes solenoid.
③—Check capacity of motor by using a 500 ampere meter and a carbon pile rheostat to control voltage. Apply volts listed across motor with armature locked. Current should be as listed.
④—Except heavy duty starter.
⑤—Heavy duty starter.

ALTERNATOR & REGULATOR SPECIFICATIONS

Year	Alternator					Regulator						
	Model	Field Current @ 80° F. 12 Volts	Cold Output @ 14 Volts		Rated Hot Output Amperes	Model	Field Relay			Voltage Regulator		
			Amperes @ 2000 R.P.M.	Amperes @ 5000 R.P.M.			Air Gap	Point Gap	Closing Voltage	Air Gap	Point Gap	Volts @ 125° F.
1974	1117225	4.1–4.5	—	77	75	Integral	—	—	—	—	—	—
	1117231	4.1–4.5	—	77	75	Integral	—	—	—	—	—	—
	1100497	4.4–4.9	—	36⑤	37	Integral	—	—	—	—	—	—
	1102346	4.0–4.5	—	37	42	Integral	—	—	—	—	—	—
	1102347	4.0–4.5	—	55	61	Integral	—	—	—	—	—	—
	1100573	4.0–4.5	—	37	42	Integral	—	—	—	—	—	—
	1100597	4.0–4.5	—	55	61	Integral	—	—	—	—	—	—
	1100547	4–4.5	—	37	42	Integral	—	—	—	—	—	—
	1100548	4–4.5	—	55	61	Integral	—	—	—	—	—	—
	1101022	4–4.5	—	74	80	Integral	—	—	—	—	—	—
	1102352	4–4.5	—	55	61	Integral	—	—	—	—	—	—
	1117141	4–4.5	—	110	105	Integral	—	—	—	—	—	—
	1117143	4–4.5	—	150	145	Integral	—	—	—	—	—	—
	655988	3	—	97②	105	Integral	—	—	—	—	—	—
	1100838	2.2–2.6	25	35	37	1119515	.015	.030	1.5–3.2	.067	.014	13.5–14.4
	1100842	2.2–2.6	28	40	42	1119507	.015	.030	3.8–7.2	.067	.014	13.5–14.4
	1100842	2.2–2.6	28	40	42	1119515	.015	.030	1.5–3.2	.067	.014	13.5–14.4
	1100849	2.2–2.6	33	58	61	1119507	.015	.030	3.8–7.2	.067	.014	13.5–14.4
	1100849	2.2–2.6	33	58	61	1119515	.015	.030	1.5–3.2	.067	.014	13.5–14.4
	1117756	4.1–4.6	20①	55②	62	9000590⑥	.012	.020	7–8	—	—	13.7–14.3
	1117781	4.1–4.6	20①	55②	62	9000590⑥	.012	.020	7–8	—	—	13.7–14.3
	1117782	4.1–4.6	20①	55②	62	1116374	—	—	4.5–6.5	—	—	13.6–14.3
	1117782	4.1–4.6	20①	55②	62	1116378	—	—	2–3	—	—	13.6–14.3
	1117787	4.1–4.6	20①	55②	62	1116374	—	—	4.5–6.5	—	—	13.6–14.3
	1100934	4.4–4.5	—	36⑤	37	Integral	—	—	—	—	—	—
1974	1100080	—	54	73	80	Integral	—	—	—	—	—	—
	1100081	—	48	61	65	Integral	—	—	—	—	—	—
	1102382	4.0–4.5	—	37	42	Integral	—	—	—	—	—	—
	1102383	4.0–4.5	—	55	61	Integral	—	—	—	—	—	—
	1100575	4.0–4.5	—	50	55	Integral	—	—	—	—	—	—
1974–75	1100545	4.0–4.5	—	31	32	Integral	—	—	—	—	—	—
	1100546	4.0–4.5	—	50	55	Integral	—	—	—	—	—	—
	1100559	4.0–4.5	—	31	32	Integral	—	—	—	—	—	—
	1100560	4.0–4.5	—	50	55	Integral	—	—	—	—	—	—
1974–76	1101016	4.4–4.9	—	76	80	Integral	—	—	—	—	—	—
1975	1100497	4.4–4.9	—	33	37	Integral	—	—	—	—	—	—
	1100575	4.0–4.5	—	50	55	Integral	—	—	—	—	—	—
	1100597	4.0–4.5	—	55	61	Integral	—	—	—	—	—	—
	1102346	4.0–4.5	—	38	42	Integral	—	—	—	—	—	—
	1102347	4.0–4.5	—	55	61	Integral	—	—	—	—	—	—
	1102483	4.0–4.5	—	33	37	Integral	—	—	—	—	—	—
	1102493	4.0–4.5	—	38	42	Integral	—	—	—	—	—	—
	1102500	4.0–4.5	—	50	55	Integral	—	—	—	—	—	—
	1102856	4.0–4.5	—	33	37	Integral	—	—	—	—	—	—
1975–76	1102497	4.0–4.5	30	57	61	Integral	—	—	—	—	—	—
	1102499	4.0–4.5	25	38	42	Integral	—	—	—	—	—	—
	1102943	2.2–2.6	25	35	37	1119515	—	—	—	—	—	—
	1102944	2.2–2.6	28	40	42	1119515	—	—	—	—	—	—
	1100081	4.4–4.9	45	57	65	Integral	—	—	—	—	—	—
	1102945	2.2–2.6	33	58	61	1119507	—	—	—	—	—	—
	1117225	4.1–4.5	—	77	75	Integral	—	—	—	—	—	—
	1117782	4.1–4.6	20⑦	55⑦	62	1116374	—	—	—	—	—	—

Continued

ALTERNATOR & REGULATOR SPECIFICATIONS—Continued

Year	Alternator					Regulator						
	Model	Field Current @ 80° F. 12 Volts	Cold Output @ 14 Volts		Rated Hot Output Amperes	Model	Field Relay			Voltage Regulator		
			Amperes @ 2000 R.P.M.	Amperes @ 5000 R.P.M.			Air Gap	Point Gap	Closing Voltage	Air Gap	Point Gap	Volts @ 125° F.
	1117231	4.1–4.5	—	77	75	1116378	—	—	—	—	—	—
	1117481	3.6–4.3	68	86	90	Integral	—	—	—	—	—	—
	1102498	2.2–2.6	33	58	61	Integral	—	—	—	—	—	—
	1101029	4.4–4.9	55	76	80	Integral	—	—	—	—	—	—
	1117147	4.0–4.5	80	110	105	Integral	—	—	—	—	—	—
	711925⑧	2.8	81	105	105	Integral	—	—	—	—	—	—
1976	1102394	4.0–4.5	—	33	37	Integral	—	—	—	—	—	—
	1102478	4.0–4.5	—	51	55	Integral	—	—	—	—	—	—
	1102479	4.0–4.5	—	51	55	Integral	—	—	—	—	—	—
	1102480	4.0–4.5	—	57	61	Integral	—	—	—	—	—	—
	1102489	4.0–4.5	—	38	42	Integral	—	—	—	—	—	—
	1102490	4.0–4.5	—	57	61	Integral	—	—	—	—	—	—
	1102491	4.4–4.9	—	33	37	Integral	—	—	—	—	—	—
	1102841	4.0–4.5	—	38	42	Integral	—	—	—	—	—	—
1977–78	1102499	4.0–4.5	25	38	42	Integral	—	—	—	—	—	—
	1102497	4.0–4.5	30	57	61	Integral	—	—	—	—	—	—
	1101029	4.4–4.9	55	76	80	Integral	—	—	—	—	—	—
	1101020	4.4–4.9	63	97	100	Integral	—	—	—	—	—	—
	711925⑧	2.8	81	105	105	Integral	—	—	—	—	—	—
	2009386⑧	5–6	85	130	130	Integral	—	—	—	—	—	—
1977–80	711925⑧	2.8	81	105⑤	105	Integral	—	—	—	—	—	—
	1100080	4.4–4.9	52	72	80	Integral	—	—	—	—	—	—
	1100081	4.4–4.9	46	58	65	Integral	—	—	—	—	—	—
	1101016	4–4.5	—	76	80	Integral	—	—	—	—	—	—
	1101020	4.4–4.9	63	97	100	Integral	—	—	—	—	—	—
	1101028	4–4.5	—	76	80	Integral	—	—	—	—	—	—
	1101029	4.4–4.9	55	76	80	Integral	—	—	—	—	—	—
	1101030	4.4–4.9	55	76	80	Integral	—	—	—	—	—	—
	1102394	4–4.5	—	33	37	Integral	—	—	—	—	—	—
	1102480	4–4.5	—	57	61	Integral	—	—	—	—	—	—
	1102485	4–4.5	—	38	42	Integral	—	—	—	—	—	—
	1102486	4–4.5	—	57	61	Integral	—	—	—	—	—	—
	1102491	4–4.5	—	33	37	Integral	—	—	—	—	—	—
	1102497	4–4.5	30	57	61	Integral	—	—	—	—	—	—
	1102499	4–4.5	25	38	42	Integral	—	—	—	—	—	—
	1102841	4–4.5	—	38	42	Integral	—	—	—	—	—	—
	1102886	4–4.5	—	57	61	Integral	—	—	—	—	—	—
	1102887	4–4.5	—	38	42	Integral	—	—	—	—	—	—
	1102888	4–4.5	—	57	61	Integral	—	—	—	—	—	—
	1102889	4–4.5	—	33	37	Integral	—	—	—	—	—	—
	1103046	4–4.5	32	60	63	Integral	—	—	—	—	—	—
	1117241	4.1–4.5	—	77	75	Integral	—	—	—	—	—	—
	1117244	4.1–4.5	—	77	75	Integral	—	—	—	—	—	—
	1117481	3.6–4.3	68②	86④	90	Integral	—	—	—	—	—	—
	1117732	3.6–4.3	72②	90④	90	Integral	—	—	—	—	—	—
	2009386⑧	5–6	85	130③	130	Integral	—	—	—	—	—	—
	5148785⑧	—	15⑨	75③	75	Integral	—	—	—	—	—	—
1980	1101042	4–4.6	55	70	70	Integral	—	—	—	—	—	—
	1101044	4–4.5	—	—	70	Integral	—	—	—	—	—	—
	1101066	4–4.5	—	—	70	Integral	—	—	—	—	—	—
	1101071	4–4.5	—	—	70	Integral	—	—	—	—	—	—
	1103043	4–4.5	—	—	42	Integral	—	—	—	—	—	—
	1103044	4–4.5	—	—	63	Integral	—	—	—	—	—	—

Continued

ALTERNATOR & REGULATOR SPECIFICATIONS—Continued

Year	Alternator					Regulator						
	Model	Field Current @ 80° F. 12 Volts	Cold Output @ 14 Volts		Rated Hot Output Amperes	Model	Field Relay			Voltage Regulator		
			Amperes @ 2000 R.P.M.	Amperes @ 5000 R.P.M.			Air Gap	Point Gap	Closing Voltage	Air Gap	Point Gap	Volts @ 125° F.
	1103069	4–5	32	42	42	Integral	—	—	—	—	—	—
	1103070	4–5	—	60	63	Integral	—	—	—	—	—	—
	1103085	4–4.5	—	—	55	Integral	—	—	—	—	—	—
	1103088	4–4.5	—	—	55	Integral	—	—	—	—	—	—
	1103091	4–4.5	—	—	63	Integral	—	—	—	—	—	—
	1103092	4–4.5	—	—	55	Integral	—	—	—	—	—	—
	1103100	4–4.5	—	—	55	Integral	—	—	—	—	—	—
	1103102	4–4.5	—	—	63	Integral	—	—	—	—	—	—
	1103118	4–4.5	—	—	37	Integral	—	—	—	—	—	—
	1103122	4–4.5	—	—	63	Integral	—	—	—	—	—	—
	1103161	4–4.5	—	—	37	Integral	—	—	—	—	—	—
	1103162	4–4.5	—	—	37	Integral	—	—	—	—	—	—
	1103169	4–4.5	—	—	63	Integral	—	—	—	—	—	—
	1117180	6.5–7.5	54	85	85	Integral	—	—	—	—	—	—
1981–82	1100110	—	—	—	42	Integral	—	—	—	—	—	—
	1101016	4–4.5	—	76	80	Integral	—	—	—	—	—	—
	1101028	4–4.5	—	76	80	Integral	—	—	—	—	—	—
	1102394	4–4.5	—	33	37	Integral	—	—	—	—	—	—
	1102480	4–4.5	—	57	61	Integral	—	—	—	—	—	—
	1102485	4–4.5	—	38	42	Integral	—	—	—	—	—	—
	1102486	4–4.5	—	57	61	Integral	—	—	—	—	—	—
	1102491	4–4.5	—	33	37	Integral	—	—	—	—	—	—
	1102841	4–4.5	—	38	42	Integral	—	—	—	—	—	—
	1102886	4–4.5	—	57	61	Integral	—	—	—	—	—	—
	1102887	4–4.5	—	38	42	Integral	—	—	—	—	—	—
	1102888	4–4.5	—	57	61	Integral	—	—	—	—	—	—
	1102889	4–4.5	—	33	37	Integral	—	—	—	—	—	—
	1103161	—	—	—	37	Integral	—	—	—	—	—	—
	1103162	—	—	—	37	Integral	—	—	—	—	—	—
1981–83	706177⑧	5–6	130	—	130	Integral	—	—	—	—	—	—
	711925⑧	2.8	81	105	105	Integral	—	—	—	—	—	—
	1100080	4.4–4.9	52	72	80	Integral	—	—	—	—	—	—
	1100081	4.4–4.9	46	58	65	Integral	—	—	—	—	—	—
	1100091	4.4–4.9	63	97	100	Integral	—	—	—	—	—	—
	1100094	4.4–4.9	45	57	65	Integral	—	—	—	—	—	—
	1100095	4.4–4.9	55	76	80	Integral	—	—	—	—	—	—
	1100096	4.4–4.9	63	97	100	Integral	—	—	—	—	—	—
	1101041	4.0–4.6	55	70	70	Integral	—	—	—	—	—	—
	1101042	4.0–4.6	55	70	70	Integral	—	—	—	—	—	—
	1101058	4.4–4.9	55	76	80	Integral	—	—	—	—	—	—
	1101059	4.4–4.9	63	97	100	Integral	—	—	—	—	—	—
	1103069	4.0–5.0	32	60	63	Integral	—	—	—	—	—	—
	1103070	4.0–5.0	32	60	63	Integral	—	—	—	—	—	—
	1117241⑧	4.1–4.5	—	77	75	Integral	—	—	—	—	—	—
	1117244⑧	4.1–4.5	—	77	75	Integral	—	—	—	—	—	—
	1117732⑧	3.6–4.3	72	90	90	Integral	—	—	—	—	—	—
	2009386⑧	5.0–6.0	85	130	130	Integral	—	—	—	—	—	—
	5148785⑧	—	—	75	75	Integral	—	—	—	—	—	—
1982	1101044	4.0–4.5	—	—	70	Integral	—	—	—	—	—	—
	1101066	4.0–4.5	—	—	70	Integral	—	—	—	—	—	—
	1101071	4.0–4.5	—	—	70	Integral	—	—	—	—	—	—
	1103043	4.0–4.5	—	—	42	Integral	—	—	—	—	—	—

Continued

ALTERNATOR & REGULATOR SPECIFICATIONS—Continued

Year	Alternator					Regulator						
	Model	Field Current @ 80° F. 12 Volts	Cold Output @ 14 Volts		Rated Hot Output Amperes	Model	Field Relay			Voltage Regulator		
			Amperes @ 2000 R.P.M.	Amperes @ 5000 R.P.M.			Air Gap	Point Gap	Closing Voltage	Air Gap	Point Gap	Volts @ 125° F.
	1103044	4.0–4.5	—	—	63	Integral	—	—	—	—	—	—
	1103085	4.0–4.5	—	—	55	Integral	—	—	—	—	—	—
	1103088	4.0–4.5	—	—	55	Integral	—	—	—	—	—	—
	1103091	4.0–4.5	—	—	63	Integral	—	—	—	—	—	—
	1103092	4.0–4.5	—	—	55	Integral	—	—	—	—	—	—
	1103118	4.0–4.5	—	—	37	Integral	—	—	—	—	—	—
	1103169	4.0–4.5	—	—	63	Integral	—	—	—	—	—	—
1983	1100200	4.0–4.5	—	—	78	Integral	—	—	—	—	—	—
	1100203	—	—	—	37	Integral	—	—	—	—	—	—
	1100204	—	—	—	37	Integral	—	—	—	—	—	—
	1100207	—	—	—	66	Integral	—	—	—	—	—	—
	1100208	—	—	—	66	Integral	—	—	—	—	—	—
	1100209	—	—	—	66	Integral	—	—	—	—	—	—
	1100217	—	—	—	78	Integral	—	—	—	—	—	—
	1100225	—	—	—	37	Integral	—	—	—	—	—	—
	1100226	4.0–4.5	—	—	37	Integral	—	—	—	—	—	—
	1100228	4.0–4.5	—	—	37	Integral	—	—	—	—	—	—
	1100229	—	—	—	42	Integral	—	—	—	—	—	—
	1100230	4.0–4.5	—	—	42	Integral	—	—	—	—	—	—
	1100231	—	—	—	42	Integral	—	—	—	—	—	—
	1100237	4.0–4.5	—	—	55	Integral	—	—	—	—	—	—
	1100239	4.0–4.5	—	—	55	Integral	—	—	—	—	—	—
	1100241	—	—	—	66	Integral	—	—	—	—	—	—
	1100242	—	—	—	66	Integral	—	—	—	—	—	—
	1100247	4.0–4.5	—	—	63	Integral	—	—	—	—	—	—
	1100259	—	—	—	78	Integral	—	—	—	—	—	—
	1100263	4.0–4.5	—	—	78	Integral	—	—	—	—	—	—
	1100264	4.0–4.5	—	—	78	Integral	—	—	—	—	—	—
	1100270	4.0–4.5	—	—	78	Integral	—	—	—	—	—	—
	1100293	—	—	—	85	Integral	—	—	—	—	—	—
	1100297	4.0–4.5	—	—	63	Integral	—	—	—	—	—	—
	1100300	4.0–4.5	—	—	63	Integral	—	—	—	—	—	—
	1101063	—	—	—	80	Integral	—	—	—	—	—	—
	1101064	—	—	—	80	Integral	—	—	—	—	—	—
	1105022	4.0–4.5	—	—	78	Integral	—	—	—	—	—	—
	1105025	4.0–4.5	—	—	63	Integral	—	—	—	—	—	—
	1105041	4.0–4.5	—	—	78	Integral	—	—	—	—	—	—

①—At 1100 R.P.M.
②—At 2500 R.P.M.
③—At 6000 R.P.M.
④—At 6500 R.P.M.
⑤—At 7000 R.P.M.
⑥—Requires field relay 1115481.
⑦—If the output is checked without a regulator, the output is 5–10 amperes higher than the value given.
⑧—Leece-Neville alternator.
⑨—At 600 R.P.M.

VALVE SPECIFICATIONS

Engine	Year	Valve Lash		Valve Angles		Valve Springs		Valve Stem Clearance		Stem Diameter, Std.	
		Int.	Exh.	Seat	Face	Installed Height	Pressure Lbs. @ In.	Intake	Exhaust	Intake	Exhaust
V6-200	1978–79	1 Turn①		46	45	1.70	200 @ 1.25	.0010–.0027	.0010–.0027	.3410–.3417	.3410–.3417
V6-229	1980–83	1 Turn①		46	45	1.70	200 @ 1.25	.0010–.0027	.0010–.0027	.3410–.3417	.3410–.3417
V6-231	1978	Hydraulic②		45	45	1.727	168 @ 1.327	.0015–.0032	.0015–.0032	.3405–.3412	.3405–.3412
V6-231	1979–81	Hydraulic②		45	45	1.727	③	.0015–.0035	.0015–.0032	.3401–.3412	.3405–.3412
V6-231	1982	Hydraulic②		45	45	1.727	220 @ 1.34	.0015–.0035	.0015–.0032	.3401–.3412	.3405–.3412
V6-231	1983	Hydraulic②		45	45	1.727	182 @ 1.34	.0015–.0035	.0015–.0032	.3401–.3412	.3405–.3412
6-250	1974–77	1 Turn①		46	45	1.66	186 @ 1.27	.0010–.0027	.0015–.0032	.3410–.3417	.3410–.3417
6-250	1978–80	1 Turn①		46	45	1.66	172 @ 1.26	.0010–.0027	.0015–.0032	.3410–.3417	.3410–.3417
6-250	1981–83	1 Turn①		46	45	1.66	175 @ 1.26	.0010–.0027	.0015–.0032	.3410–.3417	.3410–.3417
V8-267	1979–82	1 Turn①		46	45	1.70	200 @ 1.25	.0010–.0027	.0010–.0027	.3410–.3417	.3410–.3417
6-292	1974–77	1 Turn①		46	45	1.69	179 @ 1.30	.0010–.0027	.0015–.0032	.3410–.3417	.3410–.3417
6-292	1978–83	1 Turn①		46	45	1.66	175 @ 1.26	.0010–.0027	.0015–.0032	.3410–.3417	.3410–.3417
V6-305C	1974	.012H	.018H	④	④	—	200 @ 1.495	.0015–.0030	.0020–.0035	.3407–.3412	.3402–.3407
V8-305	1976–77	1 Turn①		46	45	⑤	⑥	.0010–.0027	.0010–.0027	.3410–.3417	.3410–.3417
V8-305	1978–83	1 Turn①		46	45	1.70	200 @ 1.25	.0010–.0027	.0010–.0027	.3410–.3417	.3410–.3417
V8-350	1974–75	1 Turn①		46	45	⑤	⑦	.0010–.0027	.0010–.0027	.3410–.3417	.3410–.3417
V8-350	1976–77	1 Turn①		46	45	⑤	⑥	.0010–.0027	.0010–.0027	.3410–.3417	.3410–.3417
V8-350	1978–83	1 Turn①		46	45	1.70	200 @ 1.25	.0010–.0027	.0010–.0027	.3410–.3417	.3410–.3417
V8-366	1974–83	1 Turn①		46	45	1.80	220 @ 1.40	.0010–.0027	.0012–.0029	.3715–.3722	.3713–.3720
V6-379	1974	.012H	.018H	30	30	—	200 @ 1.495	.0015–.0030	.0019–.0036	.3725–.3730	.4344–.4351
V8-400	1974–77	1 Turn①		46	45	⑤	⑦	.0010–.0027	.0010–.0027	.3410–.3417	.3410–.3417
V8-400	1978–80	1 Turn①		46	45	⑤	⑥	.0010–.0027	.0010–.0027	.3410–.3417	.3410–.3417
V8-403	1977–78	Hydraulic②		⑧	⑨	1.67	187 @ 1.27	.0010–.0027	.0015–.0032	.3425–.3432	.3420–.3427
V8-427	1974–83	1 Turn①		46	45	1.80	220 @ 1.40	.0010–.0027	.0014–.0027	.3715–.3722	.3713–.3720
V6-432	1974	.012H	.018H	②	②	1 59/64	204 @ 1.92	.0015–.0030	.0019–.0036	.3725–.3730	.4344–.4351
V8-454	1974–79	1 Turn①		46	45	1.88	300 @ 1.38	.0010–.0027	.0012–.0029	.3715–.3722	.3715–.3722
V8-454	1980–83	1 Turn①		46	45	1.80	220 @ 1.40	.0010–.0027	.0012–.0029	.3715–.3722	.3715–.3722
V8-455	1974–77	Hydraulic②		⑧	⑨	1.62	187 @ 1.27	.0010–.0027	.0015–.0032	.3425–.3432	.3420–.3427
V6-478M	1974	.012H	.018H	30	30	1 59/64	204 @ 1.92	.0015–.0030	.0025–.0040	.3725–.3730	.4340–.4345

①—Turn rocker arm stud nut until all lash is eliminated, then tighten nut the additional turns listed.
②—No adjustment.
③—Intake, 164 @ 1.34; exhaust, 182 @ 1.34.
④—Intake, 30°; exhaust, 45°.
⑤—Intake, 1.70; exhaust, 1.61.
⑥—Intake, 200 @ 1.25; exhaust, 200 @ 1.16.
⑦—Intake, 200 @ 1.25; exhaust, 189 @ 1.20.
⑧—Intake, 45°; exhaust, 31°.
⑨—Intake, 44°; exhaust, 30°.

PISTONS, PINS, RINGS, CRANKSHAFT & BEARINGS SPECIFICATIONS

Engine	Year	Piston Clearance	Ring End Gap①		Piston Pin Diam.	Rod Bearings		Main Bearings			
			Comp.	Oil		Shaft Diameter	Bearing Clearance	Shaft Diameter	Bearing Clearance	Thrust Bearing No.	Shaft End Play
V6-200	1978–79	.0007–.0017	.010	.015	.9271	2.0986–2.0998	.0013–.0035	②	③	4	.002–.006
V6-229	1980–83	.0007–.0017	.010	.015	.9271	2.0986–2.0998	.0013–.0035	②	③	4	.002–.006
V6-231	1978	.0008–.0020	.010	.015	.9392	2.2487–2.2495	.0005–.0026	2.4995	.0003–.0017	2	.004–.008
V6-231	1979–83	.0008–.0020	.013	.015	.9392	2.2487–2.2495	.0005–.0026	2.4995	.0003–.0018	2	.003–.009
6-250	1974–75	.0005–.0015	.010	.015	.9271	1.999–2.000	.0007–.0027	2.2983–2.2993	.0003–.0029	7	.002–.006
6-250	1976–77	④	.010	.015	.9271	1.999–2.000	.0007–.0027	2.2983–2.2993	.0003–.0029	7	.002–.006
6-250	1978–83	.0010–.0020	.010	.015	.9271	1.999–2.000	.0010–.0026	2.2979–2.2994	⑤	7	.002–.006
V8-267	1979–82	.0007–.0017	.010	.015	.9271	2.0986–2.0998	.0013–.0035	②	③	5	.002–.006
6-292	1974–77	.0026–.0036	.010	.015	.9271	2.099–2.100	.0007–.0027	2.2983–2.2993	.0008–.0034	7	.002–.006
6-292	1978–83	.0026–.0036	.010	.015	.9271	2.099–2.100	.0010–.0026	2.2979–2.2994	⑤	7	.002–.006
V6-305C	1974	.0027–.0035	⑥	—	⑦	2.8112–2.8122	.0015–.0035	⑧	⑨	3	.003–.008
V8-305	1976	.0007–.0017	.010	.015	.9271	2.099–2.100	.0013–.0035	②	⑩	5	.002–.006
V8-305	1977	.0007–.0017	.010	.015	.9271	2.099–2.100	.0013–.0035	②	③	5	.002–.006
V8-305	1978–83	.0007–.0017	.010	.015	.9271	2.0986–2.0998	.0013–.0035	②	③	5	.002–.006
V8-350⑫	1974–76	.0007–.0013	⑪	.015	.9271	2.099–2.100	.0013–.0035	②	③	5	.002–.006
V8-350⑫	1977	.0007–.0017	.010	.015	.9271	2.099–2.100	.0013–.0035	②	③	5	.002–.006
V8-350⑫	1978–83	.0007–.0017	.010	.015	.9271	2.0986–2.0998	.0013–.0035	②	③	5	.002–.006
V8-350⑬	1974–83	.0016–.0026	⑪	.015	.9271	2.099–2.100	.0013–.0035	②	③	5	.002–.006
V8-366	1974–83	.0030–.0040	.010	.010	.9896	2.199–2.200	.0014–.0030	⑭	⑮	5	.006–.010
V6-379	1974	.0031–.0039	⑯	—	⑦	2.8112–2.8122	.0015–.0035	⑧	⑨	3	.003–.008
V8-400	1974–75	.0014–.0020	.010	.015	.9271	2.099–2.100	.0013–.0035	⑰	③	5	.002–.006
V8-400	1976–77	.0014–.0024	.010	.015	.9271	2.099–2.100	.0013–.0035	⑰	③	5	.002–.006
V8-400	1978–80	.0014–.0024	.010	.015	.9271	2.0988–2.0998	.0013–.0035	⑰	③	5	.002–.006
V8-403	1977	.0010–.0020	.010	.015	.9805	2.1238–2.1248	.0004–.0033	⑱	⑲	3	.0035–.0135
V8-427	1974–83	.0030–.0040	.010	.010	.9896	2.199–2.200	.0014–.0030	⑭	⑮	5	.006–.010
V6-432	1974	.0033–.0041	⑳	—	⑦	2.8112–2.8122	.0015–.0035	⑧	⑨	3	.003–.008
V8-454	1974–76	.0018–.0028	.010	.010	.9896	2.1985–2.1995	.0009–.0025	㉑	㉒	5	.006–.010
V8-454	1977–79	.0014–.0024	.010	.015	.9896	2.1985–2.1995	.0009–.0025	㉑	㉒	5	.006–.010
V8-454	1980–83	.0030–.0040	.010	.015	.9896	2.199–2.200	.0009–.0025	㉓	㉒	5	.006–.010
V8-455	1974–77	.0010–.0020	.010	.015	.9805	2.4988–2.4998	.0004–.0033	2.9993–3.0003	㉔	3	.004–.008
V6-478M	1974	.0039–.0045	㉕	.017	㉖	2.8112–2.8122	.0015–.0035	⑧	⑨	3	.003–.008

①—Fit rings in tapered bores to clearance listed in tightest portion of ring travel.

②—Front, 2.4484–2.4493 inch; intermediate, 2.4481–2.4490 inch; rear, 2.4479–2.4488 inch.

③—Front, .0008–.0020 inch; intermediate, .0011–.0023 inch; rear, .0017–.0032 inch.

④—El Camino & Sprint, .0010–.0020 inch; Series 10–30 & 1500–3500, .0005–.0015 inch.

⑤—Nos. 1, 2, 3, 4, 5, 6—.0010–.0024 inch; No. 7, .0016–.0035 inch.

⑥—Top ring .017 inch; 2nd ring .015 inch; 3rd ring, .015 inch.

⑦—Red color code, 1.2403–1.2404 inch; no color code, 1.2402–1.2403 inch.

⑧—Nos. 1, 2, 3—3.1237–3.1247 inch; No. 4, 3.1229–3.1239 inch.

⑨—Nos. 1, 2, 3—.0023–.0039 inch; No. 4, .0031–.0047 inch.

⑩—No. 1, .0019–.0031 inch; Nos. 2, 3, 4—.0013–.0025 inch; No. 5, .0023–.0033 inch.

⑪—Top ring, .010 inch; 2nd ring, .013 inch.

⑫—Caballero, El Camino, Sprint & Series 10–30 & 1500–3500.

⑬—Medium duty trucks.

⑭—Nos. 1, 2, 3, 4—2.7481–2.7490 inch; No. 5, 2.7473–2.7483 inch.

⑮—Nos. 1, 2, 3, 4—.0013–.0025 inch; No. 5, .0029–.0045 inch.

⑯—Top & 2nd ring, .022 inch; 3rd ring, .015 inch.

⑰—Nos. 1, 2, 3, 4—2.6484–2.6493 inch; No. 5, 2.6479–2.6488 inch.

⑱—No. 1, 2.4988–2.4998 inch; Nos. 2, 3, 4, 5—2.4985–2.4995 inch.

⑲—Nos. 1, 2, 3, 4—.0005–.0021 inch; No. 5, .0015–.0031 inch.

⑳—Top & 2nd ring, .024 inch; 3rd ring, .015 inch.

㉑—No. 1, 2.7485–2.7494 inch; Nos. 2, 3, 4—2.7481–2.7490 inch; No. 5, 2.7478–2.7488 inch.

㉒—Nos. 1, 2, 3, 4—.0013–.0025 inch; No. 5, .0024–.0040 inch.

㉓—Nos. 1, 2, 3, 4—2.7481–2.7490 inch; No. 5, 2.7476–2.7486 inch.

㉔—Nos. 1, 2, 3, 4—.0005–.0021 inch; No. 5, .0020–.0034 inch.

㉕—Top & 2nd ring, .026 inch; 3rd ring, .017 inch.

㉖—Red color code, 1.2399–1.2400 inch; green color code, 1.2404–1.2405 inch.

CHEVROLET & GMC—Exc. S/T 10 & 15

ENGINE TIGHTENING SPECIFICATIONS

Engine	Year	Spark Plug Ft. Lbs.	Cylinder Head Bolts Ft. Lbs.	Intake Manifold Ft. Lbs.	Exhaust Manifold Ft. Lbs.	Rocker Arm Stud Ft. Lbs.	Rocker Arm Cover Ft. Lbs.	Connecting Rod Cap Bolts Ft. Lbs.	Main Bearing Cap Bolts Ft. Lbs.	Flywheel to Crankshaft Ft. Lbs.	Vibration Damper or Pulley Ft. Lbs.
V6-200	1978	20	65	30	20	—	45(1)	45	70	60	60
V6-200	1979	22	65	30	20	—	45(1)	45	70	60	60
V6-229	1980–83	22	65	30	20	—	45(1)	45	70	60	60
V6-231	1978–83	15	80	45	25	30(2)	4	40	100	60	225
6-250(3)	1974–76	15	95	35	30	—	45(1)	35	65	60	—
6-250(4)	1975–77	15	95	—	30(5)	—	45(1)	35	65	60	—
6-250	1978–83	17–27	95(6)	—	30	—	45(1)	35	65	60	—
V8-267	1979–82	22	65	30	20	—	45(1)	45	70	60	60
6-292	1974–76	15	95	35	30	—	45(1)	40	65	110	60
6-292	1977	15	95	40	45	—	45(1)	40	65	110	60
6-292	1978–83	17–27	95(6)	40	45	—	45(1)	40	65	110	60
V6-305C	1974	30–34	(7)(8)	20–25	15–20	20–25(2)	3–5	55–65(8)	(8)(9)	100–110	(10)
V8-305(11)	1976–78	15	65	30	20	—	45(1)	45	70	60	60
V8-305(11)	1979–83	22	65	30	20	—	45(1)	45	70	60	60
V8-305(12)	1977	15	65	30	20	—	45(1)	45	80	60	60
V8-305(12)	1978–82	17–27	65	30	20	—	45(1)	45	80	60	60
V8-350(11)	1974–78	15	65	30	20(13)	—	45(1)	45	70(14)	60	60
V8-350(11)	1979–83	22	65	30	20(13)	—	45(1)	45	70	60	60
V8-350(12)	1974–76	15	65	30	20(13)	—	45(1)	45	70	60	60
V8-350(12)	1977	15	65	30	20(13)	—	45(1)	45	80(15)	60	60
V8-350(12)	1978–82	17–27	65	30	20(13)	—	45(1)	45	80(15)	60	60
V8-350(16)	1974–79	13–17	60–70	25–35	18–22(13)	—	3–5	42–47	70–80(17)	55–65	70–80
V8-350(16)	1980–81	20	65	30	20(13)	—	4	45	80(15)	65	65
V8-350(16)	1982–83	15	65	30	20(3)	—	45(1)	45	80(15)	65	65
V8-366	1974–79	13–17	75–85	25–35	25–30	45–55	3–5	(18)	100–110	60–70	80–90
V8-366	1980–81	20	80	30	17	50	4	50	105	65	85
V8-366	1982–83	15	80	30	20	50	50(1)	50	105	65	85
V6-379	1974	15–20	(7)(8)	20–25	15–20	20–25(2)	3–5	55–65(8)	(8)(9)	100–110	(10)
V8-400	1974–76	15	65	30	20	—	45(1)	45	70	60	60
V8-400	1977	15	65	30	20	—	45(1)	45	80(15)	60	60
V8-400	1978–80	17–27	65	30	20	—	45(1)	45	80(15)	60	60
V8-403	1977–78	25	130(8)	40(8)	25(19)	25	—	42	(20)	60	200–310
V8-427	1974–79	13–17	75–85	25–35	25–30	45–55	3–5	(18)	100–110	60–70	80–90
V8-427	1980–81	20	80	30	17	50	4	50	105	65	85
V8-427	1982–83	15	80	30	20	50	50(1)	50	105	65	85
V6-432	1974	15–20	(7)(8)	20–25	15–20	20–25(2)	3–5	55–65(8)	(8)(9)	100–110	(10)
V8-454	1974–77	15	80	30	20	50	50(1)	50	110	65	85
V8-454	1978–83	17–27	80	30	20	50	50(1)	50	110	65	85
V8-455	1974–77	(21)	85(8)	40(8)	25	25(19)	7	42	120	60	160 Min.
V6-478M	1974	15–20	(7)(8)	25–30(8)	15–20	20–25(2)	3–5	55–65(8)	(8)(9)	100–110	(20)

(1)—Inch Lbs.
(2)—Rocker arm shaft.
(3)—Engines less integral intake manifold.
(4)—Engines with integral intake manifold.
(5)—Outer bolts, 20 ft. lbs.
(6)—Torque left hand front bolt to 85 ft. lbs.
(7)—7/16 inch bolts, 60–65 ft. lbs.; 1/2 inch bolts, 90–100 ft. lbs.
(8)—Clean & dip entire bolt in engine oil before tightening to obtain a correct specified torque reading.
(9)—Except 1/2 inch rear bolts, 170–180 ft. lbs.; 1/2 inch rear bolts, 90–100 ft. lbs.
(10)—Hub to crankshaft bolt less cone, 180–200 ft. lbs.; damper or hub to crankshaft bolt with cone, 240–260 ft. lbs.
(11)—Caballero, El Camino & Sprint models.
(12)—Series 10–30 & 15–3500.
(13)—Inside bolts, 30 ft. lbs.
(14)—Outer bolts on engines with four bolt caps, 65 ft. lbs.
(15)—Intermediate (2, 3, 4) outer bolts, 70 ft. lbs.
(16)—Medium duty trucks.
(17)—Intermediate (2, 3, 4) outer bolts, 60–70 ft. lbs.
(18)—7/16 inch bolts, 67–73 ft. lbs.; 3/8 inch bolts, 45–55 ft. lbs.
(19)—Rocker arm pivot bolt.
(20)—Nos. 1, 2, 3, 4—80 ft. lbs.; No. 5, 120 ft. lbs.
(21)—Models less H.E.I., 35 ft. lbs.; models with H.E.I., 25 ft. lbs.

LIGHT DUTY MODEL WHEEL ALIGNMENT SPECIFICATIONS

Year	Model	Caster Deg.	Camber Deg.	Toe-In Inch
CABALLERO, EL CAMINO & SPRINT				
1974	Man. Steer.	−1/2 to −1 1/2	①	0–1/8
	Power Steer.	−1/2 to +1/2	①	0–1/8
1975–76	Man. Steer.	+1/2 to +1 1/2	①	0–1/8
	Power Steer.	+1 1/2 to +2 1/2	①	0–1/8
1977	Man. Steer.	+1/2 to +1 1/2	①	0–1/8
	Power Steer.	②	①	0–1/8
1978–83	Man. Steer.	+1/2 to +1 1/2	0 to +1	1/16–3/16
	Power Steer.	+2 1/2 to +3 1/2	0 to +1	1/16–3/16
SERIES 10 THRU 30 & 1500 THRU 3500				
1974	C, G, P	③	+1/4	3/16
	K	③	+1 1/2④	3/16⑤
1975–78	C, G, P	③	③	3/16
	K	③	③	0
	Motor Home (32)	③	③	5/16
1979–80	C, P	③	+.2	3/16
	G-10, 20/1500, 2500	③	+.5	3/16
	G-30/3500	③	+.2	3/16
	K-10, 20/1500, 3500	③	+1④	0
	K-30/3500	③	+.5④	0
	Motor Home (32)	③	+.2	5/16
1981	C, P	③	+.2	3/16
	G-10, 20/1500, 2500	③	+.5	3/16
	G-30/3500	③	+.2	3/16
	K-10, 20/1500, 2500	③	+1④	3/16
	K-30/3500	③	+.5④	3/16
	Motor Home (32)	③	+.2	5/16
1982–83	C-10/1500	③	+.7	3/16
	C-20, 30/2500, 3500	③	+.2	3/16
	G-10, 20/1500, 2500	③	+.5	3/16
	G-30/3500	③	+.2	3/16
	K-10, 20/1500, 2500	③	+1④	3/16
	K-30/3500	③	+.5④	3/16
	P	③	+.2	3/16
	Motor Home (32)	③	+.2	5/16

①—Left, +1/2 to +1 1/2; right, 0 to +1.
②—Belted tires, +1/2 to +1 1/2; radial tires, +1 1/2 to +2 1/2.
③—Refer to Front Suspension section of this chapter.
④—No adjustment provision.
⑤—On K-10/1500 models w/full time four wheel drive, toe-in 0 inch.

GMC MOTOR HOME & TRANSMODE WHEEL ALIGNMENT SPECIFICATIONS

Year	Model	Caster Deg.	Camber Deg.	Toe-Out Inch
1974–78	All	+2	①	1/8

①—Left, +3/4; right, +1/2.

MEDIUM & HEAVY DUTY MODEL WHEEL ALIGNMENT SPECIFICATIONS

Axle Model	Axle Rating (lbs.)	Caster Deg.	Camber Deg.	Toe-In Inch	Kingpin Angle Deg.
1974–78 P-40-45 & MEDIUM DUTY CONVENTIONAL					
F-050①	5,000	+2–+3	+1–+2	1/8–1/4	7 1/6
F-055①	5,500	+2–+3	+1–+2	1/8–1/4	7 1/6
F-070③	7,000	+2–+3	+1–+2	1/8–1/4	7 1/6
F-090①	9,000	+2–+3	④	1/8–1/4	⑤
F-120②	9,000	+2–+3	−1/4–+3/4	1/8–1/4	⑤
F-120③	12,000	+2–+3	−1/4–+3/4	1/8–1/4	⑤
1974–78 STEEL TILT CAB					
F-070③	7,000	+3 1/2–+4 1/2	+1–+2	1/8–1/4	7 1/6
F-090①	9,000	+3 1/2–+4 1/2	④	1/8–1/4	⑤
F-120②	9,000	+3 1/2–+4 1/2	−1/4–+3/4	1/8–1/4	⑤
F-120③	12,000	+3 1/2–+4 1/2	−1/4–+3/4	1/8–1/4	⑤
1974–78 HEAVY DUTY CONVENTIONAL					
F-070②	7,000	+9/10–+1 9/10	+1–+2	1/8–1/4	7 1/16
F-090①	9,000	+1/2–+1 1/2	④	1/8–1/4	⑤
F-120②	9,000	+1/2–+1 1/2	−1/4–+3/4	1/8–1/4	⑤
F-120③	12,000	+1/2–+1 1/2	−1/4–+3/4	1/8–1/4	⑤
FL-931②	16,000	−1/10––+9/10	⑥	1/8–1/4	⑤
1974–78 ALUMINUM TILT					
FL-120②	12,000	+1/2–+1 1/2	−1/4–+3/4	1/8–1/4	⑤
FL-931②	16,000	+1/2–+1 1/2	⑥	1/8–1/4	⑤
1979–82 ALL					
F-050	5,000	⑧	+1–+2	⑨	7 1/6°
F-070D	⑪	⑧	+1–+2	⑨	7 1/6°
F-070	7,000	⑧	+1–+2	⑨	7 1/6°
F-090	9,000	⑧	④	⑨	⑤
F-120D	⑩	⑧	−1/4–+3/4	⑨	5 3/4°
F-120	12,000	⑧	−1/4–+3/4	⑨	5 3/4°
FL-931	18,000	⑧	⑦	⑨	⑤
FF-933	12,000	⑧	+1/2–+1	⑨	⑤
1983 ALL					
F-050	5,000	⑫	+1–+2	1/32–5/32	7 1/6°
F-070D	6,000	⑫	+1–+2	1/32–5/32	7 1/6°
F-070	7,000	⑫	+1–+2	1/32–5/32	7 1/6°
F-070	7,500	⑫	+1–+2	1/32–5/32	7 1/6°
F-090	9,000	⑫	④	1/32–5/32	⑤
F-120D	9,000	⑫	−1/4–+3/4	1/32–5/32	5 3/4°
F-120	12,000	⑫	−1/4–+3/4	1/32–5/32	5 3/4°
F-931	18,000	⑫	⑦	1/32–5/32	⑤
F-933	12,000	⑫	+1/2–+1	1/32–5/32	⑤

①—With hydraulic brakes.
②—With air brakes.
③—With air or hydraulic brakes.
④—Left side −1/4°–+3/4°; right side −3/4°–+1/4°.
⑤—Left side 5 3/4°; right side 6 1/4°.
⑥—Left side +3/4°; right side +1/4°.
⑦—Left side +23/60–+17/60°; right side −7/60°–+37/60°.
⑧—Forward control chassis, +2°–+3°; school bus chassis, +1 1/2°–+2 1/2°; medium duty steel conventional cab chassis, +1 1/2°–+2 1/2°; medium duty steel tilt cab chassis, +3 1/2°–+4 1/2°; heavy duty steel conventional cab chassis, +1°–+2°; heavy duty aluminum conventional cab chassis, −1/2°–+1/2°; heavy duty aluminum tilt cab chassis, −1/2°–+1/2°.
⑨—1979 all models and 1980 medium duty models, toe-in 1/8″–1/4″ except models with steel belted radials, toe-in 0″. 1980 heavy duty models; toe-in 1/32″–5/32″ for all models regardless of tire construction. 1981–82 all models, toe-in 1/32″–5/32″ regardless of tire construction.
⑩—12,000 lb. axle rated at 9,000 lbs.
⑪—7,000 lb. axle rated at 6,000 lbs.
⑫—P4 models, +2° to +3°; C5, 6, 7, & P6 models, +1 1/2° to +2 1/2°; N9 models less Ross Integral Steering, −1/2° to +1/2°; N9 models with Ross Integral Steering, +2 1/2° to +3 1/2°; D9 models, −1/2° to +1/2°.

Engine Section

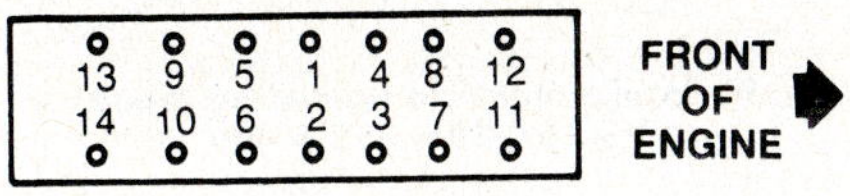

Fig. 1 Cylinder head tightening sequence. 6-250 & 292

CYLINDER HEAD, REPLACE

NOTE: A new exhaust valve part No. 336771 has been released for service on all 1974 V8-427 truck engines. This valve is of the sodium filled/hollow stem type previously utilized in the GMC V6.

CAUTION: Sodium cooled valves must not be discarded with other scrap metal. If a sodium-cooled valve is accidentally broken, the sodium will react violently with water, resulting in fire and explosion due to the chemical reaction. Also, serious burns will result if sodium or its oxide comes in contact with the skin.

6-250, 292

1. Remove necessary access covers depending on vehicle being serviced.
2. Drain cooling system and remove air cleaner and air compressor.
3. Disconnect choke cable, accelerator pedal rod at bellcrank on manifold, and fuel and vacuum lines at carburetor.
4. Disconnect exhaust pipe at manifold flange, then remove manifolds and carburetor as an assembly.
5. Disconnect wire harness from temperature sending unit and coil, leaving harness clear of clips on rocker arm cover.
6. Disconnect radiator hose at water outlet housing and battery ground strap at cylinder head.
7. Remove spark plugs. On 6-cylinder engines, remove coil.

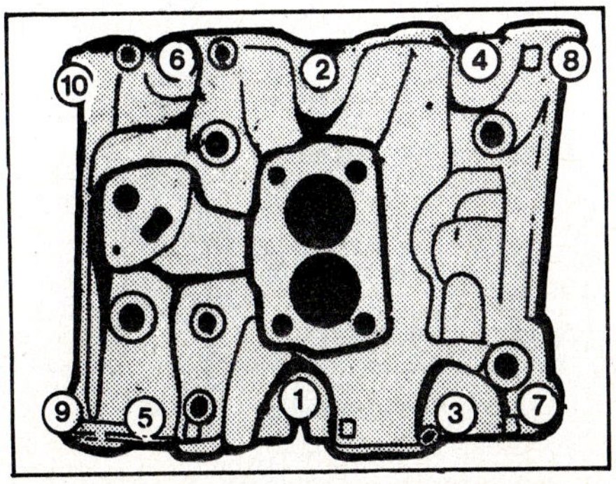

Fig. 5 Intake manifold tightening sequence. V6-231

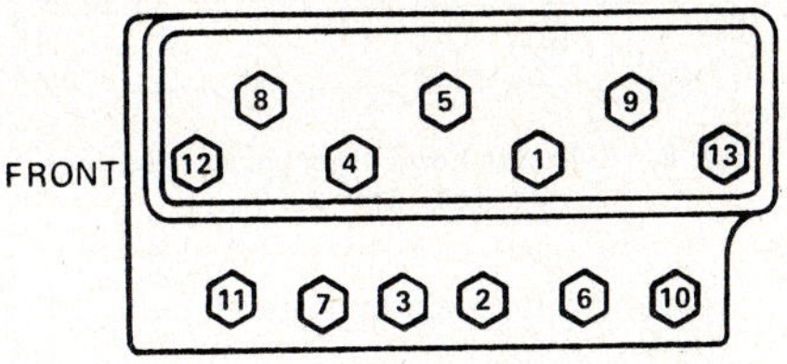

Fig. 2 Cylinder head tightening sequence. V6-200 & 229

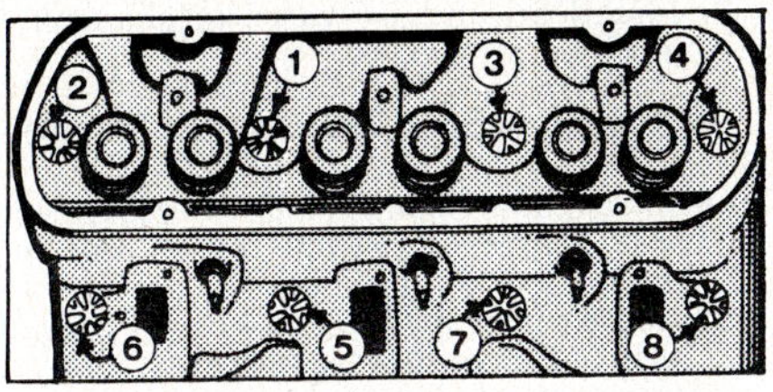

Fig. 4 Cylinder head tightening sequence. V6-231

8. Remove rocker arm cover. Back off rocker arm nuts, pivot rocker arms to clear push rods and remove push rods.
9. Unfasten and remove cylinder head.
10. Reverse removal procedure to install and tighten head bolts to the specified torque and in the sequence shown in Fig. 1.

V6-200, 229

1. Drain cooling system and engine block.
2. Remove intake and exhaust manifolds.
3. Remove alternator lower mounting bolt and position alternator aside.
4. If equipped with A/C, remove compressor and forward mounting bracket and position aside.
5. Remove rocker arm cover, rocker arms and push rods.

NOTE: Keep rocker arm, rocker arm balls and push rods in order so they can be installed in the same position.

6. Remove cylinder head bolts and cylinder head.
7. Reverse procedure to install. Tighten cylinder head and intake manifold bolts in

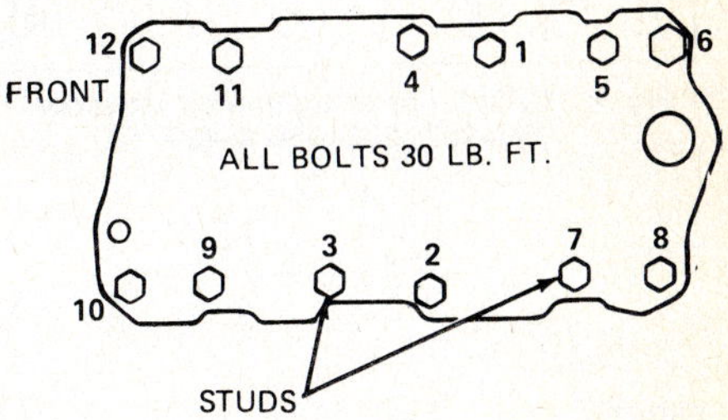

Fig. 3 Intake manifold tightening sequence. V6-200 & 229

sequence shown in Figs. 2 and 3.

V6-231

1. Drain coolant and disconnect battery.
2. Remove intake manifold.
3. When removing right cylinder head, remove Delcotron and/or A/C compressor with mounting bracket and move out of the way. *Do not disconnect hoses from A/C compressor.*
4. When removing left cylinder head, remove oil dipstick, power steering pump and move out of the way with hoses attached.
5. Disconnect exhaust manifold from head to be removed.
6. Remove rocker arm shaft and lift out push rods.
8. Remove cylinder head.
9. Reverse procedure to install and tighten bolts gradually and evenly in the sequence shown in Fig. 4.

NOTE: When installing intake manifold, refer to Fig. 5 for bolt tightening sequence.

V6-305C, 379, 432, 478

After installing cylinder head, run engine until normal operating temperature is reached. Then tighten cylinder head bolts in the sequence shown in Fig. 6 to the torque given in the *Engine Tightening* table. After head is properly tightened, adjust valve lash to the recommended clearance.

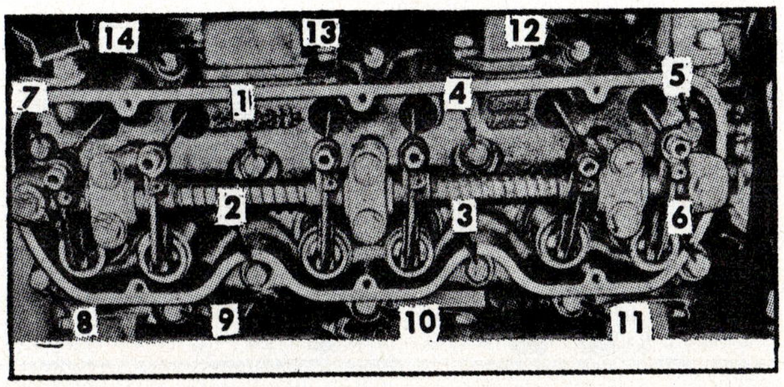

Fig. 6 Cylinder head tightening sequence. V6-305C, 379, 432 & 478

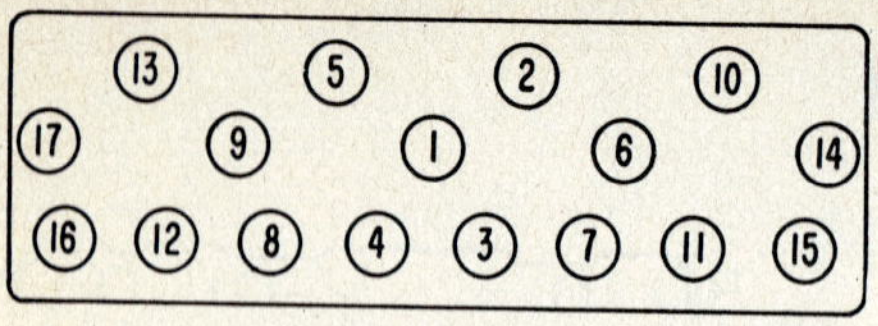

Fig. 7 Cylinder head tightening sequence. V8-267, 305, 350 & 400

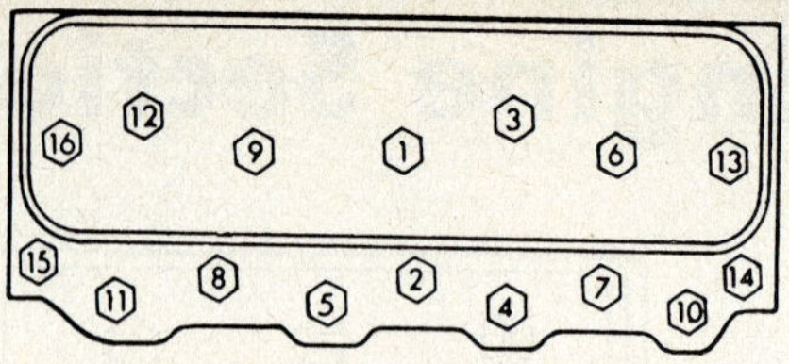

Fig. 8 Cylinder head tightening sequence. V8-366, 427 & 454

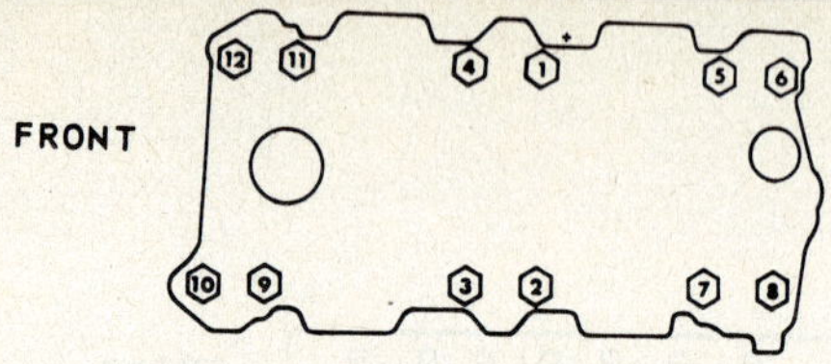

Fig. 9 Intake manifold tightening sequence. V8-267, 305, 350 & 400

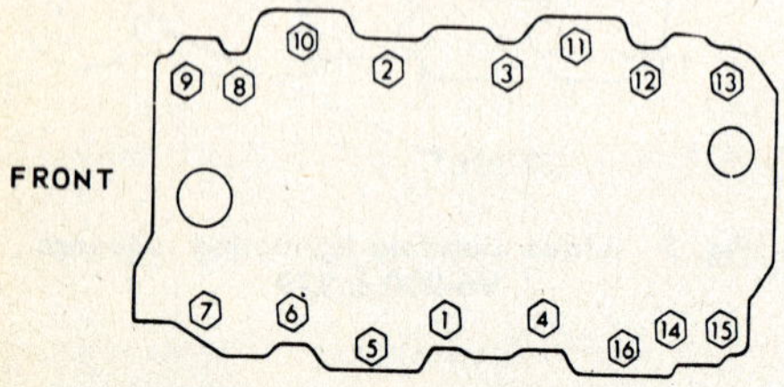

Fig. 10 Intake manifold tightening sequence. V8-366, 427 & 454

V8-267, 305, 350, 366, 400, 427, 454

1. Remove intake and exhaust manifolds and air compressor.
2. Remove valve mechanism.
3. Drain coolant from block.
4. Unfasten and remove cylinder head.

Installation:

The gasket surfaces on both head and block must be clean of any foreign material and free of nicks or heavy scratches. Cylinder bolt threads in the block and threads on the head bolts must be clean as dirt will affect bolt torque.

1. On engines using a steel gasket, coat both sides of a new gasket with a good sealer. One method of applying the sealer that will assure the proper coat is with the use of a paint roller. Too much sealer may hold the gasket away from the head or block.

CAUTION: Use no sealer on engines using a composition steel asbestos gasket.

2. Place gasket in position over the dowel pins with the head up.
3. Carefully guide cylinder head into place over dowel pins and gasket.
4. Coat threads of cylinder head bolts with sealing compound and install bolts finger tight.
5. Tighten each bolt a little at a time in the sequence shown in Figs. 7 and 8 until the specified torque is reached (see *Engine tightening Specifications* chart).
6. Install and tighten intake manifold bolts in sequence shown in Figs. 9 and 10.
7. Install and adjust valve mechanism as outlined further on.

V8-403, 455 GMC Motor Home

Some cylinder head gaskets are coated with a special lacquer to provide a good seal once the parts have warmed up. Do not use any additional sealer on such gaskets. If the gasket does not have this lacquer coating, apply suitable sealer to both sides.

Tighten cylinder head bolts a little at a time in three steps in the sequence shown in the illustrations. Final tightening should be to the torque specifications listed in the *Engine Tightening* table.

1. Drain radiator and cylinder block.
2. Disconnect spark plug wires and remove intake manifold.
3. Loosen exhaust pipe clamp at muffler then remove exhaust manifold bolts and position exhaust manifolds away from cylinder heads.
4. Remove valve cover (loosen or remove any accessory brackets that interfere).
5. Remove ground strap from right cylinder head.
6. Remove exhaust manifold, rocker arms and push rods from head to be removed.
7. Reverse removal procedure to install head, and tighten bolts in the sequence shown in Fig. 11.

VALVE ARRANGEMENT

Front to Rear

6-250, 292	E-I-I-E-E-I-I-E-E-I-I-E
V6-200, 229	
V6-231	E-I-I-E-I-E
V6-305C, 379, 432, 478	E-I-I-E-I-E
V8-267, 305, 350, 400	E-I-I-E-E-I-I-E
V8-366, 427, 454	E-I-E-I-E-I-E-I
V8-403, 455	I-E-I-E-E-I-E-I

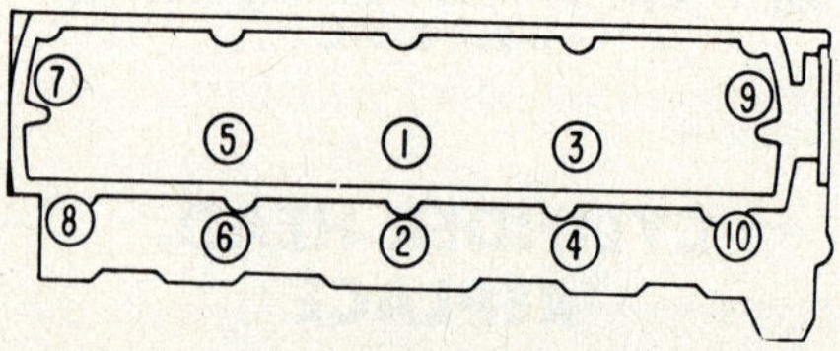

Fig. 11 Cylinder head tightening sequence V8-403 & 455

VALVES, ADJUST

V6-305C, 379, 432, 478

With engine at operating temperature and rocker arm covers removed, turn adjusting nut on each rocker arm until proper clearance is obtained between valve stem and rocker arm. For proper clearance see *Tune-Up and Valve Specifications* table.

6-250, 292, V6-200, 229, V8-267, 305, 350, 400, 427 & 454

Hydraulic Lifters

NOTE: After the engine has been thoroughly warmed up the valves may be adjusted with the engine shut off as follows: With engine in position to fire No. 1 cylinder the following valves may be adjusted: V6-200 and 229, Exhaust 1-5-6, intake 1-2-3. 6-250 and 292, Exhaust 1-3-5, intake 1-2-4. V8's, Exhaust 1-3-4-8, intake 1-2-5-7. Then crank the engine one more complete revolution which will bring No. 4 cylinder on V6-200 and 229, engine and No. 6 cylinder on 6-250, 292 and V8 engines, to the firing position at which

Fig. 12 Valve rotator & retaining components

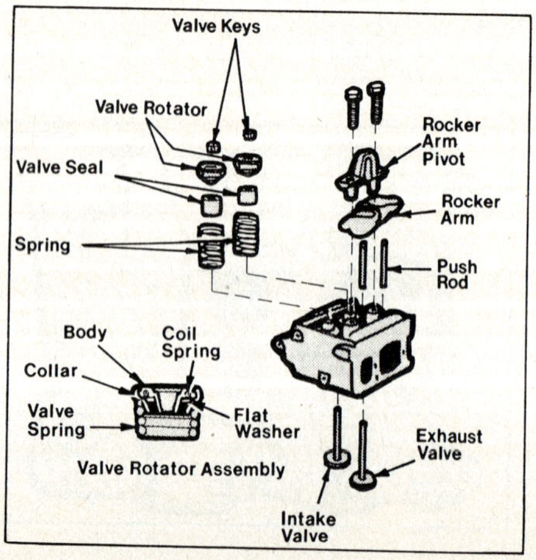

Fig. 13 Cylinder head exploded view. V8-403 & 455

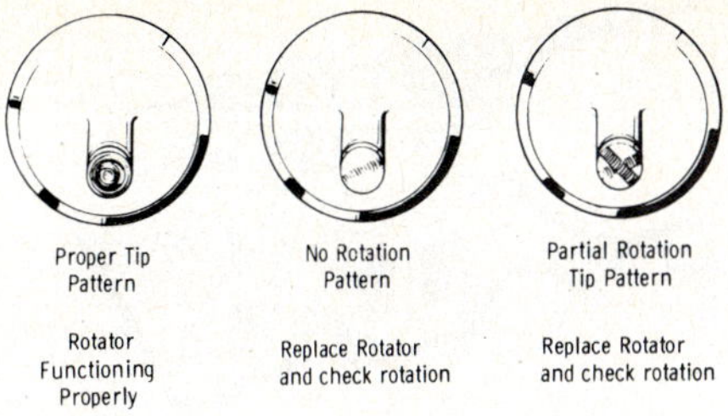

Fig. 14 Checking valve stem tips for valve rotator malfunction. V8-403 & 455

time the following valves may be adjusted: V6-200 and 229, Exhaust 2-3-4, intake 4-5-6. 6-250 and 292, Exhaust 2-4-6, intake 3-5-6. V8's, Exhaust 2-5-6-7, intake 3-4-6-8.

The following procedure, performed with the engine running should be done only in case readjustment is required.

1. After engine has been warmed up to operating temperature, remove valve cover and install a new valve cover gasket.
2. With engine running at idle speed, back off valve rocker arm nut until rocker arm starts to clatter.
3. Turn rocker arm nut down slowly until the clatter just stops. This is the zero lash position.
4. Turn nut down 1/4 additional turn and pause 10 seconds until engine runs smoothly. Repeat additional 1/4 turns, pausing 10 seconds each time, until nut has been turned down the number of turns listed in the *Valve Specifications Chart* from the zero lash position.

NOTE: This preload adjustment must be done slowly to allow the lifter to adjust itself to prevent the possibility of interference between the intake valve head and top of piston, which might result in internal damage and/or bent push rods. Noisy lifters should be replaced.

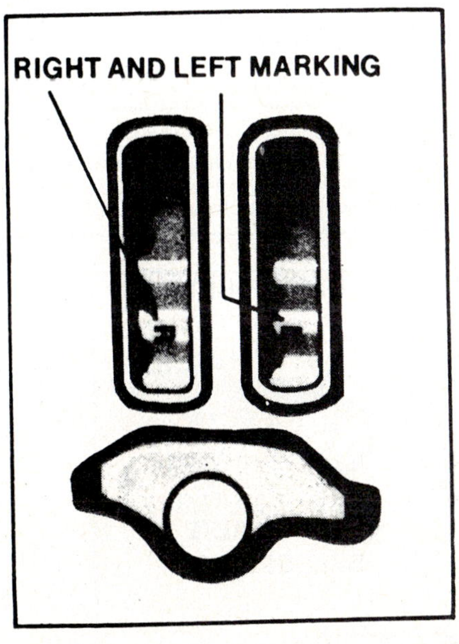

Fig. 17 Service rocker arm identification. V6-231

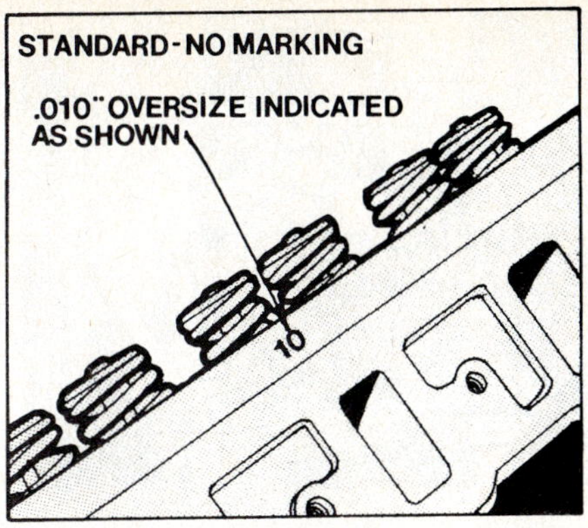

Fig. 15 Valve guide bore markings. V8-403 & 455

Mechanical Lifters

With the engine warmed to operating temperature as described above, turn rocker arm stud nut or adjusting screw as required to obtain the clearances given in the *Valve Specifications* table.

NOTE: To be sure that the valves and lifters function perfectly, it is strongly recommended that a vacuum gauge be used when making the adjustment on hydraulic lifter jobs. With vacuum gauge connected and engine idling, tighten each rocker arm nut until the *highest and steadiest vacuum is indicated on the gauge.*

VALVE ROTATORS

All Engines Except V8-403, 455

The parts involved in the rotating mechanism, Fig. 12, are a special spring seat retainer, a pair of flat half-moon keys, a close fitting cap located on the valve stem and a specially constructed valve stem.

In order to accommodate valve expansion, the valve lash must be maintained. When camshaft rotation causes this lash or clearance to be taken up, the cap on the valve stem causes the valve keys to raise the spring retainer, removing the load on the valve springs from the valve *before* the valve is moved from its seat. A clearance of .002–.004" should be maintained between the end face of the valve stem and cap. This is the distance the spring retainer is moved *before* the valve is moved. The slow valve rotating motion is caused by the vibration of the valve, the flowing of exhaust gases around the valve head, and a slight rotating motion imparted to the valve by the valve spring.

Whenever valve work is being done, a check for clearance should be made by holding the assembled valve and rotating mechanism in the hands. With one hand, press down on the valve cap and turn the valve with the other hand. If the valve turns readily, it indicates that there is clearance present.

A special gauge is available to check this clearance. If the gauge indicates no clearance, grind off the end of the valve stem. If the clearance is too great, grind off the open end of the valve cap.

V8-403, 455

The rotator operates on a Sprag clutch principle utilizing the collapsing action of coil

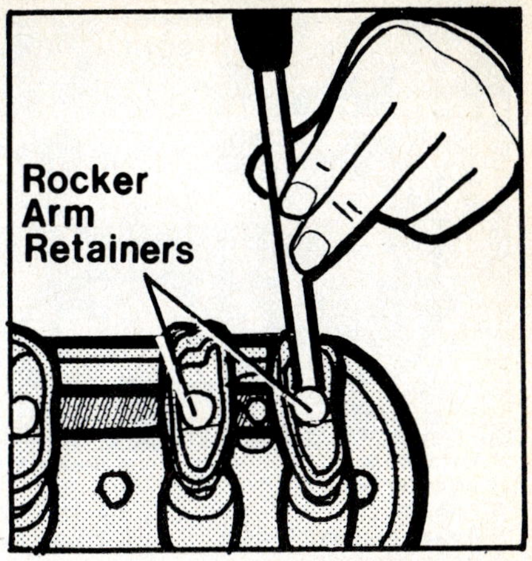

Fig. 16 Removing rocker arm nylon retainer. V6-231

spring to give rotation to the rotor body which turns the valve, Fig. 13.

To check rotator action, draw a line across rotator body and down the collar. Operate engine at 1500 rpm, rotator body should move around collar. Rotator action can be in either direction. Replace rotator if no movement is noted.

When servicing valves, valve stem tips should be checked for improper wear pattern which could indicate a defective valve rotator, Fig. 14.

VALVE GUIDES

Except V8-403, 455

CAUTION: Some V8-427 engines are equipped with exhaust valves filled with sodium for cooling purposes. When handling these valves, follow these precautions:

1. Never cut open an old valve and be extremely careful when removing a broken valve from an engine. Otherwise, the metallic sodium in the valve may contact and burn the skin.
2. Bury discarded sodium-cooled valves in some inaccessible place to prevent other persons from handling them. *Never immerse one of these valves in water.*

Valve guides in these engines are an integral part of the head and, therefore, cannot be removed. For service, guide holes can be reamed oversize to accommodate one of several service valves with oversize stems.

Check the valve stem clearance of each valve (after cleaning) in its respective valve

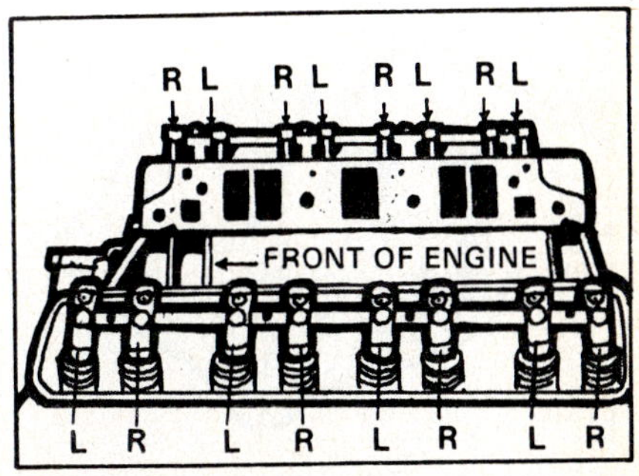

Fig. 18 Service rocker arm installation. V6-231

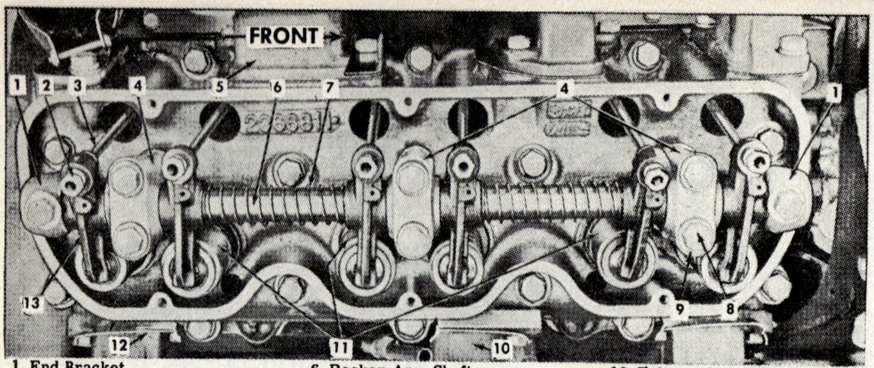

1 End Bracket
2 Adjusting Screw
3 Push Rod
4 Rocker Arm Shaft Bracket
5 Intake Manifold
6 Rocker Arm Shaft
7 Spring
8 Bracket Bolt
9 Flat Washer
10 Exhaust Manifold
11 Shields on Intake Valves
12 Exhaust Manifold Bolt Lock
13 Rocker Arm

Fig. 19 Rocker arms & shafts installed. V6-305C, 379, 432 & 478

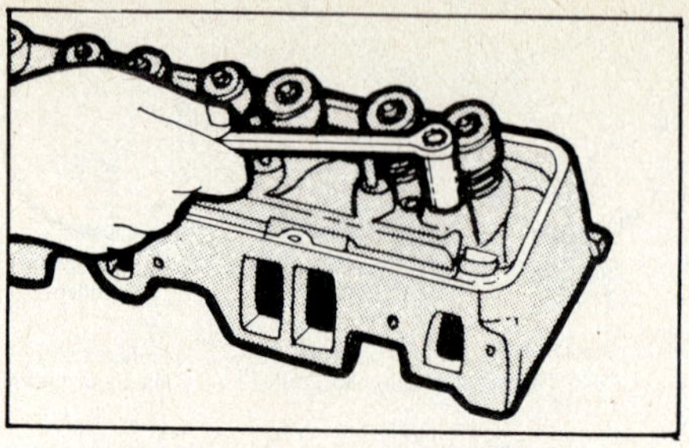

Fig. 20 Rocker arm stud removal. 6-250, 292, V6-200, 229, V8-267, 305, 350, 366, 400, 427 & 454

guide. If the clearance exceeds the service limits of .004″ on the intake or .005″ on the exhaust, ream the valve guide to accommodate the next oversize diameter valve stem.

Select the reamer for the smallest oversize which will provide a clean straight bore through the valve guide. After reaming, a new seat should be cut into the head to assure perfect seating of the new valve.

V8-403, 455 Motor Home

Valve stem guides are not replaceable, due to being cast in place. If valve guide bores are worn excessively, they can be reamed oversize.

If a standard valve guide bore is being reamed, use a .003″ or .005″ oversize reamer. For the .010″ oversize valve guide bore, use a .013″ oversize reamer. If too large a reamer is used and the spiraling is removed, it is possible that the valve will not receive the proper lubrication.

NOTE: Occasionally a valve guide will be oversize as manufactured. These are marked on the cylinder head as shown in Fig. 15. If no markings are present, the guide bores are standard. If oversize markings are present, any valve replacement will require an oversize valve. Service valves are available in standard diameters as well as .003″, .005″, .010″ and .013″ oversize.

ROCKER ARM SERVICE

V6-231

A nylon retainer is used to retain the rocker arm. Break them below their heads with a chisel, Fig. 16, or pry out with channel locks. Production rocker arms can be installed in any sequence since the arms are identical.

Replacement rocker arms for all engines are identified with a stamping, right (R) and left (L), Fig. 17 and must be installed as shown in Fig. 18.

V6-305C, 379, 432, 478

Referring to Fig. 19, assemble rocker arms, springs and brackets on rocker arm shaft, then install end brackets. Lubricate all moving parts with engine oil.

Place steel washer on each bracket bolt, then locate rocker arm shaft and bracket assembly on cylinder head and start bracket bolts into tapped holes. Back off adjusting screws, then tighten bracket bolts evenly while holding adjusting screws in respective push rod sockets. *Use special bolt at position (1) (Fig. 19) to assure correct position of shaft (6).*

NOTE: Check length of special bolt. If 2 7/8 inch bolt is used install two flat washers under bolt. If 2 3/4 inch bolt is used install one flat washer under bolt.

Check clearance between shaft end brackets and rocker arms. Clearance should be approximately .030 inch. Adjust clearance by loosening end bracket bolts and relocate brackets as allowed by clearance in bracket bolt holes. Tighten shaft bracket bolts to specified torque as listed under "Engine Tightening Specifications".

Tighten cylinder head to specified torque and then adjust valve clearance.

V8-403, 455 Motor Home

1. Remove valve cover.
2. Remove flanged bolts, rocker arm pivot and rocker arms, Fig. 13.
3. When installing rocker arm assemblies, lubricate wear surfaces with suitable lubricant. Torque flanged bolts to 25 ft. lbs.

ROCKER ARM STUDS

6-250, 292, V6-200, 229, V8-267, 305, 350, 366, 400, 427 & 454

Rocker arm studs that have damaged threads may be replaced with standard studs. Loose studs should be replaced with .003″ or .013″ oversize studs which are available for replacement.

Remove the old stud by placing a suitable spacer over the stud, Fig. 20. Install a nut and flat washer on the stud and pull out the stud by turning the nut. After reaming the hole for an oversize stud, coat the press-fit area of the new stud with rear axle lubricant. Install the stud by driving it in until it protrudes from the head the same distance as the other studs, Fig. 21.

NOTE: On V8-366, 427 & 454 the rocker arm studs are threaded into the cylinder head. Coat threads on cylinder head end of stud before assembling to head.

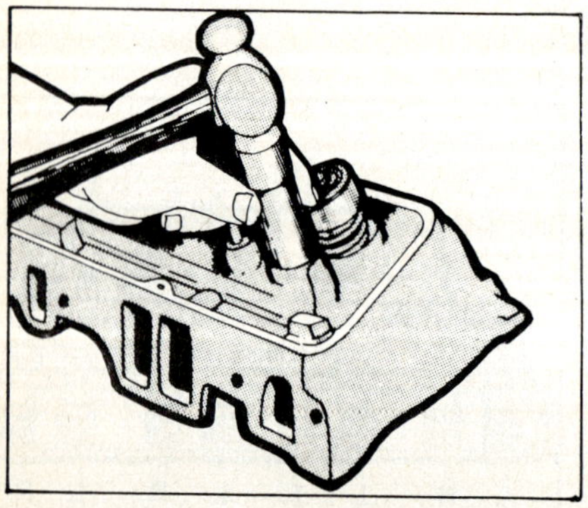

Fig. 21 Rocker arm stud installation. 6-250, 292, V6-200, 229, V8-267, 305 & 400

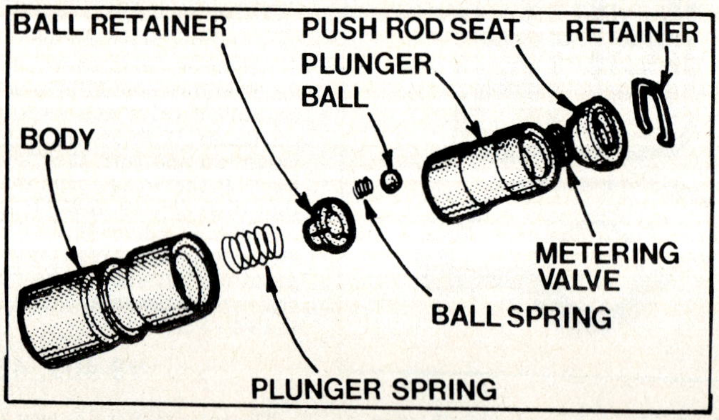

Fig. 22 Hydraulic valve lifter (Typical)

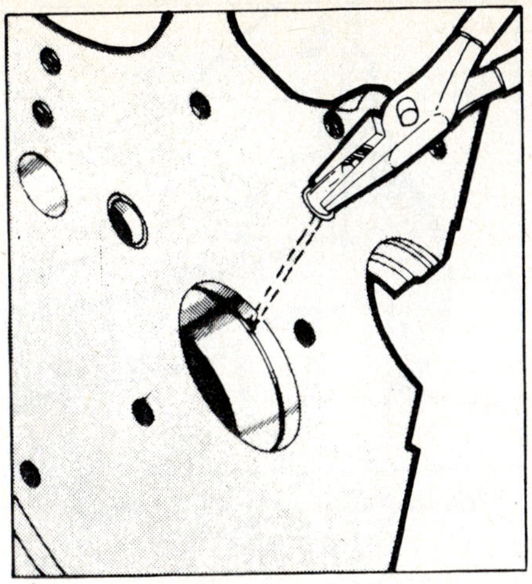

Fig. 23 Timing gear oil nozzle removal. 6-250 & 292

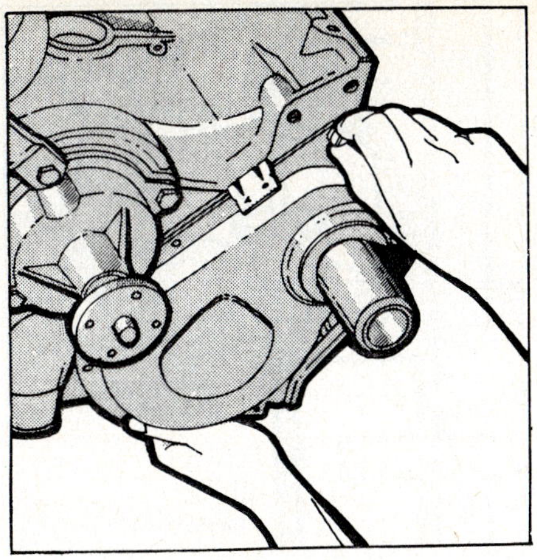

Fig. 24 Timing case cover installation. 6-250 & 292

Fig. 25 Timing case cover installation. V6-231

PUSH RODS

NOTE: When a replacement push rod has a paint stripe at one end, this painted end must be installed in contact with the rocker arm. To provide durability a hardened insert is incorporated in the rocker arm end of these push rods.

HYDRAULIC LIFTERS

The hydraulic valve lifters, Fig. 22, are simple in design, readjustments are not necessary and servicing them requires that care and cleanliness be exercised in the handling of parts.

Disassemble & Assemble

1. Hold plunger down with push rod and, using a small screw driver or awl, remove plunger retainer.
2. Remove parts from lifter body.
3. Clean all parts and inspect for damage. If any parts are damaged, the entire lifter assembly should be replaced. The inertia valve in the plunger for rocker arm lubrication should move when the plunger is shaken.
4. To reassemble, invert plunger and set ball into hole in plunger. Place ball check valve retainer over ball and on plunger. Place check valve retainer spring over retainer. Assemble body over plunger assembly. Turn assembly over and install push rod seat. Compress plunger with push rod and install retainer.
5. Compress plunger to open oil holes and fill plunger with engine oil. Work plunger up and down and refill.

TIMING CASE COVER

6-250, 292

1. To remove cover, remove radiator.
2. Remove vibration damper.
3. Remove the two oil pan to front cover attaching screws.
4. Remove front cover to cylinder block attaching screws.
5. Pull cover slightly forward; then, using a sharp knife, cut oil pan front seal flush with cylinder block at both sides of cover.
6. Remove cover and attached portion of oil pan front seal.
7. Pry oil seal out of cover with a large screwdriver. Install new seal with open side of seal inside of cover and drive or press seal into place.
8. If oil nozzle is to be replaced, remove it with pliers as shown in Fig. 23. Drive new nozzle in place, using a suitable light plastic or rubber hammer.
9. Clean gasket surfaces.
10. Cut tabs from new oil pan front seal, then install seal on cover, pressing tips into cover holes.
11. Install a suitable centering tool over end of crankshaft.
12. Coat gasket with light grease and stick it in position on cover.
13. Apply a 1/8 inch bead of RTV sealer at joint of oil pan and cover.
14. Install cover over centering tool, Fig. 24, and install cover screws, tightening them to 6 to 8 ft. lbs. *It is important that the centering tool be used to align the cover so the vibration damper installation will not damage the seal and position seal evenly around damper hub surface.*

V6-200, 229, V8-267, 305, 350, 366, 400, 427, 454

1. Remove vibration damper and water pump.
2. Remove cover retaining screws and cover.
3. Clean gasket surface of block and timing case cover.
4. Remove any excess oil pan gasket material that may be protruding at the oil pan to engine block junction.
5. Apply a thin bead of RTV #1052366 sealer or equivalent to the joint formed at oil pan and block.
6. Coat new gasket with sealer and position it on cover, then install cover to oil pan seal on cover and coat bottom of seal with engine oil.
7. Position cover on engine and loosely install the upper bolts.
8. Tighten screws alternately and evenly while pressing downward on cover so that dowels are aligned with holes in cover. Do not force cover over dowels as cover can be distorted.
9. Install remaining cover screws, vibration damper and water pump.

V6-231

1. Drain cooling system and remove radiator.
2. Remove fan, pulleys and belts.
3. Remove crankshaft pulley and reinforcement.
4. If equipped with power steering, remove any pump bracket bolts attached to timing chain cover and loosen and remove any other bolts necessary that will allow pump and brackets to be moved out of the way.
5. Remove fuel pump.
6. Remove Delcotron and brackets.
7. Remove distributor cap and pull spark plug wire retainers off brackets on rocker arm cover. Swing distributor cap with wires attached out of the way. Disconnect distributor primary lead.
8. Remove distributor. *If chain and sprockets are not to be disturbed, note position of distributor rotor for installation in the same position.*
9. Loosen and slide clamp on thermostat bypass hose rearward.
10. Remove bolts attaching chain cover to block.
11. Remove two oil pan-to-chain cover bolts and remove cover.
12. Reverse procedure to install, noting data shown in Fig. 25.

NOTE: Remove the oil pump cover and pack the space around the oil pump gears completely full of vaseline. There must be no air space left inside the pump. Reinstall the cover using a new gasket. This step is very important as the oil pump may lose its prime whenever the pump, pump cover or timing chain cover is disturbed. If the pump is not packed it may not begin to pump oil as soon as the engine is started.

V6-305C & 379

To remove cover, block crankshaft and remove retaining bolt and washer at front

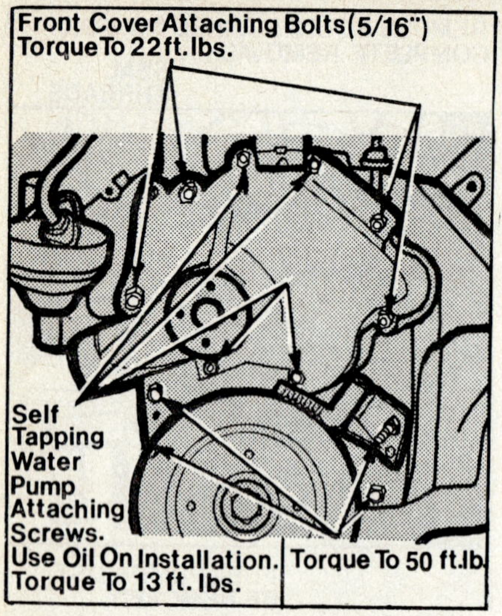

Fig. 26 Timing case cover bolt installation. V8-403 & 455

side of crankshaft pulley. Assemble a suitable puller and remove pulley. Remove bolts attaching front cover to cylinder block and jar cover loose with a lead hammer.

V6-432 & 478M

To remove the cover, drain the cooling system and disconnect lower radiator hose. Disconnect external water by-pass at the intake manifold. Drop the oil pan. Take off radiator, shroud, fan blades, pulleys, vibration damper and fuel pump. Unfasten the cover from the block and lift it off.

V8-403 & 455

NOTE: When it becomes necessary to replace the cover oil seal, the cover need not be removed.

---

1. Drain cooling system and disconnect

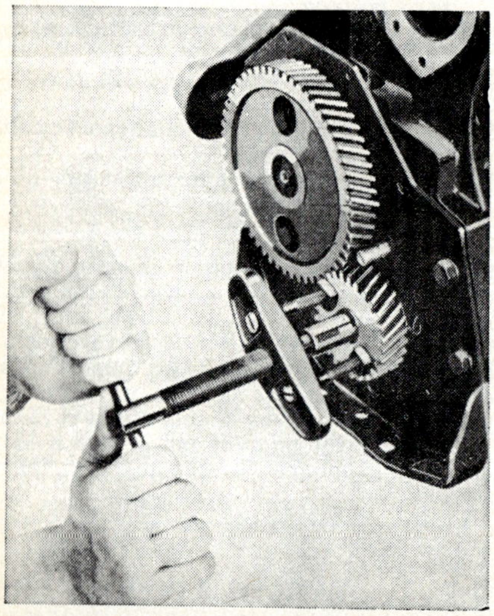

Fig. 29 Using puller to remove crankshaft gear. 6-250 & 292

Fig. 27 Camshaft gear removal. 6-250 & 292

heater hose, by-pass hose and both radiator hoses.
2. Remove all belts, fan and pulley, crankshaft pulley and pulley hub.
3. Remove oil pan.
4. Remove cover, timing, pointer and water pump assembly, Fig. 26.

TIMING GEARS

6-250 & 292

When necessary to install a new camshaft gear, the camshaft will have to be removed as the gear is a pressed fit on the shaft. The camshaft is held in position by a thrust plate which is fastened to the crankcase by two capscrews which are accessible through two holes in the gear web.

Use an arbor press to remove the gear and when doing so, a suitable sleeve, Fig. 27, should be employed to support the gear properly on its steel hub.

Before installing a new gear, assemble a new thrust plate on the shaft and press the gear on just far enough so that the thrust plate has practically no clearance, yet is free to turn. The correct clearance is from .001" to .005". Fig. 28.

The crankshaft gear can be removed by utilizing the two tapped holes in conjunction with a gear puller, Fig. 29.

When the timing gears are installed, be sure the punch-marks on both gears are in mesh, Fig. 30. Backlash between the gears should be from .004" to .006", Fig. 31. Check the run-out of the gears, Fig. 32, and if the camshaft gear run-out exceeds .004" or the crank gear run-out is in excess of .003", remove the gear (or gears) and examine for burrs, dirt or some other fault which may cause the run-out. If these conditions are not the cause, replace the gear (or gears).

V6-432, 478M

NOTE: When necessary to install a new camshaft gear, the camshaft will have to be removed as the gear is a pressed fit on the shaft.

For correct valve timing, the "X" marks on the camshaft and idler gears should be aligned, Fig. 33, and the "O" marks on the idler and crankshaft gears should be aligned.

Fig. 28 Checking camshaft end play. 6-250 & 292

TIMING CHAIN

V6-200, 229, V8-267, 305, 350, 366, 400, 427 & 454

1. Remove timing chain cover as outlined previously.
2. Remove crankshaft oil slinger.
3. Crank engine until "O" marks on sprockets are in alignment, Fig. 34.
4. Remove three camshaft-to-sprocket bolts.
5. Remove camshaft sprocket and timing chain together. Sprocket is a light press fit on camshaft for approximately 1/8". If sprocket does not come off easily, a light blow with a plastic hammer on the lower edge of the sprocket should dislodge it.
6. If crankshaft sprocket is to be replaced, remove it with a suitable gear puller. Install new sprocket, aligning key and keyway.
7. Install chain on chamshaft sprocket. Hold sprocket vertical with chain hanging below and shift around to align the "O" marks on sprockets, Fig. 34.

NOTE: The valve timing marks shown in Fig. 34, do not indicate TDC compression stroke for No. 1 cylinder, which is used during distributor installation. If distributor was removed, install timing chain and sprockets, aligning timing marks, Fig. 34, then rotate engine until No. 1 cylinder is on compression and camshaft timing mark is 180° from valve timing position shown in Fig. 34 and install distributor.

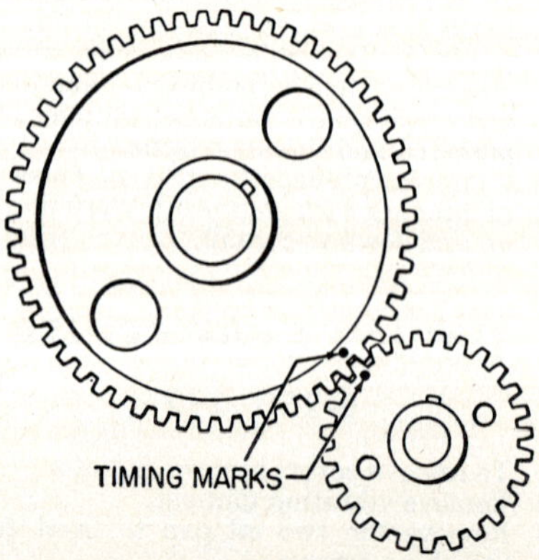

Fig. 30 Valve timing marks. 6-250 & 292

Fig. 31 Checking timing gear backlash. 6-250 & 292

Fig. 32 Checking timing gear runout. 6-250 & 292

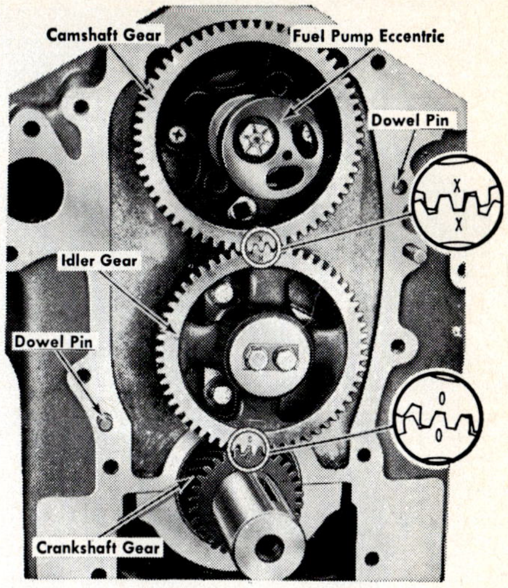

Fig. 33 Valve timing marks. V6-432 & 478M

8. Align dowel in camshaft with dowel hole in sprocket and install sprocket on camshaft. *Do not attempt to drive sprocket on camshaft as welch plug at rear of engine can be dislodged.*
9. Draw sprocket onto camshaft, using the three mounting bolts. Tighten to 20 ft. lbs. torque.
10. Lubricate timing chain and install cover.

V6-231

1. With the timing case cover removed as outlined previously, temporarily install the vibration damper bolt and washer in end of crankshaft.
2. Turn crankshaft so sprockets are positioned as shown in Fig. 35. Use a sharp rap on a wrench handle to start the vibration damper bolt out without disturbing the position of the sprockets.
3. Remove oil pan and oil slinger.
4. Remove camshaft distributor drive gear and fuel pump eccentric.
5. Use two large screwdrivers to alternately pry the camshaft sprocket then the crankshaft sprocket forward until the camshaft sprocket is free. Then remove camshaft sprocket and chain, and crankshaft sprocket off crankshaft.
6. To install, assemble chain on sprockets and slide sprockets on their respective shafts with the "O" marks on the sprockets lined up as shown.
7. Complete the installation in the reverse order of removal. Torque sprocket bolts to 22 ft. lbs.

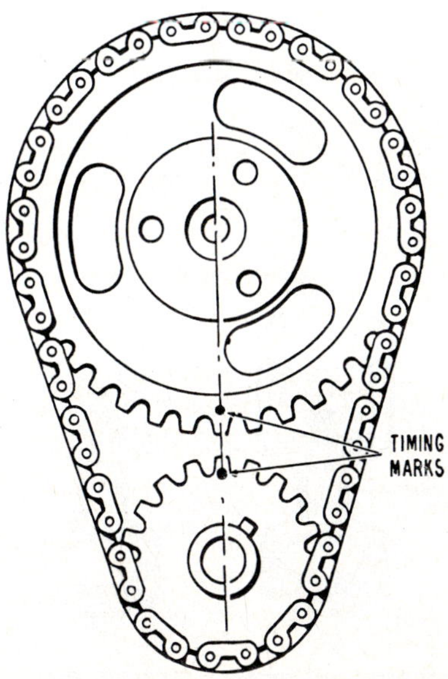

Fig. 34 Valve timing marks. V6-200, 229, V8-267, 305, 350, 366, 400, 427 & 454

V6-305C & 379

When chain is properly installed, a straightedge aligned with the center of the crankshaft and camshaft will also be aligned with the timing marks as shown in Fig. 36. Install fuel pump eccentric at front side of camshaft sprocket. Special washer and bolt (1 and 12, Fig. 36, are used to retain eccentric and sprocket. Tighten bolt to a torque of 60–70 ft. lbs.

V8-403 & 455

1. After removing front cover, remove fuel pump eccentric, oil slinger, crankshaft sprocket, chain and camshaft sprocket.
2. Install camshaft sprocket, crankshaft sprocket and timing chain together, aligning timing marks as shown in Fig. 37.

NOTE: The valve timing marks shown in Fig. 37, do not indicate TDC compression stroke for No. 1 cylinder, which is used during distributor installation. If distributor was removed, install timing chain and sprockets, aligning timing marks, Fig. 37, then rotate engine until No. 1 cylinder is on compression and camshaft timing mark is 180° from valve timing position shown in Fig. 37 and install distributor.

---

3. Install fuel pump eccentric with flat side rearward. Then install oil slinger and replace front cover, Fig. 26.

CAMSHAFT, REPLACE

6-250 & 292

It is recommended that the engine be removed from the vehicle for camshaft removal. Remove valve train components, engine front cover, fuel pump and distributor. Remove camshaft thrust plate screws and pull camshaft from cylinder block.

V6-200, 229, V8-267, 305, 350, 366, 400, 427 & 454

To remove the camshaft, remove valve lifters, fuel pump and its push rod, radiator, timing chain and camshaft sprocket. Install two $\frac{5}{16}''$-18 $\times$ 4'' bolts in two camshaft bolt holes. Using these bolts as a puller, remove camshaft from engine.

NOTE: When a new camshaft is installed, make certain that it is the correct part for the type of valve mechanism in the engine. Use of a camshaft designed for an engine with hydraulic valve lifters, or vice-versa, will result in extremely rough and noisy engine operation.

V6-231

NOTE: If engine is in the vehicle, the radiator, grille and A/C components will have to be

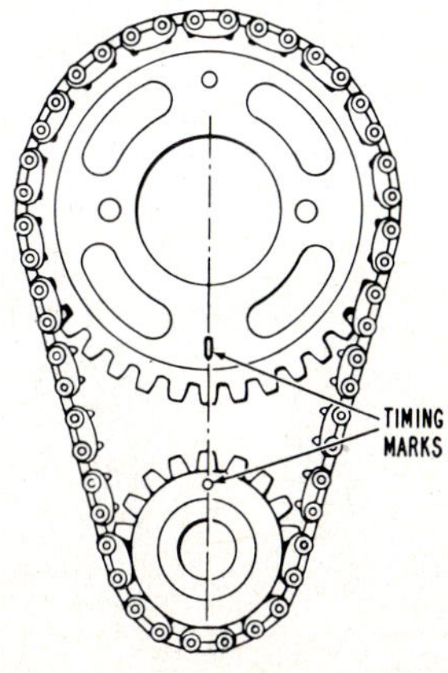

Fig. 35 Valve timing marks. V6-231

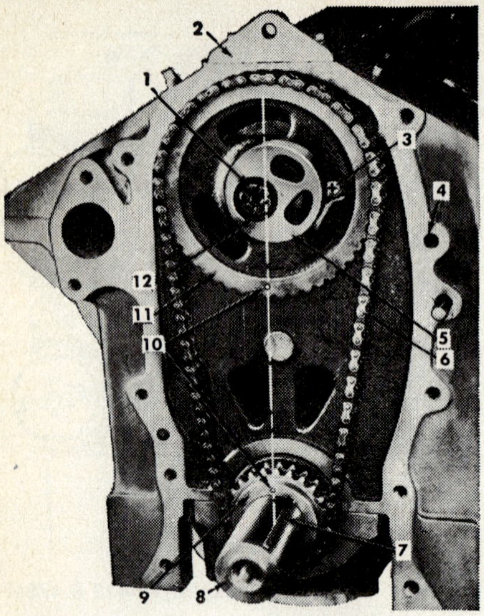

1 Bolt
2 Cap
3 Thrust Plate Screw
4 Dowel Hole
5 Fuel Pump Eccentric
6 Timing Chain
7 Pulley Hub Key
8 Crankshaft
9 Crankshaft Sprocket
10 Timing Marks
11 Camshaft Sprocket
12 Eccentric Clamp Washer (Special)

Fig. 36 Valve timing marks. V6-305C & 379

removed. If engine is out of car, proceed as follows:

1. To remove camshaft, remove intake manifold, rocker arm shaft assemblies, push rods and valve lifters.
2. Remove timing chain and sprockets.
3. Slide camshaft out of engine, using care not to mar the bearing surfaces.

V6-305C, 379, 432 & 478M

Camshaft is supported on bushing type bearings and all journals are pressure lubricated. An oil trough just below camshaft holds oil into which the cam lobes dip as the camshaft rotates.

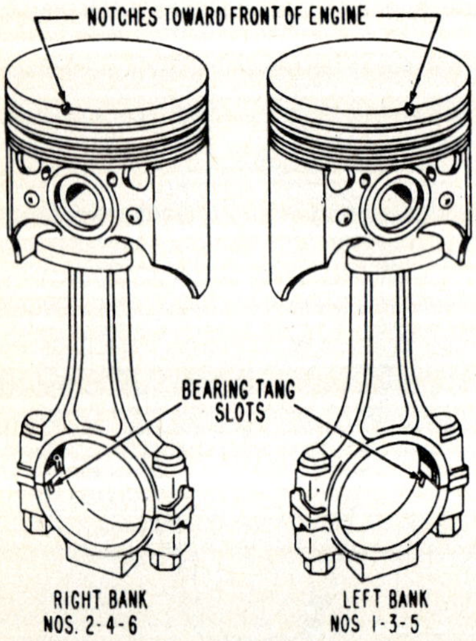

Fig. 40 Piston and rod assembly. V6-200 & 229

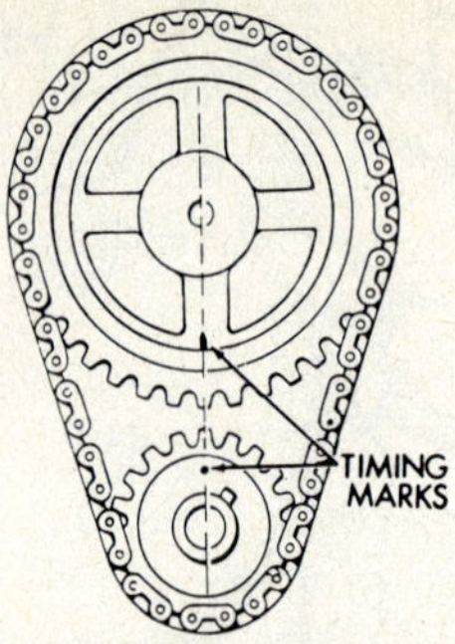

Fig. 37 Valve timing marks. V8-403 & 455

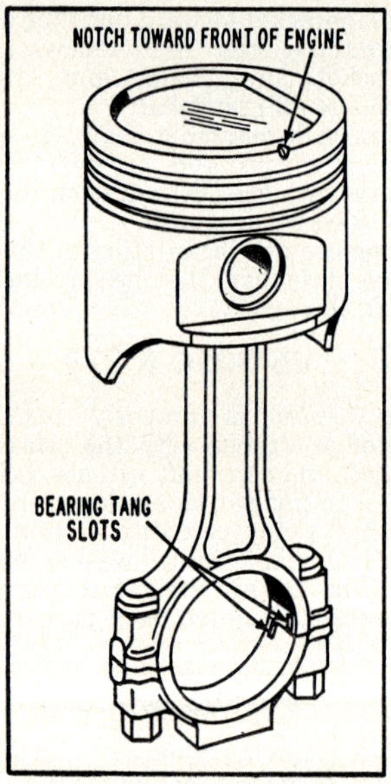

Fig. 39 Piston & rod assembly. 6-292

To remove the camshaft, remove thrust plate attaching bolts at front end of camshaft. Pull camshaft forward to remove from cylinder block. Remove key from keyway in camshafta, then remove thrust plate and gear spacer.

To install camshaft, apply engine oil to camshaft journals and cam lobes, then install camshaft in block. If spacer has been removed, install spacer with chamfer toward bearing journal and drive key into keyway. Install thrust plate. Sprocket is installed together with chain.

V8-403 & 455

1. Remove engine from vehicle.
2. Remove oil pan, crankshaft pulley and harmonic balancer and engine front cover.
3. Disconnect spark plug cables, then remove distributor cap and valve covers.
4. Remove intake manifold, rocker arms, push rods and valve lifters.

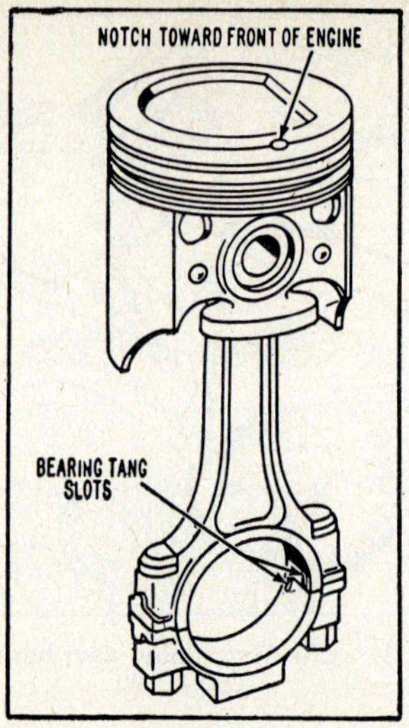

Fig. 38 Piston & rod assembly. 6-250

NOTE: Identify parts so they can be reinstalled in their original locations.

5. Remove fuel pump eccentric, camshaft gear, oil slinger and timing chain.
6. Remove camshaft by carefully sliding it out of engine.

CAUTION: Use care when removing camshaft to avoid damaging bearings.

7. Reverse procedure to install.

PISTON & ROD, ASSEMBLE

Assemble pistons to connecting rods as shown in Figs. 38 through 46.

PISTONS

Exc. V6-231

A .001" oversize piston is available for service use so that proper clearances can be obtained for slightly worn cylinder bores

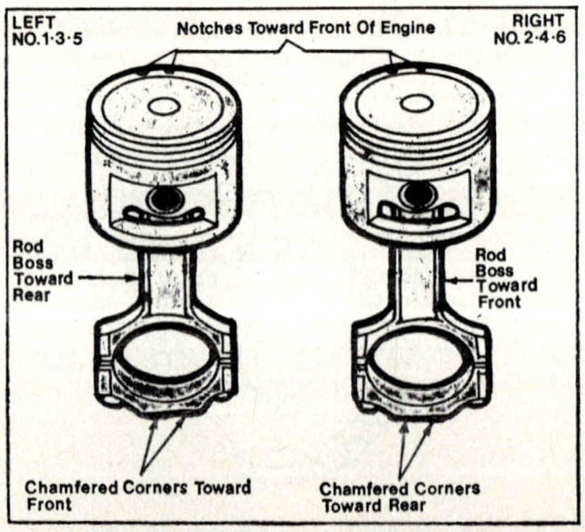

Fig. 41 Piston & rod assembly. V6-231

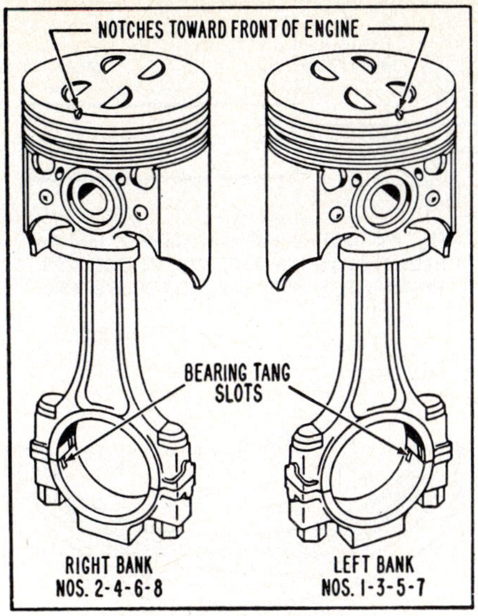

Fig. 42 Piston & rod assembly. V8-267, 305, 350 & 400

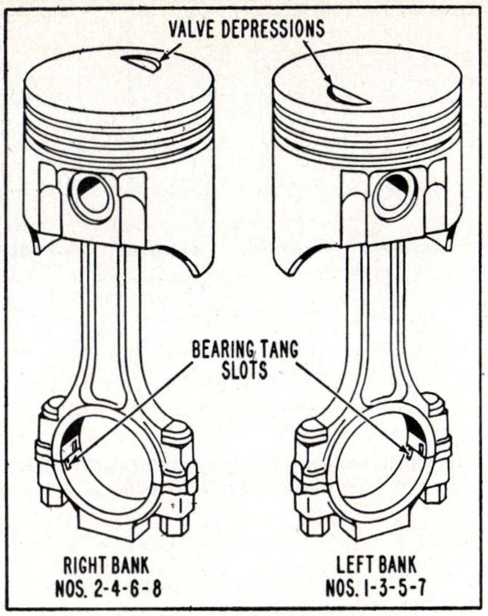

Fig. 43 Piston & rod assembly. V8-366 & 427

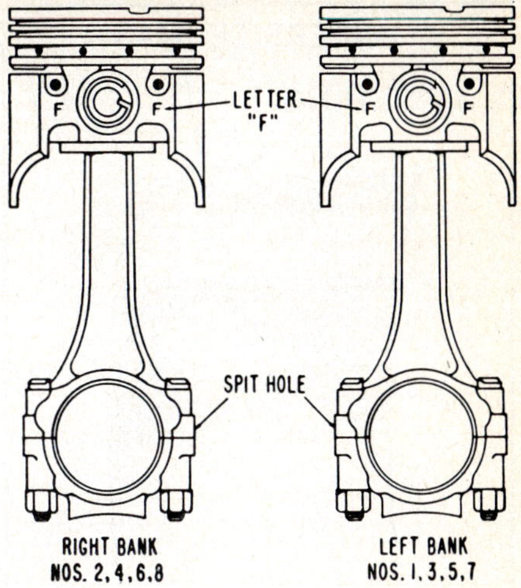

Fig. 44 Piston & rod assembly. V8-403

requiring only light honing to clean up the bores. In addition, oversizes of .020", .030" and .040" are available. If the cylinders have less than .005" taper or wear they can be reconditioned with a hone and fitted with the .001" oversize piston.

V6-231

Pistons are available in standard sizes and oversizes of .005, .010 and .030 inch.

ROD BEARINGS

6-250, 292, V6-200, 229, V8-267, 305, 350, 366, 400, 427 & 454

Connecting rod bearing inserts are available in standard size and undersizes of .001", .002", .010" and .020". The bearings can be replaced without removing the rod assembly by removing the cap and replacing the upper and lower halves of the bearing.

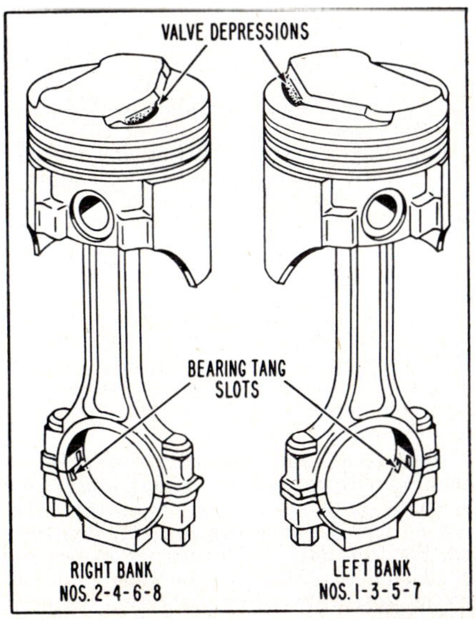

Fig. 45 Piston & rod assembly. V8-454

V6-231

Rod bearings are available in standard sizes and undersizes of .001, .002 and .010 inch. Refer to Fig. 47, for undersize markings on bearing.

V8-403 & 455

Rod bearings are available in standard sizes and undersizes of .001, .002, .005, .010, .012 and .020 inch.

MAIN BEARINGS

6-250, 292

NOTE: The rear main bearing journal has no oil hole drilling. To remove the upper bearing half (bearing half with oil hole) proceed as follows after cap is removed:

1. Use a small drift punch and hammer to start the bearing rotating out of the block.
2. Use a pair of pliers (tape jaws) to hold the bearing thrust surface to the oil slinger and rotate the crankshaft to pull the bearing out.
3. To install, start the bearing (side not notched) into side of block by hand, then use pliers as before to turn bearing half into place.
4. The last 1/4" movement may be done by holding just the slinger with the pliers or tap in place with a drift punch.

Main bearings are available in standard and undersizes of .001, .002, .009, .010 and .020 inch.

V6-200, 229, V8-267, 305, 350, 366, 400, 427 & 454

Shell type bearings are used, and if worn more than .004" should be replaced. No attempt should be made to shim, file or otherwise take up worn bearings.

Main bearings are available in standard and undersizes of .001, .002, .009, .010 and .020 inch.

V6-231

Main bearings are available in standard sizes and undersizes of .001, .002, and .010 inch. Refer to Fig. 47 for undersize markings.

V8-403 & 455

Main bearings are available in standard sizes and undersizes of .0005, .001, .0015, .002, .010 and .020".

NOTE: Main bearing clearances not within specifications must be corrected by the use of selective upper and lower shells. Fig. 48 illustrates the undersize identification marking on the bearing tang.

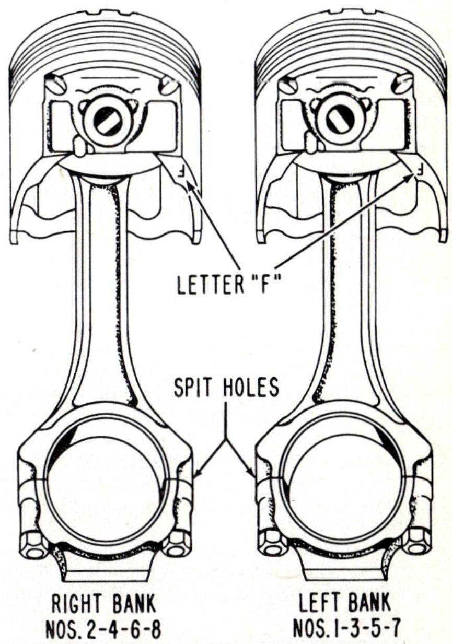

Fig. 46 Piston & rod assembly. V8-455

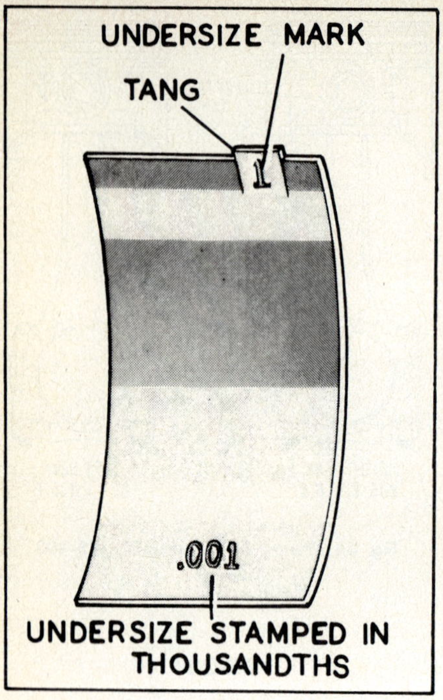

Fig. 47 Location of main & rod bearing undersize markings. V6-231

CRANKSHAFT REAR OIL SEAL, REPLACE

6-250, 292, V6-200, 229, V8-267, 305, 350, 366, 400, 427 & 454

NOTE: These engines are equipped with helix type rear seal. A seal starting tool, Fig. 49, must be used to prevent the upper seal half from coming into contact with the sharp edge of the block.

When necessary to correct an oil leak due to a defective seal, always replace the upper and lower seal halves as a unit, Fig. 50. *When installing either half, lubricate the lip portion only with engine oil, keeping oil off the parting line surface as this is treated with glue.* Always clean crankshaft surface before installing a new seal.

1. To replace the lower seal, remove seal from groove in bearing cap, using a small screwdriver to pry it out.
2. Insert new seal and roll it in place with finger and thumb.
3. To replace the upper seal (with engine in car) use a small hammer and tap a brass pin punch on one end of the seal until it protrudes far enough to be removed with pliers.
4. Position tip of tool, Fig. 49, between crankshaft and seal seat in cylinder block.
5. Position seal between crankshaft and tip of tool with seal bead contacting tip of tool. Ensure oil seal lip is facing toward front of engine.
6. Roll seal around crankshaft, using tool as a "Shoehorn" to protect seal bead from sharp corner of seal seat surface in cylinder block.

NOTE: Tool must remain in position until seal is properly seated with both ends flush with block.

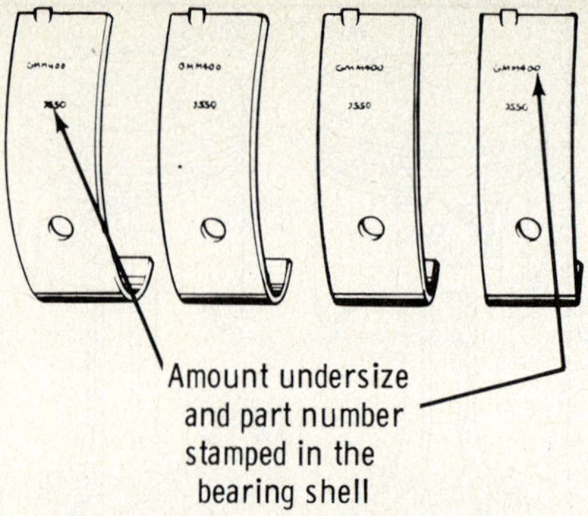

Fig. 48 Location of main bearing undersize markings. V8-403 & 455

7. Remove tool, using care not to dislodge seal.
8. Install new seal into bearing cap with tool as outlined previously.
9. Install bearing cap with sealant applied to the cap to case interface. Do not apply sealant to seal ends. Torque rear main bearing cap bolts to specifications as listed in the "Engine Tightening Specification Chart".

V6-231

Since the braided fabric seal used on these engines can be replaced only when the crankshaft is removed, the following repair procedure is recommended.

1. Remove oil pan and bearing cap.
2. Drive end of old seal gently into groove, using a suitable tool, until packed tight. This may vary between 1/4 and 3/4 inch depending on amount of pack required.
3. Repeat previous step for other end of seal.
4. Measure and note amount that seal was driven up on one side. Using the old seal removed from bearing cap, cut a length of seal the amount previously noted plus 1/16 inch.
5. Repeat previous step for other side of seal.
6. Pack cut lengths of seal into appropriate side of seal groove. A guide tool, J-21526-1, and packing tool, J-21526-2, may be used since these tools have been machined to provide a built-in stop.
7. Install new seal in bearing cap.

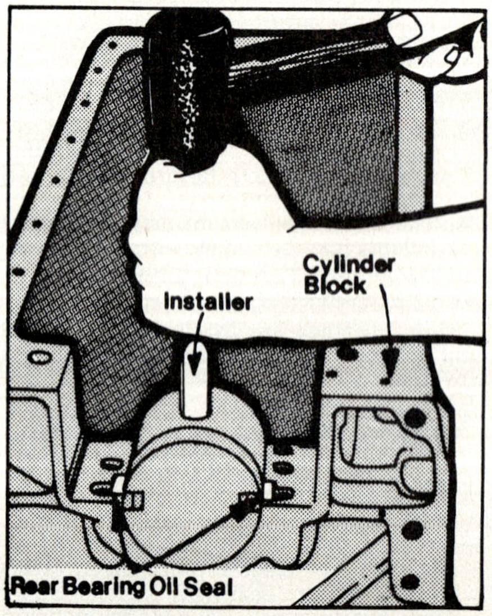

Fig. 51 Crankshaft rear oil seal installation. V6-305C, 379, 432 & 478M

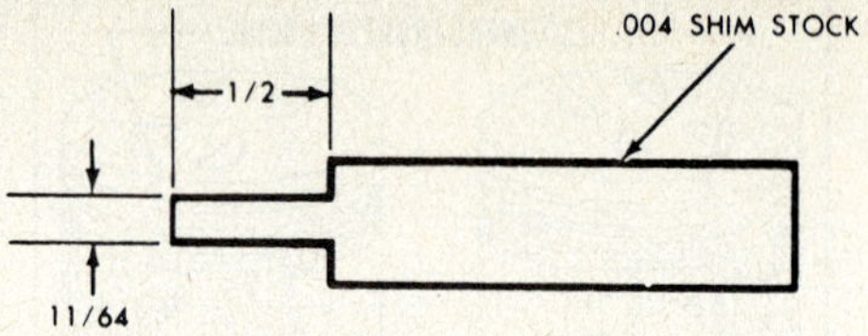

Fig. 49 Fabricated seal starting tool for helix type seal. 6-250, 292, V6-200, 229, V8-267, 305, 350, 366, 400, 427 & 454

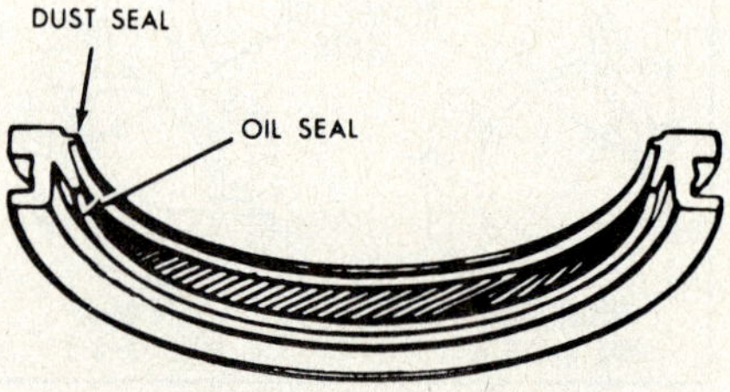

Fig. 50 Crankshaft rear oil seal. 6-250, 292, V6-200, 229, V8-267, 305, 350, 366, 400, 427 & 454

V6-305C, 379, 432 & 478M

Crankshaft rear bearing oil seal consists of two pieces of special packing. One piece is installed in the groove in the rear bearing cap and the other piece is installed in a similar groove in the cylinder block, Fig. 51.

Position the rear bearing seal in the groove in the cylinder block. Lay an improvised mandrel, Fig. 52, in the bearing bore and strike the mandrel with a hammer to drive the seal into the groove. Install the seal in the bearing cap in a similar manner. Using a sharp knife, cut off both ends of each seal which project out of the grooves. When cutting off the ends of the seals, do not leave frayed ends which would prevent proper seating of the bearing cap if the ends should extend between cap and cylinder block.

When installing rear bearing cap side seals, immerse the seals in diesel fuel oil prior to installation, and the exposed end of each side seal will protrude approximately .045", Fig. 53. *Do not cut off protruding ends of side seals.*

V8-403 & 455

To replace upper rear main bearing seal, the crankshaft must be removed. After removing crankshaft use tool No. BT23-18 to install seal, Fig. 54.

Lower rear main bearing seal can be installed with crankshaft in place and engine in vehicle. Remove oil pan, then the rear main bearing cap. Remove bearing and seal from cap. Position new seal in groove in bearing cap. Using tool BT23-18, hammer seal into groove. Rotate tool slightly, then cut seal ends flush with bearing cap surface. Pack seal end fibers away from edges, using a screw driver.

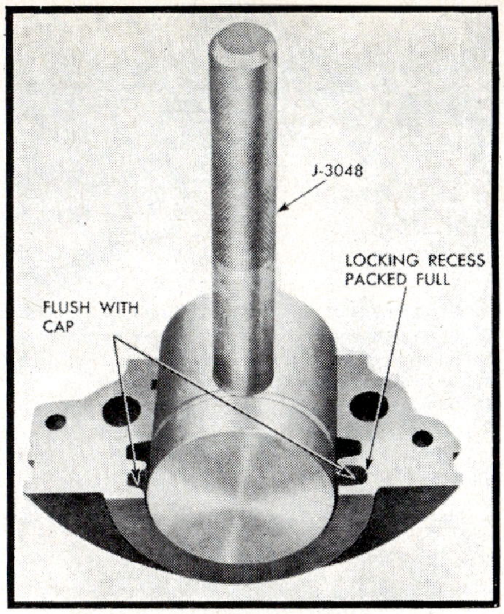

Fig. 52 Crankshaft rear oil seal to rear main bearing cap installation. V6-305C, 379, 432 & 478M

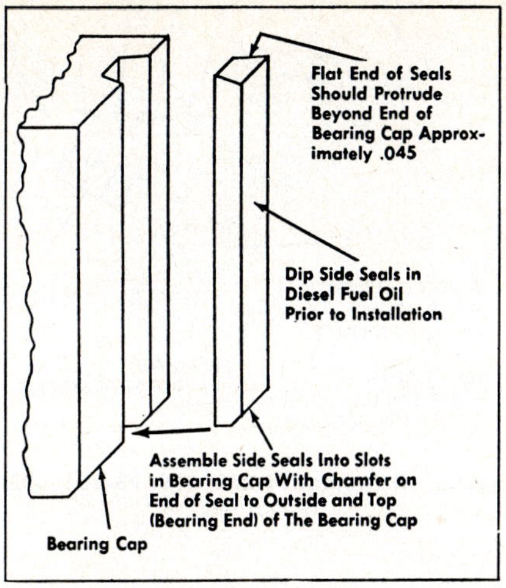

Fig. 53 Rear main bearing cap side seal installation. V6-305C, 379, 432 & 478M

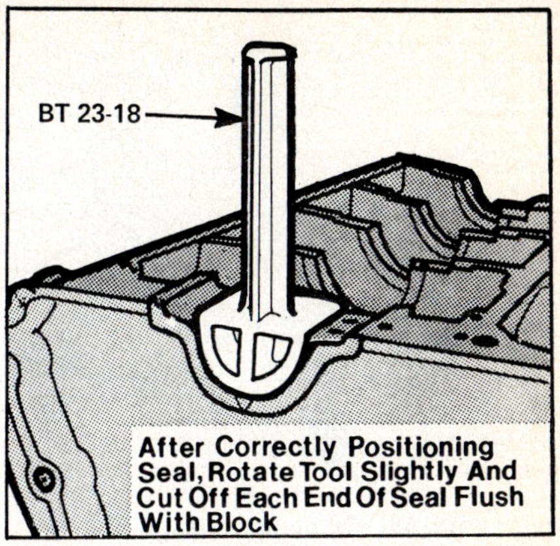

Fig. 54 Crankshaft rear seal installation. V8-403 & 455

Install bearing in cap, then install cap, lubricate bolt threads and torque to 120 ft. lbs. Install oil pan and lower flywheel cover.

OIL PAN, REPLACE

6-250, 292

1974–75

1. Disconnect battery ground cable.
2. Remove starter motor.
3. On CS series, remove engine front mount bolts (frame bracket-to-mount). Using a suitable jack with a flat piece of wood to prevent damage to oil pan, raise engine enough to insert 2″ × 4″ wood blocks between engine mount and frame brackets.
4. Drain engine oil, remove oil pan, gaskets and seals.
5. Reverse procedure to install.

1976–83 Exc. 1978–83 'G' Models

1. Disconnect battery ground cable.
2. Raise and support vehicle. Drain oil pan and remove starter.
3. Remove flywheel splash shield of converter underpan as applicable.
4. Remove front engine mount through bolts, raise front of engine and reinstall bolts, then lower engine.
5. Remove oil pan bolts and oil pan.
6. Reverse procedure to install.

1978–83 'G' Models

1. Disconnect battery ground cable. Remove engine cover.
2. Remove air cleaner and studs. Remove fan shroud and radiator upper support brackets.
3. Raise and support vehicle.
4. On models equipped with a manual transmission:
   a. Disconnect clutch cross shaft from motor mount bracket.
   b. Remove transmission to bell housing upper bolt.
   c. Remove transmission rear mount bolts and install two 7/16 × 3 inch bolts.
   d. Raise transmission and install 2 inch block between mount and crossmember.
5. On all models, remove starter and drain oil.
6. Remove front engine mount through bolts, raise engine and install wood blocks between engine mounts and cross member brackets.
7. Remove flywheel splash shield or converter cover as applicable.
8. Remove oil pan bolts and lower pan.
9. Reverse procedure to install.

V6-200, 229, V8-267, 305, 350, 366, 400, 427 & 454

1. Drain engine oil and remove dipstick and tube.
2. If equipped remove exhaust crossover pipe.
3. With automatic transmission, remove converter housing under pan.
4. Remove starter brace and inboard bolt and swing starter aside.
5. On K models, remove strut rods if equipped.
6. On 1980–83 El Camino and Caballero with V6-229, remove upper half of fan shroud. Loosen left engine front mount through bolt and remove right bolt. Raise engine and reinstall bolt. Do not tighten.
7. Remove oil pan and discard gaskets and seals.
8. Reverse procedure to install

V6-231

1. Support vehicle on hoist and drain oil.
2. Remove flywheel cover and exhaust crossover pipe.
3. Remove oil pan bolts and allow pan to drop.

NOTE: To remove oil pan, front wheels must be turned to the left, also the crankshaft may have to be turned to provide clearance.

4. Reverse procedure to install.

NOTE: Some 1981–83 engines use R.T.V. silicone sealer instead of a cork gasket for oil pan to crankcase sealing. When replacing the oil pan or sealing material on these engines, either R.T.V. sealer or a cork gasket can be used during reassembly. If R.T.V. sealer is used, the pan rail and block sealing surfaces should be cleaned thoroughly and a 1/4 inch bead of sealant applied evenly to the pan rail, avoiding any breaks or gaps in the sealer during application.

V6-305C, 379, 432 & 478M

1. Drain crankcase, then remove oil pan attaching bolts and stud nut.
2. Lower oil pan from cylinder block, then scrape old pan gasket from oil pan mating surfaces.
3. Reverse procedure to install.

V8-403 & 455

1. Remove transmission and final drive from vehicle.
2. Drain oil from engine, then disconnect relay tie rod from idler arm and relay lever and steering shock absorber from crossmember bracket.
3. Remove left hand lower venturi bracket and disconnect power steering pump.
4. Remove the four front support bolts and front motor mount bolts and position support aside.
5. Remove flywheel, then remove oil pan bolts.
6. Raise engine about 1 inch and remove oil pan.
7. Reverse procedure to install.

OIL PUMP

6-250, 292, V6-200, 229, V8-267, 305, 350, 366, 400, 427 & 454

Oil Pump, Replace

1. Remove oil pan as described previously.
2. On 6-250 engines, remove the two flange mounting bolts, pickup pipe bolt and then remove pump.
3. On all except 6-250 and 292 engines, remove pump to rear main bearing cap bolt and remove pump and extension shaft.
4. Reverse procedure to install. Make sure that installed position of oil pump screen is with bottom edge parallel to oil pan rails on all except 6-250 and 292 engines.

1 Pressure Regulator Valve
2 Pressure Regulator Spring
3 Retaining Pin
4 Screws
5 Pump Cover
6 Cover Gasket
7 Idler Gear
8 Drive Gear and Shaft
9 Pump Body
10 Pickup Screen and Pipe

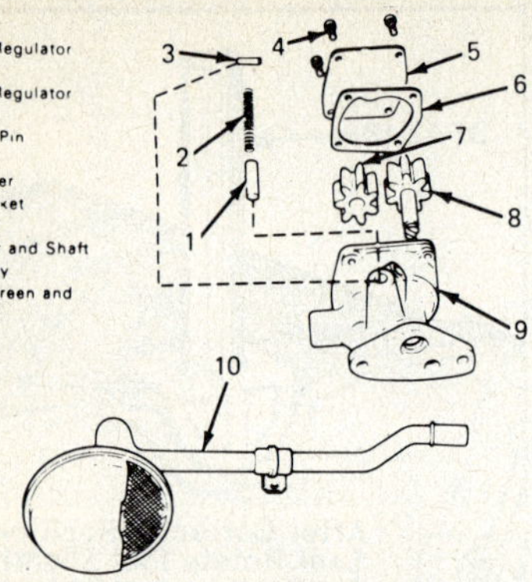

Fig. 55 Oil pump. 6-250 & 292

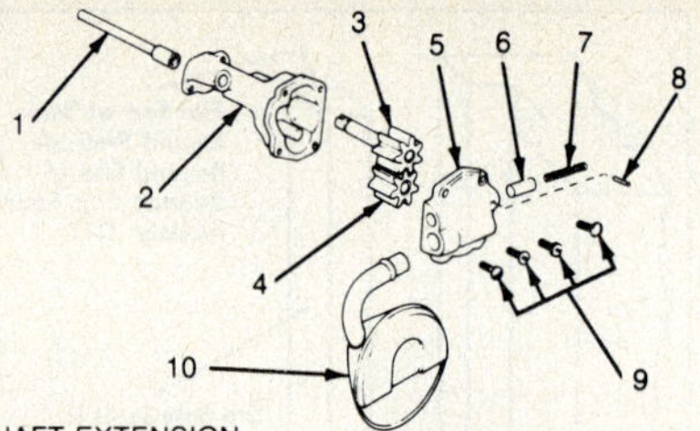

1. SHAFT EXTENSION
2. PUMP BODY
3. DRIVE GEAR AND SHAFT
4. IDLER GEAR
5. PUMP COVER
6. PRESSURE REGULATOR VALVE
7. PRESSURE REGULATOR SPRING
8. RETAINING PIN
9. SCREWS
10. PICKUP SCREEN AND PIPE

Fig. 56 Oil pump. V6-200, 229, V8-267, 305, 350 & 400

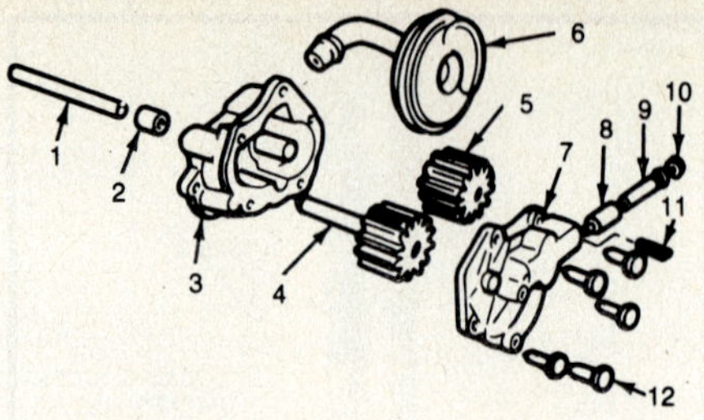

1. Shaft Extension
2. Shaft Coupling
3. Pump Body
4. Drive Gear and Shaft
5. Idler Gear
6. Pickup Screen and Pipe
7. Pump Cover
8. Pressure Regulator Valve
9. Pressure Regulator Spring
10. Washer
11. Retaining Pin
12. Screws

Fig. 57 Oil pump. V8-366, 427 & 454

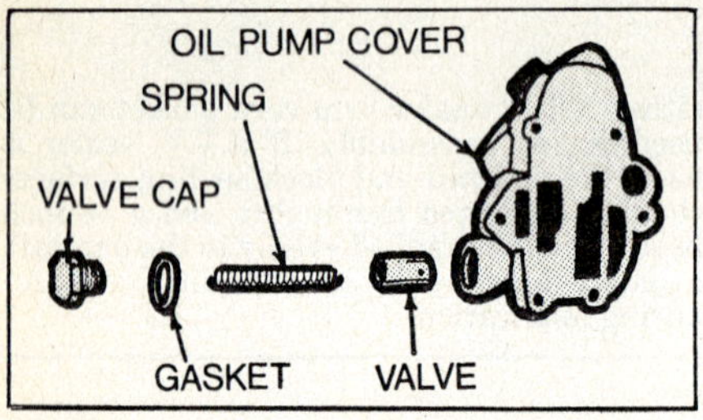

Fig. 58 Oil pump cover & bypass valve. V6-231

Oil Pump Service

1. Remove oil pump as described previously.
2. Remove pump cover screws and pump cover. On 6-250 and 292 engine oil pump, remove gasket, Figs. 55, 56 and 57.
3. Mark gear teeth so they can be reassembled with same teeth indexing, then remove drive gear, idler gear and shaft.
4. Remove pressure regulator valve retaining pin, pressure regulator valve and related parts.
5. If pickup screen and pipe require replacement, mount pump in a soft-jawed vise and extract pipe from pump.
6. Wash all parts in cleaning solvent and dry with compressed air.
7. Inspect pump body and cover for cracks and excessive wear.
8. Inspect pump gears for damage or excessive wear.
9. Check drive gear shaft for looseness in pump body.
10. Inspect inside of pump cover for wear that would allow oil to leak past the ends of the gears.
11. Inspect pickup screen and pipe assembly for damage to screen, pipe or relief grommet.
12. Check pressure regulator valve for fit in pump housing.
13. Reverse procedure to assemble. Turn drive shaft by hand to check for smooth operation.

NOTE: The pump gears and body are not serviced separately. If the pump gears or body are damaged or worn, the pump assembly should be replaced. Also, if the pick-up screen and pump assembly was removed, it should be replaced with a new one as loss of the press fit condition could result in an air leak and loss of oil pressure.

V6-231

Removal & Disassembly

1. To remove pump, take off oil filter.
2. Disconnect wire from oil pressure indicator switch in filter by-pass valve cap (if so equipped).
3. Remove screws attaching oil pump cover to timing chain cover. Remove cover and slide out pump gears. Replace any parts not serviceable.

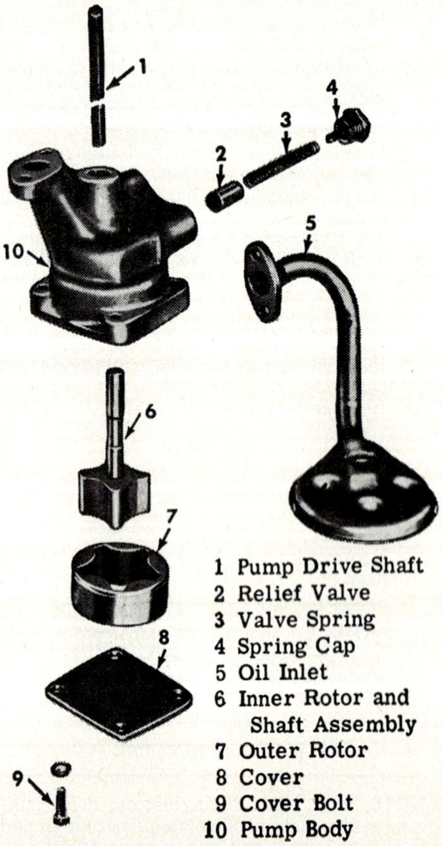

1 Pump Drive Shaft
2 Relief Valve
3 Valve Spring
4 Spring Cap
5 Oil Inlet
6 Inner Rotor and Shaft Assembly
7 Outer Rotor
8 Cover
9 Cover Bolt
10 Pump Body

Fig. 60 Oil pump. V6-305C & 379

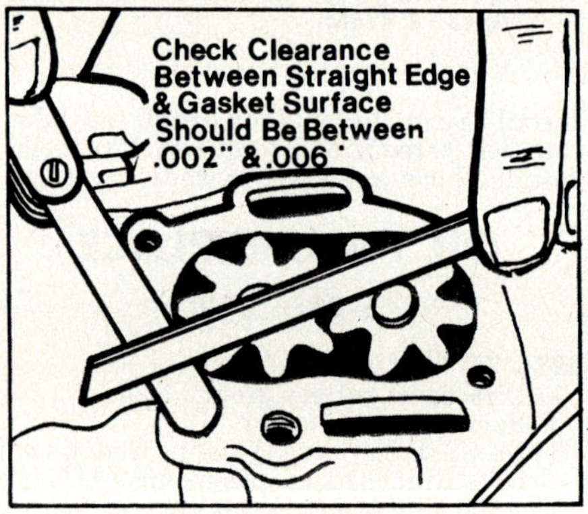

Fig. 59 Checking clearance between oil pump gear & housing gasket surface. V6-231

4. Remove oil pressure relief valve cap, spring and valve, Fig. 58. Remove oil filter by-pass valve cap, spring and valve. Replace any parts of valve not serviceable.
5. Check relief valve in its bore in cover. Valve should have no more clearance than an easy slip fit. If any perceptible side shake can be felt, the valve and/or cover should be replaced.
6. The filter by-pass valve should be flat and free of nicks and scratches.

Assembly & Installation

1. Lubricate and install pressure relief valve and spring in bore of pump cover. Install cap and gasket. Torque cap to 30–35 ft. lbs.
2. Install filter by-pass valve flat in its seat in cover. Install spring, cap and gasket. Torque cap to 30–35 ft. lbs.
3. Install pump gears and shaft in pump body section of timing chain cover to check gear end clearance. Check clearance as shown in Fig. 59. If clearance is less than .0020", check timing chain cover for evidence of wear.
4. If gear end clearance is satisfactory, remove gears and pack gear pocket *full* of vaseline, not chassis lube.
5. Reinstall gears so vaseline is forced into every cavity of gear pocket and between teeth of gears. *Unless pump* is packed

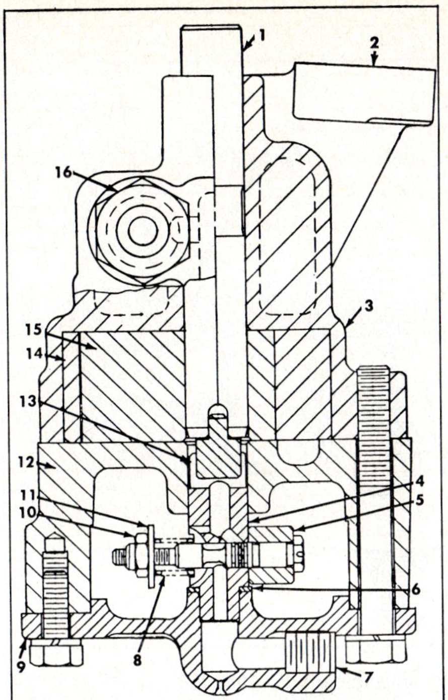

1 Shaft and Inner Rotor Assembly
2 Pump Outlet
3 Pump Body
4 Governor Spinner Shaft
5 Governor Spinner Valve
6 Shaft Thrust Washer
7 Outlet of Governor Diaphragm Pressure
8 Spinner Valve Spring
9 Governor Spinner Cover
10 Valve Adjusting Nut
11 Washer
12 Rotor Cover and Spinner Housing
13 Coupling
14 Outer Rotor
15 Inner Rotor
16 Pressure Relief Valve Cap TPM-6761

Fig. 61 Oil pump. V6-432 & 478M

with vaseline, it may not *prime itself when engine is started.*

6. Install cover and tighten screws alternately and evenly. Final tightening is 10–15 ft-lbs. torque. Install filter on nipple.

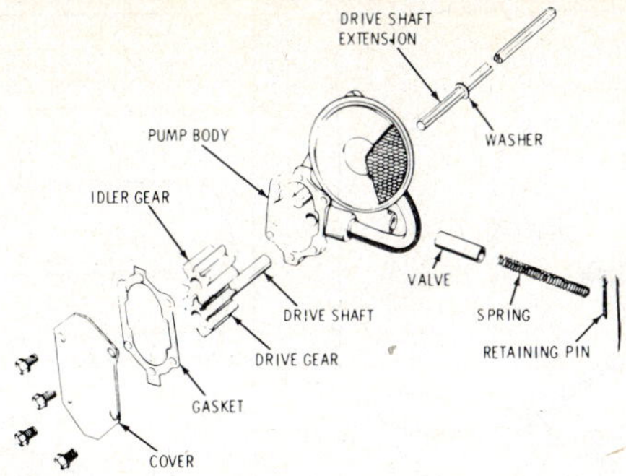

Fig. 62 Oil pump. V8-403 & 455

V6-305C & 379

After removing the oil pan, remove oil inlet pipe and screen from pump. Remove two bolts attaching oil pump to cylinder block, then remove pump and drive shaft. Disassemble pump as suggested by Fig. 60, then inspect parts as follows:

1. Inspect ends of drive shaft which engage distributor and oil pump shafts. Replace shaft if excessively worn or damaged.
2. Check fit of inner rotor shaft in body. If excessive wear is indicated at either shaft or body, replace worn parts.
3. Place outer rotor in body and insert inner rotor and shaft in operating position. Check clearance between outer rotor and body and clearance between edge of inner rotor lobe and outer rotor with feeler gauge; if clearance exceeds .006″, replace worn parts.
4. With rotors in pump body, place straightedge across rotors and body; then with feeler gauge measure clearance between rotors and straightedge, which should be .0011–.0034″.
5. Use straightedge to check rotor side of pump cover for flatness. Also look for grooves and other evidence of wear on cover. Replace cover if not in good condition.

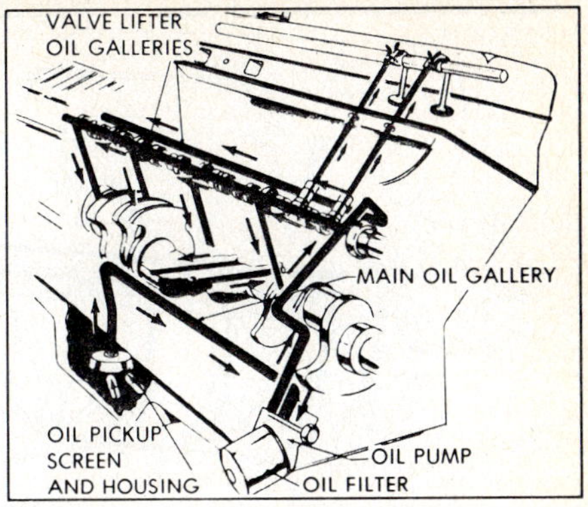

Engine oiling system. V6-231

V6-432 & 478M

After removing the oil pan and oil pump, disassemble the pump as follows, referring to Fig. 61.

Remove bolts that secure spinner valve cover, rotor cover and spinner housing to pump body and separate parts. Remove outer rotor and shaft and inner rotor from pump body. Remove relief valve spring cap, then remove spring and valve.

Inspection of Parts—Inner rotor and shaft assembly and outer rotor were furnished for service in matched sets. With rotors in pump body, place a straightedge across rotors and body and measure clearance with feeler gauge. Clearance should be .0011–.0064″.

Use feeler gauge to check rotor side of rotor cover and spinner housing for flatness. Also look for grooves and other evidence of wear on pump cover.

Clean relief valve parts and check fit of valve in bore of pump. Valve must slide into bore smoothly.

Check fit of spinner valve shaft in cover and spinner valve housing in cover. If excessive clearance is evident, replace parts as necessary. Inspect thrust washer for wear. Clean passages through spinner shaft and passages including bleed hole in spinner cover.

Soak spinner valve and shaft in solvent, then note fit of valve in shaft. It should be possible to push valve endwise, and spring should move valve to closed position. If valve

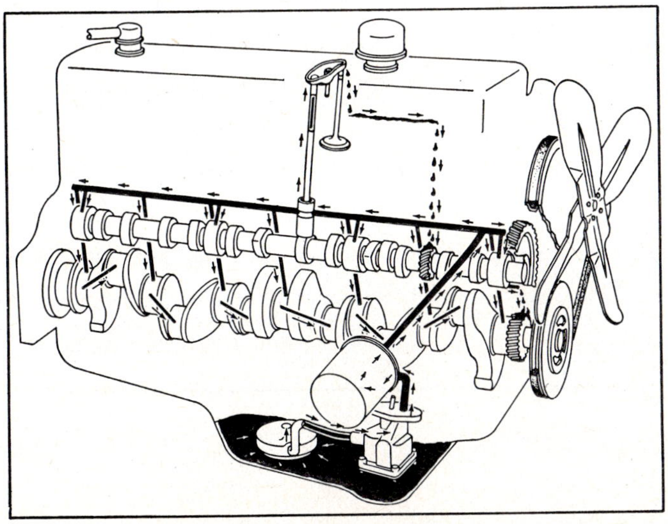

Engine oiling system. 6-250 & 292

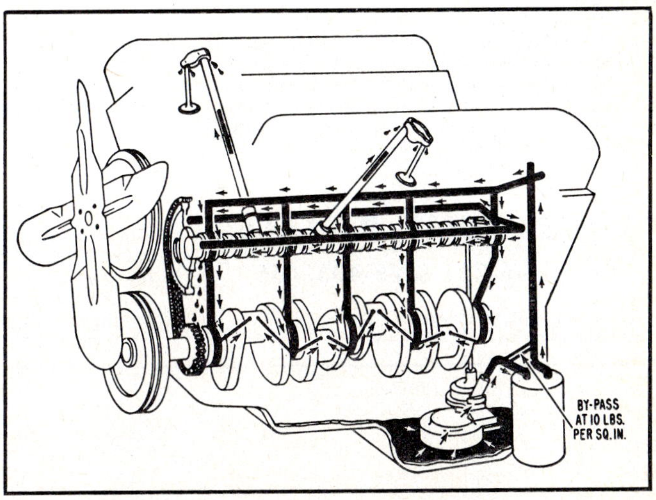

Engine oiling system. V8-267, 305, 350 & 400

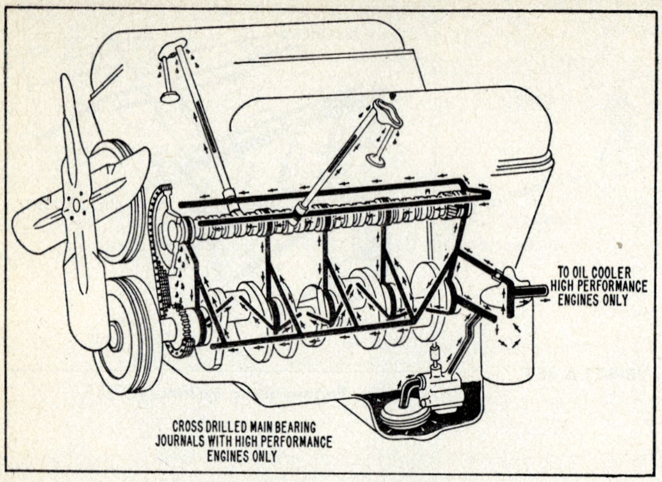

Engine oiling system. V8-366, 427 & 454

sticks or appears to be loose in shaft, obtain a new preadjusted valve assembly, for use when reassembling pump unit. *Since special equipment is required for testing spinner valve operation, the valve adjustment should not be disturbed.*

Assembling Pump & Spinner Valve

1. Install pressure relief valve parts in pump body in following order: valve, spring and spring cap. Tighten cap securely.
2. Lubricate shaft and inner rotor with engine oil and insert shaft into pump body.
3. Install outer rotor in body, then try turning inner rotor and shaft. Shaft must rotate without drag for six or more revolutions.
4. Install rotor cover and spinner housing and retain with two short self-locking bolts.
5. With oil pump inverted, drop coupling into slot in end of pump shaft. Set spinner valve and shaft in place at coupling, then place thrust washer on shaft and install governor spinner cover.

V8-403 & 455

1. Remove oil pan and pump baffle. Remove attaching screws and remove pump and drive shaft extension.
2. To service the pump, refer to Fig. 62.
3. To install, insert the drive shaft extension through the opening in the block until the shaft mates into the distributor drive gear. Position pump onto rear main bearing cap and torque the attaching bolts to 35 ft-lbs.
4. Install oil pump baffle and pan.

WATER PUMP, REPLACE

Except V6-231 & V8-403, 455

1. Drain cooling system.
2. Remove fan belt and fan.
3. Disconnect all hoses from water pump.
4. Remove pump attaching bolts and remove water pump. On 6-250 engines, pull pump straight out of block to prevent damage to impeller.
5. Reverse procedure to install.

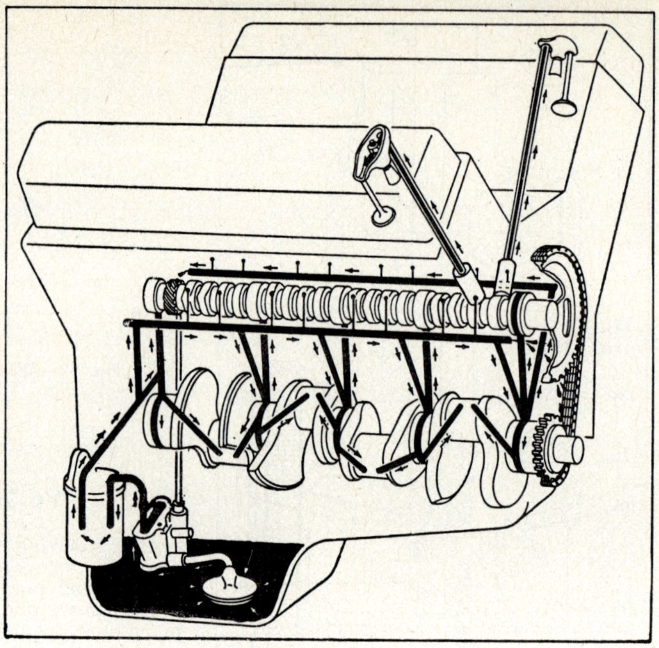

Engine oiling system. V8-403, 455

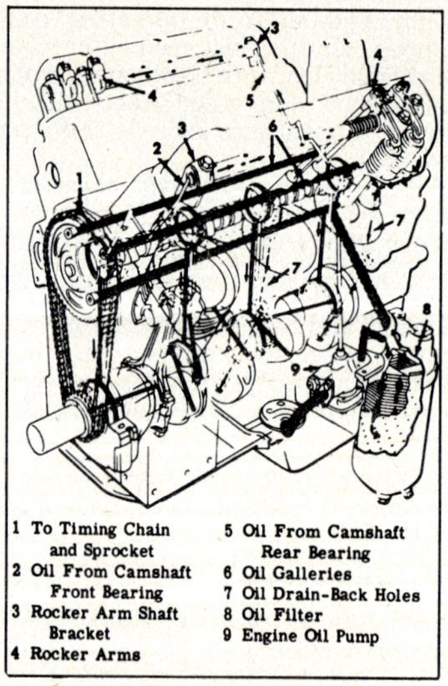

Engine oiling system.
V6-305C, 379, 432 & 478M

V6-231

1. Disconnect battery ground cable, then drain cooling system.
2. Remove drive belts, then remove fan and pulley from water pump shaft hub.
3. Disconnect radiator hose and heater hose from water pump fittings.
4. Remove bolts attaching water pump to timing case cover, then remove water pump assembly.
5. Reverse procedure to install.

V8-403 & 455

1. Drain cooling system, then loosen all belts, disconnect bypass hose and remove heater hose from water pump.
2. Raise and properly support vehicle, then disconnect lower radiator hose from water pump.
3. Remove venturi ring seal strap, then fold venturi to shroud seal forward and over shroud.
4. Remove the four nuts retaining clutch assembly to water pump, then position fan and clutch assembly in the fan shroud, using care to avoid damaging the radiator.
5. Remove venturi ring and water pump pulley.
6. Disconnect power steering pump and left hand upper venturi ring bracket.
7. Remove water pump bolts and remove water pump.

Carburetor Adjustments

ROCHESTER MONOJET M, MV, ME ADJUSTMENT SPECIFICATIONS

See Tune Up Chart for Hot Idle Speeds.

Year	Carb. Part No. ①	Initial Idle Mix Screw Turns Open	Float Level	Metering Rod	Idle Vent	Fast Idle Off Car	Choke Rod	Vacuum Break	Unloader	Fast Idle R.P.M.
1974	7044021	—	19/64	.080	—	—	.275	.350	.500	1800
	7044022	—	19/64	.080	—	—	.245	.300	.500	1800
	7044321	—	19/64	.080	—	—	.300	.375	.500	1800
	7044025	—	1/4	.070	—	—	.245	.300	.521②	2400
	7044026	—	1/4	.070	—	—	.275	.350	.521②	2400
	7044023, 33	—	1/16	—	—	—	.080	.130	.375	2000
	7044024, 34	—	1/16	—	—	—	.080	.130	.375	2200
	7044323	—	1/16	—	—	—	.080	.130	.375	2000
	7044324	—	1/16	—	—	—	.080	.130	.375	2200
	7044433	—	1/16	—	—	—	.080	.130	.375	2000
	7044434	—	1/16	—	—	—	.080	.130	.375	2200
	7044014	—	.295	.079	—	—	.230	.275	.500	1800
	7044017	—	.295	.072	—	—	.275	.350	.500	1800
	7044314	—	.295	.073	—	—	.245	.300	.500	1800
1974–76	7044011, 12	—	1/4	.070	—	—	.150	—	—	—
1975	7045002	—	11/32	.080	—	—	.260	③	.325	1800
	7045003, 5	—	11/32	.080	—	—	.275	③	.325	1800
	7045004	—	11/32	.080	—	—	.245	④	.325	1800
	7045008	—	—	—	—	—	—	—	—	—
	7045012	—	11/32	.080	—	—	.160	⑪	.215	—
	7045013	—	11/32	.080	—	—	.275	⑫	.275	—
	7045024, 28	—	1/8	—	—	—	.080	.100	.375	2000
	7045025, 29	—	1/8	—	—	—	.080	.100	.375	2200
	7045302	—	11/32	.080	—	—	.245	④	.275	1800
	7045303	—	11/32	.080	—	—	.275	⑥	.275	1800
	7045304	—	11/32	.080	—	—	.245	③	.325	1800
	7045305	—	11/32	.080	—	—	.275	⑤	.325	1800
	7045314	—	11/32	.080	—	—	.230	⑬	.275	—
1976	17056002	—	11/32	.080	—	—	.130	⑦	.335	—
	17056003	—	11/32	.080	—	—	.145	⑧	.335	—
	17056004	—	11/32	.080	—	—	.130	⑦	.335	—
	17056006	—	1/4	.080	—	—	.130	.165	.270	—
	17056007	—	1/4	.070	—	—	.130	.165	.275	—
	17056008	—	1/4	.070	—	—	.150	.190	.275	—
	17056009	—	1/4	.070	—	—	.150	.190	.275	—
	17056012	—	11/32	.084	—	—	.090	⑭	.265	—
	17056013	—	11/32	.082	—	—	.090	⑭	.265	—
	17056014	—	11/32	.080	—	—	.090	⑭	.265	—
	17056015	—	11/32	.080	—	—	.090	⑭	.265	—
	17056016	—	11/32	.080	—	—	.115	⑮	.265	—
	17056018	—	11/32	.084	—	—	.090	⑭	.265	—
	17056302	—	11/32	.080	—	—	.155	⑨	.325	—
	17056303	—	11/32	.080	—	—	.180	⑩	.325	—
	17056308	—	1/4	.070	—	—	.150	.190	.275	—
	17056309	—	1/4	.070	—	—	.150	.190	.275	—
1977	17057001	—	3/8	.080	—	—	.125	.150	.325	2100
	17057002	—	3/8	.080	—	—	.110	.135	.325	2100
	17057004	—	3/8	.080	—	—	.110	.135	.325	2100
	17057005	—	3/8	.080	—	—	.125	.180	.325	2100
	17057006	—	5/16	.070	—	—	.150	.180	.275	2400

Continued

ROCHESTER MONOJET M, MV, ME ADJUSTMENT SPECIFICATIONS—Continued

See Tune Up Chart for Hot Idle Speeds.

Year	Carb. Part No. ①	Initial Idle Mix Screw Turns Open	Float Level	Metering Rod	Idle Vent	Fast Idle Off Car	Choke Rod	Vacuum Break	Unloader	Fast Idle R.P.M.
	17057007	—	5/16	.070	—	—	.150	.180	.275	2400
	17057008	—	5/16	.065	—	—	.150	.180	.275	2400
	17057009	—	5/16	.065	—	—	.150	.180	—	2400
	17057010	—	3/8	.080	—	—	.110	.180	.325	2100
	17057011	—	1/4	.065	—	—	.150	—	—	2400
	17057013	—	3/8	.070	—	—	.100	.125	.325	2000
	17057014	—	3/8	.070	—	—	.085	.125	.325	2000
	17057015	—	3/8	.070	—	—	.100	.125	.325	2000
	17057016	—	3/8	.070	—	—	.095	.015	.325	2000
	17057018	—	3/8	.070	—	—	.085	.120	.325	2000
	17057020	—	3/8	.070	—	—	.085	.125	.325	2000
	17057302	—	3/8	.080	—	—	.110	.135	.275	2100
	17057303	—	3/8	.090	—	—	.125	.150	.325	2100
	17057308	—	5/16	.065	—	—	.150	.180	.275	2400
	17057309	—	5/16	.065	—	—	.150	.180	.275	2400
	17057310	—	3/8	.070	—	—	—	—	—	1800
	17057312	—	3/8	.070	—	—	—	—	—	1800
	17057314	—	3/8	.070	—	—	.100	.110	.225	1800
	17057318	—	3/8	.070	—	—	.100	.110	.225	1800
1978	17058006	—	5/16	.070	—	—	.275	.275	.520	2400
	17058007	—	5/16	.070	—	—	.275	.275	.520	2400
	17058008	—	5/16	.065	—	—	.275	.275	.520	2400
	17058009	—	5/16	.065	—	—	.275	.275	.520	2400
	17058013	—	3/8	.080	—	—	.190	.200	.500	2000
	17058014	—	5/16	.100	—	—	.180	.200	.500	2100
	17058020	—	5/16	.100	—	—	.180	.200	.500	2100
	17058021	—	5/16	—	—	—	.200	.250	.450	2100
	17058022	—	5/16	—	—	—	.200	.250	.450	2100
	17058024	—	5/16	—	—	—	.200	.250	.450	2100
	17058081	—	5/16	—	—	—	.200	.250	.450	2100
	17058082	—	5/16	—	—	—	.200	.250	.450	2100
	17058084	—	5/16	—	—	—	.200	.250	.450	2100
	17058308	—	5/16	.065	—	—	.275	.275	.520	2400
	17058309	—	5/16	.065	—	—	.275	.275	.520	2400
	17058310	—	5/16	.080	—	—	.190	.250	.600	2100
	17058311	—	5/16	.100	—	—	.190	.250	.600	2100
	17058312	—	5/16	.080	—	—	.190	.250	.600	2100
	17058313	—	5/16	.100	—	—	.190	.250	.600	2100
	17058314	—	3/8	.100	—	—	.190	.245	.400	2000
	17058320	—	5/16	.080	—	—	.190	.250	—	—
	17058322	—	5/16	.080	—	—	.190	.250	.600	—
	17058323	—	5/16	.100	—	—	.190	.250	.600	—
	17058358	—	5/16	.065	—	—	.275	.275	.520	2400
	17058359	—	5/16	.065	—	—	.275	.275	.520	2400
1979	17058009	—	5/16	.065	—	—	.275	.325	.520②	2400
	17058011	—	5/16	.065	—	—	.150	—	—	2400
	17059009	—	3/8	.090	—	—	.275	.400	.520②	2400
	17059012	—	5/16	.095	—	—	.180	.200	.500	—
	17059013	—	3/8	.100	—	—	.180	.200	.400	1800
	17059014	—	3/8	.100	—	—	.180	.200	.400	2000
	17059020	—	3/8	.100	—	—	.180	.200	.400	2000
	17059023	—	5/16	.095	—	—	.180	.200	.500	—
	17059309	—	3/8	.090	—	—	.275	.400	.520②	2400
	17059314	—	3/8	.100	—	—	.190	.245	.400	2000
	17059359	—	3/8	.090	—	—	.275	.400	.520②	2400

Continued

ROCHESTER MONOJET M, MV, ME ADJUSTMENT SPECIFICATIONS—Continued

See Tune Up Chart for Hot Idle Speeds.

Year	Carb. Part No. ①	Initial Idle Mix Screw Turns Open	Float Level	Metering Rod	Idle Vent	Fast Idle Off Car	Choke Rod	Vacuum Break	Unloader	Fast Idle R.P.M.
1979–80	17058011	—	5/16	.065	—	—	.150	—	—	2400
1980	17080009	—	11/32	.090	—	—	.275	.400	.520	—
	17080309	—	11/32	.090	—	—	.275	.400	.520	—
	17080359	—	11/32	.090	—	—	.275	.400	.520	—
1981–82	17081009	—	11/32	.090	—	—	.275	.400	.520	—
	17081309	—	11/32	.090	—	—	.275	.400	.520	—
	17081329	—	11/32	.090	—	—	.275	.400	.520	—
1983	All	—	11/32	.090	—	—	.275	.400	.520	2400

①—Located on tag attached to carburetor.
②—Measured at top of choke blade.
③—Primary .300; auxiliary .250.
④—Primary .300; auxiliary .150.
⑤—Primary .350; auxiliary .250.
⑥—Primary .350; auxiliary .170.
⑦—Primary .165; auxiliary .265.
⑧—Primary .180; auxiliary: wide open.
⑨—Primary .190; auxiliary: wide open.
⑩—Primary .225; auxiliary: wide open.
⑪—Primary .200; auxiliary .215.
⑫—Primary .350; auxiliary .312.
⑬—Primary .275; auxiliary .312.
⑭—Primary .110; auxiliary .215.
⑮—Primary .140; auxiliary: wide open.

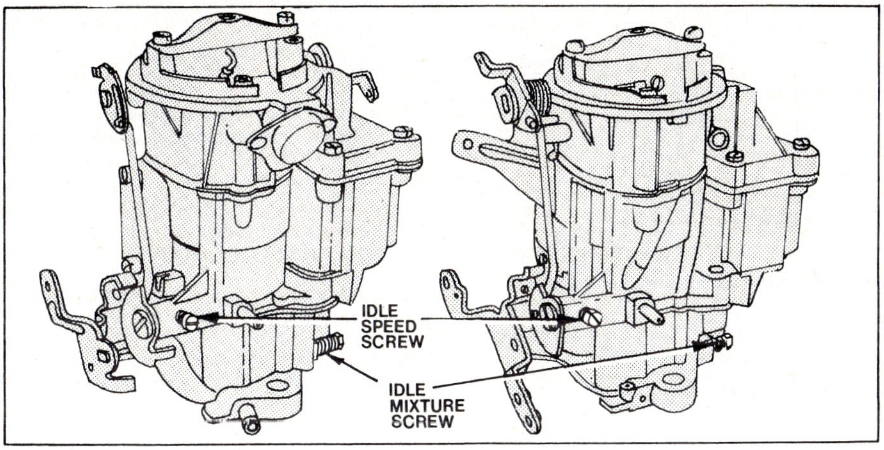

Fig. 1 Rochester Monojet carburetor. Model M

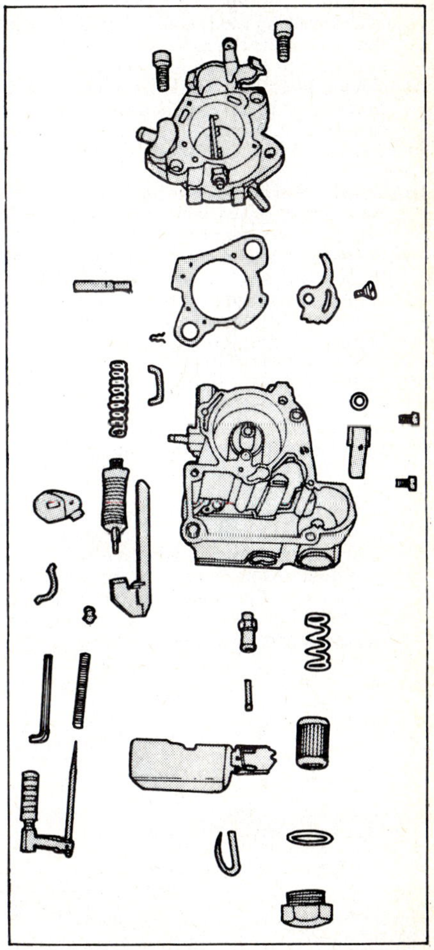

Fig. 2 Float bowl & throttle body assembly. Model M

MODELS M, MV & ME

Description

This carburetor, Fig. 1, is a single barrel unit with a triple venturi in conjunction with a plain tube nozzle. The fuel flow in the main metering system is controlled by mechanical and vacuum means.

A conventional idling system is used in conjunction with the main metering system. A hot idle compensator is available, where necessary, to maintain smooth engine idle during periods of extreme hot engine operation. The off-idle discharge port is of a vertical slot design which gives good transition between idle and main metering system operations.

Other features of the Monojet include an aluminum throttle body for decreased weight and improved heat transfer, and a thick throttle body-to-bowl insulator gasket to keep excessive engine heat from the float bowl. The carburetor has internally balanced venting through a vent hole in the air horn. Fig. 2 shows an exploded view of the float bowl and throttle body parts.

M & MV Adjustments

Float Level, Adjust, Fig. 3

1. Hold float retaining pin firmly in place and float arm against top of float needle.
2. With adjustable T-scale, measure distance from top of float (at toe) to float bowl gasket surface (gasket removed). Measurement should be made at a point 1/16" in from end of flat surface at float toe (not on radius).
3. Bend float pontoon up or down at float arm junction to adjust.

Metering Rod, Adjust, Fig. 4

1. To check adjustment, back out idle adjusting screw to be sure throttle valve is closed. Rotate fast idle cam so that cam follower is not contacting cam (if installed).
2. With throttle valve completely closed, apply pressure to hanger directly above power piston and hold piston down against its stop.
3. Holding downward pressure on power pis-

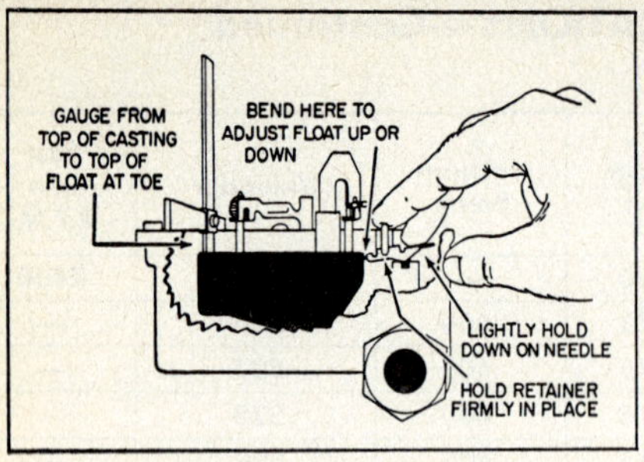

Fig. 3 Float adjustment. Model M & MV

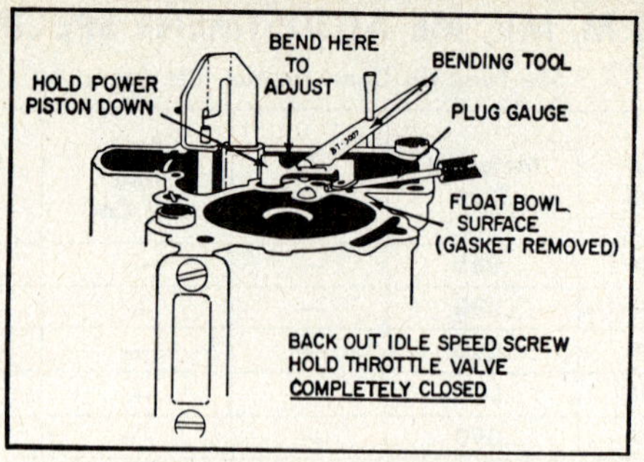

Fig. 4 Metering rod adjustment. Model M & MV

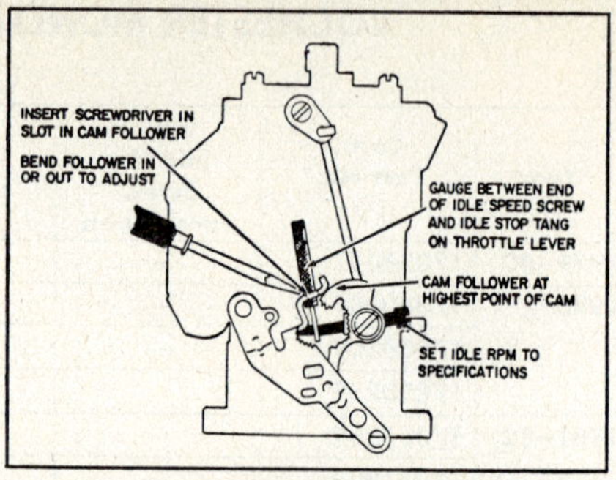

Fig. 5 Fast idle adjustment (off vehicle). Model M & MV

ton, swing metering rod holder over flat surface of bowl casting unit metering rod rests lightly against inside edge of bowl.
4. Use specified gauge of size listed and insert between bowl casting and lower surface of metering rod holder against metering rod gauge; there should be a sliding fit between surfaces as shown.
5. To adjust, carefully bend metering rod holder up or down at point shown.
6. Install air horn gasket on float bowl by carefully sliding slit portion of gasket over metering rod holder. Then align gasket with dowels provided on top of bowl casting and press gasket firmly in place.
7. Before installing air horn, check operation of entire mechanism, metering rod and accelerator pump for free operation from closed to wide open position.

Initial Idle Speed Screw Setting

1. Hold choke valve wide open so fast idle cam follower clears cam.
2. Adjust idle speed screw until it just contacts lever tang.
3. Adjust idle speed screw an additional 1¼ turns with automatic transmission and 2 turns with manual transmission.

Fast Idle, Adjust, Fig. 5

1. Set normal engine idle speed initially by turning idle speed screw 1½ turns in from closed throttle valve position.
2. Place fast idle cam follower tang on highest step of cam.
3. With tang held against cam, check clearance between end of slow idle screw and idle stop tang on throttle lever.
4. To adjust, insert end of screwdriver in slot provided in fast idle cam follower tang and bend inward (toward cam) or outward to obtain specified dimension.

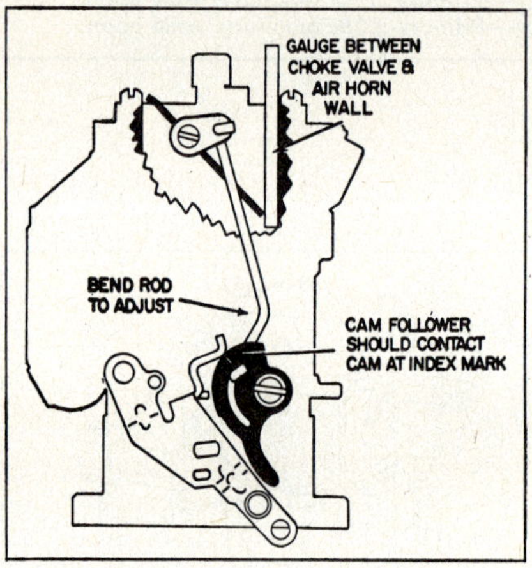

Fig. 6 Choke rod adjustment. Model M & MV

Choke Rod, Adjust, Fig. 6

1. With fast idle adjustment completed, hold fast idle cam follower firmly against cam. Index mark on side of cam should be lined up with contact point of cam follower tang.
2. Rotate choke valve toward closed position by applying force to choke lever.
3. Bend choke rod at point indicated to give specified opening between lower edge of choke valve (at center of valve) and inside air horn wall.

Vacuum Break Adjustment, Fig. 7

1. Use outside vacuum source to seat diaphragm.
2. Place gauge between lower edge of choke valve and air horn wall.
3. Bend rod to adjust.

Unloader Adjustment, Fig. 8

1. Hold down on choke valve with rod in end of slot.
2. Hold throttle valve wide open.
3. Position gauge between lower edge of choke valve and air horn wall.
4. Bend tang to adjust.

Choke Coil Rod Adjustment, Fig. 9

1. Hold choke valve closed.
2. Pull upward on coil rod to end of travel.
3. Bottom of rod end which slides into hole in choke lever should be even with top of hole.
4. Bend choke coil rod at bend to adjust.
5. Connect coil rod to choke lever and install retaining clip.
6. Ensure complete freedom of operation from full open to full closed choke.

On Vehicle Adjustments

Final slow idle, fast idle and idle vent settings should be made on the vehicle using a

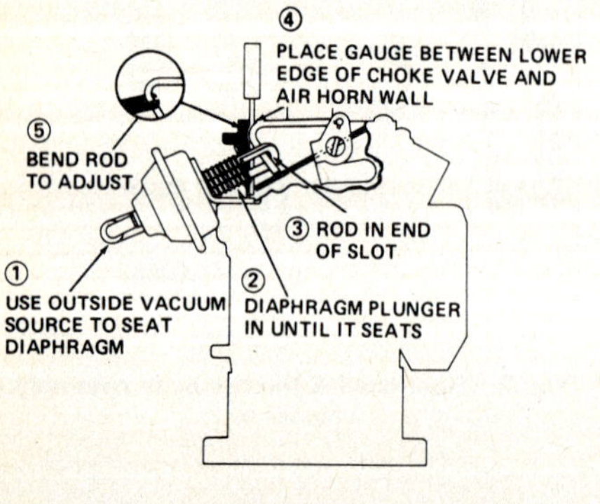

Fig. 7 Vacuum break adjustment. Model M & MV

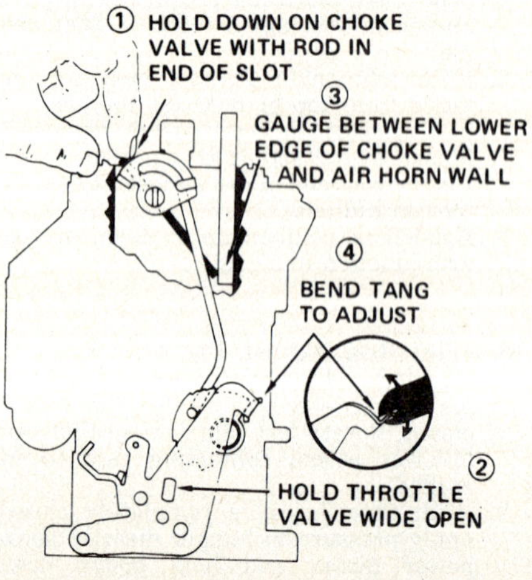

Fig. 8 Unloader adjustment. Model M & MV

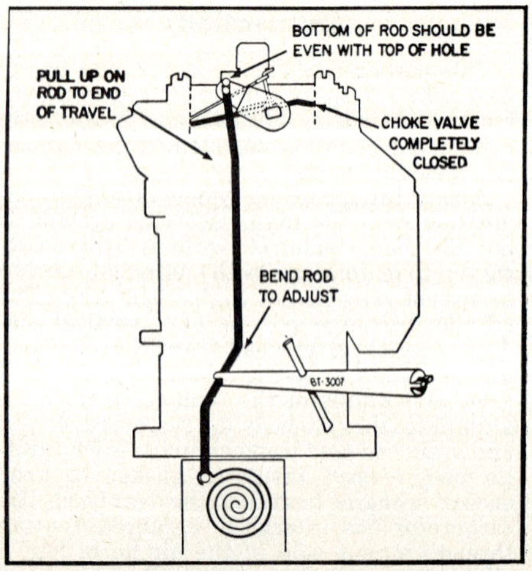

Fig. 9 Choke coil adjustment. Model M & MV

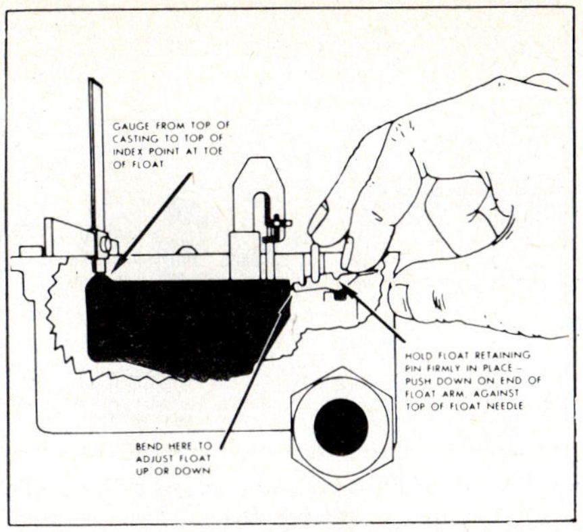

Fig. 10 Float level adjustment. Model ME

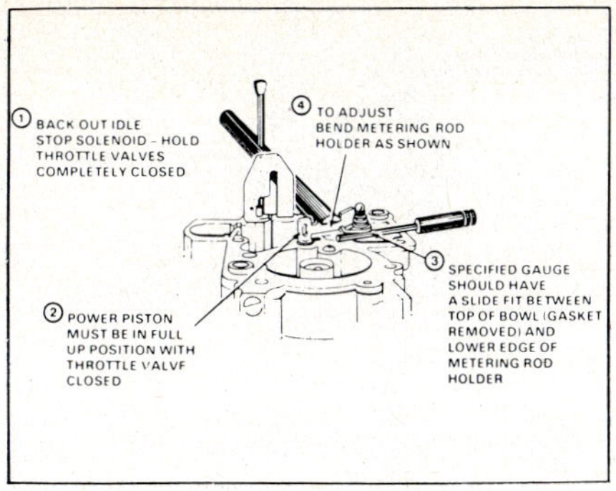

Fig. 11 Metering rod adjustment. Model ME

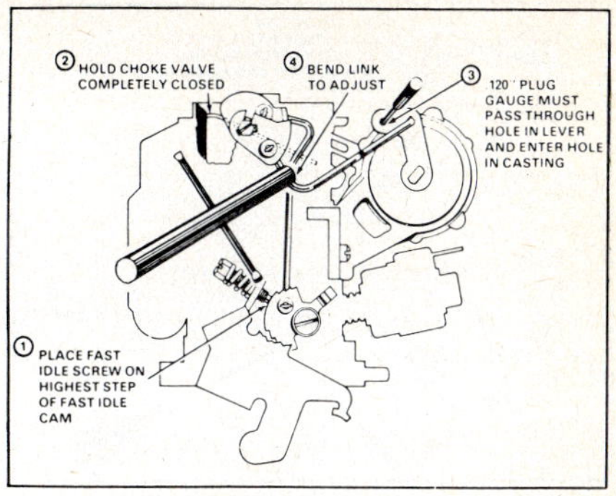

Fig. 12 Choke coil adjustment. Model ME

tachometer. Idle speed and mixture settings must be made with air cleaner on. Warm engine to normal operating temperature; thermostatic air cleaner valve must be in full open position. Slow idle setting must follow procedure shown on decal, which is to adjust mixture screw in to obtain a 20 rpm drop (lean roll), then adjust screw out 1/4 turn.

ME Adjustments

Float Level, Adjust, Fig. 10

1. Hold float retaining pin firmly in place and float arm against top of float needle by pushing downward on float arm at point between needle seat and hinge pin as shown.
2. With adjustable T-scale, measure distance from top of float at toe to float bowl gasket surface (gasket removed). Measurement should be made at a point 1/16" in from end of flat surface at float toe (not on radius).
3. Bend float pontoon up or down at float arm junction to adjust.

Metering Rod Adjustment, Fig. 11

1. Remove metering rod by holding throttle valve wide open. Push downward on metering rod against spring tension, then slide metering rod out of slot in holder and remove from main metering jet.
2. To check adjustment, back out slow idle screw and rotate fast idle cam so that fast idle cam follower is not contacting steps on cam.
3. With throttle valve completely closed, apply pressure to top of power piston and hold piston down against its stop.
4. While holding downward pressure on power piston, swing metering rod holder over flat surface of bowl casting next to carburetor bore.
5. Use specified size drill and insert between bowl casting sealing bead and lower surface of metering rod holder. Drill should have a slide fit between both surfaces as shown.
6. To adjust, carefully bend metering rod holder up or down at point shown.
7. After adjustment, install metering rod.

Choke Coil Lever Adjustment, Fig. 12

1. With fast idle adjusting screw on fast idle cam high step, close choke valve.
2. Insert a .120 inch gauge pin through hole in lever and into casting. If holes do not align, bend link to adjust.

Fast Idle Adjustment, Fig. 13

1. Set normal engine idle speed.
2. Place fast idle cam follower tang on highest step of cam.
3. To adjust, insert screwdriver in slot provided in fast idle cam follower tang and bend inwards (towards cam) or outward to obtain specified dimension.

Choke Rod Adjustment, Fig. 14

1. With fast idle adjustment made, place fast idle cam follower on second step of fast idle cam and hold firmly against the rise to the high step.
2. Rotate choke towards direction of closed choke by applying force to choke coil lever.
3. Bend choke rod at point shown to give specified opening between upper edge of choke valve (at center of valve) and inside air horn wall.

Vacuum Break Adjustment, Fig. 15

1. With fast idle screw on fast idle cam high step, seat diaphragm using an outside vacuum source. Cover purge hole with a piece of tape.
2. Close choke valve to compress plunger bucking spring and seat plunger stem.
3. Place specified gauge between upper edge of choke valve and air horn wall. Bend link to adjust.

Unloader Adjustment, Fig. 16

1. With throttle valve wide open, hold choke valve closed with specified gauge between upper edge of choke valve and air horn wall.
2. If distance is not within specifications, bend unloader tang to adjust.

1977–79 Choke Coil Adjustment, Fig. 17

1. Install electric choke assembly into housing, ensuring coil tang contacts lower side of coil lever pick-up arm.
2. Place fast cam follower on high step of cam.
3. Rotate cover and coil assembly counterclockwise until choke valve just closes, then align cover index point with the index mark on housing on 1977 78 and two notches rich on 1979.
4. Install and tighten cover retainers and screws.

NOTE: Do not install a choke cover gasket between the choke assembly and housing since the ground contact for the choke is provided by a metal plate at the rear of the choke assembly.

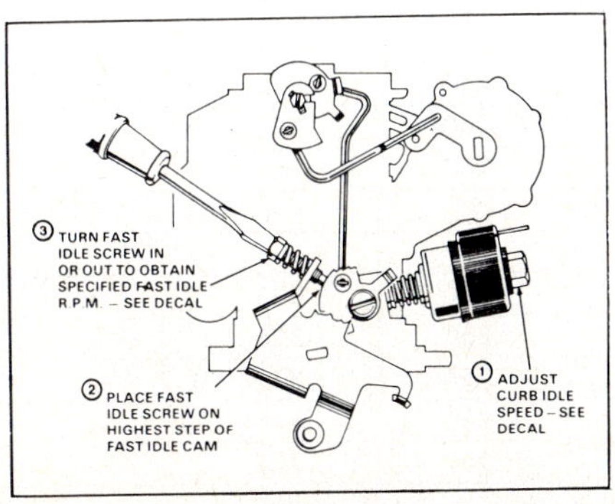

Fig. 13 Fast idle adjustment. Model ME

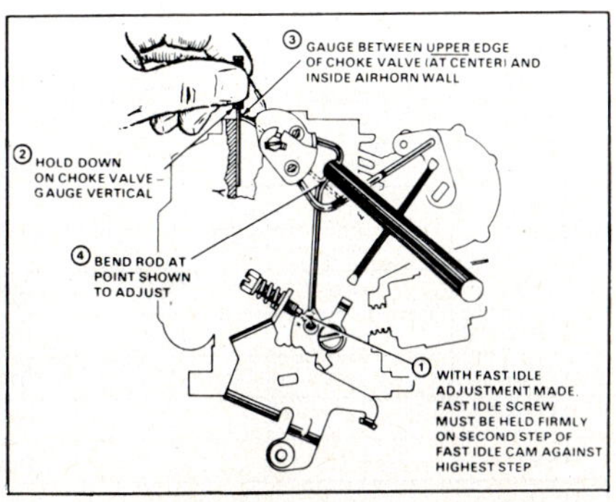

Fig. 14 Choke rod adjustment. Model ME

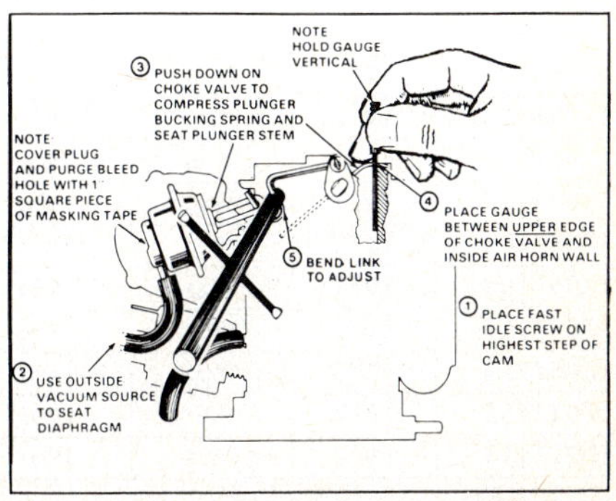

Fig. 15 Vacuum break adjustment. Model ME

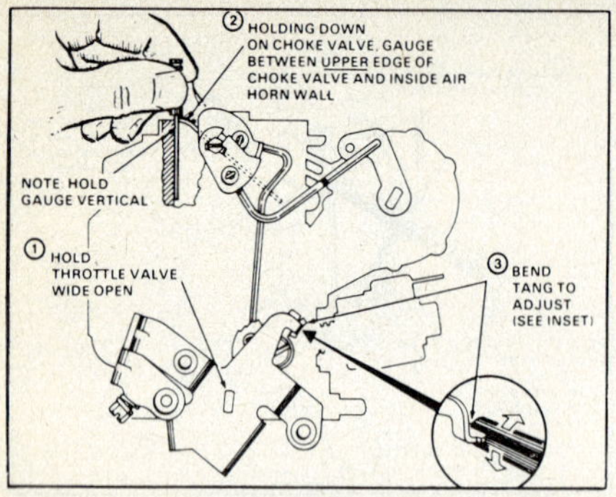

Fig. 16 Unloader adjustment. Model ME

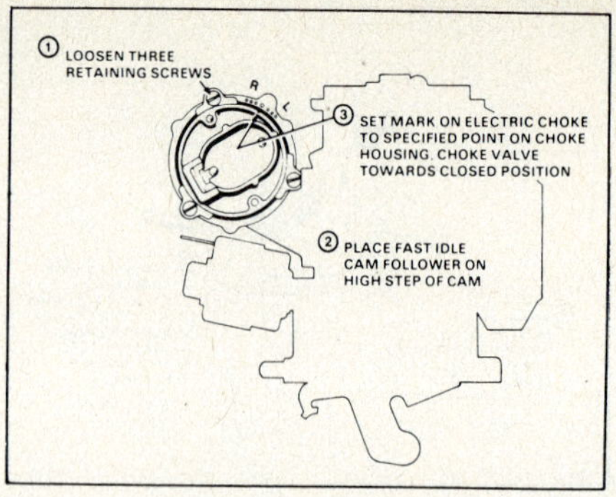

Fig. 17 Choke coil adjustment. Model ME

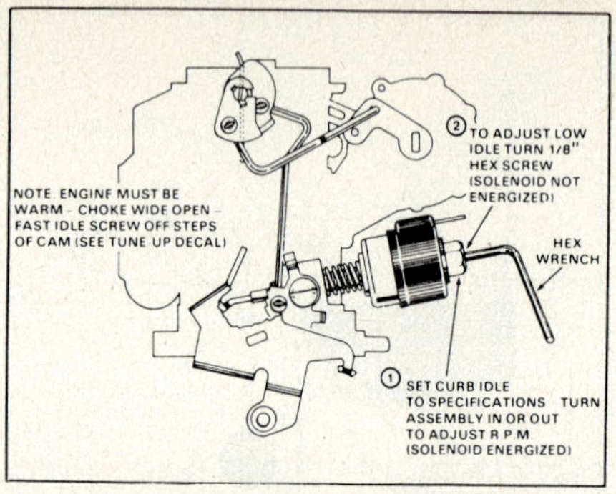

Fig. 18 Low idle & curb idle speed adjustment. Model ME

Low Idle & Curb Idle Speeds Adjustment, Fig. 18

NOTE: On all models except vehicles equipped with automatic transmission and air conditioning, the idle solenoid wire will be color-coded tan. On vehicles equipped with automatic transmission and air conditioning, the solenoid wire is color-coded green with a double white stripe. The tan wire will be left disconnected.

1. The engine should be at normal operating temperature with the air cleaner on, choke open and air conditioning off. Apply parking brake and chock wheels.
2. Disconnect and plug PCV hose at vapor canister and the vacuum advance hose from distributor.
3. On all models except equipped with automatic transmission and air conditioning, rotate solenoid body to adjust curb idle speed. Then, disconnect solenoid electrical connector. With automatic transmission in Drive or manual transmission in Neutral, turn 1/8 inch hex screw in end of solenoid to adjust low idle speed. Connect solenoid electrical connector.
4. On models equipped with automatic transmission and air conditioning, turn solenoid hex screw inward until fully bottomed. Rotate solenoid body to obtain 950 RPM. Then, rotate hex screw to obtain curb idle RPM.
5. On all models, unplug and connect PCV vapor canister hose and vacuum advance hose.

ROCHESTER 2G, 2GC, 2GV CARBURETOR ADJUSTMENT SPECIFICATIONS

See Tune-Up Chart for Hot Idle Speeds.

★Located on Tag attached to or stamped on Carburetor.

Carb. Part No. ★	Year	Float Level	Float Drop	Pump Rod	Idle Vent	Vacuum Break	Automatic Choke	Choke Rod	Choke Unloader	Fast Idle Speed
7044113	1974	19/32	1 9/32	1 9/32	—	.140	—	.200	.250	1600
7044114	1974	19/32	1 9/32	1 3/16	—	.130	—	.245	.325	1600
7044123	1974	19/32	1 9/32	1 9/32	—	.140	—	.200	.250	1600
7044124	1974	19/32	1 9/32	1 3/16	—	.130	—	.245	.325	1600
7044133	1974–78	19/32	1 9/32	1 9/16	—	—	—	—	—	—
7044134	1974–78	19/32	1 9/32	1 7/16	—	—	—	—	—	—
7044434	1974–76	19/32	1 9/32	1 7/16	—	—	—	—	—	—
7045115	1975	21/32	31/32	1 5/8	—	.130	—	.400	.350	—
7045116	1975	21/32	31/32	1 5/8	—	.130	—	.400	.350	—
7045123	1975	21/32	31/32	1 5/8	—	.130	—	.400	.350	—
7045124	1975	21/32	31/32	1 5/8	—	.130	—	.400	.350	—
17056115	1976	21/32	1 9/32	1 11/16	—	.130	Index	.260	.325	—
17056116	1976	21/32	1 9/32	1 11/16	—	.130	1 Rich	.260	.325	—
17056117	1976	9/16	1 3/16	1 21/32	—	.130	Index	.260	.325	—
17056119	1976	9/16	1 3/16	1 21/32	—	.130	Index	.260	.325	—
17056123	1976	21/32	1 9/32	1 11/16	—	.130	Index	.260	.325	—
17056124	1976	21/32	1 9/32	1 11/16	—	.130	1 Rich	.260	.325	—
17056137	1977	19/32	1 9/32	1 21/32	—	.190	Index	.260	.325	—
17056433	1977	19/32	1 9/32	1 9/16	—	—	—	—	—	—
17056434	1977	19/32	1 9/32	1 7/16	—	—	—	—	—	—
17057108	1977	19/32	1 9/32	1 21/32	—	.130	Index	.260	.325	—
17057110	1977	19/32	1 9/32	1 21/32	—	.130	Index	.260	.325	—
17057111	1977	19/32	1 9/32	1 21/32	—	②	Index	.260	.325	—
17057112	1977	19/32	1 9/32	1 21/32	—	②	Index	.260	.325	—
17057113	1977	19/32	1 9/32	1 21/32	—	②	Index	.260	.325	—

Continued

ROCHESTER 2G, 2GC, 2GV CARBURETOR ADJUSTMENT SPECIFICATIONS—Continued

See Tune-Up Chart for Hot Idle Speeds.

★Located on Tag attached to or stamped on Carburetor.

Carb. Part No. ★	Year	Float Level	Float Drop	Pump Rod	Idle Vent	Vacuum Break	Automatic Choke	Choke Rod	Choke Unloader	Fast Idle Speed
17057114	1977	19/32	1 9/32	1 21/32	—	②	Index	.260	.325	—
17057121	1977	19/32	1 9/32	1 21/32	—	②	Index	.260	.325	—
17057123	1977	19/32	1 9/32	1 21/32	—	②	Index	.260	.325	—
17057408	1977	21/32	1 9/32	1 21/32	—	③	1/2 Lean	.260	.325	—
17057410	1977	21/32	1 9/32	1 21/32	—	③	1/2 Lean	.260	.325	—
17057412	1977	21/32	1 9/32	1 21/32	—	③	1/2 Lean	.260	.325	—
17057414	1977	21/32	1 9/32	1 21/32	—	③	1/2 Lean	.260	.325	—
17058102	1978	15/32	1 9/32	1 17/32	—	④	Index	.260	.325	—
17058103	1978	15/32	1 9/32	1 17/32	—	④	Index	.260	.325	—
17058104	1978	15/32	1 9/32	1 21/32	—	⑤	Index	.260	.325	—
17058105	1978	15/32	1 9/32	1 21/32	—	④	Index	.260	.325	—
17058107	1978	15/32	1 9/32	1 17/32	—	⑤	Index	.260	.325	—
17058108	1978	19/32	1 9/32	1 21/32	—	④	Index	.260	.325	—
17058109	1978	15/32	1 9/32	1 17/32	—	④	Index	.260	.325	—
17058110	1978	19/32	1 9/32	1 21/32	—	④	Index	.260	.325	—
17058111	1978	19/32	1 9/32	1 17/32	—	④	Index	.260	.325	—
17058112	1978	19/32	1 9/32	1 21/32	—	④	Index	.260	.325	—
17058113	1978	19/32	1 9/32	1 17/32	—	④	Index	.260	.325	—
17058114	1978	19/32	1 9/32	1 21/32	—	④	Index	.260	.325	—
17058116	1978	19/32	1 9/32	1 17/32	—	④	Index	.260	.325	—
17058118	1978	19/32	1 9/32	1 17/32	—	④	Index	.260	.325	—
17058123	1978	19/32	1 9/32	1 17/32	—	④	Index	.260	.325	—
17058126	1978	19/32	1 9/32	1 17/32	—	⑤	Index	.260	.325	—
17058127	1978	19/32	1 9/32	—	—	.190	Index	.260	.325	—
17058128	1978	19/32	1 9/32	1 17/32	—	⑤	Index	.260	.325	—
17058404	1978	1/2	1 9/32	1 21/32	—	①	1/2 Lean	.260	.325	—
17058405	1978	1/2	1 9/32	1 21/32	—	①	1/2 Lean	.260	.325	—
17058408	1978	21/32	1 9/32	1 21/32	—	①	1/2 Lean	.260	.325	—
17058410	1978	21/32	1 9/32	1 21/32	—	①	1/2 Lean	.260	.325	—
17058412	1978	21/32	1 9/32	1 21/32	—	①	1/2 Lean	.260	.325	—
17058414	1978	21/32	1 9/32	1 21/32	—	①	1/2 Lean	.260	.325	—
17058120	1979	11/16	1 9/32	1 15/32	—	—	—	—	—	—
17059119	1979	17/32	1 9/32	1 15/32	—	—	—	—	—	—
17059120	1979	17/32	1 9/32	1 15/32	—	—	—	—	—	—
17059126	1979	5/8	1 9/32	1 21/32	—	—	—	—	—	—
17059127	1979	17/32	1 9/32	1 15/32	—	—	—	—	—	—
17059129	1979	5/8	19/32	1 21/32	—	.130	—	.245	.250	—
17059420	1979	17/32	1 9/32	1 15/32	—	—	—	—	—	—
17059423	1979	5/8	1 9/32	1 21/32	—	—	—	—	—	—
17059424	1979	17/32	1 9/32	1 15/32	—	—	—	—	—	—
17080127	1980–81	5/8	1 9/32	1 21/32	—	—	—	—	—	—
17080129	1980–81	5/8	1 9/32	1 21/32	—	.130	—	.245	.250	—
17080420	1980–81	5/8	1 9/32	1 21/32	—	—	—	—	—	—
17080423	1980–81	5/8	1 9/32	1 21/32	—	—	—	—	—	—
17080424	1980–81	5/8	1 9/32	1 21/32	—	—	—	—	—	—
7044134	1980–81	11/16	1 9/32	1 7/16	—	—	—	—	—	—
17080120	1980–82	5/8	1 9/32	1 21/32	—	—	—	—	—	—
7044133	1980–83	11/16	1 9/32	1 9/16	—	—	—	—	—	—
17058120	1980–83	11/16	1 9/32	1 7/16	—	—	—	—	—	—
17080126	1980–83	5/8	1 9/32	1 21/32	—	—	—	—	—	—
17082129	1982	5/8	1 9/32	1 21/32	—	.130	—	.245	.250	—

Continued

ROCHESTER 2G, 2GC, 2GV CARBURETOR ADJUSTMENT SPECIFICATIONS—Continued

See Tune-Up Chart for Hot Idle Speeds.

★Located on Tag attached to or stamped on Carburetor.

Carb. Part No. ★	Year	Float Level	Float Drop	Pump Rod	Idle Vent	Vacuum Break	Automatic Choke	Choke Rod	Choke Unloader	Fast Idle Speed
17082420	1982	5/8	19/32	1 21/32	—	—	—	—	—	—
17082423	1982	5/8	19/32	1 21/32	—	—	—	—	—	—
17080120	1983	5/8	19/32	1 15/32	—	—	—	—	—	—
17080420	1983	5/8	19/32	1 15/32	—	—	—	—	—	—

①—.140, reset to .160 at first tune up.
②—Before 22500 miles, .130; above 22500 miles, .160.
③—Before 22500 miles, .140; above 22500 miles, .160.
④—.130, reset to .150 at first tune up.
⑤—.130, reset to .160 at first tune up.

MODEL 2G, 2GC, 2GV, FIG. 1

Float Level, Fig. 2

With air horn inverted and air horn gasket installed, measure distance from top of air horn gasket to lower edge (sharp edge or lip) of float at outer end of float pontoon. To adjust, bend float arm at rear as shown in inset.

Float Drop, Fig. 3

With air horn assembly held upright and float suspended freely, measure dimension from air horn gasket to bottom of float pontoon. Adjust to specified dimension by bending tang that contacts seat at rear of float arm.

Pump Rod, Fig. 4

Back out idle stop screw and completely close throttle valve in bore. Place gauge on top of air horn ring. Bend pump rod at lower angle to obtain specified dimension to top of pump rod.

Adjust Idle Vent, 1974

NOTE: Always make pump adjustment before checking and adjusting idle vent valve.

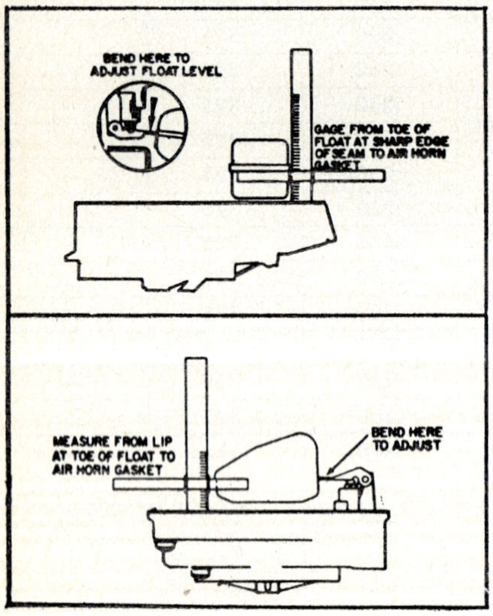

Fig. 2 Float level adjustment. Upper view for brass float & lower view for nitrophyl float

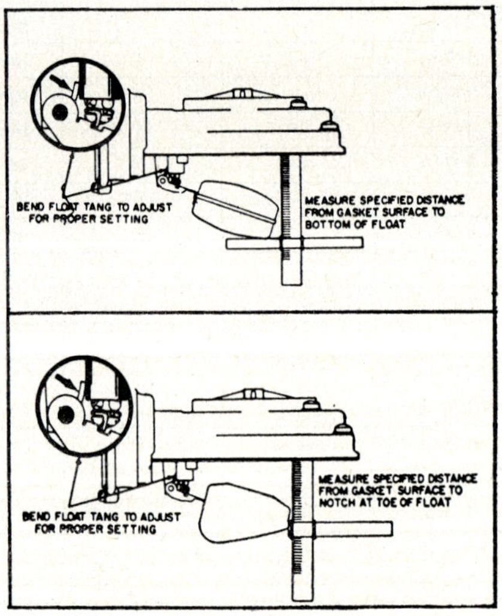

Fig. 3 Float drop adjustment. Upper view for brass float & lower view for nitrophyl float

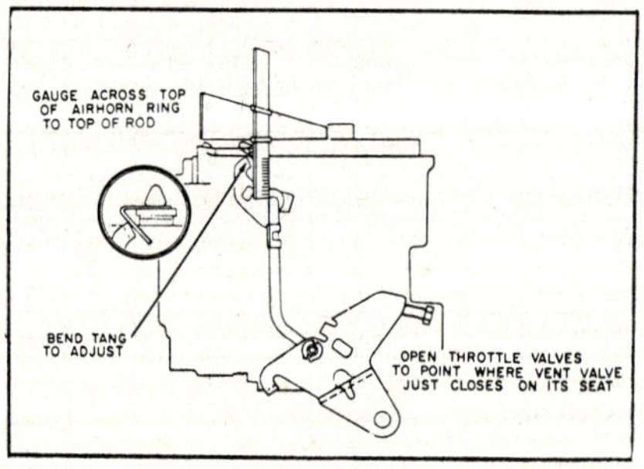

Fig. 4 Pump rod adjustment. 2G & 2GC models

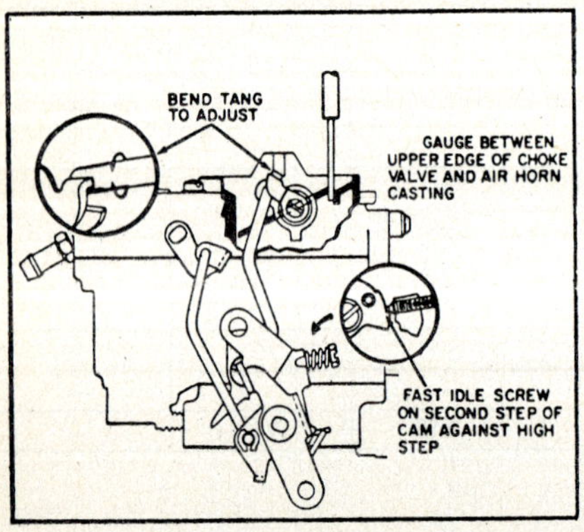

Fig. 5 Choke rod (fast idle) adjustment

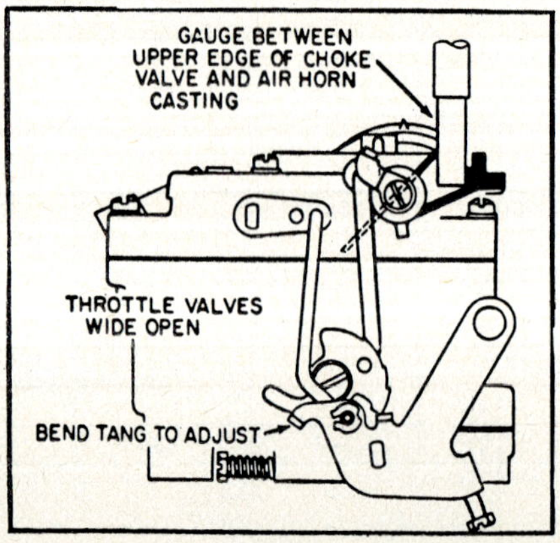

Fig. 6 Choke unloader adjustment

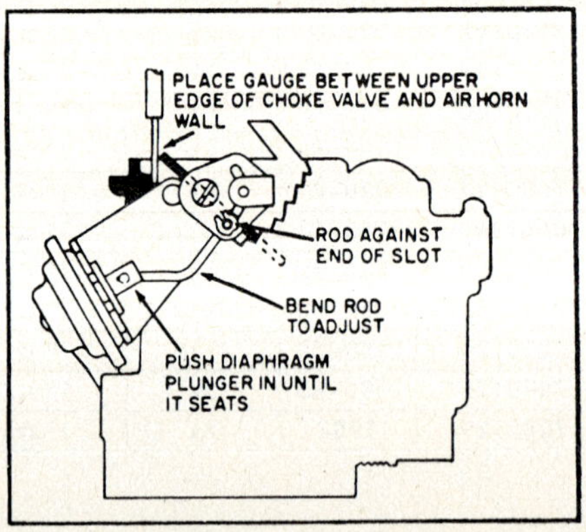

Fig. 7 Choke vacuum break adjustment

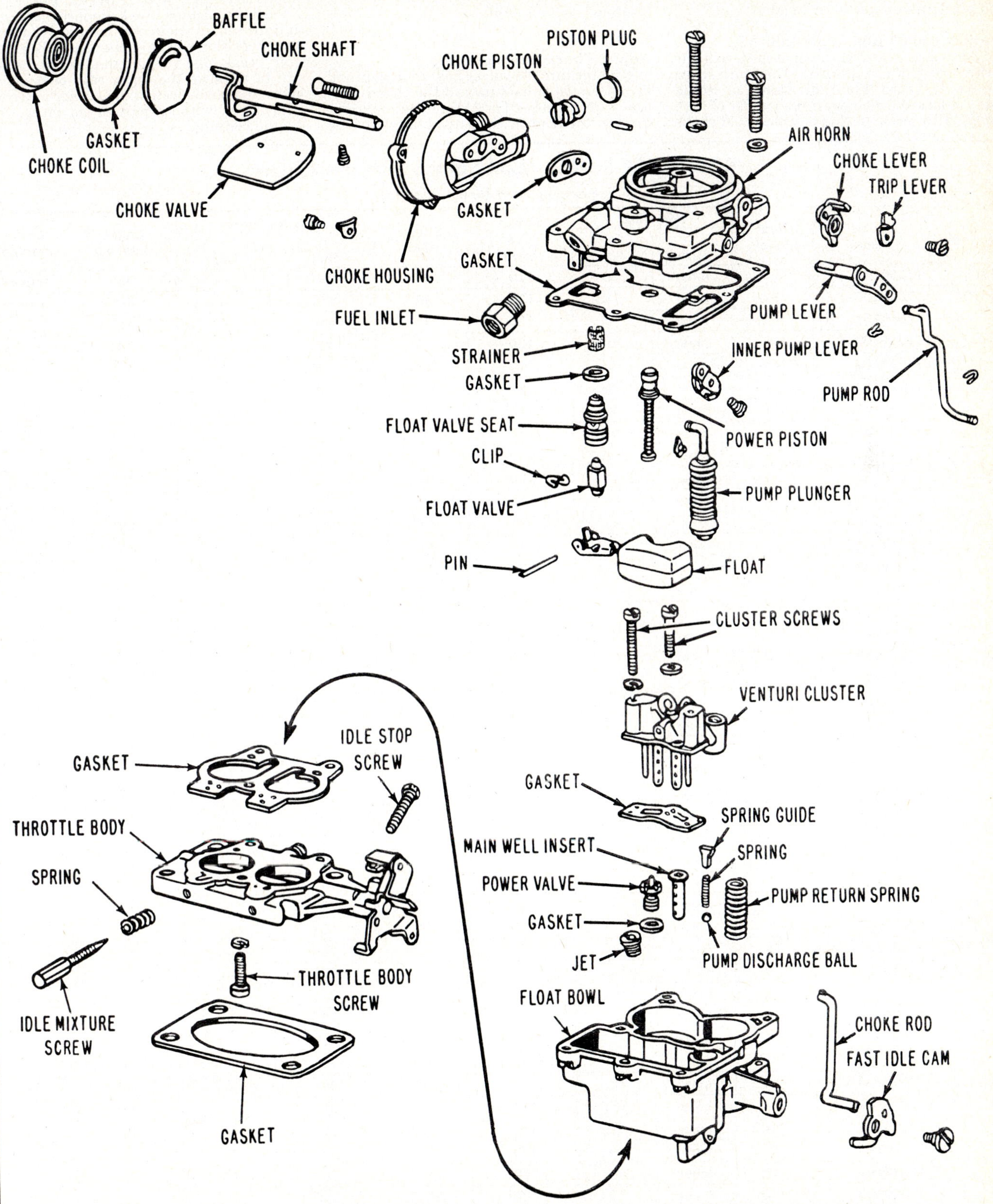

Fig. 1 Exploded view of Rochester 2GC carburetor (Typical)

With choke held in fully open position and idle screw adjusted to specifications on low step of cam, measure opening of idle vent valve and seat at its widest point. Bend tang on accelerator outer lever to adjust.

Adjust Choke Rod, (Fast Idle) Fig. 5

Turn stop screw in until it just contacts bottom step of fast idle cam. Then turn screw in one full turn. Place idle screw on second step of fast idle cam against shoulder of high step. With screw in this position, hold choke valve toward closed position with a rubber band and check clearance between upper edge of choke valve and air horn wall. Adjust to specified dimension by bending tang on choke lever and collar assembly.

Adjust Choke Unloader, Fig. 6

With the throttle valves held wide open and the choke valve held toward the closed position with a rubber band, bend the unloader tang on the throttle lever to obtain specified clearance between the upper edge of the choke valve and air horn wall.

Adjust Choke Vacuum Break, Fig. 7

Push the vacuum break diaphragm plunger in until seated. Hold the choke valve toward closed position with a rubber band. Bend vacuum break line so that the specified gauge will fit between the upper edge of the choke valve and inside wall of the air horn. Recheck adjustment with carburetor on engine and engine running.

ROCHESTER MODEL M2M SERIES SPECIFICATIONS

See tune up chart for hot idle speeds

Year	Carb. Model	Float Level	Pump Rod		Choke Coil Lever	Choke Rod	Vacuum Break		Choke Unloader	Choke Setting
			Hole	Adj.			Front	Rear		
1978	17058133	1/4	Inner	9/32	.120	46°	24°	—	46°	Index
1979	17059100	15/32	Inner	9/32	.120	38°	29°	—	46°	1 Lean
	17059101	15/32	Inner	9/32	.120	38°	29°	—	46°	1 Lean
	17059102	15/32	Inner	9/32	.120	38°	29°	—	46°	1 Lean
	17059103	15/32	Inner	9/32	.120	38°	29°	—	46°	1 Lean
	17059130	1/4	Inner	9/32	.120	38°	27°	—	38°	Index
	17059131	1/4	Inner	3/8	.120	38°	27°	—	38°	1 Lean
	17059132	1/4	Inner	9/32	.120	38°	27°	—	38°	Index
	17059133	1/4	Inner	9/32	.120	38°	27°	—	38°	1 Lean
	17059138	1/4	Inner	1/4	.120	38°	28°	—	38°	1 Lean
	17059139	1/4	Inner	1/4	.120	38°	28°	—	38°	1 Lean
	17059140	1/4	Inner	1/4	.120	38°	28°	—	38°	1 Lean
	17059141	1/4	Inner	1/4	.120	38°	28°	—	38°	1 Lean
	17059142	15/32	Inner	9/32	.120	38°	29°	—	46°	1 Lean
	17059143	15/32	Inner	9/32	.120	38°	29°	—	46°	1 Lean
	17059144	15/32	Inner	9/32	.120	38°	29°	—	46°	1 Lean
	17059145	15/32	Inner	9/32	.120	38°	29°	—	46°	1 Lean
	17059180	5/16	Inner	1/4	.120	24.5°	19°	17°	38°	2 Rich
	17059190	5/16	Inner	1/4	.120	24.5°	19°	17°	38°	2 Rich
	17059191	5/16	Inner	1/4	.120	24.5°	19°	17°	38°	2 Rich
	17059196	5/16	Inner	1/4	.120	24.5°	23°	21°	42°	1 Rich
	17059491	5/16	Inner	1/4	.120	24.5°	23°	21°	42°	1 Rich
	17059492	5/16	Inner	1/4	.120	24.5°	23°	21°	42°	2 Rich
	17059496	5/16	Outer	3/8	.120	24.5°	21°	30°	38°	1 Rich
	17059498	5/16	Inner	1/4	.120	24.5°	23°	21°	42°	1 Rich
1980	17080100	7/16	Inner	9/32	.120	38°	29°	—	38°	—
	17080102	7/16	Inner	9/32	.120	38°	29°	—	38°	—
	17080108	3/8	Inner	5/16	.120	20°	25°	—	38°	—
	17080110	3/8	Inner	5/16	.120	20°	25°	—	38°	—
	17080130	3/8	Inner	5/16	.120	20°	25°	—	38°	—
	17080131	3/8	Inner	5/16	.120	20°	25°	—	38°	—
	17080132	3/8	Inner	5/16	.120	20°	25°	—	38°	—
	17080133	3/8	Inner	5/16	.120	20°	25°	—	38°	—
	17080138	3/8	Inner	5/16	.120	20°	25°	—	38°	—
	17080140	3/8	Inner	5/16	.120	20°	25°	—	38°	—
	17080142	7/16	Inner	9/32	.120	38°	29°	—	38°	—
	17080143	7/16	Inner	9/32	.120	38°	29°	—	38°	—
	17080144	7/16	Inner	9/32	.120	38°	29°	—	38°	—
	17080145	7/16	Inner	9/32	.120	38°	29°	—	38°	—
	17080190	9/32	Inner	1/4	.120	24.5°	22°	20°	38°	—
	17080191	11/32	Inner	1/4	.120	24.5°	18°	18°	38°	—
	17080192	9/32	Inner	1/4	.120	24.5°	22°	20°	38°	—
	17080195	9/32	Inner	1/4	.120	24.5°	19°	17°	38°	—
	17080197	9/32	Inner	1/4	.120	24.5°	19°	17°	38°	—
	17080490	5/16	—	—	.120	24.5°	30°	28°	38°	—

Continued

ROCHESTER MODEL M2M SERIES SPECIFICATIONS—Cont'd

See tune up chart for hot idle speeds

Year	Carb. Model	Float Level	Pump Rod		Choke Coil Lever	Choke Rod	Vacuum Break		Choke Unloader	Choke Setting
			Hole	Adj.			Front	Rear		
	17058492	5/16	—	—	.120	24.5°	30°	28°	38°	—
	17058496	5/16	—	—	.120	24.5°	21°	30°	38°	—
	17058498	5/16	—	—	.120	24.5°	21°	30°	38°	—
1981	17080185, 187	9/32	Inner	1/4	.120	24.5°	19°	14°	38°	—
	17080191	9/32	Inner	1/4	.120	24.5°	18°	18°	38°	—
	17080491	5/16	—	—	.120	24.5°	21°	35°	38°	—
	17080496, 498	5/16	—	—	.120	24.5°	21°	33°	38°	—
	17081101, 103	13/32	Inner	5/16	.120	38°	25°	—	38°	—
	17081130, 131	3/8	Inner	—	.120	20°	25°	—	38°	—
	17081132, 133	3/8	Inner	—	.120	20°	25°	—	38°	—
	17081138, 140	3/8	Inner	—	.120	20°	25°	—	40°	—
	17081142, 143	13/32	Inner	5/16	.120	38°	25°	—	38°	—
	17081144, 145	13/32	Inner	5/16	.120	38°	25°	—	38°	—
	17081191, 192	5/16	—	—	.120	24.5°	28°	24°	38°	—
	17081194	5/16	—	—	.120	24.5°	21°	24°	38°	—
	17081196	5/16	—	—	.120	24.5°	28°	24°	38°	—
	17081197	5/16	—	—	.120	18°	28°	24°	38°	—
	17081198①	3/8	—	—	.120	24.5°	28°	24°	38°	—
	17081198②	3/8	—	—	.120	18°	28°	24°	38°	—
	17081199	3/8	—	—	.120	18°	28°	24°	38°	—
	17081496	3/8	—	—	.120	18°	28°	24°	38°	—
1982	17082130	3/8	—	—	.120	20°	27°	—	38°	—
	17082132	3/8	—	—	.120	20°	27°	—	38°	—
	17082138	3/8	—	—	.120	20°	27°	—	38°	—
	17082140	3/8	—	—	.120	20°	27°	—	38°	—
	17082497	3/8	—	—	.120	24.5°	28°	24°	32°	—
1983	17083190	5/16	—	—	.1206	18°	28°	24°	32°	—
	17083192	5/16	—	—	.1206	18°	28°	24°	32°	—
	17083193	5/16	—	—	.1206	18°	24°	28°	27°	—

①—Models with code letters "AH" stamped on fast idle cam.

②—Models with code letter "T" stamped on fast idle cam.

MODEL M2M SERIES

Description

The M2M Series (M2ME, M2MC) Duel Jet carburetor is a two barrel, single stage unit, incorporating the design features of the primary side of the Quadrajet (four barrel) carburetor. The triple venturi stack up, plus the smaller 1 3/8 inch bores results in good fuel metering control during all phases of operation.

The main metering system has a separate main well for each main nozzle for good fuel flow through the venturi.

An adjustable part throttle screw is used in the float bowl to aid in controlling fuel mixtures for good emission control. This screw is factory preset and should not be adjusted in service. However, if it becomes necessary to replace the float bowl, the new service float bowl will include the adjustable part throttle screw which has been preset.

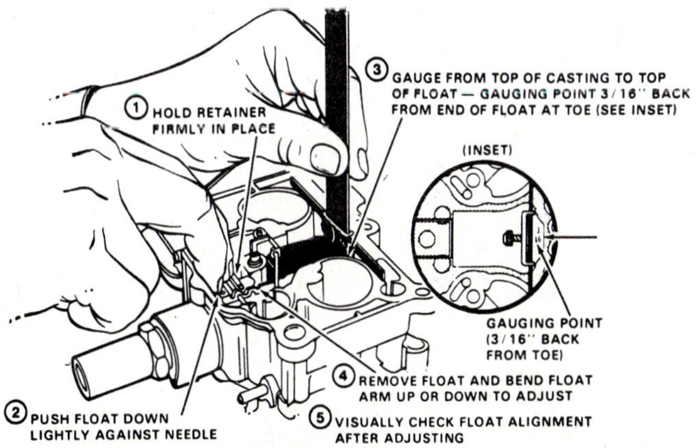

Fig. 1 Float level adjustment. Model M2M Series

Adjustments

Float Level, Adjust, Fig. 1

With adjustable T-scale, measure from top of float bowl gasket surface (gasket removed) to top of float at toe (locate gauging point 3/16" back from toe). Adjust as directed in the illustration to the dimension listed in the *Specifications Chart*. Make sure retaining pin is held firmly in place and tang of float is seated on float needle.

Pump Rod, Adjust, Fig. 2

With throttle valves completely closed and pump rod in specified hole in pump lever, measure from top of choke valve wall (next to vent stack) to top of pump stem. Dimension should be as listed in the *Specifications Chart*. To adjust, bend pump lever as required.

Choke Coil Lever, Adjust, Fig. 3

With thermostatic coil assembly removed, push upward on coil tang until choke valve closes. Insert guage specified in the *Specifica-*

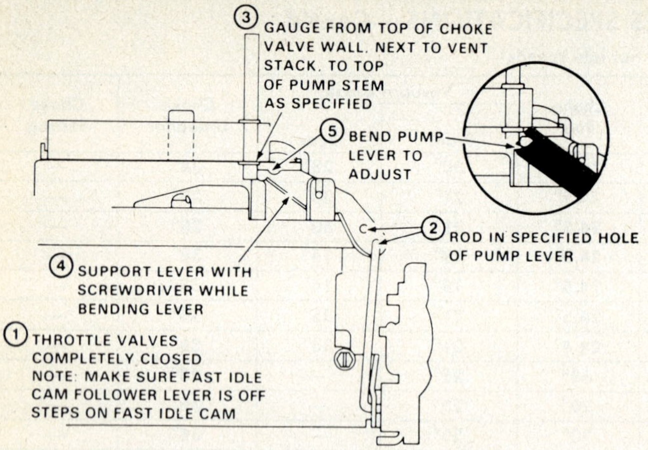

Fig. 2 Pump rod adjustment. Model M2M Series

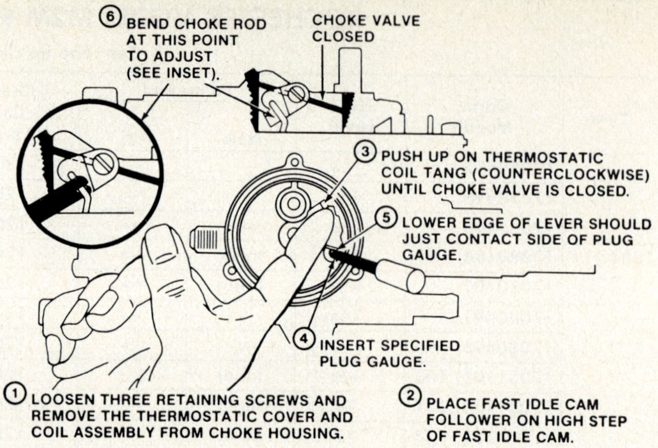

Fig. 3 Choke coil lever adjustment. Model M2M Series

tions Chart into choke housing hole. The lower edge of choke coil lever should just contact gauge. To adjust, bend choke rod as required.

Choke Rod, Adjust, Fig. 4

With fast idle adjustment made, and cam follower on second step of fast idle cam and against the high step, push upward on choke coil lever until choke valve closes. Dimension between upper edge of choke valve and air horn wall should be as specified in the *Specifications Chart*. Adjust by bending tang on intermediate choke lever.

Vacuum Break Adjustment (Gauge Method), Fig. 5

Rotate degree scale until zero is opposite pointer, then with choke valve completely closed, place gauge on top of choke valve and rotate bubble until it is centered. Refer to *Specifications Chart* and rotate gauge to degrees specified. Using an external vacuum source, seat choke vacuum diaphragm, then remove choke coil and rotate inside coil lever counterclockwise. To make final adjustment, turn screw in or out until bubble is centered.

NOTE: On delay models with air bleed, re-

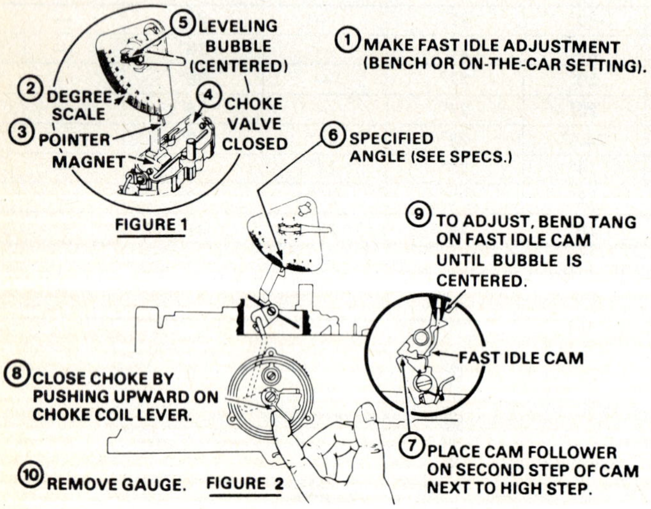

Fig. 4 Choke rod adjustment. Model M2M Series

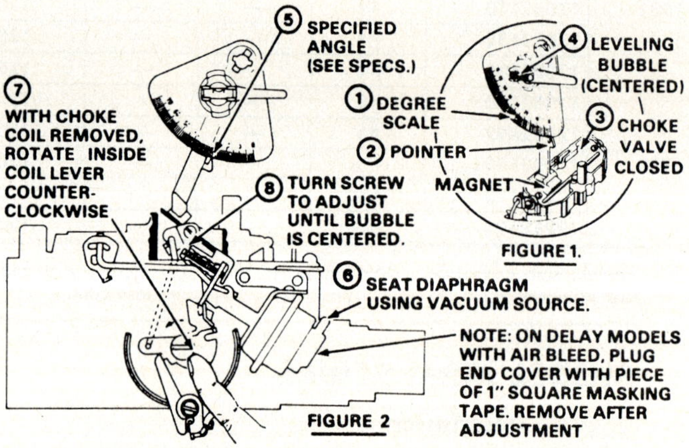

Fig. 5 Vacuum break adjustment. Model M2M Series

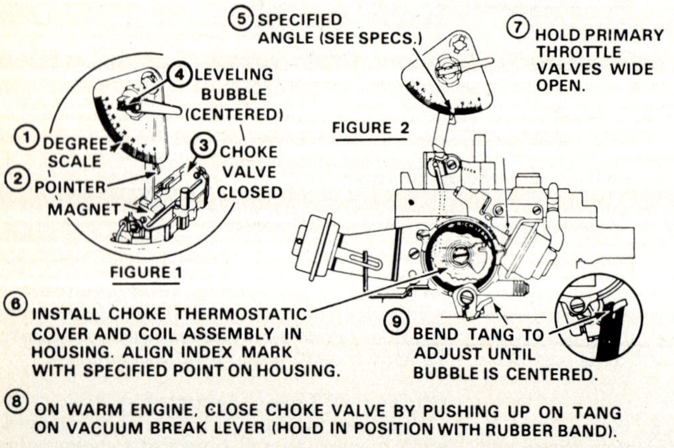

Fig. 6 Choke unloader adjustment. Model M2M Series

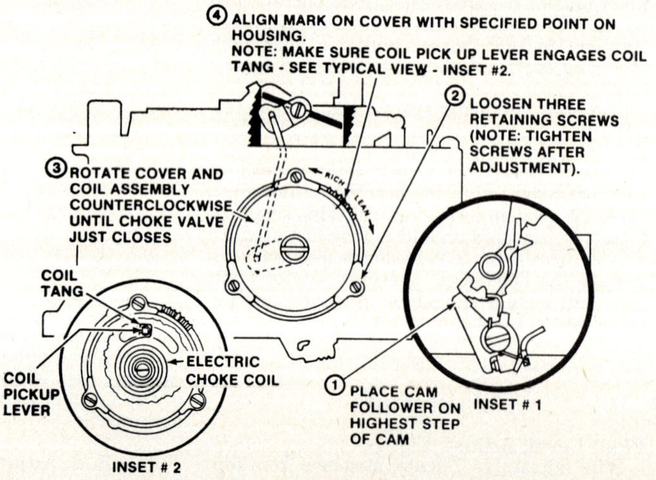

Fig. 7 Choke coil adjustment. Model M2M Series

move rubber cover over filter element and plug small bleed hole in vacuum tube with tape. On models with air bleed in end cover, plug cover with 1 inch masking tape. Remove tape after adjustment.

Choke Coil, Adjust, Fig. 6

With choke valve held closed by means of a rubber band on vacuum break lever, open throttle valves fully. With valves in this position, dimension between lower edge of choke valve and air horn wall should be as specified. To adjust, bend tang on fast idle lever.

Choke Coil, Adjust, Fig. 7

Place the fast idle cam follower on the highest step of the fast idle cam. Rotate choke cover and coil assembly counterclockwise until the choke valve just closes and the index point on cover aligns with the specified index point on the choke housing.

ROCHESTER 2SE ADJUSTMENT SPECIFICATIONS

See Tune Up chart for curb and fast idle speeds.

Year	Carb. Production No.	Float Level	Accel. Pump	Choke Coil Lever	Choke Rod	Vacuum Break		Air Valve Rod	Choke Setting	Unloader	Secondary Lockout
						Primary	Secondary				
1979	17059640	1/8"	9/16"	.085	17°	20°	37°	.040	1 Rich	49°	.025
	17059641	1/8"	9/16"	.085	17°	23.5°	37°	.040	1 Rich	49°	.025
	17059642	1/8"	9/16"	.085	17°	20°	37°	.040	1 Rich	49°	.025
	17059643	1/8"	9/16"	.085	17°	23.5°	37°	.040	1 Rich	49°	.025
	17059740	1/8"	9/16"	.085	17°	20°	37°	.040	1 Rich	49°	.025
	17059741	1/8"	9/16"	.085	17°	20°	37°	.040	1 Rich	49°	.025
	17059742	1/8"	9/16"	.085	17°	20°	37°	.040	1 Rich	49°	.025
	17059764	1/8"	9/16"	.085	17°	20°	37°	.040	1 Rich	49°	.025
	17059765	1/8"	9/16"	.085	17°	23.5°	37°	.040	2 Rich	49°	.025
	17059767	1/8"	9/16"	.085	17°	23.5°	37°	.040	1 Rich	49°	.025
1980	17059774	3/16"	15/32"	.085	18°	19°	—	2°	—	32°	.025
	17059775	—	—	.085	—	—	—	—	—	—	.025
	17059776	3/16"	15/32"	.085	18°	19°	—	2°	—	32°	.025
	17059777	—	—	.085	—	—	—	—	—	—	.025
	17080621	1/8"	9/16"	.085	17°	22°	35°	2°	—	41°	.025
	17080622	1/8"	9/16"	.085	17°	22°	35°	2°	—	41°	.025
	17080623	1/8"	9/16"	.085	17°	22°	35°	2°	—	41°	.025
	17080626	1/8"	9/16"	.085	17°	22°	35°	2°	—	41°	.025
	17080674	3/16"	1/2"	.085	18°	19°	—	2°	—	32°	.025
	17080675	3/16"	1/2"	.085	18°	21°	—	2°	—	32°	.025
	17080676	3/16"	1/2"	.085	18°	19°	—	2°	—	32°	.025
	17080677	3/16"	1/2"	.085	18°	21°	—	2°	—	32°	.025
	17080720	1/8"	9/16"	.085	17°	20°	35°	2°	—	41°	.025
	17080721	1/8"	9/16"	.085	17°	23.5°	35°	2°	—	41°	.025
	17080722	1/8"	9/16"	.085	17°	20°	35°	2°	—	41°	.025
	17080723	1/8"	9/16"	.085	17°	23.5°	35°	2°	—	41°	.025
1981	17081621	3/16"	5/8"	.085	15°	26°	38°	1°	—	38°	—
	17081622	3/16"	5/8"	.085	15°	26°	38°	1°	—	38°	—
	17081623	3/16"	5/8"	.085	15°	26°	38°	1°	—	38°	—
	17081624	3/16"	5/8"	.085	15°	26°	38°	1°	—	38°	—
	17081625	3/16"	5/8"	.085	15°	26°	38°	1°	—	38°	—
	17081626	3/16"	5/8"	.085	15°	26°	38°	1°	—	38°	—
	17081627	3/16"	5/8"	.085	15°	26°	38°	1°	—	38°	—
	17081629	3/16"	5/8"	.085	15°	24°	34°	1°	—	41°	—
	17081630	3/16"	5/8"	.085	15°	26°	38°	1°	—	38°	—
	17081633	3/16"	5/8"	.085	15°	26°	38°	1°	—	38°	—
	17081720	3/16"	5/8"	.085	15°	30°	37°	1°	—	41°	—
	17081721	3/16"	5/8"	.085	15°	30°	37°	1°	—	41°	—
	17081725	3/16"	5/8"	.085	15°	30°	37°	1°	—	41°	—
	17081726	3/16"	5/8"	.085	15°	30°	37°	1°	—	41°	—
	17081727	3/16"	5/8"	.085	15°	30°	37°	1°	—	41°	—
1982	17082334	3/16"	—	.085	15°	26°	38°	1°	—	42°	—
	17082335	3/16"	—	.085	15°	26°	38°	1°	—	42°	—

Continued

ROCHESTER 2SE ADJUSTMENT SPECIFICATIONS—Continued

See Tune Up chart for curb and fast idle speeds.

Year	Carb. Production No.	Float Level	Accel. Pump	Choke Coil Lever	Choke Rod	Vacuum Break		Air Valve Rod	Choke Setting	Unloader	Secondary Lockout
						Primary	Secondary				
	17082336	3/16	—	.085	15°	26°	38°	1°	—	42°	—
	17082337	3/16	—	.085	15°	26°	38°	1°	—	42°	—
	17082338	3/16	—	.085	15°	26°	38°	1°	—	42°	—
	17082339	3/16	—	.085	15°	26°	38°	1°	—	42°	—
	17082341	3/16	—	.085	15°	30°	37°	1°	—	42°	—
	17082342	3/16	—	.085	15°	30°	37°	1°	—	42°	—
	17082344	3/16	—	.085	15°	30°	37°	1°	—	42°	—
	17082345	3/16	—	.085	15°	30°	37°	1°	—	42°	—
	17082431	3/16	—	.085	15°	24°	38°	1°	—	42°	—
	17082433	3/16	—	.085	15°	24°	38°	1°	—	42°	—
	17082480	3/16	—	.085	15°	26°	38°	1°	—	42°	—
	17082481	3/16	—	.085	15°	26°	38°	1°	—	42°	—
	17082482	3/16	—	.085	15°	23°	38°	1°	—	42°	—
	17082483	3/16	—	.085	15°	26°	38°	1°	—	42°	—
	17082484	3/16	—	.085	15°	26°	38°	1°	—	42°	—
	17082485	3/16	—	.085	15°	26°	38°	1°	—	42°	—
	17082486	3/16	—	.085	15°	28°	38°	1°	—	42°	—
	17082487	3/16	—	.085	15°	28°	38°	1°	—	42°	—
	17082488	3/16	—	.085	15°	28°	38°	1°	—	42°	—
	17082489	3/16	—	.085	15°	28°	38°	1°	—	42°	—

MODEL 2SE

Adjustments

The Varajet model 2SE, Fig. 1, is a two barrel, two stage, down draft design carburetor. Aluminum die castings are used for the air horn, float bowl and throttle body. A heat insulator gasket is used between the throttle body and float bowl to reduce heat transfer to the float bowl.

The primary stage has a triple venturi, with a small 35mm bore, resulting in good fuel metering control during idle and part throttle operation. The secondary stage has a 46mm bore, providing sufficient air capacity for engine power requirements. An air valve is used in the secondary stage with a single tapered metering rod.

The float chamber is internally vented through a vertical vent cavity in the air horn. The float chamber is also externally vented through a tube in the air horn. A hose connects this tube directly to a vacuum operated vapor vent valve located in the vapor canister. When the engine is not running, the canister vapor vent valve is open, allowing fuel vapor from the float chamber to pass into the canister where the vapor is stored until normally purged.

An adjustable part throttle screw is used in the float bowl to aid emission control. This screw is factory pre-set and a plug is installed to prevent further adjustment or fuel leakage. The plug should not be removed or the screw setting disturbed. If float bowl replacement is required, the service float bowl will include a factory pre-set and plugged adjustable part throttle screw.

A hot idle compensator is used on some models and is located in the air horn. The opening and closing of the hot idle compensator valve is controlled by a bi-metal strip that is calibrated to a specific temperature. When the valve opens, additional air is allowed to bypass the throttle valves and enter the intake manifold to prevent rough idle during periods of hot engine operation.

The idle mixture screw is recessed in the throttle body and is sealed with a hardened steel plug to prevent alteration of the factory pre-set mixture setting. The plug should not be removed and the mixture screw readjusted unless required by major carburetor overhaul or throttle body replacement.

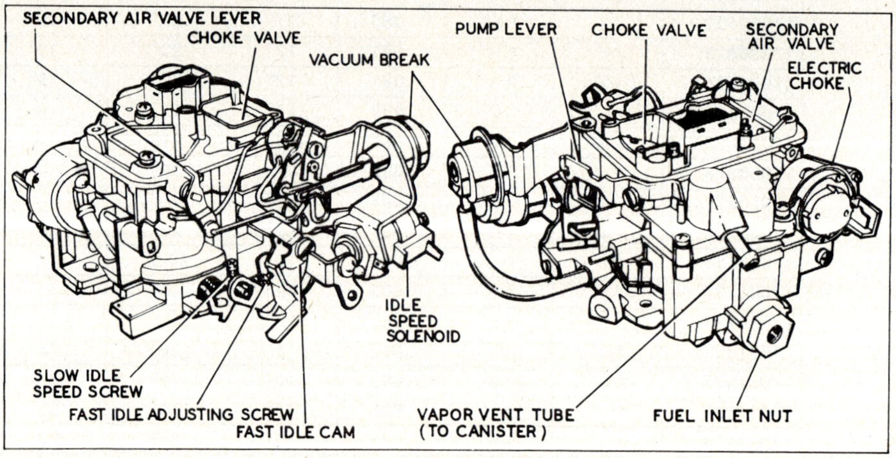

Fig. 1 Varajet 25E carburetor

Float Level Adjustment

Fig. 2

1. Hold float retainer firmly in place and push float lightly against needle.
2. With an adjustable T-scale measure distance between float bowl gasket surface (gasket removed) and float toe.
3. To adjust, remove float and bend float arm.
4. Check float alignment after adjustment.

Pump Adjustment

NOTE: The pump adjustment should not be altered from the specified setting.

Fig. 3

1. Close throttle valves and ensure that fast idle screw is not contacting fast idle cam steps.
2. Measure distance between air horn casting surface and top of pump stem.
3. To adjust, remove pump lever retaining screw and the pump lever. Place the lever in a vise and bend lever end. Do not bend

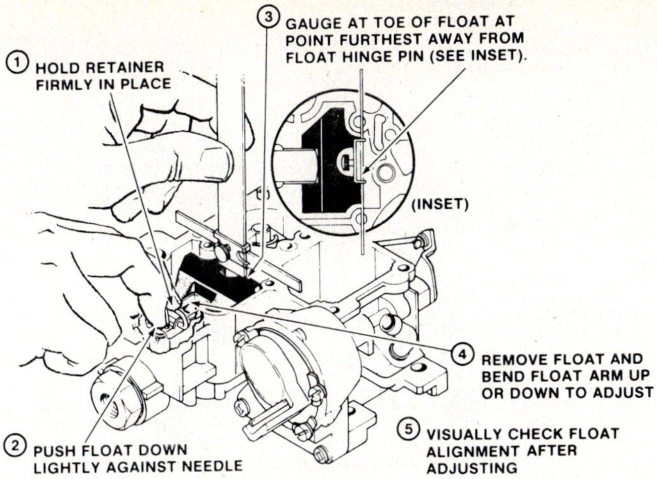

Fig. 2 Float level adjustment

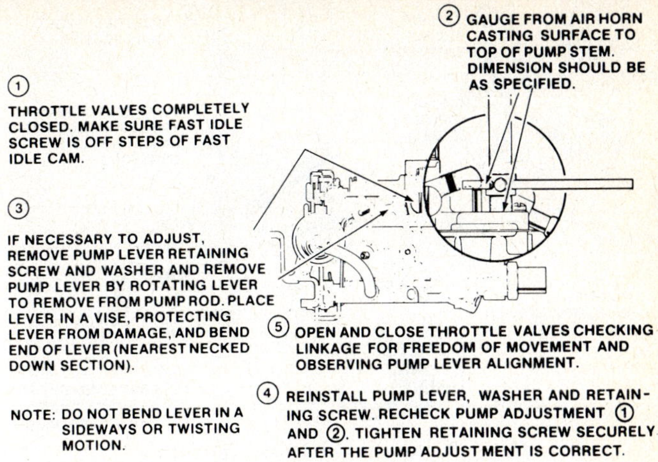

Fig. 3 Pump adjustment

lever in a twisting motion.

4. Install pump lever and retaining screw. Recheck pump adjustment.
5. Operate throttle valves to check for freedom of movement and pump lever alignment.

Fast Idle Adjustment

Fig. 4

1. Adjust curb idle speed, if necessary.
2. Place fast idle screw on highest step of cam.
3. Turn fast idle screw to obtain specified fast idle speed.

Choke Coil Lever Adjustment

Fig. 5

1. Loosen choke cover retaining screws, then the cover and coil assembly from choke housing. If a riveted choke cover is used, remove choke cover and coil assembly as outlined in the choke cover retainer kit.
2. Place fast idle screw on high step of cam.
3. Push the intermediate choke lever until the choke valve closes.
4. Insert specified gauge into hole provided in choke housing. The edge of the lever should just contact gauge.
5. To adjust, bend intermediate choke rod.

Choke Rod Adjustment

NOTE: The choke coil lever and fast idle adjustments must be made before performing this adjustment.

Fig. 6

1. Rotate degree scale until zero is opposite

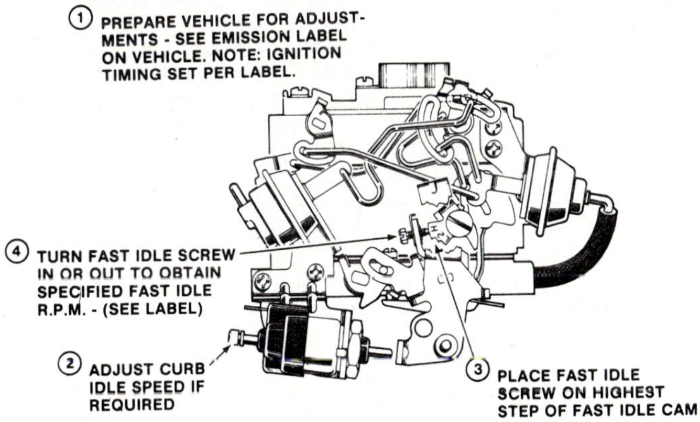

Fig. 4 Fast idle adjustment

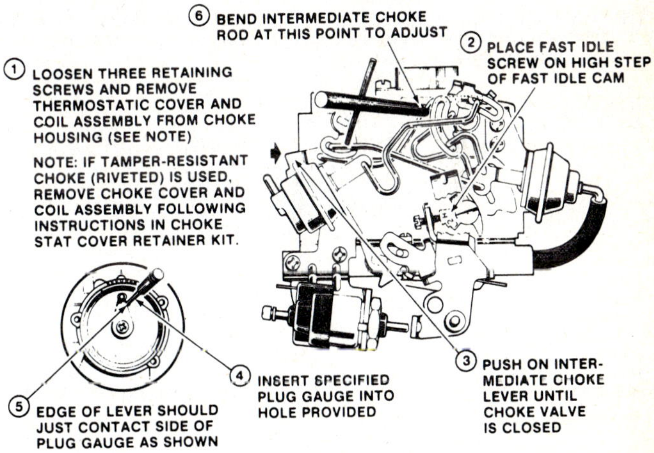

Fig. 5 Choke coil lever adjustment

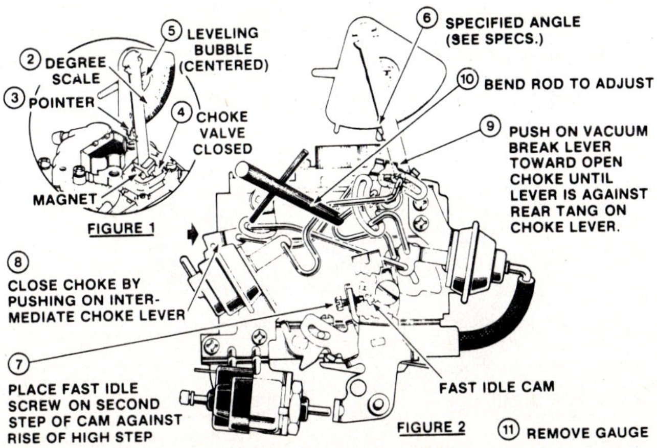

Fig. 6 Choke rod adjustment

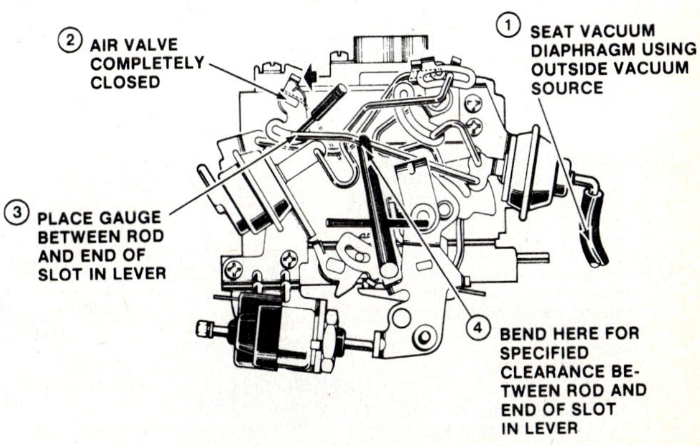

Fig. 7 Air valve rod adjustment

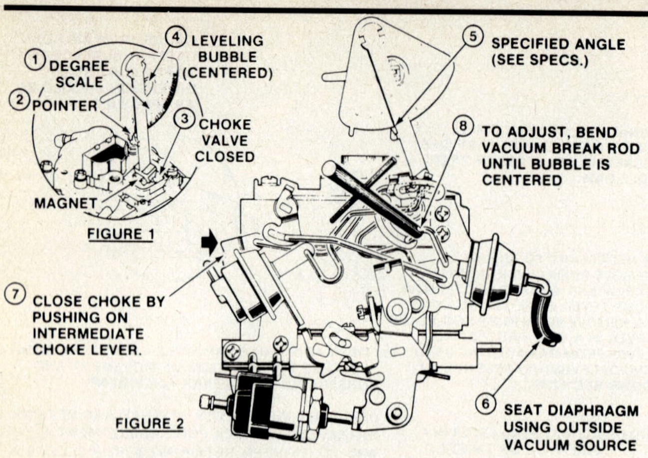

Fig. 8 Primary vacuum break adjustment

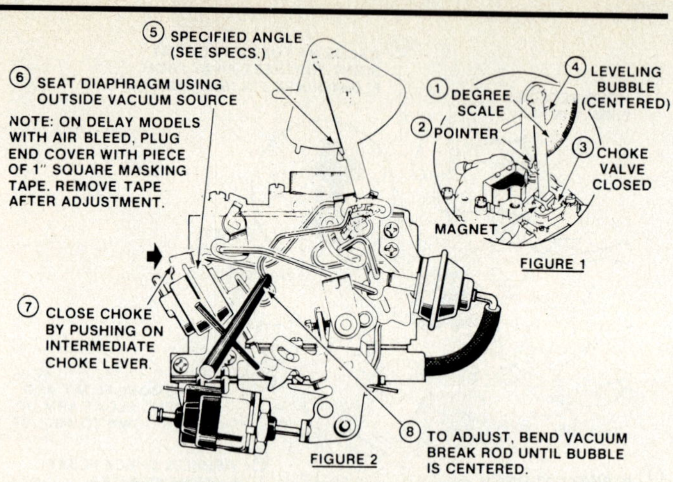

Fig. 9 Secondary vacuum break adjustment

pointer, then with choke valve completely closed, place magnet on top of choke valve and rotate bubble until centered.
2. Rotate scale so specified degree for adjustment is opposite pointer.
3. Place fast idle screw on second step of cam against shoulder of high step.
4. Close choke by pushing intermediate choke lever.
5. Push vacuum break lever toward open choke position until lever contacts rear tang on choke lever.
6. To adjust, bend fast idle cam rod until bubble is centered.

Air Valve Rod Adjustment

Fig. 7
1. Seat vacuum diaphragm with an external vacuum source.
2. Close air valve completely.
3. Place specified gauge between rod and end of lever in slot.
4. To adjust, bend air valve rod.

Primary Vacuum Break Adjustment

Fig. 8
1. Rotate degree scale until zero is opposite pointer, then with choke valve completely closed, place magnet on top of choke valve and rotate bubble until centered.
2. Rotate degree scale so specified degree for adjustment is opposite pointer.
3. Seat vacuum diaphragm with an external vacuum source.
4. Hold choke valve toward closed position by pushing on intermediate choke lever.
5. To adjust, bend vacuum break rod as shown until bubble is centered.

Secondary Vacuum Break Adjustment

Fig. 9
1. Rotate degree scale until zero is opposite pointer, then with choke valve completely closed, place magnet on top of choke valve and rotate bubble until centered.
2. Rotate scale so specified degree for adjustment is opposite pointer.
3. Seat vacuum diaphragm with an external vacuum source.
4. Hold choke valve toward closed position by pushing on intermediate choke lever. Ensure that bucking spring, if used, is compressed and seated.
5. To adjust, bend vacuum break rod as shown until bubble is centered.

Choke Setting

Fig. 10
1. Loosen choke cover retaining screws.
2. Place fast idle screw on high step of cam.
3. Align scribe line on choke cover with specified point on housing. Rotate cover toward closed choke position.
4. Tighten choke cover retaining screws.

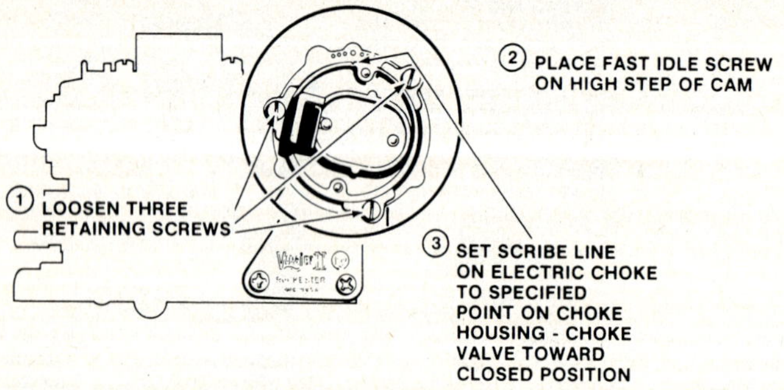

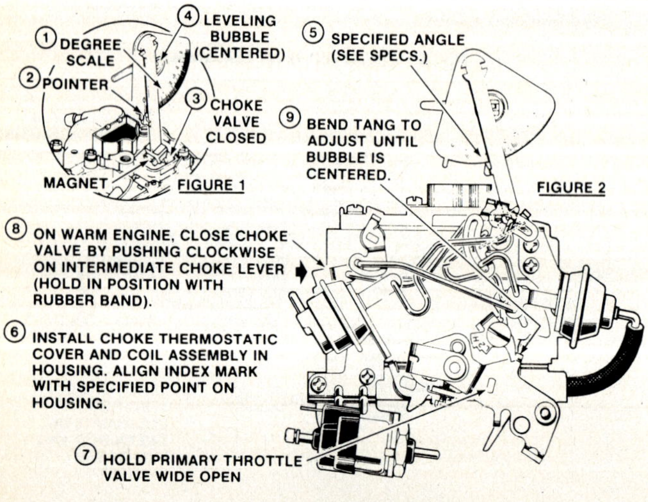

Fig. 11 Unloader adjustment

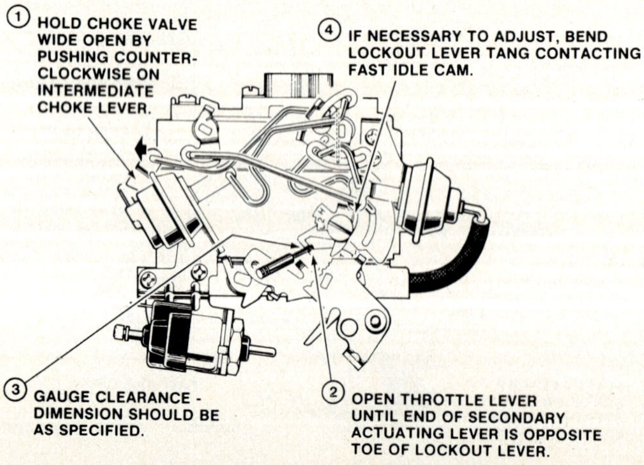

Fig. 12 Secondary lockout adjustment

Unloader Adjustment

Fig. 11

1. Rotate degree scale until zero is opposite pointer, then with choke valve completely closed, place magnet on top of choke valve and rotate bubble until centered.
2. Rotate degree scale so specified degree for adjustment is opposite pointer.
3. With choke setting properly adjusted, hold primary throttle valve wide open.
4. On warm engines, close choke valve by pushing on intermediate choke lever and hold in position with a rubber band.
5. To adjust, bend tang on throttle lever until bubble is centered.

Secondary Lockout Adjustment

Fig. 12

1. Hold choke valve wide open by pulling on intermediate choke lever.
2. Position throttle lever until end of secondary actuating lever is opposite toe of lockout lever.
3. Insert specified gauge between throttle lever and secondary lockout lever toe.
4. To adjust, bend lockout lever tang contacting fast idle cam.

ROCHESTER QUADRAJET 4MC & 4MV ADJUSTMENT SPECIFICATIONS

See Tune Up Chart for hot idle speeds.

Carb. Model	Year	Float Level	Pump Rod		Idle Vent	Choke Coil Lever	Fast Idle (Bench)	Choke Rod	Vacuum Break	Air Valve Dash-pot	Choke Unloader	Air Valve Lockout	Secondary Metering Rods	Air Valve Valve Spring Wind-Up
			Hole	Adj.										
7043254	1974	1/4	Inner	3/8	—	.120	2 Turns	.230	.215	.050	.300	.035	—	3/4
7044201	1974	3/8	Inner	13/32	—	—	2 Turns	.430	.250	.015	.450	—	—	7/16
7044202	1974	1/4	Inner	13/32	—	—	—	.430	.230	—	.450	—	—	7/8
7044203	1974	1/4	Inner	13/32	—	—	—	.430	.230	—	.450	—	—	7/8
7044206	1974	1/4	Inner	13/32	—	—	2 Turns	.430	.230	.015	.450	—	—	7/8
7044207	1974	1/4	Inner	13/32	—	—	2 Turns	.430	.230	.015	.450		—	7/8
7044208	1974	1/4	Inner	13/32	—	—	2 Turns	.430	.230	.015	.450	—	—	1 Turn
7044209	1974	1/4	Inner	13/32	—	—	2 Turns	.430	.230	.015	.450	—	—	1 Turn
7044210	1974	1/4	Inner	13/32	—	—	2 Turns	.430	.230	.015	.450	—	—	1 Turn
7044211	1974	1/4	Inner	13/32	—	—	2 Turns	.430	.230	.015	.450	—	—	1 Turn
7044212	1974	.375	Inner	13/32	—	—	—	.430	.230	—	.450	—	—	7/16
7044213	1974	11/32	Inner	13/32	—	—	—	.430	.215	—	.450	—	—	7/8
7044214	1974	11/32	Inner	13/32	—	—	—	.430	.215	—	.450	—	—	7/8
7044215	1974	11/32	Inner	13/32	—	—	—	.430	.215	—	.450	—	—	7/8
7044216	1974	11/32	Inner	13/32	—	—	—	.430	.215	—	.450	—	—	7/8
7044217	1974	.375	Inner	13/32	—	—	—	.430	.230	—	.450	—	—	7/16
7044218	1974	1/4	Inner	13/32	—	—	—	.430	.215	—	.450	—	—	7/8
7044219	1974	1/4	Inner	13/32	—	—	—	.430	.215	—	.450	—	—	7/8
7044221	1974	3/8	Inner	13/32	—	—	2 Turns	.430	.250	.015	.450	—	—	7/16
7044223	1974	.375	Inner	13/32	—	—	—	.430	.220	—	.450	—	—	7/8
7044224	1974	11/32	Inner	13/32	—	—	—	.430	.215	—	.450	—	—	7/8
7044225	1974	3/8	Inner	13/32	—	—	2 Turns	.430	.220	.015	.450	—	—	7/16
7044226	1974	1/4	Inner	13/32	—	—	2 Turns	.340	.230	.015	.450	—	—	3/4
7044227	1974	.375	Inner	13/32	—	—	—	.430	.220	—	.450	—	—	7/8
7044500	1974	.375	Inner	13/32	—	—	—	.430	.250	—	.450	—	—	7/16
7044502	1974	1/4	Inner	13/32	—	—	—	.430	.230	—	.450	—	—	7/8
7044503	1974	1/4	Inner	13/32	—	—	—	.430	.230	—	.450	—	—	7/8
7044505	1974	3/8	Inner	—	—	—	2 Turns	.430	.230	.015	.450	—	—	7/16
7044506	1974	1/4	Inner	13/32	—	—	2 Turns	.430	.230	.015	.450	—	—	7/8
7044507	1974	1/4	Inner	13/32	—	—	2 Turns	.430	.230	.015	.450	—	—	7/8
7044508	1974	3/8	Inner	13/32	—	—	2 Turns	.430	.250	.015	.450	—	—	7/16
7044509	1974	1/4	Inner	13/32	—	—	2 Turns	.430	.230	.015	.450	—	—	7/8
7044512	1974	.375	Inner	13/32	—	—	—	.430	.215	—	.450	—	—	7/8
7044513	1974	11/32	Inner	13/32	—	—	—	.430	.215	—	.450	—	—	7/8
7044514	1974	11/32	Inner	13/32	—	—	—	.430	.215	—	.450	—	—	7/8
7044515	1974	11/32	Inner	13/32	—	—	—	.430	.215	—	.450	—	—	7/8
7044516	1974	11/32	Inner	13/32	—	—	—	.430	.215	—	.450	—	—	7/8
7044517	1974	.375	Inner	13/32	—	—	—	.430	.230	—	.450	—	—	7/16
7044518	1974	1/4	Inner	13/32	—	—	—	.430	.215	—	.450	—	—	7/8
7044519	1974	1/4	Inner	13/32	—	—	—	.430	.215	—	.450	—	—	7/8
7044520	1974	.375	Inner	13/32	—	—	—	.430	.250	—	.450	—	—	7/16
7044526	1974	1/4	Inner	13/32	—	—	2 Turns	.430	.230	.015	.450	—	—	3/4
7045212	1975	3/8	Inner	.275	—	—	—	.430	.225	.015	.450	—	—	7/16

Continued

ROCHESTER QUADRAJET 4MC & 4MV ADJUSTMENT SPECIFICATIONS—Continued

See Tune Up Chart for hot idle speeds.

Carb. Model	Year	Float Level	Pump Rod		Idle Vent	Choke Coil Lever	Fast Idle (Bench)	Choke Rod	Vacuum Break	Air Valve Dash-pot	Choke Unloader	Air Valve Lockout	Secondary Metering Rods	Air Valve Valve Spring Wind-Up
			Hole	Adj.										
7045213	1975	11/32	Inner	.275	—	—	—	.430	.210	.015	.450	—	—	7/8
7045214	1975	11/32	Inner	.275	—	—	—	.430	.215	.015	.450	—	—	7/8
7045215	1975	11/32	Inner	.275	—	—	—	.430	.215	.015	.450	—	—	7/8
7045216	1975	11/32	Inner	.275	—	—	—	.430	.210	.015	.450	—	—	7/8
7045217	1975	3/8	Inner	.275	—	—	—	.430	.225	.015	.450	—	—	7/16
7045225	1975	11/32	Inner	.275	—	—	—	.430	.200	.015	.450	—	—	3/4
7045229	1975	15/32	Inner	.275	—	—	—	.430	.200	.015	.450	—	—	3/4
7045254	1975–76	1/4	Inner	5/16	—	.120	3 Turns	.230	.215	—	.300	.030	—	3/4
7045554	1975–76	1/4	Inner	5/16	—	.120	3 Turns	.230	.215	—	.300	.030	—	3/4
7045583	1975	11/32	Inner	.275	—	—	—	.430	.230	.015	.450	—	—	7/8
7045584	1975	11/32	Inner	.275	—	—	—	.430	.230	.015	.450	—	—	7/8
7045585	1975	11/32	Inner	.275	—	—	—	.430	.230	.015	.450	—	—	7/8
7045586	1975	11/32	Inner	.275	—	—	—	.430	.230	.015	.450	—	—	7/8
7045588	1975	11/32	Inner	.275	—	—	—	.430	.230	.015	.450	—	—	7/8
7045589	1975	11/32	Inner	.275	—	—	—	.430	.230	.015	.450	—	—	3/4
17056212	1976–77	3/8	Inner	9/32	—	—	—	.290	.155	.015	.295	—	—	3/4
7045213	1976	11/32	Inner	9/32	—	—	—	.290	.145	.015	.295	—	—	7/16
7045214	1976	11/32	Inner	9/32	—	—	—	.290	.145	.015	.295	—	—	7/8
7045215	1976	11/32	Inner	9/32	—	—	—	.290	.145	.015	.295	—	—	7/8
7045216	1976	11/32	Inner	9/32	—	—	—	.290	.145	.015	.295	—	—	7/8
17056217	1976–77	3/8	Inner	9/32	—	—	—	.290	.155	.015	.295	—	—	7/16
7045225	1976	11/32	Inner	9/32	—	—	—	.290	.138	.015	.295	—	—	3/4
7045229	1976–77	11/32	Inner	9/32	—	—	—	.290	.138	.015	.295	—	—	3/4
7045583	1976–77	11/32	Inner	9/32	—	—	—	.290	.155	.015	.295	—	—	7/8
7045584	1976	11/32	Inner	9/32	—	—	—	.290	.155	.015	.295	—	—	7/8
7045585	1976–77	11/32	Inner	9/32	—	—	—	.290	.155	.015	.295	—	—	7/8
7045586	1976–77	11/32	Inner	9/32	—	—	—	.290	.155	.015	.295	—	—	7/8
7045588	1976	11/32	Inner	9/32	—	—	—	.290	.155	.015	.295	—	—	7/8
7045589	1976	11/32	Inner	9/32	—	—	—	.290	.155	.015	.295	—	—	3/4
17057213	1977	11/32	Inner	9/32	—	—	—	.220	.115	.015	.205	—	—	3/4
17057215	1977	11/32	Inner	9/32	—	—	—	.220	.115	.015	.205	—	—	7/8
17057216	1977	11/32	Inner	9/32	—	—	—	.220	.115	.015	.205	—	—	7/8
17057229	1977	11/32	Inner	9/32	—	—	—	.220	.110	.015	.205	—	—	7/8
17057514	1977	11/32	Inner	9/32	—	—	—	.220	.120	.015	.225	—	—	7/8
17057525	1977	11/32	Inner	9/32	—	—	—	.220	.120	.015	.225	—	—	3/4
17057529	1977	11/32	Inner	9/32	—	—	—	.220	.110	.015	.205	—	—	7/8
17058212	1978	7/16	Inner	9/32	—	—	—	②	③	.015	⑦	—	—	7/8
17058213	1978	15/32	Inner	9/32	—	—	—	②	④	.015	⑦	—	—	7/8
17058215	1978	15/32	Inner	9/32	—	—	—	②	④	.015	⑦	—	—	7/8
17058229	1978	15/32	Inner	9/32	—	—	—	②	④	.015	⑦	—	—	7/8
17058513	1978	15/32	Inner	9/32	—	—	—	②	③	.015	⑥	—	—	7/8
17058514	1978	15/32	Inner	9/32	—	—	—	②	③	.015	⑥	—	—	7/8
17058515	1978	15/32	Inner	9/32	—	—	—	②	③	.015	⑥	—	—	7/8
17058525	1978	7/16	Inner	9/32	—	—	—	②	③	.015	—	—	—	3/4
17058529	1978	15/32	Inner	9/32	—	—	—	②	⑤	.015	—	—	—	7/8

①—.245 with auto. trans.
②—With angle gauge set to 42°.
③—With angle gauge set to 26.5°.
④—With angle gauge set to 22°.
⑤—With angle gauge set to 25°.
⑥—With angle gauge set to 43°.
⑦—With angle gauge set to 40°.

MODEL 4MC & 4MV

The primary side of the carburetor has small bores and a triple venturi for fine fuel control in the idle and economy ranges. The secondary side has large bores and an air valve for high air capacity, Fig. 1.

See Figs. 2 through 14 for adjustment procedures and the specifications chart for adjustments specifications.

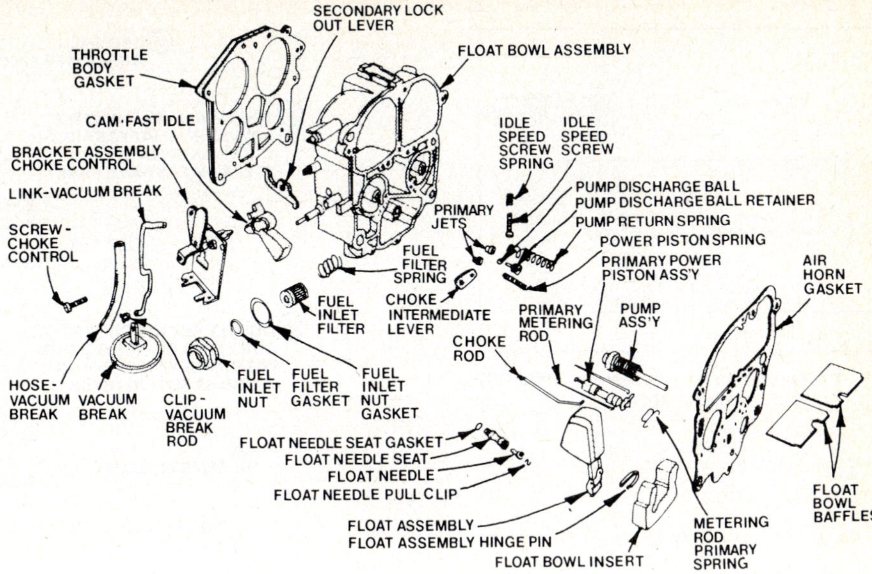

Fig. 1 Rochester Model 4MV carburetor float bowl components

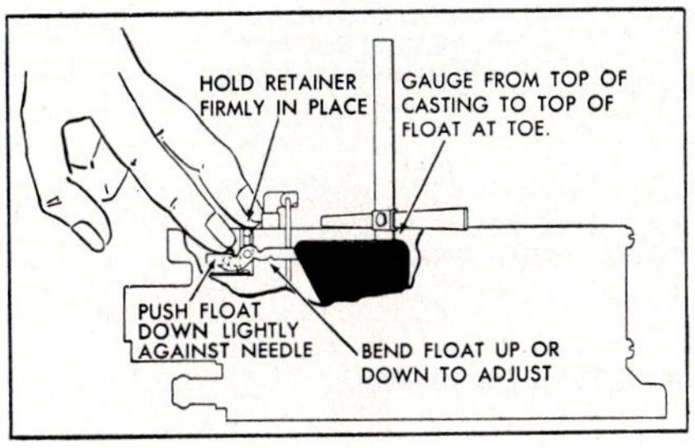

Fig. 2 Float level adjustment. 4MV

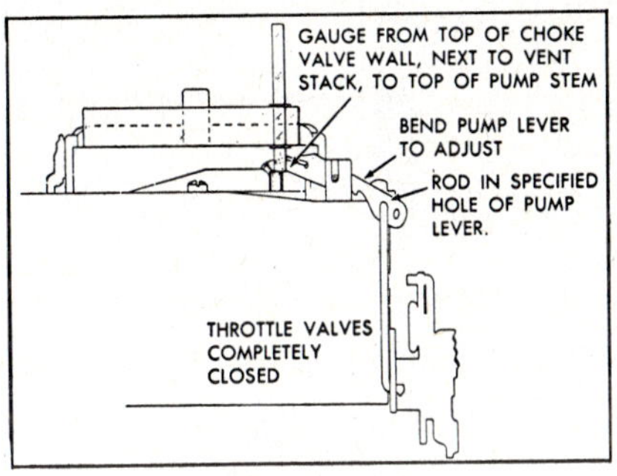

Fig. 3 Pump rod adjustment. 4MV

Fig. 5 Preliminary Fast Idle adjustment. 4MV

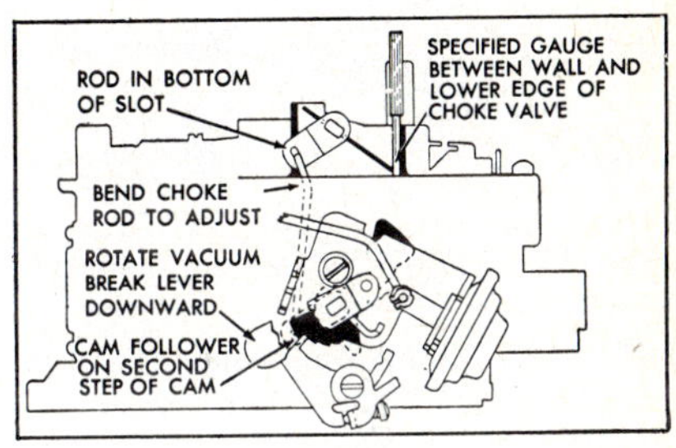

Fig. 6 Choke rod adjustment. 1973–77 Model 4MV

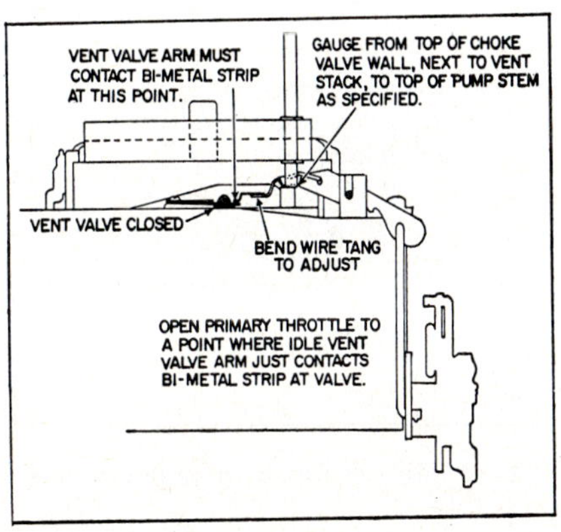

Fig. 4 Idle vent adjustment. 4MV

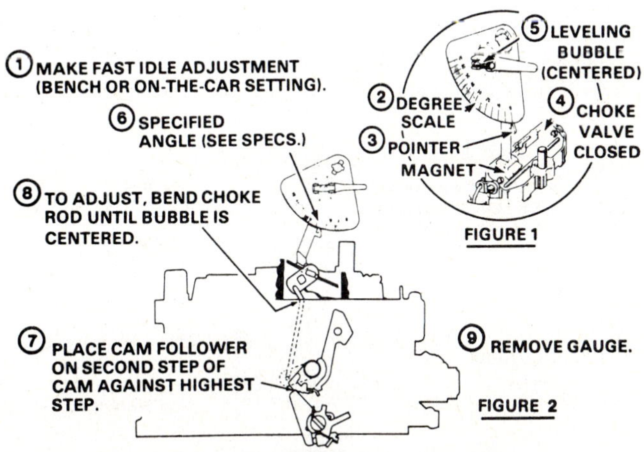

Fig. 7 Choke rod adjustment. 1978 Model 4MV

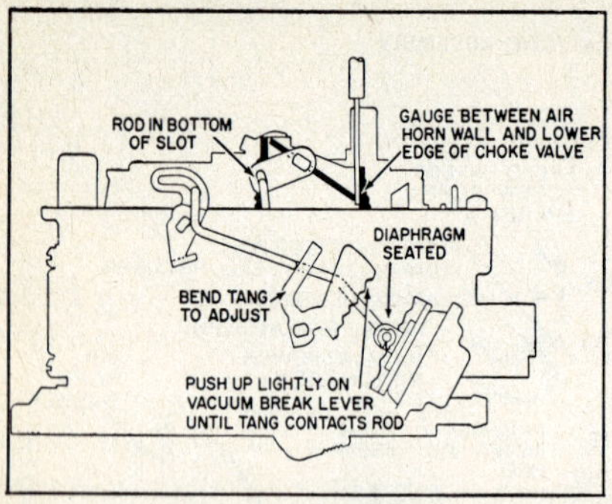

Fig. 8 Vacuum brake adjustment. 1973–77 Model 4MV

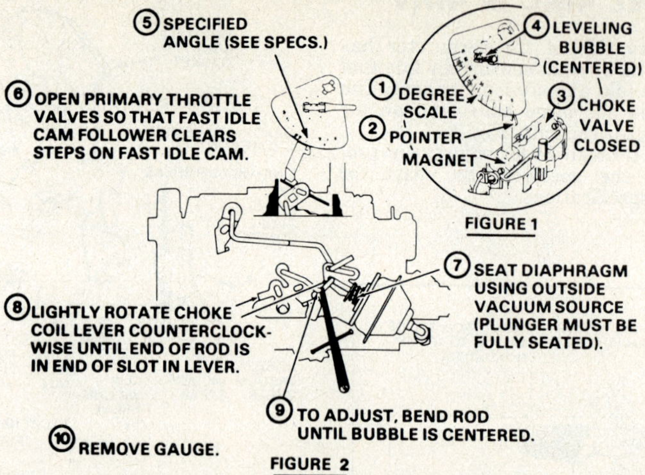

Fig. 9 Vacuum break adjustment. 1978 Model 4MV

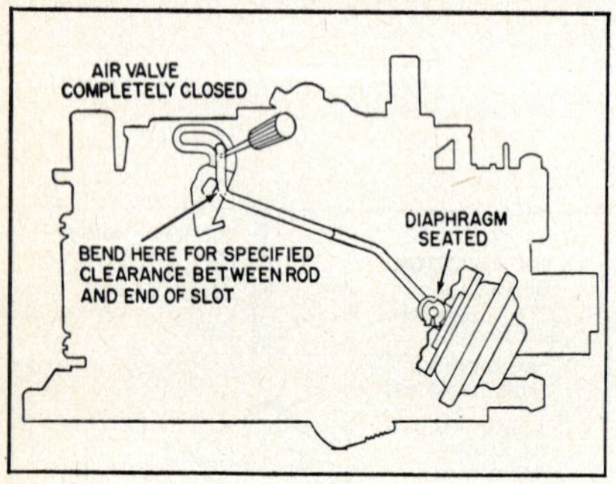

Fig. 10 Air valve dashpot adjustment. 4MV

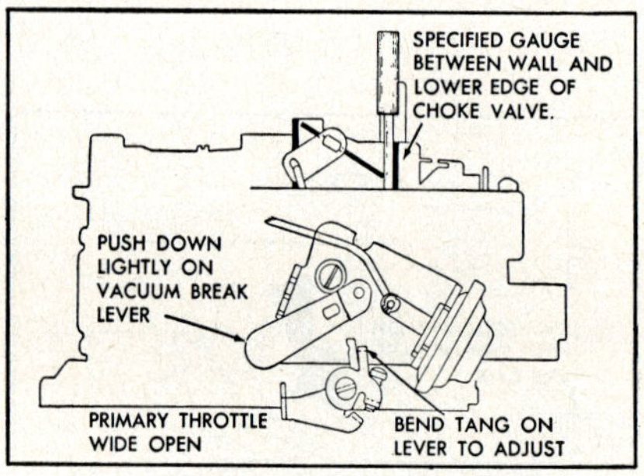

Fig. 11 Unloader adjustment. 1973–77 Model 4MV

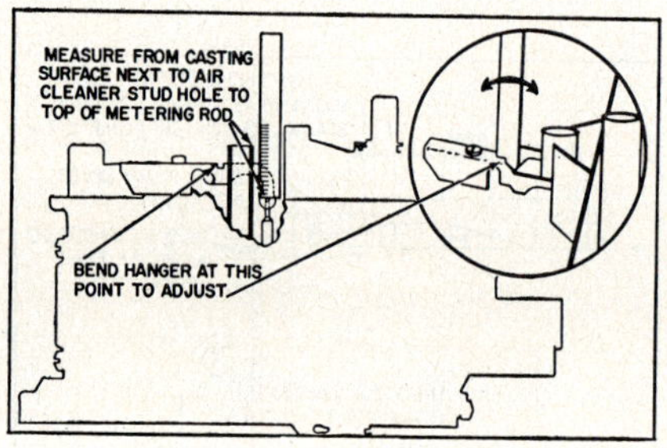

Fig. 13 Secondary metering rod adjustment. Model 4MV

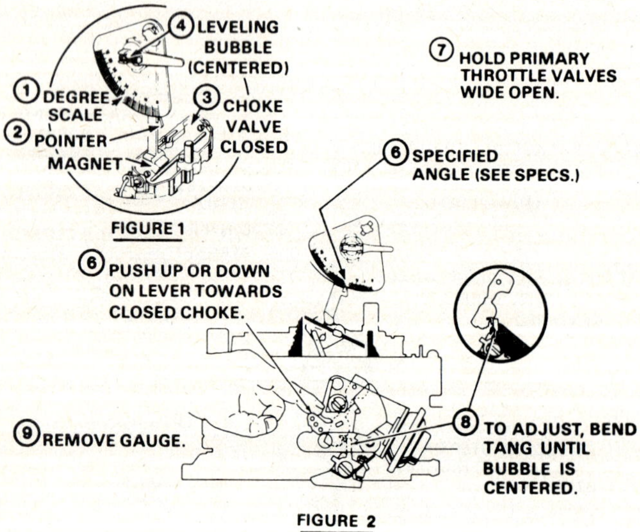

Fig. 12 Unloader adjustment. 1978 Model 4MV

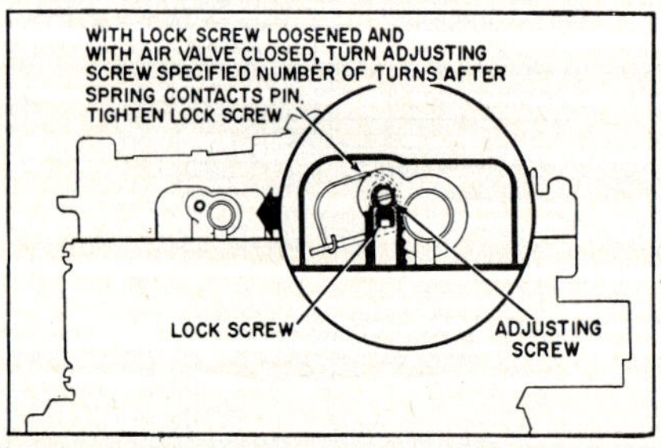

Fig. 14 Air valve spring adjustment. 4MV

QUADRAJET M4M SERIES ADJUSTMENT SPECIFICATIONS

See Tune Up Chart for hot idle speeds.

Year	Carb. Model	Float Level	Pump Rod		Bowl Vent	Fast Idle (Bench)	Choke Coil Lever	Choke Rod	Vacuum Break		Air-Valve Dash-Pot	Choke Setting	Choke Unloader	Air-Valve Valve Spring Wind-Up
			Hole	Adj.					Front	Rear				
1975	7045200	17/32	Inner	.275	—	3 Turns	.120	.300	.200	.550	.015	Index	.325	9/16
	7045202	15/32	Inner	.275	—	3 Turns	.120	.300	.180	.170	.015	—	.325	7/8
	7045203	15/32	Inner	.275	—	3 Turns	.120	.300	.180	.170	.015	—	.325	7/8
	7045206	15/32	Inner	.275	—	3 Turns	.120	.300	.180	.170	.015	Index	.325	7/8
	7045207	15/32	Inner	.275	—	3 Turns	.120	.300	.180	.170	.015	Index	.325	7/8
	7045208	15/32	Inner	.275	—	3 Turns	.120	.300	.180	.170	.015	Index	.325	7/8
	7045209	15/32	Inner	.275	—	3 Turns	.120	.300	.180	.170	.015	Index	.325	7/8
	7045210	15/32	Inner	.275	—	3 Turns	.120	.300	.180	.170	.015	Index	.325	7/8
	7045211	15/32	Inner	.275	—	3 Turns	.120	.300	.180	.170	.015	Index	.325	7/8
	7045218	15/32	Inner	.275	—	3 Turns	.120	.325	.180	.170	.015	—	.325	3/4
	7045219	15/32	Inner	.275	—	3 Turns	.120	.325	.180	.170	.015	—	.325	3/4
	7045220	17/32	Inner	.275	—	3 Turns	.120	.300	.200	.550	.015	—	.325	9/16
	7045222	15/32	Inner	.275	—	3 Turns	.120	.300	.180	.170	.015	Index	.325	7/8
	7045223	15/32	Inner	.275	—	3 Turns	.120	.300	.180	.170	.015	Index	.325	7/8
	7045224	15/32	Inner	.275	—	3 Turns	.120	.325	.180	.170	.015	Index	.325	3/4
	7045228	15/32	Inner	.275	—	3 Turns	.120	.325	.180	.170	.015	Index	.325	3/4
	7045502	15/32	Inner	.275	—	3 Turns	.120	.300	.180	.170	.015	Index	.325	7/8
	7045503	15/32	Inner	.275	—	3 Turns	.120	.300	.180	.170	.015	Index	.325	7/8
	7045504	15/32	Inner	.275	—	3 Turns	.120	.300	.180	.170	.015	Index	.325	7/8
	7045506	15/32	Inner	.275	—	3 Turns	.120	.300	.180	.170	.015	Index	.325	7/8
	7045507	15/32	Inner	.275	—	3 Turns	.120	.300	.180	.170	.015	Index	.325	7/8
	7045512	17/32	Inner	.275	—	3 Turns	.120	.300	.180	.550	.015	—	.325	9/16
	7045517	17/32	Inner	.275	—	3 Turns	.120	.300	.180	.550	.015	—	.325	9/16
1976	17056200	13/32	Inner	9/32	—	3 Turns	.120	.190	.240	.160	.015	Index	.270	7/8
	17056202	13/32	Inner	9/32	—	3 Turns	.120	.325	.185	—	.015	2 Lean	.325	7/8
	17056203	13/32	Inner	9/32	—	3 Turns	.120	.325	.185	—	.015	3 Lean	.325	7/8
	17056206	13/32	Inner	9/32	—	3 Turns	.120	.325	.185	—	.015	2 Lean	.325	7/8
	17056207	13/32	Inner	9/32	—	3 Turns	.120	.325	.185	—	.015	3 Lean	.325	7/8
	17056208	①	Inner	9/32	—	3 Turns	.120	.325	.185	—	.015	2 Lean	.325	7/8
	17056209	①	Inner	9/32	—	3 Turns	.120	.325	.185	—	.015	3 Lean	.325	7/8
	17056210	13/32	Inner	9/32	—	3 Turns	.120	.325	.185	—	.015	2 Lean	.325	1
	17056211	13/32	Inner	9/32	—	3 Turns	.120	.325	.170	—	.015	2 Lean	.325	1
	17056218	5/16	Inner	9/32	—	3 Turns	.120	.325	.185	—	.015	2 Lean	.325	7/8
	17056219	5/16	Inner	9/32	—	3 Turns	.120	.325	.185	—	.015	3 Lean	.325	7/8
	17056221	7/16	Inner	9/32	—	3 Turns	.120	.300	—	.160	.015	2 Rich	.325	7/8
	17056226	13/32	Inner	9/32	—	3 Turns	.120	.325	.185	—	.015	2 Lean	.325	3/4
	17056228	13/32	Inner	9/32	—	3 Turns	.120	.325	.185	—	.015	2 Lean	.325	7/8
	17056502	13/32	Inner	9/32	—	3 Turns	.120	.325	.185	—	.015	2 Lean	.325	7/8
	17056503	13/32	Inner	9/32	—	3 Turns	.120	.325	.185	—	.015	2 Lean	.325	7/8
	17056506	13/32	Inner	9/32	—	3 Turns	.120	.325	.185	—	.015	2 Lean	.325	7/8
	17056507	13/32	Inner	9/32	—	3 Turns	.120	.325	.185	—	.015	2 Lean	.325	7/8
	17056508	①	Inner	9/32	—	3 Turns	.120	.325	.185	—	.015	2 Lean	.325	7/8
	17056509	①	Inner	9/32	—	3 Turns	.120	.325	.185	—	.015	2 Lean	.325	7/8
	17056512	7/16	Inner	9/32	—	3 Turns	.120	.325	.185	—	.015	Index	.275	7/8
	17056517	7/16	Inner	9/32	—	3 Turns	.120	.325	.185	—	.015	Index	.275	7/8
	17056518	5/16	Inner	9/32	—	3 Turns	.120	.325	.185	—	.015	2 Lean	.325	7/8
	17056519	5/16	Inner	9/32	—	3 Turns	.120	.325	.185	—	.015	2 Lean	.325	7/8
	17056528	13/32	Inner	9/32	—	3 Turns	.120	.325	.185	—	.015	2 Lean	.325	3/4
1977	17057202	15/32	Inner	9/32	—	—	.120	.325	.160	—	.015	2 Lean	—	—
	17057203	15/32	Inner	9/32	—	—	.120	.325	.160	—	.015	3 Lean	.280	7/8
	17057204	15/32	Inner	9/32	—	—	.120	.325	.160	—	.015	2 Lean	.280	7/8
	17051210	15/32	Inner	9/32	—	—	.120	.325	.180	—	.015	2 Lean	.280	1

Continued

QUADRAJET M4M SERIES ADJUSTMENT SPECIFICATIONS—Continued

See Tune Up Chart for hot idle speeds.

Year	Carb. Model	Float Level	Pump Rod		Bowl Vent	Fast Idle (Bench)	Choke Coil Lever	Choke Rod	Vacuum Break		Air-Valve Dash-Pot	Choke Setting	Choke Unloader	Air-Valve Valve Spring Wind-Up
			Hole	Adj.					Front	Rear				
	17057211	15/32	Inner	9/32	—	—	.120	.325	.180	—	.015	3 Lean	.280	1
	17057218	15/32	Inner	9/32	—	—	.120	.325	.160	—	.015	2 Lean	.280	7/8
	17057219	7/16	Inner	9/32	—	—	.120	.325	.165	—	.015	3 Lean	.280	7/8
	17057221	3/8	Inner	9/32	—	—	.120	.325	—	.160	.015	2 Rich	.325	7/8
	17057222	7/16	Inner	9/32	—	—	.120	.325	.160	—	.015	2 Lean	.280	7/8
	17057228	13/32	Inner	9/32	—	—	.120	.325	.180	—	.015	2 Lean	.280	1
	17057254	15/32	Inner	9/32	—	3 Turns	.120	.100	.135	—	.030	3 Rich	.220	1/2
	17057502	15/32	Inner	9/32	—	—	.120	.325	.165	—	.015	2 Lean	.280	7/8
	17057503	15/32	Inner	9/32	—	—	.120	.325	.165	—	.015	1 Lean	.280	7/8
	17057504	15/32	Inner	9/32	—	—	.120	.325	.165	—	.015	2 Lean	.280	7/8
	17057510	15/32	Inner	9/32	—	—	.120	.325	.180	—	.015	2 Lean	.280	7/8
	17057512	7/16	Inner	9/32	—	—	.120	.285	.165	—	.015	Index	.240	7/8
	17057517	7/16	Inner	9/32	—	—	.120	.285	.165	—	.015	Index	.240	7/8
	17057518	7/16	Inner	9/32	—	—	.120	.325	.165	—	.015	2 Lean	.280	7/8
	17057519	7/16	Inner	9/32	—	—	.120	.325	.165	—	.015	1 Lean	.280	7/8
	17057522	7/16	Inner	9/32	—	—	.120	.325	.165	—	.015	2 Lean	.280	7/8
	17057528	15/32	Inner	9/32	—	—	.120	.325	.180	—	.015	2 Lean	.280	7/8
	17057559	15/32	Inner	9/32	—	3 Turns	.120	.100	.135	—	.030	3 Rich	.220	1/2
	17057582	15/32	Outer	9/32	—	—	.120	.325	.180	—	.015	2 Lean	.280	7/8
	17057584	15/32	Outer	9/32	—	—	.120	.325	.180	—	.015	2 Lean	.280	7/8
	17057586	7/16	Outer	9/32	—	—	.120	.325	.180	—	.015	2 Lean	.280	7/8
	17057588	7/16	Outer	9/32	—	—	.120	.325	.180	—	.015	2 Lean	.280	7/8
1978	17058201	15/32	Inner	9/32	—	—	.120	46°	28.5°	—	.015	3 Lean	42°	7/8
	17058202	15/32	Inner	9/32	—	—	.120	.314	②	—	.015	2 Lean	.277	7/8
	17058203	15/32	Inner	9/32	—	—	.120	.314	②	—	.015	2 Lean	.277	7/8
	17058204	15/32	Inner	9/32	—	—	.120	.314	②	—	.015	2 Lean	.277	7/8
	17058210	15/32	Inner	9/32	—	—	.120	.314	③	—	.015	2 Lean	.277	—
	17058211	15/32	Inner	9/32	—	—	.120	.314	③	—	.015	2 Lean	.277	—
	17058218	7/16	Inner	9/32	—	—	.120	46°	27°	—	.015	2 Lean	42°	7/8
	17058219	7/16	Inner	9/32	—	—	.120	46°	28.5°	—	.015	3 Lean	42°	7/8
	17058222	7/16	Inner	9/32	—	—	.120	46°	27°	—	.015	2 Lean	42°	7/8
	17058228	15/32	Inner	9/32	—	—	.120	.314	③	—	.015	2 Lean	.277	7/8
	17058254	15/32	Inner	9/32	—	3 Turns	.120	19°	24°	—	.030	3 Rich	35°	1/2
	17058500	3/8	Inner	9/32	—	—	.120	46°	28°	—	.015	3 Lean	42°	7/8
	17058501	3/8	Inner	9/32	—	—	.120	46°	28°	—	.015	3 Lean	42°	7/8
	17058502	15/32	Inner	9/32	—	—	.120	.314	④	—	.015	2 Lean	.277	7/8
	17058503	15/32	Inner	9/32	—	—	.120	46°	30°	—	.015	1 Lean	42°	7/8
	17058504	15/32	Inner	9/32	—	—	.120	.314	④	—	.015	2 Lean	.277	7/8
	17058506	15/32	Inner	9/32	—	—	.120	46°	30°	—	.015	1 Lean	42°	7/8
	17058508	15/32	Inner	9/32	—	—	.120	46°	30°	—	.015	1 Lean	42°	7/8
	17058509	15/32	Outer	11/32	—	—	.120	46°	30°	—	.015	2 Lean	42°	7/8
	17058510	15/32	Outer	11/32	—	—	.120	46°	30°	—	.015	2 Lean	42°	7/8
	17058512	13/32	Inner	9/32	—	—	.120	46°	30°	—	.015	Index	40°	7/8
	17058518	15/32	Inner	9/32	—	—	.120	46°	30°	—	.015	1 Lean	42°	7/8
	17058519	15/32	Inner	9/32	—	—	.120	46°	30°	—	.015	1 Lean	42°	7/8
	17058520	3/8	Inner	9/32	—	—	.120	46°	28°	—	.015	3 Lean	42°	7/8
	17058521	3/8	Inner	9/32	—	—	.120	46°	28°	—	.015	3 Lean	42°	7/8
	17058522	15/32	Inner	9/32	—	—	.120	46°	30o	—	.015	1 Lean	42°	7/8
	17058523	15/32	Inner	9/32	—	—	.120	46°	30°	—	.015	1 Lean	42°	7/8
	17058524	15/32	Inner	9/32	—	—	.120	46°	30°	—	.015	1 Lean	42°	7/8
	17058527	15/32	Inner	9/32	—	—	.120	46°	30°	—	.015	1 Lean	42°	7/8

Continued

QUADRAJET M4M SERIES ADJUSTMENT SPECIFICATIONS—Continued

See Tune Up Chart for hot idle speeds.

Year	Carb. Model	Float Level	Pump Rod		Bowl Vent	Fast Idle (Bench)	Choke Coil Lever	Choke Rod	Vacuum Break		Air-Valve Dash-Pot	Choke Setting	Choke Unloader	Air-Valve Valve Spring Wind-Up
			Hole	Adj.					Front	Rear				
	17058528	15/32	Inner	9/32	—	—	.120	46°	30°	—	.015	1 Lean	42°	7/8
	17058559	15/32	Inner	9/32	—	3 Turns	.120	19°	24°	—	.030	3 Rich	36.5°	1/2
	17058582	15/32	Inner	9/32	—	—	.120	.314	③	—	.015	2 Lean	.277	7/8
	17058584	15/32	Inner	9/32	—	—	.120	.314	③	—	.015	2 Lean	.277	7/8
	17058586	15/32	Outer	11/32	—	—	.120	46°	30°	—	.015	2 Lean	42°	7/8
	17058588	15/32	Outer	11/32	—	—	.120	46°	30°	—	.015	2 Lean	42°	7/8
1979	17059061	15/32	Inner	9/32	—	—	.120	46°	—	23°	.015	Index	42°	7/8
	17059062	15/32	Inner	1/4	—	—	.120	38°	28°	—	.015	1 Lean	38°	7/8
	17059064	15/32	Inner	1/4	—	—	.120	38°	28°	—	.015	1 Lean	38°	7/8
	17059065	15/32	Inner	9/32	—	—	.120	46°	—	23°	.015	1 Lean	42°	7/8
	17059066	15/32	Inner	9/32	—	—	.120	46°	—	23°	.015	1 Lean	42°	7/8
	17059068	15/32	Inner	9/32	—	—	.120	46°	—	23°	.015	1 Lean	42°	7/8
	17059069	15/32	Inner	9/32	—	—	.120	46°	—	23°	.015	1 Lean	42°	7/8
	17059076	15/32	Inner	9/32	—	—	.120	46°	—	23°	.015	1 Lean	42°	7/8
	17059077	15/32	Inner	9/32	—	—	.120	46°	—	23°	.015	1 Lean	42°	7/8
	17059201	15/32	Inner	9/32	—	—	.120	46°	—	23°	.015	Index	42°	7/8
	17059205	15/32	Inner	9/32	—	—	.120	46°	—	23°	.015	1 Lean	42°	7/8
	17059206	15/32	Inner	9/32	—	—	.120	46°	—	23°	.015	1 Lean	42°	7/8
	17059207	15/32	Inner	1/4	—	—	.120	38°	27°	—	.015	2 Lean	38°	7/8
	17059208	15/32	Inner	9/32	—	—	.120	46°	—	23°	.015	1 Lean	42°	7/8
	17059209	15/32	Inner	9/32	—	—	.120	46°	—	23°	.015	1 Lean	42°	7/8
	17059212	3/8	Inner	9/32	—	—	.120	46°	24°	—	.015	1 Rich	40°	3/4
	17059213	3/8	Inner	9/32	—	—	.120	37°	23°	—	.015	1 Rich	40°	1
	17059215	3/8	Inner	9/32	—	—	.120	37°	23°	—	.015	1 Rich	40°	1
	17059216	15/32	Inner	1/4	—	—	.120	38°	27°	—	.015	1 Lean	38°	7/8
	17059217	15/32	Inner	1/4	—	—	.120	38°	27°	—	.015	1 Lean	38°	7/8
	17059222	15/32	Inner	9/32	—	—	.120	38°	28°	—	.015	2 Lean	38°	7/8
	17059226	15/32	Inner	9/32	—	—	.120	46°	—	23°	.015	1 Lean	42°	7/8
	17059227	15/32	Inner	9/32	—	—	.120	46°	—	23°	.015	1 Lean	42°	7/8
	17059229	3/8	Inner	9/32	—	—	.120	37°	23°	—	.015	1 Rich	40°	1
	17059363	15/32	Inner	9/32	—	—	.120	46°	—	26°	.015	1 Lean	42°	7/8
	17059366	15/32	Inner	9/32	—	—	.120	46°	—	26°	.015	1 Lean	42°	7/8
	17059368	15/32	Inner	9/32	—	—	.120	46°	—	26°	.015	1 Lean	42°	7/8
	17059377	15/32	Inner	9/32	—	—	.120	46°	—	23°	.015	1 Lean	42°	7/8
	17059378	15/32	Inner	9/32	—	—	.120	46°	—	23°	.015	1 Lean	42°	7/8
	17059501	3/8	Inner	9/32	—	—	.120	46°	28°	—	.015	3 Lean	42°	7/8
	17059502	15/32	Inner	1/4	—	—	.120	38°	28°	—	.015	2 Lean	38°	7/8
	17059503	15/32	Inner	9/32	—	—	.120	46°	—	26°	.015	1 Lean	42°	7/8
	17059504	15/32	Inner	1/4	—	—	.120	38°	28°	—	.015	2 Lean	38°	7/8
	17059506	15/32	Inner	9/32	—	—	.120	46°	—	26°	.015	1 Lean	42°	7/8
	17059508	15/32	Inner	9/32	—	—	.120	46°	—	26°	.015	1 Lean	42°	7/8
	17059509	15/32	Outer	11/32	—	—	.120	46°	30°	—	.015	2 Lean	42°	7/8
	17059510	15/32	Outer	11/32	—	—	.120	46°	30°	—	.015	2 Lean	42°	7/8
	17059512	3/8	Inner	9/32	—	—	.120	46°	24°	—	.015	1 Rich	40°	3/4
	17059513	3/8	Inner	9/32	—	—	.120	37°	23°	—	.015	1 Rich	40°	1
	17059515	3/8	Inner	9/32	—	—	.120	37°	23°	—	.015	1 Rich	40°	1
	17059520	3/8	Inner	9/32	—	—	.120	46°	28°	—	.015	3 Lean	42°	7/8
	17059521	3/8	Inner	9/32	—	—	.120	46°	28°	—	.015	3 Lean	42°	7/8
	17059527	15/32	Inner	9/32	—	—	.120	46°	—	23°	.015	1 Lean	42°	7/8
	17059528	15/32	Inner	9/32	—	—	.120	46°	—	23°	.015	1 Lean	42°	7/8
	17059529	3/8	Inner	9/32	—	—	.120	37°	23°	—	.015	1 Rich	40°	1

Continued

QUADRAJET M4M SERIES ADJUSTMENT SPECIFICATIONS—Continued

See Tune Up Chart for hot idle speeds.

Year	Carb. Model	Float Level	Pump Rod		Bowl Vent	Fast Idle (Bench)	Choke Coil Lever	Choke Rod	Vacuum Break		Air-Valve Dash-Pot	Choke Setting	Choke Unloader	Air-Valve Valve Spring Wind-Up
			Hole	Adj.					Front	Rear				
	17059582	15/32	Outer	11/32	—	—	.120	38°	33°	—	.015	1 Lean	46°	7/8
	17059584	15/32	Outer	11/32	—	—	.120	38°	33°	—	.015	1 Lean	46°	7/8
	17059586	15/32	Outer	11/32	—	—	.120	46°	30°	—	.015	2 Lean	42°	7/8
	17059588	15/32	Outer	11/32	—	—	.120	46°	30°	—	.015	2 Lean	42°	7/8
1980	17080201	15/32	Inner	9/32	—	—	—	46°	—	23°	.025	—	42°	7/8
	17080202	13/32	Inner	1/4	—	3	.120	20°	27°	—	.025	—	38°	7/8
	17080204	13/32	Inner	1/4	—	3	.120	20°	27°	—	.025	—	38°	7/8
	17080205	15/32	Inner	9/32	—	—	—	46°	—	—	.025	—	42°	7/8
	17080206	15/32	Inner	9/32	—	—	—	46°	—	23°	.025	—	42°	7/8
	17080207	13/32	Inner	1/4	—	—	—	20°	27°	—	.025	—	38°	7/8
	17080211	13/32	Inner	9/32	—	—	—	20°	28°	—	.025	—	38°	1
	17080212	3/8	Inner	9/32	—	—	.120	46°	24°	30°	.025	—	40°	3/4
	17080213	3/8	Inner	9/32	—	—	.120	37°	23°	30°	.025	—	40°	1
	17080215	3/8	Inner	9/32	—	—	.120	37°	23°	30°	.025	—	40°	1
	17080224	15/32	Inner	9/32	—	—	.120	46°	—	23°	.025	—	42°	7/8
	17080225	15/32	Inner	9/32	—	—	.120	46°	—	23°	.025	—	42°	7/8
	17080226	15/32	Inner	9/32	—	—	.120	46°	—	23°	.025	—	42°	7/8
	17080227	15/32	Inner	9/32	—	—	.120	46°	—	23°	.025	—	42°	7/8
	17080228	13/32	Inner	9/32	—	—	.120	20°	25°	—	.025	—	38°	1
	17080229	3/8	Inner	9/32	—	—	.120	37°	23°	30°	.025	—	40°	1
	17080240	3/16	inner	9/32	—	—	.120	14.5°	16°	16°	.025	—	30°	9/16
	17080242	13/32	Inner	9/32	—	—	.120	14.5°	15°	18°	.025	—	35°	9/16
	17080243	3/16	Inner	9/32	—	—	.120	14.5°	16°	16°	.025	—	30°	9/16
	17080274	7/16	Outer	11/32	—	—	—	16°	24°	32°	.025	—	35°	5/8
	17080290	15/32	Inner	9/32	—	—	—	46°	—	26°	.025	—	42°	7/8
	17080291	15/32	Inner	9/32	—	—	—	46°	—	26°	.025	—	42°	7/8
	17080292	15/32	Inner	9/32	—	—	—	46°	—	26°	.025	—	42°	7/8
	17080293	15/32	Inner	9/32	—	—	—	46°	—	26°	.025	—	42°	7/8
	17080295	15/32	Inner	9/32	—	—	—	46°	—	23°	.025	—	42°	7/8
	17080297	15/32	Inner	9/32	—	—	—	46°	—	23°	.025	—	42°	7/8
	17080502	19/32	—	3/8	—	—	.120	20°	26°	30°	.025	—	38°	7/8
	17080503	15/32	Inner	9/32	—	—	—	46°	—	26°	.025	—	42°	7/8
	17080504	19/32	—	3/8	—	3	.120	20°	26°	30°	.025	—	38°	7/8
	17080505	15/32	Inner	9/32	—	—	—	46°	—	26°	.025	—	42°	7/8
	17080506	15/32	Inner	9/32	—	—	—	46°	—	26°	.025	—	42°	7/8
	17080508	15/32	Inner	9/32	—	—	—	46°	—	26°	.025	—	42°	7/8
	17080509	15/32	Outer	11/32	—	—	—	46°	30°	—	.025	—	42°	7/8
	17080510	15/32	Outer	11/32	—	—	.120	46°	30°	—	.025	—	42°	7/8
	17080512	3/8	Inner	9/32	—	—	.120	46°	24°	30°	.025	—	40°	3/4
	17080513	3/8	Inner	9/32	—	—	.120	37°	23°	30°	.025	—	40°	1
	17080515	3/8	Inner	9/32	—	—	.120	37°	23°	30°	.025	—	40°	1
	17080523	15/32	Inner	9/32	—	—	.120	46°	—	23°	.025	—	42°	7/8
	17080524	15/32	Inner	9/32	—	—	.120	46°	—	23°	.025	—	42°	7/8
	17080525	15/32	Inner	9/32	—	—	.120	46°	—	23°	.025	—	42°	7/8
	17080526	15/32	Inner	9/32	—	—	.120	46°	—	23°	.025	—	42°	7/8
	17080527	15/32	Inner	9/32	—	—	.120	46°	—	23°	.025	—	42°	7/8
	17080528	15/32	Inner	9/32	—	—	.120	46°	—	23°	.025	—	42°	7/8
	17080529	3/8	Inner	9/32	—	—	.120	37°	23°	30°	.025	—	40°	1
	17080540	3/8	—	3/8	—	—	.120	14.5	19°	23°	.025	—	38°	9/16
	17080542	3/8	—	3/8	—	—	.120	14.5	19°	23°	.025	—	38°	9/16
	17080543	3/8	—	3/8	—	—	.120	14.5	19°	23°	.025	—	38°	9/16

Continued

QUADRAJET M4M SERIES ADJUSTMENT SPECIFICATIONS—Continued

See Tune Up Chart for hot idle speeds.

Year	Carb. Model	Float Level	Pump Rod		Bowl Vent	Fast Idle (Bench)	Choke Coil Lever	Choke Rod	Vacuum Break		Air-Valve Dash-Pot	Choke Setting	Choke Unloader	Air-Valve Valve Spring Wind-Up
			Hole	Adj.					Front	Rear				
1981	17080212	3/8	Inner	9/32	—	—	.120	46°	24°	30°	.025	—	40°	3/4
	17080213	3/8	Inner	9/32	—	—	.120	37°	23°	30°	.025	—	40°	1
	17080215	3/8	Inner	9/32	—	—	.120	37°	23°	30°	.025	—	40°	1
	17080298	3/8	Inner	9/32	—	—	.120	37°	23°	30°	.025	—	40°	1
	17080507	3/8	Inner	9/32	—	—	.120	37°	23°	30°	.025	—	40°	1
	17080512	3/8	Inner	9/32	—	—	.120	46°	24°	30°	.025	—	40°	3/4
	17080513	3/8	Inner	9/32	—	—	.120	37°	23°	30°	.025	—	40°	3/4
	17081200	15/32	Inner	9/32	—	—	.120	46°	24°	23°	.025	—	42°	7/8
	17081201	15/32	Inner	9/32	—	—	.120	46°	23°	23°	.025	—	42°	7/8
	17081205	15/32	Inner	9/32	—	—	.120	46°	23°	23°	.025	—	42°	7/8
	17081206	15/32	Inner	9/32	—	—	.120	46°	23°	23°	.025	—	42°	7/8
	17081220	15/32	Inner	9/32	—	—	.120	46°	23°	23°	.025	—	42°	7/8
	17081226	15/32	Inner	9/32	—	—	.120	46°	24°	23°	.025	—	42°	7/8
	17081227	15/32	Inner	9/32	—	—	.120	46°	24°	23°	.025	—	42°	7/8
	17081290	13/32	Inner	9/32	—	—	.120	46°	23°	24°	.025	—	42°	7/8
	17081291	13/32	Inner	9/32	—	—	.120	46°	23°	24°	.025	—	42°	7/8
	17081292	13/32	Inner	9/32	—	—	.120	46°	23°	24°	.025	—	42°	7/8
	17081506	13/32	Inner	9/32	—	—	.120	46°	23°	36°	.025	—	36°	7/8
	17081508	13/32	Inner	9/32	—	—	.120	46°	23°	36°	.025	—	36°	7/8
	17081524	13/32	Outer	5/16	—	—	.120	46°	25°	36°	.025	—	38°	7/8
	17081526	13/32	Outer	5/16	—	—	.120	46°	25°	36°	.025	—	38°	7/8
1982	17080212	3/8	Inner	9/32	—	—	.120	46°	24°	30°	.025	—	40°	3/4
	17080213	3/8	Inner	9/32	—	—	.120	37°	23°	30°	.025	—	40°	1
	17080215	3/8	Inner	9/32	—	—	.120	37°	23°	30°	.025	—	40°	1
	17080298	3/8	Inner	9/32	—	—	.120	37°	23°	30°	.025	—	40°	1
	17080507	3/8	Inner	9/32	—	—	.120	37°	23°	30°	.025	—	40°	1
	17080512	3/8	Inner	9/32	—	—	.120	46°	24°	30°	.025	—	40°	3/4
	17080513	3/8	Inner	9/32	—	—	.120	37°	23°	30°	.025	—	40°	3/4
	17082213	3/8	Inner	9/32	—	—	.120	37°	23°	30°	.025	—	40°	1
	17082220	13/32	Inner	9/32	—	—	.120	46°	24°	34°	.025	—	39°	7/8
	17082221	13/32	Inner	9/32	—	—	.120	46°	24°	34°	.025	—	39°	7/8
	17082222	13/32	Inner	9/32	—	—	.120	46°	24°	34°	.025	—	39°	7/8
	17082223	13/32	Inner	9/32	—	—	.120	46°	24°	34°	.025	—	39°	7/8
	17082224	13/32	Inner	9/32	—	—	.120	46°	24°	34°	.025	—	39°	7/8
	17082225	13/32	Inner	9/32	—	—	.120	46°	24°	34°	.025	—	39°	7/8
	17082226	13/32	Inner	9/32	—	—	.120	46°	24°	34°	.025	—	39°	7/8
	17082227	13/32	Inner	9/32	—	—	.120	46°	24°	34°	.025	—	39°	7/8
	17082230	13/32	Inner	9/32	—	—	.120	46°	26°	36°	.025	—	39°	7/8
	17082231	13/32	Inner	9/32	—	—	.120	46°	26°	36°	.025	—	39°	7/8
	17082234	13/32	Inner	9/32	—	—	.120	46°	26°	36°	.025	—	39°	7/8
	17082235	13/32	Inner	9/32	—	—	.120	46°	26°	36°	.025	—	39°	7/8
	17082290	13/32	Inner	9/32	—	—	.120	46°	24°	34°	.025	—	39°	7/8
	17082291	13/32	Inner	9/32	—	—	.120	46°	24°	34°	.025	—	39°	7/8
	17082292	13/32	Inner	9/32	—	—	.120	46°	24°	34°	.025	—	39°	7/8
	17082293	13/32	Inner	9/32	—	—	.120	46°	24°	34°	.025	—	39°	7/8
	17082506	13/32	Inner	9/32	—	—	.120	46°	23°	36°	.025	—	39°	7/8
	17082508	13/32	Inner	9/32	—	—	.120	46°	23°	36°	.025	—	39°	7/8
	17082513	13/32	Inner	9/32	—	—	.120	37°	23°	30°	.025	—	40°	3/4
	17082524	13/32	Outer	5/16	—	—	.120	46°	25°	36°	.025	—	39°	7/8
	17082526	13/32	Outer	5/16	—	—	.120	46°	25°	36°	.025	—	39°	7/8

①—Needle seat with groove at upper edge, 5/16-inch; Needle seat w/o groove at upper edge, 7/16-inch.

②—Below 22500 miles, .157 in.; above 22500 miles, .179 in.

③—Below 22500 miles, .179 in.; above 22500 miles, .203 in.

④—Below 22500 miles, .164 in.; above 22500 miles, .187 in.

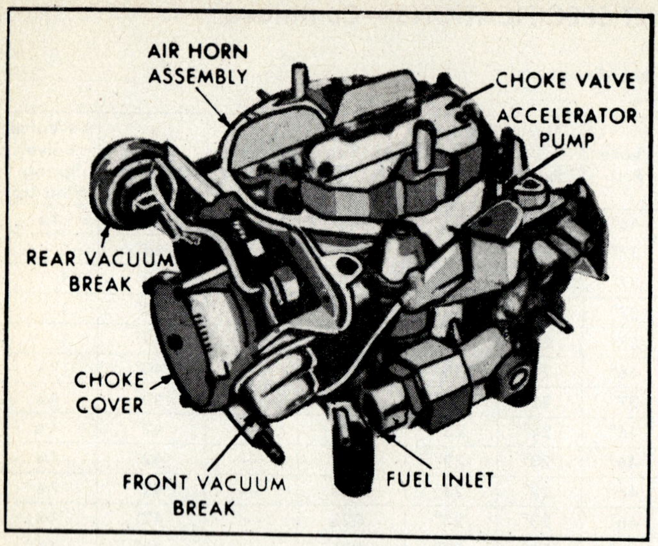

Fig. 1 Quadrajet M4M series carburetor (Typical)

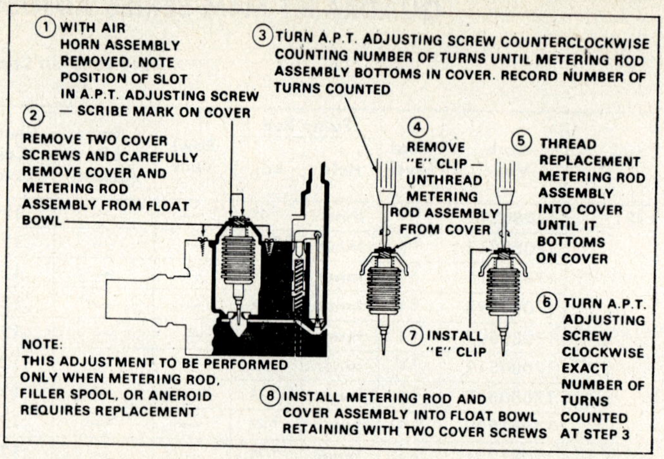

Fig. 3 Quadrajet M4M series A.P.T. metering rod adjustment

MODEL M4M

The M4M Series (M4MC, M4ME) Quadrajet carburetor, Fig. 1 is similar to the Quadrajet 4MV unit. However, some modifications have been made to the basic Quadrajet design.

An Adjustable Part Throttle (A.P.T.) feature incorporates an adjustable metering rod assembly operating in a fixed jet. On some models, a barometric pressure-sensitive aneroid (Bellows) is an integral part of the A.P.T. metering rod assembly. This provides a close tolerance control of fuel flow to the main metering system, thereby controlling air/fuel ratios during part throttle operation.

Some units use a multiple stage power enrichment system with two power pistons. One piston is an auxiliary power piston with a single metering rod operating in a fixed jet. The primary power piston with two metering rods operates in replaceable metering jets. This system provides sensitive control of the air/fuel ratio during light engine power requirements while providing richer mixtures during moderate to heavy engine loads.

All units use a bowl mounted choke housing with a thermostatic coil assembly. Also, a dual vacuum break system is used to improve cold engine warm-up and drive-away performance.

Float Level Adjustment

Fig. 2—With an adjustable T-scale, measure distance from top of float bowl gasket surface, with gasket removed, to top of float at a point 1/16 inch back from toe. Dimension should be as specified in the M4M Series Specification Chart. To adjust, bend float arm.

A.P.T. Metering Rod Adjustment

NOTE: This adjustment is performed only when the metering rod, filler spool or aneroid is replaced.

Fig. 3—Mark position of metering rod assembly adjusting screw on cover, then remove cover screws and lift metering rod assembly from float bowl. Rotating the adjusting screw counter-clockwise, count number of turns required to bottom the threaded metering rod assembly in the cover. Note this number for metering rod adjustment.

Remove "E" clip retainer from rod end, then using a small screwdriver, rotate slotted rod clockwise to disengage rod assembly from cover. Use caution when removing the rod assembly since it is spring loaded.

Install tension spring on replacement rod assembly, then thread the assembly into the cover until bottomed. Rotate the adjusting clockwise exactly the same number of turns from adjusting screw alignment mark noted previously. Install "E" clip in rod assembly groove, ensuring clip is securely locked. Place cover and rod assembly in float bowl, aligning cover assembly tab with float bowl tab nearest the fuel inlet nut, then install cover attaching screws.

Pump Rod Adjustment

Fig. 4—With throttle valves completely closed and pump rod in specified hole, measure distance from top of choke valve wall (next to vent stack) to top of pump stem. Dimension should be as specified in the M4M Series Specification Chart. To adjust, bend pump lever.

Bowl Vent Valve Adjustment

Fig. 5—Remove bowl vent valve cover and gasket. With throttle valves completely closed, rotate curb idle screw inward until it contacts throttle lever stop tang, then an additional 1½ turns. Insert the gauge specified in the M4M Series Specification Chart between curb idle screw and throttle lever stop tang. The bowl vent valve should now be closed, then remove the gauge and the valve should open. To adjust, pry out plug located next to pump plunger shaft, then turn bowl vent valve, actuating arm Allen head adjusting screw.

Fast Idle Adjustment

Fig. 6—With cam follower on highest step of cam, turn fast idle screw out until primary throttle valve is completely closed. Turn screw in to contact lever, then turn screw in the additional turns listed in the M4M Series Specification Chart.

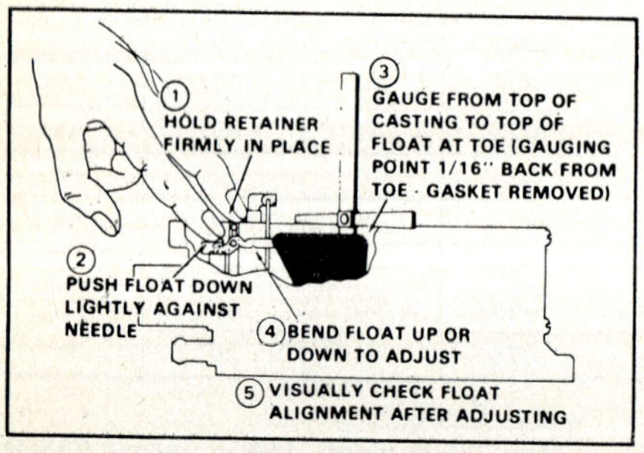

Fig. 2 Quadrajet M4M series float level adjustment

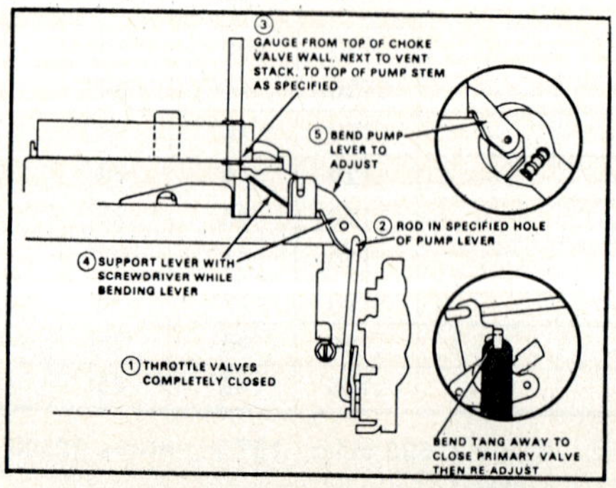

Fig. 4 Quadrajet M4M series pump rod adjustment

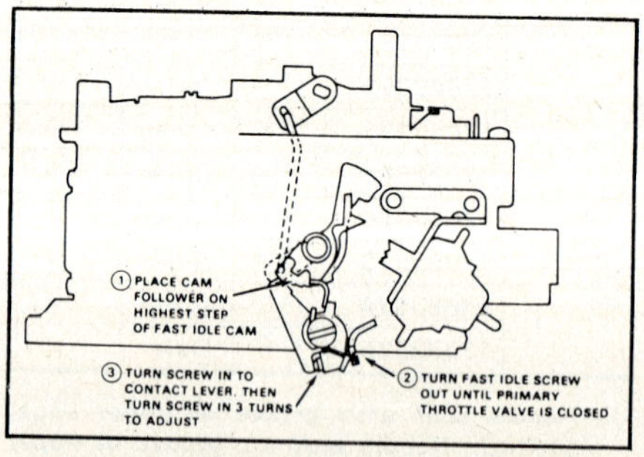

Fig. 6 Quadrajet M4M series fast idle adjustment (Bench)

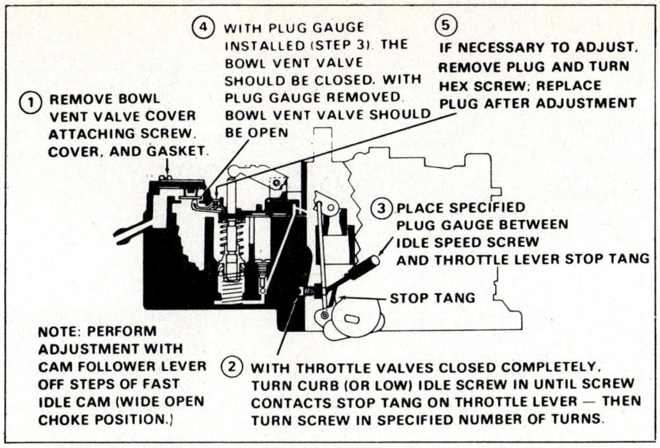

Fig. 5 Quadrajet M4M series bowl vent adjustment

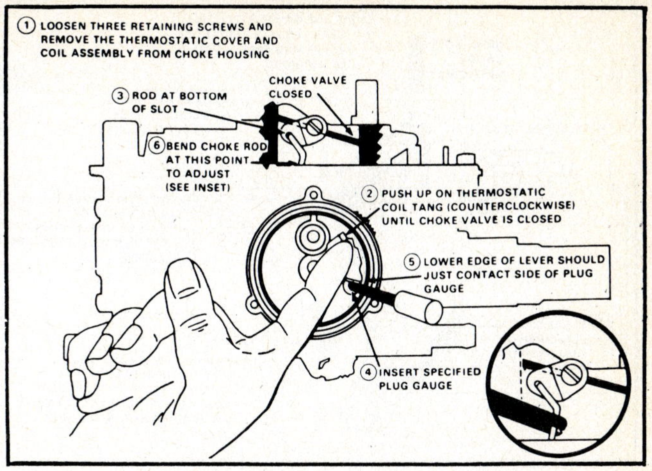

Fig. 7 Quadrajet M4M series choke coil lever adjustment

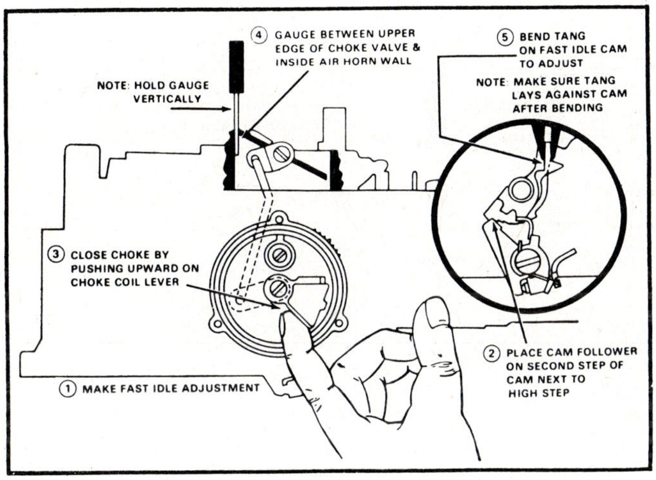

Fig. 8 Quadrajet M4M series choke rod adjustment

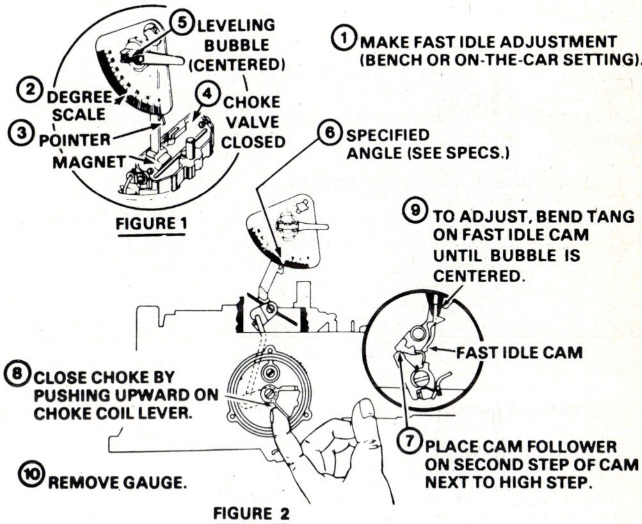

Fig. 9 Choke rod adjustment (angle gauge method). M4M series

Choke Coil Lever Adjustment

Fig. 7—With thermostatic coil assembly removed from choke housing, push upward on coil tang to close choke valve. Insert gauge specified in the M4M Series Specification Chart into hole in choke housing. The choke coil lever should contact the gauge. If not, bend choke rod to adjust.

Choke Rod Adjustment

Plug Gauge Method

Fig. 8—With the fast idle adjustment performed, and the cam follower on the second step of the fast idle cam, next to high step, close choke valve by pushing upward on choke lever. Measure dimension between the upper edge of the choke valve and air horn wall. Dimension should be as specified in the M4M Series Specification Chart. To adjust, bend fast idle cam tang.

Angle Gauge Method

Fig. 9—Rotate degree scale until zero is opposite pointer, then with choke valve completely closed, place magnet on top of choke valve and rotate bubble until it is centered. Refer to specifications and rotate guage to degree specified. Place cam follower on second step of cam next to high step. Close choke by pushing upward on choke coil lever. To adjust bend tang on fast idle cam until bubble is centered, then remove gauge.

Front Vacuum Break Adjustment

Plug Gauge Method

Fig. 10—Place cam follower on highest step of fast idle cam, then using an external vacuum source, seat diaphragm. Push choke coil lever upward until vacuum break lever tang contacts vacuum break plunger stem tang. Measure dimension between upper edge of choke valve and air horn wall. Dimension should be as specified in the M4M Series Specification Chart. To adjust, turn vacuum break plunger adjusting screw.

Angle Gauge Method

Fig. 11—Rotate degree scale until zero is opposite pointer. With choke valve completely closed, place magnet squarely on top of choke valve, then rotate bubble until it is centered. Refer to specifications and rotate scale so that degree specified is opposite pointer. Seat choke vacuum diaphragm using a vacuum source. Hold choke valve towards the closed position by pushing counter clockwise on inside coil lever. To adjust, turn screw in or out until bubble is centered, then remove gauge.

Rear Vacuum Break Adjustment

Fig. 12—Place cam follower on highest step of fast idle cam, then using an external vacuum source, seat diaphragm. Push upward on choke coil lever toward closed choke position until stem is seated. With choke rod in bottom of slot, insert gauge specified in the M4M Series Specification Chart between upper edge of choke valve and air horn wall on all units except those used on V8-454 engines. On V8-454 units, the choke valve should be wide open (fully vertical position). To adjust, bend vacuum break rod.

Air Valve Dashpot Adjustment

Fig. 13—Using an external vacuum source, seat choke vacuum diaphragm. Then, with the air valve completely closed, place gauge

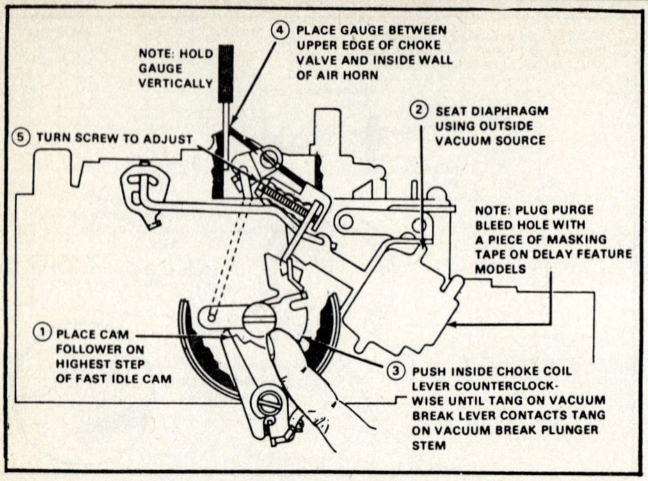

Fig. 10 Quadrajet M4M series front vacuum break adjustment

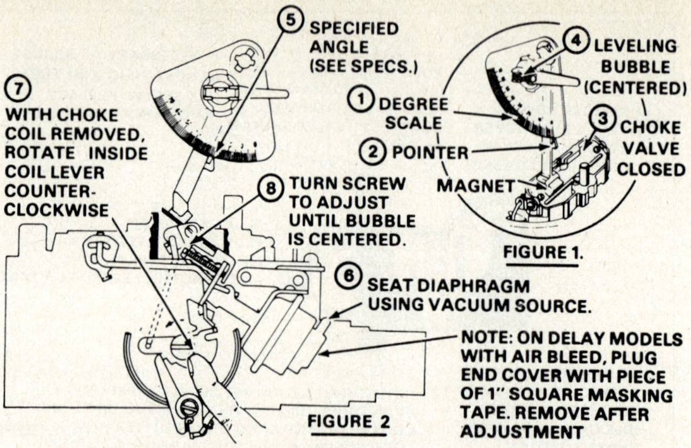

Fig. 11 Front vacuum break adjustment (angle gauge method M4M series

specified in the M4M Series Specification Chart between air valve rod and end of slot in air valve lever. To adjust, bend rod at air valve end.

Choke Coil Adjustment

Fig. 14—Install choke coil and cover assembly. Place fast idle cam follower on highest step of fast idle cam. Rotate coil and cover assembly counter-clockwise until choke just closes, then align index mark on choke cover with the specified index mark on choke housing. Refer to the M4M Series Specification Chart.

Choke Unloader Adjustment

Plug Gauge Method
Fig. 15—With the choke coil adjustment performed, close choke, valve and open throttle valve. To close choke valve on a warm engine, push upward on tang of intermediate choke lever that contacts the fast idle cam and hold in place with the rubber band. Measure dimension between upper edge of choke valve and air horn wall. To adjust, bend tang on fast idle lever.

Angle Gauge Method
Fig. 16—Rotate degree scale until zero is opposite pointer, then with choke valve completely closed, place magnet squarely on top of choke valve. Rotate bubble until it is centered. Refer to specifications and rotate scale until specified degree for adjustment is opposite the pointer. Install choke thermostatic cover and coil assembly in housing. Align index mark with specified point on housing. Hold primary throttle valve wide open. Close choke valve by pushing up on tang on vacuum break lever. To adjust bend tang on fast idle lever until bubble is centered, then remove gauge.

Secondary Throttle Valve Lockout Adjustment

Lockout-Pin Clearance
Fig. 17—With choke and throttle valves fully closed, measure clearance between lockout-pin and lockout lever. Maximum clearance should not exceed .015 inch. To adjust, bend lockout-pin.

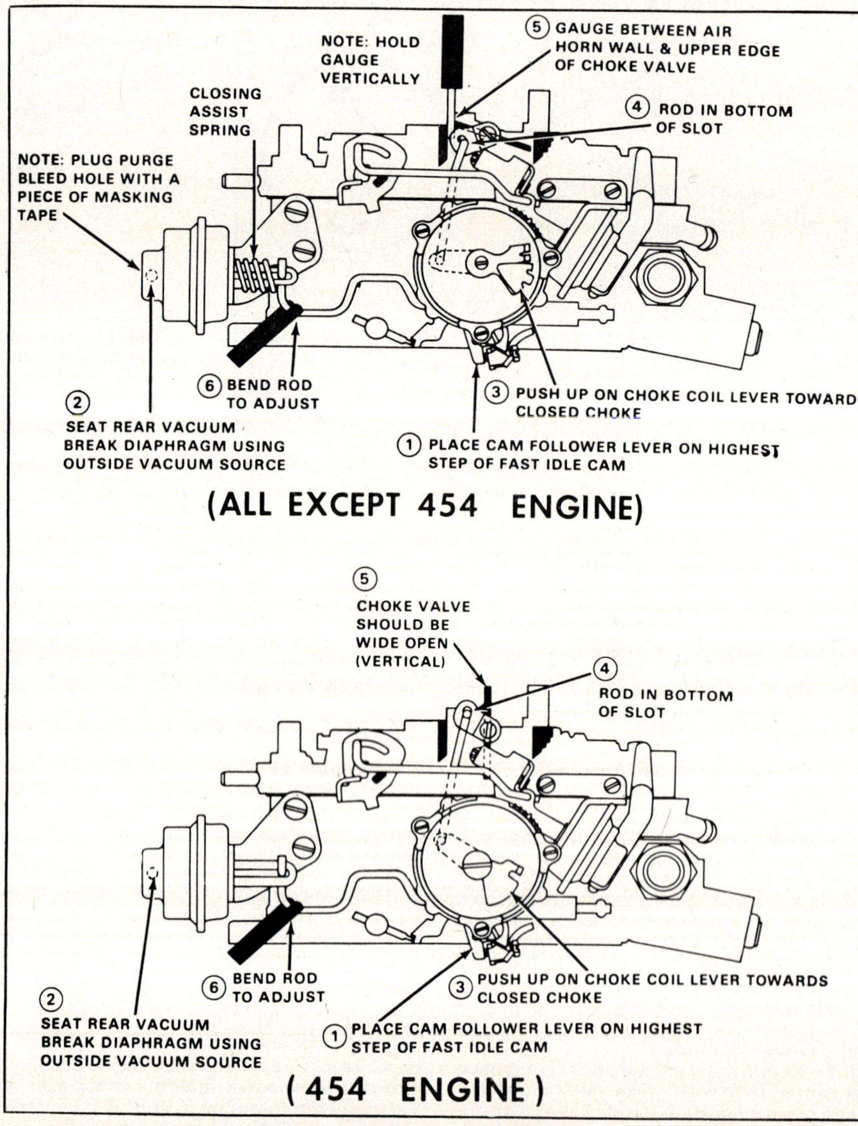

Fig. 12 Quadrajet M4M series rear vacuum break adjustment

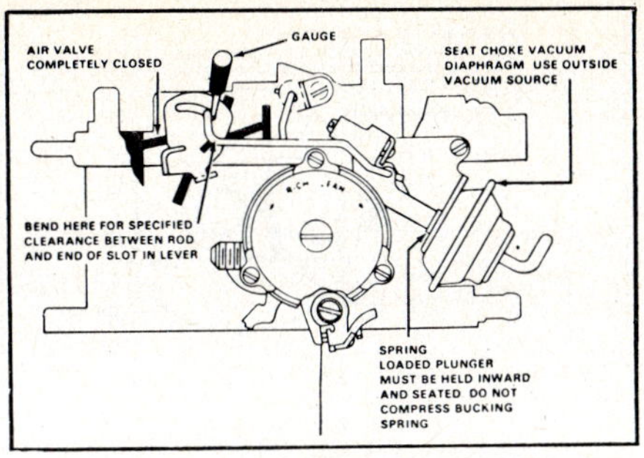

Fig. 13 Quadrajet M4M series air-valve dashpot adjustment

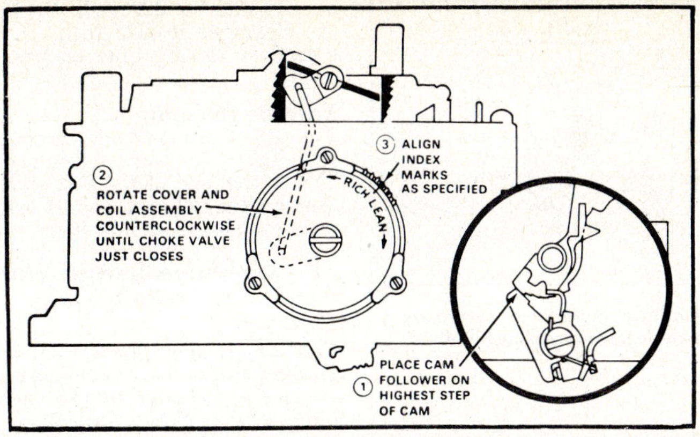

Fig. 14 Quadrajet M4M series automatic choke coil adjustment

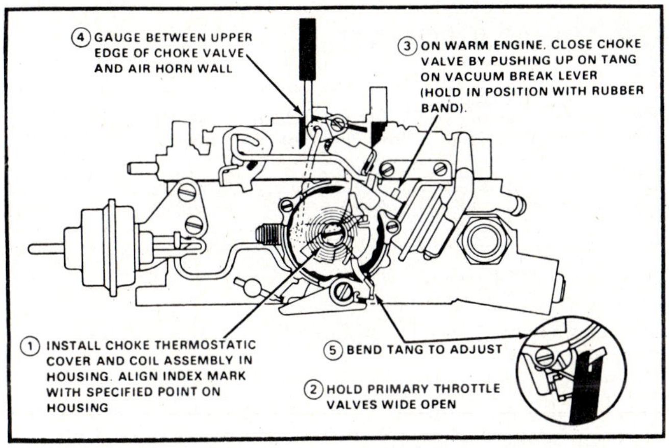

Fig. 15 Quadrajet M4M series choke unloader adjustment

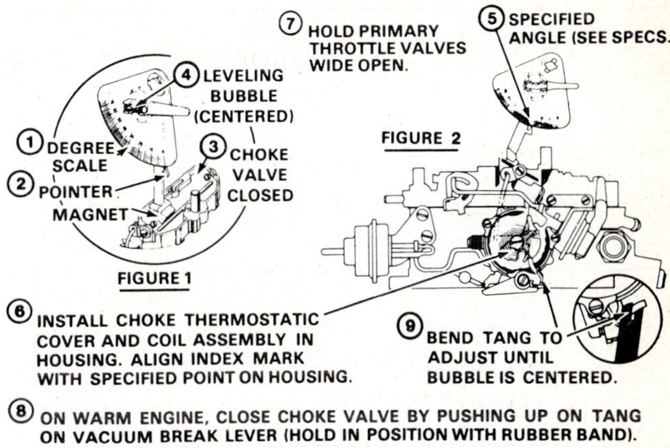

Fig. 16 Unloader adjustment (Angle Gauge Method). M4M series

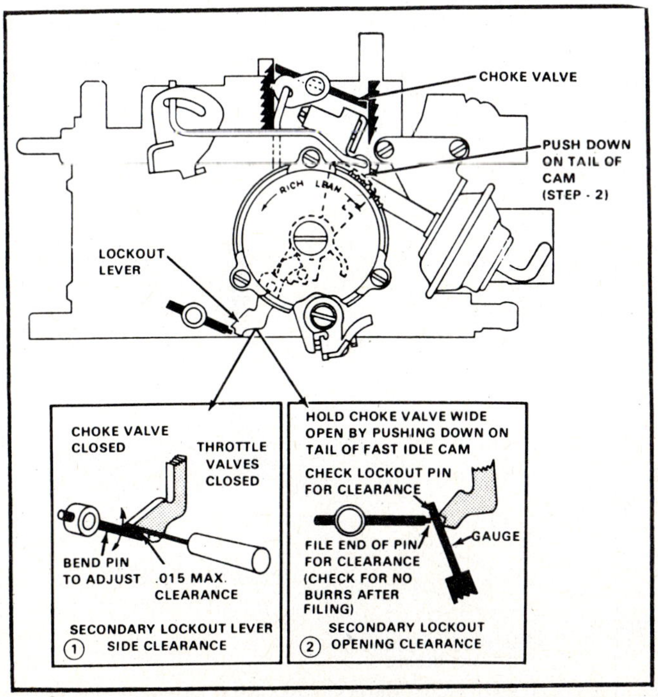

Fig. 17 Quadrajet M4M series secondary throttle valve lockout adjustment

Opening Clearance

Fig. 17—Push downward on fast idle cam tail to open choke valve, then push lock-out lever counter-clockwise so upper end of lever contacts the round pin in fast idle cam. Measure clearance between lockout-pin and lockout lever toe. Clearance should be .015 inch. To obtain power clearance, file metal from end of lockout-pin.

Secondary Throttle Valves Adjustment

Throttle Closing

Fig. 18—With curb idle speed set, hold choke valve open with cam follower lever off fast idle

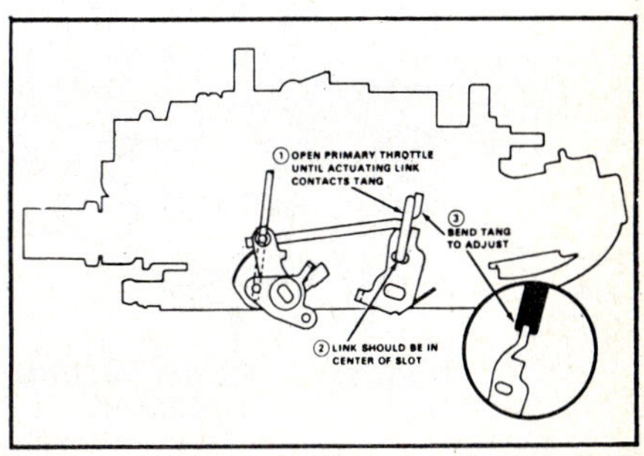

Fig. 18 Quadrajet M4M series secondary throttle valve closing adjustment

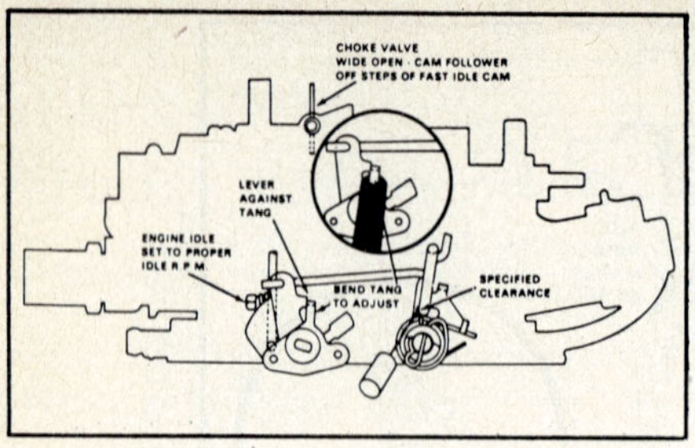

Fig. 19 Quadrajet M4M series secondary throttle valve opening adjustment

cam steps. Measure clearance between forward edge of slot in the secondary throttle valve pick-up lever and the secondary actuating rod. On all units except Pontiac, clearance should be .020 inch. On Pontiac units, the clearance is .025 inch. To adjust, bend secondary closing tang on primary throttle lever.

Throttle Opening

Fig. 19—Open primary throttle lever until link lightly contacts secondary lever tang. The bottom end of the link should be in the center of the slot in secondary lever. To adjust, bend tang on secondary lever.

Air Valve Spring Wind-Up Adjustment

Fig. 20—Remove front vacuum break diaphragm unit and air valve dashpot rod. Loosen lock screw and rotate spring adjusting screw counter-clockwise until air valve is partially open. Hold air valve closed, then rotate adjusting screw clockwise specified number of turns after the spring contacts pin. Refer to the M4M Series Specification Chart. Tighten lock screw and install air valve dashpot rod and the front vacuum break diaphragm unit.

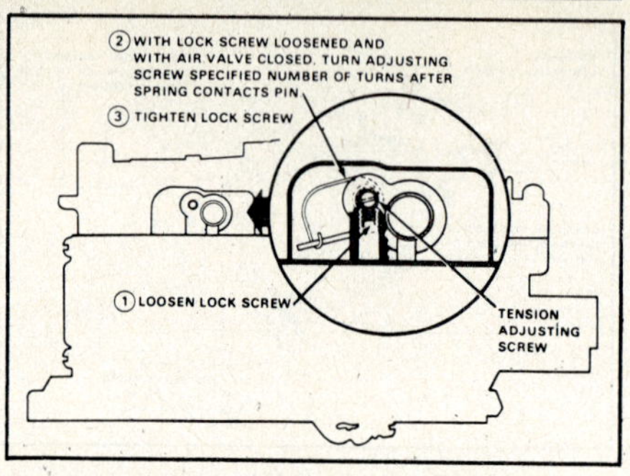

Fig. 20 Quadrajet M4M series air-valve spring wind-up adjustment

QUADRAJET E4M SERIES CARBURETOR ADJUSTMENTS

Year	Carb. Production No.	Float Level	Pump Rod		Fast Idle (Bench) Turns	Choke Coil Lever	Choke Rod	Vacuum Break		Air-Valve Dash Pot	Choke Setting	Choke Unloader	Air-Valve Valve Spring Wind-up
			Hole	Adj.				Front	Rear				
1982	17082202	11/32	—	—	4 1/2	.120	20°	—	—	.015	—	38°	7/8
	17082204	11/32	—	—	4 1/2	.120	20°	—	—	.015	—	38°	7/8

E4M SERIES ADJUSTMENTS

The E4M series (E4MC and E4ME) carburetors are used on vehicles equipped with the C4 (Computer Controlled Catalytic Converter) System or Computer Control Command System, Fig. 1. An electrically operated mixture control solenoid is mounted in the float bowl and is used to control air and fuel metered to idle and main metering systems of the carburetor. Fuel metering is controlled by two special stepped primary metering rods, operating in removable jets and positioned by a plunger in the solenoid which is controlled by an electrical signal from the Electronic Control Module. Air metering to the idle system is controlled by an idle air bleed valve located in the air horn, which follows movement of the mixture control solenoid plunger to control the amount of bleed air into the idle system. A throttle position sensor mounted in the float bowl, is used to electrically signal the Electronic Control module of various throttle position changes.

NOTE: The choke cover is retained to the choke housing by three pop rivets. With float bowl and throttle body properly supported, carefully align a No. 21 drill on pop rivet head. Drill only enough to remove the rivet heads, then using a small hammer and drift, drive remainder of rivet from choke housing. A service kit is available for choke cover installation.

Float Level Adjustment

Fig. 2—With an adjustable T-scale, measure distance from top of float bowl gasket surface, with gasket removed, to top of float at a point 3/16 inch back from toe. Dimension should be as specified in the Specification Chart. To adjust, bend float arm.

NOTE: On units equipped with Computer Controlled Catalytic Converter (C-4) System or Computer Command Control System, adjust float if level varies more than 1/16 inch from specified setting as follows:

a. If float level is too high, hold retainer firmly in place and push downward on

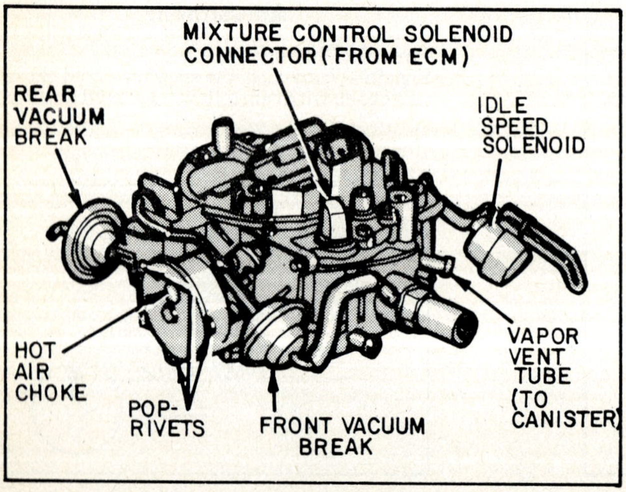

Fig. 1 Quadrajet M4M series carburetor (typical)

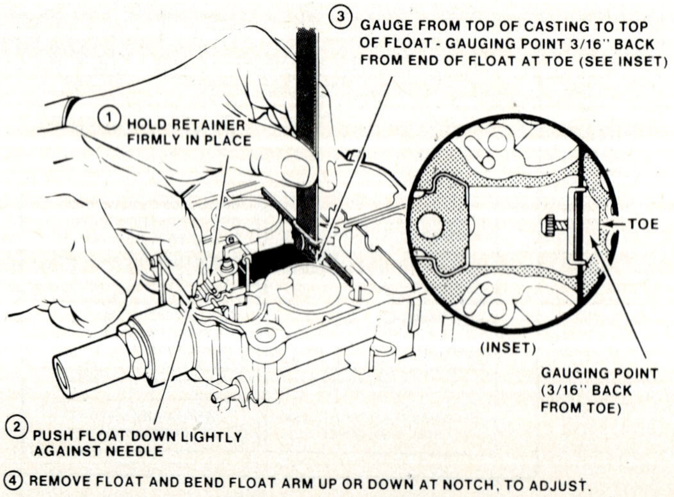

Fig. 2 Float level adjustment

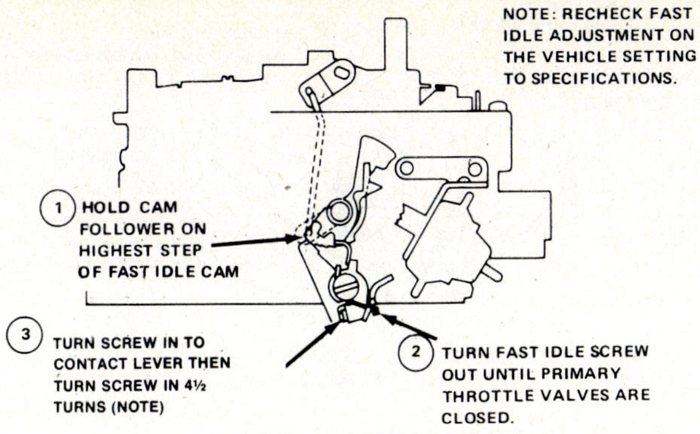

Fig. 3 Fast idle adjustment

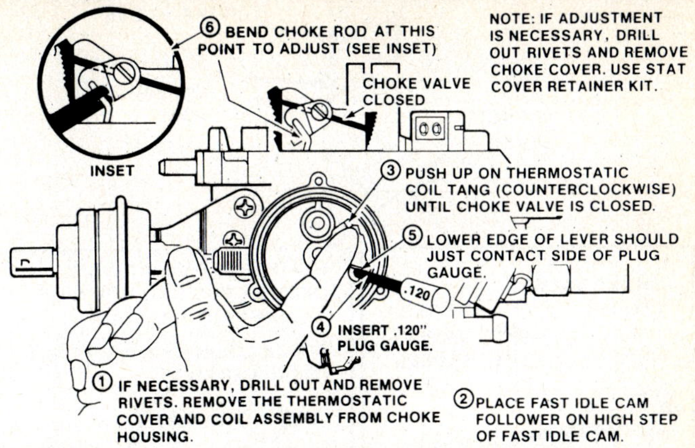

Fig. 4 Choke coil lever adjustment

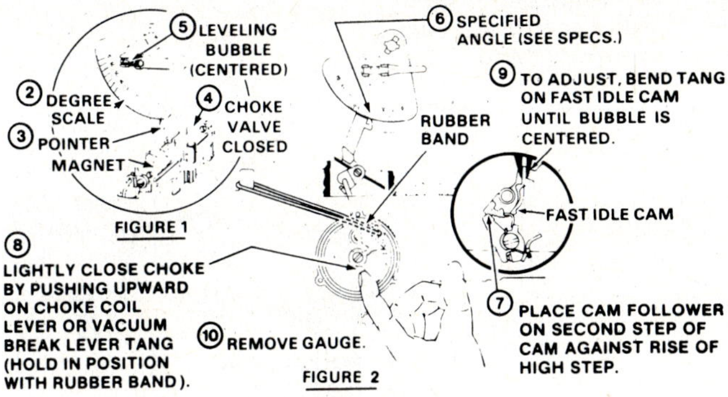

Fig. 5 Choke rod adjustment

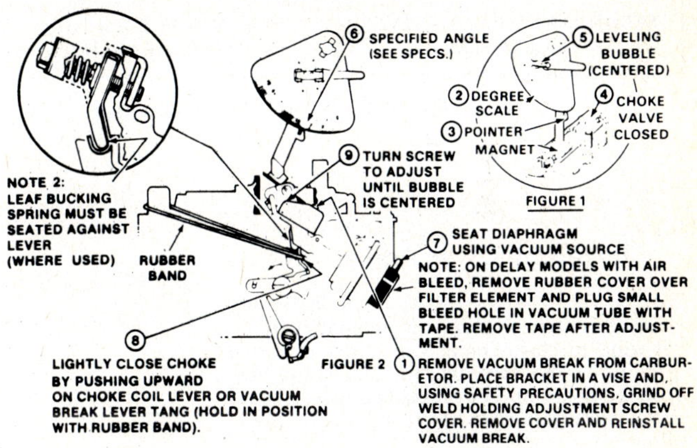

NOTE 1: MAKE CHOKE COIL LEVER ADJUSTMENT AND FAST IDLE ADJUSTMENT (BENCH OR ON-THE-CAR SETTING). DO NOT REMOVE RIVETS AND CHOKE COVER TO PERFORM THIS ADJUSTMENT. USE RUBBER BAND ON VACUUM BREAK LEVER TANG TO HOLD CHOKE VALVE CLOSED (STEP 8).

Fig. 6 Front vacuum break adjustment

center of float pontoon to obtain correct setting.

b. If float level is too low, lift out metering rods, then remove solenoid connector attaching screws. Turn lean mixture solenoid screw clockwise, noting number of turns, until screw is lightly bottomed in float bowl. Turn screw counter-clockwise and remove screw, then lift solenoid and connector from float bowl. Remove float and bend float arm to adjust. After adjusting, check float alignment. Reverse procedure to install components removed. When installing lean mixture screw, ensure that screw is backed out of float bowl the exact number of turns as noted during removal.

Fast Idle Adjustment

Fig. 3—With cam follower on highest step of cam, turn fast idle screw out until primary throttle valve is completely closed. Turn screw in to contact lever, then turn screw in the additional turns listed in the Specification Chart.

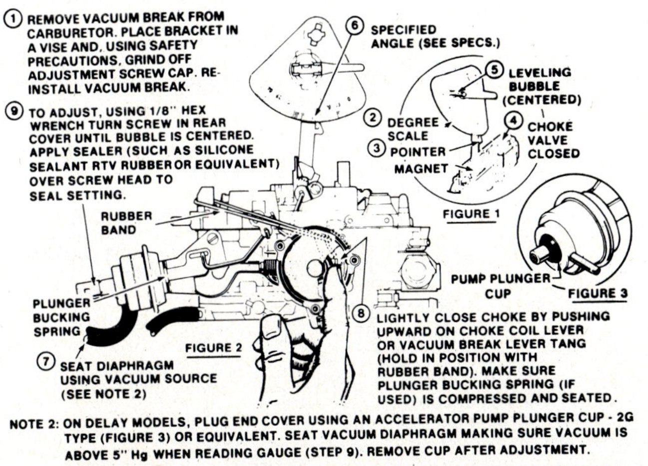

NOTE 2: ON DELAY MODELS, PLUG END COVER USING AN ACCELERATOR PUMP PLUNGER CUP - 2G TYPE (FIGURE 3) OR EQUIVALENT. SEAT VACUUM DIAPHRAGM MAKING SURE VACUUM IS ABOVE 5" Hg WHEN READING GAUGE (STEP 9). REMOVE CUP AFTER ADJUSTMENT.

NOTE 1: MAKE CHOKE COIL LEVER ADJUSTMENT AND FAST IDLE ADJUSTMENT. DO NOT REMOVE RIVETS AND CHOKE COVER TO PERFORM THIS ADJUSTMENT. USE RUBBER BAND ON VACUUM BREAK LEVER TANG TO HOLD CHOKE VALVE CLOSED (STEP 8).

Fig. 7 Rear vacuum break adjustment

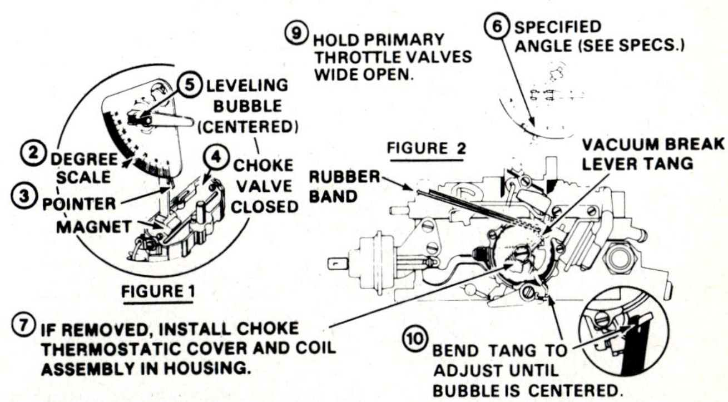

8 CLOSE CHOKE VALVE BY PUSHING UP ON TANG ON VACUUM BREAK LEVER (HOLD IN POSITION WITH RUBBER BAND).

Fig. 8 Choke unloader adjustment

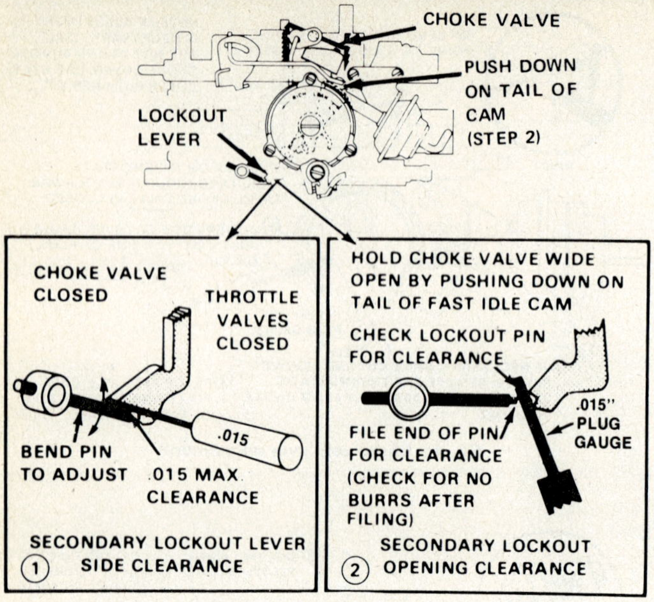

Fig. 9 Secondary throttle valve lockout adjustment

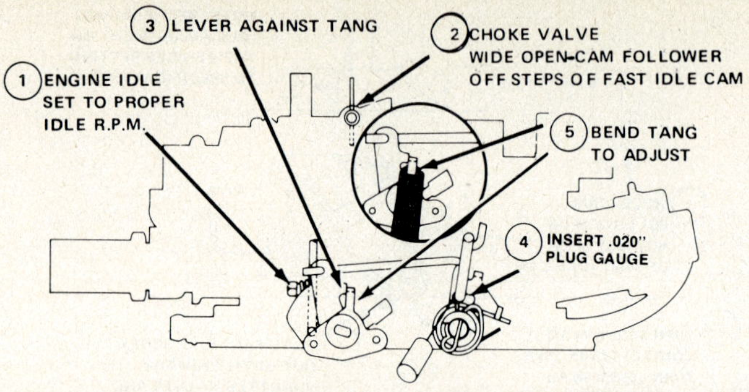

Fig. 10 Secondary throttle valve closing adjustment

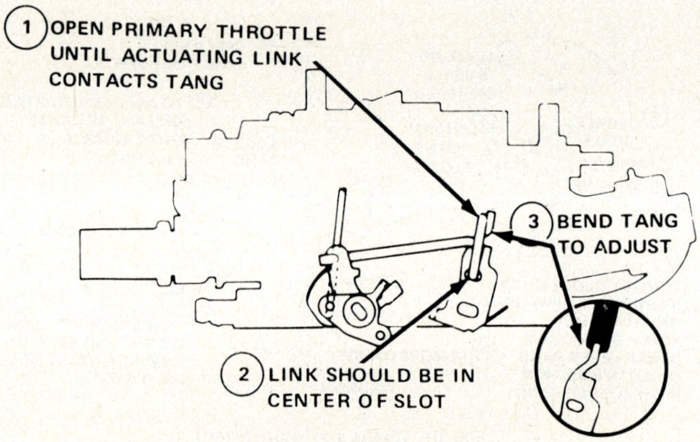

Fig. 11 Secondary throttle valve opening

Choke Coil Lever Adjustment

Fig. 4—With thermostatic coil assembly removed from choke housing, push upward on coil tang to close choke valve. Insert gauge specified in the Specification Chart into hole in choke housing. The choke coil lever should contact the gauge. If not, bend choke rod to adjust.

Choke Rod Adjustment

NOTE: Do not remove rivets and choke cover to perform this adjustment. Use a rubber band on the vacuum break lever tang to hold the choke valve in the closed position.

Fig. 5—Perform fast idle adjustment before making this adjustment. Rotate degree scale until zero is opposite pointer, then with choke valve completely closed, place magnet squarely on top of choke valve and rotate bubble until centered. Refer to Specification Chart for proper setting. Rotate scale so the specified degree for adjustment is opposite pointer. Place cam follower on second step of cam next to high step. Close choke by pushing upward on choke coil lever. To adjust, bend tang on fast idle cam until bubble is centered. Remove gauge.

Front Vacuum Adjustment

NOTE: Do not remove rivets or choke cover to place choke valve in the closed position. Use a rubber band on vacuum break lever tang to hold choke valve in the closed position.

Fig. 6—Rotate degree scale until zero is opposite pointer, then with choke valve completely closed, place magnet squarely on top of choke valve and rotate bubble until centered. Refer to Specification Chart to obtain proper setting for adjustment. Rotate scale so the specified degree for adjustment is opposite pointer. Seat choke diaphragm using an external vacuum source. On some models, it will be necessary to plug air bleed with a piece of masking tape. Hold choke valve toward closed position, pushing counter-clockwise on inside coil lever. To adjust, rotate screw until bubble is centered. Remove gauge.

Rear Vacuum Break Adjustment

Fig. 7—Rotate degree scale until zero is opposite pointer, then with choke valve completely closed, place magnet squarely on top of choke

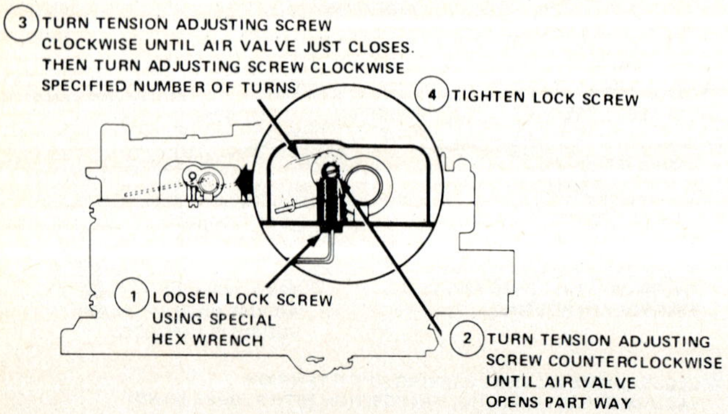

Fig. 12 Air valve spring wind-up adjustment

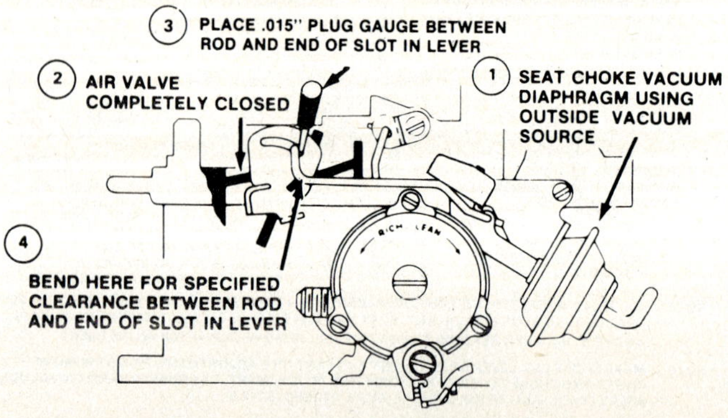

Fig. 13 Air valve rod adjustment

valve and rotate bubble until centered. Refer to Specification Chart for proper setting for adjustment. Rotate scale so specified degree for adjustment is opposite pointer. Seat choke diaphragm using an external vacuum source. On some models, it will be necessary to plug air bleed with a piece of masking tape. Hold choke valve toward closed position, pushing counter-clockwise on inside coil lever. On units equipped with an adjusting screw, using a 1/8 in. hex wrench, turn adjusting screw in rear cover until bubble is centered.

Choke Unloader Adjustment

Fig. 8—Rotate degree scale until zero is opposite pointer, then with choke valve completely closed, place magnet squarely on top of choke valve and rotate bubble until centered. Refer to Specification Chart for proper setting for adjustment. Rotate scale so specified degree for adjustment is opposite pointer. Install choke cover and coil assembly and align index mark with specified point on housing. Hold primary throttle valve wide open and close choke valve by pushing upward on vacuum break lever tang. Hold in position with a rubber band. To adjust, bend tang on fast idle lever until bubble is centered. Remove gauge.

Secondary Throttle Valve Lockout Adjustment

Lockout-Pin Clearance

Fig. 9—With choke and throttle valves fully closed, measure clearance between lockout-pin and lockout lever. Maximum clearance should not exceed .015 inch. To adjust, bend lockout-pin.

Opening Clearance

Fig. 9—Push downward on fast idle cam tail to open choke valve, then push lockout lever counter-clockwise so upper end of lever contacts the round pin in fast idle cam. Measure clearance between lockout-pin and lockout lever toe. Clearance should be .015 inch. To obtain proper clearance, file metal from end of lockout-pin.

Secondary Throttle Valves Adjustment

Throttle Closing

Fig. 10—With curb idle speed set, hold choke valve open with cam follower lever off fast idle cam steps. Measure clearance between forward edge of slot in the secondary throttle valve pick-up lever and the secondary actuating rod. Clearance should be .020 inch. To adjust, bend secondary closing tang on primary throttle lever.

Throttle Opening

Fig. 11—Open primary throttle lever until link lightly contacts secondary lever tang. The bottom end of the link should be in the center of the slot in secondary lever. To adjust, bend tang on secondary lever.

Air Valve Spring Wind-Up Adjustment

Fig. 12—Remove front vacuum break diaphragm unit and air valve dashpot rod. Loosen lock screw and rotate spring adjusting screw counter-clockwise until air valve is partially open. Hold air valve closed, then rotate adjusting screw clockwise specified number of turns after the spring contacts pin. Refer to the Specification Chart. Tighten lock screw and install air valve dashpot rod and the front vacuum break diaphragm unit.

Air Valve Dashpot Adjustment

Fig. 13—Using an external vacuum source, seat choke vacuum diaphragm. Then, with the air valve completely closed, place gauge specified in the Specification Chart between air valve rod and end of slot in air valve lever. To adjust, bend rod at air valve end.

HOLLEY 4150 CARBURETOR ADJUSTMENTS

Year	Carb. Part No. ①	Float Level (Dry)	Float Level (Wet)	Pump Lever Clearance	Bowl Vent Clearance	Fast Idle	
						Bench (R.P.M.)	On Truck (R.P.M.)
1973	R-6510A, 11A	④	③	.015	.060	.035	2100
	R-6521A, 22A	④	③	.015	.060	.035	2100
1974	R-6742A, 43A	④	③	.015	.060	.038	2200
	R-6744A, 45A	④	③	.015	.060	.038	2200
1975–76	R-6928A	⑤	③	.015	—	.038	2200
	R-6929A	⑤	③	.015	—	.038	2200
	R-6930A	⑤	③	.015	—	.031	2200
	R-6931A	⑤	③	.015	—	.031	2200
	R-7264A	⑤	③	.015	—	.038	2200
	R-7266A	⑤	③	.015	—	.038	2200
1977	R-7700A	②	③	.015	—	.038	2200
	R-7701A	②	③	.015	—	.031	2200
	R-7703A	②	③	.015	—	.038	2200
	R-7704A	②	③	.015	—	.031	2200
1978	R-7923A	②	③	.015	.060	.031	2200
	R-7924A	②	③	.015	.060	.031	2200
	R-7927A	②	③	.015	.060	.031	2200
1979	R-8278A	②	③	.015	.060	.031	2200
	R-8279A	②	③	.015	.060	.031	2200
	R-8280A	②	③	.015	—	.031	2200
	R-8281A	②	③	.015	—	.031	2200
	R-8282A	②	③	.015	.060	.031	2200
	R-8283A	②	③	.015	—	.031	2200
	R-8443A	②	③	.015	—	.031	2200
	R-8444A	②	③	.015	.060	.031	2200
1980–81	R8850-A	②	③	.015	—	.031	2200
	R8851-A	②	③	.015	—	.031	2200
	R8854-A	②	③	.015	—	.031	2200
	R8855-A	②	③	.015	—	.031	2200
	R8856-A	②	③	.015	—	.031	2200
	R8857-A	②	③	.015	—	.031	2200

Continued

HOLLEY 4150 CARBURETOR ADJUSTMENTS—Continued

Year	Carb. Part No. ①	Float Level (Dry)	Float Level (Wet)	Pump Lever Clearance	Bowl Vent Clearance	Fast Idle	
						Bench	On Truck (R.P.M.)
1980–82	R8848-A	②	③	.015	—	.031	2200
	R8849-A	②	③	.015	—	.031	2200
	R8852-A	②	③	.015	—	.031	2200
	R8853-A	②	③	.015	—	.031	2200
1982	R-9714	②	③	.015	—	.031	2200
	R-9715	②	③	.015	—	.031	2200
	R-9716	②	③	.015	—	.031	2200
	R-9717	②	③	.015	—	.031	2200
1983	All	②	③	—	—	.031	2200

①—Located on tag attached to carburetor or on casting.
②—Primary .194; secondary .213.
③—Use sight plug in fuel bowl as outlined in text.
④—With bowl inverted, adjust float parallel to bowl floor.
⑤—Primary .197; secondary .166.

HOLLEY 4150

The Holley four barrel carburetor, Fig. 1 is a two stage carburetor consisting of eight sub-assemblies. The sub-assemblies are as follows: throttle body, main body, primary and secondary fuel bowls, primary and secondary metering bodies, secondary throttle operating assembly and the governor assembly. The secondary throttle operating assembly controls the secondary stage throttle plates. A vacuum signal to the spring loaded vacuum diaphragm assembly determines the position of the throttle plates.

The governor incorporated on this carburetor provides a better means of controlling engine speed. The throttle lever controls the engine until the governor speed is reached, at this time the governor adjusts the throttle plates to maintain this speed under the various loadings. A clutch arrangement on the throttle body allows the manual control below governing speeds.

Dry Float Setting, Fig. 2.

1. Invert fuel bowl allowing float to drop to the fully closed position.
2. Using specified gauge pin, check float level setting from lower edge of float bowl to lower edge of float.
3. To adjust float level, turn adjustable needle seat nut. Lock down adjustment with screw.

NOTE: Final adjustment should be made on the vehicle.

Fuel Level Setting

1. Remove air cleaner then remove fuel level sight plugs.
2. With parking brake on and transmission in neutral, start engine and allow it to idle.
3. With car on a level surface, the fuel level should be on a level with the threads at bottom of sight plug port (plus or minus 1/32").
4. If necessary to adjust, loosen inlet needle lock screw and turn adjusting nut clockwise to lower on counterclockwise to raise fuel level, then tighten lock screw.

NOTE: 1/6 turn of adjusting nut equals approximately 1/16" fuel level change.

Secondary Throttle Stop Screw Adjustment, Fig. 3

1. Back off on adjustment screw until throttle plates are fully closed.
2. Turn adjustment screw until it just touches the throttle lever and turn one half more to position the valves.

Vent Valve Adjustment, Fig. 4.

1. Back off idle speed screw until throttle plates are fully closed. (Choke valve open and throttle arm off idle screw).
2. Check clearance between choke valve and seat.
3. Bend air vent valve rod to adjust.
4. Turn idle screw in until contact is made with throttle lever, then turn screw in 1½ additional turns for preliminary idle speed setting.

Fast Idle Cam Adjustment, Fig. 5

1. Adjust fast idle screw to provide specified clearance between screw head and fast idle cam.

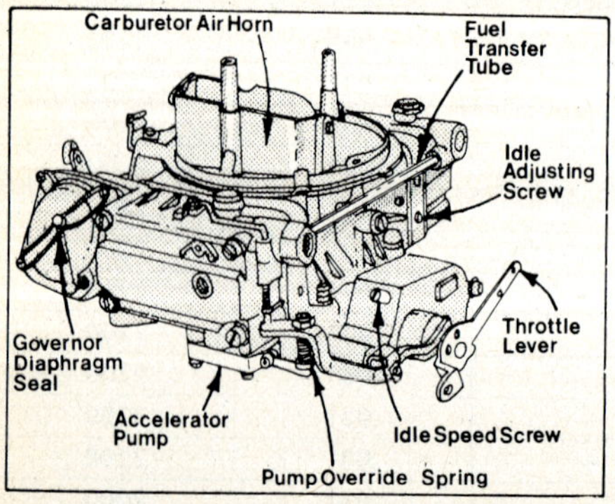

Fig. 1 Holley model 4150G carburetor.

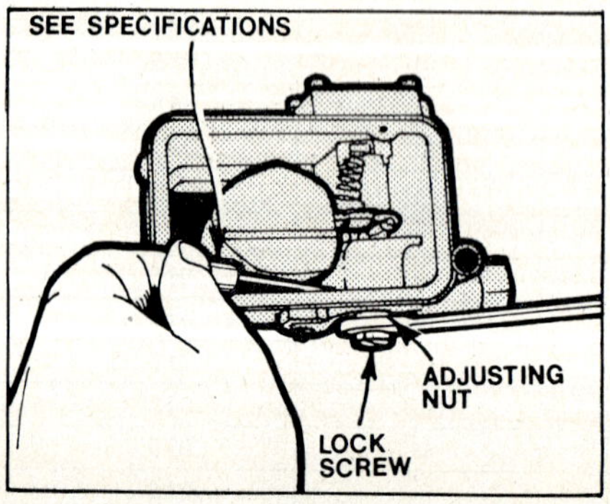

Fig. 2 Float adjustment (dry). Holley 4150

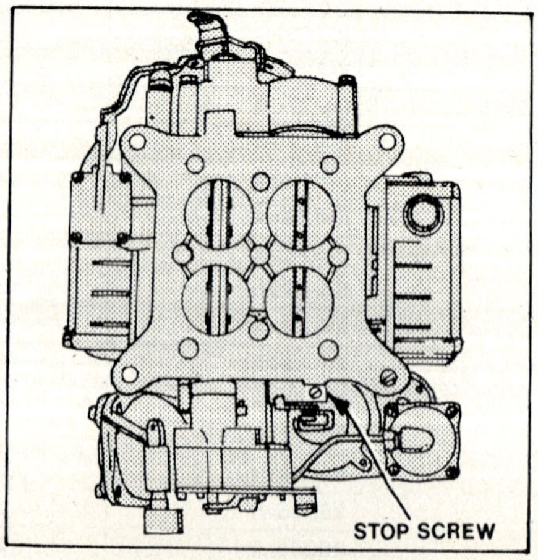

Fig. 3 Secondary throttle stop screw adjustment. Holley 4150

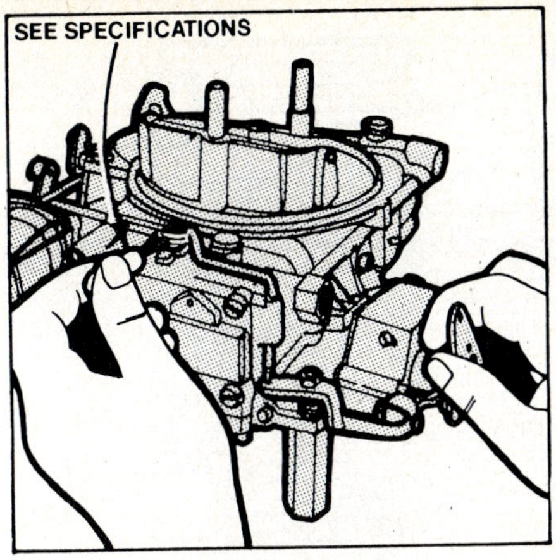

Fig. 4 Vent valve adjustment. Holley 4150

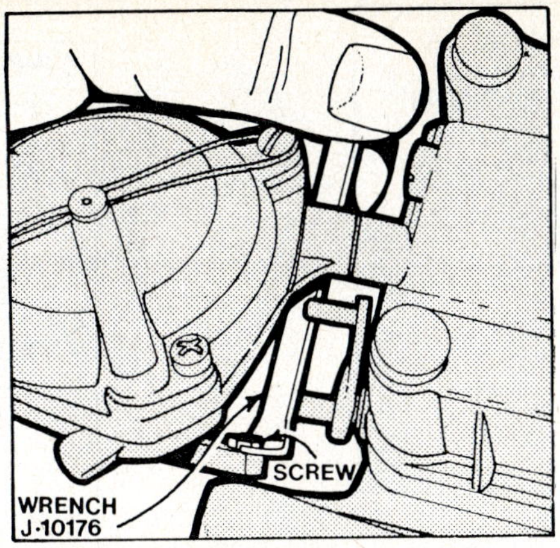

Fig. 5 Fast idle adjustment. Holley 4150

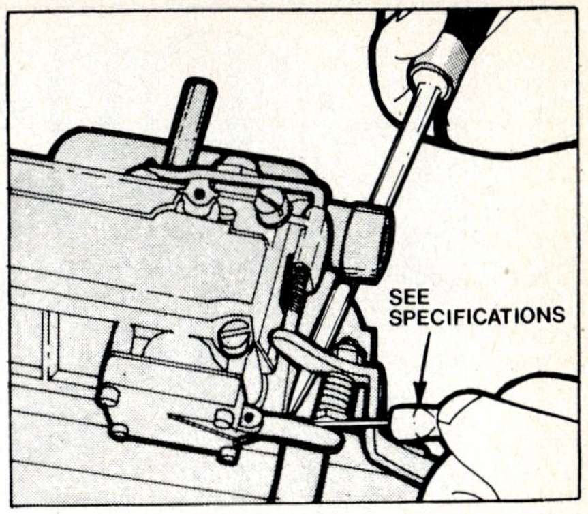

Fig. 6 Accelerator pump adjustment. Holley 4150

Adjust Accelerator Pump, Fig. 6

1. Hold throttle lever in wide open position with a rubber band and hold pump lever fully compressed (down), then measure clearance between spring adjusting nut and arm of pump lever.
2. Clearance should be as specified; adjust by turning nut or screw as required while holding opposite end. (The pump operating lever is not threaded.)
3. After adjustment is made, rotate throttle lever to fully closed and partly open again. Any movement of the throttle lever should be noticed at operating lever spring end, indicating correct pump tip-in.

STROMBERG CARBURETOR ADJUSTMENT SPECIFICATIONS

Year	Engine Application	Carb. Type	Carb. Model	Idle Mix. Screw Turns Open	Hot Idle Speed Eng. R.P.M. Std. Trans.	Hot Idle Speed Eng. R.P.M. Auto. Trans.	Float Level (Dry)	Float Coat	Pump Setting	Idle Vent Setting	Choke Setting
1974	V6-305C, 379	WW	23-264	—	①	①	7/32	—	—	—	—
	V6-432, 478M	WWC	23-265	—	525	525	7/32	—	—	—	—

①—V6-305C, 600 R.P.M.; V6-379, 575 R.P.M.

STROMBERG WW & WWC UNITS, FIGS. 1 & 2

Float Level

Remove carburetor from engine and disassemble, then invert main body to allow float to rest on needle valve and measure distance between top rib of float and top of fuel bowl, Fig. 3.

Pump Adjustment, WW models

Measure distance between top of pump link and air horn with throttle fully closed, Fig. 4. Distance should be .97 inch. Adjust by bending rod at point shown in Fig. 4.

Pump Adjustment, WWC models

Measure total distance at point at top of pump lever travels when throttle lever is moved from closed to wide open position, Fig. 5. When necessary to adjust lever travel, bend pump rod at upper end.

Fast Idle Adjustment, Fig. 6

Check position of cam with choke valve fully open. Fast idle screw must clear cam when throttle is fully closed. If there is interference, bend link to allow fast idle cam to drop below fast idle screw with throttle plate closed and choke open. To adjust fast idle properly, locate end of fast idle screw flush with inner surface of lever, then turn screw IN 7½ turns as shown.

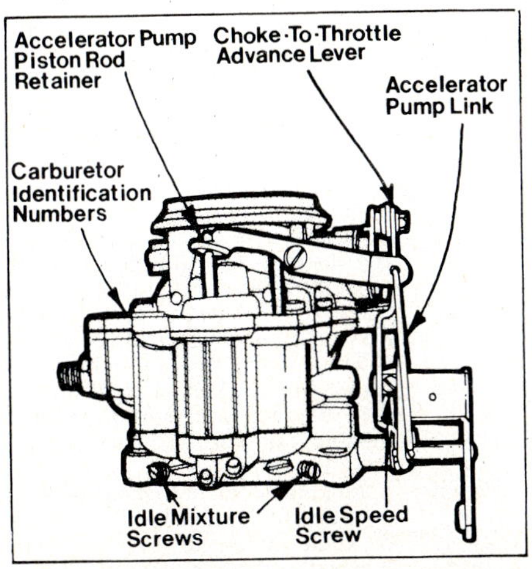

Fig. 1 Stromberg WW Carburetor

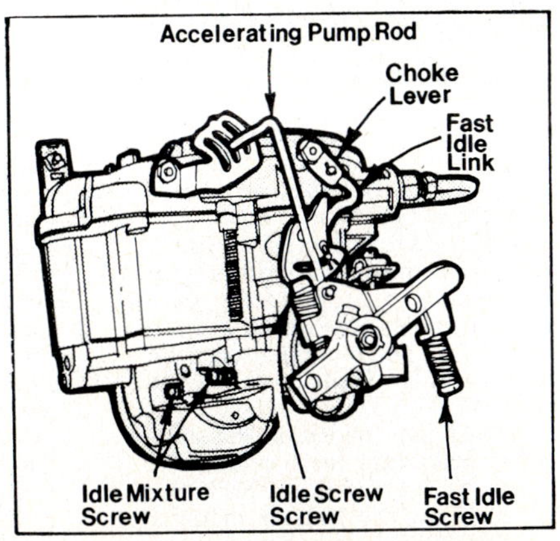

Fig. 2 Stromberg WWC Carburetor

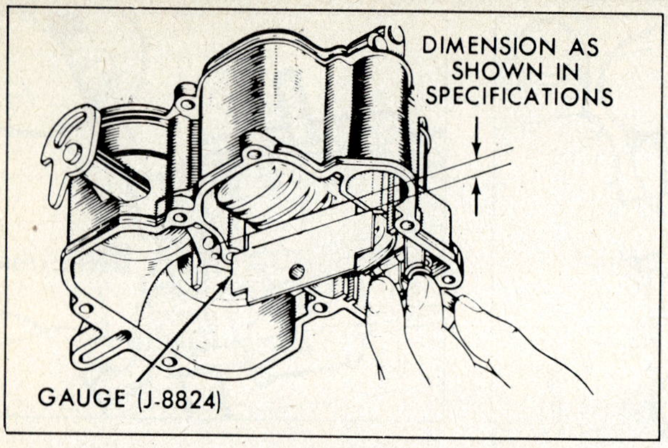

Fig. 3 Float level check. Stromberg

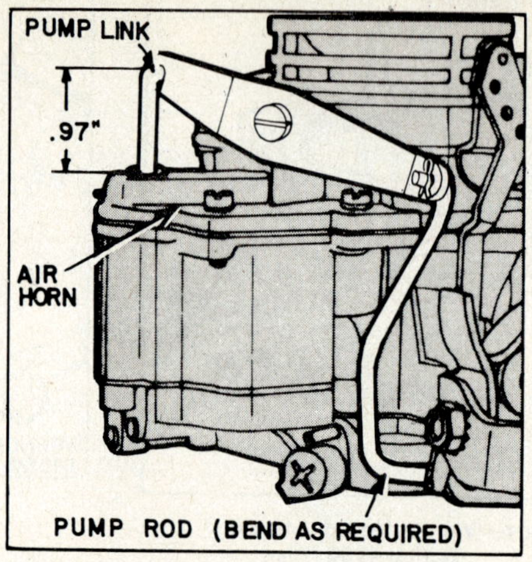

Fig. 4 Pump adjustment. WW Carburetor

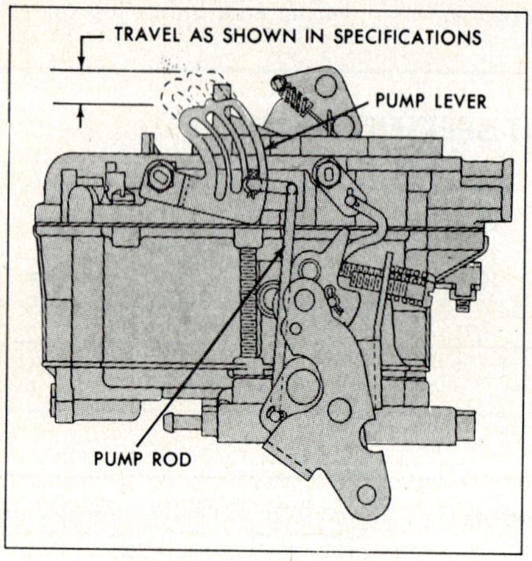

Fig. 5 Pump adjustment. WWC Carburetor

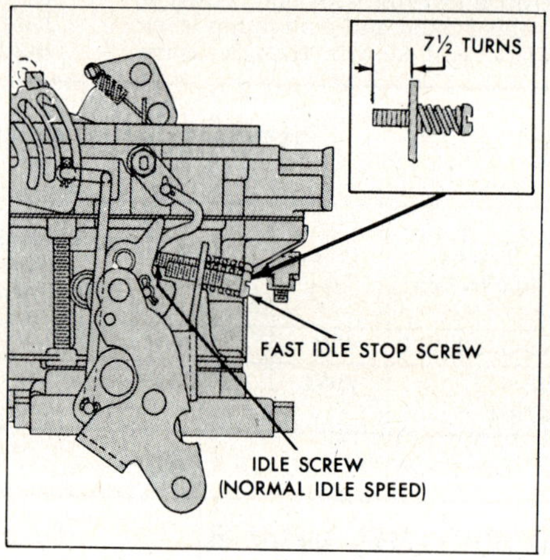

Fig. 6 Fast idle adjustment. WWC Carburetor

Clutch Section

HYDRAULIC CLUTCH CONTROLS

Fig. 1 shows a dual master cylinder, one side for clutch operation and the other for brake. A single clutch master cylinder is used in conjunction with air brakes.

When the clutch pedal is depressed, brake fluid is forced from the master cylinder, Fig. 1, into the slave cylinder, Fig. 2, and the slave cylinder push rod moves the clutch release lever. When the clutch pedal is released, fluid from the slave cylinder feeds back to the master cylinder.

A "bleeder" fitting is provided on the slave cylinder and the procedure for bleeding the system is similar to that used on the brake system. Fluid level in the master cylinder should be maintained ½" below the top of the cylinder.

Maintenance

To maintain proper clutch operation, the system should be checked and serviced periodically as follows:

1. Maintain proper level of hydraulic fluid in master cylinder. At least once a year drain and flush entire clutch system and refill with new fluid.
2. Inspect entire clutch system regularly for fluid leakage. Leakage must be corrected immediately.
3. Make sure return spring at slave cylinder is not weak or broken.
4. Check and adjust if necessary the clutch pedal linkage as outlined below.
5. If the clutch pedal action is springy or spongy, it is an indication that the clutch hydraulic system needs bleeding.

Bleeding System

The need for bleeding air from the system is generally indicated by a springy, spongy pedal action. The presence of air in the system is the result of low fluid level in the main cylinder, or if some part of the system has been opened. A bleeder valve is provided at the slave cylinder alongside of clutch.

The clutch master cylinder is accessible under the hood on conventional and "B" models, inside the cab and under dash on "L" models, and through the access door on "SP" models.

The plain end of bleeder hose can be slipped over end of bleeder valve. And the system can be bled by either the pressure or manual method in the same manner as when bleeding hydraulic brakes.

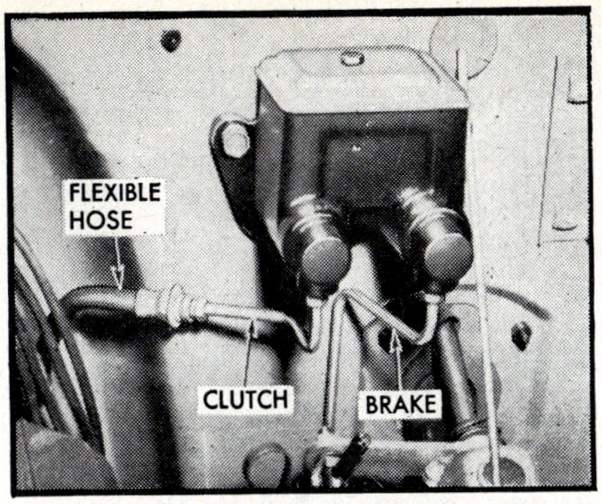

Fig. 1 Dual Master cylinder for clutch and brake operation. (typical)

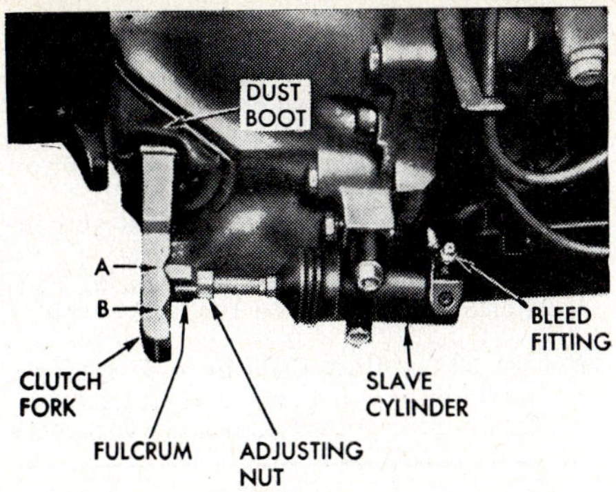

Fig. 2 Slave cylinder and clutch fork. Be sure to note which fulcrum is used ("A" or "B") before disassembly

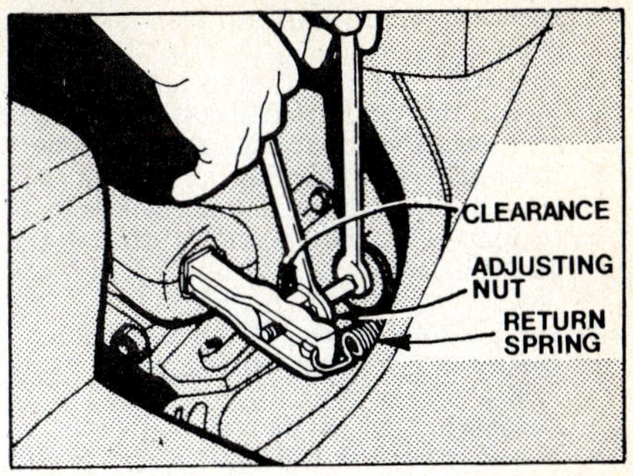

Fig. 3 Adjusting slave cylinder push rod. 1973–77

Clutch Pedal, Adjust

Two adjustments are necessary: pedal total travel, and pedal free travel. The master cylinder must be filled to 1/2" below the top of the reservoir and the system must be free of air.

A slight reduction in pedal free travel at high engine rpm is normal; it is the result of centrifugal action on the clutch release lever weights. Do not change adjustment to obtain a definite free pedal travel dimension.

Slave Cylinder Push Rod, Adjust

1. Disconnect clutch release fork retracting spring at side of clutch. *Do not apply pressure on clutch pedal while spring is disconnected.*
2. Bottom the slave cylinder push rod and piston by pushing the slave cylinder push rod to the front of the vehicle as far as it will go, Fig. 3.
3. Pull clutch fork toward rear of vehicle until the throwout bearing touches the pressure plate release levers.
4. Adjust clearance between slave cylinder push rod fulcrum ("V" block), Fig. 2, by adjusting the nut behind to the clearance given below between fulcrum pivot and clutch fork. This clearance can be checked with the shank of the drill size indicated.

1974 V6305C, 379, 432, 478M	1/2"
1974–79 V8	13/32"
1974–79 L6	11/32"
1980–82 L6-292	13/32"
1980–82 V8-350, 366	5/16"
1980–82 V8-427	1/2"

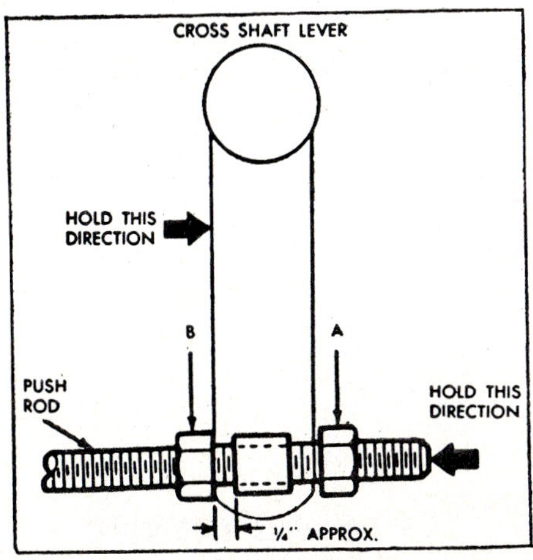

Fig. 4 Clutch pedal free travel adjustment. 1973–82 with typical mechanical linkage

Master Cylinder Push Rod, Adjust

Before making adjustment, pull master cylinder boot back on push rod. Visually check to see that piston is firmly seated against piston stop. If it isn't, push rod is too long and should be shortened.

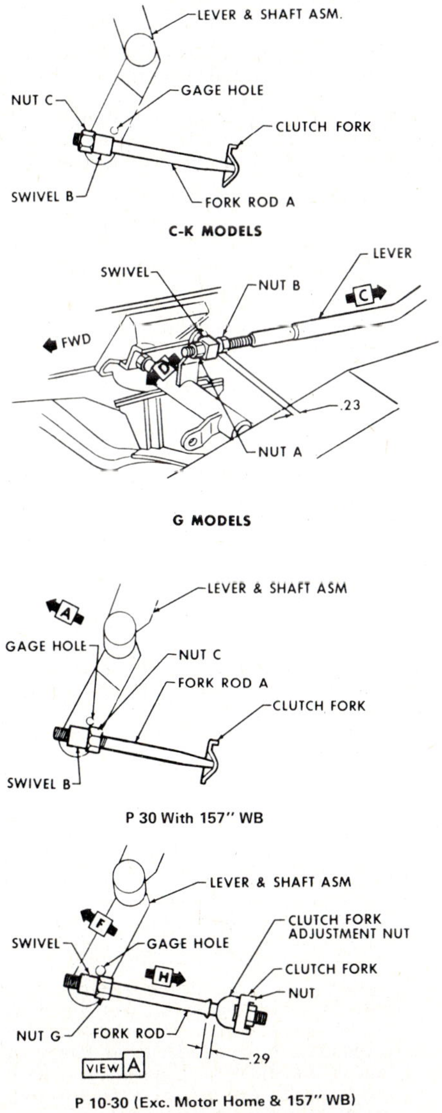

Fig. 5 Clutch pedal adjustment. 1973–82 C-K-P models

Conventional and "B" Models—Loosen lock nut on master cylinder push rod and turn hex portion of push rod as necessary to provide 1/16–1/4 inch free travel before push rod contacts master cylinder piston. Tighten lock nut on push rod.

CLUTCH PEDAL, ADJUST

Mechanical Linkage

Light Duty Models

G Models

1. Disconnect clutch fork return spring at fork.
2. Loosen nut "A", Fig. 4, and back off from swivel about 1/2".
3. Hold clutch fork push rod against fork to move clutch release bearing against clutch fingers (push rod will slide through swivel at cross-shaft).
4. Adjust nut "B" to obtain 3/16 to 1/4" clearance between nut "B" and swivel.
5. Release push rod, connect return spring and tighten nut "A" to lock swivel against nut "B."
6. Check free pedal travel which should be 1 1/4 to 1 1/2 inch.

C, K and P Models Except 157" Wheel Base P30/3500 & Motor Home, Fig. 5.

1. Disconnect return spring at clutch fork.
2. Position clutch lever firmly against rubber bumper on brake pedal bracket.
3. Push outer end of fork rearward until release bearing lightly contacts pressure plate.
4. Loosen locknut and adjust rod length so that swivel slips freely into gauge hole.

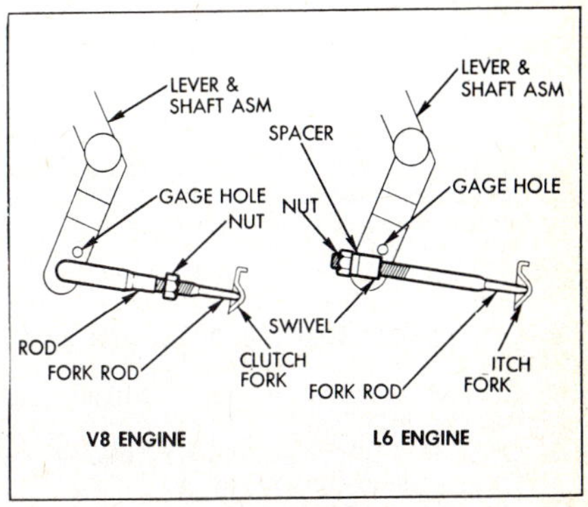

Fig. 6 Clutch pedal adjustment. 1974–81 Caballero, El Camino & Sprint

Increase length until all lash is removed from linkage.
5. Reinstall rod into lower hole on lever. Install two washers and cotter pin. Tighten locknut without changing rod length.
6. Reinstall return spring and check pedal travel. Pedal travel should be 1⅜ to 1⅝ inch on C-K models; and 1¼ to 1½ inch on P models.

157 Inch Wheel Base P30/3500, Fig. 5.

1. Disconnect return spring from clutch fork.
2. Loosen nut "G" at swivel.
3. Move clutch fork rod against fork to eliminate play between release bearing and clutch.
4. Position clutch lever against rubber stop on brake pedal bracket.
5. Rotate fork rod until a clearance of ¼ to ⅝ inch is obtained between shoulder on fork rod and the adjustment nut.
6. Tighten nut "G" against swivel and install return spring.
7. Check free pedal clearance at pedal. Pedal clearance should be 1⅜ to 1⅝ inch.

1974–81 El Camino, Caballero & Sprint

1. Disconnect return spring at clutch fork.
2. Rotate clutch lever and shaft assembly until pedal is against rubber bumper on dash brace.
3. Push outer end of clutch fork rearward until throwout bearing lightly contacts pressure plate fingers.
4. Install push rod in gauge hole and increase length until all lash is removed, Fig. 6.
5. Remove swivel or rod from gauge hole and insert into lower hole on lever. Install retainer and tighten lock nut using care not to change rod length.
6. Install return spring and check clutch pedal free travel. Free travel should be ¾ to 1 5/16 in.

Manual Shift Transmission Section

NOTE: For Clark, Fuller, New Process, Spicer and Warner transmissions used on Chevrolet and GMC trucks, see the "name" chapter for service procedures and lubricant capacities.

TRANSMISSION, REPLACE

3-Speed

1. Raise and support vehicle, drain transmission lubricant.
2. Remove driveshaft. On models equipped with two piece driveshaft, the center bearing retaining bolts must be removed before removing driveshaft.
3. Disconnect speedometer cable, then transmission shift linkage.
4. On some models it may be necessary to disconnect the parking brake lever and controls.
5. Disconnect back up switch wiring, and any other cable or wiring.
6. Position a suitable jack under transmission, slightly raise transmission and remove crossmember.
7. Place a suitable support under engine and lower transmission slightly to equalize load between transmission jack and engine support.
8. Remove transmission to clutch housing retaining bolts. Wire clutch throwout bearing lever in position to prevent the throwout bearing from falling off the lever.
9. Pull transmission assembly straight back out of the clutch housing.

NOTE: Do not allow the transmission main shaft to hang on the clutch driven disc as serious damage may occur to the clutch assembly.

10. Lower transmission to floor.
11. To reinstall transmission, lightly lubricate the splined portion of the transmission mainshaft. To ease installation, install two transmission guide pins in the clutch housing temporarily, then reverse procedure to install.

4 & 5 Speed Light & Medium Duty

1. Remove access cover from cab floor, then remove shift lever.
2. On C.O.E. models, remove bellows and transmission control tower.
3. Disconnect speedometer cable at rear of transmission case.
4. Position a suitable transmission jack under transmission and secure transmission to jack, slightly raise transmission and remove rear mounts or crossmember (if equipped).
5. Remove driveshaft. On models equipped with two piece driveshaft, the center bearing retaining bolts must be removed first; then support with a suitable jack.
6. On conventional models, remove hand brake lever and linkage at transmission.
7. Disconnect transmission strut rod (if used).
8. Remove clutch housing underpan.
9. Remove the two top transmission mounting bolts and replace with guide pins.
10. Remove lower mounting bolts.
11. With jack under transmission, move unit straight back and out.
12. Install in reverse order.

All Heavy Duty

The procedure to remove the transmission from the above model trucks is dependent generally upon the type of cab, type of body, and lifting equipment available in the shop. Operations other than those included in this procedure may be necessary if the vehicle has special equipment such as power take-off and controls.

The required operations will be obvious upon visual inspection of the vehicle. Therefore, the following instructions should be used as a guide in replacing a transmission.

1. Inside cab, remove seals and/or cover plates from floor above transmission.
2. If gearshift lever tower is removable, remove it. Some C.O.E. models have the gearshift lever clamped to a stub lever at the transmission cover. In such instances gearshift lever may be removed from stub lever instead of removing the complete tower assembly.
3. Remove the hand brake lever and bracket when those items are mounted on the transmission cover. On some models it is necessary to detach the auxiliary transmission control lever bracket which is mounted on the main transmission.
4. Disconnect propeller shaft at flange at rear of transmission. If a separate auxiliary transmission is used, it may be necessary to remove the complete propeller shaft between the transmission and auxiliary unit to provide sufficient clearance to move transmission away from engine. On long wheelbase trucks with propeller shaft center bearing, remove center bearing bolts to allow movement of propeller shafts to provide clearance.
5. Disconnect clutch control linkage and, except when an auxiliary transmission is used, disconnect speedometer cable housing from connector on mainshaft rear bearing cap. Disconnect engine ground strap if attached to transmission case or clutch housing.
6. Disconnect strut rod from rear of transmission.
7. Drain lubricant from transmission and if power take-off is used, remove unit and controls.
8. Inspect to determine if other equipment or brackets must be removed to permit removal of transmission. On some trucks the exhaust pipe or support brackets must be removed.
9. *On trucks equipped with Fuller transmissions, the air pressure must be exhausted from the air system and the air line must be disconnected from the pressure regulator valve before transmission can be removed. Also the feed wire to the circuit switch must be disconnected.*
10. Place transmission dolly or jack under transmission and adjust to carry weight of transmission. Then remove the rear support (if used). If there is sufficient clearance, the rear support may remain on transmission during removal. However, it will be necessary to disconnect support from brackets at frame.
11. Remove bolts which attach clutch housing to flywheel housing and move transmission straight back from the engine, using care to keep the main drive gear shaft in alignment with the clutch disc hub.
12. If the transmission is to be disassembled, remove the clutch release mechanism and remove the hand brake mechanism and rear support members if these parts are installed at the rear of the transmission.

1976–83 TREMEC 3 SPEED

For service procedures, refer to the Ford

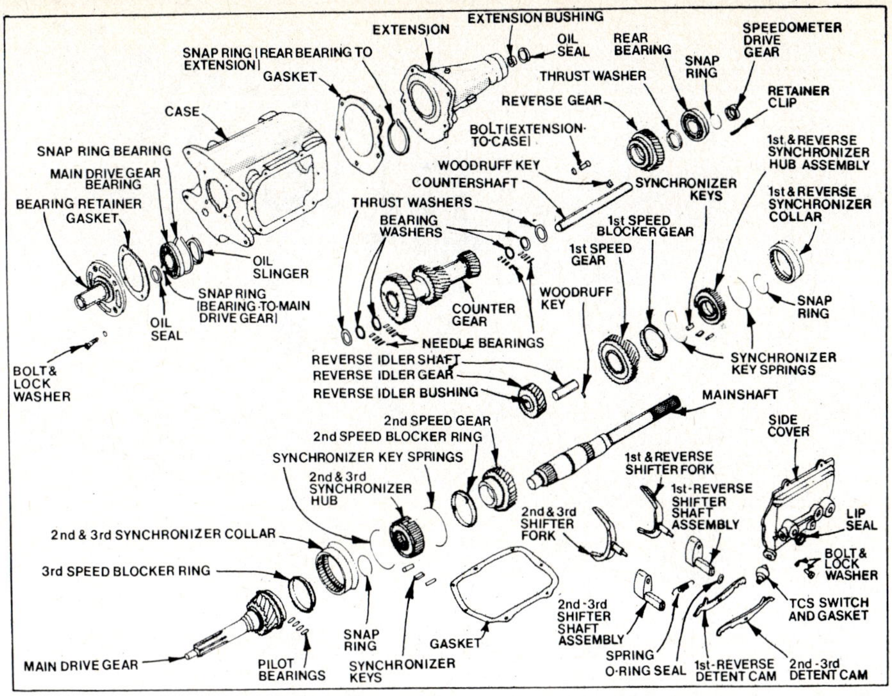

Fig. 1 1974–83 Muncie three-speed transmission. Exploded

Chapter, Transmission Section under "3 SPEED TRANS., 1974–83".

1982–83 89mm NEW PROCESS 4 SPEED

For service procedures refer to the "Dodge & Plymouth Truck Chapter, Manual Shift Transmission Section".

1974–83 MUNCIE 3 SPEED

Disassemble, Fig. 1

1. Remove side cover.
2. Remove clutch gear retainer and gasket.
3. Remove clutch gear bearing-to-gear stem snap ring, then remove bearing by pulling outward on clutch gear until a screwdriver can be inserted between large snap ring and case to complete removal. The bearing is slip fit on the gear and into the case bore.
4. Remove extension to case attaching bolts.
5. Rotate extension to left until groove in extension housing flange lines up with reverse idler shaft. Using a drift, drive reverse idler shaft out of gear and case.
6. Remove clutch gear, mainshaft and extension assembly together through rear case opening. Remove reverse idler gear from case.
7. Remove clutch gear from mainshaft.
8. Expand snap ring in extension housing, tap on end of mainshaft and remove extension from mainshaft.

Assemble

1. Load double row of roller bearings and thrust washers into each end of countergear. Use heavy grease to hold them in place.
2. Place countergear in case and install countergear shaft and woodruff key from rear of case.

NOTE: Be sure countershaft picks up both thrust washers and that the tangs are aligned in the notches in the case.

3. Position reverse idler in case.
4. Expand snap ring in extension housing and assemble mainshaft and extension.
5. Load rollers into clutch gear bore and assemble 3rd gear blocker ring onto gear with clutching teeth toward gear.
6. Install clutch gear assembly onto mainshaft. Do not install bearing over gear stem as yet.

NOTE: Be sure notches in blocker ring align with keys in the synchronizer assembly.

7. Place gasket on rear of case, and assemble the entire mainshaft-extension to the case as a unit.

NOTE: Be sure clutch gear engages the teeth of the countergear anti-lash plate and that the oil slinger is in place on the clutch gear.

8. Rotate extension and install reverse idler shaft and Woodruff key. Install extension-to-case bolts.
9. With large snap ring on bearing, install clutch gear bearing onto stem of gear and into case bore. Install small snap ring.
10. Install clutch gear bearing retainer with oil return hole at bottom.
11. Install side cover.

SAGINAW 3 SPEED

This transmission, Fig. 2, is fully synchronized in that all forward speed changes are accomplished with synchronizer sleeves. The synchronizers permit quicker shifts, greatly reduce gear clash, and permit down shifting from third to second between 40 and 20 mph, and from second to first below 20 mph.

Disassemble

1. Drain lubricant.
2. Remove side cover and gasket.
3. Remove front bearing retainer.
4. Remove front bearing snap ring, pull main drive gear out of case as far as possible and remove bearing.
5. Remove rear bearing retainer-to-case bolts.
6. Remove reverse idler shaft-to-gear "E" ring, Fig. 3.
7. From front of case, remove rear bearing retainer and mainshaft assembly.
8. Remove main drive gear, 14 needle rollers and 3rd speed gear blocking ring from mainshaft.
9. Expand snap ring and remove rear bearing retainer, Fig. 4.
10. Remove countershaft through rear of case, Fig. 5, and remove two tanged thrust washers.
11. Use a long brass drift or punch and drive reverse idler shaft and key through rear of case.
12. Remove reverse idler gear tanged thrust washer.

Disassemble Mainshaft

1. Referring to Fig. 6, remove 2–3 synchronizer sleeve.
2. Remove rear bearing snap ring.
3. Using a ram or arbor press, remove rear bearing, spring washer, thrust washer and reverse gear.
4. Remove speedometer drive gear.
5. Remove first speed synchronizer snap ring. Support first gear on a press and remove first gear and its synchronizer.
6. Remove 2–3 synchronizer snap ring and press off synchronizer and second speed gear.

Synchronizers

NOTE: The synchronizer hubs and sliding sleeves are a selected assembly and should be kept together as originally assembled. The keys and springs may be replaced if worn or broken.

1. Mark hub and sleeve so they can be reassembled in same position.
2. Remove sleeve from hub.
3. Remove keys and springs from hub.
4. Place three keys and two springs in position (one on each side of hub) so all three keys are engaged by both springs, Fig. 7. The tanged end of each synchronizer spring should be installed in different key cavities on either side of hub. Slide sleeve onto hub, aligning marks made before disassembly.

NOTE: A chamfer around outside of synchronizer hub splines identifies end that must be opposite shift fork slot in sleeve, Fig. 8.

Mainshaft, Assembly

1. Install second speed and synchronizer on mainshaft. Using a ram or arbor press, press 2–3 synchronizer (with chamfer

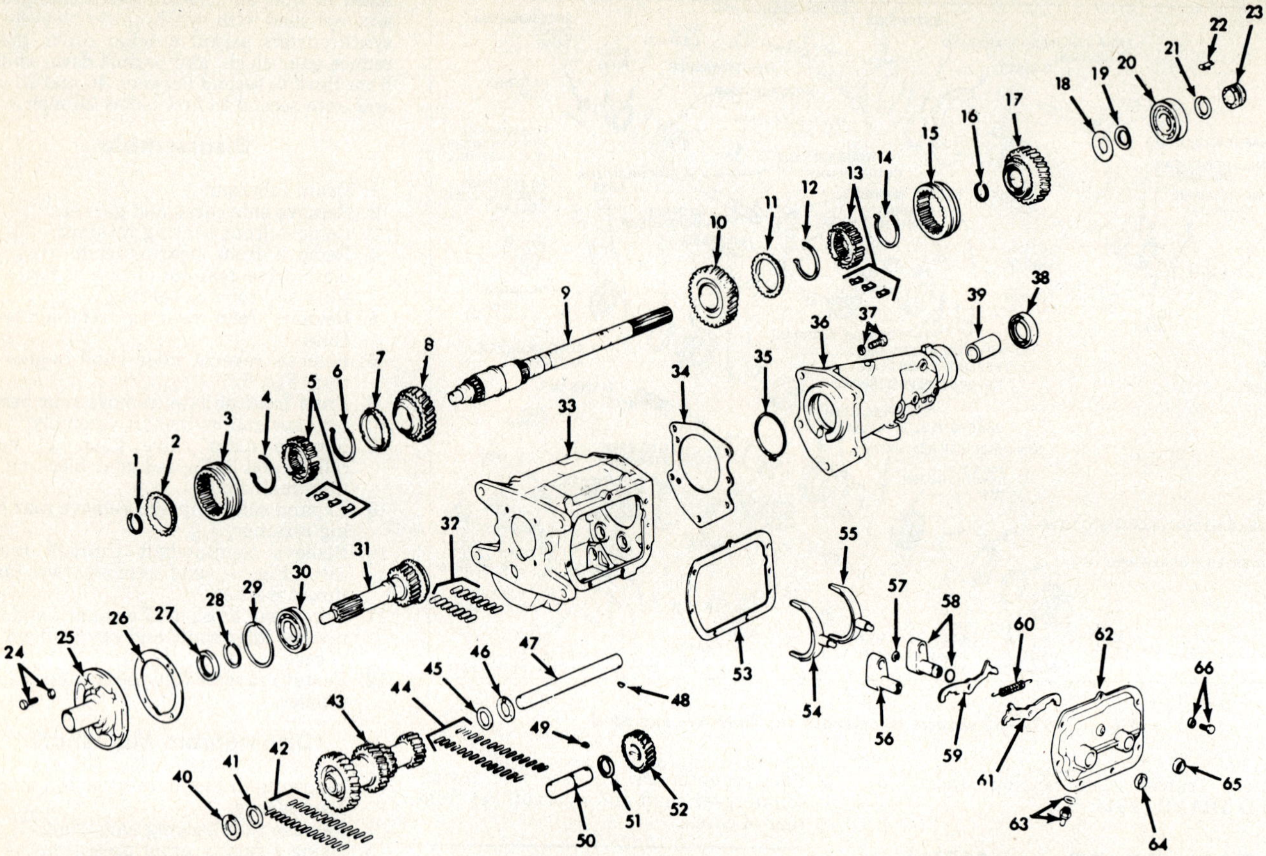

1. Snap Ring
2. Synchronizer Ring
3. 2-3 Synchronizer Sleeve
4. Synchronizer Key Spring
5. Synchronizer Hub and Keys
6. Synchronizer Key Spring
7. Synchronizer Ring
8. Second Gear
9. Main Shaft
10. First Gear
11. Synchronizer Ring
12. Synchronizer Key Spring
13. Synchronizer Hub and Keys
14. Synchronizer Key Spring
15. 1-2 Synchronizer Sleeve
16. Snap Ring
17. Reverse Gear
18. Thrust Washer
19. Waved Washer
20. Rear Bearing
21. Snap Ring
22. Speedometer Gear Clip
23. Speedometer Drive Gear
24. Bearing Retainer Bolts and Washers (4)
25. Front Bearing Retainer
26. Bearing Retainer Gasket
27. Bearing Retainer Oil Seal
28. Snap Ring
29. Bearing Snap Ring
30. Front Bearing
31. Drive Gear
32. Pilot Bearings
33. Case
34. Extension to Case Gasket
35. Rear Bearing to Extension Retaining Ring
36. Rear Extension
37. Extension to Case Retaining Bolts and Washers
38. Rear Extension Bushing
39. Rear Seal
40. Thrust Washer
41. Spacer
42. Countergear Shaft Bearings
43. Countergear
44. Countergear Shaft Bearings
45. Spacer
46. Thrust Washer
47. Countergear Shaft
48. Countergear Shaft Key
49. Idler Shaft Key
50. Reverse Idler Shaft
51. Snap Ring
52. Reverse Idler Gear
53. Side Cover Gasket
54. 2-3 Shift Fork
55. 1-Rev Shift Fork
56. 2-3 Shifter Shaft
57. Retaining "E" Ring
58. 1-Rev Shifter Shaft with "O" Ring
59. 2-3 Detent Cam
60. Detent Cam Spring
61. 1-Rev Detent Cam
62. Shift Cover
63. TCS Switch and Gasket
64. Shifter Shaft Seal
65. Shifter Shaft Seal
66. Shift Cover Bolts and Washers

Fig. 2 1974–83 Saginaw three speed transmission exploded view

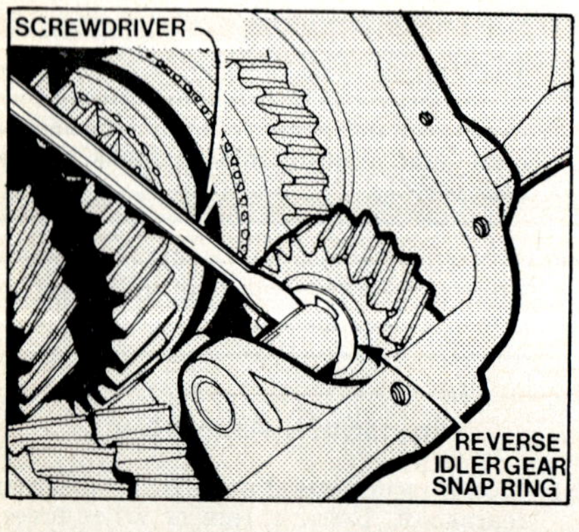

Fig. 3 Removing reverse idler "E" ring

Fig. 4 Removing rear bearing retainer after spring snap ring as shown

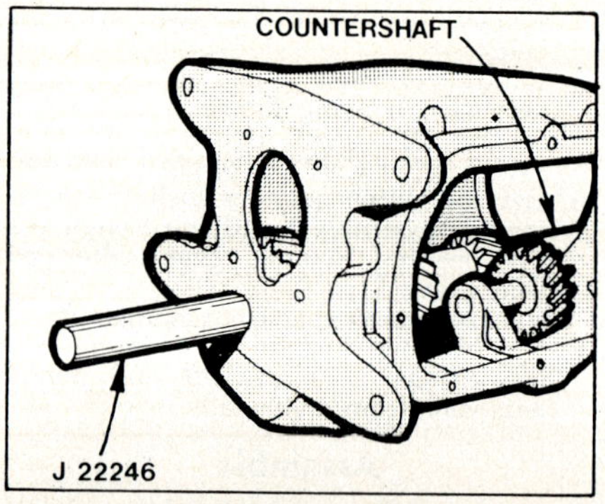

Fig. 5 Removing countershaft, using aligning arbor to hold needle bearings in place

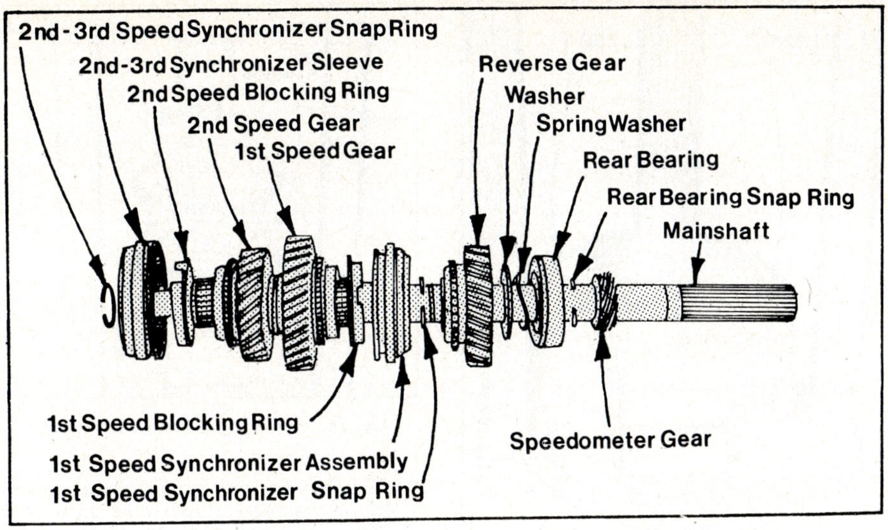

Fig. 6 Mainshaft and related parts assembled loosely to show location of parts

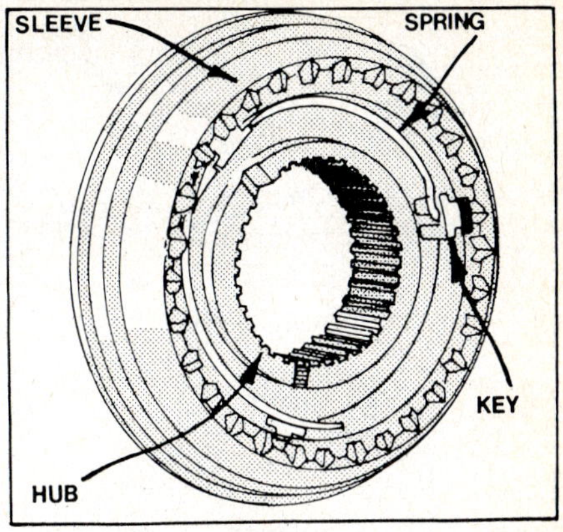

Fig. 7 Synchronizer assembly

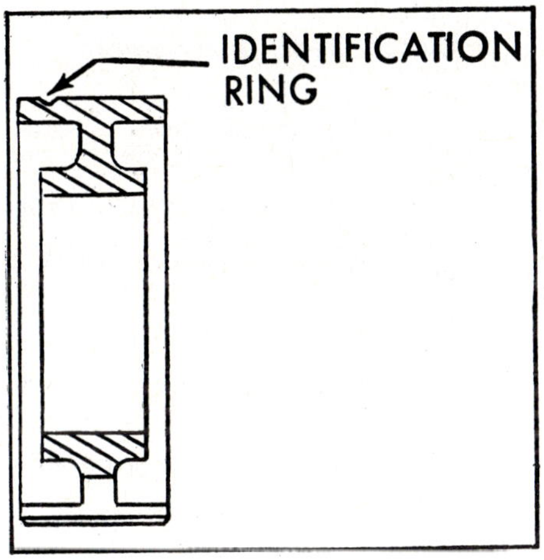

Fig. 8 Identification chamfer around synchronizer hub

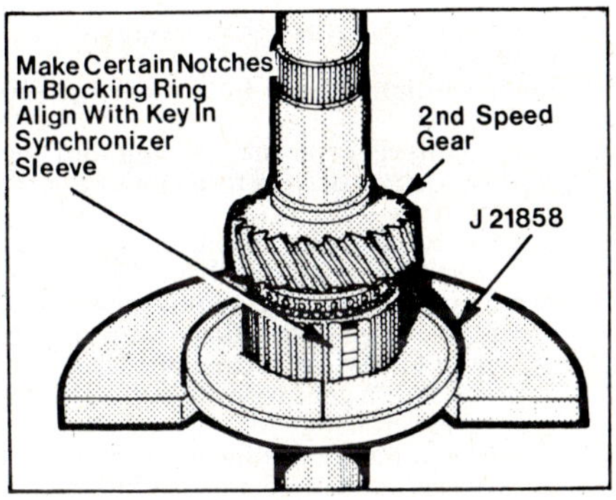

Fig. 9 Installing second speed gear

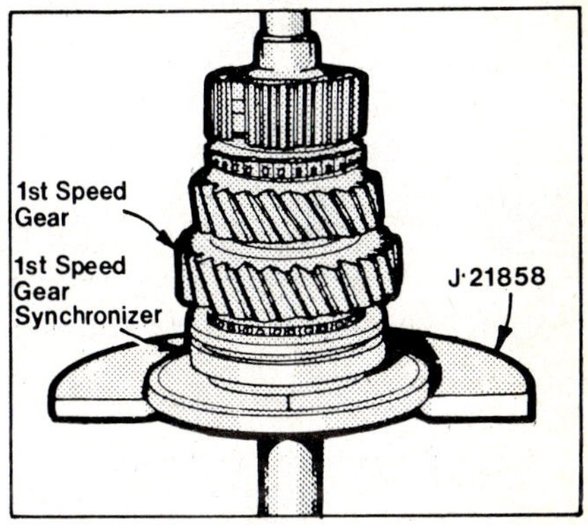

Fig. 10 Installing first speed gear

toward rear) onto mainshaft, Fig. 9.

2. Install first speed gear and synchronizer on mainshaft, Fig. 10, and install snap ring. *Make certain notches are in blocking ring align with keys.*
3. Install rear bearing retainer. Spread spring washer and rear bearing. Grooves on bearing must be toward reverse gear.
4. Install speedometer drive gear on mainshaft. Press gear until its rear face is 6 inches from rear bearing, Fig. 11.
5. Install 2–3 synchronizer sleeve.

Assemble Transmission

1. Install countergear-to-case thrust washers. Install countergear into case from rear. Make certain woodruff key is in place. Note that antirattle gear is riveted to countergear in four places and is not serviced separately, Fig. 12.
2. Install reverse idler gear tanged steel thrust washer. Install idler gear, shaft and woodruff key. *Reverse idler gear snap ring will be installed after installation of mainshaft.*
3. Install rear bearing retainer. Spread snap ring in retainer to allow snap ring to drop around rear bearing. Press on end of mainshaft until snap ring engages groove in rear bearing.
4. Install 14 needle rollers in main drive gear pocket, using grease to hold them in place. Assemble third gear. Pilot main drive gear and blocking ring over front of mainshaft. Make certain notches in blocking ring align with keys in 2–3 synchronizer.
5. Install rear bearing retainer-to-case gasket.
6. Install rear bearing retainer and mainshaft assembly into case. Torque bearing retainer-to-case bolts to 35–55 ft. lbs. torque.

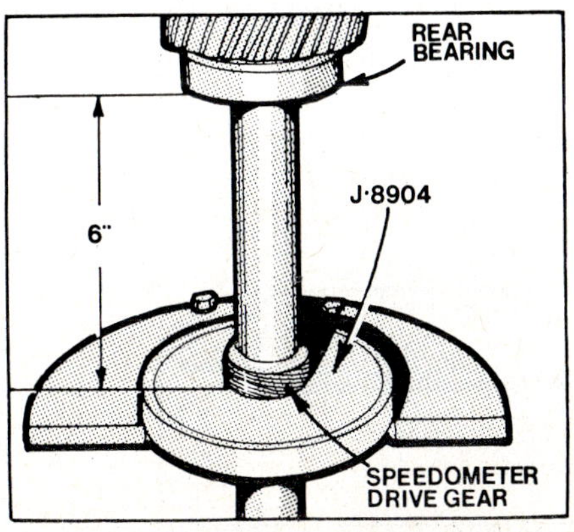

Fig. 11 Installing speedometer drive gear (6" from bearing)

Fig. 12 Anti-rattle gear is riveted to countergear

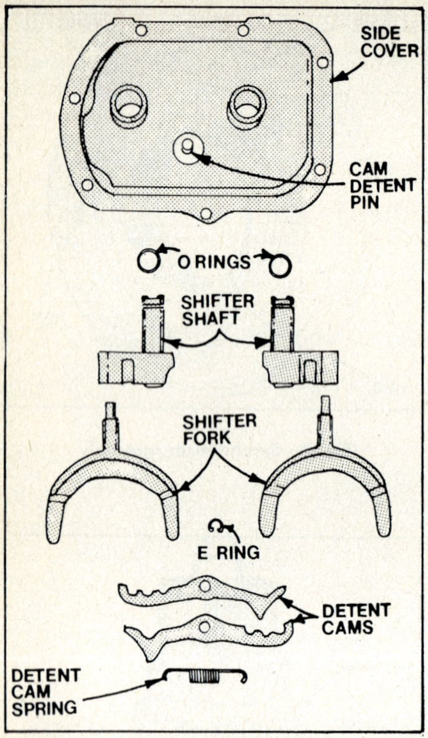

Fig. 13 Disassembled view of side cover

7. Install bearing on main drive gear. Outer snap ring groove must be toward front of gear. Install snap ring.
8. Install front bearing retainer and gasket.
9. Install reverse idler gear "E" ring.
10. If repairs are required to the side cover, refer to Fig. 13.
11. Install side cover gasket. Place transmission gears in neutral and install side cover. Install bolts and tighten evenly.

MUNCIE 4 SPEED, 1974–83

Disassemble, Fig. 14

1. Mount transmission in suitable holding fixture and remove transmission cover. If required, insert two 5/16 × 18 screws in cover flange threaded holes and turn evenly to raise cover dowel pins from case.

NOTE: Move reverse shifter fork so that reverse idler gear is partially engaged before attempting to remove cover. Forks must be positioned so rear edge of the slot in the reverse fork is in line with the front edge of the slot in the forward forks as viewed through tower opening.

---

2. Place transmission in two gears at once to lock gears and remove universal joint flange nut, universal joint from flange and brake drum assembly.
3. Remove parking brake and brake flange plate assembly on models so equipped.
4. Remove rear bearing retainer and gasket.
5. Slide speedometer drive gear off mainshaft.

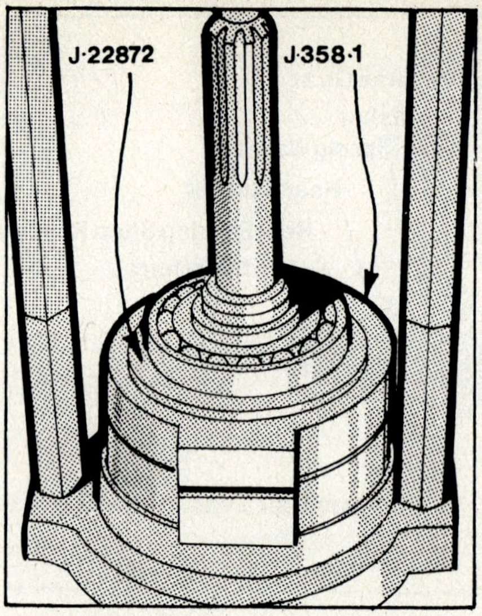

Fig. 15 Removing clutch gear bearing

6. Remove clutch gear bearing retainer and gasket.
7. Remove countergear front bearing cap and gasket.
8. Pry countergear front bearing out by inserting screw drivers into groove at cast slots in case.
9. Remove countergear rear bearing retaining rings from shaft and bearing. With a suitable tool remove countergear rear bearings. This will allow countergear assembly to rest on bottom of case.
10. Remove clutch gear bearing outer race to case retaining ring.
11. Remove clutch gear and bearing by tapping gently on bottom side of clutch gear shaft and prying directly opposite against the case and bearing snap ring groove at the same time. Remove 4th gear synchronizer ring.

CAUTION: Index cut out section of clutch gear in down position with countergear to obtain clearance for removing clutch gear.

---

12. Remove rear mainshaft bearing snap ring and using a suitable tool remove bearing from case. Slide 1st speed gear thrust washer off mainshaft.
13. Raise rear of mainshaft assembly and push rearward in case bore, then swing front end up and lift from case. Remove synchronizer cone from shaft.

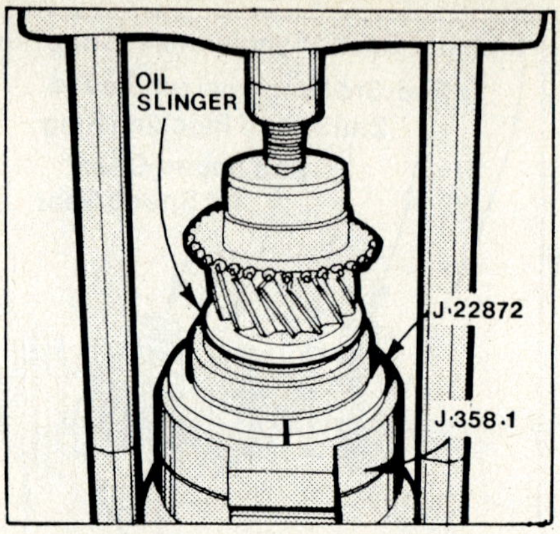

Fig. 16 Replacing clutch gear bearing

14. Slide reverse idler gear rearward and move countergear rearward until front end is free of case, then lift to remove from case.
15. To remove reverse idler gear, drive reverse idler gear shaft out of case from front to rear using a drift. Remove reverse idler gear from case.

Clutch Gear, Disassemble

1. Removing mainshaft pilot bearing rollers from clutch gear and remove roller retainer. Do not remove snap ring o inside of clutch gear.
2. Remove snap ring securing bearing on stem of clutch gear.
3. Using an arbor press, press gear and shaft out of bearing, Fig. 15.

Clutch Gear, Inspect

1. Wash all parts in cleaning solvent.
2. Inspect roller bearings for pits and galling.
3. Inspect bearing diameter in shaft recess for galling.
4. Inspect gear teeth for excessive wear.
5. Inspect clutch shaft pilot for excessive wear.
6. Re-oil bearing, then rotate clutch gear bearing slowly and check for roughness.

Clutch Gear, Assemble

1. Press bearing and new oil slinger onto clutch gear shaft, Fig. 16. Slinger should be located flush with bearing shoulder on clutch gear, Fig. 17.
2. Install snap ring to secure bearing on clutch gear shaft.
3. Install bearing retainer ring in groove on O.D. of bearing.

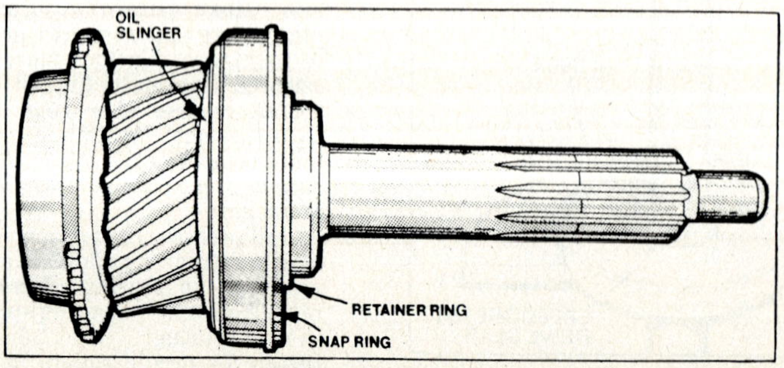

Fig. 17 Clutch gear assembly

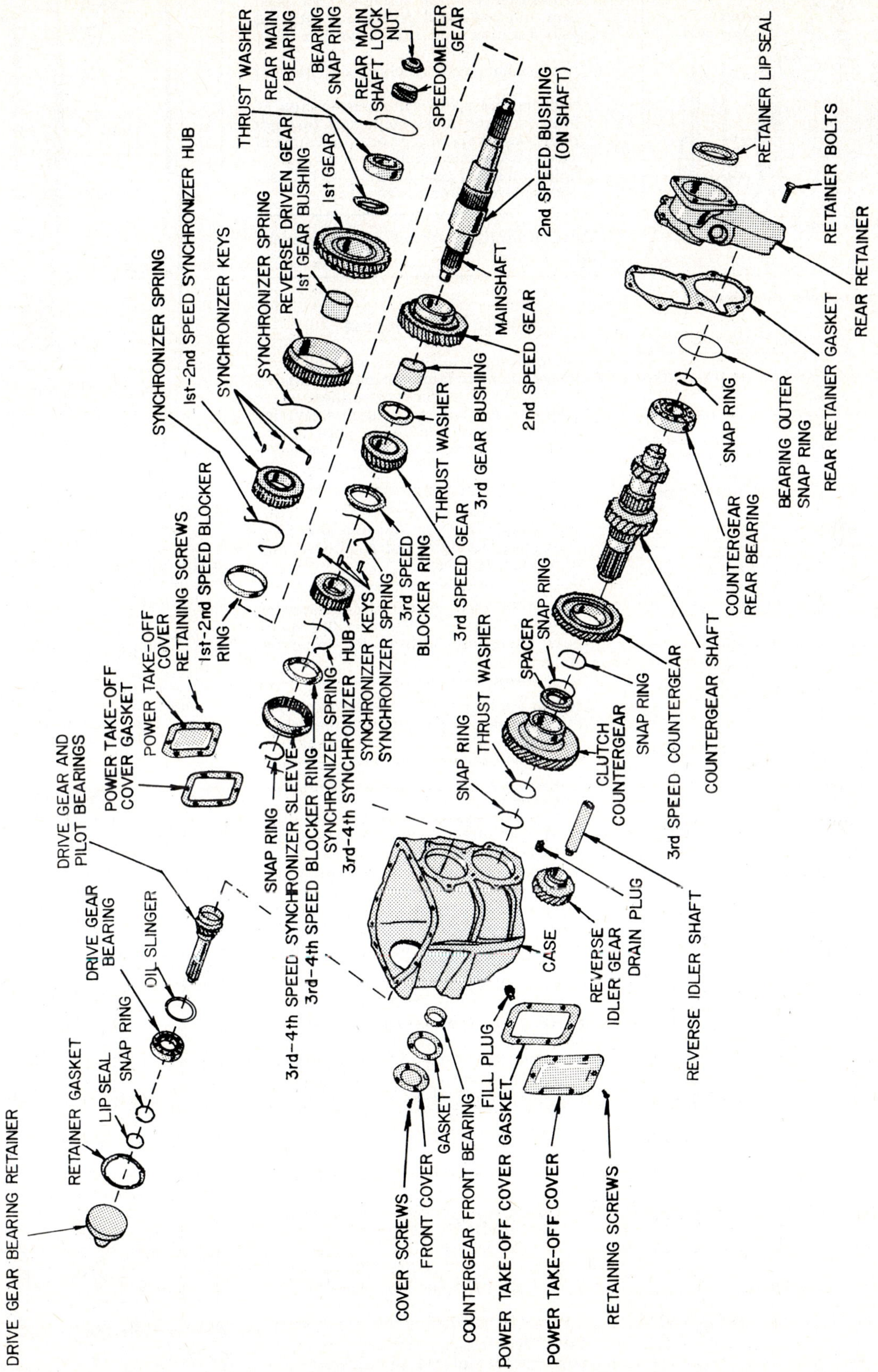

Fig. 14 GM CH & SM465 four speed synchromesh transmission. 1974–83

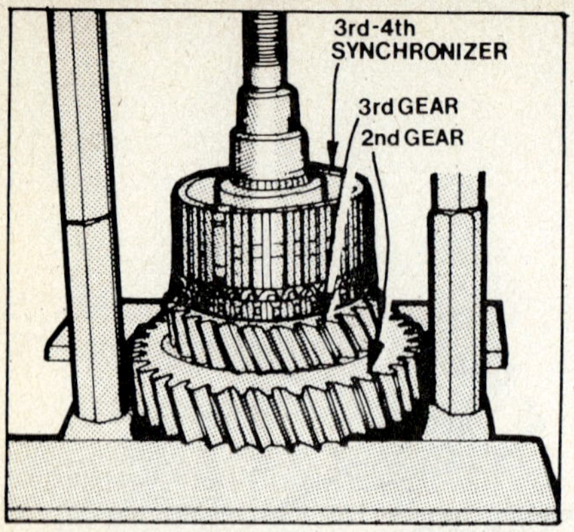

Fig. 18 Disassembly of mainshaft

CAUTION: Bearing must turn freely after it is installed on shaft.

4. Install snap ring on I.D. of mainshaft pilot bearing bore in clutch gear.
5. Apply a small amount of grease to bearing surface in shaft recess, install transmission mainshaft pilot roller bearings and install retainer.

NOTE: This roller bearing retainer holds bearings in position and in final transmission assembly is pushed forward into recess by mainshaft pilot.

Mainshaft, Disassemble

1. Remove first speed gear.
2. Remove snap ring in front of 3rd-4th synchronizer assembly.

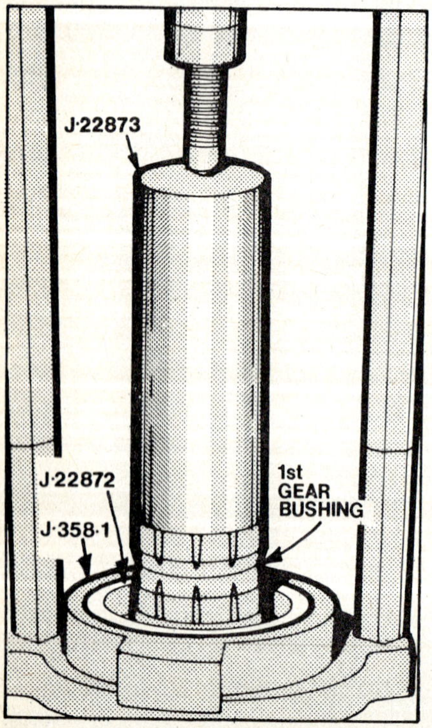

Fig. 20 Installing 1st speed gear bushing

3. Remove reverse driven gear.
4. Press behind 2nd speed gear to remove 3rd-4th synchronizer assembly, 3rd speed gear and 2nd speed gear along with 3rd gear bushing and thrust washer, Fig. 18.
5. Remove 2nd speed synchronizer ring and keys.
6. Support 2nd speed synchronizer hub at front face and press mainshaft through removing 1st speed gear bushing and 2nd speed synchronizer hub.
7. Split 2nd speed gear bushing with chisel and remove bushing from shaft.

CAUTION: Exercise care not to damage mainshaft.

Mainshaft, Inspect

1. Wash all parts in cleaning solvent.
2. Inspect mainshaft for scoring or excessive wear at thrust surfaces or splines.
3. Inspect clutch hub and clutch sleeve for excessive wear and make sure sleeve slides freely on clutch hub. Also check fit of clutch hub on mainshaft splines.

NOTE: 3rd and 4th speed clutch sleeve should slide freely on 3rd and 4th speed clutch hub but clutch hub should be snug fit on shaft splines.

4. Inspect 3rd speed gear thrust surfaces for excessive scoring and inspect 3rd speed gear mainshaft bushing for excessive wear.

NOTE: 3rd speed gear must be a running fit on mainshaft bushing and mainshaft bushing should be a press fit on shaft.

5. Check 2nd speed thrust washer for excessive scoring.
6. Inspect 2nd speed gear for excessive wear at thrust surface. Check synchronizer springs for looseness or breakage.
7. Inspect 2nd speed gear synchronizing ring for excessive wear.
8. Inspect bronze synchronizer cone on 2nd speed gear for excessive wear or damage. Also inspect clutch gear synchronizer cone and 3rd speed gear synchronizer cone for excessive wear or damage.

NOTE: 1st and reverse sliding gear must be sliding fit on synchronizer hub and must not have excessive radial and circumferential play. If sliding gear is not free on hub, inspect for burrs which may have rolled up on front ends of half-tooth internal splines and remove by honing.

9. Inspect all gear teeth for excessive wear.

Mainshaft, Assemble

1. Press 2nd speed bushing onto mainshaft until it bottoms against shoulder, Fig. 19.

CAUTION: 1st, 2nd and 3rd speed gear bushings are sintered iron. Exercise care when installing.

2. Press 1st and 2nd speed synchronizer hub onto mainshaft until it bottoms against

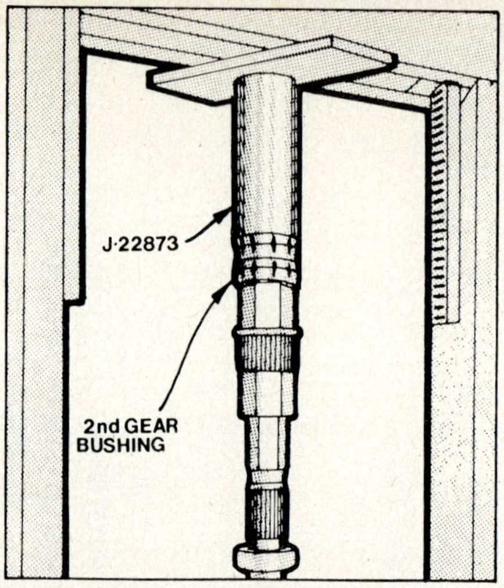

Fig. 19 Installing 2nd speed gear bushing

shoulder with annulus toward rear of shaft.

3. Install 1st and 2nd synchronizer keys and springs.
4. Press 1st speed gear bushing onto mainshaft until it bottoms against hub, Fig. 20.
5. Install synchronizer blocking ring and 2nd speed gear onto mainshaft and against synchronizer hub. Index synchronizer key slots with keys in synchronizer hub.
6. Install 3rd speed gear thrust washer onto mainshaft with tang or thrust washer in slot on shaft and against 2nd speed gear bushing. Then press 3rd speed gear bushing onto mainshaft until it bottoms against thrust washer, Fig. 21.

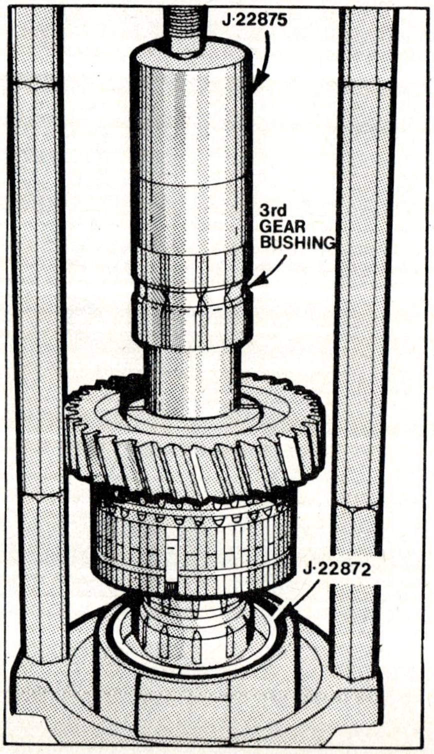

Fig. 21 Installing 3rd speed gear bushing

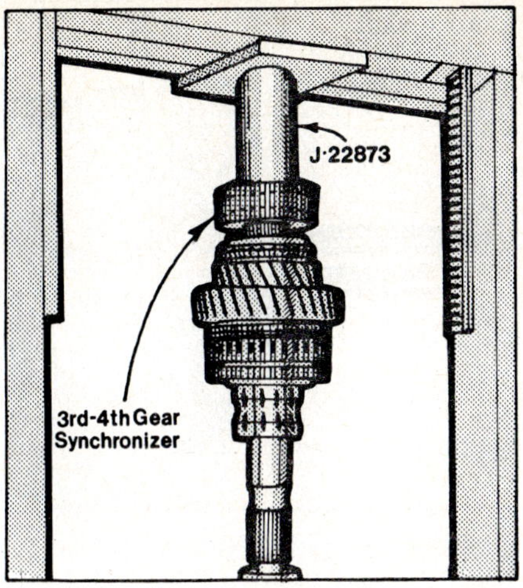

Fig. 22 Installing 3rd and 4th gear synchronizer

7. Install 3rd speed gear synchronizer blocker ring and 3rd speed gear onto mainshaft, against 3rd speed gear thrust washer.
8. Index synchronizer ring key slots with synchronizer assembly keys and drive 3rd and 4th synchronizer assembly onto mainshaft and against 3rd speed gear bushing with thrust face toward 3rd speed gear. Retain synchronizer assembly with snap ring, Fig. 22.
9. Install reverse driver gear with fork groove toward rear.
10. Install 1st speed gear onto mainshaft and against 1st and 2nd synchronizer hub. Install 1st speed gear thrust washer.

Countershaft, Disassemble

NOTE: As of 1980, a change was incorporated into the Muncie 4-speed transmission. This change consists of elimination of 1 spacer and 2 snap rings and includes a part called spacer-clutch countergear, part No. 14020811. The parts eliminated are those designated as 33, 34 and 35 in Fig. 14. The new spacer may also be used on earlier transmissions and should reduce possibility of 3rd speed countergear becoming loose and over-running retaining ring.

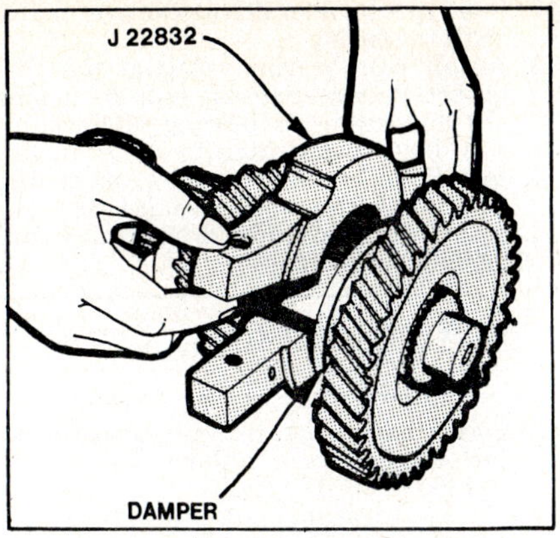

Fig. 23 Positioning Tool J-22832 on countershaft

1. Remove front countergear snap ring and thrust washer. Discard snap ring.
2. Install Tool J-22832 on countershaft, open side to damper, Fig. 23. Support assembly in arbor press and press countershaft out of clutch countergear assembly.
3. Remove clutch countergear rear retaining ring. Discard snap ring.
4. Remove 3rd speed countergear snap ring. Discard snap ring.
5. Position assembly on an arbor press and press shaft from 3rd speed countergear.

Countershaft, Assemble

1. Position 3rd speed countergear and shaft in arbor press and press gear onto shaft.

NOTE: Install gear with marked surface toward front of shaft.

2. Install new 3rd speed countergear snap ring.
3. Install new clutch countergear rear snap ring as follows: Install Tool J-22830, on end of shaft and position snap ring on tool. Using Tool J-22873, push down on snap ring until it engages groove on shaft. Using snap ring pliers, carefully expand

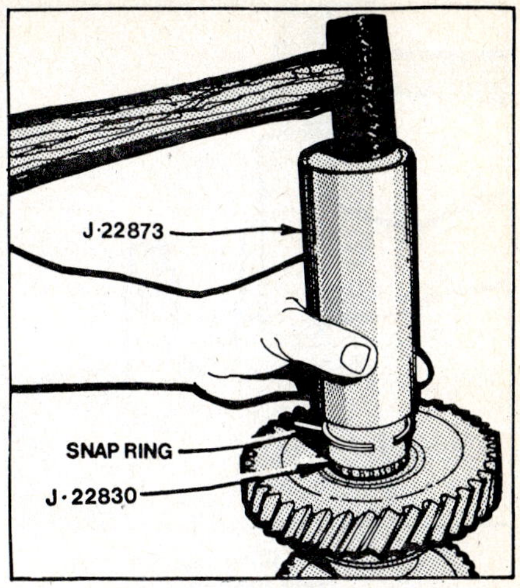

Fig. 24 Installing countergear snap ring

ring until it just slides onto splines, then push ring down on shaft until it engages groove on shaft, Fig. 24.
4. Position clutch countergear on shaft and press countergear onto shaft against snap ring, Fig. 25.
5. Install clutch countergear thrust washer and front snap ring.

Transmission, Assemble

1. Lower countergear into case until it rests on bottom.
2. Place reverse idler gear in transmission case with gear teeth toward the front. Install idler gear shaft from rear to front, being careful to have slot in end of shaft facing down. Shaft slot face must be at least flush with case.
3. Install mainshaft assembly into case with rear of shaft protruding out rear bearing hole in case. Position Tool J-22874-5 in clutch gear case opening and engaging front mainshaft. Rotate case onto front end, Fig. 26.

NOTE: Install 1st speed gear thrust washer on shaft if not previously installed.

4. Install snap ring on bearing O.D. and position rear mainshaft bearing on shaft. Using Tool J-22874-1 drive bearing onto

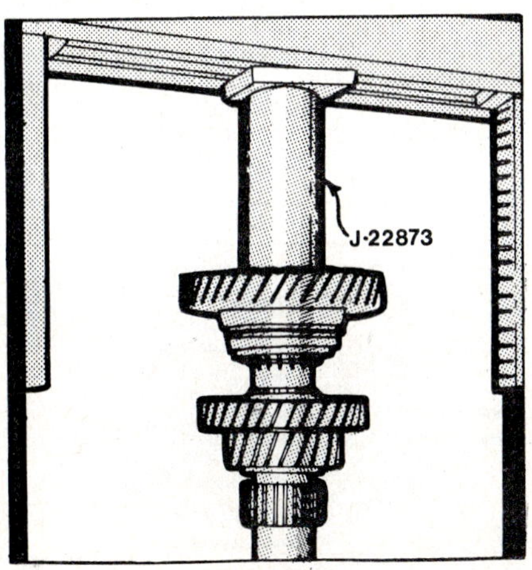

Fig. 25 Installing clutch countergear

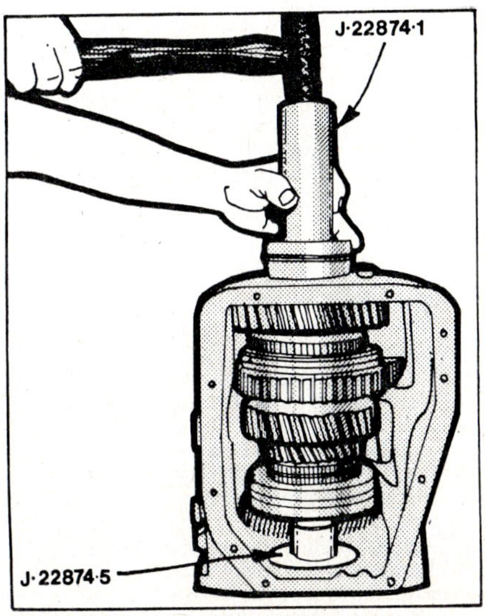

Fig. 26 Installing mainshaft rear bearing

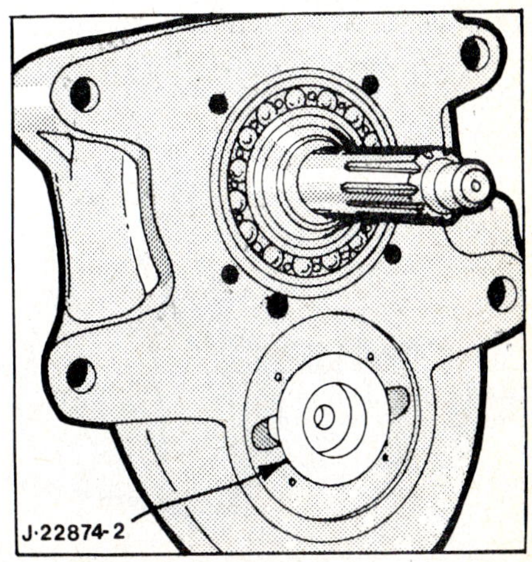

Fig. 27 Countergear front support tool

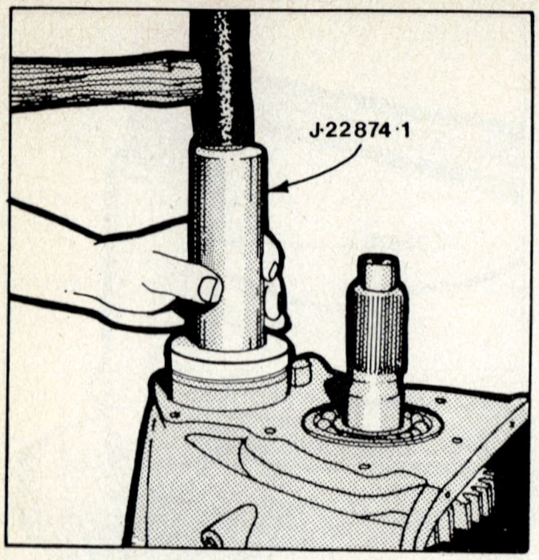

Fig. 28 Installing countergear rear bearing

shaft and into case. Align tangs on snap ring with lube slot of case and slot in tool before driving bearing. Rotate case and remove Tool J-22874-5, Fig. 26.

5. Install synchronizer cone on pilot end of mainshaft and slide rearward to clutch hub.

CAUTION: Make sure three cut-out sections of 4th speed synchronizer cone align with three clutch keys in clutch assembly.

6. Install snap ring on clutch gear bearing O.D. Index cut-out portion of clutch gear teeth to obtain clearance over countershaft drive gear teeth and install clutch gear assembly into case. Raise mainshaft to get clutch gear started and tap bearing outer race with plastic tip hammer.
7. Install clutch gear bearing retainer using a new gasket. Tighten bolts to 15–18 ft. lbs.
8. Install Tool J-22874-2 in countergear front bearing opening in case to support countergear and rotate case onto front end, Fig. 27.
9. Install snap ring on countergear rear bearing O.D., position bearing on countergear and using Tool J-22874-1, drive bearing into place. Rotate case, install snap ring on countershaft at rear bearing and then remove tool, Fig. 28.
10. Tap countergear front bearing assembly into case.
11. Install countergear front bearing cap and gasket.
12. Slide speedometer drive gear over mainshaft to bearing.
13. Install rear bearing retainer and new gasket. Be sure snap ring ends are in lube slot and cut-out in bearing retainer. Install bolts and tighten to 15–18 ft. lbs. Install brake backing plate on models equipped with propeller shaft brake.
14. Install parking brake drum and/or universal joint flange.

NOTE: Apply light coat of oil to seal surface.

15. Lock transmission in two gears at once. Install universal joint flange locknut and tighten to 90–120 ft. lbs.
16. Move all transmission gears to neutral except the reverse idler gear which should be engaged approximately 3/8 of an inch (leading edge of reverse idler gear taper lines up with front edge of 1st speed gear). Install cover assembly. Shifting forks must slide into their proper positions on clutch sleeves and reverse idler gear. Forks must be positioned as in removal.
17. Install cover bolts and tighten to 20–25 ft. lbs.

SAGINAW 4 SPEED

Transmission Disassembly

1. Remove side cover assembly and shift forks, Fig. 29. Remove damper assembly which is bolted to extension housing.
2. Remove clutch gear bearing retainer.
3. Remove clutch gear bearing to gear stem snap ring, then remove bearing by pulling clutch gear outward until a screwdriver can be inserted between large snap ring and case to complete removal, Fig. 30. Do not remove clutch gear. The bearing is a slip fit on the gear and into the case bore.
4. Remove extension to case bolts and remove clutch gear, mainshaft and extension assembly through rear case opening. Remove clutch gear and blocker ring from mainshaft.
5. Expand extension housing to rear mainshaft bearing snap ring and remove extension, Fig. 31.
6. Using a dummy shaft, drive countershaft and woodruff key out through rear of case, Fig. 32. Dummy shaft will hold roller bearings in position within countergear bore. Remove countergear.
7. Remove reverse idler gear stop ring and, using a long drift, drive idler shaft and woodruff key out through rear of case.

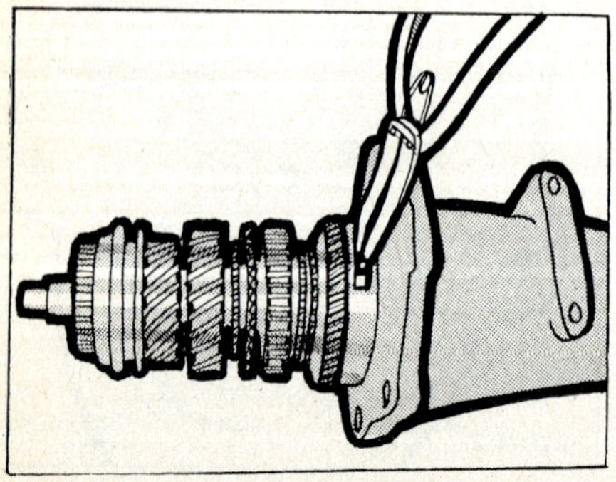

Fig. 31 Removing extension housing

Disassemble Mainshaft

1. Remove snap ring and press 3–4 synchronizer clutch assembly, 3rd speed blocker ring and 3rd gear off mainshaft, Fig. 33.
2. Depress speedometer gear retaining clip and remove gear.
3. Remove rear bearing to mainshaft snap ring, Fig. 34, support 1st gear with press plates and press on rear of mainshaft to remove 1st speed gear, thrust, washer, spring washer and rear bearing, Fig. 35.
4. Remove 1–2 sliding clutch hub snap ring, Fig. 36, support 2nd speed gear and press clutch assembly, 2nd speed blocker ring and gear from mainshaft.

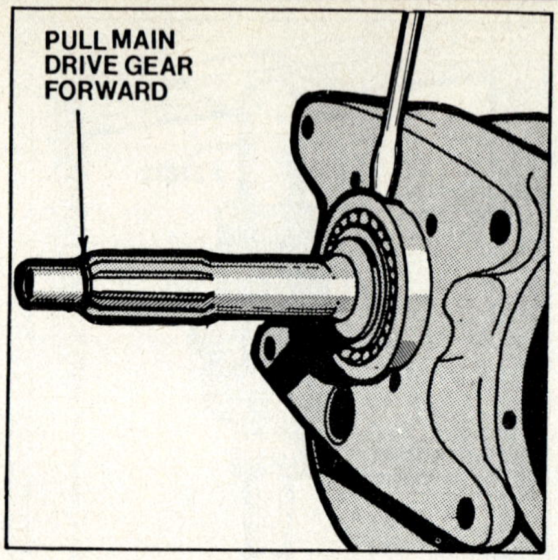

Fig. 30 Removing clutch gear bearing

Clutch Keys & Springs

NOTE: The clutch hubs and sleeves are a selected assembly and should be kept together as originally assembled, but the keys and springs may be replaced separately.

1. Mark hub and sleeve so they can be matched upon reassembly.
2. Push hub from sliding sleeve and remove keys and springs.
3. Install the three keys and two springs so all three keys are engaged by both springs. The tanged end of each spring should be installed into different key cavities on either side. Slide sleeve onto hub, aligning marks made before disassembly.

NOTE: A groove around the outside of the hub identifies the end that must be opposite the fork slot in the sleeve when assembled. This groove indicates the end of the hub with a .070″ greater recess depth.

Assemble Mainshaft

With front of mainshaft up:

1. Install 3rd speed gear with clutching teeth upward.
2. Install blocker ring with clutching teeth downward over cone of gear. All blocker rings in this unit are identical.
3. Install 3–4 synchronizer assembly with fork slot downward and press it onto mainshaft until it bottoms. *Be sure the notches of the blocker ring align with the keys of the synchronizer assembly.*

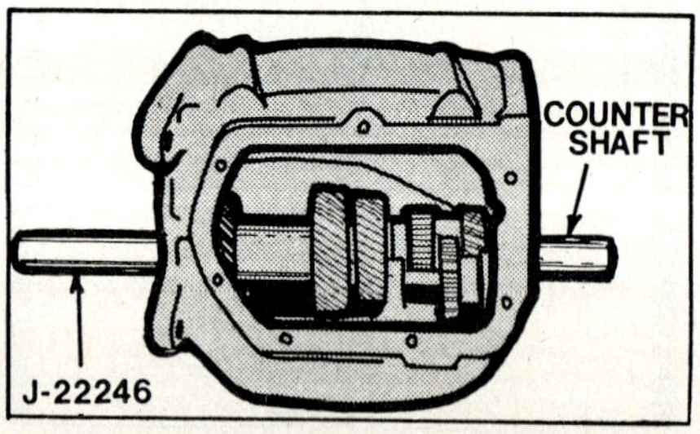

Fig. 32 Removing countershaft

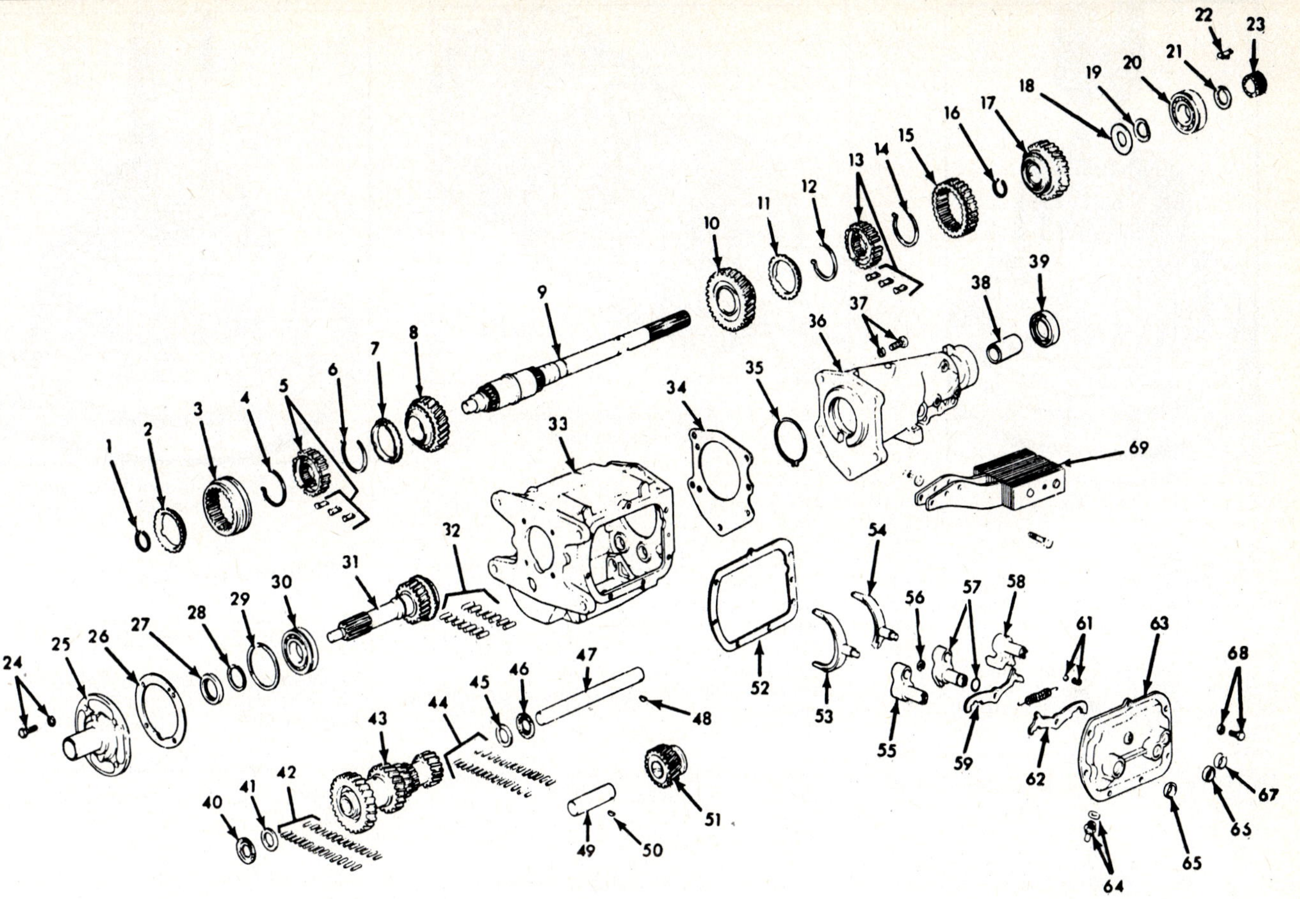

1. Snap ring, Hub to Shaft
2. Synchronizer Ring
3. 3-4 Synchronizer Sleeve
4. Synchronizer Key Spring
5. Synchronizer Hub and Keys
6. Synchronizer Key Spring
7. Synchronizer Ring
8. Third Gear
9. Main Shaft
10. Second Gear
11. Synchronizer Ring
12. Synchronizer Key Spring
13. Synchronizer Hub and Keys
14. Synchronizer Key Spring
15. 1-2 Synchronizer Sleeve & Rev. Gear
16. Snap Ring, Hub to Shaft
17. First Gear
18. Thrust Washer
19. Waved Washer
20. Ring Bearing
21. Snap Ring, Bearing to Shaft
22. Speedometer Gear Clip
23. Speedometer Drive Gear
24. Bearing Retainer Bolts and Washers (4)
25. Front Bearing Retainer
26. Bearing Retainer Gasket
27. Bearing Retainer Oil Seal
28. Snap Ring
29. Bearing Snap Ring
30. Front Bearing
31. Drive Gear
32. Pilot Bearings
33. Case
34. Extension to Case Gasket
35. Rear Bearing to Extension Retaining Ring
36. Rear Extension
37. Extension to Case Retaining Bolts and Washers
38. Rear Extension Bushing
39. Rear Seal
40. Thrust Washer
41. Spacer
42. Countergear Shaft Roller Bearings
43. Countergear
44. Countergear Shaft Roller Bearings
45. Spacer
46. Thrust Washer
47. Countergear Shaft
48. Countergear Shaft Key
49. Reverse Idler Shaft
50. Idler Shaft Key
51. Reverse Idler Gear
52. Side Cover Gasket
53. 3-4 Shift Fork
54. 1-2 Shift Fork
55. 3-4 Shifter Shaft
56. Retaining "E" Ring
57. 1-2 Shifter Shaft with "O" Rings
58. Reverse Shifter Shaft
59. 3-4 Detent Cam
60. Detent Cam Spring
61. Reverse Detent Ball & Spring
62. 1-2 Detent Cam
63. Shift Cover
64. TCS Switch and Gasket
65. Shifter Shaft Seal
66. Shifter Shaft Seal
67. Shift Cover Bolts and Washers
68. Shift Cover Attaching Bolts and Lock Washers
69. Damper Assembly

Fig. 29 Saginaw 4 speed transmission (typical). 1974–81 El Camino

4. Install synchronizer hub to mainshaft snap ring. Both synchronizer snap rings are identical.

With rear of mainshaft upward:

5. Install 2nd speed gear with clutching teeth upward.
6. Install a blocker ring with clutching teeth downward over cone of gear.
7. Press 1–2 synchronizer assembly onto mainshaft with fork slot downward. *Be sure notches in blocker ring align with keys of synchronizer assembly.*
8. Install synchronizer hub snap ring.
9. Install a blocker ring with notches downward so they align with the synchronizer keys.
10. Install 1st gear with clutching teeth downward. Install 1st gear thrust washer and spring washer.
11. Press rear bearing onto mainshaft. Install snap ring.
12. Install speedometer drive gear and clip.

Transmission Assembly

1. Load a row of roller bearings and a thrust washer at each end of the countergear, Fig. 37. Use heavy grease to hold them in place.
2. Install countergear through case rear opening with a tanged thrust washer at each end and install countershaft and woodruff key from rear of case. *Be sure countershaft picks up both thrust washers and that the tangs are aligned with their notches in the case.*
3. Install reverse idler gear, shaft and woodruff key from rear of case.
4. Expand extension housing snap ring and assemble extension housing over mainshaft.
5. Load roller bearings into clutch gear bore, using heavy grease to hold them in place, place blocker ring on gear cone with teeth toward gear, and install gear and ring onto mainshaft. Do not install clutch gear bearing at this time. *Be sure notches in blocker ring align with syn-*

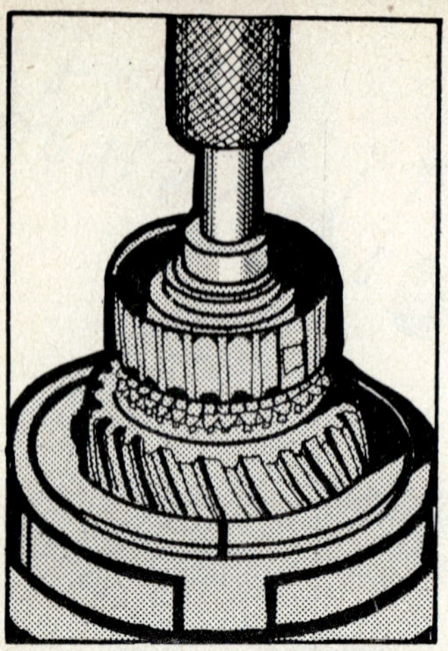

Fig. 33 Removing 3-4 synchronizer assembly

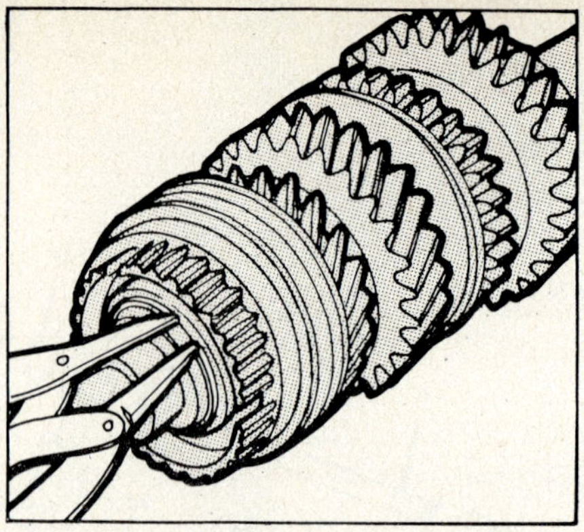

Fig. 34 Removing rear bearing to mainshaft snap ring

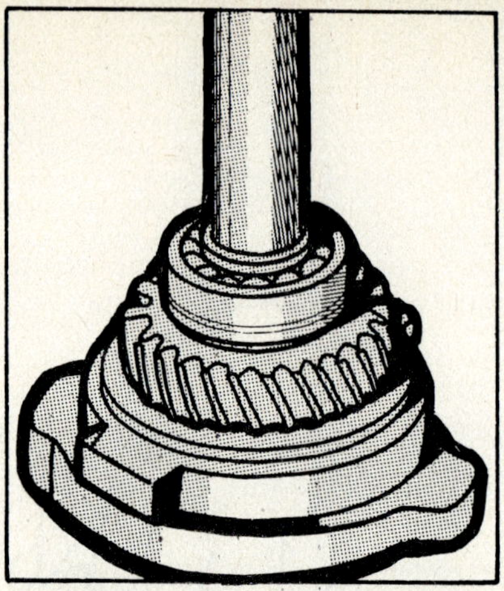

Fig. 35 Removing rear bearing and first speed gear

chronizer keys..

6. Using new gasket, install mainshaft and extension assembly through rear opening in case. Use sealing cement on bottom bolt.
7. Install large outer snap ring on clutch gear bearing and install bearing onto gear and into case bore. Install gear stem snap ring and bearing retainer.

NOTE: The retainer oil hole should be at the bottom.

8. With transmission in neutral, install cover assembly. *Be sure the shift forks are properly aligned in their grooves in the synchronizer sleeves before attempting to tighten cover bolts.*

WARNER 4 SPEED, 1975–81

Transmission Disassemble

1. Position transmission in 2nd gear, then remove side cover, gasket and both shift forks, Fig. 38.
2. Remove reverse shifter lock pin, Fig. 39, pull shifter shaft outward, disengaging lever from reverse gear.
3. Remove extension housing bolts and tap housing with a mallet to loosen, then pull extension housing rearward until idler shaft clears idler gear. Turn housing to left thereby freeing reverse shift fork from collar and remove housing.
4. Remove snap ring, then slide speedometer gear from mainshaft, remove second speedometer gear snap ring, Fig. 40.
5. Remove reverse gear from mainshaft, then slide rear reverse idler gear from rear bearing retainer, Fig. 41.
6. Remove lock bolt from rear bearing retainer and drive locating pin from bearing retainer into case.
7. Rotate rear bearing retainer counterclockwise, providing access to rear of countershaft.
9. From front of case, using a dummy shaft, tap countershaft and key out the rear of case, Fig. 42.
9. Using a mallet, tap rear bearing retainer rearward, and remove mainshaft through rear of transmission.
10. Remove bearing rollers and washer from main drive gear bore.
11. Remove front reverse idler gear and thrust washer from case.
12. Remove fourth gear synchronizing ring from main drive gear.
13. Remove main drive gear bearing retainer from case, remove snap ring and washer from main drive gear, then press gear from bearing, Fig. 43.
14. Tap drive gear bearing from case, remove countergear assembly and 2 washers. Drive locating pin from case.

NOTE: It is not necessary to remove countershaft and main gear assemblies if only main drive gear bearing is to be replaced. This is accomplished by using tools J-22912 and J-8433-1, Fig. 44.

Disassemble Mainshaft

1. From rear of mainshaft, remove snap ring and slide rear bearing retainer from mainshaft, Fig. 45.
2. Remove rear bearing snap ring and washer.
3. From front of mainshaft, remove snap ring, then remove 3–4 synchronizer assembly and 3rd gear, Fig. 38.
4. Support forward face of 2nd gear and press mainshaft from rear then remove the rear bearing, thrust washer, 1st gear and bushing, 1–2 synchronizer and 2nd gear.

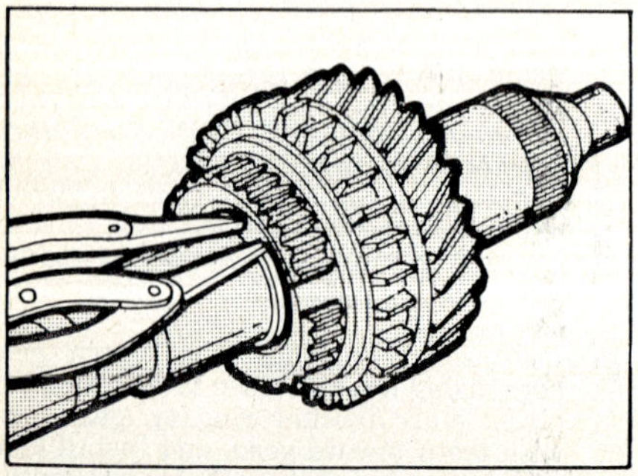

Fig. 36 Removing first and second synchronizer snap ring

Disassemble Countergear

Remove dummy shaft, bearings rollers, spacers and tubular spacer, Fig. 46.

Disassemble Side Cover

1. Remove forks from shifter shafts, Fig. 47.

NOTE: Forks are identical and are interchangeable.

2. Slowly push shifter shafts into side cover until detent balls fall from cover, then remove shifter shafts.

NOTE: The 3–4 shifter shaft has a detent.

3. Remove interlock sleeve, poppet spring and lock pin.
4. Replace shifter shaft seals if necessary.

Reverse Idler Shaft, Replace

1. Remove thrust washer, use punch to drive welch plug and pin into idler shaft, Fig. 48, until idler shaft can be removed.

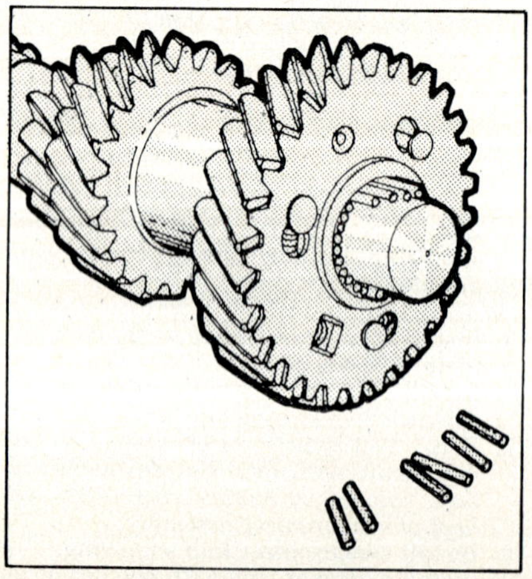

Fig. 37 Loading roller bearings into countergear

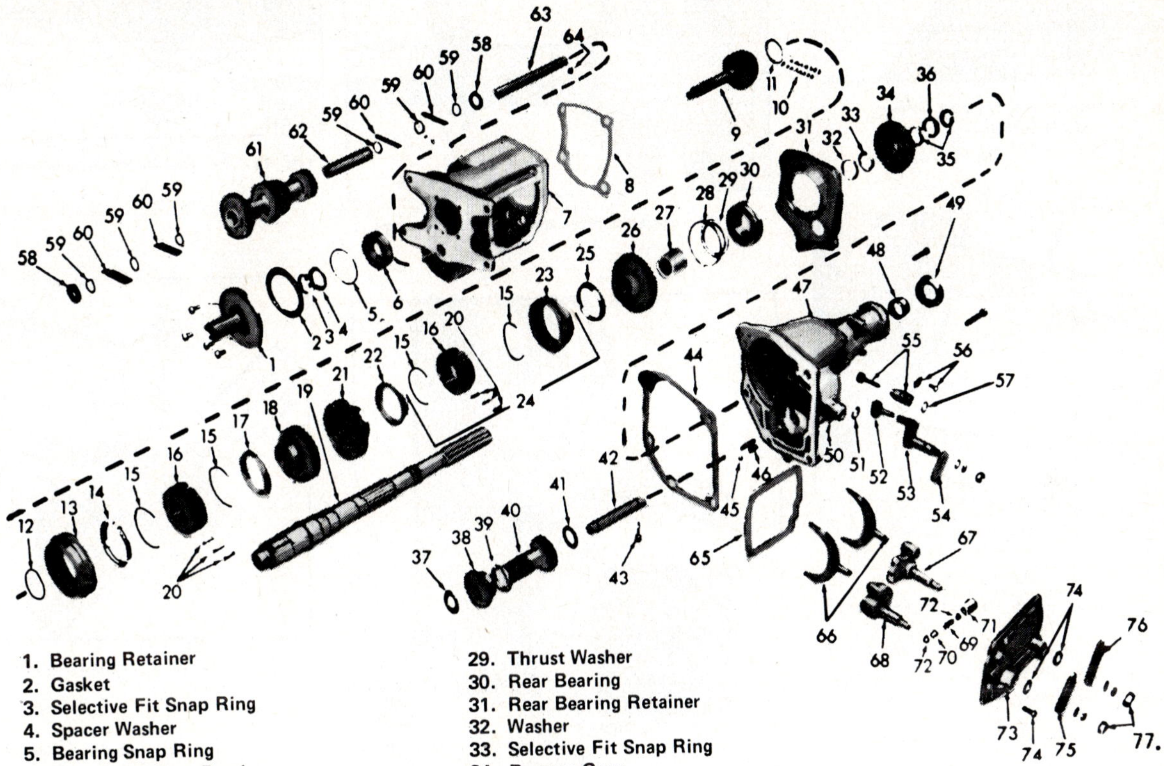

1. Bearing Retainer
2. Gasket
3. Selective Fit Snap Ring
4. Spacer Washer
5. Bearing Snap Ring
6. Main Drive Gear Bearing
7. Transmission Case
8. Rear Bearing Retainer Gasket
9. Main Drive Gear
10. Bearing Rollers (16)
11. Washer
12. Snap Ring
13. Third and Fourth Speed Clutch Sliding Sleeve
14. Fourth Speed Gear Synchronizing Ring
15. Clutch Key Spring
16. Clutch Hub
17. Third Speed Gear Synchronizing Ring
18. Third Speed Gear
19. Mainshaft
20. Clutch Keys (3)
21. Second Speed Gear
22. Second Speed Gear Synchronizing Ring
23. First and Second Speed Clutch Sliding Sleeve
24. First and Second Speed Clutch Assembly
25. First Speed Gear Synchronizing Ring
26. First Speed Gear
27. First Speed Gear Sleeve
28. Rear Bearing Snap Ring
29. Thrust Washer
30. Rear Bearing
31. Rear Bearing Retainer
32. Washer
33. Selective Fit Snap Ring
34. Reverse Gear
35. Snap Ring
36. Speedometer Drive Gear
37. Reverse Idler Front Thrust Washer (Flat)
38. Reverse Idler Gear (Front)
39. Snap Ring
40. Reverse Idler Gear (Rear)
41. Thrust Washer (Tanged)
42. Reverse Idler Shaft
43. Reverse Idler Shaft Lock Pin and Welch Plug
44. Rear Bearing Retainer To Case Extension Gasket
45. Reverse Shifter Shaft Detent Ball
46. Reverse Shifter Shaft Ball Detent Spring
47. Case Extension
48. Extension Bushing
49. Rear Oil Seal
50. Reverse Shifter Shaft Lock Pin
51. Reverse Shifter Shaft Lip Seal
52. Reverse Shift Fork
53. Reverse Shifter Shaft and Detent Plate
54. Reverse Shifter Lever
55. Speedometer Driven Gear and Fitting
56. Retainer and Bolt
57. "O" Ring Seal
58. Washer (Tanged)
59. Spacer (.050")
60. Bearing Rollers (28)
61. Countergear
62. Countergear Roller Spacer
63. Countershaft
64. Countershaft Woodruff Key
65. Gasket
66. Forward Speed Shift Forks
67. First and Second Speed Gear Shifter Shaft and Detent Plate
68. Third and Fourth Speed Gear Shifter Shaft and Detent Plate
69. Poppet Spring
70. Interlock Pin
71. Interlock Sleeve
72. Detent Balls
73. Transmission Side Cover
74. Lip Seals
75. Third and Fourth Speed Shifter Lever
76. First and Second Speed Shifter Lever
77. Lever Attaching Nuts

Fig. 38 Warner four speed transmission. 1975—81 Caballero, El Camino & Sprint

Fig. 39 Removing reverse shifter shaft lock pin

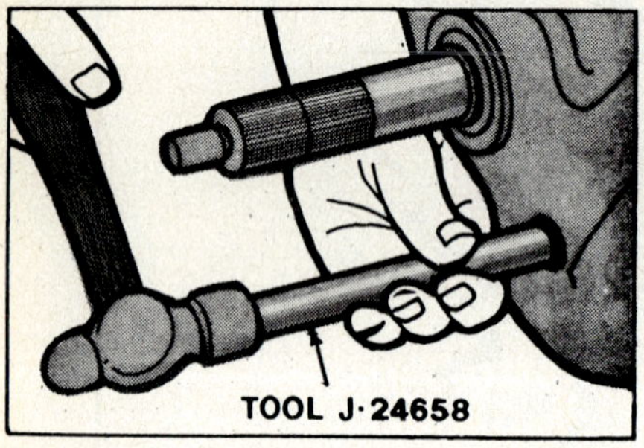

Fig. 42 Removing countershaft

2. Position idler shaft in extension housing.
3. Align hole in shaft with hole in extension housing, drive in pin and install welch plug with sealer.

Reverse Shifter & Seal, Replace

1. Remove shifter fork, tap shaft into extension housing until detent ball falls from housing, then remove shaft and detent spring.
2. Remove "O" ring and discard.
3. Install detent spring, then position "O" ring on shaft.
4. Position ball on spring, then while pressing ball downward, slide shaft into place and rotate shaft until ball is seated.
5. Install shift fork.

Extension Housing Seal & Bushing, Replace

1. Pry out oil seal, then drive out bushing, Fig. 49.
2. Press in new bushing using tool No. J-21465-17 and J-8092.
3. Coat bushing and seal with transmission oil, then using tool No. J-21359 install oil seal, Fig. 50.

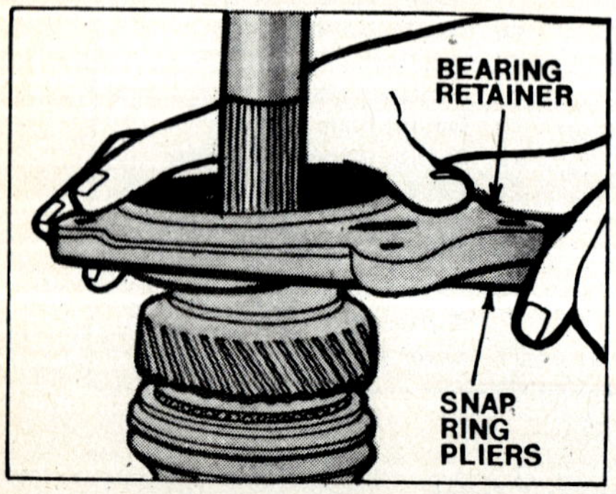

Fig. 45 Removing and installing rear bearing retainer

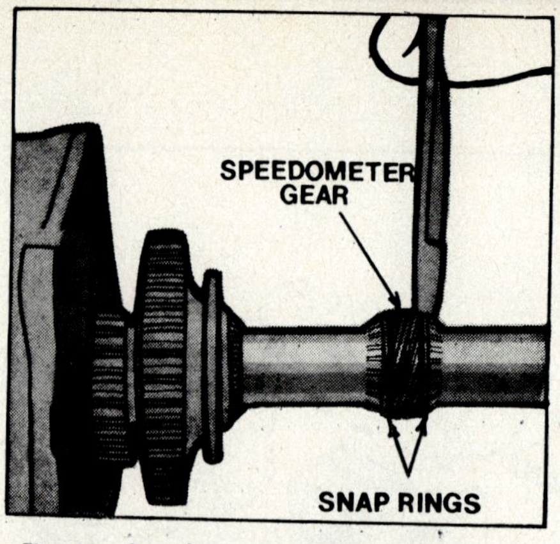

Fig. 40 Speedometer gear and snap rings

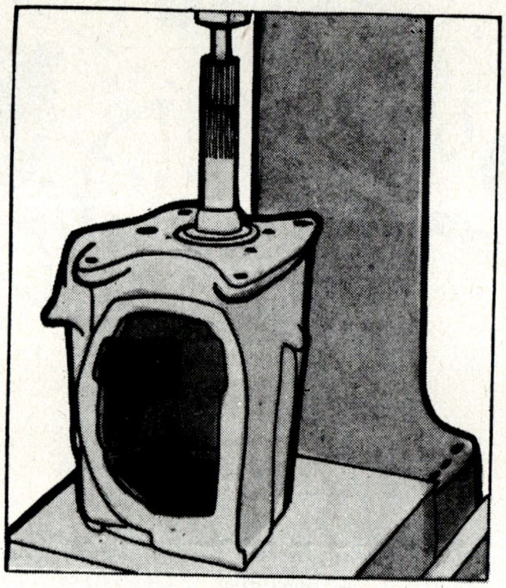

Fig. 43 Pressing drive gear from bearing

Drive Gear Bearing Oil Seal, Replace

Pry seal with screw driver. Position seal in bore and tap in with flat plate, Fig. 51.

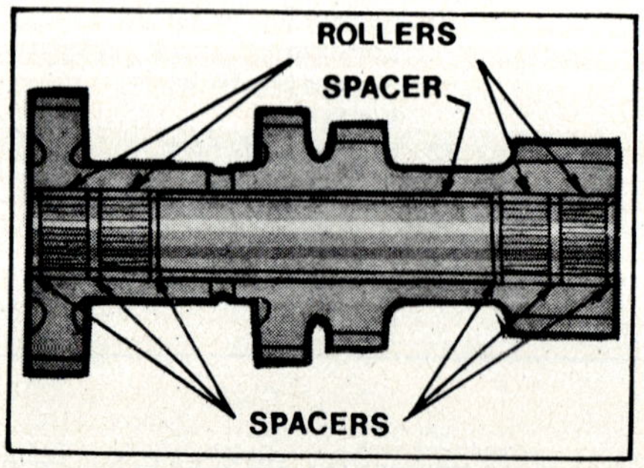

Fig. 46 Countergear assembly

Fig. 41 Removing and installing reverse gear

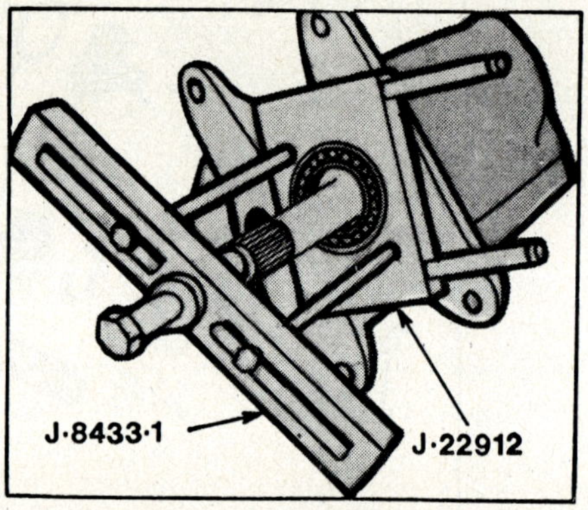

Fig. 44 Main drive gear bearing removal

Synchronizers, Overhaul

NOTE: The synchronizer clutch hub and sliding sleeve are replaced as an assembly.

1. Mark clutch hub and sliding sleeve so they can be installed in the same position, Fig. 52.
2. Press clutch hub from sliding sleeve, remove clutch keys and springs.
3. Install clutch keys and springs, align marks, install sliding sleeve on clutch hub.

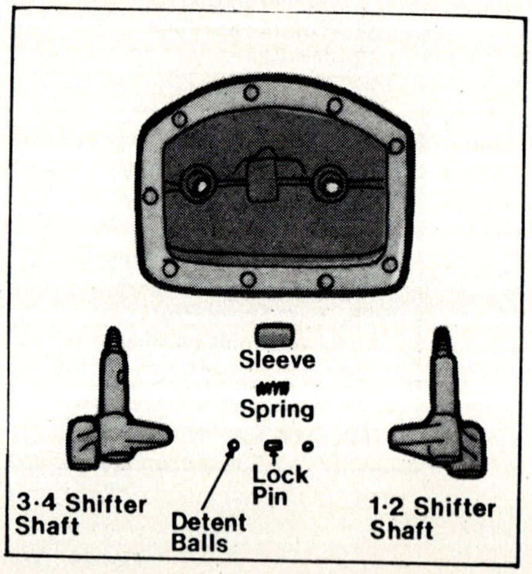

Fig. 47 Side cover assembly

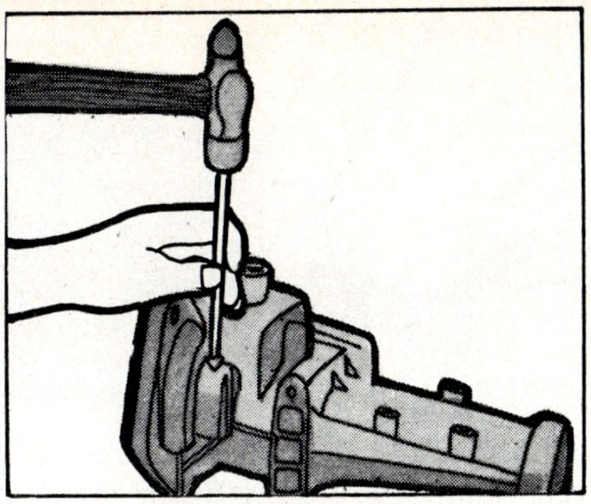

Fig. 48 Removing roll pin from reverse idler shaft

Fig. 49 Removing and installing extension housing bushing

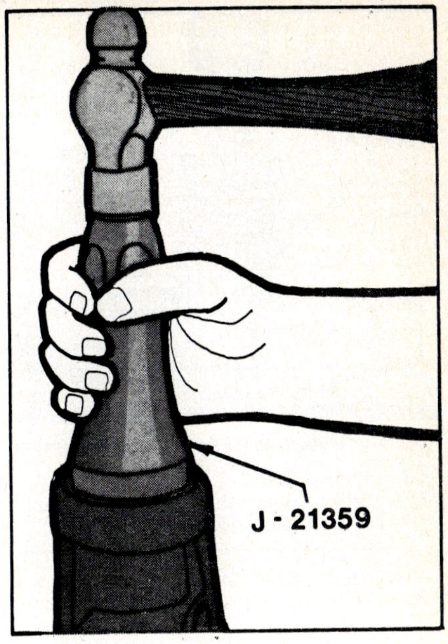

Fig. 50 Installing extension housing oil seal

Assemble Side Cover

NOTE: When assembling transmission always use new gaskets, oil seals and snap rings.

1. Install seals in shifter shaft bores, if removed.
2. Install 3–4 shifter shaft in bore, position in neutral, then install detent ball, interlock sleeve, poppet spring and lock pin, Fig. 47.
3. Position detent ball in other end of interlock sleeve, depress ball and install 1–2 shifter shaft.
4. Position 1–2 shifter shaft in 2nd gear and install shift forks.

Countergear Assemble & Install

1. Position dummy shaft in countergear bore and install tubular spacer.
2. Install spacer in bore. Using heavy grease to hold rollers in place, install 28 rollers, Fig. 46.
3. Install second spacer, 28 additional rollers and third spacer.
4. Install rollers and spacers in other end of countergear using above procedure.
5. Install thrust washers in each end of transmission case with tangs inserted in case recesses, using grease to hold washers in place.
6. Insert countershaft through rear of case, in turn forcing dummy shaft out through front of case. Install woodruff key on countershaft; tap countershaft until flush with rear of case.

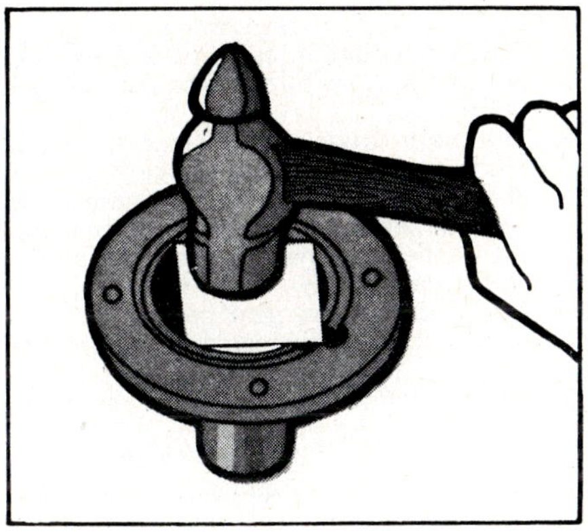

Fig. 51 Installing drive gear bearing retainer oil seal

Countergear End Play Check

Attach dial indicator, Fig. 53; check end play. If end play is greater than .025 in. new selective thrust washers must be installed.

Assemble Mainshaft

1. Front rear of mainshaft and install 2nd gear with hub facing rearward.
2. Position synchronizing rings on 1–2 synchronizer, then install synchronizer on mainshaft with hub facing forward.
3. Press 1st gear bushing on mainshaft until 2nd gear, 1–2 synchronizer and bushing are against mainshaft shoulder, Fig. 54.
4. Install 1st gear and inner race with gear hub facing front, then press rear bearing on mainshaft with snap ring groove facing front, Fig. 55.
5. Install spacer and selective thickness snap ring.

NOTE: Use the thickest snap ring that will fit in mainshaft groove.

6. Install 3rd gear with hub facing front and the 3rd gear synchronizing ring with notches facing front.

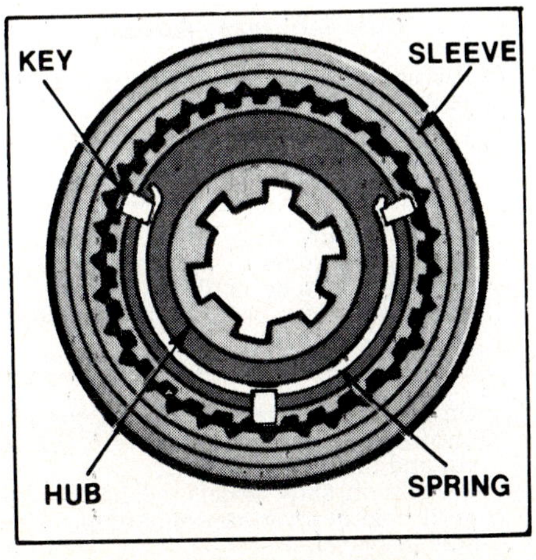

Fig. 52 Synchronizer assembly

Fig. 53 Countergear end play check

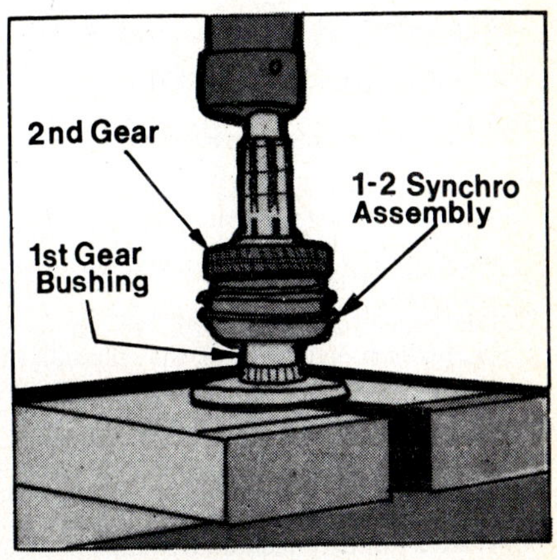

Fig. 54 Installing 2nd gear, 1–2 synchronizer assembly and 1st gear bushing

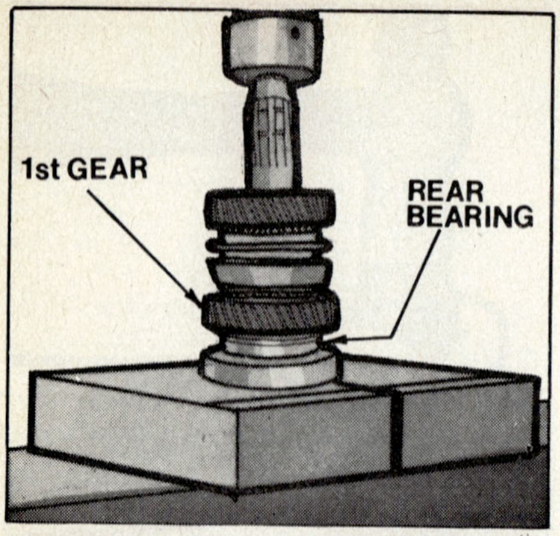

Fig. 55 Installing 1st gear and rear bearing

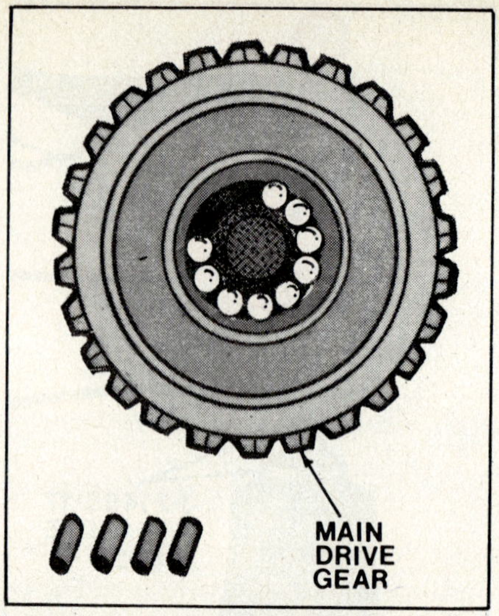

Fig. 56 Installing bearing rollers into main drive gear bore

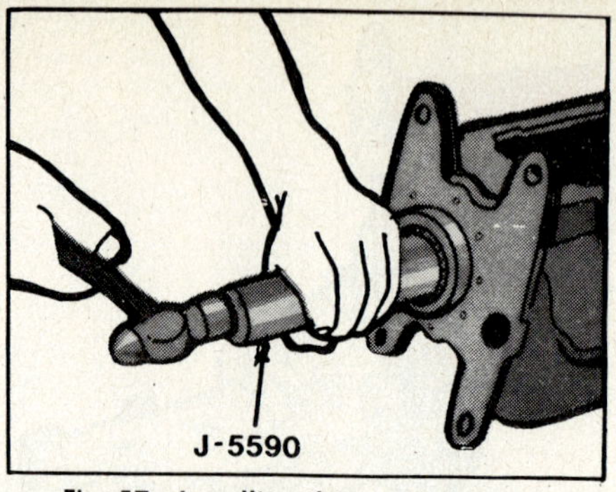

Fig. 57 Installing drive gear bearing

7. Install 3rd and 4th gear clutch assembly with the taper facing front.

NOTE: Ensure that keys in hub mate with notches in 3rd gear synchronizing ring.

8. Install selective thickness snap ring. Use thickness snap ring that will fit mainshaft groove.
9. Install locating pin in rear bearing retainer. Place bearing retainer over rear of mainshaft; spread retainer snap ring and align with groove in rear bearing, Fig. 55.
10. Using heavy grease, install 16 rollers and flat washer into main drive gear bore, Fig. 56.
11. Position 4th gear synchronizer ring on 3–4 synchronizer, aligning ring notches with hub keys.
12. Install front drive gear on main shaft.

Transmission Assemble

1. Install drain plug in transmission case.
2. Install thrust washer and front section of reverse idler gear with hub facing rear and position gasket on rear of transmission case.
3. Position synchronizers in neutral and install mainshaft in case. Install retainer locating pin and torque retainer lock bolt to 25 ft. lbs.
4. Tap main drive gear bearing into case. Using suitable tool, pull drive gear shaft forward, and tap bearing again until bearing bottoms on main drive gear and the washer and snap ring can be installed on main gear drive shaft, Fig. 57.
5. Install main drive gear bearing retainer and gasket, apply sealer to bolts and torque to 18 ft. lbs.
6. Install rear reverse idler gear, engaging splines with front section.
7. Install reverse sliding gear with hub facing rear, Fig. 41, then install the speedometer gear and two snap rings on mainshaft, Fig. 40.
8. Position gasket on rear bearing retainer and, install thrust washer on reverse idler shaft in extension housing.
9. Rotate reverse shifter shaft until fork is in full forward position. With synchronizer sleeves in neutral, slide extension housing over mainshaft. Guide reverse idler shaft into reverse idler gears, engaging fork on collar, then position shifter shaft in neutral.
10. Rotate reverse shifter shaft clockwise and push extension housing forward, aligning rear bearing retainer locating pin with extension housing. Apply sealer to bolts and torque long bolts to 40 ft. lbs., short bolts to 25 ft. lbs.
11. Install reverse shifter shaft lock pin.
12. Shift transmission into 2nd gear. The front synchronizer sleeve should be in neutral and the rear sleeve in forward position to engage 2nd gear synchronizer teeth.
13. Position slide cover and gasket to transmission case and index shift forks to shift collars, then install and torque side cover bolts to 18 ft. lbs.

Automatic Transmission Section

TURBO HYDRA-MATIC "200", "200C"

Transmission, Replace

1977–83 Caballero, El Camino & Sprint

1. Disconnect battery ground cable.
2. Remove air cleaner and dipstick. Then disconnect detent cable at bracket and carburetor.
3. On models equipped with A/C, remove heater core cover screws, then disconnect wiring connector and position heater core cover with hoses aside.
4. Raise and support vehicle.
5. Remove propeller shaft, then disconnect speedometer cable, oil cooler lines, case electrical connector and shift linkage.
6. Place a suitable jack under transmission and remove transmission rear support to frame side rail bolts.
7. Disconnect catalytic converter bracket from transmission support. Disconnect rear exhaust pipe from catalytic converter, and front pipe from manifold and remove from vehicle.
8. Remove torque converter cover and converter-to-flexplate bolts.
9. Lower transmission until jack is barely supporting it and remove transmission-to-engine bolts.
10. Raise transmission to its normal position. Place a jack under engine and remove transmission rearward from engine and lower from vehicle.

NOTE: Keep front of transmission upward to prevent torque converter from falling out.

11. Reverse procedure to install.

Shift Linkage, Adjust

1. Position shifter assembly in "NEUTRAL", Figs. 1 and 2.
2. Position transmission shift lever in "NEUTRAL"

NOTE: Obtain "NEUTRAL" position by moving transmission lever clockwise to maximum detent (PARK). Then move lever two detent positions counterclockwise (NEUTRAL).

3. On console shift models, adjust the rod until hole in the rod aligns with shifter pin and install rod on pin.
4. On column shift models, adjust the rod until hold in lever assembly aligns with hole in swivel.
5. On console shift models, install washer

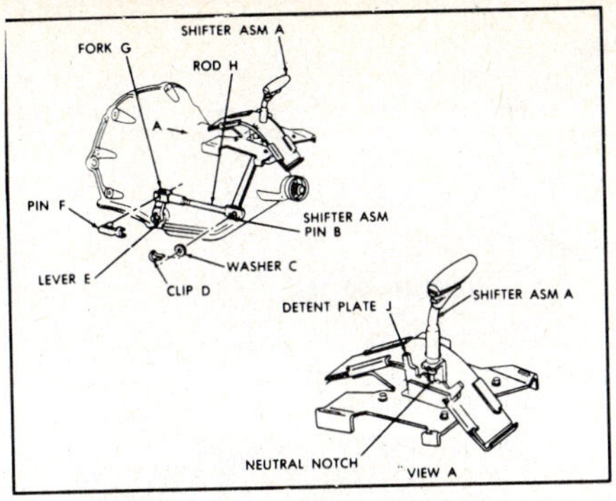

Fig. 1 Shift linkage adjustment. THM 200 with console shift

3. Set transmission shift lever in neutral position by rotating lever clockwise to park position, then two detents counterclockwise.
4. Without exerting force in either direction on rod or equalizer lever, tighten swivel clamp screw, then torque to 20 ft. lbs.
5. Check shift linkage operation for proper drive positions, then ensure the engine only starts in the "P" and "N" positions. If necessary, repeat procedure.

NOTE: If the shift linkage is improperly adjusted, an internal leak can occur at the manual valve leading to a clutch or band failure.

---

6. If necessary, adjust shift indicator pointer cable.

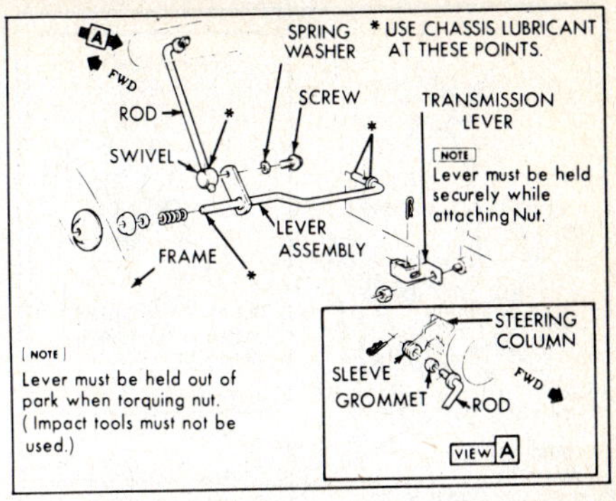

Fig. 2 Shift linkage adjustment. THM 200 with column shift

and clip. On column shift models, install swivel screw with spring washer.

TURBO HYDRA-MATIC "250", "250C"

Transmission, Replace

For transmission replacement, refer to the Turbo Hydra-Matic 350, 350C procedure.

Shift Linkage, Adjust

Console Shift

1. Place shifter in "N" (NEUTRAL) position of the detent plate, Fig. 3.
2. Place transmission lever in "N" (NEUTRAL) position by moving lever clockwise to maximum detent "P" (PARK), then counterclockwise two (2) positions (NEUTRAL-N).
3. Install the pin and lock onto the fork with lever in "N". Adjust the rod length until hole in rod aligns with pin or shifter assembly, install rod onto pin.
4. Install washer and clip.

Column Shift

1. Loosen swivel clamp set screw, Fig. 4.
2. Place shift lever in neutral detent. Do not use the shift indicator to locate the neutral detent, instead place shift lever in park position and count down two detents.

TURBO HYDRA-MATIC "350", "350C"

Transmission, Replace

(Except 1974–83 Four Wheel Drive models)

1. Place truck on hoist and drain oil.
2. Disconnect vacuum modulator line and speedometer cable at transmission.
3. Disconnect manual control lever rod and detent cable from transmission.
4. Disconnect propeller shaft from transmission.
5. Install suitable lift equipment and attach on transmission.
6. Disconnect engine rear mount on transmission extension and remove transmission support crossmember. Remove engine rear mount to bracket nut, washer and thru bolt on GA-10, 20 models.
7. Remove converter underpan and mark flywheel-converter relationship for proper assembly, then remove flywheel to converter bolts.
8. Support engine at oil pan rail with a jack of suitable brace to support engine weight when transmission is removed.
9. Lower rear of transmission slightly so upper transmission housing to engine bolts can be reached using a universal socket and a long extension. Remove upper bolts.
10. Remove remaining transmission housing to engine bolts and remove transmission by moving it slightly to the rear and downward.

CAUTION: Keep front of transmission upward to prevent converter from falling out.

1974–83 Four Wheel Drive models

1. Disconnect battery negative cable and the detent cable from carburetor, remove transmission dipstick and drain transmission.
2. Remove transfer case shift lever knob and boot, then raise vehicle.
3. Remove flywheel cover and converter to flywheel bolts.
4. Disconnect shift linkage, speedometer cable, vacuum modulator line, line to filler tube clip, detent cable to filler tube strap and oil cooler lines.

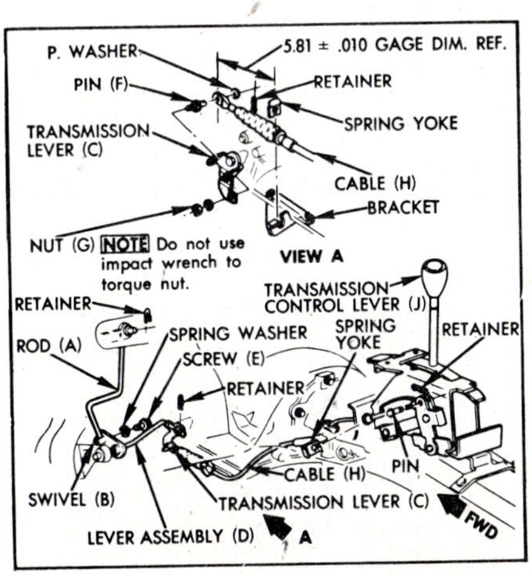

Fig. 3 THM 250C & 350 shift linkage. 1974–81 Caballero, El Camino & Sprint console shift

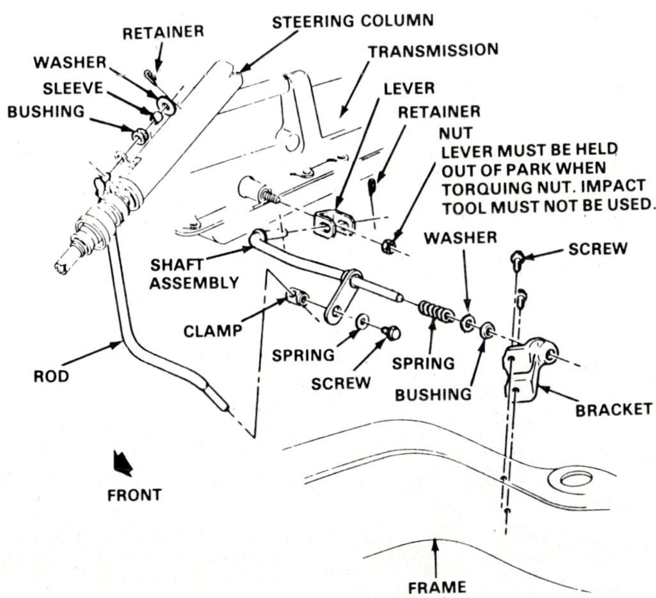

Fig. 4 Shift linkage adjustment. THM 250C with column shift

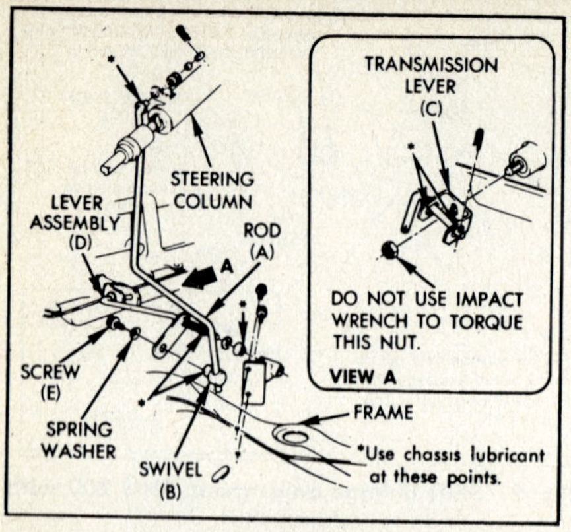

Fig. 5 THM 350 shift linkage. 1974–81 CK models, Caballero, El Camino & Sprint column shift & 1974–83 P models

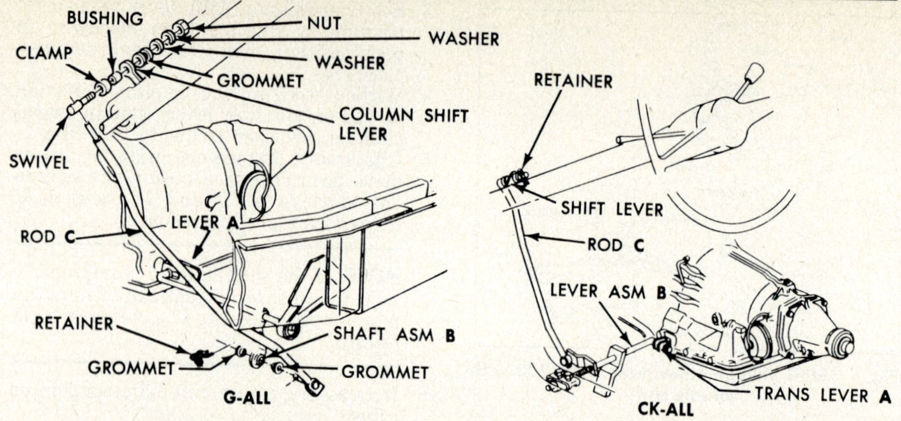

Fig. 6 THM 350C & 700-R4 shift linkage. 1982–83 CK & G models with column shift

5. Remove crossover pipe to manifold bolts.
6. Remove transfer case adapter to cross-member bolts, raise engine slightly and remove crossmember.
7. Disconnect rear propeller shaft from transfer case, front propeller shaft from front axle, parking brake cable and exhaust system.
8. Support transmission and transfer case with a suitable jack, then remove transfer case to frame bolts and engine to transmission bolts.
9. Remove transmission and transfer case assembly, then separate transfer case from transmission.

Shift Linkage, Adjust

1974–81 Sprint & Caballero Console Shift

1. Loosen swivel screw so rod is free to move in swivel, Fig. 3.
2. Place transmission control lever in Drive and loosen pin in transmission lever, so it moves in the slot.
3. Move transmission lever counterclockwise to L1 detent and then three detents clockwise to Drive position. Tighten nut in transmission lever to 20 ft. lbs.
4. Place transmission control lever in Park and ignition switch in the lock position and pull lightly against lock stop, then tighten swivel screw to 20 ft. lbs. Check for proper operation.

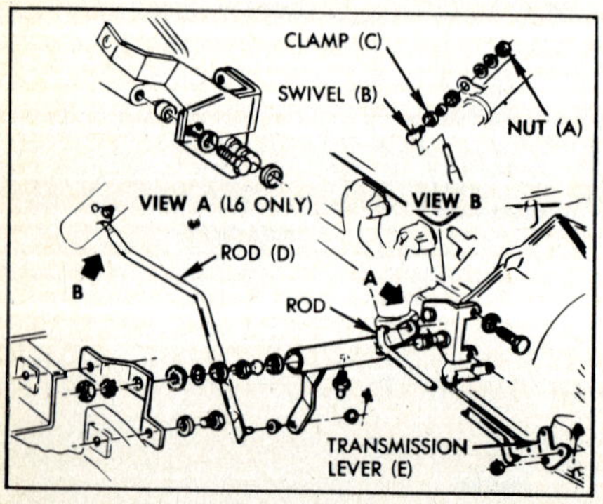

Fig. 7 THM 350 shift linkage. 1974 G Models

1974–83 "CK" Models, Caballero, El Camino & Sprint Column Shift, "P" Models & 1982 "G" Models, Figs. 5 & 6

Adjustments are made as previously described under TH-200 and 250C. Gearshift linkage adjustments for 1977–83 Caballero, El Camino and Sprint.

1974–81 "G" Models

1. Loosen swivel nut at steering column lever to allow free movement of swivel and clamp on rod.
2. Place transmission lever in Neutral position by moving lever clockwise to Park detent then clockwise two detent positions to Neutral.
3. Place selector lever in Neutral as determined by mechanical stop. Do not use indicator as reference.
4. On 1974 G models, torque swivel nut to 18 ft. lbs., Fig. 7.
5. On 1975–81 G models, hold steering column lever against Neutral stop (Park position side) and torque swivel nut to 18 ft. lbs., Fig. 8.

TURBO HYDRA-MATIC "400", "475"

Transmission, Replace

1. Remove battery ground cable and release parking brake.
2. Raise vehicle on hoist and remove drive shaft.
3. Disconnect speedometer cable, electrical lead to case connector, vacuum line at modulator and oil cooler pipes and shift control linkage.
4. Support transmission with suitable jack and disconnect rear mount from frame crossmember. Remove cross-member.
5. Remove converter under pan, remove converter-to-flywheel bolts.
6. Loosen exhaust pipe to manifold bolts approximately 1/4" and lower transmission until jack is barely supporting it.
7. Remove transmission to engine mounting bolts and remove oil filler tube at transmission.
8. Raise transmission to its normal position, support engine with jack and slide transmission rearward from engine and lower it away from vehicle.

NOTE: Use suitable converter holding tool or be sure to keep rear of transmission lower than front to prevent the converter from sliding off.

Shift Linkage, Adjust

NOTE: Do not use shift indicator pointer as a reference to position the selector lever. When adjusting linkage, the pointer is adjusted last.

1. Place selector lever in Drive (D) position as determined by the transmission detent.
2. Loosen adjustment swivel at the mast jacket and rotate the transmission lever so that it contacts the Drive stop in the steering column.
3. Tighten the swivel and recheck adjustment.
4. Readjust indicator pointer if necessary to agree with the transmission detent positions.
5. Readjust neutral safety switch if necessary so the engine will start only in the Park and Neutral positions.

TURBO HYDRA-MATIC 425

Transmission, Replace

1. Disconnect battery ground cable, then remove engine access cover.

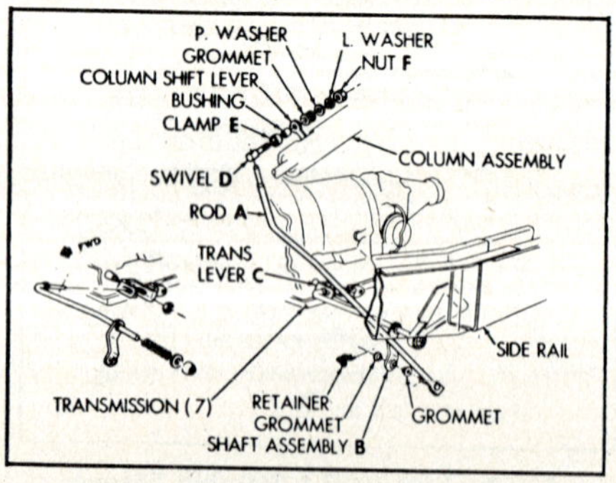

Fig. 8 THM 350 shift linkage. 1975–81 G Models

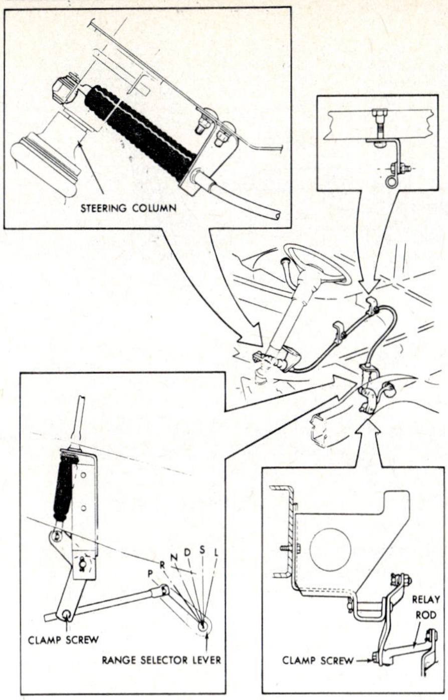

Fig. 9 THM 425 shift linkage. 1974–78 GMC Motorhome & Transmode

2. Support engine using engine holding fixture J-24603-01 or equivalent.
3. Raise vehicle, then remove starter motor.
4. Remove flywheel inspection cover, then disconnect manual linkage from transmission.
5. Disconnect speedometer cable.
6. Disconnect transmission cooler lines at transmission and cap lines and fittings.
7. Disconnect detent wire connector from detent switch and vacuum line from modulator.
8. Disconnect right axle shaft from output shaft flange.
9. Remove right hand output shaft and support assembly.
10. Disconnect left hand axle shaft from output shaft flange.
11. Remove three torque converter to flywheel attaching bolts.
12. Remove three transmission to rear support bracket attaching bolts, then remove two rear motor mount to support bracket attaching bolts.
13. Position a suitable transmission jack under transmission and secure transmission to jack using a safety chain.
14. Remove six transmission converter housing to engine attaching bolts.
15. Move transmission and final drive assembly rearward approximately 1/2 to 3/4 inch, then lower transmission and final drive assembly from vehicle.
16. Reverse procedure to install.

Shift Linkage, Adjust

1. Position shift lever in Neutral.
2. Loosen relay rod clamp screw, then position range selector lever in Neutral position, Fig. 9.
3. With shift lever against Neutral stop, tighten relay clamp to 20 ft. lbs.
4. Check transmission shift linkage for proper operation.

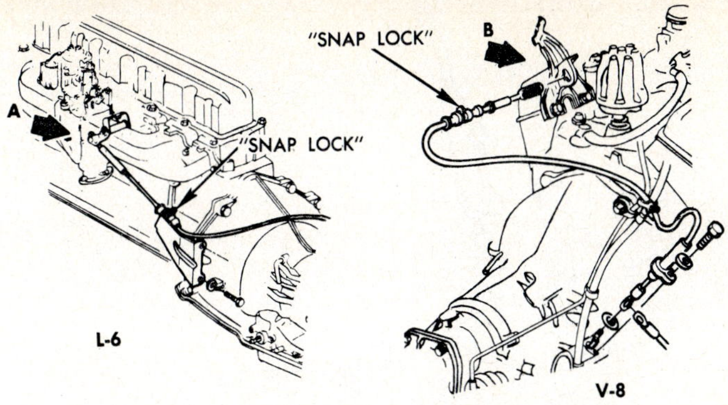

Fig. 10 Detent cable adjustment. 1974–80 models except THM 400, 425, 475 & diesel engine equipped models

TURBO HYDRA-MATIC 700-R4

Transmission, Replace

1. Remove air cleaner assembly, disconnect T.V. cable at its upper end. Remove transmission oil dipstick and bolt holding dipstick tube, if accessible.
2. Raise and support vehicle and remove driveshaft.
3. Disconnect speedometer cable, shift linkage, and all electrical leads at transmission as well as any clips that retain the leads to the transmission case.
4. Remove flywheel cover, mark flywheel and torque converter to maintain original balance. Remove torque converter to flywheel bolts and/or nuts.
5. On gasoline engine vehicles, disconnect catalytic converter support bracket.
6. If used, remove floor pan reinforcement if it interferes with removal or installation of driveshaft.
7. Remove transmission support to transmission mount bolt and transmission support to frame bolts and insulators, if used.
8. Position a suitable jack under transmission, raise transmission slightly, and slide transmission support rearward.
9. Lower transmission to gain acess to oil cooler lines and T.V. cable attachments, disconnect oil cooler lines and T.V. cable, and cap all openings.
10. Support engine with suitable tool, remove transmission to engine bolts, and disconnect transmission assembly.
11. Install torque converter holding tool J-21366 or equivalent, and remove transmission assembly from vehicle.
12. Reverse procedure to install.

Manual Linkage Adjustment

1. Place steering column shift lever in neutral.
2. Place transmission lever in neutral.
3. Assemble clamp, spring, screw to equalizer lever and control rod, Fig. 6.
4. Hold clamp flush against equalizer lever and finger-tighten clamping screw against rod, being careful not to exert any force in either direction on rod or equalizer lever.
5. Torque screw to specification.

T.V. & DETENT CABLE, ADJUST

1974–80 All Exc. THM 400, 425, 475 & Diesel Engine Models

1. Release snap lock on cable housing, Fig. 10.
2. Rotate carburetor throttle control lever to wide open throttle position.
3. Push snap lock closed.
4. Return throttle to closed position.
5. Cycle throttle a few times while observing cable operation for any binds or improper operation.
6. Road test vehicle.

1981–83 All Exc. THM 400, 425, 475 & Diesel Engine Models

1. With engine "Off", depress and hold detent cable metal lock tab located on cable housing near carburetor, Fig. 11.
2. Move cable adjusting slider through fitting in direction opposite of carburetor lever until slider stops against fitting.
3. Release metal lock tab.
4. Rotate carburetor throttle control lever to wide open throttle position. Cable will automatically adjust to the proper length. Release carburetor lever.
5. Road test vehicle to ensure proper operation of transmission. Repeat procedure if necessary.

1978–81 Diesel Engine Models

1. Remove pump rod from lever assembly.
2. Depress and hold detent cable metal lock tab located on cable housing near pump, Fig. 12.
3. Move cable adjusting slider through fitting in direction away from lever assembly until slider stops against fitting. Release metal lock tab.
4. Rotate lever assembly to its full throttle stop position to automatically adjust cable to the proper length. Release lever assembly.
5. Reconnect pump rod to lever assembly.
6. Road test vehicle to ensure proper operation of transmission. Repeat procedure if necessary.

1982–83 Diesel Engine Models

1. Depress and hold metal lock tab located on cable housing, Fig. 13.

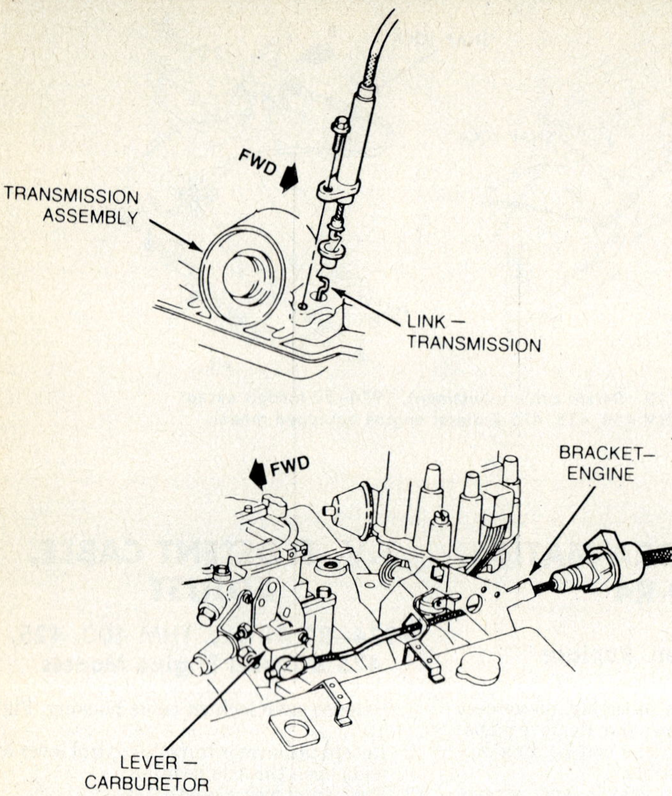

Fig. 11 T.V. & detent cable adjustment. 1981–83 models except THM 400, 475 & diesel engine equipped models

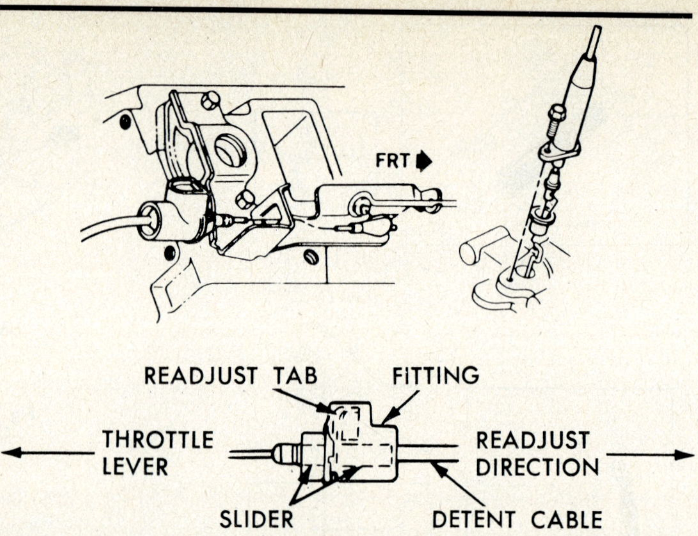

Fig. 13 T.V. & detent cable adjustment. 1982–83 diesel engine equipped models

2. Move cable adjusting slider through fitting in direction away from lever assembly until slider stops against fitting. Release metal lock tab.
3. Rotate lever assembly to its full throttle stop position to automatically adjust cable to the proper length. Release lever assembly.
4. Road test vehicle to ensure proper operation of transmission. Repeat procedure if necessary.

DETENT DOWNSHIFT SWITCH, ADJUST

Turbo Hydra-matic 400, 425 & 475

1. Working underneath dashboard in the area of the accelerator pedal, press down-shift switch plunger as far forward as possible.
2. Depress the accelerator to wide open throttle position. The switch will automatically adjust itself to the proper position.
3. Road test vehicle to ensure proper operation of transmission. Repeat procedure if necessary.

ALLISON AT-540, AT-545, MT-640, MT-643, MT-650, MT-653 & MT-654

Transmission, Replace

1. Disconnect battery ground cable and remove spark plugs so that engine can be turned manually.
2. Remove dipstick and drain transmission fluid.
3. Disconnect oil cooler lines from transmission and disconnect shift cable from transmission shift lever.
4. Disconnect vacuum modulator line, back-up light switch connector, speedometer and parking brake linkage.
5. Remove propeller shaft.
6. Remove the six flex plate-to-converter retaining nuts through access hole in transmission. Rotate flywheel as necessary to gain access to the nuts.
7. Support transmission using a transmission jack (750 lbs. minimum rating), and secure transmission to jack with a safety chain.

NOTE: Position jack so that transmission oil pan will not support weight of transmission.

8. Support engine and remove transmission to crossmember bolts, then raise engine to remove weight from rear engine mounts.
9. Remove transmission to flywheel housing bolts.

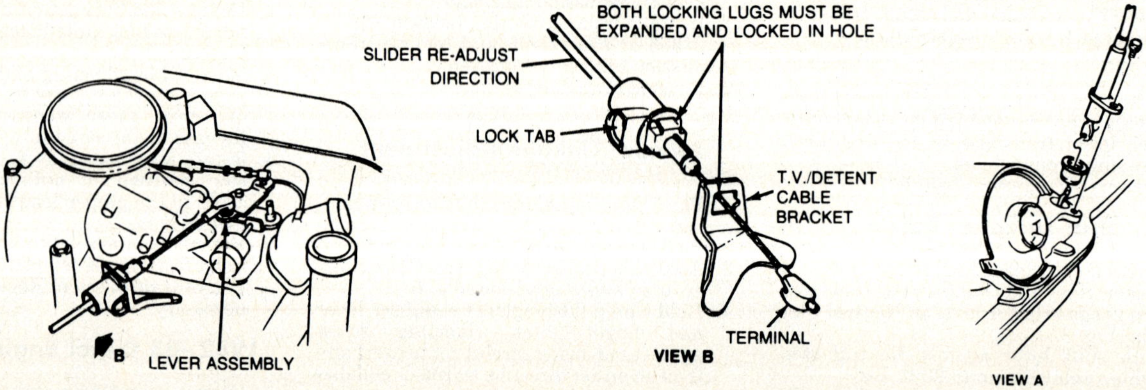

Fig. 12 Detent cable adjustment. 1978–81 diesel engine equipped models

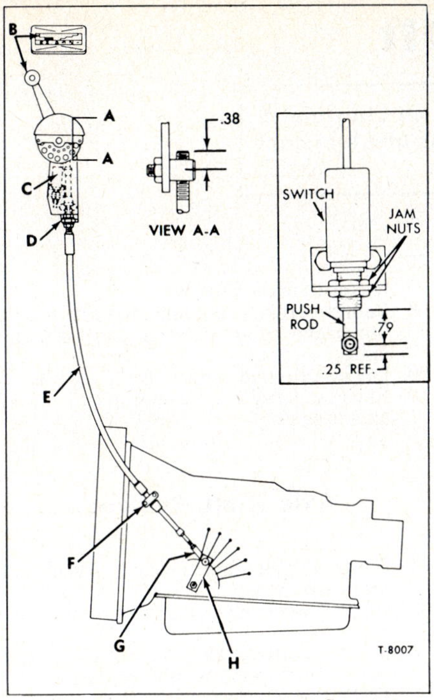

Fig. 14 Manual shift linkage. 1974–79 Allison AT-540

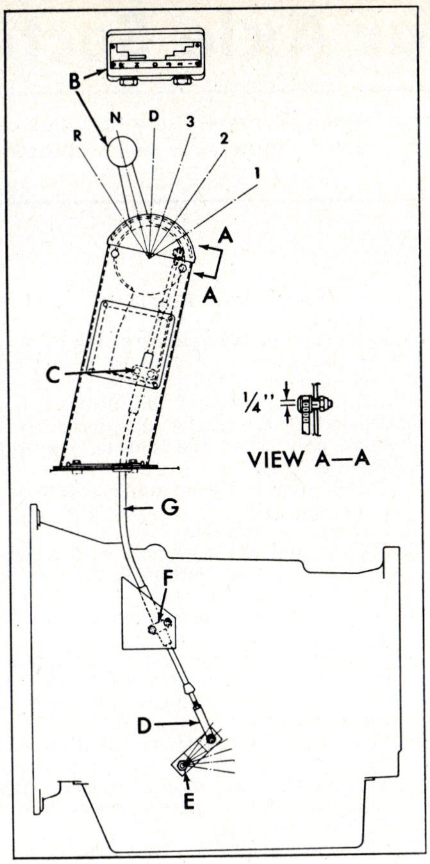

Fig. 15 Manual shift linkage. 1974–79 Allison MT-640, 650

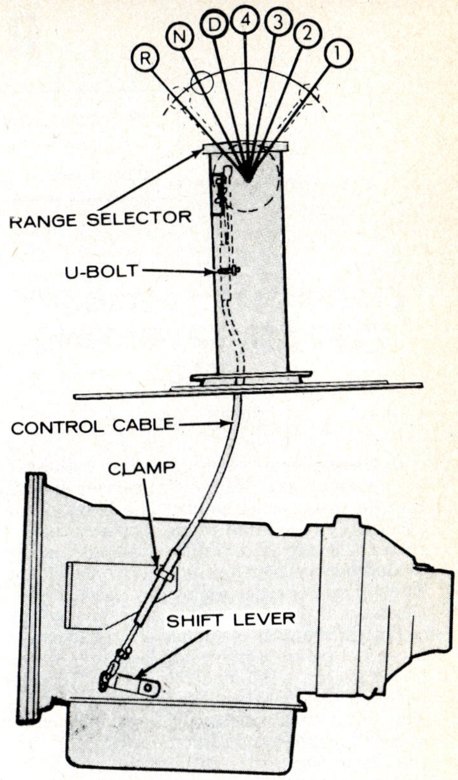

Fig. 16 Manual shift linkage (Typical). 1974–79 Allison MT-643, 653, 654 models & all 1980–83 models

---

NOTE: Inspect transmission and surrounding area to make sure that no lines, hose or wires will interfere with transmission removal.

---

10. Move transmission rearward, then lower and carefully remove transmission.

NOTE: Keep rear of transmission lower than the front so as not to lose converter.

---

11. Reverse procedure to install.

Selector Linkage, Adjust

1974–79 Models Exc. MT-654

1. Disconnect clevis from transmission shift lever and the control cable from transmission bracket. Also, remove cable retainer clips from underside of cab.
2. Remove range selector cover screws and lift range selector out to facilitate cable attachment inspection at selector lever and hanger.
3. Check cable at trunnion for dimension specified in View A-A, Figs. 14 and 15. This dimension is necessary to allow for proper cable length at clevis. If an adjustment is needed, remove cable from range selector lever and hanger assembly as follows:
   a. Remove trunnion locknut and disconnect cable from hanger anchor point.
   b. Mark location of trunnion, then remove trunnion with attached cable core from lever.

NOTE: Trunnion must be installed in original position hole since unsafe vehicle operation may result.

---

4. Loosen trunnion jam nut and rotate trunnion to obtain proper dimension, View A-A, Figs. 14 and 15. Install trunnion in original hole and tighten locknut.

NOTE: On 1977–79 models, the dimension in View A-A, Figs. 14 and 15, should be ½ inch.

---

5. Connect cable to hanger anchor point and torque nuts to 3–5 ft. lbs.
6. Install range selector cover and torque screws to 3–5 ft. lbs. and install the cable retainer clips to cab underside.
7. Connect cable to transmission bracket and tighten screws.
8. Place transmission shift lever in reverse position which is at the end of full counter-clockwise travel of lever and the range selector lever against stop in reverse position.
9. Loosen clevis jam nut and rotate clevis until clevis pin can be inserted through clevis and transmission shift lever holes. Then, rotate clevis one full turn clockwise to allow for cable backlash.
10. Connect clevis to lever with pin and a new cotter pin, however, do not spread cotter pin at this time.
11. Shift range selector lever through all ranges. The transmission detents should fully engage just before range selector lever contacts the stops incorporated in range selector cover.
12. Spread cotter pin and tighten jam nut against clevis.
13. Check adjustment for proper operation.

1980–83 Models & All MT-654 Models

NOTE: Before making adjustment, ensure controls are not damaged and operate smoothly.

---

1. Place range selector lever against stop in "Reverse" position, Fig. 16.
2. Disconnect clevis from transmission lever.
3. Move transmission shift lever fully counterclockwise to "Reverse".
4. Loosen clevis jam nut and adjust clevis until holes on clevis and transmission lever are aligned, then install clevis pin. Adjust clevis ½ turn at a time until pin enters freely.
5. Turn clevis one full turn clockwise to allow for cable backlash.
6. Connect clevis to lever and secure with a new cotter pin.
7. Check for proper operation.

Rear Axle Section

For Eaton, Spicer and Timken (Rockwell) driving axles used on Chevrolet & GMC trucks, see the "name" chapters for service procedures and lubricant capacities

CHEVROLET SALISBURY TYPE SEMI-FLOATING AXLE

1974–83

In these rear axles, Figs. 1 and 2, the rear axle housing and differential carrier are cast into an integral assembly. The drive pinion assembly is mounted in two opposed tapered roller bearings. The pinion bearings are preloaded by a spacer behind the front bearing. The pinion is positioned by a washer between the head of the pinion and the rear bearing.

The differential is supported in the carrier by two tapered roller side bearings. These bearings are preloaded by spacers located between the bearings and carrier housing. The differential assembly is positioned for proper ring gear and pinion backlash by varying these spacers. The differential case houses two side gears in mesh with two pinions mounted on a pinion shaft which is held in place by a lock pin. The side gears and pinions are backed by thrust washers.

Remove & Replace

Construction of the axle assembly is such that service operations may be performed with the housing installed in the vehicle or with the housing removed and installed in a holding fixture. The following procedure is necessary only when the housing requires replacement.

1. Raise vehicle and place stand jacks under frame side rails.
2. Remove rear wheels.
3. Support rear axle assembly with a suitable jack so that tension is relieved in springs and tie rod.
4. Disconnect tie rod at axle housing bracket.
5. Remove trunnion bearing "U" bolts from rear yoke, split universal joint, position propeller shaft to one side and tie it to frame side rail.
6. Remove axle "U" bolt nuts and allow control arm and shock absorbers to hang freely so that they do not interfere with axle. *On K Series trucks, remove spacer from axle housing.*
7. Disconnect hydraulic brake hose at connector on axle housing.
8. Remove brake drum and disconnect parking brake cable at lever and at flange plate.
9. Lower axle and remove from vehicle.
10. Reverse foregoing procedure to install axle assembly.

Axle Shaft, Replace

1. Raise vehicle and remove wheel and brake drum.
2. Drain lube from carrier and remove cover.
3. Remove differential pinion shaft lock screw and remove differential pinion shaft, Fig. 3.
4. Pull flanged end of axle shaft toward center of vehicle and remove "U" lock from button end of shaft.
5. Remove axle shaft from housing, being careful not to damage seal.
6. Reverse foregoing procedure to install the axle shaft.

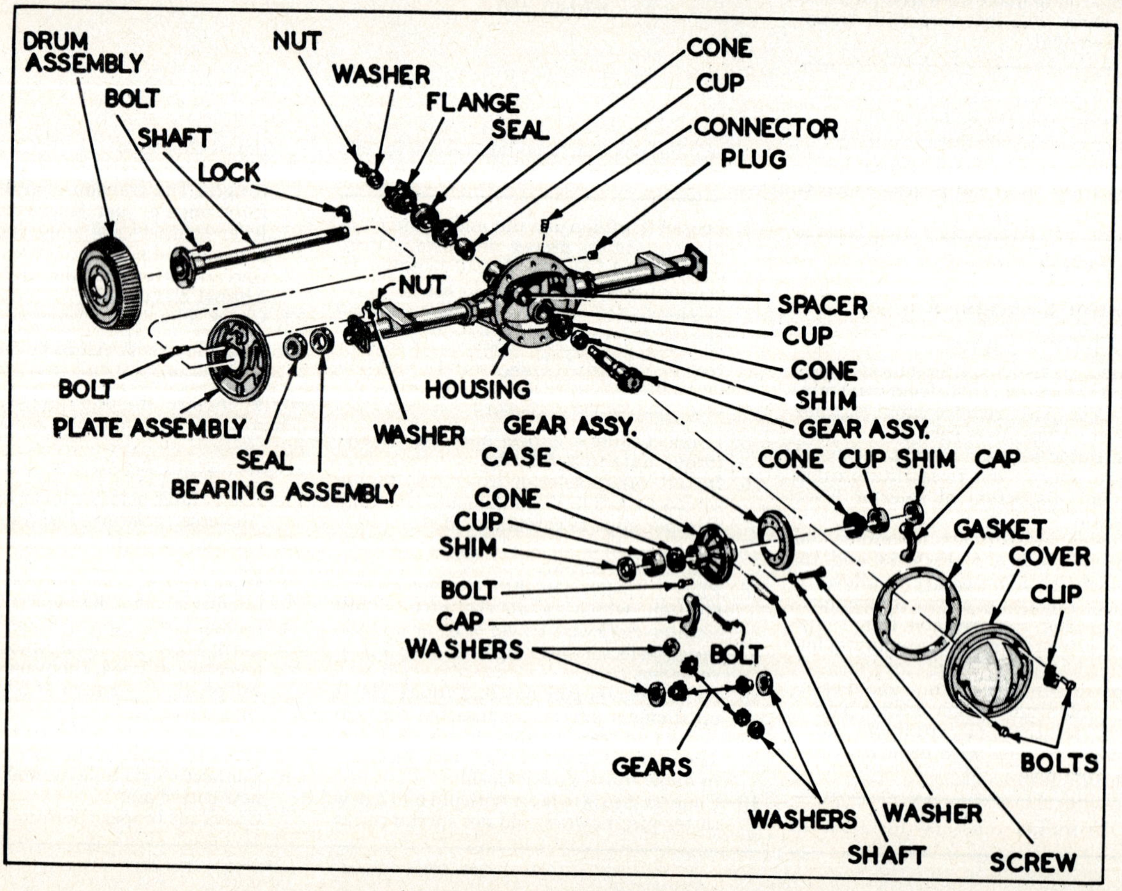

Fig. 1 Salisbury type semi-floating rear axle. 1974–83. Typical

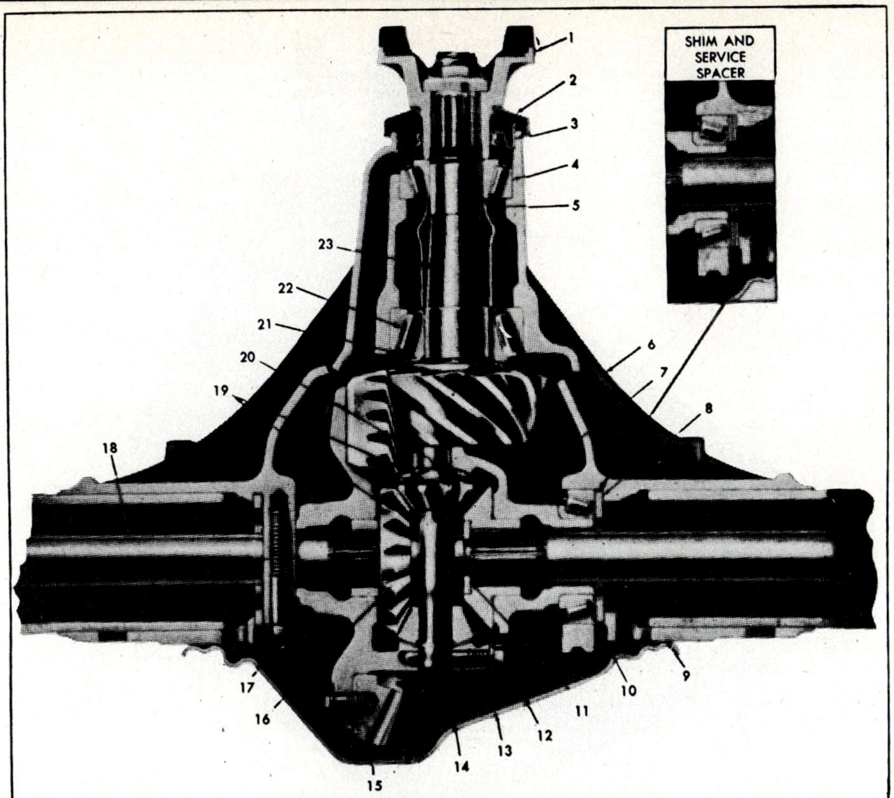

1. Companion Flange
2. Deflector
3. Pinion Oil Seal
4. Pinion Front Bearing
5. Pinion Bearing Spacer
6. Differential Carrier
7. Differential Case
8. Shim (A) with Service Shim
9. Gasket
10. Differential Bearing
11. "C" Lock
12. Pinion Shaft Lock Bolt
13. Cover
14. Pinion Shaft
15. Ring Gear
16. Side Gear
17. Bearing Cap
18. Axle Shaft
19. Thrust Washer
20. Differential Pinion
21. Shim
22. Pinion Rear Bearing
23. Drive Pinion

Fig. 2 Semi floating rear axle of 3300 and 3500 lbs. rated capacity used on 1974–83

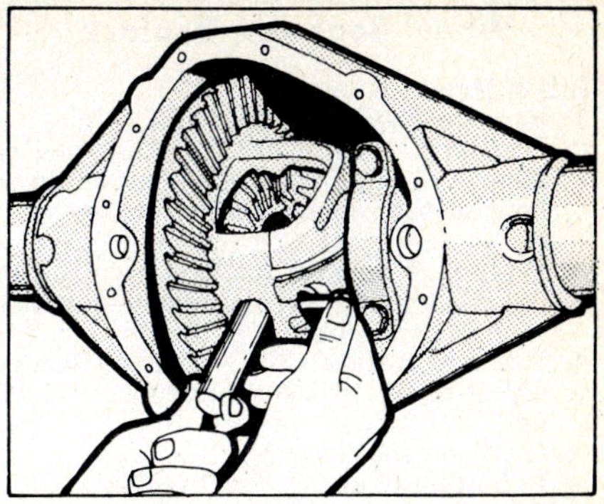

Fig. 3 Removing differential pinion shaft

3. Thoroughly clean both axle shaft flange and end of wheel hub. *Any lubricant on these surfaces tends to loosen axle shaft flange bolts.*
4. Place a new gasket over axle shaft and position axle shaft in housing so that shaft splines enter differential side gear. Position gasket so that holes are in alignment and install flange-to-hub attaching bolts. Torque bolts to 90 ft. lbs. on 1974–75 models; and 115 ft. lbs. on 1976–83 models.

NOTE: To prevent lubricant from leaking through flange holes apply Permatex to bolt threads. Use care in the amount of sealer applied. If the application is too heavy, the sealer may be forced out as the bolt is installed and may destroy the sealing effect of the flange-to-hub gasket.

11,000 & 15,000 Lb. Axles

1. Remove hub cap and install the tool shown in Fig. 7 in tapped hole in shaft flange.
2. Install slide hammer and remove axle shaft.
3. Remove all old gasket material from hub and hub cap. Clean shaft flange and mating surfaces in wheel hub.
4. Install axle shaft so that flange spines index into hub splines. Tap shaft into position.
5. Install new hub cap-to-hub gasket, position cap to hub and install attaching bolts. Torque bolts to 15 ft. lbs.

17,000 Lb. Axles

1. Remove axle shaft flange-to-hub retaining nuts.
2. Strike center of shaft flange with a lead hammer to loosen flange and dowels.
3. Remove tapered dowels from studs and pull axle shaft from housing. It may be necessary to spread dowels to permit removal.
4. Clean old gasket material from wheel hub and axle shaft flange and install a new gasket over hub studs and against hub.
5. Install axle shaft so that splines are aligned with differential side gear and flange holes index over hub studs.
6. Install tapered dowel over each hub stud. Install and tighten 1/2 inch stud nuts to 50–60 ft. lbs. and 5/8 inch stud nuts to 90–100 ft. lbs.
7. See that clearance exists between face of stud nuts and axle shaft flange. If no clearance exists, it indicates excessive wear at studs, tapered dowels or flange holes. Replace worn part if necessary.

Wheel Bearing/Oil Seal, Replace

1. Remove axle shaft as previously described.
2. Use a suitable puller to remove bearing and oil seal.
3. Lubricate wheel bearing and cavity between seal lips with wheel bearing lubricant before installation.
4. Reverse procedure to install.

CHEVROLET FULL FLOATING AXLES

1974–83 Light Duty Type

The straddle mounted drive pinion is supported at the front by two opposed tapered roller bearings. The pinion year roller bearing consists of an outer race and roller assembly and a precision ground diameter on the pinion pilot functions as an inner race.

On units with 10½ inch ring gear, Fig. 4, side bearing preload and ring gear to pinion backlash are controlled by side bearing adjusting rings threaded into the carrier. Pinion depth is controlled by a shim located between the pinion bearing retainer assembly and axle housing.

Medium and Heavy Duty Types, Figs. 5 & 6

These axles have a straddle-mounted drive pinion which is supported at the rear by a straight roller bearing. The pinion front bearing consists of a double row ball on the medium duty axles and two opposed tapered roller bearings on the heavy duty models. Selective shims placed between front bearing and its seat in the carrier housing provide a means of adjusting pinion depth to attain correct ring gear and pinion tooth contacts.

The differential is a four pinion type. Thrust washers are used between the side gears and case and also between differential pinions and case.

A thrust pad mounted on the end of an adjusting screw threaded into the carrier housing limits deflection of the ring gear under high torque conditions.

Axle Shaft, Replace

7200 Lb. Axles with 10½ inch Ring Gear

1. Remove bolts that attach axle shaft flange to wheel hub.
2. On 1974–83 units with 10½ inch ring gear, tap on flange with a rawhide mallet to loosen axle shaft, then remove axle shaft by twisting shaft with locking pliers.

Wheel Bearings, Replace

Hub & Drum, Remove

1. Remove wheel and axle shaft.
2. Disengage tang of nut lock from slot or flat of lock nut, then remove lock nut from housing tube, using appropriate tool, Fig. 8.
3. Disengage tang of nut lock from slot or flat of adjusting nut and remove nut lock from housing tube.
4. Use appropriate tool, Fig. 8, to remove adjusting nut from housing tube. *On 5200 and 7200 axles, remove thrust washer from housing tube.*
5. Pull hub and drum straight off axle housing, using care on 11,000, 15,000 and 17,000 lb. axles to avoid dropping outer bearing inner race and roller assembly.

NOTE: On 15,000 lb. axle, remove inner bearing race and inner bearing oil seal from axle housing and discard seal.

Bearing Cup, Replace

Replace inner cup (all axles) and outer bearing cup for 17,000 lb. axles as follows:

1. Cut a suitable length of 1/2" steel bar stock for press-out tool as shown in Fig. 9.
2. Place appropriate press-out tool behind bearing cup, index tool in notches provided, and press out cup with an arbor press.
3. Position bearing cup in hub with thick edge of cup toward shoulder in hub. Then, using applicable cup installer, press cup into hub until it seats on hub shoulder. Make certain cup is not cocked and that it is fully seated against shoulder.

Outer Bearing, Replace

Replace outer bearing on all axles except 17,000 lb. axle as follows. The inner bearing must be removed before attempting to replace the outer bearing.

1. Using a punch of suitable length, tap bearing outer race away from bearing retainer ring, Fig. 10. Then remove retaining ring from hub, Figs. 10 and 11.
2. Press bearing from hub, Figs. 12 and 13.
3. Place axle shaft spacer (11,000 and 15,000 lb. axles) and inner race and roller assembly in hub with larger O.D. of roller assembly toward outer end of hub. Then position bearing cup in hub with thin edge of cup toward outer end of hub. Press cup into hub until clearance is obtained for retainer ring installation.
4. Install retainer ring. Then press bearing cup into positive contact with retainer ring.

NOTE: The bearing cup-to-retainer ring seating procedure is essential to assure that an accurate wheel bearing adjustment will be obtained, and that the adjustment will not loosen up during subsequent vehicle operation.

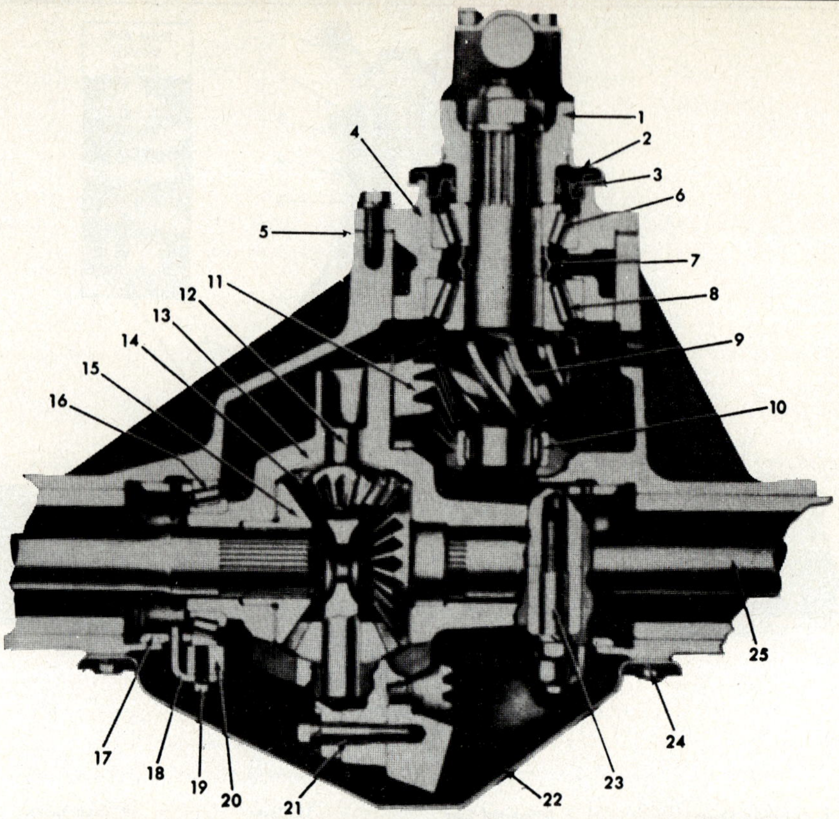

1. Companion Flange
2. Oil Deflector
3. Oil Seal
4. Bearing Retainer
5. Shim
6. Pinion Front Bearing
7. Collapsible Spacer
8. Pinion Rear Bearing
9. Drive Pinion
10. Straddle Bearing
11. Ring Gear
12. Differential Spider
13. Differential Case
14. Differential Pinion
15. Differential Side Gear
16. Side Bearing
17. Side Bearing Adjusting Nut
18. Adjusting Nut Retainer
19. Retainer Screw
20. Bearing Cap
21. Case-to-Ring Gear Bolt
22. Differential Cover
23. Bearing Cap Bolt
24. Cover Screw
25. Axle Shaft

Fig. 4 Full floating rear axle

Wheel Bearings, Adjust

Before checking wheel bearing adjustment, make sure brakes are fully released and do not drag. Check bearing play by grasping tire at top and pulling back and forth or by using a pry bar under tire. If bearings are properly adjusted, movement of brake drum in relation to brake flange plate will be barely noticeable and wheel will turn freely. If movement is excessive, adjust bearings as follows:

1. Remove axle shaft and raise vehicle until wheel is free to rotate.
2. Disengage tang of nut lock from lock nut and remove them from axle housing tube.
3. Using appropriate tool, tighten nut to specifications given below, at the same time rotating hub to make sure that all bearing surfaces are properly seated. Then back off adjusting nut one flat or notch.
   a. On 5200 and 7200 axles, tighten adjusting nut to 50–60 ft. lbs. and back off nut 1/8 turn to align nearest slot with short tang on nut lock.
   b. On 11,000 lb. axle, tighten adjusting nut to 75–100 ft. lbs. and back off nut 1/8 turn to align nearest slot with short tang on nut lock.
   c. On 15,000 lb. axle, tighten adjusting nut to 75–100 ft. lbs. and back it off to align nearest slot with short tang on nut lock.
   d. On 17,000 lb. axle, tighten adjusting nut to 65 ft. lbs. and back it off 1/16" to 1/4 turn to align slot with short tang on nut lock.
   e. On axles with 10 1/2 inch ring gear, torque adjusting nut to 50 ft. lbs.
4. Install nut lock.
5. Make sure wheel turns freely. Then install lock nut and tighten to specifications given below. Bend tang of nut lock over flat or slot of lock nut.
   a. On 5200 and 7200 lb. axles, tighten lock nut to 175 ft. lbs.
   b. On 11,000 and 15,000 lb. axles, tighten lock nut to a minimum of 250 ft. lbs.
   c. On 17,000 lb. axles, tighten lock nut to 135 ft. lbs.
   d. On axles with 10 1/2 inch ring gear, torque locknut to 65 ft. lbs.
6. Lower vehicle and install axle shaft.

Differential Carrier

Major operations on the differential carrier assembly may be performed without removing the rear axle assembly from the vehicle. After removing the axle shafts, remove the differential carrier assembly from the axle housing.

To facilitate adjusting of pinion depth in the ring gear, there are four shims available for service use. They are .012, .015, .018 and .021".

If the original ring gear and pinion are to be used, it is advisable to replace the same thickness of shims in the carrier housing counter-

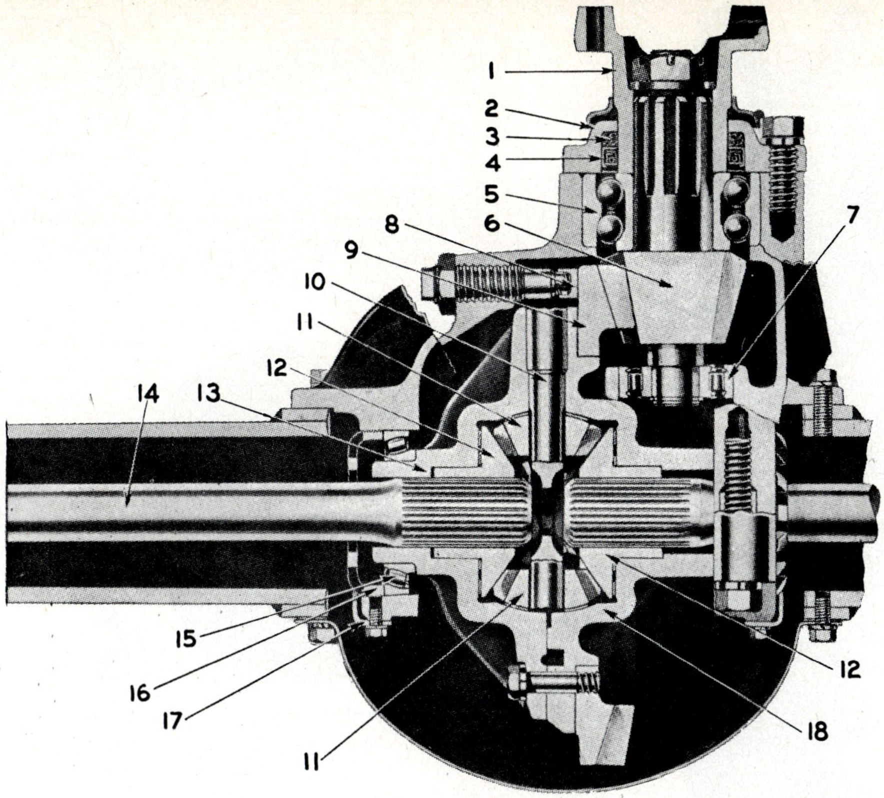

1. Universal Joint Yoke
2. Pinion Bearing Retainer and Oil Seal
3. Oil Seal Packing
4. Oil Seal
5. Front Pinion Bearing
6. Drive Pinion
7. Rear Pinion Bearing
8. Ring Gear Thrust Pad
9. Ring Gear
10. Differential Spider
11. Differential Pinion (Spider) Gear
12. Differential Side Gear
13. Differential Case—Left Half
14. Axle Shaft
15. Differential Bearing
16. Differential Bearing Adjusting Nut
17. Adjusting Nut Lock
18. Differential Case—Right Half

Fig. 5 Full-floating rear axle of 11,000 lbs. rated capacity

bore that were removed.

If a new ring gear and pinion are used, one .021" shim should be used as a standard starting setup.

Checking Pinion Depth

NOTE: Details on tooth patterns and method of checking ring gear and pinion backlash are given in the Timken and Eaton axle chapters. Backlash between ring gear and pinion in Chevrolet axles are .003–.006" on 5200 lb. axles and .005–.008" on all others.

1. Coat the ring gear teeth lightly with red lead or prussian blue. Then turn the pinion shaft several revolutions in both directions.
2. If the pinion depth is correct, the tooth pattern will be centered on the pitch line and toward the toe of the ring gear.
3. If the pattern is below the pitch line on the ring gear teeth, the pinion is too deep and it will be necessary to remove the pinion assembly and increase the shim thickness between pinion bearing and carrier.
4. If the pattern is above the pitch line on the ring gear teeth, the pinion is too shallow and it will be necessary to remove the pinion and decrease the shim thickness between pinion bearing and carrier.
5. Changing the pinion depth will make some change in backlash; therefore, it will be necessary to readjust the backlash.

Ring Gear Thrust Pad, Adjust

After the differential carrier is assembled and properly adjusted, install the thrust pad and tighten the adjusting screw until the bronze tip engages the back of the ring gear while the gear is rotated. Back off the screw 1/12 turn and tighten lock nut, Fig. 14. Make sure screw does not turn during the locking process.

GMC SINGLE SPEED AXLE

These axles, being of the full floating type, enables removal of the axle shafts without removing the truck load or jacking up the rear axle. The drive pinion is stradle mounted, being supported at the rear end by a straight roller bearing and at the front end by a double row ball bearing, Fig. 15, or opposed tapered roller bearings, Fig. 16.

Axle Shaft, Replace

Dowel Drive Type

1. Remove nuts attaching axle shaft flange to wheel hub.
2. Strike center of axle flange with a suitable hammer to loosen flange and split tapered dowels from hub studs. Remove split tapered dowels from studs.
3. Pull axle shaft from housing. Discard gasket.
4. Reverse procedure to install. Lubricate small end of axle shaft and install into housing using a new gasket. Torque 1/2 inch nuts to 50–60 ft. lbs. and 5/8 inch nuts to 90–110 ft. lbs.

NOTE: Ensure that clearance exists between nut and flange. Fig. 17. If no clearance is noted, replace worn parts as required.

Spline Drive Type

1. Remove hub cap retaining cap screws and hub cap.
2. Install a slide hammer adapter into tapped hole in axle flange.
3. Attach slide hammer onto adapter and remove axle shaft from housing, Fig. 18.
4. Reverse procedure to install. Lubricate small end of axle shaft and install into housing using a new gasket. Torque axle flange cap screws on 1974–76 models to 15–20 ft. lbs. 1977–81 models to 11–18 ft. lbs.

Wheel Bearings, Replace

1. Raise and support rear axle.
2. Remove wheel and axle shaft.
3. Remove hub and drum. Figs. 19 and 20.

NOTE: On models equipped with spline drive type axle shafts, pull hub from axle housing using care not to drop outer wheel bearing.

4. Remove inner bearing and oil seal.
5. On models equipped with dowel type axle shafts, remove outer bearing race using suitable drift.
6. On models equipped with spline drive axle shafts, tap on outer race to relieve tension on bearing snap ring. Remove snap ring, then, using a suitable drift, drive on axle shaft spacer to remove outer bearing assembly from hub.
7. Reverse procedure to install. Refer to "Wheel Bearings, Adjust" for adjustment.

Wheel Bearings, Adjust

1. Raise and support rear axle.
2. Remove axle shaft as described in "Axle Shaft, Replace". Release brakes.
3. Remove bearing lock nut and nut lock or lock ring.
4. On models equipped with tapered bearing, torque adjusting nut to 50 ft. lbs. while rotating wheel, then back off 1/4 to 1/3 turn to obtain .001 to .007 inch end play. Then proceed as follows:
   a. On models equipped with type shown in section A of Fig. 21, align flat on adjusting nut with nearest lip of nut lock. Ensure that wheel spins freely, then install nut lock and lock nut. Torque lock nut to 135 ft. lbs. and bend one lip of nut lock over a flat of each nut.
   b. On models equipped with type shown in section B of Fig. 21, align nearest slot in adjusting nut with short tang on nut lock and bend tang into slot of nut. Install lock nut with slots outward and torque to 135 ft. lbs. Bend tangs of nut lock into slots of both nuts.

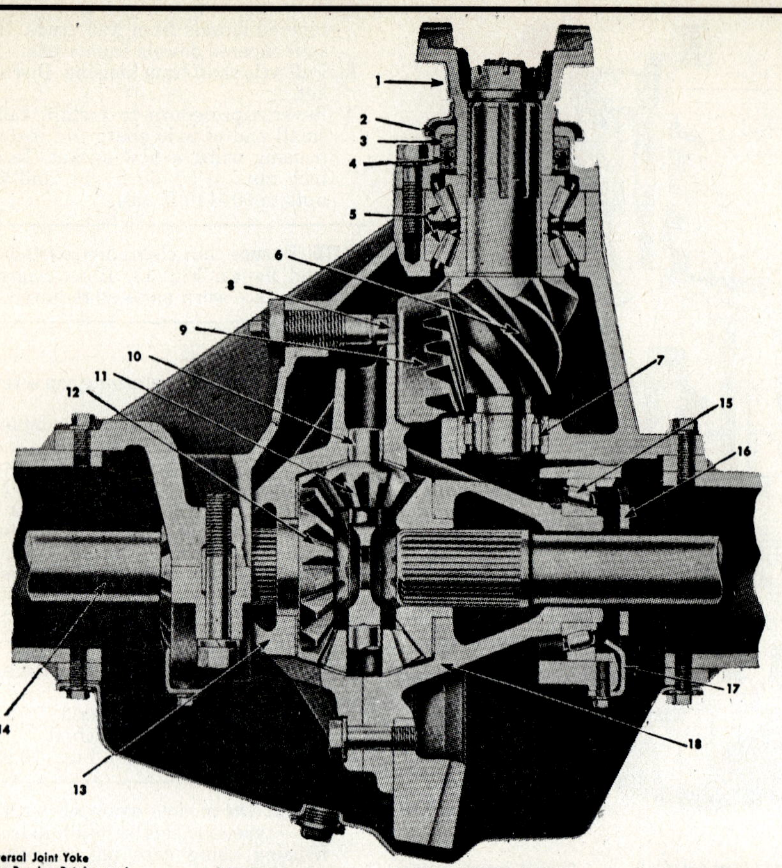

1. Universal Joint Yoke
2. Pinion Bearing Retainer and Oil Seal
3. Oil Seal Packing
4. Oil Seal
5. Pinion Bearing Assembly
6. Drive Pinion
7. Rear Pinion Bearing
8. Ring Gear Thrust Pad
9. Ring Gear
10. Differential Spider
11. Differential Pinion (Spider) Gear
12. Differential Side Gear
13. Differential Case (Left Half)
14. Axle Shaft
15. Differential Bearing
16. Differential Bearing Adjusting Nut
17. Adjusting Nut Lock
18. Differential Case (Right Half)

Fig. 6 Full-floating rear axle 13,500, 15,000 & 17,000 lbs. rated capacity

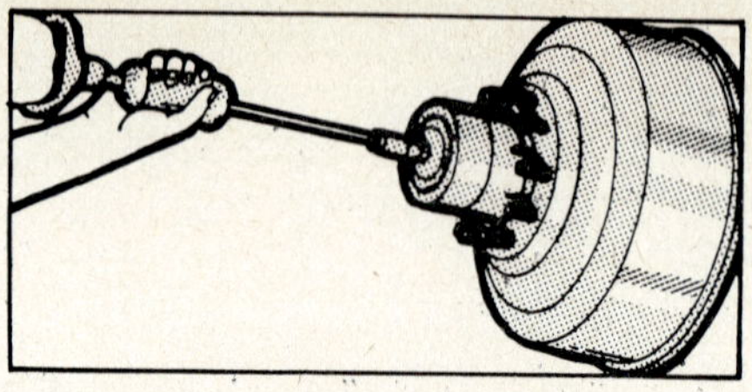

Fig. 7 Removing axle shaft. 11,000 & 15,000 lb. axles

Fig. 8 Removing bearing adjusting nut lock nut

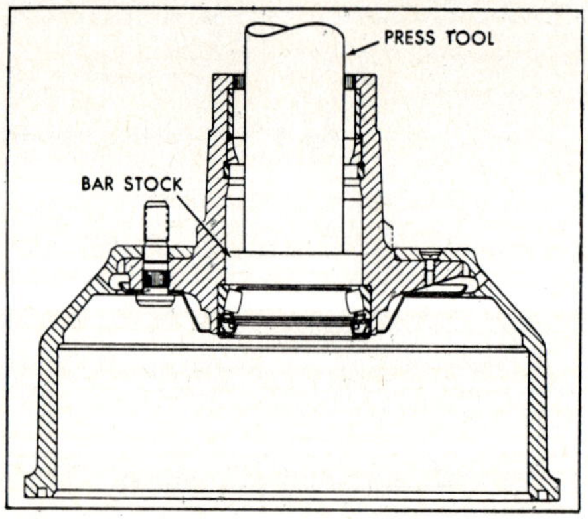

Fig. 9 Removing hub inner bearing cup

c. On models equipped with type shown in section C of Fig. 21, align dowel pin of adjusting nut with nearest hole in lock ring. Install lock nut and torque to 250–300 ft. lbs.

NOTE: Lock ring can be turned on either side to align hole with least movement of adjusting nut.

5. On models equipped with barrel bearing, torque adjusting nut to 75–100 ft. lbs. while rotating wheel back and forth to seat bearings. Back off adjusting nut a minimum amount to obtain as close to zero end play as possible. The wheel should rotate freely. Align the nearest slot in adjusting nut with short tang on nut lock and bend tang into slot of nut. Install lock nut with slots facing outward and torque to 250 ft. lbs. Bend tangs of nut lock into slots of both nuts.

Rear Axle, Replace

1. Raise vehicle from floor and support with stand jacks under frame side rails.
2. Remove rear wheels.
3. Disconnect propeller shaft from rear axle.
4. Swing propeller shaft to one side and tie to frame side rail.
5. Disconnect hydraulic or air brake line connection at rear axle housing.
6. Disconnect hand brake cables from cross shaft levers and remove cables from clamps on frame, if equipped.

Fig. 10 Removing bearing retainer. 5200 & 7200 lb. axles

Fig. 11 Removing bearing retainer ring 11,000 & 15,000 lb. axles

Fig. 12 Removing hub outer bearing. 5200 & 7200 lb. axles

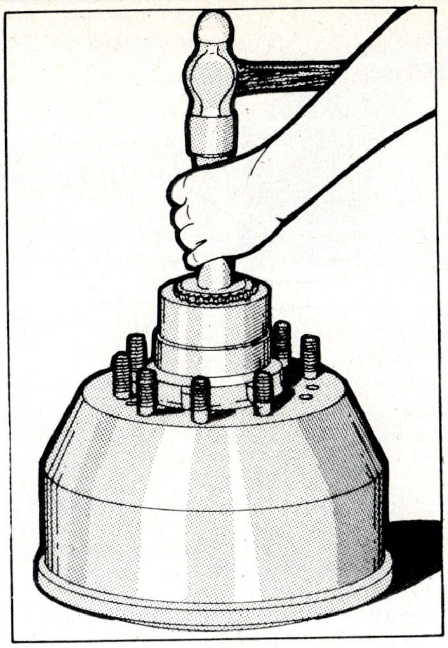

Fig. 13 Removing hub outer bearing. 11,000 & 15,000 lb. axles

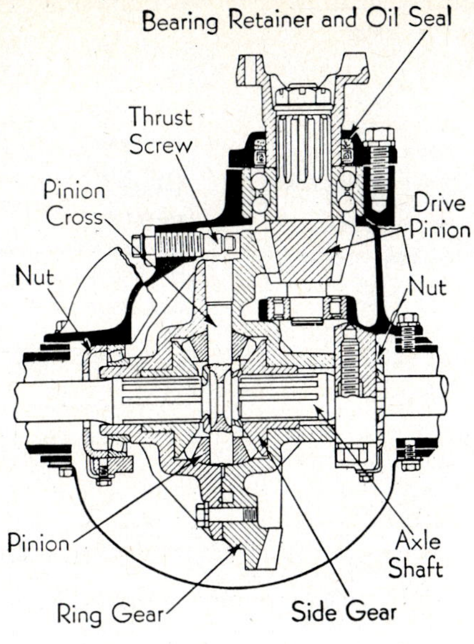

Fig. 15 Sectional view of differential carrier. GMC single speed with double row ball bearing

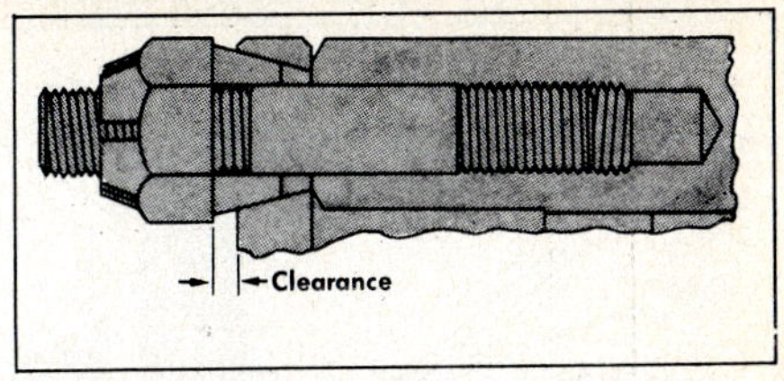

Fig. 17 Axle shaft flange to stud nut clearance. Dowel drive type axle shaft

Fig. 18 Removing axle shaft. Spline drive type axle shaft

7. Disconnect shock absorber links or eyes from anchor plates or anchor bracket, if equipped.
8. On all models, while supporting axle assembly with hydraulic jack, remove spring "U" bolts and lower axle assembly to floor.
9. Reverse procedure to install.

SERVICE NOTE

The following operations may be performed without raising the truck or removing the axle assembly from the vehicle.

Differential Carrier, Replace

1. Drain lubricant, remove axle housing cover and axle shafts.
2. Disconnect propeller shaft from rear axle.
3. Swing propeller shaft to one side and tie to frame side rail.
4. Remove bolts or nuts which attach carrier to axle housing. Support carrier with floor jack and roll it from under truck.
5. Reverse procedure to install. If equipped with cap screws, torque to 80 ft. lbs. If equipped with stud nuts, torque to 150 ft. lbs.

Differential Carrier, Disassemble

1. Loosen ring gear thrust pad adjusting screw and remove thrust pad.
2. Remove differential bearing adjuster locks, bearing cap screws and bearing caps.

NOTE: Mark bearing cap to carrier position to help reposition during assembly.

---

3. Remove differential and ring gear from carrier housing.
4. Remove pinion oil seal retainer, then the pinion and bearing assembly from carrier using suitable drift.

NOTE: Refer to Fig. 16, then remove pinion adjustment shims, making note of the number of total thickness of shims removed.

---

5. Remove pinion nut from shaft and press oil seal retainer and yoke from shaft.

Pinion Oil Seal, Replace

1. Remove old seal and packing (if equipped), from retainer.
2. Soak new seal and packing (if equipped), in light engine oil until saturated.

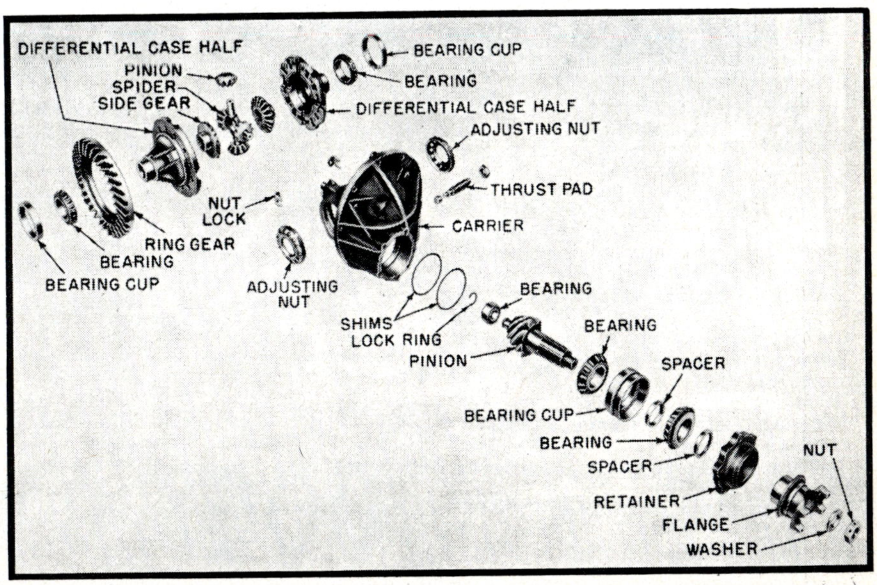

Fig. 16 Exploded view of differential carrier assembly. GMC single speed with tapered roller bearings

Fig. 14 Ring gear thrust pad adjustment

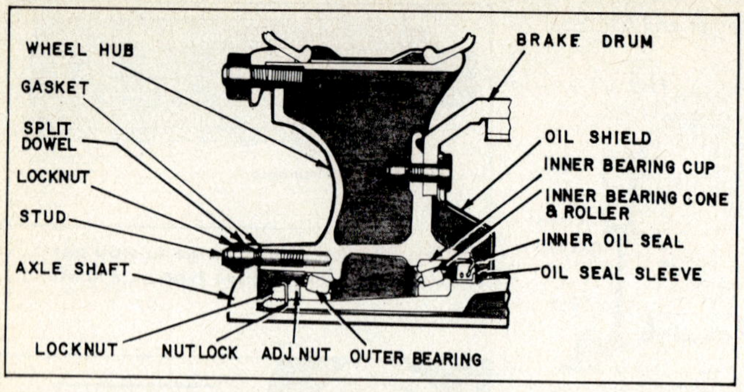

Fig. 19 Rear wheel hub and bearing sectional view. Dowel drive type axle shaft

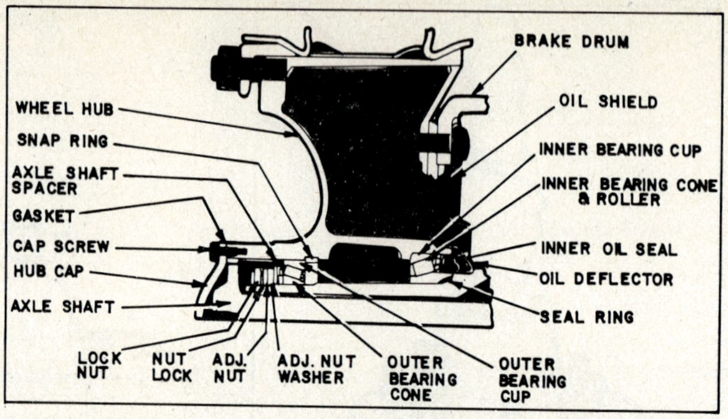

Fig. 20 Rear wheel hub and bearing sectional view. Spline drive type axle shaft

3. Clean seal bore of retainer.
4. Install felt packing (if equipped), into retainer, then lightly coat outer surface of oil seal with a suitable sealer and install into retainer with lip of seal facing toward inner side of retainer.

Pinion Bearing, Replace

1. Remove rear pinion bearing snap ring if equipped, and press bearing from shaft.
2. Press front pinion bearing from shaft. *On models with tapered roller bearings, the outer race, two inner race and roller assemblies and spacer should always be installed as a group as the spacer is preselected to give proper pinion bearing adjustment.*
3. On models with tapered roller bearings, press one new inner race and roller assembly on the pinion shaft, using a 2" pipe, applying pressure against the inner race. Large diameter of bearing race should be toward pinion. Place spacer and outer race on shaft and press on other inner roller. Small diamter of this race should be toward pinion.
4. On models with ball bearing, press new double row pinion bearing on shaft, using a piece of 2" pipe, applying pressure to inner race. Thus bearing must be installed so that extended portion of inner race is toward the back of pinion gear teeth.
5. On all models, press rear pinion bearing on shaft, using driver to apply pressure to inner race. Chamfered side of inner race should be toward pinion.
6. Install lock ring, if required.
7. Position pinion bearing oil seal retainer over hub of drive flange.
8. On all models, install drive flange and bearing retainer with washer and nut.
9. Tighten pinion shaft nut to a torque load of 220 ft. lbs. Lock nut with cotter pin.

Differential, Disassemble

1. Check differential case to make sure two halves are marked so they may be reassembled in the same position.
2. Remove attaching bolts and separate case from cover and remove pinion gears, spider and side gears.
3. Tap back of ring gear with soft hammer to remove gear from case.
4. Pull off differential bearings.

Differential, Assemble

1. Install differential bearings.
2. Install two guide pins to ring gear diametrically opposite each other. These pins can be made from 1/2"-20 × 2" capscrews with heads cut off and slotted for use of screwdriver.
3. Start guide pins through holes in case flange and tap ring gear on case.
4. Place differential pinions on spider. Then assemble side gears and pinions to left half or cover of differential.
5. Assemble ring half to left half, being sure to line up marks on both halves.
6. Install ring gear bolts and washers and tighten evenly until ring gear is flush with case flange.
7. Remove guide pins and install remaining bolts. Then torque all bolts to 150–170 ft. lbs. on 1974–78 models exc. H-110, 165 ft. lbs. on 1979–81 models exc. H-110, 100–110 ft. lbs. on 1973–78 models H-110, and 118 ft. lbs. on 1979–81 model H-110.

Differential Carrier, Assemble

1. Install differential bearings.
2. Install pinion assembly into carrier with new gasket. Torque pinion bearing securing bolts to 165 ft. lbs., 1974–78 models; 95 ft. lbs., on 1979–81 model H110; 98 ft. lbs., model H135; 93 ft. lbs., model H150; and 165 ft. lbs., model H170.
3. Install differential assembly in carrier and install adjusting nuts.
4. Install differential bearing caps, making sure marks on caps line up with marks on carrier.
5. Install bearing cap bolts and tighten until lock washers just flatten out.
6. Adjust ring gear to pinion backlash and pinion depth as described in "Ring Gear and Pinion, Adjust" and "Pinion Depth, Adjust".
7. Torque differential bearing cap bolts to 205 ft. lbs. on 1974–78 all models and 1979–81 H110 models. On 1979–81 H135, H150 and H170 models, bearing cap screws should be torqued to the point that the helical spring washers are flat.

Ring Gear & Pinion, Adjust

1. Loosen right hand adjusting nut and tighten left hand adjusting nut.
2. Continue tightening left adjusting nut until all backlash between ring gear and pinion is removed.
3. Back off left adjusting nut approximately two notches to a point where notch in nut is aligned with lock.
4. Tighten, right adjusting nut firmly to force differential and planet assembly in solid contact with left adjusting nut. Loosen right adjusting nut until it is free of its bearing, then retighten snugly against bearing. Tighten right nut from one to two notches more to a locking position. This method provides for the proper pre-load on the bearings.
5. Mount a dial indicator on the housing and check the backlash between ring gear and pinion which should be .005" to .008". Whenever backlash is more than .008", loosen right adjusting nut one notch and tighten left nut one notch. If backlash is less than .005", loosen left nut one notch and tighten right nut one notch. When backlash has been adjusted, tighten bearing cap bolts securely. Again check backlash and then install adjuster locks.

Pinion Depth, Adjust

NOTE: On models equipped with ball bearings, pinion depth is not adjustable.

1. Paint ring gear teeth with a mixture of red lead and oil to produce a contact pattern.
2. Rotate pinion through several revolutions, applying pressure to the ring gear to develop a definite contact pattern.
3. Pinion depth is controlled by the shim pack between the pinion bearing seal retainer and the carrier.

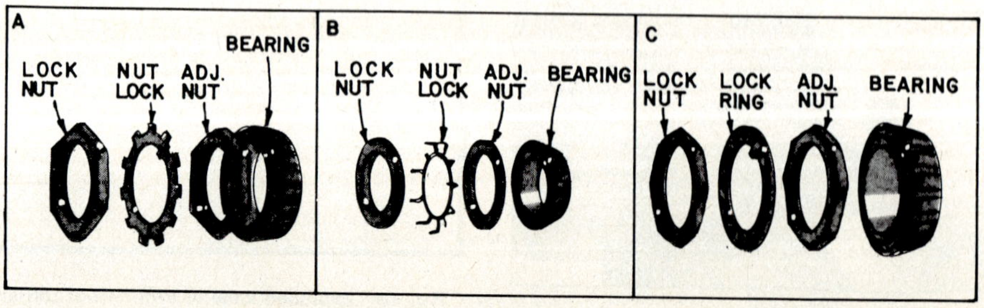

Fig. 21 Rear wheel bearing adjustment nuts. Full floating axle

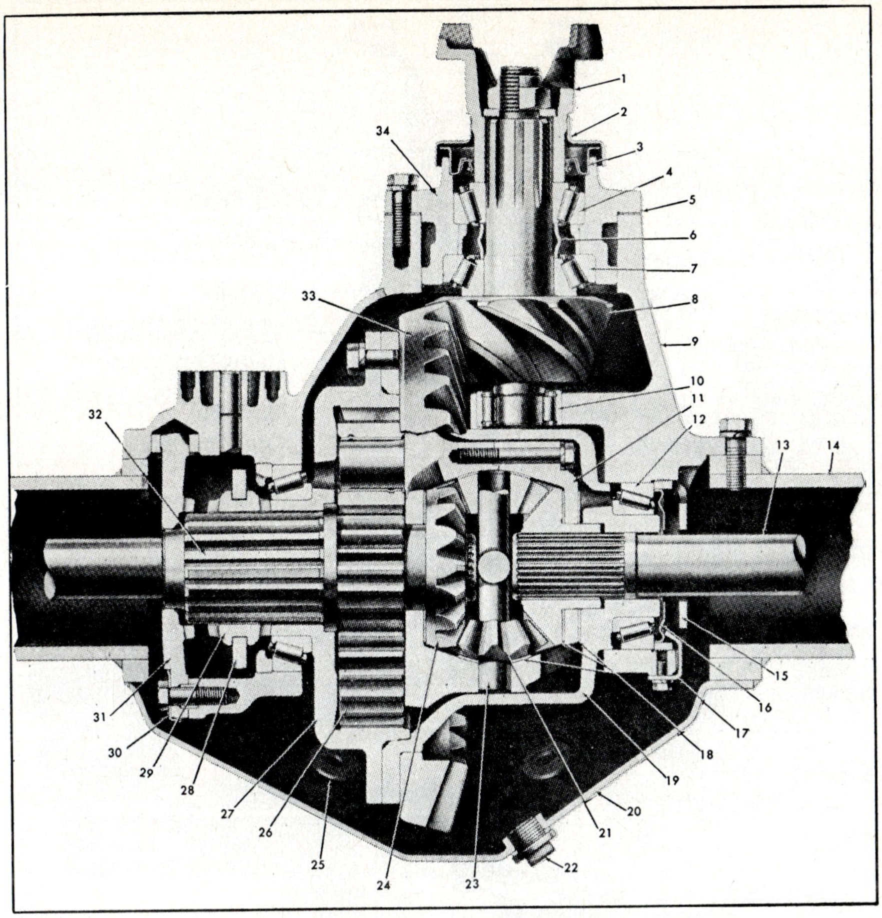

1 Pinion Flange	10. Pinion Rear Bearing	18. Thrust Washers	26. Planet Gears (4)
2 Dust Deflector	11 Differential and Planet Support	19 Differential and Planet Support Case Cover	27 Differential and Planet Support Case
3 Pinion Oil Seal	12. Differential Bearing (Right)	20 Axle Housing Cover (Welded)	28. Shift Yoke
4 Pinion Front Bearing	13. Axle Shaft	21 Differential Pinion	29 Shift Sleeve
5. Pinion Adjusting Shim	14. Axle Housing	22. Filler Plug	30. Anchor Bolt Lock
6. Pinion Bearing Preload Spacer	15 Differential Bearing Adjusting Nut (Right)	23. Differential Pinion Shaft	31 Anchor
7 Pinion Intermediate Bearing	16. Oil Baffle	24 Differential Side Gear (Left)	32. Sun Gear
8. Pinion	17. Adjusting Nut Lock (Right)	25. Magnetic Chip Collector	33. Ring Gear
9. Differential Carrier			34. Pinion and Bearing Retainer

Fig. 22 Chevrolet two-speed axle

4. Changing pinion depth will cause some change in ring gear backlash. After adjusting pinion depth, recheck backlash.

Ring Gear Thrust Pad, Adjust

Inspect the bronze tip of ring gear thrust pad and replace if worn. Install the thrust pad and tighten the screw until the bronze tip engages the back of the ring gear while rotating the gear. Back off the screw 1/12 of a turn and tighten lock nut to 135 ft. lbs. Make sure the screw does not turn during the locking process. This adjustment provides .005 to .007" clearance between ring gear and thrust pad.

CHEVROLET TWO-SPEED AXLE

This unit, Fig. 22, is available in 15,000, 17,000 and 18,000 lb. capacity. In low gear, torque is transmitted to the differential case through four planetary pinions. In high range, only the straddle mounted drive pinion and the ring gear operate to produce the high range reduction, the planet and sun gears being locked to revolve with the ring gear.

The electric axle shift unit is mounted on the differential carrier. This unit contains a reversible electric motor, which when energized, shifts the axle into the desired range through mechanism contained in the unit.

Differential Carrier Removal

1. Loosen housing attaching cap screws and drain lubricant.
2. Remove axle shafts and electric or vacuum lines.
3. Disconnect drive shaft and swing it to one side.
4. Remove all carrier attaching cap screws except two at the top. Loosen the top two but leave them installed to prevent the carrier from falling.
5. Tap along the outer edge of the carrier with a soft faced hammer to break carrier loose.
6. Support carrier with a suitable jack, remove remaining two cap screws and remove carrier.

Disassembly

1. Remove drive pinion nut and washer. Discard nut and use a new one when reassembling.
2. Remove flange.
3. Remove pinion bearing retainer and pinion. Remove shims from between carrier and retainer. Save shims for readjustment during reassembly.
4. Remove shifter assembly from carrier.
5. Remove shift yoke anchor.
6. Using a drift, drive out shifter yoke shaft.
7. Slide shifter yoke and sleeve to the left, removing the sleeve and yoke from the end of the sun gear.
8. Remove differential bearing adjusting nut lock from each nut and remove locks from both sides.
9. Remove two retaining bolts and withdraw oil trough from carrier.
10. Remove bearing caps and adjusting nuts by tapping on bosses of caps until free of dowels. Remove the differential and planet assembly from the housing.
11. Disassemble pinion.
12. Mark ring gear and differential and planet support case and case cover for alignment by scribing lines on case, cover and gear.
13. Remove cover retaining bolts and remove cover.
14. Mark or tag all gears so they can be reassembled in the same relative position then remove the sun and planet gears.
15. Lift the differential and planet support assembly from the support case.
16. Remove planet support-to-cover bolts and separate the two.
17. Mark or tag all parts for reassembly in the same relative position and remove spider and pinion gears and thrust washers.
18. Remove side gears and thrust washers from each side of the housing.
19. Remove bearings.

Reassembly

Reverse the disassembly procedure to assemble the unit, being careful to align all reference marks made during disassembly.

Ring Gear & Pinion, Adjust

1. Loosen right differential adjusting nut and tighten left adjusting nut.
2. Tighten left adjusting nut until the ring contacts the pinion (zero lash).

NOTE: Do not force gears into contact so as to bind them.

3. Back off left adjusting nut approximately two notches to a point where nut lock and nut are aligned. Install nut and lock.
4. Tighten right adjusting nut firmly to force the differential and planet assembly into solid contact with the left adjusting nut. Loosen right adjusting nut until it is free from contact with its bearing, then retighten until nut contacts bearing. Tighten right adjusting nut 1 to 2 additional notches for old bearings and 2 to 3 for new bearings, to a position where nut and lock are aligned and install lock.

NOTE: At this point, the bearings are properly preloaded. Any further adjust-

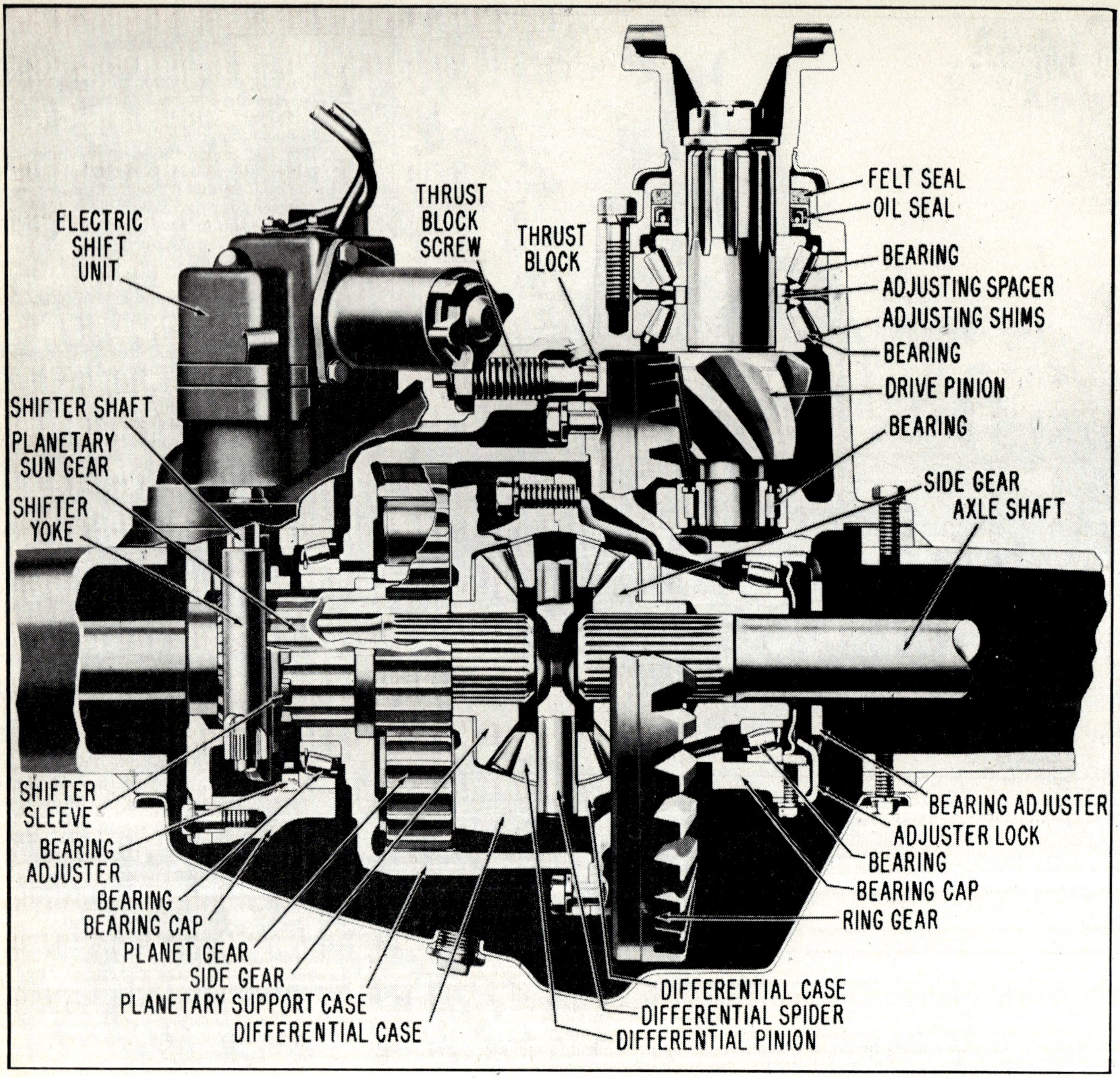

Fig. 23 GMC two-speed axle (Typical)

ment of the ring gear in either direction without a corresponding adjustment of the opposite adjusting nut, will change the bearing preload.

5. With a dial indicator mounted on the housing, measure ring and pinion backlash. Limits are .003" to .012" with .005" to .008" preferred.

Pinion Depth

1. Paint ring gear teeth with a mixture of red lead and oil to produce a contact pattern.
2. Rotate pinion through several revolutions, applying pressure to the ring gear to develop a definite contact pattern.
3. Pinion depth is controlled by the shim pack between the pinion bearing retainer and the carrier.
4. Changing pinion depth will cause some change in ring gear backlash. After adjusting pinion depth, recheck backlash.

Final Assembly

To install the unit, reverse the removal procedure, being careful to adequately support.

GMC TWO SPEED AXLE

This axle, Fig. 23, is a planetary type which is shifted by the driver as operating conditions require. Movement of the control valve shift lever causes the shift chamber at the axle housing to select the desired speed range. The axle shafts and differential carrier can be removed for repair without removing the axle assembly from the vehicle.

Disassemble

1. Remove axle shafts.
2. Place jack under carrier and remove carrier-to-housing cap screws.
3. Break carrier loose with mallet and work it out of housing.
4. Place carrier in repair stand.
5. Remove anchor from carrier.
6. Remove shifter assembly.
7. Drive out shifter yoke shaft.
8. Slide shifter yoke and sleeve to the left, removing sleeve and yoke from end of sun gear.

9. Remove differential bearing adjusting lock from each side.
10. Remove oil trough.
11. Remove thrust screw and pad.
12. Remove differential bearing caps and adjusters.
13. Remove differential and planet assembly from housing.
14. Mark ring gear and both halves of planetary support case by marking lines on case and gear.
15. Remove drive gear from planetary support case and cover.
16. Lift off planet support case.
17. Remove thrust washer used between planet cover and differential.
18. Remove sun gear and four planet gears.
19. Mark or tag all gears so they can be reassembled exactly as removed.
20. Separate differential case halves. Case halves and spider are stamped on all four sides with matching numbers to assist in assembly.
21. Remove side gears and thrust washers from each case half.
22. Remove differential side bearings.
23. Loosen bolts attaching pinion bearing oil seal retainer to carrier. As bolts are loosened they will contact deflector flange on U-joint flange.
24. Tap universal joint yoke with mallet, driving pinion, bearings, and retainer assembly gradually from carrier housing while continuing to loosen attaching bolts.
25. Remove shims from carrier housing and note the amount required in back of the double thrust bearing.
26. Remove rear pinion bearing from carrier housing.
27. Separate U-joint yoke from pinion.
28. Remove pinion bearing oil seal retainer and seal from yoke.
29. Press off double thrust bearing.

Reassembly

Before axle is reassembled, make sure that all oil passages are thoroughly clean. Apply a thin coating of differential lubricant to all thrust and bearing surfaces to prevent scoring of parts when vehicle is first placed in service.

Pinion Assembly

1. Press bearing inner race on pinion.
2. Press double row thrust bearing on pinion.
3. Install pinion rear bearing in differential carrier.
4. Install original shim pack in carrier. Grease shims thoroughly. If original ring gear and pinion are reinstalled, use shims that were removed. If new gears are used, start with one .015″ and .018″ shim (total .022″).
5. Position retaining cap screws and lock washers in oil seal retainer and install retainer on propeller shaft yoke.
6. Install new washer and nut and tighten securely.
7. Install pinion assembly into carrier and tap on yoke nut until cap screws can be started. Tighten cap screws alternately and evenly until pinion assembly is in place.

Differential & Planetary Assembly

1. Assemble differential parts into planetary half of case, then place plain half of case over assembled parts. *Be sure that all parts are assembled as indicated by marks stamped or tagged at time of disassembly. Also see that side gear with longest hub is on bevel gear side.*
2. Fasten differential case halves and tighten securely.
3. Install bearing cones at planetary support case cover.
4. Place ring gear on wood blocks with gear teeth down and install several guide pins in ring gear bolt holes. Then position planetary case cover over guide pins with marks in alignment.
5. Place a thrust washer in planetary support case cover, then install the assembled differential in case cover.
6. Install the four planet gears on differential case with long hub of gears away from case.
7. Install large planetary support case over the assembled parts with marks in alignment.
8. Fasten two halves of planetary support case and bevel gear together.

Differential & Planetary Installation

1. Locate bearing races on each side, then install differential and planetary assembly in differential carrier.
2. Install differential bearing adjusting nuts on each side. See that oil baffle is in place in right adjusting nut.
3. Install differential bearing adjusting caps and dowels at each side.
4. Install cap bolts and lock washers and tighten snugly or until bearing adjusting nuts can just be turned.

Ring Gear & Pinion, Adjust

1. Loosen right hand adjusting nut and tighten left hand adjusting nut.
2. Continue tightening left adjusting nut until all backlash between ring gear and pinion is removed.
3. Back off left adjusting nut approximately two notches to a point where notch in nut is aligned with lock.
4. Tighten right adjusting nut firmly to force differential and planet assembly in solid contact with left adjusting nut. Loosen right adjusting nut until it is free of its bearing, then retighten snugly against bearing. Tighten right nut from one to two notches more to a locking position. This method provides for the proper pre-load on the bearings.
5. Mount a dial indicator on the housing and check the backlash between ring gear and pinion which should be .005″ to .008″. Whenever backlash is more than .008″, loosen right adjusting nut one notch and tighten left nut one notch. If backlash is less than .005″, loosen left nut one notch and tighten right nut one notch. When backlash has been adjusted, tighten bearing cap bolts securely. Again check backlash and then install adjuster locks.

Pinion Depth, Adjust

1. Coat ring gear teeth lightly with red lead or prussian blue. Then turn the pinion several revolutions in both directions while applying pressure to ring gear to create a load.
2. Examine the pattern on the ring gear. If the pinion depth is correct, the tooth pattern will be centered on the pitch line and toward the toe of the ring gear.
3. If the pattern is below the pitch line the pinion is too deep and it will be necessary to remove the pinion and increase the shim thickness between the pinion bearing retainer and differential carrier housing.
4. If the pattern is above the pitch line on the ring gear teeth, the pinion is too shallow and it will be necessary to decrease the shim thickness.
5. Checking the pinion depth will make some change in the backlash; therefore, it may be necessary to readjust the backlash.
6. Tighten the bearing cap screws, recheck backlash and install adjusting nut locks.
7. Install oil trough.

Ring Gear Thrust Pad Installation

1. Inspect the bronze tip on ring gear thrust pad and replace if worn.
2. Install thrust pad and tighten until pad contacts ring gear. Then back off 30° to obtain a clearance of .005″ to .007″. Tighten lock nut, making sure screw does not turn.

Shifter Assembly Installation

1. Install shifter yoke on sleeve and install assembled parts on sun gear.
2. Install shifter lever and shaft through adapter and housing, then into the shifter yoke, mating the two parts carefully to avoid damage to the splines. The shaft must be free to avoid binding when the shifter assembly is installed.
3. Position shift anchor at left of differential carrier with anchor knob in bushing and opposite end secured to bearing cap with cap screw and lock washer.

Differential Carrier Installation

1. With new gasket in place, place carrier on roller jack, using a suitable wood block as a support.
2. Use a long drift to align assembly and install it in axle housing.
3. Install and tighten cap screws securely.
4. Use a new gasket and install electric shift units. (For service on the Electric shift unit, see the *Eaton Electric Shift section of the Rear Axles Chapter.*)

Front Wheel Drive

For service procedure on the transfer case used with four-wheel drive trucks, see the Transfer Case Chapter

FRONT AXLE

Description

The front axle is a Spicer unit equipped with steering knuckles, Figs. 1 and 2. To permit the turning and driving of the front wheels, the axle is equipped with steering knuckles and yoke and trunnion type universal joints.

The 1977–83 K30/K3500 series incorporate king pins, while 1973–83 K10/K1500 and K20/K2500 series incorporate ball joints.

BALL JOINT ADJUSTMENT

Ball joint adjustment is generally necessary when there is excessive play in steering, irregular tire wear or persistent loosening of the tie rods.

1. Raise vehicle and support with jack stands placed just inside of front springs.
2. Disconnect drag link and tie rods to permit independent movement of each steering knuckle.
3. On 1974–76 models:
   a. Apply a torque wrench to one of the steering knuckles retaining stud nuts and check torque required to turn steering knuckle. Maximum torque required should be 20 ft. lbs.

   NOTE: Steering knuckle should turn freely without vertical end play.

   b. If torque required is too low, remove upper ball joint cotter pin and locknut, then torque adjusting sleeve to 50 ft. lbs. and reinstall locknut and cotter pin.
   c. Recheck torque required to turn steering knuckle. If torque is still too low, retorque adjusting sleeve to 60 ft. lbs. as described previously.
   d. Recheck torque required to turn steering knuckle. If torque is still too low, replace parts as necessary to bring torque within specifications.
4. On 1977–83 models:
   a. Using a spring scale attached to tie rod mounting hole, determine amount of pull at right angles, required to keep knuckle assembly turning from a straight ahead position.
   b. If pull exceeds 25 lbs., remove upper ball stud nut, and loosen adjusting sleeve as required. Tighten ball stud nut and recheck turning effort.

FRONT AXLE, REPLACE

1. Disassemble propeller shaft from front axle differential.
2. Raise front of truck and support it with jacks behind front springs.
3. Disconnect drag link from steering arm.
4. Disconnect brake hoses from frame fittings.

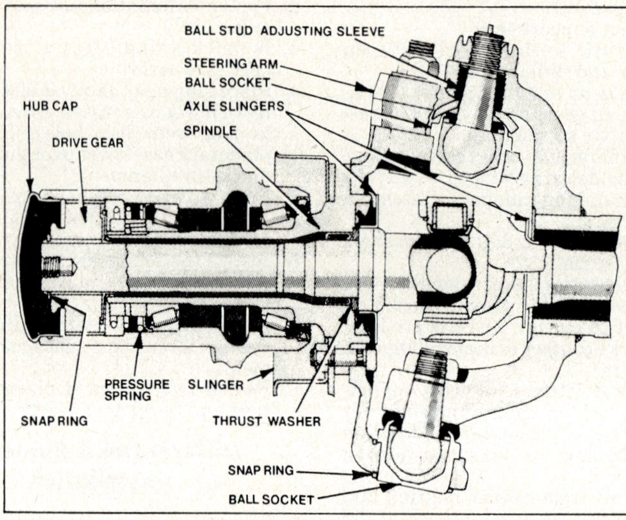

Fig. 1 Sectional view of front driving axle. 1974–83 Series K10/K1500, K20/K2500

5. Disconnect shock absorbers from spring clamp plate or axle brackets.
6. Separate axle from truck springs.
7. Roll front axle from under truck.
8. Install in reverse order of removal.

HUB AND WHEEL BEARING, REPLACE

NOTE: On models with Free-Wheeling Hubs, refer to "Front Wheel Locking Hubs" chapter for locking hub removal.

1. Raise and support vehicle.
2. On models with locked hubs, Figs. 1 & 2:
   a. Remove hub cap and snap ring.
   b. Remove drive gear. On K10/K1500 and K20/K2500 models, remove pressure spring. Place hand over drive gear and pry out using a screwdriver.
3. On all models, remove wheel bearing outer lock nut, lock ring and wheel bearing inner adjusting nut using suitable tool.
4. Remove disc brake caliper and position out of the way.
5. Remove hub-and-disc assembly, outer wheel bearing and spring retainer plate:
   a. Remove inner wheel bearing and oil seal from hub using suitable drift.
   b. Remove inner and outer bearing cups from hub using suitable drift.
6. Reverse procedure to install. Refer to "Wheel Bearings, Adjust" for wheel bearing adjustment.

AXLE SHAFT, REPLACE

1. Remove hub and wheel bearing as previously described.
2. Remove spindle, attaching nuts, then lightly tap spindle from steering knuckle.
3. Remove thrust washer, then pull axle shaft from axle, Fig. 3.
4. Reverse procedure to install. Torque spindle nuts to 45 ft. lbs. on 1974–76 K10/K1500 and K20/K2500; 25 ft. lbs. on 1977–78 K10/K1500 and K20/K2500; 60 ft. lbs. on 1977–81 K30/K3500; and 33 ft. lbs. on 1979–83 K10/K1500 and K20/K2500.

AXLE JOINT REPAIRS

1. Remove lock rings from trunnion bearings.
2. Support shaft yoke in a suitable vise.
3. With a brass drift and a hammer, tap on one end of trunnion bearing to drive the opposite bearing from yoke. Then, support other side of yoke and drive out remaining bearing.
4. Remove trunnion. Inspect and replace bearings if necessary, then lubricate with wheel bearing grease.
5. Replace trunnion and press bearings into yoke and over trunnion hubs far enough to install lock rings.
6. Gently tap yoke to seat bearings against lock rings.

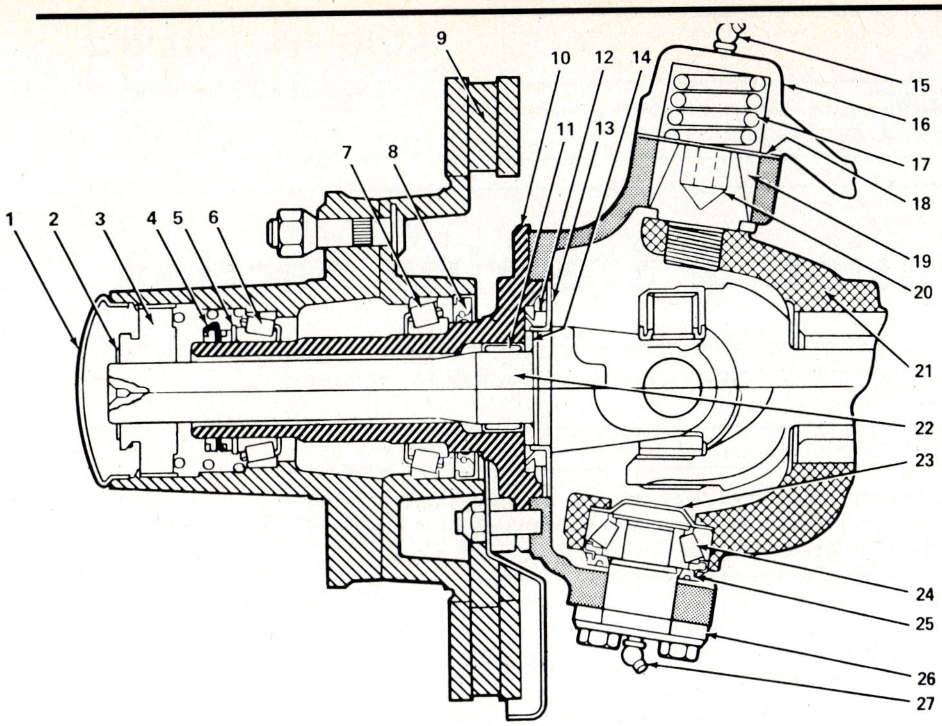

1. HUB CAP
2. SNAP RING
3. HUB DRIVE GEAR
4. ADJUSTING NUT ASM
   LOCK NUT
   LOCK TANG
   ADJUSTING NUT
5. WASHER
6. OUTER WHEEL BEARING
7. INNER WHEEL BEARING
8. SEAL
9. HUB-AND-DISC ASM
10. SPINDLE
11. SPINDLE BEARING
12. SEAL
13. DEFLECTOR
14. SPACER
15. LUBE FITTING
16. UPPER BEARING CAP
17. PRESSURE SPRING
18. GASKET
19. BUSHING, KING-PIN
20. KING-PIN
21. YOKE
22. OUTER AXLE SHAFT
23. GREASE RETAINER
24. LOWER BEARING
25. SEAL
26. BEARING CAP
27. LUBE FITTING

Fig. 2 Sectional view of front driving axle. 1977–83 K30/K3500 series

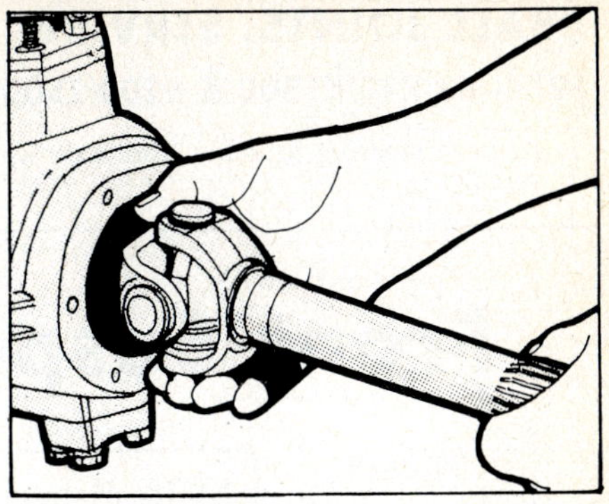

Fig. 3 Removing & installing axle shaft and U-joint

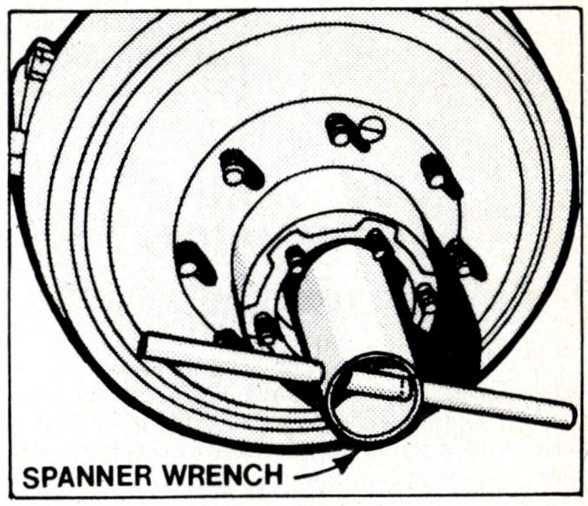

Fig. 4 Adjusting wheel bearing

WHEEL BEARINGS ADJUST

1. Remove dust cap or free wheeling hub, lockwasher and outer locknut.
2. Using suitable spanner wrench, torque inner adjusting nut to 50 ft.lbs. while rotating hub. Back off adjusting nut and retorque to 35 ft.lbs. while rotating hub, Fig. 4.
3. Back off adjusting nut 3/8 turn on 1974–83 models.
4. Assemble lockwasher by turning nut to nearest hole in lockwasher, then install and torque outer locknut to 50 ft. lbs. for all models except 1977–78 K30/K3500 series 1979–80, 80 ft. lbs. & 1981–83, 183 ft. lbs.
5. On 1977–81 K30/K3500 models, assemble lockwasher and locknut, then torque locknut to 65 ft. lbs. and bend one ear of lockwasher at least 30° over adjusting nut and another ear at least 60° over locknut.
6. On 1981–82 K10/K1500 and K20/K2500 models and 1982–83 K30/K3500 models, assemble the drag sleeve retainer washer over the axle shaft against the bearing adjusting nut. The tang on the inside diameter of this washer is assembled in the keyway of the spindle. Assemble and tighten the outer locknut to 160–205 ft. lbs.

NOTE: Hub assembly should have .001 to .010 inch end play.

7. Install dust cap or free wheeling hub.

STEERING KNUCKLE, REPLACE

1974–83 with Ball Joints

1. Remove hub and spindle.
2. Disconnect tie rod from steering arm.

CAUTION: If steering arm is removed from knuckle, discard the self-locking nuts and replace with new ones during installation. Tighten nuts to 90 ft. lbs.

3. Remove cotter pin from upper ball joint stud nut, and remove retaining nuts from both ball joints.
4. Separate stud from yoke using suitable ball joint fork at lower joint and repeat at upper joint is required.

CAUTION: Do not remove upper ball joint adjusting sleeve from yoke unless ball joint is being replaced. If it is necessary to loosen the sleeve to remove the knuckle, only loosen the sleeve two threads. The non-hardened threads in the yoke can be easily damaged by the adjusting sleeve.

5. Reverse procedure to install.
6. Torque lower ball joint to 70 ft. lbs., adjusting sleeve to 50 ft. lbs., and upper ball joint to 100 ft. lbs. Tighten tie-rod nut to 45 ft. lbs.

1977–83 with King Pins

1. Remove hub and spindle.
2. Remove upper king pin cap by alternately loosening nuts to unload the compression spring. Withdraw the spring and gasket.
3. Remove the lower bearing cap and king pin.
4. Remove upper king pin tapered bushing and remove and knuckle from yoke. Remove felt king pin seal.
5. Remove upper king pin from yoke using a large breaker bar.
6. Drive lower king pin bearing cup, cone, grease retainer, and seal through bottom of knuckle using suitable tool.
7. Reverse procedure to install.
8. Torque upper king pin to 500–600 ft. lbs., and upper and lower bearing cap nuts to 70–90 ft. lbs.

BALL JOINTS, SERVICE

1974–83 K10/K1500 & K20/K2500

1. Remove steering knuckle from axle assembly.
2. Remove lower ball joint snap ring.

NOTE: Lower ball joint must be removed before upper ball joint can be serviced.

KING PINS, SERVICE

1977–82 K30/K3500

Refer to "Steering Knuckle, Replace" for king pin service.

Front Suspension & Steering Section

NOTE: Manual and power steering gear repairs are covered in separate chapters. See Index.

NOTE: On 1979 G-Series and all 1980–83 models, steering linkage relay rod assemblies will use crimped nuts. Whenever a crimped nut is loosened or removed, it must be replaced with a prevailing torque nut, Fig. 1.

COIL SPRING SUSPENSION

1974–83

This suspension, Fig. 2, consists of upper and lower control arms pivoting on steel threaded bushings on upper and lower control arm shafts which are bolted to the front suspension crossmember and bracket assembly. These control arms are connected to the steering knuckle through pivoting ball joints. A coil spring is located between the lower control arm and a formed seat in the suspension crossmember, thus the lower control arm is the load-carrying member. The double-acting shock absorbers are also attached to the lower control arms and connect with the frame to the rear of the suspension on the upper end.

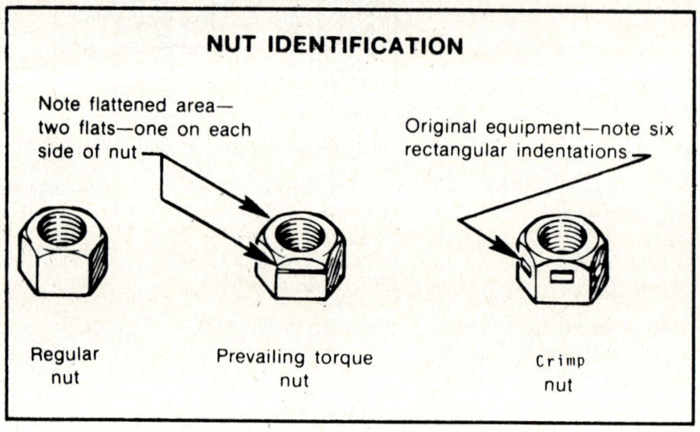

Fig. 1 Crimp nut identification

Wheel Alignment

EXCEPT CABALLERO, EL CAMINO & SPRINT

Caster and camber adjustments are made by means of shims located between upper control arm shaft and mounting bracket attached to suspension crossmember. A series of convex and concave spacers with flat opposite sides are used. These spacers allow a positive cross shaft-to-bracket attachment regardless of the number of shims used. Shims may be changed at either front or rear to vary caster, or at both points to vary camber.

Caster

1. Measure frame angle at "B", Fig. 3.
2. Check caster on alignment machine.
3. Add angle "B" and caster angle to determine *frame corrected angle.*
4. Measure dimension "A" Fig. 3.
5. Using dimension "A", and Caster Specification Chart, find the recommended caster angle.
6. The frame corrected angle should correspond to the recommended angle on the chart within plus or minus 1/4 degree. Make necessary changes to bring caster angle within limits.

Camber

1. Measure camber on alignment machine.
2. Measure dimension "A", Fig. 3.
3. Using dimension "A" and chart for appropriate vehicle, find recommended camber

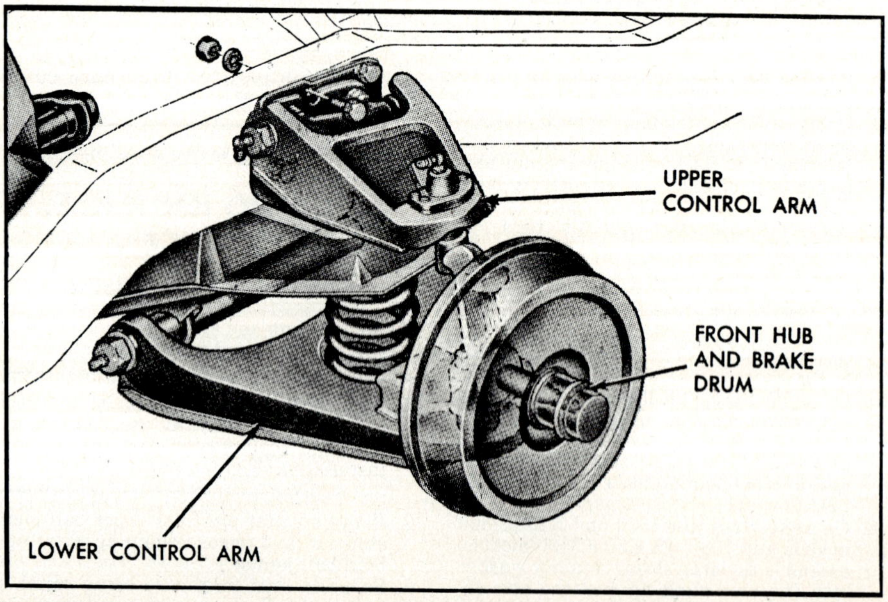

Fig. 2 Coil spring suspension. 1974–83 (typical)

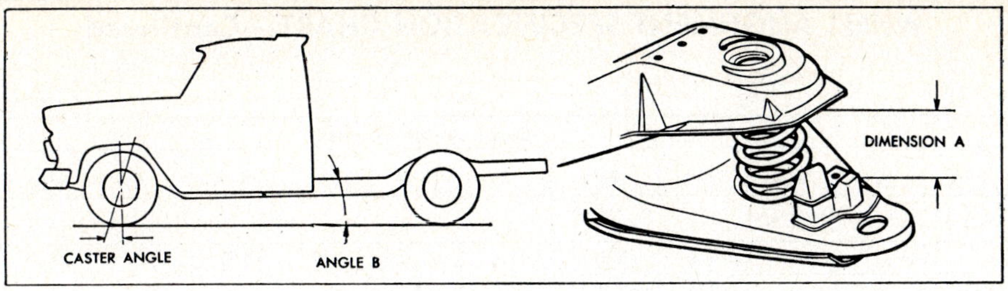

Fig. 3 Caster-camber adjustment. 1974–83 coil spring suspension

WHEEL ALIGNMENT SPECIFICATION CHART

Dimension "A" in Inches Fig. 3	Caster Angle Degrees														
	2½	2¾	3	3¼	3½	3¾	4	4¼	4½	4¾	5	5¼	5½	5¾	6
1974															
CA 10	—	—	+2	+1½	+1¼	+1	+¾	+½	+¼	0	−½	—	—	—	—
CA 20-30	+1½	+1¼	+1	+¾	+½	+¼	0	−¼	−½	−¾	−1	—	—	—	—
GA 10-30	+2¼	+2	+1½	+1¼	+1	+¾	+½	+¼	0	−¼	−½	—	—	—	—
K10-20	—	—	—	—	—	+4①	—	—	—	—	—	—	—	—	—
P10	—	—	—	+1½	+1¼	+1	+¾	+½	+¼	0	−¼	−½	−¾	−1	−1
P20-30	—	—	—	—	+1¼	+1	+¾	+½	+¼	0	−¼	−½	−¾	−1	−1
P30 (Motor Home)	—	—	+5	+4¾	+4½	+4¼	+4	+3¾	+3½	+3¼	+3	+2¾	+2¾	+2½	+2¼
1975–76															
C10	—	—	+2	+1½	+1¼	+1	+¾	+½	+¼	0	−½	—	—	—	—
C20-30	+1½	+1¼	+1	+¾	+½	+¼	0	−¼	−½	−¾	−1	—	—	—	—
K10-20	—	—	—	—	—	+4①	—	—	—	—	—	—	—	—	—
G10-20-30	+2¼	+2	+1½	+1¼	+1	+¾	+½	0	−¼	−½	—	—	—	—	—
P10-20-30	+2½	+2¼	+2	+1¾	+1½	+1	+¾	+½	+¼	0	−¼	—	—	—	—
Motor Home	+5¾	+5½	+5	+5	+4½	+4¼	+4	+4	+3½	+3¼	+3	—	—	—	—
1977															
C10	—	—	+2	+1½	+1¼	+1	+¾	+½	+¼	0	−½	—	—	—	—
C20-30	+1½	+1¼	+1	+¾	+½	+¼	0	−¼	−½	−¾	−1	—	—	—	—
K10-20-30	—	—	—	—	—	+8①	—	—	—	—	—	—	—	—	—
G10-20-30	+2¼	+2	+1½	+1¼	+1	+¾	+½	+¼	0	−¼	−½	—	—	—	—
P10-20-30	+2½	+2¼	+2	+1¾	+1½	+1	+¾	+½	+¼	0	−¼	—	—	—	—
Motor Home	+5¾	+5½	+5	+5	+4½	+4¼	+4	+4	+3½	+3¼	+3	—	—	—	—
1978															
C10	—	—	+2	+1½	+1¼	+1	+¾	+½	+¼	0	−½	—	—	—	—
C20-30	+1½	+1¼	+1	+¾	+½	+¼	0	−¼	−½	−¾	−1	—	—	—	—
K10-20-30	—	—	—	—	—	+8①	—	—	—	—	—	—	—	—	—
G10-20	+3¼	+3	+2¾	+2½	+2½	+2¼	+2	+2	+1¾	+1½	+1½	—	—	—	—
G30	+2¼	+2	+1½	+1¼	+1	+¾	+½	+¼	0	−¼	−½	—	—	—	—
P10-20-30	+2½	+2¼	+2	+1¾	+1½	+1	+¾	+½	+¼	0	−¼	—	—	—	—
Motor Home	+5¾	+5½	+5	+5	+4½	+4¼	+4	+4	+3½	+3¼	+3	—	—	—	—

Dimension "A" in Inches Fig. 3	Caster Angle Degrees														
	1½	1¾	2	2¼	2½	2¾	3	3¼	3½	3¾	4	4¼	4½	4¾	5
1979															
C10	—	—	—	—	+2.4	+2.1	+1.8	+1.5	+1.2	+1	+.7	+.5	+.2	+.1	+.3
C20, 30	—	—	—	—	+1.5	+1.2	+.9	+.6	+.3	+.1	0	−.1	−.7	−1.0	−1.2
G10, 20	+2.9	+2.7	+2.3	+2.2	+2	+1.8	+1.6	+1.4	+1.3	+1.1	+.9	+.7	—	—	—
G30	+3.4	+3.0	+2.7	+2.4	+2.1	+1.8	+1.5	+1.3	+1.0	+.7	+.4	+.2	—	—	—
P10	—	—	—	—	+2.3	+2.0	+1.7	+1.5	+1.2	+.9	+.6	+.4	+.1	−.1	−.3
P20, 30	—	—	+2.9	+2.6	+2.3	+2.0	+1.7	+1.4	+1.2	+.9	+.6	+.4	+.2	+.1	—
Motor Home②	—	—	—	—	+5.5	+5.3	+5.0	+4.7	+4.4	+4.1	+3.8	+3.6	+3.3	+3.1	+2.9

Continued

WHEEL ALIGNMENT SPECIFICATION CHART—Continued

Dimension "A" in Inches Fig. 3	Caster Angle Degrees														
	1½	1¾	2	2¼	2½	2¾	3	3¼	3½	3¾	4	4¼	4½	4¾	5
1980–81															
C10	—	—	—	—	+2.4	+2.1	+1.8	+1.5	+1.2	+1.0	+.7	+.5	+.2	+.1	+.3
C20-30	—	—	—	—	+1.5	+1.2	+.9	+.6	+.3	+.1	0	+.1	+.7	−1.0	−1.2
K10-20-30①	—	—	—	—	—	—	—	—	—	—	—	—	—	—	—
G10-20	+3.5	+3.3	+3.1	+2.9	+2.7	+2.6	+2.4	+2.2	+2.1	+1.9	+1.8	+1.6	—	—	—
G30	+2.8	+2.5	+2.2	+1.9	+1.6	+1.3	+1.0	+.7	+.5	+.2	0	−.2	—	—	—
P10	—	—	—	—	+2.3	+2.0	+1.7	+1.5	+1.2	+.9	+.6	+.4	+.1	−.1	−.3
P20-30③	—	—	—	—	+2.9	+2.6	+2.3	+2.0	+1.7	+1.4	+1.2	+.9	+.6	+.4	+.2
Motor Home	—	—	—	—	+5.5	+5.3	+5.0	+4.7	+4.4	+4.1	+3.8	+3.6	+3.1	+2.9	—

Dimension "A" in Inches Fig. 3	Caster Angle Degrees														
	1½	1¾	2	2¼	2½	2¾	3	3¼	3½	3¾	4	4¼	4½	4¾	5
1982–83															
C10	—	—	—	—	+3.6	+3.4	+3.1	+2.8	+2.6	+2.4	+2.0	+1.8	+1.5	+1.2	+1.0
C20-30	—	—	—	—	+1.5	+1.2	+0.9	+0.6	+0.3	+0.1	0	−0.1	−0.7	−1.0	−1.2
K10-20-30①	—	—	—	—	—	—	—	—	—	—	—	—	—	—	—
G10-20	+3.5	+3.3	+3.1	+2.9	+2.7	—	+2.4	+2.2	+2.1	+1.9	+1.8	+1.6	—	—	—
G30	+2.8	+2.5	+2.2	+1.9	+1.6	—	+1.0	+0.7	+0.5	+0.2	0	−0.2	—	—	—
P10	—	—	—	—	+2.3	+2.0	+1.7	+1.5	+1.2	+0.9	+0.6	+0.4	+0.1	−0.1	−0.3
P20-30③	—	—	+2.9	+2.6	+2.3	+2.0	+1.7	+1.4	+1.2	+0.9	+0.6	+0.4	+0.2	+0.1	—
Motor Home	—	—	—	—	+5.5	+5.3	+5.0	+4.7	+4.4	+4.1	+3.8	+3.6	+3.3	+3.1	+2.9

①—No provision for resetting.
②—With hydroboost brake system, add .3°. With tandem rear wheels, subtract .4°.
③—With heavy duty front disc brakes or 4 wheel disc brakes, add .3°. With tandem rear wheels, subtract .4°.

Dimension "A" in Inches Fig. 3	Camber Angle Degrees														
	2½	2¾	3	3¼	3½	3¾	4	4¼	4½	4¾	5	5¼	5½	5¾	6
1975–76															
C10-20-30	—	—	—	—	—	+¼	—	—	—	—	—	—	—	—	—
K10-20	—	—	—	—	—	+1½①	—	—	—	—	—	—	—	—	—
G10-20-30	—	—	—	—	—	+¼	—	—	—	—	—	—	—	—	—
P10	0	0	+¼	+¼	+¼	+¼	+¼	0	0	0	−¼	−½	−¾	—	—
P20-30	0	0	+¼	+¼	+¼	+¼	+¼	+¼	0	0	−¼	+½	−¾	—	—
Motor Home	0	0	+¼	+¼	+¼	+¼	0	0	0	−¼	−½	−¾	−1	—	—
1977–78															
C10-20-30	—	—	—	—	—	+¼	—	—	—	—	—	—	—	—	—
K10-20-30	—	—	—	—	—	①	—	—	—	—	—	—	—	—	—
G10-20-30	—	—	—	—	—	+¼	—	—	—	—	—	—	—	—	—
P10	0	0	+¼	+¼	+¼	+¼	+¼	0	0	0	−¼	−½	−¾	—	—
P20-30	0	0	+¼	+¼	+¼	+¼	+¼	+¼	0	0	−¼	−½	−¾	—	—
Motor Home	0	0	+¼	+¼	+¼	+¼	0	0	0	−¼	−½	−¾	−1	—	—

①—No provision for resetting.

angle.

4. If angle in Step 1 does not correspond to the recommended angle in the chart within plus or minus 1/4 degree, make necessary changes.

CABALLERO, EL CAMINO & SPRINT

Caster and camber adjustments are made by means of shims between the upper control arm inner support shaft and the support bracket attached to the frame. Shims may be added, subtracted or transferred to change the readings as follows.

Caster, Adjust

Transfer shims from front to rear or rear to front. The transfer of one shim to the front bolt from the rear bolt will decrease positive caster. On shim (1/32") transferred from the rear bolt to the front bolt will change caster about 1/2 degree.

Camber, Adjust

Change shims at both the front and rear of the shaft. Adding an equal number of shims at both front and rear of the support shaft will decrease positive camber. One shim (1/32") at

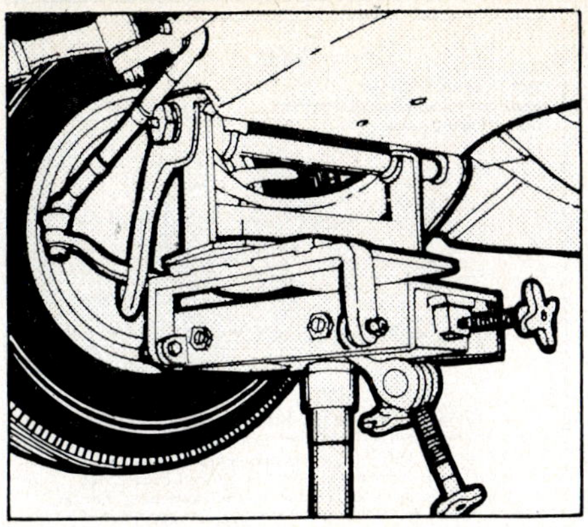

Fig. 4 Coil spring removal. 1974–83 (typical)

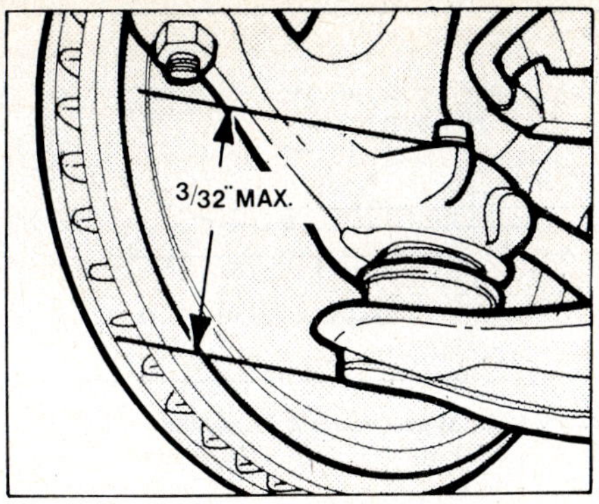

Fig. 5 Checking lower ball joint for wear

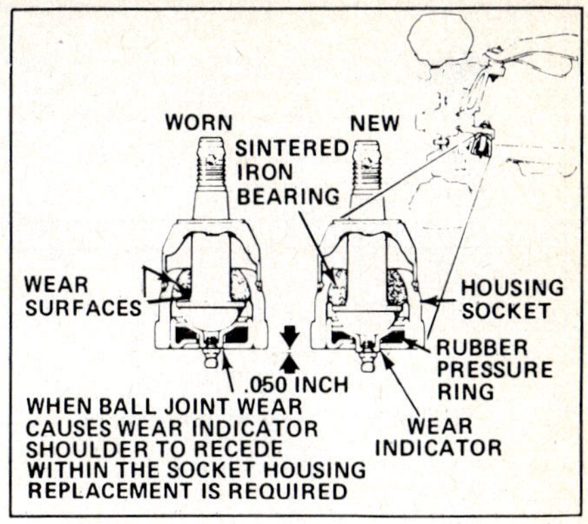

Fig. 6 Lower ball joint wear indicator

each location will move camber approximately 1/5 degree.

Coil Springs, Replace

1. Raise front of vehicle and place jack stands under frame, allowing control arms to hang free.
2. Disconnect shock absorber and stabilizer bar from lower control arm.
3. Install tool J-23028 or similar tool onto a suitable jack, Fig. 4, place jack under cross-shaft so cross-shaft seats in grooves of tool.
4. Place a safety chain through coil spring and lower control arm, then raise jack, relieving tension from coil spring.
5. Remove cross-shaft to crossmember "U" bolts, lower jack slowly and remove spring.
6. Reverse procedure to install.

Ball Joint Inspection

Upper Ball Joint

1. Raise vehicle and support on frame to allow control arms to hang free.
2. Remove wheel assembly.
3. Support lower control arm with adjustable jackstand and disconnect upper ball stud from steering knuckle.
4. Reinstall nut on ball stud and measure torque required to rotate stud with torque wrench. Specified torque for a new ball joint is 3–10 lbs. for Caballero, El Camino and Sprint and 5–10 for all other models. If torque readings are excessively high or low, replace ball joint.

Lower Ball Joint

NOTE: Lower ball joints are a loose fit when not connected to the steering knuckle. Wear may be checked without disassembling the ball stud as follows:

1. Support weight of control arms at wheel hub and drum.
2. Measure distance between tip of ball stud and tip of grease fitting below ball joint, Fig. 5.
3. Move support to control arm to allow wheel hub and drum to hang free. Measure distance as in Step 2. If difference in measurements exceeds 1/16 inch for Caballero, El Camino and Sprint and 3/32 inch for all other models, ball joint is worn and should be replaced.

NOTE: Caballero, El Camino and Sprint models have lower ball joints that incorporate wear indicators. Refer to Fig. 6 to check for wear.

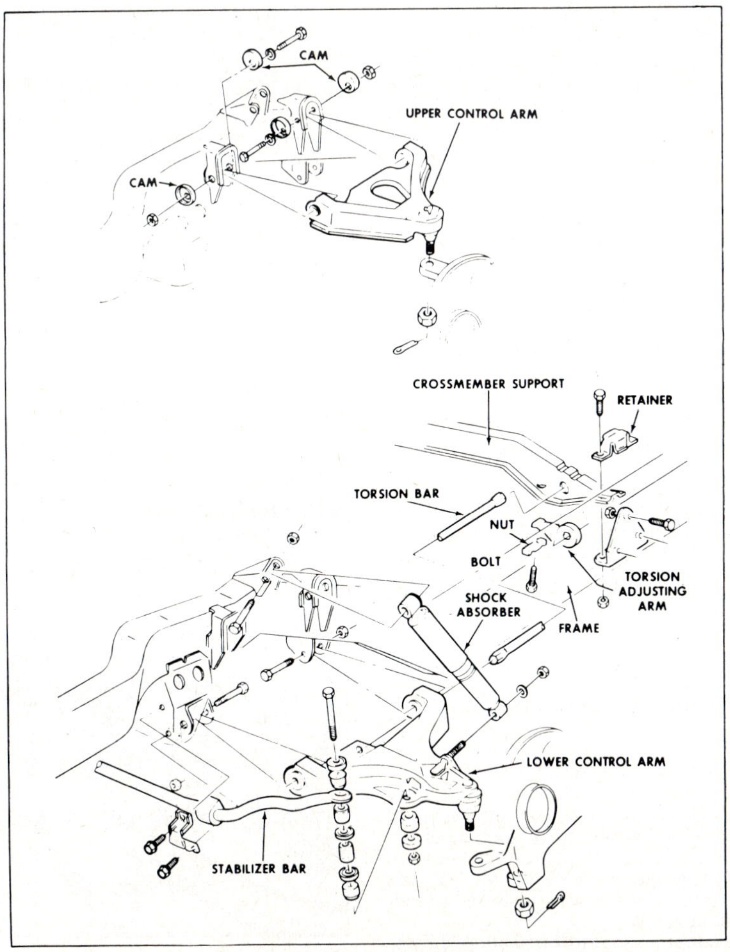

Fig. 7 Torsion bar front suspension. 1974–78 GMC Motor Home & TransMode

TORSION BAR SUSPENSION

1974–78 GMC Motor Home

The front suspension consists of control arms, stabilizer bar, shock absorbers and right and left torsion bars. The front end of the torsion bar is attached to the lower control arm, and the rear of the torsion bar is mounted in an adjustable arm at the torsion bar crossmember. The riding height of the vehicle is controlled by this adjustment, Fig. 7.

Wheel Alignment, Adjust

Riding Height

The riding height must be checked and adjusted if necessary before checking and

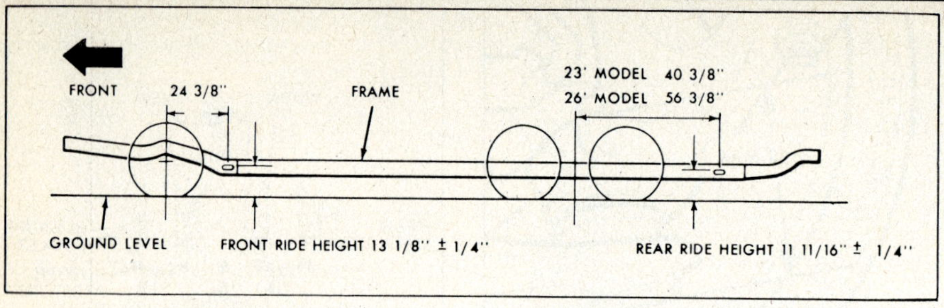

Fig. 8 Riding height dimensions. 1974–78 GMC Motor Home & TransMode

adjusting front wheel alignment. Before checking riding height, make sure that the vehicle is parked on a level surface and that tire pressure is to specification.

To check riding height, take measurements from top of oval hole to level surface, Fig. 8. If adjustments are required, proceed as follows:

CAUTION: Do not attempt to adjust riding height using the adjust bolt only. The adjusting bolt will turn, but will strip the threads. Tool J-22517-02 must be used. This tool will adjust the riding height without loading the adjusting bolt.

---

1. Install tool J-22517-02 with pin of tool aligned in hole in crossmember, and seat center screw in dimple of torsion bar adjusting arm.
2. To raise vehicle, turn tool until proper height is obtained, then turn adjusting bolt until it contacts adjusting arm and remove tool.
3. To lower vehicle, raise adjusting arm from contact with adjusting bolt, then loosen adjusting bolt and lower adjusting arm with tool until proper riding height is obtained. Turn adjusting bolt until it contacts adjustment arm and remove tool.

Camber

1. Loosen nuts on upper control arm cam bolts.
2. Turn front cam bolt (inboard or outboard) to correct for 1/2° of incorrect setting.
3. Turn rear cam bolt (same way front bolt was turned) to correct for remaining 1/2° of incorrect setting.
4. Torque front and rear cam nuts to 80 to 95 ft. lbs. while holding bolts with another wrench so that camber is not changed.
5. Recheck adjustment. Do not readjust camber unless it exceeds specifications.

NOTE: Remove any weld splatter which may be on the cam adjustment surface. Weld splatter in this area will affect front end alignment.

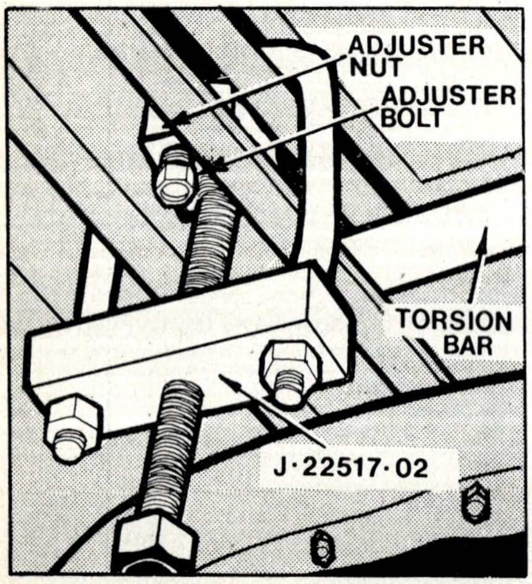

Fig. 10 Torsion bar removal tool installation. 1974–78 GMC Motor Home & TransMode

Caster

1. Loosen front and rear cam nuts while holding bolts with another wrench to avoid changing camber.
2. Turn front cam bolt so that camber changes 1/4 of the desired amount of caster to be corrected.
3. Turn rear cam bolt so that camber now returns to corrected position.
4. Recheck caster setting.
5. Torque cam nuts to 90 ft. lbs., while holding bolt with another wrench so that final setting will not be changed.

Toe-In

1. Loosen tie rod adjusting sleeves clamp bolts.

NOTE: If the torque required to remove nut from bolt (from initial breakaway) exceeds 7 ft. lbs. discard the nuts and bolts. Apply penetrating oil between clamp and tube and rotate clamps until they rotate freely. Install new nuts and bolts to assure proper retention of parts at the specified torque.

---

2. With steering wheel set in straight ahead position, turn tie rod adjusting sleeves as necessary to obtain proper toe-in adjustment at curb load.
3. After adjustment has been completed, make sure that number of threads exposed at each end of sleeve are equal and that tie rod end housings are at right angles to steering arm. Position inner and outer tie rod clamps as shown in Fig. 9, and torque nuts to 20 ft. lbs.

Torsion Bars, Replace

Removal

1. Raise vehicle on a two post hoist.
2. Remove the nuts and center screw from tool J-22517-02, then position tool over crossmember with pin of tool in crossmember hole, Fig. 10. Install the nuts and counter screw on tool. Apply chassis grease to center screw threads and rounded end of screw.
3. Turn center screw until it seats in dimple of torsion bar adjusting arm, then remove torsion bar adjusting bolt and nut.

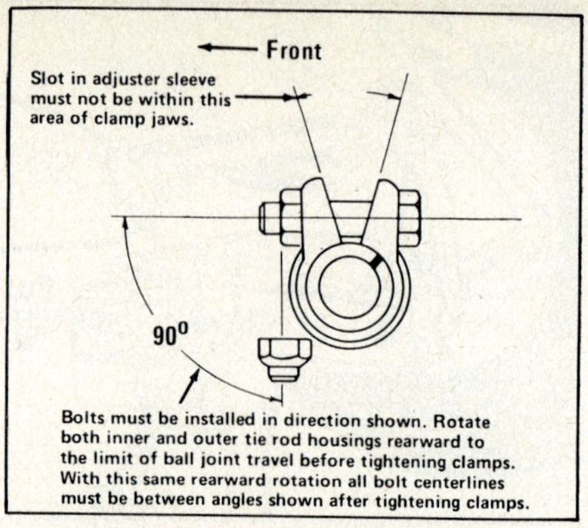

Fig. 9 Positioning tie rod clamp. 1974–78 GMC Motor Home & TransMode

---

NOTE: Count and record number of turns required to remove adjusting bolt. The number of turns will be used upon installation to obtain original riding height.

---

4. Turn center screw of tool until tension on torsion bar is completely removed, then remove tool.
5. Repeat above procedure on opposite torsion bar.
6. Remove bolts and retainer from torsion bar crossmember at frame.
7. Disconnect exhaust pipe hanger from crossmember, then loosen pipe saddle U-clamp and slide hanger rearward.
8. Move crossmember rearward until torsion bars are free and adjusting arms are removed.
9. Move torsion bar crossmember sideways to the extreme left, then move crossmembers upward and outward until opposite end clears exhaust pipe.
10. Remove torsion bars, and mark as to right or left to insure proper installation.

Installation

1. Install torsion bars in their proper locations.

NOTE: New torsion bars are stamped on one end with an "L" for left side or an "R" for right side.

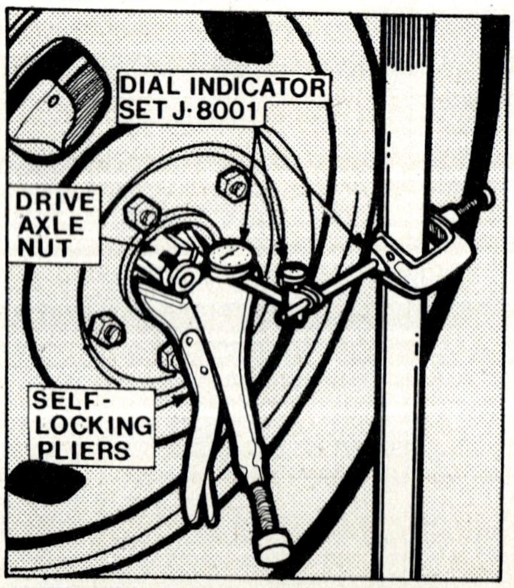

Fig. 11 Ball joint vertical check. 1974–78 GMC Motor Home & TransMode

2. Install crossmember insulators, then install crossmember positioned about two inches rearward of its normal position.
3. Raise torsion bars to align with hole in crossmember, then move torsion bars forward so that torsion bars rest on edge of hole.
4. Insert torsion bar adjusting arm into crossmember and position it so that arm will engage torsion bar and the end of arm will be 1¾–2 inches below crossmember, then tap crossmember forward enough to engage bar into arm.
5. Repeat step 4 on opposite torsion bar.
6. Place crossmember in its normal position. Torsion bar should be through and flush with rear face of adjusting arm. If not, repeat steps 4 and 5, and pulling bar slightly out from lower control arm.
7. Install retainer over each insulator and torque nut to 10 ft. lbs.
8. Connect exhaust pipe hanger to crossmember and torque saddle and U-clamp nuts to 15 ft. lbs.
9. Place tool J-22517-02 over crossmember with pin of tool in hole in crossmember. Install the two nuts and center screw.
10. Turn center screw until adjusting arm is in position to allow installation of adjusting nut, then install adjusting nut and turn adjusting bolt the same number of turns as recorded in step 3 to obtain the original riding height.
11. Turn center screw until torsion bar tension is completely removed, then remove tool.
12. Repeat steps 9 thru 11 on opposite torsion bar.
13. Lower vehicle and check riding height. If necessary, readjust riding height as described previously.

Ball Joints, Check

1. Raise vehicle and place floor stands under left and right lower control arms as close as possible to each lower ball joint.

NOTE: Vehicle must be stable and not rock on floor stands.

2. Position a dial indicator on spindle nut to measure vertical travel, Fig. 11.
3. Use a pry bar between outer constant velocity joint outer race and lower control arm and pry up and down. Reading on dial indicator should not exceed .125 inch.

Ball Joints, Replace

Lower Ball Joint

1. Remove steering knuckle as described further on.
2. Using a ¼ inch drill bit, drill side rivets to ½ inch depth. Then using a ½ inch drill bit, drill same rivets just deep enough to remove rivet heads.
3. Using a hammer and punch, drive out rivets.
4. Using a ⅝ inch drill bit, drill top rivet just enough to remove rivet head.
5. Using a hammer and punch, drive center rivet of ball joint out of control arm.
6. Install new ball joint with nuts facing toward side of control arm. Torque bolts to 25 ft. lbs. and top nut to 45 ft. lbs. and top nut to 45 ft. lbs. Stake top nut at four locations.
7. Install knuckle as described further on.
8. Check clearance from ball joint nut to drive axle outer joint. If there is no clearance, grind off nut, but not to exceed 1/16 inch.

Upper Ball Joint

1. Raise and support vehicle under lower control arms and remove wheel.
2. Remove upper ball joint stud cotter pin and nut, then disconnect brake hose clip from stud.
3. Using a hammer and brass drift, drive on steering knuckle until upper ball joint stud is disengaged from knuckle.
4. Raise control arm up, then using a ⅛ inch drill bit, drill rivet heads to a depth of ⅜ inch.

NOTE: If necessary, insert a block of wood between frame and control for support.

5. Using a ½ inch drill bit, drill off rivet heads.

CAUTION: Do not drill into control arm.

6. Using a hammer and punch, drive out rivets and remove ball joint.
7. Install new ball joint into control arm with bolts installed from the top. Torque nuts to 20 ft. lbs.
8. Install ball joint stud into knuckle and position brake hose clip over stud.
9. Install ball joint stud nut and torque to 40–60 ft. lbs., then install cotter pin.

NOTE: Do not back off nut to install cotter pin. If necessary, tighten nut to align hole in ball joint stud with slot in nut. Also make sure to bend cotter pin so as to avoid interference with constant velocity joint seal.

Steering Knuckle, Replace

Removal

1. Raise and properly support vehicle.
2. Remove wheel, hub and disc assembly.
3. Remove upper ball joint stud nut and disconnect brake line hose clip from stud.
4. Using a brass drift and hammer, strike steering knuckle to loosen upper ball joint stud.
5. Remove cotter pin and nut from tie rod end, then using a suitable puller, separate tie rod end from steering knuckle.
6. Remove cotter pin and nut from lower ball joint stud, then using a suitable tool, separate ball joint from steering knuckle and remove steering knuckle.

Installation

1. Position steering knuckle onto lower ball joint and install but not tighten nut.
2. Attach tie rod end to steering knuckle and install but do not tighten nut.
3. Install upper ball joint stud into steering knuckle, then place brake hose clip onto stud and install but do not tighten nut.
4. Torque upper ball joint nut to a minimum of 40 ft. lbs., and lower ball joint nut to a minimum of 100 ft. lbs. Tighten nuts as necessary to allow installation of new cotter pins. Do not back off nut.

CAUTION: Cotter pin on upper ball joint must be bent up so as to prevent interference with constant velocity joint seal.

5. Torque tie rod end nut to 40–50 ft. lbs., then install cotter pin. If necessary, tighten nut to allow installation of cotter pin. Do not back off nut.
6. Install hub and disc assembly and wheel.

Upper Control Arm, Replace

Removal

1. Raise vehicle and place floor stands under each lower control arm.
2. Remove shock absorber upper retaining bolt, then remove upper ball joint cotter pin and nut and disconnect brake hose clamp from ball joint stud.
3. Using a hammer and brass drift, drive on steering knuckle until upper ball joint is disengaged from knuckle.
4. Remove upper control arm cam assemblies and remove control arm by guiding shock absorber through access hole.

Installation

1. Guide upper control arm over shock absorber and install bushing ends into frame bracket.
2. Install cam assemblies, Fig. 7, then install ball joint stud into steering knuckle and install brake hose clip onto ball joint stud.
3. Install ball joint stud nut and torque to a minimum of 40 ft. lbs., Tighten nut as necessary to allow installation of new cotter pin. Do not loosen nut.

CAUTION: Cotter pin must be bent up so as not to interfere with constant velocity joint seal.

4. Install shock absorber upper retaining bolt.
5. Install wheel and lower vehicle.
6. Check front end alignment and adjust as necessary.

Lower Control Arm, Replace

Remove

1. Raise and support vehicle and remove wheel.
2. From tool J-22517-02, remove the two nuts and center screw, then place tool over crossmember support. Align pin of tool with hole in crossmember, then install the two nuts and screw on tool. Turn center screw until seated in dimple of torsion bar adjusting arm.
3. Remove torsion bar adjusting bolt and nut.

NOTE: Count and record number of turns required to remove adjusting bolt. The number of turns will be used upon installation to obtain original riding height.

4. Turn center screw of tool J-22517-02 until tension on torsion bar is completely removed then remove tool and torsion bar.
5. Disconnect shock absorber and stabilizer link from lower control arm and remove drive axle nut.
6. Remove cotter pin and nut from lower ball joint, then using suitable tool, separate lower ball joint from steering knuckle.
7. Remove bolts from lower control arm to frame and remove torsion bar.
8. Push inboard on drive axle and pull outward on knuckle to gain clearance, and remove lower control arm from knuckle.

Installation

1. Install lower control arm, making sure that shock absorber is guided onto lower control arm shock absorber, then guide ball joint stud into knuckle and install but do not tighten stud nut.
2. Install lower control arm to frame brack-

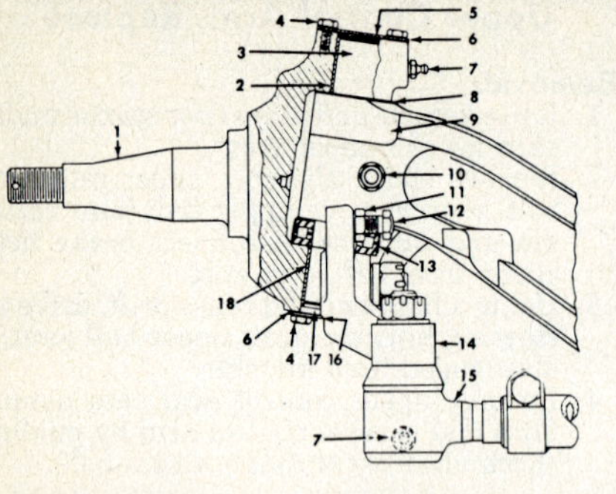

1 Steering Knuckle Spindle
2 Upper Bushing
3 Kingpin
4 Cap Screw
5 Upper Kingpin Bearing Cup
6 Kingpin Bearing Cup Gasket
7 Lubrication Fitting
8 Shim
9 Axle Center
10 Draw Key
11 Stop Screw Nut
12 Stop Screw
13 Thrust Bearing
14 Lower Steering Arm
15 Tie Rod End Assy.
16 Lower Kingpin Bearing Cap
17 Spacer - Steering Knuckle Bushing
18 Lower Bushing

Fig. 12 Steering Knuckle with single draw key. Typical F-050 and F-055 model axles

et bolts. Install and torque nuts to 90 ft. lbs.

3. Torque ball joint stud nut to 100 ft. lbs., then install new cotter pin. Tighten nut as necessary to install cotter pin. Do not back off nut.
4. Install shock absorber nut and stabilizer link nut.
5. Coat both ends of torsion bar with chassis grease, then push front end of torsion all the way into control arm.
6. Insert adjusting arm into crossmember and position it 1¾–2 inches below centerline of crossmember. Slide torsion bar rearward until flush with rear face of adjusting arm.
7. Install tool J-22517-02 making sure that pin of tool is in hole in crossmember. Turn center screw of tool until adjusting arm is in position to permit installation of adjusting nut.

1 Cap Screw
2 Cap (Upper)
3 Kingpin
4 Bushing (Upper)
5 Steering Knuckle
6 Bushing (Lower)
7 Cap (Lower)
8 Upper Grease Seal (Rounded Edge Up)
9 Shims
10 Axle Center
11 Draw Key (Short)
12 Thrust Bearing Assembly
13 Lower Grease Seal (Rounded Edge Down)
14 Draw Key (Long)

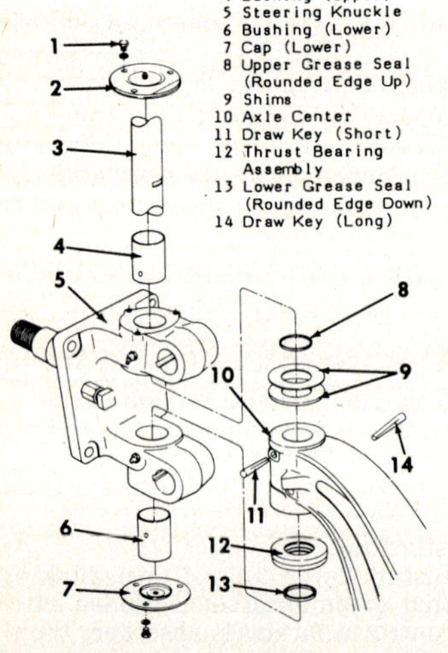

Fig. 14 Steering Knuckle with two draw keys. Typical F-120 (late model) axle

8. Apply chassis grease to adjusting bolt, then install and turn adjusting bolt the number of turns recorded in step 3 to obtain original riding height.
9. Loosen center screw of tool until tension is completely removed from tool and remove tool.
10. Install wheel and lower vehicle.
11. Check riding height and adjust if necessary as described previously.

I-BEAM FRONT AXLE

Front Axle, Replace

All Models

1. Raise and support vehicle at frame. Position floor jack under center of axle beam.
2. Remove wheels and hubs. Disconnect brake lines at axle, then the drag link from steering arm.
3. Remove U-bolt nuts and washer if equipped, and remove U-bolts.

NOTE: Note position of shims, spacer blocks, shock absorber brackets and dowel pins for proper reassembly.

4. Lower axle and remove from under vehicle.
5. Reverse procedure to install.

Steering Knuckle, Replace

All Models

1. Raise and support front axle.
2. Remove front wheels, hubs and bearings.
3. Remove brake components and disconnect tie rod from steering arm.
4. Remove steering arm from knuckle, if removable.
5. On models with single draw key, Figs. 12 and 13:
   a. Remove draw key nut and washer. Thread nut onto draw key enough to protect key threads and drive on key to loosen. Remove key from knuckle using suitable drift.
   b. Remove upper and lower king pin caps and gaskets.
   c. Drive king pin out of knuckle using suitable brass drift. Remove knuckle, shims, thrust bearing and O-ring.
6. On models with two draw keys, Fig. 14:
   a. Remove upper and lower king pin dust cap.
   b. Remove lower expansion plug retainer and expansion plug (if equipped).

   NOTE: If plug will not come out, it will come out when king pin is removed.

   c. Remove both keys using suitable drift against small end of key.

   NOTE: Keys were originally installed from opposite sides of center.

   d. Drive pin out of knuckle using suitable brass drift. Remove knuckle, shims and thrust bearing.
7. On models with tapered king pin, Figs. 15 and 16:
   a. Remove upper dust cap and gasket.
   b. Remove lower expansion plug retainer and plug.

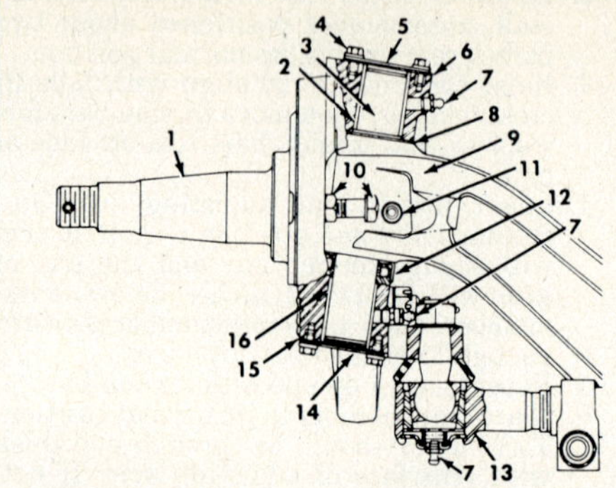

1 Steering Knuckle Spindle
2 Kingpin Bushing (Upper)
3 Kingpin
4 Cap Screw
5 Kingpin Bearing Cap (Upper)
6 Gasket
7 Lube Fitting
8 Shim
9 Axle Center
10 Stop Bolt and Lock Nut
11 Draw Key
12 Thrust Bearing
13 Tie Rod End Assy.
14 Kingpin Bearing Cap (Lower)
15 Gasket
16 Kingpin Bushing (Lower)

Fig. 13 Steering Knuckle with single draw key. Typical F-070 model axle

   NOTE: If plug will not come out, it will come out when king pin is removed.

   c. Remove cotter pin, king pin nut and steel spacer.
   d. Drive pin out of knuckle using suitable brass drift. Remove knuckle, thrust washer and spacing washer from axle. Remove sleeve bushing.
8. Reverse procedure to install.

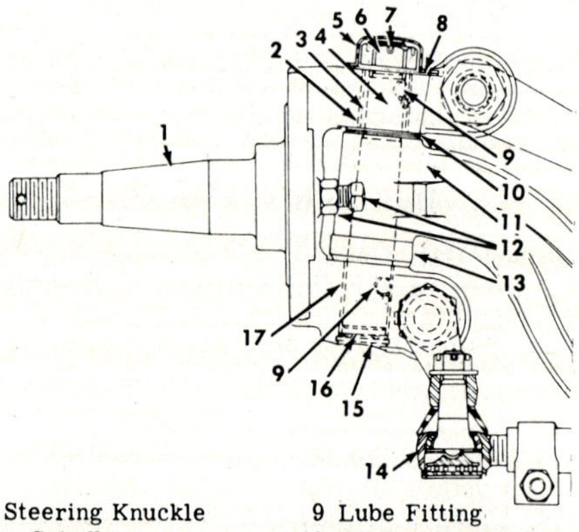

1 Steering Knuckle Spindle
2 Steering Knuckle Bushing
3 Kingpin Bushing (Upper)
4 Kingpin
5 Dust Cap
6 Kingpin Nut
7 Cotter Pin
8 Gasket
9 Lube Fitting
10 Washers or Shims
11 Axle Center
12 Stop Bolt and Lock Nut
13 Thrust Bearing
14 Tie Rod Assembly
15 Plug Retainer
16 Expansion Plug
17 Kingpin Bushing (Lower)

Fig. 15 Steering Knuckle with tapered king pin. Typical F-090 model axle

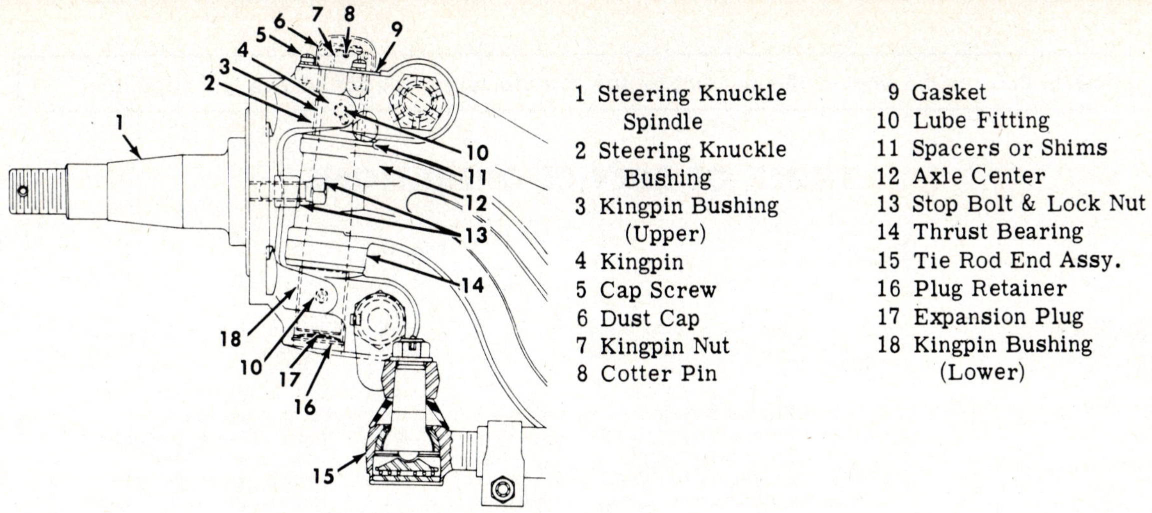

Fig. 16 Steering Knuckle with tapered king pin. Typical F-110 and F-120 (early model) axles

STEERING GEAR, REPLACE

1974–83 All Exc. GMC Motor Home & TransMode

1. Place front wheels in straight ahead position.
2. On C-K and Caballero, El Camino and Sprint models, remove flexible coupling to steering shaft bolts. On P models, remove lower universal joint pinch bolts and mark position of universal yoke to wormshaft.
3. On all models, mark position of pitman arm to pitman shaft and remove pitman shaft nut or pinch bolt, then pitman arm from shaft with a puller.
4. Unfasten steering gear and remove.
5. On C-K models, remove flexible coupling pinch bolt, then flexible coupling from steering gear wormshaft.
6. Reverse procedure to install.

1974–78 GMC Motor Home & TransMode

1. Disconnect power steering hoses from steering gear.
2. Remove pitman arm from pitman shaft using suitable puller.
3. Loosen steering shaft yoke cinch bolt.
4. Remove steering gear to frame rail attaching bolts, then the steering gear.
5. Reverse procedure to install. Torque steering gear mounting bolts to 70 ft. lbs. and pitman shaft nut to 160–210 ft. lbs.

CHEVROLET S/T 10 & GMC S/T 15

NOTE: Refer to the front of this manual for the manufacturers special service tool suppliers.

INDEX OF SERVICE OPERATIONS

GENERAL ENGINE SPECIFICATIONS

Year	Engine Model	Carb. Type	Bore & Stroke	Comp. Ratio	Net Horsepower @ R.P.M.	Torque Ft. Lbs. @ R.P.M.	Normal Oil Press. Lbs.
1982	4-119/1.9L	2 Bore	3.43 × 3.23	8.4	82 @ 4600	101 @ 3000	57
	V6-173/2.8L	2 Bore	3.50 × 3.00	8.5	110 @ 4800	145 @ 2400	30–50
1983	4-119/1.9L	2 Bore	3.42 × 3.23	8.4	82 @ 4600	101 @ 3000	57
	4-121/2.0L	2 Bore	3.50 × 3.15	9.3	82 @ 4600	108 @ 2400	45
	V6-173/2.8L	2 Bore	3.50 × 3.00	8.5	110 @ 4800	145 @ 2100	50–65

TUNE UP SPECIFICATIONS

The following specifications are published from the latest information available. This data should be used only in the absence of a decal affixed in the engine compartment.

★ When using a timing light, disconnect vacuum hose or tube at distributor and plug opening in hose or tube so idle speed will not be affected.

● When checking compression, lowest cylinder must be within 70 percent of highest.

▲ Before removing wires from distributor cap, determine location of the No. 1 wire in cap, as distributor position may have been altered from that shown at the end of this chart.

Year & Engine	Spark Plug		Ignition Timing BTDC①★				Curb Idle Speed②		Fast Idle Speed		Fuel Pump Pressure
	Type	Gap	Firing Order Fig. ▲	Man. Trans.	Auto. Trans.	Mark Location	Man. Trans.	Auto. Trans.	Man. Trans.	Auto. Trans.	
1982											
4-119	R43XLS	.035	A	6°	—	Damper	—	—	—	—	—
V6-173	R43CTS	.045	B	—	—	Damper	—	—	—	—	6–7½
1983											
4-119	R42XLS	—	A	—	—	Damper	—	—	—	—	—
4-121	R42CTS	—	C	—	—	Damper	—	—	—	—	—
V6-173	R43CTS	—	B	—	—	Damper	—	—	—	—	6–7½

①—BTDC—Before top dead center.
②—Idle speed on man. trans. vehicles is adjusted in Neutral & on auto. trans. equipped vehicles is adjusted in Drive unless otherwise specified. Where two idle speeds are listed, the higher speed is with the A/C or idle solenoid energized.

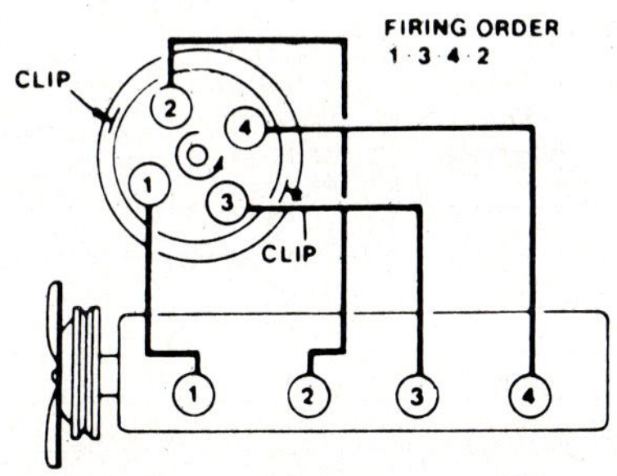

Fig. A

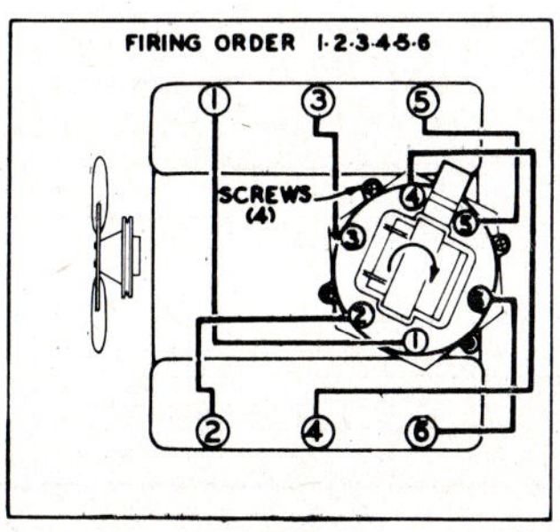

Fig. B

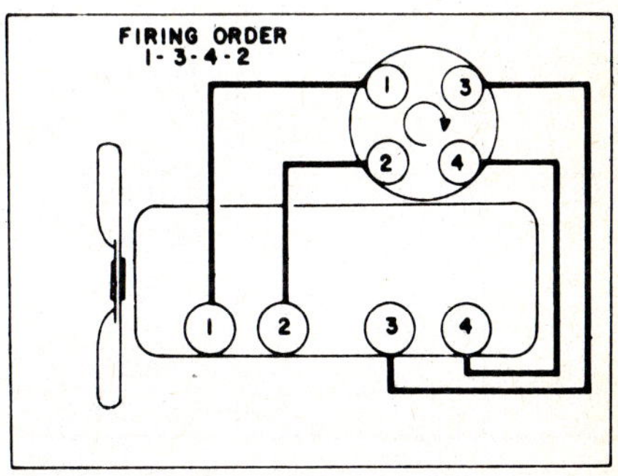

Fig. C

CHEVROLET S/T 10 & GMC S/T 15

VALVE SPECIFICATIONS

Year	Engine	Valve Lash		Valve Angles		Valve Spring Installed Height	Valve Spring Pressure Lbs. @ In.	Stem Clearance		Stem Diameter	
		Int.	Exh.	Seat	Face			Intake	Exhaust	Intake	Exhaust
1982–83	4-119	.006C	.010C	45	45	①	②	.0009–.0022	.0015–.0031	—	—
	V6-173	1½ Turn③		46	45	1.57	195 @ 1.181	.0010–.0027	.0010–.0027	.3410–.3420	.3410–.3420
1983	4-121	1½ Turn③		46	45	1.60	182 @ 1.33	.0011–.0025	.0013–.0029	—	—

①—Valve spring free length, inner 1.7835 inch; outer 1.8465 inch.

②—Valve spring tension, inner 18–21 @ 1.516; outer 32–37 @ 1.614

③—Turn rocker arm stud nut until all lash is eliminated, then tighten nut the additional turn listed.

PISTONS, PINS, RINGS, CRANKSHAFT & BEARINGS SPECIFICATIONS

Year	Engine	Piston Clearance	Ring End Gap①		Wrist Pin Diameter	Rod Bearings		Main Bearings			
			Comp.	Oil		Shaft Diameter	Bearing Clearance	Shaft Diameter	Bearing Clearance	Thrust Bearing No.	Shaft End Play
1982	4-119	.0018–.0026	②	.008	—	1.929	.0007–.0030	2.205	.0008–.0025	3	.002–.009
	V6-173	.0017–.0027	.010	.020	.9054	1.9983–1.9993	.0014–.0035	2.4937–2.4946	.0017–.0030	3	.002–.007
1983	4-119	.0018–.0026	②	.008	—	1.929	.0007–.0030	2.205	.0008–.0025	3	.002–.009
	4-121	.0006–.0016	.010	—	.9054	1.9983–1.9993	.0010–.0031	③	④	4	.002–.008
	V6-173	.0007–.0017	.010	.020	.9054	1.9983–1.9993	.0014–.0037	⑤	⑥	3	.002–.007

①—Fit rings in tapered bores for clearance listed in tightest portion of ring travel.

②—Top ring, .012 inch; second ring, .008 inch.

③—Nos. 1, 2, 3, 4; 2.4945–2.4954. No. 5; 2.4937–2.4946.

④—Nos. 1, 2, 3, 4; .0006–.0018. No. 5; .0014–.0027.

⑤—Nos. 1, 2, 4; 2.4937–2.4946. No. 3; 2.4930–2.4941.

⑥—Nos. 1, 2, 4; .0016–.0032. No. 3; .0021–.0033.

ENGINE TIGHTENING SPECIFICATIONS★

★Torque specifications are for clean and lightly lubricated threads only. Dry or dirty threads produce increased friction which prevents accurate measurement of tightness.

Year	Engine	Spark Plugs Ft. Lbs.	Cylinder Head Bolts Ft. Lbs.	Intake Manifold Ft. Lbs.	Exhaust Manifold Ft. Lbs.	Rocker Arm Shaft Bracket Ft. Lbs.	Rocker Arm Cover Ft. Lbs.	Connecting Rod Cap Bolts Ft. Lbs.	Main Bearing Cap Bolts Ft. Lbs.	Flywheel to Crankshaft Ft. Lbs.	Vibration Damper or Pulley Ft. Lbs.
1982–83	4-119	18–22	72	—	16	16	4	43	72	76	87
	V6-173	7–15	65–75	20–25	22–28	43–49①	6–9	34–40	63–74	45–55	66–84
1983	4-121	7–19	65–75	22–29	19–29	43–49①	3	34–42	63–77	45–59	66–88

①—Rocker arm stud

DISTRIBUTOR SPECIFICATIONS

Year	Engine	Distributor Number	Rotation ①	Breaker Gap (New)	Dwell Angle Deg.	Breaker Arm Spring Tension	Centrifugal Advance Degrees @ R.P.M. of Distributor		Vacuum Advance	
							Advance Starts	Full Advance	Inches of Vacuum to Start Plunger	Max. Adv. Dist. Deg. @ Vacuum
1982	4-119	94230979	CC	—	—	—	—	—	—	—
		94230990	CC	—	—	—	—	—	—	—
		94240905	CC	—	—	—	—	—	—	—
	V6-173	1110581	C	—	—	—	—	—	—	—
		1110582	C	—	—	—	—	—	—	—
		1110583	C	—	—	—	—	—	—	—

①—As viewed from top. C—clockwise CC—counterclockwise

WHEEL ALIGNMENT SPECIFICATIONS

Year	Truck Model	Caster Deg.	Camber Deg.	Toe-In (Inches)	King Pin Angle Deg.
1982–83	S-10, 15	+1.5° to +2.5°	+.3° to +1.3°	①	—

①—Toe-in, +15°.

ALTERNATOR & REGULATOR SPECIFICATIONS

Year	Alternator					Regulator						
	Model	Field Current @ 80°F. 12 volts	Cold Output @ 14 volts		Rated Hot Output Amperes	Model	Field Relay			Voltage Regulator		
			Amperes @ 2000 R.P.M.	Amperes @ 5000 R.P.M.			Air Gap	Point Gap	Closing Voltage	Air Gap	Point Gap	Volts @ 125°F.
1982	1100140	—	—	—	37	Integral	—	—	—	—	—	—
	1100146	—	—	—	63	Integral	—	—	—	—	—	—
	1100201	—	—	—	37	Integral	—	—	—	—	—	—
	1100202	—	—	—	63	Integral	—	—	—	—	—	—
1983	1100204	—	—	—	37	Integral	—	—	—	—	—	—
	1100207	—	—	—	66	Integral	—	—	—	—	—	—
	1100209	—	—	—	78	Integral	—	—	—	—	—	—
	1100227	—	—	—	37	Integral	—	—	—	—	—	—
	1100249	—	—	—	66	Integral	—	—	—	—	—	—
	1100273	—	—	—	78	Integral	—	—	—	—	—	—
	1100275	—	—	—	66	Integral	—	—	—	—	—	—
	1100276	—	—	—	78	Integral	—	—	—	—	—	—
	1105185	—	—	—	37	Integral	—	—	—	—	—	—

STARTER MOTOR SPECIFICATIONS

Year	Engine	Starter Model	Brush Spring Tension Oz.①	Free Speed Test			Resistance Test	
				Amperes	Volts	R.P.M.	Amperes	Volts
1982	4-119	—	—	—	—	—	—	—
	V6-173	1109535	56.5	50–80②	10.6	7500–11400	—	—
1983	4-119	—	—	—	—	—	—	—
	4-121	109561	56.5	50–75	—	6000–11900	—	—
	V6-173	109535	56.5	45–70	—	7000–11900	—	—

①—Minimum
②—Includes solenoid.

4-119 Engine Section

ENGINE, REPLACE

1. Remove hood and disconnect battery ground cable.
2. Raise vehicle.
3. Remove rear transmission mount nuts.
4. Support transmission and remove crossmember at frame and top transmission bolts.
5. Reinstall crossmember and remove transmission support.
6. Remove upper starter motor retaining nut.
7. Remove engine to transmission bolts.
8. On models with automatic transmission, remove torque converter cover bolts, cover and torque converter bolts.
9. Disconnect exhaust pipe
10. Disconnect starter motor wires.
11. Remove motor mount bolts and disconnect fuel lines.
12. Remove transmission cooler lines at oil pan.
13. Lower vehicle and drain coolant.
14. Disconnect any necessary wires and hoses.
15. Remove power steering pump and/or A/C compressor, if so equipped, and position aside.
16. Loosen alternator.
17. Disconnect front heater hose.
18. Remove upper half fan shroud and remove fan assembly.
19. Disconnect radiator hoses at engine and transmission cooler lines from radiator.
20. Remove radiator.
21. Disconnect accelerator cable and spring.
22. Disconnect detent cable.
23. Support transmission with a floor jack.
24. Install lift device and remove engine.
25. Reverse procedure to install.

ENGINE MOUNTS, REPLACE

Front

1. Disconnect battery ground cable.
2. Remove air cleaner duct and upper fan shroud.
3. Remove engine mount nuts and retaining wire.
4. With vehicle raised, raise engine to take weight off mounts.
5. Remove engine mount to engine bracket.
6. Remove engine mount using tool J-25510, or equivalent, Fig. 1.
7. Reverse procedure to install.

Rear

1. Disconnect battery ground cable.
2. Remove mount nut and bolts.
3. Support, and slightly raise, transmission at extension housing and remove engine mount, Fig. 1.
4. Reverse procedure to install.

CYLINDER HEAD, REPLACE

1. Remove cam cover.
2. Remove EGR pipe clamp at rear of cylinder head.
3. Raise and support vehicle.
4. Disconnect exhaust pipe at manifold.
5. Lower vehicle, drain cooling system and disconnect heater hoses.
6. Remove A/C compressor and/or power steering pump, and position aside.
7. Disconnect accelerator linkage and fuel line at carburetor, all necessary electrical connections, spark plug wires and necessary vacuum lines.
8. Remove camshaft until No. 4 cylinder is in firing position. Remove distributor cap and mark rotor to housing relationship, then remove distributor.
9. Remove fuel pump.
10. Lock shoe on timing chain automatic adjuster in fully retracted position by depressing the adjuster lock lever with a screwdriver or equivalent in direction as indicated, Fig. 2.
11. Remove timing sprocket to camshaft bolt. Remove sprocket and fuel pump drive cam from the camshaft. Do not remove the sprocket from the chain.
12. Disconnect AIR hose and check valve at

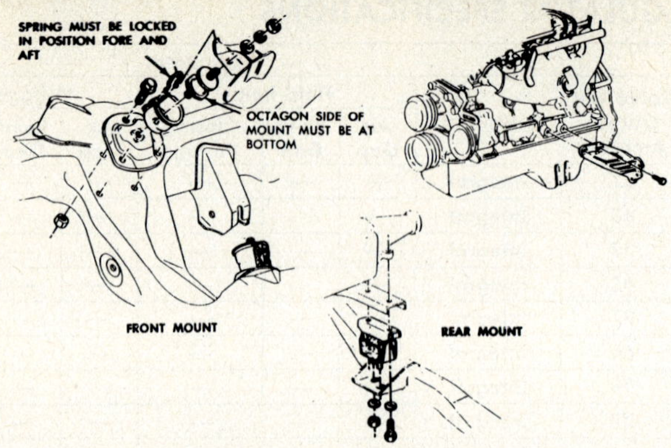

Fig. 1 Engine mount

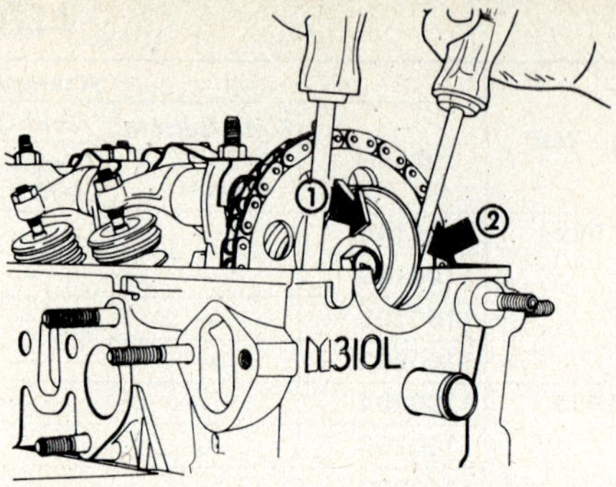

Fig. 2 Depressing adjusting lock lever

air manifold.

13. Remove cylinder head to timing cover bolts.
14. Remove cylinder head bolts using tool J-24239-01, or equivalent. Remove outer bolts first.
15. With an assistant, remove cylinder head, intake and exhaust manifold as an assembly.
16. Clean all gasket material from cylinder head and block surfaces.

NOTE: Gasket surfaces on both head and block must be clean of all foreign matter and free of nicks or heavy scratches. Cylinder bolt threads in the block and threads on the bolts must be cleaned, as dirt will affect bolt torque.

---

17. Reverse procedure to install and tighten head bolts to specifications in sequence, Fig. 3.

INTAKE MANIFOLD, REPLACE

1. Drain engine cooling system.

NOTE: Before removing intake manifold, check that the cooling system is completely drained. Any coolant left in the engine will flow into the cylinder when the manifold is removed.

---

2. Remove air cleaner assembly.
3. Disconnect radiator upper hose from intake manifold.
4. Disconnect vacuum hose from manifold.
5. Disconnect heater hoses from manifold and from connector under the dashboard.
6. Disconnect accelerator control cable from carburetor.
7. Disconnect distributor vacuum hose from distributor and the thermo-unit wiring at the connector.
8. Disconnect automatic choke and solenoid wiring at connectors.
9. Disconnect PCV hose from the rocker arm cover.
10. Disconnect oil level gage guide tube fixing bolt at the intake manifold.
11. Disconnect the EGR pipe from the EGR valve adapter and the AIR vacuum hose from the 3-way joint.
12. Remove the eight attaching nuts and remove the intake manifold.
13. Using a new gasket, reverse procedure to install.

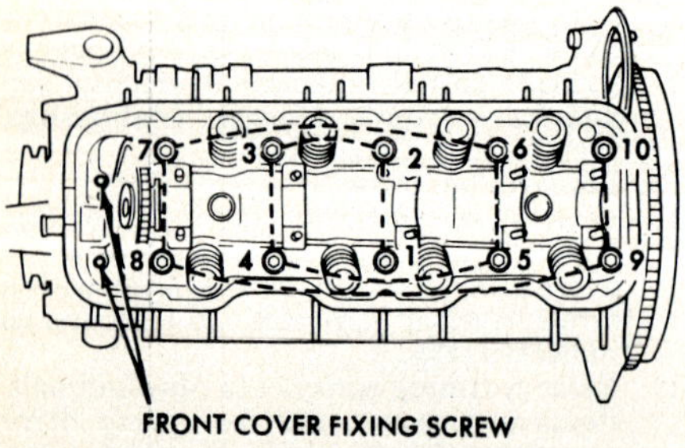

Fig. 3 Cylinder head hold-down bolt torque sequence

EXHAUST MANIFOLD, REPLACE

1. Disconnect battery ground cable.
2. Raise and support vehicle.
3. Disconnect exhaust pipe from exhaust manifold.
4. Disconnect E.G.R pipe at exhaust manifold.
5. Lower vehicle, remove bolts attaching air cleaner and loosen clamp bolts.
6. Remove hot air hose with the air cleaner slightly raised.
7. Remove the four attaching bolts and remove the manifold cover.
8. Remove the seven mounting nuts from the manifold and remove the exhaust manifold.
9. Using a new gasket, reverse procedure to install. Torque manifold stud nuts to 16 ft. lbs. in a sequence beginning with the inner nuts and working outward.

ROCKER ARMS & SHAFTS

Removal

1. Remove cam cover.
2. Alternately loosen rocker arm shaft bracket nuts and remove nuts from brackets.
3. Remove spring from rocker arm shaft, then the rocker arm brackets and arms.

Installation

1. Lubricate rocker arms, shafts and valve stems with engine oil.
2. Install the longer rocker shaft on the exhaust valve side and the shorter rocker shaft on the intake valve side, Fig. 4, with the aligning marks facing front.
3. Install rocker arm shaft brackets and rocker arms on the rocker arm shafts with the cylinder number on the upper face of the brackets facing toward front of engine.
4. Align the mark on the No. 1 rocker arm shaft bracket with the mark on the rocker arm shafts.
5. The exhaust side rocker arm shaft should project a greater distance from the face of the No. 1 rocker shaft bracket outer face than the intake side rocker arm shaft when the rocker arm shaft stud holes are aligned with the rocker arm shaft bracket stud holes.
6. Position rocker arm shaft springs between rocker arm shaft bracket and rocker arm.
7. Ensure punch mark on rocker arm shaft is facing upward. Install rocker arm shaft bracket assembly onto cylinder head studs. Align mark on camshaft with mark on No. 1 rocker arm shaft bracket.
8. Torque rocker arm shaft stud nuts to 16 ft. lbs.

NOTE: Hold the rocker arm springs in position while torquing the stud nuts to prevent spring damage.

---

9. Adjust valves and install cam cover.

VALVES, ADJUST

1. Before adjusting valve clearances, check the rocker arm shaft bracket nuts for looseness and retighten as necessary.
2. Bring either No. 1 or No. 4 piston to top dead center on the compression stroke by turning the crankshaft to align timing mark on crankshaft pulley with pointer.

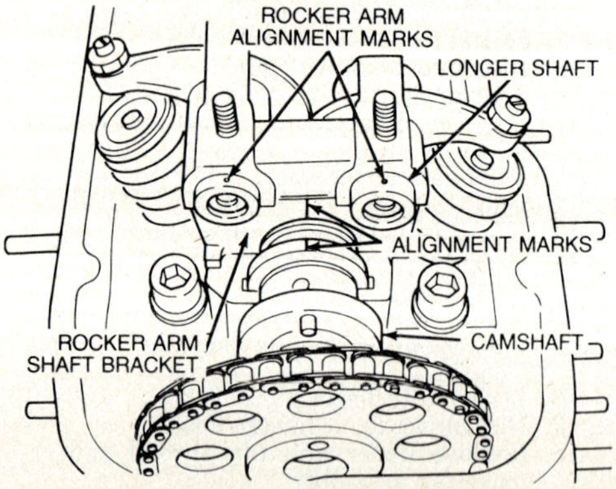

Fig. 4 Rocker arm shaft installation

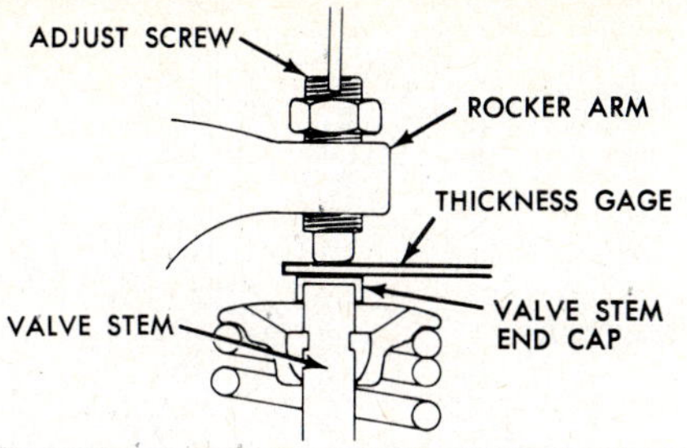

Cylinder no. / Valve	1	2	3	4
Intake	o	o	●	●
Exhaust	o	●	o	●

Note: o When piston in No. 1 cylinder is at TDC on compression stroke.
● When piston in No. 4 cylinder is at TDC on compression stroke.

Fig. 5 Valve adjustment

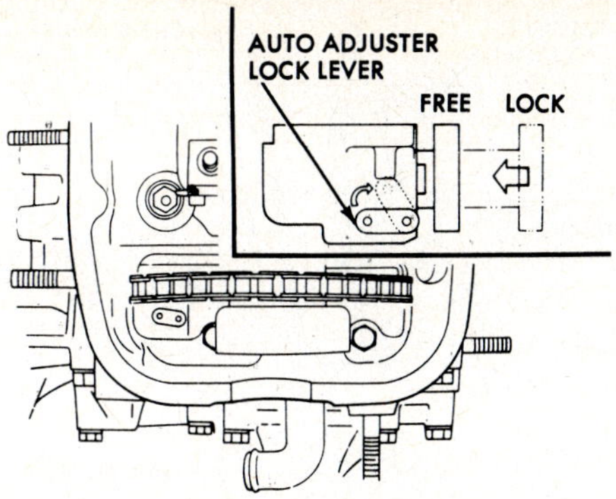

Fig. 6 Locking timing chain adapter

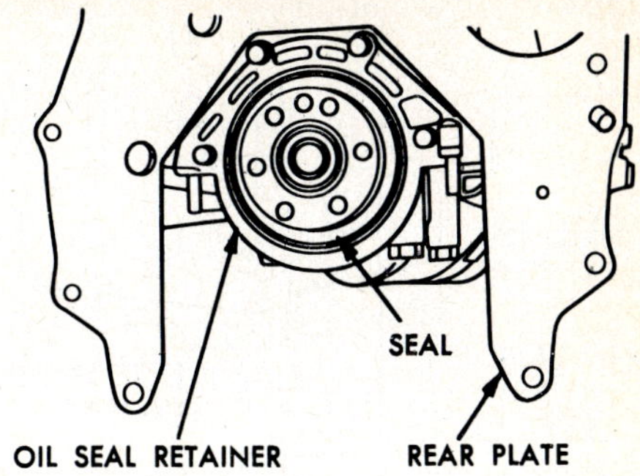

Fig. 7 Rear main oil seal

3. Hold crankshaft in above position and adjust clearance of the valves indicated, Fig. 5. to specifications. Take measurement at the clearance between rocker arm and valve stem.
4. Turn crankshaft one full turn and adjust clearance of remaining valves to specifications.

VALVE ARRANGEMENT

Front to Rear

E-I-I-E-E-I-I-E

CAMSHAFT, REPLACE

Removal

1. Remove cam cover.
2. Rotate camshaft until No. 4 cylinder is in firing position. Remove distributor cap and mark rotor to housing position, then remove distributor.
3. Remove the fuel pump.
4. Lock the shoe on timing chain automatic adjuster in fully retracted position by depressing the adjuster lock lever with a screwdriver or equivalent in direction as indicated, Figs. 2 and 6. After locking the automatic adjuster, check that the chain is loose.
5. Remove timing sprocket to camshaft bolt. Remove sprocket and fuel pump drive cam from camshaft. Do not remove sprocket from chain.
6. Remove rocker arm, shaft and bracket assembly.
7. Remove camshaft assembly.

Installation

1. Lubricate camshaft and journals of cylinder head with engine oil.
2. Install camshaft assembly.
3. Install rocker arm, shaft and bracket assembly.
4. Align mark on the No. 1 rocker arm shaft bracket with the mark on the camshaft. Also, ensure that the crankshaft pulley groove is aligned with TDC mark on front cover.
5. Assemble timing sprocket to camshaft by aligning sprocket with pin on camshaft. Use caution not to remove chain from sprocket.
6. Install fuel pump drive cam and sprocket retaining bolt and washer. Remove the half-moon seal at front of head, then torque retaining bolt to 60 ft. lbs. Then install the half-moon seal in cylinder head.
7. Install distributor.
8. Release lock by depressing shoe on timing chain automatic adjuster and check timing chain for proper tension.
9. Check that distributor rotor and mark on distributor housing are aligned with No. 4 piston firing position in distributor cap, then install distributor. Timing mark on crank pulley should also be aligned with TDC mark on front cover.
10. Install distributor cap.
11. Install cam cover.

REAR MAIN SEAL

1. Remove starter motor and position aside.
2. Remove transmission.
3. On vehicles equipped with manual transmission, remove clutch cover and pressure plate assembly, then the flywheel and flywheel cover.
4. On vehicles equipped with automatic transmission, remove flex plate.
5. Remove seal from retainer, Fig. 7.
6. Install new seal using tool J-22928-A, Fig. 8. Fill clearance between lips of seal with suitable grease and lubricate seal lip with clean engine oil.

FRONT OIL SEAL, REPLACE

1. Disconnect battery ground cable and drain cooling system.
2. Disconnect radiator hoses and remove radiator assembly.
3. Remove drive belts and engine fan.
4. Remove crankshaft pulley center bolt and the pulley and balancer assembly.
5. Pry seal from timing cover with a suitable screwdriver.
6. Install new seal with tool J-26587.

FRONT COVER, REPLACE

Removal

1. Disconnect battery ground cable.
2. Remove oil pan as described under "Oil Pan, Replace".
3. Remove cylinder head.
4. Remove oil pickup tube from oil pump.
5. Remove harmonic balancer and the AIR pump drive belt.
6. On air conditioned vehicles, remove A/C compressor and position aside. Remove compressor mounting brackets.

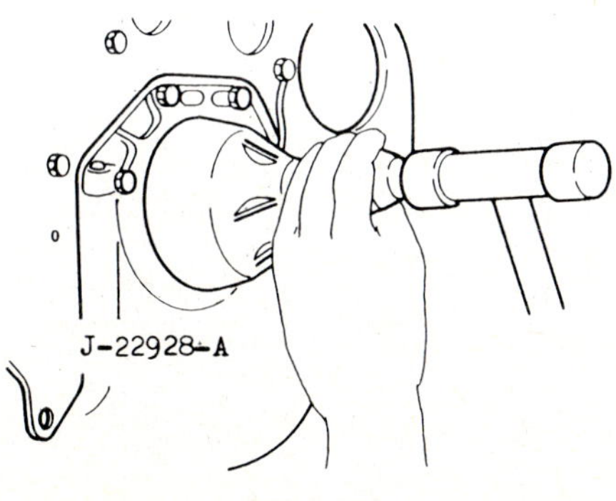

Fig. 8 Installing rear main oil seal

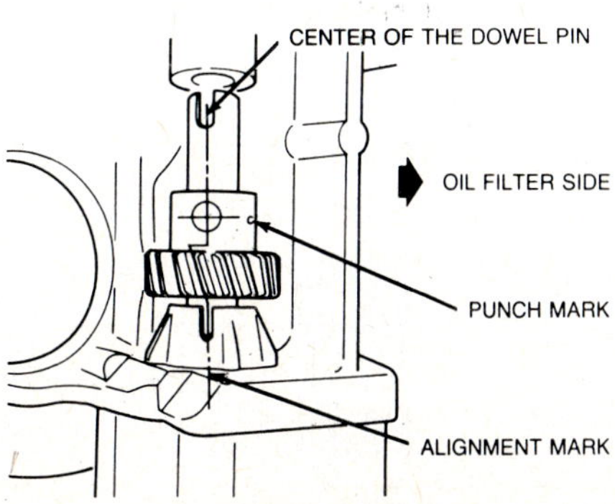

Fig. 9 Oil pump alignment

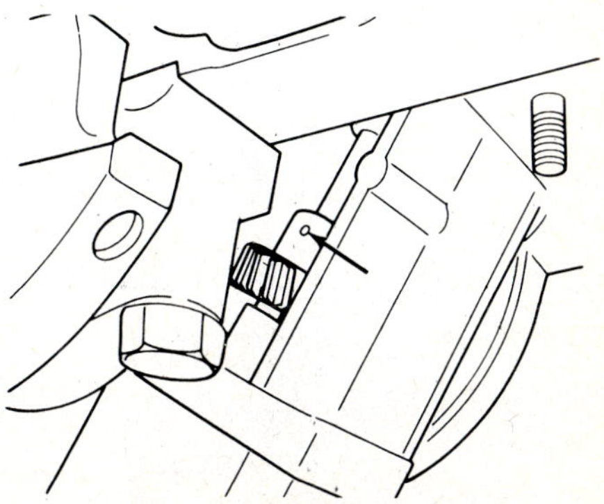

Fig. 10 Checking alignment

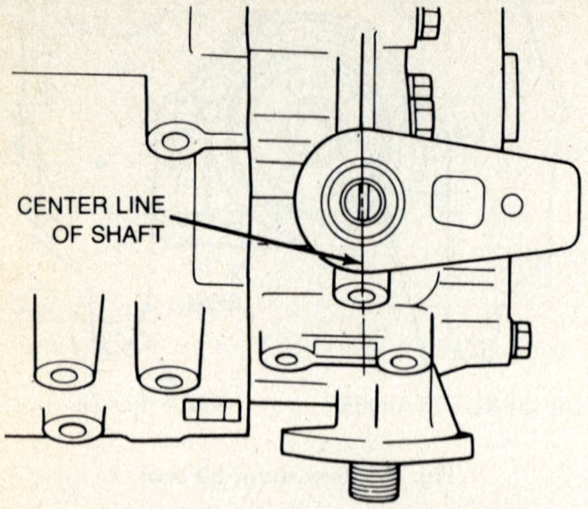

Fig. 11 Oil pump shaft alignment

7. Remove distributor assembly.
8. Remove front cover attaching bolts and the front cover.

Installation

1. Install new gasket onto cylinder block.
2. Align oil pump drive gear punch mark with oil filter side of cover, Fig. 9, then align the center of dowel pin with alignment mark on oil pump case.
3. Rotate crankshaft until Nos. 1 and 4 cylinders are at top dead center.
4. Install front cover by engaging pinion gear with oil pump drive gear on crankshaft.
5. Check that punch mark on oil pump drive gear is turned to the rear, viewed through the clearance between front cover and cylinder block, Fig. 10.
6. Check that slit at the end of oil pump shaft is parallel with front face of cylinder block and is offset forward, Fig. 11.
7. Install front cover and tighten front cover bolts.
8. Reverse steps 1 thru 7 under "Front Cover, Replace".
9. Check engine timing.
10. Check for leaks.

TIMING CHAIN, REPLACE

1. Remove front cover assembly. Refer to steps under "Front Cover, Replace".
2. Lock automatic adjuster shoe in fully retracted position.
3. Remove timing chain from crankshaft sprocket, Fig. 12.
4. Remove "E" clip and the chain adjuster and tensioner, Fig. 12. Inspect components for wear or damage. Also, inspect chain guide and oil jet. View "A", Fig. 13.
5. Install timing sprocket and pinion gear with groove side toward front cover. Align key grooves with key on crankshaft, then drive into position with tool J-26587.
6. Turn crankshaft so key is turned toward cylinder head side with Nos. 1 and 4 pistons at top dead center.
7. Install timing chain by aligning mark plate on chain with the mark on crankshaft timing sprocket. The side of the chain with mark plate is located on the front side and the side of chain with the most links between mark plates is located on the chain guide side, Fig. 14.
8. Install the camshaft timing sprocket so marked side of sprocket faces forward and the triangular mark aligns with the chain mark plate.

NOTE: Keep the timing chain engaged with camshaft timing sprocket until the camshaft timing sprocket is installed on camshaft.

9. Install automatic chain adjuster, then release lock by depressing the adjusting shoe.
10. Install front cover assembly as outlined previously.

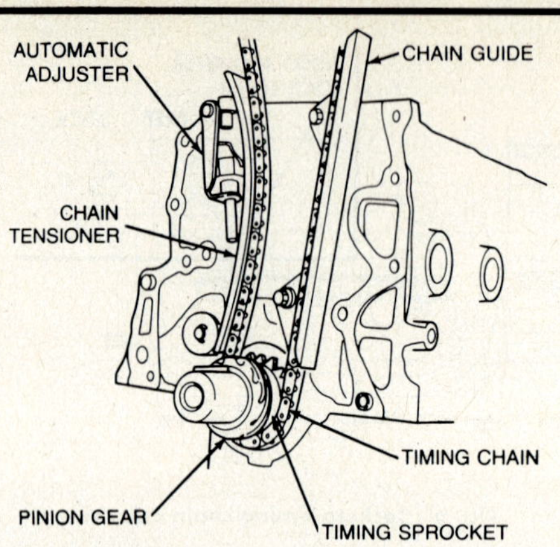

Fig. 12 Timing chain removal

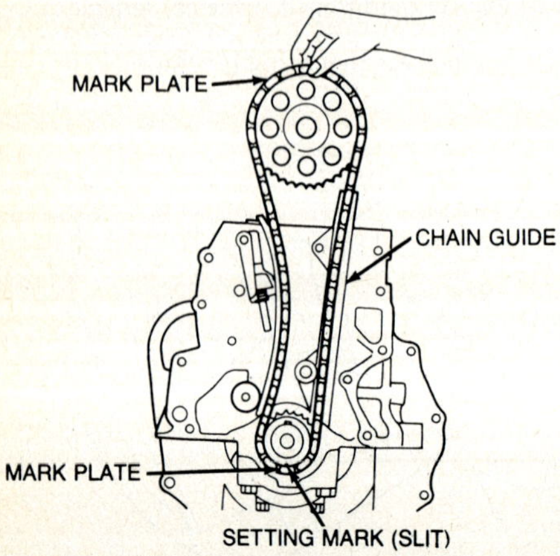

Fig. 14 Timing chain alignment

PISTONS & RODS

Assemble piston to rod so the mark on the piston is facing toward front of engine, Fig. 15, and the face of the connecting rod with the cylinder number mark is facing toward the starter side of the engine.

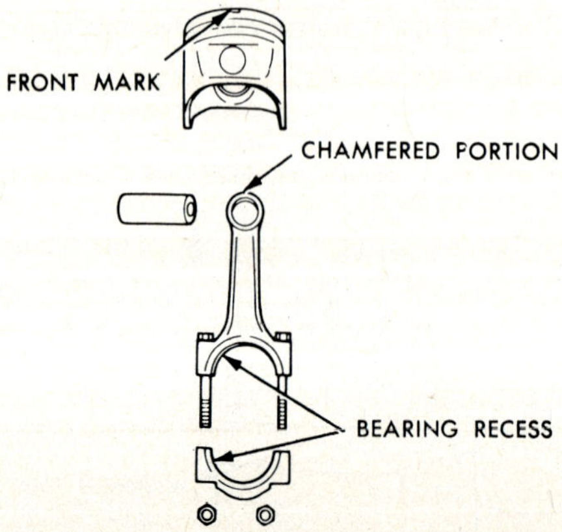

Fig. 15 Piston and connecting rod assembly

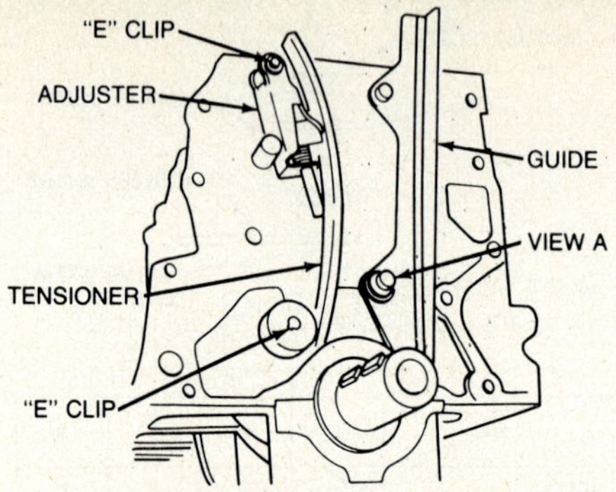

Fig. 13 Timing chain guide, tensioner and adjuster

Pistons are available in oversizes of .020 and .030 inch.

OIL PAN, REPLACE

Removal

1. Remove engine as outlined under "Engine, Replace".
2. Remove nuts and bolts attaching oil pan to cylinder block and remove oil pan.
3. Remove oil level gage guide tube from the intake manifold and oil pan.

Installation

1. Apply a thin coat of Permatex No. 2, or equivalent to locations indicated by arrows, Fig. 16.
2. Install new oil pan gasket aligning holes and install oil pan on cylinder block.
3. Install nuts and bolts and torque evenly to specifications.

NOTE: Check edge of gasket to check proper positioning. If projection of gasket edge beyond the oil pan flange is uneven, remove and reinstall, Fig. 17.

4. Install engine into vehicle as outlined under "Engine, Replace".

OIL PUMP

Removal

1. On all models, remove cam cover, distributor assembly and oil pan.
2. Remove bolt securing oil pick-up tube to engine block and the tube from the oil pump.
4. Remove rubber hose and relief valve assembly from the pump.

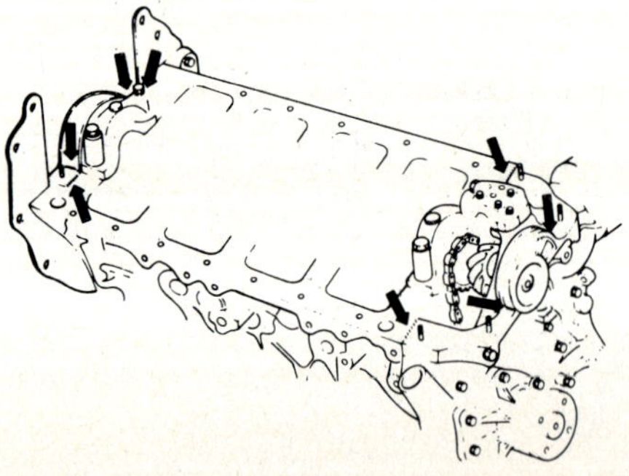

Fig. 16 Sealer locations on oil pan gasket surface

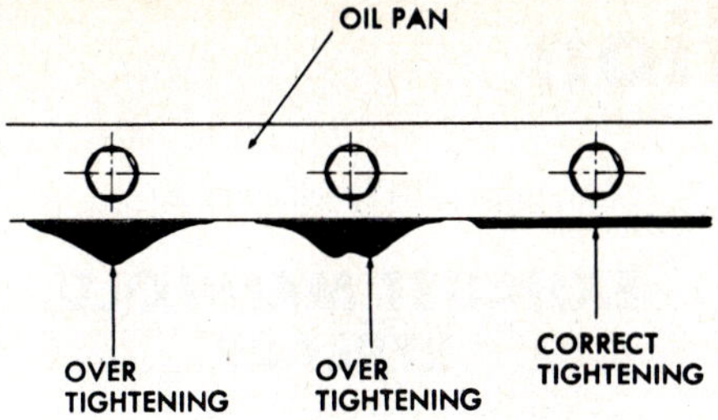

Fig. 17 Oil pan gasket installation

Service

1. Measure clearance between drive rotor and driven rotor with a feeler gauge, Fig. 18. If clearance is greater than .0079 inch, replace oil pump assembly.
2. Measure clearance between driven rotor and inner wall of pump body, Fig. 18. If clearance is greater than .0098 inch, replace oil pump assembly.
3. Measure clearance between drive rotor, driven rotor and oil pump cover with a feeler gauge and a straight edge, Fig. 18. If clearance is greater than .0079 inch, replace oil pump assembly.
4. Determine clearance between drive shaft and drive shaft hole in pump cover. Measure the inside diameter of the shaft hole in the cover and the diameter of the drive shaft. Subtract the drive shaft diameter from the inside diameter of the shaft hole. If the clearance is greater than .0098 inch, replace oil pump.
5. Inspect components for wear or damage.

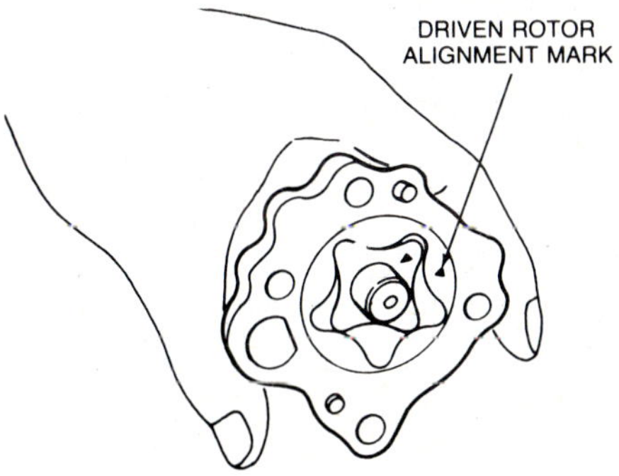

Fig. 19 Aligning drive and driven rotors

Installation

1. Align mark on camshaft with mark on No. 1 rocker arm shaft bracket. Align notch on crankshaft pulley with "O" mark on front cover. When these two sets of marks are properly aligned, No. 4 piston is at top dead center, compression stroke.
2. Install driven rotor with alignment mark aligned with mark on drive rotor. Fig. 19.
3. Engage oil pump drive gear with pinion gear on crankshaft so alignment mark on drive gear is turned rearward and approximately 20° from the crankshaft in a clock-wise direction, Fig. 20.
4. Ensure oil pump drive gear is turned rear-ward as viewed from clearance between front cover and cylinder block and the slit at the end of the oil pump drive shaft is parallel with the front face of the cylinder block and is offset forward as viewed through the distributor fitting hole, Fig. 11.
5. Install oil pump cover and mounting bolts.
6. Install relief valve assembly and rubber hose on cover.
7. Connect oil pick-up tube to rubber hose and secure tube to cylinder block.
8. Install oil pan and cam cover.
9. Install distributor so boss on shaft is fitted into slit at end of oil pump drive shaft.
10. Refill engine oil and check for leaks.

WATER PUMP, REPLACE

1. Disconnect battery ground cable.
2. Remove lower cover.
3. Drain cooling system.
4. On models less A/C, remove engine fan.
5. On models with A/C:
   a. Remove air pump and alternator mounting bolts and the fan and air pump drive belt.
   b. Remove engine fan and pulley assembly.
   c. Remove fan set plate and fan pulley.
6. On all models, remove water pump attaching bolts and the water pump.
7. Reverse procedure to install.

FUEL PUMP, REPLACE

1. Disconnect battery ground cable.
2. Remove distributor.
3. Disconnect fuel pump inlet and outlet hoses.

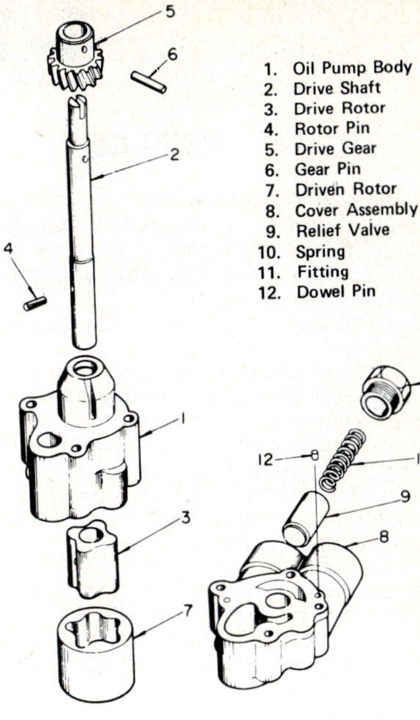

Fig. 18 Oil pump assembly

4. Remove engine lift hook.
5. Remove fuel pump and gasket.
6. Reverse procedure to install.

NOTE: Rotate camshaft to "down stroke" position before installing pump. Lubricate rod with engine oil.

7. Adjust timing and check for leaks.

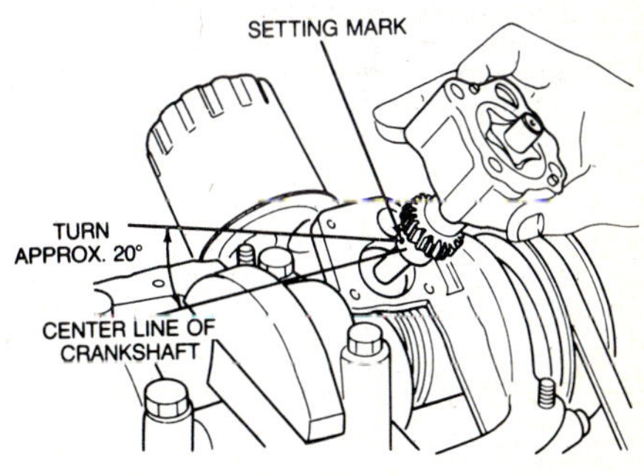

Fig. 20 Oil pump installation

4-121 Engine Section

ENGINE, REPLACE

1. Disconnect battery cables.
2. Scribe alignment marks on hood and hood hinge, then remove hood.
3. Drain cooling system, then remove radiator hose from radiator.
4. Remove upper fan shroud, then the radiator and fan.
5. Disconnect heater hoses from engine.
6. Disconnect wiring harness at bulkhead, then remove air cleaner.
7. Disconnect accelerator cable from carburetor, then the fuel line from frame.
8. Disconnect battery ground cable from engine block, then remove the ground strap from bulkhead.
9. Raise and support vehicle.
10. Remove clutch bell crank, then disconnect exhaust pipe from manifold.
11. Remove motor mount through bolt, left hand body mount bolts and left air dam bolts.
12. Raise body to gain access to upper bell housing bolts, then remove the bolts.
13. Lower body, then remove flywheel dust shield.
14. Lower vehicle and support transmission with suitable jack.
15. If equipped, disconnect power steering pump and/or A/C compressor and position aside.
16. Install suitable lifting device, then remove engine from vehicle.
17. Reverse procedure to install.

ENGINE MOUNTS, REPLACE

1. Disconnect battery ground cable.
2. Remove upper fan shroud.
3. Raise and support vehicle.
4. Remove engine mount through bolt. On left mount, remove mount-to-engine upper bracket attaching nuts.
5. Raise front of engine and remove mount-to-engine attaching bolts, then the mount.

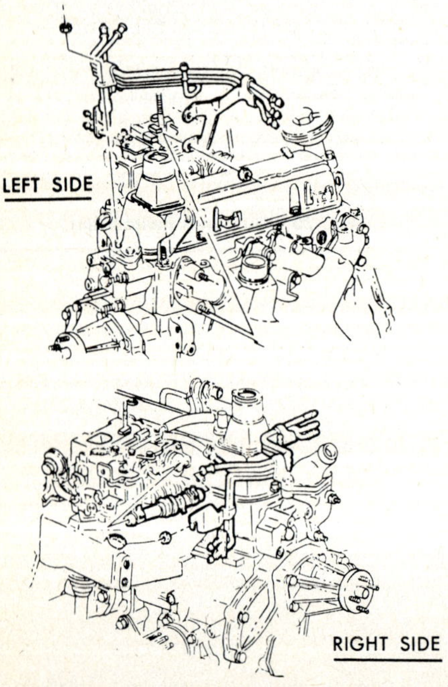

Fig. 1 Fuel vapor canister harness

NOTE: Raise engine only enough to provide sufficient clearance to remove mount. Check for interference between rear of engine and cowl panel which could cause damage to the distributor or EGR.

---

6. Reverse procedure to install. Torque mount-to-engine attaching bolts to 29–39 ft. lbs. and mount-to-frame bolt to 35–47 ft. lbs.

CYLINDER HEAD, REPLACE

1. Disconnect battery ground cable.
2. Remove air cleaner and drain cooling system.
3. Raise and support vehicle.
4. Disconnect exhaust pipe from exhaust manifold, then lower vehicle.
5. Disconnect accelerator linkage and any necessary wires and vacuum lines.
6. Remove fuel vapor canister harness lines, Fig. 1.
7. Remove distributor cap and distributor. Mark position of distributor for reference during installation.
8. Remove rocker arm cover, then the rocker arms and push rods.
9. Remove radiator and heater hoses.
10. Remove upper fan shroud, then the fan.
11. Remove air management valve, then the air pump and upper AIR bracket.
12. Remove fuel line from fuel pump.
13. Remove cylinder head attaching bolts, then the cylinder head. Disconnect wires from rear of head.
14. Reverse procedure to install. Coat cylinder head gasket and bolts with suitable sealant and install bolts finger tight. Torque bolts to specifications in sequence shown in Fig. 2.

INTAKE MANIFOLD, REPLACE

1. Disconnect battery ground cable.
2. Remove air cleaner and distributor cap.
3. Raise and support vehicle.
4. Remove middle right bell housing-to-engine block attaching bolt, then disconnect the wiring harness.
5. Remove distributor hold down nut and clamp, then disconnect vacuum line and primary wires from coil.
6. Remove fuel pump attaching bolts and position pump aside, then lower vehicle.
7. Disconnect accelerator cable, fuel inlet line and any necessary wires and vacuum lines.
8. Remove carburetor, then drain cooling system.
9. Disconnect fuel vapor canister harness lines, Fig. 1.
10. Disconnect hoses and wires from manifold.
11. Remove intake manifold attaching bolts, then the manifold.
12. Reverse procedure to install. Clean mating surface and install a new gasket. Torque bolts to specifications.

EXHAUST MANIFOLD, REPLACE

1. Disconnect battery ground cable.
2. Remove air cleaner.
3. Raise and support vehicle.
4. Remove exhaust pipe from manifold.
5. Remove AIR hose and pipe bracket bolt, then the dipstick tube bracket.
6. Remove exhaust manifold attaching bolts, then the manifold.
7. Reverse procedure to install. Torque attaching bolts to specifications.

ROCKER ARM STUDS

Rocker arm studs that have stress cracks or damaged threads should be replaced. If threads in cylinder head are damaged, the head can be retapped and a helical type insert installed.

VALVES, ADJUST

1. Crank engine until mark on crankshaft pulley is aligned with "O" mark on timing tab. Check to ensure engine is in the No. 1 cylinder firing position by placing fingers on No. 1 cylinder rocker arms as mark on pulley comes near "O" mark on timing tab. If valves are not moving, the engine is in the No. 1 firing position. If valves are moving, engine is on No. 4 cylinder firing position and should be rotated one revolution to reach the No. 1 cylinder firing position.
2. With engine in No. 1 cylinder firing position, adjust the following valves: exhaust, 1 and 3: intake 1 and 2. To adjust valves, back off adjusting nut until lash is felt at push rod, then tighten nut until all lash is removed. This can be determined by rotating push rod while tightening adjusting nut. When all lash has been removed, turn adjusting nut in an additional 1½ turns.

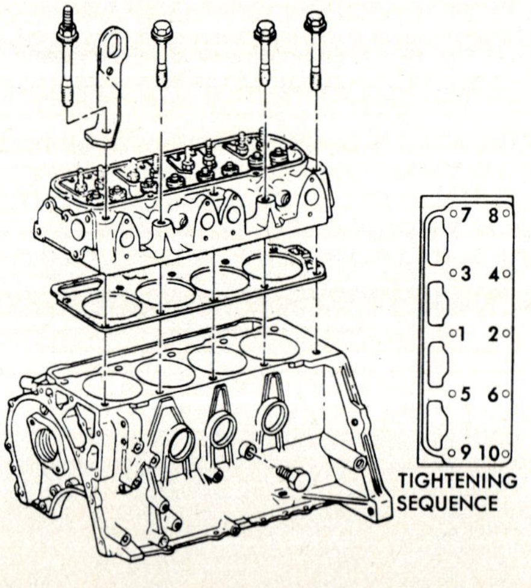

Fig. 2 Cylinder head bolt tightening sequence

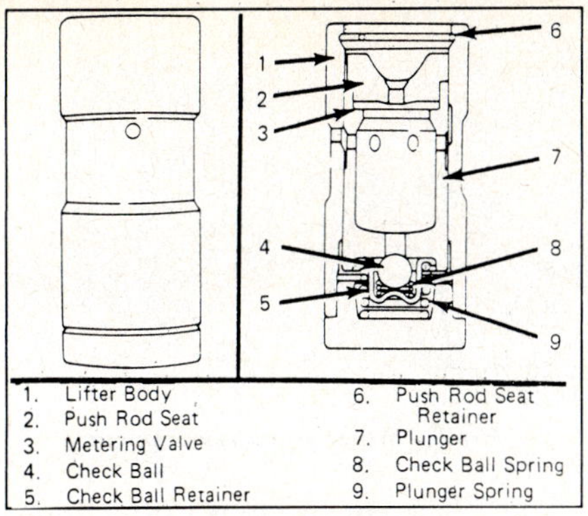

Fig. 3 Sectional view of hydraulic valve lifter

3. Crank one full revolution until mark on crankshaft pulley and "O" mark are aligned. This is the No. 4 cylinder firing position. With engine in this position, adjust the following valves: exhaust, 2 and 4: intake 3 and 4.
4. Install rocker arm cover, then start engine and check timing and idle speed.

VALVE ARRANGEMENT

E-I-I-E-E-I-I-E

VALVE LIFT SPECS.

Year	Intake	Exhaust
1983	.393	.393

VALVE TIMING

Intake Opens Before TDC

Year	Degrees
1983	32

VALVE GUIDES

Valve guides are an integral part of the cylinder head and are not removable. If valve stem clearance becomes excessive, the valve guide should be reamed to the next oversize and the appropriate oversize valves installed. Valves are available in .003, .006 and .012 inch oversizes.

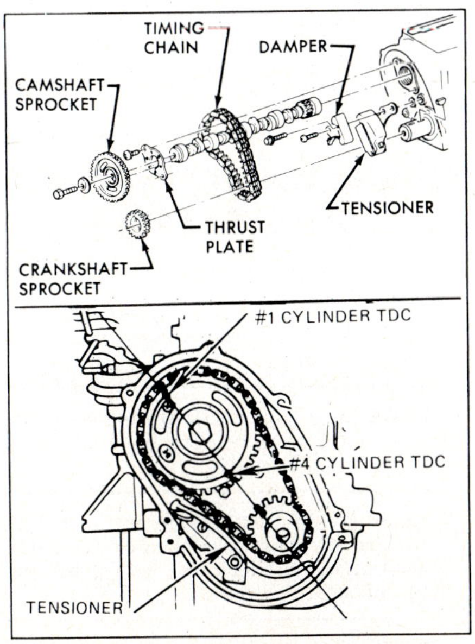

Fig. 5 Valve timing marks

VALVE LIFTERS, REPLACE

1. Remove rocker arm cover and push rod.
2. Remove rocker arm studs and push rod guide.
3. Using tool J-29834, or equivalent, remove valve lifter.
4. Reverse procedure to install. Coat foot of valve lifters with Molykote, or equivalent. Following installation, adjust valve as described under "Valves, Adjust".

VALVE LIFTERS, SERVICE

Disassembly

1. Depress plunger with a push rod, then remove push rod seat retainer using a small screwdriver, Fig. 3.
2. Remove push rod seat, metering valve, plunger, ball check valve assembly and plunger spring.
3. Pry ball retainer loose from plunger and remove ball check valve and spring.

Cleaning & Inspection

1. Clean all parts in suitable solvent.
2. Inspect all lifter components for wear or damage. If any one part is worn or damaged, that entire assembly must be replaced.
3. If lifter body wall is worn, inspect cylinder block lifter bore.
4. If bottom of lifter is worn, inspect camshaft lobe.
5. If push rod seat is worn, inspect push rod.

Assembly

1. Position check ball on small hole in bottom of plunger, then install spring and ball retainer.
2. Install plunger spring over ball retainer, then slide lifter body over spring and plunger. Make sure oil holes in lifter body and plunger are aligned.
3. Fill assembly with SAE 10 oil, then depress plunger with a 1/8 inch drift pin. Insert a 1/16 inch drift pin through both oil holes to hold plunger down against spring tension.
4. Remove the 1/8 inch drift pin and refill assembly with SAE 10 oil, then install the metering valve and push rod seat.
5. Install push rod seat retainer, then push down on the push rod seat and remove the 1/16 inch drift pin from oil holes.

ENGINE FRONT COVER, REPLACE

1. Disconnect battery ground cable.
2. Drain cooling system.
3. Remove upper fan shroud, accessory drive belts, fan and pulley.
4. Remove radiator and heater hoses.
5. Remove water pump as described under "Water Pump, Replace."
6. Remove crankshaft pulley retaining bolts, then the pulley.
7. Remove crankshaft hub using tool J-24420, or equivalent.

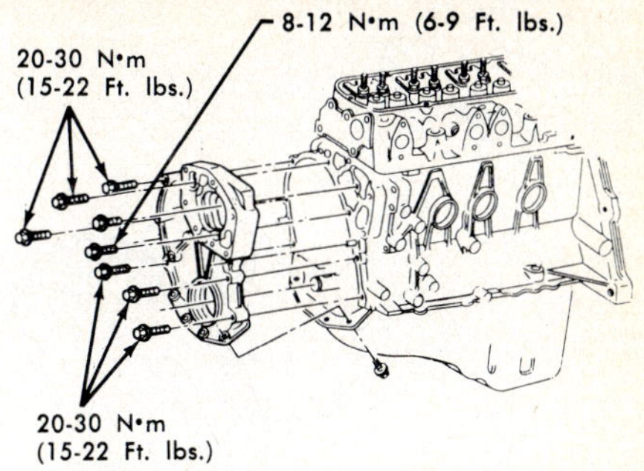

Fig. 4 Engine front cover replacement

8. Remove front cover attaching bolts, then the front cover, Fig. 4.
9. Reverse procedure to install. Apply a continuous bead of sealant 1052357, or equivalent to front cover sealing surface and 1052366, or equivalent to oil pan surface of front cover.

TIMING CHAIN, REPLACE

1. Remove engine front cover as previously described.
2. Align timing marks on camshaft and crankshaft sprockets, then remove timing chain tensioner as described under "Timing Chain Tensioner, Replace".
3. Remove camshaft sprocket and timing chain, Fig. 5.
4. If crankshaft sprocket is to be replaced, remove sprocket using a suitable puller.
5. Reverse procedure to install. Note the following:
   a. Lubricate thrust surface of sprockets with Molykote, or equivalent.
   b. Align timing marks on sprockets.
   c. Align dowel in camshaft with dowel hole in camshaft sprocket.
   d. Lubricate timing chain with clean engine oil.
   e. Torque camshaft sprocket attaching bolts to 66–88 ft. lbs.

TIMING CHAIN TENSIONER, REPLACE

1. Remove engine front cover as previously described.
2. Remove tensioner attaching bolts, then the tensioner.
3. To install tensioner, position tangs of tool J-33875 under tensioner sliding blocks, then pull tool to compress tensioner spring.

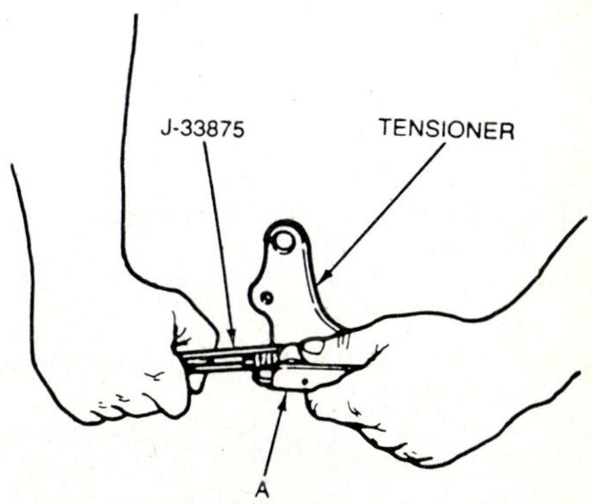

Fig. 6 Compressing timing chain tensioner spring

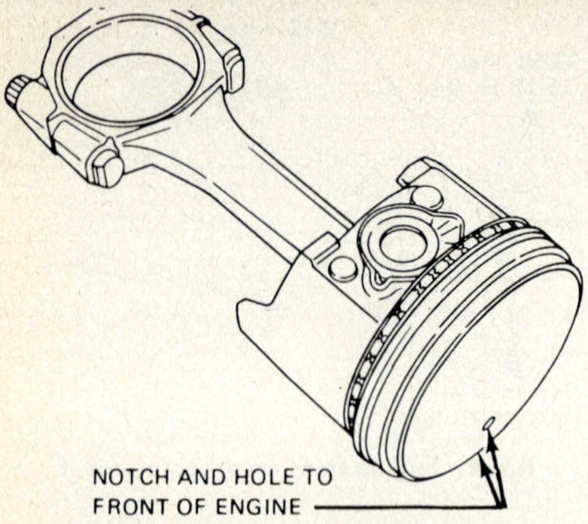

Fig. 7 Piston & rod assembly

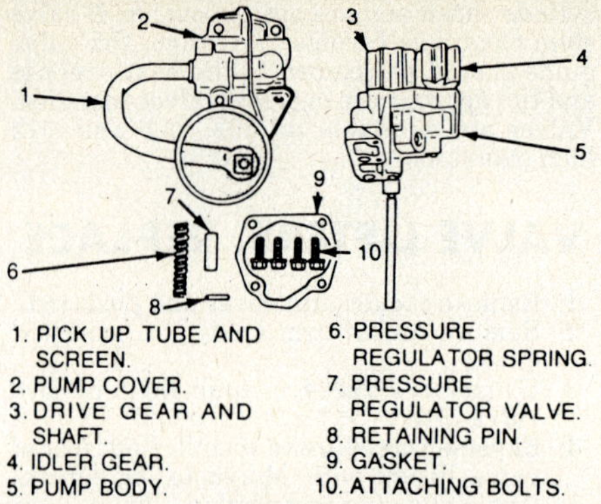

Fig. 8 Oil pump assembly

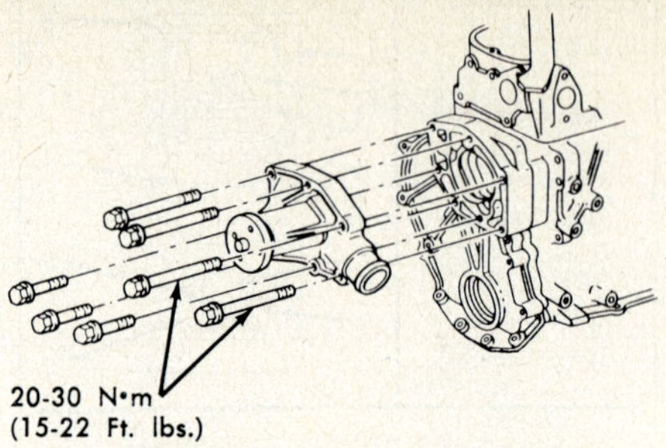

Fig. 9 Water pump replacement

4. While compressing tensioner spring, insert a cotter pin or other suitable tool into hole "A", Fig. 6, to hold spring in the compressed position. Remove tool from tensioner.
5. Install tensioner on engine, then remove cotter pin holding spring in the compressed position.

CAMSHAFT, REPLACE

1. Remove rocker arm cover, then the timing chain and camshaft sprocket as previously described.
2. Drain cooling system, then remove radiator.
3. Mark position of rotor to distributor body for assembly reference.
4. Raise and support vehicle.
5. Remove fuel pump, then the distributor hold down nut and clamp.
6. Lower vehicle, then remove distributor.
7. Remove rocker arm studs and push rod guides.
8. Remove valve lifter, then the camshaft.
9. Reverse procedure to install. Align timing marks on crankshaft and camshaft sprockets, Fig. 5.

PISTONS & RODS, ASSEMBLE

Install piston to rod with notch and hole on piston facing toward front of engine, Fig. 7.

MAIN & ROD BEARINGS

Main bearings are available in standard size and undersizes of .016 and .032 inch. Connecting rod bearings are available in standard size and undersizes of .005 and .010 inch.

OIL PAN, REPLACE

4 × 2

1. Remove engine as described under "Engine, Replace".
2. Drain engine oil, then remove oil pan attaching bolts and the oil pan.
3. Reverse procedure to install. Also, refer to step 12 under 4 × 4 oil pan replacement procedure.

4 × 4

1. Disconnect battery ground cable.
2. Remove starter motor front brace bolt.
3. Remove engine mount through bolts.
4. Raise and support vehicle.
5. Remove splash shield, then the brake and fuel line clip retaining bolts.
6. Remove crossmember attaching bolts, then the crossmember.
7. Drain crankcase, then remove starter motor and position aside.
8. Disconnect steering damper from frame.
9. Mark location of pitman arm, then disconnect pitman arm and steering gear from frame.
10. Disconnect front axle from frame, then the front propeller shaft from front differential. Slide differential forward.
11. Remove oil pan attaching bolts, then raise engine slightly and remove the oil pan.
12. Reverse procedure to install. Note the following:
    a. Apply a thin coat of RTV sealant to both ends of new rear oil pan seal prior to installing seal in rear main bearing cap. The sealant must not extend beyond tabs of seal.
    b. Apply a continuous bead of RTV sealant on oil pan side rails. The bead of sealant should circle inboard at each bolt hole location.
    c. Do not apply sealant to rear oil pan seal mating surface.
    d. Apply RTV sealant to oil pan-to-front cover mating surface. This sealant must meet the two side rail beads.
    e. Torque oil pan-to-cover bolts to 6–9 ft. lbs., oil pan-to-side rail bolts to 4–9 ft. lbs. and oil pan-to-rear cover bolts to 11–17 ft. lbs. All bolts must be torqued before sealant dries.

OIL PUMP SERVICE

Removal

1. Drain crankcase, then remove oil pan as previously described.
2. Remove pump to rear main bearing attaching bolt, then remove pump and extension shaft.

Disassemble

1. Remove four pump cover to body attaching bolts, then remove cover, idler and drive gears and shaft, Fig. 8.

NOTE: Place alignment marks on oil pump drive and idler gear teeth so they can be installed in the same position.

2. Remove pressure regulator valve retaining pin, spring and the valve from pump body.

Inspection

Inspect pump components and should any of the following conditions exist, the oil pump assembly should be replaced.

1. Inspect pump body, gears and cover for cracks or excessive wear.
2. Check drive gear shaft for looseness in housing.
3. Check inside of pump cover for wear that would allow oil to leak past ends of gears.
4. Check oil pickup screen assembly for damage to screen or pickup tube.
5. Check pressure regulator valve for fit in pump body.

Assemble

1. Install a replacement pickup screen and tube assembly, if removed. Position pump in a soft jawed vise then apply sealer to end of tube and tap into position using tool No. J8369 and a plastic hammer. Use care not to damage inlet screen and tube assembly when installing on pump housing.
2. Place pressure regulator valve, spring and retaining pin into pump body, then install drive gear and shaft.
3. Install idler gear into pump body, then the pump cover gasket, Fig. 8.
4. Install pump cover and cover retaining bolts, then torque bolts to 6–9 ft. lbs.

Installation

1. Align oil pump extension shaft to distributor drive gear socket and pump housing with dowels on cap, then install shaft retainer and pump assembly.
2. Install oil pump assembly retaining bolt to rear main bearing cap and torque bolt to 26–38 ft. lbs.
3. Install oil pan as previously described.

REAR MAIN BEARING OIL SEAL, REPAIR

1. Remove oil pan and oil pump as described previously.
2. Remove rear main bearing cap, then the upper and lower seal. Clean all oil from seal channel.

NOTE: It may be necessary to loosen No. 2, 3 and 4 main bearing bolts to install the new upper seal.

3. Apply a thin coat of oil to entire sealing surface of rubber seal, then roll seal into position in cylinder case. Position sealing lip inboard.

NOTE: Rotate crankcase to aid in seal installation. Also, protect the seal O.D.

with a piece of shim stock during installation.

4. Lubricate other half of seal, then install it in main bearing cap.
5. Place a piece of plastic gauging material on rear main bearing or journal, then install rear main bearing cap and torque to 70 ft. lbs.
6. Remove rear main bearing cap and check bearing clearance. If not within specifications, correct as necessary.
7. Clean plastic gauge material from journal and bearing, then lubricate the bearing with clean oil.
8. Apply a 1/32 inch bead of sealant 1052357, or equivalent, to bearing cap between rear main seal and the oil pan rear seal groove. Avoid getting any sealant on rear main seal or in the drain slot.
9. Apply a light coat of clean engine oil to crankshaft surface that will contact the seal, then install rear main bearing cap and torque to 70 ft. lbs.

NOTE: If any other main bearing bolts were loosened, retorque to specifications.

10. Install oil pump and oil pan, then run engine and check for oil leaks.

WATER PUMP, REPLACE

1. Disconnect battery ground cable.
2. Remove accessory drive belts.
3. Remove upper fan shroud, then drain cooling system.
4. Remove radiator and heater hoses.
5. Remove water pump attaching bolts, then the water pump, Fig. 9.
6. Reverse procedure to install. Apply a narrow bead of sealant 1052357, or equivalent, to sealing surfaces. Torque attaching bolts to 15–22 ft. lbs.

BELT TENSION DATA

	New Lbs.	Used Lbs.
Air Cond.	145	65–100
Alternator	130	50–80
Power Steer.	130	50–80
Air Pump	130	50–80

FUEL PUMP, REPLACE

1. Disconnect battery ground cable.
2. Remove distributor, then disconnect inlet and outlet lines from fuel pump.
3. Remove engine lift hook, then the fuel pump attaching bolts and fuel pump.
4. Reverse procedure to install. Install pump with camshaft on down stroke.
5. Adjust timing and check for fuel leaks.

V6-173 Engine Section

ENGINE REPLACE

4 × 2

1. Disconnect battery ground cable.
2. Raise and support vehicle.
3. Remove propeller shaft, then disconnect speedometer cable.
4. On vehicles equipped with automatic transmission, disconnect torque converter clutch connector.
5. Disconnect shift linkage.
6. Remove crossmember to transmission mount nuts and crossmember to frame bolts.
7. Disconnect exhaust pipe at manifold.
8. Remove catalytic converter rear attaching bolts.
9. Remove transmission mount to transmission attaching bolts, then the mount.
10. Support engine and transmission using a suitable jack, then remove the crossmember.
11. Remove exhaust manifold flange to catalytic converter with support.
12. On vehicles equipped with automatic transmission, remove strut bolts at dust cover and the dust cover to cylinder case bolts.
13. Remove starter motor.
14. Lower vehicle, then remove hood and the air cleaner assembly.
15. Drain cooling system, then disconnect radiator hoses from engine.
16. On vehicles equipped with A/C, remove A/C compressor and position aside.
17. Remove upper half of fan shroud, then disconnect heater hoses from engine, cooler lines from radiator and the coolant bottle hose.
18. Remove fan and radiator.
19. On vehicles equipped with power steering, disconnect pump and position aside.
20. Disconnect all necessary hoses and wires.
21. Install suitable engine lifting device, then remove motor mount through bolts.
22. On vehicles equipped with A/C, remove compressor bracket and install lift hook.
23. Raise engine slightly and disconnect throttle valve cable and any remaining wires.
24. Raise engine and transmission and remove from vehicle.
25. Reverse procedure to install.

4 × 4

1. Remove hood.
2. Disconnect battery ground cable.
3. Raise and support vehicle.
4. Remove rear transmission mount nuts.
5. Support transmission and remove crossmember at frame and top transmission bolts.
6. Install crossmember and remove transmission support.
7. Remove engine to transmission bolts.
8. Remove torque converter cover.
9. Disconnect exhaust.
10. Disconnect starter motor wires.
11. Remove motor mount bolts.
12. Disconnect fuel lines.
13. Remove transmission cooler lines at oil pan.
14. Lower vehicle.
15. Drain coolant.
16. Disconnect necessary wires and hoses.
17. Remove power steering pump and air conditioning compressor and lay aside.
18. Loosen alternator.
19. Disconnect front heater hose.
20. Remove upper half fan shroud.
21. Remove fan assembly.
22. Disconnect radiator hoses at engine.
23. Disconnect transmission cooler lines.
24. Remove radiator.
25. Disconnect accelerator cable and spring.
26. Disconnect detent cable.
27. Support transmission using a suitable floor jack.
28. Install lift device and remove engine.
29. Reverse procedure to install.

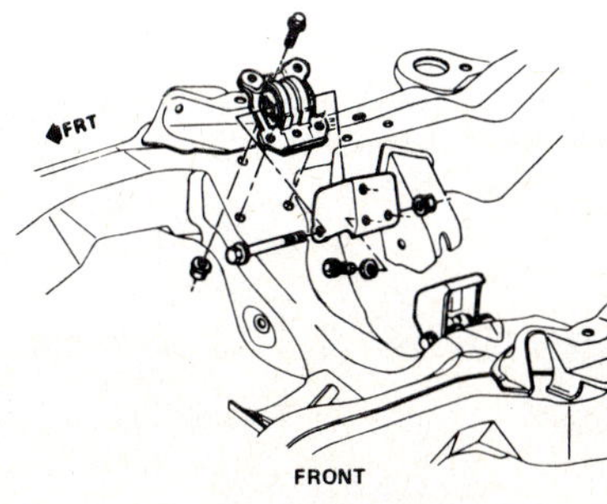

Fig. 1 Engine mount installation

ENGINE MOUNTS, REPLACE

1. Disconnect battery ground cable.
2. Remove top half of fan shroud.
3. Raise and support vehicle.
4. Remove mount through bolt, then raise front of engine and remove mount, Fig. 1.

NOTE: Do not raise engine any more than is needed to provide sufficient clearance. Also, check for interference between rear of engine and cowl panel which could damage the distributor.

5. Reverse procedure to install.

CYLINDER HEAD, REPLACE

1. Remove intake manifold.
2. Raise vehicle.
3. Drain engine block.
4. Disconnect exhaust pipe.
5. If removing left cylinder head, remove dipstick tube attachment.
6. If removing right cylinder head, remove generator bracket.
7. Lower vehicle.
8. Loosen rocker arm until able to remove push rod.
9. Remove head bolts and cylinder head.

NOTE: The gasket surfaces on both head

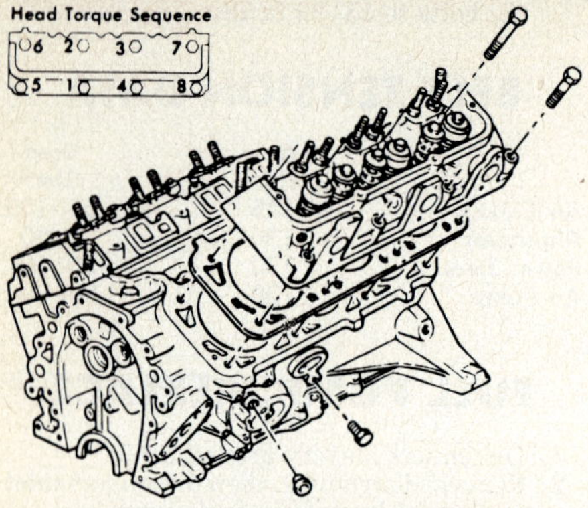

Fig. 2 Cylinder head installation

and cylinder case deck must be clean of any foreign matter and free of nicks or heavy scratches. Cylinder bolt threads in the case and threads on the cylinder head bolts must be clean. Dirt will affect bolt torque.

10. Reverse procedure to install.

NOTE: Coat cylinder head bolt threads with sealer, #1052080 or equivalent, and install bolts. Torque bolts in proper sequence to specifications, Fig. 2.

INTAKE MANIFOLD, REPLACE

1. Disconnect battery ground cable.
2. Remove air cleaner.
3. Drain coolant.
4. Remove wires and hoses at carburetor.
5. Disconnect fuel line at carburetor.
6. Disconnect cables.
7. Remove Air Management hose.
8. Disconnect Emission Canister hoses. Remove pipe bracket (front valve cover).
9. Remove left valve cover.
10. Remove Air Management bracket.
11. Remove right valve cover.
12. Remove upper radiator hose.
13. Disconnect heater hose. Remove A/C belt and rotate A/C compressor.
14. Disconnect coolant switches.
15. Remove manifold bolts and nuts.
16. Remove manifold. Discard manifold gaskets and remove loose RTV from front and rear ridges of cylinder case.
17. Reverse procedure to install.

NOTE: Ensure that no oil or water is present on surface when new RTV is applied. Place a (3/16") bead of RTV, #1052366 or equivalent, on each ridge. Install manifold retaining bolts and nuts and torque in the sequence shown in Fig. 3 to 23 ft. lbs.

EXHAUST MANIFOLDS, REPLACE

1. Disconnect battery ground cable.
2. Raise and support vehicle, then disconnect exhaust pipe.
3. If removing left manifold, remove rear manifold bolts and nut, then lower vehicle. Disconnect air management hoses and wires, remove power steering bracket, if equipped, and remove manifold.

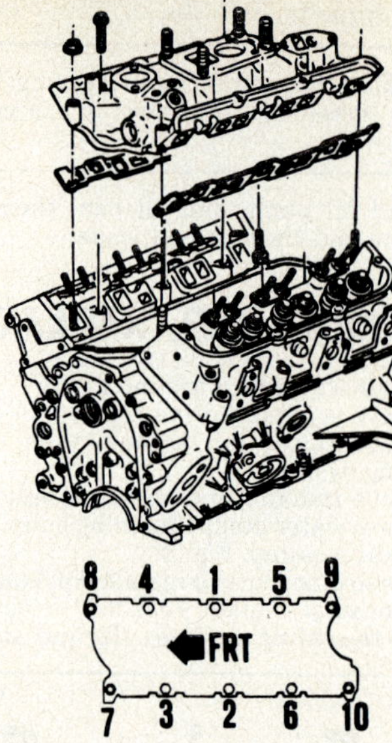

Fig. 3 Intake manifold

4. If removing right manifold, lower vehicle, remove manifold attaching bolts, disconnect air management hose and remove manifold.
5. Reverse procedure to install.

NOTE: Clean mating surfaces on manifold and head, then install manifold in position and torque manifold bolts to 25 ft. lbs.

ROCKER ARM STUDS

Rocker arm studs that are cracked or have damaged threads can be replaced. If threads in cylinder head are damaged or stripped, the head can be retapped and a helical type insert added. When installing a new rocker arm stud, torque stud to 43 to 49 ft. lbs.

VALVES, ADJUST

1. Crank engine until mark on torsional damper is aligned with TDC mark on timing tab. Check to ensure engine is in the No. 1 cylinder firing position by placing fingers on No. 1 cylinder rocker arms as mark on damper comes near TDC mark on timing tab. If valves are not moving, the engine is in the No. 1 firing position. If valves move as damper mark nears TDC mark on timing tab, engine is in the No. 4 cylinder firing position and should be rotated one revolution to reach the No. 1 cylinder firing position.
2. With engine in the No. 1 cylinder firing position, adjust the following valves: Exhaust- 1, 2, 3; Intake- 1, 5, 6. To adjust valves, back off adjusting nut until lash is felt at push rod, then tighten adjusting nut until all lash is removed. This can be determined by rotating the push rod while tightening the adjusting nut. When all lash has been eliminated, turn adjusting nut the additional number of turns listed in the Valve Specifications.
3. Crank engine one revolution until mark on torsional damper and TDC mark are again aligned. This is the No. 4 cylinder firing position. With engine in this position, the following valve can be adjusted: Exhaust- 4, 5 & 6; Intake- 2, 3 & 4.
4. Install rocker arm covers, then start engine and check timing and idle speed.

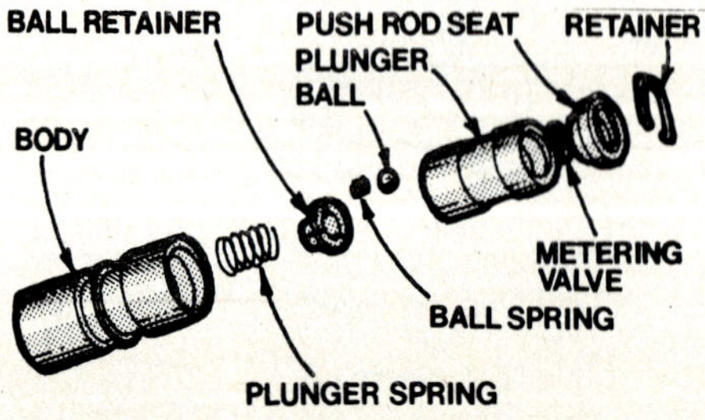

Fig. 5 Hydraulic valve lifter

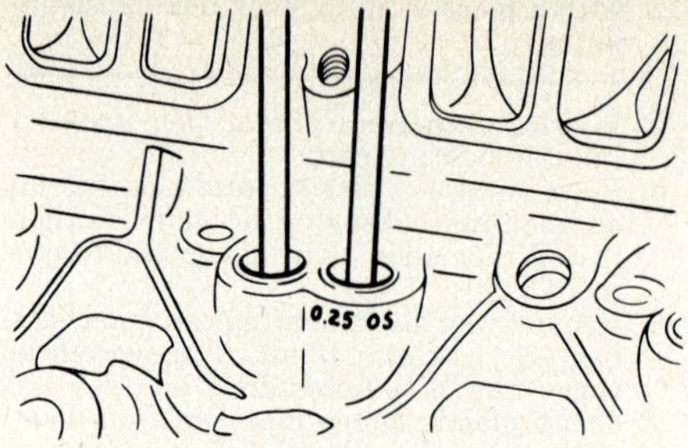

Fig. 4 Oversize valve lifter marking

VALVE GUIDES

Valve guides are an integral part of the cylinder head and are not removable. If valve stem clearance becomes excessive, the valve guide should be reamed to the next oversize and the appropriate oversize valves installed. Valves are available in .0035", .0155" and .0305" oversizes.

VALVE LIFTERS

NOTE: Some engines will be equipped with both standard and .25 mm. oversize valve lifters. The cylinder case will be marked where the oversize valve lifters are installed with a daub of white paint and .25 mm. O.S. will be stamped on the valve lifter boss, Fig. 4.

Failure of a hydraulic valve lifter, Fig. 5, is generally caused by an inadequate oil supply or dirt. An air leak at the intake side of the oil pump or too much oil in the engine will cause air bubbles in the oil supply to the lifters causing them to collapse. This is a probable cause of trouble if several lifters fail to function, but air in oil is an unlikely cause of failure of a single unit.

Valve lifters can be removed after removing rocker arm covers, intake manifold, rocker stud nuts, rocker arm balls, rocker arms and push rods.

VALVE ARRANGEMENT

E-I-I-E-I-E

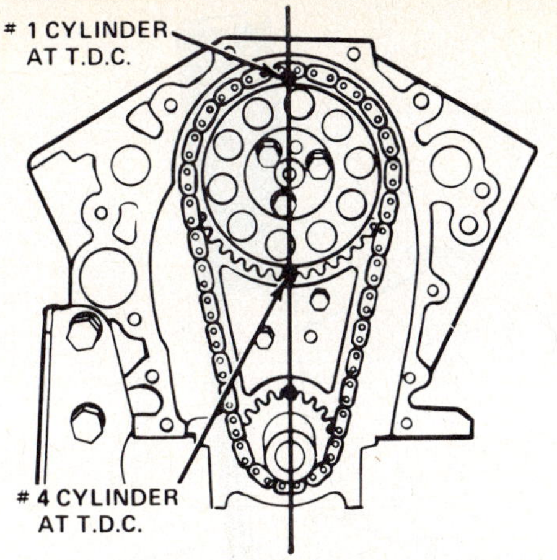

Fig. 6 Valve timing marks

VALVE TIMING

Intake Opens Before TDC

Engine	Year	Degrees
V6-173	1982–83	7°

VALVE LIFT SPECS.

Engine	Year	Intake	Exhaust
V6-173	1982	.349	.393

CAMSHAFT, REPLACE

1. Remove valve lifters and engine front cover as described previously.
2. Remove fuel pump and push rod.
3. Remove timing chain and sprocket as described under "Timing Chain, Replace".
4. Withdraw camshaft from engine, using care not to damage camshaft bearings.
5. Reverse procedure to install. When installing timing chain, align valve timing marks as shown in Fig. 6.

TIMING CHAIN, REPLACE

1. Remove front cover as described under "Front Cover, Replace".
2. Place No. 1 piston at top dead center with marks on camshaft and crankshaft sprockets aligned, Fig. 6.
3. Remove camshaft sprocket bolts, then remove sprocket and timing chain. If sprocket does not come off easily, tap lower edge of sprocket with a plastic mallet.
4. Install timing chain on camshaft sprocket. Hold sprocket vertically with chain hanging down and align marks on camshaft and crankshaft sprockets.
5. Align dowel pin hole in sprocket with dowel pin on camshaft, then install sprocket on camshaft.
6. Using camshaft sprocket attaching bolts, draw sprocket on camshaft. Torque bolts to 15 to 20 ft. lbs.
7. Lubricate timing chain with engine oil, then install front cover as outlined previously.

FRONT COVER, REPLACE

1. Remove water pump as described under "Water Pump, Replace".
2. If equipped, remove A/C compressor and bracket and position aside.
3. Remove torsional damper, then disconnect lower radiator hose from front cover and the heater hose from water pump.
4. Remove front cover attaching bolts, then the front cover.
5. Reverse procedure to install. Apply a continuous bead of suitable sealant to front cover mating surface.

NOTE: Torque M8 × 1.25 bolts to 13–18 ft. lbs., or M10 × 1.5 bolts to 20–30 ft. lbs.

PISTONS & RODS, ASSEMBLE

Assemble pistons to connecting rods as shown in Fig. 7.

OIL PAN, REPLACE

4 × 2

1. Disconnect battery ground cable.
2. Remove engine as described under "Engine, Replace".
3. Remove oil pan.
4. Reverse procedure to install.

NOTE: Before installing oil pan, thoroughly clean all sealing surfaces, then apply a 1/8 inch bead of RTV sealant, or equivalent, to the entire oil pan sealing flange.

4 × 4

1. Disconnect battery ground cable and remove dipstick.
2. Raise and support vehicle.
3. Remove drive belt splash shield, front axle shield and transfer case shield.
4. Disconnect fuel lines.
5. Remove No. 2 crossmember and the catalytic converter hanger bolts.
6. Disconnect exhaust pipes from manifold and slide exhaust rearward.
7. Disconnect front propeller shaft from drive pinion.

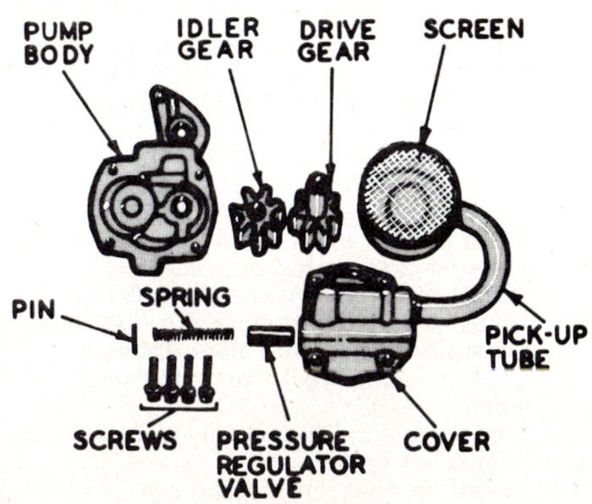

Fig. 8 Oil pump disassembled

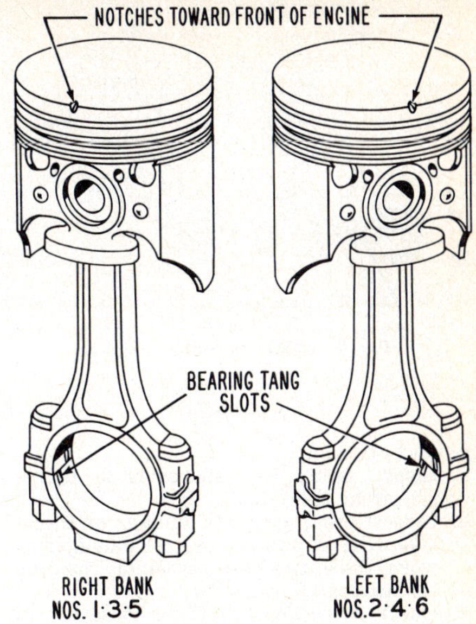

Fig. 7 Piston and rod assembly

8. Disconnect engine braces from flywheel cover and loosen braces at the block.
9. Remove flywheel cover.
10. Remove starter motor retaining bolts and position starter aside.
11. Disconnect steering shock absorber from frame bracket, then remove steering gear bolts.
12. Mark position of idler arm for reference during installation, then remove attaching bolts.
13. Pull steering gear and linkage forward, then remove differential housing attaching bolts and move the housing forward.
14. Remove motor mount bolts, then drain oil pan and remove oil pan bolts.
15. Install suitable engine lifting equipment, raise engine and remove oil pan.
16. Reverse procedure to install.

NOTE: Before installing oil pan, thoroughly clean all sealing surfaces, then apply a 1/8 inch bead of RTV sealant, or equivalent to entire oil pan sealing flange.

OIL PUMP SERVICE

Removal

1. Remove oil pan as described under "Oil Pan, Replace".
2. Remove pump to rear main bearing cap bolt and remove pump and extension shaft.

Disassembly

1. Remove pump cover attaching bolts and pump cover, Fig. 8.
2. Mark drive and idler gear teeth so they can be installed in the same position, then remove idler and drive gear and shaft from pump body.
3. Remove pin, spring and pressure regulator valve from pump cover.
4. If pick-up tube and screen assembly are to be replaced, mount pump cover in a soft jawed vise and remove pick-up tube from cover. Do not remove screen from pick-up tube, these components are serviced as an assembly.

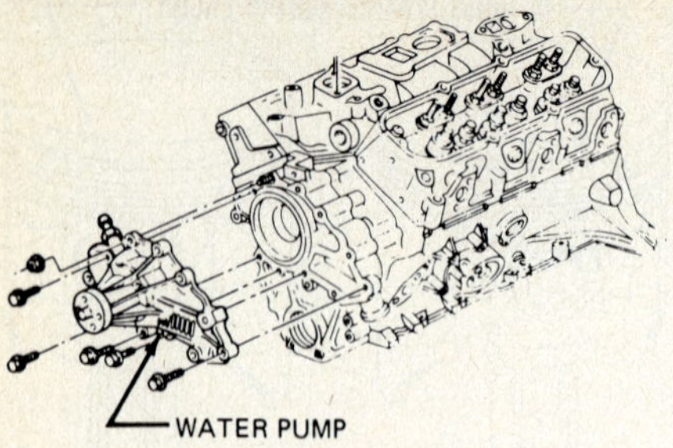

Fig. 9 Water pump removal

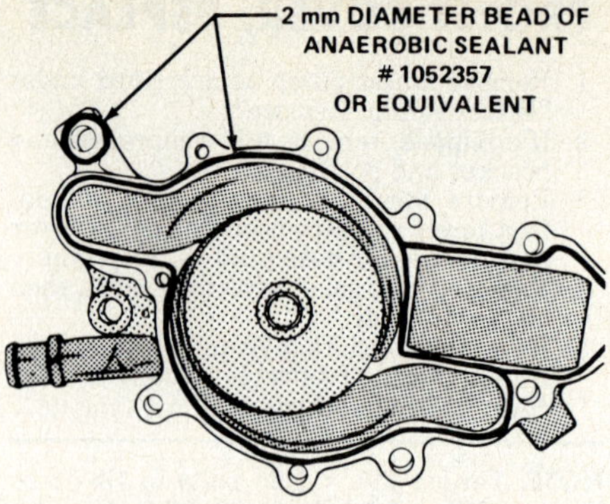

Fig. 10 Water pump sealer application

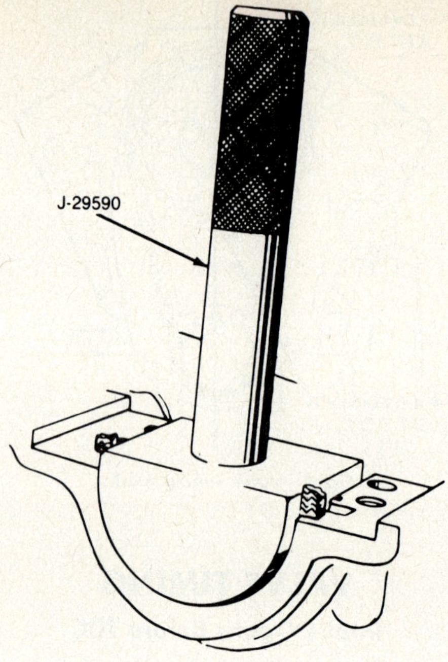

Fig. 11 Rear main bearing oil seal installation

Inspection

1. Inspect pump body and cover for excessive wear and cracks.
2. Inspect pump gear for damage or excessive wear. If pump gears are damaged or worn, the entire pump assembly must be replaced.
3. Check drive gear shaft for looseness in pump body.
4. Inspect pump cover for wear that would allow oil to leak past gear teeth.
5. Inspect pick-up tube and screen assembly for damage.
6. Check pressure regulator valve for fit in pump cover.

Assembly

1. If pick-up tube and screen were removed, apply sealer to end of pick-up tube, then mount pump cover in a soft jawed vise and using tool No. J-8369, tap pick-up tube into position using a plastic mallet.

NOTE: Whenever the pick-up tube and screen assembly has been removed, a new pick-up tube and screen assembly should be installed. Use care when installing pick-up tube and screen assembly so that tube does not twist, shear or collapse. Loss of a press fit condition could result in an air leak and a loss of oil pressure.

2. Install pressure regulator valve, spring and pin, Fig. 8.
3. Install drive gear and shaft in pump body.
4. Align marks made during disassembly, then install idler gear.
5. Install pump cover gasket, cover and attaching bolts. Torque bolts to 6 to 9 ft. lbs.
6. Rotate pump drive shaft by hand and check pump for smooth operation.

Installation

1. Assemble pump and extension shaft with retainer to rear main bearing cap, aligning top end of hexagon extension shaft with hexagon socket on lower end of distributor shaft.
2. Install pump to rear main bearing cap bolt.
3. Install oil pan as described under Oil Pan, Replace.

FUEL PUMP, REPLACE

1. Disconnect battery ground cable.
2. Disconnect inlet and outlet hoses from fuel pump.
3. Remove fuel pump attaching bolts, then the fuel pump and gasket.
4. Reverse procedure to install.

NOTE: Before installing pump, rotate camshaft to "down stroke" position. Also, when installation is completed, start engine and check for fuel leaks.

WATER PUMP, REPLACE

1. Disconnect battery ground cable.
2. Drain cooling system then remove the fan and the heater hose and drive belts from the pump.
3. Remove water pump attaching bolts and nut, then the pump, Fig. 9.
4. Reverse procedure to install.

NOTE: Prior to installing water pump, thoroughly clean sealing surfaces, then apply a 3/32 inch bead of sealant 1052357, or equivalent, to the pump sealing surface, Fig. 10. Also, coat bolt threads with sealant 1052080, or equivalent.

REAR MAIN BEARING OIL SEAL REPAIR

1. Remove oil pan and oil pump as described previously.
2. Remove rear main bearing cap.
3. Using tool No. J-29114-2, gently drive upper seal into groove approximately 1/4 in.
4. Repeat step 3 for other end of seal.
5. Measure the amount that was driven in on one side and add 1/16 in. Using a suitable cutting tool, cut this length from the oil rear main bearing cap lower seal using the main bearing cap as a guide. Repeat this step for the other end of seal.
6. Place piece of cut seal into groove of seal installer tool guide No. J-29114-1 and install tool guide onto engine block.
7. Using seal packing tool No. J29114-2, drive piece of seal into block. Drive seal in until packing tool contacts machined stop.
8. Remove tool guide and repeat steps 6 and 7 for other end of seal.
9. Install new seal in bearing cap.
10. Cut ends of seal flush with cap using tool J-29590, Fig. 11.
11. Place a piece of plastic gauging material on rear main journal, then install rear main bearing cap and torque to 70 ft. lbs.
12. Remove rear cap and check plastic gauge for bearing clearance. If clearance is not within specifications, recheck seal ends for fraying which may prevent cap from fully seating and repair as necessary.
13. Clean plastic gauge from journal and bearing.
14. Apply a thin film of sealant No. 1052357, or equivalent, to rear cap, then install cap. Use care not to allow sealant to contact the seal and bearing.

BELT TENSION DATA

	New Lbs.	Used Lbs.
Air Cond.—		
1982	146	67
1983	145	65–100
Air Pump—		
1982	135	67
1983	130	50–80
Alternator—		
1982	146	67
1983	130	50–80
Power Steering—		
1982	135	67
1983	130	50–80

Carburetor Section

ROCHESTER VARAJET 2SE & E2SE ADJUSTMENT SPECIFICATIONS

Year	Carb. Production No.	Float Level	Accel. Pump	Choke Coil Lever	Choke Rod	Vacuum Break		Air Valve Rod	Choke Setting	Unloader	Secondary Lockout
						Primary	Secondary				
1982	17082348	7/16"	—	—	22°	30°	32°	1°	—	40°	—
	17082349	7/16"	—	—	22°	30°	32°	1°	—	40°	—
	17082350	7/16"	—	—	22°	30°	32°	1°	—	40°	—
	17082351	7/16"	—	—	22°	30°	32°	1°	—	40°	—
	17082353	7/16"	—	—	22°	30°	35°	1°	—	40°	—
	17082355	7/16"	—	—	22°	30°	35°	1°	—	40°	—
	17082356	13/32"	—	.085	22°	25°	30°	1°	—	30°	.025
	17082357	13/32"	—	.085	22°	25°	32°	1°	—	30°	.025
	17082358	13/32"	—	.085	22°	25°	30°	1°	—	30°	.025
	17082359	13/32"	—	.085	22°	25°	32°	1°	—	30°	.025
1983	17083348	7/16"	—	.085	22°	30°	32°	1°	—	40°	—
	17083349	7/16"	—	.085	22°	30°	32°	1°	—	40°	—
	17083350	7/16"	—	.085	22°	30°	32°	1°	—	40°	—
	17083351	7/16"	—	.085	22°	30°	32°	1°	—	40°	—
	17083352	7/16"	—	.085	22°	30°	35°	1°	—	40°	—
	17083353	7/16"	—	.085	22°	30°	35°	1°	—	40°	—
	17083354	7/16"	—	.085	22°	30°	35°	1°	—	40°	—
	17083355	7/16"	—	.085	22°	30°	35°	1°	—	40°	—
	17083360	7/16"	—	.085	22°	30°	32°	1°	—	40°	—
	17083361	7/16"	—	.085	22°	28°	32°	1°	—	40°	—
	17083362	7/16"	—	.085	22°	30°	32°	1°	—	40°	—
	17083363	7/16"	—	.085	22°	28°	32°	1°	—	40°	—
	17083364	7/16"	—	.085	22°	30°	35°	1°	—	40°	—
	17083365	7/16"	—	.085	22°	30°	35°	1°	—	40°	—
	17083366	7/16"	—	.085	22°	30°	35°	1°	—	40°	—
	17083367	7/16"	—	.085	22°	30°	35°	1°	—	40°	—
	17083391	13/32"	—	.085	28°	30°	35°	1°	—	38°	—
	17083392	13/32"	—	.085	28°	30°	35°	1°	—	38°	—
	17083393	13/32"	—	.085	28°	30°	35°	1°	—	38°	—
	17083394	13/32"	—	.085	28°	30°	35°	1°	—	38°	—
	17083395	13/32"	—	.085	28°	30°	35°	1°	—	38°	—
	17083396	13/32"	—	.085	28°	30°	35°	1°	—	38°	—
	17083397	13/32"	—	.085	28°	30°	35°	1°	—	38°	—

DESCRIPTION

The Varajet models 2SE and E2SE, Figs, 1 and 2, are two barrel, two stage, down draft design carburetors. Aluminum die castings are used for the air horn, float bowl and throttle body. A heat insulator gasket is used between the throttle body and float bowl to reduce heat transfer to the float bowl.

The primary stage has a triple venturi, with a small 35mm bore, resulting in good fuel metering control during idle and part throttle operation. The secondary stage has a 46mm bore, providing sufficient air capacity for engine power requirements. An air valve is used in the secondary stage with a single tapered metering rod.

The float chamber is internally vented through a vertical vent cavity in the air horn. The float chamber is also externally vented through a tube in the air horn. A hose connects this tube directly to a vacuum operated vapor vent valve located in the vapor canister. When the engine is not running, the canister vapor vent valve is open, allowing fuel vapor from the float chamber to pass into the canister where the vapor is stored until normally purged.

An adjustable part throttle screw is used in the float bowl to aid emission control. This screw is factory pre-set and a plug is installed to prevent further adjustment or fuel leakage. The plug should not be removed or the screw setting disturbed. If float bowl replacement is required, the service float bowl will include a factory pre-set and plugged adjustable part throttle screw.

A hot idle compensator is used on some models and is located in the air horn. The opening and closing of the hot idle compensator valve is controlled by a bi-metal strip that is calibrated to a specific temperature. When the valve opens, additional air is allowed to bypass the throttle valves and enter the intake manifold to prevent rough idle during periods of hot engine operation.

The idle mixture screw is recessed in the throttle body and is sealed with a hardened steel plug to prevent alteration of the factory pre-set mixture setting. The plug should not be removed and the mixture screw readjusted unless required by major carburetor overhaul or throttle body replacement.

The E2SE carburetor includes special design features for use with the Computor Controlled Catalytic Converter (C4) System or the Computer Command System. An electrically operated mixture control solenoid mounted in the air horn, controls air and fuel metered to the idle and main metering systems of the carburetor. The plunger located at the end of the solenoid is submerged in fuel in the fuel chamber of the float bowl. This plunger is controlled by an electrical signal from the Electronic Control Module (ECM). The Electronic Control Module responding to signals from the oxygen sensor in the exhaust and other engine operating condition signals,

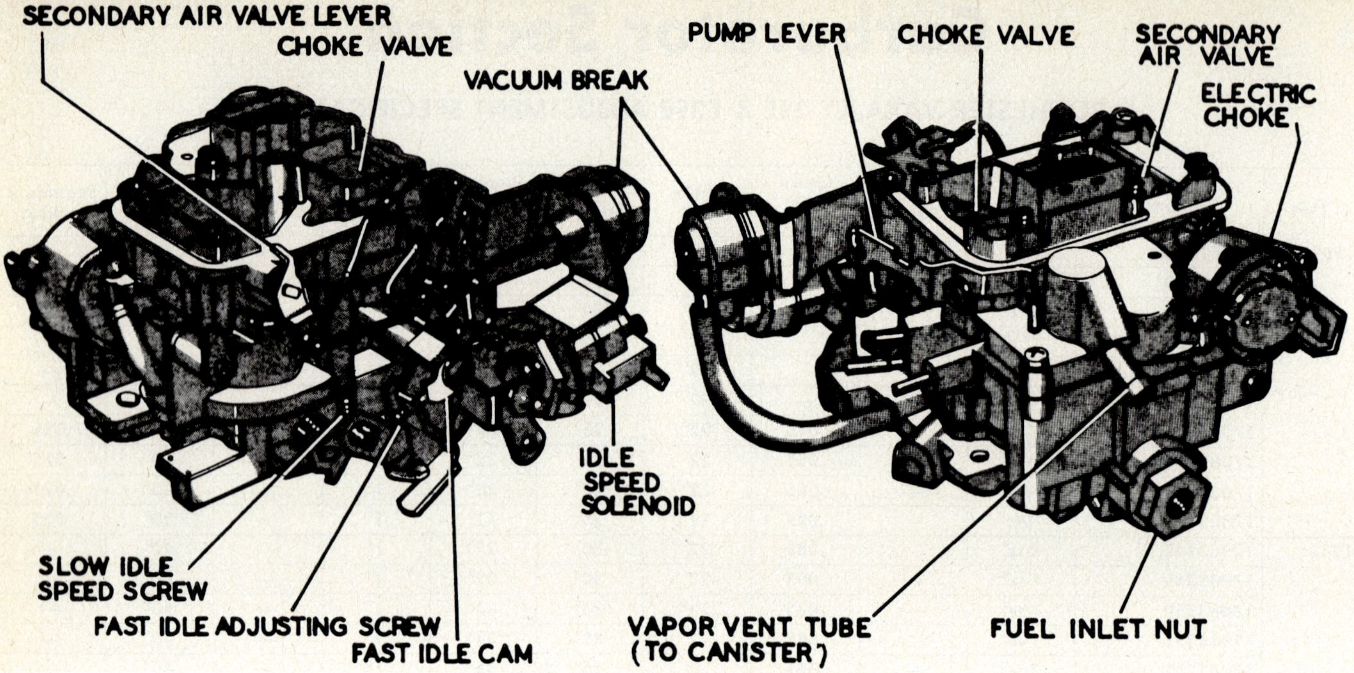

Fig. 1 Varajet 2SE Carburetor

energizes the solenoid to move the plunger down to the lean position or de-energizes the solenoid to move the plunger up to the rich position to control fuel delivery to the idle and main metering systems. When the plunger is in the lean position, fuel metering is controlled by a lean mixture screw located in the float bowl. When the plunger is in the rich position, the additional fuel is metered to the main fuel well through a rich mixture screw located at the end of the fuel supply channel in the float bowl. Air metered to the idle system is controlled by the up and down movement of the mixture control solenoid plunger. The plunger increases or decreases air supplied to the idle system which is further metered by the idle air bleed screw. The plunger cycles up and down approximately 10 times per second, controlling air and fuel mixtures.

CAUTION: Use care not to remove the special friction reducing coating applied to the primary and secondary throttle shafts, the secondary actuating lever and lockout lever. A special graphite compound is also applied to the secondary throttle bore and valve.

ADJUSTMENTS

Float Level Adjustment, Fig. 3

1. Hold float retainer firmly in place and push float lightly against needle.
2. With an adjustable T-scale, measure distance between float bowl gasket surface (gasket removed) and float toe.
3. To adjust, remove float and bend float arm.
4. Check float alignment after adjustment.

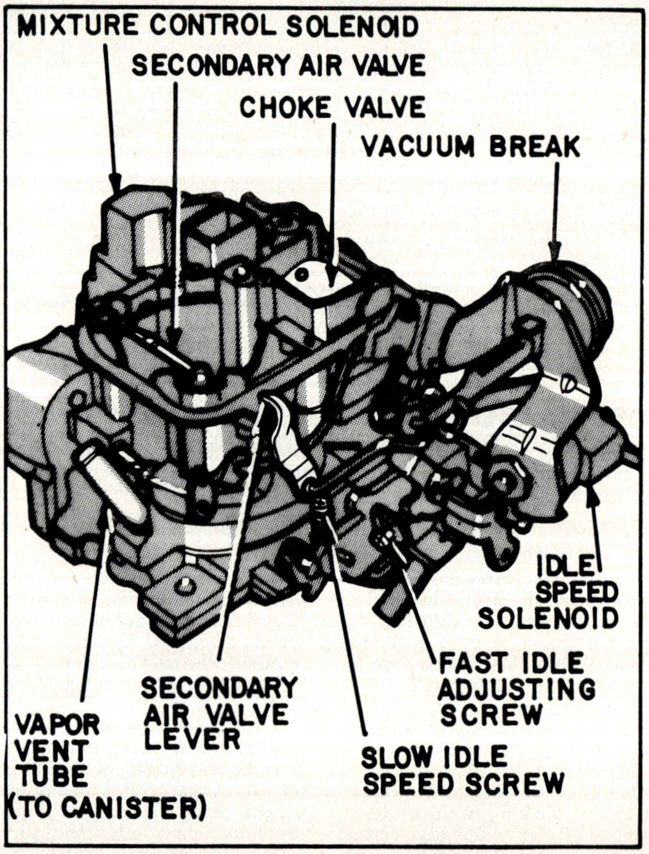

Fig. 2 Varajet E2SE Carburetor

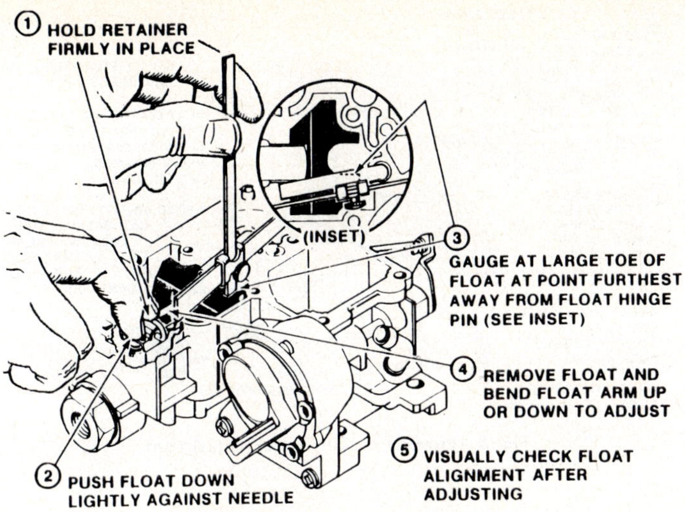

Fig. 3 Float level adjustment

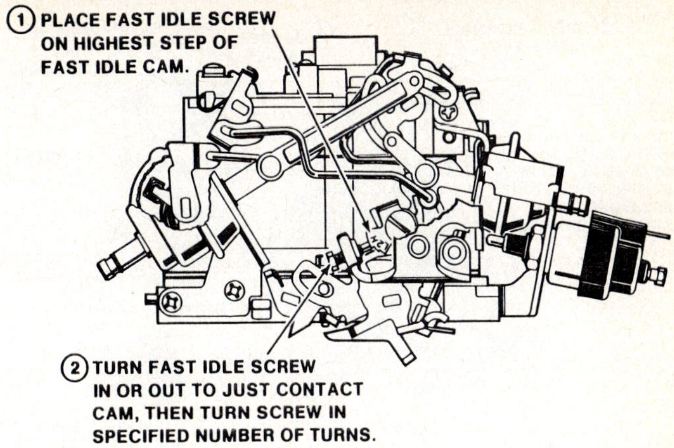

Fig. 4 Fast idle adjustment

Fast Idle Adjustment, Fig. 4

1. Adjust curb idle speed, if necessary.
2. Place fast idle screw on highest step of cam.
3. Turn fast idle screw to obtain specified fast idle speed.

Choke Coil Lever Adjustment, Fig. 5

1. Loosen choke cover retaining screws, then the cover and coil assembly from choke housing. If a riveted choke cover is used, remove choke cover and coil assembly as outlined in the choke cover retainer kit.
2. Place fast idle screw on high step of cam.
3. Push the intermediate choke lever until the choke valve closes.
4. Insert specified gauge into hole provided in choke housing. The edge of the lever should just contact gauge.
5. To adjust, bend intermediate choke rod.

Choke Rod Adjustment, Fig. 6

NOTE: The choke coil lever and fast idle adjustments must be made before performing this adjustment.

1. Rotate degree scale until zero is opposite pointer, then with choke valve completely closed, place magnet on top of choke valve and rotate bubble until centered.
2. Rotate scale so specified degree for adjustment is opposite pointer.
3. Place fast idle screw on second step of cam against shoulder of high step.
4. Close choke by pushing intermediate choke lever.
5. Push vacuum break lever toward open choke position until lever contacts rear tang on choke lever.
6. To adjust, bend fast idle cam rod until bubble is centered.

Air Valve Rod Adjustment, Fig. 7

1. Rotate gauge scale until pointer is opposite zero.
2. With air valve completely closed, place magnet on top of air valve and rotate bubble until centered.
3. Rotate scale so that specified degree for adjustment is opposite pointer.
4. Seat vacuum diaphragm using an external vacuum source.
5. Rotate air valve toward the open position by applying light pressure to the air valve shaft.
6. To adjust, bend air valve rod until bubble is centered.

Primary Vacuum Break Adjustment, Fig. 8

NOTE: Before performing adjustment procedure, remove primary vacuum break from carburetor and position bracket in vise, then grind off adjusting screw cap and reinstall vacuum break.

1. Rotate degree scale until zero is opposite pointer, then with choke valve completely closed and fast idle screw on high step of fast idle cam, place magnet squarely on top of choke valve and rotate bubble until it is centered.
2. Rotate scale so that specified degree for adjustment is opposite pointer.
3. Seat choke vacuum diaphragm using a vacuum source with over 5 inches Hg of

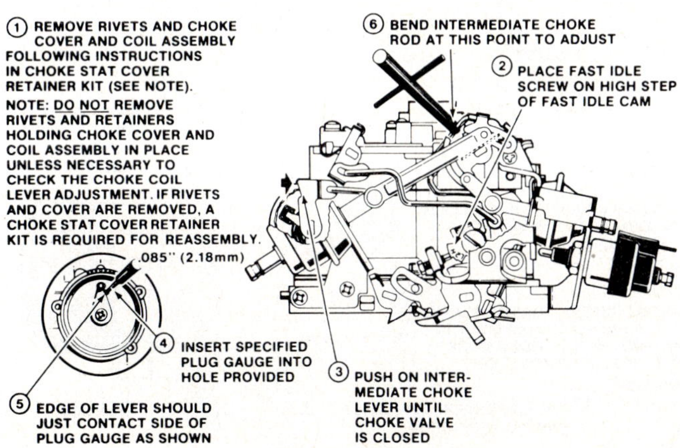

Fig. 5 Choke coil lever adjustment

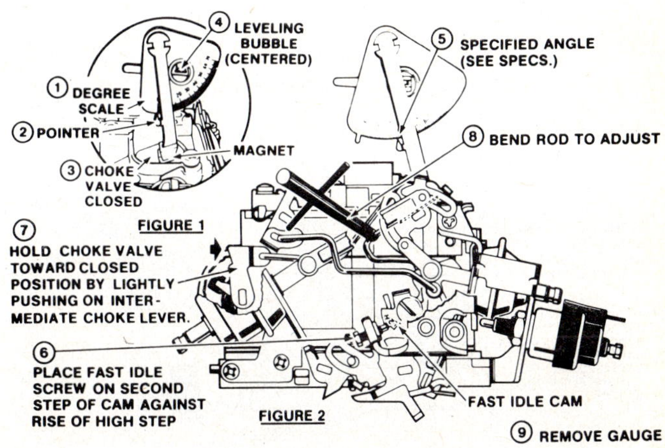

Fig. 6 Choke rod adjustment

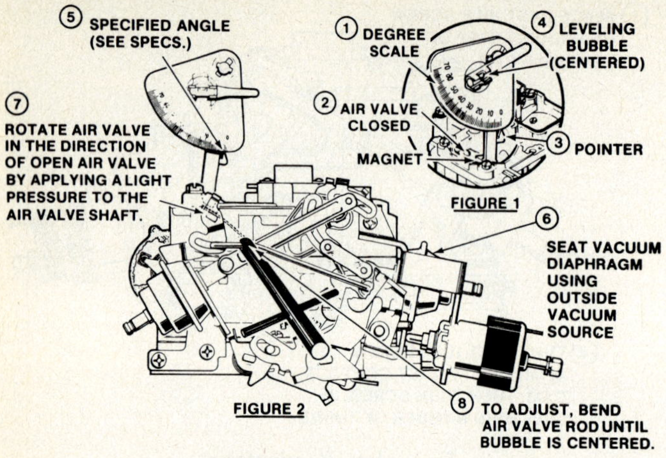

Fig. 7 Air valve rod adjustment

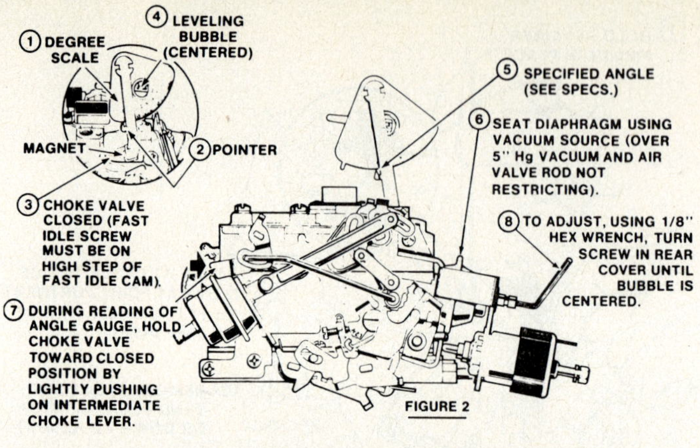

Fig. 8 Primary vacuum break adjustment

vacuum. Check to ensure that air valve rod is not restricting the vacuum diaphragm from being seated. It may be necessary to bend air valve rod to obtain a slight clearance between rod and end of slot in air valve lever. If the air valve rod adjustment is disturbed, refer to the Air Valve Rod Adjustment procedure after completing the primary vacuum break adjustment.

4. Hold choke valve toward the closed position by lightly pushing on intermediate lever and note angle gauge reading.
5. If adjustment is necessary, use a 1/8 inch hex wrench to rotate adjusting screw in rear cover until bubble is centered.
6. After completing adjustment, apply a suitable sealer over adjusting screw head.

Secondary Vacuum Break Adjustment, Fig. 9

NOTE: Before performing adjustment procedure, remove secondary vacuum break from carburetor and position bracket in a vise, then grind off adjusting screw cap and reinstall vacuum break.

1. Rotate degree scale so that zero is opposite pointer, then with choke valve completely closed and fast idle screw on high step of fast idle cam, place magnet on top of choke valve and rotate until bubble is centered.
2. Rotate scale so that specified degree for adjustment is opposite pointer.
3. Seat choke diaphragm using a vacuum source with over 5 inches Hg of vacuum.
4. Hold choke valve toward the closed position by lightly pushing on intermediate choke lever and note angle gauge reading. When noting reading check to ensure that vacuum diaphragm is seated.
5. Rotate adjusting screw in rear cover until bubble is centered. After completing adjustment, apply a suitable sealer over adjusting screw head.

Choke Setting

NOTE: The choke cover is retained to the choke housing by rivets. To remove rivets, carefully align a No. 21 drill on rivet head and drill only enough to remove rivet head. After removing rivet heads and retainers, drive remainder of rivet out using a suitable punch. A choke cover installation kit containing the proper rivets and retainers is available.

1. Loosen choke cover retaining screws.
2. Place fast idle screw on high step of cam.
3. Align scribe line on choke cover with specified point on housing. Rotate cover toward closed choke position.
4. Tighten choke cover retaining screws.

Unloader Adjustment, Fig. 10

1. Rotate degree scale until zero is opposite pointer, then with choke valve completely closed, place magnet on top of choke valve and rotate bubble until centered.
2. Rotate degree scale so specified degree for adjustment is opposite pointer.
3. With choke setting properly adjusted, hold primary throttle valve wide open.
4. On warm engines, close choke valve by pushing on intermediate choke lever and hold in position with a rubber band.
5. To adjust, bend tang on throttle lever until bubble is centered.

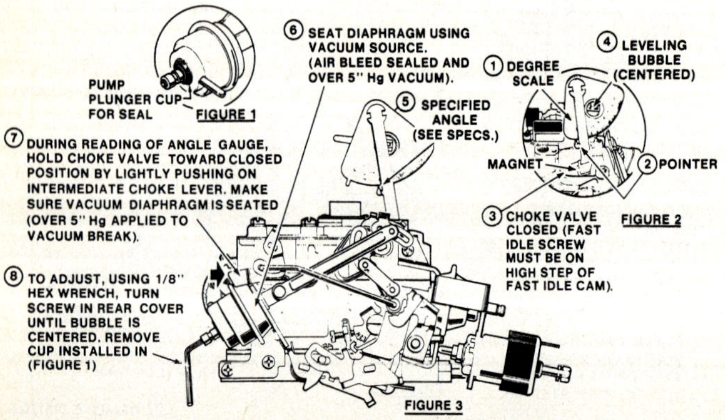

Fig. 9 Secondary vacuum break adjustment

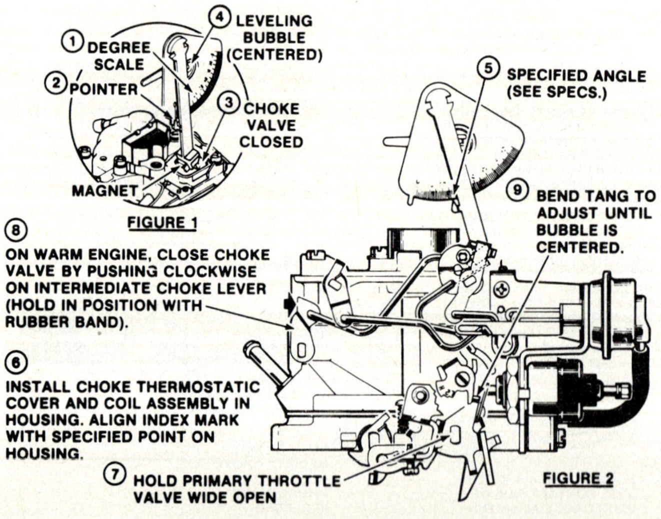

Fig. 10 Unloader adjustment

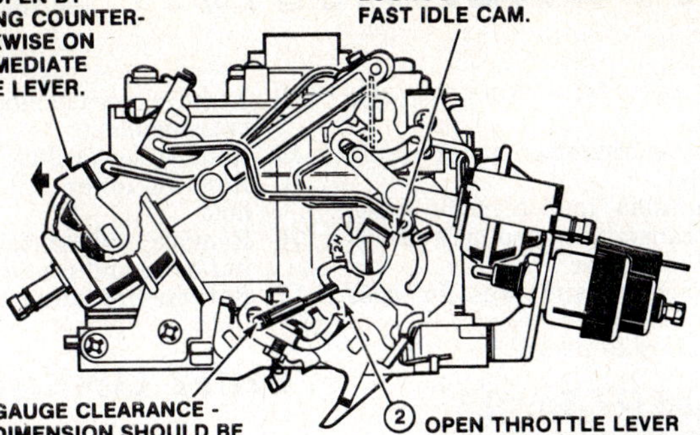

Fig. 11 Secondary lockout adjustment

Secondary Lockout Adjustment, Fig. 11

1. Hold choke valve wide open by pulling on intermediate choke lever.
2. Position throttle lever until end of secondary actuating lever is opposite toe of lockout lever.
3. Insert specified gauge between throttle lever and secondary lockout lever toe.
4. To adjust, bend lockout lever tang contacting fast idle cam.

HITACHI MODEL DCH340 CARBURETOR

Description

This carburetor is a two barrel, down draft type comprised of a low-speed, or primary side and a high-speed, or secondary side, both integrated into a single unit.

Adjustments

Float Level, Adjust, Fig. 1

If fuel level is within the mark on window of float chamber, with engine stationary, a normal fuel level is indicated. If fuel level is outside the line, adjust by bending float seat. The effective stroke of the needle valve should be about 0.059 in. To check stroke, hold carburetor bottom side up and fully raise the float. Normal needle valve stroke is indicated by a clearance of 0.059 in. between valve stem and float seat. If necessary, adjust needle valve seat by bending the float stopper.

Primary Throttle Valve, Adjust, Fig. 2

Check that the primary throttle valve is opened by fast idle adjusting screw to an angle of 16°, with manual transmission or 18°, with automatic transmission when choke valve is completely closed. To check opening angle of primary throttle valve, close choke valve completely and measure clearance between throttle valve and wall of throttle valve chamber at center part of throttle valve. Standard clearances are 0.050 to 0.059 in. with manual transmission and 0.059 to 0.069 with automatic transmission. If needed, throttle valve opening angle may be adjusted with the fast idle adjusting screw. Before measuring the clearance, be certain to turn throttle stop screw completely in.

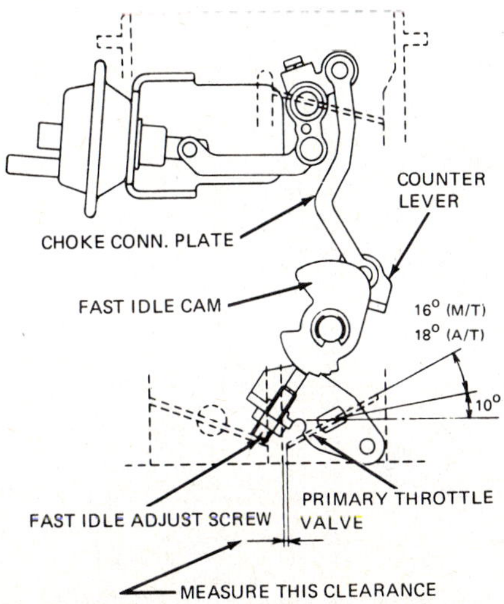

Fig. 2 Primary throttle valve adjustment

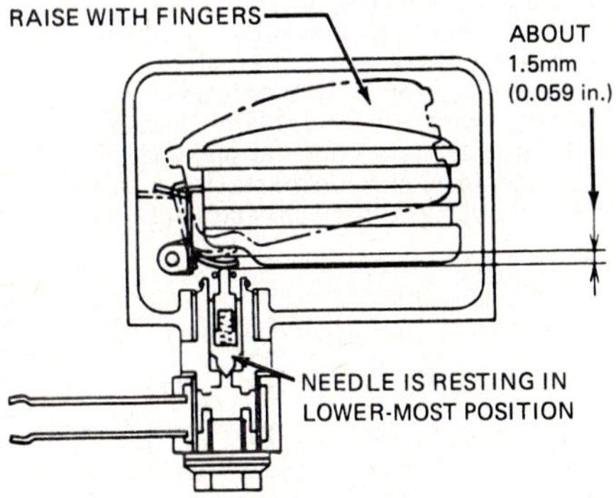

Fig. 1 Float level adjustment

Linkage, Adjust, Fig. 3

With primary throttle valve opened to an angle of 47 degrees, the adjust plate is brought into contact with portion (A) of kick lever. With primary throttle valve opened further, the return plate is pulled away from stopper (B), allowing secondary throttle valve to open. To measure secondary throttle valve opening point, measure clearance between primary throttle valve and wall of throttle chamber at center of throttle valve when adjust plate is brought into contact with point (A) of kick lever. Standard clearance (G2) is 0.24 to 0.30 in., and may be adjusted by bending portion (A) of kick lever.

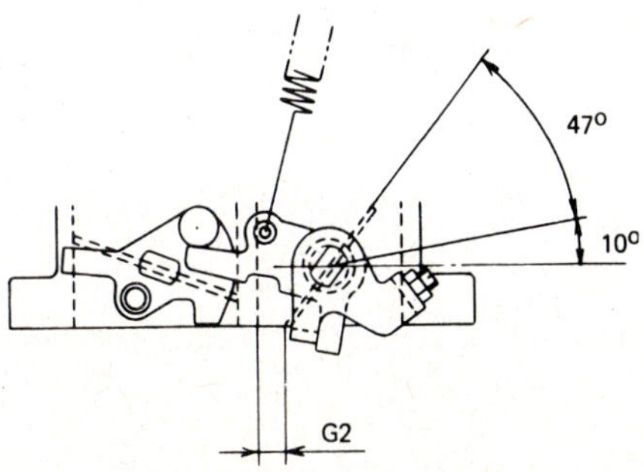

Fig. 3 Linkage adjustment

Kick Lever, Adjust, Fig. 4

Turn out throttle adjustment screw until primary side throttle valve is completely closed. Loosen lock nut on kick lever screw and turn screw until contact is made with return plate, then tighten lock nut.

Fast Idle, Adjust

Automatic choke fase idle is adjusted by opening angle of throttle valve, not by engine speed. Adjusted throttle valve opening at first step of fast idle cam is 16 degrees ± 1 with manual transmission and 18 degrees ± 1 with automatic transmission. Engine fast idle speed should be about 3,200 rpm after engine is warmed up and distributor vacuum line, idle compensator and EGR valve are disconnected and plugged.

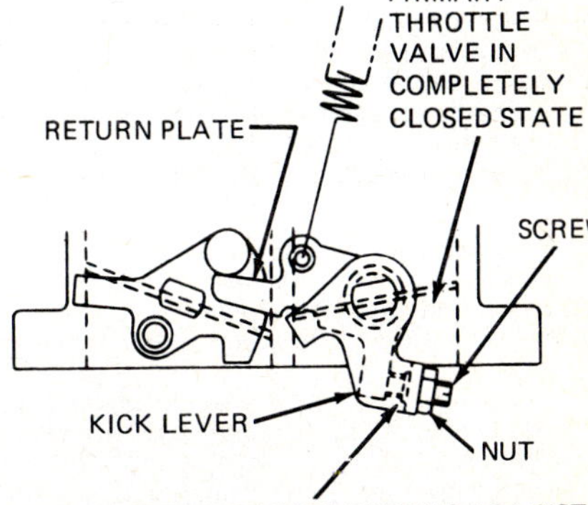

Fig. 4 Kick lever adjustment

Clutch & Manual Transmission Section

CLUTCH, ADJUST

1. Clutch is adjusted by lifting clutch pedal up to allow mechanism to adjust cable. Depress pedal several times to set pawl into mesh with detent teeth.
2. Check clutch linkage for lost motion caused by loose or worn swivels, deflection of mounting brackets or damaged cordon shaft.

CLUTCH, REPLACE

1. Disconnect clutch fork cable and remove transmission assembly as outlined under "Transmission, Replace".
2. Remove flywheel housing.
3. Slide clutch fork from ball stud and remove fork from dust boot.
4. Install Tool J-33169 (V-6) or J-33034 (4 cyl.) to support clutch assembly.

NOTE: Look for "X" mark on flywheel and on clutch cover or white painted letter on clutch cover. If marks are not evident, mark flywheel and clutch cover for assembly.

5. Loosen clutch-to-flywheel attaching bolts evenly, one turn at a time, until spring pressure is released. Then remove bolts, clutch and pressure plate assembly.
6. Clean pressure plate and flywheel mating surfaces and inspect flywheel for defects. Replace or repair as needed.
7. Reverse procedure to install.

TRANSMISSION, REPLACE

4 Speed (77mm) & 5 Speed

4 × 2

1. Disconnect battery ground cable.
2. Remove attaching screws and slide shift lever boot up shift lever.
3. With transmission in neutral, remove shift lever.
4. Raise and support vehicle.
5. Remove propeller shaft.
6. Disconnect speedometer cable and electrical connectors at transmission.
7. Disconnect clutch cable at transmission.
8. Support transmission and remove transmission mount attaching bolts.
9. Remove catalytic converter hanger.
10. Remove crossmember attaching bolts and crossmember.
11. Remove dust cover bolts.
12. Remove transmission-to-engine attaching bolts and remove transmission.
13. Reverse procedure to install.

4 × 4

1. Shift transfer case to "4 Hi".
2. Disconnect battery ground cable.
3. Raise and support vehicle.
4. Remove skid plate, then drain transfer case.
5. Mark relationship between transfer case front output shaft yoke and propeller shaft for reference during installation, then disconnect front propeller shaft from transfer case.
6. Mark relationship between rear axle yoke and propeller shaft for reference during installation, then remove rear propeller shaft.
7. Disconnect speedometer cable, vacuum harness and shift lever from transfer case.
8. Remove catalytic converter hanger bolts at the converter.
9. Raise transmission and transfer case, then remove transmission mount and catalytic converter hanger.
10. Lower transmission and transfer case from vehicle.
11. Reverse procedure to install.

4-Speed (77.5mm)

1. Disconnect battery ground cable.
2. Remove upper starter motor retaining nut.
3. Remove attaching screws and slide shift lever boot up shift lever.
4. Remove shift lever attaching bolts at transmission.
5. Disconnect electrical connection and clip at transmission shift tower.
6. Raise and support vehicle.
7. Remove propeller shaft.
8. Disconnect exhaust pipe at manifold.
9. Disconnect speedometer cable and electrical connector at transmission.
10. Disconnect clutch cable at transmission.
11. With transmission supported, remove transmission mount attaching bolts.
12. Remove catalytic converter hanger.
13. Remove crossmember attaching bolts and crossmember.
14. Remove lower dust cover bolts.
15. Remove lower starter motor attaching bolt.
16. Remove transmission-to-engine attaching bolts and remove transmission.
17. Reverse procedure to install.

SHIFT LINKAGE, ADJUST

The shift mechanism does not require adjustment and may be serviced independently of the transmission.

4 SPEED (77mm) TRANSMISSION

Transmission, Disassemble, Fig. 1

1. Drain transmission lubricant and thoroughly clean outside of transmission assembly.
2. Remove roll pin that attaches offset lever to shift rail, Fig. 1A.
3. Remove retaining bolts from extension housing, then separate extension housing from transmission case and remove hous-

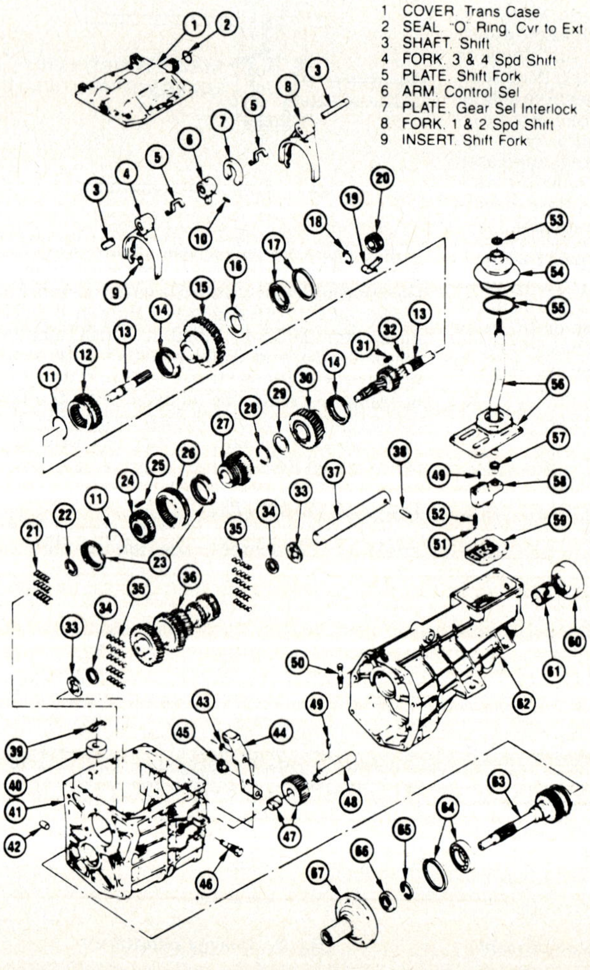

1 COVER. Trans Case
2 SEAL. "O" Ring. Cvr to Ext
3 SHAFT. Shift
4 FORK. 3 & 4 Spd Shift
5 PLATE. Shift Fork
6 ARM. Control Sel
7 PLATE. Gear Sel Interlock
8 FORK. 1 & 2 Spd Shift
9 INSERT. Shift Fork
10 PIN. Roll
11 SPRING. Syn
12 GEAR. Rev Sliding
13 SHAFT. W 1 & 2 Spd Syn
14 RING. 1 & 2 Syn Blocking
15 GEAR. 1st
16 WASHER. 1st Spd Gear Thrust
17 BEARING. Main Shaft. W Snap Ring
18 RING. Rr Brg-to Output Shf Ret
19 CLIP. Speedo Drive Gear
20 GEAR. Speedo Drive
21 ROLLER. Main Shaft
22 RING. Syn Ret
23 RING. 3 & 4 Syn Blocking
24 HUB. 3 & 4 Syn
25 KEY. 3 & 4 Syn
26 SLEEVE. 3 & 4 Syn
27 GEAR. 3rd
28 RING. 2nd Spd Gr Thr Wa Ret
29 WASHER. 2nd Spd Thrust
30 GEAR. 2nd
31 KEY. 1 & 2 Syn
32 PIN. 1st Spd Gr Thr Wa Ret
33 WASHER. Counter Gear Thrust
34 SPACER. Counter Gear Rlr
35 ROLLER. Counter Gear
36 GEAR. Counter
37 SHAFT. Counter Gear
38 PIN. Spring
39 NUT. Spring
40 MAGNET
41 CASE
42 PLUG. Fill & Drain
43 RING. Rev Rly Lvr Ret
44 LEVER. Rev Relay
45 FORK. Rev Shift Lvr
46 PIN. Rev Shift Lvr Pivot
47 GEAR. Rev Idler. W Bushing
48 SHAFT. Rev Idler Gear
49 PIN. Spr
50 VENTILATOR. Ext
51 BALL. Steel
52 SPRING. Detent
53 RETAINER. Cont Lvr Boot
54 BOOT. Cont Lvr
55 RETAINER. Cont Lvr Boot
56 CONTROL. Trans Lvr & Hsg
57 SLEEVE. Shift Lvr Damper
58 LEVER. Offset Shift
59 PLATE. Detent & Guide
60 SEAL. Ext Rear Oil
61 BUSHING. Extension Housing
62 HOUSING. Extension
63 GEAR. Main Drive
64 BEARING. Main Drive Gear. W Snap Ring
65 RING. Main Dr Gr Brg to Shf Ret
66 SEAL. Main Drive Gear Brg Oil
67 RETAINER. Main Drive Gear Brg

Fig. 1 Exploded view of 4 speed 77mm manual transmission

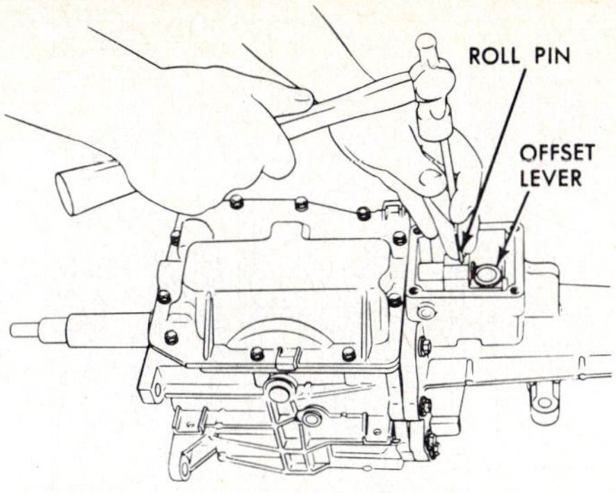

Fig. 1A Removing offset lever roll pin. 4 speed 77mm transmission

ing and offset lever as an assembly.

4. Remove detent ball and spring from offset lever, then remove roll pin from extension housing or offset lever, Fig. 2.
5. Remove attaching bolts from transmission shift cover and remove the shift cover.
6. Remove clip which retains reverse lever to reverse lever pivot bolt, then remove the bolt and remove reverse lever and fork as an assembly, Fig. 3.
7. Scribe position of front bearing cap to transmission case, then remove front bearing cap bolts and bearing cap.
8. Remove snap rings from front drive gear bearing, then, using appropriate tools, Fig. 4, remove the front bearing.
9. Remove snap rings from rear bearing and mainshaft, then, using appropriate tools, Fig. 5, remove the bearing.

NOTE: New front and rear bearings must be installed when assembling the transmission.

10. Remove drive gear from mainshaft and transmission case, then remove mainshaft from transmission case.
11. Remove roll pin retaining reverse idler gear shaft from transmission case and remove idler gear and shaft from case, Fig. 6.
12. Using loading tool J-26624, remove countershaft from rear of case, then remove countershaft gear and loading tool as an assembly from the case together with the thrust washers.

Sub-assembly Service

Mainshaft, Disassemble

1. Scribe alignment mark on third-fourth synchronizer hub and sleeve for reference during assembly, then remove snap ring and remove third-fourth synchronizer

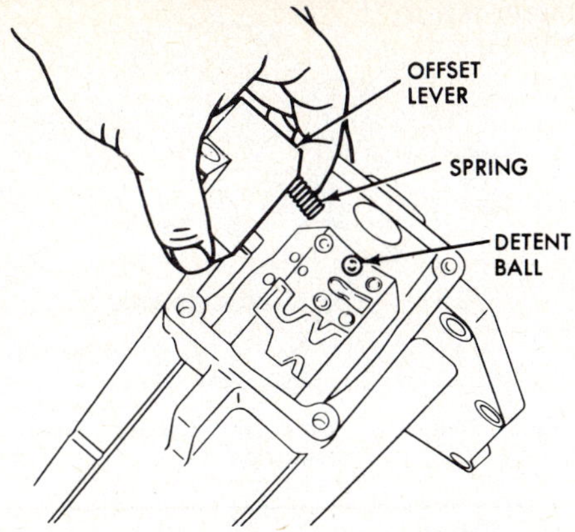

Fig. 2 Removing detent ball and spring. 4 speed 77mm transmission

assembly from mainshaft, Fig. 7.

2. Slide third gear off mainshaft, then remove snap ring from second gear and remove tabbed thrust washer, second gear and blocker ring from mainshaft, Fig. 8.
3. Remove first gear thrust washer and roll pin from mainshaft, then remove first gear and blocker ring from mainshaft.
4. Scribe alignment mark on first-second gear synchronizer hub and sleeve for reference during assembly, then remove synchronizer springs and keys from first-second sleeve and remove sleeve from shaft.

NOTE: The first-second hub and mainshaft are assembled as a unit and no attempt must be made to separate them.

5. Remove loading tool, roller bearings, spacers and thrust washers from the countershaft gear.

Drive Gear, Disassemble

1. Remove roller bearings from cavity of drive gear and wash parts in a suitable cleaning solvent.
2. Inspect gear teeth and drive shaft pilot for wear.

Cleaning & Inspection

1. Thoroughly wash transmission using a suitable cleaning solvent and inspect the case for cracks.
2. Clean magnetic disc at bottom of transmission case and check front and rear faces of case for burrs. Dress any burrs off with a fine mill file.
3. Inspect all drive gear and countergear bearing rollers and replace as necessary.

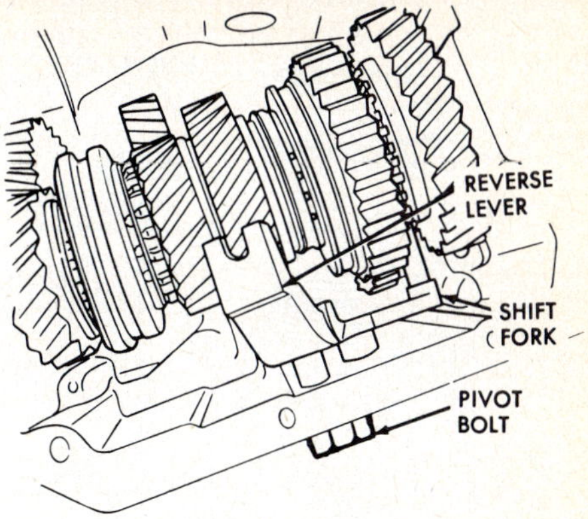

Fig. 3 Removing reverse lever and shift fork. 4 speed 77mm transmission

4. Inspect countershaft and reverse idler shaft and all spacers and replace as necessary.
5. Inspect all gears for wear or damage and replace as necessary.

Synchronizer Keys & Springs, Replace

NOTE: The synchronizer hubs and sliding sleeves are a selected assembly and must be kept together as originally assembled, however, the keys and springs may be replaced if worn or broken.

1. Mark relation of hub and sleeve for reference during assembly.
2. Push sliding sleeve from hub and remove keys and springs.
3. Install sliding sleeve on hub, aligning the marks made previously.
4. Place a blocker ring on side of hub and sleeve and install keys and a spring. Then place blocker ring on opposite side of hub and sleeve and install remaining spring.

Extension Housing Oil Seal & Bushing, Replace

1. Pry seal out with screwdriver, then drive bushing out of housing using tools J-8092 and J-23062-14, or equivalent.
2. Install new bushing in housing using same tools as for removal.
3. Coat outer edge of new seal with suitable sealing cement, then install seal using Tool J-21426, or equivalent and coat inside edge of seal with transmission lubricant.

Drive Gear Bearing Retainer Oil Seal, Replace

1. Pry seal out with screwdriver.
2. Using Tool J-23096, or equivalent, install new seal into retainer until it bottoms

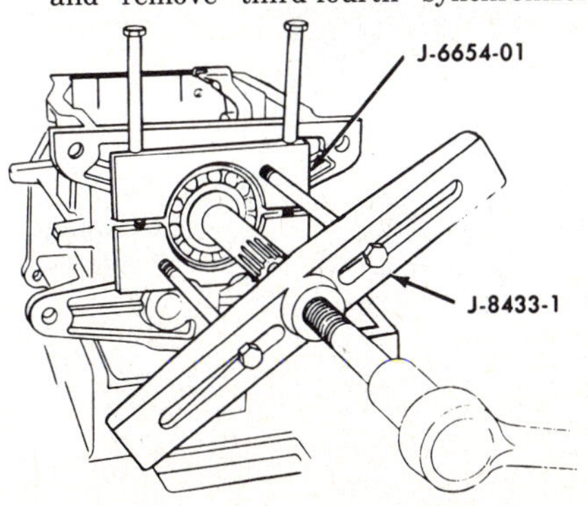

Fig. 4 Removing front bearing. 4 speed 77mm transmission

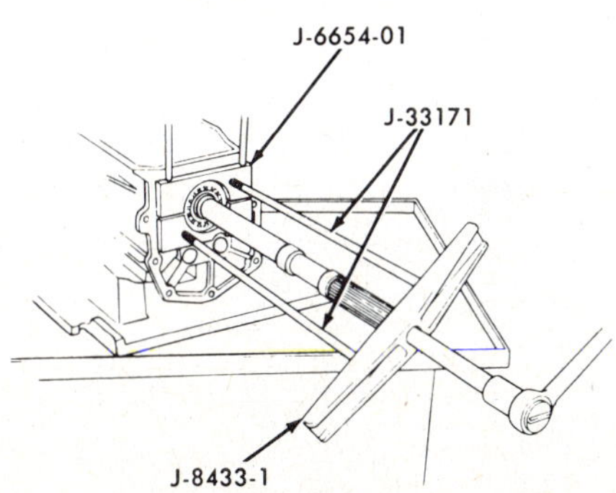

Fig. 5 Removing rear bearing. 4 speed 77mm transmission

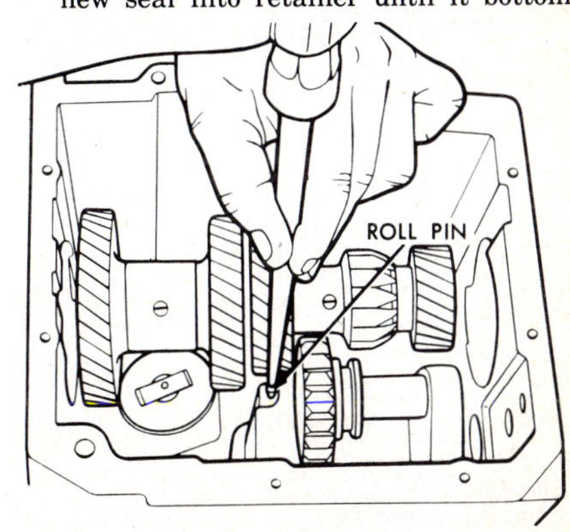

Fig. 6 Removing reverse idler gear shaft. 4 speed 77mm transmission

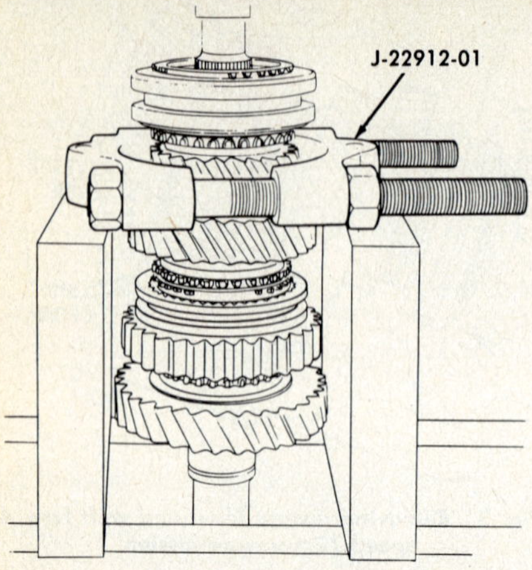

Fig. 7 Removing third-fourth synchronizer assembly. 4 speed 77mm transmission

out, then coat inner edge of seal with transmission lubricant.

Transmission Cover, Disassemble

1. Place selector arm plates and shift rail in neutral position, then rotate shift rail until selector arm disengages from selector arm plates and roll pin is accessible.
2. Remove selector arm roll pin, Fig. 9, then remove shift rail, shift forks, selector arm plates, selector arm, interlock plate and roll pin.
3. Pry out shift cover to extension housing O-ring with a screwdriver.
4. Remove nylon inserts and selector arm plates from shift forks, noting their position for reference during assembly.

Transmission Cover, Assemble

1. Install nylon inserts and selector arm plates in shift forks, then install shift rail plug after coating edges with suitable sealer.
2. Coat shift rail and rail bores with light weight grease and install shift rail in cover so that it is flush with inside edge of cover.
3. Install first-second shift fork in cover with fork offset facing rear of cover and push shift rail through fork.

NOTE: The first-second shift fork is the larger of the two forks.

4. Install selector arm and C-shaped interlock plate in cover and insert shift rail through arm.

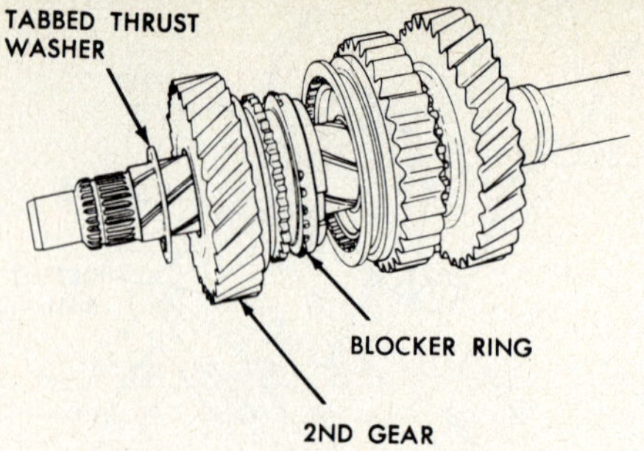

Fig. 8 Removing second gear. 4 speed 77mm transmission

NOTE: The widest part of interlock plate must face away from cover, and the selector arm roll pin hole must face downward and toward rear of cover.

5. Position third-fourth shift fork in cover with fork offset facing rear of cover.

NOTE: Third-fourth shift fork selector arm plate must be under first-second shift fork selector arm plate.

6. Push shift rail through third-fourth shift fork and into front bore in cover, then rotate shift rail until selector arm plate at foward end of rail faces away from, but is parallel to cover.
7. Align roll pin holes in selector arm and shift rail and install roll pin.

NOTE: Roll pin must be flush with surface of selector arm to prevent pin and selector arm from contacting each other during shifts.

8. Install a new shift cover to extension housing seal and coat the seal with transmission lubricant.

Drive Gear, Assemble

Coat roller bearings and drive gear bearing bore with suitable lightweight grease and install roller bearings into bore of drive gear.

Mainshaft, Assemble

1. Coat mainshaft and gear bores with transmission lubricant, then install first-second synchronizer sleeve on mainshaft, aligning marks made during disassembly.
2. Install synchronizer keys and springs into the first-second synchronizer sleeve. Engage tang end of springs into same

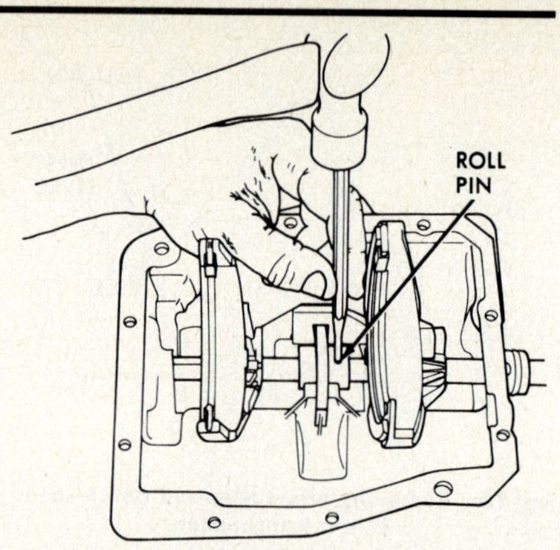

Fig. 9 Removing selector arm roll pin. 4 speed 77mm transmission

synchronizer key, but position open ends of springs so that they face away from each other, Fig. 10.

3. Place blocking ring on first gear and install gear and ring on mainshaft.

NOTE: Ensure that synchronizer keys engage notches in first gear blocking ring.

4. Install first gear roll pin in mainshaft, then place blocking ring on second gear and install gear and ring on mainshaft, ensuring synchronizer keys engage notches in second gear blocking ring.
5. Install second gear thrust washer and snap ring, ensuring that thrust washer tab is engaged in mainshaft notch.
6. Check second gear end play using a feeler gage. End play should be .004–.014 inch. If end play exceeds .014 inch, replace thrust washer and snap ring and inspect synchronizer hub for excessive wear.
7. Place blocking ring on third gear and install gear and ring on mainshaft, then install third-fourth synchronizer sleeve on hub, aligning marks made during disassembly.
8. Install synchronizer keys and springs in third-fourth synchronizer sleeve. Engage tang end of each spring in same key, but position open ends of springs so that they face away from each other.
9. Install third-fourth synchronizer assembly on mainshaft with machined groove in hub facing forward, then install snap

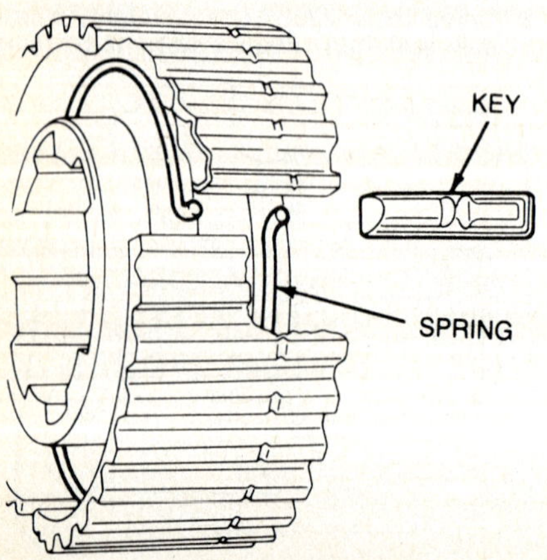

Fig. 10 Installing synchronizer spring. 4 speed 77mm transmission

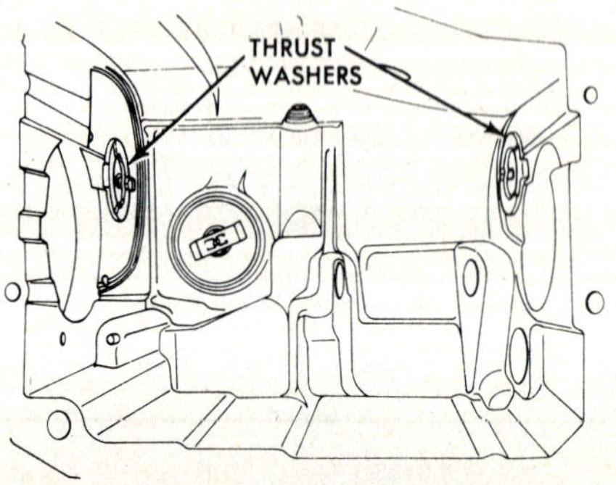

Fig. 11 Installing countershaft thrust washers. 4 speed 77mm transmission

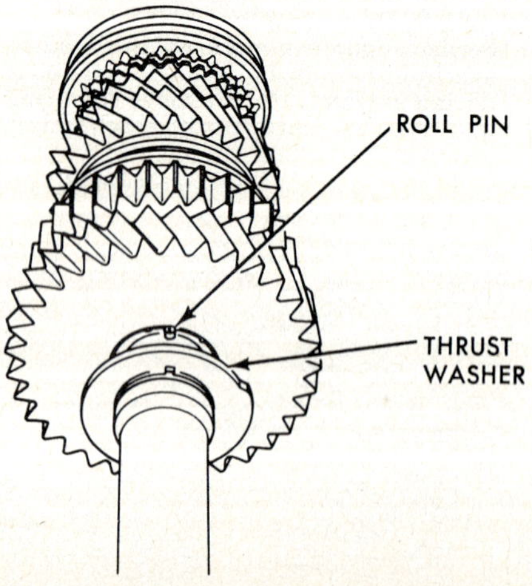

Fig. 12 Installing first gear thrust washer. 4 speed 77mm transmission

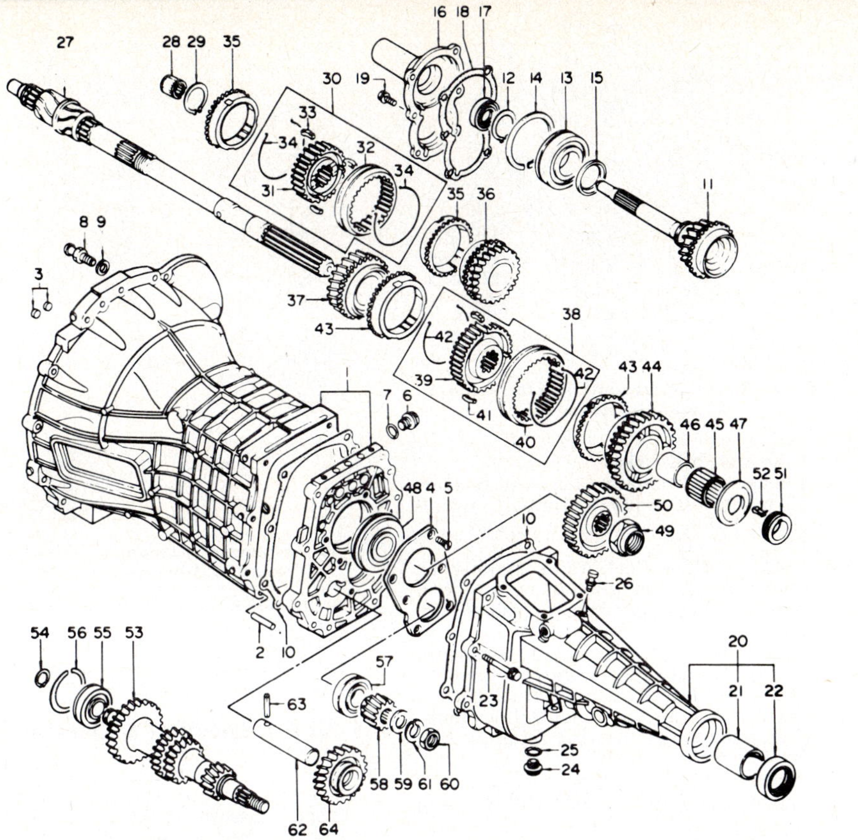

1. Transmission case and center support
2. Dowel
3. Plug
4. Rear bearing retainer
5. Screw
6. Oil filler plug
7. Gasket
8. Ball stud
9. Washer
10. Gasket
11. Drive gear shaft
12. Retaining ring
13. Bearing
14. Retaining ring
15. Spacer
16. Front bearing retainer
17. Oil seal
18. Gasket
19. Bolt and spring washer
20. Extension housing
21. Bushing
22. Rear oil seal
23. Bolt, plain washer and spring washer
24. Oil drain plug
25. Gasket
26. Ventilator
27. Mainshaft
28. Needle roller bearing
29. Retaining ring
30. 3rd/4th sychronizer
31. Clutch hub
32. Sleeve
33. Insert
34. Spring
35. Blocker ring
36. 3rd gear
37. 2nd gear
38. 1st/2nd synchronizer
39. Clutch hub
40. Sleeve
41. Insert
42. Spring
43. Blocker ring
44. 1st gear
45. Needle roller bearing
46. Bearing collar
47. Thrust washer
48. Bearing
49. Nut
50. Reverse gear
51. Speedometer drive gear
52. Clip
53. Counter shaft
54. Retaining ring
55. Bearing
56. Retaining ring
57. Bearing
58. Reverse gear
59. Plain washer
60. Nut
61. Spring washer
62. Reverse idler shaft
63. Spring pin
64. Reverse idler gear

Fig. 13 4 Speed 77.5mm transmission exploded view

ring on mainshaft, ensuring that synchronizer keys are engaged in notches in third gear blocker ring.

10. Measure third-fourth synchronizer end play following procedure in step No. 6 above.
11. Install Tool J-26624, or equivalent, into countershaft gear, then, using a suitable lightweight grease, lubricate roller bearings and install into bores at front and rear of countershaft gear. Install roller bearing retainers on Tool J-26624.

Assemble Transmission

1. Coat countershaft gear thrust washers with suitable grease and install washers in case, Fig. 11.
2. Position countershaft gear in case and install countershaft from rear of case, ensuring that thrust washer stays in place.
3. Position reverse idler gear in case with shift lever groove facing rear of case and install reverse idler shaft from rear of case, then install roll pin in shaft and center pin in shaft.
4. Install mainshaft assembly into case, being careful not to change position of synchronizer assemblies.
5. Install fourth gear blocking ring in third-fourth synchronizer sleeve, ensuring that synchronizer keys engage in notches in blocker ring, then install drive gear into case and engage with mainshaft.
6. Position mainshaft first gear against rear of case, then start new front bearing onto drive gear. Align bearing with bearing bore in case and drive bearing onto drive gear and into case using Tool J-25234, or equivalent, then install front bearing snap rings.
7. Apply a 1/8 inch bead of RTV sealant #732, or equivalent on case mating surface of front bearing cap and install bearing cap aligning marks made during disassembly. Apply non-hardening sealer on attaching bolts, install bolts, and torque to 15 ft. lbs.
8. Install first gear thrust washer with oil groove facing first gear on mainshaft, aligning slot in washer with first gear roll pin, Fig. 12.
9. Position new rear bearing on mainshaft and align bearing with bearing bore in case, then drive bearing into case using Tool J-25234, or equivalent, and install rear bearing snap rings.
10. Install speedometer gear and retaining clip on mainshaft.
11. Apply non-hardening sealer to threads of reverse lever pivot bolt and start bolt into case. Engage reverse lever fork in reverse idler gear and reverse lever on pivot bolt, tighten bolt to 20 ft. lbs., and install retaining clip.
12. While rotating drivegear and mainshaft gears, check if blocker rings stick on gears. Rings can be released by carefully prying them off the cones.
13. Apply a 1/8 inch bead of RTV sealant #732, or equivalent, on cover mating surface of transmission, then place reverse lever in neutral and position cover on case.
14. Align cover to case by installing two dowel type bolts first, then install remaining cover bolts and torque to 10 ft. lbs.

NOTE: The offset lever to shift rail roll pin hole must be in a vertical position after cover installation.

15. Apply a 1/8 inch bead of RTV sealant #732, or equivalent on extension housing to transmission case mating surface, then place extension housing over mainshaft in a position where shift rail is in shift cover opening.
16. Install detent spring in offset lever and place ball in neutral guide plate detent position. Apply pressure on offset lever, slide offset lever onto shift rail and seat extension housing to transmission case.
17. Install extension housing retaining bolts and torque to 25 ft. lbs.
18. Align hole in offset lever and shift rail and install roll pin, then fill transmission to proper level with lubricant.

4 SPEED (77.5mm) TRANSMISSION, FIG. 13

Disassembly

1. Disconnect retaining springs from bearing side and remove bearing, boot, and clutch fork, Fig. 14.
2. Drain transmission, then remove bearing retainer bolts, bearing, gasket, Fig. 15, and ball stud.
3. Remove speedometer driven gear, back-up light switch, shift cover retaining bolts, shift cover, and gasket.
4. Remove rear extension retaining bolts, rear extension, and gasket.
5. Remove speedometer driven gear from mainshaft, Fig. 16.
6. Using suitable tool, drive pin from reverse shift block while supporting shaft end with bar or block of wood, Fig. 17.
7. Remove reverse block retaining bolts and remove reverse shifter shaft, shift block, shift fork, and reverse gear as an assembly, Fig. 18.

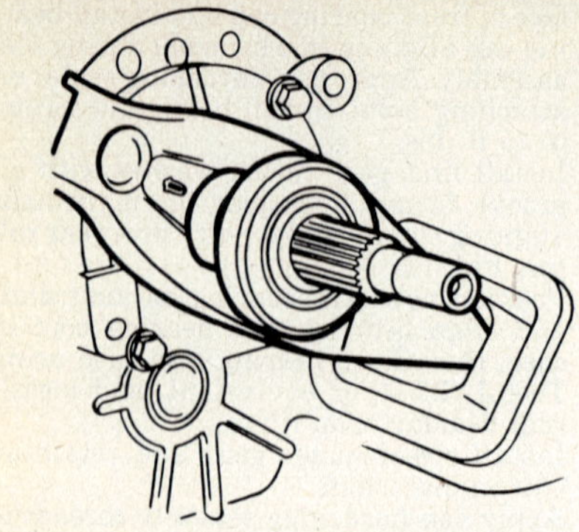

Fig. 14 **Removing clutch fork and bearing. 4 speed 77.5mm transmission**

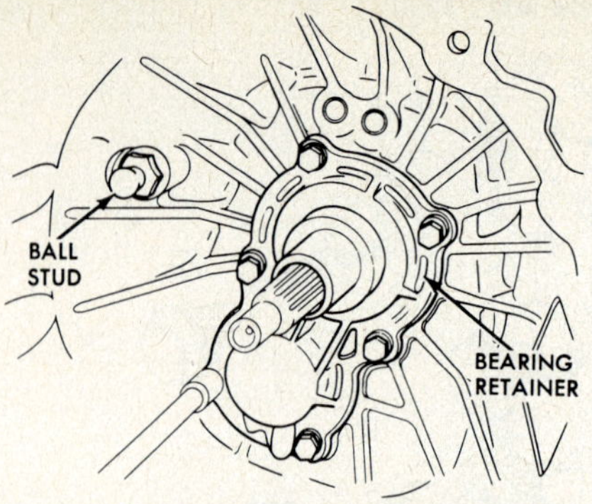

Fig. 15 **Removing bearing retainer. 4 speed 77.5mm transmission**

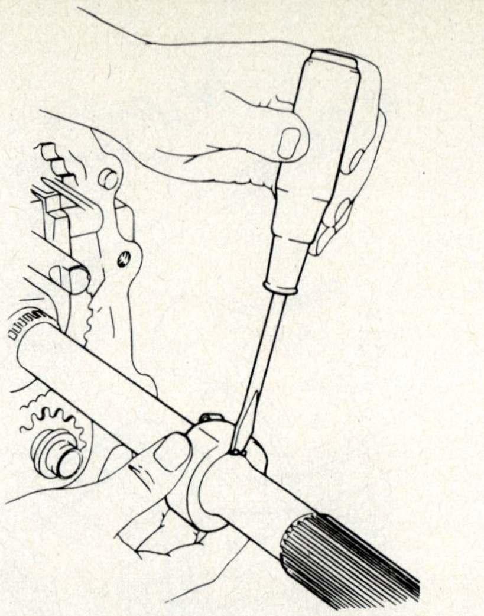

Fig. 16 **Removing speedometer driven gear. 4 speed 77.5mm transmission**

8. Remove retaining ring from drive gear shaft bearing outer race, then remove retaining ring from countershaft front bearing outer race, Fig. 19.
9. Remove center support assembly from transmission case, being careful not to hit front end of drive gear shaft while removing transmission case.
10. Using suitable tool, drive pins from 1st/2nd & 3rd/4th shift forks while supporting ends of shafts with bar or block of wood, Fig. 20.
11. Loosen retaining bolts and remove plate, gasket, and springs.
12. Place the three shifter shafts in neutral position and remove reverse shifter shaft, 1st/2nd shifter shaft, and 3rd/4th shifter shaft in that order, being careful not to lose interlock pins or detent balls.
13. Remove three detent balls, two inter-lock pins, and shift forks.
14. Engage synchronizers to prevent rotation of mainshaft and temporarily install transmission case to center support, Fig. 21.
15. Using suitable tool, raise staking on mainshaft rear nut and remove nut, then remove reverse gear from mainshaft, Fig. 22.
16. Remove nut, spring washer, plain washer, and reverse gear from countershaft, then remove case from center support.
17. Place synchronizers in neutral positions, then remove rear bearing retainer from center support, Fig. 23.
18. Slide rear bearing outer race rearward by moving countershaft fore and aft and, using suitable tool, remove outer race, Fig. 24.
19. Remove countershaft and drive gear shaft, then remove 4th blocker ring and needle roller bearing.

Unit Inspection & Overhaul

1. Thoroughly wash transmission case inside and out with suitable solvent, then inspect case for cracks.
2. Check front and rear faces for burrs and

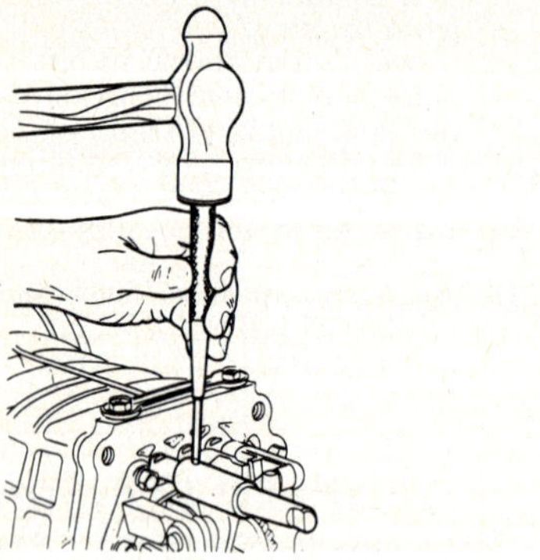

Fig. 17 **Removing reverse shift block pin. 4 speed 77.5mm transmission**

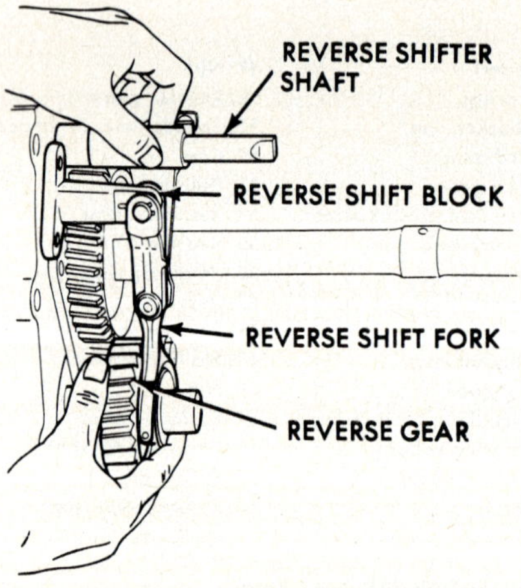

Fig. 18 **Removing reverse shifter assembly. 4 speed 77.5mm transmission**

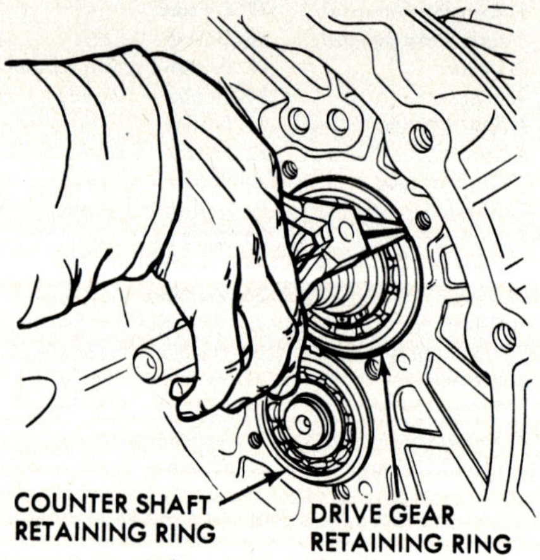

Fig. 19 **Removing drive gear retaining ring. 4 speed 77.5mm transmission**

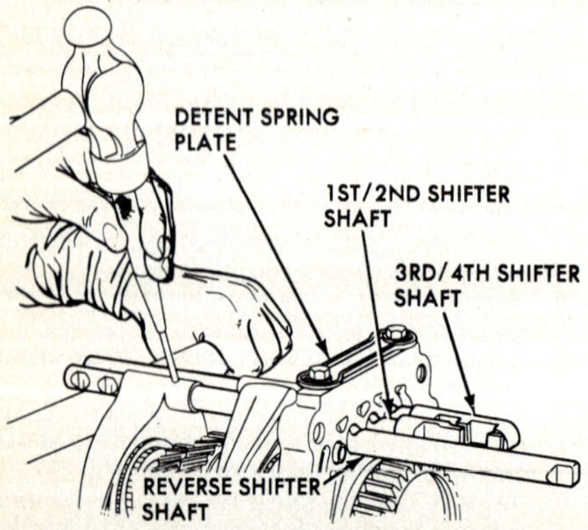

Fig. 20 **Removing pin from shifter shafts. 4 speed 77.5mm transmission**

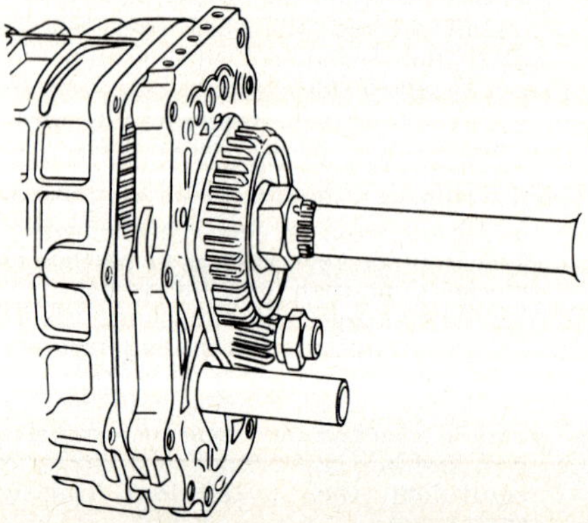

Fig. 21 **Temporarily install center support. 4 speed 77.5mm transmission**

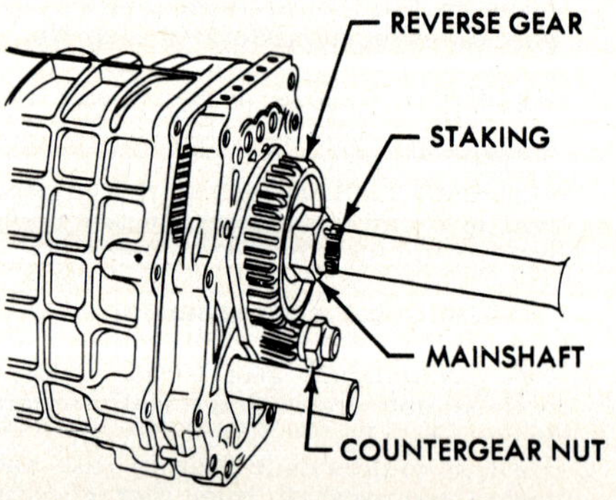

Fig. 22 **Removing reverse gear. 4 speed 77.5mm transmission**

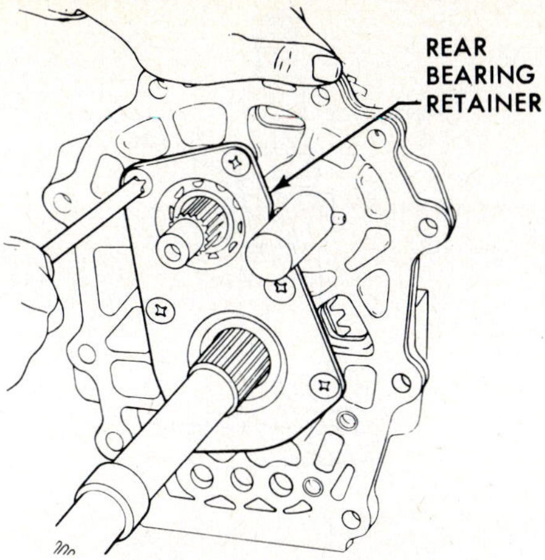

Fig. 23 Removing rear bearing retainer. 4 speed 77.5mm transmission

correct if necessary.

3. Inspect all main drive gear and countergear roller bearings and replace as required, then inspect countershaft and reverse idler shaft together and replace if necessary. Replace all worn spacers.
4. Wash front and rear ball bearings in suitable solvent and blow out bearings with compressed air.

NOTE: Do not allow bearings to spin. Turn them slowly by hand. Spinning bearings may damage race and balls.

---

5. Lubricate bearings and check for roughness by slowly turning race by hand.
6. Inspect all gears for excessive wear, chips, or cracks and replace if damaged or worn.
7. Check oil seal contact area on drive gear shaft and replace gear if pitted, rusted, or cracked.
8. Inspect interlock levers for cracking at detent and clearance notches at either end of levers.
9. Inspect shift forks for wear and inspect fork shaft bores in shift lever.

Mainshaft

1. Using tool J-22912-01 or equivalent at

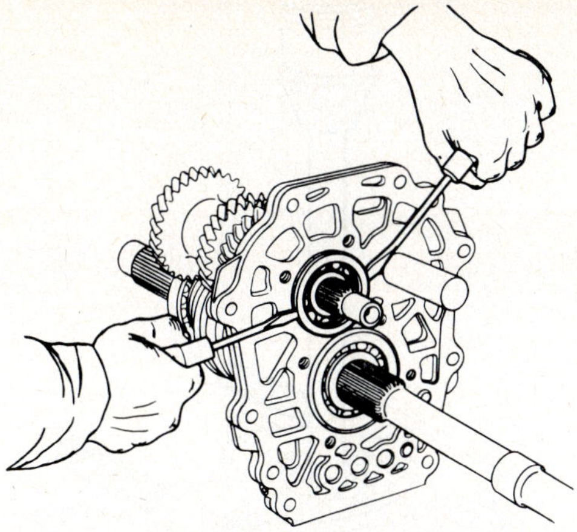

Fig. 24 Removing countershaft outer race. 4 speed 77.5mm transmission

rear face of second gear on mainshaft, remove mainshaft from center support using an arbor press, then remove 2nd gear, 1st/2nd synchronizer, 1st and 2nd blocker rings, 1st gear needle roller bearing, bearing collar, and thrust washer. Remove mainshaft rear bearing from center support, Fig. 25.

2. Remove snap ring from front of mainshaft. Set blocks to rear face of 3rd gear and remove mainshaft using an arbor press, then remove 3rd gear, 3rd blocker ring, and 3rd/4th synchronizer, Fig. 26.
3. Install 3rd gear and 3rd gear blocker ring to front of mainshaft.
4. Install 3rd/4th synchronizer assembly to mainshaft using arbor press. Plates should be fitted against parts, so that force will be exerted on boss of synchronizer hub, Fig. 27.
5. Install retaining ring to front of mainshaft, then install 2nd gear and 2nd blocker ring to rear of mainshaft.
6. Install 1st/2nd synchronizer assembly to mainshaft using arbor press. Plates should be fitted against parts, so that force will be exerted on boss of synchronizer hub, Fig. 28.
7. Install 1st gear bearing collar to main-

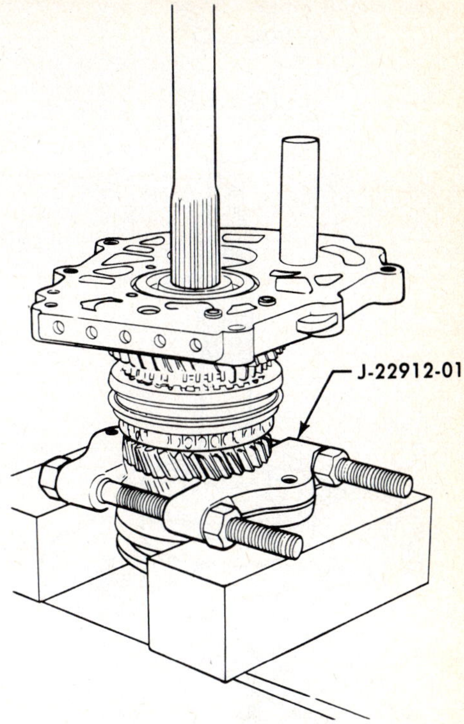

Fig. 25 Removing mainshaft from center support. 4 speed 77.5mm transmission

shaft using tool J-33036 or equivalent, and an arbor press. Install 1st blocker ring, needle roller bearing, and 1st gear, then install thrust washer with oil grooves turned toward gear.

8. Install reverse idler shaft into center support with suitable tool, after making certain spring pin is fitted to shaft.

Synchronizer Keys & Springs

NOTE: Synchronizer hubs and sliding sleeves are a selected assembly and should be kept together as originally assembled, but the keys and springs may be replaced if worn or broken.

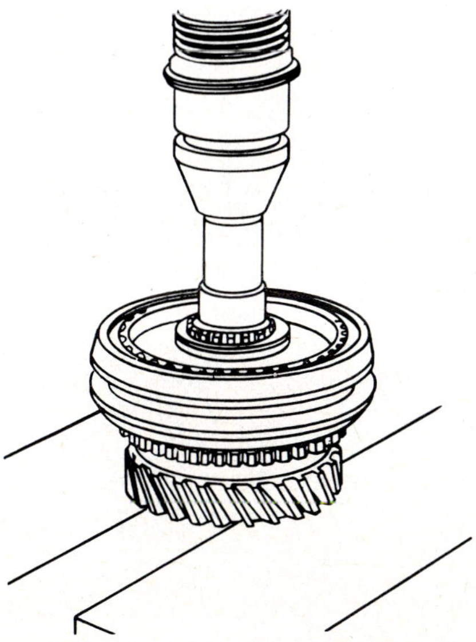

Fig. 26 Removing mainshaft from 3rd gear. 4 speed 77.5mm transmission

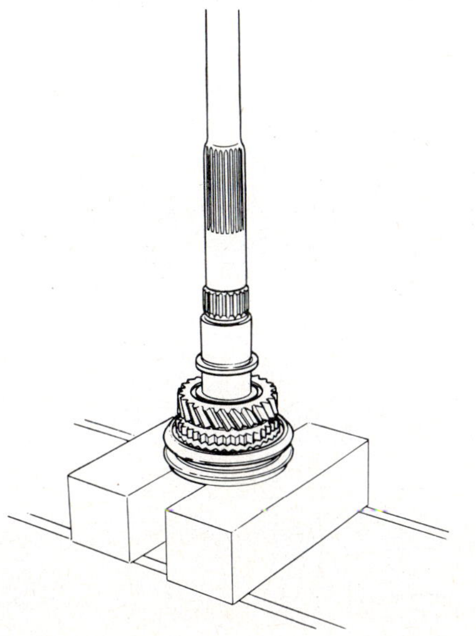

Fig. 27 Installing 3rd gear. 4 speed 77.5mm transmission

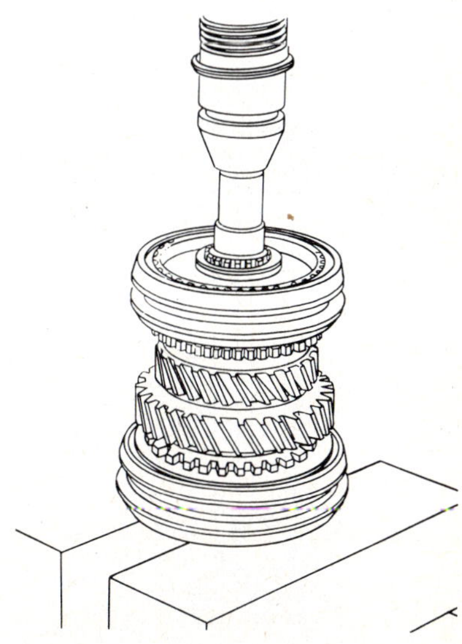

Fig. 28 Installing second gear. 4 speed 77.5mm transmission

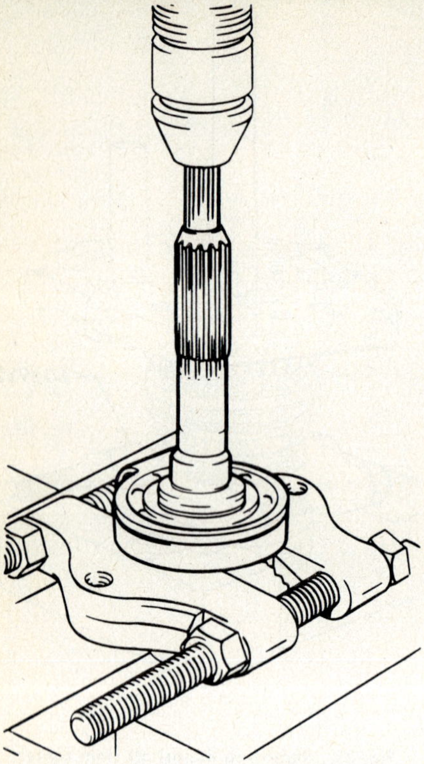

Fig. 29 Removing drive gear bearing. 4 speed 77.5mm transmission

1. Mark relation of hub and sleeve for reassembly reference.
2. Push hub from sliding sleeve, allowing keys to fall free and simplifying spring removal.
3. Place keys in position and hold in place while sliding sleeve onto hub, aligning reference marks.
4. Place the two springs in position on either side of hub so that all three keys are engaged by both springs.

Five Gear Shaft Bearing

1. Remove retaining ring from driveshaft, then remove bearing using an arbor

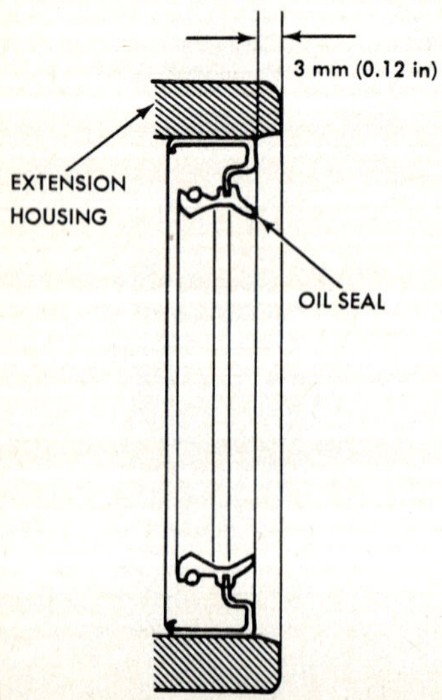

Fig. 32 Installing rear extension seal. 4 speed 77.5mm transmission

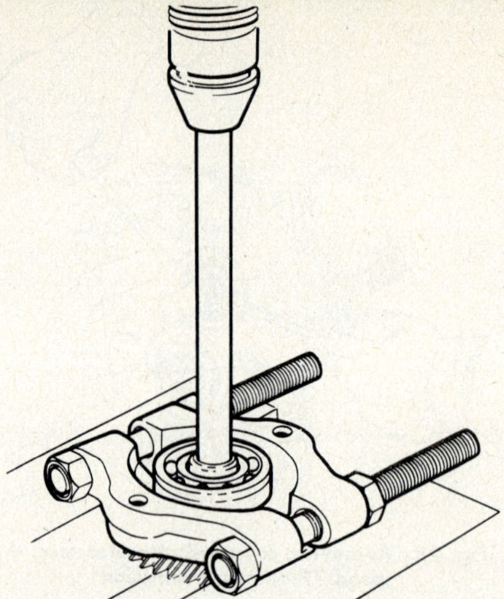

Fig. 30 Removing countergear front bearing. 4 speed 77.5mm transmission

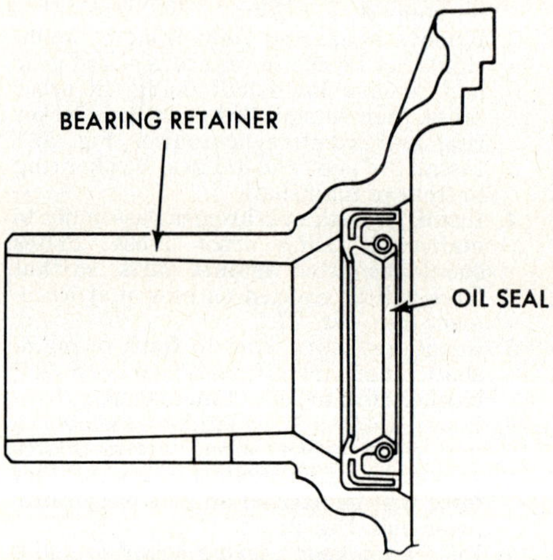

Fig. 33 Installing front bearing retaining seal. 4 speed 77.5mm transmission

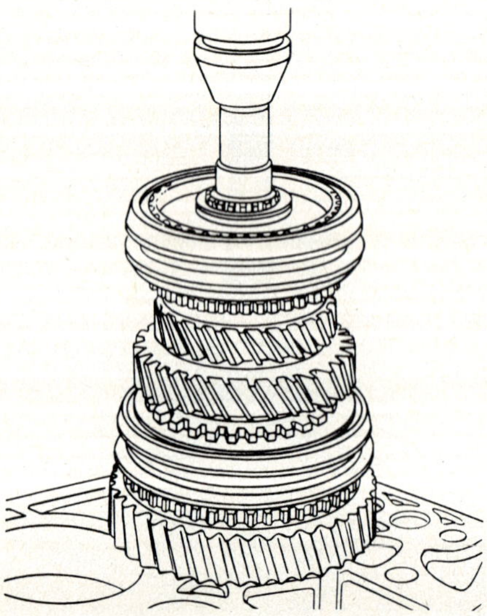

Fig. 34 Installing mainshaft to center support. 4 speed 77.5mm transmission

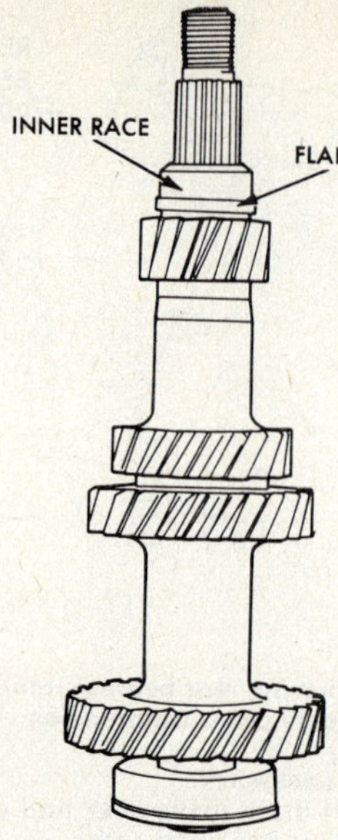

Fig. 31 Removing countergear inner race. 4 speed 77.5mm transmission

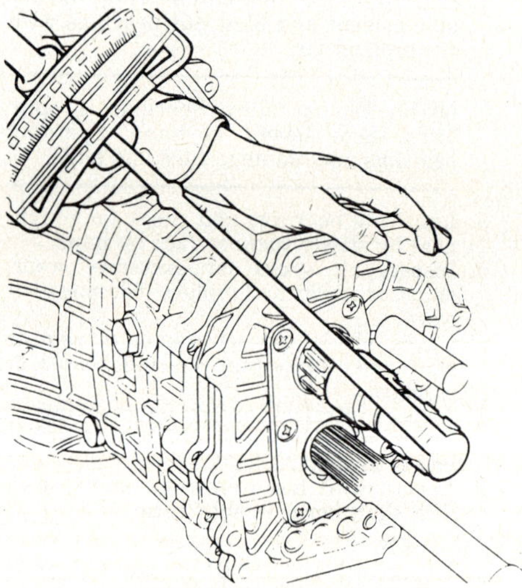

Fig. 35 Torquing countergear nut. 4 speed 77.5mm transmission

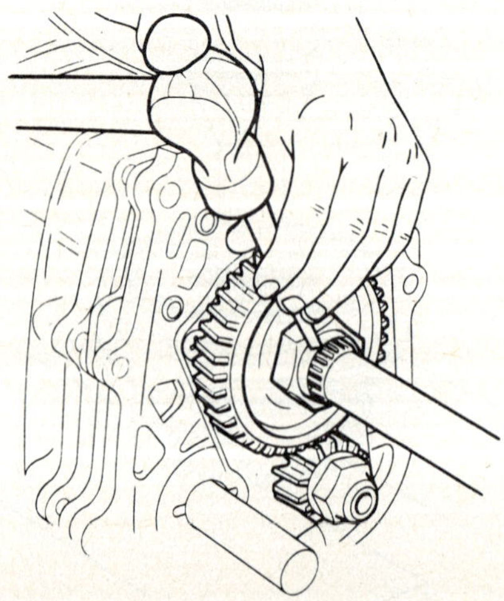

Fig. 36 Staking nut on mainshaft. 4 speed 77.5mm transmission

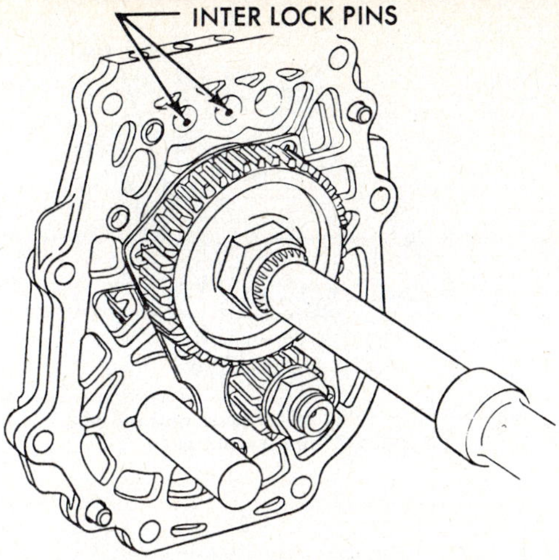

Fig. 37 Installing interlock pins. 4 speed 77.5mm transmission

press, Fig. 29.
2. Check to be sure spacer is fitted to shaft.
3. Using suitable tool and arbor press, install bearing to shaft so that snap ring groove is turned away from gear.
4. Install retaining ring to shaft.

Countergear Shaft

1. Remove retaining ring from countershaft.
2. Remove bearing using press, Fig. 30.
3. Using suitable tool and press, install bearing to shaft so that snap ring groove is turned away from gear.
4. Install retaining ring to shaft.

Rear Bearing Inner Race

1. Remove inner race using press, Fig. 31.
2. Using suitable tool and press, install bearing inner race to shaft so that flanged face of race is turned toward gears.

Extension Oil Seal

1. Remove oil seal by prying outward with suitable tool.
2. Apply thin coat of suitable sealer to outer circumference of oil seal and install into housing using tool J-33035 and hammer so that its face is recessed from rear end of housing by .12 in., Fig. 32.

Front Bearing Retainer Oil Seal

1. Remove oil seal by prying outward.
2. Drive new oil seal into retainer using tool J-26540, Fig. 33.

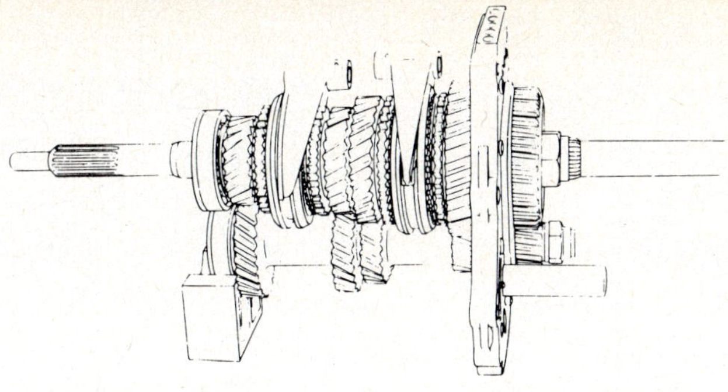

Fig. 38 Installing shift forks. 4 speed 77.5mm transmission

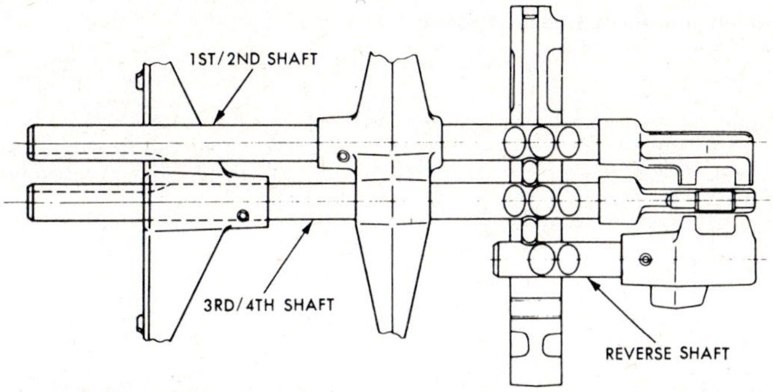

Fig. 39 Installing shifter shafts. 4 speed 77.5mm transmission

Transmission Assemble

1. Being sure snap ring is fitted to mainshaft rear bearing, install bearing into center support with snap ring side turned rearward.
2. Using an arbor press, install mainshaft into bearing which has been fitted into center support. When installing mainshaft, care should be taken so as not to damage reverse block dowels fitted to center support, Fig. 34.
3. Install 4th blocker ring, needle roller bearing, and drive gear shaft.
4. Install countershaft to center support, then install rear bearing outer race to countershaft from rear side of center support.
5. Install bearing retainer to center support. Clean threaded hole in center support, apply suitable sealer to screw threads, install screws, and torque to 15 ft. lbs.
6. Engage synchronizers in lock and place center support in transmission case temporarily.
7. Install reverse gear, washer, and spring washer to rear end of countershaft and torque nut to 80 ft. lbs., Fig. 35. Install reverse gear so that teeth with rounded edges are turned rearward.
8. Install reverse gear to mainshaft with rounded teeth turned rearward and torque nut to 95 ft. lbs., then stake nut to mainshaft, Fig. 36.
9. Remove case from center support.
10. Lubricate and install two interlock pins to center support, Fig. 37.
11. Noting direction of installation, install shift fork to synchronizers, Fig. 38.
12. Install 3rd/4th shifter shaft to center support and shift fork, then bring shaft into neutral position. Install 1st/2nd shifter shaft and reverse shifter shaft to center support, Fig. 39.
13. Supporting end of shaft with piece of wood, drive spring pin to 1st/2nd and 3rd/4th shift forks.
14. Install three detent balls and springs into holes in center support, then install gasket and plate and torque bolts to 15 ft. lbs.

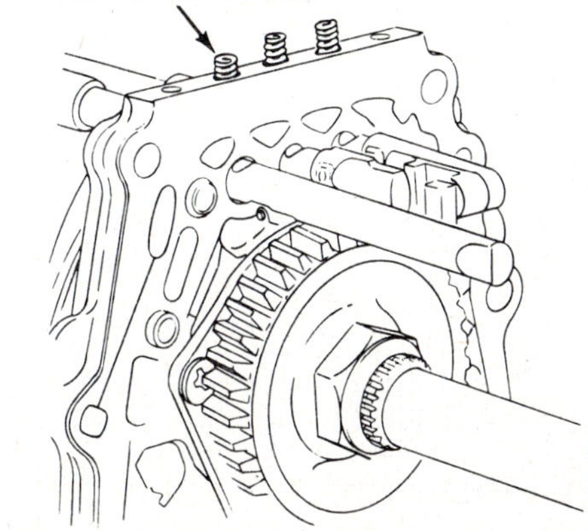

Fig. 40 Installing detent balls. 4 speed 77.5mm transmission

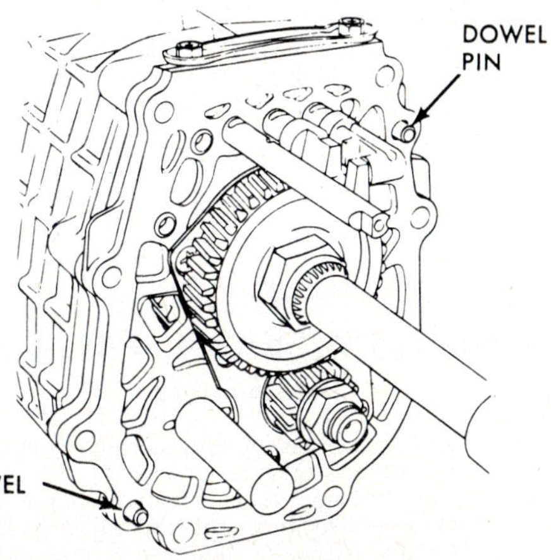

Fig. 41 Installing center support to case. 4 speed 77.5mm transmission

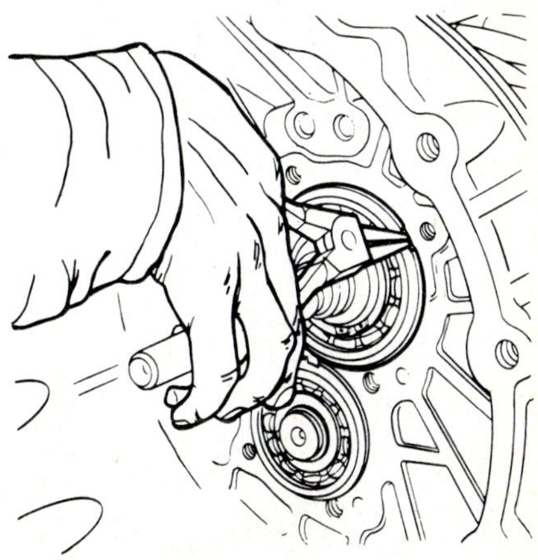

Fig. 42 Installing retaining ring. 4 speed 77.5mm transmission

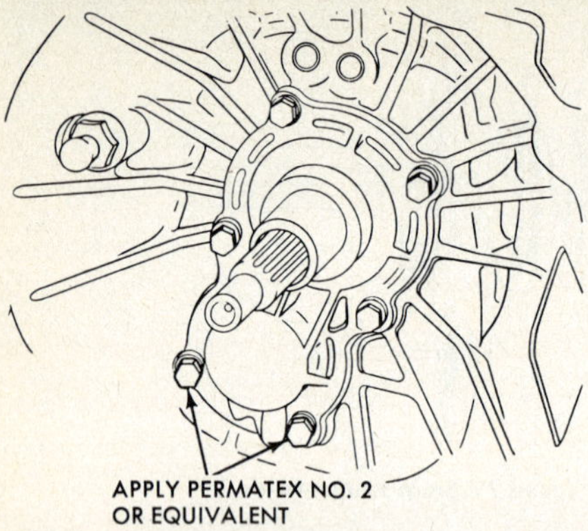

Fig. 43 Installing reverse shift block. 4 speed 77.5mm transmission

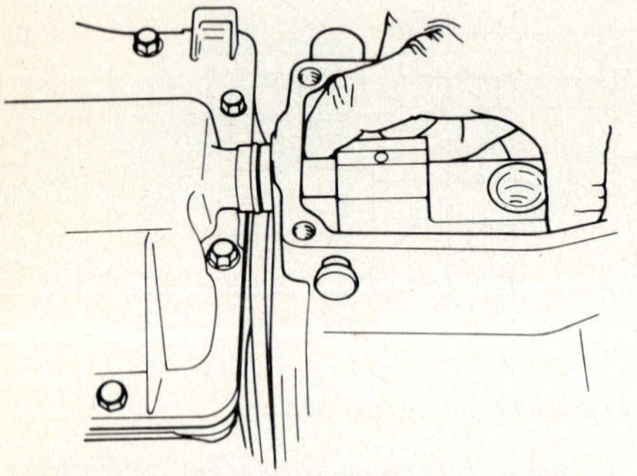

Fig. 47 Removing extension housing from transmission case. 5 speed 77mm transmission

The reverse spring is shorter than the others, Fig. 40.

15. Install gasket to transmission case and install center support by aligning it with dowels on case, Fig. 41. Avoid hitting support or case during installation.
16. Install retaining rings to drive gear shaft and countershaft bearing outer races, Fig. 42.
17. Assemble reverse shift block, reverse block, and reverse shift fork using pins and snap rings.
18. Install shift block assembly to reverse shift fork, then insert shift fork into groove in reverse idler gear. Install gear to idler shaft and reverse shift block to shifter shaft, Fig. 43.
19. Torque reverse block retaining bolts to specifications.

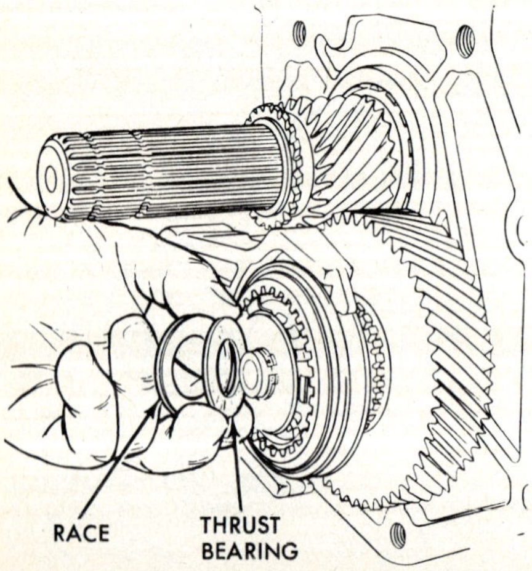

Fig. 50 Removing thrust bearing and race. 5 speed 77mm transmission

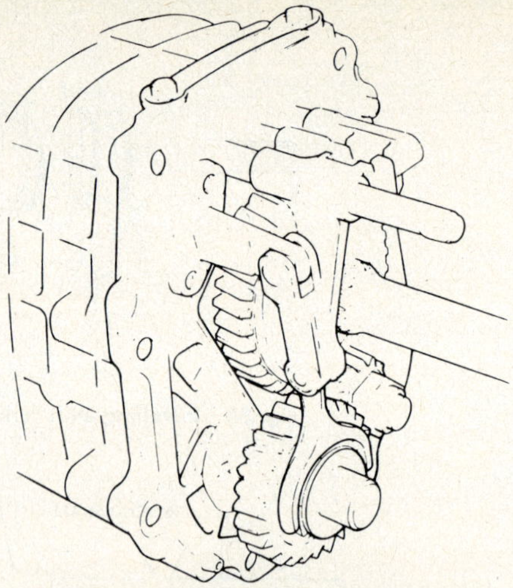

Fig. 44 Installing front bearing retainer. 4 speed 77.5mm transmission

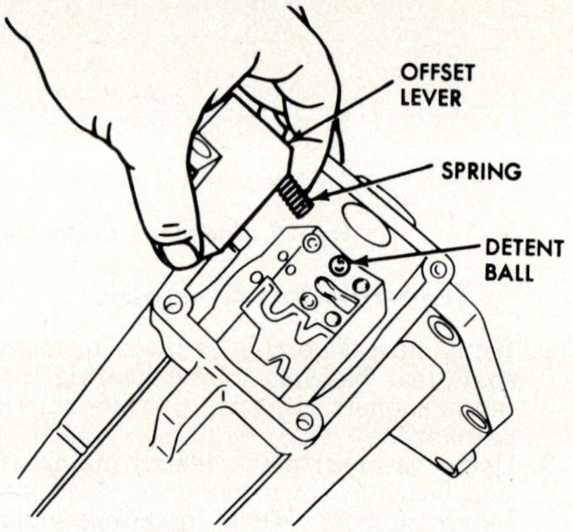

Fig. 48 Removing detent ball and spring. 5 speed 77mm transmission

20. Secure reverse shift block to shifter shaft by driving spring pin with suitable tool.
21. Insert clip end into hole in mainshaft and install speedometer driven gear.
22. Install gasket to center support, then install extension housing to support by aligning it with dowels and torque bolts to 30 ft. lbs.
23. Install back-up light switch.
24. Install gasket to upper face of extension housing and install shifter cover.
25. Install speedometer driven gear and ball stud.
26. Install gasket to transmission case, then install front bearing retainer, Fig. 44. Apply suitable sealer to bolt threads on lower side before installation and torque bolts to 15 ft. lbs.
27. Install bearing, boot, and clutch fork.
28. Install gearshift lever to shifter cover temporarily and check for smooth and correct operation.

5 SPEED (77mm) TRANSMISSION, FIG. 45

Disassembly

1. Drain transmission and thoroughly clean exterior of transmission case.
2. Using suitable tools, remove roll pin attaching offset lever to shift rail, Fig. 46.
3. Remove extension housing to transmission case bolts and remove housing and offset lever as an assembly, Fig. 47.

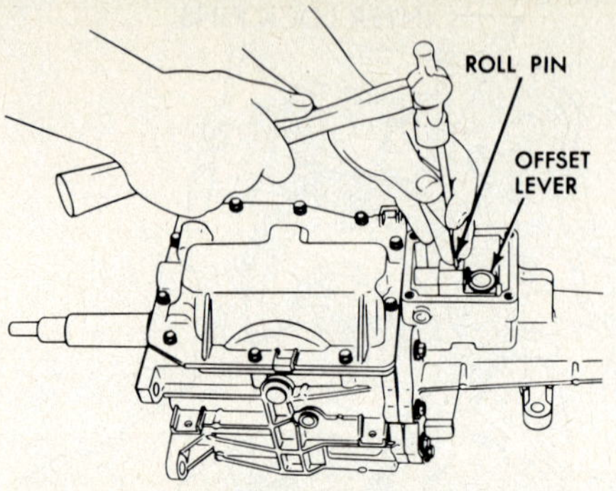

Fig. 46 Removing offset lever and roll pin. 5 speed 77mm transmission

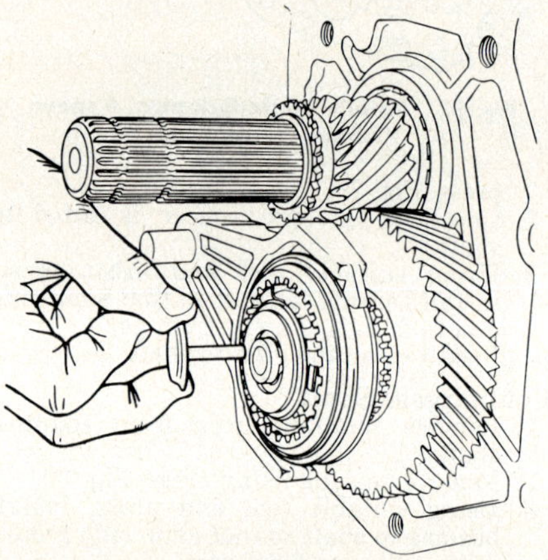

Fig. 49 Removing plastic funnel. 5 speed 77mm transmission

NOTE: Do not attempt to remove offset lever while extension housing is still bolted in place. The lever has a positioning lug engaged in the housing detent plate which prevents moving the lever far enough for removal.

4. Remove detent ball and spring from offset lever and remove roll pin from extension housing or offset lever, Fig. 48.
5. Remove plastic funnel, thrust bearing race, and thrust bearing from rear of countershaft, Figs. 49 & 50. These parts may be found inside extension housing.
6. Remove bolts attaching transmission cov-

Fig. 51 Removing transmission cover. 5 speed 77mm transmission

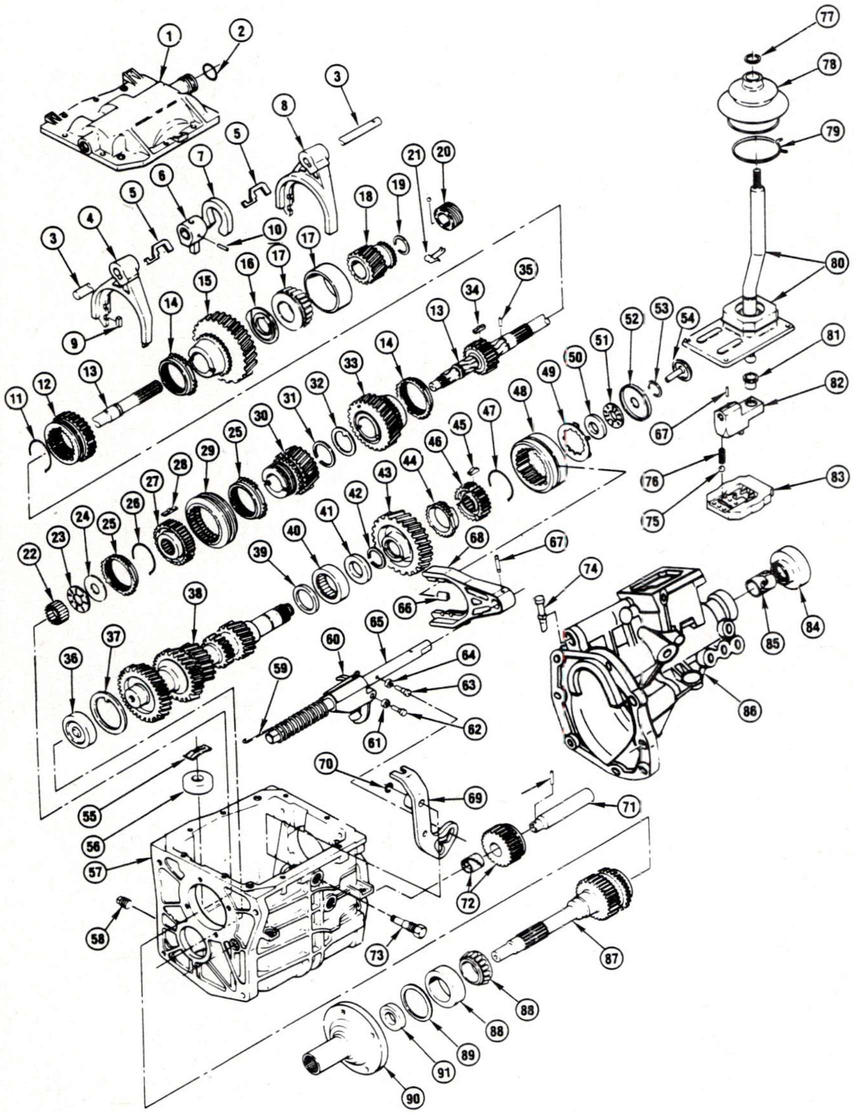

1. COVER, Trans
2. SEAL, "O" Ring, Cvr to Ext.
3. SHAFT, Shift
4. FORK, 3rd & 4th Shift
5. PLATE, Shift Fork
6. ARM, Control Selector
7. PLATE, Gear Sel Intlk
8. FORK, 1st & 2nd Shift
9. INSERT, Shift Fork
10. PIN, Roll
11. SPRING, Syn
12. GEAR, Rev Sldg
13. SHAFT, Output, W/1 & 2 Syn
14. RING, 1 & 2 Syn Blkg
15. GEAR, 1st Speed
16. WASHER, 1st Spd Gr Thrust
17. BEARING, Rear
18. GEAR, 5th Spd Drvn
19. RING, Snap
20. GEAR, Speedo Dr
21. CLIP, Speedo Dr Gr
22. BEARING, Main Shf Rir
23. BEARING, Main Dr Gr Thr Ndl
24. RACE, Main Dr Gr Ghr Brg
25. RING, 3 & 4 Syn
26. SPRING, 3 & 4 Syn
27. HUB, 3 & 4 Syn
28. KEY, 3 & 4 Syn
29. SLEEVE, 3 & 4 Syn
30. GEAR, 3rd Speed
31. RING, Snap
32. WASHER, 2nd Spd Gr Thr
33. GEAR, 2nd Speed
34. KEY, 1 & 2 Syn
35. PIN, 1st Spd Gr Thr Wa Ret
36. BEARING, Cntr Gr Frt
37. WASHER, Cntr Gr Frt Thr
38. GEAR, Counter
39. SPACER, Counter Gr Brg Frt
40. BEARING, Cntr Gr Rr
41. SPACER, Counter Gr Brg Rr
42. RING, Snap
43. GEAR, 5th Spd Drive
44. RING, 5th Syn
45. KEY, 5th Syn
46. HUB, 5th Syn
47. SPRING, 5th Syn
48. SLEEVE, 5th Syn
49. RETAINER, 5th Syn Key
50. RACE, 5th Syn Thr Brg Frt
51. BEARING, 5th Syn Ndl Thr
52. RACE, 5th Syn Thr Brg Rr
53. RING, Snap
54. FUNNEL, Trans Oiling
55. NUT, Magnet
56. MAGNET
57. CASE, Trans
58. PLUG, Fill & Drain
59. SPRING, Rev Lock
60. FORK, Rev Shift
61. ROLLER, Fork
62. PIN, Rev Fork
63. PIN, Shift Rail
64. ROLLER, Rail pin
65. RAIL, 5th & Rev Shft
66. INSERT, Shift Fork
67. PIN, Roll
68. FORK, 5th Shift
69. LEVER, 5th & Rev Relay
70. RING, Rev. Relay Lever Ret
71. SHAFT, Rev Idler Gr
72. GEAR, Rev Idler (Incl Bshg)
73. PIN, 5th Spd Shft Lvr Piv
74. VENTILATOR, Ext
75. BALL, Steel
76. SPRING, Detent
77. RETAINER, Cont Lvr Boot
78. BOOT, Cont Lvr
79. RETAINER, Cont Lvr Boot Lwr
80. CONTROL, Trans Lvr & Hsg
81. SLEEVE, Shft Lvr Dmpr
82. LEVER, Offset Shift
83. PLATE, Detent & Guide
84. SEAL, Ext Rr Oil
85. BUSHING, Extension Housing
86. HOUSING, Extension
87. GEAR, Main Drive
88. BEARING, Front
89. SHIM, Brg Adj
90. RETAINER, Drive Gr Brg
91. SEAL, Drive Gr Brg Oil

Fig. 45 Exploded view of 5 speed 77mm manual transmission

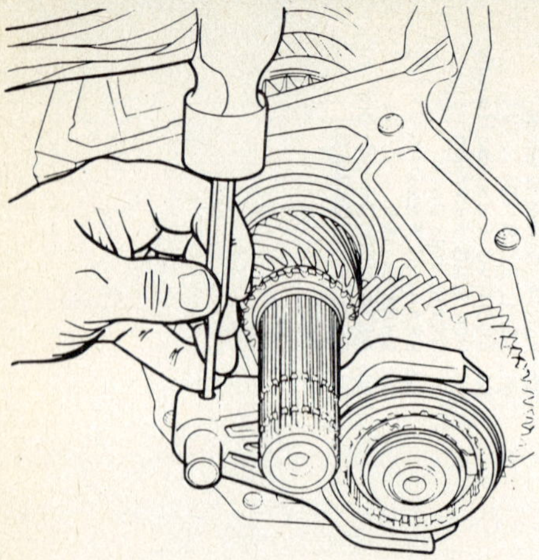

Fig. 52 Removing fifth gear shift fork roll pin. 5 speed 77mm transmission

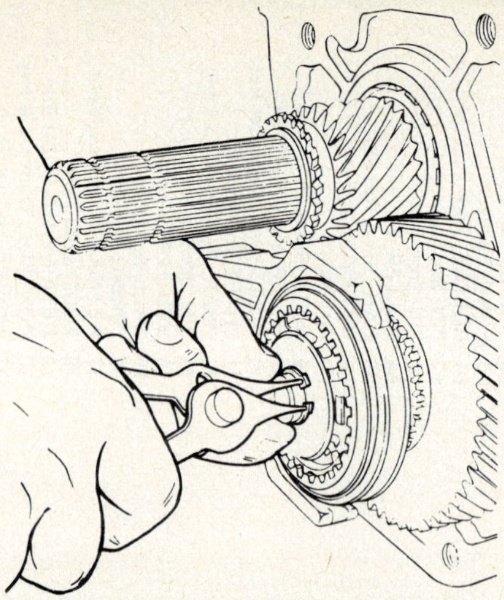

Fig. 53 Removing fifth gear synchronizing snap ring. 5 speed 77mm transmission

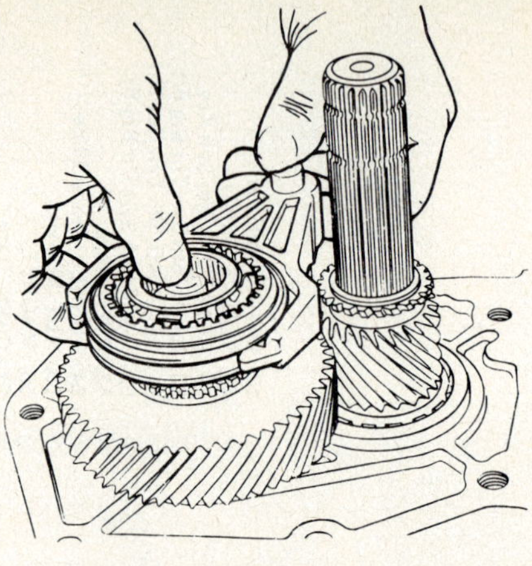

Fig. 54 Removing fifth gear synchronizer. 5 speed 77mm transmission

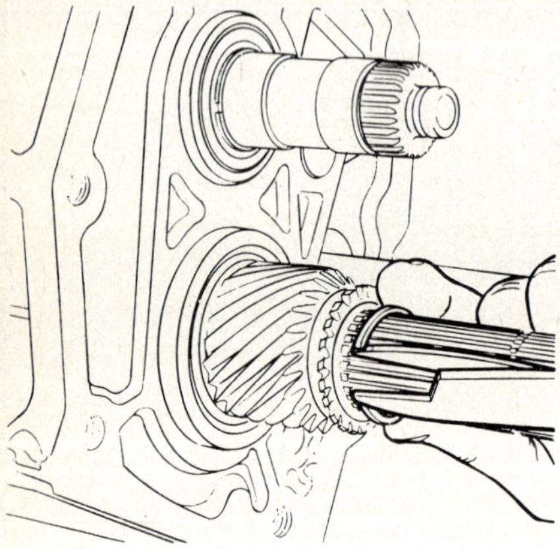

Fig. 55 Removing fifth speed gear snap ring. 5 speed 77mm transmission

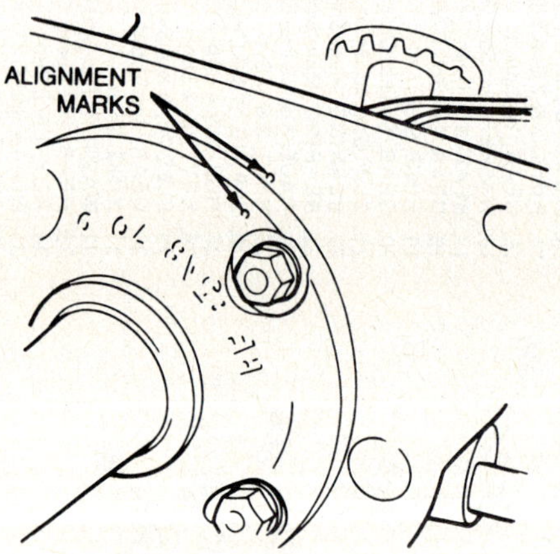

Fig. 56 Marking bearing cap to case alignment. 5 speed 77mm transmission

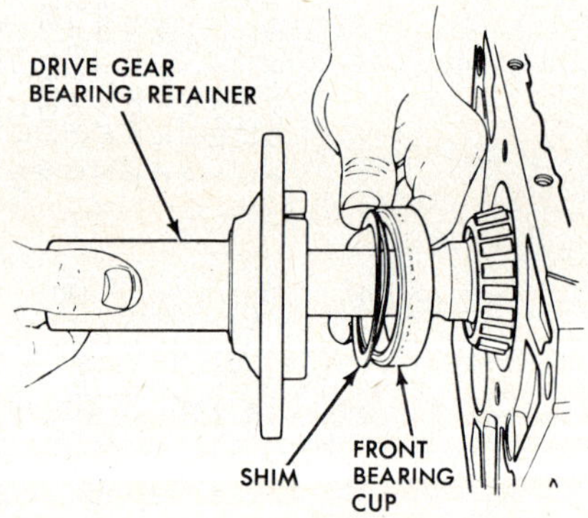

Fig. 57 Removing bearing cap and shims. 5 speed 77mm transmission

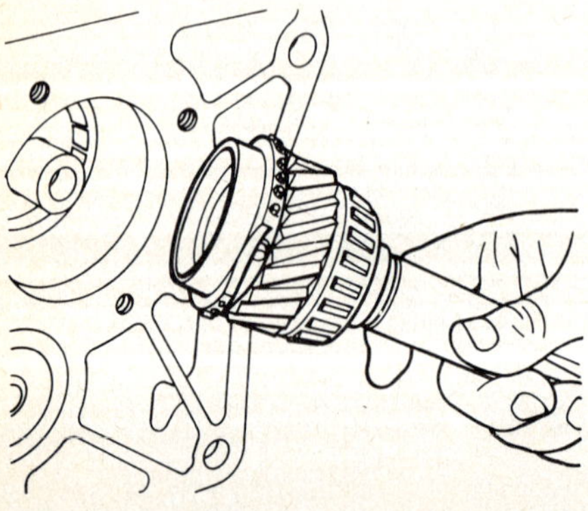

Fig. 58 Removing drive gear. 5 speed 77mm transmission

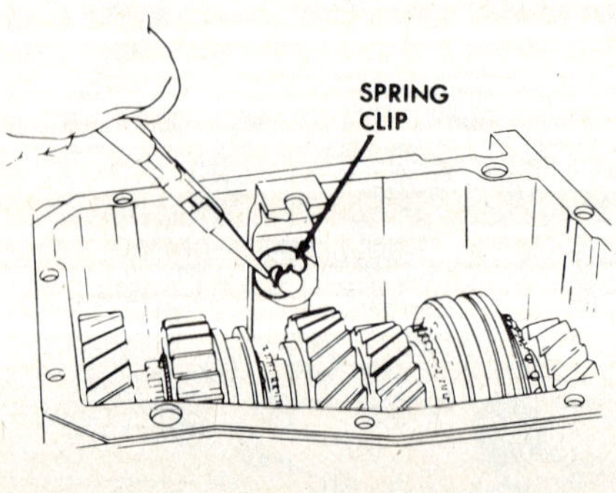

Fig. 59 Removing reverse lever retaining clip. 5 speed 77mm transmission

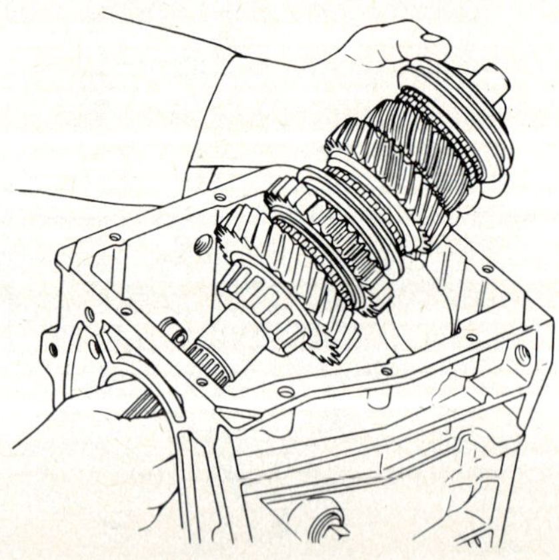

Fig. 60 Removing mainshaft. 5 speed 77mm transmission

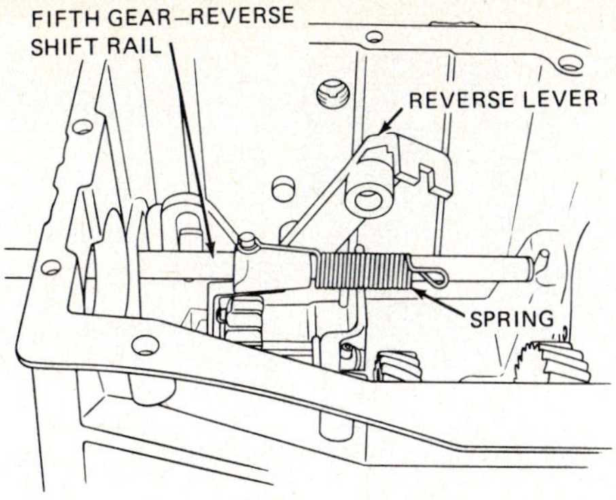

Fig. 61 Removing fifth gear reverse shift rail. 5 speed 77mm transmission

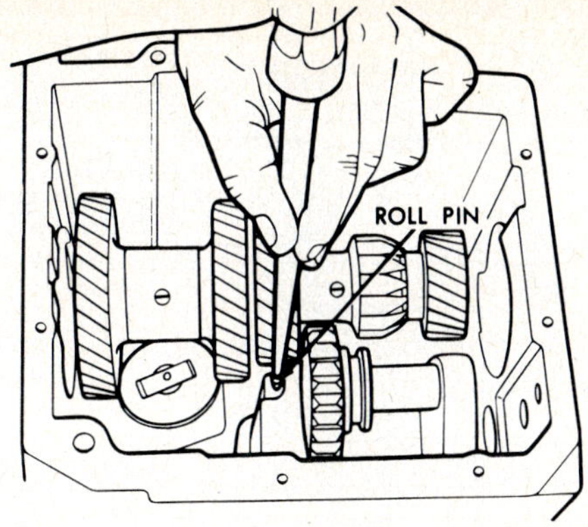

Fig. 62 Removing reverse idler gear shaft. 5 speed 77mm transmission

er and shift fork assembly and remove cover, Fig. 51. Two of the transmission cover attaching bolts are alignment type dowel bolts. Note location of these bolts for assembly reference.

7. Using suitable tools, drive roll pin from fifth gearshift fork while supporting end of shaft with a block of wood, Fig. 52.
8. Remove fifth synchronizer gear snap ring, shift fork, fifth gear synchronizer sleeve, blocking ring, and fifth speed driven gear from rear of countershaft, Figs. 53 & 54.
9. Remove snap ring from fifth speed driven gear, Fig. 55.
10. Mark bearing cap and case for assembly reference, Fig. 56.
11. Remove front bearing cap bolts and cap, then remove front bearing race and end play shims from front bearing cap, Fig. 57.
12. Turn drive gear until flat surface faces countershaft and remove drive gear from transmission case, Fig. 58.
13. Remove reverse lever "C" clip and pivot bolt, Fig. 59.
14. Remove mainshaft rear bearing race, tilt mainshaft assembly upward, and remove assembly from transmission case, Fig. 60.
15. Unhook overcenter link spring from front of transmission case.
16. Turn fifth gear-reverse shift rail to disengage rail from reverse lever assembly, then remove shift rail from rear of transmission case, Fig. 61.
17. Remove reverse lever and fork assembly from transmission case.
18. Using suitable tool, drive roll pin from forward end of reverse idler shaft, then remove rubber "O" ring and gear from transmission case, Fig. 62.
19. Remove rear countershaft snap ring and spacer, Fig. 63.
20. Insert suitable tool through drive gear opening in front of transmission case and, using arbor press, carefully press countershaft rearward to remove rear countershaft bearing, Fig. 64.
21. Move countershaft assembly rearward, tilt countershaft upward, and remove from case. Remove countershaft front thrust washer and rear bearing spacer.
22. Remove countershaft front bearing from transmission case, using arbor press.

Sub-assembly Service

Mainshaft, Disassemble

1. Remove thrust bearing washer from front end of mainshaft, then scribe alignment mark on third-fourth synchronizer hub and sleeve for reference during assembly.
2. Remove third-fourth synchronizer blocking ring, sleeve, hub and third gear as an assembly from mainshaft, Fig. 65.
3. Remove snap ring, tabbed thrust washer and second gear from mainshaft, then remove fifth gear using Tool J-22912-01 and arbor press, and slide rear bearing off mainshaft.
4. Remove first gear thrust washer, roll pin, first gear and synchronizer ring from mainshaft, Fig. 66.
5. Scribe mark on first-second synchronizer hub and sleeve for reference during assembly, then remove synchronizer spring and keys from first-reverse sliding gear and remove gear from mainshaft hub.

NOTE: The first-second-reverse hub and mainshaft are assembled and machined as a matched set and no attempt should be made to separate them.

Drive Gear, Disassemble

1. Remove bearing race, thrust bearing and roller bearings from cavity of drive gear, Fig. 67.
2. Remove bearing from drive gear using Tool J-22912-01 and arbor press, Fig. 68.
3. Wash parts in suitable cleaning solvent, then inspect gear teeth and drive pilot shaft for wear or damage and replace as necessary.

Cleaning & Inspection

1. Thoroughly wash transmission using a suitable cleaning solvent and inspect case for cracks.
2. Clean magnetic disc at bottom of transmission case and check front and rear faces of case for burrs. Dress any burrs off with a fine mill file.
3. Inspect all drive gear bearing rollers, reverse idler shaft and all spacers for wear or damage and replace as necessary.
4. Inspect all gears for wear or damage and replace as necessary.
5. Check that clutch sleeves slide freely on their hubs.
6. Wash front and rear bearings thoroughly with a suitable cleaning solvent and dry them with compressed air.

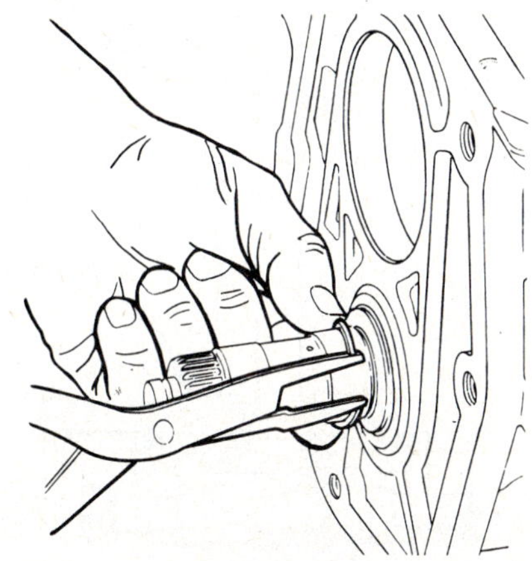

Fig. 63 Remove countershaft snap ring and spacer. 5 speed 77mm transmission

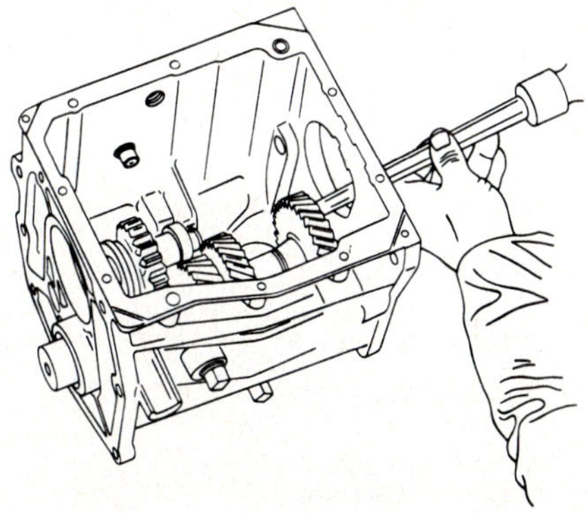

Fig. 64 Removing rear countershaft bearing. 5 speed 77mm transmission

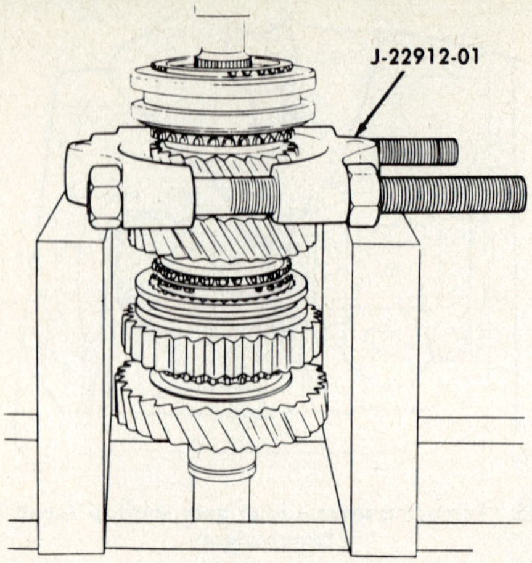

Fig. 65 Removing third-fourth synchronizer. 5 speed 77mm transmission

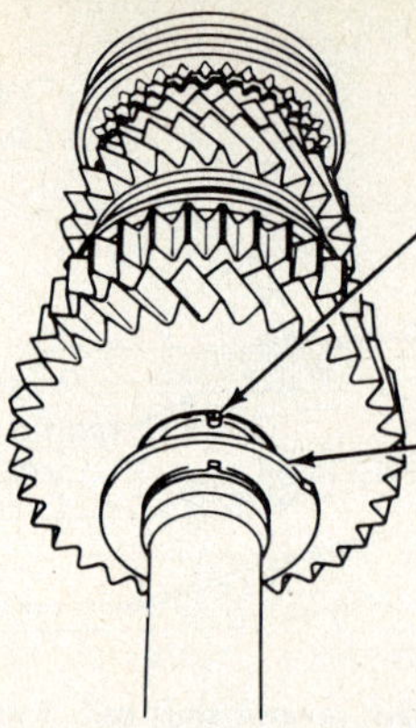

Fig. 66 Removing and installing first gear thrust washer and roll pin. 5 speed 77mm transmission

NOTE: When drying bearings, slowly turn them by hand. To avoid damaging race and balls, do not allow bearings to spin.

Synchronizer Keys & Springs, Replace

NOTE: The synchronizer hubs and sliding sleeves are a selected assembly and must be kept together as originally assembled, however, the keys and springs may be replaced if worn or broken.

1. Mark relation of hub and sleeve for reference during assembly.
2. Push sliding sleeve from hub and remove keys and springs.
3. Place a blocker ring on side of hub and sleeve and install keys and a spring, then place blocker ring on opposite side of hub and sleeve and install remaining spring.

Extension Housing Oil Seal & Bushing, Replace

Refer to four speed 77mm transmission section for replacement procedure.

Drive Gear Bearing Retainer Oil Seal, Replace

Refer to four speed 77mm transmission section for replacement procedure.

Transmission Cover, Disassemble

Refer to four speed 77mm transmission section for disassembly procedure, and refer to Fig. 69.

Transmission Cover, Assemble

Refer to four speed 77mm transmission section for assembly procedure, and refer to Fig. 69.

Drive Gear, Assemble

1. Using Tool J-22912-01 and an arbor press, install bearing on drive gear.
2. Coat roller bearings and drive gear bearing bore with grease and install roller bearings into bore of drive gear.
3. Install thrust bearing and race in drive gear.

Mainshaft, Assemble

1. Coat mainshaft and gear bores with transmission lubricant, then install first-second synchronizer sleeve on hub, aligning marks made during disassembly.
2. Install first-second synchronizer keys and springs, Fig. 70.
3. Install blocker ring, second gear, tabbed thrust washer and second gear retaining snap ring on mainshaft, ensuring that washer tab is properly seated in mainshaft notch, Fig. 71.
4. Install blocker ring and first gear on mainshaft, then install first gear roll pin and thrust washer, Fig. 66.
5. Install rear bearing on mainshaft, then install fifth speed gear on mainshaft using Tool J-22912-01 and arbor press and install snap ring.
6. Install third gear, third-fourth synchronizer assembly and thrust bearing on mainshaft.

NOTE: Synchronizer hub offset must face forward.

Assemble Transmission

1. Coat countershaft front bearing bore with Loctite 601, or equivalent, then install front countershaft bearing flush with facing of case using an arbor press.
2. Coat countershaft tabbed thrust washer with grease and install washer so that tab engages depression in case.
3. Tip transmission on end and install countershaft in front bearing bore.
4. Install countershaft rear bearing spacer, then coat countershaft rear bearing with grease and install bearing, Fig. 72.

NOTE: When correctly installed, the bearing will extend .125 inch beyond the case surface.

5. Position reverse idler gear in case with shift lever groove facing rear of case, then install reverse idler shaft from rear of case and install roll pin in idler shaft.

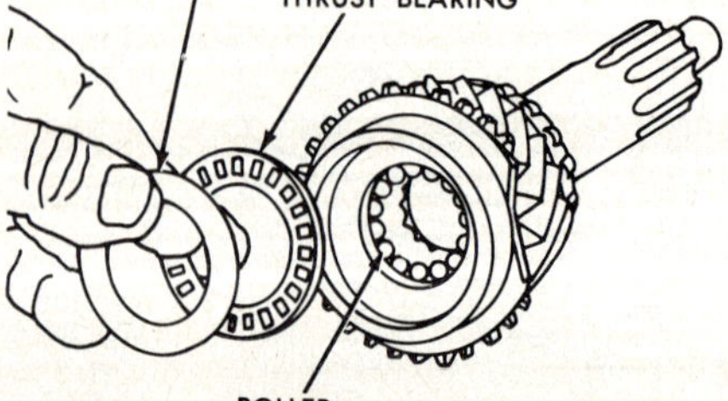

Fig. 67 Removing drive gear thrust bearing and race. 5 speed 77mm transmission

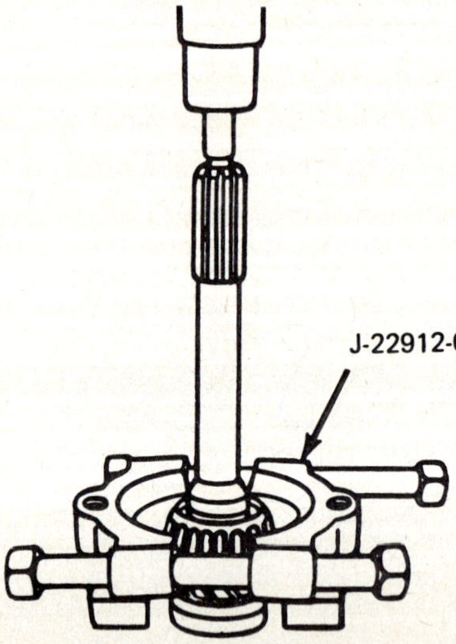

Fig. 68 Removing drive gear bearing. 5 speed 77mm transmission

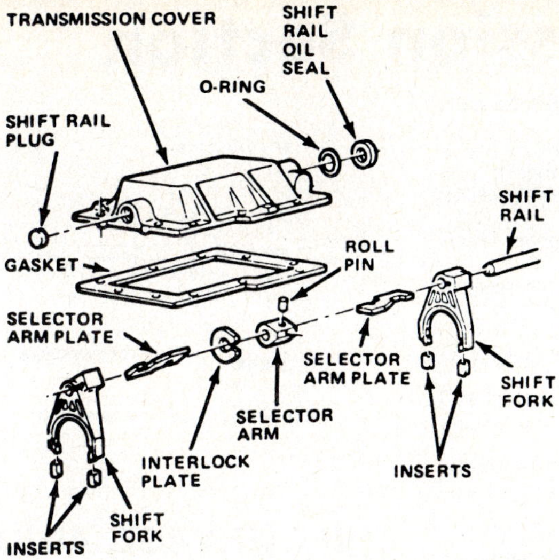

Fig. 69 Transmission cover. 5 speed 77mm transmission

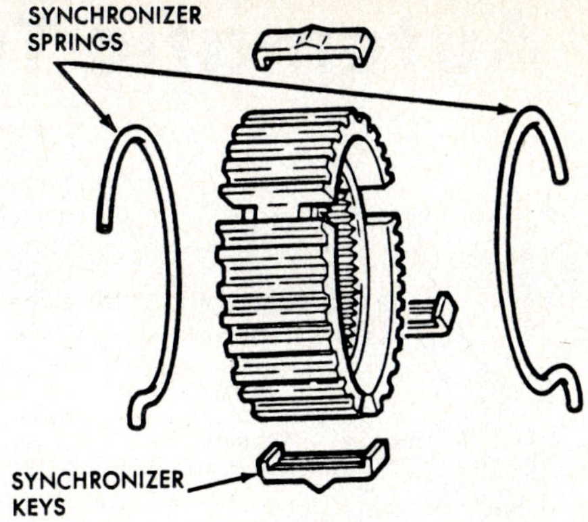

Fig. 70 Installing synchronizer keys and springs. 5 speed 77mm transmission

6. Install mainshaft and rear mainshaft bearing race in transmission case, then install drive gear in case and engage in third-fourth synchronizer sleeve and blocker ring.
7. Install front bearing race and temporarily install front bearing cap.

NOTE: Do not install shims in front bearing cap at this time.

8. Install fifth speed-reverse lever, pivot bolt and retaining clip, ensuring that reverse lever fork is engaged in reverse idler gear.

NOTE: Coat bolt threads with non-hardening sealer before installing.

9. Install countershaft rear bearing spacer and snap ring, then install fifth speed gear on countershaft.
10. Insert fifth speed-reverse rail in rear of case and install in to reverse fifth speed lever, then connect spring to front of case.

NOTE: To simplify engagement with lever, rotate rail during installation.

11. Position fifth gear shift fork on fifth gear synchronizer assembly, then install synchronizer on countershaft and shift fork on shift rail, ensuring that roll pin hole in shift fork and shift rail are aligned.
12. Support fifth gear shift rail and fork on a block of wood and install roll pin.
13. Install thrust race against fifth speed synchronizer hub and install snap ring, then install thrust bearing against race on countershaft and coat bearing and race with petroleum jelly.
14. Install lipped thrust race over needle-type thrust bearing, then install plastic funnel into hole in end of countershaft gear.
15. Temporarily install extension housing and attaching bolts, then turn transmission case on end and mount a dial indicator on extension housing, Fig. 73. Check end play of mainshaft and record reading. Mainshaft bearings must have a preload of .001–.005 inch. To set preload, use a shim pack measuring .001–.005 inch greater than reading of dial indicator.
16. Remove front bearing cap and race, install shims and reinstall bearing race.
17. Apply a 1/8 inch bead of RTV sealant #732, or equivalent, on case mating surface of front bearing cap, then install bearing cap, referring to alignment marks made during disassembly, and torque bolts to 15 ft. lbs.
18. Remove extension housing, then move shift forks on transmission cover and synchronizer sleeves inside transmission to the neutral position.
19. Apply a 1/8 inch bead of RTV sealant #732, or equivalent, to cover mating surface of transmission, then lower cover onto case while aligning shift forks and synchronizer sleeves. Center cover and install the two dowel bolts, then install remaining bolts and torque to 10 ft. lbs.

NOTE: The offset lever to shift rail roll pin hole must be in a vertical position after cover installation.

20. Apply a 1/8 inch bead of RTV sealant #732, or equivalent on extension housing to transmission case mating surface, then install extension housing over mainshaft and shift rail until shift rail just enters shift cover opening.
21. Install detent spring into offset lever and place steel ball into neutral guide plate detent.
22. Position offset lever on steel ball and apply pressure to offset lever and to extension housing against transmission case.
23. Install extension housing and bolts and torque bolts to 25 ft. lbs., then install and align roll pin in offset lever and shift rail.
24. Fill transmission to proper level with lubricant.

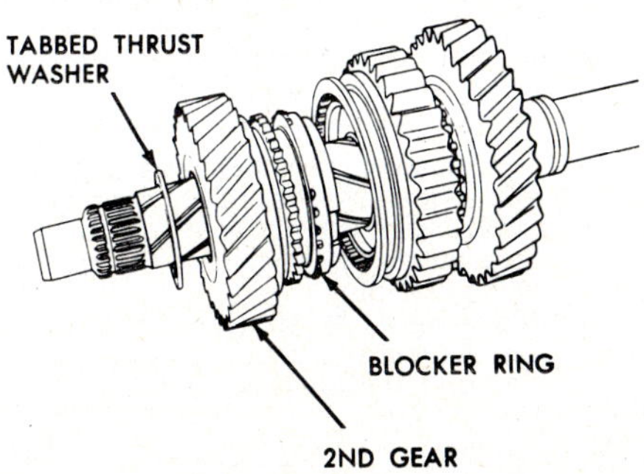

Fig. 71 Installing second gear and thrust washer. 5 speed 77mm transmission

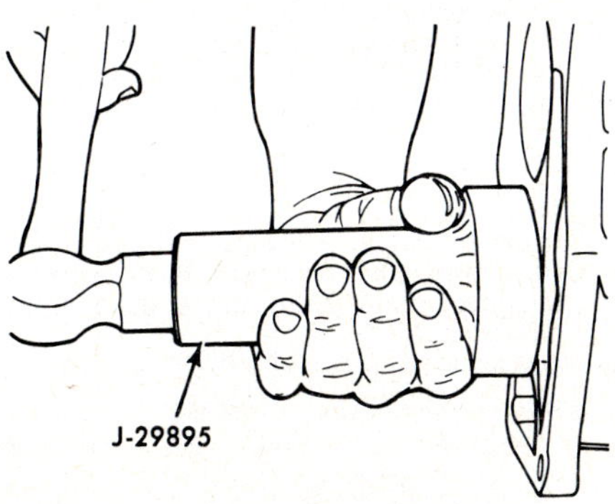

Fig. 72 Installing rear countershaft bearing. 5 speed 77mm transmission

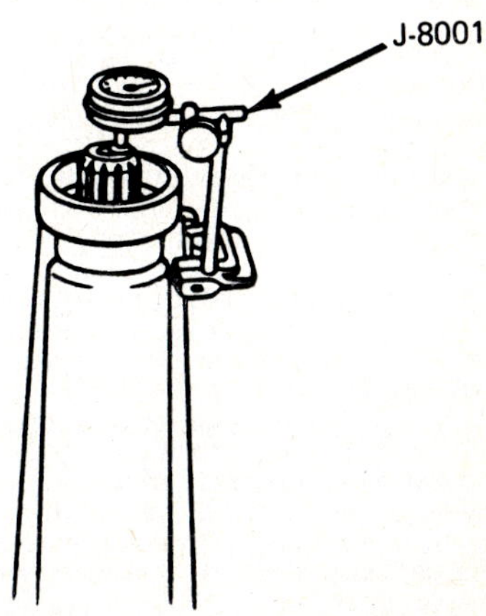

Fig. 73 Measuring mainshaft end play. 5 speed 77mm transmission

Automatic Transmission Section

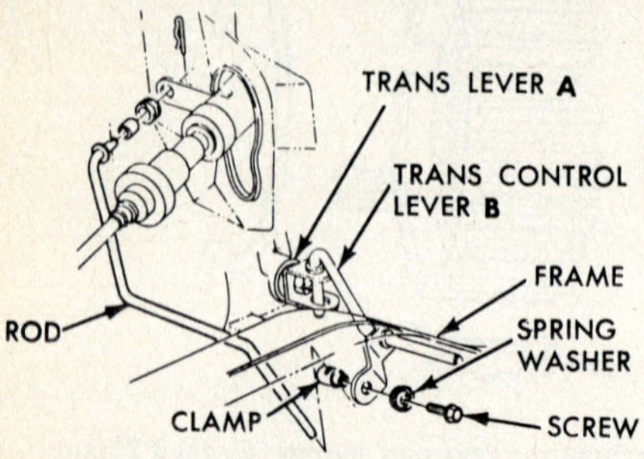

Fig. 1 Shift linkage adjustment

TURBO HYDRA-MATIC 200C

Refer to the "Turbo Hydra-Matic 200, 200C" section in the "Automatic Transmission" chapter of this manual for description and trouble shooting of this transmission.

TURBO HYDRA-MATIC 700-R4

Refer to the "Turbo Hydra-Matic 700 R4" section in the "Automatic Transmission" chapter of this manual for description and troubleshooting of this transmission.

TRANSMISSION, REPLACE

4 × 2

1. Open hood and place fender covers on both fenders.
2. Remove air cleaner assembly.
3. Disconnect T.V. cable at its upper end.
4. On vehicles equipped with 4-119 engine, remove starter motor upper retaining nut.
5. Raise and support vehicle.
6. Remove propeller shaft.
7. Disconnect speedometer cable at the transmission.
8. Disconnect shift linkage at transmission.
9. Disconnect all electrical leads at the transmission and any clips that retain the leads to the transmission case.
10. If equipped, remove transmission support brace attaching bolts at converter cover.
11. Remove converter cover and mark flywheel and torque converter to maintain original balance.
12. Remove exhaust crossover pipe and catalytic converter attaching bolts and remove as an assembly from vehicle.
13. Remove torque converter cover, then mark relationship between flywheel and torque converter for reference during installation.
14. Remove torque converter to flywheel bolts and/or nuts.
15. Disconnect the catalytic converter support bracket.
16. Position a transmission jack under the transmission and raise the transmission slightly.
17. Remove transmission support to transmission mount bolt and transmission support to frame bolts (and insulators if used).
18. Slide the transmission support rearward.
19. Lower the transmission to gain access to the oil cooler lines and T.V. cable attachments.
20. Disconnect the oil cooler lines and T.V. cable. Cap all openings.
21. Support engine with a suitable tool and remove the transmission to the engine bolts.
22. Disconnect the transmission assembly, being careful not to damage any cables, lines or linkage.
23. Install torque converter holding tool J-21366 and remove the transmission assembly from the vehicle.
24. To install, reverse the removal procedure, and note the following:
    a. Before installing the flex plate to converter bolts, make certain that the weld nuts on the converter are flush with the flex plate and the converter rotates freely by hand in this position.
    b. Hand start the bolts and tighten finger tight, then torque to 35 ft. lbs. This will insure proper converter alignment. Install new oil filler tube before installing tube.

4 × 4

1. Shift transfer case to "4 Hi".
2. Disconnect battery ground cable.
3. Raise and support vehicle.
4. Remove skid plate, then drain lubricant from transfer case.
5. Mark relationship between transfer case front output shaft yoke and propeller shaft for reference during installation, then disconnect front propeller shaft from transfer case.
6. Mark relationship between rear axle yoke and propeller shaft for reference during installation, then remove rear propeller shaft.
7. Disconnect speedometer cable, vacuum harness and shift lever from transfer case.
8. Remove catalytic converter hanger bolts at the converter.
9. Raise transmission and transfer case, then remove transmission mount and catalytic converter hanger.
10. Lower transmission and transfer case from vehicle, then remove attaching bolts and separate transfer case from transmission.

NOTE: In order to remove the upper left transfer case attaching bolt, it is necessary to remove the shift lever bracket mounting bolts from the transfer case adapter.

11. Reverse procedure to install.

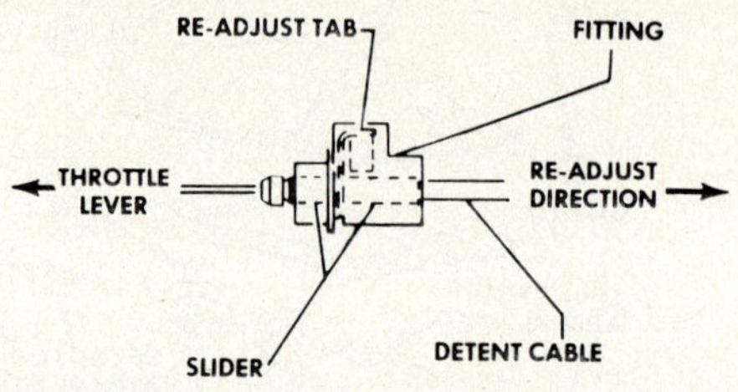

Fig. 2 T.V. cable adjustment

SHIFT LINKAGE, ADJUST

1. Position the steering column shift lever in neutral gate notch.
2. Set transmission lever (A) in neutral detent, Fig. 1.
3. Assemble clamp spring washer and screw to transmission control lever (B) and control rod, Fig. 1.
4. Hold clamp flush against transmission control lever (B) and finger tighten clamping screw against rod. No force should be exerted in either direction on the rod or transmission control lever (B) while tightening the clamping screw.
5. Tighten screw securely.

T.V. CABLE, ADJUST

Gasoline Engine

1. Depress and hold metal readjust tab. Move slider back through fitting in direction away from throttle body until slider stops against fitting, Fig. 2.
2. Release metal readjust tab.
3. Open carburetor lever to "full throttle stop" position to automatically adjust cable and release carburetor lever.
4. Check cable for proper operation. If sticking or binding occurs, proceed with the following:
    a. Remove the oil pan and inspect the throttle lever and bracket assembly.

NOTE: Check that the T.V. exhaust valve lifter rod is not distorted and not binding in the control valve assembly or spacer plate.

    b. T.V. exhaust check ball must move up and down as the lifter does. Also ensure lifter spring holds the lifter rod up against the bottom of the control valve assembly.
    c. Make sure T.V. plunger is not stuck. Inspect transmission for correct throttle lever to cable link, Fig. 3.

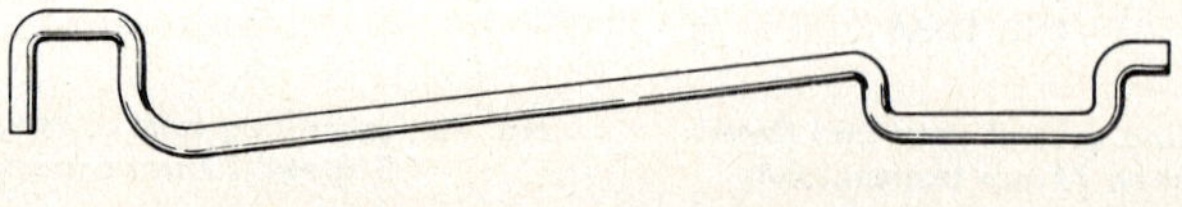

Fig. 3 Throttle lever to cable link

Diesel Engine

1. On vehicles equipped with cruise control, remove cruise control rod.
2. Disconnect T.V. cable from throttle assembly, then loosen lock nut on pump rod and shorten several turns.
3. Rotate lever assembly to full throttle stop and hold in this position.
4. Adjust pump rod until injection pump lever contacts full throttle stop.
5. Release the lever assembly and tighten pump rod lock nut, then remove pump rod from the lever assembly.
6. Reconnect T.V. cable to throttle assembly, then depress and hold metal readjust tab. Move slider back through fitting in direction away from lever assembly until slider stops against fitting.
7. Release the readjust tab, then rotate lever assembly to full throttle stop and release the lever.
8. Reconnect pump rod and, if equipped, the cruise control throttle rod.
9. On vehicles equipped with cruise control, adjust the servo throttle rod to minimum slack position, then install clip into the servo bail in first free hole closest to the bellcrank.

Rear Axle Section

AXLE SHAFT, REPLACE

1. Raise vehicle and remove wheel and brake drum.
2. Drain lube from carrier and remove cover.
3. Remove differential pinion shaft lock screw and remove differential pinion shaft, Fig. 1.
4. Push flanged end of axle shaft toward center of vehicle and remove "C" lock from button end of shaft.
5. Remove axle shaft from housing, being careful not to damage seal.
6. Reverse procedure to install.

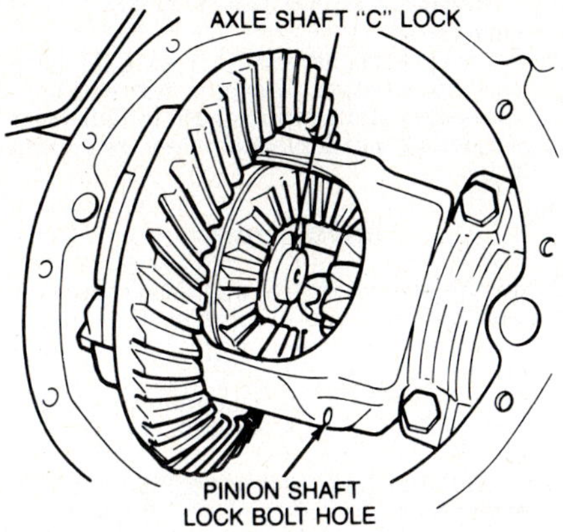

Fig. 1 Removing differential pinion shaft.

WHEEL BEARING/OIL SEAL, REPLACE

1. Remove axle shaft as previously described.
2. Use a suitable puller to remove bearing and oil seal.
3. Lubricate wheel bearing and cavity between seal lips with wheel bearing lubricant before installation.
4. Reverse procedure to install.

PINION FLANGE & SEAL, REPLACE

1. Raise and support vehicle and remove both rear wheels and brake drums.
2. Mark relation between drive shaft and pinion flange for assembly reference, then disconnect rear joint and support drive shaft out of the way. If joint bearings are not retained by a strap, use a piece of tape to keep them in place.
3. Measure preload with an inch pound torque wrench and record the results. This reading indicates combined pinion bearing, carrier bearing, axle bearing and seal preload.
4. Remove pinion flange nut and washer, then the flange.
5. Drive oil seal out of carrier. Use care not to damage the carrier.
6. Install new seal using tool J-23911.
7. Apply sealer No. 1050169, or equivalent, to pinion flange mating surface and seal lip of the seal, then install pinion flange. Tighten nut finger tight.
8. Hold pinion flange with tool J-8614-01 and tighten nut a little at a time until preload is 3–5 inch lbs. more than reading obtained in step 3. Rotate drive pinion several revolutions between tightenings to set the rollers.
9. Connect drive shaft to rear axle pinion flange.
10. Install brake drums and wheels, then check and adjust fluid level as necessary.

AXLE ASSEMBLY, REPLACE

NOTE: Construction of the axle assembly is such that service operations may be performed with the housing installed in the vehicle or with the housing removed and installed in a holding fixture. The following procedure is necessary only when the housing requires replacement.

1. Raise vehicle and place jack stands under frame side rails. Attach a hoist to the axle assembly.
2. Disconnect shock absorbers from axle.
3. Mark relation between drive shaft and pinion flange for assembly reference, then disconnect drive shaft and position aside.
4. Remove brake line junction block attaching bolt from axle housing, then disconnect brake lines from junction block.
5. Disconnect upper control arms from axle housing, then lower the axle assembly on hoist and remove springs.
6. Remove rear wheels and drums, then the axle cover.
7. Remove axle shaft as described under "Axle Shaft, Replace".
8. Disconnect brake lines from axle housing clips, then remove the backing plates.
9. Disconnect lower control arms from axle housing, then lower axle and remove from vehicle.
10. Reverse procedure to install.

Front Wheel Drive Section

NOTE: For service procedures on the transfer case used with four-wheel drive trucks, refer to the transfer case chapter.

DESCRIPTION

The front suspension is independent with torsion bars attached to the lower control arms. The front differential is mounted to the frame, with universal joints mounted on the inner and outer ends of the axle shafts. The front axle utilizes a center disconnect system which makes shifting in or out of four-wheel drive possible at any vehicle speed. When the transfer case is shifted, a vacuum diaphragm locks or unlocks the center disconnect. This system replaces automatic-locking front hubs.

TUBE & SHAFT ASSEMBLY, REPLACE

Removal

Fig. 1

1. Disconnect battery ground cable.
2. Disconnect shift cable from vacuum actuator. Disengage locking spring, then push actuator diaphragm in to release cable, Fig. 2.
3. Unlock steering wheel so linkage is free to move.
4. Raise vehicle and place jack stands under frame side rails.
5. Remove front wheels, drive belt shield and the axle skid plate, if equipped.
6. Support right hand lower control arm with a suitable jack, then disconnect right upper ball joint and remove support so that control arm hangs free.
7. Disconnect right hand drive axle shaft from tube assembly, Fig. 3. Insert a drift through opening in top of brake caliper and into corresponding vane of brake rotor to prevent axle from turning.
8. Disconnect four-wheel drive indicator light electrical connector from switch.
9. Remove the three shift cable and switch housing-to-carrier attaching bolts and pull housing out to gain access to cable locking spring. Do not remove cable coupling nut unless cable is being replaced.
10. Disconnect shift cable from fork shaft by lifting spring over slot in shift fork.
11. Remove tube bracket bolts from frame and tube assembly attaching bolts from carrier.
12. Remove tube assembly from axle. Use care not to allow sleeve, thrust washers, connector and output shaft to fall from carrier or be damaged when removing tube.

Installation

1. Install sleeve, thrust washers, connector and output shaft in carrier. The thrust washer notch must align with tab on washer, Fig. 4.
2. Coat the tube to carrier mating surface with Locktite 514 sealant, or equivalent.
3. Install tube and shaft assembly to carrier using only one bolt installed finger tight at the one o'clock position. Pull assembly down, then install cable and switch housing and four remaining bolts. Torque bolts to 30–40 ft. lbs.
4. Install the two tube to frame attaching bolts and torque to 45–60 ft. lbs.
5. Check four wheel drive mechanism for proper operation by inserting tool J-33799 into shift fork and checking for rotation of axle shaft, Fig. 5.
6. Remove tool, then install shift cable switch housing, by sliding cable through into fork shaft hole.
7. Connect four wheel drive indicator light electrical connector to switch.
8. Support and raise right hand lower control arm using a suitable jack and connect upper ball joint.
9. Install right hand drive axle to axle tube. Install one bolt first, then rotate axle to install five remaining bolts. Torque bolts to 53–63 ft. lbs.
10. Install front axle skid plate, if equipped, drive belt shield and front wheels. Torque skid plate bolts to 20–28 ft. lbs.
11. Connect shift cable to vacuum actuator by pushing cable end into actuator shaft hole.

DIFFERENTIAL CARRIER, REPLACE

1. Raise vehicle and place jack stands under frame side rails.
2. Remove tube and shaft assembly as described under "Tube & Shaft Assembly, Replace".
3. Remove steering stabilizer to frame attaching bolt.
4. Mark position of steering idler arm for reference during installation, then remove idler arm-to-frame attaching bolts.
5. Push steering linkage towards front of vehicle, then remove axle vent hose from carrier fitting.
6. Disconnect left hand drive axle shaft from carrier. When removing bolts, insert a drift through opening in top of brake caliper and into corresponding vane of brake rotor to prevent axle from turning.
7. Disconnect front propeller shaft, Fig. 6.
8. Remove carrier-to-frame attaching bolts, then the carrier.
9. Reverse procedure to install. Torque carrier attaching bolts to 30–40 ft. lbs. and axle shaft-to-carrier bolts to 53–63 ft. lbs.

DIFFERENTIAL CARRIER BUSHING, REPLACE

1. Remove tube and shaft assembly as described under "Tube & Shaft Assembly, Replace".
2. Remove differential carrier as described under "Differential Carrier, Replace".
3. Remove bushing from carrier ear using tool J-33791, Fig. 7.
4. Install new bushing using tool J-33791, Fig. 7. Ensure tool J-33791-3 is positioned properly between bushing and carrier ear to prevent bushing from being pressed in too far.

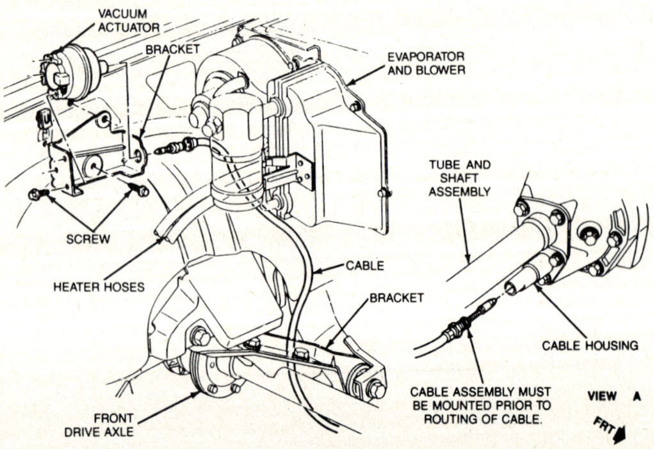

Fig. 2 Vacuum actuator

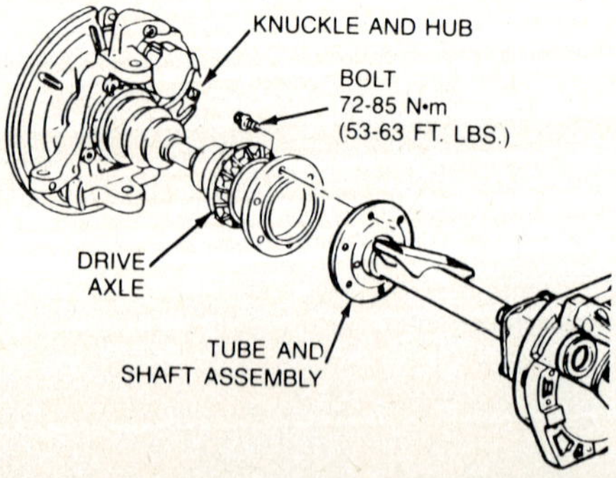

Fig. 3 Drive axle & tube assembly

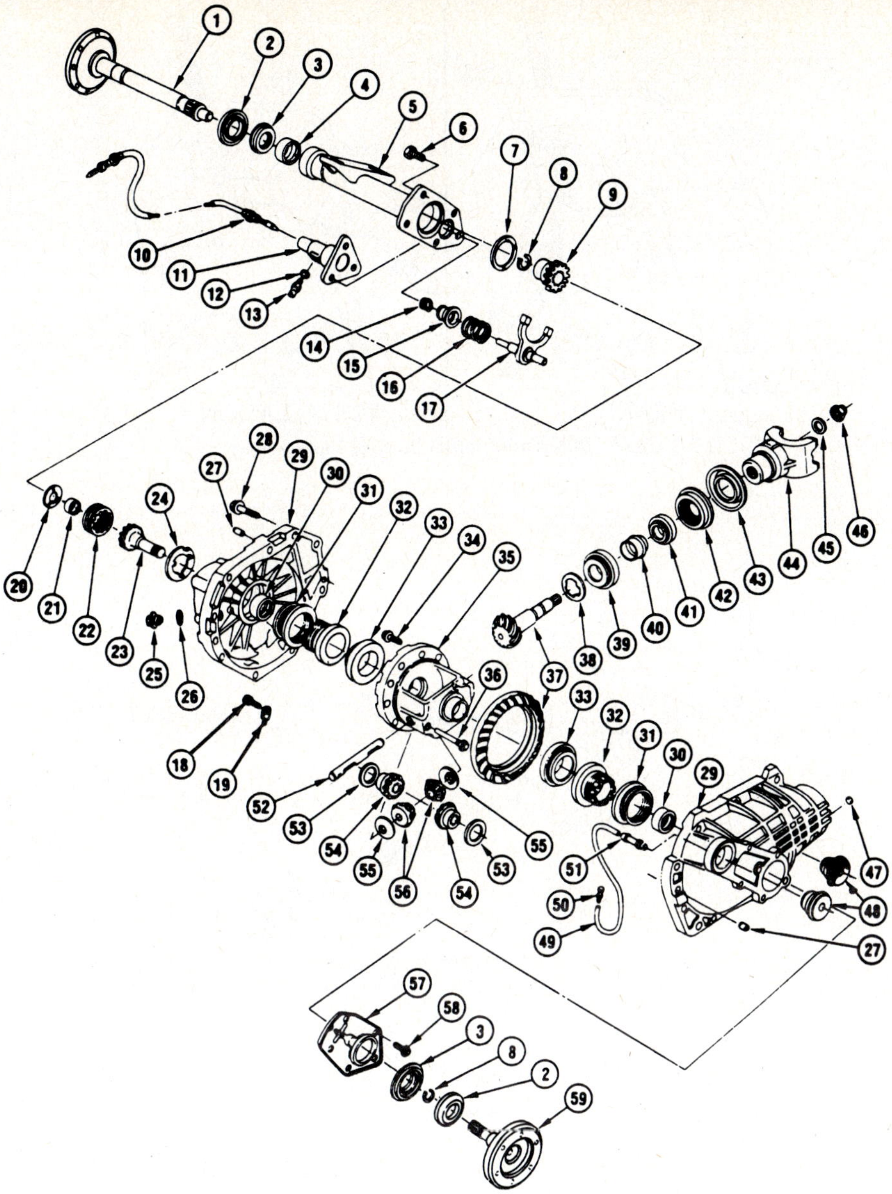

1. SHAFT, Differential Output — Right Hand (Including #2, 8)
2. DEFLECTOR, Differential Output Shaft
3. SEAL, Differential Output Shaft Tube
4. BEARING, Differential Output Shaft Tube
5. TUBE, Differential Carrier Output Shaft
6. BOLT, Tube to Carrier
7. WASHER, Carrier Thrust
8. RING, Shaft Retainer
9. CONNECTOR, Differential Carrier
10. CABLE, Differential Lock
11. HOUSING, Differential Shifter Cable
12. GASKET, Four Wheel Drive Indicator Switch
13. SWITCH, Four Wheel Drive Indicator
14. SPRING, Differential Shifter
15. SEAL, Differential Shifter
16. SPRING, Differential Shifter
17. SHAFT, w/Fork, Differential Shifter
18. BOLT, Hex Flange Head (M6 x 1 x 14)
19. LOCK, Differential Adjuster
20. WASHER, Differential Output Shaft Thrust
21. BEARING, Differential Output Shaft Pilot
22. SLEEVE, Differential
23. SHAFT, Differential Output — Left Hand
24. WASHER, Differential Thrust
25. PLUG, Differential Carrier (M20 x 1.5)
26. WASHER, Differential Carrier Plug
27. PIN, Differential Carrier Alignment
28. BOLT, Differential Carrier (10 Required)
29. CARRIER, Differential (Including #27, 28, 47, 51) (Carrier halves not serviced separately)
30. BEARING, Differential Output Shaft
31. INSERT, Differential Carrier Bearing Adjuster
32. SLEEVE, Differential Carrier Bearing Adjuster
33. BEARING, Differential Side
34. BOLT, Hypoid Drive Gear to Case
35. CASE, Differential
36. SCREW, Differential Pinion Shaft Lock
37. GEAR KIT, Ring and Pinion (Gear Kits also Include #34, 40)
38. SHIM KIT
39. BEARING, Pinion Inner
40. SPACER, Pinion Bearing
41. BEARING, Pinion Outer
42. SEAL, Differential Carrier Pinion
43. DEFLECTOR, Differential Carrier Flange
44. FLANGE, Prop Shaft Pinion (Including #43)
45. WASHER, Prop Shaft Pinion Flange
46. NUT, Prop Shaft Pinion Flange
47. PLUG, Cup Expansion (3/8")
48. BUSHING, Differential Carrier
49. HOSE, Bulk (5/16" ID)
50. VENTILATOR, Front Axle
51. CONNECTOR, Front Axle Vent Hose
52. SHAFT, Differential Pinion
53. WASHER, Differential Side Gear Thrust
54. GEAR, Differential Side
55. WASHER, Differential Carrier Pinion Thrust
56. GEAR, Differential Pinion
57. COVER, Differential Carrier
58. BOLT, Differential Carrier Cover
59. SHAFT, Differential (Including #2, 8)

Fig. 1 Exploded view of front drive axle

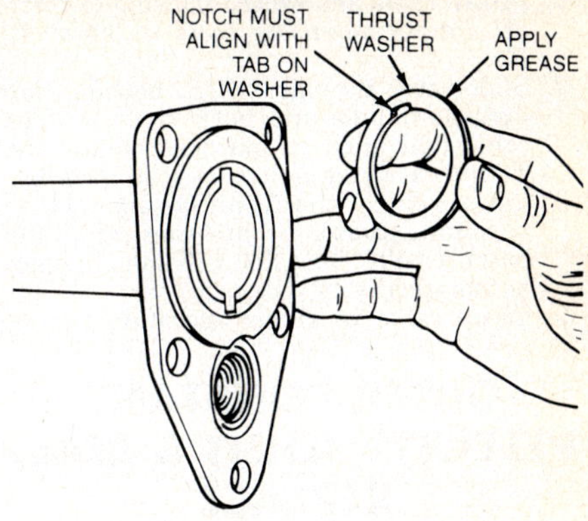

Fig. 4 Thrust washer installation

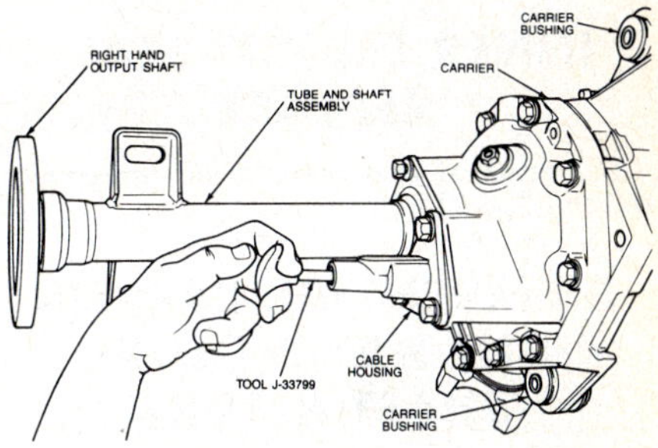

Fig. 5 Checking operation of four wheel drive mechanism

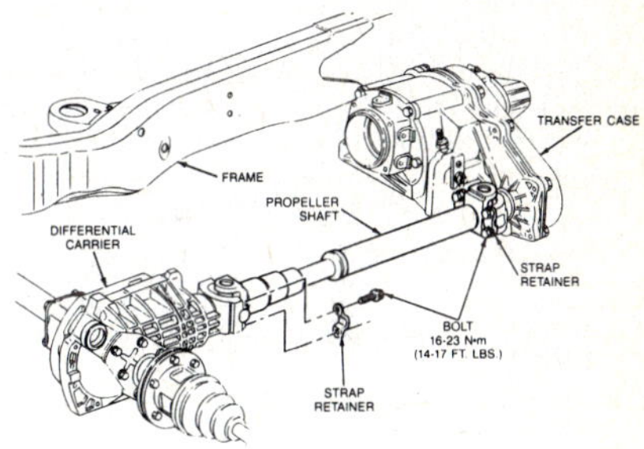

Fig. 6 Front propeller shaft

DIFFERENTIAL OUTPUT SHAFT PILOT BEARING, REPLACE

1. Remove bearing using tool J-34011.
2. Install new bearing using tool J-33842.

SHIFT CABLE, REPLACE

1. Remove shift cable from vacuum actuator by disengaging locking spring and pushing actuator diaphragm in to release cable. Compress cable locking fingers with pliers, then remove cable from bracket.
2. Raise and support vehicle.
3. Remove switch housing mounting bolts and pull housing away from axle tube flange to gain access to cable locking spring.
4. Disconnect cable from shaft fork by lifting spring over slot in shift fork, then unscrew cable from housing and remove from vehicle.
5. Install cable following proper routing, Fig. 2.

6. Install cable and switch housing to carrier. Torque mounting bolts to 30–40 ft. lbs.
7. Slide cable through switch housing into fork shaft hole and push cable in. The cable will snap in place automatically. Torque coupling nut to 71–106 inch lbs.
8. Lower vehicle, then connect shift cable to vacuum actuator by pressing cable into bracket hole. The cable will snap in place automatically.
9. Check cable for proper operation.

OUTER CONSTANT VELOCITY JOINT & SEAL, REPLACE

For removal and installation procedures, refer to Figs. 8, 9, and 10.

INNER TRI-POT SEAL, REPLACE

For removal and installation procedures, refer to Fig. 11.

DIFFERENTIAL CARRIER RIGHT HAND OUTPUT SHAFT & TUBE

Disassemble

1. Remove output shaft from tube by tapping inside of flange with a rubber mallet, Fig. 12.
2. Pry tube seal from tube, then remove bearing from tube using tool J-29369-2.
3. Drive differential shift cable housing seal out of tube using a suitable punch.

Assemble

1. Install output shaft tube bearing using tool J-33844. When bearing is properly installed, the tool will be flush with tube.
2. Install tube seal using tool J-33893. Flange of seal must be flush with tube outer surface.
3. Install output shaft into tube by tapping flange with a rubber mallet.
4. Install differential shift cable housing seal using tool J-33799.

DIFFERENTIAL CARRIER

Disassemble, Fig. 1

1. Remove differential output shaft and thrust washer as described under "Tube & Shaft Assembly, Replace".
2. Remove left hand output shaft. Use a screwdriver to pry between shaft flange and carrier head bolt and, at the same time, tap shaft flange with a rubber mallet.
3. Remove left hand side cover from carrier, then remove seal from cover.
4. Remove the ten bolts holding carrier cover halves together, then insert screwdriver in slots, Fig. 13, and pry cover halves apart.
5. Remove differential case from carrier.

NOTE: For service procedures on the differential case, refer to "Differential Case, Disassemble".

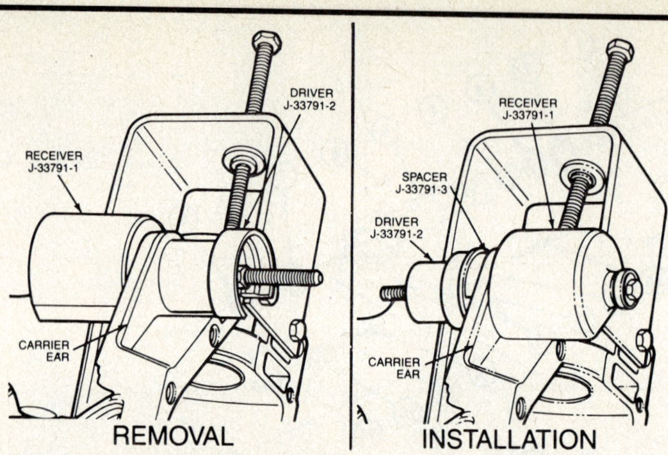

Fig. 7 Differential carrier bushing replacement

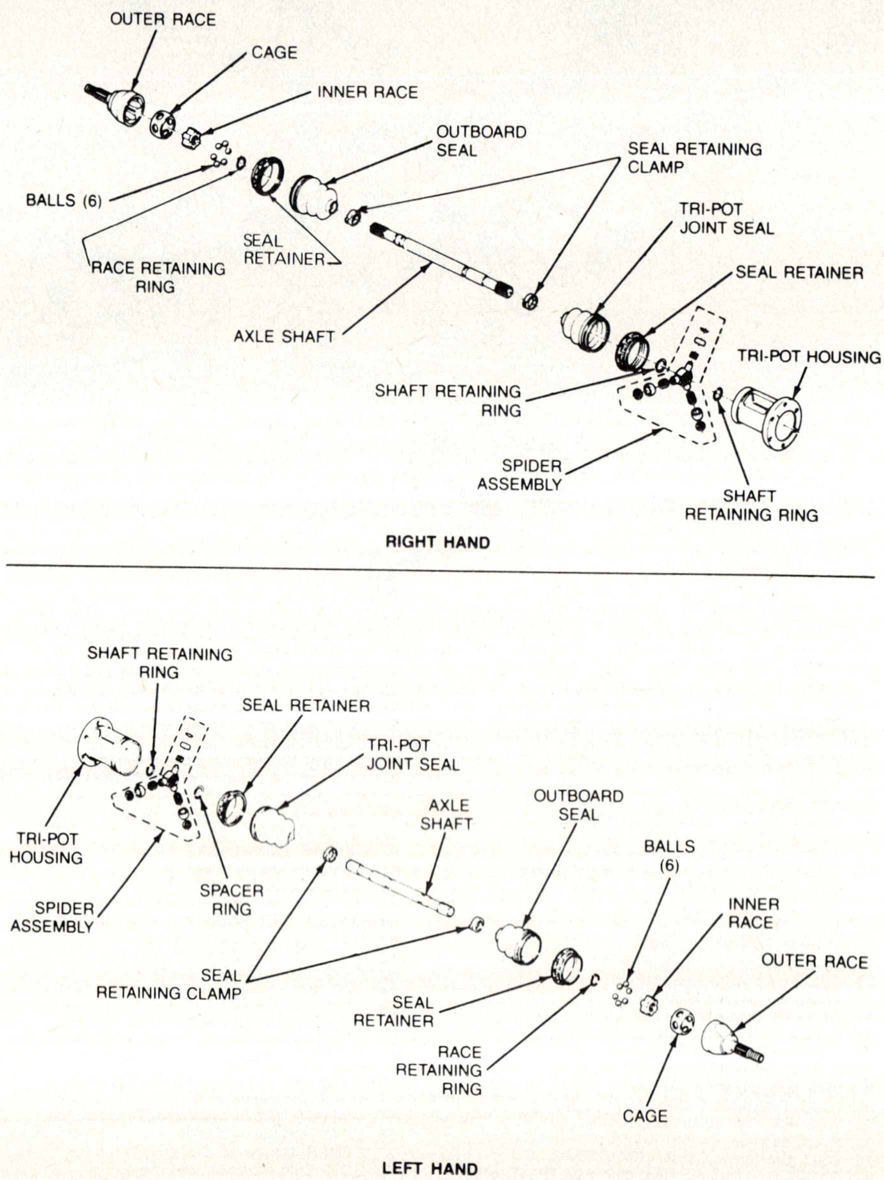

Fig. 8 Exploded view of front axle

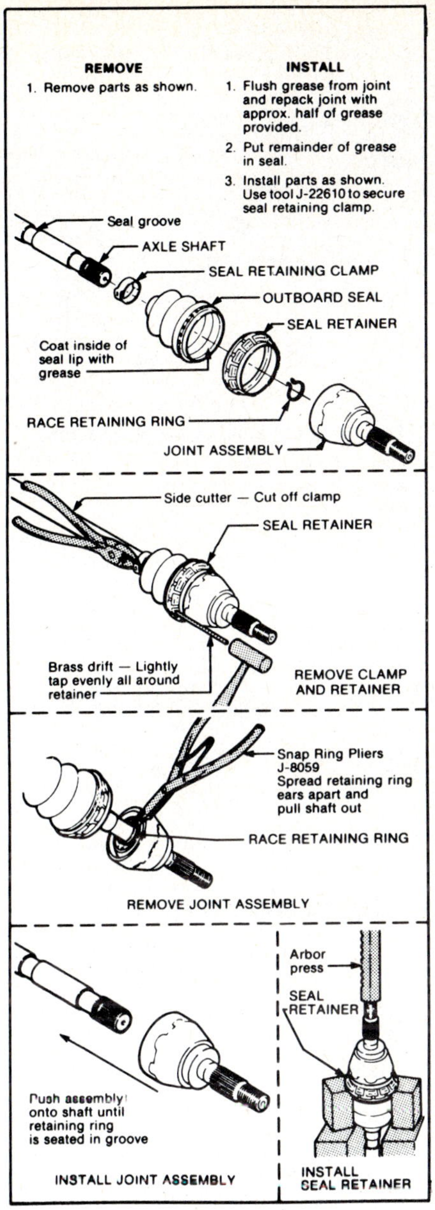

Fig. 9 Outer constant velocity joint seal removal & installation

6. Remove the two side bearing adjusting locks.
7. Remove both differential side bearing races by using tool J-33792 to turn adjusting sleeve until race is pushed out of carrier.
8. Remove pinion flange nut using tool J-8614-01, then install pinion fixture tool J-33837, Fig. 14.
9. Remove pinion from carrier by turning bolt on tool J-33837-1, Fig. 14, then remove pinion flange.
10. Install tool J-33837-6 on tool J-33837-1, then remove outer pinion bearing, pinion bearing race and pinion seat.
11. Install tool J-33837-6 on tool J-33837 and remove inner pinion bearing race from carrier.
12. Remove inner drive pinion using tool J-22912-01.

Assemble

NOTE: Lubricate all bearings, races and seals with axle lubricant before installing in differential.

1. Install outer pinion bearing race, then the inner race using tool J-33837-4 until races are fully seated in carrier.
2. Set pinion depth and install correct shim on drive pinion as described under "Setting Pinion Depth".
3. Install inner pinion bearing on pinion using tool J-33785.
4. Install a new collapsible spacer or pinion, then position the pinion assembly in carrier.
5. Install outer pinion bearing in carrier, then install bearing seal using tool J-33782.
6. Install pinion flange, washer and nut, then install holder tool J-8614-01 on flange and tighten nut until no end play is detectable.
7. Check preload and set to 5–10 inch lbs. using an inch-pound torque wrench.

NOTE: When setting preload, tighten pinion nut only a little at a time and recheck setting frequently. If preload specifications are exceeded, the collapsible spacer will be compressed beyond limits and will have to be replaced.

8. Rotate pinion several times to make sure that bearings have been seated, then recheck preload and correct as necessary.
9. Install output shaft bearings and adjusting sleeves in inserts in carrier and tighten finger tight.
10. Install differential side bearing race into carrier using tool J-23423-A. If differential output shaft bearings or adjusting sleeve inserts must be replaced, proceed as follows:
    a. Remove differential output shaft bearing from adjusting sleeve using tool J-21551.
    b. Install new bearings into adjusting sleeves using tool J-33788.
    c. Install sleeve and bearing into insert and tighten finger tight, then install complete assembly into carrier using tool J-23423-A.
11. Adjust differential backlash as described under "Differential Backlash, Adjust".
12. Install left hand carrier cover seal using tool J-33839. Support cover when installing seal to avoid bending the cover.
13. Apply sealer No. 1052366, or equivalent, to carrier mating surface, then install cover. Torque bolts to 15–20 ft. lbs.

SETTING PINION DEPTH

NOTE: The drive pinions used are "nominal" or "zero" pinions and are not marked. The thickness of the shim used will be equal to the dial indicator gauge reading.

1. Lubricate inner and outer pinion bearings with axle lubricant, then install tool J-33838 while holding bearings in place, Fig. 15.
2. Set preload of pinion bearings to 5–10 inch lbs. by tightening mounting bolt on tool while holding end of tool shaft with a wrench.
3. Zero the dial indicator, then install it on tool J-33838, Fig. 15. Push dial indicator down until needle rotates approximately three turns clockwise and tighten the dial indicator in this position.
4. Position button of tool J-33838 in differential bearing bore, Fig. 15, then rotate tool back and forth until lowest point of bore is indicated on dial indicator.
5. Zero the dial indicator, then repeat rocking action of tool to verify zero setting.
6. When zero setting has been obtained and verified, move tool button out of bearing bore and record the dial indicator reading. Select shim equal to this reading.
7. Install shim on drive pinion, then install pinion in carrier as described under "Differential Carrier, Assemble".

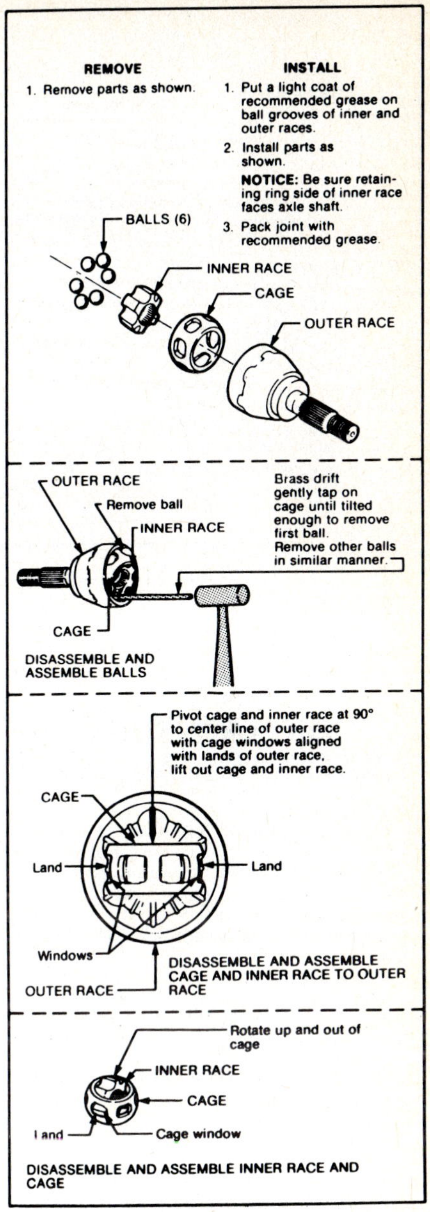

Fig. 10 Outer constant velocity joint disassembly & assembly

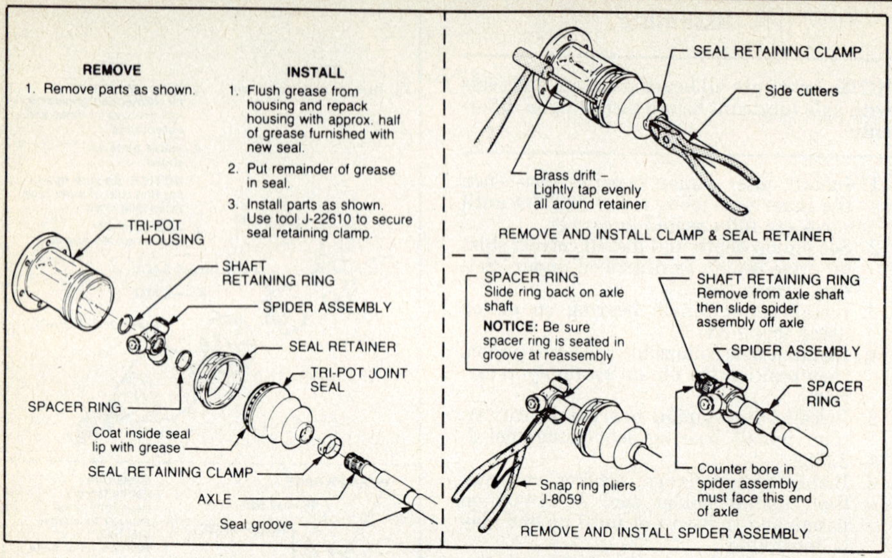

Fig. 11 Inner tri-pot seal removal & installation

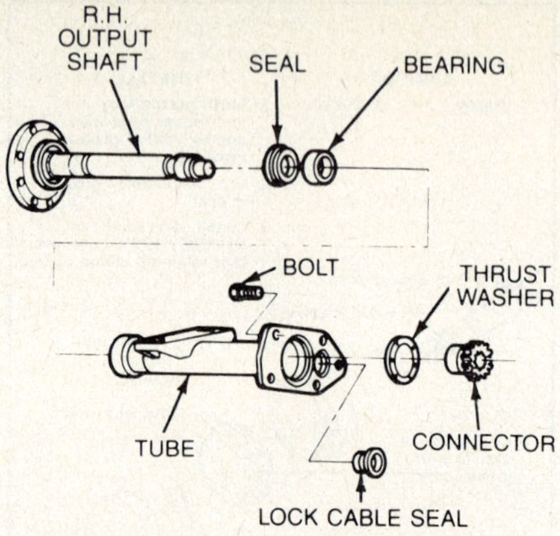

Fig. 12 Right hand output shaft & tube

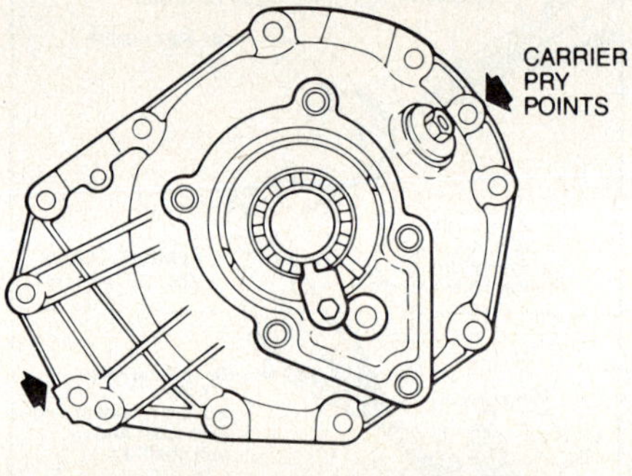

Fig. 13 Differential carrier pry points

DIFFERENTIAL BACKLASH, ADJUST

1. Turn left hand adjusting sleeve in, toward differential case, until backlash is felt between ring and pinion gears.
2. Clean carrier half mating surfaces, then install halves together without using any sealer. Install four case bolts, Fig. 16, and torque to 37 ft. lbs.

NOTE: If carrier halves will not make complete contact, back out the right hand adjusting sleeve.

3. Tighten right hand adjusting sleeve using tool J-33792 until no backlash is present (approximately 100 ft. lbs.), then tighten left hand adjusting sleeve to approximately 100 ft. lbs.
4. Mark relation of adjusting sleeves to carrier halves so that the notches in adjusting sleeves can be counted when turned.
5. Turn right hand adjusting sleeve out two notches and left hand sleeve in one notch using tool J-33792.
6. Rotate axle case several times to seat bearings, then mount dial indicator.
7. Install a small button on indicator stem and position so that contact can be made near heel end of tooth angle. This will provide an accurate backlash reading. Backlash should be .003–.010 inch, with a preferred reading of .005–.007 inch.
8. If backlash is not within specifications, readjust adjusting sleeves as necessary. Do not install adjusting sleeve locks at this time.
9. Mark position of right hand adjusting sleeve, then loosen sleeve to release contact between right hand side bearing and differential case.
10. Remove four carrier bolts and separate the carrier halves.
11. Apply sealer No. 1052357, Locktite 514 or equivalent to mating surface of one carrier half, then assemble two halves together with all ten attaching bolts. Torque bolts to 30–40 ft. lbs.
12. Set right hand adjusting sleeve in position according to mark made previously, then install both adjusting sleeve locks and torque to 62–88 inch lbs.

DIFFERENTIAL CASE

Disassemble

1. Remove side bearings using tool J-22912-01, then remove pinions, side gears and thrust washers from case. Mark relation between side gears and case for reference during assembly.
2. If ring gear is being replaced, drive it off the case using a brass drift and hammer. Do not pry between ring gear and case.

Assemble

NOTE: Prior to assembling the differential case, lubricate all parts with axle lubricant.

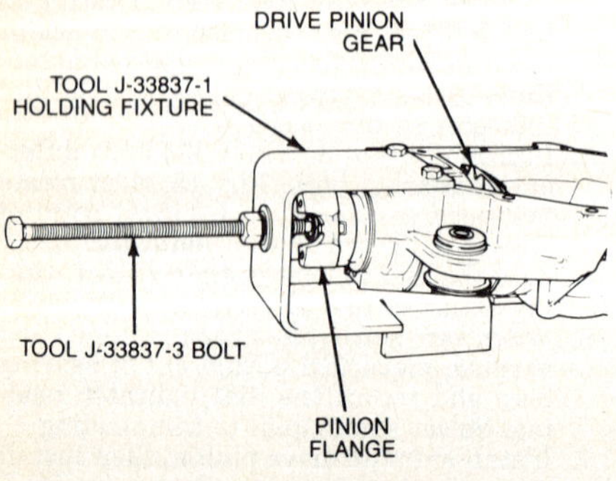

Fig. 14 Drive pinion removal

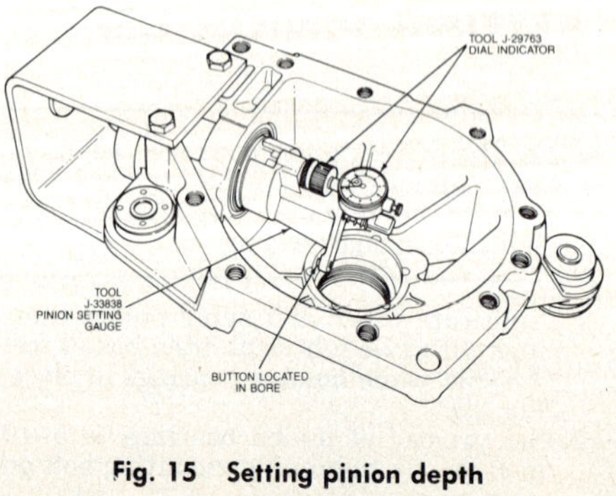

Fig. 15 Setting pinion depth

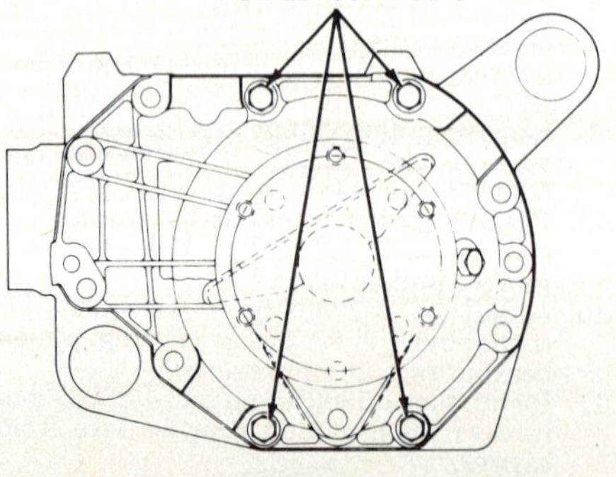

Fig. 16 Case assembly for backlash adjustment

1. Install thrust washers on side gear hubs, then install side gears in case.
2. Install one pinion, without thrust washer, between side gears and turn gears until pinion is directly opposite the loading opening in case.
3. Install other pinion between side gears so that pinion shaft holes are in line. Rotate gears to make sure that holes in pinions will line up with holes in case. If holes line up, turn pinions back toward loading opening just enough to permit installation of pinion thrust washers.
4. Check that ring gear and case mating surfaces are clean and free of burrs, then thread two bolts into opposite ends of ring gear and install gear on case.
6. Install new ring gear retaining bolts and torque alternately to 52–66 ft. lbs.
7. Install case side bearings using tool J-33790.

Front Suspension & Steering Section

WHEEL ALIGNMENT

NOTE: Before adjusting caster and camber angles, raise and release bumper twice to allow vehicle to return to it's normal height.

Caster and camber adjustments are made by means of shims inserted between upper control arm shaft and frame bracket, Fig. 1 Add, subtract, or transfer shims to change readings as follows:

Caster, Adjust

Transfer shims from front to rear or rear to front. The transfer of one shim from rear to front bolt will decrease positive caster.

Camber, Adjust

Change shims at both front and rear of shaft. Adding an equal number of shims at both front and rear of cross shaft will decrease positive camber. On 1983 "T" series trucks, the cams cannot be rotated more than 270°. Cam lobe cannot be positioned full outboard ±45°.

NOTE: To adjust caster and camber, loosen upper control arm shaft to frame nuts, add or subtract shims as required and retorque nuts. A normal shim pack will leave at least two threads of bolt exposed beyond the nut. The difference between front and rear shim packs must not exceed 0.40 inches.

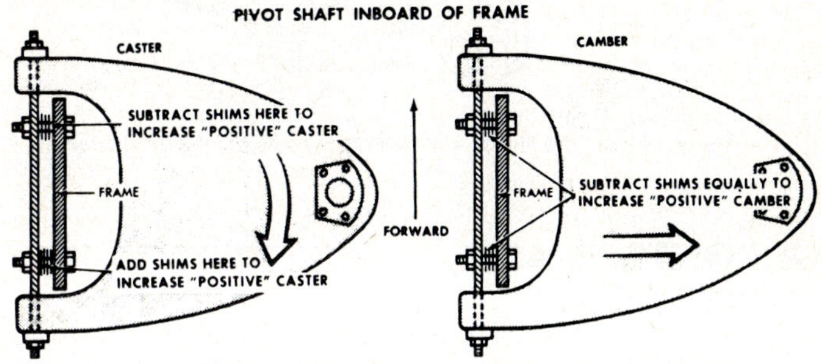

Fig. 1 Caster and camber adjustment

TOE-IN, ADJUST

NOTE: Toe-in may be adjusted by changing length of tie rods by means of a threaded sleeve. When tie rods are mounted ahead of steering knuckle then must be decreased in length to increase toe-in.

To adjust, loosen clamp bolts at each end of steering tie rod adjustable sleeves. With steering wheel in straight ahead position, turn tie rod adjusting sleeves to obtain proper adjustment. After adjusting, check that number of threads showing on each end of sleeve are equal and than the tie rod end housings are at the right angles to steering arm. Position tie rod clamps and sleeves, Fig. 2, and torque nuts to 15 ft. lbs.

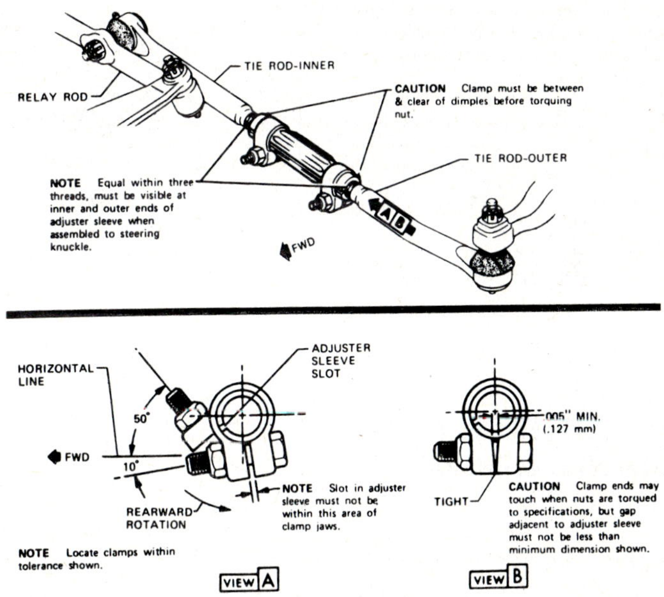

Fig. 2 Tie rod clamp and sleeve positioning

COIL SPRING, REPLACE

4 × 2

1. Raise and support vehicle.
2. Remove the two shock absorber screws and push shock up through control arm and into spring.
3. Support vehicle so that control arms hang free.
4. Place Tool J-23028, or equivalent, into position cradling the inner bushings, Fig. 3. Tool should be supported by a suitable jack.
5. Remove stabilizer to lower control arm attachment.
6. Raise the jack to remove tension on lower control arm pivot bolts, then install a chain around spring and through control arm and remove the nuts and bolts.
7. Lower control arm by slowly lowering jack.
8. With all pressure removed from spring, remove safety chain and spring.

NOTE: Do not apply force to lower control arm and ball joint to remove spring. Proper maneuvering of spring will allow for easy removal.

9. Reverse procedure to install.

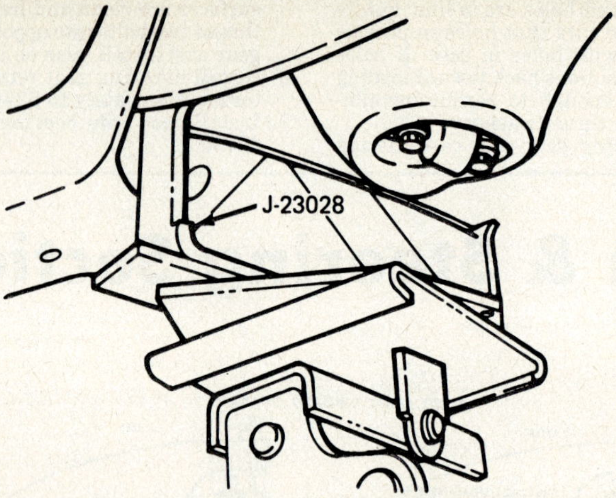

Fig. 3 Removing spring with adapter J-23028. 4 × 2 models

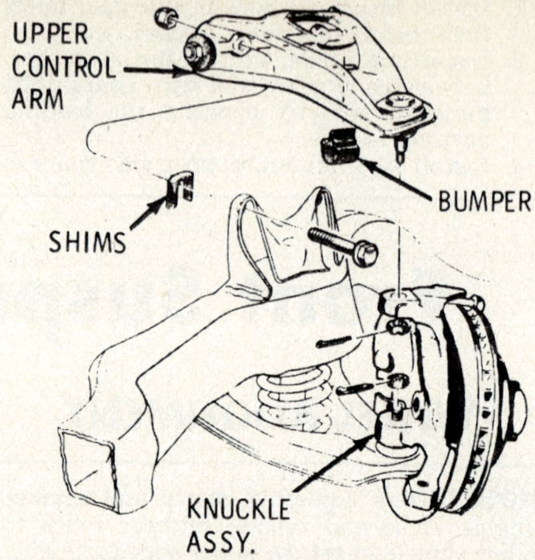

Fig. 4 Upper control arm replacement. 4 × 2 models

UPPER CONTROL ARM, REPLACE

4 × 2

1. Remove retaining nuts and shims from control arm, Fig. 4. Note location of shims for assembly reference.
2. Raise and support vehicle. Support lower control arm using a suitable jack.
3. Remove front wheel, then loosen upper ball joint from steering knuckle as described under "Upper Ball Joint, Replace".
4. Support hub assembly to prevent damage to brake line when removing control arm.
5. Remove upper control arm attaching bolts, then the control arm.
6. Reverse procedure to install. Torque retaining nuts to 45 ft. lbs.

4 × 4

Refer to Fig. 5 for upper control arm replacement procedure on four-wheel drive vehicles.

UPPER CONTROL ARM BUSHING, REPLACE

4 × 2

1. Remove upper control arm as described previously.
2. Remove nuts from end of pivot shaft, then remove bushings, Fig. 6.
3. Install bushings by installing pivot shaft in control arm and pressing new bushings into control arm, Fig. 7. Torque pivot shaft nuts to 85 ft. lbs. with weight of vehicle resting on wheels.

NOTE: Both bushings must be installed .48–.52 inch from face of control arm to bushing outer sleeve.

4 × 4

Refer to Fig. 5 for upper control arm bushing replacement procedure on four-wheel drive vehicles.

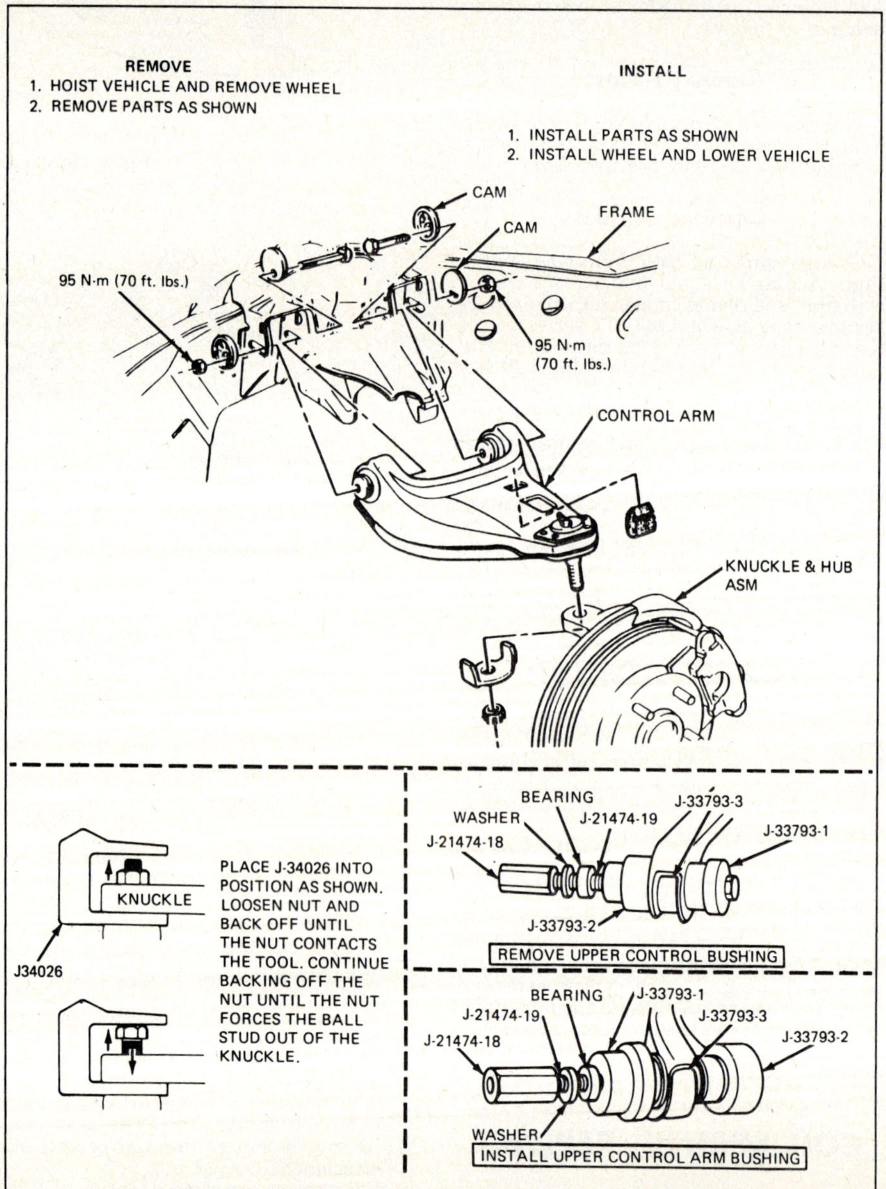

Fig. 5 Upper control arm & bushing replacement. 4 × 4 models

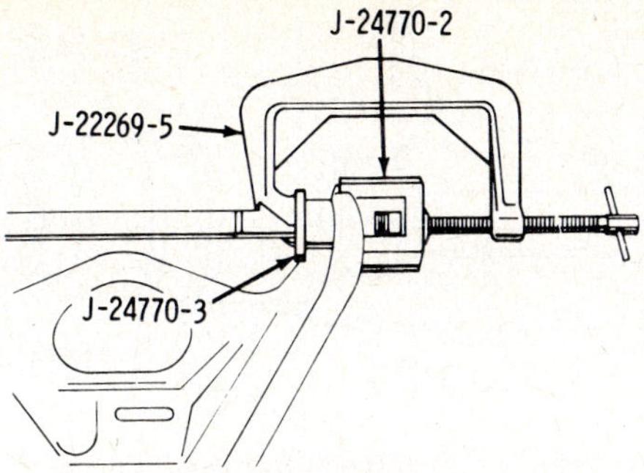

Fig. 6 Upper control arm bushing removal. 4 × 2 models

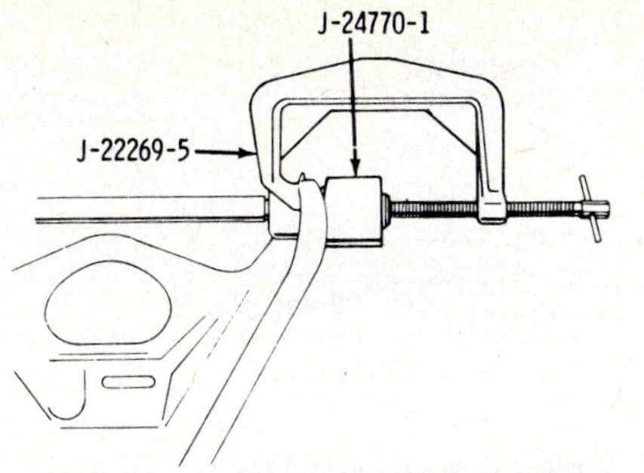

Fig. 7 Upper control arm bushing installation. 4 × 2 models

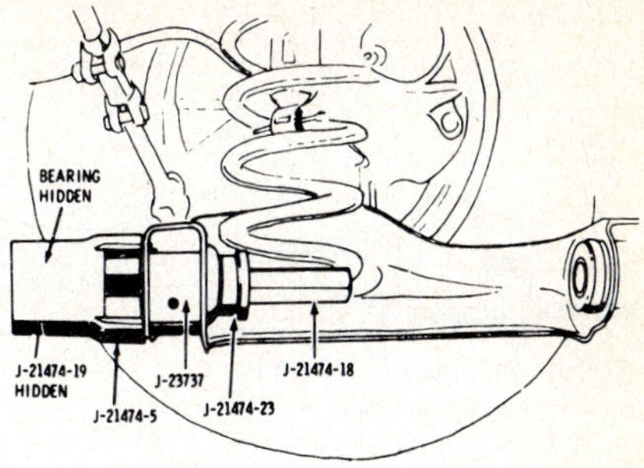

Fig. 9 Lower control arm front bushing removal. 4 × 2 models

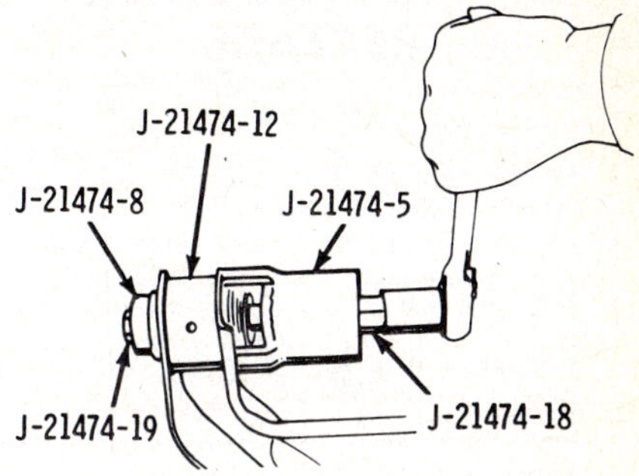

Fig. 10 Lower control arm rear bushing removal. 4 × 2 models

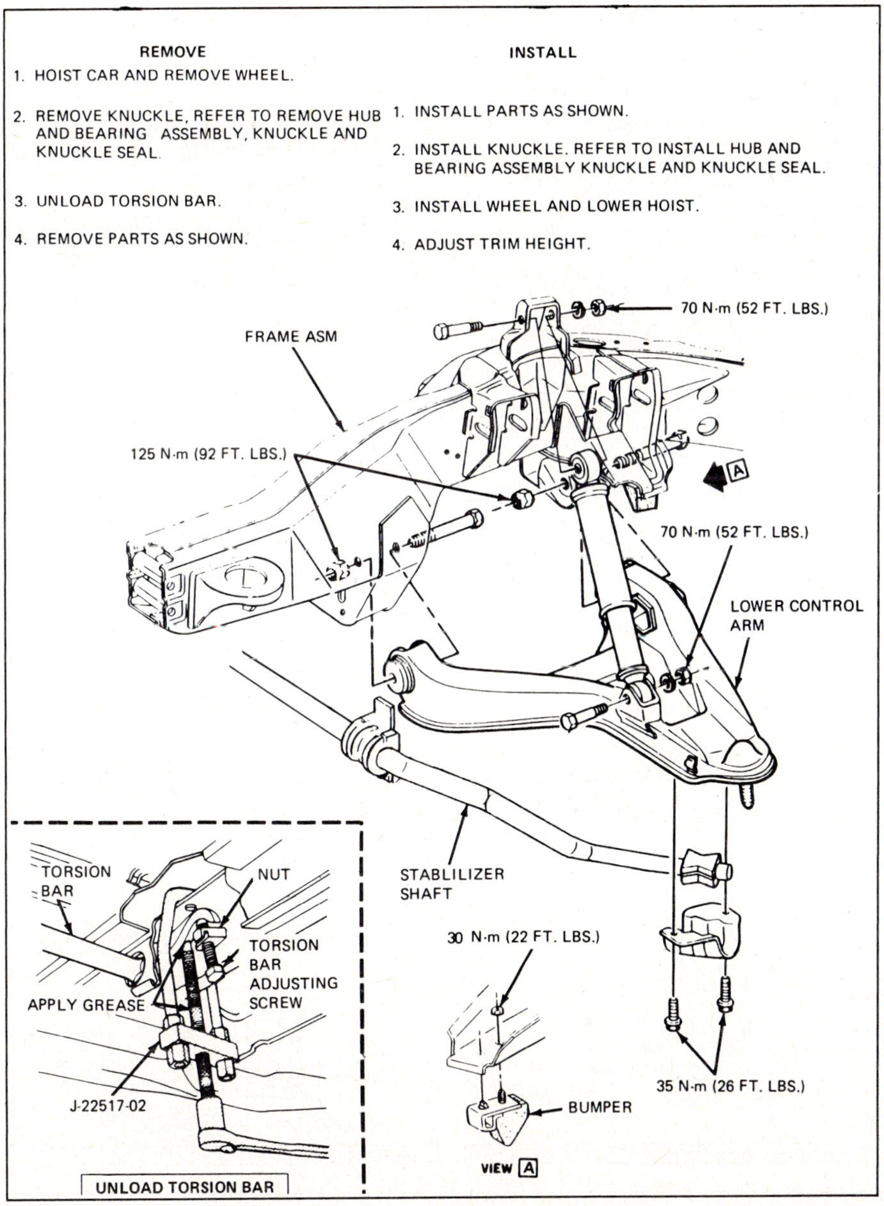

Fig. 8 Lower control arm replacement. 4 × 4 models

LOWER CONTROL ARM, REPLACE

4 × 2

1. Remove coil spring as described previously.
2. Remove lower ball joint stud, then remove lower control arm through opening in splash shield.
3. Reverse procedure to install.

4 × 4

Refer to Fig. 8 for lower control arm replacement procedure on four-wheel drive vehicles.

LOWER CONTROL ARM BUSHINGS, REPLACE

4 × 2

1. Drive bushing flare down flush with rubber of front bushing, then remove front and rear bushings from control arm, Figs. 9 and 10.
2. Install front bushing, Fig. 11, then flare the bushing, Fig. 12.
3. Install rear bushings, Fig. 13, then the lower control arm as described previously.

4 × 4

Refer to Fig. 14 for lower control arm bushing replacement procedure on four-wheel drive vehicles.

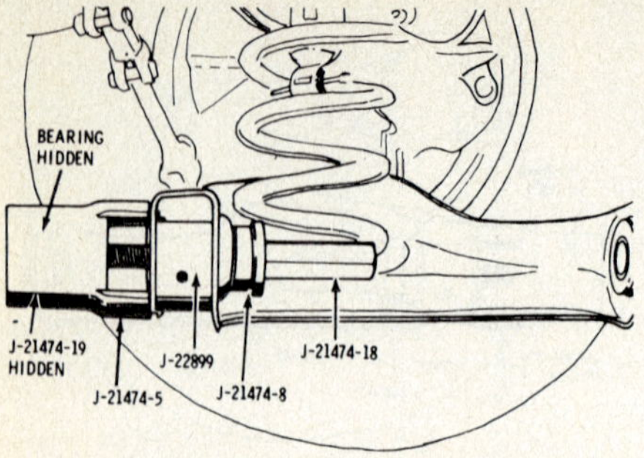

Fig. 11 Lower control arm front bushing installation. 4 × 2 models

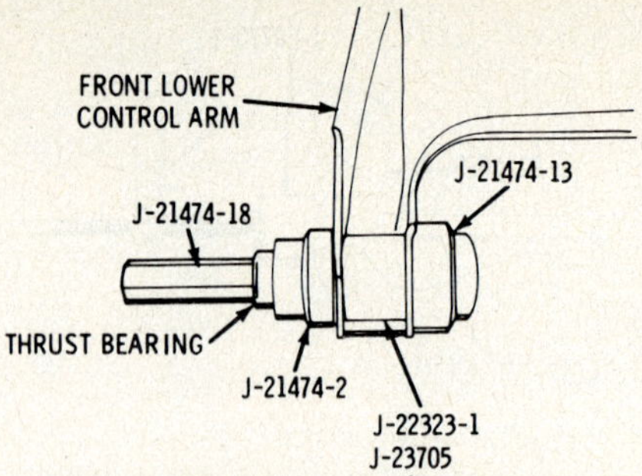

Fig. 12 Flaring lower control arm front bushing. 4 × 2 models

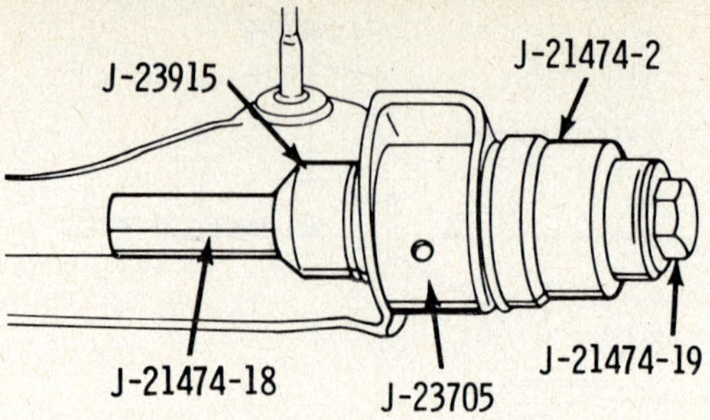

Fig. 13 Lower control arm rear bushing installation. 4 × 2 models

LOWER BALL JOINT, REPLACE

4 × 2

1. Raise vehicle and support with jack stands under frame side rails.
2. Remove front wheel, then support control arm spring seat with a suitable jack.
3. Remove cotter pin and stud nut, then break ball joint loose from steering knuckle using tool J-23742, or equivalent.

NOTE: Inspect and clean the tapered hole in steering knuckle. If hole is out of round, or damaged in any way, the steering knuckle must be replaced.

4. Guide lower control arm out of opening in splash shield, then position a wooden block between frame and upper control arm to block knuckle assembly out of the way.
5. Remove grease fittings, then press ball joint from lower control arm.
6. Reverse procedure to install. Torque stud nut to 90 ft. lbs.

4 × 4

Refer to Fig. 15 for lower ball joint replacement procedure on four-wheel drive vehicles.

UPPER BALL JOINT, REPLACE

4 × 2

1. Raise vehicle and support at lower control arm with suitable jacks.
2. Remove wheel, then remove cotter pin and stud nut from ball joint.
3. Break stud loose from steering knuckle using tool J-23742, or equivalent. Support knuckle assembly to avoid damaging brake line.
4. With control arm in raised position, drill rivets 1/4 inch deep with a 1/8 inch drill, then drill off rivet heads with a 1/2 inch drill.
5. Punch out rivets, then remove ball joint.
6. Reverse procedure to install. Replace rivets with attaching bolts and nuts. Torque attaching nuts to 8 ft. lbs. and the stud nut to 65 ft. lbs.

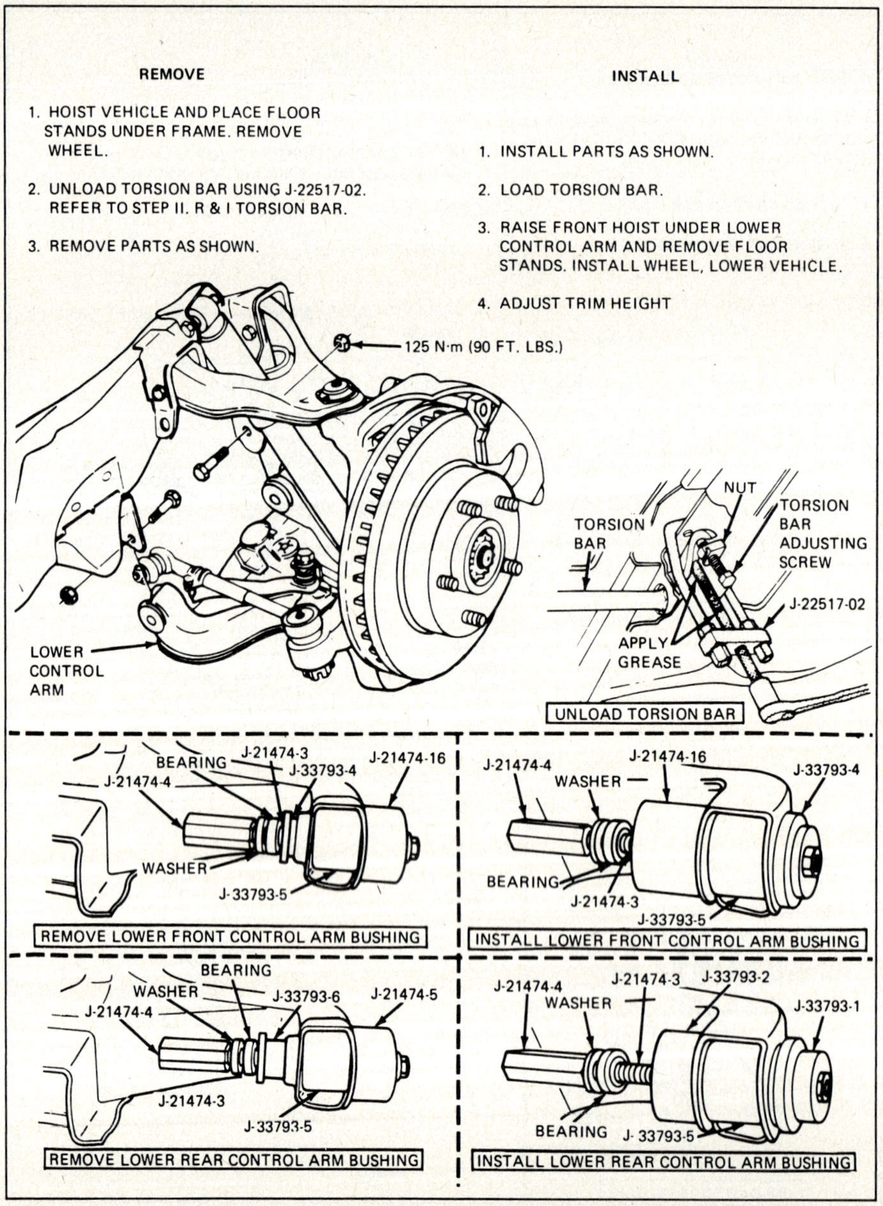

Fig. 14 Lower control arm bushing replacement. 4 × 4 models

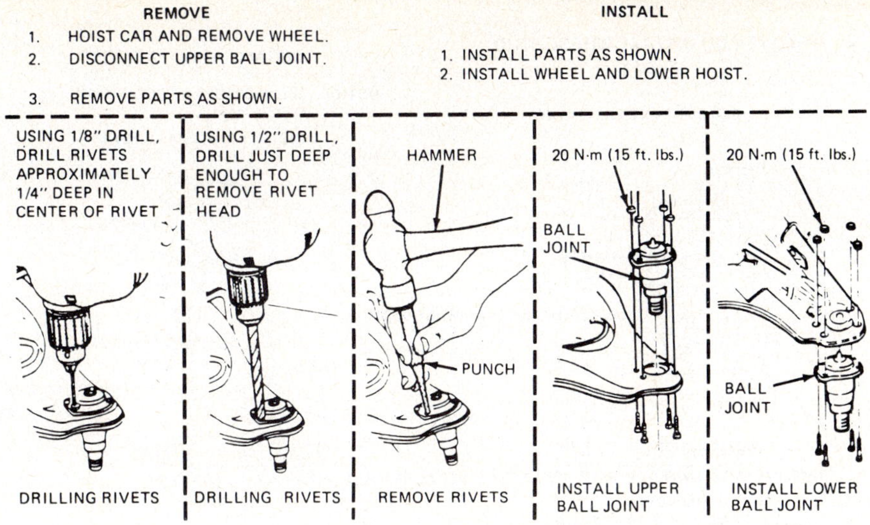

Fig. 15 Ball joint replacement. 4 × 4 models

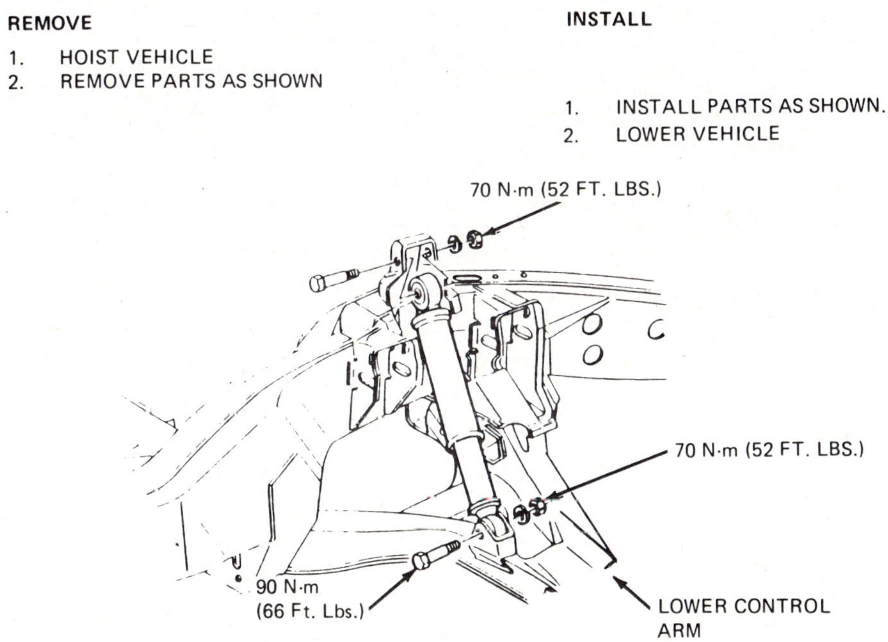

Fig. 16 Shock absorber replacement. 4 × 4 models

4 × 4

Refer to Fig. 15 for upper ball joint replacement procedure on four-wheel drive vehicles.

SHOCK ABSORBER, REPLACE

4 × 2

1. Raise and support vehicle.
2. Hold shock upper stem from turning with a suitable wrench and remove nut, retainer and grommet.
3. Remove lower shock pivot bolts, then remove shock absorber from vehicle.
4. Reverse procedure to install. Torque upper attaching nut to 8 ft. lbs. and lower attaching bolts to 20 ft. lbs.

4 × 4

Refer to Fig. 16 for shock absorber replacement procedure on four-wheel drive vehicles.

STEERING KNUCKLE, REPLACE

4 × 2

1. Raise and support vehicle. Do not position jack under lower control arm.
2. Remove brake caliper, then the hub and rotor assembly.
3. Remove splash shield-to-knuckle attaching bolts.
4. Remove tie rod end from knuckle using tool J-6627, or equivalent, then remove knuckle seal.
5. Remove ball studs from knuckle using tool J-23742, or equivalent.
6. Support lower control arm with a suitable jack, then raise upper control arm to disengage ball joint stud from knuckle.
7. Remove steering knuckle from lower ball joint stud.
8. Reverse procedure to install. Torque splash shield attaching bolts to 10 ft. lbs. When installing tie rod end, install tool J-29193 and torque to 15 ft. lbs., then remove tool and torque to 40 ft. lbs.

4 × 4

Refer to Fig. 17 for steering knuckle replacement procedure on four-wheel drive vehicles.

STABILIZER BAR, REPLACE

4 × 2

1. Raise and support vehicle.
2. Remove bolt, retainers, grommets and spacers from each side of stabilizer linkage.
3. Remove bracket to frame or body bolts, then the stabilizer bar, rubber bushings and brackets, Fig. 18.
4. Reverse procedure to install. Torque stabilizer link nut to 13 ft. lbs. and bracket bolts to 24 ft. lbs.

4 × 4

Refer to Fig. 19 for stabilizer bar replacement procedure on four-wheel drive vehicles.

TORSION BAR, REPLACE

4 × 4

Refer to Fig. 20 for torsion bar replacement procedure on four-wheel drive vehicles.

BALL JOINT INSPECTION

4 × 2

Upper Ball Joint

NOTE: Before checking ball joints, wheel bearings must be properly adjusted.

1. Raise vehicle and position floor stands under right and left lower control arms near each lower ball joint.
2. Position dial indicator against wheel rim, Fig. 21.
3. Shake wheel, Fig. 21, and read gage. Horizontal deflection should not exceed .125 in.
4. If reading exceeds .125 in., or if ball stud has been disconnected from knuckle assembly and any looseness is detected or stud is loose, replace ball joint.

Lower Ball Joint

The lower ball joint has a visual wear indicator, Fig. 22. To check, vehicle weight must rest on wheels to properly load ball joints.

4 × 4

Refer to Fig. 23 for ball joint inspection procedure on four-wheel drive vehicles.

STEERING GEAR, REPLACE

1. Disconnect battery ground cable and remove coupling shield if so equipped.
2. Remove retaining nuts, lock washers and bolts at steering coupling to steering shaft flange.
3. Remove pitman arm nut and washer from pitman shaft and mark relation of arm position to shaft, then remove pitman arm with Tool J-6632, or equivalent.
5. Remove screws securing steering gear to frame and remove gear from vehicle.
6. Reverse procedure to install.

4. REMOVE AND INSTALL WHEEL STUD

REMOVE

1. REMOVE PARTS AS SHOWN.

INSTALL

NOTICE: SEE "NOTICE" AT BEGINNING OF SECTION.

1. INSTALL PARTS AS SHOWN.

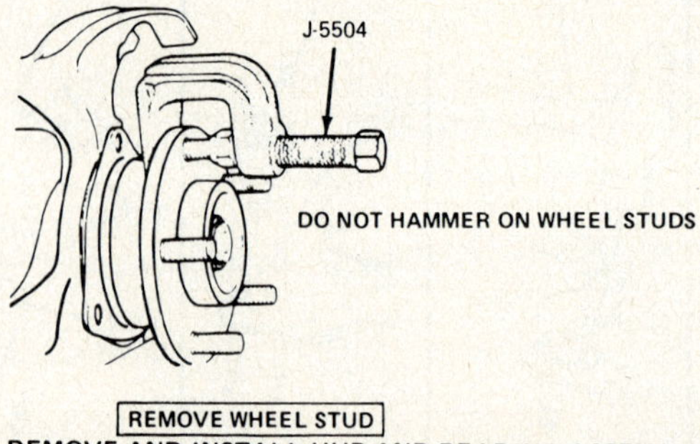

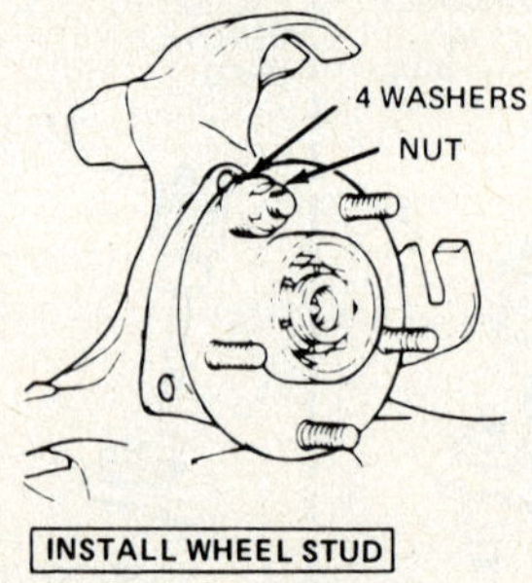

5. REMOVE AND INSTALL HUB AND BEARING ASSEMBLY, KNUCKLE AND SEAL

REMOVE

1. REMOVE PARTS AS SHOWN.

INSTALL

1. INSTALL PARTS AS SHOWN.

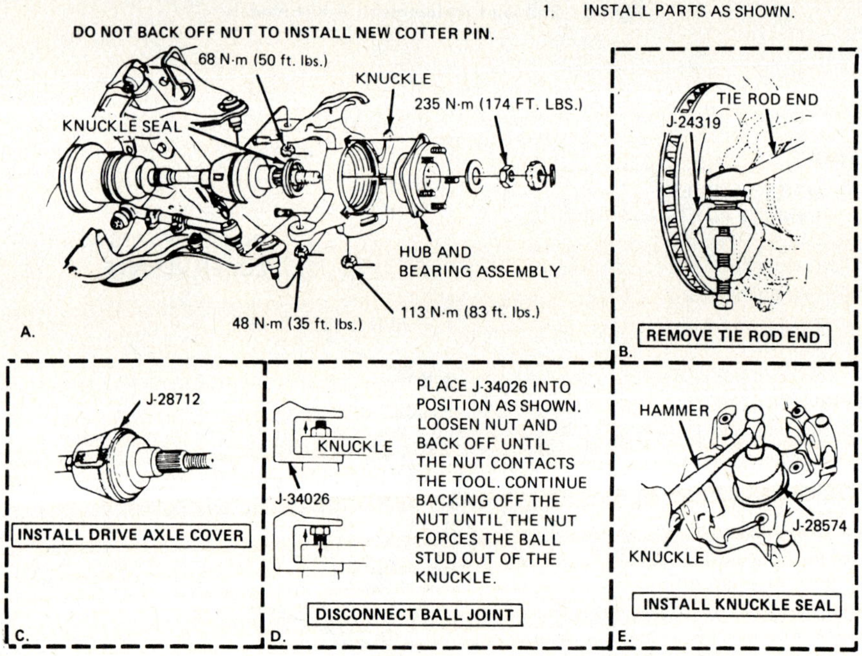

Fig. 17 Steering knuckle replacement. 4 × 4 models

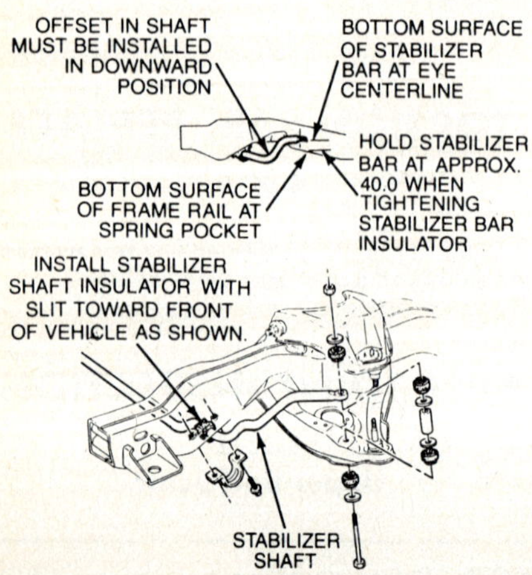

Fig. 18 Stabilizer bar replacement. 4 × 2 models

REMOVE

1. HOIST VEHICLE
2. REMOVE PARTS AS SHOWN.

INSTALL

1. INSTALL PARTS AS SHOWN.
2. LOWER VEHICLE.

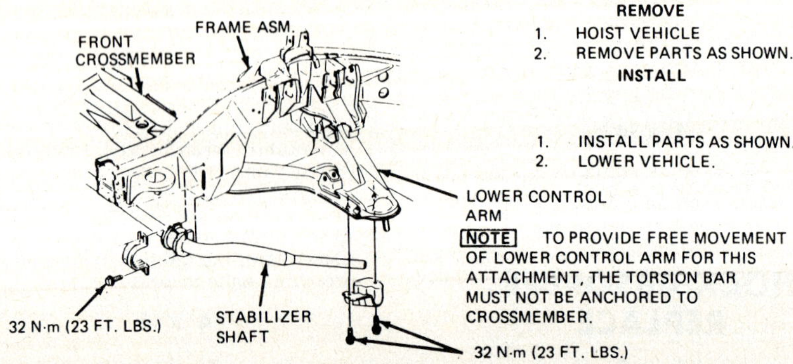

Fig. 19 Stabilizer bar replacement. 4 × 4 models

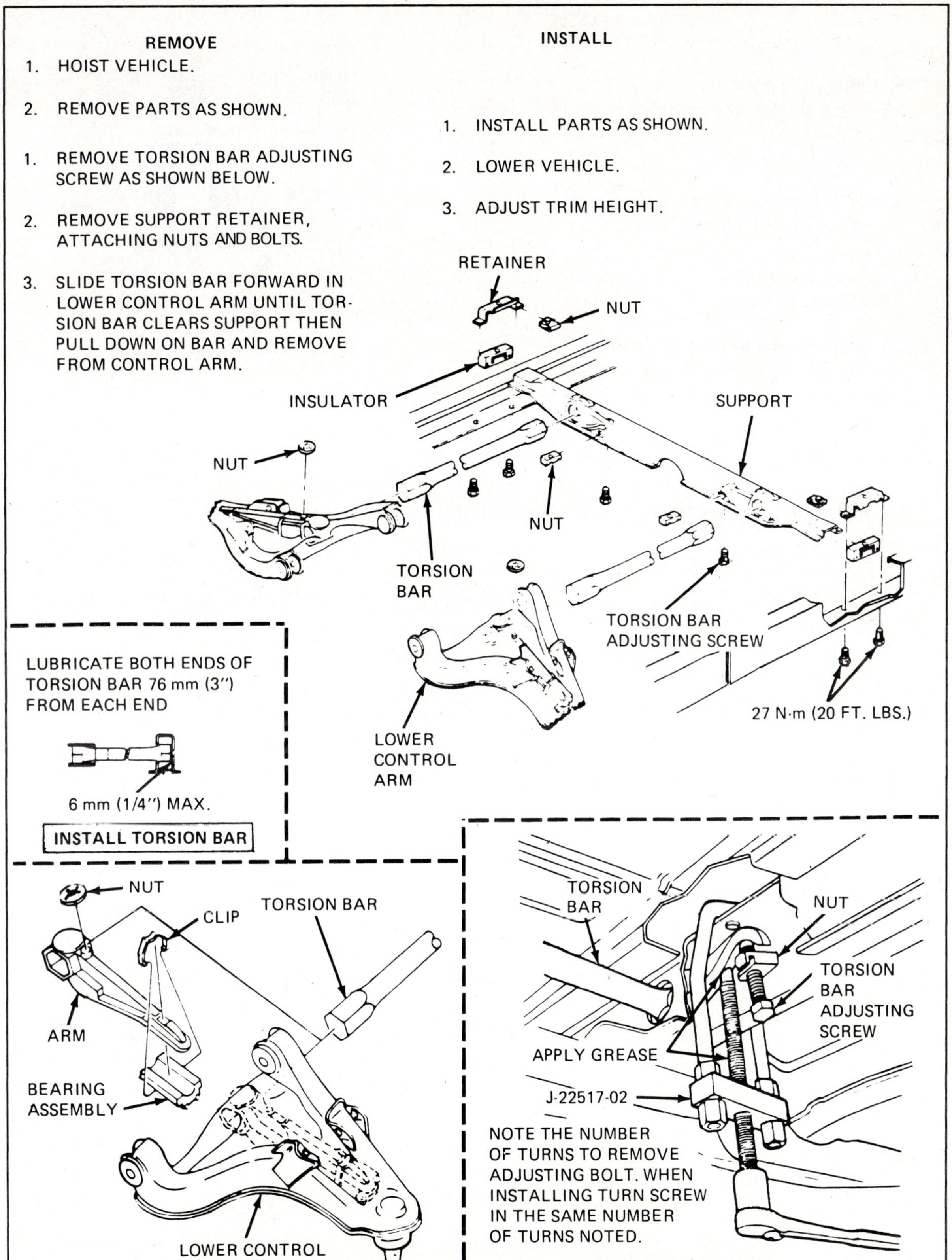

Fig. 20 Torsion bar replacement. 4 × 4 models

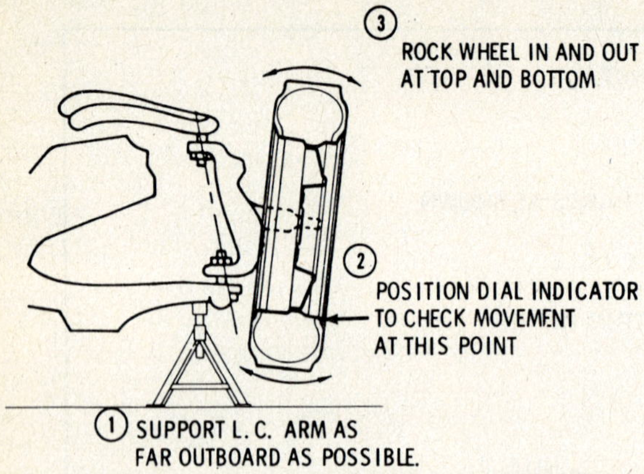

Fig. 21 Checking upper ball joint. 4 × 2 models

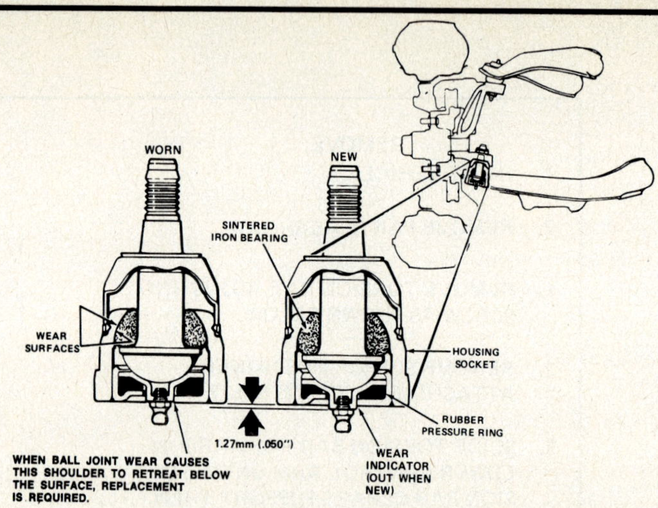

Fig. 22 Lower ball joint wear indicator. 4 × 2 models

1. CHECK BALL JOINT AS SHOWN.
2. IF DIAL INDICATOR READING EXCEEDS 3.2 mm (.125") OR IF BALL STUD IS DISCONNECTED FROM KNUCKLE AND ANY LOOSENESS IS DETECTED OR BALL STUD CAN BE TWISTED IN ITS SOCKET WITH FINGERS, REPLACE BALL JOINT.

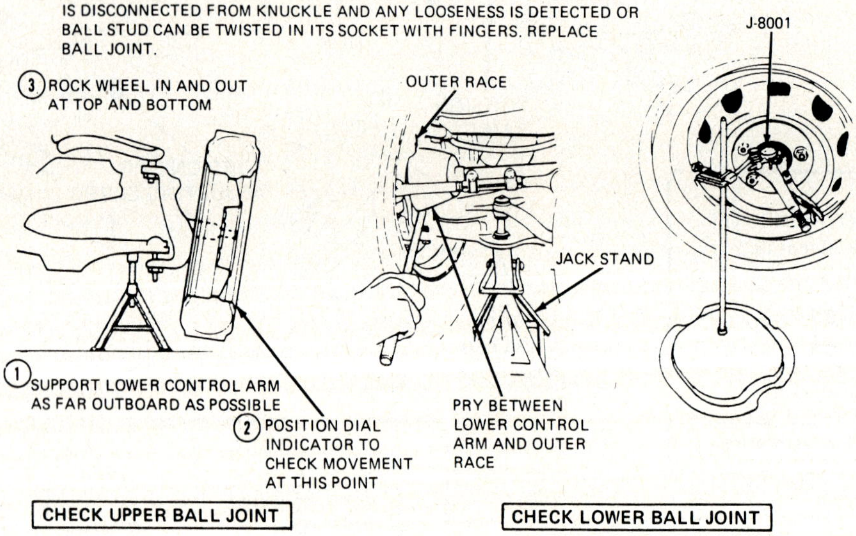

Fig. 23 Checking ball joints. 4 × 4 models

DODGE & PLYMOUTH
(Exc. Rampage & Scamp)

NOTE: Refer to the front of this manual for the manufacturers special service tool suppliers.

INDEX OF SERVICE OPERATIONS

MODEL INDEX & ENGINE APPLICATION

Model	Year	Engine Make	Standard Engine Model	Crankcase Refill Capacity Qts.	Cooling System Capacity Qts.
1974–83					
Compact, Sportsman & Ram Wagon					
B100	1974–78	Own	6-225	5	13
B100	1979–80	Own	6-225	5	12
B100	1974–78	Own	V8-318	5	17
B150	1981–83	Own	6-225	5	12
B200	1974–78	Own	6-225	5	13
B200	1980	Own	6-225	5	12
B200	1974–78	Own	V8-318	5	17
B200	1979	Own	V8-318	5	16
B250	1981–83	Own	6-225	5	12
B250	1981–83	Own	V8-318	5	16
B300	1979	Own	6-225	5	12
B300	1974–78	Own	V8-318	5	17
B300	1979–80	Own	V8-318	5	16
B350	1981–83	Own	V8-318	5	16
Dodge Ramcharger					
AW100	1974–78	Own	6-225	5	13
AW100	1974–78	Own	V8-318	5	17
AW100	1979–80	Own	V8-318	5	16
AW150	1981–83	Own	V8-318	5	16
AD100	1974–78	Own	6-225	5	13
AD100	1979–80	Own	6-225	5	12
AD100	1975–78	Own	V8-318	5	17
AD150	1981–83	Own	V8-318	5	16
Plymouth Trail Duster					
PW100	1974–78	Own	6-225	5	13
PW100	1974–78	Own	V8-318	5	17
PW100	1979–80	Own	V8-318	5	16
PW150	1981	Own	V8-318	5	16
PD100	1974–78	Own	6-225	5	13
PD100	1979–80	Own	6-225	5	12
PD100	1974–78	Own	V8-318	5	17
PD150	1981	Own	V8-318	5	16
Plymouth Voyager					
PB100	1974–78	Own	6-225	5	13
PB100	1979–80	Own	6-225	5	12
PB100	1974–78	Own	V8-318	5	17
PB150	1981–83	Own	6-225	5	12
PB200	1974–78	Own	6-225	5	13
PB200	1980	Own	6-225	5	12
PB200	1974–78	Own	V8-318	5	17
PB200	1979	Own	V8-318	5	16
PB250	1981–83	Own	6-225	5	12
PB250	1981–83	Own	V8-318	5	16
PB300	1974–78	Own	6-225	5	13
PB300	1974–78	Own	V8-318	5	17
PB300	1979–80	Own	V8-318	5	16
PB350	1981–83	Own	V8-318	5	16
Tradesman & Ram Van					
B100	1974–78	Own	6-225	5	13
B100	1979–80	Own	6-225	5	12
B100	1974–78	Own	V8-318	5	17
B150	1981–83	Own	6-225	5	12
B200	1974–78	Own	6-225	5	13
B200	1980	Own	6-225	5	12
B200	1974–78	Own	V8-318	5	17
B200	1979	Own	V8-318	5	16
B250	1981–83	Own	6-225	5	12
B300	1974–78	Own	6-225	5	13
B300	1979–80	Own	6-225	5	12
B300	1974–78	Own	V8-318	5	17
B300	1979–80	Own	V8-318	5	16
B350	1981–83	Own	V8-318	5	16
Conventional Cab					
D100	1974–78	Own	6-225	5	13
D100	1979	Own	6-225	5	12
D100	1974–78	Own	V8-318	5	17
D100	1979	Own	V8-318	5	16
W100	1974–78	Own	6-225	5	13
W100	1974–78	Own	V8-318	5	17
D150	1978	Own	6-225	5	13
D150	1978	Own	V8-318	5	17
D150	1979–83	Own	6-225	5	12
D150	1979–83	Own	V8-318	5	16
W150	1978	Own	6-225	5	13
W150	1978	Own	V8-318	5	17
W150	1979–83	Own	6-225	5	12
W150	1979–83	Own	V8-318	5	16
D200	1974–78	Own	6-225	5	13
D200	1979–80	Own	6-225	5	12
D200	1974–78	Own	V8-318	5	17
D200	1979–80	Own	V8-318	5	16
D250	1981–83	Own	6-225	5	12
D250	1981–83	Own	V8-318	5	16
W200	1974–78	Own	6-225	5	13
W200	1974–78	Own	V8-318	5	17
W200	1979–80	Own	V8-318	5	16
W200	1979–80	Own	V8-360	5	14½

Continued

MODEL INDEX & ENGINE APPLICATION—Continued

Model	Year	Engine Make	Standard Engine Model	Crankcase Refill Capacity Qts.	Cooling System Capacity Qts.
W250	1981–83	Own	V8-318	5	16
D300	1974–78	Own	6-225	5	14
D300	1974–78	Own	V8-318	5	17
D300	1979–80	Own	V8-360	5	14½
D350	1981–83	Own	V8-318	5	16
D350	1981–83	Own	V8-360	5	14.5
W300	1974–78	Own	6-225	5	14
W300	1974–78	Own	V8-318	5	17
W300	1979–80	Own	V8-360	5	14½
W350	1981–83	Own	V8-360	5	14.5
W350	1982–83	Own	V8-318	5	16
D400	1979–80	Own	V8-360	5	14½
D450	1981	Own	V8-360	5	16
W400	1977–78	Own	V8-360	5	16
W400	1979–80	Own	V8-360	5	14½
W450	1981	Own	V8-360	5	16
D500	1974	Own	6-225	5	14
D500	1974–78	Own	V8-318	6	19
D600	1974–78	Own	V8-318	6	19
W600	1974–78	Own	V8-318	6	19
D700	1975	Own	V8-361	8	26
D-700	1976–78	Own	V8-360	6	21
D-800	1974–78	Own	V8-361	8	26

LCF Cab

Model	Year	Engine Make	Standard Engine Model	Crankcase Refill Capacity Qts.	Cooling System Capacity Qts.
C800	1974–75	Own	V8-361	8	27
CN800	1974	Catpl'r	1100	—	—
CNT800	1974	Catpl'r	1100	—	—
CN800	1975	Catpl'r	1145	—	—
CNT800	1975	Catpl'r	1145	—	—
CT800	1974–75	Own	V8-361	8	27
CN900	1974–75	Cummins	NH-230	20	41
CNT900	1974–75	Cummins	NH-230	20	41
CN950	1974–75	Cummins	Super 250	28	45½
CNT950	1974–75	Cummins	Super 250	28	45½

Tilt Cab

Model	Year	Engine Make	Standard Engine Model	Crankcase Refill Capacity Qts.	Cooling System Capacity Qts.
LS1000	1974–75	Cummins	NH-230	20	44
LT1000	1974–75	Cummins	NH-230	20	44

Motor Home

Model	Year	Engine Make	Standard Engine Model	Crankcase Refill Capacity Qts.	Cooling System Capacity Qts.
M300	1975	Own	V8-318-3	6	21
M300	1976–81	Own	V8-360-3	6	21
MB300	1976–78	Own	V8-360	5	16
MB300	1979	Own	V8-360	5	15½
MB300	1980–81	Own	V8-360	5	14½
MB300	1974–75	Own	V8-318	5	17
RM300	1974	Own	V8-318-3	6	21
MBL300	1975–77	Own	V8-318	5	21
MBL300	1978	Own	V8-318	5	18
MBL300	1979	Own	V8-360	5	15½
MBL300	1980	Own	V8-360	5	14½
MBL400	1979	Own	V8-360	5	15½
MBH300	1977	Own	V8-318	5	17
MBH300	1975–77	Own	V8-360	5	16
MBH300	1978	Own	V8-318	5	18
MBH300	1979	Own	V8-360	5	15½
MBH300	1980	Own	V8-360	5	14½
MBH400	1979	Own	V8-360	5	15½
MBH400	1980	Own	V8-360	5	14½
RM350	1974	Own	V8-318-3	6	21
RM400	1974	Own	V8-440-3	6	23
M400	1975	Own	V8-318-3	6	21
M400	1976–81	Own	V8-360-3	6	21
MB400	1976–78	Own	V8-360	5	17
MB400	1979	Own	V8-360	5	15½
MB400	1980–81	Own	V8-360	5	14½
M500	1975–79	Own	V8-440-3	6	23
M500	1980–81	Int'l	V8-446	6	24½
M600	1977–79	Own	V8-440-3	6	23
M600	1980–81	Int'l	V8-446	6	24½

School Bus

Model	Year	Engine Make	Standard Engine Model	Crankcase Refill Capacity Qts.	Cooling System Capacity Qts.
S600	1974–78	Own	V8-318	6	19
S700	1975	Own	V8-361	8	26
S700	1976–78	Own	V8-360	6	21

Kary Van

Model	Year	Engine Make	Standard Engine Model	Crankcase Refill Capacity Qts.	Cooling System Capacity Qts.
CB300	1974–78	Own	6-225	5	13
CB300	1974–78	Own	V8-318	5	17
CB300	1979	Own	V8-360	5	14½
CB300	1980–81	Own	V8-360	5	14½
CB400	1979	Own	V8-360-3	6	14½
CB400	1980–81	Own	V8-360-3	6	14½
D500	1974–78	Own	V8-318	6	19
D600	1974–78	Own	V8-318	6	19
D700	1975	Own	V8-361	8	26
D700	1976–78	Own	V8-360	6	21

DODGE & PLYMOUTH—Exc. Rampage & Scamp

ENGINE IDENTIFICATION

ENGINE NUMBER LOCATION: The engine serial number location is located on a machined surface on the cylinder block. The engine serial number prefix denotes the series letter, followed by engine model, followed by serial number.

GENERAL ENGINE SPECIFICATIONS

Year	Engine Code	Engine Model	Carb. Type	Bore & Stroke	Comp. Ratio	Horsepower @ R.P.M.	Torque Ft. Lbs. @ R.P.M.	Governed Speed R.P.M.	Normal Oil Pressure Lbs.
1974	225-1	6-225①	1 Bore	3.40 × 4.125	8.4	105 @ 4000	180 @ 2000	3600	50
	225-1	6-225②	1 Bore	3.40 × 4.125	8.4	110 @ 4000	175 @ 2000	3600	50
	318-1	V8-318①	2 Bore	3.91 × 3.31	8.6	155 @ 4000	255 @ 2400	3800	50–70
	318-1	V8-318②	2 Bore	3.91 × 3.31	8.6	150 @ 4000	235 @ 2400	3800	50–70
	318-3	V8-318	2 Bore	3.91 × 3.31	7.8	140 @ 4000	250 @ 2000	3700	60–80
	360	V8-360①③	2 Bore	4.00 × 3.58	8.4	185 @ 4000	295 @ 2400	3900	50–70
	360	V8-360①④	2 Bore	4.00 × 3.58	8.4	210 @ 4000	285 @ 2400	3900	50–70
	360	V8-360②	2 Bore	4.00 × 3.58	8.4	175 @ 4000	285 @ 2400	3900	50–70
	361-3	V8-361	2 Bore	4.12 × 3.38	7.49	140 @ 3600	260 @ 2000	3600	60–80
	361-4	V8-361	2 Bore	4.12 × 3.38	7.48	155 @ 3200	295 @ 2000	3600	60–80
	400	V8-400①	2 Bore	4.342 × 3.38	8.2	185 @ 4000	305 @ 2400	—	45–65
	400	V8-400②	2 Bore	4.342 × 3.38	8.2	175 @ 4000	305 @ 2400	—	45–65
	413-2	V8-413	2 Bore	4.188 × 3.75	7.54	165 @ 3200	325 @ 2000	3600	60–80
	413-3	V8-413	4 Bore	4.188 × 3.75	7.54	180 @ 3200	335 @ 2000	3600	60–80
	440-1	V8-440①③	4 Bore	4.32 × 3.75	8.12	230 @ 4000	350 @ 3200	—	45–65
	440-1	V8-440①④	4 Bore	4.32 × 3.75	8.12	225 @ 4000	345 @ 3200	—	45–65
	440-3	V8-440①	4 Bore	4.32 × 3.75	8.2	235 @ 4000	340 @ 2400	—	45–65
1975	225-1	6-225①	1 Bore	3.40 × 4.125	8.4	95 @ 3600	175 @ 2000	3600	30–80
	225-1	6-225②	1 Bore	3.40 × 4.125	8.4	105 @ 3600	175 @ 2000	3600	30–80
	318-1	V8-318①	2 Bore	3.91 × 3.31	8.6	150 @ 4000	255 @ 2000	3900	30–80
	318-1	V8-318②③	2 Bore	3.91 × 3.31	8.6	150 @ 4000	230 @ 2400	3900	30–80
	318-1	V8-318②④	2 Bore	3.91 × 3.31	8.6	155 @ 4000	245 @ 2000	3900	30–80
	318-3	V8-318	2 Bore	3.91 × 3.31	7.8	140 @ 4000	250 @ 2000	3800	30–80
	360	V8-360	2 Bore	4.00 × 3.58	8.4	175 @ 4000	285 @ 2400	3800	30–80
	361-4	V8-361	2 Bore	4.12 × 3.375	7.5	155 @ 3200	295 @ 2000	3600	30–80
	413	V8-413	4 Bore	4.188 × 3.75	7.5	180 @ 3200	335 @ 2000	3600	30–80
	440	V8-440	4 Bore	4.32 × 3.75	8.2	235 @ 4000	340 @ 2400	3600	30–80
1976	225-1	6-225⑤	1 Bore	3.40 × 4.125	8.4	100 @ 3600	170 @ 2000	3600	30–80
	225-1	6-225⑥	1 Bore	3.40 × 4.125	8.4	105 @ 3600	175 @ 2000	3600	30–80
	318-1	V8-318⑤	2 Bore	3.91 × 3.31	8.58	150 @ 4000	255 @ 2000	3900	30–80
	318-1	V8-318③⑥	2 Bore	3.91 × 3.31	8.58	150 @ 4000	230 @ 2400	3900	30–80
	318-1	V8-318④⑥	2 Bore	3.91 × 3.31	8.58	155 @ 4000	245 @ 2000	3900	30–80
	318-3	V8-318	2 Bore	3.91 × 3.31	7.8	140 @ 4000	250 @ 2000	3800	30–80
	360-1	V8-360⑤	2 Bore	4.00 × 3.58	8.4	170 @ 4000	280 @ 2400	3800	30–80
	360-1	V8-360③⑥	2 Bore	4.00 × 3.58	8.4	175 @ 4000	285 @ 2400	3800	30–80
	360-1	V8-360④⑥	2 Bore	4.00 × 3.58	8.4	185 @ 4000	290 @ 2400	3800	30–80
	360-3	V8-360	2 Bore	4.00 × 3.58	8.4	160 @ 4000	260 @ 2000	3800	30–80
	361-4	V8-361	2 Bore	4.12 × 3.38	7.48	150 @ 3200	285 @ 2000	3600	60–80
	400-1	V8-400	2 Bore	4.342 × 3.38	8.2	165 @ 4000	290 @ 2400	—	30–80
	413-3	V8-413	4 Bore	4.188 × 3.75	7.54	175 @ 3200	325 @ 2000	3600	60–80
	440-1	V8-440	4 Bore	4.32 × 3.75	8.2	220 @ 4000	290 @ 2400	—	30–80
	440-3	V8-440	4 Bore	4.32 × 3.75	8.2	225 @ 4000	330 @ 2400	—	30–80
1977	225	6-225⑤	1 Bore	3.40 × 4.125	8.4	100 @ 3600	170 @ 1600	3600	30–80
	225-1	6-225⑥	2 Bore	3.40 × 4.125	8.4	115 @ 3600	175 @ 2000	3600	30–80
	318-1	V8-318⑤	2 Bore	3.91 × 3.31	8.58	150 @ 4000	255 @ 2000	3900	30–80
	318-1	V8-318③⑥	2 Bore	3.91 × 3.31	8.58	150 @ 4000	230 @ 2400	3900	30–80

Continued

GENERAL ENGINE SPECIFICATIONS—Continued

Year	Engine Code	Engine Model	Carb. Type	Bore & Stroke	Comp. Ratio	Horsepower @ R.P.M.	Torque Ft. Lbs. @ R.P.M.	Governed Speed R.P.M.	Normal Oil Pressure Lbs.
	318-1	V8-318④⑥	2 Bore	3.91 × 3.31	8.58	150 @ 4000	240 @ 2000	3900	30–80
	318-3	V8-318②	2 Bore	3.91 × 3.31	7.8	140 @ 4000	250 @ 2000	3800	30–80
	8-360-1	V8-360⑤	2 Bore	4.00 × 3.58	8.4	170 @ 4000	280 @ 2400	3800	30–80
	8-360-1	V8-360③⑥	2 Bore	4.00 × 3.58	8.4	175 @ 4000	285 @ 2400	3800	30–80
	8-360-1	V8-360④⑥	2 Bore	4.00 × 3.58	8.4	170 @ 4000	265 @ 2400	3800	30–80
	8-360-3	V8-360③⑥	2 Bore	4.00 × 3.58	8.4	160 @ 4000	260 @ 2400	3800	30–80
	8-360-3	V8-360④⑥	2 Bore	4.00 × 3.58	8.4	155 @ 3300	255 @ 2400	3800	30–80
	8-361-4	V8-361②	2 Bore	4.12 × 3.38	7.5	150 @ 3200	285 @ 2000	3600	60–80
	8-400-1	V8-400⑤	2 Bore	4.342 × 3.38	8.2	170 @ 4000	300 @ 2000	—	30–80
	8-400-1	V8-400⑥	2 Bore	4.342 × 3.38	8.2	165 @ 4000	290 @ 2400	—	30–80
	8-413-3	V8-413②	4 Bore	4.18 × 3.75	7.5	175 @ 3200	325 @ 2000	3600	60–80
	8-440-1	V8-440③⑥	4 Bore	4.32 × 3.75	8.2	220 @ 4000	330 @ 2400	—	30–80
	8-440-1	V8-440④⑥	4 Bore	4.32 × 3.75	8.2	220 @ 4000	320 @ 2400	—	30–80
	8-440-3	V8-400③⑥	4 Bore	4.32 × 3.75	8.2	225 @ 4000	330 @ 2400	—	30–80
	8-440-3	V8-440④⑥	4 Bore	4.32 × 3.75	8.2	225 @ 4000	320 @ 2400	—	30–80
1978	225-1	6-225⑤	2 Bore	3.40 × 4.125	8.4	115 @ 3600	175 @ 1600	—	30–70
	225-1	6-225⑥	2 Bore	3.40 × 4.125	8.4	115 @ 3600	175 @ 2000	—	30–70
	318-1	V8-318⑤	2 Bore	3.91 × 3.31	8.6	145 @ 4000	250 @ 2000	—	30–80
	318-1	V8-318⑥	2 Bore	3.91 × 3.31	8.6	150 @ 4000	230 @ 2400	—	30–80
	360-1	V8-360⑤	2 Bore	4.00 × 3.58	8.5	160 @ 4000	280 @ 2000	—	30–80
	360-1	V8-360⑥	2 Bore	4.00 × 3.58	8.5	175 @ 4000	275 @ 2400	—	30–80
	360-1	V8-360④	4 Bore	4.00 × 3.58	8.0	—	—	—	30–80
	360-3	V8-360②	2 Bore	4.00 × 3.58	8.0	160 @ 3600	225 @ 2400	—	30–80
	400-1	V8-400⑤	2 Bore	4.342 × 3.38	8.2	170 @ 4000	300 @ 2000	—	30–80
	400-1	V8-400⑥	2 Bore	4.342 × 3.38	8.2	165 @ 4000	290 @ 2400	—	30–80
	440-1	V8-440⑥	4 Bore	4.32 × 3.75	8.2	220 @ 4000	330 @ 2400	—	30–80
	440-3	V8-440②	4 Bore	4.32 × 3.75	8.1	255 @ 4000	330 @ 2400	—	30–80
1979	225-1	6-225⑤	1 Bore	3.40 × 4.125	8.4	110 @ 3600	175 @ 1600	—	30–70
	225-1	6-225⑥	1 Bore	3.40 × 4.125	8.4	115 @ 3600	175 @ 2000	—	30–70
	225-1	6-225④	2 Bore	3.40 × 4.125	8.4	110 @ 3600	—	—	30–70
	318-1	V8-318⑤	2 Bore	3.91 × 3.31	8.7	150 @ 4000	250 @ 2000	—	30–80
	318-1	V8-318⑥	2 Bore	3.91 × 3.31	8.7	150 @ 4000	230 @ 2400	—	30–80
	318-1	V8-318④⑤	4 Bore	3.91 × 3.31	8.7	150 @ 3600	—	—	30–80
	318-1	V8-318④⑦	4 Bore	3.91 × 3.31	8.7	155 @ 3800	—	—	30–80
	360-1	V8-360⑤	2 Bore	4.00 × 3.58	8.6	160 @ 3600	290 @ 2000	—	30–80
	360-1	V8-360⑥	2 Bore	4.00 × 3.58	8.6	180 @ 3600	270 @ 2000	—	30–80
	360-1	V8-360⑤	4 Bore	4.00 × 3.58	8.6	215 @ 4000	300 @ 3200	—	30–80
	360-1	V8-360⑥	4 Bore	4.00 × 3.58	8.6	175 @ 3600	280 @ 2000	—	30–80
	360-3	V8-360⑥	2 Bore	4.00 × 3.58	8.0	180 @ 3600	260 @ 2400	—	30–80
	360-1	V8-360④⑦	4 Bore	4.00 × 3.58	8.6	175 @ 3600	—	—	30–80
	360-1	V8-360④⑥	4 Bore	4.00 × 3.58	8.6	180 @ 3600	—	—	30–80
	440-1	V8-440⑥	4 Bore	4.32 × 3.75	8.2	220 @ 4000	320 @ 4000	—	30–80
	440-3	V8-440⑥	4 Bore	4.32 × 3.75	8.1	225 @ 2400	320 @ 2400	—	30–80
1980	225	6-225⑤	1 Bore	3.40 × 4.125	8.4	95 @ 3600	170 @ 1600	—	30–80
	225	6-225④⑦	1 Bore	3.40 × 4.125	8.4	90 @ 3600	150 @ 2000	—	30–80
	318	V8-318⑤	2 Bore	3.91 × 3.31	8.7	135 @ 4000	240 @ 2000	—	30–80
	318	V8-318⑤	2 Bore	3.91 × 3.31	8.7	140 @ 4000	240 @ 2400	—	30–80
	318	V8-318⑦	4 Bore	3.91 × 3.31	8.7	160 @ 4000	245 @ 2000	—	30–80
	318	V8-318④⑥	4 Bore	3.91 × 3.31	8.0	155 @ 4000	240 @ 2000	—	30–80
	360-1	V8-360⑤	4 Bore	4.00 × 3.58	8.6	170 @ 4000	270 @ 2000	—	30–80
	360-1	V8-360⑦	4 Bore	4.00 × 3.58	8.6	170 @ 4000	270 @ 2000	—	30–80
	360-1	V8-360⑤	4 Bore	4.00 × 3.58	8.6	205 @ 4000	295 @ 3200	—	30–80
	360-1	V8-360⑤⑦	4 Bore	4.00 × 3.58	8.6	175 @ 3600	260 @ 2800	—	30–80
	360-1	V8-360⑥	4 Bore	4.00 × 3.58	8.2	180 @ 3600	270 @ 2000	—	30-80
	360-3	V8-360⑥	4 Bore	4.00 × 3.58	8.0	180 @ 4000	260 @ 2400	—	30–80

Continued

GENERAL ENGINE SPECIFICATIONS—Continued

Year	Engine Code	Engine Model	Carb. Type	Bore & Stroke	Comp. Ratio	Horsepower @ R.P.M.	Torque Ft. Lbs. @ R.P.M.	Governed Speed R.P.M.	Normal Oil Pressure Lbs.
	360-3	V8-360⑥	4 Bore	4.00 × 3.58	8.0	210 @ 4000	285 @ 3200	—	30–80
	360-3	V8-360④⑥	4 Bore	4.00 × 3.58	8.0	175 @ 3600	260 @ 2800	—	30–80
	360-3	V8-360④⑥	4 Bore	4.00 × 3.58	8.0	200 @ 4000	280 @ 3200	—	30–80
	446⑧	V8-446⑥	4 Bore	4.125 × 4.18	8.0	230 @ 3600	360 @ 2800	—	15–50
1981–82	225	6-225③⑤	1 Bore	3.40 × 4.125	8.4	95 @ 3600	170 @ 1600	—	30–70
	225	6-225④⑦	1 Bore	3.40 × 4.125	8.4	90 @ 3600	165 @ 1200	—	30–70
	318	V8-318③⑤	2 Bore	3.91 × 3.31	8.6	140 @ 3600	240 @ 2400	—	30–80
	318	V8-318③⑤	2 Bore	3.91 × 3.31	8.6	135 @ 4000	240 @ 2000	—	30–80
	318	V8-318③⑤	4 Bore	3.91 × 3.31	8.6	170 @ 4000	245 @ 2000	—	30–80
	318	V8-318⑥	4 Bore	3.91 × 3.31	8.6	160 @ 4000	220 @ 3200	—	30–80
	318	V8-318④⑦	4 Bore	3.91 × 3.31	8.6	160 @ 4000	245 @ 2000	—	30–80
	360	V8-360③⑤	4 Bore	4.00 × 3.58	8.5	175 @ 4000	260 @ 2000	—	30–80
	360	V8-360④⑤	4 Bore	4.00 × 3.58	8.5	180 @ 4000	260 @ 2000	—	30–80
	360	V8-360④⑥	4 Bore	4.00 × 3.58	8.5	170 @ 4000	265 @ 2400	—	30–80
	360	V8-360③⑥	4 Bore	4.00 × 3.58	8.5	180 @ 3600	270 @ 2000	—	30–80
1983	225	6-225	1 Bore	3.40 × 4.125	8.4	—	—	—	30–70
	318	V8-318	2 Bore	3.91 × 3.31	8.6	—	—	—	30–80
	318	V8-318	4 Bore	3.91 × 3.31	8.6	—	—	—	30–80
	360	V8-360	4 Bore	4.00 × 3.58	8.5	—	—	—	30–80

①—Light duty.
②—Heavy duty.
③—Exc. Calif.
④—Calif.
⑤—Light duty emissions.
⑥—Heavy duty emissions.
⑦—Medium duty emissions.
⑧—See International chapter for service on this engine.

TUNE UP SPECIFICATIONS

The following specifications are published from the latest information available. This data should be used only in the absence of a decal affixed in the engine compartment.

★ When using a timing light, disconnect vacuum hose or tube at distributor and plug opening in hose or tube so idle speed will not be affected.

● When checking compression, lowest cylinder must be within 80 percent of highest.

▲ Before removing wires from distributor cap, determine location of the No. 1 wire in cap, as distributor position may have been altered from that shown at the end of this chart.

Year & Engine	Spark Plug		Ignition Timing BTDC①★				Curb Idle Speed②		Fuel Pump Pressure
	Type	Gap	Firing Order Fig. ▲	Man. Trans.	Auto Trans.	Mark Location	Man. Trans.	Auto. Trans.	
1983									
6-225, 1 Bbl Exc. Calif.	560PR㉙	.035	A	12°	16°	Damper	600	650N	3–4½
6-225, 1 Bbl Calif.	560PR㉙	.035	A	12°	16°	Damper	750	750N	3–4½
6-225, 2 Bbl Exc. Calif.	560PR㉙	.035	A	12°	—	Damper	700	—	3–4½
V8-318, 2 Bbl Exc. Calif.	65PR㉙	.035	B	12°	12°	Damper	750	750N	4¾–6¼
V8-318, 2 Bbl Calif.	65PR㉙	.035	B	16°	16°	Damper	700	700N	4¾–6¼
V8-318, 2 Bbl High Alt.	65PR㉙	.035	B	16°	16°	Damper	700	700N	4¾–6¼
V8-318, 4 Bbl Exc. Calif.⑤	65PR㉙	.035	B	12°	16°	Damper	750	750N	4¾–6¼
V8-318, 4 Bbl Calif.⑤	65PR㉙	.035	B	12°	—	Damper	750	—	4¾–6¼
V8-318, 4 Bbl④	65PR㉙	.035	B	8°	8°	Damper	750	750N	4¾–6¼
V8-360, 4 Bbl Exc. Calif.④	65PR㉙	.035	B	4°	4°	Damper	700	700N	4¾–6¼
V8-360, 4 Bbl Calif.④	65PR㉙	.035	B	10°	10°	Damper	750	750N	4¾–6¼
1982									
6-225, Exc. Calif.	560PR㉙	.035	A	12°	16°	Damper	600	600N	3½–5
6-225, Calif.	560PR㉙	.035	A	12°	16°	Damper	800	800N	3½–5
6-225-2, 2 Bbl	560PR㉙	.035	A	12°	—	Damper	700	—	3½–5
V8-318, 2 Bbl Exc. Calif.	64PR㉙	.035	B	12°	12°	Damper	750	750N	5–7
V8-318, 4 Bbl	64PR㉙	.035	B	12°	16°	Damper	750	750N	5–7
V8-318, 4 Bbl④	64PR㉙	.035	B	8°	8°	Damper	750	750N	5–7

Continued

TUNE UP SPECIFICATIONS—Continued

The following specifications are published from the latest information available. This data should be used only in the absence of a decal affixed in the engine compartment.

★ When using a timing light, disconnect vacuum hose or tube at distributor and plug opening in hose or tube so idle speed will not be affected.

● When checking compression, lowest cylinder must be within 80 percent of highest.

▲ Before removing wires from distributor cap, determine location of the No. 1 wire in cap, as distributor position may have been altered from that shown at the end of this chart.

Year & Engine	Spark Plug		Ignition Timing BTDC①★				Curb Idle Speed②		Fuel Pump Pressure
	Type	Gap	Firing Order Fig. ▲	Man. Trans.	Auto Trans.	Mark Location	Man. Trans.	Auto. Trans.	
1982—Continued									
V8-360, 4 Bbl Exc. Calif.④	65PR㉙	.035	B	4°	4°	Damper	700	700N	5–7
V8-360, 4 Bbl Calif.④	65PR㉙	.035	B	10°	10°	Damper	750	750N	5–7
1981									
6-225 Exc. Calif.	560PR㉙	.035	A	12°	16°	Damper	600	600N	3½–5
6-225 Calif.	560PR㉙	.035	A	12°	16°	Damper	800	800N	3½–5
V8-318 2 Barrel Carb.	64PR㉙	.035	B	10°	16°	Damper	650	650N	5–7
V8-318 4 Barrel Carb. Exc. Calif.⑤	64PR㉙	.035	B	—	16°	Damper	—	750N	5–7
V8-318 4 Barrel Carb. Calif.⑤㉚	64PR㉙	.035	B	12°	16°	Damper	750	750N	5–7
V8-318 4 Barrel Carb. Calif.⑤㉛	64PR㉙	.035	B	16°	16°	Damper	700	700N	5–7
V8-318 4 Barrel Carb.④	64PR㉙	.035	B	12°	12°	Damper	750	750N	5–7
V8-360-1 Exc. Calif.⑤	65PR㉙	.035	B	12°	16°	Damper	600	625N	5–7
V8-360-1 Calif.⑤㉜	65PR㉙	.035	B	12°	16°	Damper	750	750N	5–7
V8-360-1 Calif.⑤㉝	65PR㉙	.035	B	16°	16°	Damper	700	700N	5–7
V8-360-1④	65PR㉙	.035	B	—	4°	Damper	—	700N	5–7
V8-360-3㉞	73PR㉙	.035	B	—	4°	Damper	—	700N	5–7
V8-360-3㉟	73PR㉙	.035	B	—	10°	Damper	—	750N	5–7
1980									
6-225 Exc. Calif.	RBL16Y㉘	.035	A	12°	12°	Damper	600	600N	3½–5
6-225 Calif.	RBL16Y㉘	.035	A	12°	12°	Damper	800	800N	3½–5
V8-318 2 Barrel Carb.	RN11Y㉘	.035	B	12°	12°	Damper	600	600N	5–7
V8-318 4 Barrel Carb.③	RN11Y㉘	.035	B	6°	6°	Damper	750	750N	5–7
V8-318 4 Barrel Carb.④	RN11Y㉘	.035	B	8°	8°	Damper	750	750N	5–7
V8-360-1 4 Barrel Carb. Exc. Calif.⑤	RN12Y㉘	.035	B	12°	12°	Damper	650	650N	5–7
V8-360-1 4 Barrel Carb. Exc. Calif.④	RN12Y㉘	.035	B	4°	4°	Damper	700	700N	5–7
V8-360-1 4 Barrel Carb. Calif.	RN12Y㉘	.035	B	10°	10°	Damper	750	750N	5–7
V8-360-3 4 Barrel Carb. Exc. Calif.	RF10㉘	.035	B	4°	4°	Damper	700	700N	5–7
V8-360-3 4 Barrel Carb. Calif.⑥	RF10㉘	.035	B	10°	10°	Damper	750	750N	5–7
V8-360-3 4 Barrel Carb. Calif.⑦	RF10㉘	.035	B	4°	4°	Damper	700	700N	5–7
MV-446	RBN13Y㉘	.035	⑧	5°⑨	5°⑨	Damper	525/575	625/675N	4½–5¾
1979									
6-225 Exc. Calif.	RBL16Y㉘	.035	A	12°	12°	Damper	675	675N	3½–5
6-225 Calif.	RBL16Y㉘	.035	A	8°	8°	Damper	800	800N	3½–5
V8-318 Exc. Calif.	RN12Y㉘	.035	B	12°	12°	Damper	680	680N	5–7

Continued

TUNE UP SPECIFICATIONS—Continued

The following specifications are published from the latest information available. This data should be used only in the absence of a decal affixed in the engine compartment.

★ When using a timing light, disconnect vacuum hose or tube at distributor and plug opening in hose or tube so idle speed will not be affected.

● When checking compression, lowest cylinder must be within 80 percent of highest.

▲ Before removing wires from distributor cap, determine location of the No. 1 wire in cap, as distributor position may have been altered from that shown at the end of this chart.

Year & Engine	Spark Plug		Ignition Timing BTDC①★				Curb Idle Speed②		Fuel Pump Pressure
	Type	Gap	Firing Order Fig. ▲	Man. Trans.	Auto Trans.	Mark Location	Man. Trans.	Auto. Trans.	
1979—Continued									
V8-318 Calif.⑩	RN12Y㉘	.035	B	6°	8°	Damper	750	750N	5–7
V8-318 Calif.③	RN12Y㉘	.035	B	6°	6°	Damper	750	750N	5–7
V8-360 2 Barrel Carb. Exc. Calif.⑪	RN12Y㉘	.035	B	10°	10°	Damper	750	750N	5–7
V8-360-3 2 Barrel Carb.⑫	RF10㉘	.035	B	—	TDC	Damper	—	⑬	5–7
V8-360 4 Barrel Carb. Exc. Calif.⑤	RN12Y㉘	.035	B	—	4°	Damper	—	750N	5–7
V8-360 4 Barrel Carb. Calif.③	RN12Y㉘	.035	B	10°	10°	Damper	750	750N	5–7
V8-360 4 Barrel Carb.④	RN12Y㉘	.035	B	4°	4°	Damper	700	700N	5–7
V8-440⑪	OJ11Y㉘	.035	C	—	8°	Damper⑭	—	700N	5–7
V8-440-3⑫	OB69Y㉘	.035	C	—	8°	Damper⑭	—	700N	5–7
1978									
6-225 Exc. Calif.⑩	RBL16Y㉘	.035	A	12°	12°	Damper	750	750N	3½–5
6-225 Exc. Calif.⑮	RBL16Y㉘	.035	A	TDC	TDC	Damper	700	700N	3½–5
6-225 Calif.	RBL16Y㉘	.035	A	8°	8°	Damper	750	750N	3½–5
V8-318 Exc. Calif.⑩	RN11Y㉘	.035	B	12°	12°	Damper	750	750N	5–7
V8-318 Exc. Calif.⑮	RN11Y㉘	.035	B	2°⑯	2°⑯	Damper	750	750N	5–7
V8-318 Calif.	RN11Y㉘	.035	B	12°	12°	Damper	750	750N	5–7
V8-360 2 Barrel Carb. Exc. Calif.⑩	RN12Y㉘	.035	B	6°	6°	Damper	750	750N	5–7
V8-360 2 Barrel Carb. Exc. Calif.⑮	RN12Y㉘	.035	B	4°	4°	Damper	750	750N	5–7
V8-360 2 Barrel Carb. Calif.④	RN12Y㉘	.035	B	TDC	TDC	Damper	700	700N	5–7
V8-360-3 2 Barrel Carb.	RF10㉘	.035	B	—	TDC	Damper	—	⑬	5–7
V8-360 4 Barrel Carb. Exc. Calif.	RN12Y㉘	.035	B	4°	4°	Damper	800	800N	5–7
V8-360 4 Barrel Carb. Calif.	RN12Y㉘	.035	B	6°	6°	Damper	750	750N	5–7
V8-400-1	OJ11Y㉘	.035	C	2°	2°	Damper	700	700N	5–7
V8-440-1 Exc. Calif.	OJ11Y㉘	.035	C	8°	8°	Damper	700	700N	5–7
V8-440-1 Calif.	OJ11Y㉘	.035	C	—	8°	Damper⑭	—	⑰	5–7
V8-440-3	OBL9Y㉘	.035	C	—	8°	Damper⑭	—	700N	5–7
1977									
6-225 Exc. Calif.⑩	RBL15Y㉘	.035	A	2°	2°	Damper	750	750N	3½–5
6-225 Calif.⑩	RBL15Y㉘	.035	A	TDC	2°⑯	Damper	750	750N	3½–5
6-225 Exc. Calif.⑮	RBL11Y㉘	.035	A	TDC	TDC	Damper	700	700N	3½–5
V8-318⑩	RN11Y㉘	.035	B	2°	⑱	Damper	750	750N	5–7
V8-318-1 Exc. Calif.⑮	RN11Y㉘	.035	B	2°⑯	2°⑯	Damper	750	750N	5–7
V8-318-1 Calif.⑮	RN11Y㉘	.035	B	TDC	TDC	Damper	700	700N	5–7
V8-318-3	RF10㉘	.035	B	TDC	TDC	Damper	700	700N	5–7
V8-360⑩	RN12Y㉘	.035	B	—	6°	Damper⑲	—	700N	5–7
V8-360-1 Exc. Calif.⑮	RN12Y㉘	.035	B	⑳	⑳	Damper⑲	750	750N	5–7
V8-360-1 Calif.⑮	RN12Y㉘	.035	B	—	TDC	Damper⑲	—	700D	5–7

Continued

TUNE UP SPECIFICATIONS—Continued

The following specifications are published from the latest information available. This data should be used only in the absence of a decal affixed in the engine compartment.

★ When using a timing light, disconnect vacuum hose or tube at distributor and plug opening in hose or tube so idle speed will not be affected.

● When checking compression, lowest cylinder must be within 80 percent of highest.

▲ Before removing wires from distributor cap, determine location of the No. 1 wire in cap, as distributor position may have been altered from that shown at the end of this chart.

Year & Engine	Spark Plug		Ignition Timing BTDC[1]★				Curb Idle Speed[2]		Fuel Pump Pressure
	Type	Gap	Firing Order Fig. ▲	Man. Trans.	Auto Trans.	Mark Location	Man. Trans.	Auto. Trans.	
1977—Continued									
V8-360-3	RF10[28]	.035	B	TDC	TDC	Damper[19]	[13]	[13]	5–7
V8-361-4	RN6[28]	.035	C	2½°	2½°	Damper	700	700D	6–7½
V8-400-1	RJ11Y[28]	.035	C	2°	2°	Damper[14]	700	700D	5–7
V8-413-3	RN6[28]	.035	[21]	[22]	[22]	Damper[14]	700	700D	6–7½
V8-440-1	RJ11Y[28]	.035	C	8°	8°	Damper[14]	700	700D	5–7
V8-440-3	BL9Y[28]	.035	C	8°	8°	Damper[14]	700	700D	5–7
1976									
6-225 Exc. Calif.[10]	BL11Y[28]	.035	A	2°	2°	Damper	800	750N	3½–5
6-225 Calif.[10]	BL11Y[28]	.035	A	TDC	TDC	Damper	800	750N	3½–5
6-225-1[15]	BL11Y[28]	.035	A	TDC	TDC	Damper	700	700N	3½–5
V8-318 Exc. Calif.[10]	N11Y[28]	.035	B	2°	2°	Damper	750	750N	5–7
V8-318 Calif.[10]	N11Y[28]	.035	B	TDC	TDC	Damper	750	750N	5–7
V8-318-1[15]	N11Y[28]	.035	B	2°[16]	2°[16]	Damper	750	750N	5–7
V8-318-3 Exc. Calif.	F10[28]	.035	B	TDC	TDC	Damper	700	700N	5–7
V8-318-3 Calif.	F10[28]	.035	B	2°	2°	Damper	700	700N	5–7
V8-360 Exc. Calif.[10]	N12Y[28]	.035	B	—	6°	Damper	700	700N	5–7
V8-360 Calif.[10]	N12Y[28]	.035	B	—	4°	Damper	700	700N	5–7
V8-360-1 Exc. Calif.[15]	N12Y[28]	.035	B	TDC	TDC	Damper	700	700N	5–7
V8-360-1 Calif.[15]	N12Y[28]	.035	B	4°	4°	Damper	700	700N	5–7
V8-360-3	F10[28]	.035	B	TDC	TDC	Damper	[13]	[13]	5–7
V8-361-4	N6[28]	.035	C	2½°	2½°	Damper	700	700N	6–7½
V8-400	J11Y[28]	.035	C	—	8°	Damper[14]	700	700N	5–7
V8-413-3	N6[28]	.035	[21]	[22]	[22]	Damper	700	700N	6–7½
V8-440-1	J11Y[28]	.035	C	8°	8°	Damper[14]	700	700N	5–7
V8-440-3	BL9Y[28]	.035	C	8°	8°	Damper[14]	700	700N	5–7
1975									
6-225[10]	BL11Y[28]	.035	A	TDC	TDC	Damper	800	750N	3½–5
6-225-1[15]	BL11Y[28]	.035	A	TDC	TDC[23]	Damper	700	700N	3½–5
V8-318 Exc. Calif.[10]	N11Y[28]	.035	B	2°	2°	Damper	750	750N	5–7
V8-318 Calif.[10]	N11Y[28]	.035	B	TDC	TDC	Damper	750	750N	5–7
V8-318-1 Exc. Calif.[15]	N11Y[28]	.035	B	2°[16]	2°[16]	Damper	750	750N	5–7
V8-318-1 Calif.[15]	N11Y[28]	.035	B	TDC	TDC	Damper	700	700N	5–7
V8-318-3 Exc. Calif.	F10[28]	.035	B	TDC	TDC	Damper	700	700N	5–7
V8-318-3 Calif.	F10[28]	.035	B	2°	2°	Damper	700	700N	5–7
V8-360-1 Exc. Calif.	N12Y[28]	.035	B	TDC	TDC	Damper	750	750N	5–7
V8-360-1 Calif.	N12Y[28]	.035	B	4°	4°	Damper	700	700N	5–7
V8-360-3	F10[28]	.035	B	TDC	TDC	Damper	[13]	[13]	5–7
V8-361-4	N6[28]	.035	C	2½°	2½°	Damper	700	700N	6–7½
V8-413-3	N6[28]	.035	[21]	[22]	[22]	Damper	700	700N	6–7½
V8-440-1	J11Y[28]	.035	C	8°	8°	Damper	700	700N	5–7
V8-440-3	BL9Y[28]	.035	C	8°	8°	Damper	700	700N	5–7
1974									
6-225	N11Y[28]	.035	A	TDC[24]	TDC[24]	Damper	800	750N	3½–5
V8-318-1	N11Y[28]	.035	B	TDC[25]	TDC[25]	Damper	750	750N	5–7

Continued

TUNE UP SPECIFICATIONS—Continued

The following specifications are published from the latest information available. This data should be used only in the absence of a decal affixed in the engine compartment.

★ When using a timing light, disconnect vacuum hose or tube at distributor and plug opening in hose or tube so idle speed will not be affected.

● When checking compression, lowest cylinder must be within 80 percent of highest.

▲ Before removing wires from distributor cap, determine location of the No. 1 wire in cap, as distributor position may have been altered from that shown at the end of this chart.

Year & Engine	Spark Plug		Ignition Timing BTDC①★				Curb Idle Speed②		Fuel Pump Pressure
	Type	Gap	Firing Order Fig. ▲	Man. Trans.	Auto Trans.	Mark Location	Man. Trans.	Auto. Trans.	
V8-318-3	E10㉘	.035	B	TDC	TDC	Damper	700	700N	5–7
V8-360 Exc. Calif.⑩	N12Y㉘	.035	B	2½°	5°	Damper	750	700N	5–7
V8-360 Calif.⑩	N12Y㉘	.035	B	—	2½°	Damper	—	㉖	5–7
V8-360⑮	N12Y㉘	.035	B	TDC	TDC	Damper	750	750N	5–7
V8-361	N6㉘	.035	C	2½°	2½°	Damper	700	700N	6–7½
V8-400⑩	J13Y㉘	.035	C	7½°	7½°	Damper	750	750N	5–7
V8-400⑮	J11Y㉘	.035	C	2½°	2½°	Damper	750	750N	5–7
V8-413-2	N6㉘	.035	㉑	2½°	—	Damper	700	—	6–7½
V8-413-3	N6㉘	.035	㉑	5°	—	Damper	700	—	6–7½
V8-440⑩	J11Y㉘	.035	C	10°㉗	10°㉗	Damper	700	700N	5–7
V8-440-1	J11Y㉘	.035	C	7½°	7½°	Damper	700	700N	5–7
VS-440-3	BL9Y㉘	.035	C	7½°	7½°	Damper	700	700N	5–7

①—B.T.D.C.—Before top dead center.
②—N: Neutral.
③—Medium duty emission, GVWR 6001 to 8500 lbs.
④—Heavy duty emissions, GVWR 8501 lbs. & above.
⑤—Light duty emissions, GVWR, 8500 lbs & under.
⑥—Exc. models with carburetor No. TQ-9261S.
⑦—Models with carburetor No. TQ-9261S.
⑧—Cylinder numbering (front to rear), right bank, 1-3-5-7; left bank, 2-4-6-8. Firing order, 1-2-7-3-4-5-6-8.
⑨—There are two timing mark plates located on the engine front cover. If the upper timing plate is to be used to check timing, connect timing light to No. 1 cylinder. If the lower timing plate is used to check timing, connect timing light to No. 7 cylinder.
⑩—Light duty emissions, GVWR 6000 lbs. & under.
⑪—Except motor home chassis.
⑫—Motor home chassis.
⑬—Except Calif., 750N RPM; California, 700N RPM.
⑭—On B series, Voyager & motor home chassis models, timing mark located on torque converter.
⑮—Heavy duty emissions, GVWR 6001 lbs. & above.
⑯—A.T.D.C.—After top dead center.
⑰—Medium duty emissions, GVWR 6001 to 8500 lbs., 750N RPM; heavy duty emissions, GVWR 8501 lbs. & above, 700N RPM.
⑱—Except high altitude, 2° BTDC; high altitude, 6° BTDC.
⑲—On B series & Voyager models w/air pump, timing mark located on torque converter.
⑳—Models equipped with Holley 2210 carburetor, TDC; models equipped with Holley 2245 carburetor, 4° BTDC.
㉑—To determine proper firing order diagram it is necessary to observe rotor rotation. Use Fig. C if rotation is counter clockwise & Fig. D if clockwise. This is caused by the fact that some engines use timing gears while others use a timing chain.
㉒—Except school bus, 5° BTDC; TDC.
㉓—California vehicles built after September 11, 1975, set at 2° ATDC.
㉔—Distributor #3755056, 2½°
㉕—Distributor #3755201, 2½°
㉖—Series 100–800, 750 RPM; except Series, 100–800, 850 RPM.
㉗—California, 5° BTDC.
㉘—Champion
㉙—Mopar.
㉚—Models w/distributor No. 4111501 for man. trans., No. 4145602 for auto. trans.
㉛—Models w/distributor No. 4145753.
㉜—Models w/distributor No. 4145364 for man. trans., No. 4145604 for auto. trans.
㉝—Models w/distributor No. 4145350.
㉞—Models less catalytic converter.
㉟—Models w/catalytic converter.

Continued

TUNE UP NOTES—Continued

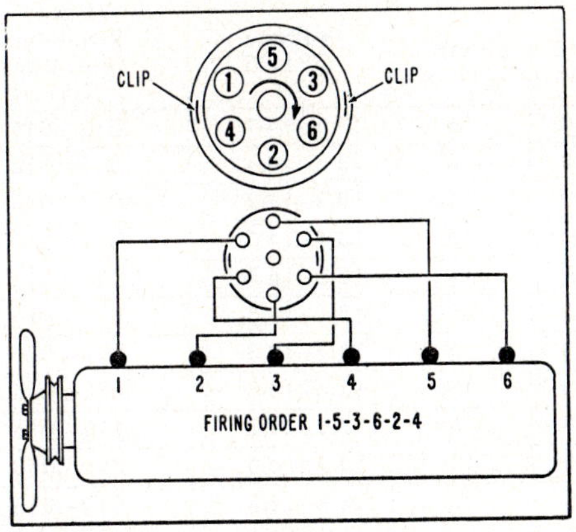

Fig. A

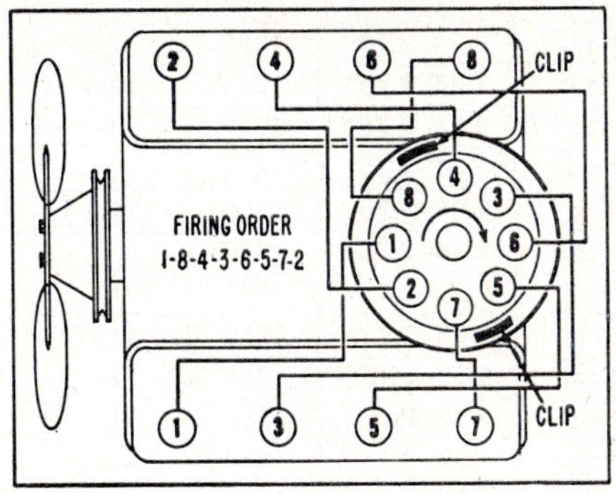

Fig. B

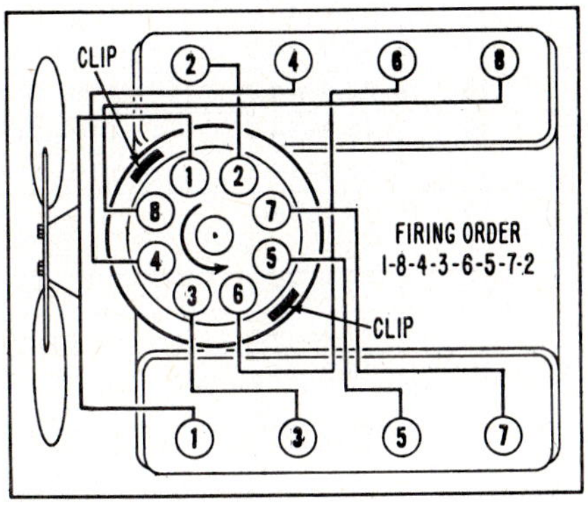

Fig. C

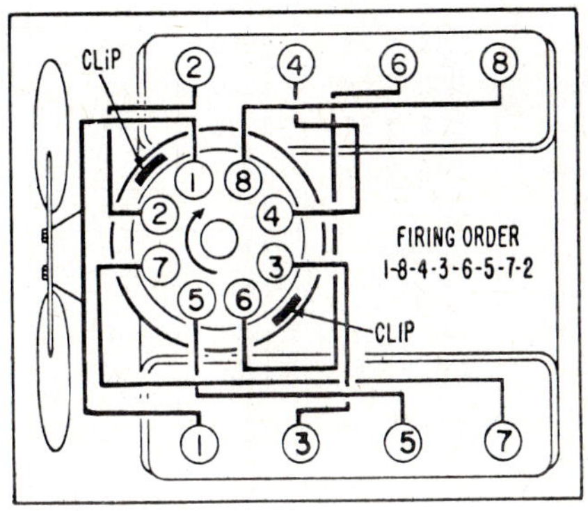

Fig. D

CHRYSLER ALTERNATOR AND REGULATOR SPECIFICATIONS

Unit Number	Ground Polarity	Field Coil Draw Amperes ②	Current Output			Operating Voltage			Voltage Regulator Point Gap	Regulator Armature Air Gap
			Engine R.P.M.	Amperes ③	Volts	Engine R.P.M.	Volts	Voltage @ 117° ①		
2095060	Neg.	2.38–2.75	1250	35	14.6	1250	15	13.5–14.1	.012–.016	.048–.052
2095425	Neg.	2.38–2.75	1250	40	14.6	1250	15	13.5–14.1	.012–.016	.048–.052
2095661	Neg.	2.38–2.75	1250	30	14.6	1250	15	13.5–14.1	.012–.016	.048–.052
2095663	Neg.	2.38–2.75	1250	34	14.6	1250	15	13.5–14.1	.012–.016	.048–.052
2095666	Neg.	2.38–2.75	1250	40	14.6	1250	15	13.5–14.1	.012–.016	.048–.052
2095667	Neg.	2.38–2.75	1250	40	14.6	1260	15	13.5–14.1	.012–.016	.048–.052
2098256	Neg.	2.38–2.75	1250	33.5	15	1250	15	13.5–14.1	.012–.016	.048–.052
2098262	Neg.	2.38–2.75	1250	32.5	15	1250	15	13.5–14.1	.012–.016	.048–.052
2098265	Neg.	2.38–2.75	1250	33.5	15	1250	15	13.5–14.1	.012–.016	.048–.052
2098268	Neg.	2.38–2.75	1250	32.5	15	1250	15	13.5–14.1	.012–.016	.048–.052
2098290	Neg.	2.38–2.75	1250	36.0	15	1250	15	13.5–14.1	.012–.016	.048–.052
2098330	Neg.	2.38–2.75	1250	39.0	15	1250	15	13.5–14.1	.012–.016	.048–.052
2098830	Neg.	2.38–2.75	1250	35.5	15	1250	15	13.5–14.1	.012–.016	.048–.052
2098840	Neg.	2.38–2.75	1250	35.5	15	1250	15	13.3–14.2	.012–.016	.048–.052
2098850	Neg.	2.38–2.75	1250	44	15	1250	15	13.3–14.3	.010–.018	.032–.042

Continued

CHRYSLER ALTERNATOR AND REGULATOR SPECIFICATIONS—Continued

Unit Number	Ground Polarity	Field Coil Draw Amperes ②	Current Output			Operating Voltage			Voltage Regulator Point Gap	Regulator Armature Air Gap
			Engine R.P.M.	Amperes ③	Volts	Engine R.P.M.	Volts	Voltage @ 117° ①		
2098855	Neg.	2.38–2.75	1250	33.5	15	1250	15	13.5–14.1	.012–.016	.048–.052
2098860	Neg.	2.38–2.75	1250	32.5	15	1250	15	13.5–14.1	.012–.016	.048–.052
2098865	Neg.	2.38–2.75	1250	37	15	1250	15	13.3–14.3	.012–.016	.048–.052
2098870	Neg.	2.38–2.75	1250	36	15	1250	15	13.4–14.3	.012–.016	.048–.052
2098875	Neg.	2.38–2.75	1250	32.5	15	1250	15	13.5–14.1	.012–.016	.048–.050
2098880	Neg.	2.38–2.75	1250	36	15	1250	15	13.4–14.3	.012–.016	.048–.052
2444897	Neg.	2.38–2.75	1250	51	15	1250	15	13.3–14.3	.012–.016	.048–.052
2444898	Neg.	2.38–2.75	1250	51	15	1250	15	13.3–14.3	.012–.016	.048–.052
2444899	Neg.	2.38–2.75	1250	51	15	1250	15	13.3–14.3	.012–.016	.048–.052
2444900	Neg.	2.38–2.75	1250	51	15	1250	15	13.3–14.3	.012–.016	.048–.052
2642537	Neg.	2.38–2.75	1250	34.5	15	1250	15	13.3–14.3	.010–.018	.032–.042
2875724	Neg.	2.38–2.75	1250	44	15	1250	15	13.3–14.3	.010–.018	.032–.042
2875844	Neg.	2.38–2.75	1250	44	15	1250	15	13.3–14.3	.010–.018	.032–.042
3000010	Neg.	2.38–2.75	1250	51	15	1250	15	13.3–14.3	.010–.018	.032–.042
3438165	Neg.	2.38–2.75	1250	44	15	1250	15	13.3–14.3	.010–.018	.032–.042
3438168	Neg.	2.38–2.75	1250	51	15	1250	15	13.3–14.3	.010–.018	.032–.042
3438173	Neg.	2.38–2.75	1250	44	15	1250	15	13.3–14.3	.010–.018	.032–.042
3438174	Neg.	3.38–2.75	1250	51	15	1250	15	13.3–14.3	.010–.018	.032–.042
3438178	Neg.	2.38–2.75	1250	44	15	1250	15	13.3–14.3	.010–.018	.032–.042
3438180	Neg.	2.38–2.75	1250	51	15	1250	15	13.3–14.3	.010–.018	.032–.042
Red Tag	Neg.	2.5–3.7⑤	1250	40	15	1250	15	13.3–14④	—	—
Bronze Tag	Neg.	4.5–6.5	1250	40	15	1250	15	13.3–13.9④	—	—
Green Tag	Neg.	2.5–3.7⑤	1250	47	15	1250	15	13.3–14④	—	—
Blue Tag	Neg.	2.5–3.7⑤	1250	57	15	1250	15	13.3–14④	—	—
Natural Tag⑥	Neg.	2.5–3.7⑤	1250	57	15	1250	15	13.3–14④	—	—
Yellow Tag⑥	Neg.	4.5–6.5	1250	57	15	1250	15	13.3–13.9④	—	—
Black Tag	Neg.	2.5–3.7⑤	1250	62	15	1250	15	13.3–14④	—	—
Yellow Tag⑦	Neg.	4.75–6.0⑨	900	72	13	900	13	13.3–13.9④	—	—
Violet Tag	Neg.	4.5–6.5	1250	40	15	1250	15	13.3–13.9④	—	—
Natural Tag⑧⑦	Neg.	4.75–6.0	900	72	13	900	13	13.3–13.9④	—	—

①—For each 10° rise in temperature subtract .04 volt; for each 10° drop add .04 volt. Temperature is checked with a thermometer 2 inches from installed voltage regulator cover.

②—Current draw at 12 volts while turning rotor shaft by hand.

③—If output is low, stator or rectifier is shorted.

④—Rated @ 140°.

⑤—1976–77, 4.5–6.5 amps.

⑥—60 amp. alternator.

⑦—1977–78 100 amp. & 1979–83 117 Amp. alternator.

⑧—1978.

⑨—1979–83, 4.5–6.5 amps.

DELCO REMY ALTERNATOR & REGULATOR SPECIFICATIONS

Year	Alternator					Regulator						
	Model	Field Current @ 80° F. 12 Volts	Cold Output @ 14 Volts		Rated Hot Output Amperes	Model	Field Relay			Voltage Regulator		
			Amperes @ 2000 R.P.M.	Amperes @ 5000 R.P.M.			Air Gap	Point Gap	Closing Voltage	Air Gap	Point Gap	Volts @ 125° F.
1974	1117089	6.0–6.6	31①	58②	60	9000590	.010	.015–.025	7–9	—	—	13.7–14.3
	1117756	4.14–4.62	20③	55②	62	9000590	.010	.015–.025	7–9	—	—	13.7–14.3
	1100072	4.4–4.9	45④	57⑤	65	—	—	—	—	—	—	—
	1100073	4.4–4.9	54④	73⑤	80	—	—	—	—	—	—	—
	1117090	6.0–6.6	30③	89②	85	9000590	.010	.015–.025	7–9	—	—	13.7–14.3
	1117790	6.0–6.6	24③	80②	85	9000590	.010	.015–.025	7–9	—	—	13.7–14.3

①—At 1000 R.P.M.

②—At 2500 R.P.M.

③—At 1100 R.P.M.

④—At 2000 R.P.M.

⑤—At 5000 R.P.M.

STARTING MOTOR SPECIFICATIONS

Model Number ①	Year	Brush Spring Tension, Oz.	No Load Test			Torque Test		
			Amperes	Volts	R.P.M.	Amperes	Volts	Torque Ft. Lbs.
1114088	1974	80	140–190	9	4000–7000	700	1.5	15
2642692	1974	32–36	78	11	3800	370	4	8.5
1113167	1974	80	50–70	9	3500–5500	230	1.5	6
1114058	1974	80	140–190	9	4000–7000	700	1.5	15
1113691	1974	—	130–160	9	5000–7000	—	—	—
1114085	1974	—	120–150	9	3000–4500	—	—	—
1114128	1974	—	140–190	9	4000–7000	—	—	—
1114161	1974	—	140–190	9	4000–7000	—	—	—
3656575	1974	32–36	90	11	3700–4200	475–550	4	—
3656650	1974	32–36	90	11	1925–2600	400–450	4	—
3755900	1974–77	32–36	90	11	3700–4200	475–550	4	—
2642692	1977–78	32–36	78	11	3800	310–445	4	—
4091975	1978–79	32–36	90	11	3700	475–550	4	—
4111855	1980–83	32–36	90	11	3700	475–550	4	—

①—Stamped on plate riveted to housing.

DISTRIBUTOR SPECIFICATIONS

Distributor Part No. ①	Rotation ②	Breaker Gap	Dwell Angle Deg.	Breaker Arm Spring Tension	Centrifugal Advance Degrees @ R.P.M. of Distributor		Vacuum Advance	
					Advance Starts	Full Advance	Inches of Vacuum To Start Plunger	Max. Adv. Dist. Deg. @ Vacuum
2095270	C	.020	40–45	17–21	1 @ 450	11 @ 2200	4.9	10 @ 12
2095843	C	.017	27–32	17–21	1 @ 600	9 @ 2000	6	11 @ 12
2095844	CC	.017	27–32	17–21	2 @ 425	14 @ 1750	7	11 @ 12
2095845	CC	.017	27–32	17–21	2 @ 425	14 @ 1950	6	11 @ 12
2095846	CC	.017	27–32	17–21	1 @ 440	14 @ 2100	6	11 @ 12
2095957	C	.020	36–42	17–21	2 @ 440	8 @ 1550	5	12 @ 16
2095974	C	.020	40–45	17–21	2 @ 475	14 @ 2200	6	12 @ 12
2095976	C	.020	40–45	17–21	1 @ 475	12 @ 2300	6	7 @ 13
2098665	C	.020	40–45	17–20	1 @ 525	14 @ 2200	8.5	8 @ 10
2098670	C	.020	40–45	17–20	1 @ 560	12 @ 2500	10.5	7 @ 13
2098604	CC	.017	28–33	17–20	2 @ 425	13 @ 1600	9.2	11 @ 12
2098675	C	.020	40 45	17–20	4 @ 475	14 @ 2200	8.5	8 @ 10
2098687	CC	.017	28–33	17–20	2 @ 425	13 @ 2000	9.2	11 @ 12
2098692	C	.020	40–45	17–20	1 @ 540	10 @ 2000	8.5	8 @ 10
2444254	C	.020	40–45	17–20	1 @ 560	12 @ 2500	10.5	7 @ 13
2444255	C	.020	40–45	17–20	1 @ 525	14 @ 2200	9.2	11 @ 12
2444256	C	.020	40–45	17–20	3 @ 475	14 @ 2200	8.5	8 @ 10
2444258	C	.017	28–32	17–20	1 @ 480	11 @ 2300	9	12 @ 16
2444275	CC	.017	28–33	17–20	4 @ 425	13 @ 1600	9.2	11 @ 12
2444276	CC	.017	28–33	17–20	3 @ 425	13 @ 2000	9.2	11 @ 12
2444277	C	.020	36–42	17–21	1 @ 440	8 @ 1550	10	12 @ 16
2444291	C	.017	28–33	17–20	1 @ 450	6 @ 2000	9.2	11 @ 12
2444292	CC	.017	28–33	17–20	2 @ 425	14 @ 1750	9.2	11 @ 12
2444295	C	.020	40–45	17–20	1 @ 540	9 @ 2000	8.5	8 @ 10
2444648	C	.020	40–45	17–20	1 @ 475	12 @ 2200	6	7 @ 13
2444671	C	.020	40–45	17–20	1 @ 475	12 @ 2500	6	8 @ 10
2642234	C	.017	28–32	17–20	1 @ 450	12 @ 1750	6.5	13 @ 13
2642238	C	.017	28–32	17–20	1 @ 475	10 @ 1750	6.5	13 @ 13

Continued

DISTRIBUTOR SPECIFICATIONS—Continued

Distributor Part No. ①	Rotation ②	Breaker Gap	Dwell Angle Deg.	Breaker Arm Spring Tension	Centrifugal Advance Degrees @ R.P.M. of Distributor		Vacuum Advance	
					Advance Starts	Full Advance	Inches of Vacuum To Start Plunger	Max. Adv. Dist. Deg. @ Vacuum
2642244	CC	.017	28–32	17–20	1 @ 450	12 @ 2150	6	11 @ 13
2642329	C	.020	40–45	17–20	0–7½ @ 475	17½ @ 2200	7.1	8½ @ 10
2642346	C	.017	28–32	17–20	3 @ 500	17 @ 1900	8	13 @ 15
2642349	C	.020	40–45	17–20	2 @ 500	20 @ 2500	6	11 @ 12
2642352	C	.020	40–45	17–20	3 @ 475	20 @ 2500	6	8 @ 10
2642354	C	.020	40–45	17–20	2 @ 500	17 @ 2200	6	7 @ 13
2642356	C	.017	28–32	17–20	4 @ 500	17 @ 1500	6	13 @ 13
2642371	C	.017	28–32	17–20	2 @ 475	14 @ 2000	6	11 @ 12
2642373	CC	.017	28–32	17–20	4 @ 500	20 @ 2200	5.5	14 @ 14
2642531	CC	.017	28–32	17–20	2 @ 550	17 @ 2200	6	14 @ 16
2642718	C	.017	28–32	17–20	2 @ 525	11 @ 2350	8	13 @ 15
2642721	C	.017	28–32	17–20	1 @ 475	13 @ 2250	8	13 @ 15
2642724	C	.017	28–32	17–20	2 @ 475	19 @ 2350	8	13 @ 16
2642727	CC	.017	28–32	17–20	1 @ 450	11 @ 2150	10	13 @ 16
2642752	C	.020	40–45	17–20	1 @ 475	9 @ 2250	6	8 @ 10
2642755	C	.020	40–45	17–20	0–4½ @ 475	14½ @ 2200	7.1	8.5 @ 10
2642758	C	.020	40–45	17–20	0–2½ @ 525	14½ @ 2200	7.5	10 @ 10½
2642786	C	.020	40–45	17–20	0–3¾ @ 500	19 @ 2200	7.1	11½ @ 12
2642792	C	.020	40–45	17–20	1 @ 500	13 @ 2400	8	7 @ 16
2642795	C	.020	40–45	17–20	2 @ 475	13 @ 2400	6	7 @ 10
2642810	CC	.017	28–32	17–20	2 @ 550	15 @ 2200	12	14 @ 16
2642941	C	.020	40–45	17–20	0–5¾ @ 525	16½ @ 2200	7.1	8½ @ 10
2642949	CC	.017	28–32	17–20	3 @ 525	15 @ 2300	12	14 @ 16
2875199	C	.020	40–45	17–20	0–3¾ @ 500	19 @ 2200	7.1	8½ @ 10
2875202	C	.020	40–45	17–20	0–6¾ @ 500	16 @ 2200	7.1	8½ @ 10
2875206	C	.020	40–45	17–20	0 @ 425	21 @ 2200	6	7 @ 13
2875338	C	.017	28–33	17–20	1 @ 475	14 @ 2250	8	13 @ 15
2875332	C	.020	40–45	17–20	0–2 @ 475	10½ @ 2250	7.1	8½ @ 10
2875340	C	.017	28–33	17–20	0 @ 425	12 @ 2350	8	13 @ 15
2875342	C	.017	28–33	17–20	0 @ 425	19 @ 2350	9	10 @ 15
2875346	CC	.017	28–32	17–20	0 @ 350	14 @ 1750	7	11 @ 12
2875348	CC	.017	28–33	17–20	0 @ 350	13 @ 1600	7	11 @ 12
2875350	CC	.017	28–33	17–20	0 @ 350	13 @ 2000	7	11 @ 12
2875352	CC	.017	28–33	17–20	0 @ 425	19 @ 2200	6.5	10 @ 13
2875354	CC	.017	28–33	17–20	0 @ 425	16 @ 2250	5	8 @ 10
2875364	C	.020	40–45	17–20	0 @ 425	14 @ 2000	8	6 @ 13
2875366	C	.020	40–45	17–20	0 @ 425	14 @ 2000	6	6 @ 8
2875822	C	.020	42–47	17–20	1 @ 550	14 @ 2000	10	7.75 @ 16
2875826	C	.020	42–47	17–20	1 @ 550	14 @ 2000	7	7.75 @ 10
2875838	C	.020	42–47	17–20	1–4 @ 550	12 @ 1900	7.5	8.5 @ 10
2875796	C	.017	30–35	17–20	1 @ 550	19 @ 2400	10.5	10.75 @ 15
2875800	C	.017	30–35	17–20	1 @ 500	14 @ 2150	10.5	11 @ 15
2875804	C	.017	30–35	17–20	½–3½ @ 500	17 @ 2250	9.5	13.5 @ 15
2875653	CC	.016	28–32	17–21	1–4½ @ 500	14.5 @ 1600	7	11.5 @ 12
2875770	CC	.017	30–35	17–21	1–5 @ 500	17 @ 1750	7	11.5 @ 12
2875876	CC	.017	30–35	17–21	1–5 @ 500	17 @ 1750	7	11.5 @ 12
2875742	CC	.017	30–35	17–20	1 @ 550	23 @ 2350	8.5	13.5 @ 13.5
2875747	CC	.017	30–35	17–20	1 @ 500	19 @ 2300	8.5	13.5 @ 13.5
2875655	CC	.016	28–32	17–21	1–4½ @ 500	14.5 @ 1600	7	11.5 @ 12
2875654	CC	.016	28–32	17–21	½–3½ @ 500	14.5 @ 2000	7	11.5 @ 12
2875966	CC	.016	28–32	17–21	½–3½ @ 500	14.5 @ 2000	7½	11.5 @ 12
2959339	CC	.016	28–32	17–21	0 @ 420	12 @ 1600	7	8 @ 11¼

Continued

DISTRIBUTOR SPECIFICATIONS—Continued

Distributor Part No. ①	Rotation ②	Breaker Gap	Dwell Angle Deg.	Breaker Arm Spring Tension	Centrifugal Advance Degrees @ R.P.M. of Distributor		Vacuum Advance	
					Advance Starts	Full Advance	Inches of Vacuum To Start Plunger	Max. Adv. Dist. Deg. @ Vacuum
2959396	CC	.016	25–29	17–21	0 @ 320	8 @ 1600	5	8 @ 11.2
3438225	C	.017	30–34	17–20	1 @ 550	16 @ 2100	12	10.5 @ 15
3438227	C	.017	30–34	17–20	1 @ 550	12 @ 2100	10.5	9.75 @ 15
3438237	C	.020	41–46	17–20	0 @ 525	14 @ 1600	7	7.75 @ 10
3438255	C	.017	30–34	17–20	1 @ 550	16 @ 2100	10.5	9.75 @ 15
3498087	CC	.016	28–32	17–21	0–2 @ 340	8 @ 1600	7	8 @ 11¼
3656252	C	.020	41–46	17–20	½–4½ @ 550	14 @ 2000	10	7½ @ 15
3656257	C	.020	41–46	17–20	½–4½ @ 550	14 @ 2000	7	7½ @ 9½
3656260	C	.020	41–46	17–20	½–3½ @ 700	14 @ 2000	10	7½ @ 15
3656266	C	.020	41–46	17–20	½–3½ @ 700	14 @ 2000	7	7½ @ 9½
3656269	C	.020	41–46	17–20	1–4 @ 550	11 @ 1900	7½	8½ @ 10
3656272	C	.017	30–34	17–20	1½–5½ @ 550	16 @ 2100	10	10½ @ 15
3656275	C	.017	30–34	17–20	0–4 @ 600	16 @ 2100	10	10½ @ 15
3656287	C	.017	30–34	17–20	½–3½ @ 550	14 @ 2200	10	10½ @ 16
3656294	C	.017	30–34	17–20	½–4 @ 550	16 @ 2200	10	10½ @ 15
3656329	CC	.017	30–34	17–20	1–5 @ 650	14 @ 2000	10½	10½ @ 15½
3656359	CC	.017	30–35	17–20	1–6 @ 550	16 @ 1750	7	11½ @ 12
3656365	CC	.016	28–32	17–21	½–4 @ 550	14 @ 2000	7	11.5 @ 12
3656373	CC	.017	30–35	17–20	1–6 @ 550	16 @ 1750	7	11½ @ 12
3656375	CC	.016	28–32	17–21	1–6 @ 550	14 @ 1600	7	11½ @ 12
3656377	CC	.016	28–32	17–21	1–5 @ 550	14 @ 1600	7	11½ @ 12
3656379	CC	.016	28–32	17–21	½–4 @ 550	14 @ 2000	7	11.5 @ 12
3656667	C	—	—	—	½–3½ @ 550	16 @ 2200	9½	10 @ 14
3656672	C	—	—	—	½–3½ @ 500	14 @ 2200	10	9½ @ 15
3656678	CC	—	—	—	0–3½ @ 500	16 @ 1700	7	10.5 @ 11½
3656762	C	—	—	—	0–2 @ 500	14 @ 2150	10	9½ @ 14½
3656763	C	—	—	—	0–3½ @ 500	16 @ 2050	9	12 @ 15
3656780	C	—	—	—	0–3½ @ 500	15½ @ 1950	9	12½ @ 15
3656791	CC	—	—	—	½–3½ @ 550	15 @ 1950	10½	10½ @ 15½
3656830	CC	—	—	—	1–5 @ 550	16 @ 1600	9½	9.5 @ 13½
3656839	CC	—	—	—	1–4 @ 550	13½ @ 1900	9½	9½ @ 13½
3656872	CC	—	—	—	1–4½ @ 550	14 @ 2000	9½	9½ @ 13½
3656876	CC	—	—	—	1½–4½ @ 550	14½ @ 1650	7½	11.5 @ 12
3656877	CC	—	—	—	1½–4½ @ 550	14½ @ 1650	7½	11.5 @ 12
3656878	CC	—	—	—	1–3½ @ 550	14 @ 1900	9½	9.5 @ 13½
3755037	C	—	—	—	½–4½ @ 550	14 @ 2000	9½	14 @ 15
3755042	C	—	—	—	½–4½ @ 550	14 @ 2000	7¼	10 @ 11
3755056	C	—	—	—	1–4½ @ 550	14 @ 2200	11	8 @ 12½
3755062	C	—	—	—	½–4½ @ 550	14 @ 2000	10	8 @ 15
3755071	C	—	—	—	½–4½ @ 550	14 @ 2000	7	7½ @ 9½
3755075	C	—	—	—	½–2½ @ 500	12 @ 1900	7	5½ @ 9½
3755150	CC	—	—	—	1½–4½ @ 650	14 @ 2000	10½	10½ @ 15½
3755157	CC	—	—	—	½–3½ @ 650	12 @ 2000	10½	10½ @ 15½
3755160	C	—	—	—	0–3½ @ 500	15½ @ 2000	9½	10 @ 14
3755201	C	—	—	—	½–3 @ 550	14 @ 2150	10	9½ @ 15
3755205	C	—	—	—	1–3½ @ 550	14 @ 1900	7½	10.5 @ 13½
3755467	C	—	—	—	½–3½ @ 550	14 @ 2000	9	10 @ 15½
3755470	C	—	—	—	½–3½ @ 550	14 @ 2000	7	10 @ 11½
3755475	C	—	—	—	½–4 @ 550	14 @ 2000	7	12 @ 12½
3755518	CC	—	—	—	½–3½ @ 650	12 @ 2000	8	11 @ 14
3755520	CC	—	—	—	1–5 @ 550	16 @ 1600	9½	9.5 @ 13½
3755522	CC	—	—	—	½–3 @ 650	10 @ 2000	8	11 @ 14

Continued

DISTRIBUTOR SPECIFICATIONS—Continued

Distributor Part No. ①	Rotation ②	Breaker Gap	Dwell Angle Deg.	Breaker Arm Spring Tension	Centrifugal Advance Degrees @ R.P.M. of Distributor		Vacuum Advance	
					Advance Starts	Full Advance	Inches of Vacuum To Start Plunger	Max. Adv. Dist. Deg. @ Vacuum
3755820	C	—	—	—	1–3½ @ 700	13½ @ 2400	9	9½ @ 13
3755821	C	—	—	—	1–4½ @ 600	16 @ 2050	9	12½ @ 15½
3755825	C	—	—	—	½–3½ @ 650	14 @ 2000	9	10 @ 15½
3755841	C	—	—	—	1–4½ @ 650	14 @ 2000	7	12 @ 12½
3755681	CC	—	—	—	1–4.5 @ 650	16 @ 2150	8	11 @ 14
3874082	C	—	—	—	1–4½ @ 600	14 @ 2200	7	10 @ 11½
3874090	C	—	—	—	1½–5½ @ 550	14 @ 2200	7	12 @ 12½
3874101	CC	—	—	—	½–3 @ 600	12 @ 2000	8	11 @ 14
3874115	C	—	—	—	1–3.5 @ 600	12 @ 2000	7	12 @ 12.5
3874302	C	—	—	—	1½–5½ @ 550	14 @ 2200	9	12½ @ 15½
3874323	C	—	—	—	½–3 @ 550	14 @ 2150	11	8 @ 12½
3874596	CC	—	—	—	1½–3½ @ 600	10 @ 2200	10	11 @ 15
3874598	C	—	—	—	1–4½ @ 600	11½ @ 2300	7	10 @ 11½
3874672	—	—	—	—	—	—	—	—
3874876	C	—	—	—	1½–3½ @ 650	4.5 @ 2200	8	12 @ 12½
3874887	CC	—	—	—	1–3.2 @ 650	12 @ 2400	7	8 @ 11½
4091282	C	—	—	—	.8–4.7 @ 500	13½ @ 2400	7	8 @ 13½
4091661	C	—	—	—	1–3 @ 550	13 @ 2000	7	8 @ 15
4091821	CC	—	—	—	½–2½ @ 600	14 @ 2000	10	10½ @ 15½
4091902	C	—	—	—	0–1 @ 650	7 @ 2150	7	12 @ 15
4111487	C	—	—	—	4–6 @ 700	11.5 @ 2200	7	12 @ 15
4111501	C	—	—	—	0–1.5 @ 700	10 @ 2200	7	12 @ 15
4111783	CC	—	—	—	1–2.5 @ 550	13.5 @ 2100	10	9.5 @ 15
4111947	C	—	—	—	1.5–3.5 @ 550	8.5 @ 2200	7	12 @ 15
4111950	C	—	—	—	0.5–2.5 @ 650	10 @ 2200	7	12 @ 15
4145364	C	—	—	—	1.0–3.0 @ 500	8 @ 2100	6	12 @ 13
4145602	C	—	—	—	.5–2.5 @ 900	7.5 @ 2200	6	12 @ 13
4145751	C	—	—	—	—	—	—	—
4145753	C	—	—	—	—	—	—	—

①—Plate on housing.

②—C: Clockwise; CC: Counterclockwise.

PISTON, PIN, RING, CRANKSHAFT & BEARING SPECIFICATIONS

Year	Engine Model ②	Wristpin Diameter	Piston Clearance, Inch ①	Ring End Gap In. (Minimum)		Crankpin Diameter, Inch	Rod Bearing Clearance, Inch	Main Bearing Journal Diameter, Inch	Main Bearing Clearance, Inch	End Thrust On Bearing No.	Shaft End Play
				Comp.	Oil						
1974	6-225	.9008	.0015	.010	.015	2.1865–2.1875	.0005–.002	2.7495–2.7505	.0005–.002	3	.002–.007
1975	6-225	.9008	.0015	.010	.015	2.1865–2.1875	.0005–.0015	2.7495–2.7505	.0005–.0015	3	.002–.007
1976	6-225	.9008	.0015	.010	.015	2.1865–2.1875	.0005–.0025	2.7495–2.7505	.0005–.002	3	.002–.009
1977–78	6-225	.9008	.0015	.010	.015	2.1865–2.1875	.0005–.0015	2.7495–2.7505	.0005–.0015	3	.002–.009
1979–83	6-225	.9008	.0015	.010	.015	2.1865–2.1875	.0010–.0022	2.7495–2.7505	.0010–.0025	3	.0035–.0095
1974	8-318	.9842	.0015	.010	.015	2.124–2.125	.0005–.0025	2.4995–2.5005	.0005–.0025	3	.002–.007
1975	8-318	.9842	.0015	.010	.015	2.124–2.125	.0002–.0022	2.4995–2.5005	.0005–.0015	3	.002–.007
1976	8-318	.9842	.0015	.010	.015	2.124–2.125	.0005–.0025	2.4995–2.5005	.0005–.002	3	.002–.009
1977–78	V8-318-1	.9842	.0015	.010	.015	2.124–2.125	.0002–.0022	2.4995–2.5005	.0005–.0015	3	.002–.007
1979	V8-318	.9842	.0015	.010	.015	2.124–2.125	.0005–.0025	2.4995–2.5005	.0005–.0020	3	.002–.007
1980	V8-318	.9842	.0015	.010	.015	2.124–2.125	.0005–.0025	2.4995–2.5005	⑥	3	.002–.007
1981–83	V8-318	.9842	.0015	.010	.015	2.124–2.125	.0005–.0022	2.4995–2.5005	⑥	3	.002–.007
1974–76	LA318-3	.9842	.0015	.010	.015	2.1235–2.1245	.001–.002	2.4995–2.5005	.001–.002	3	.002–.007
1977–78	V8-318-3	.9842	.0015	.010	.015	2.1235–2.1245	.001–.002	2.4995–2.5005	.0005–.0015	3	—

Continued

PISTON, PIN, RING, CRANKSHAFT & BEARING SPECIFICATIONS—Continued

Year	Engine Model ②	Wristpin Diameter	Piston Clearance, Inch ①	Ring End Gap In. (Minimum)		Crankpin Diameter, Inch	Rod Bearing Clearance, Inch	Main Bearing Journal Diameter, Inch	Main Bearing Clearance, Inch	End Thrust On Bearing No.	Shaft End Play
				Comp.	Oil						
1974	8-360	.9842	.0015	.010	.015	2.124–2.125	.0005–.0025	2.8095–2.8105	.0005–.0025	3	.002–.007
1975	8-360	.9842	.0015	.010	.015	2.124–2.125	.0005–.0025	2.8095–2.8105	.0005–.0015	3	.002–.007
1976	8-360	.9842	.0015	.010	.015	2.124–2.125	.0005–.0025	2.8095–2.8105	.0005–.002	3	.002–.009
1977–78	V8-360	.9842	.0015	.010	.015	2.124–2.125	.0005–.0025	2.8095–2.8105	.0005–.0015	3	.002–.009
1979	V8-360	.9842	.0015	.010	.015	2.124–2.125	.0005–.0025	2.8095–2.8105	.0005–.0020	3	.002–.009
1980	V8-360	.9842	.0015	.010	.015	2.124–2.125	.0005–.0025	2.8095–2.8105	⑥	3	.002–.009
1981–83	V8-360	.9842	.0015	.010	.015	2.124–2.125	.0005–.0022	2.8095–2.8105	⑥	3	.002–.009
1974–76	8-361	1.0936	.0015	.013	.013	2.374–2.375	.001–.002	2.6245–2.6255	.0015–.0025	3	.002–.007
1977–78	V8-361-4	1.0936	.0015	—	—	2.374–2.375	.001–.002	2.6245–2.6255	.0015–.0025	3	.002–.009
1976	8-400	1.0936	.0013	.013	.015	2.375–2.376	⑤	2.6245–2.6255	.0005–.002	3	.002–.007
1977–78	V8-400	1.0936	.0013	.013	.015	2.376	⑤	2.6245–2.6255	.0005–.002	3	.002–.009
1974	8-413	1.0936	.0015	.013	.013	2.374–2.375	.001–.002	2.7495–2.7505	.0015–.0025	3	.002–.007
1975–76	8-413	1.0936	.0013	—	—	2.374–2.375	.001–.002	2.7495–2.7505	.0015–.0025	3	.002–.007
1977–78	V8-413-3	1.0936	.0013	—	—	2.374–2.375	.001–.002	2.7495–2.7505	.0015–.0025	3	.002–.009
1974	440-1	1.0941	.0013	—	—	2.374–2.375	.001–.0025	2.7495–2.7505	.0005–.002	3	.002–.007
1975	440-1	1.0941	.0013	.013	.015	2.374–2.375	.0005–.0015	2.7495–2.7505	.0020–.0022	3	.002–.007
1976	8-440	1.0936	.0013	.013	.015	2.375–2.376	.0005–.003	2.7495–2.7505	.0005–.002	3	.002–.007
1977–78	V8-440	1.0936	.0013	.013	.015	2.376	.0005–.0030	2.7495–2.7505	.0005–.002	3	.002–.009
1979	V8-440	1.0936	.0013	.013	.015	2.375–2.376	.0005–.0030	2.7495–2.7505	.0001–.0022	3	.002–.009
1974–76	440-3	1.0936	.0013	.013	.015	2.374–2.375	.001–.0025	2.7495–2.7505	.0005–.002	3	.002–.007
1980	V8-446③	1.020	.0017	.013	.013	2.498–2.499	.0011–.0036	3.1228–3.1236	.0010–.0036	3	.0025–.0085
1974	8-478③	1.3118	.004	.013	.013	2.623–2.624	.002–.004	3.123–3.124	.0015–.0045	3	.004–.009
1974	8-549③	1.3118	.004	.013	.013	2.623–2.624	.002–.004	3.123–3.124	.0015–.0045	3	.004–.009

①—Using a 1/2" wide feeler gauge, a pull of 6 to 9 lbs. should be required to pull feeler gauge past piston. Pistons removed from above.

②—Engine size—cubic inch displacement and number of cylinders.

③—See International chapter for service on this engine.

④—Clearance measured at piston pin centerline.

⑤—2 bbl. carb. .0005–.0025; 4 bbl. carb. .0005–.003.

⑥—No. 1, .0005–.0015; Nos. 2, 3, 4 & 5, .0005–.0020.

VALVE SPECIFICATIONS

Engine Model	Year	Valve Lash		Valve Angles		Valve Springs		Valve Stem Clearance		Stem Diameters, Std.	
		Int.	Exh.	Seat	Face	Installed Height	Pressure Lbs. @ In.	Intake	Exhaust	Intake	Exhaust
6-225	1974	.010H	.020H	45	①	1 11/16	144 @ 1 5/16	.001–.003	.002–.004	.372–.373	.371–.372
	1975	.012H	.024H	45	②	⑪	⑦	.001–.003	.002–.004	.372–.373	.371–.372
6-225-1	1976	.010H	.020H	45	45	1 11/16	144 @ 1 5/16	.001–.003	.002–.004	.372–.373	.371–.372
6-225	1977–78	.010H	.020H	45	②	1 59/64	144 @ 1 5/16	.001–.003	.002–.004	.372–.373	.371–.372
6-225-1	1977–78	.010H	.020H	45	②	㉑	⑦	.001–.003	.002–.004	.372–.373	.371–.372
6-225	1979–80	.010H	.020H	45	②	1 21/32	144 @ 1 5/16	.001–.003	.002–.004	.372–.373	.371–.372
6-225	1981–83	Zero	Zero	45	②	1 11/16	144 @ 1 5/16	.001–.003	.002–.004	.372–.373	.371–.372
LA-318-1	1974	Zero	Zero	45	②	1 11/16	177 @ 1 5/16	.001–.003	.002–.004	.372–.373	.371–.372
	1975	Zero	Zero	45	45	1 21/32	⑫	.017⑱	.017⑱	.372–.373	.371–3.72
8-318-1	1976	Zero	Zero	45	②	1 11/16	177 @ 1 5/16	.001–.003	.002–.004	.372–.373	.371–.372
LA-318-3	1974	Zero	Zero	45	45	1 11/16	⑬	.001–.003	.002–.004	.372–.373	.371–.372
8-318-3	1974–76	Zero	Zero	45	45	⑯	⑰	.017⑱	.017⑱	.3715	.371–.372

Continued

VALVE SPECIFICATIONS—Continued

Engine Model	Year	Valve Lash		Valve Angles		Valve Springs		Valve Stem Clearance		Stem Diameters, Std.	
		Int.	Exh.	Seat	Face	Installed Height	Pressure Lbs. @ In.	Intake	Exhaust	Intake	Exhaust
8-318-1	1977–78	Zero	Zero	45	45	1 21/32	⑫	.017⑱	.017⑱	.372–.373	.371–.372
8-318-3	1977–78	Zero	Zero	45	45	㉒	177 @ 1 15/16	.017⑱	.017⑱	.372–.373	.371–.372
8-318	1978–83	Zero	Zero	45	45	㉓	㉔	.001–.003	.002–.004	.372–.373	.371–.372
8-360	1974	Zero	Zero	45	②	1 21/32	193 @ 1 1/4	.001–.003	.002–.004	.372–.373	.371–.372
	1975	Zero	Zero	45	②	1 11/16	177 @ 1 5/16	.001–.003	.002–.004	.372–.373	.371–.372
8-360	1976	Zero	Zero	45	②	1 11/16	183 @ 1 5/16	.001–.003	.002–.004	.372–.373	.371–.372
8-360	1977–78	Zero	Zero	45	45	1 21/32	177 @ 1 5/16	.017⑱	.017⑱	.372–.373	.371–.372
8-360	1978–79	Zero	Zero	45	45	㉓	㉔	.001–.003	.002–.004	.372–.373	.371–.372
8-360	1980–83	Zero	Zero	45	45	㉓	㉕	.001–.003	.002–.004	.372–.373	.371–.372
8-361	1974–76	Zero	Zero	45	45	⑧	⑨	.001–.003	.003–.005	.372–.373	.433–.434
8-400	1974	Zero	Zero	45	45	1 55/64	200 @ 1 7/16	.001–.0027	.001–.0027	.372–.373	.371–.372
8-400	1976–78	Zero	Zero	45	45	1 55/64	200 @ 1 7/16	.0011–.0028	⑲	.3723–.373	⑳
8-413-2, 3	1974–76	Zero	Zero	45	45	⑧	⑨	.001–.003	.003–.005	.372–.373	.433–.434
8-413-1	1974	Zero	Zero	45	45	⑤	⑥	.001–.003	.003–.005	.372–.373	.433–.434
8-440-1	1974	Zero	Zero	45	45	1 55/64	246 @ 1 23/64	.0016–.0033	.0015–.0032	.3723–.3730	.3718–.3725
	1975	Zero	Zero	45	45	⑭	⑮	.017⑱	.017⑱	.3718–.3725	.3718–.3725
8-440-1	1976–78	Zero	Zero	45	45	1 55/64	200 @ 1 7/16	.0011–.0028	⑲	.3723–.373	⑳
8-440-3	1974–75	Zero	Zero	45	45	⑭	⑮	.0016–.0033	.0015–.0032	.3718–.3725	.371–.372
8-440	1979	Zero	Zero	45	45	1 55/64	200 @ 1 7/16	.0011–.0028	⑲	.3723–.3730	⑳
8-446③	1980	Zero	Zero	45	45	—	188 @ 1.429	.00115–.00285	.00165–.00235	.37215–.37285	.37165–.37235
8-478③	1974	Zero	Zero	④	④	⑩	115 @ 1 21/32	.0015–.004	.0025–.005	.434–.435	.433–.434
8-549③	1974	Zero	Zero	④	④	⑩	115 @ 1 21/32	.0015–.004	.0025–.005	.434–.435	.433–.434

①—Intake 45, exhaust 47.
②—Intake 45, exhaust 43.
③—See International Chapter for service on this engine.
④—Intake 15°, exhaust 45°.
⑤—Intake 1 55/64", exhaust 1 47/64".
⑥—Intake 180 @ 1 15/32", exhaust 115 @ 1 3/8".
⑦—Intake 144 @ 1 5/16", exhaust 185 @ 1 5/32".
⑧—Intake 1 55/64", exhaust 1 3/4".
⑨—Intake 180 @ 1 15/32", exhaust 175 @ 1 21/64".
⑩—Inner spring 83 @ 1 17/32"; outer spring 115 @ 1 21/32".
⑪—Intake 1 11/16", exhaust 1 9/16".
⑫—Intake 185 @ 1 1/4", exhaust 192 @ 1 9/32".
⑬—Intake 177 @ 1 5/16", exhaust 185 @ 1 5/32".
⑭—Intake 1 55/64", exhaust 1 47/64".
⑮—Intake 200 @ 1 7/16", exhaust 205 @ 1 5/16".
⑯—Intake 1 11/16", exhaust 1 31/64".
⑰—Intake 177 @ 1 5/16", exhaust 187 @ 1 1/2".
⑱—Wobble method.
⑲—Hot end, .0021–.0038; Cold end, .0011–.0028.
⑳—Hot end, .3713–.372; Cold end, .3723–.373.
㉑—Intake, 1 59/64"; exhaust, 1 7/8".
㉒—Seat to retainer, 1 21/32"; with rotators, 1 31/64".
㉓—Intake or exhaust without rotator; 1 11/16", Exhaust with rotator; 1 33/64".
㉔—Intake or exhaust without rotator; 177 @ 1 5/16", exhaust with rotator; 187 @ 1 5/64".
㉕—Intake, 177 @ 1 5/16; exhaust with rotator, 189 @ 1 1/16.

ENGINE TIGHTENING SPECIFICATIONS★

★ Torque specifications are for clean and lightly lubricated threads only. Dry or dirty threads produce increased friction which prevents accurate measurement of tightness.

Engine	Spark Plugs Ft. Lbs.	Cylinder Head Bolts Ft. Lbs.	Intake Manifold Ft. Lbs.	Exhaust Manifold Ft. Lbs.	Rocker Arm Shaft Bracket Ft. Lbs.	Rocker Arm Cover Ft. Lbs.	Connecting Rod Cap Bolts Ft. Lbs.	Main Bearing Cap Bolts Ft. Lbs.	Flywheel to Crankshaft Ft. Lbs.	Vibration Damper or Pulley Ft. Lbs.
6-225	②	70	240③	10	25	40①③	45	85	55	④
8-318	⑥	⑦	⑧	⑨	200 ③	40①③	45	85	55	100
8-360	⑥	⑦	⑭	⑨	210③⑬	40①③	45	85	55	100
8-361	30	70	⑩	30	25	40③	45	85	55	135

Continued

ENGINE TIGHTENING SPECIFICATIONS★—Continued

★ Torque specifications are for clean and lightly lubricated threads only. Dry or dirty threads produce increased friction which prevents accurate measurement of tightness.

Engine	Spark Plugs Ft. Lbs.	Cylinder Head Bolts Ft. Lbs.	Intake Manifold Ft. Lbs.	Exhaust Manifold Ft. Lbs.	Rocker Arm Shaft Bracket Ft. Lbs.	Rocker Arm Cover Ft. Lbs.	Connecting Rod Cap Bolts Ft. Lbs.	Main Bearing Cap Bolts Ft. Lbs.	Flywheel to Crankshaft Ft. Lbs.	Vibration Damper or Pulley Ft. Lbs.
8-400	30	70	40⑫	30	25	40③	45	85	55	135
8-413	30	70	⑩	30	25	40③	45	85	55	135
8-440	⑪	70	45	30	25	40③	45	85	55	135
8-446⑤	15	105	—	18	20	6	41	95	—	90
8-549⑤	30	100	—	—	90	—	70	110	100	—

①—1974–80, 40 in. lbs.; 1981–83, 80 in. lbs.
②—1974, 30 ft. lbs.; 1975–83, 10 ft. lbs.
③—In. Lbs.
④—Press fit.
⑤—See International Chapter for service on this engine.
⑥—14MM plugs, 30; 18MM plugs, 20.
⑦—1974–77, 95 ft. lbs., 1978–83 105 ft. lbs.
⑧—1974–75 & 1977, 35 ft. lbs.; 1976 & 1978–83, 40 ft. lbs.
⑨—Screw 20 ft. lbs. & nut 15 ft. lbs.
⑩—1974, 40 ft. lbs.; 1975–77, 45 ft. lbs.
⑪—V8-440-1, 30 ft. lbs.; V8-440-3, 10 ft. lbs.
⑫—1977–78, 45 ft. lbs.
⑬—1977–83, 200 in. lbs.
⑭—1974–77, 35 ft. lbs.; 1978–83, 40 ft. lbs.

WHEEL ALIGNMENT SPECIFICATIONS

Model	Caster, Deg.	Camber, Deg.	Toe-In, Inch	Kingpin Inclination, Deg.
1974 SERIES				
AW-100, PW-100	+3	+1½	0-1/8	—
B100, B200, B300	0-+1⑦	0-+½	1/16-1/8	—
MBH300, MBL300, CB300	0-+1⑦	0-+½	1/16-1/8	—
PB100, PB200, PB300	0-+1⑦	0-+½	1/16-1/8	—
M300, RM300	+1½②⑧	+2	1/16-3/16	—
RM350, M375, RM400	+1②⑨	+2	1/16-3/16	—
D100, D200, D300	+½	0-+½	0-1/8	—
P300	+3②⑧	+2	1/16-3/16	7
P400, D500, D600⑩	+1②⑨	+2	0-1/8	7
S600⑪	+1②⑨	+2	0-1/8	7
S600, D600, D800	+1½②⑧	+1	1/16-3/16	5½
C800, CT800	0-+1⑥	+1	0-1/8	5½
CNT800, CN800	0-+1⑥	+1	0-1/8	5½
CN900, CNT900	0-+1⑥	+1	0-1/8	5½
LS1000, LT1000	0-+1⑥	+1	0-1/8	5½
CN950, LN950	0-+1⑥	+1	0-1/8	5½
1975 SERIES				
AD-100, PD-100⑫	+½⑦	+¼	1/16-1/8	—
AW-100, PW-100⑬	+3	+1½	0	—
B-100, B-200, B-300	+½⑦	+¼	1/16-1/8	—
PB-100, PB-200, PB-300	+½⑦	+¼	1/16-1/8	—
CB-300, MB-300	+½⑦	+¼	1/16-1/8	—
M-300, M-400, M-500⑭	+3①⑮	+2	1/16-3/16	7
M-300, M-400, M-500	+2½①⑯	+2	0-1/8	7
D-100, D-200, D-300	+½①	+½	1/16-1/8	—
W-100, W-200⑰	+3②	+1½	3/32-11/32	8½
W-200, W-300⑭	+3②	+½	0-1/4	8½
W-600⑱	—	—	—	—
W-600⑲	+2½②	+¾	0-1/8	8
D-500, D-600	+2½①	+2	0-1/8	7
D-600, S-600, D-700, S-700⑳	—	+1	1/16-3/16	7
D-600, S-600, D-700, S-700, D-800	+1½	+1	0-1/8	5½
D-800③	+½	+1	0-1/8	5½
1976 SERIES				
AD-100, PD-100	+½	+¼	1/8	—
AW-100, PW-100 ⑬	+3	+1½	0	—
B-100, B-200, B-300	+½⑦	+½	1/8	—
CB-300	+½⑦	+½	1/8	—
MB-300, MB-400	+2¼	+½	1/8	—
M-300, M-400, M-500	+5½	+2	1/8	—
D-100, D-200, D-300	+½	+¼	1/8	—
W-100, W-200⑬④	+3	+½	0	—
W-200, W-300⑬⑭	+3	+½	0	—
W-600⑬	+3	—	1/8	—
D-500, D-600, D-700 D-800, S-600, S-700	+2	+1	1/8	—

Continued

WHEEL ALIGNMENT SPECIFICATIONS—Continued

Model	Caster, Deg.	Camber, Deg.	Toe-In, Inch	Kingpin Inclination, Deg.
1977–78 SERIES				
AD-100, PD-100⑫	+1/2	+1/4	1/8	—
AW-100, PW-100⑬	+3	+1/2	0	—
B-100, B-200, B-300	+1/2⑦	+1/2	1/8	—
PB-100, PB-200, PB-300	+1/2⑦	+1/2	1/8	—
CB-300, MB-300	+1/2⑦	+1/2	1/8	—
MB-400	+2 1/2	+1/2	1/8	—
M-300, M-400, M-500	+5	+2	1/16	—
M-600	+6.41	+1	1/16	—
D-100, D-200, D-300, D-400	+1/2	+1/4	1/8	—
W-100, W-200⑬④	+3	+1 1/2	0	—
W-200, W-300⑬⑭	+3	+1/2	0	—
W-600⑬⑱	+3	+1 1/2	1/16	—
W-600⑬⑲	+3	+1	1/16	—
D-500, D-600, D-700, D-800, S-600, S-700	+2.37	+1	1/16	—
1979 SERIES				
AD-100, PD-100	+1/2	+1/4	1/8	—
AW-100, PW-100⑬	—	—	—	—
B-100, B-200, B-300	+1/2⑦㉒	+1/2㉓	1/8㉔	—
PB-100, PB-200, PB-300	+1/2⑦㉒	+1/2㉓	1/8㉔	—
CB-300, CB-400	+1/2⑦	+1/2	1/8	—
MB-300, MB-400	+1/2⑦	+1/2	1/8	—
M-300	+3	+2	1/8	7
M-400, M-500	+2 1/2	+1	1/16	7
M-600	+2	+1	1/16	7
D-100, D-150, D-200, D-300, D-400	+1/2⑤	+1/4	1/8	—
W-150, W-200②⑬	+3	+1 1/2	0	8 1/2
W-200, W-300, W-400②⑬⑭	+3	+1/2	0	8 1/2
1980 SERIES				
AD-100, PD-100	+1/2⑤	+1/4	1/8	—
AW-100, PW-100⑬	—	—	—	—
B-100, B-200, B-300	+2 1/4㉒	+1/2㉓	1/8㉔	—
PB-100, PB-200, PB-300	+2 1/4㉒	+1/2㉓	1/8㉔	—
CB-300, CB-400	+2 1/4	+1/2	1/8	—
MB-300, MB-400	+2 1/4	+1/2	1/8	—
M-300	—	—	—	—
M-400, M-500	—	—	—	—
M-600	—	—	—	—
D-100, D-150, D-200, D-300, D-400	+1/2⑤	+1/4	1/8	—
W-150, W-200	+3②	+1 1/2	0	8 1/2
W-200, W-300, W-400②⑬⑭	+3②	+1/2	0	8 1/2
1981 SERIES				
AD-150	+1/2	+1/4	1/8	—
B-150, B-250, B-350	+2 1/4	+1/2	1/8	—
CB-350, CB-450	+2 1/4	+1/2	1/8	—
D-150, D-250, D-350, D-450	+1/2	+1/4	1/8	—
MB-250, MB-350, MB-450	+2 1/4	+1/2	1/8	—
PB-150, PB-250, PB-350	+2 1/4	+1/2	1/8	—
PD-150	+1/2	+1/4	1/8	—
W-150, W-250	+3	+1 1/2	0	8 1/2
1982 SERIES				
AD-150, D-150, D-250, D-350	+1/2	+1/4	1/8	—
B-150, B-250, B-350	+2 1/4	+1/2	1/8	—
PB-150, PB-250, PB-350	+2 1/4	+1/2	1/8	—
W-150, W-250, W-350	+3	+1 1/2	1/8	8 1/2
W-250, W-350㉑	+3	+1/2	1/8	8 1/2
1983 SERIES				
AD-150, D-150, D-250, D-350	+1/2	+1/2	1/8	—
B-150, B-250, B-350	+2 1/4	+1/2	1/8	—
PB-150, PB-250, PB-350	+2 1/4	+1/2	1/8	—
W-150, W-250, W-350	+2	+1	1/4	8 1/2

①—Caster should be checked with vehicle loaded. If caster cannot be checked when vehicle is loaded the no load caster is approximately 1° to 2° for manual steering and 5° to 6° for power steering. If vehicle wanders caster should be increased. If steering effort is very high, especially when taking corners, caster should be decreased.
②—No load.
③—9000 lb. axle.
④—3500, 3800 & 4000 lb. axles.
⑤—Loaded.
⑥—Power Steering 4 to 5°.
⑦—Power Steering +2 1/4°.
⑧—Power Steering 4 1/2°.
⑨—Power Steering +4°.
⑩—All D600 except 240" wheel base.
⑪—197" and 221" wheel base.
⑫—Exc. four wheel drive.
⑬—Four wheel drive.
⑭—4500 lb. axle.
⑮—Power steering +6°.
⑯—Power steering +5 1/2°.
⑰—3500 lb. axle.
⑱—6000 lb. axle.
⑲—7500 lb. axle.
⑳—5500 lb. axle.
㉑—Models with 135" W.B. and 149" W.B.
㉒—Heavy front axle load applications, +1 1/2.
㉓—Heavy front axle load applications, +5/8.
㉔—Heavy front axle load applications, 0.

Engine Section

NOTE: Refer to International Chapter for service procedures on V8-446, 478 & 549 engines.

CYLINDER HEAD, REPLACE

V8 Engines

Service Note

Some 1979 and all 1980–83 V8-318 engines have cylinder head bolt holes drilled through the block into the water jacket in certain locations, Fig. 3A. On 1979 models, this type cylinder block is identified by engine numbers beginning with 9M3180702 or 4104230-318. If engine number cannot be determined, refer to Fig. 3A and insert a screwdriver into the head bolt holes. If the screwdriver can be inserted at least two inches into a hole, that hole is open to the water jacket. Cylinder head bolts in these locations must have sealer 4057989 or equivalent applied to the threads to prevent engine coolant leakage. Be sure old sealer is cleaned from the threads before applying new sealer. Cylinder head bolt torque on these engines has been reduced to 95 ft. lbs.

Before an engine is disassembled for valve grinding or major repair work, the operation of the exhaust valve rotators, if so equipped, should be checked by removing the cylinder head covers and observing exhaust valve rotation with the engine running. While the direction of rotation is not important, there should be a large difference in rotation speeds between 1,000 and 3,000 RPM. Improperly functioning rotators should be replaced.

Before installing cylinder heads, clean gasket surfaces of the cylinder block and cylinder heads. Check mating surfaces for flatness using a straightedge and feeler gauge. Use new gaskets and check that all holes are properly punched through. After coating gaskets with a suitable sealer, install cylinder heads ensuring that dowel pins and holes are aligned. Install and torque cylinder head bolts in sequence shown in Figs. 1 and 2. Refer to "Engine Tightening Specifications."

Service Note

V8-318 & 360 U.S. Built Engines

The intake manifold attaching bolts on this engine are tilted upward about 30 degrees at an angle to the manifold-to-cylinder head gasket face. The purpose of this design is to provide more effective sealing at the cylinder block end gaskets. If the intake manifold is removed the installation should be such that the bolt tightening is done evenly and in the sequence shown in Fig. 3.

With gaskets in place start all bolts, leaving them loose. Run bolts 1 through 4 down so the heads just touch manifold. Then tighten these four bolts to 25 ft. lbs. torque. After checking to see that gaskets are properly seated at all surfaces, tighten remaining bolts to 25 ft. lbs. Finally torque all bolts in sequence shown to 35 ft. lbs. on 1974–77 models. On 1978–83 models, torque bolts to 40 ft. lbs.

Before an engine is disassembled for valve grinding or major repair work, the operation of the exhaust valve rotators, if so equipped, should be checked by removing the cylinder head cover and observing exhaust valve rotation with the engine running. While the direction of rotation is not important, there should be a large difference in rotation speeds between 1,000 and 3,000 RPM. Improperly functioning rotators should be replaced.

Before installing cylinder head, clean gasket surface of the cylinder block and cylinder head. Check mating surfaces for flatness using a straightedge and feeler gauge. Use a new gasket and check that all holes are properly punched through. After coating gasket with a suitable sealer, install cylinder head ensuring that dowel pins and holes are aligned. Install and tighten cylinder head bolts in sequence, Fig. 4. Repeat procedure and torque bolts in sequence to torque load listed in "Engine Tightening Specifications".

Installation of intake and exhaust manifolds should be performed as follows to avoid breakage due to overtightening and/or expansion:

1. Loosen bolts and nut holding intake manifold to exhaust manifold. After coating with a suitable sealer, position manifold to cylinder head gasket on cylinder head.
2. Install intake and exhaust manifold assembly to cylinder head with cup side of conical washers against manifolds. Snug nuts to approx. 20 inch lbs.
3. Torque intake to exhaust manifold nut to 240 inch lbs. then torque intake to exhaust manifold bolts to 200 inch lbs.
4. Starting from the center and working outward, torque intake and exhaust manifold nuts on cylinder head to 120 in. lbs.

ROCKER ARM SERVICE

NOTE: *When disassembling rocker arms, place all parts on the work bench in their proper sequence to insure correct assembly.*

Clean all sludge and gum formation from the inside and outside of the shafts. Clean oil holes and passages in the rocker arms and shafts. Inspect the shafts for wear.

V8-318 & 360

To provide proper lubrication of the rocker arms, the rocker shafts have a small notch machined at one end. Install rocker arm and shaft assemblies with notch on end of rocker shaft pointing to centerline of engine and toward front of engine on the left bank and to the rear on the right bank. If rocker arms are removed from shaft, care must be taken to ensure proper reassembly. Some exhaust rocker arms have a relieved area on the underside for rotator clearance. Refer to Fig. 5 for proper positioning of rocker arms. Note placement of long stamped steel retainers in the number two and four positions between the rocker arms.

V8-361, 400, 413, 440

The rocker arms are of stamped steel and are arranged on one rocker arm shaft, per cylinder head. The push rod angularity tends to force the pairs of rocker arms toward each oth-

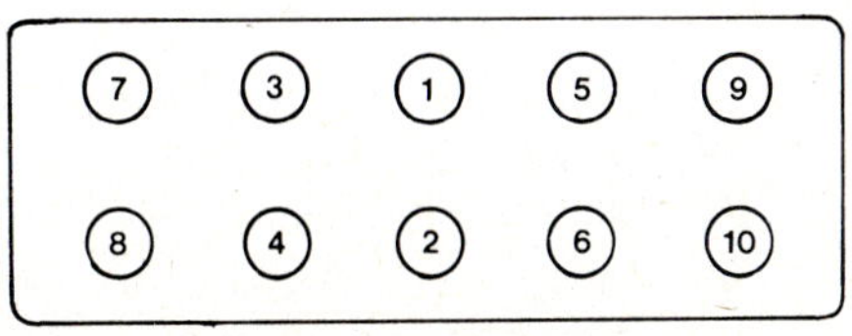

Fig. 1 Cylinder head tightening sequence. V8-318 & 360

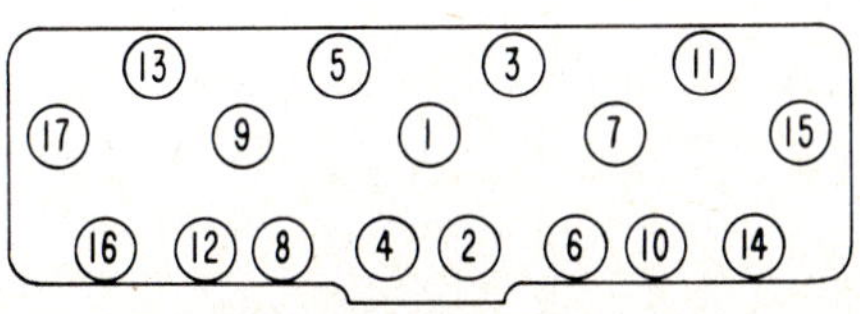

Fig. 2 Cylinder head tightening sequence. V8-361, 400, 413 & 440

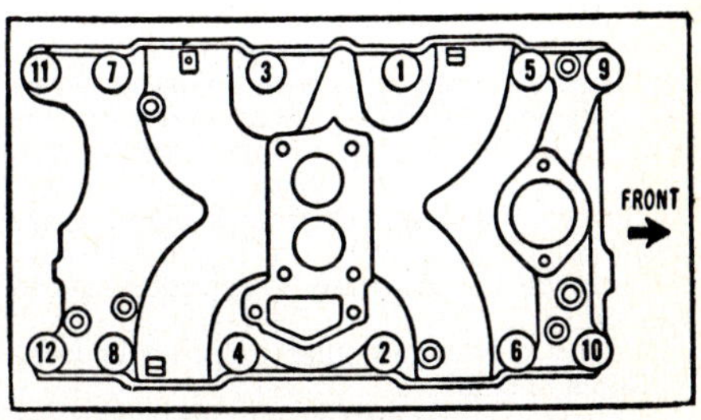

Fig. 3 Intake manifold tightening sequence. V8-318 & 360

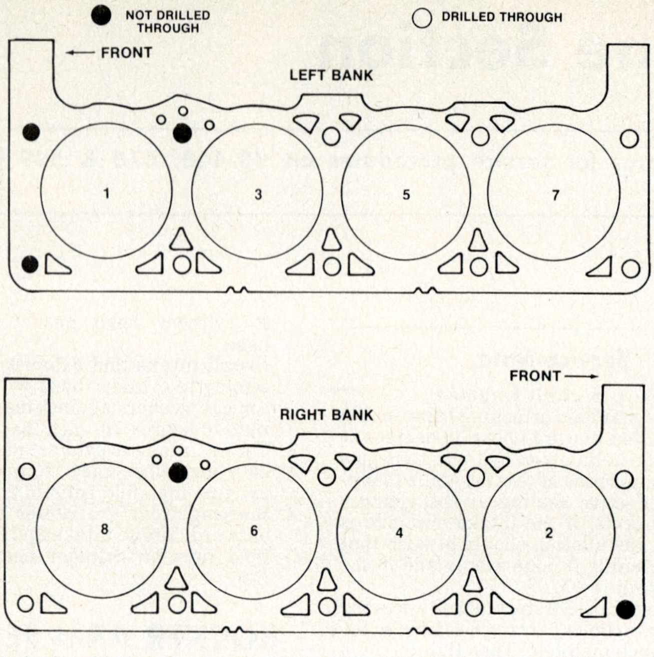

Fig. 3A Cylinder head bolt hole identification. Some 1979 & all 1980–83 Chrysler Corp. V8-318 engines

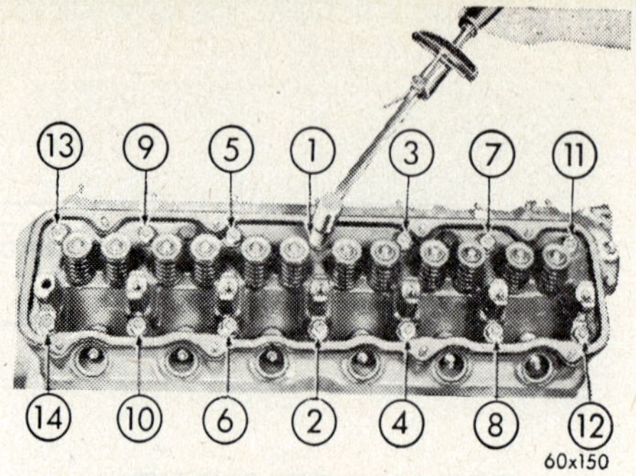

Fig. 4 Head tightening sequence. 6-225

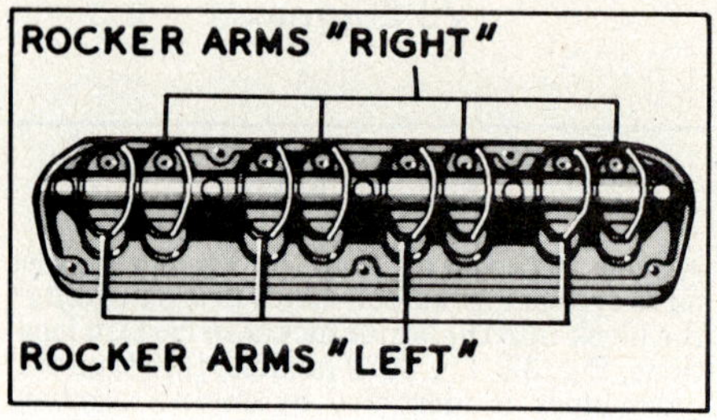

Fig. 5 Rocker arm and shaft assembly installed. V8-318 & 360

er where oilite spacers carry the side thrust at the rocker arms, Fig. 6.

1. Install rocker shafts so that the 3/16" diameter rocker arm lubrication holes point downward into rocker arm, and so that the 15 degree angle of these holes point outward toward valve end of rocker arm, Fig. 7. The 15 degree angle is determined from the center line of the bolt holes through the shaft which are used to attach the shaft assembly to the cylinder head.
2. On all engines, install rocker arms and shaft assembly, making sure to install long stamped steel retainers in No. 2 and 4 positions.

NOTE: Use extreme care in tightening the bolts so that valve lifters have time to bleed down to their operating length. Bulged lifter bodies, bent push rods and permanent noisy operation may result if lifters are forced down too rapidly.

3. Installation should be as shown in Fig. 8.

6-225

Positioning rocker arms on rocker shaft requires flat and/or oil hole on rocker shaft to be on top and pointing to front of engine. This ensures proper lubrication of rocker assembly. Install rocker arms with adjusting screw (if equipped), to right side of engine in this order: One rocker arm and one spacer, then two rocker arms and a spacer. Continue this sequence until all rocker arms and spacers are positioned, Fig. 9.

Install rocker assembly to cylinder head in this order:

1. Place a bolt and small retainer in front hole of shaft.
2. Place a bolt and the one wide retainer through the center hole of shaft with six rocker arms on each side of center.

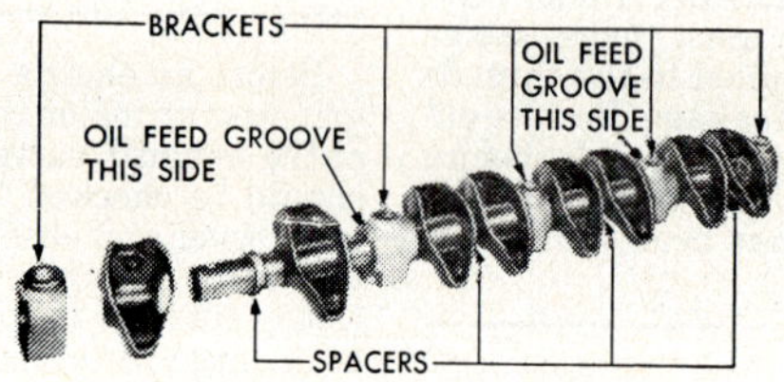

Fig. 6 Rocker arms and shaft. V8-361, 400, 413, 440 (Typical)

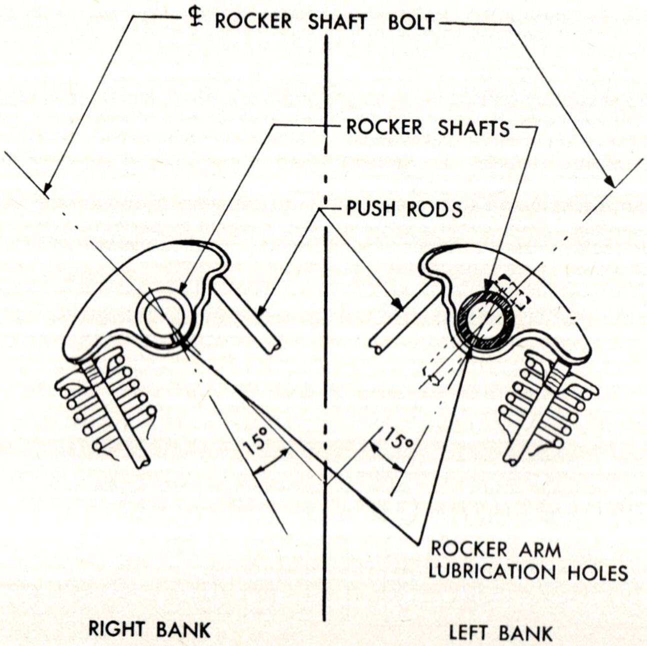

Fig. 7 Rocker arm shaft installation, V8-361, 400, 413 & 440

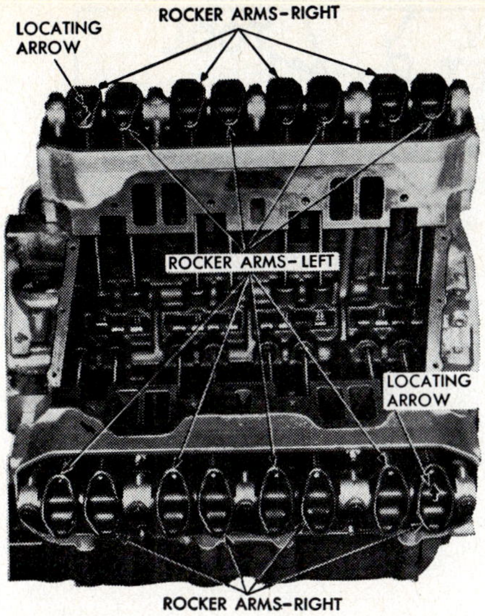

Fig. 8 Rocker arm installed. V8-361 & 413 (Typical)

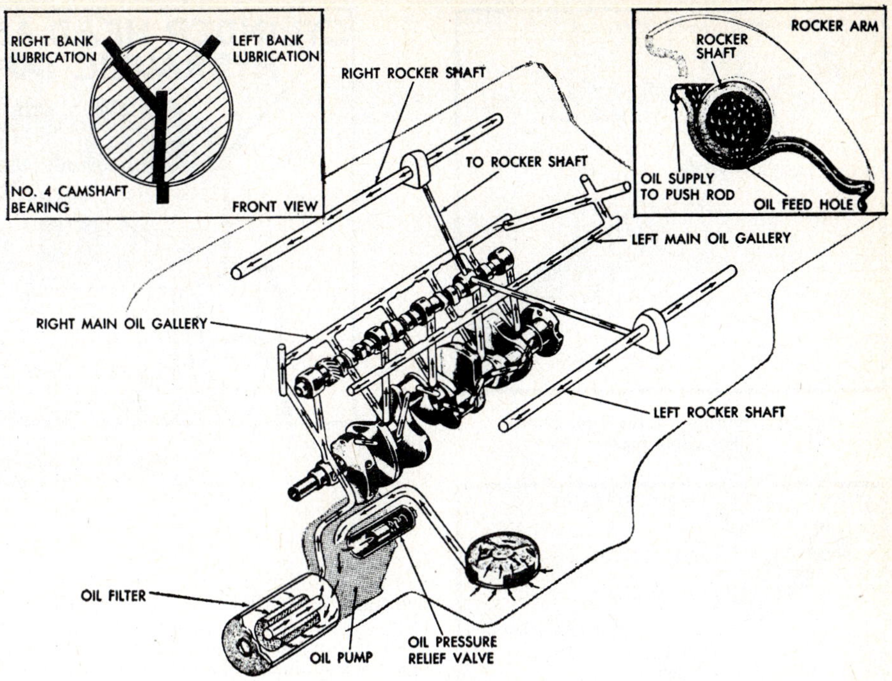

Engine lubrication. V8-361, 400, 413 & 440 (Typical)

3. Insert remaining bolts and retainers, separating the six pairs of rocker arms. Place long or special bolt and retainer in rear hole of shaft.
4. Locate assembly on cylinder head and position rocker arm adjusting screws in pushrods.
5. Tighten bolts finger tight, bringing retainers in contact with shaft between rocker arms. Retainers must not seat on extended bushing of rocker arm. Torque bolts to specifications listed in "Engine Tightening Specifications".
6. On engines equipped with adjustable rocker arms, temporarily set valve lash cold: intake .012 inch (0.304 mm), exhaust .028 inch (0.711 mm). Start and run engine at 550 RPM idle until normal operating temperature is reached. To reduce oil splatter during warm-up, temporarily fit cylinder head cover over rocker assembly.
7. On engines equipped with adjustable rocker arms, allow engine to reach normal operating temperature, then adjust valve lash to specifications. On all engines, complete installation of remaining parts removed.

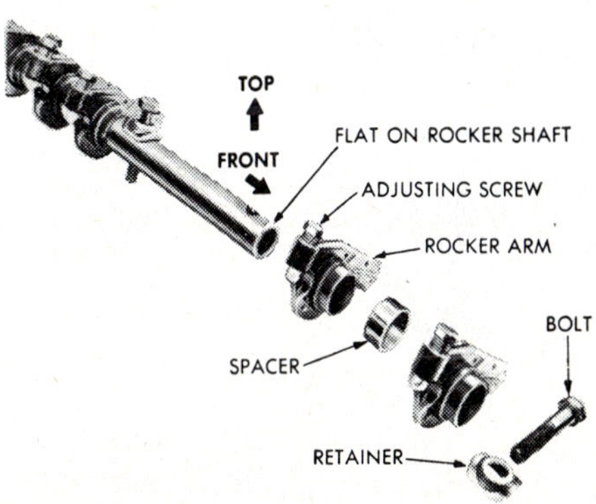

Fig. 9 Rocker arms and shaft. 6-225

VALVES, ADJUST

6-225 1974–83

NOTE: On 1981–83 6-225 engines valve adjustment has been eliminated through the use of hydraulic tappets similar in design to those used in the 8 cylinder engines, except the oil enters through the push rod socket.

If the cylinder head has been removed it is a good practice to make an initial valve adjustment before starting the engine. Make two chalk marks spaced 120 degrees apart (1/3 of circumference) on the vibration damper so that with the timing mark, the damper is divided into thirds. With the crankshaft at TDC, temporarily set the intake valve lash for No. 1 cylinder at .012 inch and the exhaust at .028 inch. Repeat the procedure for the remaining valves, turning the crankshaft 1/3 turn in the direction of normal rotation while adjusting the valves in the firing order sequence of 1-5-3-6-2-4. For final adjustment, start and run engine at 550 RPM idle until normal operating temperature is reached. Then adjust valve lash to clearance listed in "Valve Specifications".

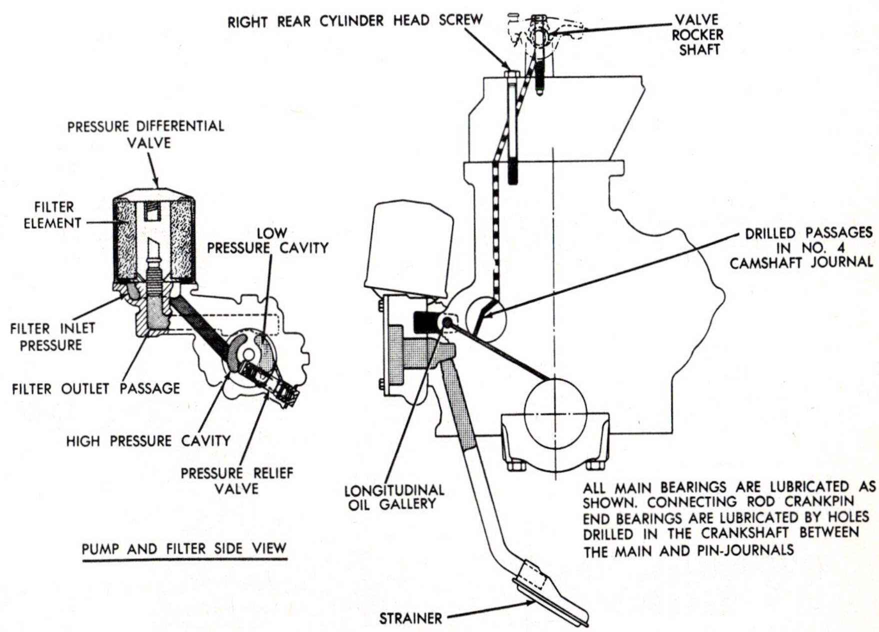

Lubrication diagram 6-225 (Typical)

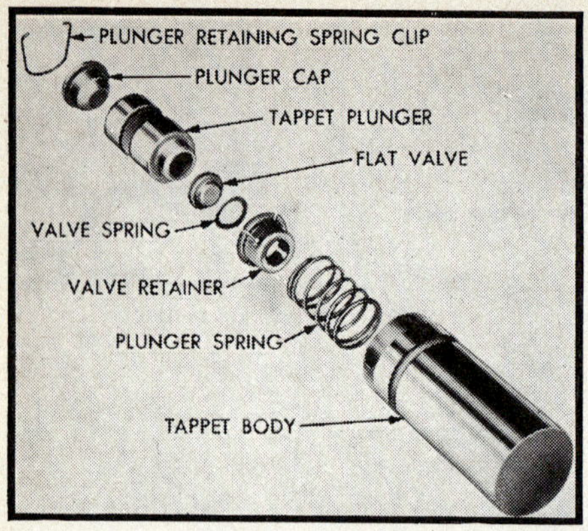

Fig. 10 Hydraulic valve lifter of the flat valve type

Fig. 12 Removing stuck hydraulic lifter with special tool. Typical

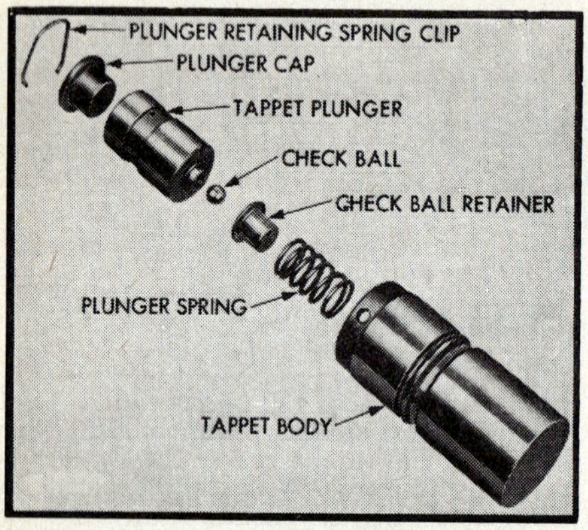

Fig. 11 Hydraulic valve lifter of the ball check type (Typical)

VALVE GUIDES

V8-318, 360, 361, 400, 413, 440 & 6-225

These engines do not have removable valve guides. The valves operate in guide holes bored in the cylinder head. Valves with oversize valve stems are available for service replacement when necessary to ream the guide holes.

Standard production reaming of both intake and exhaust valves is .374–.375". When reaming guides for oversize valve stems do not attempt to ream from standard to .030" oversize. If necessary to ream to that size, use the step procedure of .005", .015" and .030". This must be done in order to maintain a true relationship of the guide to the valve seat. The following chart indicates reamer size and valve stem size.

Reamer Oversize	Valve Stem Size
.005"	.379–.380"
.015"	.389–.390"
.030"	.404–.405"

If an "I" or an "E" is stamped on the cylinder head at the rocker cover gasket boss, all intake or exhaust valves are oversize in that bank, depending upon which letter is stamped.

HYDRAULIC VALVE LIFTERS

Figs. 10 and 11 illustrate the type of hydraulic valve lifters used. Before disassembling any part of the engine to check for noise, check the oil pressure at the gauge and the oil level in the oil pan. The oil level in the pan should *never be above the "full"* mark on the dipstick, *nor below* the "add oil" mark. Either of the two conditions could be responsible for noisy lifters.

Lifter, Replace

Worn valve guides or cocked springs are sometimes mistaken for noisy lifters. Determine which lifter is noisy. If the application of side thrust on the valve spring fails to noticeably reduce the noise, the lifter is probably faulty and should be removed for inspection. Removal of stuck lifters requires a special tool, Fig. 12. When installing hydraulic lifters in the engine, fill them with light engine oil to avoid excessive time required to quiet them during initial operation of engine.

TIMING CHAIN COVER

6-225

1. To remove cover, drain cooling system and remove radiator and fan.
2. Remove vibration damper with a puller.
3. Loosen oil pan bolts to allow clearance and remove chain case cover.
4. Reverse above procedure to install cover.

V8 Engines

To remove cover, first drain cooling system, remove radiator, fan belt and water pump assembly. Then, remove pulley from vibration damper and after removing bolt and washer, use puller to remove damper. Next, remove fuel lines and fuel pump. Loosen oil pan bolts and after removing front bolt at each side, timing cover may be removed. Reverse procedure to install cover, using new oil seal.

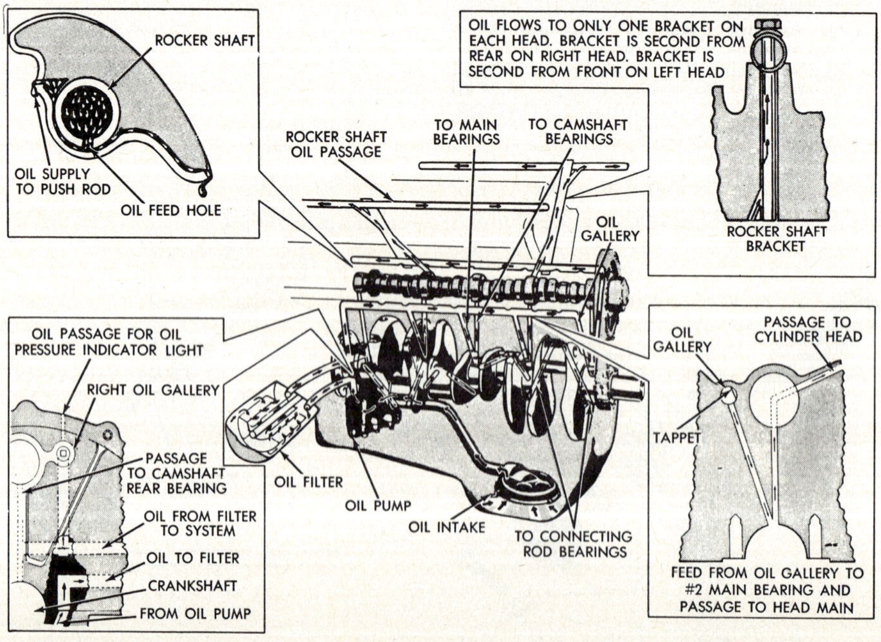

Engine lubrication. V8-318 & 360 (Typical)

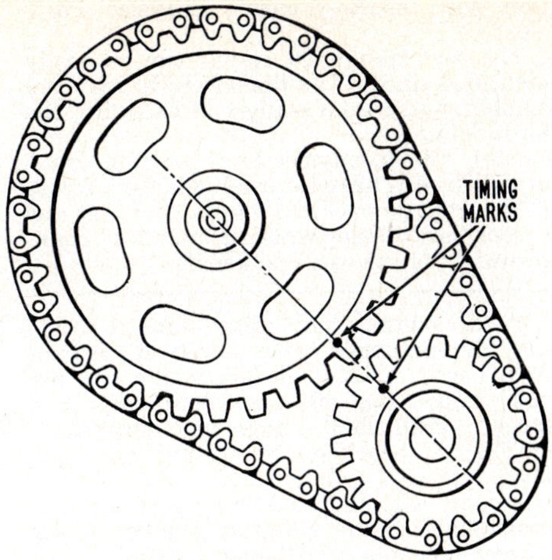

Fig. 13 Valve timing. All six cylinder engines

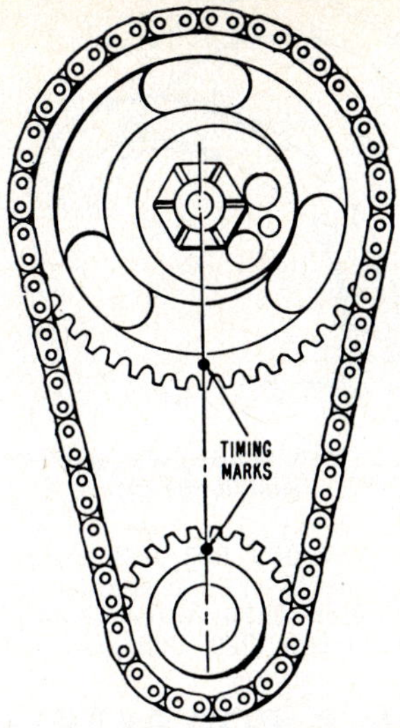

Fig. 14 Valve Timing. V8 Engines

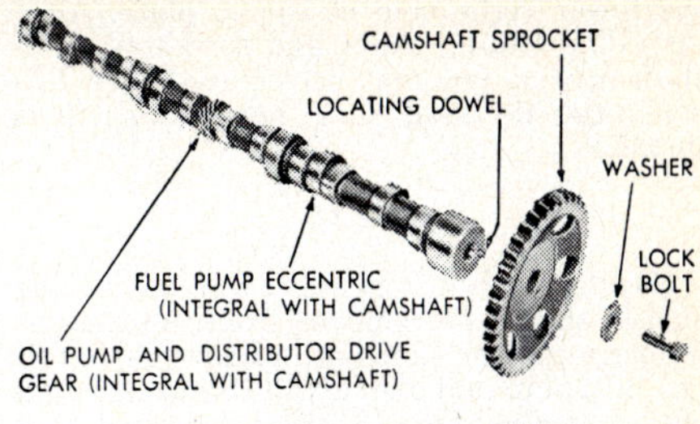

Fig. 15 Camshaft assembly. 6-225

TIMING CHAIN

6-225

1. After removing timing chain cover as outlined above, remove camshaft sprocket lock bolt, then remove timing chain and sprocket.
2. Clean all parts and inspect chain for broken or damaged links. Inspect sprockets for cracks and chipped, worn or damaged teeth.
3. Turn crankshaft so sprocket timing mark is toward and directly in line with centerline of camshaft.
4. Temporarily install camshaft sprocket. Rotate camshaft to position sprocket timing mark toward and directly in line with centerline of crankshaft; then remove camshaft sprocket.
5. Place chain on crankshaft sprocket and position camshaft sprocket in chain so sprocket can be installed with timing marks aligned and without moving camshaft, Fig. 13.
6. Install remaining components in reverse order of removal.

V8 Engines

To install chain and sprockets, lay both camshaft and crankshaft sprockets on bench. Position sprockets so that the timing marks are next to each other. Place chain on both sprockets, then push sprockets apart as far as the chain will permit. Use a straightedge to form a line through the exact centers of both gears. The timing marks must be on this line.

Slide the chain with both sprockets on the camshaft and crankshaft at the same time, then recheck the alignment, Fig. 14.

NOTE: Use Tool C-3509 to prevent camshaft from contacting welch plug in rear of engine block. Remove distributor and oil pump-distributor drive gear. Position tool against rear side of camshaft gear and attach with distributor retainer plate bolt.

CAMSHAFT & BEARINGS

V8 Engines

If camshaft bearings are to be replaced, the engine should be removed from the chassis and the crankshaft taken out in order that any chips or foreign matter may be removed from the oil passages.

With camshaft removed, drive out rear welch plug. A special tool is required to remove and install camshaft bearings. This tool pulls out all the bearings at once and is also used to install new bearings.

When installing new bearings, be sure oil holes in bearings line up with oil holes in cylinder block. Install new welch plug, with sealer applied, in the rear of block and carefully insert properly lubricated camshaft to within 2 inches of rear welch plug. Position and secure Tool C-3509 with tongue behind distributor drive gear so that camshaft is prevented from pushing out welch plug. Install camshaft thrust plate and chain oil tab making sure tang of tab is correctly positioned in

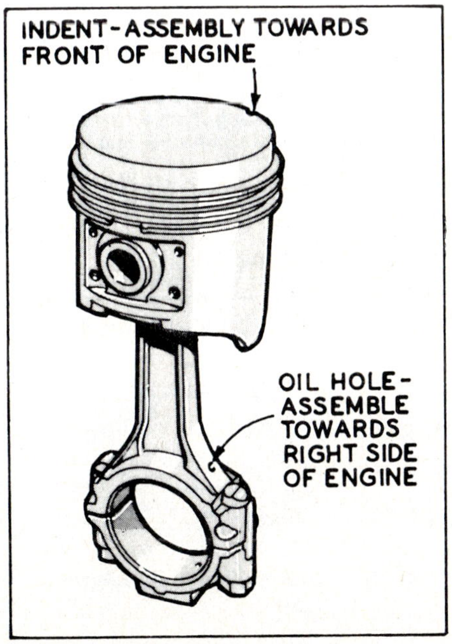

Fig. 16 Piston & rod assembly. 1974–75 6-225 & 1976–77 6-225 with forged crankshaft

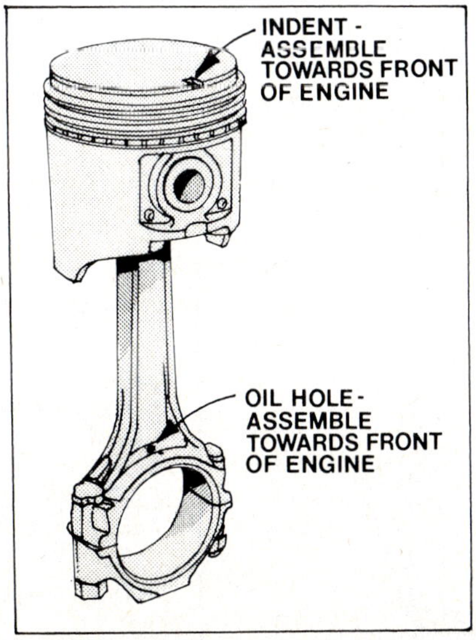

Fig. 17 Piston & rod assembly. 1976–77 6-225 with cast iron crankshaft & 1978–83 6-225

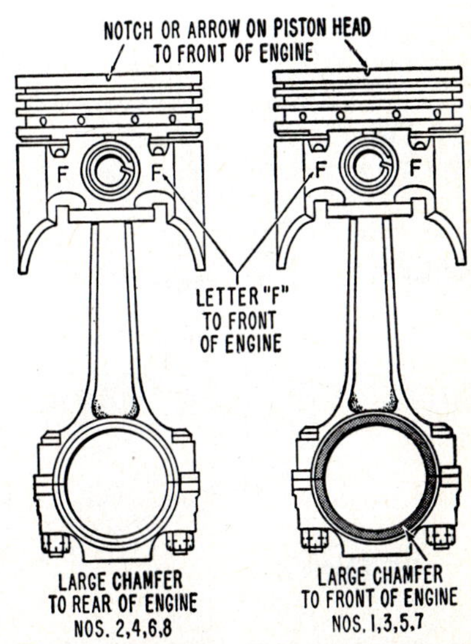

Fig. 18 Correct assembly of pistons and rods. V8 engines

the lower right hole of thrust plate. After installing timing chain and sprockets, camshaft holding tool may be removed. New lifters must be used if a new camshaft is installed.

6-225

If camshaft bearings are to be replaced, the engine should be removed from the chassis. Before removing camshaft, remove valve lifters, oil pump and distributor. Slide camshaft out of engine, Fig. 15, then drive out rear welch plug. A special tool is required to remove and install camshaft bearings. When installing new bearings, make certain oil holes in bearings line up exactly with corresponding drilled oil passages in cylinder block. Install a new, well sealed welch plug at the rear of block. Carefully insert properly lubricated camshaft. New lifters must be used if a new camshaft is installed.

PISTON & ROD, ASSEMBLE

6-225

Piston and rod assemblies should be installed as shown in Figs. 16 and 17.

All V8 Engines

When installing piston and rod assemblies in the cylinders, the compression ring gaps should be diametrically opposite one another and not in line with the oil ring gap. The oil ring expander gap should be toward the outside of the "V" of the engine. The oil ring gap should be turned toward the inside of the engine "V".

Immerse the piston head and rings in clean engine oil and, with a suitable piston ring compressor, insert the piston and rod assembly into the bore. Tap the piston down into the bore, using the handle of a hammer.

Assemble and install the pistons and rods as shown in Fig. 18.

PISTONS, PINS & RINGS

Pistons are available in standard sizes and the following oversizes: 1974–77 V8, .005, .020, .040"; 1974–77 6 cyl., .005, .020, .040, .060"; 1978–83 6 cyl., .020"; 1978–83 V8, .020".

Pins are available in the following over-

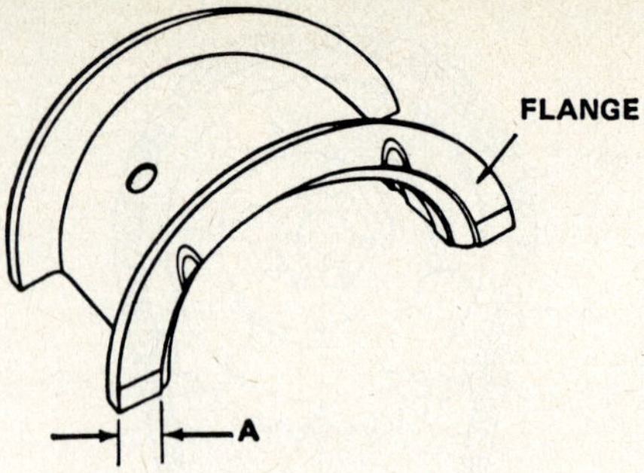

Fig. 20 Main bearing with small thrust wall flange (A)

sizes: V8-318 & 360, .003, .008". Oversize pins are not available on other engines.

Rings are available in the following oversizes: std. to .009, .020–.029, .040–.049".

MAIN & ROD BEARINGS

Main bearings are furnished in standard sizes and the following undersizes: .001, .002, .003, .010, .012".

Rod bearings are furnished in standard sizes and the following undersizes: .001, .002, .003, .010, .012".

NOTE: Some 1974 V8-400 and V8-440 engines were built with low thrust wall cylinder blocks. These engines require a small thrust wall or flange (No. 3) crankshaft main bearing. The dimension "A" in Fig. 20, must be noted if the thrust bearing (No. 3) is to be replaced.

Engines using the small thrust wall (flange) bearing, Fig. 20, can be identified by the yellow "X" painted on the engine identification pad and also engine will be date coded 8-12 on the pad.

All other 1974 V8-400 and V8-440 engines require the large thrust wall bearing.

NOTE: 1974–76 V8-400 and 440 use either forged or cast crankshafts. These engines require that a matching torque converter, short block, damper and connecting rods be used.

If replacement of the crankshaft, torque converter, crankshaft damper or short engine is required, it is important that matching parts are used otherwise severe engine vibration will result. (Consult Chrysler Parts Dept.).

The cast crankshaft engine can be easily identified since it has the letter "E" stamped on the engine numbering pad following the built date.

Also, V8-440 engines built starting January 1975, will have external balance weights at the front damper and torque converter and a clock face will appear on the number 1 counterweight of the cast crankshaft.

6-225 engines built after June 1, 1976 may be equipped with either a cast or forged crankshaft. The following parts are not interchangeable between cast and forged crankshaft engines, crankshaft, crankshaft bearings, connecting rods and bearings and cylinder block. On models equipped with manual transmission, cast crankshaft engine vibration damper is not interchangeable with forged crankshaft engine and requires that a crankshaft screw and washer be used and torqued to 135 ft. lbs. On models equipped with automatic transmission, vibration damper and torque converter are interchangeable between cast and forged crankshaft engines.

CRANKSHAFT OIL SEAL, REPLACE

6-225

Replacement seals are of two piece rubber type composition which make possible the replacement of upper rear seal without removing crankshaft. Both halves must be used. After removing oil pan, rear main bearing cap and seal retainer, pry lower rope seal from retainer with small screwdriver. Screw a special tool into upper rope seal and carefully pull to remove seal while rotating crankshaft. Wipe crankshaft surface clean, then oil lightly before installing new upper seal. After oiling seal lip, hold seal with paint stripe to rear tightly against crankshaft with thumb. Carefully slide seal into groove in block making sure sharp edge does not shave or nick seal. Crankshaft may be rotated to ease seal into groove but sealing lip must not be damaged. Install other half of seal into lower seal retainer again with paint stripe to rear, then install rear main bearing cap and torque to specifications. Before installing seal retainer, apply a small amount of gasket sealer to mating surface of retainer but not on seal ends or seal lip, Fig. 21. Install retainer and torque to 30 ft. lbs.

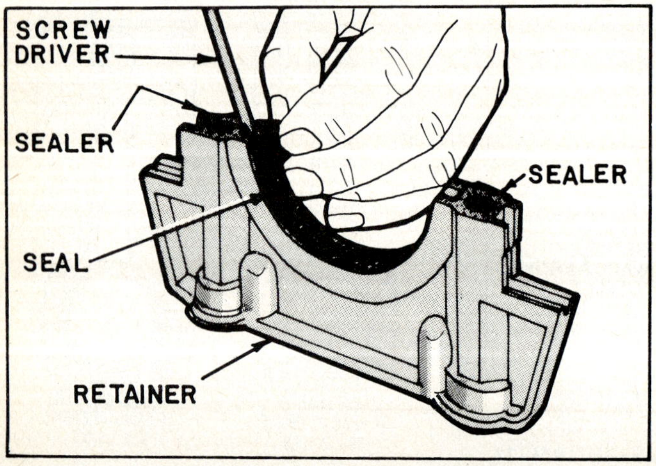

Fig. 21 Lower oil seal and retainer. 6-225

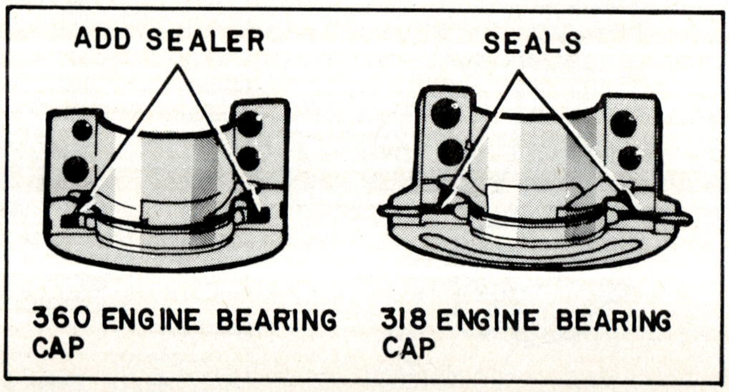

Fig. 22 Rear main bearing caps. V8-318 & 360

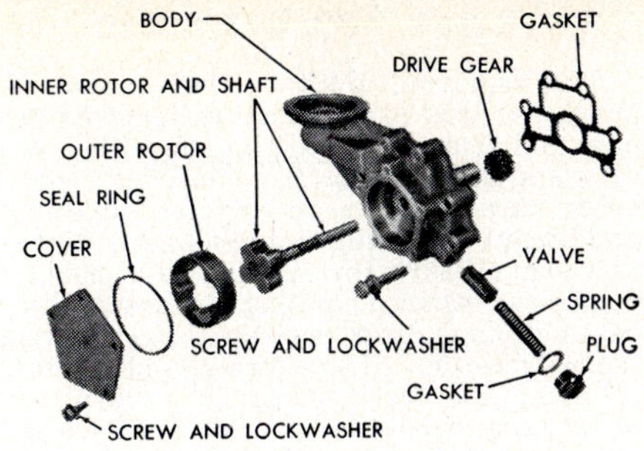

Fig. 23 Oil pump. 6-198 & 225

V8-318 & 360

Replacement of rear main bearing oil seals is similar to procedure given above for 6 cylinder engines. A seal retainer is not found on these engines; lower half of seal is installed into groove in rear main bearing cap.

The 318 engine has capseals in addition to lower seal secured by rear main bearing cap. Cap seal with yellow paint is installed, narrow sealing edge up, into right side with bearing cap in engine position. Cap seals must be flush with shoulder of bearing cap to prevent oil leakage.

The 360 engine requires sealer to be applied adjacent to rear main bearing oil seal as cap seals are not used, Fig. 22. After applying sealer, quickly assemble rear main bearing cap to block and torque to specifications.

V8-361, 400, 413, 440

Replacement seals are of two piece rubber type composition which make possible the replacement of upper rear seal without removing crankshaft. Both halves must be used. After removing oil pan, seal retainer and rear main bearing cap, pry lower rope seal and side seals from lower seal retainer with small screwdriver. Thoroughly clean seal retainer grooves. Screw a special tool into upper rope seal and carefully pull to remove seal while rotating crankshaft. Wipe crankshaft surface clean, then oil lightly before installing new upper seal. After oiling seal lip, hold seal with paint stripe to rear tightly against crankshaft with thumb. Carefully slide seal into groove in block making sure sharp edge does not shave or nick seal. Crankshaft may be rotated to ease seal into groove but sealing lip must not be damaged. Install other half of seal with paint stripe to rear into seal retainer. Install rear main bearing cap and torque to specifications. The side seals used with the rear seal retainer should be installed in the retainer and then into the engine as quickly as possible. These seals are made from a material that expands rapidly when oiled. Apply sealer to mating face of seal retainer but not to side seals or main bearing seal ends, then oil side seals with mineral spirits or diesel fuel. Install seals in seal retainer grooves then quickly install retainer and torque to 30 ft. lbs. Failure to pre-oil seals will result in an oil leak.

OIL PAN, REPLACE

V8-318, 360, 400, & 440 Except 1974–78 B Series & Voyager

1. Drain engine oil.
2. On 318 and 360 engines, remove torque converter housing support bracket and disconnect cross-over pipe from both manifolds. On Sport Utility vehicles, remove the left engine to transmission support only.
3. Remove the oil pan attaching screws and oil pan.
4. Reverse the procedure to install. Clean the oil pan in solvent and inspect the oil strainer alignment. Use a new pan gasket set and add a drop of sealant 4026070 or an equivalent at corners of rubber and cork.

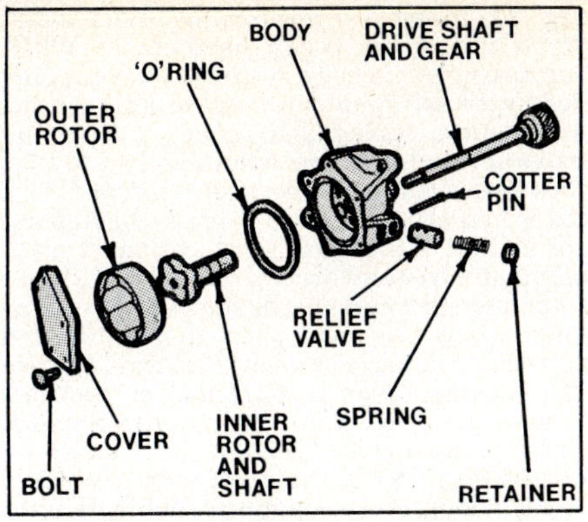

Fig. 24 Oil pump. V8-318 & 360

1974–78 B Series & Voyager V8-318 & 360

1. Disconnect the battery ground cable, then remove dipstick, engine cover, and air cleaner.
2. Disconnect throttle linkage at rear of engine and either clutch or automatic transmission linkage.
3. Raise the engine slightly and support with a suitable fixture installed in the vehicle.
4. Raise the vehicle and drain the oil. Remove the starter motor and oil filter.
5. Remove the propeller shaft and engine rear mount.
6. Remove transmission from vehicle.
7. Remove clutch assembly and flywheel or driveplate from crankshaft.
8. Raise rear of engine approximately 2 inches.

NOTE: To avoid damage to radiator, position fan to provide maximum clearance between blades and core.

---

9. Position crankshaft to clear oil pan for removal. Maximum clearance is obtained when crankshaft flange cut-out is at 3 o'clock position.
10. Remove oil pan screws and lower oil pan far enough to reach and turn the oil pick-up tube and strainer slightly toward right side to clear pan for removal. Remove pan.
11. Reverse procedure to install. Clean pan in solvent and install new gasket adding a drop of sealer 4026070 or equivalent at corners of rubber and cork.

1974–78 B Series & Voyager V8-400 & 440

1. Disconnect battery ground cable and remove engine cover.
2. Remove carburetor and install lifting fixture to intake manifold.
3. Install an engine support inside vehicle.
4. Raise vehicle and remove idle arm brackets from frame.
5. Disconnect strut bars from lower control arms.
6. Remove shock absorbers and motor mount insulator upper nuts.
7. Mark frame and crossmember on both sides for installation alignment. Remove crossmember outboard retaining bolts and replace with 4 1/2 inch long bolts to enable re-alignment.
8. Place floor stands under crossmember and remove inboard crossmember retaining bolts.
9. Carefully lower crossmember with floorstands. Drain oil and remove dipstick and tube.
10. Remove oil pan.
11. Reverse procedure to install. Check front end alignment and correct if necessary.

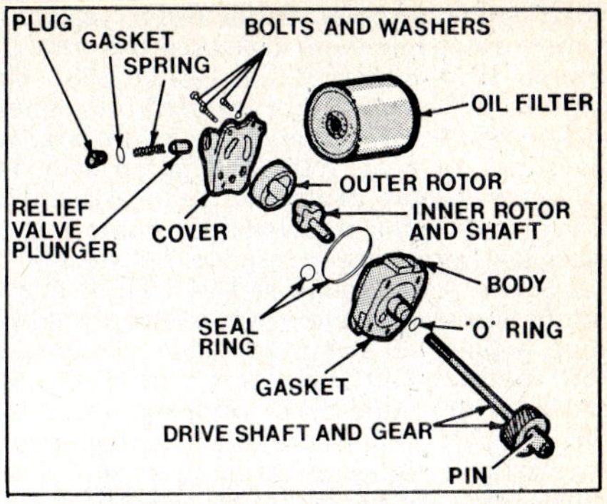

Fig. 25 Oil pump. V8-361, 383, 400, 413 & 440

1974–83 6-225

1. Disconnect battery ground cable, then remove engine oil dipstick, dipstick tube, engine cover, and air cleaner.
2. Raise and support front of vehicle, then drain crankcase and remove engine to transmission strut, if equipped.
3. On models equipped with automatic transmission, remove torque converter inspection cover.
4. Remove frame reinforcement if so equipped.
5. On 1974–78 vans, raise the engine slightly and support with suitable fixture. Remove propeller shaft and engine rear support. If equipped with automatic transmission, remove transmission from vehicle.
6. Remove clutch assembly and flywheel or driveplate from crankshaft flange. Raise rear of engine approximately six inches.
7. Remove oil pan screws and position the crankshaft to permit clearance for oil pan removal. Remove oil pan.
8. Reverse the procedure to install. On California engines, left side gasket is of high temperature type. Apply 1/8 inch diameter drop of sealer 4026070 or equivalent to four corners of rubber seal and cork gasket.

OIL PUMP REPAIRS

6-225

Remove pump from side of engine by first unbolting oil pump cover, Fig. 23. Outer rotor will drop out when cover is removed. Do not allow rotor to be damaged by dropping it. Oil pump can now be unbolted and disassembled for inspection. Press off drive gear, supporting gear to keep load off aluminum body. Inner rotor and shaft can now be removed. Remove oil pressure relief valve plug and lift out spring and plunger. Clean all parts thoroughly. If mating surface of oil pump cover is scratched or grooved, replace pump assembly. Lay a straightedge across pump cover surface. If a .0015 inch feeler gauge can be inserted between cover and straightedge, replace pump assembly. If outer rotor thickness measures .649 inch on 1974–80 or .825 inch or less on 1981–83 engines, and the diameter is 2.469 inches or less, replace shaft and both rotors.

Shaft and both rotors should also be replaced if inner rotor thickness measures .649 inch or less on 1974–80, or .825 or less on 1981–83 engines. With outer rotor inserted into pump body, press rotor to one side with fingers. If clearance between rotor and pump body is .014 inch or more, replace oil pump assembly. With inner rotor inserted into pump body, place a straightedge across face between bolt holes. If a feeler gauge of .004 inch or more can be inserted between rotors and straightedge, replace oil pump assembly. Shaft and both rotors should be replaced if tip clearance between inner and outer rotor exceeds .010 inch. Inspect oil pressure relief valve plunger for scoring and free operation in its bore. The relief valve spring has a free length of 2¼ inches and should test between 22.3 and 23.3 lbs. when compressed to 1 19/32 inch. Replace spring that fails to meet specifications.

Using new parts as required, assemble pump except for outer rotor and cover. Install oil pump on engine then wet inside of pump with oil to assure priming. Install outer rotor and cover using new seal ring.

V8-318 & 360

Removal of oil pump for servicing requires oil pan to be removed and oil pump unbolted from rear main bearing cap. With pump removed, disassemble pressure relief valve by pulling out cotter pin and drilling a 1/8 inch hole into center of relief valve retainer cap, Fig. 24. Insert a self-threading sheet metal screw into cap and secure head in vise. While supporting pump body, remove cap by tapping body with soft hammer. Discard retainer cap and remove spring and relief valve. Relief valve spring has a free length of 2 1/32 to 2 3/64 inch and should test between 16.2 and 17.2 lbs. when compressed to 1 11/32 inch. Replace spring that fails to meet specifications. Unbolt oil pump cover and discard oil seal ring. Inner rotor and shaft can now be removed as well as outer rotor. Clean all parts thoroughly and inspect for damage or wear. If mating surface of oil pump cover is scratched or grooved, replace pump assembly. Lay a straightedge across pump cover surface. If a .0015 inch feeler gauge can be inserted between cover and straightedge, replace pump assembly. If outer rotor thickness measures .825 inch or less on 318 engine and 1974–76 360 engine, or less than .943 inch on 1977–83 360 engine, or the diameter is 2.469 inches or less, replace the outer rotor. If inner rotor measures .825 inch or less on 318 engine and 1974–76 360 engine, or less than .943 on 360 engine, replace inner rotor and shaft assembly. With outer rotor inserted into pump body, press rotor to one side with fingers. If clearance between rotor and pump body is .014 inch or more, replace oil pump assembly. With inner rotor inserted into pump body, place a straightedge across face between bolt holes. If a feeler gauge of .004 inch can be inserted between rotors and straightedge, replace oil pump assembly. Shaft and both rotors should be replaced if tip clearance between inner and outer rotor exceeds .010 inch. Using new parts as required, assemble pump. Prime pump before installation by filling rotor cavity with engine oil.

V8-361, 400, 413 & 440

After removing oil filter element unbolt oil pump and filter base assembly. Remove filter base and discard oil seal ring, Fig. 25. Inner rotor and shaft can now be removed as well as outer rotor. Remove pressure relief valve plug and lift out the spring and relief valve plunger. Clean all parts thoroughly and inspect for damage or wear. If mating surface of filter base (oil pump cover) is scratched or grooved, replace filter base. Lay a straight edge across oil pump filter base surface. If a .0015 inch feeler gauge can be inserted between base and straightedge, filter base should be replaced. If outer rotor thickness measures .943 inch or less and diameter measures 2.469 inches or less, replace shaft and both rotors. If inner rotor thickness measures .943 inch or less, replace shaft and both rotors. With outer rotor inserted into pump body, press rotor to one side with fingers. If clearance between rotor and pump body is .014 inch or more, replace pump body. With inner rotor inserted into pump body, place a straightedge across face between bolt holes. If a feeler gauge of .004 inch or more can be inserted between rotors and straightedge, replace pump body. If the top clearance between inner and outer rotor is .010 inch or more, replace shaft and both rotors. Inspect oil pressure relief valve plunger for scoring and free operation in its bore. The relief valve spring has a free length of 2¼ inches and should test between 22.3 and 23.3 lbs. when compressed to 1 19/32 inch. Replace spring that fails to meet specifications.

Using new parts as required, assemble pump and fill rotor cavity with engine oil to prime. Before bolting oil pump to engine, install new "O" ring seal on pilot.

Carburetor Section

MODEL BBS ADJUSTMENTS

Float Level Adjustment

Fig. 1

Remove both check balls, then invert main body so that only weight of floats is forcing needle against seat. If proper gauge is not available, measure from surface of fuel bowl to crown of each float at center. Float setting should be as listed in the "BBS Specifications Chart". If adjustment is necessary, hold floats on bottom of bowl and bend lip of float lever in or out until correct setting is obtained. Do not allow lip to push against needle as the synthetic rubber tip can be compressed sufficiently to cause a false setting which will affect correct level of fuel in bowl. When float is set correctly, float lip should be perpendicular to or slanted less than 10 degrees away from needle.

Pump & Bowl Vent Adjustment 1974 With C.A.P. & C.A.S.

Fig. 2

This adjustment automatically adjusts the accelerator pump as well. The procedure is as follows:

1. Back off idle speed adjusting screw. Open choke valve so that when throttle valve is closed the fast idle adjusting screw will not contact fast idle cam.
2. Be sure pump operating rod is in center hole in throttle lever and that bowl vent clip on pump stem is in center groove.
3. Close throttle valve tightly. It should be possible to insert a drill of the specified size between bowl vent and air horn.
4. If an adjustment is necessary, bend pump operating rod at the lower angle as required to obtain the correct bowl vent opening.

NOTE: This is an important adjustment since too much lift at the bowl vent will result in considerable loss in low speed fuel economy. If the pump operating rod is moved to either the short or long stroke position, a corresponding change must be made in the location of the bowl vent clip, and the amount of lift of the bowl rechecked and adjusted.

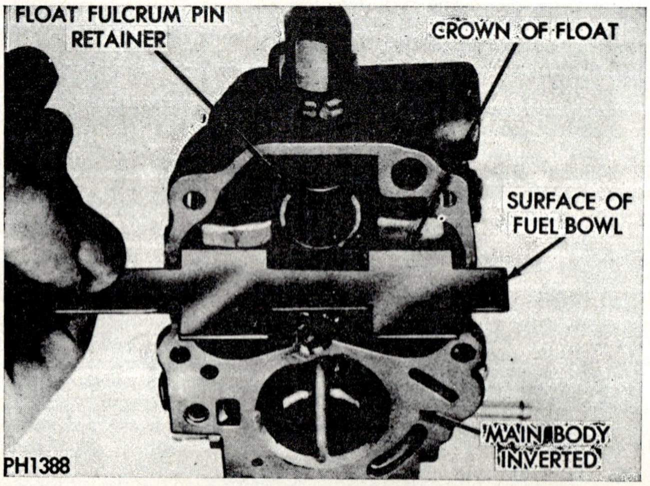

Fig. 1 Checking float level. BBS and BBD carburetors.

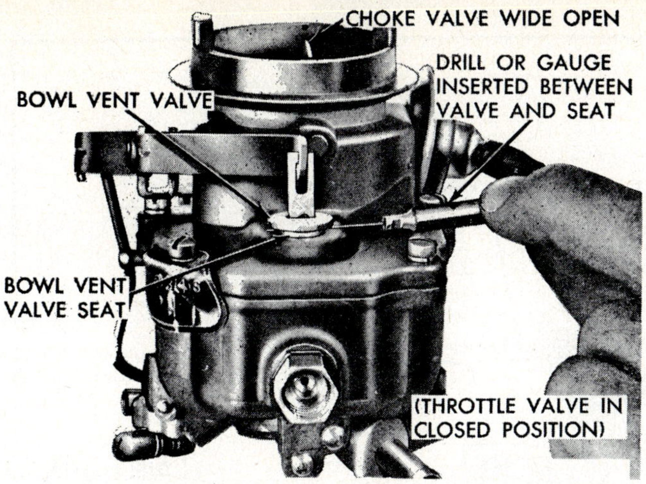

Fig. 2 Checking bowl vent opening. BBS and BBD carburetors

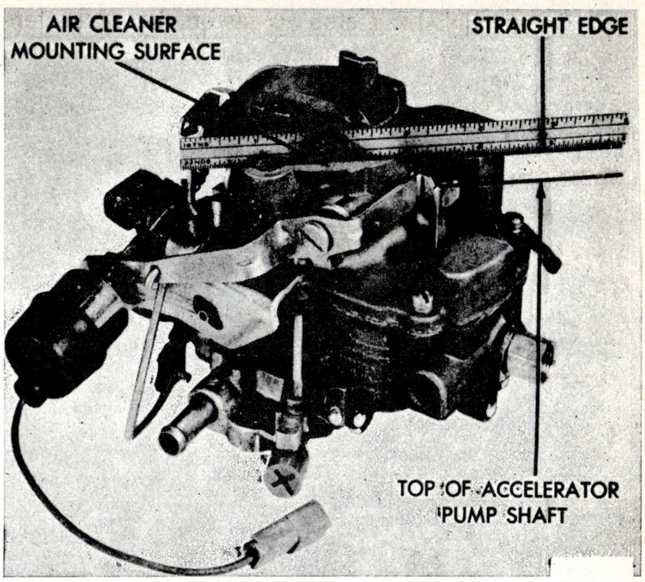

Fig. 3 Checking accelerator pump travel. BBS carburetors with E.C.S.

Pump and Bowl Vent Adjustments E.C.S.

On the E.C.S. BBS carburetor, it is necessary to check or set the accelerator pump travel before checking the bowl vent valve opening.

Accelerator Pump Travel Adjustment

1. Be sure the accelerator pump rod is in outer hole (long stroke) of throttle lever.
2. Close throttle valve to curb idle.
3. Using a straight edge placed on flat surface (air cleaner mounting surface), measure distance between straight edge and top of accelerator pump plunger, Fig. 3. Refer to *BBS Specification Chart.*
4. If an adjustment is necessary, bend accelerator pump rod at angle until correct pump travel has been obtained.

Bowl Vent Valve Adjustment

This adjustment to be made after accelerator pump adjustment.

1. With throttle valve at curb idle, measure distance from top of casting to top of bowl vent valve stem, Fig. 4. Refer to *BBS Specifications Chart.*
2. If an adjustment is necessary, bend lower tang on bowl vent valve operating lever (at pivot) until correct opening has been obtained.

Choke Unloader Adjustment

Fig. 5—Hold throttle valve in wide open position. Insert a drill of the specified size between upper edge of choke valve and inner wall of air horn. With a finger lightly pressing against choke valve, a slight drag should be felt as the drill is being withdrawn. If an adjustment is necessary, bend unloader tang to throttle lever as required.

Fast Idle Cam Position

Fig. 6

1. With fast idle speed adjusting screw contacting the second highest step on the fast idle cam, move choke valve toward closed position with light pressure on choke shaft lever.
2. Insert specified drill between choke valve and wall of air horn. If an adjustment is necessary, bend fast idle rod at upper angle until correct valve opening has been obtained.

Vacuum Kick Adjustment

Fig. 7

The choke diaphragm adjustment controls

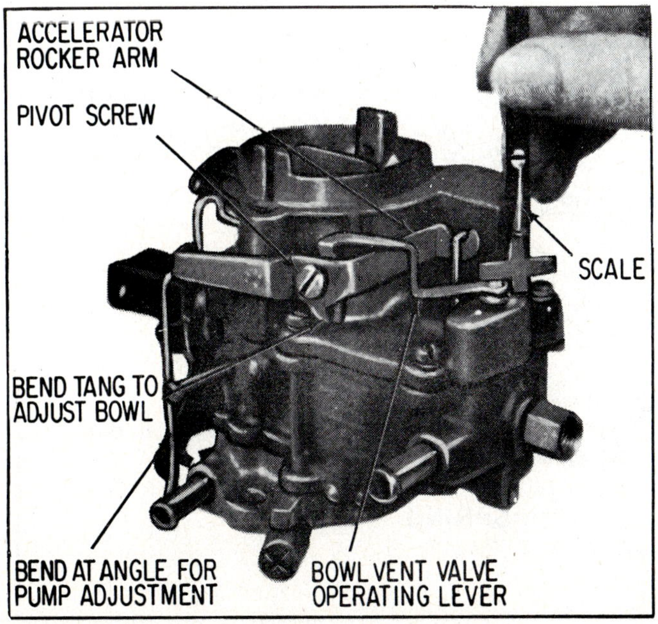

Fig. 4 Bowl vent adjustment. BBS carburetors with E.C.S.

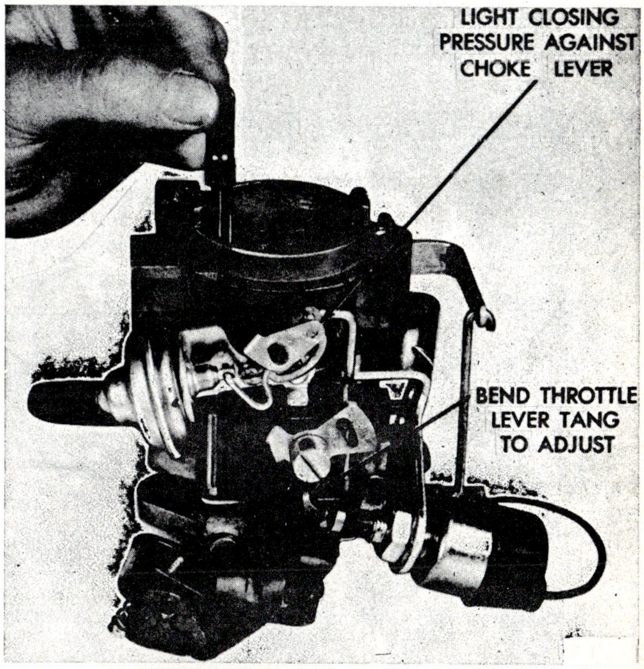

Fig. 5 Checking choke unloader setting (wide open kick). BBS carburetors

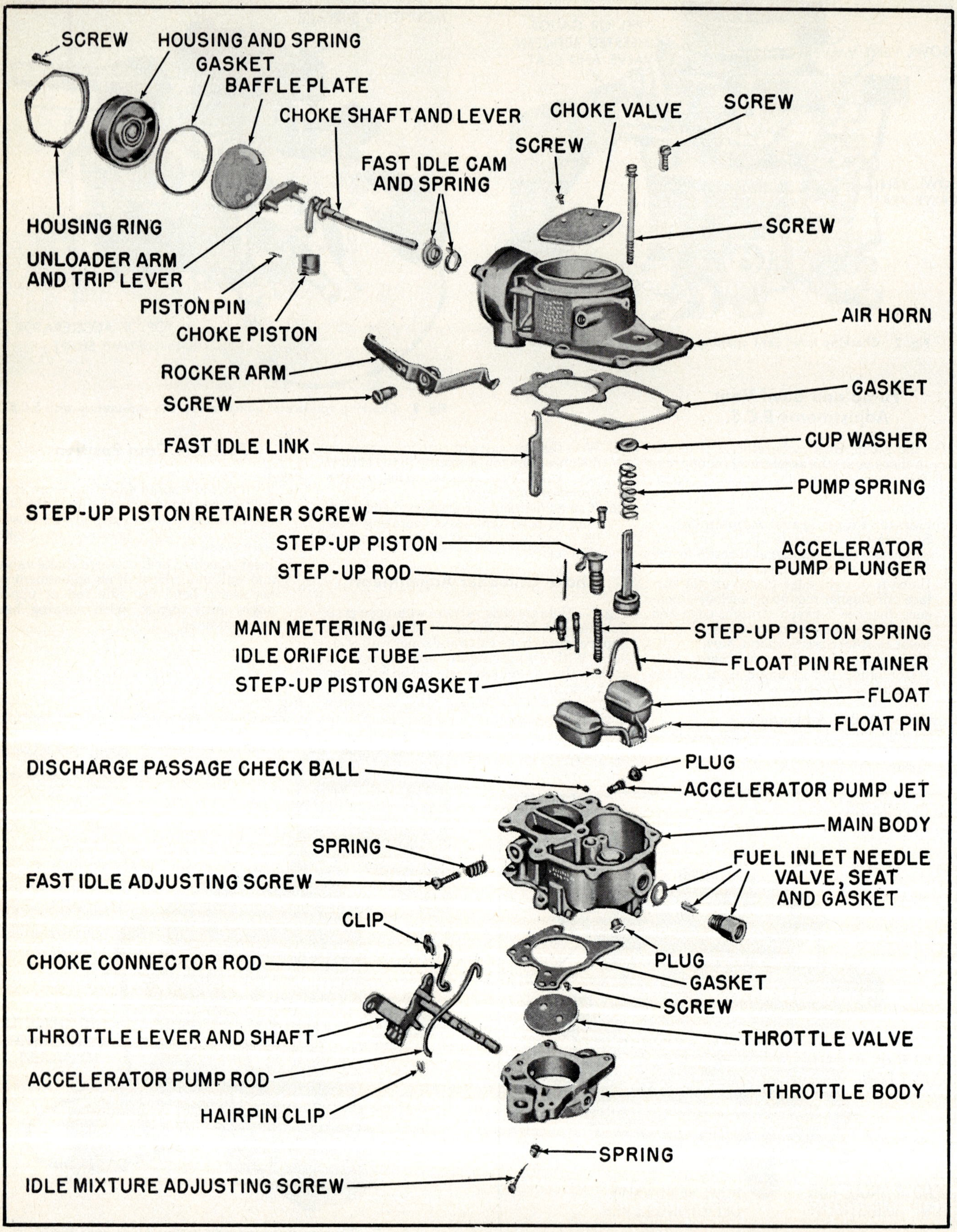

Carter BBS single barrel unit with built-in automatic choke

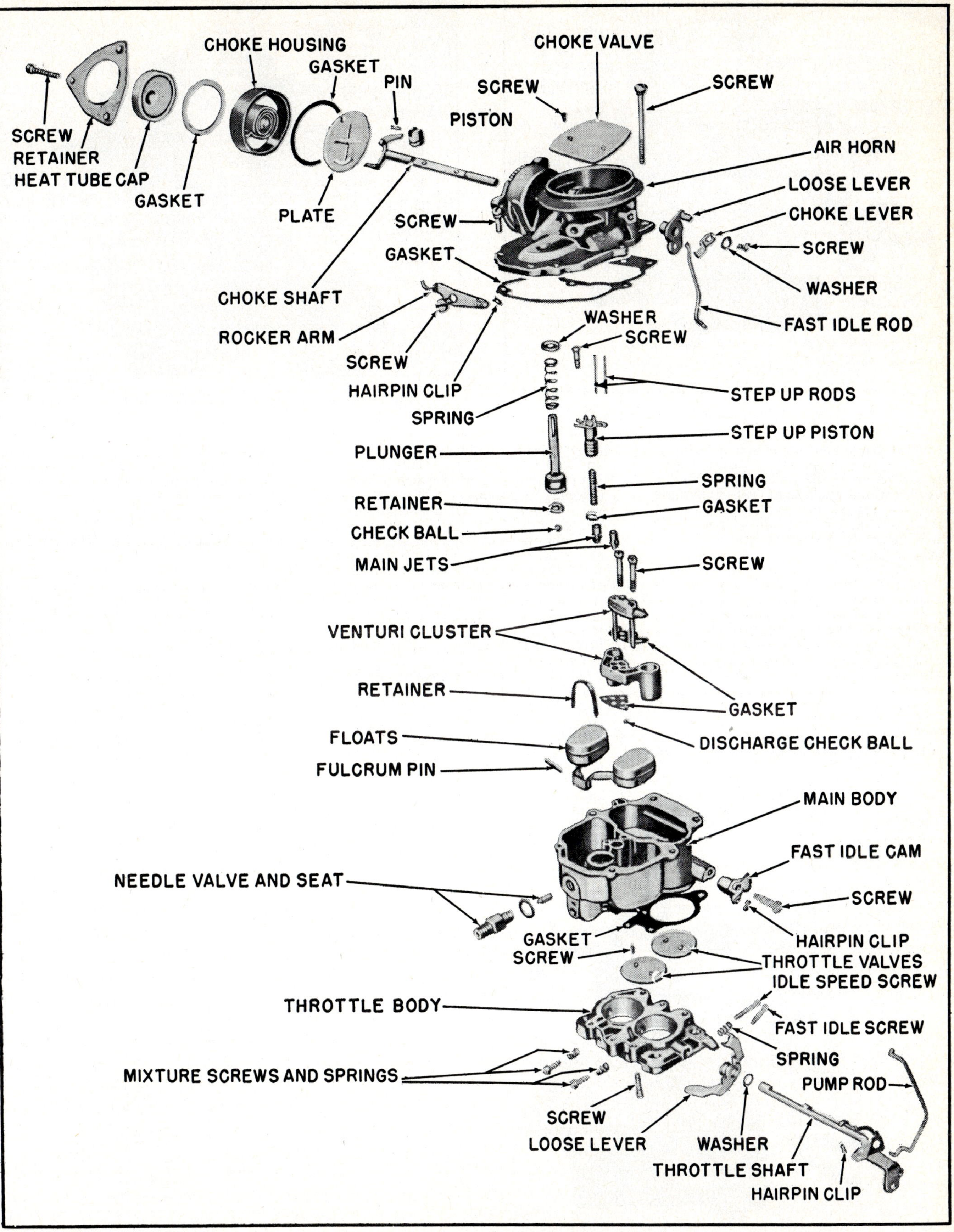

Carter BBD two-barrel unit with built-in automatic choke

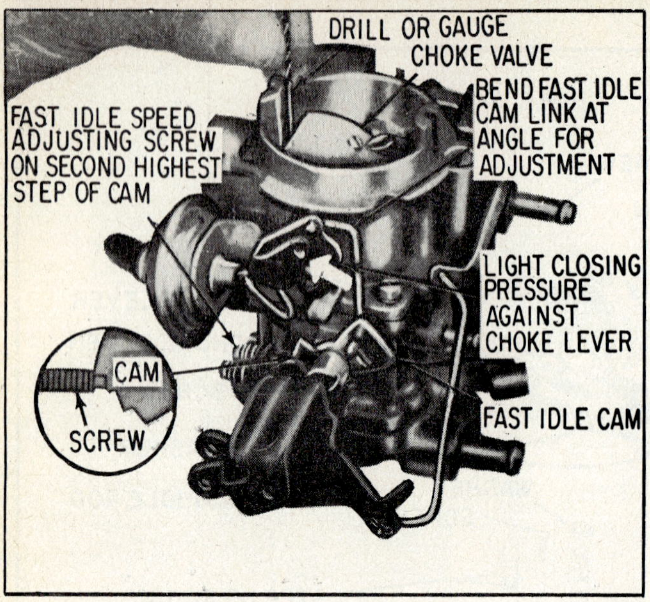

Fig. 6 Fast idle cam position. BBS carburetors

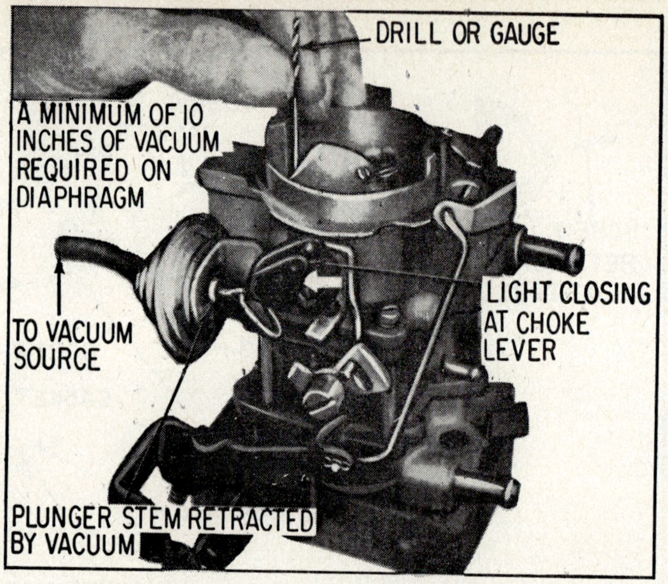

Fig. 7 Vacuum kick adjustment. BBS and BBD carburetors

the fuel delivery while the engine is running. It positions the choke valve within the air horn by action of the linkage between the choke shaft and diaphragm. The diaphragm must be energized to measure the vacuum kick adjustment. Use a vacuum source, or vacuum supplied by another vehicle.

Adjust as follows:

1. If adjustment is to be made with the engine running, back off the fast idle speed screw until choke can be closed to the kick position with engine at curb idle. Note the number of screw turns required so that fast idle can be returned to original adjustment. If an auxiliary vacuum source is to be used, open throttle valve (engine not running) and move choke to closed position. Release throttle first, then release choke.
2. When using an auxiliary vacuum source, disconnect vacuum hose from carburetor and connect it to hose from vacuum supply with a small length of tube to act as a fitting. Removal of hose from diaphragm may require forces which change the system. Apply a vacuum of 10 inches or more of mercury.
3. Insert specified drill between choke valve and wall of air horn. Apply sufficient closing pressure on lever to which choke rod attaches to provide a minimum stroke valve opening without distortion of diaphragm link. Note that cylindrical stem of diaphragm will extend as internal spring is compressed. This spring must be fully compressed for proper measurement of vacuum kick adjustment.
4. An adjustment will be necessary if a slight drag is not obtained as drill is being removed. Shorten or lengthen diaphragm link to obtain correct choke opening. Length changes should be made carefully by bending (opening or closing) the bend provided in the diaphragm link. *Do not apply twisting or bending force to diaphragm.*
5. Reinstall vacuum hose on correct carburetor fitting. Return fast idle screw to its original location if disturbed as suggested in Step 1.
6. Check the adjustment as follows: With no vacuum applied to diaphragm, *choke valve should move freely between open and closed positions.* If movement is not free, examine linkage for misalignment or interferences caused by bending operation. Repeat adjustment if necessary to provide proper link operation.

Spring Staged Choke Adjustment

Fig. 8

The spring staged choke is a device incorporated in the choke mechanism that limits the choke blade closing torque when cranking the engine at temperatures below zero. Thus the spring staging of the choke is a better match for the engine's starting mixture requirements at low temperatures.

To check the spring staged choke for correct operating clearance, proceed as follows:

1. Push on hub lever with finger at closed choke position. A small opening should exist between shaft and hub levers as indicated.
2. Using a drill or gauge, measure the opening which should be from .010 to .040".
3. If adjustment is necessary, bend hub lever tang until correct opening is obtained.

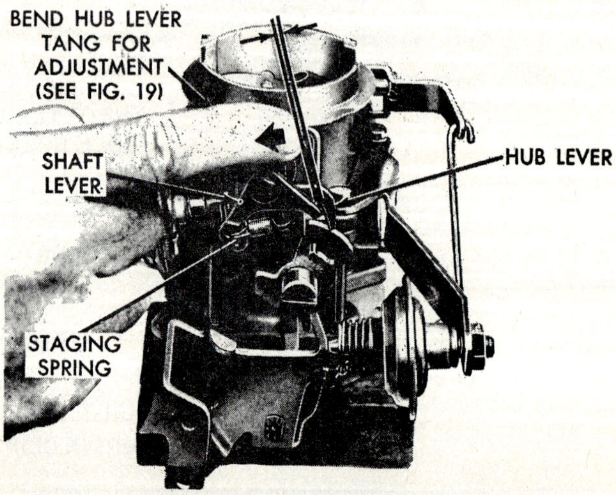

Fig. 8 Spring staged choke adjustment. BBS carburetors

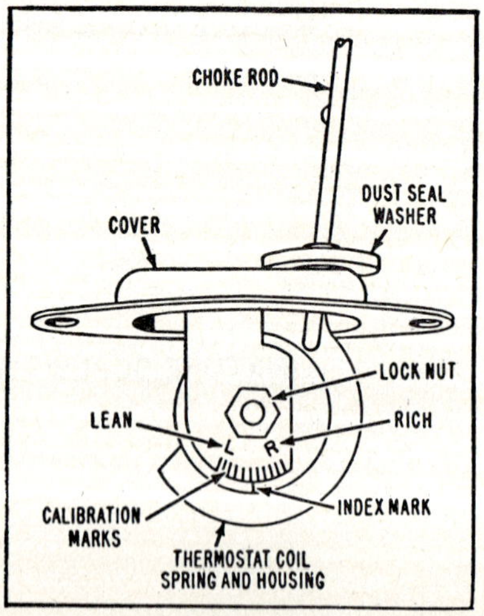

Fig. 9 Well type choke setting. BBS and BBD carburetors

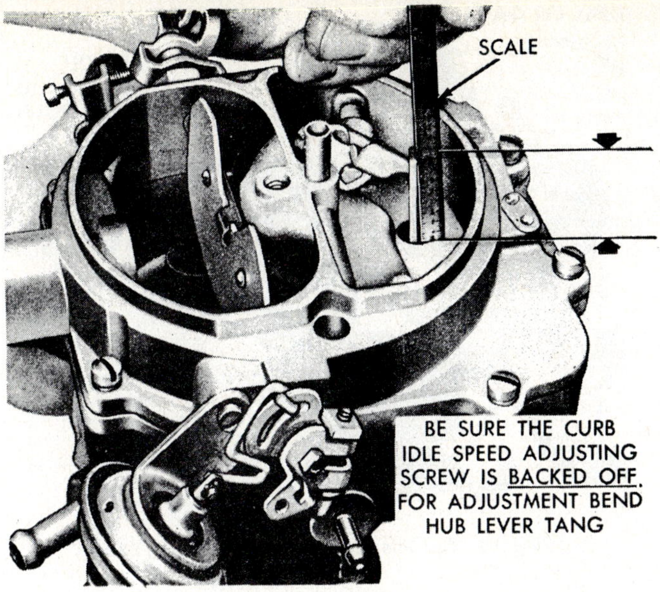

Fig. 10 Checking accelerator pump travel. 1974 BBD except 1 1/4" bore carburetors

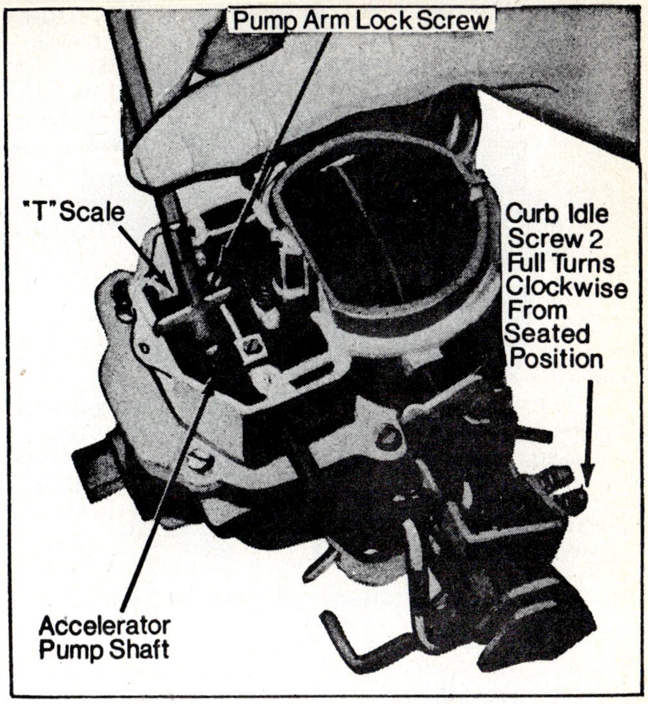

Fig. 11 Accelerator pump setting. 1974 1 1/4" E.C.S. units & 1975–79 & 1982–83 all

Choke Adjustment

Fig. 9—Loosen mounting post lock nut and turn mounting post with screwdriver until index mark on disc is positioned as listed in the *BBS Specifications Chart.* Hold in this position with screwdriver and tighten lock nut.

NOTE: Screwdriver may be held in a vise so that one hand may be used to support housing while tightening nut. After adjustment is completed and coil housing and rod and carburetor are installed on engine, lift cover disc and open and close choke valve manually to see if connector rod clears sides of hole in housing cover without binding. If rod does not clear housing cover without binding, replace with a new unit since connecting rod cannot be bent without affecting calibration.

Dashpot Adjustment

Clean Air Carburetors

To adjust the dashpot, have the curb idle speed and mixture properly adjusted, and install a tachometer. Position throttle lever so that actuating tab on lever is contacting stem of dashpot but not depressing it. The tachometer should read 1000 rpm if the setting is correct. If not correct, screw dashpot in or out as required, then tighten lock nut on dashpot against the bracket.

MODEL BBD ADJUSTMENTS

Float Level Adjustment

Fig. 1—With carburetor body inverted so that weight of floats ONLY is forcing needle against its seat, use a T-scale or the tool shown, and check the float level from surface of fuel bowl to crown of each float at center.

If an adjustment is necessary, hold floats on bottom of bowl and bend float lip as required to give the specified dimension.

CAUTION: When bending the float lip, do not allow the lip to push against the needle as the synthetic rubber tip (if used) can be compressed sufficiently to cause a false setting which will affect correct level of fuel in bowl. After being compressed, the tip is very slow to

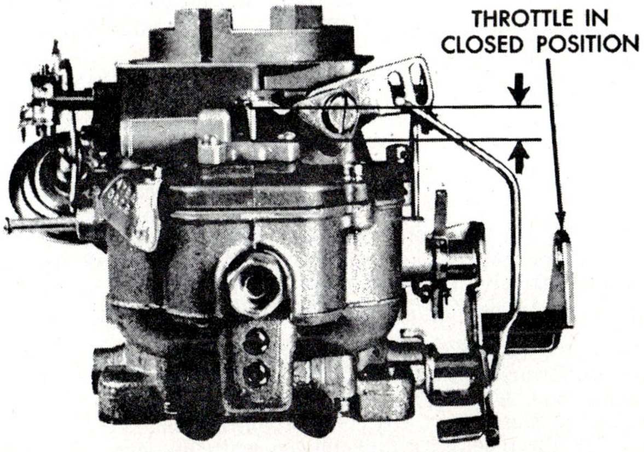

Fig. 12 Checking bowl vent opening. BBD 1 1/4" E.C.S. units

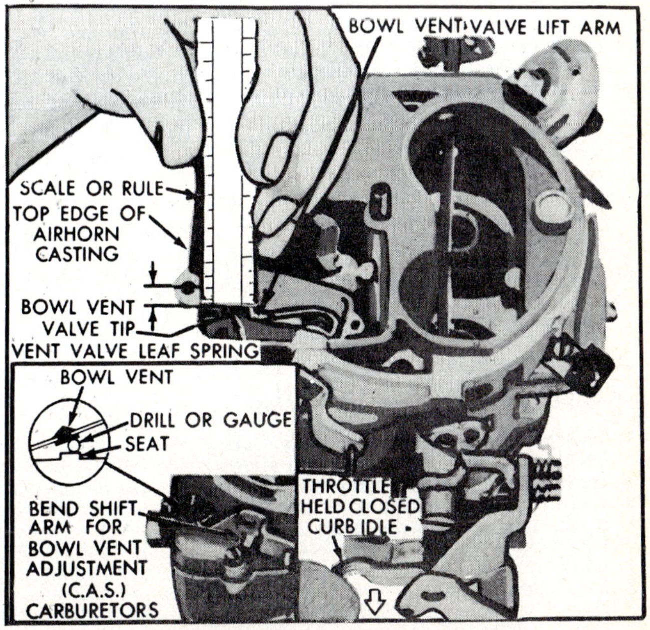

Fig. 13 Measuring bowl vent valve opening. BBD 1 1/2" carburetors

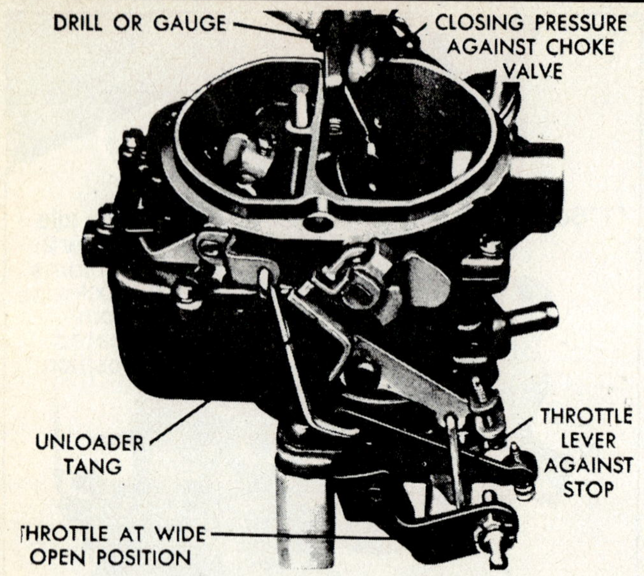

Fig. 14 Choke unloader setting. BBD carburetors

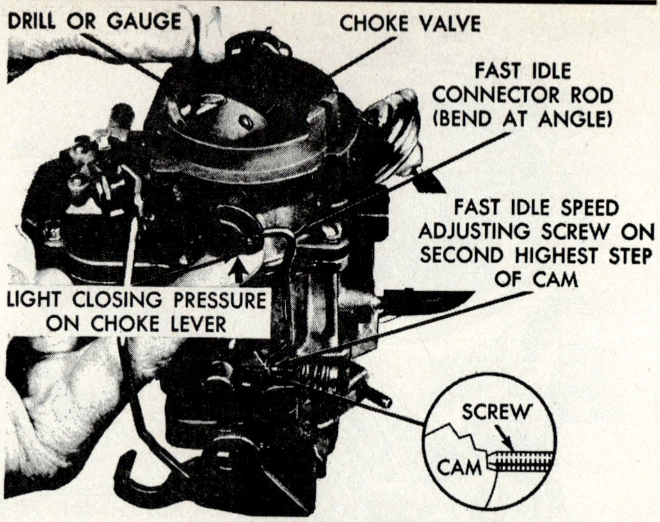

Fig. 15 Fast idle cam position adjustment. BBD 1/4" bore carburetors

recover its original shape.

Accelerator Pump

1974 except 1¼" Bore Units, Fig. 10

1. Back off idle adjusting screw. Open choke valve so that fast idle cam allows throttle valves to fully close. Be sure that pump connector rod is installed in outer hole of throttle lever.
2. With throttle valves closed tightly, measure distance between top of air horn and end of pump plunger shaft. If the dimension is not as specified, bend pump connector rod at the angle on the rod until correct setting is obtained.

1974 1¼" Bore Units & 1975–79 & 1982–83 All, Fig. 11

1. Back off curb idle adjusting screw, completely closing throttle valve, then open choke valve, allowing throttle valves to seat in bores. Ensure accelerator pump "S" link is located in outer hole or pump arm.
2. Turn curb idle adjusting screw until screw contacts, stop, then rotate screw two additional turns.
3. Measure distance between air horn surface and top of accelerator pump shaft. Refer to BBD Specifications Chart.
4. Adjust by loosening pump arm adjusting screw and rotating sleeve until proper dimension is obtained. Tighten adjusting screw.

Accelerator Pump and Bowl Vent

C.A.S. Equipped 1¼" Bore Units, Figs. 1 & 2

1. Back off idle speed adjusting screw. Open choke valve so that fast idle cam allows throttle valves to close completely.
2. Be sure pump operating rod is in medium stroke hole in throttle lever, and that bowl vent clip on pump stem is on center notch.
3. With throttle valves closed tightly, it should be just possible to insert the specified gauge or drill size between bowl vent and its seat.
4. If an adjustment is necessary, bend pump operating rod at the angle. On CAP carburetors, bend pump operating rod to give the height of pump plunger stem above bowl cover as specified in the table.

E.C.S. Equipped 1¼" Bore Units, Fig. 12

1. Back off idle speed adjusting screw to completely close throttle valves. Open choke valve so that fast idle cam allows throttle valves to seat in bores.
2. Be sure accelerator pump operating rod is in medium stroke hole in throttle lever.
3. Close throttle valves tightly. Measure distance between top of vent valve plastic housing and end of pump plunger stem.
4. To adjust pump travel, bend accelerator pump operating rod at lower angle. Refer to BBD Specifications Chart.

NOTE: This is an important adjustment, since too much lift at the bowl vent will result in considerable loss in low speed fuel economy.
Remember that if the pump operating rod is moved to either the short or long stroke position, a corresponding change must be made in the location of the bowl vent clip, and the amount of lift of the bowl vent rechecked.

Bowl Vent Adjustment

1½" Bore Units with Separate Pump, Fig. 13

1. Open choke valve so that fast idle cam allows valves to close, curb idle.
2. Be sure that pump operating rod is in long stroke hole in throttle lever. Remove bowl vent valve cover if not previously done.
3. On E.C.S. units, close throttle valves tightly. Using a narrow ruler, measure the distance from top of bowl vent valve (rubber tip) to top of air horn casting. Refer to Specifications Chart. If an adjustment is necessary, bend bowl vent lift arm, using a suitable tool, until correct opening has been obtained. On C.A.S. units, with the throttle valves closed, (curb idle), there should be specified clearance between bowl vent valve and seat on air horn. Measure at outermost or largest dimension with a drill shank. If an adjustment is necessary, bend vent valve lifter arm until correct clearance has been obtained.

NOTE: Do not bend bowl vent valve leaf spring during bending operation or improper vent valve operation will result.

4. Install bowl vent valve cover and secure with attaching screws.

Choke Unloader Adjustment

Fig. 14—The choke unloader is a mechanical device to partially open the choke valve at wide open throttle. It is used to eliminate choke enrichment during engine cranking. Engines that have been flooded or stalled by excessive choke enrichment can be cleared by the use of the unloader. Adjust as follows:

1. Hold throttle valve in wide open position. Insert the specified drill size between upper edge of choke valve and inner wall of air horn.
2. With a finger lightly pressing against choke valve, a slight drag should be felt as the drill is being withdrawn.
3. If an adjustment is necessary, bend unloader tang on throttle lever until specified opening has been obtained.

Fast Idle Cam Position

Figs. 15 & 16

1. With fast idle adjusting screw contacting second highest step on fast idle cam, move choke valve toward closed position with light pressure on choke shaft lever.
2. Insert the specified size drill between choke valve and air horn wall. An adjustment will be necessary if a slight drag is not obtained as drill is being removed.

Choke Vacuum Kick Adjustment

Figs. 7 & 17

The choke diaphragm adjustment controls the fuel delivery while the engine is running. It positions the choke valve within the air horn by action of the linkage between choke

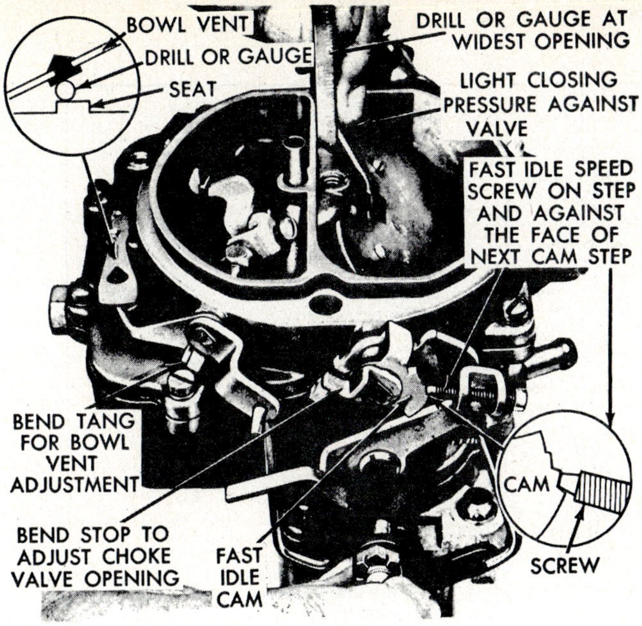

Fig. 16 Fast idle cam position adjustment. BBD except 1¼" bore carburetors

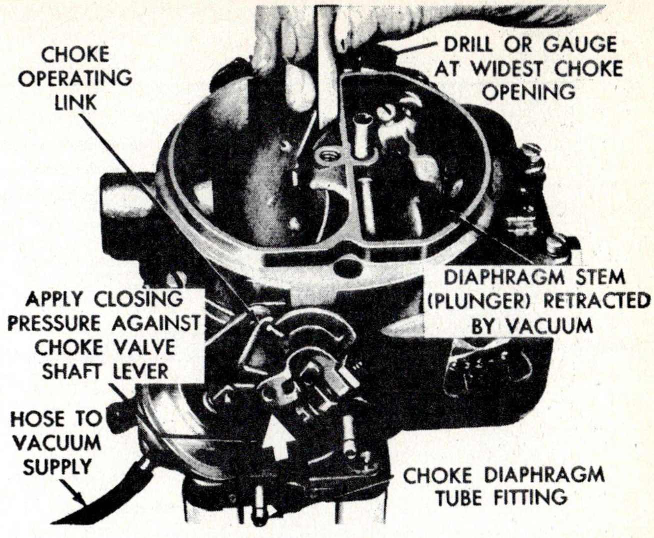

Fig. 17 Choke vacuum kick setting. BBD except 1¼" bore carburetors

shaft and diaphragm. The diaphragm must be energized to measure the vacuum kick adjustment. Use either a distributor test machine with a vacuum source, or vacuum supplied by another vehicle.

1. If adjustment is to be made with engine running, disconnect fast idle linkage to allow choke to close to kick position with engine at curb idle. If an auxiliary vacuum source is to be used, open throttle valves (engine not running) and move choke to closed position. Release throttle first, then release choke.
2. When using an auxiliary vacuum source, disconnect vacuum hose from carburetor and connect it to hose from vacuum supply with a small length of tube to act as a fitting. Removal of hose from diaphragm may require forces which damage the system. Apply a vacuum of 15 or more inches of mercury.
3. Insert the specified drill size between choke valve and wall of air horn. Apply sufficient closing pressure on lever to which choke rod attaches to provide a minimum choke valve opening without distortion of diaphragm link. Note that the cylindrical stem of diaphragm will extend as internal spring is compressed. This spring must be fully compressed for proper measurement of vacuum kick adjustment.
4. An adjustment will be necessary if a slight drag is not obtained as drill is being removed. Shorten or lengthen diaphragm link to obtain correct choke opening. Length changes should be made carefully by bending (opening or closing) the bend provided in the diaphragm link. *Do not apply twisting or bending force to diaphragm.*
5. Reinstall vacuum hose on correct carburetor fitting. Return fast idle linkage to its original condition if it has been disturbed as in Step 1.
6. Check as follows: With no vacuum applied to diaphragm, choke valve should move freely between open and closed positions. If movement is not free, examine linkage for misalignment or interferences caused by bending operation. Repeat adjustment if necessary.

Step-up Piston Adjustment

1974 1¼" Units & 1975–79 & 1982–83 All, Fig. 18

1. Adjust gap in step-up piston to .030 inch on 1974 models and .035 inch on 1975 models by turning allen head screw on top of piston, then install step-up piston assembly into air horn bore, making certain that metering rods are positioned in metering jets.

NOTE: On 1976 units, do not remove or adjust allen head screw on top of vacuum

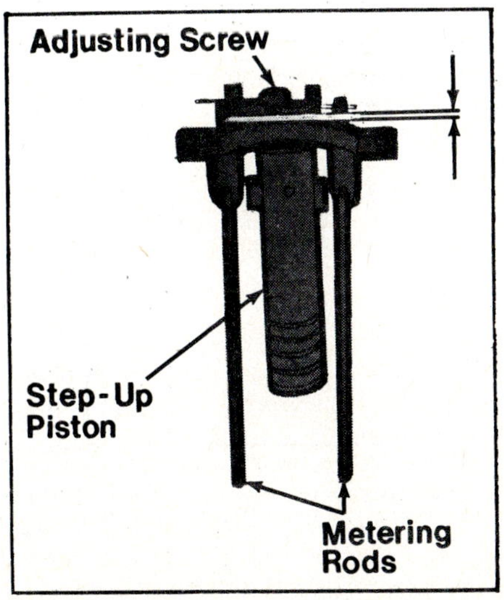

Fig. 18 Step up piston adjustment

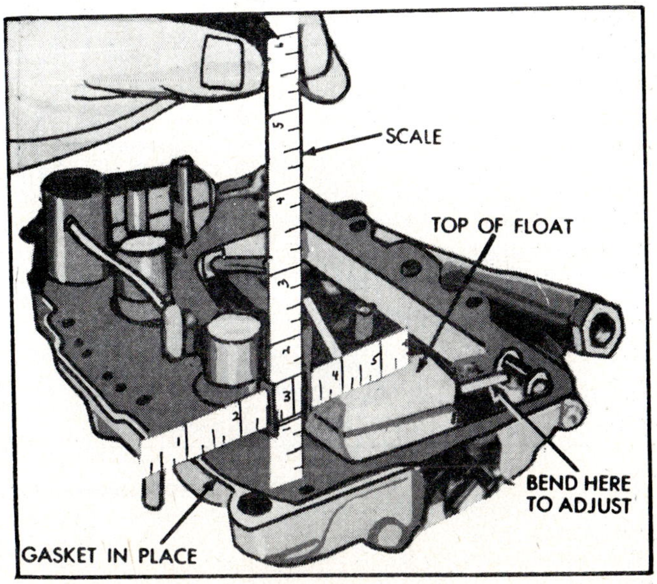

Fig. 19 TQ float setting

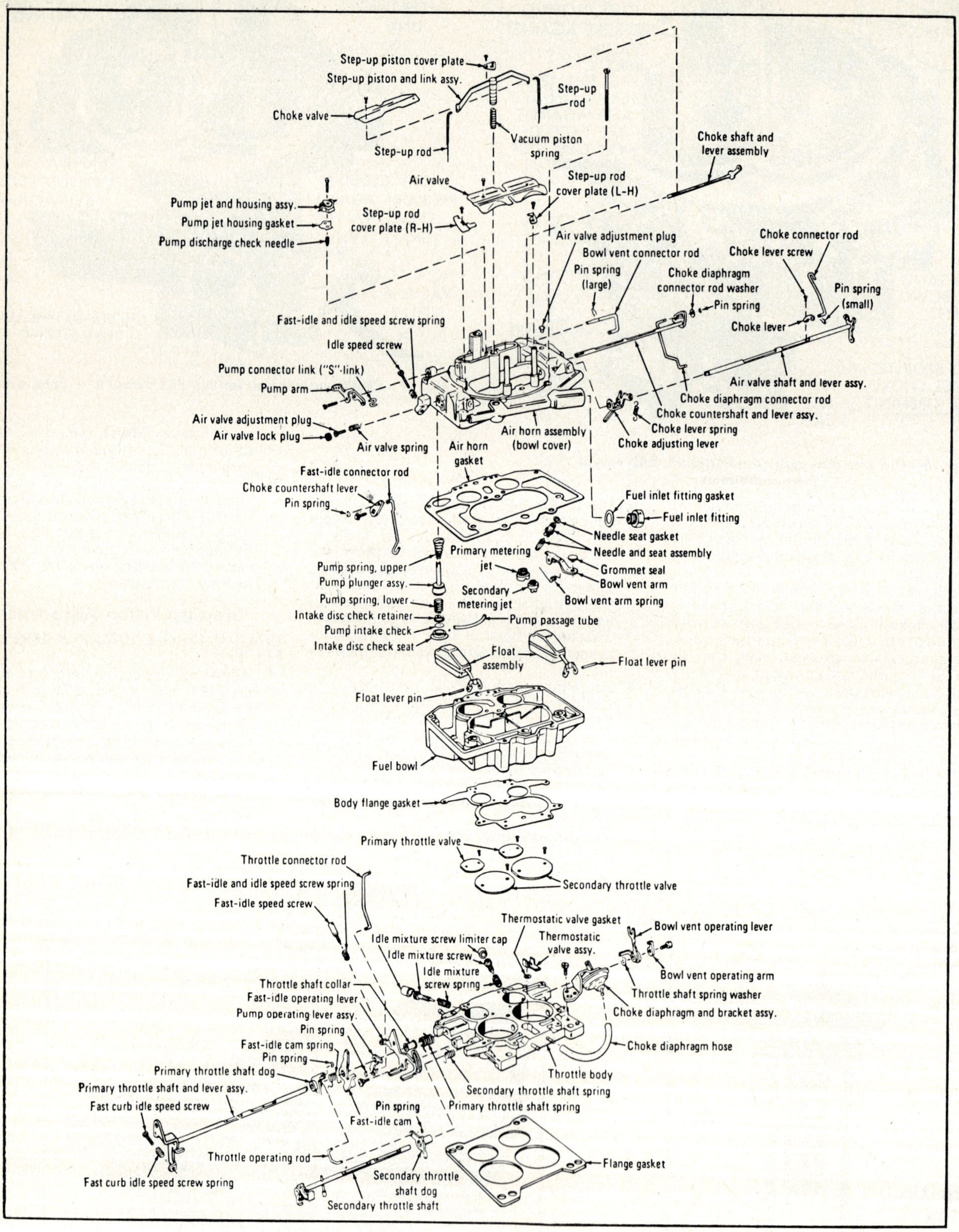

Carter TQ model exploded view

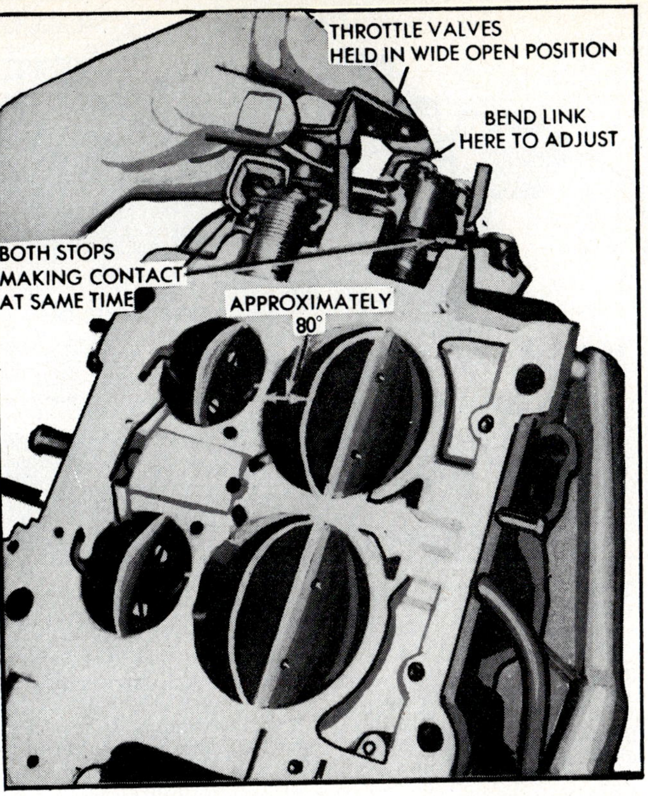

Fig. 20 TQ secondary throttle adjustment

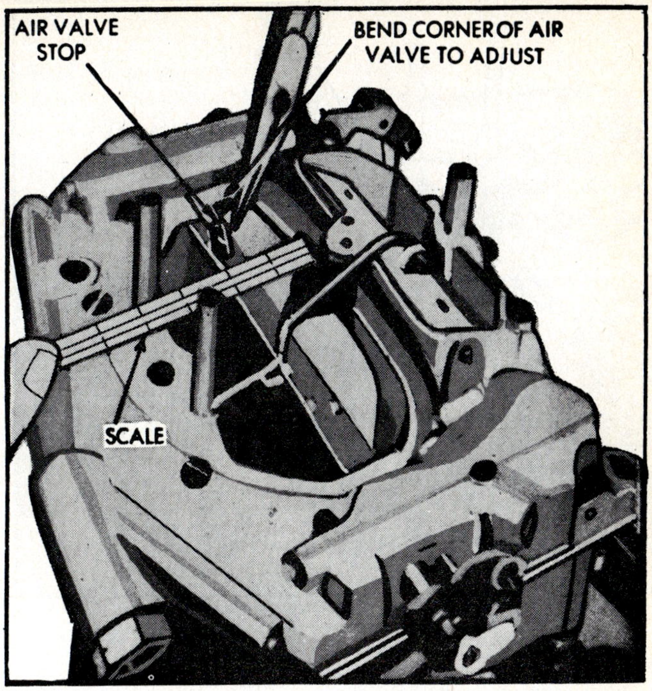

Fig. 21 TQ secondary air valve opening

piston.

2. Back off curb idle screw until throttle valves are completely closed. Count number of turns so that screw can be returned to its original setting.
3. Fully depress piston and while applying moderate pressure on rod lifter, tighten rod lifter screw.
4. Release piston and rod lifter and reset curb idle screw.

Well-Type Choke Setting

Fig. 9—Loosen mounting post lock nut and turn mounting post with screwdriver until index mark on disc is positioned as listed in the *BBD Specifications Chart*. Hold in this position with screwdriver and lock nut.

NOTE: Screwdriver may be held in vise so that one hand may be used to support housing while tightening nut. After adjustment is completed and coil housing, rod and carburetor are installed on engine, lift cover disc and open and close choke valve manually to see if connector rod clears sides of hole in housing cover without binding. If rod does not clear housing cover without binding, replace with a new unit since connecting rod cannot be bent without affecting calibration.

Dashpot Adjustment

Manual Transmission Only

To adjust the dashpot, have the curb idle speed and mixture properly adjusted, and install a tachometer. Position throttle lever so that actuating tab on lever is contacting stem of dashpot but not depressing it. Tachometer should read 2500 rpm if the setting is correct. If not correct, screw dashpot in or out as required, then tighten lock nut on dashpot against the bracket.

CARTER TQ ADJUSTMENTS

Float Setting

Fig. 19 With bowl cover inverted, gasket installed and floats resting on seated needle, the dimension of each float from bowl cover gasket to bottom side of float should be as shown in the specifications chart. If adjustment is necessary, bend lever as shown but never allow lip of float lever to be pressed against needle when adjusting.

Secondary Throttle Linkage

Fig. 20 Block choke valve in wide open position and invert carburetor. Slowly open the primary throttle valves until it is possible to measure between lower edge of primary valve and its bore. When dimension is as shown in the specifications chart, the secondary valves should just start to open. If necessary to adjust, bend rod until correct dimension is obtained.

Secondary Air Valve Opening

Fig. 21

1. With air valve in closed position, the opening along air valve at its long side must be at its maximum and parallel with air horn gasket surface.
2. With air valve wide open, the opening of the air valve at the short side and air horn must be as shown in the specifications chart. The corner of air valve is notched for adjustment. Bend the corner with a pair of pliers to give proper opening.

Secondary Air Valve Spring Tension

Fig. 22—Loosen air valve lock plug and allow air valve to position itself wide open. With a long screwdriver that will enter center of tool C-4152 positioned on air valve adjustment plug, turn plug counterclockwise until air valve contacts stop lightly, then turn additional turn as specified in the specifications chart. Hold plug with screwdriver and tighten lock plug securely with tool C-4152.

NOTE: On 1980 light and medium duty cycle models and all 1981–83 models, use tool C-4152-B.

Accelerator Pump Stroke

1974

Fig. 23—Move choke valve wide open to release fast idle cam. Back off idle speed adjusting screw until throttle valves are seated in bores. Be sure throttle connector rod is in center hole of pump arm. Close throttle valve tightly and measure distance between top of bowl cover and end of plunger shaft. Dimension should be as shown in the specifications chart. Bend throttle connector rod at lower angle to adjust.

1975–80 & 1983

Fig. 24—***First Stage.*** Ensure throttle connector rod is in center hole on three hole pump arm or inner hole on two hole pump arm. With idle adjusting screw adjusted to the specified curb idle speed, measure distance from air horn surface to top of accelerator pump plunger. If equipped with an idle stop solenoid the ignition switch must be in the on position. Dimension should be as shown in the specifications chart. Bend throttle connector rod at lower angle to adjust.

Second Stage. With choke in the open position, open throttle until secondary lockout latch is just applied. Plunger downward travel stops at this point. Measure distance from air horn surface to top of accelerator pump plunger. Dimension should be as shown in the specification chart. Bend tang on throttle to adjust.

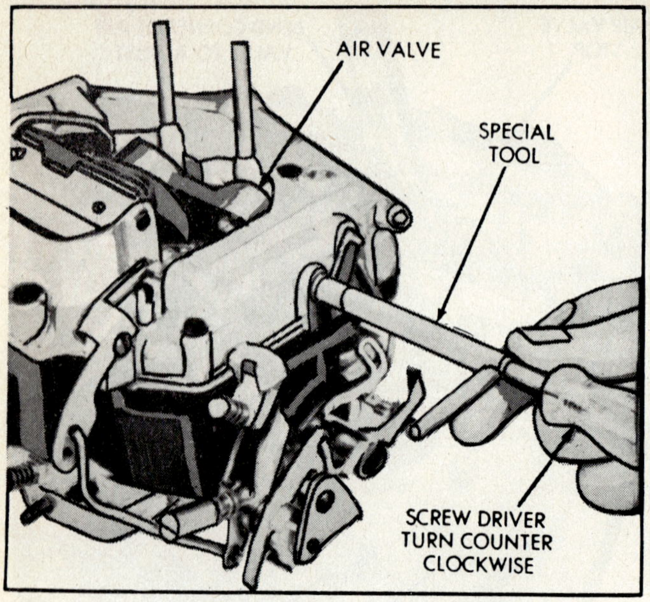

Fig. 22 TQ secondary air valve spring tension

Fig. 23 TQ accelerator pump stroke

1981–82

Accelerator pump stroke is determined by measurement of the accelerator pump plunger height above the air horn surface at curb idle.

1. Be sure throttle connector rod is in the specified hole of the pump arm.
2. Use a scale to measure height of the accelerator pump plunger stem (top) at curb idle. Pump heights are shown in specifications.
3. Adjust plunger height by bending the throttle connector rod in the proper area.

Choke Control Lever

Fig. 25—Remove the tamper resistant cover, if equipped, by drilling out the two rivets. Place carburetor on a flat surface and disconnect choke diaphragm rod at diaphragm. Close choke by pushing on choke lever with throttle partly open. Measure vertical distance between top of rod hole in control lever and base of carburetor (flat surface). Dimension should be as shown in the specifications chart. Adjust by bending link connecting the two choke shafts. On 1983 light duty cycle models, adjust by loosening lever mounting screw and rotating lever to correct position.

If choke control lever adjustment is changed, the vacuum kick, fast idle cam position and choke unloader adjustments must also be reset.

Choke Diaphragm Connector Rod

Fig. 26—Remove tamper resistant cover, if equipped, by drilling out the two rivets. Apply a vacuum of 15 or more inches of Mercury to diaphragm to fully depress diaphragm stem. An auxiliary source like a distributor test machine can be used for this purpose. With air valve closed, adjust connector rod to give .040" clearance between air valve and stop.

If choke diaphragm connector rod adjustment is changed, the vacuum kick adjustment must also be reset.

Vacuum Kick Adjustment

Fig. 27—With engine running, back off fast idle speed screw until choke can be closed to kick position at idle. Note number of screw turns so fast idle can be turned back to original adjustment. Insert specified gauge between long side (lower edge) of choke valve and the air horn wall. Apply sufficient pressure on choke control lever to provide a minimum choke valve opening. The spring connecting the control lever to the adjustment lever must be fully extended for proper adjustment. Bend tang as shown to change contact with end of diaphragm rod. Do not adjust diaphragm rod. A slight drag should be felt as gauge is being removed.

Fast Idle Cam & Linkage

Fig. 28—With fast idle adjusting screw on

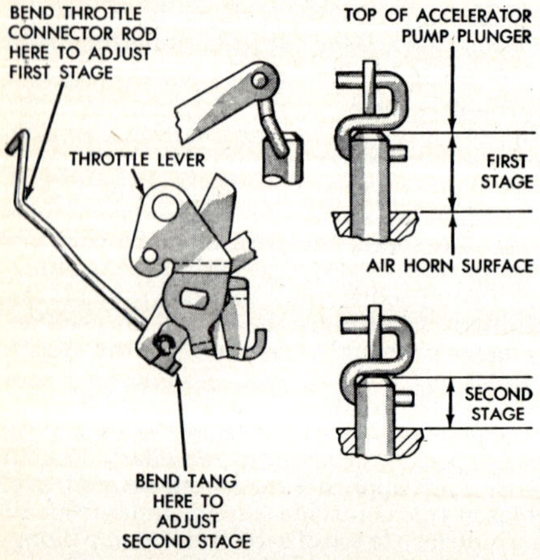

Fig. 24 TQ staged accelerator pump adjustment

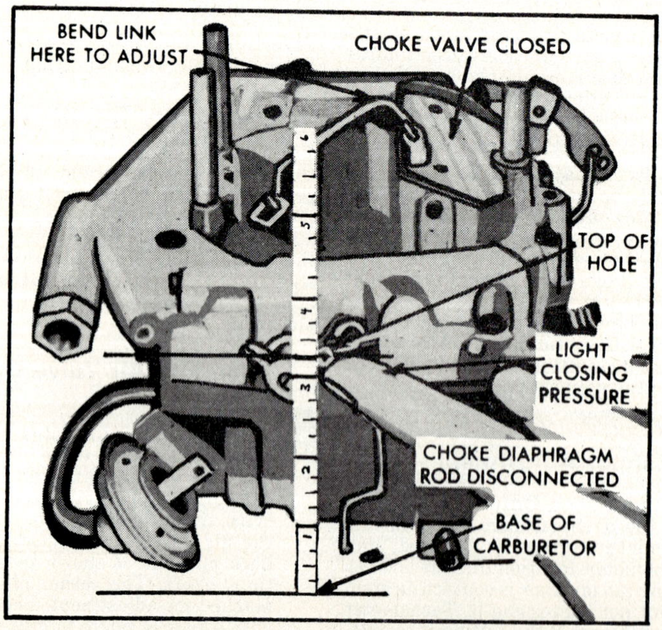

Fig. 25 TQ choke control lever adjustment

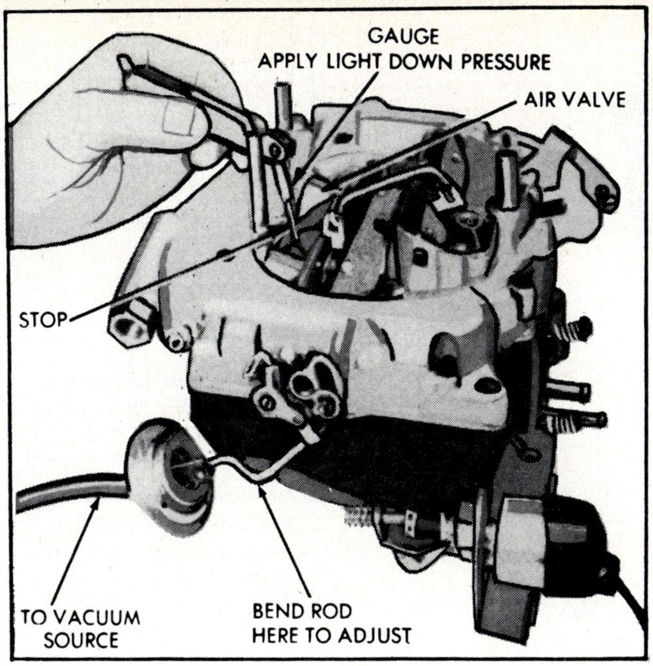

Fig. 26 TQ choke diaphragm connector rod

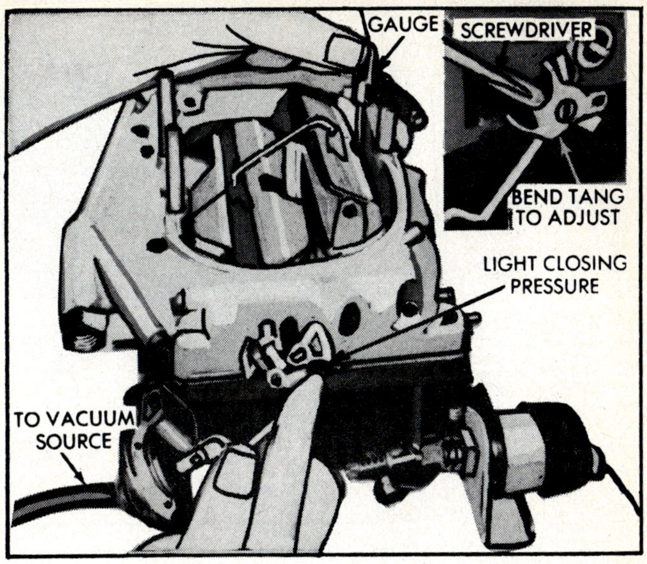

Fig. 27 TQ vacuum kick adjustment

second highest step of fast idle cam, move choke valve towards closed position with light pressure on fast idle cam lever. Clearance between choke valve and air horn wall should be .100 inch on 1974–83 units. Federal 318 engines should have a clearance of .130 inch while California 318 and Federal light duty 360 engines require .100 inch. California light duty 360 engines should have a clearance of .120 inch and all heavy duty 360 engines require .130 inch. If adjustment is necessary, bend fast idle connector rod at angle.

Choke Unloader Adjustment

Fig. 29—Hold throttle valves in wide open position and insert specified drill between long side (lower edge) of choke valve and inner wall of air horn. With finger lightly pressing against choke valve control lever, a slight drag should be felt as drill is withdrawn. Refer to the specification chart for proper drill size or dimension. Adjust by bending tang on fast idle control lever.

Secondary Throttle Lockout

Fig. 30—Move choke control lever to open choke position. Measure clearance between lockout lever and stop. Clearance should be .060–.090 inch. Adjust by bending tang on fast idle control lever.

Bowl Vent Valve Adjustment

Figs. 31 & 31A—Remove bowl vent valve checking hole plug in bowl cover. With throttle valves at curb idle, insert a narrow ruler down through hole. Allow ruler to rest lightly on top of valve. Dimension should be .812 inch. Adjust by bending bowl vent operating lever at notch. Install a new plug.

Units with a solenoid bowl vent diaphragm should be tested by removing hose and applying an auxiliary vacuum source of at least 15 inches of Mercury to diaphragm. Viewed through air horn vent tube, valve should move down when vacuum is applied; if it does not, replace leaking diaphragm assembly. Turn ignition switch on and remove auxiliary vacuum source. Valve should remain in down position; if it does not, solenoid or related wiring is defective.

Fast Idle Speed Cam

1974–81

NOTE: On 1974–75 units, plug vacuum fit-

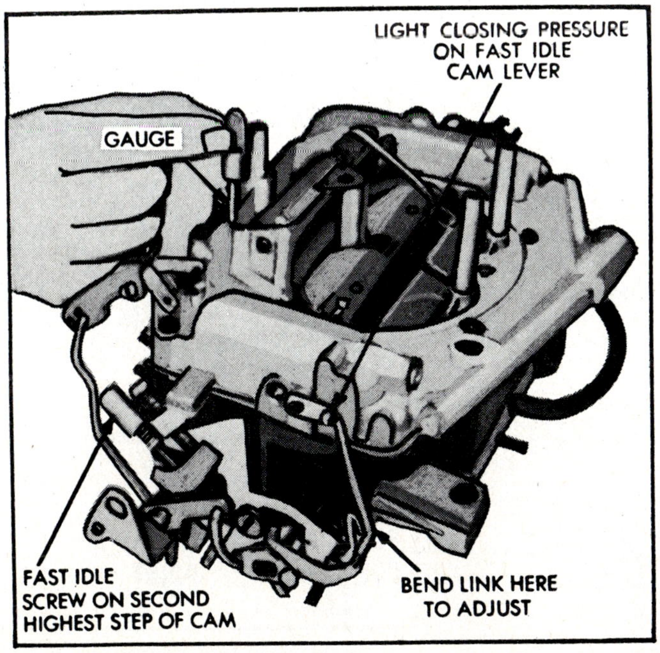

Fig. 28 TQ fast idle cam & linkage adjustment

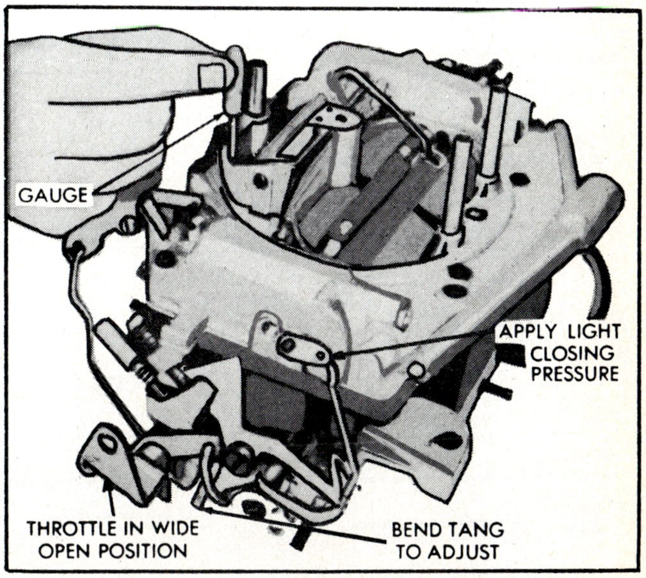

Fig. 29 TQ choke unloader adjustment

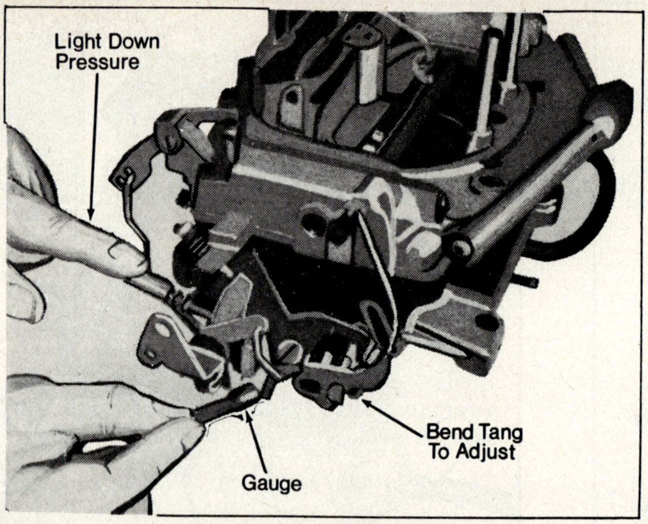

Fig. 30 TQ secondary throttle lockout

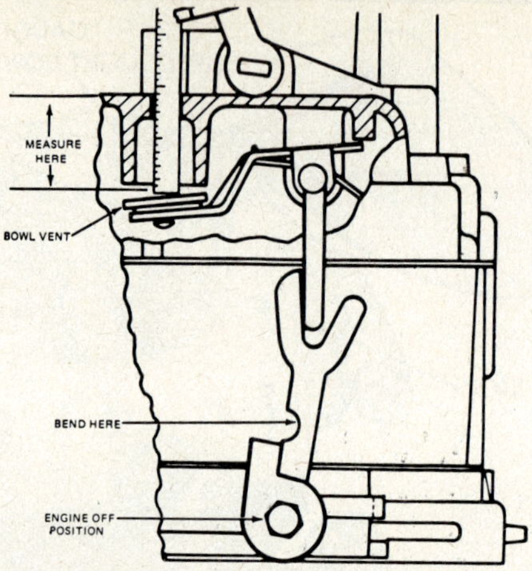

Fig. 31 TQ bowl vent valve adjustment

tings to heated air control and OSAC valves.

Fig. 32—With engine off, transmission in neutral and parking brake set, open throttle and close choke. Close throttle to place fast idle speed screw on highest speed step. Turn screw until it drops to second highest speed step. Start engine and determine stabilized speed. Turn fast idle speed screw to obtain speed listed in specifications chart.

1982–83
Fig. 32—Disconnect and plug vacuum hose at EGR valve and, on 1983 models, at distributor. Disconnect and plug hose from carburetor at heated air temperature sensor and at the OSAC valve, if equipped. Remove air cleaner, then disconnect and plug control hose from canister. Remove PCV valve, then start and run engine until normal operating temperature is reached. Open throttle and place fast idle adjustment screw on second highest step of fast idle cam. With choke fully open, turn fast idle screw to obtain speed listed in specifications chart. Return engine to idle and reposition fast idle adjusting screw on second highest step of cam to verify fast idle speed. Stop engine, then unplug and reconnect disconnected vacuum hoses and install PCV valve and air cleaner.

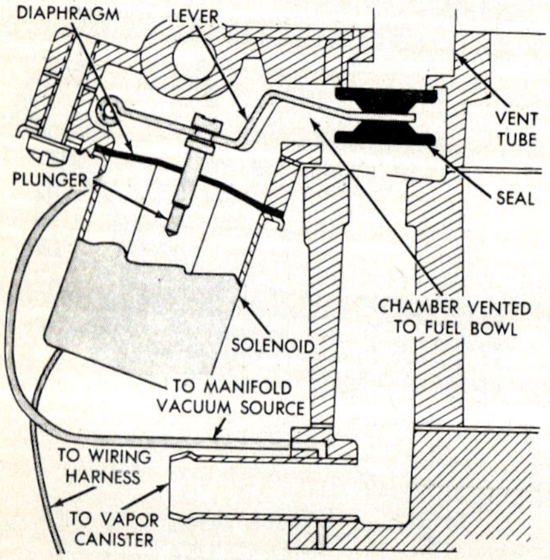

Fig. 31A TQ solenoid bowl vent valve adjustment

HOLLEY EXC. 1945, 2210, 2245, 2280 & 6145 ADJUSTMENTS

Float Level

Model 1920, Fig. 33
With carburetor inverted, slide float gauge into position and check setting on "touch" leg of gauge. Float should just touch gauge. Reverse gauge and check the "no touch" leg. Float should just clear gauge. To adjust, bend float tab that touches head of fuel inlet needle.

NOTE: Do not allow float tab to contact float needle head while making the adjustment as the rubber tip of the needle can be compressed, giving a false reading.

Models 2300, 4150 & 4160
With fuel bowl inverted, turn adjusting nut in or out until float lever is parallel to floor of float bowl. Tighten lock bolt.

Fuel Level

Model 1920, Fig. 34
With engine running and vehicle on a level floor, measure fuel level through economizer diaphragm opening. Using a 6" scale with a depth gauge as shown, measure distance from machined surface of opening to the exact fuel level. If necessary to adjust, proceed as outlined for float level.

Models 2300, 4150 & 4160, Fig. 35
With engine running, remove the sight plug from the fuel bowl and turn adjusting nut up or down until fuel just dribbles out of the sight hole. After correct level has been obtained, tighten the lock screw while holding the adjusting nut.

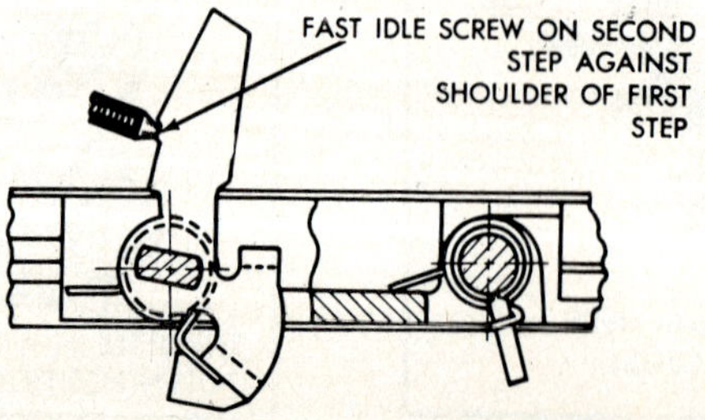

Fig. 32 TQ fast idle speed cam position

Fast Idle Cam Position

Model 1920, Fig. 36
With fast idle speed adjusting screw contacting second highest step on fast idle cam, move choke valve toward closed position with light pressure on choke shaft lever. Insert specified drill size between choke valve and air horn wall. If an adjustment is necessary, bend fast idle link at lower angle until correct valve opening is obtained.

NOTE: When the correct fast idle position cam adjustment has been made, the choke unloader (wide open kick) adjustment has also been obtained. No further adjustment is required.

Model 2300, Fig. 37
After setting curb idle speed and with manual control of choke in wide open position, turn fast idle speed screw to obtain clearance of .035 inch between cam and screw head.

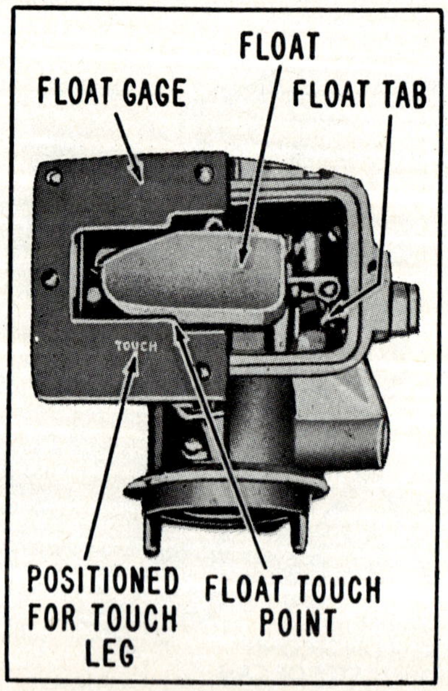

Fig. 33 Float level adjustment. Holley 1920

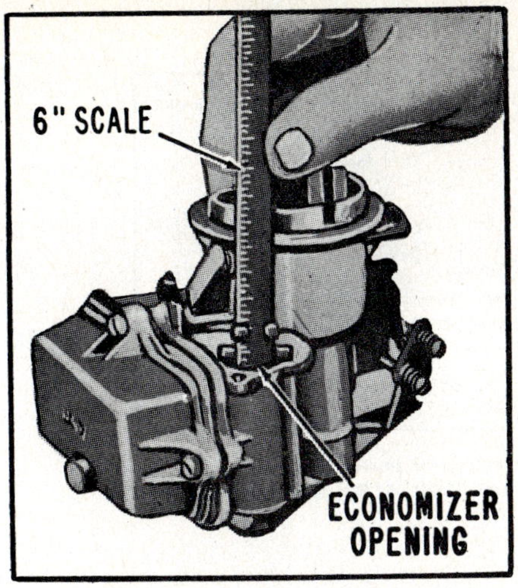

Fig. 34 Checking fuel level. Holley 1920

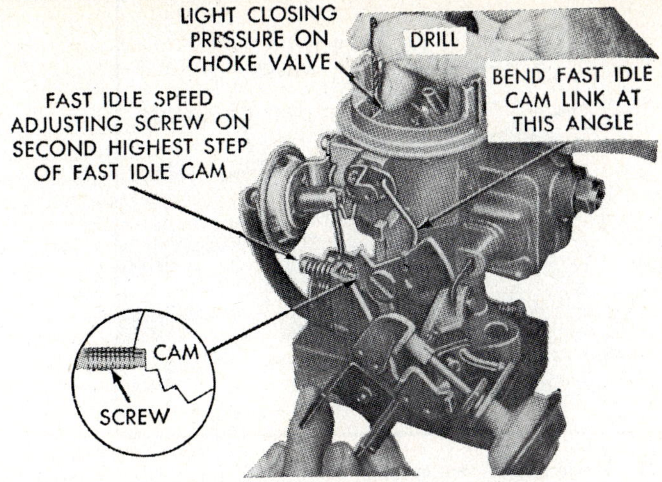

Fig. 36 Fast idle cam position. Holley 1920

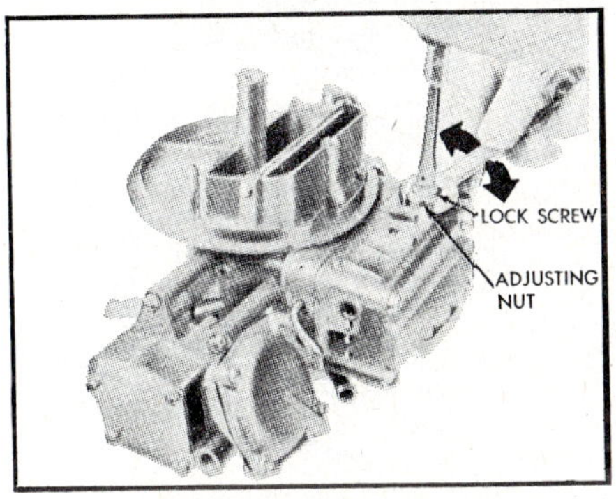

Fig. 35 Adjusting fuel level. Holley 2300, 4150 & 4160

Models 4150 & 4160, Fig. 38

Insert specified gauge between top of fast idle cam and choke housing post and apply closing pressure to choke valve. If slight drag is not felt when gauge is withdrawn, adjust by bending choke linkage as indicated.

Bowl Vent

Model 1920, Fig. 39

With throttle valve at curb idle speed, it should be possible to insert a drill of the size specified between bowl vent and the seat. If an adjustment is necessary, bend vent rod at horizontal portion until correct clearance has been obtained. Be sure vent rod does not bind in guide after adjusting.

Dashpot Setting

Model 1920, C.A.P. UNITS, Fig. 40

With curb idle speed and mixture properly set and a tachometer installed, position throttle lever so that actuating tab on lever is contacting stem of dashpot but not depressing it. The tachometer should read 2500 rpm if the setting is correct.

If necessary to adjust the setting, screw the dashpot in or out as required. When the desired setting is obtained, tighten the lock nut on the dashpot against the bracket.

Accelerator Pump Lever Setting

Models 2300, 4150 & 4160, Fig. 41

With throttle valves wide open and pump lever held down, it should be possible to insert a .015" gauge between adjusting nut and lever. Adjust pump override screw to obtain correct clearance.

HOLLEY 1945 & 6145 ADJUSTMENTS

Dry Float Setting

Fig. 42—Hold float fulcrum retaining pin in position and invert carburetor bowl. Place a straight edge across surface of bowl, contacting float toes. Remove straight edge and measure distance float dropped from surface of fuel bowl. Refer to the specifications chart. Adjust by bending float tang to obtain proper dimension.

Fast Idle Cam Position Adjustment

Fig. 43—With fast idle speed adjusting screw contacting second highest step on fast idle cam, move choke valve toward closed position with light pressure on choke shaft lever. Insert specified gauge between top of choke valve and wall of air horn. Refer to the specifications chart. An adjustment will be necessary if a slight drag is not obtained as drill shank is being removed. Adjust by bending fast idle link at lower angle, until correct valve opening has been obtained.

Choke Vacuum Kick Adjustment

NOTE: Test can be made on or off vehicle.

Fig. 44—If adjustment is to be made with engine running, back off fast idle speed screw until choke can be closed to the kick position with engine at curb idle. (Note number of screw turns required so that fast idle can be returned to original adjustment). If an auxiliary vacuum source is to be used, open throttle valve (engine not running) and move choke to closed position. Release throttle first, then release choke.

When using an auxiliary vacuum source, disconnect vacuum hose from carburetor and

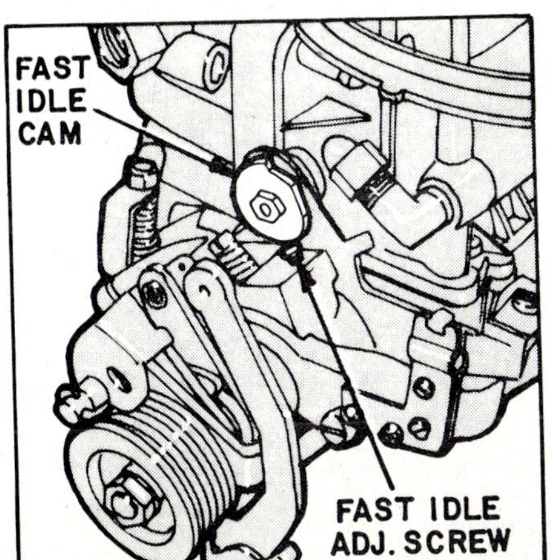

Fig. 37 Fast idle adjustment. Holley 2300

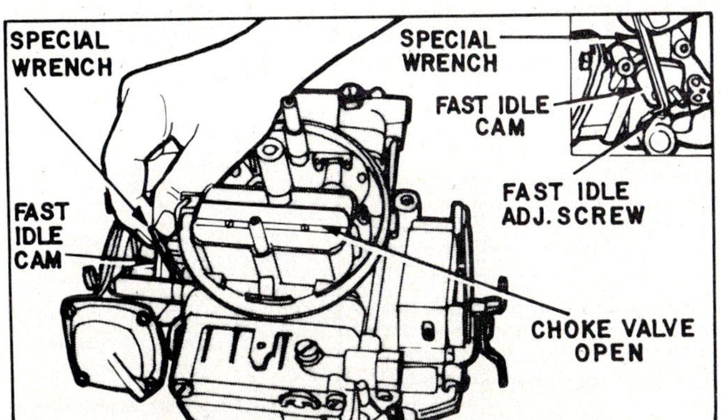

Fig. 38 Fast idle adjustment. Holley 4150

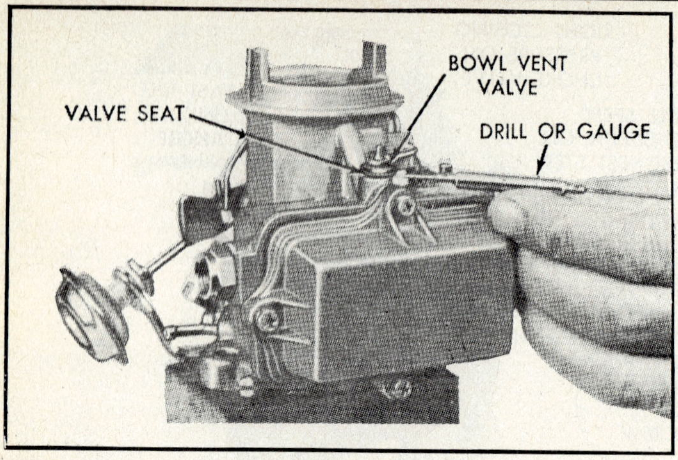

Fig. 39 Checking bowl vent opening. Holley 1920

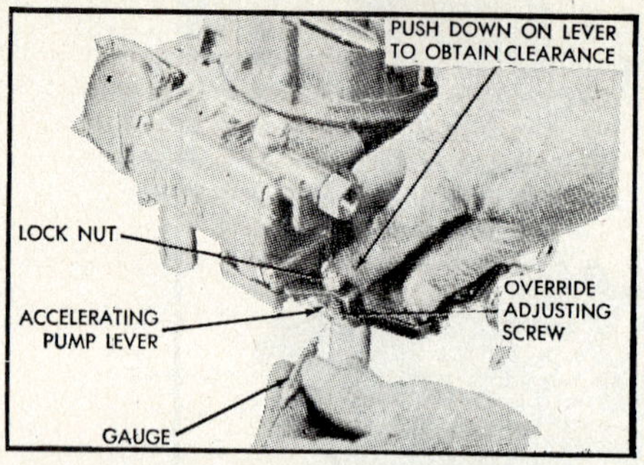

Fig. 41 Accelerator pump lever clearance. Holley 2300, 4150, & 4160

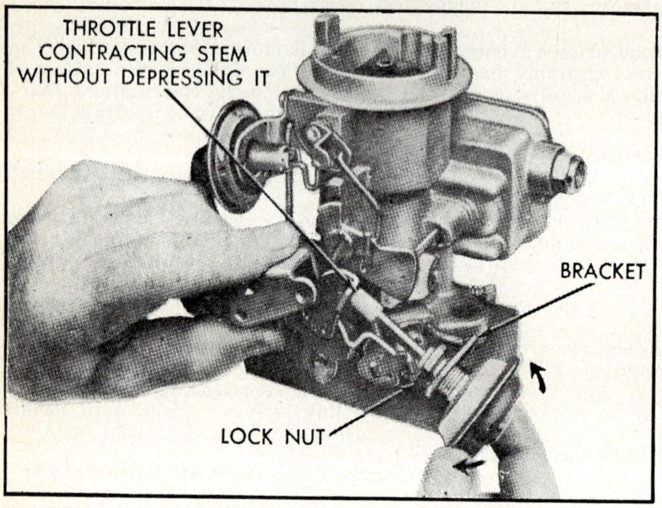

Fig. 40 Adjusting dashpot on C.A.P. units. Holley 1920

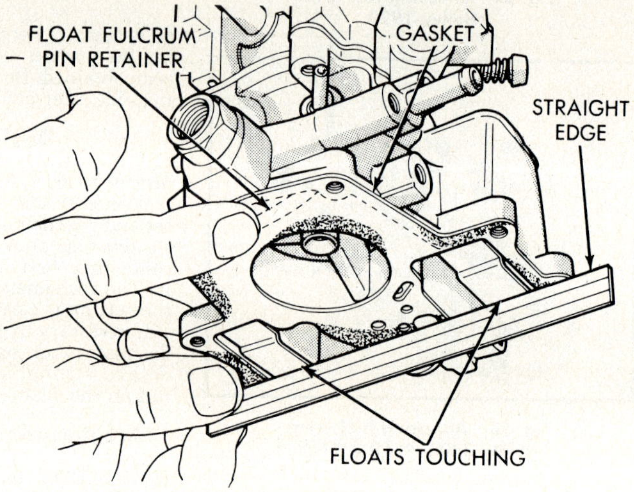

Fig. 42 Measuring float level. 1945 & 6145 carburetors

connect it to hose from vacuum supply with a small length of tube to act as a fitting. Removal of hose from diaphragm may require forces which damage the system. Apply a vacuum of 15 or more inches of mercury.

Insert gauge between top of choke valve and wall of air horn. Refer to the specifications chart. Apply sufficient closing pressure on lever to which choke rod attaches to provide a minimum choke valve opening without distortion of diaphragm link.

NOTE: The cylindrical stem of diaphragm extends as the internal spring is compressed. This spring must be fully compressed for proper measurement of vacuum kick adjustment.

Adjustment is necessary if slight drag is not obtained when removing the gauge. Shorten or lengthen diaphragm link to obtain correct choke valve opening. Length changes should be made by carefully opening or closing the U-bend provided in the link. Improper bending causes contact between the U-section and the diaphragm assembly.

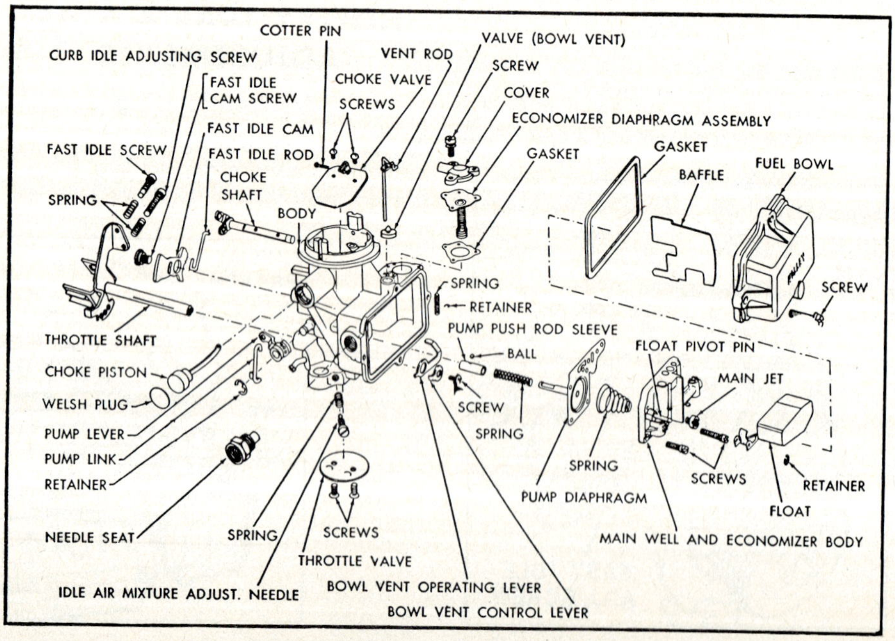

Holley Model 1920 single barrel carburetor

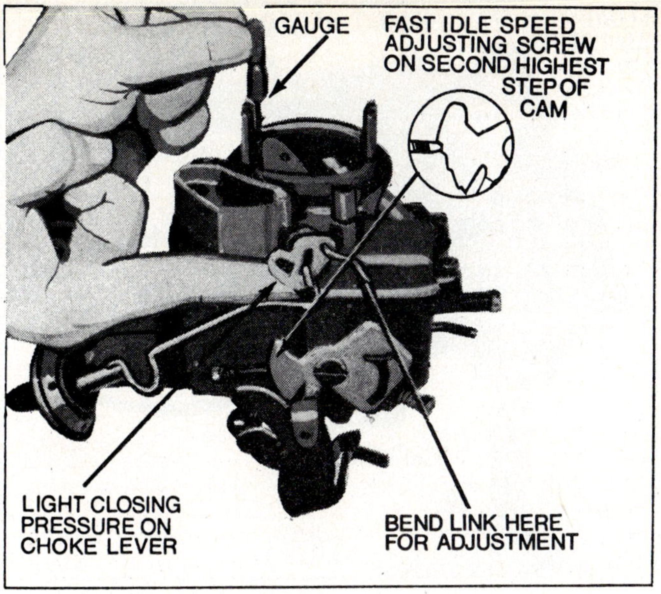

Fig. 43 Fast idle cam position adjustment. 1945 & 6145 carburetors

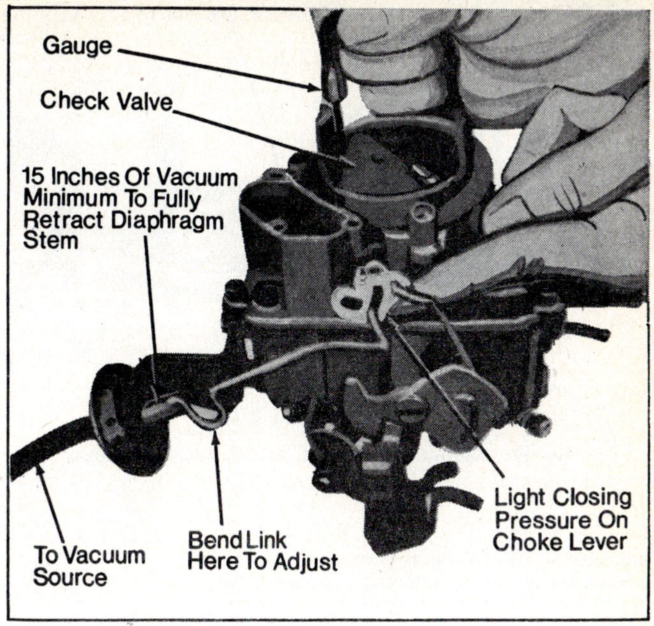

Fig. 44 Choke vacuum kick adjustment. 1945 & 6145 carburetors. Typical

NOTE: Do not apply twisting or bending force to diaphragm.

After completion of adjustment, reinstall vacuum hose on correct carburetor fitting. Return fast idle screw to its original location if disturbed. Make following check. With no vacuum applied to diaphragm, the choke valve should move freely between open and closed positions. If movement is not free, examine linkage for misalignment or interferences caused by bending operation.

NOTE: On 1982–83 models, adjustment is made by means of a 5/64 inch allen wrench inserted into vacuum diaphragm.

Choke Unloader (Wide Open Kick) Adjustment

Fig. 45—With throttle valves in wide open position, insert drill gauge between upper edge of choke valve and inner wall of air horn. Refer to the specifications chart. With a finger lightly pressing against shaft lever, a slight drag should be felt as drill is being withdrawn. Adjust by bending unloader tang on throttle lever until correct opening has been obtained.

Accelerator Pump Setting

Fig. 46—With throttle at curb idle position, measure distance from vacuum passage casting to outer edge of hole in pump operating rod. Refer to the specifications chart. Adjust by bending link between throttle lever and pump operating rod.

HOLLEY 2210 & 2245 ADJUSTMENTS

Float Adjustment

Invert air horn so that weight of float only is forcing needle against seat. Measure the clearance between top of float and float stop, Fig. 47. Be sure drill gauge is perfectly level when measuring. Refer to the specifications chart. Adjust by bending float lip toward or away from needle, using a narrow blade screwdriver, Fig. 48, until correct clearance of setting has been obtained.

Float Drop Adjustment

Check float drop, by holding air horn in an

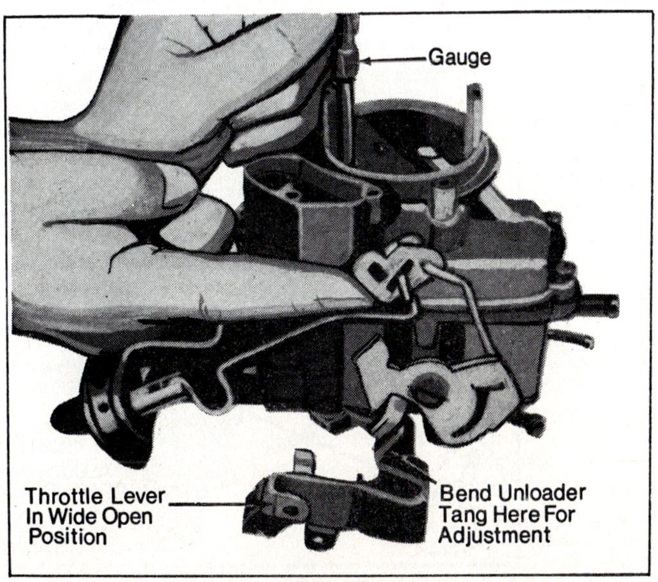

Fig. 45 Choke unloader adjustment. 1945 & 6145 carburetors

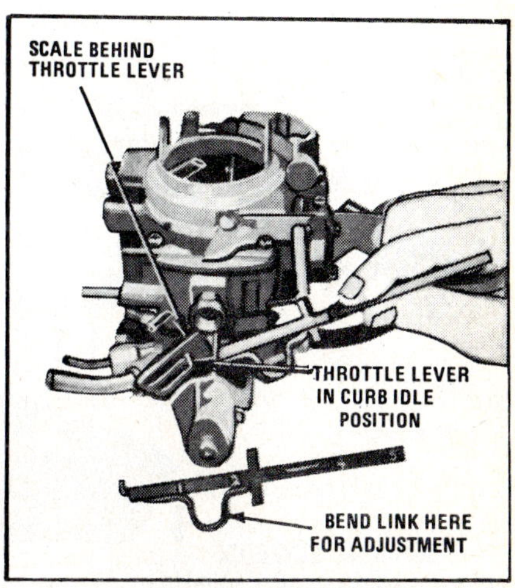

Fig. 46 Accelerator pump adjustment. 1945 & 6145 carburetors. Typical

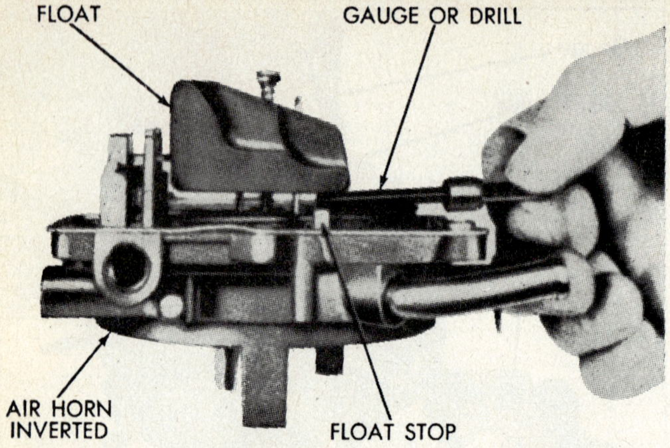

Fig. 47 Measuring float level. 2210 & 2245 carburetors

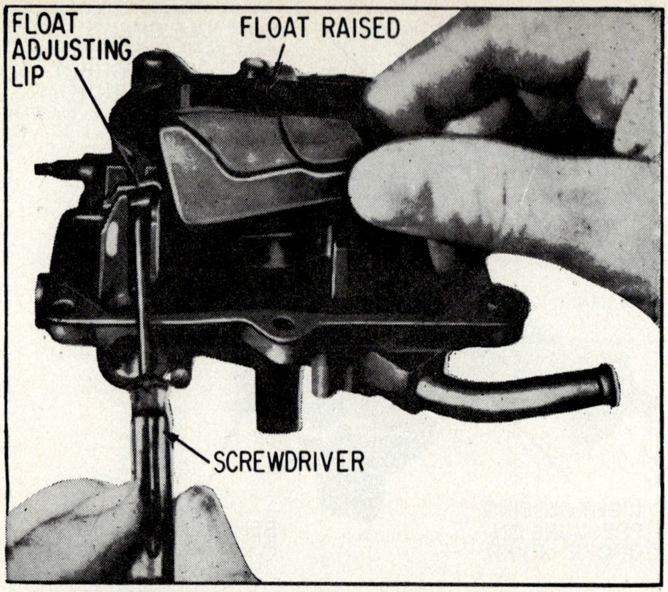

Fig. 48 Float adjustment. 2210 & 2245 carburetors

upright position. The bottom edge of float should be parallel to underside surface of air horn, Fig. 49. Adjust by bending tang on float arm until parallel surfaces have been obtained.

Fast Idle Cam Position Adjustment

Fig. 50—With fast idle speed adjusting screw contacting second highest step on fast idle cam, move choke valve toward closed position with light pressure on choke shaft lever. Insert specified gauge between top of choke valve and wall of air horn. Refer to the specifications chart. An adjustment will be necessary if a slight drag is not obtained as drill shank is being removed. Adjust by bending fast idle link at angle, until correct valve opening has been obtained.

Choke Vacuum Kick Adjustment

NOTE: Test can be made on or off vehicle.

Fig. 51—If adjustment is to be made with engine running, back off fast idle speed screw until choke can be closed to the kick position with engine at curb idle. (Note number of screw turns required so that fast idle can be returned to original adjustment). If an auxiliary vacuum source is to be used, open throttle valve (engine not running) and move choke to closed position. Release throttle first, then release choke.

When using an auxiliary vacuum source, disconnect vacuum hose from carburetor and connect it to hose from vacuum supply with a small length of tube to act as a fitting. Removal of hose from diaphragm may require forces which damage the system. Apply a vacuum of 10 or more inches of mercury on 1974–76, or 15 or more inches of mercury on 1977–83 models.

Insert gauge between top of choke valve and wall of air horn. Refer to the specifications chart. Apply sufficient closing pressure on lever to which choke rod attaches to provide a minimum choke valve opening without distortion of diaphragm link.

NOTE: The cylindrical stem of diaphragm extends as the internal spring is compressed. This spring must be fully compressed for proper measurement of vacuum kick adjustment.

Adjustment is necessary if slight drag is not obtained when removing the gauge. Shorten or lengthen diaphragm link to obtain correct choke valve opening. Length changes should be made by carefully opening or closing the U-bend provided in the link. Improper bending causes contact between the U-section and the diaphragm assembly.

NOTE: Do not apply twisting or bending force to diaphragm.

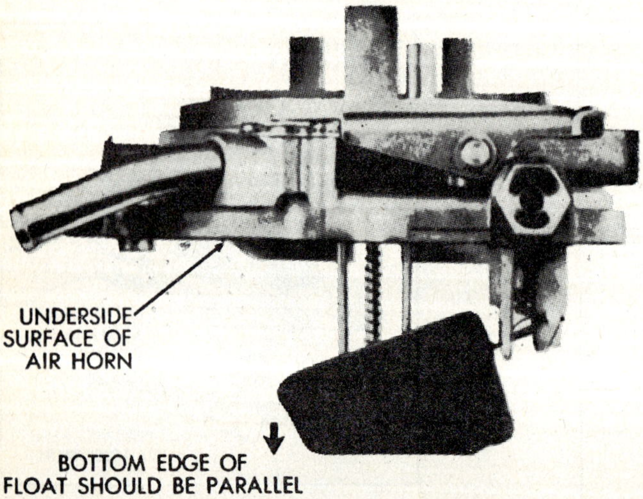

Fig. 49 Checking float drop. 2210 & 2245 carburetors

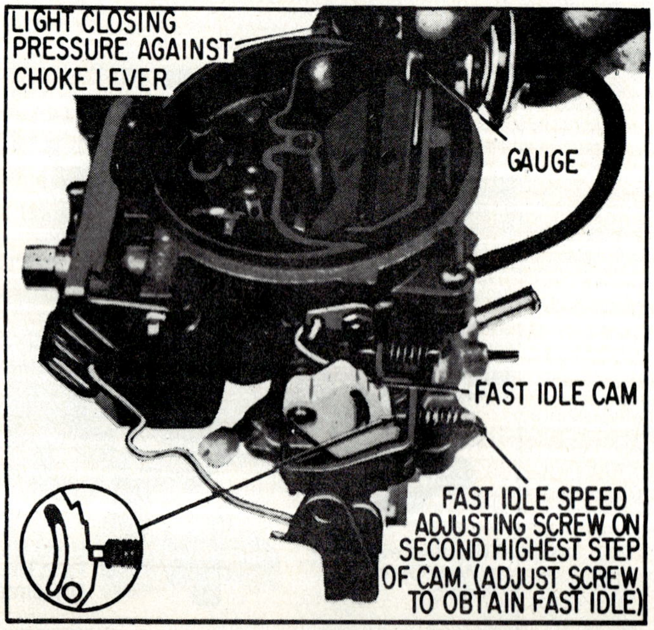

Fig. 50 Fast idle cam position adjustment. 2210 & 2245 carburetors

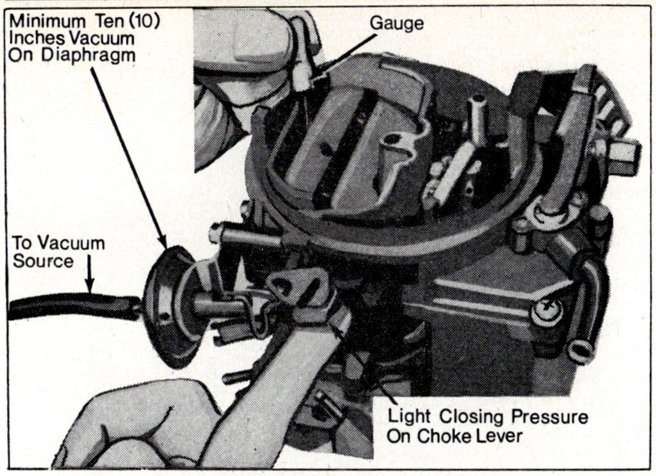

Fig. 51 Vacuum kick adjustment. 2210 & 2245 carburetors

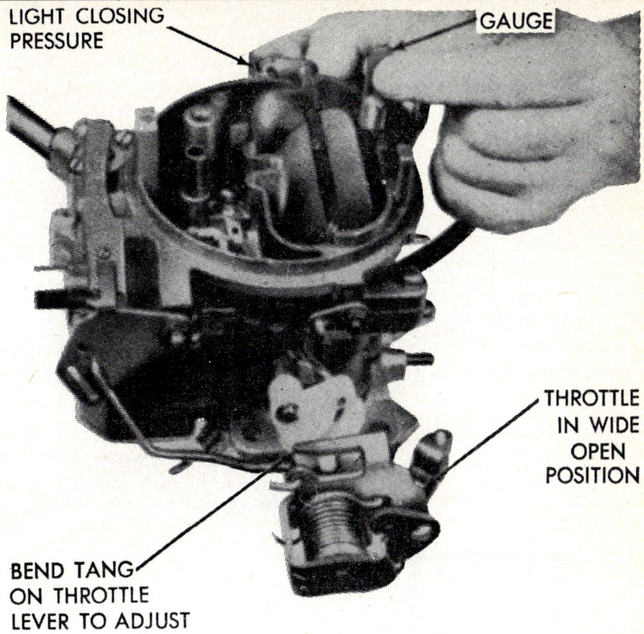

Fig. 52 Choke unloader adjustment. 2210 & 2245 carburetors

After completion of adjustment, reinstall vacuum hose on correct carburetor fitting. Return fast idle screw to its original location if disturbed. Make following check. With vacuum applied to diaphragm, the choke valve should move freely between open and closed positions. If movement is not free, examine linkage for misalignment or interferences caused by bending operation.

Choke Unloader (Wide Open Kick) Adjustment

Fig. 52—With throttle valves in wide open position, insert drill gauge between upper edge of choke valve and inner wall of air horn. Refer to the specifications chart. With a finger lightly pressing against shaft lever, a slight drag should be felt as drill is being withdrawn. Adjust by bending unloader tang on throttle lever until correct opening has been obtained.

Accelerator Pump Adjustment

Fig. 53—Back off curb idle speed adjusting screw. Open choke valve so that fast idle cam allows throttle valves to be completely seated in bores. Be sure that pump connector rod is installed in correct slot of accelerator pump rocker arm. First slot for manual transmissions next to attaching screw.

Close throttle valves tightly. Measure the distance between top of air horn and end of plunger shaft. Refer to the specifications chart. Adjust pump travel by bending pump operating rod, at loop of rod, until correct setting has been obtained.

Bowl Vent Valve Clearance Adjustment

Fig. 54—With the throttle valves at curb idle, it should be possible to insert an inch gauge between the bowl vent valve plunger stem and operating rod. Refer to the specifications chart. Adjust by bending the tang on pump lever to change arc of contact with throttle lever, until correct clearance has been obtained.

MODEL 2280

This carburetor, Fig. 55, uses four basic metering systems. The basic idle system pro-

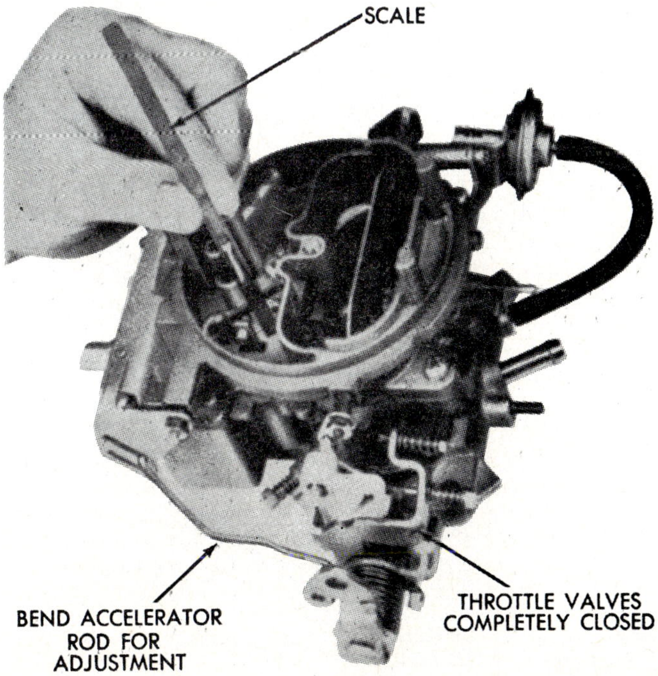

Fig. 53 Accelerator pump adjustment. 2210 & 2245 carburetors

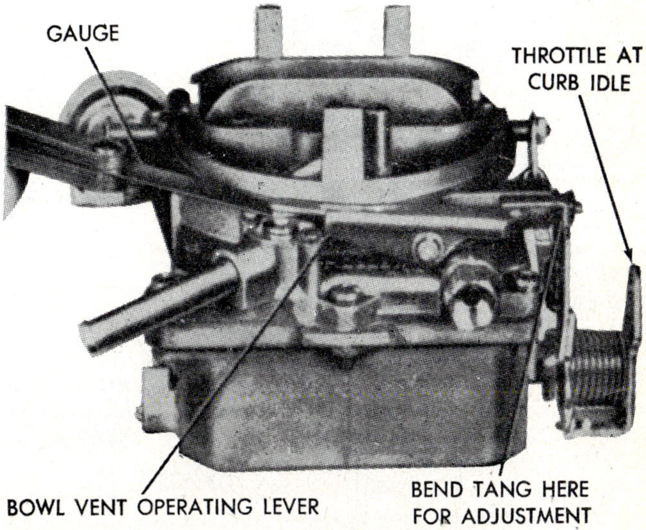

Fig. 54 Bowl vent adjustment. 2210 & 2245 carburetors

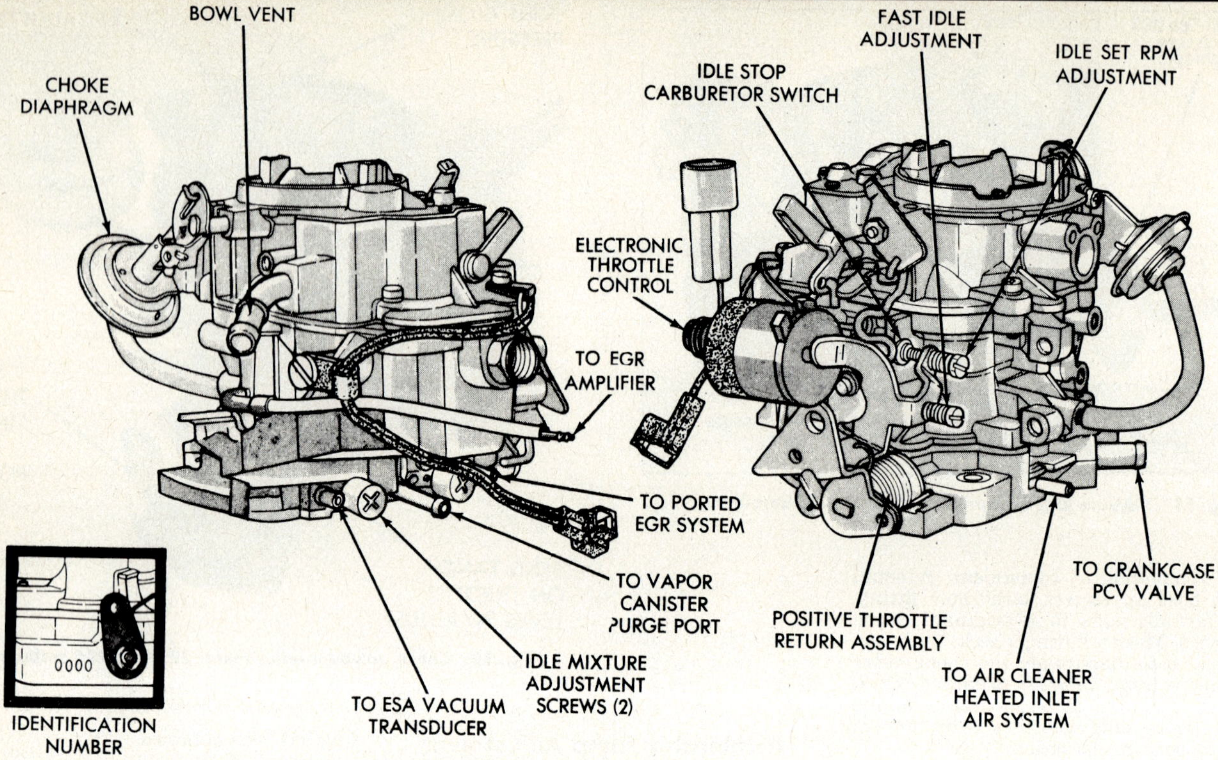

Fig. 55 Holley model 2280 two barrel carburetor

vides the proper air/fuel mixture for idle and low speed operations. The accelerator pump system provides additional fuel during acceleration. The main metering system provides the proper air/fuel mixture during normal cruising conditions. The power enrichment system, combining a mechanical and vacuum operated power valve, provides a richer mixture when higher engine power is required.

In addition to the four basic systems, there is a fuel inlet system which supplies fuel to the four basic systems and a choke system which temporarily enriches the mixture to aid in starting and running a cold engine.

Float Adjustment

Fig. 56—Invert main body so the weight of the floats only is forcing the needle against the seat. Hold finger against hinge pin retainer to fully seat in the float hinge cradle. Using a suitable scale, measure distance between the float bowl surface and the toe of each float. To adjust, bend the float tang. If necessary, bend either float arm to equalize individual float positions.

Accelerator Pump Adjustment

1980

Fig. 57—Remove bowl vent cover plate and vent valve spring. Do not dislodge vent valve lever retainer. Ensure that accelerator pump rod is in the inner hole of the pump operating lever and the throttle is at curb idle. Place a

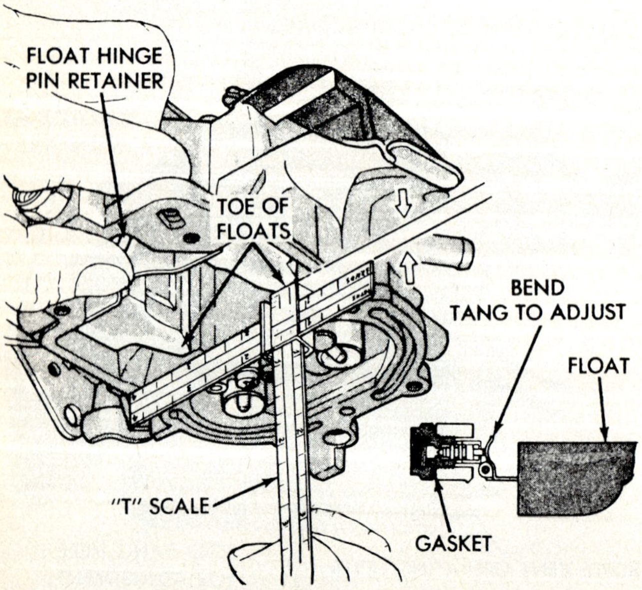

Fig. 56 Float adjustment. 2280 carburetor

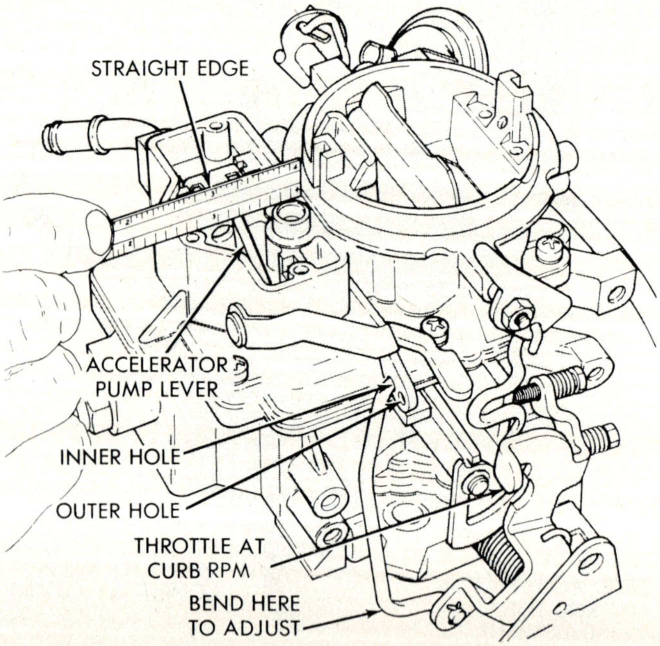

Fig. 57 Accelerator pump adjustment. 1980 2280 carburetor

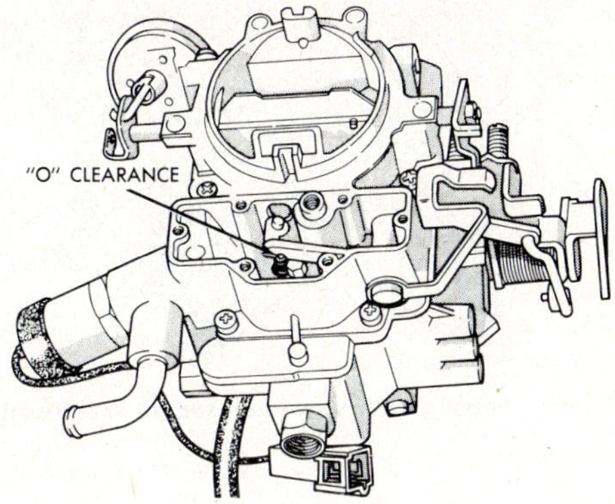

Fig. 57A Accelerator pump adjustment. 1981–83 2280 carburetor

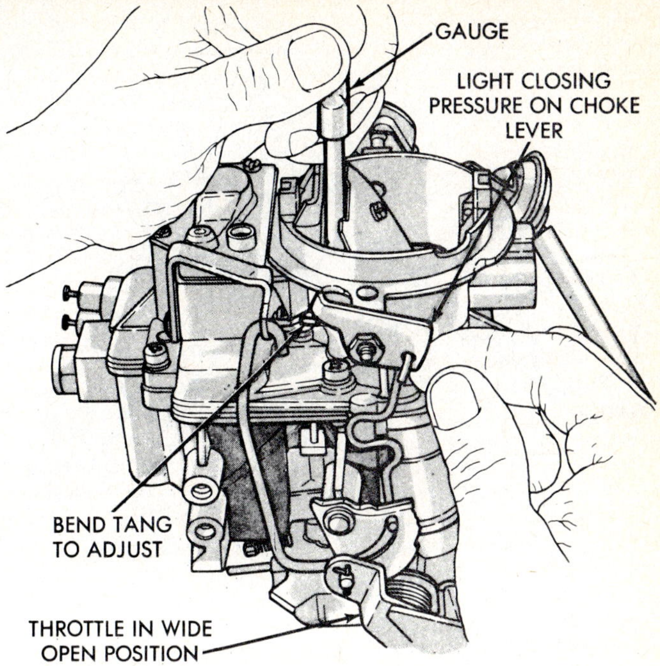

Fig. 58 Choke unloader adjustment. 2280 carburetor

straight edge on bowl vent cover surface of air horn over accelerator pump lever. Adjust until lever surface is flush with air horn surface by bending the accelerator pump connector rod. Install vent valve lever spring and bowl vent cover plate.

NOTE: If this adjustment is changed, the bowl vent and mechanical power valve adjustments must be reset.

1981–83
Fig. 57A—Remove bowl vent cover plate and gasket. With all pump links and levers installed, adjust accelerator pump cap nut for zero clearance between pump lever and cap nut. Check that wide open throttle can be reached without binding, then install gasket and bowl vent cover plate.

Choke Unloader Adjustment

Fig. 58—Hold throttle valves in wide open position. Lightly press finger against control lever to move choke valve toward closed position. Insert specified gauge between top of choke valve and air horn wall. To adjust, bend tang on accelerator pump lever.

Choke Vacuum Kick Adjustment

Fig. 59—Open throttle, close choke, then close throttle to trap fast idle cam at closed choke position. Apply approximately 15 inches of vacuum to diaphragm with an external vacuum source. Apply closing pressure on choke lever to completely compress spring in diaphragm stem without distorting linkage. The cylindrical stem of the diaphragm extends to a stop as the spring compresses. Insert specified gauge between top of choke valve and air horn wall. On 1980–81 models, adjust by bending diaphragm link at U-bend. On 1982–83 models, adjust by inserting a 5/64 inch allen wrench into diaphragm and turning as necessary. Check for free movement between the open and adjusted positions.

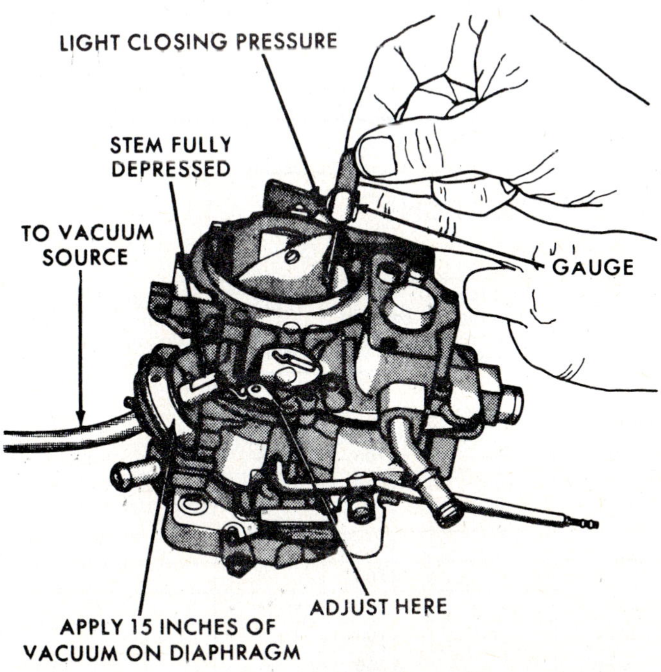

Fig. 59 Choke vacuum kick adjustment. 2280 carburetor

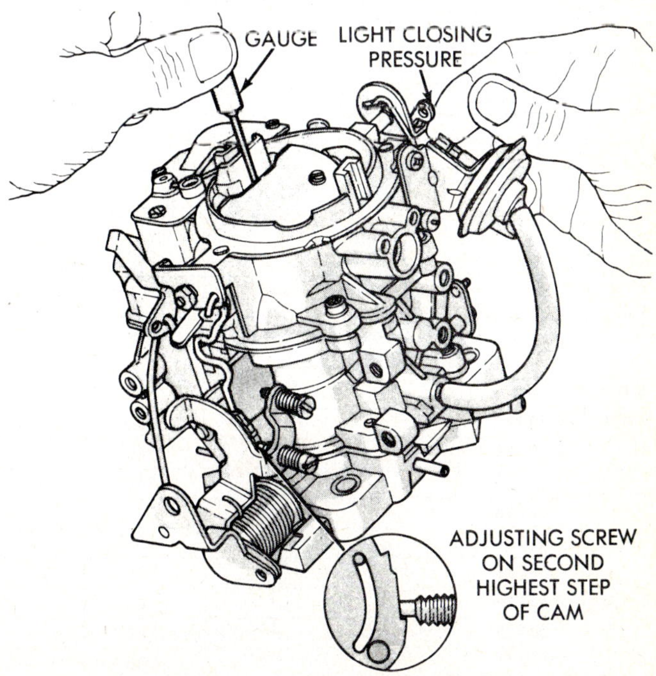

Fig. 60 Fast idle cam position adjustment. 2280 carburetor

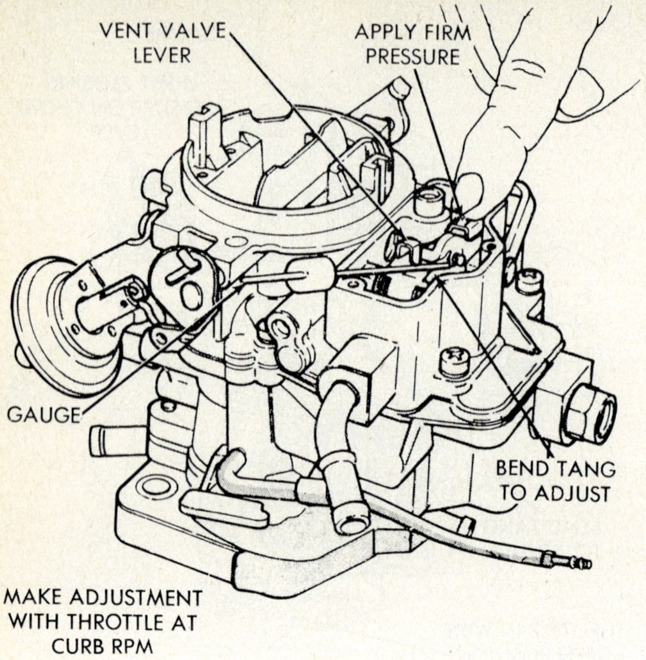

Fig. 61 Bowl vent valve adjustment. 2280 carburetor

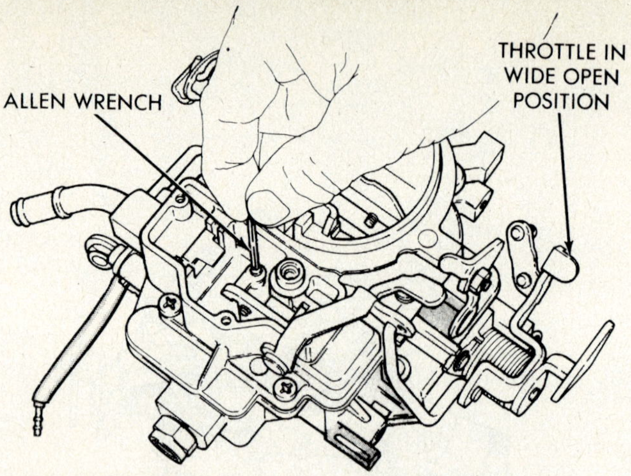

Fig. 62 Mechanical power valve adjustment. 2280 carburetor

Fast Idle Cam Position Adjustment

Fig. 60—With fast idle speed adjusting screw contacting second highest step of fast idle cam, move choke valve toward closed position with light pressure on choke shaft lever. Insert specified gauge between top of choke valve and air horn wall. To adjust, bend the U-bend in fast idle connector link to obtain proper setting.

Bowl Vent Valve Adjustment

Fig. 61—Remove bowl vent cover plate and vent valve lever spring. Do not dislodge vent valve lever retainer. With throttle at curb idle, press downward on vent valve lever where spring seats. Measure distance between contact surfaces of vent valve tang and vent valve lever. Adjust by bending end of vent valve lever. Install vent valve lever spring and bowl vent cover plate.

Mechanical Power Valve Adjustment

Fig. 62—Remove bowl vent cover plate. On 1980 models, remove vent valve lever spring and retainer, then the vent valve lever and pivot pin. Hold throttle in wide open position. Insert a 5/64 inch Allen wrench in mechanical power valve adjustment screw. Push screw downward and release to determine if clearance exists. Turn screw clockwise until zero clearance is obtained. Adjust by turning screw one turn counter-clockwise. On 1980 models, install vent valve lever, pivot pin and retainer, then the vent valve lever spring. On all models, install the bowl vent cover plate.

CARTER BBS ADJUSTMENT SPECIFICATIONS

See Tune Up Chart for hot idle speed.

Year	Model	Float Level	Pump Setting	Bowl Vent	Choke Unloader	Fast Idle Cam Position	Fast Idle Speed RPM	Choke Vacuum Kick	Automatic Choke Setting
1974	7044S	1/4	—	—	3/16	.076	2000	.110	Fixed
	7045S	1/4	—	—	3/16	.076	1800	.075	Fixed

①—Pump is automatically set when bowl vent is properly adjusted.

CARTER BBD ADJUSTMENT SPECIFICATIONS

See Tune Up Chart for hot idle speed.

Year	Carb. Model	Idle Mix. Screws Turns Open	Float Setting	Pump Setting	Bowl Vent Adjust	Fast Idle Linkage Adjust	Fast Idle Throttle Valve Clearance	Fast Idle R.P.M.	Choke Unloader Adjust	Choke Diaphragm Linkage Adjust	Choke Setting
BBD MODELS											
1974	6221S	—	1/4	.280	—	.095	—	1700	1/4	—	—
	6222S	—	1/4	.280	—	.095	—	1700	1/4	—	—
	6224S	—	1/4	.340	—	—	—	1900	—	—	—
	6228S	—	1/4	.340	—	—	—	1600	—	—	—
	6229S	—	1/4	.390	—	—	—	1600	—	—	—
	6316S	—	1/4	.280	—	.095	—	1700	1/4	—	—
	6317S	—	1/4	.280	—	.095	—	1700	1/4	—	—
	6343S	—	1/4	.280	—	.095	—	1700	1/4	—	—
	6344S	—	1/4	.280	—	.095	—	1700	1/4	—	—

Continued

CARTER BBD ADJUSTMENT SPECIFICATIONS—Continued

See Tune Up Chart in truck chapters for hot idle speed.

Year	Carb. Model	Idle Mix. Screws Turns Open	Float Setting	Pump Setting	Bowl Vent Adjust	Fast Idle Linkage Adjust	Fast Idle Throttle Valve Clearance	Fast Idle R.P.M.	Choke Unloader Adjust	Choke Diaphragm Linkage Adjust	Choke Setting
	6363S	—	1/4	.310	—	.095	—	1600	1/4	—	—
	6364S	—	1/4	.310	—	.095	—	1700	1/4	—	—
	6365S	—	1/4	.340	—	—	—	1900	—	—	—
	6367S	—	1/4	.200	—	.095	—	1800	1/4	—	2 Rich
	6368S	—	1/4	.310	—	—	—	1600	—	—	—
	6610S	—	1/4	1/2	—	.095	—	1700	.28	.150	—
	6611S	—	1/4	1/2	—	.095	—	1500	.28	.110	—
	6613S	—	1/4	1/2	—	.095	—	1600	.28	.110	—
	8008S	—	1/4	1/2	—	.095	—	1700	.28	.150	—
1974–76	6537S	—	1/4	1/2	—	.095	—	1500	.31	.110	—
	6540S	—	1/4	1/2	—	.095	—	1700	.25	—	—
	6585S	—	1/4	1/2	—	—	—	1900	—	—	—
	6586S	—	1/4	1/2	—	—	—	1900	—	—	—
1974–77	6536S	—	1/4	1/2	—	.095	—	1700	.28	.150	—
1975–76	8013S	—	1/4	1/2	—	.070	—	1500	.31	.130	—
	8014S	—	1/4	1/2	—	.070	—	1500	.31	.110	—
	8016S	—	1/4	1/2	—	.070	—	1500	.31	.130	—
	8020S	—	1/4	1/2	—	.070	—	1500	.28	.130	—
	8024S	—	1/4	1/2	—	.070	—	1500	.28	.130	—
	8025S	—	1/4	1/2	—	.070	—	1500	.31	.070	—
	8026S	—	1/4	1/2	—	.070	—	1500	.31	.110	—
1977	8081S	—	1/4	1/2	—	.070	—	1500	.310	.150	—
	8082S	—	1/4	1/2	—	.070	—	1500	.280	.130	—
	8085S	—	1/4	1/2	—	.070	—	1500	.280	.130	—
	8108S	—	1/4	1/2	—	.070	—	1500	.310	.150	—
	8110S	—	1/4	1/2	—	.070	—	1500	.280	.095	—
	8112S	—	1/4	1/2	—	.070	—	1500	.310	.110	—
	8113S	—	1/4	1/2	—	.070	—	1500	.310	.130	—
	8115S	—	1/4	1/2	—	.070	—	1500	.280	.130	—
	8121S	—	1/4	1/2	—	.070	—	1500	.310	.070	—
	8146S	—	1/4	1/2	—	.070	—	1500	.310	.070	—
	8147S	—	1/4	1/2	—	.095	—	1700	.280	.150	—
1978	8146S	—	1/4	1/2	—	.070	—	1600	.280	—	—
	8147S	—	1/4	1/2	—	.070	—	1500	.310	.070	—
	8149S	—	1/4	1/2	—	.070	—	1500	.280	.130	—
	8151S	—	1/4	1/2	—	.070	—	1500	.280	.130	—
	8152S	—	1/4	1/2	—	.070	—	1500	.280	.130	—
	8154S	—	1/4	1/2	—	.070	—	1500	.310	.130	—
	8156S	—	1/4	1/2	—	.070	—	1500	.310	.110	—
	8176S	—	1/4	1/2	—	.070	—	①	.310	.070	—
	8180S	—	1/4	1/2	—	.070	—	1500	.280	.150	—
1979	8210S	—	1/4	1/2	—	.070	—	1400	.280	.110	—
	8211S	—	1/4	1/2	—	.070	—	1500	.280	.110	—
	8214S	—	1/4	1/2	—	.070	—	1400	.280	.110	—
	8215S	—	1/4	1/2	—	.070	—	1600	.280	.110	—
	8230S	—	1/4	1/2	—	.070	—	1600	.250	.190	—
	8232S	—	1/4	1/2	—	.070	—	1500	.280	.110	—
	8249S	—	1/4	1/2	—	.070	—	1400	.280	.110	—
1982	8348S	—	1/4	1/2	—	.070	—	1600	.310	—	—
	8352S	—	1/4	1/2	—	.070	—	1600	.310	—	—
1983	8358S	—	1/4	.470	—	.070	—	1400	.280	—	—
	8359S	—	1/4	.470	—	.070	—	1400	.280	—	—
	8371S	—	1/4	.470	—	.070	—	1600	.280	—	—
	8374S	—	1/4	.470	—	.070	—	1400	.280	—	—

①—Automatic transmission, 1700 RPM; manual transmission, 1600 RPM.

CARTER TQ ADJUSTMENT SPECIFICATIONS

See Tune Up Chart in truck chapters for hot idle speeds.

Year	Carb. Model	Float Setting	Secondary Throttle Linkage	Secondary Air Valve Opening	Secondary Air Valve Spring	Pump Travel @ Curb Idle	Pump Travel @ Secondary Pick-up	Choke Control Lever (Off Vehicle)	Choke Vacuum Kick	Choke Unloader	Fast Idle R.P.M.
1974	9017S	②	①	1/2	1 1/4	1/2	—	3 3/8	.160	.310	1700
	9022S	②	①	1/2	1 1/4	1/2	5/16	3 3/8	.210	.310	1800
	9025S	②	①	1/2	1 1/4	1/2	5/16	3 3/8	.160	.310	1700
1974–76	9036S	②	①	1/2	1 1/4	1/2	5/16	3 3/8	.160	.310	1700
1974–77	6545S	②	①	1/2	1 1/4	1/2	5/16	3 3/8	.160	.310	1700
1975–76	9034S	29/32	①	1/2	1 1/4	1/2	5/16	3 3/8	.160	.310	1700
	9035S	29/32	①	1/2	1 1/4	1/2	5/16	3 3/8	.160	.310	1700
1977	9096S	29/32	①	1/2	1 1/4	1/2	5/16	3 3/8	.100	.310	1700
1978	9116S	29/32	①	1/2	1 1/4	1/2	5/16	3 3/8	.100	.500	1600
	9117S	29/32	①	1/2	1 1/4	1/2	5/16	3 3/8	.100	.310	1700
	9118S	29/32	①	1/2	1 1/4	1/2	5/16	3 3/8	.160	.310	1700
	9123S	29/32	①	1/2	2	1/2	5/16	3 3/8	.150	.500	1600
	9124S	29/32	①	1/2	2	1/2	5/16	3 3/8	.100	.500	1600
	9125S	29/32	①	1/2	2	1/2	5/16	3 3/8	.150	.500	1700
	9126S	29/32	①	1/2	2	1/2	5/16	3 3/8	.100	.500	1600
	9149S	29/32	①	1/2	1 1/2	1/2	5/16	3 3/8	.100	.500	1600
	9173S	29/32	①	1/2	1 1/4	1/2	5/16	3 3/8	.150	.310	1500
1978–79	9150S	29/32	①	1/2	2	1/2	5/16	3 3/8	.100	.500	1400
	9151S	29/32	①	1/2	2	1/2	5/16	3 3/8	.160	.500	1400
1979	9207S	29/32	①	7/16	2	31/64	23/64	3.30	.150	.310	1600
	9208S	29/32	①	7/16	2	31/64	23/64	3.30	.150	.310	1600
	9209S	29/32	①	7/16	2	31/64	23/64	3.30	.150	.310	1600
	9210S	29/32	①	1/2	2	31/64	23/64	3.30	.150	.310	1600
	9211S	29/32	①	1/2	1 1/2	31/64	23/64	3.30	.100	.500	1400
	9212S	29/32	①	1/2	1 1/2	31/64	23/64	3.30	.100	.500	1400
	9223S	29/32	①	1/2	3	11/32	9/64	3.30	.100	.500	1600
	9224S	29/32	①	1/2	3	5/16	3/16	3.30	.100	.500	1600
	9225S	29/32	①	1/2	3	5/16	3/16	3.30	.100	.500	1600
	9226S	27/32	①	1/2	2 5/8	31/64	23/64	3.30	.100	.310	1750
	9228S	29/32	①	1/2	3	11/32	3/16	3.30	.100	.500	1600
	9229S	29/32	①	1/2	3	11/32	3/16	3.30	.100	.500	1600
	9247S	29/32	①	7/16	1 1/2	31/64	23/64	3.30	.100	.500	1400
	9248S	29/32	①	7/16	1 1/2	31/64	23/64	3.30	.100	.500	1400
1980	9251S	29/32	①	7/16	2 1/4	.340	.190	3 3/8	.150	.310	1600
	9252S	29/32	①	7/16	2 1/4	.340	.190	3 3/8	.150	.310	1600
	9254S	29/32	①	3/8	2 1/2	.340	.190	3 3/8	.150	.310	1500
	9261S	29/32	①	7/16	2	.340	.190	3 3/8	.130	.310	1600
	9265S	29/32	①	7/16	2 1/2	.340	.140	3 3/8	.150	.310	1600
	9279S	29/32	①	3/8	2	.340	.190	3 3/8	.130	.310	1800
	9281S	29/32	①	3/8	2	.340	.190	3 3/8	.180	.310	1800
	9288S	29/32	①	1/2	2 3/4	.340	.190	3 5/16	.120	.310	1600
	9292S	29/32	①	7/16	2	.340	.190	3 3/8	.130	.310	1600
	9296S	29/32	①	3/8	2 1/2	.340	.140	3 3/8	.110	.310	1500
	9300S	29/32	①	3/8	2	.340	.190	3 3/8	.180	.310	1800
1981	9311S	29/32	①	27/64	2 1/2	.340	—	3 3/8	.150	.310	1500
	9314S	29/32	①	7/16	2 1/2	.340	—	3 3/8	.150	.310	1500
	9329S	29/32	①	27/64	2 1/2	.340	—	3 3/8	.130	.310	1600

Continued

CARTER TQ ADJUSTMENT SPECIFICATIONS—Continued

See Tune Up Chart in truck chapters for hot idle speeds.

Year	Carb. Model	Float Setting	Secondary Throttle Linkage	Secondary Air Valve Opening	Secondary Air Valve Spring	Pump Travel @ Curb Idle	Pump Travel @ Secondary Pick-up	Choke Control Lever (Off Vehicle)	Choke Vacuum Kick	Choke Unloader	Fast Idle R.P.M.
	9330S	29/32	①	27/64	2 1/2	.340	—	3 3/8	.130	.310	1500
	9331S	29/32	①	7/16	2 1/2	.340	—	3 3/8	.110	.310	1700
	9332S	29/32	①	7/16	2 1/2	.340	—	3 3/8	.110	.310	1700
	9357S	29/32	①	27/64	2	.340	—	3 3/8	.130	.310	—
	9358S	29/32	①	7/16	2	.340	—	3 3/8	.180	.310	1700
	9359S	29/32	①	7/16	2	.340	—	3 3/8	.130	.310	1500
1982	9342S	29/32	①	27/64	2 1/2	.340	.190	—	.130	.310	1600
	9375S	29/32	①	27/64	2	.340	.190	—	.130	.310	1800
	9376S	29/32	①	27/64	2	.340	.190	—	.130	.310	1700
	9379S	29/32	①	27/64	2	.340	—	—	.130	.310	1500
1983	9342S	29/32	①	27/64	2 1/2	.340	.190	—	.130	.310	1600
	9375S	29/32	①	3/8	2 1/2	.340	.190	—	.130	.310	1800
	9376S	29/32	①	3/8	2	.340	.190	—	.130	.310	1700
	9379S	29/32	①	7/16	2	.340	—	—	.130	.310	1500

①—Adjust link so primary & secondary stops both contact at same time.

②—Brass float, 1 inch; cellular plastic float, 29/32 inch.

HOLLEY ADJUSTMENT SPECIFICATIONS

See Tune Up Chart in truck chapters for hot idle speeds.

Year	Carb. Model	Carb. Part No.	Idle Mix. Screw Turns Open	Float Level Dry	Fuel Level Wet	Vacuum Kick	Fast Idle Cam Position	Bowl Vent Valve	Fast Idle Speed	Pump Setting
1974	1920	R6593A	—	See Text	See Text	.100	.065	.015	2000	—
	1920	R6594A	—	See Text	See Text	.100	.065	.015	1700	—
	1920	R6595A	—	See Text	See Text	.100	.065	.015	2000	—
	1920	R6596A	—	See Text	See Text	.100	.065	.015	1700	—
	2210	R6452A	—	.180	See Text	.150	.110	.015	1900	—
	2210	R6454A	—	.180	See Text	.150	.110	.015	1800	—
	2210	R6472A	—	.180	See Text	.150	.110	.015	1800	—
	2210	R6484A	—	.180	See Text	.160	.110	.015	1900	—
	2210	R6485A	—	.180	See Text	.160	.110	.015	1900	—
	2210	R6486A	—	.180	See Text	.110	.110	.015	1800	—
	2210	R6487A	—	.180	See Text	.160	.110	.015	1800	—
	2210	R6488A	—	.180	See Text	.110	.110	.015	1800	—
	2210	R6575A	—	.180	See Text	.150	.110	.015	1900	—
	2300	R6277A	1	See Text	See Text	—	.035	—	—	.015②
	2300	R6278A	1	See Text	See Text	—	.035	—	—	.0625②
	4160C	R6495A	1	See Text	See Text	.070	#65 Drill	.015	1800	①
	1945	R6725A	—	See Text	See Text	.140	.080	—	1600	2 7/32
	1945	R6921A	—	See Text	See Text	.140	.080	—	1600	2 7/32
	1945	R6875A	—	See Text	See Text	.090	.080	—	1800	2 11/32
	2210	R6764A	—	.180	See Text	.150	.110	—	1700	.270②
	2210	R-6765A	—	.180	See Text	.150	.110	—	1800	.250②
	2210	R6775A	—	.180	See Text	.150	.110	—	1800	—
	2210	R-6886A	—	.180	See Text	.150	.110	—	1800	.250②
	2210	R-7052A	—	.180	See Text	.160	.110	—	1800	.250②
	2245	R6762-1A	—	.180	See Text	.150	.110	—	1700	.255②⑤
	2245	R6860A	—	.180	See Text	.150	.110	—	1800	.255②⑤

Continued

HOLLEY ADJUSTMENT SPECIFICATIONS—Continued

Year	Carb. Model	Carb. Part No.	Idle Mix. Screw Turns Open	Float Level Dry	Fuel Level Wet	Vacuum Kick	Fast Idle Cam Position	Bowl Vent Valve	Fast Idle Speed	Pump Setting
	2245	R-6990A	—	.180	See Text	.150	.110	.015	1600	.255②⑤
	2300	R-6769A	1	See Text	See Text	—	.035	—	—	.015⑥
	2300	R-6770A	1	See Text	See Text	—	.035	—	—	.015⑥
	4150	R-6771A	1	See Text	See Text	—	.066	—	—	.015②
	4150	R-7138A	—	See Text	See Text	—	.035	—	—	.015②
1975–76	1945	R-7074A	—	3/64	—	.110	.080	—	1600	2 7/32
	1945	R-7076A	—	3/64	—	.090	.080	—	1700	2 21/64
	1945	R-7078A	—	3/64	—	.110	.080	—	1600	2 7/32
	1945	R-7079A	—	3/64	—	.090	.080	—	1700	2 21/64
	1945	R-7080A	—	3/64	—	.110	.080	—	1600	2 7/32③
	1945	R-7081A	—	3/64	—	.090	.080	—	1700	2 21/64④
	1945	R-7082A	—	3/64	—	.110	.080	—	1600	2 7/32③
	1945	R-7083A	—	3/64	—	.090	.080	—	1700	2 21/64④
	2210	R-6764A	—	.180	See Text	.150	.110	—	1700	.320②
	2210	R-6765A	—	.180	See Text	.150	.110	—	1800	.310②
	2245	R-7088A	—	.180	See Text	.150	.110	—	1600	.310⑤
	2245	R-7089A	—	.180	See Text	.150	.110	—	1600	.310⑤
	2245	R-7090A	—	.180	See Text	.150	.110	—	1600	.310⑤
	2245	R-7091A	—	.180	See Text	.150	.110	—	1600	.310⑤
	2245	R-7092A	—	.180	See Text	.150	.110	—	1600	.310⑤
	2245	R-7103A	—	.180	See Text	.150	.110	—	1600	.310⑤
	2245	R-7187A	—	.180	See Text	.150	.110	—	1600	.310⑤
	2245	R-7188A	—	.180	See Text	.150	.110	—	1600	.310⑤
	2300	R-6769-1A	—	See Text	See Text	—	.035	—	—	.015⑥
	2300	R-7137A	—	See Text	See Text	—	.035	—	—	.015⑥
	4150	R-6771A	—	See Text	See Text	—	.035	—	—	.015②
	4150	R-7138A	—	See Text	See Text	—	.035	—	—	.015②
1977	1945	R-7815A	—	3/64	—	.110	.080	—	1600	2 7/32
	1945	R-7816A	—	3/64	—	.110	.080	—	1700	2 21/64
	1945	R-7847A	—	3/64	—	.110	.080	—	1600	2 7/32
	1945	R-7848A	—	3/64	—	.110	.080	—	1700	2 21/64
	1945	R-7849A	—	3/64	—	.110	.080	—	1600	2 21/64
	2210	R-6764A	—	.180	—	.150	.110	—	1700	.310⑤
	2210	R-6886A	—	.180	—	.150	.110	—	1600	.310⑤
	2210	R-6886-1A	—	.180	—	.150	.110	—	1400	.310②⑤
	2210	R-7870A	—	.180	—	.150	.110	—	1800	.310⑤
	2245	R-7088A	—	.180	—	.150	.110	—	1600	.310⑤
	2245	R-7091A	—	.180	—	.150	.110	—	1600	.310⑤
	2245	R-7092A	—	.180	—	.150	.110	—	1600	.310⑤
	2245	R-7103A	—	.180	—	.150	.110	—	1600	.310⑤
	2245	R-7403A	—	.180	—	.150	.110	.025	1600	.310②⑤
	2245	R-7697A	—	.180	—	.150	.110	—	1600	.310⑤
	2245	R-7698A	—	.180	—	.150	.110	—	1600	.310⑤
	2245	R-7871A	—	.180	—	.130	.110	—	1600	.310⑤
	2245	R-8036A	—	.180	—	.150	.110	—	1600	.310②⑤
1977–78	2300	R-6769-1A	—	See Text	See Text	—	.035	—	—	.015②
	2300	R-7137A	—	See Text	See Text	—	.035	—	—	.015②
	4150	R-6771A	—	See Text	See Text	—	.035	—	—	.015②
	4150	R-7138A	—	See Text	See Text	—	.035	—	—	.015②
1978	2245	R-6886A	—	.180	—	.150	.110	—	1600	—
	2245	R-6886-1A	—	.180	—	.150	.110	—	1600	.310②⑤
	2245	R-7088A	—	.180	—	.150	.110	—	1600	—
	2245	R-7871A	—	.180	—	.130	.110	—	1600	.310②⑤
	2245	R-8026A	—	.180	—	.150	.110	.025	1600	.310②⑤
	2245	R-8028A	—	.180	—	.150	.110	—	1600	—
	2245	R-8135A	—	.180	—	.150	.110	.025	1800	.260②⑤

Continued

HOLLEY ADJUSTMENT SPECIFICATIONS—Continued

Year	Carb. Model	Carb. Part No.	Idle Mix. Screw Turns Open	Float Level Dry	Fuel Level Wet	Vacuum Kick	Fast Idle Cam Position	Bowl Vent Valve	Fast Idle Speed	Pump Setting
	2245	R-8453A	—	.180	—	.150	.110	.025	1700	.260②⑤
	2245	R-8484A	—	.180	—	.150	.110	—	1600	—
1979	1945	R-8593A	—	See Text	See Text	.100	.080	.0625	1600	2 7/32⑥
	1945	R-8594A	—	See Text	See Text	.100	.080	.0625	1600	2 21/64⑦
	2245	R-8597A	—	.200	—	.110	.110	.025	1600	.290②⑤
	2245	R-8598A	—	.200	—	.110	.110	.025	1600	.290②⑤
	2245	R-8925A	—	.200	—	.110	.110	.025	1600	.290②⑤
1980	1945	R-8720A	—	See Text	—	.130	.090	.060	1600	—
	1945	R-8721A	—	See Text	—	.130	.090	.060	1700	1.61
	1945	R-8978A	—	See Text	—	.130	.080	.060	1600	1.70
	1945	R-8979A	—	See Text	—	.130	.080	.060	1600	1.70
	2280	R-8999A	—	1/32	—	.130	.070	.030	1600	.030⑧
	2280	R-9000A	—	1/32	—	.130	.070	.030	1600	.030⑧
	2280	R-9001A	—	5/16	—	.150	.070	.030	1600	.030⑧
	2280	R-9209A	—	1/32	—	.150	.070	.030	1600	.030⑧
1981	1945C	R-9131A	—	⑨	—	.130	.090	⑩	1600	1.61
	1945C	R-9132A	—	⑨	—	.130	.090	⑩	1600	1.61
	1945C	R-9152A	—	⑨	—	.130	.080	⑩	1800	1.70
	1945C	R-9153A	—	⑨	—	.130	.080	⑩	1800	1.70
	2280	R-9135A	—	9/32	—	.130	.070	⑩	1500	⑪
	2280	R-9136A	—	9/32	—	.110	.070	⑩	1500	⑪
	2280	R-9151A	—	9/32	—	.110	.070	⑩	1500	⑪
1982	1945	R-9132A	—	⑨	—	.130	.090	⑩	1600	1.61
	1945	R-9153A	—	⑨	—	.130	.080	⑩	1800	1.70
	1945	R-9762A	—	⑨	—	.130	.090	⑩	1600	1.61
	1945	R-9765A	—	⑨	—	.130	.080	⑩	1800	1.70
	2280	R-9491A	—	9/32	—	.140	.052	⑩	1500	⑪
	2280	R-9493A	—	9/32	—	.140	.052	⑩	1500	⑪
1983	1945	R-40055A	—	⑨	—	.130	.080	—	1600	1.70
	1945	R-40056A	—	⑨	—	.130	.090	—	1600	1.61
	2280	R-9499A	—	9/32	—	.140	.052	—	1500	⑪
	2280	R-9951A	—	9/32	—	.140	.052	—	1500	⑪
	6145	R-40029A	—	⑫	—	.150	.090	—	1600	1.70
	6145	R-40030A	—	⑫	—	.150	.090	—	1600	1.61

①—Set automatic choke at 2 Rich.
②—Slot #1.
③—Slot #2.
④—Slot #3.
⑤—From closed throttle.
⑥—Hole position #1.
⑦—Hole position #2.
⑧—At idle.
⑨—Flush with top of bowl gasket ± 1/32".
⑩—2 Way electric.
⑪—Flush with top of bowl vent casting.
⑫—Flush with top of main body casting to .050" above.

Clutch & Manual Transmission Section

NOTE: For Clark, Fuller, New Process, Spicer & Warner transmissions used on Dodge & Plymouth trucks, see the "name" chapter for service procedures and lubricant capacities.

CLUTCH PEDAL, ADJUST (MECHANICAL LINKAGE)

1974–83 All Except 1974–77 "B" Series

Adjust fork rod by turning self-locking adjusting nut to provide 3/32 inch free movement at end of fork. This movement will provide 1 inch free play at pedal.

1974–77 "B" Series

Shorten or lengthen clutch release fork rod by turning adjusting nut until there is .187 inch free movement at the clutch fork outer

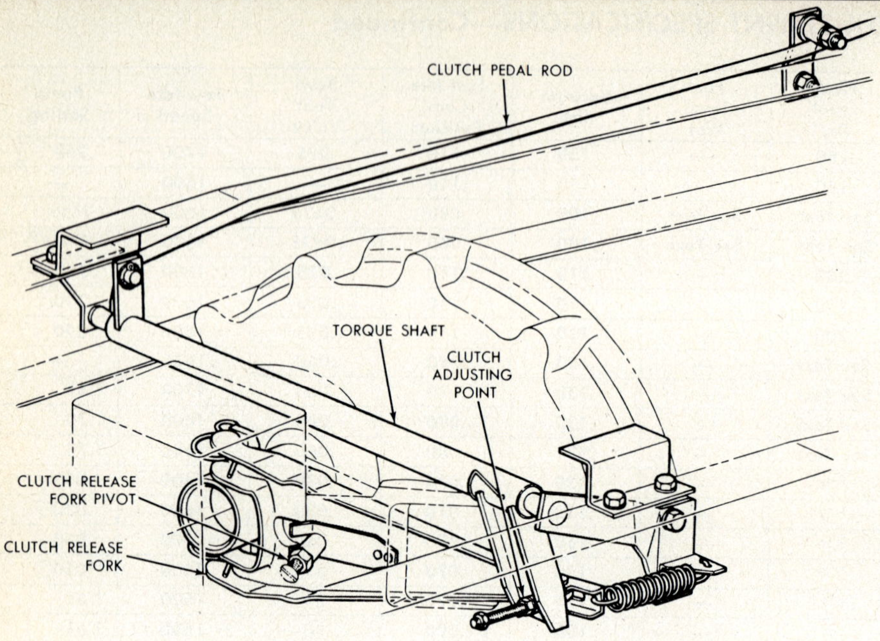

Fig. 1 Clutch linkage & adjusting rod. Medium duty Tilt Cab

end. This will give the necessary 1 to 1½ inch free play at pedal.

1974 Tilt Cab

Adjust fork rod by turning adjusting nut until there is 3/16 inch free play at clutch fork outer end, Fig. 1. This will give the necessary 1 inch free play at pedal.

1975 L, LN, LNT–800 Series W/14 Inch Clutch

Release bearing free play of .069 inch as measured at clutch release shaft lever adjusting screw yields 1 to 1½ inches free play at pedal.

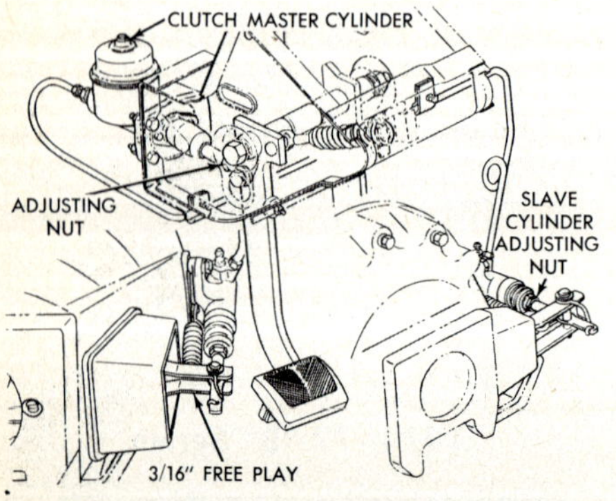

Fig. 2 Hydraulic clutch adjusting points. (Typical)

HYDRAULIC CLUTCH

A hydraulic master cylinder and slave cylinder replaces the mechanical linkage. The slave cylinder is connected to and operates the clutch release fork. The master cylinder and slave cylinder are serviced in a similar manner to a brake master cylinder.

When the master cylinder is installed on the dash panel and hydraulic line connected, connect the push rod end to the clutch pedal and adjust as follows:

1. Adjust clutch release over center spring adjusting end nut counterclockwise (to increase tension) until clutch pedal will not return to normal position with 3/4" open end wrench.
2. Turn adjusting end nut clockwise (to reduce tension) approximately 1½ to 2 turns or until good pedal return is achieved.

Adjust master cylinder push rod so that there is .010" free play between the push rod and master cylinder piston when the piston is in the fully returned position and when the clutch pedal is in its normal at rest position.

Clutch Adjustment

1. Inside the cab, adjust clutch pedal free play to 3/16", which is equal to .010" clearance between push rod and master cylinder piston.
2. On Light and Medium Duty models, at the slave cylinder, adjust clutch linkage free play at the clutch fork pivot pin to 3/16", Fig. 2.
3. On Heavy Duty models, adjust the linkage free play at the clutch fork shaft lever to Dimension "A", Fig. 3

NOTE: On 1974 Heavy Duty models, Dimension "A", Fig. 3 is .120 inch.

4. Inside the cab, check clutch pedal reserve which should be a minimum of 1". Pedal reserve is that portion of pedal travel remaining after free play and complete clutch disengagement.

NOTE: If there is not enough pedal reserve, bleed the clutch hydraulic system to make sure there is no air in the system.

CLUTCH, REPLACE

Single Disc-Type

1. Remove the transmission.
2. Remove the clutch housing pan.
3. Remove clutch fork, bearing and sleeve assembly.
4. Mark clutch cover and flywheel to ensure proper balance when clutch is installed.
5. Remove clutch cover retaining screws and pull the pressure plate clear of flywheel and, while supporting pressure plate, slide the disc from between flywheel and pressure plate. On some models it may be necessary to rotate the pressure plate to the point of maximum clearance. Then allow plate assembly to drop while tilting the top edge of the clutch cover back to clear the frame crossmember.

Two Plate Type

1. After transmission has been removed, insert two 3/4" blocks of wood between release bearing housing and spring plate. This facilitates removal of clutch-to-flywheel screws and protects the pull-back springs from excessive tension.
2. Remove clutch retaining screws and remove pressure plate and flywheel assembly.
3. Remove rear driven disc, intermediate plate and front driven disc.
4. Remove four drive pins from flywheel. Allen head screws are used to retain the drive pins. They must be removed to clear drive pins.

Installation

NOTE: Heavy duty models use a permanently

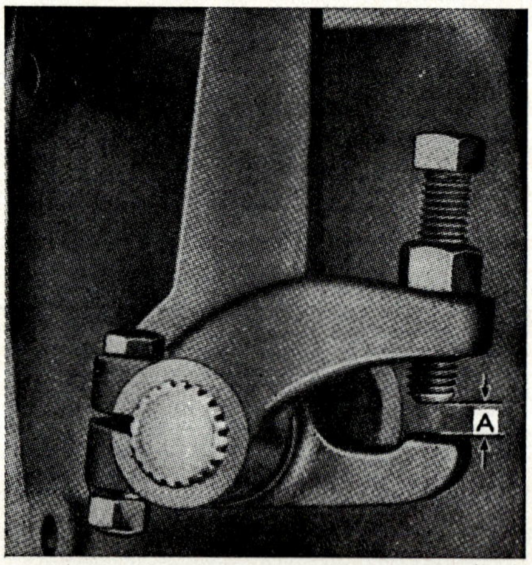

Fig. 3 Heavy duty clutch release bearing adjustment

lubricated pilot bearing. New pilot bushings should be soaked in oil before installing. Pilot bushing requires a grease which will stay in place during high temperature operation. A small amount of grease should be placed in front of bushing and inner surface should be lightly coated.

Clean the surfaces of the flywheel and pressure plate, making certain no oil or grease remains on these parts. Hold the cover plate and disc in place and insert a special clutch aligning tool or a spare clutch shaft through the hub of the disc and into the pilot bearing in the crankshaft. Bolt the clutch cover loosely to the flywheel, being sure that punch marks previously made are lined up.

To avoid distortion of the clutch cover, tighten the cover bolts a few turns each in progression until all are tight.

Guide the transmission into place, using care to see that the driven disc is not bent. Use a floor jack to support the transmission so that the clutch shaft may be guided through the driven disc safely. Finally, adjust the clutch pedal free travel.

TRANSMISSION, REPLACE

Three Speed Units

Disconnect propeller shaft, speedometer cable, speedometer pinion sleeve, gearshift control rods and hand brake cable. Remove transmission-to-clutch housing bolts and pull transmission straight back until input shaft clears clutch disc and release levers before lowering transmission out of truck. Do not disengage clutch or otherwise move clutch or clutch disc unless it is to be removed for service.

Four Speed Overdrive

1. Remove gear shift components, then drain fluid from transmission.
2. Disconnect propeller shaft at rear universal joint, mark parts for reinstallaton in same position, then carefully pull shaft yoke out of transmission housing.
3. Disconnect speedometer cable and back-up light switch.
4. Install engine lifting fixture and raise engine slightly.
5. Using a suitable jack, support transmission, then remove center crossmember.
6. Remove transmission to clutch housing bolts, then slide transmission rearward until drive pinion shaft clears clutch disc.
7. Lower transmission and remove from vehicle.

New Process 4-Speed

1. Remove transfer case, if equipped, as described under "Transfer Case, Replace."
2. Disconnect back-up light switch.
3. Install engine support fixture, then, using a suitable jack, support transmission and remove transmission crossmember.

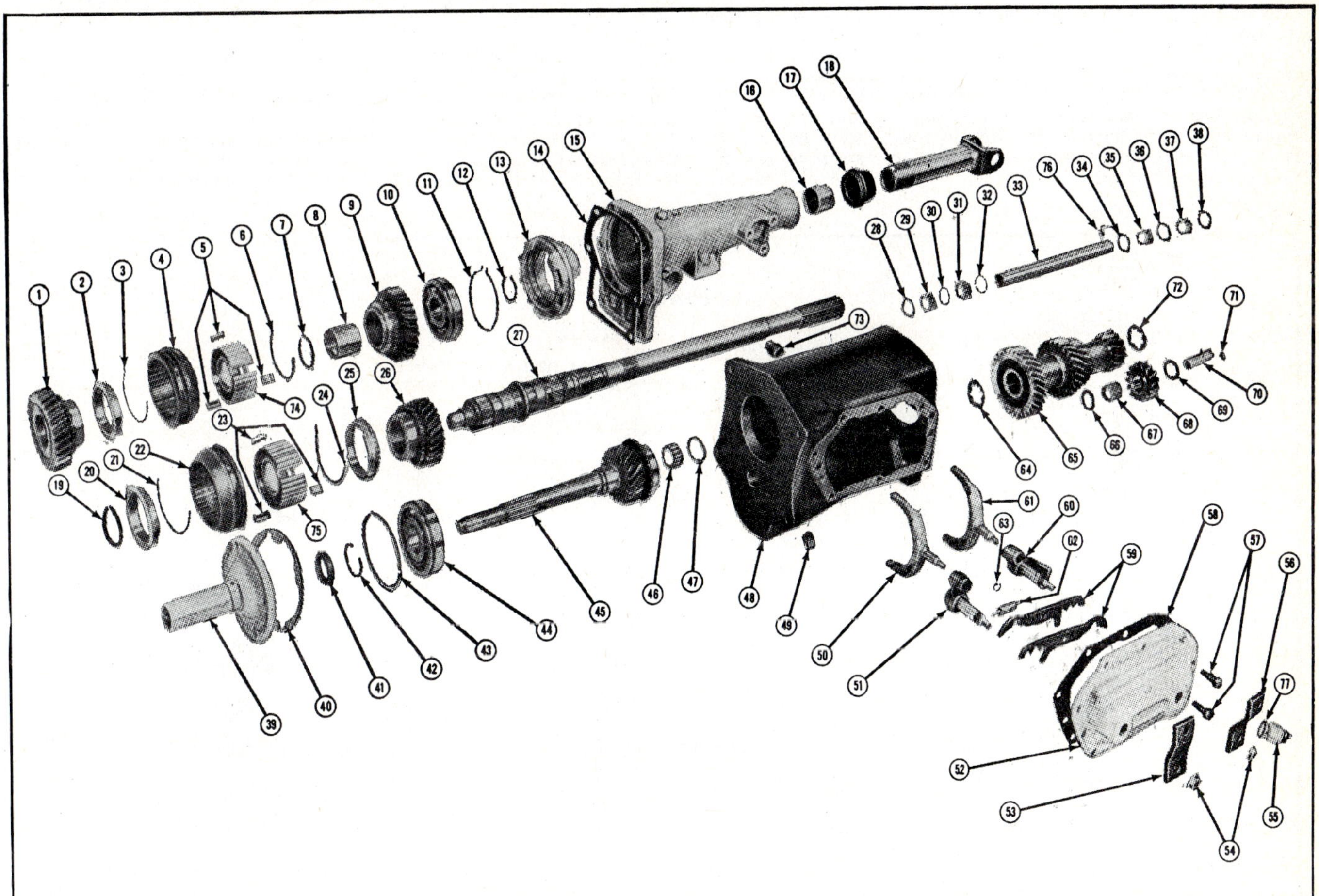

1. Gear, First
2. Ring
3. Spring
4. Sleeve
5. Struts (3)
6. Spring
7. Snap Ring
8. Bushing
9. Gear, Reverse
10. Bearing
11. Snap Ring
12. Snap Ring
13. Retainer
14. Gasket
15. Extension
16. Bushing
17. Seal
18. Yoke
19. Snap Ring
20. Ring
21. Spring
22. Sleeve
23. Struts (3)
24. Spring
25. Ring
26. Gear, Second
27. Shaft, Output
28. Washer
29. Roller
30. Washer
31. Roller
32. Washer
33. Countershaft
34. Washer
35. Roller
36. Washer
37. Roller
38. Washer
39. Retainer
40. Gasket
41. Seal
42. Snap Ring
43. Snap Ring
44. Bearing
45. Pinion, Drive
46. Roller
47. Snap Ring
48. Case
49. Plug, Drain
50. Fork
51. Lever
52. Housing
53. Lever
54. Nut Locking
55. Switch
56. Lever
57. Bolt
58. Gasket
59. Lever, Interlock
60. Lever
61. Fork
62. Spring
63. Snap Ring
64. Washer
65. Gear, Countershaft
66. Washer
67. Roller
68. Gear, Idler
69. Washer
70. Shaft
71. Key
72. Washer
73. Plug, Filler
74. Gear, Clutch
75. Gear, Clutch
76. Key
77. Gasket

Fig. 4 A-230 fully synchronized three-speed transmission

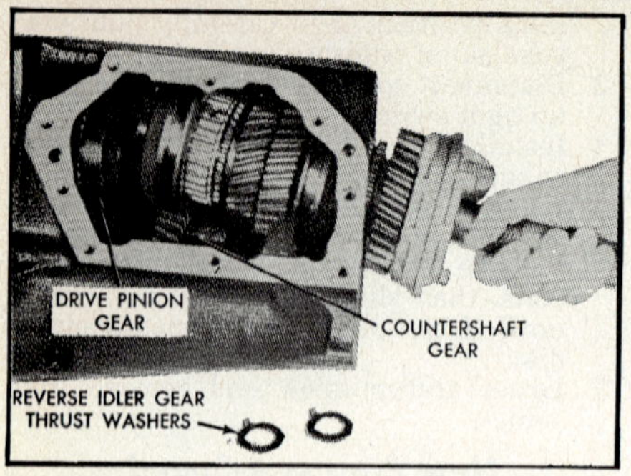

Fig. 5 Removing mainshaft assembly

4. Remove transmission to clutch housing attaching bolts.
5. Slide transmission rearward until drive pinion shaft clears clutch disc, then lower transmission and remove from vehicle.

New Process 5-Speed

Remove gearshift lever, floor mat and floor cover over transmission. Drop center bearing (if equipped). Place jack under transmission and remove transmission from clutch housing. Pull transmission with jack straight back about 6 inches. Then lower jack slightly. Move transmission with jack to the left so that clutch shaft will clear clutch housing. Lower transmission out of truck.

Clark & Spicer Units

Remove floor mat and transmission access plate. Disconnect speedometer cable. On models with hand brake assembly, disconnect hand brake lever to hand brake rod or cable. Disconnect clutch control linkage at transmission bell housing. Disconnect propeller shaft and center bearing, if equipped.

If equipped with auxiliary transmission, disconnect transmission-to-auxiliary transmission companion flange. Place transmission jack under transmission and remove clutch housing-to-transmission cap screws. Move transmission back until clutch shaft is clear of clutch housing. Lower transmission from truck.

Spicer Auxiliary Units

Drain transmission lubricant. Shift each shaft to a forward position. Remove clevis pins and then place shafts to neutral position. Disconnect universal joint companion flange next to front of auxiliary transmission. Disconnect emergency brake rod and speedometer cable. *If it is difficult to disconnect speedometer cable, remove brake band.*

Mark sliding yoke and drive shaft with a center punch for proper installation. Position portable hydraulic jack under transmission and remove transmission support bolts. *Observe position of mounting components to assist in proper installation.* Lower transmission and pull sliding yoke from propeller shaft

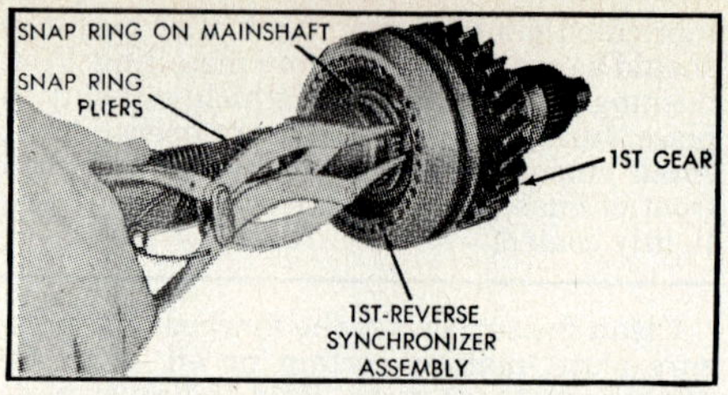

Fig. 6 Removing 1st reverse synchronizer snap ring

spline. Then remove transmission from truck.

FULLY SYNCHRONIZED 3 SPEED

NOTE: For service procedures on the A–390 model transmission, refer to the "3 SPEED TRANS., 1974–76" in the Ford chapter.

MODEL A–230

Transmission, Disassemble, Fig. 4

1. Shift transmission into 2nd gear for shift fork clearance and remove side cover and shifter assembly.
2. Remove front bearing retainer.

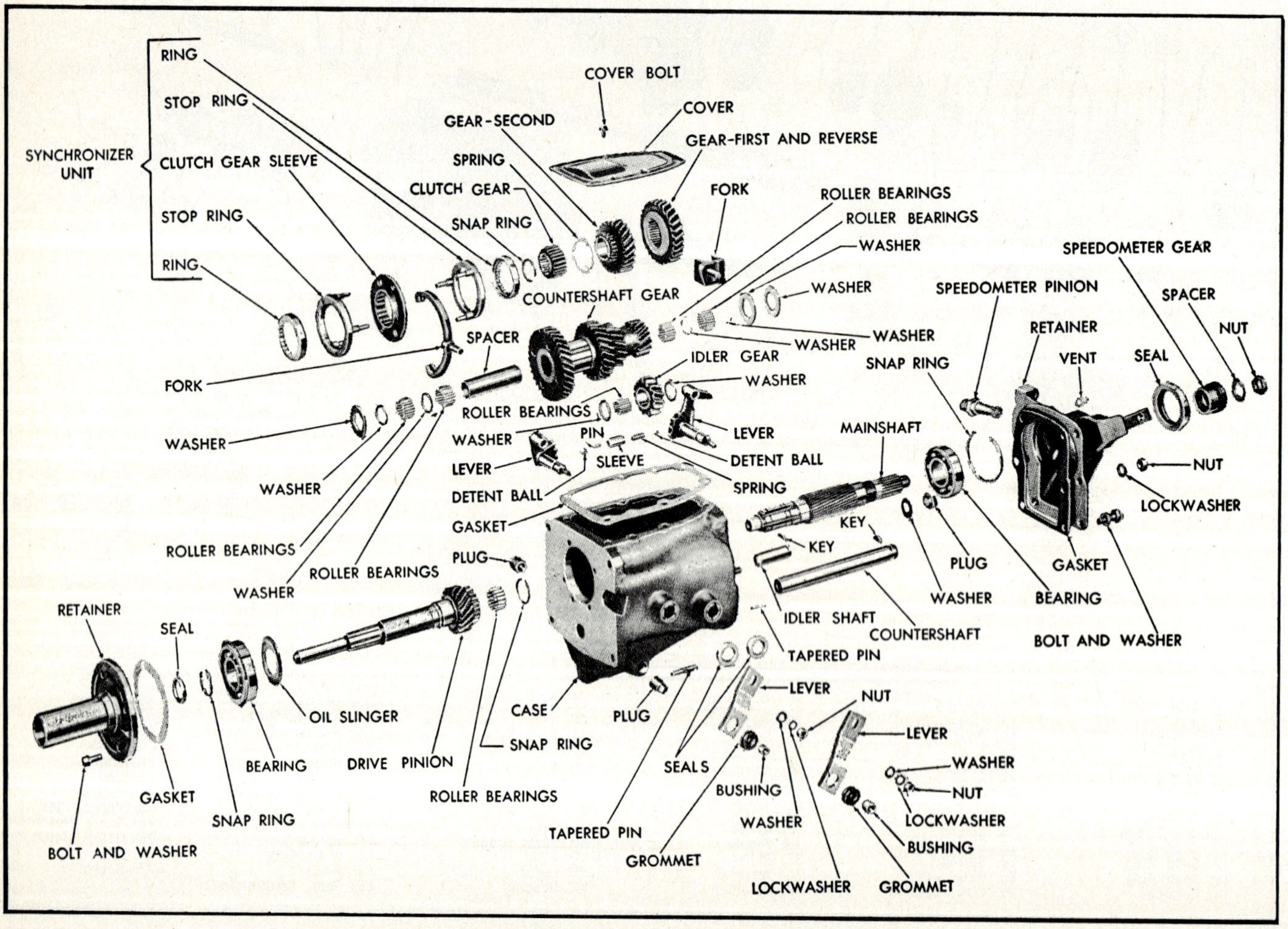

Fig. 7 A-250 three speed transmission 1974–76 (typical)

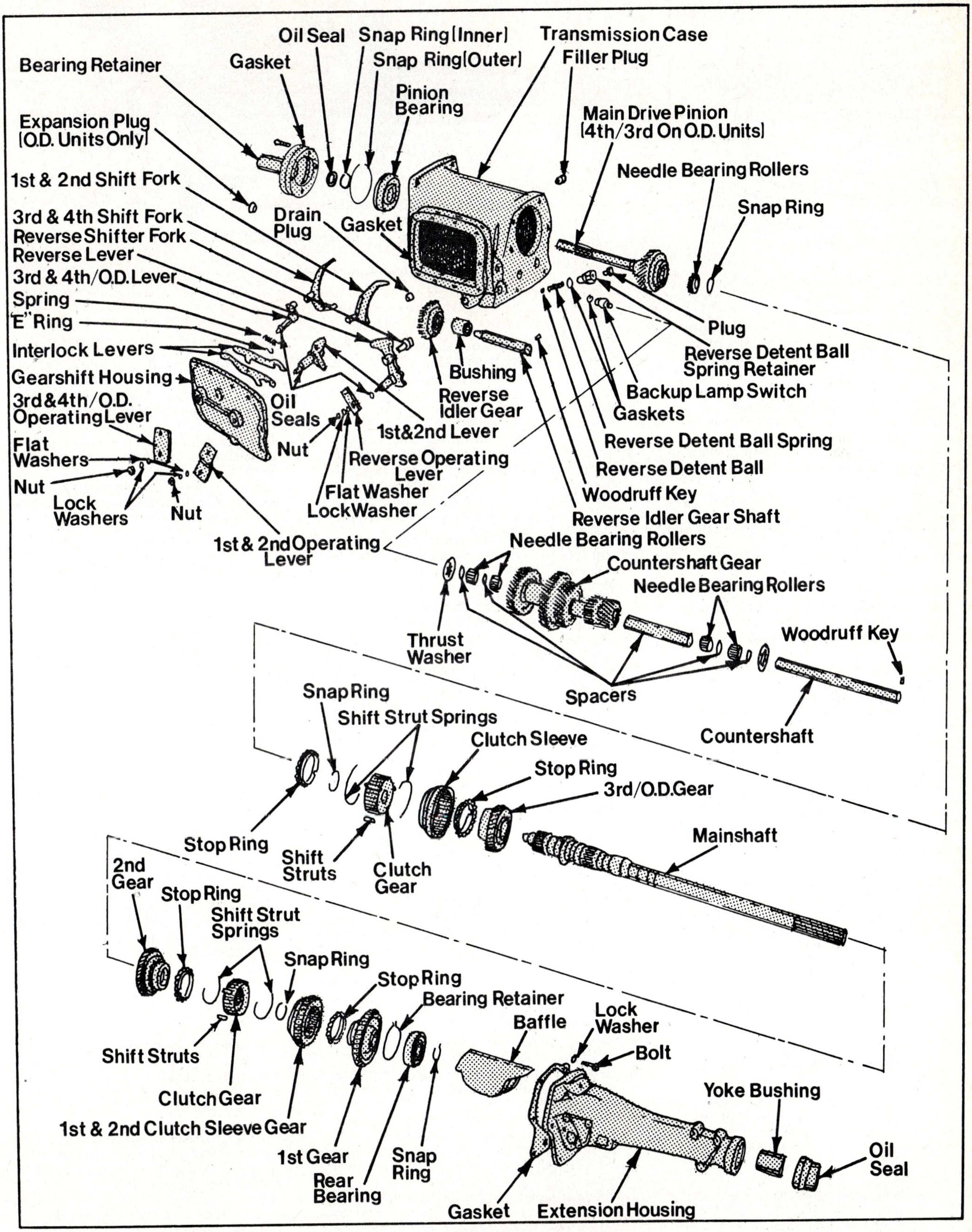

Fig. 8 Four speed overdrive transmission

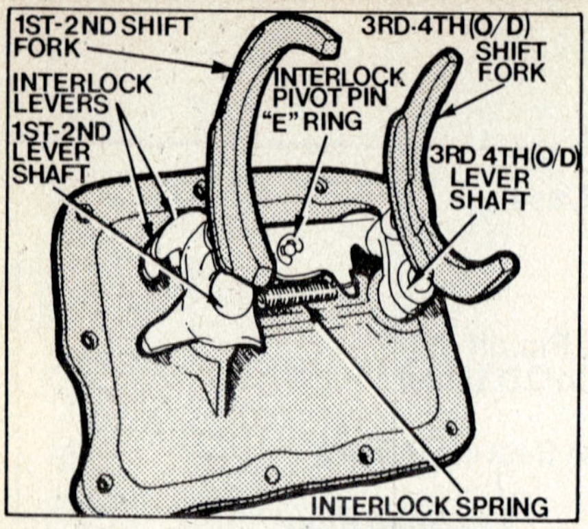

Fig. 9 Gearshift housing assembly. Four speed units

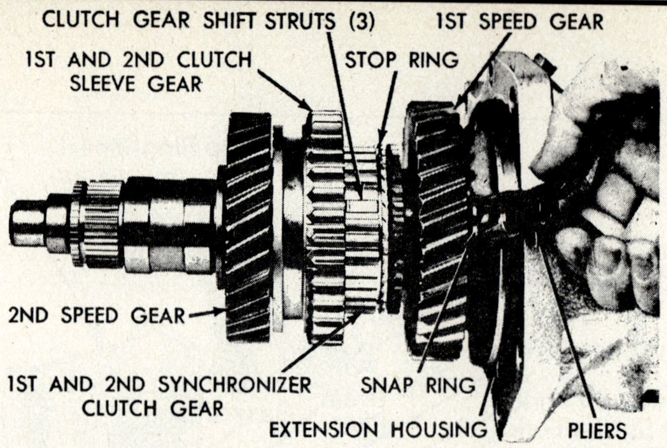

Fig. 10 Compressing center bearing snap ring. Four speed units

3. Tap drive pinion forward with brass drift as far as possible to provide clearance for mainshaft removal.
4. Rotate cut away part of second gear next to countergear for mainshaft removal clearance. Shift 2–3 synchronizer sleeve forward.
5. Remove speedometer gear.
6. Remove rear extension housing.
7. Using dummy shaft, push reverse idler shaft and key out of case.
8. Remove idler gear with dummy shaft in place to retain rollers. Remove thrust washers.
9. Remove mainshaft through rear case opening, Fig. 5
10. Using dummy shaft to retain rollers, tap countershaft out rear of case and lower countergear to bottom of case to permit removal of drive pinion.
11. Remove snap ring from pinion bearing outer race, drive pinion into case and remove through rear case opening.
12. Remove countergear through rear case opening.

Mainshaft

1. Remove 2–3 synchronizer clutch gear retaining ring from front of mainshaft.
2. Slide 2–3 synchronizer and 2nd gear stop ring off of shaft. Remove 2nd gear.
3. Spread snap ring in mainshaft bearing retainer and slide retainer off bearing race.
4. Remove snap ring securing bearing to mainshaft.
5. Support front side of the reverse gear in press and press bearing off shaft. *When bearing clears shaft, do not allow parts to drop through.*
6. Remove 1st-Reverse synchronizer retaining ring and remove synchronizer assembly from mainshaft, Fig. 6
7. Remove 1st gear and stop ring.
8. Reverse procedure to assemble mainshaft.

Transmission, Assemble

1. Using dummy shaft to hold rollers in place and heavy grease to hold thrust washers, carefully place countergear assembly in bottom of case. *Do not finish installation until drive pinion is installed.*
2. Load rollers and retaining ring in drive pinion bore and install drive pinion through rear case opening and into case bore. Install large snap ring on bearing and install front bearing retainer.
3. Align countergear with its shaft bore and install countershaft through gear, driving dummy shaft out as countershaft is installed. Install countershaft key.
4. Carefully tap drive pinion forward to provide mainshaft installation clearance.
5. With 2–3 synchronizer sleeve fully forward and cut out on 2nd gear turned so it is toward countershaft, insert mainshaft assembly through rear case opening.

NOTE: If installation is correct, the bearing retainer will bottom in the case without force. If not, check for out of position strut, roller or stop ring.

---

6. Using dummy shaft to hold rollers, install reverse idler in case and install idler shaft and key.
7. Install rear extension housing, speedometer gear and with transmission shifted into 2nd gear, install cover and shifter assembly.

MODEL A–250

Transmission, Disassemble, Fig. 7

1. Referring to Fig. 7, remove input shaft retainer.
2. Remove extension housing.
3. Remove transmission cover.
4. Remove clutch shaft bearing retainer.
5. Grasp clutch shaft with hand and pull assembly out of case. *Be careful not to bind inner synchronizer ring on clutch teeth.*
6. With transmission in reverse, remove outer center bearing snap ring, using a hook or a flat blade, then partially remove mainshaft.
7. Cock mainshaft; then remove synchronizer sleeve, rings and 2–3 shift fork.
8. Remove clutch gear snap ring, using snap ring pliers. Slide clutch gear off end of mainshaft.
9. Slide second speed gear, stop ring and synchronizer spring off mainshaft.
10. Remove low-reverse sliding gear and shift fork as mainshaft is withdrawn from case.
11. Using a feeler gauge, check end play of cluster gear, which should be .004" to .012". This measurement will determine if new thrust washers are to be installed at reassembly.
12. Drive countershaft rearward and out of case. Remove key from end of shaft.
13. Lift cluster gear and thrust washers out of case. Disassemble cluster gear by removing needle bearings and spacer.
14. Using a suitable drift, drive reverse idler gear shaft towards rear and out of case. Remove key from shaft.
15. Lift out reverse idler gear, thrust washers and needle bearings out of case. Remove needle bearings from gear.

Assembly Notes

In addition to some outward visual differences, the late model (1975–76) A-250 transmission differs from the early model in that two sets of roller bearings are used to support the countershaft gear. Early model countershaft gears were supported by four sets of roller bearings.

Reassembling the transmission is accomplished by reversing the disassembly procedure. However, observe the following:

1. When installing the reverse idler gear, have the bevelled ends of the gear teeth forward. Raise the idler gear slightly to align with shaft; then drive shaft into case, through thrust washer and gear, until end of shaft is approximately 1/64" below surface of case.
2. Cluster gear thrust washers are available in two sizes marked "A" and "B". Make a selection to give .004" to .012" total end play of cluster gear. Make sure tabs on thrust washers slide into grooves in case. Countershaft should be installed until approximately 1/64" below surface of case.
3. When installing mainshaft, have offset of low-reverse fork to rear. Engage fork in low-reverse gear, then position in case by shifting in reverse.
4. After sliding the synchronizer clutch gear on mainshaft and down against second speed gear, select a snap ring of the correct thickness and install. *Snap ring should eliminate end play and must be a snug fit.*
5. Check clearance between clutch gear and second speed gear, which should be .003" to .008". *End play in excess of .008" will permit gear "jump out."*
6. After mainshaft is properly positioned, select a snap ring that will be a snug fit at the rear bearing.

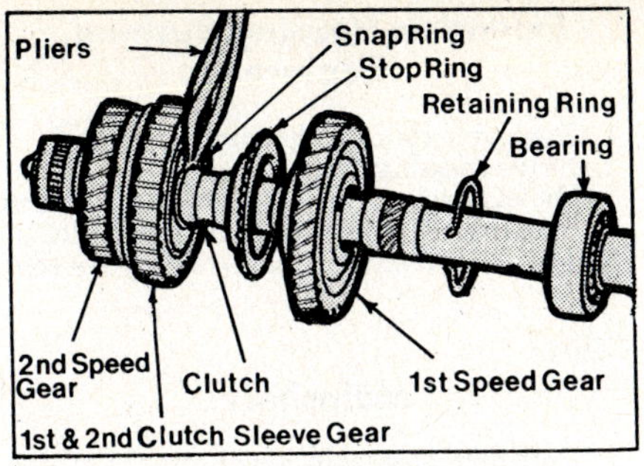

Fig. 11 Removing clutch gear snap ring. Four speed units

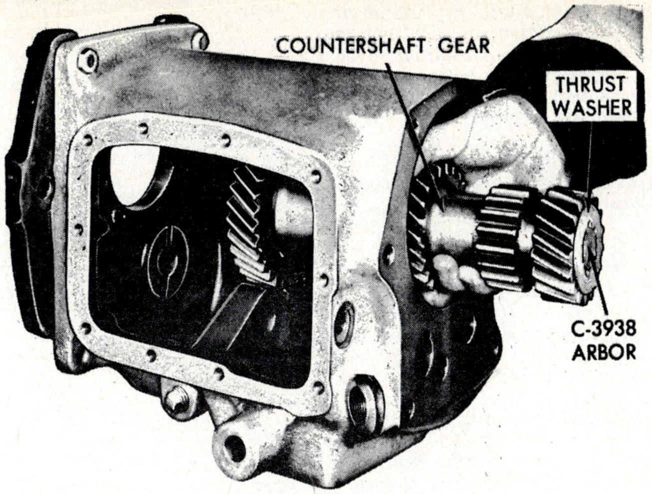

Fig. 12 Replacing countergear. Four speed units

FOUR SPEED & FOUR SPEED OVERDRIVE

TRANSMISSION, DISASSEMBLE

Gearshift Housing

1. Remove reverse shift lever from shaft, Fig. 8, then the gearshift housing attaching bolts, Fig. 9.
2. With levers in neutral detent position, pull housing away from case. Remove forks from sleeves.
3. If oil leakage is present around gearshift lever shafts or the interlock levers are cracked, perform the following:
   a. Remove nuts and the shift operating levers from shafts.
   b. Remove gearshift lever shafts from housing.
   c. Remove "E" ring from interlock lever pivot pin, then the interlock levers and spring from housing.

Extension Housing & Mainshaft

Exc. Overdrive Units

1. Remove bolt and retainer securing speedometer pinion adapter in extension housing, then the adapter and pinion from housing.
2. Remove extension housing attaching bolts.
3. Slide third-fourth synchronizer sleeve slightly forward, slide reverse idler gear to center of shaft, then using a mallet, tap extension housing rearward. Slide extension housing and mainshaft assembly from case.
4. Remove snap ring retaining third-fourth synchronizer clutch gear and sleeve assembly, then slide the assembly from mainshaft.
5. Slide third speed gear and stop ring from mainshaft.
6. Compress snap ring retaining mainshaft ball bearing in extension housing and pull mainshaft assembly and bearing from extension housing, Fig. 10.
7. Remove mainshaft bearing retaining snap ring. Remove bearing by installing steel plates on the first speed gear, then press or drive mainshaft through bearing. Use caution not to damage gear teeth.
8. Remove bearing, bearing retainer ring, first speed gear and first speed stop ring from shaft.
9. Remove first-second clutch gear and sleeve assembly snap ring, Fig. 11, then slide assembly and the second speed gear from shaft.
10. Measure countershaft gear end play by inserting a feeler gauge between thrust washer and gear. End play should be .015–.029 in. If not, install new thrust washers when reassembling transmission.

Overdrive Units

1. Remove bolt and retainer securing speedometer pinion adapter in extension housing, then the adapter and pinion from housing.
2. Remove extension housing attaching bolts.
3. Rotate extension housing on output shaft to expose the rear of countershaft. Clearance is provided on extension housing flange to permit one bolt to be reinstalled to hold the extension housing in the inverted position for access to countershaft.
4. Drill a hole in the countershaft expansion plug at front of case. Then, working through hole, push countershaft rearward until woodruff key is exposed and remove key. Push countershaft forward against expansion plug and, using a brass drift, tap countershaft forward until expansion plug is driven from case.
5. With a dummy shaft, drive countershaft through rear of case, using caution so the countershaft washers do not fall out of position. Lower cluster gear to bottom of case.
6. Rotate extension housing back to normal position.
7. Remove drive pinion bearing retainer attaching bolts, then slide retainer and

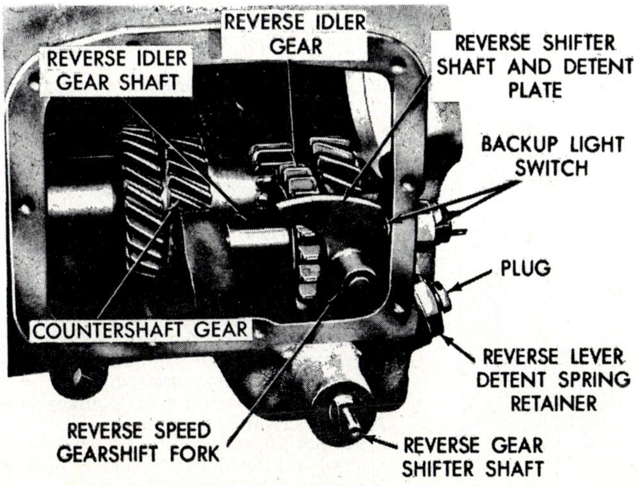

Fig. 13 Reverse gearing and countergear. Four speed units

Fig. 14 Removing reverse gear fork and lever. Four speed units

gasket from pinion shaft. Pry pinion or seal from retainer. Use caution not to nick or scratch seal bore in retainer or the surface on which the seal bottoms.

8. With a brass drift, drive the pinion and bearing assembly through the front of case.
9. Perform steps 3 through 9 as outlined under "Extension Housing & Mainshaft, Except Overdrive Units" procedure.

Drive Pinion & Countershaft Gear

1. Remove pinion bearing inner snap ring, then press pinion from bearing.
2. Remove snap ring and roller bearings from drive pinion cavity.
3. Remove countershaft gear from bottom of case, Fig. 12
4. Remove dummy shaft, needle bearings, thrust washers and spacer from countershaft gear.

Reverse Gear, Lever & Fork

1. Remove reverse gearshift lever detent spring retainer, gasket, plug and detent ball spring from rear of case, Fig. 13.
2. Press reverse idler gear shaft from case and remove woodruff key.
3. If oil leakage is present around the reverse gearshift lever shaft, remove any burrs from shaft. Carefully push reverse gearshift lever shaft inward and remove from case, Fig. 14. Remove detent ball from bottom of case, then the shift fork from shaft and detent plate.

TRANSMISSION, ASSEMBLE

Reverse Gear, Lever & Fork

1. Install new "O"-ring on reverse gearshift lever shaft. Lubricate lever shaft and install lever shaft into case bore, then the reverse fork in lever.
2. Install reverse detent spring retainer and gasket and torque to 50 ft. lbs. Insert ball and spring, then install and torque gasket and plug to 24 ft. lbs.
3. Place reverse idler gearshaft in position in end of case and drive into case to position reverse idler gear on the protruding end of shaft with shift slot facing toward rear and engine slot with reverse shift fork.
4. With reverse idler gear properly positioned, drive reverse idler gearshaft into case and install woodruff key. Ensure shaft is flush with surface of case.
5. Install and torque back-up light switch and gasket to 15 ft. lbs. if removed.

Countershaft Gear & Drive Pinion

1. Lubricate countershaft gear bore and install roller bearing spacer with dummy shaft into gear. Center the spacer and dummy shaft.
2. Install 19 rollers, a spacer ring, 19 more rollers and spacer ring into gear.
3. Lubricate and install new thrust washers, if needed, on dummy shaft with tanged side facing toward case boss.
4. Place countershaft gear assembly on bottom of case, Fig. 12
5. Press drive pinion bearing onto pinion shaft with outer snap ring groove facing front of transmission. Fully seat the bearing against gear shoulder.

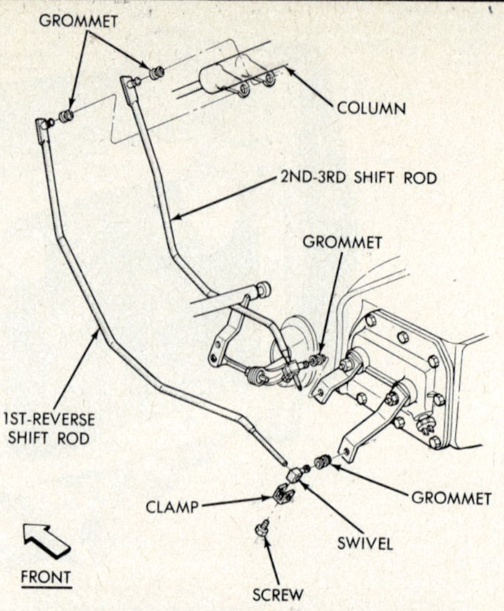

Fig. 15 Disassembled view of gearshift linkage. Except B-Series & Voyager w/A-230 transmission

6. Install new bearing retaining snap ring.
7. With pinion shaft held in a soft jawed vise, install 16 roller bearings in shaft cavity. Coat bearings with heavy grease and install bearing retaining snap ring in groove.
8. On all except overdrive units:
   a. Install drive pinion and bearing assembly through rear of case and position in front bore. Tap into place with a mallet and install outer snap ring in bearing groove.
   b. Insert countershaft in bore at rear of case and raise the countershaft gear until the teeth mesh with drive pinion gear. Ensure thrust washers remain properly positioned.
   c. Align dummy shaft with case bores, then push countershaft into gear. Tap shaft into case until shaft end is flush with case surface. Remove dummy shaft.
9. On all units, install new oil seal in retainer bore.

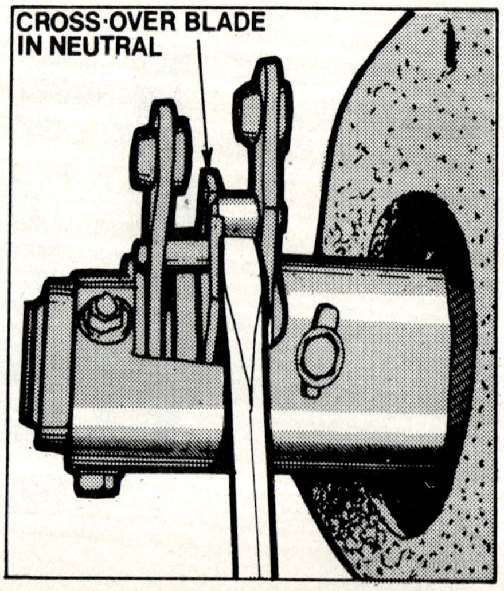

Fig. 16 Gearshift lever adjustments. Three speed transmission

Extension Housing Bushing, Replace

1. Remove extension housing yoke seal.
2. Drive bushing from housing.
3. Drive new bushing into housing, aligning bushing oil hole with housing oil slot.
4. Install new extension housing yoke seal.

Mainshaft

1. Partially assemble synchronizer components as follows:
   a. Place a stop ring flat on bench, then the clutch sleeve and gear.
   b. Place struts into slots and snap a strut spring into place with the tang inside one strut.
   c. Turn assembly over and install second strut spring with the tang in a different strut.
2. Slide second speed gear onto mainshaft with synchronized cone facing toward rear against shoulder.
3. Slide first-second synchronizer with stop ring lugs indexed in slots onto mainshaft against second gear cone and secure with new snap ring. Slide next stop ring onto shaft and index lugs into clutch hub slots.
4. Slide first speed gear with synchronizer cone facing forward into position against clutch sleeve gear.
5. Install mainshaft bearing retaining ring, then the mainshaft rear bearing. Press bearing into place and install new snap ring to retain bearing.
6. Install partially assembled mainshaft into extension housing and compress bearing retaining ring so mainshaft ball bearing can move inward and bottom against thrust shoulder in extension housing. Release and seat ring in groove in the housing.
7. On all except overdrive units:
   a. Slide third speed gear onto mainshaft with synchronized cone facing forward, then the third gear stop ring.
   b. Install third-fourth synchronizer clutch gear assembly on mainshaft with shift fork slot facing rearward against third speed gear.
   c. Install retaining snap ring, then, with grease, position front stop ring over clutch gear, indexing ring lugs with shift struts.
8. On overdrive units:
   a. Slide overdrive gear onto mainshaft with synchronizer cone facing forward, then the overdrive gear stop ring.
   b. Install third-overdrive synchronizer clutch gear assembly on mainshaft with shift fork slot facing rearward against overdrive gear.
   c. Install retaining snap ring then, with grease, position front stop ring over clutch gear, indexing ring lugs with shift struts.
9. On all units, coat new extension housing to case gasket with grease, then place gasket on housing.
10. Slide reverse idler gear to center of the shaft, then move the third-fourth synchronizer sleeve or the third-overdrive synchronizer sleeve forward.
11. On all except overdrive units, move drive pinion forward to provide maximum clearance for mainshaft pilot end.
12. On all units, slowly insert mainshaft assembly into case, tilting the shaft as necessary to clear idler and cluster gears.

On all except overdrive units, engage the mainshaft with the pilot rollers in the drive pinion gear.

13. On all except overdrive units:
    a. Place third-fourth synchronizer sleeve in neutral position.
    b. If all components are positioned properly the extension housing will bottom on the case without force. If not, check position of struts, pinion rollers and stop rings.
    c. Install and torque extension housing attaching bolts to 50 ft. lbs.
    d. Install drive pinion bearing retainer and gasket. Coat attaching bolt threads with a suitable sealing compound, then install and torque bolts to 30 ft. lbs.
14. On overdrive units:
    a. Place third-overdrive synchronizer sleeve in neutral position.
    b. Rotate extension housing on mainshaft to expose countershaft rear and install one extension housing attaching bolt, holding housing in inverted position and preventing housing from moving rearward.
    c. Install drive pinion and bearing assembly through front of case and into front bore. Install outer snap ring in bearing groove. Tap lightly into position with a mallet. If all components are positioned properly, the bearing outer snap ring will bottom onto case face without force. If not, check position of struts, pinion rollers and stop rings.
    d. Turn transmission upside down while holding countershaft gear assembly to prevent damage.
    e. Lower countershaft gear assembly into position with teeth meshed with drive pinion gear. Ensure thrust washers are positioned properly with the tangs aligned with case slots.
    f. Start countershaft into bore at rear of case and push until approximately halfway into case, then install woodruff key. Push shaft forward until end is flush with case surface. Remove dummy shaft.
    g. Properly align extension housing on case and install and torque extension housing attaching bolts to 50 ft. lbs.
    h. Install drive pinion bearing retainer and gasket. Coat attaching bolt threads with a suitable sealing compound, then install and torque bolts to 30 ft. lbs.
    i. Install new expansion plug in countershaft bore at front of case.

Gearshift Housing

1. Install interlock levers on pivot pin and retain with "E" ring, Fig. 9. Use pliers to install spring on interlock lever hangers.
2. Lubricate and install new "O" rings on shift lever shafts. Lubricate housing bores and push shafts into proper bore.
3. Install operating levers and torque retaining nuts to 18 ft. lbs. Ensure the third-overdrive lever points downward.
4. Rotate shift fork bores to neutral position and install the third-overdrive shift fork in proper bore and under both interlock levers.
5. Position both synchronizer sleeves in neutral and place first-second shift fork in the first-second synchronizer sleeve groove.
6. Lay transmission on its right side and place gearshift housing gasket on case with grease to hold in position.
7. Lower gearshift housing into place, guiding the third-overdrive shift fork into synchronizer groove, and lead the first-second shift fork shaft into the first-second shift lever bore. Raise interlock lever against its spring tension with a screwdriver, allowing first-second shift fork shaft to slip under the levers. The gearshift housing should now seat against case gasket.
8. Install gearshift housing attaching bolts finger tight and shift transmission through all gears to insure proper operation.

NOTE: Eight gearshift housing attaching bolts are shoulder bolts used to accurately locate the mechanism on the transmission. One bolt shoulder is longer and acts as a dowel, passing through the cover and into the case at center of rear flange. Two bolts are of standard design, located at the lower rear of cover.

9. Torque gearshift housing attaching bolts to 15 ft. lbs.
10. The reverse shift lever and the first-second shift lever incorporate cam surfaces which mate in the reverse position to lock the first-second lever, fork and synchronizer in the neutral position. To check, place transmission in reverse, then, while turning input shaft, move first-second lever in each direction. If input shaft lock

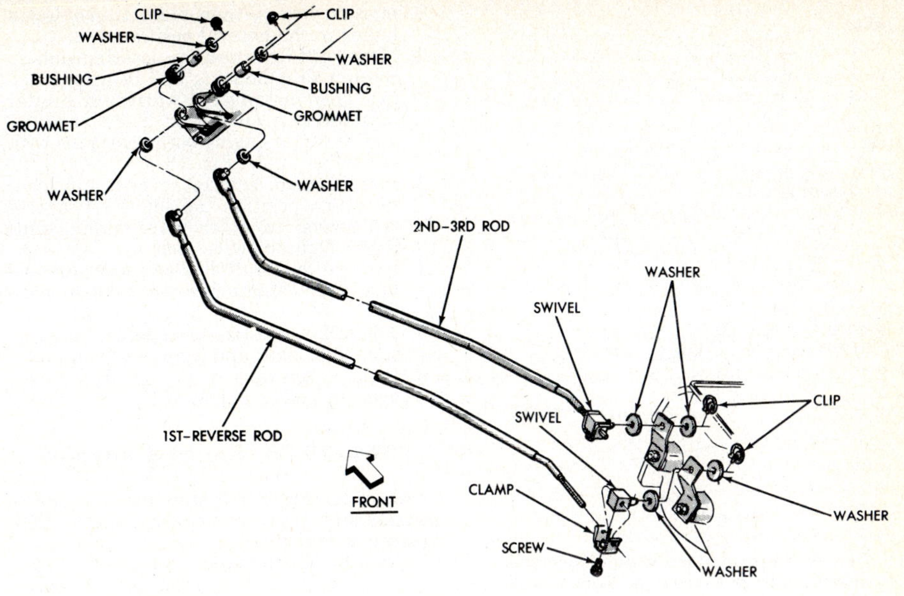

Fig. 17 Disassembled view of column shift linkage. B-Series & Voyager w/A-230 transmission

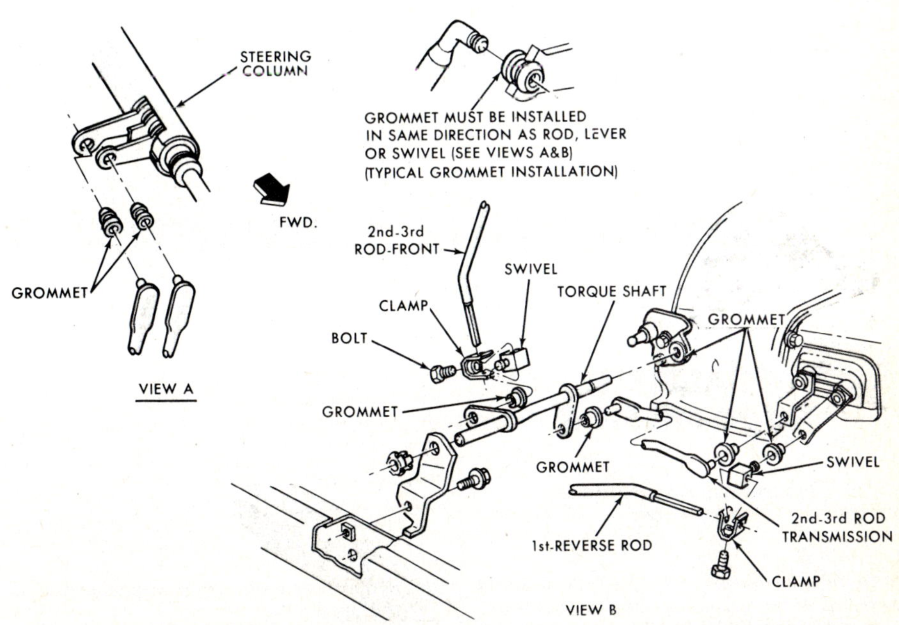

Fig. 18 Disassembled view of column shift linkage. B-Series & Voyager w/A-390 transmission

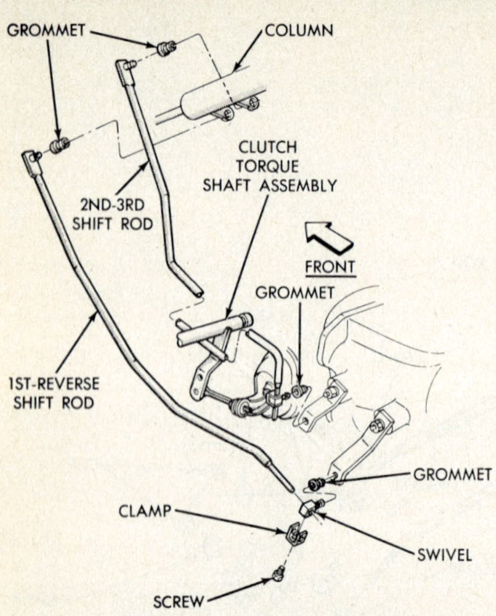

Fig. 19 Disassembled view of gearshift linkage. Except B-Series & Voyager w/A-390 transmission

becomes difficult to turn, the synchronizer is partially engaged, caused by excessive cam clearance. Install a new first-second shift lever of different size. If insufficient cam clearance exists, it is difficult or impossible to shift into reverse.

11. Lubricate reverse shaft and install operating lever, then torque attaching nut to 18 ft. lbs.
12. Install speedometer drive pinion gear and adapter, ensuring range number (indicating number of teeth) is in the six o'clock position.

GEARSHIFT, ADJUST

THREE-SPEED

Except A-390 Units

1974–79 Exc. "B" Series & Voyager

1. Disconnect shift rod swivels from transmission levers, then place transmission levers in the neutral position.
2. Move shift lever to align locating slots in column shift housing and bearing housing. Then install a screwdriver or similar tool between cross-over blade and 2–3 shift lever, engaging both lever pins with cross-over blade, Fig. 16.
3. Place transmission 1st reverse shift lever in reverse position and adjust 1st reverse rod swivel to enter transmission shift lever. Tighten swivel bolt.
4. Remove gearshift housing locating tool and place column lever in neutral position.
5. Adjust 2–3 rod swivel to enter transmission shift lever and tighten swivel bolt.
6. Remove screwdriver or tool from cross-over and check adjustment.

1974–79 "B" Series & Voyager

Adjust length of 2–3 shift rod to properly position shift lever on steering column, then proceed as follows:

1. Install both shift rods and secure at each lever with a clip, then loosen swivel clamp bolts, Fig. 17.
2. Place transmission 2–3 shift lever in 3rd position and place steering column shift lever into a position approximately five degrees above the horizontal position, then tighten swivel clamp bolt.
3. Shift transmission into neutral and install a screwdriver or other suitable tool between cross-over blade and 2–3 lever at steering column, engaging both lever pins and cross-over blade, Fig. 16.
4. Place transmission 1st-reverse lever in neutral position and tighten swivel clamp bolt.
5. Remove screwdriver or tool from cross-over and check adjustment.

A-390 Units

1. Loosen both shift rod swivels, Fig. 18 and 19.
2. Place transmission shift levers in neutral (middle detent) position.
3. Move shift lever to align locating slots in bottom of steering column shift housing and bearing housing. Install a suitable tool into slot.

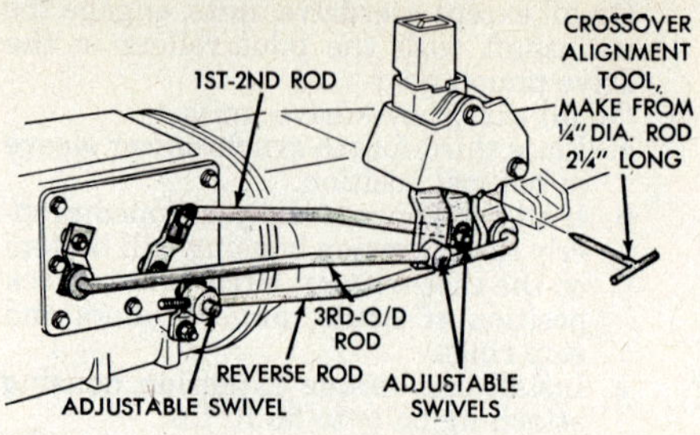

Fig. 20 Gearshift linkage adjustment. Four-speed overdrive

4. Place a screwdriver or other suitable tool between cross-over blade and 2–3 lever at steering column so both lever pins are engaged by the cross-over blade, Fig. 16.
5. Tighten both shift rod swivel bolts.
6. Remove gearshift housing locating tool.
7. Remove tool from cross-over blade at steering column.
8. Shift through gears to check adjustment and cross-over smoothness.
9. Check for proper operation of steering column lock in reverse. With proper adjustment, the ignition should lock in the reverse position only.

FOUR-SPEED OVERDRIVE

1. Install gear shift lever aligning tool, Fig. 20, to hold levers in neutral cross-over position.
2. With all rods removed from transmission shift levers, place levers in neutral detent position.
3. Adjust threaded shift rod levers so they enter transmission levers freely without any rearward or forward movement.

NOTE: Start with 1–2 shift rod, it may be necessary to pull clip at shifter end to rotate this rod.

4. Install washers and clips, then remove aligning tool and check linkage operation.

Automatic Transmission Section

LOADFLITE TRANSMISSION

Transmission, Replace 1974–83

The transmission and converter must be removed as an assembly, otherwise the converter drive plate, pump bushing and oil seal will be damaged. The drive plate will not support a load; therefore, none of the weight of the transmission should be allowed to rest on the plate during removal. Also, on four-wheel drive models, the transfer case must be removed before removing the transmission. Proceed as follows:

1. Disconnect battery ground cable.
2. On some models it may be necessary to lower exhaust system to provide clearance.
3. Remove engine to transmission struts, if equipped.
4. Disconnect and plug oil cooler lines at transmission.
5. Remove starter motor, oil cooler line bracket and converter access cover.
6. Using a socket wrench on vibration damper bolt, rotate engine clockwise to bring converter drain plug to bottom.
7. Remove plug and drain converter, then loosen transmission oil pan and drain transmission.
8. Mark converter and drive plate to aid in assembly.
9. To provide access for removal of converter to drive plate attaching bolts, rotate engine clockwise using a socket wrench on vibration damper bolt.

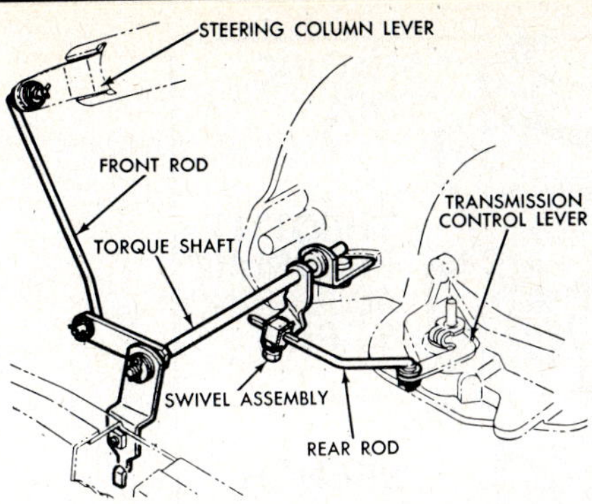

Fig. 1 Gearshift linkage. Loadflite (Typical)

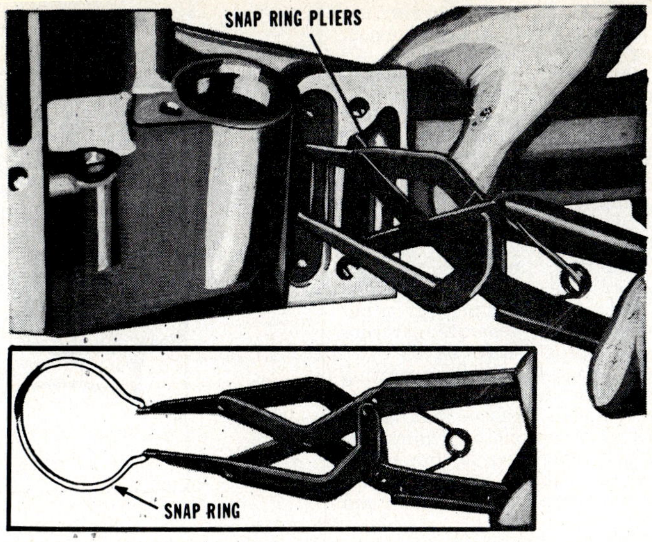

Fig. 2 Removing or installing extension housing snap ring

10. Disconnect propeller shaft at rear universal joint, then carefully pull shaft assembly out of extension housing.
11. Disconnect wire connector from back-up light and neutral start switch.
12. Disconnect gearshift rod and torque shaft assembly from transmission.
13. Disconnect throttle rod from lever at left side of transmission.
14. Remove oil filler tube and disconnect speedometer cable.
15. Support engine at rear, then using a transmission jack raise transmission slightly to relieve load on supports.
16. Remove crossmember, then remove converter housing to engine attaching bolts.
17. Carefully work transmission and converter off engine block dowels and disengage converter hub from end of crankshaft.

NOTE: Attach a small "C" clamp to edge of converter housing to hold converter in place during transmission removal.

18. Lower transmission and remove assembly from vehicle.
19. Remove "C" clamp and carefully slide converter assembly from transmission.

Gearshift Linkage, Adjust

1. Place gear shift lever in Park position.
2. Move transmission control lever all the way to rear (in Park detent), Fig. 1.
3. Adjust rear rod swivel assembly so that swivel can be freely installed in transmission control lever with no load in either direction on linkage.
4. On 1974–78 models torque the lock bolt to 125 in. lbs. On 1979–83 models torque the bolt to 90 in. lbs., then check adjustment.

Extension Housing & Parking Lock Control Rod 1974–83

1. Mark parts for reassembly and remove propeller shaft.
2. Remove speedometer pinion and adapter assembly, then drain about two quarts of fluid from transmission.
3. Remove extension housing to crossmember bolts, then raise transmission with jack and remove crossmember.
4. Remove extension housing to transmission bolts. On console shift models, remove torque shaft lower bracket to extension housing bolts.

NOTE: In the following step, the gearshift must be in "low" therefore positioning the parking lock control rod rearward so it can be disengaged or engaged with the parking lock sprag.

5. Remove two screws, plate and gasket from bottom of extension housing mounting pad, then spread snap ring from output shaft bearing, Fig. 2, and carefully tap extension housing off output shaft bearing.
6. Slide extension housing off shaft to remove parking sprag and spring. Remove snap ring and slide reaction plug and pin assembly out of housing, Fig. 3.
7. To replace parking lock control rod, refer to "Valve Body."

Valve Body

1. Drain transmission and remove oil pan.
2. Loosen clamp bolts and remove throttle and gear selector levers from manual lever.
3. Remove neutral safety switch and oil filter.
4. Place a drain pan under transmission and remove the ten valve body to transmission bolts. Hold valve body in place while removing bolts.
5. Carefully lower valve body while pulling it forward to disengage parking control rod.

NOTE: It may be necessary to rotate output shaft to permit parking control rod to clear sprag.

6. Remove accumulator piston and spring from transmission case. Inspect piston for nicks, scores and wear. Inspect spring for distortion. Inspect rings for freedom in piston grooves and wear or breakage. Replace parts as necessary.

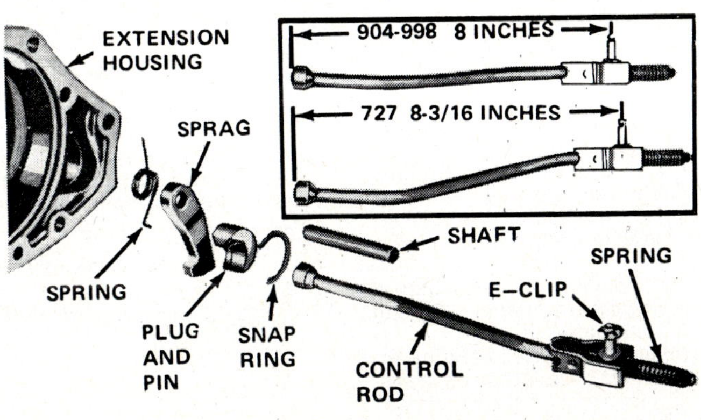

Fig. 3 Parking lock components

Output Shaft Oil Seal

1. Mark propeller shaft to aid in reassembly and remove propeller shaft being careful not to scratch or nick surface on sliding spline yoke.
2. Using a screwdriver, or tool C-3985, and hammer, drive between extension housing and seal and remove seal.
3. Position new seal and drive it into extension housing using tool C-3995 or C-3972.
4. Carefully install yoke into housing, then align marks made at removal and install propeller shaft.

Governor

1. Remove extension housing, then remove

output shaft bearing rear snap ring and remove bearing. On 727 Series, remove remaining snap ring from shaft.
2. Remove snap ring, Fig. 4, from weight end of governor valve shaft and remove valve shaft from governor body.
3. Remove large snap ring from weight end of governor housing, and lift out weight assembly.
4. Remove snap ring from inside governor weight and remove inner weight and spring from outer weight.
5. Remove snap ring from behind governor housing, then slide governor housing and parking brake sprag assembly off output shaft. If necessary, separate governor housing from sprag (4 screws).
6. The primary cause of governor operating failure is due to a sticking governor valve or weights. Rough surfaces may be removed with crocus cloth. Thoroughly clean all parts and check for free movement before assembly.
7. Reverse above operations to assemble and install governor.

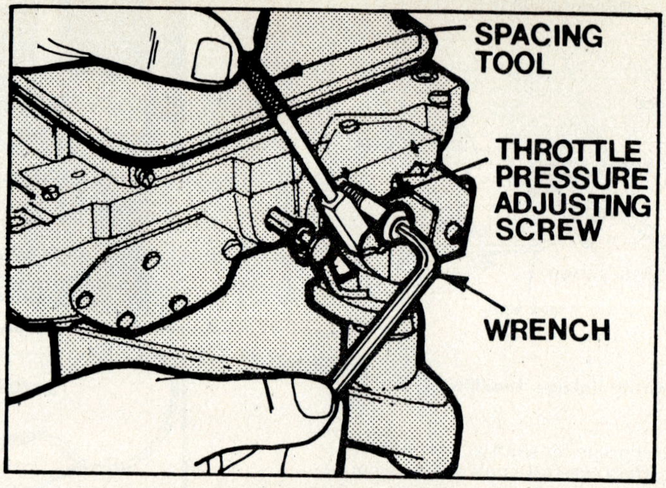

Fig. 5 Throttle pressure adjustment. New Process A-345

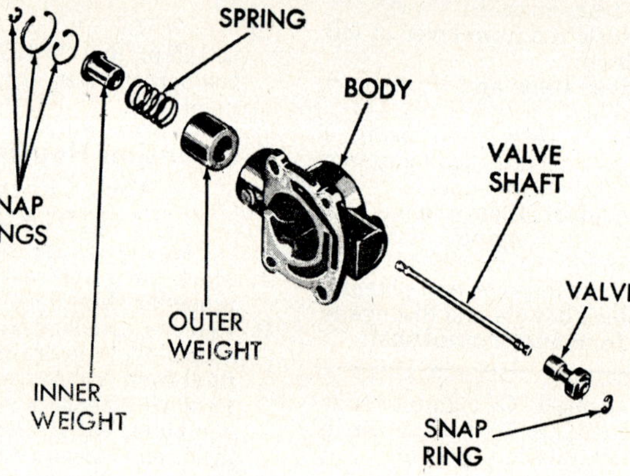

Fig. 4 Governor disassembled

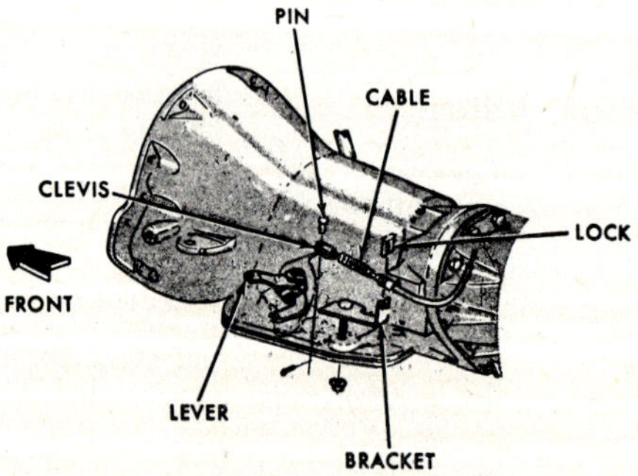

Fig. 6 Gearshift linkage adjustment. New Process A-345

NEW PROCESS A-345

Transmission, Replace

NOTE: The compounder, transmission and converter must be removed as an assembly. Do not allow weight of transmission to rest on drive plate during removal, otherwise damage to drive plate, pump bushing and oil seal may result.

1. Remove converter access cover.
2. Connect remote control starter switch to starter solenoid so that engine can be rotated from under vehicle, then disconnect high tension lead from ignition coil.
3. Rotate engine using remote starter switch to bring drain plug to 6 o'clock position and drain converter, compounder and transmission.
4. Position a suitable jack under transmission oil pan, raise transmission slightly to relieve weight from mountings.
5. Support rear of engine using a suitable support fixture.
6. Mark converter and drive plate to aid in assembly.

NOTE: Crankshaft flange, converter and inner and outer bolt hole circles of drive plate have one bolt hold offset, so these components can be installed in their original position.

7. Rotate engine using remote starter switch to locate two converter to drive plate attaching bolts at 5 and 7 o'clock positions. Remove both bolts, then rotate engine and remove remaining two bolts.

NOTE: Do not use screwdriver or similar tool to rotate converter or drive plate otherwise drive plate may become distorted. Do not engage starter if drive plate is not attached to converter with at least one bolt or if transmission case to engine attaching bolts have been loosened.

8. Disconnect battery ground cable, then remove starter assembly and transmission oil filler tube.
9. Disconnect oil cooler lines, then plug fittings and case openings.
10. Disconnect wire connector from backup light and neutral start switch.
11. Disconnect gearshift cable, speedometer cable and throttle rod from transmission.
12. Remove crossmember, then disconnect propeller shaft at transmission yoke and tie shaft to frame side rail.
13. Remove converter housing to engine attaching bolts, then carefully work transmission and converter assembly rearward off engine block dowels and disengage converter hub from end of crankshaft.

NOTE: Attach a small "C" clamp to edge of converter housing to hold converter in place during transmission removal.

14. Lower transmission and remove from vehicle.
15. Remove "C" clamp from edge of converter housing, then carefully slide converter out of housing.

Line Pressure, Adjust

Incorrect throttle pressure setting will

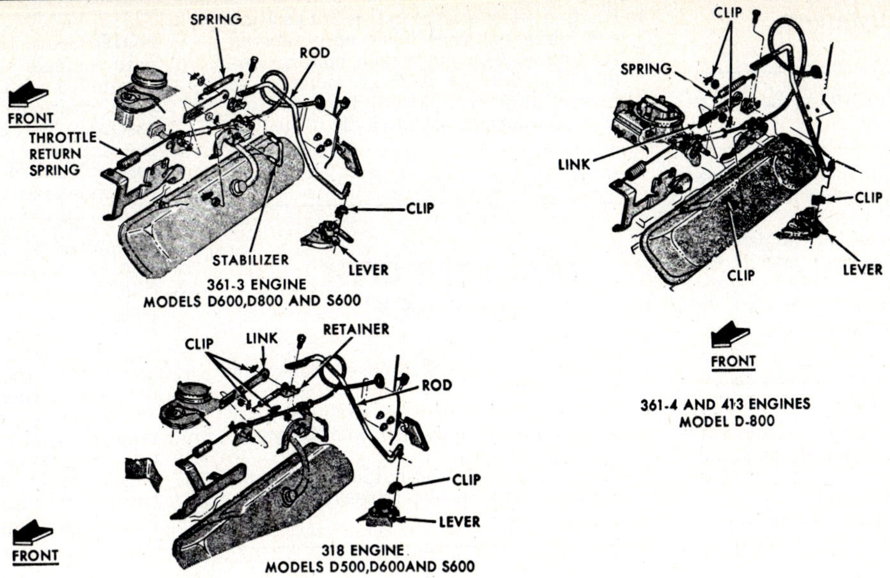

Fig. 7 Throttle linkage adjustment. New Process A-345

cause incorrect line pressure readings even though line pressure adjustment is correct. Always inspect and correct throttle pressure adjustment before adjusting line pressure. Line pressure may be adjusted by turning adjusting screw with an allen wrench. Turning adjusting screw counterclockwise increases pressure and clockwise decreases pressure. One complete turn of adjusting screw changes closed throttle line pressure approximately 1⅔ psi.

Throttle Pressure, Adjust

NOTE: If malfunction is evident, the adjustment should be measured, as throttle pressures cannot be accurately tested.

1. Insert tool C-4233 between throttle lever can and kickdown valve, Fig. 5.
2. Push in on tool and compress kickdown valve against its spring so throttle valve is completely bottomed inside valve body.
3. Exert force to compress spring and turn throttle lever stop screw until head of screw touches throttle lever tang with throttle lever cam touching tool and throttle valve bottomed. Be sure adjustment is made with spring fully compressed and valve bottomed in valve body.

Gearshift Control Linkage, Adjust

1. Disconnect clevis from transmission lever, Fig. 6, then place gear shift selector in neutral and move control lever on transmission two detent positions from rear to neutral.
2. Align clevis hole in transmission lever, then install clevis pin and secure with cotter pin.
3. Recheck adjustment as follows:
   a. Detent position must be close enough to gate stops in neutral and drive to assure that gear shift lever will not remain out of detent position when it is placed against gate and then released.
   b. Engine must start only when gearshift is in neutral. If not, recheck adjustment.
   c. Make certain that vehicle does not move rearward with engine on and gearshift in neutral.

Throttle Linkage, Adjust

1. Operate accelerator to check for throttle opening and closing. If any binding occurs, lubricate linkage and check cable routing or pedal interference with carpet.
2. Disconnect choke rod or block choke valve in full open position, then release fast idle cam by opening throttle and return carburetor to curb idle.
3. Hold transmission throttle lever forward against its stop while adjusting linkage, making certain there is no other force exerted against lower end of rod.
4. Remove spring from stabilizer and retainer and loosen bolt retaining upper end of slotted rod of retainer, Fig. 7.
5. Lightly push rearward on rod, making certain not to disturb throttle lever.
6. Tighten bolt at retainer and install spring at stabilizer and retainer, then connect choke rod or remove blocking and check linkage for correct operation.

ALLISON AT540, MT640 & 650

Transmission, Replace

NOTE: Transmission and converter must be removed as an assembly.

1. Disconnect battery ground cable.
2. On models with gasoline engine, remove spark plugs so engine can be turned manually. On models with diesel engine, ensure fuel stop knob is pulled all the way out.
3. Disconnect oil filler tube and allow transmission oil to drain.

NOTE: After transmission oil has drained, plug oil pan openings to prevent entry of dirt.

4. Disconnect oil cooler lines, then plug fittings and connections.
5. Disconnect range selector cable from shift lever and vacuum line from modulator.
6. Disconnect speedometer shaft fitting from adapter at rear of transmission.
7. Disconnect propeller shaft from transmission, then disconnect mechanical parking brake linkage at right side of transmission, if equipped.
8. Remove six bolts attaching flex plate assembly to converter cover.

NOTE: Through opening in flywheel cover, use a pry bar to manually rotate flywheel to position flex plate to converter attaching bolts.

9. Support transmission using a suitable jack. Position jack so that oil pan is not supporting weight of transmission. Place safety chain around top of transmission, then attach chain to each side of jack.
10. Support engine at rear, then remove transmission case to crossmember attaching bolts. Raise engine slightly to relieve weight from rear mounting.
11. Remove transmission case to flywheel housing attaching bolts, then inspect transmission and surrounding area to ensure no lines, cables or wires will interfere with transmission removal.
12. Position a jack or sling relative to transmission center of gravity, then separate transmission from engine and lower transmission and converter assembly from vehicle.

NOTE: When removing transmission, keep rear of transmission lower than front to prevent converter from falling out of housing. Ensure converter does not separate from transmission during removal. As soon as transmission is clear of mountings, install a retaining strap across front of converter housing to hold converter in place.

Linkage Adjustments

NOTE: The following general procedure is applicable to most vehicles:

1. The shift selector lever should move freely and have a positive detent feel in each position.
2. The shift selector should be adjusted so that the shift tower stops match the transmission detents.
3. With shift linkage correctly adjusted, the shift linkage lever retaining pin should move freely in each range.

CAUTION: Proper installation of external selector lever is important for proper transmission operation. Excessive torque applied to external selector retaining nut may damage internal selector lever.

4. Position selector lever to D or 3rd position and place external lever onto tapered area of selector shaft, then making certain that flats on shaft are adjacent to flats on lever, install retaining nut. Hold lever and torque nut to 15–20 ft. lbs.

NOTE: Do not exceed 20 ft. lbs.

TRANSFER CASE, REPLACE

1. Remove skid plate, if equipped, and drain transfer case.
2. Disconnect speedometer cable and, if equipped with a parking brake assembly, parking brake cable.
3. Disconnect input and output shafts and position shafts aside, properly supported.

NOTE: Some 1978–81 models equipped with a single piece propeller shaft were assembled with the propeller shaft slip yoke positioned toward the rear axle assembly. When servicing these vehicles, the slip yoke should be positioned as assembled in production whether at the rear axle or at the transfer case.

4. Disconnect shift rods and de-clutch rod from transfer case.
5. Support transfer case with a suitable jack.
6. If transfer case is supported by crossmember, or a support bracket, remove crossmember or the support bracket bolts, then lower transfer case from vehicle.
7. If transfer case is secured to an adapter, remove transfer case to adapter mounting bolts. Slide transfer case rearward to disengage front input spline and lower transfer case from vehicle.
8. Reverse procedure to install.

Rear Axle Section

NOTE: For Eaton, Spicer and Timken (Rockwell) driving axles used on Dodge & Plymouth trucks, see the "name" chapters for service procedures and lubricant capacities

DODGE SEMI-FLOATING AXLE ALL EXC. 8 3/8, 9 1/4

In these axles, Figs. 1 & 2, the drive pinion is mounted in two tapered roller bearings, preloaded by a spacer and/or shims.

The differential is supported by two tapered roller side bearings. In the removable type, the differential assembly is positioned for proper ring and pinion back lash and bearing preload by two threaded ring nuts between the bearings and the mounting pedestals. In the integral type, the differential assembly is positioned by shims located between the bearing cone and the hub shoulder of the differential case.

Removal

1. With wheels removed, remove drum retaining clips and brake drum.
2. Through hole in axle shaft flange, remove five nuts from shaft retainer plate. The right hand shaft with adjuster in retainer plate will also have a lock under one of the studs that will be removed at this time.
3. Pull axle shaft assembly with tool.
4. The inner seal should be replaced at this time. Remove seal using Tool C-637, Fig. 3. Install a new seal using Tool C-839, Fig. 4.

Disassembly

1. Install axle shaft in a vise and slide sleeve

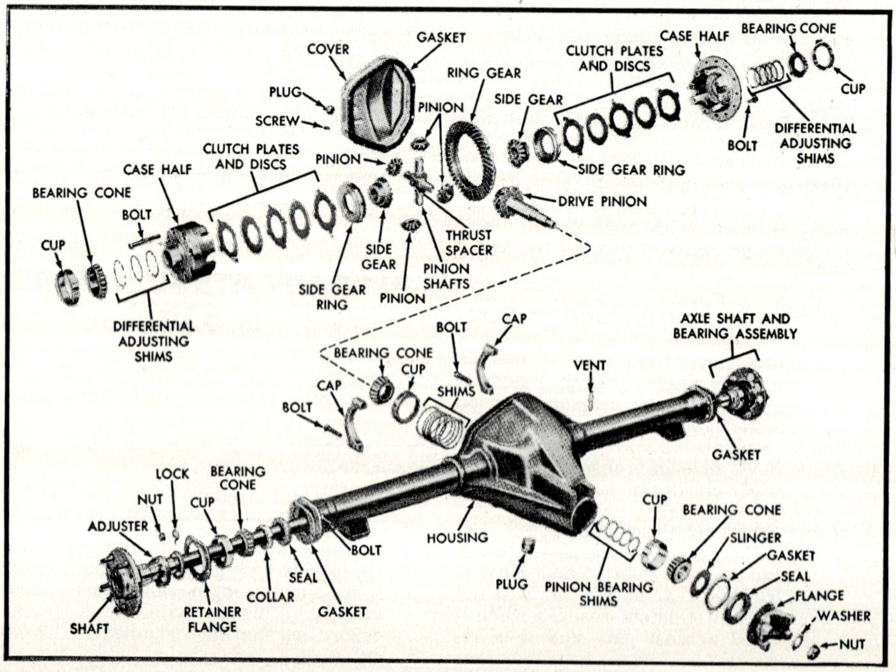

Fig. 2 9¾" rear axle assembly. Integral carrier

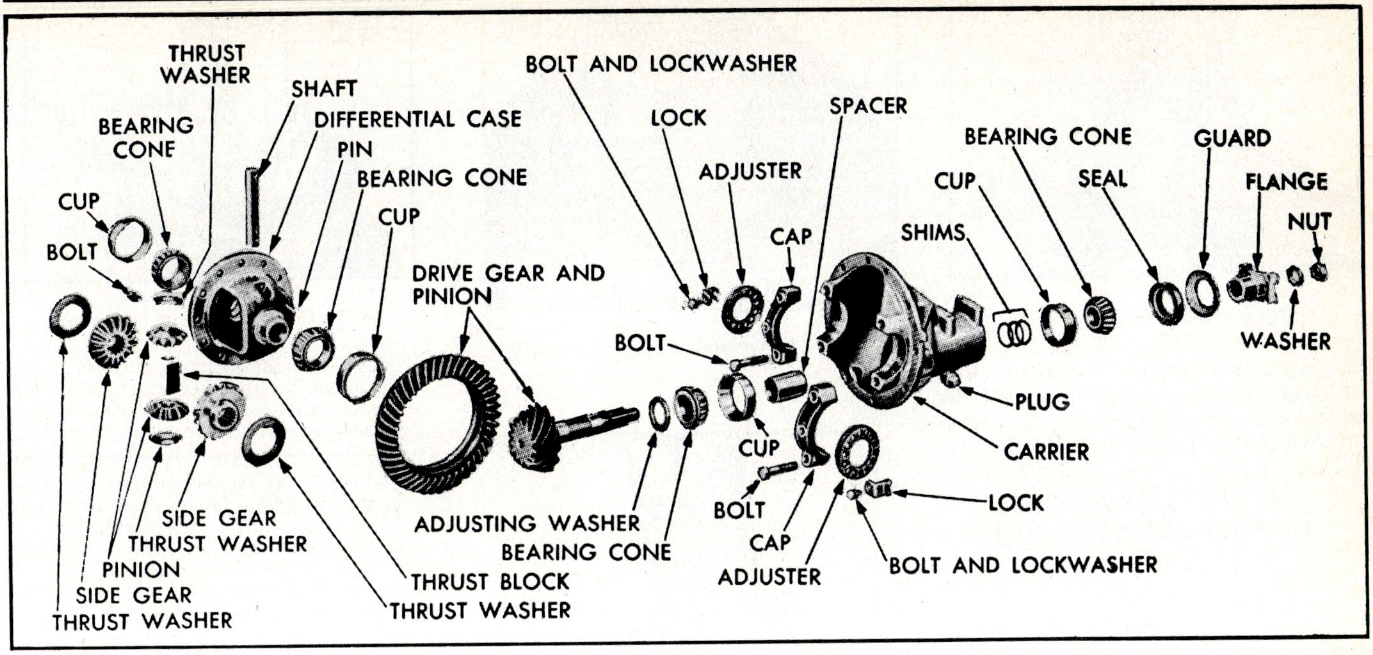

Fig. 1 Differential carrier assembly used with semi-floating axle. Removable type

SP5041 over seal surface of shaft and notch collar, Fig. 5. Slide collar off shaft.

2. Remove roller retainer flange by cutting it off lower edge with a chisel, Fig. 6.
3. Grind a section off flange of inner bearing cone, Fig. 7.
4. Remove bearing rollers from section of flange that was ground away, Fig. 8.
5. Pull roller retainer down as far as possible and cut it with a pair of side cutters and remove from the assembly, Fig. 9.
6. Remove roller bearing cup and protective sleeve from shaft.
7. To avoid scuffing seal journal when bearing cone is being removed, it should be protected by a single wrap of .002" shim stock held in place by a rubber band, Fig. 10.
8. Remove bearing cone with tool set. Tighten bolts of tool alternately until cone is removed, Fig. 11.
9. Remove seal in bearing retainer plate and replace with a new seal.

Assembly

1. Install retainer plate and seal on axle shaft.
2. Install new axle shaft bearing cup, cone and collar on shaft, using tool shown in Fig. 12. Tighten bolts of tool alternately until bearing and collar are seated properly.
3. Inspect axle shaft seal journal for scratches and polish with #600 crocus cloth if necessary.
4. Lubricate wheel bearings.

Installation

1. Install new gasket on left end of housing, add brake backing plate and a second new gasket. Both gaskets are the same on this installation.
2. Slide left axle shaft assembly into position carefully to avoid damaging the new inner seal. Install and tighten retainer plate nuts to 30–35 ft.-lbs.
3. Repeat Step 1 for right side of axle housing.

Axle Shaft End Play

When setting end play both rear wheels must be off the ground, otherwise a false end play setting will occur.

1. Using a dial indicator mounted on a stud or right axle shaft as shown in Fig. 13, turn the adjuster clockwise until there is zero end play. Back off adjuster counterclockwise four notches to establish an axle shaft end play of .013–.023".
2. Tap end of left axle shaft lightly with a plastic mallet to seat right wheel bearing cup against adjuster, and rotate axle shaft several revolutions so that a true end play reading is indicated.
3. Remove one retainer plate nut and install adjuster lock. If tab on lock does not mate with notch in adjuster, turn adjuster slightly until it does. Install nut and torque to 30–35 ft. lbs.
4. Recheck axle shaft end play. If not within prescribed limits, repeat adjustment procedure.
5. Remove dial indicator and install brake drum, drum retaining clips and wheel.

8 3/8 & 9 1/4 INCH INTEGRAL CARRIER TYPE

These axle assemblies, Fig. 14, are of the integral carrier housing hypoid gear type with the centerline of the drive pinion mounted below the centerline of the ring gear. The drive pinion is supported by two preloaded taper roller bearings and the front and rear pinion bearing cones are pressed on the pinion stem. The front and rear pinion bearing cups are pressed against a shoulder that is

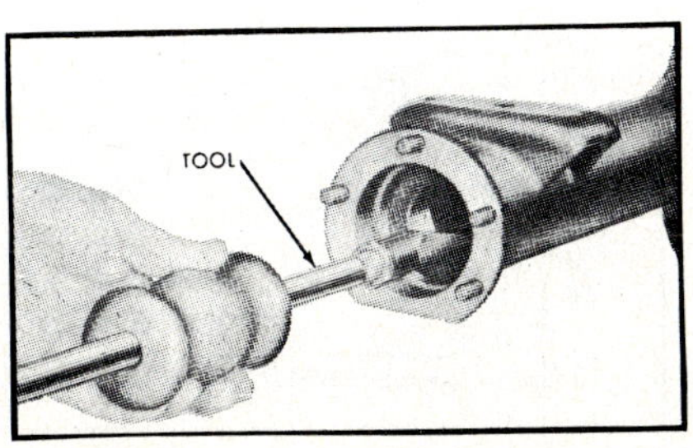

Fig. 3 Removing inner oil seal

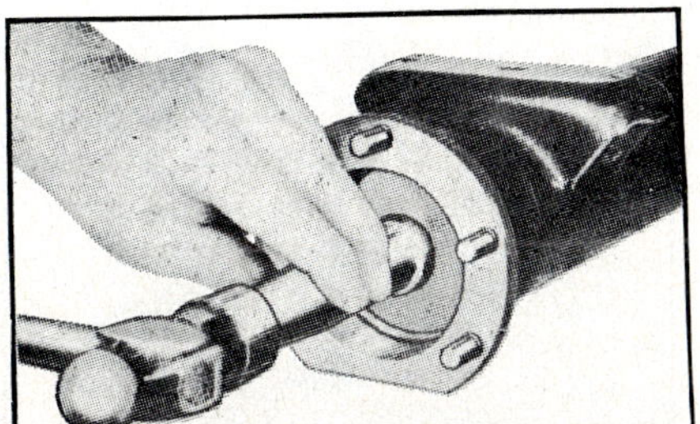

Fig. 4 Installing inner oil seal

Fig. 5 Notching bearing retainer collar

Fig. 6 Removing roller retainer

recessed within the carrier casting. Drive pinion depth of mesh adjustment is controlled by installing metal shims between the rear pinion bearing cup and carrier casting.

AXLE SHAFTS & BEARINGS REPLACE

1. Raise and support vehicle and remove brake drum.
2. Clean area around housing cover, then loosen housing cover and allow lubricant to drain and remove cover.
3. Turn differential case until pinion shaft lock screw is accessible and remove lock screw and pinion shaft, Fig. 15.
4. Push axle shaft inward and remove C-washer locks from axle shaft, Fig. 16, then pull axle shaft from housing being careful not to damage axle shaft bearing.

NOTE: Inspect axle shaft bearing surfaces for signs of spalling or pitting. If any of these conditions exist, both shaft and bearing should be replaced. Normal bearing contact on shaft should be a dull gray and may appear lightly dented.

5. Using tools C-4167 and C-637, remove axle shaft bearing and seal from axle housing, Fig. 17. If bearing shows no sign of excessive wear or damage, it can be reused along with a new seal. Never reuse an axle shaft seal.

NOTE: Remove any burrs that may be present in housing bearing shoulder, as bearing could become cocked during installation.

6. Using tools C-4198 and C-4171, Fig. 18, install bearing making sure it does not become cocked. Drive bearing until it bottoms against shoulder.

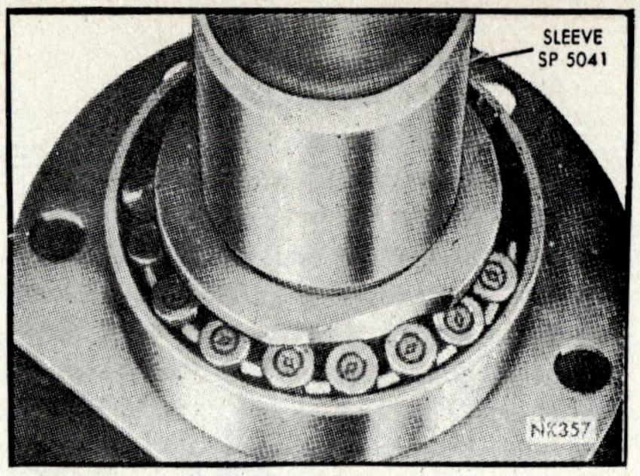

Fig. 7 Flange ground off inner bearing

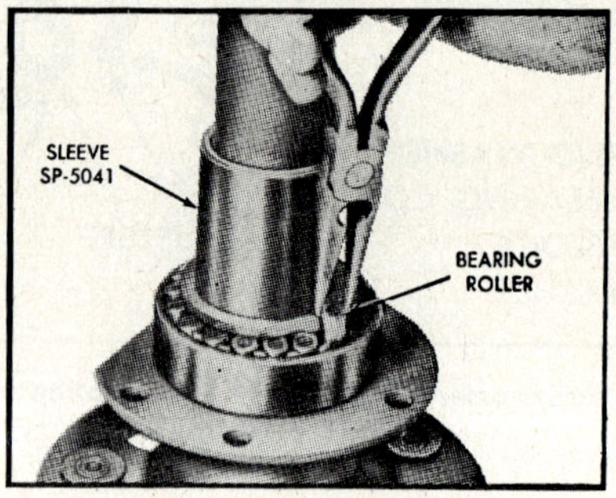

Fig. 8 Removing bearing rollers

NOTE: Do not use seal to position or bottom bearing as this will damage seal.

7. Using tool C-4130, install axle shaft bearing seal until outer flange of tool bottoms against housing flange face. This will position seal to the proper depth.
8. Reverse disassembly procedure to reassemble axle.

DIFFERENTIAL CASE SERVICE

Removal & Disassembly

1. Remove axle shafts as described previously.

NOTE: On 1979–83 9¼ inch units, prior to removing axle shafts, measure axle shaft end play by inserting a feeler gauge between end of each axle shaft and the differential pinion shaft. If end play measures less than .005 inch, measure side gear clearance by inserting two feeler

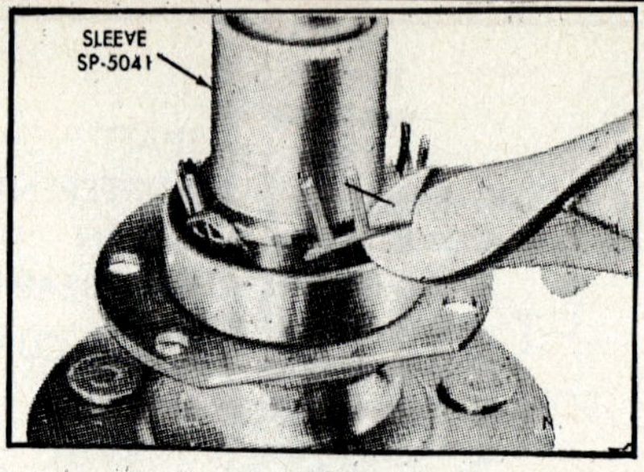

Fig. 9 Cutting out bearing retainer

gauges of equal thickness at the side gear thrust surface, one above and one below the side gear hub. If clearance measures more than .012 inch, replace side gears.

2. Scribe marks on propeller shaft universal joint and pinion flange, then disconnect propeller shaft from differential yoke and secure in an upright position to avoid damage to front universal.
3. Measure and record side play and runout for use during reassembly as follows:
   a. Measure differential side play by positioning a screwdriver or pinch bar between left side of axle housing and differential case, Fig. 19. There should be no side play when differential case is pryed sideways.
   b. Measure drive gear runout by mounting a dial indicator at right angles to back face of drive gear, then rotate drive gear several complete revolutions and note reading. Mark drive gear and differential case at point of maximum runout, this marking will facilitate in determining differential case runout. If total indicated runout is more than .005 inch, the differential case may be damaged. Test for case runout as described further on.
   c. Using an inch pound torque wrench, measure and record pinion bearing preload.
4. Remove drive pinion nut, then using appropriate holding tool, remove pinion flange.
5. Using a screwdriver and hammer, drive pinion oil seal from carrier.
6. Mark axle housing and differential bearing caps for proper relocation during reassembly, Fig. 20, then remove adjuster locks from bearing cap and loosen but do not remove bearing caps.
7. Using tool C-4164 inserted through axle shaft tube, loosen hex adjuster, Fig. 21.
8. Hold differential assembly in place, then using extreme caution, remove bearing caps, adjusters and differential assembly.

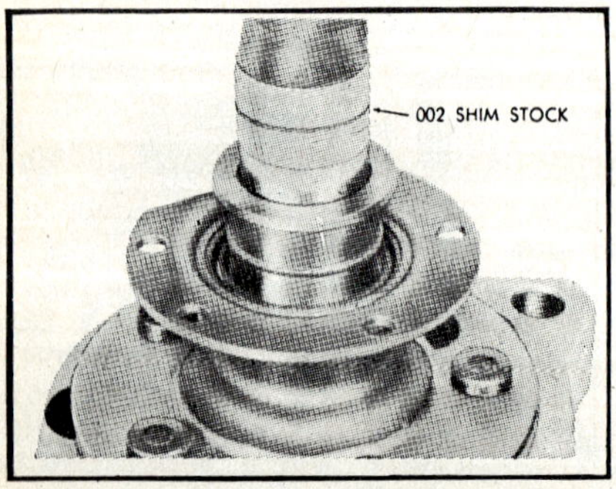

Fig. 10 Seal journal protection

Fig. 11 Removing bearing cone

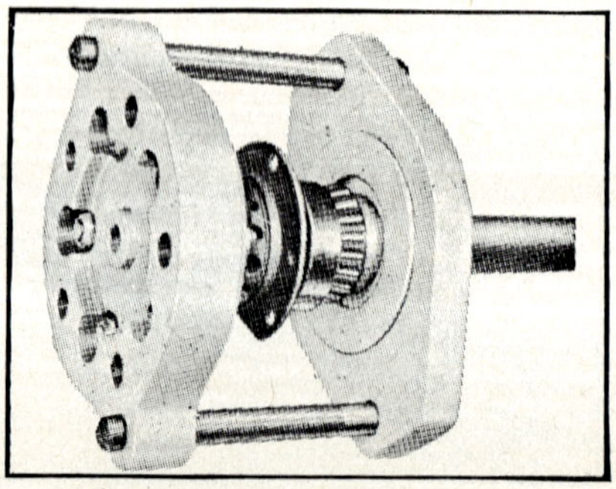

Fig. 12 Installing bearing and collar

Fig. 13 Measuring axle shaft end play

NOTE: Differential bearing cups and threaded adjusters must be kept with their respective bearing cones.

9. If necessary to remove drive pinion or front pinion bearing cone, drive pinion stem rearward and out of bearing.

NOTE: This procedure will result in damage to bearing rollers and cups, and both bearing cone and cup must be replaced with new parts. Discard collapsible spacer.

10. Using a brass drift and hammer, remove front and rear bearing cups from housing, then remove shim from behind rear bearing cup and record thickness. Discard damaged shim.
11. Using tool C-293-P and adapters C-293-42 for 8⅜ inch units, or C-293-37 for 9¼ inch units, remove rear bearing cone from drive pinion.

NOTE: Make certain that adapters are located so as not to pull on bearing cage.

12. Mount differential case in a soft jawed vise, then remove and discard drive gear bolts. Note that bolts are left hand threaded. Using a plastic hammer or brass drift, tap drive gear loose and remove from differential case pilot.
13. If drive gear runout exceeded .005 inch in previous check, differential case flange runout should be checked. Install differential case and respective bearing cups and adjusters in housing.
14. Install bearing caps and lightly tighten cap bolts, then using tool C-4164, Fig. 21, tighten both hex adjusters until all side play in bearings has been eliminated.
15. Mount a dial indicator on housing so that pointer on indicator squarely contacts drive gear surface of differential case flange between outer edge of flange and drive gear bolts.
16. Rotate differential case several revolutions and note total indicated runout. If reading is in excess of .003 inch, differential case should be replaced.

NOTE: If runout does not exceed .003 inch, sometimes it is possible to reduce runout by positioning drive gear 180° from point of maximum runout when reassembling drive gear on differential case.

17. Remove pedestal cap bolts and remove differential case assembly from carrier.
18. Rotate differential side gears until differential pinions appear at differential case window and remove pinions, differential side gears and thrust washers.
19. Using tool C-293-P and adapter plates C-293-44 for 8⅜ inch units or C-293-47 for the 9¼ inch units and plug SP-3289 for 8⅜ inch units, or C-293-3 for the 9¼ inch units, remove differential bearing cones.

Assembly & Installation

1. Lubricate all components with rear axle lubricant.

Fig. 16 Removing "C" lock washers

2. Install thrust washers on differential side gears, then position gears in counterbores of differential case.

NOTE: On 1979–83 9¼ inch units, if axle shaft end play measured during disassembly was more than .005 inch, replace thrust washers with the thickest set of thrust washers that will reduce the axle shaft end play to less than .005 inch.

3. Position thrust washers on both differential pinion gears, then mesh the pinion gears making sure pinion gears are exactly 180° apart.
4. Rotate side gears until pinion gears and washers are aligned with differential pinion shaft holes in case.

NOTE: The contact surface of the drive gear and differential case and flange must be clean with all burrs removed.

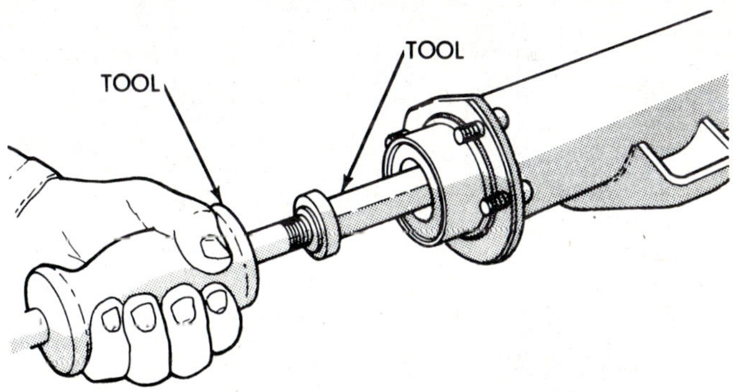

Fig. 17 Removing axle shaft bearing and seal

Fig. 15 Removing differential pinion shaft lock pin

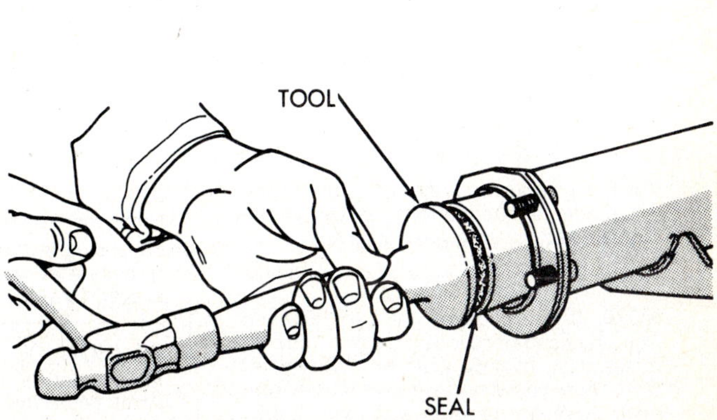

Fig. 18 Installing axle shaft oil seal

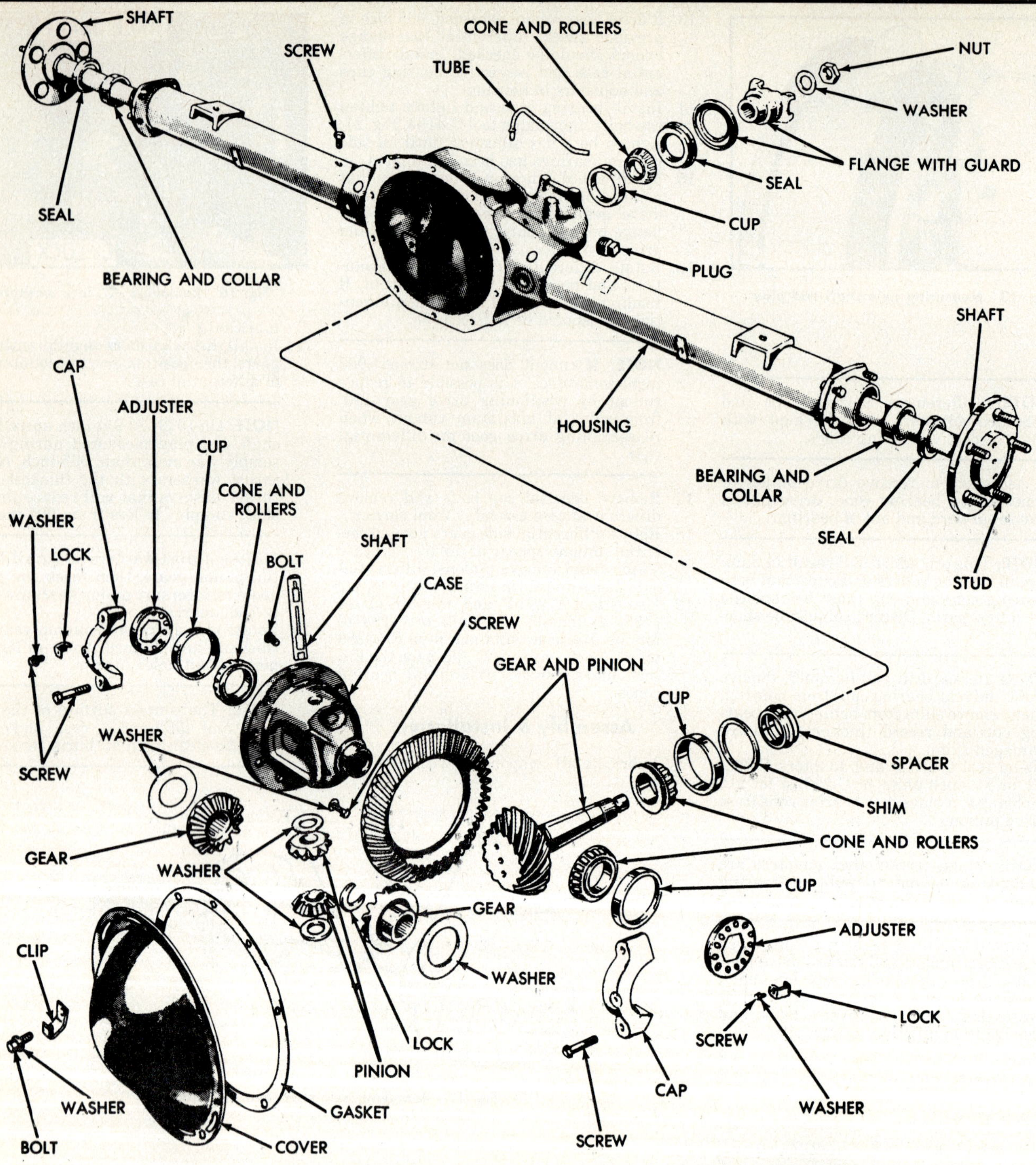

Fig. 14 8³/₈" & 9¹/₄" integral carrier rear axle assembly. The 8³/₈ inch unit is equipped with a vent tube mounted on the carrier housing, while the 9¹/₄ inch unit has a vent located on the left leg of the housing

5. With an Arkansas stone, relieve the sharp edge of the chamfer on the inside diameter of the ring gear, Fig. 22.

NOTE: If sharp edges are not removed from inside diameter of ring gear, during installation of ring gear on differential case, these edges will remove metal from pilot diameter of case and will imbed between differential case flange and gear. This will cause improper seating of ring gear.

---

6. Heat ring with a heat lamp or immersing in hot water or oil, with temperatures not exceeding 300° F.

CAUTION: Do not use a torch. It is advisable to use pilot studs equally spaced in three positions to align gear with case.

---

7. Install unit in a soft jawed vise and alternately torque new drive gear screws (left hand threads) to 70 ft. lbs.
8. Position each differential bearing cone on hub of differential case with taper away from drive gear and using tools C-4107 (for 8³/₈ inch units) or C-4213 and C-4171 (for 9¹/₄ inch units) install bearing cones. An arbor press may be used in conjunction with installing tool.

CAUTION: Never exert pressure against

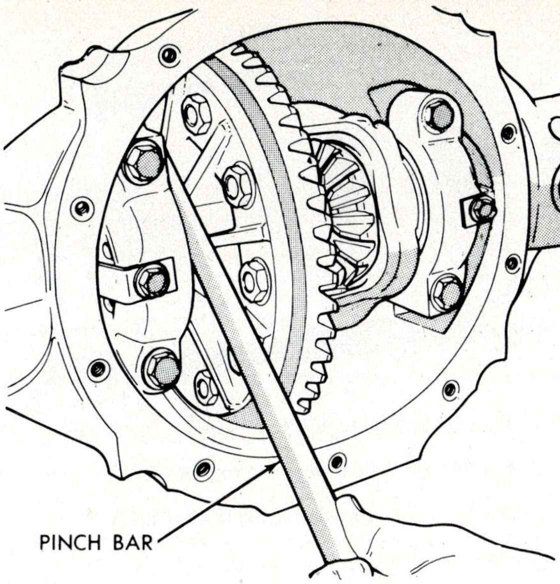

Fig. 19 Checking differential side play

Fig. 20 Marking bearing caps and housing for proper identification

bearing cage, since bearing damage will result.

---

9. Install drive pinion bearing cups and using setting gauge tools and adapter sets, Figs. 23 and 24, determine pinion depth of mesh.
10. Start both drive pinion bearing cups into axle housing bores making certain they are not cocked.
11. Assemble spacer SP-5408 for 8³/₈ inch units or SP-6017 for 9¹/₄ inch units over tool SP-5385 for 8³/₈ inch units or SP-526 for 9¹/₄ inch units followed by the rear pinion bearing cone and insert through rear side of axle carrier.
12. Install front pinion bearing cone over spacer SP-5382 for 8³/₈ inch units, or SP-6022 for the 9¹/₄ inch units and position onto stem of tool followed by compression sleeve SP-3194-B for 8³/₈ inch units, or SP-535-A for the 9¹/₄ inch units, centralizing washer SP-534 and compression nut SP-3193 for 8³/₈ inch units or SP-533 for the 9¹/₄ inch units.
13. While holding compression sleeve from turning with tool C-3281, tighten nut, thereby drawing pinion bearing cups into axle housing bearing cup bores. Permit tool to turn several revolutions while tightening to prevent hardening of bearing cups or cones.

NOTE: The position of the drive gear pinion with respect to the drive gear (depth of mesh) is determined by the location of the bearing cup shoulders in the carrier and by the portion of pinion behind the rear bearing.

---

14. With main tool remaining in axle housing, loosen nut SP-3193 for 8³/₈ inch units, or SP-533 for the 9¹/₄ inch units. Lubricate front and rear pinion bearings with rear axle lubricant and retighten nut to obtain 10–30 inch lbs. of preload for 8³/₈ inch units, or 15–25 inch lbs. for the 9¹/₄ inch units. Rotate pinion several complete revolutions to align and lubricate bearing rollers.
15. Install gauge block SP-5383 for 8³/₈ inch units, or SP-6020 for the 9¹/₄ inch units on the end of main tool and tighten set screw.
16. Position cross bore arbor SP-5380 for 8³/₈ inch units, or SP-6018 for 9¹/₄ inch units into axle housing differential bearing seats. Center arbor to maintain an approximate equal distance at both ends. Position bearing caps and bolts on carrier pedestals and insert a piece of .002 inch shim stock between arbor and each cap. Torque cap bolts to 10 ft. lbs.
17. Using a feeler gauge, select a rear pinion bearing mounting shim which will fit between cross bore and arbor and gauge. This fit must be snug but not too tight. Shims are available in increments of .001 inch from .061–.078 inch for 8³/₈ inch units, or .020–.038 inch for 9¹/₄ inch units. To select a shim pack for installation, read the markings on the end of the pinion head (−0, −1, −2, +1, +2 etc). When marking is minus, add that amount to the thickness of shim pack selected. When marking is plus, subtract that amount. Apply other pinion markings in a similar manner.
18. Remove arbor and tool from axle housing.
19. On 8³/₈ inch units, remove rear pinion bearing cup from casting.
20. Position shim selected previously on pinion followed by rear pinion bearing

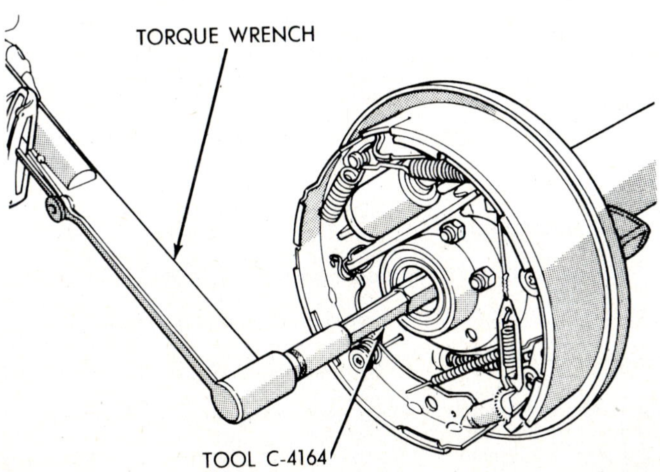

Fig. 21 Loosening or tightening hex adjuster

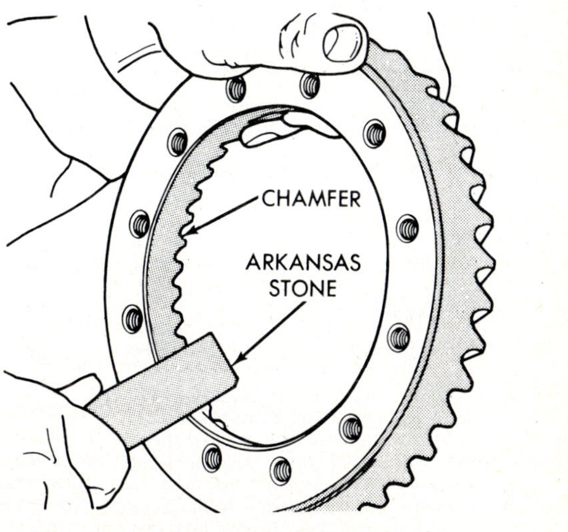

Fig. 22 Relieving sharp edge on chamfer ring gear

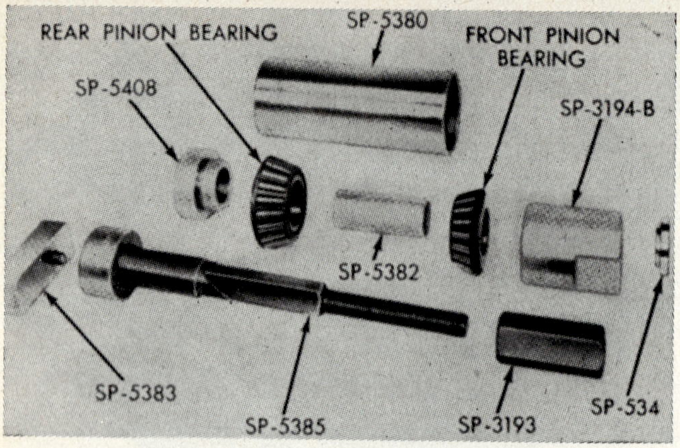

Fig. 23 8⅜ inch axle pinion setting tools

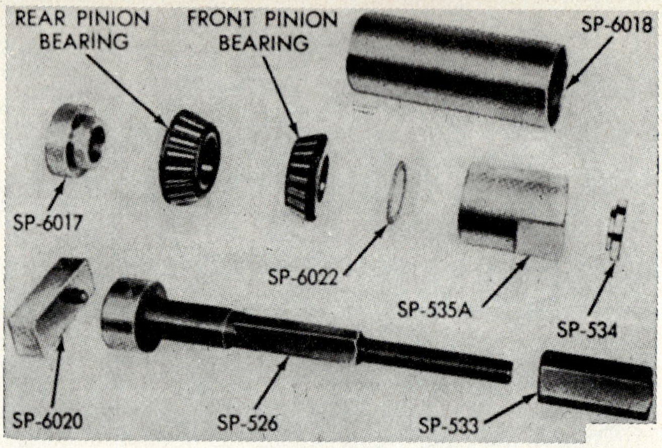

Fig. 24 9¼ inch axle pinion setting tools

cone, making certain that pinion head contact surface, bearing cone and shim are clean. Using tool DD-955 on all 8⅜ inch units and 1974–78 9¼ inch units, or C-3095A on 1979–83 9¼ inch units, press bearing onto pinion stem.

21. Lubricate front and rear pinion bearing cones with rear axle lubricant and insert drive pinion and bearing assembly through axle housing. Install new collapsible spacer over pinion stem and position front pinion bearing cone over pinion stem.
22. Install companion flange using tool C-3718 for 8⅜ inch units, DD-999 for 1974–78 9¼ inch units, or C-496 on 1979–83 9¼ inch units and holding tool C-3281. This is necessary in order to properly install front pinion bearing cone on pinion due to interference fit.

CAUTION: During installation of front pinion bearing cone, be careful not to collapse spacer.

23. Remove tool and flange from pinion stem.
24. Using tool C-4076 for 8⅜ inch units, or either C-3980 or C-4109 for the 9¼ inch units install drive pinion oil seal. Seal is properly seated when seal flange contacts housing flange face.

NOTE: Outside diameter of seal is coated with a special sealer so no sealing compound is required.

25. While supporting pinion in carrier, reinstall companion flange using tools previously described.
26. Remove tools and install Belleville washer with convex side of washer out, then the pinion nut.
27. Using holding tool on universal flange, tighten pinion nut to remove end play in pinion, while rotating pinion to insure proper bearing roller seating.
28. Remove holding tool and rotate pinion several revolutions in both directions to permit bearing rollers to seat.
29. Torque pinion nut to 210 ft. lbs. and while rotating pinion, measure pinion bearing preload. The correct bearing preload specifications are 20–35 inch lbs. for new bearings with the pinion nut torqued to 210 ft. lbs. minimum. If when rebuilding axle and the original rear pinion bearing and a new front pinion bearing are used, the correct bearing preload is 10 inch lbs. more than the reading at the time of disassembly with a minimum of 210 ft. lbs. of torque on the pinion nut. If correct preload cannot be obtained at 210 ft. lbs., continue tightening nut in small increments and checking until proper preload is obtained. Bearing preload should be uniform during complete revolution. A preload that varies during rotation indicates a binding condition which must be corrected. The assembly is unacceptable if final pinion nut torque is below 210 ft. lbs. or pinion bearing preload is not within specifications.

CAUTION: The pinion nut should not be backed off to decrease pinion bearings preload. If the desired preload is exceeded for any reason, a new collapsible spacer must be installed and the nut retightened until proper preload is obtained.

30. Coat differential bearing cones, cups and adjusters with axle lubricant and carefully position the entire assembly into housing.
31. Install differential bearing cups on their respective sides, using the marks made during disassembly.
32. Install bearing cap and torque top bolts to 10 ft. lbs. and tighten bottom bolts finger tight until head is just seated on bearing cap.

Differential Bearing Preload & Drive Gear Backlash

The following precautions must be observed when adjusting differential bearing preload and drive gear backlash.

a. Permissible variation is .003 inch. This variation is actually allowable runout. Therefore it is important to index the gears engaging same teeth during all backlash measurements.
b. Also maintain specified adjuster torque to obtain accurate differential bearing preload and drive gear backlash measurements.

NOTE: Excessive torque will produce high bearing load and cause premature bearing failures. Inadequate torque will not support the drive gear load properly, will lead to free play of the differential assembly and excessive drive gear noise.

CAUTION: Differential bearing cups will not always move directly with the adjusters, therefore to insure accurate adjustment changes and to maintain gear mesh index, bearings must be seated by vigorously moving the drive pinion a half turn in each direction five to ten times each time adjusters are moved.

1. Using tool C-4164 turn each hex adjuster, Fig. 23 until bearing free play is eliminated with some drive gear backlash (approximately .010 inch) between drive gear and pinion. Now seat bearings.
2. Install dial indicator and position contact point against drive side of gear tooth. To find point of minimum backlash, check backlash at four positions at approximately 90 degree intervals around drive gear. Rotate gear to position of least backlash. Mark the index so that all backlash readings will be taken with the same teeth in mesh.
3. Loosen right adjuster and tighten left adjuster until the backlash is .003–.004 inch with each adjuster torqued to 10 ft. lbs. Again seat bearings. Torque differential bearing cap bolts to 100 ft. lbs. Using tool C-4164 torque right hex adjuster to 70 ft. lbs. for 8⅜ inch units and 75 ft. lbs. for the 9¼ inch units. Again seat bearings. Continue to tighten right adjuster and seat rollers until torque remains constant at 70 ft. lbs. for 8⅜ inch units and 75 ft. lbs. for the 9¼ inch units. Measure backlash. If backlash does not measure between .006–.008 inch, increase torque on right adjuster and seat bearings until correct backlash is obtained. Torque left adjuster to 70 ft. lbs. for 8⅜ inch units and 75 ft. lbs. for the 9¼ inch units. Seat bearing until torque remains constant.

NOTE: If all previous steps have been properly performed, initial reading on left adjuster will be about 70 ft. lbs. for 8⅜ inch models and 75 ft. lbs. for the 9¼ inch units. If it is considerably less, complete procedure must be repeated.

4. When adjustments are completed, install adjuster locks. Be sure lock finger is engaged in adjuster hole on 8⅜ inch units, or that lock teeth are engaged in adjuster threads on 9¼ inch units. Torque lock screws to 90 inch lbs.

Front Wheel Drive Section

NOTE: For overhaul procedure on the transfer case used with front-wheel drive trucks, see the Transfer Case chapter.

MODEL 44F

The front axle drive shaft outer end connects through a splined slip joint drive flange to the wheel hub. The inner and outer axle shafts are connected through universal joints which revolve in the steering knuckles. Some service parts replacement can be made without complete removal of the axle assembly. However, it is generally desirable to remove the assembly for complete overhaul.

AXLE ASSEMBLY

Remove and Replace

1. Secure brake pedal in the "up" position, then raise vehicle.
2. Disconnect front drive shaft at drive pinion yoke.
3. Disconnect drag link at steering knuckle arm (left side only).
4. Disconnect and plug brake line at frame crossmember.
5. Disconnect shock absorbers at lower mounts, then disconnect sway bar assembly from spring plates, if equipped.
6. Remove nuts and washers from "U" bolt spring clips, then remove axle from vehicle.
7. Reverse removal procedure to install axle assembly. Lubricate all fittings, free brake pedal and bleed brakes.
8. Lower vehicle and check front wheel alignment.

Front Drive Axle & Joint Repairs

1. Remove wheel assembly.
2. Remove grease cap from drive flange.
3. Remove bolts and snap ring; then, using a suitable puller, remove drive flange.
4. Remove outer lock nut.
5. Remove nut lock and front wheel bearing adjusting nuts.
6. Remove hub and brake drum.
7. Unfasten brake support and disconnect brake line. If axle is on vehicle, attach brake support with wire to front spring, thus making it unnecessary to disconnect brake lines.
8. Tap end of spindle with soft-head hammer to remove spindle.
9. Inspect spindle for wear and damage.
10. Inspect spindle axle flanged bushing for wear, Fig. 1. If a new bushing is to be installed it requires no reaming. It is so designed that when it is pressed into the spindle, it will be compressed to provide proper running clearance.
11. Remove axle shaft assembly.
12. Disassemble front axle as suggested by Fig. 1.
13. Reverse the removal procedure to install the front axle shaft and joint assembly.

Steering Knuckle Repairs

1. Remove wheel, hub, drum, brake support, spindle and axle shaft, Fig. 2.
2. Disconnect tie rod ends from steering knuckles.
3. Disconnect drag link from steering knuckle arm.
4. Remove steering knuckle felt and oil seal retainers, Fig. 3.
5. Remove lower bearing cap.
6. Unscrew nuts from top of steering knuckle arm. Remove arm, pivot and shims.
7. Remove knuckle housing, felt and oil seal.
8. Remove upper and lower bearing cones.

Assembling Knuckle

1. Place felt and oil seal over end of axle housing.
2. Place original shims on knuckle upper studs, Fig. 3.
3. Install steering knuckle arm and tighten retaining nuts to 35–55 ft. lbs torque.
4. Coat bronze cone with short fiber grease. Insert cone in steering knuckle and install it on knuckle arm pivot. Make sure cone is properly positioned on key.
5. Slide knuckle over yoke, entering upper cone in its cup.
6. Pack lower bearing cone with short fiber grease; tilt bottom of knuckle outward and insert lower cone.
7. Position knuckle over lower cone and install lower bearing cap. Tighten capscrews to 30–50 ft. lbs torque.
8. Check preload of knuckle bearings by hooking a spring scale in the tie rod socket hole, and pulling on scale at right angles to the axle. Preload should be 9 to 25 lbs. when in motion without seal attached to Spicer 44F axle. Adjust preload by adding or removing shims under knuckle arm.
9. Attach oiled felt and oil seal to knuckle housing with retainers and screws.
10. Connect tie rod ends and drag link.
11. Install axle shaft, spindle and brake support. Torque to 40 ft. lbs.
12. Install hub, drum and wheel and adjust front wheel bearings as outlined below.
13. Fill steering knuckle joint housing with S.A.E. 140 oil to level of filler plug opening.
14. Check toe-in and adjust if required at the tie rod end. Clamp should be at the bottom for Models W100 and W200.
15. Check turning radius as excessive angle may cause damage to joint assembly. Adjust to 29 degrees turning angle by breaking weld at stop screw lock nut. Secure nut by spot weld after making adjustment.

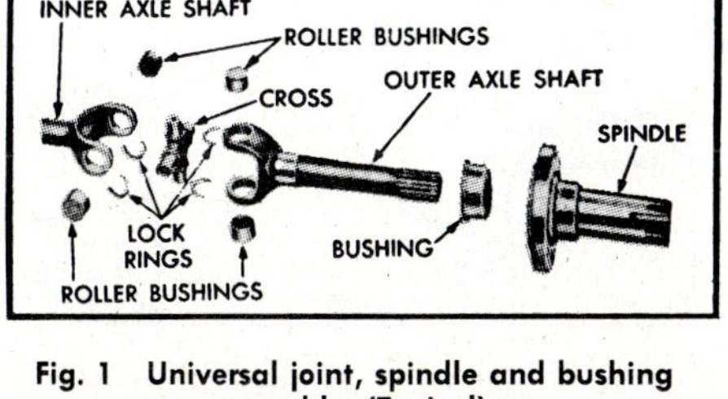

Fig. 1 Universal joint, spindle and bushing assembly. (Typical)

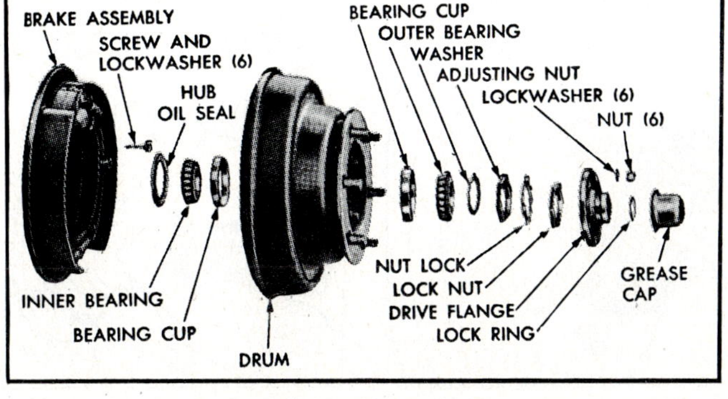

Fig. 2 Brake hub and drum assembly Models W100, W200, W300

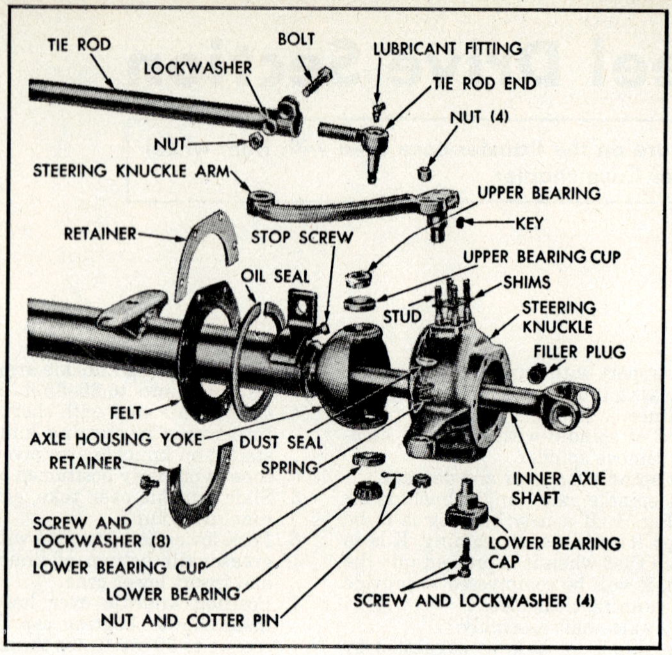

Fig. 3 Spicer 44F axle and steering knuckle assembly. Models W100, W200, Ramcharger & Trailduster. (Typical)

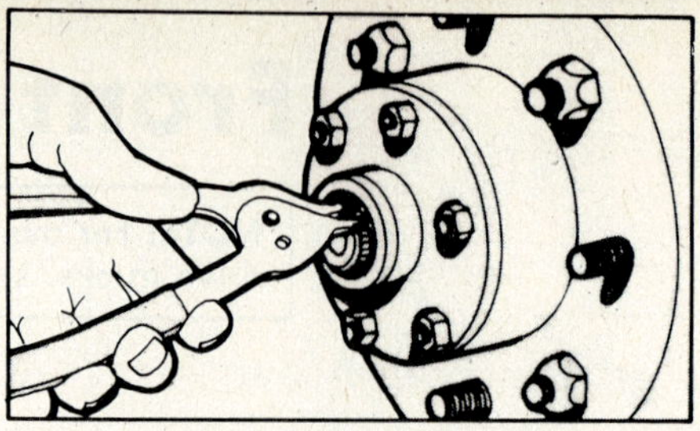

Fig. 4 Removing snap ring. 44 FBJ Units w/drum brakes

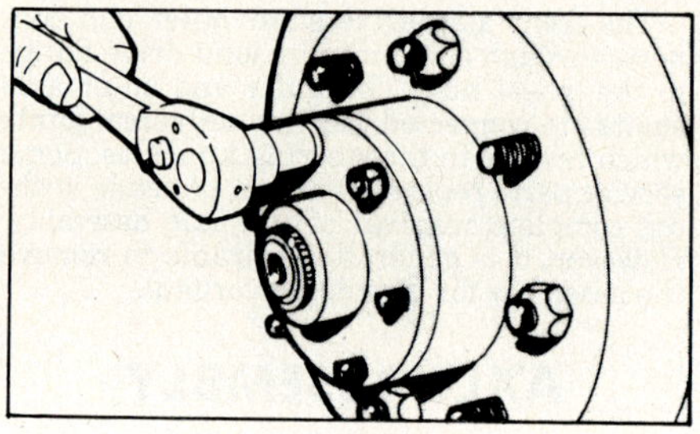

Fig. 5 Removing flange nuts. 44 FBJ Units w/drum brakes

Wheel Bearings, Adjust

Tighten bearing adjusting nut while rotating wheel slowly in both directions. Continue tightening until some binding is evident then back off 1/8 turn. Engage lock, then tighten lock nut securely and bend down tabs on lockwasher.

44FBJ, UNITS WITH DRUM BRAKES

Upper or Lower Ball Sockets

Removal

1. Raise and support truck.
2. Remove wheel.
3. Remove hub cap and snap ring, Fig. 4.
4. Remove flange nuts and lock washers, Fig. 5. Remove flange and discard gasket.
5. Using suitable tool, unscrew outer lock nut and remove lock ring, inner lock nut, bearing and brake drum, Fig. 6.
6. Remove nuts from brake support plate then slide brake assembly off spindle and secure out of the way. Do not allow assembly to hang by hydraulic brake hose.
7. Remove spindle, Fig. 7, and bronze spacer from axle shaft.
8. On left side only, remove tie rod from steering knuckle and disconnect steering link from knuckle arm. Unscrew upper steering knuckle arm nuts then remove coned washers and arm, Fig. 8.
9. Remove cotter pin and nut from upper ball socket. Unscrew and discard nut from lower ball socket. Do not reuse lower nut.
10. Remove lower ball socket snap ring, Fig. 9.
11. After securing adapter to knuckle face, Fig. 10, position puller (Tool P-833) and step plate (Tool C-1144) under lower ball socket, Fig. 11, and remove knuckle from yoke.
12. Using suitable tools press ball sockets from yoke and knuckle and discard sockets.
13. Unscrew and discard sleeve from yoke.
14. Clean components thoroughly and inspect. Replace as necessary parts before assembly.

Installation

1. Install upper and lower ball sockets using suitable tools and ensure they are not cocked. Check clearance between sockets and knuckle with .0015 inch feeler gauge. Feeler gauge must not enter at minimum area of contact. Install snap ring on lower ball socket.
2. Using a suitable tool, screw new sleeve into yoke leaving two threads exposed, Fig. 12.
3. Position boots on ball sockets, then assemble knuckle to yoke and finger tighten new nut on lower ball socket.

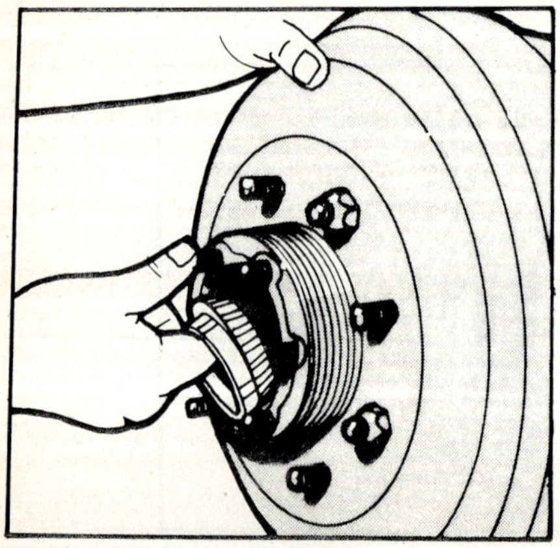

Fig. 6 Removing bearing cone. 44 FBJ Units w/drum brakes

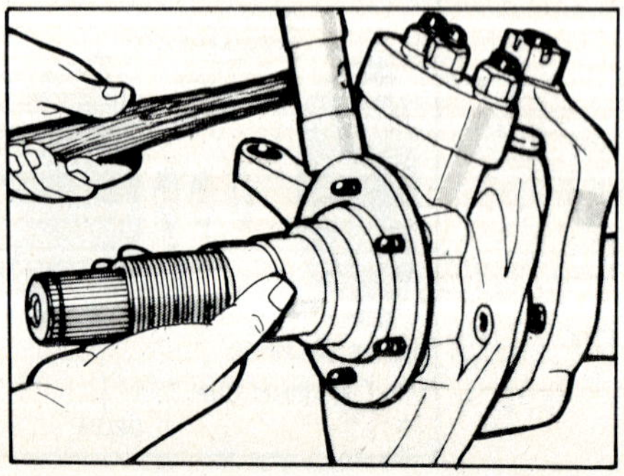

Fig. 7 Removing spindle from knuckle. 44 FBJ Units w/drum brakes

Fig. 8 Removing steering knuckle arm from knuckle. 44 FBJ Units w/drum brakes

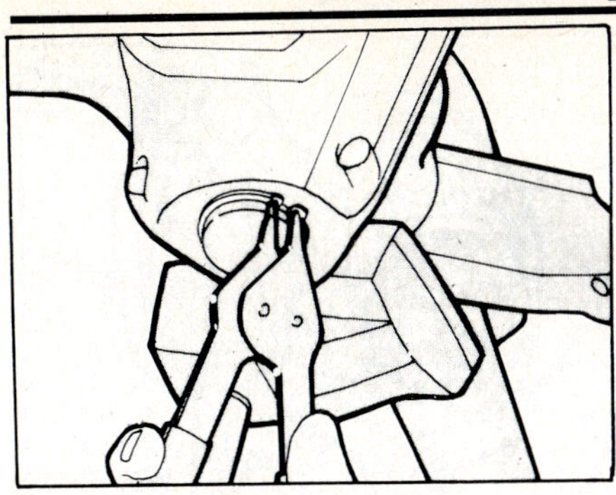

Fig. 9 Lower ball socket snap ring removal. 44 FBJ Units w/drum brakes

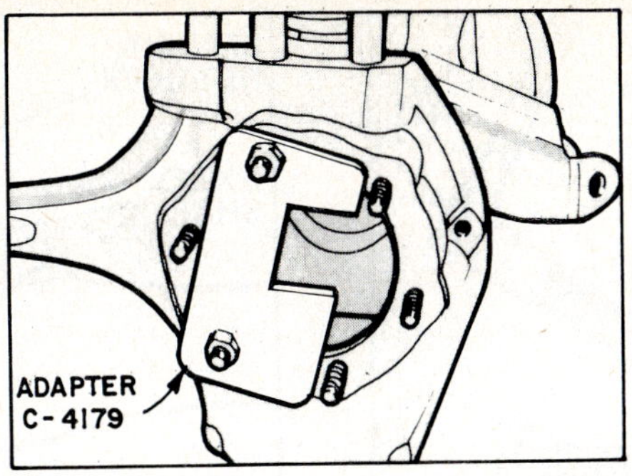

Fig. 10 Adapter tool secured to knuckle. 44 FBJ Units w/drum brakes

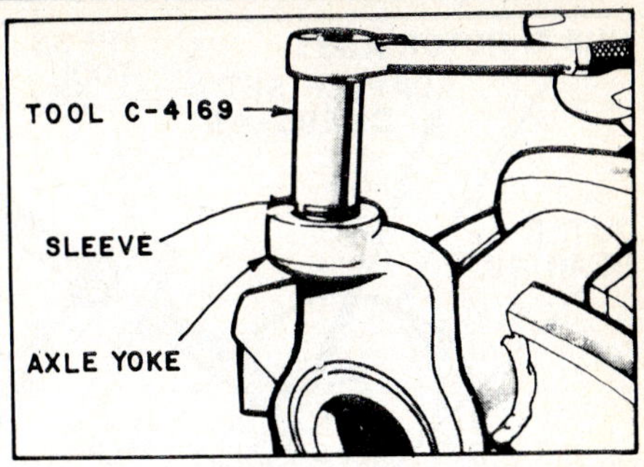

Fig. 12 Yoke sleeve installed. 44 FBJ Units w/drum brakes

4. Seat ball sockets by positioning and tightening proper tools, Fig. 13. Before loosening tools, torque lower socket nut to 80 ft. lbs.
5. After removing tools, tighten adjusting sleeve to 50 ft. lbs. and install upper socket nut. Torque nut to 100 ft. lbs. and install cotter pin, if slot in nut and hole in stud are aligned. If not, tighten nut and install pin. Do not loosen.
6. Secure tie rod end to knuckle and torque nut to 45 ft. lbs. On left side only, secure steering knuckle arm to knuckle and torque nuts to 90 ft. lbs. Then assemble steering link to knuckle arm and torque nut to 45 ft. lbs.
7. Carefully slide axle shaft into housing so as not to damage differential seal at side gears. Position bronze spacer on shaft with chamfer inboard and position spindle against knuckle, Fig. 14.
8. Secure brake support plate to knuckle then position brake drum and bearing on spindle and adjust wheel bearings as outlined below.
9. Install new gasket and drive flange. Torque flange nuts to 40 ft. lbs., then install snap ring and hub cap.
10. Install wheel then lower truck and test operation.

Wheel Bearings, Adjust

Tighten bearing adjusting nut while rotating wheel slowly in both directions. Torque nut to 50 ft. lbs. then loosen 1/4 to 1/3 turn and install locknut. Torque outer nut to 50 ft. lbs. Final bearing adjustment should be from .001 inch to .010 inch end play.

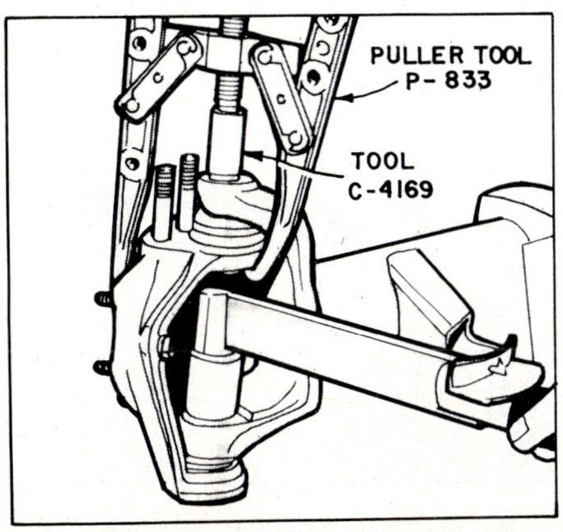

Fig. 13 Tightening lower ball socket nut. 44 FBJ Units w/drum brakes

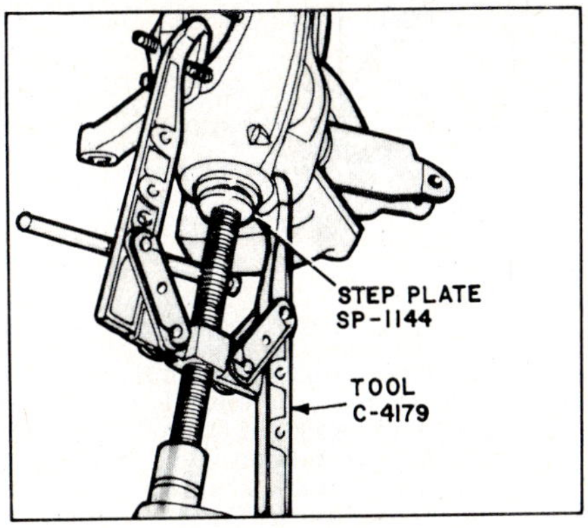

Fig. 11 Using puller to separate knuckle from yoke. 44 FBJ Units w/drum brakes

44FBJ, UNITS WITH DISC BRAKES

NOTE: On units with locking hubs refer to the chapter on locking hubs for disassembly and inspection.

Axle Assembly

Remove & Replace

1. Secure brake pedal in the up position then raise vehicle.
2. Disconnect front drive shaft at drive pinion yoke.
3. Disconnect drag link at steering knuckle arm (left side only).
4. Disconnect and plug brake line at frame crossmember.
5. Disconnect shock absorbers at lower mounts then disconnect sway bar link assembly, if equipped, from spring plates.
6. Remove nuts and washers from "U" bolt spring clips, then remove axle from vehicle.
7. Reverse procedure to install axle assembly. Lubricate all fittings, free brake pedal and bleed brakes.
8. Lower vehicle and check front wheel alignment.

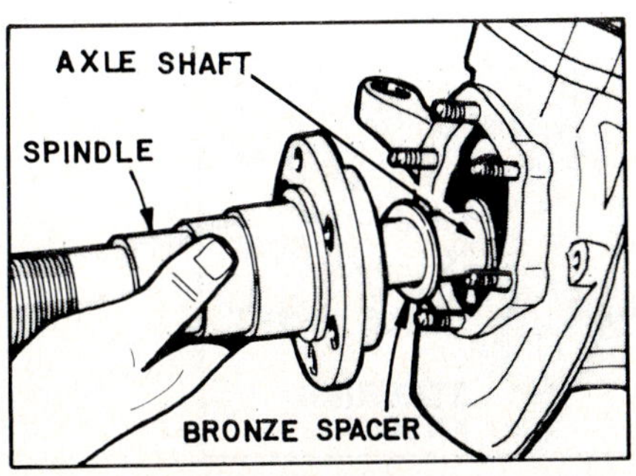

Fig. 14 Installing spindle & spacer to knuckle. 44 FBJ Units w/drum brakes

Servicing Rotor, Hub or Bearings

EXCEPT 1980–83 MODELS W/LOCKING HUBS

Removal & Inspection

1. Remove axle shaft cotter pin and loosen outer axle shaft nut.
2. Raise vehicle then remove wheel assembly.
3. After removing caliper retainer and anti-rattle spring assembly, slide caliper out and away from rotor. Hang caliper out of the way. Do not allow caliper to hang by

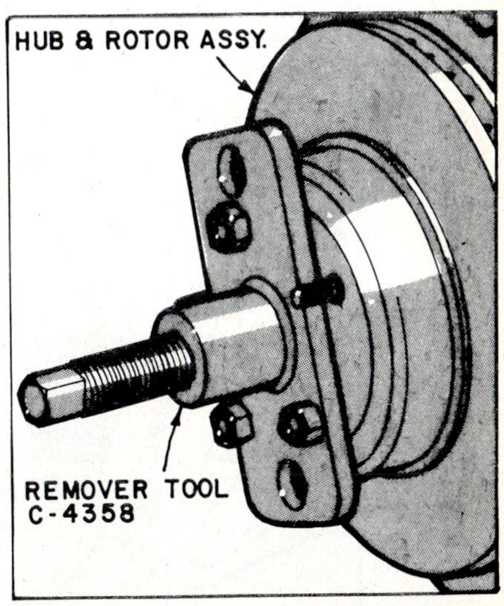

Fig. 15 Using puller to remove hub & rotor. 44 FBJ Units w/disc brakes. Except 1980–83 models w/locking hubs

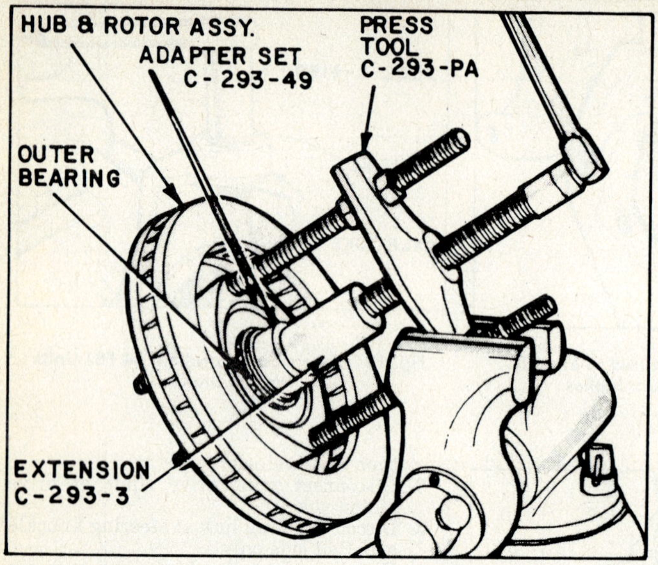

Fig. 16 Removing outer bearing cone. 44 FBJ Units w/disc brakes. Except 1980–83 models w/locking hubs

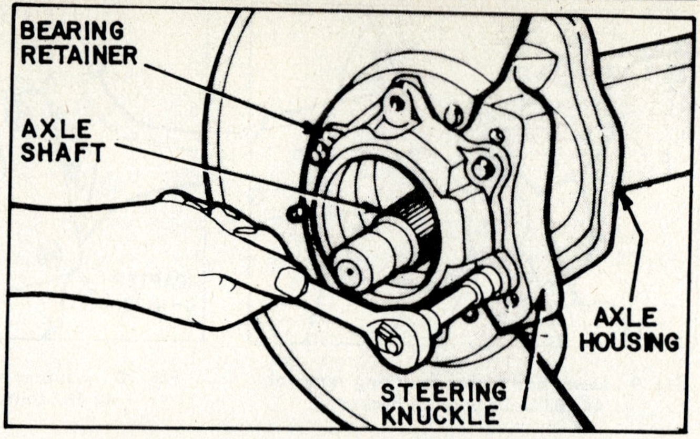

Fig. 17 Removing bearing retainer. 44 FBJ Units w/disc brakes. Except 1980–83 models w/locking hubs

hydraulic brake hose. Remove inboard brake pad.

4. Remove outer axle shaft nut and washer. Secure a suitable puller to wheel studs and tighten main screw of tool to remove hub and rotor assembly, Fig. 15.
5. Remove puller from hub and rotor assembly.
6. Assemble Modified Bearing Press, Tool C-293-PA, Extension, Tool C-293-3, and Adapters, Tool No. C-293-49, to hub and rotor assembly and position in a vise. Pull outer bearing cone from hub and rotor and discard outer seal, Fig. 16.
7. Remove the six retainer bolts and retainer, Fig. 17, then remove brake caliper adapter from knuckle, if required.
8. Place a pry bar behind inner axle shaft yoke and push bearings out of knuckle, Fig. 18.
9. Remove and discard "O" ring from knuckle, if equipped, then carefully remove axle shaft assembly.

Examine inner seal surface and knuckle bore for wear or damage and replace knuckle if required. If outer axle shaft seal surface is grooved, repair as follows:

a. Measure in from yoke shoulder of axle approximately 3/8 inch. Use a center punch and stake at 1/4 inch intervals around circumference of shaft. This will size shaft and ensure a tight fit of inner seal slinger, Fig. 19.
b. Proper bearing clamp should be checked by installing bearing cups and spacer into knuckle bore and bolting bearing retainer to knuckle, Figs. 20 and 21. If a .004 inch feeler gauge can not be inserted between knuckle and retainer at the six places midway between retainer mounting ears, the knuckle must be replaced, Fig. 22. The brake dust shield may have to be removed to complete this check. If knuckle is serviceable, remove retainer, bearing cups and spacer and install dust shield if removed.

Assembly & Installation

1. Apply RTV sealer or equivalent to seal surface of axle shaft.
2. Using driver, Tool No. C-4398-1, install seal slinger with lip toward splines onto outer axle shaft, Figs. 23 and 24.
3. Carefully insert axle shaft into housing so as not to damage differential seal at side gears. After sliding axle shaft completely in, wedge a pry bar through "U" joint to retain shaft, Fig. 25.
4. Install seal cup using adapter, Tool C-4398-2, and driver, Tool No. C-4398-1. Use a small amount of wheel bearing grease on adapter face to hold cup in position, then drive up until bottomed in knuckle, Figs. 26 and 27. Do not remove tool at this time.
5. Using a suitable tool, install new outer seal in retainer plate, then locate retainer plate over hub of rotor.
6. Thoroughly pack wheel bearings with Multi-Purpose grease and press outer bearing onto hub using Tool No. C-4246-A and adapters, Fig. 28. Remove tool and place grease coated outer bearing cup over bearing cone followed by spacer, grease coated inner bearing cup and inner bearing cone. Again use Tool No. C-4246-A and adapters to press components into position, Fig. 28. Remove tool.
7. Apply a 1/4 inch bead of RTV sealer to retainer face on the chamfer, Fig. 29. This replaces "O"-ring discarded during disassembly.
8. Carefully remove seal installing tool from knuckle bore so that outer axle shaft remains centered. If shaft is moved, be sure that lip seal is still riding inside cup.

Fig. 18 Using pry bar to remove bearing from knuckle. 44 FBJ Units w/disc brakes. Except 1980–83 models w/locking hubs

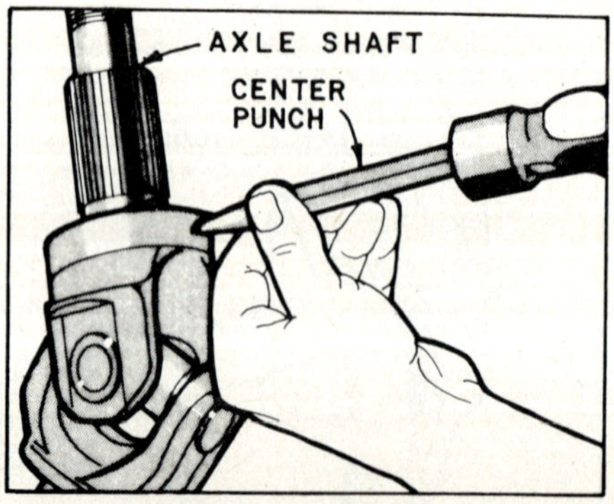

Fig. 19 Sizing axle shaft seal surface. 44 FBJ Units w/disc brakes. Except 1980–83 models w/locking hubs

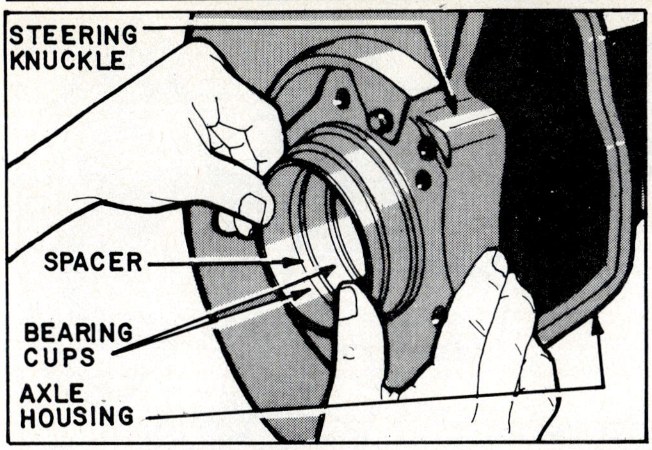

Fig. 20 Installing bearing cups before checking bearing clamp. 44 FBJ Units w/disc brakes. Except 1980–83 models w/locking hubs

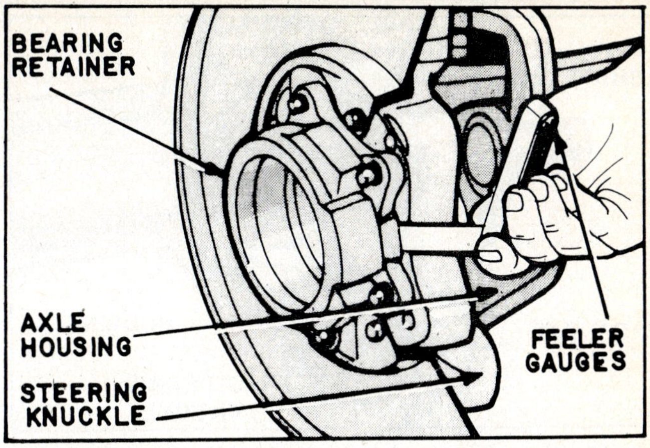

Fig. 22 Checking bearing clamp. 44 FBJ Units w/disc brakes. Except 1980–83 models w/locking hubs

Correct if necessary.

9. Before assembling hub and rotor to knuckle, position bearing retainer in hub so that grease fitting is facing forward, if equipped. Using a criss-cross method, torque retainer plate bolts to 30 ft. lbs.

NOTE: Bearing retainers that have a grease fitting must be positioned on knuckle so that fitting is facing directly forward, Fig. 30.

10. Install brake adapter and remove pry bar from "U" joint.
11. Install axle shaft washer and nut. Torque nut to 100 ft. lbs. and continue to tighten nut until next slot in nut aligns with help in axle shaft. Install cotter pin.
12. Through the access hole in hub, lubricate fitting in bearing retainer with Multi-Purpose grease until grease flows through new inner seal. Seal may be viewed at the "U" joint area. Spin hub several times and lubricate fitting again. Grease must flow from 1/2 of seal diameter.
13. Replace brake caliper assembly and wheel, then lower truck and test operation.

1980–83 MODELS W/LOCKING HUBS

Removal & Inspection

1. Remove locking hub assembly.
2. Raise and support vehicle, then remove wheel assembly.
3. Remove caliper retainer and anti-rattle assemblies.
4. Remove caliper from disc and support caliper to prevent damage to brake line.
5. Remove inboard brake shoe.

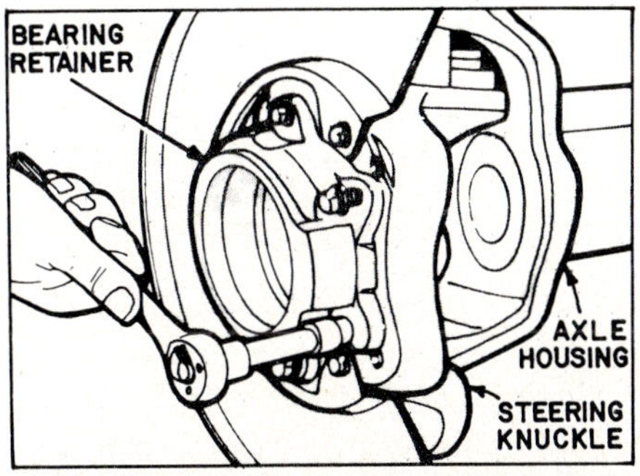

Fig. 21 Installing bearing retainer to check bearing clamp. 44 FBJ Units w/disc brakes. Except 1980–83 models w/locking hubs

6. Using tool C-4170, remove wheel bearing adjusting lock nut, then the washer and adjusting nut.
7. Remove rotor assembly. The outer wheel bearing and retainer spring plate will slide out as rotor is removed.
8. Pry inner wheel bearing grease seal from hub, then remove bearing cone and the inner and outer bearing cups.

Assembly & Installation

1. Install inner and outer bearing cups in rotor.
2. Lubricate and install inner bearing, then install the grease seal using tools D-155 and C-4171.
3. Install rotor, then the outer wheel bearing.
4. Install inner lock nut and torque to 50 ft. lbs., using tool C-4170, to seal bearings.
5. Loosen inner lock nut, then retorque to 30–40 ft. lbs. while rotating hub. Back off lock nut 135° to 150°. Set retaining washer in position by turning nut so that the pin pressed into lock nut will engage nearest hole in washer.
6. Install outer lock nut and torque to 50 ft. lbs. End play after final bearing adjustment should be .001–.010 inch.
7. Install inboard brake shoe on adapter with shoe flanges in adapter slots, then install caliper assembly.

NOTE: Use care not to pull dust boot out of its grooves while sliding piston and boot over the inboard shoe.

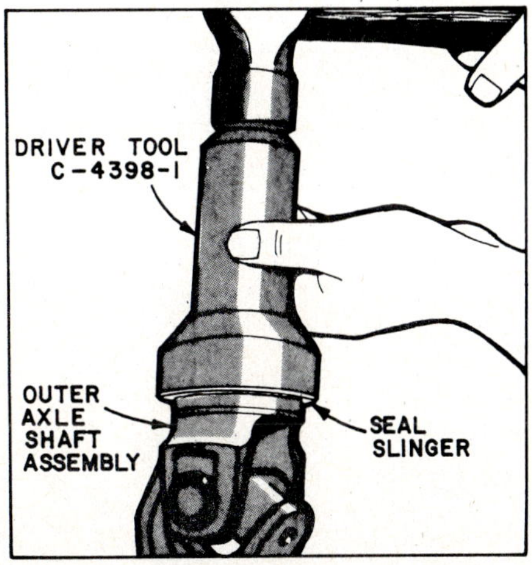

Fig. 23 Installing seal slinger on outer axle shaft. 44 FBJ Units w/disc brakes. Except 1980–83 models w/locking hubs

8. Install anti-rattle springs and retaining clips and torque to 180 inch lbs.

NOTE: The inboard shoe anti-rattle spring must be installed on top of the retainer spring plate.

9. Install wheel assembly, then lower truck and test operation.

Steering Knuckle & Ball Joint

EXCEPT 1980–83 MODELS W/LOCKING HUBS

Removal & Disassembly

1. Remove axle shaft cotter pin and loosen outer axle shaft nut.
2. Raise vehicle, then remove wheel assembly.
3. After removing caliper retainer and anti-rattle spring assembly, slide caliper out and away from rotor. Hang caliper out of the way. Do not allow caliper to hang by hydraulic brake hose. Remove inboard brake pad.
4. Remove outer axle shaft nut and washer and through access hole in rotor assembly, remove the six bearing retainer bolts.
5. Secure a suitable puller to wheel studs and tighten main screw of tool to remove hub, rotor, bearings, retainer and outer seal as an assembly. Remove puller from rotor.
6. Remove brake caliper adapter from knuckle, then remove and discard "O" ring from knuckle, if equipped.
7. Carefully pull axle shaft assembly out and remove seal and slinger from shaft.

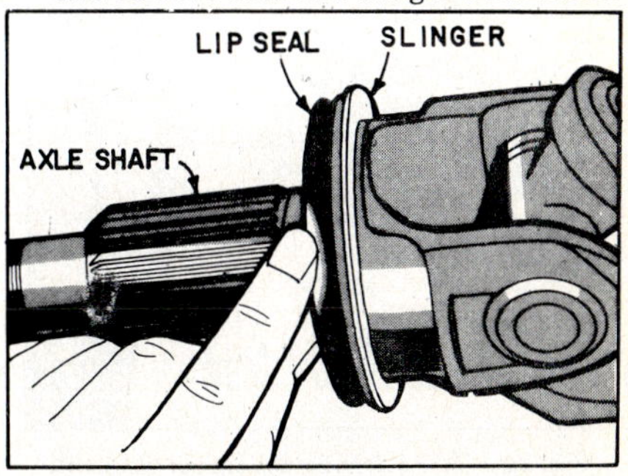

Fig. 24 Installing lip seal on outer axle shaft. 44 FBJ Units w/disc brakes. Except 1980–83 models w/locking hubs

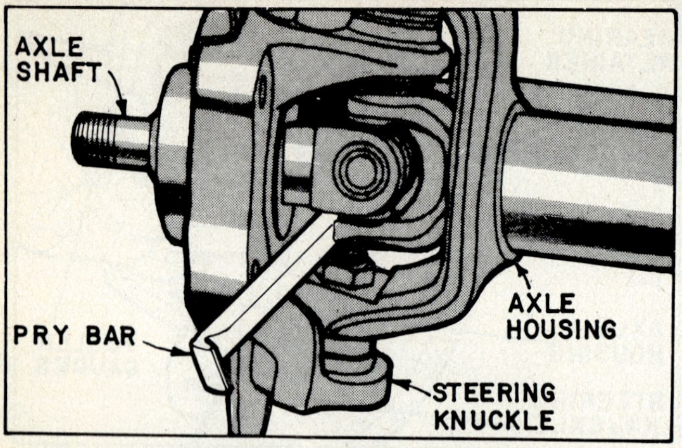

Fig. 25 Using pry bar to retain axle shaft. 44 FBJ Units w/disc brakes. Except 1980–83 models w/locking hubs

8. Disconnect tie rod from steering knuckle using a suitable tool, Fig. 31, so as not to damage seal.
9. On left side only, disconnect drag link from steering knuckle arm again using a suitable tool to avoid seal damage, Fig. 32.
10. Remove nuts from steering knuckle arm on left side only. Tap arm to loosen tapered dowels. Remove dowels and arm.
11. Remove cotter pin from upper ball joint nut then remove upper and lower ball joint nuts. Discard lower nut.
12. Separate steering knuckle from axle housing yoke using a brass drift and a hammer, then, using a suitable tool, remove and discard sleeve from upper ball joint yoke on axle housing.
13. Secure steering knuckle upside down in a vise and remove snap ring from lower ball joint.
14. Use proper tools to press upper and lower ball joints from steering knuckle. Replace ball joints if any looseness or end play exists.

Assembly & Installation

1. Secure steering knuckle right side up in a vise and press upper and lower ball joints into position using proper tools. Install snap ring on lower joint and new boots on both joints.
2. Thread new sleeve into ball joint yoke on axle housing ensuring that two threads are exposed at the top.

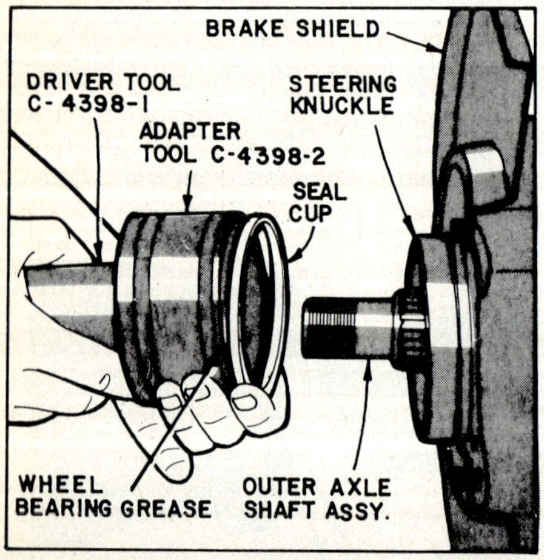

Fig. 26 Using driving & adapter to install seal cup into knuckle. 44 FBJ Units w/disc brakes. Except 1980–83 models w/locking hubs

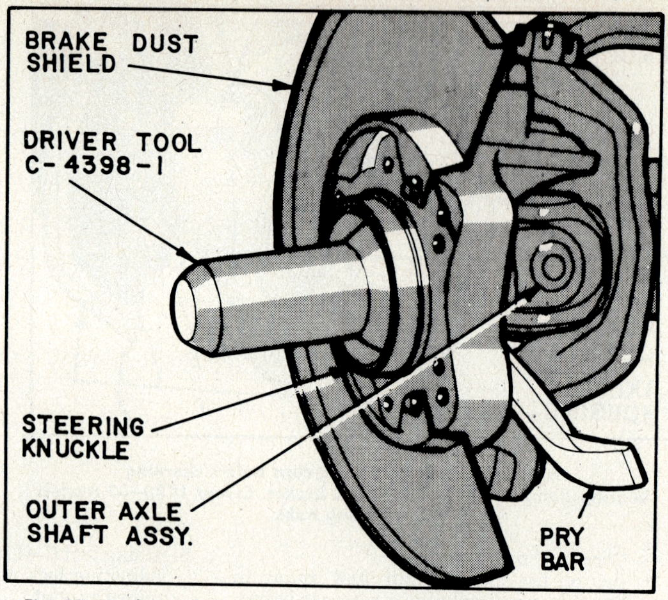

Fig. 27 Tools in place after seal cup is installed. 44 FBJ Units w/disc brakes. Except 1980–83 models w/locking hubs

3. Position steering knuckle on axle housing yoke, then install a new lower ball joint nut and torque to 80 ft. lbs.
4. Use Tool No. C-4169 and a torque wrench to torque sleeve in upper ball joint yoke to 40 ft. lbs, then install upper ball joint nut and torque to 100 ft. lbs. Install cotter pin if slot in nuts and hole in stud align. If not, tighten, do not loosen nut to align.
5. On left side only, position steering knuckle arm over studs on steering knuckle. Install tapered dowels and nuts, then torque nuts to 90 ft. lbs. Secure drag link to steering knuckle arm and torque nut to 60 ft. lbs.
6. Secure tie rod end to steering knuckle.

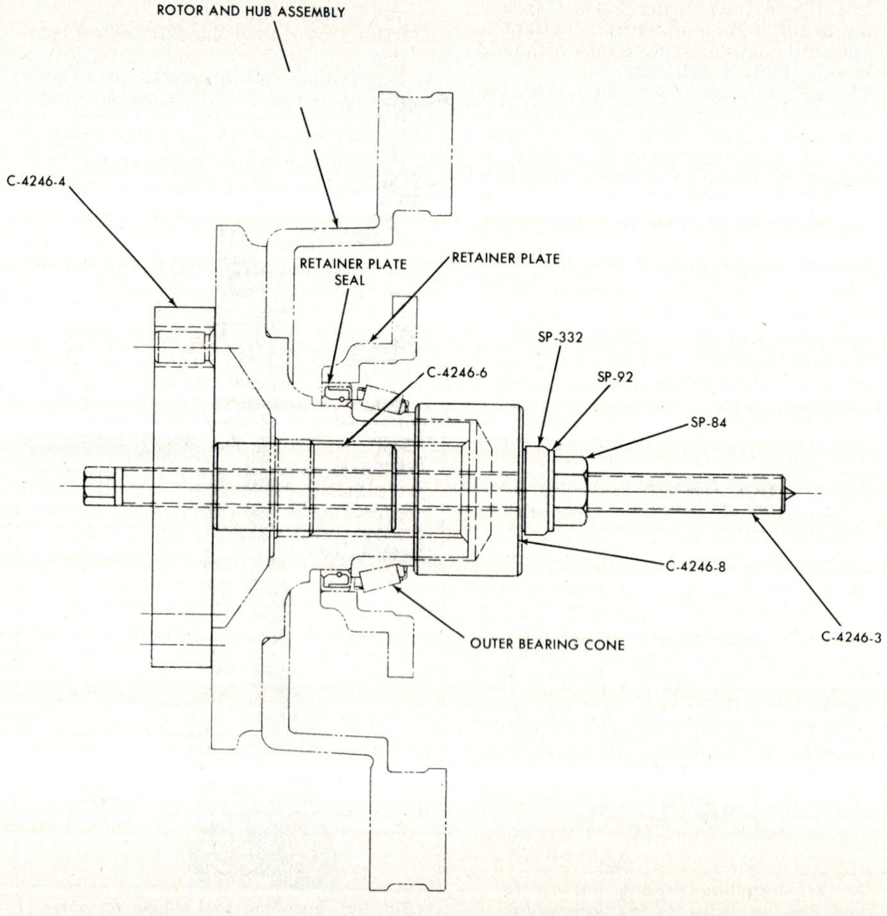

Fig. 28 Tools positioned to press outer bearing onto hub. 44 FBJ Units w/disc brakes. Except 1980–83 models w/locking hubs

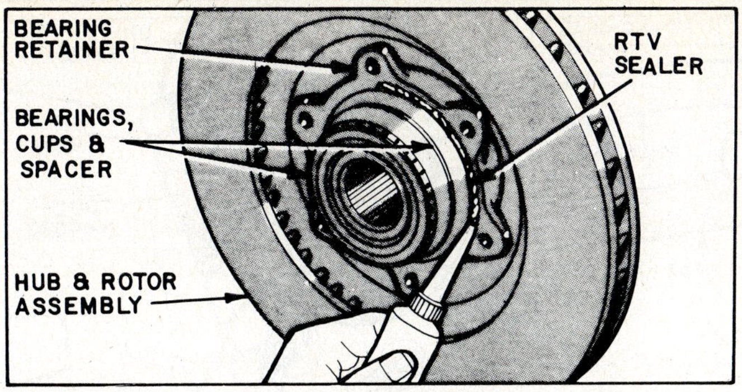

Fig. 29 Forming a seal on bearing retainer face. 44 FBJ Units w/disc brakes. Except 1980–83 models w/locking hubs

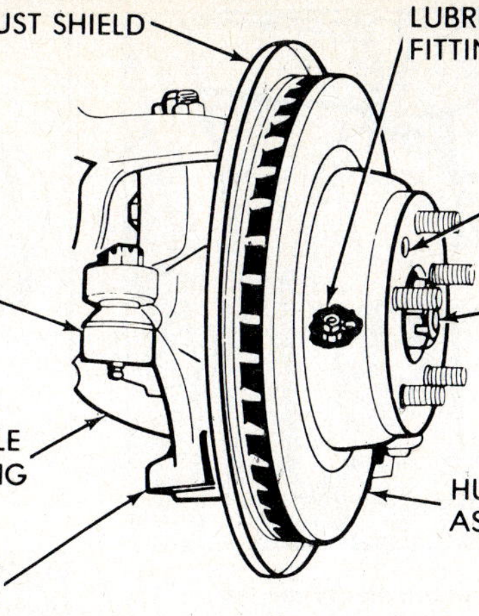

Fig. 30 Bearing retainer positioned with grease fitting facing forward. 44 FBJ Units w/disc brakes. Except 1980–83 models w/locking hubs

Torque nut to 45 ft. lbs.
7. Install brake dust shield, if removed.
8. Inspect outer axle shaft seal surface for grooving. If surface is grooved, repair as described in "Servicing Rotor Hub or Bearings".
9. Apply RTV sealer to seal surface of axle shaft then using driver, C-4398-1, install seal slinger with lip toward splines onto outer axle shaft, Figs. 23 and 24.
10. Carefully insert axle shaft into housing so as not to damage differential seal at side gears. After sliding axle shaft completely in, wedge a pry bar through "U" joint to retain shaft, Fig. 25.
11. Install seal cup using adapter C-4398-2, and driver C-4398-1. Use a small amount of wheel bearing grease on adapter face to hold cup in position, then drive cup until bottomed in knuckle, Figs. 26 and 27. Do not remove tool at this time.
12. Apply a 1/4 inch bead of RTV sealer to retainer face on the chamfer, Fig. 29. This replaces the "O"-ring discarded during disassembly.
13. Carefully remove seal installing tool from knuckle bore so that outer axle shaft remains centered. If shaft is moved, ensure that lip seal is still riding inside cup. Correct if necessary.
14. Before installing hub, rotor, retainer and bearing assembly on knuckle, position bearing retainer in hub so that grease fitting is facing forward, if equipped. Use a criss-cross method and torque retainer plate bolts to 30 ft. lbs.

NOTE: Bearing retainers that have a grease fitting must be positioned on knuckle so that fitting is facing directly forward, Fig. 30.

15. Install brake adapter and remove any pry bar from "U" joint.
16. Install axle shaft washer and nut. Torque nut to 100 ft. lbs. and continue to tighten nut until next slot in nut aligns with hole in axle shaft. Install cotter pin.
17. Through the access hole in hub, lubricate fitting in bearing retainer with Multi-Purpose grease until grease flows through new inner seal. Seal may be viewed through "U" joint area. Spin hub several times and lubricate fitting again. Grease must flow from 1/2 of seal diameter.
18. Replace brake caliper assembly and wheel, then lower vehicle and test operation.

1980–83 MODELS W/LOCKING HUBS

Removal & Disassembly

1. Remove locking hub assembly.
2. Raise vehicle and remove wheel assembly.
3. Remove caliper retainer and anti-rattle spring assemblies.
4. Remove caliper from disc and support caliper to prevent damage to brake line.
5. Remove the inboard brake pad and disc.
6. Remove the caliper adapter from knuckle.
7. Remove the six nuts and washers from spindle to steering knuckle attaching bolts.
8. Remove brake splash shield.
9. Tap spindle lightly with soft faced hammer to free from steering knuckle.
10. Upon removal, examine bronze spacer between needle bearing and shaft joint assembly. If wear is evident, replace.
11. Clamp spindle in vise avoiding bearing carrying surfaces. Remove needle bearing grease seal.
12. Using a suitable puller remove inner axle needle bearings.
13. Carefully remove axle shaft, axle seal and stone shield.
14. Remove the tie rod from the steering knuckle.
15. On the left side only, remove drag link from steering knuckle.
16. On the left side only, remove the nuts and cone washers from steering knuckle arm. Tap the steering knuckle arm to free knuckle. Remove arm.
17. Remove cotter pin from upper ball joint nut. Remove upper and lower ball joint nut. Discard lower nut.
18. Using a brass drift and hammer separate steering knuckle from axle housing yoke. Remove sleeve from upper ball joint using tool C-4169. Discard sleeve.
19. Install steering knuckle in vise and remove snap ring from lower ball joint with suitable snap ring pliers.
20. Using tool C-4212-1 and adapter set C-4288, press lower ball joint from steering knuckle.
21. Reposition tool, and press upper ball joint from steering knuckle.

NOTE: Replace ball joints if any looseness or end play exists.

Cleaning & Inspection

1. Clean all parts using a suitable solvent.
2. Blow dry parts using compressed air.
3. Inspect all parts for cracks, wear, chips, burrs and distortion.
4. Replace any parts not suitable for further service.

Assembly & Installation

1. Position steering knuckle right side up in a vise. Using tool C-4212-1 and adapter set C-4288; press the lower ball joint into position.
2. Using the same tool and adapter set as above, install the upper ball joint.
3. Install new boots on the ball joints and remove the steering knuckle from the vise.
4. Screw a new sleeve into the upper ball joint yoke leaving two threads showing at the top.
5. Position steering knuckle on axle housing yoke and torque new lower ball joint nut to 80 ft. lbs.
6. Using tool C-4169 and a torque wrench torque sleeve in upper ball joint to 40 ft. lbs. Install upper ball joint and torque to 100 ft. lbs. Align slot in nut with hole in stud and install cotter pin.

NOTE: Do not loosen to align.

7. On left side only, position steering knuckle arm over studs on steering knuckle. Install cone washers and nuts and torque to 90 ft. lbs.
8. Install drag link on steering knuckle arm and torque to 60 ft. lbs. Install cotter pin.
9. Install tie rod end to steering knuckle. Torque nut to 45 ft. lbs. and install cotter pin.
10. Install lip seal on stone shield with lip toward axle shaft spline.
11. Carefully insert axle shaft into housing. Avoid damaging differential seal at side gears.
12. Using tools D-122 and C-4171, install new needle bearings into spindle.
13. Fill seal area with NLG1 grease or equivalent, and install seal with tools D-155 and C-4171.
14. Install new bronze spacer on axle shaft, install spindle and brake splash shield.
15. Install new nuts and torque to 25–35 ft. lbs.
16. Mount braking disc assembly and outer wheel bearing cone onto spindle.
17. Install wheel bearing nuts and locking washer.
18. Install locking hub assembly.
19. Install brake adapter and torque attaching bolts to 85 ft. lbs.
20. Install the assembled brake caliper assembly.
21. Install the wheel and torque the nuts to 110 ft. lbs. Lower vehicle and test operation.

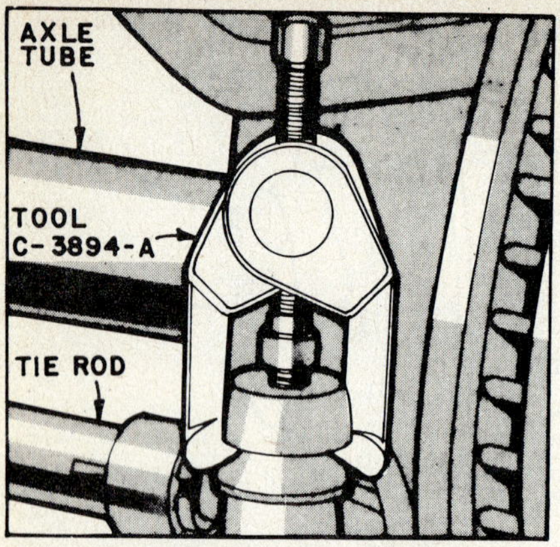

Fig. 31 Using puller to remove tie rod. (Typical)

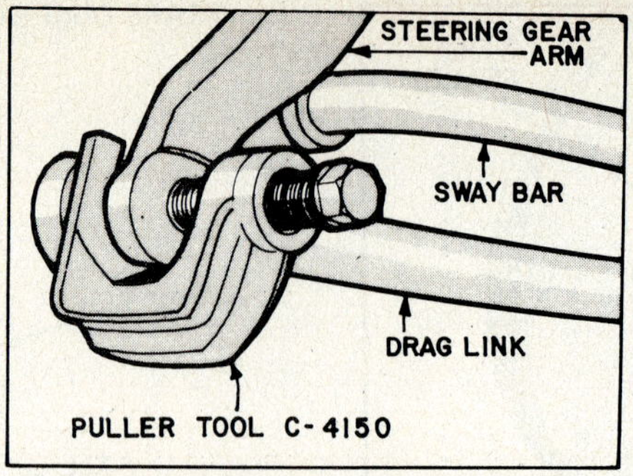

Fig. 32 Using puller to remove drag link from steering arm. (Typical)

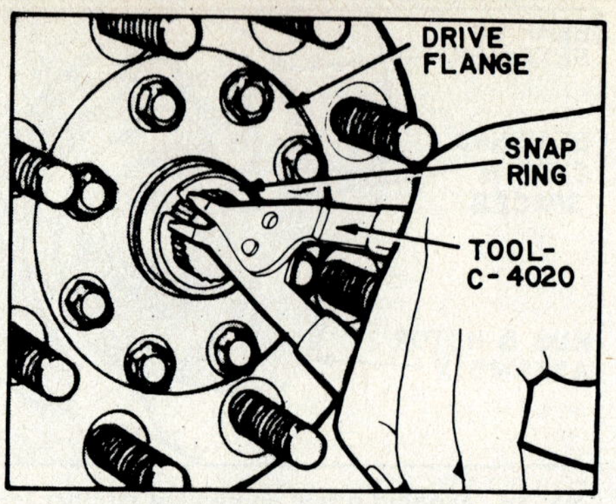

Fig. 33 Removing snap ring. 60 Front Axle

60 FRONT AXLE

Axle Shaft, Ball Joint & Steering Knuckle Repairs

Removal & Disassembly

1. Block brake pedal in "Up" position.
2. Raise and support vehicle.
3. Remove wheel.
4. Remove brake caliper from adapter and, using a piece of wire, suspend caliper. Do not hang caliper by brake hose. The inner brake pad will remain on adapter.
5. Remove hub cap and snap ring, Fig. 33.
6. Remove flange nuts and lock washers.
7. Remove drive flange and discard gasket, or remove locking hub, if equipped.
8. Straighten tang on lock ring, then remove outer lock nut, lock ring, inner lock nut and outer bearing, Fig. 34. Slide hub and rotor assembly from spindle.
9. Remove inner brake pad from adapter.
10. Remove nuts and washers securing brake splash shield, brake adapter and spindle to steering knuckle.
11. Remove spindle from knuckle. Slide inner and outer axle shaft with bronze spacer, seal and slinger from axle, Fig. 35.
12. Remove cotter key and nut from tie rod. Disconnect tie rod from steering knuckle.
13. On left side only, remove cotter key and nut from drag link. Disconnect drag link from steering knuckle arm. Also, remove nuts and upper knuckle cap. Discard gasket. Remove spring and upper socket sleeve.
14. Remove capscrews from lower knuckle cap and free cap from knuckle and housing.

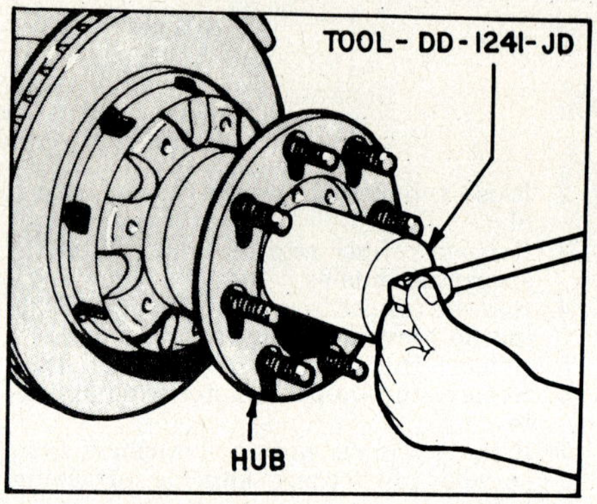

Fig. 34 Removing hub lock nut. 60 Front Axle

15. To remove knuckle from housing, swing outward at bottom, then lift up and off upper socket pin.
16. Using a suitable tool, loosen and remove upper socket pin, then the seal, Fig. 36.
17. Press lower ball socket from axle housing with suitable tools.
18. Referring to Fig. 1, disassemble shaft.

Assembly & Installation

1. Lubricate lower ball socket assembly with suitable grease.
2. With suitable tools, press seal and lower bearing cup into axle housing. Then, press lower bearing and seal into axle housing.
3. Using a suitable tool and torque wrench, install and torque upper socket pin to 500–600 ft. lbs, Fig. 36. Install seal over socket pin.
4. Place steering knuckle over socket pin. Fill lower socket cavity with suitable grease. Work lower knuckle cap into place on knuckle and housing. Install capscrews and torque to 70–90 ft. lbs.
5. Lubricate upper socket pin with suitable grease. Align upper socket sleeve in keyway of steering knuckle and slide into position, Fig. 37.
6. Install new gasket over upper steering knuckle studs. Place spring over sleeve. Install cap on left side steering knuckle arm. Install nuts and torque to 70–90 ft. lbs.
7. On left side only, attach drag link to steering knuckle arm and install and torque nut to 60 ft. lbs.
8. Connect tie rod to steering knuckle and install and torque nut to 45 ft. lbs. Install cotter key.
9. Referring to Fig. 1, assemble shaft.
10. Slide axle shaft into position. Place bronze spacer on axle shaft with chamfer facing toward universal joint, Fig. 35.
11. Install spindle, brake adapter and splash shield. Install and torque nuts to 50–70 ft. lbs.
12. Place inner brake pad on adapter.
13. Install hub and rotor assembly onto spindle. Install outer bearing and inner lock nut. Using tool DD-1241-JD and tool C-3952, torque lock nut to 50 ft. lbs. to seat bearings. Then, back off locknut and retorque to 30–40 ft. lbs. while rotating hub and rotor. Back off nut 135 to 150 degrees. Assemble lock ring and outer lock nut. Torque lock nut to 65 ft. lbs. minimum. Bend one tang of lock ring over inner lock nut and another tang over outer lock nut. Bearing end play should be .001–.010 inch.

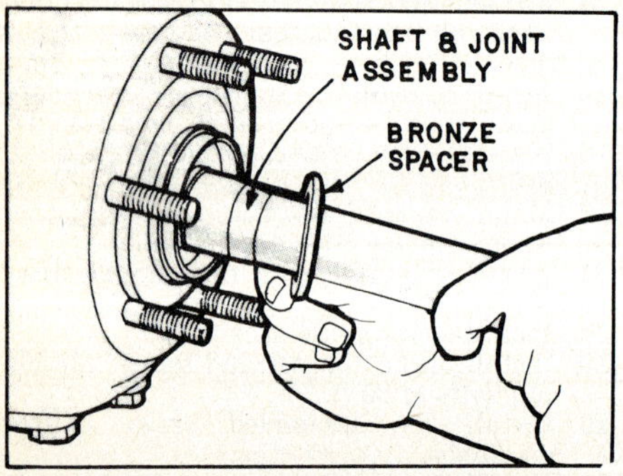

Fig. 35 Removing or replacing axle shaft. 60 Front Axle

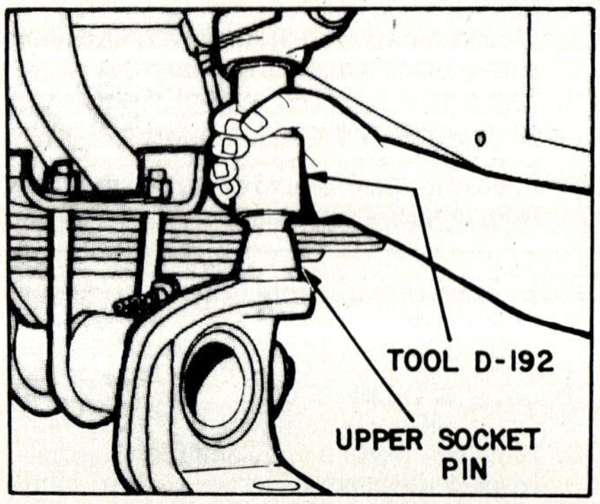

Fig. 36 Removing or replacing upper socket pin. 60 Front Axle

Fig. 37 Correct alignment of upper socket pin to steering knuckle. 60 Front Axle

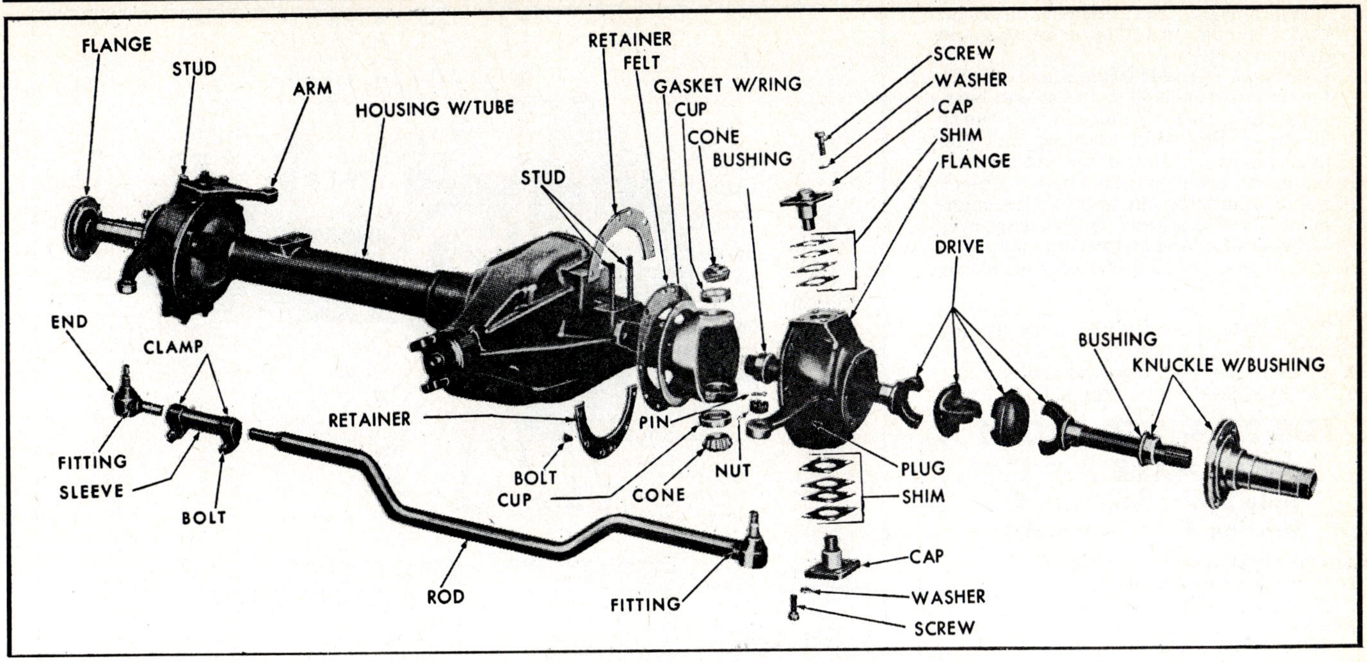

Fig. 38 Spicer 70F axle and steering knuckle assembly. (Typical)

14. Install new gasket on hub, then the drive flange, lock washers and nuts. Torque nuts to 30–40 ft. lbs. Install snap ring and hub cap, or locking hub, if equipped, Fig. 33.
15. Position caliper on adapter and torque allen screw to 12–18 ft. lbs.
16. Install wheel and lower vehicle.
17. Remove block from brake pedal.

Wheel Bearing Adjusting Procedure

1. Raise and support truck.
2. Remove outer drive flange snap ring, Fig. 33.
3. Remove flange nuts and lock washers, then remove flange and discard gasket.
4. Straighten tang on lock ring and using suitable tool, remove outer lock nut and lock ring, Fig. 34.
5. To seat bearings, use proper tools to torque inner lock nut to 50 ft. lbs. Then loosen lock nut and retorque to 30 to 40 ft. lbs. while rotating hub. Back off lock nut 135° to 150°. Install lock ring and torque outer lock nut to at least 65 ft. lbs. Bend one tang of lock ring over outer lock nut and one over inner nut. Final bearing adjustment should be .001 to .010 inch.
6. Install drive flange using new gasket then install snap ring.
7. Lower truck and test operation.

70F FRONT AXLE

Axle Shaft, Ball Joint & Steering Knuckle Repairs

Removal & Disassembly

1. Block brake pedal in "Up" position.
2. Raise and support vehicle.
3. Remove wheel.
4. Remove flange nuts and lock washers.
5. Remove drive flange and discard gasket.
6. Straighten tang on lock ring. Then, remove outer lock nut, lock ring, inner lock nut and outer bearing. Slide hub and drum assembly from spindle.
7. Remove brake support plate attaching nuts and support assembly aside with a piece of wire. Do not hang assembly by the brake hose.
8. Remove spindle from steering knuckle by tapping with a mallet to loosen.
9. Inspect spindle for wear and damage. Replace bushing, if necessary. The flanged bushing requires no reaming. When pressed into spindle, it compresses to provide the correct running clearance.
10. Remove axle shaft and joint assembly from axle, Fig. 38.
11. Disconnect tie rod.
12. On left side only, disconnect drag link from steering knuckle arm.
13. Remove capscrews from rear of steering knuckle. Remove the two retainer plates. Discard felt seal and oil seal.
14. Remove capscrews from bottom bearing cap. Pry cap from knuckle and record shim thickness for use at assembly.
15. Lift steering outward and upward off housing to remove. Be sure to retain the lower bearing when removing steering knuckle.
16. Remove capscrews from upper bearing cap.
17. Place steering knuckle upside down in a vise. With a suitable drift and hammer, separate cap or steering knuckle arm from knuckle and bronze bushing. Discard woodruff key. Record thickness of shims for use at assembly.
18. Remove bearing caps from spherical ball on housing. Remove bronze bushing from ball yoke.
19. Referring to Fig. 1, disassemble axle shaft.

Assembly & Installation

1. Lubricate all components with suitable grease.
2. Install bronze bushing into ball yoke and bearing cups into spherical ball on housing with a hammer and a suitable installation tool.
3. Place steering knuckle in a vise.
4. Install .060 inch shim pack on upper cap or steering knuckle. Assemble steering knuckle. Install and torque capscrews to 80–85 ft. lbs.
5. Invert position of steering knuckle in vise.
6. Install new woodruff key. Then, using a hammer and suitable installation tool, tap bronze bearing onto king pin of cap or steering knuckle arm until properly seated. Coat bronze bearing with suitable grease. Pack lower bearing cone with suitable grease.
7. Install new felt seal and oil seal over spherical ball.
8. Assemble steering knuckle to housing. Install lower bearing cone, shim pack, cap and capscrews. Torque capscrews to 80–85 ft. lbs.
9. Place a torque wrench on one of the upper bearing cap capscrews and record torque required to rotate knuckle through complete revolution, not the starting torque. The torque reading should be 20–30 ft. lbs. If preload is too tight, correct by adding shims. If preload is too loose, correct by removing shims. Shims are available in thicknesses of .003, .005, .010 and .030 inch. Note that preload corrections are made to the shim pack under the lower knuckle cap only.
10. Install new oil seal on steering knuckle with split in seal at the top center of the knuckle. Position felt seal against oil seal followed by the two retainer plates. Install and torque capscrews to 20–25 ft. lbs.
11. Connect tie rod to steering knuckle. Install and torque nut to 45–75 ft. lbs. Install cotter key.
12. On left side only, connect drag link to steering knuckle arm. Install and torque nut to 45–75 ft. lbs. Install cotter key.
13. Assemble axle shaft, referring to Fig. 1.
14. Slide axle shaft into axle.
15. Place spindle on steering knuckle followed by the brake support plate assembly. Torque retaining nuts to 85–135 ft. lbs.

16. Install hub and drum assembly on grease coated spindle. Install outer bearing and inner lock nut.
17. Using tool DD-1241-JD and tool C-3952, torque lock nut to 50 ft. lbs. to seat bearings. Then, back off nut and retorque to 30–40 ft. lbs. while rotating hub and drum assembly. Back off nut 1/4 turn maximum. Assemble lock ring and outer lock nut. Torque lock nut to 50 ft. lbs. minimum. Bend one tang of lock ring over inner lock nut and another tang over outer lock nut. Bearing end play should be .001–.010 inch.
18. Install new gasket on hub, then the drive flange, lock washers and nuts. Torque nuts to 30–40 ft. lbs.
29. Install wheel and lower vehicle.
20. Remove block from brake pedal.

FDS 75 & 750 FRONT AXLE

Axle Shaft, Ball Joint & Steering Knuckle Repairs

Removal & Disassembly, Fig. 39.

1. Raise and support vehicle.
2. Drain axle lubricant.
3. Remove capscrews securing hub cap to drive flange, then the hub cap.
4. Remove snap ring from end of outer drive shaft.
5. Remove capscrews from drive flange, then the flange from axle shaft and hub.
6. Straighten adjusting lock nut. Remove jam nut, adjusting lock nut, wheel bearing adjusting nut, outer wheel bearing and wheel assembly complete with the hub and drum assembly from spindle.
7. Disconnect and plug hydraulic brake lines. Remove the eight brake support attaching capscrews. Slide oil slinger and brake support from spindle.
8. Remove spindle attaching screw from top of spindle flange. Pull spindle straight out from steering knuckle and outer drive shaft.
9. Pull axle shaft assembly from axle housing.
10. Disconnect tie rod and steering link from steering knuckle.
11. Remove four upper sleeve to steering knuckle capscrews. Disassemble knuckle upper sleeve from steering knuckle. Pull upper sleeve from steering knuckle bore. Upper knuckle bearing cone will remain on the sleeve and is removed at the same time. Remove shim pack and reuse, if not damaged, at assembly.
12. Remove four lower sleeve to steering knuckle capscrews. Disassemble knuckle lower sleeve from steering knuckle. Pull lower sleeve from steering knuckle bore. The lower bearing cone will remain on the sleeve and is removed at the same time. Remove shim pack and reuse, if not damaged, at assembly.
13. Pull knuckle assembly from axle housing. With a suitable tool, remove upper and lower knuckle oil seals from bores in axle housing.
14. Using a suitable sleeve, approximately the same diameter as the socket plug, tap out upper and lower bearing caps and socket plugs.
15. Referring to Fig. 1, disassemble axle shaft.

Assembly & Installation

1. Apply a coat of silastic gasket material or equivalent to mating surface on concave side of socket plugs. Seat plugs with hand pressure in upper and lower bores of axle housing. Remove excess material from plugs and housing.
2. With a suitable sleeve, press upper and lower bearing cups in bores in axle housing. Install upper and lower seals into axle housing, then position steering knuckle over end of axle housing.
3. If bearing cones were removed from upper and lower sleeves, use a suitable diameter sleeve to press cones into position on sleeves.
4. Assemble shim pack to lower knuckle sleeve. The shim pack must not exceed .010 inch thickness. Press sleeve and shim pack through bottom bores of steering knuckle until sleeve bottoms. Install and torque capscrews to 210 ft. lbs.
5. Assemble shim pack to upper knuckle sleeve. Press sleeve and shim pack through upper bores of knuckle and housing until sleeve bottoms. Install and torque capscrews to 210 ft. lbs.
6. Measure and record distance between top grease fitting in upper knuckle sleeve and the center of the steering arm ball stud. The distance is measured perpendicular to the centerline of the knuckle.
7. Wrap a soft wire or cord around steering arm ball stud and attach a spring scale to wire or cord. Pull spring scale to rotate knuckle. Note number of pounds required to rotate knuckle. Use rotating torque not the starting torque. Rotating torque should be between 11 and 15 pounds. To determine proper preload use the following formulas:

$$\text{Minimum preload} = \frac{\text{11 ft. lbs} \times \text{12 inches}}{\text{Grease Fitting to Ball Stud Dimension}}$$

$$\text{Maximum preload} = \frac{\text{15 ft. lbs.} \times \text{12 inches}}{\text{Grease Fitting to Ball Stud Dimension}}$$

To increase bearing preload, remove shims. To decrease bearing preload, add shims. The adjustment is made at the upper sleeve shim pack.

8. Assemble axle shaft, referring to Fig. 8.
9. Slide axle shaft assembly into axle housing. Position spindle against knuckle mating surface and rotate spindle until countersunk hole in spindle flange aligns with mating hole at top of knuckle. This will align all capscrew holes. Install spindle position screw and torque to 25 ft. lbs. Install felt seal on spindle.
10. Connect tie rod and steering link. Torque tie rod nut to 90 ft. lbs. Install cotter keys. If holes in shafts and slots in nuts are not aligned, continue to tighten nut until cotter key installation is possible.
11. Position brake assembly and oil slinger over spindle. Install and torque capscrews to 70 ft. lbs. Connect hydraulic lines.
12. Place wheel and hub and drum assembly over spindle. Install outer wheel bearing and adjusting nut. Torque nut to 50 ft. lbs. while rotating assembly in both directions to seat bearings. Back off adjusting nut 1/4 to 1/3 turn. Install adjusting lock nut and jam nut. Torque jam nut to 125 ft. lbs. and bend nut lock over nuts.
13. Coat inside hub mating surface with silastic gasket material or equivalent. Coat internal splines of drive flange with wheel bearing grease.
14. Install flange and torque capscrews to 70 ft. lbs.
15. Install snap ring in outer drive shaft groove.
16. Coat inside drive flange mating surface of hub cap with silastic gasket material or equivalent. Assemble cap to drive flange and torque capscrews to 50 ft. lbs.
17. Fill axle assembly with recommended lubricant.
18. Bleed and adjust brakes.
19. Lower vehicle and test operation.

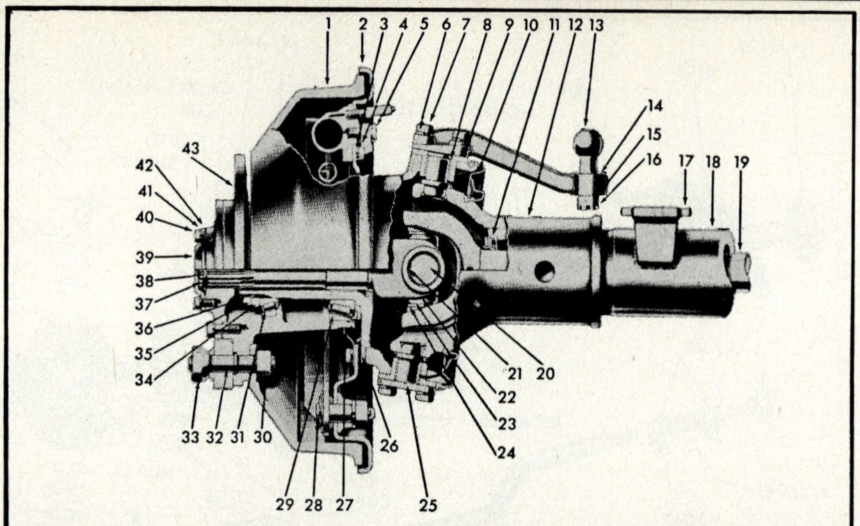

1—Drum
2—Backing Plate
3—Backing Plate Lockwasher
4—Backing Plate Nut
5—Brake Cylinder
6—Steering Arm Capscrew
7—Steering Arm Capscrew
8—Trunnion Pin Bushing
9—Trunnion Pin Bushing Retainer
10—Steering Knuckle Boot
11—Oil Seal Assembly
12—Steering Knuckle Socket
13—Steering Ball
14—Steering Arm
15—Steering Arm Nut
16—Steering Arm Cotter
17—Spring Seat
18—Housing Tube
19—Cardon Joint Axlé Shaft
20—Cardon Joint Needle Bearing Cap
21—Bearing Cap Retaining Ring
22—Cardon Joint Cross
23—Needle Bearing Retainer
24—Needle Bearing Cap
25—Lower Steering Knuckle Plate
26—Steering Knuckle Spindle
27—Oil Slinger
28—Anchor Pin
29—Inner Cup and Cone
30—Hub and Drum Inner Stud Nut
31—Outer Cup and Cone
32—Hub and Drum Stud
33—Outer Hub and Drum Stud Nut
34—Wheel Bearing Jam Nut
35—Jam Nut Retainer
36—Wheel Bearing Adjusting Nut
37—Outer Cardon Joint Axle Shaft
38—Axle Shaft Retaining Ring
39—Hub Cap
40—Hub Cap Capscrew
41—Hub Cap Lockwasher
42—Drive Flange
43—Hub

Fig. 39 Detail of housing assembly

Front Axle Section

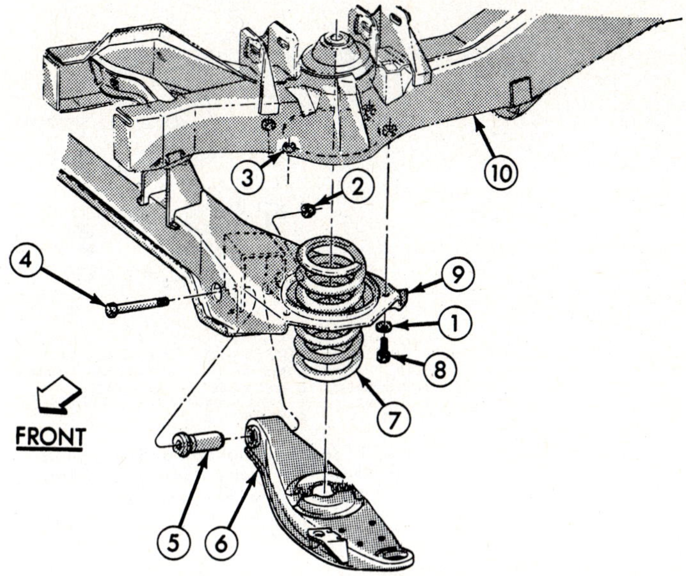

1. Coned Washer
2. Nut
3. Weld Nut
4. Bolt
5. Lower Arm Pivot Bushing
6. Lower Control Arm Assembly
7. Coil Spring
8. Bolt
9. Crossmember Assembly
10. Longitudinal Rail

Fig. 1 Lower control arm and coil spring. "B" Series & Voyager. 1974–78

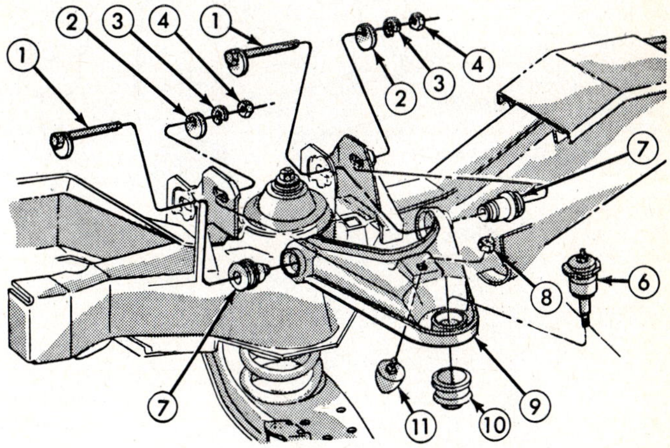

1. Cam and Bolt Assembly
2. Cam
3. Lock Washer
4. Nut
6. Ball Joint Assembly
7. Bushing Assembly
8. Lock Nut
9. Upper Control Arm
10. Upper Ball Joint Assembly
11. Bumper Assembly

Fig. 2 Upper control arm, "B" Series & Voyager. 1974–78

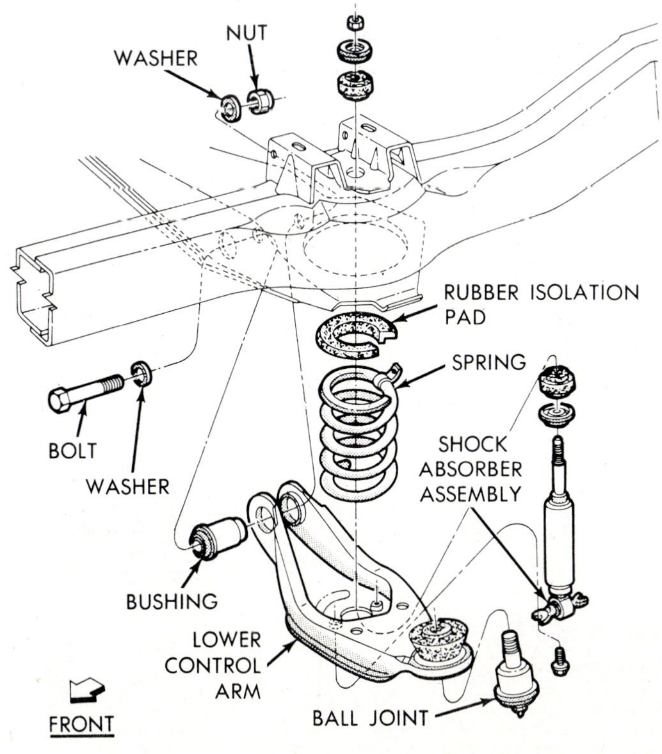

Fig. 3 Lower control arm and coil spring. "B" Series & Voyager. 1979–83

INDEPENDENT FRONT SUSPENSION

Description

These vehicles are equipped with a coil spring front suspension system, Figs. 1 through 6. The upper control arms are mounted on longitudinal rails and the lower control arms are mounted on a removable crossmember. Both control arms have replaceable bushings on the inner ends and ball joints on the outer ends. The upper control arms also control caster and camber adjustments through eccentric pivot bolts at the inner ends.

Coil Spring, Replace

1. Raise vehicle on hoist and position jack stands under extreme front ends of longitudinal rails. Lower hoist and remove wheel and tire.
2. On models with disc brakes, remove the caliper retainer and anti-rattle spring assemblies. Remove the caliper, but do not allow it to hang by brake line. Remove inboard brake pad.
3. On 1977–83 Ramcharger, Trailduster, "D" series and "W" series models, remove grease cap, cotter key, lock nut, adjusting nut, washer and outer bearing, then carefully slide rotor off of steering knuckle. Remove splash shield.
4. Remove shock absorber and upper shock bushing and sleeve. Disconnect sway bar at link if equipped and remove strut.
5. Install spring compressor and remove cotter keys and ball joint nuts.
6. Install ball joint breaker to free lower ball joint and remove breaker.
7. Slowly loosen the coil spring compressor until all tension is relieved and remove the spring.

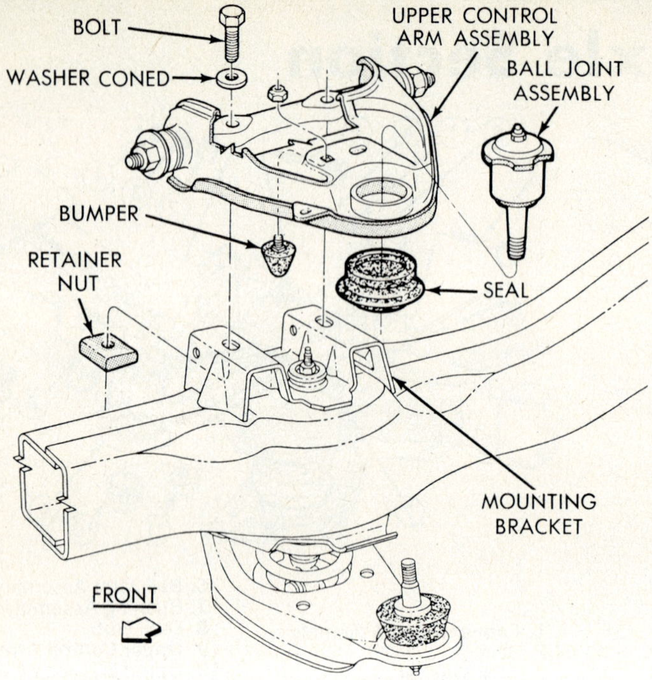

Fig. 4 Upper control arm. "B" Series & Voyager. 1979–83

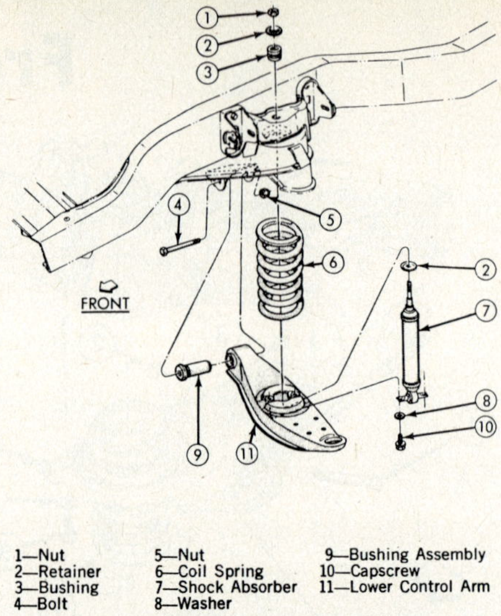

Fig. 5 Lower control arm and coil spring. "D" Series, "W" Series, Ramcharger & Trail Duster except four wheel drive.

Lower Ball Joint

1. Remove coil spring as described previously.
2. Unfasten and remove lower control arm.
3. Remove ball joint seal and using a suitable press and sleeve, press ball joint from control arm.

Upper Ball Joint

1. Place jack under outer end of lower control arm and raise vehicle.
2. Remove wheel and tire.
3. Remove ball joint nuts and free upper ball joint using ball joint breaker.
4. Using tool C-3561 unscrew ball joint from control arm.

Wheel Bearings, Adjust

Torque adjusting nut to 41–54 ft. lbs. on 1974–75 models, or 30–40 ft. lbs. on 1976–83 models, while rotating wheel. Stop rotation and loosen nut, then finger tighten adjusting nut and install lock nut and cotter pin. Final bearing adjustment should be from .0001 to .003 inch end play. Clean dust cap, then coat inside with wheel bearing grease but do not fill. Install cap.

5,000, 5,500, 7,000 OR 9,000 LB. FRONT AXLE

Wheel Bearings, Adjust

Torque adjusting nut to 50 ft. lbs. while rotating wheel. Loosen adjusting nut 1/4 to 1/3 turn and install lock ring. Nut should not be more than finger tight. Install lock washer and lock nut, torque to 100–150 ft. lbs., then bend down tabs on lock washer.

Steering Knuckle & Bushings

Removal

1. After removing the wheel and hub, secure brake pedal and remove brake support plate.
2. Secure brake assembly out of the way. Do not allow assembly to hang by hydraulic brake hose.
3. Remove steering knuckle arm from steering knuckle, then remove pivot pin locking screw or pin. Some models have two locking screws.
4. Remove upper pivot pin oil seal plug and drive pivot pin down, pushing lower seal plug from assembly.
5. Pivot thrust bearing is also removed at this time.

NOTE: 9000 lb. axles are equipped and serviced with bronze pivot pin bushings. Lighter axles are equipped with Delrin or Zytel bushings which can distort if steam cleaning equipment is used to clean axle. Service replacement bushings can be the "plastic" type or bronze. Bronze replacement bushings are recommended where severe service is anticipated.

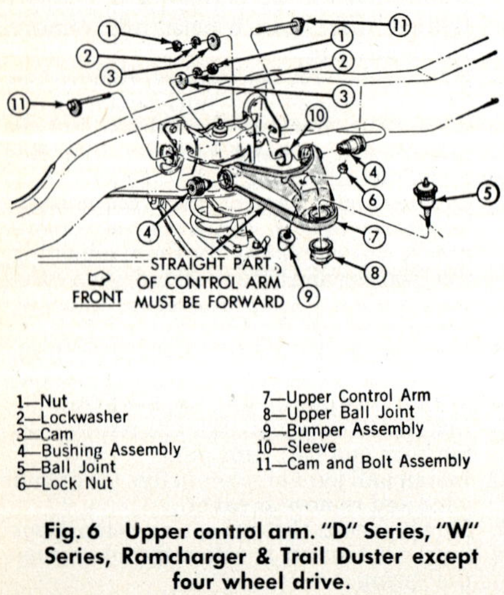

Fig. 6 Upper control arm. "D" Series, "W" Series, Ramcharger & Trail Duster except four wheel drive.

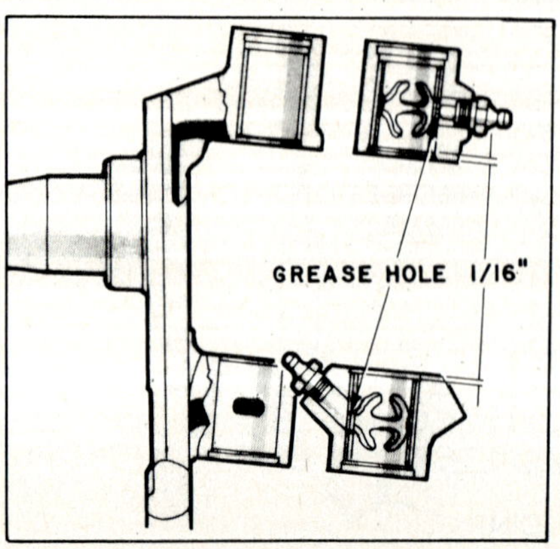

Fig. 7 Pivot pin bushings aligned & installed. (Typical)

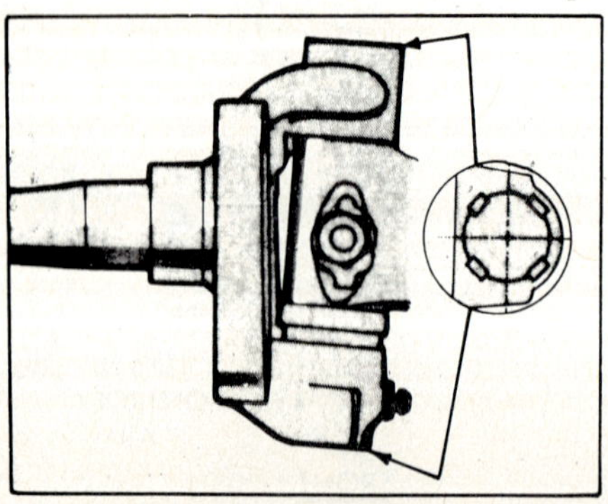

Fig. 8 Grease seal plugs installed & staked. (Typical)

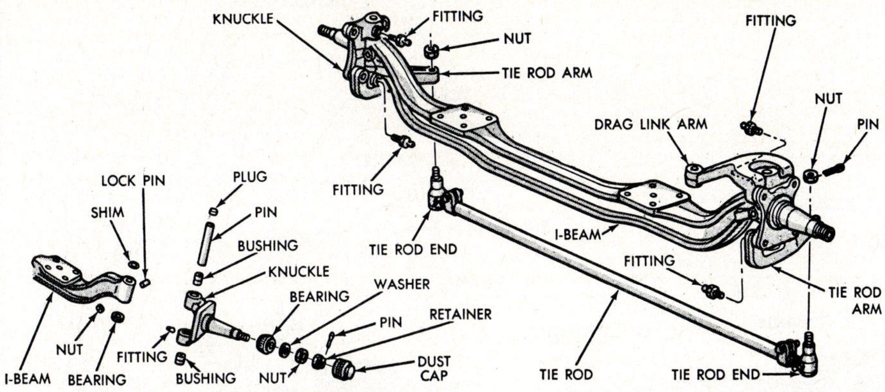

Reverse Elliot 6000 lb. front axle

Pivot Pin Bushings

Replace

1. Bronze bushings should be removed and installed using a press or other suitable tool to prevent bushing damage.
2. Ensure grease hole in bushing aligns with grease fitting hole, Fig. 7.
3. Line ream both bronze bushings to insure pivot pin fit. Plastic type bushings do not require line reaming.
4. Fill grease grooves in bushing with Multi-Purpose grease, then check pivot pin for wear or damage.
5. Rough edges of expansion plug mating surfaces should be cleaned for better sealing.
6. Position steering knuckle and pivot pin bushing on axle, then install pivot pin and pivot pin lock.
7. Install expansion plug seals at top and bottom of axle and stake at four points, Fig. 8.
8. Install and tighten steering knuckle arm and brake support plate.
9. Lubricate all fittings, then install hub assembly and wheel, free brake pedal and test operation.

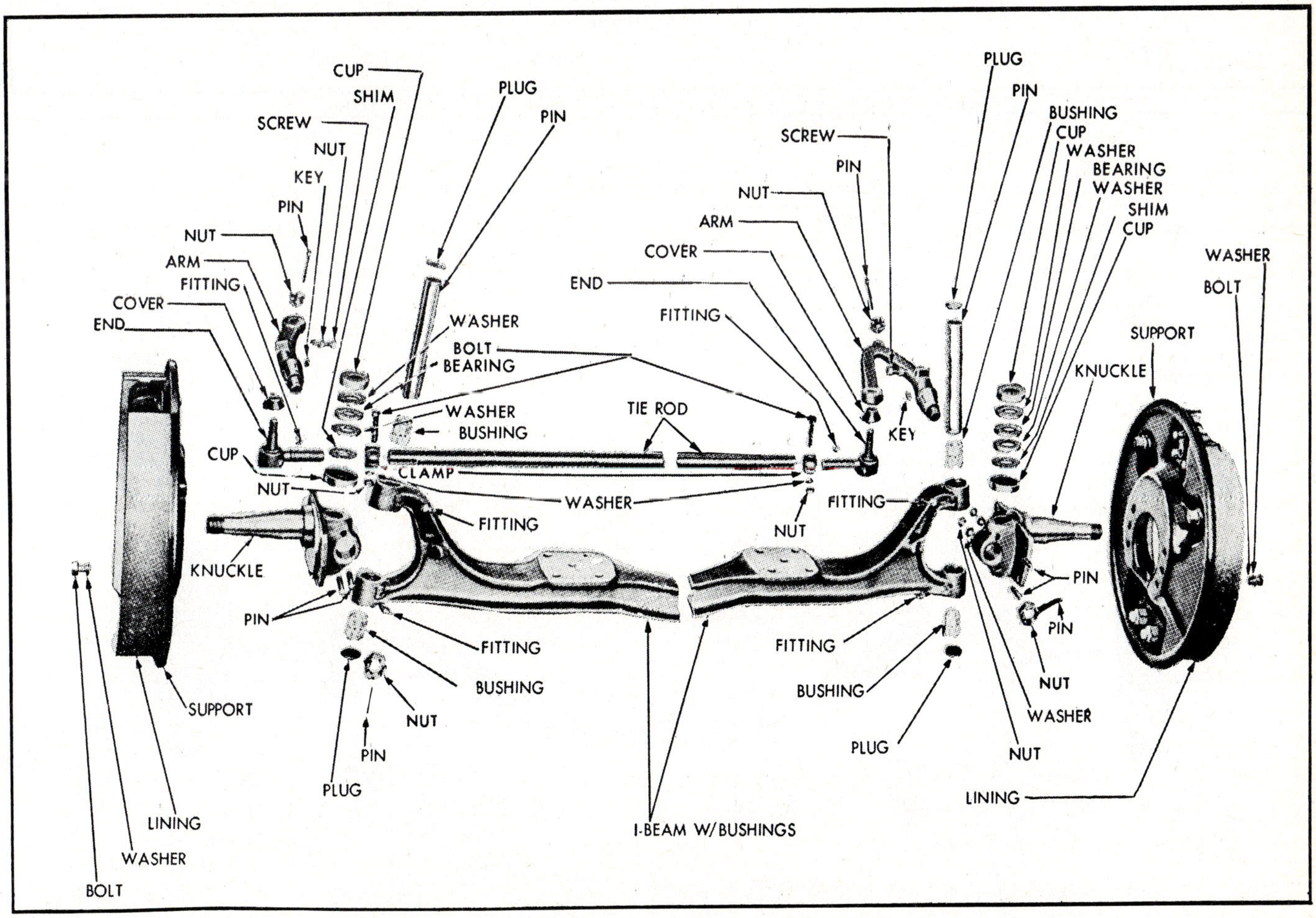

Elliot type 4000 lb. & 5000 lb. front axle

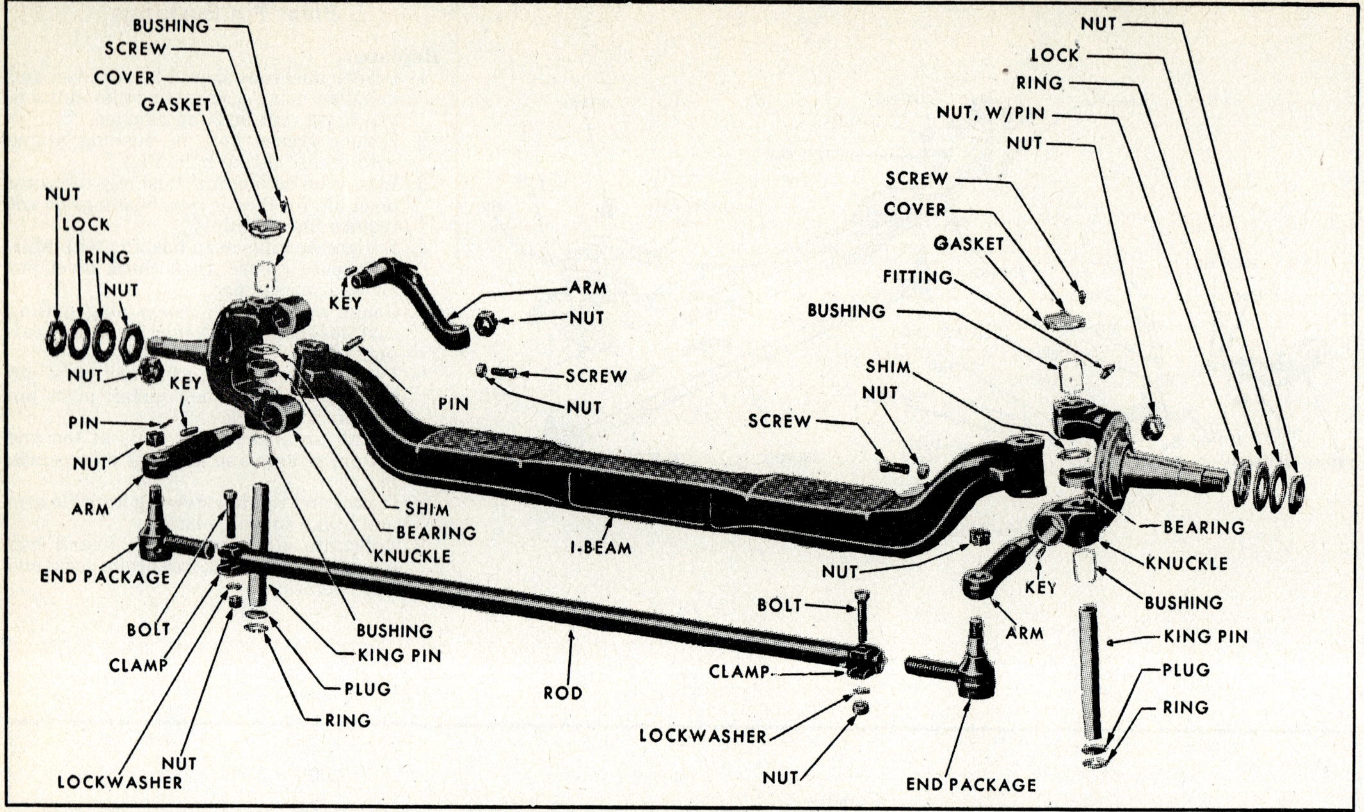

Front axle. 1974–76 Series 500 through 1000 & 1973–76 Series 100–1000

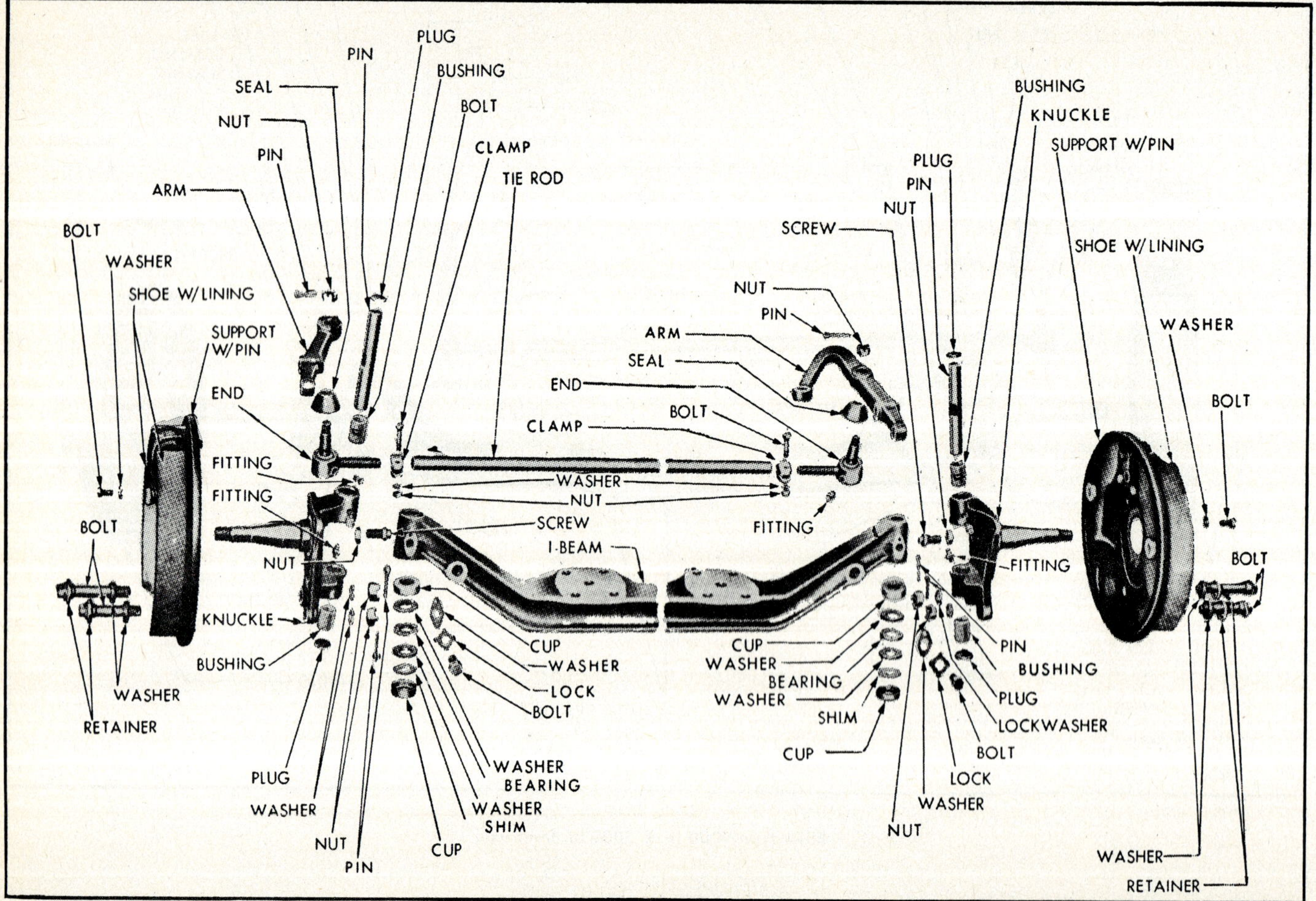

Reverse Elliot 2800 lb. front axle

DODGE RAMPAGE & PLYMOUTH SCAMP

INDEX OF SERVICE OPERATIONS

NOTE: Refer to the front of this manual for the manufacturers special service tool suppliers.

DODGE RAMPAGE & PLYMOUTH SCAMP

MODEL INDEX & ENGINE APPLICATION

Model	Year	Engine Make	Standard Engine Model	Crankcase Refill Capacity Qts.	Cooling System Capacity Qts.
Rampage	1982–83	OWN	4-135	4	9
Scamp	1983	OWN	4-135	4	9

GENERAL ENGINE SPECIFICATIONS

Year	Engine	Carb.	Bore & Stroke	Comp. Ratio	Horsepower @ R.P.M.	Torque Ft. @ R.P.M.	Normal Oil Pressure
1982	4-135	2 Bbl.	3.44 × 3.62	8.5:1	84 @ 4800	111 @ 2400	45–50①
1983	4-135	2 Bbl.	3.44 × 3.62	9:1	94 @ 5200	117 @ 3200	50①

①—Pounds per square inch.

TUNE UP SPECIFICATIONS

The following specifications are published from the latest information available. This data should be used only in the absence of a decal affixed in the engine compartment.

★ When using a timing light, disconnect vacuum hose or tube at distributor and plug opening in hose or tube so idle speed will not be affected.

● When checking compression, lowest cylinder must be within 80 percent of highest.

▲ Before removing wires from distributor cap, determine location of the No. 1 wire in cap, as distributor position may have been altered from that shown at the end of this chart.

Year & Engine	Spark Plug		Ignition Timing BTDC①★				Curb Idle Speed		Fast Idle Speed		Fuel Pump Pressure
	Type	Gap	Firing Order Fig. ▲	Man. Trans.	Auto. Trans.	Mark Location	Man. Trans.	Auto. Trans.②	Man. Trans.	Auto. Trans.	
1982											
4-135	65 PR	.035	A	12°	12°	Flywheel	900	900N	—	—	4.5 to 6
1983											
4-135 Exc. Calif. & High Alt.	65 PR	.035	A	10°③	10°③	Flywheel	775	900N	1300	1500	4.5–6
4-135 Calif.	65 PR	.035	A	10°③	10°③	Flywheel	775	900N	1400	1500	4.5–6
4-135 High Alt.	65 PR	.035	A	6°③	6°③	Flywheel	900	900N	1350	1275	4.5–6

①—B.T.D.C.—Before top dead center.

②—N: Neutral

③—Basic Timing set with vacuum line disconnected—do not readjust curb idle.

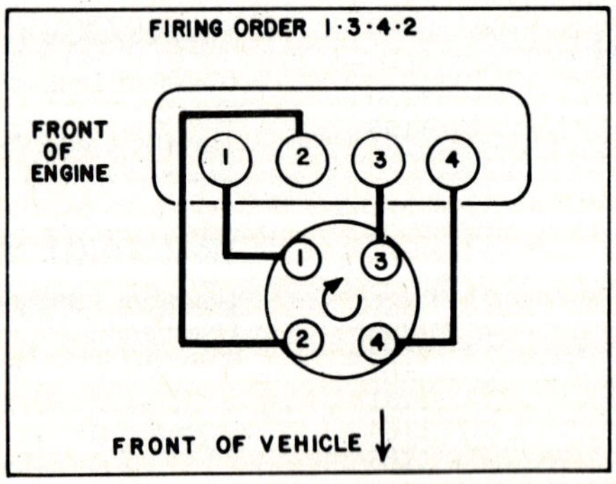

Fig. A

ALTERNATOR & REGULATOR SPECIFICATIONS

Unit Number	Ground Polarity	Field Coil Draw Amperes ①	Current Output			Operating Voltage		
			Engine R.P.M.	Amperes②	Volts	Engine R.P.M.	Volts	③
Yellow Tag	Neg.	4.5 to 6.5	1250	45	15	1250	15	13.3 to 13.9
Brown Tag	Neg.	4.5 to 6.5	1250	56	15	1250	15	13.3 to 13.9

①—@ 12 volts.

②—If current is low, stator or rectifier is shorted.

③—Voltage range @ 140° F.

STARTING MOTOR SPECIFICATIONS

Engine	Model	Ident. No.③	Brush Spring Tension Oz.	Free Speed Test		
				Amps.④	Volts	R.P.M.⑤
4-135①	Bosch	5213045	—	47	11	6600
4-135②	Bosch	5213395	—	47	11	6600
4-135	Nippondenso	5213645	—	47	11	6600

①—Auto. Trans.
②—Manual Trans.
③—Number located on plate riveted to starter housing.
④—Maximum current drawn.
⑤—Minimum speed.

DISTRIBUTOR SPECIFICATIONS

Distributor Part No.①	Centrifugal Advance Degree @ R.P.M.			Vacuum Advance	
	Advance Starts	Intermediate Advance	Full Advance	Start Dist. Degree @ Inches of Vacuum	Maximum Dist. Degree
5206975	—	—	—	—	—

①—Part No. stamped on distributor housing.

PISTON, PIN, RING, CRANKSHAFT & BEARING SPECIFICATIONS

Year	Engine Model	Wristpin Diameter	Piston Clearance, Inch	Ring End Gap, Inch (Minimum)		Crankpin Diameter, Inch	Rod Bearing Clearance, Inch	Main Bearing Journal Diameter, Inch	Main Bearing Clearance, Inch	End Thrust on Bearing No.	Shaft End Play
				Comp.	Oil						
1982–83	4-135	.9008	.0005–.0015	.011	.015	1.968–1.969	.0008–.0034	2.362–2.363	.0003–.0031	3	.003 to .007

VALVE SPECIFICATIONS

Year	Engine Model	Valve Lash		Valve Angle		Valve Spring		Valve Stem Clearance		Stem Diameter, Std.	
		Int.	Exh.	Seat	Face	Installed Height	Pressure Lbs. @ In.	Intake	Exhaust	Intake	Exhaust
1982	4-135	Hydraulic①	Hydraulic①	45°	45°	1.65	175 @ 1.22	.0009–.0026	.0028–.0044	.3124	.3103
1983	4-135	Hydraulic①	Hydraulic①	45°	45°	1.65	175 @ 1.22	.0009–.0026	.0030–.0047	.3124	.3103

①—No adjustment.

ENGINE TIGHTENING SPECIFICATIONS★

★ Torque specifications are for clean and lightly lubricated threads only. Dry or dirty threads produce increased friction which prevents accurate measurement of tightness.

Year	Engine Model	Spark Plug Ft. Lbs.	Cylinder Head Bolts Ft. Lbs.	Intake Manifold Inch Lbs.	Exhaust Manifold Inch Lbs.	Camshaft Cover Inch Lbs.	Connecting Rod Cap Bolts Ft. Lbs.	Main Bearing Cap Bolts Ft. Lbs.	Flywheel to Crankshaft Ft. Lbs.	Crankshaft Pulley Ft. Lbs.
1982–83	4-135	26	45①	200	200	105	40①	30①	—	—

①—Turn torque wrench an additional 1/4 turn after the specified torque has been achieved.

WHEEL ALIGNMENT SPECIFICATIONS

Model	Caster Angle Degrees	Camber Angle Degrees	Toe-out Inch
Rampage & Scamp	—	+0.3°	1/16″ ± 1/16″

Engine Section

ENGINE REPLACE

1. Disconnect battery ground cable.
2. Scribe alignment marks on hood and hood hinge, then remove hood.
3. Drain cooling system, then disconnect radiator hoses at radiator and engine.
4. Remove radiator and fan shroud, then remove air cleaner.
5. If equipped with A/C, remove compressor from the mounting bracket and position aside with the hoses attached.
6. Remove power steering pump from mounting bracket and position aside with the hoses attached, if equipped.
7. Drain crankcase and remove the oil filter.
8. Disconnect the wire connectors at the alternator, carburetor and engine, Fig. 1.
9. Disconnect fuel line, heater hose and the accelerator cable, Fig. 2.
10. Remove alternator from mounting bracket and position aside.
11. On models equipped with manual transmission, disconnect clutch cable, then remove transmission lower cover.
12. Disconnect exhaust pipe at the manifold, then remove the starter motor, Fig. 3.
13. On models equipped with automatic transmission, remove transmission case lower cover and place alignment marks on the flex plate and torque converter. Remove converter to flex plate attaching screws. Attach a C-clamp to the front lower position of the housing so as to retain the torque converter in the housing when removing the engine.
14. On all models, install a suitable transmission holding fixture and attach a suitable lifting device, Fig. 4.
15. Remove right hand inner splash shield, then disconnect ground strap, Fig. 5.
16. Remove right hand engine mount through bolts, Fig. 6.
17. Remove transmission case-to-engine-block attaching bolts.
18. Remove front engine mount bolt, Fig. 7.
19. Carefully lift engine from vehicle.
20. Reverse procedure to install.

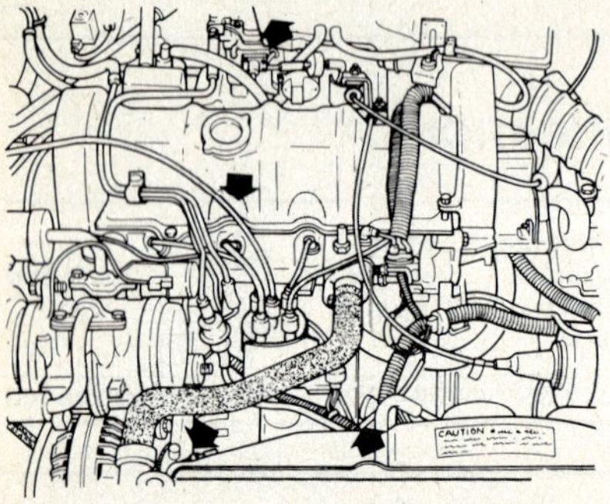

Fig. 1 Position of electrical connectors

TIMING SPROCKETS & OIL SEAL

Alternator Belt Removal

1. Disconnect battery ground cable.
2. Loosen alternator locking screw, then loosen adjusting screw and remove the alternator belt.
3. Reverse procedure to install.

Alternator & Compressor Mounting Bracket Removal, Fig. 8

1. Remove power steering locking screw, Fig. 9.
2. Remove pivot bolt and nut, then remove drive belt.
3. Reverse procedure to install.

Power Steering Pump Mounting Bracket Removal

1. Remove power steering pump from bracket, then remove bracket from engine, Fig. 9.
2. Reverse procedure to install.

Fig. 2 Fuel line, heater hose and accelerator connections

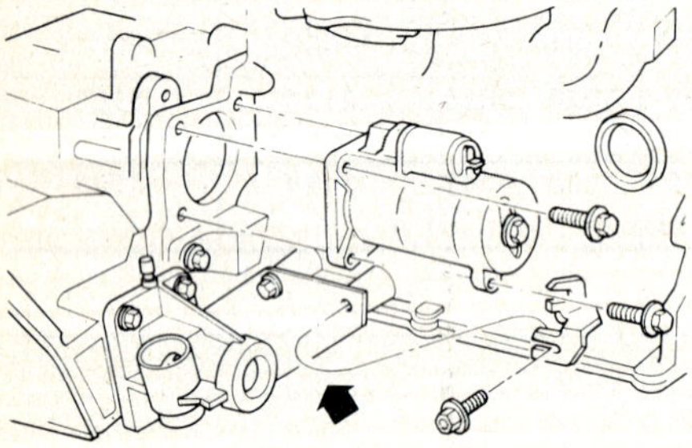

Fig. 3 Starter motor removal

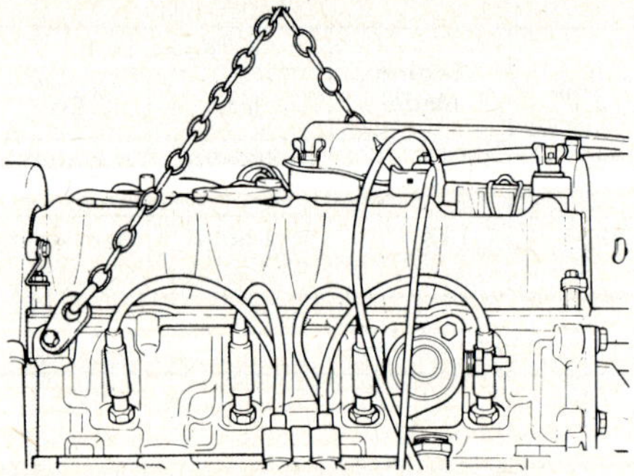

Fig. 4 Engine lifting device attached

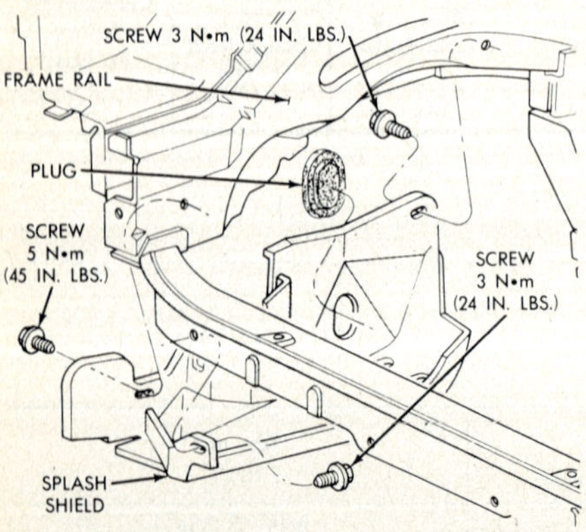

Fig. 5 Right inner splash shield

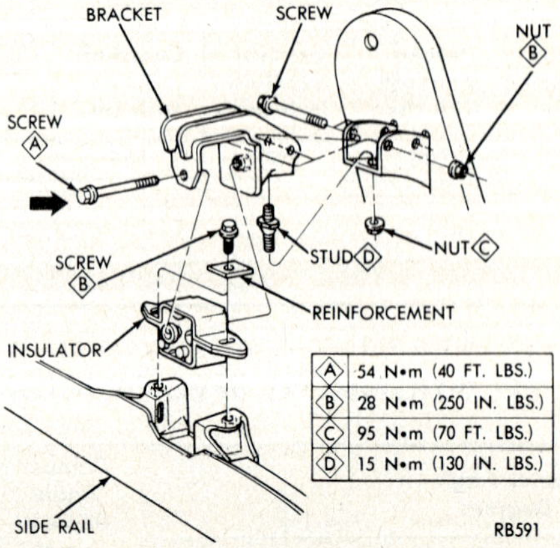

Fig. 6 Position of right engine mount through bolt (arrow)

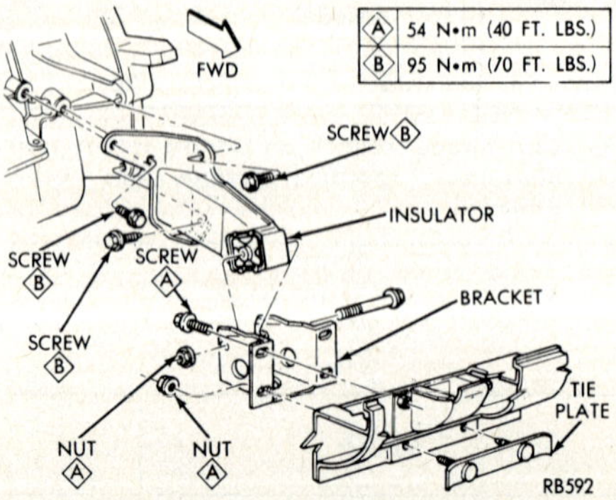

Fig. 7 Front engine mount

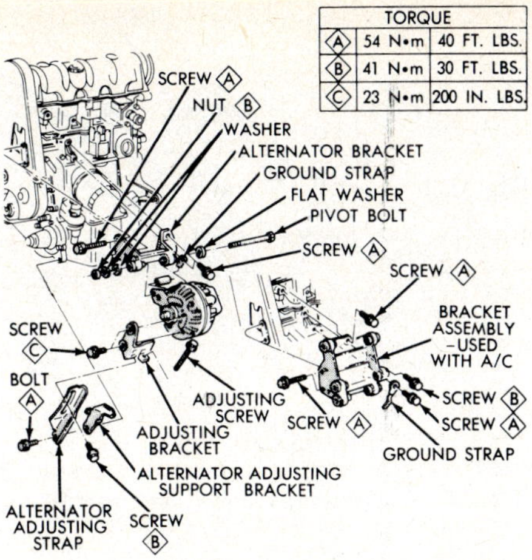

Fig. 8 Alternator & compressor mounting bracket removal

Crankshaft Pulley & Water Pump Pulley Removal

1. Remove screws retaining water pump pulley to pump shaft, Fig. 10.
2. Remove bolts retaining crankshaft pulley.
3. Raise and support front of vehicle, then remove right inner splash shield and remove crankshaft pulley.
4. Reverse procedure to install.

Timing Belt Cover Removal

1. Remove nuts securing timing belt cover to the cylinder head, Fig. 11.
2. Remove screws securing the cover to the cylinder head, then remove both halves of the timing belt cover.
3. Position a suitable jack under engine, then remove right hand engine mount-bolt and raise engine slightly.
4. Loosen timing belt tensioner, then remove timing belt.
5. Reverse procedure to install.

NOTE: On 1983 models, torque all attaching hardware to 40 inch lbs.

Crankshaft Sprocket Removal

1. With the timing belt removed from engine, remove the crankshaft sprocket bolt.

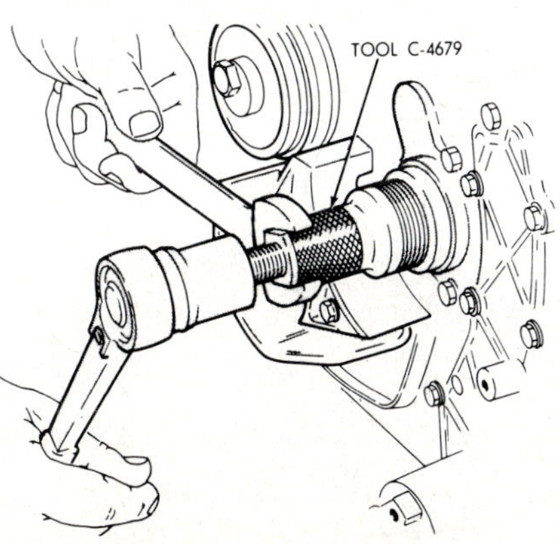

Fig. 12 Crankshaft intermediate shaft & camshaft oil seal removal

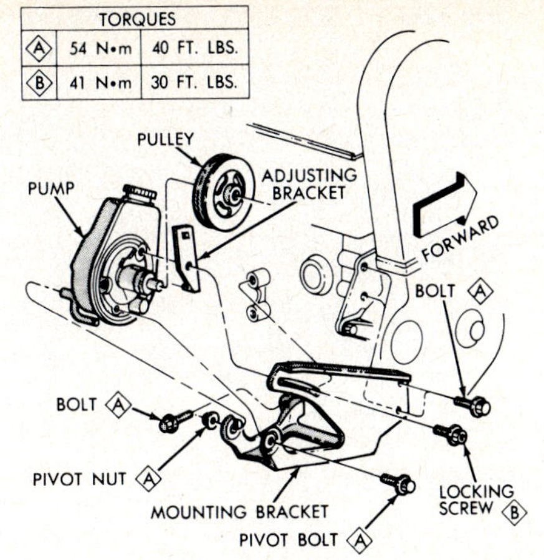

Fig. 9 Power steering pump mounting bracket

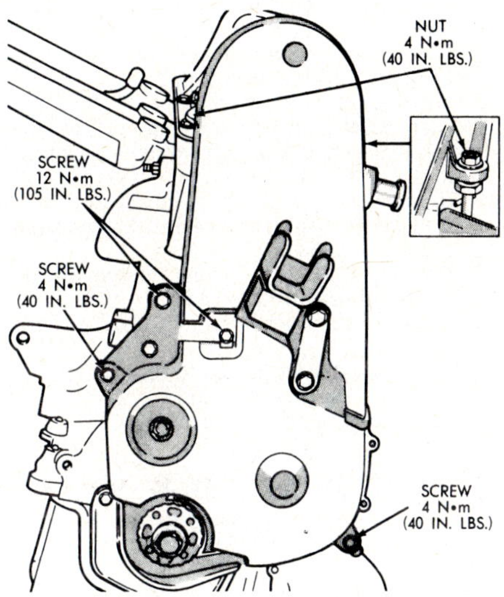

Fig. 11 Timing belt cover removal

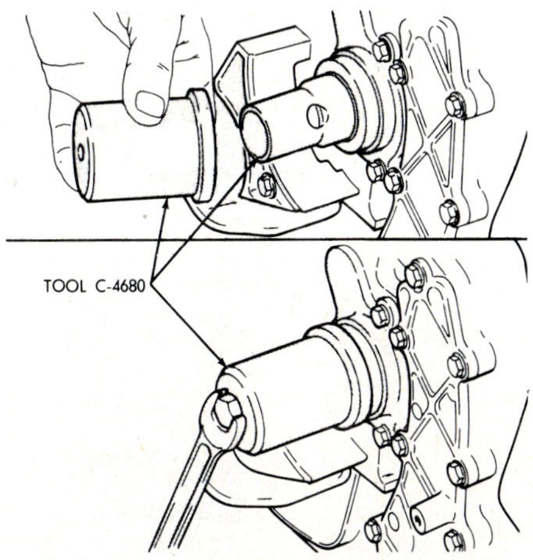

Fig. 13 Crankshaft intermediate shaft & camshaft oil seal installation

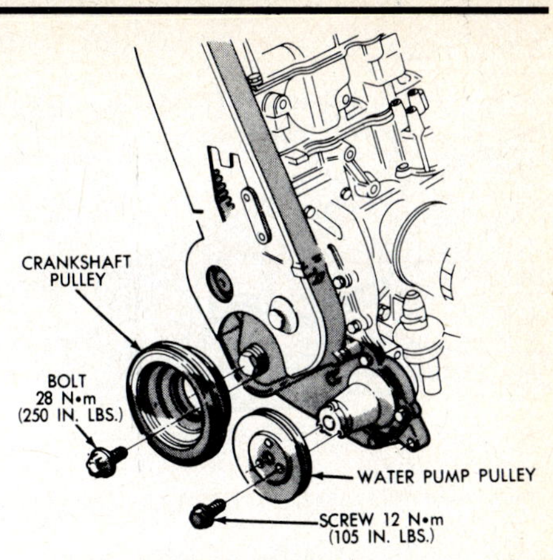

Fig. 10 Crankshaft and water pump pulley removal

2. Remove crankshaft sprocket using a suitable puller.

Crankshaft, Intermediate Shaft & Camshaft Oil Seal Service

Refer to Figs. 12 and 13, for removal and installation of crankshaft, intermediate shaft or camshaft seals.

Crankshaft & Intermediate Shaft Timing

1. Rotate crankshaft and intermediate shaft until markings on sprockets are aligned, Fig. 14.

Camshaft Timing

1. Rotate camshaft until arrows on hub are aligned with No. 1 camshaft cap to cylinder head line. Small hole must be located along vertical centerline.
2. Install timing belt. Refer to "Adjusting Drive Belt Tension" described elsewhere, for proper drive belt adjustment.
3. Rotate crankshaft two full revolutions and recheck timing.

NOTE: Do not allow oil or solvents to contact the timing belt, since they will deteriorate the rubber and cause tooth slippage.

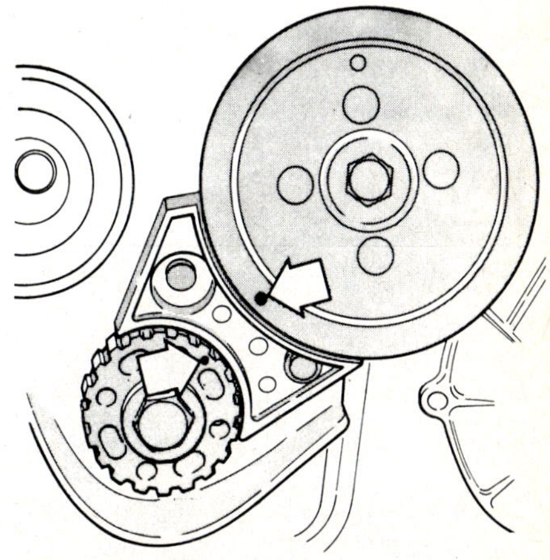

Fig. 14 Aligning crankshaft & intermediate shaft timing marks

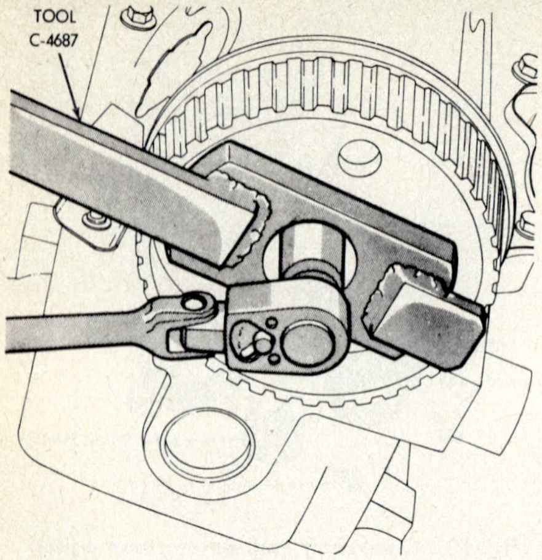

Fig. 15 Camshaft & intermediate shaft replacement

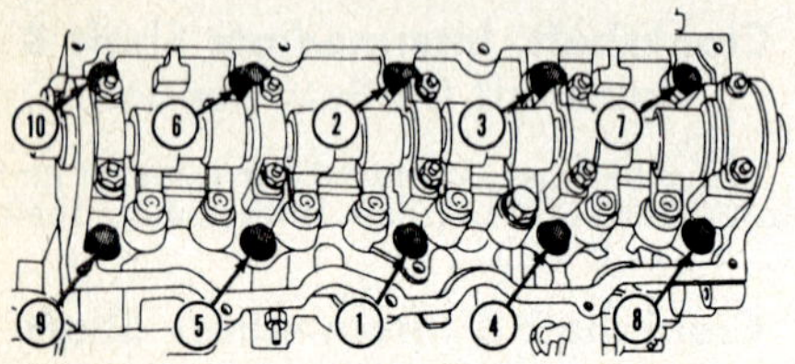

Fig. 18 Cylinder head bolt tightening sequence

Camshaft & Intermediate Shaft Removal & Installation

Refer to Fig. 15, for removal and installation of camshaft and intermediate shaft sprocket.

Adjusting Drive Belt Tension

1. Remove spark plugs, then rotate crankshaft to TDC position.
2. Using a suitable tool, loosen tensioner lock nut, Fig. 16.
3. Reset tension so that belt tensioning tool's axis is within 15° of horizontal.
4. Rotate crankshaft two revolutions in a clockwise direction and position at TDC, then tighten tensioner lock nut.

INTAKE & EXHAUST MANIFOLD

1. Disconnect battery ground cable and drain coolant system.

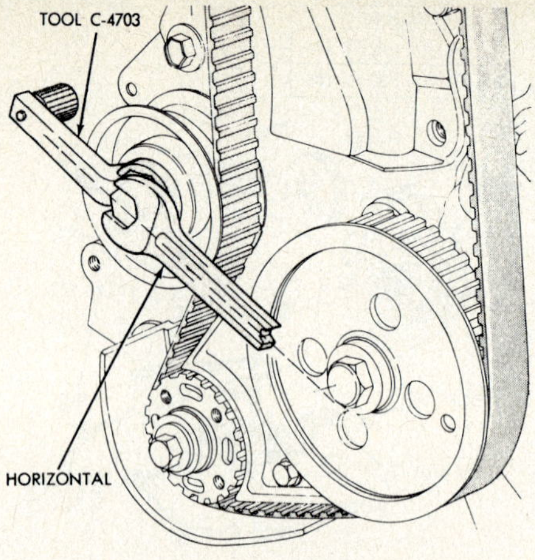

Fig. 16 Adjusting drive belt tension

2. Remove air cleaner and disconnect all vacuum and fuel lines, and electrical connectors from carburetor.
3. Disconnect throttle linkage, and remove power steering pump drive belt.
4. Disconnect power brake vacuum hose from carburetor, if equipped.
5. Disconnect hose from water crossover, then raise and support vehicle and disconnect exhaust pipe from exhaust manifold.
6. Remove power steering pump and position aside, then remove intake manifold support bracket.
7. Remove EGR tube, then remove intake manifold retaining screws.
8. Lower vehicle and remove intake manifold.
9. Remove exhaust manifold retaining nuts, then the exhaust manifold.
10. Reverse procedure to install.

CYLINDER HEAD & VALVE ASSEMBLY

Removing & Installing Valve Springs

Cylinder Head-Off Engine

1. Using a suitable tool to hold valve in the head, compress valve springs enough to remove and install valve stem locks, Fig. 17.

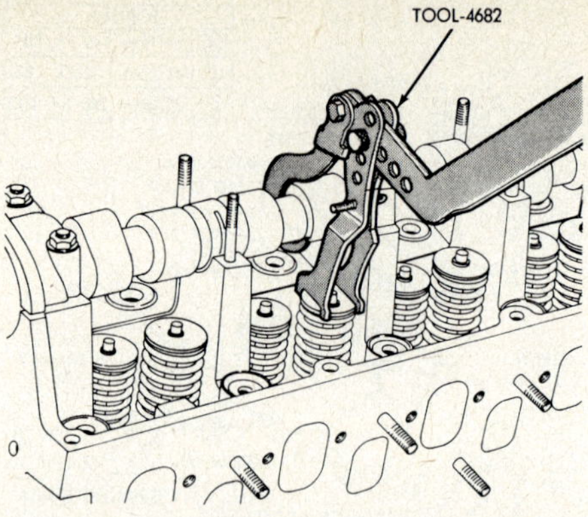

Fig. 17 Valve spring removal and installation

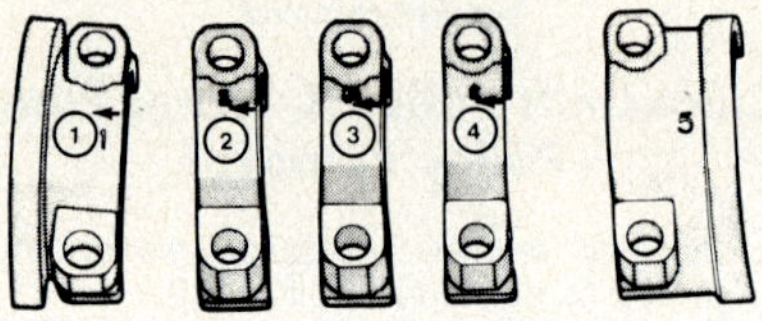

Fig. 19 Camshaft bearing cap installation

Cylinder Head-On Engine

1. Rotate crankshaft until piston is at TDC on compression stroke.
2. Apply 90–120 psi of compressed air into spark plug hole of valve being removed.
3. Using suitable tool, compress valve enough to remove valve stem locks.
4. Remove valve spring and spring seat.
5. Remove valve seal.

Cylinder Head Bolt Tightening Sequence

Refer to Fig. 18 for cyinder head bolt tightening sequence.

Camshaft Bearing Caps

1. With caps removed from engine, check oil holes for obstructions.
2. With caps aligned in proper sequence, make sure arrow on caps 1, 2, 3 and 4 point toward timing belt, Fig. 19.

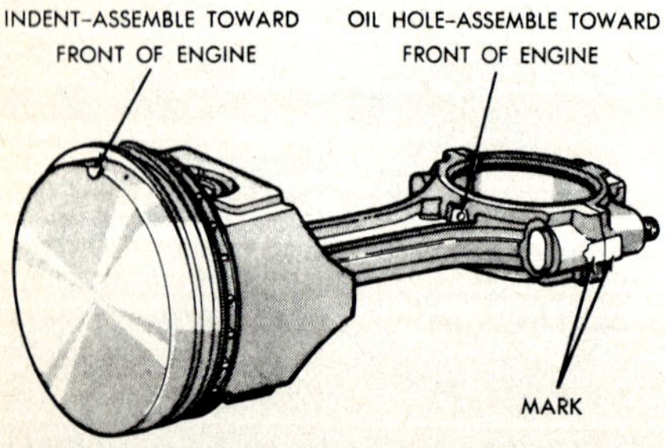

Fig. 20 Piston & connecting rod assembly

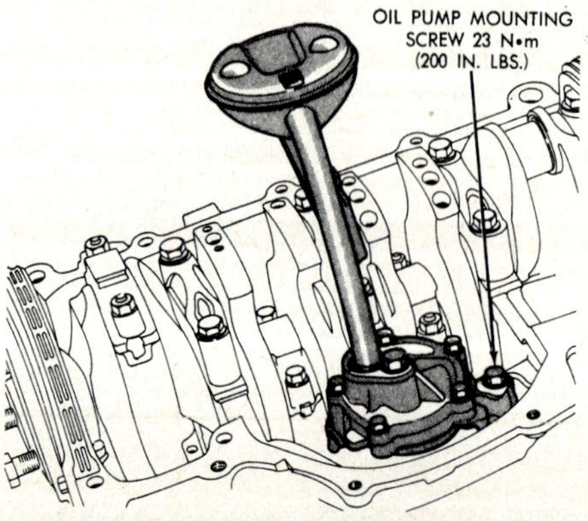

Fig. 21 Oil pump assembly

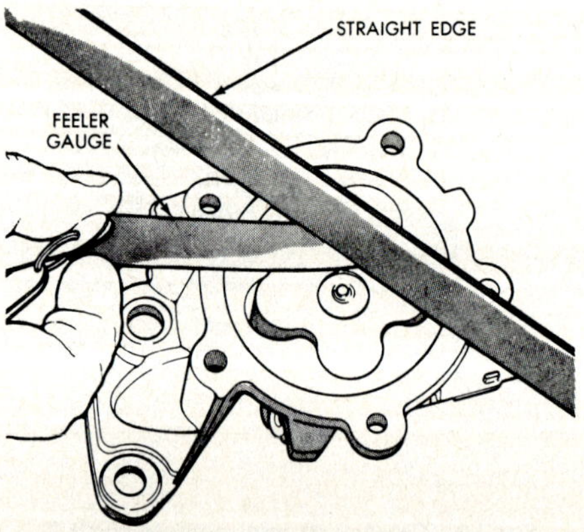

Fig. 22 Checking oil pump end play

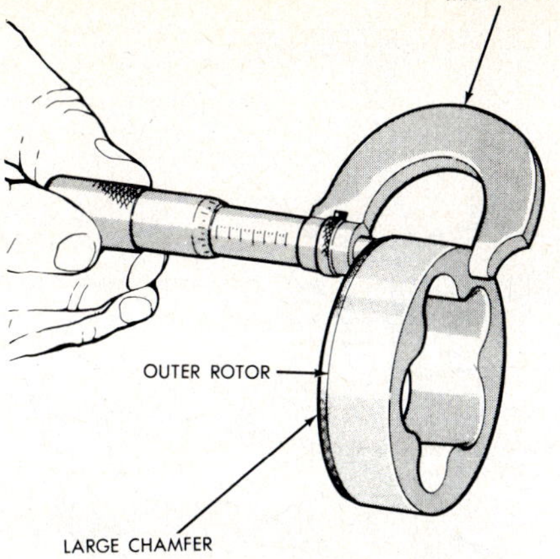

Fig. 23 Measuring oil pump outer rotor thickness

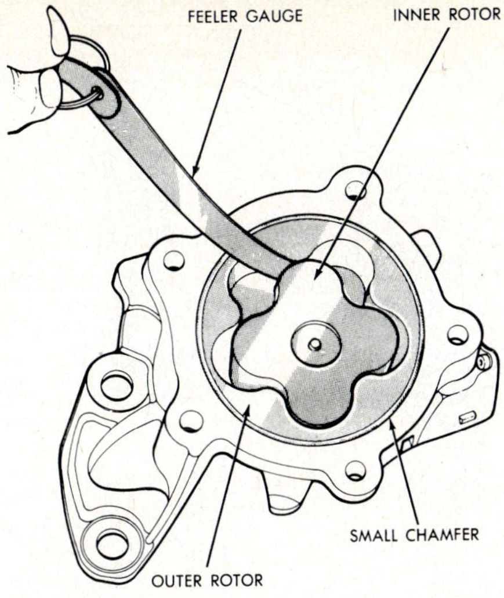

Fig. 24 Measuring clearance between oil pump rotors

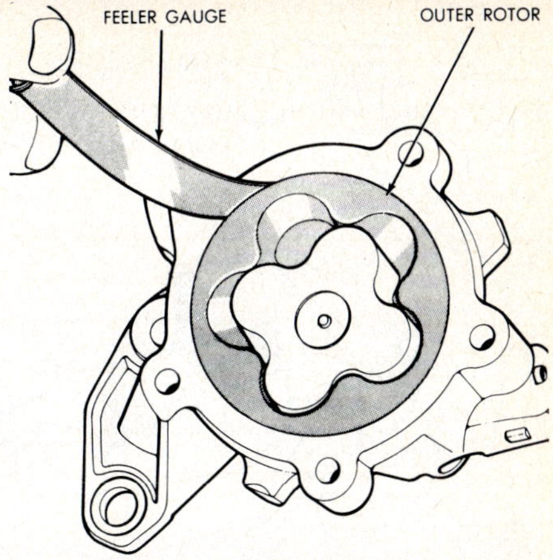

Fig. 25 Measuring oil pump outer rotor clearance

3. Apply suitable sealant to No. 1 and 5 bearing caps.
4. Install caps before installing camshaft seals, then torque cap bolts to 165 inch lbs.

PISTON & ROD ASSEMBLY

When installing the piston and rod assembly, the indentation on the top of piston must face the timing belt side of the engine, Fig. 20. The oil hole on connecting rod must also face the timing belt side of the engine and be on the same side as the indented mark on the piston.

ENGINE LUBRICATION SYSTEM

Oil Pump Assembly

1. With oil pan removed, remove the screw securing oil pump to the cylinder block, Fig. 21.

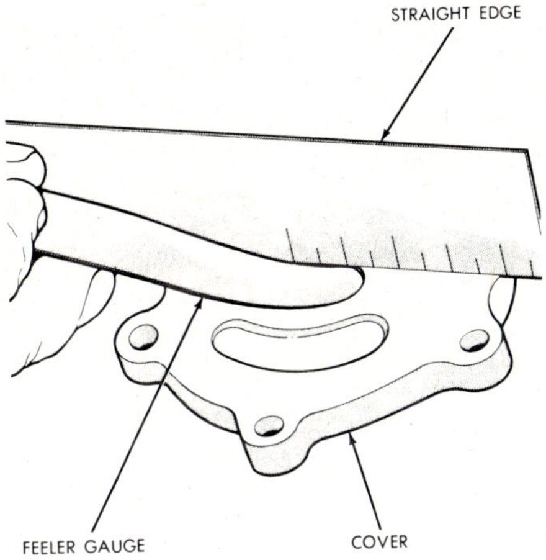

Fig. 26 Measuring oil pump cover clearance

2. Reverse procedure to install. Torque oil pump attaching screw to 200 inch lbs.

Oil Pump Service

1. Measure the following oil pump clearances:
   a. End play, Fig. 22. End play should be .001–.006 inch on 1982 vehicles, or .001–.004 inch on 1983 vehicles.
   b. Outer rotor thickness, Fig. 23. Thickness should be .825 inch, minimum. Install outer rotor with chamfered edge in pump body.
   c. Clearance between rotors, Fig. 24, should be .010 inch, maximum.
   d. Outer rotor clearance, Fig. 25. Clearance should be .014 inch, maximum.
   e. Oil pump cover clearance, Fig. 25, should be no greater than .015 inch on 1982 vehicles, or .003 inch on 1983 vehicles.
   f. Oil pressure relief valve spring length should be 1.95 inches.

Carburetor Section

Holley Model 5220 Adjustment Specifications

Year	Model No.	Float Level	Float Drop	Acc Pump Hole No.	Choke Vacuum Kick	Choke Setting
1982	R-9582A	.480	1.875	3	.060	—
	R-9583A	.480	1.875	3	.060	—
	R-9584A	.480	1.875	3	.060	—
	R-9585A	.480	1.875	3	.060	—
1983	R-40020A	.480	1.875	3	.055	—
	R-40022A	.480	1.875	3	.055	—

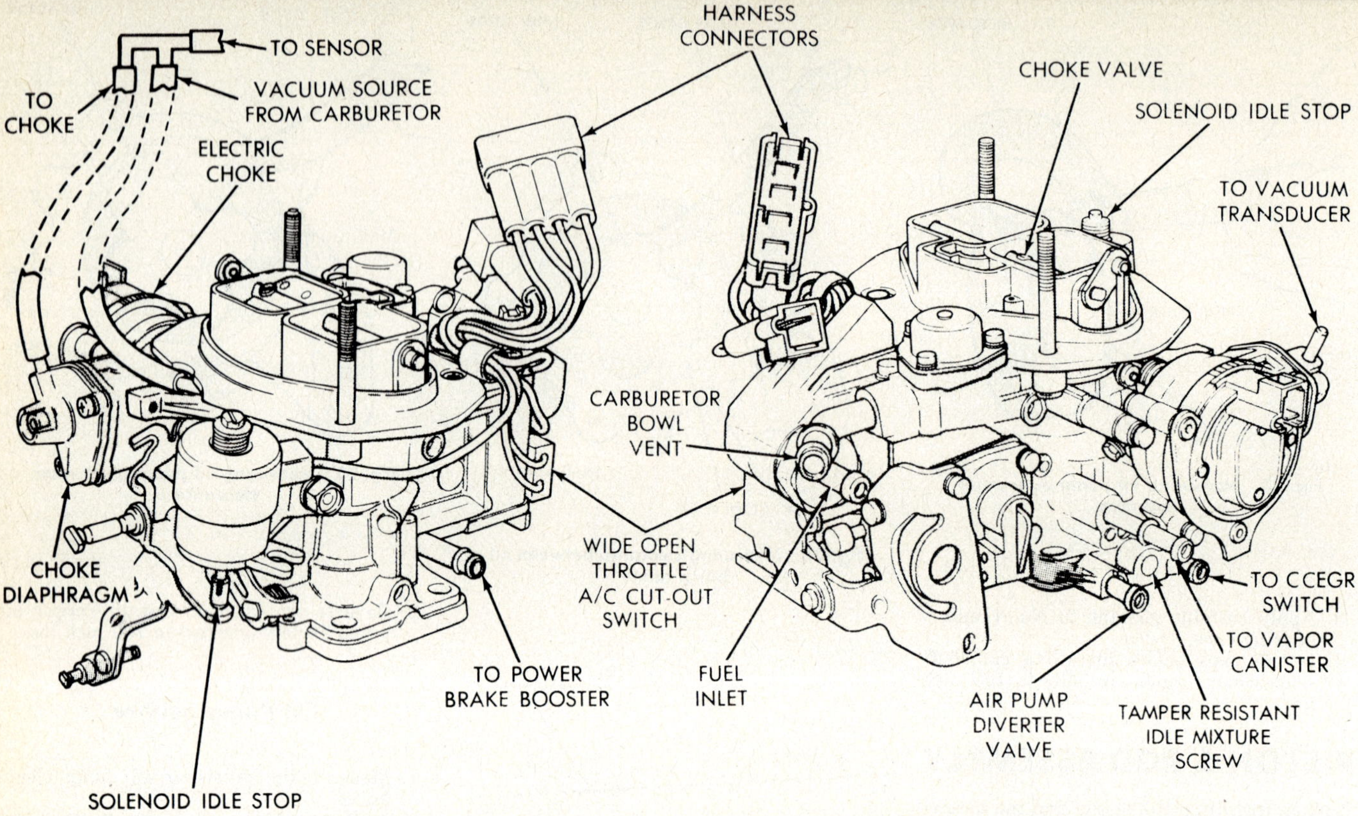

Fig. 1 Holley model 5220 carburetor. Vehicle equipped with air conditioning (typical)

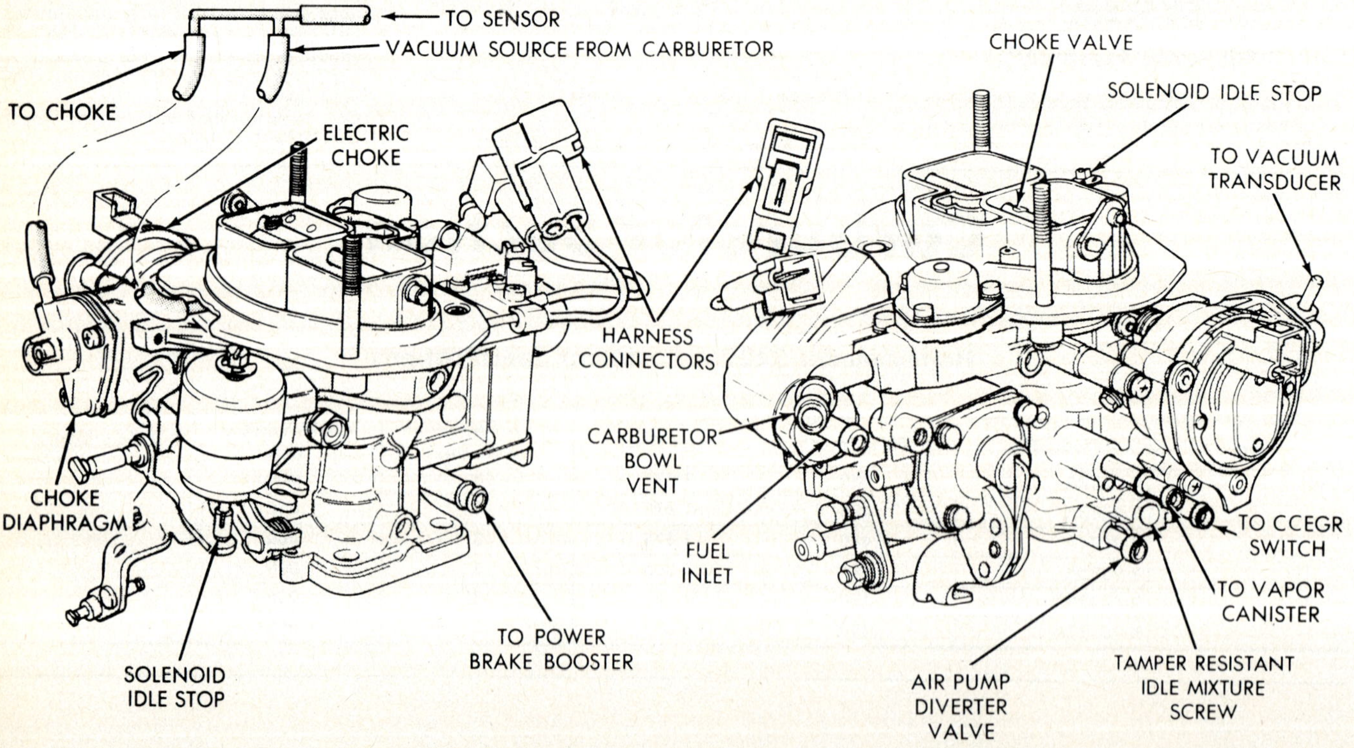

Fig. 2 Holley model 5220 carburetor. Vehicle less air conditioning (typical)

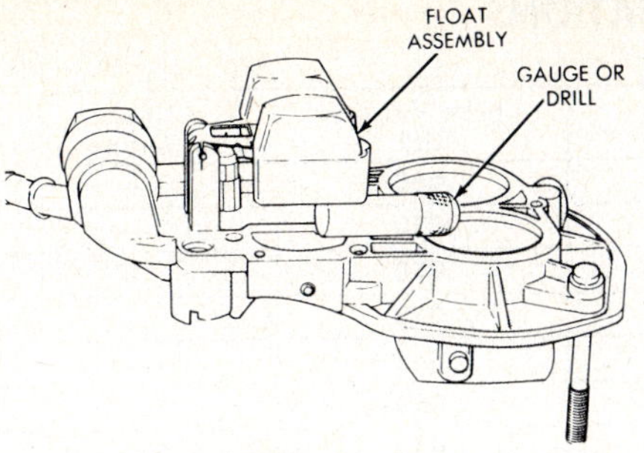

Fig. 3 Measuring float level. 5220 carburetor

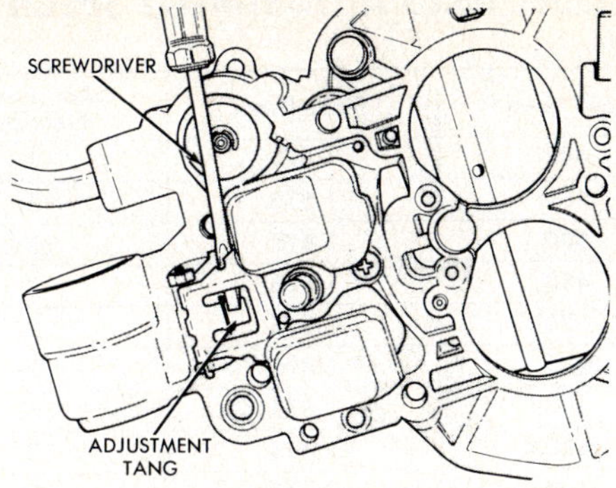

Fig. 4 Adjusting float level. 5220 carburetor

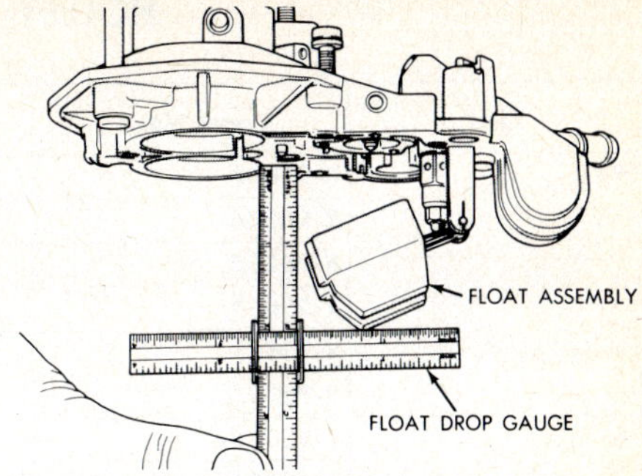

Fig. 5 Measuring float drop. 5220 carburetor

HOLLEY MODEL 5220 DESCRIPTION

The Holley model 5220, Figs. 1 and 2, is a staged dual venturi carburetor. The primary bore or venturi is smaller than the secondary bore. The secondary stage is mechanically operated by linkage connecting the primary and secondary throttle levers. The primary stage includes a curb idle and transfer system, diaphragm-type accelerator pump system, main metering system and power enrichment system. The secondary stage includes a main metering system and power system. Both the primary and secondary venturi draw fuel from a common fuel bowl. The electric automatic choke has a bimetal two-stage heating element. The carburetor also has an electronic solenoid and a vacuum-operated bowl vent.

FLOAT LEVEL ADJUSTMENT

Figs. 3 & 4—Invert air horn and insert specified gauge between float and air horn. To adjust, use a small screwdriver to bend tang.

FLOAT DROP ADJUSTMENT

Figs. 5 & 6—Using a suitable depth gauge, measure float drop. To adjust, use a small screwdriver to bend tang.

CHOKE VACUUM KICK ADJUSTMENT

Fig. 7—Open throttle, close choke, then close throttle to trap fast idle system in closed choke position. Using an external vacuum source, apply 15 inches of vacuum to choke diaphragm. Apply closing pressure to position choke at smallest opening without distorting linkage. An internal spring will compress to a stop inside choke system. Insert specified gauge between upper edge of choke valve and air horn wall at primary throttle end of carburetor. To adjust, rotate allen head screw in center of diaphragm housing.

FAST IDLE SPEED ADJUSTMENT

Fig. 8—Remove top of air cleaner and eliminate EGR signal. Cap or plug all disconnected vacuum fittings. Do not disconnect vacuum hose to spark control computer. Use a jumper wire to ground idle stop switch. Turn A/C "Off." Disconnect engine cooling fan at radiator and complete the circuit at plug with a jumper wire to energize fan. Apply parking brake, place transmission in Neutral with engine "Off," open throttle slightly and place lowest speed step of fast idle cam under adjusting screw. Start engine and determine stabilized speed. The choke valve must be fully open. If engine continues to rise, the idle stop switch is not grounded properly. To adjust speed, turn speed screw to obtain proper RPM. Reposition screw on cam after each adjustment to provide proper throttle closing torque and screw position.

CHOKE SETTING ADJUSTMENT

Loosen choke coil cap retaining screws and rotate cap assembly to align mark on cap with specified mark on housing. Tighten choke coil cap retaining screws.

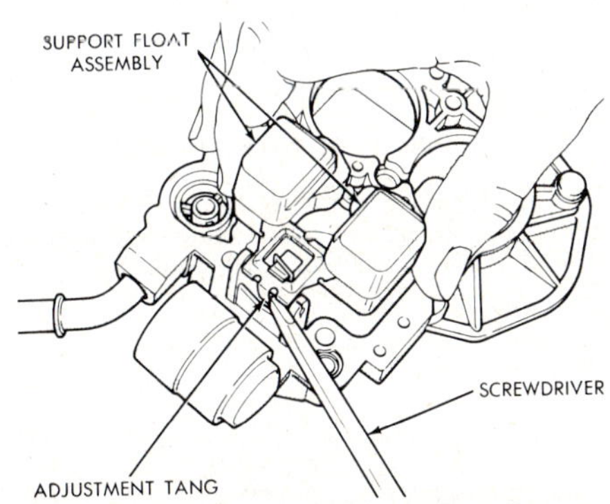

Fig. 6 Adjusting float drop. 5220 carburetor

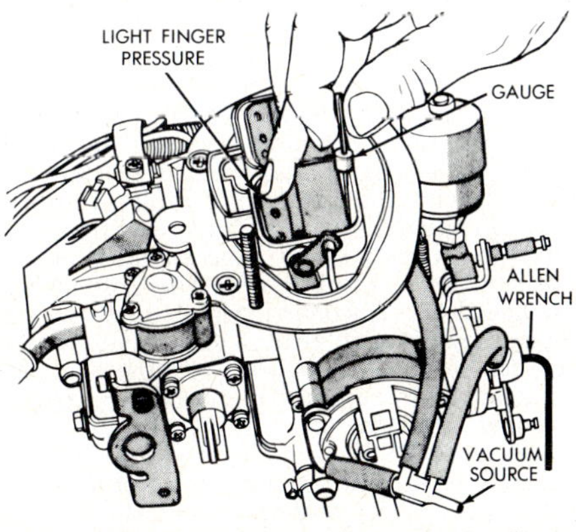

Fig. 7 Choke vacuum kick adjustment. 5220 carburetor

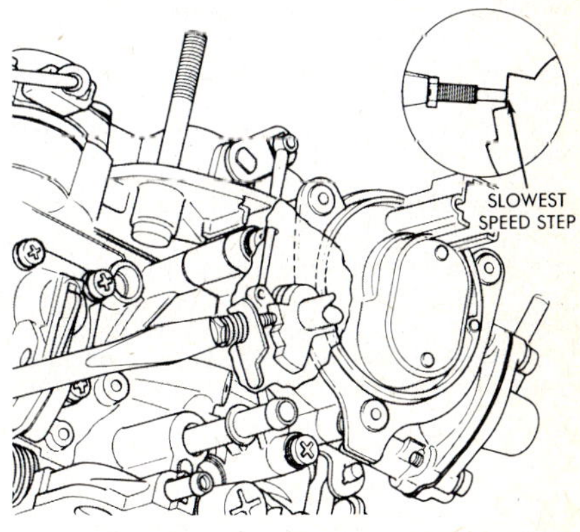

Fig. 8 Fast idle speed adjustment. 5220 carburetor (typical)

HOLLEY MODEL 6520 ADJUSTMENT SPECIFICATIONS

Year	Model No.	Float Level	Float Drop	Acc. Pump Hole No.	Choke Vacuum Kick	Choke Setting
1982	R-9503A	.480	1.875	3	.060	—
	R-9504A	.480	1.875	3	.070	—
	R-9505A	.480	1.875	3	.070	—
	R-9506A	.480	1.875	3	.070	—
	R-9507A	.480	1.875	3	.085	—
	R-9508A	.480	1.875	3	.085	—
1983	R-40003A	.480	1.875	3	.070	—
	R-40004A	.480	1.875	3	.065	—
	R-40005A	.480	1.875	3	.080	—
	R-40006A	.480	1.875	3	.080	—
	R-40010A	.480	1.875	3	.065	—
	R-40014A	.480	1.875	3	.080	—

HOLLEY MODEL 6520

The Holley model 6520, Fig. 1, is a staged dual venturi electronic feedback carburetor. The primary bore is smaller than the secondary bore. The secondary stage is mechanically operated by linkage connecting the primary and secondary throttle levers. The primary stage includes a curb idle and transfer system, diaphragm type accelerator pump system, main metering system and a fuel regulator solenoid responsive to oxygen sensor. The secondary stage includes a main metering system and power system. Both the primary and secondary venturi draw fuel from a common fuel bowl. The electric automatic choke incorporates a two stage heating element. On manual transmission carburetors, only one choke valve is used, while on automatic transmission carburetors, two choke valves are used.

The 6250 electronic feedback carburetor also incorporates a duty cycle solenoid which provides a limited regulation of air-fuel ratio in response to electrical signals from the spark control computer. The solenoid meters the main fuel system and operates in parallel with a conventional fixed main metering jet. When there is no electrical signal applied to the solenoid, the valve spring pushes upward through the main system fuel valve, fully uncovering the solenoid controlled main metering orifice so that the richest condition exists within the carburetor for any given air-flow. When the electrical signal is applied to the solenoid, the field windings are energized, thus causing the armature to move the push rod and main system valve downward against the valve spring. This movement will continue until the main system valve bottoms against the main system valve seat. In this position the solenoid controlled main metering orifice is fully sealed so that the leanest condition exists within the carburetor for any given airflow. This condition will remain unchanged until the signal from the spark control computer to the solenoid is switched off. The main system fuel may be regulated between richest and leanest limits by controlling the amount of time that the solenoid is in the power on position. Under normal operating conditions, 12 volts at a frequency of 10 Hz is applied to the field windings. By controlling the duration of the voltage signal, the power on time to total time, referred to as the duty cycle, is established.

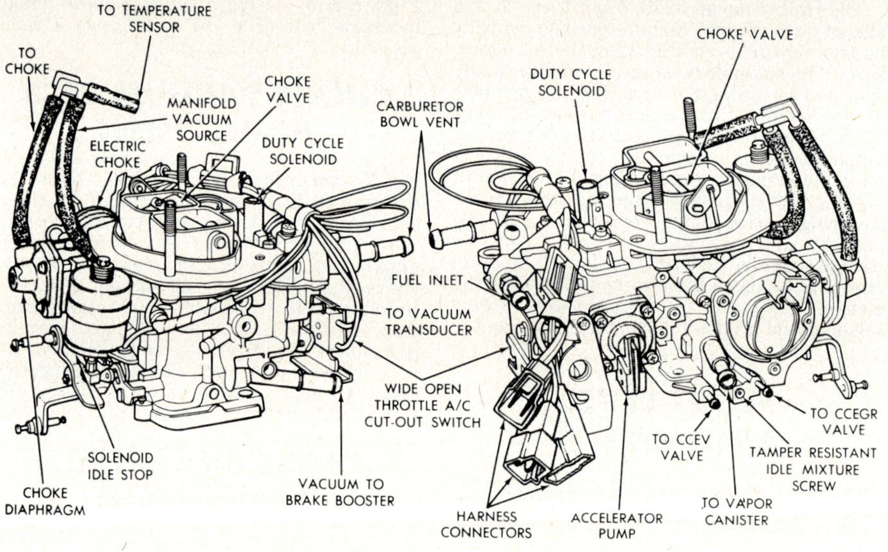

Fig. 1 Holley Model 6520 Carburetor

Float Level Adjustment

Figs. 2 & 3—Invert air horn and insert specified gauge between float and air horn. To adjust, use a small screwdriver to bend tang.

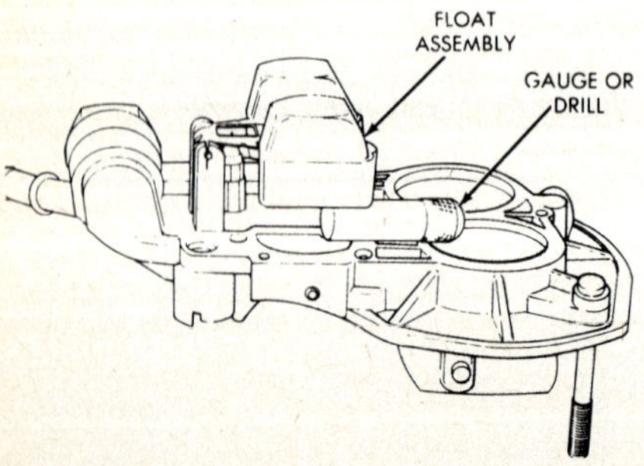

Fig. 2 Measuring float level. 6520 carburetor

Float Drop Adjustment

Figs. 4 & 5—Using a suitable depth gauge measure float drop. To adjust, use a small screwdriver to bend tang.

Choke Vacuum Kick Adjustment

Fig. 6—Open throttle and close choke valve, then close throttle to retain choke valve in the closed position. Disconnect vacuum hose from

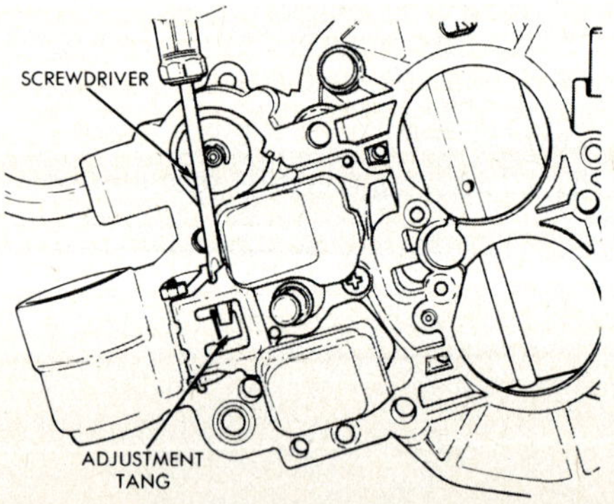

Fig. 3 Adjusting float level. 6520 carburetor

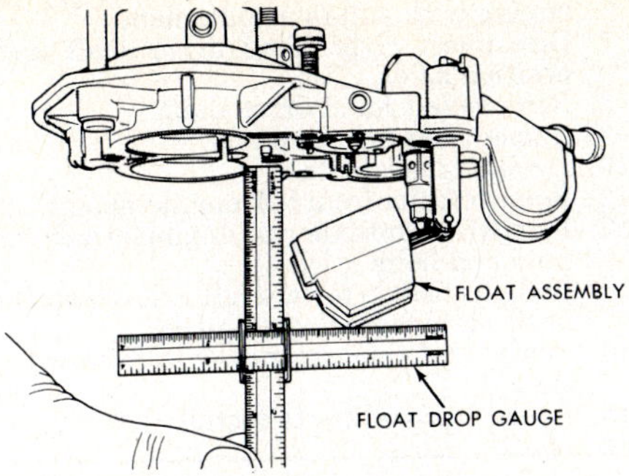

Fig. 4 Measuring float drop. 6520 carburetor

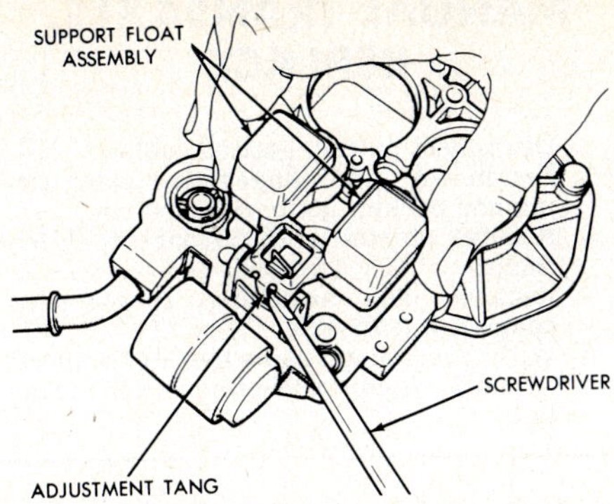

Fig. 5 Adjusting float drop. 6520 carburetor

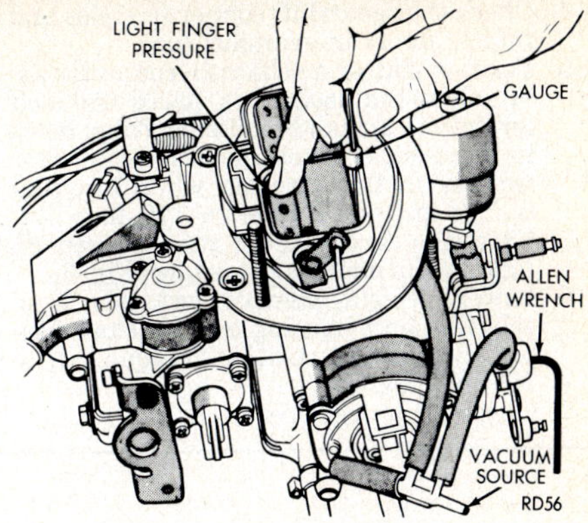

Fig. 6 Adjusting choke vacuum kick. 6520 carburetor

carburetor and connect an external vacuum source. Apply 15 inches Hg. of vacuum or more. Apply sufficient closing pressure to choke valve, using care not to distort linkage. Note that an internal spring will compress to a stop within the choke system. Insert the specified gauge in center area between top of choke valve and air horn wall at primary throttle end of carburetor. Adjust clearance by rotating the allen head screw in center of diaphragm housing.

Clutch & Manual Transaxle Section

CLUTCH, ADJUST

The clutch release cable, Fig. 1, on these models cannot be adjusted. When the cable is properly routed, the spring between the clutch pedal and positioner adjuster will hold the clutch cable in the proper position. An adjuster pivot is used to hold release cable in place to ensure complete clutch release when the clutch pedal is depressed.

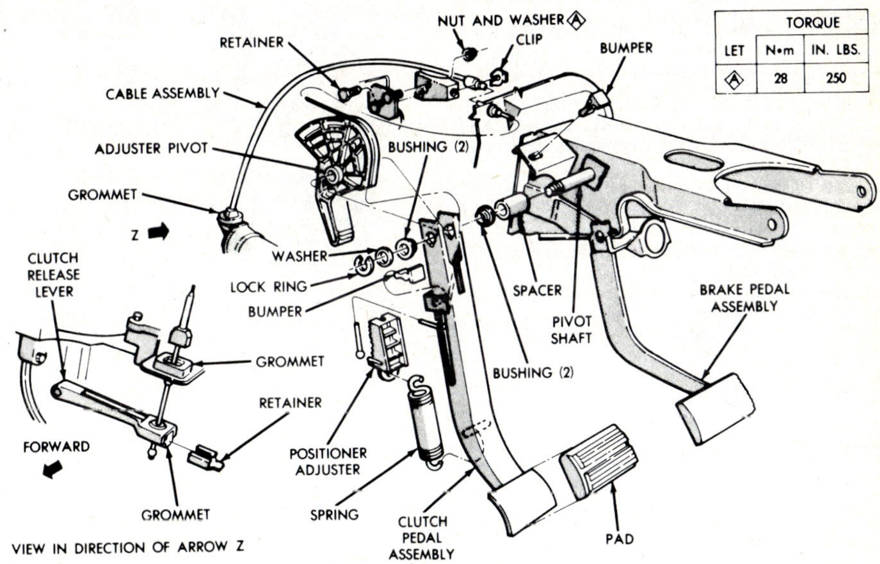

Fig. 1 Clutch cable routing

CLUTCH, REPLACE

1. Remove transmission as outlined under "Manual Transaxle, Replace" procedure.
2. Mark relationship between clutch cover and flywheel for reference during reassembly, then insert suitable clutch disc aligning tool through clutch disc hub.
3. Gradually loosen clutch cover attaching bolts, then remove pressure plate and cover assembly and disc from flywheel.
4. Remove clutch release shaft and slide release bearing assembly off input shaft seal retainer. Remove fork from release bearing thrust plate.
5. Reverse procedure to install. Align reference marks made during reassembly. Using clutch disc alignment tool, install disc, plate and cover to flywheel and torque bolts to 21 ft. lbs.

GEARSHIFT LINKAGE, ADJUST

1. Remove lock pin from transaxle selector shaft housing, Fig. 2.
2. Reverse lock pin, so long end is facing downward, and insert pin into same threaded hole while pushing selector shaft into selector housing.
3. Raise and support vehicle, then loosen clamp bolt that secures gearshift tube to gearshift rod.

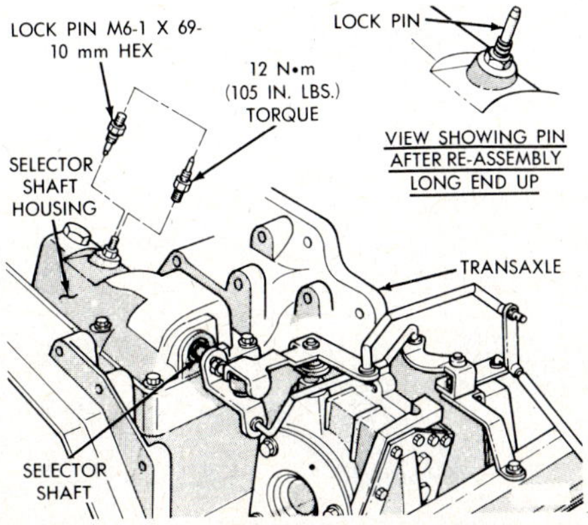

Fig. 2 Lock pin removal & installation

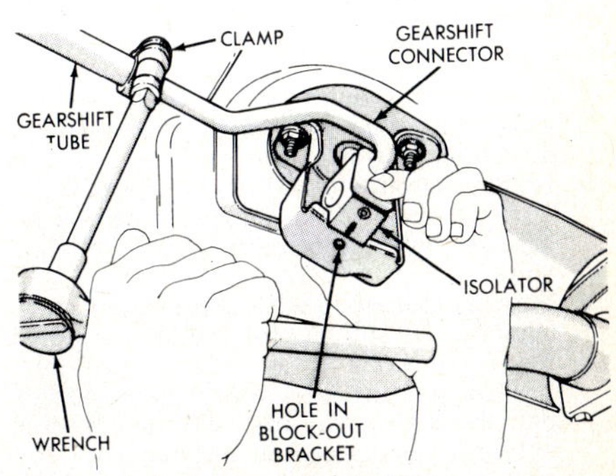

Fig. 3 Adjusting gear shift linkage

4. Check that gearshift connector slides and rotates freely in gearshift tube.
5. Position shifter mechanism connector assembly so that isolater is spaced .050 inch away from upstanding flange. Align holes in block-out bracket, then hold connector isolater in this position while torquing nut on gearshift tube to 170 inch lbs. Excessive force should not be used on linkage during this procedure, Fig. 3.
6. Lower vehicle, remove lock pin from selector shaft housing and reinstall lock pin in reversed position. Torque pin to 105 inch lbs.
7. Check for proper operation.

MANUAL TRANSAXLE, REPLACE

1. Disconnect battery ground cable.
2. Disconnect shift linkage rods, starter wiring and backup lamp switch wiring.
3. Remove starter and disconnect clutch cable.
4. Remove bolt attaching speedometer adapter to transaxle.
5. With speedometer cable housing connected, pull adapter and pinion from transaxle.
6. Raise and support vehicle. Also, support engine with suitable equipment.
7. Disconnect right hand drive shaft and position aside.
8. Remove left hand drive shaft.
9. Remove left splash shield.
10. Drain transaxle fluid.
11. Remove bolts from left engine mount.
12. Remove transaxle to engine-attaching nuts and bolts.
13. Slide transaxle toward left side of vehicle until mainshaft clears clutch.
14. Lower and remove transaxle from vehicle.
15. Reverse procedure to install.

Torqueflite A-413 Automatic Transaxle

TRANSAXLE, REPLACE

NOTE: The transaxle and converter must be removed as an assembly.

1. Disconnect battery ground cable.
2. Disconnect transaxle shift control and throttle cables from transaxle and position aside.
3. On 1982 models, disconnect upper oil cooler hose. On 1983 models, disconnect both upper and lower oil cooler hoses.
4. Support engine with suitable engine lifting equipment.
5. Remove upper bell housing bolts.
6. Raise and support vehicle, then remove front wheels and left splash shield.
7. On 1982 models, remove differential cover.
8. Remove speedometer adapter, cable and pinion as an assembly, Fig. 1.
9. Remove sway bar and both lower ball joint to steering knuckle bolts, then, using a suitable pry bar, remove lower ball joint.
10. Remove drive shaft from hub.
11. On 1982 models, rotate both drive shafts until circlip ends are visible. Compress circlip and pry drive shaft out of side gear, Fig. 2.
12. Remove both drive shafts.
13. Remove dust cover, then mark position of torque converter on drive plate and remove torque converter mounting bolts. Remove plug from access hole in right splash shield to rotate crankcase for bolt removal.
14. Disconnect Neutral/Park safety switch electrical connector and, on 1982 models, disconnect lower oil cooler hose.
15. Remove engine mount bracket from front crossmember, front mount insulator through-bolt and remaining bell housing bolts.
16. Support transaxle with a suitable jack.
17. Remove left side engine mount.
18. Move transaxle away from engine and lower from vehicle. It may be necessary to pry transaxle away from engine between extension housing and engine block for proper clearance.
19. Reverse procedure to install. When in-

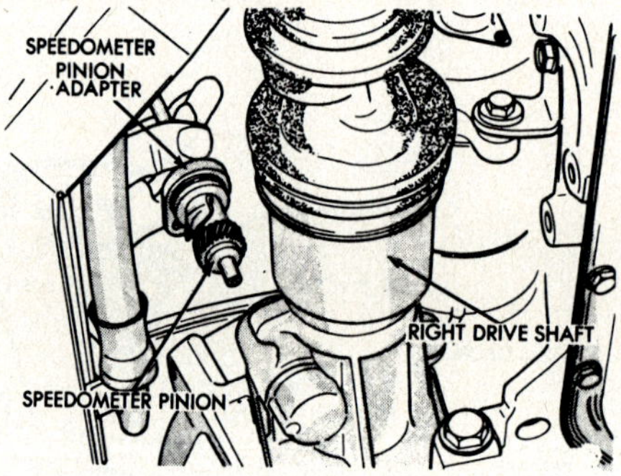

Fig. 1 Removing speedometer adapter, cable and pinion

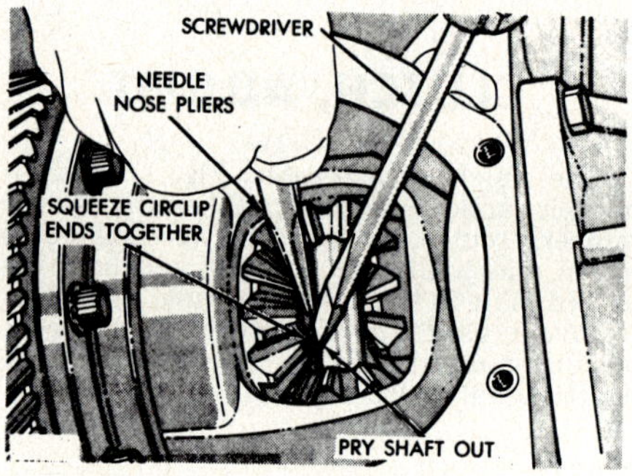

Fig. 2 Prying drive shaft out of side gear. 1982

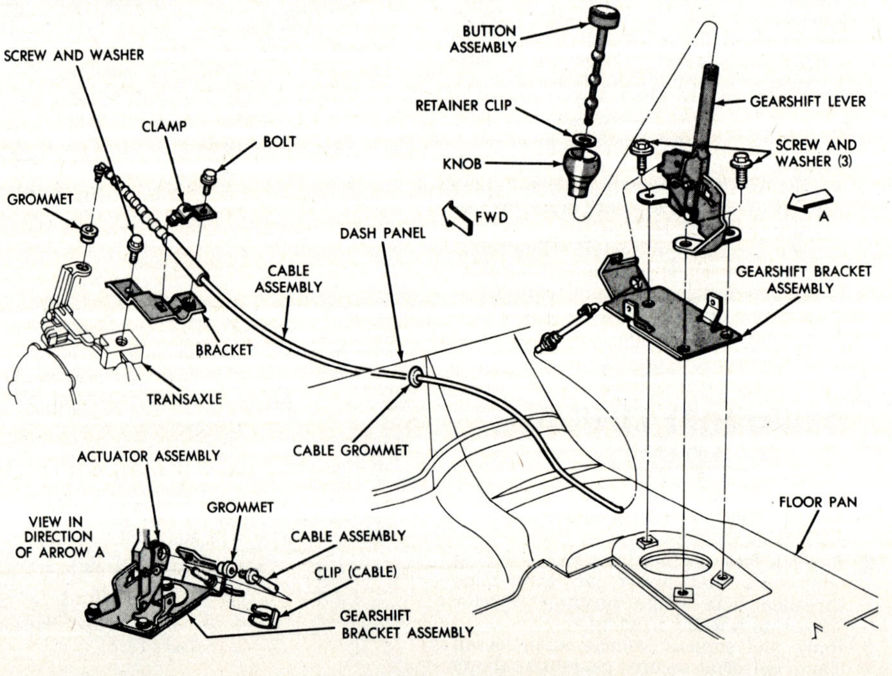

Fig. 3 Gearshift linkage

stalling differential cover on 1982 models, apply a 1/8 inch bead of RTV sealant to cover.

GEARSHIFT LINKAGE, ADJUST

NOTE: When linkage cable must be disconnected from levers which use plastic grommets as retainers, the grommets must be replaced.

1. Place selector lever in "P" position.
2. Pull shift lever, Fig. 3, to front detent (Park) position with approximately 10 lbs. of pressure.
3. While maintaining pressure on shift lever, torque lock nut bolt to 90 inch lbs. and check adjustment.

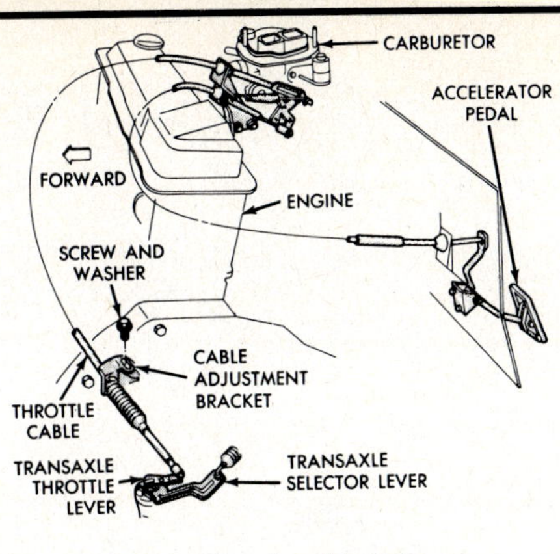

Fig. 4 Throttle cable linkage

THROTTLE CABLE, ADJUST

1. Perform adjustment with engine at operating temperature, otherwise ensure carburetor is not on fast idle cam by disconnecting choke.
2. Loosen adjusting bracket lock screw, Fig. 4.
3. Hold transmission lever rearward against internal stop and tighten adjusting bracket lock screw to 105 inch lbs.

Front Suspension & Steering Section

DESCRIPTION

These vehicles use a MacPherson-type front suspension with the vertical shock absorber struts attached to the upper fender reinforcement and the steering knuckle, Fig. 1. The lower control arms are attached inboard to a crossmember and outboard to the steering knuckle through a ball joint to provide lower steering knuckle position. During steering maneuvers, the strut and steering knuckle rotate as an assembly.

The drive shafts are attached inboard to the transaxle output drive flanges and outboard to the driven wheel hub.

WHEEL ALIGNMENT

Caster

The caster angle on these vehicles cannot be adjusted.

Camber

To adjust camber, loosen the cam and through bolts, Fig. 2. Rotate the upper cam bolt to move the top of the wheel in or out to achieve the specified camber angle. Torque cam bolts to 45 ft. lbs., then advance bolts an additional 1/4 turn (90 degrees).

Toe-In

To adjust toe-in, center the steering wheel and hold in position with a suitable tool. Loosen the tie rod lock nuts and rotate the rod, Fig. 3, to adjust toe-in to specifications. Use care not to twist the steering gear rubber boots. Torque the tie rod lock nuts to 55 ft. lbs. (75 Nm). Adjust position of steering gear rubber boots. Remove steering wheel holding tool.

STRUT DAMPER ASSEMBLY, REPLACE

Removal

1. Raise and support vehicle, then remove front wheels.
2. Mark position of camber adjusting cam, then remove the camber adjusting bolt and through bolt and the brake hose to damper bracket retaining screw, Fig. 4.
3. Remove strut damper to fender shield mounting nut and washer assemblies.
4. Remove strut damper from vehicle.

Installation

1. Position strut assembly into fender reinforcement, then install retaining nuts and washers and torque to 20 ft. lbs.
2. Position steering knuckle and washer plate to strut, then install upper cam and lower through bolts.
3. Install brake hose retainer on damper, then index cam bolt to alignment mark made during removal.
4. Position a 4-inch or larger, C-clamp on steering knuckle and strut, Fig. 5. Tighten clamp just enough to eliminate any looseness between strut and knuckle. Check alignment of marks made during removal, then tighten bolts to 45 ft. lbs. plus an additional 1/4 turn beyond the specified torque.
5. Remove C-clamp, then install wheel and tire assembly.

COIL SPRING, REPLACE

1. Remove strut damper assembly as outlined previously.
2. Using a suitable tool, compress coil spring.
3. Remove strut rod nut while holding strut rod to prevent rotation.
4. Remove the mount assembly, Fig. 6.
5. Remove coil spring from strut damper.

1. FRONT SUSPENSION CROSSMEMBER
2. FRONT PIVOT BOLT
3. LOWER CONTROL ARM
4. SWAY ELIMINATOR SHAFT ASSEMBLY
5. LOWER ARM BALL JOINT ASSEMBLY
6. STEERING GEAR
7. TIE ROD ASSEMBLY
8. DRIVE SHAFT
9. STEERING KNUCKLE
10. STRUT DAMPER ASSEMBLY
11. COIL SPRING
12. UPPER SPRING SEAT
13. REBOUND STOP
14. UPPER MOUNT ASSEMBLY
15. JOUNCE BUMPER
16. DUST SHIELD

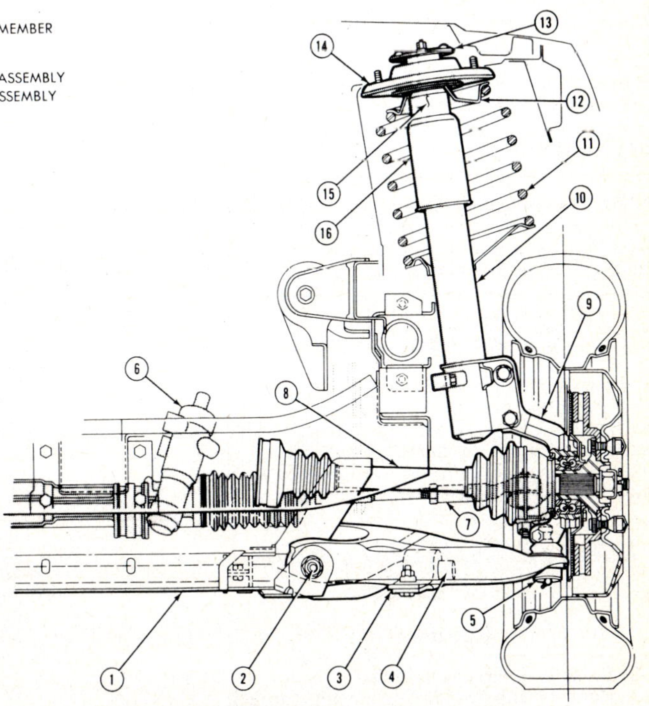

Fig. 1 Front suspension

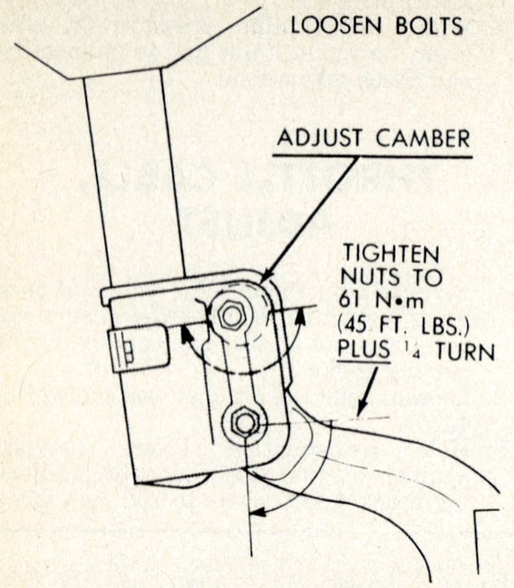

Fig. 2 Chamber adjustment

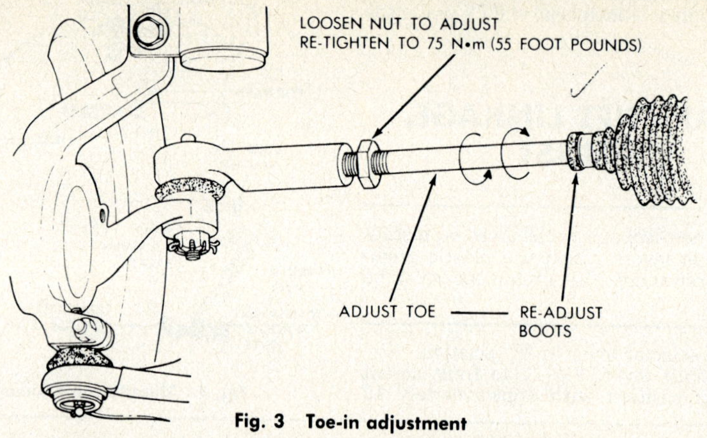

Fig. 3 Toe-in adjustment

6. Inspect mount assembly for deterioration of rubber isolator, retainers for cracks and distortion and bearings for binding.
7. Install the bumper dust shield assembly.
8. Install spring and seat, upper spring retainer, bearing and spacer, mount assembly and the rebound bumper, retainer and rod nut upper.

NOTE: Position the spring retainer alignment notch parallel to the damper lower attaching brackets.

---

9. Torque strut rod nut to 60 ft. lbs. (81 Nm). Do not release spring compressor before torquing nut.
10. Remove spring compressor.

BALL JOINTS

The ball joint is pressed into the lower control arm. On these models, the ball joint can be pressed from the lower control arm using a 1$^{1}/_{16}$ inch deep socket and tool No. C-4699-2. When pressing ball joint into lower control arm, use tools Nos. C-4699-1 and C-4699-2. Install ball joint seal using a 1½ socket and tool No. C-4699-2.

NOTE: On some models the ball joint is welded to the lower control arm. On these models, the ball joint and lower control arm must be replaced as an assembly.

Checking Ball Joints

With weight of vehicle resting on wheel and tire assembly, attempt to move grease fitting with fingers, Fig. 7. Do not use a tool or added force to attempt to move grease fitting. If grease fitting moves freely, ball joint is worn and should be replaced.

LOWER CONTROL ARM, REPLACE

Removal

1. Raise and support vehicle.
2. Remove the front inner pivot through bolt, the rear stub strut nut, retainer and bushing and the ball joint to steering knuckle clamp bolt, Fig. 8.
3. Separate the ball joint from the steering knuckle by prying between the ball stud retainer and the lower control arm.

NOTE: Pulling the steering knuckle "Out" from vehicle after releasing from ball joint can separate inner C/V joint.

---

4. Remove sway bar to control arm nut and reinforcement and rotate control arm over sway bar. Remove rear stub strut bushing, sleeve and retainer.

Installation

1. Install retainer, bushing and sleeve on stub strut.
2. Position control arm over sway bar and install rear stub strut and front pivot into crossmember.
3. Install front pivot bolt and loosely assemble nut, Fig. 8.
4. Install stub strut bushing and retainer and loosely assemble nut.
5. Place sway bar bracket stud through control arm and install retainer and nut. Torque nut to 22 ft. lbs. (30 Nm).
6. Install ball joint stud into steering knuckle, then the clamp bolt. Torque bolt to 50 ft. lbs. (67 Nm).
7. Lower vehicle and with suspension support vehicle torque front pivot bolt to 105 ft. lbs. (142 Nm) and the stub strut nut to 70 ft. lbs. (94 Nm).

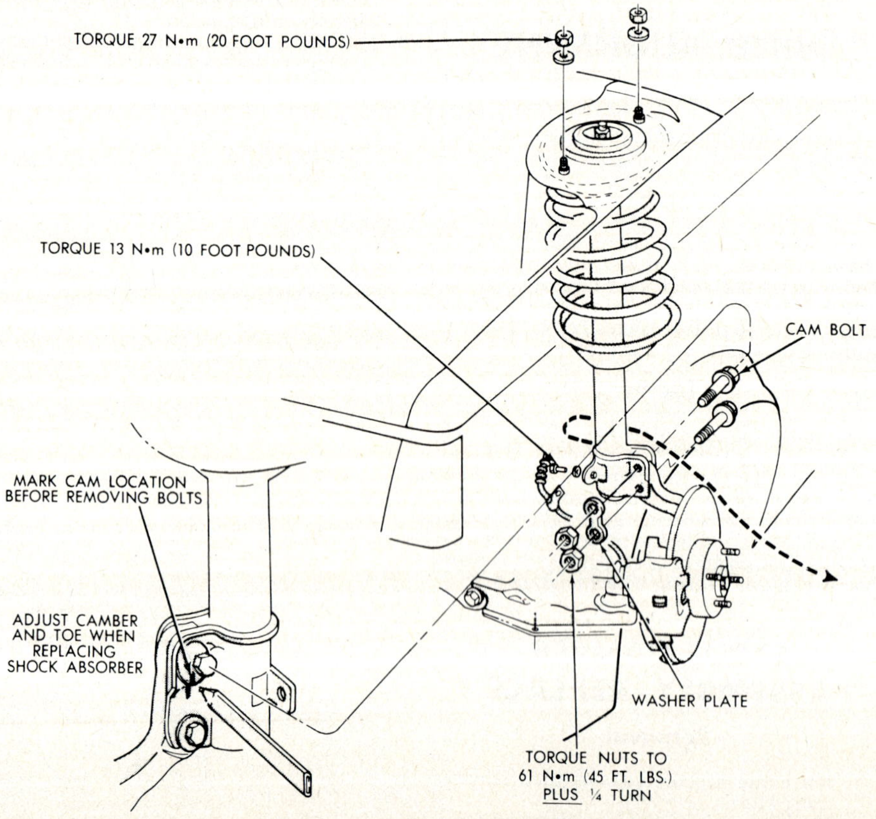

Fig. 4 Strut damper replacement

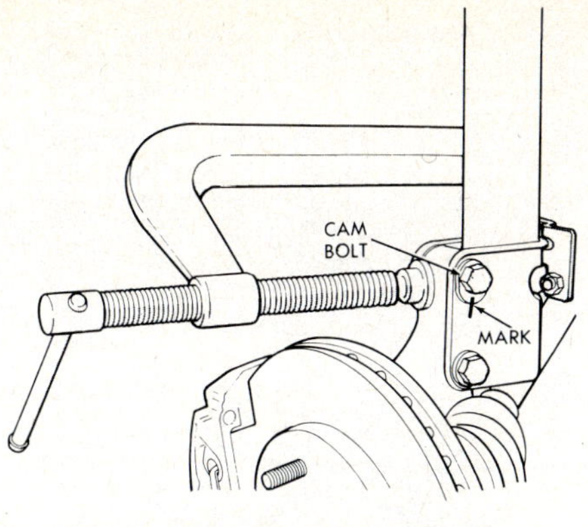

Fig. 5 Strut damper installation

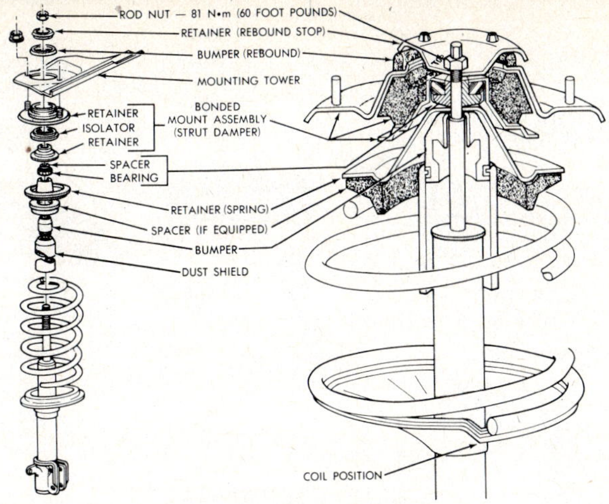

Fig. 6 Strut damper assembly

STEERING KNUCKLE, REPLACE

Removal

1. Remove cotter pin and nut lock.
2. Loosen hub nut with brakes applied, Fig. 9.

NOTE: The nub and driveshaft are splined together through the knuckle (bearing) and retained by the hub nut.

3. Raise and support vehicle, then remove front wheel.
4. Remove hub nut. Ensure that the splined driveshaft is free to separate from spline in hub during knuckle removal. A pulling force on the shaft can separate the inner C/V joint. Tap lightly with a brass drift, if required.
5. Disconnect the tie rod end from steering arm with a suitable puller.
6. Disconnect brake hose retainer from strut damper.
7. Remove clamp bolt securing ball joint stud into steering knuckle and brake caliper adapter screw and washer assemblies.
8. Support caliper with a piece of wire. Do not hang by brake hose.
9. Remove rotor.
10. Mark position of camber cam upper adjusting bolt and loosen both bolts.
11. Support steering knuckle and remove cam adjusting and through bolts. Move upper knuckle "neck" from strut damper bracket and lift knuckle from ball joint stud.

NOTE: Support driveshaft during knuckle removal. Do not permit driveshaft to hang after separating steering knuckle from vehicle.

Installation

1. Place steering knuckle on lower ball joint stud and the driveshaft through hub.
2. Position upper "neck" of knuckle into strut damper bracket and install cam and through bolts. Place cam in original position. Place a 4-inch or larger, C-clamp on strut and steering knuckle, then tighten clamp just enough to eliminate looseness between knuckle and strut. Check to ensure that cam alignment marks made during remove are aligned, then tighten bolts to 45 ft. lbs. plus an additional 1/4 turn beyond specified torque. Remove C-clamp.
3. Install and torque ball joint to steering knuckle clamp bolt to 50 ft. lbs. (68 Nm).
4. Install tie rod end into steering arm and torque nut to 35 ft. lbs. (47 Nm). Install cotter pin.
5. Install rotor.
6. Install caliper over rotor and position adapter to steering knuckle. Install adapter to knuckle bolts and torque to 85 ft. lbs. (115 Nm).
7. Attach brake hose retainer to strut damper and torque screw to 10 ft. lbs. (13 Nm).
8. a. Install and hub nut.
   b. With brakes applied, torque hub nut to 180 ft. lbs. (245 Nm).
   c. Install nut lock and new cotter pin, Fig. 9.

HUB & BEARING, REPLACE

Removal

1. Remove steering knuckle as outlined pre-

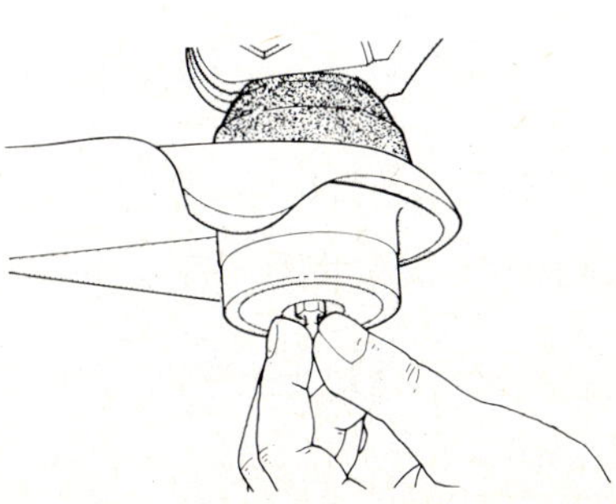

Fig. 7 Checking ball joint for wear

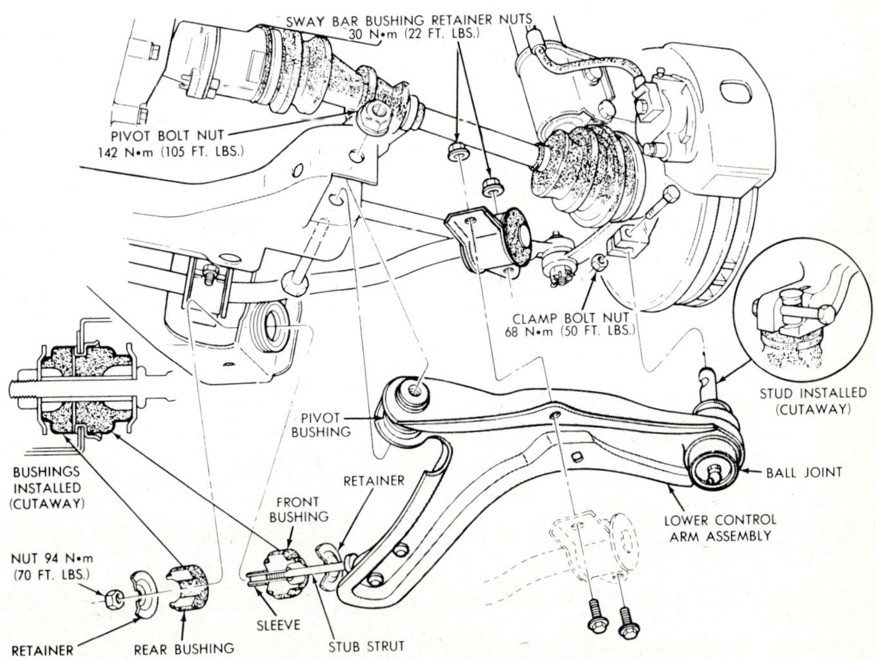

Fig. 8 Lower control arm assembly

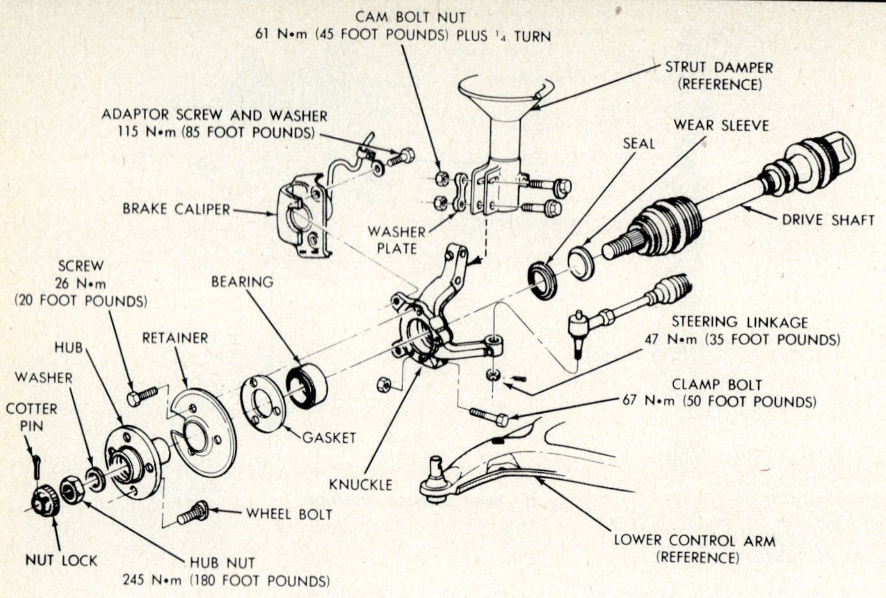

Fig. 9 Steering knuckle assembly

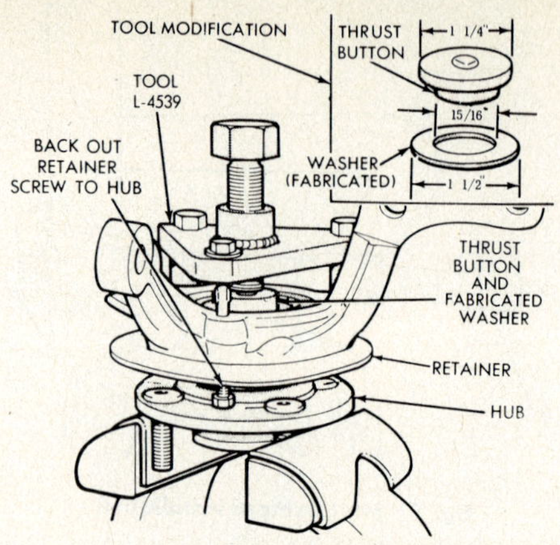

Fig. 10 Hub removal

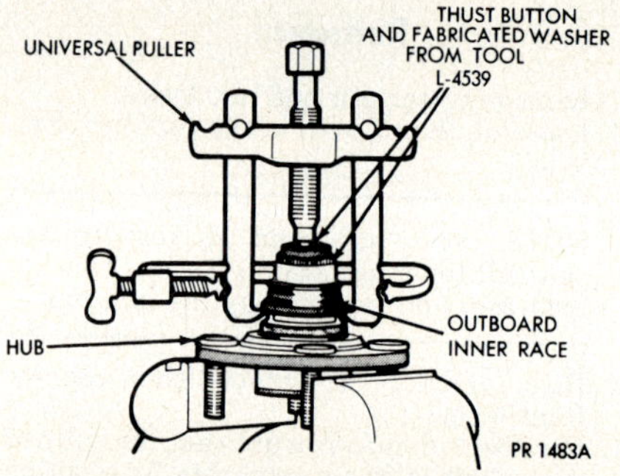

Fig. 11 Removing outboard inner race

viously.

2. Remove hub with a suitable puller or tool L-4539, Fig. 10.

NOTE: The bearing inner races will separate and the outer race will remain in the hub.

---

3. Remove bearing outer race from hub with a suitable puller, Fig. 11.
4. Remove bearing retainer.
5. Remove bearing from knuckle with a suitable press and socket, Fig. 12.

Installation

1. Press new bearing into knuckle with a suitable press and installer tool, Fig. 13.
2. Install bearing retainer. Torque retaining screws to 20 ft. lbs. (27 Nm).
3. Press hub into bearing with a suitable press and socket, Fig. 14.
4. Install steering knuckle as outlined previously.

DRIVESHAFTS, REPLACE

Removal

NOTE: On early 1982 models, the inboard C/V joints have stub shafts splined into the differential side gears and are retained with circlips, Fig. 15. The circlip "Tangs" are located on a machined surface on the inner end of the stub shafts and are removed and installed with the shaft. On late 1982 and 1983 models, the driveshafts are spring loaded and are retained to the side gears by constant spring pressure provided by the spring contained in the C/V joints, Fig. 16.

---

1. On early 1982 models, drain transaxle differential unit and remove cover.
2. If removing the right hand driveshaft, the speedometer pinion must be removed prior to driveshaft removal, Fig. 17.
3. On early 1982 models, rotate driveshaft to expose circlip tangs, Fig. 18. Using needle nose pliers, compress circlip tangs while prying shaft into side gear splined cavity, Fig. 19. The circlip will be compressed in the cavity with the shaft.
4. Remove clamp bolt securing ball joint clamp bolt to steering knuckle, then, separate ball joint stud from steering knuckle. Do not damage ball joint or C/V joint boots.
5. Separate outer C/V joint splined shaft from hub by holding C/V housing while moving knuckle hub assembly away from C/V joint.

NOTE: Do not damage slinger on outer C/V joint. Do not attempt to remove, repair or replace.

---

6. Support assembly at C/V joint housings and remove by pulling outward on the inner C/V joint housing. Do not pull on the shaft.

NOTE: If removing left hand driveshaft assembly, the removal may be aided by

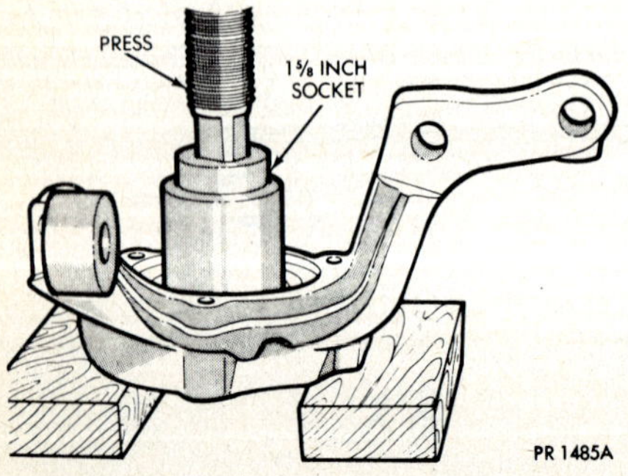

Fig. 12 Removing bearing from knuckle

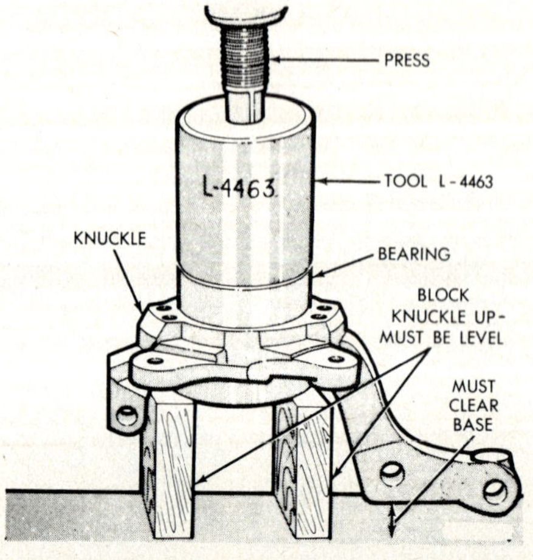

Fig. 13 Installing bearing into knuckle

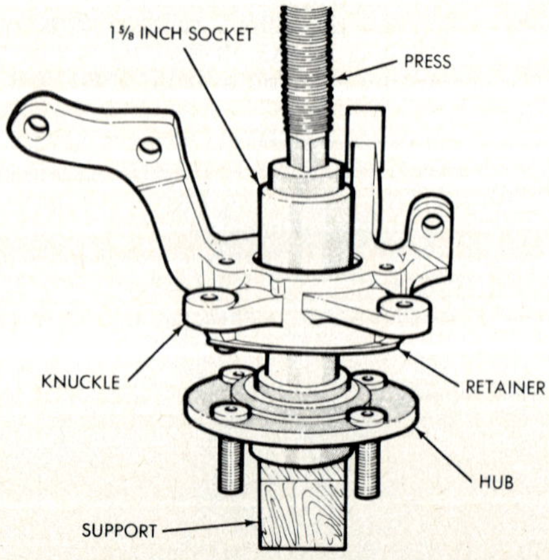

Fig. 14 Installing hub into knuckle

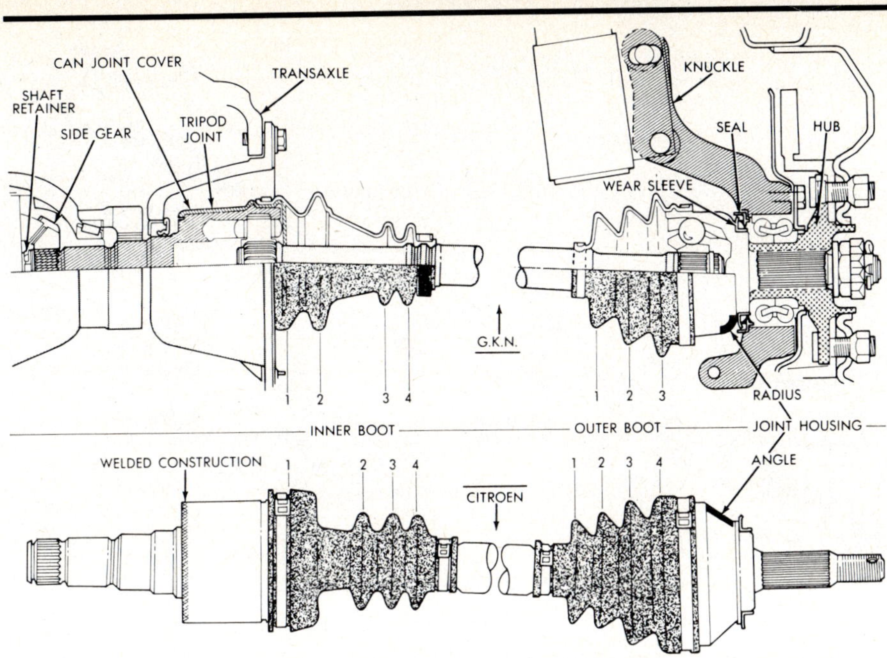

Fig. 15 Driveshaft assembly. Early 1982

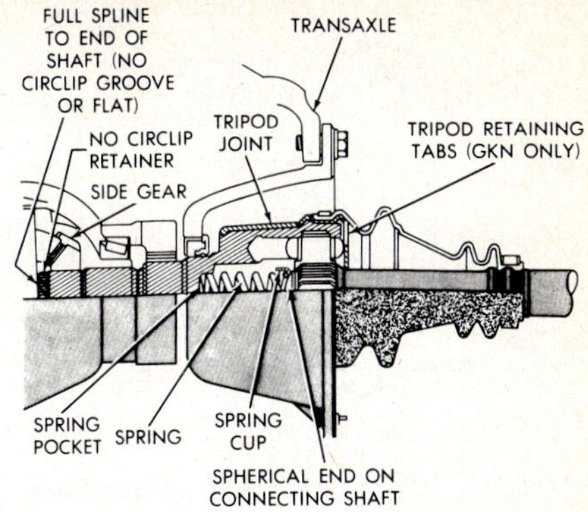

Fig. 16 Driveshaft assembly. Late 1982 & 1983

inserting a screwdriver blade between the differential pinion shaft and carefully prying against the end face of stub.

7. Remove driveshaft assembly from vehicle.

Installation

NOTE: On early 1982 models, install new circlips on inner joint shaft before installation, Fig. 20.

1. On early 1982 units, be sure tangs on circlips are aligned with flattened end of shaft before inserting shaft into transaxle. If not, jamming or component damage may result.
2. Hold inner joint assembly at housing while aligning and guiding the inner joint spline into transaxle.
3. While holding the inner joint housing, quickly thrust the shaft into the differential. This will complete the lock-up of the driveshaft to the axle side gear.

NOTE: On early 1982 models, inspect circlip positioning in side gears to verify lock-up.

4. Push knuckle/hub assembly out and install splined outer C/V joint shaft into hub.
5. Install knuckle assembly on ball joint stud.
6. Install and torque clamp bolt to 50 ft. lbs. (68 Nm).
7. Install speedometer pinion, Fig. 17.
8. On early 1982 models, apply a 1/16 inch bead of silicone sealant, part number 4026070, to differential cover sealing surface and mating surface of transaxle case after both have been properly cleaned and inspected.
9. On early 1982 models, install differential cover and torque retaining screws to 165 inch lbs. (19 Nm).
10. Fill differential to bottom of filler plug hole with Dexron automatic transmission fluid.
11. Install washer and hub nut. Torque hub nut to 180 ft. lbs. (245 Nm). Install nut lock and cotter pin.
12. If, after attaching driveshaft assembly in vehicle the inboard boot appears collapsed or deformed, vent the inner boot by inserting a round tipped, small diameter rod between the boot and shaft. As venting occurs, the boot will return to the normal shape.

INNER CONSTANT VELOCITY JOINT SERVICE

Disassemble, Fig. 21

1. Remove clamp from boot and discard.
2. On Citroen units, separate tripod from housing by slightly deforming retaining ring at three locations, Fig. 22. If necessary, cut retaining ring from housing and install replacement retaining ring with inner flange rolled, then stake into machined groove with a suitable punch.

NOTE: When removing tripod from housing, secure rollers. After tripod has been

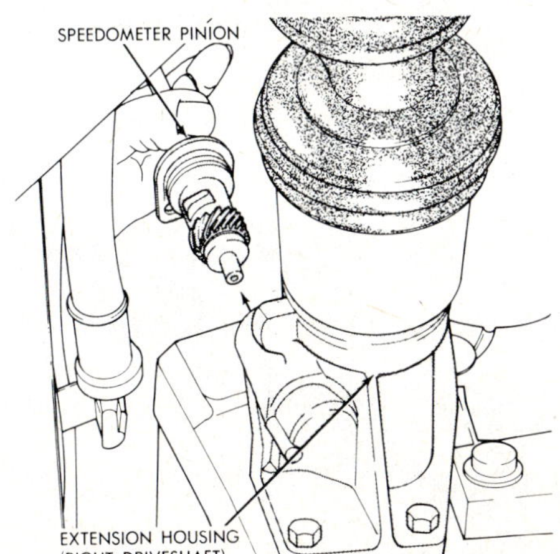

Fig. 17 Speed pinion replacement

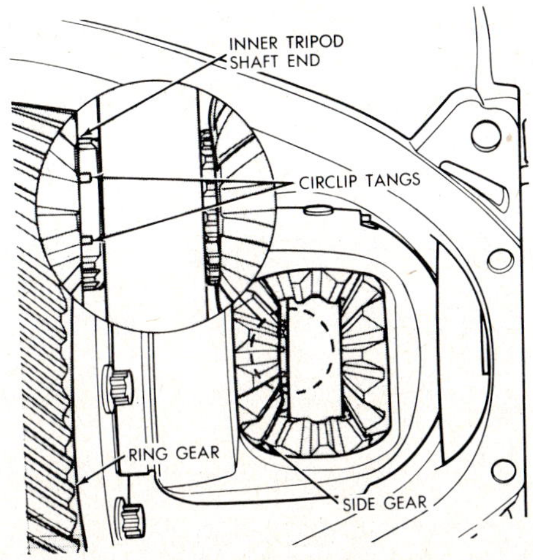

Fig. 18 Circlips exposed. Early 1982 models

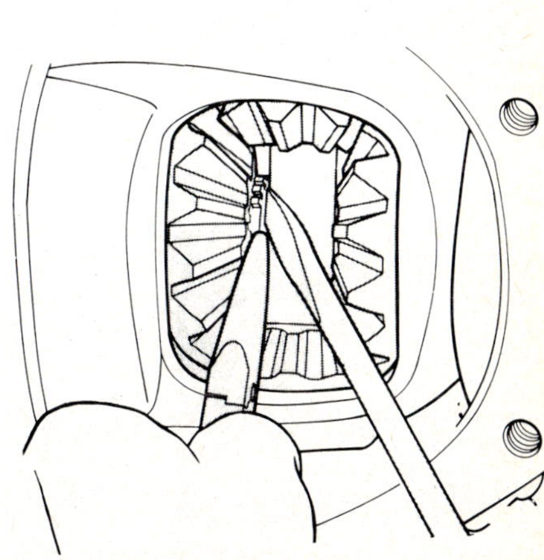

Fig. 19 Compressing circlips. Early 1982 models

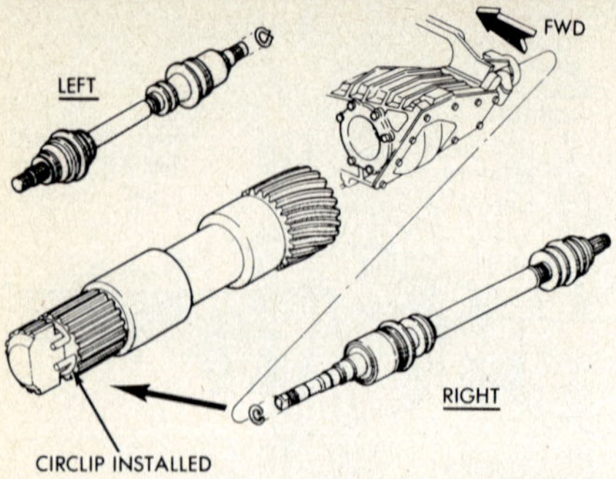

Fig. 20 Circlips installation. Early 1982 models

removed, secure assembly with tape.

3. On early 1982 G.K.N. nonspring loaded units, slide tripod from housing, Fig. 22. On late 1982 and 1983 G.K.N. spring loaded units, bend tabs on joint cover using needle nose pliers, then remove tripod from housing.
4. Remove snap ring from end of shaft, then remove tripod using brass punch.

Inspection

Remove grease from assembly and inspect bearing race and tripod components for wear and damage and replace as necessary. On late 1982 and 1983 spring loaded joints inspect spring, spring cup and spherical end of connecting shaft for wear and damage and replace as necessary.

NOTE: Components of spring loaded and nonspring loaded inner C/V joints cannot be interchanged.

Assemble

1. Slide small end of boot over shaft. On tubular type shafts, align boot lip with mark on shaft outer diameter. On solid type shafts, position small end of boot in groove on shaft.
2. Place rubber clamp over groove on boot.
3. Install tripod on shaft with non-chamfered face of tripod body facing shaft retainer groove.
4. Lock tripod assembly on shaft by installing retaining ring in shaft groove.
5. Distribute two packets of special grease provided with boot and clamp kit in boot. On 1983 Citroen units, use only 2/3 tube of grease.
6. On early 1982 models distribute one packet of grease in housing before positioning housing over tripod. On Citroen units, reform retainer ring. On all units, secure boot to housing with boot clamp.
7. On late 1982 and 1983 models distribute one packet of grease in housing, then position spring with spring cup attached to exposed end, into spring pocket. Place a small amount of grease on spring cup, then position housing over tripod. On Citroen units, reform retainer ring. On G.K.N. units, bend retaining tabs. On all units, position boot over boot groove in housing, then install clamp.

NOTE: When installing housing, check to insure that spring remains in pocket and centered in housing. Also ensure that spring cup contacts spherical end of connecting shaft.

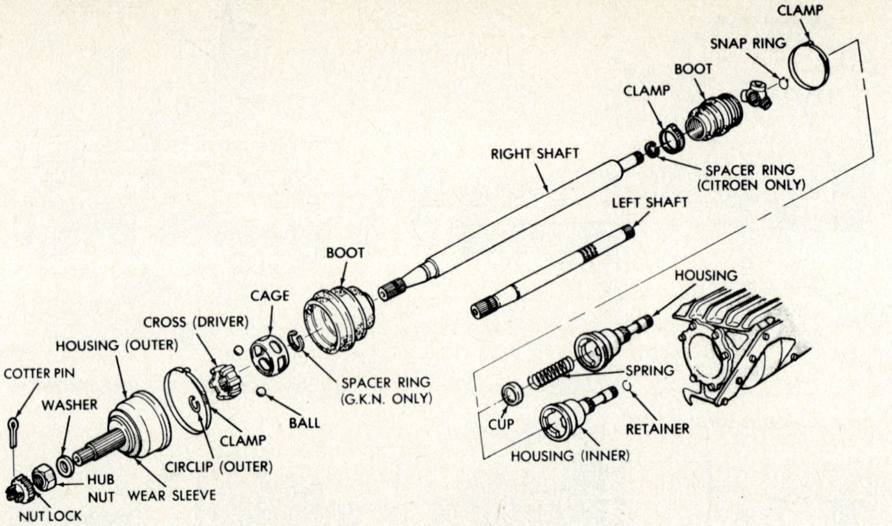

Fig. 21 Driveshaft components

OUTER CONSTANT VELOCITY JOINT SERVICE

Disassembly

Figs. 21 & 24

1. Cut boot clamps from boot and discard.
2. Clean grease from joint.
3. Support shaft in a soft-jawed vise. Support the outer joint, and tap with a mallet to dislodge joint from internal circlip installed in a groove at the outer end of the shaft, Fig. 25. Do not remove slinger from housing.
4. Remove circlip from shaft groove and discard, Fig. 26.
5. Unless the shaft requires replacement, do not remove the heavy lock ring from shaft, Fig. 26.
6. If constant velocity joint was operating satisfactorily and grease does not appear contaminated, proceed to "Assembly" procedure, step 7.
7. If the constant velocity joint is noisy or badly worn, replace entire unit. The repair kit will include boot, clamps, circlip and lubricant. Clean and inspect the joint outlined in the following steps.
8. Clean surplus grease and mark relative position of inner cross, cage and housing with a dab of paint.
9. Hold joint vertically in a soft jawed vise.

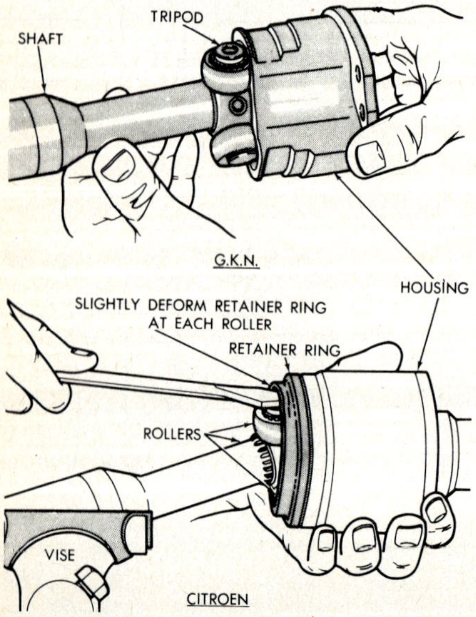

Fig. 22 Removing 3 ball tripod housing. Inner C/V joint

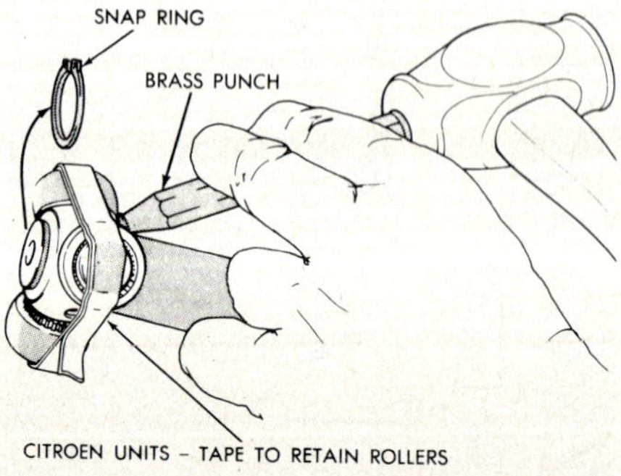

Fig. 23 Removing snap ring & tripod inner C/V joints

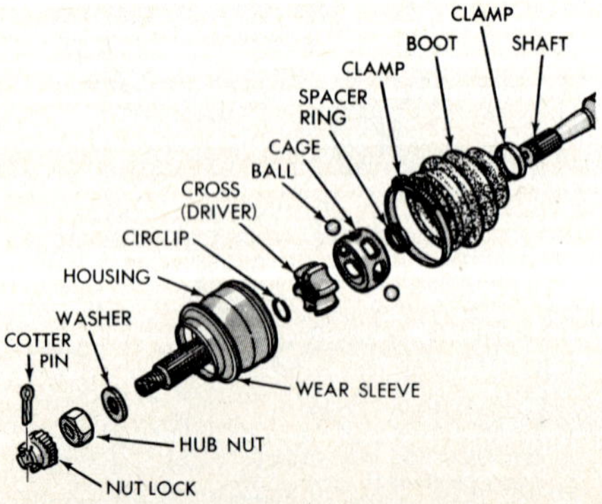

Fig. 24 Outer C/V joint disassembled

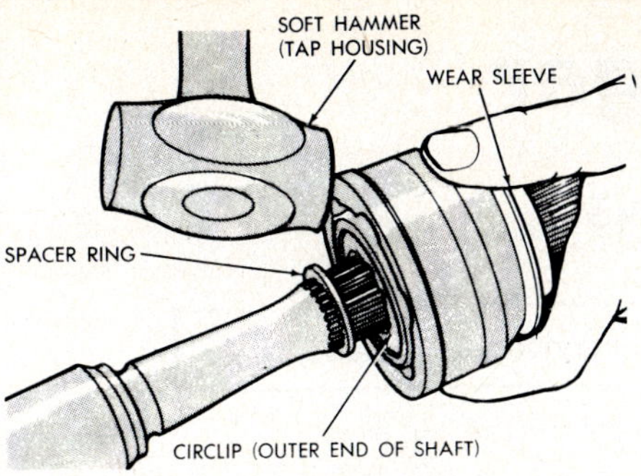

Fig. 25 Removing joint from shaft. Outer C/V joints

10. Press downward on one side of the inner race to tilt cage and remove ball from opposite side, Fig. 27. If joint is tight, use a hammer and a brass drift to tap inner race. Do not strike the cage. Repeat this step until all six balls are removed. A screwdriver may be used to pry the balls loose.
11. Tilt the cage assembly vertically and position the two opposing, elongated cage windows in area between ball grooves. Remove cage and inner race assembly by pulling upward from the housing, Fig. 28.
12. Rotate inner cross 90 degrees to cage and align one of the race spherical lands with an elongated cage window. Raise land into cage window and remove inner race by swinging outward, Fig. 29.

Inspection

1. Check housing ball races for excessive wear.
2. Check splined shaft and nut threads for damage.
3. Inspect the balls for pitting, cracks, scouring and wearing. Dulling of the surface is normal.
4. Inspect cage for excessive wear on inner and outer spherical surfaces, heavy brinelling of cage, window cracks and chipping.
5. Inspect inner race (Cross) for excessive wear or scoring of ball races.
6. If any of the defects listed in steps 1 through 5, are found, replace the C/V assembly as a unit.

NOTE: Polished areas in races (Cross and housing) and on cage spheres are normal and do not indicate a need for joint replacement

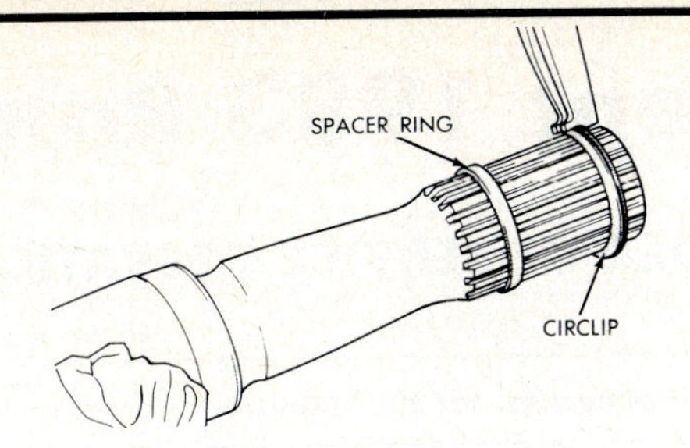

Fig. 26 Circlip removal. Outer C/V joints

unless they are suspected of causing noise and vibration.

Assembly

Figs. 21 & 24

1. If removed, position wear sleeve on joint housing, then tap sleeve onto housing, using tool No. C-4698.
2. Lightly oil components, then align marks made during disassembly.
3. Align one of the inner race lands with elongated window of cage, then insert race into cage and pivot 90°.
4. Align elongated cage windows with housing land, then pivot cage 90°. The curved side of the elongated cage windows and inner race counterbore should face outward from joint.
5. Lubricate ball races with one packet of grease from kit.
6. Tilt cage and inner race assembly and insert balls.
7. With shaft supported in a soft-jawed vise, install boot. On G.K.N. units, slip small clamp over spacer ring and shaft.
8. Slide small end of boot over spacer ring and shaft, then position boot end in machined groove.

NOTE: On Citroen units, position vent sleeve under boot at clamp area.

9. Install snap ring on shaft. When installing use care not to overexpand snap ring.
10. Position joint housing on shaft, then engage by tapping sharply with a soft-faced mallet.
11. Check to ensure that snap ring is properly seated, by attempting to pull joint from shaft.
12. Locate large end of boot over housing.
13. On G.K.N. units, secure boot clamps using tool No.C-4124. On Citroen units, secure boot clamps using tool C-4653.

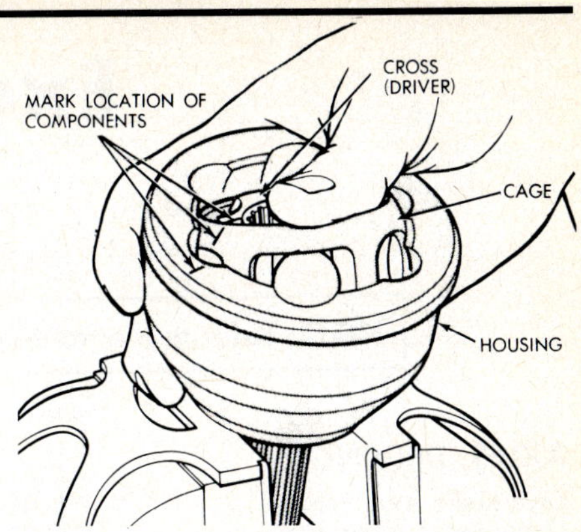

Fig. 27 Ball removal. Outer C/V joints

NOTE: G.K.N. units use steel clamps, while Citroen units use a strap-type clamp.

RACK & PINION STEERING GEAR, REPLACE

NOTE: These vehicles are equipped with manual steering gears supplied by two different manufacturers, Burman and Cam Gears. The replacement rubber boots for these gears are not interchangeable.

The Burman gear can be identified by the name "Burman" cast on the housing and is visible from under the hood.

The Cam Gear steering gear has no identification markings.

1. Raise and support vehicle, then remove front wheels.
2. Remove tie rod ends with a suitable puller.
3. Remove splash shields and boot seal shields.
4. Drive out lower roll pin attaching pinion shaft to lower universal joint.
5. Support front suspension crossmember with a suitable jack.
6. On power steering units, disconnect hoses from steering gear.
7. Disconnect tie rod ends from steering knuckles.
8. On all models, remove bolts attaching steering gear to front suspension crossmember. Loosen crossmember from vehicle frame.
9. Remove steering gear from left side of vehicle.
10. Reverse procedure to install.

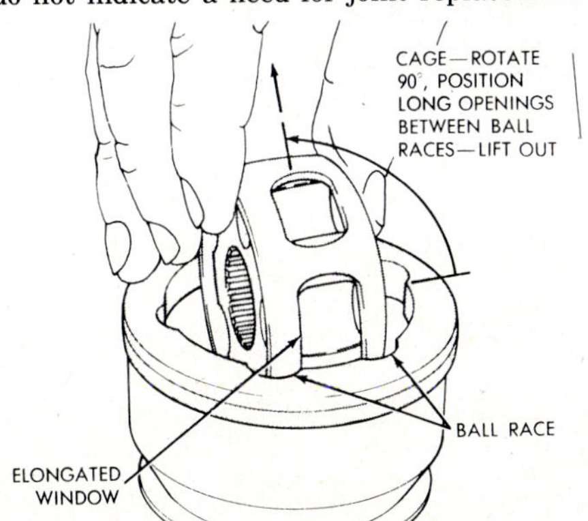

Fig. 28 Cage & cross assembly removal. Outer C/V joints

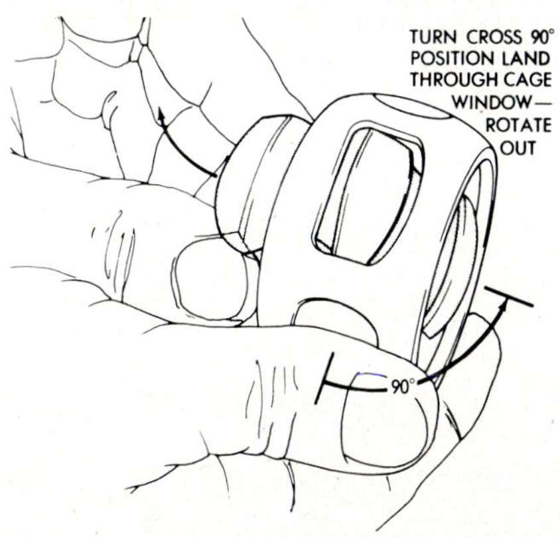

Fig. 29 Removing cross from cage. Outer C/V joints

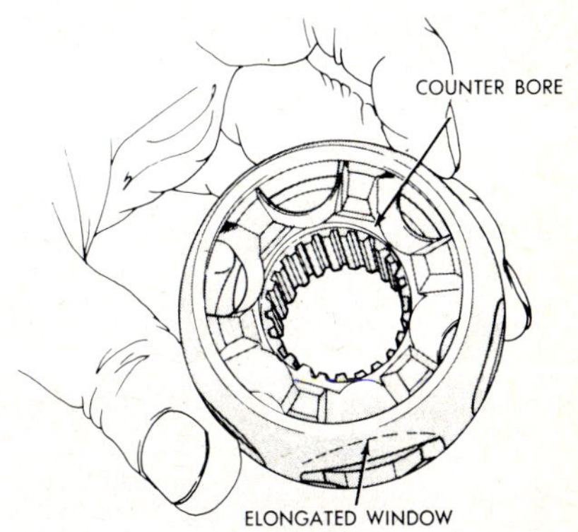

Fig. 30 Cage & cross assembly. Outer C/V joints

FORD
Except 1983 Ranger

INDEX OF SERVICE OPERATIONS

NOTE: Refer to the front of this manual for the manufacturers special service tool suppliers.

MODEL INDEX & ENGINE APPLICATION

CONVENTIONAL GASOLINE MODELS

Model	Year	Engine Make	Standard Engine Model	Crankcase Refill Capacity, Qts.	Cooling System Capacity, Qts.⑳
Bronco	1974	Own	6-200	6	9.2
	1975–77	Own	V8-302	5	14.7
	1978–79	Own	V8-351	5	22
	1980–83	Own	6-300	5	13
	1980	Own	V8-302	5	13
Econoline	1974	Own	6-240	4	—
	1975–79	Own	6-300	5	15
	1980–83	Own	6-300	5	16
	1980	Own	V8-302	5	16
Ranchero	1974	Own	6-250	4	10
	1974	Own	V8-302	4	15
	1975–76	Own	V8-351	4	15.9
	1977–79	Own	V8-302	4	14.3
	1978	Own	V8-351	4	15.4
F-100 (4×4)	1974	Own	6-240	5	14
	1975	Own	6-300	5	14
	1976	Own	V8-360	5	22.3
(4×2)	1974	Own	6-240	4	13
	1974	Own	V8-360	5	19.6
	1974–79	Own	6-300	5	14½
	1980–82	Own	6-300	5	13
	1982–83	Own	V6-232	5	11
F-150 (4×2)	1975–79	Own	6-300	5	14.4
	1980–83	Own	6-300	5	13
(4×4)	1978–79	Own	V8-351	5	15
	1980–83	Own	6-300	5	13
	1980	Own	V8-302	5	13
F-250 (4×2)	1974	Own	6-240	4	14
	1975–79	Own	6-300	5	14.4
	1980–83	Own	6-300	5	13
(4×4)	1974	Own	6-240	5	14
	1975–79	Own	6-300	5	14.4
	1978–79	Own	V8-351	5	20
	1980–83	Own	6-300	5	13
	1980	Own	V8-302	5	13
F-350	1974	Own	6-300	5	14½③
	1975–79	Own	6-300	5	16.7④
	1975–76	Own	V8-360	5	22.3
	1977–79	Own	V8-351	5	22
(4×2)	1980–83	Own	6-300	5	13
(4×4)	1980–83	Own	V8-351	5	15
F-500	1974–75	Own	6-300	5	19
	1976–77	Own	6-300	5	18
F-600	1974–75	Own	6-300	6	19
	1976–79	Own	6-300	5	18
	1980–83	Own	V8-370	8	23.5
(4×4)	1974–75	Own	V8-330	8	26
	1976–78	Own	V8-330	8	24.8
	1979	Own	V8-370	8	24.5
F-700	1974–75	Own	V8-361	8	⑦
	1976–78	Own	V8-361	8	⑧
	1979–83	Own	V8-370	8	23.5
F-750	1974	Own	V8-361	8	24①
	1975	Own	V8-361	8	⑦
	1976–79	Own	V8-361	8	⑧
F-800	1979	Own	V8-429	8	23.5
	1980	Own	V8-429	8	23.8
	1981–83	Own	V8-370	8	24
F-880	1975–79	Own	V8-475	9	46

TILT CAB GASOLINE MODELS

Model	Year	Engine Make	Standard Engine Model	Crankcase Refill Capacity, Qts.	Cooling System Capacity, Qts.⑳
C-600	1974–77	Own	6-300	6	22.5
	1978	Own	V8-330	6	28
	1979	Own	V8-370	8	20.5
	1980–83	Own	V8-370	8	22
C-700	1974	Own	V8-361	8	25
	1975	Own	V8-361	8	29①
	1976–78	Own	V8-361	8	28①
	1979	Own	V8-370	8	20.5
	1980–81	Own	V8-370	8	24
	1982–83	Own	V8-370	8	22.5
C-750	1974	Own	V8-361	8	28①
	1975	Own	V8-361	8	29①
	1976–78	Own	V8-361	8	28①
C-800	1974	Own	V8-361	8	28①
	1975	Own	V8-361	8	29①
	1976–78	Own	V8-361	8	28①
	1979	Own	V8-429	8	20.8
	1980	Own	V8-370	8	23.8
	1981–83	Own	V8-370	8	22.5
C-900	1974–75	Own	V8-401	9	52②
	1976–79	Own	V8-477	9	⑪
	1980	Own	V8-477	9	52
	1981	Own	V8-477	9	⑩

TANDEM AXLE GASOLINE MODELS

Model	Year	Engine Make	Standard Engine Model	Crankcase Refill Capacity, Qts.	Cooling System Capacity, Qts.⑳
CT-800	1974	Own	V8-361	8	28①
	1975	Own	V8-389	8	29①
	1976	Own	V8-361	8	28①
	1977–78	Own	V8-389	8	⑨
	1979	Own	V8-370	8	27.3
	1980	Own	V8-370	8	28.4
	1981	Own	V8-370	9	27.5
CT-900	1974–75	Own	V8-401	9	52②
	1976–79	Own	V8-477	9	⑪
	1980	Own	V8-477	9	52
	1981	Own	V8-477	9	⑩

TANDEM AXLE DIESEL MODELS

Model	Year	Engine Make	Standard Engine Model	Crankcase Refill Capacity, Qts.	Cooling System Capacity, Qts.⑳
CT-800	1982–83	Detroit Diesel	8.2L-T	12	27

SCHOOL BUS MODELS

Model	Year	Engine Make	Standard Engine Model	Crankcase Refill Capacity, Qts.	Cooling System Capacity, Qts.⑳
B-500	1974–75	Own	6-300	5	19

Continued

MODEL INDEX & ENGINE APPLICATION—Continued

Model	Year	Engine Make	Standard Engine Model	Crankcase Refill Capacity, Qts.	Cooling System Capacity, Qts.⑳
	1976–78	Own	6-300	5	18
B-600	1974–75	Own	6-300	6	19
	1976–77	Own	6-300	6	18
	1978	Own	V8-330	8	24
	1979–83	Own	V8-370	8	23.5
B-700	1974	Own	V8-330	8	25
	1975	Own	V8-330	5	⑦
	1976–77	Own	V8-330	5	⑫
	1978	Own	V8-361	8	⑭
	1979–83	Own	V8-370	8	23.5
B-750	1974	Own	V8-361	8	24①
	1975	Own	V8-361	5	⑦
	1976–77	Own	V8-361	5	⑧

PARCEL DELIVERY GASOLINE MODELS

Model	Year	Engine Make	Standard Engine Model	Crankcase Refill Capacity, Qts.	Cooling System Capacity, Qts.⑳
P-350	1974–76	Own	6-300	5	18
P-400	1974–76	Own	6-300	5	18
P-450	1975–76	Own	V8-360	5	—
P-500	1974–78	Own	6-300	5	18
P-550	1975–76	Own	V8-390	5	—
P-600	1975–79	Own	6-300	5	18.1

MOTOR HOME MODELS

Model	Year	Engine Make	Standard Engine Model	Crankcase Refill Capacity, Qts.	Cooling System Capacity, Qts.⑳
M-400	1974–75	Own	V8-360	5	22
M-450	1974–76	Own	V8-360	5	22
M-500	1974–76	Own	V8-390	5	23

LOUISVILLE GASOLINE MODELS

Model	Year	Engine Make	Standard Engine Model	Crankcase Refill Capacity, Qts.	Cooling System Capacity, Qts.⑳
LN-600	1974	Own	6-300	6	19
	1975	Own	V8-330	8	25
	1976–78	Own	V8-330	8	⑤
	1979	Own	V8-370	8	25
	1980–81	Own	V8-370	8	26.4
	1982–83	Own	V8-370	8	26
LN-700	1974–75	Own	V8-361	8	25
	1976–78	Own	V8-361	8	⑥
	1979	Own	V8-370	8	25
	1980	Own	V8-370	8	⑮
	1981–83	Own	V8-370	8	⑯
LN-750	1974	Own	V8-361	8	25
	1975	Own	V8-361	8	25
	1976–78	Own	V8-361	8	⑥
L-800	1974–75	Own	V8-361	8	25.5
	1976	Own	V8-361	8	⑨
	1977	Own	V8-330	8	⑫
	1978	Own	V8-389	8	24.5
	1979	Own	V8-370	8	23.5
	1980–81	Own	V8-370	8	26.8
	1982–83	Own	V8-370	8	25
LN-800	1974–75	Own	V8-361	8	25.5
	1976–77	Own	V8-361	8	⑨
	1978	Own	V8-389	8	24.5
	1979	Own	V8-370	8	23.5
	1980–81	Own	V8-370	8	26.8
	1982–83	Own	V8-370	8	25
LT-800	1974	Own	V8-361	8	25½
	1975	Own	V8-361	8	25½
	1976	Own	V8-361	8	⑨
	1977–78	Own	V8-389	8	⑨
	1979	Own	V8-370	8	23.5
	1980–81	Own	V8-370	8	26.8
	1982–83	Own	V8-370	8	25
LNT-800	1974	Own	V8-361	8	25½
	1975	Own	V8-361	8	25½
	1976	Own	V8-361	8	⑨
	1977–78	Own	V8-389	8	⑨
	1979	Own	V8-370	8	24.5
	1980–81	Own	V8-370	8	26.8
	1982–83	Own	V8-370	8	25
LTS-800	1974	Own	V8-361	8	25½
	1975	Own	V8-361	8	25½
	1976	Own	V8-361	8	⑨
	1977–78	Own	V8-475	9	⑬
	1979	Own	V8-429	9	22.5
	1980	Own	V8-429	9	26.8
	1981	Own	V8-370	8	26
	1982	Own	V8-429	8	—
LT-880	1974–75	Own	V8-475	9	47
	1976–78	Own	V8-475	9	45.2
LNT-880	1974–75	Own	V8-475	9	47
	1976–78	Own	V8-475	9	45.2
L-900	1974–75	Own	V8-401	9	47
	1976–81	Own	V8-477	9	47
LN-900	1974–75	Own	V8-401	9	47
	1976–81	Own	V8-477	9	47
LT-900	1974–75	Own	V8-401	9	47
	1976–77	Own	V8-477	9	52
	1978–81	Own	V8-477	9	⑰
LNT-900	1974–75	Own	V8-401	9	47
	1976–77	Own	V8-477	9	52
	1978–81	Own	V8-477	9	⑰
LTS-900	1974–75	Own	V8-401	9	47
	1976–77	Own	V8-477	9	52
	1978–81	Own	V8-477	9	⑰

LOUISVILLE DIESEL MODELS

Model	Year	Engine Make	Standard Engine Model	Crankcase Refill Capacity, Qts.	Cooling System Capacity, Qts.⑳
LN-6000	1974	Catpl'r	V-150	12	25
LN-7000	1974	Catpl'r	3208⑱	12	25
	1975–80	Catpl'r	3208⑱	10	42
	1981	Catpl'r	3208⑱	14	40.5
	1982–83	Catpl'r	3208⑱	20	40.5
L-8000	1974	Catpl'r	3208⑱	12	25
	1975	Cummins	V-555	—	28
	1975–80	Catpl'r	3208⑱	10	38
	1981	Catpl'r	3208⑱	14	39
	1982–83	Catpl'r	3208⑱	20	39
LN-8000	1974	Catpl'r	3208⑱	12	25
	1975	Cummins	V-555	—	28
	1975–80	Catpl'r	3208⑱	10	38
	1981	Catpl'r	3208⑱	14	39
	1982–83	Catpl'r	3208⑱	20	39

Continued

MODEL INDEX & ENGINE APPLICATION—Continued

Model	Year	Engine Make	Standard Engine Model	Crankcase Refill Capacity, Qts.	Cooling System Capacity, Qts.⑳
LT-8000	1974	Catpl'r	3208⑱	12	25
	1975–80	Catpl'r	3208⑱	10	38
	1981	Catpl'r	3208⑱	14	39
	1982–83	Catpl'r	3208⑱	20	39
LTS-8000	1974	Catpl'r	3208⑱	12	25
	1975–80	Catpl'r	3208⑱	10	40
	1981	Catpl'r	3208⑱	14	38
	1982–83	Catpl'r	3208⑱	20	39
LNT-8000	1974	Catpl'r	3208⑱	12	25
	1975–80	Catpl'r	3208⑱	10	38
	1981	Catpl'r	3208⑱	14	39
	1982–83	Catpl'r	3208⑱	20	39
CL-9000	1978–82	Cummins	NTC-290	28	48
CL-9000	1983	Cummins	NTC-300	36	48
CLT-9000	1978–82	Cummins	NTC-290	28	48
CLT-9000	1983	Cummins	NTC-300	36	48
L-9000	1974–75	Cummins	NH-230	24	47
	1975	Cummins	NTC-250	28	43
	1976–77	Cummins	NH-230	28	32
	1978–80	Cummins	NTC-230	28	52
	1981–83	Cummins	PT-240	28	50
LN-9000	1974–75	Cummins	NH-230	24	47
	1975	Cummins	NTC-250	28	43
	1976–77	Cummins	NH-230	28	32
	1978–80	Cummins	NTC-230	28	52
	1981–83	Cummins	PT-240	28	50
LT-9000	1974	Cummins	NH-230	24	47
	1975	Cummins	NH-230	28	43
	1976–77	Cummins	NH-230	28	32
	1978–80	Cummins	NTC-230	28	46
	1981–83	Cummins	PT-240	28	50
LTS-9000	1974	Cummins	NH-230	24	47
	1975	Cummins	NH-230	28	43
	1976–77	Cummins	NH-230	28	32
	1978–80	Cummins	NTC-230	28	46
	1981–83	Cummins	PT-240	28	50
LNT-9000	1974	Cummins	NH-230	24	47
	1975	Cummins	NH-230	28	43
	1976–77	Cummins	NH-230	28	32
	1978–80	Cummins	NTC-230	28	46
	1981–83	Cummins	PT-240	28	50

FORD DIESEL MODELS

C-6000	1974	Catpl'r	V-150	12	25

Model	Year	Engine Make	Standard Engine Model	Crankcase Refill Capacity, Qts.	Cooling System Capacity, Qts.⑳
F-6000	1974	Catpl'r	V-150	12	25
B-7000	1974–79	Catpl'r	3208⑱	12	25
C-7000	1974	Catpl'r	3208⑱	12	25
	1975	Catpl'r	3208⑱	12	32
	1976–81	Catpl'r	3208⑱	10	44
	1982–83	Catpl'r	3208⑱	20	43
F-7000	1974	Catpl'r	3208⑱	12	25
	1975–79 & 83	Catpl'r	3208⑱	10	41
C-8000	1974	Catpl'r	3208⑲	12	25
	1975–80	Catpl'r	3208⑱	10	46
	1981	Catpl'r	3208⑱	10	44
	1982–83	Catpl'r	3208⑱	20	43
CT-8000	1974	Catpl'r	V-200	12	25
	1975	Catpl'r	3208⑱	12	32
	1976–79	Catpl'r	3208⑱	10	46
	1980	Catpl'r	3208⑲	10	46
	1981	Catpl'r	3208⑱	14	43
	1982–83	Catpl'r	3208⑱	20	43
W-9000	1974	Cummins	NH-230	20	49
	1975	Cummins	NTC-290	28	49
	1976–77	Cummins	NTC-290	28	48
WT-9000	1974	Cummins	NH-230	20	49
	1975	Cummins	NTC-290	28	49
	1976–77	Cummins	NTC-290	28	48

①—Add 1½ qts. if equipped with Transmatic transmission.
②—Add 7 qts. if equipped with Transmatic transmission.
③—Add 2 qts. if equipped with dual rear tires.
④—Single Rear with Manual Transmission 14.4 quarts.
⑤—Man. Trans. 20.8 quarts; Auto. Trans. 26.7 quarts.
⑥—Man. Trans. 24 quarts; Auto. Trans. 26.7 quarts.
⑦—Man. Trans. 25 quarts; Auto. Trans. 27.6 quarts.
⑧—Man. Trans. 24.8 quarts; Auto. Trans. 26.8 quarts.
⑨—Man. Trans. 21.8 quarts; Auto. Trans. 26.6 quarts.
⑩—Man. Trans. 51.5 quarts; Auto. Trans. 58 quarts.
⑪—Man. Trans. 53.2 quarts; Auto. Trans. 59 quarts.
⑫—Man. Trans. 24.5 quarts; Auto. Trans. 27 quarts.
⑬—Man. Trans. 45.2 quarts; Auto. trans. 46.9 quarts.
⑭—Man. trans. 24 quarts; Auto. trans. 30 quarts.
⑮—Man. Trans. 27.2 quarts; Auto. Trans. 35.6 quarts.
⑯—Man. Trans. 26 quarts; Auto. trans. 34.5 quarts.
⑰—Man. trans. 45 quarts; Auto. trans. 47 quarts.
⑱—175 H.P.
⑲—200 H.P.
⑳—Cooling system capacities are for models equipped with standard engines, tramsmissions and cooling systems.

FORD—Exc. 1983 Ranger

GENERAL ENGINE SPECIFICATIONS

Year	Engine Code On Truck Rating Plate	Engine Model	Carb. Type	Bore & Stroke	Comp. Ratio	Horsepower @ R.P.M.	Torque Ft. Lbs. @ R.P.M.	Governed Speed R.P.M. No Load	Normal Oil Pressure Lbs.
1974	T	6-200	1 Bore	3.68 × 3.13	8.3	85 @ 3800	150 @ 1800	—	35
	A	6-240	1 Bore	4.00 × 3.18	7.9	108 @ 3800	182 @ 2400	4000	40–50
	L	6-250	1 Bore	3.68 × 3.91	8.0	92 @ 3200	234 @ 1600	—	35–55
	B	6-300 (2)	1 Bore	4.00 × 3.98	7.9	126 @ 3400	234 @ 1600	4000	40–50
	B	6-300 (3)	1 Bore	4.00 × 3.98	7.9	126 @ 3400	272 @ 1800	4000	40–50
	(27)	V8-302	2 Bore	4.00 × 3.00	8.0	145 @ 4000	239 @ 2600	—	50–65
	C	V8-330	2 Bore	3.88 × 3.50	7.4	(7)	—	3900	35–55
	D	V8-330 (3)	2 Bore	3.88 × 3.50	7.4	(8)	(9)	3900 (29)	(28)
	Q	V8-351	2 Bore	4.00 × 3.50	7.9	163 @ 4200	278 @ 2000	—	50–70
	Q	V8-351	4 Bore	4.00 × 3.50	7.9	254 @ 5600	292 @ 3400	—	50–70
	Y	V8-360	2 Bore	4.05 × 3.50	8.0	(10)	(11)	—	45–55
	E	V8-361 (3)	2 Bore	4.05 × 3.50	7.2	(12)	(13)	3800	45–70
	H	V8-390	2 Bore	4.05 × 3.78	8.2	(16)	(17)	—	45–55
	M	V8-390	4 Bore	4.05 × 3.78	8.2	(18)	(19)	—	45–55
	F	V8-391 (3)	4 Bore	4.05 × 3.79	7.2	(20)	(21)	3800	45–70
	S	V8-400	2 Bore	4.00 × 4.00	8.0	171 @ 3400	329 @ 2000	—	50–70
	H	V8-401 (26)	4 Bore	4.125 × 3.75	7.3	171 @ 3400	274 @ 2800	3600 (5)	52–62
	J	V8-460	4 Bore	4.36 × 3.850	8.0	(22)	(23)	—	35–65
	J	V8-475	4 Bore	4.50 × 3.75	7.2	(24)	(25)	—	52–62
	K	V8-477 (26)	4 Bore	4.50 × 3.75	7.2	198 @ 3200	344 @ 2600	3400 (5)	52–62
	L	V8-534 (26)	4 Bore	4.50 × 4.20	7.3	213 @ 3000	393 @ 1800	3200 (5)	52–62
1975	B	6-300	1 Bore	4.00 × 3.98	7.9	126 @ 3400	234 @ 1600	—	40–50
	B	6-300 (3)	1 Bore	4.00 × 3.98	7.9	126 @ 3400	234 @ 1600	—	40–50
	G	V8-302	2 Bore	4.00 × 3.00	8.0	135 @ 3600	222 @ 2000	—	50–65
	C	V8-330	2 Bore	3.87 × 3.50	7.4	137 @ 3200	272 @ 1800	—	30–55
	D	V8-330 (3)	2 Bore	3.87 × 3.50	7.4	137 @ 3600	250 @ 2400	—	(28)
	M	V8-351	2 Bore	4.00 × 3.50	8.0	—	—	—	—
	Y	V8-360	2 Bore	4.05 × 3.50	—	155 @ 3800	263 @ 2000	—	45–55
	E	V8-361 (3)	4 Bore	4.05 × 3.50	7.2	170 @ 3600	284 @ 2400	—	45–70
	F	V8-389 (3)	4 Bore	4.05 × 3.78	7.2	180 @ 3600	302 @ 2700	—	—
	H	V8-390	2 Bore	4.05 × 3.78	—	157 @ 3600	291 @ 1800	—	45–55
	M	V8-390	4 Bore	4.05 × 3.78	—	208 @ 4500	324 @ 2800	—	45–55
	S	V8-400	2 Bore	4.00 × 4.00	8.0	—	—	—	—
	H	V8-401 (26)	4 Bore	4.125 × 3.75	7.3	171 @ 3400	274 @ 2800	—	52–62
	J	V8-460	4 Bore	4.36 × 3.85	8.0	239 @ 4200	374 @ 2600	—	35–65
	J	V8-475	4 Bore	4.50 × 3.75	7.2	203 @ 3400	341 @ 2600	—	—
	K	V8-477 (26)	4 Bore	4.50 × 3.75	7.2	197 @ 3200	341 @ 2600	—	52–62
	L	V8-534 (26)	4 Bore	4.50 × 4.20	7.3	212 @ 3000	392 @ 1800	—	52–62
1976	B	6-300 (30)	1 Bore	4.00 × 3.98	7.3	(32)	223 @ 1600	—	40–50
	B	6-300 (31)	1 Bore	4.00 × 3.98	7.3	(33)	(34)	—	40–50
	B	6-300 (30)	1 Bore	4.00 × 3.98	8.0	(35)	(36)	—	40–50
	B	6-300 (31)	1 Bore	4.00 × 3.98	8.0	(37)	(38)	—	40–50
	B	6-300 (3) (30)	1 Bore	4.00 × 3.98	7.9	126 @ 3400	234 @ 1600	—	40–50
	B	6-300 (3) (31)	1 Bore	4.00 × 3.98	7.9	127 @ 3600	223 @ 2400	—	40–50
	G	V8-302 (30)	2 Bore	4.00 × 3.00	8.0	(39)	(40)	—	50–65
	G	V8-302 (31)	2 Bore	4.00 × 3.00	8.0	(41)	(42)	—	50–65
	C	V8-330 (4) (30)	2 Bore	3.87 × 3.50	7.4	137 @ 3200	272 @ 1800	—	35–55
	C	V8-330 (4) (31)	2 Bore	3.87 × 3.50	7.4	144 @ 3200	261 @ 2400	—	35–55
	D	V8-330 (6) (30)	2 Bore	3.87 × 3.50	7.4	137 @ 3600	250 @ 2400	3600	(28)
	D	V8-330 (6) (31)	2 Bore	3.87 × 3.50	7.4	141 @ 3200	255 @ 2400	3600	(28)
	H	V8-351 (30)	2 Bore	4.00 × 3.50	8.1	(43)	(44)	—	50–65
	H	V8-351 (31)	2 Bore	4.00 × 3.50	8.1	(45)	(46)	—	50–65
	H	V8-351W (30) (47)	2 Bore	4.00 × 3.50	8.0	154 @ 3400	286 @ 1800	—	50–70
	H	V8-351M (47)	2 Bore	4.00 × 3.50	8.0	152 @ 3800	274 @ 1600	—	50–70

Continued

GENERAL ENGINE SPECIFICATIONS—Continued

Year	Engine Code On Truck Rating Plate	Engine Model	Carb. Type	Bore & Stroke	Comp. Ratio	Horsepower @ R.P.M.	Torque Ft. Lbs. @ R.P.M.	Governed Speed R.P.M. No Load	Normal Oil Pressure Lbs.
	Y	V8-360(30)	2 Bore	4.05 × 3.50	8.0	(48)	(49)	—	45–55
	Y	V8-360(31)	2 Bore	4.05 × 3.50	8.0	(50)	(51)	—	45–55
	E	V8-361(30)	4 Bore	4.05 × 3.50	7.2	170 @ 3600	284 @ 2400	3600	45–70
	E	V8-361(31)	4 Bore	4.05 × 3.50	7.2	166 @ 3600	278 @ 2600	3600	45–70
	F	V8-389(30)	4 Bore	4.05 × 3.78	7.2	178 @ 3400	316 @ 2400	3600	45–70
	F	V8-389(31)	4 Bore	4.05 × 3.78	7.2	175 @ 3400	307 @ 2600	3600	45–70
	H	V8-390(30)	2 Bore	4.05 × 3.78	8.2	156 @ 3400	292 @ 2000	—	45–55
	H	V8-390(31)	2 Bore	4.05 × 3.78	8.2	147 @ 3200	288 @ 2000	—	45–55
	M	V8-390(30)	4 Bore	4.05 × 3.78	8.2	195 @ 4000	312 @ 2400	—	45–55
	M	V8-390(30)(52)	4 Bore	4.05 × 3.78	8.2	190 @ 4000	292 @ 2600	—	45–55
	M	V8-390(31)(52)	4 Bore	4.05 × 3.78	8.2	183 @ 3800	300 @ 2400	—	45–55
	S	V8-400(47)	2 Bore	4.00 × 4.00	8.0	180 @ 3800	336 @ 1800	—	50–70
	J	V8-460(30)	4 Bore	4.36 × 3.85	8.0	(53)	(54)	—	35–65
	J	V8-460(30)(52)	4 Bore	4.36 × 3.85	8.0	230 @ 4000	(55)	—	35–65
	J	V8-460(31)(52)	4 Bore	4.36 × 3.85	8.0	(56)	(57)	—	35–65
	J	V8-475(30)	4 Bore	4.5 × 3.75	7.2	203 @ 3400	341 @ 2600	3400	52–62
	J	V8-475(31)	4 Bore	4.5 × 3.75	7.2	215 @ 3400	354 @ 2600	3400	52–62
	K	V8-477(30)	4 Bore	4.5 × 3.75	7.2	197 @ 3200	341 @ 2600	3200	52–62
	K	V8-477(31)	4 Bore	4.5 × 3.75	7.2	208 @ 3200	354 @ 2600	3200	52–62
	L	V8-534(30)	4 Bore	4.5 × 5.2	7.3	212 @ 3000	392 @ 1800	3000	52–62
	L	V8-534(31)	4 Bore	4.5 × 5.2	7.3	218 @ 3000	298 @ 1800	3000	52–62
1977	B	6-300(1)	1 Bore	4.00 × 3.98	8.9	(59)	(60)	—	40–60
	B	6-300(58)(31)	1 Bore	4.00 × 3.98	8.0	(37)	(38)	—	40–60
	B	6-300(58)(30)	1 Bore	4.00 × 3.98	8.0	(61)	(62)	—	40–60
	B	6-300(3)(30)	1 Bore	4.00 × 3.98	7.9	126 @ 3400	234 @ 1600	—	40–60
	G	V8-302(1)	2 Bore	4.00 × 3.00	8.4	(63)	(64)	—	40–60
	C	V8-330(4)(30)	2 Bore	3.88 × 3.50	7.4	137 @ 3200	272 @ 1800	—	35–75
	D	V8-330(6)(30)	2 Bore	3.88 × 3.50	7.4	137 @ 3600	250 @ 2400	3600(86)	35–75
	H	V8-351W(1)	2 Bore	4.00 × 3.50	8.3	(65)	(66)	—	40–65
	H	V8-351M(1)(30)	2 Bore	4.00 × 3.50	8.0	(67)	(68)	—	50–75
	H	V8-351M(1)(31)	2 Bore	4.00 × 3.50	8.0	(69)	(70)	—	50–75
	H	V8-351W(58)	2 Bore	4.00 × 3.50	8.3	(71)	(72)	—	40–65
	H	V8-351M(58)	2 Bore	4.00 × 3.50	8.0	(73)	(74)	—	50–75
	H	V8-351W(47)	2 Bore	4.0 × 3.50	8.3	135 @ 3200	291 @ 1600	—	40–65
	H	V8-351M(47)	2 Bore	4.0 × 3.50	8.0	161 @ 3600	(83)	—	50–75
	—	V8-359	2 Bore	4.05 × 3.50	—	—	—	3300(88)	35–75
	E	V8-361(30)	4 Bore	4.05 × 3.50	7.2	170 @ 3600	284 @ 2400	3600(87)	35–75
	E	V8-361(31)	4 Bore	4.05 × 3.50	7.2	166 @ 3600	284 @ 2400	3600(87)	35–75
	F	V8-389(30)	4 Bore	4.05 × 3.78	7.2	178 @ 3400	316 @ 2400	—	35–75
	F	V8-389(31)	4 Bore	4.05 × 3.78	7.2	175 @ 3400	316 @ 2400	—	35–75
	F	V8-391	4 Bore	4.05 × 3.79	—	—	—	3800	35–75
	S	V8-400(1)	2 Bore	4.00 × 4.00	8.0	(75)	(76)	—	50–75
	S	V8-400(58)	2 Bore	4.00 × 4.00	8.0	(77)	(78)	—	50–75
	S	V8-400(47)	2 Bore	4.00 × 4.00	8.0	(84)	(85)	—	50–75
	J	V8-460(30)	4 Bore	4.36 × 3.85	8.0	(79)	(80)	—	40–65
	J	V8-460(31)	4 Bore	4.36 × 3.85	8.0	(81)	(82)	—	40–65
	J	V8-460(47)	4 Bore	4.36 × 3.85	8.0	216 @ 4000	366 @ 2600	—	40–65
	J	V8-475	4 Bore	4.50 × 3.75	7.2	212 @ 3400	378 @ 2200	3400(88)	35–60
	K	V8-477	4 Bore	4.50 × 3.75	7.2	206 @ 3200	380 @ 2200	3400(88)	35–60
	L	V8-534	4 Bore	4.50 × 4.20	7.3	218 @ 3100	436 @ 2100	3400(88)	35–60
1978	B	6-300(89)	1 Bore	4.00 × 3.98	8.9	—	—	—	40–60
	B	6-300(90)	1 Bore	4.00 × 3.98	8.0	—	—	—	40–60
	B	6-300(91)(31)	1 Bore	4.00 × 3.98	7.9	126 @ 3400	234 @ 1600	—	40–60

Continued

GENERAL ENGINE SPECIFICATIONS—Continued

Year	Engine Code On Truck Rating Plate	Engine Model	Carb. Type	Bore & Stroke	Comp. Ratio	Horsepower @ R.P.M.	Torque Ft. Lbs. @ R.P.M.	Governed Speed R.P.M. No Load	Normal Oil Pressure Lbs.
	G	V8-302	2 Bore	4.00 × 3.00	8.4	—	—	—	40–60
	D	V8-330 (92)	2 Bore	3.87 × 3.50	7.4	(93)	(94)	3600 (86)	—
	D	V8-330 (95)	2 Bore	3.87 × 3.50	7.4	(96)	(94)	3600 (86)	—
	D	V8-330 (97)	2 Bore	3.87 × 3.50	7.4	(98)	(99)	3600 (86)	—
	H	V8-351W	2 Bore	4.00 × 3.50	8.3	—	—	—	40–65
	H	V8-351M	2 Bore	4.00 × 3.50	8.0	—	—	—	50–75
	E	V8-361	2 Bore	4.05 × 3.50	7.2	(100)	(101)	3600 (87)	—
	E	V8-361 (102)	4 Bore	4.05 × 3.50	7.2	(103)	(104)	3600 (87)	—
	E	V8-361 (105)	4 Bore	4.05 × 3.50	7.2	175 @ 3600	284 @ 2400	3600 (87)	—
	E	V8-361 (106)	4 Bore	4.05 × 3.50	7.2	(107)	282 @ 2400	3600 (87)	—
	E	V8-361 (108)	4 Bore	4.05 × 3.50	7.2	167 @ 3600	280 @ 2400	3600 (87)	—
	F	V8-389 (109)	4 Bore	4.05 × 3.78	7.2	(110)	(111)	3600 (87)	—
	F	V8-389 (112)	4 Bore	4.05 × 3.78	7.2	(113)	313 @ 2400	3600 (87)	—
	F	V8-389 (108)	4 Bore	4.05 × 3.78	7.2	175 @ 3400	311 @ 2400	3600 (87)	—
	F	V8-389 (105)	4 Bore	4.05 × 3.78	7.2	180 @ 3400	318 @ 2400	3600 (87)	—
	S	V8-400	2 Bore	4.00 × 4.00	8.0	—	—	—	50–75
	J	V8-460	4 Bore	4.36 × 3.85	8.0	—	—	—	40–65
	J	V8-475 (30)	4 Bore	4.5 × 3.75	7.2	(114)	(115)	3400	52–62
	K	V8-477 (30)	4 Bore	4.5 × 3.75	7.2	(116)	(117)	3200	52–62
	L	V8-534 (30)	4 Bore	4.5 × 4.20	7.3	(118)	(119)	3200	52–62
1979	B	6-300 (30) (2)	1 Bore	4.00 × 3.98	7.9	126 @ 3600	234 @ 1800	—	40–60
	B	6-300 (31) (2)	1 Bore	4.00 × 3.98	7.9	—	—	—	40–60
	K	6-300 (3) (30)	1 Bore	4.00 × 3.98	7.9	117 @ 3600	227 @ 2400	—	40–60
	K	6-300 (3) (31)	1 Bore	4.00 × 3.98	7.9	—	—	—	40–60
	G	V8-302	2 Bore	4.00 × 3.00	8.4	—	—	—	40–60
	H	V8-351W	2 Bore	4.00 × 3.50	—	—	—	—	40–65
	H	V8-351M	2 Bore	4.00 × 3.50	—	—	—	—	50–75
	C	V8-370	2 Bore	4.05 × 3.59	8.0	184 @ 3600 (120)	302 @ 2800 (120)	3600	40–65
	M	V8-370	4 Bore	4.05 × 3.59	8.0	204 @ 3600 (120)	311 @ 2800 (120)	3600	40–65
	S	V8-400	2 Bore	4.00 × 4.00	—	—	—	—	50–75
	C	V8-429	4 Bore	4.36 × 3.59	8.0	233 @ 4000 (120)	369 @ 2700 (120)	4000	40–65
	J	V8-460	2 Bore	4.36 × 3.85	—	—	—	—	40–65
	A	V8-460	4 Bore	4.36 × 3.85	—	—	—	—	40–65
	J	V8-475	4 Bore	4.5 × 3.75	7.2	(114)	(115)	3400	35–60
	K	V8-477	4 Bore	4.5 × 3.75	7.2	(116)	(117)	3200	35–60
	L	V8-534	4 Bore	4.5 × 4.20	7.3	(118)	(119)	3200	35–60
1980	E	6-300	1 Bore	4.00 × 3.98	8.0	(121)	(122)	—	40–60
	F	V8-302	2 Bore	4.00 × 3.00	8.4	(123)	(124)	—	40–60
	G	V8-351M	2 Bore	4.00 × 3.50	8.0	(125)	(126)	—	50–75
	W	V8-351W	2 Bore	4.00 × 3.50	8.3	132 @ 3000	266 @ 1200	—	40–65
	C	V8-370	2 Bore	4.05 × 3.59	8.0	154 @ 3600	274 @ 2200	3600	40–65
	A	V8-370	4 Bore	4.05 × 3.59	8.0	(127)	(128)	3600	40–65
	H	V8-370	2 Bore	4.05 × 3.59	8.0	(129)	(130)	3600	40–65
	H	V8-370	4 Bore	4.05 × 3.59	8.0	(131)	(132)	3600	40–65
	Z	V8-400	2 Bore	4.00 × 4.00	8.0	(133)	(134)	—	50–75
	K	V8-429	4 Bore	4.36 × 3.59	8.0	(135)	343 @ 2700	4000	40–65
	L	V8-460	4 Bore	4.36 × 3.85	8.0	212 @ 4000	339 @ 2400	—	40–65
	M	V8-477	4 Bore	4.50 × 3.75	7.2	218 @ 3200	393 @ 2300	3200	35–60
	—	V8-534	4 Bore	4.50 × 4.20	7.3	227 @ 3200	432 @ 2100	3200	35–60
1981	E	6-300	1 Bore	4.00 × 3.98	8.9	(136)	(137)	—	40–60
	D	V8-255	2 Bore	3.68 × 3.00	8.2	118 @ 3400	206 @ 2000	—	40–60
	F	V8-302	2 Bore	4.00 × 3.00	8.4	(138)	(139)	—	40–60
	G	V8-351	2 Bore	4.00 × 3.50	8.3	(140)	(141)	—	40–65

Continued

GENERAL ENGINE SPECIFICATIONS—Continued

Year	Engine Code On Truck Rating Plate	Engine Model	Carb. Type	Bore & Stroke	Comp. Ratio	Horsepower @ R.P.M.	Torque Ft. Lbs. @ R.P.M.	Governed Speed R.P.M. No Load	Normal Oil Pressure Lbs.
	H	V8-370	2 Bore	4.05 × 3.59	8.0	(142)	(143)	3600	40–65
	H	V8-370	4 Bore	4.05 × 3.59	8.0	(144)	(145)	3600	40–65
	Z	V8-400	2 Bore	4.00 × 4.00	8.0	153 @ 3200	296 @ 1600	—	50–75
	K	V8-429	4 Bore	4.36 × 3.59	8.0	(135)	343 @ 2700	4000	40–65
	L	V8-460	4 Bore	4.36 × 3.85	8.0	212 @ 4000	339 @ 2400	—	40–65
	M	V8-477	4 Bore	4.50 × 3.75	7.2	218 @ 3200	393 @ 2300	3200	52–62
	P	V8-534	4 Bore	4.50 × 4.20	7.3	227 @ 3200	432 @ 2100	3200	52–62
1982	3	V6-232	2 Bore	3.81 × 3.39	8.65	109 @ 3600	184 @ 1600	—	40–60
	E	6-300	1 Bore	4.00 × 3.98	8.9	(146)	(147)	—	40–60
	D	V8-255	2 Bore	3.68 × 3.00	8.2	118 @ 3400	206 @ 2000	—	40–60
	F	V8-302	2 Bore	4.00 × 3.00	8.4	(148)	(149)	—	40–60
	W	V8-351	2 Bore	4.00 × 3.50	8.3	(140)	(141)	—	40–65
	H	V8-370	2 Bore	4.05 × 3.95	8.0	(142)	(143)	3600	40–65
	H	V8-370	4 Bore	4.05 × 3.95	8.0	(144)	(145)	3600	40–65
	Z	V8-400	2 Bore	4.00 × 4.00	8.0	153 @ 3200	296 @ 1600	—	50–75
	K	V8-429	4 Bore	4.36 × 3.95	8.0	(135)	343 @ 2300	4000	40–65
	L	V8-460	4 Bore	4.36 × 3.85	8.0	212 @ 4000	339 @ 2400	—	40–65
1983	3	V6-232	2 Bore	3.81 × 3.39	8.7	109 @ 3600	184 @ 1600	—	40–60
	Y	6-300	1 Bore	4.00 × 3.98	8.4	(150)	(151)	—	40–60
	F	V8-302	2 Bore	4.00 × 3.00	8.4	139 @ 3400	250 @ 2000	—	40–60
	G	V8-351	2 Bore	4.00 × 3.50	8.3	(152)	(153)	—	40–65
	H	V8-370	2 Bore	4.05 × 3.59	8.0	(142)	(143)	3600	40–65
	H	V8-370	4 Bore	4.05 × 3.59	8.0	(144)	(145)	3600	40–65
	K	V8-429	4 Bore	4.36 × 3.59	8.0	(135)	343 @ 2700	4000	40–65
	L	V8-460	4 Bore	4.36 × 3.85	8.0	202 @ 4000	331 @ 2200	—	40–65

(1)—Under 6000 lbs. GVW emissions standard.
(2)—Light duty.
(3)—Heavy duty.
(4)—Medium duty.
(5)—Auto. Trans. 200 R. P. M. higher.
(6)—Extra heavy duty.
(7)—F-500 137 @ 3200; B-500 138 @ 3200.
(8)—F & LN Series Man. Trans. 137 @ 3600, C6 Auto. Trans. 135 @ 3600, Auto. Trans. except C6 131 @ 3600; B-Series Man. Trans. 138 @ 3600, C6 Auto. Trans. 136 @ 3600, Auto. Trans. except C6 132 @ 3600; C Series Man. Trans. 140 @ 3600, Auto. Trans. 128 @ 3800.
(9)—F, LN & B Series Man. Trans. and C6 Auto. Trans. 250 @ 2400, Auto. Trans. except C6 247 @ 2400; C Series Man. Trans. 253 @ 2400, Auto. Trans. 246 @ 2400.
(10)—F-100 4 × 2 Man. Trans. 157 @ 4000, Auto. Trans. 151 @ 3800; F-100 4 × 4 Man. Trans. 153 @ 4000, Auto. Trans. 148 @ 3800; F-250, 350 @ 3600.
(11)—F-100 4 × 2 Man. Trans. 265 @ 2200, Auto Trans. 151 @ 3800; F-100 4 × 4 Man. Trans. 262 @ 2200, Auto. Trans. 256 @ 2200; F-250, 350, 263 @ 2000.
(12)—F-LN-600, 750 Man. Trans. 153 @ 3600, Auto. Trans. 147 @ 3400; B Series & CT-800 Man. Trans. 154 @ 3600; B Series & C-600, 800 Auto. Trans. 148 @ 3400; C-600, 800 Man. Trans. 156 @ 3600; L-Line 800 158 @ 3600.
(13)—F-LN-600, 750 & B Series Man. Trans. 278 @ 2400, Auto. Trans. 275 @ 2400; C-600, 800 & CT-800 Man. Trans. 279 @ 2400; C-600, 800 Auto. Trans. 273 @ 2400; L-Line 800 280 @ 2400.
(14)—F-100 Auto. Trans. 257 @ 2000; F-250-350 Exc. Calif. 148 @ 3800; F-250-350 Calif. 145 @ 3600.
(15)—F-100 Auto. Trans. 161 @ 3600; F-250-350 Exc. Calif. 264 @ 2200; F-250-350 Calif. 263 @ 2000.
(16)—F-100 Auto. Trans. 160 @ 3600; F-250, 350 159 @ 3400.
(17)—F-100 Auto. Trans. 295 @ 2000; F-250, 350 296 @ 2000.
(18)—F-250, 350 Man. Trans. 208 @ 4500, Auto. Trans. 206 @ 4600.
(19)—F-250, 350 Man. Trans. 324 @ 2800, Auto. Trans. 320 @ 2800.
(20)—F-LN-750 Man. Trans. 180 @ 3600, Auto. Trans. 174 @ 3400; B Series Man. Trans. 181 @ 3600; B Series & C-750, 800 Auto. Trans. 175 @ 3400; C-750, 800 Man. Trans. 179 @ 3600; CT-800 177 @ 3600: L-Line 800 182 @ 3600.
(21)—F-LN-750 & B Series Man. Trans. 320 @ 2400, Auto. Trans. 317 @ 2400; C-750, 800 Man. Trans. 318 @ 2400, Auto. Trans. 315 @ 2400; CT-800 317 @ 2400: L-Line 800 182 @ 3600.
(22)—Ranchero 216 @ 4600; F-100 238 @ 4200: F-250, 350 239 @ 4200.
(23)—Ranchero 356 @ 2600; F-100 380 @ 2600: F-250, 350 374 @ 2600.
(24)—LT-880 203 @ 3400: LTN-880 197 @ 3400.
(25)—LT-880 341 @ 2600: LTN-880 336 @ 2600.
(26)—Super Duty.
(27)—Ranchero engine code F, all except Ranchero G.
(28)—With velocity governor 35–55 p.s.i., without velocity governor 45–70 p. s. i.
(29)—Auto. Trans. 100 R. P. M. lower.
(30)—Except California.
(31)—California.
(32)—E-100: 101 @ 3000: F-100: 100 @ 3000.
(33)—E-100: 102 @ 3000: F-100: 100 @ 3000.
(34)—E-100: 225 @ 1600: F-100: 221 @ 1600.
(35)—E-150-350: 118 @ 3400; F-150-350: 114 @ 3400: P-350-500; 117 @ 3600.
(36)—E-150-350: 226 @ 1800; F-150-350; 222 @ 1600: P-350-500: 227 @ 1800.
(37)—E-150-350: 124 @ 3600; F-150-350: 118 @ 3600; P-350-500:124 @ 3400.
(38)—E-150-350: 217 @ 2200; F-150-350: 205 @ 2200; P-350-500:222 @ 2200.
(39)—Bronco: 121 @ 3600; F-100: 130 @ 3800.
(40)—Bronco: 217 @ 1800; F-100: 222 @ 2000.
(41)—Bronco: 122 @ 3600; F-100: 134 @ 3600.
(42)—Bronco: 215 @ 1800; F-100: 222 @ 2400.
(43)—E-100: 156 @ 4000; E-150-350: 178 @ 4000.
(44)—E-100: 262 @ 2200: E-150-350: 270 @ 2600.
(45)—E-100: 156 @ 4000; E-150-350: 176 @ 4000.
(46)—E-100: 263 @ 2200; E-150-350: 266 @ 2400.
(47)—Ranchero.
(48)—F-100: 145 @ 3800; F-150-350: 145 @ 3600.
(49)—F-100: 264 @ 2200; F-150-350: 263 @ 2000.
(50)—F-100: 145 @ 3800; F-150-350: 148 @ 3600.
(51)—F-100: 250 @ 2200; F-150-350: 251 @ 2200.
(52)—Noise Legislated area engines.

Continued

GENERAL ENGINE SPECIFICATIONS—NOTES—Continued

(53)—E-250-350: 237 @ 4000; F-150-350: 245 @ 4200.
(54)—E-250-350: 365 @ 2600; F-150-350: 371 @ 2600.
(55)—E-250-350: 359 @ 2600; F-150-350: 362 @ 2600.
(56)—E-250-350: 230 @ 4000; F-150-350: 234 @ 4000.
(57)—E-250-350: 360 @ 2600; F-150-350: 362 @ 2600.
(58)—Over 6000 lbs. GVW emissions standards.
(59)—F-100: 119 @ 3200; E-100 exc. Calif: 122 @ 3200; E-100 Calif. 123 @ 3200.
(60)—F-100: 252 @ 1600; E-100 exc. Calif: 252 @ 1600; E-100 Calif. 253 @ 1600.
(61)—F-150-350: 114 @ 3200; E-150-350: 120 @ 3400; P-500 117 @ 3600.
(62)—F-150-350: 227 @ 1800; E-150-350: 229 @ 1400; P-500 227 @ 1800.
(63)—F-100: 136 @ 3600; Bronco 133 @ 3600; Ranchero: 130 @ 3400.
(64)—F-100: 245 @ 2000; Bronco exc. Calif. 236 @ 1800: Bronco Calif 235 @ 1800; Ranchero: 243 @ 1800.
(65)—Exc. Calif.: 141 @ 3200; Calif. auto. trans.: 143 @ 3200; Calif. Man. Trans. 152 @ 3200.
(66)—Exc. Calif.: 286 @ 1400; Calif. auto trans.: 287 @ 1600; Calif. Man. Trans.: 286 @ 2000.
(67)—Non-noise legislated 162 @ 3800; Noise legislated man. trans. 161 @ 3800; Noise legislated auto. trans. 165 @ 3800.
(68)—Non-noise legislated 282 @ 2000; Noise legislated man. trans. 281 @ 2000; Noise legislated auto. trans. 282 @ 2000.
(69)—Auto. Trans. 165 @ 3800; Man. Trans., 161 @ 3800.
(70)—Auto. Trans. 282 @ 2000; Man. Trans. 281 @ 2000.
(71)—Exc. Calif. 147 @ 3400; Calif. 143 @ 3200.
(72)—Exc. Calif. Non-noise legislated 276 @ 1800; Exc. Calif. Noise legislated 275 @ 1800; Calif. 272 @ 2000.
(73)—Exc. Calif. Non-noise legislated 163 @ 3800; Exc. Calif., Noise legislated 162 @ 3800: Calif. 161 @ 3800.
(74)—Exc. Calif. Non-noise legislated 267 @ 2200: Exc. Calif. Noise legislated 266 @ 2200: Calif. 265 @ 2200.
(75)—Non-noise legislated 175 @ 3600; Noise legislated 177 @ 3600.
(76)—Non-noise legislated 326 @ 2000; Noise legislated 325 @ 2000.
(77)—Exc. Non-noise legislated: 167 @ 3600; Non-noise legislated 169 @ 3600.
(78)—Exc. Non-noise legislated: 302 @ 2200; Non-noise legislated 303 @ 2200.
(79)—Non-noise legislated (F-150-350) 245 @ 4200; Noise legislated (F-150-350 & E-250/350) 230 @ 4000: Non-noise legislated (E-250/350) 237 @ 4000.
(80)—Non-noise legislated (F-150-350) 371 @ 2600; Noise legislated (F-150-350) 362 @ 2600: Non-noise legislated (E-250/350) 365 @ 2600 Noise legislated (E-250/350) 359 @ 2600.
(81)—F-150-350: 234 @ 4000; E-250/350; 230 @ 4000.
(82)—F-150-350: 362 @ 2600: E-250/350: 360 @ 2600.
(83)—Exc. Calif. 285 @ 1800; Calif. 286 @ 1800.
(84)—Exc. Calif. 173 @ 3800; Calif. 181 @ 4000.
(85)—Exc. Calif. 326 @ 1600: Calif. 323 @ 1600.
(86)—3800 W/C-6 Auto. Trans.
(87)—3400 W/AT-540 Trans.
(88)—400 W/C6 Auto. trans.
(89)—E100 & F100.
(90)—E150-350, F150-350 & P600.
(91)—F600 & B600.
(92)—F-LN Series.
(93)—Auto. trans., exc. C6, 131 @ 3600; C6, 135 @ 3800: manual trans.,137 @ 3600.
(94)—Auto. trans., exc. C6, 247 @ 2400; C6, 250 @ 2400: manual trans., 250 @ 2400.
(95)—B series.
(96)—Auto. trans., exc. C6, 132 @ 3600; C6, 136 @ 3800: manual trans., 138 @ 3600.
(97)—C series.
(98)—Auto. trans., exc. C6, 128 @ 3600; C6, 139 @ 3800: manual trans.,exc. with velocity governor, 140 @ 3600; with velocity governor 140 @ 3200.
(99)—Auto. trans., exc. C6, 251 @ 2400; C6, 253 @ 2400: manual trans. exc. with velocity governor, 253 @ 2400: with velocity governor, 275 @ 1800.
(100)—F-B600/B700, 149 @ 3800; C600, 144 @ 3800.
(101)—F-B600/B700, 278 @ 2400; C600, 274 @ 2400.
(102)—F/B/LN600-700 series.
(103)—Auto. trans., 163 @ 3400: manual trans., exc. Calif., 170 @ 3600: Calif., 166 @ 3600.
(104)—Auto. trans., 280 @ 2400: manual trans., exc. Calif., 280 @ 2400; Calif., 284 @ 2400.
(105)—L-line 800 series.
(106)—C600-800 series.
(107)—Auto. trans., 168 @ 3400: manual trans., 169 @ 3600.
(108)—CT800 series.
(109)—F/B/LN700, F800 series.
(110)—Auto trans., 171 @ 3400: manual trans., exc. Calif., 178 @ 3400; Calif., 175 @ 3400.
(111)—Auto. trans., 309 @ 2400: manual trans., 316 @ 2400.
(112)—C700-800 series.
(113)—Auto. trans., 176 @ 3400: manual trans. 177 @ 3400.
(114)—F-C 800, 213 @ 3400; L-LT-LTS 800, 212 @ 3400; LN-LNT 800, 210 @ 3400; CT 800, 211 @ 3400.
(115)—F 800, 377 @ 2300; L-LT-LTS 800, 380 @ 2200; LN-LNT 800, 375 @ 2200; C 800, 383 @ 2200; CT 800, 382 @ 2200.
(116)—L-LT-LTS-LN-LNT 900, 209 @ 3200; C 900, 210 @ 3200; CT 900, 208 @ 3200.
(117)—L-LT-LTS 900, 380 @ 2200; LN-LNT 900, 379 @ 2200; C 900, 383 @ 2200; CT 900, 382 @ 2200.
(118)—L-LT-LTS 900, 220 @ 3100; LN-LNT 900, 218 @ 3100; C 900, 221 @ 3100; CT 900, 219 @ 3100.
(119)—L-LT-LTS 900, 437 @ 2100; LN-LTN 900, 436 @ 2100; C 900, 440 @ 2100; CT 900, 439 @ 2100.
(120)—Gross SAE.
(121)—Bronco & F-100-250 exc. Calif., 119 @ 3200; E-100-250 exc. Calif., 115 @ 3200; Bronco, F-100-250 & E-150-250 Calif., 116 @ 3200; E-350 & F-350, 118 @ 3200.
(122)—Bronco & F-150-250 exc. Calif., 243 @ 1200; E-150-250 exc. Calif., 241 @ 1200; Bronco, F-100-250 & E-150-250 Calif., 244 @ 1200; E-350 & F-350, 238 @ 1200.
(123)—Bronco & F-100-250 exc. Calif., 137 @ 3600; E-100-250 exc. Calif., 138 @ 3600; Bronco, F-150 (4 × 4), F-250 & E-100-250, 136 @ 3600; F-100-150 (4 × 2), 133 @ 3400.
(124)—Bronco & F-100-250 exc. Calif., 239 @ 1800; E-100-250 exc. Calif., 242 @ 1800; Bronco, F-150 (4 × 4), F-250 & E-100-250 Calif., 235 @ 1800; F-100-150 (4 × 2), 235 @ 2000.
(125)—Bronco & F-150-250 exc. Calif., 138 @ 3400; Bronco & F-150-250, Calif., 135 @ 3200; E-250 133 @ 3200; E-350 & F-350, 142 @ 3400.
(126)—Bronco & F-150-240 exc. Calif., 263 @ 2000; Bronco & F-150-250 Calif., 259 @ 1600; Bronco, E-250, 133 @ 3200; E-350 & F-350, 251 @ 2400.
(127)—Exc. Calif., 171 @ 3600; Calif., 177 @ 3600.
(128)—Exc. Calif., 284 @ 2600; Calif., 293 @ 2600.
(129)—LN series, 164 @ 3600; C series, 163 @ 3600.
(130)—LN series, 278 @ 2200; C series, 274 @ 2200.
(131)—L series-800 Calif., 181 @ 3600; exc. Calif., 180 @ 3600; C & CT series, Calif., 180 @ 3600; exc. Calif., 186 @ 3600.
(132)—L series-800 Calif., 288 @ 2600; exc. Calif., 289 @ 2600; C & CT series, Calif., 288 @ 2600; exc. Calif., 289 @ 2600.
(133)—F-250 (4 × 4), 136 @ 1200; E-250, 309 @ 1200; F-350 & E-350, 153 @ 3200.
(134)—F-250 (4 × 4), 310 @ 1200; E-250, 309 @ 1200; F-350 & E-350, 296 @ 1600.
(135)—F, B, L, C & CT models, exc. L series with RT-613 or RT-6613 trans., 223 @ 4000; L series with RT-613 or RT-6613 trans., 212 @ 3400.
(136)—Bronco & F100-250 exc. Calif., 122 @ 3000; E-100-250 exc. Calif., 119 @ 3000; F-100-250, E-100-250 & F-250 under 8500 lbs. Calif., 116 @ 3200; F-250 over 8500 lbs., E & F-350, 118 @ 3200.
(137)—Bronco & F-100-250 exc. Calif., 255 @ 1400; E-100-250 exc. Calif., 253 @ 1200; F-100-250, E-100-250 & F-250 under 8500 lbs. Calif., 250 @ 1200; F-250 over 8500 lbs., E & F-350, 238 @ 1200.
(138)—Bronco & F-100-250, 133 @ 3400; E-100-250 exc. Calif., 130 @ 3400; E-100-250 Calif., 132 @ 3600.
(139)—Bronco & F-100-250 exc. Calif., 233 @ 2000; E-100-250 exc. Calif., 230 @ 1400; Bronco, F-100-250 & E-100-250 Calif., 239 @ 1400.
(140)—E-100-250 exc. Calif., 144 @ 3200; E-250-350, 157 @ 3400; Bronco, F-150-250 & F-250 under 8500 lbs. exc. Calif., 136 @ 3000; Bronco, F & E-150-250 & F-250 under 8500 lbs. Calif., 139 @ 3200; F-250 over 8500 lbs. & F-350, 142 @ 3400.
(141)—E-100-250 exc. Calif., 269 @ 1200; E-250-350, 271 @ 2400; Bronco, F-150-250 & F-250 under 8500 lbs. exc. Calif., 262 @ 1600; Bronco, F & E-150-250 & F-250 under 8500 lbs. Calif., 279 @ 1400; F-250 over 8500 lbs. & F-350 251 @ 240.
(142)—F&B series, 154 @ 3600; LN series, 164 @ 3600; C series, 163 @ 3600.
(143)—F, B & C series, 274 @ 2200; LN series, 278 @ 2200.
(144)—F, B & LN series exc. Calif., 177 @ 3600; L series-800 exc. Calif., 187 @ 3600; C & CT series exc. Calif., 186 @ 3600; F, B & LN series Calif., 171 @ 3600; L series-800 Calif., 181 @ 3600; C & CT series Calif., 180 @ 3600.
(145)—F, B & LN series exc. Calif., 293 @ 2600; L series-800 exc. Calif., 299 @ 2600; C & CT series exc. Calif., 298 @ 2600; F, B & LN series Calif., 284 @ 2600; L series-800 Calif., 289 @ 2600; C & CT series Calif., 288 @ 2600.

Continued

GENERAL ENGINE SPECIFICATIONS—NOTES—Continued

(146)—Bronco, F-100-250 & E-100-250 under 8500 lbs. exc. Calif., 119 @ 3000; F-100-250 under 8500 lbs. & E-100-250 Calif., 116 @ 3000; F-250 over 8500 lbs. & F&E-350, 117 @ 3200.

(147)—Bronco, F-100-250 & E-100-250 under 8500 lbs. exc. Calif., 253 @ 1200; F-100-250 under 8500 lbs. & E-100-250 Calif., 250 @ 1200; F-250 over 8500 lbs. & F&E-350, 234 @ 1200.

(148)—Bronco, F-100-250 exc. Calif., 132 @ 3400; E-100-250 exc. Calif., 131 @ 3600; Bronco F-100-250, E-100-250 Calif., 132 @ 3600.

(149)—Bronco, F-100-250 exc. Calif., 232 @ 1800; E-100-250 exc. Calif., 231 @ 1400; Bronco, F-100-250, E-100-250 Calif., 239 @ 1400.

(150)—120 @ 3200 exc. F-250 over 8500 lbs. & F, E-350; F-250 over 8500 lbs. & F, E-350 models, 119 @ 3200.

(151)—250 @ 1600 exc. F-250 over 8500 lbs. & F, E-350; F-250 over 8500 lbs. & F, E-350 models, 230 @ 2200 exc. Calif., 220 @ 2200.

(152)—Bronco, F-150-F-250 under 8500 lbs., 139 @ 3200; F-250 over 8500 lbs. & F-350, 147 @ 3200; E-100, E-150 & E-250 van, 145 @ 3200; E-250 wagon & E-350, 150 @ 3200.

(153)—Bronco, F-150 & 250 under 8500 lbs., 278 @ 1400; F-250 over 8500 lbs. & F-350, 276 @ 2000; E-100-150 & E-250 vans, 270 @ 1400; E-250 wagons & E-350, 279 @ 2000.

TUNE UP SPECIFICATIONS

The following specifications are published from the latest information available. This data should be used only in the absence of a decal affixed in the engine compartment.

★ When using a timing light, disconnect vacuum hose or tube at distributor and plug opening in hose or tube so idle speed will not be affected.

● When checking compression, lowest cylinder must be within 75 percent of highest.

▲ Before removing wires from distributor cap, determine location of the No. 1 wire in cap, as distributor position may have been altered from that shown at the end of this chart.

Year & Engine	Spark Plug		Ignition Timing BTDC(1)★				Curb Idle Speed(2)		Fuel Pump Pressure
	Type	Gap	Firing Order Fig. ▲	Man. Trans.	Auto Trans.	Mark Location	Man. Trans.	Auto. Trans.	
1983									
V6-232	AGSP-52	.044	E	2°	10°	Damper	700/850	650/750D	6–8
6-300 Exc. High Alt.(74)	BSF-42	.044	(62)	6°(94)	10°	Damper	500/700	500/600	5–7
6-300 High Alt(74)	BSF-42	.044	(62)	10°	14°	Damper	500/700	500/600	5–7
6-300 Exc. Calif.(75)	BSF-42	.044	(62)	10°	10°	Damper	700	600D	5–7
6-300 Calif.(75)	BSF-42	.044	(62)	12°(95)	12°	Damper	500/700	550D	5–7
V8-302 Exc. Calif. & High Alt.	ASF-42	.044	A	8°	8°	Damper	700/800	600/675D	6–8
V8-302 High Alt.	ASF-42	.044	A	12°	12°	Damper	700/800	600/675D	6–8
V8-302 Calif.	ASF-42	.044	A	—	—	Damper	—	575D	6–8
V8-351 Exc. High Alt.(74)	ASF-42	.044	B	—	10°	Damper	750	550/625	6–8
V8-351 High Alt.(74)	ASF-42	.044	B	—	14°	Damper	—	550/625	6–8
V8-351(75)	ASF-42	.044	B	8°	8°	Damper	700	650D	6–8
V8-370 2 Bbl.	ASF-32	.044	(63)	6°	6°	Damper	600	600D	(96)
V8-370 4 Bbl. Exc. Calif.	ASF-32	.044	(63)	6°	6°	Damper	600	600D	(96)
V8-370 4 Bbl. Calif.	ASF-32	.044	(63)	2°	2°	Damper	600	600D	(96)
V8-429	ASF-32	.044	(63)	6°	6°	Damper	600	600D	(96)
V8-460 Exc. Calif.	ASF-42	.044	(63)	8°	8°	Damper	800	650D	6–8
V8-460 Calif.	ASF-42	.044	(63)	6°	6°	Damper	800	600D	6–8
1982									
V6-232	AGSP-52	.044	E	10°	12°	Damper	650/750	550/600D	
6-300 Exc. High Alt.(74)	BSF-42	.044	(62)	6°	10°	Damper	500/600		
6-300 High Alt.(74)	BSF-42	.044	(62)	10°	14°	Damper	500/600	500/550D	
6-300 Exc. Calif.(75)	BSF-42	.044	(62)	12°	12°	Damper	500/700	500/550D	
6-300(75)	BSF-42	.044	(62)	10°	—	Damper	500/700	—	
V8-255	ASF-42	.044	A	8°	8°	Damper	750	625/700D	
V8-302 Exc. Calif. & High Alt.	ASF-42	.044	A	8°	8°	Damper	700	575D	
V8-302 Calif.(90)	ASF-42	.044	A	—	—	Damper	—	575/650D	
V8-302 High Alt.	ASF-42	.044	A	12°	12°	Damper	750	575/650D	
V8-351(76)	ASF-42	.044	B	—	10°	Damper	—	550/625D	

Continued

TUNE UP SPECIFICATIONS—Continued

The following specifications are published from the latest information available. This data should be used only in the absence of a decal affixed in the engine compartment.

★ When using a timing light, disconnect vacuum hose or tube at distributor and plug opening in hose or tube so idle speed will not be affected.

● When checking compression, lowest cylinder must be within 75 percent of highest.

▲ Before removing wires from distributor cap, determine location of the No. 1 wire in cap, as distributor position may have been altered from that shown at the end of this chart.

Year & Engine	Spark Plug		Ignition Timing BTDC(1)★				Curb Idle Speed(2)		Fuel Pump Pressure
	Type	Gap	Firing Order Fig. ▲	Man. Trans.	Auto Trans.	Mark Location	Man. Trans.	Auto. Trans.	
1982—Continued									
V8-351(77)	ASF-42	.044	B	—	6°	Damper	—	525/650D	
B8-351 F & E 150, 250 Calif.(90)	ASF-42	.044	B	—	—	Damper	—	600	
V8-351 E-100-250 High Alt.	ASF-42	.044	B	—	14°	Damper	—	550/625D	
V8-370 2 Bbl.	ASF-32	.044	(63)	6°	6°	Damper	600	600D	
V8-370 4 Bbl.	ASF-32	.044	(63)	6°	6°	Damper	600	600D	
V8-370 4 Bbl. Calif.	ASF-32	.044	(63)	2°	2°	Damper	600	600D	
V8-400	ASF-32	.044	B	6°	6°	Damper	600	500/600D	
V8-429	ASF-32	.044	(63)	6°	—	Damper	600	—	
V8-460	ASF-32	.044	A	8°	—	Damper	650	—	
1981									
6-300 Exc. E & F-350	BSF-42	.044	(62)	6°	10°	Damper	600/700(89)	550D(89)	5–7
6-300 E & F-350 Exc. Calif.	BSF-42	.044	(62)	12°	12°	Damper	500/700	500/550D	5–7
6-300 F-350 Calif.	BSF-42	.044	(62)	10°	—	Damper	500/700	—	5–7
V8-255	ASF-42	.044	A	4°	10°	Damper	750(89)	575/650D(89)	6–8
V8-302	ASF-42	.044	A	8°	8°	Damper	700(89)	575/650D(89)	6–8
V8-351W Exc. Calif. & High Alt.(10)(90)	ASF-42	.044	B	—	10°	Damper	—	550/625D	6–8
V8-351W Exc. Calif.(10)(91)	ASF-42	.044	B	—	6°	Damper	—	525/600D	6–8
V8-351W Calif.	ASF-42	.044	B	—	—	Damper	—	600D(89)	6–8
V8-351W High Alt.	ASF-42	.044	B	—	8°	Damper	—	550/625D	6–8
V8-351M(6)(15)(92)	ASF-42	.044	B	10°	6°	Damper	650	550/625D	6–8
V8-351M(6)(15)(93)	ASF-52	.044	B	10°	10°	Damper	600	500/600D	6–8
V8-370 2 Barrel Carb.	ASF-32	.044	(63)	6°	6°	Damper	600	600D	6–8(24)
V8-370 4 Barrel Carb. Exc. Calif.	ASF-32	.044	(63)	6°	6°	Damper	600	600D	6–8(24)
V8-370 4 Barrel Carb. Calif.	ASF-32	.044	(63)	2°	2°	Damper	600	600D	6–8(24)
V8-400 Exc. Calif.	ASF-52	.044	B	6°	3°	Damper	500/600	500/600D	6–8
V8-400 Calif.	ASF-52	.044	B	6°	6°	Damper	600	500/600D	6–8
V8-429	ASF-32	.044	(63)	6°	6°	Damper	600	600D	6–8(24)
V8-460	ASF-42	.044	A	—	8°	Damper	650	650D	6–8
V8-477, 534	BYSF-31-4	.040	(64)	10°	10°	Damper	600	600D	6–8(24)
1980									
6-300 Bronco, E & F100, 150, 250 Exc. Calif.	BSF-42	.044	(62)	6°	10°	Damper	600/700	550D	5–7
6-300 Bronco, E & F100, 150, 250 Calif.	BSF-42	.044	(62)	6°	(3)(6)	Damper	500/700	550D	5–7
6-300 E & F350 Exc. Calif.	BSF-42	.044	(62)	12°	12°	Damper	500/700	550D	5–7
6-300 F350 Calif.	BSF-42	.044	(62)	10°	10°	Damper	500/700	550D	5–7
V8-302 E100, 150 Exc. Calif.	ASF-42	.044	A	8°	10°	Damper	700	575D	6–8
V8-302 E100 Calif.	ASF-42	.044	A	—	6°	Damper	—	650D	6–8
V8-302 E150 Calif.	ASF-42	.044	A	—	(6)(7)	Damper	—	600D	6–8

Continued

TUNE UP SPECIFICATIONS—Continued

The following specifications are published from the latest information available. This data should be used only in the absence of a decal affixed in the engine compartment.

★ When using a timing light, disconnect vacuum hose or tube at distributor and plug opening in hose or tube so idle speed will not be affected.

● When checking compression, lowest cylinder must be within 75 percent of highest.

▲ Before removing wires from distributor cap, determine location of the No. 1 wire in cap, as distributor position may have been altered from that shown at the end of this chart.

Year & Engine	Spark Plug		Ignition Timing BTDC(1)★				Curb Idle Speed(2)		Fuel Pump Pressure
	Type	Gap	Firing Order Fig. ▲	Man. Trans.	Auto Trans.	Mark Location	Man. Trans.	Auto. Trans.	
1980—Continued									
V8-302 E250 Calif.	ASF-42	.044	A	—	10°	Damper	—	575D	6–8
V8-302 F100	ASF-42	.044	A	6°	8°	Damper	700	575D	6–8
V8-302 F150 Exc. Calif.(4)	ASF-42	.044	A	2°	8°	Damper	550/800	575D	6–8
V8-302 F150 Calif.(4)	ASF-42	.044	A	(8)	8°	Damper	(9)	650D	6–8
V8-302 F150 & Bronco Exc. Calif.(5)	ASF-42	.044	A	8°	(6)(11)	Damper	700	575D	6–8
V8-302 F150 & Bronco Calif.(5)	ASF-42	.044	A	4°	(6)(12)	Damper	750	575D	6–8
V8-302 F250 Exc. Calif.(4)	ASF-42	.044	A	(13)	8°	Damper	700	575D	6–8
V8-302 F250 Calif.(4)	ASF-42	.044	A	—	10°	Damper	—	575D	6–8
V8-302 F250(5)	ASF-42	.044	A	8°	8°	Damper	700	575D	6–8
V8-351W E100, 150 Exc. Calif.(6)(10)(14)	ASF-42	.044	B	—	6°	Damper	—	525/650D	6–8
V8-351W E100, 150 Exc. Calif.(6)(10)(16)(17)	ASF-42	.044	B	—	16°	Damper	—	500/600D	6–8
V8-351W E100, 150 Exc. Calif.(6)(10)(16)(18)	ASF-42	.044	B	—	14°	Damper	—	500/600D	6–8
V8-351W E100, 150 Calif.(6)(10)(19)	ASF-42	.044	B	—	8°	Damper	—	500/600D	6–8
V8-351W E100, 150 High Alt.(6)(10)(20)	ASF-42	.044	B	—	14°	Damper	—	500/600D	6–8
V8-351W E100, 150(6)(10)(21)	ASF-42	.044	B	—	10°	Damper	—	500/600D	6–8
V8-351W E100, 150(6)(10)(22)	ASF-42	.044	B	—	8°	Damper	—	500/600D	6–8
V8-351M E250, 350(6)(15)(23)	ASF-42	.044	B	—	6°	Damper	—	550/625D	6–8
V8-351M E250, 350(6)(15)(78)	ASF-42	.044	B	—	4°	Damper	—	550/625D	6–8
V8-351M F100-350 & Bronco(6)(15)(79)	ASF-52	.044	B	10°(80)	10°(80)	Damper	600	500/600D	6–8
V8-351M F100-350 & Bronco(6)(18)(81)	ASF-42	.044	B	16°(80)	14°(80)	Damper	650	550/625D	6–8
V8-351M F100-350 & Bronco(6)(15)(82)	ASF-42	.044	B	10°	12°(80)	Damper	650	550/625D	6–8
V8-351M F100-350 & Bronco(6)(15)(83)	ASF-42	.044	B	10°	8°(80)	Damper	550/800	550/625D	6–8
V8-351M F100-350 & Bronco(6)(15)(84)	ASF-42	.044	B	—	6°(80)	Damper	—	550/625D	6–8
V8-351M F100-350 & Bronco(6)(15)(85)	ASF-42	.044	B	—	4°(80)	Damper	—	550/625D	6–8
V8-370 2 Barrel Carb.	ASF-32	.044	(63)	6°	6°	Damper	600	600D	6–8(24)
V8-370 4 Barrel Carb. Exc. Calif.	ASF-32	.044	(63)	6°	6°	Damper	600	600D	6–8(24)
V8-370 4 Barrel Carb. Calif.	ASF-32	.044	(63)	2°	2°	Damper	600	600D	6–8(24)
V8-400 E250, 350	ASF-42	.044	B	—	4°	Damper	—	550/625D	6–8
V8-400 F100-350 Exc. Calif.(6)(86)	ASF-52	.044	B	6°(80)	3°(80)	Damper	600	500/600D	6–8
V8-400 F100-350(6)(87)	ASF-52	.044	B	6°(80)	6°(80)	Damper	600	500/600D	6–8
V8-400 F100-350(6)(88)	ASF-42	.044	B	—	4°(80)	Damper	—	550/625D	6–8

Continued

TUNE UP SPECIFICATIONS—Continued

The following specifications are published from the latest information available. This data should be used only in the absence of a decal affixed in the engine compartment.

★ When using a timing light, disconnect vacuum hose or tube at distributor and plug opening in hose or tube so idle speed will not be affected.

● When checking compression, lowest cylinder must be within 75 percent of highest.

▲ Before removing wires from distributor cap, determine location of the No. 1 wire in cap, as distributor position may have been altered from that shown at the end of this chart.

Year & Engine	Spark Plug		Ignition Timing BTDC(1)★				Curb Idle Speed(2)		Fuel Pump Pressure
	Type	Gap	Firing Order Fig. ▲	Man. Trans.	Auto Trans.	Mark Location	Man. Trans.	Auto. Trans.	
1980—Continued									
V8-429	ASF-32	.044	(63)	6°	6°	Damper	600	600D	6–8(24)
V8-460	ASF-32	.044	A	—	8°	Damper	—	650D	6–8
V8-477, 534	BYSF-31-4	.040	(64)	10°	10°	Damper	600	600D	6–8(24)
1979									
6-300 E & F-100, 150 & 250	(25)	.044	(62)	6°	10°	Damper	500/700	500/550D	5–7
6-300 E & F-350	(25)	.044	(62)	12°	12°	Damper	500/700	500/550D	5–7
6-300 FB & P-600	(26)	.044	(62)	12°	—	Damper	500/700	—	4–6(24)
V8-302 Ranchero Exc. Calif.	(27)	.050	A	—	8°	Damper	—	600/675D	6–8
V8-302 Ranchero Calif.	(27)	.050	A	—	12°	Damper	—	600/675D	6–8
V8-302 E-100, 150, 250	(28)	.044	A	4°	6°	Damper	700	550/600D(29)	6–8
V8-302 F-100, 150 Exc. Calif.	(28)	.044	A	6°	8°	Damper	700	550/600D	6–8
V8-302 F-100 Calif.	(28)	.044	A	—	8°	Damper	—	550/650D	6–8
V8-302 F-150 Calif.	(28)	.044	A	—	6°	Damper	—	550/600D	6–8
V8-351W Ranchero(10)	(27)	.050	B	—	15°	Damper	—	600/650D	6–8
V8-351W E-100(10)	(28)	.044	B	—	10°	Damper	—	500/600D	6–8
V8-351W E-150 Exc. Calif.(10)	(28)	.044	B	4°	10°	Damper	500/800	500/600D	6–8
V8-351W E-150 Calif.(10)	(28)	.044	B	—	8°	Damper	—	500/650D	6–8
V8-351W E-250(10)	(28)	.044	B	4°	12°	Damper	500/800	500/600D	6–8
V8-351W E-350 & F-800(10)	(28)	.044	B	—	6°	Damper	—	525/650D	6–8
V8-351M Ranchero Exc. Calif.(15)	(27)	.050	B	—	12°	Damper	—	600/650D	6–8
V8-351M Ranchero Calif.(15)	(27)	.050	B	—	14°	Damper	—	600/650D	6–8
V8-351M F-100 Exc. Calif.(15)	(28)	.044	B	10°	6°	Damper	650	550D	6–8
V8-351M F-100 Calif.(15)	(28)	.044	B	—	10°	Damper	—	550D	6–8
V8-351M F-150, 250 & Bronco(15)	(28)	.044	B	(30)	6°	Damper	650	550D	6–8
V8-351M F-350(15)	(28)	.044	B	6°	12°	Damper	600	500/600D	6–8
V8-370 2 Barrel Carb.	(31)	.044	(63)	6°	—	Damper	600	—	6–8(24)
V8-370 4 Barrel Carb.	(31)	.044	(63)	2°	—	Damper	600	—	6–8(24)
V8-400 F-150, 250 & Bronco	(28)	.044	B	10°	6°	Damper	650	550D	6–8
V8-400 F-350	(28)	.044	B	3°	3°	Damper	600	600D	6–8
V8-429	(31)	.044	(63)	8°	—	Damper	600	—	6–8(24)
V8-460 E-250	(28)	.044	A	—	14°	Damper	—	650/800D	6–8
V8-460 E-350	(28)	.044	A	—	8°	Damper	—	650D	6–8
V8-475, 477 & 534	(32)	.040	(64)	10°	—	Damper	600	—	6–8(24)
1978									
6-300 E-100	BSF-42	.044	(62)	6°	(6)(33)	Damper	700	550D	5–7
6-300 F-100	BSF-42	.044	(62)	6°	10°	Damper	700	550D	5–7

Continued

TUNE UP SPECIFICATIONS—Continued

The following specifications are published from the latest information available. This data should be used only in the absence of a decal affixed in the engine compartment.

★ When using a timing light, disconnect vacuum hose or tube at distributor and plug opening in hose or tube so idle speed will not be affected.

● When checking compression, lowest cylinder must be within 75 percent of highest.

▲ Before removing wires from distributor cap, determine location of the No. 1 wire in cap, as distributor position may have been altered from that shown at the end of this chart.

Year & Engine	Spark Plug		Ignition Timing BTDC①★				Curb Idle Speed②		Fuel Pump Pressure
	Type	Gap	Firing Order Fig. ▲	Man. Trans.	Auto Trans.	Mark Location	Man. Trans.	Auto. Trans.	
1978—Continued									
6-300 E & F-150, 250 Exc. Calif.	BSF-42	.054	㊷	10°	10°	Damper	600	600D	5–7
6-300 E & F-150, 250 Calif.	BSF-42	.044	㊷	6°	8°	Damper	700	550D	5–7
6-300 E & F-350	BSF-42	.054	㊷	10°	10°	Damper	600	600D	5–7
6-300 F, B & P-600	㉞	.054	㊷	10°	—	Damper	600	—	4–6
V8-302 Ranchero	ASF-52	.050	A	—	14°	Damper	—	600D	6–8
V8-302 F-100 Exc. Calif. & High Alt.	ASF-42	.044	A	6°	8°	Damper	700	600D	6–8
V8-302 F-150 Exc. Calif.	ASF-42	.044	A	6°	6°	Damper	750	600D	6–8
V8-302 F-100, 150 Calif.㉟	ASF-42	.044	A	6°	8°	Damper	750	650D	6–8
V8-302 F-150 Calif.㊱	ASF-42	.044	A	—	10°	Damper	—	650D	6–8
V8-302 F-100 High Alt.	ASF-42	.044	A	12°	10°	Damper	800	550D	6–8
V8-330 Exc. Calif.㊲	BSF-31	.030	㊸	6°	—	Damper	550	—	5–7㉔
V8-330 Exc. Calif.㊳	BSF-31	.030	㊸	6°	6°	Damper	600	550D	5–7㉔
V8-330 Calif.㊲	BSF-31	.030	㊸	4°	—	Damper	650	—	5–7㉔
V8-330 Calif.㊳	BSF-31	.030	㊸	8°	—	Damper	650	—	5–7㉔
V8-351W Ranchero⑩	ASF-52	.050	B	—	14°	Damper	—	600/675D	6–8
V8-351W Bronco⑩	ASF-42	.044	B	8°	14°	Damper	650	650D	6–8
V8-351W E-100 Exc. Calif. & High Alt.⑩	ASF-42	.044	B	8°	14°	Damper	600/900	500/600D	6–8
V8-351W E-100 Calif.⑩	ASF-42	.044	B	—	10°	Damper	—	500/700D	6–8
V8-351W E-100 High Alt.⑩	ASF-42	.044	B	—	12°㊴	Damper	—	550D	6–8
V8-351W E-150, 250, 350 Exc. Calif.⑩	㊵	.044	B	6°	8°	Damper	650	550D	6–8
V8-351W E-150, 250 Calif.⑩	ASF-42	.044	B	4°	8°	Damper	650	550D	6–8
V8-351W E-350 Calif.⑩	㊵	.044	B	—	8°	Damper	—	650D	6–8
V8-351M Ranchero Exc. Calif. & High Alt.⑮	ASF-52	.050	B	—	9°	Damper	—	600/675D	6–8
V8-351M Ranchero Calif.⑮	ASF-52	.050	B	—	14°	Damper	—	600/675D	6–8
V8-351M Ranchero High Alt.⑮	ASF-52	.050	B	—	12°㊴	Damper	—	650D	6–8
V8-351M F-100 Exc. Calif. & High Alt.⑮	ASF-42	.044	B	8°	10°	Damper	750	600D	6–8
V8-351M F-100 Calif.⑮	ASF-42	.044	B	—	8°	Damper	—	650D	6–8
V8-351M F-100 High Alt.⑮	ASF-42	.044	B	—	14°	Damper	—	650D	6–8
V8-351M F-150, 250, 350, Bronco Exc. Calif.⑮	ASF-42	.044	B	6°	14°	Damper	650	650D	6–8
V8-351M F-150, 250, Bronco Calif.⑮	ASF-42	.044	B	6°	8°	Damper	800	650D	6–8
V8-351M F-350 Calif.⑮	ASF-42	.044	B	10°	10°	Damper	650	650D	6–8
V8-361 Exc. Calif.	BSF-31	.030	㊸	4°	4°	Damper	650	650D	5–7㉔
V8-361 Calif.	BSF-31	.030	㊸	6°	6°	Damper	650	650D	5–7㉔
V8-391	BSF-31	.030	㊸	4°	4°	Damper	650	650D	5–7㉔

Continued

TUNE UP SPECIFICATIONS—Continued

The following specifications are published from the latest information available. This data should be used only in the absence of a decal affixed in the engine compartment.

★ When using a timing light, disconnect vacuum hose or tube at distributor and plug opening in hose or tube so idle speed will not be affected.

● When checking compression, lowest cylinder must be within 75 percent of highest.

▲ Before removing wires from distributor cap, determine location of the No. 1 wire in cap, as distributor position may have been altered from that shown at the end of this chart.

Year & Engine	Spark Plug		Ignition Timing BTDC①★				Curb Idle Speed②		Fuel Pump Pressure
	Type	Gap	Firing Order Fig. ▲	Man. Trans.	Auto Trans.	Mark Location	Man. Trans.	Auto. Trans.	
1978—Continued									
V8-400 Ranchero Exc. Calif. & High Alt.㉟	ASF-52	.050	B	—	13°	Damper	—	575/650D	6–8
V8-400 Ranchero Exc. Calif. & High Alt.㊱	ASF-52	.050	B	—	14°	Damper	—	600/675D	6–8
V8-400 Ranchero Calif.	ASF-52	.050	B	—	14°	Damper	—	600/675D	6–8
V8-400 Ranchero High Alt.	ASF-52	.050	B	—	8°㊴	Damper	—	650D	6–8
V8-400 F-100	ASF-42	.044	B	—	6°	Damper	—	600D	6–8
V8-400 F-150, 250, 350, Bronco Exc. Calif.	ASF-42	.044	B	12°	12°	Damper	650	650D	6–8
V8-400 F-150, 250, Bronco Calif.	ASF-42	.044	B	—	8°	Damper	—	650D	6–8
V8-400 F-350 Calif.	ASF-42	.044	B	—	10°	Damper	—	650D	6–8
V8-460 Exc. Calif.	ASF-42	.044	A	—	12°	Damper	—	650D	㊿65
V8-460 E-250, F-150, 250 Calif.	ASF-42	.044	A	—	14°	Damper	—	600D	(65)
V8-460 E & F-350 Calif.	ASF-42	.044	A	—	12°	Damper	—	650D	(65)
V8-477, 534	BYSF-31-4	.040	(64)	10°	10°	Damper	600	600D	—
1977									
6-300 E & F-100	BSF-42	.044	(62)	6°	6°	Damper	700	550D	5–7
6-300 E & F-150, 250, 350	BSF-42	.044	(62)	10°	10°	Damper	600	600D	5–7
6-300 P-500	BSF-42	.044	(62)	10°	—	Damper	600	—	4–6
6-300 C-600	BSF-31	.032	(62)	10°	—	Damper	600	—	4–6
V8-302 Ranchero	ARF-52	.050	A	—	2°	Damper	—	500/600D	6–8
V8-302 Bronco Exc. High Alt.	ASF-42	.044	A	⑥㊶	8°	Damper	850	550/650D	6–8
V8-302 Bronco High Alt.	ASF-42	.044	A	8°	8°	Damper	800	550D	6–8
V8-302 E-100 Exc. Calif. & High Alt.	ASF-42	.044	A	㊷	8°	Damper	850	650D	6–8
V8-302 E-100 Calif.	ASF-42	.044	A	6°	—	Damper	800	—	6–8
V8-302 E-100 High Alt.	ASF-42	.044	A	8°	8°	Damper	800	550D	6–8
V8-302 F-100 Exc. Calif. & High Alt.	ASF-42	.044	A	⑥㊸	⑥㊹	Damper	850	650D	6–8
V8-302 F-100 Calif.	ASF-42	.044	A	6°	—	Damper	800	—	6–8
V8-302 F-100 High Alt.	ASF-42	.044	A	8°	8°	Damper	800	550D	6–8
V8-302 F-150	ASF-42	.044	A	6°	6°	Damper	750	650D	6–8
V8-330	BSF-31	.030	(63)	6°	6°	Damper	600	550D	5–7㉔
V8-351W Ranchero⑩	ARF-52	.050	B	—	4°	Damper	—	625D	6–8
V8-351W E-100 Exc. Calif.⑩	ASF-42	.044	B	㊷	12°	Damper	㊷	550D	6–8
V8-351W E-100 Calif.⑩	ASF-42	.044	B	8°	14°	Damper	900	700D	6–8
V8-351W E-150, 250, 350⑩	ASF-34	.044	B	6°	8°	Damper	650	550D	6–8
V8-351M Ranchero Exc. Calif. & High Alt.⑮㉟	ARF-52	.050	B	—	⑥㊺	Damper	—	650D	6–8
V8-351M Ranchero Exc. Calif. & High Alt.⑮㊱	ARF-52	.050	B	—	8°	Damper	—	500/650D	6–8

Continued

TUNE UP SPECIFICATIONS—Continued

The following specifications are published from the latest information available. This data should be used only in the absence of a decal affixed in the engine compartment.

★ When using a timing light, disconnect vacuum hose or tube at distributor and plug opening in hose or tube so idle speed will not be affected.

● When checking compression, lowest cylinder must be within 75 percent of highest.

▲ Before removing wires from distributor cap, determine location of the No. 1 wire in cap, as distributor position may have been altered from that shown at the end of this chart.

Year & Engine	Spark Plug		Ignition Timing BTDC(1)★				Curb Idle Speed(2)		Fuel Pump Pressure
	Type	Gap	Firing Order Fig. ▲	Man. Trans.	Auto Trans.	Mark Location	Man. Trans.	Auto. Trans.	
1977—Continued									
V8-351M Ranchero Calif.(15)	ARF-52-6	.060	B	—	(6)(46)	Damper	—	600D	6–8
V8-351M Ranchero High Alt.(15)	ARF-52	.050	B	—	12°	Damper	—	650D	6–8
V8-351M F-100 Exc. Calif. & High Alt.(15)	ASF-42	.044	B	10°	(6)(47)	Damper	800	(6)(48)	6–8
V8-351M F-100 Calif.(15)	ASF-42	.044	B	—	8°	Damper	—	650D	6–8
V8-351M F-100 High Alt.(15)	ASF-42	.044	B	—	10°	Damper	—	650D	6–8
V8-351M F-150, 250, 350 Exc. Calif.(15)	ASF-42	.044	B	12°	(6)(49)	Damper	650	650D	6–8
V8-351M F-150, 250, 350 Calif.(15)	ASF-42	.044	B	—	(6)(50)	Damper	—	650D	6–8
V8-361, F, B, C-600 & B-700 Exc. Calif.(6)(51)	BSF-31	.030	(63)	—	6°	Damper	—	550D	5–7(24)
V8-361F, B, LN, C-600, 700, 750 Exc. Calif.(6)(52)	BSF-31	.030	(63)	—	4°	Damper	—	650D	5–7(24)
V8-361F, B, LN, C-600, 700, 750 Calif.	BSF-31	.030	(63)	—	6°	Damper	—	650D	5–7(24)
V8-361L, LN, LT, LNT, LTS, C, CT-800 Exc. Calif.	BSF-31	.030	(63)	—	4°	Damper	—	650D	5–7(24)
V8-361L, LN, LT, LNT, LTS, C, CT-800 Calif.	BSF-31	.030	(63)	—	6°	Damper	—	650D	5–7(24)
V8-391	BSF-31	.030	(63)	4°	4°	Damper	650	650D	5–7(24)
V8-400 Ranchero Exc. Calif. & High Alt.	ARF-52	.050	B	—	(6)(53)	Damper	—	(6)(54)	6–8
V8-400 Ranchero Calif.	ARF-52-6	.060	B	—	6°	Damper	—	500/600D	6–8
V8-400 Ranchero High Alt.	ARF-52	.050	B	—	10°	Damper	—	650D	6–8
V8-400 F-100	ASF-42	.044	B	—	6°	Damper	—	650D	6–8
V8-400 F-150, 250, 350, Bronco	ASF-42	.044	B	12°	12°	Damper	650	650D	6–8
V8-460 E & F-150, 250, 350	ASF-42	.044	A	—	12°	Damper	—	650D	(65)
V8-477, 534	BYSF-31-4	.040	(64)	—	10°	Damper	—	600D	—
1976									
6-300 E & F-150, 250, 350	BTRF-42	.044	(62)	10°	10°	Damper	600	600D	4–6
6-300 P-350, 400, 500	BTRF-42	.044	(62)	10°	10°	Damper	600	600D	4–6
6-300F, B-500, 600 & C-600 Exc. Calif.(55)	BRF-31	.030	(62)	10°	10°	Damper	600	600D	4–6
6-300 F, B-500, 600 Calif.	BRF-31-4	.044	(62)	10°	10°	Damper	600	600D	4–6
V8-302	ARF-42	.044	C	(42)	(42)	Damper	(42)	(42)	6–8
V8-330(56)	BRF-31	.030	(63)	6°	6°	Damper	600	550D	5–7(24)
V8-351W Ranchero(10)	ARF-42	.044	D	—	12°	Damper	—	650D	6–8
V8-351W E-150, 250, 350(6)(10)(57)	ARF-32	.044	D	14°	14°	Damper	650	650D	5–7
V8-351W E-150, 250, 350(6)(10)(58)	ARF-32	.044	D	12°	12°	Damper	650	650D	5–7
V8-351W E-150, 250, 350(6)(10)(59)	ARF-42	.044	D	14°	14°	Damper	650	650D	5–7

Continued

TUNE UP SPECIFICATIONS—Continued

The following specifications are published from the latest information available. This data should be used only in the absence of a decal affixed in the engine compartment.

★ When using a timing light, disconnect vacuum hose or tube at distributor and plug opening in hose or tube so idle speed will not be affected.

● When checking compression, lowest cylinder must be within 75 percent of highest.

▲ Before removing wires from distributor cap, determine location of the No. 1 wire in cap, as distributor position may have been altered from that shown at the end of this chart.

Year & Engine	Spark Plug		Ignition Timing BTDC(1)★				Curb Idle Speed(2)		Fuel Pump Pressure
	Type	Gap	Firing Order Fig. ▲	Man. Trans.	Auto Trans.	Mark Location	Man. Trans.	Auto. Trans.	
1976—Continued									
V8-351M Ranchero(15)	ARF-42	.044	D	—	8°	Damper	—	650D	6–8
V8-360 Exc. Calif.	BRF-42	.044	(63)	6°	6°	Damper	650	550D	5–7
V8-360 Calif.(6)(60)	BRF-42	.044	(63)	12°	12°	Damper	600	600D	5–7
V8-360 Calif.(6)(61)	BRF-42	.044	(63)	4°	4°	Damper	600	600D	5–7
V8-361 Exc. Calif.(56)	BRF-31	.030	(63)	4°	4°	Damper	600	600D	5–7
V8-361 Calif.(56)	BRF-31	.030	(63)	6°	6°	Damper	650	650D	5–7(24)
V8-390 F-100	BRF-42	.044	(63)	—	12°	Damper	—	650D	5–7
V8-390 F-150, 250, 350 Exc. Calif.	BRF-42	.044	(63)	14°	16°	Damper	650	650D	5–7
V8-390 F-150, 250, 350 Calif.	BRF-42	.044	(63)	14°	14°	Damper	650	650D	5–7
V8-391(56)	BRF-31	.030	(63)	4°	4°	Damper	650	650D	5–7(24)
V8-400 Ranchero Exc. Calif.	ARF-42	.044	D	—	12°	Damper	—	650D	6.7–8.7
V8-400 Ranchero Calif.	ARF-42	.044	D	—	12°	Damper	—	625D	6.7–8.7
V8-460 Ranchero Exc. Calif.	ARF-52	.044	C	—	10°	Damper	—	650D	6.7–8.7
V8-460 Ranchero Calif.	ARF-52	.044	C	—	14°	Damper	—	650D	6.7–8.7
V8-460 F-150, E & F-250, 350	ARF-42	.044	C	—	12°	Damper	—	650D	(65)
V8-477, 534 Exc. Calif.(56)	BRF-31	.030	(64)	6°	6°	Damper	550	550D	—
V8-477, 534 Calif.(56)	BRF-31	.030	(64)	12°	12°	Damper	550	550D	—
1975									
6-300	BTRF-42	.044	(62)	12°	12°	Damper	700	550D	4–6
V8-302	ARF-42	.044	C	(42)	(42)	Damper	(42)	(42)	(66)
V8-330(56)	BRF-31	.030	(63)	(42)	(42)	Damper	(42)	(42)	5–7
V8-351W Ranchero(10)	ARF-42	.044	D	—	12°	Damper	—	650D	5–7
V8-351M Ranchero(15)	ARF-42	.044	D	—	12°	Damper	—	650D	5–7
V8-360 F-100	BRF-42	.044	(63)	6°	(42)	Damper	850	650D	5–7
V8-390 F-100	BRF-42	.044	(63)	—	6°	Damper	—	650D	5–7
V8-391(56)	BRF-31	.030	(63)	(42)	(42)	Damper	(42)	(42)	5–7
V8-400 Ranchero	ARF-42	.044	D	—	12°	Damper	—	650D	5–7
V8-401(56)	BRF-31	.030	(64)	(42)	(42)	Damper	(42)	(42)	—
V8-460 Ranchero	ARF-42	.044	C	—	14°	Damper	—	650D	5.7–7.7
V8-477(56)	BRF-31	.030	(64)	(42)	(42)	Damper	(42)	(42)	—
V8-534(56)	BRF-31	.030	(64)	(42)	(42)	Damper	(42)	(42)	—
1974									
6-200, 250	BRF-82	.034	(62)	(42)	(42)	Damper	(42)	(42)	4½–5½
6-240	BRF-42	.034	(62)	(42)	(42)	Damper	(42)	(42)	4–6
6-300 Light Duty Exc. Calif.	BRF-42	.034	(62)	(42)	(42)	Damper	(42)	(42)	4–6
6-300 Light Duty Calif.	BTRF-42	.044	(62)	(42)	(42)	Damper	(42)	(42)	4–6
6-300 Heavy Duty	BRF-31	.030	(62)	(42)	(42)	Damper	(42)	(42)	4–6
V8-302 Ranchero(67)	BRF-42	.034	C	6°	6°	Damper	800	625D	4–6
V8-302 Ranchero(68)	BRF-42	.054	C	6°	6°	Damper	800	625D	4–6
V8-302 Exc. Ranchero(67)	BRF-42	.034	C	(42)	(42)	Damper	(42)	(42)	(69)
V8-302 Exc. Ranchero(68)	BRF-42	.054	C	(42)	(42)	Damper	(42)	(42)	(69)

Continued

TUNE UP SPECIFICATIONS—Continued

The following specifications are published from the latest information available. This data should be used only in the absence of a decal affixed in the engine compartment.

★ When using a timing light, disconnect vacuum hose or tube at distributor and plug opening in hose or tube so idle speed will not be affected.

● When checking compression, lowest cylinder must be within 75 percent of highest.

▲ Before removing wires from distributor cap, determine location of the No. 1 wire in cap, as distributor position may have been altered from that shown at the end of this chart.

Year & Engine	Spark Plug		Ignition Timing BTDC(1)★				Curb Idle Speed(2)		Fuel Pump Pressure
	Type	Gap	Firing Order Fig. ▲	Man. Trans.	Auto Trans.	Mark Location	Man. Trans.	Auto. Trans.	
1974—Continued									
V8-330(56)	BRF-31	.030	(63)	(42)	(42)	Damper	(42)	(42)	5–6(24)
V8-351W Ranchero(10)(67)	BRF-42	.034	D	—	6°	Damper	—	600D	5–6
V8-351W Ranchero(10)(68)	BRF-42	.034	D	—	6°	Damper	—	600D	5–6
V8-351W Exc. Ranchero(10)	BRF-42	.034	D	(42)	(42)	Damper	(42)	(42)	5–6
V8-351C 2 Barrel Carb. Ranchero(67)(70)	ARF-42	.044	D	—	10°	Damper	—	600D	5–6
V8-351C 2 Barrel Carb. Ranchero(68)(70)	ARF-42	.044	D	—	10°	Damper	—	600D	5–6
V8-351CJ 4 Barrel Carb. Ranchero(70)(71)	ARF-42	.044	D	(42)	(42)	Damper	(42)	(42)	5–6
V8-351C Exc. Ranchero(70)	ARF-42	.044	D	(42)	(42)	Damper	(42)	(42)	5–6
V8-360 Exc. Calif.(56)	BRF-42	.034	(63)	(42)	(42)	Damper	(42)	(42)	5–6
V8-360(56)	BRF-42	.034	(63)	(42)	(42)	Damper	(42)	(42)	5–6
V8-360(68)	BRF-42	.054	(63)	(42)	(42)	Damper	(42)	(42)	5–6
V8-361(56)	BRF-31	.030	(63)	(42)	(42)	Damper	(42)	(42)	5–6(24)
V8-390 Exc. Calif.	BRF-42	.044	(63)	(42)	(42)	Damper	(42)	(42)	5–6
V8-390	BRF-42	.034	(63)	(42)	(42)	Damper	(42)	(42)	5–6
V8-391	BRF-31	.031	(63)	(42)	(42)	Damper	(42)	(42)	5–6(24)
V8-400 Ranchero	ARF-42	.044	D	—	6°	Damper	—	625D	5–6
V8-400 Exc. Ranchero	ARF-42	.044	D	(42)	(42)	Damper	(42)	(42)	5–6
V8-401	BRF-31	.030	(64)	(42)	(42)	Damper	(42)	(42)	—
V8-460 Ranchero	ARF-52	.054	C	—	14°	Damper	—	650D	5–6
V8-460 Exc. Ranchero(72)	ARF-52	.044	C	(42)	(42)	Damper	(42)	(42)	5–6
V8-460 Exc. Ranchero(73)	ARF-42	.044	C	(42)	(42)	Damper	(42)	(42)	5–6
V8-477	BRF-31	.030	(64)	(42)	(42)	Damper	(42)	(42)	—
V8-534	BRF-31	.030	(64)	(42)	(42)	Damper	(42)	(42)	—

(1)—B.T.D.C.—Before top dead center.
(2)—Idle speed on man. trans. vehicles is adjusted in Neutral & on auto. trans. vehicles is adjusted in Drive unless otherwise specified. Where two idle speeds are listed, the higher speed is with the A/C or idle solenoid energized.
(3)—Except calibration code 0-52S-R10, 10° BTDC; calibration code 0-52S-R10, 6° BTDC.
(4)—Except 4 wheel drive models.
(5)—4 wheel drive models.
(6)—Refer to engine calibration code on engine identification label, located at rear of left valve cover on V8 engines, on front of valve cover on 6 cylinder engines. The calibration code is located on the label after the engine code number and is preceded by the letter C and the revision code is located below the calibration code and is preceded by the letter R.
(7)—Calibration code 0-54T-R0, 12° BTDC; calibration code 0-54T-R10, 8° BTDC.
(8)—Models w/3 spd. or 4 spd. overdrive man. trans., 2° BTDC; models w/4 spd. man. trans., 4° BTDC.
(9)—Models w/3 spd. or 4 spd. overdrive man. trans., 550/800 RPM; models w/4 spd. man. trans., 750 RPM.
(10)—Windsor engine.
(11)—Calibration code 0-54D-R0, 14° BTDC; 0-54D-R11, 12° BTDC; 0-54F-R0 & 0-54M-R0, 8° BTDC.
(12)—Calibration code 0-54M-R0, 8° BTDC; 0-54R-R0, 10° BTDC.
(13)—Models w/3 spd. or 4 spd. overdrive man. trans., 6° BTDC; models w/4 spd. man. trans., 8° BTDC.
(14)—Calibration code 7-76J-R11.
(15)—Modified engine.
(16)—Except high altitude.
(17)—Calibration codes 0-64A-R0, R10 & R11.
(18)—Calibration codes 0-64B-R0 & R10.
(19)—Calibration codes 0-64T-R0, R10 & R11.
(20)—Calibration code 0-64B-R11.
(21)—Calibration codes 0-64G-R0, R10 & R11.
(22)—Calibration codes 0-64H-R0, R10 & R11.
(23)—Calibration codes 0-60D-R0 & R10.
(24)—Mechanical fuel pump.
(25)—BSF-42 or BRF-42.
(26)—BRF-31-4 or BSF-31-4.
(27)—ASF-52 or ARF-52.
(28)—ASF-42 or ARF-42.
(29)—E-100 California models w/2.75 rear axle ratio, set at 500/650D.
(30)—Man. trans., except Calif. models w/3.50 or 3.54 axle ratio 10° BTDC; California models w/3.50 or 3.54 axle ratio, 8° BTDC.
(31)—ASF-32 or ARF-32.
(32)—BYSF-31-4 or BYRF-31-4.
(33)—Except Calibration code 8-52L-R12, 10° BTDC; calibration code 8-52L-R12, 12° BTDC.
(34)—BSF-31-5 or BSF-42.
(35)—Models equipped w/Thermactor pump.
(36)—Models less Thermactor pump.
(37)—Early 1978 models.
(38)—Late 1978 models.
(39)—With dual mode switch disconnected from ignition module.
(40)—ASF-42 or ASF-34.
(41)—Calibration code 7-53H-R0, set at 8° BTDC; calibration code 7-53H-R10, set at 4° BTDC.
(42)—Must refer to engine decal due to running production changes.
(43)—Calibration codes 7-53G-R0, 7-53G-R10 & 7-53G-R11, set at 6° BTDC; calibration code 7-53G-R17, set at 2° BTDC.
(44)—Calibration code 7-54K-R0, set at 8° BTDC; calibration code 7-54K-R15, set at 4° BTDC.
(45)—Calibration code 7-14A-R1, set at 10° BTDC; calibration codes 7-14A-R10, 7-14B-R10 & 7-14C-R11 set at 9° BTDC.

Continued

TUNE UP NOTES—Continued

(46)—Calibration codes 7-14N-R0 & 7-14N-R12, set at 8° BTDC; calibration codes 7-14N-R13 & 7-14N-R17, set at 6° BTDC.
(47)—Calibration codes 7-60G-R0 & 7-60G-R10, set at 10° BTDC; calibration code 7-60G-R11, set at 6° BTDC.
(48)—Except Calibration code 7-60G-R0, set at 600D RPM; calibration code 7-60G-R0, set at 650D RPM.
(49)—Calibration code 7-72-R0, set at 12° BTDC; calibration code 7-72-R10, set at 14° BTDC.
(50)—Calibration code 7-72J-R0, set at 12° BTDC; calibration codes 7-72J-R1 & 7-72J-R11, set at 10° BTDC.
(51)—Calibration code 7-86C-R0.
(52)—Calibration code 5-86A-R0.
(53)—Calibration codes 7-17B-R1, 7-37A-R0 & 7-37A-R1, 8° BTDC; 7-17B-R15, 12° BTDC; 7-17B-R16, 10° BTDC; 7-17C-R1, 6° BTDC.
(54)—Except calibration codes, 7-37A-R0 & 7-37A-R1, set at 600D RPM; calibration codes 7-37A-R0 & 7-37A-R1, 650D RPM.
(55)—Breaker point gap, .024"; dwell angle, 33°–39°.
(56)—Breaker point gap, .017"; dwell angle, 24°–30
(57)—Calibration codes except Calif., man. trans. 5-75A-R4, auto. trans. 5-76A-R6; California, man. trans. 5-75J-R6, auto. trans. 5-76J-R8.
(58)—Calibration codes, except Calif., man. trans. 5-75A-R6, auto. trans. 5-76A-R8; California, man. trans. 5-75J-R7, auto. trans. 5-76J-R9.
(59)—Calibration codes, except Calif., man. trans. 5-75A-R3, auto. trans. 5-76A-R5; California, man. trans. 5-75J-R5, auto. trans. 5-76J-R7.
(60)—Calibration codes, man. trans. 5-83J-R3, R4; auto. trans. 5-84J-R7, R8.
(61)—Calibration codes, man. trans. 5-83J-R2; auto. trans. 5-84J-R6.
(62)—Cylinder numbering front to rear, firing order 1-5-3-6-2-4.
(63)—Cylinder numbering (front to rear), right bank 1-2-3-4, left bank 5-6-7-8; firing order 1-5-4-2-6-3-7-8.
(64)—Cylinder numbering (front to rear), right bank 1-2-3-4, left bank 5-6-7-8; firing order 1-5-4-8-6-3-7-2.
(65)—E series, 6–8 psi.; F series, 5–7 psi.
(66)—Ranchero, 5–7 psi.; F & Bronco, except Calif., 4–6 psi.; California, 5–7 psi.
(67)—Models w/breaker point type ignition, point gap, .017 in.; dwell angle, 26–30°.
(68)—Models w/breakerless ignition.
(69)—F-100 & Bronco except Calif., 4–6 psi; E series & California F-100 & Bronco, 5–6 psi.
(70)—Cleveland engine.
(71)—Man. trans. models, breaker point gap, both sets .020 in.; dwell angle both sets operating 32–35°. Auto. trans. models, breaker point gap .017 in.; dwell angle, 24–30°.
(72)—Models w/G.V.W.R. 6000 lbs. & under.
(73)—Models w/G.V.W.R. 6000 lbs. & above.
(74)—Light duty models.
(75)—Heavy duty models.
(76)—E-100, 150 exc. Calif. or high alt. carb. calibration code 1-64H-R2.
(77)—E-250 exc. Calif. or high alt. carb. & E-350.
(78)—Calibration code 0-62D-R10.
(79)—Calibration codes, man. trans. 9-71J-R10; auto. trans. 9-72J-R11.
(80)—On early 1980 Bronco & F Series, a sight hole located at the upper right hand corner of the fan shroud is used for viewing the pointer when adjusting ignition timing. It is recommended that the timing be set 2° less than the valve observed on the engine damper through the sight hole, due to the angle employed when viewing the pointer.
(81)—Calibration codes, man. trans. 0-59C-R0 & R10; auto. trans. 0-60A-R0 & R10.
(82)—Calibration codes, man. trans. 0-59G-R0 & R10, 0-59H-R0 & R10 & 0-59J-R0 & R10; auto. trans. 0-60B-R0 & R10 & 0-60C-R0 & R10.
(83)—Calibration codes, man. trans. 0-59S-R0; auto. trans. 0-60G-R0, 0-60H-R11 & 0-60H-R13.
(84)—Calibration codes 0-60H-R0, 0-60-H-R12, 0-60K-R0, R10, R11, R12 & O-60J-R0.
(85)—Calibration code 0-60L-R10.
(86)—Calibration codes, man. trans. 9-73J-R11; auto. trans. 9-74J-R11.
(87)—Calibration codes, man. trans. 9-7EJ-R12; auto. trans. 9-74J-R12.
(88)—Calibration code 0-62L-R0.
(89)—If mileage on vehicle is less than 100 mi., set idle speed 50 RPM less than specified.
(90)—Models with G.V.W.R. less than 8500 lbs.
(91)—Models with G.V.W.R. 8500 lbs. & above.
(92)—Except calibration codes 9-71J-R10, R11, & 9-72J-R11 & R12.
(93)—Calibration codes 9-71J-R10, R11, & 9-72J-R11 and R12.
(94)—Except calibration code 3-51P-R00, 10°.
(95)—Except calibration code 9-77J-R12, 10°.
(96)—Mechanical fuel pump, 6–8 psi. Electric fuel pump, minimum flow @ 12 volts & 2 psi 1 pint within 20 seconds.

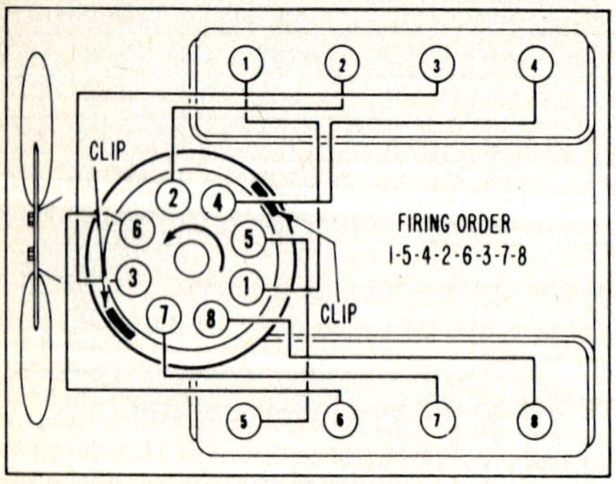

Fig. A

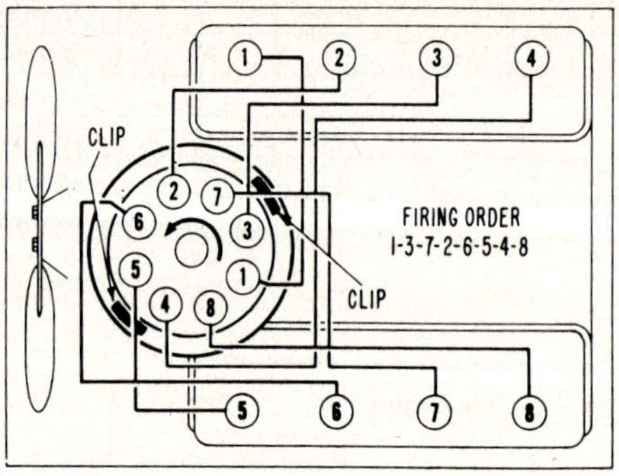

Fig. B

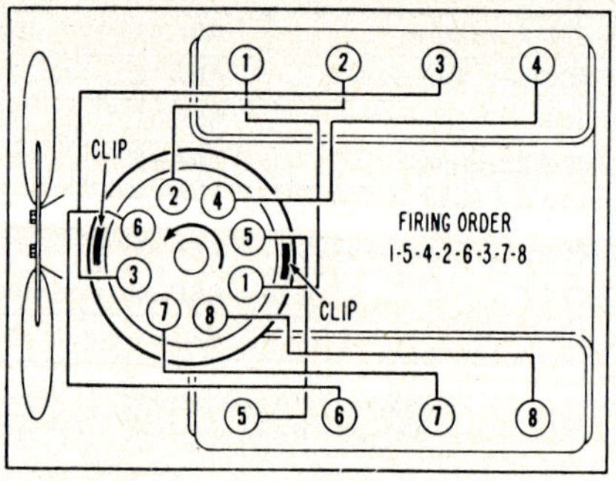

Fig. C

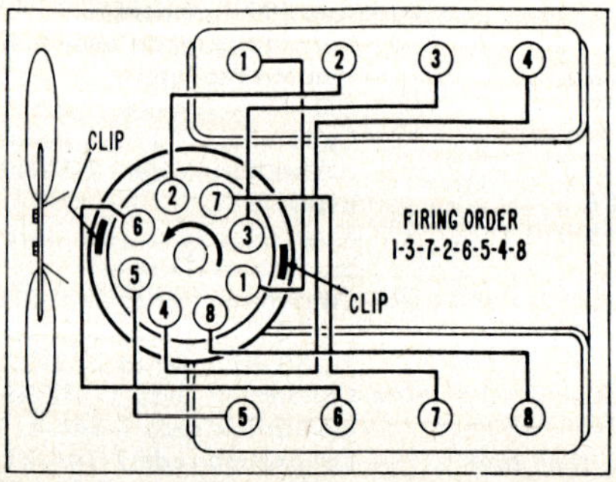

Fig. D

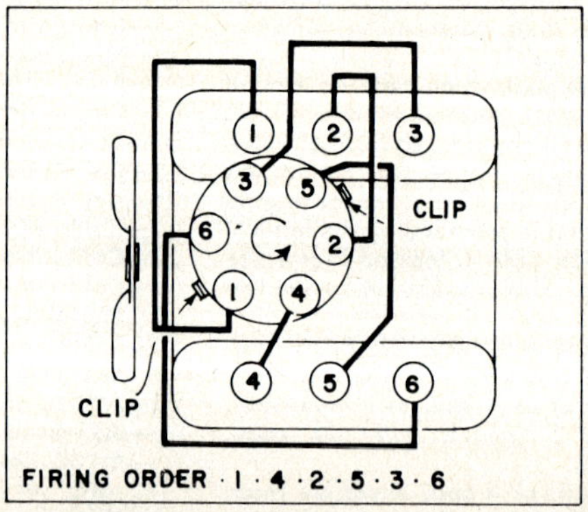

Fig. E

DISTRIBUTOR SPECIFICATIONS—Continued

Year	Engine Model	Distributor Number	Breaker Gap, In.	Dwell Angle, Deg.	Breaker Arm Spring Tension Oz.	Centrifugal Advance Deg. @ Dist. R.P.M.		Vacuum Advance Deg. @ In. of Mercury	
						Advance Starts	Maximum Advance	Advance Starts	Maximum Advance
1974	6-200	D3BF-12127-CA	—	—	—	3/4 @ 500	12 1/2 @ 2000	5/8 @ 5	7 1/4 @ 20
	6-240	D3UF-12127-BA	—	—	—	1/4 @ 500	10 @ 2000	1/2 @ 5	7 1/4 @ 20
	6-240	D3UF-12127-PA	—	—	—	1/2 @ 500	10 1/2 @ 2000	4 @ 5	5 1/2 @ 20
	6-240	D3UF-12127-RA	—	—	—	1/4 @ 500	10 @ 2000	4 @ 5	5 1/2 @ 20
	6-250	D3DF-12127-FA	—	—	—	3/4 @ 500	10 1/2 @ 2000	1 1/2 @ 5	9 1/4 @ 20
	6-250	D30F-12127-RA	—	—	—	3/4 @ 500	10 @ 2000	2 @ 5	9 1/4 @ 20
	6-250	D4DE-12127-LA	—	—	—	3/4 @ 500	9 1/2 @ 2000	3/4 @ 5	9 1/4 @ 20
	6-250	D4DE-12127-NA	—	—	—	3/4 @ 500	10 @ 2000	3/4 @ 5	9 1/4 @ 20
	6-250	D4DE-12127-RA	—	—	—	3/4 @ 500	10 1/2 @ 2000	2 @ 5	9 1/4 @ 20
	6-300	D3TF-12127-GA	—	—	—	1/2 @ 500	7 1/4 @ 2000	5/8 @ 5	9 1/4 @ 20
	6-300	D4TE-12127-EA	—	—	—	1/2 @ 500	7 1/4 @ 2000	5/8 @ 5	9 1/4 @ 20
	8-302	D3BF-12127-BA	—	—	—	1/2 @ 500	12 @ 2000	1/2 @ 5	11 1/2 @ 20
	8-302	D3BF-12127-DA	—	—	—	5/8 @ 500	13 1/2 @ 2000	1/2 @ 5	11 1/2 @ 20
	8-302	D30F-12127-DA	—	—	—	1/2 @ 500	11 @ 2000	5/8 @ 5	5 1/4 @ 20
	8-302	D30F-12127-HB	—	—	—	3/4 @ 500	13 1/2 @ 2000	3/4 @ 5	11 1/4 @ 20
	8-302	D3UF-12127-EA	—	—	—	5/8 @ 500	14 1/2 @ 2000	3/4 @ 5	9 1/4 @ 20
	8-302	D3UF-12127-FA	—	—	—	3/4 @ 500	14 @ 2000	3/4 @ 5	13 1/4 @ 20
	8-302	D3UF-12127-GA	—	—	—	1/2 @ 500	11 1/2 @ 2000	1/4 @ 5	5 1/4 @ 20
	8-302	D3UF-12127-JA	—	—	—	3/4 @ 500	10 1/2 @ 2000	1/2 @ 5	7 1/4 @ 20
	8-302	D3UF-12127-KA	—	—	—	3/4 @ 500	11 @ 2000	1/2 @ 5	7 1/4 @ 20
	8-302	D4BE-12127-FA	—	—	—	1/4 @ 500	11 1/2 @ 2000	1/4 @ 5	5 1/4 @ 20
	8-302	D4BF-12127-EA	—	—	—	1/4 @ 500	13 1/2 @ 2000	1/4 @ 5	11 1/2 @ 20
	8-302	D4TE-12127-PA	—	—	—	1/4 @ 500	11 1/2 @ 2000	1/2 @ 5	11 1/2 @ 20
	8-302	D4TE-12127-VA	—	—	—	1/4 @ 500	10 @ 2000	1/4 @ 5	11 1/2 @ 20
	8-302	D4UE-12127-FA	—	—	—	1/4 @ 500	11 @ 2000	1/2 @ 5	11 1/2 @ 20
	8-302	D4UE-12127-HA	—	—	—	1/4 @ 500	12 1/2 @ 2000	3/4 @ 5	9 1/2 @ 20
	8-330	D3HF-12102-BA	—	—	—	1/2 @ 500	14 @ 2000	2 @ 5	5 1/2 @ 20
	8-330	D3TF-12127-EA	—	—	—	3/4 @ 500	12 1/2 @ 2000	5/8 @ 5	9 1/4 @ 20
	8-330	DETF-12102-AA	—	—	—	1/4 @ 500	14 @ 2000	2 @ 5	5 1/2 @ 20
	8-351C	D3ZF-12127-GA	—	—	—	1/2 @ 500	11 @ 2000	1/2 @ 5	12 1/2 @ 20
	8-351C	D4AE-12127-AA	—	—	—	1/4 @ 500	10 1/2 @ 2000	1/2 @ 5	12 1/2 @ 20
	8-351CJ	D30F-12127-FA	—	—	—	1/2 @ 500	13 @ 2000	1 1/2 @ 5	5 1/4 @ 20
	8-351CJ	D30F-12127-GA	—	—	—	1/2 @ 500	13 1/2 @ 2000	1 3/4 @ 5	5 1/2 @ 20
	8-360	D3TF-12127-HA	—	—	—	5/8 @ 500	14 @ 2000	2 @ 5	5 1/4 @ 20
	8-360	D3TF-12127-PA	—	—	—	3/4 @ 500	17 1/2 @ 2000	1 1/4 @ 5	9 1/4 @ 20
	8-360	D3TF-12127-UA	—	—	—	3/4 @ 500	17 1/2 @ 2000	3/4 @ 5	9 1/4 @ 20
	8-360	D4TE-12127-AA	—	—	—	—	—	—	—
	8-360	D4TE-12127-TA	—	—	—	1/2 @ 500	16 1/2 @ 2000	5 @ 5	7 1/4 @ 20
	8-360	D4TE-12127-YA	—	—	—	1/4 @ 500	13 1/2 @ 2000	3 7/8 @ 5	5 1/4 @ 20
	8-361	D3TF-12102-CA	—	—	—	1 1/4 @ 500	14 @ 2000	1 3/4 @ 5	5 1/2 @ 20
	8-361	D3TF-12102-FA	—	—	—	1 1/4 @ 500	14 @ 2000	1 3/4 @ 5	5 1/2 @ 20
	8-361	D4TE-12102-AA	—	—	—	1 1/4 @ 500	14 @ 2000	1 3/4 @ 5	5 1/2 @ 20
	8-390	D3TF-12127-HA	—	—	—	5/8 @ 500	14 @ 2000	2 @ 5	5 1/2 @ 20
	8-390	D3TF-12127-VA	—	—	—	3/4 @ 500	11 @ 2000	1/2 @ 5	5 1/2 @ 20
	8-390	D3TF-12127-XA	—	—	—	1/2 @ 500	15 1/4 @ 2000	1 1/4 @ 5	13 1/4 @ 20
	8-390	D4TE-12127-UA	—	—	—	1/4 @ 500	10 1/2 @ 2000	1 @ 5	7 1/2 @ 20
	8-391	D3TF-12102-BA	—	—	—	1/4 @ 500	14 @ 2000	1/4 @ 5	7 1/2 @ 20
	8-391	D3TF-12102-GA	—	—	—	1/4 @ 500	14 @ 2000	1/4 @ 5	7 1/2 @ 20
	8-400	D4OE-12127-CA	—	—	—	1/4 @ 500	10 1/2 @ 2000	2 1/4 @ 5	13 1/2 @ 20
	8-460	D4TE-12127-HA	—	—	—	1/4 @ 500	11 1/4 @ 2000	5/8 @ 5	11 1/4 @ 20
	8-460	D4TE-12127-NA	—	—	—	1/4 @ 500	8 1/2 @ 2000	3/4 @ 5	9 1/2 @ 20
	8-460	D4VE-12127-CA	—	—	—	1/4 @ 500	10 1/2 @ 2000	5/8 @ 5	11 1/4 @ 20
	8-401, 475, 477, 534	D3HF-12127-AA	—	—	—	1 1/4 @ 500	14 @ 2000	1/4 @ 5	9 1/2 @ 20
1975–83	No longer available								

Continued

DISTRIBUTOR SPECIFICATIONS—Continued

①—Dual vacuum diaphragm.
②—Single vacuum diaphragm.
③—Transistor ignition, .020″.
④—Transistor ignition, 22–24°.
⑤—With Exhaust Emission System.
⑥—Light duty.
⑦—Except California.
⑧—California.
⑨—With Improved Combustion (IMCO).
⑩—Medium Duty.
⑪—Heavy Duty.

VALVE SPECIFICATIONS

Engine Model	Year	Valve Lash①		Valve Angles		Valve Springs		Valve Stem Clearance		Stem Diameter, Std.	
		Int.	Exh.	Seat	Face	Installed Height	Pressure Lbs. @ In.	Intake	Exhaust	Intake	Exhaust
6-200	1974	.079–.129		45	44	1 19/32	⑥	.0008–.0025	.0010–.0027	.3100–.3107	.3098–.3105
6-232	1982–83	.088–.189		45	44	1.74	215 @ 1.79	.0010–.0027	.0015–.0032	.3716–.3423	.3411–.3418
6-240	1974	.100–.150		45	44	⑩	⑪	.0010–.0027	.0010–.0027	.3416–.3423	.3416–.3423
6-250	1974	.095–.145		45	44	1 19/32	150 @ 1.18	.0008–.0025	.0010–.0027	.3100–.3107	.3098–.3105
V8-255	1981–82	.123–.173		45	44	⑱	⑨	.0010–.0027	.0015–.0032	.3416–.3423	.3411–.3418
6-300 L.D.	1974–75	⑮		45	44	⑩	⑪	.0010–.0027	.0010–.0027	.3416–.3423	.3416–.3423
6-300 L.D.	1976–77	.125–.175		45	44	⑩	⑫	.0010–.0027	.0010–.0027	.3416–.3423	.3416–.3423
6-300 H.D.	1974–83	⑮		45	44	⑩	⑪	.0010–.0027	.0010–.0027	.3416–.3423	.3416–.3423
8-302	1974	.100–.150		45	44	⑩	⑪	.0010–.0027	.0010–.0027	.3416–.3423	.3416–.3423
8-302	1975	.115–.130		45	44	⑱	200 @ 1.31	.0010–.0027	.0015–.0032	.3416–.3423	.3411–.3418
8-302	1976	.115–.165		45	44	⑱	200 @ 1.31	.0010–.0027	.0015–.0032	.3416–.3423	.3411–.3418
8-302	1977–83	.096–.165		45	44	⑱	⑨	.0010–.0027	.0015–.0032	.3416–.3423	.3411–.3418
8-330 M.D.	1974	.080–.130		45	44	1 53/64	189 @ 1.42	.0010–.0027	.0020–.0037	.3711–.3718	.3701–.3708
8-330 M.D.	1975–76	.105–.155		45	44	1 53/64	189 @ 1.42	.0010–.0027	.0020–.0040	.3711–.3718	.3701–.3708
8-330 M.D.	1977	.105–.155		45	44	1 53/64	220 @ 1.38	.0010–.0027	.0020–.0037	.3711–.3718	.3701–.3708
8-330 H.D.	1974–76	.105–.155		45	44	1 43/64	185 @ 1.24	.0010–.0027	.0020–.0040	.3711–.3718	.4338–.4348
8-330 H.D.	1977–79	.105–.155		45	44	㉒	㉓	.0010–.0027	.0020–.0040	.3711–.3718	.4338–.4348
8-351C-2V	1974	.100–.150		45	44	1 53/64	226 @ 1.39	.0010–.0027	.0015–.0032	.3416–.3423	.3411–.3418
8-351C-4V	1974	.100–.150		45	44	1 53/64	285 @ 1.32	.0010–.0027	.0015–.0032	.3416–.3423	.3411–.3418
8-351-M	1975–78	.125–.175		45	44	1 53/64	226 @ 1.39	.0010–.0027	.0015–.0032	.3416–.3423	.3411–.3418
8-351M	1979–81	.125		45	44	⑦	226 @ 1.39	.0010–.0027	.0015–.0032	.3416–.3423	.3411–.3418
8-351W	1975–76	.131–.181		45	44	1 51/64	200 @ 1.34	.0010–.0027	.0015–.0032	.3416–.3423	.3411–.3418
8-351W	1977–83	.123–.173		45	44	㉗	⑬	.0010–.0027	.0015–.0032	.3416–.3423	.3411–.3418
8-359, 361	1977	.105–.155		45	44	㉒	㉓	.0010–.0027	.0020–.0040	.3711–.3718	.4338–.4348
8-360	1974–76	.119–.169㉑		45	44	1 53/64	220 @ 1.38	.0010–.0027	.0015–.0032	.3711–.3718	.3706–.3713
8-360 F-100	1974–76	.119–.169㉑		45	44	㉒	㉓	.0010–.0027	.0015–.0032	.3711–.3718	.3706–.3713
8-361	1978	.105–.155		45	44	㉒	㉓	.0010–.0027	.0020–.0040	.3711–.3718	.4338–.4348
8-361, 391	1974–76	.105–.155		45	44	1 43/64	185 @ 1.24	.0010–.0027	.0020–.0040	.3711–.3718	.4338–.4348
8-370③	1979	.100–.150		45	44	④	⑳	.0010–.0027	.0020–.0037	.3711–.3718	.3701–.3708
8-370③	1980–83	.125–.175		45	44	④	⑧	.0010–.0027	.0020–.0037	.3711–.3718	.3701–.3708
8-370, 429⑭	1979	.100–.150		45	44	1.72	⑳	.0010–.0027	.0016–.0035	.3711–.3718	.3703–.3712
8-370, 429⑭	1980–83	.125–.175		45	44	1 23/32	⑯	.0010–.0027	.0015–.0035	.3711–.3718	.3703–.3712
8-389	1975–76	.105–.155		45	44	1 43/64	185 @ 1.24	.0010–.0027	.0020–.0040	.3711–.3718	.4338–.4348
8-389, 391	1977–78	.105–.155		45	44	⑰	⑲	.0010–.0027	.0020–.0040	.3711–.3718	.4338–.4348
8-390	1974–75	.119–.169㉑		45	44	1 53/64	220 @ 1.38	.0010–.0027	.0015–.0032	.3711–.3718	.3706–.3713
8-390 F-100	1974–75	.119–.169㉑		45	44	㉒	㉓	.0010–.0027	.0015–.0032	.3711–.3718	.3706–.3713
8-400	1974–76	.100–.150㉔		45	44	1 53/64	226 @ 1.39	.0010–.0027	.0015–.0032	.3416–.3423	.3411–.3418
8-400	1977–78	—		45	44	1 53/64	226 @ 1.39	.0010–.0027	.0015–.0032	.3416–.3423	.3411–.3418
8-400	1979–82	.175		45	44	⑦	226 @ 1.39	.0010–.0027	.0015–.0032	.3416–.3423	.3411–.3418
8-401	1974–76	.020H②	.020H②	45	44	1 45/64	180 @ 128	.0010–.0026	.0024–.0040	.4349–.4358	.4335–.4344
8-460	1974–83	.100–.150㉘		45	44	1 13/16	253 @ 1.33⑤	.0010–.0027	.0010–.0027	.3416–.3423	.3416–.3423
8-475	1974–79	.020H②	.020H②	45	44	1 45/64	180 @ 1.28	.0010–.0026	.0024–.0040⑯	.4349–.4358	.4335–.4344
8-477, 534	1974–80	.020H②	.020H②	45	44	1 45/64	180 @ 1.28	.0010–.0026	.0024–.0040⑯	.4349–.4358	.4335–.4344

Continued

VALVE SPECIFICATIONS—Continued

①—On engines with hydraulic lifters, clearance specified is at valve stem tip with lifter collapsed.
②—Mechanical lifters.
③—2 barrel carb.
④—Intake 1 23/32; Exhaust 1 25/32
⑤—1977–83 229 @ 1.33
⑥—1974–75 Intake 150 @ 1.18; Exhaust 150 @ 1.22.
⑦—Intake 1 53/64; Exhaust 1 45/64.
⑧—Intake 216 @ 1.26; Exhaust 216 @ 1.32.
⑨—1977–78, Intake 200 @ 1.31; Exhaust 200 @ 1.20; 1979, Intake 202 @ 1.36; Exhaust 200 @ 1.20; 1980–83, Intake 204 @ 1.36; Exhaust 200 @ 1.20.
⑩—Intake 1 45/64; Exhaust 1 37/64.
⑪—Intake 197 @ 1.30; Exhaust 192 @ 1.18.
⑫—Intake 197 @ 1.30; Exhaust 192 @ 1.80.
⑬—1977–78, Intake 200 @ 1.34; Exhaust 200 @ 1.20; 1979–83, Intake 200 @ 1.36; Exhaust 200 @ 1.20
⑭—4 barrel carb.
⑮—1974, .100–.150; 1975–83, .125–.175.
⑯—Intake 230 @ 1.26; Exhaust, exc. V8-370 with 2 barrel carb, 230 @ 1.32; V8-370 with 2 barrel carb, 216 @ 1.32.
⑰—Intake 1 43/64; Exhaust 1 45/64.
⑱—Intake 1 11/16; Exhaust 1 19/32.
⑲—Intake 220 @ 1.38; Exhuast 180 @ 1.28.
⑳—Intake 230 @ 1.26; Exhaust 230 @ 1.32.
㉑—1975, .144–.184.
㉒—Intake 1 53/64; Exhaust 1 43/64.
㉓—Intake 220 @ 1.38; Exhaust 185 @ 1.24.
㉔—1975, .125–.175.
㉕—Intake 30°; Exhaust 45°.
㉖—Intake 29°; Exhaust 44°.
㉗—Intake 1 25/32; Exhaust 1 19/32.
㉘—1974, .075–.125.

PISTON, PIN, RING, CRANKSHAFT & BEARING SPECIFICATIONS

Year	Engine Model	Wristpin Diameter	Piston Clearance, Inch①	Ring End Gap, In. (Minimum)		Crank Pin Diameter, Inch	Rod Bearing Clearance, Inch	Main Bearing Journal Diameter, In.	Main Bearing Clearance, Inch	Shaft End Play
				Comp.	Oil					
1974	6-200	.9122	.0013–.0021	.008	.015	2.123–2.124	.0008–.0024	2.248–2.249	.0008–.0024	.004–.008
	6-240	.9751	.0014–.0022	.010	.015	2.1228–2.1236	.0008–.0024	2.3982–2.3990	.0005–.0015	.004–.008
	6-250	.9121	.0013–.0021	.008	.015	2.1232–2.1240	.0008–.0024	2.3982–2.3990	.0008–.0024	.004–.008
	6-300	.9751	.0014–.0022	.010	.015	2.1228–2.1236	.0009–.0027	2.3982–2.3990	.0009–.0026	.004–.008
	6-300 H.D.	.9751	.0014–.0022	.010	.015	2.1228–2.1236	.0009–.0027	2.3982–2.3990	.0009–.0028	.004–.008
	V8-302	.9121	.0018–.0026	.010	.015	2.1228–2.1236	.0008–.0026	2.2482–2.2490	⑤	.004–.008
	V8-330	.9751	.0025–.0033	.010	.015	2.4380–2.4388	.001–.0029	2.7484–2.7492	.001–.0029	.004–.010
	V8-330 H.D.	.9751	.0025–.0033	.010	.015	2.4377–2.4385	.001–.0029	2.7484–2.7492	.001–.0029	.004–.010
	8-351C	.9121	.0014–.0022	.010	.015	2.3103–2.3111	.0008–.0026	2.7484–2.7492	.0009–.0026	.004–.008
	V8-360, 390	.9751	.0015–.0023	④	.015	2.4380–2.4388	.0008–.0025	2.7484–2.7492	.0005–.0025	.004–.010
	V8-361, 391	.9751	.0025–.0033	④	.015	2.4377–2.4388	.001–.0029	2.7479–2.7487	.001–.0029	.004–.010
	8-400	.9752	.0014–.0022	.010	.015	2.3103–2.3111	.0011–.0026	2.9994–3.0002	.0011–.0028	.004–.008
	V8-401, 475, 477, 534	1.2200	.0028–.0034	②	.013	2.7092–2.7100	.0017–.0035	3.1246–3.1254	.0018–.0036	.004–.008
	V8-460	1.0400	.0022–.0030	.010	.015	2.4992–2.5000	.0008–.0028	2.9994–3.0002	⑥	.004–.008
1975	6-300	.9752	.0014–.0022	.010	.015	2.1228–2.1236	.0008–.0015	2.3982–2.3990	.0008–.0015	.004–.008
	8-302	.9122	.0018–.0026	.010	.015	2.1228–2.1236	.0008–.0015	2.2482–2.2490	⑧	.004–.008
	8-330	.9752	.0025–.0033	.010	.015	2.4380–2.4388	.0008–.0015	2.7484–2.7492	.0005–.0015	.004–.010
	8-351W	.9122	.0018–.0026	.010	.015	2.3103–2.3111	.0008–.0015	2.9994–3.0002	.0008–.0015	.004–.008
	8-351M	.9752	.0014–.0022	.010	.015	2.3103–2.3111	.0008–.0015	2.9994–3.0002	.0008–.0015	.004–.008
	8-360	.9752	.0015–.0023	④	.015	2.4380–2.4388	.0008–.0015	2.7484–2.7492	.0005–.0015	.004–.010
	8-361	.9752	.0025–.0033	④	.015	2.4377–2.4385	.0008–.0015	2.7479–2.7492	.0005–.0015	.004–.010
	8-390	.9752	.0015–.0023	④	.015	2.4380–2.4388	.0008–.0015	2.7484–2.7492	.0005–.0015	.004–.010
	8-391	.9752	.0025–.0033	④	.015	2.4377–2.4388	.0008–.0015	2.7479–2.7492	.0005–.0015	.004–.010
	8-400	.9752	.0014–.0022	.010	.015	2.3103–2.3111	.0008–.0015	2.9994–3.0002	.0008–.0015	.004–.008
	8-401, 477, 534	1.2200	.0028–.0034	②	.013	2.7092–2.7100	.0017–.0025	3.1246–3.1254	.0018–.0025	.004–.008
	8-460	1.0401	.0014–.0022	.010	.015	2.4992–2.5000	.0008–.0015	2.9994–3.0002	⑨	.004–.008
1976	6-300	.9752	.0014–.0022	.010	.015	2.1228–2.1236	.0008–.0015	2.3982–2.3990	.0008–.0015	.004–.008
	8-302	.9121	.0018–.0026	.010	.015	2.1228–2.1236	.0008–.0015	2.2482–2.2490	.0008–.0015	.004–.008
	8-330	.9752	.0025–.0033	.010	.015	2.4380–2.4388	.0008–.0015	2.7484–2.7492	.0005–.0015	.004–.010
	8-351W	.9121	.0018–.0026	.010	.015	2.3103–2.3111	.0008–.0015	2.9994–3.0002	.0008–.0015	.004–.008
	8-351M	.9752	.0014–.0022	.010	.015	2.3103–2.3111	.0008–.0015	2.9994–3.0002	.0008–.0015	.004–.008
	8-360	.9752	.0015–.0023	④	.015	2.4380–2.4388	.0008–.0015	2.7484–2.7492	.0005–.0015	.004–.010
	8-361	.9752	.0025–.0033	④	.015	2.4377–2.4385	.0008–.0015	2.7479–2.7487	.0005–.0015	.004–.010
	8-389	—	—	—	—	—	—	2.7479–2.7487	—	—
	8-390	.9752	.0015–.0023	④	.015	2.4380–2.4388	.0008–.0015	2.7484–2.7492	.0005–.0015	.004–.010

Continued

PISTON, PIN, RING, CRANKSHAFT & BEARING SPECIFICATIONS—Continued

Year	Engine Model	Wristpin Diameter	Piston Clearance, Inch①	Ring End Gap, In. (Minimum)		Crank Pin Diameter, Inch	Rod Bearing Clearance, Inch	Main Bearing Journal Diameter, In.	Main Bearing Clearance, Inch	Shaft End Play
				Comp.	Oil					
	8-391	.9752	.0025–.0033	④	.015	2.4377–2.4385	.0008–.0015	2.7479–2.7487	.0005–.0015	.004–.010
	8-400	.9752	.0014–.0022	.010	.015	2.3103–2.3111	.0008–.0015	2.9994–3.0002	.0008–.0015	.004–.008
	8-401, 477, 534	1.2200	.0028–.0034	②	.013	2.7092–2.100	.0017–.0025	3.1246–3.1254	.0018–.0025	.004–.008
	8-460	1.0401	.0014–.0022	.010	.015	2.4992–2.5000	.0008–.0015	2.9994–3.0002	.0008–.0015	.004–.008
1977	6-300	.9752	.0014–.0022	.010	.010	2.1228–2.1236	.0008–.0015	2.3982–2.3990	.0008–.0015	.004–.008
	8-302	.9122	.0018–.0026	.010	.010	2.1228–2.1236	.0008–.0015	2.2482–2.2490	⑧	.004–.008
	8-330	.9752	.0025–.0033	.010	.010	⑩	.0008–.0015	2.7484–2.7492	.0008–.0015	.004–.010
	8-351W	.9122	.0022–.0030	.010	.010	2.3103–2.3111	.0008–.0015	2.9994–3.0002	.0008–.0015	.004–.008
	8-351M	.9750	.0014–.0022	.010	.010	2.3103–2.3111	.0008–.0015	2.9994–3.0002	.0008–.0015	.004–.008
	8-359	.9752	.0025–.0033	④	.015	2.4377–2.4385	.0008–.0015	2.7479–2.7487	.0008–.0015	.004–.010
	8-361	.9752	.0025–.0033	④	.015	2.4377–2.4385	.0008–.0015	2.7479–2.7487	.0008–.0015	.004–.010
	8-389	.9752	.0025–.0033	④	.015	2.4377–2.4385	.0008–.0015	2.7479–2.7487	.0008–.0015	.004–.010
	8-391	.9752	.0025–.0033	④	.015	2.4377–2.4385	.0008–.0015	2.7479–2.7487	.0008–.0015	.004–.010
	8-400	.9750	.0014–.0022	.010	.010	2.3103–2.3111	.0008–.0015	2.9994–3.0002	.0008–.0015	.004–.008
	8-460	1.0401	.0014–.0022	.010	.010	2.4992–2.5000	.0008–.0015	2.9994–3.0002	.0008–.0015	.004–.008
	8-475, 477, 534	1.2200	.0028–.0034	②	.013	2.7092–2.7100	.0019–.0027	3.1246–3.1254	.0019–.0025	.004–.008
1978	6-300	.9752	.0014–.0022	.010	.010	2.1228–2.1236	.0008–.0015	2.3982–2.3990	.0008–.0015	.004–.008
	8-302	.9122	.0018–.0026	.010	.010	2.1228–2.1236	.0008–.0015	2.2482–2.2490	.0008–.0015	.004–.008
	8-330	.9752	.0025–.0033	.010	.010	2.4377–2.4385	.0008–.0015	2.7484–2.7492	.0008–.0015	.004–.010
	8-351M	.9750	.0014–.0022	.010	.015	2.3103–2.3111	.0008–.0015	2.9994–3.0002	.0008–.0015	.004–.008
	8-351W	.9122	.0022–.0030	.010	.010	2.3103–2.3111	.0008–.0015	2.9994–3.0002	.0008–.0015	.004–.008
	8-361	.9752	.0025–.0033	④	.015	2.4377–2.4385	.0008–.0015	2.7479–2.7487	.0008–.0015	.004–.010
	8-389	.9752	.0025–.0033	④	.015	2.4377–2.4385	.0008–.0015	2.7479–2.7487	.0008–.0015	.004–.010
	8-391	.9752	.0025–.0033	④	.015	2.4377–2.4385	.0008–.0015	2.7479–2.7487	.0008–.0015	.004–.010
	8-400	.9750	.0014–.0022	.010	.015	2.3103–2.3111	.0008–.0015	2.9994–3.0002	.0008–.0015	.004–.008
	8-460	1.0401	.0022–.0030	.010	.010	2.4992–2.5000	.0008–.0015	2.9994–3.0002	.0008–.0015	.004–.008
	8-475, 477, 534	1.2200	.0028–.0034	②	.013	2.7092–2.7100	.0019–.0027	3.1246–3.1254	.0019–.0025	.004–.008
1979	6-300	.9752	.0014–.0022	.010	.010	2.1228–2.1236	.0008–.0015	2.3982–2.3990	.0008–.0015	.004–.008
	8-302	.9122	.0018–.0026	.010	.015	2.1228–2.1236	.0008–.0015	2.2482–2.2490	⑧	.004–.008
	8-351W	.9122	.0022–.0030	.010	.015	2.3103–2.3111	.0008–.0015	2.9994–3.0002	.0008–.0015	.004–.008
	8-351M	.9750	.0014–.0022	.010	.010	2.3103–2.3111	.0008–.0015	2.9994–3.0002	.0008–.0015	.004–.008
	8-370	1.0401	.0018–.0039	.010	—	2.4992–2.5000	.0008–.0015	2.9994–3.0002	.0008–.0026	.004–.008
	8-400	.9750	.0014–.0022	.010	.010	2.3103–2.3111	.0008–.0015	2.9994–3.0002	.0008–.0015	.004–.008
	8-429	1.0401	.0019–.0037	.013	.013	2.4992–2.5000	.0008–.0015	2.9994–3.0002	.0008–.0026	.004–.008
	8-460	1.0401	.0022–.0030	.010	.010	2.4992–2.5000	.0008–.0015	2.9994–3.0002	.0008–.0015	.004–.008
	8-475, 477, 534	1.2200	.0028–.0034	②	.013	2.7092–2.7100	.0019–.0027	3.1246–3.1254	.0019–.0028	.004–.008
1980–81	8-477, 534	1.2200	.0028–.0034	②	.013	2.7092–2.7100	.0019–.0027	3.1246–3.1254	.0019–.0028	.004–.008
1980–83	6-300	.9752	.0014–.0022	.010	.010⑦	2.1228–2.1236	.0008–.0015	2.3982–2.3990	.0008–.0015	.004–.008
	8-302	.9122	.0018–.0026	.010	.010	2.1228–2.1236	.0008–.0015	2.2482–2.2490	⑧	.004–.008
	8-351W	.9122	③	.010	.010	2.3103–2.3111	.0008–.0015	2.9994–3.0002	.0008–.0015	.004–.008
	8-351M	.9752	.0014–.0022	.010	.010	2.3103–2.3111	.0008–.0015	2.9994–3.0002	.0008–.0015	.004–.008
	8-370	1.0401	.0018–.0039	⑪	⑪	2.4992–2.5000	.0008–.0015	2.9994–3.0002	.0008–.0015	.004–.008
	8-400	.9752	.0014–.0022	.010	.010	2.3103–2.3111	.0008–.0015	2.9994–3.0002	.0008–.0015	.004–.008
	8-429	1.0401	.0019–.0037	.013	.013	2.4992–2.5000	.0008–.0015	2.9994–3.0002	.0008–.0015	.004–.008
	8-460	1.0401	.0022–.0030	.010	.010	2.4992–2.5000	.0008–.0015	2.9994–3.0002	.0008–.0015	.004–.008
1981–82	8-255	.9122	.0018–.0026	.010	.010	2.1228–2.1236	.0008–.0015	2.2482–2.2490	⑧	.004–.008
1982–83	6-232	.9122	.0014–.0022	.010	.015	2.4266–2.4274	.001–.0014	2.5190–2.5198	.001–.0014	.004–.008

Continued

PISTON, PIN, RING, CRANKSHAFT & BEARING SPECIFICATIONS—Continued

①—Measured at bottom of skirt.
②—Top ring .018", others .015".
③—1980–81, .0022–.0030; 1982–83, .0018–.0026.
④—Top ring, .015", others .010".
⑤—No. 1 .0001–.0020", others .0005–.0024".
⑥—No. 1 .0004–.0020", others .0012–.0028".
⑦—Exc. 1983; 1983, .015.
⑧—No. 1, .0001–.0015"; others, .0005–.0015".
⑨—No. 1, .0004–.0015"; others, .0009–.0015".
⑩—Medium duty engines 2.4380–2.4388; Heavy duty engines 2.4377–2.4385.
⑪—2 barrel carb, comp. .010; oil. .010. 4 barrel carb, comp. .013; oil .013.

ENGINE TIGHTENING SPECIFICATIONS★

★ Torque specifications are for clean and lightly lubricated threads only. Dry or dirty threads produce increased friction which prevents accurate measurement of tightness.

Year	Engine Model	Spark Plug Ft-Lbs.	Cylinder Head Ft-Lbs.	Intake Manifold Ft-Lbs.	Exhaust Manifold Ft-Lbs.	Rocker Shaft Support Ft-Lbs.	Rocker Cover Ft-Lbs.	Conn. Rod Cap Ft-Lbs.	Main Bearing Caps Ft-Lbs.	Flywheel To Crankshaft Ft-Lbs.	Damper or Pulley Ft-Lbs.
1974	6-250	15–20	70–75	—	13–18	30–35	3–5	21–26	60–70	75–85	85–100
	6-200	15–20	70–75	—	13–18	30–35	3–5	19–24	60–70	75–85	85–100
	6-240, 300	15–20	70–75	28–33	23–28	⑪	4–7	40–45	65–72	75–85	130–150
	8-302	15–20	65–72	17–25	12–16	17–23	3–5	19–24	60–70	75–85	70–90
	8-330 MD	15–20	80–90	40–45	18–24	40–45	4–7	40–45	95–105	75–85	70–90
	8-330 HD, 361, 391	15–20	80–90	40–45	18–24	40–45	4–7	40–45	95–105	75–85	150–175
	V8-351	15–20	105–112	③	12–22	—	3–5	40–45	⑤	75–85	70–90
	8-360, 390	15–20	80–90	40–45	12–18	40–45	4–7	40–45	95–105	75–85	70–90
	V8-400	15–20	95–105	⑥	12–16	—	3–5	40–45	⑤	75–85	70–90
	8-401, 477, 534	15–20	170–180	23–28	23–28	—	4–7	60–65	150–165	100–110	130–175
	V8-429	15–20	130–140	25–30	28–33	—	5–6	40–45	95–105	75–85	70–90
	8-460	15–20	130–140	25–30	28–33	②	5–6	40–45	95–105	75–85	70–90
1975	6-250	15–25	70–75	—	18–24	30–35	3–5	21–26	60–70	75–85	85–100
	6-300	15–20	70–85	①	23–28	⑪	4–7	40–45	60–70	75–85	130–150
	8-302	10–15	65–72	23–25	18–24	⑦	3–5	19–24	60–70	75–85	70–90
	8-330	15–20	90	40–45	18–24	40–45	4–7	40–45	95–105	75–85	130–150
	8-351W	10–15	105–112	23–25	18–24	⑦	3–5	40–45	95–105	75–85	70–90
	8-351M	10–15	95–105	⑥	18–24	④	3–5	40–45	⑤	75–85	70–90
	8-360, 390	15–20	90	40–45	12–18	40–45	4–7	40–45	95–105	75–85	130–150
	8-361, 391	15–20	90	40–45	18–24	40–45	4–7	40–45	95–105	75–85	150–175
	8-389	15–20	90	40–45	18–24	40–45	4–7	40–45	95–105	75–85	150–175
	8-400	10–15	95–105	⑥	18–24	④	3–5	40–45	⑤	75–85	70–90
	8-401	15–25	170–180	—	22–32	—	5–7	60–65	150–165	100–110	130–175
	8-460	10–15	130–140	22–37	28–33	②	5–6	40–45	95–105	75–85	70–90
	8-477, 534	15–25	170–180	—	22–32	—	5–7	60–65	150–165	100–110	130–175
1976	6-300	15–20	70–85	①	23–28	⑪	4–7	40–45	60–70	75–85	130–150
	V8-302, 351	10–15	65–70	23–25	18–24	17–23	3–5	19–24	60–70	75–85	70–90
	V8-330, 361	15–20	90	40–45	18–24	40–45	4–7	40–45	95–105	75–85	130–150
	V8-360	15–20	90	40–45	12–18	40–45	4–7	40–45	95–105	75–85	150–175
	V8-390	15–20	90	40–45	12–18	40–45	4–7	40–45	95–105	75–85	130–150
	V8-391	15–20	90	40–45	18–24	40–45	4–7	40–45	95–105	75–85	150–175
	V8-460	—	130–140	—	28–33	—	5–6	40–45	95–105	75–85	70–90
	V8-401, 477, 534	15–25	170–180	—	22–32	—	5–7	60–65	150–165	100–110	35–50
1977	6-300	15–25	70–85	22–32	28–33	⑪	4–7	40–45	60–70	75–85	130–150
	8-302	10–15	65–72	23–25	18–24	⑦	3–5	19–24	60–70	75–85	70–90
	8-330	15–25	90	40–45	18–24	40–45	4–7	40–45	95–105	75–85	⑨
	8-351W	10–15	105–112	23–25	18–24	⑦	3–5	40–45	95–105	75–85	70–90
	8-351M	10–15	95–105	⑧	18–24	④	3–5	40–45	95–105	75–85	70–90
	8-359	15–25	90	40–45	18–24	40–45	4–7	40–45	95–105	75–85	⑨
	8-361	15–25	90	40–45	18–24	40–45	4–7	40–45	95–105	75–85	⑨
	8-389	15–25	90	40–45	18–24	40–45	4–7	40–45	95–105	75–85	⑨

Continued

ENGINE TIGHTENING SPECIFICATIONS—Continued

★ Torque specifications are for clean and lightly lubricated threads only. Dry or dirty threads produce increased friction which prevents accurate measurement of tightness.

Year	Engine Model	Spark Plug Ft-Lbs.	Cylinder Head Ft-Lbs.	Intake Manifold Ft-Lbs.	Exhaust Manifold Ft-Lbs.	Rocker Shaft Support Ft-Lbs.	Rocker Cover Ft-Lbs.	Conn. Rod Cap Ft-Lbs.	Main Bearing Caps Ft-Lbs.	Flywheel To Crankshaft Ft-Lbs.	Damper or Pulley Ft-Lbs.
	8-391	15–25	90	40–45	18–24	40–45	4–7	40–45	95–105	75–85	⑨
	8-400	10–15	95–105	⑧	18–24	④	3–5	40–45	95–105	75–85	70–90
	8-460	10–15	130–140	22–23	28–33	④	5–6	40–45	95–105	75–85	70–90
	8-475, 477 534	10–15	170–180	25–32	22–32	—	5–7	60–65	150–165	100–110	130–175
1978	6-300	15–25	70–85	22–32	28–33	⑪	4–7	40–45	60–70	75–85	130–150
	8-302	10–15	65–72	23–25	18–24	⑩	3–5	19–24	60–70	75–85	70–90
	8-330	15–25	90	40–45	18–24	40–45	4–7	40–45	95–105	75–85	⑨
	8-351M	10–15	105	⑧	18–24	④	3–5	40–45	95–105	75–85	70–90
	8-351W	10–15	105–112	23–25	18–24	⑩	3–5	40–45	95–105	75–85	70–90
	8-361	15–25	90	40–45	18–24	40–45	4–7	40–45	95–105	75–85	⑨
	8-389	15–25	90	40–45	18–24	40–45	4–7	40–45	95–105	75–85	⑨
	8-391	15–25	90	40–45	18–24	40–45	4–7	40–45	95–105	75–85	⑨
	8-400	10–15	105	⑧	18–24	④	3–5	40–45	95–105	75–85	70–90
	8-460	10–15	130–140	22–32	28–33	④	5–6	40–45	95–105	75–85	70–90
	8-475, 477, 534	15–25	170–180	25–32	22–32	—	5–7	60–65	150–165	100–110	130–175
1979	6-300	15–25	70–85	22–32	28–33	⑪⑫	4–7	40–45	60–70	75–85	130–150
	8-302	10–15	65–72	23–25	18–24	④	3–5	19–24	60–70	75–85	70–90
	8-351W	10–15	105–112	23–25	18–24	④	3–5	40–45	95–105	75–85	70–90
	8-351M, 400	10–15	95–105	⑧	18–24	④	3–5	40–45	95–105	75–85	70–90
	8-370, 429	5–10	130–140	22–32	28–33	④	5–6	45–50	95–105	75–85	150–175
	8-460	5–10	130–140	22–32	28–33	④	5–6	40–45	95–105	75–85	70–90
	8-475, 477, 534	—	—	25–32	—	—	—	—	—	—	—
1980–83	6-300	10–15	85	22–32	28–33	⑪⑫⑬	4–7	40–45	60–70	75–85	130–150
	8-255, 302	10–15	65–72	23–25	18–24	④	3–5	19–24	60–70	75–85	70–90
	8-351W	10–15	105–112	23–25	18–24	④	3–5	40–45	95–105	75–85	70–90
1980–83	8-351M, 400	10–15	95–105	⑧	18–24	④	3–5	40–45	95–105	75–85	70–90
	8-460	5–10	130–140	22–32	28–32	④	5–6	40–45	95–105	75–85	70–90
	8-370, 429	10–15	130–140	22–32	28–32	④	5–7	45–50	95–105	75–85	150–175
	8-477, 534	10–15	170–180	25–32	22–32	170–180	5–7	60–65	150–165	100–110	130–175
1982–83	V6-232	17–22	74	18	15–22	26	3–5	31–36	65–81	54–64	93–121

①—Intake man. to exhaust man., 28–33; Intake man. to cyl. head, 23–28.
②—Rocker arm bolt, 18–25 ft. lbs.
③—1/4" bolts 6–9 ft. lbs; 3/8" bolts 22–23 ft. lbs.; 5/16 bolts 25 ft. lbs.
④—Rocker arm bolts to cylinder head 18–25.
⑤—3/8" bolts 35–45 ft. lbs.; 1/2" bolts 95–105 ft. lbs.
⑥—5/16" 21–25 ft. lbs., 3/8" 27–33 ft. lbs.
⑦—Rocker arm nut to cylinder head bolt, 17–23 ft. lbs.
⑧—3/8" 22–32 ft. lbs.; 5/16" bolts 19–25 ft. lbs.
⑨—5/8" bolts 130–150 ft. lbs.; 3/4" bolts 150–175 ft. lbs.
⑩—Early 1978 with positive stop nut, 17–23 ft. lbs. Late 1978 with fulcrum and bolt, 18–25 ft. lbs.
⑪—Breakaway torque, 4.5–15 ft. lbs. Refer to "Valves, Adjust" procedure
⑫—Rocker arm stud nut torque, 17–23 ft. lbs. Refer to "Valves, Adjust" procedure.
⑬—On 1982–83 models the rocker arm bolt to cylinder head 18–25.

STARTING MOTOR SPECIFICATIONS

Year	Starter Type	Ampere Draw Normal Load	Engine Cranking Speed R.P.M.	Minimum Stall Torque @ 5 volts Ft. Lbs.	Maximum head Amperes	No Load Ampere @ 12 Volts	Brushes		
							Length Inch	Wear Limit Inch	Spring Tension Ounces
1974	Prestolite	200	—	17.2	525	60	.47	.25	45–53
1976–80	Prestolite	200	—	17.2	525	80	.47	.25	45–53
1974–77	Ford Solenoid Actuated	180–210	140–170	—	—	70	.50	.25	40
1974	Delco Positive Engagement	150①	—	15②	700	140	.75	.25	45
	Delco Inertia	200	—	17.5	525	60	.75	.25	24
1974–79	Ford 4″ Diameter	150–200	180–250	9	460	70	.50	.25	40
	Ford 4½″ Diameter	150–180	150–290	15.5	670	70	.50	.25	40③
1975–80	Delco Positive Engagement	150①	—	15②	900	180	.75	.25	45
	Delco Inertia	200	—	17.5	525	60	.75	.25	24
1980–83	Ford 4″ Diameter	150–200	180–250	—	—	70	.50	.25	40
	Ford 4½″ Diameter	150–180	150–290	—	—	80	.50	.25	80

①—Diesel engines: 650–750 amperes. ②—At 1.5 volts. ③—1978–79, 80 ounces.

ALTERNATOR & REGULATOR SPECIFICATIONS

Year	Unit Make	Rated Output @ 15 Volts Amperes	Field Coil Draw @ 12 Volts Amperes	Regulation							
				Voltage Regulation @ 75° F.	Current Limiter		Voltage Limiter		Field Relay		
					Amperes	Air Gap	Contact Gap	Air Gap	Contact Gap	Air Gap	Closing Voltage
1974–75	Motorcraft (Purple)	38	2.5	—	—	—	—	—	—	—	—
	Motorcraft (Orange)	42	2.9	—	—	—	—	—	—	—	—
	Motorcraft (Red)	55	2.9	—	—	—	—	—	—	—	—
	Motorcraft (Green)	61	2.9	—	—	—	—	—	—	—	—
	Motorcraft (Black)	70	2.9	—	—	—	—	—	Sealed	Sealed	2.0–4.2
	Motorcraft (Red)	90	2.9	—	—	—	—	—	Sealed	Sealed	2.0–4.2
	Leece-Neville	70	2.9	—	—	—	—	—	Sealed	Sealed	2.0–4.2
	Leece-Neville	105	2.9	—	—	—	—	—	Sealed	Sealed	2.0–4.2
1976	Motorcraft (Orange)	40	2.9	—	—	—	—	—	—	—	—
	Motorcraft (Green)	60	2.9	—	—	—	—	—	—	—	—
	Motorcraft (Black)	70	2.9	—	—	—	—	—	—	—	—
	Motorcraft (Red)	90	2.9	—	—	—	—	—	—	—	—
	Leece-Neville	70	2.9	—	—	—	—	—	—	—	—
	Leece-Neville	105	2.9	—	—	—	—	—	—	—	—
1977	Motorcraft (Orange)	40	2.9	—	—	—	—	—	—	—	—
	Motorcraft (Green)	60	2.9	—	—	—	—	—	—	—	—
	Motorcraft (Black)	70	2.9	—	—	—	—	—	—	—	—
	Motorcraft (Red)	90	2.9	—	—	—	—	—	—	—	—
	Leece-Neville	70	2.9	—	—	—	—	—	—	—	—
	Leece-Neville	105	2.9	—	—	—	—	—	—	—	—
1978	Motorcraft (Orange)	40	2.9	13.5–14.3	—	—	—	—	—	—	—
	Motorcraft (Green)	60	2.9	13.5–14.3	—	—	—	—	—	—	—
	Motorcraft (Green)⑥	60	4.0	13.8–14.6	—	—	—	—	—	—	—

Continued

ALTERNATOR & REGULATOR SPECIFICATIONS—Continued

Year	Unit Make	Rated Output @ 15 Volts Amperes	Field Coil Draw @ 12 Volts Amperes	Regulation							
				Voltage Regulation @ 75° F.	Current Limiter		Voltage Limiter		Field Relay		
					Amperes	Air Gap	Contact Gap	Air Gap	Contact Gap	Air Gap	Closing Voltage
	Motorcraft (Black)	70	2.9	13.5–14.3	—	—	—	—	—	—	—
	Motorcraft (Red)	90	2.9	13.5–14.3	—	—	—	—	—	—	—
	Motorcraft (Red)⑥	90	4.0	13.8–14.6	—	—	—	—	—	—	—
	Leece-Neville	70	2.9	—	—	—	—	—	—	—	—
	Leece-Neville	105	2.9	—	—	—	—	—	—	—	—
1979	Motorcraft (Orange)⑥	40	4.0	13.8–14.6	—	—	—	—	—	—	—
	Motorcraft (Green)⑥	60	4.0	13.8–14.6	—	—	—	—	—	—	—
	Motorcraft (Black)⑥	65	4.0	13.8–14.6	—	—	—	—	—	—	—
	Motorcraft (70 Amp)⑥	70	4.0	13.8–14.6	—	—	—	—	—	—	—
	Motorcraft (100 Amp)⑥	100	4.0	13.8–14.6	—	—	—	—	—	—	—
	Leece-Neville	70	2.9	—	—	—	—	—	—	—	—
	Leece-Neville	105	2.9	—	—	—	—	—	—	—	—
	Leece-Neville	130	5.7	—	—	—	—	—	—	—	—
1980	Leece-Neville	70	2.9	—	—	—	—	—	—	—	—
	Leece-Neville	105	2.9	—	—	—	—	—	—	—	—
	Leece-Neville	130	5.7	—	—	—	—	—	—	—	—
	Motorcraft (100 Amp)⑥	100	4	—	—	—	—	—	—	—	—
	Delco	75	—	—	—	—	—	—	—	—	—
	Delco	90	—	—	—	—	—	—	—	—	—
1980–81	Motorcraft (Orange)⑥	40	4	—	—	—	—	—	—	—	—
	Motorcraft (Green)⑥	60	4	—	—	—	—	—	—	—	—
	Motorcraft (Black)⑥	65	4	—	—	—	—	—	—	—	—
	Motorcraft (70 Amp)⑥	70	4	—	—	—	—	—	—	—	—
1982–83	Leece-Neville	70	2.9	—	—	—	—	—	—	—	—
	Leece-Neville	90	2.9	—	—	—	—	—	—	—	—
	Leece-Neville	105	2.9	—	—	—	—	—	—	—	—
	Leece-Neville	130	5.7	—	—	—	—	—	—	—	—
	Motorcraft (Orange)⑥	40	4	—	—	—	—	—	—	—	—
	Motorcraft (Green)⑥	60, 65	4	—	—	—	—	—	—	—	—
	Motorcraft (Black)⑥	70, 75	4	—	—	—	—	—	—	—	—
	Motorcraft (Red)⑥	100	4	—	—	—	—	—	—	—	—

①—Lower contacts 0.1 to 0.3 below (listed) upper contact voltage.
②—At 125° F.
③—Econoline.
④—Units w/electro mechanical regulator.
⑤—Units w/transistorized regulator.
⑥—Solid state alternator, these units are marked with blue ink on front of pulley. The electronic voltage regulator used with these units is color coded black for systems w/ warning indicator lamp & blue for systems w/ammeter.

WHEEL ALIGNMENT SPECIFICATIONS

Truck Model	Caster Deg.	Camber Deg.	Toe-In Inch	Kingpin Angle Deg.
1974–75 MODELS				
Bronco	+3 1/2	+1 1/2 (9)	5/32	8 1/2
Econoline E300, 350	+5	+2	1/32	—
Ranchero 1974	+2	(15)	1/8	—
Ranchero 1975	+4	(16)	1/8	—
E100, 150, 200, 250	+4 1/2	+2	1/32	—
F100 (4×4)	+3 1/2	+1 1/2 (9)	1/8	8 1/2
F250 (4×4)	+4	+1 1/2 (9)	1/8	7 1/2
F600 (4×4)	+3	+1	1/8	—
F100, 250	+4 1/2	+1	3/32	—
F-350	+4 1/2	+1 1/2	3/32	—
P350, 3500	+4 1/2	+5/8	1/8	4 1/2
P400, 500, 4000, 5000	+3 1/2	+1/2	1/8	4 1/2
F-B-500 thru 750 (13)	+3 1/2	+5/8	1/4	—
F-B-500 thru 750 (14)	+3 1/2	(10)	3/8	—
C500 thru C750 (6)	+3	+5/8	1/4	—
C750 (8)	+3	(10)	3/8	—
C-CT6000, 7000	+3	+5/8	1/4	—
C-CT9000, 12000	+3	(10)	3/8	—
C-CT15000	+3	+5/8	3/8	—
L Ser. Exc. LN500-750 (6)	+3 3/4	+5/8	1/4	—
L Ser. Exc. LN500-750 (8)	+4	+5/8	3/8	—
L-W-WT Ser. 12000 Conv.	+3	+5/8	1/4	—
LN500-750	+4 1/2	+5/8	1/4	—
W-WT9000	+3 1/2	(10)	1/4	—
Center Point Steering	+1/2	(11)	7/16	—
1976 MODELS				
Ranchero	+4	(16)	1/8	—
Bronco	+4 1/2	+1 1/2	5/32	8 1/2
Econoline E-350	+5	+2	1/8	—
E-100, 150, 260	+4 1/2	+2	1/8	—
F-100, 250 (4×2)	+4 1/2	+1	3/32	—
F-350 (4×2)	+4 1/2	+1 1/2	3/32	—
F-100, 150 (4×4)	+4 1/2	+1 1/2	5/32	8 1/2
F-250 (4×4)	+4	+1 1/2	5/32	7 1/2
P-350	+4 1/2	+5/8	1/8	4 1/2
P-400, 500	+3 1/2	+1/2	1/8	4 1/2
F-600 (4×4)	+3	+1	1/8	—
F-B-500 thru 750 (13)	+3 1/2	+5/8	1/4	—
F-B-500 thru 750 (14)	+3 1/2	(10)	3/8	—
C-600 thru 750 (6)	+3	+5/8	1/4	—
C-600 thru 750 (8)	+3	(10)	3/8	—
LN-600-750 (13)	+4 1/2	+5/8	1/4	—
LN-600-750 (14)	+4 1/2	(10)	3/8	—
L Ser 800-9000 (6)	+3 3/4	+5/8	1/4	—
L Ser 800-9000 (18)	(29)	+5/8	3/8	—
C Series 800 thru 8000	+3	(10)	(19)	—
W-Wt-9000	+3	(10)	1/4	—
1977 MODELS				
F-250 (20)	4	1	5/32	—
F-150 (20)(21)	3 1/2	1 1/2	3/16	—
P-500	3 1/2	1/2	1/8	—
F-500-600, LN-600 (22)(32)	(24)	5/8	1/4	—
B-500, F-C-B-LN-600-750, LT-LNT-800, LT-880, F-C-LN-6000, F-B-C-7000, L800-900, LN-800 (23)(6)	(25)	5/8	1/4	—
F-600-700, B-LN-600-700, LN-F-B-750, F-B-LN-7000, F-880 (22)(33)	(26)	5/8	3/8	—
F-C-L-LN-750-7000, LT-880, CT-LN-LNT-800, C-L-LT-LTS-800-900 (23)(14)	(27)	(17)	3/8	—
F-LN-750-7000, F-880 (22)(14)	(29)	5/8	3/8	—
L-LT-LN-LNT-800, L-LN-LNT,8000, LNT-9000, L-LN-LNT-900, C-7000, F-LT-LNT-880, C-600-700-750 (22)(33)	(28)	5/8	3/8	—
L-LT-LTS-LN-LNT-800-900-8000-9000, LT-LNT-880, C-750-800-900-7000-8000, CT-800-900-8000 (22)(14)	(30)	5/8	3/8	—
L-LT-LTS-LN-LNT-C-CT-800-900-L-LT-LTS-LN-LNT-8000-9000, LNT-LT-880, C-8000, CT-8000, W-WT-9000 (22)(34)	(1)	5/8	1/4	—
C-900, CT-800-900, CT-8000, LT-800-900, LTS-800-900, LT-LTS-8000-9000, LN-LNT-800-900-8000-9000 (22)(37)	(32)	5/8	3/8	—
F-100/150 (4×2), F-250, 6200-6900 GVW	6 1/2	1	1/8	—
F-250, 7800-8000 GVW, F-250 Supercab (4×2)	5 7/8	1 5/8	1/8	—
F-350, F-250, F-350 Crew CAB, F-250, 8100 GVW S.C.	6 1/8	1 7/8	1/8	—
E-100/150	3 7/8	1/2	1/8	—
E-250/350	6 1/2	7/8	1/8	—
1978 MODELS				
F-150 (20)(21)	+4	+1 1/2	3/32	8 1/2
F-250 (20)	+8	+1 1/2	3/32	8 1/2
P-600	+3 1/2	+1/2	1/8	4 1/2
F-600, LN-600 (23)(32)	(24)	+5/8	1/4	—
B-600, F-C-B-LN-600-700, LT-LNT-800, LT-800, F-C-LN-6000, F-B-C-7000, L-800-900, LN-800 (23)(6)	(25)	+5/8	1/4	—
F-600-700, B-LN-600-700, LN-F-B-700, F-B-LN-700, F-800 (22)(33)	(26)	5/8	3/8	—
F-C-L-LN-700-7000, LT-800, C-L-LT-LTS-800-900, CT-LN-LNT-800 (23)(14)	(27)	(17)	3/8	—
L-LT-LN-LNT-800, L-LN-LNT-8000, LNT-9000, L-LN-LNT-900, F-LT-LNT-800, C-600-700, C-7000 (22)(33)	(28)	5/8	3/8	—
F-LN-700, 7000, F-800 (22)(14)	(29)	5/8	3/8	—

Continued

WHEEL ALIGNMENT SPECIFICATIONS—Continued

Truck Model	Caster Deg.	Camber Deg.	Toe-In Inch	Kingpin Angle Deg.
L-LT-LTS-LN-LNT-800-900-8000-9000, LT-LNT-800, C-700-800-900-7000-8000, CT-800-900-8000㉒⑭	㉚	5/8	3/8	—
L-LT-LTS-LN-LNT-C-CT-800-900, L-LT-LTS-LN-LNT-8000-9000, LNT-LT-800, CT-8000, C-8000㉒㉞	①	5/8	1/4	—
C-900, CT-800-900, CT-8000, LT-800-900, LTS-LN-LNT-8000-9000, LNT-LT-800, CT-8000, C-8000㉒㊲	⑦	5/8	1/4	—
CL-CLT-9000㉒	3 1/4	5/8	1/4	—
F-100/150/250/350	②	②	②	—
E-100/150/250/350	②	②	②	—

1979 MODELS

Truck Model	Caster Deg.	Camber Deg.	Toe-In Inch	Kingpin Angle Deg.
F-250-350⑳	+4	+1 1/2	3/32	8 1/2
F-150③	+4	+1 1/2	3/32	8 1/2
F-150⑳㉑	+8	+1 1/2	3/32	8 1/2
P-600	+3 1/2	+1/2	1/8	4 1/2
F-600, LN-600㉓㉜	㉔	5/8	1/4	—
B-600, F-C-B-LN-600-700, LT-LNT-800, F-C-LN-6000, F-B-C-7000, L-800, 900, LN-800㉓⑥	㉕	5/8	1/4	—
F-600-700, B-LN-600-700, LN-F-B-700, F-B-LN-7000, F-800㉒㉝	㉖	5/8	3/8	—
F-C-L-LN-700-7000, C-L-LT-LTS-800-900, CT-LN-LNT-800㉓⑭	㉗	⑰	3/8	—
L-LN-LNT-800-8000, LT-800, LNT-9000, L-LN-LNT-900, F-LT-LNT-800, C-600-700, C-7000㉒㉝	㉘	5/8	3/8	—
F-LN-700-7000, F-800㉒⑭	㉙	5/8	3/8	—
LT-LTS-LN-LNT-800-900-8000-9000, LT-LNT-800, C-700-800-900-7000-8000, CT-800-900-8000㉒⑭	㉚	5/8	3/8	—
LT-LTS-LN-LNT-C-CT-800-900, L-LT-LTS-LN-LNT-8000-9000, LNT-LT-800, C-CT-8000, W-WT-9000㉒㉞	①	5/8	1/4	—
C-900, CT-800-900, CT-8000, LT-800, 900, LTS-LN-LNT-8000-9000, LNT-LT-800, C-8000, CT-8000㉒㊲	⑦	5/8	3/8	—
CL-CLT-9000㉒	3 1/4	5/8	1/4	—
F-100-150-250-350	②	②	②	—
E-100-150-250-350	②	②	②	—

1980 MODELS

Truck Model	Caster Deg.	Camber Deg.	Toe-In Inch	Kingpin Angle Deg.
F-100-150-250-350	②	②	②	8
E-100-150-250-350	②	②	②	—
F-150⑳㉑	②	②	②	13
F-250-350⑳	②	②	②	13
F-600, LN-600㉓㉜	㉔	+5/8	1/4	—
F-C-B-LN-600-700, LT-LNT-800, F-C-LN-6000, C-7000, L-800-900, LN-800㉓⑥	㉕	+5/8	1/4	—
F-B-LN-600-700, LN-7000, F-800㉒㉝	㉖	+5/8	3/8	—
F-C-L-LN-700-7000, C-L-LT-LTS-800-900, CT-LN-LNT-800㉓⑭	㉗	⑰	3/8	—
F-L-LT-LN-LNT-800, LNT-9000, L-LN-LNT-8000, L-LN-LNT-900, C-600-700, C-7000㉒㉝	㉘	5/8	3/8	—
F-LN-700, F-800, LN-7000㉒⑭	㉙	5/8	3/8	—
L-LT-LTS-LN-LNT-800-900-8000-9000, C-700-800-900-7000-8000, CT-800-900-8000㉒⑭	㉚	5/8	3/8	—
L-LT-LTS-LN-LNT-C-CT-800-900, L-LT-LTS-LN-LNT-8000-9000, C-CT-8000, W-WT-9000㉒㉞	①	5/8	1/4	—
C-900, LT-CT-800-900, CT-8000, LTS-LN-LNT-8000-9000, LNT-LT-800, C-CT-8000㉒㊲	⑦	5/8	3/8	—
CL-CLT-9000㉒	3 1/4	5/8	1/4	—

1981 MODELS

Truck Model	Caster Deg.	Camber Deg.	Toe-In Inch	Kingpin Angle Deg.
F-100-150-250-350	②	②	②	8
E-100-150-250-350	②	②	②	—
F-150, Bronco (4×4)	②	②	②	13
F-250-350 (4×4)	②	②	②	13
F-600, LN-600㉓㉜	㉔	5/8	3/16	—
F-C-B-LN-600-700-7000, LT-LNT-800, L-800-900, LN-800㉓⑥	㉕	5/8	3/16	—
F-B-LN-700, 7000, F-800㉒㉝	㉖	5/8	3/16	—
L-LT-LN-LNT-800, L-LN-LNT-8000, LNT-9000, L-LN-LNT-900, C-600-700, C-7000㉒㉝	㉚	5/8	3/16	—
F-B-LN-700, 7000, F-800㉒⑭	㉙	5/8	3/16	—
C-L-LT-LTS-800-900, CT-LN-LNT-800㉓⑭	㉗	⑰	3/16	—
L-LT-LTS-LN-LNT-800-900-8000-9000, C-700-800-900-7000-8000, CT-800-900-8000㉒⑭	㉚	5/8	3/16	—
L-LT-LTS-LN-LNT-C-CT-800-900, L-LT-LTS-LN-LNT-8000-9000, C-8000, CT-8000㉒㉞	①	5/8	1/16	—
C-900, CT-800-900, CT-8000, LT-800-900, LTS-LN-LNT-8000-9000, LNT-LT-800, C-8000㉒㊲	㉛	5/8	3/16	—
CL-9000, CLT-9000	3 1/4	5/8	1/16	—

Continued

WHEEL ALIGNMENT SPECIFICATIONS—Continued

Truck Model	Caster Deg.	Camber Deg.	Toe-In Inch	Kingpin Angle Deg.
1982 MODELS				
F-100-150-250-350	②	②	②	8
E-100-150-250-350	②	②	②	—
F-150⑳㉑	②	②	②	13
F-250-350-⑳	②	②	②	13
F-600, LN-600㉓㉜	㉔	5/8	1/16	—
F-C-B-LN-600-700㉓⑥	㉕	5/8	1/16	—
F-B-700, F-800, L-800㉒㉝	㉖	5/8	1/16	—
C-600-700, C-7000, L-800-8000㉒㉝	㉘	5/8	1/16	—
LN-600-700-7000-800-8000, F-B-700, F-800㉒⑭	㉙	5/8	1/16	—
C-800, CT-800, L-LN-LT-LTS-LNT-800-8000㉓⑭	㉗	⑰	1/16	—
C-700-800-7000-8000, CT-800-8000, L-LT-LTS-LN-LNT-800-900-8000-9000㉒⑭	㉚	5/8	1/16	—
C-CT-800, C-8000, CT-8000, L-LN-LT-LTS-LNT-800-8000-9000㉒㉞	①	5/8	1/16	—

Truck Model	Caster Deg.	Camber Deg.	Toe-In Inch	Kingpin Angle Deg.
CT-800, C-CT-8000, LT-LTS-LNT-8000-9000, LTL-900㉒㊲	㉛	5/8	1/16	—
1983 MODELS				
F-100-150-250-350	②	②	②	—
F-100-150-250-350	②	②	②	—
F-150-250-350⑳㉑	②	②	②	—
F-600㉓㉜	4	5/8	1/16	—
F-C-B-600-700㉓⑥	㉕	5/8	1/16	—
F-B-700, F-800㉒㉝	3	5/8	1/16	—
C-600-700, C-7000㉒㉝	㉘	5/8	1/16	—
F-B-700, F-800㉒⑭	3	5/8	1/16	—
C-800, CT-800⑭㉓	3	⑰	1/16	—
C-700-800, C-7000-8000, CT-800-8000⑭㉒	㉘	5/8	1/16	—
F-FT-800, F-FT-8000	㉟	⑩	1/16	—
C-CT-800, C-CT-8000㉒㉞	①	㊱	1/16	—
CT-800, CT-8000, LT-800, C-8000㉒㊲	3	5/8	1/16	—

①—C series w/9.5 frame & L-W series w/2° wedge, 3°; L series w/drop frame 3½°; C series w/9.76 frame, 4½°.
②—Refer to Front End Section of this chapter.
③—Super cab.
④—Power Steering 7½°.
⑤—Power Steering 6½°.
⑥—With 6000 and 7000 lb. capacity axle.
⑦—C series w/16000 axle, 3°; all w/16000-18000-20000 axles exc. C series, 3½°.
⑧—With 9000 and 12000 Lb. capacity axle.
⑨—Designed into axle and not adjustable.
⑩—Left side +3/8, right side −1/8.
⑪—Left side +1/2, right side zero.
⑫—Except center point steering.
⑬—With 5000–7000 lb. capacity axle.
⑭—With 9000 lb. capacity axle.
⑮—Left side 1½°, right side +1/8°.
⑯—Left side +½°, right side +¼°.
⑰—C series LH, 3/8°; C series RH, −1/8°; exc. C series, 5/8°.
⑱—With 9000, 12000, 16000, 18000, 20000 lb. capacity axles.
⑲—With 7000 lb. capacity axle 1/4; With 9000 lb. capacity axle and above, 3/8.
⑳—4-wheel drive.
㉑—Bronco.
㉒—Air brakes.
㉓—Hydraulic brakes.
㉔—F-B series, 3½°; LN series, 4½°.
㉕—F-B series, 3½°; LN series medium 4½°; L-LN series heavy 3¾°; C series, 3°.
㉖—F-B series, 3°; L series, 3½°.
㉗—Exc. C series, 4°; C series, 3°.
㉘—C series w/9.5 frame, 2°51′; C series w/9.76 frame, 4½°; L series exc. drop frame, 5°; LN-LNT series w/drop frame, 3½°.
㉙—F-B series, 3°; LN series, 3½°.
㉚—L series exc. drop frame, 5°; LN-LNT w/drop frame, 3½°; C series w/9.5 frame, 2°51′; C series w/9.76 frame, 4½°.
㉛—C series w/16000–18000 axles, 3; all w/16000–18000–20000 axles exc. C series, 3½.
㉜—With 5000 lb. or 5500 lb. capacity axle.
㉝—With 7000 lb. or 7500 lb. capacity axle.
㉞—With 9500 lb. or 12000 lb. capacity axle.
㉟—With 9500 lb. capacity axle, 5¼; with 12000 lb. capacity axle, 5½.
㊱—Left side +3/8; right side +1/8.
㊲—With 16000 lb., 18000 lb. or 20000 lb. capacity axle.

Engine Section

CYLINDER HEAD, REPLACE

V8-401, 477, 534

1. Remove intake manifold.
2. Remove push rod cover and rocker arm covers.
3. Disconnect muffler inlet pipes at exhaust manifolds.
4. Release spring tension on rocker arms by loosening adjusting screws.
5. Remove rocker arm shaft and oil baffle plate. *Because five cylinder head bolts also serve as pedestals for the rocker arm shaft, each time the rocker arm assembly is removed, a new cylinder head gasket must be installed.*
6. Remove push rods, keeping them in proper sequence so they may be installed in original location.
7. Remove remaining cylinder head bolts and lift off head.
8. Install heads in the reverse order of removal and tighten bolts in the sequence shown in Fig. 1 to the torque given in the *Engine Tightening Chart.*

V8-370, 429, 460

1. Remove intake manifold and disconnect exhaust pipe from exhaust manifold.
2. Loosen air conditioner drive belt (if equipped), then loosen alternator bolts and remove the bolt attaching alternator bracket to cylinder head.
3. If equipped with air conditioning, shut off compressor service valves and remove valves and hoses from compressor then remove compressor and place out of way. Remove compressor upper mounting bracket from cylinder head.
4. If not equipped with air conditioning, remove bolts attaching power steering reservoir bracket to left cylinder head and position reservoir and bracket out of the way.
5. On V8-370, 460 and V8-429 engines, remove rocker arm bolts, rocker arms, oil deflectors, fulcrums and push rods.

NOTE: Keep all parts in sequence so that they can be reinstalled in their original locations.

6. Remove cylinder head bolts taking note as to the length of each bolt and remove the cylinder head.

NOTE: If necessary to loosen cylinder head gasket seal, pry at forward corners of cylinder head and against casting provided on cylinder block.

7. Install heads in the reverse order of removal and torque cylinder head bolts in three steps and in the sequence shown in Fig. 1. Final tightening should be to the torque listed in the Engine Tightening Table. Torque intake manifold bolts to torque spec given in Engine Tightening Table and in sequence shown in Fig. 1A.

V8-330, 360, 361, 389, 390, 391

1. Remove intake manifold and carburetor as a unit.
2. Disconnect exhaust manifolds at exhaust pipes.
3. If left head is being removed, remove ignition coil.
4. Unfasten and lift off head.
5. Install heads in the reverse order of removal and torque cylinder head bolts in three steps and in the sequence shown in Fig. 1. Final tightening should be to the torque listed in the Engine Tightening Specifications chart. Torque intake manifold bolts to torque specs given in Engine Tightening Specifications chart and in sequence shown in Fig. 1B.

V8-255, 302, V8-351 & 400

1. Remove intake manifold and rocker arm covers.
2. If removing right side cylinder head on 1974–78 models with V8-302 and 351W engines, loosen alternator adjusting arm and remove alternator mounting bracket and spacer. Swing alternator downward. On 1979–83 models with V8-302 and 351W engines, loosen alternator and air pump mounting bracket and swing alternator downward. On Bronco and Econoline models, remove ignition coil and air cleaner inlet duct from right side cylinder head. On models with 351M and 400 engines, remove air cleaner inlet duct and ground wire from right side cylinder head.
3. If removing left cylinder head, remove accelerator shaft assembly from front of cylinder head on all 1974–78 vehicles equipped with V8-302 and 351 engines. On 1979–83 models equipped with V8-302 and 351 engines, remove bolts securing air conditioner compressor bracket to cylinder head. On all vehicles equipped with V8-351M and 400 engines, remove power steering pump from cylinder head and remove drive belt from pump pulley. Position pump aside without disconnecting any hoses. On all vehicles equipped with V8-351M and 400 engines and air conditioning, isolate and remove compressor.
4. Disconnect exhaust pipes from exhaust manifolds.
5. Loosen rocker arm stud nuts or bolts and remove push rods in sequence so that they can be installed in their original locations.
6. On V8-302, and 1980–82 V8-351 and 400, remove exhaust valve stem caps.
7. On 1979–83 E100 thru E350 series equipped with V8-255, 302 and 351W engines, remove bolts securing thermactor air supply manifold to rear of cylinder head, then disconnect hose at air pump. Remove hose pump valve and air supply manifold as an assembly. On F100 thru F350 series and Bronco, disconnect thermactor air supply hoses at check valves and plug check valves. On 1980–83 E150 thru E350 series equipped with V8-351M and 400 engines, it may be necessary to remove forward bolt on left side of head as a unit with the head, due to lack of clearance with body flange.
8. Remove the cylinder head bolts and cylinder head.
9. Install cylinder heads in the reverse order of removal and torque cylinder head bolts

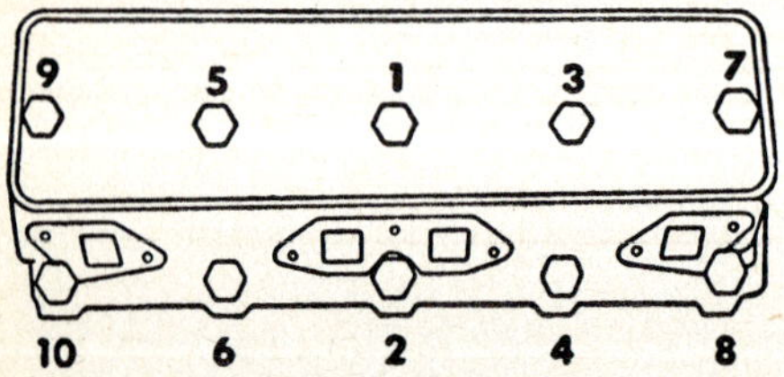

Fig. 1 Cylinder head tightening sequence. V8 engines

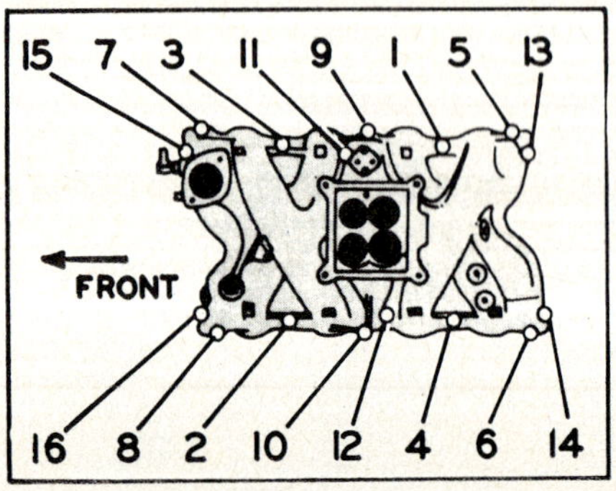

Fig. 1A Intake manifold tightening sequence. V8-370, 429 & 460

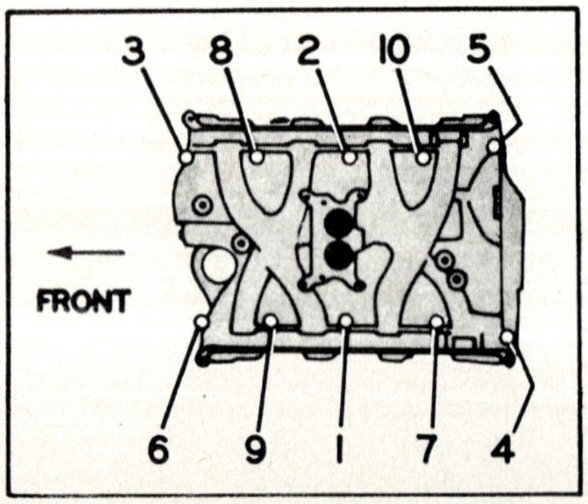

Fig. 1B Intake manifold tightening sequence. V8-330, 360, 361, 389, 390, 391

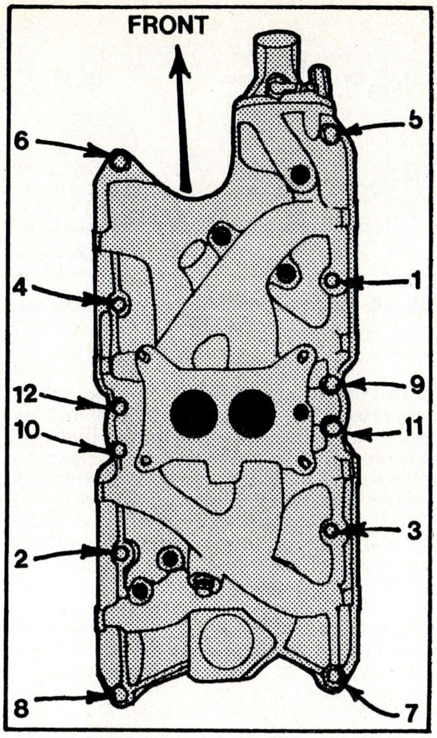

Fig. 1C Intake manifold tightening sequence. V8-255, 302 & 1976–83 V8-351W

in three steps and in sequence shown in Fig. 1. Final tightening should be to the torque listed in the Engine Tightening Table. Torque intake manifold bolts to torque specs given in Engine Tightening Table and in sequence shown in Figs. 1C, 1D and 1E.

6-240, 300

1. Drain cooling system. On Econoline, remove right front seat and engine compartment cover.
2. Remove air cleaner.
3. Remove PCV valve from rocker arm cover, and disconnect vent hose at intake manifold inlet tube.
4. Disconnect hoses, wires, carburetor linkage, fuel and vacuum lines as necessary for removal of head.
5. Disconnect exhaust pipe from manifold.
6. Unfasten and position coil to one side.
   a. If equipped with Thermactor, remove air tube and vacuum hose from anti-backfire and intake manifold. Disconnect air pump outlet hoses at air manifold and at anti-backfire valve. Disconnect air pump to air cleaner hose at the pump. Remove the air pump and air manifold assembly from the cylinder head. Remove the air cleaner and mounting bracket assembly, exhaust manifold heat shield, air pump mounting bracket and anti-backfire valve and bracket assembly.
7. Remove valve rocker arm cover.
8. Loosen rocker arm stud nuts so that rocker arms can be rotated to one side.
9. Remove valve push rods, keeping them in proper sequence so they may be installed in their original locations.
10. Remove cylinder head bolts and use a lifting sling to remove cylinder head with manifolds attached.
11. Reverse procedure to install and tighten head bolts in three steps and in the sequence shown in Fig. 2 to the torque listed in the *Engine Tightening Table*. Torque intake manifold bolts to torque specs given in Engine Tightening Table and in sequence shown in Fig. 2A.

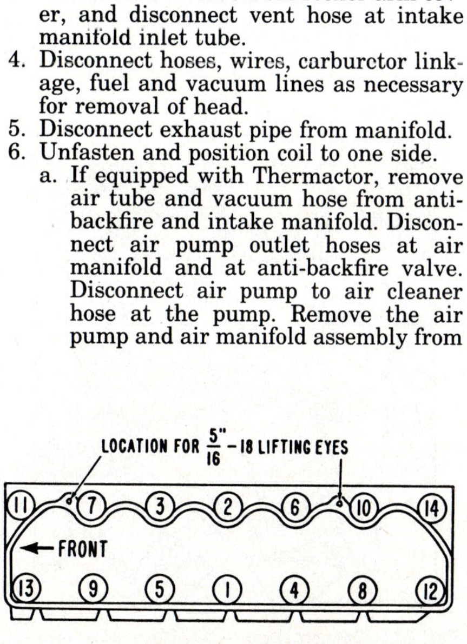

Fig. 2 Cylinder head tightening sequence. 6-240, 300

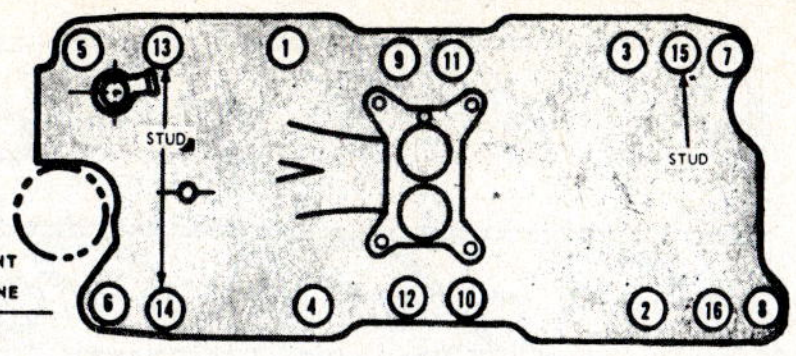

Fig. 1D Intake manifold tightening sequence. 1974–75 V8-351W

6-200 & 250

1. Drain cooling system.
2. Remove air cleaner.
3. Disconnect upper radiator hose at engine.
4. Disconnect exhaust pipe from manifold.
5. Disconnect accelerator retracting spring choke control cable and accelerator rod at carburetor.
6. Disconnect fuel line and distributor vacuum line at carburetor.
7. Disconnect coolant lines at carburetor spacer.
8. Disconnect distributor vacuum line at distributor and carburetor fuel line at the fuel pump and remove lines as an assembly.
9. Disconnect plug wires at the spark plugs and temperature sending unit wire at the sending unit.
10. Remove PCV valve from rocker arm cover and disconnect vent hose at intake manifold inlet tube.
11. If equipped with thermactor, remove hoses from Thermactor system as necessary for accessibility.
12. Remove rocker arm cover.
13. Remove the rocker arm shaft support bolts and remove rocker shaft assembly.
14. Remove valve push rods, keeping them in proper sequence so they may be installed in their original locations.
15. Remove cylinder head bolts and remove the cylinder head.
16. Reverse procedure to install head and tighten head bolts in three steps and in the sequence shown in Fig. 3 to the torque listed in the *Engine Tightening Table*.

V6-232

1. Disconnect battery ground cable, then drain cooling system.
2. Remove air cleaner, air intake duct and heat tube.

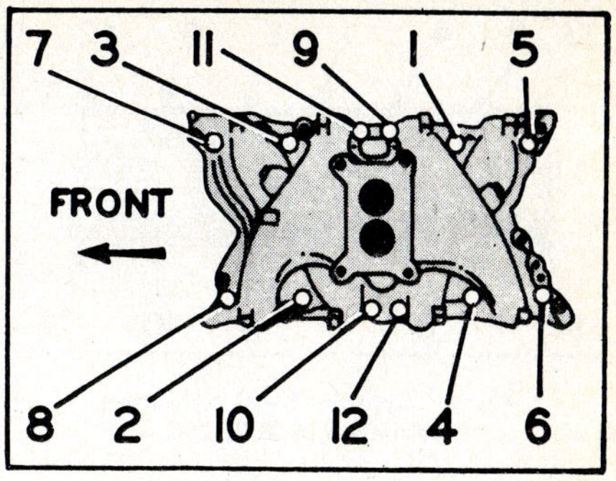

Fig. 1E Intake manifold tightening sequence. V8-351M & 400

3. Loosen accessory drive belt idler and remove drive belt.
4. If left cylinder head is to be removed, proceed as follows:
   a. Remove oil filler cap.
   b. If equipped with power steering, remove pump bracket attaching bolts, then remove pump and bracket assembly and position pump aside with hoses attached.
   c. If equipped with A/C, remove compressor bracket attaching bolts, then position compressor and bracket assembly aside with refrigerant lines attached.
5. If right cylinder head is to be removed, proceed as follows:
   a. Remove Thermactor diverter valve and hose assembly.
   b. Remove accessory drive belt idler, then remove alternator.
   c. Remove Thermactor pump pulley, then remove Thermactor pump.
   d. Remove alternator mounting bracket.

NOTE: On models equipped with Tripminder, the fuel supply line from the fuel pump to the fuel sensor will have to be disconnected to gain access to the upper alternator bracket bolt.

   e. Remove PCV valve.
6. Remove intake manifold and exhaust manifolds.
7. Remove rocker arm cover attaching screws, then loosen cover by using a putty knife under cover flange and remove cover. Do not use excess force when loosening rocker arm cover as cover may become damaged.
8. Loosen rocker arm fulcrum bolt enough to allow rocker arms to be rotated to one side, then remove push rods.

NOTE: Tag push rods so they can be installed in the same position.

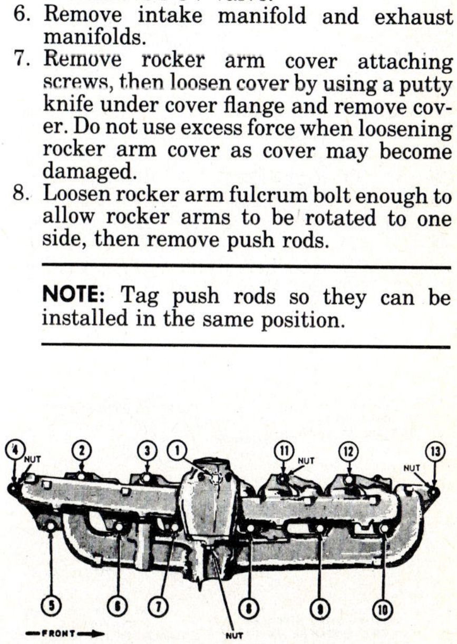

Fig. 2A Intake & exhaust manifold tightening sequence. 6-240 & 300

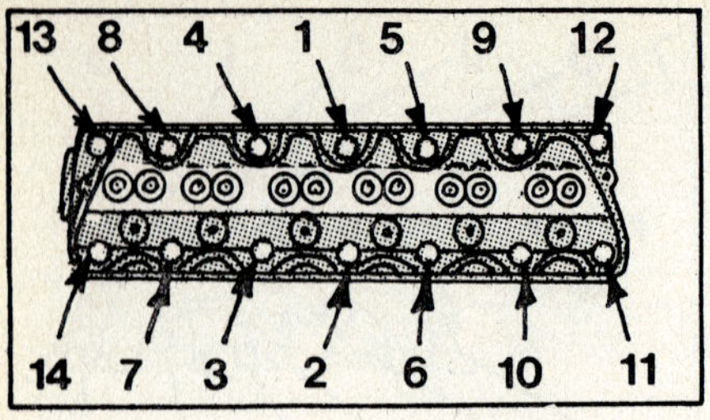

Fig. 3 Cylinder head tightening sequence. 6-200, 250

9. Remove cylinder head attaching bolts, then remove cylinder head and gasket.
10. Reverse procedure to install. Apply a thin coating of pipe sealant D8AZ-19558-A or equivalent to the shorter cylinder head bolts which are installed on the exhaust manifold side of the cylinder head. Do not apply pipe sealant to long bolts which are installed on the intake manifold side of the cylinder head. Tighten cylinder head bolts in four steps to torque listed under Engine Tightening Specifications using sequence shown in Fig. 3A. Loosen cylinder bolts approximately 2 to 3 turns, then retighten bolts in four steps to specified torque in sequence, Fig. 3A. When installing intake manifold, tighten mounting bolts to torque specified under Engine Tightening specifications in sequence shown in Fig. 3B.

NOTE: Before installing intake manifold, apply a 1/8 inch bead of silicone rubber sealer D6AZ-19562-B or equivalent at mating surfaces of intake manifold cylinder heads and cylinder block. Fig. 3C. Also apply a 1/8 inch bead of sealer to the outer end of each intake manifold seal for the full width of the seal.

VALVE ARRANGEMENT

Front to Rear

6-240, 300	E-I-E-I-E-I-E-I-E-I-E-I
6-200, 250	E-I-I-E-I-E-E-I-E-I-I-E
V6-232 Right	I-E-I-E-I-E
V6-232 Left	E-I-E-I-E-I
V8-255, 302 Left Bank	E-I-E-I-E-I-E-I
V8-255, 302 Right Bank	I-E-I-E-I-E-I-E
V8-330, 361, 391	E-I-E-I-I-E-I-E
V8-351, 400 Right Bank	I-E-I-E-I-E-I-E
V8-351, 400 Left Bank	E-I-E-I-E-I-E-I
V8-360, 389, 390	E-I-E-I-I-E-I-E
V8-401, 477, 534	E-I-E-I-I-E-I-E
V8-370, 429, 460 Left Bank	E-I-E-I-E-I-E-I
V8-370, 429, 460 Right Bank	I-E-I-E-I-E-I-E

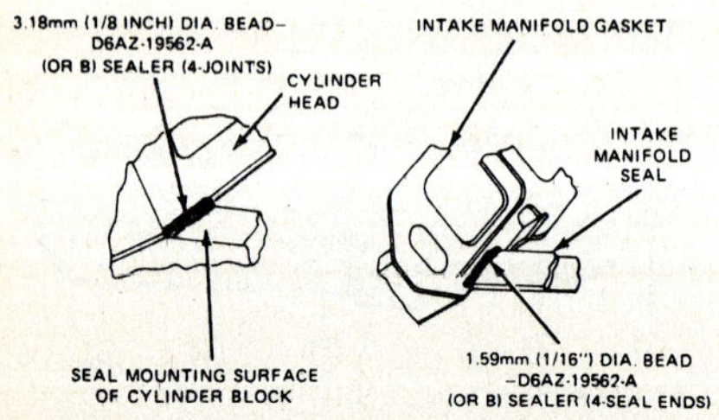

Fig. 3C Applying RTV sealer for intake manifold installation. V6-232

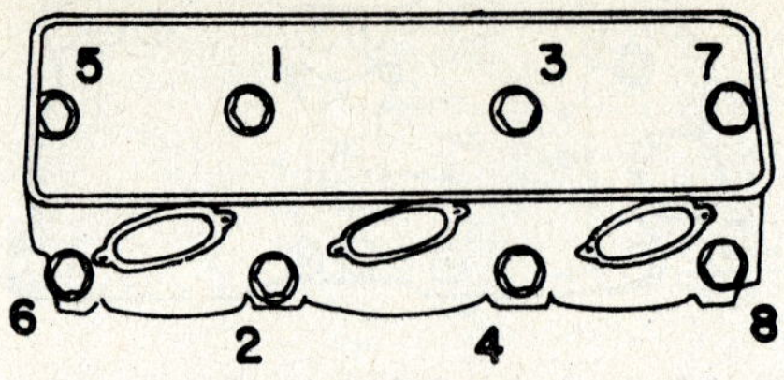

Fig. 3A Cylinder head tightening sequence. V6-232

VALVES, ADJUST

6-200, 250

Rotate crankshaft to place No. 1 piston at TDC of its compression stroke and adjust the following valves:

No. 1 Intake	No. 3 Exhaust
No. 1 Exhaust	No. 4 Intake
No. 2 Intake	No. 5 Exhaust

Rotate crankshaft one full revolution to place No. 6 piston at TDC compression and adjust the remaining valves:

No. 2 Exhaust	No. 5 Intake
No. 3 Intake	No. 6 Intake
No. 4 Exhaust	No. 6 Exhaust

Adjustable Rocker Arms

Turn rocker arm adjusting screw to obtain proper valve adjustment as listed in *Valve Specifications* table.

NOTE: If the torque required to tighten the self-locking adjusting screw is less than 3 ft. lbs., install a new standard or .002" oversize screw. If unable to obtain a minimum torque of 7 ft. lbs. with the oversize screw, replace the rocker arm and adjusting screw assembly.

Non-Adjustable Rocker Arms

The push rod of the non-adjustable rocker arm has a spherical push rod socket instead of an adjusting screw. The push rods are available in standard size, .060" undersize and .060" oversize.

Valve clearance is checked by compressing the push rod end of the rocker arm until the valve lifter plunger is completely bottomed. While holding the lifter in the fully collapsed position, check the available clearance between the rocker arm and valve stem. If the clearance is less than .067", install undersize push rod. If clearance is more than .200", install an oversize push rod.

6-240, 300

1974–78

NOTE: Before adjusting valves, check breakaway torque (torque required to turn nut in a counterclockwise direction) of stud nut. If breakaway torque is less than 4.5–15 ft. lbs., replace the stud nut. If breakaway torque is still not within specifications, replace the stud.

Starting with No. 1 cylinder, position each piston at TDC compression stroke as described previously. With No. 1 piston at TDC compression stroke, adjust intake and exhaust valve clearance for that cylinder by loosening the stud nut until there is clearance between the push rod and rocker arm, then tighten stud nut just enough to remove all rocker arm clearance. This may be determined by rotating and/or moving the push rod as the stud nut is tightened. When clearance has been eliminated, tighten stud nut one additional turn.

Follow the firing order and repeat the above procedure for the remaining valves with piston of cylinder at TDC compression stroke.

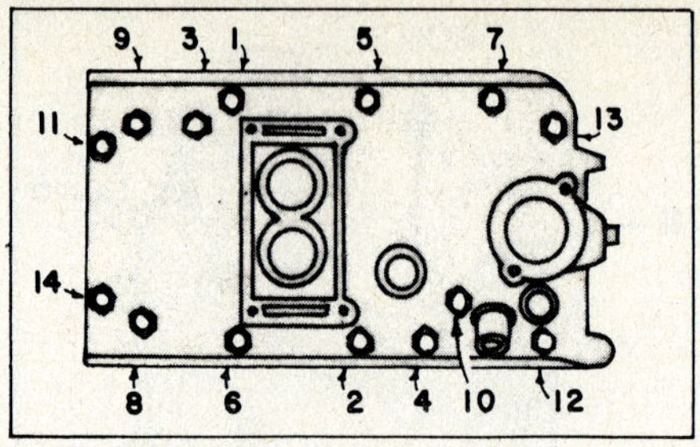

Fig. 3B Intake manifold tightening sequence. V6-232

1979–83

NOTE: Before adjusting valves, check breakaway torque (torque required to turn nut in a counterclockwise direction) of stud nut. If breakaway torque is less than 4.5–15 ft. lbs., replace the stud nut. If breakaway torque is still not within specifications, replace the stud nut. Also, for these engines, a .060 inch shorter push rod or .060 inch longer push rod is available to compensate for dimensional changes in the valve mechanism and provide for valve adjustment.

Starting with No. 1 cylinder, position each piston at TDC compression stroke as described previously. With No. 1 piston at TDC compression stroke, install and torque stud nut to 17–23 ft. lbs. Using a suitable tool, Fig. 3D, slowly apply pressure to bleed down the lifter until the plunger is completely bottomed. While holding the lifter in this position, check clearance between rocker arm and

Fig. 3D Compressing valve lifter with tool to check clearance on engines with hydraulic valve lifters

valve. If clearance is not as specified in the VALVE SPECIFICATIONS CHART, install a longer or shorter push rod as necessary to bring clearance within specifications.

Follow the firing order and repeat the above procedure for the remaining valves with piston of cylinder at TDC compression stroke.

V6-232

A .060 inch longer or a .060 inch shorter push rod is available to compensate for dimensional changes in the valve train. If clearance is less than specified, the .060 inch shorter push rod should be used. If clearance is more than the maximum specified, the .060 inch longer push rod should be used.

Using an auxiliary starter switch crankshaft until No. 1 cylinder is at TDC compression stroke, then compress valve lifter using tool T82C-6500-A or equivalent, Fig. 3D. At this point, the following valves can be checked:

No. 1 Intake	No. 3 Intake
No. 1 Exhaust	No. 4 Exhaust
No. 2 Exhaust	No. 6 Intake

After clearance on these valves has been checked, rotate crankshaft until No. 5 cylinder is at TDC compression stroke (1 revolution of crankshaft), and then compress valve lifter using tool No. T82C-6500-A or equivalent. Fig. 3D, and check the following valves:

No. 2 Intake	No. 5 Intake
No. 3 Exhaust	No. 5 Exhaust
No. 4 Intake	No. 6 Exhaust

All V8's With Hydraulic Lifters

To provide a means to compensate for dimensional changes in the valve train and provide for valve adjustment, .060 inch shorter or longer push rods are available. If the valve clearance is less than the minimum, the .060 inch shorter push rod should be used. If the clearance is more than the maximum, the longer push rod should be used. To check the valve clearance, proceed as follows:

1974–83 V8-255, 302, 351, 400 & 460, Exc. 1975 V8-351W

1. Mark crankshaft pulley at three locations, with number 1 location at TDC timing mark (end of compression stroke), number 2 location one half turn (180°) clockwise from TDC and number 3 location three quarter turn clockwise (270°) from position number 2.
2. Turn crankshaft to number 1 location, then compress valve lifter using tool T71P-6513-A or equivalent, Fig. 3D, and check the clearance on the following valves:

V8-255, 302, 460

No. 1 Intake	No. 1 Exhaust
No. 7 Intake	No. 5 Exhaust
No. 8 Intake	No. 4 Exhaust

V8-351 & 400

No. 1 Intake	No. 1 Exhaust
No. 4 Intake	No. 3 Exhaust
No. 8 Intake	No. 7 Exhaust

3. Turn crankshaft to number 2 location, then compress valve lifter using tool T71P-6513-A or equivalent, Fig. 3D, and check the clearance on the following valves:

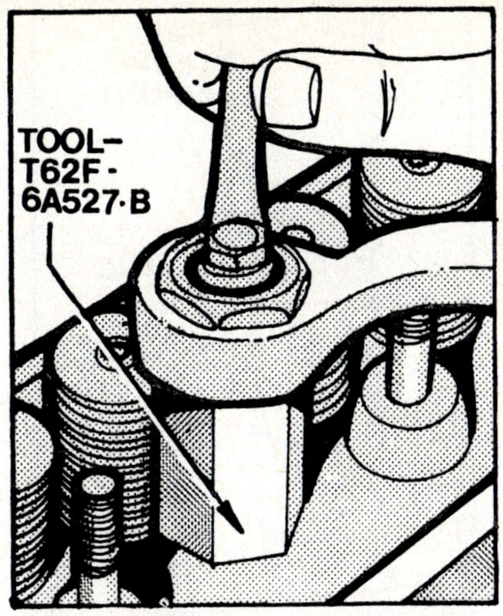

Fig. 4 Rocker arm stud removal. 6-240, 300, 1974–77 & Early 1978 V8-302, 1975–77 & Early 1978 V8-351W

V8-255, 302, 460

No. 5 Intake	No. 2 Exhaust
No. 4 Intake	No. 6 Exhaust

V8-351 & 400

No. 3 Intake	No. 2 Exhaust
No. 7 Intake	No. 6 Exhaust

4. Turn crankshaft to number 3 location, then compress valve lifter using tool T70P-6513-A or equivalent, Fig. 3D, and check the clearance on the following valves:

V8-255, 302, 460

No. 2 Intake	No. 7 Exhaust
No. 3 Intake	No. 3 Exhaust
No. 6 Intake	No. 8 Exhaust

V8-351 & 400

No. 2 Intake	No. 4 Exhaust
No. 5 Intake	No. 5 Exhaust
No. 6 Intake	No. 8 Exhaust

All V8-330, 360, 361, 389, 390, 391 & 1975 V8-351W

1. Turn the engine until number 1 piston is at TDC and at the end of its compression stroke and check the clearance on the following valves after compressing the valve lifter using tool T58P-6565-A, Fig. 3D, or equivalent:

No. 1 Intake	No. 1 Exhaust
No. 3 Intake	No. 4 Exhaust
No. 7 Intake	No. 5 Exhaust
No. 8 Intake	No. 8 Exhaust

2. Turn the crankshaft one complete turn (360°) to position number 6 cylinder at TDC and at the end of its compression stroke and check the clearance on the following valves after compressing valve lifter using tool T58P-6565-A, Fig. 3D, or equivalent:

No. 2 Intake	No. 2 Exhaust
No. 4 Intake	No. 3 Exhaust
No. 5 Intake	No. 6 Exhaust
No. 6 Intake	No. 7 Exhaust

1979–83 V8-370 & 429

1. Rotate engine until number 1 piston is at TDC and at the end of the compression stroke. Check clearance of the following valves after compressing valve lifter, using tool T58P-6565-A or equivalent, Fig. 3D:

No. 1 Intake	No. 1 Exhaust
No. 3 Intake	No. 4 & No. 5 Exhaust
No. 7 & No. 8 Intake	No. 8 Exhaust

2. Rotate crankshaft one revolution (360°) clockwise from No. 1 position and check the following valves:

No. 2 Intake	No. 2 Exhaust
No. 4 & No. 5 Intake	No. 3 Exhaust
No. 6 Intake	No. 6 Exhaust
	No. 7 Exhaust

V8's With Mechanical Lifters

Make three chalk marks on the vibration damper. Space the marks 90 degrees apart (1/4 of circumference) so that with the timing mark, the damper is divided into four equal parts. Rotate the crankshaft until No. 1 piston is near TDC at the end of the compression stroke and adjust the following valves:

No. 1 intake and exhaust
No. 2 intake
No. 4 exhaust
No. 5 exhaust
No. 7 intake

Rotate the crankshaft 180 degrees (1/2 turn) which puts No. 4 piston on TDC. Then adjust the following valves:

No. 4 intake
No. 5 intake
No. 6 exhaust
No. 8 exhaust

Rotate the crankshaft 270 degrees (3/4 turn), which puts No. 3 piston on TDC. Then adjust the following valves:

No. 2 exhaust
No. 3 intake and exhaust
No. 6 intake
No. 7 exhaust
No. 8 intake

VALVE GUIDES

Valve guides in these engines are an integral part of the head and, therefore, cannot be removed. For service, guides can be reamed oversize to accommodate one of three service valves with oversize stems (.003", .015" and .030").

Check the valve stem clearance of each valve (after cleaning) in its respective valve guide. If the clearance exceeds the service limits of .004" of the intake or .005" on the exhaust, ream the valve guides to accommodate the next oversize diameter valve.

ROCKER ARM STUD

6-240, 300, & V8-302, 351W Exc. Late 1978 & 1979–83 V8-302 & 351W

If necessary to replace a rocker arm stud, a rocker arm stud kit is available and contains a stud remover, a stud installer, and two reamers, one .006" and the other .015".

Rocker arm studs that are broken or have damaged threads may be replaced with standard studs. Loose studs in the head may be

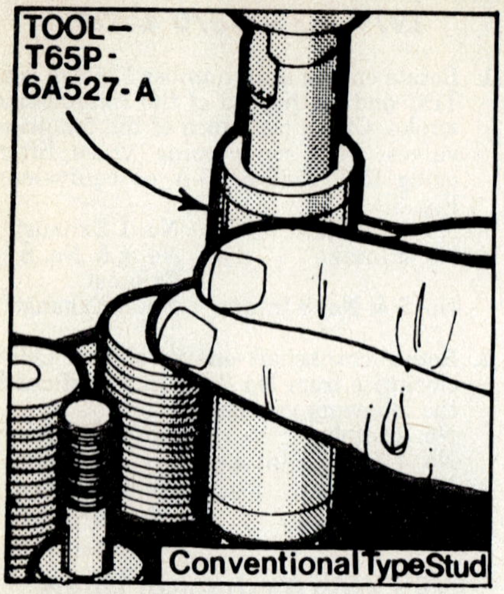

Fig. 5 Rocker arm stud installation. 6-240 300

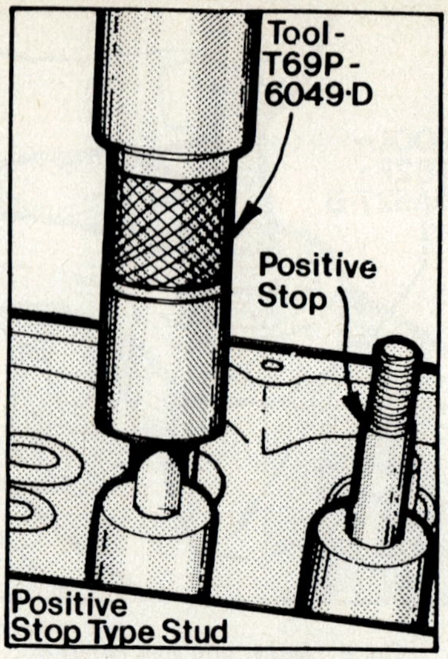

Fig. 6 Rocker arm stud installation. 1974–77 & Early 1978 V8-302, 1975–77 & Early 1978 V8-351W

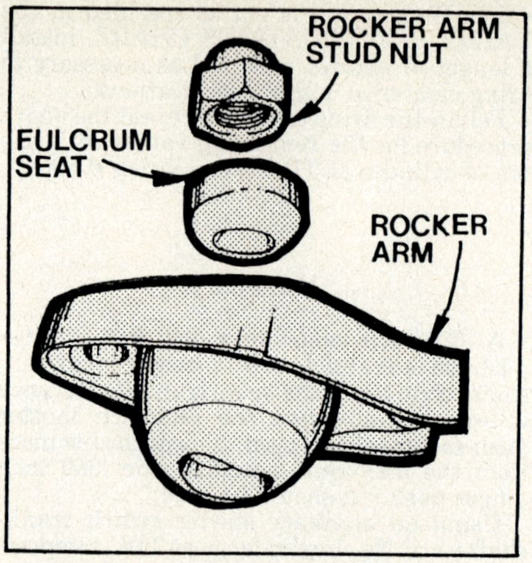

Fig. 7 Rocker arm assembly. 6-240, 300, 1974–77 & Early 1978 V8-302, 1975–77 & Early 1978 V8-351W

replaced with .006", .010" or .015" oversize studs which are available for service.

When going from a standard size stud to a .015" oversize stud, always use a .006" reamer and a .010" reamer before finish reaming with a .015" reamer.

If a stud is broken off flush with the stud boss, use an easy-out to remove the broken stud, following the instructions of the tool manufacturer.

Remove stud with tool T62F-6A527-B, Fig. 4. On 6-240-300, install new stud with tool T65P-6A527-A, Fig. 5. On 1974–77 and early 1978 V8-302 and V8-351W, install new stud with tool T69P-6049-D, Fig. 6.

ROCKER ARM, REPLACE

6-240, 300 & V8-302, 351W Exc. Late 1978 & 1979–83 V8-302 & 351W

These engines use a rocker arm stud and nut. To disassemble, remove rocker arm nut, fulcrum seat and rocker arm, Fig. 7. Inspect condition of nut, Fig. 7A, and replace if necessary.

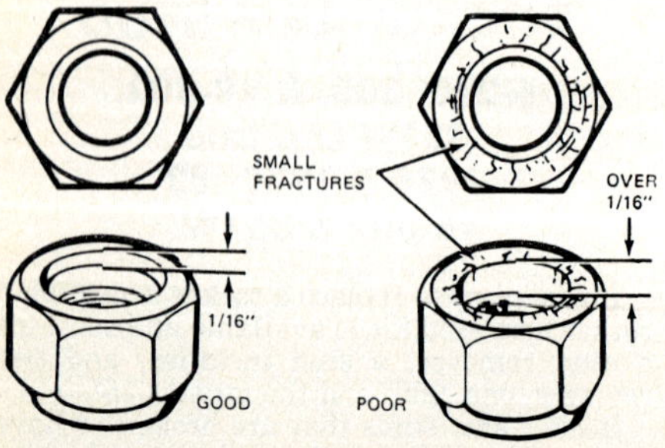

Fig. 7A Inspection of rocker arm stud nut.

ROCKER ARM & FULCRUM BOLT

V6-232, V8-255, Late 1978 & 1979–83 V8-302 & 351W, 1979–83 V8-429 & All V8-351C, 370, 400 & 460

The rocker arm is supported by a fulcrum bolt which fits through the fulcrum seat and threads into the cylinder head. To disassemble, remove the bolt, oil deflector, fulcrum seat and rocker arm, Figs. 8 and 8A.

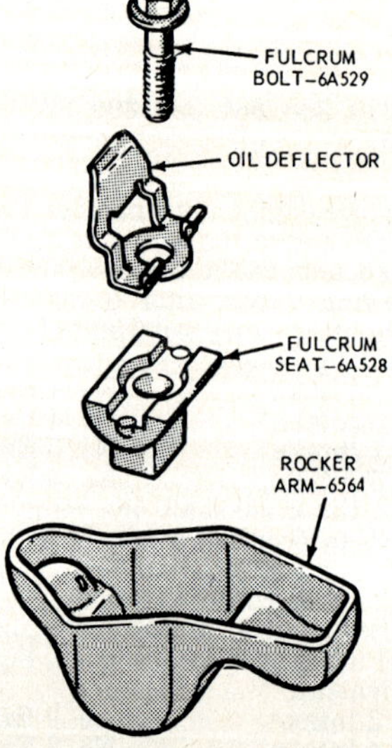

Fig. 8 Rocker arm assembly. V8-351C, 351M, 370, 400, 460, 1979–83 429

ROCKER ARMS & SHAFTS

To disassemble the rocker arms, remove cotter pins from each end of the shaft and remove the flat washers and spring washers. Slide rocker arms, springs and supports off shaft, being sure to identify all parts so they can be assembled in the same position.

If it is necessary to remove the plugs from each end of the shaft, drill or pierce one plug, then insert a steel rod through the plug and knock out the plug on the opposite end. Working from the open end, knock out the remaining plug.

Assemble the rocker arms and related parts as indicated by Figs. 9 to 11.

HYDRAULIC VALVE LIFTERS

Hydraulic Lifters, Fig. 12 Replace

6-200 & 250

1. Remove cylinder head assembly.

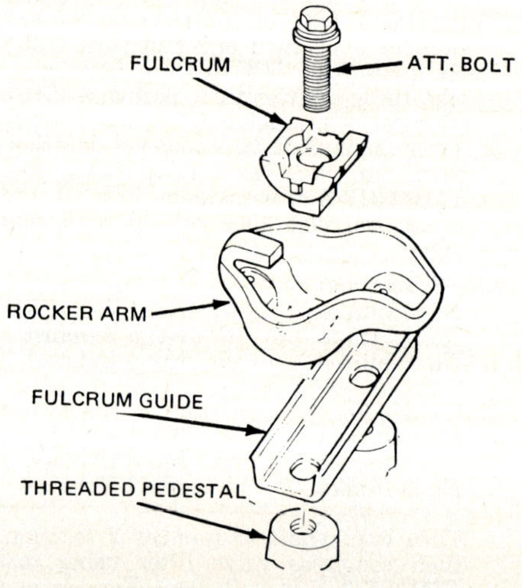

Fig. 8A Rocker arm assembly. Late 1978 & 1979–83 V8-255, 302 & 351W

2. Using a magnet, remove and install one lifter at a time.

NOTE: If lifters are stuck in their bores by excessive varnish or gum, it may be necessary to use a plier-type tool to remove them. Rotate the lifter back and forth to loosen the gum and varnish which may have formed on the lifter.

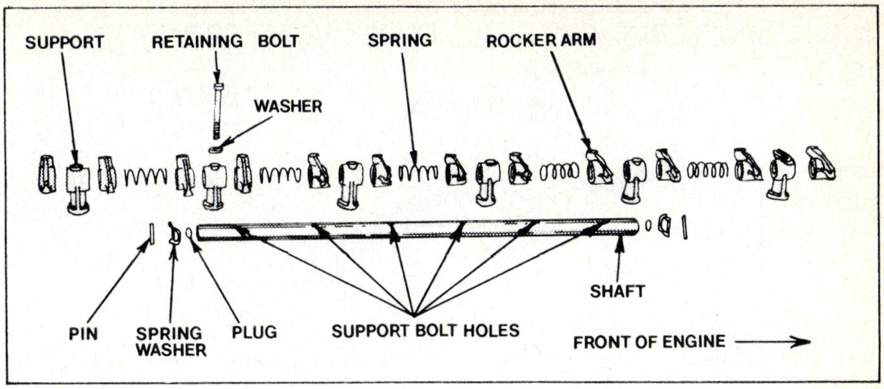

Fig. 9 Rocker arm shaft assembly. 6-200, 250

6-240, 300

1. Remove valve rocker arm cover.
2. Remove valve push rod cover.
3. Loosen rocker arm stud nuts until rocker arms can be disengaged from push rods.
4. Remove push rods, keeping them in a rack so they may be installed in their original location.
5. Remove valve lifters, using a magnet rod. Place lifters in a rack so they may be installed in their original location.

ALL V8's WITH ROCKER ARM SHAFTS

1. Remove intake manifold.
2. Remove rocker arms and shafts.
3. Remove push rods, keeping them in a rack in sequence so they may be installed in their original location.
4. Remove valve lifters with a magnet rod and place them in a rack in sequence so they may be installed in original location.
5. Reverse procedure to install.

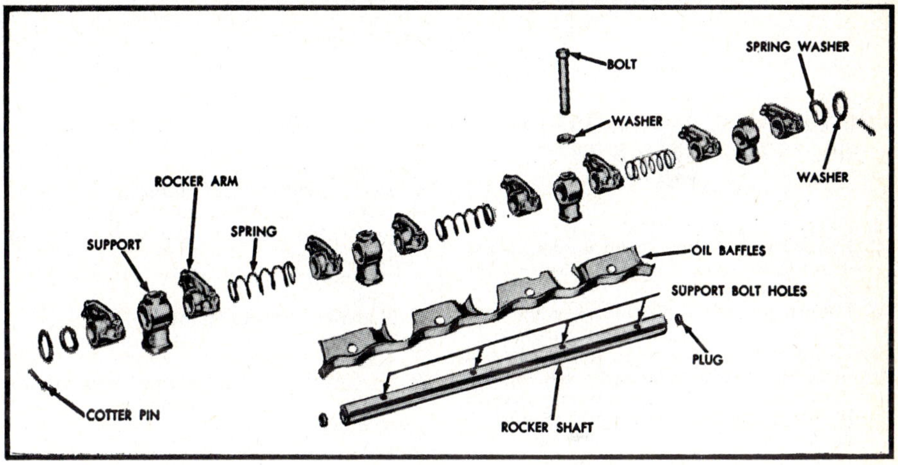

Fig. 10 Layout of rocker arm parts. V8-330, 360, 361, 389, 391

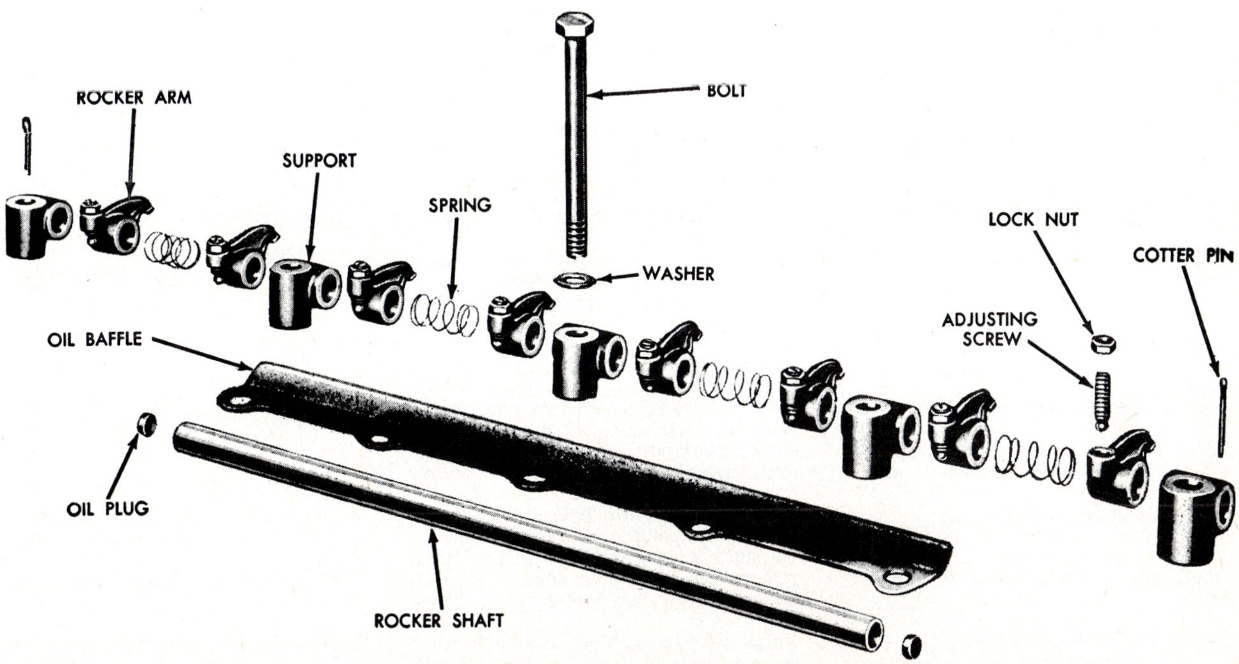

Fig. 11 Layout of rocker arm parts. V8-401, 477, 534 engines

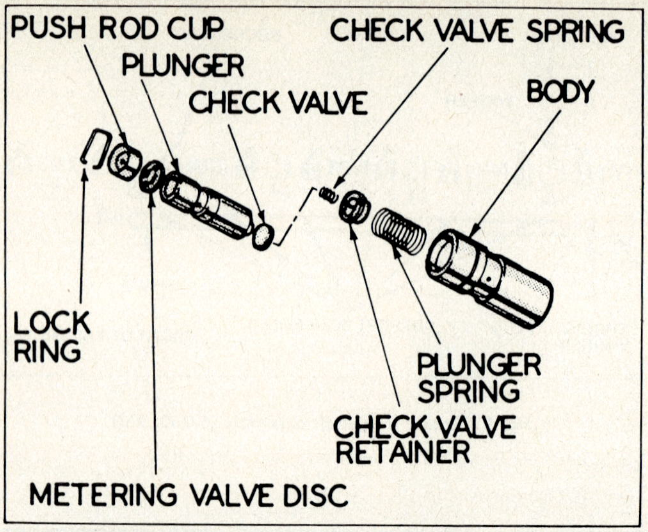

Fig. 12 Hydraulic valve lifter

V6-232, V8-255, 302, 351, 370, 400, 1979–83 429 & 1979–83 460

1. Remove intake manifold.
2. Remove rocker arm covers. Loosen rocker arm stud nuts or bolts and rotate rocker arms to the side.
3. Remove push rods in sequence so they can be installed in their original bores.
4. Using a magnet, remove the lifters and place them in a numbered rack so they can be installed in their original bores. If lifters are stuck in their bores, it may be necessary to use a plier-type tool to remove them.
5. The internal parts of each lifter are matched sets. Do not intermix parts.

TIMING CASE COVER

V8-255, 302, 351 & 400

1. Remove air cleaner and intake duct assembly.
2. Drain cooling system and crankcase.
3. Remove radiator, fan, pulley and drive belts.
4. Disconnect fuel pump outlet line from pump. Remove fuel pump retaining bolts and lay the pump to one side.
5. Remove crankshaft pulley and vibration damper.
6. Remove oil pan to cylinder front cover retaining bolts.
7. Remove cylinder front cover and water pump as an assembly.

1974–75 6-240, 6-300

1. Drain cooling system and crankcase.
2. On N Series, remove air cleaner and grille.
3. Remove shroud and radiator.
4. Remove fan, drive belts and pulley. It may be necessary to remove air compressor belt and/or power steering pump belt.
5. Remove vibration damper.
6. Remove oil dipstick and oil pan.
7. Remove oil pump screen and inlet tube.
8. Remove timing gear cover.

1976–83 6-300

1. Drain cooling system.
2. Remove shroud and radiator.
3. Remove alternator adjusting arm bolt, loosen drive belt and swing the adjusting arm aside.
4. Remove fan, drive belts, spacer and pulleys.
5. Remove vibration damper.
6. Remove the front oil pan and front cover attaching screw.
7. Remove timing case cover and discard gasket. Replace crankshaft oil seal with new one.
8. Reverse procedure to install.

V6-232

1. Disconnect battery ground cable, then drain cooling system.
2. Remove air cleaner and air intake duct.
3. Remove fan shroud attaching screws and fan and fan clutch attaching bolts, then remove fan and clutch assembly and shroud.
4. Loosen accessory drive belt idler, then remove drive belt and water pump pulley.
5. On models equipped with power steering, remove pump bracket attaching bolts, then position pump aside with hoses attached.
6. On models equipped with A/C, remove compressor front support bracket.
7. Disconnect coolant bypass hose and heater hose at water pump and upper radiator hose at thermostat housing.
8. Disconnect ignition coil secondary wire from distributor cap, then remove distributor cap with ignition wires attached.
9. With No. 1 cylinder at TDC compression stroke, mark position of rotor to distributor housing and position of distributor housing to front cover.
10. Remove distributor hold down clamp, then lift distributor from front cover.
11. On models equipped with Tripminder, remove fuel flow meter support bracket. Do not remove flow meter or disconnect fuel lines.
12. Raise and support front of vehicle, then remove crankshaft pulley using a suitable puller.

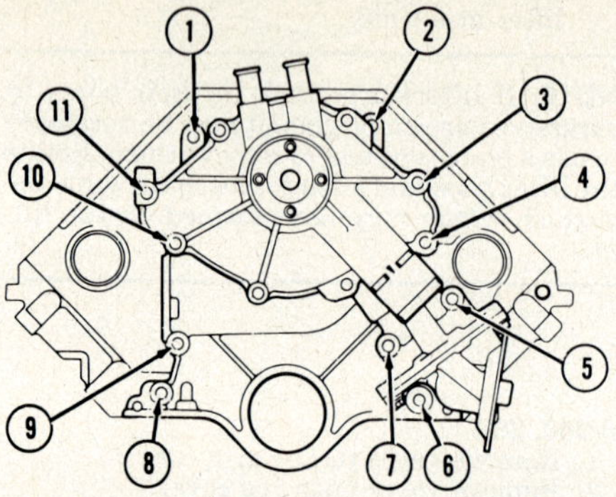

Fig. 12A Front cover attaching bolt location. V6-232

13. Remove fuel pump shield, then disconnect fuel pump to carburetor fuel line at fuel pump.
14. Remove fuel pump attaching bolts and position aside with fuel hose attached.
15. Remove oil filter, then disconnect lower radiator hose from water pump.
16. Remove oil pan as described under Oil Pan, Replace.
17. Lower vehicle and remove front cover attaching bolts, Fig. 12A.

NOTE: One of the front cover attaching bolts is located behind the oil filter adapter. Also tag bolts as they are removed so that they can be installed at the same location.

18. Remove ignition timing indicator, then remove front cover and water pump as an assembly.
19. Remove camshaft thrust button and spring from camshaft.
20. Reverse procedure to install. Before installing front cover attaching bolts at location 10, Fig. 12A, coat threads of bolt with pipe sealant D8AZ-19558-A or equivalent. Also lubricate camshaft thrust button with polyethylene grease before installing. Torque front cover attaching bolts to 15 to 22 ft. lbs.

NOTE: If a replacement front cover is to be installed, the water pump, oil pump, oil filter adapter and intermediate shaft must be removed from the front cover to be replaced and reinstalled on the replacement front cover. Also it may be necessary to rotate crankshaft 180° from the No. 1 cylinder TDC location to position fuel pump eccentric for fuel pump installation. When installing distributor, No. 1 cylinder must be at TDC position and marks made during removal must be aligned.

1974–78 V8-330, 360, 361, 390, 391

1. Drain cooling system and crankcase.
2. Remove air cleaner.
3. Disconnect battery ground cable.
4. Disconnect distributor vacuum line.
5. Disconnect necessary water hose.
6. If equipped, disconnect transmission oil cooler lines.
7. Remove radiator and support. If equipped with automatic radiator shutter, leave it attached to radiator.

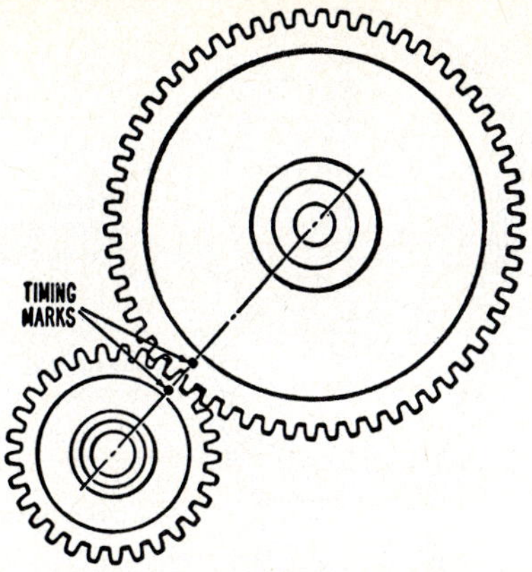

Fig. 13 Valve timing. 6-240, 300

8. On high fan installation, remove fan and drive belt.
9. Remove power steering pump and position it to one side, leaving hoses attached.
10. If equipped, remove air compressor.
11. Remove alternator.
12. Remove water pump and fan.
13. On C Series, remove fan from vibration damper.
14. Remove vibration damper.
15. Remove fuel pump and lay it aside with flexible fuel line attached.
16. Remove crankshaft sleeve.
17. Remove front cover, alternator support bracket and adjusting arm bracket, and engine front support bracket.

1974–80 V8-401, 475, 477, 534

1. Drain cooling system and crankcase.
2. Remove fan, shroud, radiator and shutter.
3. Remove alternator, power steering pump and air compressor drive belts.
4. On F, T, L & LT Series, remove front engine support bolts, then lift front of engine about one inch with a floor jack to allow removal of vibration damper.
5. Remove vibration damper.
6. Remove air compressor idler pulley and disconnect alternator adjusting arm at front cover.
7. Remove water pump.
8. Remove engine front support bracket and oil level dipstick.
9. Remove timing cover attaching bolts, then using a thin bladed knife between cover and cylinder block; cut oil pan gasket flush with cylinder block and remove timing gear cover.

V8-460

1. Drain cooling system and crankcase.
2. Remove radiator fan and shroud, then disconnect upper and lower hoses and oil cooler lines from radiator. Remove radiator upper support and radiator.
3. Loosen alternator attaching bolts and the air conditioning idler pulley, then remove drive belts and water pump pulley.
4. Remove bolts attaching compressor support, water pump and compressor and remove compressor support (if used).
5. Remove crankshaft pulley, then using a puller, remove vibration damper and woodruff key.
6. Loosen by-pass hose from water pump and disconnect heater return hose from water pump.

Fig. 14 Valve timing. Typical of V8 engines with timing gears

7. Remove fuel pump and plug fuel lines to prevent entry of dirt.
8. Remove cylinder front cover bolts, then using a thin blade knife, cut oil pan seal between cover and block and remove cover.

V8-370, 429

1. Drain cooling system and crankcase.
2. Remove air cleaner.
3. Disconnect battery ground cable.
4. Disconnect distributor vacuum line.
5. Disconnect necessary water hose.
6. If equipped, disconnect transmission oil cooler lines.
7. Remove radiator and support. If equipped with automatic radiator shutter, leave it attached to radiator.
8. On high fan installation, remove fan and drive belt.
9. Remove power steering pump and position it to one side, leaving hoses attached.
10. If equipped, remove air compressor.
11. Remove alternator.
12. Remove water pump and fan.
13. On C Series, remove fan from vibration damper.
14. Remove vibration damper.
15. Remove fuel pump and lay it aside with flexible fuel line attached.
16. Remove front cover and front mount to engine and oil pan attaching bolts. Raise engine slightly, then using a thin bladed knife, cut oil pan gasket flush with cylinder block to oil pan junction before removing the cover and front mount from block.

6-200, 250

1. Remove any front sheet metal that would interfere with the removal of the front cover.
2. Drain the cooling system and the crankcase.
3. Disconnect the radiator upper hose at the outlet housing and the lower hose at the water pump.
4. Remove the radiator, drive belt, fan and pulley.
5. If equipped with thermactor system, remove the air pump drive belt.

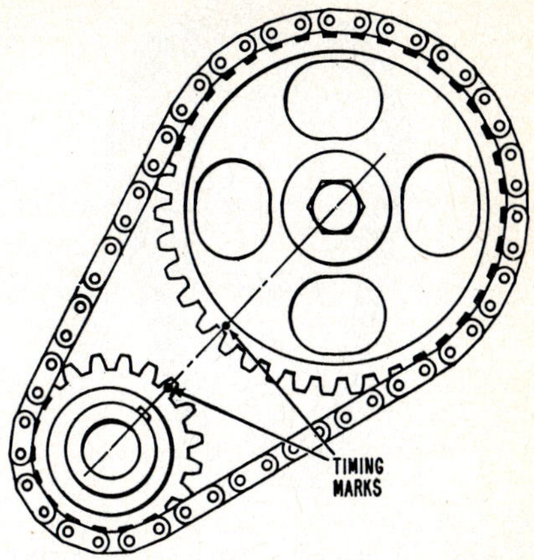

Fig. 15 Valve timing marks. 6-170, 200, 250

6. If so equipped, remove the accessory drive pulley.
7. Remove the vibration damper.
8. Remove the front cover and the oil pan.

TIMING GEARS

6-240, 6-300

CAUTION: When the crankshaft and camshaft lose their timing relationship through removal of timing gears, interference may occur between crankshaft and cam lobes. Therefore, to prevent possible damage to camshaft lobes, do not rotate crankshaft or camshaft without timing gears installed.

Camshaft Metal Gear

1. With front cover removed, crank engine until timing gear marks are aligned as shown in Fig. 13.
2. Install gear puller and remove gear.
3. Be sure key and spacer are properly installed. Align keyway with key and install gear on camshaft.
4. Check backlash between crank and cam gear, using a dial indicator. Hold gear firmly against block while making check. Backlash should be between .002" and .004".

Camshaft Fiber Gear

1. With timing gear cover removed, remove crankshaft oil slinger and camshaft.
2. Press gear off camshaft and remove thrust plate spacer and key.
3. Install camshaft in engine and install spacer and thrust plate on camshaft.
4. Insert key in keyway.
5. Position gear on camshaft with timing marks aligned, Fig. 13. Be sure gear and spacer are tight against shoulder on camshaft and that thrust plate can be moved freely. Install thrust plate screws.
6. Check gear backlash as outlined above; it should be .002–.004".

Crankshaft Gear

1. After removing gear cover, remove oil slinger from crankshaft.
2. Use a suitable puller to remove gear.
3. Install key in crankshaft keyway.
4. Install crankshaft gear and oil slinger.

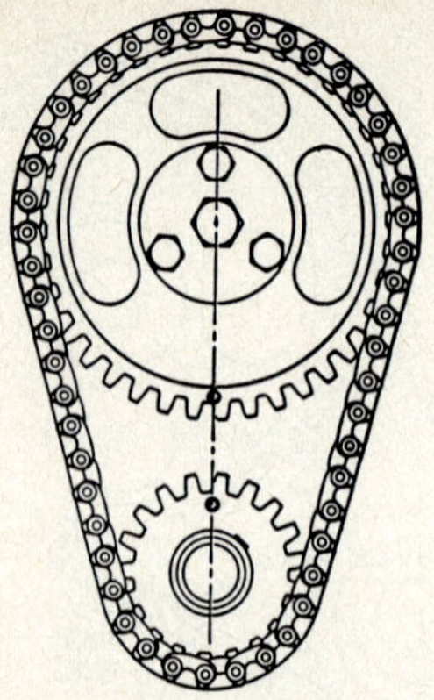

Fig. 16 Valve timing. Typical of V8 engines with timing chain

All V8-401, 477, 534

The camshaft gear is a slip-on fit and is retained on the end of the camshaft by a lock plate and three cap screws.

To remove, drain cooling system and crankcase. Remove oil pan, radiator, water pump and front cover. Crank engine until timing marks are aligned as shown in Fig. 14. Remove cap screws and lock plate and take off camshaft gear.

When installing the gear, be sure to have the timing marks lined up as shown.

TIMING CHAIN

6-200, 250, V6-232, V8-255, 302, 351, 400

After removing the cover as outlined above, remove the crankshaft front oil slinger. Crank the engine until the timing marks are aligned as shown in Figs. 15, 16 and 16A. Remove camshaft sprocket retaining bolt, washer and fuel pump eccentric. Slide sprockets and chain forward and remove them as an assembly. Reverse the order of the foregoing procedure to install the chain and sprockets, being sure the timing marks are properly aligned.

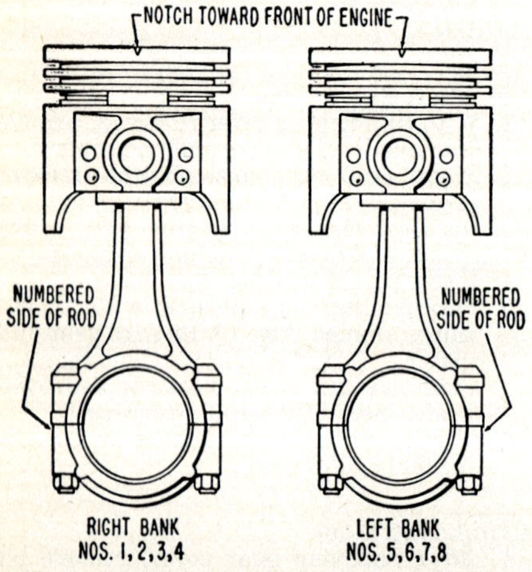

Fig. 17 Piston and rod assembly. V8-390 Light Duty & V8-255, 302, 351, 400, 460 & 1979–80 370 & 429

V8-330, 360, 361, 370, 389, 390, 391, 429, 460

To install the chain, align the slotted camshaft spacer with the dowel hole in the camshaft. Position the sprockets and chain on the camshaft and crankshaft. Be sure the timing marks on sprockets are positioned as shown in Fig. 16. Install fuel pump eccentric and camshaft sprocket cap screw.

CAMSHAFT, REPLACE

Six Cylinder

NOTE: On some Econoline models, it is necessary to remove the engine and transmission from the vehicle. Ensure that adequate clearance exists for camshaft removal with the engine in the vehicle.

1. Disconnect battery ground cable.
2. Drain cooling system and oil pan.
3. Remove radiator, valve lifters & front cover. On 1974 vehicles, remove oil pan and oil pump.
4. Disconnect outlet lines at fuel pump and position fuel pump aside.
5. Remove distributor.
6. Rotate crankshaft to align timing marks, Figs. 13 and 15.
7. Remove camshaft thrust plate screws and the camshaft. Use caution not to damage the camshaft lobes when removing camshaft.
8. Reverse procedure to install.

V6-232

1. Remove radiator.
2. If equipped with air conditioning, remove condensor.
3. Remove grille, intake manifold and tappets, then remove front cover and timing chain.
4. Remove oil pan and remove camshaft through front of engine, being careful not to damage bearing surfaces.
5. Reverse procedure to install.

All V8's Exc. V8-401, 477 & 534

1. On V8-370, 1979–83 V8-429 and Econoline models with V8-255, 302, 351, and 460, remove grille.
2. Drain cooling system, disconnect all hoses and lines from radiator and remove radiator.
3. Remove rocker arm covers, loosen rocker arms and remove pushrods in sequence.
4. Remove intake manifold and remove valve lifters in sequence.
5. Remove cylinder front cover and timing chain as described previously.
6. Remove camshaft thrust plate and carefully remove camshaft by pulling toward front of engine.

NOTE: Use caution to avoid damaging the camshaft bearings.

V8-401, 477 & 534

1. Disconnect battery ground cable.
2. Remove intake manifold and front cover.

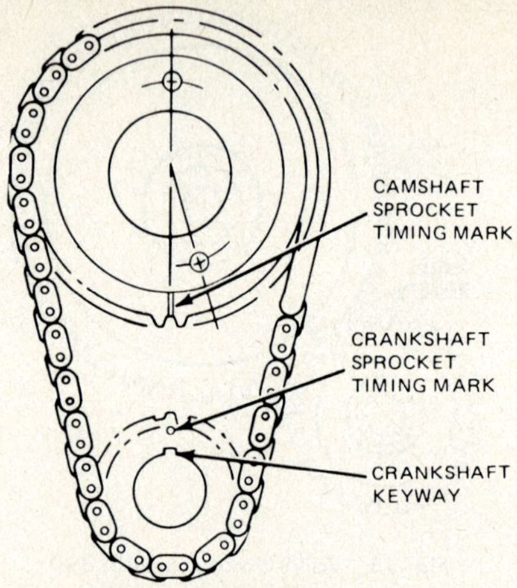

Fig. 16A Timing marks aligned for correct valve timing. 1982 V6-232

3. Rotate crankshaft to follow firing order and loosen valve adjusting screws. Then, slide rocker arm to one side and remove push rods.
4. Remove distributor, governor and push rod cover.
5. Remove valve lifters.
6. Remove crankshaft front oil slinger.
7. Align timing marks, Fig. 14.
8. Remove camshaft thrust plate attaching bolts, then the gear, spacer, thrust plate and camshaft as an assembly.
9. Reverse procedure to install.

PISTON & ROD, ASSEMBLE

Lubricate all parts with light engine oil. Position the connecting rod in the piston and push the pin into place, Figs. 17 to 20B. Insert new piston pin retainers (when used) by spiralling them into the piston with the fingers. Do not use pliers.

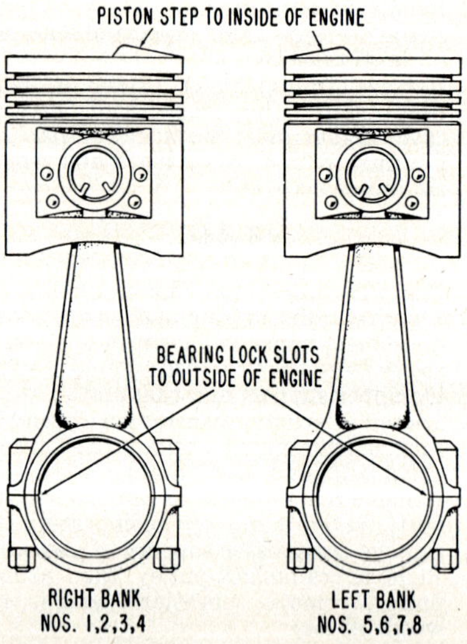

Fig. 18 Assembly of pistons and rods. V8-401, 477, 534

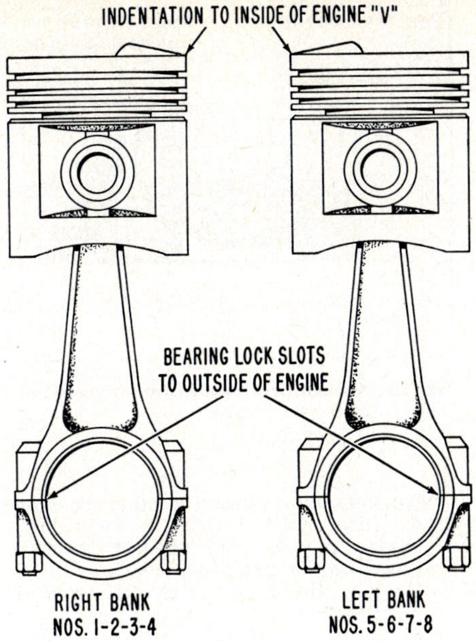

Fig. 19 Piston and rod assembly V8-330, 360, 361, 389, 390 Heavy Duty, 391

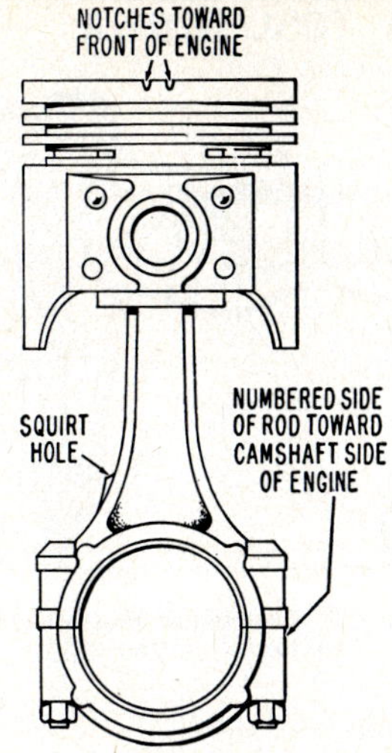

Fig. 20 Piston and rod assembly. 6-200, 250, 240, 300

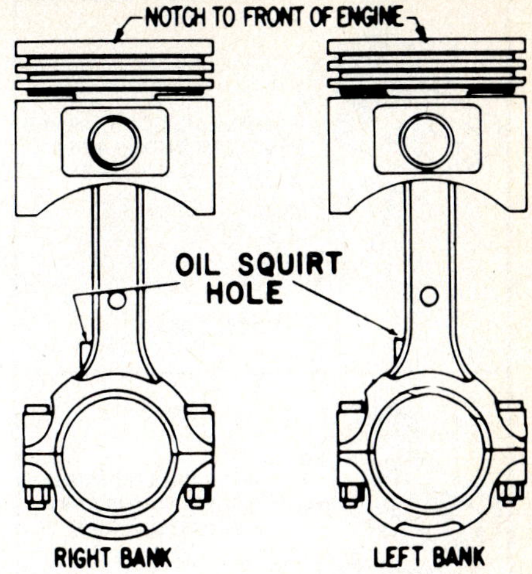

Fig 20A Piston & assembly. V6-232

CRANKSHAFT REAR OIL SEAL

A new rubber split-lip rear crankshaft oil seal is released for service. This seal can be installed without removal of the crankshaft and also eliminates the necessity of seal installation tools.

1. Remove oil pan.
2. Remove rear main bearing cap.
3. Loosen remaining bearing caps, allowing crankshaft to drop down about 1/32".
4. Remove old seals from both cylinder block and rear main bearing cap. Use a brass rod to drift upper half of seal from cylinder block groove. Rotate crankshaft while drifting to facilitate removal.
5. Carefully clean seal groove in block with a brush and solvent. Also clean seal groove in bearing cap. Remove the oil seal retaining pin from the bearing cap if so equipped. *The pin is not used with the split-lip seal.*
6. Dip seal halves in clean engine oil.
7. Carefully install upper seal half in its groove with undercut side of seal toward front of engine, Fig. 21, by rotating it on shaft journal of crankshaft until approximately 3/8" protrudes below the parting surface. *Be sure no rubber has been shaved from outside diameter of seal by bottom edge of groove.*
8. Retighten main bearing caps and torque to specifications.
9. Install lower seal in main bearing cap with undercut side of seal toward front of engine, and allow seal to protrude about 3/8" above parting surface to mate with upper seal upon cap installation.
10. Apply suitable sealer to parting faces of cap and block, Fig. 22. Install cap and torque to specifications.

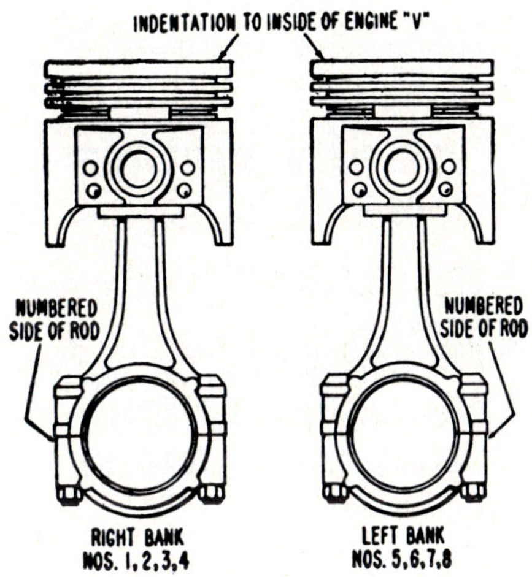

Fig. 20B Piston and rod assembly. 1981–83 V8-370 & 429

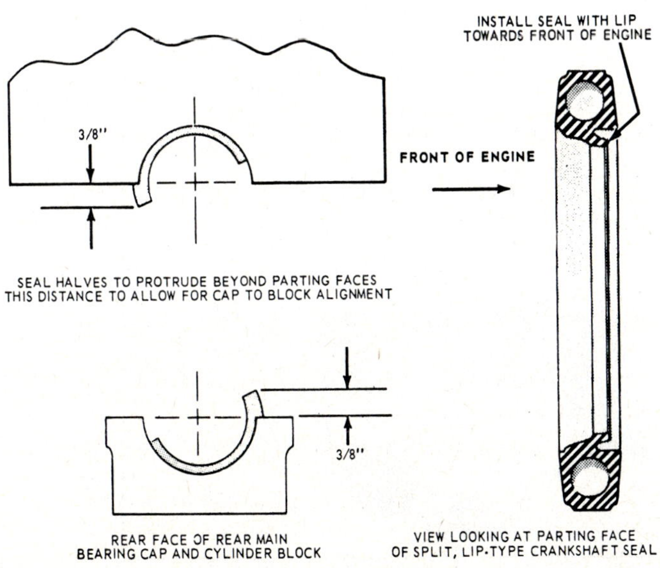

Fig. 21 Crankshaft rear oil seal installation. All exc. 6-240, 300

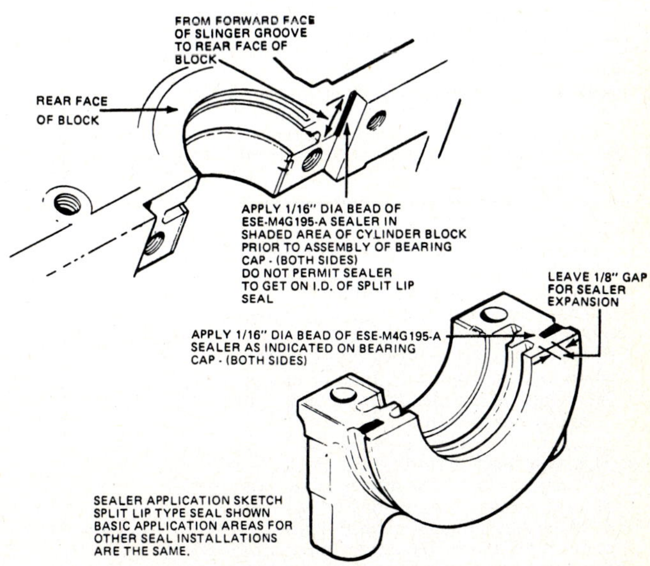

Fig. 22 Applying sealer to crankshaft rear oil seal. All exc. 6-240 & 300

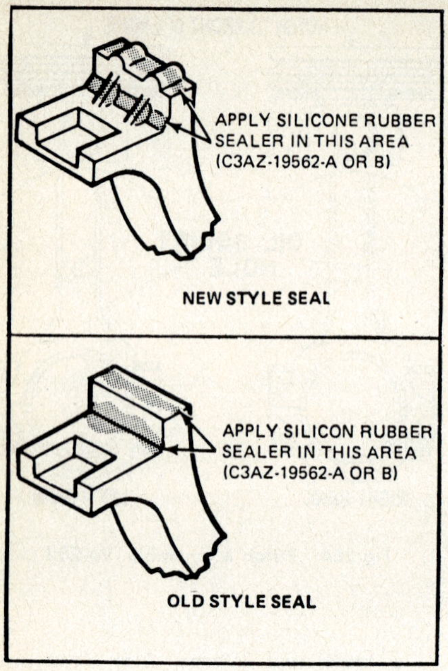

Fig. 23 Revised crankshaft rear oil seal

NOTE: If difficulty is encountered in installing the upper half of the seal in position, lightly lap (sandpaper the side of the seal opposite the lip side using a medium grit paper). After sanding, the seal must be washed in solvent, then dipped in clean engine oil prior to installation.

SERVICE BULLETIN

A revised crankshaft rear oil seal has been released for service. This new seal may be received when ordering an oil pan gasket kit and is installed in the same manner as described previously, Fig. 23.

6-240, 300

NOTE: If crankshaft rear oil seal replacement is the only operation being performed, it can be done in the vehicle. If the oil seal is being replaced in conjunction with a rear main bearing replacement, the engine must be removed from the vehicle. To replace the seal only, proceed as follows:

1. Remove starting motor.
2. Disconnect transmission from engine and slide it back. On manual shift transmission, remove clutch assembly.
3. Remove flywheel and engine rear cover plate.
4. Use an awl to punch two holes in crankshaft rear oil seal. Punch holes on opposite sides of crankshaft and just above bearing cap-to-cylinder block split line. Insert a sheet metal screw in each hole.
5. Use two large screwdrivers or pry bars and pry against both screws at the same time to remove seal. It may be necessary to place small blocks of wood against cylinder block to provide a fulcrum point for pry bars. Use caution to avoid scratching or otherwise damaging crankshaft oil seal surfaces.

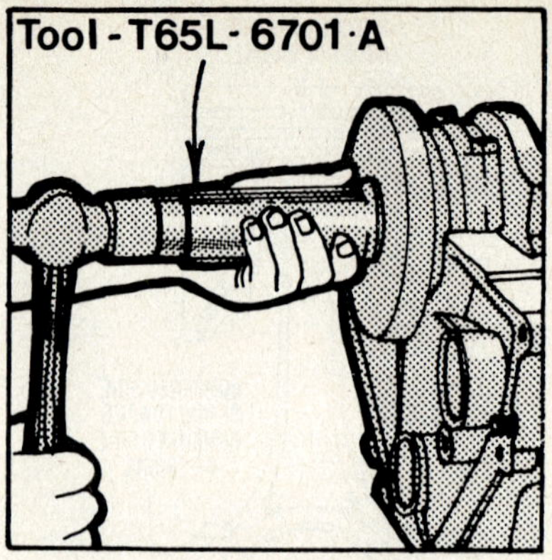

Fig. 24 Crankshaft rear oil seal installation. 6-240, 300

Installation

1. Clean oil seal recess in cylinder block and rear main bearing cap.
2. Coat new oil seal and crankshaft with a light film of engine oil.
3. Start seal in recess and install it until it is fully seated in seal recess, Fig. 24.
4. Be sure seal was not damaged during installation and reverse the procedure of removal to complete the operation.

OIL PAN, REPLACE

6-200 & 250 Exc. Econoline

1. Drain the crankcase.
2. Remove the dipstick and flywheel housing inspection cover.
3. Remove oil pan and gasket and the oil pump inlet tube and screen assembly.
4. Reverse the procedure to install.

6-240, 300 Exc. 1976–83 Econoline

1. Drain crankcase.
2. Drain cooling system and remove radiator and starter. Remove engine front mounting nuts and raise engine with a transmission jack and wood block. Place one-inch wood blocks between front engine mounts and support brackets. Lower engine and remove jack.
3. Remove oil pan and inlet tube retaining bolts. Remove inlet tube and screen from oil pump and let it lay in oil pan. Then remove oil pan.
4. Reverse procedure to install, Fig. 25.

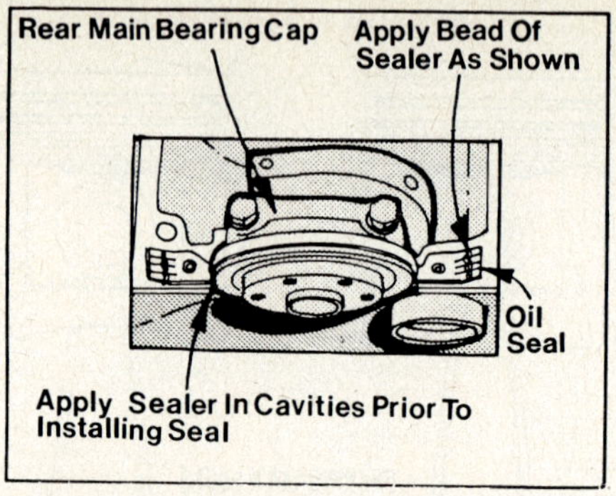

Fig. 25 Oil pan rear seal installation. 6-240, 300

V6-232

1. Disconnect ground cable from battery.
2. Remove air cleaner assembly.
3. Remove oil dipstick from engine.
4. Raise the front of vehicle.
5. Drain engine oil and replace drain plug. Remove the oil filter.
6. Disconnect muffler pipes from exhaust manifolds then disconnect pipes from the converter, remove pipes.
7. Disconnect transmission shift linkage at the transmission.
8. Disconnect transmission cooler lines at the radiator.
9. Remove nuts attaching the engine supports to the chassis brackets.
10. Raise engine as high as possible, and place wood blocks between the engine support and chassis brackets. Remove the jack.

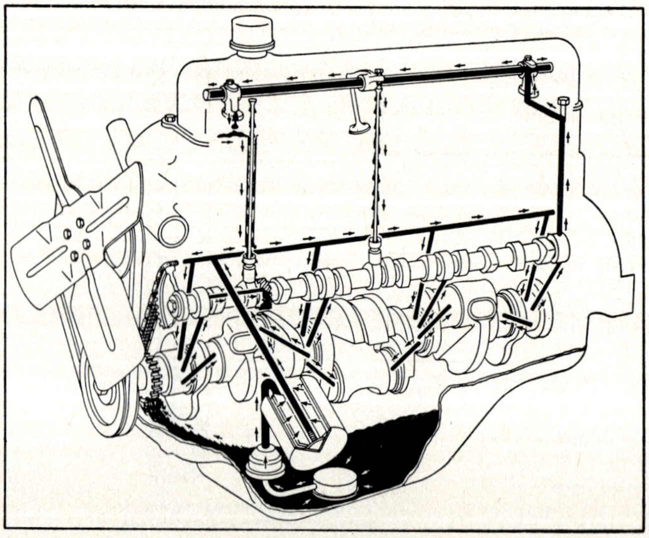

Engine oiling diagram. 6200, 250

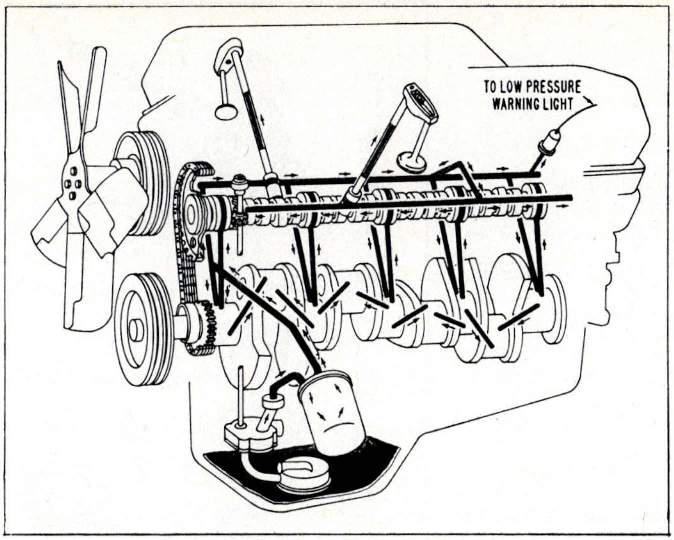

Engine oiling system. 6-240, 6-300

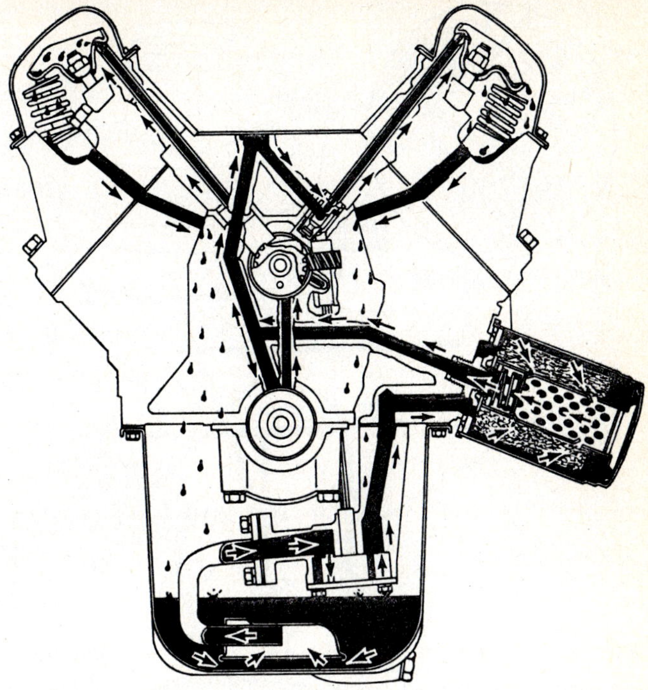

Engine oiling system. V8-302, 351, 400

11. Remove oil pan attaching bolts and oil pan.
12. Reverse procedure to install. Clean oil pan and sealing surface thoroughly and replace old gasket. Tighten attaching bolts to 15–22 ft. lbs.

1976–83 Econoline with 6-300

1. Remove engine cover, air cleaner and carburetor.
2. On vehicles equipped with air conditioning, discharge refrigerant system, then remove compressor.
3. Remove EGR valve if equipped, then on E250-HD and E350 models disconnect thermactor check valve inlet hose and remove check valve.
4. Remove upper radiator hose, fan shroud and automatic transmission filler tube.
5. Remove nuts securing exhaust pipe to manifold.
6. Raise and support vehicle, then disconnect and plug fuel pump inlet hose. Remove alternator heat shield from engine support nuts.
7. Remove power steering return line clip from crossmember, then disconnect lower radiator hose and transmission cooler lines.
8. Remove starter, then raise engine and place 3 inch blocks under engine mounts. Remove oil pan dipstick tube from pan.
9. Remove oil pan bolts, then the pick-up screen and tube from oil pump. Remove the pan.
10. Reverse procedure to install, Fig. 25.

1974 V8-302 Econoline

1. Remove oil dipstick attaching bolts and position it out of the way.
2. Drain crankcase and remove oil filter.
3. Disconnect steering rod at idler arm.
4. Remove nuts and washers retaining front engine supports to crossmember.
5. Position support jack under damper and raise engine as required.
6. Remove nuts retaining engine support crossmember to side rails and frame. Remove engine support crossmember.
7. Position engine support tools to side rails and front engine supports. Lower support jack.
8. Remove oil pan.

Bronco V8-302 & 351W, Exc. 1980–83

1. Remove air cleaner and duct.
2. Remove oil level dipstick and tube.
3. Remove oil pan attaching bolts and remove oil pan.

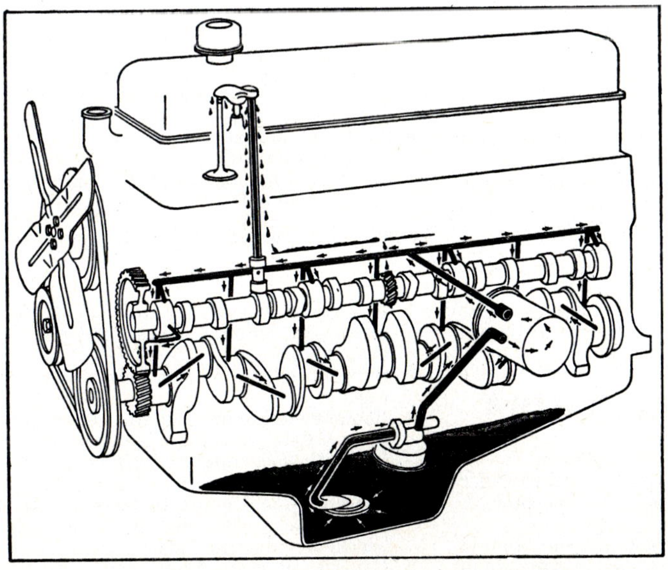

Engine lubrication V8-330, 332, 360, 361, 390, 391

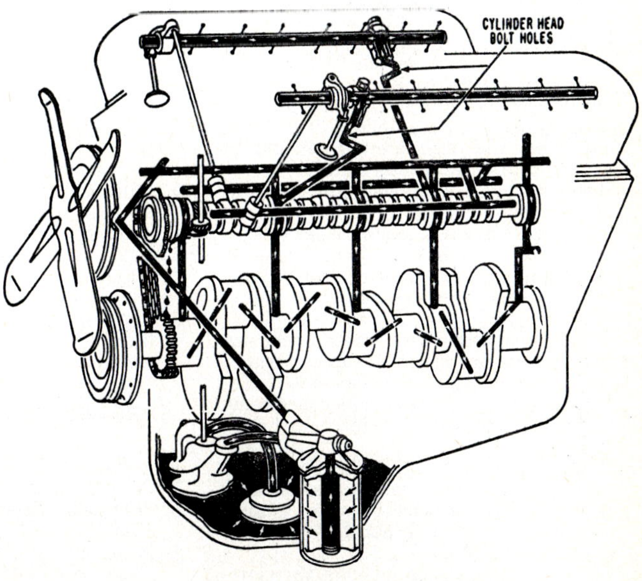

Engine oiling system. V8-429, 460

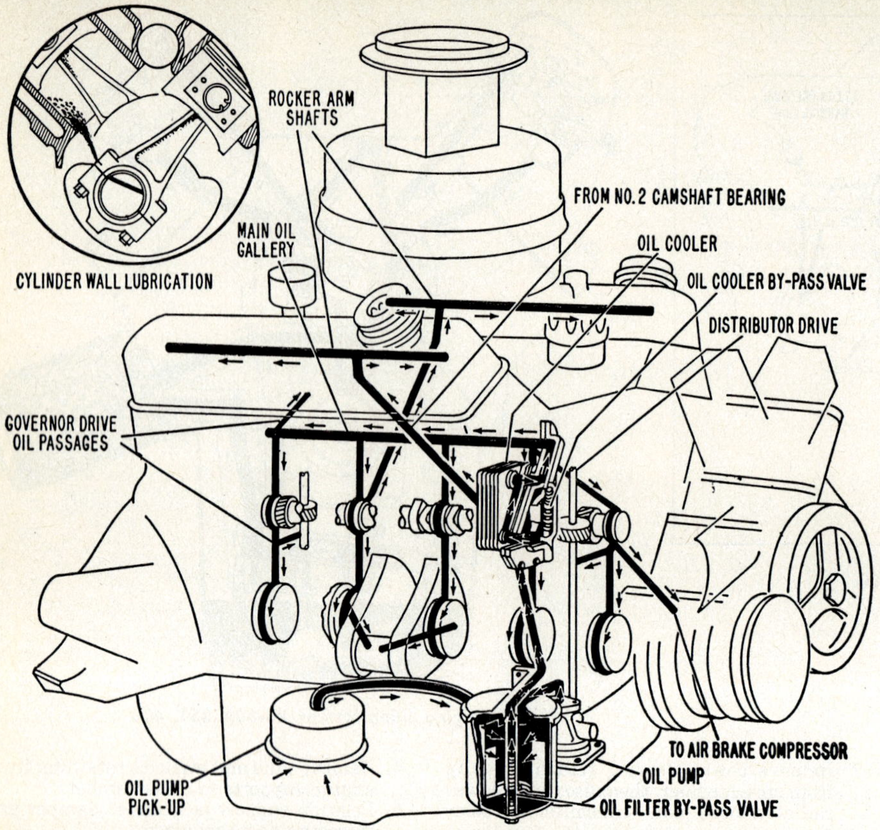

Engine lubrication. V8-401, 477, 534 engines

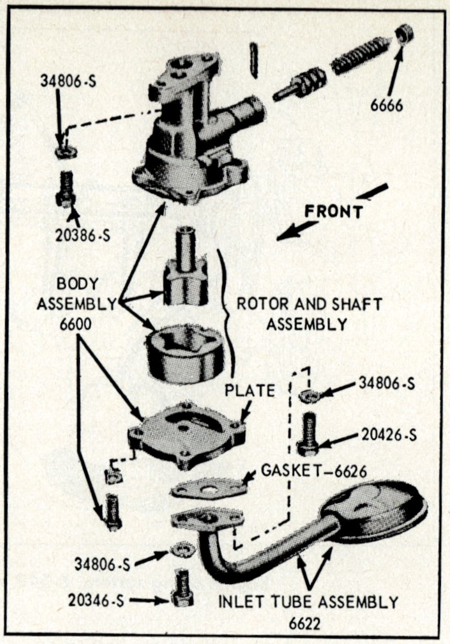

Fig. 26 Oil pump. 6-200 & 250

1975–83 Econoline V8-255, 302 & 351W

1. Disconnect battery ground cable.
2. Remove engine cover, air cleaner and drain cooling system.
3. If used, disconnect power steering pump and air conditioning pump from their mounts and position the units aside.
4. Remove fan shroud bolts and position shroud over fan, then remove oil filler tube and oil dipstick tube and disconnect upper and lower radiator hoses.
5. Raise and support vehicle and remove alternator splash shield.
6. Disconnect fuel line from pump and exhaust pipes from manifolds, then drain crankcase and remove engine mount to chassis bracket retaining nuts.
7. If equipped with automatic transmission, disconnect oil cooler lines from radiator and remove dipstick and oil filler tube.
8. Disconnect shift linkage from transmission, then remove center driveshaft support and remove driveshaft from transmission.
9. Remove rear crossmember retaining bolts and remove braces, then support transmission and remove crossmember. Remove mount from crossmember then reinstall crossmember and install bolts finger tight.
10. Remove right exhaust pipe heat deflector, then raise engine as far as possible and insert wooden blocks between engine mounts and chassis brackets to support engine.
11. Remove oil pan attaching bolts, then lower oil pan onto crossmember and remove oil pump pickup tube attaching bolts and lower the tube into the oil pan and remove the oil pan.

F100-F350, & 1980–83 Bronco V8-255, 302, 351W, 351M & 400 & 1977–79 Econoline V8-351M & 400

1. Remove the oil level dipstick and tube, then remove the fan shroud attaching bolts and position shroud over the fan.
2. Remove engine mounts to chassis bracket retaining nuts.
3. If equipped with automatic transmission, disconnect the oil cooler line from the left side of the radiator.
4. On all vehicles equipped with V8-351M and 400 engines, remove starter.
5. Raise engine and place wooden blocks between the engine mounts and chassis brackets.
6. Drain the oil pan, then remove the oil pan attaching bolts and lower the oil pan onto the crossmember.
7. Remove the pick tube attaching bolts then lower the pickup tube into the oil pan and remove the oil pan.

1980–83 Econoline V8-351M & 400

1. Perform steps 1 thru 8 as described under "1975–83 Econoline V8-255, 302 & 351W".
2. Position suitable jack under oil pan and place 1¾ inch block of wood between pan and jack. Raise engine 2 inches.

NOTE: Do not raise engine more than 2 inches, as damage to the air conditioner system will result.

3. Position wood blocks under engine mounts, then lower engine onto mounts.
4. Remove oil pan from vehicle.

V8-401, 477, 534

To remove pan, drain oil and remove oil level dipstick. Unfasten and remove oil pan.

V8-330, 360, 361, 370, 389, 390, 391, 1979–83 429

1. To remove pan and oil pump, drain crankcase and remove oil level dipstick and tube.
2. Unfasten and lower oil pan to axle. Position crankshaft so that counterweight will clear pan.
3. Remove oil pump and inlet tube retaining bolts. Then place pump, screen and intermediate drive shaft in the pan.
4. Remove pan and pump.
5. Reverse removal procedure to install. First place pan on axle. Install pump and screen. Then attach oil pan.

1974–79 V8-460 except 1975–79 Econoline

1. Disconnect battery ground cable, then disconnect radiator shroud and place it over the fan.
2. Support vehicle on hoist and drain crankcase.
3. Remove engine support thru bolts, then place a floor jack under front edge of oil pan with wood block between oil pan and jack.
4. Raise engine enough to place 1¼ inch wood blocks between insulators and brackets and remove floor jack.
5. Remove oil filter, then remove oil pan bolts and oil pan.

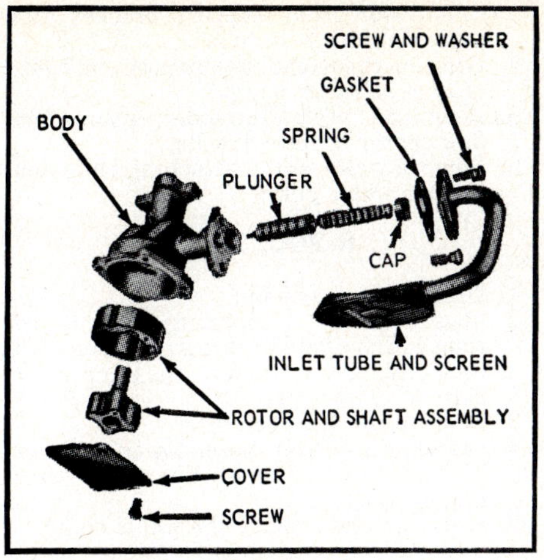

Fig. 27 Oil pump. 6-240, 300

NOTE: It may be necessary to rotate crankshaft to obtain clearance between the crankshaft throws and oil pan.

V8-460 1975–79 Econoline & 1980–83 All

1. Remove engine cover, then disconnect battery ground cable and drain coolant system.
2. Remove air cleaner assembly, then disconnect throttle and transmission linkages from carburetor.
3. Disconnect all remaining lines, hoses and wiring from carburetor and, on vehicles with power brakes, disconnect brake booster vacuum hose.
4. Remove radiator, then remove power steering pump and position aside.
5. On vehicles equipped with air conditioning, position lines at rear of compressor downward to clear dash, or remove lines if necessary.
6. Raise and support vehicle, then remove front engine mount attaching bolts and engine oil dipstick tube.
7. Remove oil filler tube and bracket, then drain crankcase and remove oil filter.
8. Remove muffler inlet pipe, then disconnect manual and kickdown linkages from transmission.

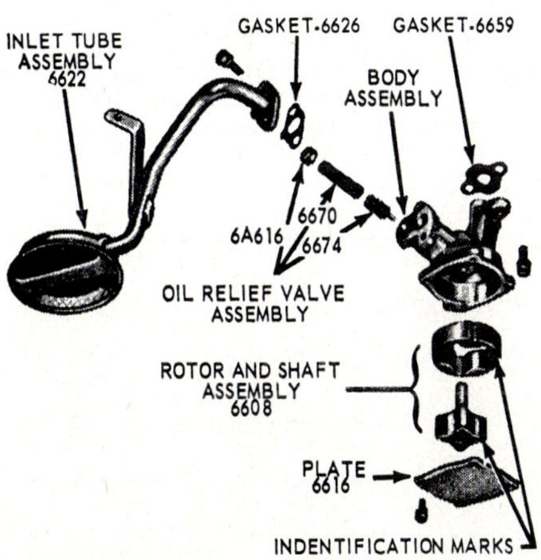

Fig. 30 Oil pump. V8-255, 302, 351W

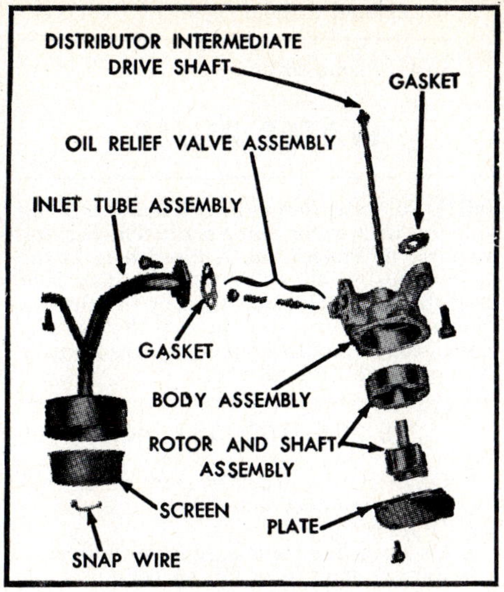

Fig. 28 Oil pump. V8-330, 360, 361, 390, 391

9. Remove driveshaft from vehicle. Position suitable jack under engine oil pan with block of wood placed between jack and pan. Raise engine 4 inches, then position wood blocks under engine mounts and allow engine to rest on blocks.
10. Remove oil pan from vehicle.
11. Reverse procedure to install.

OIL PUMP, REPLACE

6-200, 250

1. Remove oil pan as outlined previously.
2. Remove oil pump bolts and remove the pump, gasket and intermediate drive shaft.
3. To install, prime the pump with engine oil.
4. Position the intermediate drive shaft into the distributor socket.
5. Position a new gasket on the pump housing and insert the drive shaft into the pump.
6. Install the pump and shaft as an assembly.

NOTE: Do not force the pump if it will not seat readily. The drive shaft hex may be misaligned with the distributor shaft. To align, rotate the intermediate shaft into a new position.

7. Install oil pan.

6-240, 6-300

1. After removing oil pan as outlined previously, unfasten and remove oil pump.
2. Prime pump with engine oil.
3. Coat a new pump gasket with oil-resistant sealer and place it on pump.
4. Install pump and oil pan.

V6-232

NOTE: On these engines the oil pump is contained within the front cover.

1. If necessary remove oil filter.

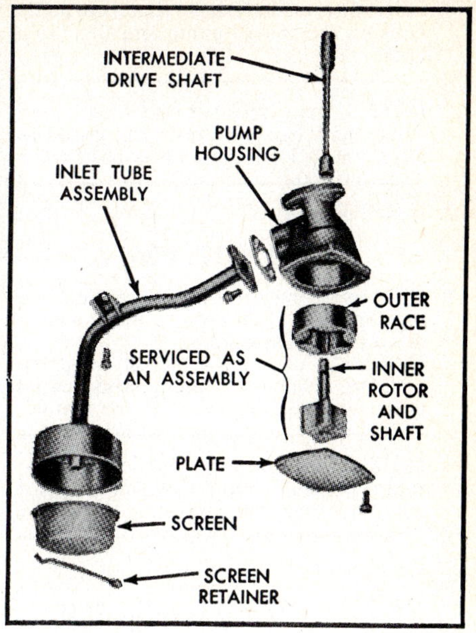

Fig. 29 Oil pump. V8-401, 477, 534

2. Remove oil pump cover attaching bolts and remove cover.
3. Lift pump gears off pocket in front cover.
4. Remove cover gasket. Discard gasket.
5. If necessary, remove pump gears from cover.
6. Reverse procedure to install. If gears have been removed, pack gear pocket with petroleum jelly. Do not use chassis lubricants.

V8-330, 360, 361, 370, 390, 391, 1979–83 429

1. Oil pump is removed together with oil pan as outlined previously.
2. To install, position a new pump inlet tube gasket on pump and install tube.
3. Prime pump with engine oil.
4. Place pump in oil pan and position pan on crossmember.
5. Place a new pump gasket on engine.

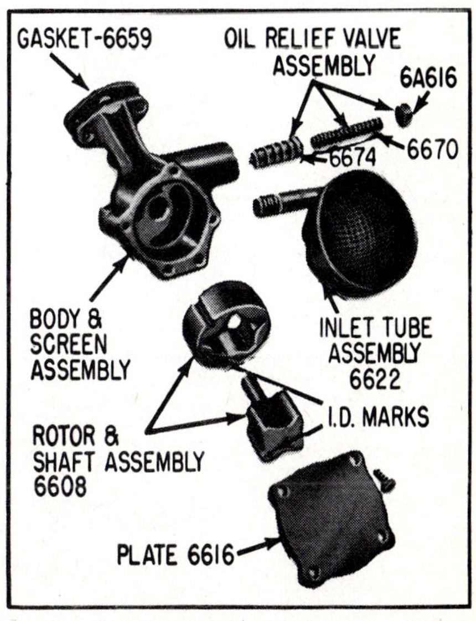

Fig. 31 Oil pump assembly. V8-351C, 351M & 400

6. Insert intermediate drive shaft into pump housing and install pump and shaft as a unit.

NOTE: If pump will not seat readily, the drive shaft hex may be misaligned. To align, rotate shaft into a new position.

7. Install oil pan.

V8-401, 477, 534

1. After removing oil pan, remove pump and pick-up tube. Then remove intermediate drive shaft from engine.
2. To install, place new gasket on pump housing, prime the pump and install pump, shaft and pick-up tube as a unit.

NOTE: If the pump will not seat readily, rotate the shaft into a new position as it may have been misaligned.

3. Install oil pan.

V8-255, 302, 351, 400, 460

1. Remove oil pan as described previously.
2. Remove oil pump bolts and oil pump.
3. To install, prime oil pump with engine oil and apply sealant on gasket.
4. Insert distributor intermediate shaft making certain that it is properly seated and install oil pump.

NOTE: Do not force pump into place if it will not seat readily, as the intermediate shaft may be misaligned with distributor shaft. To align, rotate intermediate shaft until pump can be seated without applying force.

5. Install oil pan.

OIL PUMP, SERVICE

Rotor Type Pump

1. To disassemble, Figs. 26 thru 31, remove cover, inner rotor and shaft assembly and outer race. Remove staking marks at relief valve chamber cap. Insert a self-threading sheet metal screw of proper diameter into oil pressure relief valve chamber cap and pull cap out of chamber. Then remove spring and plunger.
2. To assemble, install pressure relief valve plunger, spring and a new cap. Stake cap in place. Install outer race, inner rotor and shaft.

NOTE: Be sure identification dimple mark on outer race is facing outward and on the same side as dimple on rotor. Inner rotor and shaft and outer race are furnished only as a unit. One part should not be replaced without replacing the other.

3. Install pump cover.

WATER PUMP

NOTE: All water pumps have a sealed bearing integral with water pump shaft. The bearing requires no lubrication. A bleed hole in the pump housing allows water that may leak past the seal to be thrown out by the slinger. This is not a lubrication hole. If the pump is damaged and requires repair, replace it with a new pump or a rebuilt one.

V8-255, 302, 351, 400

1. On Econoline models, remove air cleaner and intake duct assembly.
2. Drain cooling system.
3. On Econoline models, remove radiator.
4. Remove drive belts, fan, spacer and water pump pulley.
5. On Bronco and F-100 models, remove fan shroud if so equipped, and disconnect the lower radiator hose.
6. Loosen alternator pivot bolt and bolt retaining alternator adjusting arm to water pump.
7. On Thermactor engines, remove air pump mounting bolts.
8. Disconnect heater by-pass hoses.
9. Remove bolts securing water pump to cylinder front cover. Remove water pump.

V8-460

1. Drain cooling system.
2. Remove fan and shroud and loosen the power steering pump attaching bolts.
3. If equipped with air conditioning, remove the compressor attaching bolts and secure compressor to left fender brace.
4. Remove power steering attaching bolts and place pump aside.
5. If equipped with air conditioning, remove compressor brackets.
6. Disconnect heater, by-pass and radiator lower hose from water pump.
7. Remove water pump attaching bolts and remove water pump.

V6-232

1. Drain cooling system.
2. Remove air cleaner assembly and air intake duct.
3. Remove fan shroud attaching screws and fan/clutch assembly attaching bolts, then remove fan/clutch assembly and shroud.
4. Loosen accessory drive idler and remove drive belt and water pump pulley.
5. If equipped with power steering, remove attaching bolts from pump mounting bracket, then, with hoses attached, place pump/bracket assembly aside.
6. If equipped with air conditioning, remove compressor front support bracket and leave compressor in place.
7. Disconnect coolant by-pass hose and heater hose at water pump.
8. If equipped with tripminder, remove fuel flow sensor support bracket.
9. Remove water pump attaching bolts and remove water pump.

6-200, 250

1. Drain cooling system.
2. Loosen alternator adjusting arm bolt and remove drive belt. If unit is equipped with an air compressor, remove the compressor belt.
3. Remove the fan and pulley.
4. Disconnect heater hose, lower radiator hose and radiator supply line at the water pump.
5. Unfasten and remove the water pump.

6-240, 6-300

1. To remove pump, drain cooling system and remove alternator drive belt, and air compressor belt (if used).
2. Remove fan and pulley.
3. Disconnect hose as required.
4. Unfasten and remove pump.

V8-330 MD

1. Drain cooling system and remove alternator drive belt.
2. Remove fan, spacer and pulley.
3. Disconnect hose as required.
4. Unfasten and remove pump.

V8-330 HD, 360, 361, 370, 390, 391, 429

1. Drain cooling system and remove drive belts.
2. On engines with water pump mounted fans, remove fan, spacer and pulley. Let fan rest down in shroud. On high fan engines, remove fan and bracket. On low fan engines, remove fan from crankshaft.
3. If equipped with air compressor, unfasten and position compressor on frame rail. Remove compressor mounting bracket.
4. If equipped with power steering, remove steering pump and position it to one side, leaving hoses attached.
5. Disconnect hose as required.
6. Unfasten and remove pump.

V8-401, 477, 534

1. Drain cooling system and disconnect hose as required.
2. Remove fan belts.
3. On F and T Series trucks, remove fan and pulley and lower them into bottom of radiator shroud. Unfasten and remove the pump. On C and H Series trucks, unfasten and remove pump, pulley and gaskets as a unit.

Carburetor Section

1974–83 CARTER YF, YFA ADJUSTMENT SPECIFICATIONS

See Tune Up Chart for hot idle speed.

Year	Carb. Model	Float Level	Fast Idle Cam Setting	Dechoke or Unloader Setting	Pulldown Setting	Choke Setting
1974	D4TE-YC	3/8	.110	.280	.290	Index
	D4TE-ZA	3/8	.175	.280	.230	1 Lean
	D4TE-ABA	3/8	.120	.120	.260	Index
	D4UE-KB	3/8	.110	.280	.290	Index
	D4UE-LA	3/8	.175	.280	.230	1 Lean
	D4UE-NA	3/8	.175	.280	.290	Index
1975	D4HE-AA	3/8	—	—	—	—
	D4TE-AGA	3/8	—	—	—	—
	D4TE-AUA	3/8	—	—	—	—
	D5TE-ADA	3/8	.110	.280	.290	Index
	D5TE-AFA	3/8	—	—	—	—
	D5TE-AGA	3/8	.110	.280	.230	1 Rich
	D5TE-AJA	3/8	—	—	—	—
	D5TE-AKA	3/8	.110	.280	.290	Index
	D5TE-ALA	3/8	.110	.280	.290	Index
	D5TE-AMA	3/8	.110	.280	.230	1 Rich
	D5TE-ANA	3/8	.110	.280	.230	1 Rich
	D5TE-APA	3/8	.110	.280	.290	Index
1976	D5TE-AFA	3/8	—	—	—	—
	D5TE-AFB	3/8	—	—	—	—
	D5TE-AGA	3/8	.110	.280	.230	1 Rich
	D5TE-AGB	23/32	.110	.280	.230	1 Rich
	D5TE-AJA	3/8	—	—	—	—
	D5TE-AJB	3/8	—	—	—	—
	D5TE-APA	3/8	.110	.280	.290	Index
	D5TE-APB	23/32	.110	.280	.290	Index
	D5TE-CAB	23/32	.110	—	.280	Index
	D5UE-AAB	23/32	.110	.280	.290	Index
	D5UE-EA	3/8	.110	.280	.290	Index
	D5UE-FA	3/8	.110	.280	.290	Index
	D6TE-DA	23/32	.110	.280	.290	Index
	D6TE-HA	23/32	.110	.280	.290	Index
	D6TE-KA	23/32	.110	.280	.290	Index
	D6TE-ZA	23/32	.110	.280	.290	Index
	D6UE-FA	23/32	.110	.280	.290	Index
	D6UE-MA	23/32	.110	.280	.290	Index
1977	D7TE-MA	25/32	.110	.280	.290	Index
	D7TE-PA	25/32	.110	.280	.290	Index
1978	D8TE-AAA	23/32	.110	.280	.230	1 Rich
	D8TE-BUA	25/32	.140	.280	.230	Index
	D8TE-BVA	25/32	.140	.280	.230	Index
	D8TE-BWA	25/32	.140	.280	.230	Index
	D8TE-CDA	23/32	—	—	—	—
	D8TE-CKB	25/32	.140	.280	.230	Index
	D8TE-CNA	25/32	.140	.280	.230	Index
	D8TE-UA	23/32	—	—	—	—
	D8UE-AAA	25/32	.140	.280	.230	Index
	D8UE-EA	25/32	—	—	—	—
	D8UE-ZA	25/32	.140	.280	.230	Index
	D2UE-EA	25/32	—	—	—	—

Continued

1974–83 CARTER YF, YFA ADJUSTMENT SPECIFICATIONS—Continued

See Tune Up Chart for hot idle speed.

Year	Carb. Model	Float Level	Fast Idle Cam Setting	Dechoke or Unloader Setting	Pulldown Setting	Choke Setting
1979	D9TE-KA	11/16	.140	.280	.230	Index
	D9TE-ZB	11/16	.140	.280	.230	Index
	D9UE-FA	11/16	.140	.280	.230	Index
	D9TE-NA	11/16	.140	.280	.230	Index
	D9TE-LA	11/16	.140	.280	.230	Index
	D9TE-MA	11/16	.140	.280	.230	Index
	D9TE-CA	11/16	.140	.280	.290	Index
	D9TE-VA	11/16	.140	.280	.290	Index
	D9TE-BA	11/16	—	—	—	—
	D9TE-AA	11/16	—	—	—	—
1980	E0TE-ABA	11/16	.140	.280	.290	—
	E0TE-ACA	11/16	.140	.280	.320	—
	E0TE-AEA	11/16	.140	.280	.230	—
	E0TE-AFA	11/16	.140	.280	.230	—
	E0TE-AHA	11/16	.140	.280	.230	—
	E0TE-AKA	11/16	.140	.280	.230	—
	E0TE-ALA	11/16	.140	.280	.230	—
	E0TE-ATA	11/16	.140	.280	.320	—
	E0TE-CA	11/16	.140	.280	.230	—
	E0TE-FA	11/16	.140	.280	.290	—
	E0UE-GA	11/16	.140	.280	.230	—
	E0UE-LA	11/16	.140	.280	.230	—
	ET0E-ARA	11/16	.140	.280	.320	—
	E0UE-KA	11/16	.140	.280	.230	—
1981	D5TE-AGB	—	.375	.110	.280	1 Rich
	D9TE-CA	—	.69	.140	.280	Index
	D9TE-VA	—	.69	.140	.280	Index
	E0TE-AMA	—	.69	.140	.280	Index
	E0TE-FA	—	.69	.140	.280	Index
	E1TE-ANA	—	.78	.140	.280	Index
	E1TE-ARA	—	.78	.140	.280	Index
	E1TE-AUA	—	.78	.140	.280	2 Rich
	E1TE-AZA	—	.78	.140	.330	2 Rich
	E1TE-DA	—	.78	.140	.280	Index
	E1TE-EA	—	.78	.140	.280	Index
	E1TE-GA	—	.78	.140	.330	2 Rich
	E1TE-TA	—	.78	.140	.280	Index
	E1TE-UA	—	.78	.140	.280	Index
	E1TE-VA	—	.78	.140	.280	2 Rich
	E1TE-ANB	—	.78	.140	.280	Index
	E1TE-EB	—	.78	.140	.280	Index
1982	E2TE-BZA,BVA	.78	—	.280	.270	Index
	E2TE-AMA	.78	—	.280	.230	Index
	E2UE-EA	.78	—	.280	.230	Index
	E2TE-CEA	.78	—	.330	.320	2 Rich
	E2TE-JA	.78	—	.330	.320	2 Rich
	E2TE-YA	.78	—	.280	.300	Index
	E2TE-AAA	.78	—	.280	.300	Index
	E2TE-MA	.78	—	.280	.300	2 Rich
	E2TE-ANA	.78	—	.280	.300	2 Rich
	E2TE-KA	.78	—	.330	.320	2 Rich
	E2UE-DA	.78	—	.330	.320	2 Rich

Continued

1974–83 CARTER YF, YFA ADJUSTMENT SPECIFICATIONS—Continued

See Tune Up Chart for hot idle speed.

Year	Carb. Model	Float Level	Fast Idle Cam Setting	Dechoke or Unloader Setting	Pulldown Setting	Choke Setting
1983	D5TE-AGB	3/8	.110	.280	.290	—
	E0TE-FB	.69	.140	.280	.290	—
	E0TE-AMB	.69	.140	.280	.290	—
	E2TE-ZA	.69	.140	.280	.290	—
	E3TE-FA	.78	.140	.330	.320	—
	E3TE-GA	.78	.140	.330	.320	—
	E3TE-YA	.78	.140	.280	.270	—
	E3TE-ZA	.78	.140	.280	.300	—
	E3TE-AAA	.78	.140	.280	.300	—
	E3TE-ABA	.78	.140	.280	.300	—
	E3TE-AFA	.78	.140	.280	.270	—
	E3TE-AGA	.78	.140	.280	.270	—
	E3TE-AHA	.78	.140	.280	.270	—
	E3TE-AJA	.78	.140	.280	.270	—
	E3TE-AKA	.78	.140	.280	.300	—
	E3TE-ALA	.78	.140	.330	.320	—
	E3TE-AMA	.78	.140	.280	.270	—
	E3TE-ANA	.78	.140	.280	.270	—
	E3TE-APA	.78	.140	.280	.270	—
	E3TE-ARA	.78	.140	.280	.300	—
	E3TE-ASA	.78	.140	.280	.270	—
	E3TE-BDA	.78	.140	.330	.320	—
	E3TE-BKA	.78	.140	.330	.320	—
	E3TE-BNA	.78	.140	.330	.320	—
	E3TE-BRA	.78	.140	.280	.300	—

1974–83 MODEL 2100, 2150-2V ADJUSTMENT SPECIFICATIONS

See Tune Up Chart for hot idle speeds.

Year	Carb. Model	Float Level (Dry)	Fuel Level (Wet)	Pump Setting Hole No.	Choke Plate Clearance (Pull down)	Fast Idle Cam Linkage Clearance	Dechoke Clearance	Dashpot Setting	Choke Setting
1974	D4AE-DA	7/16	13/16	#2A	—	—		—	1 Rich
	D4AE-EA	7/16	13/16	#2A	—	—	—	—	3 Rich
	D4AE-FA	7/16	13/16	#3A	—	—	—	—	3 Rich
	D4AE-GA	7/16	13/16	#3A	—	—	—	—	3 Rich
	D4AE-HB	7/16	13/16	#3A	—	—	—	—	3 Rich
	D4BE-FA	1/2	29/32	#3	—	—	—	—	2 Rich
	D4DE-LA	7/16	13/16	#2	—	—	—	—	2 Rich
	D4DE-NB	7/16	13/16	#2	—	—	—	—	3 Rich
	D4ME-BA	7/16	13/16	#3A	—	—	—	—	3 Rich
	D4OE-CA	—	—	—	—	—	—	—	—
	D4OE-FA	7/16	13/16	#2A	—	—	—	—	3 Rich
	D4OE-GA	7/16	13/16	#2	—	—	—	—	3 Rich
	D4OE-JA	—	—	—	—	—	—	—	—
	D4TE-EA	7/16	13/16	#3	—	—	—	—	2 Rich
	D4TE-GA	31/64	7/8	#3	—	—	—	—	1 Lean
	D4TE-LA	7/16	13/16	#2	—	—	—	—	2 Rich
	D4TE-TA	31/64	7/8	#3	—	—	—	—	Index
	D4TE-AZA	7/16	13/16	#2	—	—	—	—	2 Rich

Continued

1974–83 MODEL 2100, 2150-2V ADJUSTMENT SPECIFICATIONS—Continued

See Tune Up Chart for hot idle speeds.

Year	Carb. Model	Float Level (Dry)	Fuel Level (Wet)	Pump Setting Hole No.	Choke Plate Clearance (Pull down)	Fast Idle Cam Linkage Clearance	Dechoke Clearance	Dashpot Setting	Choke Setting
	D4TE-BAA	31/64	7/8	#3	—	—	—	—	1 Rich
	D4TE-CAA	7/16	13/16	#2	—	—	—	—	2 Rich
	D4TF-AA	1/2	29/32	#3	—	—	—	—	2 Rich
	D4UE-AA	7/16	13/16	#3	—	—	—	—	2 Rich
	D4UE-BA	7/16	13/16	#2	—	—	—	—	2 Rich
	D4UE-FA	7/16	13/16	#3	—	—	—	—	3 Rich
	D4UE-GA	7/16	13/16	#3	—	—	—	—	2 Rich
	D4UE-HA	7/16	13/16	#2	—	—	—	—	2 Rich
	D4UE-JA	7/16	13/16	#2	—	—	—	—	2 Rich
	D4UE-UA	7/16	13/16	#2	—	—	—	—	2 Rich
	D4UF-AA	7/16	13/16	#3	—	—	—	—	2 Rich
1975	D5AE-AA	7/16	13/16	#3	.125	—	—	—	3 Rich
	D5AE-EA	7/16	13/16	#3	.125	—	—	—	3 Rich
	D5DE-HA	7/16	13/16	#3	.140	—	—	—	3 Rich
	D5DE-UA	7/16	13/16	#2	.140	—	—	—	3 Rich
	D5ME-BA	7/16	13/16	#2	.125	—	—	—	3 Rich
	D5ME-FA	7/16	13/16	#2	.125	—	—	—	3 Rich
	D5OE-AA	7/16	13/16	#2	.140	—	—	—	3 Rich
	D5OE-BA	7/16	13/16	#3	.125	—	—	—	3 Rich
	D5OE-CA	7/16	13/16	#3	.125	—	—	—	3 Rich
	D5OE-DA	7/16	13/16	#2	.140	—	—	—	3 Rich
	D5OE-GA	7/16	13/16	#2	.125	—	—	—	3 Rich
	D5TE-LA	31/64	7/8	#2	.135	—	—	—	Index
	D5TE-PA	31/64	7/8	#2	.160	—	—	—	3 Rich
	D5TE-VA	31/64	7/8	#3	.179	—	—	—	2 Rich
	D5TE-YD	31/64	7/8	#3	.179	—	—	—	3 Rich
	D5TE-ZA	31/64	7/8	#4	.179	—	—	—	3 Rich
	D5TE-AAD	31/64	7/8	#3	.179	—	—	—	3 Rich
	D5TE-ABA	7/16	13/16	#4	—	—	—	—	—
	D5TE-ACB	7/16	13/16	#4	—	—	—	—	—
	D5TE-ASA	31/64	7/8	#4	.179	—	—	—	Index
	D5TE-ATA	31/64	7/8	#4	.179	—	—	—	Index
	D5TE-AUB	31/64	7/8	#3	.179	—	—	—	2 Rich
	D5TE-BCA	31/64	7/8	#2	.179	—	—	—	2 Rich
	D5TE-BCB	31/64	7/8	#2	.179	—	—	—	2 Rich
	D5TE-BDA	31/64	7/8	#2	.179	—	—	—	2 Rich
	D5TE-BDB	31/64	7/8	#2	.179	—	—	—	2 Rich
	D5TE-BEA	31/64	7/8	#3	.179	—	—	—	2 Rich
	D5TE-BFA	31/64	7/8	#3	.179	—	—	—	2 Rich
	D5TE-BFB	31/64	7/8	#3	.179	—	—	—	2 Rich
	D5TE-BGA	31/64	7/8	#3	.179	—	—	—	3 Rich
	D5TE-BHA	31/64	7/8	#2	.135	—	—	—	Index
	D5TE-BJA	31/64	7/8	#2	.160	—	—	—	2 Rich
	D5UE-BA	31/64	7/8	#3	.153	—	—	—	1 Rich
	D5UE-DC	31/64	7/8	#3	.153	—	—	—	3 Rich
	D5UE-JD	31/64	7/8	#3	.160	—	—	—	3 Rich
	D5UE-KD	31/64	7/8	#3	.160	—	—	—	3 Rich
	D5UE-ZA	31/64	7/8	#3	.153	—	—	—	1 Rich
1976	D5TE-PA	31/64	7/8	#2	.160	—	—	—	3 Rich
	D5TE-VA	31/64	7/8	#3	.179	—	—	—	2 Rich
	D5TE-YA	—	—	—	—	—	—	—	—
	D5TE-YF	31/64	7/8	#3	.140	.160	—	—	2 Rich

Continued

1974–83 MODEL 2100, 2150-2V ADJUSTMENT SPECIFICATIONS—Continued

See Tune Up Chart for hot idle speeds.

Year	Carb. Model	Float Level (Dry)	Fuel Level (Wet)	Pump Setting Hole No.	Choke Plate Clearance (Pull down)	Fast Idle Cam Linkage Clearance	Dechoke Clearance	Dashpot Setting	Choke Setting
	D5TE-ZA	31/64	7/8	#3	.179	—	—	—	3 Rich
	D5TE-ZB	31/64	7/8	#3	.140	.160	—	—	2 Rich
	D5TE-AAA	—	—	—	—	—	—	—	—
	D5TE-AAC	31/64	7/8	#4	.160	—	—	—	3 Rich
	D5TE-AAF	31/64	7/8	#3	.140	.160	—	—	3 Rich
	D5TE-ABA	7/16	13/16	#4	—	—	—	—	—
	D5TE-ACB	7/16	13/16	#4	—	—	—	—	—
	D5TE-ASA	31/64	7/8	#4	.179	—	—	—	Index
	D5TE-ATA	31/64	7/8	#4	.179	—	—	—	Index
	D5TE-AUA	—	—	—	—	—	—	—	—
	D5TE-AUB	31/64	7/8	#3	.179	—	—	—	2 Rich
	D5TE-BCA	31/64	7/8	#2	.179	—	—	—	2 Rich
	D5TE-BCB	31/64	7/8	#2	.179	—	—	—	2 Rich
	D5TE-BDA	31/64	7/8	#2	.179	—	—	—	2 Rich
	D5TE-BDB	31/64	7/8	#2	.179	—	—	—	2 Rich
	D5TE-BEA	31/64	7/8	#3	.179	—	—	—	2 Rich
	D5TE-BEB	31/64	7/8	#3	.140	.160	—	—	2 Rich
	D5TE-BFA	31/64	7/8	#3	.179	—	—	—	2 Rich
	D5TE-BFB	31/64	7/8	#3	.179	—	—	—	2 Rich
	D5TE-BGA	31/64	7/8	#3	.179	—	—	—	3 Rich
	D5TE-BMA	31/64	7/8	#2	.140	.160	—	—	Index
	D5UE-AA	31/64	7/8	#2	.160	.180	—	—	3 Rich
	D5UE-BA	31/64	7/8	#3	.153	—	—	—	1 Rich
	D5UE-CA	1/2	7/8	#2	.160	—	—	—	3 Rich
	D5UE-DA	—	—	—	—	—	—	—	—
	D5UE-DC	31/64	7/8	#3	.153	—	—	—	3 Rich
	D5UE-JA	—	—	—	—	—	—	—	—
	D5UE-JD	31/64	7/8	#3	.160	—	—	—	3 Rich
	D5UE-KA	—	—	—	—	—	—	—	—
	D5UE-KD	31/64	7/8	#3	.160	—	—	—	3 Rich
	D5UE-LA	31/64	7/8	#2	.160	.180	—	—	3 Rich
	D5UE-MA	31/64	7/8	#2	.160	.180	—	—	3 Rich
	D5UE-ZA	31/64	7/8	#3	.153	—	—	—	1 Rich
	D6AE-HA	7/16	13/16	#2	.160	.180	—	—	3 Rich
	D6AE-JA	7/16	13/16	#2	.160	.180	—	—	2 Rich
	D6AE-KA	7/16	13/16	#2	.160	.180	—	—	2 Rich
	D6ME-AA	7/16	13/16	#2	.160	.180	—	—	3 Rich
	D6ME-BA	7/16	13/16	#2	.160	.180	—	—	2 Rich
	D6OE-AA	7/16	13/16	#3	.160	.180	—	—	3 Rich
	D6TE-FA	31/64	7/8	#2	.140	.160	—	—	Index
	D6TE-GA	31/64	7/8	#2	.140	.160	—	—	3 Rich
	D6TE-GB	—	—	—	—	—	—	—	—
	D6TE-RA	31/64	7/8	#2	.179	—	—	—	2 Rich
	D6TE-SA	31/64	7/8	#3	.179	—	—	—	2 Rich
	D6TE-TA	31/64	7/8	#3	.179	—	—	—	2 Rich
	D6TE-VA	31/64	7/8	#2	.135	—	—	—	2 Rich
	D6TE-YA	31/64	7/8	#2	.135	—	—	—	3 Rich
	D6TE-YB	31/64	7/8	#2	.135	—	—	—	3 Rich
	D6TE-AAC	31/64	7/8	#4	.160	—	—	—	3 Rich
	D6UE-HA	—	—	—	—	—	—	—	—
	D6UE-JA	31/64	7/8	#2	.180	.200	—	—	3 Rich
	D6UE-NA	—	—	—	—	—	—	—	—

Continued

1974–83 MODEL 2100, 2150-2V ADJUSTMENT SPECIFICATIONS—Continued

See Tune Up Chart for hot idle speeds.

Year	Carb. Model	Float Level (Dry)	Fuel Level (Wet)	Pump Setting Hole No.	Choke Plate Clearance (Pull down)	Fast Idle Cam Linkage Clearance	Dechoke Clearance	Dashpot Setting	Choke Setting
	D6WE-AA	7/16	13/16	#2	.160	.180	—	—	3 Rich
	D6WE-BA	7/16	13/16	#2	.160	.180	—	—	2 Rich
1977	D7AE-ACA	7/16	13/16	#2	.156	.170	—	—	Index
	D7AE-AHA	7/16	13/16	#3	.179	.189	—	—	Index
	D7AE-AKA	7/16	13/16	#3	.179	.189	—	—	Index
	D7BE-YA	7/16	13/16	#2	.147	.167	—	—	1 Rich
	D7OE-LA	7/16	13/16	#4	.169	.189	—	—	3 Rich
	D7OE-RA	3/4	3/4	#3	.167	.187	—	—	2 Rich
	D7OE-TA	7/16	13/16	#3	.185	.205	—	—	2 Rich
	D7TE-ABA	31/64	7/8	#2	.140	—	—	—	3 Rich
	D7TE-ACA	31/64	7/8	#2	.140	—	—	—	3 Rich
	D7TE-ADA	31/64	7/8	#3	.140	—	—	—	3 Rich
	D7TE-AFA	31/64	7/8	#2	.140	—	—	—	3 Rich
	D7UE-TA	7/16	13/16	#3	.170	—	—	—	3 Rich
	D7UE-YA	7/16	13/16	#2	.170	—	—	—	3 Rich
	D7UE-ZC	7/16	13/16	#3	.170	—	—	—	Index
	D7UE-ACA	31/64	7/8	#3	.170	—	—	—	1 Rich
	D7UE-ADA	31/64	7/8	#4	.170	—	—	—	2 Rich
	D7UE-AEA	31/64	7/8	#4	.170	—	—	—	1 Rich
	D7UE-ANA	31/64	7/8	#3	.170	—	—	—	2 Rich
	D7TE-AHA	31/64	7/8	#3	.160	—	—	—	Index
	D7TE-AKA	31/64	7/8	#3	.160	—	—	—	Index
	D7TE-ALA	31/64	7/8	#3	.160	—	—	—	1 Rich
	D7TE-AMA	31/64	7/8	#4	.160	—	—	—	3 Rich
	D7TE-ANA	31/64	7/8	#4	.160	—	—	—	3 Rich
	D7TE-APA	31/64	7/8	#4	.160	—	—	—	3 Rich
	D7TE-ARA	31/64	7/8	#4	.160	—	—	—	3 Rich
	D7TE-BEA	31/64	7/8	#3	.160	—	—	—	Index
	D7TE-AKA	31/64	7/8	#3	.160	—	—	—	Index
	D7TE-AUA	31/64	7/8	#4	.160	—	—	—	3 Rich
	D7TE-AYA	31/64	7/8	#4	.160	—	—	—	3 Rich
	D7TE-AZA	31/64	7/8	#4	.160	—	—	—	3 Rich
1978	D8AE-JA	3/8	3/4	#3	.167	—	—	—	3 Rich
	D8BE-ACA	7/16	3/4	#4	.155	—	—	—	2 Rich
	D8BE-ADA	7/16	13/16	#2	.110	—	—	—	3 Rich
	D8BE-AEA	7/16	13/16	#2	.110	—	—	—	4 Rich
	D8BE-AFA	7/16	13/16	#2	.110	—	—	—	4 Rich
	D8DE-HA	1 19/32	13/16	#3	.157	—	—	—	Index
	D8FE-BA	1 19/32	13/16	#2	.128	—	—	—	2 Rich
	D8KE-EA	1 19/32	13/16	#2	.135	—	—	—	3 Rich
	D8ME-BA	7/16	13/16	#2	.135	—	—	—	Index
	D8OE-BA	3/8	3/4	#3	.167	—	—	—	3 Rich
	D8OE-EA	1 19/32	13/16	#2	.136	—	—	—	Index
	D8OE-HA	7/16	13/16	#3	.180	—	—	—	2 Rich
	D8SE-CA	1 19/32	13/16	#3	.150	—	—	—	2 Rich
	D8SE-DA	7/16	13/16	#3	.147	—	—	—	3 Rich
	D8SE-EA	7/16	13/16	#3	.147	—	—	—	3 Rich
	D8SE-FA	3/8	13/16	#3	.147	—	—	—	3 Rich
	D8SE-GA	3/8	13/16	#3	.147	—	—	—	3 Rich
	D8TE-AA	31/64	7/8	#4	.150	—	—	—	—
	D8TE-BA	31/64	7/8	#3	.140	—	—	—	Index
	D8TE-DA	31/64	7/8	#4	.130	—	—	—	Index

Continued

1974–83 MODEL 2100, 2150-2V ADJUSTMENT SPECIFICATIONS—Continued

See Tune Up Chart for hot idle speeds.

Year	Carb. Model	Float Level (Dry)	Fuel Level (Wet)	Pump Setting Hole No.	Choke Plate Clearance (Pull down)	Fast Idle Cam Linkage Clearance	Dechoke Clearance	Dashpot Setting	Choke Setting
	D8TE-GA	31/64	7/8	#2	.130	—	—	—	3 Rich
	D8TE-LA	31/64	7/8	#3	.130	—	—	—	Index
	D8TE-ARA	31/64	7/8	#3	.145	—	—	—	Index
	D8TE-ATA	31/64	7/8	#3	.145	—	—	—	2 Rich
	D8TE-BEA	31/64	7/8	#2	.200	—	—	—	3 Rich
	D8TE-BJA	31/64	7/8	#3	.175	—	—	—	3 Rich
	D8TE-BLA	31/64	7/8	#4	.180	—	—	—	2 Rich
	D8TE-BNA	31/64	7/8	#3	.130	—	—	—	3 Rich
	D8TE-CCA	31/64	7/8	#2	.130	—	—	—	3 Rich
	D8TE-CGA	31/64	7/8	#2	.130	—	—	—	1 Rich
	D8TE-CHA	31/64	7/8	#2	.130	—	—	—	3 Rich
	D8TE-CPA	31/64	7/8	#3	.185	—	—	—	3 Rich
	D8UE-DA	31/64	7/8	#3	.185	—	—	—	3 Rich
	D8UE-GA	31/64	7/8	#2	.205	—	—	—	1 Rich
	D8UE-HA	31/64	7/8	#2	.215	—	—	—	Index
	D8UE-KA	7/16	13/16	#3	.185	—	—	—	Index
	D8UE-MA	31/64	7/8	#2	.215	—	—	—	Index
	D8UE-SA	31/64	7/8	#3	.180	—	—	—	3 Rich
	D8UE-VA	31/64	7/8	#4	.145	—	—	—	Index
	D8WE-DA	7/16	13/16	#4	.143	—	—	—	1 Rich
	D8WE-TA	3/8	3/4	#4	.135	—	—	—	Index
	D8ZE-UA	3/8	3/4	#4	.135	—	—	—	Index
	D84E-EA	3/8	3/4	#3	.167	—	—	—	3 Rich
1979	D9TE-CRA	31/64	7/8	2	.145	—	—	—	3 Rich
	D9TE-BSB	31/64	7/8	2	.136	—	—	—	1 Rich
	D9TE-BUB	31/64	7/8	3	.140	—	—	—	3 Rich
	D9TE-BTB	31/64	7/8	2	.145	—	—	—	3 Rich
	D9TE-ALB	31/64	7/8	3	.145	—	—	—	Index
	D9TE-AMB	31/64	7/8	3	.145	—	—	—	Index
	D9TE-BPB	31/64	7/8	3	.150	—	—	—	3 Rich
	D9TE-BRB	31/64	7/8	3	.150	—	—	—	3 Rich
	D9TE-CPA	31/64	7/8	3	.145	—	—	—	Index
	D9TE-CTA	31/64	7/8	3	.130	—	—	—	3 Rich
	D9TE-CSA	31/64	7/0	3	.130	—	—	—	3 Rich
	D9TE-BNB	31/64	7/8	2	.145	—	—	—	Index
	D9TE-BJB	31/64	7/8	2	.135	—	—	—	Index
	D8TE-BJA	31/64	7/8	3	.160	—	—	—	3 Rich
	D9UE-DB	7/16	13/16	2	.200	—	—	—	Index
	D9UE-LB	31/64	7/8	3	.140	—	—	—	3 Rich
	D9UE-GA	7/16	13/16	2	.200	—	—	—	Index
	D9UE-KA	31/64	7/8	4	.175	—	—	—	1 Rich
	D9TE-DCB	31/64	7/8	2	.145	—	—	—	2 Rich
	D9UE-EB	7/16	13/16	2	.190	—	—	—	Index
	D9DE-RA	7/16	13/16	2	.125	—	—	—	3 Rich
	D9DE-RB	7/16	13/16	2	.125	—	—	—	3 Rich
	D9DE-SA	7/16	13/16	2	.125	—	—	—	3 Rich
	D9OE-CB	7/16	13/16	3	.132	—	—	—	3 Rich
	D9OE-DB	7/16	13/16	3	.132	—	—	—	3 Rich
	D9AE-AHA	7/16	13/16	3	.147	—	—	—	3 Rich
	D9AE-AJA	7/16	13/16	3	.147	—	—	—	3 Rich
	D9WE-JA	7/16	13/16	3	.150	—	—	—	2 Rich
	D9WE-MB	7/16	13/16	3	.132	—	—	—	1 Rich

Continued

1974–83 MODEL 2100, 2150-2V ADJUSTMENT SPECIFICATIONS—Continued

See Tune Up Chart for hot idle speeds.

Year	Carb. Model	Float Level (Dry)	Fuel Level (Wet)	Pump Setting Hole No.	Choke Plate Clearance (Pull down)	Fast Idle Cam Linkage Clearance	Dechoke Clearance	Dashpot Setting	Choke Setting
	D9WE-NB	7/16	13/16	3	.132	—	—	—	1 Rich
1980	EOTE-BEA	—	13/16	#2	.140	—	.25	—	—
	EOTE-BGA	—	13/16	#2	.135	—	.25	—	—
	EOTE-BHA	—	13/16	#2	.135	—	.25	—	—
	EOTE-BRA	—	7/8	#3	.128	—	.115	—	—
	EOTE-BSA	—	7/8	#2	.140	—	.115	—	—
	EOTE-BYA	—	7/8	#2	—	—	.115	—	—
	EOTE-CFA	—	13/16	#2	.128	—	.25	—	—
	EOTE-CLA	—	7/8	#2	.140	—	.200	—	—
	EOTE-CVA	—	7/8	#2	.105	—	.25	—	—
	EOTE-CYA	—	13/16	#2	.140	—	.25	—	—
	EOTE-CZA	—	13/16	#2	.140	—	.25	—	—
	EOTE-DBA	—	7/8	#3	.130	—	.25	—	—
	EOTE-DCA	—	7/8	#3	.130	—	.25	—	—
	EOTE-DDA	—	7/8	#3	.128	—	.115	—	—
	EOTE-EAA	—	13/16	#2	.128	—	.25	—	—
	EQUE-AA	—	13/16	#3	.105	—	.25	—	—
	EQUE-ABA	—	13/16	#2	.140	—	.25	—	—
	E0UE-AAA	—	13/16	#3	.195	—	.25	—	—
	E0TE-BLA	—	7/8	#3	.148	—	.25	—	—
	E0TE-BFA	—	7/8	#3	.148	—	.25	—	—
	E0TE-BZA	—	7/8	#3	.148	—	.25	—	—
	E0TE-CCA	—	7/8	#3	.159	—	.25	—	—
	E0TE-CBA	—	7/8	#3	.159	—	.25	—	—
	E0TE-EDA	—	7/8	#4	.155	—	.25	—	—
	E0TE-DGA	—	7/8	#4	.155	—	.25	—	—
	E0TE-EEA, EFA	—	7/8	#4	.160	—	.25	—	—
	E0UE-NA	—	13/16	#3	.105	—	.25	—	—
	E0UE-PA, RA	—	13/16	#4	.185	—	.25	—	—
	E0UE-SA, TA	—	13/16	#3	.195	—	.25	—	—
	E0UE-VA	—	13/16	#4	.185	—	.25	—	—
	E0TE-DEA	—	7/8	#4	.170	—	.25	—	—
	E0TE-DFA	—	7/8	#4	.180	—	.25	—	—
	E0TE-ECA	—	7/8	#4	.175	—	.25	—	—
1981	E1TE-BFA	.484	.875	#2	.140	—	.200	—	V-Notch
	E1TE-BGA	.484	.875	#3	.148	—	.250	—	V-Notch
	E1TE-BHA	.484	.875	#2	.140	—	.200	—	3 Rich
	E1TE-BJA	.484	.875	#3	.148	—	.250	—	V-Notch
	E1TE-BSA	.484	.875	#3	.130	—	.250	—	V-Notch
	E1TE-BTA	.484	.875	#3	.130	—	.250	—	V-Notch
	E1TE-BVA	.438	.810	#2	.130	—	.200	—	V-Notch
	E1TE-BYA	.484	.875	#4	.175	—	.250	—	V-Notch
	E1TE-BZA	.484	.875	#4	.180	—	.250	—	V-Notch
	E1TE-CAA	.484	.875	#4	.175	—	.250	—	V-Notch
	E1TE-CBA	.484	.875	#4	.180	—	.250	—	V-Notch
	E1TE-CCA	.484	.875	#4	.155	—	.250	—	V-Notch
	E1TE-CDA	.484	.875	#4	.155	—	.250	—	V-Notch
	E1TE-CEA	.484	.875	#4	.160	—	.200	—	V-Notch
	E1TE-CFA	.484	.875	#4	.160	—	.200	—	V-Notch
	E1TE-CHA	.484	.875	#3	.148	—	—	—	—
	E1TE-CKA	.438	.810	#2	.125	—	.200	—	V-Notch
	E1TE-CLA	.438	.810	#2	.125	—	.200	—	V-Notch

Continued

1974–83 MODEL 2100, 2150-2V ADJUSTMENT SPECIFICATIONS—Continued

See Tune Up Chart for hot idle speeds.

Year	Carb. Model	Float Level (Dry)	Fuel Level (Wet)	Pump Setting Hole No.	Choke Plate Clearance (Pull down)	Fast Idle Cam Linkage Clearance	Dechoke Clearance	Dashpot Setting	Choke Setting
	E1TE-CMA	.438	.810	#2	.125	—	.200	—	V-Notch
	E1TE-CNA	.438	.810	#2	.125	—	.200	—	V-Notch
	E1TE-CPA	.438	.810	#2	.125	—	.200	—	V-Notch
	E1TE-CRA	.438	.810	#2	.125	—	.200	—	V-Notch
	E1TE-CSA	.438	.810	#2	.125	—	.200	—	V-Notch
	E1UE-GA	.438	.810	#2	.130	—	.200	—	V-Notch
	E1UE-HA	.484	.875	#2	.125	—	.200	—	V-Notch
	E1UE-KA	.484	.875	#3	.180	—	.250	—	V-Notch
	E1VE-CA	.484	.875	#3	.120	—	.200	—	V-Notch
	E1VE-FA	.484	.875	#3	.120	—	.200	—	Index
	E1VE-JA	.484	.875	#3	.120	—	.200	—	V-Notch
1982	E1UE-JA	31/64	.875	#3	.120	—	.200	—	V-Notch
	E2TE-AYA	7/16	.810	#2	.130	—	.200	—	V-Notch
	E2TE-BAA	7/16	.810	#2	.125	—	.200	—	V-Notch
	E2TE-BBA	7/16	.810	#2	.125	—	.200	—	V-Notch
	E2TE-BEA	7/16	.810	#2	.130	—	.200	—	V-Notch
	E2TE-BFA	7/16	.810	#2	.125	—	.200	—	V-Notch
	E2TE-BGA	31/64	.875	#4	.180	—	.250	—	V-Notch
	E2TE-BHA	31/64	.875	#4	.180	—	.250	—	V-Notch
	E2TE-BJA	31/64	.875	#4	.175	—	.250	—	V-Notch
	E2TE-BKA	31/64	.875	#4	.175	—	.250	—	V-Notch
	E2TE-BLA	7/16	.810	#2	.125	—	.200	—	V-Notch
	E2TE-BMA	7/16	.810	#2	.125	—	.250	—	V-Notch
	E2TE-BNA	7/16	.810	#2	.125	—	.200	—	V-Notch
	E2TE-CFA	7/16	.810	#2	.125	—	.250	—	V-Notch
	E2TE-CGA	7/16	.810	#2	.125	—	.200	—	V-Notch
	E2TE-CJA	7/16	.810	#2	.130	—	.200	—	V-Notch
	E2TE-CKA	7/16	.810	#2	.120	—	.200	—	V-Notch
	E2UE-FA	31/64	.875	#3	.120	—	.200	—	V-Notch
	E2UE-JA	31/64	.875	#2	.130	—	.200	—	V-Notch
	E2UE-KA	31/64	.875	#2	.120	—	.200	—	V-Notch
1983	E2TE-BPA	31/64	.875	#3	.130	—	.250	—	V-Notch
	E2TE-BRA	31/64	.875	#3	.130	—	.250	—	V-Notch
	E2UE-FA	31/64	.875	#3	.120	—	.200	—	V-Notch
	E2UE-KA	31/64	.875	#2	.120	—	.200	—	V-Notch
	E2UE-AKA	31/64	.875	#3	.180	—	.250	—	V-Notch
	E2UE-ANA	31/64	.875	#3	.180	—	.250	—	V-Notch
	E3UE-BA	31/64	.875	#2	.120	—	.200	—	V-Notch
	E3UE-CA	31/64	.875	#3	.120	—	.200	—	V-Notch
	E3UE-DA	31/64	.875	#3	.180	—	.250	—	V-Notch
	E3UE-EA	31/64	.875	#3	.180	—	.250	—	V-Notch
	E3TE-AUA	7/16	.810	#3	.142	—	.200	—	V-Notch
	E3TE-AVA	7/16	.810	#3	.149	—	.250	—	V-Notch
	E3TE-AYA	7/16	.810	#4	.137	—	.250	—	V-Notch
	E3TE-AZA	31/64	.875	#3	.130	—	.250	—	V-Notch
	E3TE-BAA	31/64	.875	#3	.130	—	.250	—	V-Notch
	E3TE-BBA	—	.810	#3	.125	—	.250	—	V-Notch
	E3TE-BCA	—	.810	#3	.125	—	.250	—	V-Notch
	E3TE-BEA	7/16	.810	#3	.149	—	.200	—	V-Notch
	E3TE-BFA	—	.810	#3	.125	—	.250	—	V-Notch
	E3TE-BGA	—	.810	#3	.125	—	.250	—	V-Notch

Continued

1974–83 MODEL 2100, 2150-2V ADJUSTMENT SPECIFICATIONS—Continued

See Tune Up Chart for hot idle speeds.

Year	Carb. Model	Float Level (Dry)	Fuel Level (Wet)	Pump Setting Hole No.	Choke Plate Clearance (Pull down)	Fast Idle Cam Linkage Clearance	Dechoke Clearance	Dashpot Setting	Choke Setting
	E3TE-BHA	7/16	.810	#3	.152	—	.200	—	V-Notch
	E3TE-BJA	7/16	.810	#4	.157	—	.200	—	V-Notch
	E3TE-BLA	7/16	.810	#3	.157	—	.200	—	V-Notch
	E3TE-BMA	7/16	.810	#4	.150	—	.200	—	V-Notch
	E3TE-BPA	7/16	.810	#3	.157	—	.200	—	V-Notch

HOLLEY 2300-2V ADJUSTMENT SPECIFICATIONS

See Tune Up Chart for hot idle speeds.

Year	Model	Float Level (Dry)	Float Level (Wet)	Pump Cam Position	Pump Override Spring Adj.	Choke Setting
1974–78	D4TE-AHA	①	②	#2	.015	—
	D4TE-AJA	①	②	#2	.015	—
	D4TE-AKA	①	②	#2	.015	—
	D4TE-AMA	①	②	#2	.015	—
1978	D8TE-AJA	③	②	#2	.015	—
	D8TE-BPA	③	②	#2	.015	—
	D8TE-CEA	③	②	#2	.015	—
1979	D9TE-CUA	①	②	#2	.015	—
1979–80	D9TE-APA	①	②	#2	.015	—
	D9TE-ABC	①	②	#2	.015	—
1980	E0TE-PA	—	②	#1	—	—
	E0TE-BCA	—	②	#1	—	—
1981	D9TE-ABC	—	②	#2	—	④
	E0TE-BCA	—	②	#1	—	④
1981–82	D9TE-APA	—	②	#2	—	④
	E0TE-PA	—	②	#1	—	④
1983	D9TE-APA	—	②	—	—	④
	E0TE-PA	—	②	—	—	④
	E2TE-NA	—	②	—	—	④
	E2TE-AGA	—	②	—	—	④
	E2TE-DPA	—	②	#1	—	⑤

①—Parallel with float bowl floor (inverted).
②—Lower edge of sight plug.
③—Parallel with float bowl floor (inverted) plus one turn clockwise.
④—Choke pulldown setting, .35–.40 inch.
⑤—Choke pulldown setting, .21–.23 inch.

HOLLEY MODEL 4150, 4160, 4180 & 4190-4V ADJUSTMENT SPECIFICATIONS

See Tune Up Chart for hot idle speeds.

Year	Carb. Model	Float Level (Dry)	Float Level (Wet)	Pump Cam Position	Pump Override Spring Adjustment	Pulldown Setting	Choke Unloader	Choke Setting
1974	D4HE-DA	①	②	#1	.015	—	—	—
	D4HE-CA	①	②	#1	.015	—	—	—
	D4HE-BA	—	②	#1	—	—	—	—
	D4TE-ARA	①	②	#1	.015	—	.315	Index
	D4TE-ASA	①	②	#1	.015	—	.315	Index
	D4TE-ANA	—	②	#2	—	—	—	—
1974–75	D4HE-BA	①	②	#1	.015	—	—	—
	D4HE-CA	①	②	#1	.015	—	—	—
	D4HE-DA	①	②	#1	.015	—	—	—
1975	D5TE-BA	①	②	#2	.015	—	—	—

Continued

HOLLEY MODEL 4150, 4160, 4180 & 4190-4V ADJUSTMENT SPECIFICATIONS—Continued

See Tune Up Chart for hot idle speeds.

Year	Carb. Model	Float Level (Dry)	Float Level (Wet)	Pump Cam Position	Pump Override Spring Adjustment	Pulldown Setting	Choke Unloader	Choke Setting
	D5TE-CB	①	②	#2	.015	—	—	—
	D5TE-AA	①	②	#2	.015	—	—	—
	D5TE-BRA	①	②	#2	.015	—	—	—
	D5TE-DA	①	②	#1	.015	.180	.315	Index
	D5TE-EA	①	②	#1	.015	.180	.315	Index
	D5TE-FA	①	②	#1	.015	.180	.315	Index
	D5TE-GA	①	②	#1	.015	.180	.315	Index
	D5HE-BA	①	②	#1	.015	—	—	—
	D5HE-CA	①	②	#1	.015	—	—	—
1975–76	D4HE-CA	①	②	#1	.015	—	—	—
	D4HE-DA	①	②	#1	.015	—	—	—
	D5TE-AA	①	②	#2	.015	—	—	—
	D5TE-BA	①	②	#2	.015	—	—	—
	D5TE-BRA	①	②	#2	.015	—	—	—
	D5TE-CB	①	②	#2	.015	—	—	—
	D5HE-BA	①	②	#1	.015	—	—	—
	D5HE-CA	①	②	#1	.015	—	—	—
	D5TE-DA	①	②	#1	.015	.180	.315	Index
	D5TE-EA	①	②	#1	.015	.180	.315	Index
	D5TE-FA	①	②	#1	.015	.180	.315	Index
	D5TE-GA	①	②	#1	.015	.180	.315	Index
1976	D6HE-CB	①	②	#1	.015	—	—	—
	D6HE-DB	①	②	#1	.015	—	—	—
	D5TE-FB	①	②	#1	.015	.200	.315	2 Lean
	D5TE-GB	①	②	#1	.015	.190	.315	2 Lean
1976–77	D5HE-BA	①	②	#1	.015	—	—	—
	D5HE-CA	①	②	#1	.015	—	—	—
	D5TE-BRA	①	②	#2	.015	—	—	—
	D5TE-AA	①	②	#2	.015	—	—	—
	D5TE-CB	①	②	#2	.015	—	—	—
1977	D6HE-DA	—	②	#1	—	—	—	—
	D6HE-CA	—	②	#1	—	—	—	—
	D4TE-EAA	—	②	#2	—	—	—	—
	D4TE-CA	—	②	#2	—	—	—	—
	D5TE-B	—	②	#2	—	—	—	—
	D7HE-CA	—	②	#1	—	—	—	—
	D7HE-RA	—	②	#1	—	—	—	—
1977–78	D5TE-BRA	①	②	#2	.015	—	—	—
	D5TE-CB	①	②	#2	.015	—	—	—
1978	D8TE-AGA	③	②	#2	.015	—	—	—
	D8TE-AHA	③	②	#2	.015	—	—	—
	D8TE-BRA	③	②	#2	.015	—	—	—
	D8TE-BSA	③	②	#2	.015	—	—	—
	D8HE-AA	③	②	#1	.015	—	—	—
	D8HE-CA	③	②	#1	.015	—	—	—
	D8HE-EA	①	②	#1	.015	.375	—	—
	D8HE-BA	③	②	#1	.015	—	—	—
	D8HE-DA	③	②	#1	.015	—	—	—
	D8HE-FA	①	②	#1	.015	.375	—	—
	D8TE-BZA	①	②	#2	.015	—	—	—
	D8TE-CBA	①	②	#2	.015	—	—	—
	D6HE-CB	①	②	#1	.015	.375	—	—
	D6HE-DB	①	②	#1	.015	.375	—	—
1979	D9HE-DA	①	②	#1	—	—	—	—
	D9HE-CA	①	②	#1	.015	—	—	—

Continued

HOLLEY MODEL 4150, 4160, 4180 & 4190-4V ADJUSTMENT SPECIFICATIONS—Continued

See Tune Up Chart for hot idle speeds.

Year	Carb. Model	Float Level (Dry)	Float Level (Wet)	Pump Cam Position	Pump Override Spring Adjustment	Pulldown Setting	Choke Unloader	Choke Setting
	D9HE-FA	①	②	#1	.015	—	—	—
	D9HE-EA	①	②	#1	.015	—	—	—
	D9TE-BKA	①	②	#1	.015	.210	.315	5 Rich
	D9TE-AGA	①	②	#1	.015	.210	.315	5 Rich
	D9TE-AEA	①	②	#1	.015	.210	.315	5 Rich
	D9TE-ACC	①	②	#1	.015	—	—	—
	D9TE-AZA	①	②	#1	.015	—	—	—
	D9TE-AYB	①	②	#1	.015	—	—	—
	D9TE-AHD	①	②	#1	.015	—	—	—
1980	D9TE-AHE	—	②	#2	—	—	—	—
	D9TE-EBA	—	②	#2	—	—	—	—
	D9TE-ETA	—	②	#1	—	—	—	—
	D9TE-EUA	—	②	#1	—	—	—	—
	E0TE-JA	—	②	#1	—	—	—	—
	E0TE-RA	—	②	#1	—	—	—	—
1981	D9HE-CA	—	②	#1	—	—	—	—
	D9HE-EA	—	②	#1	—	—	—	—
	D9TE-BKA	—	②	#1	—	.210	—	5 Rich
	D9TE-AHE	—	②	#1	—	.210	—	—
	D9TE-ETA	—	②	#1	—	.210	—	—
	E0TE-JA	—	②	#1	—	.210	—	—
	E0TE-MA	—	②	#1	—	.210	—	—
1981–82	D9TE-EBA	—	②	#1	—	.210	—	—
	D9TE-EUA	—	②	#1	—	.210	—	—
	E0TE-RA	—	②	#1	—	.210	—	—
	E0TE-SA	—	②	#1	—	.210	—	—
	E1UE-RA	—	②	#1	—	.210	—	—
	E2TE-AGA	—	②	#1	—	.210	—	—
	E2TE-AJA	—	②	#1	—	.210	—	—
	E2TE-AKA	—	②	#1	—	.210	—	—
	E2TE-NA	—	②	#1	—	.210	—	—
1983	E2TE-CMA	—	②	#1	—	—	—	—
	E2TE-CNA	—	②	#1	—	—	—	—
	E2TE-CPA	—	②	#1	—	—	—	—
	E2TE-CRA	—	②	#1	—	—	—	—
	E2TE-CSA	—	②	#1	—	—	—	—
	E2TE-CTA	—	②	#1	—	—	—	—
	E2TE-CUA	—	②	#1	—	—	—	—
	E2TE-CVA	—	②	#1	—	—	—	—
	E3TE-PC	—	②	#1	—	.220	.315	3 Rich
	E3TE-RC	—	②	#1	—	.220	.315	3 Rich
	E3TE-SB	—	②	#1	—	.220	.315	3 Rich
	E3TE-TB	—	②	#1	—	.220	.315	3 Rich

①—Parallel with float bowl floor (inverted).

②—Lower edge of sight plug.

③—Parallel with float bowl floor (inverted) plus one turn clockwise.

1974–78 MODEL 4300/4350-4V ADJUSTMENT SPECIFICATIONS

See Tune Up Chart for hot idle speed.

Year	Carb. Model	Float Level (Dry)	Pump Setting (Hole No.)	Choke Plate Clearance (Pulldown)	Fast Idle Cam Linkage Setting	Auxiliary Inlet Valve Setting	Metering Rod Setting	Dechoke Clearance	Dashpot Setting	Choke Setting
1974	D4AE-NA	3/4	#1	.220	.200	1/16	—	—	—	Index
	D4TE-ATA	13/16	#1	.220	.180	1/16	—	—	—	Index
	D4TE-BB	3/4	#1	.220	.200	1/16	—	—	—	Index
	D4VE-AB	3/4	#1	.220	.200	1/16	—	—	—	Index
1975–76	D5AE-CA, DA	31/32	Inner	②	.160	—	.120	.300	—	2 Rich
	D5TE-ARC	15/16	#2	.160	.170	1/16	①	.300	—	Index
	D5TE-ARD	1.00	#2	.160	.170	.030	①	.300	—	Index
	D5TE-BBA	15/16	#2	.160	.170	1/16	①	.300	—	Index
	D5TE-BBB	59/64	#2	.160	.170	.060	①	.300	—	Index
	D5TE-BBC	1.00	#2	.160	.170	.030	①	.300	—	Index
	D5UE-NA	15/16	#2	.160	.170	1/16	①	.300	—	Index
	D5UE-NB	59/64	#2	.160	.170	.060	①	.300	—	Index
	D5UE-NC	1.00	#2	.160	.170	.030	①	.300	—	Index
	D5UE-SA	15/16	#2	.160	.170	1/16	①	.300	—	Index
	D5UE-SB	1.00	#2	.160	.170	.030	①	.300	—	Index
	D5VE-AD, BA	15/16	#1	②	.160	1/16	①	.300	—	2 Rich
1976	D6AE-CA, FA	1.00	#2	③	.140	.030	①	.300	—	2 Rich
	D6AE-DA	31/32	#2	④	.160	.030	①	.300	—	2 Rich
	D6UE-KA, LA	1.00	#2	.160	.170	—	—	.300	—	Index
	D6TE-NA, UA	1.00	#2	.160	.170	—	—	.300	—	Index
1977	D7UE-AFA	1.00	#2	.160	.170	—	—	.300	—	Index
	D7UE-AGA	1.00	#2	.160	.170	—	—	.300	—	Index
	D7TE-BJA	1.00	#2	.160	.170	—	—	.300	—	Index
	D7TE-BLA	1.00	#2	.160	.170	—	—	.300	—	Index
1978	D8TE-AKA	1.00	#3	.160	.160	—	—	.300	—	Index
	D8UE-AA	1.00	#3	.160	.160	—	—	.300	—	Index
	D7UE-ASA	1.00	#3	.160⑤	.160	—	—	.300	—	Index
	DTUE-ASA	1.00	#3	.160⑤	.160	—	—	.300	—	Index
	D8UE-AA	1.00	#3	.160	.160	—	—	.300	—	Index
	D8TE-AMA	1.00	#3	.160	.170	—	—	.300	—	Index
	D8UE-CA	1.00	#3	.160	.170	—	—	.300	—	Index

①—Check before disassembling and reset to original setting.
②—Initial setting, .160″; delayed setting, .190″.
③—Initial setting, .140″; delayed setting, .190″.
④—Initial setting, .160″; delayed setting, .210″.
⑤—Delayed setting, .200″.

MODEL 7200VV ADJUSTMENT SPECIFICATIONS

See Tune Up Chart for hot idle speed.

Year	Carb. Model (Code 9510)	Float Level (Dry)	Float Drop	Control Vacuum Regulator	Venturi Valve Limiter	Fast Idle Cam	Choke Setting
1981–82	E1TE-ABA	1.040	1.460	.250	③	①	Index
	E1TE-AHA	1.040	1.460	.250	②	①	Index
	E1TE-YA	1.040	1.460	.250	③	①	Index
	E1TE-ZA	1.040	1.460	.250	②	①	Index
1982	E2TE-CAA, CBA	1.040	1.460	.250	③	①	Index
	E2TE-CCA, CDA	1.040	1.460	.250	②	①	Index
1983	E2TE-CCB	1.040	1.46	.250	②	①	Index
	E2TE-CCC	1.040	1.46	.250	②	①	Index
	E2TE-CDB	1.040	1.46	.250	②	①	Index
	E2TE-CDD	1.040	1.46	.250	②	①	Index

①—.36 on 2nd stop
②—Limiter adjusting screw, .48–.52; limiter stop screw, .99–1.01.
③—Limiter adjusting screw, .39–.41; limiter stop screw, .99–1.01.

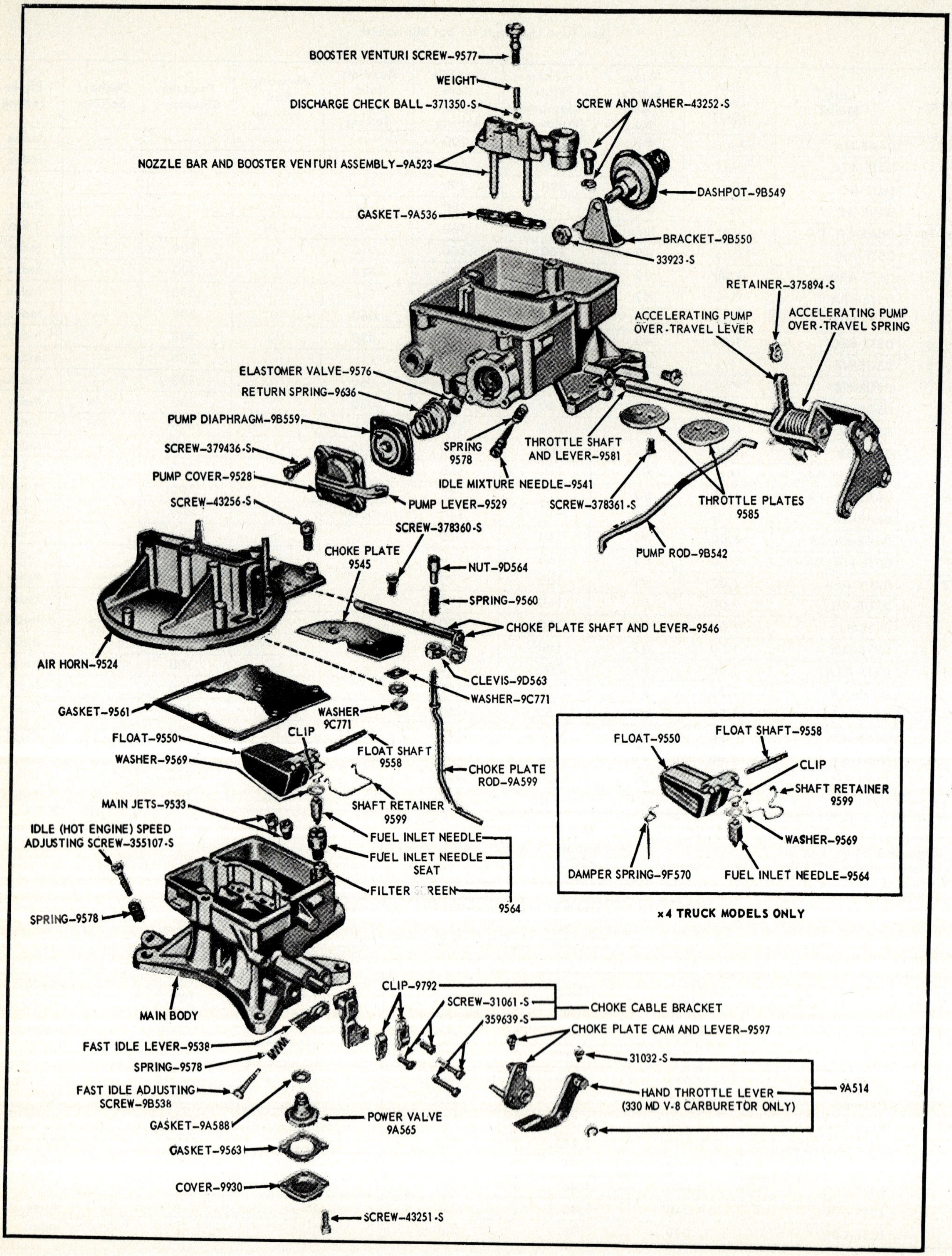

Ford Motorcraft Model 2100, 2150-2V (Typical)

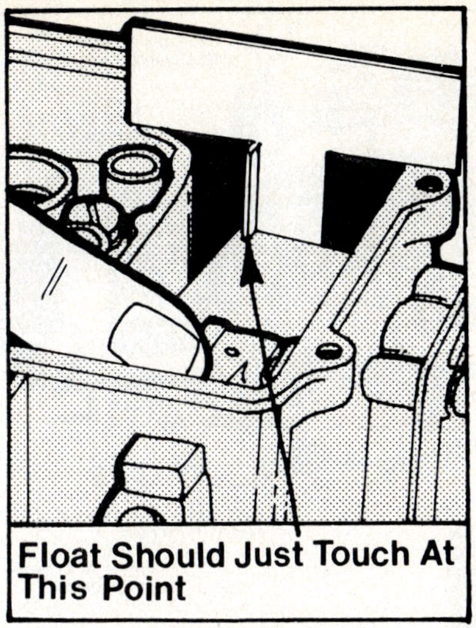

Fig. 1 Float adjustment. Ford Motorcraft 2100, 2150-2V

FORD MOTORCRAFT CARBURETOR ADJUSTMENTS

Model 2100, 2150-2V

Float Adjustment (Dry), Fig. 1

The dry float fuel level adjustment is a preliminary adjustment only. The final adjustment must be made after the carburetor is mounted on the engine.

Remove the air horn. With float raised and fuel inlet needle seated, check distance between top surface of main body and top surface of float. Take measurement at a point 1/8" from the free end of the float and 1/8" (5/16" metal float) in from side of float adjacent to inside wall of air horn.

If cardboard gauge is used, place gauge in corner of the enlarged end section of fuel bowl as shown. The gauge should touch float near the end but not on the end radius.

Depress the float tab to seat the fuel inlet needle. The float height is measured from the gasket surface of the main body with gasket removed. If necessary, bend float tab to bring setting within specified limits. This should provide the correct fuel level setting.

Fuel Level (Wet), Fig. 2

1. Operate engine to normalize engine temperatures, and place vehicle on a flat surface as near level as possible. Stop the engine.
2. Remove air cleaner and carburetor air horn assembly and gasket.
3. Temporarily place air horn gasket in position on carburetor main body and start engine. Let engine idle for a few minutes, then remove air horn and gasket to provide accessibility to float assembly.
4. While engine is idling, use a standard depth scale to measure vertical distance from top machined surface of carburetor main body to level of fuel in bowl. Measurement must be made at least 1/4" away from any vertical surface to insure an accurate reading, because the surface of the fuel is higher at the edges than in the center.
5. If an adjustment is required, stop the engine and bend the float tab as necessary to bring the fuel level within specifications.

Accelerating Pump Adjustment, Figs. 3, & 4

The primary throttle shaft lever (overtravel lever) has 4 holes on 1974–80 models or three holes on 1981–83 models and the accelerating pump link has 2 or 4 holes to control the pump stroke for various atmospheric temperatures, operating conditions and specific engine applications.

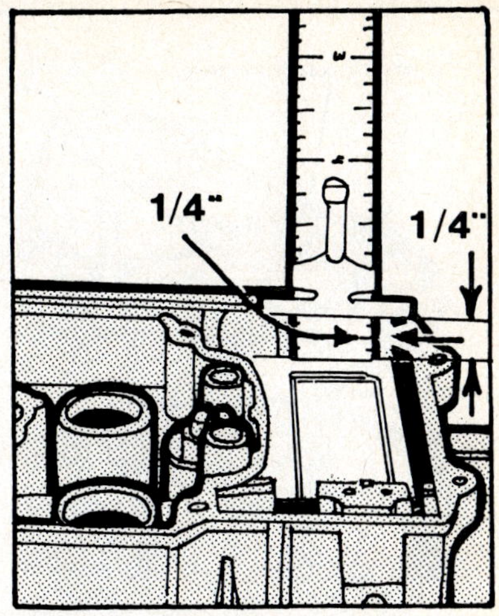

Fig. 2 Fuel level adjustment. Ford Motorcraft 2100, 2150-2V

NOTE: The stroke should not be changed from the specified setting.

---

1. To release rod from retainer clip, press tab end of clip toward rod. Then, at the same time, press rod away from clip until it is disengaged.
2. Position clip over specified hole in overtravel lever. Press ends of clip together and insert operating rod through clip and lever. Release clip to engage rod.

Choke Plate Clearance, Manual Choke, Fig. 5

Remove air cleaner. Pull choke cam and lever to full-choke position. Insert the specified drill or gauge between downward side of choke plate and air horn wall. The gauge or drill will open the choke plate against the pull-down spring tension. Adjust the choke operating (pull-down) rod adjusting nut so it just contacts plastic swivel.

Choke Plate Clearance, 1974 All

Since the choke plate pulldown is set in production by means of an air-fuel meter; no spe-

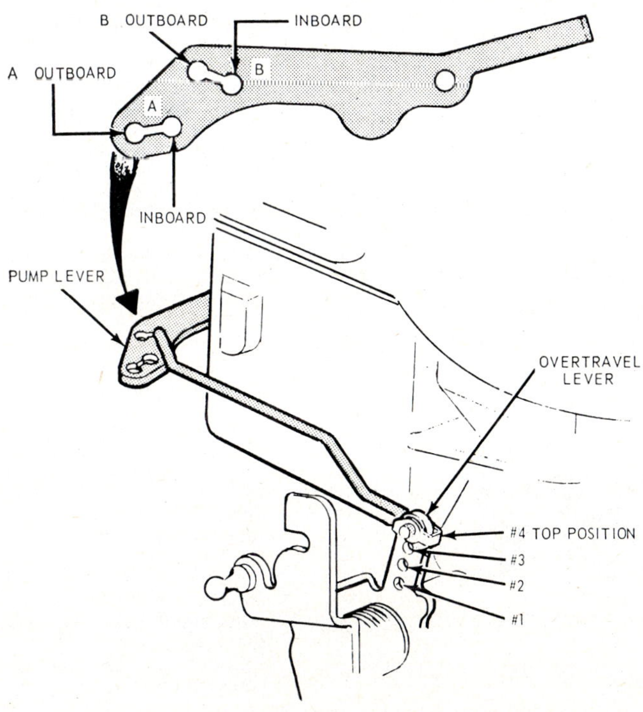

Fig. 3 Pump stroke adjustment, 1974–80. Ford Motorcraft 2100, 2150-2V

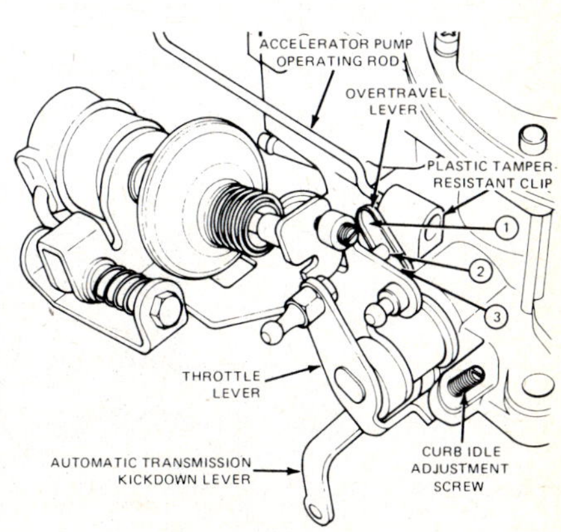

Fig. 4 Pump stroke adjustment. 1981–83 Motorcraft 2150-2V

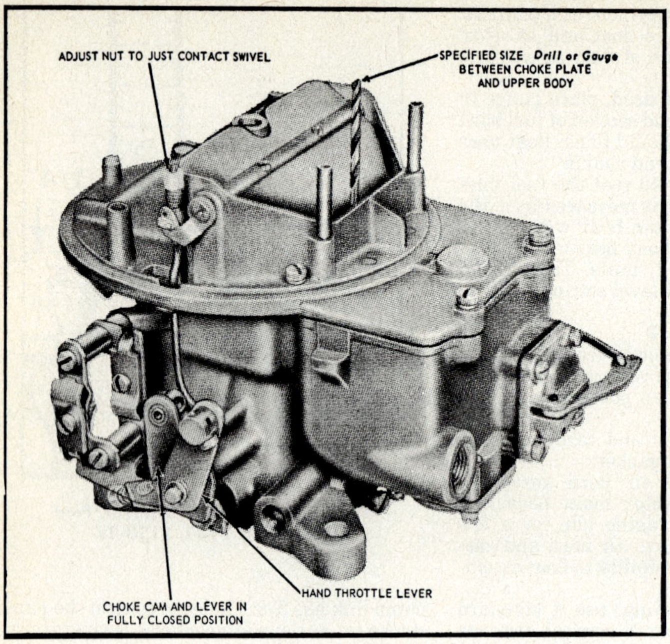

Fig. 5 Choke plate clearance adjustment, manual choke. Ford Motorcraft 2100-2V

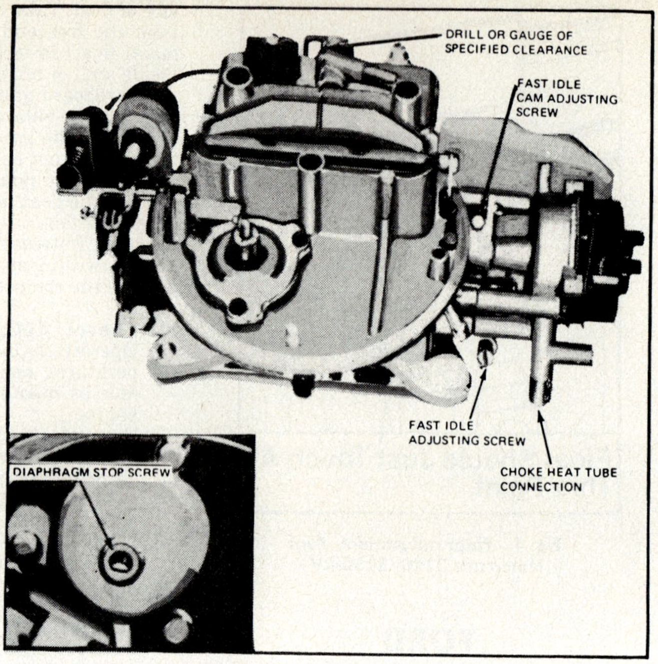

Fig. 6 Choke plate clearance, 1974 All. Ford Motorcraft 2100-2V

cific clearance is indicated for pulldown adjustment. If the vehicle shows indication of leanness during cold starting, decrease the clearance between the choke plate and the air horn wall by 0.020 inch. If the engine shows signs of an overrich condition during cold starting, increase the pulldown clearance by 0.020 inch. If additional adjustment is required always make the adjustments in steps of 0.020 inch. If the original pulldown adjustment is lost, set the clearance between the choke plate and the air horn to 0.160 inch. Then adjust as required in steps on 0.020 inch, Fig. 6.

Choke Plate Clearance, 1975–83 All

1. Set throttle on fast idle cam top step, Fig. 8, then loosen choke thermostatic housing retaining screws and set housing 90° in rich direction.
2. Activate pulldown motor by manually forcing pulldown control diaphragm link in direction of applied vacuum or by applying vacuum to external vacuum tube.
3. Check clearance between choke plate and center of carburetor air horn wall nearest fuel bowl, Fig. 7. Refer to the Specifications Chart. If clearance is not as specified, reset by adjusting diaphragm stop on end of choke pulldown diaphragm.

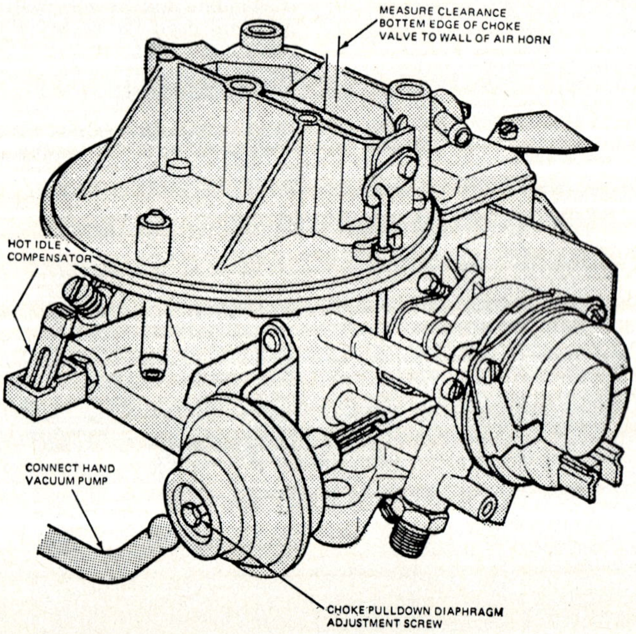

Fig. 7 Choke plate clearance. 1975–83 All. Ford Motorcraft 2150-2V

Fast Idle Cam Clearance, 1974

1. Loosen the choke thermostatic spring housing retainer screws and set the housing 90 degrees in the rich direction.
2. Position the fast idle speed screw at the kickdown step of the fast idle cam. The kickdown step is identified on most units by a V stamped on the cam, Fig. 8. When a two-piece fast idle lever is used, a tang on the top lever will align with the V mark on the cam, Fig. 8.
3. Be sure the cam is at the kickdown position while checking or adjusting the fast idle cam clearance. Check clearance between the lower edge of the choke plate and the air horn wall, Fig. 9. Refer to the *Specifications Chart.* Adjust clearance by turning the fast idle cam clearance adjusting screw clockwise to increase and counterclockwise to decrease the clearance.
4. Set the choke thermostatic spring housing to specifications, Fig. 12. Adjust the anti-stall dashpot, idle speed and fuel mixture.

Fast Idle Cam Clearance, 1975–83

1. Open and close throttle slightly and observe where fast idle screw contacts cam. The fast idle screw should be set to approximate high cam speed and should strike the second highest step on fast idle cam.
2. After adjusting fast idle cam with fast idle screw, allow choke plate to close. Choke plate should close tightly in air horn.
3. If choke plate is loose, an additional adjustment must be made to pulldown diaphragm rod assembly, by loosening lock nut and tightening turnbuckle until choke plate is tight on airhorn.

Automatic Choke Valve Tension

Turn thermostatic spring cover against spring tension until index mark on cover is

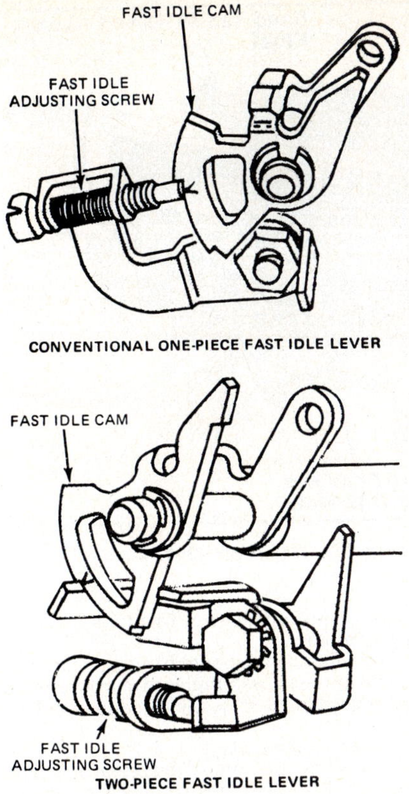

Fig. 8 Fast idle cam assemblies. Ford Motorcraft 2100, 2150-2V

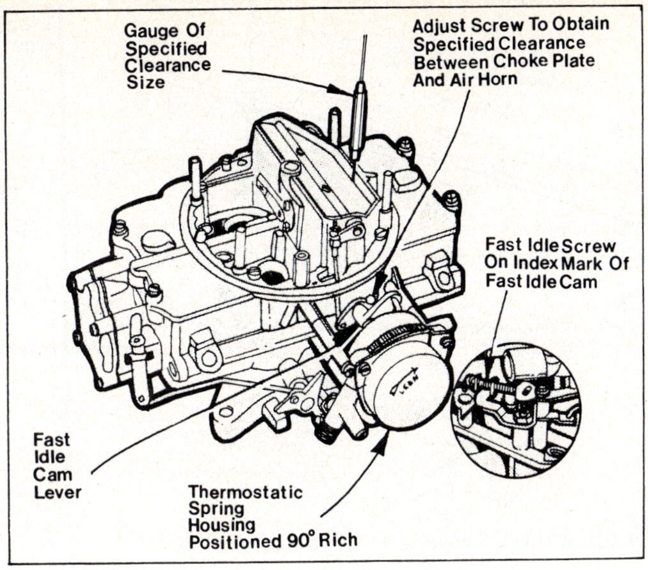

Fig. 9 Fast idle cam linkage adjustment. Ford Motorcraft 2100, 2150-2V

aligned with mark specified in the *Specifications Chart* on choke housing, Fig. 10.

Fast Idle Speed, Figs. 8 & 11

The fast (cold engine) idle speed is adjusted with the hand throttle control rod pushed fully inward to prevent contact of the hand (manual) throttle lever with the fast idle pick-up lever during the adjustment procedure.

With the hot idle speed properly adjusted, manually rotate the choke cam and lever until the fast idle adjusting screw rests on the index mark (arrow) on the cam. Turn fast idle adjusting screw in or out as required so that screw just contacts the cam at the index mark to obtain the specified fast idle rpm.

Anti-Stall Dashpot, Fig. 12

With air cleaner removed, and with engine idle speed and mixture properly adjusted, loosen anti-stall dashpot lock nut. Hold throttle in closed position and depress plunger with a screwdriver. Check clearance between throttle lever and plunger tip with a feeler gauge of the specified clearance dimension. Then turn dashpot in its bracket in a direction to provide the specified clearance.

Model 4300-4V

Float Setting, Figs. 13 & 14

1. Adjust gauge to specified height.
2. Insert gauge into air horn outboard holes as shown.
3. Check clearance and alignment of float pontoons to gauge. Both pontoons should just touch gauge for proper setting. Align pontoons if necessary by slightly twisting pontoons.
4. If it is necessary to adjust float clearance, bend primary needle tab downward to raise float and upward to lower float.

NOTE: To raise float, insert open end of bending tool to *right* side of float lever tab and between needle and float hinge. Raise float lever off needle and bend tab downward.

To lower float, insert open end of bending tool to *left* side of float lever tab, between needle and float hinge. Support float lever and bend tab upward.

Auxiliary Valve Setting, Fig. 15

1. Turn the air horn assembly upright allowing the float to hang freely.
2. Push up on the float until the primary fuel inlet needle lightly contacts its seat.
3. While holding the float in this position, measure the clearance between the float lever auxiliary tab and the auxiliary inlet valve plunger.
4. Adjust by bending the tab. Refer to the *Specifications Chart.*

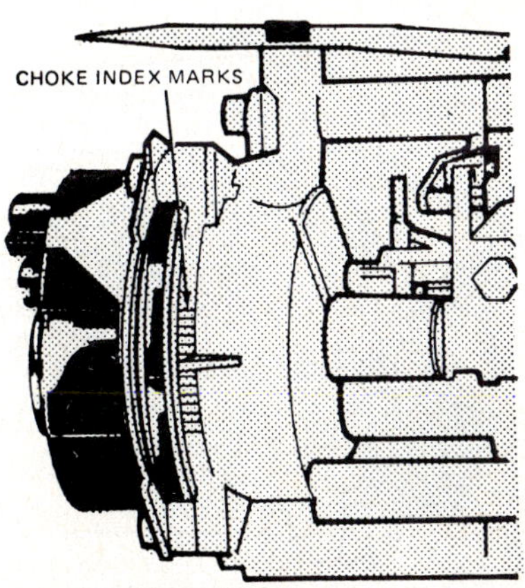

Fig. 10 Automatic choke adjustment. Ford Motorcraft units

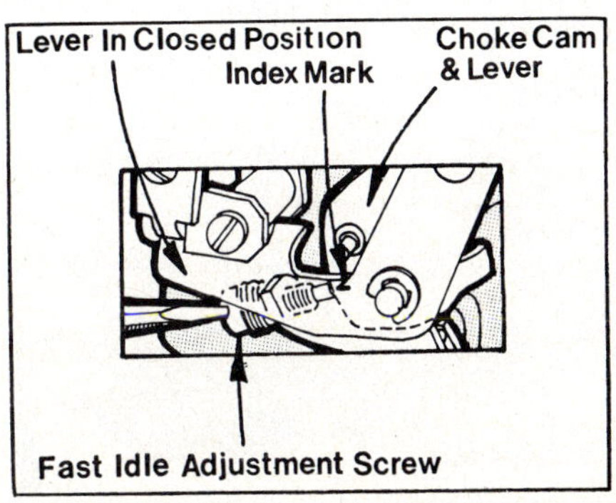

Fig. 11 Fast idle speed adjustment. Ford Motorcraft 2100, 2150-2V (Typical)

Fig. 12 Dashpot adjustment. Typical of all Ford applications

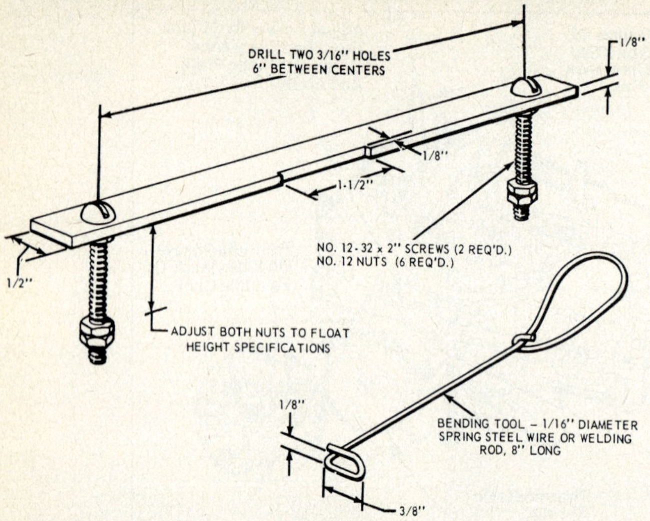

Fig. 13 Float setting gauge & bending tool. Ford Motorcraft 4300, 4350-4V

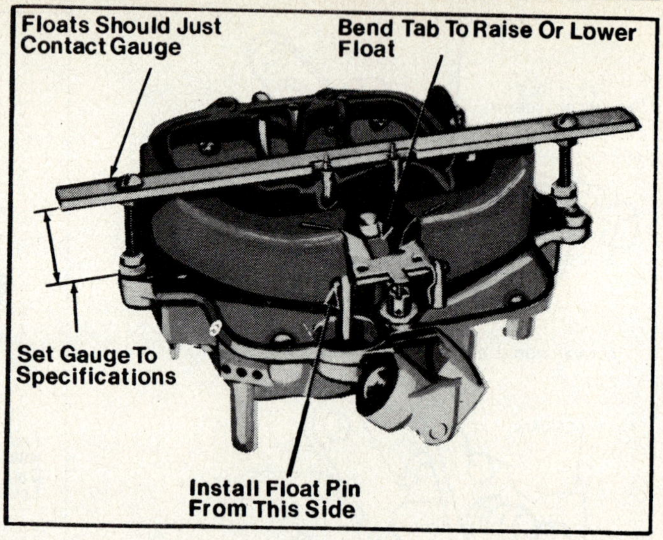

Fig. 14 Float setting. Ford Motorcraft 4300, 4350-4V

NOTE: To measure this clearance on a police fleet carburetor with a semi-articulating float, the float assembly must be positioned so that the bottoms of both pontoons are same distance from gasket surface of air horn when primary fuel inlet is seated.

Fast Idle Speed Adjustment

The fast idle adjusting screw contacts one edge of the fast idle cam. Each position on the fast idle cam permits a slower idle rpm as engine temperature rises and choking is reduced.

NOTE: Make certain the curb idle speed and mixture are adjusted to specification before attempting to set the fast idle speed.

1. With the engine operating temperature normalized (hot), air cleaner removed and the tachometer attached, manually rotate the fast idle cam until the fast idle adjusting screw rests on the center step of the cam.
2. Adjust by turning the fast idle adjusting screw inward or outward as required. Refer to the *Specifications Chart.*

Accelerator Pump Stroke Adjustment, Fig. 16

The accelerator pump stroke has been calibrated to inject a pre-determined quantity of fuel into the air stream and help keep the exhaust emission level of the engine within the specified limits. The additional holes provided for pump stroke adjustment are for adjusting the stroke for specific engine applications.

NOTE: The stroke should not be changed from the specified setting.

If the pump stroke has been changed from the specified setting, refer to the following instructions to correct the stroke to specification. Before adjusting the accelerator pump stroke, measure the height of the pump piston. Adjust by bending the pump control rod to correct the piston stem height. Refer to the *Specifications Chart.*

Automatic Choke Setting, Fig. 10

1. If carburetor is installed on engine, loosen choke heat tube nut.
2. Loosen choke cover retaining screws.
3. Rotate choke cover clockwise to reduce choking action or counterclockwise to increase choking action.
4. Tighten choke cover screws and choke heat tube nut.

Staged Choke Vacuum Control Adjustment

This adjustment should be made only if the control unit has been replaced, carburetor has been overhauled or a choke adjustment has been made.

1. Close choke plate.
2. Back off the adjusting sleeve locknut until it is clear of the adjusting sleeve.
3. Turn the adjusting sleeve inward until the choke plate just begins to move. Scribe a mark on the upper flat of the sleeve, then back off the sleeve one complete turn.
4. Tighten the locknut firmly against the sleeve to set the adjustment.

NOTE: Do not turn the diaphragm stem.

Dechoke Clearance, Fig. 17

1. Open and hold throttle plate to wide open position.
2. Rotate choke plate toward closed position until pawl on fast idle speed lever contacts fast idle cam.
3. Check clearance between lower edge of choke plate and air horn wall.
4. Adjust clearance to specifications by bending pawl on fast idle speed lever forward to increase (backward to decrease) clearance.

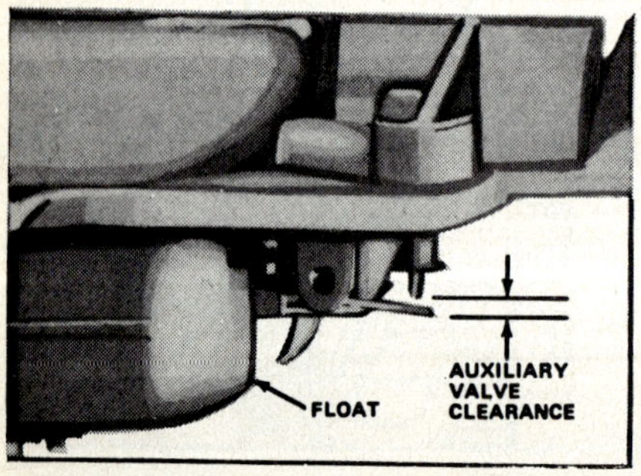

Fig. 15 Auxiliary inlet valve adjustment. Ford Motorcraft 4300, 4350-4V

Fig. 16 Pump stroke adjustment. Ford Motorcraft 4300-4V

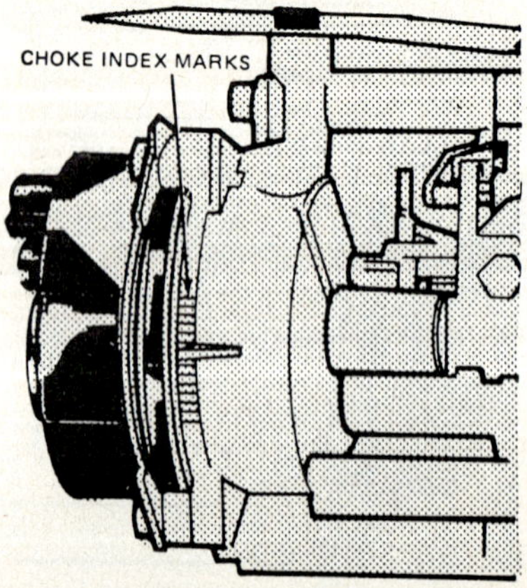

Fig. 17 Automatic choke setting. Ford Motorcraft 4300-4V

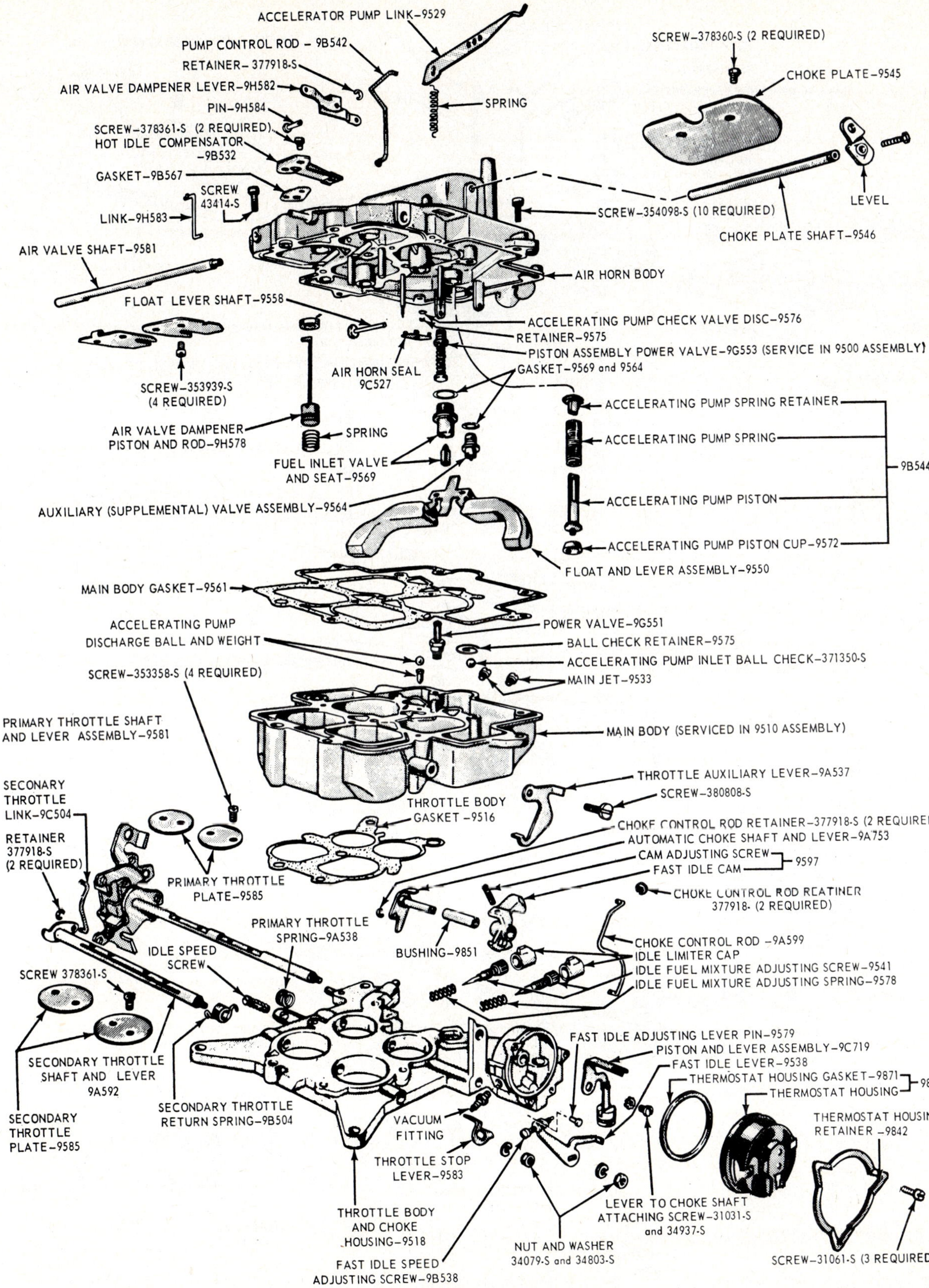

Ford Motorcraft Model 4300-A, 4V (Typical)

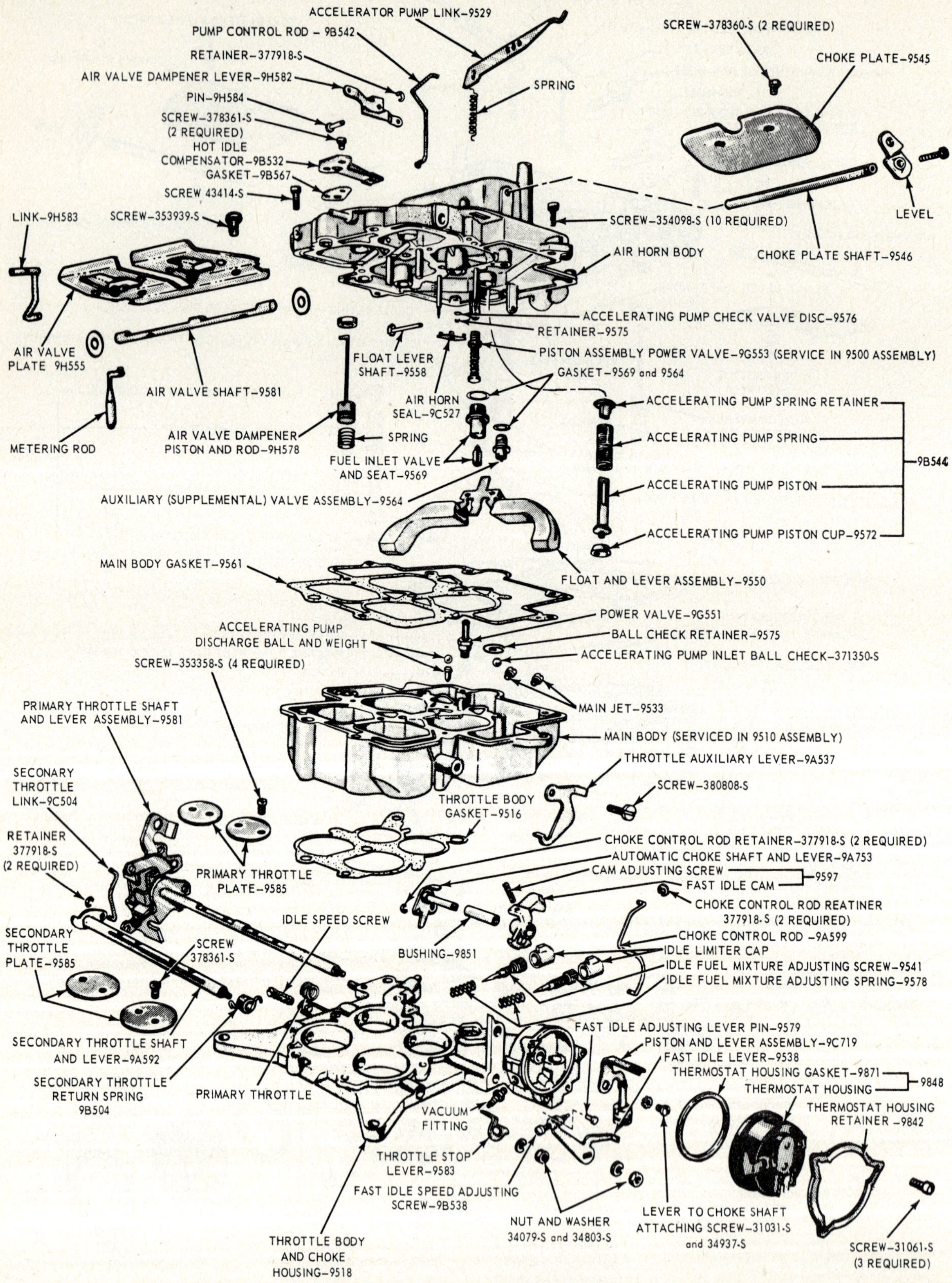

Ford Motorcraft Model 4300-D, 4V (Typical)

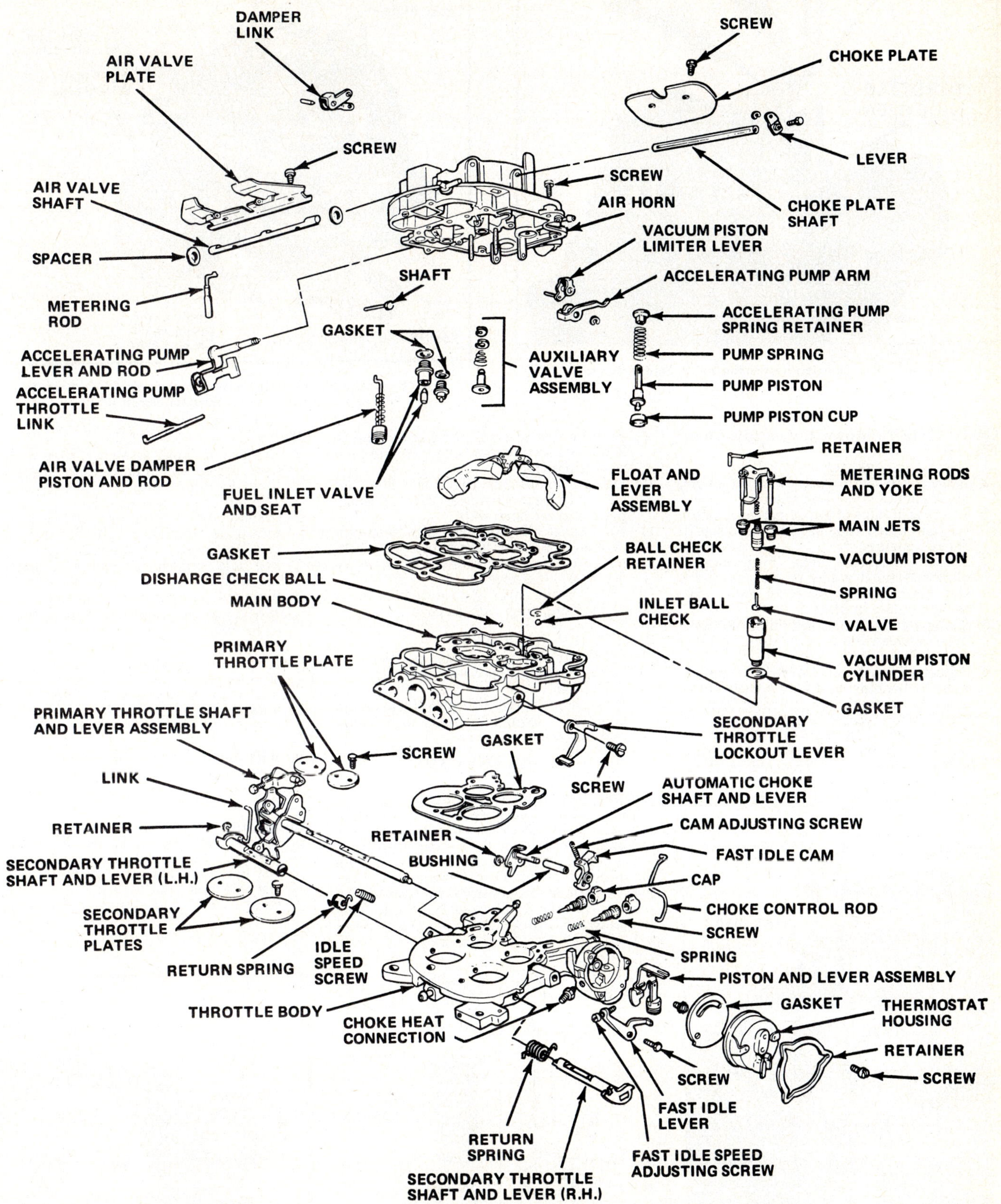

Ford Motorcraft 4350-4V (Typical)

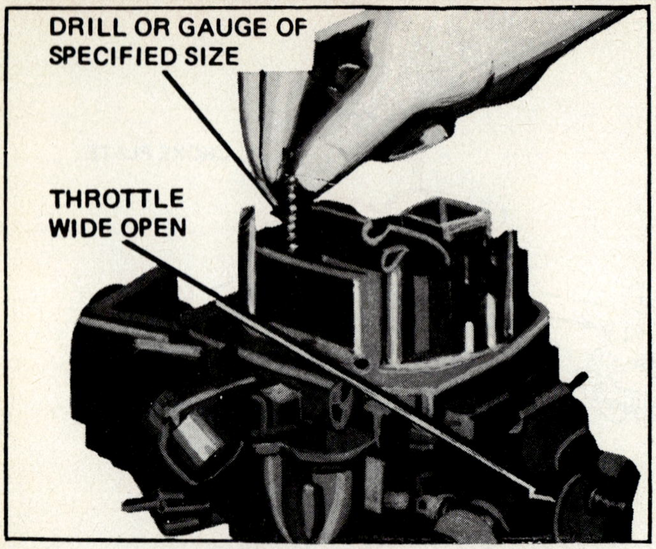

Fig. 17 Dechoke clearance adjustment. Ford Motorcraft 4300, 4350-4V

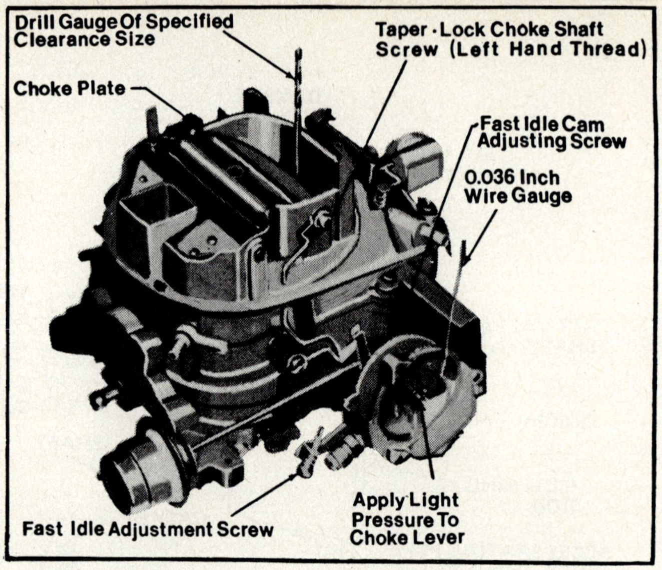

Fig. 18 Choke plate pulldown, Ford Motorcraft 4300, 4350-4V. Fast idle cam adjustment, Ford Motorcraft 4300-4V

Choke Plate Pulldown & Fast Idle Cam, Fig. 18

1. Remove choke cover.
2. Bend a .036" wire gauge at a 90° angle, approximately 1/8" from the end.
3. Insert bent edge of gauge between piston slot and upper edge of right hand slot in choke housing.
4. Rotate automatic choke lever counterclockwise until gauge is snug in piston slot. Exert light pressure on choke lever to hold gauge in place.
5. Using a gauge pin, check pulldown clearance between lower edge of choke plate and air horn wall.
6. Adjust pulldown clearance to specifications by bending adjusting arm on choke shaft lever. Bend downward to increase (upward to decrease) clearance.
7. Remove gauge and install choke cover loosely so it can rotate. Be sure thermostatic spring end is engaged in choke lever slot.
8. Rotate choke cover to 90° rich.
9. Position fast idle adjusting screw end to kickdown step on fast idle cam and hold in this position.
10. Using a gauge pin, check fast idle cam clearance between lower edge of choke plate and air horn wall.
11. Adjust fast idle cam clearance to specifications by turning adjusting screw clockwise to increase (counterclockwise to decrease) clearance.
12. Install choke cover and rotate it to specified setting, then tighten cover.

Bowl Vent Valve, Fig. 19

1. Set throttle plates in closed position.
2. Check clearance between vent valve and valve seat.
3. If clearance is not .070 inch, bend end of vent valve lever downward to decrease and upward to increase.

Model 4350-4V

Accelerator Pump Adjustment

The piston-to-shaft pin position which is the only adjustment, is pre-set to deliver the correct amount of fuel for the engine on which it is installed and should not be changed from the specified setting. Do not attempt to adjust the accelerator pump stroke by turning the vacuum limiter level adjusting nut. This adjustment is pre-set and changing it could affect drive.

1. Remove air horn assembly, then disconnect accelerator pump from operating lever by depressing spring and sliding arm out of pump shaft slot.
2. Disassemble spring and nylon keeper retaining adjustment pin. Refer to the specifications chart. If pin is not in specified hole, remove it, reposition shaft in specified hole and reinstall pin, Fig. 20.
3. Slide nylon retainer over pin and position spring on shaft; then compress spring on shaft and install pump on arm.

Vent Valve Adjustment

1. With engine at curb idle speed and solenoid throttle positioner (if used) in extended position, adjust vent valve so that arm is just touching valve stem, Fig. 21.
2. With engine off, observe that vent operating linkage pushes down on vent valve stem.

Initial Choke Pulldown

1. Remove choke housing assembly, then open throttle about half-way and position fast idle adjusting screw on high stem of fast idle cam.
2. Bend a .036" wire gauge at 90° angle,

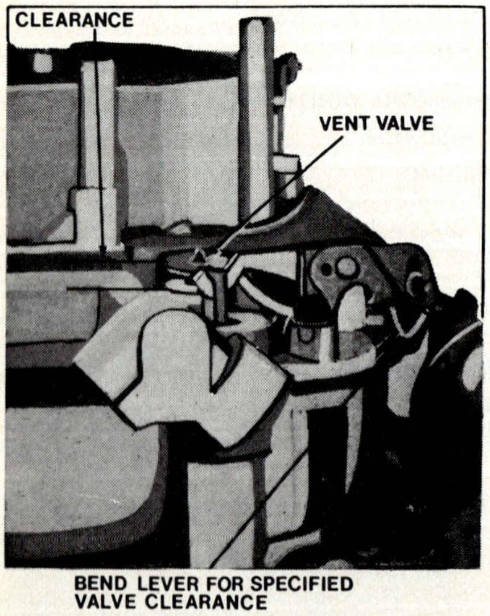

Fig. 19 Bowl vent valve adjustment. Ford Motorcraft 4300-4V

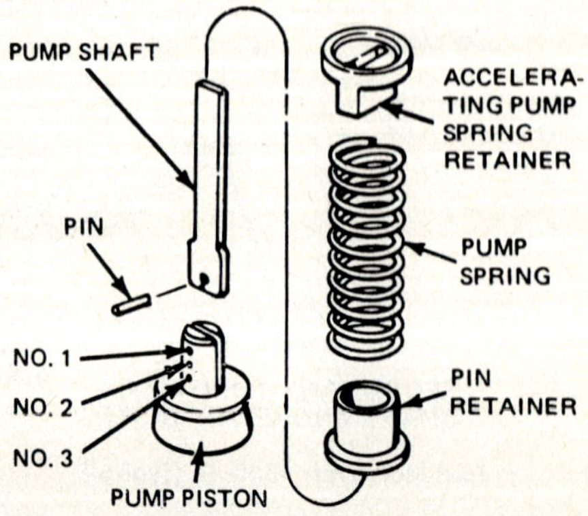

Fig. 20 Accelerator pump stroke adjustment. Ford Motorcraft 4350-4V

Fig. 21 Mechanical fuel vent valve adjustment. Ford Motorcraft 4350-4V

about 1/8" from end, then insert bent end of gauge between choke piston slot and upper edge of right hand slot in choke housing, Fig. 18
3. Rotate choke lever counterclockwise until gauge is snug in piston slot, then apply light force to lever to hold gauge and move top of choke rod away from carburetor while moving bottom of rod toward carburetor to remove end play from linkage.
4. Using a drill gauge or pin of specified size, check clearance between lower edge of choke plate and wall, Fig. 18. Refer to the Specifications Chart.
5. Adjust clearance to specifications, by turning lock screw on choke plate shaft 3 full turns in a clockwise direction. Pry choke lever from shaft to break taper lock. Choke lever should rotate freely on choke shaft.
6. With drill gauge and .036" wire gauge positioned as above, tighten screw on choke shaft.
7. Install choke housing assembly.

Delayed Choke Pulldown

The delayed choke pulldown opens the choke to a wider setting about 6 to 18 seconds after the engine is started.
1. With throttle set on fast idle cam, note position of index marks on choke housing, then loosen retaining screws and rotate cap 90° in closing (rich) direction.
2. Connect a vacuum source of 14–18 inches Hg. to vacuum supply port of delayed choke pulldown diaphragm and check clearance between lower edge of choke plate and wall, Fig. 22. Refer to the Specifications Chart.
3. Adjust clearance to specifications by turning stop screw on delayed choke pulldown diaphragm, Fig. 25.

Fast Idle Top Step Pulloff
1. Operate delayed pulldown diaphragm manually or by applying 14 to 18 inches Hg.
2. Position fast idle screw on top step fast idle cam with choke plate closed. If necessary, rotate choke housing cap to close choke plate.
3. Observe that fast idle speed adjusting screw drops to second step of cam as delayed choke pulldown diaphragm is operated, Fig. 24.

Dechoke (Unloader) Clearance
1. Hold throttle plate fully open and apply pressure on choke valve toward close position.
2. Measure clearance between lower edge of choke valve and air horn wall, Fig. 17. Refer to the Specifications Chart.
3. Adjust clearance to specifications by bending unloader tang which contacts fast idle cam forward to increase or rearward to decrease clearance.

NOTE: Do not bend tang downward from a horizontal plane. After adjustment, make certain that there is at least .070 inch clearance between unloader tang and choke housing with throttle fully open.

Float Setting

To simplify float setting, refer to Fig. 13 for construction of an adjustable float gauge and float tab bending tool.
1. Adjust gauge to specified height and insert gauge into air horn, Fig. 14.
2. Check clearance and alignment of floats to gauge. Refer to the Specifications Chart. Both floats should just touch gauge. If necessary, align floats by twisting floats slightly.
3. To raise float setting, insert open end of bending tool to right side of float lever tab between needle and float hinge. Raise float lever off needle and bend tab downward.
4. To lower float setting, insert open end of bending tool to left side of float lever tab between needle and float hinge. Support float lever and bend tab upward.
5. If above gauge is not available, invert upper body and measure distance from top of float to gasket surface of upper body. Adjust alignment and height of floats with bending tool as described previously.

Vacuum Piston & Metering Rods Adjustment
1. Remove air horn assembly, then depress metering rod hanger and turn vacuum piston adjusting screw and metering rod adjusting screws until metering rod hanger is fully seated against vacuum cylinder top face, Fig. 25.
2. Hold metering rod hanger in full downward position and turn each metering rod adjusting screw clockwise until the metering rods are now properly positioned in relation to vacuum piston.
3. Turn vacuum piston adjusting screw clockwise until specified clearance between metering rod hanger and vacuum cylinder top face is obtained.

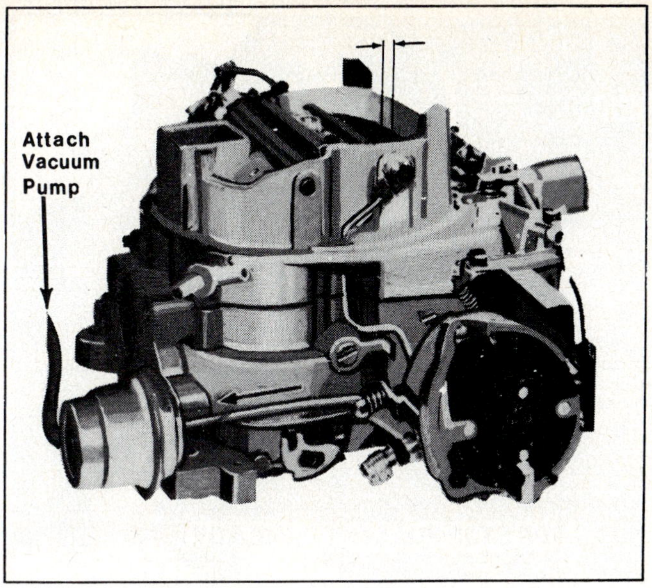

Fig. 22 Choke plate delayed pulldown adjustment. Ford Motorcraft 4350-4V

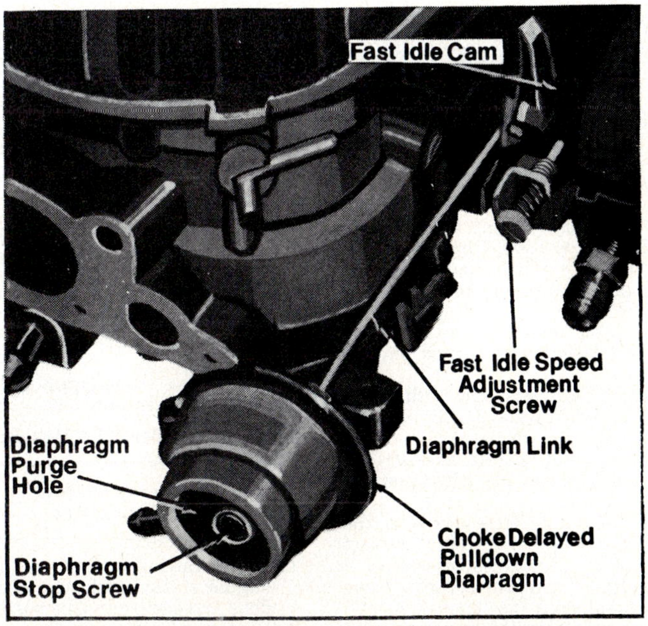

Fig. 23 Choke plate delayed pulldown stop screw adjustment. Ford Motorcraft 4350-4V

Fig. 24 Fast idle top step pulldown adjustment. Ford Motorcraft 4350-4V

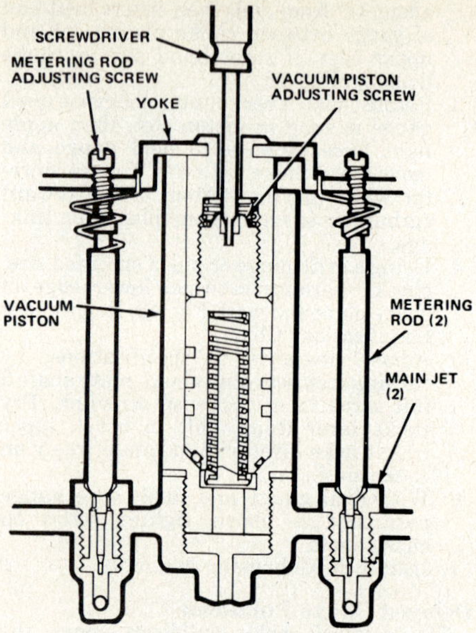

Fig. 25 Metering rod adjustment. Ford Motorcraft 4350-4V

Auxiliary Valve Setting

Check auxiliary valve clearance as shown in Fig. 15. If necessary to adjust clearance, use bending tool, Fig. 13.

Fast Idle Cam

1. Loosen choke cover screws and rotate cover to align index marks on cover and housing. Then, rotate cover an additional 90° counterclockwise and tighten attaching screws.
2. Place fast idle speed adjusting screw on kickdown (center) step of fast idle cam. Fully close choke plate and check clearance between air horn and lower edge of choke plate. Adjust by turning fast idle cam adjusting screw, ensuring adjusting screw remains on kickdown step during adjustment.
3. Loosen choke cover screws and rotate cover to specified index mark and tighten cover screws.

Model 7200 VV Adjustments, Figs. 26 thru 28

Carburetor Vent Setting

This adjustment is performed on vehicles equipped with solenoid switch vent. Remove air cleaner, then disconnect bowl vent lines from carburetor and thermal valve. Check for proper operation of bowl vent solenoid by connecting suitable vacuum pump to vent line. Turn ignition to On position, then apply vacuum. If vacuum is maintained, solenoid is operating properly. If not, solenoid is leaking and check as follows:

1. Place hand on switch vent assembly while ignition switch is turned on and off.
2. If clicking is heard and felt with hand, solenoid is activating and deactivating as voltage is applied.
3. If solenoid does not operate as described in step 2, disconnect electrical connector.
4. Check for broken or loose wires to the solenoid and for proper terminal contact in the connector. Repair as necessary.
5. Connect voltmeter and check that voltage is applied to solenoid. If solenoid is receiving voltage, check for binding or plugged vent valve and clean or replace as necessary.

Fuel Level

1. Remove upper body assembly, then remove upper body gasket. Install new gasket prior to performing adjustment.
2. Invert upper body, then position fuel level gauge, Fig. 29, onto cast surface of upper body and measure vertical distance between cast surface of upper body and bottom of float.

NOTE: Do not place gauge on gasket.

3. To adjust, bend float operating lever away from fuel inlet needle to decrease setting and toward needle to increase setting. Check and adjust float drop.

NOTE: Float pontoon must be parallel with gasket.

Float Drop

1. Position upper body assembly in upright position, then place gauge, Fig. 30, against cast surface of upper body and measure vertical distance between cast surface of upper body and bottom of float.

NOTE: Do not place gauge on gasket.

2. To adjust, bend stop tab on float lever away from hinge pin to increase setting and toward hinge pin to decrease setting.
3. Install upper body assembly.

Venturi Valve Wide Open Throttle Opening Adjustment

PRE ADJUSTMENT CHECKS

NOTE: These checks are performed with carburetor installed on engine.

1. With transmission lever in Park and engine static, push accelerator pedal to floor and secure in this position.
2. Remove air cleaner, then push venturi valve lightly toward closed position and check gap between valve and air horn, Fig. 31. This is WOT limiter opening and must be .48–.52 inch.
3. Position venturi valve at wide open stop position and check gap between valve and air horn, Fig. 32. This is WOT maximum opening and must be .99–1.01 inch.
4. If above specifications are incorrect, carburetor must be removed from vehicle and adjustments performed on bench.

ADJUSTMENT PROCEDURE

1. Remove carburetor from engine, then remove expansion plug covering venturi valve limiter adjustment screw, Fig. 33.
2. Remove WOT adjustment screw and spring with 5/32 inch allen wrench, Fig. 33.
3. Position throttle plates in wide open position, then lightly push venturi valve toward closed position and check gap between valve and air horn, Fig. 31. Using 5/64 inch allen wrench, rotate venturi valve limiter adjustment screw until gap of .48–.52 inch is obtained, Fig. 34.
4. Install WOT adjustment screw and spring.
5. Position throttle plates to wide open position, then lightly push venturi valve to wide open stop and check gap between valve and air horn, Fig. 32. Gap should be .99–1.01 inch.
6. Rotate WOT adjustment screw until specification is obtained, Fig. 35.
7. Install new expansion plug, then install carburetor onto engine.

CARTER CARBURETOR ADJUSTMENTS

Model YF, YFA

Float Adjustment, Fig. 36

Invert the air horn assembly, and check the clearance from the top of the float to the bottom of the air horn with the float level gauge. Hold the air horn at eye level when gauging the float level. The float arm (lever) should be resting on the needle pin. Do not load the needle when adjusting the float. Bend the float arms as necessary to adjust the float level (clearance). Do not bend the tab at the end of the float arm. It prevents the float from striking the bottom of the fuel bowl when empty.

Pump Adjustment

With throttle valve seated in bore of carburetor, press down on upper end of diaphragm shaft until it reaches its bottom position. The metering rod arm should now contact the pump lifter link at the outer end nearest the springs. Adjust by bending the pump connector link at its lower angle.

Metering Rod Adjustment, Fig. 37

Back out the idle speed adjusting screw until the throttle plate is closed tight in the throttle bore. Press down on upper end of diaphragm shaft until diaphragm bottoms in vacuum chamber. Metering rod should contact bottom of metering rod well, and metering rod should contact lifter link at the outer end, nearest the springs and at supporting lug. For models not equipped with metering rod adjusting screw, adjust by bending lip of metering rod arm to which metering rod is attached, up or down as required. For models equipped with a metering rod adjusting screw, turn the adjusting screw until metering rod just bottoms in the body casting. For final adjustment turn metering rod adjusting screw in (clockwise) one additional turn.

Fast Idle Cam Linkage Adjustment, Fig. 38

Position fast idle screw on second step of fast idle cam and against shoulder of highest step. Using a drill of specified size, check clearance between lower edge of choke plate and bore. Refer to the Specifications Chart. To adjust, bend choke connector rod as required.

Choke Unloader Adjustment, Fig. 39

With throttle valve held wide open and choke valve held toward closed position with a rubber band, there should be the clearance listed in the *Specifications Chart* between lower edge of choke valve and inner air horn wall. Adjust by bending unloader lug on choke trip lever.

Choke Plate Pulldown Adjustment

Bend a 0.026 in. diameter wire gauge at a 90 degree angle approximately 1/8-inch from one end. Insert the bent end of the gauge between the choke piston slot and the right hand slot in choke housing. Rotate the choke piston lever counterclockwise until gauge is snug in the piston slot. Exert a light pressure on choke piston lever to hold the gauge in place, then use a drill with a diameter equal to the specified edge of choke plate and carburetor bore to check clearance, Fig. 40.

To adjust the choke plate pulldown clearance, bend the choke piston lever as required to obtain specified setting.

NOTE: When bending the lever, be careful not to distort the piston link. Install the choke thermostatic spring housing and gasket. Set the housing to specifications.

Dechoke Adjustment

Hold the throttle plate fully open and close the choke plate as far as possible without forcing it. Use a drill of specified diameter to check the clearance between choke plate and air horn, Fig. 41. If clearance is not within specification, adjust by bending arm on choke trip lever of the throttle lever. Bending the arm downward will decrease the clearance, bending it upward will increase the clearance.

If the choke plate clearance and fast idle cam linkage adjustment was performed with the carburetor on the engine, adjust the engine idle speed and fuel mixture. Adjust dashpot (if so equipped).

Dashpot Adjustment

With the engine idle speed and mixture properly adjusted, the engine at normal operating temperature, loosen the antistall dashpot lock nut, Fig. 41. Hold the throttle in the curb idle position and depress the dashpot plunger. Measure the clearance between the throttle lever and plunger tip. Turn the antistall dashpot to provide 7/64" ± 1/64" clearance between the tip of the plunger and the throttle lever. Tighten the locknut to secure the adjustment.

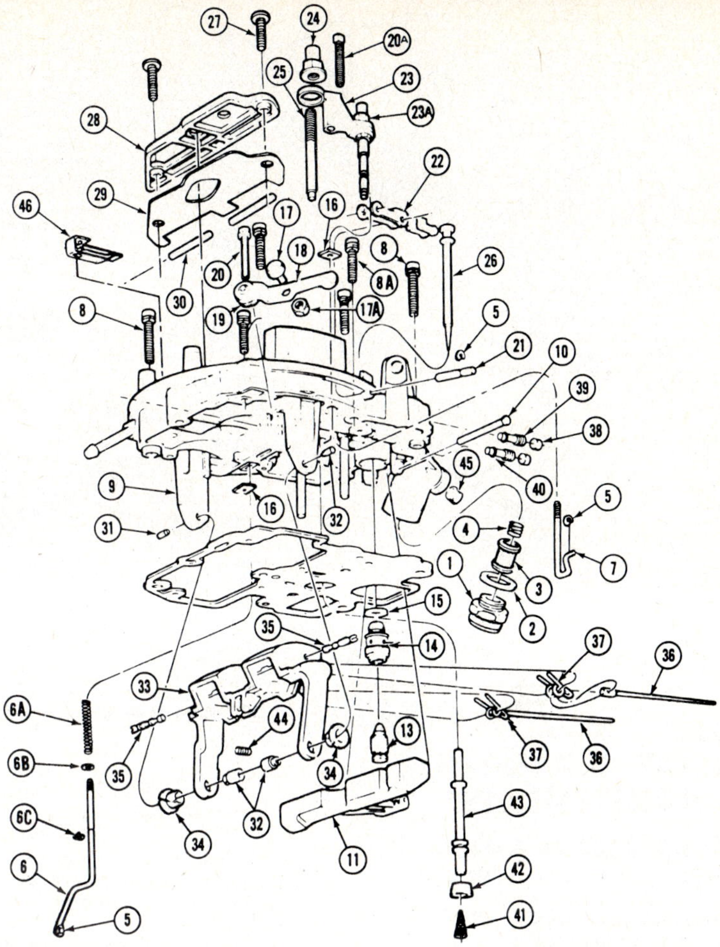

1. FUEL INLET FITTING	15. FUEL INLET SEAT GASKET	30. ROLLER BEARING
2. FUEL INLET FITTING GASKET	16. DUST SEAL	31. VENTURI VALVE PIVOT PLUG
3. FUEL FILTER	17. SCREW	32. VENTURI VALVE PIVOT PIN
4. FUEL FILTER SPRING	17A. NUT	33. VENTURI VALVE
5. 1/8 RETAINING E-RING	18. ACCELERATOR PUMP LINK AND SWIVEL ASSEMBLY	34. VENTURI VALVE PIVOT PIN BUSHING
6. ACCELERATOR PUMP ROD	19. ACCELERATOR PUMP SWIVEL (PART OF 18)	35. METERING ROD PIVOT PIN
6A. SPRING—OVERTRAVEL	20. NUT — NYLON ADJUSTING	36. METERING ROD
6B. WASHER	20A. CHOKE TRAVEL STOP ADJUSTING SCREW	37. METERING ROD SPRING
6C. CLIP	21. CHOKE HINGE PIN	38. CUP PLUG
7. CHOKE CONTROL ROD	22. COLD ENRICHMENT ROD LEVER	39. MAIN METERING JET ASSEMBLY
8. SCREW (2) 8-32 X .88	23. COLD ENRICHMENT ROD SWIVEL	40. "O" RING
8A. SCREW (5) 8-32 X .75	23A. COLD ENRICHMENT ROD ADJUSTING NUT	41. ACCELERATOR PUMP RETURN SPRING
9. UPPER BODY	24. CONTROL VACUUM REGULATOR ADJUSTING NUT	42. ACCELERATOR PUMP CUP
10. FLOAT HINGE PIN	25. CONTROL VACUUM REGULATOR	43. ACCELERATOR PUMP PLUNGER
11. FLOAT ASSEMBLY	26. COLD ENRICHMENT ROD	44. VENTURI VALVE LIMITER ADJUSTING SCREW
12. FLOAT BOWL GASKET	27. SCREW (2) 8-32 X .75	45. PIPE PLUG
13. FUEL INLET VALVE	28. VENTURI VALVE COVER PLATE	46. HOT IDLE COMPENSATOR
14. FUEL INLET SEAT	28A. PLUG	
	29. GASKET	

Fig. 26 Upper body. Motorcraft 7200 VV

Automatic Choke Adjustment

Loosen choke cover retaining screws and turn choke cover so that line or index mark on cover lines up with the specified mark listed in the *Specifications Chart* on choke housing.

Fuel Bowl Vent Adjustment

1. With engine at operating temperature and idle RPM set to specification, open throttle lever so that there is clearance between throttle lever actuating lever and fuel bowl vent rod, Fig. 42.
2. Close throttle lever to idle set position and measure travel of fuel bowl vent rod at point A. Travel of vent rod at point A should be .100–.150 inch.
3. Bend throttle actuating lever at point indicated in Fig. 42 to adjust as necessary.

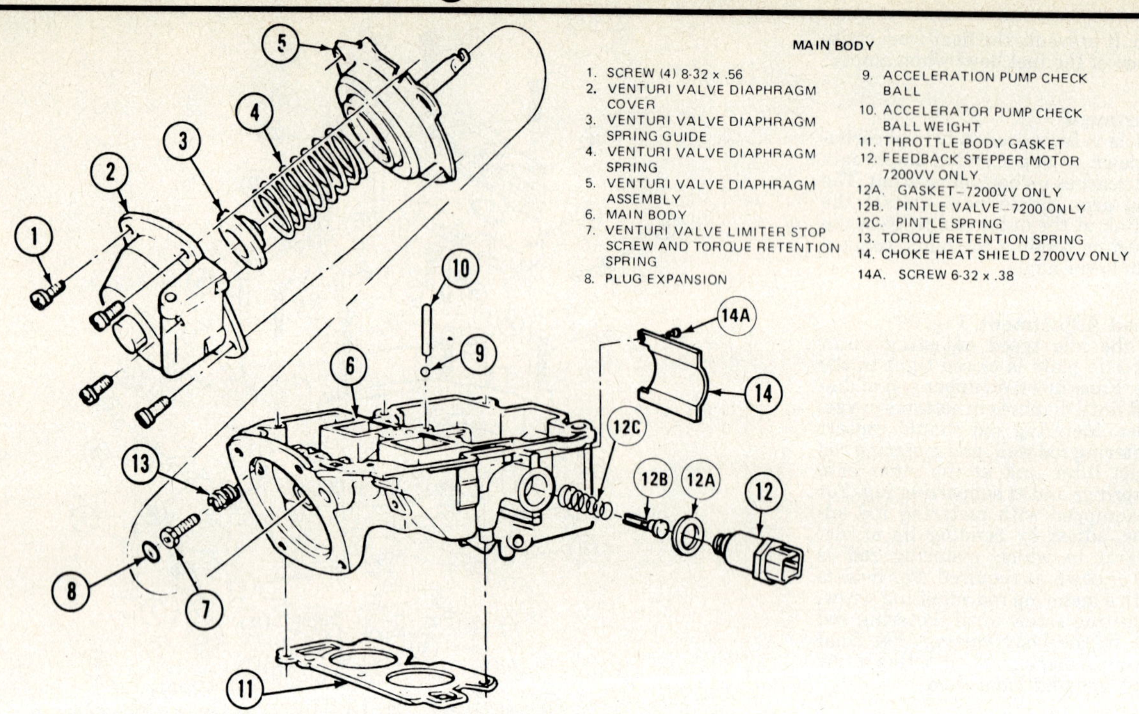

Fig. 27 Main body. Motorcraft 7200 VV

HOLLEY CARBURETOR ADJUSTMENTS

Models 2300-2V & 4150, 4160, 4180-4V & 4190-4V

Float Adjustment (Dry), Fig. 43

The dry float adjustment is a preliminary fuel level adjustment only. The final adjustment must be performed after the carburetor is installed on the engine.

With fuel bowl and float removed, adjust the float so that it is parallel to the fuel bowl with the top of the fuel bowl inverted. On four-barrel units, adjust both sides in the same manner.

Fuel Level (Wet) Adjustment, Fig. 44

1. Operate engine to normalize engine temperatures and place the vehicle on a flat surface as near level as possible. Stop engine and remove air cleaner.
2. Check fuel lever in each bowl separately. Place a suitable container below the fuel level sight plug in the fuel bowl to collect any spillover of fuel. Carefully remove sight plug and gasket. On a four-barrel unit, check each fuel bowl separately. The fuel level within the bowl should be at the lower edge of the sight plug opening (plus or minus 1/16").
3. If the fuel level is too high, drain the bowl and refill it. Check the fuel level again before altering the float setting. Install the bolt removed for draining and sight plug and start engine to fill the fuel bowl. After the engine has stabilized, stop the engine and check the fuel level.
4. If the fuel level is still too high, it should first be lowered below specifications and then raised until it is just at the lower edge of the sight plug opening. To adjust the fuel level, proceed as follows:
5. With engine stopped, loosen lock screw just enough to allow rotation of the adjusting nut underneath. *Do not loosen lock screw or attempt to adjust the fuel level with the sight plug removed and engine running because the pressure in the line will spray fuel out and cause a fire.*
6. Turn adjusting nut approximately 1/2 turn IN to lower the fuel below the sight plug opening (1/6 turn, depending upon direction of rotation, will raise or lower the float at the sight plug opening 3/64"). Tighten lock screw and install sight plug. Start engine and check.

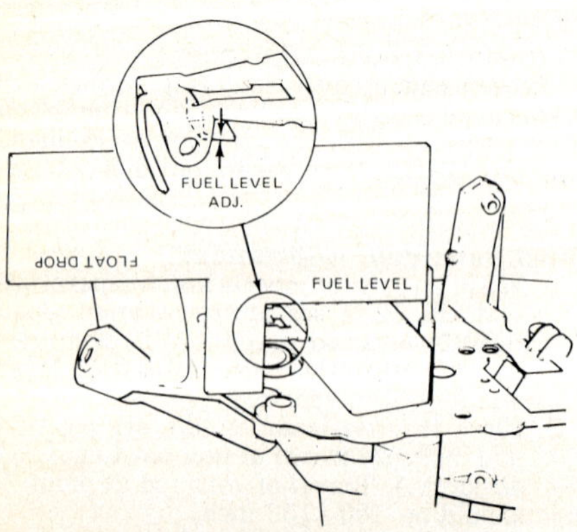

Fig. 29 Fuel level adjustment. Motorcraft 7200 VV

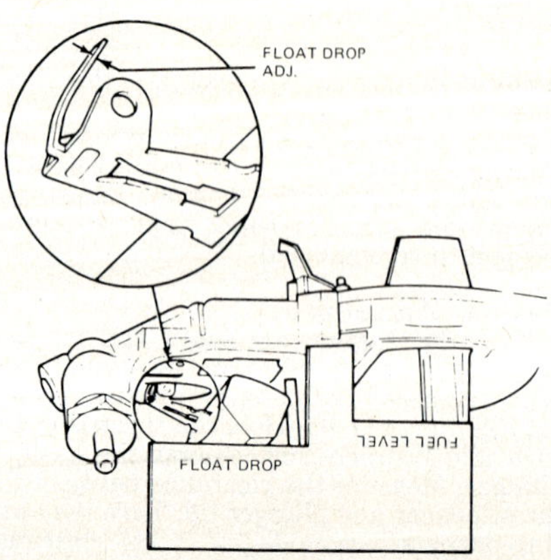

Fig. 30 Float drop adjustment. Motorcraft 7200 VV

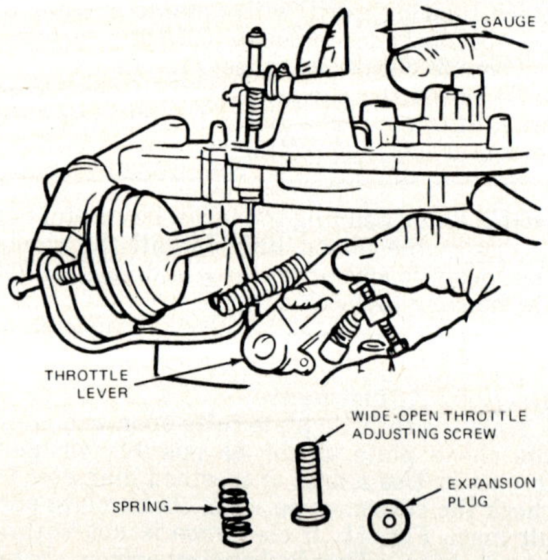

Fig. 31 Checking gap between venturi valve & air horn. Motorcraft 7200 VV

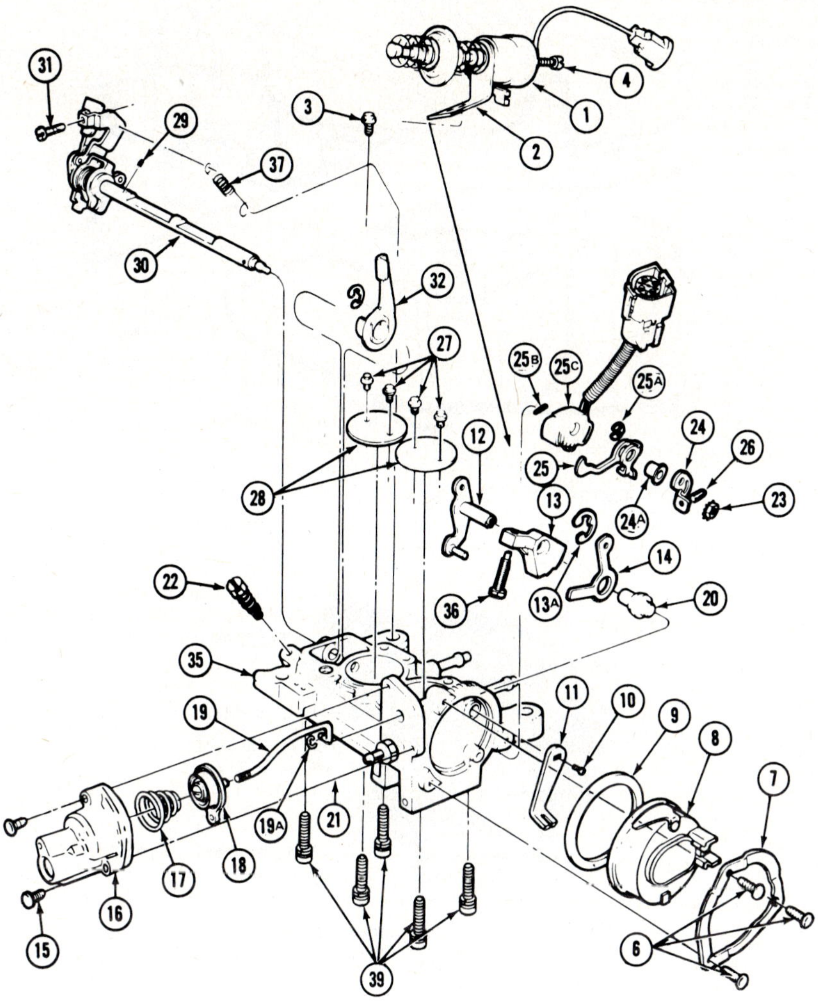

1.	THROTTLE RETURN CONTROL DEVICE.	22.	CURB IDLE ADJUSTING SCREW (TSP OFF).
2.	THROTTLE RETURN CONTROL DEVICE BRACKET.	23.	RETAINING NUT, 10-32
3.	MOUNTING SCREW, 10-32 X .50	24.	FAST IDLE ADJUSTING LEVER.
4.	ADJUSTING SCREW (TSP ON)	24A.	NYLON WASHER AND BUSHING ASSEMBLY.
5.	(NOT APPLICABLE)	25.	FAST IDLE LEVER.
6.	SCREW (3), 8-32 X .50	25A	LARGE E-CLIP.
7.	CHOKE THERMOSTATIC HOUSING RETAINER.	25B.	ROLL PIN.
8.	CHOKE THERMOSTATIC HOUSING.	25C.	THROTTLE POSITION SENSOR (7200VV ONLY).
9.	CHOKE THERMOSTATIC HOUSING GASKET.	26.	FAST IDLE ADJUSTING SCREW.
10.	SCREW, 6-32 X .50	27.	THROTTLE PLATE SCREWS (4).
11.	CHOKE THERMOSTATIC LEVER.	28.	THROTTLE PLATES.
12.	CHOKE SHAFT LEVER AND PIN ASSEMBLY	29.	VENTURI VALVE LIMITER STOP PIN.
13.	FAST IDLE CAM. 13A. LARGE E-CLIP	30.	THROTTLE SHAFT ASSEMBLY.
14.	FAST IDLE INTERMEDIATE LEVER	31.	TRANSMISSION KICKDOWN ADJUSTING SCREW.
15.	SCREW (2), 8-32 X .75	32.	VENTURI VALVE LIMITER LEVER AND BUSHING ASSEMBLY.
16.	CHOKE CONTROL DIAPHRAGM COVER.	33.	E-CLIP.
17.	CHOKE CONTROL DIAPHRAGM SPRING.	34.	(NOT APPLICABLE)
18.	CHOKE CONTROL DIAPHRAGM ASSEMBLY.	35.	THROTTLE BODY.
19.	CHOKE CONTROL DIAPHRAGM ROD.	36.	FAST IDLE CAM ADJUSTING SCREW.
19A.	CLIP	37.	TRANSMISSION KICKDOWN LEVER RETURN SPRING (IF EQUIPPED)
20.	CHOKE HOUSING BUSHING.	38.	(NOT APPLICABLE)
21.	CHOKE HEAT TUBE (IF EQUIPPED)	39.	SCREW (5) 8-32 X .75

Fig. 28 Throttle body. Motorcraft 7200 VV

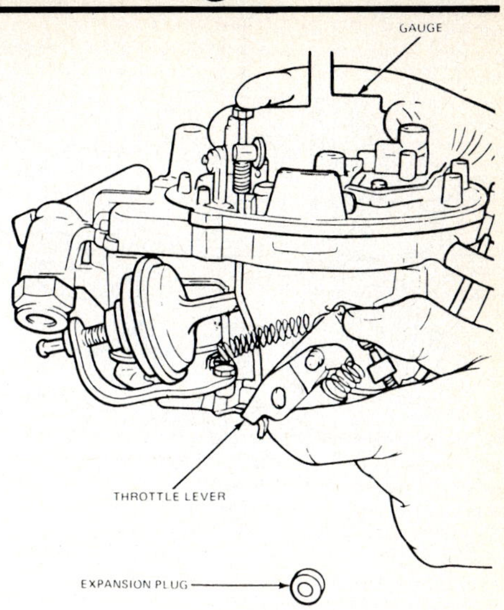

Fig. 32 Checking venturi valve maximum opening. Motorcraft 7200 VV

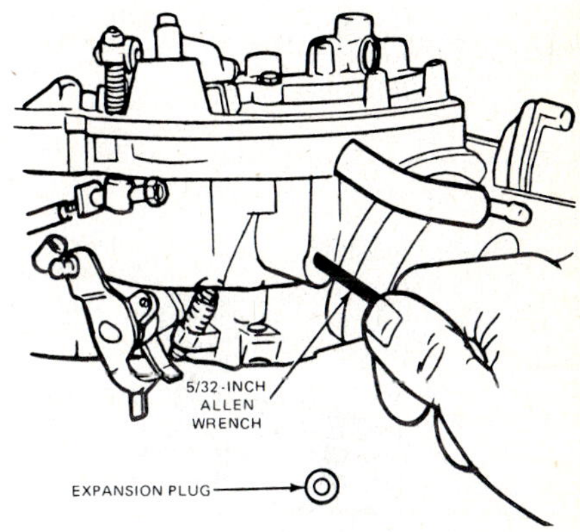

Fig. 33 Removing WOT adjusting screw. Motorcraft 7200 VV

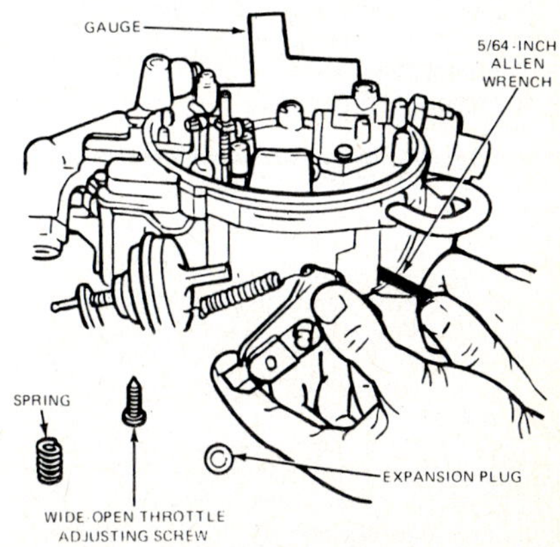

Fig. 34 Adjusting venturi valve limiter. Motorcraft 7200 VV

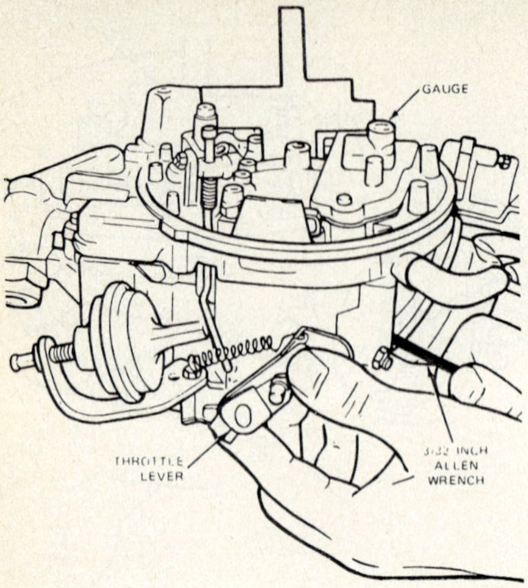

Fig. 35 Adjusting venturi valve maximum opening clearance. Motorcraft 7200 VV

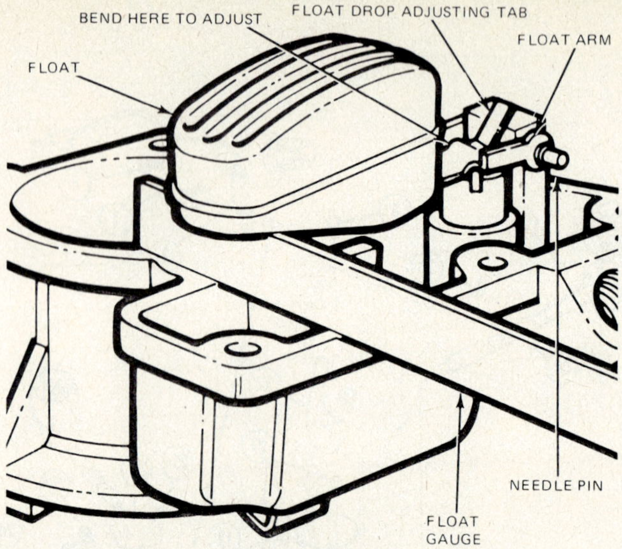

Fig. 36 Float level adjustment. Carter YF, YFA

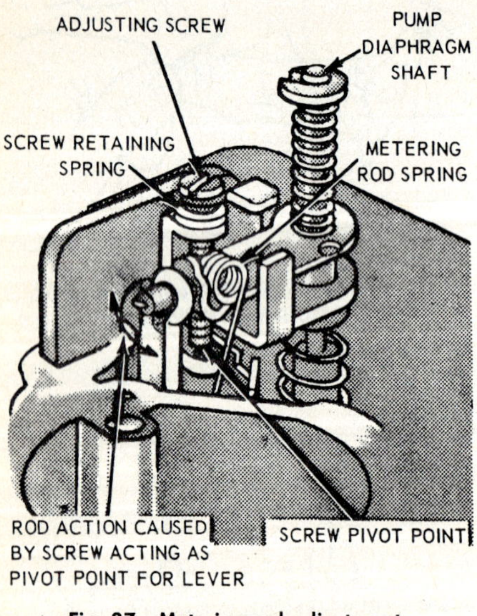

Fig. 37 Metering rod adjustment. Carter YF, YFA

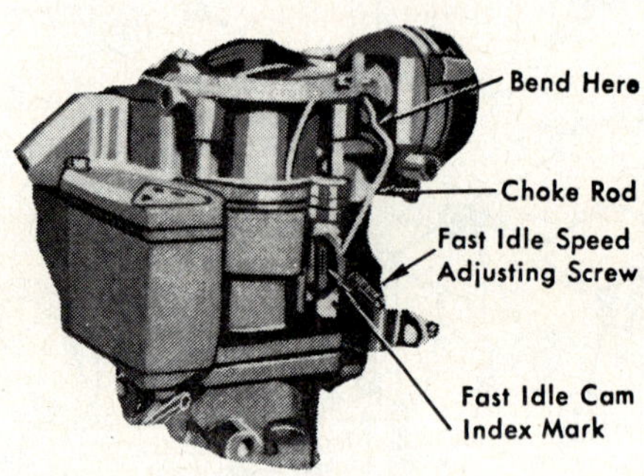

Fig. 38 Fast idle cam linkage adjustment, Carter YF, YFA

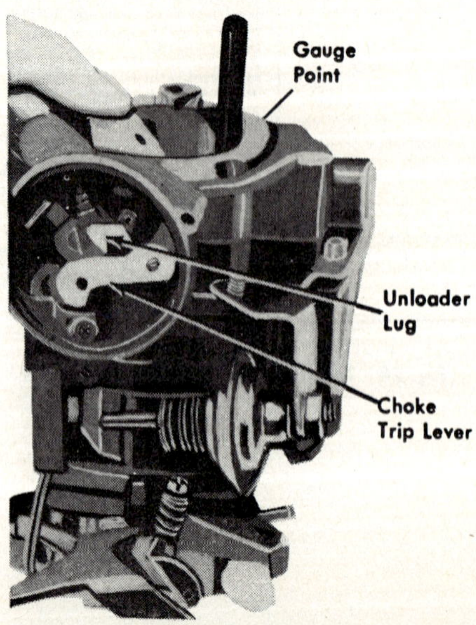

Fig. 39 Choke unloader adjustment. Carter YF

Fig. 40 Choke plate pulldown adjustment. Carter YF & YFA

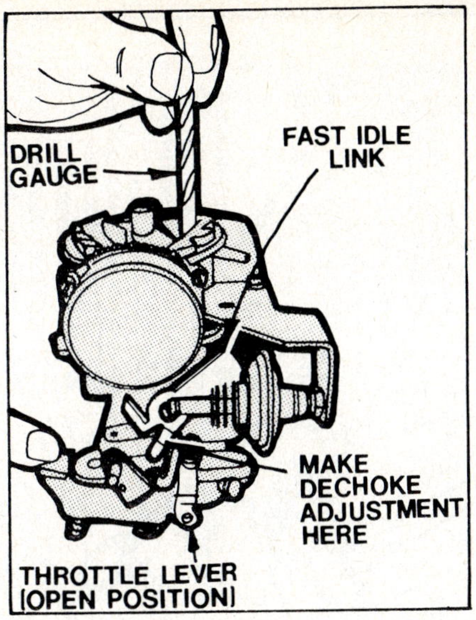

Fig. 41 Dechoke adjustment. Carter YF, YFA

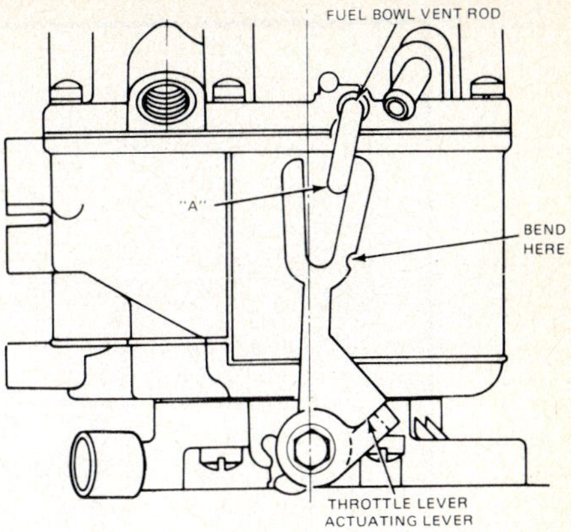

Fig. 42 Fuel bowl vent adjustment. Carter YFA

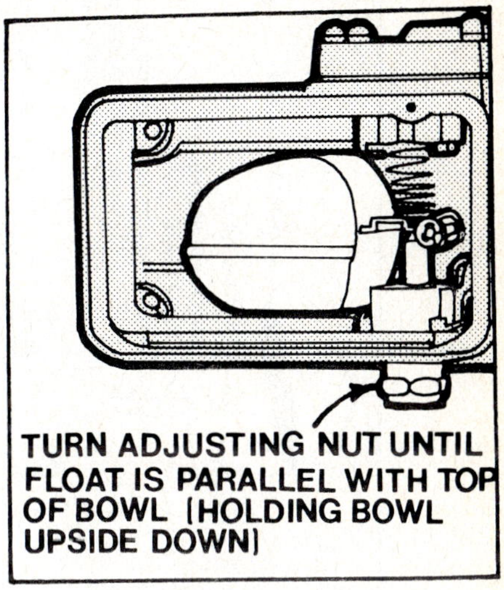

Fig. 43 Float adjustment. Holley 2300-2V & 4150, 4160, 4180-4V & 4190-4V

7. If adjustment is not satisfactory, loosen lock screw and turn adjusting nut in increments of 1/6 turn or less until correct fuel level is achieved. After each adjustment, tighten lock screw, install sight plug, start engine and, after fuel is stabilized, check the fuel level at the sight plug opening.

Accelerating Pump Level, Fig. 45

1. Using a feeler gauge and with throttle plates wide open, there should be the specified clearance (minimum) between pump operating lever adjustment screw and pump arm when pump arm is depressed manually.
2. If adjustment is required, loosen lock nut and turn adjusting screw in to increase clearance and out to decrease clearance. One-half turn of screw is equal to approximately .015".
3. When proper adjustment has been made, hold screw in position with a wrench and tighten lock nut.

Accelerating Pump Stroke

To satisfy the requirements in various climates, the pump discharge can be adjusted. The bottom hole (No. 2) in the pump cam and throttle lever provides the maximum pump discharge for extreme cold weather, and the top hole (No. 1) provides the minimum pump discharge for warm weather operation.

If a change in the adjustment is required, make certain the proper hole in plastic pump cam, located behind throttle lever, is properly indexed with the numbered hole in the throt-

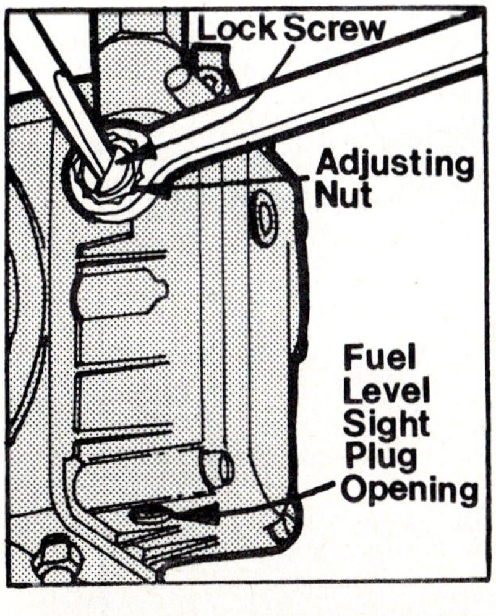

Fig. 44 Fuel level adjustment. Holley 2300-2V & 4150, 4160 4160 4180-4V & 4190-4V

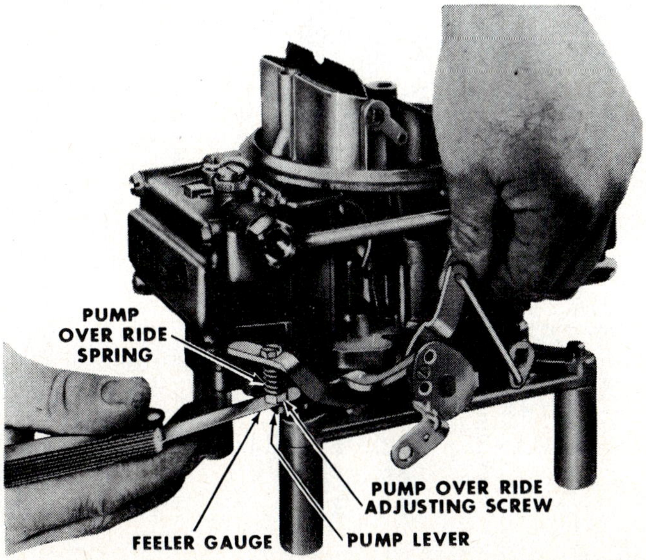

Fig. 45 Accelerator pump lever adjustment. Holley 2300-2V & 4150, 4160, 4180-4V & 4190-4V (Typical)

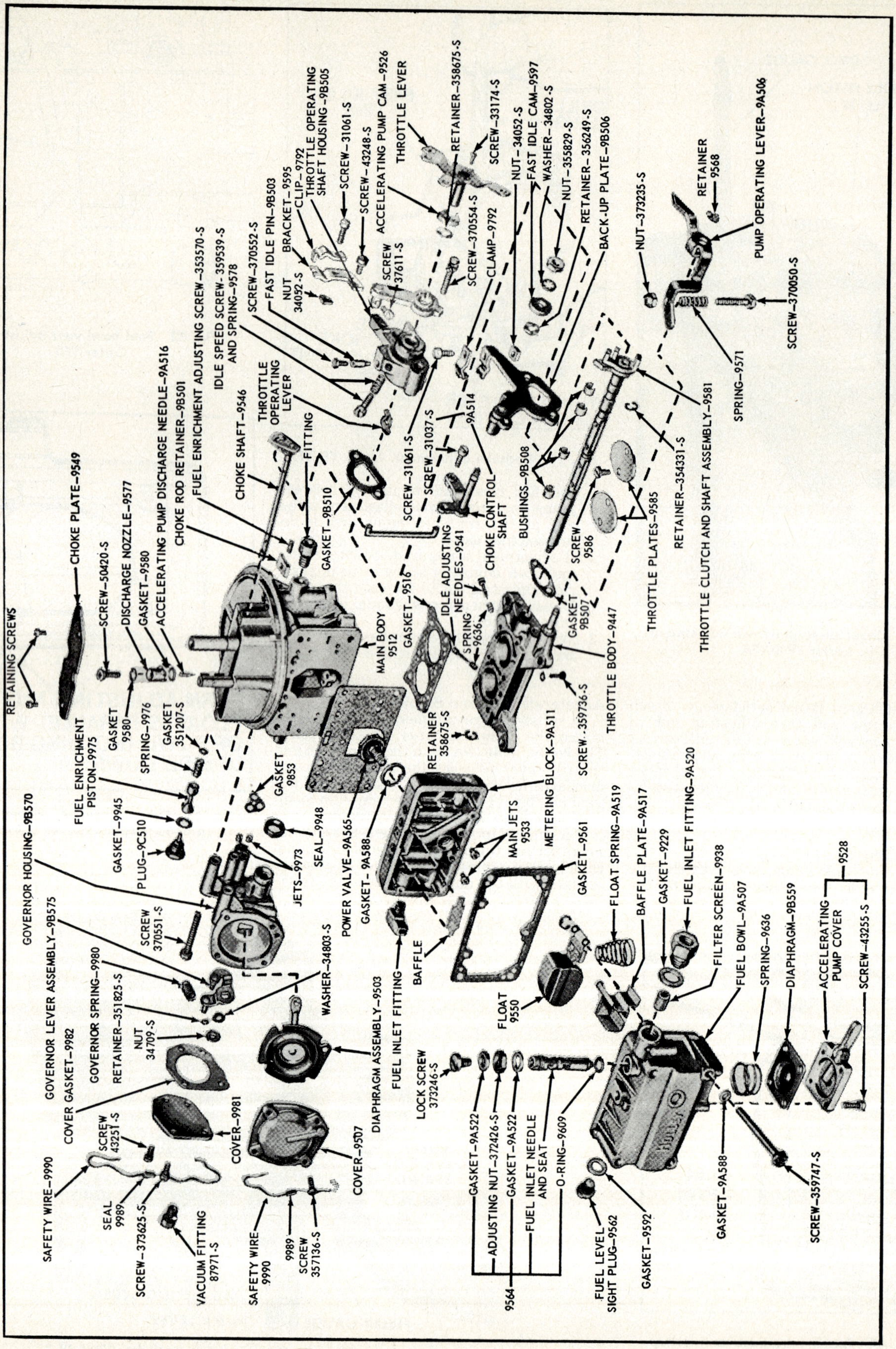

Holley Model 2300-2V (Typical)

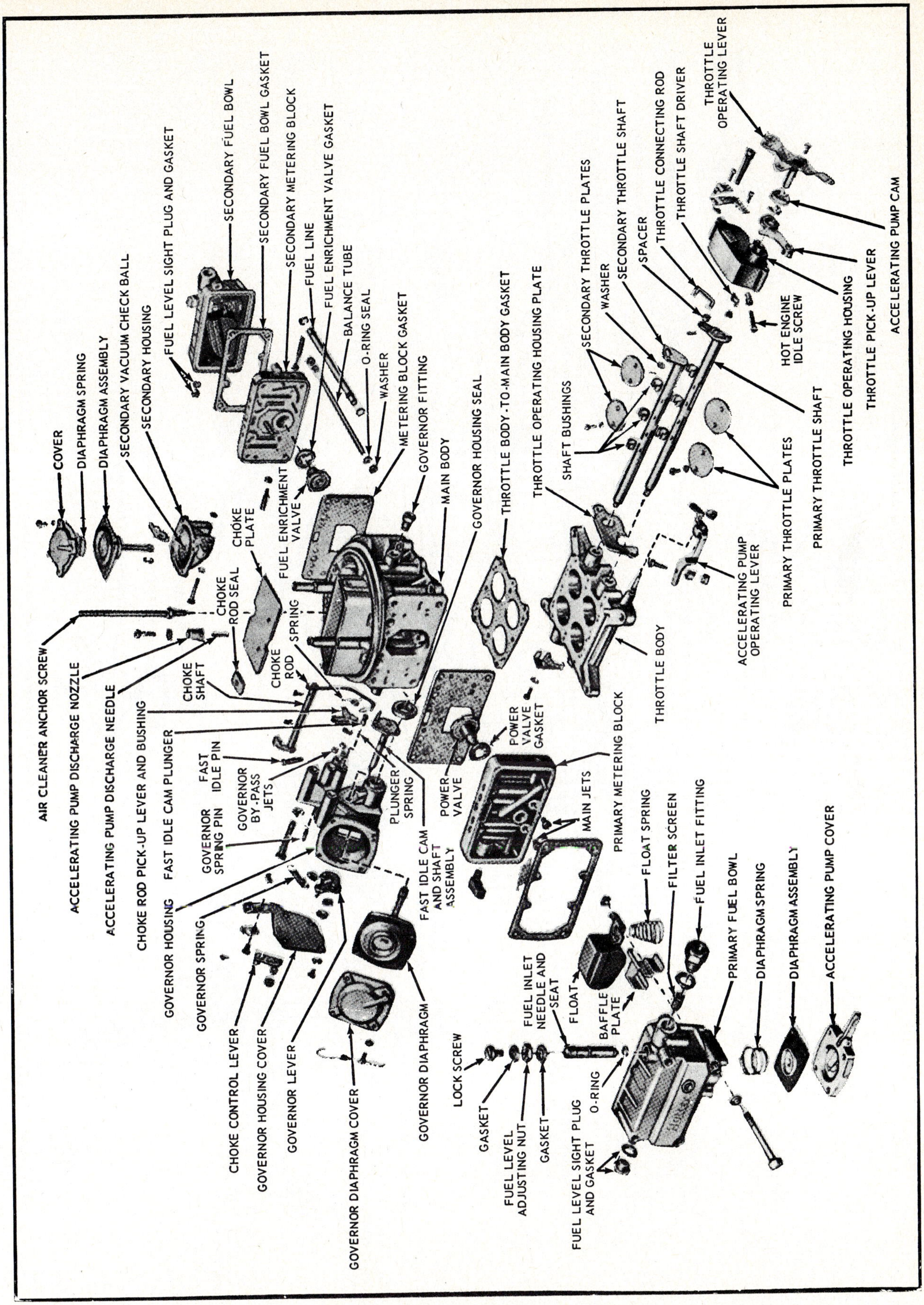

Holley Model 4150-4V (Typical)

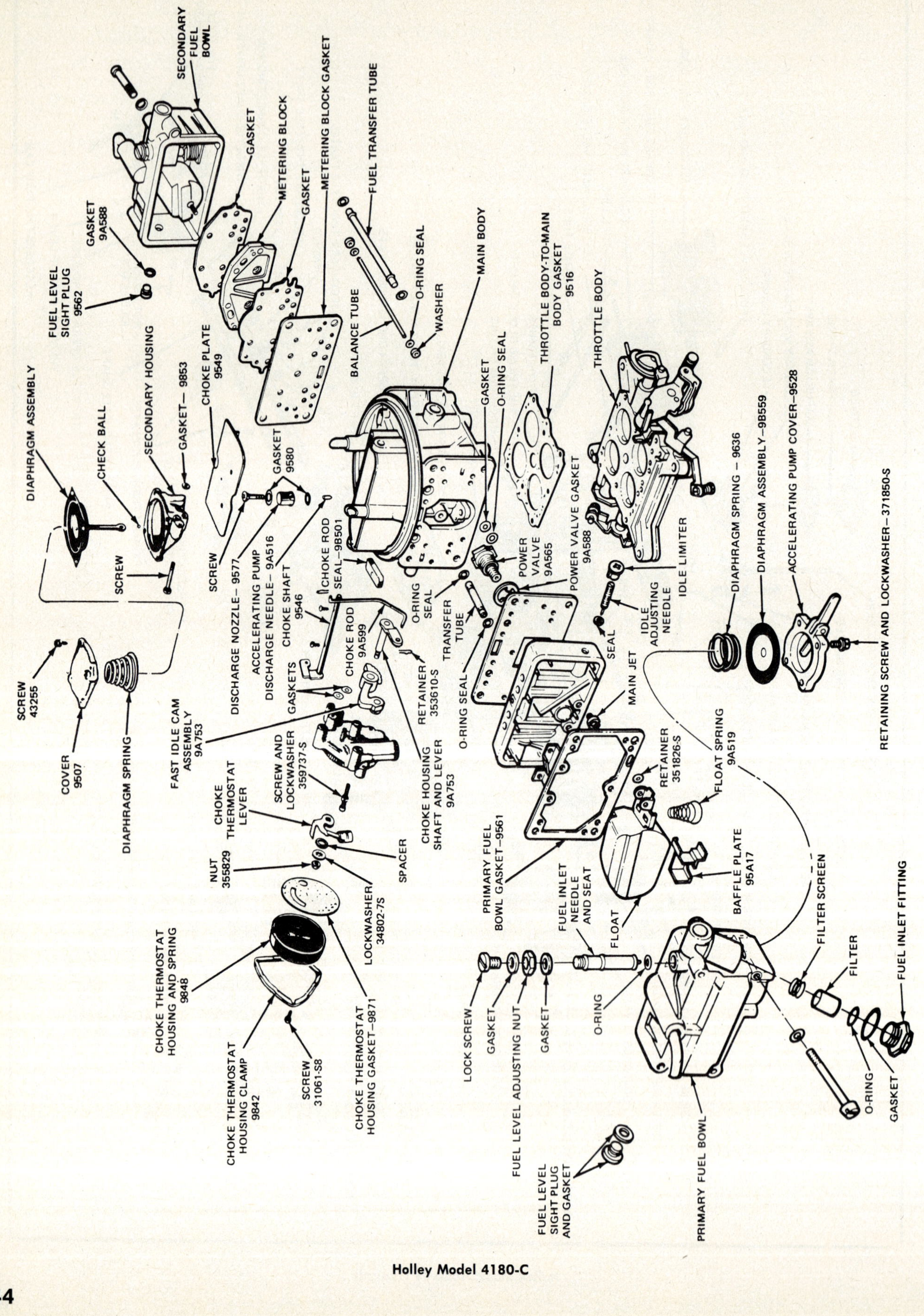

Holley Model 4180-C

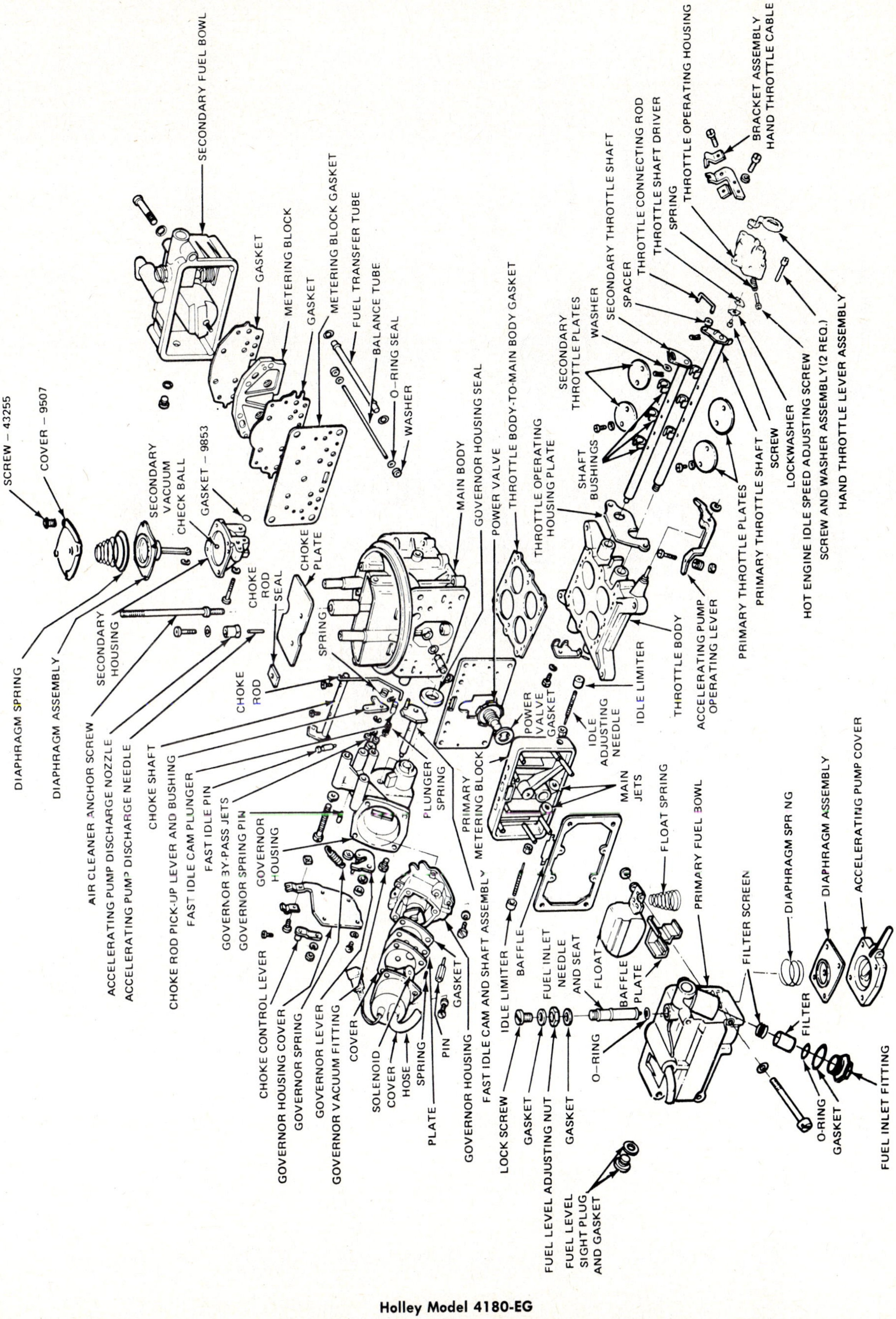

Holley Model 4180-EG

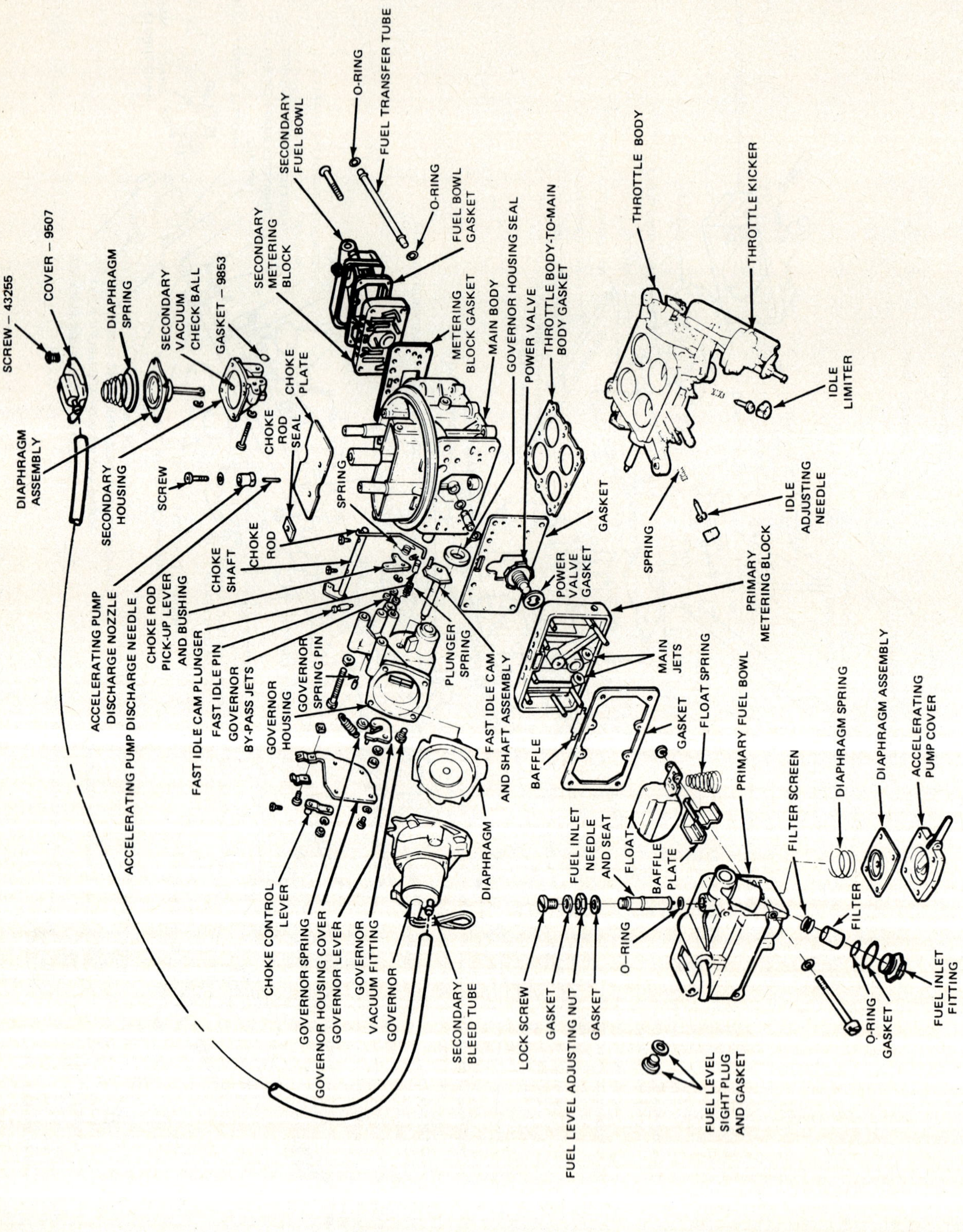

Holley Model 4190-EG

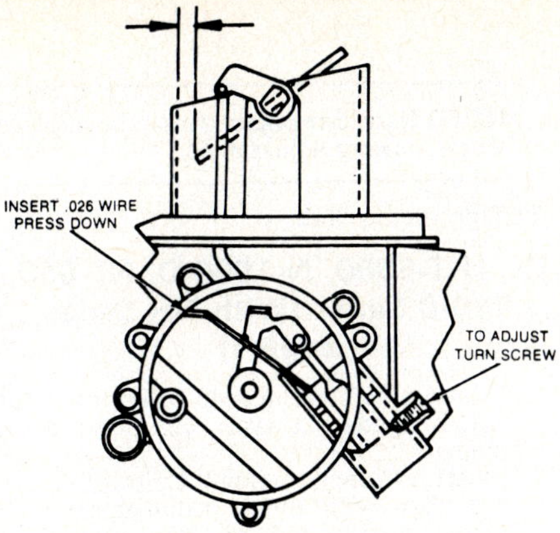

Fig. 46 Choke plate pulldown adjustment. Holley 4180 units

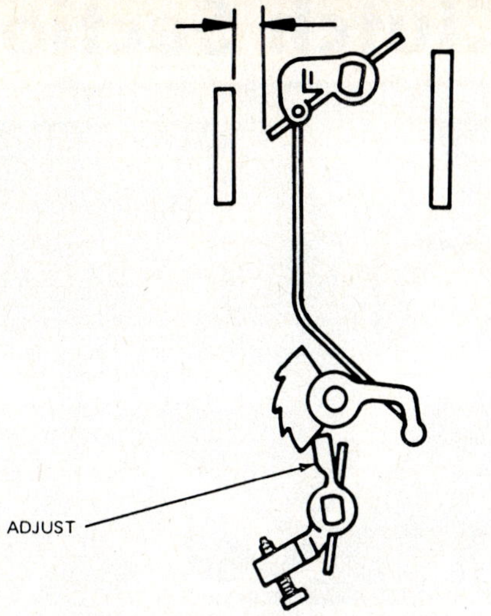

Fig. 47 Dechoke adjustment. Holley 4180 units

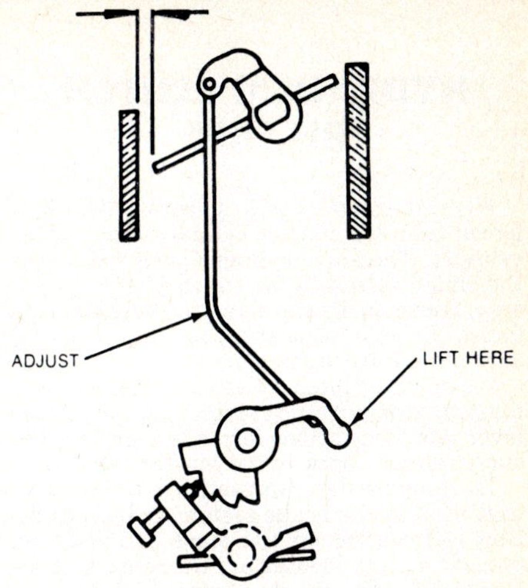

Fig. 48 Fast idle cam adjustment.

tle lever before installing the retaining screw.

Secondary Throttle Plate Adjustment, 4150, 4160, 4180 & 4190 Units

With secondary throttle plates closed, rotate secondary throttle shaft lever adjusting screw counter-clockwise until the secondary throttle plates seat in the bores. On all units rotate adjusting screw clockwise until it contacts the secondary lever, then an additional 1/4 turn on 4190 & 4150 units or 1/2 turn on 4160 units.

Choke Plate Pulldown, 4160 Units

1. Remove choke thermostatic spring housing from carburetor.
2. Block throttle half open so the fast idle cam does not contact the adjusting screw.
3. With a suitable tool, depress choke piston down against the adjusting screw.
4. Check clearance between front edge of choke plate and the air horn wall. The clearance should be as follows: No. D4TE-ARA, .150 inch; No. D4TE-ASA, .180 inch; Nos. D5TE-EA, GA, .190 inch.
5. To adjust clearance, rotate adjusting screw clockwise to decrease clearance or counter-clockwise to increase clearance.
6. Install thermostatic spring housing on carburetor and rotate housing counter-clockwise to align the index mark on spring housing with center index mark on choke housing.

Choke Plate Pulldown, 4180 Units

1. Remove thermostat housing, gasket and retainer.
2. Insert a piece of wire, Fig. 46, into choke piston bore to move the piston down against the stop screw. Maintain light closing pressure on choke plate and measure the gap between the lower edge of the choke plate and air horn wall.
3. To adjust, remove the putty covering the adjustment screw and turn clockwise to decrease or counter-clockwise to increase gap setting.
4. Reinstall choke thermostatic housing, gasket and retainer. Do not tighten screws until Fast Idle Cam Set adjustment has been performed.

Dechoke Clearance Adjustment, 4160 Units

1. With throttle held in the wide open position, rotate choke plate toward closed position until pawl on fast idle speed lever contacts fast idle cam.
2. Check clearance between upper edge of choke plate and air horn wall. Clearance should be as follows: D4TE-ARA, ASA, .300 inch; D5TE-EA, GA, .315 inch.
3. To adjust clearance, bend the pawl on the fast idle speed lever forward to increase clearance or rearward to decrease clearance.

Dechoke Clearance, 4180 Units

1. Hold throttle in wide open throttle position.
2. Apply a light closing pressure on the choke plate and measure the gap between the lower edge of the choke plate and air horn wall.
3. To adjust, bend the pawl on the fast idle lever, Fig. 47, forward to increase or backward to decrease the clearance.

Fast Idle Cam Set, 4180 Units

1. Rotate choke cap 45 degrees counter-clockwise to close the choke plate. Tighten screws.
2. Open and close throttle to place fast idle screw on top step of cam, Fig. 48.
3. Place a pulldown gauge between lower edge of choke plate and air horn wall, then open and close the throttle to allow fast idle cam to drop.
4. Press upward on fast idle cam. There should be little or no movement indicating that the fast idle screw is on the kick-down step of the cam, against the first step.
5. To adjust, use bending tool and bend the choke control rod to assure proper fast idle location on the cam, Fig. 48.
6. Loosen screws and reset the choke thermostatic housing to the proper index mark and retighten screws.

Decel Throttle Modulator, 4180 & 4190 Units

1. With engine at normal operating temperature, apply 19 inches Hg. of vacuum to decel throttle modulator.
2. Loosen decel throttle modulator lock nut and rotate modulator to achieve 1800 rpm.

Choke Thermostatic Spring Adjustment, 4160 & 4180 Units

1. Loosen thermostatic spring housing clamp retaining screws.
2. Rotate spring housing to align index mark on spring housing with center index mark on choke housing.

Clutch Section

HYDRAULIC CLUTCH RELEASE

When the clutch pedal is depressed, fluid is forced from the master cylinder into a slave cylinder. The slave cylinder push rod moves the clutch release lever which in turn pivots on a trunnion in the flywheel housing, and forces the release bearing against the pressure plate release fingers.

When the clutch pedal is released, the master cylinder return spring and the release lever retracting spring force the fluid from the slave cylinder back to the master cylinder.

The fluid used in this system is the same as that used in the brake system. A bleeder fitting is installed on the slave cylinder and procedures for bleeding are the same as those for brakes. The master cylinder fluid level should be maintained at ½″ below the top of the master cylinder reservoir.

CLUTCH PEDAL, ADJUST

MECHANICAL RELEASE

Bronco & Econoline

NOTE: For 1975–83 Econoline & 1978–83 Bronco models, refer to 1974–83 under "F-100 thru 350 Series".

1. Measure total travel of pedal. Travel should be within specifications shown below:

   Bronco
   1974–77 .... 6¾–7″

   Econoline
   1974 .... 7½–7¾″
2. If total pedal travel is not as specified in step 1, adjust clutch pedal stop:
   a. On Bronco models, move clutch pedal and bumper as necessary to obtain desired total pedal travel.
   b. On Econoline models, loosen clutch pedal bumper eccentric bolt and rotate bumper until desired pedal travel is obtained.
3. Check clutch pedal free travel. Travel should be within specifications shown below:

   Bronco
   1974–83 .... ¾–1½″

   Econoline
   1974 .... ¾–1½″
4. On Bronco and 1974 Econoline models, if clutch pedal free travel is not as specified in step 3, loosen locknut and adjust pushrod as necessary until desired clutch pedal free play is obtained.

F-100 thru 350

1974–83

NOTE: Clutch pedal total travel is not adjustable on 1974–83 vehicles.

1. Check clutch pedal free travel by measuring distance from clutch pedal pad to steering wheel rim, then depress clutch pedal until free travel is diminished and take a second measurement. If the difference between the two measurements is less than ¾ inches on 1974–79 vehicles or ½ inch on 1980–83 vehicles, the clutch linkage must be adjusted.
2. Loosen and back-off the two jam nuts several turns, then adjust the first jam nut until a free travel measurement of 1½ inches is obtained at the clutch pedal.
3. While holding the second jam nut, tighten the first jam nut against the second one.
4. Recheck free pedal travel and readjust as necessary.

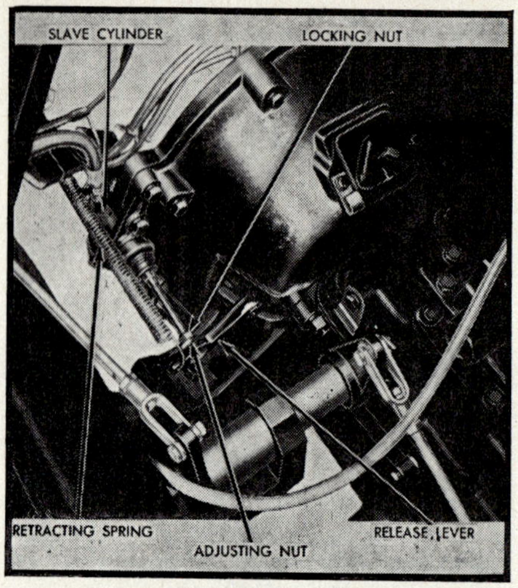

Fig. 1 Slave cylinder push rod adjustment

Ranchero

1. Disconnect clutch return spring from release lever.
2. Loosen release lever rod lock nut.
3. Move release lever rearward until release bearing lightly contacts clutch pressure plate release fingers.
4. Adjust adapter length until adapter seats in release lever pocket.
5. Insert the proper feeler gauge (see below) against back face of rod adapter, then tighten lock nut finger tight against feeler gauge.

   1974 .... .194″
6. Remove feeler gauge. Hold lock nut in position and tighten adapter against nut.
7. Install release spring and check for free travel of pedal which is ⅞″ to 1⅛″.

F, B & LN, N 500-750 Series

1. Adjust clutch pedal stop to obtain total pedal travel of 7⅞ inches on 1974–83 except LN, L and N series, or 10½ inches on 1974–83 LN, L and N Series.
2. Adjust clutch pedal free travel by backing off jam nut and adjusting clearance between spherical nut and release lever. Clearance on gasoline engine models should be ¼ inch, and clearance on diesel engine models should be ⅛ inch.

NOTE: Retracting spring must be in place when making adjustment.

3. Tighten jam nut.

LN, LNT-8000, N-1000-D, NT-850, 950-D Series (With NH Series Engine)

1. Adjust clutch pedal stop to obtain total clutch pedal travel of 8 15/32–8 23/32 inches.
2. Insert a ¼ inch alignment pin into relay rod, then with clutch pedal again stop, adjust vertical rod length to allow free installation of retaining bolt.
3. Install ¼ inch alignment pins into relay lever and clutch equalizer shaft, then adjust relay to clutch equalizer shaft to allow free installation of retaining bolt.
4. On vehicles equipped with Spicer 6000, 8000, 8312 and Fuller 5H74 and R96 transmission, adjust equalizer shaft to release lever push rod to obtain a clearance of ¼–5/16 inch between edge of push rod and bottom of groove in clutch release lever. On vehicles equipped with Fuller T-905 or RT-910, adjust clearance to 7/32–15/64 inch. On vehicles equipped with Clark 390 or Spicer CM 5000 or CM 6000 transmissions, adjust clearance to 1⅞–2⅛ inches.

HYDRAULIC RELEASE CLUTCH PEDAL, ADJUST

Before making any adjustments, be sure that the system is free of air and that the master cylinder fluid is at the specified level.

F, N, NT, T Series

1. Adjust pedal stop to permit total pedal travel of 8¼″.
2. Set clutch pedal initial free travel to 3/16″ to ⅜″ by adjusting the eccentric bolt. This measurement is taken at the centerline of the pedal pad.
3. If necessary, readjust pedal stop to maintain initial specified free travel.
4. To adjust slave cylinder on gasoline engines without SAE flywheel housing, disconnect release lever retracting spring and push slave cylinder push rod toward front of vehicle as far as possible, then pull release lever back until release bearing contacts pressure plate release fingers. Adjust clearance between slave cylinder push rod spherical nut and release lever to ¼ inch on all except super duty engines, and 13/64 inch on super duty engines. Tighten jam nut and reconnect retracting spring.

NOTE: As a result of the above adjustment, free pedal travel should be 2 inches.

5. To adjust slave cylinder on super duty engines with SAE flywheel housing, disconnect release lever retracting spring, then loosen clamp bolt of external release lever and push slave cylinder push rod toward front of vehicle as far as possible. Hold push rod against internal stop, then loosen push rod jam nut and adjust wing nut to obtain 8⅞ inches as measured

from flat end of slave cylinder casting to angled edge of wing nut. Tighten jam nut, then loosen outer adjusting arm set screw jam nut and rotate arm clockwise to eliminate release bearing lash. Adjust set screw to obtain 1/4 inch clearance between edge of wing nut and notch on external release lever. Tighten set screw and reconnect retracting spring.

6. On vehicles with diesel engines, adjust slave cylinder by pushing push rod against piston and piston against internal cylinder stop. Disconnect retracting spring, then with release lever held against bearing retainer, loosen jam nut and adjust wing nut on push rod to obtain 1/4 inch clearance. Tighten jam nut and reconnect retracting spring.

NOTE: Do not adjust slave cylinder to perform subsequent clutch adjustments.

If clearance between push rod wing nut and release lever is reduced to 3/32 inch, internal clutch adjustment must be performed.

C Series

1. Adjust eccentric bolt and pedal stop to give 5/32″ to 1/4″ of initial free travel before master cylinder push rod contacts piston. This measurement is taken at centerline of pedal pad.
2. Disconnect release lever retracting spring. Push slave cylinder push rod the full distance of its travel toward front of truck. Pull release lever toward rear of truck until release bearing touches release fingers.
3. Loosen locknut and adjust conical nut or wing nut until there is 1/4 inch clearance between release lever and conical nut or wing nut on all except 1973–83 models with 13 inch clutch, or 13/64 inch on all 1973–83 models with 13 inch clutch.

H and HT Series

1. Adjust clutch pedal height to 10 3/8″ by moving bumper and its bracket.
2. By means of eccentric bolt, adjust initial free pedal travel 1/8″ to 1/4″.
3. Disconnect release lever retracting spring. Push slave cylinder push rod the full distance of its travel forward, and move release lever to the rear until release bearing touches release fingers.
4. Adjust clearance between slave cylinder push rod spherical nut and release lever to 3/16″ to obtain final free travel. As a result of this adjustment, there should be final free travel of approximately 1 1/2″. Tighten jam nut.

P SERIES

1974–76 350, 400

NOTE: Pedal free play is controlled by the master cylinder and is not adjustable. To adjust slave cylinder, proceed as follows:

1. Bottom slave cylinder push rod against the internal stop.
2. Push clutch release lever back until release bearing contacts clutch fingers.
3. Adjust conical nut to obtain 1/4 inch clearance between release lever and conical nut.
4. Tighten jam nut against conical nut.

1974–79 500, 600

1. With retracting spring attached and pedal against stop, adjust master cylinder push rod to obtain 1/4 inch free pedal play, then tighten jam nut.
2. Bottom slave cylinder push rod against internal stop, then push clutch release lever back until release bearing contacts clutch fingers.
3. Loosen jam nut and adjust conical nut to obtain 1/4 inch clearance between release lever and conical nut.
4. Tighten jam nut against conical nut.

CLUTCH, REPLACE

1. Remove release lever rod. Remove transmission from clutch housing. Remove release bearing and hub from release lever.
2. On all models, mark the clutch assembly and flywheel so the parts may be installed in the same relative position.
3. Loosen the clutch-to-flywheel attaching bolts a little at a time until the tension on the clutch springs is relieved. Then unscrew the bolts completely and remove the clutch assembly and driven disc.

Installation

1. Coat pilot bearing bore in flywheel with a small amount of wheel bearing grease. *Avoid using too much grease as it may be thrown on the clutch disc when the clutch revolves.*
2. Install pilot bearing in flywheel. On 12″ clutch, install the bearing with its shielded side toward the rear.
3. Install clutch pilot tool in clutch disc and position disc on flywheel so that the pilot tool enters the pilot bearing.
4. Position pressure plate and cover assembly on flywheel and install the retaining cap screws. If the original clutch is being installed, be sure to align the marks made prior to removal.
5. Position the release bearing and hub on the release lever. Install the transmission. Position release bearing and hub on the release lever. Install adjusting nut and lock nut on release rod. Connect the clutch pedal retracting spring to the release lever and check the clutch pedal free travel.

Transmission Section

> **NOTE:** For Clark, Fuller, New Process, Spicer and Warner transmissions used on Ford trucks, see the "name" chapter for service procedures and lubricant capacities

TRANSMISSION IDENTIFICATION

1974

Code	Mfr. Name[1]	Mfr. Model[1]
A	New Process	435
B	Spicer	8516
B	Spicer	P8516
C	Ford	3 Spd.
C	Fuller	RT-610
D	Clark	387-V
E	Warner	T-19G
E	Fuller	5H74
F	Fuller	5HA74
F	Warner	T-18
G	Ford	C4 Auto.
G	Warner	T-87G
G	Spicer	SST 6 + 1
H	Allison	MT650
J	Fuller	RTO-9513
J	Ford	FMX Auto.
J	Spicer	6453
K	Spicer	6453A
L	Allison	AT540
M	Clark	285V
N	Spicer	6352
O	New Process	542FL
O	Fuller	T-905B
P	Warner	T-19
P	Ford	C6 Auto.
P	Fuller	RT-613
Q	Spicer	5652
R	Spicer	8716
S	Spicer	5756B
T	New Process	542-FO
T	Fuller	TRO-9509B
U	Spicer	6852G
V	Fuller	RT-910
W	Spicer	6352B
X	Fuller	T-905A
Y	Allison	MT41
Z	Allison	MT40
1	Spicer	8552A
1	Allison	MT42
2	Clark	282V
3	Fuller	RT-906
4	Clark	280VO
5	Fuller	RTO-910
6	Fuller	RTO-915
7	Clark	385V
8	Allison	MT640
9	Fuller	RT-915
9	New Process	542-FD

1975

Code	Mfr. Name[1]	Mfr. Model[1]
A	New Process	435
A	Spicer	SST 10
B	Spicer	RP8516-3B
C	Ford	3 Spd.
C	Fuller	RT-610
D	Clark	387V
E	Spicer	RT12510
E	Warner	T19
F	Spicer	RTO-12-513
F	Warner	T18
G	Ford	Auto.
G	Spicer	SST 6 + 1
H	Allison	MT650
J	Fuller	RTO-9513
J	Ford	FMX Auto.
K	Spicer	6453
L	Allison	AT540
M	Clark	285V
N	Spicer	6352
O	New Process	542FL
O	Fuller	T-905B
P	Warner	T19
P	Fuller	RT-613
Q	Spicer	5652
R	Spicer	P8716-3A
S	Spicer	5756-B
T	New Process	542FO
T	Fuller	TRO-9509-B9
U	Spicer	6852G
V	Fuller	RT-910
W	Spicer	6352B
X	Fuller	T-905A
1	Spicer	8552A
2	Clark	282V
3	Fuller	RT-906
4	Clark	280VO
4	Fuller	RT-9513
5	Fuller	RTO-910
6	Fuller	RTO-915
7	Clark	385V
8	Allison	MT640
9	Fuller	RT-915
9	New Process	542FDJ

1975 Auxiliary Units

Code	Mfr. Name[1]	Mfr. Model[1]
2	Spicer	5831D
3	Spicer	7231B
4	Spicer	7231D
5	Spicer	8341C
6	Spicer	8031B
8	Spicer	7041J

1976

Code	Mfr. Name[1]	Mfr. Model[1]
A	New Process	435
A	Spicer	SST10
B	Spicer	RP8516-3B
C	Ford	3 Spd.
C	Fuller	RT-610
D	Clark	387V
E	Fuller	RT12510
E	Warner	T19
F	Fuller	RTO-12-513
F	Warner	T18
G	Ford	Auto.
G	Spicer	SST1007-2F
G	Warner	T87G
H	Allison	MT650
J	Fuller	RTO-9513
J	Ford	FMX Auto.
K	Spicer	6453
L	Allison	AT540
M	Clark	285V
N	Spicer	6352
O	New Process	542FL
O	Fuller	T-905B
P	Warner	T19
P	Fuller	RT-613
P	Ford	C6 Auto.
Q	Spicer	5652
Q	Fuller	RT-1110
Q	Fuller	RT-1110
R	Spicer	RP8716-3A
S	Spicer	5756-B
T	New Process	542-FO
T	Fuller	RTO-9509-B9
U	Spicer	6852G
V	Fuller	RT-910
W	Spicer	6352B
X	Fuller	RT-910
Z	Clark	397V
1	Spicer	8552A
2	Clark	282V
3	Fuller	RT-906
4	Clark	280-VO
4	Fuller	RT-9513
5	Fuller	RTO-910
6	Fuller	RTO-915
6	Clark	390V
7	Clark	385V
8	Allison	MT640
9	Fuller	RT-915
9	New Process	542FD

1976 Auxiliary Units

Code	Mfr. Name[1]	Mfr. Model[1]
2	Spicer	5831D
3	Spicer	7231B
4	Spicer	7231D
5	Spicer	8341C
6	Spicer	8031B
8	Spicer	7041

1977

Code	Mfr. Name[1]	Mfr. Model[1]
A	New Process	435
A	Spicer	SST10
B	Spicer	CM6052A
C	Ford	3 Spd.
C	Fuller	RT610
D	Spicer	CM5252
E	Fuller	RT12510
F	Warner	T18
F	Fuller	RT012513
G	Spicer	SST10072A
G	Ford	Auto
H	Allison	MT650
J	Fuller	RT09513
J	Ford	Auto
K	Spicer	6052B
L	Allison	AT540
M	Clark	285
N	Spicer	CM6052
N	Spicer	6352
O	New Process	542FL
O	Fuller	T905B
P	Warner	T19
P	Fuller	RT613
Q	Spicer	CM5052A
Q	Spicer	5652
Q	Spicer	CM6052A
Q	Fuller	RT1110
R	Spicer	CM6052C
S	Spicer	CM6052B
S	Spicer	5656B
T	New Process	542FO
T	Fuller	RT09509B
U	Spicer	6852G
V	Fuller	RT910

TRANSMISSION IDENTIFICATION—Continued

Code	Mfr. Name①	Mfr. Model①
X	Fuller	T905A
Y	Spicer	RP85163A
Z	Clark	397
Z	Ford	Auto
2	Clark	282
3	Fuller	RT906
4	Clark	280
4	Fuller	RT9513
5	Fuller	RT0910
6	Clark	390
7	Spicer	CM5052
8	Allison	MT640
8	Fuller	RT9509A
9	New Process	542

1977 Auxiliary Units

Code	Mfr. Name①	Mfr. Model①
4	Spicer	7231D
5	Spicer	R8341-2
8	Spicer	7041

1978

Code	Mfr. Name①	Mfr. Model①
A	New Process	435
A	Spicer	SST-1010
B	Ford	4 Spd. O/D
B	Spicer	CM-6052A
C	Ford	3 Spd.
C	Fuller	RT-610
D	Spicer	CM-5252A
E	Fuller	RT-12510
F	Fuller	RTO-12513
F	Warner	4 Spd.
G	Spicer	SST-1007-2A
G	Ford	C6
H	Allison	MT-650
I	Ford	C6
J	Fuller	RT-12510
K	Spicer	CM-6052B
L	Allison	AT-540
M	Clark	285V
O	New Process	542-FL
P	Fuller	RT-613
P	Warner	T-19
Q	Fuller	RT-1110
S	Spicer	CM-6052C
T	Fuller	RTOF-12513
T	New Process	542-FO
U	Ford	C6
V	Fuller	RT-910
W	Ford	C4
X	Fuller	T-905A
X	Ford	FMX
Y	Spicer	RP-85163A
Z	Clark	397
2	Clark	282V
3	Fuller	RT-906
4	Fuller	RT-9513
4	Clark	280-VO
5	Fuller	RTO-910
6	Clark	390
7	Spicer	CM-5052A
8	Allison	MT-640
9	Fuller	RTO-958LL
9	New Process	542-FD

1978 Auxiliary Units

Code	Mfr. Name①	Mfr. Model①
4	Spicer	7231D
7	Spicer	R-8341-2
8	Spicer	7041

1979

Code	Mfr. Name①	Mfr. Model①
A	New Process	435
B	Ford	4 Spd. O/D
B	Spicer	CM-6052B
C	Ford	3 Spd.
C	Fuller	RT-610
D	Spicer	CM-5252A
E	Fuller	RT-12510
F	Warner	4 Spd.
F	Fuller	RT-12513
G	Ford	Auto. Trans.
G	Clark	282-VHD
H	Allison	MT-653
J	Fuller	RTO-9513
K	Spicer	CM-6052C
L	Allison	AT-540
M	Clark	285
M	Clark	285V
N	Clark	285-VHD
O	Fuller	T-905B
O	New Process	542-FL
P	Warner	T-19
P	Fuller	RT-613
Q	Fuller	RT-1110
S	Spicer	CM-6052C
T	Warner	T-19
U	Ford	C6
U	Spicer	SST-10073A
V	Fuller	RT-910
W	Ford	C4
X	Ford	C6
X	Ford	MFX
X	Fuller	T-905A
Y	Spicer	RP-85163A
Z	Clark	397
2	Clark	282
2	Clark	282V
3	Fuller	RT-906
4	Clark	280-VO
4	Fuller	RT-9513
5	Fuller	RTO-910
6	Clark	390
7	Spicer	CM-5052A
8	Allison	MT-643
8	Fuller	RT-9509A
9	New Process	5542-F

1979 Auxiliary Units

Code	Mfr. Name①	Mfr. Model①
3	Spicer	7231D
4	Spicer	7231D
5	Spicer	7041
6	Spicer	R-8341L
7	Spicer	R-8341L
8	Spicer	7041

1980
F-100-350, E-100-350, Bronco, F & B-600

Code	Mfr. Name①	Mfr. Model①
A	New Process	4 Spd.
B	Ford	4 Spd. O/D
C	Ford	3 Spd.
D	Spicer	5052A
E	Spicer	5252A
F	Warner	4 Spd.
G	Ford	Auto. Trans.
J	Ford	Auto. Trans.
K	Ford	Auto. Trans.
L	Allison	AT-540
M	Clark	285-V
O	New Process	542-FL
P	Warner	T-19
W	Clark	282 VHD
X	Clark	285 VHD
X	Ford	Auto. Trans.
2	Clark	282-V
9	New Process	542-FD

700 Thru 9000 Series & N600-C600 Parcel

Code	Mfr. Name①	Mfr. Model①
AA	Allison	AT-540
8	Allison	MT-640
AB	Allison	MT-643
H	Allison	MT-650
AC	Allison	MT-653
CA	Clark	280
4	Clark	280-VO
CB	Clark	282
CC	Clark	282 VHD
CD	Clark	285
CE	Clark	285 VHD
CF	Clark	390
CG	Clark	397
FA	Fuller	RT-610
FB	Fuller	RT-613
FC	Fuller	T-905A
FD	Fuller	T-905B
FE	Fuller	RT-906
FF	Fuller	RT-910
FG	Fuller	RTO-910
FH	Fuller	RTO-958LL
FI	Fuller	RT-1110
FJ	Fuller	RT-9509A
FK	Fuller	RT-9513
FL	Fuller	RTO-9513
FM	Fuller	RT-12510
FN	Fuller	RTO-12513
T	Fuller	RTOF-12513
NA	New Process	NP-435
NB	New Process	NP-542-FD
NC	New Process	NP-542-FL
T	New Process	NP-542-FO
G	Spicer	SST-1007-2A
SA	Spicer	SST-1007-3A
A	Spicer	SST-1010
SB	Spicer	CM-5052A
SC	Spicer	CM-5252A
SD	Spicer	CM-6052A
SE	Spicer	CM-6052B
SF	Spicer	CM-6052C
SG	Spicer	RP-85163-A
WA	Warner	T-19

1981
F100-350, E100-350, Bronco, F & B-600

Code	Mfr. Name①	Mfr. Model①
A	New Process	4 Spd.
B	Ford	4 Spd. O/D
C	Ford	3 Spd.
D	Spicer	5052A
E	Spicer	5252A
F	Warner	4 Spd.
G	Ford	Auto. Trans.
J	Ford	Auto. Trans.
K	Ford	Auto. Trans.
L	Allison	AT 540
O	New Process	542 FL
P	Warner	T-19
T	Ford	Auto. Trans. O/D
W	Clark	282 VHD
X	Clark	285 VHD
2	Clark	282 V
9	New Process	542-FD

700 Thru 9000 Series & N600-C600 Parcel

Code	Mfr. Name①	Mfr. Model①
AA	Allison	AT-540
AB	Allison	MT-643
AC	Allison	MT-653
CA	Clark	280
CB	Clark	282
CC	Clark	282-VHD
CD	Clark	285
CE	Clark	285-VHD

TRANSMISSION IDENTIFICATION—Continued

Code	Mfr. Name①	Mfr. Model①
CF	Clark	390
CG	Clark	397
FA	Fuller	RT-610
FB	Fuller	RT-613
FC	Fuller	T-905A
FD	Fuller	T-905B
FE	Fuller	RT-906
FF	Fuller	RT-910
FG	Fuller	RTO-910
FH	Fuller	RTO-958LL
FI	Fuller	RT-1110
FJ	Fuller	RT-9509A
FK	Fuller	RT-9513
FL	Fuller	RTO-9513
FM	Fuller	RT-12510
FN	Fuller	RTO-12513
NA	New Process	NP-435
NB	New Process	NP-542-FD
NC	New Process	NP-542-FL
SA	Spicer	SST-1007-3A
SB	Spicer	CM-5052A
SC	Spicer	CM-5252A
SD	Spicer	CM-6052A
SE	Spicer	CM-6052B
SF	Spicer	CM-6052C
SG	Spicer	RP-85163A
WA	Warner	T-19

1980–81 Auxiliary Units

Code	Mfr. Name①	Mfr. Model①
3	Spicer	7231D
4	Spicer	7231D
5	Spicer	7041
6	Spicer	R-8341L
7	Spicer	R-8341L
8	Spicer	7041

1982–83 F100-350, E100-350, Bronco, F & B-600

Code	Mfr. Name①	Mfr. Model①
A	New Process	4 Spd.
B	Ford	4 Spd. O/D
C	Ford	3 Spd.
D	Spicer	5052A
E	Spicer	5252A
F	Warner	4 Spd.
G	Ford	Auto. Trans.
K	Ford	Auto. Trans.
L	Allison	AT-545
M	Clark	285-V
O	New Process	542-FL
P	Warner	T-19
T	Ford	Auto. Trans. O/D
W	Clark	282 VHD
W②	Ford	Auto. Trans.
X	Clark	285 VHD
2	Clark	282-V
9	New Process	542-FD

700 Thru 9000 Series & N600-C600 Parcel

Code	Mfr. Name①	Mfr. Model①
AA	Allison	AT-545
AB	Allison	MT-643
AC	Allison	MT-653
CA	Clark	280
CB	Clark	282
CC	Clark	282 VHD
CD	Clark	285
CE	Clark	285 VHD
CF	Clark	390
CG	Clark	397
DC	Fuller	RT-11608
FA	Fuller	11608LL
FB	Fuller	RTO-11607L
FC	Fuller	T-905A
FD	Fuller	RTO-11607LL
FE	Fuller	RT-11609A
FF	Fuller	RT-910
FG	Fuller	RTO-11609B
FH	Fuller	RT-0958LL
FI	Fuller	RT-1110
FJ	Fuller	RT-9509
FL	Fuller	RTO-9513
FM	Fuller	RT-12510
FN	Fuller	RTO-12513
FP	Fuller	RT-12509
FQ	Fuller	RT-11509
FR	Fuller	RTO-9509B
FS	Fuller	RT-6610
FT	Fuller	RT-6613
FU	Fuller	T-11605A
FV	Fuller	RTO-12515
FW	Fuller	RT-9508
FX	Fuller	RTO-1157DDL
FY	Fuller	RTO-1157DL
NA	New Process	435
NB	New Process	542-FD
NC	New Process	542-FL
NF	Fuller	RT-14609A
PD	Fuller	RT-14608
SA	Spicer	SST-1007-3A
SB	Spicer	CM-5052A
SC	Spicer	CM-5252A
SD	Spicer	CM-6052A
SE	Spicer	CM-6052B
SF	Spicer	CM-6052C
SH	Spicer	SST-1062C
SJ	Spicer	SST-1107-2A
WA	Warner	T-19
1F	Fuller	RTO-11608LL
1G	Fuller	RT-11610
1H	Fuller	RTO-J14610
3J	Fuller	RTO-11613
3K	Fuller	RTO-14613
7A	Spicer	1207-3A
7B	Spicer	1372-A
52	Fuller	RT-11615
53	Fuller	RTO-14615

①—See name chapter for repair procedures.
②—Light duty only.

3 SPEED TRANS., 1974–83

Disassembly

1. Remove transmission cover, Fig. 1.
2. Remove extension housing. To prevent mainshaft from following housing (with resultant loss of needle bearings) tap end of mainshaft while withdrawing housing.
3. Remove front bearing retainer.
4. Remove filler plug from right side of case. Then working through plug opening, drive roll pin out of case and countershaft with small punch, Fig. 2.
5. Hold counter gear with a hook and, with a dummy shaft, push countershaft out rear of case until counter gear can be lowered to bottom of case, Fig. 3.
6. Pull main drive gear forward until gear contacts case, then remove large snap ring.

NOTE: On some models, it is necessary to move gear forward to provide clearance when removing mainshaft assembly. On other models, the drive gear is removed from front of case.

7. Remove snap ring and slide speedometer drive gear off mainshaft. Remove lock ball from shaft.
8. Remove snap ring and remove mainshaft rear bearing from shaft and case, Fig. 4.
9. Place both shift levers in neutral (central) position.
10. Remove set screw that retains detent springs and plugs in case. Remove one spring and plug. Fig. 5.
11. Remove low-reverse set screw and slide shift rail out through rear of case.
12. Rotate low-reverse shift fork upward and lift it from case.
13. Remove 2-3 set screw and rotate 2-3 shift rail 90 degrees with pliers.
14. Lift interlock plug from case with a magnet rod, Fig. 5.
15. Tap inner end of 2-3 shift rail to remove expansion plug from front of case. Remove shift rail.
16. Remove 2-3 detent plug and spring from detent bore.
17. Rotate 2-3 shift fork upward and lift from case.
18. Lift mainshaft assembly out through top of case.
19. On some transmissions, push main drive gear into case until bearing is free of bore, then lift gear and bearing through top of case.
20. Working through front bearing opening, drive reverse idler gear shaft out through rear of case with a drift, Fig. 6.
21. Lift reverse idler gear and two thrust washers from case.
22. Lift counter gear and thrust washers from case, Fig. 7. Be careful not to allow dummy shaft and needle bearings to fall out of gear.
23. Remove countershaft-to-case retaining pin and any needle bearings that may have fallen into case.
24. Unfasten and lift shift levers off shafts. Slide each lever and shaft out of case. Discard O-ring seal from each shaft, Fig. 8.
25. Remove snap ring from front end of mainshaft and remove synchronizers, gears and related parts, Fig. 9.
26. If main drive gear is to be disassembled, refer to Fig. 10.

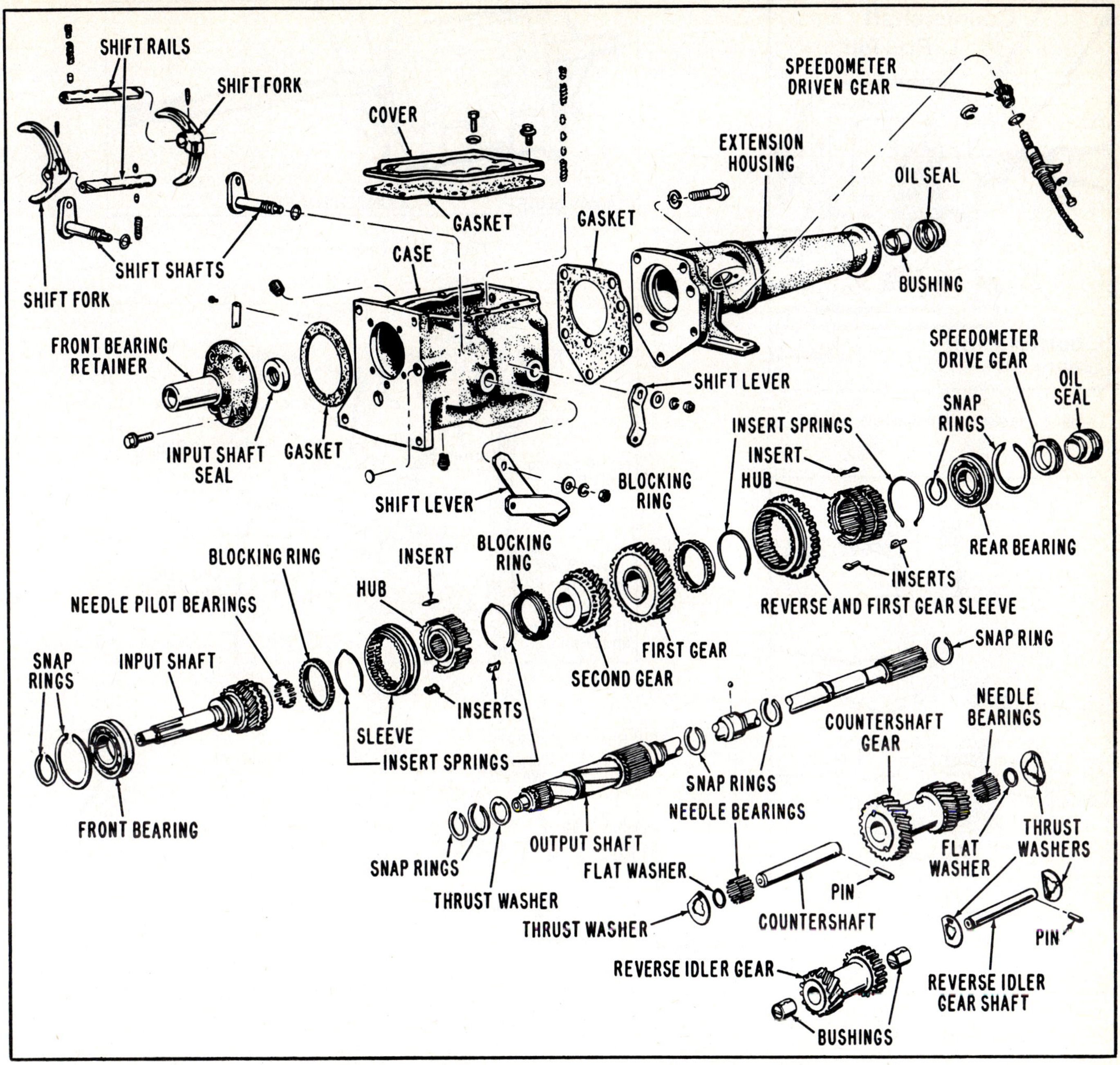

Fig. 1 Exploded view of three-speed transmission. 1974–83

Assembly

Counter Gear

1. Coat bore at each end of counter gear with grease.
2. Hold appropriate dummy shaft in gear and install 25 needle bearings and a retainer washer in each end of gear, Fig. 7.
3. Install counter gear, thrust washers and countershaft in case.
4. Place transmission case in vertical position and check end play with a feeler gauge as shown in Fig. 11. If end play exceeds .018″, replace thrust washers as required to obtain .004–.018″ end play.
5. Once end play has been established, remove countershaft with dummy shaft.
6. Allow counter gear assembly to remain in case.

Reverse Idler Gear

1. Install idler gear, thrust washers and shaft in case.
2. Make sure that thrust washer with flat side is at the web end, and that the spur gear is toward the rear of case, Fig. 6.
3. Check reverse idler gear end play in same manner and to the same clearance as the counter gear. If end play is within limits of .004–.018″ leave gear in case.

Low-Reverse Synchronizer

1. Install an insert spring, Fig. 12, in groove of low-reverse synchronizer hub. Make sure spring covers all insert grooves.
2. Start hub in sleeve, making sure that alignment marks are properly indexed.
3. Position three inserts in hub, making sure that small end is over spring and that shoulder is on inside of hub.
4. Slide sleeve and reverse gear onto hub until detent is engaged.
5. Install other insert spring in front of hub to hold inserts against hub.

Second-Third Synchronizer

1. Install one insert spring, Fig. 13, into a

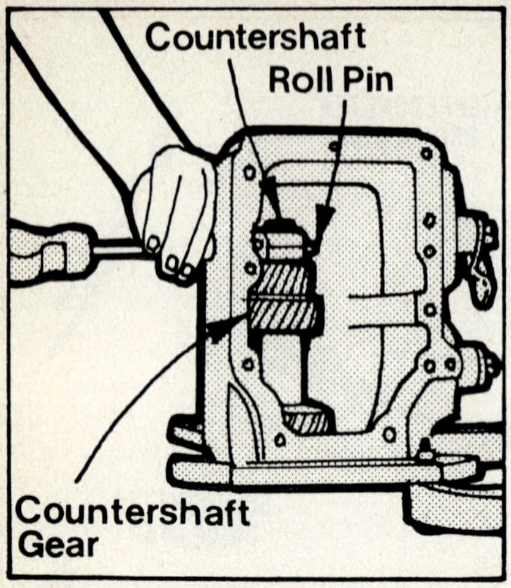

Fig. 2 Removing countershaft roll pin

Fig. 3 Removing countershaft

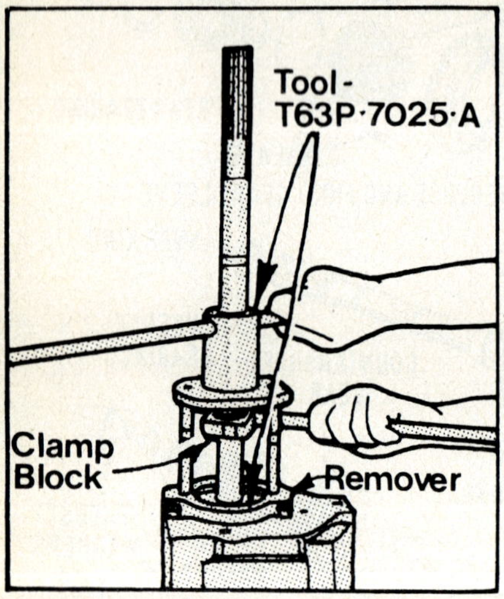

Fig. 4 Removing mainshaft bearing

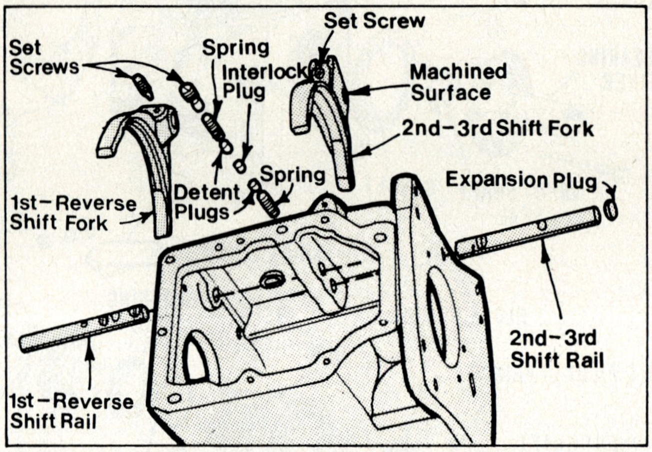

Fig. 5 Shift rods & forks, disassembled

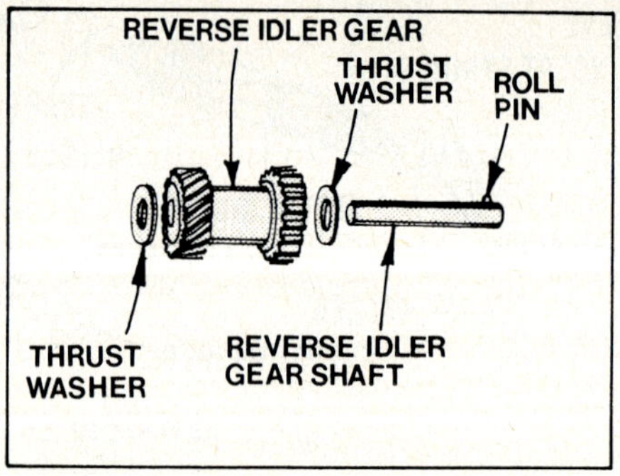

Fig. 6 Reverse idler gear, disassembled

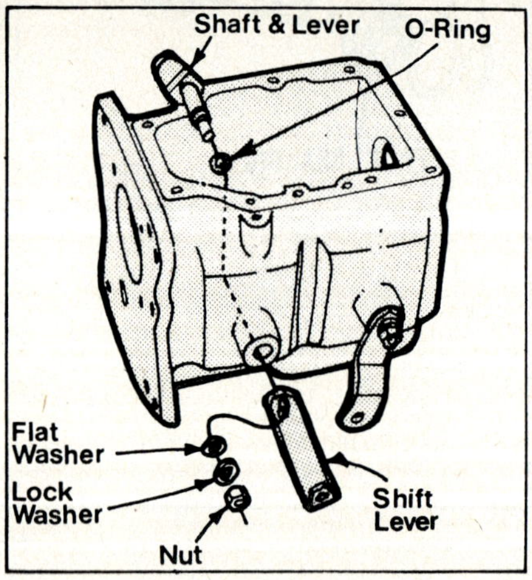

Fig. 8 Shift lever & related components

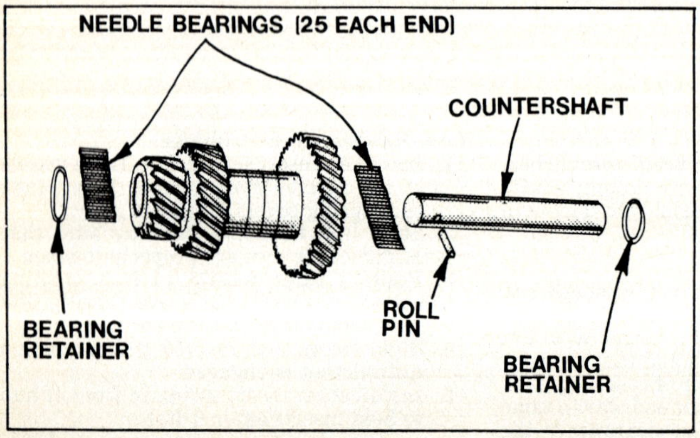

Fig. 7 Countergear, disassembled

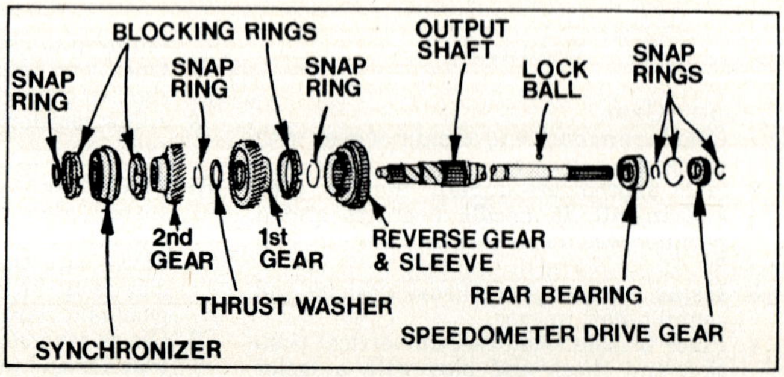

Fig. 9 Mainshaft, disassembled

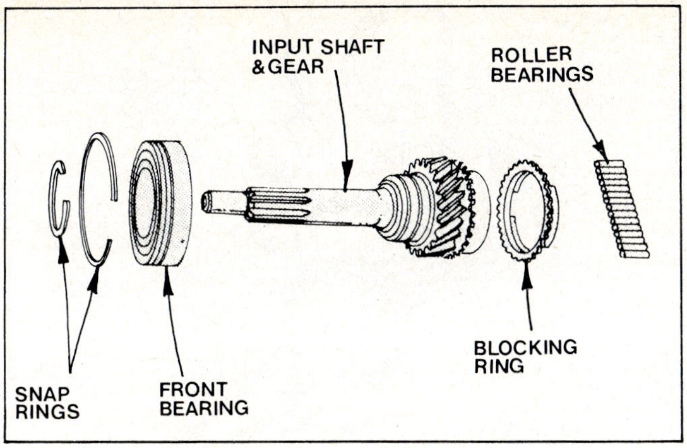

Fig. 10 Main drive gear, disassembled

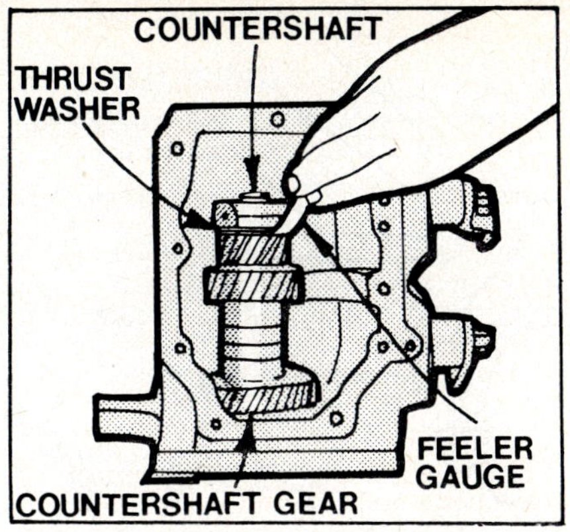

Fig. 11 Checking countergear end play

groove of synchronizer hub, making sure that all three insert slots are fully covered.

2. With alignment marks on hub and sleeve aligned, start hub into sleeve.
3. Place three inserts on top of retaining spring and push assembly together.
4. Install remaining insert spring so that spring ends cover same slots as do other spring. Do not stagger springs.
5. Place a synchronizer blocking ring in each end of sleeve.

Main Drive Line

1. Lubricate mainshaft splines and machined surfaces with transmission lube.
2. Slide low-reverse synchronizer, Fig. 9, onto mainshaft with teeth end of gear facing toward rear of shaft. Secure in place with snap ring.
3. Coat tapered machined surface of low gear with grease. Place blocking ring on greased surface.
4. Slide low gear onto mainshaft with blocking ring toward rear of shaft. Rotate gear as necessary to engage the three notches in blocking ring with synchronizer inserts. Secure low gear with thrust washer and snap ring.
5. Coat tapered machined surface of 2nd gear with grease and slide blocking ring onto it. Slide 2nd gear with blocking ring and 2-3 synchronizer onto mainshaft. Tapered machined surface of 2nd gear must be toward front of shaft. Make sure that notches in blocking ring engage synchronizer inserts. Secure synchronizer with snap ring.
6. Install new O-ring on each of two shift lever shafts, Fig. 8. Lubricate shafts with transmission lube and install them in case. Secure each lever on its shaft with a flat washer, lock washer and nut.
7. Coat bore of main drive gear shaft with thick coat of grease. Install 15 needle bearings in gear pocket, Fig. 10.
8. Position drive gear assembly into case.
9. Place a detent spring and plug in case, Fig. 5. Place 2-3 shift fork in synchronizer groove. Rotate fork into position and install 2-3 shift rail. It will be necessary to depress detent plug to enter rail in bore. Move rail inward until detent plug engages center (neutral) notch. Secure fork to shaft with set screw.
10. Install interlock plug in case. If 2-3 shift rail is in neutral position, top of interlock will be slightly lower than surface of low-reverse shift rail bore.
11. Place low-reverse shift fork in groove of synchronizer. Rotate fork into position and install low-reverse shift rail. Move rail inward until center (neutral) notch is aligned with detent bore. Secure fork to shaft with set screw.
12. Install remaining detent plug and spring. Secure spring with slotted head set screw. Turn set screw in until head is flush with case.
13. Install new expansion plug in case.
14. Install large snap ring on drive bearing. Work drive gear into case and onto mainshaft until snap ring is seated against case. Make sure that needle bearings do not fall out of place and that notches in blocking ring engage inserts in synchronizer.

NOTE: The foregoing applies only to transmissions used with V8-390 engines. On all other models, the drive gear is installed through front of case with bearing in place on shaft.

---

15. Position new front bearing retainer gasket on case. Place bearing retainer on case, making sure oil return groove is at the bottom. Install and tighten attaching screws to 30–36 ft-lbs.
16. Install large snap ring on mainshaft rear bearing. Place bearing on mainshaft with snap ring end toward rear of shaft. Press bearing into place and secure with snap ring.
17. Hold speedometer drive gear lock ball in detent and slide gear into place. Secure gear with snap ring.

Final Assembly

1. Place transmission in vertical position. Working through drain hole in bottom of case, align bore of counter gear and thrust

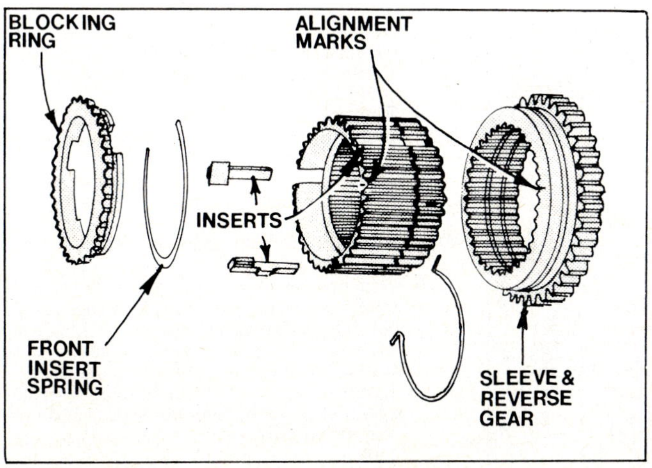

Fig. 12 Low-reverse synchronizer, disassembled

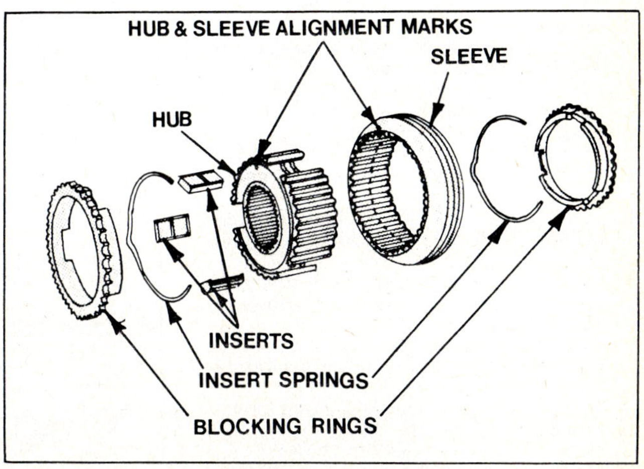

Fig. 13 Second-high synchronizer, disassembled

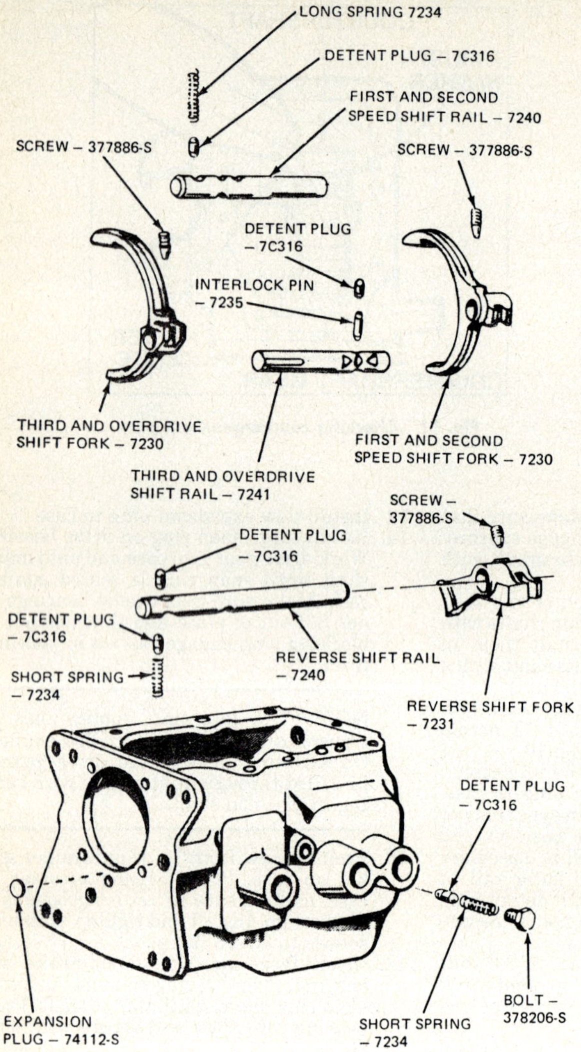

Fig. 13A Shift forks & rails, disassembled

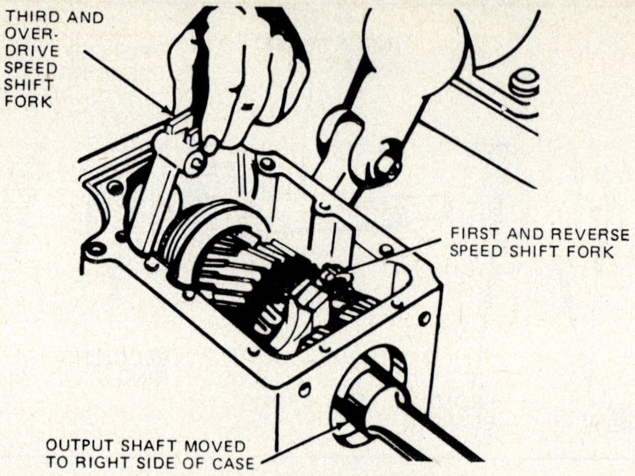

Fig. 13B Removing shift forks

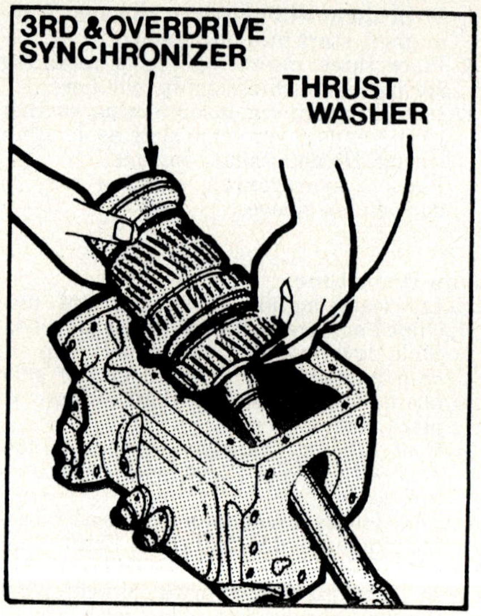

Fig. 13C Replacing output shaft assembly

washers with bore of case, using a screwdriver.

2. Working through rear of case, push dummy shaft out with countershaft. Before countershaft is completely inserted, make sure the hole that accommodates roll pin is aligned with hole in case.
3. Working through lubricant filler hole, install roll pin (Fig. 2) in case and countershaft.
4. Install filler and drain plugs, making sure magnetic plug is installed in bottom of case.
5. Install extension housing with new gasket and torque attaching cap screws to 42–50 ft-lbs.
6. Place transmission in gear, pour lubricant over entire gear train while rotating input and output shafts.
7. Install cover with new gasket and torque attaching screws to 14–19 ft-lbs. on 1974–76 models, and 20–25 ft-lbs. on 1977–83 models. Coat gasket and cover screws with sealer.

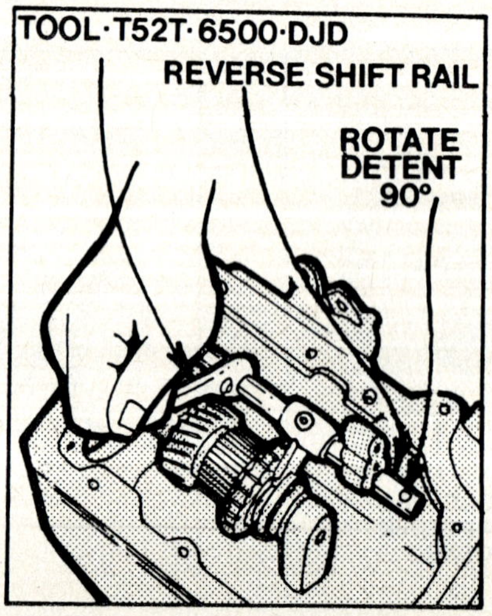

Fig. 13D Replacing reverse shift rail

FOUR SPEED OVERDRIVE TRANS.

Disassembly

1. Remove the lower extension housing to case attaching screw to drain lubricant.
2. Remove cover attaching screws and the cover from case.
3. Remove the long spring retaining detent plug in case, then, using a magnet, remove detent plug.
4. Remove extension housing to case screws and the extension housing.
5. Remove input shaft bearing retainer attaching screws and slide retainer from input shaft.
6. Support countershaft gear with a wire hook. Working from front of case, push countershaft out rear of case with a dummy shaft. Lower countershaft gear to bottom of case and remove the wire hook.
7. Remove set screw from first and second speed shift fork, then slide first and second speed shift rail out rear of case, Fig. 13A.
8. With a magnet, remove interlock detent from between the first and second and the

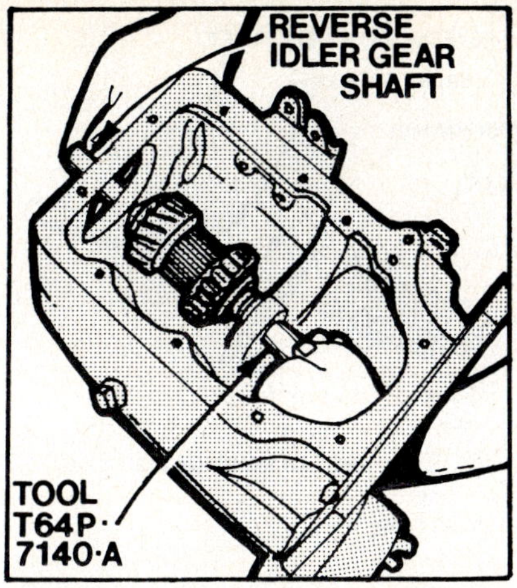

Fig. 13E Removing reverse idler gear shaft

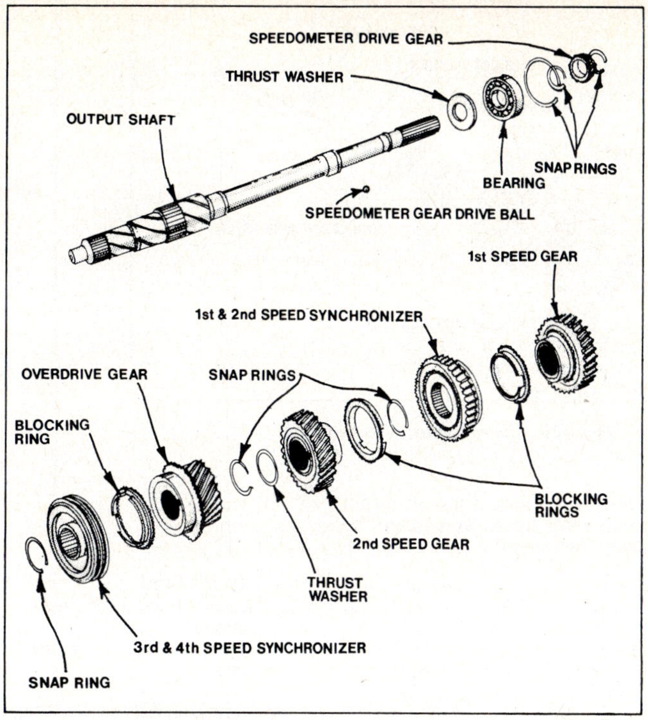

Fig. 13F Output shaft disassembled

third and overdrive shift rails.
9. Shift transmission into overdrive position. Remove set screw from the third and overdrive shift fork. Remove the side detent bolt, detent plug and spring. Rotate the third and overdrive speed shift rail 90 degrees and tap rail out front of case with a suitable punch and hammer.
10. With a magnet, remove interlock plug from top of case.
11. Remove snap ring securing speedometer drive gear to output shaft, then slide gear from shaft. Remove speedometer gear drive ball.
12. Remove output shaft bearing retaining snap ring and the snap ring from outside diameter of output shaft bearing.
13. Remove output shaft bearing.
14. Remove input shaft bearing retaining snap ring and the snap ring from outside diameter of input shaft bearing.
15. Remove input shaft bearing from shaft and case, then the input shaft and blocking from front of case.
16. Position output shaft to right side of case and remove shift forks, Fig. 13B.
17. Lift output shaft assembly from case, Fig. 13C, support the thrust washer and first speed gear to prevent the washer and gear from sliding from shaft.
18. Remove reverse gear shift fork set screw, rotate reverse shift rail 90 degrees, Fig. 13D, slide shift rail out rear of case and lift the reverse shift fork from case.
19. With a magnet, remove reverse detent plug and spring with a magnet.
20. Remove reverse idler gear shaft with a dummy shaft and lower reverse idler gear to bottom of case, Fig. 13E.
21. Lift countershaft gear and thrust washers from bottom of case, using caution not to drop bearings or the dummy shaft from countershaft gear.
22. Lift reverse idler gear shaft from case, using caution not to drop bearings or the dummy shaft from reverse idler gear.

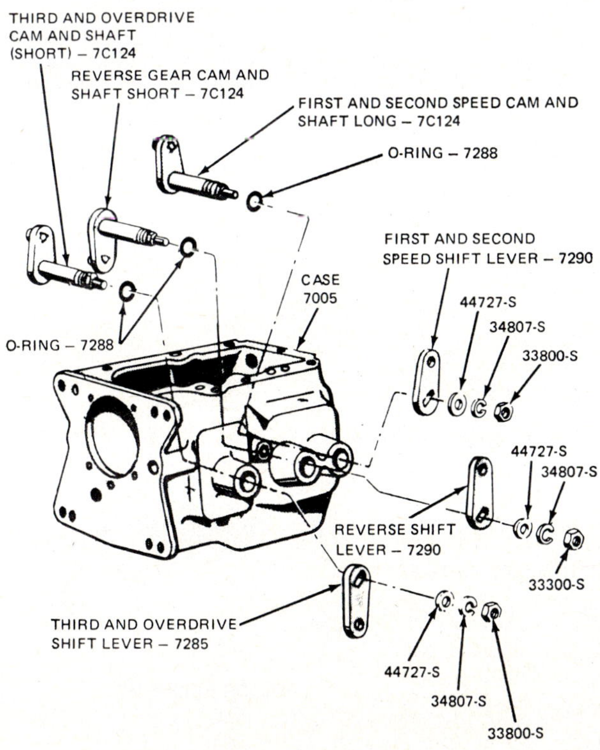

Fig. 13G Cams, shafts & levers

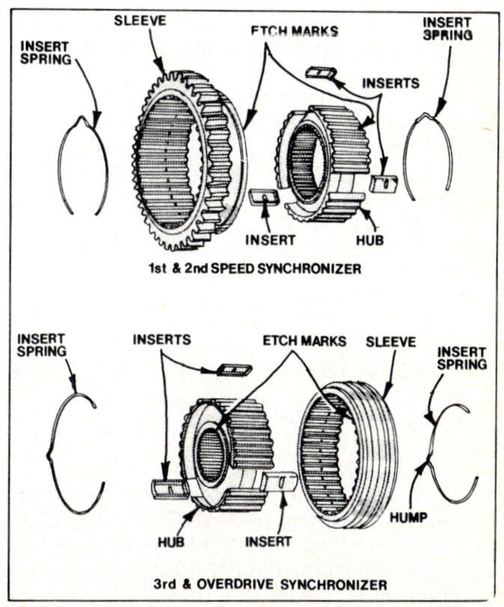

Fig. 13H Synchronizer assemblies, disassembled

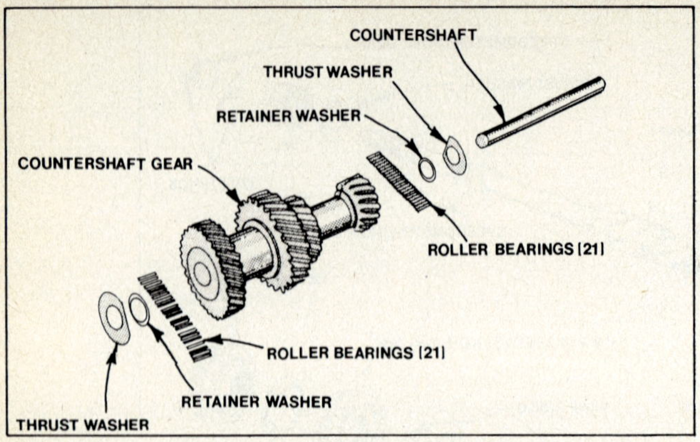

Fig. 13I Countergear, disassembled

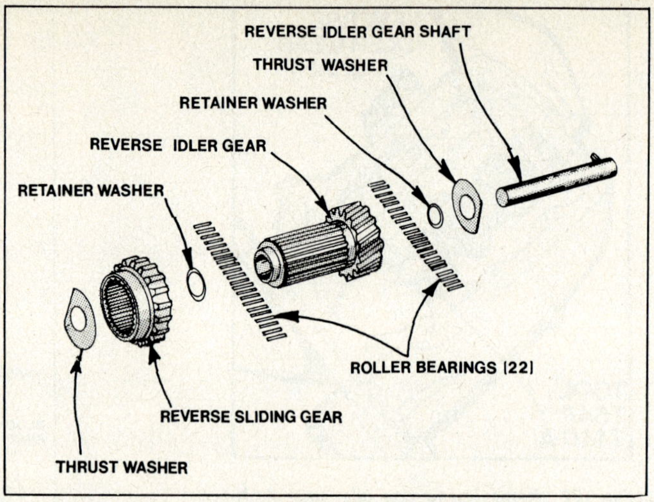

Fig. 13J Reverse idler gear, disassembled

23. Remove snap ring from front of output shaft and slide the third and overdrive synchronizer, blocking and gear from output shaft, Fig. 13F.
24. Remove the next snap ring and the second speed gear thrust washer, gear and blocking ring from output shaft.
25. Remove the next snap ring and the first speed gear thrust washer, gear and blocking ring from rear of output shaft.

Sub-Assembly Service

Cam & Shaft Seals

NOTE: To facilitate reassembly, note position of cams and shafts assemblies and levers before removal from transmission case.

1. Remove three shift levers, Fig. 13G.
2. Remove three cams and shafts from inside of case.
3. Remove O-ring from each cam and shaft.
4. Dip new O-rings into gear lubricant and install them on cam and shafts.
5. Slide each cam and shaft into its respective bore in case.
6. Secure each shift lever.

Synchronizers

1. Place alignment marks between hub and sleeve. Push synchronizer hub from sleeve, Fig. 13H.
2. Separate inserts and springs from hubs. Do not mix parts from one synchronizer to another.
3. Position hub in sleeve, being sure that alignment marks are properly indexed.
4. Place three inserts into hub. Install insert springs, making sure that irregular surface (hump) is seated in one of inserts. Do not stagger springs.

Countershaft Gear Bearings

1. With unit disassembled, coat bore in each end of countergear with grease, Fig. 13I.
2. Hold a suitable dummy shaft in gear and insert 21 rollers and a retainer washer in each end of gear.

Reverse Idler Gear Bearings

1. With unit disassembled, Fig. 13J, coat bore at each end of gear with grease.
2. Hold a suitable dummy shaft in gear and insert 22 rollers and retainer washer at each end of gear.
3. Install sliding gear on reverse idler gear, making sure that shift fork groove is toward front.

Input Shaft Seal

1. Remove seal from input shaft bearing retainer.
2. Lubricate sealing surface and install new seal.

Transmission Assembly

1. Lubricate countershaft gear thrust surfaces and position a thrust washer at each end of case. Place countershaft gear, dummy shaft and roller bearings in the case.
2. Place case in the vertical position, align gear bore and thrust washers with case bores and install countershaft.
3. Place case in the horizontal position and, with a feeler gauge, check countershaft gear end play which should be .004–.018 inch. If not, replace thrust washers.
4. Reinstall dummy shaft in countershaft gear and place gear at bottom of case.
5. Lubricate reverse idler gear thrust surfaces and place the two thrust washers in position. Install reverse idler gear, sliding gear, dummy shaft and roller bearings in place, ensuring the sliding gear shift fork groove is facing toward front of case. Align gear bore and thrust washers with case bores and install the reverse idler shaft.
6. With a feeler gauge, check reverse idler gear end play which should be .004–.018 inch. If not, replace thrust washers. If end play is within limits, do not remove gear and shaft.
7. Install reverse gear shift rail detent spring and detent plug into case. Holding reverse shift fork in place on the reverse idler sliding gear, install the shift rail from rear of case. Install shift rail to fork set screw.
8. Place first and second speed synchronizer onto output shaft with shift fork groove facing toward rear of shaft. The synchronizer hub is installed with the teeth end of the gear facing toward rear of output shaft.
9. Place blocking ring on second speed gear, then slide the gear onto front of output shaft, ensuring synchronizer inserts engage the blocker ring notches.
10. Install second speed gear thrust washer and snap ring.
11. Slide overdrive gear onto shaft with coned surface facing toward front of out-

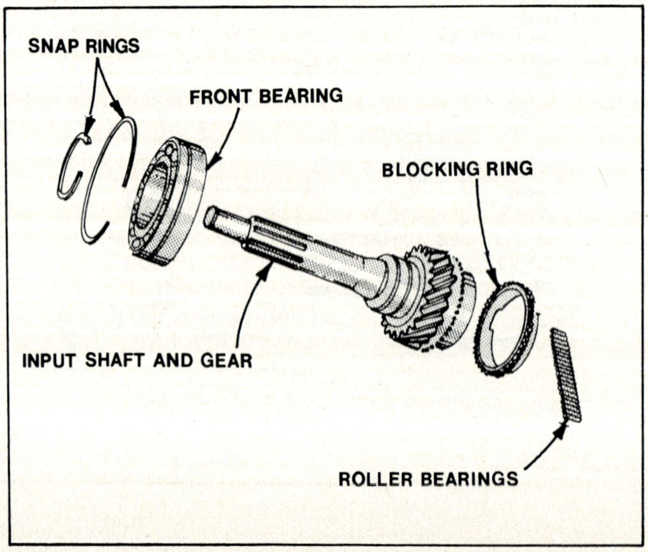

Fig. 14 Input shaft, disassembled

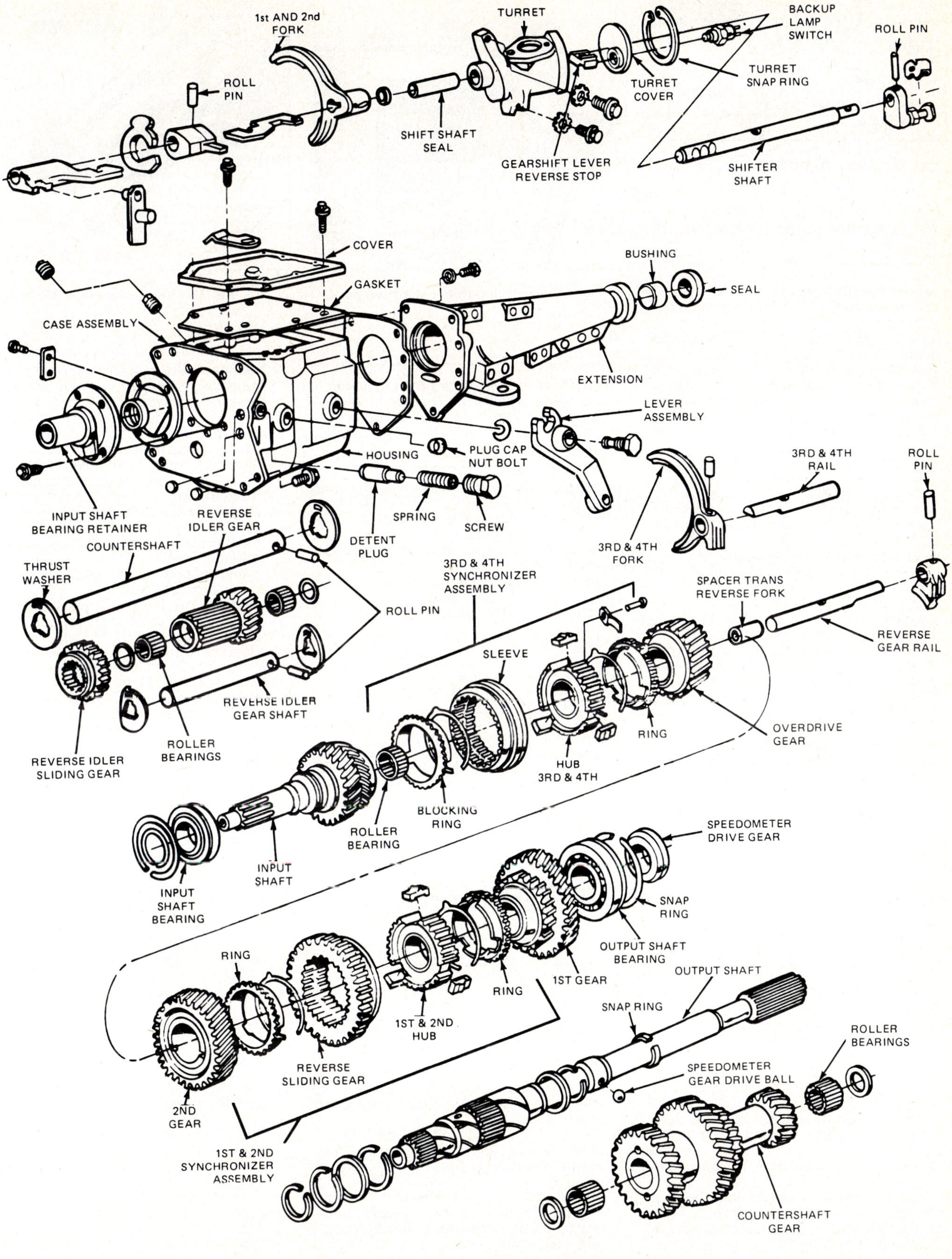

Fig. 15 Exploded view of Ford single rail 4-speed overdrive transmission

Fig. 16 Removing detent plug, spring & screw

put shaft, then install blocking ring on gear.

12. Slide third and overdrive gear synchronizer onto shaft, ensuring synchronizer inserts engage blocking ring notches and the thrust surface is facing toward overdrive gear.
13. Install snap ring at front of output shaft.
14. Place blocking ring on first speed gear, then slide first speed gear onto rear of output shaft, ensuring synchronizer inserts engage blocking ring notches.
15. Install the heavy thrust washer at rear of output shaft.
16. Support thrust washer and first speed gear to prevent washer and gear from sliding off shaft and lower assembly into case, Fig. 13C.
17. Install first and second speed shift fork and the third and overdrive speed shift fork, rotating forks into place.
18. Install detent spring and plug into reverse shift rail, then place rail in neutral position.
19. Lubricate third and overdrive shift rail interlock pin and install in shift rail.
20. Align third and overdrive shift fork with shift rail bores and slide shift rail into position, ensuring the three detents are facing toward outside of case. Place front synchronizer into neutral position and install the third and overdrive shift fork set screw. Place synchronizer in neutral position and install the third and overdrive shift rail detent plug, spring and bolt into left side of case. Install detent plug into case bore.
21. Align first and second speed shift fork with case bores and slide shift rail into place, then install set screw.
22. Lubricate input gear bore and install 15 roller bearings into bore.

NOTE: Apply a thin coat of lubricant since a heavy application may plug the lubricant holes and restrict lubrication of the bearings.

23. Place front blocking ring in the third and overdrive synchronizer.
24. Place a dummy bearing on rear of output shaft to support and align shaft in case.
25. Install input shaft gear into case, ensuring output shaft pilot enters roller bearings in input gear pocket.
26. Place input shaft bearing on input shaft and press bearing onto shaft and into case bore, Fig. 14.
27. Install input shaft and bearing snap rings, Fig. 14.
28. Place new gasket on input shaft bearing retainer, apply sealer to attaching bolts and install and torque attaching bolts to 19–25 ft. lbs.
29. Remove dummy bearing from output shaft and install output shaft bearing onto shaft and press into case bore.

Fig. 17 Removing shifter shaft roll pin

NOTE: Before pressing bearing onto output shaft, ensure the bearing is aligned with the case bore and the countershaft is not interfering with the output shaft assembly.

30. Install output shaft and bearing snap rings.
31. Place transmission in the vertical position. Align countershaft gear bore and thrust washers with case bore and install countershaft.
32. Install extension housing on case with a new gasket, apply sealer to attaching screws and install and torque to 42–50 ft. lbs.
33. Install filler plug.
34. Pour transmission lubricant over gear train while rotating input shaft.
35. Place each shift fork in all positions to ensure proper operation.
36. Install remaining detent plug in case and the long spring to secure detent plug.
37. Install cover with a new gasket, apply sealer to attaching screws and install and torque attaching screws to 20–25 ft. lbs.
38. Apply sealer to the third and overdrive shift rail plug bore and install new expansion plug.

SINGLE RAIL FOUR SPEED OVERDRIVE TRANS.

Disassembly

NOTE: Refer to Fig. 15, for exploded view of transmission

1. Remove transmission cover retaining screws, and remove cover and gasket.
2. Remove screw, detent spring and detent plug from case, Fig. 16. A magnetic rod may have to be used to remove spring and detent plug.
3. Drive roll pin from shifter shaft, Fig. 17, then remove back-up light switch, snap ring and dust cover from rear of extension housing.
4. Remove shifter shaft from turret assembly, Fig. 18, then remove extension housing retaining screws, and extension housing and gasket.
5. Remove speedometer gear retaining snap ring, then slide speedometer gear off shaft and remove drive ball.

Fig. 18 Removing shifter shaft

6. Remove output shaft bearing to shaft retaining snap ring, then using two screwdrivers, pry out bearing, Fig. 19.
7. From front of case, push countershaft out of rear of case with a dummy shaft. Lower countershaft to bottom of case.
8. Remove input shaft bearing retainer bolts, then slide retainer and gasket off input shaft.
9. Remove input shaft bearing to shaft retaining snap ring, then remove input shaft bearing from input shaft and transmission case.
10. Remove input shaft, blocking ring and roller bearings from case.
11. Remove overdrive shift pawl, gear selector interlock plate and 1–2 gearshift selector arm plate.
12. Remove roll pin from 3rd-overdrive shift fork, Fig. 20, then remove 3rd overdrive shift rail, Fig. 21.
13. Remove first, second and third overdrive speed shift fork.
14. Lift countershaft gear and thrust washers from case. Use care to avoid dropping the bearings or dummy shaft from countershaft gear.
15. Remove snap ring from front of output shaft, then slide third and overdrive synchronizer blocking ring and gear off the shaft.
16. Remove the next snap ring and thrust washer, first speed gear and blocking ring from rear of shaft.
17. Lift countershaft gear thrust washers and roller bearings from case. Use care to avoid dropping the bearings or dummy shaft from countershaft gear.
18. From front of case, drive the reverse gear shaft out of case.
19. Remove reverse idler gear, thrust washers and roller bearings. Use care to avoid dropping bearing.
20. Remove retaining clip, reverse gearshift relay lever and reverse gear selector fork pivot pin.
21. Remove overdrive shift control link assembly, then remove shift shaft seal from rear of case.
22. Remove expansion plug from front of case.

Assembly

1. Reverse disassembly procedure to assemble.
2. The transmission mainshaft bearing rollers, extension housing bushing, reverse idler bearing rollers and countershaft gear bearing rollers must be lubri-

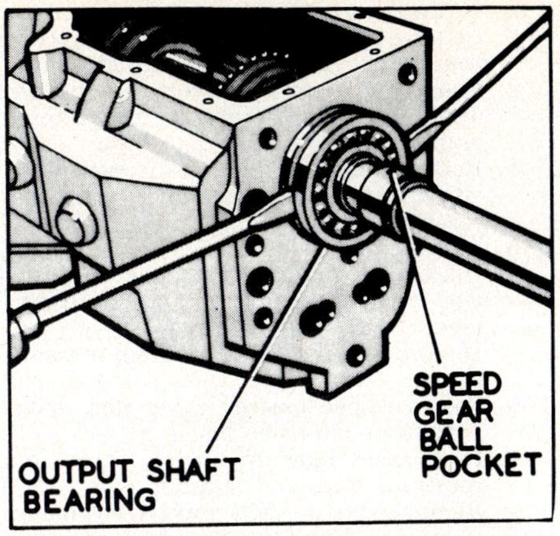

Fig. 19 Removing output shaft bearing

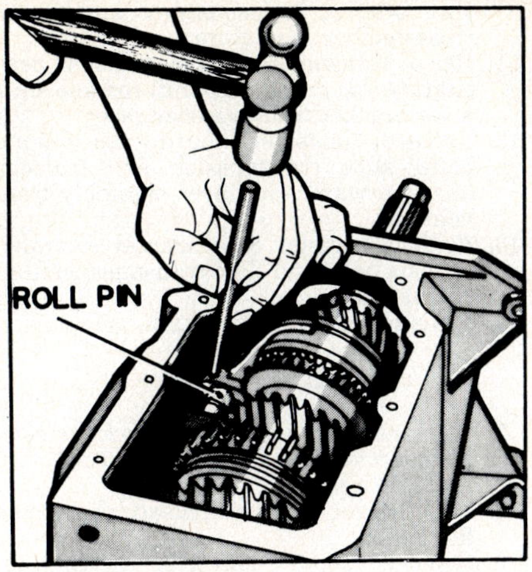

Fig. 20 Removing 3rd-overdrive shift fork roll pin

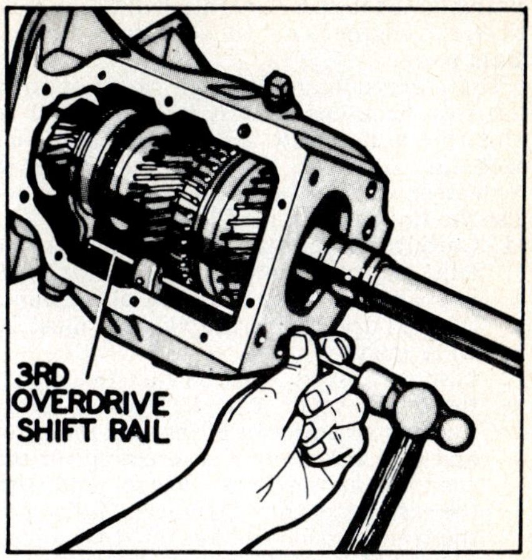

Fig. 21 Removing 3rd-overdrive shift rail

cated with ESW-M1C109-A or equivalent at each location. The low gear and 2nd and overdrive gear journals are to be lubricated with ESP-M2C83-G transmission oil or equivalent. All other internal components of the transmission are to be lubricated with ESP-M2C83-C transmission oil or equivalent.

3. The intermediate and high rail welch plug must be seated firmly and must not protrude above the front face of case, nor must it be recessed more than .06 inch below front face of case.
4. The end play of the 1st, 2nd and overdrive gears after their assembly on the output shaft must be checked using a suitable gauge and must meet the following specifications:
   a. With 1st gear thrust washer clamped against the shoulder on the output shaft, the 1st gear end play should be .005–.024 inch.
   b. 2nd gear end play should be .003–.021 inch.
   c. Overdrive gear end play should be .009–.023 inch.
   d. Countershaft gear end play after its assembly into case between thrust washers must be .004–.018.

NOTE: If dimensions are not within specifications, replace components as necessary.

5. When gearshift selector arm plate is firmly seated in the first-second shift fork plate slot, the transmission shifter shaft must pass freely through the first-second shift fork shifter shaft bore without indication of binding.
6. Torque specifications for the bolts or parts are as follows:

Bolt or Part	Torque (Ft. Lbs.)
Back-up Lamp Switch	8–12
Bearing Retainer	11–25
Case Access Cover	20–25
Detent Bolt	10–15
Extension Housing	42–50
Filler Plug	10–20
Reverse Gear Fork Pivot Pin	15–25
Turret Assembly	8–12

TILT CAB SHIFT LINKAGE

1974–83 Adjustment

1. Shift transmission into neutral and loosen both turnbuckle nuts, Fig. 22.
2. Rotate turnbuckle until shift lever is located 6¼ inch from instrument panel (view Z), Fig. 23.
3. Rotate rear universal joint counterclockwise (looking rearward from front of truck) as far as possible.

NOTE: Spring plunger in transmission must be fully compressed on CM50, CM60, QM60, 280, 380, 390, 5000 and 6000 model transmissions.

4. Position shift arm at 2½° from vertical (angle "B"), Fig. 23, then tighten turnbuckle nuts.
5. Adjust shift rod to position shift lever ball 7 11/16 inch from steering column (view Y), Fig. 14B, then tighten jam nuts on shift rod. Threads on shift rod must be centered within 1/8 inch.

FORD AUTOMATIC TRANS.

Linkage Adjustments

Before making any checks or adjustments of the transmission linkage check the performance of the engine. The transmission is

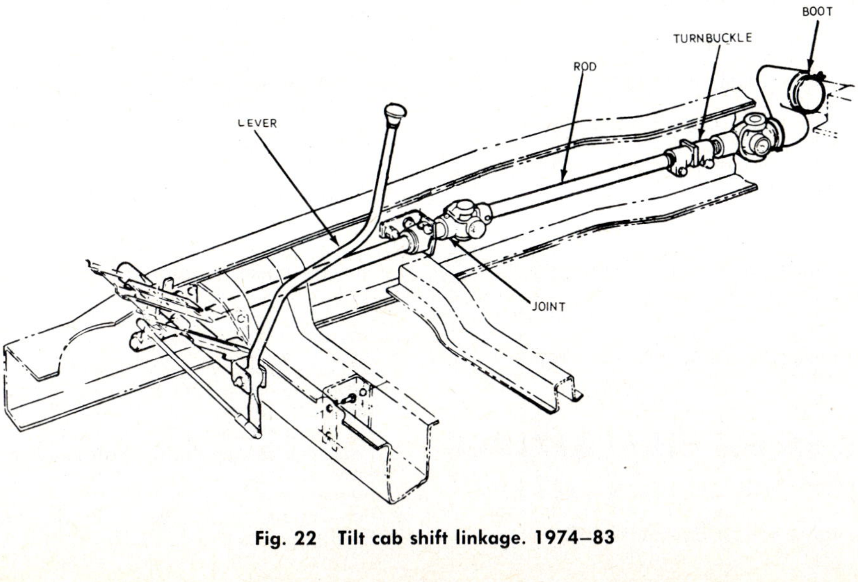

Fig. 22 Tilt cab shift linkage. 1974–83

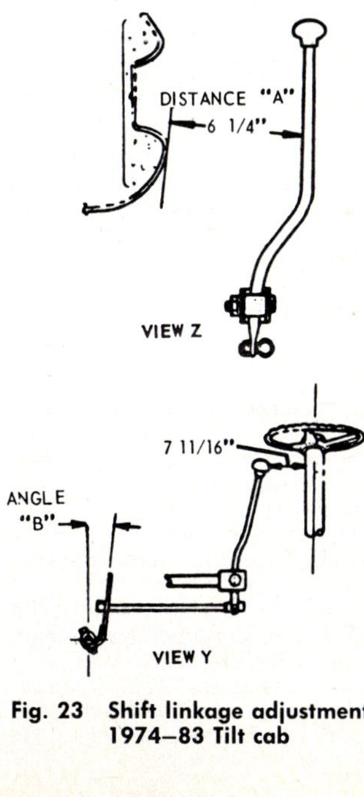

Fig. 23 Shift linkage adjustments. 1974–83 Tilt cab

often blamed for poor operation of the vehicle when the engine is not tuned to deliver maximum power.

There are so many linkage variations and intricate hook-ups that it is impractical to illustrate and go into specific procedures for each one. However, there are two things to remember when it becomes necessary to adjust the linkage.

1. Carburetor throttle linkage must be so adjusted that when the accelerator pedal is pushed down to the limit of its travel the carburetor throttle valve(s) must be wide open.
2. Manual linkage must be adjusted so that the hand lever can be moved freely through all positions of the shift indicator and that the pointer is directly over the positions as they are selected. And with the carburetor throttle valve(s) fully open the transmission linkage must be adjusted so that the TV lever on the transmission will move to the limit of its travel.

C4 CRUISEOMATIC, REPLACE

Exc. Bronco, Econoline & Ranchero

1. Drain transmission and converter.
2. Remove drive shaft.
3. Disconnect oil cooler lines from transmission.
4. Remove downshift lever return spring from low-reverse servo cover.
5. Disconnect shift rods from transmission levers.
6. Disconnect neutral start switch wiring harness and back-up wires at connectors, then retaining clips and retainer.
7. Remove speedometer gear from extension housing.
8. Remove four converter-to-flywheel bolts.
9. Remove starter.
10. Disconnect vacuum line from diaphragm unit.
11. Support transmission with a jack.
12. Remove crossmember from under extension housing.
13. Remove transmission-to-engine bolts.
14. Move transmission away from engine. Then lower and remove transmission from under truck.
15. Reverse procedure to install.

Bronco

1. Remove fan shroud and raise vehicle.
2. Remove transfer case shield, if used, and drain transmission, converter and transfer case.
3. Disconnect rear drive shaft from transfer case and remove front drive shaft from vehicle.
4. Disconnect exhaust system and position aside.
5. Remove speedometer gear from transfer case.
6. Disconnect oil cooler lines, shift linkage and neutral start switch wires from transmission.
7. Remove starter.
8. Disconnect vacuum lines from modulator and retaining clip.
9. Remove crossmember-to-transfer case adapter and crossmember-to-frame side support attaching bolts.
10. Raise transmission and transfer case with a suitable jack and remove left side support bracket-to-frame bolts, side support bracket, crossmember and upper crossmember insulators.
11. Raise transmission and transfer case slightly and disconnect shift rod and shift lever bracket from transfer case.
12. Remove converter housing to engine bolts, slide transmission and transfer case rearward and lower assembly from vehicle.
13. Remove transfer case adapter to transmission bolts and slide transmission from transfer case.
14. Reverse procedure to install.

Econoline

1. Working from inside the truck, remove engine cover and disconnect neutral start switch wires.
2. On V8 units, remove air cleaner heat tube flex hose.
3. Remove upper converter housing-to-engine bolts.
4. On V8, disconnect exhaust pipe from manifold.
5. Raise vehicle and drain transmission and converter.
6. Remove converter cover and remove converter-to-flywheel bolts.
7. Disconnect drive shaft.
8. Remove transmission filler tube.
9. Remove starter.
10. Install suitable engine support bar.
11. Disconnect vacuum and cooler lines, shift linkage and speedometer cable.
12. Support transmission with suitable jack and remove crossmember.
13. Remove remaining converter housing-to-engine bolts and lower transmission away from vehicle.
14. Reverse procedure to install.

Ranchero

1. Disconnect neutral safety switch wires in engine compartment.
2. Raise vehicle and remove converter cover at low front side of converter housing.
3. Drain converter.
4. Remove propeller shaft.
5. Remove vacuum line hose from transmission vacuum unit. Disconnect vacuum line from clip.
6. Remove two extension housing to crossmember bolts.
7. Remove speedometer cable from extension housing.
8. Disconnect exhaust pipe from manifold.
9. Remove parking brake cable from equalizer lever.
10. Loosen transmission oil pan bolts and drain oil at one corner of pan.
11. Disconnect cooler lines from the transmission.
12. Remove the manual and kick down linkage rods from transmission shift levers.
13. Where necessary, disconnect the neutral start switch wires.
14. Remove starter.
15. Remove fluid.
16. Remove four converter-to-flywheel nuts.
17. Support transmission with jack.
18. Remove crossmember.
19. Remove converter housing-to-engine bolts.
20. Lower transmission and remove from vehicle.

C5 SELECT SHIFT REPLACE

1982 F100 & F150 & 1983 F100

1. Raise vehicle on suitable lift.
2. Drain fluid from transmission.
3. Remove converter drain plug access cover from lower end of converter housing.
4. Remove the four converter-to-flywheel attaching nuts. Rotate the converter to gain access to the nuts.
5. Rotate converter to gain access to converter drain plug. Drain converter then reinstall plug
6. Remove drive shaft.
7. Disconnect manual and downshift linkage rods from transmission control levers.
8. Disconnect oil cooler lines from transmission.
9. Remove speedometer cable and driven gear from extension housing.
10. Disconnect back-up switch wires from retaining clips and retainer.
11. Remove starter from converter housing.
12. Remove vacuum line hose from transmission vacuum unit. Disconnect vacuum line from retaining clip.
13. Using a transmission jack to support the transmission remove two engine rear support crossmember-to-frame attaching bolts.
14. Remove two engine rear support-to-extension housing attaching bolts.
15. Raise transmission and remove rear support. Remove six converter housing-to-engine attaching bolts.
16. Move converter and transmission assembly away from the engine. Lower the transmission and remove it from under vehicle.
17. Reverse procedure to install.

C6 CRUISEOMATIC, REPLACE

1974–83 Exc. Ranchero & 1975–83 Econoline

1. Remove two upper converter housing-to-engine bolts and oil filler tube-to-cylinder head bolt.
2. Raise vehicle and drain transmission.
3. Disconnect drive shaft.
4. Disconnect speedometer cable, shift linkage, oil cooler lines and vacuum line.
5. Remove starter and driveshaft.

NOTE: On F-150-250 models with four wheel drive and Bronco, remove the transfer case.

6. Remove four converter-to-flywheel bolts.
7. Support transmission with a suitable jack and remove both crossmembers.
8. Secure transmission to jack, remove remaining converter-to-engine bolts and lower transmission away from vehicle.
9. Reverse procedure to install.

1974–78 Ranchero

NOTE: On models with the neutral safety switch wire harness connected at the dash panel, disconnect the harness before raising the vehicle.

1. Working from engine compartment, remove the two bolts retaining the fan shroud to the radiator. On 1975–78 models, disconnect the battery.
2. Raise vehicle and drain transmission and converter.
3. Remove drive shaft and starter.
4. Remove four converter to flywheel attaching points.

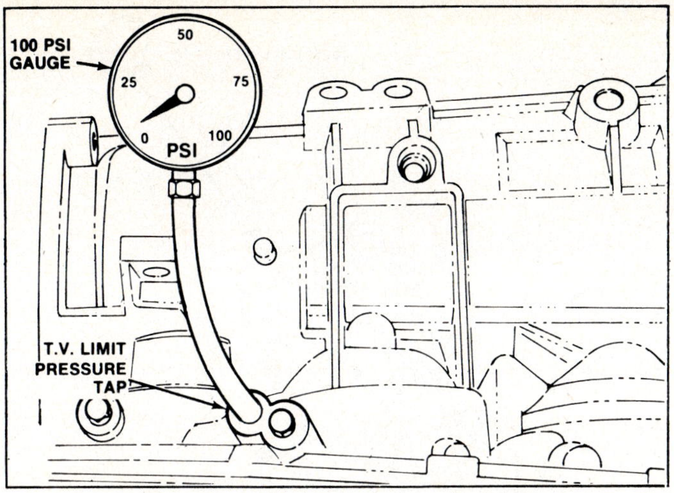

Fig. 24 Gauge connections to throttle valve pressure tap

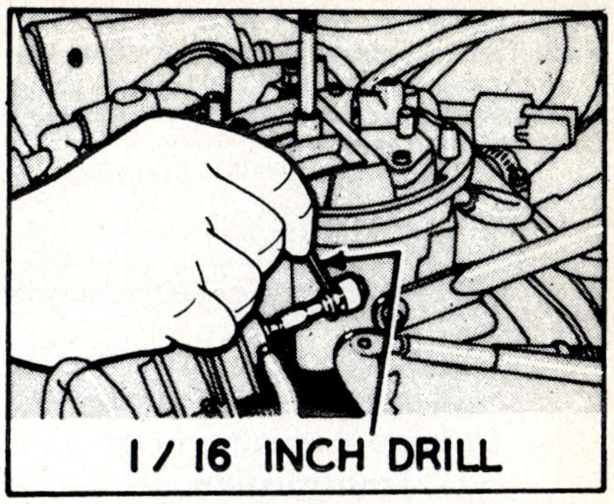

Fig. 25 Positioning throttle lever

5. Disconnect parking brake front cable from equalizer.
6. Disconnect speedometer cable and transmission linkage.
7. Where necessary, disconnect muffler inlet pipes from exhaust manifold.
8. Support the transmission with a suitable jack, remove parking brake rear cables from the equalizer and remove the crossmember and rear engine support.
9. Lower transmission and remove oil cooler lines, vacuum line and transmission oil filler tube.
10. Secure the transmission to the jack with the chain, remove the converter housing to cylinder block bolts and carefully move the transmission away from the engine, at the same time lowering it to clear the underside of the vehicle.

1975–83 Econoline

1. Working from inside the truck, remove engine cover and disconnect neutral start switch wires.
2. On V8 units, remove air cleaner heat tube flex hose.
3. Remove upper converter housing-to-engine bolts.
4. On V8, disconnect exhaust pipe from manifold.
5. Raise vehicle and drain transmission and converter.
6. Remove converter cover and remove converter-to-flywheel bolts.
7. Disconnect drive shaft.
8. Remove transmission filler tube.
9. Remove starter.
10. Install suitable engine support bar.
11. Disconnect vacuum and cooler lines, shift linkage and speedometer cable.
12. Support transmission with suitable jack and remove crossmember.
13. Remove remaining converter housing-to-engine bolts and lower transmission away from vehicle.
14. Reverse procedure to install.

HEAVY DUTY CRUISEOMATIC

1974–77 Exc. Ranchero

Transmission & Converter

1. Drain converter.
2. Disconnect oil filler pipe.
3. On 1974–77 models, wedge the converter in place.
4. Remove starter.
5. Disconnect oil cooler hoses (if used).
6. Disconnect linkage from transmission.
7. Remove hose from vacuum unit.
8. Disconnect speedometer cable and remove drive shaft.
9. Remove both rear engine support bolts.
10. With transmission jack raise engine and transmission high enough to remove engine rear supports.
11. Lower engine against floor stand or engine support bar so that converter housing is clear of crossmember when all weight is off transmission jack.
12. Remove remaining converter housing-to-engine bolts.
13. Where necessary, loosen and drop the exhaust system to allow converter to clear exhaust pipe.
14. Remove flywheel-to-converter bolts.
15. Move transmission rearward and lower it, leaving flywheel attached to crankshaft. If additional clearance is needed, tilt rear of assembly upright slightly and to the rear enough to allow removal of flywheel-to-crankshaft bolts.
16. Move assembly to rear and remove it.
17. Reverse procedure to install.

1974–77 Ranchero

NOTE: On some models, it is necessary to remove the two upper converter housing to engine bolts before raising vehicle.

1. Raise vehicle and place on jack stands, then drain transmission oil pan.
2. Remove converter access cover and remove converter drain plug.
3. Remove converter to flywheel attaching nuts, reinstall converter drain plugs and converter housing access cover to hold converter in place when transmission is removed.
4. Remove starter and propeller shaft.
5. Where necessary, disconnect exhaust pipes from manifold.
6. Disconnect oil cooler lines, speedometer cable, vacuum hose and manual downshift linkage.
7. Disconnect TRS switch wire, if equipped.
8. Support transmission with suitable jack and remove crossmember.

NOTE: On some models, it is necessary to remove the engine rear support to transmission bolts before crossmember removal.

9. Lower transmission and remove filler tube and dipstick.
10. Remove converter housing-to-engine bolts. On 1974–77 models, move transmission and jack rearward and lower away from vehicle.
11. Reverse procedure to install.

AUTOMATIC OVERDRIVE TRANSMISSION

Refer to the "Automatic Transmission" chapter elsewhere in this manual for service procedures not contained in this section.

ADJUSTMENTS

Throttle Valve Control Linkage

1. Check that idle is set to specifications.
2. Position carburetor throttle lever on its idle stop. Place the shift lever in Neutral and set the parking brake.
3. With engine off, rotate the linkage lever adjusting screw fully counter-clockwise.
4. Rotate adjusting screw clockwise until a thin shim or a piece of paper fits snugly between end of screw and throttle lever.

NOTE: Do not apply any load on levers with tools or hands while checking gap.

5. Rotate adjusting screw clockwise an additional four turns. Two turns is minimum permitted.
6. If it is not possible to turn in the adjusting screw at least two additional turns or if there is insufficient screw adjusting capacity to obtain an initial gap, refer to "Linkage Adjustment At Transmission."

Idle Speed

NOTE: When an adjustment of idle speed of more than 50 RPM is required, the adjustment screw on linkage lever at the carburetor should also be readjusted.

IDLE SPEED CHANGE	CHANGE ON LINKAGE LEVER ADJUSTING SCREW
Less than 50 RPM	No change required
50–100 RPM increase	1½ turns counterclockwise
50–100 RPM decrease	1½ turns clockwise
100–150 RPM increase	2½ turns counterclockwise
100–150 RPM decrease	2½ turns clockwise

Linkage Adjustment at Transmission

NOTE: Linkage lever adjustment screw has limited adjustment capacity. If it is not possible to make the adjustments using this screw, the length of the TV control rod assembly must be readjusted using the following procedure.

1. Set the idle speed to specification.
2. With engine off, position carburetor throttle lever against idle stop. Place selector in neutral and set parking brake.
3. Set linkage lever adjustment screw at its approximate mid-range.
4. If a new TV control rod assembly is being installed, connect the rod to the linkage lever at the carburetor.
5. Raise and support vehicle.
6. Loosen nut on sliding stud on TV control rod assembly. Remove any corrosion from the control rod and free up the stud so that it slides freely on control rod. Insert pin into transmission lever grommet.
7. Push up the lower end of the control rod to insure that the linkage lever at the carburetor is firmly against the throttle lever. Release rod. Rod must stay up.
8. Force control lever on the transmission against its internal stop with a force of approximately five pounds, then tighten stud nut. Do not release the force on the lever until the nut is tightened.

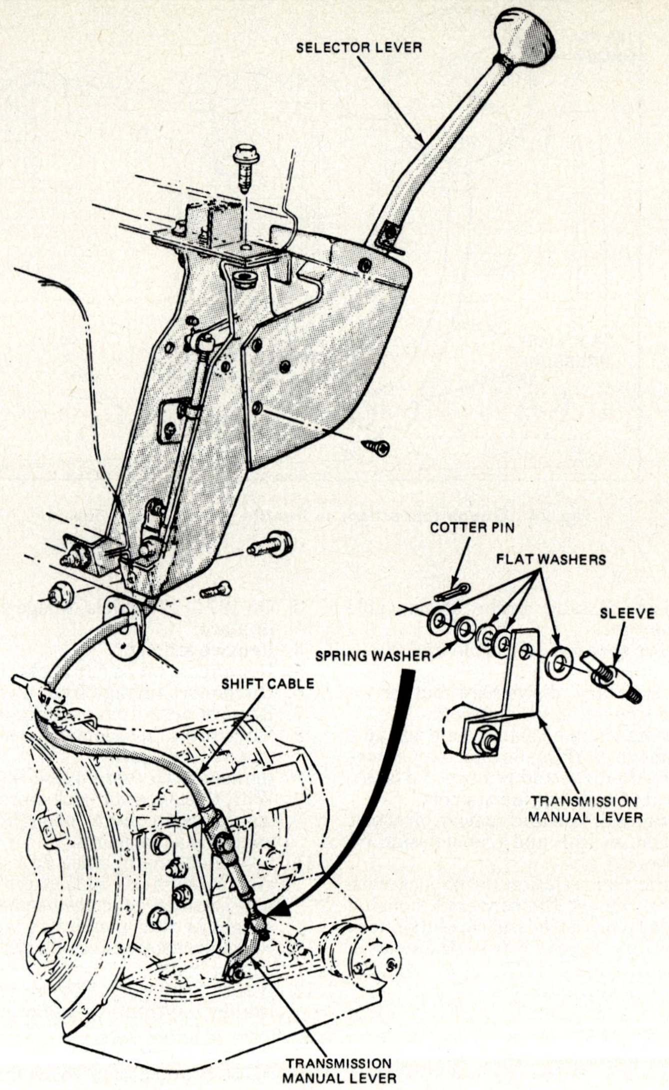

Fig. 26 AT-540, MT-640 & MT-650 range selector linkage (typical)

Throttle Valve Pressure

NOTE: This procedure is used to check and or adjust the throttle valve control linkage using the throttle valve control pressure. Idle speed should be set to specifications.

1. Install a suitable pressure gauge to the throttle valve pressure tap on the transmission, Fig. 24.
2. Place the selector in Neutral and disconnect idle kicker solenoid. Set parking brake and remove air cleaner.
3. Allow engine to reach operating temperature with throttle valve in off fast idle.

NOTE: Do not make checks in Park.

4. Read the pressure gauge and slowly pull back TV rod at carburetor, causing a gap to occur between the throttle valve linkage lever and the throttle lever. The pressure should remain near zero and then rapidly increase to approximately 15 to 30 psi, with only a slight increase in gap. The point where pressure increases rapidly is the breakpoint.
5. Breakpoint gap should be between 3/16 and 1/4 inches on 1981 vehicles, or 5/32 to 7/32 inch on 1982 vehicles. However, a gap between 1/16 to 5/16 is acceptable. Gap may be gauged using standard sized drills inserted between lever adjustment screw and throttle lever, Fig. 25.
6. If breakpoint gap is less than 1/16 inch linkage is set too short. Correct by backing out linkage lever adjustment screw.
7. If breakpoint gap is greater than 5/16 inch, linkage is set too short. Correct by turning in the linkage lever adjustment screw.

NOTE: If the correct throttle valve breakpoint cannot be obtained using above methods, it may be necessary to adjust the throttle valve control rod length at the transmission as outlined under "Linkage Adjustment at Transmission."

TRANSMISSION, REPLACE

1. Raise and support vehicle.
2. Drain transmission fluid from pan. After fluid is drained, install pan.
3. Remove converter drain plug access cover from lower end of converter housing. Rotate engine to gain access to converter drain plug. Remove drain plug, drain fluid from converter, then replace drain plug.
4. Remove converter to flywheel attaching nuts, then the driveshaft from vehicle.
5. Disconnect battery cable from starter motor, then remove starter motor. Disconnect neutral start switch electrical connector.
6. Remove rear mount to crossmember bolts and crossmember to frame bolts.
7. Remove engine rear support to extension housing bolts.
8. Disconnect manual linkage from transmission, then remove bolts securing bellcrank bracket to converter housing.
9. Raise transmission with suitable jack and remove crossmember.
10. Lower transmission slightly and disconnect oil cooler lines and speedometer cable from transmission.
11. Remove filler tube and dipstick from transmission.
12. With transmission secured to jack, remove converter housing to cylinder block attaching bolts. Move transmission

and converter assembly rearward, then lower transmission and remove from under vehicle.
13. Reverse procedure to install.

AT-540 & MT-640, 650

For a detailed description on the operation and servicing of this transmission see the *Allison AT-540 & MT-640, 650 Chapters* elsewhere in this manual. The following material deals with linkage adjustments and removal of the unit from the truck.

Range Selector Linkage, Adjust

AT-540

1. With engine stopped, position transmission selector lever in "R" position.
2. Disconnect shift cable sleeve at transmission manual lever, Fig. 26.
3. Shift transmission manual lever to "R" (all the way forward and upward).
4. With transmission manual lever in "R" position, adjust sleeve so that it freely enters into hole of manual lever.
5. Connect sleeve to manual lever with flat washers, spring washer and cotter pin.
6. Check operation in all ranges. It may be necessary to readjust sleeve slightly to obtain detent position in all drive ranges.

MT-640, 650

When linkage is correctly adjusted, the pin which engages the shift lever linkage at the transmission should move freely in each range. Proper installation of the external selector lever is imperative to the function of the transmission. Excessive torquing of the selector lever nut may damage the internal selector lever.

Refer to AT-540 procedure for selector linkage adjustment.

Transmission, Replace 1974–83

1. Remove right and left door sill scuff plates and transmission access cover.
2. On one piece drive shaft models, disconnect drive shaft from pinion flange and yoke at parking brake drum.
3. On two piece drive shaft models, disconnect the coupling shaft from yoke at parking brake drum, then remove center support bearing bracket. Place aside forward end of drive shaft.
4. On all models, disconnect parking brake linkage and remove parking brake drum from output shaft flange.
5. Remove parking brake handle and brackets from transmission case.
6. Remove filler tube and drain transmission.
7. Disconnect speedometer cable, oil cooler lines, shift cable, vacuum modulator line and remove modulator and shift cable clamp bracket from transmission.
8. Remove flywheel housing dust shield and converter to flywheel bolts.
9. Remove four upper and two lower converter housing to flywheel housing bolts.
10. Remove engine rear support lower insulators and loosen engine front support bolts.
11. Support transmission with a suitable jack and with a second jack, raise rear of engine, relieving weight from crossmember.
12. Remove engine rear support upper insulators and brackets from crossmember.
13. Remove the remaining converter housing to flywheel housing bolts and move transmission assembly rearward until converter housing contacts metal body. Then, tilt rear of transmission upward to clear crossmember and lower transmission from vehicle.
14. Reverse procedure to install.

Rear Axle Section

NOTE: For Eaton, Spicer and Timken (Rockwell) driving axles used on Ford Trucks, see the "name" chapter for service procedures and lubricant capacities

DANA/SPICER INTEGRAL HOUSING TYPE AXLE

This rear axle, Fig. 1, is a semi-floating type, while the axle shown in Fig. 1A is of the full-floating type. The drive pinion is of the over-hung design, mounted on pre-loaded taper roller bearings. Sealing of the pinion shaft is accomplished at the front end by a spring-loaded leather seal bearing on the companion flange, which is splined and secured to the pinion shaft by a nut.

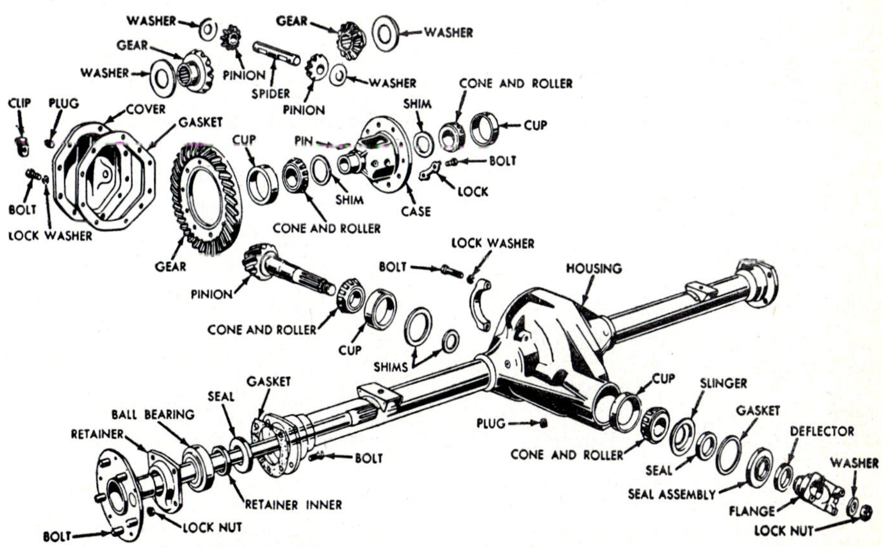

Fig. 1 Semi-floating integral housing hypoid rear axle

Axle Shaft, Bearing, Oil Seal, Replace

Full-Floating Axle

1. Set parking brake and loosen axle shaft bolts.
2. Raise and support rear of vehicle so that axle is parallel with floor.
3. Remove axle shaft bolts.
4. Remove axle shaft and discard gasket.
5. Pry out locking wedge with screwdriver.
6. Raise wheel until weight is removed from wheel bearing.
7. Remove wheel bearing adjustment nut.
8. Remove outer bearing cone and pull wheel assembly from axle.
9. Drive inner bearing cone and seal out of wheel hub with brass drift.
10. Inspect bearings and races and replaced if worn or damaged. Drive races from wheel hub with brass drift.
11. Lubricate bearing cone and roller assemblies and place inner assemblies into wheel hub. Install new hub inner seal.
12. Wrap spindle threads with tape, then slide wheel assembly onto axle housing spindle and remove tape.
13. Install outer wheel bearing and start adjuster nut. Lower wheel to ground.
14. While rotating wheel, torque adjusting nut to 120–140 ft. lbs. Back off the nut to obtain 0.001–0.010 inch end play (1/8–3/8 turn). If nut or locking wedge are damaged or do not allow for proper end play they should be replaced.
15. Position locking wedge in key way slot, making sure it is not bottomed against

the shoulder of adjusting nut.
16. Install axle shaft and new flange gasket, then install retaining bolts and torque locking bolts to 40–50 ft. lbs.
17. Adjust brakes.
18. Remove supports and lower vehicle.

Semi-Floating Axle
1. Raise and support vehicle.
2. Remove wheel and brake drum.
3. Remove cover plate and discard gasket.
4. Remove differential pinion shaft lock screw and the shaft.
5. Remove "C" clip from button end of shaft by pushing inward on other end.
6. Pull shaft from axle tube. Do not rotate differential side gears.
7. Remove and discard oil seal from axle tube.
8. Pull bearing from axle tube.
9. Lubricate new bearing and install in axle tube.
10. Lubricate and install new oil seal.
11. Insert shaft into axle tube, making sure splines engage side gears, then install "C" clip and pull shaft outward until the clip locks.
12. Install pinion mate shaft, aligning lock pin holes, and pinion gear side washers.
13. Install new lock screw and torque to 8 ft. lbs.
14. Install cover plate and gasket. Torque bolts to 30–40 ft. lbs.
15. Install wheel and drum assembly.
16. Remove supports and lower vehicle.

Rear Axle Assembly, Replace

1. Raise vehicle from floor and support with stand jacks under frame side rails.
2. Remove rear wheels.
3. Split rear universal joint.
4. Disconnect parking brake cable from equalizer rod and unfasten brake cable brackets from frame crossmember.
5. Disconnect hydraulic brake line connection at rear axle housing.
6. Loosen and move shock absorbers out of the way.
7. While supporting axle housing with hydraulic jack, remove spring clips and lower axle assembly to the floor.
8. Reverse the foregoing procedure to install the rear axle assembly, being sure to bleed the brake system when the installation is completed.

FORD TYPE AXLE

Axle Shaft, Replace

Removal, Units with Ball Bearing
1. Remove wheel assembly.
2. Remove brake drum from flange.
3. Working through hole provided in axle shaft flange, remove nuts that secure wheel bearing retainer.
4. Pull axle shaft out of housing. If bearing is a tight fit in axle housing use a slide hammer-type puller. *Brake carrier plate must not be dislodged. Install one nut to hold the plate in place after axle shaft is removed.*

NOTE: On 1978–83 models, remove brake backing plate and secure to frame rail.

---

5. If the axle shaft bearing is to be replaced, loosen the inner retainer by nicking it deeply with a chisel in several places. The bearing will then slide off easily.
6. Press bearing from axle shaft.
7. Inspect machined surface of axle shaft and housing for rough spots that would affect sealing action of the oil seal. Carefully remove any burrs or rough spots.
8. Press new bearing on shaft until it seats firmly against shoulder on shaft.
9. Press inner bearing retainer on shaft until it seats firmly against bearing.
10. If oil seal is to be replaced, use a hook-type tool to pull it out of housing. Wipe a small amount of oil resistant sealer on outer edge of seal before it is installed.

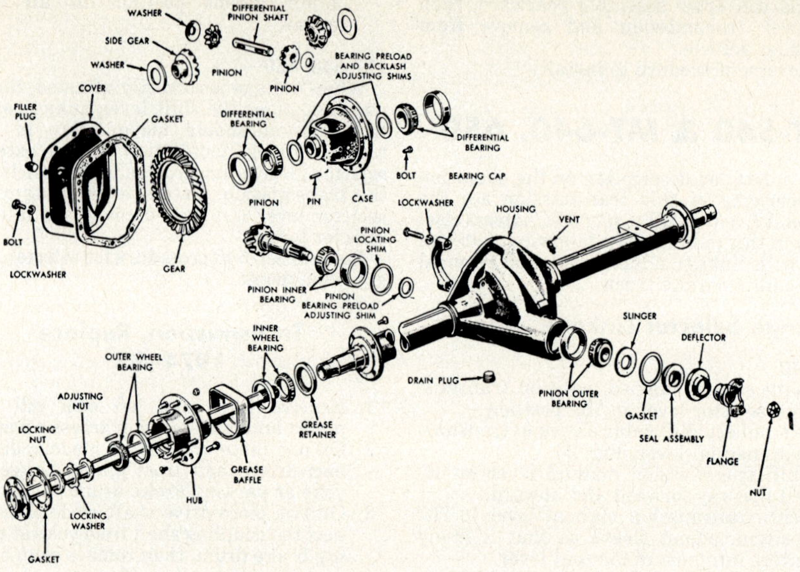

Fig. 1A Full-floating integral housing hypoid rear axle

Removal, Units with Roller Bearing
1. Remove wheel assembly.
2. Remove brake drum from flange.
3. Working through hole provided in axle flange, remove nuts securing wheel bearing retainer.
4. Pull axle shaft carefully from housing to prevent damage to outer seal rubber. Use a slide hammer type puller to remove tapered bearing cup from housing. Remove brake backing plate and secure to frame rail.
5. If the axle shaft bearing or seal is to be replaced, split the inner bearing retainer.
6. Press bearing from axle shaft.
7. Inspect machined surface of axle shaft and housing for rough spots that would affect sealing action of the oil seal. Carefully remove any burrs or rough spots.
8. Install outer retainer plate on axle shaft. Press lubricated seal and bearing onto shaft until firmly seated.

NOTE: Oil seal for rear drum brake equipped vehicles is different than that used on vehicles with rear disc brakes. Seals used with drum brakes have a gray metal colored outer rim, while seals used with disc brakes have a black or orange color.

---

9. Press new bearing retainer onto shaft until seated.

Installation
1. On 1978–83 models, place a new gasket on each side of brake carrier plate and slide axle shaft into housing. On models with roller bearing, ensure that outer seal is not fully mounted on the bearing.
2. Start the splines into the differential side gear and push the shaft in until bearing bottoms in housing.
3. Torque bearing retainer nuts 30–40 ft. lbs. on light duty axles, 50–75 ft. lbs. on heavy duty models.

FORD REMOVABLE CARRIER TYPE REAR AXLE

This rear axle, Fig. 2, is a banjo-housing hypoid in which the drive pinion is mounted 2¼" below the centerline of the drive gear. The drive pinion is straddle-mounted in that two opposed tapered roller bearings support the pinion shaft at the front of the pinion gear and a straight roller bearing supports the pinion shaft at the rear of the pinion gear.

The pinion shaft and gear are assembled in the pinion retainer which is bolted to the carrier housing. Two carrier and differential cases are used to accommodate two bearing sizes. The right and left axle shafts are not interchangeable since the left shaft is shorter than the right.

Disassembly

1. Mark one differential bearing cap and bearing support to ensure proper assembly. Remove adjusting nut locks, bearing caps and adjusting nuts. Lift differential out of carrier.
2. Remove drive gear from differential case.
3. Drive out differential pinion shaft retainer and separate the differential pinion shaft and remove gears and thrust washers.
4. Remove drive pinion retainer from carrier. Remove O-ring from retainer.
5. Remove pinion locating shim. Measure shim thickness with micrometer.
6. If the drive pinion pilot bearing is to be

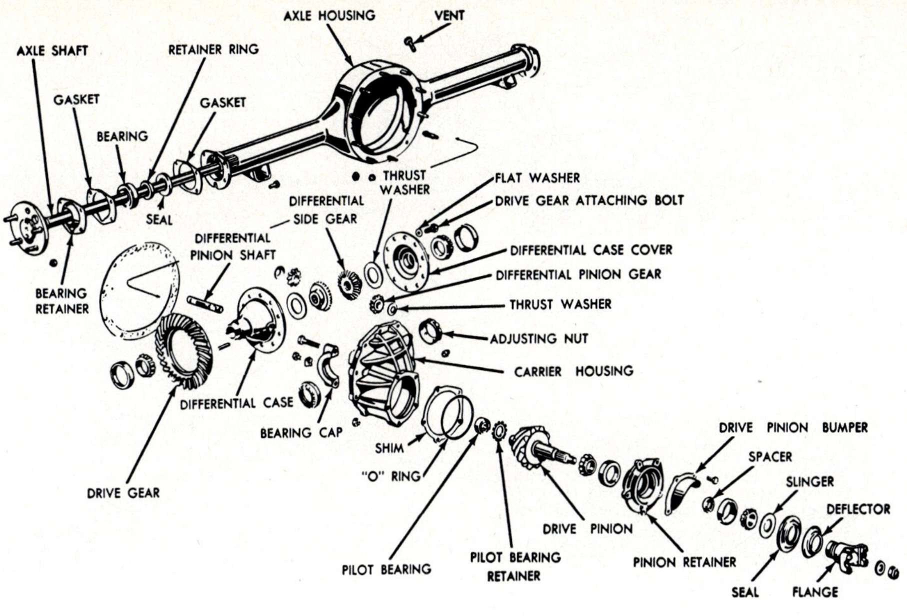

Fig. 2 Ford removable carrier rear

replaced, drive the pilot end and bearing retainer out at the same time. When installing, drive the bearing in until it bottoms. Install a new retainer with the concave side up.

7. Mount the drive pinion retainer assembly in a holding fixture and remove pinion shaft nut and flat washer. Then remove universal joint flange, oil seal and slinger.
8. Remove pinion retainer from fixture and place a protective sleeve or hose on the pinion pilot bearing surface. Then press the pinion shaft out of front bearing cone and remove spacer.
9. Remove pinion rear bearing cone.
10. Do not remove pinion bearing cups from retainer unless they are worn or damaged. The flange and pilot are machined by locating on these cups after they are installed in the bores. If new cups are to be installed, make sure they are seated in the retainer by trying to insert a .0015" feeler gauge between cup and bottom of bore.

Assembly

Differential Case—Place a side gear and thrust washer in the differential case bore. *Lubricate all parts liberally with axle lubricant during assembly.* With a soft-faced hammer, drive pinion shaft into case only far enough to retain a pinion thrust washer and pinion gear. Place the second pinion and thrust washer in position. Drive the pinion shaft into place. Be careful to line up pinion shaft retainer holes. Place second side gear and thrust washer in position and install differential case cover. Install retainer. A pinion or axle shaft spline can be inserted in side gear spline to check for free rotation of differential gears.

Insert two 7/16" × 2" bolts through differential flange and thread them 3 or 4 turns into the drive gear as a guide in aligning the drive gear bolt holes. Press or tap the drive gear into position. Install and tighten the drive gear bolts evenly and alternately across the gear to 60–65 lb. ft. torque.

If the differential bearings have been removed, use a suitable press to install them.

Pinion and Retainer—Install the pinion rear bearing cone on the pinion shaft and install a new spacer on the shaft. Place the bearing retainer on the pinion shaft, then install the front bearing cone. *As the bearing is pressed into position, rock the pinion retainer. Do not press the bearing on the shaft, until all play is removed between retainer and bearings as the spacer may be compressed too much.*

Lubricate both bearings with axle lubricant. Place oil slinger in position. Coat the outside edge of a new oil seal with an oil-resistant sealer and press it into the bearing retainer. New seals need not be soaked before installation. Install the universal joint flange. Place the flat washer over the pinion shaft and start the pinion shaft nut.

Hold the flange and tighten the pinion shaft nut until a torque required to turn the pinion is 8–12 *inch* lbs. with used bearings or 17–27 *inch* lbs. with new bearings. As the pinion shaft nut is tightened, rotate the pinion shaft frequently to allow the bearing to seat.

Shim Selection—Manufacturing tolerances in the pinion bore dimensions are in the best operating position of the gears make an adjustment shim necessary. This shim is placed between the pinion retainer and the carrier, Fig. 2. An increase in the thickness of the shim moves the pinion *away* from the drive gear. Manufacturing objectives are to make axles requiring a .0015" shim and if a new assembly is being built, a .0015" shim should be used for a tentative buildup. Shims are available in .010" to .021" thicknesses in steps of .001".

Pinions and drive gears are marked, when matched, with the same number. Following the number on the pinion is a minus (−) or plus (+) followed by a number. If the pinion is marked "−1" it indicates that a shim .001" thinner than a standard shim for this carrier is required. A minus number means the pinion should be moved closer to the drive gear and a thinner shim is required. A plus number means the pinion should be moved farther from the drive gear and a thicker shim is required. A pinion marked zero (0) is a standard pinion.

To select a shim, measure the original shim with a micrometer. Note the dimensional mark on the original pinion. Compare the mark on the original pinion with the mark on the new pinion to determine how the original shim should be modified.

For example, if the original shim is .015" and the original pinion is marked "−1", the new pinion requires a +1 shim. Therefore, the new pinion requires a .002" thicker shim, and a .017" shim should be used. If the new pinion is marked the same as the old pinion, no shim change is required.

Backlash Adjustment—The threaded nut type of differential bearing adjustment is used.

FORD INTEGRAL CARRIER TYPE REAR AXLE

This rear axle, Fig. 3, is a hypoid gear design with the centerline of the pinion set below the centerline of the ring gear. The drive pinion is of the overhung design mounted on two opposed tapered roller bearings. Pinion bearing preload is adjusted by the pinion nut and a collapsible spacer maintains the seating of the inner race.

Pinion Oil Seal, Replace

1. Raise and support vehicle. Remove rear wheels and brake drums.
2. Disconnect driveshaft from rear axle companion flange, then remove from transmission extension housing.

NOTE: Mark driveshaft end flange to axle companion flange location for proper positioning during installation.

3. Record torque required to maintain rotation of pinion through several revolutions.
4. Remove pinion nut using tool T57T-4851-B to hold companion flange from turning.
5. Mark companion flange and pinion shaft for reinstallation purposes and remove rear axle companion flange using tool T70P-4221-AB.
6. Remove seal from axle carrier.
7. Reverse procedure to install. Use tool T83T-4676-A to install a seal into carrier. Apply lubricant C1AZ-19590-B to lips of seal, then install companion flange and new pinion nut. Torque nut to 160 ft. lbs. minimum, taking frequent pinion bearing preload readings until original bearing preload (as noted in step 3) is reached. Torque driveshaft to rear axle companion flange and driveshaft to transmission housing bolts to 7–15 ft. lbs.

NOTE: If pinion preload recorded in step 3 is less than 6 inch lbs. for original bearings, or 16 inch lbs. for new bearings, reset bearing preload to 6–14 inch lbs. for original bearings or 16–29 inch lbs. for new bearings. If pinion preload recorded in step 3 is more than 14 inch lbs. for original bearings or 29 inch lbs. for new bearings, set pinion bearing preload to the specifications obtained in step 3.

CAUTION: Never back off pinion nut to reduce preload. If a reduced preload is required, a new collapsible pinion spacer is required.

Axle Shaft, Replace

1. Raise and support vehicle, then remove

rear wheel and brake drum.
2. Loosen rear housing cover screws and allow lubricant to drain, then remove cover.
3. Remove differential pinion shaft lock bolt and differential pinion shaft.
4. Move flanged end of axle shaft inward and remove C-lock from end of shaft.
5. Remove axle shaft from housing by pulling outward.
6. Reverse procedure to install. Apply locktite EOAZ-19554-B to pinion shaft lock bolt threads, then torque lock bolt to 15–30 ft. lbs.

Oil Seal and Wheel Bearing, Replace

1. Remove axle shaft as outlined in "Axle Shaft, Replace".
2. Insert tool T50T-100-A and T83T-1225-A into axle bore so tangs on tool engage bearing outer race. Remove bearing and seal as an assembly.
3. Install new bearing using tool T83T-1225-B and new seal using tool T83T-1175-A. Pack lips of seal with C1AZ-19590-B or equivalent.
4. Reinstall axle.

Differential Case

Remove

1. Remove axle shafts as outlined in "Axle Shaft, Replace".

NOTE: Before removing differential case, check ring gear runout with a suitable indicator. If ring gear runout exceeds .004 inch, differential case runout check must be performed to isolate problem. Refer to "Differential Case Runout Check".

2. Mark differential bearing caps for proper installation during assembly.

NOTE: Right and left bearing caps are not interchangeable.

3. Loosen differential bearing caps and bolts, pry differential case up until loose in bearing caps. Remove hearing caps and differential assembly from carrier.

Runout Check

1. Remove differential case from axle housing.
2. Remove ring gear bolts and press or tap ring gear off case.
3. Place differential bearing cups over cones and place differential assembly into carrier.
4. Install a .265 inch shim on left side

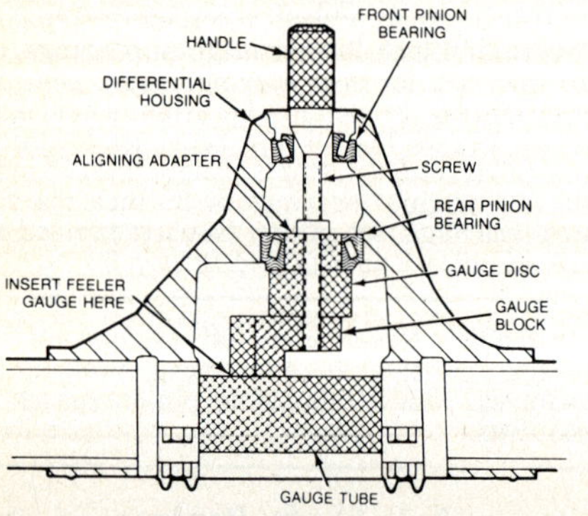

Fig. 4 Tool kit T79P-4020-A installed in axle carrier

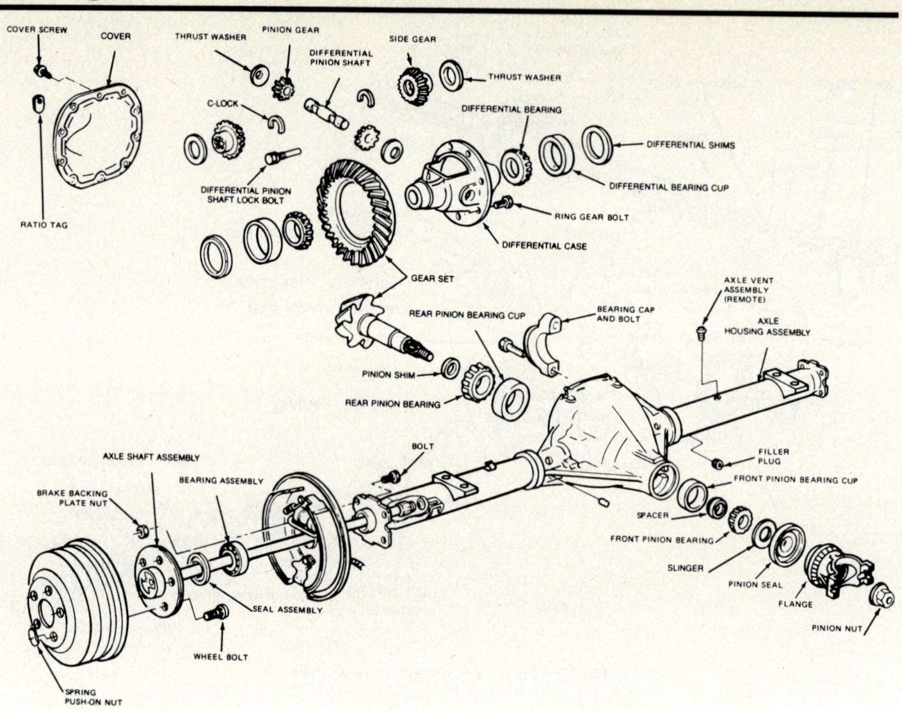

Fig. 3 Ford integral carrier type rear axle

between carrier and bearing cup and install left bearing cap and tighten bolts finger tight.
5. Install progressively larger shims on right side until shim with slight drag is installed.

NOTE: Apply pressure to left side of case to insure bearing cup is seated.

6. Install right side bearing cap and torque bearing cap bolts to 70–85 ft. lbs. Rotate assembly to ensure proper seating of bearings.
7. Check runout of differential case flange with dial indicator Tool 4201-C. If runout is within .003 maximum, install new ring and pinion gear. If runout exceeds specification, ring gear is true and trouble is either a damaged case or bearings. If bearings are not damaged, replace both case and bearings. If bearings are damaged, use tool T70P-4221-AB to remove bearings and tool T57L-4221-A to install new bearings. Check case runout again with new bearings. If runout exceeds specifications, case is damaged and should be replaced.

Install

1. Install differential bearings on case hubs using tool T57L-4221-A, then press ring gear onto case and install attaching bolts. Torque bolts to 70–85 ft. lbs.
2. Insert differential case sub-assembly into carrier with a .265 inch shim on left side.

NOTE: Pinion gear must be installed and pinion depth set before differential case is installed.

3. Install left bearing cap and tighten bolts finger tight. Install progressively larger shims on right side until shim with slight drag is selected. Install right side bearing cap and torque bearing cap bolts to 70–85 ft. lbs.
4. Check ring gear and pinion backlash with Tool 4201-C. Backlash should be .008–.015 inch. If backlash is within specifications, install bearing caps and torque bolts to 70–85 ft. lbs. If a zero backlash condition occurs, add .020 inch shim to right side and subtract .020 inch shim to left side. Recheck backlash. If backlash is not within specification, add or subtract shims as necessary. To decrease backlash, increase shims on left side and decrease shims on right side. To increase backlash, decrease shims on left side and increase shims on right side.

Drive Pinion

Remove

1. Raise and support vehicle, then remove rear wheels and brake drums.
2. Scribe reference marks on driveshaft and rear axle companion flange to aid in reassembly, then remove driveshaft.
3. Remove pinion nut using tool T57T-4851-B to hold companion flange from turning.
4. Scribe reference marks on companion flange and pinion shaft to aid in reassembly, then remove companion flange using tool T70P-4221-AB.

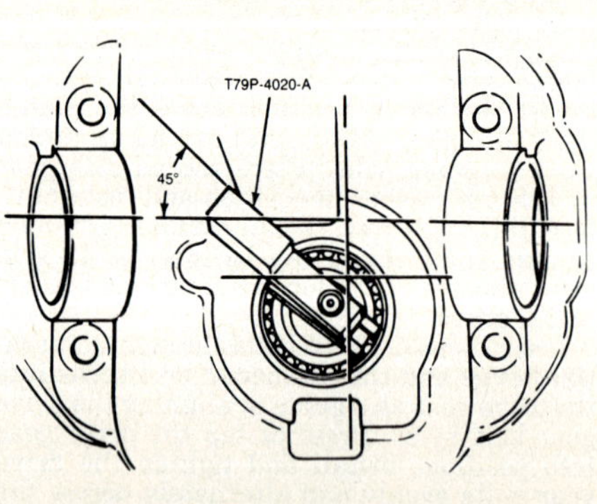

Fig. 5 Gauge block angle setting

5. Remove differential case, refer to "Differential Case, Remove".
6. Using a suitable soft faced mallet, tap pinion out of front bearing cone and remove through rear of axle housing. To remove rear bearing, use tool T71P-4621-B and suitable hydraulic press ram.

NOTE: Measure shim located under bearing cone and record thickness.

Shim Selection

1. Using Tool kit 7T9P-4020-A determine shim thickness by placing rear bearing over aligning disc and insert into rear pinion bearing cup, Fig. 4.
2. Place front bearing into front bearing cup and assemble tool handle into screw. Move assembly back and forth to seat bearings. Torque tool handle to 20 ft. lbs.

NOTE: Gauge block must be off-set 45° to obtain accurate reading, Fig. 5.

3. Place gauge tube into center of differential bore and install bearing caps. Torque cap bolts to 70–85 ft. lbs., Fig. 4.
4. Using a feeler gauge, insert blades between gauge tube and gauge block. Correct thickness will be indicated by blade that produces a slight drag. Install this thickness shim.
5. Install correct thickness shim(s) on pinion and, using tool T53T-4621-C and press, install rear pinion bearing until firmly seated.

NOTE: Pinion bearing used in step 5 must be same bearing used in final assembly.

Install

1. Install spacer up against rear bearing and place drive pinion into housing case bore.
2. Install front bearing and oil slinger on pinion shaft.
3. Install pinion oil seal. Refer to "Pinion Oil Seal, Replace" for procedure.

Pinion Bearing Cup, Replace

NOTE: Do not remove pinion bearing cups unless damaged.

1. Remove drive pinion, refer to "Drive Pinion, Remove" for procedure.
2. Tap bearings from bores with suitable drift using care not to cock bearings during removal.
3. Install new cups with tool T67P-4616-A. Check to see if cups are properly seated. If a .0015 inch feeler gauge can be inserted between cup and bottom of bore, cup is not properly seated.

Axle Housing, Replace

1. Remove four backing plate nuts from each side of axle and wire backing plate aside.
2. Remove vent hose from vent and vent from brake junction block on rear axle housing.
3. Remove brake line from clips on axle housing and remove brake junction block.

NOTE: Do not open brake lines when removing junction box.

4. Support rear axle housing on suitable jack and remove U-bolt nuts and plates.
5. Disconnect lower bolts on shock absorber mounting bracket and remove housing from vehicle.
6. Reverse procedure to install. Torque U-bolt nuts to 75–100 ft. lbs. on F-100, F-150, F-250, Bronco, E-100 and E-150, 150–210 ft. lbs. on F-350 and 150–180 ft. lbs. on E-250 and E-350. Torque shock absorber lower bolts to 40–64 ft. lbs. on all F-models and to 40–60 ft. lbs. on all E-models. Torque backing brake attaching bolts to 20–40 ft. lbs. on F-100 and F-150 and to 50–70 ft. lbs. on E-250 and F-250 with 4050 lb. axle.

NOTE: No gaskets are required on brake backing plate.

Companion Flange, Replace

Refer to "Pinion Oil Seal, Replace", for procedure.

Rear Cover, Replace

NOTE: Before removing rear cover, clean carrier around rear cover area with wire brush.

1. Loosen cover bolts and allow lubricant to drain, then remove rear cover.
2. Clean gasket mating surface of rear cover and carrier casting and apply continuous bead of sealant D6AZ-19562-B or equivalent to carrier casting on inside of bolt holes.
3. Install cover and torque bolts in crosswise pattern to 25–35 ft. lbs.
4. Fill rear axle assembly with EOAZ-19580-A lubricant or equivalent for conventional and Traction-Lok differentials. For Traction-Lok limited slip differentials add 4 oz. friction modifier C8AZ-19B546-1A. Install filler plug and torque to 25–50 ft. lbs.

Front End Section

NOTE: Refer to the Main Index for transfer case, locking hub and driving axle service.

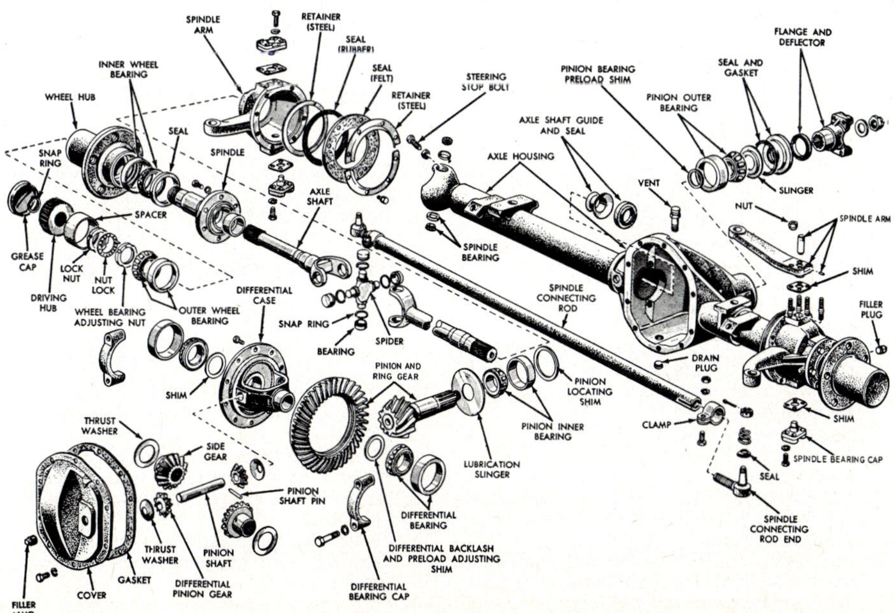

Fig. 1 Four wheel drive front axle assembly; closed spindle type (Typical)

1974–79 4×4 SERVICE

Front Axle Shaft

Closed Spindle Type

1. Raise truck on a hoist and then lower it so that the front axle is supported on a stand.
2. Referring to Fig. 1, remove hub dust cap and snap ring and slide splined driving hub from between outer axle shaft and wheel hub and drum.
3. After removing lock nut, washer and wheel bearing retainer nut from steering spindle, remove wheel, hub and drum as an assembly. The wheel outer bearing will be forced off the spindle at the same time.
4. Remove wheel inner bearing.
5. Place a drain pan under steering knuckle, and remove brake backing plate and spindle.
6. Pull outer axle shaft, universal joint and inner axle shaft from housing as an assembly.
7. Reverse procedure to install, then remove the steering knuckle plug and fill the knuckle with ½ pint of the proper lubricant.

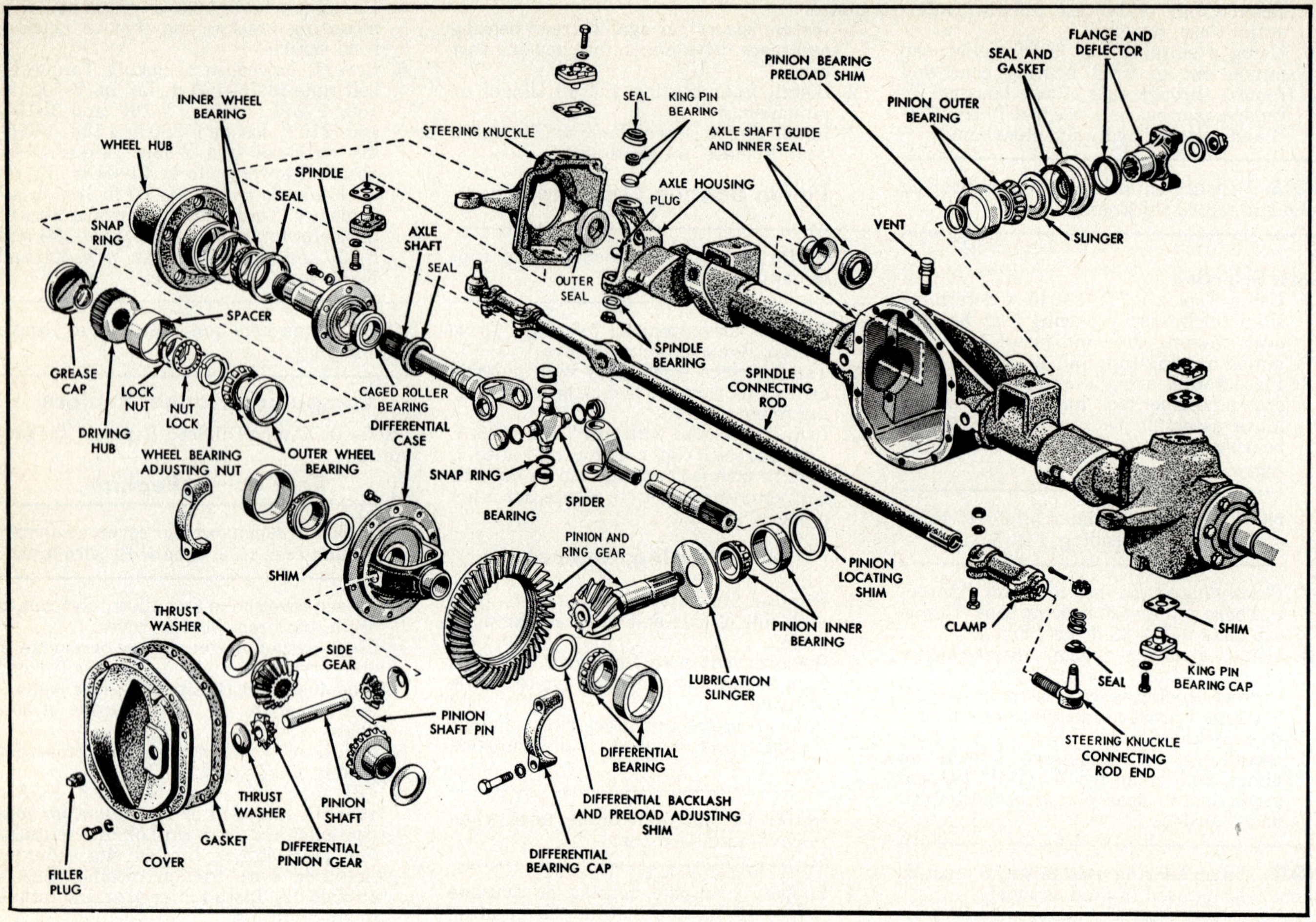

Fig. 2 Four wheel drive front axle assembly; open spindle type

Spindle Bearing

Removal

1. Referring to Fig. 1, remove the inner and outer axle shafts and U-joint from each end of the axle housing.
2. Disconnect upper spindle arm from steering drag link (left side only).
3. Disconnect spindle connecting rod (tie rod) from front spindle arm.
4. Remove oil seal retaining bolts from rear flange of steering knuckle and remove the two-piece retainer.
5. Remove upper spindle arm (left side) and integral bearing cap and shaft from steering knuckle. On the right side there is only a bearing cap and shaft assembly. Remove bearing preload shims.
6. Remove lower spindle bearing cap and shaft from steering knuckle.
7. Remove steering knuckle from axle housing yoke.
8. After removing upper and lower spindle bearings from the axle housing yoke, remove upper and lower bearing cups from yoke.
9. Remove felt, rubber and metal oil seal rings from axle housing.

Installation

Reverse the removal procedure to install the bearings. However, to mount the steering knuckle on the axle housing yoke, first fit the lower end of the knuckle over the bottom of the yoke so that the lower bearing will fit into the cup; then bring the entire knuckle into position on the yoke.

Place the bearing preload shims over the studs on the upper side of the knuckle and install the upper bearing cap and shaft. The shaft is inserted in the bearing and the cap is bolted to the upper side of the knuckle.

To check the bearing preload, place a torque wrench on one of the bearing cap nuts and pivot the knuckle on its bearings. As the knuckle is turned, a preload reading of 5 to 10 inch-pounds should register on the torque wrench. After removing or adding shims to establish correct preload, torque bearing cap nuts to 25–40 ft. lbs.

Front Axle, Open Spindle Type

Front Axle Shaft & Spindle

1. To remove, raise vehicle on a hoist.
2. Drain lubricant from axle. *This must be done to prevent lubricant flooding the outer axle housing when axle shaft is removed.*
3. Remove hub grease cap and driving hub retaining snap ring and slide splined hub from between axle shaft and wheel hub. Remove hub spacer.
4. Remove wheel bearing adjusting nut.
5. Remove wheel, hub and drum.
6. Remove inner bearing cone.
7. Remove brake carrier plate and secure it to one side to avoid damaging brake hose. Remove spindle.
8. Pull axle shaft from housing, working universal joint through bore of spindle arm. *At this point the following parts can be replaced without further disassembly of the axle: axle shaft, universal joint, axle housing outer seal, spindle bore seal, and spindle bore roller bearing.*
9. To remove spindle arm (housing), disconnect steering connecting rod end from spindle arm and remove bearing caps and spindle arm on models without ball joints. On models with ball joints, loosen the nuts from upper and lower joints and separate the knuckle from the yoke.

NOTE: Discard the lower joint nut, replace with a new one.

---

10. Install knuckle into a vise and press out ball joints with suitable tool.
11. Reverse procedure to install.

1980–83 BRONCO F150, 250, 350 (4×4) SERVICE

Front Axle Assembly, Replace

1980–83 Bronco & F150, Fig. 3

1. Raise and support vehicle. Remove front wheels.
2. Drain fluid from differential.
3. Remove brake caliper, hub assembly, brake rotor, spindle, splash shield and axle shaft.
4. Position a jack under tube and yoke assembly and remove upper coil spring retainers.
5. Lower jack and remove coil spring, cushion and lower spring seat.

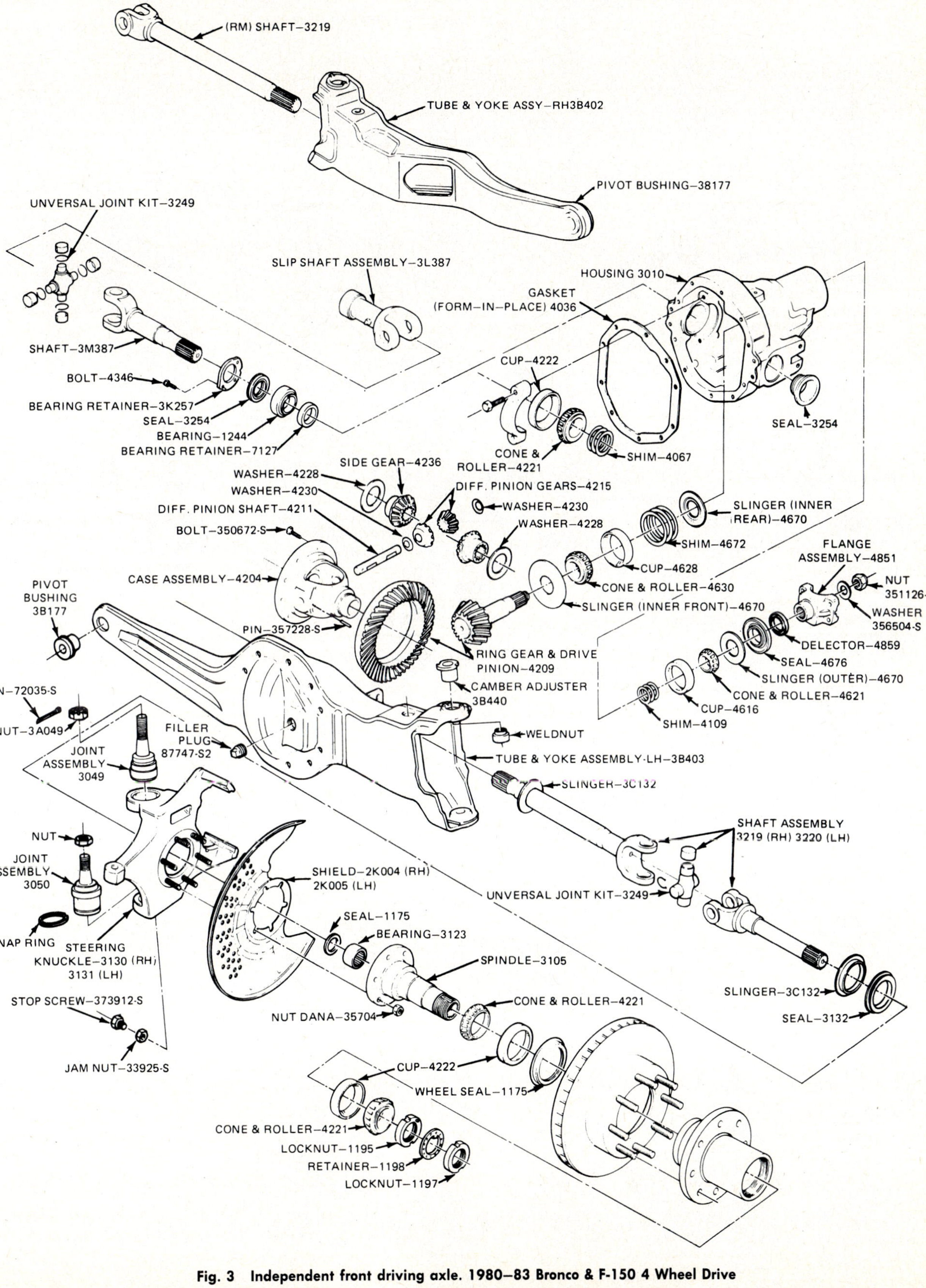

Fig. 3 Independent front driving axle. 1980–83 Bronco & F-150 4 Wheel Drive

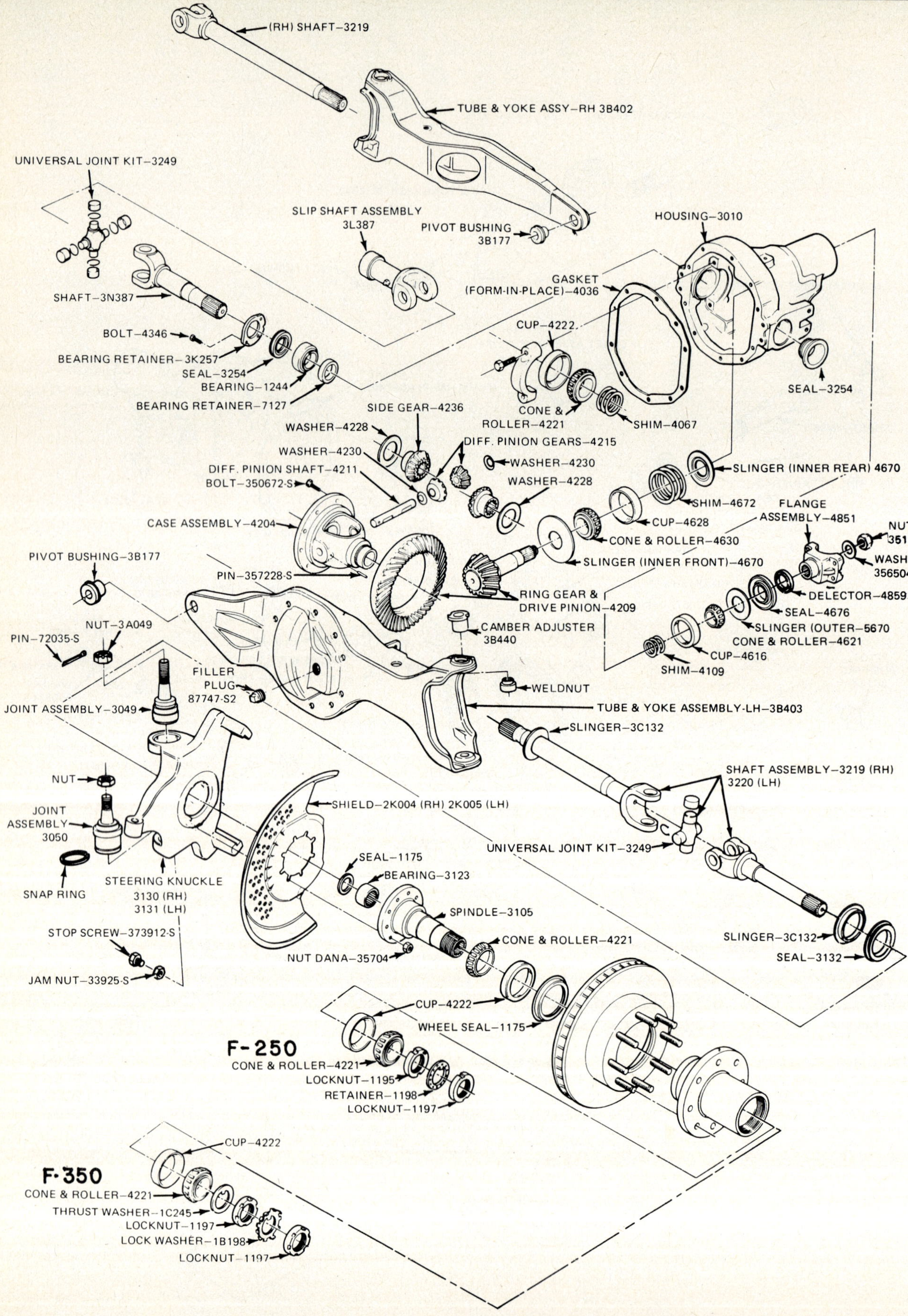

Fig. 4 Independent front driving axle. 1980–83 F-250, 350 4 Wheel Drive

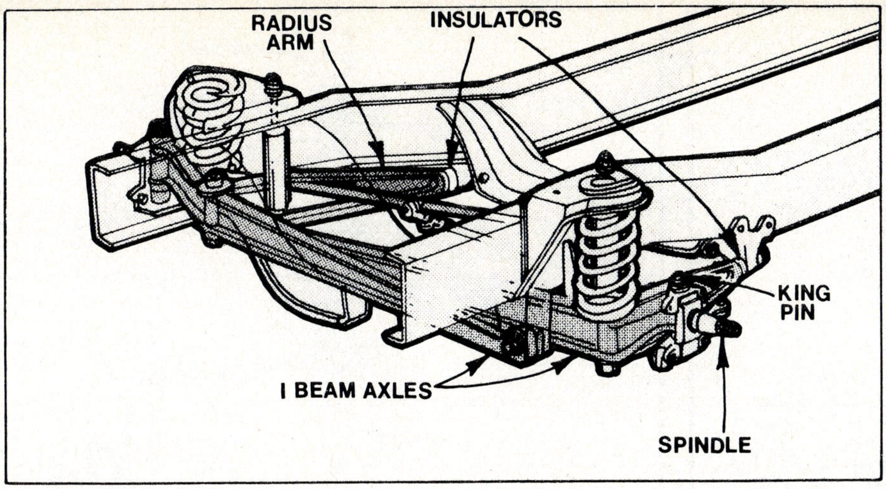

Fig. 5 Twin I-Beam front axle

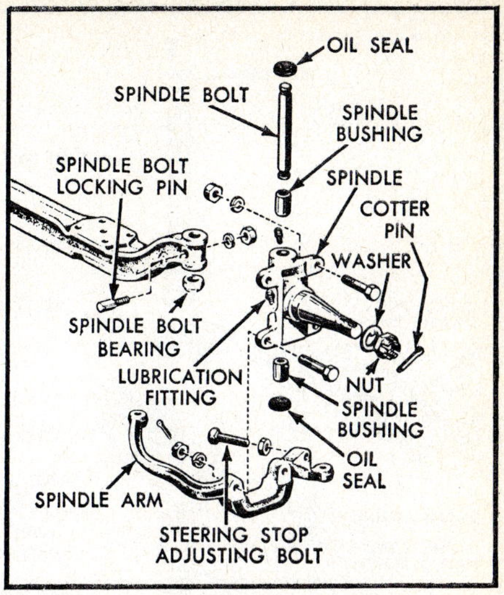

Fig. 6 Spindle used with twin I Beam front axle. Typical

NOTE: Place caliper on frame or support caliper to prevent damage to brake hose.

---

6. Remove shock absorber, then steering knuckle. Refer to "Steering Knuckle, Replace" for removal procedure.
7. Remove stud and spring seat at upper radius arm and tube and yoke assembly. Remove bolt attaching lower radius arm to tube and yoke assembly.
8. Disconnect vent tube from left tube and yoke assembly.
9. Remove right tube and yoke assembly to crossmember pivot bolt. Remove right hand tube and yoke assembly.
10. Position jack and drain pan under differential housing. Remove differential attaching bolts from housing. Remove left tube and yoke assembly to crossmember pivot bolt. Remove left tube and yoke assembly.
11. Reverse procedure to install.

1980–83 F-250 & F-350, Fig. 4

1. Raise and support vehicle. Remove front wheels.
2. Drain fluid from differential.
3. Remove brake calipers, hubs, brake rotors, spindles, splash shields and axle shafts.
4. Position jack under right tube and yoke assembly. Disconnect shock absorber and leaf spring plate from tube and yoke assembly.
5. Remove steering knuckle from tube and yoke assembly. Refer to "Steering Knuckle, Replace" for removal procedure.
6. Disconnect vent tube from left tube and yoke assembly. Remove right tube and yoke assembly.
7. Position jack under left tube and yoke assembly. Disconnect shock absorber and leaf spring plate from tube and yoke assembly.
8. Remove steering knuckle assembly from left tube and yoke assembly.
9. Position jack and drain pan under differential housing. Remove differential attaching bolts. Remove left tube and yoke assembly.
10. Reverse procedure to install.

Axle Shaft Bearing, Replace, Figs. 3 & 4

1. Remove axle shaft assembly.
2. Remove axle shaft bearing retainer plate bolts, then pull stub shaft from housing.
3. Place stub shaft in vise and drill a 1/4 inch hole 3/4 through the bearing retainer.
4. Cut the retainer using a chisel across the hole. Press bearing from stub shaft.
5. Remove seal and bearing retainer from stub shaft.
6. Reverse procedure to install. Install bearing retainer and new seal onto stub axle. Lubricate seal with wheel bearing grease. Install bearing onto axle with the large radius on inner race facing yoke of axle. Press bearing, then inner bearing retainer until fully seated. Push the seal and outer retainer away from bearing and fill the space with wheel bearing grease. Wrap a piece of tape around the space, then pull the retainer and seal against the bearing. This will force the grease into the bearing. Repeat until the grease is visible on the small end of rollers.

TWIN I-BEAM AXLE

As illustrated in Fig. 5 there are two I-beam axles, one for each front wheel. One end of each axle is attached to the spindle and a radius arm and the other end is attached to a frame bracket on the opposite side of the vehicle.

Each spindle is held in place on the axle by a spindle bolt which pivots in bushings pressed in the upper and lower ends of the spindle, Fig. 6. A thrust bearing is installed between the lower end of the axle and the spindle to support the load on the axle. A spindle arm is installed on each spindle for attachment to the steering linkage.

SINGLE I-BEAM AXLES

These axles are the I-beam type with the front wheel spindles attached to the ends of the axles, Figs. 7 and 8.

Each spindle is held in place on the axle by a spindle bolt. On 4000, 5000 and 5500 lb. axles, steel-backed, bronze bushings are pressed into the upper and lower ends of the spindle to provide bearing surfaces for the spindle bolt. Delrin bushings are used on 6000 through 11,000 lb-axles.

A thrust bearing is installed between the lower end of the axle and the spindle to support the load on the axle. A spindle arm, to which the steering linkage is attached, is installed on each spindle.

FRONT SUSPENSION

1980–83 Bronco & F-150 4 Wheel Drive

The independent front suspension on these vehicles is composed of a two piece front driving axle assembly, two coil springs and two radius arms, Fig. 9.

The front driving axle consists of two independent yoke and tube assemblies. One end of yoke and tube assembly is anchored to the frame. The other end of each yoke and tube assembly is supported by the coil spring and radius arm.

1980–83 F-250 & F-350 4 Wheel Drive

The independent front suspension on these vehicles has a two piece driving axle attached to the frame with two semi-elliptic leaf springs, Fig. 10. Each spring is clamped to the axle yoke with two U-bolts. The rear tube assembly of the spring rests in a rear hanger bracket. The front of the spring is attached to a shackle bracket.

1974–79 F-100, 150 4 Wheel Drive & Bronco

The front driving axle is attached to the frame with two coil springs, two radius arms and a track bar, Fig. 11.

The radius arms clamp around the axle tube and attach to frame brackets behind the front axle. A coil spring is attached to each radius arm directly above the axle and extends to the frame spring seat. The shock absorbers are attached to the frame and to brackets on the radius arm.

The track bar is mounted to a bracket on the left frame side member and extends to the right side of the axle housing.

1974–79 F-250, 350 4 Wheel Drive

The front suspension on these vehicles consists of a driving axle attached to the frame with two semi-elliptic leaf type springs. Each spring is clamped to the axle with U-bolts. The front of the spring rests in a front hanger bracket, while the rear of the spring is attached to a shackle bracket. The shock absorbers are attached to a frame bracket at the top

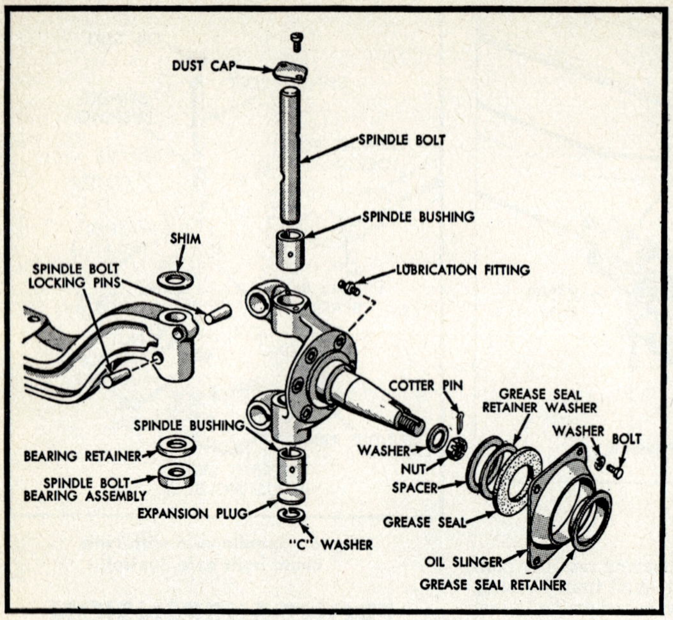

Fig. 7 Spindle used with 7000 thru 9000 lb. axles

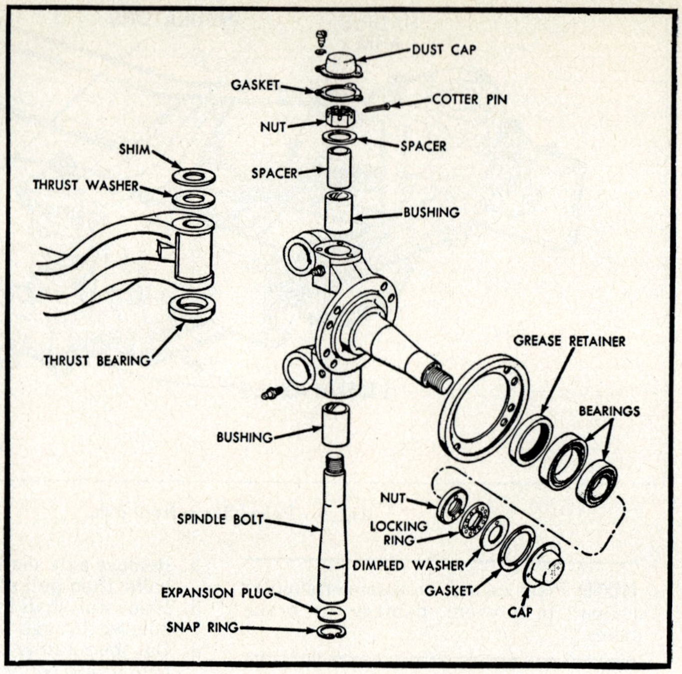

Fig. 8 Spindle used with 11,000 lb. axle

and to an axle bracket at the bottom.

RANCHERO

1974–79

The front suspension, Fig. 12, has been redesigned with the coil spring mounted on the lower arm rather than on the upper arm as in the past.

WHEEL ALIGNMENT

Prior to checking caster and camber, ensure vehicle front ride height is within 1/8 inch side to side on 1974–79 models and 1980–83 E-100-350. On 1980–83 F-100-350 models, the left side ride height should be within 0–3/8 inch higher than the right side ride height. On 1980–83 F-150-350 (4×4) and Bronco, the front ride heights should be within 3/16 inch side to side. Refer to Figs. 13, 14, 15 and 16. If the vehicle ride heights are not within specifications, redistribute load on loaded vehicles or slightly load empty vehicles on one side. If the ride heights cannot be brought within specifications, verify correct spring installation.

NOTE: The ride height variations stated above are for checking purposes only. The vehicle does not have to operate within these specifications.

Check and correct as necessary all tire inflation pressures, then check front tires for the same size, ply rating and load range. Check front wheel bearings and adjust as necessary.

If all the above checks have been made, check wheel alignment with suitable alignment equipment. Using the ride heights obtained earlier, compare caster and camber readings to those listed in the "Wheel Alignment Specifications Chart". If the caster and camber angles exceed the specifications, inspect front end for damaged suspension components. Replace as necessary.

Alignment equipment indicates a true reading only when the vehicle's frame is horizontal. Therefore, if the frame is not level (due to tire, spring or load differences), the caster angle reading must be modified to compensate for the frame angles. If the front is higher than the rear, subtract the amount of angle from the reading. If the front is lower than the rear, add the angle. To check frame angle, use a spirit protractor, and take the frame angle measurement on the lower frame flange at the flat area immediately adjacent to the rear spring front hanger.

1974–77 F-250 4×4

The caster and camber angles are designed into the axle and cannot be adjusted except for models equipped with leaf springs. On models equipped with leaf springs, caster angle can be adjusted by placing shims between the spring and spring pad.

NOTE: Do not attempt to correct caster and camber angles by bending axle using heat or mechanical means.

1974–83 F-100-350 & Econoline, 1974–79 F-100 4×4 & Bronco, 1976–83 E-100-350, 1976–79 F-150 4×4, 1978–79 F-250 4×4 & 1979 F-350 4×4

The caster and camber angles are designed into the front end and are not adjustable.

1980–83 F-150-350 4×4 & Bronco

The caster angle is designed into the axle and is not adjustable. The camber angle may be adjusted by means of mounting sleeves placed on the upper ball joint, Fig. 17. Four sleeves are available in 1/2° camber increments to allow a 3° range of adjustment from negative 1½° to positive 1½°.

1975–83 500-9000, All Series

The caster angle is controlled by the front spring pad angle and changes with varying axle loads or as the spring sags. The spring pads must parallel each other. Changes in caster angle are the result of a twisted axle or sagging spring.

If a change in caster angle is caused by a sagging spring, the caster angle can be adjusted by replacing the springs or by installing tapered wedges between the springs and spring pads. Wedges are available in different thicknesses and only one wedge should be used per side. Position thick edge of wedge towards rear of vehicle to increase caster or toward the front of the vehicle to decrease caster. When installing wedges, ensure that the front spring tie bolts extend through the wedge and into the axle pilot hole.

Camber is designed into the axle and is not adjustable.

1974 500-9000 Series

Camber can be adjusted by bending the front axle beam with tools shown in Figs. 18 and 19. Install tool as shown in Fig. 18 to increase camber and as shown in Fig. 19 to decrease camber. Bend axle between front spring pad and spindle. Do not heat axle to bend it.

For caster angle adjusting procedure refer to "1975–83 500-9000, All Series".

1974–79, Ranchero

Caster and camber can be adjusted by loosening the bolts that attach the upper suspension arm to the shaft at the frame side rail, and moving the arm assembly in or out in the elongated bolt holes. Since any movement of the arm affects both caster and camber, both factors should be balanced against one another when making the adjustment.

Install the tool with the pins in the frame holes and the hooks over the upper arm inner shaft. Tighten the hook nuts snug before loosening the upper arm inner shaft attaching bolts, Fig. 20.

Caster Adjust

1. Tighten the tool front hook nut or loosen the rear hook nut as required to increase caster to the desired angle.
2. To decrease caster, tighten the rear hook nut or loosen the front hook nut as required.

NOTE: The caster angle can be checked

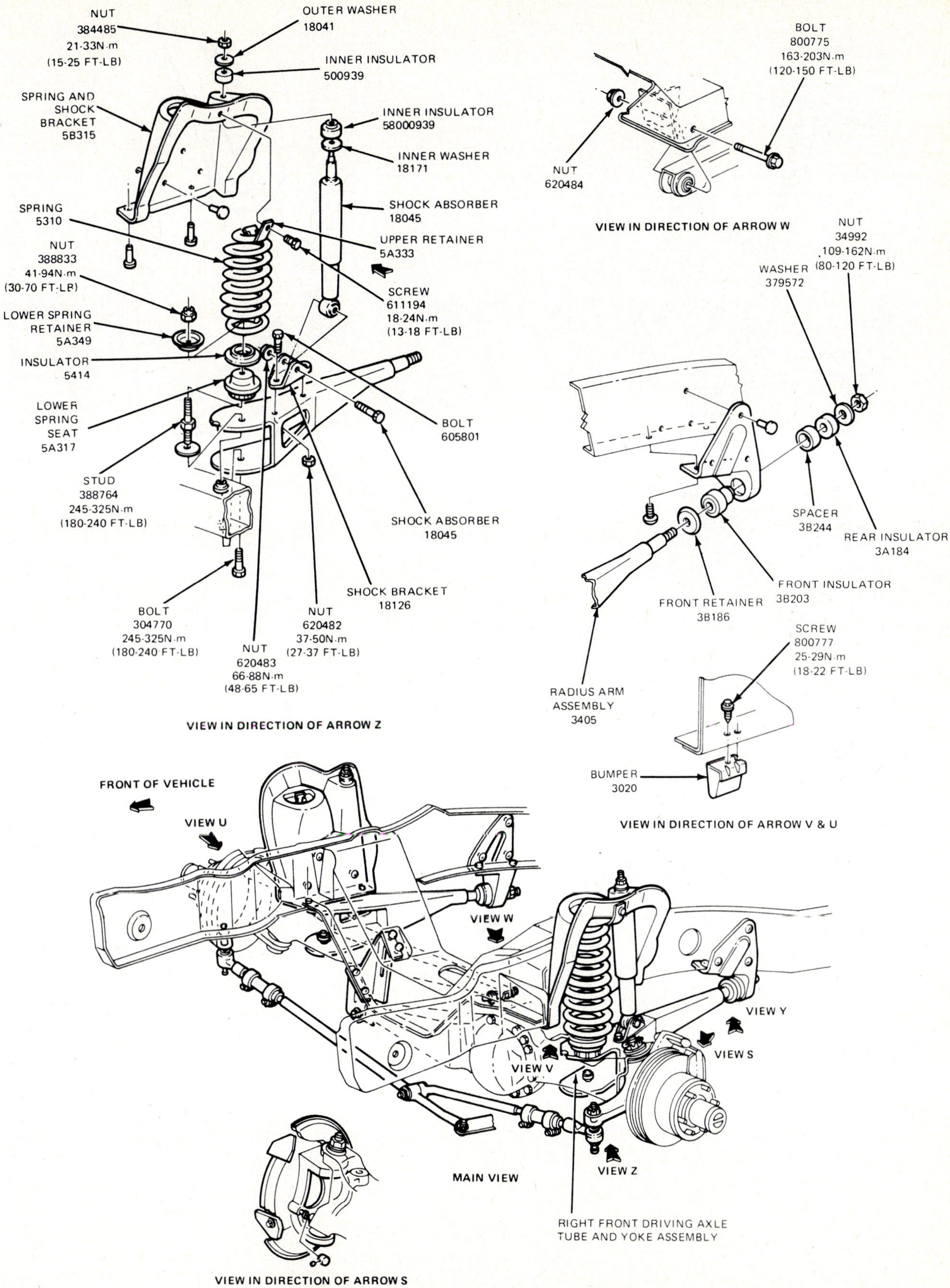

Fig. 9 Independent front suspension. 1980–83 Bronco & F-150 Four Wheel Drive

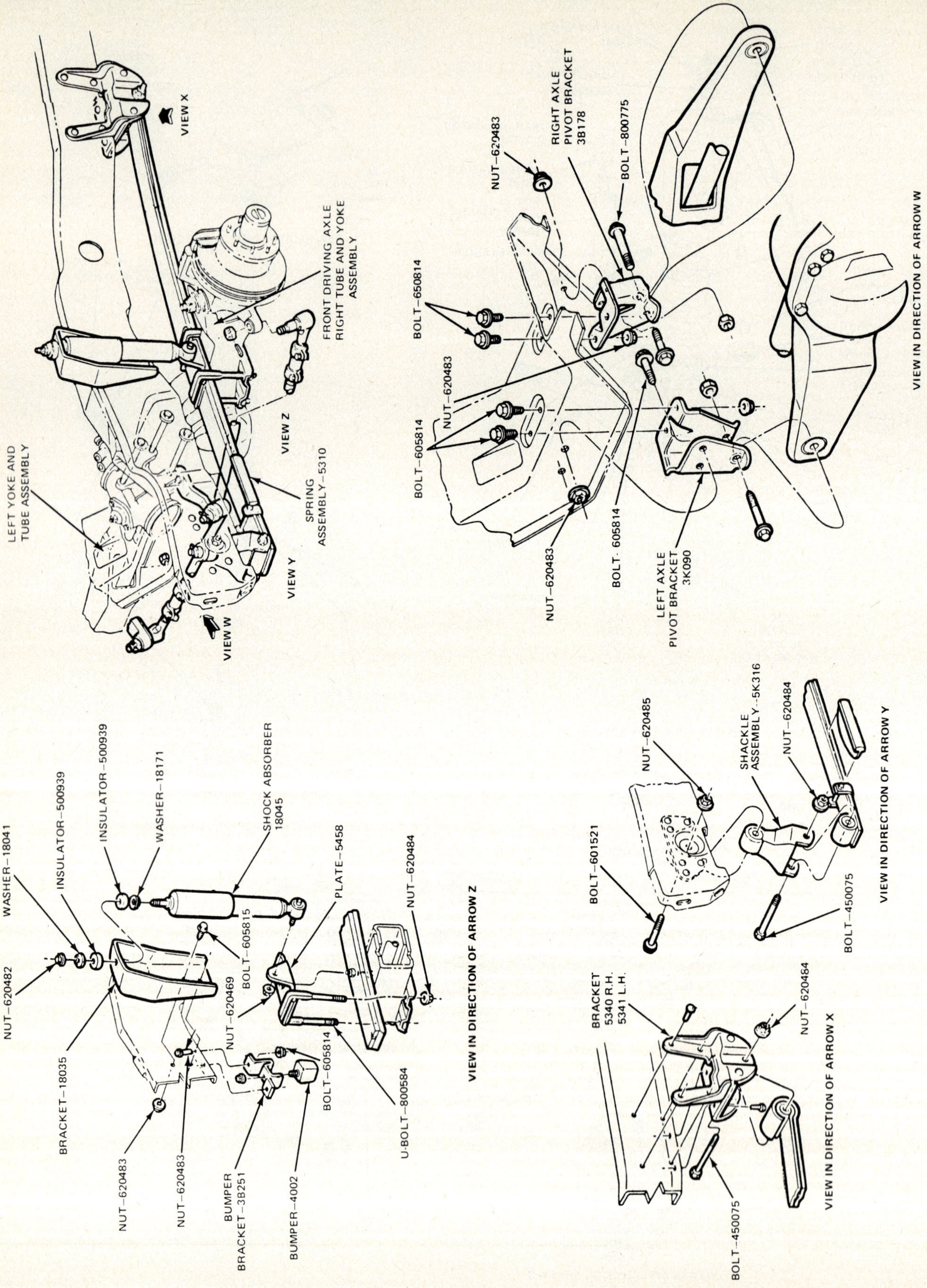

Fig. 10 Independent front suspension. 1980–83 F-250 & F-350 Four Wheel Drive

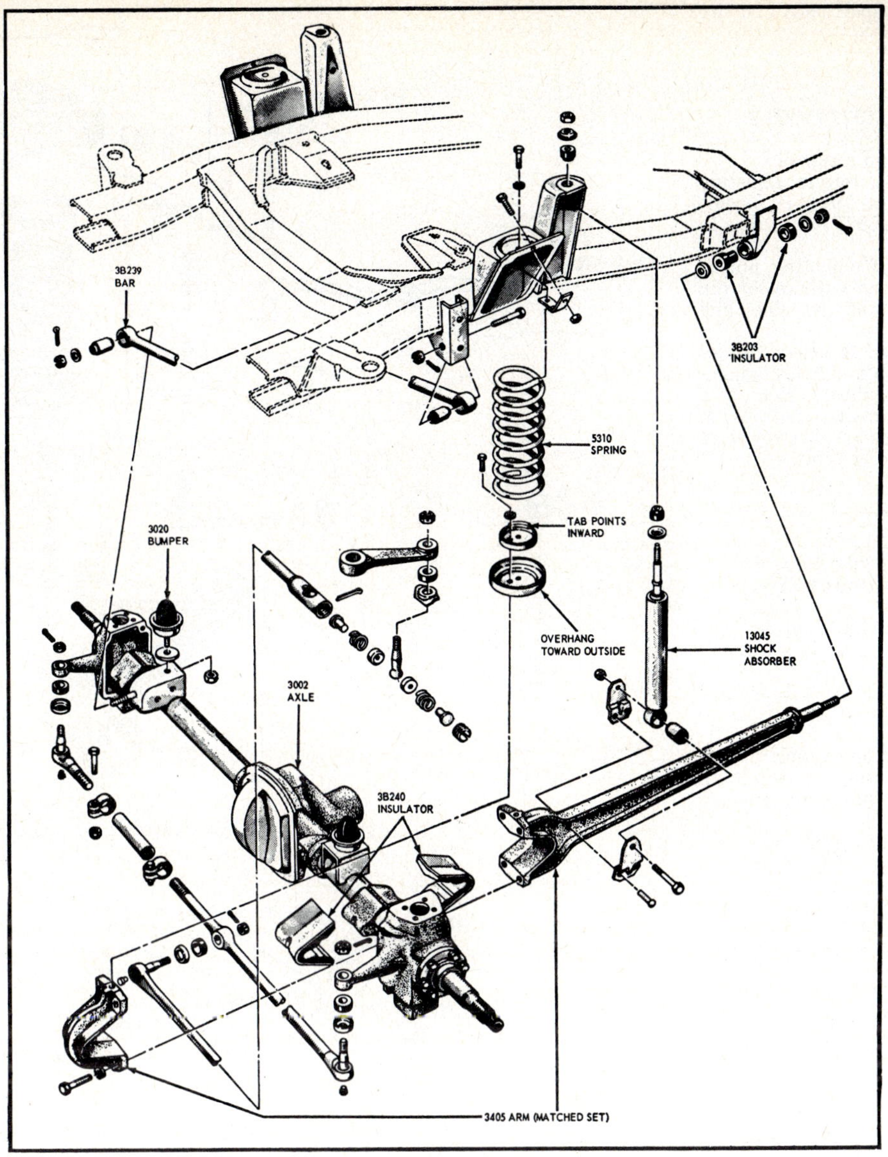

Fig. 11 Front suspension (Typical) F-100 & 150 4 wheel drive & Bronco. Less independent front suspension

without tightening the inner shaft retaining bolts.

3. Check the camber angle to be sure it did not change during the caster adjustment and adjust if necessary.
4. Tighten the upper arm inner shaft retaining bolts and remove tool.

Camber, Adjust

1. Install as previously outlined.
2. Loosen both inner shaft retaining bolts.
3. Tighten or loosen the hook nuts as necessary to increase or decrease camber.
4. Recheck caster angle.

Toe-In, Adjust

Check the steering wheel spoke position when the front wheels are in the straight-ahead position. If the spokes are not in the normal position, they can be adjusted while toe-in is being adjusted.

1. Loosen clamp bolts on each tie rod end sleeve.
2. Adjust toe-in. If steering wheel spokes are in their normal position, lengthen or shorten both rods equally to obtain correct toe-in. If spokes are not in normal position, make necessary rod adjustments to obtain correct toe-in and steering wheel spoke alignment.

WHEEL BEARINGS, ADJUST

Except 4 Wheel Drive Vehicles

Ranchero

1. With wheel rotating, torque adjusting nut to 17–25 ft. lbs.
2. Back off adjusting nut 1/2 turn and retorque nut to 10–15 inch lbs.
3. Place nut lock on nut so that castellations on lock are aligned with cotter pin hole in spindle and install cotter pin.
4. Check wheel rotation. If it rotates rough or noisily, clean or replace bearings.

1974–76 F-100 Thru F-350 & E-100 Thru E-350

1. Remove grease cap, cotter pin and nut lock.
2. While rotating wheel, torque adjusting nut to 17–25 ft. lbs. to seat bearings.
3. Install nut lock so a slot is aligned with cotter pin hole in spindle.
4. Back off adjusting nut and nut lock two slots of the nut lock and install cotter pin.
5. Check wheel rotation. If wheel rotates freely with no noticeable end play, adjustment is acceptable. If wheel is loose or if it rotates rough or noisily, the bearing cones, rollers and cups are dirty or worn and should be cleaned or replaced.
6. Install grease cap.

1977–83 F-100 Thru F-350 & E-100 Thru E-350

1. Remove wheel and tire assembly, disc brake caliper and pads, dustcap, locknut, adjusting nut, washer and cotter pin.
2. Tighten wheel adjusting nut to 22–25 ft. lbs. while rotating disc brake rotor in opposite direction.
3. Back off wheel retention nut 1/8 turn and install retainer and cotter pin without any additional movement of nut.

4. Reinstall dustcap, caliper, pads and tire and wheel assembly.

1974–76 P-350 thru 500, 1977 P-500 & 1978–79 P-600

1. Remove grease cap and cotter pin.
2. While rotating wheel, torque adjusting nut to 40–55 ft. lbs. on 1974–76 models and 40–50 ft. lbs. on 1977–79 models to seat bearings.
3. If necessary, back off adjusting nut to next slot so nut aligns with cotter pin hole, then back off nut two slots and install cotter pin.
4. Check wheel rotation. If wheel rotates freely with no noticeable end play, adjustment is acceptable. If wheel is loose, or if it rotates rough or noisily, the bearing cones, rollers and cups are dirty or worn and should be cleaned or replaced.
5. Install grease cap.

1974–77 Vehicles With 12000 Pound Center Point Axle With Double Nut & Washer

1. Back off brake adjustment, then straighten lock washer and remove lock nut and washer.
2. While rotating wheel in both directions, torque adjusting nut to 100–125 ft. lbs. to seat bearings.
3. Back off adjusting nut one complete turn, then again, while rotating wheel in both directions, retorque adjusting nut to 40–55 ft. lbs.
4. Back off adjusting nut to obtain an initial .004–.010 inch end play between hub and spindle.
5. Install outer washer and lock nut and torque to 100–125 ft. lbs.
6. Final end play should be .001–.010 inch. Readjust if necessary.
7. Bend washer outward in two places on opposite sides of lock nut, to lock outer nut in position.
8. Install gasket and grease cap and readjust brakes.

NOTE: Final adjustment must not result in a preloaded bearing.

1974–77 Vehicles With 15000 Pound Axle With Double Nut & Lock Ring

1. Back off brake adjustment, then remove grease cap and gasket.
2. While rotating wheel in both directions, torque adjusting nut to 100–125 ft. lbs. to seat bearings.
3. Back off adjusting nut one complete turn, then while again rotating wheel in both directions, retorque nut to 40–55 ft. lbs.
4. Back off adjusting nut to obtain an initial .004–.010 inch end play between wheel hub and spindle.
5. Install lock ring making sure that adjacent dowel is inserted into a lock ring hole, then install washer with dimple indexed in one of the holes in lock ring.
6. Install lock nut and torque to 100–125 ft. lbs.
7. Recheck end play. Final bearing adjustment must allow .001–.010 inch end play. Readjust if necessary.
8. Bend lock washer in two places on opposite sides to lock nut in position.
9. Install gasket and grease cap, and readjust brakes.

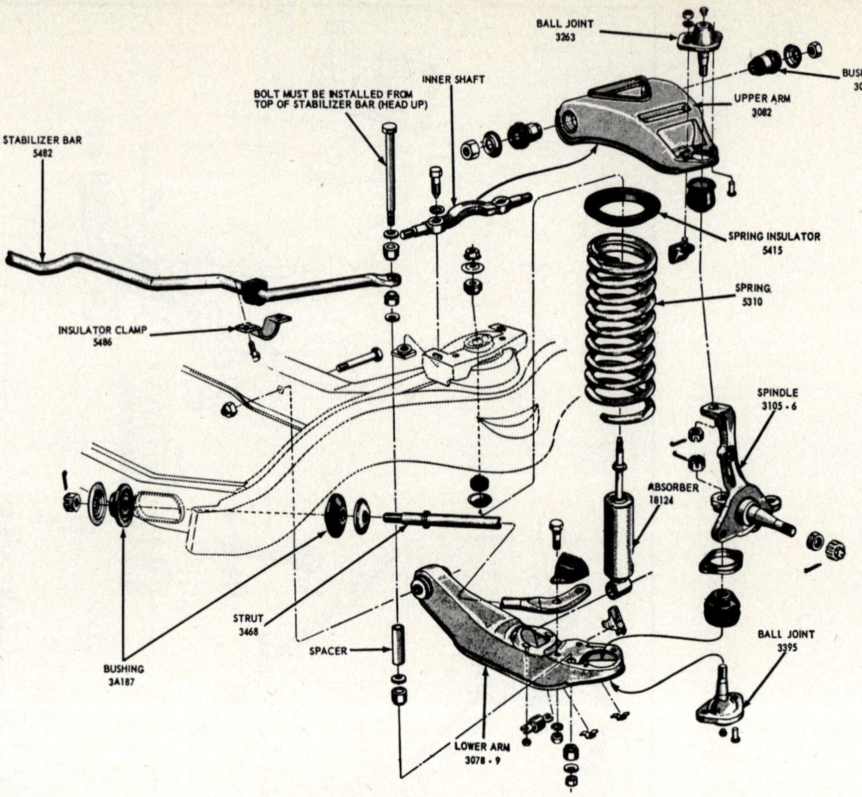

Fig. 12 Front suspension. 1974–79 Ranchero

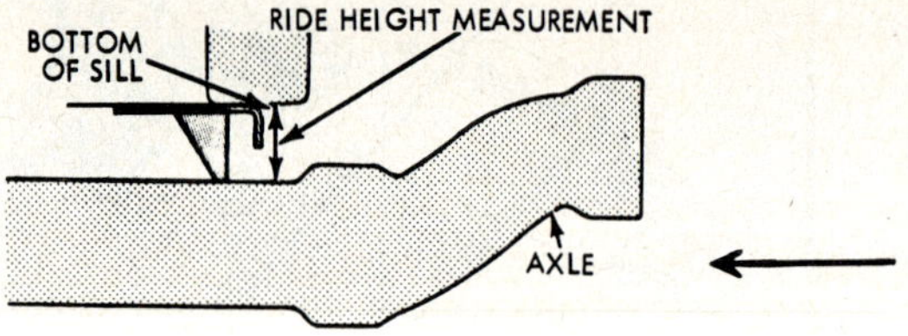

Fig. 13 Measuring riding height. 1974–76 E series

NOTE: Final adjustment must not result in a preloaded bearing.

1974–80 Vehicles With 5000–7000 Pound Axles With Single Nut & Cotter Pin, 1980–83 Vehicles With 7000, 9000 & 12000 Pound Axles With Air Brakes & F & B Models With 9000 Lb. Front Axle & Hydraulic Brakes

1. Back off brake adjustment, then remove grease cap and cotter pin.
2. While rotating wheel, torque adjusting nut to 70–100 ft. lbs. to seat bearings.
3. Back off adjusting nut one half turn, then hand tighten adjusting nut (13–17 inch lbs.) while moving top part of tire in and out.
4. If cotter pin hole lines up with slot in nut, insert a new cotter pin. If it does not, back off adjusting nut to align a slot with nearest cotter pin hole and install a new cotter pin. Adjustment should provide .001–.010 inch end play.
5. Install gasket and grease cap, then adjust brakes.

NOTE: Final adjustment must not result in a preloaded bearing.

1974–79 Vehicles With 9000 & 12000 Pound Axle With Double Nut & Lock Ring, 1980–83 9000 & 12000 Pound Axle With Double Nut, Lock Ring & Hydraulic Brakes

1. Back off brake adjustment, then remove gasket and grease cap and straight lock washer.
2. While rotating wheel in both directions, torque adjusting nut to 100–125 ft. lbs. to seat bearings.
3. Back off adjusting nut one complete turn, then while again rotating wheel in both directions, retorque adjusting nut to 50–70 ft. lbs.
4. Back off adjusting nut to obtain an initial end play of .004–.010 inch between hub and spindle.
5. Install lock ring making sure that adjusting nut dowel is inserted into lock ring, then install outer washer locating dimple in washer in one of the lock ring holes.

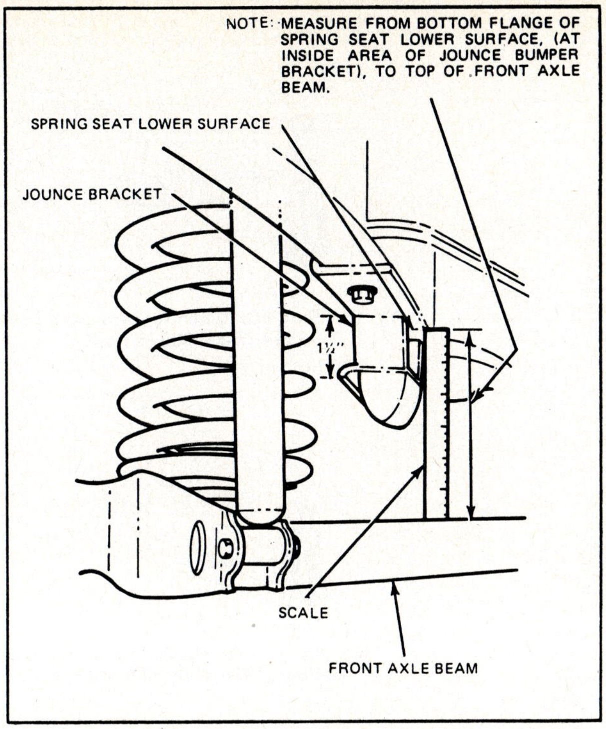

Fig. 14 Measuring riding height. 1977–83 E series

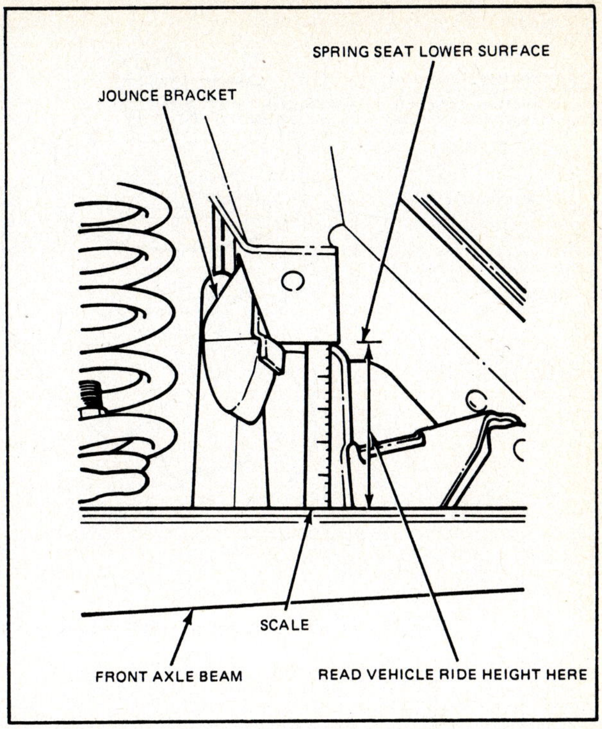

Fig. 15 Measuring riding height. 1974–79 F series

Install lock nut and torque to 100–150 ft. lbs.

6. Recheck to make sure that final bearing adjustment is .001–.010 inch end play. Readjust if necessary.
7. Bend washer outward in two places to lock outer nut in position, then install gasket and grease cap and adjust brakes.

NOTE: Final adjustment must not result in a preloaded bearing.

1974–83 16000, 18000 & 20000 Pound Axle With Single Nut & Cotter Pin

1. Back off brake adjustment, then remove grease cap and gasket.
2. While rotating wheel in both directions, torque adjusting nut to 100–125 ft. lbs. to seat bearings.
3. Back off adjusting nut one complete turn, then while again rotating wheel in both directions, retorque adjusting nut to 50–70 ft. lbs.
4. If necessary, back off adjusting nut to align a slot with next cotter pin hole.
5. Using same cotter pin hole as a reference point, back off adjusting nut to next slot to allow wheel to rotate freely. End play should be .001–.010 inch.
6. Install cotter pin, gasket and grease cap and readjust brakes.

NOTE: Final adjustment must not result in a preloaded bearing.

1981–83 Vehicles Except 9000 & 12000 Pound Axles With Hydraulic Brakes

1. Remove grease cap and cotter pin.
2. Torque adjusting nut to 50–70 ft. lbs. while rotating wheel.
3. Back off adjusting nut 1/6 turn on 5000, 6000 and 7000 lb. axles with hydraulic brakes and 16000, 18000 and 20000 lb. axles with air brakes. Back off adjusting nut 1/4 turn on 7000, 9000 and 12000 lb. axles with air brakes.
4. Install cotter pin in spindle hole. Back off nut if necessary to install cotter pin.

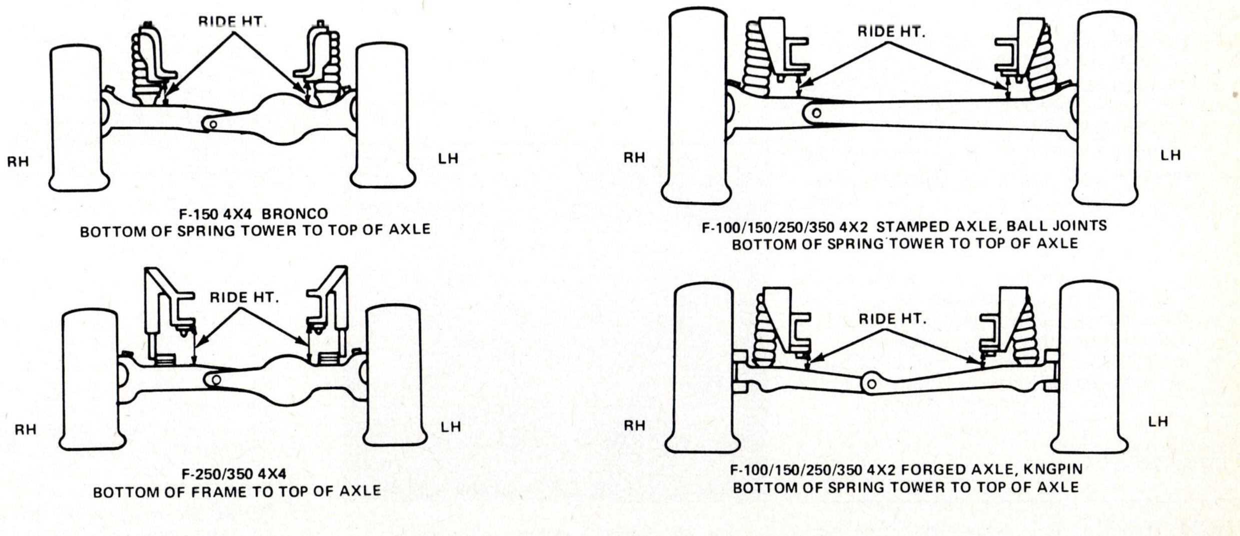

Fig. 16 Measuring riding height. 1980–83 F-100-350 & Bronco

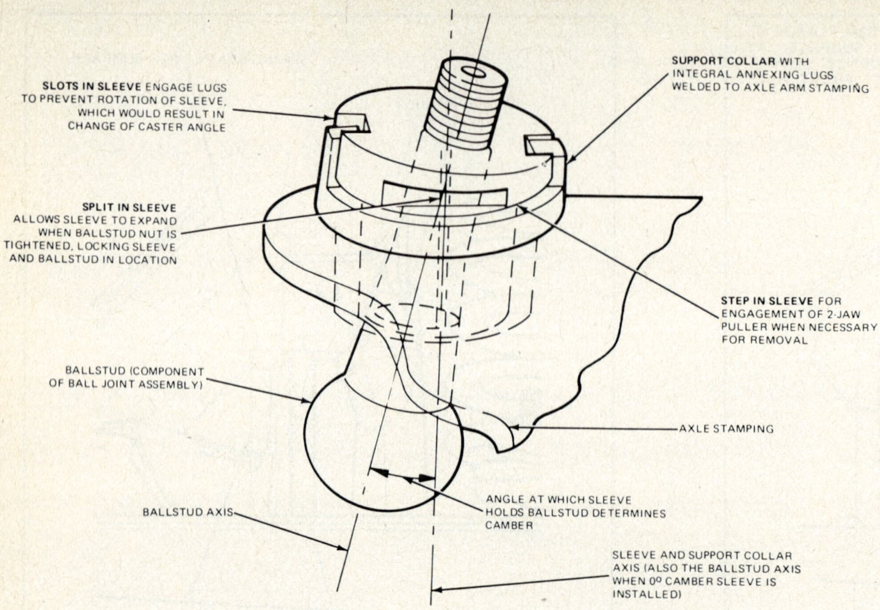

Fig. 17 Camber adjustment. 1980–83 F-150-350 & Bronco

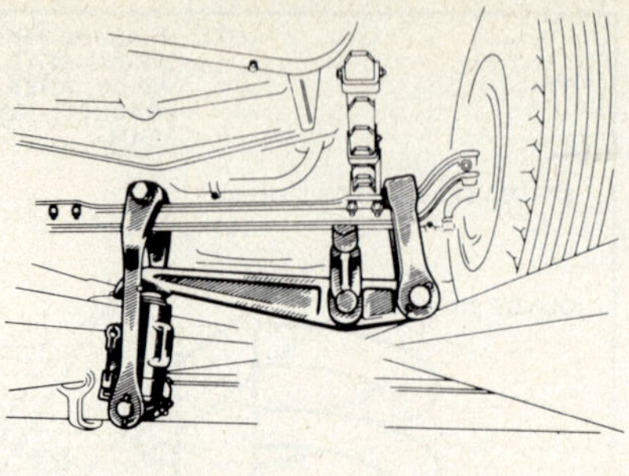

Fig. 18 Tool installation for increasing camber. 1974 500–9000, all series

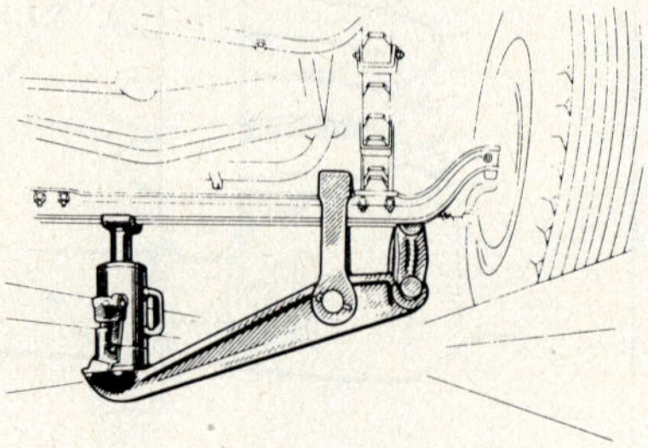

Fig. 19 Tool installation for decreasing camber. 1974 500–9000 all series

5. Check for .001–.010 inch end play and adjust, if necessary. Install grease cap and adjust brakes, if necessary.

NOTE: Final adjustment must not result in a preloaded bearing.

4 Wheel Drive Vehicles

1974–78 F-600

1. Remove outer hub cap retaining screws and cap.
2. Remove retaining lock ring, screws and splined drive plate.
3. Straighten tab of lock washer and remove lock nut and washer. Discard lock washer.
4. While rotating wheel in both directions, torque adjusting nut to 50 ft. lbs., then back off adjusting nut 1/4 to 1/3 turn.
5. Place a new lock washer against adjusting nut and apply oil to outer surface of lock washer.
6. Turn lock nut up against lock washer and torque to 100–150 ft. lbs.
7. Bend one tab of lock washer over lock nut, and bend other tab of lockwasher over adjusting nut.

NOTE: Use a blunt tool when bending tabs to avoid making any chips which could cause damage to bearing.

8. Apply silastic sealer or equivalent to front and rear mounting surfaces of splined drive plate, then position drive plate and install retaining screws.
9. Install retaining lock ring, outer hub cap and retaining screws.

1974–77 F-250

NOTE: If vehicle is equipped with free running lock-out hubs, refer to "FRONT WHEEL LOCKING HUBS" section for removal and installation.

1. Raise vehicle and support with stands.
2. Remove free-running hubs, if equipped.
3. Remove grease cap, then using tool T59T-1197-B, remove lock nut and lock ring from spindle.
4. Using tool T59T-1197-B and a torque wrench, torque bearing adjusting nut to 50 ft. lbs. while rotating wheel in both directions to seat bearings.
5. Back off adjusting nut 1/4 turn, then assemble lock ring by turning nut to nearest notch for dowel pin installation.
6. Install outer lock nut and torque to 80–100 ft. lbs. on 1974–76 models, and 50–70 ft. lbs. on 1977 models.
7. Install grease cap. Install free-running hubs, if equipped.

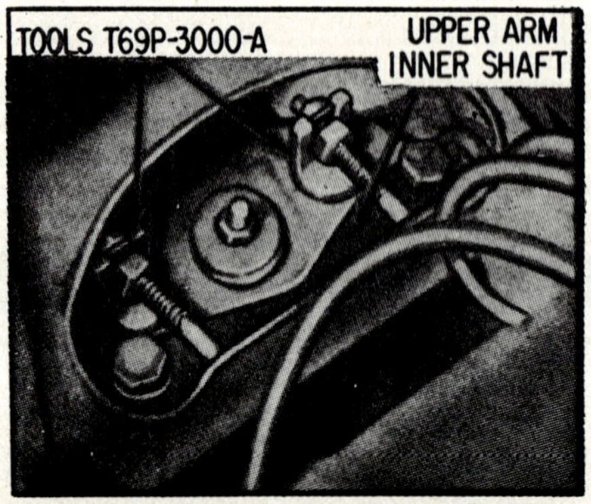

Fig. 20 Caster & camber adjustments. 1974–79 Ranchero

1974–79 Bronco, 1974–76 F-100, 1977–79 F-150, 1978–79 F-250 & 1979 F-350

NOTE: If vehicle is equipped with free running lock-out hubs, refer to "FRONT WHEEL LOCKING HUBS" section for removal and installation

1. Raise vehicle and support with stands.
2. Back off brake adjustment, if necessary.
3. Remove grease cap and hub snap ring, then remove splined driving hub and pressure spring. This may require prying slightly.
4. Remove wheel bearing lock nut and lock ring.
5. Using tool T59T-1197-B and a torque wrench, torque bearing adjusting nut to 50 ft. lbs. while rotating wheel back and forth to seat bearings.
6. Back off adjusting nut 1/4 turn, then assemble lock ring by turning nut to nearest notch for dowel pin installation.

NOTE: The dowel pin must seat in a lock ring hole for proper bearing adjustment and wheel retention.

7. Install outer lock nut and torque to 80–100 ft. lbs. on 1974–76 models, 50–70 ft. lbs. on 1977 models and 50–80 ft. lbs. on 1978–79 models. Final end play of wheel when installed on spindle should be .001–.010 inch on all except 1979 F-250 with a Dana 44-9F axle, and .001–.006 inch on 1979 F-250 with a Dana 44-9F axle.
8. Install pressure spring, driving hub, snap ring and grease cap, then adjust brakes, if necessary.

1974–82 WHEEL ALIGNMENT SPECIFICATIONS CHART

1974 F100, 250, 350

Riding Height, Inches		3	3.50	4	4.50	5	5.50	Toe In
F100, 250①	Caster	+5 1/4°	+4 1/2°	+3 1/4°	+1 7/8°	+1°	—	3/32
	Camber	+7/8°	+5/8°	+1 1/2°	+2 3/8°	+3 1/8°	+4 1/4°	
F35	Caster	—	+5 1/4°	+4 1/2°	+3 1/4°	+1 7/8°	+1°	3/32
	Camber	+7/8°	+5/8°	+1 1/2°	+2 3/8°	+3 1/8°	+4 1/4°	

1974 E100, 200 SERIES

Riding Height, Inches		3.69	3.99	4.19	4.44	4.69	4.94	5.19	5.44	5.69	5.94	Toe In
1974 E100, 200	Caster	—	+8 1/4°	—	+7 1/2°	—	+6 1/4°	—	+4 3/4°	—	+4°	3/32
	Camber	−7/8°	—	+5/8°	—	+1 1/2°	—	+2 3/8°	—	+3 1/8°	—	

1974–76 F SERIES

Riding Height, Inches		2.25	2.75	3	3.25	3.50	3.75	4	4.25	4.50	4.75	5	5.25	5.50	Toe In
1974 F100, 250①	Caster	—	—	+8 1/4°	—	+7 1/2°	—	+6 1/4°	—	+4 3/4°	—	+4°	—	—	3/32
	Camber	—	—	−3/4°	—	−1/2°	—	+1 1/22°	—	+2 3/8°	—	+3 1/4°	—	+4°	
1974 F250②, F350	Caster	—	+8 1/4°	—	+7 3/8°	—	+6 1/8°	—	+4 3/4°	—	+4°	—	—	—	3/32
	Camber	—	−3/4°	—	+1/2°	—	+1 1/2°	—	+2 1/2°	—	+3 1/4°	—	+4 1/4°	—	
1975 F100③, 250④	Caster	—	—	+8 5/8°	—	+7 3/8°	—	+6 1/4°	—	+4 7/8°	—	+3 7/8°	—	—	1/8
	Camber	—	—	−3/4°	—	+1/2°	—	+1 1/2°	—	+2 1/4°	—	+3 1/4°	—	+4 1/8°	
1975 F250⑤	Caster	—	+8 5/8°	—	+7 3/8°	—	+6 1/4°	—	+4 7/8°	—	+3 7/8°	—	—	—	1/8
	Camber	—	−3/4°	—	+1/2°	—	+1 1/2°	—	+2 1/4°	—	+3 1/4°	—	+4 1/8°	—	
1975 F100⑥, 150⑥, 250⑦	Caster	—	—	+8 5/8°	—	+7 3/8°	—	+6 1/4°	—	+4 7/8°	—	+3 7/8°	—	—	1/8
	Camber	−3/4°	—	+1/2°	—	+1 1/2°	—	+2 1/4°	—	+3 1/4°	—	+4 1/8°	—	—	
1975 F250⑧, 350	Caster	—	—	+8 5/8°	—	+7 3/8°	—	+6 1/4°	—	+4 7/8°	—	+3 7/8°	—	—	1/8
	Camber	—	—	−3/4°	—	+1/2°	—	+1 1/2°	—	+2 1/4°	—	+3 1/4°	—	+4 1/8°	
1976 F100③, 150③, 250③	Caster	—	—	—	—	+7 7/8°	—	+6 3/8°	—	+4 7/8°	—	+4 1/4°	—	—	1/8
	Camber	—	−1 1/2°	—	−5/8°	—	+5/8°	—	+1 7/8°	—	+3 1/4°	—	+4 1/4°	—	
1976 F250⑩, 350	Caster	—	—	—	—	—	—	+7 7/8° ③⑩	—	+6 3/8° ③⑩	—	+4 7/8° ③⑩	—	+4 1/4° ③⑩	1/8
	Camber	−1 1/2°	−5/8°	—	−5/8°	—	+1 7/8°	—	+3 1/4°	—	—	—	—	—	
1976 F100⑥, 150⑥	Caster	—	—	—	+7 7/8°	—	—	+6 3/8°	—	+4 7/8°	—	+4 1/4°	—	—	1/8
	Camber	—	−1 1/2°	—	−5/8°	—	+5/8°	—	+1 7/8°	—	+3 1/4°	—	+4 1/4°	—	

Continued

1974–82 WHEEL ALIGNMENT SPECIFICATIONS CHART—Continued

1975–76 E SERIES

Riding Height, Inches		2.80	3	3.30	3.50	3.80	4	4.30	4.50	4.80	5	5.30	5.50	Toe In
E100, 150	Caster	—	—	—	+8 5/8°	—	+7 3/8°	—	+6 1/4°	—	+4 7/8°	—	+3 7/8°	1/8
	Camber	—	−3/4°	—	+1/2°	—	+1 1/2°	—	+2 1/4°	—	+3 1/4°	—	+4 1/8°	
E250	Caster	—	—	+8 5/8°	—	+7 3/8°	—	+6 1/4°	—	+4 7/8°	—	+3 7/8°	—	1/8
	Camber	—	−3/4°	—	+1/2°	—	+1 1/2°	—	+2 1/4°	—	+3 1/4°	—	+4 1/8°	
E350	Caster	—	—	+8 5/8°	—	+7 3/8°	—	+6 1/4°	—	+4 7/8°	—	+3 7/8°	—	1/8
	Camber	−3/4°	—	+1/2°	—	+1 1/2°	—	+2 1/4°	—	+3 1/4°	—	+4 1/8°	—	
1976 E100, 150	Caster	—	—	—	+9°	—	+7 3/4°	—	+6 1/2°	—	+5 3/8°	—	+4 1/4°	1/8
	Camber	—	—	—	−1 1/4°	—	+1/8°	—	+3/4°	—	+1 7/8°	—	+2 3/4°	
1976 E250, 350	Caster	—	—	+6°	—	+4 3/4°	—	+3 1/2°	—	+2 3/8°	—	+1 1/4°	—	1/8
	Camber	—	—	−1 1/4°	—	−1/8°	—	+3/4	—	+1 7/8°	—	+2 3/4°	—	

1977–79 E & F SERIES

Riding Height, Inches		3.00–3.25	3.50–3.75	4.00–4.25	4.25–4.50	4.75–5.00	5.25–5.50	Toe In
1977 E100, 150	Caster	—	—	—	+3 1/4° to +5 3/4°	+2° to +4 1/2°	+3/4° to +3 1/4°	—
	Camber	—	—	—	−3/4° to +3/4°	0° to +1 3/4°	+1 1/2° to +3 1/4°	
1977–78 E250, 350	Caster	—	—	+6 1/4° to +9°	+5 3/4° to +8 1/4°	+4 1/2° to +1 1/4°	+3 1/4° to +6°	1/32
	Camber	—	—	−1° to +3/4°	−1/2° to +1 1/4°	+1/2° to +2 1/4°	+1 1/2° to +3 1/4°	
1977 F100, 150, 250⑪	Caster	+7° to 9 1/2°	+5 3/4° to +8 1/2°	+4 1/2° to +7°	+4° to +6 1/2°	—	—	3/32
	Camber	−1/2° to +1°	0° to +2 1/4°	+3/4° to +2 1/2°	+1 1/4° to +3°	—	—	
1977 F250⑫	Caster	+6 1/2° to +9°	+5 1/4° to +7 3/4°	+4° to +6 1/2°	+3 1/4° to +6°	—	—	3/32
	Camber	−1/4° to +1 1/2°	0° to +1 3/4°	+1 1/4° to +3°	+1 3/4° to +3 1/4°	—	—	
1977 F250⑬⑭, 350	Caster	—	+7° to +9 1/2°	+5 3/4° to +8 1/2°	+5 1/4° to +7 3/4°	+4° to +6 1/2°	—	3/32
	Camber	—	−1/4° to +1 1/2°	+1/2° to +2 1/4°	+3/4° to +2 1/2°	+1 3/4° to +3 1/4°	—	
1978 E100, 150	Caster	—	—	+3 3/4° to +6 1/2°	+3 1/4° to +5 3/4°	+2° to +4 1/2°	+3/4° to +3 1/4°	1/8
	Camber	—	—	−3/4° to +1/2°	−1/2° to +3/4°	+1/2° to +1 3/4°	+1 3/4° to +3 1/4°	
1978 F100, 150, 250⑪	Caster	+7 3/4° to +9°	+6 1/2° to +7 3/8°	+5° to +6 1/2°	+4 1/2° to +5 7/8°	+3 1/8° to +4 1/2°	—	3/32
	Camber	−1/2° to +1°	0° to +1 3/4°	+3/4° to +2 1/2°	+1 1/4° to +3°	+2° to +3 3/4°	—	
1978 F250⑫	Caster	+6 7/8° to +8 3/8°	+5 3/4° to +7°	+4 1/2° to +5 7/8°	+3 7/8° to +5 1/8°	+2 1/2° to +4°	—	3/32
	Camber	−1/4° to +1 1/2°	+1/2° to +2 1/4°	+1 1/4° to +3°	+1 3/4° to +3 1/4°	+2 3/8° to +4 1/8°	—	
1978 F250⑬⑭, 350	Caster	+9° to +10 1/4°	+7 5/8° to +8 5/8°	+6 3/8° to +7 1/4°	+5 3/4° to +7°	+4 1/2° to +5 7/8°	—	3/32
	Camber	−1° to +5/8°	−1/4° to +1 1/2°	+1/2° to +2 1/4°	+3/4° to +2 1/2°	+1 3/4° to +3 1/4°	—	
1979 E100, 150	Caster	—	+5 3/4° to +7 1/4°	+4 1/2° to +5 3/4°	+4° to +5 1/4°	+2 1/2° to +4°	+1 1/2° to +2 3/4°	1/32
	Camber	—	−1 1/2° to +1/4°	−1/2° to +1 1/4°	0° to +1 3/4°	+1° to +2 3/4°	+2° to +3 3/4°	
1979 E250, 350	Caster	—	+8 1/2° to +9 3/4°	+7 1/8° to +8 1/2°	+6 1/2° to +7 3/4°	+5 1/4° to +6 1/2°	+4° to +5 1/2°	1/32
	Camber	—	−1 1/2° to +1/4°	−1/2° to +1 1/4°	0° to +1 3/4°	+1° to +2 3/4°	+2° to +3 3/4°	
1979 F100, 150, 250⑮	Caster	+7 3/4° to +9°	+6 1/4° to +7 5/8°	+5 1/8° to +6 1/2°	+4 1/2° to +5 7/8°	+3 1/4° to +4 5/8°	—	3/32
	Camber	−1 1/2° to +1/8°	−3/4° to +1°	0° to +1 5/8°	+3/8° to +2°	+1 1/4° to +2 3/4°	—	
1979 F250⑨	Caster	+8 3/8° to +7°	+5 7/8° to +7 1/8	+4 1/2° to +5 7/8°	+3 7/8° to +5 1/4°	+2 5/8° to +4°	—	3/32
	Camber	−1 1/2° to +1/8°	+7/8° to −3/4°	0° to +1 5/8°	+3/8° to +2°	+1 1/4° to +2 3/4°	—	
1979 F250⑬⑭, 350	Caster	+8 7/8° to +10 3/8°	+7 3/4° to +9°	+6 3/8° to 7 3/4°	+5 3/4° to +7 1/8°	4 1/2° to +5 7/8°	—	3/32
	Camber	−1 3/4° to −1/4°	−1 1/8° to +1/2°	−1/4° to +1 1/4°	0° to +1 5/8°	+3/4° to +2 3/8°	—	

Continued

1974–82 WHEEL ALIGNMENT SPECIFICATIONS CHART—Continued

1980–82 E SERIES

Riding Height, Inches		3.25–3.50	3.50–3.75	4.00–4.25	4.25–4.50	4.50–4.75	5.00–5.25	5.25–5.50	Toe In
E100, 150	Camber	−1¾° to −¼°	−1½° to +¼°	−½° to +1¼°	0° to +1¾°	+½° to +2¼°	+1½° to +3¼°	+2° to +3¾°	⅛°
	Caster	+6¼° to +8°	+5¾° to +7¼°	+4½° to +5¾°	+4° to +5¼°	+3¼° to +4½°	+2° to +3¼°	+1½° to +2¾°	
E250, 350	Camber	−1¾° to −¼°	−1½° to +¼°	−½° to +1¼°	0° to +1¾°	+½° to +2¼°	+1½° to +3¼°	+2° to +3¾°	1/32°
	Caster	+9° to +10½°	+8½° to +9¾°	+7⅛° to +8½°	+6½° to +7¾°	+5¾° to +7°	+4⅝° to +6°	+4° to +5½°	

1980–82 F SERIES EXC. F-250/350 (4×4)

Riding Height, Inches		2.00–2.25	2.25–2.75	2.75–3.25	3.25–3.50	3.50–4.00	4.00–4.25	4.25–4.75	4.75–5.00	Toe In
F100, 150 (4×2)	Camber	—	−3°–−½°	−2°–+½°	−1¼°–+1¼°	−¼°–+2¼°	+½°–+3°	+1½°–+4°	—	3/32
	Caster	—	+6°–+10°	+5°–+9°	+4°–+8°	+3°–+7°	+2°–+6°	+1°–+5°	—	
F250/350 (4×2)	Camber	−2½°–0°	−1½°–+1°	+¾°–+1¾°	+¼°–+2¾°	+1°–+3½°	+2°–+4½°	—	—	⑯
	Caster	+5¾°–+9°	+4¾°–+8°	+3¾°–+7°	+2¾°–+6°	+1¾°–+5°	+¾°–+4°	—	—	
F150 Bronco (4×4)	Camber	—	—	−2½°–−¼°	−1¾°–+½°	−¾°–+1½°	0°–+2¼°	+1°–+3¼°	+1¾°–+4°	⑰
	Caster	—	—	+6°–+9°	+5°–+8°	+4°–+7°	+3°–+6°	+2°–+5°	+1°–+4°	

1980 F SERIES (4×4)

Riding Height, Inches		4.25–4.75	4.75–5.00	5.00–5.50	5.50–6.00	6.00–6.25	6.25–6.75	6.75–7.00	Toe In
F250-350 (4×4)	Camber	−4°–−1¼°	−2¾°–−0°	−1½°–+1¼°	−¼°–+2½°	+1°–+3¾°	+2¼°–+5°	—	2/32
	Caster	+3½°–+5¾°	+3¼°–+5½°	+3°–+5¼°	+2¾°–+5°	+2½°–+4¾°	+2¼°–+4½°	—	

1981–82 F SERIES (4×4)

Riding Height, Inches		4.25–4.75	4.75–5.00	5.00–5.50	5.50–6.00	6.00–6.25	6.25–6.75	6.75–7.00	Toe In
F250-350 (4×4)	Camber	—	−2¾°–−¼°	−1¾°–+¾°	−¾°–+1¾°	+¼°–+2¾°	+1¼°–+4°	+2½°–+5°	3/32
	Caster	—	+3°–+5°	+3⅛°–+5⅛°	+3⅛°–+5⅛°	+3¼°–+5¼°	+3⅜°–+5⅜°	+3½°–+5½°	

①—Except 1974 7500-8050 GVW.
②—1974 7500-8050 GVW.
③—Except Super Cab.
④—Except 7500-8050 GVW & 6900 GVW with V8-460 engine.
⑤—6900 GVW with V8-460 engine.
⑥—Super Cab.
⑦—Super Cab 8100 GVW.
⑧—Super Cab except 8100 GVW.
⑨—Regular Cab 7700-7900 GVW & Super Cab 6300-7800 GVW.
⑩—F250 7800-8000 GVW.
⑪—Regular Cab 6200-6900 GVW.
⑫—Regular Cab 7800-8000 GVW & Super Cab 6350-7800 GVW.
⑬—Super Cab 8100 GVW.
⑭—Super Cab with standard suspension.
⑮—6200-6800 GVW.
⑯—1980, 7/32; 1981, 3/32.
⑰—1980, 2/32; 1981, 3/32.

Continued

1983 WHEEL ALIGNMENT SPECIFICATIONS

1983 E-100 & 150

Riding Height In Inches	Caster Angle Degrees①							
	4	4 1/4	4 1/2	4 3/4	5	5 1/4	5 1/2	5 3/4
E-100, 150	+4 1/2–+6 1/2	+4–+6	+3 1/4–+5 1/4	+2 1/2–+4 1/2	+2–+4	+1 1/4–+3 1/4	+3/4–+2 3/4	+1/4–+2 1/4

①—Side to side caster readings must be within 1 1/2°, when vehicle riding height is within 1/8 inch side to side.

1983 E-250 & 350

Riding Height In Inches	Caster Angle Degrees①							
	3.8	4.05	4.3	4.55	4.8	5.05	5.3	5.55
E-250, 350	+4 1/2–+6 1/2	+4–+6	+3 1/4–+5 1/4	+2 1/2–+4 1/2	+2–+4	+1 1/4–+3 1/4	+3/4–+2 3/4	+1/4–+2 1/4

①—Side to side caster readings must be within 1 1/2°, when vehicle riding height is within 1/8 inch side to side.

1983 F-100 & 150 (4×2)

Riding Height In Inches	Caster Angle Degrees①									
	3–3.2	3.2–3.4	3.4–3.6	3.6–3.8	3.8–4	4–4.2	4.2–4.4	4.4–4.6	4.6–4.8	4.8–5
F-100, 150	+5 1/4–+7 1/4	+5–+7	+4 1/2–+6 1/2	+4 1/4–+6 1/4	+3 3/4–+5 3/4	+3 1/4–+5 1/4	+2 3/4–+4 3/4	+2 1/2–+4 1/2	+2–+4	+1 1/2–+3 1/2

①—Side to side caster readings must be within 1 1/2°, when truck frame to axle riding height is such that right hand side is from 0 to .4 inch less than left hand side.

1983 F-250 & 350 (4×2)

Riding Height In Inches	Caster Angle Degrees①								
	2.4	2.6	2.8	3	3.2	3.4	3.6	3.8	4
F-250, 350	+6 1/4–+7 1/4	+5 3/4–+7 1/4	+5 1/4–+7 1/4	+4 1/2–+6 1/2	+4–+6	+3 1/2–+5 1/2	+3–+5	+3–+4 1/2	+3–+4

①—Side to side caster readings must be within 1 1/2°, when truck frame to axle riding height is such that right hand side is from 0–.2 inch less than left hand side.

1983 F-150 & Bronco (4×4)

Riding Height In Inches	Caster Angle Degrees①								
	3.2	3.4	3.6	3.8	4	4.2	4.4	4.6	4.8
F-150, Bronco	+6–+8	+5 1/2–+7 1/2	+5–+7	+4 1/2–+6 1/2	+4–+6	+3 1/2–+5 1/2	+3–+5	+2 1/2–+4 1/2	+2–+4

①—Side to side caster readings must be within 1 1/2°, when vehicle riding height is within .16 inch side to side.

1983 F-250 & 350 (4×4)

Riding Height In Inches	Caster Angle Degrees①								
	5.2	5.4	5.6	5.8	6	6.2	6.4	6.6	6.8
F-250, 350	+3 1/16–+5 1/16	+3 1/16–+5 1/16	+3 1/8–+5 1/8	+3 3/16–+5 3/16	+3 1/4–+5 1/4	+3 5/16–+5 5/16	+3 3/8–+5 3/8	+3 7/16–+5 7/16	+3 1/2–+5 1/2

①—Side to side caster readings must be within 1 1/2°, when vehicle riding height is within .16 inch side to side.

Continued

1983 WHEEL ALIGNMENT SPECIFICATIONS—Continued

Riding Height In Inches	Camber Angle Degrees①							
	4	4 1/4	4 1/2	4 3/4	5	5 1/4	5 1/2	5 3/4
1983 E-100 & 150								
E-100, 150	−5/8—+5/8	0—+1 1/4	+3/8—+1 5/8	+7/8—+2 1/4	+1 3/8—+2 5/8	+1 7/8—+3 1/4	+2 3/8—+3 5/8	+2 7/8—+4 1/4

①—Side to side camber readings must be within 5/8°, when vehicle riding height is within 1/8 inch side to side.

Riding Height In Inches	Camber Angle Degrees①							
	3.9	4.15	4.4	4.65	4.9	5.15	5.4	5.65
1983 E-250 & 350								
E-250, 350	−5/8—+5/8	0—+1 1/4	+3/8—+1 5/8	+7/8—+2 1/4	+1 3/8—+2 5/8	+1 7/8—+3 1/4	+2 3/8—+3 5/8	+2 7/8—+4 1/4

①—Side to side camber readings must be within 5/8°, when vehicle riding height is within 1/8 inch side to side.

Riding Height In Inches	Camber Angle Degrees①									
	3—3.2	3.2—3.4	3.4—3.6	3.6—3.8	3.8—4	4—4.2	4.2—4.4	4.4—4.6	4.6—4.8	4.8—5
1983 F-100 & 150 (4×2)										
F-100, 150	−1—+1/4	−3/4—+3/4	−1/4—+1 1/4	+1/4—+1 3/4	+3/4—+2	+1—+2 1/2	+1 1/2—+3	+2—+3 1/2	+2 1/2—+4	+3—+4 1/2

①—Side to side camber readings must be within .7°, when truck frame to axle riding height is such that, right hand side is from 0—.4 inch less than left hand side.

Riding Height In Inches	Camber Angle Degrees①									
	2.6	2.8	3	3.2	3.4	3.6	3.8	4	4.2	4.4
1983 F-250 & 350 (4×2)										
F-250, 350	−1/2—+1/4	−1/2—+1/2	−1/4—+1	0—+1 1/2	+3/4—+2	+1—+2 1/4	+1 1/2—+2 3/4	+2—+3 1/4	+2 1/2—+3 1/2	+2 3/4—+3 1/2

①—Side to side camber readings must be within .7°, when truck frame to axle riding height is such that, right hand side is from 0—.2 inch less than left hand side.

Riding Height In Inches	Camber Angle Degrees①								
	3.1	3.3	3.5	3.7	3.9	4.1	4.3	4.5	4.7
1983 F-150 & Bronco (4×4)									
F-150, Bronco	−1 5/8—−1/4	−1—+1/4	−5/8—+5/8	−1/4—+1 1/4	+1/4—+1 5/8	+3/4—+2 1/4	+1 1/4—+2 5/8	+1 5/8—+3 1/4	+2 1/4—+3 1/2

①—Side to side camber readings must be within .7°, when truck frame to axle riding height is within 0—.16 inch of side to side.

Riding Height In Inches	Camber Angle Degrees①									
	5.1	5.3	5.5	5.7	5.9	6.1	6.3	6.5	6.7	6.9
1983 F-250 & 350 (4×4)										
F-250, 350	−1 5/8—−1/4	−1 1/8—+1/4	−5/8—+3/4	−1/8—+1 1/4	+3/8—+1 3/4	+7/8—+2 3/8	+1 1/2—+2 7/8	+2—+3 1/2	+2 1/2—+4	+3 1/8—+4 1/2

①—Side to side camber readings must be within .7°, when truck frame to axle riding height is within 0—.16 inch of side to side.

1980–83 Bronco, F-150, 1980–82 F-250 & 1980–81 F-350

NOTE: If vehicle is equipped with free running lock-out hubs, refer to "FRONT WHEEL LOCKING HUBS" section for removal and installation.

1. Raise vehicle and support with stands.
2. Remove hub assembly, then using tool T59T-1197-B and a torque wrench, torque bearing inner adjusting nut to 50 ft. lbs. while rotating wheel back and forth to seat bearings.
3. Back off adjusting nut 1/8 turn, then assemble lock tab and outer locknut. Torque locknut to 150 ft. lbs. Bend at least one tab into a slot on both the inner adjusting nut and outer lock nut.
4. Final end play of wheel when installed on spindle should be .001–.006 inch.
5. Install hub assembly.

1982–83 & 1983 F-250 & F-350

1. Raise vehicle and support with stands.
2. Remove hub assembly, then using front wheel bearing spanner, D78T-1197-A and torque wrench and tighten inner locknut to 50 ft. lbs. to seat bearing.
3. Back off inner locknut and retorque to 31–39 ft. lbs.
4. While rotating hub, backoff locknut 3/8 turn.
5. Assemble lockwasher and outer locknut and torque to 65 ft. lbs. Bend at least one into a slot on both the inner adjusting nut and outer lock nut.
6. Final end play of the wheel when installed on spindle should be .001–.009 inch.
7. Install hub assembly.

FRONT SPRING, REPLACE

1980–83 Bronco & F-150 4 Wheel Drive, Fig. 9

1. Raise and support vehicle.
2. Remove shock absorber to lower bracket attaching bolt.
3. Remove spring lower retainer attaching nuts from inside coil spring.
4. Remove spring upper retainer attaching screw and the upper retainer.
5. Place jack stands under frame side rails and lower axle to relieve spring tension.

NOTE: The axle must be supported when replacing spring and not be permitted to hang by the brake hose. If the length of the brake hose is not sufficient to permit spring replacement, it will be necessary to remove the disc brake caliper. Do not suspend caliper by brake hose, if removed.

6. Remove spring and lower retainer.
7. Reverse procedure to install.

1980–83 F-250 & 350 4 Wheel Drive, Fig. 10

1. Raise vehicle until weight is relieved from front spring with the front wheels still contacting floor. Support axle to prevent rotation.
2. Disconnect shock absorber lower end from U-bolt spacer, then remove U-bolts, cap and spacer.
3. Remove rear hanger bolt at rear of spring.
4. Remove front shackle bolt and remove spring.
5. Reverse procedure to install.

RADIUS ARM, REPLACE

1980–83 Bronco & F-150 4 Wheel Drive, Fig. 3

1. Raise and support vehicle under frame side rails.
2. Remove front spring as outlined under "Front Spring, Replace".
3. Remove nut attaching radius arm to frame bracket, then the radius arm rear insulator.
4. Remove bolt and stud attaching radius arm to axle.
5. Move axle forward and remove radius arm from axle, then pull radius arm from frame bracket.
6. Reverse procedure to install.

AXLE HOUSING PIVOT BUSHING, REPLACE

1980–83 Bronco, F-150 Thru F-350 4 Wheel Drive, Figs. 3 & 4

1. Raise and support vehicle, remove front wheels.
2. Remove front axle assembly from vehicle.
3. Remove pivot bushing from axle housing using suitable tool.
4. Reverse procedure to install.

STEERING KNUCKLE, REPLACE

1980–83 Bronco, F-150 Thru F-350 4 Wheel Drive, Figs. 3 & 4

1. Remove axle shaft and spindle from steering knuckle.
2. Disconnect tie rod from steering arm.
3. Remove upper ball joint nut and loosen lower ball joint nut. Strike upper ball joint stud to loosen ball joints.
4. Remove lower ball joint nut and remove knuckle from yoke tube.
5. Remove camber adjuster bushing from tube and yoke assembly.
6. Install knuckle into vise and press out ball joints using suitable tool.

NOTE: On some models, the lower ball joint is retained by a snap ring.

7. Reverse procedure to install. Install knuckle onto tube and yoke assembly, then install camber adjusting bushing with arrow pointing outward for "positive camber" or towards center of vehicle for "negative camber". Install new lower ball joint nut and torque both upper and lower ball joint nuts to 100 ft. lbs.

MANUAL STEERING GEAR, REPLACE

1974 Econoline

1. Raise vehicle on hoist and remove attaching bolt from lower half of flex coupling.
2. Disconnect Pitman arm with suitable puller.
3. While supporting steering gear, remove three attaching bolts and lower steering gear.
4. Remove Pitman arm attaching nut. Remove Pitman arm.
5. Reverse procedure to install.

1975–83 Econoline & 1980–83 F-150-350 2 Wheel Drive

1. Raise and support front of vehicle, then disconnect flex coupling from steering shaft flange by removing two attaching nuts.
2. Disconnect drag link from Pitman arm using tool No. 3290-C.
3. Remove Pitman to sector shaft attaching nut and washer, then remove Pitman arm from gear sector shaft using tool No. T64P-3590-F.
4. While supporting steering gear, remove bolts and washer attaching gear assembly to frame side rail and lower gear assembly from vehicle.
5. Reverse procedure to install.

1974–79 F-100-350 2 Wheel Drive

1. Remove bolt attaching flex coupling to steering gear. Remove brake line bracket.
2. Raise and support front of vehicle, then disconnect Pitman arm from sector shaft using tool No. T64P-3590.
3. Remove steering gear to frame attaching bolts, then remove steering gear assembly.
4. Reverse procedure to install.

1974–77 F-100 Thru 350 4 Wheel Drive & 1974–77 Bronco

1. Raise the vehicle on a hoist and remove the Pitman arm from the sector shaft.
2. Unfasten the steering gear from the frame side rail and lower vehicle.
3. Remove the flex coupling bolt and nut from the coupling clamp.
4. Loosen the clamp on the steering shaft at end of steering column and separate the coupling from the gear input shaft by pushing the steering shaft toward the column.
5. Remove the flex coupling clamp from the steering gear input shaft and remove the gear from the frame side rail.
6. Reverse procedure to install.

1978 F-150 4 Wheel Drive & 1979 Bronco

Recirculating Ball Manual Steering Gear

1. Remove bolt attaching flex housing to coupling assembly shaft. Note alignment of bolt with groove in shaft.
2. Raise and support front of vehicle, then disconnect drag link from pitman arm using tool No. 3290-C.
3. Remove pitman arm nut and washer, then remove pitman arm from sector shaft using tool No. T64P-3590-F.
4. Support steering gear, then remove steering gear to frame attaching bolts and lower steering gear assembly from vehicle.

1978–79 F-150 4 Wheel Drive & Bronco

Worm & Roller Manual Steering Gear

1. Raise and support front of vehicle, then remove pitman arm attaching nut and washer.
2. Remove pitman arm from sector shaft using tool No. T64P-3590-F.
3. Remove bolts attaching steering gear to

frame side rail, then lower front of vehicle.
4. Remove flex coupling clamp bolt and nut. Loosen clamp from coupling at end of steering column and separate coupling from steering gear input shaft by pushing steering shaft toward steering column.
5. Remove flex coupling clamp from steering gear input shaft, then remove steering gear assembly from vehicle.

1979 F-250 & 350 4 Wheel Drive

1. Remove two bolts attaching steering shaft to steering gear and column shaft.
2. Remove fan shroud attaching bolts and position shroud to right hand side of engine compartment.
3. Raise and support front of vehicle, then remove pitman arm nut and washer.
4. Disengage steering shaft coupling from steering gear shaft.
5. Remove steering gear to frame attaching bolts and lower steering gear assembly from vehicle.

F-B 500-800, 6000-7000

1. Loosen steering column lower clamp and slide clamp down.
2. Remove bolt and nut retaining pitman arm from sector shaft.
3. Remove retaining bolts and retaining strap if so equipped and remove steering gear from under vehicle.

L-N 500-900, 6000-9000, L-LT-LNT-W-WT 800-900, 8000-9000 & CL-9000

1. Remove Pitman arm.
2. Remove bolt and nut retaining U joint to steering gear shaft and slide joint off shaft.
3. Remove retaining bolts and straps if so equipped and remove steering gear.

P Series

1. Remove steering wheel.
2. Remove column bracket bolts from instrument panel.
3. Raise front of truck about 10 inches and install safety stands.
4. Remove pitman arm from sector shaft and remove steering gear retaining bolts.
5. Loosen steering column lower clamp.
6. Disconnect horn wire.
7. Move gear to left and remove it from beneath truck.
8. Reverse procedure to install.

C Series

1. Remove horn wire brush from steering column just below instrument panel bracket.
2. Remove U-joint upper nut and bolt.
3. Turn wheels to right and remove Pitman arm from sector shaft.
4. Unfasten and remove steering gear with U-joint atttached.
5. Reverse procedure to install.

Ranchero with Manual Steering Gear

1. Remove flex coupling bolts.
2. Remove Pitman arm nut and remove Pitman arm from sector shaft using a puller.
3. On vehicles with manual transmission it may be necessary to disconnect the clutch linkage and on V8 models it may be necessary to lower the exhaust system.
4. Unfasten and remove steering gear.

POWER STEERING GEAR, REPLACE

1974–82 F-100-350 With 2 Wheel Drive, 1977–82 F-150, 1978–82 F-250 & Bronco With 4 Wheel Drive & 1983 All Models

1. Disconnect pressure and return lines from gear and plug all open lines and fittings.
2. Disconnect brake line bracket attached to gear, if necessary.
3. Remove bolts securing flex coupling to gear, then raise and support vehicle and remove sector shaft nut.
4. Remove Pitman arm from sector shaft using tool T64P3590F, or equivalent. On vehicle equipped with standard transmission, remove clutch release lever spring, then support steering gear and remove three gear attaching bolts. Remove gear from vehicle.
5. Reverse procedure to install.

1974–77 W Series

1. Raise cab to full tilt position, then remove grille.
2. Mark location of power steering hoses for reference during installation, then remove hoses from gear. Plug all open lines and fittings.
3. Remove hydraulic cylinder stud from Pitman arm, then position cylinder aside.
4. Remove steering shaft U joint nut, bolt and washer, then separate joint from steering gear input shaft.
5. Remove gear to bracket retaining bolts, then remove gear from vehicle.
6. Reverse procedure to install.

1974–79 F, B, L, LN, LNT 500 thru 9000 Series & 1979–80 C Series

1. Disconnect pressure and return lines from steering gear and plug all open lines and fittings.
2. Remove Pitman arm from sector shaft.
3. Remove bolt and nut securing U joint to steering gear input shaft, then slide joint from shaft.
4. On 1974–79 F-600-750 & B-600-750 models, with manual transmission, loosen front engine mount bolt, then remove both exhaust pipe retaining bolts. Remove exhaust heat shield and clutch spring bracket. Remove transmission mount bolts, then support transmission with suitable jack and remove mounting pads. Remove clutch bracket from over steering gear, then remove gear retaining bolts. Remove floorboard section from over shift lever and emergency brake lever, then position engine and transmission assembly toward the right side to allow access for gear removal.
5. On all other vehicles, remove gear or gear bracket to frame attaching bolts, then remove gear.
6. Reverse procedure to install.

1981–82 F, B & C Series

1. Disconnect pressure and return lines from steering gear and plug all open lines and fittings.
2. Disconnect assist cylinder lines, if equipped.
3. Remove cotter pin, bolt and nut holding Pitman to sector shaft, and remove Pitman arm from sector shaft using tool T64P-3590-F or equivalent.
4. Remove bolts and nuts holding steering gear bracket to the frame side rail. Remove steering gear.
5. Reverse procedure to install.

1974–77 Bronco

1. Disconnect pressure and return lines from gear and plug all open lines and fittings.
2. Raise and support vehicle, then remove Pitman arm from sector shaft using tool T64P3590F, or equivalent.
3. Remove bolts securing gear to frame, then remove pinch bolt from flange and insulator assembly.
4. Loosen horn retaining nut, then pivot horn outward.
5. Remove steering gear, shaft and joint assembly from vehicle.
6. Remove pinch bolt and nut from shaft and joint assembly, then remove shaft and joint assembly from gear.
7. Reverse procedure to install.

1975–82 Econoline & E-150-350

1. Disconnect pressure and return lines from gear and plug all open lines and fittings.
2. Raise and support vehicle, then remove drag link from Pitman arm.
3. Remove nuts securing flex coupling to steering column shaft assembly.
4. Support gear, then remove gear attaching bolts.
5. Remove flex coupling with pinch bolt, then the coupling from the gear.
6. Remove Pitman arm from sector shaft using tool T64P3590F, or equivalent.
7. Reverse procedure to install.

Ranchero with Integral Power Steering Gear

These models use an integral power steering gear which replaces the linkage power steering used previously. Remove the gear assembly as follows:
1. Disconnect pressure and return lines from gear and plug openings to prevent entry of dirt.
2. Remove two bolts that secure flex coupling to the gear and to the column.
3. Raise vehicle and remove pitman arm with suitable puller.
4. If vehicle is equipped with synchromesh transmission, remove clutch release lever retracting spring to provide clearance to remove gear.
5. Support gear and remove three gear attaching bolts.

1983 FORD RANGER

INDEX OF SERVICE OPERATIONS

GENERAL ENGINE SPECIFICATIONS

Year	Engine CID①/Liter	Carburetor	Bore and Stroke	Compression Ratio	Net H.P. @ R.P.M.	Maximum Torque Lbs. Ft. @ R.P.M.	Normal Oil Pressure Pounds
1983	4-122, 2.0 L	YFA, 1 Bbl②	3.52 × 3.13	9.0	73 @ 4000	107 @ 2400	40–60
	4-140, 2.3 L④	YFA, 1 Bbl②③	3.78 × 3.13	9.0	79 @ 3800	124 @ 2200	40–60
	4-140, 2.3 L⑤	YFA, 1Bbl②③	3.78 × 3.13	9.0	82 @ 4200	126 @ 2200	40–60

①—Cubic Inch Displacement.
②—Carter.
③—Feedback carburetor, California only.
④—Manual Trans.
⑤—Automatic Trans.

ENGINE TIGHTENING SPECIFICATIONS★

★ Torque specifications are for clean and lightly lubricated threads only. Dry or dirty threads produce increased friction which prevents accurate measurement of tightness.

Year	Engine	Spark Plugs Ft. Lbs.	Cylinder Head Bolts Ft. Lbs.	Intake Manifold Ft. Lbs.	Exhaust Manifold Ft. Lbs.	Rocker Arm Shaft Bracket Ft. Lbs.	Rocker Arm Cover Ft. Lbs.	Connecting Rod Cap Bolts Ft. Lbs.	Main Bearing Cap Bolts Ft. Lbs.	Flywheel to Crank-shaft Ft. Lbs.	Vibration Damper or Pulley Ft. Lbs.
1983	4-122	5–10	①	14–21	②	—	6–8	③	①	56–64	100–120
	4-140	5–10	①	14–21	②	—	6–8	③	①	56–64	100–120

①—Torque bolts in sequence in two steps: Step 1 (50–60 ft. lbs.); Step 2 (80–90 ft. lbs.).
②—Torque nuts in sequence in two steps: Step 1 (5–7 ft. lbs.); Step 2 (16–23 ft. lbs.).
③—Torque bolts in sequence in two steps: Step 1 (25–30 ft. lbs.); Step 2 (30–36 ft. lbs.).

TUNE UP SPECIFICATIONS

The following specifications are published from the latest information available. This data should be used only in the absence of a decal affixed in the engine compartment.

★ When using a timing light, disconnect vacuum hose or tube at distributor and plug opening in tube or hose so idle speed will not be affected.

● When checking compression, lowest cylinder must be within 75% of the highest.

▲ Before removing wires from distributor cap, determine location of the No. 1 wire in cap, as distributor position may have been altered from that shown at the end of this chart.

Year & Engine	Spark Plug		Ignition Timing BTDC①★				Curb Idle Speed②		Fast Idle Speed		Fuel Pump Pressure
	Type	Gap	Firing Order Fig. ▲	Man. Trans.	Auto. Trans.	Mark Fig.	Man. Trans.	Auto. Trans.	Man. Trans.	Auto. Trans.	
1983											
4-122	AWSE-42	.034	A	6°	—	B	800	—	2000	—	5–7
4-140 Exc. High Alt.	AWSE-42	.034	A	6°	8°	B	800	④	2000	2000	5–7
4-140 High Alt.	AWSE-42	.034	A	10°	8°	B	850	800	2000	2000	5–7

①—Before top dead center
②—Idle speed on man. trans. vehicles is adjusted in Neutral & and on auto. trans. vehicles in Drive unless otherwise specified. Adjustment should be made with all vacuum hoses connected.
③—Transmission should be in neutral with EGR valve vacuum hose disconnected and plugged.
④—Calibration code 3-50S-R01, 750; calibration code 3-50S-R11, 800.

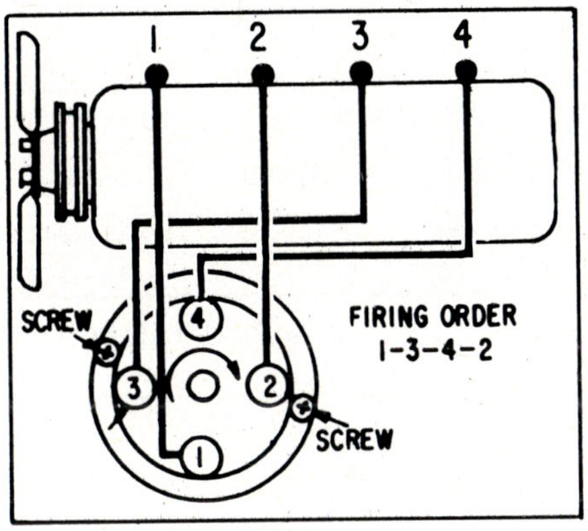

Fig. A

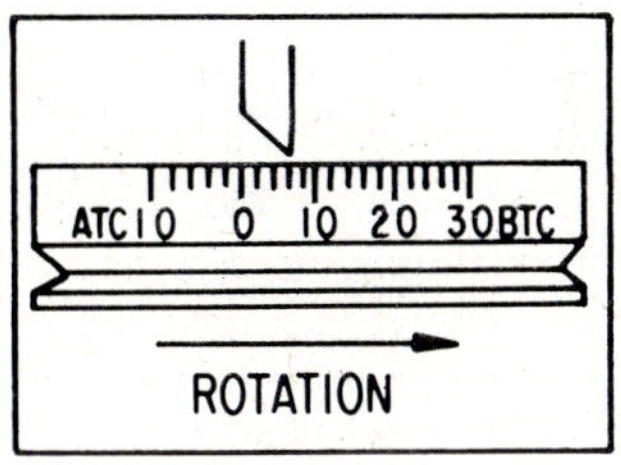

Fig. B

COOLING SYSTEM & CAPACITY DATA

Year	Model or Engine	Cooling Capacity Qts.		Radiator Cap Relief Pressure, Lbs.	Thermo. Opening Temp.	Fuel Tank Gals.	Engine Oil Refill Qts. ①	Transmission Oil			Rear Axle Oil Pints
		Less A/C	With A/C					3 Speed Pints	4 Speed Pints	Auto. Trans. Qts. ②	
1983	4-122	6.5	7.2	13	192	14.5	4	—	3	④	③
	4-140	6.5	7.2	13	192	14.5	4	—	3	④	③

①—Add 1/2 qt. with filter change
②—Approximate. Make final check with dipstick.
③—With 6 3/4" ring gear, 3.0 pts.; with 7 1/2" ring gear, 5.0 pts.
④—With C-3 transmission, 8.0 qts.; with C-5 transmission, 7.5 qts.

BRAKE SPECIFICATIONS

Year	Model	Brake Drum Inside Diameter	Wheel Cylinder Bore Diameter			Master Cylinder Bore Diameter		
			Front Disc Brake	Front Drum Brake	Rear Brake	With Disc Brakes	With Drum Brakes	With Power Brakes
1983	Ranger	9.00	2.60	—	—	—	—	—

ALTERNATOR & REGULATOR SPECIFICATIONS

Year	Make or Model	Current Rating		Field Current @ 75°F.		Voltage Regulator				Field Relay	
		Amperes	Volts	Amperes	Volts	Part No.	Voltage @ 75°F.	Contact Gap	Armature Air Gap	Armature Air Gap	Closing Voltage @ 75°F.
1983	Orange①②	40	15	4	12	—	—	—	—	—	—
	Green①②	60	15	4	12	—	—	—	—	—	—
	Black①②	65	15	4	12	—	—	—	—	—	—

①—Color of identification tag. ②—Rear terminal alternator.

STARTING MOTOR SPECIFICATIONS

Year	Starter Type	Ampere Draw Normal Load	Engine Cranking Speed R.P.M.	Minimum Stall Torque @ 5 Volts Ft. Lbs.	Maximum Head Amperes	No Load Ampere @ 12 Volts	Brushes		
							Length Inch	Wear Limit Inch	Spring Tension Ounces
1983	4" Diameter	150–200	180–250	—	—	70	.50	.25	40
	4½" Diameter	150–180	150–290	—	—	80	.50	.25	80

VALVE SPECIFICATIONS

Year	Engine	Valve Lash		Valve Angles		Valve Spring Installed Height	Valve Spring Pressure Lbs. @ In.	Stem Clearance		Stem Diameter, Standard	
		Int.	Exh.	Seat	Face			Intake	Exhaust	Intake	Exhaust
1983	4-122	.035–.055①		45	44	1.49–1.55	②	.0010–.0027	.0015–.0032	.3416–.3423	.3411–.3418
	4-140	.035–.055①		45	44	1.53–1.59	③	.0010–.0027	.0015–.0032	.3416–.3423	.3411–.3418

①—Measured at camshaft with hydraulic lifter collapsed. ②—Intake, 75 @ 1.52; Exhaust 149 @ 1.12. ③—Intake, 75 @ 1.56; Exhaust 167 @ 1.16.

PISTONS, PINS, RINGS, CRANKSHAFT & BEARINGS

Year	Engine	Piston Clearance	Ring End Gap①		Wristpin Diameter	Rod Bearings		Main Bearings			
			Comp.	Oil		Shaft Diameter	Bearing Clearance	Shaft Diameter	Bearing Clearance	Thrust on Bear. No.	Shaft End Play
1983	4-122	.0014–.0022	.010	.015	.9121	2.0462–2.0472	.0008–.0015	2.3982–2.3990	.0008–.0015	3	.004–.008
	4-140	.0014–.0022	.010	.015	.9121	2.0462–2.0472	.0008–.0015	2.3982–2.3990	.0008–.0015	3	.004–.008

①—Fit rings in tapered bores for clearance listed in tightest portion of ring travel.

REAR AXLE SPECIFICATIONS

Year	Ring Gear Diameter	Carrier Type	Ring Gear & Pinion Backlash Inch	Nominal Pinion Locating Shim, Inch	Pinion Bearing Preload				Differential Bearing Preload	Pinion Nut Torque Ft. Lbs.
					New Bearings With Seal Inch-Lbs.	Used Bearings With Seal Inch-Lbs.	New Bearings Less Seal Inch-Lbs.	Used Bearings Less Seal Inch-Lbs.		
1983	6¾"	Integral	.008–.015	—	—	8–14	16–29	—	.006	140
	7½"	Integral	.008–.015	—	—	8–14	16–29	—	.006	170

4-122 & 4-140 Engine Section

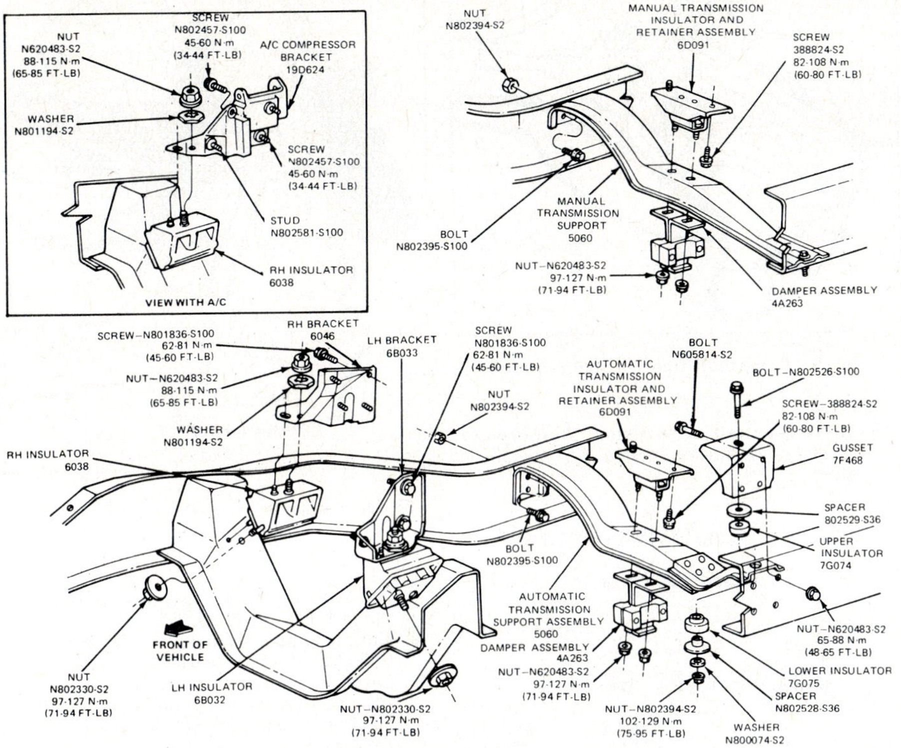

Fig. 1 Front & rear engine mounts. Typical

ENGINE MOUNTS, REPLACE

Front

1. Remove nuts from top of insulator brackets.

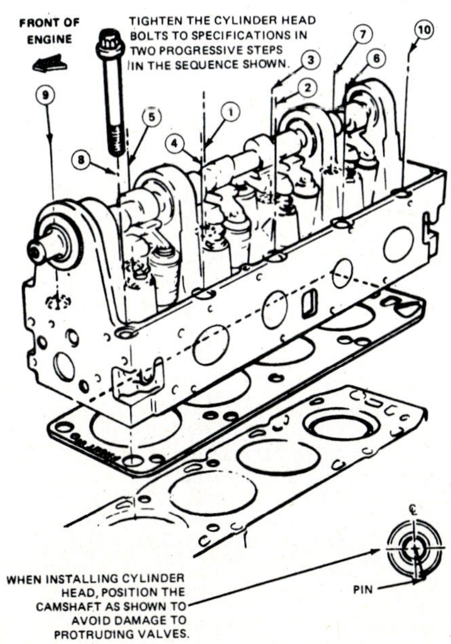

Fig. 2 Cylinder head installation

2. Raise and support vehicle, then lift engine until it clears mounts.
3. Remove attaching screws and mounts.
4. Reverse procedure to install. Refer to Fig. 1 for torque specifications.

Rear

1. Raise and support vehicle.
2. Remove the two nuts attaching rear insulator to engine support, Fig. 1.
3. Using a suitable jack and wood block placed between engine and jack, raise transmission.
4. Remove bolts attaching mount to rear of engine and remove mount.
5. Reverse procedure to install. Torque bolts to 60–80 ft. lbs. and nuts to 71–94 ft. lbs.

ENGINE, REPLACE

1. Raise hood, then mark location of hood hinges and remove hood.
2. Disconnect battery ground cable at engine and battery positive cable at battery.
3. Drain coolant from radiator, then remove air cleaner and duct assembly.
4. Remove upper and lower radiator hoses, engine fan, shroud and radiator.
5. Remove oil filler cap.
6. If equipped with air conditioning, remove compressor from mounting bracket and position it aside.
7. Disconnect heater hoses at engine, and fuel line at fuel pump.
8. Disconnect wires from starter, alternator, ignition coil, water temperature sending unit, and oil pressure sending unit.
9. If equipped with automatic transmission, disconnect transmission kickdown rod.
10. Disconnect heater hoses at engine, and fuel line at fuel pump.
11. Disconnect power brake vacuum hose.
12. Remove engine mount nuts.
13. Raise vehicle, then drain oil from crankcase and remove starter motor.
14. Disconnect exhaust pipe from manifold.
15. Remove dust cover on manual transmission vehicles, or inspection cover on automatic transmission vehicles.
16. On vehicles with manual transmission, remove flywheel housing cover lower bolts. On vehicles with automatic transmission, remove converter-to-flywheel bolts, then remove converter housing lower bolts.
17. On manual transmission vehicles, remove slave cylinder.
18. Lower vehicle, then support transmission using a floor jack.
19. Remove flywheel housing or converter housing bolts.
20. Using a suitable hoist, remove engine from vehicle.
21. Reverse procedure to install.

CYLINDER HEAD, REPLACE

1. Disconnect battery ground cable.
2. Drain cooling system, then remove air cleaner assembly.
3. Remove heater hose to rocker arm cover retaining screw.
4. Remove distributor cap and spark plug wires, then the spark plugs.
5. Disconnect vacuum hoses as necessary.
6. Remove dipstick and rocker arm cover.
7. Remove intake manifold retaining bolts.
8. Remove alternator belt and alternator bracket-to-cylinder head bolts.
9. Remove upper radiator hose.
10. Remove timing belt cover. On vehicles with power steering, remove power steering pump bracket bolts.
11. Loosen cam idler retaining bolts, then position idler in the unloaded position

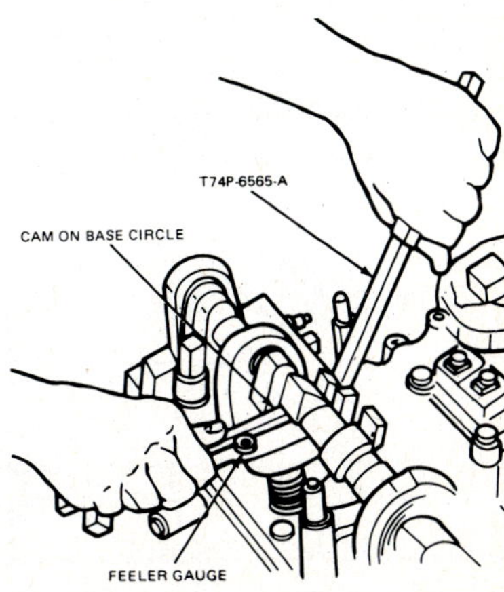

Fig. 3 Collapsing lash adjuster

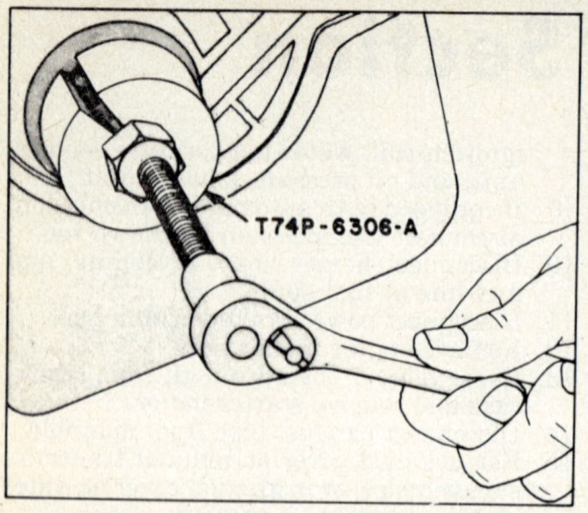

Fig. 4 Crankshaft sprocket removal

and tighten bolts.
12. Remove timing belt.
13. Remove heated air intake pipe and exhaust manifold bolts.
14. Remove timing belt idler, the two bracket bolts and idler spring from cylinder head.
15. Disconnect oil pressure sending unit lead wire.
16. Remove cylinder head bolts and cylinder head.
17. Thoroughly clean all gasket surfaces, then blow out oil out of the cylinder head bolt block holes.

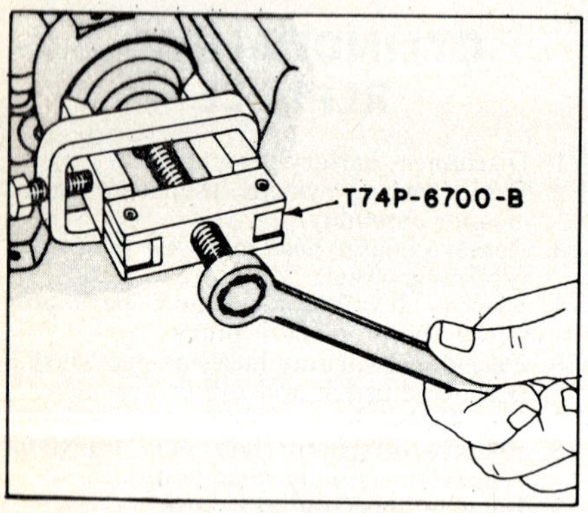

Fig. 7 Camshaft or auxiliary shaft seal removal

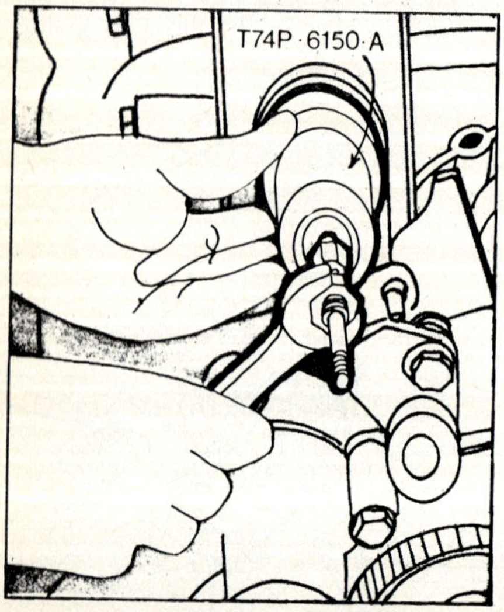

Fig. 8 Camshaft seal installation

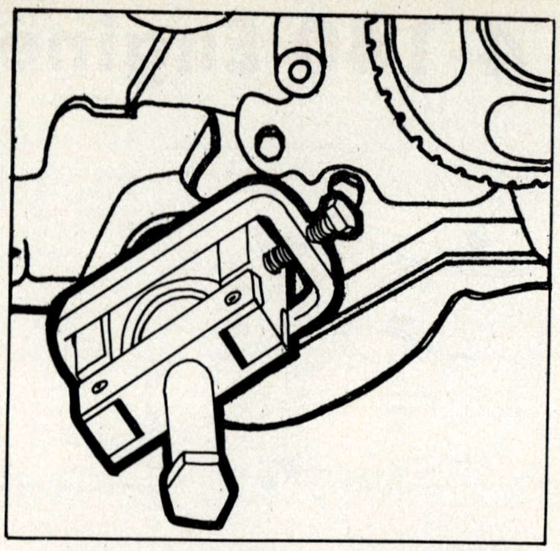

Fig. 5 Crankshaft front oil seal removal

18. Reverse procedure to install. Torque bolts in sequence shown in Fig. 2 to specifications.

VALVE ARRANGEMENT

Front to Rear

All Engines .... E-I-E-I-E-I-E-I

VALVE LIFT SPECS.

Engine	Year	Intake	Exhaust
4-122	1983	.390	.390
4-140	1983	.390	.390

VALVE GUIDES

Valve guides consist of holes bored in the cylinder head. For service the guides can be reamed oversize to accommodate valves with oversize stems of .015 inch and .030 inch.

VALVES, ADJUST

The valve lash on this engine cannot be adjusted due to the use of hydraulic valve lash adjusters. However, the valve train can be checked for wear as follows:

1. Crank engine to position camshaft with flat section of lobe facing rocker arm of valve being checked.
2. Collapse lash adjuster with tool T74P-6565A and insert correct size feeler gauge between rocker arm and camshaft lobe, Fig. 3. If clearance is not as listed in the "Valve Specifications" chart in front of this chapter, remove rocker arm and check for wear and replace as necessary. If rocker arm is found satisfactory, check valve spring assembled height and adjust as needed. If valve spring assembled height is within specifications listed in the front of this chapter, remove lash adjuster and clean or replace as necessary.

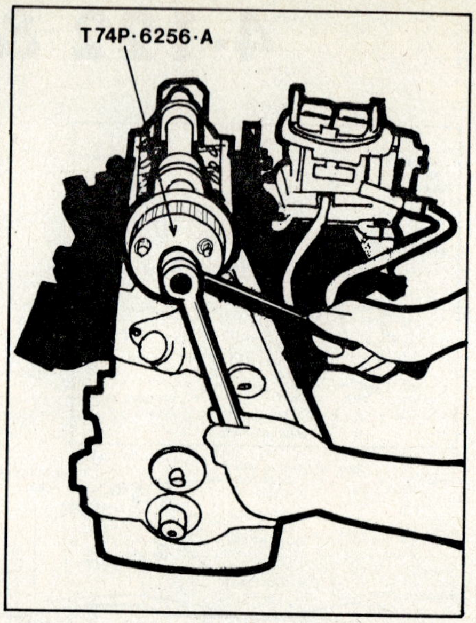

Fig. 6 Camshaft or auxiliary shaft sprocket removal

ROCKER ARM SERVICE

1. Remove rocker arm cover.
2. Rotate camshaft until flat section of lobe faces rocker being removed.
3. Collapse lash adjuster and, if necessary, valve spring and slide rocker arm over lash adjuster, Fig. 3.
4. Reverse procedure to install.

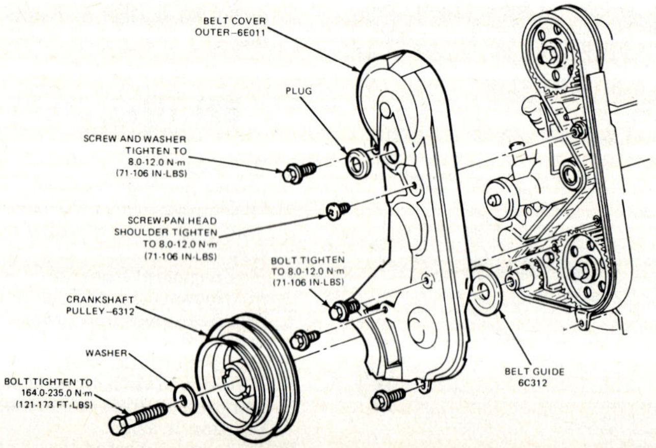

Fig. 9 Timing belt outer cover. Exploded view

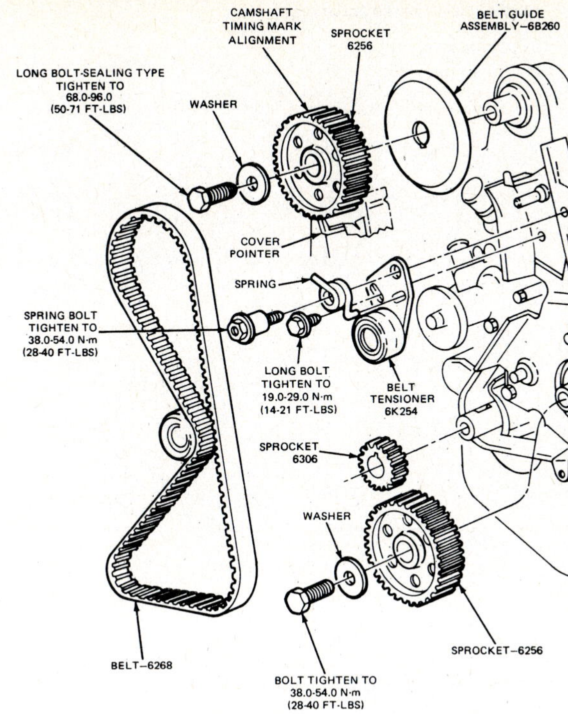

Fig. 10 Timing belt drive train

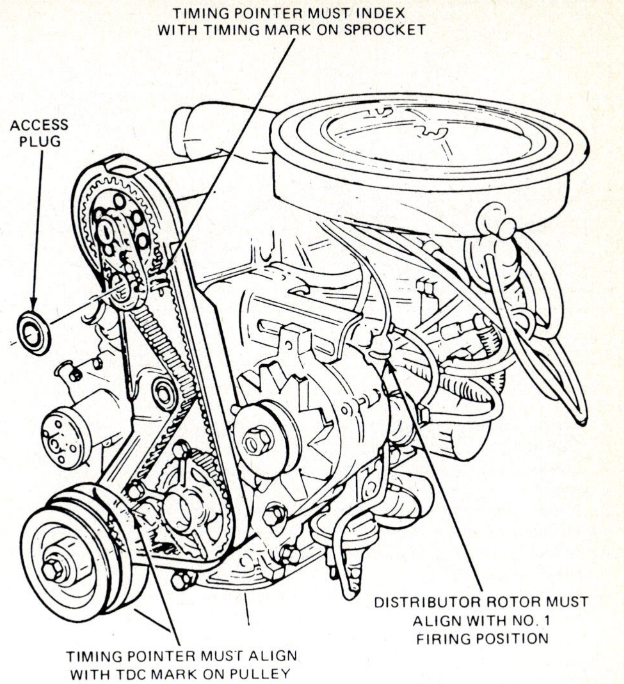

Fig. 12 Timing belt installation

LASH ADJUSTERS, REPLACE

The hydraulic valve lash adjusters can be replaced after rocker arm removal.

FRONT ENGINE SEALS, REPLACE

Crankshaft Oil Seal

1. Remove crankshaft sprocket with tool T74P-6306-A, Fig. 4.
2. Remove crankshaft oil seal with tool A7082-1A, Fig. 5.
3. Install a new crankshaft oil seal with tool T74P-6150-A.
4. Install crankshaft sprocket with tool T74P-6306-A.

Camshaft & Auxiliary Shaft Oil Seals

1. Remove camshaft or auxiliary shaft sprocket with tool T74P-6256-A, Fig. 6.
2. Remove camshaft or auxiliary shaft seal with tool T74P-6700-B, Fig. 7.
3. Install a new oil seal with tool T74P-6150-A, Fig. 8.
4. Install camshaft or auxiliary shaft sprocket with tool T74P-6256-A with center arbor removed.

TIMING BELT

Checking Belt Timing

1. Remove timing belt cover access plug, Fig. 9
2. Rotate crankshaft in engine normal direction of rotation until TDC is reached. This is obtained by aligning the crankshaft pulley mark with the TC mark on the belt cover.

NOTE: Never rotate engine against normal rotation, as this may cause the timing belt to jump time.

---

3. While looking through the belt cover access hole, ensure the timing mark on the camshaft sprocket is aligned with the pointer on the inner belt cover.
4. Remove distributor cap to ensure the distributor rotor is pointed towards the distributor cap No. 1 firing position.
5. If belt timing is satisfactory, install distributor cap and belt cover access plug. If belt timing is unsatisfactory, refer to "Adjustment or Replacement" procedure.

Adjustment or Replacement

1. Remove timing bolt outer cover, Fig. 9.
2. Loosen belt tensioner adjustment bolt, Fig. 10.
3. Using tool, T74P-6254-A positioned on tension spring rollpin, retract tensioner, then tighten adjustment screw to hold tensioner in retracted position, Fig. 11.
4. Remove crankshaft pulley and belt guide, Fig. 9.
5. Remove timing belt and inspect for signs of wear and damage. Replace as necessary.
6. Refer to Fig. 12, for proper camshaft and crankshaft sprocket position.
7. Remove distributor cap, then set distributor rotor to No. 1 firing position by turning the auxiliary shaft as necessary.
8. Install timing belt on crankshaft sprocket, then working counterclockwise, position belt on auxiliary sprocket and camshaft sprocket. Ensure timing marks do not change position.
9. Align belt on sprockets, then loosen tensioner adjustment bolt to allow tensioner to move against the belt.
10. Remove spark plugs.

NOTE: Failure to remove spark plugs, may result in timing belt jumping time during next step.

---

11. Rotate engine in normal direction of rotation for two complete turns to remove slack from belt. Tighten tensioner adjustment and pivot bolts to specifications given in Fig. 10.
12. Recheck timing mark alignment, then install crankshaft pulley and belt guide.
13. Install timing belt cover, refer to Fig. 9 for torque specifications. Install spark plugs.
14. Start engine and set ignition timing to specifications.

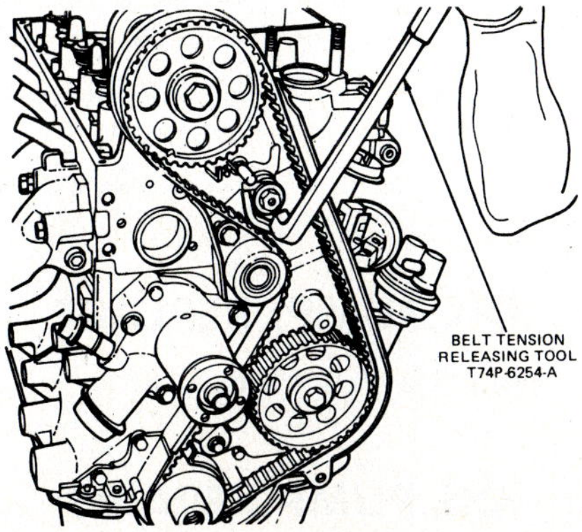

Fig. 11 Retracting belt tensioner

CAMSHAFT, REPLACE

1. Disconnect battery ground cable, then

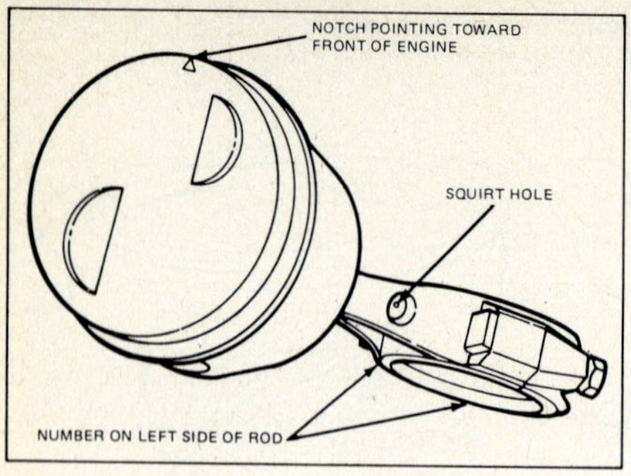

Fig. 13 Piston & rod

drain cooling system and remove air cleaner.
2. Disconnect ignition wires and position aside.
3. Disconnect any vacuum hoses interfering with camshaft removal.
4. Remove rocker arm cover, then remove alternator belt.
5. Remove alternator mounting bracket retaining bolts and position aside.
6. Remove upper radiator hose, then remove fan shroud retaining bolts and fan shroud.
7. Remove cam belt cover retaining bolts and cover.
8. If vehicle is equipped with power steering, remove power steering pump bracket.
9. Loosen idler cam retaining bolts, then position idler in the unloaded position and tighten retaining bolts.
10. Remove timing belt from cam and auxiliary pulleys.
11. Remove rocker arms as described under "Rocker Arm Service", then remove gear and seal as described under "Front Engine Seals, Replace".
12. Remove camshaft rear retainer.
13. Raise and support vehicle.
14. Remove right and left engine support bolts and nuts, then position a suitable transmission jack under engine.
15. Place a block of wood on transmission jack, raise engine as high as possible and position blocks of wood between engine mounts and chassis bracket, then remove jack.
16. Carefully remove camshaft from front of head.
17. Reverse procedure to install.

PISTON & ROD, ASSEMBLE

Assemble rod to piston with arrow on top of piston facing front of engine, Fig. 13.

CRANKSHAFT OIL SEAL

1. Remove oil pump retaining bolts if necessary.
2. Insert small sheet metal screws into seal and pull screws to remove seal.
3. Install new seal with tool T82L-6701-A, Fig. 14.

OIL PAN, REPLACE

Refer to Fig. 15 to replace oil pan.

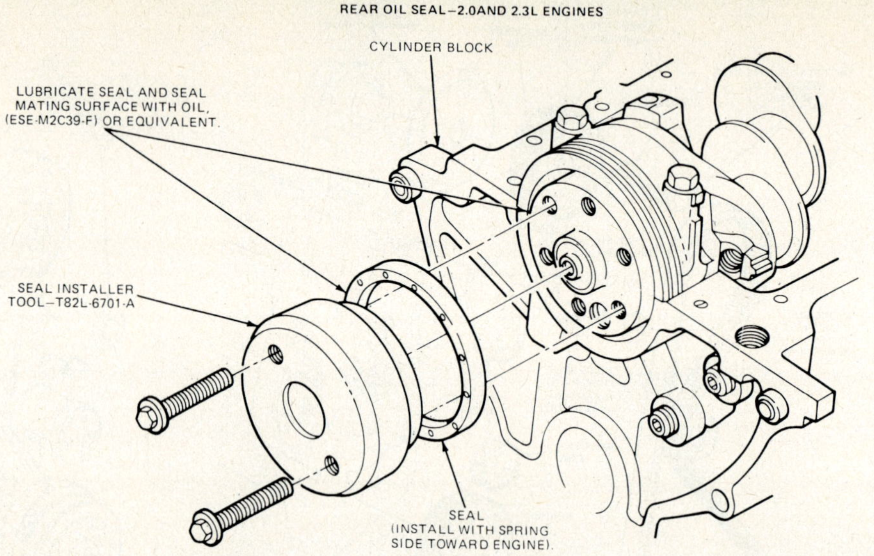

NOTE: REAR FACE OF SEAL MUST BE WITHIN 0.127mm (0.005-INCH) OF THE REAR FACE OF THE BLOCK.

Fig. 14 Crankshaft rear oil seal replacement

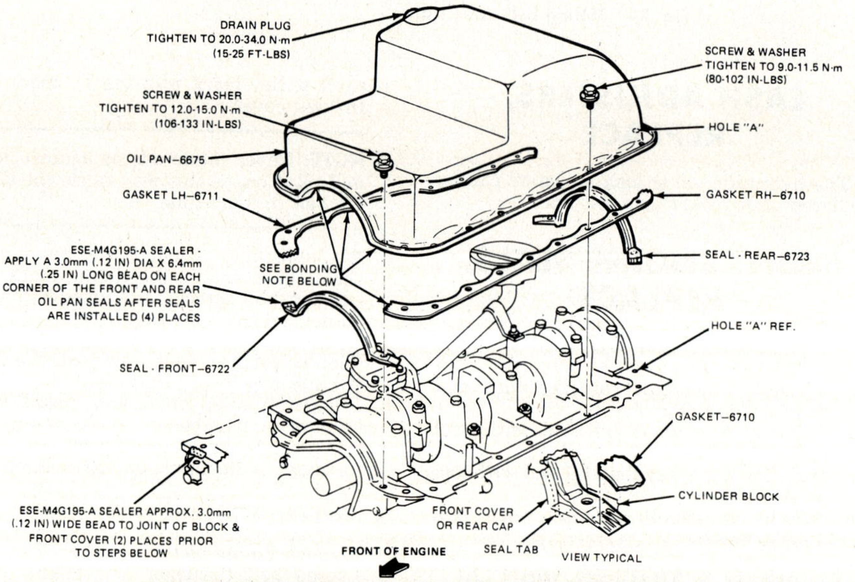

THERMAL BONDING INSTRUCTIONS - OIL PAN GASKETS TO BE BONDED SECURELY TO OIL PAN USING A THERMAL PROCESS MEETING THE REQUIREMENTS OF THE (ES-DOAE-6584-A OR EQUIVALENT) ADHESIVE COATING SPECIFICATION - IF NECESSARY IN PLACE OF THERMAL BONDING USE ADHESIVE (ESE-M2G52-A OR B OR EQUIVALENT) APPLY EVENLY TO OIL PAN FLANGE & TO PAN SIDE OF GASKETS - ALLOW ADHESIVE TO DRY PAST "WET" STAGE THEN INSTALL GASKETS TO OIL PAN.

1. APPLY SEALER AS NOTED ABOVE
2. INSTALL SEALS TO FRONT COVER & REAR BEARING CAP - PRESS SEAL TABS FIRMLY INTO BLOCK
3. INSTALL (2) GUIDE PINS
4. INSTALL OIL PAN OVER GUIDE PINS & SECURE WITH (4) BOLTS
5. INSTALL (18) BOLTS
6. TORQUE ALL BOLTS IN SEQUENCE CLOCKWISE FROM HOLE "A" AS NOTED ABOVE

Fig. 15 Oil pan replacement

OIL PUMP, REPLACE

Refer to Fig. 16 to replace oil pump.

OIL PUMP REPAIRS

1. Inspect inside of pump housing, outer race and rotor for excessive wear or scoring.
2. Inspect mating surface of pump cover for excessive wear, or scoring. Either of these conditions will necessitate replacement of pump.
3. Measure inner rotor tip clearance, Fig. 17.
4. Measure rotor end play, Fig. 18. Maximum end play is .004 inch.

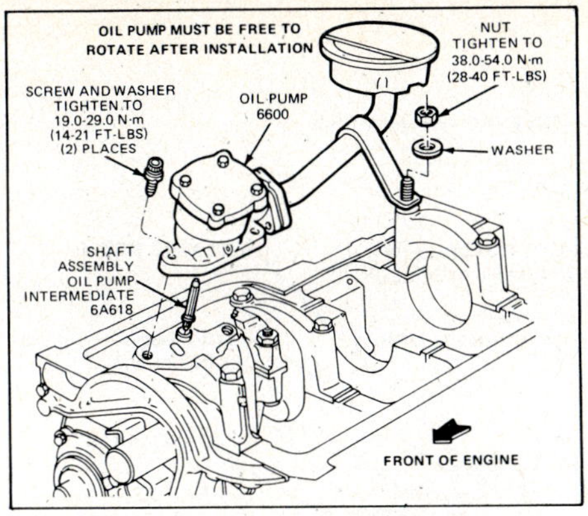

Fig. 16 Oil pump replacement

5. Check driveshaft to housing bearing clearance. This measurement should be .0015–.0030 inch.

NOTE: Individual components of oil pump are not serviced. If any part of pump requires replacement, the entire pump assembly must be replaced.

6. Check relief valve spring tension. This should be 15.2–17.2 Lbs. @ 1.20 inches.
7. Inspect relief valve piston for scores and free operation in bore.

BELT TENSION DATA

	New Ft. Lbs.	Used Ft. Lbs.
1983 Exc. 1/4 inch	160	145
1/4 inch	65	50

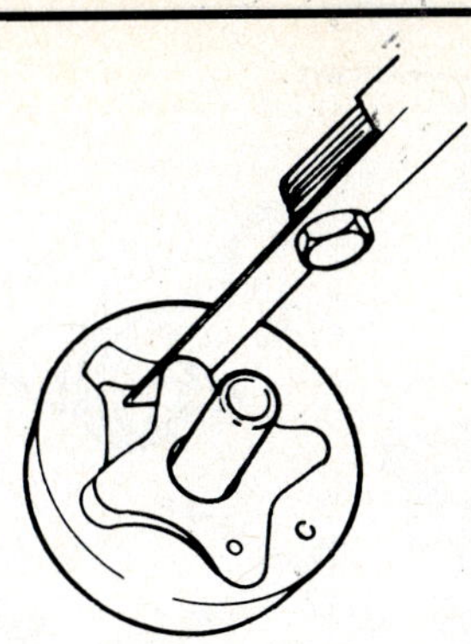

NOTE: INNER TO OUTER ROTOR TIP CLEARANCE MUST NOT EXCEED 0.25mm (.012 IN) WITH FEELER GAUGE INSERTED 13mm (1/2") MINIMUM AND ROTORS REMOVED FROM PUMP HOUSING.

Fig. 17 Checking oil pump inner rotor tip clearance

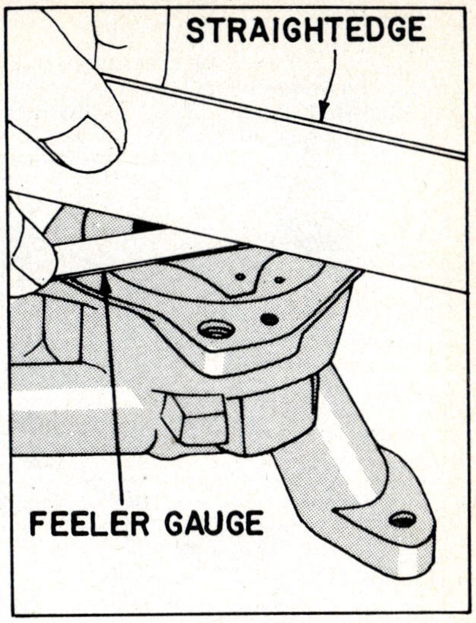

Fig. 18 Checking oil pump rotor end play

WATER PUMP, REPLACE

1. Remove timing belt outer cover.
2. Drain cooling system, then remove the two fan shroud retaining bolts and position shroud over fan.
3. Remove fan retaining bolts then remove fan and shroud assembly.
4. If equipped with air conditioning, loosen compressor adjusting idler pulley and remove drive belt.
5. If equipped with power steering, loosen power steering pump bolts, then remove alternator and power steering bolts.
6. Remove water pump pulley and vent tube to canister.
7. Disconnect heater hose at water pump, then remove cam belt cover.
8. Remove lower radiator hose from pump, then remove water pump attaching bolts and water pump.
9. Reverse procedure to install.

Carburetor Section

1983 CARTER YF, YFA ADJUSTMENT SPECIFICATIONS

See Tune Up Chart for hot idle speed

Year	Carb. Model	Float Level	Fast Idle Cam Setting	Dechoke or Unloader Setting	Pulldown Setting	Choke Setting
1983	E27E-BB	.65	.140	.220	.320	—
	E27E-CB	.65	.140	.220	.320	—
	E27E-CC	.65	.140	.220	.320	—
	E27E-EB	.65	.140	.220	.320	—
	E27E-FB	.65	.140	.220	.320	—
	E27E-GB	.65	.140	.220	.320	—
	E27E-HB	.65	.140	.220	.320	—

YFA 1-V & YFA 1-V FEEDBACK CARBURETOR ADJUSTMENTS

The YFA 1-V carburetor is used on all 4-122 and 4-140 engines except Calif. and high altitude models. The YFA 1-V feedback carburetor, Fig. 1, is used on 4-140 Calif. and high altitude models.

Float Adjustment

Fig. 2—Invert the air horn assembly, and check the clearance from the top of the float to the bottom of the air horn with the float level gauge. Hold the air horn at eye level when gauging the float level. The float arm (lever) should be resting on the needle pin. Do not load the needle when adjusting the float. Bend the float arm as necessary to adjust the float level (clearance). Do not bend the tab at the end of the float arm. It prevents the float from striking the bottom of the fuel bowl when empty.

Float Drop Adjustment

Fig. 3—Hold air horn upright and measure maximum clearance from top of float to bottom of air horn with float drop gauge. Bend tab at end of float arm to obtain specified setting listed under YFA 1-V Adjustment Specifications.

Metering Rod Adjustment

Fig. 4—Back out the idle speed adjusting screw until the throttle plate is closed tight in the throttle bore. Press down on upper end of diaphragm shaft until diaphragm bottoms in vacuum chamber. Metering rod should contact bottom of metering rod well, and metering rod should contact lifter link at the outer end nearest the springs and at supporting lug. Turn the adjusting screw until metering rod just bottoms in the body casting. For final adjustments turn metering rod adjusting screw in (clockwise) one additional turn.

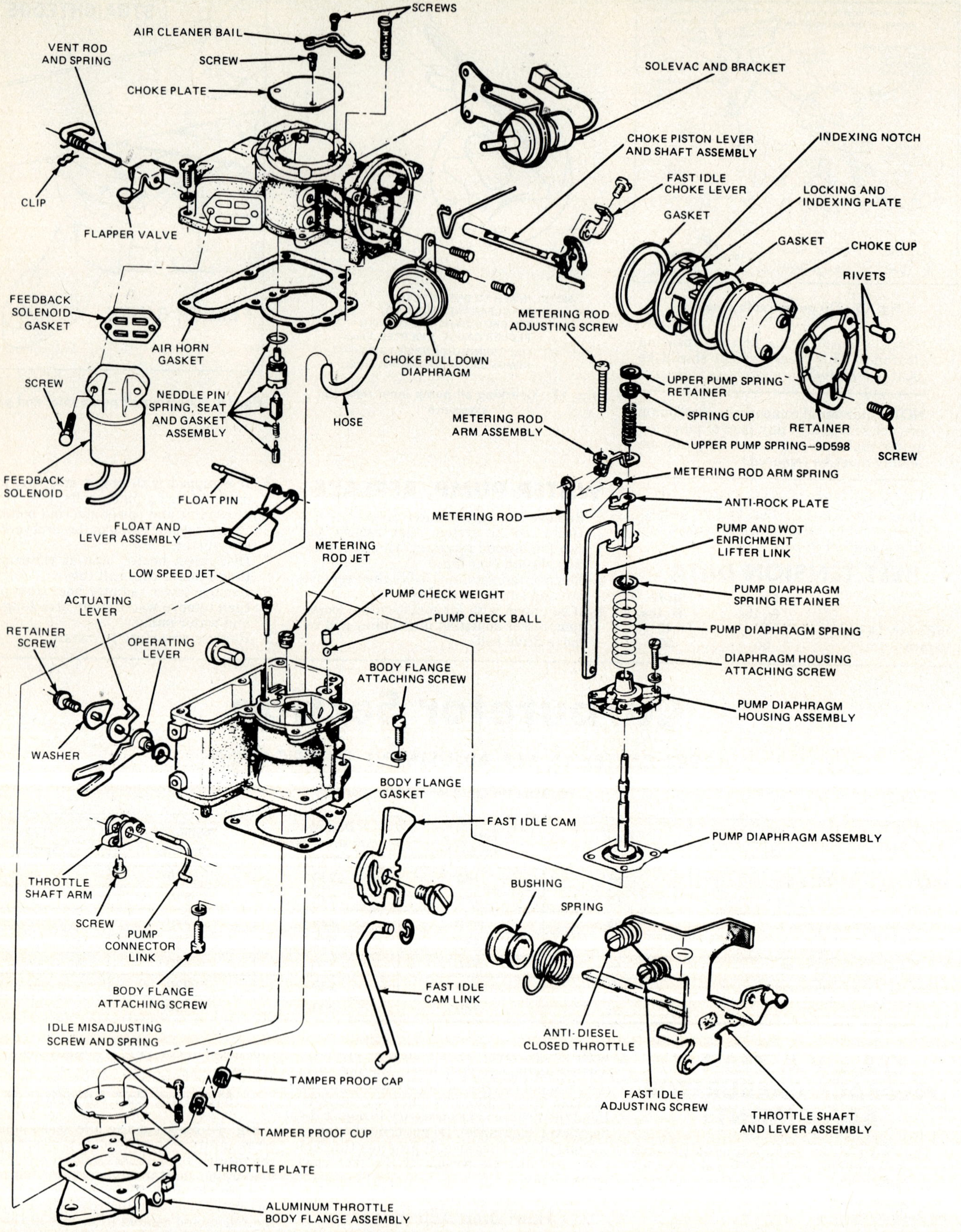

Fig. 1 Model YFA 1-V feedback carburetor (Non-feedback carburetor similar)

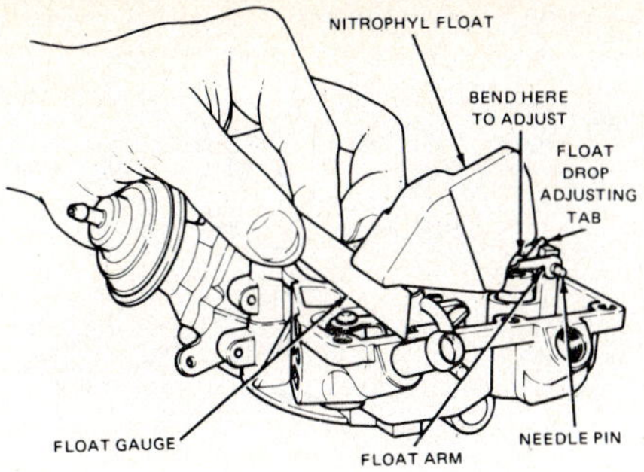

Fig. 2 YFA & YFA feedback float level adjustment

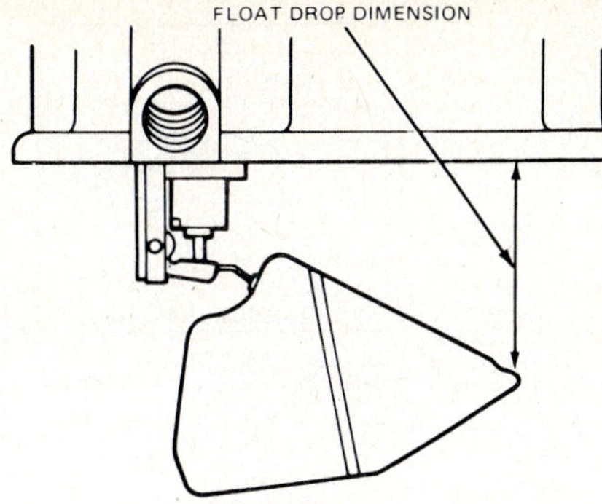

Fig. 3 YFA & YFA feedback float drop adjustment

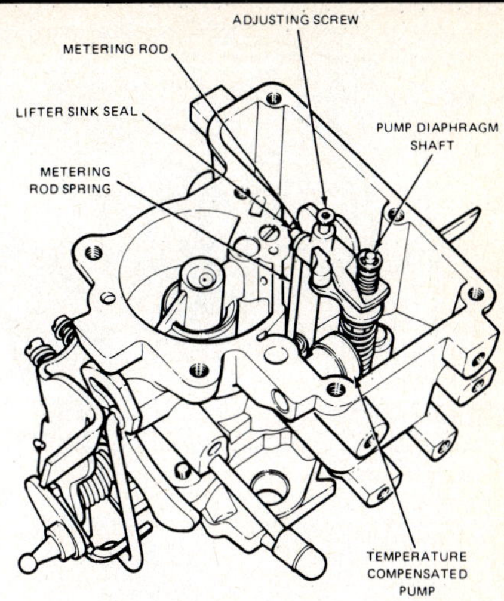

Fig. 4 YFA & YFA feedback metering rod adjustment

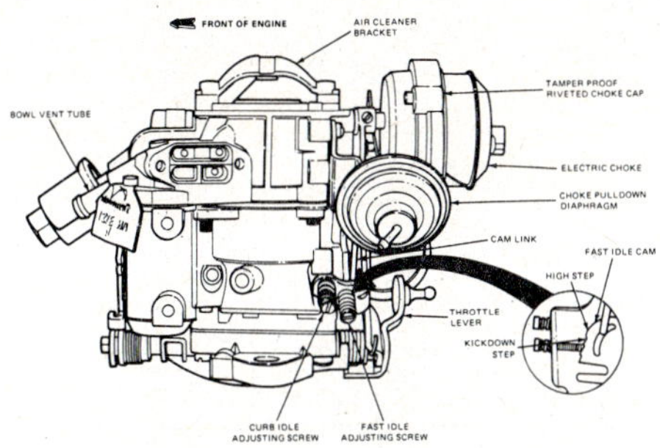

Fig. 5 YFA fast idle cam index adjustment

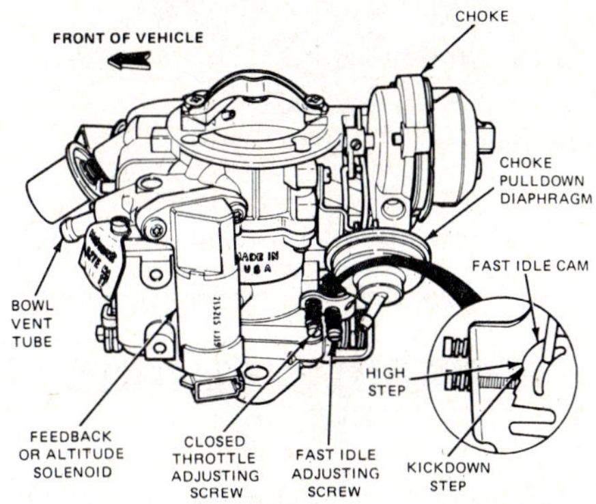

Fig. 6 YFA feedback fast idle cam index adjustment

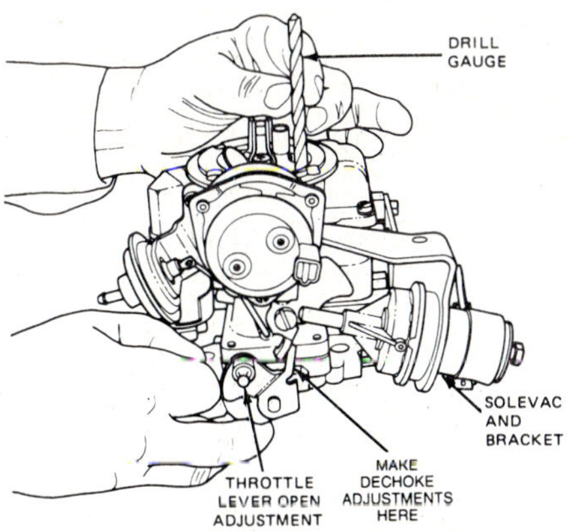

Fig. 7 YFA & YFA feedback dechoke adjustment

Fast Idle Cam Index Adjustment

Position fast idle screw on kickdown step of fast idle cam, against shoulder of high step, Figs. 5 and 6. Adjustment is made by bending fast idle cam link to specification listed in YFA 1-V Specification Chart.

Dechoke Adjustment

Hold the throttle plate fully open and close the choke plate as far as possible without forcing it. Use a drill of specified diameter to check the clearance between choke plate and air horn, Fig. 7. If clearance is not within specification, adjust by bending arm on choke trip lever of the throttle lever. Bending the arm downward will decrease the clearance, bending it upward will increase the clearance.

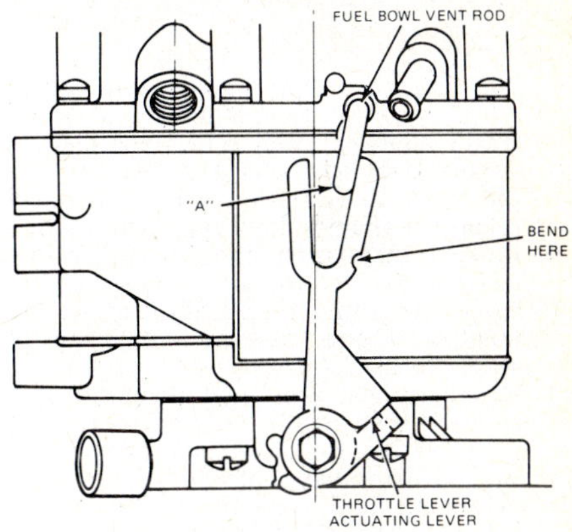

Fig. 8 YFA & YFA feedback fuel bowl vent adjustment

Fuel Bowl Vent Adjustment

With engine at operating temperature and idle RPM set to specification, open throttle lever so that there is clearance between throttle lever actuating lever and fuel bowl vent rod, Fig. 8. Close throttle lever to idle set position and measure travel of fuel bowl vent rod at point A. Travel of vent rod at point A should be .100–.150 inch. Bend throttle actuating lever at point indicated in Fig. 8 to adjust as necessary.

Clutch Section

CLUTCH, REPLACE, FIG.1

1. Raise and support vehicle, then remove dust cover from clutch housing and disconnect hydraulic clutch linkage from housing and release lever. Remove starter.
2. Remove clutch housing to engine block bolts, noting direction of bolt installation, then mark alignment of driveshaft and companion flange and remove driveshaft.
3. Remove nuts which attach transmission and insulator to #2 crossmember support, Fig. 2, then raise transmission with suitable jack and remove #2 crossmember support. Lower transmission and clutch housing.
4. Remove release lever, and hub and bearing, then mark assembled position of pressure plate and cover to flywheel.
5. Loosen pressure plate and cover attaching bolts evenly until pressure plate springs are expanded, then remove bolts.
6. Remove pressure plate and cover assembly and clutch disc from flywheel.
7. Position clutch disc on flywheel so that tool D79T-7550-A or equivalent can enter clutch pilot bearing and align disc.
8. Align pressure plate and cover assembly to flywheel and position pressure plate and cover assembly on flywheel, aligning pressure plate and disc, then install assembly to flywheel retaining bolts. Torque bolts to 15–24 ft. lbs. and remove tool.
9. Position clutch release bearing and bear-

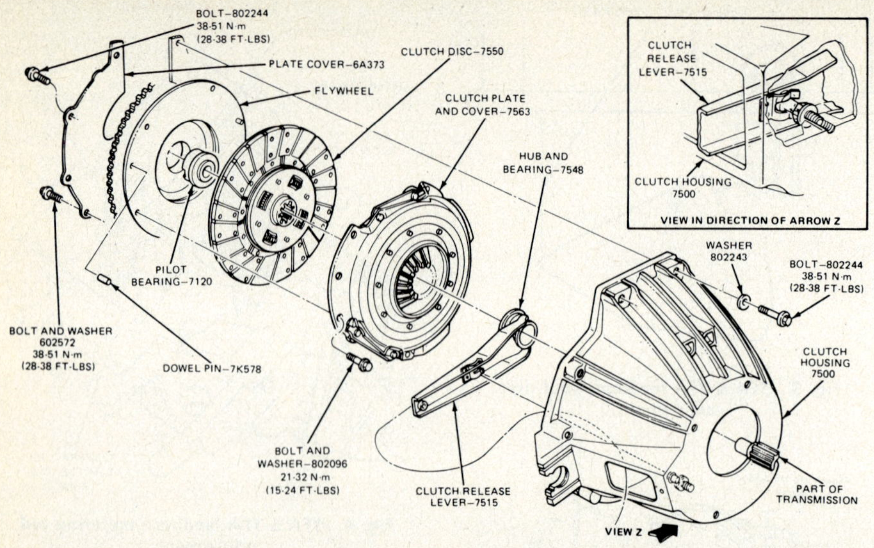

Fig. 1 Clutch Housing

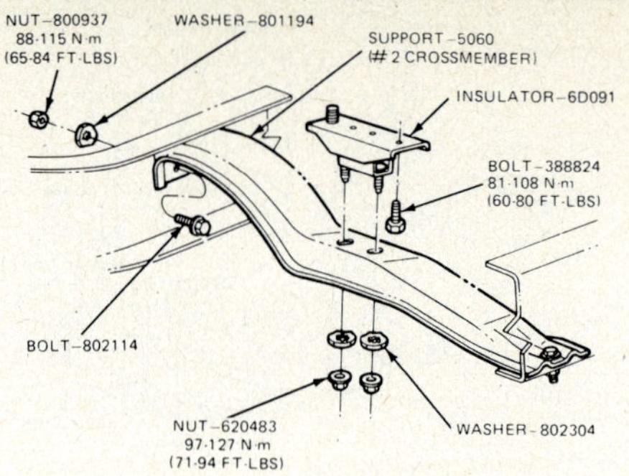

Fig. 2 Transmission Rear Support

ing hub on release lever and install release lever on release lever seat in flywheel housing, then coat release lever fingers and lever pivot ball with suitable lubricant and fill annular groove of release bearing hub with grease.

10. Raise transmission and clutch housing into position and install #2 crossmember support to frame, then install connecting nuts, bolts and washers and torque to 48–65 ft. lbs., Fig. 2.
11. Lower transmission and insulator into support, then install nuts and torque to 71–94 ft. lbs. Remove transmission jack.
12. Install driveshaft, aligning marks on driveshaft and companion flange, and torque bolts to 70–95 ft. lbs.
13. Install housing to engine block bolts in same direction as removed and torque to 28–38 ft. lbs.
14. Install hydraulic clutch linkage, on housing in position with release lever, dust shield, and starter, then lower vehicle and check clutch for proper operation.

HYDRAULIC CLUTCH, FIG. 3

The hydraulic clutch control system consists of a fluid reservoir, a master cylinder, a slave cylinder and connective tubing. The clutch master cylinder converts mechanical clutch pedal movement into hydraulic fluid movement. The slave cylinder converts hydraulic fluid movement to mechanical movement to activate the clutch release lever. The hydraulic clutch system locates the clutch pedal and provides automatic clutch adjustment. Adjustment of clutch pedal position or clutch linkage is not required.

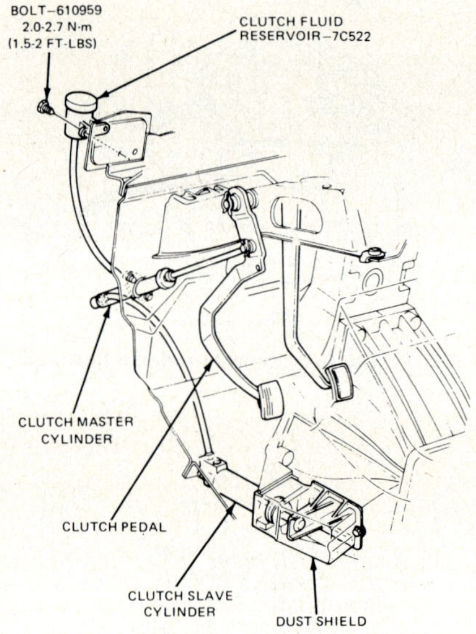

Fig. 3 Hydraulic Clutch System

4 Speed Manual Transmission Section

TRANSMISSION, REPLACE, FIG. 1

1. Disconnect battery ground cable.
2. Place gearshift lever in neutral position then remove boot retainer screws.
3. Remove retainer cover to gearshift lever retainer attaching bolts then disconnect clutch master cylinder push rod from clutch pedal.
4. Pull gearshift lever assembly, shim and bushing straight up away from the gearshift lever retainer.
5. To avoid dirt entering shift tower opening in extension housing, cover with cloth.
6. Disconnect clutch hydraulic system master cylinder push rod from clutch pedal.
7. Raise and support vehicle.
8. Disconnect driveshaft at rear axle drive flange then pull driveshaft rearward and disconnect from transmission. Install suitable plug in extension housing to prevent lubricant leakage.
9. Remove clutch housing dust shield and slave cylinder and secure to the side.
10. Disconnect starter motor and back-up lamp switch wires.
11. Place suitable jack under engine using a block of wood between jack and engine to protect the oil pan.
12. Remove speedometer cable from extension housing.
13. On four-wheel drive vehicles, remove the transfer case as described in the "Transfer Case Section".
14. Remove starter motor and position a transmission jack under transmission.
15. Remove bolts, lockwasher and flat washer attaching transmission to the engine rear plate.
16. Remove nuts and bolts attaching transmission mount to crossmember.
17. Remove nuts attaching crossmember to frame side rails and remove crossmember.
18. Lower the jack then work clutch housing from locating dowels and slide transmission rearward until input shaft spline clears clutch disc.
19. Remove transmission from vehicle.
20. Reverse procedure to install.

TRANS. DISASSEMBLY

1. Remove drain plug and drain lubricant from transmission. If necessary clean metal filings from drain plug and install.
2. Place transmission in neutral.
3. Remove four bolts attaching gearshift lever retainer to extension housing then remove gearshift lever retainer and gasket.
4. Remove bell housing to transmission case attaching nuts then remove bell housing and gasket.
5. Remove six bolts attaching extension housing to transmission case.
6. Raise control lever to the left and slide toward rear of transmission.
7. Slide extension housing off transmission being careful not to damage the oil seal.

NOTE: If necessary remove gearshift control lever end and control by removing attaching bolt. If required also remove back-up lamp switch from extension housing.

8. Remove anti-spill seal from output shaft and discard, seal is not necessary for reas-

sembly.

9. Remove snap ring from speedometer drive gear. Slide drive gear off mainshaft and remove lock ball.
10. Evenly loosen fourteen bolts securing transmission case cover to transmission case and remove cover and gasket.
11. Remove three spring cap bolts. Remove detent springs and balls from transmission using a magnet, Fig. 2.
12. Remove four blind covers-to-transmission case attaching bolts, and remove blind covers and gasket.
13. Slide reverse shift fork shaft assembly and idler gear out of transmission case.
14. Shift transmission into fourth gear. With a small drift, drive roll pin from third and fourth fork assembly. Slide third and fourth shift fork shaft out of transmission case.
15. Remove roll pin from first and second shift fork. Slide shift fork out of transmission case. Remove inter-lock pins, Fig. 2.
16. Reinstall reverse idler gear then install tool No. T77J-7025-E between fourth speed synchronizer ring and synchromesh gear on mainshaft, and shift transmission into second gear.
17. Fold back ends of lockwasher then remove locknut and washer, slide reverse idler gear off mainshaft, Fig. 3.
18. Remove key from mainshaft then remove reverse idler gear.
19. Remove snap ring from end of countershaft, and slide reverse gear off countershaft.
20. Remove four bearing retainer attaching bolts then remove bearing retainer together with reverse idler gear shaft.
21. Remove countershaft rear bearing using a suitable puller, Fig. 4.

NOTE: The retainer ring may need to be turned to position the split in the retainer ring midway between the recessed areas, before the puller is installed, Fig. 5. This will reduce the possibility of the retainer ring becoming distorted as the bearing is removed.

22. Turn forcing screw clockwise to remove bearing.

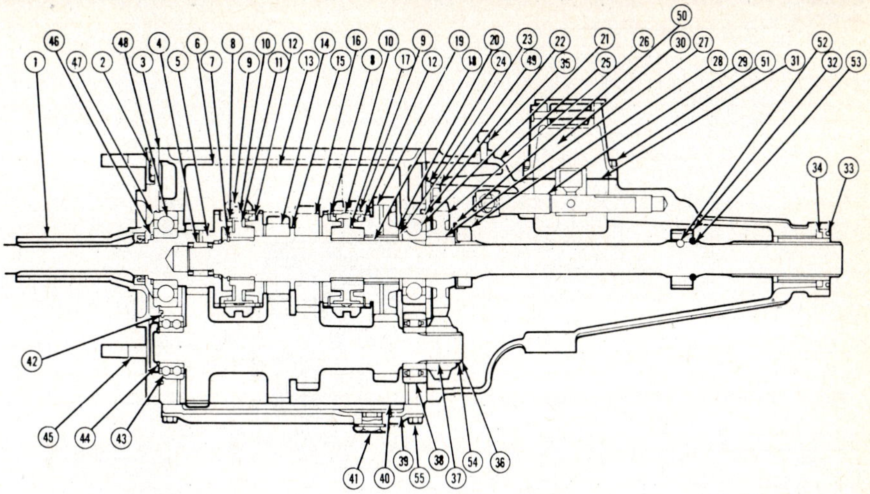

1	Clutch Adapter Plate	28	Lock Nut
2	Flange Bolt	29	Control Rod
3	Gasket	30	Control Case
4	Needle Bearing	33	Oil Seal
5	Main Drive Gear	34	Anti-Spill Seal
6	Transmission Case	35	Breather
7	Snap Ring	36	Counter Shaft Gear
8	Synchronizer Ring	37	Counter Reverse Gear
9	Clutch Hub Sleeve	38	Needle Bearing
10	Synchronizer Key	39	Under Cover
11	Clutch Hub 3rd-4th	40	Gasket
12	Synchronizer Spring	41	Magnet Plug
13	I.D. Tag	42	Adjust Shim
14	3rd-Gear	43	Ball Bearing
15	Main Shaft	44	Retaining Ring
16	2nd-Gear	45	Stud
17	Clutch Hub 1st-2nd	46	Retaining Ring
18	Gear Sleeve	47	Adjust Shim
19	1st Gear	48	Ball Bearing
20	Thrust Washer	49	Washer and Bolt
21	Ball Bearing	50	Lock Washer
22	Gasket	51	Bolt
23	Adjust Shim	52	Steel Ball
24	Bearing Stopper	53	Retaining Ring
25	Extension Housing	54	Retaining Ring
26	Reverse Gear	55	Bolt
27	Key		

Fig. 1 Four speed manual transmission, cutaway view

23. Remove mainshaft rear bearing using suitable puller, Fig. 6. Turn forcing screw clockwise to remove bearing.
24. Remove shim spacer from behind mainshaft rear bearing.
25. Remove four studs and bolts attaching front cover to case and remove front cover. Save shim on inside of the cover.
26. Remove snap ring from input shaft.
27. Remove mainshaft drive gear bearing using suitable puller, Fig. 6. Turn forcing screw clockwise to remove bearing.
28. Rotate both shift forks so that main gear train will fall to bottom of case. Remove shift forks. Rotate input shaft so that one of two flats on input shaft face upward, Fig. 7.

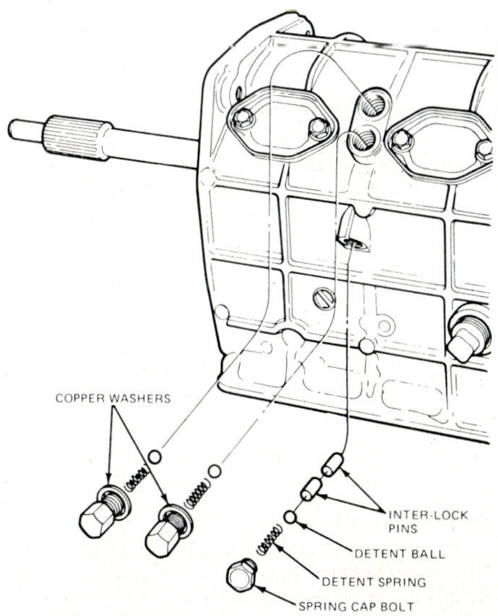

Fig. 2 Detent spring and ball removal

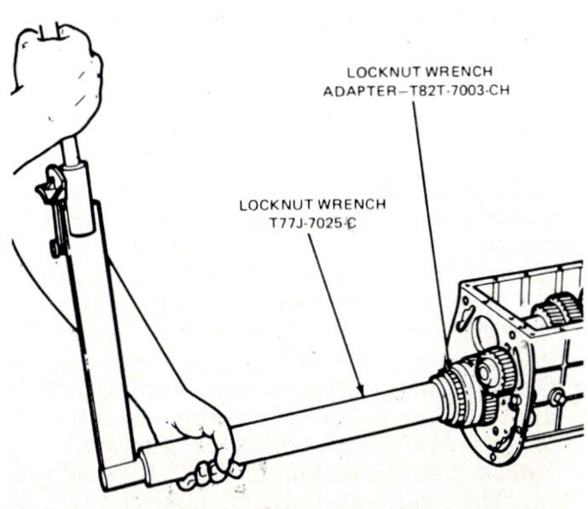

Fig. 3 Removing or installing locknut

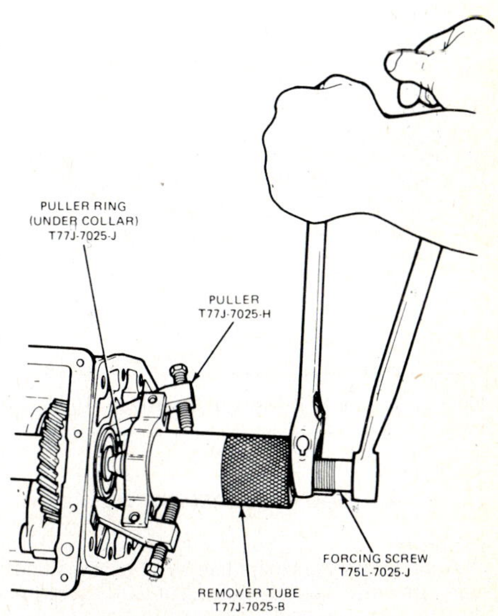

Fig. 4 Countershaft bearing removal

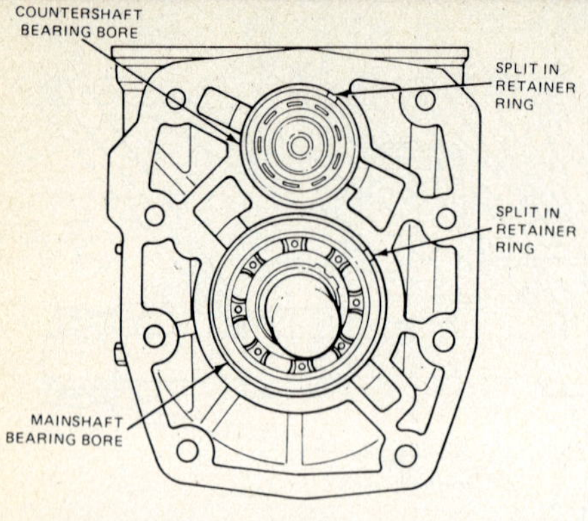

Fig. 5 Positioning of the retainer ring for removal

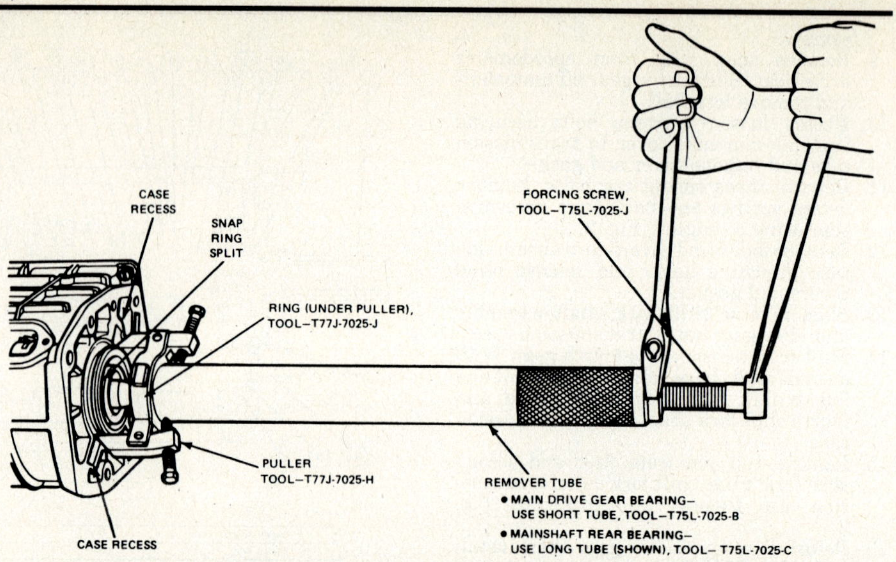

Fig. 6 Mainshaft rear bearing & mainshaft drive gear bearing removal

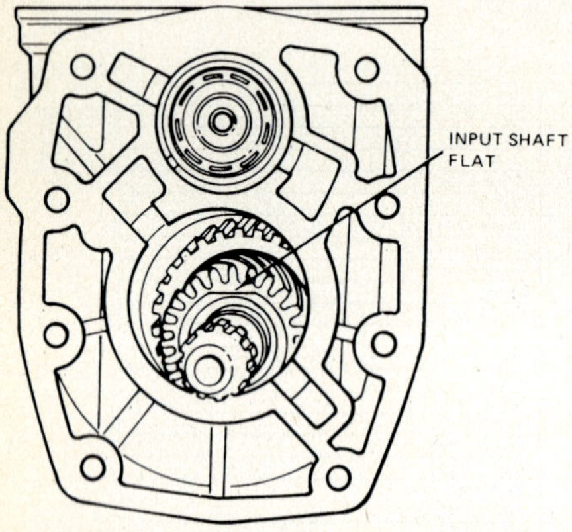

Fig. 7 Positioning input shaft flats

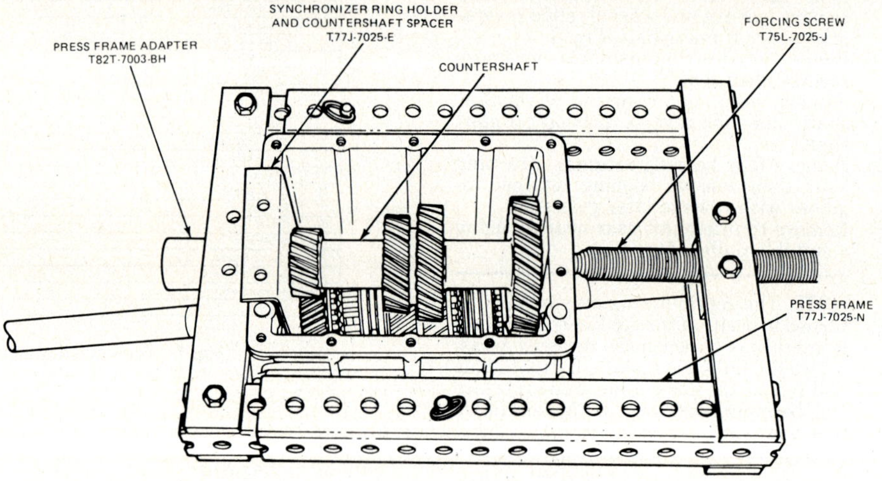

Fig. 8 Countershaft removal

29. Remove snap ring from front of countershaft.
30. Install tool No. T77J-7025-E between first gear on countershaft and rear of case, Fig. 8.
31. Install tools No. T75L-7025-J, T82T-7003-BH and T77J-7025-N, Fig. 8.
32. Turn forcing screw clockwise until countershaft contacts tool No. T77J-7025-E.
33. Remove countershaft front bearing using suitable puller. Turn forcing screw clockwise.
34. Remove shim from behind countershaft front bearing.
35. Remove input shaft from transmission case then remove synchronizer ring and caged bearing from main driveshaft.
36. Remove countershaft from transmission case.
37. Remove inner race of countershaft center bearing from countershaft, using suitable press.
38. Remove mainshaft and gear assembly from transmission case.
39. Remove snap rings from front of mainshaft. Slide clutch hubs, sleeve assembly synchronizer ring, gears and thrust washer from mainshaft, Fig. 9.

NOTE: Do not mix the synchronizer rings.

Cleaning & Inspection

Cleaning

Wash all parts, except the ball bearings and seal, in suitable solvent. Brush all foreign matter from the parts. Be careful not to damage any parts. Do not clean, wash or soak transmission seals in cleaning solvent. Dry all parts with compressed air.

Rotate the bearings in a cleaning solvent until all lubricant is removed. Hold the bearing assembly to prevent it from rotating and dry it with compressed air. Lubricate the bearings with approved transmission lubricant and wrap them in a clean, lint-free cloth or paper, until ready for use

Inspection

Inspect the transmission case and housing for cracks, worn or damaged bores, damaged threads, or any other damage that could affect the operation of the transmission. Inspect the machined mating surfaces for burrs, nicks or damage.

Inspect the front face of the case for small nicks or burrs that could cause misalignment of the transmission with the flywheel housing. Remove all small nicks or burrs with a fine stone. Inspect the bell housing for cracks. Make sure the machined mating surfaces are free from burrs, nicks or other damage.

Check the condition of the shift covers, forks, shift rails and shafts. Inspect the ball bearings for cracks or deformations and replace any roller bearings that are broken, worn or rough.

TRANS. ASSEMBLY

1. Assemble first and second synchromesh mechanism by installing clutch hub to sleeve, placing three synchronizer keys into clutch hub keys slots and installing key springs to clutch hub, Fig. 10.

NOTE: When installing key springs, the open end of the spring should be kept 120° apart, Fig. 11. This will keep the spring tension on each key uniform.

2. Assemble third and fourth synchromesh mechanisms in the same manner as step 1.
3. Place synchronizer ring on third gear and slide third gear to front of mainshaft with synchronizer ring toward front.
4. Slide third and fourth clutch hub and sleeve assembly to front of mainshaft, making sure that three synchronizer keys

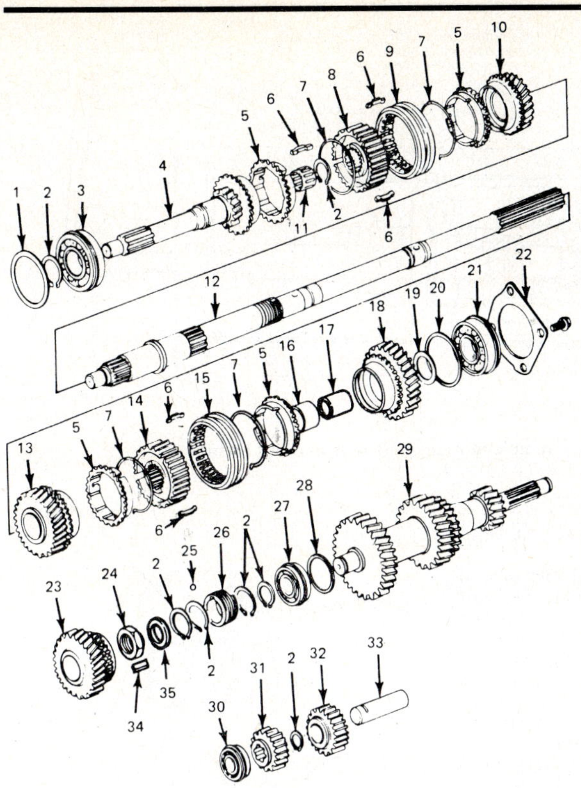

1. SHIM
2. SNAP RING
3. MAIN DRIVE SHAFT BEARING
4. MAIN DRIVE SHAFT GEAR
5. SYNCHRONIZER RING
6. SYNCHRONIZER KEY
7. SYNCHRONIZER KEY SPRING
8. 3RD-AND-4TH CLUTCH HUB
9. CLUTCH SLEEVE
10. 3RD GEAR
11. CAGED BEARING
12. MAIN SHAFT
13. 2ND GEAR
14. 1ST-AND-2ND CLUTCH HUB
15. CLUTCH SLEEVE
16. FIRST GEAR SLEEVE
17. BUSHING
18. 1ST GEAR
19. THRUST WASHER
20. SHIM
21. MAIN SHAFT FRONT BEARING
22. BEARING COVER
23. REVERSE GEAR
24. MAIN SHAFT LOCK NUT
25. LOCK BALL
26. SPEEDOMETER DRIVE GEAR
27. COUNTER SHAFT FRONT BEARING
28. SHIM
29. COUNTER SHAFT
30. COUNTER SHAFT REAR BEARING
31. COUNTER REVERSE GEAR
32. REVERSE IDLER GEAR
33. IDLER GEAR SHAFT
34. KEY
35. LOCKWASHER

Fig. 9 Shafts and gears disassembled

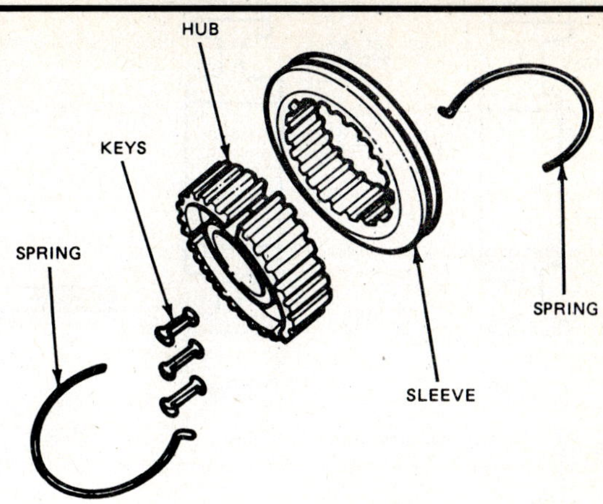

Fig. 10 Synchromesh mechanism

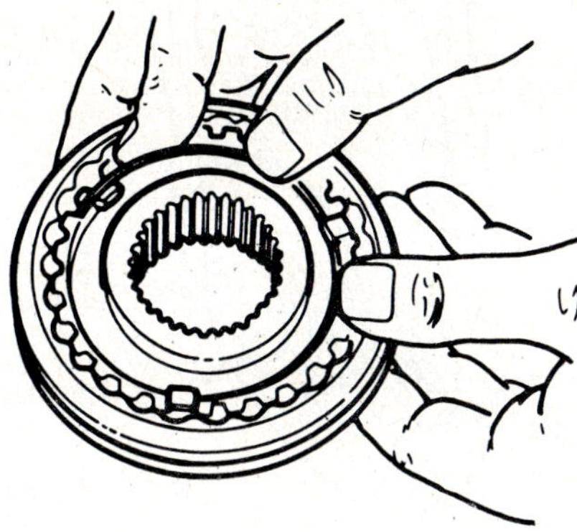

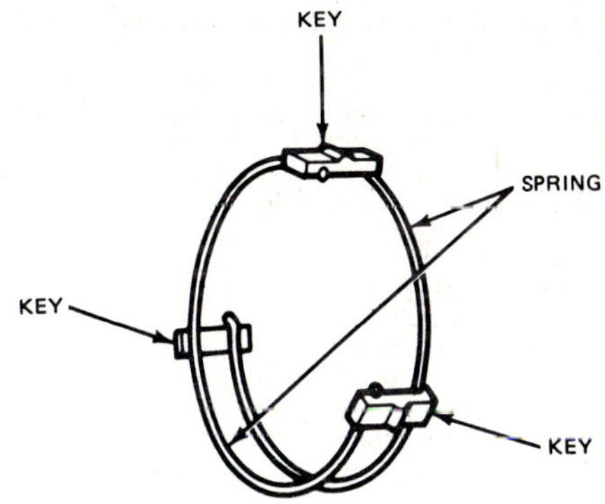

Fig. 11 Install key spring

are engaged with notches in synchronizer ring.

NOTE: The direction of the third and fourth clutch hub and assembly should be as shown in Fig. 12.

---

5. Install snap ring in front of mainshaft then place synchronizer ring on second gear and slide second gear to mainshaft with synchronizer ring toward rear of shaft.
6. Slide first and second clutch hub and sleeve assembly to mainshaft with oil grooves of clutch hub toward front of mainshaft. Make sure that three synchronizer keys in synchromesh mechanism engage with notches in second synchronizer ring.
7. Insert first gear sleeve in mainshaft then press bushing in first gear using a suitable press adapter.
8. Place synchronizer ring on first gear and slide first gear onto mainshaft with ring facing front of shaft then rotate gear as necessary to engage three notches in ring with keys.
9. Install original thrust washer on mainshaft and position mainshaft and gears assembly in case.
10. Position caged bearing in front end of mainshaft.
11. Place synchronizer ring on fourth gear input shaft, and install input shaft to front end of mainshaft that three keys engage notches in ring.
12. Position first and second shift fork and third and fourth shift fork in groove of clutch hub and sleeve.
13. Using suitable press install inner race of countershaft rear bearing onto countershaft.
14. Position countershaft gear in case, making sure that gear engages each gear of mainshaft assembly.
15. Install correct shim in mainshaft rear bearing bore, Fig. 13.
16. Position main drive gear bearing and mainshaft rear bearing into proper bearing bores. Be sure synchronizer and shifter forks have not been moved out of position.
17. Press both main drive gear and rear mainshaft bearings into case using suitable press until both bearings are seated properly, Fig. 14. Install main drive gear bearing snap ring.
18. Place shim in countershaft front bearing bore, Fig. 15.
19. Position countershaft front and rear bearings in bores and install suitable tools, Fig. 16. Turn forcing screw until bearing is properly seated. Install snap ring to secure countershaft front bearing.
20. Install bearing retainer together with reverse idler gear shaft to transmission case. Tighten the four attaching bolts.
21. Slide counter reverse gear onto countershaft with chamfer to rear then install snap ring.
22. Install key on mainshaft.

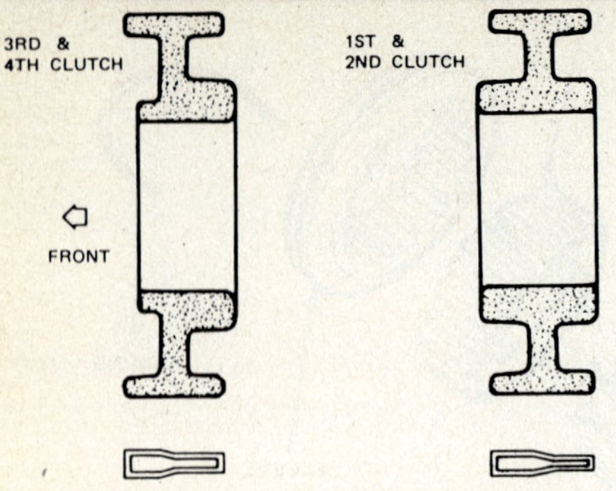

Fig. 12 Clutch hub assembly direction

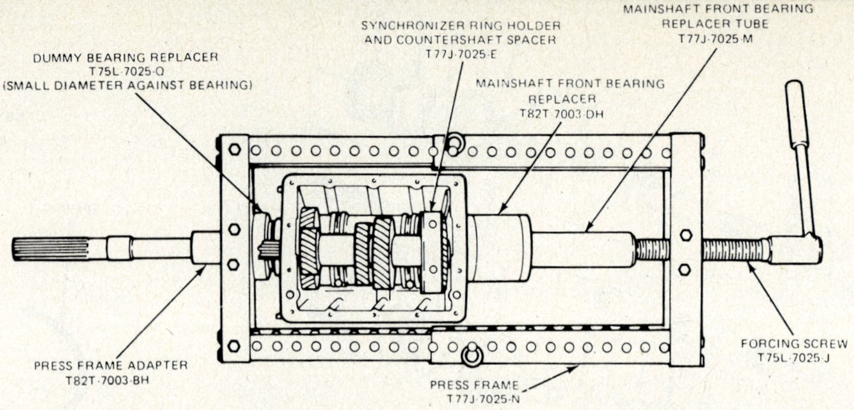

Fig. 14 Main drive gear and mainshaft rear bearing installation

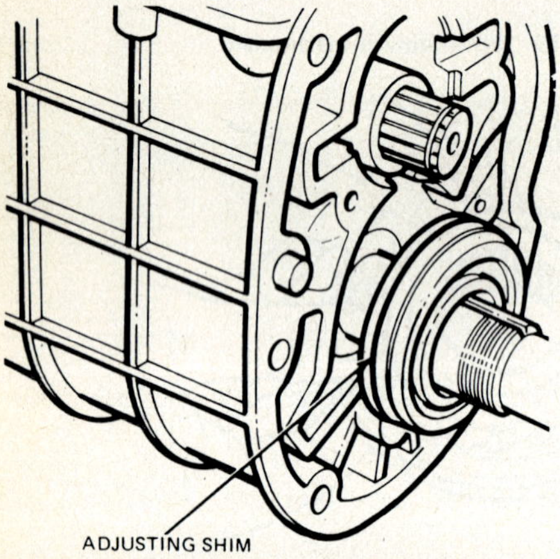

Fig. 13 Mainshaft thrust play adjusting shim

23. Slide reverse gear and lockwasher onto mainshaft. Install new locknut and hand tighten.
24. Shift into second gear and reverse gear to lock rotation of mainshaft. Torque locknut to 145–203 ft. lbs., Fig. 3.
25. Place fourth and third clutch sleeve in third gear using tool No. T77J-7025-E.
26. Check clearance between synchronizer key and exposed edge of synchronizer ring with feeler gauge, Fig. 17. If measurement is greater than 0.079 inch, change thrust washer between mainshaft rear bearing and first gear. Check clearance again; if within specifications, bend tab of lockwasher.
27. From rear of case install first and second shift fork shaft assembly then install roll pin. Stake roll pin to secure shift fork to fork shaft.
28. Insert inter-lock pin into transmission.
29. From rear of case install third and fourth shift fork shaft assembly, then install roll pin and stake to secure shift fork to fork shaft. Place transmission in neutral.
30. Use the preceding two steps to install the reverse fork shaft assembly and reverse idler gear.
31. Position three detent balls and springs into case. Place copper washers on top two bolts and install three spring cap bolts, then install two blind covers and gaskets and torque the attaching bolts to 23–34 ft. lbs.
32. Install lock ball, speedometer drive gear, and snap ring onto mainshaft.
33. Apply a thin coat of suitable sealing agent to contacting surfaces of transmission case and extension housing.
34. Position extension housing with gearshift control lever end laid down to left as far as it will go, tighten the four attaching bolts.
35. If removed, insert speedometer driven gear assembly to extension housing and secure it with bolt.
36. Check to ensure gearshift control lever operates properly.

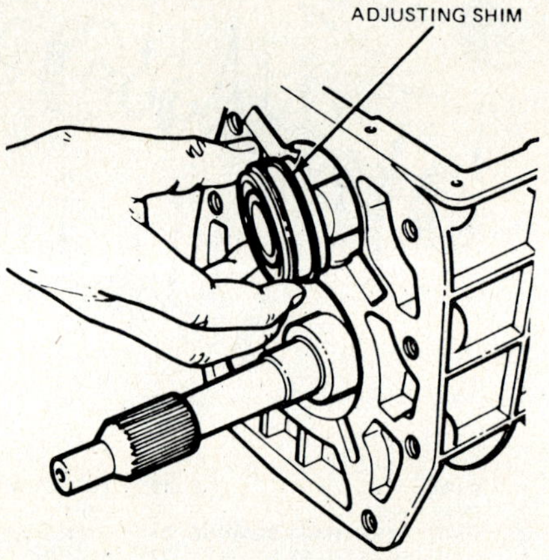

Fig. 15 Placing shim in countershaft front bearing bore

37. Install transmission case cover gasket and cover with drain plug to rear. Install and torque the fourteen attaching bolts to 23–34 ft. lbs.
38. Position gasket and gearshift lever retainer to extension housing, and tighten the four attaching bolts.
39. Install shim on second step of clutch adapter plate, then coat with suitable sealer. Install clutch adapter plate to transmission case and tighten the four bolts and studs.
40. Remove filler plug and fill transmission with 3.0 pints of suitable lubricant. Reinstall plug and torque to 18–39 ft. lbs.

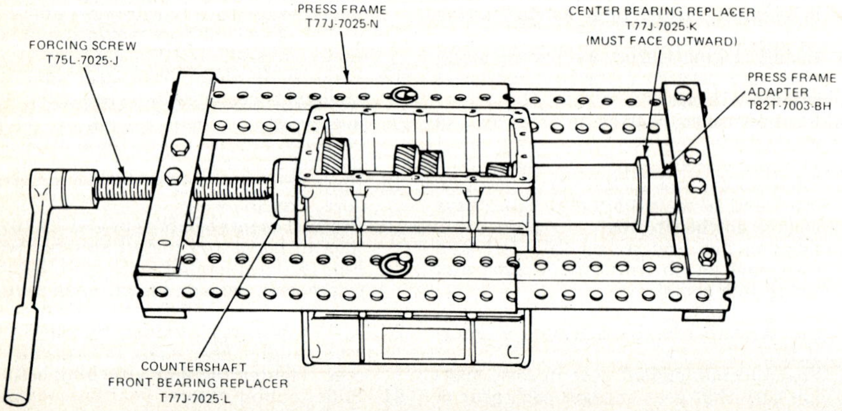

Fig. 16 Countershaft front bearing installation

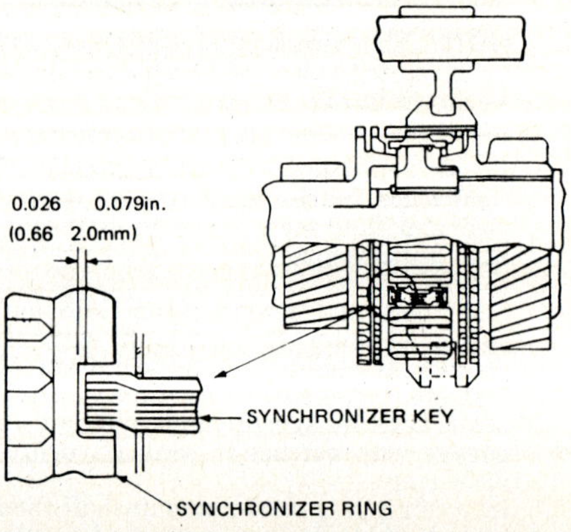

Fig. 17 Checking synchronizer key clearance

C3 Automatic Transmission Section

Refer to the Ford C4 transmission in the "Automatic Transmission" section of this manual for Description, Troubleshooting and Servicing of this transmission.

NOTE: The intermediate band only is adjusted on this transmission, Fig. 1. Prior to adjusting band, remove downshift rod from downshift lever. Also, when performing band adjustment, the adjusting screw must be backed out exactly two turns.

TRANSMISSION, REPLACE

1. Raise and support vehicle.
2. Place a drain pan under transmission oil pan, then, starting at rear, loosen attaching bolts and drain fluid from transmission. Leave two bolts in front and reinstall two bolts on rear of oil pan to temporarily hold it in place.
3. Remove converter drain plug access cover and adapter plate bolts from lower end of converter housing.
4. Remove converter-to-flywheel attaching nuts. Place a wrench on crankshaft pulley attaching bolt to rotate converter to gain access to nuts.

NOTE: Do not turn engine backwards on belt driven overhead camshaft engines.

5. Rotate converter to gain access to converter drain plug, then remove the plug. After fluid has drained from converter, reinstall drain plug and torque to 20–30 ft. lbs.
6. Remove driveshaft, then install extension housing seal replacer tool in extension housing.
7. Remove speedometer cable from extension housing, then disconnect shift rod from transmission manual lever and the downshift rod from transmission downshift lever.
8. Remove starter-to-converter housing attaching bolts and position starter motor aside.
9. Disconnect neutral start switch wires from the switch, then remove vacuum line from transmission vacuum modulator.
10. Position a suitable transmission jack under transmission and raise it slightly, then remove engine rear support-to-crossmember attaching bolts.
11. Remove crossmember-to-frame side support attaching bolts, then the crossmember insulator and support and damper.
12. Lower the transmission jack, then position a suitable jack under front of engine and raise engine to gain access to the two upper converter housing-to-engine attaching bolts.
13. Disconnect oil cooler lines from transmission. Plug all openings to prevent contamination.
14. Remove lower converter housing-to-engine attaching bolts, then the transmission filler tube.
15. Install a safety chain to hold transmission on jack, then remove two upper converter housing-to-engine attaching bolts.
16. Move transmission rearward and lower it from vehicle.
17. Reverse procedure to install. Note the following torques: converter housing to engine bolts, 28–38 ft. lbs.; filler tube bolt, 28–38 ft. lbs.; crossmember-to-frame side rail bolts, 20–30 ft. lbs.; rear engine-to-crossmember bolt, 60–80 ft. lbs.; starter bolts, 15–20 ft. lbs.; flywheel-to-converter nuts, 20–34 ft. lbs.; converter drain plug access cover and adapter plate bolts, 12–16 ft. lbs.; companion flange bolts, 70–95 ft. lbs.

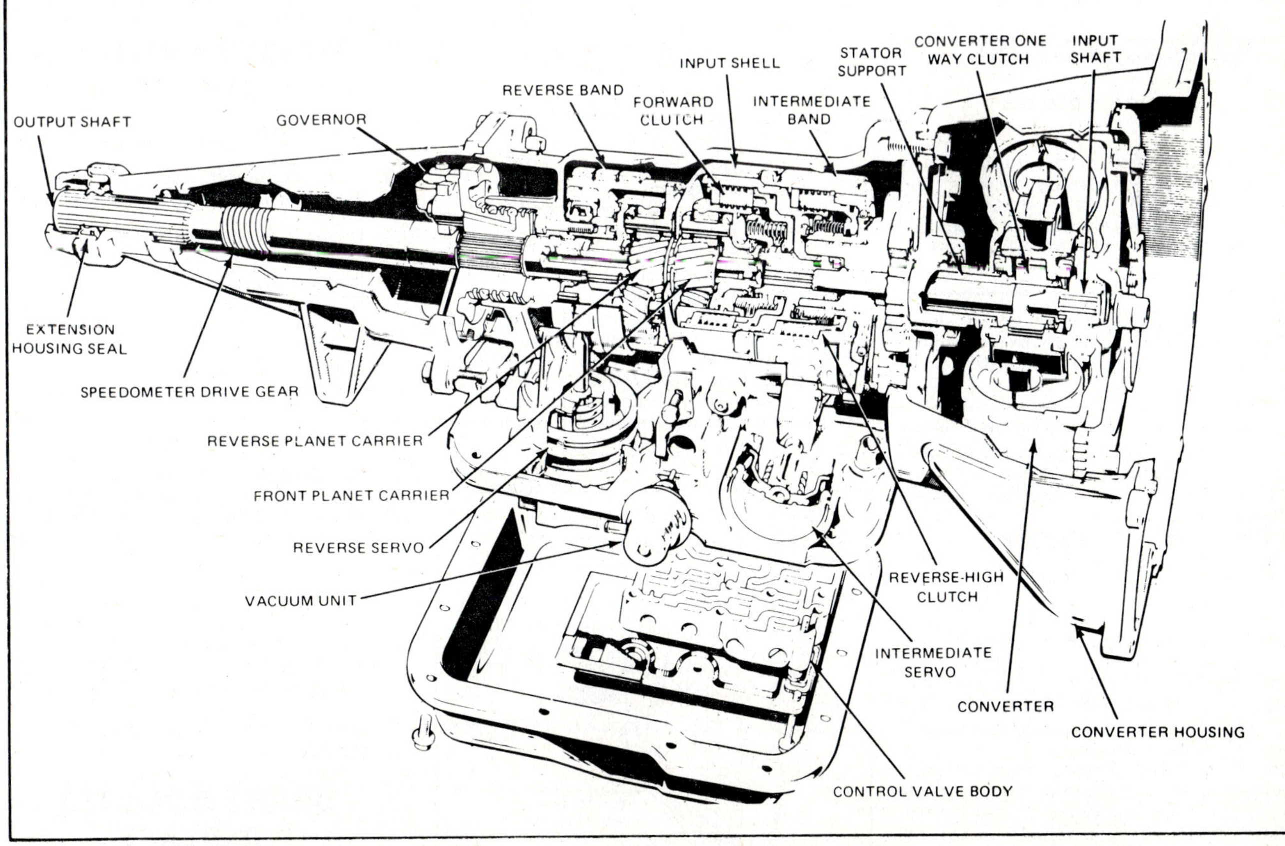

Fig. 1 C3 dual range automatic transmission

C5 Automatic Transmission Section

Refer to "Automatic Transmission" section of this manual for Description, Troubleshooting and Servicing of this transmission.

TRANSMISSION, REPLACE

1. Raise and support vehicle.
2. Place a drain pan under the transmission and lower edge of the oil pan to drain transmission.
3. Remove converter drain plug access cover from the lower end of the converter housing and remove converter-to-flywheel attaching nuts. Place a wrench on the crankshaft pulley attaching bolt to turn the converter to gain access to the remaining nuts.
4. With the wrench on the crankshaft pulley attaching bolt, turn converter to gain access to the converter drain plug and remove plug. After fluid has been drained, reinstall plug.
5. Remove driveshaft and disconnect the oil cooler lines from the transmission.
6. Disconnect manual and downshift linkage rods from the transmission control levers.
7. Remove speedometer cable and driven gear from the extension housing.
8. Disconnect back-up switch wires from the retaining clips and retainer.
9. Disconnect starter cable and remove the three starter-to-converter housing attaching bolts and remove the starter.
10. Remove vacuum line hose from the transmission vacuum unit, also disconnect vacuum line from the retaining clip.
11. Position a transmission jack to support the transmission. Install a safety chain to hold the transmission on the jack.
12. Remove the two rear engine support crossmember-to-frame attaching bolts. Then remove the engine rear support-to-extension housing attaching bolts.
13. Raise transmission and remove the rear support. Then proceed to remove the six converter housing-to-engine attaching bolts.
14. Move converter and transmission assembly away fron the engine. Lower transmission and remove it from under the vehicle.
15. Reverse procedure to install. Tighten converter housing-to-engine attaching bolts to 45 ft. lbs. Tighten converter-to-flywheel attaching nuts to 20–34 ft. lbs.

NOTE: The converter must rest squarely against the flywheel. This indicates that the converter pilot is not binding in the engine crankshaft.

Brake Section

BRAKES, ADJUST

Rear Drum Brakes

The rear brake shoes adjust automatically when the vehicle is driven forward and reverse and the brakes are applied several times sharply. Manual adjustment is required only when brake shoes are relined or replaced.

NOTE: When adjusting rear brake shoes, check parking brake cable for proper adjustment and be sure equalizer is operating properly.

1. With drums removed, clean areas where shoes contact backing plate, then apply suitable lubricant to these contact areas, being careful not to get lubricant on shoes.
2. Using brake shoe adjusting gauge or equivalent, adjust gauge to inside diameter of drum braking surface.
3. Reverse tool and adjust brake shoes until they touch gauge, being sure the gauge contact points on the shoes are parallel to vehicle with center line through center of axle. Holding automatic adjusting lever out of way, rotate adjustment screw. Oil screw threads with suitable lubricant if screw will not turn freely.
4. Install drums, drum retaining nuts, wheels and mounting nuts, then complete adjustment by applying brakes several times while driving vehicle in reverse and check brake operation by making several stops while driving forward.
5. Drum brakes may also be adjusted by raising vehicle enough to raise wheels off ground and turning adjusting screws through holes in brake backing plate.

Parking Brake

1. Adjust parking brake with service brakes adjusted and drums cold.
2. Fully depress parking brake and, holding threaded rod to prevent it from spinning, tighten equalizer nut six full turns past its original position.
3. Attach suitable cable tension gauge behind equalizer assembly toward either rear drum assembly and measure cable tension which should be 400–600 lbs. with parking brake pedal fully depressed. If tension is low, repeat steps 2 and 3.
4. Release parking brake and check for rear wheel drive.

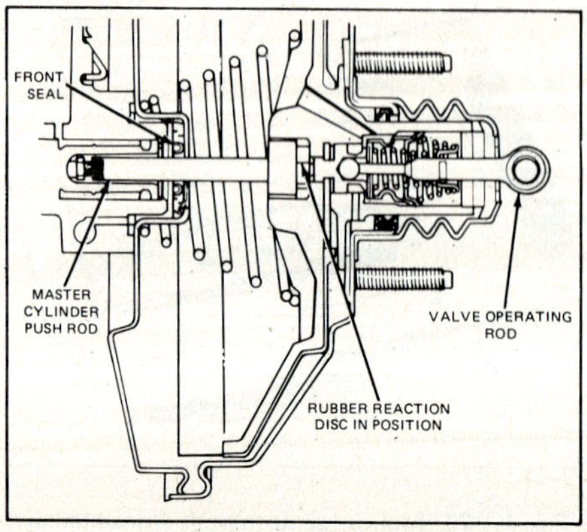

Fig. 1 Checking Reaction Disc Installation

MASTER CYLINDER, REPLACE

With Power Brakes

1. With engine off, depress brake pedal to expel vacuum from brake booster system.
2. Disconnect hydraulic lines from master cylinder.
3. Remove brake booster to master cylinder retaining nuts and remove master cylinder from brake booster.
4. Reverse procedure to install and bleed brake system, then refill master cylinder and bleed it.

Less Power Brakes

1. Disconnect wires from stoplight switch inside cab below instrument panel.
2. Remove retaining nut, shoulder bolt and spacers which secure master cylinder push rod to brake pedal assembly, then remove stoplight switch from pedal.
3. Disconnect hydraulic brake lines from master cylinder, then remove master cylinder to dash panel retaining nuts, master cylinder, and boot from master cylinder push rod.
4. Reverse procedure to install and bleed brake system, then refill and bleed master cylinder.

BRAKE BOOSTER, REPLACE

NOTE: Make sure booster rubber reaction disc is properly installed as shown in Fig. 1. if the master cylinder push rod is removed or accidentally pulled out. A dislodged disc may cause excessive pedal travel and extreme operation sensitivity. The disc is black compared to the silver colored valve plunger that will be exposed after push rod and front seal are removed. The booster unit is serviced as an assembly and must be replaced if the reaction disc cannot be properly installed and aligned, or if it cannot be located within the unit itself.

1. Disconnect stoplight switch wiring, support master cylinder from underside, and remove master cylinder to booster retaining nuts.
2. Loosen clamp which secures manifold vacuum hose to booster check valve and remove hose, then remove booster check valve.
3. Pull master cylinder off booster and support far enough away to allow removal of booster assembly.
4. Remove, from inside of cab on vehicles equipped with push rod mounted stoplight switch, retaining pin and slide stoplight switch, push rod, spacer, and bushing off brake pedal arm.
5. Remove, from engine compartment, booster to dash panel bolts.
6. Reverse procedure to install, then start engine and check brake operation.

Rear Axle Section

6¾ INCH RING GEAR AXLE

Description, Fig. 1

NOTE: This axle can be identified by the 8 cover bolts.

This rear axle is an integral design hypoid with the centerline of the pinion set below the centerline of the ring gear. The semi-floating axle shafts are retained in the housing by ball bearings and bearing retainers at axle ends.

The differential is mounted on two opposed tapered roller bearings which are retained in the housing by removable caps. Differential bearing preload and drive gear backlash is adjusted by nuts located behind each differential bearing cup.

The drive pinion assembly is mounted on two opposed tapered roller bearings. Pinion bearing preload is adjusted by a collapsible spacer on the pinion shaft. Pinion and ring gear tooth contact is adjusted by shims between the rear bearing cone and pinion gear.

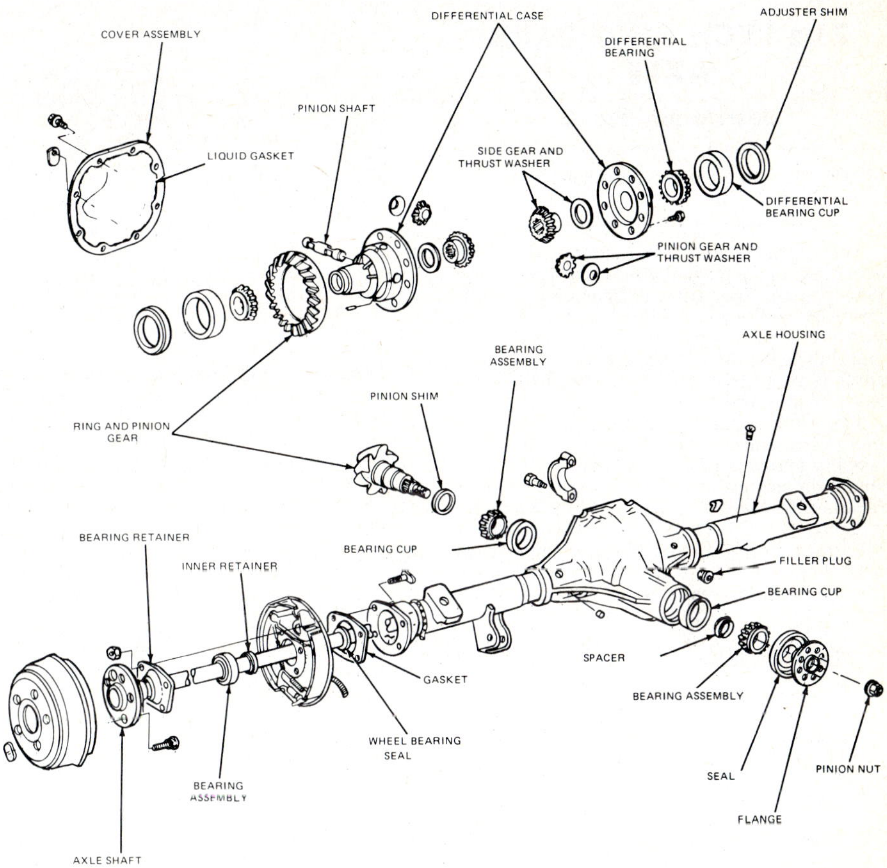

Fig. 1 Disassembled view of Ford 6¾ inch Ring Gear axle

Rear Axle Assembly, Replace

1. Raise vehicle and support at rear frame members.
2. Drain lubricant from axle.
3. Mark driveshaft and pinion flange for proper reassembly. Disconnect driveshaft from pinion flange.
4. Remove wheels, brake drums and both axle shafts as outlined in "Axle Shaft & Oil Seal, Replace".
5. Remove vent tube hose, then remove vent tube from axle housing.
6. Remove bolt securing hydraulic brake T-fitting from axle housing, then carefully remove the hydraulic brake hose from retaining clip. Carefully wire rear brake backing plates, hydraulic brake system and parking brake cables as one unit out of the way.
7. Using a suitable jack, support rear axle. Remove U-bolts, shock absorber lower mounting and spring plates.
8. Lower axle housing and remove from vehicle.
9. Reverse procedure to install, noting the following; Torque U-bolt nuts to 55–75 ft. lbs., shock absorber lower mounts to 40–60 ft. lbs., driveshaft to pinion flange bolts to 70–95 ft. lbs., rear cover attaching bolts to 25–35 ft. lbs. and rear axle shaft retainer plate nuts to 20–40 ft. lbs.

Axle Shaft & Oil Seal, Replace

1. Remove wheel and tire assembly, then remove nuts attaching brake drum to axle shaft flange and remove brake drum.
2. Working through opening in axle shaft flange, remove nuts securing axle shaft bearing retainer.
3. Using a suitable puller, pull axle shaft from housing.
4. Remove brake backing plate and attach to frame side rail with a piece of wire.
5. Using a hook type puller, remove oil seal from housing.
6. Wipe all lubricant from oil seal area of axle housing, then install oil seal using tool No. T79P-1177-A.
7. Install gasket on housing flange, then install brake backing plate.
8. Carefully slide axle shaft into housing using care not to damage oil seal, then install bearing retainer attaching nuts and torque to 20 to 40 ft. lbs.
9. Install brake drum and retaining nuts, then install wheel and tire assembly.

Rear Axle Shaft Bearings, Replace

1. Remove axle shaft as described in "Axle Shaft & Oil Seal, Replace".
2. Loosen axle inner retainer ring by nicking deeply with a cold chisel in several places and slide off shaft.
3. Using a suitable press, remove bearing

from axle.

NOTE: Do not ever attempt to use heat to make bearing removal easier, as it will weaken the axle shaft bearing journal area.

4. Position bearing retainer plate and bearing on axle, then using a suitable press, install bearing. Using the same tool, install bearing retainer ring.

NOTE: Do not press on the bearing and the retainer ring at the same time. They must be installed separately.

5. Reinstall axle in axle housing. Ensure new seal is installed.

7½ INCH RING GEAR AXLE

Description, Fig. 2

NOTE: This axle can be identified by the 10 cover bolts.

The gear set consists of a ring gear and an overhung drive pinion which is supported by two opposed tapered roller bearings. The differential case is a one-piece design with openings allowing assembly of the internal parts and lubricant flow. The differential pinion shaft is retained with a threaded bolt (lock) assembled to the case.

The roller type wheel bearings have no inner race, and the rollers directly contact the bearing journals of the axle shafts. The axle shafts do not use an inner and outer bearing retainer. Rather, they are held in the axle by means of C-locks. These C-locks also fit into a machined recess in the differential side gears within the differential case. There is no retainer bolt access hole in the axle shaft flange.

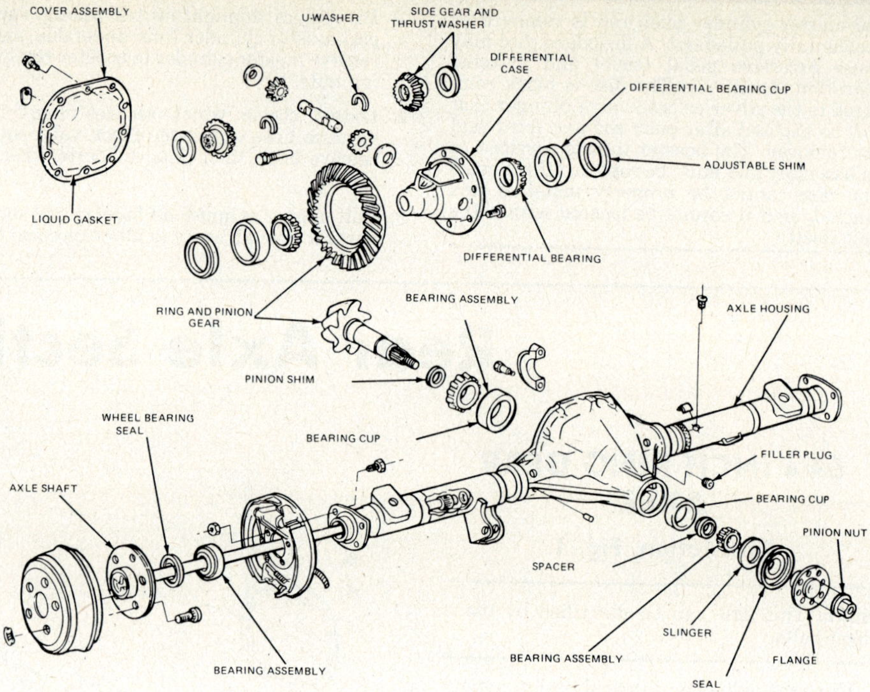

Fig. 2 Disassembled view of Ford 7½ inch Ring Gear Axle

Rear Axle Assembly, Replace

1. Raise vehicle and support at rear frame members. Remove rear wheels and brake drums.
2. Drain rear axle lubricant by loosening rear cover assembly.
3. Remove rear cover then remove differential pinion lock bolt and pinion shaft.
4. Push both axles in towards center of vehicle, then remove axle shaft C-washers.
5. Remove axles, then brake backing plate retaining nuts. Wire backing plates out of the way.
6. Disconnect rear axle housing vent.
7. Remove bolt securing hydraulic brake T-fitting from axle housing, then carefully remove the hydraulic brake hose from retaining clip. Wire hydraulic system out of the way.
8. Mark driveshaft and pinion flange for proper reassembly. Disconnect driveshaft from pinion flange.
9. Using a suitable jack, support rear axle. Remove U-bolts, shock absorber lower mounts and spring plates.
10. Lower axle housing and remove from vehicle.
11. Reverse procedure to install, noting the following; Torque U-bolt nuts to 55–75 ft. lbs., shock absorber lower mounts to 40–60 ft. lbs., pinion lock bolt to 15–22 ft. lbs., drive-shaft to pinion flange bolts to 70–95 ft. lbs. and rear cover attaching bolts to 25–35 ft. lbs., except the ratio tag bolt which is torqued 15–24 ft. lbs.

Axle Shaft, Seals & Bearings, Replace

1. Raise car on hoist and remove wheels.
2. Drain differential lubricant by loosening differential housing cover.
3. Remove brake drums.
4. Remove differential housing cover.
5. Position safety stands under rear frame member and lower hoist to allow axle to lower as far as possible.
6. Working through differential case opening, remove pinion shaft lock bolt and pinion shaft.
7. Push axle shaft(s) inward toward center of axle housing and remove C-lock(s) from housing.
8. Remove axle shaft, using extreme care to avoid contact of shaft seal lip with any portion of axle shaft except seal journal.
9. Use a hook-type puller to remove seal and bearing.
10. Reverse procedure to install, using suitable driving tools, to install seal and bearing. New seals are prepacked with lubricant and do not require oil soaking before installation. Torque pinion shaft lock bolt to 15–22 ft. lbs. and differential housing cover bolts to 25–35 ft. lbs, except the ratio tag bolt which is torqued 15–24 ft. lbs.

Front Suspension & Steering Section

4 × 2

The twin I-beam suspension consists of coil springs, I-beam axle arms, radius arms, upper and lower ball joints, spindles, tie rods, shock absorbers and an optional stabilizer bar, Fig. 1.

4 × 4

The independent front suspension used on four wheel drive models is comprised of a two piece front driving axle assembly, two coil springs and two radius arms, Fig. 2. The front driving axle consists of two independent axle arm assemblies. One end of each axle arm assembly is mounted to the frame, with the opposite end supported by the coil spring and radius arm.

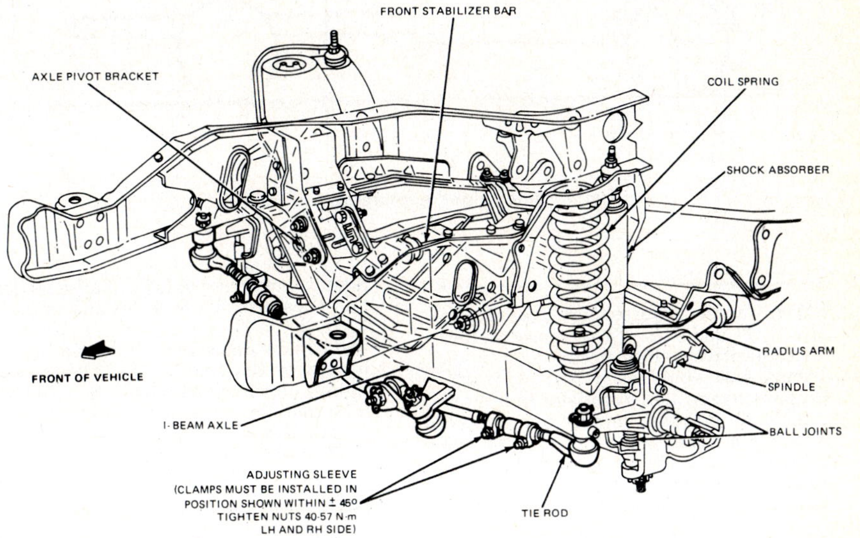

Fig. 1 Front suspension 4 × 2 models

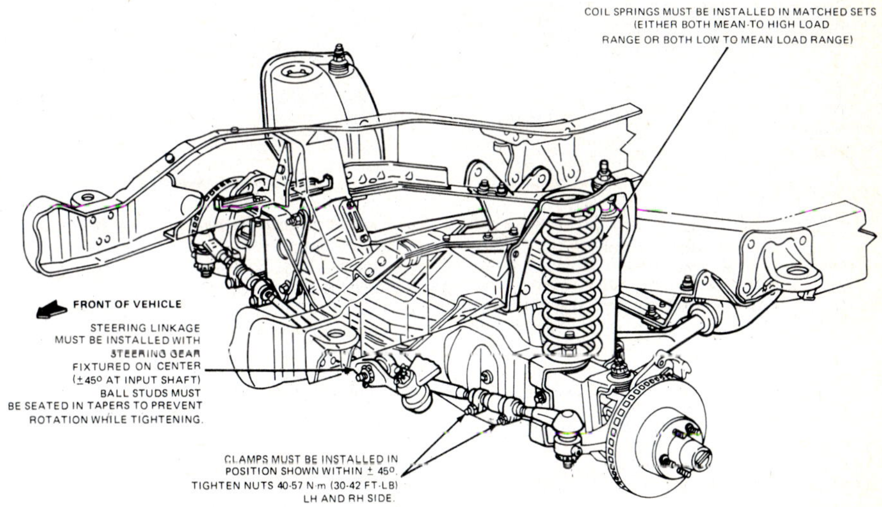

Fig. 2 Front suspension. 4 × 4 models

WHEEL ALIGNMENT

Front wheel alignment specifications and method of checking caster, camber and toe-in for Twin I-beam front axles on 2×4 models or independent front suspension on 4×4 models is determined with truck at normal operating height, provided front ride height is within specifications, and tires are inflated to specified cold pressure.

To determine riding height, measure distance between bottom of spring tower and top of axle, Figs. 3 and 4. Refer to "Wheel Alignment Specifications Chart" to determine alignment specifications for the particular riding height of truck. If riding height does not fall within specifications, it should be corrected by installing proper springs or by use of shims.

Alignment equipment indicates a true reading only when frame is horizontal. Measure left and right hand angle as shown in Fig. 5. If frame is not level (due to tire, spring or load differences), the caster angle reading must be modified to compensate for frame angles. If front is higher than rear, subtract amount of angle from reading. If front is lower than rear, add angle. Check frame angle with a spirit lever protractor and take frame angle measurement on flat area immediately adjacent to rear spring front hanger.

NOTE: Axles are not to be bent or twisted to correct caster or camber readings.

Camber Adjustment

4 × 2

1. Raise and support vehicle. Remove front wheels.
2. Remove upper ball joint nut, then remove cotter pin on lower ball stud and back nut down stud. Pop ball joint/camber adjuster taper assembly from mounting.
3. Wedge camber adjuster out of spindle with camber adjusting tool D81T-3010-B. Replace adjuster with desired camber adjuster, Fig. 6.

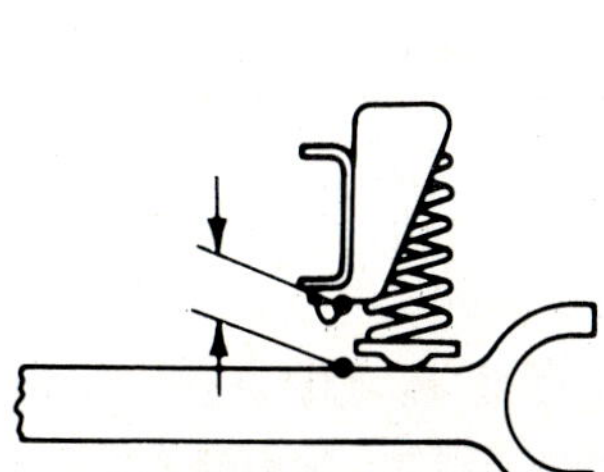

Fig. 3 Ride height measurement at front end. 4 × 2 models

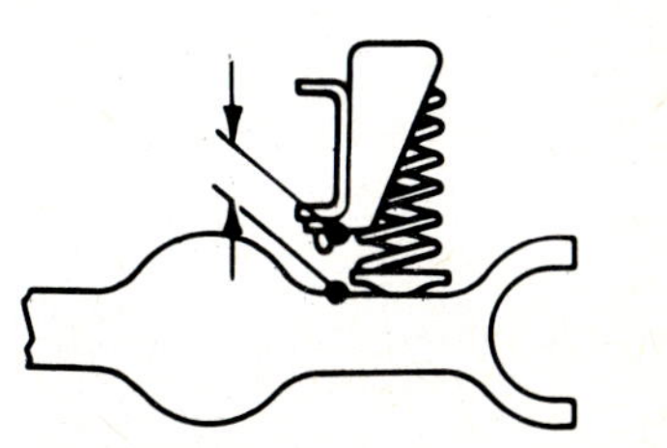

Fig. 4 Ride height measurement at front end. 4 × 4 models

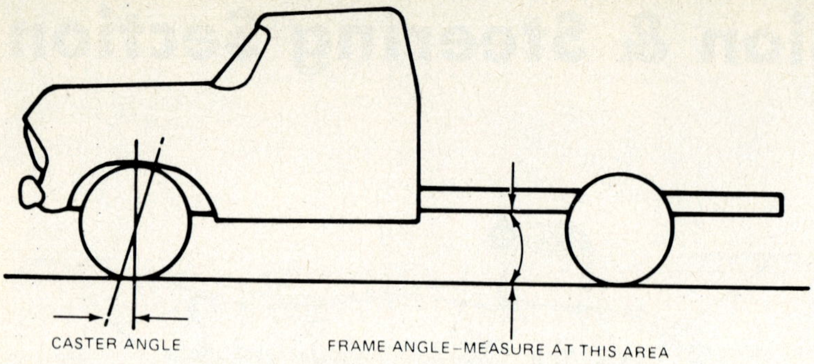

Fig. 5 Ride height measurement at frame

NOTE: Camber adjusters are available in 0°, 1/2°, 1° and 11/2°. To increase camber (more positive), align slot as follows: on driver's side, point slot to rear of vehicle. On passenger's side, point slot to front of vehicle. To decrease camber (more negative), align slot as follows: on driver's side point slot forward. On passenger's side, point slot rearward.

4. Reverse procedure to install. Remove lower ball joint stud nut and apply Locktite 242 to upper and lower ball joint studs. Hand start upper ball stud nut and torque lower ball stud nut to 35 ft. lbs. Torque upper ball stud nut to 85–110 ft. lbs. and finish torqueing lower stud nut to 104–146 ft. lbs. Install cotter pin.

4 × 4

1. Raise and support vehicle. Remove front wheels.
2. Remove upper ball joint cotter pin and nut, then back off lower ball joint nut to end of stud.
3. Break spindle loose from ball joint studs, then remove camber adjuster sleeve from spindle using tool T64P-3590-F or equivalent, if necessary, Fig. 7.
4. Install camber adjuster on top ball joint stud with arrow pointing outboard for positive camber or inboard for negative camber.

NOTE: Zero camber bushings will not have an arrow and may be turned in either direction as long as the lugs on the yoke engage the slots in the bushing.

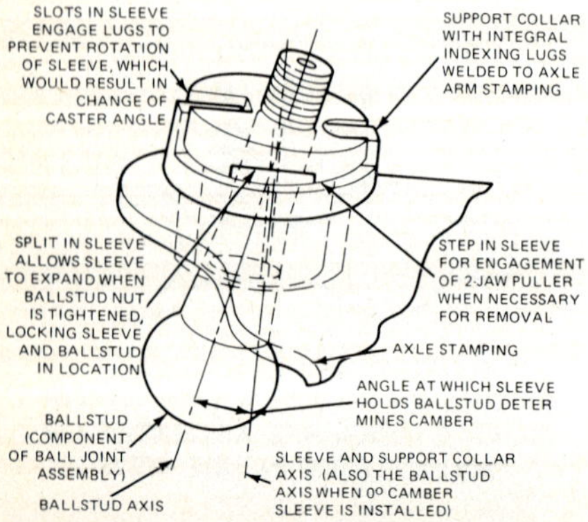

Fig. 7 Camber adjustment. 4 × 4 models

5. Remove and discard lower ball joint stud nut, then install a new nut and torque to 40 ft. lbs.
6. Install new upper ball joint stud nut and torque to 85–100 ft. lbs. Continue to tighten nut until cotter pin hole lines up with castellation on nut, then install cotter pin. Retorque lower ball joint stud nut to 95–110 ft. lbs.

NOTE: The camber adjuster will be seated at a predetermined position during the tightening sequence. Do not attempt to change this position.

7. Reinstall wheels and lower vehicle.

Caster Adjustment

Caster angle is non-adjustable on these vehicles.

Toe-In, Adjust

1. Loosen clamp bolts at each end of spindle connecting rod tube, then rotate the sleeve until correct toe alignment of 1/32 inch is obtained.
2. Center clamps between adjustment sleeve nibbs, then position bolts horizontally with clamps nuts positioned as shown in Fig. 8.
3. Recheck toe-in adjustment, then check that steering wheel spokes are properly positioned and adjust as necessary.

SHOCK ABSORBER, REPLACE

1. Remove nut and washer from shock absorber at spring seat, Figs. 9 and 10.
2. Remove nut and bolt from shock absorber at radius arm and lower shock bracket. Slightly compress shock and remove.
3. Reverse procedure to install.

LOWER & UPPER BALL JOINTS, REPLACE

4 × 2

1. Remove spindle. Remove snap rings from ball joints, Fig. 11.
2. Assemble C-Frame T74P-4635-C and receiving cup D81T-3010-A on upper ball joint. Turn forcing screw clockwise until

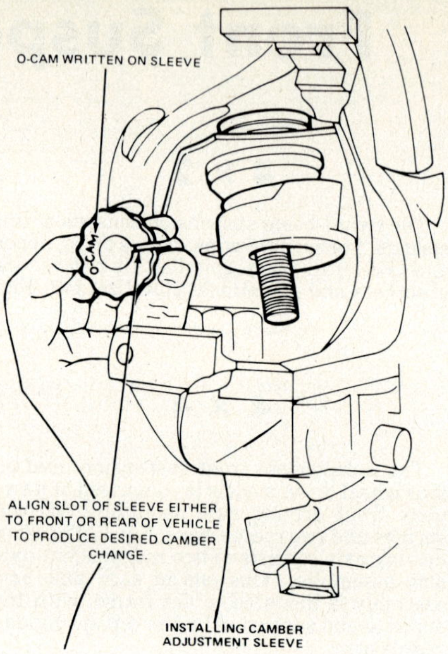

Fig. 6 Camber adjustment. 4 × 2 models

ball joint is removed. Assemble C-Frame and cup on lower ball joint and turn forcing screw clockwise until ball joint is removed.

NOTE: Always remove upper ball joint first.

3. Reverse procedure to install. Install lower ball joint first using ball joint receiver cup D81T-3010-A5 and installation cup D81T-3010-A1 inside adapter cup D81T-3010-A4. Turn forcing screw until ball joint is seated. Install snap ring onto lower ball joint. Install upper ball joint in same manner.

4 × 4

Refer to the "Front Wheel Drive Section" in this chapter for ball joint replacement procedure.

COIL SPRING, REPLACE

4 × 2

1. Raise and support front of vehicle. Support axle.

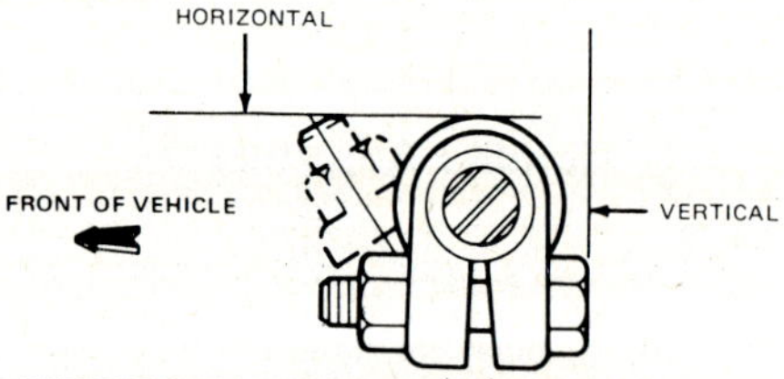

AFTER SETTING TOE, THE TWO CLAMP BOLTS/NUTS ON EACH ADJUSTING SLEEVE MUST BE POSITIONED WITHIN A LIMIT OF ± 45 DEGREES AS SHOWN WITH THE THREADED END OF THE BOLTS ON BOTH ADJUSTING SLEEVES POINTING TOWARDS THE FRONT OF THE VEHICLE.

Fig. 8 Toe-in adjustment

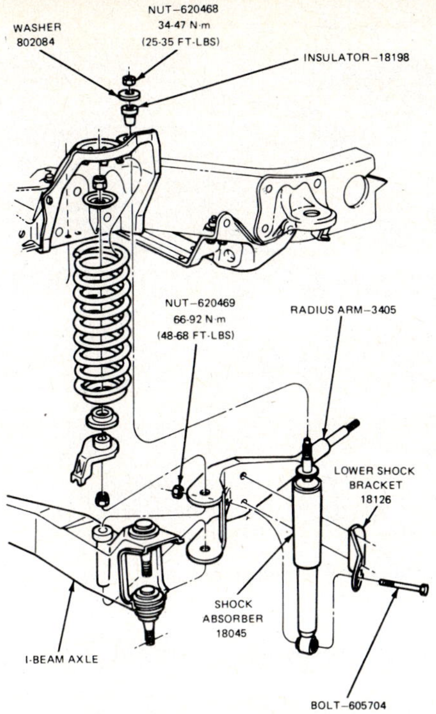

Fig. 9 Shock absorber replacement. 4 × 2 models

2. Remove lower retainer nut and lower retainer, Fig. 12. Lower axle until it hangs unsupported.
3. Insert pry bar between axles and force appropriate I-beam axle down far enough to allow spring to be lifted over bolt in lower spring seat. Rotate spring until retainer on upper spring seat is cleared. Remove spring.

NOTE: The axle must be supported when replacing spring and not be permitted to hang by the brake hose. If the length of the brake hose is not sufficient to permit spring replacement, it will be necessary to remove the disc brake caliper. Do not suspend caliper by brake hose, if removed.

---

4. Reverse procedure to install. Install bolt in axle arm and tighten nut all the way down. Install top of spring into upper spring seat and rotate into position. Raise axle until spring is seated in lower spring upper seat. Install lower retainer and nut.

4 × 4

1. Raise and support front of vehicle. Support axle with a suitable jack so the spring is compressed.
2. Remove shock absorber to radius arm attaching bolt, then remove shock absorber from bracket.
3. Remove spring lower retainer attaching nut, then the retainer.
4. Lower axle until all spring tension is released, then remove spring by rotating upper coil out of tabs in upper spring seat, Fig. 13.
5. Remove spacer and seat from spring and

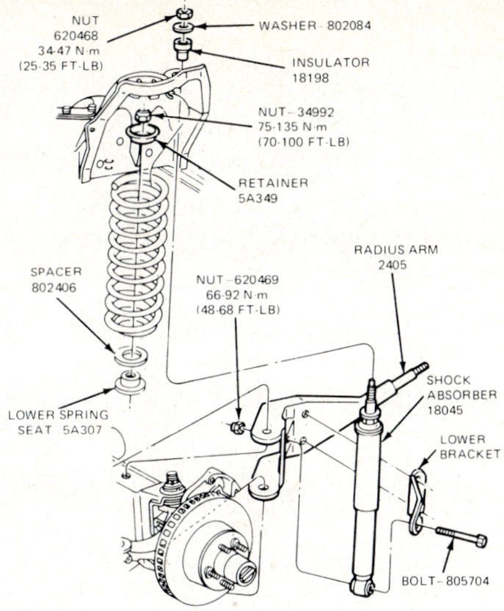

Fig. 10 Shock absorber replacement. 4 × 4 models

the stud from axle assembly, if necessary.

NOTE: The axle must be supported when replacing spring and not be permitted to hang by the brake hose. If the length of the brake hose is not sufficient to permit spring replacement, it will be necessary to remove the disc brake caliper. Do not suspend caliper by brake hose, if removed.

---

6. Reverse procedure to install.

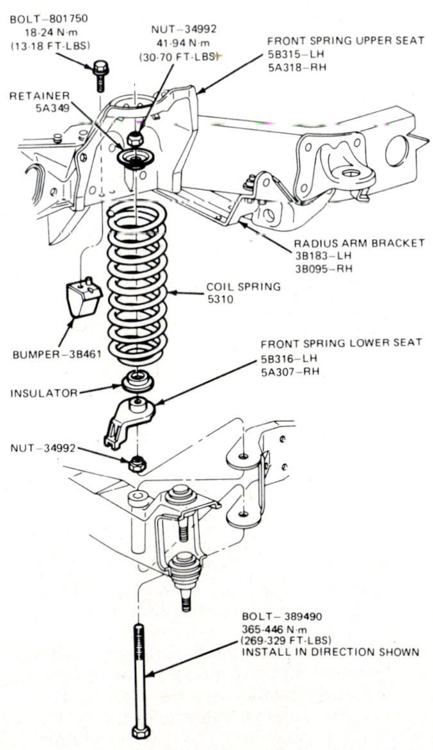

Fig. 12 Coil spring replacement. 4 × 2 models

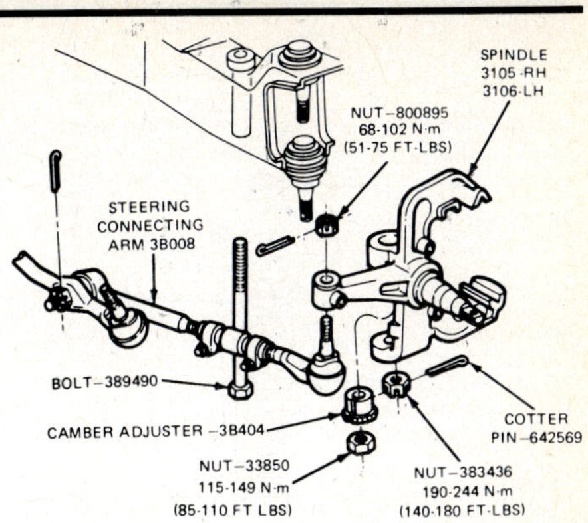

Fig. 11 Ball joint replacement. 4 × 2 models

RADIUS ARM, REPLACE

1. Raise and support front of vehicle. Support axle.
2. Disconnect lower end of shock absorber from shock lower bracket and remove front spring as described previously.
3. Loosen axle pivot bolt. Remove spring lower seat from radius arm. Remove nut and bolt from radius arm to axle and front bracket, Figs. 14 and 15.
4. Remove nut, rear washer and insulator from rear side of radius arm rear bracket and remove radius arm. Remove inner insulator and retainer from radius arm stud.
5. Reverse procedure to install.

RADIUS ARM INSULATORS

1. Loosen axle pivot bolt and upper shock absorber pivot bolt and compress shock.
2. Remove nut and washer attaching radius arm to radius arm bracket. Remove outer insulator and spacer, Fig. 14 and 15.
3. Move radius arm and axle assembly forward out of radius arm bracket. Remove inner insulator and retainer.
4. Reverse procedure to install. Torque upper shock bolt to 25–35 ft. lbs. and axle pivot bolt and nut to 120–150 ft. lbs.

FRONT I-BEAM AXLE, REPLACE

1. Remove front wheel spindle and front spring.
2. Remove spring lower seat from radius arm and bolt and nut for radius arm to front axle.
3. Remove axle-to-frame pivot bracket bolt and nut. Remove axle.
4. Reverse procedure to install.

AXLE PIVOT BRACKET, REPLACE

4 × 2

1. Remove front spring, radius arm, wheel spindle and I-beam.
2. Remove four attaching nuts and two bolts and retainer assemblies. Remove axle pivot bracket, Fig. 16

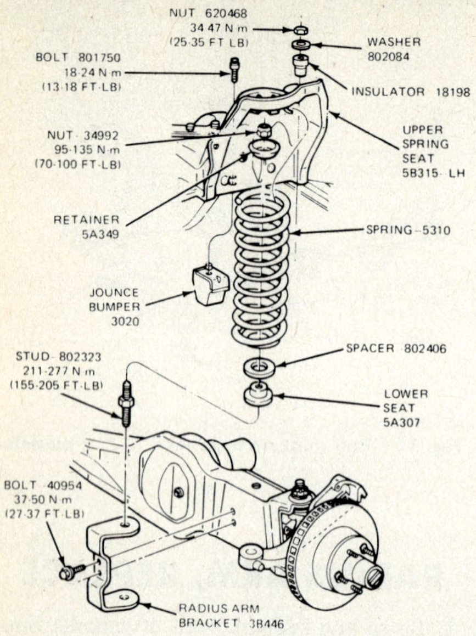

Fig. 13 Coil spring replacement. 4 × 4 models

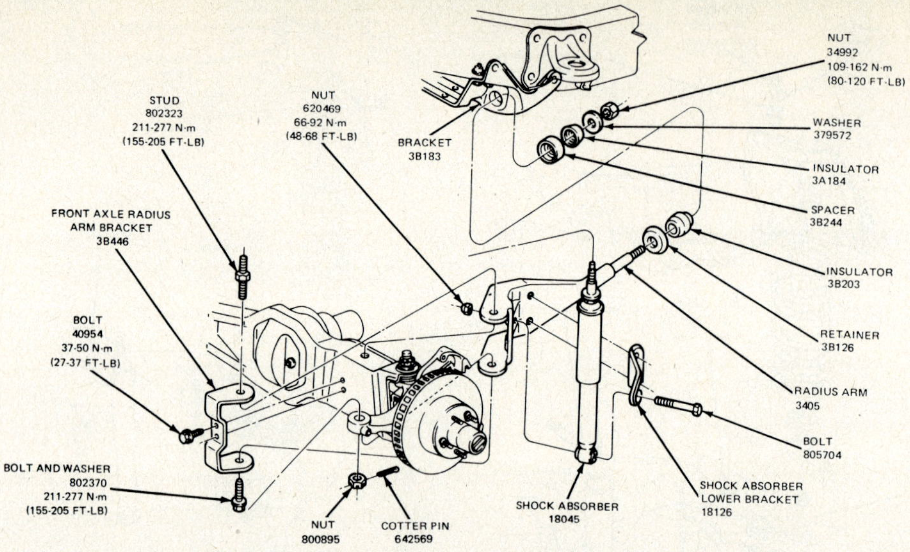

Fig. 15 Radius arm replacement. 4 × 4 models

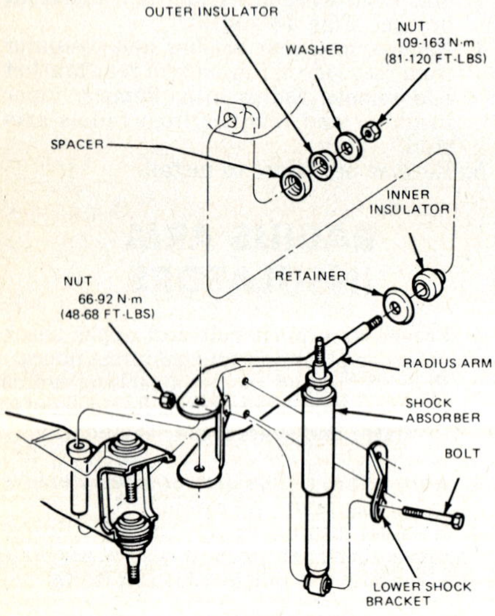

Fig. 14 Radius arm replacement. 4 × 2 models

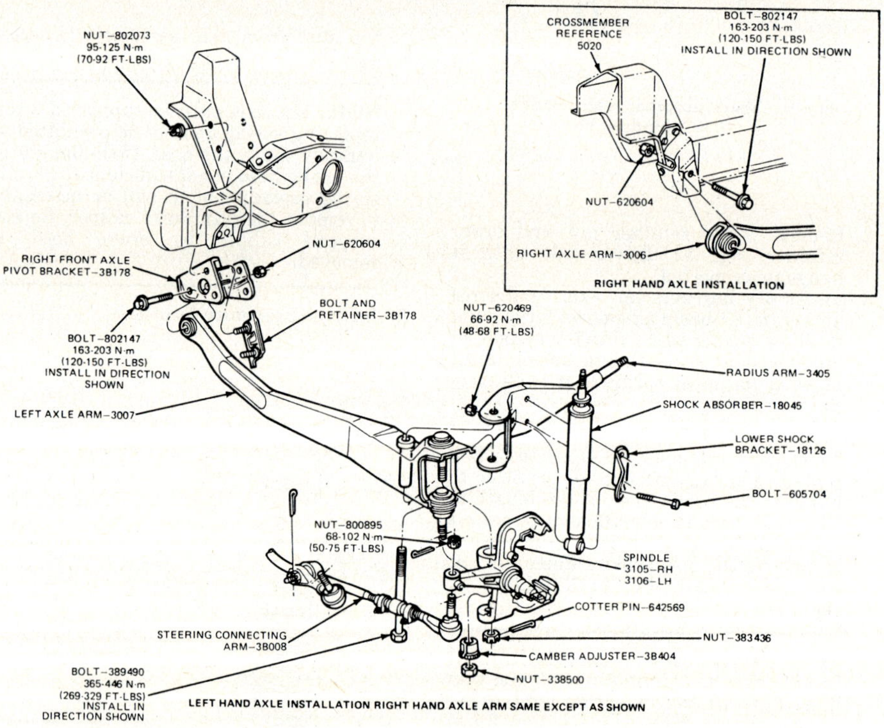

Fig. 16 Axle pivot bracket replacement. 4 × 2 models

3. Reverse procedure to install. Position axle pivot bracket to frame and install forward and rearward bolts and retainer assemblies from inside of pivot bracket out through crossmember. Loosely install four nuts on outside of crossmember (two forward and two rearward). Torque nuts to 70–92 ft. lbs.

NOTE: Use nut N8802073-S2 or install one .20 inch thick hardened washer under each nut if a standard nut is used.

4 × 4

1. Remove coil spring and radius arm as described previously.
2. Remove front drive axle as described in the "Front Wheel Drive Section" of this chapter.
3. To remove right pivot bracket, remove upper bolt and retainer, then the side bolt and retainer, Fig. 17. Remove and discard lower bolt and retainer, then remove pivot bracket from crossmember.
4. To remove left pivot bracket, drill out rivets so that mounting holes in bracket and crossmember are 9/16 inch in diameter, then remove the bracket, Fig. 18.
5. Reverse procedure to install. Refer to Figs. 17 and 18 for replacement hardware and torque to specifications.

AXLE PIVOT BUSHING, REPLACE

4 × 2

1. Remove front coil spring as outlined in this chapter.
2. For left I-beam axle, remove axle pivot bolt and nut and pull left I-beam axle down until bushing is exposed. For right I-beam axle, entire right I-beam axle must be removed. Refer to "Front I-Beam Axle, Replace."
3. Install forcing screw T78P-5638-A1, bushing remover T80T-5638-A2, spacer T82T-3006-AH and receiver cup T78P-

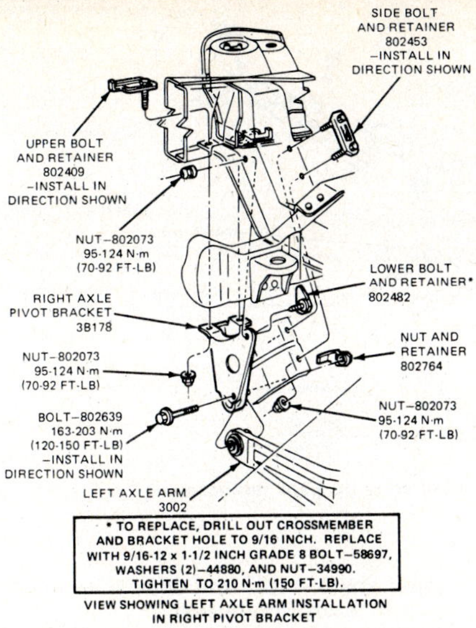

Fig. 17 Right axle pivot bracket replacement. 4 × 4 models

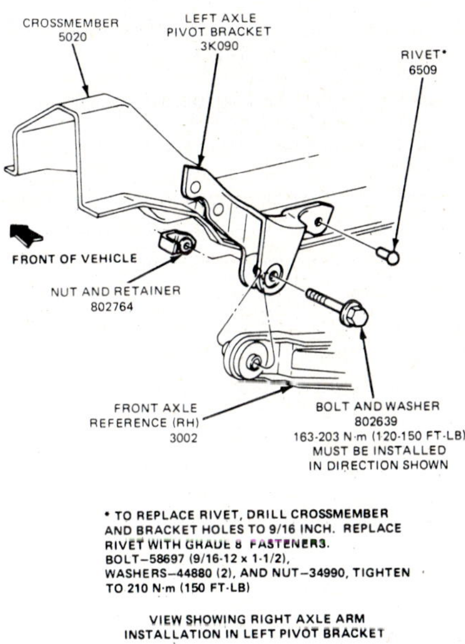

Fig. 18 Left axle pivot bracket replacement. 4 × 4 models

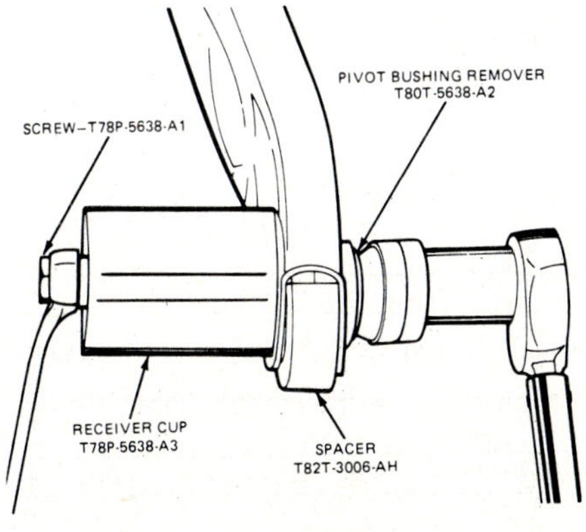

Fig. 19 Axle pivot bushing removal. 4 × 2 models

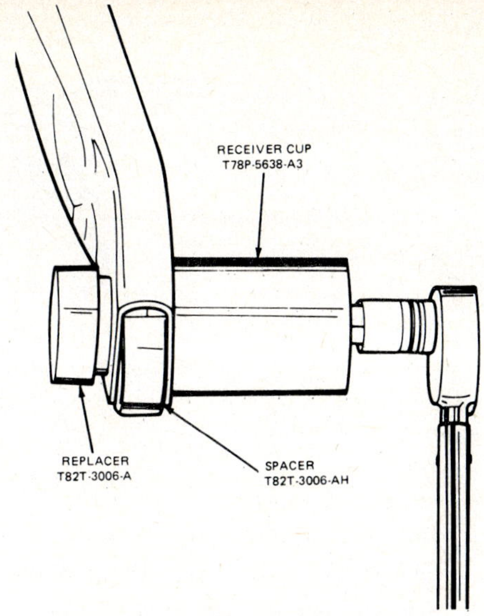

Fig. 20 Axle pivot bushing installation. 4 × 2 models

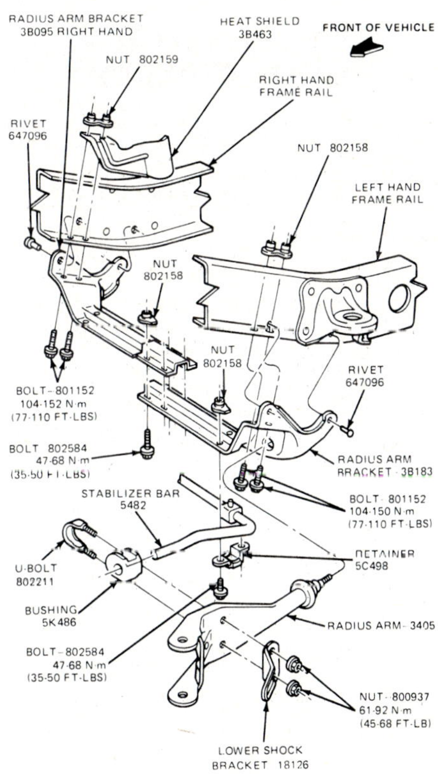

Fig. 21 Front stabilizer bar replacement

5638-A3 onto pivot bushing. Turn forcing screw and remove pivot bushing, Fig. 19.

4. Reverse procedure to install. Refer to Fig. 20. Lower vehicle and with weight on suspension torque pivot bushing and nut to 120–150 ft. lbs.

4 × 4

Refer to the "Front Wheel Drive Section" in this chapter for axle pivot bushing replacement procedure.

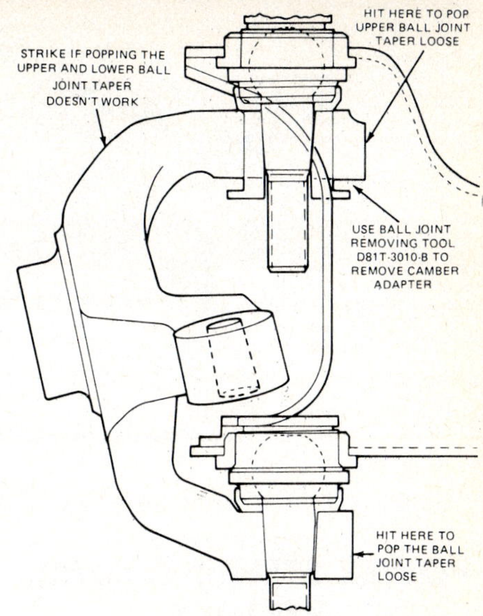

Fig. 22 Front wheel spindle replacement. 4 × 2 models

FRONT STABILIZER BAR, REPLACE

1. Remove nuts and U-bolts retaining lower shock bracket/stabilizer bar bushing to radius arm. Remove retainers and stabilizer bar and bushings, Fig. 21.
2. Reverse procedure to install.

FRONT WHEEL SPINDLE, REPLACE

4 × 2

1. Raise and support front of vehicle. Remove front wheel, then the brake caliper assembly. Support caliper with a length of wire.
2. Remove dust cap, cotter pin, nut retainer, nut, washer, outer bearing and rotor from spindle.
3. Remove inner bearng cone and seal, then the brake dust shield. Discard the seal.
4. Disconnect steering linkage from spindle by removing cotter pin and nut.
5. Remove nut from upper ball joint stud, then the cotter pin and nut from lower ball joint stud.
6. Strike lower side of spindle to pop ball joints loose from spindle, Fig. 22, then remove the spindle.
7. Reverse procedure to install. Note the following:
   a. When installing ball joint stud nuts, coat studs with Loctite 242, or equivalent.
   b. Install lower ball joint stud nut first and torque to 30 ft. lbs., then install upper ball joint stud nut and torque to 85–110 ft. lbs. Complete the tightening sequence by torquing lower ball joint stud nut to 104–106 ft. lbs.
   c. Torque steering linkage to spindle nut to 51–75 ft. lbs.

4 × 4

Refer to the "Front Wheel Drive Section" in this chapter for front wheel spindle replacement procedure.

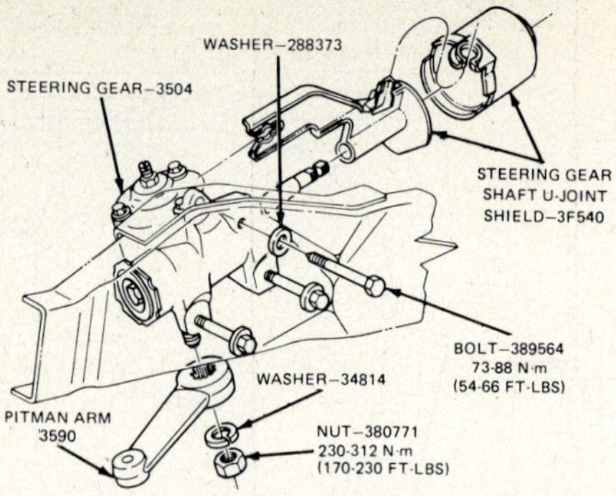

Fig. 23 Power steering gear replacement

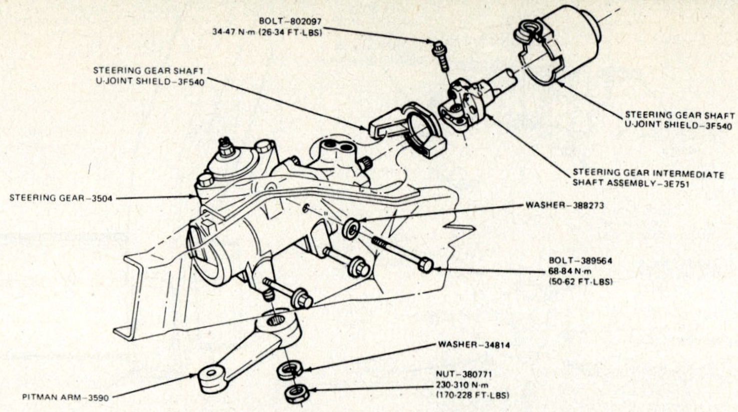

Fig. 24 Manual steering gear replacement

POWER STEERING GEAR, REPLACE

1. Disconnect pressure and return lines from steering gear and plug openings to prevent entry of dirt.
2. Remove steering gear shaft U-joint shield from flex coupling, remove bolt securing flex coupling to steering gear, and disconnect flex coupling.
3. Raise and support vehicle and remove pitman arm using a suitable puller.
4. Support steering gear and remove gear attaching bolts.
5. Work steering gear free of flex coupling and remove gear from vehicle, Fig. 23.
6. Reverse procedure to install, making sure gear is centered with index flat on input shaft pointing down, and steering wheel spokes are in horizontal position.

MANUAL STEERING GEAR, REPLACE

1. Disconnect flex coupling shield from input shaft shield and slide it to upper part of intermediate shaft.
2. Remove bolt securing flex coupling to steering gear and remove steering gear input shaft shield.
3. Raise and support vehicle and remove pitman arm using a suitable puller.
4. Remove bolts securing gear to side rail and remove steering gear from vehicle, Fig. 24.
5. Reverse procedure to install, making sure gear is centered and flat on input shaft and is aligned with flat on flex coupling.

Riding Height In Inches Fig. 3	Caster Angle Degrees①								
	2.95	3.15	3.35	3.55	3.75	3.95	4.15	4.35	4.55
1983 RANGER (4×2)									
Ranger	+5 1/8°—+8°	+4 5/8°—+7 1/2°	+4°—+7°	+3 1/2°—+6 1/2°	+3°—+6°	+2 3/8°—5 3/8°	+1 7/8°—+4 7/8°	+1 3/8°—+4 1/4°	+3/4—3 3/4°

①—Side to side caster readings must be within 1 1/2°, when vehicle riding height is within .12 inch side to side.

Riding Height In Inches Fig. 4	Caster Angle Degrees①									
	2.75	2.95	3.15	3.35	3.55	3.75	3.95	4.15	4.35	4.55
1983 RANGER (4×4)										
Ranger	+5 3/8—+8 1/8	+4 3/4—+7 5/8	−4 1/4—+7	+3 3/4—+6 1/2	+3 1/8—+5 7/8	+2 1/2—+5 3/8	+2—+4 7/8	+1 1/2—+4 3/8	+1—+3 7/8	+1—+3 3/8

①—Side to side caster readings must be within 1 1/2°, when vehicle riding height is within .16 inch side to side.

Riding Height In Inches Fig. 3	Camber Angle Degrees①							
	2.75	2.95	3.15	3.35	3.55	3.75	3.95	4.15
1983 RANGER (4×2)								
Ranger	−1—−3/4	−1—−1/4	−1—+1/8	−5/8—+5/8	−1/8—+1 1/8	+3/8—+1 5/8	+7/8—+2 1/8	+1 3/8—+2 3/4

①—Side to side camber readings must be within .7°, when truck frame to axle riding height is within .12 inch side to side.

Riding Height In Inches Fig. 4	Camber Angle Degrees①									
	2.75	2.95	3.15	3.35	3.55	3.75	3.95	4.15	4.35	4.55
1983 RANGER (4×4)										
Ranger	−1 7/8—−3/8	−1 3/8—+1/8	−7/8—+3/4	+3/8—+1 1/8	+1/8—+1 3/4	+5/8—+2 1/8	+1 1/8—+2 5/8	+1 3/4—+3 1/8	+2 1/8—+3 5/8	+2 3/4—+4

①—Side to side camber readings must be within .7°, when truck frame to axle riding height is within .16 inch side to side.

Front Wheel Drive Section

FRONT AXLE ASSEMBLY, REPLACE

1. Raise and support vehicle. Install jack stands under radius arm brackets.
2. Disconnect driveshaft from front axle yoke, then remove front wheels.
3. Remove disc brake calipers and position aside. Do not let caliper hang down on brake hose.
4. Disconnect steering linkage from spindle by removing the cotter pin and retaining nut.
5. Place a jack under axle arm assembly and compress the coil spring, then remove nut which retains bottom of spring to axle arm.
6. Lower axle until all spring tension is released, then remove the coil spring, spacer, seat and stud.

NOTE: The axle assembly must be supported when removing spring and not be permitted to hang by the brake hose. If the length of the brake hose is not sufficient to provide adequate clearance for spring removal, the disc brake caliper must be removed. If so, do not suspend caliper by brake hose.

---

7. Disconnect shock absorber from radius arm bracket, then remove the bracket and radius arm.
8. Remove pivot bolt securing right hand axle arm to crossmember.
9. Remove clamps from axle shaft slip yoke and axle shaft, then slide rubber boot over yoke.
10. Disconnect right driveshaft from slip yoke assembly, then lower jack and remove right axle arm assembly.
11. Position another jack under differential housing, then remove bolt securing left axle arm to crossmember and remove left axle arm assembly.
12. Reverse procedure to install. Note the following torques: pivot bracket bolts, 120–150 ft. lbs.; radius arm stud, 155–205 ft. lbs.; radius arm bracket bolts, 27–37 ft. lbs.; coil spring attaching nut, 70–100 ft. lbs.; shock absorber attaching nut, 48–68 ft. lbs.; ball joint attaching nut, 85–115 ft. lbs.; drive shaft to front axle yoke U-bolt nuts, 8–15 ft. lbs.

SPINDLE, SHAFT & JOINT ASSEMBLY, REPLACE, FIG. 1

1. Raise and support vehicle. Remove front wheels.
2. Remove disc brake caliper, hub locks, wheel bearings and lock nuts, then the hub, rotor and outer wheel bearing cone.
3. Remove and discard grease seal from rotor using tools 1175-AC and T50T-100-A, or equivalent.
4. Remove inner wheel bearing, then the inner and outer bearing cups using tool D78P-1225-B, or equivalent.
5. Remove spindle to steering knuckle attaching nuts, then break spindle loose from knuckle and remove splash shield.
6. Remove shaft and joint assembly from left side of vehicle by pulling assembly out of carrier.
7. Remove and discard clamp from shaft and joint assembly and the stub shaft on right side of carrier. Slide rubber boot onto stub shaft, then pull shaft and joint assembly from stub shaft splines.
8. Install spindle, on second step, in suitable soft jawed vise.
9. Remove oil seal and needle bearing from spindle using tools T50T-100-A and 1175-AC, or equivalent. Remove oil seal from shaft, if necessary.
10. Reverse procedure to install. Torque spindle to steering knuckle attaching nuts to 35–45 ft. lbs.

RIGHT HAND SLIP YOKE & STUB SHAFT ASSEMBLY, CARRIER, CARRIER OIL SEAL & BEARING, REPLACE, FIG. 1

1. Disconnect driveshaft from yoke. Position driveshaft aside so it will not interfere with carrier removal.
2. Remove both spindles, then the left and right shaft and U-joint assemblies as described previously.
3. Support carrier using a suitable jack, then remove carrier to support arm attaching bolts.
4. Remove carrier from support arm, then drain lubricant and remove carrier from vehicle.
5. Install carrier in suitable holding fixture, then rotate slip yoke and shaft assembly until open end of snap ring is exposed.
6. Remove snap ring, then the slip yoke and shaft assembly from carrier.
7. Remove oil seal and caged needle bearings simultaneously from carrier using tools T50T-100-A and D880L-100-A or equivalent. Discard seal and bearings.
8. Reverse procedure to install. Note the following:
   a. Clean mating surfaces of carrier and support arm. Apply a narrow bead of RTV sealant to mating surfaces.
   b. Install carrier attaching bolts hand tight, then torque in a clockwise or counterclockwise pattern to 40–50 ft. lbs.
   c. Torque carrier to axle arm shear bolt to 75–95 ft. lbs.
   d. Torque driveshaft to yoke U-bolt nuts to 8–15 ft. lbs.

STEERING KNUCKLE & BALL JOINTS, REPLACE, FIG. 1

1. Remove spindle, then the shaft and joint assembly as described previously.
2. If tie rod has not been removed, remove cotter pin and nut and disconnect tie rod from steering arm.
3. Remove cotter pin and loosen nut on top ball joint stud. Loosen bottom nut inside knuckle, then remove top nut.
4. Break spindle loose from ball stud, then remove camber adjuster. Mark position of adjuster for assembly reference.
5. Install steering knuckle in suitable soft jawed vise, then remove snap ring from lower ball joint socket.
6. Press ball joints from steering knuckle using C-Frame D79T-3010-A with tools D79T-3010-AE and T83T-3050-A, or equivalent.
7. Reverse procedure to install. The lower ball joint must be installed first. When installing ball joint stud nuts, first torque nut on lower stud to 40 ft. lbs., then torque nut on upper stud to 85–100 ft. lbs. Complete tightening sequence by torquing lower nut to 129–149 ft. lbs.

AXLE PIVOT BUSHING, REPLACE

1. Remove front axle assembly as described previously.
2. Remove pivot bushing from axle housing, Fig. 2.
3. Install bushing in axle housing, Fig. 3, then flare the bushing to prevent movement, Fig. 4. Complete installation by crimping bushing lip.

FRONT DRIVESHAFT, REPLACE

1. Remove nuts and U-bolts securing front driveshaft to front drive axle.
2. Remove U-joint assembly from axle yoke, then slide the splined yoke assembly out of transfer case from front output shaft and remove driveshaft assembly.
3. Reverse procedure to install. Torque attaching nuts to 8–15 ft. lbs.

DIFFERENTIAL CARRIER

Disassemble, Fig. 1

1. Remove left hand axle arm assembly as described previously.
2. Loosen carrier to axle arm retaining bolts, then drain fluid from differential.
3. Remove carrier retaining bolts, then the carrier.
4. Remove slip yoke and stub shaft as described previously, then install carrier in suitable holding fixture.
5. Clean gasket material from all mating surfaces, then remove bearing caps. Note location of markings on caps and carrier for assembly reference.
6. Remove differential case, then the bearing cups from carrier. Note which side of carrier bearing cups were removed from.
7. Support carrier with nose end up and hold end yoke with a suitable tool, then remove nut and washer from pinion shaft.
8. Remove end yoke using tool T65L-4851-B, or equivalent.

NOTE: The yoke must be replaced if the seal contact area shows any signs of wear.

---

9. Remove drive pinion, drive pinion oil

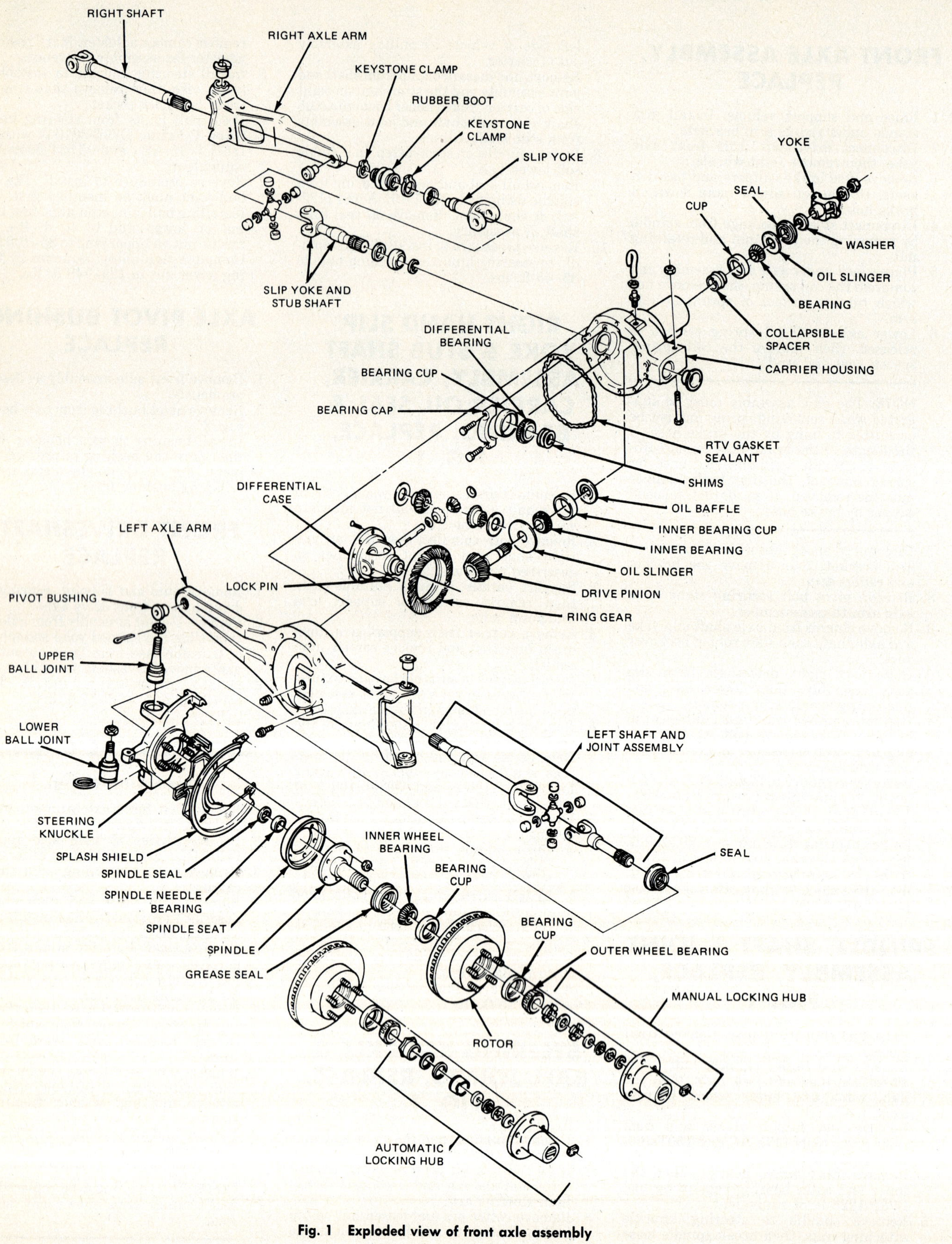

Fig. 1 Exploded view of front axle assembly

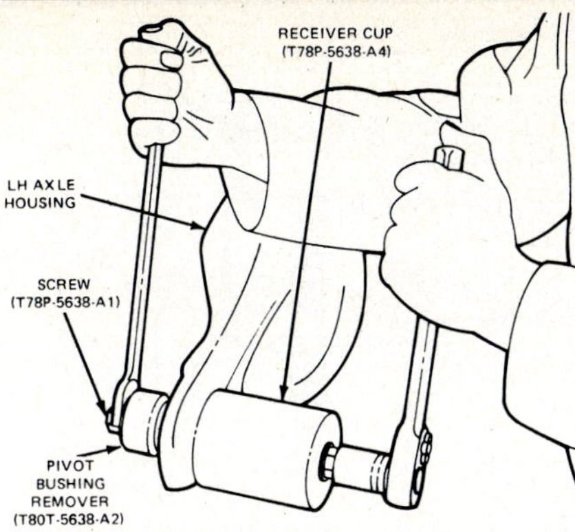

Fig. 2 Axle pivot bushing removal

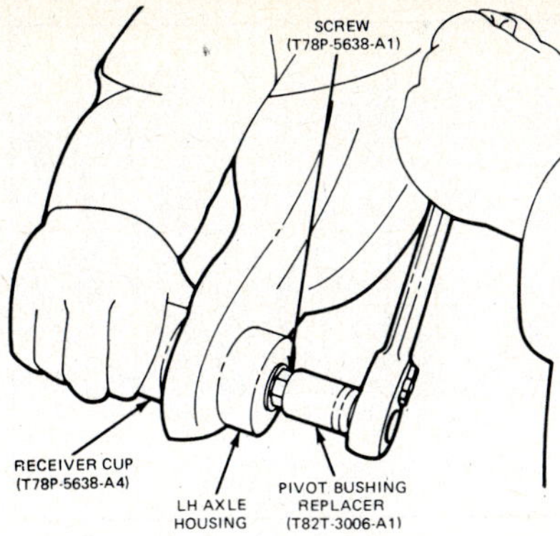

Fig. 3 Axle pivot bushing installation

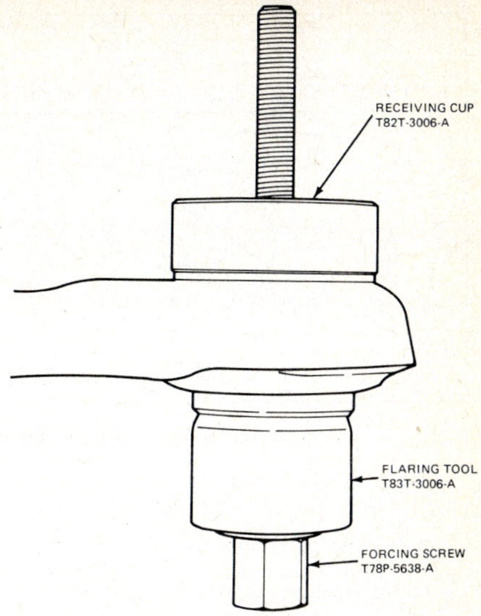

Fig. 4 Flaring axle pivot bushing

seal, outer pinion bearing cone, oil slinger and collapsible spacer from carrier. Discard oil seal and collapsible spacer.

10. Remove inner pinion bearing cup, Fig. 5, then remove oil baffle from bearing cup bore.
11. Invert carrier, then remove outer pinion bearing cup, Fig. 5.
12. Remove differential case bearings and shims. Wire shims, bearing cup and cone together and note which side of case they were removed from.
13. Install differential case in suitable soft jawed vise. Remove ring gear attaching bolts, then the ring gear from case.

NOTE: Whenever the ring gear is removed, the ring gear attaching bolts must be replaced.

Total Differential Case End Play Check

1. Install ring gear onto differential case. Torque bolts alternately and evenly to 50–60 ft. lbs.
2. Clean trunnions on differential, then install master differential bearings onto case. Remove any burrs and nicks from hubs so bearings will rotate freely.
3. Install differential case into carrier, without drive pinion. Ensure that the case moves freely in the carrier.
4. Install suitable dial indicator on differential case flange. Position tip of indicator on a flat of one ring gear bolt.
5. Move differential case toward dial indicator as far as possible, then zero the indicator with force still applied. The dial indicator should have a minimum travel of .200 inch.
6. Move differential case away from dial indicator as far as possible and record reading. This reading indicates amount of shims needed behind differential side bearings to take up total clearance between differential bearing and case.
7. Remove differential case from carrier. Do not remove master differential bearings from case at this time.

Depth Gauge Check

1. Install oil baffle into inner bearing cup bore in carrier.
2. Install inner and outer pinion bearing cups in carrier, Fig. 6.
3. Position a new rear pinion bearing over aligning adapter T76P-4020-A1, then install into pinion bearing retainer assembly.
4. Install front pinion bearing into bearing cup in carrier. The old bearing may be reused, if serviceable. Assemble handle T76P-4020-A11 onto screw T76P-4020-A9 and hand tighten.
5. Install gauge tube T76P-4020-A7 into differential bearing bore, then install bearing caps and torque to 35–40 ft. lbs. Set bearing preload by torquing handle to 20–40 inch lbs.
6. Using a feeler gauge, measure distance between gauge tube and gauge block T76P-4020-A10.
7. If there are no markings on pinion gear, select an oil slinger with a thickness equal to the dimension measured in step 6. If pinion gear is marked with a "+" reading, this amount must be subtracted from the dimension measured in step 6. For example, if pinion is marked +2, use a slinger .002 inch thinner than measured. If pinion is marked with a "–" reading, this amount must be added to the measured dimension.
8. Measure thickness of oil slinger with a micrometer to verify correct size, then install slinger on pinion and press bearing onto pinion.
9. Apply lubricant C1AZ-19590-B, or equivalent, to ends of outer pinion bearing rollers, then install outer bearing cone in outer bearing cup.
10. Install drive pinion with inner bearing cone and oil slinger into carrier.

Drive Pinion Preload & Final Depth Checks

1. With pinion installed in carrier, install outer bearing cone and oil slinger.
2. Install end yoke, washer, deflector and slinger on pinion shaft using alignment tool T80T-4000-G and flange holder T78P-4851-A.
3. Install new pinion nut and tighten to a rotating torque of 10 inch lbs. Rotate pinion several times to seal bearing.
4. Install gauge tube T76P-4020-A7 in carrier, then install the bearing caps and torque to 35–40 ft. lbs.
5. Install final check gauge block T83T-4020-F58 on top of pinion button under gauge tube, Fig. 7. Press down on ends of gauge block to make sure it is level on pinion.
6. Measure clearance between gauge tube and final check block using a feeler gauge until a slight drag is felt. Clearance should be .020 inch added to the drive pinion etching, with a tolerance of ±.002 inch. For example, a pinion with an etching of "+2" will have a .022 inch clearance and a .002 inch tolerance. Therefore, shims equalling .020–.024 inch would be required.
7. When drive pinion is set at correct depth, remove yoke using tools T78P-4851-A and T65L-4851-B. Remove yoke, nut and washer, and outer bearing cone.
8. Install a new collapsible spacer and outer bearing cone.
9. Apply lubricant C6AZ-19580-E, or equivalent, to drive pinion oil seal, then install the seal using tool T71T-3010-R. Make sure that garter spring remains in place.
10. Install yoke using tool T83T-4851. Torque nut to 175 ft. lbs.
11. Using an inch-pound torque wrench, check rotational torque of pinion. Rotational torque should be 15–35 inch lbs.

NOTE: If rotational torque is less than 15 inch lbs., tighten the pinion nut in small increments until 15–35 inch lbs. is reached. Do not tighten nut more than 225 ft. lbs.

If reading is more than 35 inch lbs., the collapsible spacer has been compressed too far and must be replaced.

Differential Case, Service

1. Install differential case in suitable vise, then drive out lock pin which retains pinion mate shaft to case.
2. Remove drive pinion mate shaft using a suitable drift pin.
3. Rotate pinion mate gears and side gears until mate gears are exposed in windows of case, then remove mate gears and spherical washers.
4. Remove side gears and thrust washers, then install lock pin in case. Set lock pin

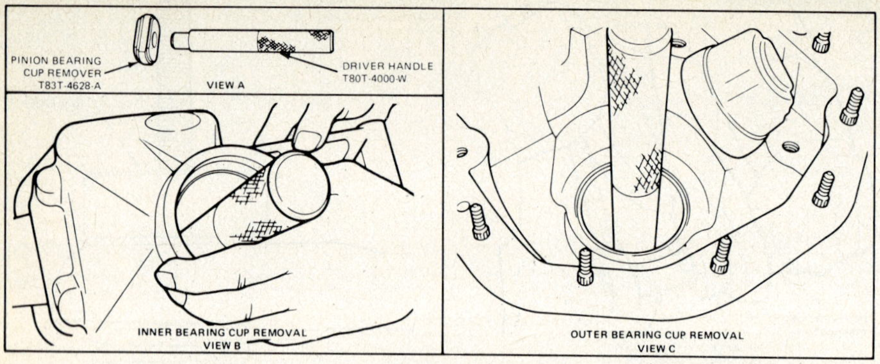

Fig. 5 Bearing cup removal

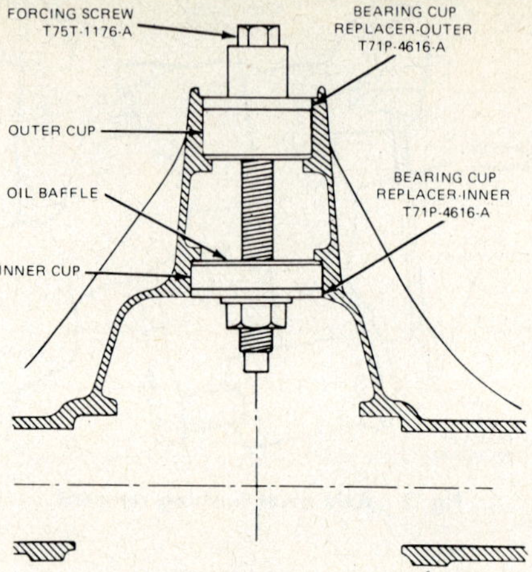

Fig. 6 Pinion bearing cup installation

in place by peening some metal of the case over pin.

5. Apply lubricant C1AZ-19590-B, or equivalent, to all gears and washers.
6. While holding side gears in position in case, install pinion mate gears and spherical washers. Rotate side gears and pinion mate gears until holes in washers and pinion mate gears line up with holes in case, then install pinion mate shaft into case. Ensure that lock pin hole in shaft lines up with lock pin holes in case.

Assemble, Fig. 1

1. Install ring gear on differential case, using new bolts. Torque bolts alternately and evenly to 50–60 ft. lbs.
2. Position differential case in carrier. Move case away from drive pinion gear until it is seated against cross bore face of carrier.
3. Install suitable dial indicator on differential so that indicator tip rests on a flat of one case bolt. Zero the dial indicator.
4. Force ring gear against pinion gear, then rock ring gear to make sure gear teeth are in contact. Move ring gear away from drive pinion gear, making sure indicator returns to zero. Repeat this procedure until reading is the same. This reading, minus .006 inch, indicates amount of shims needed between differential case and differential bearing on ring gear side.
5. Remove differential case from carrier, then the master bearings from case.
6. Install required shims, as determined in step 4, on ring gear hub of differential case.
7. Install bearing cone and bearing onto ring gear hub using tool T80T-4000-J.
8. To determine amount of shims needed on hub of drive pinion side of differential case, subtract reading obtained in step 4 from reading obtained in "Total Differential Case End Play" procedure, then add .003 inch to this amount.

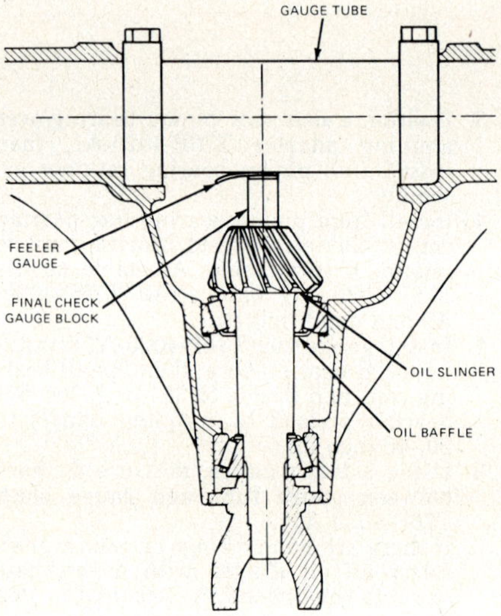

Fig. 7 Pinion depth check

9. Install required shims, as determined in step 8, on hub of drive pinion side of differential case.
10. Install bearing cone and bearing onto drive pinion hub using tool T80T-4000-J. Protect ring gear side bearing using step plate D80L-630-5, or equivalent.
11. Install bearing cone on pinion side of differential case. Protect ring gear bearing using tool T-53T-4621-C, or equivalent.
12. Install differential bearing cups on bearing cones, then install differential case into carrier. If ring gear and drive pinion have alignment marks, the marks must be aligned during assembly.
13. Remove dial indicator and spreader from case, then install bearing caps and bolts. Make sure that markings on caps correspond, in both position and direction, to markings on carrier. Torque bolts to 35–40 ft. lbs.
14. Install suitable dial indicator on differential case and measure ring gear and pinion backlash at three equally spaced points on ring gear. Backlash must measure .004–.010 inch and cannot vary more than .003 inch between measuring points. If backlash is high, move ring gear closer to pinion by moving shims to ring gear side from opposite side. If backlash is low, move ring gear away from pinion by moving shims from ring gear side to opposite side.
15. Apply a narrow bead of RTV sealant on carrier mounting face support arm mating surfaces.

NOTE: Allow the sealant at least one hour curing time after the carrier is assembled to the axle arm before filling the differential with lubricant or operating vehicle.

16. Install differential assembly onto left hand axle arm, using two guide pins. Install bolts and torque to 40–50 ft. lbs. Torque one bolt, then the bolt directly opposite it. Torque remaining bolts in a clockwise or counterclockwise direction.
17. Install carrier shear bolt and nut. Torque to 75–95 ft. lbs.

Transfer Case Section

DESCRIPTION

The Borg-Warner 13-50 transfer case, Fig. 1, is a three-piece aluminum part time unit. When actuated, the transfer case transfers power from the transmission to both front and rear axles. Lubrication is accomplished by a positive displacement oil pump which routes oil through drilled holes in the rear output shaft.

TRANSFER CASE, REPLACE

1. Raise and support vehicle.
2. Remove skid plate from frame, if equipped, then drain fluid from transfer case.
3. Disconnect four-wheel drive indicator switch electrical connector from transfer case.
4. Disconnect front driveshaft from axle input yoke.
5. Loosen clamp retaining front driveshaft boot to transfer case, then remove driveshaft and front boot assembly from transfer case front output shaft.
6. Disconnect rear driveshaft from transfer case output shaft yoke, then the speedometer driven gear from transfer case rear cover.
7. Disconnect vent hose from control lever, then remove shoulder bolt and the shifter to extension housing retaining bolt. Pull control lever out until bushing slides off shift lever pin. If necessary, remove shift lever from control lever.
8. Remove heat shield from transfer case.
9. Support transfer case with a suitable jack, then remove retaining bolts. Slide transfer case rearward and lower from vehicle.
10. Reverse procedure to install. Note the following:
    a. Torque transfer case retaining bolts to 25–35 ft. lbs. in sequence shown, Fig. 2.
    b. Torque heat shield retaining bolts to 27–37 ft. lbs.
    c. Torque shifter to extension housing retaining bolt to 31–42 ft. lbs. and shoulder bolt to 70–90 ft. lbs. The shifter retaining bolt must be tightened before the shoulder bolt.
    d. Torque speedometer driven gear attaching screw to 20–25 inch lbs., front and rear driveshaft retaining bolts to 12–15 ft. lbs., fill and drain plugs to 14–22 ft. lbs. and skid plate retaining bolts to 22–30 ft. lbs.
    e. Fill transfer case with 3 pints of Dexron II automatic transmission fluid, or equivalent.

SHIFT LEVER, REPLACE

NOTE: Removal of the shift ball is necessary only if the shift ball, boot, or lever is being replaced. If not, remove the ball, boot and lever as an assembly.

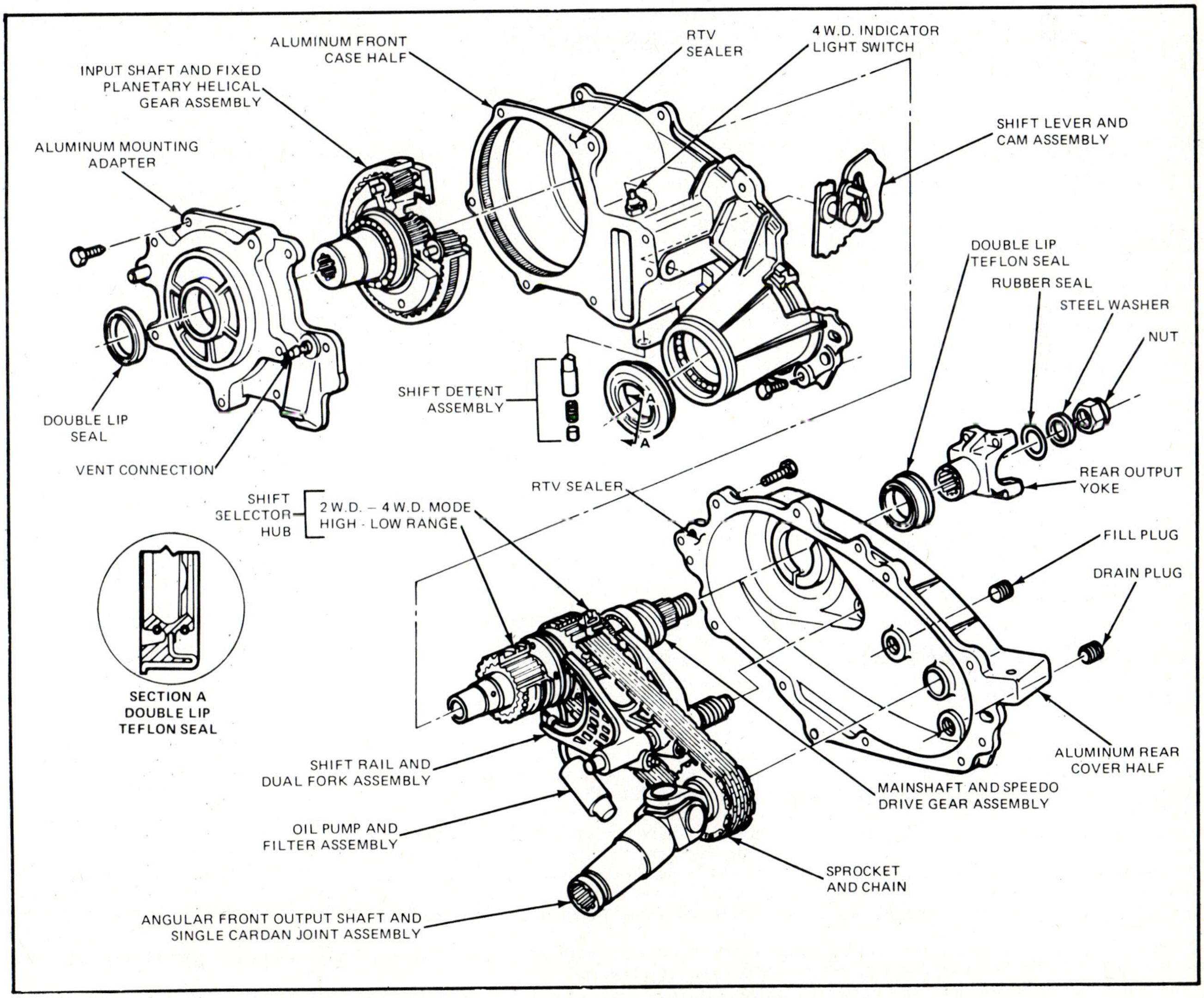

Fig. 1 Exploded view of Borg-Warner 13-50 tranfer case

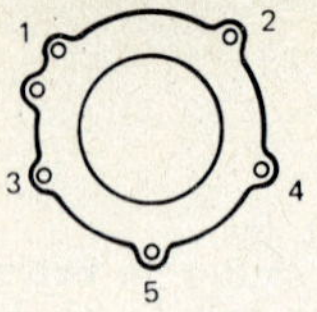

Fig. 2 Transfer case retaining bolt torque sequence

1. Remove plastic insert from shift ball, then warm ball to approximately 140–180° F and knock ball off lever with a hammer and a block of wood.
2. Remove rubber boot and pan cover, then disconnect vent hose from control lever.
3. Unscrew shift lever from control lever, then remove shifter to extension housing retaining bolts and the shoulder bolt. Remove control lever and bushings.
4. Reverse procedure to install. Note the following:
   a. Torque shifter to extension housing retaining bolt to 31–42 ft. lbs. and shoulder bolt to 70–90 ft. lbs. The shifter retaining bolt must be tightened before the shoulder bolt.
   b. Torque shift lever to 27–37 ft. lbs.
   c. Install vent assembly so the white marking on housing is in line with notch in shifter. Position upper end of vent hose inside shift lever bolts.
   d. Install shift ball to the end of knurl on upper part of shift lever.

FRONT OUTPUT SHAFT OIL SEAL, REPLACE

1. Raise and support vehicle.
2. Remove front driveshaft from axle input yoke.
3. Loosen clamp retaining front driveshaft boot to transfer case, then remove driveshaft and boot assembly from front output shaft.
4. Drain fluid from transfer case, then remove oil seal using tools 1175-AC and T50T-100-A.
5. Reverse procedure to install. Note the following:
   a. Apply lubricant C1A-19590-B, or equivalent to oil seal and front output shaft female spline
   b. Install oil seal using tools T83T-7065-B and T80T-4000-W, or equivalent.
   c. Torque front driveshaft retaining bolts to 12–15 ft. lbs. and the drain and fill plugs to 14–22 ft. lbs.

REAR OUTPUT SHAFT OIL SEAL, REPLACE

1. Raise and support vehicle.
2. Remove rear driveshaft from transfer case output shaft yoke and position aside.
3. Remove output shaft yoke, then the oil seal using tools 1175-AC and T50T-100-A.
4. Reverse procedure to install. Note the following:
   a. Apply lubricant C1AZ-19590-B, or equivalent, to oil seal.
   b. Install oil seal using tools T83T-7065-B and T80T-4000-W, or equivalent.
   c. Torque output shaft yoke retaining nut to 100–130 ft. lbs. and rear driveshaft retaining bolts to 12–15 ft. lbs.

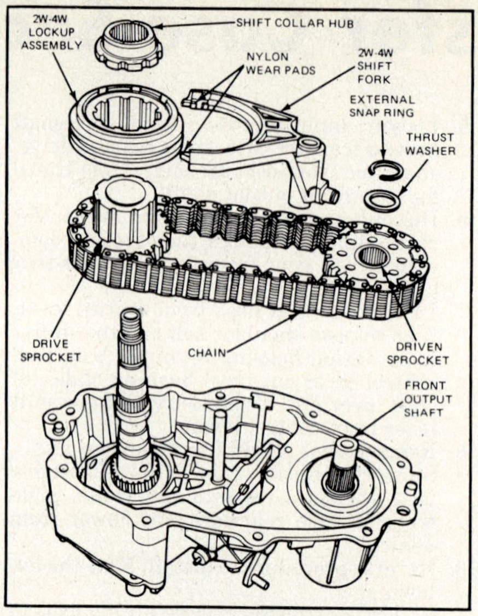

Fig. 3 Shift collar hub & 2W-4W lock-up & shift fork assemblies

TRANSFER CASE, DISASSEMBLE

1. Remove transfer case as described under "Transfer Case, Replace".
2. Drain lubricant from transfer case, then remove four-wheel indicator switch and the breather vent plug.
3. Remove rear output yoke, then the front case to rear cover attaching bolts.
4. Separate front case from rear cover, then remove all gasket material from mating surfaces.
5. If speedometer drive gear or ball bearing assembly is to be replaced, remove output shaft oil seal. The seal can be removed from inside of rear cover with a brass drift and hammer, or from outside by bending and pulling the oil seal lip. Discard oil seal, then remove speedometer drive gear assembly.
6. Remove rear output shaft ball bearing snap ring, then the bearing using tools T83T-7025-B and T50T-100-A, or equivalent.
7. Remove front output shaft caged needle bearing from rear cover using tools D80L-100-S and T50T-100-A, or equivalent.
8. Remove 2W-4W shift fork spring from rear cover.
9. Remove shift collar hub from output shaft, then the 2W-4W lock-up assembly and shift fork as an assembly, Fig. 3.
10. Remove shift fork from lock-up assembly, Fig. 3. If necessary, remove external clip, then roller bushing assembly from shift fork.
11. Remove internal snap ring from lock-up collar, then remove lock-up hub and spring from collar, Fig. 4.
13. Remove snap ring and thrust washer retaining drive sprocket to front output shaft, then remove chain, driven sprocket and drive sprocket as an assembly, Fig. 3. Remove magnet from front case, Fig. 5.
14. Remove output shaft and oil pump as an assembly.
15. Remove shift rail, high-low range shift fork and high-low shift hub from front case, Fig. 5.
16. Invert front case and remove mounting adapter retaining bolts, then remove mounting adapter, input shaft and planetary gearset as an assembly, Fig. 6.

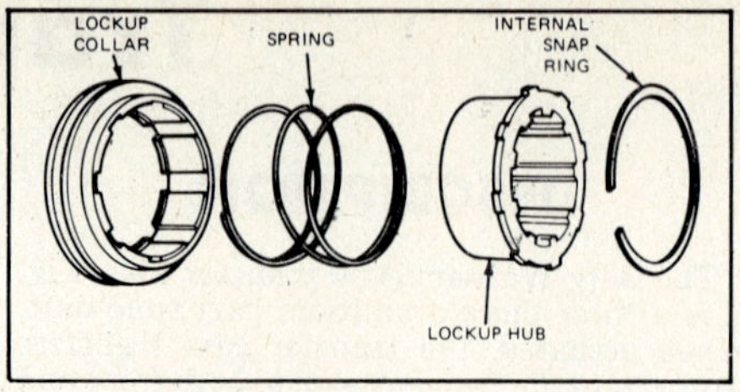

Fig. 4 2W-4W lock-up assembly

17. Press ring gear from front case. Note position of ring gear for assembly reference.
18. Remove planetary gearset and input shaft from mounting adapter by expanding large snap ring under adapter and prying under gearset with screwdrivers, Fig. 6.
19. Remove oil seal from mounting adapter using tools 1175-AC and T50T-100-A, or equivalent.
20. Remove internal snap ring from planetary carrier, then remove planetary gearset from input shaft assembly, Fig. 6.
21. Remove external snap ring from input shaft, then install input shaft assembly in suitable press and remove ball bearing using tool D79L-4621-A.
22. Remove thrust washer, thrust plate and sun gear from input shaft, Fig. 6.
23. Place shift lever in "TWO WHEEL HIGH" (2WH) position, then make an alignment mark on outside of front case at the shift lever.
24. Remove set screws from front case and shift cam, then invert front case and remove external clip.
25. Remove shift lever from front case and shift cam. Remove "O" ring from second groove in shift lever shaft.
26. Remove detent plunger and compression spring from inside of front case.
27. Remove internal snap ring and ball bearing retainer from front case. Remove bearing from retainer using tools T83T-7025-B and T80T-4000-W.
28. Remove front output shaft and U-joint assembly from front case. Remove oil seal, then the internal snap ring and ball bearing from case, Fig. 7.
29. Install front output shaft and U-joint assembly in a soft jawed vise, then remove internal snap rings which retain bearings in shaft.
30. Press bearing out of front output shaft using tool T74P-4635-C, or equivalent. Reposition tool on spider and remove opposite bearing. Repeat procedure for remaining bearings.

TRANSFER CASE, ASSEMBLE

NOTE: Prior to assembly, lubricate all transfer case components with Dexron II automatic transmission fluid, or equivalent.

1. Install output shaft in a soft jawed vise, then install a new bearing into end of shaft ear.
2. Install spider into bearing. Press bearing below snap ring groove using tool T74P-4625-C. Remove tool and install a new

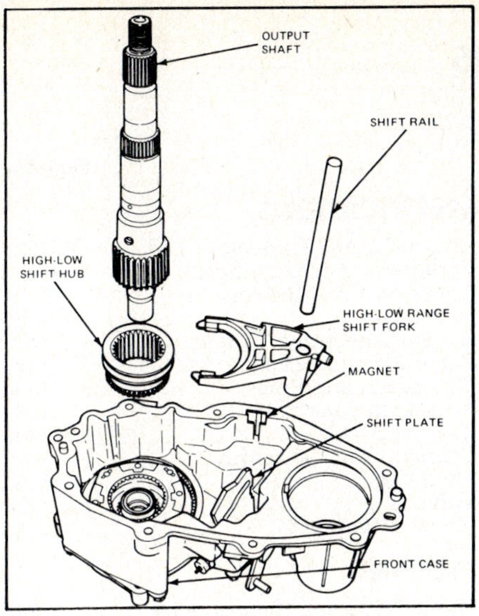

Fig. 5 High-low shift assembly, output shaft & collector magnet

internal snap ring in groove.

3. Install a new bearing into opposite end of shaft ear. Press bearing until opposite bearing contacts snap ring. Remove tool and install a new internal snap ring in groove.
4. Reposition output shaft in vise and install remaining two bearings. Check U-joint for free movement. If there is any binding, tap ears of both shafts to relieve the bind.
5. Install ball bearing into front output shaft case bore using tools T83T-7024-B and T80T-4000-W, then install internal snap ring. Make sure bearing is not cocked in the bore.
6. Install front output oil seal in front case bore.
7. Install ring gear in front case, using a suitable press.
8. Install ball bearing, then the internal snap ring into bearing retainer bore, Fig. 7.
9. Install front output shaft and U-joint assembly through front case seal, then install the ball bearing and retainer assembly and internal snap ring. The clip on bearing retainer must align with slot in front case, Fig. 7.
10. Install compression spring and detent plunger into bore from inside of case.
11. Install a new "O" ring into second groove of shift lever shaft.
12. Install shift cam in front case with "2WH" detent position over the detent plunger. While holding shift cam in position, push shift lever shaft into front case to engage shift cam, using alignment mark made during disassembly.
13. Install external clip on end of shift lever shaft.
14. Install two set screws in front case and shift cam. Torque screws to 5–7 ft. lbs.

NOTE: The set screw in the front case must be in first groove of shift lever shaft, and not bottomed against the shaft. Also, check that the shift lever can be moved to all positions.

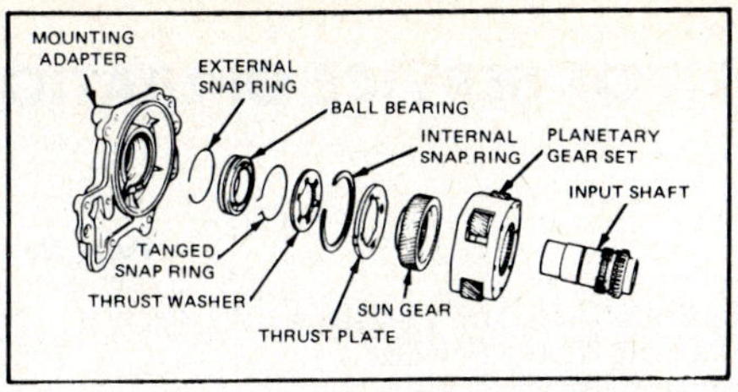

Fig. 6 Mounting adapter assembly

15. Install sun gear, thrust plate, thrust washer, ball bearing and external snap ring onto input shaft, Fig. 6.
16. Install planetary gearset onto sun gear and input shaft assembly, then the internal snap ring on planetary carrier, Fig. 6.
17. Install oil seal, then the tanged snap ring into mounting adapter.
18. Install input shaft and planetary gearset in mounting adapter. Push the assembly into adapter until it is fully seated and the snap ring engages.
19. Apply a bead of RTV sealant to front case mating surface, then install mounting adapter on front case. Torque bolts to 35–40 ft. lbs.
20. Install high-low shift hub into planetary gearset. Position high-low shift fork into inside track of shift cam and groove of the high-low shift hub, Fig. 5. Ensure that nylon wear pads are installed on shift fork and dot on pad lines up with fork hole.
21. Install shift rail through high-low fork until rail is fully seated in front case bore, Fig. 5.
22. Install oil pump cover with "TOP" marking facing front of front case. Install pump pins, flats facing up, with spring between pins, then install assembly in oil pump bore in output shaft.
23. Install oil pump body and pickup tube over shaft, then install oil pump rear body with "TOP REAR" marking facing rear of front case.
24. Install oil pump retainer and attaching bolts. Torque bolts, while rotating output shaft, to 36–40 inch lbs.
25. Install output shaft and oil pump assembly in input shaft. Ensure that external splines of output shaft engage internal splines of high-low shift hub and that the oil pump retainer and oil filter leg are in the groove and notch of front case.
26. Install collector magnet in front case, Fig. 5.
27. Install chain, drive sprocket and driven sprocket assembly over shafts. Install thrust washer on front output shaft, then the external snap ring over thrust washer.
28. Assemble 2W-4W lockup assembly, then install 2W-4W shift fork to 2W-4W lockup assembly, Fig. 3. Ensure nylon wear pads are installed on shift fork and dot on pad lines up with fork hole.

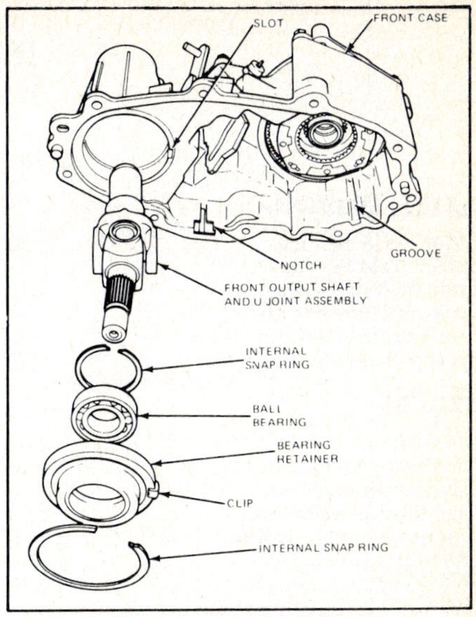

Fig. 7 Front output shaft assembly

29. Install 2W-4W lockup collar and hub assembly over output shaft and into shift rail. Install shaft, bushing and external clip on 2W-4W lockup fork.
30. Install shift collar hub on output shaft, then install caged needle bearing into rear cover using tools T83T-7127-A and T80T-4000-W, or equivalent.
31. Install ball bearing in rear cover bore using tools T83T-7025-B and T80T-4000-W, or equivalent. Make sure that bearing is not cocked in bore, then install snap ring to retain bearing in rear cover.
32. Install speedometer drive gear assembly into rear cover so that round end of speedometer gear clip faces the inside of rear cover.
33. Install oil seal in rear cover using tools T83T-7065-B and T80T-4000-W, or equivalent.
34. Install 2W-4W shift fork spring on inside boss of rear cover.
35. Apply a bead of silicone rubber sealant D6AZ-19562-A, or equivalent, to front case mating surface.
36. Install front case on rear cover, making sure the 2W-4W shift fork spring engages shift rail. Install attaching bolts, starting with rear cover bolts, and torque to 35–40 ft. lbs.
37. Install rear yoke on output shaft with rubber seal, washer and nut. Torque nut to 100–130 ft. lbs.
38. Install four-wheel drive indicator switch and torque to 8–12 ft. lbs.
39. Install breather plug and torque to 6–14 ft. lbs.
40. Install drain plug and torque to 14–22 ft. lbs.
41. Fill transfer case with 3 pints of Dexron II automatic transmission fluid, or equivalent, then install fill plug and torque to 14–22 ft. lbs.
42. Install transfer case as described under "Transfer Case, Replace".

INTERNATIONAL

INDEX OF SERVICE OPERATIONS

MODEL INDEX & ENGINE APPLICATION

ENGINE IDENTIFICATION—Engine model takes the form of a prefix to the engine serial number. For example, a V-266 engine serial number would read V-266-000000. Engine numbers are located on machined boss on block.

Truck Model or Series	Year	Standard Engine Make	Standard Engine Model	Oil Pan Refill Capacity Qts.	Cooling System Capacity Qts.
SCOUT					
Scout II	1974	Own	4-196	4	11.6
Scout II	1974	Jeep	6-258	4	12
Scout II	1975–80	Own	4-196	4	14
LOADSTAR					
1600	1974	Own	V-345	10	22
1600	1975–78	Own	V-345	—	—
1600 4×4	1974	Own	V-304	—	—
1600 4×4	1975	Own	V-345	—	—
1610FC	1978	Own	V-345	—	—
1650	1978	Own	D-150	—	—
1700	1974	Own	V-345	10	22
1700	1975–78	Own	MV-404	—	—
1700 4×4	1975–77	Own	V-345	—	—
1750	1974	Own	DV-550B	—	42
1750	1975–78	Own	D-150	12	—
1800	1974	Own	V-345	10	22
1800	1975–78	Own	MV-404	—	—
1850	1974	Own	DV-550B	—	42
1850	1975–78	Own	D-170	12	—
F-1800	1974	Own	V-345	10	22
F-1800	1975–78	Own	MV-404	—	—
F-1850	1974	Own	DV-550B	—	42
F-1850	1975–78	Own	D-170	12	42
CARGOSTAR					
CO-1610B	1974–83	Own	V-345	—	—
CO-1710B	1974, 1983	Own	V-345	—	—
CO-1710B	1975–82	Own	MV-404	—	—
CO-1750B	1981–83	Own	D-165	—	—

Continued

MODEL INDEX & ENGINE APPLICATION—Continued

Truck Model or Series	Year	Standard Engine Make	Standard Engine Model	Oil Pan Refill Capacity Qts.	Cooling System Capacity Qts.
CO-1810B	1974	Own	V-345	—	—
CO-1810B	1975–82	Own	MV-404	—	—
CO-1850B	1974	Own	DV-550B	10	—
CO-1850B	1975–79	Own	D-170	12	—
CO-1850B	1980	Own	D-180	—	—
CO-1850B	1981–83	Own	D-165	—	—
CO-1910B	1974	Own	VS-401	10	—
CO-1910B	1975–77	Own	VS-478	—	—
CO-1910B	1978–81	Own	V-537	—	—
CO-1950B	1974	Own	DV-550B	10	—
CO-1950B	1975–76	Own	D-170	12	—
CO-1950B	1977–79	Own	D-190	—	—
CO-1950B	1980–83	Own	D-180	—	—
COF-1810B	1974	Own	V-345	—	—
COF-1810B	1975–80	Own	MV-404	—	—
COF-1910B	1974	Own	VS-401	10	—
COF-1910B	1975–76	Own	VS-478	—	—
COF-1910B	1977–80	Own	V-537	—	—
COF-1950B	1974	Own	DV-550B	10	—
COF-1950B	1975–76	Own	D-170	12	—
COF-1950B	1977–79	Own	D-190	—	—
COF-1950B	1980	Own	D-180	—	—
FLEETSTAR					
F-2010A	1974	Own	RD-450	9	47.7
F-2010A	1975–77	Own	VS-478	—	53
F-2010A	1978	Own	V-537	—	—
F-2050A	1974	Own	DV-550B	—	57.3
F-2050A	1975–77	Own	D-170	—	—
F-2050A	1978	Own	D-190	—	—
F-2500A	1975	Own	D-170	12	57.3
F-2070A	1974–77	Detroit	6-71N	30	48
2010A	1974	Own	RD-450	9	47.7
2010A	1975–77	Own	VS-478	—	53
2010A	1978	Own	V-537	—	—
2050A	1974	Own	DV-550B	—	57.3
2050A	1975–77	Own	D-170	12	57.3
2050A	1978	Own	D-190	—	—
2070A	1974–77	Detroit	6-71N	30	48
TRANSTAR					
4270	1974–75	Detroit	8V-71	27	67
4270	1974–77	Detroit	8V-71NE	27	67
4270	1978	Detroit	8V-71T	—	—
4270	1978–79	Detroit	6V-92TT	—	67
4270	1980–83	Detroit	6V-92TTA	—	67
4370	1974–80	Cummins	NTC-290	34.5	54
4370	1974–77	Cummins	NH-230	32	54
4370	1981–83	Cummins	Form. 300	—	—

Truck Model or Series	Year	Standard Engine Make	Standard Engine Model	Oil Pan Refill Capacity Qts.	Cooling System Capacity Qts.
F-4270	1974–75	Detroit	8V-71	27	67
F-4270	1974–77	Cummins	NTC-290	—	—
F-4270	1978	Detroit	8V-71T	—	—
F-4270	1979–83	Detroit	6V-92TT	—	67
F-4370	1974–80	Cummins	NTC-290	34.5	54
F-4370	1981–83	Cummins	Form. 300	—	—
CO-4070B	1974–80	Cummins	NTC-290	—	55
CO-4070B	1981	Cummins	Form. 290	—	—
COF-4070B	1974–80	Cummins	NTC-290	—	55
COF-4070B	1981	Cummins	Form. 290	—	—
COF-5370	1978	Cummins	NTC-250	—	—
TRAVELALL					
100 4×2	1974	Own	V-345	—	—
100 4×4	1974	Own	V-392	—	—
150 4×2	1975	Own	V-304	—	19
150 4×4	1975	Own	V-304	—	19
200 4×2	1974	Own	V-392	—	—
200 4×4	1974	Own	V-392	—	—
200 4×2	1975	Own	V-304	—	19
200 4×4	1975	Own	V-304	—	19
PAYSTAR					
5050 4×4	1974	Own	DV-550B	—	—
5050 4×4	1975–76	Own	D-190	12	—
5050 4×4	1977–83	Own	DT-466	—	—
5070 4×4	1974–83	Detroit	6-71N	—	—
5050 6×4	1974	Own	DV-550B	—	—
5050 6×6	1974	Own	DV-550B	—	—
COF-5370 6×4	1979–81	Cummins	NTC-250	—	—
F-5050 6×4	1975–76	Own	D-190	12	—
F-5050 6×4	1977–83	Own	DT-466	—	—
F-5050 6×4	1978	Cat	3208	—	—
F-5050 6×6	1975–76	Own	D-190	12	—
F-5050 6×6	1977–83	Own	DT-466	—	—
F-5070 6×4	1974–83	Detroit	6-71N	30	—
F-5070 6×4	1980–83	Detroit	6V-92TTA	—	—
F-5070 6×6	1974–83	Detroit	6-71N	—	—
F-5050SF 6×4	1974	Own	DV-550B	—	—
F-5050SF 6×4	1975–76	Own	D-190	12	—
F-5050SF 6×4	1977	Detroit	6V-53	—	—
F-5050SF 6×4	1978–83	Own	DT-466	—	—
F-5050SF 6×4	1978–82	Cat	3208	—	—
F-5070SF 6×4	1974–83	Detroit	6-71N	30	—
S-SERIES & FC BUS CHASSIS					
1624	1979–82	Own	V-345	—	—
1654	1981	Own	D-165	—	—

Continued

MODEL INDEX & ENGINE APPLICATION—Continued

Truck Model or Series	Year	Standard Engine Make	Standard Engine Model	Oil Pan Refill Capacity Qts.	Cooling System Capacity Qts.
1723	1980–83	Own	V-345	—	—
1724	1979–83	Own	V-345	—	26
1754	1979	Own	D-150	—	—
1754	1980–83	Own	D-165	—	—
1823	1980–82	Own	MV-404	—	—
1824	1979–82	Own	MV-404	—	25
1824 4×4	1979–83	Own	V-345	—	—
1853	1980–83	Own	D-165	—	—
1854	1979	Own	D-190	—	—
1854	1980–81	Own	D-180	—	—
1854 4×4	1979	Own	D-170	—	—
1854 4×4	1980–83	Own	D-165	—	—
1924	1979–82	Own	MV-404	—	26
F-1924	1979–82	Own	MV-404	—	26
F-1924	1983	Own	V-392	—	—
F-1924 6×6	1979–82	Own	MV-404	—	26
1925	1980–82	Own	MV-404	—	—
F-1925	1979–80	Own	MV-404	—	26
1954	1979–83	Own	DT-466	—	37
F-1954	1979	Own	D-190	—	—
F1954	1980–83	Own	D-180	—	—
F-1954 6×6	1979	Own	D-190	—	—
F-1954 6×6	1980–83	Own	D-180	—	—
1955	1979	Own	D-190	—	—
1955	1980–83	Own	D-180	—	—
2125	1979–80	Own	V-537	—	—
F-2125 6×4	1979–80	Own	V-537	—	—
2155	1979–83	Own	DT-466		37
2275 4×2	1978–83	Detroit	6-71	—	—
F-2275 6×4	1978–83	Detroit	6-71	—	—
2524	1980	Own	V-537	—	—
2524 4×2	1978–79	Own	V-537	—	—
F-2524	1980	Own	V-537	—	—
F-2524 6×4	1978–79	Own	V-537	—	—
2554 4×2	1978–83	Own	DT-466	—	—
F-2554 6×4	1978–83	Own	DT-466	—	—
2574 4×2	1978–83	Detroit	6-71	—	—
F-2574 6×4	1978–83	Detroit	6-71	—	—
2575 4×2	1978–83	Detroit	6-71	—	—
F-2575 6×4	1978–83	Detroit	6-71	—	—
F-2624 6×4	1978–80	Own	V-537	—	—
F-2654 6×4	1978–83	Own	DT-466	—	—
F-2674 6×4	1978–83	Detroit	6-71	—	—

Truck Model or Series	Year	Standard Engine Make	Standard Engine Model	Oil Pan Refill Capacity Qts.	Cooling System Capacity Qts.
ALL WHEEL DRIVE MODELS					
1603	1974	Own	V-304	9	21
1603	1975–78	Own	V-345A	—	—
1610-F.C.	1978	Own	V-345	—	—
1703	1974	Own	V-304	9	21
1703	1975–78	Own	V-345A	—	—
1723	1979	Own	V-345	—	—
1803	1974–78	Own	V-345	9	21.9
1823	1979	Own	MV-404	—	26
1853	1974	Own	RD-406	9	27
1853	1977–78	Own	D-170	—	—
METRO SERIES, MULTI-STOP & MOTOR HOME CHASSIS					
MS-1210	1974	Jeep	6-258	4	15.6
MS-1210	1975	Own	V-345A	—	20
MHC-1310	1974–75	Own	V-392	—	24
MS-1510	1974	Own	V-345	—	15.6
MS-1510	1975	Own	V-345A	—	20
MHC-1510	1974–75	Own	V-392	—	24
100, 110, 120, 130 SERIES					
100	1974	Jeep	6-258	4	12
100 4×4	1974	Own	V-304	—	17
140, 150, 160 SERIES					
150	1975	Own	V-304	—	19
150 4×4	1975	Own	V-304	—	19
200, 210, and 230 SERIES					
200	1974	Jeep	6-258	4	12
200 4×4	1974	Jeep	6-258	4	—
200	1975	Own	V-304	—	19
200 4×4	1975	Own	V-304	—	19
500 SERIES					
500	1974	Jeep	6-258	4	12
500	1975	Own	V-304	—	19
XL SERIES					
CO-9670 4×2	1981–82	Detroit	6V92TTA	—	—
COF-9670 6×4	1981–82	Detroit	6V92TTA	—	—
COF-9670 Eagle	1981–82	Cummins	NTC-350	—	—

GENERAL ENGINE SPECIFICATIONS

Year	Engine Model	Carb. Type	Bore & Stroke	Comp. Ratio	Horsepower @ R.P.M.	Torque Ft. Lbs. @ R.P.M.	Comp. Pressure Cranking Speed ①	Governed Speed R.P.M.	Normal Oil Pressure Lbs.
1974	OHV-4 4-196	1 Bore	4 1/8 × 3 21/32	8.1	103 @ 4000	177 @ 2000	—	—	—
1975–80	OHV-4④ 4-196	1 Bore	4 1/8 × 3 21/32	8.02	86 @ 3800	157 @ 2200	—	—	40–50
1979–80	OHV-4③ 4-196	1 Bore	4.125 × 3.656	8.02	79 @ 3600	155 @ 2000	—	—	40–50
1979	OHV-6 6-232	1 Bore	3.75 × 3.50	8.5	120 @ 4000	202 @ 1000	—	—	32

Continued

GENERAL ENGINE SPECIFICATIONS—Continued

Year	Engine Model	Carb. Type	Bore & Stroke	Comp. Ratio	Horsepower @ R.P.M.	Torque Ft. Lbs. @ R.P.M.	Comp. Pressure Cranking Speed ①	Governed Speed R.P.M.	Normal Oil Pressure Lbs.
1974	OHV-6 6-258	1 Bore	3 3/4 × 3.895	8.0	115 @ 3800	199 @ 2000	—	—	32
1974	OHV-6 .... RD-406	2 Bore	4 3/8 × 4 1/2	7.1	160 @ 2750	336 @ 1400	143	2750	35–45
1974	OHV-6 .... RD-450	2 Bore	4 3/8 × 5	7.1	168 @ 2600	372 @ 1400	142	2600	35–45
1974	OHV-6 .... RD-501	4 Bore	4 1/2 × 5 1/4	6.8	181 @ 2600	406 @ 1400	145	2600	35–45
1974	OHV-V8 .... V-304	2 Bore	3 7/8 × 3 7/32	8.2	147 @ 3900	240 @ 2400	145	—	50
1974	OHV-V8 .... V-304	4 Bore	3 7/8 × 3 7/32	8.2	153 @ 3900	246 @ 2400	145	—	50
1975–77	OHV-V8 .... V-304	2 Bore	3 7/8 × 3 7/32	8.2	140 @ 3800	243 @ 2400	—	—	45
1978–80	OHV-V8 .... V-304A	2 Bore	3.875 × 3.219	8.19	144 @ 3600	247 @ 2400	—	—	40–50
1974	OHV-V8 .... V-345	2 Bore	3 7/8 × 3 21/32	8.1	157 @ 3800	266 @ 2400	143	3800	50
1975–80	OHV-V8 .... V-345	2 Bore	3 7/8 × 3 21/32	8.05	158 @ 3600	288 @ 2000	—	—	45
1978–80	OHV-V8 .... V-345A	2 Bore	3.875 × 3.656	8.28	163 @ 3600	292 @ 2000	—	—	40–50
1981	OHV-V8 .... V-345⑤⑥	2 Bore	3 7/8 × 3 21/32	8.28	147 @ 3600	278 @ 1900	—	—	40–50
1981	OHV-V8 .... V-345⑤⑦	2 Bore	3 7/8 × 3 21/32	8.28	162 @ 3600	286 @ 2200	—	—	40–50
1981	OHV-V8 .... V-345③	2 Bore	3 7/8 × 3 21/32	8.28	155 @ 3600	281 @ 2200	—	—	40–50
1974	OHV-V8 .... V-345	4 Bore	3 7/8 × 3 21/32	8.1	163 @ 3800	273 @ 2400	143	3800	50
1975–78	OHV-V8 .... V-345	4 Bore	3 7/8 × 3 21/32	8.05	168 @ 3800	288 @ 2000	—	3800	45
1974	OHV-V8 .... V-392	2 Bore	4 1/8 × 3 21/32	8.0	191 @ 3600	299 @ 2800	143	3600	50
1974	OHV-V8 .... V-392	4 Bore	4 1/8 × 3 21/32	8.0	194 @ 3600	308 @ 2800	143	3600	50
1975–77	OHV-V8 .... V-392	4 Bore	4 1/8 × 3 21/32	8.02	187 @ 3600	307 @ 2400	—	—	45
1975–80	OHV-V8 .... V-392	4 Bore	4 1/8 × 3 21/32	8.0	191 @ 3600	307 @ 2800	—	3600	45
1981	OHV-V8 .... V-392	4 Bore	4 1/8 × 3 21/32	8.0	185 @ 3600	302 @ 2800	—	—	40–50
1974	OHV-V8② V-400	2 Bore	4 11/64 × 3 11/16	8.25	211 @ 4000	326 @ 2800	—	—	37–75
1974	OHV-V8 .... V-401	2 Bore	4 1/8 × 3 3/4	7.7	186 @ 3400	322 @ 2400	139	3400	40–55
1974–76	OHV-V8 Short VS-401	2 Bore	4 1/8 × 3 3/4	7.7	186 @ 3400	322 @ 2400	140	3400	40–55
1977	OHV-V8 Short VS-401	2 Bore	4 1/8 × 3 3/4	7.69	180 @ 3400	339 @ 1700	—	3400	45–55
1975–80	OHV-V8 .... MV-404	2 Bore	4 1/8 × 3 3/4	8.1	188 @ 3600	311 @ 2300	—	3600	44–50
1981	OHV-V8 .... MV-404⑤	2 Bore	4 1/8 × 3 3/4	8.0	189 @ 3600	326 @ 2200	—	—	44–50
1981	OHV-V8 .... MV-404③	2 Bore	4 1/8 × 3 3/4	8.0	183 @ 3600	323 @ 2200	—	—	44–50
1975–80	OHV-V8 .... MV-404	4 Bore	4 1/8 × 3 3/4	8.1	210 @ 3600	336 @ 2800	—	3600	44–50
1981	OHV-V8 .... MV-404	4 Bore	4 1/8 × 3 3/4	8.0	206 @ 3600	336 @ 2200	—	—	44–50
1975–80	OHV-V8 .... MV-446	4 Bore	4 1/8 × 4 3/16	8.1	235 @ 3600	385 @ 2600	—	3600	44–50
1981	OHV-V8 .... MV-446	4 Bore	4 1/8 × 4 3/16	8.1	227 @ 3600	385 @ 2200	—	—	44–50
1975–76	OHV-V8 .... V-478	2Bore	4 1/2 × 3 3/4	7.6	209 @ 3400	384 @ 2200	140	3400	40–55
1974–75	OHV-V8 Short VS-478	2 Bore	4 1/2 × 3 3/4	7.6	209 @ 3400	384 @ 2200	140	3400	40–55
1977	OHV-V8 Short VS-478	2 Bore	4 1/2 × 3 3/4	7.64	234 @ 3600	409 @ 1800	—	3400	45–55
1975–80	OHV-V8 .... V-537	2 Bore	4.625 × 4.00	7.5	208 @ 3200	415 @ 2000	—	—	45–55
1975–80	OHV-V8 .... V-537	4 Bore	4.625 × 4.00	7.5	236 @ 3200	429 @ 2200	—	—	45–55
1981	OHV-V8 .... V-537	4 Bore	4.625 × 4.00	7.5	203 @ 3200	410 @ 1800	—	—	45–55
1974–76	OHV-V8 Short V-549	4 Bore	4 1/2 × 4 5/16	7.6	227 @ 3200	446 @ 2000	140	3200	40–55
1974–76	OHV-V8 Short VS-549	4 Bore	4 1/2 × 4 5/16	7.6	227 @ 3200	446 @ 2000	140	3200	40–55
1977	OHV-V8 Short VS-549	4 Bore	4 1/2 × 4 5/16	7.57	230 @ 3200	479 @ 1600	—	3200	45–55
1974–76	OHV-V8 Short FTVS-549	4 Bore	4 1/2 × 4 5/16	7.6	232 @ 3200	446 @ 2000	140	3200	40–55
1977	OHV-V8 Short FTVS-549	4 Bore	4 1/2 × 4 5/16	7.57	230 @ 3200	479 @ 1600	—	3200	45–55

①—Plus or minus 10 lbs.
②—See V8-401 in Jeep Chapter for service data.
③—Calif.
④—Exc. 1979–80 Calif.
⑤—Except Calif.
⑥—Standard engine.
⑦—Optional engine.

TUNE UP SPECIFICATIONS

The following specifications are published from the latest information available. This data should be used only in the absence of a decal affixed in the engine compartment.

★ When using a timing light, disconnect vacuum hose or tube at distributor and plug opening in hose or tube so idle speed will not be affected.

● When checking compression, lowest cylinder must be within 75 percent of highest.

▲ Before removing wires from distributor cap, determine location of the No. 1 wire in cap.

Year & Engine	Spark Plug		Ignition Timing BTDC①★				Curb Idle Speed		Fuel Pump Pressure
	Type	Gap	Firing Order ▲	Man. Trans.	Auto Trans.	Mark Location	Man. Trans.	Auto. Trans.②	
1981									
V-345 Exc. Scout⑥	RJ-10Y	.035	④	5°⑤	5°⑤	Damper	625–675	625–675N	4½–5¾
V-345 Exc. Scout⑦	RJ-10Y	.035	④	TDC⑤	TDC⑤	Damper	625–675	625–675N	4½–5¾
V-392	RJ-10Y	.035	④	TDC⑤	TDC⑤	Damper	625–675	625–675N	4½–5¾
MV-404 Exc. Calif.	RBN-13Y	.035	⑧	9°	9°	Damper	525–575	625–675N	4½–5¾
MV-404 Calif.	RBN-13Y	.035	⑧	5°	5°	Damper	525–575	625–675N	4½–5¾
MV-446 Exc. Calif.	RBN-13Y	.035	⑧	5°	5°	Damper	525–575	625–675N	4½–5¾
MV-446 Calif.	RBN-13Y	.035	⑧	1°	1°	Damper	525–575	625–675N	4½–5¾
V-537	ON-11Y	.030	⑨	7°	7°	Damper	500–550	600–650N	4½–5¾
1980									
4-196	RJ-10Y	.035	③	TDC	TDC	Damper	700	700N	4–5½
V-304	RJ-10Y	.035	④	10°⑤	10°⑤	Damper	650	650N	4½–5¾
V-345 Scout	RJ-10Y	.035	④	7½°⑤	7½°⑤	Damper	700	700N	4½–5¾
V-345 Exc. Scout⑥	RJ-6	.035	④	5°⑤	5°⑤	Damper	625–675	625–675N	4½–5¾
V-345 Exc. Scout⑦	RJ-6	.035	④	TDC⑤	TDC⑤	Damper	625–675	625–675N	4½–5¾
V-392	RJ-6	.035	④	TDC⑤	TDC⑤	Damper	625–675	625–675N	4½–5¾
MV-404 Exc. Calif.	RBN-13Y	.035	⑧	9°	9°	Damper	525–575	625–675N	4½–5¾
MV-404 Calif.	RBN-13Y	.035	⑧	5°	5°	Damper	525–575	625–675N	4½–5¾
MV-446 Exc. Calif.	RBN-13Y	.035	⑧	5°	5°	Damper	525–575	625–675N	4½–5¾
MV-446 Calif.	RBN-13Y	.035	⑧	1°	1°	Damper	525–575	625–675N	4½–5¾
V-537	ON-11Y	.030	⑨	7°	7°	Damper	500–550	600–650N	4½–5¾
1979									
4-196	RJ-10Y	.035	③	TDC	TDC	Damper	700	700N	4–5½
V-304	RJ-10Y	.035	④	TDC⑤	TDC⑤	Damper	700	700N	4½–5¾
V-345 Scout	RJ-10Y	.035	④	TDC⑤	TDC⑤	Damper	700	700N	4½–5¾
V-345 Exc. Scout⑥	RJ-6	.030	④	5°⑤	5°⑤	Damper	625–675	625–675N	4½–5¾
V-345 Exc. Scout⑦	RJ-6	.030	④	TDC⑤	TDC⑤	Damper	625–675	625–675N	4½–5¾
V-392	RJ-6	.030	④	TDC⑤	TDC⑤	Damper	625–675	625–675N	4½–5¾
MV-404	RBN-13Y	.030	⑧	9°	9°	Damper	525–575	625–675N	4½–5¾
MV-446	RBN-13Y	.030	⑧	5°	5°	Damper	525–575	625–675N	4½–5¾
V-537	RN-11Y	.030	⑨	7°	7°	Damper	500–550	600–650N	4½–5¾
1978									
4-196	RJ-10Y	.035	③	TDC	TDC	Damper	525–575	525–575N	4–5½
V-304	RJ-10Y	.035	④	TDC⑤	TDC⑤	Damper	650–700	650–700N	4½–5¾
V-345 Exc. Calif.	RJ-6	.030	④	TDC⑤	TDC⑤	Damper	650–700	650–700N	4½–5¾
V-345 Calif.⑥	RJ-6	.030	④	5°⑤	5°⑤	Damper	625–675	625–675N	4½–5¾
V-345 Calif.⑦	RJ-6	.030	④	TDC⑤	TDC⑤	Damper	625–675	625–675N	4½–5¾
V-392 Exc. Calif.	RJ-6	.030	④	TDC⑤	TDC⑤	Damper	650–700	650–700N	4½–5¾
V-392 Calif.	RJ-6	.030	④	TDC⑤	TDC⑤	Damper	625–675	625–675N	4½–5¾
MV-404	RBN-13Y	.030	⑧	9°	9°	Damper	525–575	625–675N	4½–5¾
MV-446	RBN-13Y	.030	⑧	5°	5°	Damper	525–575	625–675N	4½–5¾
V-537	RN-11Y	.030	⑨	7°	7°	Damper	500–550	600–650N	4½–5¾
1977									
4-196	RJ-10Y	.035	③	TDC	TDC	Damper	525–575	525–575N	4–5½
V-304 Exc. Calif.⑩	RJ-10Y	.035	④	TDC⑤	TDC⑤	Damper	650–700	650–700N	4½–5¾
V-304 Calif.	RJ-10Y	.035	④	5°⑤	5°⑤	Damper	600–650	600–650N	4½–5¾
V-345 Exc. Calif.⑩	RJ-6	.030	④	TDC⑤	TDC⑤	Damper	650–700	650–700N	4½–5¾

Continued

TUNE UP SPECIFICATIONS—Continued

The following specifications are published from the latest information available. This data should be used only in the absence of a decal affixed in the engine compartment.

★ When using a timing light, disconnect vacuum hose or tube at distributor and plug opening in hose or tube so idle speed will not be affected.

● When checking compression, lowest cylinder must be within 75 percent of highest.

▲ Before removing wires from distributor cap, determine location of the No. 1 wire in cap.

Year & Engine	Spark Plug		Ignition Timing BTDC①★				Curb Idle Speed		Fuel Pump Pressure
	Type	Gap	Firing Order ▲	Man. Trans.	Auto Trans.	Mark Location	Man. Trans.	Auto. Trans.②	
1977—Continued									
V-345 Calif.⑥	RJ-6	.030	④	5°⑤	5°⑤	Damper	625–675	625–675N	4½–5¾
V-345 Calif.⑦	RJ-6	.030	④	TDC⑤	TDC⑤	Damper	625–675	625–675N	4½–5¾
V-392 Exc. Calif.⑩	RJ-6	.030	④	TDC⑤	TDC⑤	Damper	650–700	650–700N	4½–5¾
V-392 Calif.	RJ-6	.030	④	TDC⑤	TDC⑤	Damper	625–675	625–675N	4½–5¾
MV-404	RBN-13Y	.030	⑧	9°	9°	Damper	525–575	625–675N	4½–5¾
MV-446	RBN-13Y	.030	⑧	5°	5°	Damper	525–575	625–675N	4½–5¾
V-537	RN-11Y	.030	⑨	7°	7°	Damper	500–550	600–650N	4½–5¾
1976									
4-196	RJ-10Y	.035	③	TDC	TDC	Damper	525–575	525–575N	4–5½
V-304⑩	RJ-6	.030	④	TDC⑤	TDC⑤	Damper	650–700	650–700N	4½–5¾
V-345 Exc. Calif.⑩	RJ-6	.030	④	TDC⑤	TDC⑤	Damper	650–700	650–700N	4½–5¾
V-345 Calif.⑩	RJ-6	.030	④	5°⑤	5°⑤	Damper	625–675	625–675N	4½–5¾
V-392 Exc. Calif.⑩	RJ-6	.030	④	TDC⑤	TDC⑤	Damper	650–700	650–700N	4½–5¾
V-392 Calif.⑩	RJ-6	.030	④	TDC⑤	TDC⑤	Damper	625–675	625–675N	4½–5¾
MV-404	RBN-4	.030	⑧	9°	9°	Damper	525–575	525–575N	4½–5¾
MV-446	RBN-4	.030	⑧	5°	5°	Damper	525–575	525–575N	4½–5¾
V, VS-478⑪	RJ-6	.030	⑨	12°	12°	Damper	500–550	500–550N	4½–5¾
V-537	RN-11Y	.030	⑨	7°	7°	Damper	550–550	500–550N	4½–5¾
V, VS-549⑪	RJ-6	.030	⑨	7°	7°	Damper	550–550	500–550N	4½–5¾
1975									
4-196	RJ-10Y	.035	③	TDC	TDC	Damper	525–575	525–575N	4–5½
V-304⑫	RJ-6	.030	④	TDC⑤	TDC⑤	Damper	650–700	650–700N	4½–5¾
V-345 Exc. Calif.⑫	RJ-6	.030	④	TDC⑤	TDC⑤	Damper	650–700	650–700N	4½–5¾
V-345 Calif.⑫	RJ-6	.030	④	5°⑤	5°⑤	Damper	625–675	625–675N	4½–5¾
V-392 Exc. Calif.⑫	RJ-6	.030	④	TDC⑤	TDC⑤	Damper	650–700	650–700N	4½–5¾
V-392 Calif.⑫	RJ-6	.030	④	TDC⑤	TDC⑤	Damper	⑬	⑬	4½–5¾
MV-404	RBN-4	.030	⑧	9°	9°	Damper	525–575	525–575N	4½–5¾
MV-446	RBN-4	.030	⑧	5°	5°	Damper	525–575	525–575N	4½–5¾
V, VS-478	RJ-6	.030	⑨	12°	12°	Damper	500–550	500–550N	4½–5¾
V-537	RN-11Y	.030	⑨	7°	7°	Damper	500–550	500–550N	4½–5¾
V, VS-549	RJ-6	.030	⑨	7°	7°	Damper	500–550	500–550N	4½–5¾
1974									
4-196	UJ10Y	.030	③	4°	4°	Damper	575	575	4–5½
6-258	N12-Y	.035	⑭	3°	3°	Damper	700	700N	4–5½
V-304	J6	.030	④	TDC⑤	TDC⑤	Damper	675	675N	4½–5¾
V304A	J6	.030	④	TDC⑤	TDC⑤	Damper	675	675N	4½–5¾
V-345	⑮	.030⑯	④	TDC⑤⑰	TDC⑤⑰	Damper	675	675N	4½–5¾
V-392	⑮	.030	④	TDC⑤⑱	TDC⑤⑱	Damper	700	700N	4½–5¾
V-400	J6	.035	④	5°	5°	Damper	700	700N	—
V-401	J6	.027	⑨	7°	7°	Damper	600	600N	4½–5¾
RD-406	J6	.030	⑭	5°	5°	Damper	525	525N	3½–5½
RD-450	J6	.030	⑭	5°	5°	Damper	500⑳	500⑳	3½–5½
V8-478	J6	.027	⑨	10°	10°	Damper	500	500N	4½–5¾
RD-501	J6	.030	⑭	5°	5°	Damper	—	—	3½–5½
V-549	J6	.027	⑨	7°⑲	7°⑲	Damper	—	—	4½–5¾

Continued

TUNE UP NOTES—Continued

①—B.T.D.C.—Before top dead center.
②—N: Neutral; D: Drive.
③—Cylinder numbering front to rear, firing order 1-3-4-2.
④—Cylinder numbering (front to rear), left bank, 1-3-5-7; right bank, 2-4-6-8. Firing order 1-8-4-3-6-5-7-2.
⑤—Timed at No. 8 cylinder, located on right bank, rear of engine.
⑥—Models equipped with Holley 2210C carburetor.
⑦—Models equipped with Holley 2300GLEG carburetor.
⑧—Cylinder numbering (front to rear), right bank, 1-3-5-7; left bank, 2-4-6-8. Firing order, 1-2-7-3-4-5-6-8.
⑨—Cylinder numbering (front to rear), left bank, 1-3-5-7; right bank, 2-4-6-8. Firing order, 1-8-7-3-6-5-4-2.
⑩—Breaker point gap, new .019″, used .016″; dwell angle, 24°–34°.
⑪—Breaker point gap, new .017″, used .014″; dwell angle, 28°–32°.
⑫—Breaker point gap, .015″; dwell angle, 28°–32°.
⑬—Models equipped with Holley 4150 G carburetor, 625–675N RPM; models equipped with Carter Thermo-Quad carburetor, 650–700N RPM.
⑭—Firing order 1-5-3-6-2-4.
⑮—Scout, 150–500, RJ10Y; 1600–1800, RJ6.
⑯—L.P.G. models .015 inch.
⑰—L.P.G. models 10°; Low compression engine 5°.
⑱—With Thermo Quad Carburetor or Low compression engine, 5°.
⑲—FTVS-549, 9°.
⑳—Carburetor List Nos. 6398, 6398-1, 6438, 525 RPM.

DISTRIBUTOR SPECIFICATIONS

Engine Application	Distributor		Rotation Rotor End	Point Gap Inch	Dwell Angle Deg.	Breaker Arm Spring Tension Ounces	Centrifugal Advance Deg. @ Dist. R.P.M.		Vacuum Advance	
	Make	Model					Advance Starts	Full Advance	Inches of Vacuum to Start Advance	Max. Adv. Dist. Deg. @ Vacuum
P.T., RD ENGINES										
	Delco-Remy	1110429	C	.019	28–35	17–21	5 @ 400	14 @ 1000	—	—
	Delco-Remy	1110479	CC	.019	31–34	17–21	1 @ 400	14 @ 2000	5 to 7	—
	Delco-Remy	1110497	CC	.019	31–34	19–23	1 @ 550	14 @ 2200	5 to 7	7½ @ 11
	Delco-Remy	1110522	CC	.016	31–34	19–23	1 @ 450	13 @ 2200	—	—
	Delco-Remy	1112276	C	.025	31–37	17–21	1 @ 300	14 @ 1350	—	—
	Delco-Remy	1112288	C	.025	31–37	17–21	1 @ 300	14 @ 1350	—	—
	Delco-Remy	1112293	C	.025	31–37	17–21	1 @ 300	14 @ 1350	—	—
	Delco-Remy	1112321	C	.025	31–37	17–21	1 @ 300	14 @ 1350	—	—
	Delco-Remy	1112324	C	.025	31–37	17–21	1 @ 250	10 @ 900	—	—
	Delco-Remy	1112341	C	.025	31–37	17–21	1 @ 300	14 @ 1350	—	—
	Delco-Remy	1112342	C	.025	31–37	17–21	1 @ 300	14 @ 1350	—	—
	Delco-Remy	1112355	CC	.025	31–37	17–21	1 @ 250	15 @ 1500	3 to 5	7½ @ 14
	Delco-Remy	1112359	C	.025	31–37	17–21	1 @ 300	14 @ 1350	—	—
	Delco-Remy	1112366	C	.025	31–37	17–21	1 @ 300	14 @ 1350	—	—
	Delco-Remy	1112370	C	.025	31–37	17–21	1 @ 300	11 @ 1600	—	—
	Delco-Remy	1112371	C	.025	31–37	17–21	1 @ 300	14 @ 1350	—	—
	Delco-Remy	1112397	C	.025	31–37	17–21	1 @ 225	14 @ 1050	3 to 5	7½ @ 14
	Delco-Remy	1112398	C	.025	31–37	17–21	1 @ 250	10 @ 900	—	—
	Delco-Remy	1112412	C	.025	31–37	17–21	1 @ 300	11 @ 1600	—	—
	Delco-Remy	1112416	CC	.019	28–35	17–21	2 @ 300	12 @ 1100	6 to 8	7½ @ 14
	Delco-Remy	1112417	C	.019	28–35	17–21	1 @ 350	14 @ 1350	—	—
	Delco-Remy	1112422	CC	.019	28–35	17–21	2 @ 300	12 @ 1100	6 to 8	7½ @ 14
	Delco-Remy	1112423	C	.019	28–35	17–21	1 @ 275	11 @ 1600	—	—
	Delco-Remy	1112424	CC	.019	28–35	17–21	1 @ 225	14 @ 1050	6 to 8	7½ @ 14
	Delco-Remy	1112429	C	.019	28–35	17–21	1 @ 300	11 @ 1600	—	—
	Delco-Remy	1112492	CC	.019	31–34	19–23	1 @ 300	11 @ 1700	—	—
	Delco-Remy	1112704	C	.019	28–35	17–21	1 @ 350	14 @ 1350	—	—
	Delco-Remy	1112705	C	.019	28–35	17–21	1 @ 300	11 @ 1600	—	—
	Delco-Remy	1112706	C	.019	28–35	17–21	1 @ 300	11 @ 1600	—	—
	Delco-Remy	1112714	C	.019	28–35	17–21	1 @ 350	14 @ 1350	—	—
	Delco-Remy	1112690	CC	.019	31–34	17–21	1 @ 375	15 @ 1350	—	—
	Delco-Remy	1112691	CC	.019	31–34	17–21	1 @ 300	12 @ 1750	—	—
	Delco-Remy	1112692	CC	.019	31–34	19–23	2 @ 350	12 @ 1750	—	—
V8 ENGINES										
V8-304 & 345	Delco-Remy	1110953	C	.016	28–32	17–21	1 @ 350	16 @ 1800	3 to 5	7 @ 10
	Delco-Remy	1110954	C	.016	28–32	17–21	1 @ 350	14 @ 2100	3 to 5	7 @ 10

Continued

DISTRIBUTOR SPECIFICATIONS—Continued

Engine Application	Distributor		Rotation Rotor End	Point Gap Inch	Dwell Angle Deg.	Breaker Arm Spring Tension Ounces	Centrifugal Advance Deg. @ Dist. R.P.M.		Vacuum Advance	
	Make	Model					Advance Starts	Full Advance	Inches of Vacuum to Start Advance	Max. Adv. Dist. Deg. @ Vacuum
V8 ENGINES—Continued										
	Delco-Remy	1110966	C	.016	28–32	17–21	1 @ 350	14 @ 2100	3 to 5	7 @ 10
	Delco-Remy	1110967	C	.016	28–32	17–21	1 @ 350	18 @ 1900	3 to 5	7 @ 10
	Delco-Remy	1111002	C	.016	28–32	17–21	1 @ 350	16 @ 1800	3 to 5	7 @ 10
	Delco-Remy	1112740	C	.016	28–32	17–21	1 @ 350	16 @ 1800	3 to 5	7 @ 10
	Delco-Remy	1112743	C	.016	28–32	17–21	1 @ 350	14 @ 2100	3 to 5	7 @ 10
	Delco-Remy	1112744	C	.016	28–32	17–21	1 @ 350	14 @ 2100	3 to 5	7 @ 10
	Delco-Remy	1112745	C	.016	28–32	17–21	1 @ 350	16 @ 1800	3 to 5	7 @ 10
	Delco-Remy	1112752	C	.016	28–32	17–21	1 @ 350	14 @ 2100	3 to 5	7 @ 10
	Delco-Remy	1112753	C	.016	28–32	17–21	1 @ 350	14 @ 2100	3 to 5	7 @ 10
	Delco-Remy	1112771	C	.016	28–32	17–21	1 @ 375	12 @ 2100	3 to 5	7 @ 10
	Delco-Remy	1112772	C	.016	28–32	17–21	1 @ 375	12 @ 2100	3 to 5	7 @ 10
	Delco-Remy	1112775	C	.016	28–32	17–21	1 @ 375	12 @ 2100	3 to 5	7 @ 10
	Delco-Remy	1112776	C	.016	28–32	17–21	1 @ 350	14 @ 2100	3 to 5	7 @ 10
	Delco-Remy	1112777	C	.016	28–32	17–21	1 @ 350	16 @ 1800	3 to 5	7 @ 10
	Delco-Remy	1112778	C	.016	28–32	17–21	1 @ 350	16 @ 1800	3 to 5	7 @ 10
V-304	Prestolite	285098-C91	—	.017	25–29	17–20	1 @ 350	14 @ 2000	5 to 7	14 @ 11
V-304, 345	Prestolite	285093-C91	—	.017	25–29	17–20	1 @ 350	14 @ 2000	5 to 7	14 @ 11
V-304, 345	Holley③	461270-C91	C	.008④	26–32	—	7.5 @ 1000	17 @ 2100	5 to 7	6 @ 15
V-304, 345	Prestolite③	484778-C91	C	.008④	26–32	—	7.5 @ 1000	17 @ 2100	5 to 7	6 @ 12
V-304, 345	Prestolite③	496004-C91	C	.008④	28–34	—	0 @ 400	15 @ 2100	3 to 5	6 @ 13
V-345	Prestolite	285096-C91	—	.017	25–29	17–20	1 @ 350	14 @ 2000	5 to 7	14 @ 11
①	Prestolite	291262-C92	—	②	25–29	17–21	1 @ 350	14½ @ 2000	5 to 7½	7 @ 11½
①	Prestolite	359206-C91	—	②	25–29	17–21	1 @ 350	14½ @ 2000	5 to 7½	7 @ 11½
V-304	Prestolite	428078-C91	—	.019	28–32	17–21	½ @ 400	13½ @ 2000	5 to 7	4 @ 8½
V-304	Prestolite	428079-C91	—	.019	28–32	17–21	½ @ 400	13½ @ 2000	5 to 7	4 @ 8½
V-304	Prestolite	428095-C91	—	.019	28–32	17–21	½ @ 400	16 @ 2000	5 to 7	4 @ 8½
V-304	Prestolite	428509-C91	—	.019	28–32	17–21	½ @ 400	13½ @ 2000	5 to 7	6 @ 10
V-304	Prestolite	448510-C91	—	.019	28–32	17–21	½ @ 400	13½ @ 2000	5 to 7	6 @ 10
V-304	Holley③	467073-C91	C	.008④	26–32	—	1 @ 600	16.8 @ 2100	5 to 7	8 @ 19
V-304	Holley③	480870-C91	C	.008④	26–32	—	2.5 @ 1000	15.5 @ 2100	3 to 5	4 @ 10
V-304	Prestolite	484834-C91	C	.016	28–32	17–22	0 @ 400	17 @ 2000	5 to 7	4 @ 12
V-304	Prestolite③	467073-C91	C	.008④	26–32	—	2 @ 600	16.8 @ 2100	5 to 7	8 @ 15
V-304, 345	Prestolite	291260-C94	—	②	25–29	17–21	1 @ 350	14½ @ 2000	5 to 7½	7 @ 11½
V-304, 345	Prestolite	359217-C91	—	②	25–29	17–21	1 @ 350	14½ @ 2000	5 to 7½	7 @ 11½
V-304, 345	Prestolite	359220-C91	—	②	25–29	17–21	1 @ 350	14½ @ 2000	5 to 7½	7 @ 11½
⑥	Prestolite	359225-C91	—	.017	27–33	17–21	½ @ 350	11 @ 1600	5½ to 7½	7 @ 11¼
⑥	Prestolite	359234-C91	—	.017	27–33	17–21	½ @ 350	11 @ 1600	5½ to 7½	7 @ 11¼
⑥	Prestolite	360514-C91	—	.019	28–32	17–21	½ @ 400	16 @ 2000	5½ to 7½	6 @ 11
⑥	Prestolite	379282-C91	—	.019	28–32	17–21	½ @ 400	14 @ 2000	5 to 7	6 @ 11
⑥	Prestolite	383148-C91	—	.019	28–32	17–21	½ @ 400	16 @ 2000	5 to 7	6 @ 11
⑥	Prestolite	383149-C91	—	.019	28–32	17–21	½ @ 400	14 @ 2000	5 to 7	6 @ 11
⑥	Prestolite	383150-C91	—	.019	28–32	17–21	½ @ 400	16 @ 2000	5 to 7	6 @ 11
⑥	Prestolite	383151-C91	—	.019	28–32	17–21	½ @ 400	14 @ 2000	5 to 7	6 @ 11
⑥	Prestolite	408129-C91	—	.019	28–32	17–21	1 @ 400	16 @ 2000	5 to 7	8 @ 11
⑥	Prestolite	408130-C91	—	.019	28–32	17–21	1 @ 400	16 @ 2000	5 to 7	8 @ 11
⑥	Prestolite	461326-C91	—	.019	28–32	17–21	1 @ 400	16 @ 2000	5 to 7	8 @ 11
⑥	Prestolite	484799-C91	—	.016	28–32	17–21	½ @ 300	14.5 @ 2000	5½ to 7½	8 @ 15
⑥	Prestolite	484831-C91	C	.016	28–32	17–21	0 @ 400	17 @ 2000	5 to 7	8 @ 15
⑥	Prestolite	487830-C91	C	.016	28–32	17–21	7.1 @ 700	16.5 @ 1800	5 to 7	8 @ 19
⑥	Prestolite	487877-C91	C	.016	28–32	17–21	7.1 @ 700	16.5 @ 1800	5 to 7	8 @ 19
⑥	Prestolite	491357-C91	C	.016	28–32	17–21	1.5 @ 500	15 @ 2200	5 to 7	6 @ 15
V-304, 345	Prestolite③	461270-C91	C	.008④	26–32	—	2 @ 550	16.7 @ 2200	5 to 7	6 @ 15

Continued

DISTRIBUTOR SPECIFICATIONS—Continued

Engine Application	Distributor Make	Distributor Model	Rotation Rotor End	Point Gap Inch	Dwell Angle Deg.	Breaker Arm Spring Tension Ounces	Centrifugal Advance Deg. @ Dist. R.P.M. Advance Starts	Centrifugal Advance Deg. @ Dist. R.P.M. Full Advance	Vacuum Advance Inches of Vacuum to Start Advance	Vacuum Advance Max. Adv. Dist. Deg. @ Vacuum
V8 ENGINES—Continued										
⑤	Prestolite	359207-C91	—	.019	28–32	17–21	½ @ 300	16 @ 1800	4¾ to 6¼	8 @ 10
⑤	Prestolite	359219-C91	—	.019	28–32	17–21	½ @ 300	16 @ 1800	4¾ to 6¼	8 @ 10
⑤	Prestolite	359222-C91	—	.019	28–32	17–21	½ @ 300	16 @ 1800	4¾ to 6¼	8 @ 10
⑤	Prestolite	361558-C91	—	.019	28–32	17–21	½ @ 300	16 @ 1800	4¾ to 6¼	8 @ 10
⑤	Prestolite	427908-C91	—	.019	28–32	17–21	½ @ 400	16 @ 2000	5 to 7	7 @ 11
⑤	Holley③	463333-C91	C	.008④	26–32	—	½ @ 300	14.5 @ 2000	5 to 7.5	8 @ 15
⑤	Holley③	463334-C91	C	.008④	26–32	—	½ @ 300	17 @ 1800	4¾ to 6¼	8 @ 15
⑤	Holley③	463335-C91	C	.008④	26–32	—	½ @ 300	14.5 @ 2000	5 to 7.5	8 @ 15
⑤	Holley③	463336-C91	C	.008④	26–32	—	½ @ 300	17 @ 1800	4¾ to 6¼	8 @ 15
⑤	Prestolite③	484804-C91	—	.008④	26–32	—	½ @ 300	17 @ 1800	4¾ to 6¼	8 @ 15
V-345	Prestolite	427964-C91	—	.019	28–32	17–21	½ @ 400	11½ @ 2000	5 to 7	4 @ 8½
V-345	Prestolite	427966-C91	—	.019	28–32	17–21	½ @ 400	11½ @ 2000	5 to 7	4 @ 8½
V-345	Prestolite	433142-C91	—	.019	28–32	17–21	½ @ 400	16 @ 2000	5 to 7	4 @ 8½
V-345	Prestolite	436869-C91	—	.019	28–32	17–21	½ @ 400	16 @ 2000	5 to 7	6 @ 11½
V-345	Prestolite	448511-C91	—	.019	28–32	17–21	½ @ 400	11½ @ 2000	5 to 7	6 @ 10
V-345	Prestolite	448512-C91	—	.019	28–32	17–21	½ @ 400	11½ @ 2000	5 to 7	6 @ 10
V-345	Holley③	448685-C91	C	.008④	26–32	—	½ @ 400	17 @ 2000	5 to 7	4 @ 12
V-345	Holley③	461271-C91	C	.008④	26–32	—	½ @ 400	12.4 @ 2000	5 to 7	6 @ 15
V-345	Holley③	461272-C91	C	.008④	26–32	—	½ @ 400	12.4 @ 2000	5 to 7	6 @ 15
V-345	Holley③	469648-C91	C	.008④	26–32	—	.9 @ 700	13.9 @ 2100	5 to 7	4 @ 12
V-345	Holley③	469649-C91	C	.008④	26–32	—	.9 @ 700	12.5 @ 2000	5 to 7	4 @ 12
V-345	Holley③	469650-C91	C	.008④	26–32	—	.9 @ 700	12.5 @ 2000	5 to 7	4 @ 12
V-345	Holley③	483423-C91	C	.008④	26–32	—	.9 @ 700	13.9 @ 2100	10 to 12	4 @ 15
V-345	Holley③	483492-C91	C	.008④	26–32	—	1 @ 600	16.8 @ 2100	—	—
V-345	Holley③	483493-C91	C	.008④	26–32	—	1 @ 600	16.8 @ 2100	—	—
V-345	Prestolite	484780-C91	C	.016	28–32	17–22	7.5 @ 1000	17 @ 2100	5 to 7	6 @ 12
V-345	Prestolite③	484785-C91	C	.008④	26–32	—	.9 @ 700	13.9 @ 2100	5 to 7	4 @ 15
V-345	Prestolite③	484786-C91	C	.008④	26–32	—	1 @ 600	16.8 @ 2100	—	—
V-345	Prestolite③	484797-C91	C	.008④	28–32	—	0 @ 313	12.5 @ 2000	5 to 7	8 @ 15
V-345	Prestolite	484835-C91	C	.016	28–32	17–22	1 @ 400	17 @ 2000	5 to 7	4 @ 12
V-345	Prestolite	484836-C91	C	.016	28–32	17–22	1 @ 600	17 @ 2000	5 to 7	6 @ 12
⑤	Prestolite③	463333-C91	C	.008④	26–32	—	½ @ 300	14½ @ 2000	5 to 7½	8 @ 15
⑤	Prestolite③	463334-C91	C	.008④	26–32	—	½ @ 300	17 @ 1800	4¾ to 6¼	8 @ 15
⑤	Prestolite③	463335-C91	C	.008④	26–32	—	½ @ 300	14½ @ 2000	5 to 7½	8 @ 15
⑤	Prestolite③	463336-C91	C	.008④	26–32	—	½ @ 300	17 @ 1800	4¾ to 6¼	8 @ 15
V-345	Prestolite③	448685-C91	C	.014④	22–28	—	½ @ 400	17 @ 2000	5 to 7	4 @ 11
V-345	Prestolite③	461271-C91	C	.008④	26–32	—	½ @ 400	12.4 @ 2000	5 to 7	6 @ 15
V-345	Prestolite③	461272-C91	C	.008④	26–32	—	½ @ 400	12.4 @ 2000	5 to 7	6 @ 15
V-345 L.P.G.	Prestolite③	463330-C91	C	.008④	26–32	—	1 @ 400	12½ @ 2000	5 to 7	8 @ 15
V-345 L.P.G.	Prestolite③	463331-C91	C	.008④	26–32	—	1 @ 400	12½ @ 2000	5 to 7	8 @ 15
V-345 L.P.G.	Prestolite③	463332-C91	C	.008④	26–32	—	1 @ 400	12½ @ 2000	5 to 7	8 @ 15
V-345 L.P.G.	Holley③	463330-C91	C	.008④	26–32	—	1 @ 400	12.5 @ 2000	5 to 7	8 @ 15
V-345 L.P.G.	Holley③	463331-C91	C	.008④	26–32	—	1 @ 400	12.5 @ 2000	5 to 7	8 @ 15
V-345 L.P.G.	Holley③	463332-C91	C	.008④	26–32	—	1 @ 400	12.5 @ 2000	5 to 7	8 @ 15
V-345 L.P.G.	Prestolite	484795-C91	C	.016	28–32	17–22	0 @ 400	12.5 @ 2000	5 to 7	8 @ 15
V-345 L.P.G.	Prestolite③	484797-C91	C	.008④	26–32	—	0 @ 400	12.5 @ 2000	5 to 7	8 @ 15
V-345, 392	Prestolite	484781-C91	C	.016	28–32	17–22	4 @ 700	12.3 @ 2100	5 to 7	6 @ 12
V-345, 392	Prestolite③	484783-C91	C	.008④	26–32	—	4 @ 700	12.3 @ 2100	5 to 7	6 @ 12
V-392	Prestolite	285092-C91	—	.017	25–29	17–20	1 @ 350	12 @ 2000	5 to 7	14 @ 11
V-392	Prestolite	285095-C91	—	.017	25–29	17–20	1 @ 350	12 @ 2000	5 to 7	14 @ 11
V-392	Prestolite	323418-C93	—	②	25–29	17–21	½ @ 350	12½ @ 2000	5 to 7	7 @ 11
V-392	Prestolite	359218-C91	—	②	25–29	17–21	½ @ 350	12½ @ 2000	5 to 7	7 @ 11
V-392	Prestolite	359221-C91	—	②	25–29	17–21	½ @ 350	12½ @ 2000	5 to 7	7 @ 11

Continued

DISTRIBUTOR SPECIFICATIONS—Continued

Engine Application	Distributor		Rotation Rotor End	Point Gap Inch	Dwell Angle Deg.	Breaker Arm Spring Tension Ounces	Centrifugal Advance Deg. @ Dist. R.P.M.		Vacuum Advance	
	Make	Model					Advance Starts	Full Advance	Inches of Vacuum to Start Advance	Max. Adv. Dist. Deg. @ Vacuum
V8 ENGINES—Continued										
V-392	Prestolite	427965-C91	—	.019	28–32	17–21	1/2 @ 400	11 1/2 @ 2000	5 to 7	7 @ 10
V-392	Prestolite	427967-C91	—	.019	28–32	17–21	1/2 @ 400	11 1/2 @ 2000	5 to 7	7 @ 10
V-392	Holley③	448686-C91	C	.008④	26–32	—	1.5 @ 500	15 @ 2200	5 to 7	6 @ 15
V-392	Holley③	461273-C91	C	.008④	26–32	—	0 @ 400	12.4 @ 2000	5 to 7	6 @ 15
V-392	Holley③	467034-C91	C	.008④	26–32	—	3.4 @ 700	14 @ 1800	5 to 7	8 @ 19
V-392	Holley③	467035-C91	C	.008④	26–32	—	3.4 @ 700	14 @ 1800	5 to 7	8 @ 19
V-392	Holley③	472203-C91	C	.008④	26–32	—	4 @ 600	15 @ 2100	3 to 5	6 @ 13
V-392	Prestolite③	484788-C91	—	.008④	26–32	—	3.4 @ 700	14 @ 1800	5 to 7	8 @ 19
V-392	Holley③	519861-C91	C	.008④	26–32	—	0 @ 400	12.4 @ 2000	5 to 7	6 @ 15
V-392	Prestolite③	448686-C91	C	.008④	26–32	—	2.7 @ 400	15 @ 2200	5 to 7	6 @ 15
V-392	Prestolite③	461273-C91	C	.008④	26–32	—	1/2 @ 400	12.4 @ 2000	5 to 7	6 @ 15
V-392	Prestolite③	467034-C91	C	.008④	26–32	—	3.6 @ 650	15 1/2 @ 2000	5 to 7	8 @ 15
V-392	Prestolite③	467035-C91	C	.008④	26–32	—	3.6 @ 650	15 1/2 @ 2000	5 to 7	8 @ 15
V-392	Prestolite③	519861-C91	C	.008④	26–32	—	1/2 @ 400	12.4 @ 2000	5 to 7	6 @ 15
V-400	Delco-Remy	1112215	—	.016	29–31	17–21	3/4 @ 400	14 1/4 @ 2000	4 to 6	8 1/4 @ 13
V-400	Delco-Remy	1112904	—	.016	29–31	17–21	3/4 @ 400	14 1/4 @ 2000	9 to 11	6 @ 15 1/2
MV-404	Prestolite③	446916-C91	CC	.008④	26–32	—	2 @ 600	12.4 @ 1900	5 to 7	4 @ 12
MV-404	Prestolite③	446920-C91	CC	.008④	26–32	—	2 @ 600	12.4 @ 1900	5 to 7	4 @ 12
MV-404	Prestolite③	446921-C91	CC	.008④	26–32	—	2 @ 600	12.4 @ 1900	5 to 7	4 @ 12
MV-404	Holley③	446916-C92	CC	.008④	26–32	—	1 @ 600	12.8 @ 1800	5 to 7	4 @ 12
MV-404	Holley③	446920-C92	CC	.008④	26–32	—	1 @ 600	12.8 @ 1800	5 to 7	4 @ 12
MV-404	Holley③	446921-C92	CC	.008④	26–32	—	1 @ 600	12.8 @ 1800	5 to 7	4 @ 12
MV-404	Prestolite③	484790-C91	CC	.008④	26–32	—	0 @ 400	12.8 @ 1800	5 to 7	4 @ 12
MV-446	Prestolite③	484792-C91	—	.008④	26–32	—	1 @ 600	12.8 @ 1800	10 to 12	4 @ 15
MV-446	Prestolite③	446917-C91	CC	.008④	26–32	—	2 @ 600	12.4 @ 1900	10 to 12	4 @ 15
MV-446	Prestolite③	446918-C91	CC	.008④	26–32	—	2 @ 600	12.4 @ 1900	10 to 12	4 @ 15
V-401, 549	Holley	1883-1	—	.016	28–32	17–20	1 @ 350	11 @ 1600	6 to 7	7 @ 11
L.P.G.	Holley	2293	—	.016	28–32	17–20	1 @ 350	7 @ 1600	5 to 6	5 @ 10
V-401, 549	Prestolite	286686-C93	—	②	25–29	17–21	1/2 @ 350	8 @ 1600	5 1/2 to 7 1/2	7 @ 11 1/2
V-401, 549	Prestolite	359223-C91	—	②	25–29	17–21	1/2 @ 350	8 @ 1600	5 1/2 to 7 1/2	7 @ 11 1/2
V-401, 549	Prestolite	359226-C91	—	②	25–29	17–21	1/2 @ 350	8 @ 1600	5 1/2 to 7 1/2	7 @ 11 1/2
V-401, 549	Prestolite	359228-C91	—	②	25–29	17–21	1/2 @ 350	8 @ 1600	5 1/2 to 7 1/2	7 @ 11 1/2
V-401, 549	Prestolite	359230-C91	—	②	25–29	17–21	1/2 @ 350	8 @ 1600	5 1/2 to 7 1/2	7 @ 11 1/2
V-401, 549	Prestolite	359231-C91	—	②	25–29	17–21	1/2 @ 350	8 @ 1600	5 1/2 to 7 1/2	7 @ 11 1/2
V-401, 549	Prestolite	359233-C91	—	②	25–29	17–21	1/2 @ 350	8 @ 1600	5 1/2 to 7 1/2	7 @ 11 1/2
V/VS-401	Prestolite	401031-C91	—	.017	28–32	17–20	1/2 @ 350	7 @ 1600	—	—
V/VS-401	Prestolite	401035-C91	—	.017	28–32	17–20	1/2 @ 350	7 @ 1600	—	—
V-478	Prestolite	286684-C93	—	②	25–29	17–21	0 @ 350	12 @ 1600	5 1/2 to 7 1/2	7 @ 11 1/4
V-478	Prestolite	359227-C91	—	②	25–29	17–21	0 @ 350	12 @ 1600	5 1/2 to 7 1/2	7 @ 11 1/4
V-478	Prestolite	359229-C91	—	②	25–29	17–21	0 @ 350	12 @ 1600	5 1/2 to 7 1/2	7 @ 11 1/4
V-478	Prestolite	359232-C91	—	②	25–29	17–21	0 @ 350	12 @ 1600	5 1/2 to 7 1/2	7 @ 11 1/4
VS-478	Prestolite	469620-C92	—	.017	27–33	17–20	0 @ 375	10 @ 1750	—	—
VS-478	Prestolite	470399-C91	—	.017	27–33	17–20	0 @ 375	10 @ 1750	—	—
V/VS-478	Prestolite	401032-C91	—	.017	27–33	17–20	0 @ 350	11 @ 1600	—	—
V/VS-478	Prestolite	401034-C91	—	.017	27–33	17–20	0 @ 350	11 @ 1600	—	—
V/VS-478	Prestolite	427199-C91	—	.017	28–32	17–20	0 @ 400	12.5 @ 1550	—	—
V/VS-478	Prestolite	427200-C91	—	.017	28–32	17–20	0 @ 400	12.5 @ 1550	—	—
V/VS-478	Prestolite	427201-C91	—	.017	28–32	17–20	0 @ 400	12.5 @ 1550	—	—
⑦	Prestolite	401030-C91	—	.017	28–32	17–20	0 @ 320	7 @ 1700	—	—
⑦	Prestolite	401033-C91	—	.017	28–32	17–20	0 @ 320	7 @ 1700	—	—
⑦	Prestolite	413377-C91	—	.017	28–32	17–20	0 @ 320	7 @ 1700	—	—
V-537	Holley③	448195-C91	C	.008④	26–32	—	.2 @ 500	8.9 @ 1700	—	—
V-537	Prestolite③	484794-C91	C	.008④	26–32	—	.2 @ 500	8.9 @ 1700	—	—

Continued

DISTRIBUTOR SPECIFICATIONS—Continued

Engine Application	Distributor Make	Distributor Model	Rotation Rotor End	Point Gap Inch	Dwell Angle Deg.	Breaker Arm Spring Tension Ounces	Centrifugal Advance Deg. @ Dist. R.P.M. Advance Starts	Full Advance	Vacuum Advance Inches of Vacuum to Start Advance	Max. Adv. Dist. Deg. @ Vacuum
FOUR CYLINDER ENGINES										
4-196	Holley	3857-P	—	.017	42–48	17–20	1 @ 300	16 @ 1800	5 to 7	11 @ 14
4-196	Prestolite③	464359-C91	C	.008④	26–32	—	5 @ 600	17.8 @ 2200	5 to 7	6 @ 15
4-196	Prestolite	484824-C91	C	.016	42–48	17–22	0 @ 300	17 @ 1800	5 to 7	8 @ 15
4-196	Holley③	464359-C91	C	.008④	26–32	—	4 @ 600	17.8 @ 2200	5 to 7	6 @ 15
4-196	Holley③	476165-C91	C	.008④	26–32	—	4 @ 600	17.8 @ 2200	3 to 5	6 @ 13
4-196	Prestolite③	484776-C91	C	.008④	28–34	—	4 @ 600	17.8 @ 2200	5 to 7	6 @ 15
4-196	Prestolite③	484777-C91	C	.008④	28–34	—	4 @ 600	17.8 @ 2200	3 to 5	6 @ 13
4-196	Prestolite③	496003-C91	C	.008④	28–34	—	2.7 @ 500	15 @ 2200	5 to 7	6 @ 19
4-196	Prestolite③	497600-C91	C	.008④	28–34	—	4 @ 600	17.8 @ 2200	5 to 7	6 @ 15

①—V-304, 345.
②—New .017″, used .014″.
③—Electronic breakerless ignition system.
④—Clearance between end of trigger wheel tooth and sensor.
⑤—V-304, 345 & 392 low compression engines.
⑥—V-304, 345, & 392
⑦—V/VS/FTV/FTVS-549

STARTING MOTOR SPECIFICATIONS

Starter Part No.	Brush Spring Tension, Oz.	Free Speed Test Amperes	Free Speed Test Volts	Free Speed Test R.P.M.	Resistance Test④ Amperes	Resistance Test④ Volts	Lock Torque Test Amperes	Lock Torque Test Volts	Lock Torque Test Torque Ft.-Lbs.
BENDIX DRIVE TYPE									
1108009	24	70①	5.65	5500②	—	—	570	13.5	3.15
1108142	24	70①	11.3	6000②	—	—	530	16	6.7
1108221	24	70①	5.0	3500②	—	—	600	22	3.0
1108312	35	58–80	10.6	6750–10500	280–320	4.0	—	—	—
1108840	36–40	115①	11.6	7000②	—	—	570	20	2.3
1113336	48	30①	23.4	6000②	—	—	200	20	6.6
1113456	28–36	50①	11.5	3600②	—	—	500	34	3.25
1113458	28–36	50①	5.7	2300②	—	—	500	31	3.0
1113460	36	35①	23.5	3500②	—	—	200	28	6.6
1114000	80	75–115	11.5	5600–7600	—	—	600	16.5	2.5
1114089	80	75–115	11.5	5600–7600	—	—	600	16.5	2.5
DYER DRIVE TYPE									
1109715	36–40	100①	23	8000②	—	—	525	32	3.0
1113708	80	60–90	23	7000–10700	—	—	500	26	3.5
1113709	80	60–90	23	7000–10700	—	—	500	26	3.5
1113781	80	60–90	23	7000–10700	—	—	500	26	3.5
1113823	80	60–90	23	7000–10700	—	—	500	26	3.5
ENCLOSED SHIFT LEVER TYPE (Standard)									
1107220	35	55–80③	9	3500–6000	300–360	3.5	—	—	—
1107678	35	50–80③	9	5500–10500	270–310	4.3	—	—	—
1107679	35	55–80③	9	3500–6000	300–360	3.5	—	—	—
1107680	35	50–80③	9	3500–10500	300–360	3.5	—	—	—
1107709	35	50–80③	9	5500–10500	270–310	4.3	—	—	—
1107710	35	55–80③	9	3500–6000	300–360	3.5	—	—	—
1107742	35	40–105③	9	3500–6500	300–360	3.5	—	—	—
1108233	—	20–30③	20	2250–2750	—	—	—	—	—
1108238	35	40–75③	20	3300–5600	385–435	15.0	—	—	—
1108278	—	50–80③	9	5500–10500	—	—	—	—	—
1108341	35	50–80③	9	5500–10500	300–360	3.5	—	—	—
1108384	35	50–80③	9	5500–10500	270–310	4.3	—	—	—
1108478	—	50–80③	9	5500–10500	—	—	—	—	—
1108515	—	50–80③	9	5500–10500	—	—	—	—	—
1108769	—	65–95③	9	7500–10500	—	—	—	—	—

Continued

STARTING MOTOR SPECIFICATIONS—Continued

Starter Part No.	Brush Spring Tension, Oz.	Free Speed Test			Resistance Test④		Lock Torque Test		
		Amperes	Volts	R.P.M.	Amperes	Volts	Amperes	Volts	Torque Ft.-Lbs.
ENCLOSED SHIFT LEVER TYPE (Standard)—Continued									
1109256	—	50–80③	9	5500–9000	—	—	—	—	—
1109356	—	60–85③	9	6800–10300	—	—	—	—	—
1109358	—	60–85③	9	6800–10300	—	—	—	—	—
1109369	—	65–95③	9	7500–10500	—	—	—	—	—
1109437	—	60–85③	9	6800–10300	—	—	—	—	—
ENCLOSED SHIFT LEVER TYPE (Heavy Duty)									
11109630	—	95–120③	20	5500–7500	—	—	—	—	—
1109646	—	95–120③	20	5500–7500	—	—	—	—	—
11109989	—	95–120③	20	5500–7500	—	—	—	—	—
1113167	80	57–70	9	3500–5500	—	—	500	22	3.4
1113172	80	57–70	9	3500–5500	—	—	500	22	3.4
1113183	—	50–70③	9	3500–5500	—	—	—	—	—
1113217	—	50–70③	9	3500–5500	—	—	—	—	—
1113218	—	50–70③	9	3500–5500	—	—	—	—	—
1113245	—	75–105③	9	5000–7000	—	—	—	—	—
1113491	—	75–105③	9	5000–7000	—	—	—	—	—
1113626	80	75–105③	9	5000–7000	—	—	600	20	2.2
1113644	80	75–105③	9	5000–7000	—	—	600	20	2.2
1113650	—	75–105③	9	5000–7000	—	—	—	—	—
1113846	—	75–95③	20	5500–7500	—	—	—	—	—
1113887	—	75–95③	20	5500–7500	—	—	—	—	—
1113655	80	75–105③	9	5000–7000	—	—	600	20	2.2
1113820	80	75–95③	20	5500–7500	—	—	500	26	3.5
1113914	80	75–95③	20	5500–7500	—	—	500	26	3.5
1113969	—	75–95③	20	5500–7500	—	—	—	—	—
1114052	80	120–150③	9	3000–4500	—	—	600	16.5	2.5
1114058	—	140–190③	9	4000–7000	—	—	—	—	—
1114064	80	120–150③	9	3000–4500	—	—	600	19	2.5
1114066	80	120–150③	9	3000–4500	—	—	600	16.5	2.5
1114070	80	140–190③	9	4000–7000	—	—	700	15	1.5
1114071	80	120–150③	9	3000–4500	—	—	600	19	2.5
1114074	80	140–190③	9	4000–7000	—	—	700	15	1.5
1114076	80	140–190③	9	4000–7000	—	—	700	15	1.5
1114085	80	120–150③	9	3000–4500	—	—	600	16.5	2.5
1114088	80	140–190③	9	4000–7000	—	—	700	15	1.5
1114089	—	105–135③	9	3000–4500	—	—	—	—	—
1114098	80	140–190③	9	4000–7000	—	—	700	15	1.5
1114101	—	120–150③	9	3000–4500	—	—	—	—	—
1114102	80	140–190③	9	4000–7000	—	—	700	15	1.5
1114107	80	120–150③	9	3000–4500	—	—	600	16.5	2.5
1114112	—	120–150③	9	3000–4500	—	—	—	—	—
1114113	80	140–190③	9	4000–7000	—	—	700	15	1.5
1114116	80	120–150③	9	3000–4500	—	—	600	16.5	2.5
1114120	80	140–190③	9	4000–7000	—	—	700	15	1.5
1114122	80	120–150③	9	3000–4500	—	—	600	16.5	2.5
1114127	—	140–190③	9	4000–7000	—	—	—	—	—
1114128	—	140–190③	9	4000–7000	—	—	—	—	—
1114134	—	140–190③	9	4000–7000	—	—	—	—	—
1114145	—	140–190③	9	4000–7000	—	—	—	—	—
1114161	—	140–190③	9	4000–7000	—	—	—	—	—
1114165	—	140–190③	9	4000–7000	—	—	—	—	—
1114168	—	120–150③	9	3000–4500	—	—	—	—	—
1114171	—	140–190③	9	4000–7000	—	—	—	—	—

Continued

STARTING MOTOR SPECIFICATIONS—Continued

Starter Part No.	Brush Spring Tension, Oz.	Free Speed Test			Resistance Test④		Lock Torque Test		
		Amperes	Volts	R.P.M.	Amperes	Volts	Amperes	Volts	Torque Ft.-Lbs.
ENCLOSED SHIFT LEVER TYPE (Heavy Duty)—Continued									
1114176	—	140–190③	9	4000–7000	—	—	—	—	—
1114180	—	120–150③	9	3000–4500	—	—	—	—	—
1114187	—	120–150③	9	3000–4500	—	—	—	—	—
1114189	—	120–150③	9	3000–4500	—	—	—	—	—
1114194	—	140–190③	9	4000–7000	—	—	—	—	—
1114196	—	120–150③	9	3000–4500	—	—	—	—	—
1114197	—	140–190③	9	4000–7000	—	—	—	—	—
1114734	—	140–190③	9	4000–7000	—	—	—	—	—
508702C91	—	130–160③	9	5000–7000	—	—	—	—	—
EXPOSED SHIFT LEVER TYPE									
1107156	24	70①	5.65	5500②	—	—	550	11	3.25
1107168	24	80①③	5.6	5500②	—	—	550	11	3.3
1107218	35	91①	10.6	3240②	330–395	3.5	—	—	—
1107689	35	112①	10.6	3240②	320–385	3.5	—	—	—
1108033	24	80①	5.67	5500②	—	—	600	14	3.0
1108042	24	80①	5.67	5500②	—	—	600	14	3.0
1108051	24	80①	5.67	5500②	—	—	600	14	3.0
1108153	24	70①	11.3	6000②	—	—	530	16	6.7
1108157	24	70①	11.3	6000②	—	—	530	16	6.7
1108161	24	70①	11.3	6000②	—	—	530	16	6.7
1108233	24	35①	23.5	2500②	—	—	265	19	19.1
1109930	24	35①	23.5	2500②	—	—	265	19	19.1
1109941	24	35①	23.5	2500②	—	—	265	19	19.1
1113001	28–36	50①	11.5	8000②	—	—	500	22	3.25
1113039	28–36	50①	5.7	5000②	—	—	500	21	3.0
1113040	28–36	50①	11.5	8000②	—	—	500	22	3.25
1113070	48	57–70③	11.5	5000–7400	—	—	500	22	3.4
1113081	48	57–70③	11.5	5000–7400	—	—	500	22	3.4
1113098	48	57–70③	11.5	5000–7400	—	—	500	22	3.4
1113108	28	32–40③	23.5	5800–6800	—	—	250	19.5	8.0
1113109	28	32–40③	23.5	5800–6800	—	—	250	19.5	8.0

①—Maximum.
②—Minimum.
③—Includes solenoid.
④—Check capacity of motor by using a 500 ampere meter and a carbon pile rheostat to control voltage. Apply volts listed across motor with armature locked. Amperage should be as listed.

ALTERNATOR & REGULATOR SPECIFICATIONS

Alternator					Regulator						
Model	Field Current @ 80°F. 12 Volts	Cold Output @ 14 V		Rated Hot Output Amperes	Model	Field Relay			Voltage Regulator		
		Amperes @ 2000 R.P.M.	Amperes @ 5000 R.P.M.			Air Gap	Point Gap	Closing Voltage	Air Gap	Point Gap	Volts @ 125° F.
1100072⑥	4.4–4.9	45	57	65	Integral	—	—	—	—	—	13.2–14.4
1100073⑥	4.4–4.9	54	73	80	Integral	—	—	—	—	—	13.2–14.4
1100080⑥	4.4–4.9	54	73	80	Integral	—	—	—	—	—	—
1100081⑥	4.4–4.9	48	61	65	Integral	—	—	—	—	—	—
1100543⑥	4–4.5	25	38	42	Integral	—	—	—	—	—	—
1100544⑥	—	—	—	37	Integral	—	—	—	—	—	13.3–15.6
1100554⑥	4–4.5	32	60	63	Integral	—	—	—	—	—	—
1100588⑥	—	—	—	61	Integral	—	—	—	—	—	13.3–15.6
1100628⑥	1.90–2.30	25	35	37	1119507	.015	.030	3.8–7.2	.067	.014	13.5–14.4
1100629⑥	1.90–2.30	28	40	42	—	—	—	—	—	—	—

Continued

ALTERNATOR & REGULATOR SPECIFICATIONS—Continued

Alternator					Regulator						
Model	Field Current @ 80°F. 12 Volts	Cold Output @ 14 V		Rated Hot Output Amperes	Model	Field Relay			Voltage Regulator		
		Amperes @ 2000 R.P.M.	Amperes @ 5000 R.P.M.			Air Gap	Point Gap	Closing Voltage	Air Gap	Point Gap	Volts @ 125° F.
1100630⑥	1.90–2.30	21	30	32	—	—	—	—	—	—	—
1100633⑥	1.90–2.30	32	50	52	1119507	.015	.030	3.8–7.2	.067	.014	13.5–14.4
1100687⑥	2.2–2.6	21	30	32	—	—	—	—	—	—	—
1100688⑥	2.2–2.6	25	35	37	—	—	—	—	—	—	—
1100689⑥	2.2–2.6	28	40	42	—	—	—	—	—	—	—
1100690⑥	2.2–2.6	32	50	55	—	—	—	—	—	—	—
1100750⑥	2.2–2.6	33	58	61	—	—	—	—	—	—	—
1102848⑥	4–4.5	25	38	42	Integral	—	—	—	—	—	—
1102849⑥	4–4.5	32	60	63	Integral	—	—	—	—	—	—
1102915⑥	4–4.5	25	38	42	Integral	—	—	—	—	—	—
1117070⑥	6.00–6.60	31①	58②	60	9000554	.014	.030	3.8–7.2	.075	—	13.2–14.2
1117078⑥	6.00–6.60	31①	58②	60	9000552	.014	.027	5.0–9.5	.075	—	13.2–14.2
1117089⑥	6.00–6.60	31①	58②	60	9000552	.014	.027	5.0–9.5	.075	—	13.2–14.2
1117090⑥	6.00–6.60	24③	80②	85	—	—	—	—	—	—	—
1117096⑥	3.60–4.00	23⑤	61②	60	—	—	—	—	—	—	—
1117115⑥	1.23–1.33	32③	98②	105	—	—	—	—	—	—	—
1117128⑥	2.22–2.40	40③	126②	130	—	—	—	—	—	—	—
1117141⑥	4–4.5	80	110	105	Integral	—	—	—	—	—	—
1117142⑥	4–4.5	110	150	145	Integral	—	—	—	—	—	—
1117143⑥	4.0–4.5	—	150	145	Integral	—	—	—	—	—	13.2–14.4
1117147⑥	4–4.5	80	110	105	Integral	—	—	—	—	—	—
1117148⑥	4–4.5	110	150	145	Integral	—	—	—	—	—	—
1117149⑥	4–4.5	110	150	145	Integral	—	—	—	—	—	—
1117225⑥	4.1–4.5	—	77	75	—	—	—	—	—	—	—
1117228⑥	4.1–4.5	—	77	75	Integral	—	—	—	—	—	—
1117235⑥	4.1–4.5	—	75	75	Integral	—	—	—	—	—	—
1117241⑥	4.1–4.5	—	77	75	Integral	—	—	—	—	—	—
1117242⑥	4.1–4.5	—	77	75	Integral	—	—	—	—	—	—
1117481⑥	3.6–4.3	—	60②	86④	Integral	—	—	—	—	—	—
1117732⑥	3.6–4.3	—	72②	90④	Integral	—	—	—	—	—	—
1117753⑥	2.10–2.40	14③	55④	55	1119500	.015	.015	6.5–8.5	.067	.016	13.5–14.4
1117754⑥	3.70–4.40	18③	62④	62	9000595	.015	.027	2.3–3.5	.075	—	13.2–14.2
1117756⑥	3.70–4.40	18③	62④	62	9000567	.014	.027	5.0–9.5	.075	—	13.2–14.2
8AL2026F⑦	1.8–2.5	—	—	37	8RD2001	—	—	—	—	—	13.3–13.95
8AL2033L⑦	1.8–2.5	—	—	42	8RD2001	—	—	—	—	—	13.3–13.95
513303-C91⑥	—	—	—	85	Integral	—	—	—	—	—	—
513542-C91⑧	2.9	—	—	105	Integral	—	—	—	—	—	—
514089-C91⑧	2.9	—	—	65	Integral	—	—	—	—	—	—
514251-C91⑧	—	—	—	130	Integral	—	—	—	—	—	—
517368-C91⑧	—	—	—	75	Integral	—	—	—	—	—	—
520443-C91⑧	2.9	—	—	105	Integral	—	—	—	—	—	—
521092-C91⑧	2.9	—	—	105	Integral	—	—	—	—	—	—
527890-C91⑧	—	—	—	145	—	—	—	—	—	—	—

①—At 1000 R.P.M.
②—At 2500 R.P.M.
③—At 1100 R.P.M.
④—At 6500 R.P.M.
⑤—At 5000 R.P.M.
⑥—Delco-Remy.
⑦—Motorola.
⑧—Leece Neville.

VALVE SPECIFICATIONS

Engine Model	Valve Lash	Valve Angles		Valve Springs		Valve Stem Clearance		Stem Diameter, Std.	
		Seat	Face	Installed Height	Pressure Lbs. @ In.	Intake	Exhaust	Intake	Exhaust
4-196	Zero①	②	②	③	188 @ 1.42	.001–.0035	.0015–.004	.372–.373	.414–.415
6-258	Zero①	④	④	—	195 @ 1 7/16	.001–.003	.001–.003	.3715–.3725	.3715–.3725
V-304	Zero①	45°	45°	31/32	188 @ 1.429	.001–.0035	.0015–.004	.372–.373	.3715–.3725
V-304A	Zero①	45°	45°	⑤	188 @ 1.429	.001–.0035	.0015–.004	.372–.373	.3715–.3725
V-345	Zero①	45°	45°	⑤	188 @ 1.429	.001–.0035	.0015–.004	.372–.373	.3715–.3725
V-392	Zero①	45°	45°	③	188 @ 1.429	.001–.0035	.0015–.004	.372–.373	.3715–.3725
V-400	Zero①	⑥	⑥	—	⑦	.001–.003	.001–.003	.3715–.3725	.3715–.3725
V-401	Zero①	⑧	⑧	—	116 @ 1.663	.001–.0035	.0025–.0045	.434–.435	.433–.434
MV-404	Zero①	45°	45°	—	188 @ 1.429	.0011–.0028	.00165–.00235	.3721–.3728	.37165–.37235
RD-406	.024–.026H	⑧	⑧	—	137 @ 1 45/64	.0015–.004	.002–.0045	.434–.435	.4335–.4345
MV-446	Zero①	45°	45°	—	188 @ 1.429	.0011–.00285	.00165–.00235	.3721–.3728	.37165–.37235
RD-450	.024–.026H	⑧	⑧	—	137 @ 1 45/64	.0015–.004	.002–.0045	.434–.435	.4335–.4345
V-478	Zero①	⑧	⑧	—	116 @ 1.663	.001–.0035	.0025–.0045	.434–.435	.433–.434
RD-501	.024–.026H	⑧	⑧	—	137 @ 1 45/64	.0015–.004	.002–.0045	.434–.435	.4335–.4345
V-537	Zero①	⑧	⑧	—	200 @ 1.39	.0016–.0034	.002–.0037	.4341–.4349	.4338–.4345
V-549	Zero①	⑧	⑧	—	116 @ 1.663	.001–.0035	.0025–.0045	.434–.435	.433–.434

①—Hydraulic lifters.
②—Intake, 30°; Exhaust 45°.
③—Intake, 31/32; Exhaust 1 11/64.
④—Intake, 30°; Exhaust 44°.
⑤—Intake, 31/32; Exhaust 1 3/32.
⑥—Intake, 29°; Exhaust 44°.
⑦—Intake, 218 @ 1.365; Exhaust 218 @ 1.183.
⑧—Intake, 15°; Exhaust 45°.

PISTON, PIN, RING, CRANKSHAFT & BEARING SPECIFICATIONS

Year	Engine Model	Wristpin Diameter	Piston Clearance, Inch	Ring End Gap, In. (Minimum)		Crank Pin Diameter, Inch	Rod Bearing Clearance, In.	Main Bearing Journal Diameter, In.	Main Bearing Clearance, Inch	Thrust On Bearing No.	Shaft End Play
				Comp.	Oil						
1974–80	4-196	1.0624	.0035	.013	.020	2.373–2.374	.0011–.0036	2.7484–2.7494	.001–.004	3	.003–.008
1974	6-258	.9306	.0009	.010	.015	2.0934–2.0955	.001–.002	2.4981–2.5001	.001–.002	3	.0015–.007
1974	RD-372	1.109	④	.025	.013	2.751–2.752	.0012–.0037	3.2495	.001–.004	7	.006–.015
1974	RD-406	1.109	④	.025	.013	2.751–2.752	.0012–.0037	2.2495–2.2505	.0013–.0043	7	.006–.015
	RD-450	1.109	④	.025	.013	2.751–2.752	.0012–.0037	3.2495–3.2505	.0013–.0043	7	.006–.015
	RD-501	1.109	④	.025	.013	2.751–2.752	.0012–.0037	3.2495–3.2505	.0013–.0043	7	.006–.015
1974–80	V-304	1.11248	.004	.010	⑤	2.373–2.374	.0009–.0034	2.7484–2.7494	.001–.004	3	.002–.010
1974–79	V-345	1.11248	.004	.010	⑤	2.373–2.374	.0009–.0034	2.7484–2.7494	.001–.004	3	.002–.010
1974–80	V-392	1.11248	.004	.013	②	2.373–2.374	.0009–.0034	2.7484–2.7494	.001–.004	3	.002–.010
1974	V-400	1.006	.0010	.010	③	2.2485–2.2464	.001–.002	2.7474–2.7489	.001–.002	3	.002–.010
1974–75	VS-401	1.3118	.002	.013	.013	2.623–2.624	.0017–.0042	3.124–3.125	.0008–.0038	3	.003–.008
1975–80	MV-404	1.200	.0012	.013	.013	2.4980–2.4990	.0011–.0036	3.1228–3.1236	.0010–.0036	3	.006–.011
1975–80	MV-446	1.200	.0012	.013	.013	2.4980–2.4990	.0011–.0036	3.1228–3.1236	.0010–.0036	3	.0025–.0085
1974–75	VS-478	1.3118	.002	.013	.013	2.623–2.624	.0017–.0042	3.124–3.125	.0008–.0038	3	.0025–.0085
1980	V-537	1.3118	.0012	①	.012	2.6230–2.6240	.0011–.0036	3.1235–3.1245	.0015–.0035	3	.006–.011
1974–77	VS-549	1.3118	.002	.013	.013	2.623–2.624	.0017–.0042	3.124–3.125	.0008–.0038	3	.006–.012
1975	FTVS-549	1.3118	.002	.013	.013	2.623–2.624	.0017–.0042	3.124–3.125	.0008–.0038	3	.006–.011

①—Top .012, Second .014.
②—No gap at ring joint. Steel rail gap .013 inch.
③—No gap at ring joint. Steel rail gap .0015 inch.
④—Top of skirt .0045 inch, bottom .002 inch.
⑤—No gap at ring joint. Steel rail gap .015".

ENGINE TIGHTENING SPECIFICATIONS★

★ Torque specifications are for clean and lightly lubricated threads only. Dry or dirty threads produce increased friction which prevents accurate measurement of tightness.

Engine	Spark Plugs Ft. Lbs.	Cylinder Head Bolts Ft. Lbs.	Intake Manifold Ft. Lbs.	Exhaust Manifold Ft. Lbs.	Rocker Arm Shaft Bracket Ft. Lbs.	Rocker Arm Cover Ft. Lbs.	Connecting Rod Cap Bolts Ft. Lbs.	Main Bearing Cap Bolts Ft. Lbs.	Flywheel to Crankshaft Ft. Lbs.	Vibration Damper or Pulley Ft. Lbs.
4-196⑤	28–30	90–100	40–45	40–45	14–16	—	45–55	75–80	45–55	100–110
4-196⑥	28–30	110	25–30	25–30	14–16	—	45–50	75–85	45–55	100–110
6-258	25–30	80–85	20–25	20–25	20–33	—	26–30	75–85	100–110	70–80
RD-406	28–30	100–110②	25–30	25–30	25–30	—	65–75	100–110	150–160	①
RD-450	28–30	100–110②	25–30	25–30	25–30	—	65–75	100–110	150–160	①
RD-501	28–30	100–110②	25–30	25–30	25–30	—	65–75	100–110	150–160	①
V-304, 345	28–30	90–100	40–45	40–45	—	—	45–55	75–85	45–55	100–110③
V-392	28–30	90–100	40–45	40–45	—	—	40–45	75–85	45–55	100–110③
V-400	28	110	43	25	—	—	38	100	105	55
V-401	28–30	90–100	—	—	80–90④	—	60–70	100–110	90–100	①
MV-404	15	100–110	—	15–21	18–22④	4–7	42–46	70–90	—	80–100
MV-446	15	100–110	—	15–21	18–22④	4–7	42–46	70–90	—	80–100
V-478	28–30	90–100	—	—	80–90	—	60–70	100–110	90–100	①
V-537	25–30	80–90	30–38	30–38	70–85	7–10	65–70	125–135	110–120	260–290
V-549	28–30	90–100	—	—	80–90	—	60–70	100–110	90–100	①

①—Pressed on.
②—Oil supply bolt 75–85 ft. lbs.
③—Pulley hub bolt.
④—Rocker arm cap screw.
⑤—Prior to engine serial No. 37396.
⑥—Engine serial No. 37396 & after.

WHEEL ALIGNMENT SPECIFICATIONS

Front Axle Model①	Caster, Degrees	Camber, Degrees	Toe-In, Inch	Kingpin Inclination, Degrees
FA-3	0	1	3/32–5/16	8½
FA-5	⑯	⑯	⑯	—
FA-6	⑯	⑯	⑯	—
FA-7	⑯	⑯	⑯	—
FA-44	0	1	3/32–5/16	8½
FA-54	0	1	1/16	8
FA-64	2	½	1/16	0
FA-71	2	③	1/16	②
FA-73	2	③	1/16	②
FA-74	3½	③	1/8	②
FA-78	2	½	0–3/8	0
FA-79	2	½	0–3/8	0
FA-101	④	③	1/16	②
FA-103	3½	③	1/8	②
FA-109	⑤	③	⑥	②
FA-109⑱	3½	③	1/16	②
FA-109⑰	2½	③	1/16	②
FA-118⑱	3½	③	1/16	②
FA-118⑰	2½	③	1/16	②
FA-127⑱	3½	③	1/16	②
FA-127⑰	2½	③	1/16	②
FA-139	⑤	③	⑥	②
FA-139⑱	3¾	③	1/16	②
FA-139⑰	2¾	③	1/16	②
FA-191	3½	−¼	1/16	0

Front Axle Model①	Caster, Degrees	Camber, Degrees	Toe-In, Inch	Kingpin Inclination, Degrees
FA-192	3½	−¼	1/16	0
FA-193	3½	−¼	1/16	0
FA-194	3½	¾	1/16	0
FA-195	3½	¾	1/16	0
FA-231	⑦	⑧	⑥	⑨
FA-232	⑩	⑧	1/16	⑨
FA-233	3½	⑧	1/16	⑨
FA-237	3½	⑧	1/16	⑨
FA-238	3½	⑧	1/16	⑨
FA-239	3½	⑧	1/16	⑨
FA-309	⑪	③	⑥	②
FA-309⑱	3½	③	1/16	②
FA-309⑰	2½	③	1/16	②
FA-329	⑫	③	⑥	②
FA-329⑱	3½	③	1/16	②
FA-329⑰	2½	③	1/16	②
FA-339	⑬	③	⑥	②
FA-339⑱	3¾	③	1/16	②
FA-339⑰	2¾	③	1/16	②
FA-340	⑭	③	1/16	②
FA-341	⑮	③	1/16	②
FA-341⑱	3½	③	1/16	②
FA-341⑰	2½	③	1/16	②
FA-342	1	③	1/16	②

Continued

WHEEL ALIGNMENT SPECIFICATIONS—Continued

(1)—Model is given on aluminized paper plate glued in truck cab.
(2)—Left 4 1/4°, right 4 1/2°.
(3)—Left 1/2°, right 1/4°.
(4)—On Loadstar & Fleetstar (4×2) models, 1°; on S-series, 1 3/4°; on Fleetstar (6×4), 4°.
(5)—On Loadstar models, 1°; on Fleetstar (4×2) models, 1 1/2°; on Fleetstar (6×4) models, 3°; on Cargostar models, 3 1/2°.
(6)—Except Cargostar, 1/16 in.; Cargostar, 1/8 in.
(7)—On conventional Transtar models; 3/4°; on Co-Transtar models, 1 1/4°, on S-series models, 3 1/4°; on Cargostar, Paystar & COF-5370, 3 1/2°; on Fleetstar (4×2), 2°; on Fleetstar (6×4), 4°.
(8)—Left 1/2°, right 0°
(9)—Left 5 3/4°, right 6 1/4°.
(10)—On S-series models, 3 1/4°; on Paystar & COF-5370, 3 1/2°; on Fleetstar, 4°.
(11)—On Fleetstar (4×2) models, 1 1/2°; on Fleetstar (6×4) models, 3°; on Cargostar, 3 1/2°.
(12)—On Loadstar & conventional Transtar, 1°; on Co-Transtar & Fleetstar (4×2), 1 1/2°; on Fleetstar (6×4), 3°; on Cargostar, 3 1/2°.
(13)—On conventional Transtar, 3/4°; on Co-Transtar, 1 1/4°; on Fleetstar (4×2), 1 1/2°; on Loadstar, 2°; on Fleetstar (6×4), 3°; on Cargostar & Paystar, 3 1/2°.
(14)—On conventional Transtar, 3/4°; on Co-Transtar, 1 1/4°; on Fleetstar (4×2), 1 1/2°; on Fleetstar (6×4), 3°.
(15)—On conventional Transtar, 1°; on Co-Transtar, 1 1/2°.
(16)—Refer to "Front Suspension & Steering Gear" section for alignment procedure.
(17)—With manual steering.
(18)—With power steering.

Engine Section

ENGINE, REPLACE

CAUTION: If the truck is equipped with LPG fuel system, be sure that the tank valves are tightly closed and all fuel is exhausted from the lines before starting any repair work on the fuel system.

Any trucks of this type which have been involved in an accident should not be brought into the shop for repair until the tanks and system is checked for possible leaks.

6-258

1. Drain cooling system and crankcase and disconnect battery ground cable.
2. Remove hood, radiator hoses and radiator.
3. Remove carburetor air cleaner and disconnect fuel line at fuel pump.
4. Disconnect exhaust pipe at manifold. Disconnect carburetor throttle and choke linkages.
5. Disconnect wiring harnesses from alternator, starter, temperature and oil pressure sending units and ignition coil.
6. Remove starter motor.
7. Disconnect power brake vacuum hose, power steering lines and freon compressor lines, if so equipped.
8. Disconnect vacuum line from rear of intake manifold.
9. Remove automatic transmission filler tube support if so equipped.
10. Attach lifting sling to engine and raise engine enough to remove load from engine mounts. Disconnect mounting brackets from crossmember.
11. Remove flywheel housing front cover. Remove five flywheel housing-to-crankcase bolts.
12. The engine must be pulled forward to clear the transmission main drive gear and clutch disc then raised and lifted out of the chassis.

NOTE: Use extreme care while removing engine to avoid damaging the clutch disc.

13. Reverse procedure to install.

V8 Engines Exc. V-537

The engine is removed as an assembly complete with accessories, less the carburetor air cleaner. The transmission is separated from the engine at the flywheel housing. Drain the cooling system; drain points are located one on each side and toward the front of the cylinder block and the lower left of the radiator. Drain engine oil. Disconnect battery ground cable and tape the end of the terminal to avoid accidental grounding.

On V8-401, 478, 549, tapped holes at the rear of the intake manifold and engine front cover are provided for lifting eye bolts.

On V8-304, 345, 392, the engine lifting fixture can be attached to the intake manifold by utilizing the carburetor mounting capscrews with flat washers. *When installing the lifting fixture, the capscrews are to be installed finger tight; then tighten the capscrews one-half additional turn.*

V-537

1. Remove front bumper if equipped with steel hood.
2. Disconnect battery ground cable and drain cooling system.
3. Remove hood and disconnect shutter cylinder air line, if equipped.
4. Remove radiator hoses, support, and core

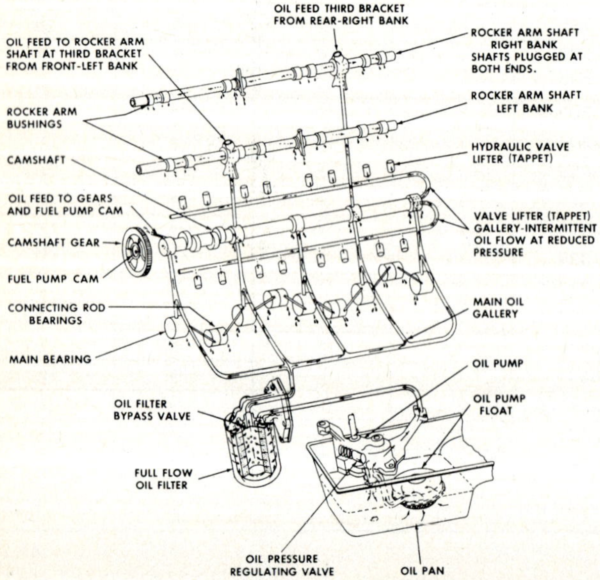

Engine oiling system. V304, V345 & V392

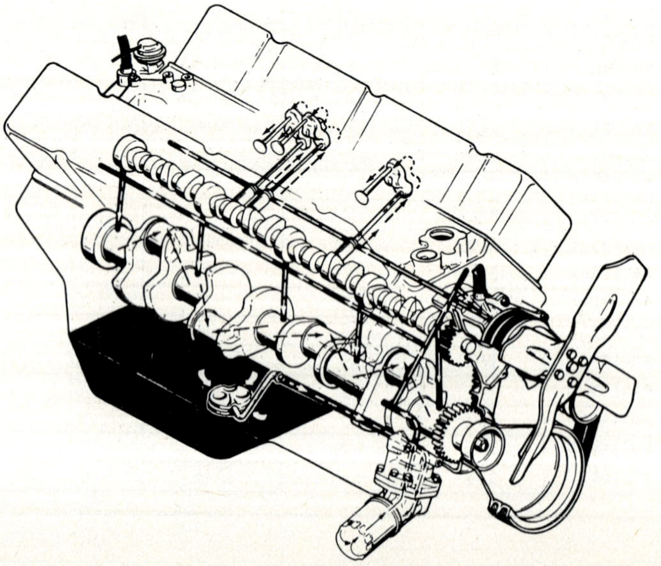
Engine oiling system. 400

as an assembly on models with fiberglass hood.

5. If equipped with steel hood, disconnect wiring at cowl and remove radiator support and core assembly with fenders and splash panels as an assembly.
6. Remove air cleaner braces and air inlet hose between air cleaner and right cylinder head.
7. Remove air inlet hose to air compressor and air cleaner with shroud hose.
8. Remove stud in carburetor and cover to prevent entry of foreign material.
9. Disconnect heater hoses, wires from alternator, coil, distributor, heat, and oil sender unit.
10. Remove engine ground strap, choke, and throttle wire.
11. Disconnect tachometer cable if equipped.
12. Remove accelerator linkage at carburetor and fuel line at fuel pump.
13. Disconnect exhaust pipe at manifold and starter wires.
14. Disconnect canister vent and purge lines at carburetor, if equipped.
15. Disconnect freon compressor wire at connector and freon lines from compressor.
16. On models equipped with air brakes, disconnect main air supply line and flexible air line at compressor.
17. On models equipped with hydraulic brakes and hydrovac, disconnect vacuum air cleaner hose at rear of intake manifold.
18. Remove distributor cap and thermostat housing.
19. Disconnect governor vacuum line and remove distributor from engine.
20. Remove brace from rear of freon compressor.
21. Install lifting plate SE-1896 to thermostat housing base make rear lifting plate locally.
22. Connect a hoist to a lifting sling installed between the lifting plates.
23. Install a support under the transmission and remove the twelve bolts securing bell housing to flywheel housing.
24. Remove the engine front and rear engine supports.
25. Pull the engine forward to clear transmission main drive gear and clutch driven disc.
26. Raise and lift out of chassis. Reverse procedure to install.

4-196 Engines

1. Drain cooling system, crankcase and disconnect battery ground cable.
2. Remove fan blade, radiator, and other parts such as radiator cross braces, heater air ducts, etc., which may interfere with engine removal.
3. Remove carburetor linkage and engine wiring.
4. Remove fuel line from pump.
5. Disconnect exhaust pipe from manifold.
6. Install lifting fixture on engine.
7. Hoist engine enough to support it while disconnecting engine front mountings.
8. Separate engine from transmission by separating transmission from flywheel housing or flywheel housing from rear of engine.
9. Pull engine forward to clear clutch from transmission main drive gear shaft. Then tilt engine upward and remove from chassis.

RD Engines

1. Drain engine oil and cooling system.
2. Disconnect upper and lower hoses.
3. Remove air cleaner.
4. Disconnect vacuum line at manifold or air line at compressor when equipped with air brakes.
5. Disconnect coil wire, starter cable, engine ground strap and instrument sender unit wires.
6. Disconnect fuel line at fuel pump.
7. Disconnect throttle control linkage and remove choke wire at carburetor.
8. Remove hood and floor boards.
9. Remove fenders and grille as a unit.
10. Unfasten and remove radiator and its support.
11. Remove engine front mounting bolts from support bracket. These are the bolts at the front cross member.
12. Support transmission with floor jack or blocking. Then remove cap screws from around bell housing.
13. Disconnect exhaust pipe from manifold.
14. Attach engine sling to front and rear right side cylinder head bolts and remove engine from chassis.
15. Reverse procedure to install.

CYLINDER HEAD, REPLACE

All Engines

Cylinder heads should be tightened down by starting from the center, working outward from side to side and to the ends in the sequence shown in Figs. 1 to 4A.

6-258

CAUTION: The cylinder head bolt located at the left front corner of the head (No. 11, Fig. 4) must have the threads sealed to prevent cooling system leakage. Permatex No. 2 or equivalent is recommended.

ROCKER ARM SERVICE

Early 6-258

The pressure supply for each rocker arm is obtained from No. 5 camshaft bearing located where the camshaft meters the flow of oil from the main lubrication gallery through a groove in the camshaft bearing surface to a gallery extending upwards to the cylinder head gas-

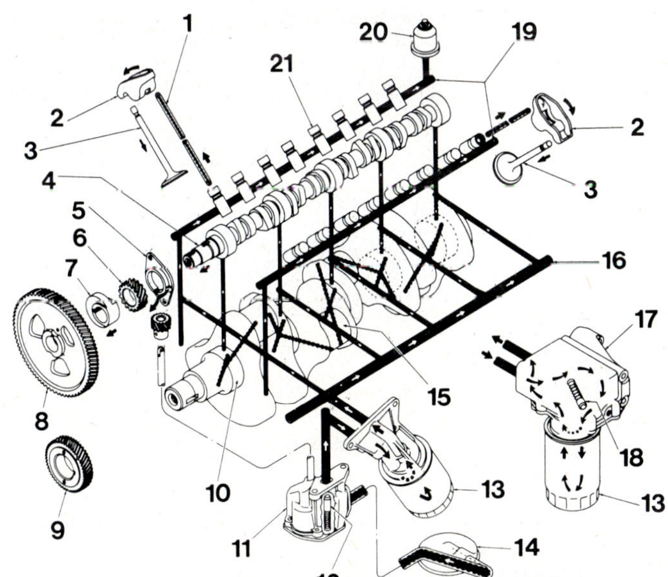

1. Push rod
2. Rocker arm
3. Valve
4. Camshaft
5. Camshaft thrust flange
6. Distributor drive gear
7. Fuel pump cam
8. Camshaft gear
9. Crankshaft gear
10. Main Bearing
11. Oil pump
12. Pressure control valve
13. Oil filter
14. Oil pump float
15. Connecting rod bearings
16. Main oil gallery
17. Oil cooler
18. Cooler by-pass valve
19. Valve lifter (tappet) oil gallery
20. Oil pressure sender unit
21. Hydraulic valve lifter (tappet)

Engine oiling system. MV404 & 446

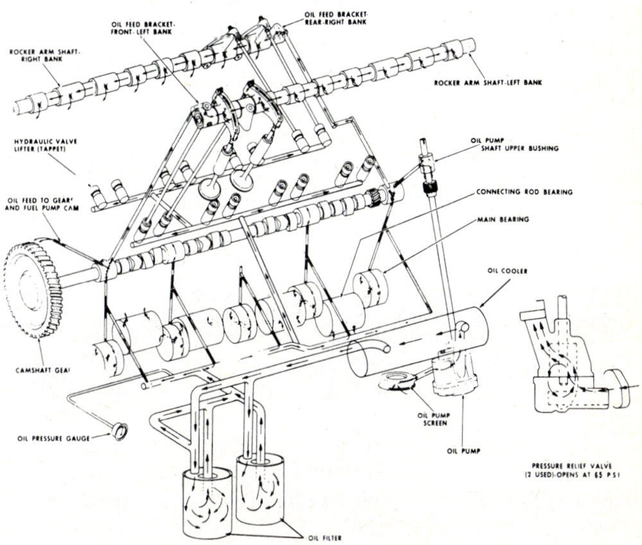

Engine oiling system. V537

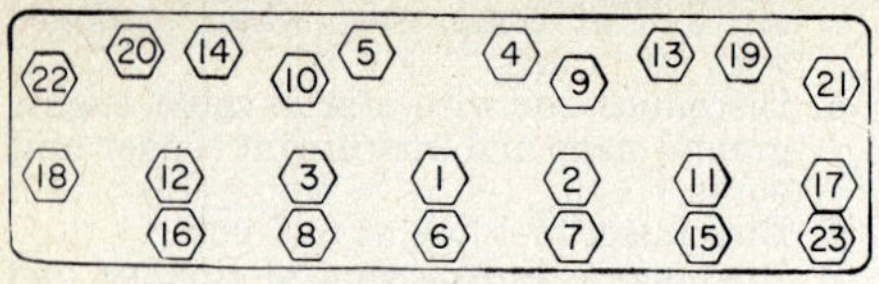

Fig. 1 Cylinder head tightening sequence. RD engines

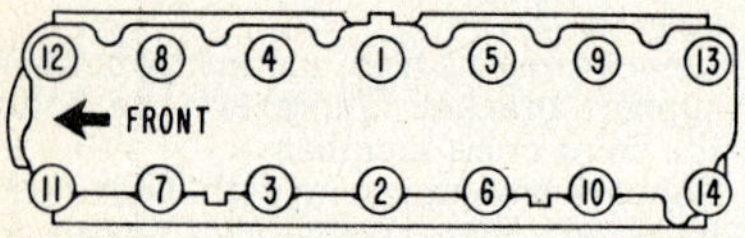

Fig. 4 Cylinder head tightening sequence. 6-258

ket surface. When installing the rocker arm shaft the oil holes must face down to the cylinder head, Fig. 5.

Late 6-258

As shown in Fig. 6, the rocker arm assembly consists of a pair of rocker arms and one rocker arm bracket for each cylinder. When disassembling, keep parts in order to assure reassembly in their original position if in satisfactory condition.

4-196 Engines

As shown in Fig. 7, the rocker arm assembly consists of the shaft, shaft supports, rocker arms and spacers. The rocker arms are lubricated by oil supplied to the hollow shaft through an oil passage in one of the supports.

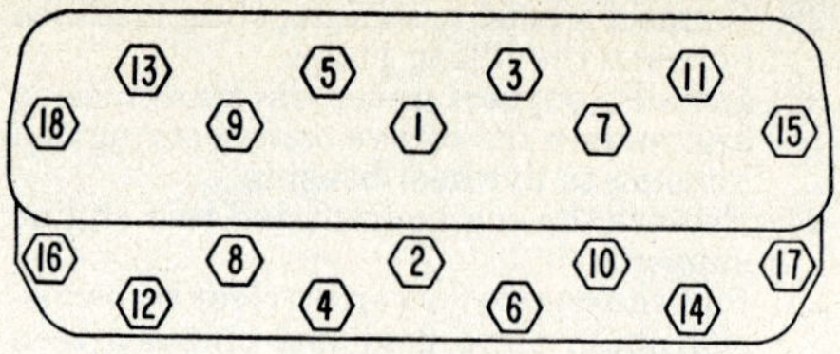

Fig. 2 Cylinder head tightening sequence. V-401, 478, 549

Formation of sludge in the oil passages of the rocker arms and shaft will restrict flow of oil to the rocker arm bushings and valves. Therefore, each time the assembly is removed, it should be disassembled and cleaned thoroughly.

When disassembling, keep parts in order to assure reassembly in their original position if in satisfactory condition.

If necessary to remove plugs from each end of shaft, drill or pierce one plug, then knock out the opposite plug with a steel rod. Working from the other end, knock out other plug.

When reassembling, lubricate rocker arm bushings with engine oil. Make sure the notches at ends of rocker arm shaft are up. Be sure to position the rocker arm support having the oil feed hole so that the oil hole will index with the oil feed hole in the cylinder head (third from rear, Fig. 7).

NOTE: Before installing push rods in 4-196 engines, rotate the crankshaft to position the crankshaft pulley hub as shown in Fig. 8. This is to prevent possible damage to engine parts if the hydraulic lifters have not leaked down to operating position.

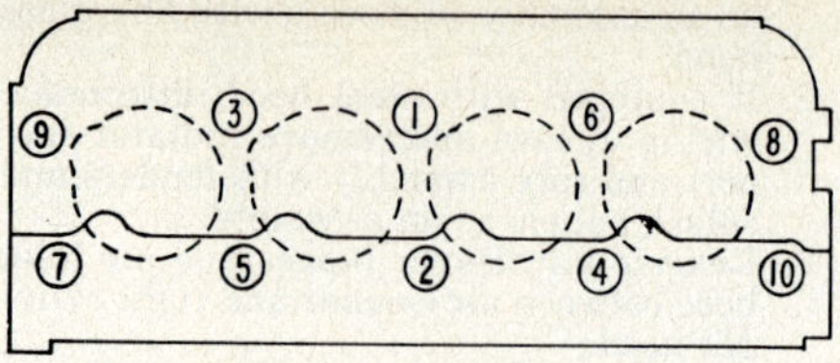

Fig. 3 Cylinder head tightening sequence. 4-196, V-304, 345, 392 & MV 404, 446

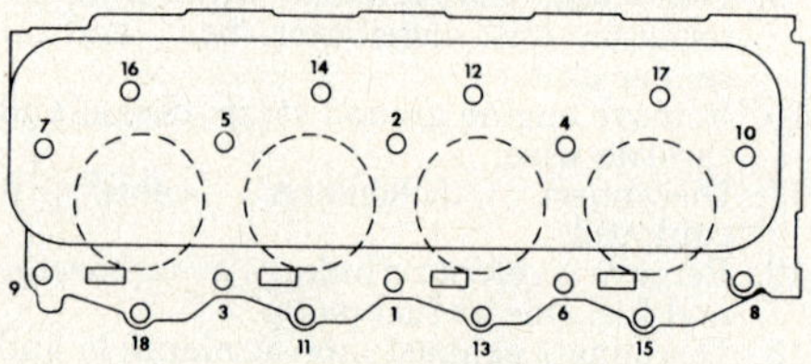

Fig. 4A Cylinder head tightening sequence. V-537

V-304, 345, 392 Engines

When assembling the rocker arms, Fig. 9, make sure the notches at the end of the rocker arm shaft are facing upward. Also make sure the rocker arm shaft support incorporating the oil feed hole is located *third from rear on right bank and third from front on left bank.*

Rocker arm shaft assemblies are interchangeable from one bank to another. Make sure rocker arms are correctly positioned to actuate the valves. Check all push rods for straightness by rolling on a flat surface. Replace any that are bent or have loose ends.

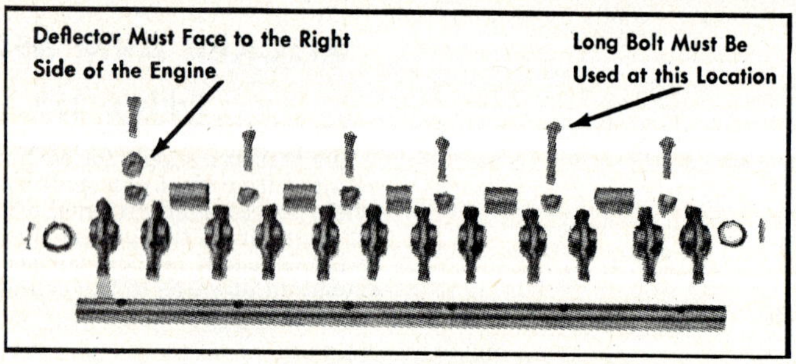

Fig. 5 Rocker arm assembly. Early 6-258

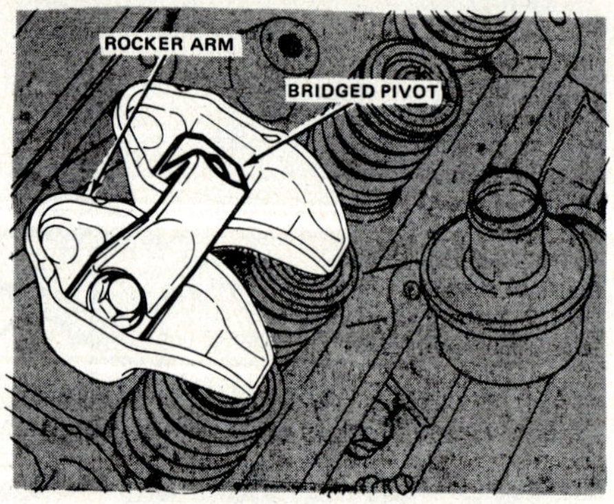

Fig. 6 Rocker arm removal. Late 6-258

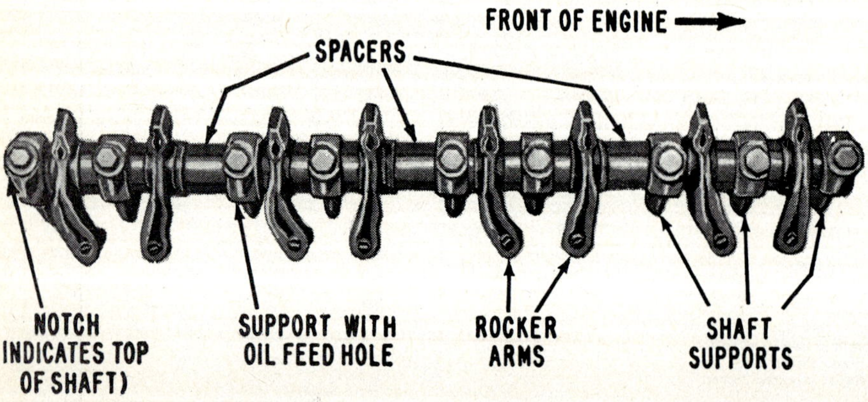

Fig. 7 Rocker arm assembly on 4-196 engines

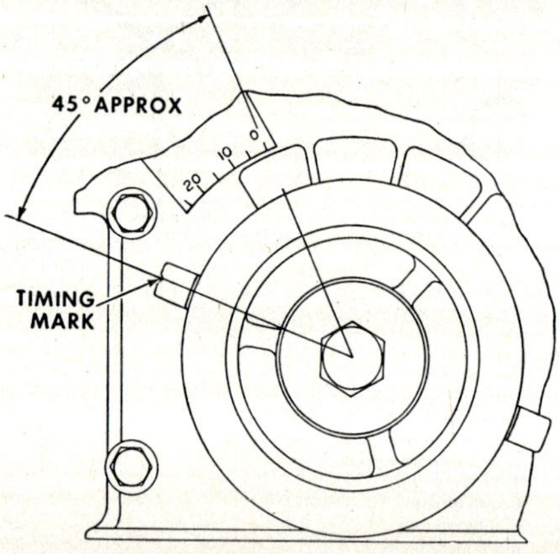

Fig. 8 Correct position of crankshaft pulley when installing push rod on 4-196

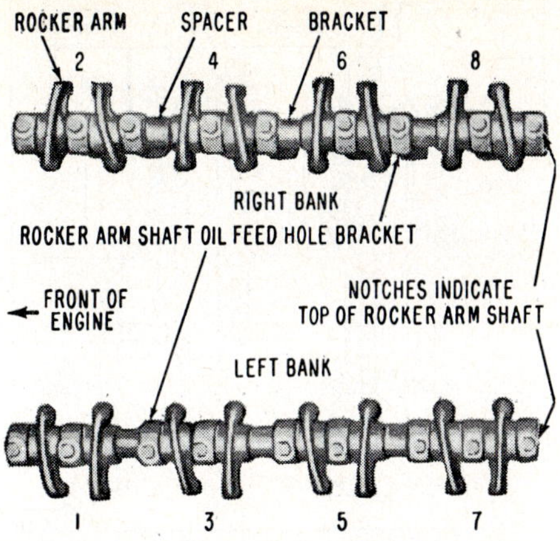

Fig. 9 Rocker arm assemblies. V-304, 345, 392

V-401, 478, 549 Engines

The rocker arm assemblies, Fig. 10, are removed from the engine as a unit. After removing the hair pin retaining rings and spring washers from the ends of the shaft, the shafts may be stripped of arms, springs and supports. Keep parts in order so they may be replaced in their original position if in satisfactory condition. Remove plugs from ends of shafts and clean all parts thoroughly, making sure all oil passages are open.

When assembling, make sure rocker arms are correctly positioned to actuate the valves. Be sure the rocker arm end supports incorporate the dowel sleeves for alignment to cylinder head.

MV-404, 446 Engines

The rocker arms on these engines are individually mounted and are retained by flange head bolts and pivot balls. Install the rocker arm components in original position.

Inspect the pivot surfaces of the rocker arms and pivot balls for signs of scuffing, pitting or excessive wear. Inspect the valve stem contact surface of the rocker arms for pitting. Replace any component found unsatisfactory.

RD Engines

The rocker arm assembly, Fig. 11, is composed of front and rear shafts joined at the center, on which are mounted the twelve rocker arms and four tension springs. The shafts are mounted on seven brackets, and are prevented from turning at the two center brackets which contain locking washers fitting into slots in the shaft. *Note that the third bolt from the rear on the left side of the cylinder head is drilled with an oil hole which supplies oil to the rocker arms.*

Inspect the component parts of the assembly as outlined for BD engines. And when new bushings are installed in the rocker arms they should be burnished and reamed to provide a clearance of .0015–.004". Ream dimension is .8745–.8760".

V-537

The rocker arm assemblies are removed as a unit. Each rocker arm is marked with an "A" or "B" for identification purposes, Fig. 11A. There are two different shapes of rocker arms used on each rocker arm assembly. Oil is fed through the support located on the rocker arm shaft at the front on the left bank and at the rear on the right bank. Rocker shafts are hollow to provide for the oil flow to each rocker arm. Keep parts in order to aid in reassembly. Remove rocker arm shaft and plugs. Clean and inspect all components for wear or damage. The valve stem contact pad of the rocker arm may be resurfaced; however pad thickness must not be reduced below .010 inches.

VALVE ROTATORS

RD Engines

The parts involved in the rotating mechanism are a special spring seat retainer, a pair of flat halfmoon keys, a close fitting cap located on the valve stem, and a specially constructed valve stem, Fig. 12.

In order to accomodate valve expansion, the valve lash must be maintained. When camshaft rotation causes this lash or clearance to be taken up, the cap on the valve stem causes the valve keys to lower the spring retainer, removing the load of the valve springs from the valve *before* the valve is raised from its seat. A clearance of .002–.006" should be maintained between the end face of the valve stem and cap, Fig. 12. This is the distance the spring retainer is lowered *before* the valve is moved. The slow valve rotating motion is caused by vibration of the valve, the flowing of exhaust gases around the valve head, and a slight rotating motion imparted to the valve by the valve spring.

Whenever valve work is done, check the clearance, using the special gauge shown in Fig. 13. If there is no clearance, grind off the end of the valve stem. If the clearance is too great, grind off the open end of the valve cap.

4-196 & V8 Engines Exc. 400, 404, 446

The "Roto Cap" valve rotator may be used on the intake valve or both valves. 4-196 and V8-304, 345 and 392 engines have the "Roto Cap" on the intake valve, Fig. 14. V8-401, 461, 478 and 549 engines, use the "Roto Cap" on both valves.

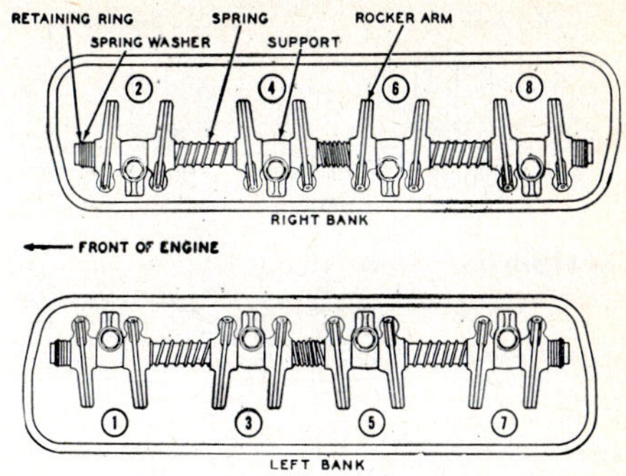

Fig. 10 Installation of rocker arms. V-401, 478, 549

VALVE GUIDES

6-258

Using a micrometer and a suitable telescope hole gauge, check the diameter of the valve stem in three places (top, center and bottom). Insert telescope hole gauge in valve guide bore, measuring at the center. Subtract the highest reading of valve stem diameter from valve guide bore center diameter to obtain valve-to-guide clearance. If clearance is not within specified limits, use the next oversize valve and ream bore to fit. Valves with oversize stems are available in .003", .015" and .030". Valve-to-guide clearance should be .001"–.003".

VALVE GUIDES, REPLACE

Clean valve guides with a suitable cleaning tool. Then check each valve guide with a "GO and NO-GO" gauge, if available; otherwise, use a new valve to check the fit. If the "NO-GO" portion of the gauge enters, the guide must be replaced.

Replacement is recommended for valve guides with diameters exceeding specifications, bell-mouthed more than .0005 inch or guides which are out-of round.

New guides should be installed so that the distance from the cylinder head to the top of the guide is as follows.

4-196 Intake	31/32"
4-196 Exhaust	1 11/64"
V-304, V 345, V-392 Intake	31/32"
V-304, V-345 Exhaust	1 3/32"
V-392 Exhaust	1 11/64"
V-401, 478, 549 Intake	1'
V-401, 478, 549 Exhaust	1 1/16"
MV-404, 446 Intake	27/32"
MV-404, 446 Exhaust	1 1/8"
RD-406, 450, 501 Intake	1 1/16"
RD-406, 450, 501 Exhaust	1"
V-537 Intake	1.04"
V-537 Exhaust	1.04"

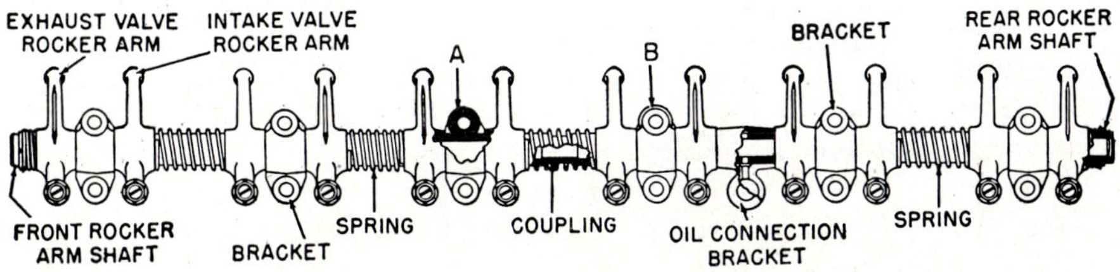

Fig. 11 Rocker arm assembly. RD engines

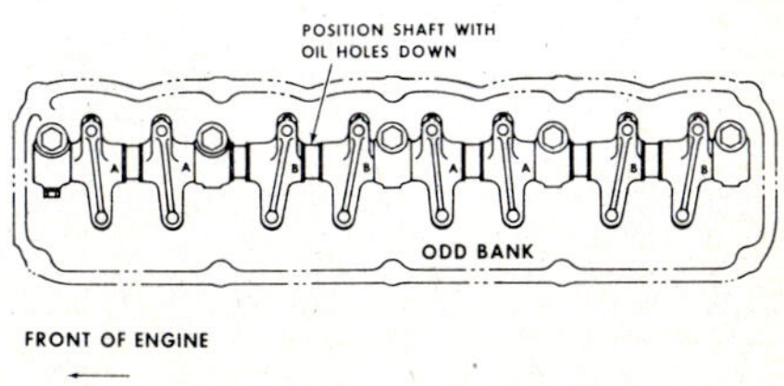

Fig. 11A Rocker arm assembly. V-537

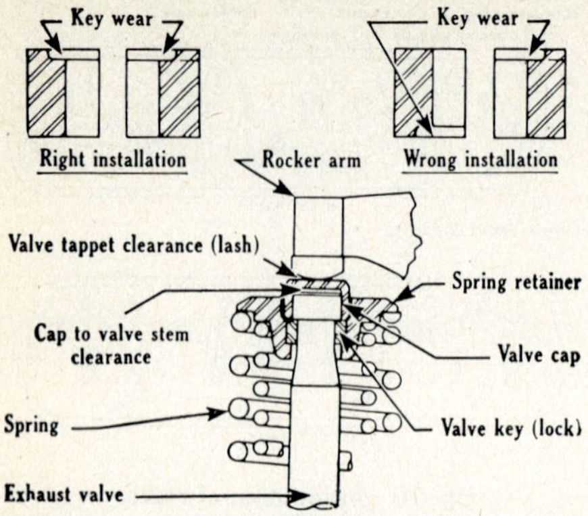

Fig. 12 Rotating exhaust valve mechanism. RD engines

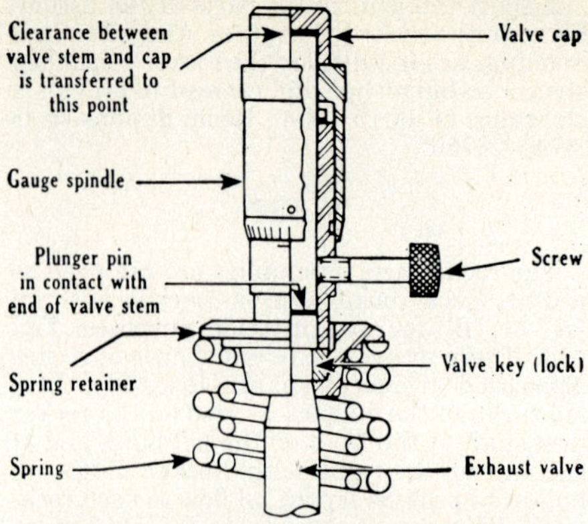

Fig. 13 Showing use of special gauge for checking clearance between valve stem and cap. RD engines

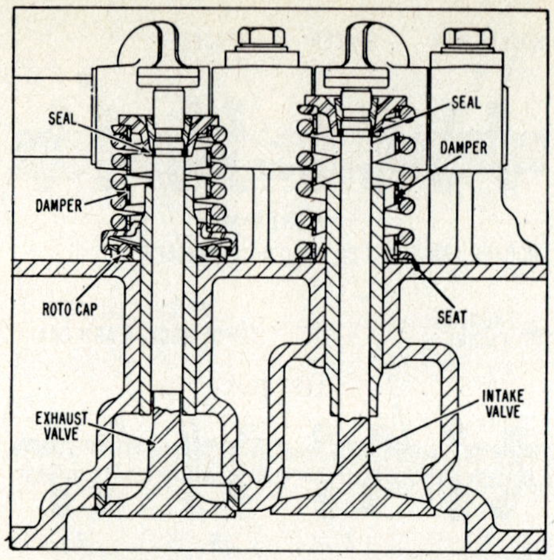

Fig. 14 Valves & related parts. 4-196 & V8-304, 345 & 392 (Typical of V8-401, 478 & 549)

When guides are properly installed, they should be reamed to the following diameters:

4-196	.376–.377″
V-266, 304, 345	.374–.3755″
V-401, 478, 549	.4365–.4380″
V-537	.4365–.4375″
MV-404, 446	.374–.375″
RD	.4365–.4380″

VALVE LIFTERS

4-196, 6-258 & V8 Engines

Hydraulic valve lifters, Figs. 15 and 16, are removed from the engine as an assembly. The lifters are removed by removing the valve cover, rocker arm and shaft assembly, then lifting out the push rods. If a lifter cannot be removed due to carbon build-up, use a carburetor solvent in the lifter bore. However, if an excessive amount of solvent is used, it will be necessary to change the engine oil. The lifters are also accessible by removal of the lifter covers.

As shown in Figs. 15 and 16, different designs of hydraulic lifters are used. When disassembling, do not mix the components of any lifter. Service kits of component parts are available for both types of lifters.

RD Engines

These lifters are of the mushroom type operating in guide holes bored in the cylinder block. To remove lifters, therefore, it is necessary to take out the camshaft and remove the lifters through the bottom of the engine.

The valve lifter bores have spring-loaded retainers in the block, Fig. 17, which are used to hold up the lifters when pulling out the camshaft. When the lifter is pushed up the retainer snaps into the groove at the top of the lifter. After the camshaft is installed, the lifters may be pushed down into position by the push rods.

TIMING GEARS OR CHAIN

When valves are correctly timed, the timing marks on the gears or sprockets should be adjacent to each other. Figs. 18 thru 21.

CAMSHAFT & BEARINGS

All Engines

Camshaft bearings should be installed in sets and care must be exercised to see that the oil holes in the bearings and the oil feed holes in the block are aligned to assure adequate lubrication.

Service Note

4-196 & V8-304, 345, 392

Camshaft bushings must be installed so as to provide running clearance between valve lifters and bushings as shown in Fig. 22.

PISTONS & RODS, ASSEMBLE

V-304, 345, 392

When assembling connecting rod to piston, the rod bearing tangs will be towards side of piston (marked "up"). It is important that piston and rod are assembled correctly so that when piston is installed in engine the chamfered side of each rod is located against the crankshaft face.

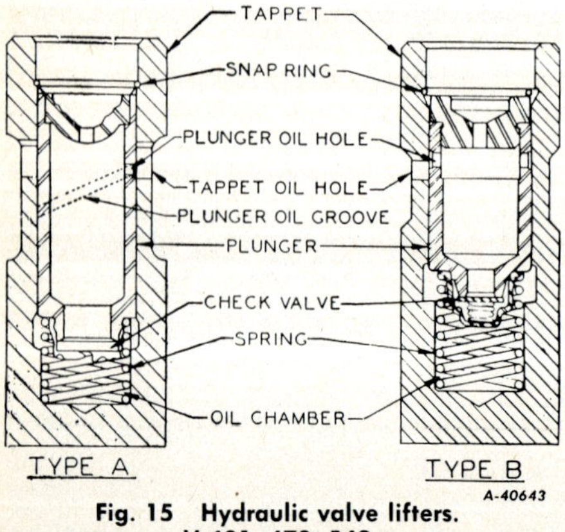

Fig. 15 Hydraulic valve lifters. V-401, 478, 549

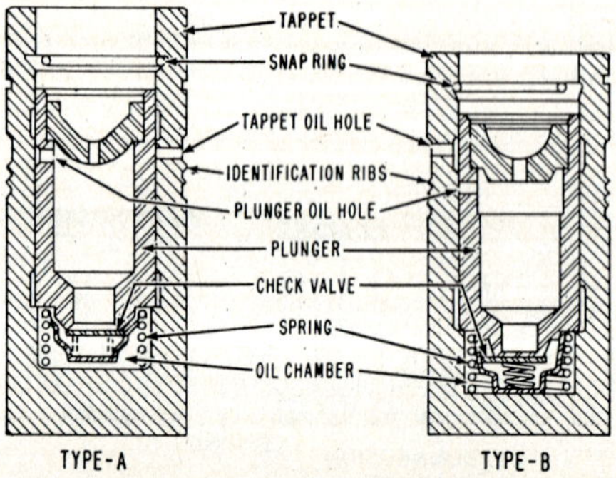

Fig. 16 Hydraulic valve lifters. V-304, 345, 392

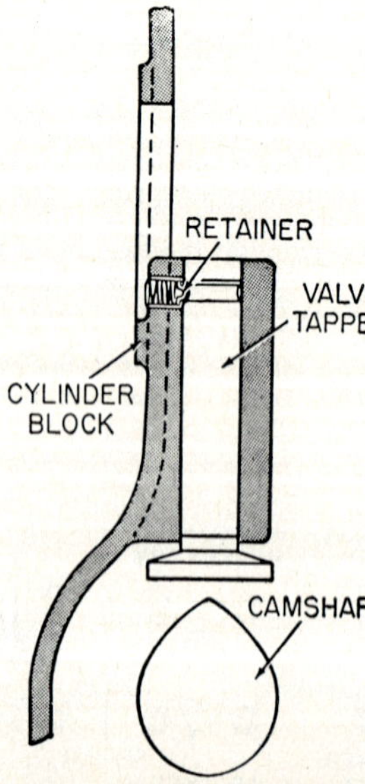

Fig. 17 Tappet and guide construction. RD engines

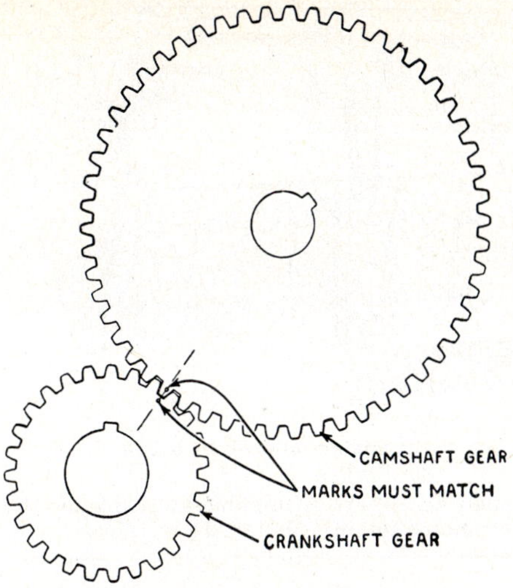

Fig. 18 Valve timing on 6 cylinder engines except 6-258

When installing piston in engine, position side of piston (marked "up") towards engine center line. Install bearing cap with corresponding number on same side as number on connecting rod. Torque connecting bolts to specifications as listed under "Engine Tightening Specifications."

CAUTION: If bolt is over tightened it must be replaced.

MV-404, 446

Before disassembling, mark piston on same side as large chamfer on connecting rod, so they can be assembled in the same position. New pistons may be installed either way on connecting rod.

When installing piston in engine the large chamfered side of each connecting rod must be located against the crankshaft face. The chamfer provides clearance at the crankshaft fillet.

V8-401, 478, 549 Engines

For correct assembly of the connecting rod to the piston, the rod bearing tangs will be toward the domed side of the piston (marked "up") for proper installation, Fig. 23. This is important in that the rod and piston must be assembled correctly so that when the piston is installed in the engine the chamfered side of each rod is located against the crankshaft face. The chamfer provides clearance at the crankshaft fillet.

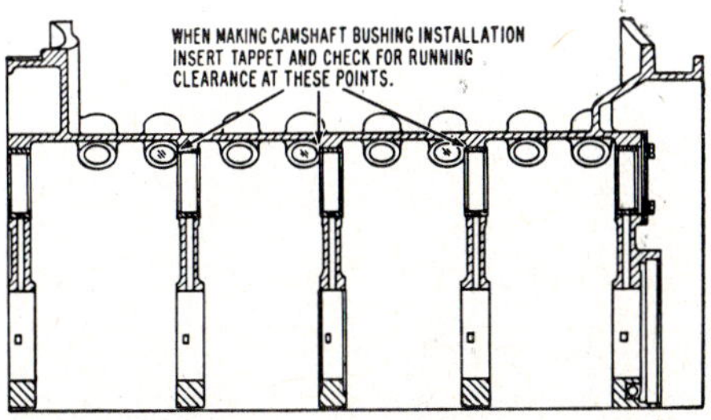

Fig. 22 Checking clearance between valve lifters and camshaft bearings. 4-196 & V8-304, 345, 392

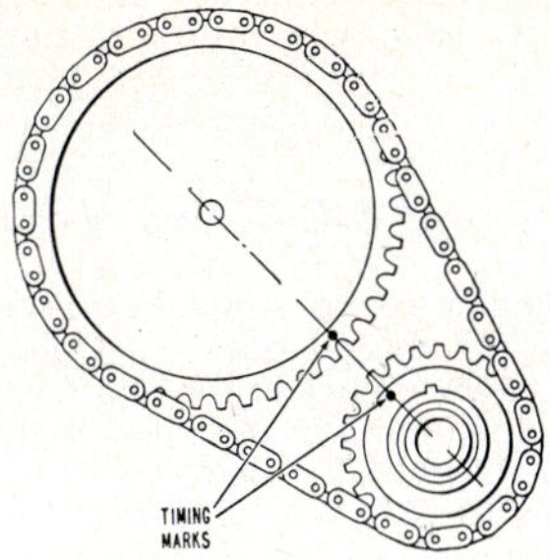

Fig. 19 Valve timing. 6-258

Place the piston in a piston vise, and with the piston heated to approximately 200°, in a piston heater or boiling water, position the rod in the piston and install the piston pin.

6-Cyl. Engines

On RD engines, the step on the piston head must be installed in the cylinder on the side opposite to the camshaft, Fig. 24.

On 6-258 engines, assemble the piston and rod so that the notch in the piston head faces the front of the engine and the numbered side of the rod faces the camshaft side of the engine, Fig. 25.

4-196 Engines

For correct assembly of the connecting rod to the piston, the rod bearing locators (bearing tangs) must be toward the dome side (marked "up") of the piston. And when installed in the cylinder, the word "up" on the piston goes toward the top (camshaft) side of the engine.

V-537

Position piston and rod assembly in cylinder bore with dome side of piston (marked up) toward centerline of engine block. Carefully

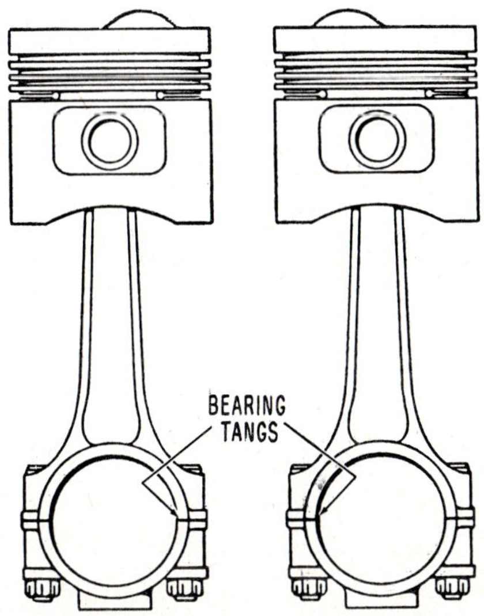

Fig. 23 Correct assembly of pistons and rods. V8-401, 478, 549 engines

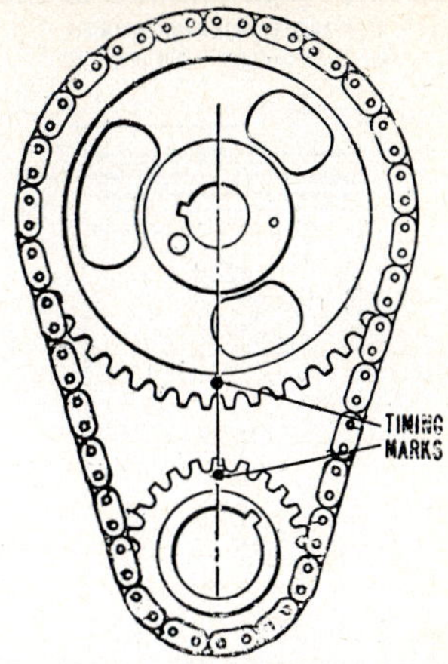

Fig. 20 Valve timing marks. V8-400

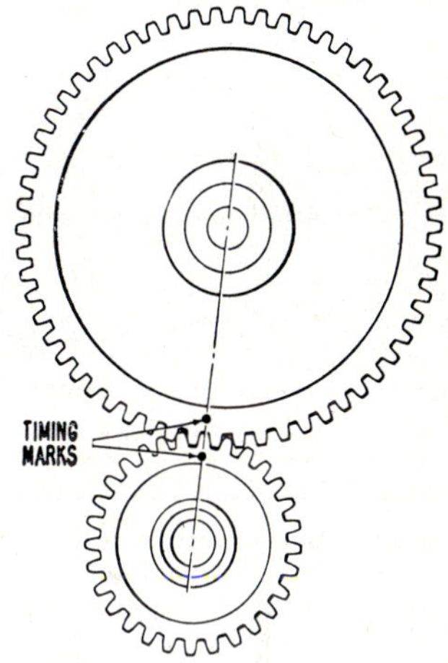

Fig. 21 Valve timing marks. V8 exc. 400

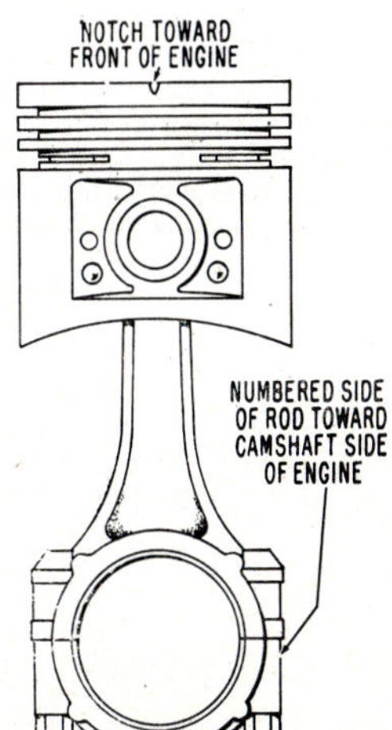

Fig. 24 Piston and rod assembly. RD engines

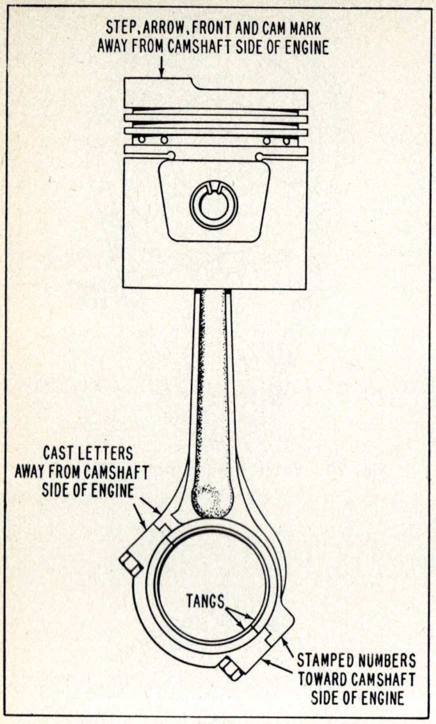

Fig. 25 Piston and rod assembly. 6-258

push piston and rod assembly through ring compressor with hammer handle into bore.

NOTE: Avoid striking cylinder bore with connecting rod.

Coat the crankshaft journal and connecting rod bearings with clean engine oil. Install connecting rod cap to rod with marked sides matching. When properly installed large chamfered side of the rod and cap will be on the fillet side of the crankpin.

CRANKSHAFT OIL SEAL, REPLACE

4-196 Engines

The rear main bearing oil seal can be replaced with the engine in the chassis by first removing the transmission, clutch and flywheel.

Position rear bearing oil seal on pilot No. SE-1942-2, then position pilot on installer No. SE-1942-1. Place installer and seal over crankshaft, drive seal into bore until flush with engine.

V8's and 6-Cyl. Exc. V-537

Crankshaft rear bearing oil seal consists of two pieces of special packing. One piece is installed in the groove in the rear bearing cap and the other piece is installed in a similar groove in the cylinder block.

Position the rear bearing seal in the groove in the cylinder block. Lay an improvised mandrel in the bearing bore and strike the mandrel with a hammer to drive the seal into the groove. Install the seal in the bearing cap in a similar manner. Using a sharp knife, cut off both ends of each seal which project out of the grooves. When cutting off the ends of the seals, do not leave frayed ends which would prevent proper seating of the bearing cap if the ends should extend between cap and cylinder block.

V-537

The rear main bearing cap oil seal can be removed with the engine in the chassis, using tool SE-2091. Screw the tool into the seal and using the slide hammer remove the seal.

OIL PUMP REPAIRS

196, V8's and 6-Cyl. Exc. V-537

Conventional gear type oil pumps, located in the crankcase, are used on all engines. After removing the oil pan, the pump may be removed from its mounting and disassembled for repairs.

Inspect the pump gears for nicks or burrs. Inspect the bottom cover for wear and dress down on a surface plate with emery if necessary. Place the bottom cover on the pump and check the drive shaft end play. If end play is excessive (more than .006"), replace the drive and idler gears.

NOTE: Whenever the oil pump is removed, the distributor will have to be taken off as well in order to reinstall the pump without damage. The reason for this is that the slots in the end of the oil pump drive shafts must match the tangs on the end of the distributor drive shaft. If an attempt is made to install the pump while the distributor is in place, it will be difficult or impossible to align the slots and tangs with the result that the pump will be damaged when its mounting bolts are tightened. Be sure to retime the ignition when the oil pump and distributor are installed.

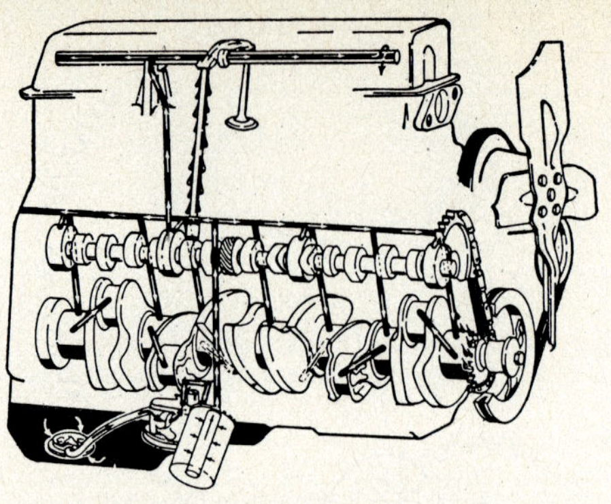

Engine lubrication. 6-258

V-537

1. Remove the mounting nuts, lockwashers, flat washers and bolts from the oil pump screen brace and clamp.
2. Remove the oil pump mounting bolts and remove the pump.

NOTE: To remove oil pump, position crankshaft rear counterweight away from the oil pump shaft.

3. Wash all the pump components and screen assembly in cleaning solvent.
4. With the cover plate and gaskets removed and the gears and shafts in place, exert pressure against the gears with thumb and push gears away from outlet side of pump. Clearance between outside diameter of gear and bore of housing should be .0007–.0027 inches.
5. Check pump shaft clearance in bore. If wear is excessive, replace pump assembly.
6. Check that backlash between pump body gears is .0005–.0065 inches. If wear is excessive, replace gears.
7. Check that body gear end clearance is within .003–.0045 inches.

NOTE: Gaskets control the clearance between pump body gear and pump cover plate.

8. Lubricate pump gears and shaft with engine oil before installation.
9. Press oil pump drive gear onto shaft 1.95 inches from shaft end, and install pump screen after pump installation.

Carburetor Section

CARBURETOR ADJUSTMENT SPECIFICATIONS

Carb. Make	Carb. Model	Carb. Part No.	Engine Application	Float Level (Dry)	Fuel Level (Wet)	Idle Mix. Screws Turns Open	Pump Link Position	Dashpot Clearance	Fast Idle Setting	Idle CO %
Carter	TQ	6550S, 6591S	V-345	1.06	—	—	Inner	⑨	.089–.109	1.5
		6551S	V-392	1.06	—	—	Inner	⑨	.089–.109	1.5
		6590S, 6592S	V-392	1.06	—	—	Inner	⑨	.089–.109	1.5
		91285	V-345	—	—	—	Inner	—	.089–.109	.1–.8
Holley	852-FFG	2691-A	RD-450	1 1/4	—	1/2	Center	—	—	—
		2409-2	V-401, VS-401	1 1/4	—	7/8–1 1/8	Center	—	—	—
		3806	V-478, VS-478	1 1/4	—	1/2–3/4	Center	—	—	—
		4329, 4330	RD-450	1 1/4	5/8	1 1/2–2	#2 Hole	—	—	3
		4378, 4438	V-401, VS-401	1 1/4	5/8	1 1/2–2	#2 Hole	—	—	3
		4679	V-401, VS-401	1 1/4	5/8	1 1/2–2	—	—	—	3
		4331, 4332	V-478, VS-478	1 1/4	5/8	1 1/2–2	#2 Hole	—	—	3
		4333, 4377	V-478, VS-478	1 1/4	5/8	1 1/2–2	#2 Hole	—	—	3
		4338	V-478, VS-478	1 1/4	5/8	1 1/2–2	—	—	—	3
		6398,-1	RD-450	1 1/4	5/8	1 1/2–2	—	—	—	3.5 max
		6438	RD-450	1 1/4	5/8	1 1/2–2	—	—	—	3.5 max
Holley	885-FFG	126583	RD-406	7/32	—	1–1 1/2	Center	—	—	—
		260612	RD-406, 450	7/32	—	1–1 1/2	Center	—	—	—
		4327	RD-406	7/32	1/2	1 1/2–2	#2 Hole	—	—	3
		4328	RD-406	7/32	1/2	1 1/2–2	#2 Hole	—	—	3
		6397,-1	RD-406	7/32	1/2	1 1/2–2	Center	—	—	2.5–3.5
		6437	RD-406	7/32	1/2	1 1/2–2	Center	—	—	2.5–3.5
		6625	RD-406	7/32	1/2	1 1/2–2	Center	—	—	3.5 max
Holley	1904	3716	4-196	See Text	—	1	—	—		—
		3842	4-196E	See Text	—	1	—	—	—	—
Holley	1920	3993-1, 4487	4-196	—	27/32⑧	3	#2 Hole	.125–.155	—	—
		4591	4-196	—	27/32⑧	3	#2 Hole	.075–.125	—	—
		4405	6-232	—	27/32⑧	3	#2 Hole	.105–.135	—	—
		4542	6-232	—	27/32⑧	3	#2 Hole	—	—	—
Holley	1940C	6442	6-258	—	11/16	—	—	.070–.090	—	1.5
		7161	4-196	—	11/16	—	—	—	2200RPM	.3–1.5
		7576	4-196	—	11/16	—	—	—	2000RPM	.5–2.0
		7771	4-196	—	11/16	—	—	—	2200RPM	.3–1.5
Holley	2140-C	864-1	RD-501	1/4	—	3/8–5/8	See Text	—	—	—
		977	RD-501	1/4	—	3/8–5/8	See Text	—	—	—
		987-2	V-549	1/4	—	1 7/8–2 1/8	See Text	—	—	—
		1348-1	V-549	1/4	—	1 7/8–2 1/8	See Text	—	—	—
		2407	V-549	1/4	—	1 7/8–2 1/8	See Text	—	—	—
		2408	V-549	1/4	—	1 7/8–2 1/8	See Text	—	—	—
		2461	V-549	1/4	—	1 7/8–2 1/8	See Text	—	—	—
		2486	V-549	1/4	—	1 7/8–2 1/8	See Text	—	—	—
Holley	2140-G	4334, 4335	RD-501	1/4	1/2	7/8	#2 Hole	—	—	3
		4334	RD-501	1/4	1/2	7/8	#2 Hole	—	—	3
		864-1	RD-501	1/4	1/2	7/8	#2 Hole	—	—	3
		977	RD-501	1/4	1/2	7/8	#2 Hole	—	—	3
		4337, 4338	V-549, VS-549	1/4	1/2	1 1/2–2	#2 Hole	—	—	3
		4339, 4340	V-549, VS-549	1/4	1/2	1 1/2–2	#2 Hole	—	—	3

Continued

CARBURETOR ADJUSTMENT SPECIFICATIONS—Continued

Carb. Make	Carb. Model	Carb. Part No.	Engine Application	Float Level (Dry)	Fuel Level (Wet)	Idle Mix. Screws Turns Open	Pump Link Position	Dashpot Clearance	Fast Idle Setting	Idle CO %
		4341	V-549, VS-549	1/4	1/2	1 1/2–2	#2 Hole	—	—	3
		4635	V-549, VS-549	1/4	1/2	1 1/2–2	#2 Hole	—	—	3
		4636	V-549, VS-549	1/4	1/2	1 1/2–2	#2 Hole	—	—	3
		4342, 4343	FTV, FTVS-549	1/4	1/2	1 1/2–2	#2 Hole	—	—	3
		4344	RD-501	1/4	1/2	7/8	#2 Hole	—	—	3
		4634, 4633	FTV-FTVS-549	1/4	1/2	1 1/2–2	#2 Hole	—	—	3
Holley	2210C	6443	V-304, 345	13/64①	1/2	—	—	.070–.090	—	1.5
		6443-1	V-304, 345	13/64①	1/2	—	—	.070–.090	—	1.5
		6776	V-304, 345	13/64①	1/2	—	—	.070–.090	—	1.5
		6620,-1,2	V-345	13/64①	1/2	—	—	—	—	2
		6674-1	V-400	13/64①	1/2	—	—	.070–.090	—	1.5
		6828	V-304, 345	.2	1/2	—	—	.070–.090	2200RPM	1.5 max
		7133	V-345	.180	1/2	—	—	—	2200RPM	.5–1.5
		7214	MV-404	.180	1/2	—	—	—	1800RPM	.5–2.5
		7217	MV-404	.180	1/2	—	—	—	1800RPM	.5–2.5
		7309	V-304, V-345	.180	1/2	—	—	—	2200RPM	2.0 max
		7657	MV-404	.180	1/2	—	—	—	1800RPM	.5–2.5
		8241	V-345	.180	1/2	—	—	—	2200RPM	.5–2.5
		8244	V-404	.180	1/2	—	—	—	1800RPM	.5–2.5
Holley	2245C	7773	V-304	.180	1/2	—	—	—	2200RPM	.1–.8
Holley	2300	1710	V-304	—	3/8③	3/4	#1 Hole	1/16	—	⑥
		2977-2	V-304, 345	—	3/8③	3	—	.090–.120	⑤	⑥
		1710-4	V-304, 345	—	3/8③	3	—	.090–.120	⑤	⑥
		3991-1	V-304, 345	—	3/8③	3	—	.090–.120	⑤	3
		4079-1	V-304, 345	—	3/8③	3	—	.090–.120	⑤	3
		4308	V-304, 345	—	3/8③	3	—	.090–.120	⑤	3
		4306	V-304, 345	—	3/8③	3	—	.090–.120	⑤	3
		4574	V-304, 345	—	3/8③	3	—	—	⑤	—
		4594	V-304, 345	—	3/8③	3	—	.090–.120	⑤	—
		4081-1	V-304, 345	—	3/8③	3	—	.090–.120	⑤	3
		4083-1	V-304, 345	—	3/8③	3	—	.090–.120	⑤	3
		6380	V-304, V-304E, V-345, V-345E	—	3/8③	3	—	.075–.105	2000RPM	—
		6393	V-304, V-304E, V-345, V-345E	—	3/8③	1.5–2	—	⑤	2000RPM	—
		6393-1	V-304, V-345	—	3/8③	1.5–2	—	—	2000RPM	—
		6394-1	V-304, V-345	—	3/8③	1.5–2	—	—	2000RPM	—
Holley	2300C	3679-3	V-304, 345	—	3/8③	3	—	.090–.120	⑤	⑥
		2520-3	V-304, 345	—	3/8③	3	—	.090–.120	⑤	⑥
		3937-2	V-304, 345	—	3/8③	3	—	.090–.120	⑤	⑥
		3936-2	V-304, 345	—	3/8③	3	—	.090–.120	⑤	⑥
Holley	2300C	4078-1	V-304, 345	—	3/8③	3	—	.090–.120	⑤	3
		4080-1	V-304, 345	—	3/8③	3	—	.090–.120	⑤	3
		4082-1	V-304, 345	—	3/8③	3	—	.090–.120	⑤	3
		4084-1	V-304, 345	—	3/8③	3	—	.090–.120	⑤	3
		4307	V-304, 345	—	3/8③	3	—	.090–.120	⑤	3
		4309	V-304, 345	—	3/8③	3	—	.090–.120	⑤	3
		4593	V-304, 345	—	3/8③	3	—	—	⑤	—
		4595	V-304, 345	—	3/8③	3	—	.090–.120	⑤	—
Holley	2300G	1706-2, 2979	V-304	—	3/8③	3	#2 Hole	—	.018–.020	—
		1707-2, 2980	V-345	—	3/8③	3	#2 Hole	—	.018–.020	—
		2979	V-304, 345	—	3/8③	3	—	—	.018–.020	⑥
		2980	V-304, 345	—	3/8③	3	—	—	.018–.020	⑥

Continued

CARBURETOR ADJUSTMENT SPECIFICATIONS—Continued

Carb. Make	Carb. Model	Carb. Part No.	Engine Application	Float Level (Dry)	Fuel Level (Wet)	Idle Mix. Screws Turns Open	Pump Link Position	Dashpot Clearance	Fast Idle Setting	Idle CO %
		2975	V-304, 345	—	3/8 ③	3	—	—	.018–.020	⑥
		4310-1	V-304, 345	—	3/8 ③	3	—	—	.018–.020	3
		4310-2	V-304, V-345	—	3/8 ③	1.5–2	—	—	.018–.020	—
		4311-1	V-304, 345	—	3/8 ③	3	—	—	.018–.020	3
		6623	V-304	—	3/8 ③	—	—	—	.018–.020	2 max
		6624-1	V-345	—	3/8 ③	—	—	—	.018–.020	1.5 max
		6801-1	V-304	—	3/8 ③	—	—	—	.018–.020	2 max
		6802	V-345	—	3/8 ③	—	—	—	.018–.020	1.5 max
		6899	V-345	—	3/8 ③	—	—	—	2000RPM	.5–1.5
		6908	V-537	—	3/8 ③	—	—	—	2000RPM	1.5–3.0
		7198	V-537	—	3/8 ③	—	—	—	2000RPM	1.0–2.0
		7213	MV-404	—	3/8 ③	—	—	—	2400RPM	.5–2.5
		7216	MV-404	—	3/8 ③	—	—	—	2400RPM	.5–2.5
		7656	V-345	—	3/8 ③	—	—	—	2000RPM	1.0–3.0
		7922, 7580	MV-404	—	3/8 ③	—	—	—	2400RPM	.5–2.0
Holley	2300EG	6801-1	V-304, V-345	—	3/8 ③	—	—	—	2200RPM	1.5 max
		7922	MV-404	—	3/8 ③	—	—	—	2400RPM	.5–2.5
		8235	MV-404	—	3/8 ③	—	—	—	2400RPM	.5–2.5
		8238	V-537	—	3/8 ③	—	—	—	2000RPM	1.5–3.0
		8242	V-345	—	3/8 ③	—	—	—	2000RPM	1.0–3.0
		8245	MV-404	—	3/8 ③	—	—	—	2400RPM	.5–2.0
		8248	V-537	—	3/8 ③	—	—	—	2000RPM	1.0–2.0
		8736	V-345	—	3/8 ③	—	—	—	2000RPM	1.0–3.0
		8741	V-537	—	3/8 ③	—	—	—	2000RPM	1.0–2.0
Holley	4150	4237	V-392	—	⑦	⑩	—	—	.025	3
		4264	V-392	—	⑦	⑩	—	.060–.090	⑪	3
		4318	V-392	—	④	4	—	—	.025	—
		4320	V-392	—	④	4	—	.060–.090	.025	—
		4572	V-392	—	⑦	3	—	—	.025	—
		4601	V-392	—	⑦	3	—	.090–.120	—	—
		4601-1, 2	V-392	—	⑦	3	—	.090–.120	⑫	—
Holley	4150C	4312	V-392	—	⑦	⑩	—	—	—	②
		4313	V-392	—	⑦	⑩	—	.060–.090	—	②
		4319	V-392	—	④	4	—	—	.025	—
		4321	V-392	—	④	4	—	.060–.090	.025	—
		4599	V-392	—	⑦	3	—	—	—	—
		4602	V-392	—	⑦	3	—	.090–.120	—	—
		6390	V-392	—	⑦	3	—	.075–.105	—	1.0–1.5
Holley	4150G	1708	V-392	—	④	3/4	.015	—	.025	—
		4323	V-392	—	④	⑩	—	—	⑪	②
		4324	V-392	—	④	⑩	—	—	⑪	②
		6803	V-392	—	④	3	—	—	.010–.015	2
		6911	V-537	—	④	—	—	—	2000RPM	1.5–3.0
		6974	V-537	—	④	—	—	—	2000RPM	1.0–2.0
		7028	V-392	—	④	—	—	—	.015–.020	.5–1.5
		7029,-1	MV-446	—	④	—	—	—	2400RPM	.5–2.5
		7215	MV-404	—	④	—	—	—	2400RPM	.5–2.5
		7218,-1	MV-404	—	④	—	—	—	2400RPM	.5–2.5
		7579	V-392	—	④	—	—	—	.015–.020	.5–1.5
		7581	MV-404	—	④	—	—	—	2400RPM	.5–2.5
Holley	4150EG	7921, 8246	MV-404	—	④	—	—	—	2400RPM	.5–2.5
		8119	MV-446	—	④	—	—	—	2400RPM	.5–2.5
		8139	V-537	—	④	—	—	—	2000RPM	1.5–3.0

Continued

CARBURETOR ADJUSTMENT SPECIFICATIONS—Continued

Carb. Make	Carb. Model	Carb. Part No.	Engine Application	Float Level (Dry)	Fuel Level (Wet)	Idle Mix. Screws Turns Open	Pump Link Position	Dashpot Clearance	Fast Idle Setting	Idle CO %
		8233	V-392	—	④	—	—	—	2000RPM	2.0 max
		8243	V-392	—	④	—	—	—	.015–.020	.5–1.5
		8247, 8740	MV-446	—	④	—	—	—	2400RPM	.5–2.5
Holley	4160C	6444	V-392	—	⑦	—	—	.070–.090	—	1.5

①—Measure from flat on float to inner edge of bowl with carburetor held upside down.
②—1974 models, 2.0.
③—At 5½ psi fuel pump pressure.
④—Primary float 3/8", secondary float 5/8".
⑤—Zero clearance.
⑥—12–13 air fuel ratio.
⑦—Primary float 1/2", secondary float 5/8".
⑧—At 4½ psi fuel pump pressure.
⑨—Manual Trans. .050"–.070"; Auto. Trans. .070"—.090".
⑩—1974 models, 3 turns.
⑪—1974 models, .010"–.015".
⑫—Man. trans., .025"; auto. trans. exc. Calif., .070"; auto. trans. Calif., .040".

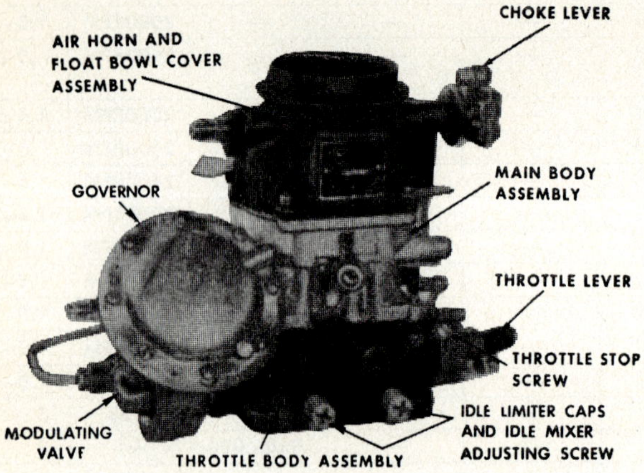

Holley Model 852-FFG carburetor

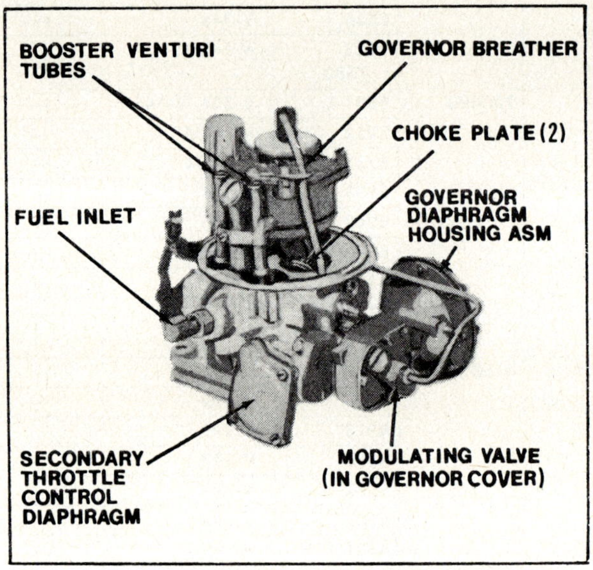

Holley Model 2140G carburetor

CARBURETOR ADJUSTMENTS

Holley 852-FFG

Float Level, Fig. 1

The float level can be set accurately when the air horn is off by gauging the distance between the bottom of the float and the flange surface of the air horn. Holding the air horn upside down (without gasket) and with the float in the closed position, measure the distance from the flange surface of the air horn to what would normally be the bottom side of the float, not the soldered seam. If the dimension is not as specified, bend the float arm up or down as required. Pushing the float toward the float chamber cover raises the level; pushing it away from the cover lowers the level.

Accelerating Pump Stroke, Fig. 2

When the engine does not accelerate properly, check the pump system for dirt. If the pump inlet check valve does not seat, fuel will return to the float bowl instead of discharging through the pump discharge passages. This can be checked by removing the main body cover and operating the pump with a small amount of fuel in the bowl. If the check ball is leaking, air or fuel will bubble back into the float bowl through the inlet hole. After cleaning this seat, extreme care should be taken when reinstalling the pump piston to be sure the piston is not damaged and that it contacts the cylinder wall. If the piston is not a snug fit against the cylinder wall, fuel will leak by the piston on acceleration, causing a weak discharge.

Holley 2140-G

Float Level, Fig. 3

To check the float level, lift the float to completely close the fuel inlet valve. Hold the float closed and place a straight edge across the float chamber so that one end of the straight edge rests on the rim of the float chamber and the other end rests on the center of the main body. Use a scale and measure the distance between the straight edge and the flat top surface of the float at a point farthest from the fuel inlet valve. Check both halves of float in this manner. If one side is lower than the other, bend that part of the float lever next to the float to correct the mis-alignment. The float level can be adjusted by bending the tab on the float lever that contacts the fuel inlet needle. Bend the tab up to lower or down to raise the float.

Accelerating Pump System

The pump system is provided with an adjustment for varying the quantity of pump discharge to meet seasonal requirements and climatic changes. The adjustment is made by changing the position of the pump link in the holes located in the throttle lever. The hole most distant from the throttle shaft is the cold weather setting since it gives the longest stroke and greatest fuel discharge. The hole nearest the throttle shaft is used in hot weather.

Holley 885-FFG

Float Level

Check and adjust the float level as outlined for the 2140G carburetor and refer to Fig. 3.

Pump Setting, Fig. 4

The accelerating pump is provided with an adjustment for varying the quantity of the pump discharge to meet seasonal requirements and climatic changes. The adjustment is made by changing the position of the pump link in the holes provided in the throttle lever. The position farthest from the throttle shaft is the cold weather setting since it gives the longest stroke and the greatest discharge. The middle position gives the average setting, and the position nearest the throttle shaft is used in hot weather.

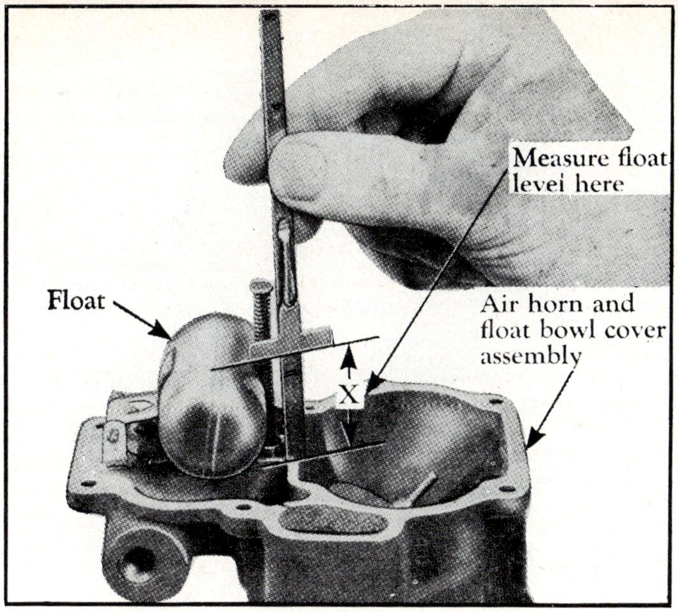

Fig. 1 Float level check. Holley 852-FFG

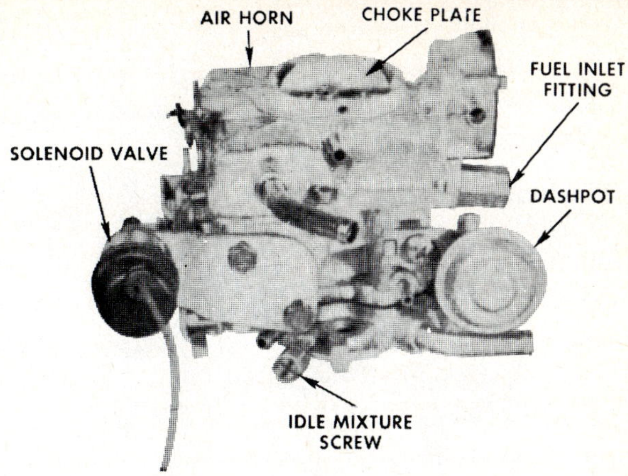

Holley Model 1940C carburetor

Holley 1904

Float Level, Fig. 5

Invert main body, allowing float to drop to closed position. Using the float gauge shown, check the setting on both the "touch," and "no touch" legs of gauge. The level of the float may be adjusted by bending the small tab in the float lever that contacts the head of the fuel inlet needle pin.

Dashpot Adjustment

The dashpot on carburetors for engines equipped with automatic transmission is to be adjusted after the idle speed and mixture settings have been completed. Close the throttle lever to the idle position. Set the dashpot adjusting screw to the clearance specified and with the dashpot diaphragm rod in fully compressed position. To assure an accurate adjustment, check the choke plate to be sure it remains fully open while setting the dashpot adjusting screw.

Holley 1940C

Float Level

With the Air Horn removed and carburetor body in an upright position, place a straight edge across the surface of the fuel bowl. The straight edge should just touch the toes of the float when the fuel inlet valve is held closed. Adjust by bending the float tang.

Fast Idle Cam Adjustment, Fig. 6

With throttle lever screw against second highest step, move choke plate toward closed position. Check clearance between choke plate and wall of air horn. Adjust by bending fast idle connector rod with a suitable tool.

Dashpot Adjustment

With engine at idle speed, depress dashpot plunger completely into dashpot. Measure clearance between plunger and throttle lever. Loosen jam nut and rotate dashpot to obtain specified clearance.

Holley 2210C

Float Level, Fig. 7

Invert air horn so weight of float forces needle against seat. Measure clearance between top of float and float stop. Adjust by bending float lip toward or away from needle using a narrow blade screw driver.

Dashpot Adjustment

With engine at idle speed, depress dashpot plunger completely into dashpot. Measure clearance between plunger and throttle lever. Loosen jam nut and rotate dashpot to obtain specified clearance.

Holley 2245C

Float Level

Invert air horn so weight of float forces needle against seat. Measure clearance between top of float and float stop. Adjust by bending float lip toward or away from needle using a narrow blade screwdriver.

Dashpot Adjustment

With engine at curb idle speed, depress plunger completely into dashpot. Measure clearnace between plunger and throttle lever pad. Loosen jam nut and rotate dashpot to obtain specified clearance.

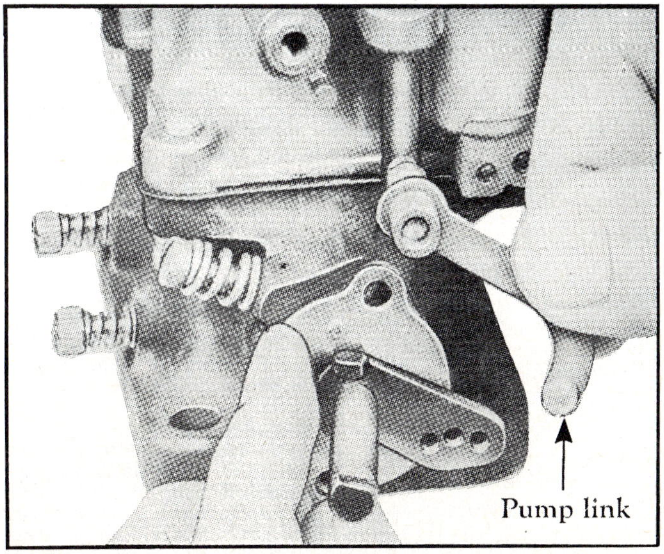

Fig. 2 Accelerating pump stroke. Holley 852-FFG

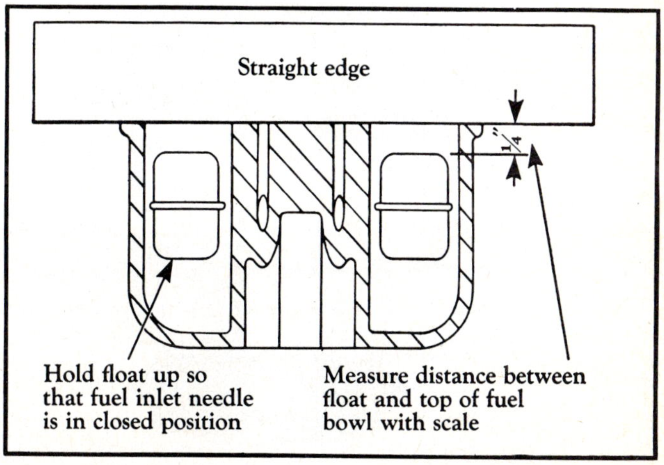

Fig. 3 Float level check. Holley 2140G and 885-FFG

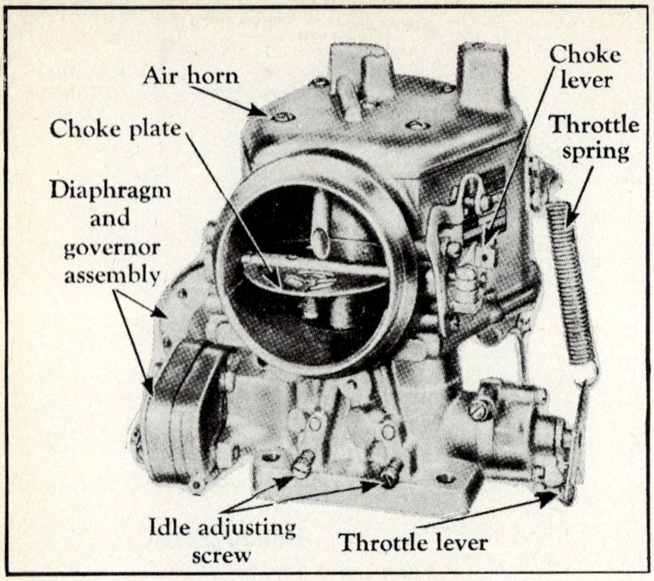

Holley Model 885-FFG carburetor

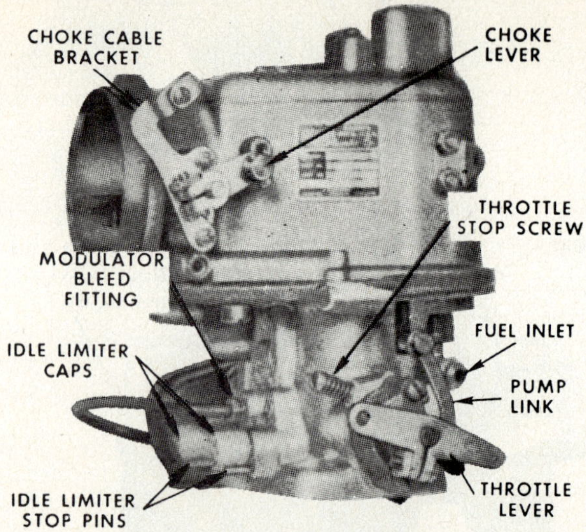

Fig. 4 Accelerating pump stroke. Holley 885-FFG

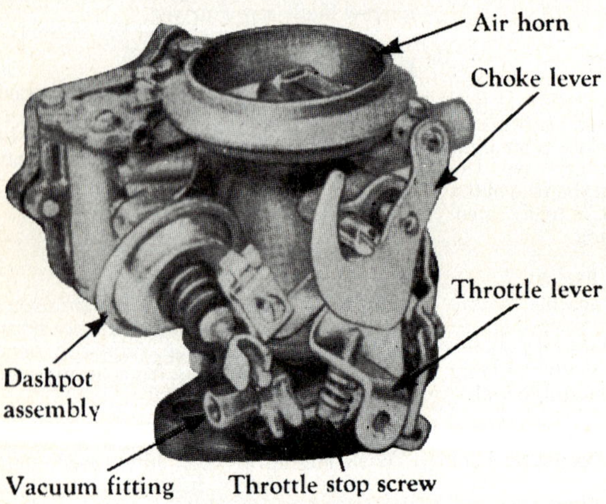

Holley 1904 carburetor used with automatic transmission

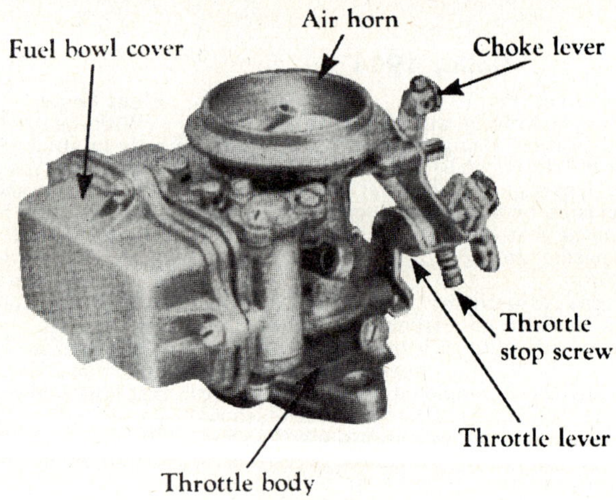

Holley Model 1904 carburetor used with standard transmission

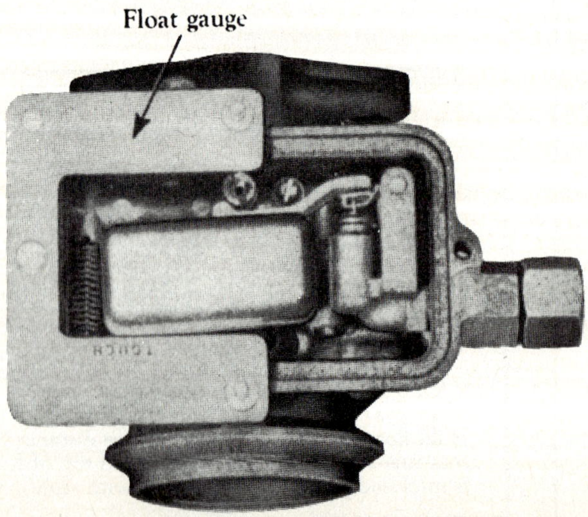

Fig. 5 Float level check. Holley 1904

Holley 2300, 2300C, 2300G

Fast Idle, Model 2300G

Adjust fast idle stud so as to have the specified clearance between stud and fast idle cam when throttle plates are set for curb idle and choke plate is fully open.

Fast Idle, Model 2300, 2300C

Set throttle lever stop screw so that screw just touches fast idle cam at curb idle.

Dashpot Adjustment

Dashpot adjustment is made only after idle speed and mixture settings have been completed. Loosen dashpot jam nut and rotate complete dashpot to obtain the specified clearance between dashpot and throttle lever.

Fuel Level, Figs. 8 and 9

Check fuel pump pressure before changing fuel level. Pump pressure should be 5½ psi.

1. With ignition off and air cleaner removed, remove fuel bowl lower mounting screw (farthest from fuel inlet). Drain fuel into a shallow pan to purge bowl of possible dirt. Reinstall bowl screw.

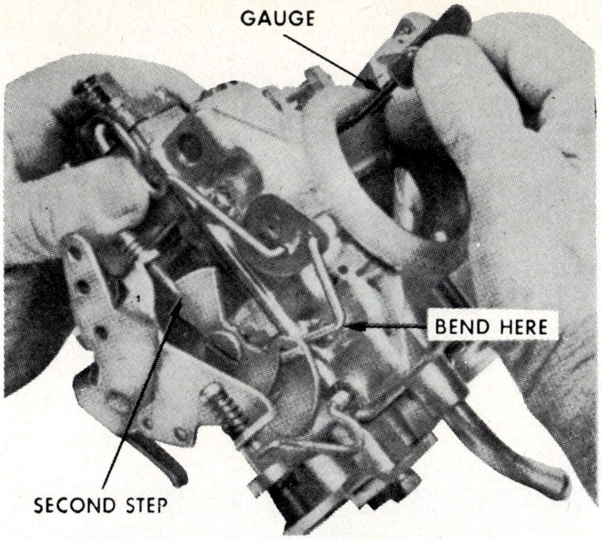

Fig. 6 Fast idle cam adjustment. Holley 1940C

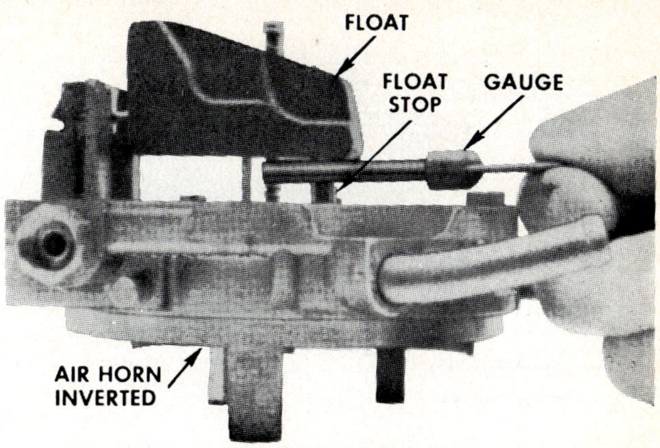

Fig. 7 Float level check. Holley 2210C

2. Start engine and remove sight plug from bowl. The fuel level should be on line with threads at bottom of sight plug hole.
3. To adjust fuel level, loosen lock bolt slightly and turn the adjusting nut clockwise to lower and counterclockwise to raise the level. Tighten lock bolt. Operate engine until fuel level is stabilized, then recheck level at sight plug hole. Repeat adjustment until fuel level is correct.

Pump Discharge, Fig. 10

The adjustment that regulates the pump discharge is located at the override spring on the pump lever. With choke wide open and throttle set for curb idle, lengthen or shorten adjusting bolt until the bolt touches the diaphragm lever and takes up the slack in the linkage. Then lengthen the adjusting bolt 1/2 turn.

Holley 4150G

Fuel Level, Fig. 11

With vehicle on a level floor, operate engine until normal operating temperature has been reached. Place a suitable container below fuel level sight plug to collect any fuel spill over. Check each fuel bowl separately.

With engine stopped, remove air cleaner, fuel sight plug and gasket and check fuel level. The level within the bowl should be at the lower edge of the sight plug opening (plus or minus 1/16").

If the fuel level is too high or low, install sight plug, drain fuel bowl by loosening one lower retaining bolt from bowl and drain fuel into a suitable container. Install bolt and sight plug, and start engine to fill fuel bowl. Check level again before altering the float setting. This will eliminate the possibility of foreign material causing a temporary flooding condition.

If the fuel level requires adjustment, loosen the lock screw on top of the fuel inlet seat and turn the adjusting nut in or out as required to correct the fuel level. A 1/6 turn of the adjusting nut equals 3/64" in fuel level. Remove sight plug and recheck the level.

Fast Idle Adjustment, Fig. 12

The fast idle speed adjusting screw is located on the governor housing. Adjust the curb idle speed and mixture, then adjust the fast idle speed.

With choke plate wide open, adjust fast idle screw to provide the clearance specified between head of screw and fast idle cam. Check fast idle rpm, using a tachometer. If the fast idle is not within the limits of 900 to 1000 rpm, turn fast idle screw clockwise to decrease engine rpm or counter-clockwise to increase the rpm.

Accelerating Pump Lever, Fig. 13

1. Using a feeler gauge and with throttle plates wide open, there should be the specified clearance (minimum) between pump operating lever adjustment screw and pump arm when pump arm is depressed manually.
2. If adjustment is required, loosen lock nut and turn adjusting screw in to increase clearance and out to decrease clearance. One-half turn of screw is equal to approximately .015".
3. When proper adjustment has been made, hold screw in position with a wrench and tighten lock nut.

Accelerating Pump Stroke

To satisfy the requirements in various climates, the pump discharge can be adjusted. The bottom hole (No. 2) in the pump cam and throttle lever provides the maximum pump discharge for extreme cold weather, and the top hole (No. 1) provides the minimum pump discharge for warm weather operation.

If a change in the adjustment is required, make certain the proper hole in plastic pump cam, located behind throttle lever, is properly indexed with the numbered hole in the throt-

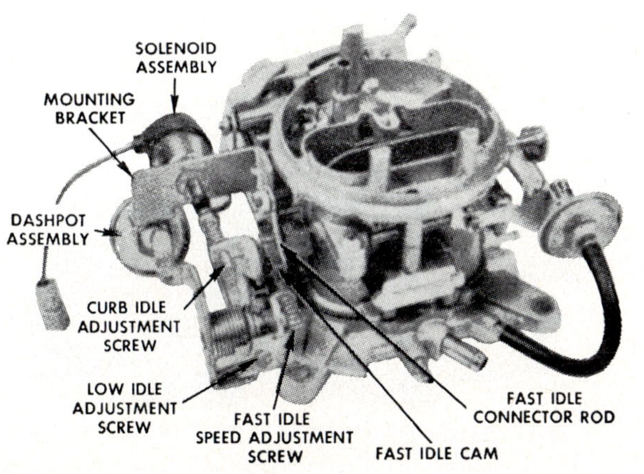

Holley Model 2210C—Early Production

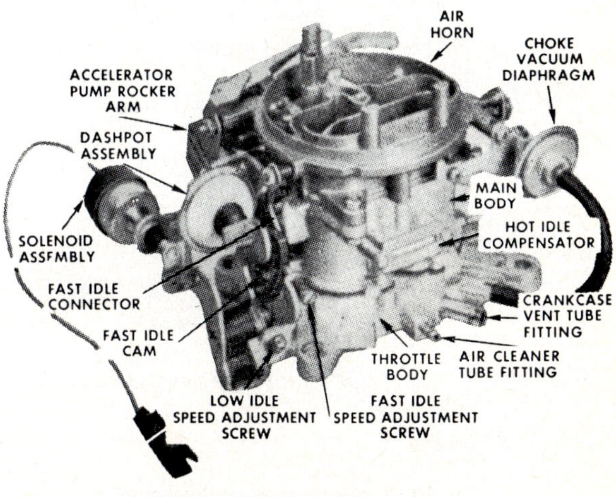

Holley Model 2210C—Late production

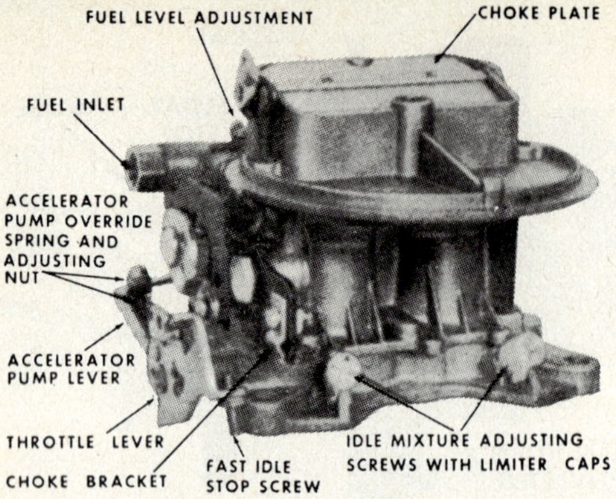

Holley Model 2300 carburetor

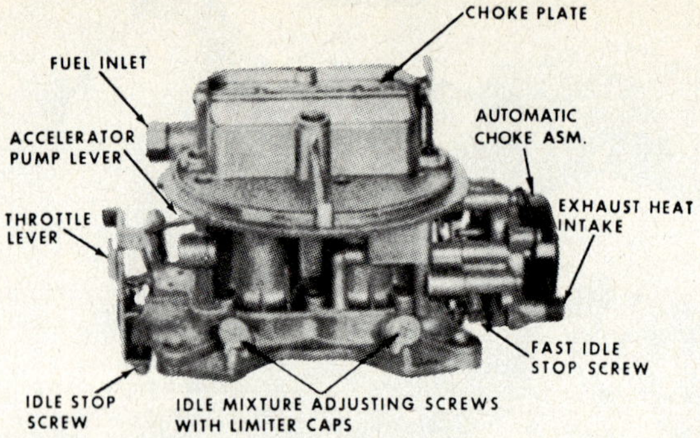

Holley Model 2300C carburetor

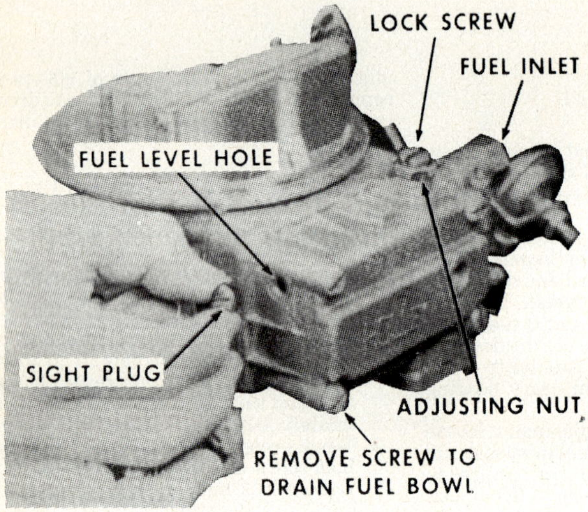

Fig. 8 Removing sight plug from fuel bowl. Holley 2300, 2300C, 2300G

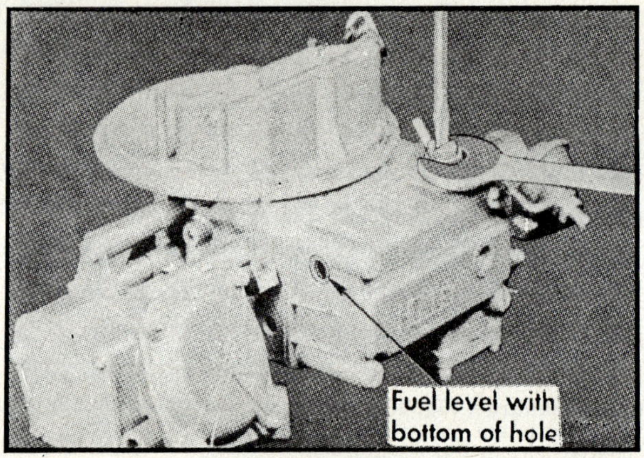

Fig. 9 Adjusting fuel level. Holley 2300, 2300C, 2300G

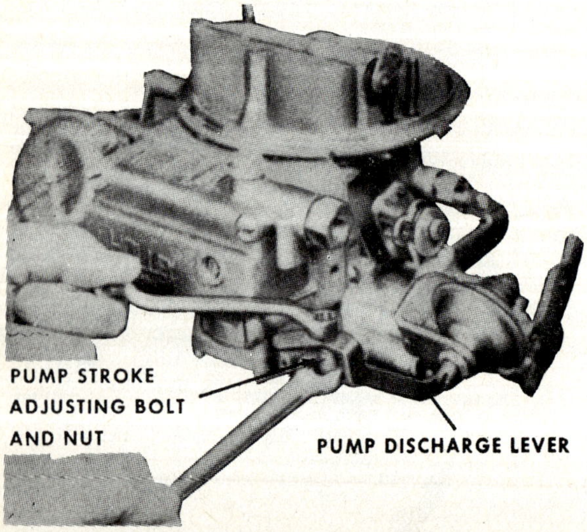

Fig. 10 Accelerating pump discharge Adjustment. Holley 2300, 2300C, 2300G

tle lever before installing the retaining screw.

Holley 4160C

Float Level, Fig. 14

With bowl in an inverted position, adjust float so that top of float is parallel to bowl floor.

Dashpot Adjustment

With engine at idle speed, depress dashpot plunger completely into dashpot. Measure clearance between plunger and throttle lever. Loosen jam nut and rotate dashpot to obtain specified clearance.

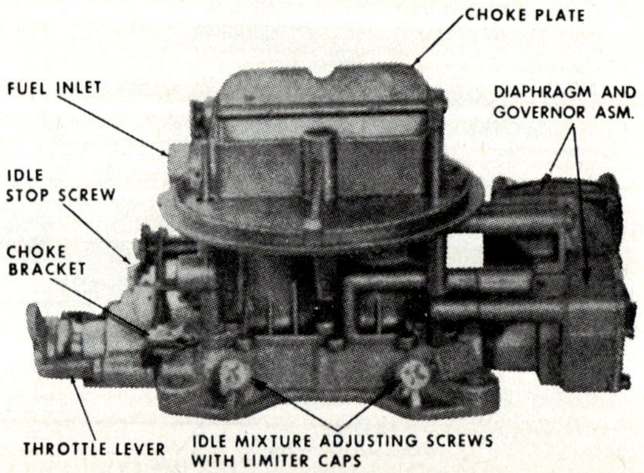

Holley Model 2300G carburetor

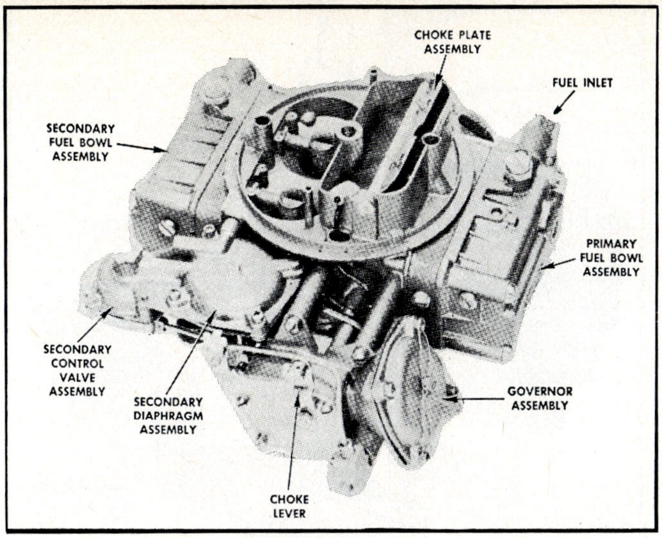

Holley Model 4150G

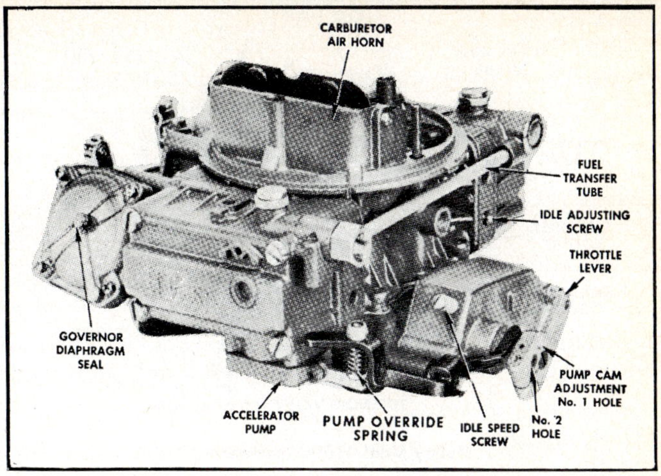

Holley Model 4150G

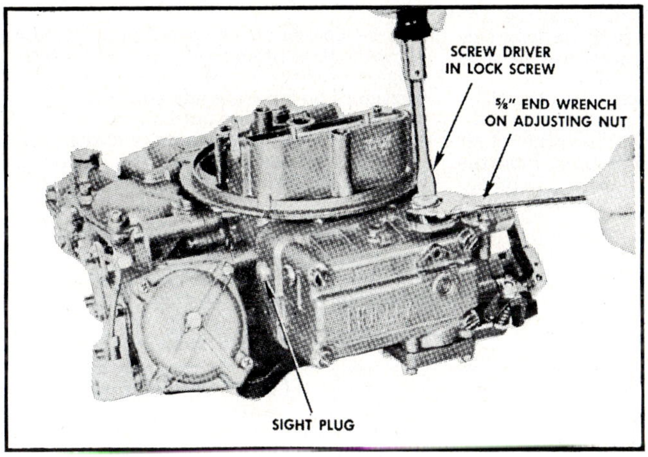

Fig. 11 Fuel level check. Holley 4150G

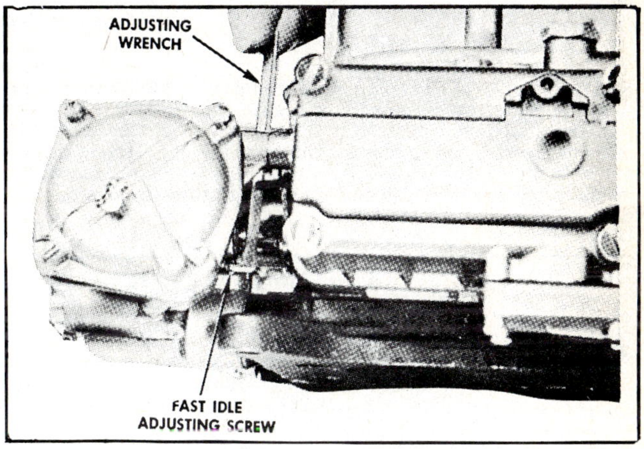

Fig. 12 Fast idle adjustment. Holley 4150G

Fig. 13 Pump lever adjustment. Holley 4150G

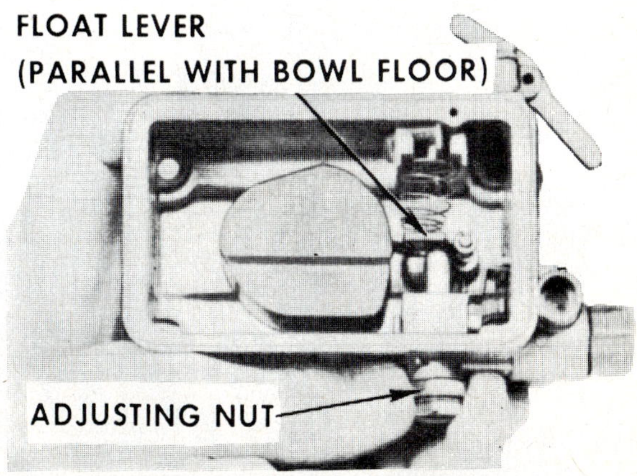

Fig. 14 Float level check. Holley 4160C

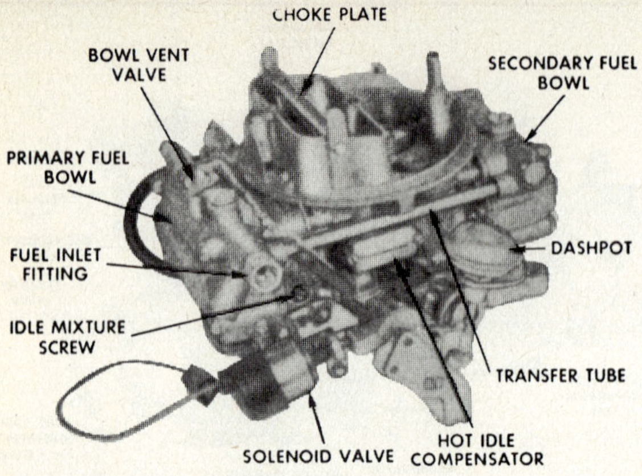

Holley Mdel 4160C carburetor

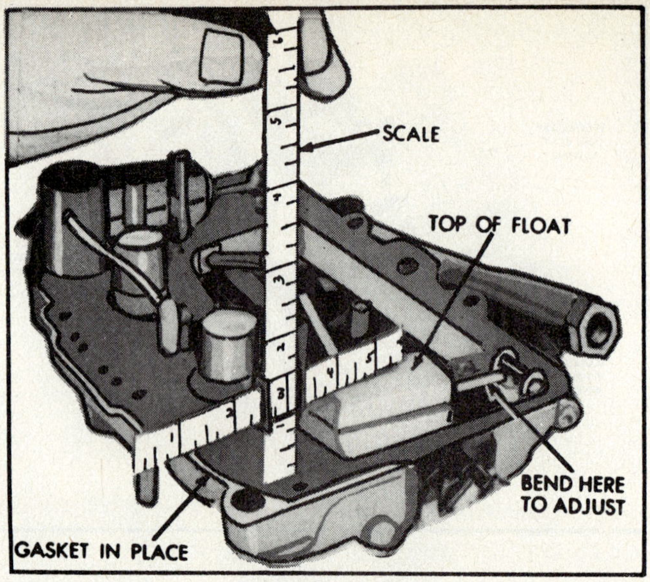

Fig. 1 TQ float setting

CARTER TQ ADJUSTMENTS

Float Setting

Fig. 1—With bowl cover inverted, gasket installed and floats resting on seated needle, the dimension of each float from bowl cover gasket to bottom side of float should be 1.06 inch.

Secondary Throttle Linkage

Fig. 2—Block choke valve in wide open position and invert carburetor. Slowly open the primary throttle valves until it is possible to measure between lower edge of primary valve and its bore. If necessary to adjust, bend rod so primary and secondary stops both contact at same time.

Secondary Air Valve Opening

Fig. 3

1. With air valve in closed position, the opening along air valve at its long side must be at its maximum and parallel with air horn gasket surface.
2. With air valve wide open, the opening of the air valve at the short side and air horn must be 31/64 inch. The corner of air valve is notched for adjustment. Bend the corner with a pair of pliers to give proper opening.

Secondary Air Valve Spring Tension

Fig. 4—Loosen air valve lock plug and allow air valve to position itself wide open. With a long screwdriver that will enter center of tool positioned on air valve adjustment plug, turn plug counterclockwise until air valve contacts stop lightly, then turn an additional 1 1/4 turns. Hold plug with screwdriver and tighten lock plug securely with tool.

Accelerator pump stroke

Fig. 5—With engine at operating temperature, choke in wide open position, idle sole-

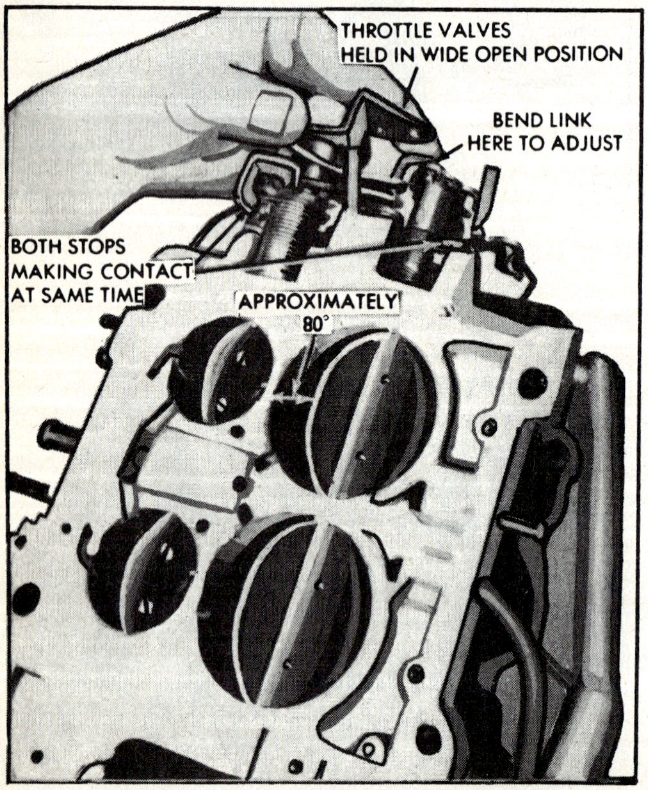

Fig. 2 TQ secondary throttle adjustment

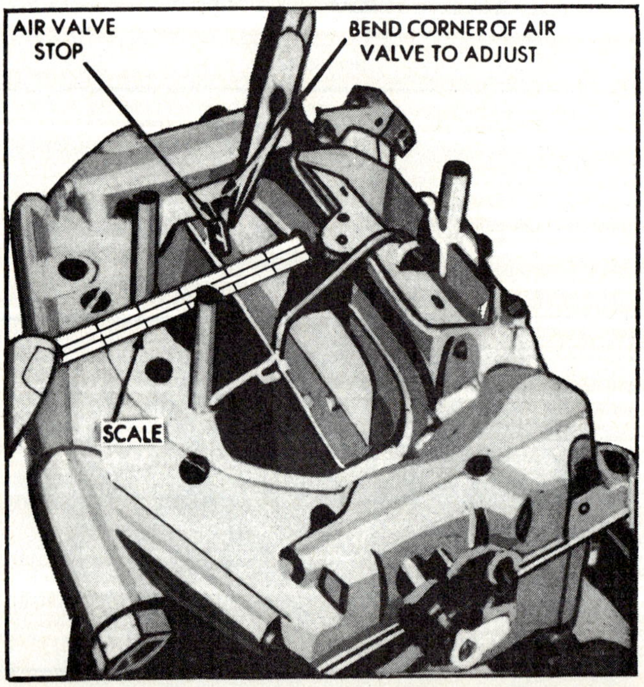

Fig. 3 TQ secondary air valve opening

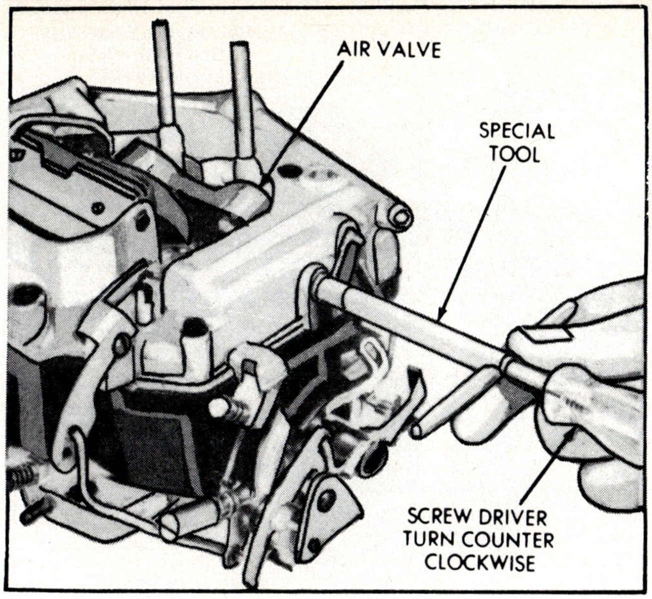

Fig. 4 TQ secondary air valve spring tension

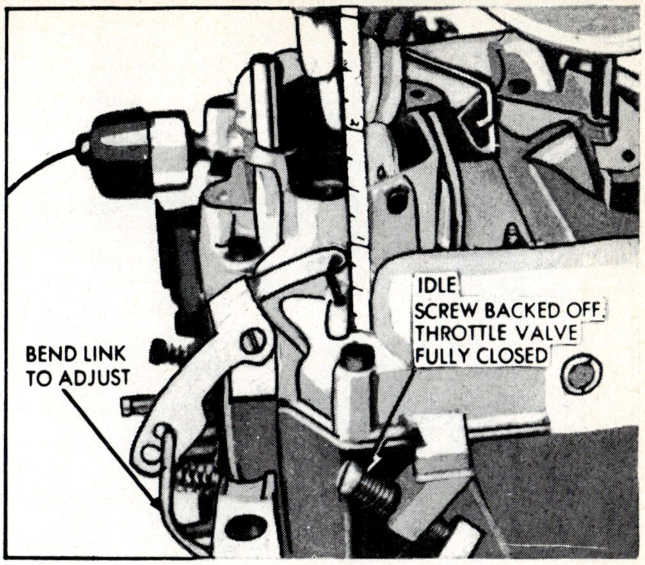

Fig. 5 TQ accelerator pump stroke

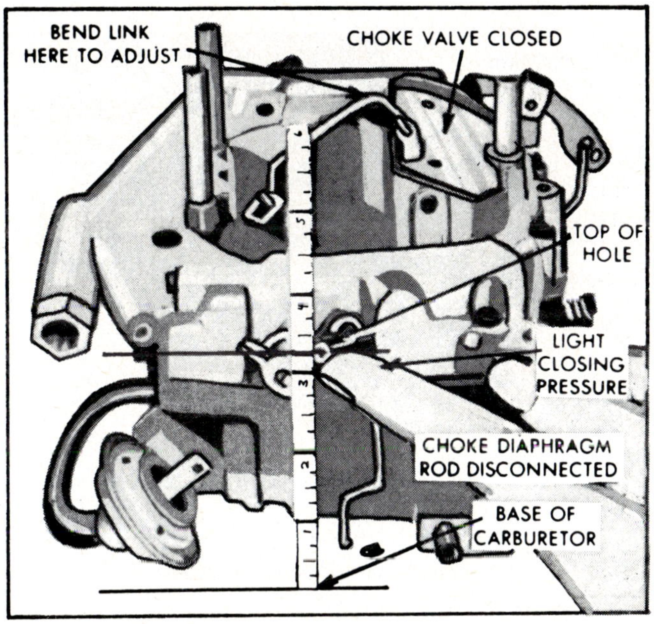

Fig. 6 TQ choke control lever adjustment

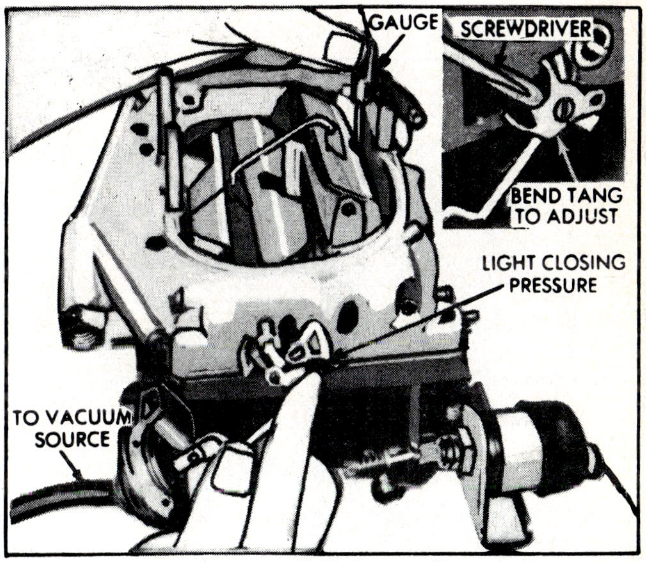

Fig. 7 TQ choke diaphragm connector rod

noid activated and engine idle speed adjusted to proper rpm, measure primary accelerator pump stroke. Measure distance between top of bowl cover to bottom of accelerator pump link. Dimension should be .328–.358 inch. Bend throttle connector rod at upper angle to adjust.

Position throttle lever just below secondary opening and measure secondary accelerator pump stroke. Measure distance between top of bowl cover and bottom of accelerator pump link. Dimension should be .120–.160 inch. Bend accelerator pump secondary pick-up tab to adjust.

Choke Control Lever

Fig. 6—Place carburetor on a flat surface. Close choke by pushing on choke lever with throttle partly open. Measure vertical distance between top of rod hole in control lever and base of carburetor (flat surface). Dimension should be 3.30 inch. Adjust by bending link connecting the two choke shafts.

Choke Diaphragm Connector Rod

Fig. 7–Apply a vacuum of 10 or more inches of Mercury to diaphragm to fully depress diaphragm stem. An auxiliary source like a distributor test machine can be used for this purpose. With air valve closed, adjust connector rod to give .020–.060 inch clearance between air valve and stop.

Vacuum Kick Adjustment

Fig. 8—With engine running, back off fast idle speed screw until choke can be closed to kick position at idle. Note number of screw turns so fast idle can be turned back to original adjustment. Insert a #35 drill between long side (lower edge) of choke valve and the air horn wall. Apply sufficient pressure on choke control lever to provide a minimum choke valve opening. The spring connecting the control lever to the adjustment lever must be fully extended for proper adjustment. Bend tang as shown to change contact with end of diaphragm rod. Do not adjust diaphragm rod. A slight drag should be felt as drill is being removed.

Fast Idle Cam & Linkage

Fig. 9—With fast idle adjusting screw on second step of fast idle cam, move choke valve toward closed position with light pressure on choke control lever. Clearance between choke valve lower edge and air horn wall should be .089–.109 inch. Adjust by bending fast idle connector rod at angle.

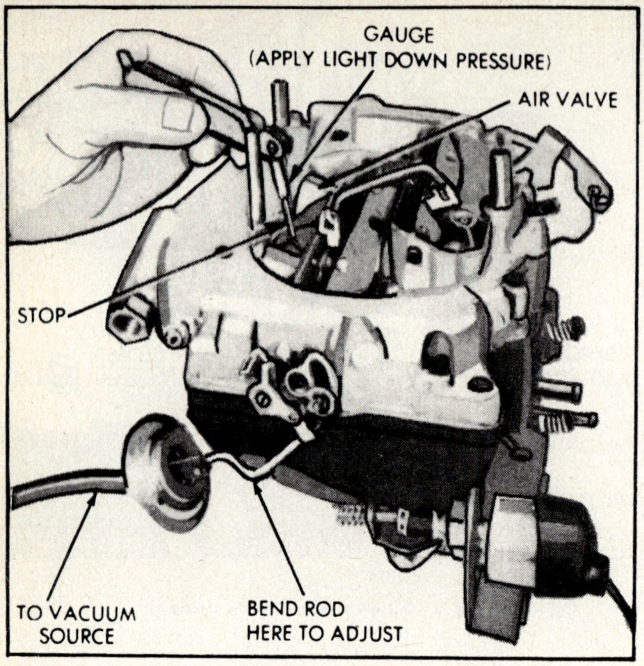

Fig. 8 TQ vacuum kick adjustment

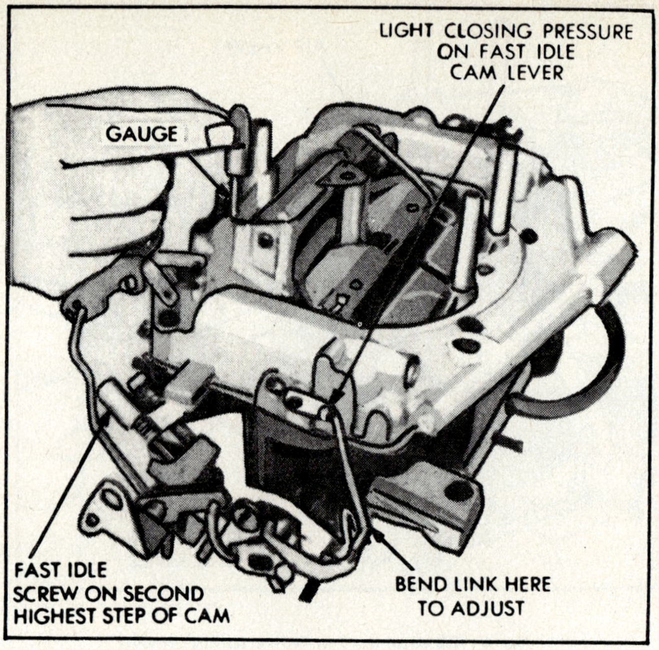

Fig. 9 TQ fast idle cam & linkage adjustment

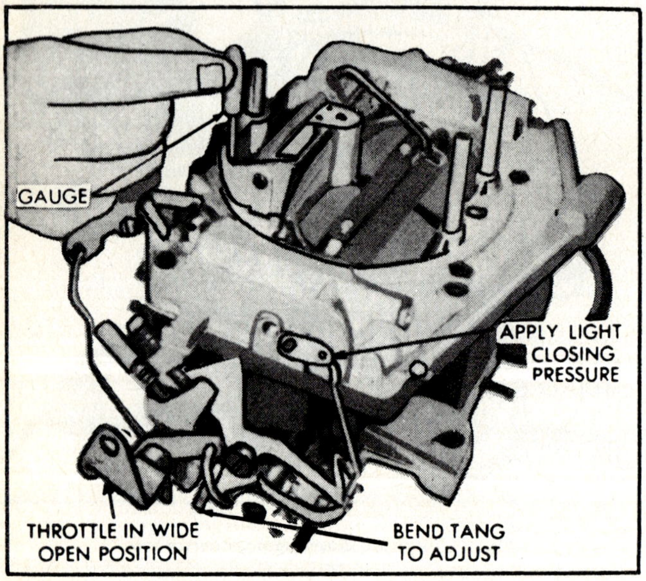

Fig. 10 TQ choke unloader adjustment

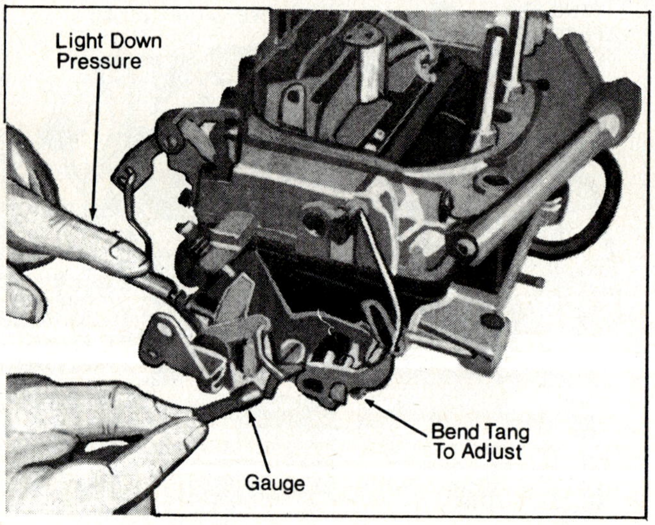

Fig. 11 TQ secondary throttle lockout

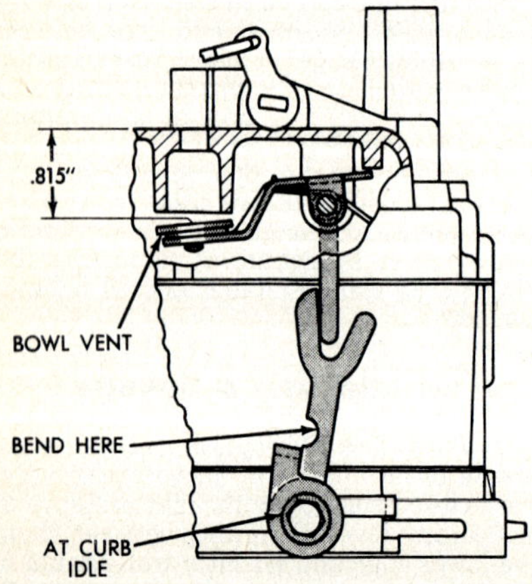

Fig. 12 TQ bowl vent valve adjustment

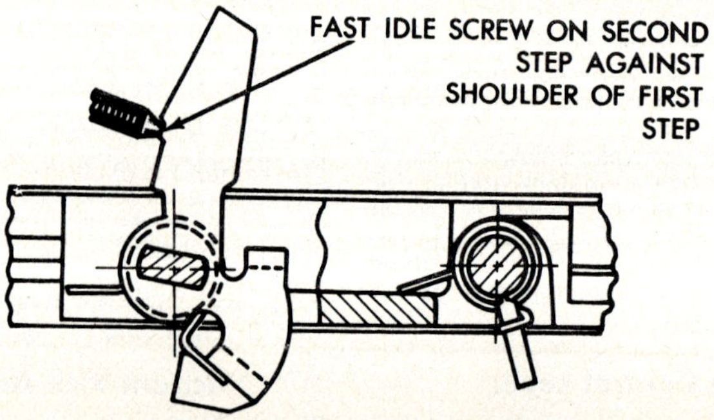

Fig. 13 TQ fast idle speed cam position

Choke Unloader Adjustment

Fig. 10—Hold throttle valves in wide open position and insert a .280–.320 inch gauge between long side (lower edge) of choke valve and inner wall of air horn. With finger lightly pressing against choke valve control lever, a slight drag should be felt as drill is withdrawn. Adjust by bending tang on fast idle control lever.

Secondary Throttle Lockout

Fig. 11—Move choke control lever to open choke position. Measure clearance between lockout lever and stop. Clearance should be .060–.090 inch. Adjust by bending tang on fast idle control lever.

Bowl Vent Valve Adjustment

Fig. 12—Remove bowl bent valve checking hole plug in bowl cover. With throttle valves at curb idle, insert a narrow ruler down through hole. Allow ruler to rest lightly on top of valve. Dimension should be .800–.830 inch. Adjust be bending bowl vent operating lever at notch. Install a new plug.

Fast Idle Speed Cam

NOTE: Disconnect and plug the EGR valve vacuum line.

Fig. 13—With engine off and transmission in Park or Neutral, open throttle slightly. Close choke valve until fast idle screw is positioned on second step of cam against shoulder of first step. Start engine and stabilize RPM, then adjust fast idle speed to 1550–1600 RPM.

Clutch Section

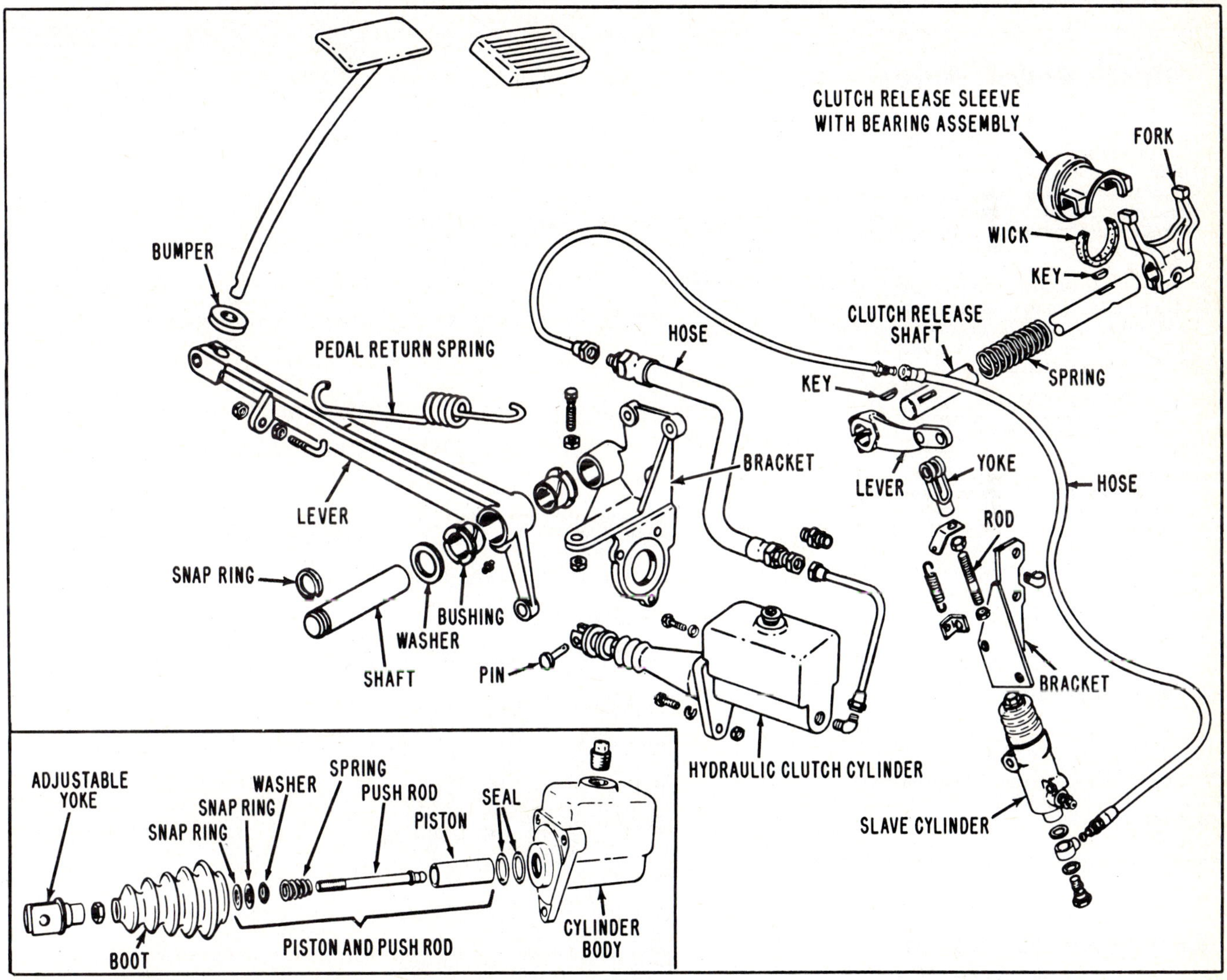

Fig. 1 Hydraulic clutch control. CM-75, 80, 110 and 1600 thru 1890 Models have suspended pedal

HYDRAULIC CLUTCH CONTROL

As shown in Figs. 1 and 2, the system consists of a pedal-operated cylinder, similar to a brake master cylinder, which transmits hydraulic fluid to the slave cylinder mounted on the transmission. The slave cylinder moves the clutch release bearing to engage and disengage the clutch. Tubing and flexible hose connect the two units.

The hydraulic cylinder is filled with high grade hydraulic brake fluid. To replenish fluid, fill the reservoir to within 3/4″ from the top of the filler neck.

Bleeding System

Any air trapped inside the system will eventually bleed itself to the reservoir. However, this is not accomplished immediately and takes place over an extended period of clutch pedal operation. For this reason a bleeding operation should be performed whenever the units have been disassembled or fluid lines disconnected. The bleeding procedure is accomplished in the same manner as for hydraulic brake systems. The bleeder screw is located on the lower end of the slave cylinder.

Adjustments

Two adjustments affect clutch operation and must be checked each time the hydraulic clutch is serviced. The first of these is clutch pedal stroke adjustment which is made at the hydraulic cylinder piston and push rod. The second adjustment is for clutch release bearing clearance. This adjustment is made at the upper end of the slave cylinder.

Clutch Pedal, Adjust

Except 1600-1890

1. With clutch pedal disconnected from hydraulic cylinder push rod, pull piston and push rod assembly its full length of stroke out of cylinder. *Be sure piston is held snugly against snap ring stop in end of cylinder.*
2. Hold clutch pedal up against pedal stop.
3. Adjust clevis yoke on end of piston push rod until yoke can be fastened to clutch pedal. Tighten jam nut on yoke. Install end pin and cotter pin.

1600 Thru 1890

1. Measure clutch pedal height from lower edge of clutch pedal lever to toe boards. (Measurement should be made at 90° to toe boards).
2. If pedal height is not 5⅝" for 1600 thru 1890 models, loosen lock nut on master cylinder push rod and adjust push rod to obtain proper pedal height.

CO-1610B thru COF-1950B Cargostar and COF-5370

The pedal stop setting is adjusted by screwing bolt in and out of master cylinder body. Pedal stop must limit travel before piston bottoms in cylinder. Clutch pedal free travel is determined by clutch release bearing clearance. Correct adjustment of release bearing will provide proper pedal free travel, Fig. 2.

Clutch Release Bearing Clearance

Except CM-75, 80, 110 1600 Thru 1890 Models, & CO-1610B thru COF-1950B

1. Remove slave cylinder push rod return spring, Fig. 3. Loosen clevis yoke lock nut on push rod. Remove cotter pin and clevis yoke end pin. Remove two hex head

Clutch Code	Setting Clutch Pedal Stop	Clutch Pedal Travel	Release Bearing Clearance
11129 11133	(3 31/32 inches)	(10 inches)	(.134 inches)①
11235 11343 11344	(3 13/32 inches)	(10 inches)	(1/8 inches)①
11345	(3 13/16 inches)	(10 inches)	*Internal Adjustments* With clutch brake 7/16 inches between release bearing housing and brake disc Without clutch brake 9/16 inches between release bearing housing and clutch cover *External Adjustments* 1/8 inch②
11249 11902	(3 31/32 inches)	(10 inches)	*Internal Adjustments* With clutch brake 1/2 inch between release bearing housing and brake disc Without clutch brake 9/16 inches between release bearing housing and clutch cover *External Adjustments* 1/8 inch②

①—Between release bearing and clutch fingers.
②—Between release shaft fork and release bearing housing.

Fig. 2 Clutch specifications for Cargostar & COF-5370 vehicles

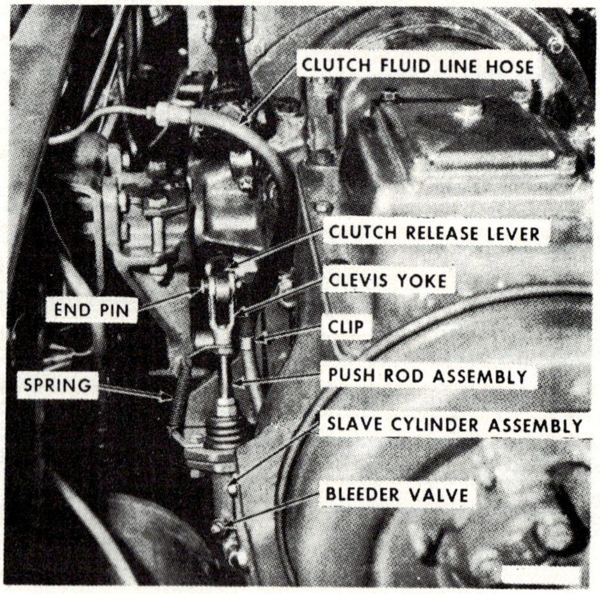

Fig. 3 Slave cylinder installation

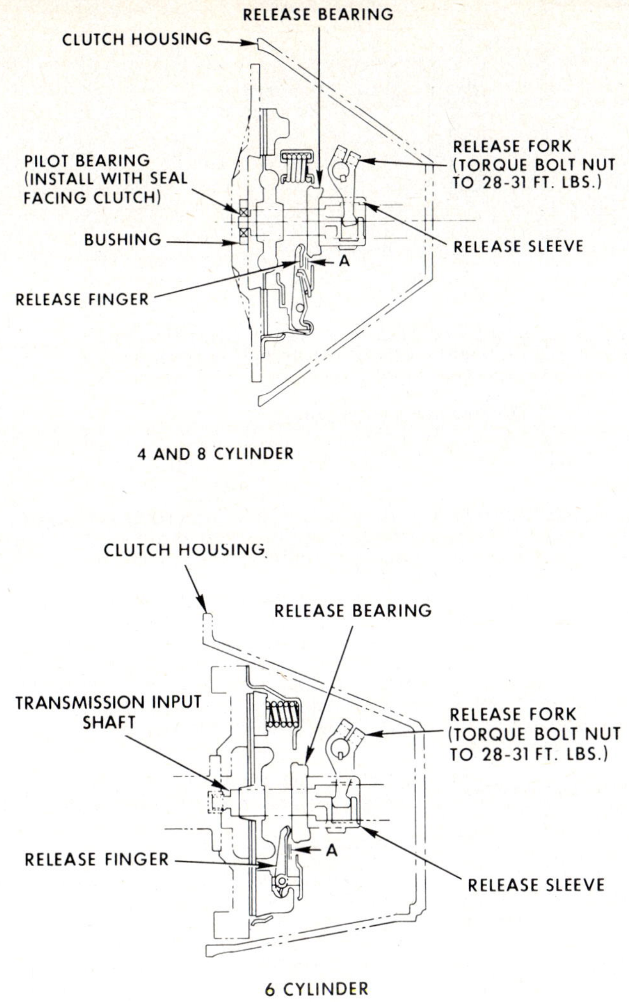

Fig. 4 Clutch release bearing clearance adjustment. Scout

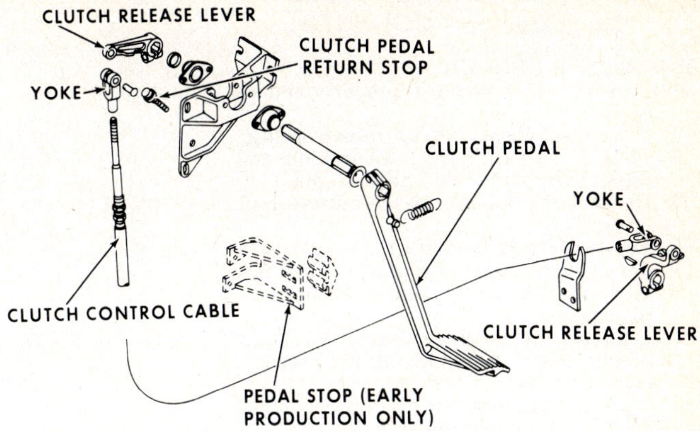

Fig. 5 Clutch control linkage

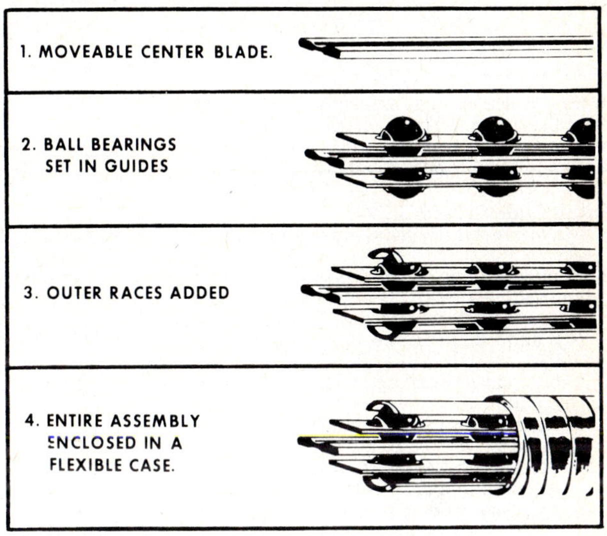

Fig. 6 Clutch control cable

mounting bolts from slave cylinder body and move cylinder away from clutch release lever.
2. Move clutch release lever upward until release bearing contacts clutch release levers. With push rod in released position, turn clevis yoke until yoke hole indexes with clutch release lever hole when slave cylinder mounting holes are aligned with mounting bracket holes. *Be certain that clutch release bearing is held against release levers while completing this step.*
3. After completing Step 2, turn clevis yoke on to push rod 2½ turns which provides approximately ⅛" clutch release bearing-to-levers clearance. Tighten lock nut on push rod to keep clevis yoke from turning while reinstalling slave cylinder.
4. Reinstall slave cylinder. Connect clevis yoke and clutch release lever with end pin. Insert cotter pin. Recheck lock nut on push rod and install push rod return spring, keeping spring clips aligned.

CM-75, 80, 110

This adjustment should follow clutch pedal adjustment.

1. Depress clutch pedal until there is the feel of the release bearing contacting the clutch release levers.
2. Adjust the pedal free movement by loosening the lock nut on the slave cylinder and turning the push rod to obtain 1½" pedal free movement.
3. Tighten lock nut and recheck pedal free movement.

1600 Thru 1890 Models

This adjustment should follow clutch pedal adjustment.

1. Loosen lock nut on clutch slave cylinder push rod. Remove cotter pin and yoke pin.
2. Move clutch release lever forward until release bearing contacts clutch release fingers. With slave cylinder push rod assembly retained in the released position (bottomed), turn yoke until yoke hole indexes with clutch release lever hole. Install yoke pin.
3. After completing step 2, remove yoke pin and shorten linkage by turning yoke 4 complete turns. Reassemble yoke pin and cotter pin and tighten lock nut on push rod assembly.

Scout II

1. Disconnect return spring and loosen clutch rod lock nut.
2. Hold clutch pedal against stop and lengthen or shorten clutch rod until there is zero clearance between the throw out bearing and clutch release fingers.
3. After obtaining zero clearance, rotate clutch two turns to obtain the desired 3/32 inch clearance between the throw out bearing and the clutch release fingers, Fig. 4, and tighten clutch rod lock nut.

CO-1610B thru COF-1950B Cargostar and COF-5370

PUSH-TYPE CLUTCH

1. Be sure clutch pedal is properly adjusted.
2. Disconnect slave cylinder return spring.
3. Loosen slave cylinder yoke locknut and remove the cotter pin and yoke pin.
4. Disconnect yoke from clutch release shaft lever.
5. Rotate the yoke to adjust push rod length until release bearing contacts clutch fingers with yoke pin in place and piston bottomed.
6. Back off the yoke five turns to obtain proper clearance and install the yoke pin and cotter pin.
7. Tighten yoke lock nut and connect slave cylinder return spring.

NOTE: Return spring must be capable of fully retracting piston to assure release bearing to clutch finger clearance is obtained. If spring is weak or missing replace.

PULL-TYPE CLUTCH

NOTE: Do not change external (slave cylinder) adjustment without checking clutch internal adjustment first.

Clutch Codes 11249, 11345

1. Make sure clutch pedal is properly adjusted.
2. Remove clutch housing inspection cover; on vehicles with clutch brake, check clearance between release bearing housing and clutch brake disc. On vehicles without clutch brake check clearance between release bearing housing and clutch cover hub.
3. If clearance must be adjusted proceed as follows:
   a. Rotate flywheel until adjusting ring is exposed and remove ring lock bolt.
   b. Pry lock free of adjusting ring, using caution as ring is spring loaded.
   c. Release clutch by blocking clutch in depressed position.
   d. Use a large screw driver to turn the clutch adjustment ring. Counterclockwise moves the release bearing toward the flywheel. Clockwise moves the bearing away from the flywheel. Each rotation notch is equivalent to .020 inches.
   e. Engage the clutch and recheck clearance. Tolerance allowable is 1/32 inches.
   f. After adjustment, install lockplate with bolt and washer and reinstall cover plate.
4. Using a 1/8 inch rod as a gauge, check the clearance between release shaft fork and contact pads of release bearing housing.

Clutch Codes 11902

1. Make sure that clutch pedal is properly adjusted.
2. Remove inspection cover and inspect clearance; on vehicles with clutch brake check between release bearing housing and clutch brake disc. On vehicles without clutch-brake check between the release bearing housing and clutch cover hub.
3. If clearance must be adjusted proceed as follows:
   a. Rotate engine until adjuster assembly is exposed and remove right hand bolt from adjuster and loosen left hand bolt one turn.
   b. Rotate adjuster upward disengaging the worm gear from the adjusting ring to allow manual adjustment. Hold the adjuster disengaged and tighten the left hand bolt.
   c. Block the pedal in the depressed position.
   d. Rotate ring to obtain specified clearance. Rotate clockwise to move bearing away and clockwise to move bearing toward flywheel.
   e. Engage clutch and recheck clearance.
   f. Loosen left hand bolt and rotate adjuster assembly downward to engage worm gear with adjusting ring teeth. Adjusting ring may have to be rotated slightly to allow worm gear to mesh.

NOTE: Make sure that adjuster actuator arm is engaged in hole in release sleeve retainer.

   g. Install right hand bolt and torque both to 30–35 ft. lbs.
4. Using a 1/8 inch rod as a gauge, check the clearance between the release shaft fork and contact pads of release bearing housing.

NOTE: If necessary, disconnect slave cylinder yoke from release shaft lever and adjust slave cylinder push rod length to obtain specified clearance between release shaft fork and release bearing housing with yoke pin in place and slave cylinder piston bottomed.

5. Install yoke pin and cotter pin and tighten yoke locknut.
6. Reinstall inspection cover.

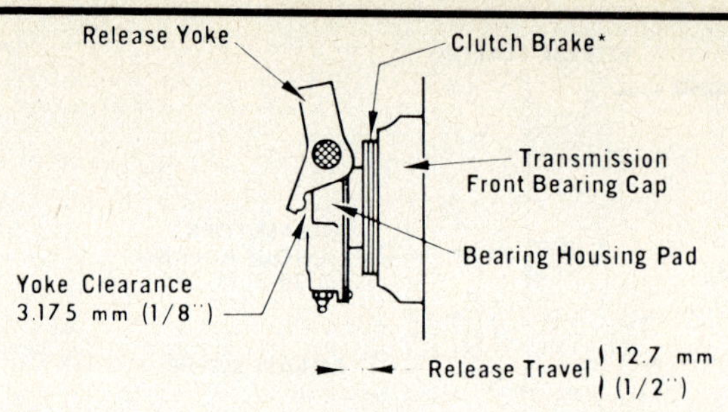

*The Clutch Brake is not used with synchronized transmissions—in which case adjust the release Bearing Housing to be approximately 14.288 mm (9/16") from the Front Bearing Cap.

Fig. 7 Release bearing clearance

CABLE CLUTCH CONTROL

The clutch control linkage consists of a suspended type pedal connected by a control cable to the clutch release shaft assembly. Each end is attached by means of a yoke and pin. The upper end is anchored through the cab floor and the lower end is anchored by a mounting bracket at the left rear engine mount, Fig. 5.

Release bearing travel is controlled by the release bearing fork which is mounted on the release shaft assembly.

The cable is a precision ball bearing remote control cable, Fig. 6. This cable is designed to bend 180° or into the shape of a "U" in either direction around the flat section of the center blade.

To assure correct clutch operation and long service life of clutch components, the release system must be kept in proper adjustment. Three factors affect proper adjustment. They are release bearing clearance, proper installation of clutch control cable, and correct clutch pedal height and free travel.

NOTE: Do not change or readjust external linkage without checking internal adjustment first.

To obtain proper clutch control operation, it is recommended that adjustment operations be performed in the following sequence:

a. Check clutch internal adjustment. (Release bearing)
b. Make sure lower end of clutch control cable is properly adjusted.
c. Check the clutch pedal height and pedal free travel adjustments.

Clutch Release Bearing Adjustment

Clutch Codes 11155, 11239, 11249, 11250, 11254

1. Remove the clutch housing inspection cover.
2. On vehicles with clutch brake, weld a piece of 1/2 inch thick stock to a rod (handle) to check the gap between release bearing and clutch brake assembly. On vehicles without clutch brake, weld a piece of 17/32 inch stock to a rod (handle) and check the gap between release bearing and clutch cover hub.
3. If adjustment is necessary proceed as follows:
   a. Rotate flywheel until adjusting ring is exposed. Remove lock bolt and pry lock free of adjusting ring. Use caution as ring is spring loaded.
   b. Block clutch in depressed position, and using a large screwdriver turn the adjustment ring. Clockwise moves the bearing toward flywheel; while counterclockwise moves the bearing away from the flywheel.
   c. Each rotation notch represents .020 inches of bearing housing movement.
   d. Engage the clutch and recheck the clearance. Allowable tolerance is 1/32 inch.
   e. When desired tolerance is obtained,

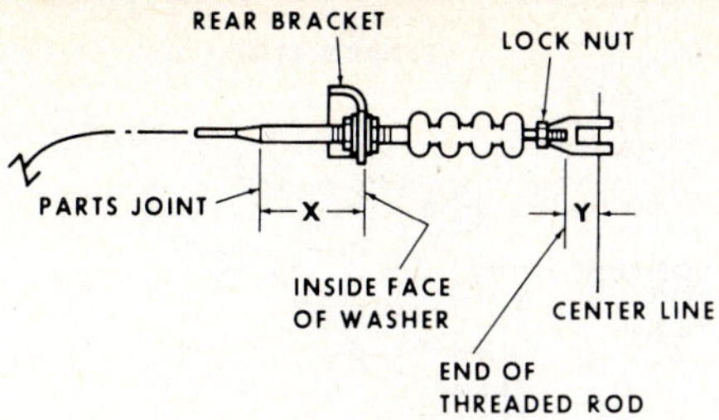

Fig. 8 Clutch cable lower end adjustment dimensions

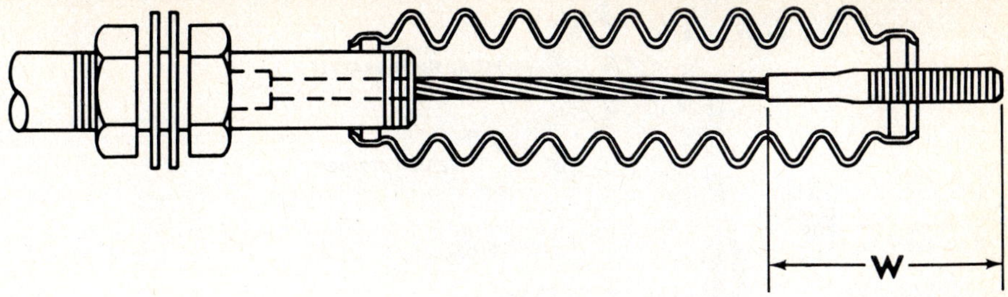

Fig. 9 Control cable rear stud length

install lockplate with bolt and washer.

f. Install clutch cover plate.

NOTE: When following are adjusted properly, release bearing, clutch pedal, and clutch control cable, clearance between contact pad of release bearing housing and release shaft yoke will be approximately 1/8 inch. Use a 1/8 inch diameter rod to check clearance.

Clutch Codes 11145 & 11346

1. Remove the clutch housing inspection cover.

NOTE: Clearance should be 1/2 inch between the release bearing housing and front brake disc. On vehicles without the clutch brake, clearance should be 9/16 inch between the release bearing housing and the front bearing cap, Fig. 7.

2. Hold or block the clutch pedal down. (Clutch disengaged)
3. Loosen release sleeve locknut using a spanner or a punch and hammer.
4. Release the clutch pedal. (Engage clutch)
5. Rotate sleeve adjusting nut to obtain proper clearance between release bearing housing and clutch brake disc or transmission front bearing cap.

NOTE: Keep light tension on release yoke or clutch pedal while turning adjusting nut. This keeps release spider snugly against release levers and prevents false clearance reading.

6. Hold or block clutch pedal down and tighten release sleeve locknut.

NOTE: Always disengage clutch while lock nut is being loosened or tightened; however engage clutch when turning release sleeve to obtain proper clearance. When release bearing, clutch pedal, and control cable are properly adjusted clearance between contact pad of release bearing housing and release shaft yoke will be approximately 1/8 inch. Use a 1/8 inch diameter rod to measure this clearance.

Clutch Pedal Adjustment

Vehicles Built Before 2/24/75 (Vehicle identification numbers below EGA-15693)

1. Check proper clutch pedal height and free travel. Rear edge of pedal should be approximately 13.6 inches from the kickpanel. Free travel should be approximately 1.7 inches.
2. Adjust pedal as follows:
   a. Remove inner portion of kickpanel to expose upper end of clutch control cable.

Engine	Dim. "W" *	Dim. "X"	Dim. "Y"
Cummins In-Line (6x4)	88.9 mm (3.5") 98.6 mm (3.88")	137.2 mm (5.4") 137.2 mm (5.4")	61.0 mm (2.4") 48.3 mm (1.9")
Cummins In-Line (4x2) 5 & 7 Deg.	88.9 mm (3.5") 98.6 mm (3.88")	114.3 mm (4.5") 116.8 mm (4.6")	61.0 mm (2.4") 50.8 mm (2.0")
Cummins In-Line KT-450	88.9 mm (3.5") 98.6 mm (3.88")	129.5 mm (5.1") 129.5 mm (5.1")	61.0 mm (2.4") 50.8 mm (2.0")
Caterpillar In-Line 3406	88.9 mm (3.5") 98.6 mm (3.88")	109.2 mm (4.3") 109.2 mm (4.3")	33.0 mm (1.3") 22.9 mm (0.9")
IH V-800 (6x4)	88.9 mm (3.5") 98.6 mm (3.88")	109.2 mm (4.3") 111.8 mm (4.4")	61.0 mm (2.4") 50.8 mm (2.0")
IH V-800 (4x2)	88.9 mm (3.5") 98.6 mm (3.88")	109.2 mm (4.3") 109.2 mm (4.3")	61.0 mm (2.4") 50.8 mm (2.0")
Cummins VT-350 (VT-903)	88.9 mm (3.5") 98.6 mm (3.88")	119.4 mm (4.7") 119.4 mm (4.7")	61.0 mm (2.4") 50.8 mm (2.0")
Detroit 8V (6x4) 3 Deg. & (4x2) 5 Deg.	88.9 mm (3.5") 98.6 mm (3.88")	109.2 mm (4.3") 91.4 mm (3.6")	61.0 mm (2.4") 50.8 mm (2.0")
Detroit 6V92T (4x2) 5 Deg.	88.9 mm (3.5") 98.6 mm (3.88")	119.4 mm (4.7") 129.5 mm (5.1")	50.8 mm (2.0") 50.8 mm (2.0")
Detroit 6V92T (6x4) 3 Deg.	88.9 mm (3.5") 98.6 mm (3.88")	119.4 mm (4.7") 129.5 mm (5.1")	43.2 mm (1.7") 43.2 mm (1.7")

* Dimension "W" varies between cable types. Use Dimensions "X" and "Y" corresponding with Dimension "W" of cable being adjusted.

Fig. 10 Clutch control cable adjustment specifications

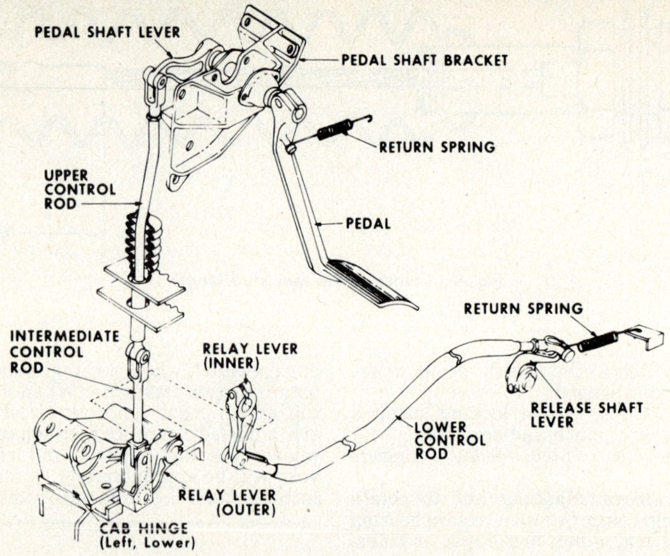

Fig. 11 Clutch control linkage

b. Disconnect control cable yoke from release lever.
c. Set pedal return stop to 1.2 inches above the bracket face to provide proper pedal height from kickpanel.
d. Adjust the cable yoke to obtain approximately 1.7 inches of pedal free travel. This will provide approximately 1/8 inch clearance between release bearing housing and release shaft yoke.
e. Check release bearing travel by depressing clutch pedal.

NOTE: Vehicles with clutch brakes should obtain clutch braking action when the clutch pedal compresses the rubber stop by approximately 1.7 inches. On vehicles without clutch brake release bearing housing should not contact transmission front bearing cap.

---

f. Operate the vehicle and test clutch and clutch brake operation.
g. Reinstall inner portion of kick panel.

Vehicles Built After 2/24/75 (Vehicle identification numbers EGA-15693 and up)

1. Check for proper clutch pedal height and free travel. Rear edge of pedal should be approximately 10.7 inches from the kick panel. Free travel should be approximately 1 inch.
2. Adjust pedal as follows:
   a. Remove inner portion of kick panel and expose upper end of clutch control cable.
   b. Disconnect cable from clutch release lever and loosen cable yoke locknut adjust-yoke as needed to obtain full clutch release (and clutch brake application) just before pedal contacts kickpanel. Tighten the yoke locknut and install cotter pin.
   c. Adjust pedal return stop to obtain approximately 1 inch of pedal free travel. This will provide approximately 10.7 inches of pedal height.
   d. Operate engine and check clutch and if equipped, clutch brake operation.
   e. Install inner portion of kick panel.

Cable Installation & Adjustment

1. Make certain that internal clutch adjustment is correct.
2. Install lower end of clutch cable housing in rear bracket.
3. Determine cable adjustment dimensions applicable to vehicle being serviced. Refer to cable adjustment specifications.

NOTE: Dimension W: Fig. 8, Length of cable threaded rear stud.
Dimension X: Fig. 8, Distance from parts joint on cable rear anchor sleeve to inside face of rear anchor nut washer.
Dimension Y: Fig. 9, Distance from end of cable to centerline of yoke connecting pin.
Refer to Fig. 10.

---

4. Tighten cable anchor nuts according to specified dimension "X". Do not allow clutch cable anchor sleeve to turn when tightening anchor nuts.
5. Install jam nut and yoke on cable rear stud according to dimension "Y". Do not tighten jam nut.
6. Remove seal nut from upper end of clutch cable.
7. Install upper end of cable through cab floor.

NOTE: Do not force cable into radius less than 7 inches.

---

8. Lower cab and align control housing in floor so wire and yoke will pull in line with housing. Maximum misalignment of yoke and anchor housing centerline is 1/2 inch.
9. Install 3/4 inch flat washer and 3/4-16 hex nut on cable. Torque nut to 10 ft. lbs.
10. Install seal nut with felt wiper and brass scraper finger tight only.
11. Place lock nut and yoke on upper end of clutch control cable. Do not tighten jam nut.
12. Make sure clutch pedal height is set properly.
13. Connect both yokes to their attachment points. Secure with cotter pins. Make sure that wire rope ends are free to turn in threads of both yokes.
14. Tighten lock nut against yoke on each end of cable, making sure that wire ends do not turn while tightening. Replace kick panel.

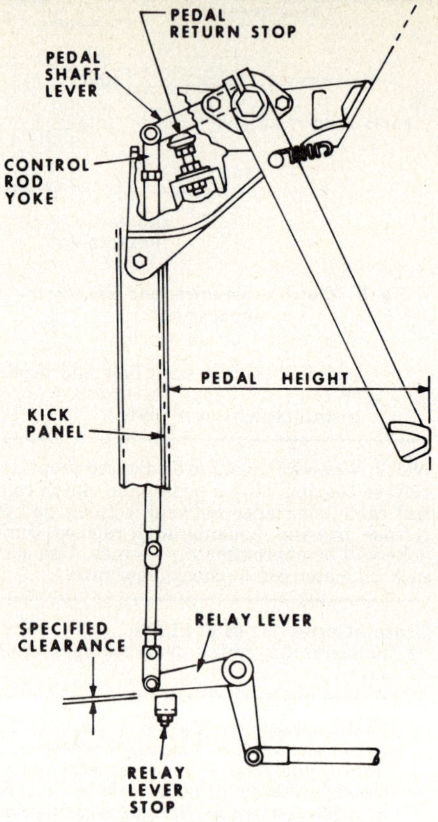

Fig. 12 Clutch pedal adjustment

ROD TYPE CLUTCH CONTROL

CO, COF-4070B (Transtar)

The clutch control linkage consists of a suspended type pedal connected by rod to a relay (bellcrank) which is connected by another rod to the clutch release shaft cover, Fig. 11.

To assure correct clutch operation and long service life the clutch and linkage must be kept in proper adjustment. Two types of adjustment must be made; clutch internal (release bearing) adjustment and external adjustment (pedal and linkage). Adjustment factors are release bearing clearance, pedal height, and release shaft yoke to release bearing clearance (clutch pedal free travel). Some clutches have a self adjusting feature which automatically compensates for wear. However, initial settings must be established by manual adjustment.

Clutch service adjustments must be performed in the following sequence:

A. Check clutch internal adjustment (release bearing).
B. Check clutch pedal adjustment.
C. Check clutch release shaft yoke-to-release bearing clearance (clutch pedal free travel).

Release Bearing Adjustment

Dana Angle Spring Clutches

1. Remove the clutch housing inspection cover.

2. Check for clearance of 9/16 inches as follows:
   a. On vehicles with clutch brake check the clearance between release bearing housing and clutch brake disc.
   b. On vehicles without clutch brake check the clearance between release bearing housing and clutch cover.

NOTE: Gauge tools may be made locally by welding a piece of stock material to a rod (handle). Refer to specifications for clearance dimensions.

3. On manual adjustment type clutches rotate engine flywheel until adjusting ring is exposed. Remove the adjusting ring lock bolt and pry the lock free of adjusting nut. Use caution, adjusting ring is spring loaded.
4. Release the clutch by blocking pedal in depressed position.
5. Using a large screwdriver, turn the clutch adjusting ring counterclockwise to move release bearing toward flywheel and clockwise to move it away.

NOTE: Rotation of each notch will move the housing .020 inches.

6. Reengage the clutch and check clearance. If proper clearance has been obtained, install lock plate with bolt and washer.
7. On self-adjusting type rotate the engine until adjuster assembly is exposed.

NOTE: Incorrect release bearing clearance may indicate adjuster mechanism is faulty. Adjust clearance and recheck after short period of service.

8. Remove right hand bolt from adjuster and loosen left hand bolt one turn.
9. Rotate the adjuster upward disengaging worm gear from adjusting ring to allow manual adjustment. Hold adjuster disengaged and tighten left hand bolt.
10. Release clutch by blocking pedal in depressed position.
11. Rotate adjusting ring to obtain proper release bearing clearance. Rotate ring clockwise to move bearing housing away from flywheel and counterclockwise to move housing toward flywheel.
11. Reengage the clutch and check the release bearing clearance.
12. Loosen the left hand bolt and rotate the adjuster downward to engage the worm gear with the adjusting ring teeth. Adjusting ring may have to be rotated slightly to allow worm gear to mesh. Make sure that actuator arm is engaged in hole in release sleeve retainer.
13. Install righthand bolt and torque both bolts to 30–35 ft. lbs. and install inspection cover.

LIPE LP, LPA Clutches

1. Remove the inspection cover.
2. On vehicles with a clutch brake check for a clearance of 3/4 inches between release housing and clutch brake disc. On vehicles without clutch brake check the clearance between release bearing housing and transmission front bearing cap.
3. On manual adjustment type LP clutches hold or block clutch pedal down, and loosen release sleeve locknut using either a spanner or hammer and punch.
4. Release clutch pedal (engage clutch).
5. Rotate sleeve adjusting nut to obtain proper clearance between release bearing housing and clutch brake disc or transmission front bearing cap.
6. Keep slight tension on release yoke or clutch pedal while turning sleeve adjusting nut. This keeps release spider against release levers preventing false clearance readings. Depress pedal a few times and then recheck clearances.

NOTE: Always disengage clutch while locknut is being loosened or tightened, however, have clutch engaged when turning release sleeve for proper clearance. On self adjuster type LPA, incorrect release bearing clearance may indicate faulty self-adjuster. Readjust clearance and then recheck after period of service.

7. On self adjusting models if clearance exceeds specified dimension, turn adjusting ratchet counterclockwise using a soft drift and a small hammer. Do not attempt to turn clockwise without releasing ratchet pawls with pawl block tool.
8. If clearance is less than specified, insert pawl block tool SE-2745 between pawl carrier and ratchet. Rotate position of tool until three pawls are released.
9. As viewed from rear of vehicle, use a soft drift and small hammer to turn ratchet clockwise.
10. Remove pawl block tool and install clutch housing inspection plate.

Clutch Pedal, Adjustment

1. Check for a clutch pedal height of 11.5 inches.
2. With pedal in full return position (at specified pedal height) check clearance between outer relay lever and lever stop.
3. To adjust proceed as follows:
   a. Remove the kickpanel to expose upper control rod adjusting yoke and loosen locknut on control rod adjusting yoke, Fig. 12.
   b. Disconnect control rod from pedal shaft lever by removing cotter pin and yoke pin.
   c. Insert spacer gauge with thickness of 1/8 inches between outer relay lever and relay lever stop.
   d. Adjust control rod to remove all slack from upper and intermediate control rod linkage.
   e. Remove spacer gauge and check for clearance between lever and lever stop, if necessary, repeat steps to obtain specified clearance.
   f. Tighten adjusting yoke locknut and install cotter pin.
   g. Reinstall kick panel. Adjust so clutch pedal clearance is .7 inches.

SCOUT

Clutch, Replace

When removing the transmission for the purpose of gaining access to the clutch or for any other reason, extreme care should be taken to support the weight of the transmission until it is completely removed so that the clutch shaft splines will clear the clutch disc hub. There is a possibility of distorting the clutch disc which will then not permit a free release of the clutch.

When installing the clutch, position the disc and clutch assembly on the flywheel, using a stub shaft (or a spare clutch shaft) to align the clutch disc. Then tighten the cover-to-flywheel capscrews gradually and evenly.

Free Pedal Travel, Adjust

The clutch pedal should have a minimum of 1½" free movement before clutch pressure is felt. If pedal free movement is less than specified, an adjustment is necessary to provide adequate bearing-to-lever clearance.

On models with turnbuckle release rod, adjustment is made by loosening release rod lock nuts on each side of turnbuckle. Looking down on turnbuckle, turn it counter-clockwise to increase or clockwise to decrease free travel.

On models with adjustable end yokes, remove yoke pins and turn yokes up or down on release rod as required.

Manual Shift Transmission Section

NOTE: For Clark, Fuller, New Process, Spicer and Warner transmissions used in International trucks, see the "name" chapter for service procedures

MODELS T1, T2, T4

Figs. 1, 2

1. Take off control cover.
2. Remove universal flange nut and pull off flange.
3. Remove speedometer driven gear and front and rear bearing retainers.
4. Drive countershaft out rearward, allowing cluster gear to lie in case.
5. Pull main drive gear out through front.
6. Remove mainshaft front pilot bearing and push mainshaft far enough to rear to permit removal of speedometer drive gear and rear bearing.
7. Lift mainshaft and gears out through top.
8. Lift out cluster gear and washers.
9. Drive reverse idler shaft out through rear and lift out gear.
10. To disassemble mainshaft, slide the low and reverse sliding gear and synchronizer unit from the shaft. Release snap ring and take off the second speed gear, bushing and thrust washer.
11. Reverse the order of the above procedure to assemble the transmission.

T2, T8 O. D. TRANS.

Disassemble, Fig. 3

1. Drain lubricant.
2. Remove companion flange.
3. Remove lockout switch.
4. Remove governor.
5. Drive out tapered pin (from underside) which holds overdrive control shaft in case. Pull shaft with lever from case.
6. Remove screws which attach overdrive case and adapter to transmission. Tap end of output shaft lightly with a soft mallet and remove overdrive case. This prevents shaft and ring gear from being removed with case and spilling the 12 clutch rollers.
7. With case removed, slide governor and speedometer drive gear from shaft.
8. Reinstall one capscrew to hold adapter to transmission.
9. Remove shift rail adapter spring from inside of overdrive case. Pry out control shaft and output shaft oil seal from case. Discard seals and replace with new ones upon reassembly.
10. Remove outer snap ring, rear bearing and inner snap ring from shaft.
11. Remove output shaft and ring gear, catching 12 clutch rollers, Fig. 4.
12. Separate ring gear and output shaft by removing large snap ring.
13. Pry retaining clip that retains roller clutch assembly (free wheeling unit) to output shaft. Then remove free wheeling cam and pinion case assembly, Fig. 5. Remove retaining clip holding the two units together and separate.

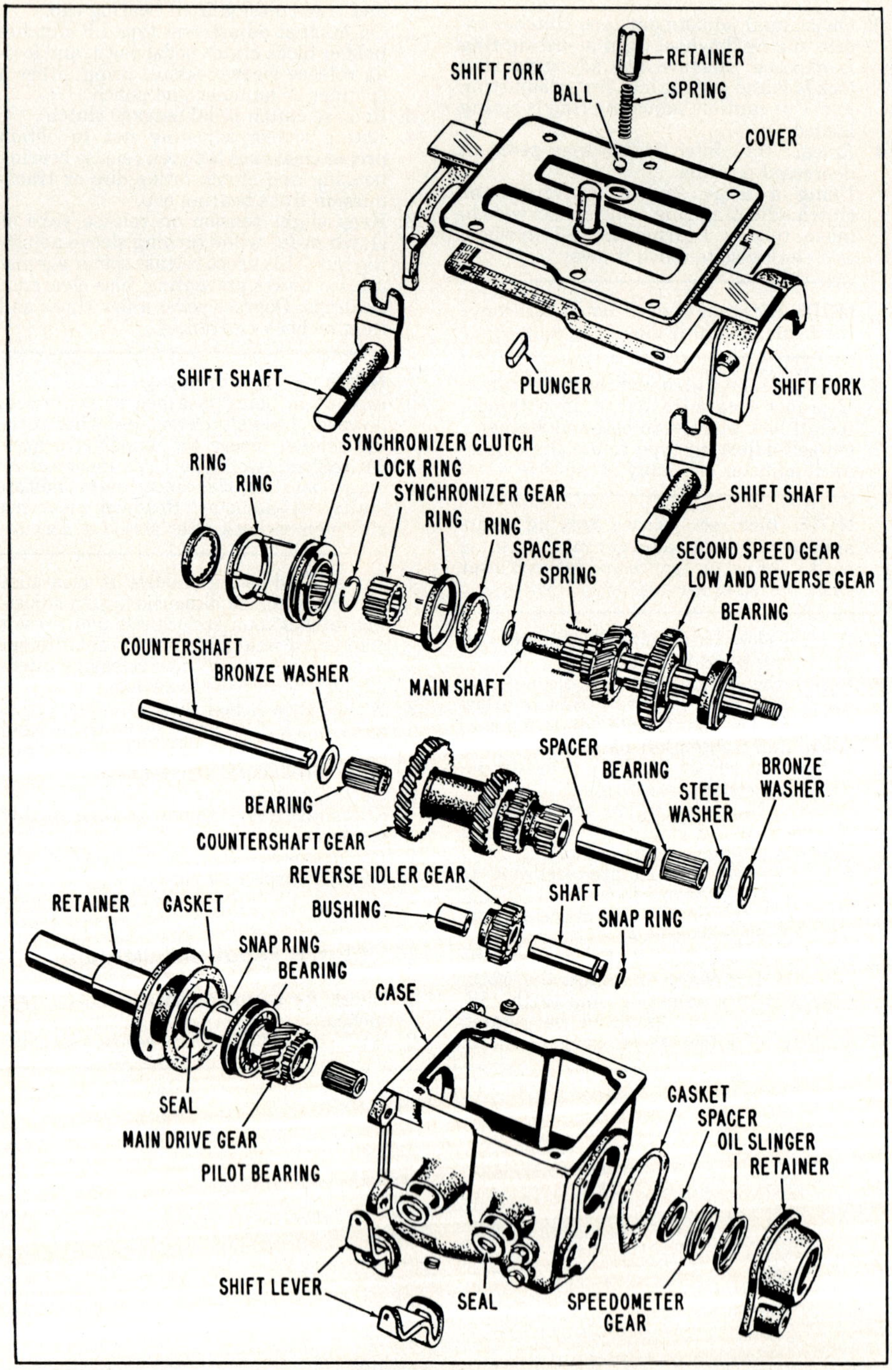

Fig. 1 Three speed transmission. Model T-1

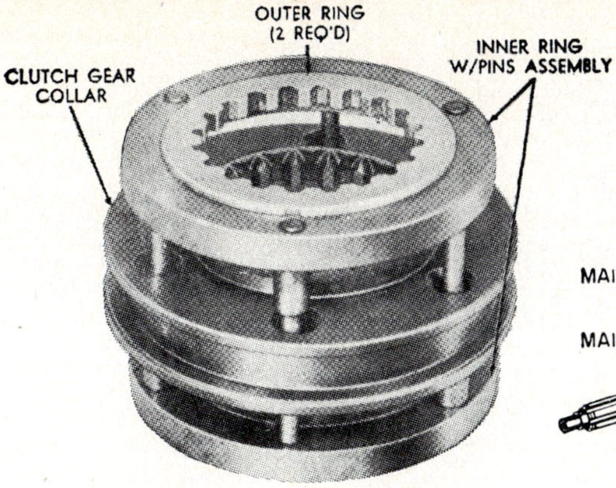

Fig. 2 On T1 and T2 transmissions the synchronizer consists of the parts shown plus the synchronizer clutch gear which is secured to the mainshaft with a snap ring

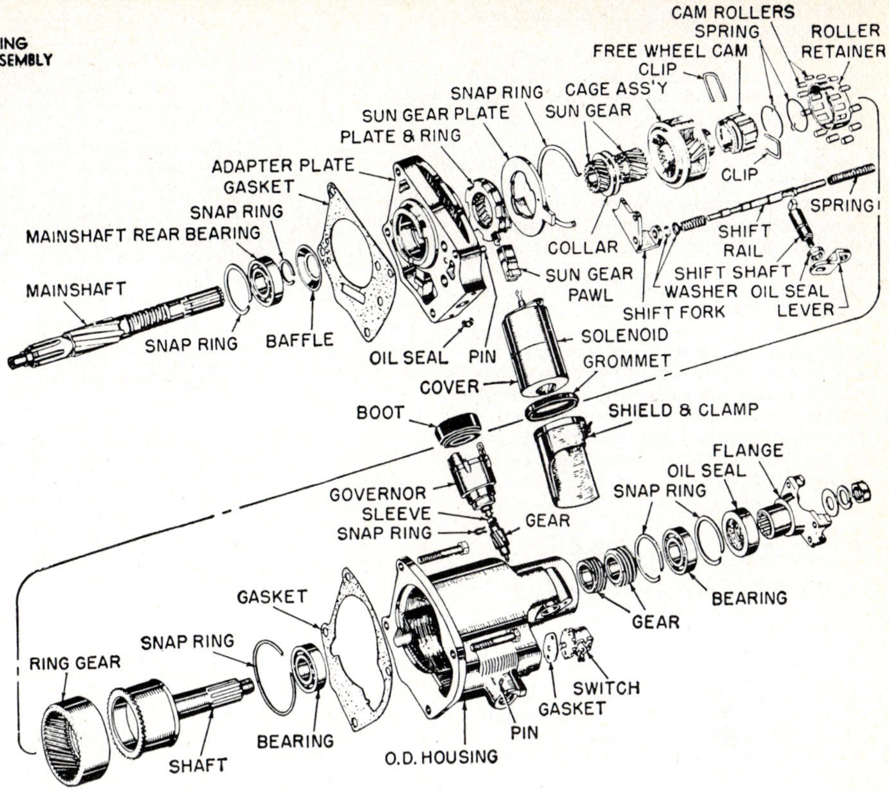

Fig. 3 T2 overdrive

14. Slide sun gear with shifting collar off end of transmission mainshaft and at the same time remove shift rail from case adapter.
15. Remove two snap rings which retain shifting collar to sun gear and separate collar and gear.
16. Remove "C" washer located on shift rail between shift fork and "C" washer cup. Then slide fork, "C" washer cup and reverse lock-up spring from shift rail.
17. Unfasten solenoid from case and turn clockwise 1/4 turn and remove from adapter.
18. Remove large snap ring located in adapter next to retainer plate with oil trough and remove parts.
19. Before further disassembly of the overdrive unit can be made, it will be necessary to partially disassemble the transmission as follows:
20. Remove transmission cover.
21. Remove main drive gear bearing retainer, noting that oil drain is located below.
22. Start main drive gear bearing out of case by tapping on outer race from inside of case. As soon as bearing is started, work it out of case by moving end of shaft from side to side.
23. Pull overdrive adapter and mainshaft toward rear of transmission case, lowering extended end of mainshaft to permit removal of synchronizer from inside of case. Remove snap ring and strip mainshaft.
24. Remove large snap ring holding bearing in adapter. Pull adapter off bearing and press bearing from mainshaft.
25. Turn transmission low and reverse shift lever shaft to its forward limits. This will provide clearance for removal of reverse lockout cam.

Assemble

1. Reverse the disassembly procedure to assemble the transmission and overdrive, observing the following:
2. Prior to assembling the overdrive unit, select correct thickness of snap ring that will secure (a) mainshaft rear bearing in overdrive case adapter (b) rear bearing to mainshaft, (c) retaining plate with oil trough in adapter, and (d) output shaft bearing in overdrive case with no perceptible end play.
3. Always install new gaskets and oil seals if unit has accumulated any extensive mileage or if damaged in removal.
4. When installing sun gear control plate and blocker ring assembly, *be sure* that the blocker ring and pawl are properly positioned. Position blocker ring so that pawl will be located between the two lugs of the blocker ring. The notched side of the pawl should be positioned so that the notch faces toward top of transmission.
5. The solenoid should be installed so that opening in end of solenoid is located to the

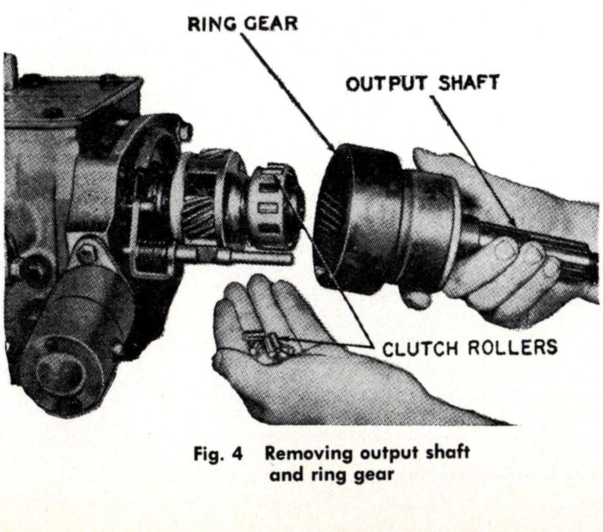

Fig. 4 Removing output shaft and ring gear

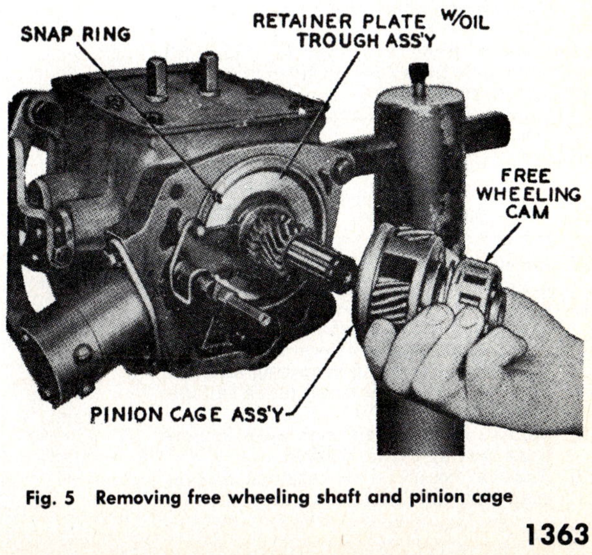

Fig. 5 Removing free wheeling shaft and pinion cage

bottom and that plunger is engaged and locked in pawl.

6. To facilitate installing output shaft and ring gear, apply heavy grease to roller clutch cage to hold the 12 rollers in position. Then, with low gear of transmission engaged, turn cage and rollers counterclockwise until the rollers are in their low positions on cam. Install output shaft and ring gear on pinion cage and roller clutch (free wheeling unit), turning the shaft to the left as the outer race encloses the rollers.
7. Prior to installing overdrive case to adapter, position shift rail retractor spring in case. Insert a pilot in case through retractor spring to assure spring alignment. While installing case, guide shift rail through retracting spring. Pilot will be forced from case. *Be sure* slotted portion in shift rail is properly positioned to take overdrive control shaft before installing case.

T-495 & T-496 TRANSMISSIONS

Transmission, Disassembly Fig. 10

1. Position shift lever in Reverse, then remove bolts from shift bar housing and lift housing and gasket from top of case.
2. Remove poppet springs and balls from shift bar housing.
3. Remove lock pins from 1st and reverse speed shift block and from shift forks.
4. Remove 4th and 5th speed shift rail and shift forks.
5. Remove interlock ball and cross pin retainer plug. Remove 1st and reverse and 2nd and 3rd speed shift forks and shift rails.
6. Lock transmission by selecting two gears and remove mainshaft flange nut, washer and flange. Remove parking brake drum and band.
7. Remove six bolts from mainshaft rear bearing retainer, then remove retainer and gasket.
8. Remove speedometer drive gear from mainshaft, then remove countershaft rear bearing nut.
9. Position mainshaft assembly towards rear of case, then remove mainshaft rear bearing and snap ring using suitable tools.
10. Remove main drive gear bearing retainer, then remove main drive gear, bearing, spacer and mainshaft pilot bearing using suitable tools.
11. Remove front synchronizer, then lift mainshaft from case by tilting as necessary to clear case.

NOTE: Countershaft slinger impeller must be in horizontal position to prevent damage to slinger impeller blades when removing mainshaft.

---

12. Remove reverse idler shaft lock plate, then remove idler shaft from case.
13. Remove reverse idler gear, bearings, spacer and thrust washer.
14. Position countershaft toward rear of case, then remove bearing snap ring and rear bearing from countershaft using suitable tools.
15. Remove countershaft from case. Remove front bearing thrust washer.
16. Remove countershaft expansion plug snap ring and expansion plug, then re-

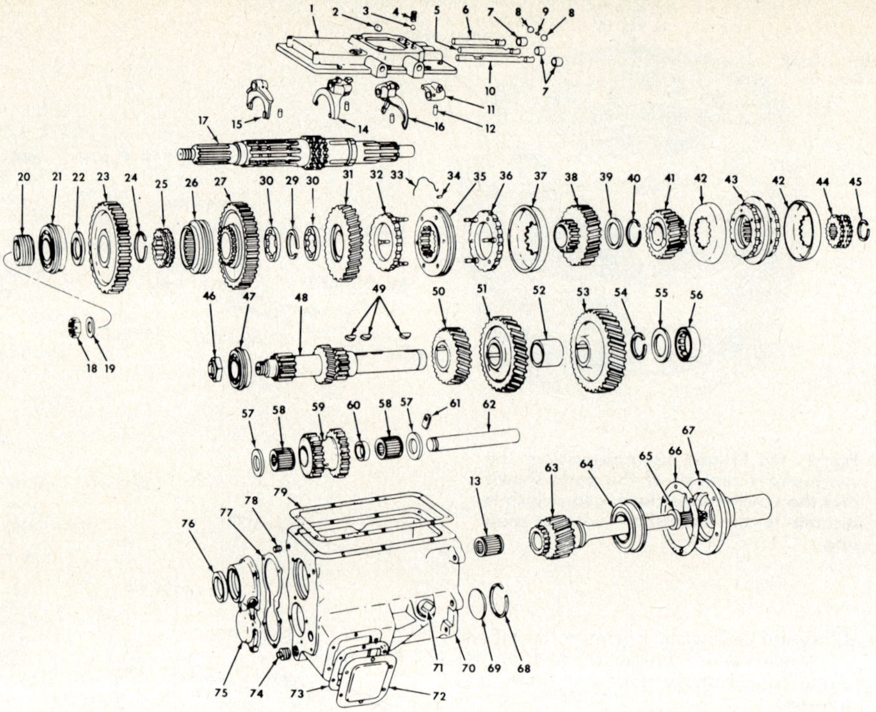

1 HOUSING, W/First and Reverse Speed Poppet
2 PLUG, Shift Bar Dust
3 BALL, Shift Bar Poppet
4 SPRING, Poppet Ball
5 BAR, Second and Third Speed Shift
6 BAR, Fourth and Drive Speed Shift
7 CUP, Shift Bar Dust
8 BALL, Shift Bar Interlock
9 PIN, Shift Bar Interlock Plunger
10 BAR, First and Reverse Speed Shift
11 BLOCK, W/Plunger and Spring, First and Reverse Shift
12 PIN, Shift Fork Set
13 BEARING, Mainshaft Pilot
14 FORK, Second and Third Speed Shift
15 FORK, First and Reverse Speed Shift
16 FORK, Fourth and Fifth Speed Shift
17 MAINSHAFT
18 NUT, Companion Flange
19 WASHER, Companion Flange Nut
20 GEAR, Speedometer Drive
21 BEARING, Mainshaft Rear
22 WASHER, Mainshaft Bearing
23 GEAR, Mainshaft First Speed
24 RING, Mainshaft First Speed Gear Snap, Select Fit
1.96 mm to 2.00 mm (.077 - .079") Thick
2.03 mm to 2.08 mm (.080 - .082") Thick
2.11 mm to 2.16 mm (.083 - .085") Thick
2.18 mm to 2.24 mm (.086 - .088") Thick
2.26 mm to 2.31 mm (.089 - .091") Thick
25 HUB, Clutch
26 CLUTCH, Sliding
27 GEAR, Mainshaft Reverse
28 SEE ITEM 30
29 RING, Mainshaft Reverse Speed Gear Snap
30 WASHER, Mainshaft Reverse and Second Speed Gear Thrust, Select Fit
3.89 mm to 3.94 mm (.153 - .155") Thick
3.96 mm to 4.01 mm (.156 - .158") Thick
4.04 mm to 4.09 mm (.159 - .161") Thick
4.12 mm to 4.17 mm (.162 - .164") Thick
31 GEAR, Mainshaft Second Speed, W/Washer
32 RING, W/Pins, Synchronizer Unit Inner Rear
33 SPRING, Sliding Clutch
34 PIN, Spring
35 CLUTCH, W/Pins and Springs, Sliding
36 RING, W/Pins, Synchronizer Unit Inner Forward
37 RING, Synchronizer Unit Outer Stop
38 GEAR, Mainshaft Third Speed
39 WASHER, Thrust, Select Fit
3.02 mm to 3.09 mm (.119 - .122") Thick
3.12 mm to 3.18 mm (.123 - .125") Thick
3.23 mm to 3.30 mm (.127 - .130") Thick
3.33 mm to 3.41 mm (.131 - .134") Thick
40 RING, External Snap
41 GEAR, Mainshaft Fourth Speed
42 RING, Outer Stop
43 SYNCHRONIZER, Front, Assembly
44 GEAR, Mainshaft Clutch
45 RING, Mainshaft Snap, Select Fit
2.18 mm to 2.24 mm (.086 - .088") Thick
2.26 mm to 2.31 mm (.089 - .091") Thick
2.34 mm to 2.39 mm (.092 - .094") Thick
2.41 mm to 2.46 mm (.095 - .097") Thick
46 NUT, Countershaft Rear Bearing
47 BEARING, Countershaft Rear
48 COUNTERSHAFT
49 KEY, Woodruff
50 GEAR, Countershaft Third Speed
51 GEAR, Countershaft Overdrive
52 SPACER, Countershaft Gear
53 GEAR, Countershaft Drive
54 RING, Countershaft Gear Snap
55 WASHER, Countershaft Front Bearing
56 BEARING, Countershaft Front
57 WASHER, Thrust
58 BEARING, Reverse Idler Gear
59 GEAR, Reverse Idler
60 NOT USED
61 LOCK, Reverse Idler Gear Shaft
62 SHAFT, Reverse Idler Gear
63 GEAR, Main Drive
64 BEARING, Main Drive Gear
65 SEAL, Bearing Retainer Oil
66 GASKET, Bearing Retainer
67 RETAINER, W/Seal, Main Drive Gear Bearing
68 RING, Expansion Plug Snap
69 PLUG, Countershaft Front Bearing Expansion
70 CASE, Transmission
71 PLUG, Pipe, Sq. Hd. 1-1/4, Filler
72 COVER, PTO Opening
73 GASKET, PTO Opening
74 PLUG, Pipe Countersunk 3/4, Drain
75 RETAINER, W/Seal and Bushing, Mainshaft Rear Bearing
76 SEAL, Mainshaft Oil
77 GASKET, Mainshaft Rear Bearing Retainer
78 PLUG
79 GASKET, Shift Bar Housing Mounting

Fig 6. T-495 & T-496 transmission, exploded view

move countershaft front roller bearing from case.

Sub-Assembly Service

MAINSHAFT

Disassembly

1. Remove 4th and 5th speed synchronizer assembly from mainshaft.
2. Remove snap ring and mainshaft clutch gear.
3. Remove 4th speed gear, 3rd speed gear snap ring, thrust washer and 3rd speed gear.
4. Remove 2nd and 3rd speed synchronizer assembly, 1st speed gear thrust washer and 1st speed gear.
5. Remove 1st and reverse gear sliding clutch snap ring and sliding clutch.
6. Remove clutch hub snap ring, clutch hub, reverse gear and thrust washer.
7. Remove 2nd speed snap ring and 2nd speed gear.

Assembly

1. Lubricate all friction surfaces with suitable grease.
2. Assemble 2nd and 3rd speed synchronizer assembly. Ensure wider inner stop ring is positioned to face 2nd gear cavity.
3. Install 3rd speed gear onto mainshaft and select a thrust washer that will allow clearance between snap ring and thrust washer of 3rd speed gear of .002–.006 inch (.05–.15 mm). Thrust washers are available in the following thicknesses: .119–.122 inch (3.02–3.09 mm), .123–.126 inch (3.12–3.18 mm), .127–.130 inch (3.23–3.30 mm), .131–.134 inch (3.33–3.41 mm).
4. Install 3rd speed gear snap ring and 4th speed gear with oil slots facing third speed gear.
5. Install mainshaft clutch gear with oil slots facing 4th speed gear.
6. Install mainshaft snap ring. Select a snap ring of proper thickness that will provide a tight fit and will allow 4th speed gear to rotate on mainshaft after snap ring is installed. Snap rings are available in the following thicknesses: .086–.088 inch (2.18–2.24 mm), .089–.091 inch (2.26–2.31 mm), .092–.094 inch (2.34–2.39 mm) and .095–.097 inch (2.41–2.46 mm).
7. Install 2nd speed gear and selective thrust washer.
8. Check 2nd speed gear end play. With mainshaft supported in suitable vise, insert feeler gauges between 2nd speed gear and forward section of mainshaft. Rotate 2nd speed gear one full turn to locate gear high spot. Install selective thrust washer to obtain .002–.006 inch (.05–.15 mm) end play. Thrust washers are available in the following thicknesses: .153–.155 inch (3.89–3.94 mm), .156–.158 inch (3.96–4.01 mm), .159–.161 inch (4.04–4.09 mm) and .162–.164 inch (4.12–4.17 mm).
9. Temporarily install 1st and reverse clutch hub onto mainshaft and install select snap ring into mainshaft groove. Snap rings are available in the following thicknesses: .077–.079 inch (1.92–2.00 mm), .080–.082 inch (2.03–2.08 mm), .083–.085 inch (2.11–2.16 mm), .086–.088 inch (2.18–2.24 mm) and .089–.091 inch (2.26–2.31 mm). Select snap ring should provide a tight fit in mainshaft groove.

NOTE: Select snap ring and 1st and reverse clutch hub are installed after reverse gear end play is checked.

10. Install selective thrust washer selected in step 8, reverse gear and 1st and reverse clutch hub onto mainshaft. Install selective snap ring selected in step 9 onto mainshaft and re-install feeler gauges as described in step 8. Install feeler gauge on each side of mainshaft and between 1st and reverse clutch hub and reverse gear. Rotate reverse gear one full turn until high spot between gear hubs is located. Select a thrust washer that will provide a clearance of .002–.006 inch (.05–.15 mm). Thrust washers are available in thicknesses described in step 8.
11. With 2nd speed gear and 1st and reverse speed gear feeler gauges left in position, measure 1st speed gear end play as follows: Install first speed gear and thrust washer onto mainshaft. Position suitable adapter onto mainshaft to serve as a load for 1st speed gear thrust washer. Install feeler gauge between 1st speed gear washer and 1st speed gear hub. Clearance should be .002–.006 inch (.05–.15 mm).

NOTE: If minimum .002 inch (.05 mm) clearance cannot be obtained, 1st speed gear should be replaced. Distance between 1st speed gear hub gear teeth and hub should be 1.222 ± .001 inch (31.14 ± .025 mm).

COUNTERSHAFT

Disassembly

NOTE: Remove nylon slinger prior to countershaft disassembly.

1. Remove countershaft snap ring, then remove countershaft drive gear, spacer, 4th speed gear and 3rd speed gear from countershaft using suitable tools.
2. Remove countershaft woodruff keys.

Assembly

NOTE: Install nylon slinger after countershaft is assembled.

1. Install 3rd and 4th speed gears, spacer and countershaft drive gear onto countershaft.
2. Install woodruff keys.
3. Install countershaft snap ring.
4. Install nylon slinger with raised mounting ear facing flat mounting ear.
5. Install nylon slinger onto countershaft and torque bolts to 35–40 inch lbs.

Transmission, Assembly

1. Install reverse idler gear. Ensure that 21 tooth gear is assembled so it is facing rear of case.
2. Install countershaft front roller bearing and countershaft assembly in reverse order of removal.
3. Install mainshaft into case, then install front synchronizer onto mainshaft.
4. Install main drive gear and bearing onto mainshaft.

NOTE: Early model main drive gears require use of pilot bearing spacer. Early model gear may be identified by oil hole positioned 1 1/8 inch (28.6 mm) from rear edge of gear and late model version may be identified by oil hole positioned 15/32 inch (11.9 mm) from rear edge of gear.

5. Install main drive gear bearing retainer, rear bearing and snap ring.
6. Install speedometer drive gear onto rear of mainshaft.
7. Install rear bearing retainer, then select two gears and lock transmission.
8. Install companion flange washer and nut and torque nut to 190–230 ft. lbs. (258–312 Nm).
9. Refer to "Transmission, Disassembly" and perform steps 2 thru 5 in reverse order.
9. Position mainshaft into reverse position, then place shift lever into Reverse position on shift bar housing.
10. Install shift bar housing onto transmission case.
11. Install cover and torque retaining bolts to 30–40 ft. lbs. (41–57 Nm).
12. Fill transmission with 14 pints of suitable lubricant.

T-429, T-430, T-431, T-690 & T-691 TRANSMISSIONS, FIG. 7

Transmission, Disassembly

1. Shift transmission into neutral and remove cover from case.
2. Remove main drive gear bearing assembly.
3. Remove mainshaft rear bearing retainer, gasket and speedometer drive gear.
4. Tap front end of mainshaft rearward to drive rear bearing from bore. Pull bearing off using a suitable puller.
5. Tilt front end of mainshaft assembly upward, then pull assembly forward and out of transmission case.
6. Remove reverse idler gear shaft lock screw and lock plate, then drive reverse idler gear shaft out of bore toward rear of case. Remove gear and washers from case.
7. Remove countershaft rear bearing retainer, gasket and bearing.
8. Tilt front end of countershaft upward and remove from transmission case. Remove spacer washer from front end of countershaft, then push front bearing retainer with needle bearings out of case.

Sub-Assembly Service

MAINSHAFT

Disassembly

1. Remove spacer-washer and first speed gear.
2. Remove retainer ring and first speed and reverse clutch assembly.
3. Remove reverse gear and thrust washer assembly.

NOTE: Thrust washer consists of two halves held together by a retaining clip.

4. Remove second speed gear and second and third speed gear synchronizer.
5. Remove snap ring, fourth and fifth speed synchronizer assembly, fourth speed

gear, retaining ring washer and third speed gear.

Assembly

NOTE: Lubricate all components with transmission oil before assembly. Use sealer on all capscrews.

1. Place mainshaft in soft jawed vise, forward end up.
2. Place third gear on shaft with clutching teeth facing downward, then install spacer washer and lock ring.
3. Place fourth speed gear on shaft with clutching teeth facing upward, then install front synchronizer assembly and lock ring.
4. Remove mainshaft from vise and install rear synchronizer assembly and second speed gear.
5. Place thrust washer retaining pin into recess on mainshaft. Position the two thrust washer halves on mainshaft and install thrust washer retaining ring with large diameter contacting second gear. Install reverse gear.
6. Position first and reverse clutch gear on mainshaft, then install retaining ring.
7. Position sliding clutch on clutch gear and install first gear. Place spacer washer on mainshaft.
8. Measure end play at rear synchronizer. Use two feeler gauges diametrically opposite one another between third speed outer stop ring and third gear. End play should be between .070 to .090 in. If not, install new component parts for assembly of synchronizer.

NOTE: Be sure all synchronizer components are assembled and square and gauges are inserted close to mainshaft and up on shoulder of third gear.

9. Measure end play at first gear. Place mainshaft in suitable vise with rear of shaft facing upward. Install mainshaft rear bearing, speedometer gear and rear flange and rear flange retaining nut. Torque retaining nut to 90 to 130 ft. lbs. Insert two equal size gauges diametrically opposite one another between spacer washer and first gear. A minimum of .003 to .030 in. end clearance must exist. If not, install new first gear and spacer washer.

COUNTERSHAFT

Disassembly

1. Remove snap ring and place assembly in arbor press with blocks supporting drive gear and carefully press shaft out. Repeat this step supporting fourth gear. Remove the one key.

Assembly

1. Place key in position on shaft and press gears on countershaft. Install snap ring on countershaft drive gear.

NOTE: Make sure key does not come out of position as gear is being pressed on.

MAIN DRIVE GEAR

Disassembly

1. Remove snap ring and washer holding pilot roller bearing in place and remove bearings.
2. Remove staking on main drive gear ball bearing nut and remove ball bearing assembly. Ball bearing retainer nut has left hand threads.

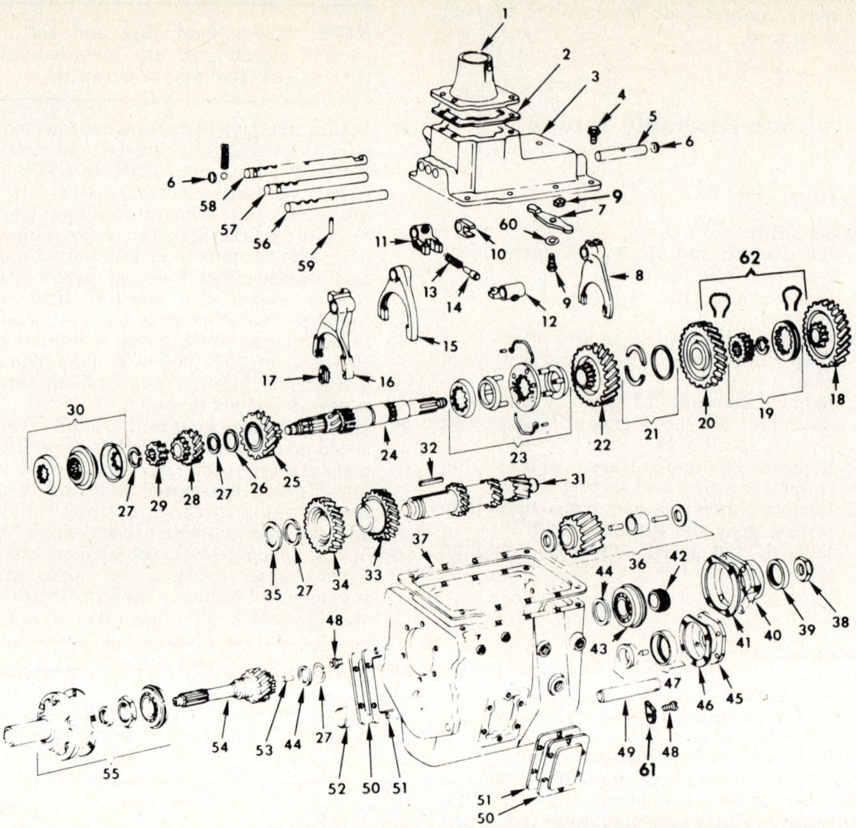

1. HOUSING, Control Lever
2. GASKET
3. COVER
4. BOLT, w/Lockwasher
*5. RAIL, First/Reverse Shift
6. PLUG
*7. PIVOT, First/Reverse Shift Rail
8. FORK, First/Reverse Shift
*9. BOLT, Pivot and Nut
10. BLOCK, Lug
11. BLOCK, Lug
12. BLOCK, Lug
13. SPRING
14. PLUNGER
15. FORK, Second/Third Shift
16. FORK, Fourth/Fifth Shift
17. RETAINER
18. GEAR, First Speed
19. CLUTCH, Assembly, Sliding
20. GEAR, Reverse
21. WASHER, Assembly Thrust
22. GEAR, Second Speed
23. SYNCHRONIZER, Second/Third Speed
24. MAINSHAFT
25. GEAR, Third Speed
26. WASHER
27. RING, Snap Selective Fit
28. GEAR, Fourth Speed
29. CLUTCH, Gear
30. SYNCHRONIZER, Fourth/Fifth Speed
31. COUNTERSHAFT
32. KEY
33. GEAR, Countershaft Fourth
34. GEAR, Countershaft Drive
35. WASHER, Thrust
36. IDLER, Assembly, Reverse
37. GASKET
38. NUT
39. OIL SEAL
40. RETAINER
41. GASKET
42. GEAR, Speedometer Drive
43. BEARING
44. WASHER, Thrust
45. RETAINER
46. GASKET
47. BEARING, Assembly
48. SET, Screw
49. SHAFT, Reverse Idler
50. COVER, PTO
51. GASKET
52. BEARING, Countershaft Front
53. BEARING, Roller (15 Used)
54. GEAR, Main Drive
55. RETAINER, Main Drive Gear
56. RAIL, Fourth/Fifth Speed Shift
57. RAIL, Second/Third Speed Shift
58. RAIL, First/Reverse Speed Shift
59. PIN
*60. WASHER, Pivot
61. PLATE, Lock, Reverse Idler Shaft
62. STOP, First and Reverse Sliding Clutch Gear

* Not with T-690 and T-691 transmissions.

Fig. 7 T-429, T-430, T-431, T-690 & T-691 transmission, exploded view

Assembly

1. Insert greased pilot rollers into bore of drive gear and install snap ring.
2. Press large bearing onto stem with shielded side of bearing next to gear.
3. Install bearing retainer nut and tighten. Nut has left hand threads.
4. Stake nut to lock in place and install snap ring on large bearing.

TRANSMISSION COVER

Disassembly

1. Place shifter forks and rails in neutral. Remove split pins from shift forks and shifter lugs.

NOTE: Split pins may be removed by using an easyout installed in a tap handle.

2. On T-429, T-430 and T-431 units, drive short, first-reverse shift rail toward rear and out of transmission cover, then remove fork.
3. On T-429, T-430 and T-431 units, remove first-reverse shift rail pivot bolt and nut and remove pivot.
4. Drive fourth-fifth and second-third speed shift rails forward and out of transmission cover. Remove first-reverse shift rail in same manner.

5. Remove four interlock balls and pin from bore through width of cover.

Assembly

1. Push first-reverse shift rail into cover bore far enough to permit installation of two lugs, poppet ball, spring, split pins and fork. Move shift rail into neutral position.
2. Apply a small amount of grease to four interlock balls and pin. Place two balls in interlock bore and move both balls toward shift rail to seat ball in neutral notch.
3. Place interlock pin into hole located in second-third speed shift rail. Install second-third speed shift rail, lug and fork. Move rail into neutral position and install remaining two balls into interlock bore.
4. Push fourth-fifth speed shift rail into cover bore and install shift fork. Move rail into neutral position.

NOTE: Steps 5, 6, and 7 do not apply to the T690 and T691 transmissions.

5. Place pivot in proper position in cover and install short first-reverse shift rail, fork and split pin.
6. Reposition pivot to mate in notches in both long and short first-reverse shift rails.
7. Install pivot bolt through pivot and transmission cover. Install nut and washer and tighten nut.
8. Install new plugs.

Transmission, Assembly

1. Install front bearing assembly into countershaft bearing bore, grease thrust washer and place in position. Install countershaft.
2. Install countershaft front bearing journal into front bearing and seat against thrust washer.
3. Install countershaft rear bearing assembly with snap ring toward gear teeth. Install gasket and retainer and tighten screws securely.
4. Check countershaft end clearance. Clearance should be between .008 to .020 in. End clearance can be adjusted by changing countershaft rear bearing retaining gasket.
5. Place bearing in reverse idler gear and position gear in case.
6. Using a brass drift and hammer, drive shaft through case and gear.

NOTE: Position lock strap slot in shaft so that lock strap capscrew can be installed.

7. Install lock strap on shaft and tighten cap screw on shaft.
8. Install mainshaft assembly in case and press rear bearing onto shaft and seat bore in case.
9. Place speedometer drive gear on shaft. Install oil seal in mainshaft rear bearing retainer. Install rear bearing retainer with gasket and torque bolts to 50 to 55 ft. lbs.
10. Install main drive gear and bearing, oil seal in main drive gear retainer and gasket and retainer. Torque main drive gear retainer bolts to 25 to 35 ft. lbs. Install U-joint flange and tighten flange nut to 90 to 130 ft. lbs.
11. Measure end play at front synchronizer. Use two feeler gauges diametrically opposite one another between fourth speed gear and outer stop ring. End play float should not exceed .095 in. Install shims between fourth speed gear and outer stop ring to adjust end play float.
12. Shift transmission and cover into neutral and lower into position on case.
13. Install both aligning screws and lockwasher thumb tight. Install remaining screws. Tighten all screws until cover is properly seated.
14. Install parking brake drum and flange assembly. Install brake band assembly.
15. Fill transmission to proper level.

TRANSMISSION IDENTIFICATION

I. H. C. Code①	I. H. C. Model	Mfr's Name	Mfr's Model	No. Speeds	Oil Pints
13001	T-1	I.H.C	T-1	3	2½
13002	T-2	I.H.C	T-2	3	3½
13004	T-4	Warner	—	3	2½
13005	T-5	Warner	T-87②	3	6
13007	T-7	Warner	T-87②	3	6
13008	T-8	Warner	—	3	3½
13010	T-10	Warner	T-9②	4	5½
13011	T-11	Warner	T-9②	4	5
13013	T-13	Warner	—	3	2¾
13014	T-14	Warner	—	3	2¾
13015	T-15	Warner	T-98②	4	7
13016	T-16	Warner	T-98②	4	7
13017	T-17, 18	New Process	435②	4	7
13018	T-18	Fuller	—	4	7
13024	T-24	—	—	3	2½
13030	T-30	Fuller	5-A-33②	5	12
13031	T-31	Fuller	5-A-330②	5	12
13032	T-32	Fuller	—	5	12
13033	T-33	Fuller	—	5	12
13034	T-34	Fuller	—	5	12
13035	T-35	Fuller	—	5	12
13036	T-36	Fuller	—	5	12
13039	T-39	Fuller	WG-11②	3	19
13040	T-40	Spicer	4652②	5	13
13041	T-41	Spicer	4753②	5	13
13044	T-44	Fuller	WG-11②	4	7
13045	T-45	Fuller	—	4	7
13049	T-49	Fuller	WG-11②	3	19
13050	T-50	Fuller	5-A-430②	5	19
13051	T-51	Fuller	5-A-43②	5	19
13053	T-53	Fuller	—	5	19
13054	T-54	Fuller	—	5	19
13057	T-57	Fuller	5W-43②	5	19
13059	T-59	Fuller	5-A-620②	5	24
13060	T-60	Fuller	5-A-620②	5	24
13061	T-61	Fuller	5-A-62②	5	24
13062	T-62	Fuller	5-C-65②	5	24
13063	T-63	Fuller	5-C-650②	5	24
13064	T-64	Fuller	5-C-650②	5	24
13068	T-68	Spicer	5652	—	—
13070	T-70	Fuller	5-C-65②	5	24
13071	T-71	Fuller	5-C-65②	5	24
13072	T-72	Fuller	5-C-65②	5	24
13073	T-73	Fuller	5-C-65②	5	26
13074	T-74	Fuller③	R-35②	7	16
13075	T-75	Fuller③	R-46②	8	17
13076	T-76	Fuller③	R-46S②	8	17
13078	T-78	Fuller	5-C-65②	5	26
13080	T-80	Fuller	10CB-650②	10	31
13081	T-81	Fuller	10CB-65②	10	31
13082	T-82	Spicer	8125②	12	28
13083	T-83	Spicer	3152	5	28
13088	T-88	Spicer	5752C	—	—
13089	T-89	Spicer	8125②	12	28
13090	T-90	Fuller	10CB-650②	10	—
13091	T-91	Fuller	10CB-650②	10	—
13104	T-104	Fuller	5-H-74②	5	24
13105	T-105	Fuller	5-HA-74②	5	24
13106	T-106	Fuller	5-H-75②	5	24
13107	T-107	Fuller	5-HA-75②	5	24
13109	T-109	Allison	SA-1126J	6	—
13110	T-110	Fuller	MT-41②	6	19
13111	T-111	Fuller	MT-42②	6	19
13112	T-112	Fuller	MT-42②	6	19

Continued

TRANSMISSION IDENTIFICATION—Continued

I. H. C. Code①	I. H. C. Model	Mfr's Name	Mfr's Model	No. Speeds	Oil Pints
13121	T-121	Fuller	6352②	6	17
13122	T-122	Fuller	MT-42	6	19
13126	T-126	Fuller	5C-720	5	22
13128	T-128	Fuller	RT-915②	15	28
13129	T-129	Fuller	RT-910②	10	25
13130	T-130	Fuller	—	10	39
13132	T-132	Fuller	—	5	24
13135	T-135	Fuller	—	5	24
13136	T-136	Fuller	—	5	24
13137	T-137	Fuller	—	5	24
13194	T-194	Fuller	RT-12509	—	—
13197	T-197	Fuller	TO-955 All	—	—
13204	T-204	Spicer	6352B	—	—
13208	T-208	Fuller	5C-720②	5	24
13209	T-209	Fuller	5C-720②	5	24
13210	T-210	Fuller	5C-720②	5	24
13211	T-211	Fuller	5C-72②	5	24
13212	T-212	Fuller	5C-72②	5	24
13214	T-214	Fuller	5W-74②	5	24
13215	T-215	Fuller	5CA-720②	5	24
13216	T-216	Fuller	5CA-720②	5	24
13217	T-217	Fuller	5CA-720②	5	24
13218	T-218	Fuller	5CA-720②	5	24
13219	T-219	Fuller	5WA-74②	5	24
13220	T-220	Fuller	5A-1120②	5	29
13223	T-223	Spicer	6352②	5	17
13227	T-227	Fuller③	R-660②	10	—
13228	T-228	Fuller③	R-63②	10	30
13230	T-230	Fuller③	R-96C②	10	39
13231	T-231	Fuller③	R-960C②	10	39
13233	T-233	Fuller	R-96A②	10	39
13234	T-234	Fuller	R-960A②	10	39
13252	T-252	Fuller	—	5	24
13298	T-298	Spicer	6352G	—	—
13316	T-316	Fuller	TO-955 All	—	—
13320	T-320	Spicer	6853C	—	—
13331	T-331	IH	T-331	3	3
13331	T-331	Warner	T-15D	—	—
13332	T-332	IH	T-332	3	3
13332	T-332	Warner	T-15D	—	—
13345	T-345	Fuller	RTO-910②	10	25
13346	T-346	Fuller	RTO-915②	15	28
13347	T-347	Fuller	T-905A	—	—
13348	T-348	Fuller	T-905B	—	—
13362	T-362	Fuller	TO-905C	—	—
13365	T-365	Fuller③	R-46②	8	—
13366	T-366	Fuller	—	10	39
13367	T-367	Fuller	—	10	39
13401	T-401	Fuller	5CW-62②	5	24
13402	T-402	Fuller	5CW-62②	5	24
13403	T-403	Fuller	5CW-62②	5	24
13404	T-404	Fuller	5CW-620②	5	24
13407	T-407	Chrysler	Loadflite	3	19
13411	T-411	Fuller	5CW-65②	5	26
13412	T-412	Fuller	5CW-65②	5	24
13413	T-413	Fuller	5CW-65②	5	26
13414	T-414	Fuller	5CW-650②	5	26
13418	T-418	—	—	4	7
13419	T-419	—	—	4	7
13422	T-422	Fuller	RT-613	—	—
13424	T-424	Fuller	RTO-613	—	—
13425	T-425	New Process	4590	4	8
13427	T-427	Warner	T-19	4	—
13428	T-428	Warner	T-19A	4	—
13429	T-429	IH	T-429	5	13½
13430	T-430	IH	T-430	5	13½
13431	T-431	IH	T-431	5	13½
13445	T-445	—	—	4	7
13448	T-448	Fuller	RT-610	—	—
13451	AT-451	Allison	SA-1241-A	4	—
13454	AT-454	Allison	SA-1241-A	4	—
13456	T-456	Spicer	1007-3A	—	—
13457	T-457	Spicer	1211-3A	—	—
13459	T-459	Fuller	RTO-9513	—	—
13460	T-460	Fuller	RTOO-9513	—	—
13462	T-462	Fuller	RT-9513	—	—
13463	T-463	Fuller	T-955 All	—	—
13464	AT-464	Allison	SA-1317-A	4	—
13465	AT-465	Allison	SA-1317-A	5	—
13466	T-466	Fuller	TO-955 All	—	—
13468	T-468	Fuller	RTO-12513	—	—
13470	T-470	Fuller	RTO-958 LL	—	—
13471	T-471	Fuller	RT-9509A	—	—
13474	AT-474	Allison	SA-1270-B	4	—
13475	AT-475	Allison	SA-1270-B	5	—
13478	T-478	Spicer	RP-8516-3A	—	—
13480	T-480	Spicer	1010-2A	—	—
13482	T-482	Spicer	1252A	—	—
13486	AT-486	Allison	SA-1270-B	5	—
13487	T-487	Fuller	RT-12510	—	—
13488	AT-488	Allison	SA-1317-A	4	—
13489	AT-489	Allison	SA-1317-A	4	—
13491	T-491	Spicer	1252A	—	—
13493	T-493	Fuller	12515	—	—
13494	T-494	IH	T-494	5	12
13495	T-495	IH	T-495	5	12
13496	T-496	IH	T-496	5	12
13501	AT-501	Spicer	5831②	3 Aux.	10
13510	AT-510	Fuller	2-A-45②	2 Aux.	10
13522	AT-522	Spicer	7231②	3 Aux.	8
13523	AT-523	Spicer	7231②	3 Aux.	8
13524	AT-524	Spicer	6041②	4 Aux.	8
13527	AT-527	Fuller	3-H-65②	3 Aux.	6½
13528	AT-528	Fuller	3-H-65②	3 Aux.	6½
13529	AT-529	Fuller	3-A-65②	3 Aux.	6½
13530	AT-530	Fuller	3-A-65②	3 Aux.	6½
13536	AT-536	Spicer	7041	—	—
13539	AT-539	Spicer	8031②	3 Aux.	12
13540	AT-540	Spicer	8031②	3 Aux.	12
13546	AT-546	Spicer	8341②	4 Aux.	12
13548	AT-548	IH	AT-548	4	12

Continued

TRANSMISSION IDENTIFICATION—Continued

I. H. C. Code①	I. H. C. Model	Mfr's Name	Mfr's Model	No. Speeds	Oil Pints
13551	AT-551	Spicer	8345②	5 Aux.	12
13552	AT-552	Spicer	R-8341	—	—
13601	AT-601	Fuller	2A-92	2 Aux.	12
13610	AT-610	—	3A-92	3 Aux.	12
13610	AT-610	—	3B-92	3 Aux.	12
13620	AT-620	Fuller	3T-92	3 Aux.	24
13625	AT-625	Fuller	4M-112	4 Aux.	21
13630	AT-630	Spicer	8035②	3 Aux.	12
13690	T-690	IH	T-690	5	13½
13691	T-691	IH	T-691	5	13½

①—Code number is listed on Vehicle Inspection Card attached to left sun visor.
②—See the Stock Transmissions Chapter for repairs on this unit.
③—Roadranger transmission.

Rear Axle Section

NOTE: For Eaton and Timken (Rockwell) driving axles used on International trucks, see the "name" chapter for service procedures.

I. H. C. Code①	I. H. C. Model	Mfr's Name②	Axle Type③	Oil Pints
14001	RA-1	Spicer	S. R.	3
14002	RA-2	Spicer	S. R.	2
14003	RA-3	Spicer	S. R.	3
14004	RA-4	Spicer	S. R.	2.4
14005	RA-5	Timken	S. R.	6
14006	RA-6	Timken	S. R.	6
14009	RA-9	Spicer	S. R.	3
14010	RA-10	Timken	S. R.	6
14011	RA-11	Timken	S. R.	6
14012	RA-12	Spicer	S. R.	2.4
14013	RA-13	Spicer	S. R.	2.4
14014	RA-14	Spicer	S. R.	2.4
14015	RA-15	Timken	S. R.	6½
14016	RA-16	Spicer	S. R.	5½
14017	RA-17	Spicer	S. R.	5½
14018	RA-18	Spicer	S. R.	3
14019	RA-19	Spicer	S. R.	3
14020	RA-20	Timken	S. R.	6½
14021	RA-21	Spicer	S. R.	2
14023	RA-23	Spicer	S. R.	3
14024	RA-24	Spicer	S. R.	3
14025	RA-25	Timken	S. R.	9½
14026	RA-26	Timken	S. R.	8
14028	RA-28	Spicer	S. R.	3
14030	RA-30	Timken	S. R.	9½
14031	RA-31	Timken	S. R.	8
14032	RA-32	Timken	S. R.	8
14033	RA-33	Timken	S. R.	8
14035	RA-35	Timken	S. R.	12
14039	RA-39	Timken	S. R.	12
14043	RA-43	Timken	S. R.	24
14044	RA-44	Timken	S. R.	24
14045	RA-45	Timken	S. R.	22
14047	RA-47	Timken	S. R.	31½
14048	RA-48	Timken	S. R.	31½
14050	RA-50	Timken	S. R.	21
14051	RA-51	Spicer	S. R.	2½
14053	RA-53	Spicer	S. R.	6
14054	RA-54	Spicer	S. R.	6
14055	RA-55	Timken	S. R.	24
14057	RA-57	Timken	S. R.	31
14060	RA-60	Timken	S. R.	23
14063	RA-63	Spicer	S. R.	5¾
14065	RA-65	Timken	S. R.	31
14067	RA-67	Timken	S. R.	28
14070	RA-70	Timken	S. R.	30
14071	RA-71	Timken	S. R.	28
14072	RA-72	Timken	S. R.	—
14073	RA-73	Timken	S. R.	34
14074	RA-74	Timken	S. R.	—
14075	RA-75	Eaton	S. R.	34
14076	RA-76	Timken	S. R.	—
14078	RA-78	Timken	S. R.	34
14083	RA-83	Spicer	S. R.	6
14084	RA-84	Spicer	S. R.	6
14120	RA-120	Eaton	2 Sp.	13
14125	RA-125	Eaton	2 Sp.	11
14126	RA-126	Eaton	2 Sp.	13
14127	RA-127	Eaton	2 Sp.	13
14128	RA-128	Eaton	2 Sp.	13
14130	RA-130	Eaton	2 Sp.	11
14131	RA-131	Eaton	2 Sp.	11
14135	RA-135	Eaton	2 Sp.	16
14136	RA-136	Eaton	2 Sp.	20
14138	RA-138	Eaton	2 Sp.	20
14140	RA-140	Timken	2 Sp.	26④
14143	RA-143	Eaton	2 Sp.	16
14145	RA-145	Eaton	2 Sp.	18
14146	RA-146	Eaton	2 Sp.	22
14147	RA-147	Eaton	2 Sp.	22
14149	RA-149	Eaton	2 Sp.	20
14150	RA-150	Eaton	2 Sp.	22
14152	RA-152	Eaton	2 Sp.	28½
14155	RA-155	Eaton	2 Sp.	24
14157	RA-157	Eaton	2 Sp.	29½
14158	RA-158	Eaton	2 Sp.	34
14160	RA-160	Timken	2 Sp.	29
14162	RA-162	Timken	2 Sp.	28

Continued

REAR AXLE IDENTIFICATION—Continued

I. H. C. Code①	I. H. C. Model	Mfr's Name②	Axle Type③	Oil Pints
14163	RA-163	Eaton	2 Sp.	22④
14165	RA-165	Timken	2 Sp.	29④
14167	RA-167	Timken	2 Sp.	32½
14170	RA-170	Timken	2 Sp.	34④
14172	RA-172	Timken	2 Sp.	31
14175	RA-175	Timken	2 Sp.	37④
14178	RA-178	Timken	2 Sp.	37④
14180	RA-180	Eaton	2 Sp.	32
14181	RA-181	Eaton	2 Sp.	32
14184	RA-184	Eaton	2 Sp.	13
14186	RA-186	Eaton	2 Sp.	22
14187	RA-187	Eaton	2 Sp.	13
14189	RA-189	Eaton	2 Sp.	20
14190	RA-190	Timken	2 Sp.	35
14191	RA-191	Timken	2 Sp.	60
14192	RA-192	Eaton	2 Sp.	28
14196	RA-196	Eaton	2 Sp.	29½
14197	RA-197	Eaton	2 Sp.	29½
14199	RA-199	Eaton	2 Sp.	29½
14245	RA-245	Eaton	D. R.	22
14246	RA-246	Eaton	D. R.	22
14250	RA-250	Eaton	D. R.	22
14252	RA-252	Eaton	D. R.	28.5
14255	RA-255	Eaton	D. R.	35
14270	RA-270	Timken	D. R.	36④
14272	RA-272	Timken	D. R.	31
14273	RA-273	Timken	D. R.	34
14275	RA-275	Timken	D. R.	38④
14276	RA-276	Timken	D. R.	38
14277	RA-277	Timken	D. R.	38
14286	RA-286	Eaton	D. R.	22
14290	RA-290	Timken	D. R.	34
14292	RA-292	Eaton	D. R.	31
14301	RA-301	Eaton	T. S. R.	11 Each⑤
14302	RA-302	Eaton	S. R.	18
14303	RA-303	Eaton	2 Sp.	18
14304	RA-304	Eaton	D. R.	㉓
14305	RA-305	Eaton	T. S. R.	14 Each⑤
14306	RA-306	Eaton	3 Sp.	24
14310	RA-310	Eaton	T. S. R.	⑥
14312	RA-312	Eaton	T. S. R.	⑭
14313	RA-313	Eaton	T. 2 Sp.	⑭
14314	RA-314	Eaton	T. D. R.	⑭
14315	RA-315	Timken	T. D. R.	⑮
14316	RA-316	Timken	T. D. R.	⑮
14317	RA-317	Timken	T. D. R.	28 Each
14318	RA-318	Timken	T. S. R.	⑧
14319	RA-319	Timken	T. S. R.	⑩
14320	RA-320	Eaton	T. S. R.	⑨
14323	RA-323	Eaton	T. S. R.	⑭
14324	RA-324	Eaton	T. 2 Sp.	⑭

I. H. C. Code①	I. H. C. Model	Mfr's Name②	Axle Type③	Oil Pints
14325	RA-325	Eaton	T. D. R.	⑭
14328	RA-328	Timken	T. S. R.	⑪
14329	RA-329	Timken	T. D. R.	—
14330	RA-330	Timken	T. D. R.	22 Each⑦
14331	RA-331	Timken	T. D. R.	22 Each
14332	RA-332	Timken	T. D. R.	⑬
14333	RA-333	Timken	T. S. R.	⑫
14334	RA-334	Timken	T. S. R.	⑫
14335	RA-335	Timken	T. D. R.	22 Each
14336	RA-336	Timken	T. D. R.	22 Each⑦
14341	RA-341	Timken	T. S. R.	⑯
14346	RA-346	Timken	T. S. R.	⑪
14348	RA-348	Timken	T. D.	—
14349	RA-349	Timken	T. D.	—
14351	RA-351	Timken	T. S. R.	⑰
14355	RA-355	Timken	T. S. R.	⑰
14356	RA-356	Eaton	T. D. R.	⑭
14360	RA-360	Eaton	T. 3 Sp.	⑭
14361	RA-361	Eaton	T. 3 Sp.	⑭
14364	RA-364	Rockwell	T. D. R.	⑲
14367	RA-367	Rockwell	T. D. R.	20
14368	RA-368	Rockwell	T. D. R.	⑳
14370	RA-370	Timken	T. D. R.	28 Each
14371	RA-371	Timken	T. D. R.	28 Each⑦
14372	RA-372	Timken	T. D. R.	28 Each
14373	RA-373	Timken	T. D. R.	28 Each⑦
14375	RA-375	Eaton	T. D. R.	22 Each⑱
14383	RA-383	Timken	T. D.	—
14386	RA-386	Timken	T.S.R.	—
14387	RA-387	Timken	T.S.R.	—
14388	RA-388	Rockwell	T.S.R.	㉕
14390	RA-390	Eaton	T.S.R.	㉑
14391	RA-391	Eaton	2 Sp.	㉑
14392	RA-392	Eaton	D.R.	㉑
14393	RA-393	Eaton	3 Sp.	㉑
14395	RA-395	Eaton	D.R.	24
14399	RA-399	Eaton	T.D.	㉔
14446	RA-446	Eaton	S.R.	㉓
14448	RA-448	Eaton	2 Sp.	㉒
14449	RA-449	Eaton	D.R.	㉒
14459	RA-459	Rockwell	T.S.R.	—
14460	RA-460	Rockwell	T.D.R.	⑳
14461	RA-461	Rockwell	T.S.R.	—
14462	RA-462	Rockwell	T.D.R.	⑲
14487	RA-487	Eaton	S.R.	㉓
14488	RA-488	Eaton	2 Sp.	㉓
14489	RA-489	Eaton	D.R.	㉓
14490	RA-490	Eaton	S.R.	㉑
14491	RA-491	Eaton	2 Sp.	㉑
14492	RA-492	Eaton	D.R.	㉑
14495	RA-495	Eaton	T.D.R.	24 Each

①—Code number is listed on Vehicle Specification Card attached to left sun visor.

②—See Rear Axles Chapter for repairs on Timken and Eaton Axles.

③—S. R.: Single reduction. D. R.: Double reduction. 2 Sp.: Two speed. T. S. R.: Tandem drive single reduction. T. D. R.: Tandem drive double reduction. T. 2 Sp.: Tandem drive 2 speed. T. 3 Sp.: Tandem drive 3 speed.

④—Add 1 pint to pinion cage when new or reconditioned drive unit is installed.

⑤—Power divider 6½ pints.

⑥—Forward axle 28, rearward axle 33, power divider 3.

⑦—Interaxle differential 2 pints.

⑧—Forward axle 27, rearward 24, interaxle differential 2.

⑨—Forward axle 20, rearward 21, power divider 2.

⑩—Forward axle 31, rearward 32, interaxle differential 2.

⑪—Forward axle 31½, rearward 32, interaxle differential 2.

Continued

REAR AXLE IDENTIFICATION NOTES—Continued

⑫—Forward axle 34, rearward 31, interaxle differential 2.
⑬—Forward axle 34, rearward 36, interaxle differential 2.
⑭—Forward axle 24, rearward 22.
⑮—Forward axle 26, rearward 28, interaxle differential 2.
⑯—Forward axle 25, rearward 26, interaxle differential 1.
⑰—Forward axle 23, rearward 24, interaxle differential 1.
⑱—Power divider 6 pints.
⑲—Forward axle 46, rearward 44; interaxle differential 2.
⑳—Forward axle 30, rearward 28, interaxle differential 2.
㉑—Forward axle 27, rearward 20.
㉒—Forward axle 43, rearward 34, interaxle differential 1.
㉓—Forward axle 24; rearward 18.
㉔—Forward axle 32, rearward 28, power drive 2.
㉕—Forward axle 33, rearward 32.

SPICER AXLE

Integral Housing Type

In as much as the axle tubes are pressed into the differential carrier to form a one-piece housing, the rear axle assembly must be removed from the chassis when it becomes necessary to overhaul the unit, Fig. 1.

AXLE SHAFT, REPLACE

1. Raise vehicle and support on jack stands. Remove wheels.
2. Remove brake drum with a suitable puller.
3. Disconnect brake lines at wheel cylinders.
4. Remove bolts securing seal retainer, bearing retainer, backing plate and shims. Retain shim for assembly.
5. Using a suitable puller, remove axle shaft from axle housing.

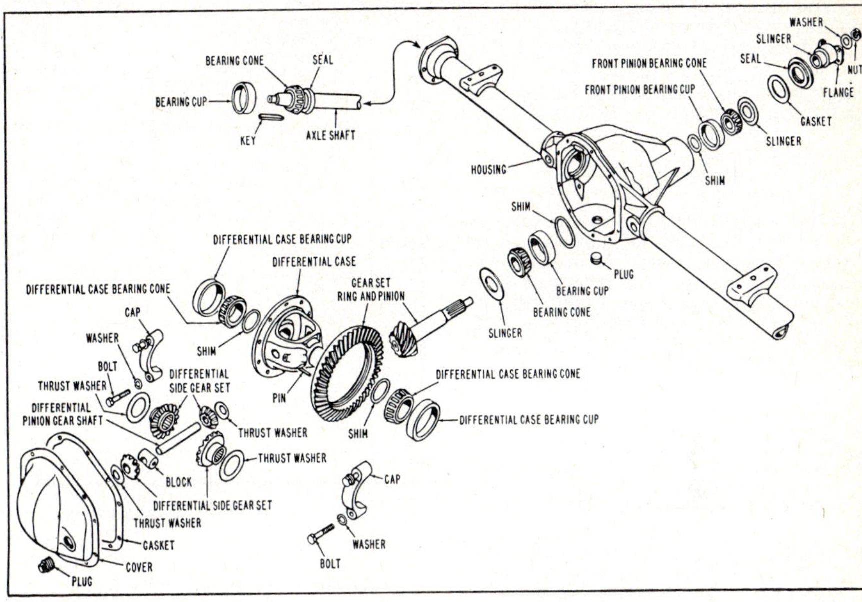

Fig. 1 Disassembled view of Spicer rear axle

AXLE ASSEMBLY, REPLACE

1. Raise vehicle and support on jack stands. Remove wheels.
2. Disconnect brake lines and cables from housing and backing plate, then shock absorbers from lower mountings and propeller shaft from carrier.
3. Support axle with a suitable jack and remove leaf spring "U" bolts.
4. Lower axle assembly from vehicle.

Front Wheel Drive

SCOUT FRONT DRIVING AXLE

The differential carrier assembly is the same as the rear driving axle except that the front unit is full floating. The axle shafts can be removed without dismantling the steering knuckles.

Axle Shaft & Universal Joint

Removal

1. Referring to Figs. 1 and 2, remove hub cap and wheel.
2. Remove grease cap and snap ring.
3. Remove drive flange and gasket.
4. Remove brake drum.
5. Remove wheel bearing adjusting nut and washer.
6. Remove wheel hub with wheel bearing, taking care not to damage oil seal.
7. Disconnect brake hose at backing plate.
8. Remove six backing plate and wheel spindle bolts and washers.
9. Remove wheel bearing spindle with bushing.
10. Pull axle shaft and universal joint from axle housing.

Installation

1. Clean dirt from all parts.
2. Enter shaft and joint assembly into housing. Enter spline end of axle shaft into differential and push into place.
3. Install wheel bearing spindle with bushing.
4. Install backing plate and brake hose.
5. Pack wheel bearings with lubricant and install.
6. Assemble driving flange, wheel hub and drum.
7. Install wheel bearing lockwasher and adjusting nut. Tighten adjusting nut until there is a slight drag on bearings when hub is turned, then back off 1/6th of a turn.
8. Install lockwasher and lock nut; tighten nut and bend lip of lockwasher over lock nut.
9. Install snap ring, grease cap, wheel and hub cap.

Steering Knuckle Bearings

Replacement of steering knuckle bearings or bearing cups on the kingpin requires removal of the tie rod, brake drum, hub, wheel bearing, axle shaft, spindle and steering knuckle. The following procedure should be followed when removing the steering knuckle:

1. Referring to Figs. 1 and 2, remove studs and tapered capscrews holding kingpin bearing lower cap.
2. Remove studs and tapered capscrews holding kingpin bearing upper cap and shims.
3. Remove eight oil seal retaining capscrews.
4. Remove upper and lower steering knuckle bearing cone and cup.
5. Remove steering knuckle and seals.

Installation: When reinstalling the steering knuckle, sufficient shims must be installed under the upper bearing cap so the proper tension may be maintained on the knuckle bearing. No shims are used on the lower bearing cap.

Shims are available in thicknesses of .005", .010", and .030". Install one each of the three thicknesses over the studs on top of the knuckle. Then proceed as follows:

1. Install wheel bearing cups and cones, steering arm, bearing cap, bolts, lockwashers and nuts.
2. Check tension of bearings by hooking a spring scale around the steering arm ball.

3. Pull the scale at right angle to the steering arm. Take the reading when the arm is moving.
4. Remove or add shims until the load is approximately 25–35 pounds without the oil seal assembly in position.

Steering Knuckle Oil Seal

To remove the seal, remove the eight capscrews holding the steering knuckle oil seal in place and remove the seal.

Before installing the new seal, examine the spherical surface of the axle for scores or scratches that might damage the new seal. Roughness of any kind should be smoothed with fine emery cloth.

Lubrication

The front axle universal joints are lubricated by a constant supply of lubricant in the steering knuckle housings. A level plug for checking the lubricant level is located at the rear of each housing. Every 1,000 miles the level should be checked. The recommended capacity for each steering knuckle housing is 38 oz. Use viscous chassis lubricant.

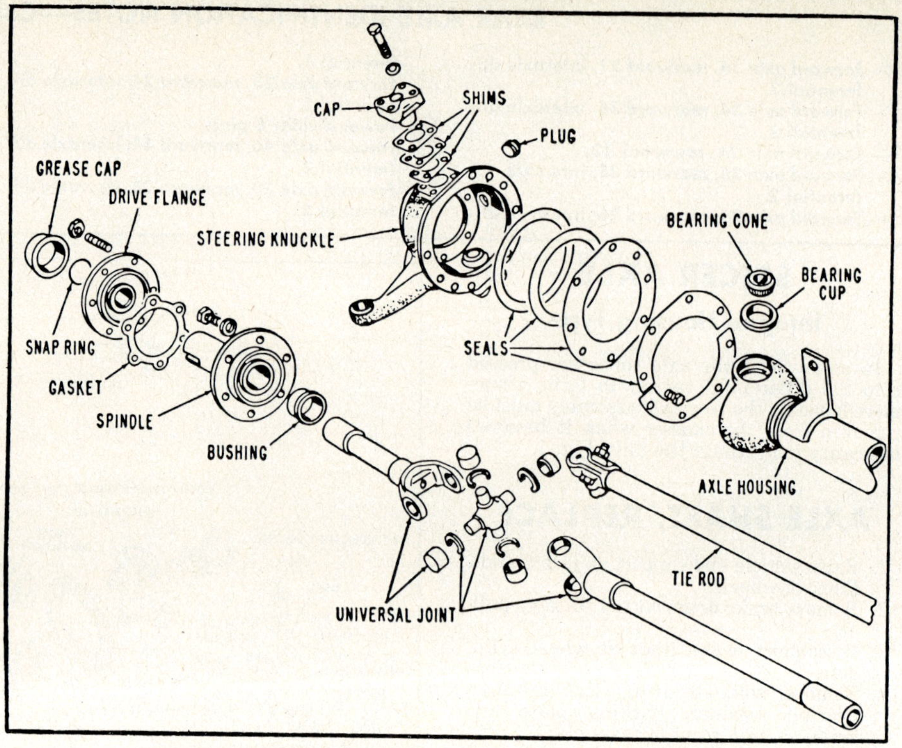

Fig. 1 Exploded view of 4 × 4 front axle shaft and universal joint

MODELS FA-15, 16, 18, 19, 25, 52, 54

The axles covered in this section are front driving units of the Spicer, type incorporating hypoid gears and spherical steering knuckles. The axle shaft assemblies are full floating and may be removed without disassembling the steering knuckles.

Two different means of transmitting driving torque from the axle shaft to the wheel are employed in these axles. One type uses a drive flange having a splined ID to mate with the axle shaft splines. The flange is bolted to the wheel hub, thus transmitting driving torque from the axle shaft to the wheel hub, Fig. 2.

The second type has a drive gear splined both internally and externally, Fig. 3. The internal splines mate with the axle shaft splines and the external splines index with the splines on the inside of the wheel hub. Driving torque is carried from the axle shaft to the wheel hub through the splined drive gear.

NOTE: The different designs outlined above necessitate two different designs in front wheel locking hubs. For locking hub service, see the *Front Wheel Locking Hub Chapter.*

Axle Shaft & U-Joint

Axles Having Drive Flange, Fig. 2

1. Remove wheel and grease cap and snap ring from end of axle shaft.
2. Remove drive flange. If equipped with locking hubs, take out capscrews and remove clutch body. Lift off clutch body, holding it erect so as not to let drive pins fall out of body.
3. Remove hub body. Loosen setscrew and unscrew drag shoe from spindle.
4. Remove brake drum.
5. Remove wheel hub with wheel bearing.
6. Remove backing plate and wheel spindle bolts. Support backing plate to prevent damage to brake hose if hose is not disconnected.
7. Remove wheel spindle with bushing.
8. Pull out axle shaft and U-joint assembly.
9. Reverse procedure to install.

Axles Having Drive Gear, Fig. 3

1. Remove wheel. Lightly tap alternately around edge of hub cap until hub cap is removed.
2. If axle is equipped with locking hubs, remove setscrews securing hub clutch assembly to wheel hub. Drive pins may fall out of hub clutch when separated from wheel hub. Be certain to replace them upon reassembly.
3. Remove retaining ring from wheel hub if equipped with locking hubs.
4. Remove snap ring from axle shaft.

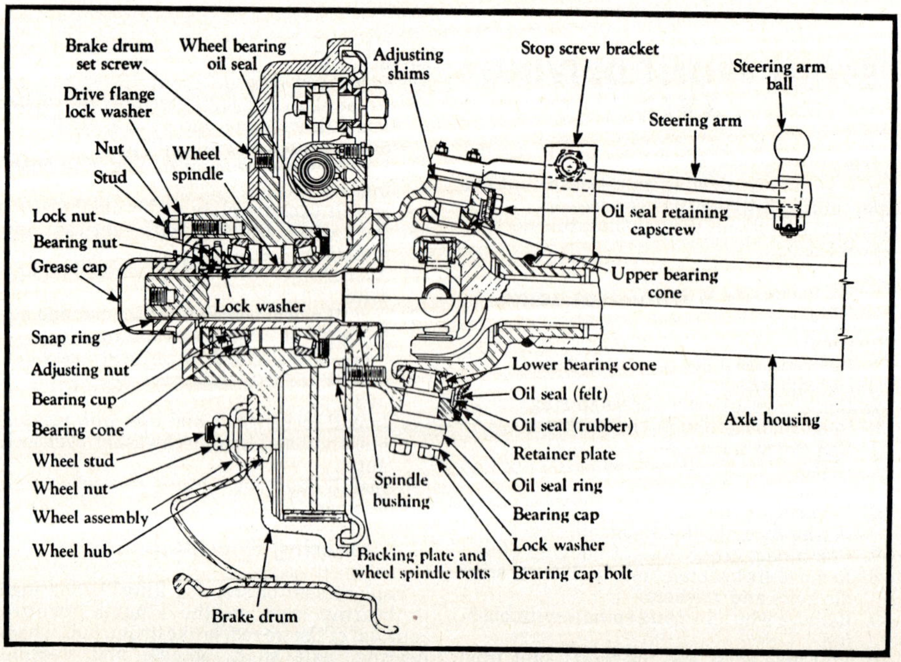

Fig. 2 Front wheel end assembly having drive flange

5. Pull drive gear out of wheel hub. If difficulty is encountered in removing drive gear, use an offset screwdriver and insert it into groove in drive gear and withdraw gear. If necessary, move wheel backward and forward to aid removal of gear.
6. Remove retaining ring and locking hub body (if equipped).
7. Remove wheel bearing outer nut and slide lock ring off axle shaft. Remove wheel bearing inner nut.
8. Pull drive gear spacer out of wheel hub.
9. Remove brake drum and slide wheel hub off spindle.
10. Remove grease guard and gasket from backing plate.
11. Unfasten spindle and backing plate from steering knuckle. Pull spindle with bushing off axle shaft.
12. Pull axle shaft and U-joint out of axle housing.
13. Reverse procedure to install.

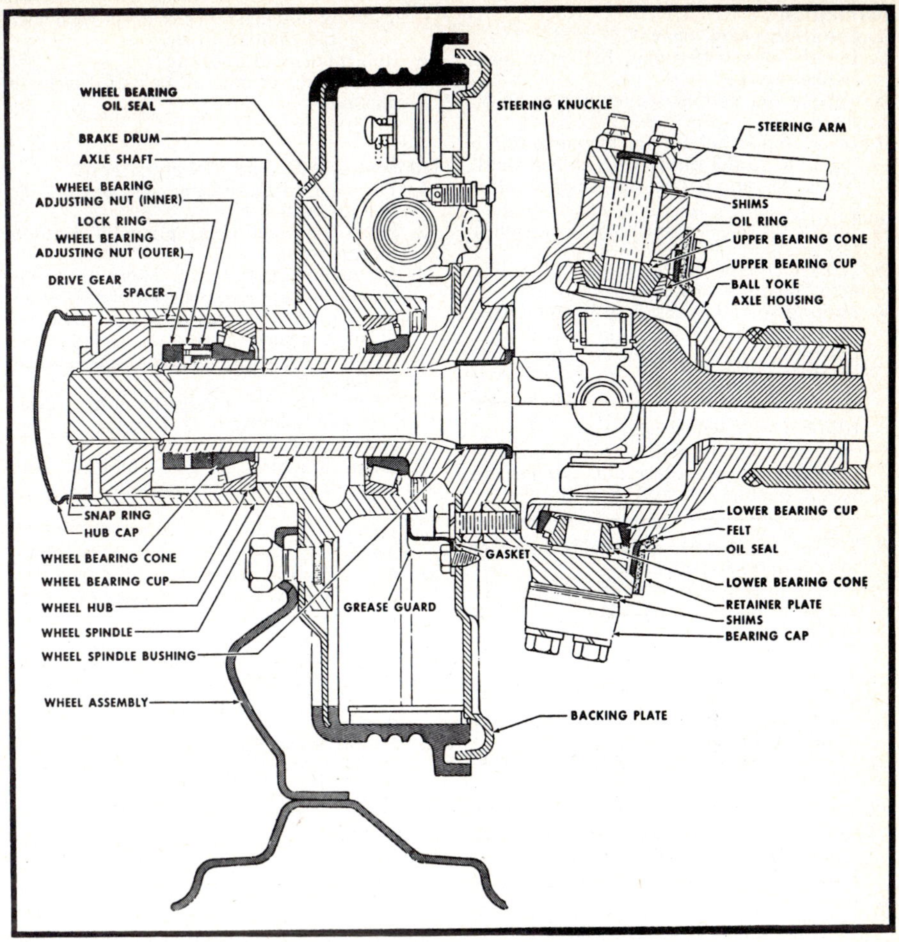

Fig. 3 Front wheel end assembly having disc gear

MODELS FA-3, 13, 43, 44, 45, 46, 62, & 63

Axle Shaft & Steering Knuckle Figs. 4 & 5

Removal

1. Loosen but do not remove wheel stud nuts.
2. Raise vehicle and back off brake shoe adjusting screw.
3. Remove wheel stud nuts and wheel.
4. Remove hub and drum flange nuts and lock washers.
5. Straighten washer to permit locking nut removal. Remove nut and washer.
6. Remove adjusting nut and bearing thrust washer.
7. Remove outer wheel bearing.
8. Remove hub and drum from wheel spindle.
9. Check condition of wheel seal, bearings, and races.
10. Remove nuts from backing plate bolts and remove plate.
11. Tap spindle lightly with soft faced hammer to separate.
12. Check bronze spacer with spindle and axle shaft joint assembly for wear.
13. Remove cotter pin and loosen tie rod nut. Tap on nut with raw-hide hammer to break stud loose from steering arm.
14. Remove nut and disconnect tie-rod.
15. Remove cotter pin from upper ball socket and then nut.
16. Remove nut from lower socket and discard.
17. Install adapter plate SE-2385 to the backing plate screws and tighten nuts finger tight. Locate the puller and plate. Tighten the nuts. Remove ball socket snap ring.
18. In event that top ball socket remains in yoke, remove by hitting on top of stud with rawhide hammer. Discard socket.
19. Remove bottom ball socket using tools SE-1336-16 and SE-1336-18. Tighten adjusting screw until tension is felt then rap top of adjusting screw.
20. Remove threaded sleeve.

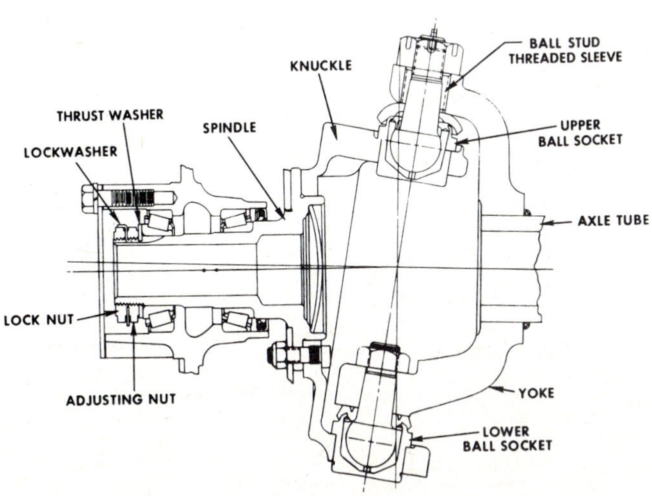

Fig. 4 FA-3 Front axle 4 × 2 wiht 40° steer

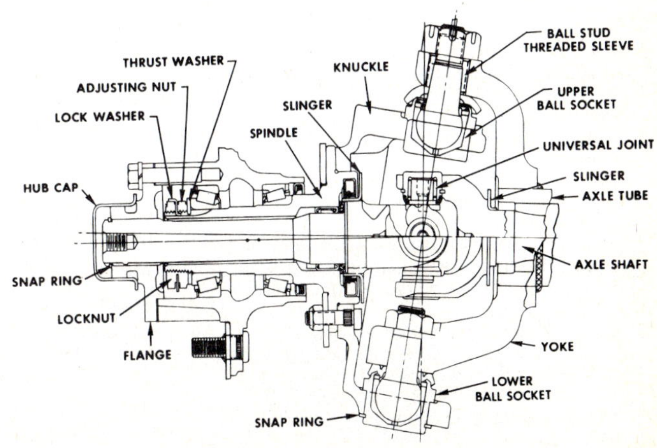

Fig. 5 FA-13 Front axle 4 × 4 with 40° steer

Installation

1. Install threaded sleeve.
2. Install yoke with upper ball joint positioned in yoke.
3. Torque the bottom socket nut to 80 ft. lbs.
4. Install top socket nut and torque to 100 ft. lbs. Line up slot in nut with hole in shaft. Never loosen to line up. Install cotter pin.
5. Install tie rod to steering arm of knuckle.
6. Loosen clamps on tie rod to set toe-in adjustments at 1/16 inches ± 1/16 inches.
7. Making sure that the drop in the tie rod tube is down following the adjustment, assemble inner protective slinger to shaft.

NOTE: Outer wheel bearing spindle nut or similar adapter will serve as a guide to assemble slinger. Place nut (adapter) in vise and slinger on end of shaft.

8. Tap on shaft with a rawhide hammer until slinger is seated.
9. Install outer protective slinger on shaft.

NOTE: Wheel spindle can serve as guide to assemble slinger.

10. Place spindle in vise, using caution to avoid bearing diameters.
11. Place slinger on shaft, and tap with rawhide hammer until seated.

NOTE: Use caution to avoid damaging seal diameter of slinger.

12. Place spindle in vise, using caution to avoid damaging bearing surfaces.
13. Install needle roller bearings using a driver and rawhide hammer. Recess bearings .025–.035 inches to prevent bearing contact with bronze spacer.
14. Install grease seal into spindle flush with spindle face.
15. After assembly of needle bearing and oil seal, pack needle bearing and lip of seal with grease.
16. Assemble axle shaft joint assembly, bronze spacer and spindle to knuckle.
17. Install brake assembly.

MODELS FA-64, 78

Axle Shaft & Universal Joint

Disassembly

1. Place axle shaft in a vise. Grasp shaft end of universal joint and pull while rapping back side of joint with a soft faced hammer.
2. Remove lock ring from axle shaft end and discard. Use new ring for assembly.
3. Place universal joint in a vise equipped with soft jaws with outer race bell upward.
4. Tilt inner race in outer race until one ball can be removed. Remove all balls using this method. A soft faced hammer may be used to aid disassembly.
5. Roll universal joint cage at a right angle to universal joint outer race bell with the two elongated openings in cage aligned with opposite teeth of outer race bell.
6. Remove cage and inner race from outer race bell.
7. Separate inner race from cage by turning inner race at right angle to cage and aligning notched tooth of inner race with elongated opening in cage.
8. Roll inner race from cage.
9. Clean and examine all parts for wear or damage.

Oil Seal and Bearing Replacement

1. To replace axle shaft oil seal pry out old seal and discard.
2. Install new seal making sure it contacts counterbore.
3. To remove upper and lower trunnion bearing cups, drive cups out of axle housing bores with driver and drift inserted through opposite bore.
4. Using care not to nick or scar mating surfaces, install cups.
5. To remove the spindle bushing, insert puller in bushing bore. Remove bushing and discard.
6. Install bushing in bore, making sure bushing contacts counter bore.

Assembly

1. Assemble the inner race and cage by indexing notched tooth of inner race with elongated opening in cage and rolling inner race into cage.
2. Align elongated openings of cage with opposite teeth of outer race bell and lower inner race and cage assembly into outer race bell.
3. Tilt inner race in outer race until one ball can be inserted, continue until all balls are inserted.

NOTE: Prelubrication of components and a soft faced hammer may be used to aid inner race movement.

4. Install new lock ring on axle shaft end and place axle shaft in vise.
5. Place universal joint on top of axle shaft.
6. Index end of axle shaft in splined inner race. Tap end of joint shaft with soft hammer to collapse lock ring, securing assembly.
7. Pack universal joint bell with lubricant.

MODELS FA-138, 184, 186, 191, 192, & 193

Disassembly

Axle Shafts & U-Joint

1. Block rear wheels of vehicle.
2. Raise vehicle and block up axle securely under spring seats.
3. Drain lubricant.
4. Remove hub cap from drive flange, Fig. 6.
5. Remove outer driveshaft snap ring.
6. Remove drive flange from wheel hub.

NOTE: Because of gasket material under hub cap and drive flange, it may be necessary to tap off these parts with a soft faced hammer.

7. Straighten adjusting nut lock from over bearing jam nut.
8. Remove hub, drum, and wheel from axle, using tire to grip and pull. It may be necessary to use a mallet on inside of wheel to remove hub from spindle.

NOTE: Outer bearing cone is loose and will drop off spindle when hub is removed.

9. Disassemble wheel from hub to facilitate hub bearing and seal removal if required.

NOTE: Before removing bearing cups and seal, it may be desirable to first disassemble the brake drum from hub. If bearing cup removal is required proceed with following.

10. If hub and spindle seal is still seated in the hub, remove by using a long screwdriver or other pointed tool to pry loose. Replace with new seal at reassembly.
11. Remove inner and outer bearing cups from hub using a press and sleeve or suitable bearing puller.

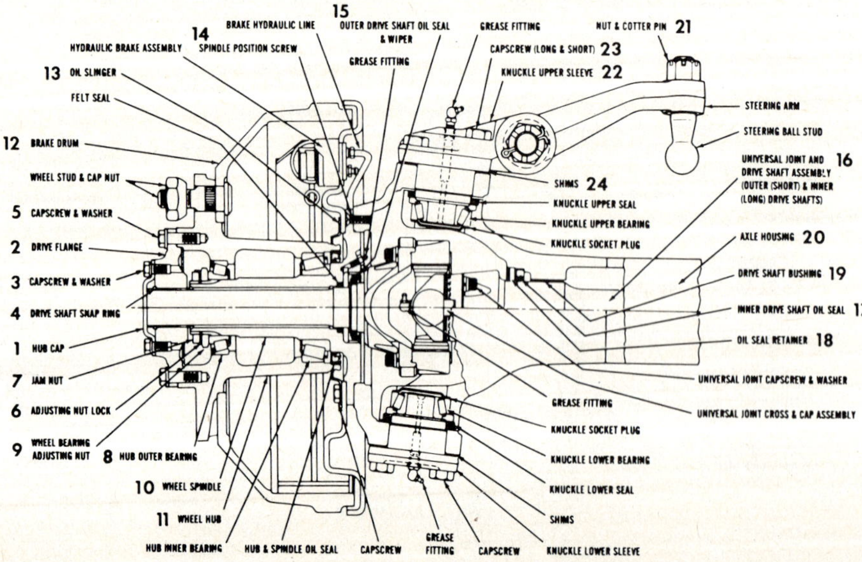

Fig. 6 FA-138, 184, 186, 191, 192, & 193 Front axle

12. Remove wheel studs from brake drum and hub by placing hub and drum assembly in press. Position open end of hub downward and press out studs.
13. Disconnect all brake lines at brake assembly.
14. Disassemble hub oil slinger, brake assembly, and wheel spindle from steering knuckle by removing eight attaching screws. Oil slinger and brake assembly may be removed at this time.
15. Remove spindle positioning screw, at top of spindle attaching flange, and pull spindle straight out from knuckle and outer driveshaft.

NOTE: Spindle felt seal and retainer assembly and outer driveshaft oil seal wiper will remain seated in spindle. Remove only if required using a suitable pointed tool.

U-Joints

1. Remove universal joint and driveshaft assembly from the axle by pulling complete assembly straight out of carrier and housing. If no maintenance is required set aside for reassembly.
2. Disassemble universal joint only if required, by removing eight yoke to cross and cap assembly capscrews and washers, four capscrews per yoke.

NOTE: It is not recommended that the cross and cap assembly be disassembled. Replace old parts with new preassembled unit.

3. Remove inner driveshaft oil seal, seal retainer, and bushing from axle housing, if required.
4. Pull out seal retainer and set aside for reassembly.
5. Remove oil seal by pulling it out of housing and using suitable puller remove bushing.

NOTE: It is not necessary to disassemble the steering knuckle to remove inner driveshaft oil seal or bushing. However, if knuckle must be removed from axle, it will be easier to disassemble these parts after knuckle removal.

Steering Knuckle

NOTE: Steering arm and tie rod assemblies may be serviced individually without removing complete steering knuckle. If it is necessary to disassemble both knuckle and steering arm from axle, steering arm may be disassembled from knuckle upper sleeve on bench.

1. Disconnect vehicle steering linkage from steering arm ball stud. It is not necessary to remove steering arm from knuckle at this time.
2. Remove tie rod from knuckle arm by removing cotter pin and nut.
3. Push tie rod through knuckle arm. It may be necessary to use a soft faced mallet to tap out end assembly from knuckle.

NOTE: It is not necessary to disassemble both tie rod ends unless both knuckle assemblies are being serviced at the same time.

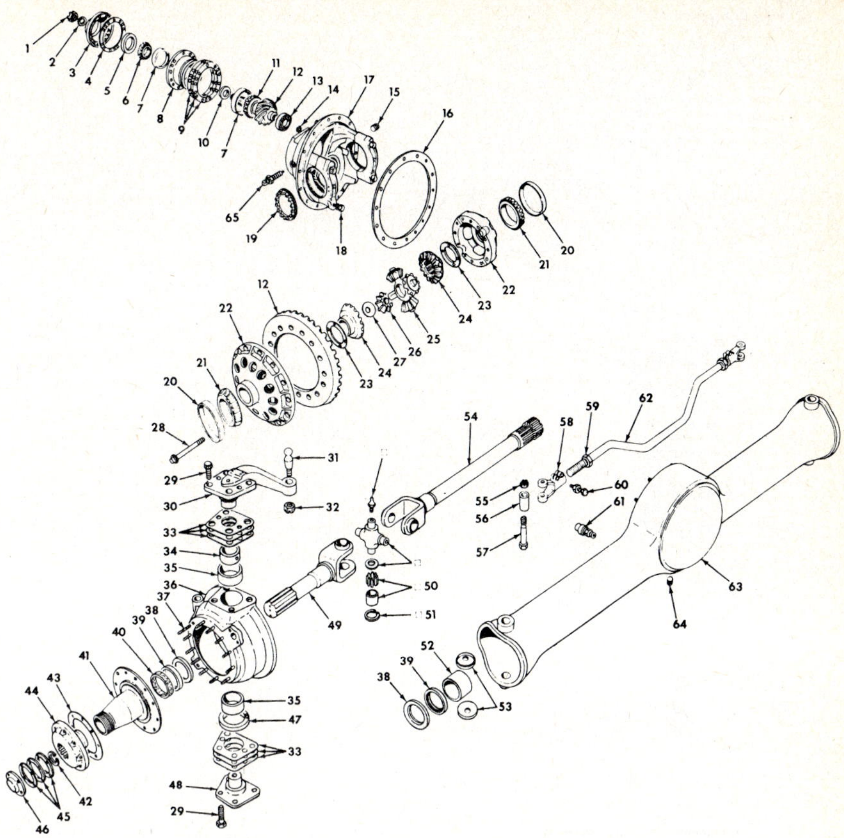

1	NUT, Companion Flange	34	SLEEVE, Steering Arm
2	WASHER, Companion Flange	35	BUSHING, Steering Arm
3	RETAINER, w/Seal, Pinion Bearing	36	HOUSING, w/Bushing
4	SHIM, Pinion Bearing Cage	37	STUD, Steering Knuckle Flange
5	SEAL, Oil	38	WASHER, Drive Shaft Oil Retainer
6	BEARING, Pinion Rear Cone	39	SEAL, Drive Shaft Oil
7	BEARING, Pinion Cup	40	SEAL, Drive Shaft Oil
8	CAGE, w/Cups, Pinion Bearing	41	KNUCKLE, w/Seal and Retainer
9	GASKET, Cage to Carrier	42	RING, Drive Shaft Snap
10	SPACER, Pinion Bearing	43	GASKET, Drive Flange
11	BEARING, Pinion Front Cone	44	FLANGE, Drive
12	GEAR SET, Ring and Pinion	45	SHIM, Hub Cap Spacer
13	BEARING, Pinion Gear Pilot	46	CAP, Hub
14	RING, Pinion Gear Pilot Bearing Snap	47	WASHER, Steering Knuckle Thrust
15	PLUG, Pipe Countersunk 3/4	48	CAP, Trunnion Housing Lower
16	GASKET, Carrier to Housing	49	SHAFT, Drive Axle, Outer
17	CARRIER, w/Caps Differential	50	NOT USED
18	BOLT, Bearing Cap	51	NOT USED
19	ADJUSTER, Differential Bearing	52	BUSHING, Drive Shaft
20	BEARING, Differential Cup	53	COVER, Knuckle Pin
21	BEARING, Differential Cone	54	SHAFT, Drive Axle, Inner
22	CASE, Differential Assembly	55	NUT, Tie Rod Yoke Bolt
23	WASHER, Diff Side Gear Thrust	56	BUSHING, Tie Rod Yoke Bolt
24	GEAR, Differential Side	57	BOLT, Tie Rod Yoke
25	SPIDER, Differential	58	YOKE, Tie Rod
26	GEAR, Differential Pinion	59	NUT, Tie Rod Yoke Lock
27	WASHER, Diff Pinion Gear Thrust	60	BOLT, Tie Rod Yoke Clamp
28	BOLT, Differential Case	61	BREATHER, Axle Housing
29	BOLT, Trunnion Cap, Inner & Outer	62	ROD, Front Axle Tie
30	ARM, w/Trunnion Cap	63	HOUSING, Front Axle
31	NOT USED	64	PLUG, Pipe 3/4, Drain
32	NOT USED	65	SCREW, Pinion Adjuster
33	SHIM, Trunnion Housing Cap		

Fig. 7 FA-138 & FA-191. Entire front axle assembly

4. Disassemble knuckle upper sleeve from knuckle by first removing four sleeve to knuckle capscrews, two long and two short.
5. Pull upper sleeve from knuckle bore. Knuckle upper bearing cone will remain seated on the sleeve and will be removed at same time. Remove bearing cone from sleeve, if required, using a suitable puller.

NOTE: Shims are between sleeve attaching flange and knuckle. Remove shim and inspect. If reuseable, keep together for reassembly.

6. Disassembly procedures of knuckle lower sleeve, bearing cone and shims are same as upper sleeve. After removing both upper and lower sleeves, pull knuckle from axle housing.

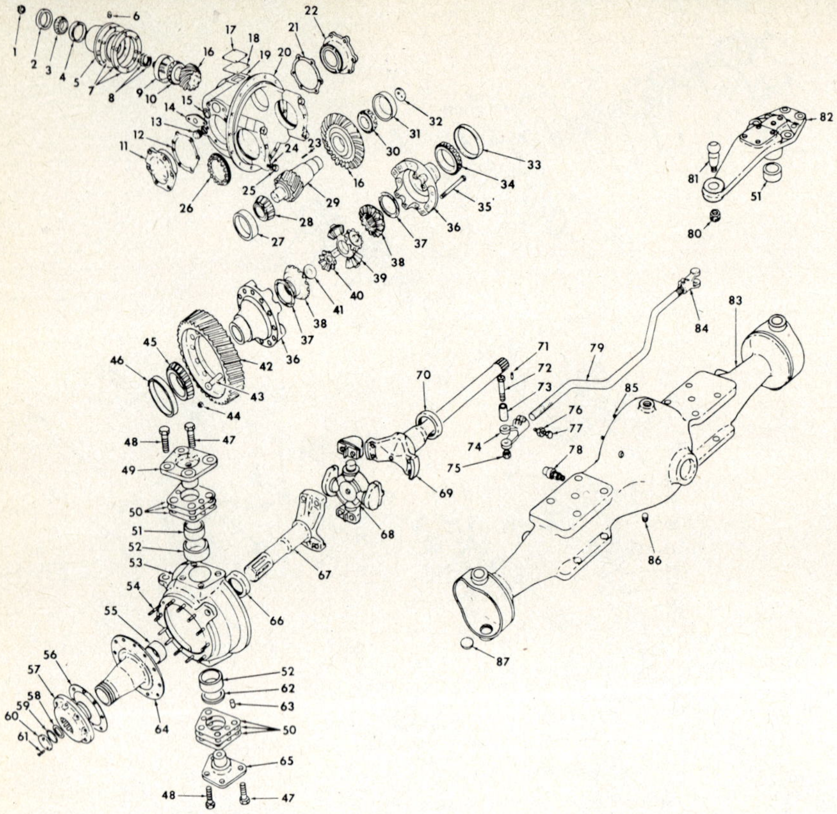

1 NUT, Companion Flange
2 SEAL, Oil
3 BEARING, Pinion Rear Cone
4 BEARING, Pinion Rear Cup
5 CAGE, w/Cups, Pinion Bearing
6 PLUG, Pinion Bearing Cage
7 SHIM, Pinion Bearing Cage
8 SPACER, Pinion Bearing
9 BEARING, Pinion Front Cup
10 BEARING, Pinion Front Cone
11 CAGE, w/Cup, Right Helical or Gear Bearing
12 SHIM, Bearing Cage
13 PLUG, Pipe
14 COVER, Shift Chamber Hole
15 GASKET, Shift Chamber Hole
16 GEAR SET, Hypoid Ring and Pinion
17 COVER, Diff. Inspection Hole
18 GASKET, Diff. Inspection Hole Cover
19 CARRIER, w/Caps and Cover, Diff.
20 GASKET, Carrier to Housing
21 SHIM, Bearing Cage
22 CAGE, w/Cup, Left Helical or Gear Bearing
23 KEY, Hypoid Ring Gear
24 WASHER, Bearing Cap Bolt
25 BOLT, Bearing Cap
26 SPACER, Differential Bearing
27 BEARING, Helical Drive Gear Right Cup
28 BEARING, Helical Drive Gear Right Cone
29 GEAR, Helical Drive
30 BEARING, Helical Drive Gear Left Cone
31 BEARING, Helical Drive Gear Left Cup
32 PLATE, Helical Drive Gear Lock
33 BEARING, Diff. Left Cup
34 BEARING, Diff. Left Cone
35 BOLT, Differential Case
36 CASE, Diff. Assy.
37 WASHER, Diff. Side Gear Thrust
38 GEAR, Diff. Side
39 SPIDER, Differential
40 GEAR, Differential Pinion
41 WASHER, Diff. Pinion Gear Thrust
42 GEAR, Helical Driven
43 RIVET, Helical Driven Gear to Case
44 NUT, Hex, Lock
45 BEARING, Differential Right Cone
46 BEARING, Differential Right Cup
47 BOLT, Trunnion Cap Inner
48 BOLT, Trunnion Cap Outer
49 NOT USED
50 SHIM, Trunnion Housing Cap
51 SLEEVE, Steering Arm
52 BUSHING, Housing Cap
53 HOUSING, Left Trunnion Yoke, Assy.
54 STUD, Knuckle to Trunnion Housing
55 BUSHING, Steering Knuckle
56 GASKET, Drive Flange to Hub
57 FLANGE, Front Wheel Drive
58 RING, Drive Shaft Snap
59 O-RING, Drive Shaft Ret. Washer
60 WASHER, Drive Shaft Retainer
61 SCREW, Drive Shaft Washer
62 WASHER, Bearing Thrust
63 NOT USED
64 KNUCKLE, w/Bushing, Steering
65 CAP, Trunnion Housing Lower
66 SEAL, Axle Shaft Oil
67 SHAFT, Axle Outer
68 KIT, Axle Shaft U-Joint
69 SHAFT, Axle Inner
70 RETAINER, U-Joint Shaft w/Seal
71 PIN, Tie Rod Yoke Bolt
72 BOLT, Tie Rod Yoke to Trunnion Housing
73 BUSHING, Tie Rod Yoke
74 YOKE, Tie Rod Right
75 NUT, Tie Rod Yoke Bolt
76 NUT, Tie Rod Yoke Bolt
77 BOLT, Tie Rod Yoke
78 BREATHER, Axle Housing, Assy.
79 ROD, Tie
80 NOT USED
81 NOT USED
82 ARM, Steering, Assy.
83 HOUSING, w/Seals, Bushings, Pins and Studs, Axle
84 YOKE, Tie Rod Left
85 STUD, Axle Housing to Diff. Carrier
86 PLUG, Pipe Sq-Socket
87 PLUG, Expansion

Fig. 8 FA-184, 186, 192, & 193. Entire front axle assembly

7. Remove upper and lower knuckle oil seals from their bores in the axle housing using screwdriver or other suitable pointed tool.
8. Remove both upper and lower knuckle bearing cones and socket plugs from their bores in axle housing. Remove from inside of axle housing using suitable sleeve having a diameter approximately the same as the plug bore. Tap out the plug and bearing cup at the same time.

Assembly

Steering Knuckle

Reassembly of knuckle to housing is done in reverse manner of disassembly, except for the following:

1. If inner driveshaft bushing, oil seal, and seal retainer were removed from housing reinstall these parts at this time before assembling knuckle components. Coat seal retainer O.D. with "Permatex" before installing.

NOTE: Do not use excessive pressure to install seal or tap seal after it has bottomed. This will crush the seal retainer and seal damage will result.

---

2. If knuckle socket plugs were removed from axle housing, reassemble by first applying a coat of silastic gasket material around one mating surface only, plug or inside shoulder of plug bore.
3. Install socket plug in bore; do not allow gasket material to set up. Position plug so concave side is facing outward and apply hand pressure to plug.
4. Wipe top of plug and bearing bore clean of any excess silastic material.
5. Install knuckle oil seals into axle housing, bearing cones onto upper and lower sleeves and the steering arm and ball stud on knuckle upper sleeve.
6. Install knuckle over end of axle housing, aligning knuckle bores with housing bores.
7. Assemble lower knuckle sleeve and shims first. Press sleeves with shims through bottom bores of knuckle and housing until sleeve bottoms. Install capscrews.
8. Assemble upper knuckle sleeve to knuckle and housing in similar manner as lower sleeve. Add enough shims between upper sleeve and knuckle between upper sleeve and knuckle to obtain 11–15 ft. lbs. of preload on knuckle bearings. Use a minimum of three shims. Use the following procedures:
   a. Assemble upper sleeve with shims through bore in knuckle and housing until sleeve bottoms. Secure sleeve with capscrews (2 long and 2 short).
   b. Measure from top grease fitting in upper knuckle sleeve (center of knuckle) to center of steering arm ball stud. Measure distance perpindicular to center line of knuckle. Note this measurement.
   c. Wrap soft wire or cord around steering arm ball stud (point of measurement) and install common pound scale to end of wire or cord. Pull knuckle around with scale. While rotating knuckle, record pound pull indicated on scale.

NOTE: Use rotating torque, not starting torque. To increase pounds pulled on scale, remove shims. To decrease bearing preload, and ponds pulled on scale, add shims.

U-Joints

1. If cross and cap assembly was found in need of repair, replace parts with complete new unit.
2. Before reassembling universal yokes to cross and cap assembly, clean all capscrews with "Locquic" and allow to dry.
3. Coat all capscrews with "Loctite" sealant and assemble universal joint. Tighten bolts and allow to dry for 40 minutes.
4. If outer driveshaft oil seal was removed, replace with new seal. Coat I.D. of seal retainer with "Permatex" and assemble onto shaft.

NOTE: Do not use excessive pressure to install.

5. Install universal joint and driveshaft assembly into axle housing and carrier.

NOTE: It may be necessary to index splines of inner shaft with splines of differential side gear for proper installation.

Spindle and Oil Slinger

Reassembly is done in reverse manner of disassembly, except for following:

1. Install spindle flange against knuckle mating surface. Turn spindle until countersunk hole in spindle flange aligns with corresponding hole in knuckle (at top).
2. Install spindle position screw and tighten.
3. Install brake assembly and oil slinger onto spindle flange and secure parts to knuckle.
3. Reconnect brake lines.

Wheel Bearings

1. If wheel bearings were removed from hub, reassemble inner and outer bearing cups into hub using a suitable sleeve and press.
2. Place inner bearing cone into position, inside hub against cup and install new hub and spindle oil seal. Coat O.D. of seal retainer with "Permatex" and tap into position until seal bottoms in bore.

NOTE: Do not use excessive pressure to install.

Wheel Bearing Adjustment

1. Pack bearings and hub cavity with wheel bearing grease even with inside diameter of bearing cups.
2. Assemble hub and drum assembly onto spindle being careful not to damage oil seal in hub. Press hub until inner bearing is seated flush on spindle.
3. Place outer bearing cone in position on spindle against cup in the hub.
4. Adjust bearings using following procedure:
   a. Install and tighten bearing adjusting nut onto spindle and against outer bearing. Torque nut to 50 ft. lbs. while hub is rotated in both directions to seat bearings.
   b. Back off adjusting nut 1/4 to 1/3 turn.
   c. Assemble nut lock (retainer) and jam nut onto spindle and against adjusting nut. Torque jam nut to 100 to 150 ft. lbs.
   d. Bend nut lock over nuts.

Drive Flange

1. Coat inside hub mating surface of drive flange with silastic gasket material.
2. Apply small amount of wheel bearing grease to internal splines of drive flange.
3. Assemble drive flange onto hub and secure with capscrews.
4. Assemble driveshaft lock ring in its groove in outer driveshaft.
5. Coat inside drive flange mating surface of hubcap with silastic gasket material.
6. Assemble hubcap to drive flange and secure with capscrews.

Tie Rod

1. If tie rod end assembly and clamp were removed, reinstall onto tie rod.
2. Insert tie rod end assembly shaft through bottom of bore in knuckle arm.
3. Secure tie rod to knuckle by installing slotted nut and cotter pin.
4. Install rim and tire to hub.

Front Suspension & Steering Gear

NOTE: Manual and power steering gear repairs are covered in separate chapters. See Index.

WHEEL ALIGNMENT

All Exc. FA-5, 6 & 7 Axles

Checking Front Suspension Height

Front suspension height should be checked before front wheels are aligned or whenever any service work is performed on the torsion bars. If suspension is not up to specifications, Fig. 1, measured between top edge of lower control arm and bottom edge of rubber bumper frame bracket as shown, the height should be checked and reset.

1. Place vehicle on alignment equipment or on a level floor.
2. Inflate tires to recommended pressures, with only the vehicle weight on the torsion bars (no passenger or unusual weight in vehicle).
3. Grasp front bumper and jounce vehicle up and down several times to settle suspension system.
4. Adjustment can be made by loosening or tightening the torsion bar retainer lever bolt, Fig. 2.

NOTE: If the difference between the two sides of the vehicle are within 1/8" of each other and are within the limits specified, they may be considered acceptable. If these values differ more than 1/8" or if one or both of them are outside the specified limits, the suspension height on both sides must be reset.

Camber, Adjust

1. Check front suspension height as described previously.
2. Referring to Fig. 2, loosen nut on camber adjusting bolt.
3. Turn camber bolt in the direction required to obtain a camber angle of one degree (plus or minus 1/2 degree).
4. Holding camber bolt, tighten nut to 81–135 ft. lbs. torque.

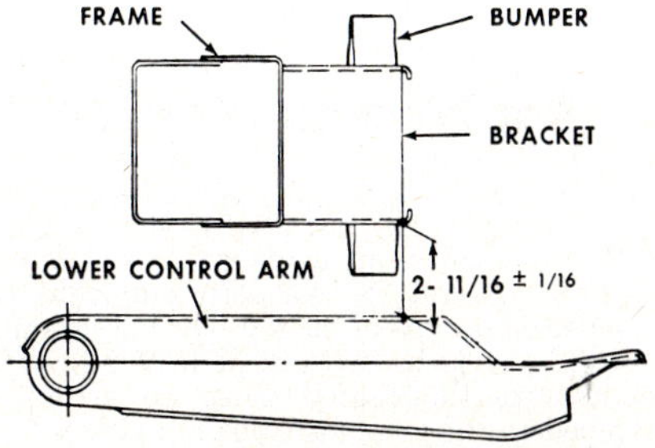

Fig. 1 Checking front suspension height

Caster, Adjust

1. Check front suspension height as described previously.
2. Loosen inner and outer nuts on threaded end of strut.
3. Adjust either the inner or outer nut to obtain a caster setting of two degrees *negative* (plus or minus 1/2 deg.).
4. After caster has been established, tighten nuts to 120–150 ft-lbs torque.

CAUTION: Make sure rubber insert washers are parallel with frame bracket when nuts are tightened.

5. Recheck caster after nuts are tightened.

Toe-In, Adjust

1. Check front suspension height as described previously.
2. Disconnect vertical link from left steering arm.
3. Position steering gear in mid-position.
4. Position left front wheel in straight ahead position.
5. Adjust vertical link so ball joint stud will enter hole in left steering arm with wheel in straight ahead position.
6. Insert ball stud in steering arm, install nut and tighten securely.
7. Recheck measurements. If toe-in is not

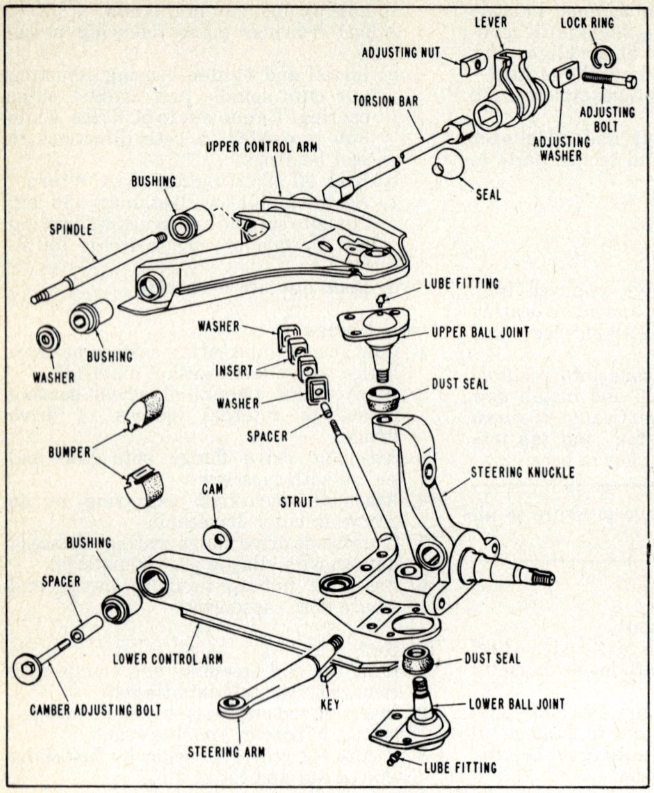

Fig. 2 Torsion bar suspension

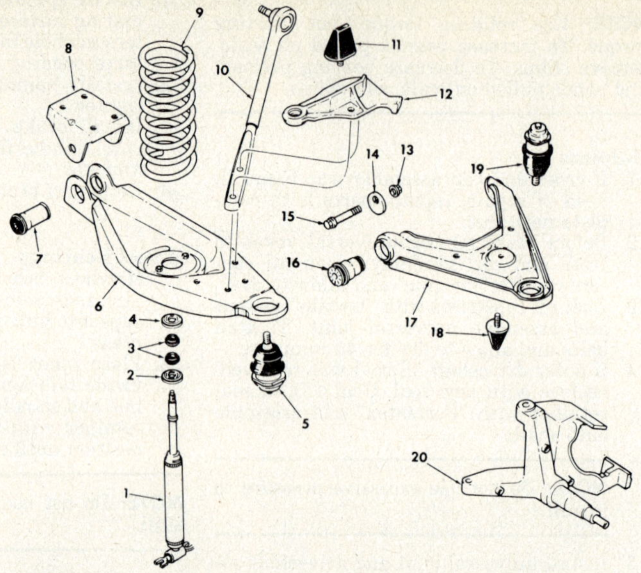

1. SHOCK ABSORBER	6. LOWER CONTROL ARM	11. BUMPER	16. BUSHING
2. RETAINER	7. BUSHING	12. WHEEL STOP BRKT.	17. UPPER CONTROL ARM
3. CUSHION	8. BRACKET	13. NUT	18. BUMPER
4. RETAINER	9. COIL SPRING	14. CASTER ADJUST CAM	19. UPPER BALL JOINT
5. LOWER BALL JOINT	10. CONTROL ARM ROD	15. BOLT	20. STEERING KNUCKLE

Fig. 3 Front suspension. FA-5, 6 and 7 models

within 1/8 to 3/16", it will be necessary to loosen clamp bolts on right vertical link sleeve and turn sleeve to obtain desired setting.

FA-5, 6 & 7 Axles

Checking Front Suspension Height

With aid of a pair of inside calipers, check the minimum gap from stop plate, welded on the underside of the suspension crossmember, to the ridge on the axle stop brackets located on the lower control arm, Fig. 3. Repeat axle gap measurement on both front wheels.

Camber, Adjust

NOTE: Always adjust camber first before adjusting caster.

1. Loosen nuts on front and rear cams of upper control arm inner pivot bolts, Fig. 3.
2. Adjust camber by simultaneously turning front and rear cam bolts clockwise to decrease positive camber and counterclockwise to increase positive camber. Refer to "FA-5, 6 & 7 Axles Wheel Alignment Specifications Chart," Fig. 3A.
3. Tighten nuts and check caster.

Caster, Adjust

1. Loosen nut on front cam of upper control arm pivot bolt, Fig. 3, and turn bolt counterclockwise to increase positive caster and clockwise to decrease positive caster. Refer to "FA-5, 6 & 7 Axles Wheel Alignment Specifications Chart," Fig. 3A.
2. Tighten nut.
3. Loosen nut on rear cam of upper control arm pivot bolt and turn bolt clockwise to increase positive caster and counterclockwise to decrease positive caster. Refer to "FA-5, 6 & 7 Axles Wheel Alignment Specifications Chart," Fig. 3A.
4. After camber and caster are adjusted to specifications, torque pivot bolt nuts to 65–75 ft. lbs.

Toe-In, Adjust

NOTE: Check all steering joints and make sure they are not binding or cocked to one side.

1. Check toe-in with pitman arm on center of travel of steering gear.
2. To adjust, loosen clamp bolt nuts and rotate turnbuckle as necessary. Refer to "FA-5, 6 & 7 Axles Wheel Alignment Specifications Chart," Fig. 3A.
3. After correct alignment has been obtained, torque clamp bolt nuts to 25–35 ft. lbs.

TORSION BAR, REPLACE

The torsion bars are not interchangeable, side to side. The left-hand bar cannot be used on the right side and vice versa. The bars are marked "L" and "R" for left and right or stamped on the end of the rod. The bars should always be installed with these letters toward the rear of the vehicle. Also an arrow is stamped on the end of the bar to indicate the direction of wind-up for applying load application.

Removal

1. Referring to Fig. 2, raise vehicle with jack under frame crossmember.
2. Release load from torsion bar by loosening retaining lever adjusting bolt.
3. Remove retainer lever adjusting bolt, washer and nut and slide retainer lever from end of torsion bar.
4. Disengage torsion bar from upper control arm and slide bar to rear to remove from vehicle.

CAUTION: When removing torsion bars, be careful not to nick or scratch them as it may create a fracture.

Installation

1. With vehicle raised off floor, position torsion bar stamped with the letter "L" in upper control arm on left side of vehicle and bar stamped with letter "R" in upper control arm on right side of vehicle. Also make sure arrow stamped on end of bar is located toward rear of vehicle.
2. Position retainer lever on end of torsion bar so adjusting bolt can just be threaded into nut.
3. Install washer on adjusting bolt and insert bolt into retainer lever and bracket.
4. Install nut on adjusting bolt and adjust height of suspension as outlined below.
5. Always check caster and camber whenever torsion bars have been removed or replaced. Also check aiming of headlamps after setting front suspension height.

COIL SPRING, REPLACE

FA-5, 6 & 7 Axles

1. Raise vehicle and position two jack stands

Axle Gap (Inch) Fig. 3	Caster, Degrees	Camber, Degrees		Toe-In, Inch
		Left	Right	
2.80	1½	⅛	−⅛	0–⅛
3.20	1	¼	0	0–⅛
3.60	½	¼	0	0–⅛
4.00	0	¼	0	0–⅛
4.40	−½	¼	0	0–⅛
4.80	−1	0	−¼	0–⅛
5.20	−1½	−¼	−½	0–⅛

Fig. 3A FA-5, 6 & 7 axles wheel alignment specifications chart

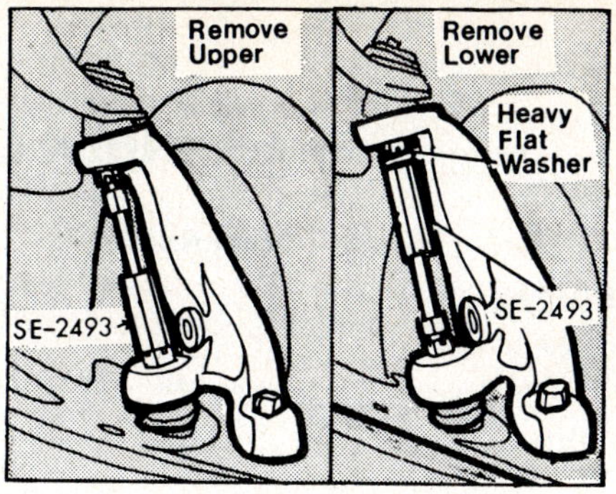

Fig. 4 Removing ball joint stud from steering knuckle

under frame.
2. Remove wheel and tire assembly.
3. Remove caliper, rotor, and brake shield from steering knuckle.

NOTE: Secure caliper to frame to prevent damage to brake hose.

4. Remove shock absorber.
5. Disconnect sway bar from axle bumper and wheel stop bracket, if equipped, then remove bumper and stop bracket from lower control arm.
6. Insert Compressor screw No. SE-2491-2 through shock absorber upper mounting hole in crossmember. Working through opening in lower control arm, assemble puller hook SE-2491-1 to compressor screw, then position puller hooks under second coil from bottom of spring.
7. Turn spring compressor until spring is unseated from lower control arm.
8. Remove cotter pins from upper and lower ball joint studs, then loosen lower ball joint stud nut approximately two turns.
9. Position tool No. SE-2493, Fig. 4, between upper and lower ball joint studs. Turn tool until a slight pressure is applied to lower ball joint stud, then tap steering knuckle to free stud.
10. Torque upper ball joint stud nut to 70–80 ft. lbs. on 100 models and 80–90 ft. lbs. on 200 models and install cotter pin.
11. Remove lower ball joint stud nut and lower control arm from steering knuckle.
12. Turn spring compressor until spring is unseated from crossmember, then remove compressor and spring.
13. Reverse procedure to install. Torque lower ball joint stud nut to 80–90 ft. lbs. on 100 models and 90–100 ft. lbs. on 200 models.

BALL JOINT, REPLACE

FA-5, 6 & 7 Axles

Upper Ball Joint

1. Raise vehicle and position two jack stands under frame.
2. Remove wheel and tire assembly and shock absorber.
3. Install spring compressor SE-2491 and puller hook SE-2491-1. Turn compressor screw until spring is unseated from lower control arm.
4. Using a suitable jack, support lower control arm at ball joint.
5. Remove cotter pins from upper and lower ball joint studs, then loosen upper ball joint stud nut approximately two turns.
6. Position tool No. SE-2493, Fig. 4, between upper and lower ball joint studs. Turn tool until a slight pressure is applied to the upper ball joint stud, then tap steering knuckle to free stud.
7. Torque lower ball joint stud nut to 80–90 ft. lbs. on 100 models and 90–100 ft. lbs. on 200 models and install cotter pin.
8. Remove upper ball joint stud nut and lift control arm from steering knuckle.

NOTE: Support upper control arm to prevent brake line from becoming damaged.

9. Using tool No. SE-2491-1, remove ball joint from control arm.
10. Install ball joint using tool No. 2491-1, thread ball joint into control arm until fully seated.
11. Pack dust cover ⅔ full with grease, then install dust cover on ball joint using tool No. SE-2495-2.
12. Place ball joint stud into steering knuckle, then install nut and cotter pin. Torque nut to 70–80 ft. lbs. on 100 models and 80–90 ft. lbs. on 200 models.
13. Using the jack, raise lower control arm until it contacts the spring.
14. Loosen spring compressor and lower jack.
15. Remove spring compressor.
16. Install shock absorber and wheel and tire assembly.

Lower Ball Joint

1. Remove spring as described under "Coil Spring Replace."
2. Remove lower control arm from frame bracket.
3. Position lower control arm in a vise.
4. Remove ball joint from control arm using tool No. 2494-2.
5. Install ball joint using tool No. 2494-2, thread ball joint into control arm until fully seated.
6. Pack dust cover ⅔ full with grease, then install dust cover on ball joint using tool No. 2495-1.
7. Remove control arm from vise and install on frame bracket.

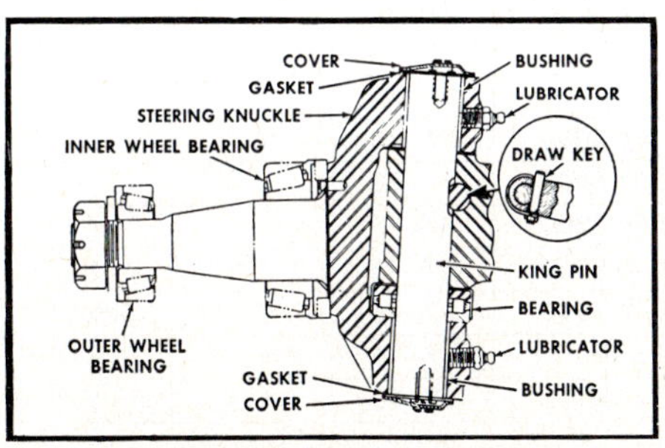

Steering knuckle using kingpin seal cap with hold-down screws

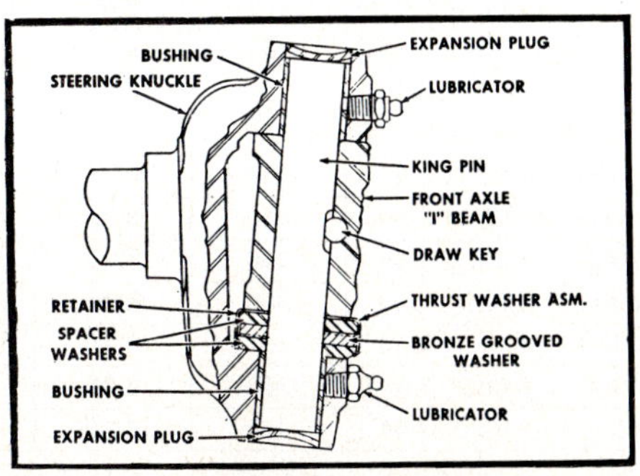

Steering knuckle using thrust bearing with bronze grooved washer

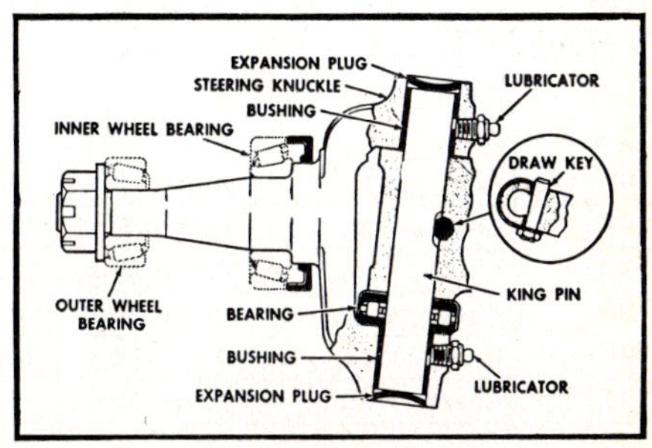

Steering knuckle with expansion type seal plugs

JEEP

INDEX OF SERVICE OPERATIONS

GENERAL ENGINE SPECIFICATIONS

Year	Model or Series	Engine Make	Engine Model	Bore & Stroke	Piston Displ. Cu. In.	Comp. Ratio	Horsepower @ R.P.M.	Normal Oil Pressure P.S.I.	Oil Pan Refill Capacity Qts. ②	Cooling System Capacity Qts. ①
JEEP										
1974–75	CJ Series	Amer. Mtrs.	6-232	3.750 × 3.500	232	8.0	100 @ 3600③	37–75	5	10½
1977–78	CJ Series	Amer. Mtrs.	6-232	3.750 × 3.500	232	8.0	—	37–75	5	10½
1974–75	Commando	Amer. Mtrs.	6-232	3.750 × 3.500	232	8.0	100 @ 3600③	37–75	5	10½
1976	CJ, Chero., Wag.	Amer. Mtrs.	6-232	3.750 × 3.500	232	8.0	90 @ 3050	37–75	5	10½
1974–75	J-100 Series	Amer. Mtrs.	6-258	3.750 × 3.895	258	8.0	110 @ 3500③	37–75	5	10½
1976	CJ, Chero., Wag.	Amer. Mtrs.	6-258	3.750 × 3.895	258	8.0	95 @ 3050	37–75	5	10½
1977–78	CJ Series	Amer. Mtrs.	6-258	3.750 × 3.895	258	8.0	—	37–75	5	10½
1977–78	Cherokee	Amer. Mtrs.	6-258	3.750 × 3.895	258	8.0	—	37–75	5	10½
1974–75	All	Amer. Mtrs.	V8-360④	4.08 × 3.44	360	8.25	175 @ 4000	37–75	4	—
1974–75	All	Amer. Mtrs.	V8-360⑤	4.08 × 3.44	360	8.25	195 @ 4400	37–75	4	—
1976	Chero., Wag.	Amer. Mtrs.	V8-360④	4.08 × 3.44	360	8.25	140 @ 3200	37–75	4	14
1976	Chero., Wag.	Amer. Mtrs.	V8-360⑤	4.08 × 3.44	360	8.25	165 @ 3500	37–75	4	14
1976	Chero., Wag.	Amer. Mtrs.	V8-401	4.16 × 3.68	401	8.25	205 @ 3500	37–75	4	14
1977–78	Chero., Wag.	Amer. Mtrs.	V8-360④	4.08 × 3.44	360	8.25	—	37–75	4	14
1977–78	Chero., Wag.	Amer. Mtrs.	V8-360⑤	4.08 × 3.44	360	8.25	—	37–75	4	14

Continued

GENERAL ENGINE SPECIFICATIONS—Continued

Year	Model or Series	Engine Make	Engine Model	Bore & Stroke	Piston Displ. Cu. In.	Comp. Ratio	Horsepower @ R.P.M.	Normal Oil Pressure P.S.I.	Oil Pan Refill Capacity Qts. ②	Cooling System Capacity Qts. ①
1977	Chero., Wag.	Amer. Mtrs.	V8-401	4.16 × 3.68	401	8.25	—	37–75	4	14
1978	Chero., Wag.	Amer. Mtrs.	V8-401	4.16 × 3.68	401	8.30	—	37–75	4	14
1974–75	CJ Series	Amer. Mtrs.	V8-304	3.75 × 3.440	304	8.4	150 @ 4200③	37–75	4	14
1976	CJ Series	Amer. Mtrs.	V8-304	3.750 × 3.44	304	8.4	120 @ 3200	37–75	4	14
1977–78	CJ Series	Amer. Mtrs.	V8-304	3.75 × 3.44	304	8.4	—	37–75	4	13
1979	All	Amer. Mtrs.	6-258	3.75 × 3.90	258	8.3	—	37–75	5	10.5
1979	All	Amer. Mtrs.	V8-304	3.75 × 3.44	304	8.4	—	37–75	4	13.0
1979	All	Amer. Mtrs.	V8-360	4.08 × 3.44	360	8.25	—	37–75	4	13.0
1980	CJ Series	Pontiac	4-151	4.00 × 3.00	151	8.2	—	37–75	3⑦	7.8
1980	All	Amer. Mtrs.	6-258	3.75 × 3.90	258	8.0	—	37–75	5	10.5
1980–81	CJ Series	Amer. Mtrs.	V8-304	3.75 × 3.44	304	8.4	—	37–75	4	13
1980	All Exc. CJ	Amer. Mtrs.	V8-360	4.08 × 3.44	360	8.25	—	37–75	4	14
1981–83	CJ Series	Pontiac	4-151	4.00 × 3.00	151	8.2	—	36–41	3⑦	7.8
1981–82	All	Amer. Mtrs.	6-258	3.75 × 3.90	258	8.0	—	37–75	5	10.5
1981–83	All Exc. CJ	Amer. Mtrs.	V8-360	4.08 × 3.44	360	8.25	—	37–75	4	14.0
1983	All	Amer. Mtrs.	6-258	3.75 × 3.90	258	9.2	—	37–75	5	10.5
JEEP TRUCKS										
1974–75	J-2000	Amer. Mtrs.	6-258	3.750 × 3.895	258	8.0	110 @ 3500③	37–75	5	10½
1974–75	All	Amer. Mtrs.	V8-360④	4.08 × 3.44	360	8.25	175 @ 4000	37–75	4	—
	All	Amer. Mtrs.	V8-360⑤	4.08 × 3.44	360	8.25	195 @ 4400	37–75	4	—
	All	Amer. Mtrs.	V8-401	4.16 × 3.68	401	8.25	215 @ 4200	37–75	4	—
1976	J-10, 20	Amer. Mtrs.	6-232	3.75 × 3.50	232	8.0	90 @ 3050	37–75	5	10½
	J-10, 20	Amer. Mtrs.	6-258	3.750 × 3.895	258	8.0	95 @ 3050	37–75	5	10½
	J-10, 20	Amer. Mtrs.	V8-360④	4.08 × 3.44	360	8.25	140 @ 3200	37–75	4	14
	J-10, 20	Amer. Mtrs.	V8-360⑤	4.08 × 3.44	360	8.25	165 @ 3500	37–75	4	14
	J-10, 20	Amer. Mtrs.	V8-401	4.16 × 3.68	401	8.25	205 @ 3500	37–75	4	14
1977–78	J-10	Amer. Mtrs.	6-258	3.750 × 3.895	258	8.0	—	37–75	5	10½
	J-10, 20	Amer. Mtrs.	V8-360④	4.08 × 3.44	360	8.25	—	37–75	4	14
	J-10, 20	Amer. Mtrs.	V8-360⑤	4.08 × 3.44	360	8.25	—	37–75	4	14
	J-10, 20	Amer. Mtrs.	V8-401	4.16 × 3.68	401	⑥	—	37–75	4	14
1979	All	Amer. Mtrs.	6-258	3.75 × 3.90	258	8.3	—	37–75	5	10.5
	All	Amer. Mtrs.	V8-304	3.75 × 3.44	304	8.4	—	37–75	4	13.0
	All	Amer. Mtrs.	V8-360	4.08 × 3.44	360	8.25	—	37–75	4	13.0
1980	All	Amer. Mtrs.	6-258	3.75 × 3.90	258	8.0	—	37–75	5	10.5
1980	All	Amer. Mtrs.	V8-360	4.08 × 3.44	360	8.25	—	37–75	4	14
1981–82	J-10	Amer. Mtrs.	6-258	3.75 × 3.90	258	8.3	—	37–75	5	10.5
1981–83	All	Amer. Mtrs.	V8-360	4.08 × 3.44	360	8.2	—	37–75	4	14.0
1983	All	Amer. Mtrs.	6-258	3.75 × 3.90	258	9.2	—	37–75	5	10.5

①—With heater.
②—Add 1 qt. with filter change.
③—Ratings are NET—as installed in vehicle.
④—2 Bar. Carb.
⑤—4 Bar. Carb.
⑥—1977; 8.25, 1978; 8.30.
⑦—Includes ½ qt. for oil filter.

TUNE UP SPECIFICATIONS

The following specifications are published from the latest information available. This data should be used only in the absence of a decal affixed in the engine compartment.

★ When using a timing light, disconnect vacuum hose or tube at distributor and plug opening in hose or tube so idle speed will not be affected.

● When checking compression, lowest cylinder must be within 80 percent of highest.

▲ Before removing wires from distributor cap, determine location of the No. 1 wire in cap, as distributor position may have been altered from that shown at the end of this chart.

Year & Engine	Spark Plug		Ignition Timing BTDC①★				Curb Idle Speed②		Fuel Pump Pressure
	Type	Gap	Firing Order Fig. ▲	Man. Trans.	Auto. Trans.	Mark Location	Man. Trans.	Auto. Trans.	
1983									
4-151	R44TSX⑰	.060	E	12°	—	Damper	900	—	6½–8
6-258 Exc. Calif	RFN14LY⑱	.035	A	6°	6°	Damper	600㉖	500㉗	5–6½㉚
6-258 Calif.	RFN14LY⑱	.035	A	6°	6°	Damper	650㉖	550㉗	5–6½㉚
6-258 High Alt.	RFN14LY⑱	.035	A	13°	13°	Damper	700㉘	650㉙	5–6½㉚
V8-360	RN12Y⑱	.035	B	10°㉔	10°㉕	Damper	600	600	5–6½
1982									
4-151 Exc. Calif.	R44TSX⑰	.060	E	12°㉓	—	Damper	900	—	6½–8
4-151 Calif.	R44TSX⑰	.060	E	8°	—	Damper	900	—	6½–8
6-258 CJ Model Exc. Calif	RFN14LY⑱	.035	A	8°	8°	Damper	650④	550⑤	4–5
6-258 CJ Model Calif.	RFN14LY⑱	.035	A	15°	15°	Damper	650④	550⑤	4–5
6-258 Exc. CJ Models	RFN14LY⑱	.035	A	15°㉔	15°㉕	Damper	600④	500⑤	5–6½
V8-360	RN12Y⑱	.035	B	10°㉔	10°㉕	Damper	600	600	5–6½
1981									
4-151 Exc. Calif.	R44TSX⑰	.060	E	10°	12°	Damper	900	500/700D	6½–8
4-151 Calif.	R44TSX⑰	.060	E	10°	10°	Damper	900	500/700D	6½–8
6-258 Exc. Calif.	RFN14LY⑱	.035	A	8°	8°	Damper	650④	550D⑤	4–5
6-258 Calif.	RFN14LY⑱	.035	A	4°	6°	Damper	650④	550D⑤	4–5
V8-304	N12Y⑱	.033	B	8°⑥	10°	Damper	600⑦	600D	5–6½
V8-360	N12Y⑱	.033	B	10°	10°	Damper	600	600D	5–6½
1980									
4-151	R44TSX⑰	.060	E	12°	—	Damper	900	—	6½–8
6-258 CJ Models Exc. Calif.	③⑱	.035	A	8°	10°	Damper	700	500/600D	4–5
6-258 CJ Models Calif.	N13L⑱	.035	A	6°	8°	Damper	700	500/600D	4–5
6-258 Exc. CJ Models	N14LY⑱	.035	A	8°	8°	Damper	700	700D	4–5
V8-304 Exc. Calif.	N12Y⑱	.035	B	8°⑥	10°	Damper	700	500/600D	5–6½
V8-304 Calif.	N12Y⑱	.035	B	5°	5°	Damper	700	500/600D	5–6½
V8-360	N12Y⑱	.035	B	8°	8°	Damper	800	600D	5–6½
1979									
6-258 CJ Models	N13L	.035	A	6°	4°	Damper	700	600D	4–5
6-258 Exc. CJ Models	N13L	.035	A	8°	8°	Damper	600	600D	4–5
V8-304	N12Y	.035	B	5°	8°	Damper	⑧	600D	5–6½
V8-360	N12Y	.035	B	8°	8°	Damper	800	600D	5–6½
1978									
6-232 Exc. High Alt.	N13L	.035	A	5°	—	Damper	850	—	4–5
6-232 High Alt.	N13L	.035	A	10°⑨	—	Damper	600	—	4–5
6-258 1 Barrel Carb.⑩	N13L	.035	A	3°	8°	Damper	850	550D	4–5
6-258 1 Barrel Carb. Calif.	N13L	.035	A	8°	8°	Damper	850	550D	4–5
6-258 1 Barrel Carb. High Alt.	N13L	.035	A	10°⑨	10°⑨	Damper	600	550D	4–5
6-258 2 Barrel Carb.	N13L	.035	A	6°	6°	Damper	650	550D	4–5
V8-304	N12Y	.035	B	5°	⑪	Damper	750	700D	5–6½
V8-360 Exc. Calif.	N12Y	.035	B	5°	8°	Damper	750	700D	5–6½
V8-360 Calif.	N12Y	.035	B	—	10°	Damper	—	700D	5–6½
V8-401	N12Y	.035	B	5°	8°	Damper	750	700D	5–6½

Continued

TUNE UP SPECIFICATIONS—Continued

The following specifications are published from the latest information available. This data should be used only in the absence of a decal affixed in the engine compartment.

★ When using a timing light, disconnect vacuum hose or tube at distributor and plug opening in hose or tube so idle speed will not be affected.

● When checking compression, lowest cylinder must be within 80 percent of highest.

▲ Before removing wires from distributor cap, determine location of the No. 1 wire in cap, as distributor position may have been altered from that shown at the end of this chart.

Year & Engine	Spark Plug		Ignition Timing BTDC①★				Curb Idle Speed②		Fuel Pump Pressure
	Type	Gap	Firing Order Fig. ▲	Man. Trans.	Auto. Trans.	Mark Location	Man. Trans.	Auto. Trans.	
1977									
6-232 Exc. High Alt.	N12Y	.035	C	5°	—	Damper	850	—	4–5
6-232 High Alt.	N12Y	.035	C	10°⑫	—	Damper	600	—	4–5
6-258 1 Barrel Carb.⑩	N12Y	.035	C	3°	8°	Damper	850	550D	4–5
6-258 1 Barrel Carb. Calif.	N12Y	.035	C	6°	8°	Damper	850	700D	4–5
6-258 1 Barrel Carb. High Alt.	N12Y	.035	C	10°⑬	10°⑫	Damper	600	550D	4–5
6-258 2 Barrel Carb.	N12Y	.035	C	6°	6°	Damper	650	550D	4–5
V8-304	RN12Y	.035	D	5°	⑪	Damper	700	700D	5–6½
V8-360, 401	RN12Y	.035	D	5°	8°	Damper	750	700D	5–6½
1976									
6-232	N12Y	.035	C	8°	—	Damper	600	—	4–5
6-258 Exc. Calif.	N12Y	.035	C	6°	8°	Damper	⑭	550D	4–5
6-258 Calif.	N12Y	.035	C	6°	8°	Damper	600	700D	4–5
V8-304 Exc. Calif.	N12Y	.035	D	5°	10°	Damper	750	700D	5–6½
V8-304 Calif.	N12Y	.035	D	5°	5°	Damper	750	700D	5–6½
V8-360, 401 Exc. Calif.	N12Y	.035	D	5°	8°	Damper	750	700D	5–6½
V8-360, 401 Calif.	N12Y	.035	D	5°	5°	Damper	750	700D	5–6½
1975									
6-232	N12Y	.035	C	5°	—	Damper	⑮	—	4–5
6-258	N12Y	.035	C	3°	3°	Damper	⑯	550D	4–5
V8-304	N12Y	.035	D	5°	—	Damper	750	—	5–6½
V8-360, 401	N12Y	.035	D	2½°	2½°	Damper	750	700D	5–6½
1974									
6-232⑲	N12Y	.035	C	5°	—	Damper	600	—	4–5½
6-258⑲	N12Y	.035	C	3°	3°	Damper	600	550D⑳	4–5½
V8-304㉑	N12Y	.035	D	5°	—	Damper	750	—	4–6½
V8-360㉑	N12Y	.035	D	5°	5°	Damper	750	700D⑳	4–6½
V8-401㉑	N12Y	.035	D	㉒	㉒	Damper	650	600D⑳	4–6½

①—B.T.D.C.—Before top dead center.

②—Idle speed on man. trans. vehicles is adjusted in Neutral & on auto. trans. equipped vehicles is adjusted in Drive unless otherwise specified. When two idle speeds are listed, the higher speed is with the A/C or idle solenoid energized.

③—Man. trans., use N14LY; auto. trans., use N13L.

④—With holding solenoid energized, 750 RPM; with vacuum actuator energized, 900 RPM.

⑤—With holding solenoid energized, 650D RPM; with vacuum actuator energized, 800D RPM.

⑥—On hilly terrain, set at 12° BTDC.

⑦—On hilly terrain, set at 700 RPM.

⑧—Except Calif., 700 RPM; California, 750 RPM.

⑨—At altitudes below 4000 ft., set at 5° BTDC.

⑩—Except Calif. & high altitude.

⑪—Except Calif., 10° BTDC; California, 5° BTDC.

⑫—At altitudes below 4000 ft., set at 8° BTDC.

⑬—At altitudes below 4000 ft., set at 6° BTDC.

⑭—Except CJ models, 650 RPM; CJ models, 600 RPM.

⑮—Except Calif., 700 RPM; California, 600 RPM.

⑯—Except Calif., less EGR 700 RPM, with EGR 650 RPM; California, 600 RPM.

⑰—AC.

⑱—Champion.

⑲—Breaker point gap, .016 inch; dwell angle, 31–34°.

⑳—Disconnect TCS wires at solenoid vacuum valve.

㉑—Breaker point gap, .016 inch; dwell angle, 29–31°.

㉒—Except heavy duty, 5° BTDC; heavy duty, 2½° BTDC.

㉓—On hilly terrain, set at 17° BTDC.

㉔—On hilly terrain, set at 19° BTDC.

㉕—On hilly terrain, set at 21° BTDC.

㉖—With holding solenoid energized, 750 RPM; with vacuum actuator energized, 950 RPM.

㉗—With holding solenoid energized, 650 RPM; with vacuum actuator energized, 750 RPM.

㉘—With holding solenoid energized, 950 RPM; with vacuum actuator energized, 1000 RPM.

㉙—With holding solenoid energized, 750 RPM; with vacuum actuator energized, 850 RPM.

㉚—Except CJ models. On CJ models, 4–5.

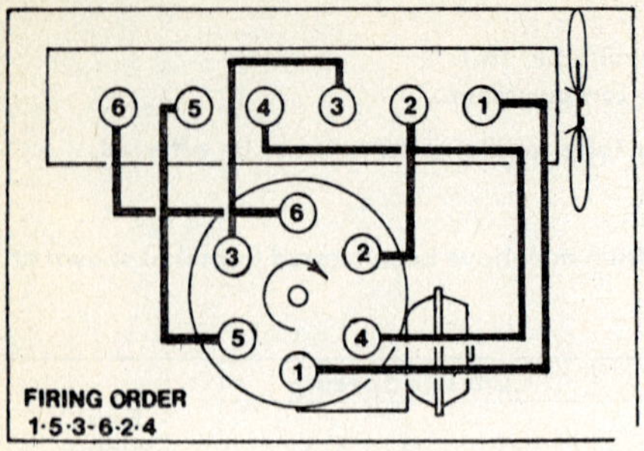

Fig. A

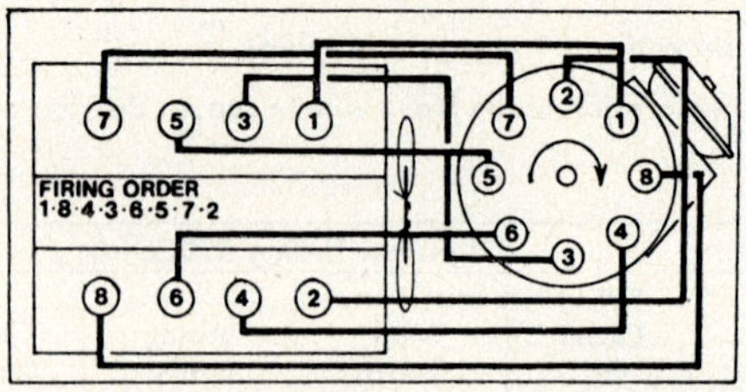

Fig. B

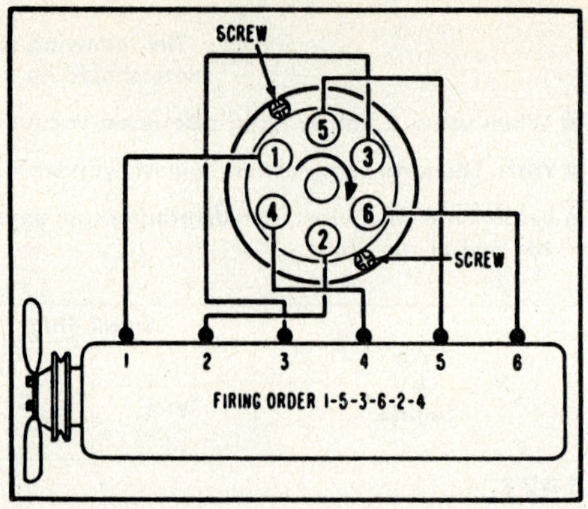

Fig. C

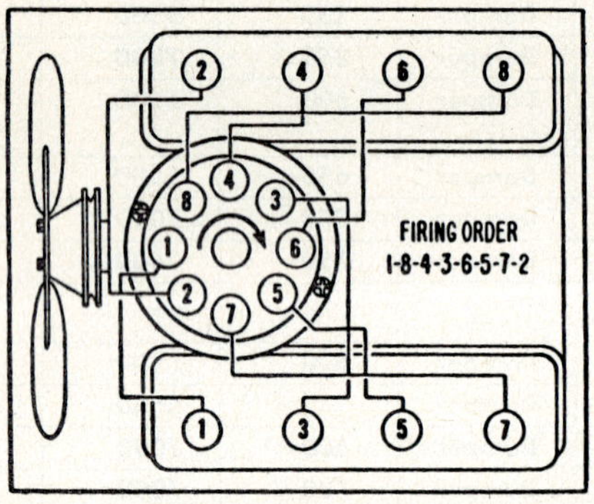

Fig. D

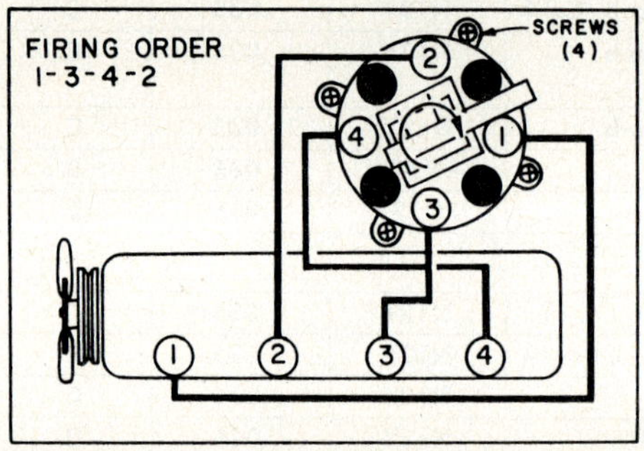

Fig. E

DISTRIBUTOR SPECIFICATIONS

★NOTE: If advance is checked on the vehicle, double the R.P.M. and degrees to get crankshaft figures.

Year	Engine	Distributor Part No.	Rotation	Breaker Gap	Dwell Angle	Breaker Arm Spring Tension	Centrifugal Advance Degrees @ R.P.M. of Distributor★		Vacuum Advance	
							Advance Starts	Full Advance	Inches of Vacuum To Start Plunger	Max. Adv. Dist. Deg. @ Vacuum
1974	6-232, 258	1110529	C	.016	31–34	17–21	0–2 @ 500	14 @ 2300	5	9 @ 13
	V8-360	1112112	C	.016	29–31	17–21	0–1¾ @ 500	14 @ 2000	5	8.25 @ 12¾
	V8-304	1112179	C	.016	29–31	17–21	0–1 @ 400	17 @ 2200	5	8.25 @ 12¾
	V8-360	1112215	C	.016	29–31	17–21	0–1½ @ 400	16 @ 2200	4	8.25 @ 13
	V8-360, 401	1112215	C	.016	29–31	17–21	0–1½ @ 400	16 @ 2200	4	8.25 @ 13
1975	6-232, 258	3224968	C	—	—	—	0–2 @ 600	15 @ 2250	5–7	9 @ 13
	6-232, 258	3224969	C	—	—	—	0–2 @ 500	14 @ 2250	5–7	9 @ 13
	6-258	3224969	C	—	—	—	0–2 @ 500	14 @ 2250	5–7	9 @ 13
	V8-304	3224965	C	—	—	—	0–2 @ 500	17 @ 2200	5–7	7½ @ 12½
	V8-360, 401	3224746	C	—	—	—	0–4.5 @ 500	16 @ 2300	6	8½ @ 12.7
	V8-360, 401	3225295	C	—	—	—	0–2 @ 500	14 @ 2000	5–7	17 @ 12½
1976	6-232, 258	3227331	C	—	—	—	0–.4 @ 500	6.25 @ 2200	5–7	8 @ 12.6
	V8-304	3228263	C	—	—	—	0–2.8 @ 500	12 @ 2200	4–6	7.5 @ 12.7
	V8-304	3228264	C	—	—	—	0–2.4 @ 500	14.5 @ 2200	5–7	7.5 @ 12.7
	V8-360, 401	3228263	C	—	—	—	0–2.8 @ 500	12 @ 2200	4–6	7.5 @ 12.7
	V8-360, 401	3228265	C	—	—	—	0–2.7 @ 500	8.5 @ 2200	4–6	7.5 @ 12.7
1977	Six	3229719	C	—	—	—	0–1.8 @ 500	11 @ 2200	5–7	9 @ 12.7
	V8-304	3228264	C	—	—	—	0–2.4 @ 500	15½ @ 2200	5–7	8½ @ 11.4
	V8-304	3228263	C	—	—	—	0–2.8 @ 500	13 @ 2200	4–6	8½ @ 11.7

Continued

DISTRIBUTOR SPECIFICATIONS—Continued

★NOTE: If advance is checked on the vehicle, double the R.P.M. and degrees to get crankshaft figures.

Year	Engine	Distributor Part No.	Rotation	Breaker Gap	Dwell Angle	Breaker Arm Spring Tension	Centrifugal Advance Degrees @ R.P.M. of Distributor★		Vacuum Advance	
							Advance Starts	Full Advance	Inches of Vacuum To Start Plunger	Max. Adv. Dist. Deg. @ Vacuum
	V8-360	3228263	C	—	—	—	0–2.8 @ 500	13 @ 2200	4–6	8 1/2 @ 11.7
	V8-360	3228265	C	—	—	—	0–2.7 @ 500	9 1/2 @ 2200	4–6	8 1/2 @ 11.7
	V8-401	3228265	C	—	—	—	0–2.7 @ 500	9 1/2 @ 2200	4–6	8 1/2 @ 11.7
1978	Six	3232434	C	—	—	—	0 @ 400	9 1/2 @ 2200	3–7	8 @ 13
	V8-304	3230443	C	—	—	—	0 @ 400	12 @ 2200	2–5	8 @ 12.7
	V8-304	3231340	C	—	—	—	0 @ 400	14 1/2 @ 2200	2–7	8 @ 12.2
	V8-360	3230443	C	—	—	—	0 @ 400	12 @ 2200	2–5	8 @ 12.7
	V8-360	3231341	C	—	—	—	0 @ 300	8 @ 2200	2–5	8 @ 12.7
	V8-401	3231341	C	—	—	—	0 @ 300	8 @ 2200	2–5	8 @ 12.7
1979	6-258	3231915	C	—	—	—	0 @ 450	9 1/2 @ 2200	2	12 @ 12
	6-258	3232434	C	—	—	—	9 @ 500	9 @ 2200	3	8 @ 13
	V8-304	3231340	C	—	—	—	0 @ 450	14 1/2 @ 2200	3	8 @ 12
	V8-304	3233959	C	—	—	—	0 @ 600	13 @ 2200	3	16 @ 12
	V8-304	3234693	C	—	—	—	0 @ 500	13 @ 2200	2	12 @ 13
	V8-360	3233174	C	—	—	—	0 @ 450	8 1/2 @ 2200	3	12 @ 13
1980	V8-304	3231340	C	—	—	—	0 @ 425	16 @ 2200	2	9 1/2 @ 12
1980–81	4-151	1110560	C	—	—	—	0 @ 475	8 @ 2100	3 1/2–4 1/2	11 @ 9 1/2
	4-151	1110561	C	—	—	—	0 @ 475	8 @ 2100	2 1/2–3 1/2	11 @ 9 1/2
	6-258	3235141	C	—	—	—	0 @ 450	7 1/2 @ 2200	3	13 1/2 @ 15 3/4
	V8-304	3237198	C	—	—	—	0 @ 650	11 @ 2300	2 1/2	12 1/2 @ 12
	V8-304	3237199	C	—	—	—	0 @ 500	11 1/2 @ 2200	2 1/2	17 1/2 @ 12
	V8-360	3233174	C	—	—	—	0 @ 400	9 1/2 @ 2200	1 3/4	13 @ 11
1982	4-151	1110598	C	—	—	—	0 @ 400	8 @ 2000	2	11 @ 7
	6-258	3241334	C	—	—	—	0 @ 400	6 1/2 @ 1000	2 1/2	12 1/2 @ 17
	6-258	3241333	C	—	—	—	0 @ 550	7 1/2 @ 1025	2 1/2	13 @ 17 1/2
	V8-360	3233174	C	—	—	—	0 @ 425	9 1/2 @ 2100	1 1/2	13 @ 11
1983	4-151	1103527	C	—	—	—	0 @ 400	8 1/2 @ 2050	2 3/4	10 1/2 @ 7
	4-151	1110598	C	—	—	—	0 @ 400	8 @ 2000	3 1/2	9 1/2 @ 8 1/2
	6-258	3242409	C	—	—	—	0 @ 550	6 @ 1000	2 1/2	13 @ 17 1/2
	V8-360	3233174	C	—	—	—	0 @ 425	9 1/2 @ 2200	1 1/2	13 @ 11

STARTING MOTOR SPECIFICATIONS

Year	Engine Model	Starter Number	Brush Spring Tension Oz.	Free Speed Test			Resistance Test		Lock Torque Test		
				Amps.	Volts	R.P.M.	Amps.	Volts	Amps.	Volts	Torque Ft. Lbs.
1974–77	All	—	40	65	12	9250	—	—	600	3.4	13.0
1978	6-232, 258 & V8-304	—	40	67	12	9356	—	—	—	—	—
	V8-360 & 401	—	40	77	12	9600	—	—	—	—	—
1979–80	Exc. 4–151	—	40	77	12	9600	—	—	—	—	—
1980	4–151	1109526	—	45–70	9	7000–11,900	—	—	—	—	—
1981	4–151	1109526	—	45–70	9	7000–11,900	—	—	—	—	—
	Exc. 4–151	—	40	77	12	8900–9600	—	—	—	—	—
1982–83	4-151	1109526	—	45–70	9	7000–11,900	—	—	—	—	—
	Exc. 4-151	—	—	67	12	7380–9356	—	—	—	—	—

ALTERNATOR SPECIFICATIONS

Year	Alternator							Regulator			
	Make	Model	Ground Polarity	Rated Output		Field Current		Model ②	Regulator Test @ 120°F.		
				Amperes	Volts	Amperes ①	Volts		Ampere Load	Altern. R.P.M.	Volts
1974	Motorola	8AL2025F	Negative	37	15	1.8–2.5	12	8RD-2001	—	—	—
1975	Motorola	—	Negative	37	—	1.8–2.5	12	8RH-2003	—	—	—
	Motorola	—	Negative	51	—	1.8–2.5	12	8RH-2003	—	—	—
	Motorola	—	Negative	62	—	1.8–2.5	12	8RH-2003	—	—	—
	Delco-Remy	—	Negative	37	—	—	—	1116387	—	—	—
	Delco-Remy	—	Negative	55	—	—	—	1116387	—	—	—
	Delco-Remy	—	Negative	63	—	—	—	1116387	—	—	—
1976–77	Delco-Remy	—	Negative	37	—	③	12	1116387	—	—	—
	Delco-Remy	—	Negative	63	—	③	12	1116387	—	—	—
	Motorcraft	—	Negative	40	—	2.5–3	—	—	10	—	13.7–14.5
	Motorcraft	—	Negative	60	—	2.5–3	—	—	10	—	13.7–14.5
1978–80	Delco-Remy	—	Negative	37	—	4–5	—	1116387	—	—	—
	Delco-Remy	—	Negative	63	—	4–5	—	1116387	—	—	—
1981–83	Delco	—	Negative	42	—	4–5	—	1116387	—	—	—
	Delco	—	Negative	63	—	4–5	—	1116387	—	—	—
	Delco	—	Negative	70	—	4–5	—	1116387	—	—	—
	Delco	—	Negative	85	—	4–5	—	1116387	—	—	—

①—Excessive current draw indicates shorted field windings; no current draw indicates an open winding.

②—Regulator is a sealed assembly, requiring no adjustments.

③—4–4.5 amps @ 80° F.

PISTONS, PINS, RINGS, CRANKSHAFT & BEARINGS

Year	Engine Model	Piston Clearance	Ring End Gap①		Wristpin Diam-eter	Rod Bearings		Main Bearings			
			Comp.	Oil		Shaft Diameter	Bearing Clearance	Shaft Diameter	Bearing Clearance	Thrust on Bear. No.	Shaft End Play
1974–75	6-232	.0009–.0017	.010	.010	.9307	2.0934–2.0955	.001–.003	2.4986–2.5001	.001–.003	3	.0015–.0065
1974–75	6-258	.0009–.0017	.010	.010	.9307	2.0934–2.0955	.001–.003	2.4986–2.5001	.001–.003	3	.0015–.0065
1974–75	8-304	.0010–.0018	.010	.010	.9311	2.0934–2.0955	.001–.003	②	.001–.003	3	.003–.008
1974–75	V8-360	.0012–.0020	.010	.015	.9311	2.0934–2.0955	.001–.003	②	.001–.003	3	.003–.008
1974–75	V8-401	.0010–.0018	.010	.015	1.0011	2.2464–2.2485	.001–.003	②	.001–.003	3	.003–.008
1976–78	6-232	.0009–.0017	.010	.010	.9307	2.0934–2.0955	.0010–.0025	2.4986–2.5001	.001–.003	3	.0015–.0065
1976–80	6-258	.0009–.0017	.010	.010	.9307	2.0934–2.0955	.0010–.0025	2.4986–2.5001	.001–.003	3	.0015–.0065
1976–81	V8-304	.0010–.0018	.010	.010	.9311	2.0934–2.0955	.001–.003	②	④	3	.003–.008
1976–81	V8-360	.0012–.0020	.010	.015	.9311	2.0934–2.0955	.001–.003	②	④	3	.003–.008
1976–78	V8-401	.0010–.0018	.010	.015	1.0011	2.2464–2.2485	.001–.003	②	④	3	.003–.008
1980–83	4-151	.0017–.0041	.010	.015	.927	2.000	.0005–.0026	2.300	.0005–.0022	5	.0035–.0085
1981	6-258	.0009–.0017	.010	.010	.9307	2.0934–2.0955	.0010–.0025	2.4986–2.5001	③	3	.0015–.0065
1982–83	6-258	.0009–.0017	.010	.010	.9307	2.0934–2.0955	.0010–.0025	2.4996–2.5001	.001–.0025	3	.0015–.0065
	V8-360	.0012–.0020	.010	.015	.9311	2.0934–2.0955	.001–.003	②	④	3	.003–.008

①—Fit rings in tapered bores for clearance listed in tightest portion of ring travel.

②—Rear, 2.7464–2.7479"; others 2.7474–2.7489".

③—No. 1, .0005–.0026; No. 2, 3, 4, 5, & 6, .0005–.0030; No. 7 .0011–.0035

④—Rear, .002–.004"; others .001–.003".

VALVE SPECIFICATIONS

Year	Engine Model	Valve Lash Int.	Valve Lash Exh.	Valve Angles Seat	Valve Angles Face	Valve Spring Installed Height	Valve Spring Pressure Lbs. @ In.	Stem Clearance Intake	Stem Clearance Exhaust	Stem Diameter Intake	Stem Diameter Exhaust
1974–78	6-232	Zero	Zero	⑤	②	1 13/16⑨	195 @ 1 7/16⑧	.001–.003	.001–.003	.3715–.3725	.3715–.3725
1974–78	6-258⑥	Zero	Zero	⑤	②	1 13/16⑨	195 @ 1 7/16⑧	.001–.003	.001–.003	.3715–.3725	.3715–.3725
1977–80	6-258⑦	Zero	Zero	⑤	②	1 13/16	④	.001–.003	.001–.003	.3715–.3725	.3715–.3725
1974–80	V8-304	Zero	Zero	⑤	②	1 13/16	213 @ 1 23/64	.001–.003	.001–.003	.3715–.3725	.3715–.3725
1974–80	V8-360	Zero	Zero	⑤	②	1 13/16	213 @ 1 23/64	.001–.003	.001–.003	.3715–.3725	.3715–.3725
1974–78	V8-401	Zero	Zero	⑤	②	1 13/16	213 @ 1 23/64	.001–.003	.001–.003	.3715–.3725	.3715–.3725
1980–81	4-151	Zero	Zero	46	45	1.66	172 @ 1.25	.0010–.0027	.0010–.0027	.3418–.3425	.3418–.3425
1981	6-258	Zero	Zero	⑤	②	①	③	.001–.003	.001–.003	.3715–.3725	.3715–.3725
1981	V8-304	Zero	Zero	⑤	②	1.78	211 @ 1.356	.001–.003	.001–.003	.3715–.3725	.3715–.3725
1981	V8-360	Zero	Zero	⑤	②	1.78	211 @ 1.356	.001–.003	.001–.003	.3715–.3725	.3715–.3725
1982–83	4-151	Zero	Zero	46°	45°	1.66	172 @ 1.25	.0010–.0027	.0010–.0027	.3418–.3425	.3418–.3450
1982–83	6-258	Zero	Zero	⑤	②	1.78	188 @ 1.41	.001–.003	.001–.003	.3715–.3725	.3715–.3725
1982–83	V8-360	Zero	Zero	⑤	②	1.78	211 @ 1.35	.001–.003	.001–.003	.3715–.3725	.3715–.3725

①—Intake valve 1.625, exhaust valve 1.786.
②—Intake 29°, exhaust 44°.
③—Intake 188–202 lbs. @ 1.411", exhaust 210–226 lbs. @ 1.188".
④—1977; 204 @ 1 23/64", 1978–80; 195 @ 1 13/32".
⑤—Intake 30°, exhaust 44 1/2°.
⑥—1 Barrel carb.
⑦—2 Barrel carb.
⑧—With valve rotators, 218 @ 1 3/16".
⑨—With valve rotators, 1 5/8".

ENGINE TIGHTENING SPECIFICATIONS★

★ Torque Specifications are for clean and lightly lubricated threads only. Dry or dirty threads produce increased friction which prevents accurate measurement of tightness.

Year	Engine	Spark Plugs Ft. Lbs.	Cylinder Head Bolts Ft. Lbs.	Intake Manifold Ft. Lbs.	Exhaust Manifold Ft. Lbs.	Rocker Arm Shaft Bracket Ft. Lbs.	Rocker Arm Cover Ft. Lbs.	Connecting Rod Cap Bolts Ft. Lbs.	Main Bearing Cap Bolts Ft. Lbs.	Flywheel to Crank-shaft Ft. Lbs.	Vibration Damper or Pulley Ft. Lbs.
1975–76	6-232, 258	22–33	95–115	18–28	18–28	16–26①	42–58③	30–35	75–85	95–120	48–64
1977–80	6-232, 258	22–33	95–115	18–28②	18–28②	16–26①	42–58③	30–35	75–85	95–120	70–90
1974	V8-304, 360	22–33	100–120	37–47	20–30	16–26①	42–58③	26–30	90–105	95–120	48–64
1974–76	V8-401	22–33	100–120	37–47	20–30	16–26①	48–58③	35–40	90–105	95–120	48–64
1975–76	V8-304, 360	22–33	100–120	37–47	20–30	16–26①	42–58③	30–35	90–105	95–120	48–64
1977–78	V8-401	22–33	100–120	37–47	20–30	16–26①	48–58③	35–40	90–105	95–120	80–100
1977–79	V8-304, 360	22–33	100–120	37–47	20–30	16–26①	42–58③	30–35	90–105	95–120	80–100
1980	4-151	7–15	81–103	34–40	34–40	17–23	77–83③	27–33	62–68	65–71	157–163
1980–83	V8-304, 360	25–30	100–120	37–47	②	16–26①	42–58③	30–35	90–105	95–115	80–100
1982	4-151	7–15	81–103	34–40	36–42	17–23	4–10	27–33	62–68	65–71	157–163
	6-258	7–15	80–90	18–28	18–28	16–26	25–31 in-lbs	30–35	65–70	95–115	70–90
1983	4-151	7–15	81–103	21–30	36–42	17–23	6–9	27–33	59–70	65–71	133–192
	6-258	7–15	80–90	18–28	18–28	16–26	25–31 in-lbs	30–35	75–85	95–115	70–90

①—Rocker arm cap screw.
②—Center two bolts; 20–30 ft. lbs.; outer four bolts, 12–18 ft. lbs.
③—Inch pounds.

WHEEL ALIGNMENT SPECIFICATIONS

Model	Caster, Deg.	Camber, Deg.	Toe in, inch	Kingpin Inclination Deg.
CJ & DJ Series	+3°①	+1½	3/64–3/32	8½
J-Series	+4°①	+1½	3/64–3/32	8½
Commando	+3°	+1½	3/64–3/32	7½

Model	Caster, Deg.	Camber, Deg.	Toe in, inch	Kingpin Inclination Deg.
Cherokee	+4°①	+1½	3/64–3/32	8½
Wagoneer	+4°①	+1½	3/64–3/32	8½

①—1981–82, +6°

Engine Section

CYLINDER HEAD, REPLACE

4-151

1. Disconnect battery ground cable, then remove cylinder head cover.
2. Drain coolant from radiator and engine block.
3. Disconnect exhaust pipe from exhaust manifold.
4. Disconnect vacuum hoses. Tag vacuum hoses so they can be installed in the same position.
5. Remove alternator and position aside, then disconnect fuel line from carburetor.
6. Disconnect upper radiator and rear heater hoses from cylinder head.
7. Remove power steering pump and bracket and position aside with hoses attached, if equipped.
8. Remove dipstick, then remove rocker arm assemblies and push rods.

NOTE: Tag rocker arms and push rods so they can be installed in the same position.

9. Remove cylinder head attaching bolts, then remove cylinder head by inserting a pry bar into alternator bracket and prying upward.
10. Reverse procedure to install. Tighten cylinder head bolts to torque listed in Engine Tightening Specification in sequence shown in Fig. 1.

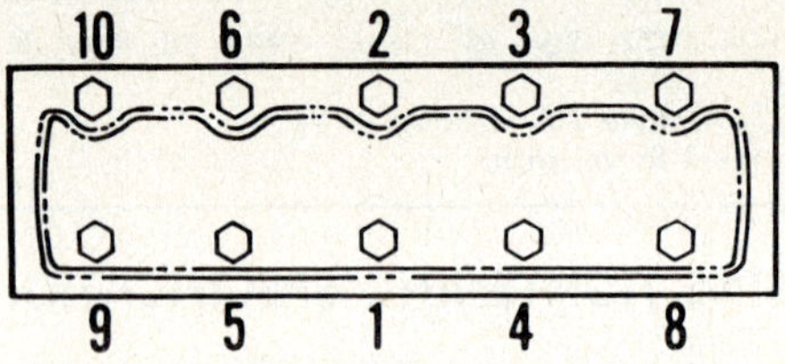

Fig. 1 Cylinder head tightening sequence. 4-151 engine

6-232, 258

1. Drain cooling system, then disconnect upper radiator hose and remove valve cover.
2. On early 1974 engines, remove rocker arm and shaft assembly.
3. On late 1974 and 1975–83 engines, remove bridged pivot and rocker arm assemblies.

NOTE: Loosen each bridged pivot bolt one turn at a time to avoid breaking the bridge pivot. Also keep bridged pivots and rocker arms in same order as removed, so that they can be reinstalled in their original locations.

4. Remove push rods.

NOTE: Keep push rods in same order as removed, so that they can be installed in their original locations.

5. Remove intake and exhaust manifold bolts.
6. Disconnect power steering and air pumps and brackets, then lay them aside. Do not disconnect hoses.
7. If equipped with air conditioning, remove idler bracket and loosen alternator belt, then remove compressor bracket bolts and set compressor aside.
8. Disconnect spark plug wires and temperature sending lead, then remove spark plugs and ignition coil and bracket assembly.
9. Remove cylinder head bolts and cylinder head.
10. Reverse removal procedure to install. Torque cylinder head bolts to the specification given in the Engine Tightening Chart and in the sequence shown in Fig. 2.

CAUTION: The cylinder head bolt located at the left front corner of the head (No. 11, Fig. 2) must have the threads sealed to prevent coolant leakage. Permatex No. 2 or equivalent is recommended.

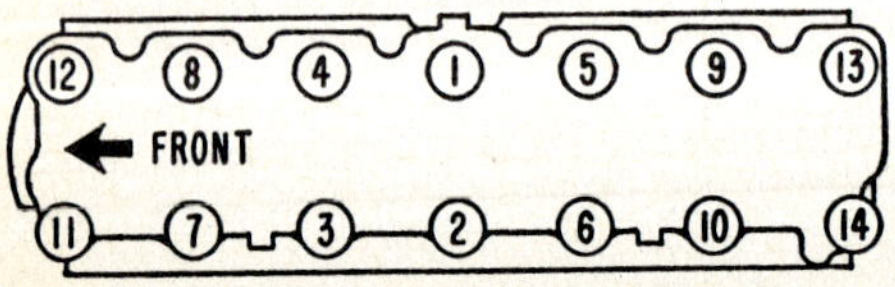

Fig. 2 Cylinder head bolt tightening sequence. 6-232, 258

V8-304, 360, 401

1. Remove air cleaner and drain cooling system, then if equipped with Air Guard emission system, disconnect air hose from injection manifold.
2. Disconnect all hoses and electrical leads from intake manifold.
3. Remove valve covers, bridged pivot assemblies, rocker arms and push rods.

NOTE: Loosen each bridged pivot bolt one turn at a time to avoid breaking the bridged pivot. Also, keep rocker arms, bridged pivot and push rods in same order as removed so that they can be reinstalled in their original locations.

4. Disconnect spark plug wires and remove spark plugs.
5. Remove intake manifold and disconnect exhaust manifolds from cylinder head.
6. Loosen all drive belts, then if equipped with air conditioning, remove compressor bracket attaching bolts and set compressor aside.
7. Disconnect alternator support brace, power steering mounting bracket and Air Guard pump (if used) mounting bracket from cylinder head.
8. Remove cylinder head bolts and cylinder head.
9. Reverse removal procedure to install. First torque the cylinder head bolts to 80 ft. lbs. and in the sequence shown in Fig. 3, then finally torque the cylinder head bolts to the specifications given in the "Engine Tightening" chart.

NOTE: The cylinder block has two locating dowels on each bank to assist in lining up and holding the cylinder head and gasket in position during installation.

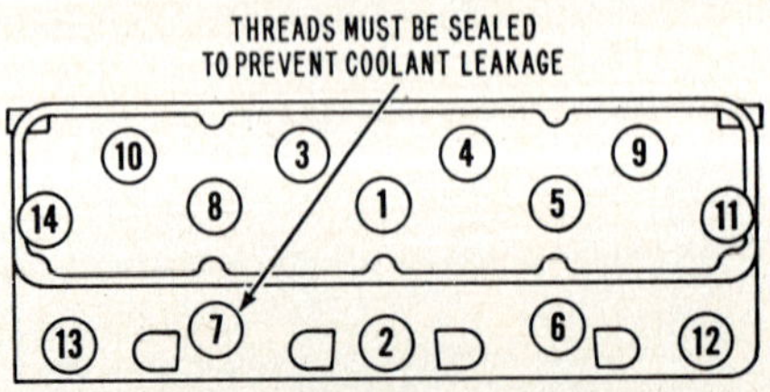

Fig. 3 Cylinder head tightening sequence. V8-304, 360, 401

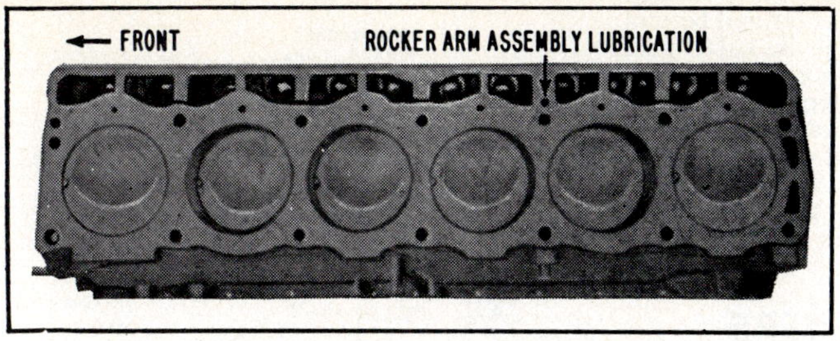

Fig. 4 Rocker arm lubrication gallery. Early 1974 6-232, 258

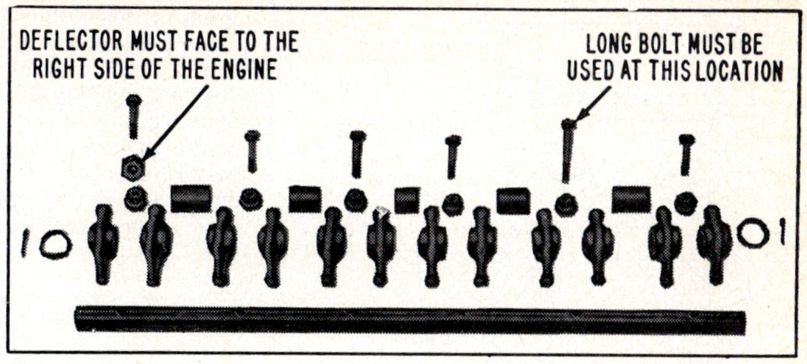

Fig. 5 Rocker arm assembly. Early 1974 6-232, 258

CAUTION: The No. 7 bolt shown in Fig. 3, second from front on the left bank, must have the threads sealed to prevent coolant leakage. Permatex No. 2 or equivalent is recommended.

ROCKER ARM SERVICE

4-151

To remove rocker arm assembly, proceed as follows:

1. Remove cylinder head cover.
2. Remove rocker arm nut and ball.
3. Remove rocker arm.
4. Reverse procedure to install. Clean all parts using a suitable cleaning solvent. Blow dry parts using compressed air. Inspect the pivot contact surface of each rocker arm, pivot ball and push rod assembly. Replace any part that has been found unserviceable.

6-232, 258 Engines

On early 1974 engines, the pressure supply for each rocker arm is obtained from No. 3 camshaft bearing location where the camshaft meters the flow of oil from the main lubricating gallery through a groove in the camshaft bearing surface to a gallery extending upwards to the cylinder head gasket surface, Fig. 4. On late 1974 and 1975–83 232, 258, oil is supplied through the valve lifters and hollow push rods. On early 1974 232 and 258, install the rocker arm shaft with oil holes facing down toward the cylinder head, Fig. 5. The late 1974 and 1975–83 232, 258 engines have the intake and exhaust rocker arms pivoting on a bridged pivot assembly which is secured to the cylinder head by two cap screws, Fig. 6.

Fig. 6 Rocker arms, push rod & pivot assembly. Late 1974 & 1975–83 6-232, 258 & all V8's.

NOTE: The rocker arms used on late 1974 engines (beginning with build date code 704 (A or E) 04, have been redesigned with the oil hole on the push rod end relocated, pointing away from the valve assemblies allowing for improved oil control, Fig. 7.

The exhaust valve train on certain engines, has been modified to incorporate valve rotators to improve valve durability.

NOTE: When replacing rocker arms on these engines, use only the cast type rocker arms, as they are designed to accommodate the valve rotators.

V8-304, 360, 401

These engines have the intake and exhaust rocker arms pivoting on a bridged pivot assembly which is secured to the cylinder head by two cap screws, Fig. 6.

The push rods are hollow, serving as oil galleries for lubricating each individual rocker arm assembly. Prior to installing, the push rods should be cleaned thoroughly, inspected for wear and deposits which may restrict the flow of oil to the rocker arm assembly.

The push rods also serve as guides to maintain correct rocker arm to valve stem relationship; therefore, a contact pattern on the push rods where they contact the cylinder head is normal.

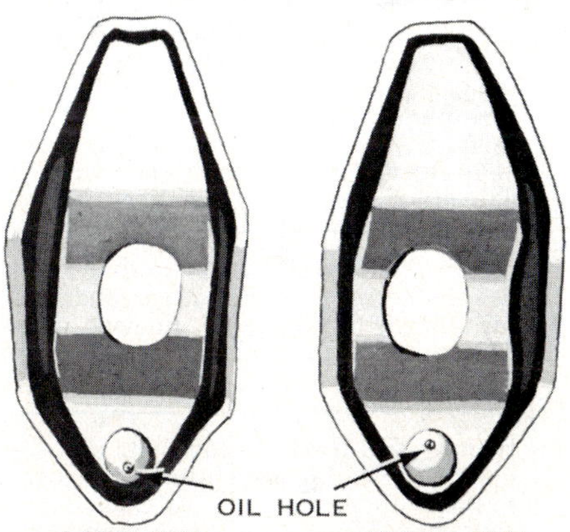

Fig. 7 1974 rocker arm identification

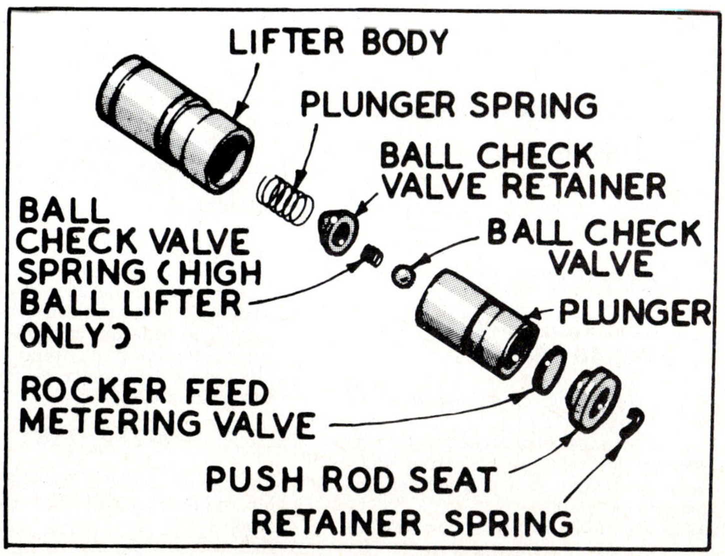

Fig. 8 Valve lifter disassembled view. Typical

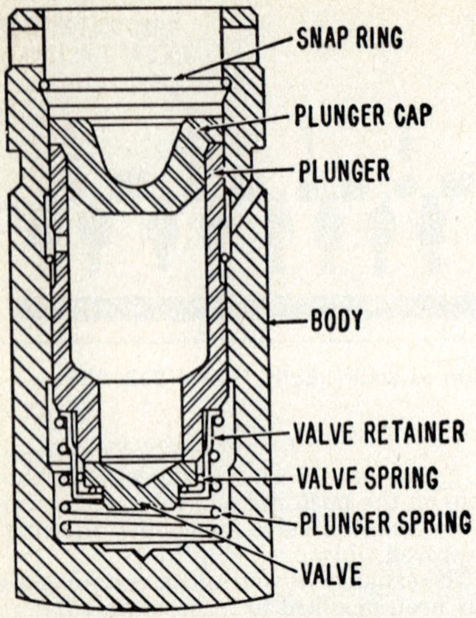

Fig. 9 Hydraulic valve lifter

Lubrication to each rocker arm is supplied by the corresponding hydraulic valve lifter. A metering system located in each valve lifter consists of a stepped lower surface on the push rod cap that contacts a flat plate, causing a restriction, Fig. 8. The restriction meters the amount of oil flow through the push rod cap, hollow push rod, and upper valve train components. A loss of lubrication to the rocker arm could be caused by a restricted or plugged push rod or a defective hydraulic valve lifter.

CAUTION: Correct installation of push rods in these engines is critical and more than normal care must be taken upon installation. When placing the push rods through the guide hole in the cylinder head, it is important that the push rod end is inserted in the plunger cap socket. It is possible that the push rod may seat itself on the edge of the plunger cap which will restrict valve lifter rotation and lubrication to rocker arms.

It is recommended that, just prior to installation of the cylinder head covers, the engine be operated and the supply of lubrication to each rocker arm be visually inspected. If inspection reveals that an individual rocker arm is not being supplied with lubrication, the push rod and/or valve lifter must be inspected to determine the cause.

VALVE ARRANGEMENT

Front to Rear

All V8s . . . . E-I-I-E-E-I-I-E
6-232, 258 . . . . E-I-I-E-I-E-E-I-E-I-I-E

VALVE GUIDES

6-232, 258, V8-304, 360, 401 Engines

Excessive valve stem-to-guide clearance will cause lack of power, rough idling and noisy valves, and may cause valve breakage. Insufficient clearance will result in noisy and sticky functioning of valves and disturb engine smoothness of operation.

Valve stem-to-guide clearances are listed in the *Engine Valve Specifications* table. By using a micrometer and a suitable telescope hole gauge, check the diameter of the valve stem in three places (top, center and bottom). Insert telescope hole gauge in valve guide bore, measuring at the center. Subtract the highest reading of valve stem diameter from valve guide bore center diameter to obtain valve-to-guide clearance. If clearance is not within specified limits, use the next oversize valve and ream bore to fit. Valves with oversize stems are available in .003", .015" and .030".

VALVE LIFTERS

4-151 Engines

To replace valve lifters, remove cylinder head cover, intake manifold and push rod cover. Loosen rocker arm nuts, then rotate rocker arms so push rods can be removed. Remove push rod and valve lifter. Adjustable pliers with taped jaws may be used to remove lifters that are stuck due to varnish, carbon, etc.

All Exc. 4-151

Valve lifters may be removed from their bores after removing the cylinder head. A suitable lifter removing tool or adjustable pliers with taped jaws may be used to remove inaccessible or stuck lifters. Fig. 9 illustrates the type of lifter used.

TIMING CASE COVER

4-151

1. Disconnect battery ground cable.
2. Remove crankshaft pulley hub, then remove alternator bracket.
3. Remove fan and shroud attaching nuts, then loosen drive belts and remove fan and shroud.
4. Remove oil pan to timing case cover attaching screws, then pull cover slightly forward to permit cutting of oil pan front seal.
5. Using a suitable cutting tool, cut oil pan front seal flush with engine block at both sides of cover.
6. Remove front cover with attached portion of oil pan front seal.
7. Clean timing case cover and cylinder block gasket surfaces.
8. Cut tabs from a replacement oil pan front oil seal.
9. Install seal on timing case cover, inserting tips into holes provided in cover.
10. Coat gasket with gasket sealer, then position gasket on cover.
11. Apply a 1/8 inch bead of RTV sealer to joint surface formed at oil pan and cylinder block.
12. Position centering tool J-23042 into timing case cover oil seal, then position cover on block and install and partially tighten two oil pan to timing case cover attaching screws.
13. Install timing case cover to cylinder block attaching screws.
14. Tighten all timing case cover attaching screws, then remove centering tool.
15. Install alternator bracket, crankshaft pulley hub, fan and shroud.
16. Tighten all drive belts, then connect battery ground cable.

6-232, 258

1. Remove drive belts, fan and hub assembly, damper pulley and the vibration damper.
2. Remove oil pan-to-timing case cover screws and the cover-to-cylinder block screws.

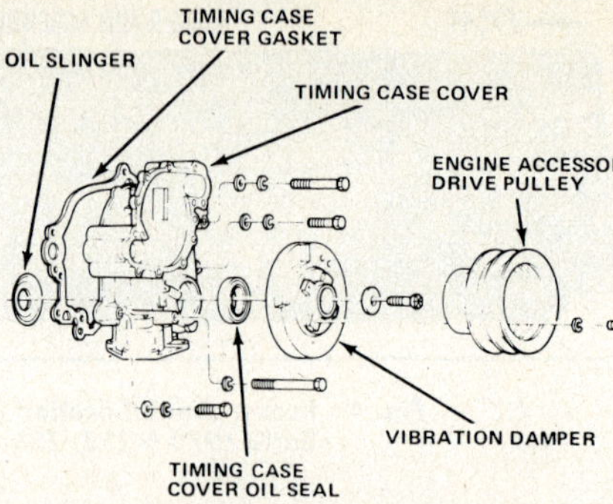

Fig. 10 Timing chain cover assembly V8-304, 360, 401

3. Raise timing case cover enough to detach the retaining nibs of the oil pan seal from the bottom side of the cover. This prevents the seal end tabs pulling away from the tongues of the oil pan gasket which would cause an oil leak requiring oil pan removal to correct.
4. Remove timing case cover and gasket from engine.
5. Cut off oil pan seal end tabs flush with front face of cylinder block and remove seal.
6. Coat both sides of new timing case cover gasket with suitable sealer and position gasket on cylinder block.
7. Cut off tabs of a new pan seal. Coat seal end tabs with suitable sealer and position seal on timing case cover.
8. Install timing case cover on engine, then the cover-to-block and oil pan-to-cover screws.
9. Torque the timing case cover-to-cylinder block screws to 5 ft. lbs. and the oil pan-to-timing case cover screws to 11 ft. lbs.
10. Install vibration damper, pulley, fan and hub assembly and the drive belts.

When installing the cover it is important that the cover be properly aligned when installing the vibration damper to prevent damage to the oil seal. This is accomplished by leaving the cover-to-block cap screws loose until the vibration damper has been partially installed. Then tighten the cover screws.

V8-304, 360, 401

The timing chain cover is a die casting incorporating an oil seal at the vibration damper hub, Fig. 10. The crankshaft front seal is installed from the back side of the cover; therefore, it is necessary to remove the cover when replacement of the seal is required. To remove cover, proceed as follows:

1. Drain cooling system completely.
2. Remove lower radiator hose and by-pass hose from cover.
3. Remove distributor, fuel pump, drive belts, fan and hub assembly and vibration damper, using a suitable puller.

NOTE: It is not necessary to disconnect power steering or discharge air conditioning system (if equipped). Remove units from their mounting brackets and place them aside.

4. Remove two front oil pan bolts and the eight hex head bolts retaining the cover to the cylinder block.
5. Pull cover forward until free from locating dowel pins.
6. Remove used seal and clean seal bore and gasket surface of cover.

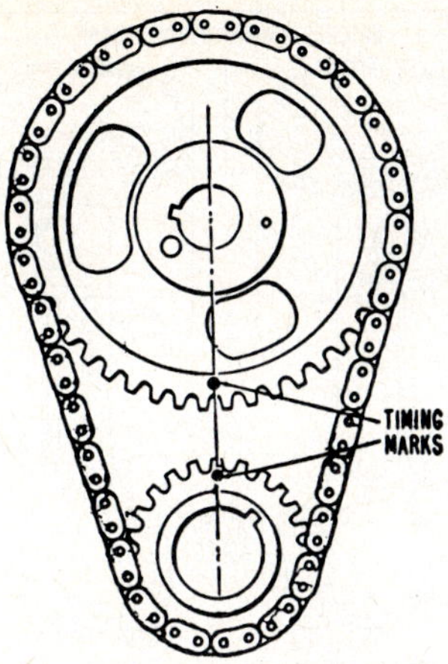

Fig. 11 Valve timing. V8 engine

7. Apply sealing compound to outer surface of seal and a film of Lubriplate or equivalent to seal lips. Drive seal into cover bore until seal contacts outer flange of cover.

Installation of Cover

1. Prior to installation of cover, remove lower dowel pin from cylinder block.
2. Using a sharp knife or razor blade, cut oil pan gasket flush with cylinder block on both sides of oil pan.
3. Cut corresponding pieces of gasket from the replacement oil pan gasket set. Cement gasket to cover. Install replacement Neoprene oil pan seal into cover and align cork gasket tabs to the pan seal.
4. Apply a strip of sealing compound to both the cut-off oil pan gaskets at the oil pan to cylinder block location.
5. Replace cover in position, install oil pan bolts in cover, tighten evenly and slowly until cover aligns with upper dowel. Then install lower dowel through cover. Drive dowel in corresponding hole in cylinder block.
6. Install cover attaching bolts and torque to 20–30 ft-lbs.

TIMING CHAIN

6-232, 258, V8-304, 360, 401

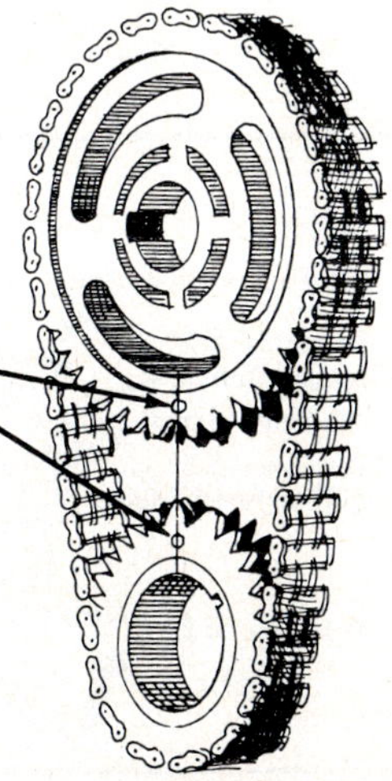

Fig. 13 Valve timing. V8-360, 401 with double row timing chain

When installing a timing chain, see that the timing marks on the sprockets are in line as shown in Figs. 11, 12 and 13.

NOTE: All V8-360 and 401 heavy duty engines built after April 22, 1974 use a new double row roller type timing chain to improve timing chain durability, Fig. 13. This new timing chain is available as a service replacement item and should be installed when timing chain failures occur to eliminate recurrence of the problem. Installation of the new timing chain requires the use of a new heavy duty camshaft sprocket and crankshaft sprocket.

TIMING GEARS

4-151

The camshaft is driven by an iron crankshaft gear meshed with a fiber gear attached to the camshaft, Fig. 14. To remove the fiber timing gear requires the removal of the camshaft. Position camshaft and timing gear assembly in suitable press plate and press camshaft out of gear using appropriate size socket and arbor press.

NOTE: Thrust plate must be properly aligned with camshaft woodruff key when pressing shaft out of gear or damage to thrust plate may result.

Reverse procedure to install timing gear. End clearance of thrust plate should be .0015–.0050 inch. If clearance is less than .0015 inch, replace spacer ring. If clearance is more than .0050 inch, replace thrust plate.

CAMSHAFT

4-151 Engines

To remove camshaft, remove timing case cover as outlined previously and proceed as follows:

1. Remove air cleaner.

WARNING: Do not remove block drain plugs or loosen radiator draincock with system under pressure. Scalding can occur from the hot coolant.

2. Drain radiator.
3. Disconnect radiator hoses at radiator. Refer to radiator removal procedure and remove radiator.
4. Remove camshaft thrust plate screws in camshaft gear.
5. Refer to the appropriate removal procedures and remove distributor, oil and fuel pumps.
6. Remove camshaft and gear assembly, withdraw assemblies through front of block. Secure shaft carefully when removing to prevent camshaft bearing damage.
7. Reverse procedure to install, aligning timing marks, Fig. 14. Thoroughly clean and inspect all parts. Any component found to be unserviceable should be replaced.

6-232, 258

To remove camshaft, remove all valve lifters. Take off timing chain and sprockets. Remove fuel pump, distributor and distributor oil pump drive gear. When withdrawing camshaft from engine, use care to avoid damage to the camshaft bearings.

When installing the timing chain, be sure to align the timing marks as shown in Fig. 12.

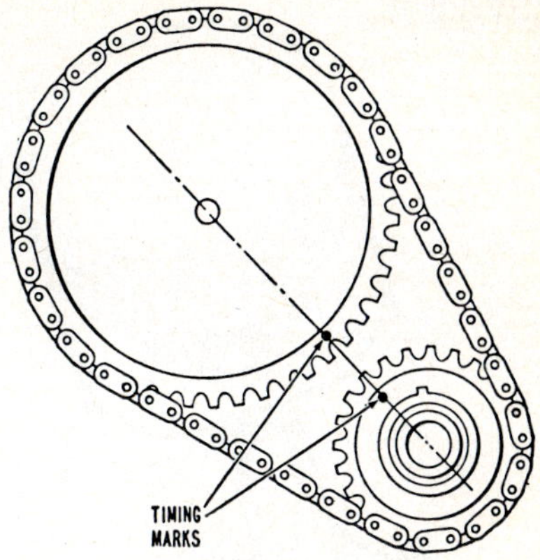

Fig. 12 Valve timing. 6-232, 258

V8-304, 360, 401

1. Disconnect battery ground cable.
2. Disconnect transmission cooler lines at radiator if so equipped.
3. Remove radiator.
4. Remove distributor, wires and coil.
5. Remove intake manifold and carburetor as an assembly.
6. Remove cylinder head covers, loosen rocker arms and remove push rods and lifters.
7. Dismount power steering pump.
8. Remove fan and hub, fuel pump and heater hose at water pump.
9. Remove alternator.
10. Remove vibration damper and pulley and lower radiator hose at water pump.
11. Remove timing chain cover, distributor-oil pump drive gear, fuel pump eccentric sprockets and chain.
12. Remove hood latch support bracket upper retaining screws and move bracket as required to allow removal of camshaft.

When installing the timing chain, be sure to align the timing marks as shown in Fig. 11.

PISTONS & RODS, ASSEMBLE

4-151 Engines

Assemble piston to rod with notch on piston facing toward front of engine and the raised

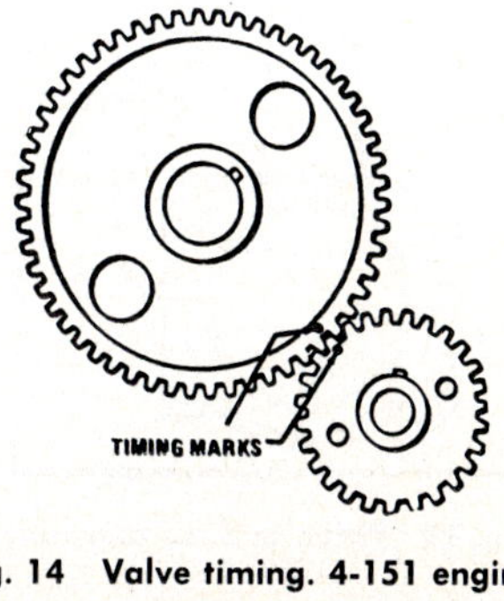

Fig. 14 Valve timing. 4-151 engine

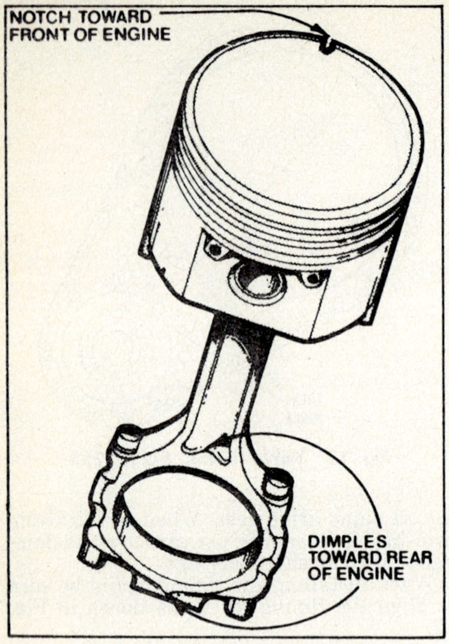

Fig. 15 Piston & rod assembly. 4-151

notch side of rod at bearing end facing toward rear of engine, Fig. 15.

V8-304, 360, 401

Rods and pistons should be assembled and installed as shown in Figs. 16 and 17.

6-232, 258 Engines

Pistons are marked with a depression notch on the top perimeter, Fig. 18. When installed in the engine this notch must be toward the front of the engine. Always assemble rods and caps with the cylinder numbers facing the camshaft side of engine.

PISTONS, PINS & RINGS

Pistons are furnished in standard sizes and oversizes of .002, .005, .010 and .020″.

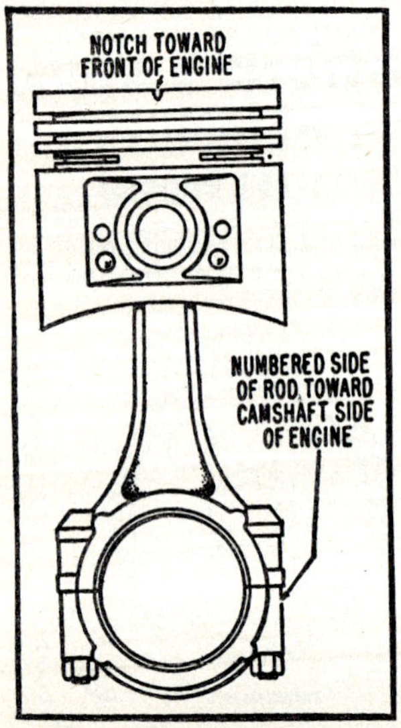

Fig. 18 Piston and rod assembly. 6-232, 258

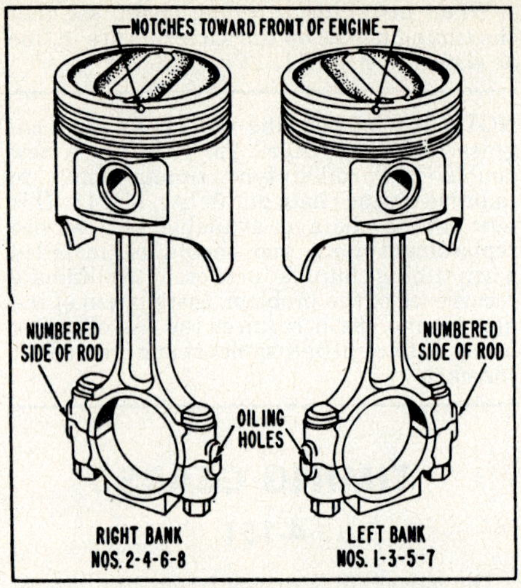

Fig. 16 Piston and rod assembly V8-401

Piston pins are furnished in oversizes of .003 and .005″.

Piston rings are available in .020″ oversizes.

MAIN & ROD BEARINGS

Both main and rod bearings are supplied in undersizes of .001, .002, .010 and .012″.

CRANKSHAFT REAR OIL SEAL

4-151 Engines

The rear main bearing oil seal is a one piece unit and can be removed or installed without removing the oil pan or the crankshaft.

1. Disconnect battery ground cable and raise and support vehicle.
2. Remove transmission and transfer case as a unit. Refer to Transmission, Replace.
3. Remove starter motor.
4. On manual transmissions only, remove inspection plate from flywheel housing.
5. On manual transmissions only, remove hydraulic clutch slave cylinder from flywheel housing.
6. On all models, remove flywheel/drive plate housing from engine.
7. On manual transmissions, remove clutch pressure plate and disc assembly by loosening bolts 1/4 turn in equal amounts until pressure is relieved.
8. On all models, mark flywheel/drive plate location to ensure correct assembly.
9. Remove flywheel/drive plate.
10. Using a small screwdriver, pry out rear main oil seal, using care not to damage seating groove or crankshaft.
11. Center replacement seal over crankshaft with lip facing toward front of engine. Using a soft hammer, tap around perimeter of seal until it seats in groove taking care to prevent seal from binding on crankshaft and not seating properly.
12. Install flywheel/drive plate on crankshaft and torque to 68 ft. lbs.

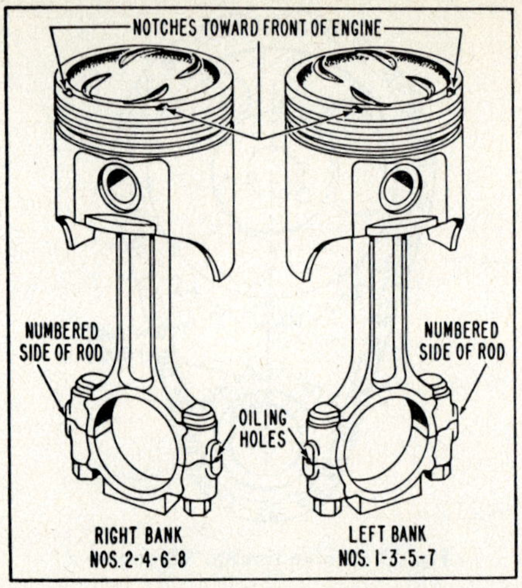

Fig. 17 Piston and rod assembly. V8-304 & V8-360 std.

13. On manual transmissions only, install clutch pressure plate and disc assembly. Alternately tighten bolts 1/4 turn at a time until assembly is flush against flywheel. Torque bolts to 18 ft. lbs.

NOTE: Using alignment tool J-5824-01 or equivalent align clutch disc prior to tightening bolts.

14. Install flywheel/drive plate housing and torque bolts to 35 ft. lbs.
15. On manual transmissions only, install inspection plate on flywheel housing, and install hydraulic clutch slave cylinder on flywheel housing. Torque bolts to 18 ft. lbs.
16. Install transmission/transfer case assembly. Refer to Transmission, Replace.
17. Install starter motor, lower vehicle, and connect battery ground cable.

6-232, 258, V8-304, 360, 401 Engines

1. To replace the seal, Fig. 19, remove oil pan and scrape oil pan surfaces clean.

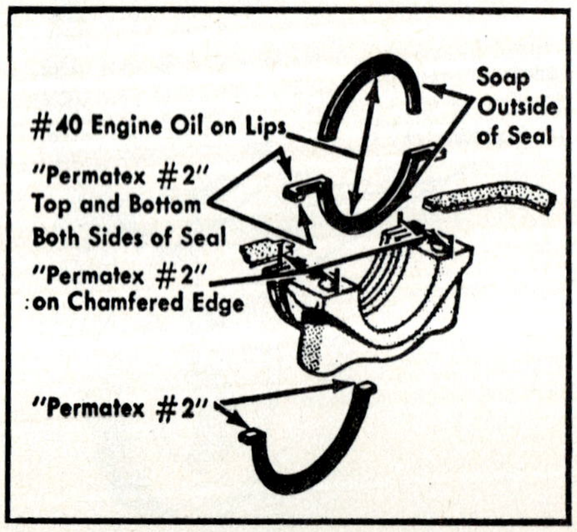

Fig. 19 Rear main bearing oil seal. 6-232, 258, V8-304, 360, 401

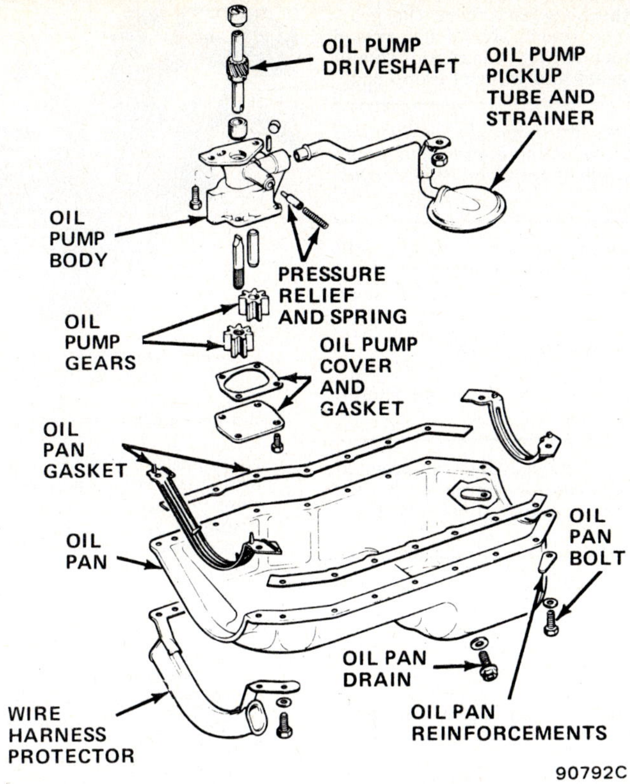

Fig. 20 Oil pan & pump. 4-151 engine

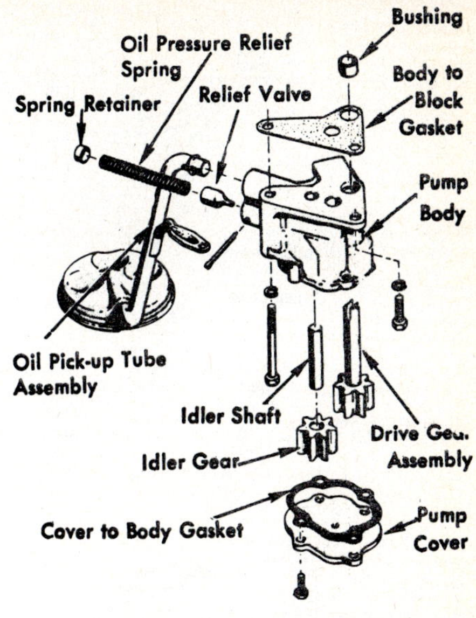

Fig. 21 Oil pump. 6-232, 258

2. Remove rear main bearing cap.
3. Remove and discard oil seals.
4. Clean cap thoroughly.
5. Loosen all remaining main bearing cap screws.
6. With a brass drift and hammer, tap upper seal until sufficient seal is protruding to permit pulling seal out completely with pliers.
7. Wipe seal surface of crankshaft clean, then oil lightly.
8. Coat back surface of upper seal with soap, and lip of seal with No. 40 engine oil.
9. Install upper seal into cylinder block. *Lip of seal must face to front of engine.*
10. Coat cap and cylinder block mating surface portion of seal with Permatex No. 2 or equivalent, being careful not to apply sealer on lip of seal.
11. Coat back surface of lower seal with soap, and lip of seal with No. 40 engine oil. Place into cap, seating seal firmly into seal recess in cap.
12. Place Permatex No. 2 or equivalent on both chamfered edges of rear main bearing cap.
13. Install main bearing and cap. On 6 cylinder engines torque cap to 65–70 ft. lbs. On V-8 engines torque cap to 100 ft. lbs.
14. Cement oil pan gasket to cylinder block with tongue of gasket at each end coated with Permatex or equivalent before installing into rear main bearing cap at joint of tongue and oil pan front neoprene seal.
15. Coat oil pan rear seal with soap. Place into recess of rear main bearing cap, making certain seal is firmly and evenly seated.
16. Install oil pan and tighten drain plug securely.

OIL PUMP

4-151

Engine lubrication is accomplished with a gear type pump that pumps engine oil from the oil pan sump through the full flow oil filter and into an oil passage that runs along the right side of the block and intersects the hydraulic valve tappet bores. Oil from this passage is directed to the crankshaft and camshaft bearings. Oil reaches the rocker arms through holes in the hydraulic valve tappets which force oil up through the tubular push rods to the rocker arms.

A bypass valve in the oil filter mounting boss and a pressure regulator valve in the oil pump body ensure the proper flow of oil. The full flow oil filter is mounted on the right side of the engine.

Removal

1. Drain oil and remove oil pan.
2. Remove flange mounting bolts and nut from main bearing cap bolt and remove pump and pickup assembly as a unit.

Disassembly

1. Remove oil pump cover, idler gear and drive gear and shaft, Fig. 20.
2. Remove pressure regulator valve and parts.

NOTE: The oil pickup pipe is attached during factory assembly and should not be disturbed.

3. The complete pump assembly should be replaced if any of the following conditions are observed during inspection:
   a. Pump body is cracked or shows excessive wear.

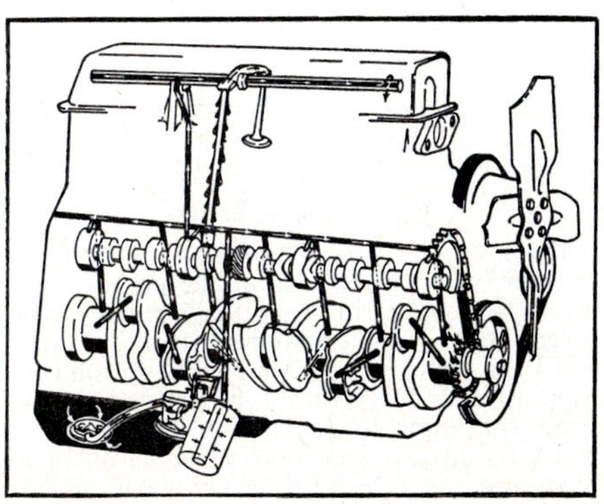

Engine oiling system. Early 1974 6-232, 258

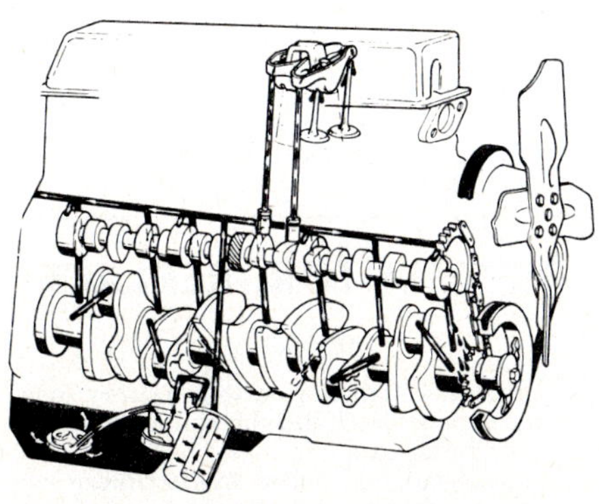

Engine oiling system. Late 1974 & 1975–83 6-232, 258

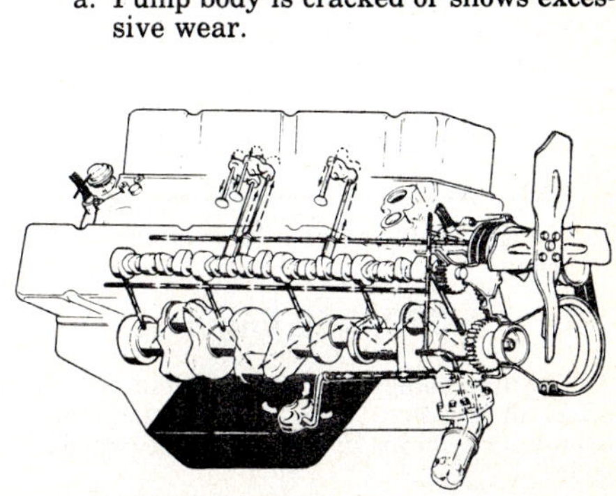

Engine oiling system. V8-304, 360, 401

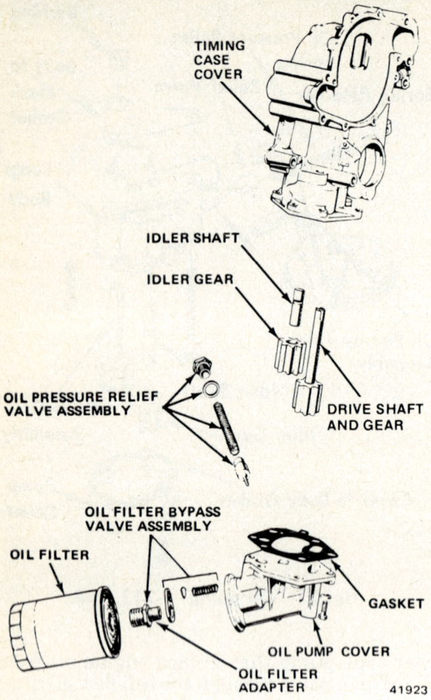

Fig. 22 Oil pump. V8-304, 360, 401

b. Pump gears have cracks, excessive wear, and damage.
c. Shaft is loose in housing.
d. Inside of cover is worn enough to allow oil to leak past end of gears.
e. Oil pickup assembly has damaged strainer screen or relief grommet.

NOTE: Remove any debris found on strainer screen surface.

---

f. Pressure regulator valve plunger does not fit properly in body.

Assembly

1. Place drive gear and shaft in oil pump body.
2. Install idler gear so smooth side of gear will face cover.
3. Install cover and torque attaching screws to 9 ft. lbs. Make sure that shaft turns freely.
4. Install regulator valve plunger, spring, retainer, and pin.

Installation

1. Position oil pump gear shaft tang so it aligns with oil pump drive shaft slot. Install oil pump to block positioning flange over oil pump drive shaft lower bushing. Do not use gasket. Torque bolts to 18 ft. lbs.

NOTE: If oil pump does not slide easily into place, remove shaft and relocate slot.

---

2. Install oil pan with new gaskets and seals.

6-232, 258 Engines

Oil pump removal or replacement will not affect distributor timing as the distributor drive gear remains in mesh with the camshaft gear.

Upon disassembly of the oil pump, Fig. 21 locate a straight edge across the pump body and gears in the body and check the gear-to-cover clearance, which should not exceed .004". A clearance of .008" maximum should exist between gears and walls of pump body.

NOTE: *The pump cover should be installed with the pump out of the engine and pump checked for freedom of operation before installation.*

The oil pressure relief valve, which is built into the pump, is not adjustable, the correct pressure being built into the relief valve spring.

V8-304, 360, 401

The oil pump is driven by the distributor drive shaft, Fig. 22.

NOTE: Oil pump gear removal or replacement will not affect distributor timing as the distributor drive gear remains in mesh with the camshaft gear.

After removal of the oil pump cover and gasket, place a straight edge across the pump body and gears in their operational cavity and check the gear to end clearance with a feeler gauge, Fig. 23. The clearance should be .002–.006 in.

If the gear end clearance is less than specified, replace the timing chain cover and the gear and shaft assemblies.

Measure between the gear tooth and the pump body inner wall directly opposite the point of gear mesh, Fig. 23. Rotate the gears to check each tooth for .0005–.0025 in. clearance. If the gear to body clearance is more than specified, replace the idler gear, idler shaft and drive gear assembly.

NOTE: To ensure self-priming of the oil pump, the pump must be filled with oil prior to installation.

WATER PUMP, REPLACE

Remove the fan belt and blades, unfasten the pump from the cylinder block and lift it off. On some models, it may be necessary to loosen the radiator core and pull it forward in order to remove the water pump assembly.

NOTE: It will be necessary to remove some accessories and the mounting brackets and position aside to gain access to water pump.

CAUTION: 1981–83 California six cylinder engines with a single serpentine drive belt have reverse rotating water pump and viscous fan drive assemblies. The words "REVERSE" stamped on the cover of the viscous drive and inner side of the fan, and "REV" cast into the water pump body identify such components. Do not use components designed for conventional drive belts.

DISTRIBUTOR, REPLACE

4-151

1. Disconnect ignition switch battery lead, tachometer lead and coil connectors from distributor cap.
2. Turn securing latches on distributor cap counterclockwise, and remove cap.
3. Remove hose from vacuum advance unit.

Fig. 23 Checking oil pump gear end clearance

4. Remove clamps and screw from distributor.
5. Note position of rotor, pull distributor up until rotor stops turning, and again note rotor position.
6. If engine was cranked after distributor was removed, use following procedure to install distributor:
   a. Remove No. 1 spark plug, place finger over spark plug hole and crank engine slowly until pressure is felt on finger. At this time piston is at TDC of its compression stroke.
   b. Align timing marks on vibration damper at 0° on engine timing indicator.
   c. Turn rotor to point between No. 1 and No. 3 spark plug towers on distributor cap.
   d. Install distributor lead, ignition lead, distributor cap and spark plug wires.

6-232, 258, V8-304, 350, 360, 401

1. Disconnect distributor primary wire from coil terminal.
2. Remove distributor cap and rotor. *Mark position of rotor arm on distributor housing so distributor can be installed in same position.*
3. Remove vacuum line from distributor.
4. Remove distributor hold-down clamp.
5. Note relative position of distributor in block, then work it out of the engine.

Installation

1. Turn rotor about 1/8 of a turn counterclockwise past the mark previously placed on the distributor housing.
2. Push the distributor down into the block with the housing in the normal "installed" position. *On gear-driven distributors, it may be necessary to move the rotor slightly to start gear into mesh with camshaft gear, but rotor should line up with mark when distributor is down in place.*
3. Tighten distributor clamp screw snugly and connect vacuum line, primary wire to coil, and install cap.

V8 Service Note

If the engine was disturbed while the distributor was removed from the engine, first crank the engine to bring No. 1 piston up on its compression stroke and continue cranking until the timing mark is adjacent to the timing indicator. Then rotate the distributor cam until the rotor is in position to fire No. 1 cylinder. Install the distributor as outlined above and set the ignition timing as directed.

Carburetor Section

MOTORCRAFT 2100, 2150 CARBURETOR ADJUSTMENTS

Year	Carb. Part No.①	Float Level (Dry)	Float Level (Wet)	Choke Plate Pulldown Clearance	Pump Setting	Fast Idle Cam	Fast Idle Speed (on Car)	Choke Unloader Clearance	Dashpot	Choke Setting
1974	4DM2, 4DMJ2	23/32	25/32	.130	—	.130	1600④	1/4	9/64	2 Rich
	4RA2	23/32	25/32	.140	—	.130	1600④	1/4	—	2 Rich
	4RHD2	3/8	3/4	.140	—	.130	1600④	1/4	9/64	2 Rich
1975	5DM2, 5DM2J	.400	.78	.130	—	.130	1600④	1/4	.095	2 Rich
	5RHA2	.555	.93	.140	—	.130	1600④	1/4	—	2 Rich
	5RHM2	.555	.93	.140	—	.130	1600④	1/4	—	2 Rich
1976	6DA2J	.555	.930	⑤	—	⑥	1600③	.250	—	1 Rich
	6DM2	.555	.930	⑤	—	⑥	1600③	.250	.093	1 Rich
	6DM2J	.555	.930	⑤	—	⑥	1600③	.250	.093	2 Rich
	6RHA2, 6RHM2	.555	.930	⑤	—	⑥	1600③	.250	—	2 Rich
1977	6DA2J	.555	.930	.136	—	.126	1600③	.250	—	1 Rich
	6DM2	.555	.930	.132	—	.120	1600③	.250	.093	2 Rich
	6DM2J	.555	.930	.132	—	.120	1600③	.250	.093	1 Rich
	6RHA2, 6RHM2	.555	.930	.136	—	.115	1600③	.250	—	2 Rich
	7DA2A	.555	.930	.104	—	.089	1600③	.290	—	1 Rich
	7DM2A	.555	.930	.110	—	.089	1600③	.290	—	2 Rich
1978	8DM2A	.555	.930	.093	—	.078	1600③	.290	—	2 Rich
	8DM2	.555	.930	.132	—	.120	1500③	.250	—	2 Rich
	8DM2C	.555	.930	.132	—	.120	1500③	.250	—	1 Rich
	8DA2J	.555	.930	.136	—	.126	1600③	.250	—	1 Rich
	8DA2JC	.555	.930	.136	—	.126	1600③	.250	—	1 Rich
	6RHA2	.555	.930	.136	—	.115	1600③	.250	—	2 Rich
	6RHM2	.555	.930	.136	—	.115	1600③	.250	—	2 Rich
	8RHA2C	.555	.930	.136	—	.126	1600③	.250	—	2 Rich
1979	9DA2J	.555	.930	.128	—	.113	1600②	.300	—	1 Rich
	9DM2	.555	.930	.125	—	.120	1500②	.300	—	2 Rich
	9DM2C	.555	.930	.132	—	.120	1500②	.250	—	1 Rich
	9DM2H	.555	.930	.140	—	.125	1500②	.360	—	Index
	9RHA2	.555	.930	.113	—	.093	1600③	.350	—	2 Rich
	9RHM2	.555	.930	.104	—	.086	1500③	.348	—	2 Rich
1980	0DA2J	.375	.930	.128	—	.113	1600②	.300	—	1 Rich
	0DA2J2	.375	.930	.120	—	.106	1600②	.300	—	2 Rich
	0DM2A	.375	.930	.128	—	.113	1500②	.360	—	1 Rich
	0DMJ2	.375	.930	.113	—	.113	1500②	.300	—	2 Rich
	0DM2JC	.375	.930	.120	—	.106	1500②	.300	—	2 Rich
	0DHA2	.575	.930	.113	—	.086	1600②	.350	—	2 Rich
	0DHM2	.575	.930	.014	—	.081	1500②	.348	—	2 Rich
1981	DMJ2	—	.930	.125	—	.113	1500	.300	—	2 Rich
	DA2J	—	.930	.128	—	.113	1600	.300	—	1 Rich
	DM2A	—	.930	.128	—	.113	1500	.360	—	1 Rich
	RHM2	—	.930	.104	—	.081	1500	.348	—	2 Rich
	RHA2	—	.930	.113	—	.086	1600	.350	—	2 Rich
1982	RHM2	—	.930	.116	—	.076	1500	.350	—	1 Rich
	RHA2	—	.930	.116	—	.076	1600	.350	—	1 Rich
1983	2RHA2	—	.930	.116	—	.076	1600	.420	—	1 Rich
	1RHM2	—	.930	.104	—	.081	1500	.348	—	2 Rich
	1RHA2	—	.930	.113	—	.086	1600	.350	—	2 Rich

①—Located on tag attached to carburetor on casting.
②—Engine hot and EGR valve disconnected.
③—On second step of fast idle cam with TCS and EGR disconnected.
④—Second step of cam.
⑤—V8-304 with manual trans.; .132". V8-304 with auto. trans.; .136". V8-360; .136".
⑥—V8-360; .115". V8-304 with manual trans.; .120". V8-304 with auto. trans.; .126".

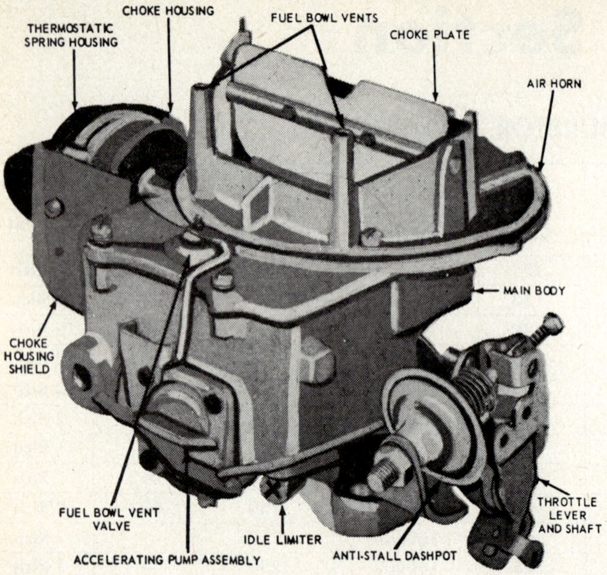

Fig. 1 Autolite/Motorcraft 2100 carburetor (Typical)

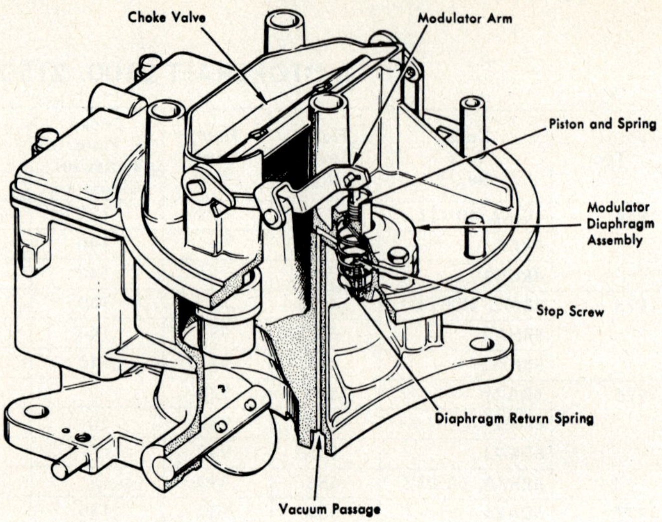

Fig. 3 Autolite/Motorcraft 2100 choke diaphragm assembly

Model 2100, 2150

Figs. 1 & 2

These carburetors have two main bodies—the air horn and throttle body. The air horn assembly, which serves as a cover for the throttle body, contains the choke plate and vents for the fuel bowl. The 1974 units use a fuel bowl vent valve.

A choke modulator assembly is incorporated in the 1974–81 2100 units, Fig. 3. The 1974–81 units have an additional port for the Exhaust Gas Recirculation (EGR) system. Model 2150-2 incorporates a high altitude compensation circuit which parallels the main carburetor intake circuit. Compensation circuit is controlled by an aneroid calibrated during factory assembly and is non-adjustable.

The throttle plate, accelerating pump, power valve and fuel bowl are in the throttle body. The choke housing is attached to the throttle body.

The two bodies each contain a main and booster venturi, main fuel discharge, accelerating pump discharge, idle fuel discharge, and a throttle plate. An antistall dashpot is attached to the carburetor when the vehicle is equipped with an automatic transmission.

Float Level Adjustment

Fig. 4—This is a preliminary adjustment; the final adjustment must be made after the carburetor is mounted on the engine.

With air horn removed, float raised and fuel inlet needle seated, measure distance between top surface of throttle body and top surface of float.

Depress the float tab to seat the fuel inlet needle. The float height is measured from the gasket surface of the throttle body with gasket removed. If the float height is not as listed in the 2100, 2150 *Specifications Chart,* bend tab on float as required to achieve the desired setting.

Fuel Level Adjustment

Fig. 5—With vehicle on a level surface, operate engine until normal temperature is reached, then stop engine and check fuel level as follows:

1. Remove carburetor air cleaner.
2. Remove air horn retaining screws and carburetor identification tag.
3. Temporarily leave air horn and gasket in position on throttle body and start engine.
4. Allow engine to idle for several minutes, then rotate air horn and remove air horn gasket to gain access to float or floats.
5. While engine is idling, use a standard depth gauge to measure vertical distance from top machined surface of throttle body to level of fuel in bowl. The measurement must be made at least 1/4″ away from any vertical surface to assure an accurate reading.
6. If the fuel level is not as listed in the 2100, 2150 *Specifications Chart,* stop the engine to avoid any fire hazard due to fuel spray when float setting is disturbed.
7. To adjust fuel level, bend float tab (contacting fuel inlet needle) upward in relation to original position to raise the fuel level, and downward to lower it.
8. Each time an adjustment is made to the float tab to alter the fuel level the engine must be started and permitted to idle for at least three minutes to stabilize the fuel level. Check fuel level after each adjustment until the specified level is achieved.
9. Assemble carburetor with a new air horn gasket. Then adjust idle speed and mixture, and anti-stall dashpot, if so equipped.

Fig. 4 Float level adjustment 2100, 2150

Accelerating Pump Adjustment

Fig. 6—The primary throttle shaft lever (overtravel lever) has 4 holes and the accelerating pump link has 2 holes to control the pump stroke for various atmospheric temperatures, operating conditions and specific engine applications.

NOTE: The stroke should not be changed from the specified setting.

1. To release rod from retainer clip, press tab end of clip toward rod. Then, at the same time, press rod away from clip until it is disengaged.
2. Position clip over specified hole in overtravel lever. Press ends of clip together and insert opening rod through clip and lever. Release clip to engage rod.

Choke Plate Clearance (Pulldown) Adjustment

1978–81 2100 Units

1. Loosen choke cover retaining screws and

1. Pivot Pin
2. Modulator Arm
3. Choke Valve Retaining Screw (2)
4. Choke Valve
5. Choke Shaft
6. Air Horn
7. Air Horn Retaining Screw (4)
8. Air Horn Gasket
9. Float Shaft Retainer
10. Float and Lever Assembly
11. Needle Retaining Clip
12. Needle and Seat Assembly
13. Needle Seat Gasket
14. Float Shaft
15. Curb Idle Adjusting Screw
16. Curb Idle Adjusting Screw Spring
17. Throttle Shaft and Lever Assembly
18. Dashpot
19. Dashpot Locknut
20. Dashpot Bracket
21. Dashpot Bracket Retaining Screw
22. Throttle Valve Retaining Screw (4)
23. Throttle Valve (2)
24. Main Jet (2)
25. Main Body
26. Pump Rod Retainer
27. Pump Rod
28. Elastomer Valve
29. Pump Return Spring
30. Pump Diaphragm
31. Pump Lever Pin
32. Pump Cover
33. Pump Lever
34. Pump Cover Retaining Screw (4)
35. Fuel Inlet Fitting
36. Power Valve Gasket
37. Power Valve
38. Power Valve Cover Gasket
39. Power Valve Cover
40. Power Valve Cover Retaining Screw (4)
41. Idle Limiter Cap (2)
42. Idle Mixture Screw (2)
43. Idle Mixture Screw Spring (2)
44. Retainer
45. Retainer
46. Fast Idle Lever Retaining Nut
47. Fast Idle Lever Pin
48. Retainer
49. Retainer
50. Fast Idle Cam Rod
51. Choke Shield
52. Choke Shield Retaining Screw (2)
53. Piston Passage Plug
54. Heat Passage Plug

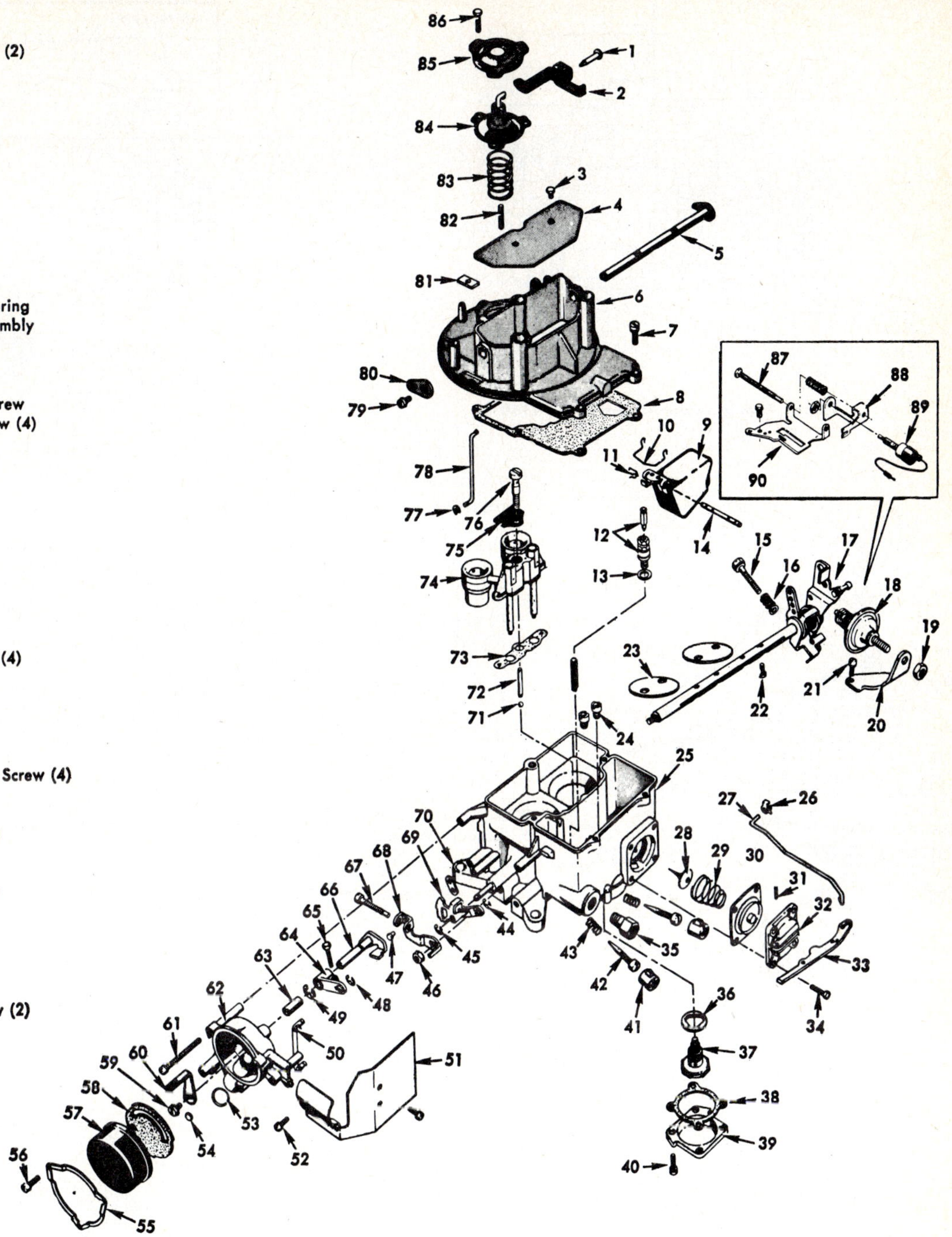

55. Choke Cover Retaining Clamp
56. Choke Cover Retaining Screw (3)
57. Choke Cover
58. Choke Cover Gasket
59. Thermostat Lever Retaining Screw
60. Thermostat Lever
61. Choke Housing Retaining Screw (3)
62. Choke Housing
63. Choke Shaft Bushing
64. Fast Idle Cam Lever
65. Fast Idle Cam Lever Adjusting Screw
66. Thermostatic Choke Shaft
67. Fast Idle Speed Adjusting Screw
68. Fast Idle Lever
69. Fast Idle Cam
70. Choke Housing Gasket
71. Pump Discharge Check Ball
72. Pump Discharge Weight
73. Booster Venturi Gasket
74. Booster Venturi Assembly
75. Air Distribution Plate
76. Pump Discharge Screw
77. Retainer
78. Choke Rod
79. Choke Lever Retaining Screw
80. Choke Lever
81. Choke Rod Seal
82. Stop Screw
83. Modulator Return Spring
84. Modulator Diaphragm Assembly
85. Modulator Cover
86. Modulator Retaining Screw (3)
87. Adjusting Screw
88. Carriage
89. Electric Solenoid
90. Mounting Bracket

Fig. 2 Exploded view of a typical 2100 two barrel carburetor

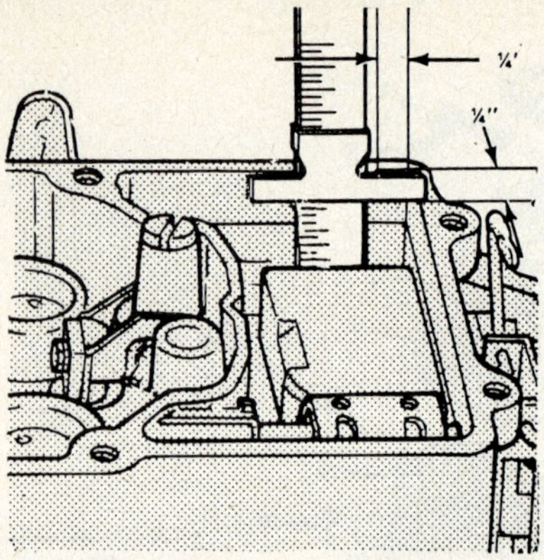

Fig. 5 Fuel level adjustment 2100, 2150

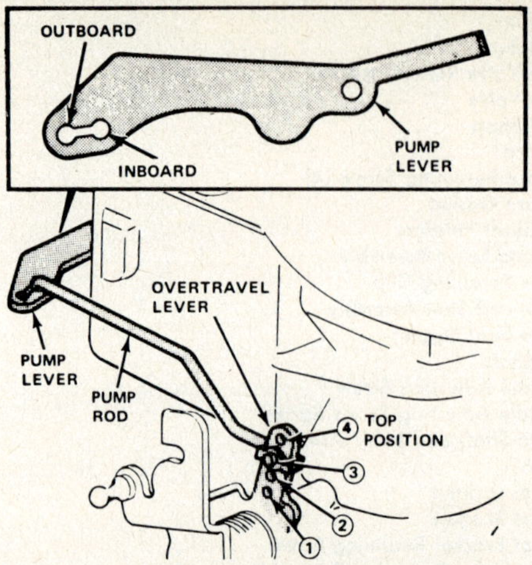

Fig. 6 Pump stroke adjusting points 2100, 2150

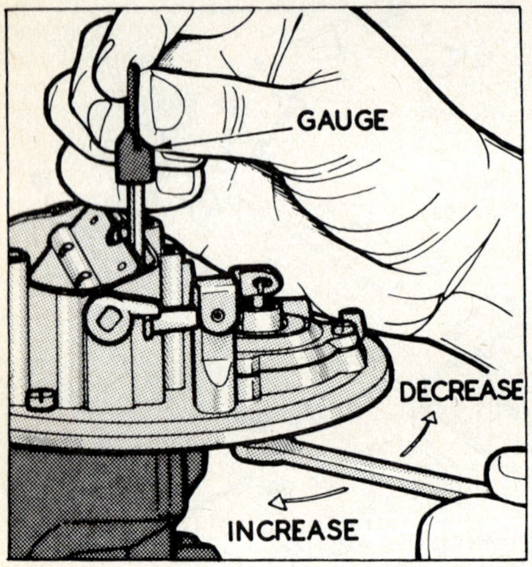

Fig. 7 Choke plate pulldown adjustment. 1978–80 2100 units

rotate cover 1/4 turn counterclockwise, then tighten one screw.
2. Disconnect choke heat inlet tube. Align fast idle speed adjusting screw with second step of fast idle cam.
3. Start engine without moving accelerator linkage. Turn fast idle cam lever adjusting screw counterclockwise three turns. Measure clearance between upper edge of choke valve and air horn wall, Fig. 7. Refer to 2100, 2150 Specification Chart.
4. Adjust by turning set screw on bottom of modulator.
5. Stop engine and connect choke heat tube. Adjust fast idle cam clearance.

1974–77 2100 Units

1. Loosen the choke cover retaining screws. Rotate the choke cover 1/4 turn counterclockwise (rich) from index and tighten the retaining screws. Disconnect the choke heat inlet tube.
2. Align the fast idle speed adjusting screw with the second step of the fast idle cam.
3. Start the engine without moving the accelerator linkage. Turn the fast idle cam lever adjusting screw out counterclockwise 3 full turns.
4. Measure the clearance between the lower edge of the choke valve and the air horn wall. Refer to *2100, 2150 Specifications Chart* for the correct setting. Adjust by grasping the modulator arm securely with a pair of pliers at point "A" and twisting the arm at point "B" with a second pair of pliers. Twist toward the front of the carburetor to increase clearance and toward the rear to decrease clearance, Fig. 8.

CAUTION: Use extreme care while twisting the modulator arm to avoid damaging the nylon piston rod of the modulator assembly.

NOTE: Connect the choke heat tube. Turn the fast idle cam lever adjusting screw in (clockwise) 3 full turns. Do not reset the choke cover until the fast idle cam linkage adjustment has been performed.

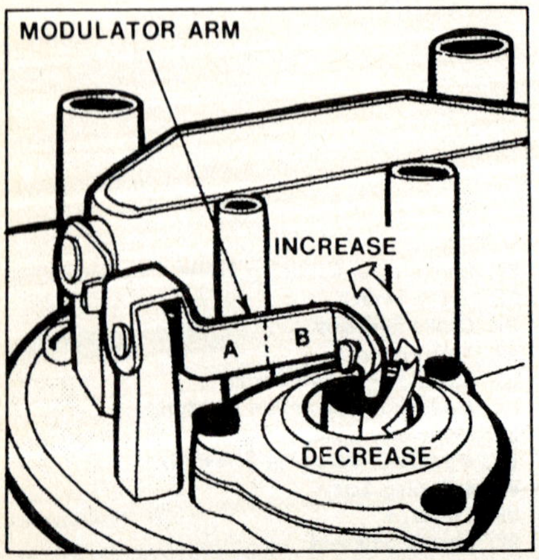

Fig. 8 Choke plate pulldown clearance 2100

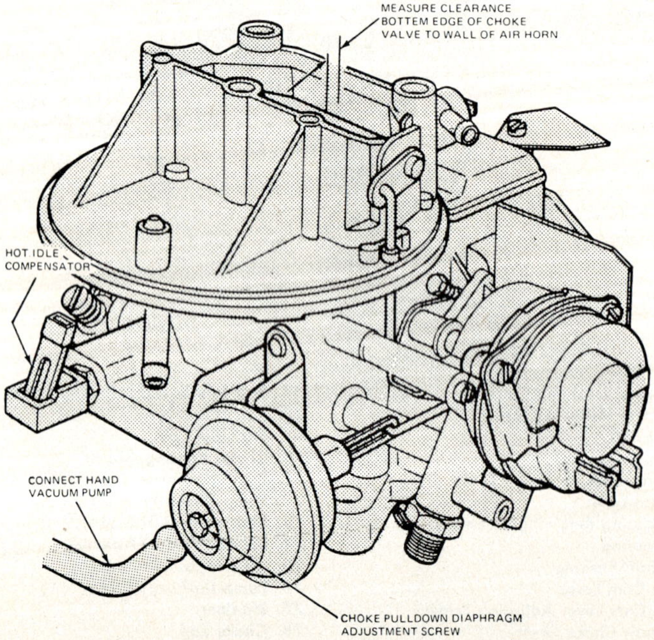

Fig. 9 Choke plate clearance adjustment. 2150

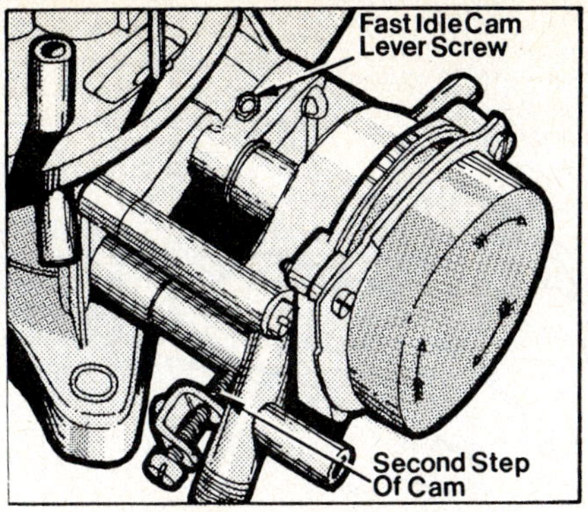

Fig. 10 Fast idle cam linkage adjustment 2100, 2150

2150 Units

1. Set throttle on fast idle cam top step, then loosen choke thermostatic housing retaining screws and set housing 90° in rich direction.
2. Activate pulldown motor by manually forcing pulldown control diaphragm link in direction of applied vacuum or by applying vacuum to external vacuum tube.
3. Check clearance between lower edge of choke plate and center of carburetor air horn wall nearest fuel bowl Fig. 9. Refer to 2100, 2150 Specifications Chart. If clearance is not as specified, reset by adjusting diaphragm stop on end of choke pull-down diaphragm.

Fast Idle Cam Clearance

1. Loosen the choke thermostatic spring housing retainer screws and set the housing 90 degrees in the rich direction.
2. Position the fast idle speed screw at the kickdown step of the fast idle cam. The kickdown step is identified on most units by a V stamped on the cam. When a two-piece fast idle lever is used, a tang on the top lever will align with the V mark on the cam.
3. Be sure the cam is at the kickdown position while checking or adjusting the fast idle cam clearance. Check clearance between the lower edge of the choke plate and the air horn wall, Fig. 10. Refer to *2100, 2150 Specifications Chart.* Adjust clearance, by turning the fast idle cam clearance adjusting screw clockwise to increase and counterclockwise to decrease the clearance.
4. Set the choke thermostatic spring housing to specifications. Adjust the anti-stall dashpot, idle speed and fuel mixture.

Fig. 11 Choke unloader adjustment. 2100, 2150

Unloader Adjustment

1. Hold throttle fully open and apply pressure on choke valve toward the closed position.
2. Measure clearance between lower edge of choke valve and air horn wall. Refer to the 2100, 2150 Specifications Chart.
3. Adjust by bending the unloader tang contracting fast idle cam, Fig. 11. Bending the tang toward the cam increases the clearance and bending the tang away from the cam decreases the clearance.

NOTE: Do not bend unloader tang downward from a horizontal plane.

4. Open the throttle until the unloader tang is directly below the fast idle cam pivot. A clearance of .070 inch must be obtained between the unloader tang and the edge of the fast idle cam, Fig. 12.

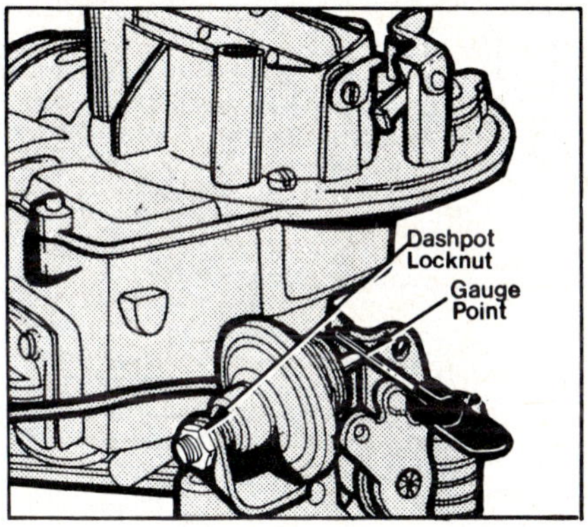

Fig. 13 Dashpot adjustment 2100, 2150

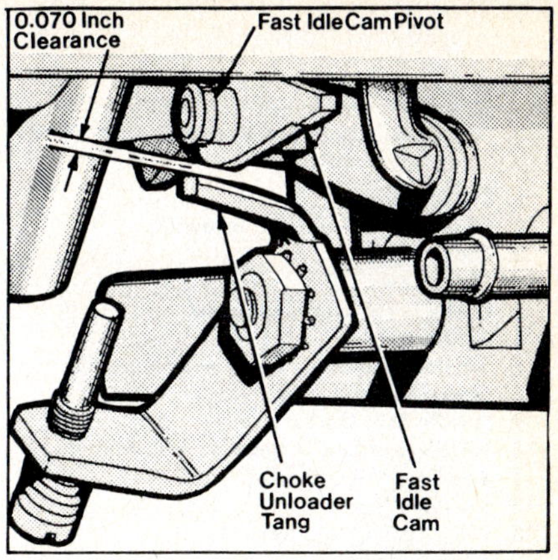

Fig. 12 Choke unloader to fast idle cam clearance. 2100, 2150

Anti-Stall Dashpot Adjustment

Fig. 13—With engine idle speed and mixture properly adjusted and with engine at normal operating temperature, loosen dashpot lock nut. Hold throttle in closed position and depress plunger with screwdriver as shown. Check clearance between throttle lever and plunger tip. If clearance is not as listed in the *2100, 2150 Specifications Chart,* turn dashpot in its bracket as required to obtain the desired clearance. Tighten lock nut.

MOTORCRAFT 4300 CARBURETOR ADJUSTMENTS

Year	Carb. Part No.①	Float Level	Choke Plate Pulldown Clearance	Pump Setting	Fast Idle Cam	Fast Idle Speed (Hot Engine)	Choke Unloader Clearance	Dashpot Setting	Choke Setting
1974	4TA4	13/16	11/64	—	.160	1600②	.325	—	2 Rich
	4THD4	13/16	11/64	—	.160	1600②	.325	9/64	2 Rich

①—Located on tag attached to carburetor on casting.

②—Second step of cam.

MODEL 4300

The model 4300, Figs. 1 and 2, is a four barrel, three piece, separately cast design consisting of air horn, main body and mounting flange. A cast-in center fuel inlet has provision for a supplementary fuel inlet system. The fuel bowl is vented by an internal balance vent and a mechanical atmospheric vent operates during idle. An idle air by-pass system is designed to provide a consistent idle and a hot idle compensator is used to help stability.

The main (primary) fuel system has booster-type venturi cast integral with the air horn, and the main venturi are cast integral with the main body. The secondary throttle plates are mechanically operated from the primary linkage.

An integral hydraulic dashpot dampens sudden movement of the air valve plates to help prevent plate flutter and erratic engine operation.

The automatic choke system consists of a standard bimetal thermostat.

An electric assist choke is incorporated within the choke cover to speed up the opening of the choke valve when the underhood air

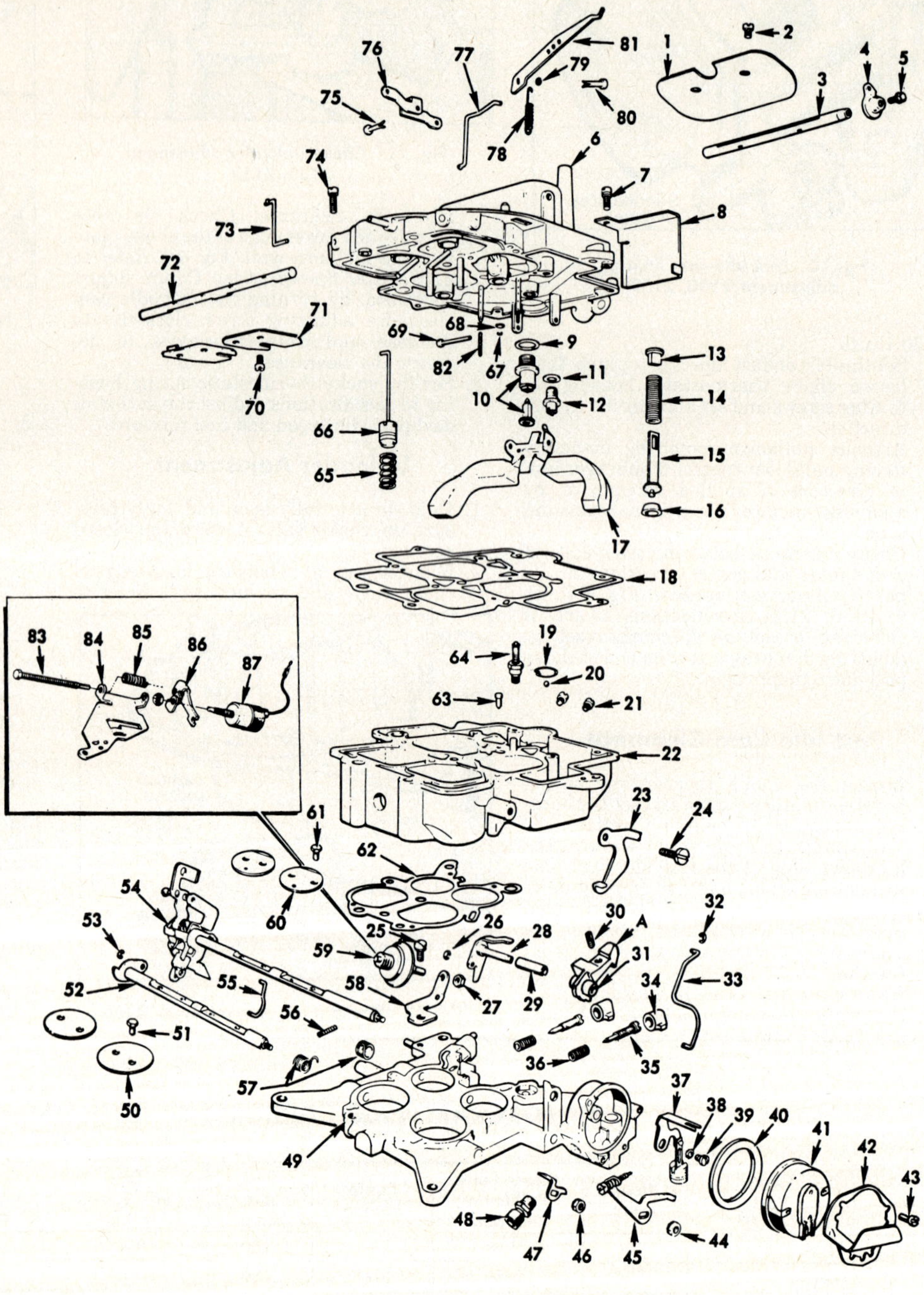

1. Choke Valve
2. Choke Valve Retaining Screw (2)
3. Choke Shaft
4. Choke Lever
5. Choke Lever Retaining Screw
6. Air Horn
7. Air Horn Retaining Screw — Phillips (10)
8. Choke Shield
9. Needle Seat Gasket
10. Needle and Seat Assembly
11. Auxiliary Valve Gasket
12. Auxiliary Valve
13. Pump Spring Retainer
14. Pump Spring
15. Pump Piston
16. Pump Rubber Cup
17. Float and Lever Assembly
18. Air Horn Gasket
19. Inlet Check Valve Retainer
20. Inlet Check Valve
21. Main Jet (2)
22. Main Body
23. Secondary Throttle Lockout Lever
24. Lockout Lever Retaining Screw
25. Dashpot Bracket Retaining Screw
26. Choke Rod Lower Retainer
27. Dashpot Locknut
28. Thermostatic Choke Shaft
29. Thermostatic Choke Shaft Bushing
30. Fast Idle Cam Adjusting Screw
31. Fast Idle Cam
32. Choke Rod Upper Retainer
33. Choke Rod
34. Idle Limiter Cap (2)
35. Idle Mixture Screw (2)
36. Idle Mixture Screw Spring (2)
37. Choke Piston and Lever Assembly
38. Choke Piston Lever Retaining Washer
39. Choke Piston Lever Retaining Screw
40. Choke Cover Gasket
41. Choke Cover (Electric)
42. Choke Cover Retaining Clamp
43. Choke Cover Retaining Screw (3)
44. Fast Idle Lever Retaining Nut
45. Fast Idle Lever
46. Secondary Stop Lever Retaining Nut
47. Secondary Throttle Stop Lever
48. Choke Heat Inlet Fitting
49. Throttle Body
50. Secondary Throttle Valve (2)
51. Throttle Valve Retaining Screw (4)
52. Secondary Throttle Shaft
53. Secondary Link Retainer
54. Primary Throttle Shaft Assembly
55. Secondary Throttle Link
56. Curb Idle Adjusting Screw
57. Throttle Return Spring
58. Dashpot Bracket
59. Dashpot
60. Primary Throttle Valve (2)
61. Throttle Valve Retaining Screw (4)
62. Throttle Body Gasket
63. Pump Discharge Needle
64. Power Valve
65. Dampener Spring
66. Dampener Piston
67. Pump Air Bleed Retainer
68. Pump Air Bleed (Viton Disc)
69. Float Pin
70. Air Valve Retaining Screw (4)
71. Air Valve (2)
72. Air Valve Shaft
73. Air Valve Link
74. Air Horn Retaining Screw — Common
75. Pivot Pin
76. Air Valve Lever
77. Pump Rod
78. Pump Return Spring
79. Pump Rod Washer
80. Pivot Pin
81. Pump Lever
82. Atomizing Screen
83. Adjusting Screw
84. Mounting Bracket
85. Adjusting Screw Spring
86. Carriage
87. Electric Solenoid

Fig. 2 Exploded view of a Motorcraft 4300 4-barrel carburetor

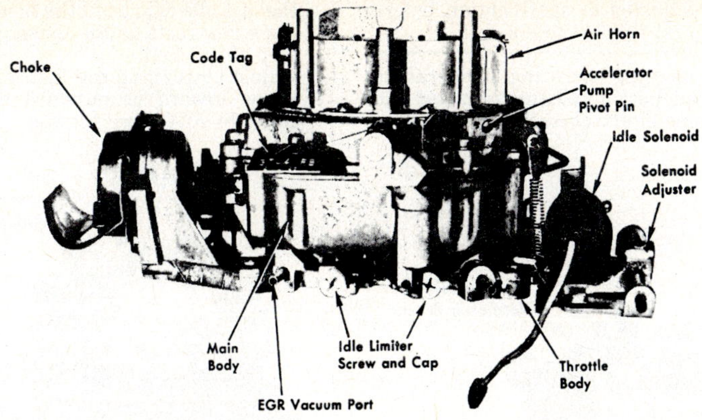

Fig. 1 Motorcraft 4300 4-barrel carburetor

temperature reaches approximately 95 degrees F.

A single fuel bowl supplies both the primary and secondary fuel systems. Pontoon-type floats are used to help stabilize fuel level during cornering and hill-climbing. The accelerator pump is of the piston type. It is located in the fuel bowl for more positive displacement and a safeguard against external leaks.

Float Setting

Fig. 3—Check clearance and alignment of float pontoons to gauge. Both pontoons should just touch gauge for proper setting. Align pontoons if necessary by slightly twisting pontoons.

Adjust the float level by bending the tab which contacts the fuel inlet needle.

Auxiliary Inlet Valve Adjustment

1. Turn the air horn assembly upright allowing the float to hang freely.
2. Push up on the float until the primary fuel inlet needle lightly contacts its seat.
3. While holding the float in this position, measure the clearance between the float level auxiliary tab and the auxiliary inlet valve plunger, Fig. 4.
4. Adjust by bending the tab. Refer to *4300 Specifications Chart.*

Accelerator Pump Stroke Adjustment

Fig. 5—The accelerator pump stroke has been calibrated to inject a pre-determined quantity of fuel into the air stream and help keep the exhaust emission level of the engine within the specified limits. The additional holes provided for pump stroke adjustment are for adjusting the stroke for specific engine applications.

NOTE: The stroke should not be changed from the specified setting.

If the pump stroke has been changed from the specified setting, refer to the following instructions to correct the stroke to specification. Before adjusting the accelerator pump stroke, measure the height of the pump piston. Adjust by bending the pump control rod to correct the piston stem height. Refer to *4300 Specifications Chart.*

Unloader Adjustment

With the throttle held fully open, apply pressure on the choke valve and measure the clearance between the lower edge of the choke valve and the air horn wall, Fig. 6. Refer to *4300 Specifications Chart.* Adjust by bending the tang on the fast idle lever which contacts the fast idle cam.

NOTE: For 1974, the unloader adjustment has been eliminated. It will be necessary to check for wide-open throttle to ensure proper unloading.

NOTE: Final unloader adjustment must be accomplished on vehicle. To make sure there is a full throttle opening, the throttle should be fully depressed with accelerator pedal to floor.

Choke Plate Pulldown Adjustment

1. Bend a .035 inch wire gauge at a 90 degree angle approximately 1/8 inch from the end.
2. Partially open the throttle and close the choke valve. This will move the fast idle cam and choke piston into the cold start position.
3. Hold the choke valve fully closed; release the throttle and the choke valve. Insert the wire gauge into the slot at the front of the choke piston passage, Fig. 7.
4. Pull the choke piston lever counterclockwise until the choke piston locks the wire gauge in place.
5. Hold the choke piston in this position and check the clearance between the lower edge of the choke valve and the air horn wall, Fig. 7. Refer to *4300 Carburetor Specifications Chart.*
6. Adjust by loosening the choke lever retaining screw and rotating the choke shaft.
7. Install choke cover and rotate it to specified setting, then tighten cover.

NOTE: The retaining screw has a left hand thread and must be turned clockwise to loosen.

Fast Idle Cam Linkage Adjustment

1. Rotate the choke cover 1/4 turn clockwise and tighten the retaining screws.
2. Operate the throttle to allow the choke valve to close completely.
3. Push down on the fast idle cam counterweight until the screw is in contact with the second step and against the shoulder of the high step.
4. Measure the clearance between the lower edge of the choke valve and the air horn wall, Fig. 8. Refer to *4300 Carburetor Specifications Chart.*
5. Adjust by turning the fast idle cam adjusting screw.
6. Loosen the choke cover retaining screw and adjust the choke. Refer to *4300 Carburetor Specifications Chart.*

Dashpot Adjustment

Set throttle to curb idle and fully depress the dashpot stem. Measure the clearance between the stem and the throttle lever Fig. 5. Refer to *4300 Specifications Chart.* Adjust by

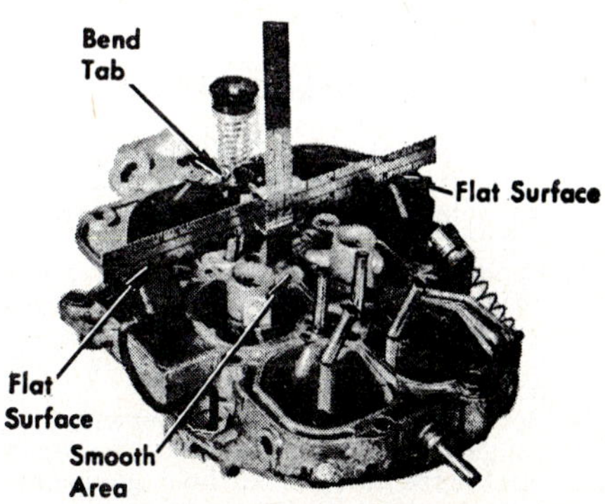

Fig. 3 Float adjustment 4300 4-barrel carburetor

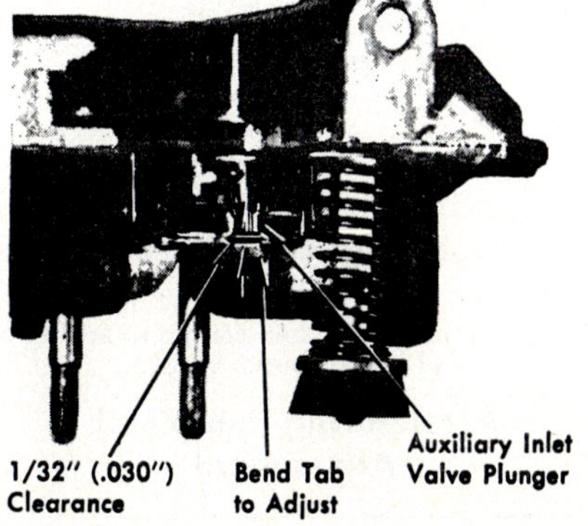

Fig. 4 Auxiliary inlet valve adjustment

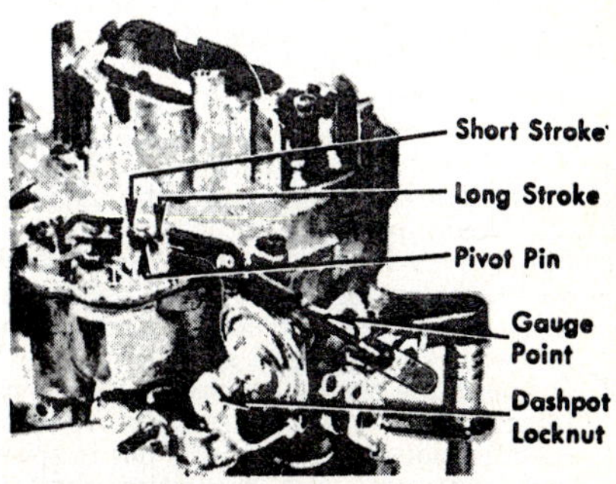

Fig. 5 Accelerator pump and dashpot adjustment

loosening the lock nut and turning the dash-pot.

Fast Idle Speed Adjustment

NOTE: Make certain the curb idle speed and mixture are adjusted to specification before attempting to set the fast idle speed.

1. With the engine operating temperature normalized (hot), air cleaner removed and the tachometer attached, manually rotate the fast idle cam until the fast idle adjusting screw rests on the kickdown (middle) step of the cam.
2. Adjust by turning the fast idle adjusting screw inward or outward as required. Refer to *4300 Specifications Chart.*

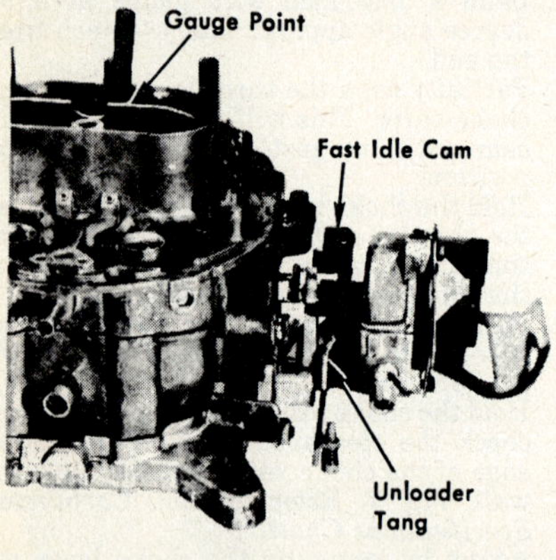

Fig. 6 Unloader adjustment

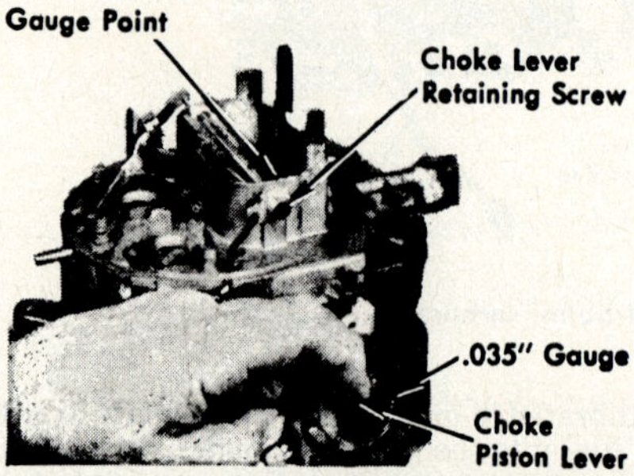

Fig. 7 Choke plate pulldown adjustment

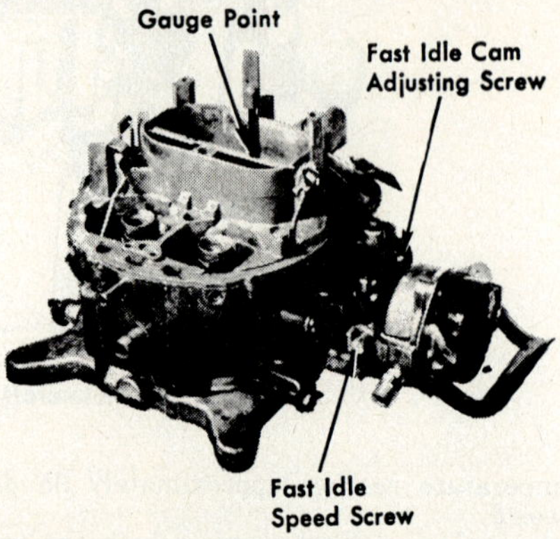

Fig. 8 Fast idle cam linkage adjustment

MOTORCRAFT 4350 CARBURETOR ADJUSTMENTS

Year	Carb. Model (Code 9510)	Float Level (Dry)	Pump Setting (Hole No.)	Choke Plate Clearance (Pulldown)	Fast Idle Cam Linkage Setting	Fast Idle Speed (Hot Engine)	Auxiliary Inlet Valve Setting	Dechoke Clearance	Metering Rod Setting	Choke Setting
1975	5THA4	.90	Lower	.140	.160	1600①	.050	.325	.120	2 Rich
	5THA4D	29/32	Lower	.135	.135	1600①	.050	.325	.120	2 Rich
	5THM4	.90	Lower	.140	.160	1600①	.050	.325	.120	2 Rich
	5THM4D	29/32	Lower	.135	.135	1600①	.050	.325	.120	2 Rich
1976	6THA4	.90	—	.135	.135	1600①	.050	.325	—	2 Rich
	6THA4C	.90	—	.135	.135	1600①	.050	.325	—	2 Rich
	6THM4	.90	—	.135	.135	1600①	.050	.325	—	2 Rich
1977–78	6THA4	.90	Middle	.135	.135	1600①	.050	.325	.117	2 Rich
	6THA4C	.90	Middle	.135	.135	1600①	.050	.325	.112	2 Rich
	6THM4	.90	Middle	.135	.135	1600①	.050	.325	.108	2 Rich

①—On second step of cam with TCS solenoid and EGR disconnected.

MODEL 4350-4V ADJUSTMENTS

Accelerator Pump Stroke Adjustment

This piston-to-shaft pin position which is the only adjustment, is pre-set to deliver the correct amount of fuel for the engine on which it is installed and should not be changed from the specified setting. Do not attempt to adjust the accelerator pump stroke by turning the vacuum limiter level adjusting nut. This adjustment is pre-set and changing it could affect driveability.

1. Remove air horn assembly, then disconnect accelerator pump from operating lever by depressing spring and sliding arm out of pump shaft slot.
2. Disassemble spring and nylon keeper retaining adjustment pin. Refer to 4350 specifications chart. If pin is not in specified hole, remove it, reposition shaft in specified hole and reinstall pin, Fig. 1.
3. Slide nylon retainer over pin and position spring on shaft, then compress spring on shaft and install pump on arm.

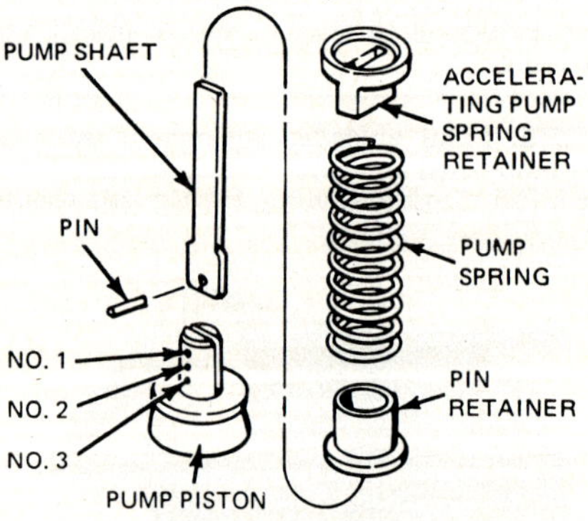

Fig. 1 Accelerating pump stroke adjustment

Accelerating Pump Link Adjustment

NOTE: The pump throttle link must be installed in the lower hole of the throttle shaft lever, Fig. 2.

1. Turn out idle speed screw until primary throttle plates are completely seated in the bores.
2. Turn accelerating pump throttle link adjusting nut until notch in pump lever is aligned with index mark on air horn, Fig. 2.

Initial Choke Pulldown

1. Remove choke housing assembly, then open throttle about half-way and position fast idle adjusting screw on high step of fast idle cam.
2. Bend a .036" wire gauge at 90° angle, about 1/8" from end, then insert bent end of gauge between choke piston slot and upper edge of right hand slot in choke housing, Fig. 3.
3. Rotate choke lever counterclockwise until gauge is snug in piston slot, then apply light force to lever to hold gauge and move top of choke rod away from carburetor while moving bottom of rod toward carburetor to remove end play from link-

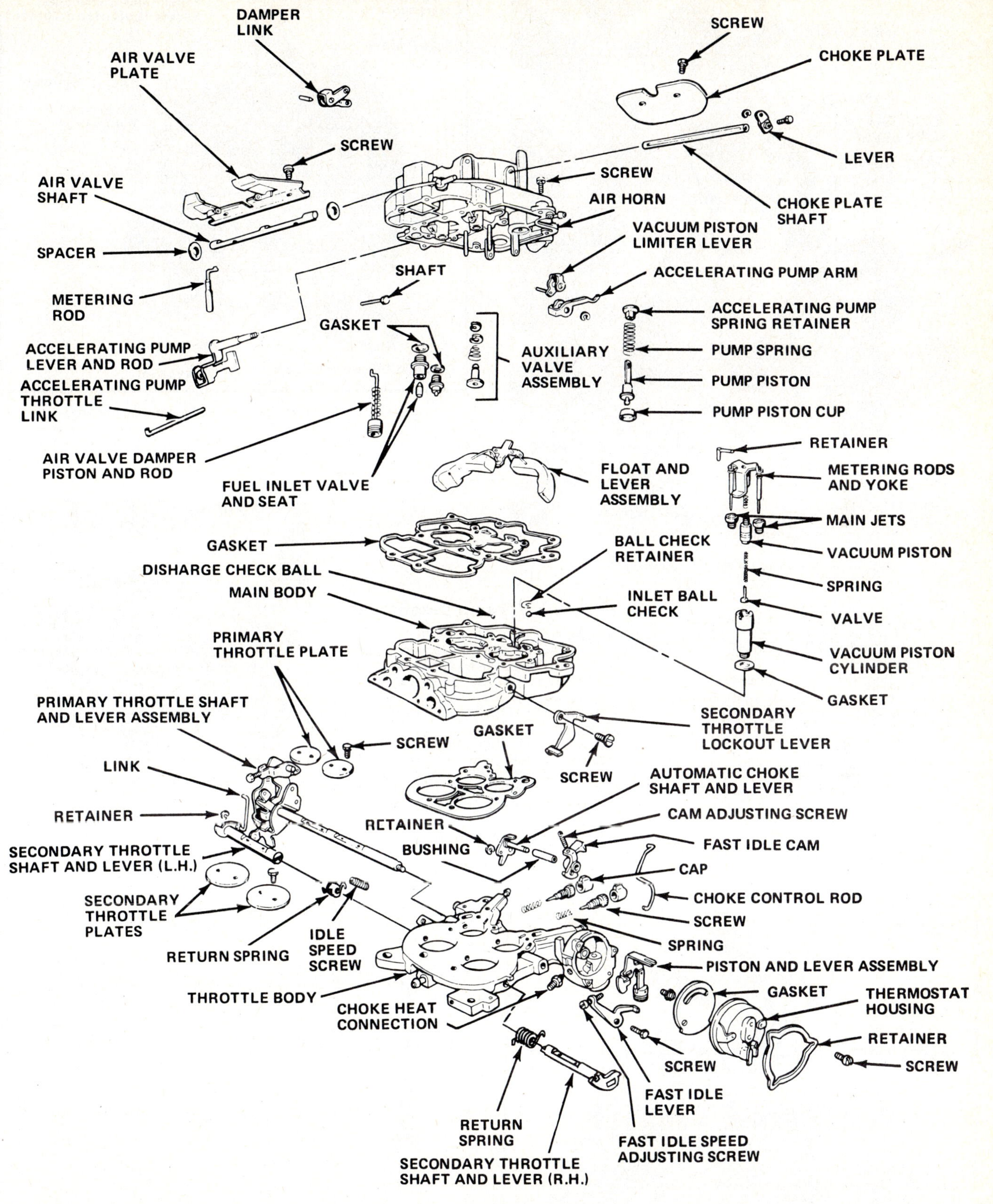

Exploded view of 4350 4V carburetor

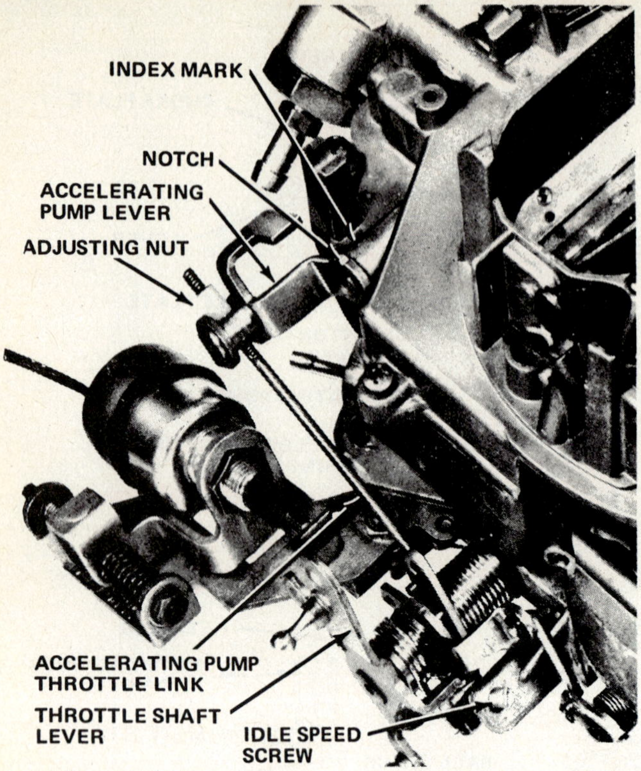

Fig. 2 Accelerating pump link adjustment

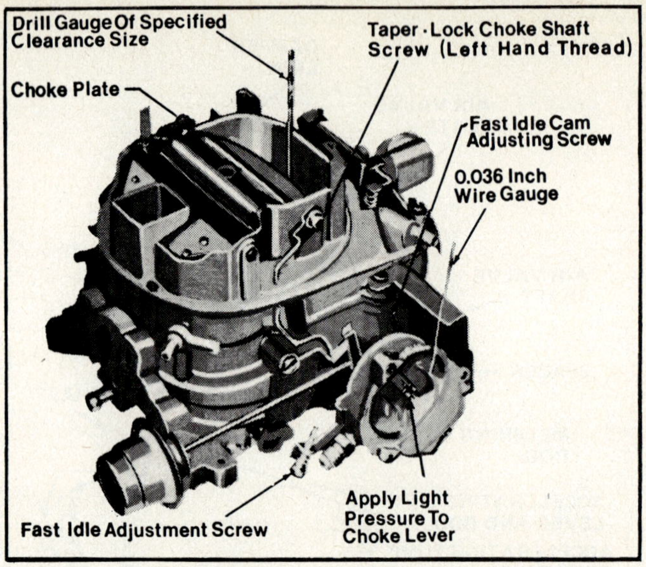

Fig. 3 Initial choke plate pulldown adjustment

age.

4. Using a drill gauge or pin of specified size, check clearance between lower edge of choke plate and wall, Fig. 3. Refer to 4350 specifications chart.
5. Adjust clearance to specifications, by turning lock screw on choke plate shaft 3 full turns in a clockwise direction. Pry choke lever from shaft to break taper lock. Choke lever should rotate freely on choke shaft.
6. With drill gauge and .036" wire gauge positioned as above, tighten, screw on choke shaft.
7. Install choke housing assembly.

Units With Vacuum Diaphragm

Initial Choke Valve Clearance

1. Loosen choke cover screws, open throttle and rotate choke cover until choke valve closes. Then, tighten one cover screw to hold cover in place.
2. Close throttle. The fast idle screw should be on top step of cam.
3. Bottom choke diaphragm on setscrew, however, do not press on links, Fig. 4.
4. Measure clearance between air horn and lower edge of choke valve, Fig. 4. Adjust by turning adjusting screw at rear of diaphragm housing.
5. Adjust fast idle cam linkage.

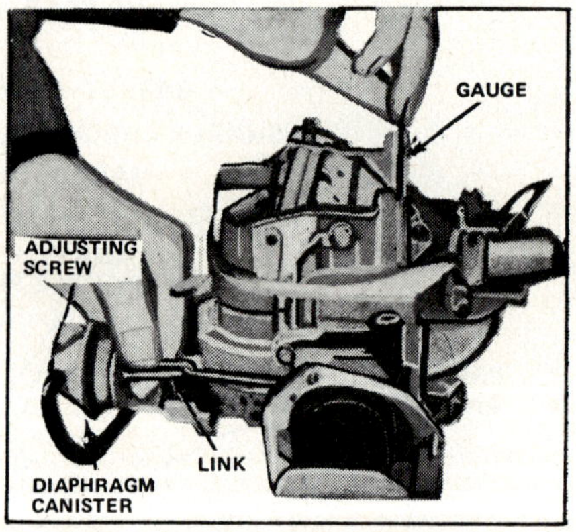

Fig. 4 Initial choke valve clearance

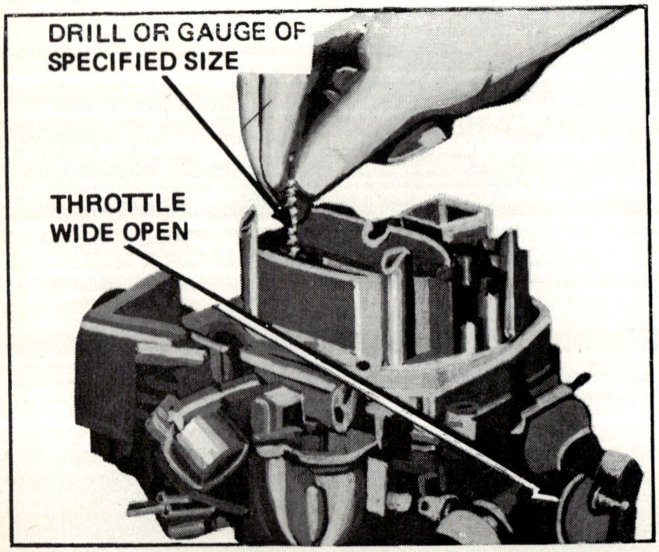

Fig. 5 Dechoke clearance adjustment

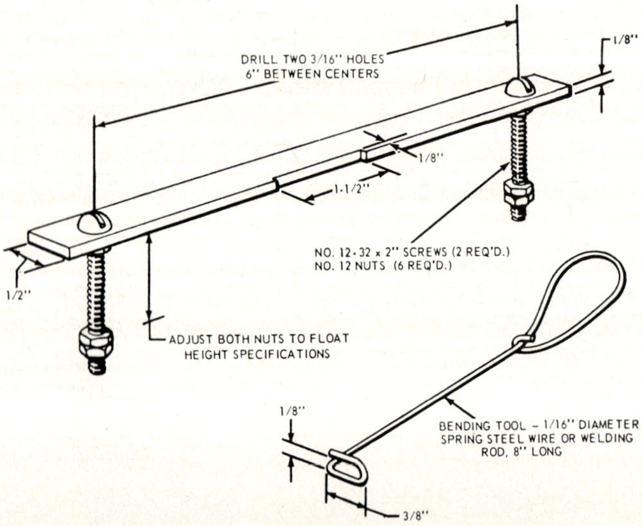

Fig. 6 Float setting gauge and bending tool

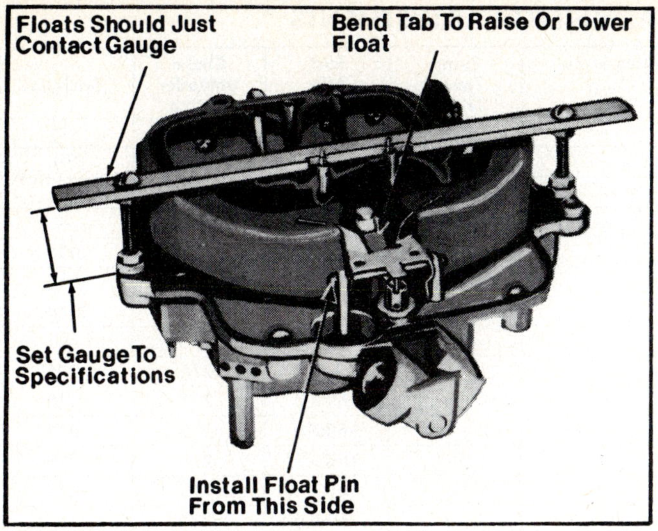

Fig. 7 Setting float level

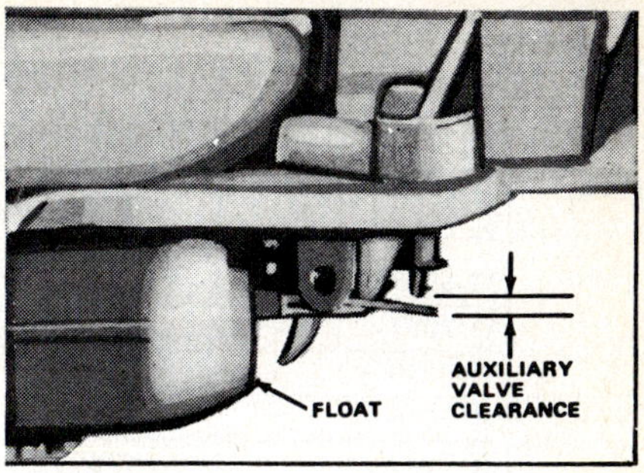

Fig. 8 Auxiliary inlet valve adjustment

6. Loosen choke cover screw and set choke cover to specifications.

Choke Plate Indexing

1. Loosen choke lever attaching screw and pry lever from choke shaft to permit choke valve rotation.

NOTE: The choke lever attaching screw has a left hand thread.

---

2. Loosen choke cover screws and rotate cover 90° counterclockwise so the choke shaft lever contacts cam adjusting screw.
3. Position fast idle screw on top step of fast idle cam. Then, back off cam adjusting screw until clearance is obtained between screw and choke shaft lever.
4. Turn cam adjusting screw until it contacts choke shaft lever, then turn 6–7 additional turns.
5. Manually close choke plate and tighten choke shaft lever attaching screw.
6. Adjust initial choke valve clearance and fast idle cam linkage.
7. Set choke cover and tighten cover screws.

Dechoke (Unloader) Clearance

1. Hold throttle plate fully open and apply pressure on choke valve toward close position.
2. Measure clearance between lower edge of choke valve and air horn wall. Fig. 5. Refer to 4350 Specifications Chart.
3. Adjust clearance to specifications by bending unloader tang which contacts fast idle cam forward to increase or rearward to decrease clearance.

NOTE: Do not bend tang downward from a horizontal plane. After adjustment, make certain that there is at least .070 inch clearance between unloader tang and choke housing with throttle fully open.

Float Setting

To simplify float setting, refer to Fig. 6 for construction of an adjustable float gauge and float tab bending tool.

1. Adjust gauge to specified height and insert gauge into air horn, Fig. 7.
2. Check clearance and alignment of floats to gauge. Refer to 4350 Specifications Chart. Both floats should just touch gauge. If necessary, align floats by twisting floats slightly.
3. To raise float setting, insert open end of bending tool to right side of float lever tab between needle and float hinge. Raise float lever off needle and bend tab downward.

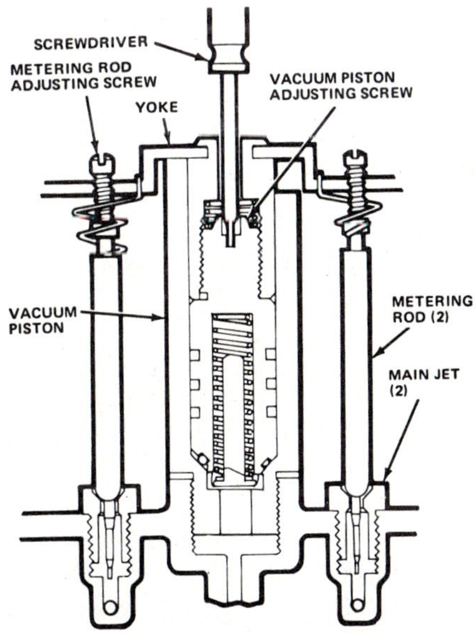

Fig. 9 Metering rod adjustment

4. To lower float setting, insert open end of bending tool to left side of float lever tab between needle and float hinge. Support float lever and bend tab upward.
5. If above gauge is not available, invert upper body and measure distance from top of float to gasket surface of upper body. Adjust alignment and height of floats with bending tool as described previously.

Auxiliary Valve Setting

Check auxiliary valve clearance as shown in Fig. 8. If necessary to adjust clearance, use bending tool, Fig. 6. Refer to 4350 Specifications Chart.

Vacuum Piston & Metering Rods Adjustment

1. Remove air horn assembly. Depress metering rod hanger and turn vacuum piston adjusting screw and metering rod adjusting screws until metering rod hanger is fully sealed against vacuum cylinder topface, Fig. 8.
2. Hold metering rod hanger in full downward position and turn each metering rod adjusting screw clockwise until the metering rod hanger starts to rise. Metering rods are now properly positioned in relation to vacuum piston.
3. Turn vacuum piston adjusting screw clockwise until specified clearance between metering rod hanger and vacuum cylinder top face is obtained.

Fast Idle Cam

1. Loosen choke cover screws and rotate cover to align index marks on cover and housing. Then, rotate cover an additional 90° counterclockwise and tighten attaching screws.
2. Place fast idle speed adjusting screw on kickdown (center) step of fast idle cam. Fully close choke plate and check clearance between air horn and lower edge of choke plate. Adjust by turning fast idle cam adjusting screw, ensuring adjusting screw remains on kickdown step during adjustment.
3. Loosen choke cover screws and rotate cover to specified index mark and tighten cover screws.

CARTER YF CARBURETOR ADJUSTMENTS

Year	Carburetor Model	Float Level	Float Drop	Idle Screw Turns Open	Pump Travel Setting	Fast Idle Setting	Choke Unloader Setting	Dashpot	Choke Setting
1974	YF-6423	.476	1.38	—	—	1600①	.275	3/32	1
	YF-6431	.476	1.38	—	—	1600①	.275	3/32	1
	YF-6511	.476	1.38	—	—	1600①	.275	3/32	1
	YF-7001	.476	1.38	—	—	1600①	.275	—	1
	YF-7029⑭	.476	1.38	—	—	1600①	.275	.095	1 Rich
1975	YF-7040	.476	1.38	—	—	1600②	.275	.075	1 Rich
	YF-7041	.476	1.38	—	—	1600②	.275	.075	1 Rich
	YF-7043	.476	1.38	—	—	1600②	.275	—	1 Rich
1976	YF-7083	.476	1.38	—	—	1600②	.275	.075	1 Rich
	YF-7084	.476	1.38	—	—	1600②	.275	.075	2 Rich
	YF-7085	.476	1.38	—	—	1600②	.275	.075	1 Rich
	YF-7088	.476	1.38	—	—	1600②	.275	.075	1 Rich
	YF-7109	.476	1.38	—	—	1600②	.275	.075	2 Rich
1977	YF-7110	.476	1.375	—	—	1600①	.275	—	2 Rich
	YF-7111	.476	1.375	—	—	1800	.275	—	2 Rich
	YF-7151	.476	1.375	—	—	1600①	.275	—	1 Rich
	YF-7153	.476	1.375	—	—	1600①	.275	—	Index
	YF-7154	.476	1.375	—	—	1600①	.275	—	1 Rich
1978	YF-7201	.476	1.375	—	—	1600②	.275	—	Index
	YF-7228	.476	1.375	—	—	1600②	.275	—	1 Rich
	YF-7230	.476	1.375	—	—	1600②	.275	—	1 Rich

①—On second step of fast idle cam.

②—On second step of fast idle cam with TCS solenoid and EGR disconnected.

MODEL YF SERIES ADJUSTMENTS

Float Adjustment

Fig. 1—Invert the air horn assembly, and check the clearance from the top of the float to the bottom of the air horn with the float level gauge. Hold the air horn at eye level when gauging the float level. The float arm (lever) should be resting on the needle pin. Do not load the needle when adjusting the float. Bend the float arm as necessary to adjust the float level (clearance). Do not bend the tab at the end of the float arm. It prevents the float from striking the bottom of the fuel bowl when empty.

Float Drop Adjustment

Fig. 2—Hold air horn upright and measure maximum clearance from top of float to bottom of air horn with float drop gauge. Bend tab at end of float arm to obtain specified setting listed under YF Adjustment Specifications.

Metering Rod Adjustment

Fig. 3—Back out the idle speed adjusting screw until the throttle plate is closed tight in the throttle bore. Press down on upper end of diaphragm shaft until diaphragm bottoms in vacuum chamber. Metering rod should contact bottom of metering rod well, and metering rod should contact lifter link at the outer end nearest the springs and at supporting lug. For models not equipped with metering rod adjusting screw, adjust by bending lip of metering rod arm to which metering rod is attached, up or down as required. For models equipped with a metering rod adjusting screw, turn the adjusting screw until metering rod just bottoms in the body casting. For final adjustments turn metering rod adjusting screw in (clockwise) one additional turn.

Fast Idle Cam Linkage Adjustment

1974–78

Position fast idle screw on second step of fast idle cam and against shoulder of highest step, Fig. 4. Using a drill of specified size, check clearance between lower edge of choke plate and bore. Refer to YF Specifications Chart. To adjust, bend choke connector rod as required.

Choke Plate Pulldown Adjustment

1974–78

Bend a 0.025 in. diameter wire gauge at a 90 degree angle approximately 1/8-inch from one end. Insert the bent end of the gauge

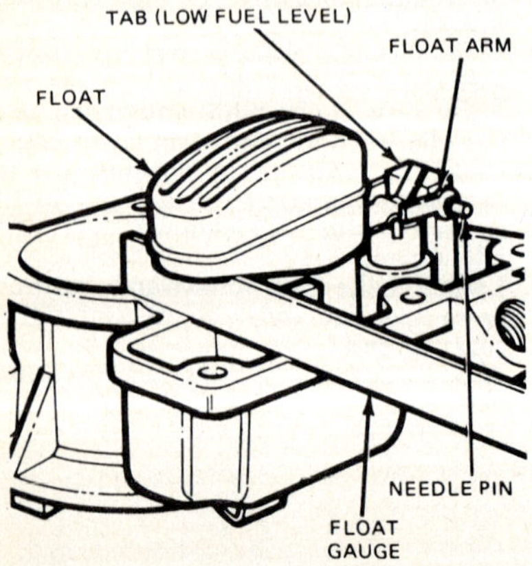

Fig. 1 YF float level adjustment

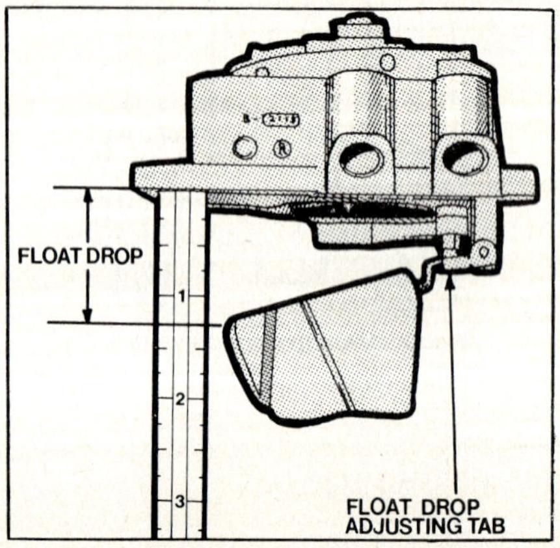

Fig. 2 YF float drop adjustment

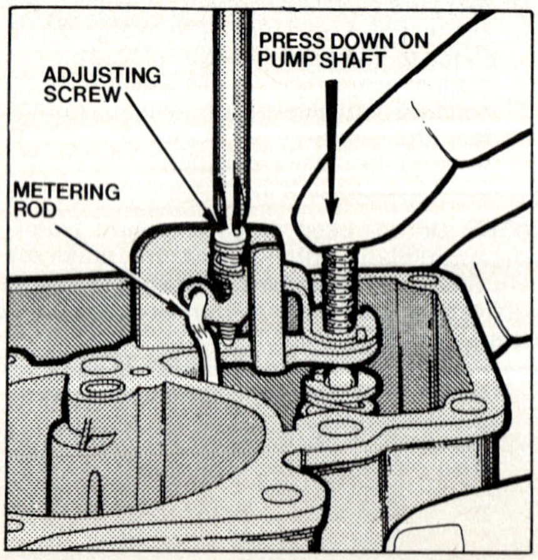

Fig. 3 YF metering rod adjustment. 1974–78

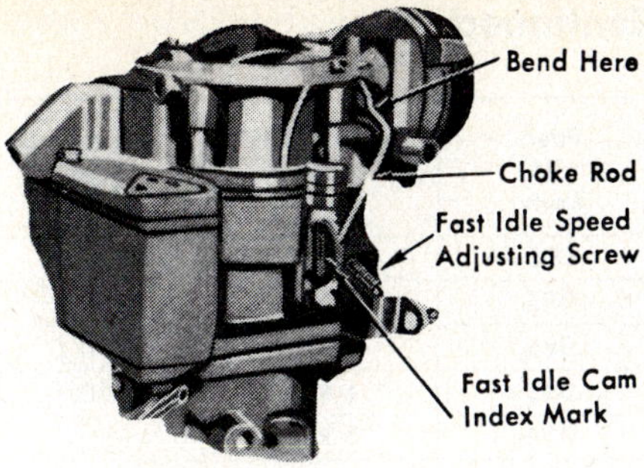

Fig. 4 YF fast idle cam linkage adjustment

between the choke piston slot and the right hand slot in choke housing. Rotate the choke piston lever counterclockwise until gauge is snug in the piston slot. Exert a light pressure on choke piston lever to hold the gauge in place, then use a drill with a diameter equal to the specified pulldown clearance between the lower edge of choke plate and carburetor bore to check clearance, Fig. 5.

To adjust the choke plate pulldown clearance, bend the choke piston lever as required to obtain specified setting.

NOTE: When bending the lever, be careful not to distort the piston link. Install the choke thermostatic spring housing and gasket. Set the housing to specifications.

Dechoke Adjustment

1974–78

Hold the throttle plate fully open and close the choke plate as far as possible without forcing it. Use a drill of specified diameter to check the clearance between choke plate and air horn, Fig. 6. If clearance is not within specification, adjust by bending arm on choke trip lever of the throttle lever. Bending the arm downward will decrease the clearance, bending it upward will increase the clearance.

If the choke plate clearance and fast idle cam linkage adjustment was performed with the carburetor on the engine, adjust the engine idle speed and fuel mixture. Adjust dashpot (if so equipped).

Fig. 5 YF choke plate pulldown adjustment 1974–78

Dashpot Adjustment

With the engine idle speed and mixture properly adjusted, the engine at normal operating temperature, loosen the anti-stall dashpot lock nut, Fig. 7. Hold the throttle in the curb idle position and depress the dashpot plunger. Measure the clearance between the throttle lever and plunger tip. Turn the anti-stall dashpot to provide 7/64" ± 1/64" clearance between the tip of the plunger and the throttle lever. Tighten the locknut to secure the adjustment.

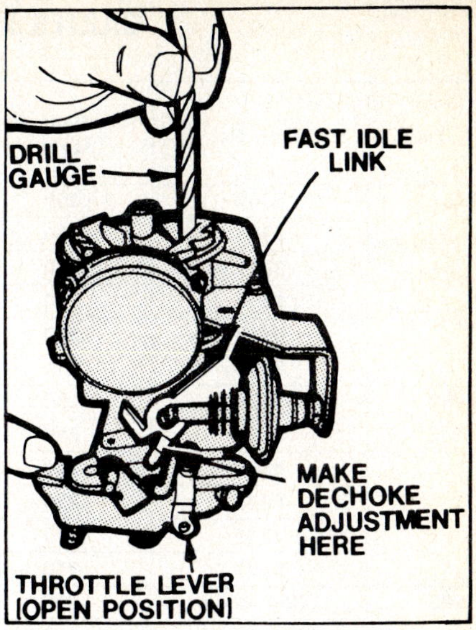

Fig. 6 YF dechoke adjustment. 1974–78

Fig. 7 YF dashpot adjustment

Automatic Choke Adjustment

Loosen choke cover retaining screws and turn choke cover so that line or index mark on cover lines up with the specified mark listed in *YF Specifications Chart* on choke housing.

CARTER BBD ADJUSTMENT SPECIFICATIONS

See Tune Up Chart for hot idle speed.

Year	Ident. No.	Float Level	Vacuum Piston Gap	Initial Choke Valve Clearance	Choke Unloader	Fast Idle Cam	Automatic Choke Setting	Fast Idle Speed
1977–78	8107	.250	.040	.128	.280	.095	2 Rich	1700①
1979	8185	.250	.035	.140	.280	.110	1 Rich	1600③
	8186	.250	.035	.150	.280	.110	1 Rich	1500③
	8187	.250	.035	.140	.280	.110	1 Rich	1600③
	8188	.250	.035	.150	.280	.110	1 Rich	1500③
	8195	.250	.035	.140	.280	.110	1 Rich	②③
	8229	.250	.035	.128	.280	.095	1 Rich	1500③
1980	8253	.250	.035	.128	.280	.095	2 Rich	1850
	8254	.250	.035	.120	.280	.086	2 Rich	1700
	8255	.250	.035	.140	.280	.093	2 Rich	④
	8256	.250	.035	.128	.280	.093	2 Rich	1850
	8257	.250	.035	.128	.280	.095	2 Rich	1700
	8277	.250	.035	.116	.280	.081	1 Rich	1700

Continued

CARTER BBD ADJUSTMENT SPECIFICATIONS—Continued

See Tune Up Chart for hot idle speed.

Year	Ident. No.	Float Level	Vacuum Piston Gap	Initial Choke Valve Clearance	Choke Unloader	Fast Idle Cam	Automatic Choke Setting	Fast Idle Speed
1981	8302	.250	.035	.140	.280	.095	1 Rich	1850③
	8303	.250	.035	.140	.280	.095	1 Rich	1700③
	8311	.250	.035	.120	.280	.085	1 Rich	1700③
	8306	.250	.035	.140	.280	.095	1 Rich	1700③
	8312	.250	.035	.140	.280	.095	1 Rich	③④
	8307	.250	.035	.140	.280	.095	1 Rich	1700③
1982	8338	.250	.035	.140	.280	.095	1 Rich	1850
	8339	.250	.035	.140	.280	.095	1 Rich	1700
	8340	.250	.035	.150	.280	.110	1 Rich	1700
	8341	.250	.035	.150	.280	.150	1 Rich	1700
	8349	.250	.035	.128	.280	.095	2 Rich	④
	8351	.250	.035	.130	.280	.095	0 Rich	1700
1983	8360, 8362	.250	.035	.140	.280	.095	1 Rich	1850
	8364, 8367	.250	.035	.140	.280	.095	1 Rich	1700

①—On second step of fast idle cam with TCS solenoid and EGR disconnected.
②—Manual trans.; 1500. Auto. trans.; 1600.
③—Engine hot with EGR valve disconnected.
④—Manual trans., 1700; auto. trans., 1850.

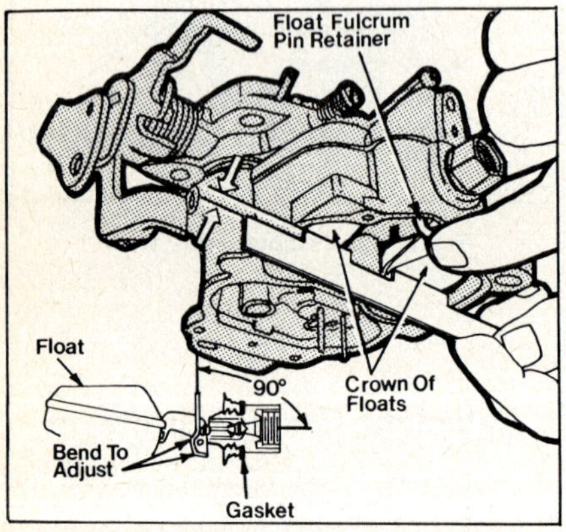

Fig. 1 BBD float level adjustment

MODEL BBD ADJUSTMENTS

Float Level Adjustment

Fig. 1—With carburetor body inverted so that weight of floats ONLY is forcing needle against its seat, use a T-scale or the tool shown, and check the float level from surface of fuel bowl to crown of each float at center.

If an adjustment is necessary, hold floats on bottom of bowl and bend float lip as required to give the specified dimension.

CAUTION: When bending the float lip, do not allow the lip to push against the needle as the synthetic rubber tip (if used) can be compressed sufficiently to cause a false setting which will affect correct level of fuel in bowl. After being compressed, the tip is very slow to recover its original shape.

Accelerator Pump

Fig. 2

1. Back off curb idle adjusting screw, completely closing throttle valve, then open choke valve, allowing throttle valves to

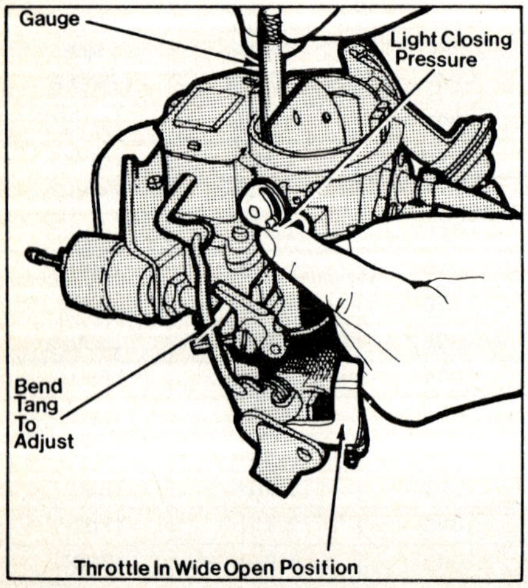

Fig. 3 BBD choke unloader adjustment

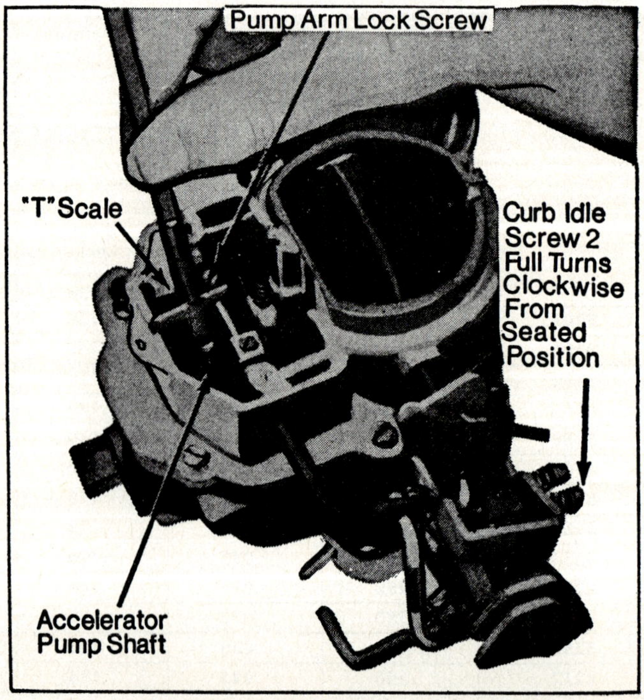

Fig. 2 BBD accelerating pump setting

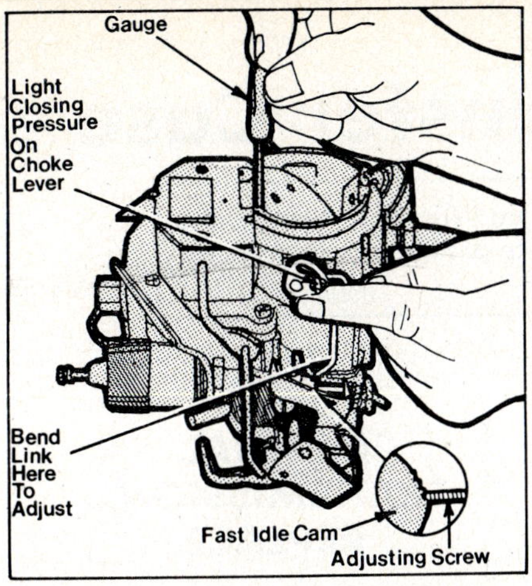

Fig. 4 BBD fast idle cam position adjustment

seat in bores. Ensure accelerator pump "S" link is located in outer hole or pump arm.
2. Turn curb idle adjusting screw until screw contacts; stop, then rotate screw two additional turns.
3. Measure distance between air horn surface and top of accelerator pump shaft. Refer to BBD Specifications Chart.
4. Adjust by loosening pump arm adjusting screw and rotating sleeve until proper dimension is obtained. Tighten adjusting screw.

Choke Unloader Adjustment

Fig. 3—The choke unloader is a mechanical device to partially open the choke valve at wide open throttle. It is used to eliminate choke enrichment during engine cranking. Engines that have been flooded or stalled by excessive choke enrichment can be cleared by the use of the unloader. Adjust as follows:
1. Hold throttle valve in wide open position. Insert the specified drill size between upper edge of choke valve and inner wall of air horn.
2. With a finger lightly pressing against choke valve, a slight drag should be felt as the drill is being withdrawn.
3. If an adjustment is necessary, bend unloader tang on throttle lever until specified opening has been obtained.

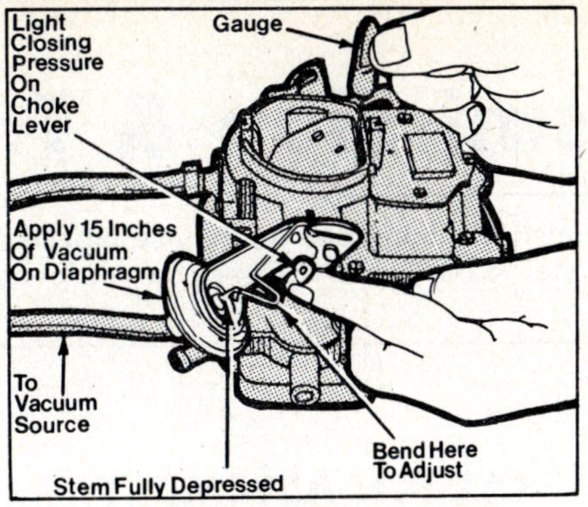

Fig. 5 BBD initial choke valve clearance adjustment

Fast Idle Cam Position

Fig. 4
1. With fast idle adjusting screw contacting second highest step on fast idle cam, move choke valve toward closed position with light pressure on choke shaft lever.
2. Insert the specified size drill between choke valve and air horn wall. An adjustment will be necessary if a slight drag is not obtained as drill is being removed.
3. Adjust by bending fast idle connector rod at lower angle.

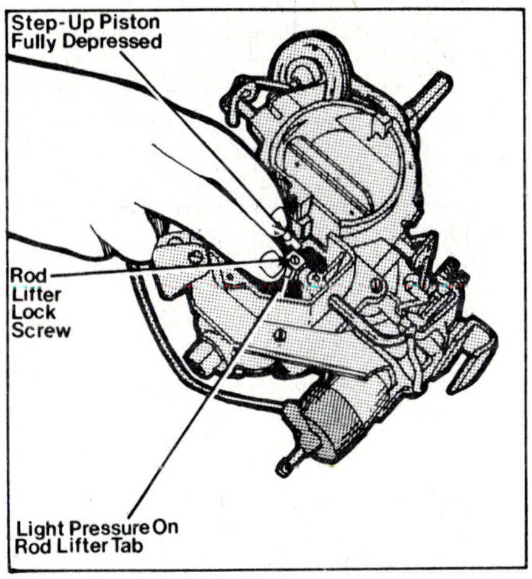

Fig. 7 BBD vacuum piston gap adjustment

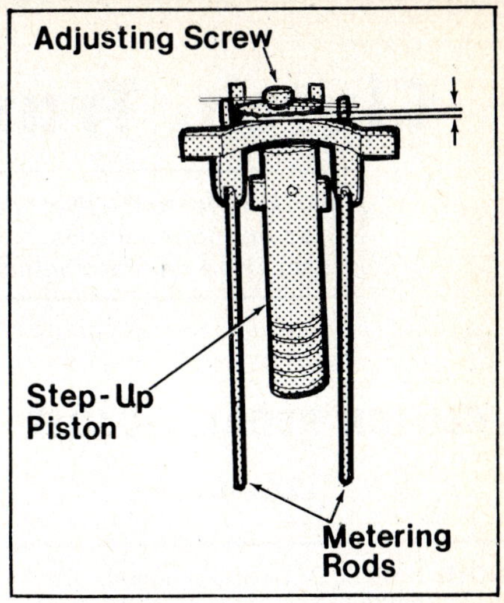

Fig. 6 BBD step-up piston adjustment

Initial Choke Valve Clearance

Fig. 5
1. Fully seat choke diaphragm by applying a minimum of 19 inches of mercury from an external vacuum source.
2. Open throttle valve slightly to place fast idle screw on high step of cam.
3. Apply light closing pressure to choke lever or rotate choke coil housing to apply closing pressure to choke.
4. Measure clearance between upper edge of choke plate and air horn wall. To adjust, bend diaphragm connector link.

Vacuum Piston Gap Adjustment

Figs. 6 & 7
1. Adjust vacuum piston gap to specifications. Rotate allen head screw on top of piston to adjust gap.
2. Install vacuum piston assembly into air horn bore, ensuring metering rods are positioned in the metering jets.
3. Back off curb idle screw until throttle valves are completely closed. Count number of turns so that screw can be returned to its original setting.
4. Fully depress piston and while applying moderate pressure on rod lifter, tighten rod lifter-screw.
5. Release piston and rod lifter and reset curb idle screw.

ROCHESTER 2SE & E2SE TWO BARREL CARB. ADJUSTMENTS

NOTE: Refer to Chevrolet Chapter for carburetor adjustments.

Year	Carburetor Number	Float Level (Inches)	Pump Setting (Inches)	Fast Idle Cam Adj.	Fast Idle Speed ± 100 RPM	Air Valve Link	Vac. Break Pri.	Unloader	Choke	Secondary Lockout	Choke Coil Lever Gauge
1980	17080685	3/16	1/2	18°	2500	2°	19°	32°	TR	.004–.012	.085
1980	17080781	7/32	1/2	18°	2500	2°	19°	32°	TR	.004–.012	.085
1981	17081790	.208	.128	25°	2400	2°	19°	32°	TR	.050–.080	.050–.080
1981	17081791	.256	.128	25°	2600	2°	19°	32°	TR	.085	.085
1981	17081796	.208	.128	25°	2400	2°	19°	32°	TR	.050–.080	.050–.080
1981	17081797	.208	.128	25°	2600	2°	19°	32°	TR	.085	.085
1982	17082380	.169	.128	18°①	2400	2°	21°	34°	TR	.050–.080	.050–.080
1982	17082381	.169	.128	18°①	2400	2°	21°	34°	TR	.050–.080	.050–.080
1982	17082389	.169	.128	18°①	2400	2°	19°	34°	TR	.050–.080	.050–.080
1983	17082380	.216	.128	18°①	2500	2°	21°	34°	TR	.050–.080	.085
1983	17083385	.138	.128	18°①	2500	2°	19°	34°	TR	.050–.080	.085

①—Second step of cam.

Clutch, Transmission & Transfer Case

NOTE: Refer to the Warner Transmission chapter for service procedures on Warner transmissions used on these vehicles. Also refer to the Transfer Case chapter for service procedures on transfer cases used on these vehicles.

CLUTCH PEDAL, ADJUST

1974–83

NOTE: 1980–83 CJ models equipped with the 4-151 engine use a hydraulic clutch mechanism which is self adjusting. Free play adjustments are not necessary and there is no provision for such an adjustment.

1. On 1974–75 models adjust bellcrank outer support bracket to obtain about 1/8 inch bellcrank end play.
2. On all models, position clutch pedal against pedal support bracket stop.
3. On Cherokee, Wagoneer and all truck models, adjust pedal to bellcrank push rod until bellcrank inner lever is parallel to front face of clutch housing and slightly forward from the vertical position.
4. Adjust clutch fork to bellcrank rod to obtain the following clutch pedal free travel: 3/8 to 5/8 inch on 1974–81 Cherokee, Wagoneer, and Truck, 1–1 1/4 inch on 1982–83 Cherokee, Wagoneer and Truck with 258 cu. inch engine, 3/8 to 5/8 on 1982–83 Cherokee, Wagoneer and Truck with 360 cu. engine, 1/2 to 3/4 inch on 1974 CJ/DJ and Commando, 7/8 to 1 inch on 1975–77 CJ and 1 to 1 1/4 inch on 1978–83 CJ.

CLUTCH MASTER CYLINDER REPLACE

1. Disconnect hydraulic line at clutch cylinder.
2. Cap hydraulic line and master cylinder opening to prevent entry of dirt.
3. Remove cotter pin and washer retaining cylinder push rod on clutch pedal and slide rod off pedal pivot.
4. Remove nuts attaching clutch cylinder to firewall and remove cylinder.
5. Reverse procedure to install. Torque cylinder attaching nuts to 11 ft. lbs.

CLUTCH SLAVE CYLINDER REPLACE

1. Raise vehicle and disconnect hydraulic line at cylinder.
2. Remove throwout lever-to-cylinder push rod retaining spring.
3. Remove slave cylinder, heat shield, throwout lever pivot, washer and seal from clutch housing.

CLUTCH ASSEMBLY, REPLACE

1974–83

Removal

1. Raise and support vehicle.
2. Remove transmission as outlined under "Transmission, Replace" procedure.
3. Remove starter motor.
4. Remove throwout bearing and clutch housing.
5. Mark position of pressure plate on flywheel for assembly reference.
6. Alternately loosen the pressure plate attaching bolts to relieve spring tension.

NOTE: The pressure plate bolts must be loosen evenly to prevent distortion of cover.

7. Remove pressure plate bolts, then the pressure plate and driven plate from flywheel.

NOTE: Place a mark on the side of the driven plate facing the flywheel for assembly reference.

8. Remove pilot bushing lubricating wick and soak wick in engine oil.

Installation

1. Lightly lubricate release lever pivots.
2. Install pilot bushing lubricating wick into crankshaft bore.
3. Install a suitable clutch alignment tool or a spare clutch shaft into driven plate hub.
4. Install assembled plate and tool on flywheel. Ensure that tool is seated fully in the pilot bushing.

NOTE: Ensure that the proper side of the driven plate faces the flywheel.

5. Place pressure plate on flywheel and over driven plate and alignment tool. Align pressure plate and flywheel using the marks made during removal. Install pressure plate attaching bolts finger tight.
6. Torque pressure plate evenly and alternately to 40 ft. lbs. except on CJ models with four cylinder engine. Torque these pressure plate bolts to 23 ft. lbs.
7. Install clutch housing and starter motor.
8. Install throwout bearing and waved washer, if used.
9. Install transmission.

MANUAL TRANSMISSION, REPLACE

1980–83 All

1. Remove transmission shift lever boot attaching screws and slide boot upward.
2. On models with SR-4, T4 or T5 transmissions, remove bolts attaching transmission shift lever housing to transmission. Lift shift lever and housing upward and out of transmission.
3. On models with T-18A transmission, remove shift lever cap, gasket, spring seat, spring and shift lever as an assembly, then remove shift lever locating pins from housing.
4. On models with T-176 transmission, press and rotate transmission shift lever retainer, then remove lever, boot, spring and seat as an assembly.
5. Raise vehicle, then place alignment marks on propeller shaft and transfer case yoke. Disconnect propeller shaft at transfer case yoke and secure to under body with wire.
6. On Cherokee and truck models, disconnect front parking brake cable at equalizer, then remove clip that retains rear cable to rear crossmember and position cable out of way.
7. Support engine by positioning a jack stand under clutch housing.
8. Remove bolts and nuts attaching rear crossmember to frame side rails and support cushion, then remove crossmember.
9. Disconnect speedometer cable, backup lamp switch wire and four wheel drive indicator switch wire.
10. Disconnect transfer case vent hose at transfer case.
11. Place alignment marks on front propeller shaft and transfer case yoke, then disconnect propeller shaft and secure to underbody with wire.
12. On CJ models, remove transfer case shifter shaft retaining nut and control link pins, then remove shifter shaft and disengage shift lever from shift control links. Slide shift lever upward into boot to position lever out of way.

NOTE: On some units, it may be necessary to unthread shifter shaft to remove.

13. On Cherokee and truck models, remove cotter pin and washer that connect link to shift lever, then disconnect link.
14. Using a suitable jack, support transmission and transfer case assembly.
15. Remove transmission to clutch housing attaching bolts, then remove transmission and transfer case assembly.

16. Reverse procedure to install. Soak pilot bushing lubricating wick in engine oil before installing. Torque transmission to clutch housing bolts to 55 ft. lbs. On CJ models, torque transfer case bolts to 30 ft. lbs., on Cherokee, Wagoneer and Truck models torque bolts to 40 ft. lbs. Torque crossmember to 30 ft. lbs. On SR-4, T4 and T5 transmissions, torque shift lever housing bolts to 10 ft. lbs.

1974–79 All

1. Remove floor lever knob, trimming, boot, floor covering and pan.
2. On three-speed transmissions, remove shift control lever housing assembly and on four-speed transmissions, remove shift cap, spring retainer, spring, shift lever and pin.
3. Remove transfer case shift lever and bracket assembly.
4. Raise vehicle, remove front propeller shaft and disconnect rear propeller shaft from transfer case.
5. Disconnect speedometer cable, backup lamp switch and TCS switch wires, and brake cable.
6. On vehicles with V8 engines, disconnect exhaust pipes from manifolds.
7. Support transmission and engine with jacks and remove frame center crossmember.
8. Remove transmission to bell housing bolts and lower transmission slightly, then slide transmission and transfer case toward rear of vehicle until clutch shaft clears bell housing.

MANUAL TRANSMISSION SHIFT MECHANISM

The shift mechanism on all manual transmission models is located within the shift control housing which also serves as the transmission top cover. The shift mechanism does not require adjustment and can be serviced separately from the transmission.

AUTOMATIC TRANSMISSION SHIFT LINKAGE ADJUSTMENT

1974–79

1. Place lever in N and raise vehicle on hoist.
2. Loosen gearshift rod trunnion enough to allow movement of rod in trunnion.
3. Place transmission outer range selector lever fully in to detent position and torque gearshift rod trunnion locknut to 9 ft. lbs.
4. Lower vehicle and check operation of lever in all ranges. Vehicle should only start in Park and Neutral.

1980–83

1. Raise vehicle and loosen shift rod trunnion jamnuts.
2. Remove lock pin retaining shift rod trunnion to bell crank and disengage trunnion and shift rod at bell crank.
3. Place lever in Park and lock steering column.
4. Move valve body lever rearward into Park detent (last rearward detent) as far as possible.
5. Make sure park lock is engaged by attempting to turn driveshaft. Shaft should not turn.
6. Adjust shift rod trunnion to obtain free pin fit in bell crank arm and tighten trunnion jamnuts.

NOTE: Shift linkage play must be eliminated in order to obtain proper adjustment. Pull down on shift rod and press up on outer bell crank to eliminate play.

7. Place lever in Park and Neutral positions and check engine starting. Engine should start only in these positions. If engine does not start or starts in other positions, adjustment in incorrect or neutral switch is defective.
8. Check steering lock for proper operation and lower vehicle.

AUTOMATIC TRANSMISSION, REPLACE

1974–79

1. Remove transmission dipstick.
2. Remove radiator shroud if equipped.
3. Raise vehicle and mark front and rear universal joints and axle yokes for assembly reference.
4. On Cherokee, Wagoneer and Truck models, remove parking brake cable jamnut and adjuster nut, and pull parking brake cable out of crossmember.
5. On Cherokee and Wagoneer models with low range reduction unit, remove unit shift lever from shift shaft and remove unit. On CJ and Truck models with low range unit, disconnect shift rod at unit shift lever and remove unit.
6. Disconnect speedometer cable, mark Emergency Drive control vacuum lines for reassembly and disconnect lines and indicator lamp wire.
7. Remove vacuum line routing bracket from rear of transfer case and disconnect detent solenoid wire at transmission case connector.
8. Remove starter and converter housing inspection cover.
9. Mark torque converter and drive plate for reassembly and remove attaching bolts.
10. Remove rear support cushion-to-crossmember attaching nuts.
11. Support transmission with jack and remove rear crossmember.
12. Disconnect transmission gearshift rod.
13. Disconnect front propeller shaft at transfer case yoke and secure shaft aside.
14. Disconnect transmission oil cooler lines at transmission case fittings and disconnect engine-to-modulator vacuum hose at modulator.
15. Place support stand under engine and remove converter housing-to-engine bolts.
16. Remove filler tube, move transmission rearward until it clears drive plate and lower transmission while holding converter in position.
17. Reverse procedure to install. Torque converter housing-to-engine bolts to 28 ft. lbs. and front propeller universal joint U-bolts to 15 ft. lbs.

1980–83

1. Disconnect fan shroud if equipped and transmission fill tube at upper bracket.
2. Raise vehicle and remove inspection cover from converter housing.
3. Remove transmission fill tube and starter.
4. Mark propeller shafts and axle yokes for assembly alignment reference. Disconnect shafts at transfer case yokes and secure aside.
5. On V8 vehicles, disconnect exhaust pipes at manifolds.
6. On 1981–83 Cherokee, Wagoneer and Truck models, drain transfer case lubricant and on all models, disconnect speedometer cable at transfer case, disconnect gearshift and throttle linkage, and disconnect wires at neutral start switch.
7. Mark converter drive plate and converter for assembly and remove bolts attaching converter to drive plate. Using ratchet handle and socket on crankshaft front pulley bolt rotate crankshaft and drive plate to gain access to drive plate bolts.
8. Support transmission-transfer case assembly using jack and remove rear crossmember.
9. Lower transmission slightly and disconnect oil cooler lines at transmission.
10. Unbolt and move transmission and converter rearward to clear engine. Lower transmission while holding converter in position.
11. Remove transfer case from transmission.
12. Reverse procedure to install. Torque converter-to-engine bolts to 28 ft. lbs. If torque converter was removed use Pump Aligning Tool J-24033 (Model 904/999) or J-24045 (Model 727) to engage rotor drive lugs and rotate tool until drilled hole in tool is vertical and remove. Rotate converter so pump drive slots in converter hub are vertical. Insert converter hub into pump and make sure drive lugs of pump inner rotor are properly engaged in drive slots of converter hub.

TRANSFER CASE

1974–75

1. Remove shift lever knob, boot and shift lever.
2. Raise and support vehicle.
3. Drain transfer case lubricant.
4. Disconnect front and rear propeller shafts from output shaft yokes.
5. Disconnect parking brake cable from equalizer.
6. Disconnect speedometer cable from transfer case.
7. Remove transfer case to transmission attaching bolts. Install one 3/8 × 4 inch dowel pin on each side of case for use as guides during installation.
8. Remove transfer case from vehicle.
9. Remove transmission to transfer case gasket and discard.
10. Install a new transmission to transfer case gasket.
11. Reverse steps 1 through 8 to install.

1976–79

1. Remove shift lever knob, trim ring and boot from transmission.
2. Remove transfer case shift levers.
3. Remove floor covering, if equipped, then the transmission access cover from floorpan.
4. Raise and support vehicle.
5. Drain transfer case lubricant. On CJ models, drain transmission lubricant.
6. Disconnect torque reaction bracket from crossmember, if equipped.
7. On CJ models, place a jack stand under clutch housing to support engine and transmission, then remove rear cross-

member.
8. On all models, disconnect front and rear propeller shafts from output shaft yokes.
9. Disconnect speedometer cable from transfer case.
10. On Cherokee and truck models, disconnect parking brake cable at equalizer and the exhaust pipe support bracket from transfer case.
11. On all models, remove transfer case to transmission attaching bolts.

NOTE: One attaching bolt is located at the front of the transfer case at the lower right corner of the transmission.

12. Remove transfer case from vehicle. Remove and discard transfer case to transmission gasket.
13. Install new transfer case to transmission gasket.
14. Reverse steps 1 through 12 to install.

1980–83

CJ Models
1. On models equipped with automatic transmission, remove shift lever knob, trim ring and boot from transfer case shift lever.
2. On models equipped with manual transmission, remove shift lever knob, trim ring and boot from transmission and transfer case shift levers.
3. On all models, remove floor covering, if equipped, then the transmission access cover from floorpan.
4. Raise and support vehicle. Drain lubricant from transfer case.
5. Place a jack stand under clutch housing to support engine and transmission, then remove rear crossmember.
6. Disconnect front and rear propeller shafts from output shaft yokes.
7. Disconnect speedometer cable from transfer case.
8. Disconnect parking brake cable at equalizer, if necessary.
9. Disconnect exhaust pipe support bracket from transfer case, if equipped.
10. Remove transfer case to transmission attaching bolts.
11. Remove transfer case from vehicle.
12. Remove and discard transfer case to transmission gasket.
13. Install new transfer case to transmission gasket.
14. Reverse steps 1 through 11 to install.

Cherokee, Wagoneer & Truck Models
1. Raise and support vehicle.
2. Drain transfer case lubricant.
3. Mark alignment of transfer case front and rear output shaft yokes and propeller shafts, then disconnect front and rear propeller shafts from output shaft yokes.
4. Disconnect speedometer cable and indicator switch wires from transfer case.
5. Support transmission with a suitable jack and remove rear crossmember.
6. Disconnect parking brake cable guide from pivot, located at right frame rail, if necessary.
7. Remove bolts attaching exhaust pipe support bracket to transfer case, if necessary.
8. Remove transfer case to transmission attaching bolts.
9. Remove transfer case from vehicle.
10. Remove transfer case to transmission gasket and discard.
11. Install new transfer case to transmission gasket.
12. Reverse steps 1 through 9 to install.

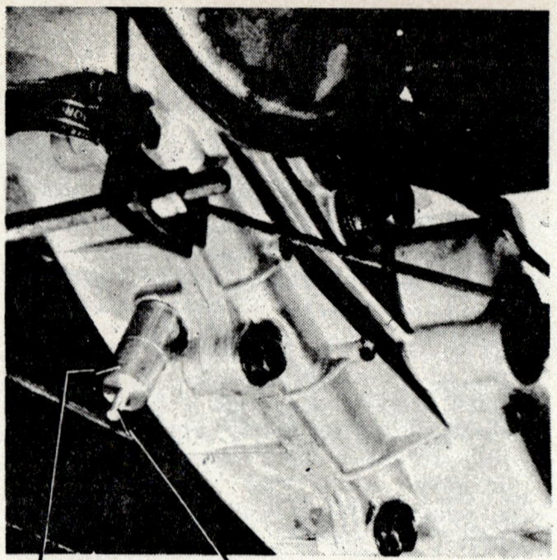

Fig. 1 Warner Quadra-Trac, drive chain tension check

WARNER QUADRA-TRAC

1974–79

Drive Chain Tension Check

Drain lubricant from transfer case and reinstall drain plug. Remove chain inspection plug and install chain tension gauge into hole. If chain tension is satisfactory, plunger will protrude past end of tool, Fig. 1. If plunger is flush or is recessed within tool, the drive chain has lost tension, requiring replacement.

Transfer Case, Removal
1. Disconnect front and rear propeller shafts from transfer case, mark universal joints for proper alignment during installation.
2. Remove exhaust system support bracket from transfer case.
3. Disconnect diaphragm control vacuum hoses, lock-out indicator switch wire and speedometer cable from transfer case.
4. Disconnect parking brake cable guide from frame right side pivot.
5. On CJ-7 models, place a support under clutch housing and remove rear crossmember.
6. Remove two transfer case to transmission bolts entering from front and the two bolts entering from rear and install guide pins in the upper holes.
7. Slide transfer case rearward until unit clears output shaft and guide pins, then remove assembly from vehicle.
8. Reverse procedure to install.

Reduction Unit, Removal
1. Raise vehicle on hoist.
2. Loosen bolts attaching reduction unit to transfer case and move unit rearward to drain oil from unit.
3. Loosen control lever cable retaining bolt and remove control cable from clamp bracket and control lever.
4. Remove reduction unit to transfer case bolts and move reduction unit rearward to clear transmission output shaft and remove unit from vehicle.
5. Reverse procedure to install.

TRANSFER CASE SHIFT LINKAGE, ADJUST

1. Disconnect adjustable link at transfer case.
2. On dual range transfer cases, move both transfer case shift rods out to the last detent position (maximum forward position). On single range cases, move the single shift rod to the first detent position (maximum rearward position). Make sure detents are fully seated.
3. Position transfer case shift lever knob in the forward position with 3/4" clearance between knob and floor pan.
4. With shift rods and lever knob positioned as above, adjust and install adjustable link.
5. Test operation in all positions. If necessary, readjust to eliminate any interference between shift lever knob and floor pan.

1980–81 SR-4 4 SPEED TRANSMISSION

Transmission, Disassembly, Fig. 2

1. Remove transfer case from transmission, then drain transmission fluid.
2. Remove offset lever from shift rail, then remove adapter housing from transmission.
3. Remove shift control housing and gasket from transmission. Note position of two dowel type bolts for reference during assembly.
4. Remove spring clip securing reverse lever to reverse lever pivot bolt, then remove reverse lever pivot bolt, reverse lever and reverse lever fork as an assembly.
5. Scribe alignment marks on front bearing cap and transmission case for reference during assembly, then remove bearing cap and gasket.
6. Remove snap rings from front and rear bearings, then remove front bearing clutch shaft using suitable tool. Remove clutch shaft from case.
7. Remove rear bearing from output shaft using suitable puller.
8. Remove output shaft and gear train as an assembly. Check that synchronizer sleeves do not separate from hubs during removal.
9. Remove reverse idler gear and shaft from rear of case, then remove countershaft from rear of case using tool J26624 or equivalent.
10. Remove countershaft gear and tool used to remove countershaft as an assembly, then remove countershaft gear thrust washers and all clutch shaft pilot bearings from case.
11. Remove needle bearing retainer and 50 needle bearings from countershaft gear.

Output Shaft Service

Disassembly
1. Place reference marks on 3rd-4th synchronizer hub and sleeve for reference during assembly.
2. Remove output shaft snap ring, then remove 3rd-4th synchronizer assembly.
3. Remove 3rd gear, 2nd gear snap ring, thrust washer, 2nd gear and blocking ring.
4. Remove 1st gear thrust washer and roll

pin, then remove 1st gear and blocking ring. Thrust washer has oil groove and roll pin has locating slot on one side. This side must face 1st gear when assembled.

5. Place alignment marks on 1st-2nd synchronizer sleeve and output shaft hub for reference during assembly.
6. Remove insert spring and inserts from 1st-2nd sleeve, then remove sleeve from output shaft hub.

NOTE: Do not attempt to remove 1st-2nd-reverse hub from output shaft. Hub and shaft are machined and matched as an assembly.

Assembly

1. Lubricate output shaft and gear bores with transmission fluid.
2. Install 1st-2nd synchronizer sleeve on output shaft hub using reference marks made during disassembly.
3. Install three 1st-2nd synchronizer inserts and two insert springs into 1st-2nd synchronizer sleeve. Engage tang end of each insert spring into same synchronizer insert but position open ends of springs so that they face away from each other.
4. Install blocking ring onto 1st gear, then install gear and ring onto output shaft. Ensure that synchronizer inserts engage notches in 1st gear blocking ring.
5. Install 1st gear roll pin into output shaft, then install gear and ring onto output shaft. Ensure that sharp edge of washer faces outward and that tab engages output shaft notch.
6. Measure 2nd gear end play by inserting feeler gauge between gear and thrust washer. End play should be .004–.014 inch. If end play is greater than .0014 inch, replace thrust washer and snap ring and inspect synchronizer hub for damage or wear.

NOTE: If any output shaft gear must be replaced, the countershaft gear must also be replaced to maintain proper gear mesh and avoid excessive noise during operation.

---

7. Install blocking ring onto 3rd gear, then install gear and ring onto output shaft.
8. Install 3rd-4th synchronizer sleeve onto 3rd-4th synchronizer hub and align marks made during disassembly.
9. Refer to step 3 for assembly of synchronizer.
10. Measure 3rd-4th synchronizer end play using feeler gauge inserted between output shaft snap ring and 3rd-4th synchronizer hub. End play should be .004–.014 inch.
11. If end play is greater than .014 inch, replace snap ring and inspect synchronizer hub for damage and wear.

Shift Control Housing, Fig. 3

Disassembly

1. Remove detent plug, spring and plunger.
2. Position selector arm plates and shift rail in Neutral. Rotate shift rail counterclockwise until selector arm disengages from selector arm plates and selector arm roll pin is accessible.
3. Position shift rail rearward until selector arm contacts 1st-2nd shift fork.
4. Remove selector arm roll pin, then remove shift rail. Remove shift forks, selector arm plates, selector arm, roll pin and interlock plate.

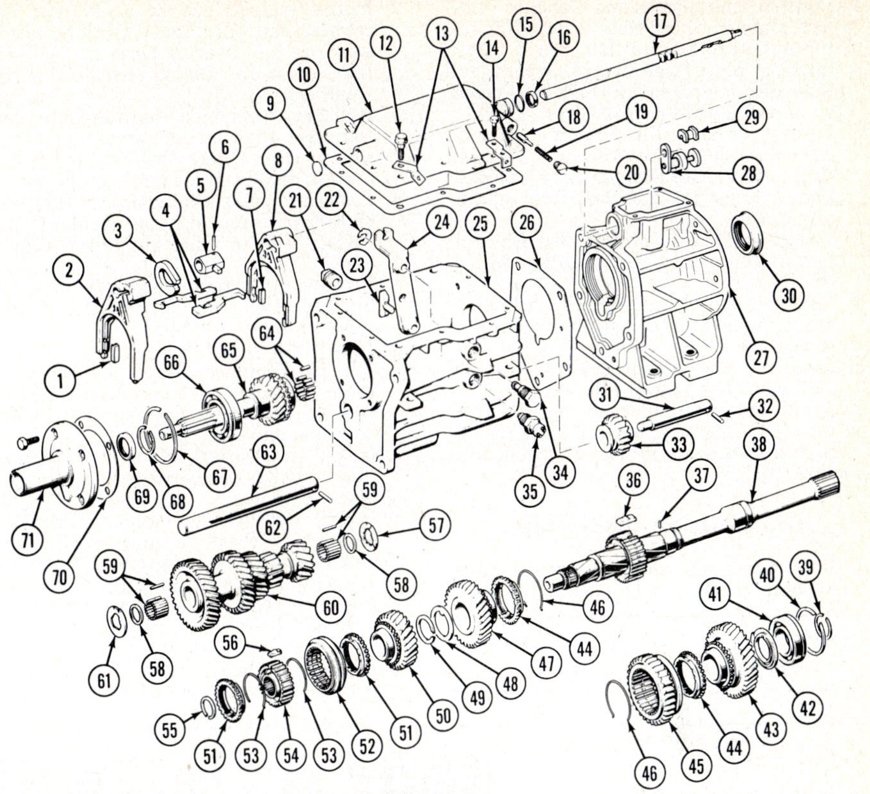

1. THIRD – FOURTH SHIFT INSERT
2. THIRD – FOURTH SHIFT FORK
3. SELECTOR INTERLOCK PLATE
4. SELECTOR ARM PLATE (2)
5. SELECTOR ARM
6. SELECTOR ARM ROLL PIN
7. FIRST – SECOND SHIFT FORK INSERT
8. FIRST – SECOND SHIFT FORK
9. SHIFT RAIL PLUG
10. TRANSMISSION COVER GASKET
11. TRANSMISSION COVER
12. TRANSMISSION COVER DOWEL BOLT (2)
13. CLIP
14. TRANSMISSION COVER BOLT (8)
15. SHIFT RAIL O-RING SEAL
16. SHIFT RAIL OIL SEAL
17. SHIFT RAIL
18. DETENT PLUNGER
19. DETENT SPRING
20. DETENT PLUG
21. FILL PLUG
22. REVERSE LEVER PIVOT BOLT C-CLIP
23. REVERSE LEVER FORK
24. REVERSE LEVER
25. TRANSMISSION CASE
26. GASKET
27. ADAPTER HOUSING
28. OFFSET LEVER
29. OFFSET LEVER INSERT
30. EXTENSION HOUSING OIL SEAL
31. REVERSE IDLER SHAFT
32. REVERSE IDLER SHAFT ROLL PIN
33. REVERSE IDLER GEAR
34. REVERSE LEVER PIVOT BOLT
35. BACKUP LAMP SWITCH
36. FIRST – SECOND SYNCHRONIZER INSERT (3)
37. FIRST GEAR ROLL PIN
38. OUTPUT SHAFT AND HUB ASSEMBLY
39. REAR BEARING RETAINING SNAP RING
40. REAR BEARING LOCATING SNAP RING
41. REAR BEARING
42. FIRST GEAR THRUST WASHER
43. FIRST GEAR
44. FIRST – SECOND SYNCHRONIZER BLOCKING RING (2)
45. FIRST – REVERSE SLEEVE AND GEAR
46. FIRST – SECOND SYNCHRONIZER INSERT SPRING (2)
47. SECOND GEAR
48. SECOND GEAR THRUST WASHER (TABBED)
49. SECOND GEAR SNAP RING
50. THIRD GEAR
51. THIRD – FOURTH SYNCHRONIZER BLOCKING RING (2)
52. THIRD – FOURTH SYNCHRONIZER SLEEVE
53. THIRD – FOURTH SYNCHRONIZER INSERT SPRING (2)
54. THIRD – FOURTH SYNCHRONIZER HUB
55. OUTPUT SHAFT SNAP RING
56. THIRD – FOURTH SYNCHRONIZER INSERT (3)
57. COUNTERSHAFT GEAR REAR THRUST WASHER (METAL)
58. COUNTERSHAFT NEEDLE BEARING RETAINER (2)
59. COUNTERSHAFT NEEDLE BEARING (50)
60. COUNTERSHAFT GEAR
61. COUNTERSHAFT GEAR FRONT THRUST WASHER (PLASTIC)
62. COUNTERSHAFT ROLL PIN
63. COUNTERSHAFT
64. CLUTCH SHAFT ROLLER BEARINGS (15)
65. CLUTCH SHAFT
66. FRONT BEARING
67. FRONT BEARING LOCATING SNAP RING
68. FRONT BEARING RETAINING SNAP RING
69. FRONT BEARING CAP OIL SEAL
70. FRONT BEARING CAP GASKET
71. FRONT BEARING CAP

Fig. 2 1980–81 SR-4 4 speed transmission exploded view.

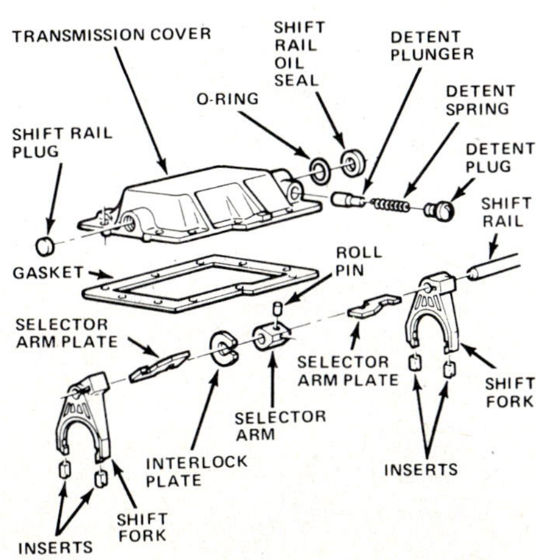

Fig. 3 Shift control housing exploded view.

5. Remove shift rail oil seal and O-ring, then remove shift rail plug.
6. Remove nylon inserts and selector arm plates from shift forks. Note position of inserts and plates for reference during assembly.

Assembly

1. Install nylon inserts and selector arm plates into shift forks, then install shift rail plug.
2. Lubricate shift rail and shift rail bores with vaseline and install shift rail into cover until end of rail is flush with inside edge of cover.
3. Position 1st-2nd shift fork in cover with fork offset facing rear of cover, then install shift rail through fork.

NOTE: 1st-2nd shift fork is larger of two forks.

---

4. Position selector arm and C shaped interlock plate in cover, then insert shift rail through arm. Ensure that widest part of interlock plate faces away from cover and that selector arm roll pin hole faces downward and toward rear of cover.
5. Position 3rd-4th shift fork in cover with fork offset facing rear of cover. 3rd-4th shift fork selector arm plate must be positioned under 1st-2nd shift fork selector arm plate.
6. Install shift rail through 3rd-4th shift fork and into front shift rail bore in cover.
7. Rotate shift rail until selector arm plate at rail forward end faces away from, but is parallel to, cover.
8. Align roll pin holes in selector arm and shift rail, then install roll pin.
9. Install detent plunger, spring and plug. Install O-ring in groove of shift rail oil seal. Install shift rail oil seal using suitable tools.

Transmission, Assembly

1. Lubricate countershaft gear thrust washers with vaseline, then position in case.

NOTE: Install plastic washer at front of case and metal washer at rear.

2. Install tool J26624 or equivalent into countershaft gear, then install 50 needle bearings into bearing bores at front and rear of gear. Install needle bearing retainers.
3. Position assembled countershaft gear into case, then install countershaft from rear of case. Check that thrust washers are not disturbed during installation.
4. Position reverse idler gear into case with shift lever groove facing front of case. Install reverse idler shaft from rear of case.
5. Install output shaft and gear train assembly into case using caution not to disturb synchronizer assemblies.
6. Install 4th gear blocking ring into 3rd-4th synchronizer sleeve. Lubricate 15 roller bearings and clutch shaft roller bearing bore with vaseline, then install bearings into shaft bore.
7. Install clutch shaft into case and engage into 3rd-4th synchronizer sleeve and blocking ring.
8. Install front bearing using tool J22697 or equivalent. Position front bearing onto clutch shaft, then position output shaft 1st gear against rear of case. Align bearing with case bearing bore and install bearing. Install bearing retainer and snap ring.
9. Install front bearing cap oil seal in front bearing cap using tool J26625 or equivalent.
10. Install front bearing cap onto case. Check that groove in cap aligns with oil hole in case. Torque retaining bolts to 13 ft. lbs.
14. Install 1st gear thrust washer onto output shaft, then install rear bearing using J22697 or equivalent. Install retaining ring and snap ring.
15. Position reverse lever in case, then install reverse lever pivot bolt into case. Install reverse lever onto bolt, install spring clip, then torque bolt to 20 ft. lbs.

NOTE: Check that reverse lever fork engages reverse idler gear.

16. Rotate clutch shaft and output shaft gears. If blocking rings stick to gear cones separate rings from cones by prying them off with a screwdriver.
17. Replace adapter housing oil seal, if necessary, then position reverse lever in Neutral and install transmission cover gasket and cover assembly onto case. Torque bolts evenly to 10 ft. lbs. Ensure that cover dowel bolts are installed in proper location.
18. Position adapter housing gasket onto case, then install adapter housing.
19. Fill transmission with 3 pints of fluid, then torque drain plug to 23 ft. lbs.
20. Install transfer case onto transmission.

Rear Axle Section

NOTE: For unit overhaul procedure, refer to "Spicer Driving Axles" chapter.

AXLE SHAFT, REMOVE

1976–83 Semi-Floating-Tapered Shaft, Fig. 1

The hub and drum are separate units, and the hub and axle shaft are serrated to mate and fit together on the taper. Both are marked to insure correct assembly. The axle shaft and bearing assembly may be removed as follows:

1. Remove rear wheel, drum and hub, then disconnect parking brake cable at equalizer.

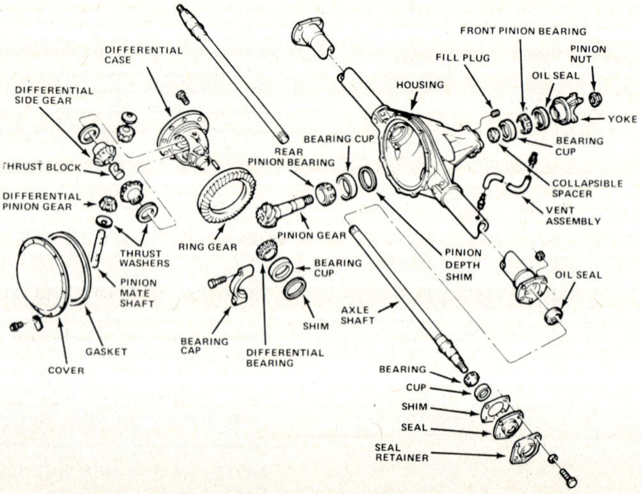

Fig. 1 Semi-Floating axle with tapered shaft. 1976–83

CAUTION: Do not use a slide hammer or knockout type puller to remove hub from axle shaft. Damage to the axle bearings or other components will result.

2. Disconnect brake tube from wheel cylinder and remove brake support plate assembly, oil seal and axle shims from axle shaft.

NOTE: Axle shaft end play shims are located on the left side only.

3. Using suitable puller, pull axle shaft and bearing from axle tube.

CAUTION: On models equipped with Trak-Lok differential, do not rotate differential unless both axle shafts are in place.

1974–83 Semi-Floating - Flanged Shaft, Remove, Fig. 2

1. Raise and support vehicle and remove rear wheels.
2. Remove drum.
3. Remove axle retainer nuts using access hole in axle shaft flange.

NOTE: On 1974–77 models, remove access hole cup plug from flange to gain access to nuts.

4. Remove axle shaft from housing using suitable slide hammer. Remove bearing cup from housing.

1974–83 Full-Floating, Remove, Fig. 3

NOTE: It is not necessary to remove rear wheels to facilitate axle shaft replacement.

1. Remove axle flange nuts, lockwashers, and split washers retaining axle shaft flange.
2. Remove axle shaft from housing.

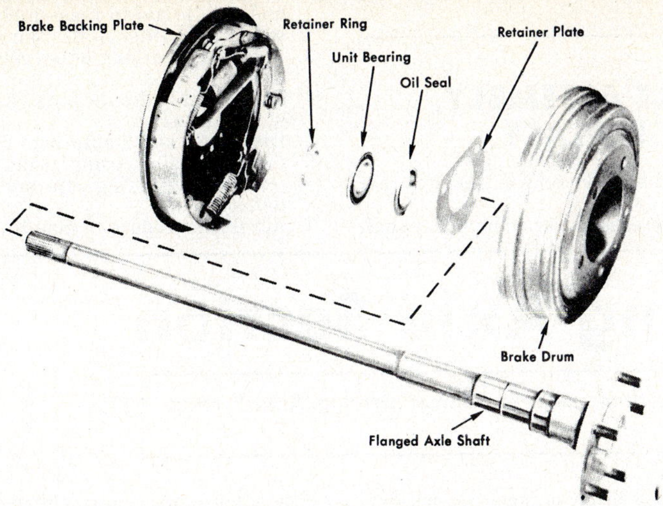

Fig. 2 Semi-Floating flanged axle shaft

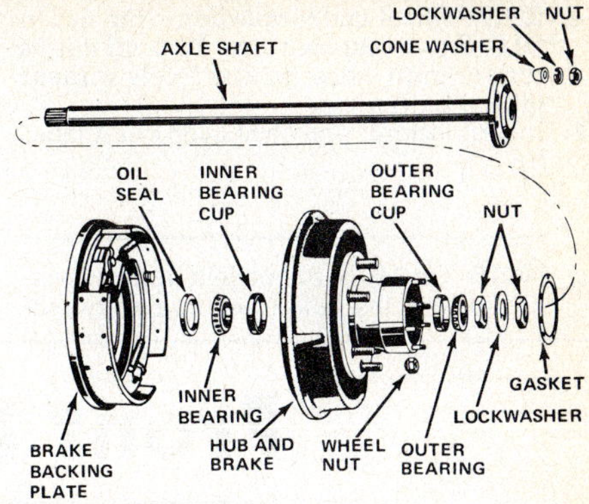

Fig. 3 Full-Floating axle with wheel bearing

BEARING OR OIL SEAL, REPLACE

Semi-Floating - Tapered Shaft, Fig. 1

Remove and install axle shaft bearing using a suitable arbor press. Remove old oil seal from axle housing bore, clean bore and install new seal.

Semi-Floating - Flanged Shaft, Fig. 2

1. Remove oil seal from housing bore, clean bore and install new seal.
2. To remove bearing, drill 1/4 inch hole in retainer ring. Hole depth should be 3/4 of the ring thickness. Do not allow drill to contact axle shaft.
3. Use a large chisel to deeply nick the retaining ring. This will enlarge the ring, or split it, allowing it to be removed.

NOTE: Be certain not to damage axle shaft.

4. Cut through outer oil seal using a hacksaw, then remove seal from shaft.
5. Using suitable press, remove bearing from axle shaft. Remove retainer plate.
6. Reverse procedure to install. When installing new bearing, pack bearing with wheel bearing grease and install with cup rib ring facing axle flange.

Full-Floating Axle, Fig. 3

1. Remove axle shaft.
2. Bend lip of lockwasher and remove locknut and lockwasher.
3. Raise and support vehicle.
4. Remove adjusting nut, outer wheel bearing and pull the wheel straight off the axle.
5. Service wheel bearings as required.
6. Reverse procedure to install. Adjust wheel bearings as described in "Wheel Bearings, Adjust".

AXLE SHAFT, INSTALL

1976–83 Semi-Floating - Tapered Shaft, Fig. 1

Replace the parts in the reverse order of their removal. If the old parts are replaced and the shims have not been disturbed, the axle shaft end play should be correct when the parts are assembled. However, if a new shaft, bearing, differential carrier or housing has been installed, it will be necessary to check the end play.

The end play can be checked when all parts have been replaced except the wheel and hub. To make this check, rap each axle shaft after the nuts are tight to be sure the bearing cups are seated. Then place a dial indicator so that its stem contacts the end of the shaft and work the shaft in and out to determine the amount of existing end play. End play should be .004 to .008 inch. If an adjustment is necessary, remove the outer oil seal and brake support and add or remove shims as required. When making this adjustment, add or subtract shims on left side of axle only.

NOTE: *The application of a bead of sealing material such as "Pliobond" or "Permatex" to the outer diameter of axle tube flange and the brake support contact area is recommended. The sealing material will be used in addition to the gasket for improved sealing.*

When installing hub onto axle, install two well lubricated thrust washers and axle shaft nut. Tighten axle shaft nut until hub is installed to the dimensions shown in Fig. 4. Remove axle shaft nut and one thrust washer. Reinstall axle shaft nut and tighten to 250 ft. lbs. If cotter pin hole is not aligned, tighten the nut to the next castellation and install cotter pin.

NOTE: Do not use an original hub on a replacement axle shaft; use a new hub. A new hub may be installed on an original axle shaft providing the serrations on the shaft are not worn or damaged. Be certain that the hub and axle shaft are punch marked to insure proper alignment on installation. A replacement hub, which is not serrated, can be installed and serrations will be cut in the hub when installed on the shaft due to the difference in

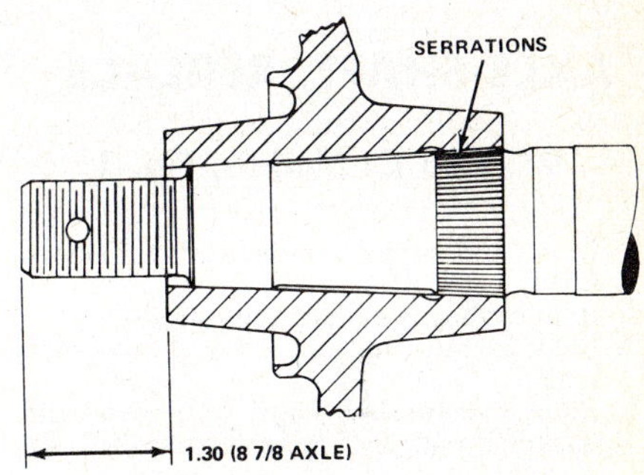

Fig. 4 Replacement hub installation measurement. 1976–83 with tapered shaft

hardness of the shaft and the hub.

1974–83 Semi-Floating - Flanged Shaft, Fig. 2

1. Lubricate outer diameter of bearing cup and install axle shaft and bearing assembly into axle housing using care not to damage oil seal.
2. Install axle shaft retainer and backing plate to housing. Torque retainer nuts to 30 ft. lbs.
3. Install cup plug into axle flange hole if required.
4. Install brake drum and wheel. Lower vehicle.

1974–83 Full-Floating Axle Shaft, Fig. 3

1. Install new gasket on hub.
2. Install axle shaft and tighten nuts securely.

NOTE: It may be necessary to rotate wheel hub to align axle shaft splines with differential splines.

WHEEL BEARING, ADJUST

Full-Floating Axle, Fig. 3

1. Remove axle shaft as described in "Axle Shaft, Service".
2. Straighten lip of lockwasher and remove locknut and lockwasher.
3. Raise and support vehicle.

4. Rotate wheel and torque adjusting nut to 50 ft. lbs. to seat bearings. Back off nut 1/6 turn or until wheel rotates freely without lateral movement.
5. Install locknut and bend over lockwasher lip.

NOTE: On 1975–83 models, torque locknut to 50 ft. lbs. before bending lockwasher lip.

AXLE ASSEMBLY, REPLACE

1. Raise and support vehicle. Remove rear wheels.
2. Scribe reference marks on propeller shaft and axle to aid in reassembly, then disconnect propeller shaft from axle.
3. Disconnect parking brake cable to equalizer.
4. Disconnect hydraulic brake hose from tee at axle housing.
5. Disconnect shock absorbers from axle.
6. Support axle housing using a suitable jack, remove spring clips and lower axle to floor.
7. Reverse procedure to install.

Front Driving Axle Section

NOTE: For unit overhaul procedure, refer to "Spicer Driving Axles" chapter.

AXLE SHAFT, REPLACE

1977–83 CJ Models, Fig. 1

1. Raise and support vehicle, then remove wheel.
2. Remove brake caliper if equipped.
3. Remove hub cap and drive flange snap ring.
4. Remove rotor hub bolts, hub cover and gasket if equipped with disc brakes. On models with drum brakes, remove axle flange bolts.
5. Remove axle flange.
6. Straighten lip of lockwasher and remove outer nut, lockwasher, inner adjusting nut and bearing lockwasher.
7. Remove outer bearing and drum/rotor assembly.

NOTE: Do not damage oil seal.

8. Remove brake support plate or caliper adapter and splash shield as equipped.
9. Remove spindle and spindle bearing, then slide out axle shaft assembly.
10. Reverse procedure to install. Torque inner adjusting nut to 50 ft. lbs. while rotating hub to seat bearings. Back off adjusting nut 1/3 turn on 1977–79 models and 1/6 turn on 1980–83 models. Install outer lockwasher and locknut. Torque locknut to 50 ft. lbs. and bend lip of washer over nut. On 1980–83 models, torque hub body bolts to 30 ft. lbs.

1974–83 Less Selec-Trac Except CJ Models, Fig. 1

1. Raise and support vehicle.
2. Remove wheel and dust cover.
3. Remove axle shaft snap ring, drive gear and pressure spring. It may be necessary to pry drive gear from shaft with a suitable tool.

NOTE: 1980–83 models with front drive hubs do not use spring cup, pressure spring, drive gear and dust cover.

4. Remove wheel bearing locknut, lock ring, and wheel bearing adjusting nut.
5. Remove brake caliper and position aside if equipped.
6. Remove hub and drum or rotor assembly. Note that the spring retainer and outer bearing will slide out as the assembly is removed.
7. Remove nuts and bolt retaining spindle and disc brake shield, if equipped.
8. Remove spindle and disc brake shield, then slide axle shaft from housing.
9. Reverse procedure to install. Torque inner wheel bearing adjusting nut to 50 ft. lbs. while rotating wheel to seat bearings, then back nut off 1/4 turn on 1974–78 models, 1/3 turn on 1979 models and 1/6 turn on 1980–83 models. Install lockwasher with inner tab aligned with keyway in spindle. Turn inner adjusting locknut until the peg engages nearest hole in lockwasher. Install outer locknut and torque to 50 ft. lbs.

NOTE: Install spring retainer with cupped side facing toward center of vehicle.

1983 Models with Selec-Trac, Exc. CJ Models, Fig. 2

NOTE: To remove right side axle shaft refer to

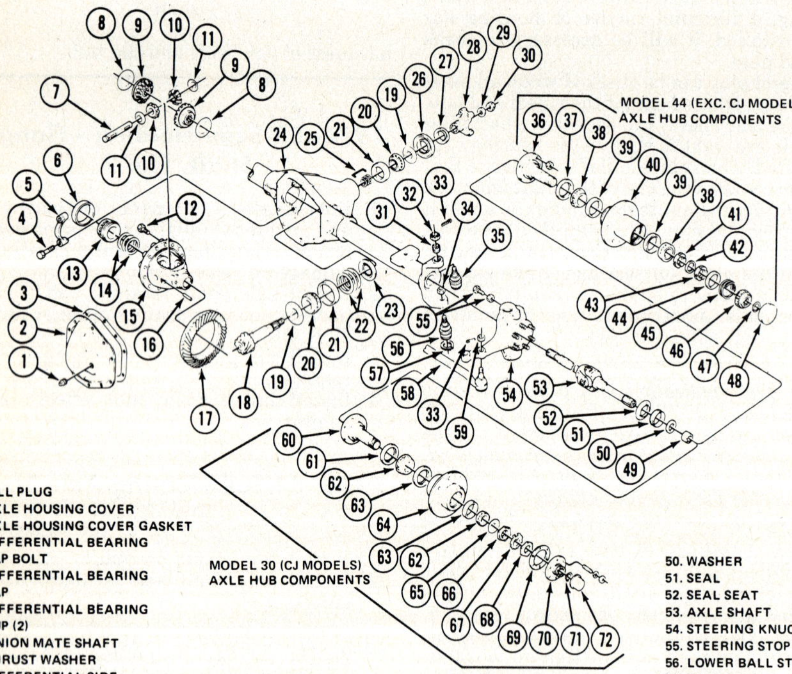

1. FILL PLUG
2. AXLE HOUSING COVER
3. AXLE HOUSING COVER GASKET
4. DIFFERENTIAL BEARING CAP BOLT
5. DIFFERENTIAL BEARING CAP
6. DIFFERENTIAL BEARING CUP (2)
7. PINION MATE SHAFT
8. THRUST WASHER
9. DIFFERENTIAL SIDE GEAR
10. DIFFERENTIAL PINION GEAR
11. THRUST WASHER
12. RING GEAR MOUNTING BOLTS
13. DIFFERENTIAL BEARING (2)
14. DIFFERENTIAL BEARING PRELOAD SHIMS
15. DIFFERENTIAL CASE
16. PINION MATE SHAFT PIN
17. RING GEAR
18. PINION GEAR
19. SLINGER
20. PINION BEARING
21. PINION BEARING CUP
22. PINION DEPTH SHIMS
23. BAFFLE
24. AXLE HOUSING
25. PINION PRELOAD SHIMS
26. OIL SEAL
27. DUST CAP
28. YOKE
29. WASHER
30. PINION NUT
31. UPPER BALL STUD SPLIT RING SEAT
32. UPPER BALL STUD NUT
33. COTTER PIN
34. LOWER BALL STUD JAMNUT
35. UPPER BALL STUD
36. SPINDLE
37. SEAL
38. BEARING
39. BEARING CUP
40. HUB
41. INNER LOCKNUT
42. WASHER
43. OUTER LOCKNUT
44. SPRING CUP
45. PRESSURE SPRING
46. DRIVE GEAR
47. SNAP RING
48. HUB CAP
49. SPINDLE BEARING
50. WASHER
51. SEAL
52. SEAL SEAT
53. AXLE SHAFT
54. STEERING KNUCKLE
55. STEERING STOP BOLT
56. LOWER BALL STUD
57. SNAP RING
58. TIE ROD
59. TIE ROD END NUT
60. SPINDLE
61. SEAL
62. BEARING
63. BEARING CUP
64. HUB
65. TABBED WASHER
66. INNER LOCKNUT
67. LOCK WASHER
68. OUTER LOCKNUT
69. GASKET
70. DRIVE FLANGE
71. SNAP RING
72. HUB CAP

Fig. 1 Front axle assembly. Open knuckle type

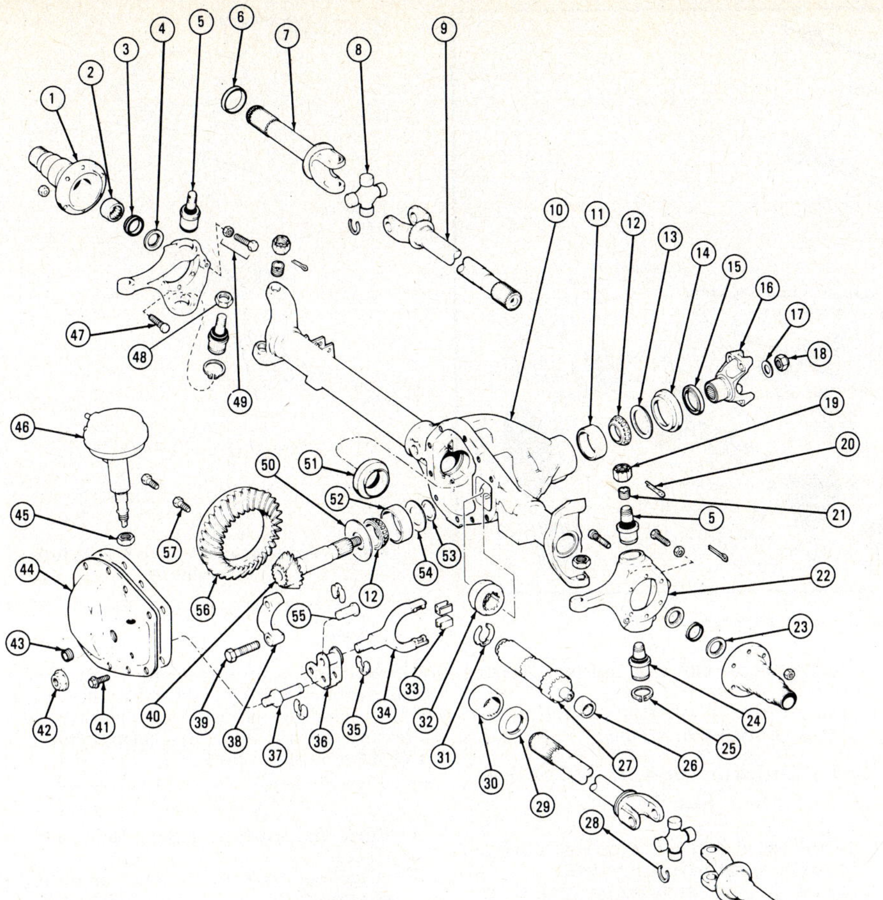

1 STEERING SPINDLE
2 SPINDLE BEARING
3 SEAL
4 SPACER
5 UPPER BALL STUD
6 OIL SEAL
7 OUTER AXLE SHAFT
8 SPIDER ASSEMBLY
9 INNER AXLE SHAFT
10 HOUSING
11 REAR PINION BEARING CUP
12 REAR PINION BEARING
13 REAR OIL SLINGER
14 REAR PINOIN BEARING OIL SEAL
15 YOKE DUST SEAL
16 YOKE
17 PINION WASHER
18 PINION NUT
19 SLOTTED NUT
20 COTTER PIN
21 SPLIT RING
22 STEERING KNUCKLE
23 RETAINING RING
24 LOWER BALL STUD
25 SNAP RING
26 STUB SHAFT BEARING
27 STUB SHAFT
28 SPIDER ASSEMBLY SNAP RING
29 AXLE SHAFT INNER OIL SEAL
30 NEEDLE BEARING
31 STUB SHAFT RETAINING CLIP
32 SHIFT COLLAR
33 SHIFT FORK PAD (2)
34 SHIFT FORK
35 C-RING
36 BELLCRANK
37 PIVOT STUD
38 BEARING CAP
39 BEARING CAP BOLT
40 PINION GEAR
41 HOUSING COVER BOLT
42 LOCK NUT (PIVOT STUD)
43 FILL PLUG
44 HOUSING COVER
45 VACUUM MOTOR RETAINER NUT
46 AXLE VACUUM MOTOR
47 STEERING KNUCKLE TO SPINDLE BOLT
48 JAM NUT
49 TURNING RADIUS STOP BOLT AND LOCK NUT
50 OIL SLINGER
51 INNER OIL SEAL
52 FRONT PINION BEARING CUP
53 FRONT PRE-LOAD SHIM
54 FRONT DEPTH SHIM
55 CLEVIS
56 RING GEAR
57 RING GEAR TO DIFFERENTIAL CASE BOLT

Fig. 2 Selec-Trac front axle assembly. Open knuckle type

"1974–83 less Selec-Trac, except CJ models" for procedure. To remove left side axle shaft refer to "1974–83 less Selec-Trac, except CJ models" steps 1–8, then use following procedure.

1. Disconnect vacuum hoses from axle shift motor, then loosen axle housing cover bolts, allowing lubricant to drain from housing.
2. Remove axle housing cover and shift motor as an assembly.
3. Remove retaining clip from inner axle shaft, then, using a suitable magnet remove inner axle shaft from axle tube. Remove shift collar from housing, Fig. 3.
4. Remove outer axle shaft bearing and seal using Tool J-26941 and J2619-01, Fig. 4.
5. Reverse procedure to install. Torque inner wheel bearing adjusting nut to 50 ft. lbs. while rotating wheel to seat bearings, then back nut off 1/6 turn. Install lockwasher with inner tab aligned with keyway in spindle. Turn inner adjusting nut until the peg engages nearest hole in lockwasher. Install outer locknut and torque to 50 ft. lbs.

NOTE: Install spring retainer with cupped side facing toward center of vehicle.

1974–76 CJ, Figs. 1 & 5

1. Remove drive flange hub cap and snap ring.
2. Remove axle flange bolts, lock washers and flat washers.
3. Have a helper apply the service brakes, then remove axle flange using suitable puller.
4. Release locking lip of lock washer and remove outer nut, lockwasher, adjusting nut and bearing lock washer.
5. Remove brake drum assembly with bearings. Use care not to damage oil seal.

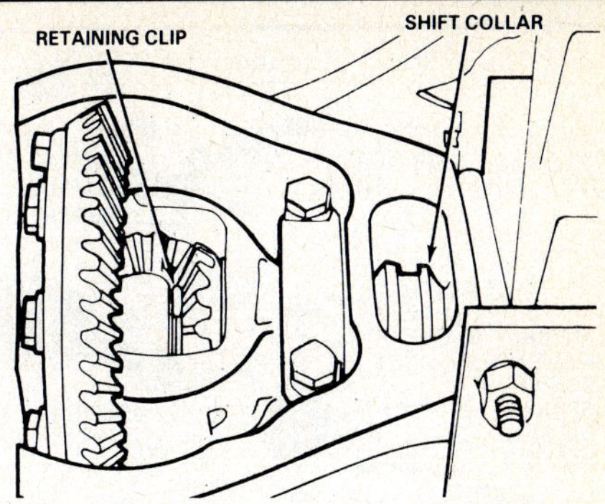

Fig. 3 Inner axle shaft retaining clip & shift collar

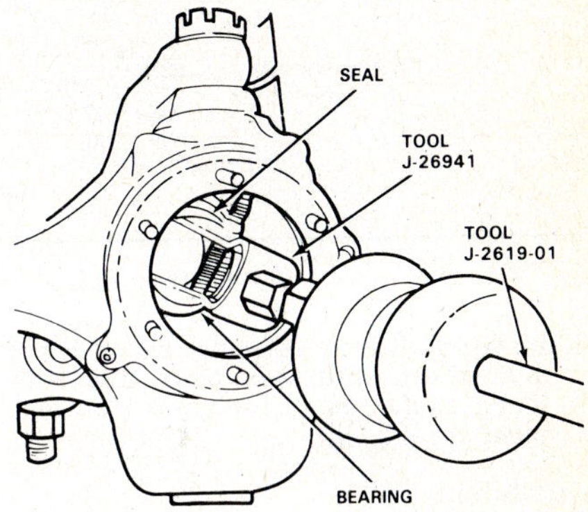

Fig. 4 Outer axle shaft bearing & seal removal

6. Disconnect brake hose between front brake line and the flexible connection, then remove brake backing plate.
7. Remove spindle and spindle bushing.
8. Remove axle shaft and universal joint assembly.
9. Reverse procedure to install. Before installing drive flange, tighten bearing adjust-nut until a slight drag on the bearings is present as the hub is rotated. Then, back off nut approximately 1/6 turn and install lock washer and nut.

AXLE SHIFT MOTOR SERVICE

1983 Models with Selec-Trac

Removal

1. Raise and support front of vehicle, then drain lubricant from front axle.
2. Disconnect vacuum lines from shift motor.
3. Remove axle housing cover, shift motor and shift fork as an assembly, Fig. 6.

Disassembly

1. Remove shift fork to pivot pin retaining E-clip, Fig. 6.
2. Disconnect shift fork pivot from shift motor shaft by threading pivot out of shaft.
3. Remove shift fork assembly from pivot pin, Fig. 7.
4. Remove shift motor retaining nut and remove shift motor from axle housing cover.
5. Remove O-ring from shift motor.

Assembly

1. Install new O-ring on motor.
2. Apply small amount of sealer around shift motor mounting hole and install shift motor and retaining nut, Fig. 6.

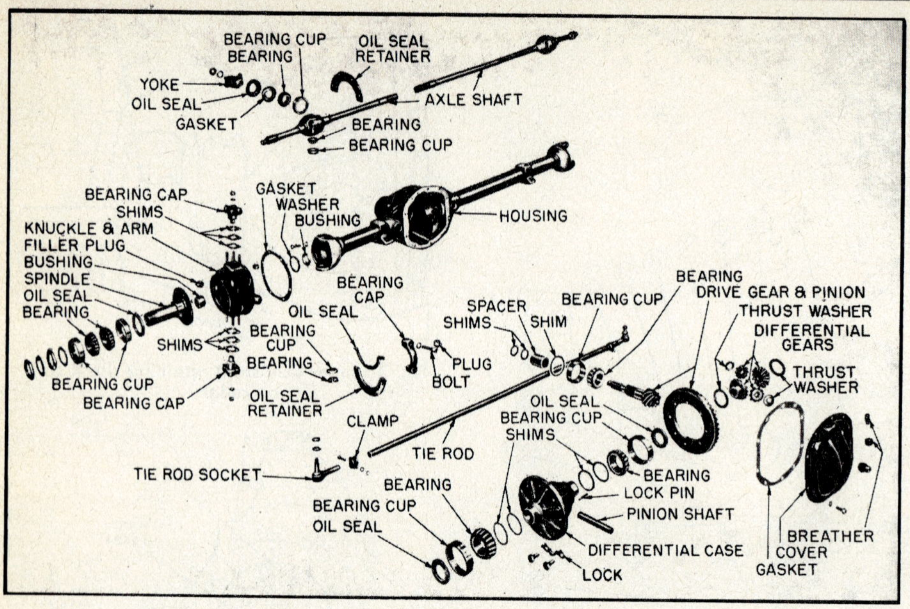

Fig. 5 Front axle assembly. Closed knuckle type

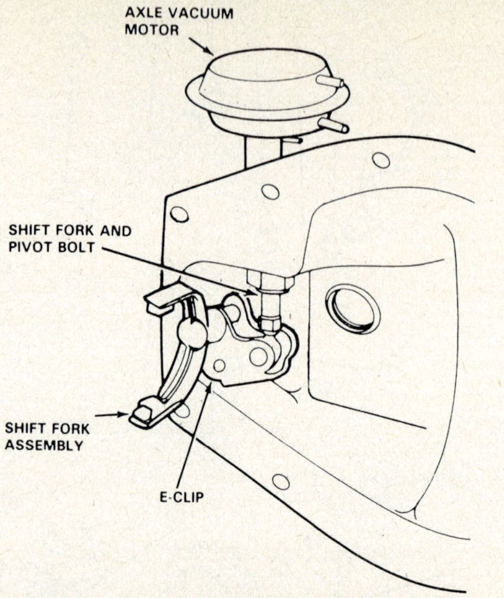

Fig. 6 Axle vacuum motor & shift fork assembly

3. Install shift fork assembly on pivot pin and thread shift fork pivot into shift motor shaft. Install E-clip on shift fork pivot pin, Fig. 6.

Installation

1. Apply RTV sealant to axle cover sealing surfaces and install on axle.

NOTE: Shift fork and fork tabs must be engaged in shift collar before installing cover bolts.

2. Torque axle housing cover bolts to 20 ft. lbs. Fill axle to bottom of fill plug hole with SAE 75 W gear lubricant and reconnect vacuum lines.

WHEEL BEARINGS, REPLACE & ADJUST

Refer to "Axle Shaft, Replace" for wheel bearing replacement and adjustment procedure.

STEERING KNUCKLE, REPLACE

Closed Knuckle Type, Fig. 5

1. Remove axle shaft as described in "Axle Shaft, Replace".
2. Remove oil seal retainer cap screws.
3. Remove the pivot pin cap screws, lock washers and pivot pins. Remove knuckle from yoke.
4. Reverse procedure to install. Check the pivot pin bearing preload as described in "Pivot Pin Bearing Preload, Adjust".

Open Knuckle Type, Removal, Fig. 1

1. Remove axle shaft as described in "Axle Shaft, Replace".
2. Disconnect tie rod from knuckle arm.
3. Remove lower ball stud jamnut and discard. Loosen upper ball stud nut until top of nut is flush with top of stud.
4. Unseat ball studs using suitable hammer. Remove upper ball stud nut and steering knuckle.
5. Remove upper ball stud split ring seat using suitable nut wrench.

Open Knuckle Type, Installation, Fig. 1

1. Install upper ball split ring seat until top of seat is flush with top of yoke.
2. Install steering knuckle on yoke then a new lower ball stud jamnut finger tight.
3. On 1974 models, position Nut Wrench W-355, Button SP-5583, Plate SP-5581 and Puller SP-5574. On 1975–80 models, position Nut Wrench J-25158, Button J-25211-3, Plate J-25211-3 and Puller J-25215. On 1981–83 models, position Nut Wrench J-23447, Button J-25211-3, Plate J-25211-1 and Puller J-25212 as shown, Fig. 8.
4. Tighten puller screw until lower ball stud is held firmly seated. Torque jamnut to 85 ft. lbs. on CJ models, and 75 ft. lbs. on all other models. Remove puller and plate.
5. Torque upper ball stud split ring seat to 50 ft. lbs.
6. Install upper ball stud nut and torque to 100 ft. lbs. Install cotter pin.
7. Install tie rod to steering knuckle. Torque tie rod end nuts to 50 ft. lbs.
8. Install axle shaft as described in "Axle Shaft, Replace".

BALL STUDS, REPLACE

1. Remove steering knuckle as described in "Steering Knuckle, Replace".
2. Remove lower ball stud snap ring.
3. Using suitable tools, press lower and upper ball studs from steering knuckle.
4. Reverse procedure to install. Install a new replacement snap ring on the lower ball stud.

BALL STUD PRELOAD MEASUREMENT

1977–83

NOTE: Ball stud preload is measured when vehicle exhibits high steering effort or slow return of the steering mechanism after turns. If this condition occurs and all other items affecting steering effort are normal, ball stud preload should be checked.

1. Raise and support vehicle. Remove front wheels.
2. Disconnect steering damper at tie rod if equipped.
3. Unlock steering column. Disconnect steering connecting rod at right-side of steering knuckle on CJ models; and on all other models, at right side of tie rod.
4. Remove cotter pin and nut attaching tie rod to right side steering knuckle.
5. Rotate both steering knuckles completely several times, working from right side of vehicle.
6. Assemble a socket and 0–50 ft. lb. torque wrench onto right tie rod attaching nut.

NOTE: Torque wrench must be positioned parallel with the steering knuckle arm.

7. Rotate steering knuckles slowly and steadily through a complete arc and measure torque required to rotate knuckles.
   a. If reading is less than 25 ft. lbs., turning effort at knuckles is normal. Check other steering components for defects or binding.
   b. If reading is greater than 25 ft. lbs., proceed to step 8.
8. Disconnect tie rod from both steering knuckles. Install a ½ × 1 inch bolt, flat washer and nut in tie rod mounting hole in each of the steering knuckles.
9. Measure torque required to rotate each steering knuckle as described previously.
   a. If torque is less than 10 ft. lbs., steering effort is within specifications.
   b. If torque is more than 10 ft. lbs., perform "Ball Stud Preload, Adjust" procedure.
10. If both steering knuckles are within specification, check for damaged or tight tie rod ends.

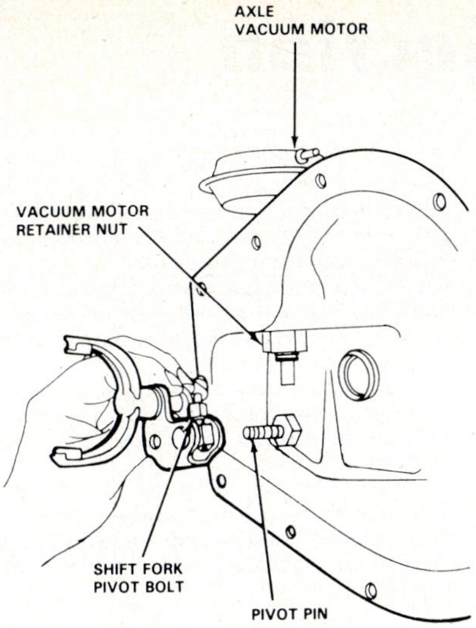

Fig. 7 Removal & installation of shift fork assembly

BALL STUD PRELOAD, ADJUST

1977–83

1. Remove front axle shafts as described in "Axle Shaft, Replace".
2. Loosen lower ball stud jamnut, then remove upper ball stud nut.
3. Unseat ball studs from yoke using suitable hammer.
4. Remove upper ball stud split ring seat and discard.
5. Remove lower ball stud jamnut and steering knuckle. Discard jamnut.
6. Clean upper ball stud split ring seat threads, lower ball stud taper in steering knuckle, threads and tapered surfaces of ball studs, and upper ball stud retaining nut threads.
7. Position steering knuckle on axle and install a new lower ball stud jamnut finger tight.
8. Install upper ball stud nut and torque 10–20 ft. lbs. to draw upper ball stud into yoke.

NOTE: Do not install upper ball stud split ring seat at this time.

9. Torque lower ball stud jamnut to 80 ft. lbs.
10. Remove upper ball stud nut and install a new split ring seat. Torque split ring seat to 50 ft. lbs. Install upper ball stud nut and torque to 100 ft. lbs. Align and install cotter pin.

NOTE: Do not loosen nut to align cotter pin holes. Tighten only enough to align the holes and install cotter pin.

11. Install front axle shafts and steering spindles loosely and measure turning effort of each steering knuckle as described in "Ball Stud Preload, Measurement".
    a. If turning effort is less than 10 ft. lbs., proceed to next step.
    b. If turning effort is more than 10 ft. lbs., replace upper and lower ball studs as described in "Ball Studs, Replace".
12. Install front axle shafts and connect tie rod to steering knuckle arms. Torque tie rod end retaining nuts to 45 ft. lbs.
13. Attach connecting rod to tie rod. Torque connecting rod nut to 60 ft. lbs. on CJ models; and 75 ft. lbs. on all other models. Install steering damper if equipped.
14. Install front wheels and lower vehicle.

PIVOT PIN BEARING PRELOAD, ADJUST

NOTE: The pivot pin bearing preload is checked before installing the oil seal.

1. Check the preload in the bearing by hooking a spring scale in the tie rod socket hole in the steering arm with the steering knuckle in the straight ahead position.
2. Make note of the scale reading when the knuckle had just started its sweep. The reading should be 12–16 lb.
3. Shims are available in the following thicknesses: .003, .005, .010, and .030 inch. Add or remove shims from under top bearing cap to obtain this reading, Fig. 2.

FRONT AXLE ASSEMBLY, REPLACE

1. Raise and support vehicle. Support at frame behind front springs.
2. Remove front wheels.
3. Disconnect front propeller shaft at axle.
4. Disconnect steering linkage from steering knuckles.
5. Disconnect shock absorbers from axle. Disconnect sway bar link bolts if equipped.
6. Disconnect brake hoses. Disconnect breather tube from axle if equipped.
7. Remove brake drums and support plates or brake calipers, hub, rotors and brake shields.
8. Support axle assembly with suitable jack.
9. On vehicles with springs mounted under the axle, remove nuts from spring clips. Remove spring clips, then disconnect springs from shackles by removing lower spring shackle bolts. Lower springs and slide axle assembly from under vehicle.
10. On vehicles with springs mounted on top of axle, remove nuts from spring to axle U-bolt clips. Remove spring clip plates and slide axle assembly from under vehicle.
11. Reverse procedure to install.

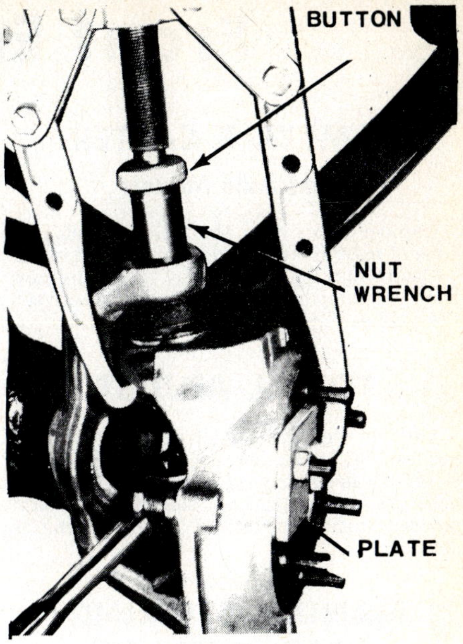

Fig. 8 Steering knuckle installation tools. (Typical)

Front End & Steering Section

CAMBER & CASTER

Solid Axle Models

Camber is set into the axle assembly at time of manufacture and cannot be adjusted.

Caster may be adjusted by installing new front end components or caster shims between the axle pad and the springs.

TOE-IN, ADJUST

Position the wheels to a true straight ahead position with the steering gear also in a straight ahead driving position. Then turn both tie rod adjusting sleeves an equal amount until the desired toe-in setting is obtained.

MANUAL STEERING GEAR, REPLACE

J-Series, Cherokee, Commando, Truck & 1974–76 CJ, DJ

1. Disconnect steering gear from column by removing the flexible coupling nuts or flex coupling pinch bolt.
2. Disconnect the connecting rod from the steering arm.
3. Remove the bolts attaching the gear to the frame and remove the gear by sliding it slightly forward and to the right and lifting it out of the engine compartment.

1977–83 Models

1. Remove intermediate shaft to worm-shaft coupling clamp bolt and disconnect intermediate shaft.
2. Remove pitman arm nut and lockwasher, then separate pitman arm from steering gear shaft using a puller.
3. On all except CJ models, remove steering gear to frame bolts and remove steering gear.
4. On CJ models:
   a. Slightly raise left side of vehicle to relieve tension on left front spring, then place support stand under frame.
   b. Remove steering gear lower bracket to frame bolts.
   c. Remove steering gear upper bracket to crossmember bolts and remove gear.
   d. Remove bolts attaching upper bracket to tie plate and lower bracket to steering gear and remove brackets.
5. Reverse procedure to install. Apply Locktite 271 or equivalent to the steering gear to frame bolts, the bracket to gear bolts and the bracket to tie plate bolt. Torque steering gear to frame bolts and bracket to gear bolts to 70 ft. lbs., the bracket to tie plate bolt to 55 ft. lbs., the coupling pinch bolt and nut to 45 ft. lbs., and pitman arm nut to 185 ft. lbs.

NOTE: If there is a slightly rough feel after the steering gear is installed, turn the steering wheel completely to the left and then to the right for 10 to 15 complete cycles.

POWER STEERING GEAR, REPLACE

1974–76

1. Disconnect pressure and return hoses from gear and plug hoses and gear ports to prevent contamination of system
2. Disconnect steering gear from column by removing the flexible coupling nuts or flex coupling pinch bolt.
3. Remove pitman arm and the steering gear to frame bolts and remove gear from vehicle.
4. Reverse procedure to install.

1977–83

1. Disconnect lines from gear. Plug lines to prevent entry of dirt and keep lines raised to avoid excessive fluid loss.
2. On all except CJ models, remove clamp bolt and nut attaching flex coupling to steering gear shaft and disconnect intermediate shaft.
3. On CJ models, remove clamp bolt and nut attaching intermediate shaft coupling to steering gear shaft and disconnect intermediate shaft.
4. Paint alignment marks on pitman shaft and pitman arm for assembly reference, then remove and discard the pitman arm nut and lockwasher. Use new nut and lockwasher.
5. Remove pitman using a puller.
6. On all except CJ models, remove steering gear to frame mounting bolts and remove steering gear.
7. On CJ models:
   a. Raise left side of vehicle to relieve tension on left front spring and place support stand under the frame.
   b. Remove the three lower steering gear mounting brackets to frame bolts.
   c. Remove the two upper steering gear mounting brackets to crossmember bolts and remove steering gear and mounting brackets.
   d. Remove mounting bracket to gear attaching bolts and remove upper and lower mounting brackets from steering gear.
8. Reverse procedure to install. Apply Locktite 271 or equivalent to mounting bracket to steering gear bolts and to steering gear to frame bolts. Torque steering gear to frame bolts to 70 ft. lbs. on all except CJ models, and 55 ft. lbs. on CJ models, the mounting bracket to steering gear bolts to 70 ft. lbs., the flex coupling clamp bolt to 30 ft. lbs., the intermediate shaft coupling to steering gear bolt to 45 lbs. and the pitman arm nut to 185 ft. lbs.

KENWORTH

INDEX OF SERVICE OPERATIONS

MODEL INDEX & ENGINE APPLICATION

ENGINE IDENTIFICATION—Kenworth identification nameplate can be found on the cab door. Stamped on the plate is the model of the engine, transmission, auxiliary transmission, rear axle and front axle. This data represents only the equipment on the vehicle when it was shipped from the factory.

Model	Engine Make	Basic Engine Model	Crank-case Refill Capacity, Qts.	Cooling System Capacity, Qts.
C500	Cummins	NTC 290	—	—
C500	Cummins	NTC 300	—	—
C500	Cummins	NTC 270	—	—
C823S-2	①	①	②	③
C823S-12	①	①	②	③
C824S-2	①	①	②	③
C824S-12	①	①	②	③
C825S-2	①	①	②	③
C825S-12	①	①	②	③
K100	Cummins	NTC 290	—	—
K100	Cummins	NTC 300	—	—
K121	①	①	②	③
K122	①	①	②	③
K123	①	①	②	③
K124	①	①	②	③
K125	①	①	②	③
L700	Detroit	6V-92TTA 270	—	—
W900	Cummins	NTC 290	—	—
W900	Cummins	NTC 300	—	—
W900S-12	Cummins	NTC 290	—	—
W921	①	①	②	③
W921S-2	①	①	②	③
W921S-12	①	①	②	③
W922	①	①	②	③
W922S-2	①	①	②	③
W922S-12	①	①	②	③
W923	①	①	②	③
W923S-2	①	①	②	③
W923S-12	①	①	②	③
W924	①	①	②	③
W924S-2	①	①	②	③
W924S-12	①	①	②	③
W925	①	①	②	③
W925S-2	①	①	②	③
W925S-12	①	①	②	③

①—Various Diesel engines are available. Consult nameplate to identify engine.
②—Capacity varies depending on engine installed.
③—Capacity varies depending on engine and radiator installed.

STARTING MOTOR SPECIFICATIONS

Unit Part No.	Rotation ①	Brush Spring Tension Oz.	No Load Test			Torque Test		
			Amperes	Volts	R.P.M.	Amperes	Volts	Torque Lb. Ft.
1113818	C	80	60–90	23	7000–10700	500	3.5	26
1113845	C	80	60–90	23	7000–10700	500	3.5	26
1113887	C	80	60–90	23	7000–10700	500	3.5	26

①—As viewed from drive end. C—Clockwise. CC—Counterclockwise.

WHEEL ALIGNMENT SPECIFICATIONS

Front Axle Model	Caster, Deg.	Camber, Deg.	Toe-In, Inch	Kingpin Inclina-tion, Deg.
1-F-1①	0	1	1/4	0
2-F-1①	2–2½	1	1/8	0
2-F-2①	1½	0	1/16	0
3-F-1①	1½	1	1/8–1/4	8
4-F-1①	1½	1	1/8–1/4	8
DCA53-4③	2–2½	1	1/8	0
DCA53-5③	2–2½	1	1/8	0
FDS-20000②	0	3/4	1/16–1/8	0
FDS-22500②	1½	3/4	1/16–1/8	8
FE-15③	2–2½	1	1/8	0
FE-18③	2–2½	1	1/8	0
F-223④	5	0	1/8–1/4	0
F-429④	3	0	1/8–1/4	0
F-3200④	6½	0	3/16–5/16	0
F-7900④	1½	1	1/8–1/4	8
FE-900⑤	2–3	1	1/16	5½
FE-970⑤	1½–1¾	1	1/16	0
FF-901⑤	2–3	1	1/16	5½
FF-921⑤	2–3	1	1/16	5½
FG-900⑤	1½	1	1/16	8
FH-901⑤	3–4	1	1/16	5½
FL-901⑤	3–4	1	1/16	5½
FU-900⑤	1½	0	1/16	0
27462⑤	1½	1	1/16	8

①—Kenworth axle.
②—Clark axle.
③—Shuler axle.
④—Wisconsin (Rockwell) axle.
⑤—Timken (Rockwell) axle.

Clutch Linkage

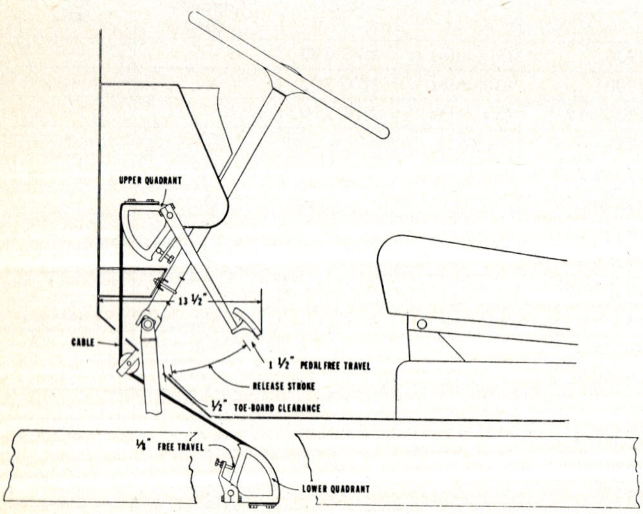

Fig. 1 Cable operated clutch adjustments

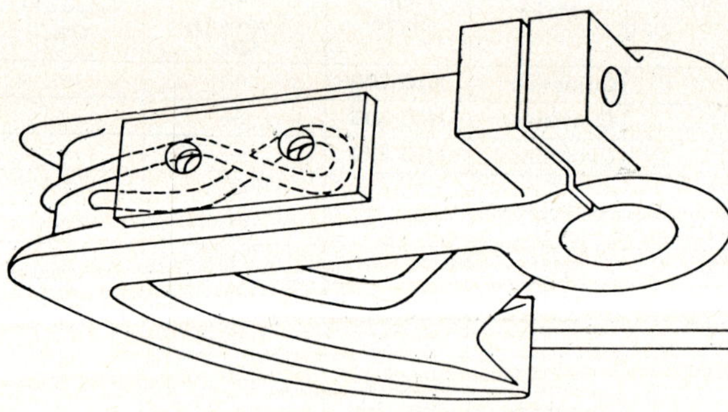

Fig. 2 Cable properly laced to quadrant

External Linkage Adjustment

1. Check the clutch pedal to be sure it is fully engaged (pedal up).
2. With the clutch in the engaged position, free travel clearance at the lever arm must not be more than 1/8" or less than 1/16".

NOTE: Use a 1/8" drill rod to establish proper clearance at adjusting setscrew, as the contact surface is not flat.

3. After setting the adjusting setscrew, check the clutch pedal to make sure there is sufficient free travel, Fig. 1.

Cable, Adjust

1. Loosen the locknut and turn adjusting setscrew so about 1/2" of setscrew protrudes on quadrant side.
2. Loosen two capscrews at either end of the cable.
3. Shorten or lengthen cable as necessary, being careful to lace cable in "figure eight" fasion, Fig. 2, at its connecting point on the quadrant.
4. Tighten capscrews.
5. Adjust setscrews so clearance is 1/8" as outlined in preceding External Linkage Adjustment.

Cable, Replace

1. Remove old cable.

NOTE: When removing old cable from lower quadrant, note the number of turns that spring is wound on the quadrant (about 1½ turns).

2. Turn adjusting setscrew so about 1/2" of it is protruding on lower quadrant side.
3. Install cable on upper quadrant being sure to lace it properly as shown in Fig. 2 and tighten capscrews.
4. Thread cable past pulley and through guide.
5. Turn lower quadrant until it is in original position and fasten cable being sure to lace it properly.
6. Take up all slack in cable and tighten capscrews.
7. Adjust clutch as outlined previously.

MACK

INDEX OF SERVICE OPERATIONS

1974–79 MODEL INDEX & ENGINE APPLICATION

Truck Model	Engine Model ①	Crank-case Refill Capacity Qts.	Cooling System Capacity Qts.
DM & DMM MODELS			
DM401	END475	15	32
DM403	END465C	15	42
DM410	EN414A	8	32
DM487	ENDT475	16	37
DM487, S	ETZ477	—	—
DM492, S	3208⑪	—	—
DM606	ET673	—	—
DM606S	ETZ673E	—	—
DM607	END673	15	44
DM607	END673E	—	49
DM609	END707	19	49
DM611	ENDT673	15	52
DM611S, SX	ETZ673C	—	—
DM611S	ET673	—	—
DM612, S, SX	ETAZ673A	—	—
DM615	END684	⑧	56
DM640	END540A	13	58
DM685	ENDT675	19	57
DM685S, SX, S	ETZ675	—	—
DM686, S, X	ENDT676	—	58
DM807	END673	15	62
DM809	END707	19	62
DM811	ENDT673	—	62
DM811	ET673	—	—
DM811SX	ETZ673C	—	—
DM812, SX	ETAZ673A	—	—
DM815	END864	⑧	62
DM819	ENDT864	⑧	64
DM821ST, SX	ETAZ1000	—	—
DM831	NH220②	—	60
DM833	Super 250②	—	—
DM837	NHC250②	—	60
DM845	NTC335②	—	64
DM861	V903②	—	—
DM863	NTC335②	—	64
DM866ST	NTA400②	—	—
DM867, SX	NTC350②	—	—
DM885	ENDT675	19	64
DM885SX	ETZ675	—	—
DM886, SX	ENDT676	—	64
DM895	ENDT865	—	66
DM897	ENDT866	—	66
DM6076, S	END673	15	52
DML821SX	ETAZ1000	—	—
DML886SX	ENDT676	—	—
DMM612, S	ETAZ673A	—	—
DMM685	ENDT675	—	—

Continued

1974–79 MODEL ENGINE & ENGINE APPLICATION—Continued

Truck Model	Engine Model ①	Crank-case Refill Capacity Qts.	Cooling System Capacity Qts.
DMM685S	ETZ675	—	—
DMM686, S	ENDT676	—	—
DMM4876	ENDT475	—	37
DMM6076	END673E	—	44
DMM6096	END707	—	—
DMM6116	ENDT673	—	52
DMM6856	ENDT675	—	58
MM6856S, SX	ETZ675	—	—
DMM6866, SX	ENDT676	—	58

F MODELS

Truck Model	Engine Model ①	Crank-case Refill Capacity Qts.	Cooling System Capacity Qts.
F607	END673	15	44
F609	END707	19	⑦
F609	END711	19	⑨
F611	ENDT673	15	④
F685	ENDT675	19	55
F707	END673P	15	44
F707	END673E	—	54
F709	END711	19	⑦
F711	ENDT673	15	54
F711T, ST	ETE673	—	—
F712T, ST	ETAZ673A	—	—
F715	END864	⑧	55
F719	ENDT864	⑧	60
F721ST, T	ETAZ1000	—	—
F723	ENDT864A	⑧	60
F726, ST	6V92TT③	—	—
F731	NH220②	—	⑤
F733	Super 250②	—	—
F737	NH250②	—	⑤
F741	NTO-6-B②	—	58
F743	V8-265②	—	48
F745	NT280②	—	52
F747	NTC290②	—	—
F747, ST	Formula 290②	—	—
F749	NHS②	—	52
F759	NHRS②	—	58
F759T, ST	VT903②	—	—
F761	V903②	—	—
F763	NTC335②	—	54
F765	NT-380②	—	58
F765	NTA370②	—	—
F767	NTC350②	—	—
F767, ST	Formula 350②	—	—
F769	8V71T③	—	—
F773	8V71N③	—	—
F773	8V71NE③	—	—
F785	ENDT675	—	55
F785ST, T	ETZ675	—	—
F786, T	ENDT676	—	55
F795	ENDT865	—	70
F797	ENDT866	—	70
F885	ENDT675	—	62
F895	ENDT865	—	70
F985	ENDT675	—	62
F995	ENDT865	—	70

FL & FS MODELS

Truck Model	Engine Model ①	Crank-case Refill Capacity Qts.	Cooling System Capacity Qts.
FL707	END673	15	48
FL709	END707	19	53
FL711	ENDT673	15	55
FL711L, LS	ENDLT673C	—	—
FL715	END864	⑧	59
FL719	ENDT864	⑧	59
FL731	NH220②	—	48
FL733	Super 250②	—	—
FL737	NHC250②	—	48
FL739	NHCT270②	—	48
FL747	NTC290L②	—	—
FL753	1674⑪	—	—
FL755	1693T⑪	—	—
FL759	VT903L②	—	—
FL759LS	ENDT865	—	—
FL761	V903L②	—	—
FL763	NTC335②	—	54
FL765	NTA370L②	—	—
FL767	NTC350L②	—	—
Fl769	8V71T③	—	—
FL773	8V71③	—	66
FL781	6-71N③	—	—
FL785	ENDT675	19	55
FL786	ENDT676	—	55
FL795	ENDT865	—	71
FL797	ENDT866	—	72
FL797LS	ENDT866	—	—
FS711	ENDLT673C	—	55
FS733	Super 250②	—	—
FS747	NTC290L②	—	—
FS753	1674⑪	—	—
FS755	1693T⑪	—	—
FS759	VT903L②	—	—
FS761	V903L②	—	—
FS673L, LS	NTC335	—	—
FS765	NTA370L②	—	—
FS767	NTC350L②	—	—
FS769	8V71T③	—	—
FS773	8V71N③	—	—
FS781	6-71N③	—	—
FS785	ENDT675	—	55
FS786	ENDT676	—	55
FS795	ENDT865	—	71
FS797	ENDT866	—	72

H, HM, HMM, MB, MC & MR MODELS

Truck Model	Engine Model ①	Crank-case Refill Capacity Qts.	Cooling System Capacity Qts.
H693	ENDT673	15	52
HMM685	ENDT675	—	—
HMM6856	ENDT675	—	58
HMM6856S	ETZ675	—	—
MB401	END475	15	38
MB402	EN402	8	36
MB403	END465	15	42
MB410	EN414A	8	32
MB478	ENDT475	16	37
MB483	6-53N③	—	—
MB487	ENDT475	15	42
MB487P, S, T	ET477	—	—
MB47P, S	ETZ477	—	—
MB491	V8-210②	—	—
MB492	3208⑪	—	—
MB605	END673	15	44
MB606	ET673E	—	—
MB607S	END673E	—	—
MB607	END673	15	49
MB609	END707	—	49
MB611	ENDT673	—	52
MB611S	ET673	—	—
MB685	ENDT675	—	56
MC487	ET477	—	—
MC487P, S	ETZ477B	—	—
MC492, S	3208⑪	—	—
MC606	ET673	—	—
MC606P, S	ETZ673E	—	—
MC685	ENDT675	—	—
MC685P, S	ETZ675	—	—
MC686, S, P	ENDT676	—	—
MR487	ET477	—	—
MR487P, S	ETZ477B	—	—
MR492, S	3208⑪	—	—
MR606	ET673	—	—
MR606P, S	ETZ673E	—	—
MR685	ENDT675	—	—
MR685P, S	ETZ675	—	—
MR686, S, P	ENDT676	—	—

R MODELS

Truck Model	Engine Model ①	Crank-case Refill Capacity Qts.	Cooling System Capacity Qts.
R401	END475	15	37
R402	EN402	8	36
R403	END465C	15	42
R410	EN414A	8	32
R487	ENDT475	15	37
R487P	ET477	—	—
R487P	ETZ477	—	—
R482	3208⑪	—	—
R492P	3208⑪	—	—
R567	END465D	15	40
R606	ET673E	—	—
R606T	ETZ673E	—	—
R607	END673P	15	⑥
R607	END673E	—	49
R609	END707	19	⑦
R609	END711	19	⑨
R611	ENDT673	15	④

Continued

1974–79 MODEL ENGINE & ENGINE APPLICATION—Continued

Truck Model	Engine Model ①	Crank-case Refill Capacity Qts.	Cooling System Capacity Qts.
R611	ENDT673C	—	57
R611T, ST	ETZ673	—	—
R612	ETAZ673	—	—
R612ST, T	ETAZ673A	—	—
R615	END864	⑧	56
R640	EN540A	13	58
R685	ENDT675	19	57
R685ST, T	ETZ675	—	—
R686	ENDT676	—	57
R686ST, T	ENDT676	—	—
R707	END673	15	58
R709	END707	19	60
R709	END711	19	60
R711	ENDT673	15	60
R711ST	Super 250②	—	—
R715	END864	⑧	62
R719	ENDT864	⑧	64
R721S, ST	ETAZ1000	—	—
R731	NH220②	—	54
R733	Super 250②	—	—
R737	NHC250②	—	54
R743	V8-265②	—	50
R747	NTC290②	—	—
R747ST	Formula 250②	—	—
R761	V903②	—	—
R763	NTC335②	—	66
R767	NTC350②	—	—
R767ST	Formula 350②	—	—
R769	8V-71T③	—	—
R773	8V71C③	—	—
R773ST	8V-71N③	—	—
R773ST	8V-71NE③	—	—
R786	ENDT676	—	—
R795	ENDT865	—	68
R797	ENDT866	—	68
RD487	ENDT475	—	37
RD487P	ETZ477	—	—
RD487P, S	ET477	—	—
RD492, S	3208⑫	—	—
RD606	ET673E	—	—
RD606S	ETZ673E	—	—
RD607	END673E	15	44
RD611	ENDT673	15	52
RD611	ENDT673C	15	57
RD611S, SX	ETZ673C	—	—
RD611P, S, SX	ET673	—	—
RD612	ETAZ673	—	—
RD612S	ETAZ673A	—	—
RD612SX	ETAZ673	—	—
RD615	END864	⑧	56
RD685	ENDT675	19	57
RD685P, S, SX	ETZ675	—	—
RD686	ENDT676	—	57
RD686S, SX	ENDT676	—	—
RD721SX	ETAZ1000	—	—
RD731	NH230②	—	66
RD733	Super 250②	—	—
RD747	NTC 290②	—	—
RD759	ENDT865	—	—
RD673	NTC335②	—	54
RD767SX	NTC350②	—	—
RD769SX	8V71T③	—	—
RD773	8V71N③	⑩	68
RD785L	ENDT675	—	—
RD795	ENDT865	—	68
RD797	ENDT866	—	68
RL456L, LS	1150⑫	—	—
RL457L, LS	1160⑫	—	—
RL483L, LS	6V-53N③	—	—
RL493L, LS	V555②	—	—
RL611	ET673	—	—
RL611L, LS	ENDLT673	—	—
RL611L, LS	ETZ673C	—	—
RL612L, LS	ETAZ673A	—	—
RL626L, LS	6V92TA③	—	—
RL673L, LS	RV-71N③	—	—
RL673L, LS	RV-71N③	—	—
RL673L	8V-71NE③	—	—
RL679L, LS	6V-71N③	—	—
RL685L, LS	ENDT675	—	—
RL685L, LS	ETZ675	—	—
RL686L, LS	ENDT676	—	—
RL711	ET673	—	—
RL711	ENDLT673C	—	55
RL711L	ETZ673C	—	—
RL712L	ETAZ673E	—	—
RL721L	ETAZ1000	—	—
RL733	Super 250②	—	—
RL747, L	NTC290L②	—	—
RL753	1674⑫	—	—
RL755	1693T⑫	—	—
RL759	VT903L②	—	—
RL761	V903L②	—	—
RL765	NTA370L②	—	—
RL767, L	NTC350L②	—	—
RL769, L	8V71T③	—	—
RL773	8V71N③	—	—
RL781	6-71N③	—	—
RL785	ENDT675	—	55
RL785L	ETZ675	—	—
RL786, L	ENDT786	—	55
RL795	ENDT865	—	71
RL797	ENDT866	—	72
RM685	ENDT675	—	—
RM685X	ETZ675	—	—
RM686	ENDT676	—	—
RM4874	END475	—	37
RM4874X	ET477	—	—
RM4874X, S	ETZ477	—	—
RM4876	ENDT475	—	37
RM6004	ENDT475	—	—
RM6074	END673E	—	44
RM6076	END673E	—	44
RM6114	ENDT673	—	52
RM6114X	ETZ673	—	—
RM6116	ENDT673	—	52
RM6854	ENDT675	—	58
RM6856, S	ENDT675	—	58
RM6864, X	ENDT676	—	58
RM6866, SX	ENDT676	—	58
RS456L	1150⑪	—	—
RS457L	1160⑪	—	—
RS483L	6V-53N③	—	—
RS493L	V555②	—	—
RS611L, LS	ENDLT673	—	—
RS611L, LS	ETZ673C	—	—
RS612L, LS	ETAZ673A	—	—
RS626L, LS	6V92TA3	—	—
RS673L, LS	8V-71N③	—	—
RS673L	NTC335②	—	—
RS679LS	6V-71N③	—	—
RS685L, LS	ETZ675	—	—
RS686LS	ENDT676	—	—
RS711	ET673	—	—
RS711	ENDLT673C	—	55
RS711L, LS	ETZ673C	—	—
RS712L	ETAZ673E	—	—
RS721L	ETAZ1000	—	—
RS733	Super 250②	—	—
RS747, L	NTC290L②	—	—
RS753	1674⑫	—	—
RS755	1693T⑫	—	—
RS759	VT903L②	—	—
RS761	V903L②	—	—
RS765	NTA370L②	—	—
RS767, L	NTC350L②	—	—
RS769, L	8V71T③	—	—
RS773	8V71N③	—	—
RS781	6-71N③	—	—
RS785	ENDT675	—	55
RS785L	ETZ675	—	—
RS786, L	ENDT676	—	55
RS795	ENDT865	—	71
RS797	ENDT866	—	72
RWL711	ETZ673C	—	—
RWL712	ETAZ673A	—	—
RWL721	ETAZ1000	—	—
RWL747	Formula 290②	—	—

Continued

1974–79 MODEL INDEX & ENGINE APPLICATION—Continued

Truck Model	Engine Model ①	Crank-case Refill Capacity Qts.	Cooling System Capacity Qts.
RWL754	3406DIT⑪	—	—
RWL766	NTC400②	—	—
RWL767	NTC350②	—	—
RWL768	KT450②	—	—
RWL770	8V92TA③	—	—
RWL785	ETZ675	—	—
RWL786	ENDT676	—	—
RWL711	ETZ673C	—	—
RWL712	ETAZ673A	—	—
RWL721	ETAZ1000	—	—
RWL747	Formula 290②	—	—
RWL754	3406DIT⑪	—	—
RWL766	NTC400②	—	—
RWS711	ETZ673C	—	—
RWS712	ETAZ673A	—	—
RWS721	ETAZ1000	—	—
RWS747	Formula 290②	—	—
RWS754	3406DIT⑪	—	—
RWS766	NTC400②	—	—
RWS767	NTC350②	—	—
RWS768	KT450②	—	—
RWS770	8V927A③	—	—
RWS785	ETZ675	—	—
RWS786	ENDT676	—	—

U MODELS

Truck Model	Engine Model ①	Crank-case Refill Capacity Qts.	Cooling System Capacity Qts.
U92T	3208⑪	—	—
U401	END475	15	32
U403	END465C	15	42
U410	EN414A	8	32
U487	ENDT475	—	37
U492T	3208⑪	—	—
U567	END465D	15	40
U606	ET673	—	—
U606T	ETZ673E	—	—
U607	END673P	15	⑥
U609	END707	19	④
U609	END711	19	⑨
U611	ENDT673	15	④
U612	ETAZ673	—	—
U612ST, T	ETAZ673A	—	—
U615	END864	⑧	56
U626	6V92TT③	—	—
U626ST, TA	6V92TA③	—	—
U661	V903②	—	—
U685	ENDT675	19	57
U685ST, T	ETZ675	—	—
U686	ENDT676	—	57
U686ST, T	ENDT675	—	—
U734ST	Formula 250②	—	—
U747ST	Formula 290②	—	—
U767ST	Formula 350②	—	—
U769ST	8V71TA③	—	—
U770ST	8V92TA③	—	—
U795	ENDT865	—	—
U797	ENDT866	—	—
U795	ENDT865	—	74
U797	ENDT866	—	74

W MODELS

Truck Model	Engine Model ①	Crank-case Refill Capacity Qts.	Cooling System Capacity Qts.
WL711	ENDT673C	—	—
WL711L, LS	ETZ673C	—	—
WL712L, LS	ETAZ673A	—	—
WL721L, LS	ETAZ1000	—	—
WL747L, LS	Formula 290②	—	—
WL754L, LS	3406 DIT⑪	—	—
WL766L, LS	NTC400②	—	—
WL767L, LS	Formula 350②	—	—
WL768L, LS	KT450②	—	—
WL770L, LS	8V92TA③	—	—
WL785	ENDT675	—	—
WL785L, LS	ETZ675	—	—
WL786L, LS	ENDT676	—	—
WL795	ENDT865	—	—
WL797	ENDT866	—	—
WS711	ENDT673C	—	—
WS711L, LS	ETZ673C	—	—
WS712L, LS	ETAZ673A	—	—
WS721L, LS	ETAZ1000	—	—
WS747L, LS	Formula 290②	—	—
WS754L, LS	3406 DIT⑪	—	—
WS766L, LS	NTC400②	—	—
WS767L, LS	Formula 350②	—	—
WS768L, LS	KT450②	—	—
WS770L, LS	V92TA③	—	—
WS785	ENDT675	—	—
WS785L, LS	ETZ675	—	—
WS786, L, LS	ENDT676	—	—
WS795	ENDT865	—	—
WS797	ENDT866	—	—

1980–83 MODEL INDEX & ENGINE APPLICATION

DM & DMM MODELS

Truck Model	Engine Model ①	Crank-case Refill Capacity Qts.	Cooling System Capacity Qts.
DM487S	E4-210	18	⑬
DM492S	3208/210⑪	—	⑬
DM606S	E6-200	⑫	⑬
DM611S,SX	E6-250	⑫	⑬
DM612S,SX	E6-315	⑫	⑬
DM685S,X,SX	EM6-237/235	⑫	⑬
DM685S,SX	EM6-237R	⑫	⑬
DM686S,SX	EM6-285	⑫	⑬
DM686S,SX	EM6-285R	⑫	⑬
DM688S,SX	E6-350	⑫	⑬
DM811SX	E6-250	⑫	⑬
DM811SX	E6-315	⑫	⑬
DM812SX	E6-315	⑫	⑬
DM821SX	EM9-400R	58.5	⑬
DM821ST,SX	EM9-400	58.5	⑬
DM836ST,SX	EM9-400R	58.5	⑬
DM836ST,SX	EM9-400	58.5	⑬
DM847ST	NTC300②	—	⑬
DM847ST	Formula 300②	—	⑬
DM867ST,SX	NTC350②	—	⑬
DM867ST	Formula 350②	—	⑬
DM867SX	NTC350②	—	⑬
DM885SX	EM6-237	⑫	⑬
DM885SX	EM6-237R	⑫	⑬
DM886SX	EM6-285	⑫	⑬
DM886SX	EM6-285R	⑫	⑬
DM888ST,SX	E6-350	⑫	⑬
DMM612S	E6-315	⑫	⑬
DMM685S	EM6-237/235	⑫	⑬
DMM686S	EM6-285	⑫	⑬
DMM6856S	EM6-237/235	⑫	⑬
DMM6866S	EM6-285	⑫	⑬

F & FM MODELS

Truck Model	Engine Model ①	Crank-case Refill Capacity Qts.	Cooling System Capacity Qts.
F712T,ST	E6-315	⑫	⑬
F726ST	6V92TTA/ 307③	—	⑬
F747ST	NTC300②	—	⑬
F747ST	Formula 300②	—	⑬
F767ST	NTC300②	—	⑬

Continued

1980–83 MODEL INDEX & ENGINE APPLICATION—Continued

Truck Model	Engine Model ①	Crank-case Refill Capacity Qts.	Cooling System Capacity Qts.
F767ST	Formula 300②	—	⑬
F770ST	8V92TA/ 435③	—	⑬
F770ST	8V92TTA/ 365③	—	⑬
F785T,ST	EM6-237	⑫	⑬
F785T,ST	EM6-237R	⑫	⑬
F786T,ST	EM6-285	⑫	⑬
F786T,ST	EM6-285R	⑫	⑬
F788T,ST	E6-350	⑫	⑬
FM712ST	E6-315	⑫	⑬
FM767ST	NTC350②	—	⑬
FM767ST	Formula 350②	—	⑬
FM785ST	EM6-237	⑫	⑬
FM785ST	EM6-237R	⑫	⑬
FM786ST	EM6-285	⑫	⑬
FM786ST	EM6-285R	⑫	⑬

MC & MR MODELS

Truck Model	Engine Model ①	Crank-case Refill Capacity Qts.	Cooling System Capacity Qts.
MC487P,S,T,ST	E4-210	18	⑬
MC492P,T	3208/175⑪	—	⑬
MC492P,S,T,ST	3208/210⑪	—	⑬
MC606P,S,T,ST	E6-200	⑫	⑬
MC611P,S,T,ST	E6-250	⑫	⑬
MC685P,S,T,ST	EM6-237/235	⑫	⑬
MC685P,S	EM6-237R	⑫	⑬
MC686P,S,T,ST	EM6-285	⑫	⑬
MC686P,S	EM6-285R	⑫	⑬
MR487P,S	E4-210	18	⑬
MR492P	3208/210⑪	—	⑬
MR606P,S	E6-200	⑫	⑬
MR611P,S	E6-250	⑫	⑬
MR685P,S	EM6-237/235	⑫	⑬
MR685P,S	EM6-237R	⑫	⑬
MR686P,S	EM6-285	⑫	⑬
MR686P,S	EM6-285R	⑫	⑬

R, RD, RL, RM & RS MODELS

Truck Model	Engine Model ①	Crank-case Refill Capacity Qts.	Cooling System Capacity Qts.
R487P	E4-210	18	⑬
R492P	3208/175⑪	—	⑬
R492P	3208/210⑪	—	⑬
R606T	E6-200	⑫	⑬
R611T,ST	E6-250	⑫	⑬
R612T,ST	E6-315	⑫	⑬
R626T,ST	6V92TTA/ 307-335③	—	⑬
R685T	EM6-225	⑫	⑬
R685T,ST	EM6-237	⑫	⑬
R685T,ST	EM6-237R	⑫	⑬
R685T,ST	EM6-250	⑫	⑬
R685T,ST	EM6-250R	⑫	⑬
R686T	EM6-275	⑫	⑬
R686T,ST	EM6-275R	⑫	⑬
R686T,ST	EM6-285	⑫	⑬
R686T,ST	EM6-285R	⑫	⑬
R686T,ST	EM6-300	⑫	⑬
R686T,ST	EM6-300R	⑫	⑬
R688T,ST	E6-325	⑫	⑬
R688T,ST	E6-325R	⑫	⑬
R688T,ST	E6-350	⑫	⑬
R688T,ST	E6-350R	⑫	⑬
R747S,ST	NTC300②	—	⑬
R747S,ST	Formula 300②	—	⑬
R767S,ST	NTC350②	—	⑬
R767S,ST	Formula 350②	—	⑬
R770S,ST	8V92TA/ 355③	—	⑬
R770S,ST	8V92TA/ 445③	—	⑬
R770S,ST	8V92TA/ 435③	—	⑬
R770S,ST	8V92TTA/ 365③	—	⑬
RD487P,S	E4-210	18	⑬
RD492S	3208/210⑪	—	⑬
RD606P	E6-200	⑫	⑬
RD611P	E6-250	⑫	⑬
RD612S,SX	E6-315	⑫	⑬
RD626S,SX	6V92TA/307- 330③	—	⑬
RD685P,S	EM6-225	⑫	⑬
RD685P,S,SX	EM6-237/235	⑫	⑬
RD685P,S,SX	EM6-237R	⑫	⑬
RD685P,S	EM6-250	⑫	⑬
RD686P,S,SX	EM6-285	⑫	⑬
RD686P,S,SX	EM6-285R	⑫	⑬
RD686P	EM6-300	⑫	⑬
RD688S	E6-325	⑫	⑬
RD688S	EM6-350	⑫	⑬
RD688SX	E6-350	⑫	⑬
RD821S,SX	EM9-400	58.5	⑬
RD821S,SX	EM9-400R	58.5	⑬
RD822S	E9-440	58.5	⑬
RD828SX	E9-440	58.5	⑬
RD866S,SX	NTC400②	—	⑬
RD867S,SX	NTC350②	—	⑬
RD886S,SX	EM6-285/283	⑫	⑬
RD888S,SX	E6-350	⑫	⑬
RL612LS	E6-315	⑫	⑬
RL626LS	6V92TA/307- 330③	—	⑬
RL685LS	EM6-237	⑫	⑬
RL685LS	EM6-250	⑫	⑬
RL685LS	EM6-250R	⑫	⑬
RL686LS	EM6-275R	⑫	⑬
RL686LS	EM6-285	⑫	⑬
RL686LS	EM6-300	⑫	⑬
RL686LS	EM6-300R	⑫	⑬
RL688LS	E6-325	⑫	⑬
RL688LS	E6-325R	⑫	⑬
RL688LS	E6-350	⑫	⑬
RL688LS	E6-350R	⑫	⑬
RM685S	EM6-237/235	⑫	⑬
RM686S,SX	EM6-285/283	⑫	⑬
RM4874X	E4-210	18	⑬
RM6114X	E6-250	⑫	⑬
RM6854X	EM6-237/235	⑫	⑬
RM6856S	EM6-237/235	⑫	⑬
RM6864X	EM6-285	⑫	⑬
RM6866S,SX	EM6-285/283	⑫	⑬
RS612L,LS	E6-315	⑫	⑬
RS626L,LS	6V92TA/307- 330③	—	⑬
RS685L,LS	EM6-250	⑫	⑬
RS685L,LS	EM6-250R	⑫	⑬
RS685LS	EM6-237	⑫	⑬
RS686L,LS	EM6-237	⑫	⑬
RS686L,LS	EM6-275	⑫	⑬
RS686L,LS	EM6-275R	⑫	⑬
RS686L,LS	EM6-285	⑫	⑬
RS686L,LS	EM6-300	⑫	⑬
RS686L,LS	EM6-300R	⑫	⑬
RS686L,LS	E6-325	⑫	⑬
RS688L,LS	E6-325	⑫	⑬
RS688L,LS	E6-325R	⑫	⑬
RS688L,LS	E6-350	⑫	⑬
RS688L,LS	E6-350R	⑫	⑬

RWL & RWS MODELS

Truck Model	Engine Model ①	Crank-case Refill Capacity Qts.	Cooling System Capacity Qts.
RWL721L,LS	EM9-400	58.5	⑬
RWL721L,LS	EM9-400R	58.5	⑬
RWL722L,LS	E9-400	58.5	⑬
RWL747L,LS	NTC300②	—	⑬
RWL747L,LS	Formula 300②	—	⑬
RWL754L,LS	3406 DITA/ 310⑪	—	⑬
RWL762L,LS	NTC475 Twin Turbo②	—	⑬
RWL762L,LS	Formula 450②	—	⑬
RWL764L,LS	KTA525②	—	⑬
RWL766L,LS	NTC400②	—	⑬
RWL767L,LS	NTC350②	—	⑬
RWL767L,LS	Formula 350②	—	⑬
RWL768L,LS	KT450②	—	⑬
RWL770L,LS	8V92TA/ 335③	—	⑬
RWL770L,LS	8V92TA/ 445③	—	⑬
RWL786L,LS	EM6-300	⑫	⑬

Continued

1980–83 MODEL INDEX & ENGINE APPLICATION—Continued

Truck Model	Engine Model ①	Crank-case Refill Capacity Qts.	Cooling System Capacity Qts.
RWL786L,LS	EM6-300R	⑫	⑬
RWL788L,LS	E6-325	⑫	⑬
RWL788L,LS	E6-325R	⑫	⑬
RWL788L,LS	E6-350	⑫	⑬
RWL788L,LS	E6-350R	⑫	⑬
U SERIES			
U606T	E6-200	⑫	⑬
U611T,ST	E6-250	⑫	⑬
U612T,ST	E6-315	⑫	⑬
U626T,ST	6V92TTA/ 270③	—	⑬
U626T,ST	6V92TA/ 335③	—	⑬
U626T,ST	6V92TTA/ 307③	—	⑬
U626T,ST	6V92TA/307-350③	—	⑬
U685T,ST	EM6-237	⑫	⑬
U685T,ST	EM6-237R	⑫	⑬
U685T,ST	EM6-250	⑫	⑬
U685T,ST	EM6-250R	⑫	⑬
U686T,ST	EM6-275R	⑫	⑬
U686T,ST	EM6-285	⑫	⑬
U686T,ST	EM6-285R	⑫	⑬
U686T,ST	EM6-300	⑫	⑬
U686T,ST	EM6-300R	⑫	⑬
U688T,ST	E6-350	⑫	⑬
U688T,ST	E6-350R	⑫	⑬
U747ST	Form. 300②	—	⑬
U747ST	NTC300②	—	⑬
U767ST	NTC300②	—	⑬
U767ST	Form. 350②	—	⑬
U770ST	8V92TA/ 435③	—	⑬
U770ST	8V92TTA/ 335③	—	⑬
W SERIES			
WL,WS712L,LS	E6-315	⑫	⑬
WL,WS721L,LS	EM9-400	58.5	⑬
WL,WS721L,LS	EM9-400R	58.5	⑬
WL,WS722L,LS	E9-400	58.5	⑬
WL,WS726L,LS	6V92TA/307-330③	—	⑬
WL,WS747L,LS	NTC300②	—	⑬
WL,WS747L,LS	Form. 300②	—	⑬
WL,WS754L,LS	3406 DITA/ 310⑪	—	⑬
WL,WS762L,LS	NTC475 Twin Turbo②	—	⑬
WL,WS762L,LS	Form. 450②	—	⑬
WL,WS764L,LS	KTA525②	—	⑬
WL,WS766L,LS	NTC400②	—	⑬
WL,WS767L,LS	NTC350②	—	⑬
WL,WS767L,LS	Form. 350②	—	⑬
WL,WS768L,LS	KT450②	—	⑬
WL,WS770L,LS	8V92TA/ 355③	—	⑬
WL,WS770L,LS	8V92TA/ 445③	—	⑬
WL,WS785L,LS	EM6-250	⑫	⑬
WL,WS785L,LS	EM6-250R	⑫	⑬
WL,WS786L,LS	EM6-275	⑫	⑬
WL,WS786L,LS	EM6-275R	⑫	⑬
WL,WS786L,LS	EM6-300	⑫	⑬
WL,WS786L,LS	EM6-300R	⑫	⑬
WL,WS788L,LS	E6-325	⑫	⑬
WL,WS788L,LS	E6-325R	⑫	⑬
WL,WS788L,LS	E6-350	⑫	⑬
WL,WS788L,LS	E6-350R	⑫	⑬

①—Mack design unless otherwise noted.
②—Cummins engine.
③—Detroit diesel engine.
④—Radiator frontal area: 700 sq. in. 52 qts; 1000 sq. in. 60 qts.
⑤—Radiator frontal area: 700 sq. in. 42 qts; 1000 sq. in. 52 qts.
⑥—Radiator frontal area: 560 sq. in. 42 qts; 1000 sq. in. 60 qts.
⑦—Radiator frontal area: 700 sq. in. 49 qts; 1000 sq. in. 60 qts.
⑧—High level 16 qts.; low level 12 qts.
⑨—Radiator frontal area; 700 sq. in. 50 qts.; 1000 sq. in. 60 qts.
⑩—High level 20 qts.; low level 16 qts., 1000 sq. in. 60 qts.
⑪—Caterpillar engine.
⑫—ESI, 33.5 qts.; ESI Plus, 58.5 qts.
⑬—Cooling system capacity will vary depending on vehicle & engine model.

STARTING MOTOR SPECIFICATIONS

Starter Type	Rotation	No Load Test	
		Amperes	R.P.M.
12 Volt Leece Neville	CW① or CCW②	95–105	4000
24 Volt Leece Neville	CW①	85–90	3800

①—Clockwise. ②—Counterclockwise.

WHEEL ALIGNMENT SPECIFICATIONS

Front Axle Model ① FA-	Caster, Degrees		Camber Degrees	Kingpin Angle Degrees
	Manual Steering	Power Steering		
532	1 to 3	4 to 6	½ to 1	7
L532	1 to 3	4 to 6	½ to 1	7
W532	1 to 3	4 to 6	½ to 1	7
5321	1 to 3	4 to 6	½ to 1	7
534,5341	1 to 3	4 to 6	½ to 1	8
535,5351	1 to 3	4 to 6	½ to 1	7
L535	1 to 3	4 to 6	½ to 1	7
535N	1 to 3	4 to 6	½ to 1	7
W536,C,D,N,W	1 to 3	4 to 6	½ to 1	7

Continued

WHEEL ALIGNMENT SPECIFICATIONS—Continued

Front Axle Model ① FA-	Caster, Degrees		Camber Degrees	Kingpin Angle Degrees
	Manual Steering	Power Steering		
FW537C,D	1 to 3	②	½ to 1	7
W537,C,D,N,W	1 to 3	4 to 6	½ to 1	7
W538C	1 to 3	4 to 6	½ to 1	7
600,601	1 to 3	4 to 6	½ to 1	3½
6001,60011	1 to 3	4 to 6	½ to 1	3½
604,6041	1 to 3	4 to 6	½ to 1	5½
6042,6043	1 to 3	4 to 6	½ to 1	5½
6044	1 to 3	4½ to 6½	½ to 1	5½
612	1 to 3	4 to 6	½ to 1	3½
W612	1 to 3	4 to 6	½ to 1	3½
616C,D	1 to 3	4 to 6	½ to 1	5½
W616D	1 to 3	4 to 6	½ to 1	5½
617C,D	1 to 3	4 to 6	½ to 1	5½
W617C,D	1 to 3	4 to 6	½ to 1	5½
618W-DS	1 to 3	4 to 6	½ to 1	③
W618W-DS	1 to 3	4 to 6	½ to 1	③
619W	1 to 3	4 to 6	½ to 1	③
W619W	1 to 3	4 to 6	½ to 1	③
701,702	1 to 3	4 to 6	½ to 1	3½
703,703C	1 to 3	4 to 6	½ to 1	3½
705	1 to 3	4 to 6	½ to 1	3½

①—Axle model number is located at right front of axle housing.

②—Ross, 1–3; Sheppard 4–6.

③—Right side, 6¼; left side 5¾.

Electrical Section

LEECE-NEVILLE 12 & 24 VOLT STARTER MOTORS

Disassembly & Assembly

1. Disconnect wires from No. 3 and No. 4 terminals of solenoid switch assembly and remove the switch.

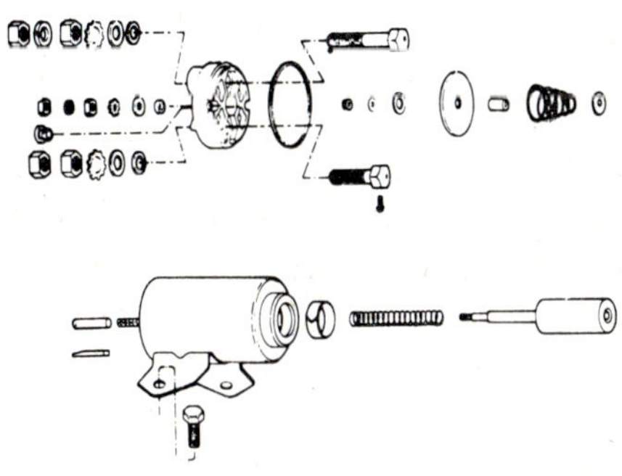

Fig. 1 Exploded view of switch assembly

NOTE: To ease removal of jumper strap retaining nut at No. 3 switch stud and the field ring, heat the stud to approximately 300° F. This will prevent damage to the plastic base which secures the stud.

2. Loosen motor field terminal stud nut several turns. Secure jam nut from turning while loosening nut stud.
3. Remove rubber plug from contact housing to gain access to timing shaft.
4. Remove the two switch-to-field ring attaching screws, then unscrew nut on end of timing shaft to release switch plunger from arm and shaft assembly.
5. Scribe alignment marks on all housings to mark their relationship to each other and to the field ring for assembly reference.
6. Remove brush opening cover band and gasket assembly, then remove brushes from holders.
7. Remove nose housing attaching screws, then the nose housing from armature shaft.
8. Remove shaft housing attaching screws, then remove housing, with arm and shaft and drive assemblies, from armature shaft.
9. Remove commutator and housing attaching screws, then remove housing and brush rigging assembly from field ring.
10. Remove shift arm and shaft-to-shift housing retaining screw and washer, then remove shaft by prying into slot provided.
11. Disassembly switch and brush rigging assemblies as shown in Figs. 1 and 2.
12. Reverse procedure to assemble.

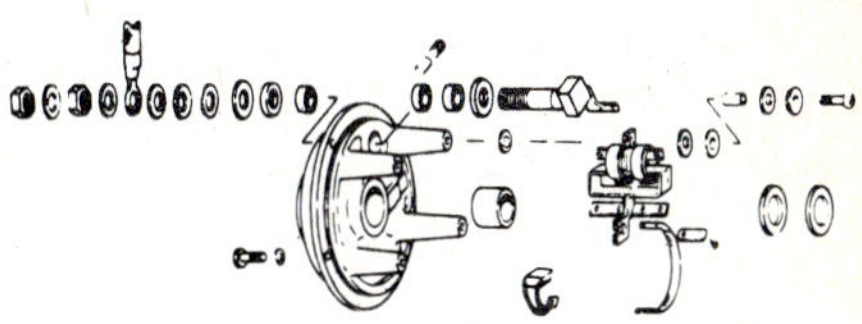

Fig. 2 Exploded view of brush rigging assembly

Clutch Section

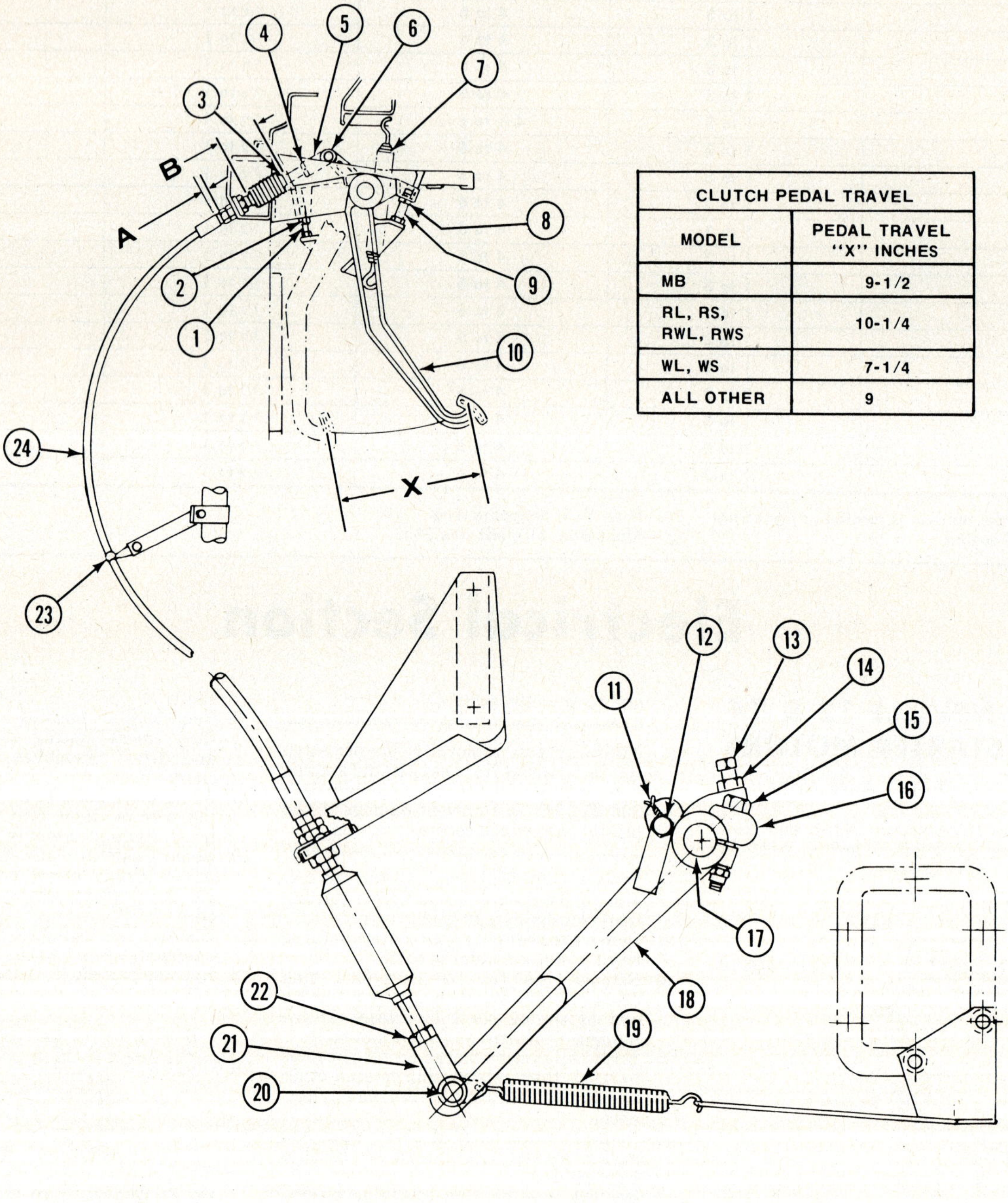

CLUTCH PEDAL TRAVEL	
MODEL	PEDAL TRAVEL "X" INCHES
MB	9-1/2
RL, RS, RWL, RWS	10-1/4
WL, WS	7-1/4
ALL OTHER	9

1. ADJUSTING BOLT
2. JAM NUT
3. CLAMP
4. CLEVIS NUT
5. CLEVIS
6. CLEVIS PIN
7. SPRING
8. JAM NUT
9. ADJUSTING BOLT
10. CLUTCH PEDAL
11. LOCKWIRE
12. LOCK BOLT
13. ADJUSTING BOLT
14. LOCKNUT
15. PINCHBOLT
16. CLUTCH RELEASE ADJUSTING LEVER
17. CLUTCH RELEASE SHAFT
18. CLUTCH RELEASE LEVER
19. SPRING
20. CLEVIS PIN
21. CLEVIS
22. NUT
23. CLAMP
24. CABLE

Fig. 1 Clutch linkage arrangement late models. Mack trans. only typical

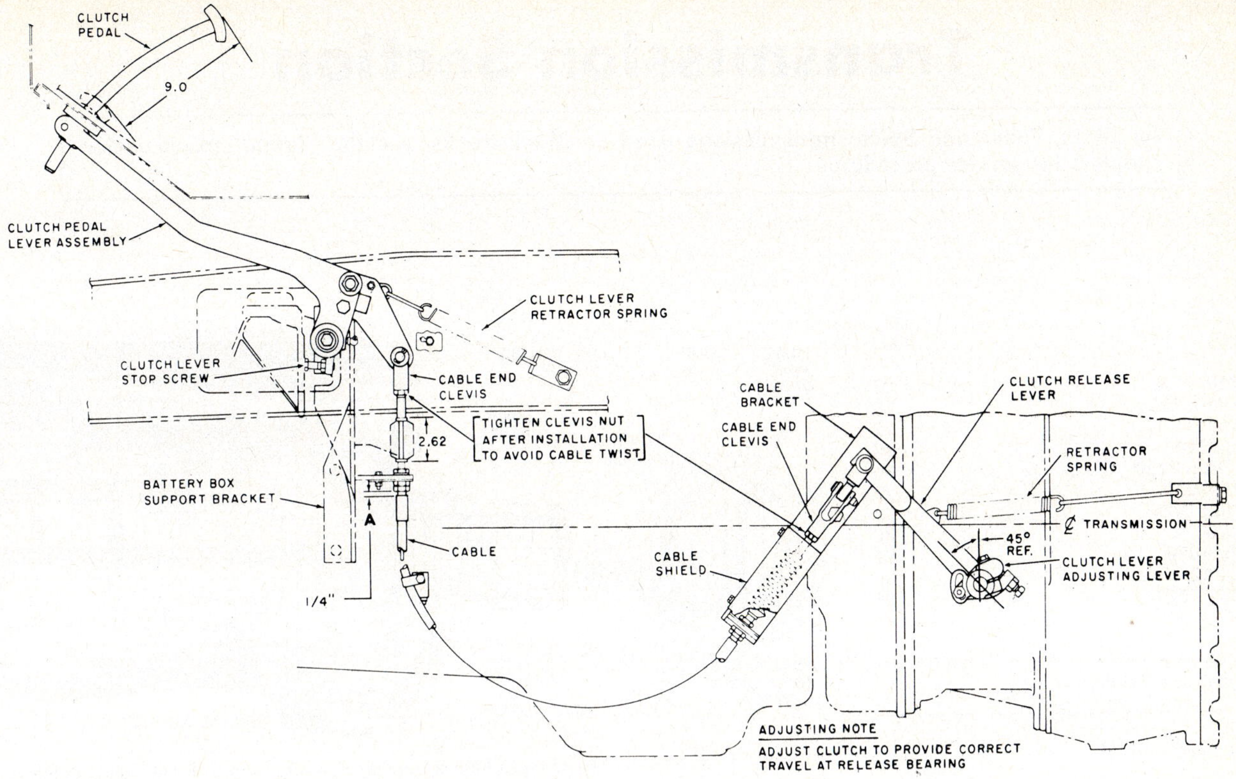

Fig. 2 Clutch linkage arrangement early models. Mack trans. only typical

CLUTCH LINKAGE ADJUSTMENT, FIGS. 1 & 2

As disc facings wear, the thrust washer moves toward the release bearing to reduce release bearing clearance. Moving release collar toward transmission by adjusting external linkage restores this clearance. Do not adjust position of release levers.

The correct release bearing clearance when clutch is engaged is 1/8 inch to 3/16 inch on CL-28, 40, 50 and 58 Series 1/16 inch to 3/16 inch on CL-47 Series, Fig. 3. To obtain full release, pedal travel must be sufficient to move the release assembly an additional distance toward the flywheel after assembly has contacted the thrust washer, Fig. 4, for correct distance.

After repeated adjustment, the release collar approaches the driving cover (a total rearward movement of approximately 1 inch) and the control linkage cannot be readjusted to provide release bearing clearance. When this occurs, disc facings are completely worn and the driven disc must be replaced. It is also important to check clearance between release levers and clutch cover at same time. When clearance at that point reaches .050 inch after making a 1/8 to 3/16 inch collar adjustment, the clutch has reached final adjustment and should be scheduled for removal at no less than .005 inch clearance or serious damage will result from overheating.

Second Stage Adjustment, CL-50 Series

The CL-50 Series two-plate clutches are designed to provide two equal stages of wear. When the first stage is completed (indicated by lack of adjustment in clutch linkage) the 12 adjusting spacers between flywheel and clutch cover must be removed. With spacers removed, the clutch assembly is moved towards the flywheel an amount approximately equal to that worn off clutch disc, thus restoring clutch, linkage, etc. to their original state. The clutch now has a second wear stage equal to the first stage. To make an adjustment, loosen all 12 cover-to-flywheel screws three full turns, then remove one screw at a time, take out the spacer it retains and reinstall the screw. When all 12 spacers have been removed, tighten cover-to-flywheel screws.

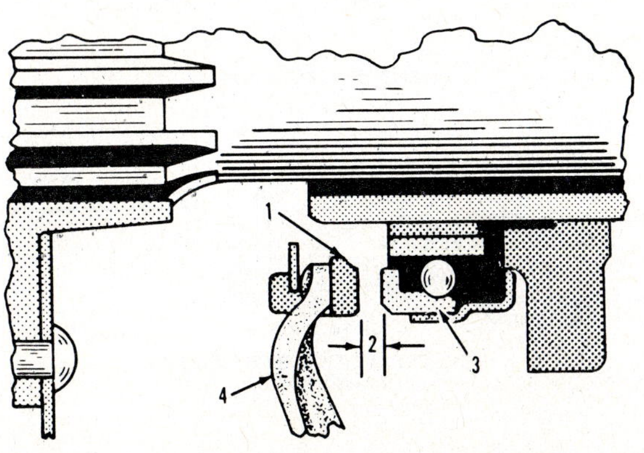

Fig. 3 Release bearing clearances

TRANSMISSION COLLAR TRAVEL			
CLUTCH	RELEASE COLLAR TRAVEL		
	FREE	WORKING	TOTAL
CL-28 & 58 SERIES	.125	.375	.500
CL-40 SERIES EXCEPT CL-404	.125	.281	.406
CL-404	.125	.437	.562
CL-47 SERIES	.062	.500	.562
CL-50 SERIES	.125	.500	.625

Fig. 4 Clutch collar clearance chart

Transmission Section

For Clark, Fuller and Spicer transmissions used on Mack trucks, see the "name" transmission chapters for service procedures

Mack Unit No.①	Oil, Pints
TR-72, 720, 7220	22
TRD-72, 720, 7220	28
TRDLG-1070	22
TRT-722, 7220	30
TRQ-72, 720, 7220	36
TRDL-107, 1070	22
TRDXL-107, 1070	28
TRDX-7220	28
TRL-1076, 1078	22
TRTL-72, 722	30
TRTXL-107, 1070	28
TRXL-107, 1071	22

①—See name plate on transmission housing for model identification.

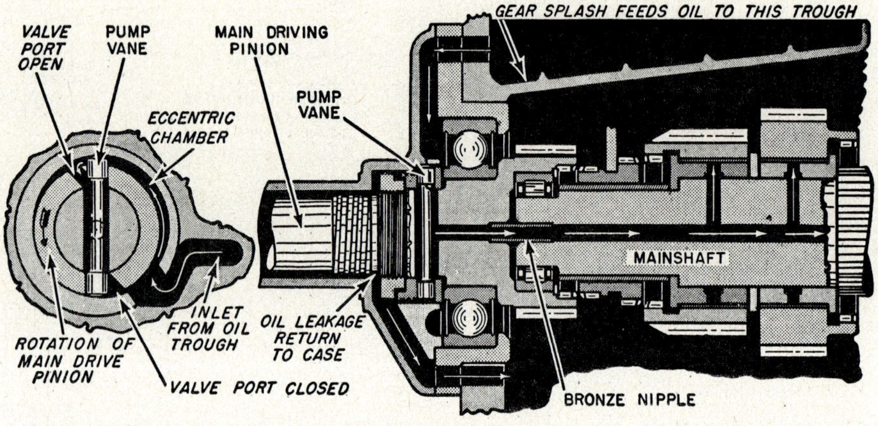

Fig. 1 Sectional view of typical Mack transmission oil pump and mainshaft drillings

GENERAL DATA

Transmission Oil Pump

All Mack-made transmissions employ constant mesh helical gears which run free on the mainshafts and are brought into selective engagement by means of sliding gear-type clutches. These free-running or floating gears run on thinshell bushings to maintain gear alignment with greater precision. However, because bushings require more positive provision for lubrication, pressure lubrication of these floating gears is provided by a built-in pump in all Mack-made transmissions except the small capacity TRL-1076 and TRL-1078 which is lubricated by gear throw-off and gear pumping action.

Pressure lubrication of these bushings is provided by oil delivered through rifle-drilled passages in the mainshafts and supplied by a simple eccentric shuttle-type pump built in the main drive pinion, Fig. 1. This pump has but one moving part—a double-ended reciprocating vane which dispenses with valves. Calculated leakage past the bronze nipple acting as a jumper tube, which conducts oil from the pump to the drilled mainshafts, supplies copious lubrication to the spigot bearings. Gravity feed oil reaches in the reverse cluster gear bearings through a rifle-drilled connecting hole in the supporting shaft.

All working and sliding parts of Mack-made transmissions are bathed in oil from gear throw-off when the engine is running; the countershaft being partially submerged in lubricant. To supply the pump, part of this gear throw-off is collected by a gutter high on the right-hand side of the case. Through a drilled connecting passage in the case, this oil is led to the eccentric space between the driving pinion shaft and its bearing cover. As the shaft turns, the pump vane reciprocates in its eccentric orbit and the vane ends pick up the oil and force it into the center passage through the bronze nipple to the mainshaft drillings.

Magnetic Cleaner Plug

Of course, clean oil in a transmission is essential to long life of gears and bearings. Minute particles of steel, dislodged as a result of wear, form an abrasive of great destructiveness and tend to circulate with the oil. This greatly increases the wear on gear teeth and bearings unless means are provided to separate them from the oil. To combat this, all Mack-made transmissions are provided with a magnetic oil cleaner plug. As viewed from the rear of the transmission, Fig. 2, the oil cleaner is located high inside the transmission case on the right-hand side above the level of the oil. It is so situated that it receives the throw-off from the upper gears. The cleaner consists of an integral open trough and

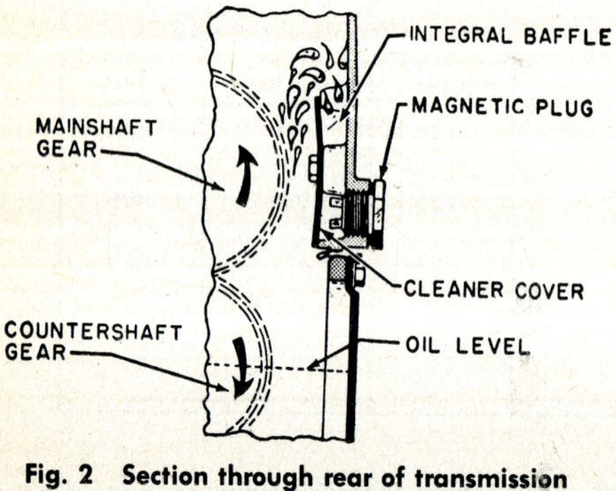

Fig. 2 Section through rear of transmission showing magnet cleaner plug

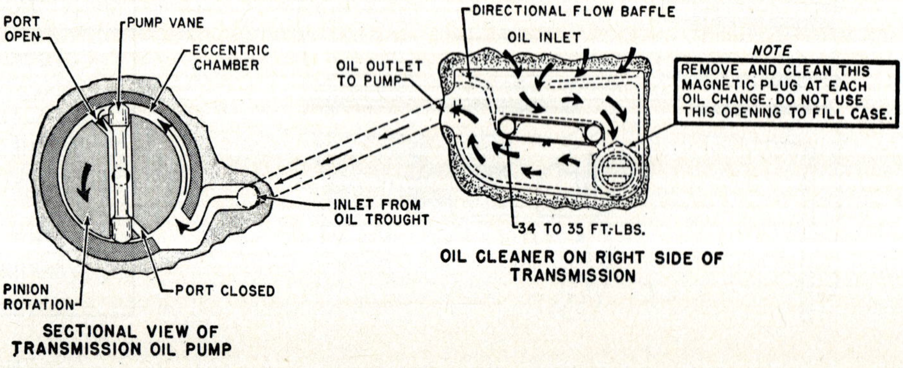

Fig. 3 Showing oil pump and oil cleaner

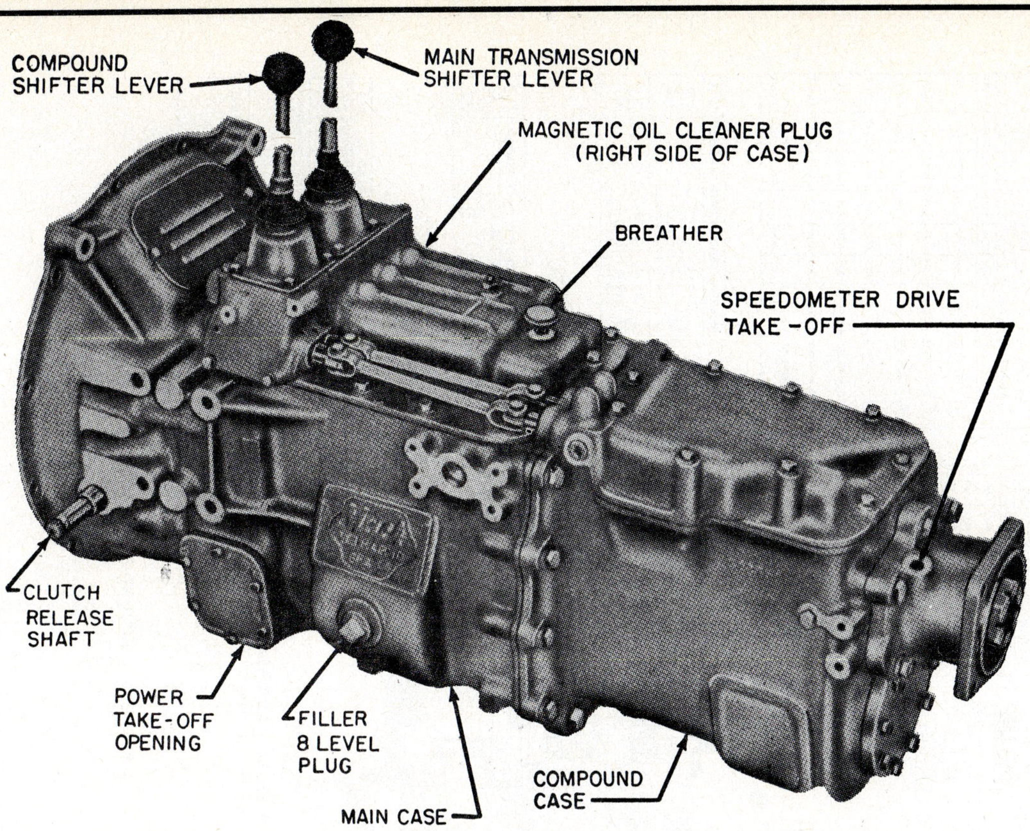

Fig. 4 Mack 20 speed TRQ-72 & 7220 transmission. Typical

slanting baffle with a removable sheet metal cover, Fig. 3. At the bottom of the slanting baffle, a tapped hole in the case accomodates a large hex plug with a powerful built-in magnet installed from outside, Fig. 2.

It is not necessary to drain the case in order to remove the plug for cleaning or for removing sludge from the bottom of the trough. The oil cleaner plug may be removed at any time without losing any oil or disturbing other parts of the transmission. When removed, it carries all foreign matter with it which is easily removed from the magnet simply by wiping it clean. Remove sludge from the sump and then replace the plug in the case.

In most Mack transmissions, magnetic drain plugs are also employed in addition to the magnetic cleaner plug to aid in extracting any magnetizeable particles from the lubricating oil.

MACK 20 SPEED TRQ-72, 7210, 7220

Disassembly, Figs. 4 and 5

Unless a complete overhaul is necessary, remove only those parts required to gain access to the faulty parts. Do not disturb parts having press fits unless replacement is necessary. In that case, use proper press setups and pullers so that usable parts are not damaged. After cleaning all transmission parts with suitable solvent, inspect each part thoroughly. Replace worn or damaged parts, staked nuts, gaskets and cotter pins to avoid subsequent transmission failures.

1. After removing transmission from vehicle and cleaning externally, mount unit in a suitable overhaul stand. Remove hand brake parts, clutch release bearing assembly with return spring, clutch release shaft and yoke, P.T.O. covers and gear shifter hand control levers.
2. Remove both case covers and lay aside for bench disassembly. Engage sliding gear and ALL sliding clutches so that the entire gearset is locked up. Remove cotter pin from drive flange nut and then remove nut. Install standard puller and remove drive flange. Remove both rear bearing covers and save countershaft rear bearing shim pack intact. Slide speedometer gear off of mainshaft.
3. Remove all case-to-case dowel bolts and capscrews. With a large enough nylon headed mallet, tap compound case and separate it and the enclosed parts from the main case. Lay enclosed parts aside.
4. Remove mainshaft rear bearing retainer capscrews, oil tube, countershaft rear bearing cone clamp plate and capscrews. Remove countershaft front bearing cover capscrews and use jackscrews to remove the cover. Remove main driving pinion cover capscrews and cover. Carefully remove oil pump vane from pinion and save. To remove pinion, tap on gear end of pinion with a nylon mallet and work assembly out of case. Lay pinion aside.
5. Remove mainshaft spigot bearing inner race retaining snap ring. With 4th and 5th speed sliding clutch against bearing inner race, tap on clutch with nylon mallet and remove both parts from mainshaft. Pry key out of mainshaft spline. With a small tool, rotate 4th speed gear thrust washer one spline tooth and slide washer off of mainshaft.
6. Butt a soft drift forward end of mainshaft and with a hammer, drive mainshaft rearward to separate rear bearing from case.
7. Carefully slide gears, sleeve, thrust washer fixed and sliding clutches from mainshaft while withdrawing mainshaft from rear of main case. Remove sleeve Woodruff key from mainshaft. Place mainshaft in a vise with soft jaws and pry retaining ring off of mainshaft rear bearing 2-piece thrust washer. Pry 2-piece thrust washer out of groove in mainshaft. Turn mainshaft over and remove oil tube, retaining snap ring and compound mainshaft spigot bearing. Remove from vise and press mainshaft out of rear bearing retainer and front cone. Remove bearing cone spacer shims and save. Remove remaining bearing cone with a segmented-type puller.
8. Remove reverse gear shaft lock and with puller 17-T-1425, withdraw gear shaft from case. Remove cluster gear, bearings and bearing spacer from case.
9. Slide countershaft rearward and partially through rear bearing opening. Install segmented puller on rear bearing cones and remove. With a soft steel drift, remove rear bearing cup from main case. When using a drift, be very careful not to damage part or case and apply alternate equal blows to opposite sides to prevent cocking of part in the case.
10. Tilt forward end of countershaft upward and remove entire assembly from case. Remove countershaft front bearing inner race and gear snap ring. For shops having a small capacity hydraulic press, press off one countershaft gear at a time. For shops

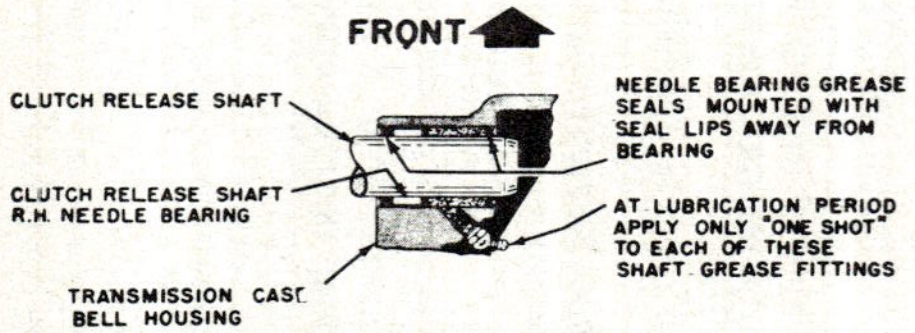

R.H. MOUNTING ARRGT. OF CLUTCH RELEASE SHAFT

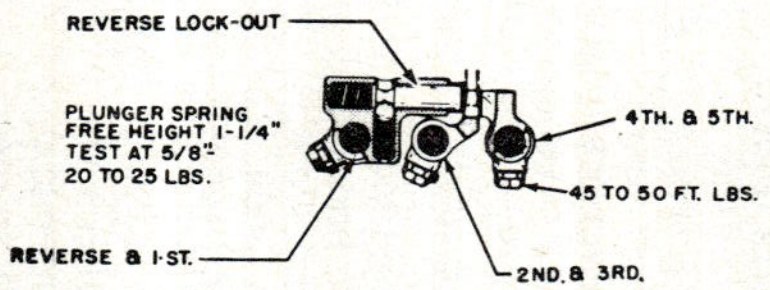

FRONT VIEW OF SHIFTER ARRANGEMENT

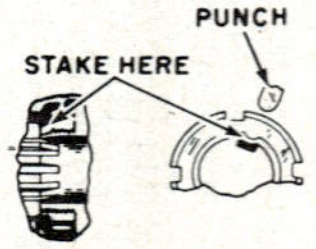

POSITIVE STAKING METHOD

USE ROUND-NOSED PUNCH TO STAKE METAL LIP OF NUT INTO MILLED SLOT IN SHAFT

MAINSHAFT BEARING ADJUSTMENT	MIN. NEW	MAX. NEW	MAX. WEAR
END-PLAY AT J	.001	.005	.015
END-PLAY AT K	.003	.007	.017

	THRUST WASHER	AVAILABLE THICKNESSES	
Y	3RD SPEED GEAR USE (1) AS REQUIRED	.271	.291
Z	1 ST. SPEED GEAR SELECT AS REQUIRED FOR TIGHT FIT IN GROOVE	.365 .390 .400 .410 .420	.370 .395 .405 .415 .425

SCREW TORQUES IN FT.-LBS.			
SCREW		LUBRICATED	
1		60	80
2		125	155
3		45	65
4		225	255

FLOATING GEAR	GEAR END-PLAY MIN. NEW	MAX. NEW	MAX. WEAR	BUSHING CLEARANCE MIN. NEW	MAX. NEW	MAX. WEAR
(A) 4-TH/5-TH SPEED	.003	.007	.015	.0036	.0046	.0076
(B) 3-RD SPEED	.004	.008	.015	.0043	.0053	.0080
(C) 2-ND SPEED	.004	.008	.015	.0043	.0053	.0080
(D) 1-ST SPEED	.006	.011	.015	.0039	.0049	.0079
(E) COMPOUND HI-SPLIT	.005	.009	.015	.0048	.0058	.0080
(F) COMPOUND LO-SPLIT	.005	.009	.015	.0043	.0053	.0080
(G) COMPOUND LO-LO SPLIT	.005	.009	.015	.0053	.0063	.0080

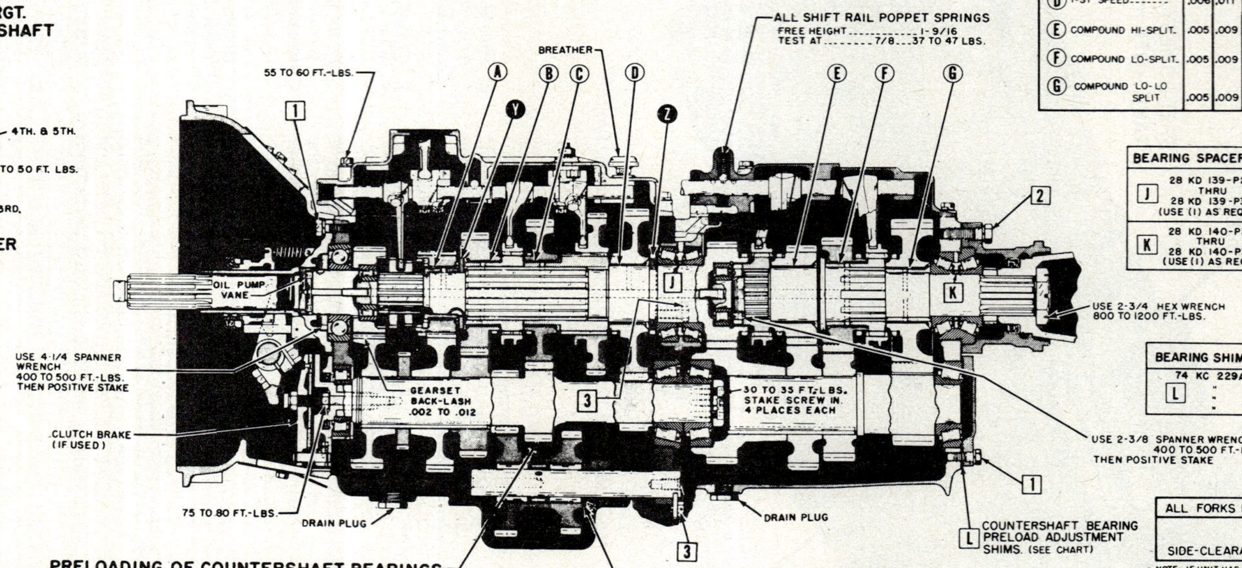

BEARING SPACER PART NUMBER	THICK-NESS
J 28 KD 139-P270 THRU 28 KD 139-P334 (USE (1) AS REQUIRED)	.270 THRU .334
K 28 KD 140-P306 THRU 28 KD 140-P361 (USE (1) AS REQUIRED)	.306 THRU .361

BEARING SHIM PART NUMBER	THICK-NESS
L 74 KC 229A-P1	.005
" P2	.007
" P3	.020
" P4	1/32

ALL FORKS IN SLIDING CLUTCHES	MIN. NEW	MAX. NEW	MAX.* WEAR
SIDE-CLEARANCE	.005	.020	.050

* NOTE- IF UNIT HAS EXPERIENCED DISENGAGEMENT SIDE CLEARANCE MUST NOT EXCEED .030 MAX.

PRELOADING OF COUNTERSHAFT BEARINGS

WRAP STRING AROUND 2ND. SPEED GEAR SEVERAL TIMES. ATTACH SPRING SCALE TO LOOPED FREE END AND WITH A STEADY PULL, ROTATE ASSEMBLY SLOWLY. ADJUST REAR BEARING SHIM PACK UNTIL ROLLING — DRAG IS:

NEW TAPERED BEARINGS...8 TO 11 LBS. PULL
USED OR RUN-IN TAPERED BEARINGS...4 TO 5-1/2 LBS.

REVERSE IDLER GEAR
END-PLAY.... .004 TO .022

RECOMMENDED TORQUES FOR CAPSCREWS NOT ILLUSTRATED

BELL HOUSING-TO-MAIN CASE CAPSCREWS — 4
MAIN CASE-TO-COMPOUND CASE DOWEL BOLT AND CAPSCREWS — 2

5-SPEED MAIN BOX GEAR SET

4-SPEED COMPOUND BOX GEAR SET

Fig. 5 Mack 20 speed TRQ-72 transmission. Model TRQ-7220 is the overdrive version

having a large capacity hydraulic press, place countershaft assembly in press, slide length of heavy tubing over countershaft integral gears to engage 3rd speed gear and remove all gears in one operation.

11. Clamp gear end of main driving pinion securely in large bench vise with soft jaws. Unstake pinion bearing nut and with a wrench, remove left-hand threaded nut. Remove ball bearing by pressing or tapping off. Turn pinion over and remove retaining snap ring and spigot bearing. Remove oil tube carefully and save.
12. Position compound case on arbor press table and press mainshaft out of case-mounted rear bearing retainer and cone. Carefully remove and save cone spacer and shims. Turn case over and remove compound mainshaft rear bearing retainer and compound countershaft rear bearing cup.
13. Press compound mainshaft out of LO-split gear, fixed clutch, LO-LO split gear, gear sleeve and rear bearing front cone. Turn shaft over and clamp in vise with soft jaws. Remove spigot bearing inner race from spigot, unstake nut and remove. Nut has left-hand threads. Slide fixed clutch, sliding clutch and HI-split gear off of shaft. Remove shaft from vise.
14. Remove bearing cup from compound countershaft constant-mesh gear. With segmented puller, remove rear bearing cone. Remove gears in hydraulic press one at a time or all in one operation, depending on capacity of press. Remove gear key from shaft.
15. Clamp flange of main case cover in bench vise and remove shifter and fork setscrews. Remove Welch plugs at shifter ends with shifter rails. Withdraw shifter rails from main case cover and at the same time remove shifters and shifter forks.

CAUTION: The spring-loaded balls over the shifter rails will fly out with CONSIDERABLE FORCE when released by withdrawing the rails. Therefore, it is advisable to hold a clean shop wiper over the holes while withdrawing the rails.

Be sure to save the balls, spring, and rail interlock pins for use at reassembly. Remove breather and toggle levers from top of cover.

16. Disassemble compound case cover in same manner. *Watch out for spring-loaded balls over rails when withdrawing rails from case.*

Inspection and Repair

1. Wash and clean all parts thoroughly in clean solvent and dry with moisture-free compressed air as damage to the bearing may result.
2. Replace bearings which are pitted and/or spalled. Replace lubricated bearings having excessive axial or radial clearances. Replace cases if outer races of bearings turn freely in case bores.
3. Replace gears having pitted or badly worn teeth. Replace gears with damaged internal or external teeth as a result of gear clashing. Floating gears having excessive radial clearance should be rebushed and bored to size.
4. Check all splines. If excessive wear is obvious, replace part or parts. Check thrust washers for scored surfaces or reduced thickness from excessive wear. Replace if either or both conditions prevail.
5. Check all gray iron parts for cracks and breaks. Case castings may be welded or brazed provided cracks do not extend into bearing bores or bolting surfaces. Replace or repair parts found to be damaged.
6. Check all other parts for wear and damage. Replace all parts that cannot be repaired or straightened. Replace all gaskets, damaged oil seals, O-rings, staked nuts and any part that shows mutilation. Replace poppet springs that have lost their tension. Clean up any threads that show mutilation.

Reassembly

NOTE: All working parts, especially the bearings, should be generously coated with SAE-30 oil while the transmission is being assembled. This will assure immediate lubrication when first starting and will prevent seizing of these parts. As moving parts are assembled, check frequently to see that they move freely.

1. Insert gear key in countershaft and press gears on one at a time. These gears have an interference fit and can be pressed on cold but for best results, the gears should be shrunk-fitted on shaft. With a heat lamp or hot oil, heat gears 270° to 300° F. for a period of not more than 1/2 hour. Oil shaft for each gear. Install gear snap ring, install front bearing inner race on spigot and retain with its snap ring.
2. Install countershaft front bearing in case. Install bearing cover and gasket. Place countershaft assembly in case with small end extending out of rear bearing opening. Install rear bearing cup in case. Install tapered roller bearing cones back-to-back on rear end of countershaft and secure with clamp plate. Position countershaft in front bearing.
3. Insert gear key in compound countershaft and press gears on one at a time. These gears have an interference fit and can be pressed on cold but for best results, the gears should be shrunk-fitted on shaft. With a heat lamp or hot oil, heat gears 270° to 300° F. for a period of not more than 1/2 hour. Oil shaft for each gear. Press bearing cup in front end of compound countershaft constant-mesh gear and bearing cone on rear end. Push countershaft rear bearing cup in compound case and install shim pack and bearing cover. Temporarily assemble compound countershaft, case-to-case gasket and compound case to the main case and countershaft.
4. Preload countershaft bearings as follows: Lubricate all bearings with SAE-30 oil. Wrap a string around countershaft 2nd-speed gear (largest integral gear) several times. Attach a spring scale to looped free end of string and with a steady pull, rotate the assembly SLOWLY. Repeat this operation and adjust thickness of rear bearing shim pack until the rolling-drag of 8 to 11 lbs.-pull is read on the scale. When correct shim pack thickness is established, remove compound case and compound countershaft from main case.
5. Position reverse cluster gear shaft so that slot in end of shaft will permit shaft lock to engage it. Start shaft into case. Assemble roller bearings and spacer in cluster gear. Position assembled gear in case and with a nylon hammer, tap shaft into case and gear. Install shaft lock and lock capscrew after tightening 60 to 65 ft.-lbs. with soft iron wire.
6. On bench, install compound mainshaft spigot roller bearing in end of mainshaft (compound driving pinion) and retain with snap ring. Place gear end of mainshaft in vise with soft metal jaws. Install rear bearing cone, bearing retainer with two cups, bearing retainer with two cups, bearing cone shims, front bearing cone and bearing split thrust washer. Adjust bearing shim pack thickness until end-play is .001 to .005" then press or drive positioning ring over split washer. Install 3rd-speed thrust washer Woodruff key in shaft.
7. Remove mainshaft from vise and start in main case through rear bearing hole. As mainshaft is advanced into working position, slide parts on mainshaft as follows: 1st-speed gear, reverse gear fixed clutch, reverse gear, 2nd-speed gear, 3rd-speed fixed and sliding clutches, 3rd speed gear, engage 3rd-speed gear thrust washer and 4th-speed gear sleeve with Woodruff key. Install 4th-speed gear on sleeve. Install 4th/5th-speed toothed thrust washer and with a small tool, move washer one tooth so that inner teeth align with mainshaft splines. Install washer key in deepened spline root and install 4th/5th-speed sliding clutch on splines and over inlaid key.
8. At this point the mainshaft rear bearing retainer should be seated in the main case. Install retainer capscrews and tighten 60 to 65 ft.-lbs., then lock heads together with soft iron wire. Install inner race of spigot bearing on front end of mainshaft and retain with snap ring.
9. Assemble mainshaft spigot bearing in gear end of main driving pinion and retain with snap ring. Turn pinion over and place in vise with soft metal jaws. Install pinion ball bearing and bearing nut. Tighten nut (left-hand thread) 400 to 500 ft.-lbs., the positive stake lip of nut to pinion in slot provided. Install oil tube in pinion gear end and install pinion assembly in transmission case. *Be very careful not to damage oil tube as it engages mainshaft.* Install oil pump vane in pinion and install bearing cover. Tighten cover capscrews 75 to 80 ft.-lbs.
10. At this point, the main case should be positioned in a vertical position with the pinion pointing downward. If overhaul stand is not used, block up bell housing so that end of pinion clears the floor.
11. Clamp compound mainshaft in vise with soft metal jaws or improvised holding device and assemble as follows: Slide HI-split floating gear and fixed clutch on front end of shaft, install clutch nut (left-hand threads) and tighten 400 to 500 ft.-lbs., then positive stake lip of nut to shaft in slot(s) provided. Install inner race of bearing on shaft spigot.
12. Slide LO-split floating gear on fixed clutch and install both parts on rear end of compound mainshaft. Install sliding clutch on fixed clutch. Install LO-LO-split gear sleeve on shaft then slide LO-LO floating gear on sleeve. Press bearing cone on shaft next to sleeve, install double-row cup, cone spacer and shims, then press rear cone on shaft. With a piece of tubing over flange splines, place assembly in a press and with a load on bearing cones, measure bearing end-play. Add or remove shims between cones until end-play is .003 to .007" as read on a dial gage,

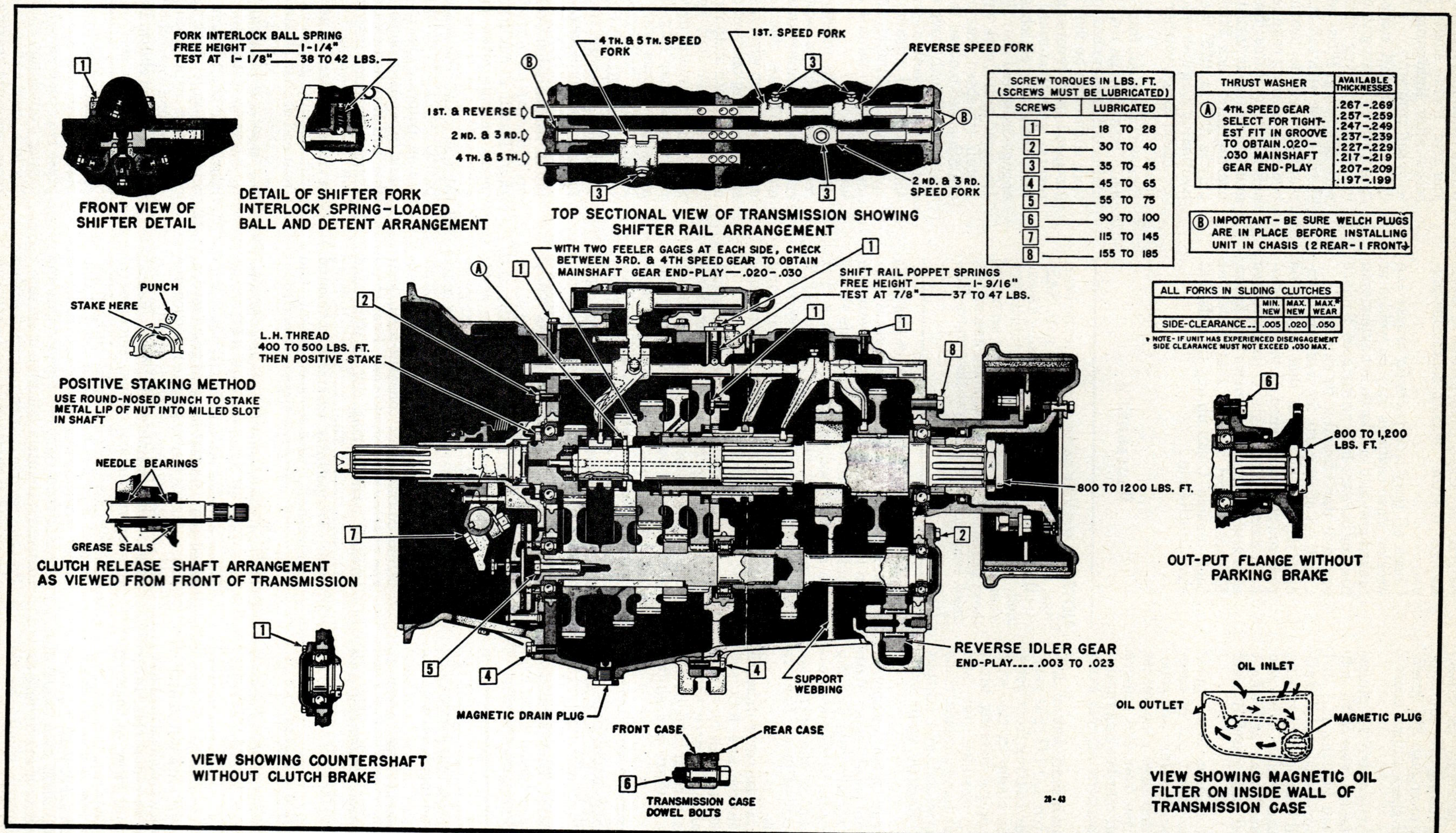

Fig. 6 Mack TRL-107 five speed triple countershaft transmission

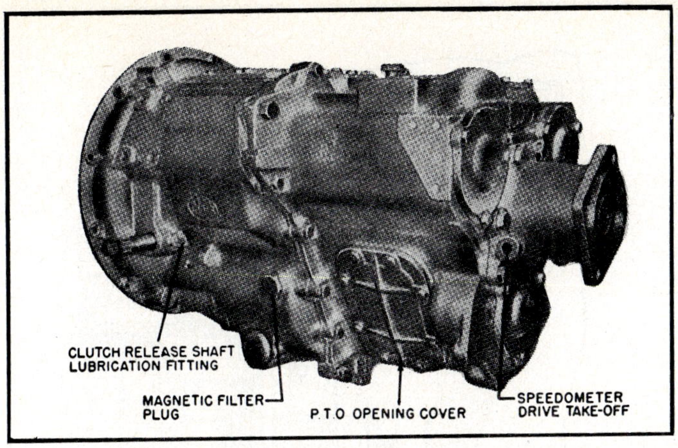

Fig. 7 Left rear view of Mack TRL-107 transmission

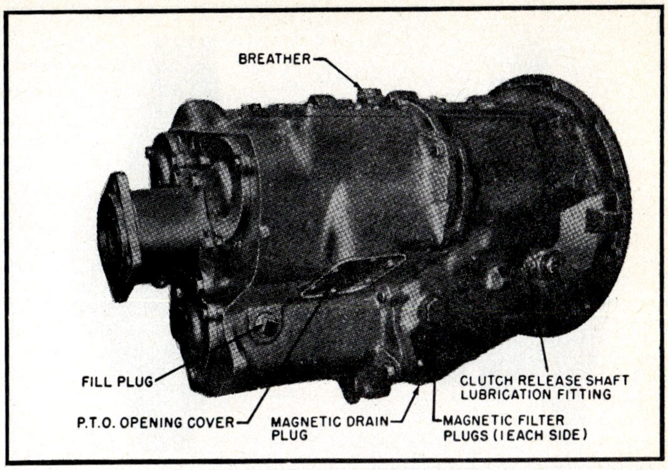

Fig. 8 Right rear view of Mack TRL-107 transmission

then remove double cup and rear bearing cone.

13. Slide HI-split clutch on its fixed clutch. Install interconnecting oil tube in end of mainshaft and carefully engage spigot of assembled compound mainshaft with oil tube and spigot bearing in end of mainshaft. *Be very careful not to damage oil tube as it engages compound mainshaft spigot.* Place compound countershaft (assembled earlier) on countershaft rear bearing cone and make sure all gears are properly meshed.
14. Paint end of main case with a good gasket sealer compound and install case-to-case gasket. Carefully lower compound case down over assembled parts and see that bearings are properly seated. Install case-to-case capscrews and dowel bolts. Tighten 90 to 100 ft.-lbs. Install double-row rear bearing cup in compound case. Install rear bearing cone and speedometer gear. Apply gasket sealer in case and gasket and install rear bearing retainer gasket. Install retainer and capscrews with lockwashers. Tighten capscrews 90 to 100 ft.-lbs. Dip drive flange oil seal in oil and install in rear bearing retainer. Install drive flange and nut. Return transmission to normal position.
15. Lock up gear train by engaging both sliding clutches of the compound gearset. Tighten drive flange nut 800 to 1200 ft.-lbs., then install nut cotter pin. Return all sliding clutches to their neutral positions.
16. Position main case cover in bench vise with soft metal jaws. Install transmission case breather and toggle shifter levers. Starting with 1st and reverse speed shifter rail, start rail in cover and shifter, with a small blunt tool compress detent ball and spring, push rail past compressed ball, through a shifter and into neutral position. Install large interlock pin next to rail just installed. Install 2nd and 3rd-speed rail in same manner. Install small interlock pin in 2nd and 3rd-speed rail then another large pin between. Install short 1st and reverse speed rail with fork then 4th and 5th-speed rail with fork and shifter. Install shifter and fork setscrews, tighten 45 to 50 ft.-lbs., then lock to shafts individually with soft iron wire. Install new Welch plugs in shifter rail openings. Remove from vise.
17. Place compound case cover in vise with soft metal jaws. Dip new rail oil seals in oil and install. Install LO-split rail in cover and fork. Install interlock pin between rails. Install HI-split rail in cover and fork. Install pipe plug in pin hole. Install fork setscrews and tighten 45 to 50 ft.-lbs. Remove cover from vise.
18. Apply gasket sealer compound to case openings and cover gaskets and install gaskets on cases. Position sliding gear and all sliding clutches in neutral position. Install both assembled covers and make sure the shifter forks engage the grooves in their mating parts. Install cover capscrews with lockwashers and tighten 55 to 60 ft.-lbs.
19. Apply sealer compound to all other cover gaskets and install covers and capscrews with lockwashers. Check for proper shifting and free rotation of gears. Install clutch release shaft, yoke bearing, return spring and any other external part that was removed for disassembly. Recheck all fasteners for correct torques and locking means. Install all plugs, then remove transmission from stand.

MACK TRL-107

Five Speed with Three Countershafts

This transmission, Figs. 6 to 9, is a triple countershaft unit providing five speeds forward and one reverse. The five speed gearset is housed in an aluminum case that is constructed in two pieces for convenience of disassembly and reassembly.

Three countershafts are equally spaced around the input shaft. By distributing the load equally between three shafts, normal deflection is reduced to a minimum. The mainshaft gears are self-centered between the three countershafts, eliminating the need for gear bushings. The mainshaft gears are connected to the mainshaft only when engaged by a sliding clutch.

As shown in Fig. 9, all gears are of the spur type design and, with the exception of the reverse speed sliding gear, are in constant mesh with the countershaft gears. Other than reverse, driving engagement for the five forward speeds is effected by means of sliding clutches. Reverse is obtained by engaging reverse sliding spur gear with the three individually mounted idler gears. Conventional forks spanning the sliding clutches and reverse gear move these parts in response to the movement of the gear shifter and hand control lever.

A single row ball bearing supports the main driving pinion. The mainshaft is supported by a roller spigot bearing at the front end and a ball bearing at the rear. The countershafts are carried by a ball bearing and roller bearing combination. Ball bearings that are snap ring type provide shaft location, while the roller bearing is located adjacent to the more heavily loaded low range gears. In constant mesh with the countershafts, each of the three reverse idler gears revolves on a roller bearing mounted on a shaft.

Lubrication

All rotating and sliding parts of the transmission are bathed in oil from gear throw off when the engine is operating. However, the sliding clutches and mainshaft gears require more positive provision for lubrication, and for this reason, pressure lubrication of the clutches and gears is provided by a simple eccentric shuttle type pump built in the main driving pinion. As the pinion rotates the pump vane reciprocates in its eccentric housing, thus forcing lubricant under pressure through rifle drilled holes in the mainshaft to the sliding clutches and mainshaft gears. To supply the pump, oil from gear throw-off is collected by a trough located above the main driving pinion and is then gravity fed to the pump.

Removal of all foreign metallic particles within the transmission accomplished by two magnetic oil filters, one on each side of the case. The filter consists of an integral open trough and baffle arrangement with a removable sheet metal cover. At the bottom of the baffle, a tapped hole in the case accommodates a large hex plug with a built-in magnet installed from outside.

The oil from gear throw-off is collected by the filter and is made to pass the magnetic plugs that pull all ferrous metal particles out of the passing oil and holds them. After passing the magnet, clean and chip-free oil that rises to the outlet near the top of the filter and drops down into the transmission case. A magnetic drain plug is provided at the bottom of the case.

When the oil is hot, as when coming in from a run, the oil level of the transmission should be checked with the vehicle on level ground. To add oil, remove plug on right side of case and fill. The level should not be above the filler plug hole. At every check or change, always remove, clean and reinstall magnetic plugs on right and left side of case.

To change oil, remove magnetic drain plug and drain oil from case while hot. If necessary, also flush case with flushing oil and drain thoroughly. Clean and replace magnetic drain plug in case. Remove fill plug and fill with

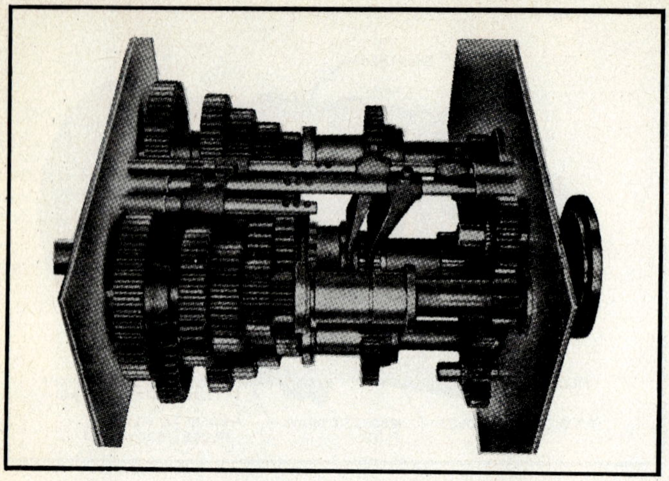

Fig. 9 Skeletal view of Mack TRL-107 gearset (in support plates)

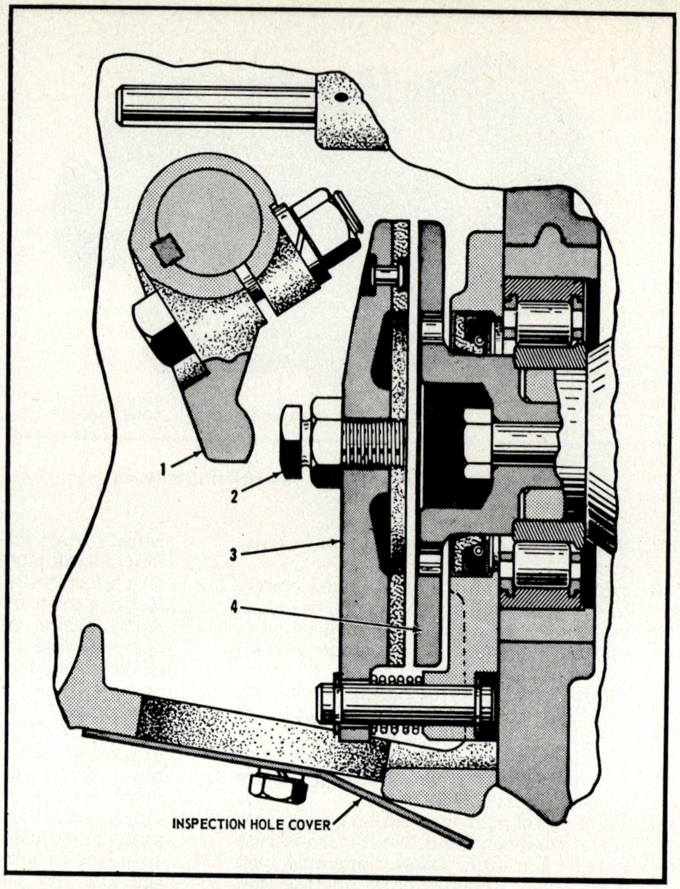

Fig. 10 Sectional view of Mack Clutch Brake. 1. Clutch throwout lever. 2. Adjusting screw. 3. Pressure plate and lining assembly. 4. Flange

mild extreme pressure SAE-90 gear oil, year-round grade, to level of filler plug hole.

Clutch Brake, Fig. 10

In any unsynchronized transmission, the rapidity with which an upshift can be made is limited by the time it takes for the free-spinning clutch disc and countershaft assemblies to slow down to the slower rotating speed of the transmission mainshaft. In order to reduce this time element, Mack Clutch Brake is employed.

The clutch brake is assembled on the transmission lower countershaft, and since the countershaft is always connected to the clutch disc through the constant mesh driving pinion, action of the brake will overcome the tendency of the discs to continue to rotate at high speed when the clutch is disengaged.

Clutch Brake, Adjust

Remove inspection hole cover on bottom of bell housing. Loosen set screw lock nut. Depress clutch pedal to within 1/2" to 1" of the end of its travel. Adjust set screw so that brake pressure plate and lining assembly firmly contacts flange. Lock set screw with lock nut and install inspection hole cover.

NOTE: It must be remembered that each time a clutch adjustment is made to provide the 1/8" to 3/16" release bearing clearance, the set screw, which pushes the clutch brake pressure plate and lining assembly, must also be adjusted to suit location of clutch throwout lever. When initial wear on clutch driven disc has been used up and the spacers are to be removed, first move the set screw in to allow sufficient clearance with the throwout lever; then remove adjusting spacers. When the clutch cover is tightened down, readjust release bearing and clutch brake properly.

TRTL-72 & TRTL-722 15 SPEED & TR722 & TRD722 5 & 10 SPEED TRANSMISSIONS

Transmission Disassembly

1. After removing transmission from vehicle and cleaning externally, mount unit on a suitable work stand. Remove hand brake components, clutch release bearing assembly with return spring clutch release shaft and yoke, P.T.O. covers and gear shifter hand control levers. On 15 speed units, remove compound low split rocker, rocker arm, pivot pin, rail and fork through top of compound case.
2. Remove bolts, shifter, shifter forks and shifter rails from covers.
3. Shift transmission into any two gears to lock up the assembly, then remove cotter pin and nut from drive flange.
4. Remove drive flange using suitable puller and both rear bearing covers.
5. Slide speedometer gear, and on 5 speed units the bearing thrust washer, from mainshaft.
6. On 10 and 15 speed units, remove case-to-case bolts. Using mallet, tap compound case to separate case and related parts from main case.
7. On 10 and 15 speed units, remove mainshaft rear bearing bolts and oil tube, then, on all units, the countershaft rear bearing cone clamp plate, main drive pinion bearing cover and countershaft front bearing cover.
8. Remove oil pump vane from main drive pinion, then the main drive pinion assembly from case. Tap on gear end with mallet, if necessary, to remove drive pinion.
9. Remove oil tube. Clamp main drive pinion gear end securely in a soft jaw vise, then unstake and remove bearing nut, ball bearing, snap ring and pilot bearing from pinion.
10. Remove mainshaft pilot bearing inner race snap ring. With fourth and fifth speed sliding clutches positioned against bearing inner race, tap on clutch with mallet and remove both fourth and fifth speed sliding clutches from mainshaft. Pry key out of mainshaft spline. Using suitable tool, rotate fourth and fifth speed gear sleeve slightly (one spline) and slide sleeve and fourth and fifth speed gear from mainshaft.
11. Using a drift and hammer, drive mainshaft rearward to separate rear bearing from case.
12. Carefully slide gears, sleeve, thrust washer, splined sleeve and sliding clutch from mainshaft while withdrawing mainshaft from rear of main case. Remove Woodruff key from mainshaft. On 5 speed units, remove mainshaft bearing and thrust washer.
13. On 10 and 15 speed units, clamp gear end of mainshaft in a soft-jaw vise, and remove thrust washer ring, first speed gear and thrust washer pieces from mainshaft. Press mainshaft from rear bearing retainer and front bearing cone.
14. On 10 and 15 speed units, remove bearing cone spacer and bearing cone from mainshaft.
15. On 10 and 15 speed units, remove snap ring and compound mainshaft pilot bearing from gear end.
16. Remove reverse gear shaft lock and using suitable puller, withdraw gear shift from case. Remove cluster gear, bearings, bearing spacer and thrust washers from case.
17. On 5 speed units, insert screws into holes tapped into countershaft rear bearing retainer and remove retainer and rear bearing cone. Remove rear bearing shim

pack, then slide countershaft rearward and partially through rear bearing opening. Remove remaining bearing cone using suitable puller.

18. On 10 and 15 speed units, slide countershaft partially rearward and through the rear bearing opening. Install puller on inside bearing cone, and remove cones.
19. On 10 and 15 speed units, remove countershaft case mounted bearing and inner race with a drift and hammer.

NOTE: When driving out the bearing and race, apply equally placed blows around the circumference of the assembly.

20. Tilt forward end of countershaft upward and remove entire assembly from case. Using a press arbor, remove bearings from countershaft.

Inspection

Inspect case and all other components of the transmission thoroughly. Check for any defects that will lead to transmission failure. If any defects are apparent, replace parts as required. Clean all serviceable parts using a suitable solvent and blow drying them using compressed air. Replace all "O"-rings and seals.

Check bushing clearance, Fig. 11. Replace bushings when clearances specified in Figs. 12 and 13 are exceeded.

Assembly

NOTE: Lubricate parts to prevent seizing of transmission components during vehicle operation. Refer to Figs. 12 and 13 for specifications during assembly.

1. Insert gear key into countershaft keyway. Using press arbor, install gears onto countershaft.

NOTE: The gears incorporate a press fit with the countershaft and should be heated using a lamp or hot oil to 270–300°F, before being pressed onto countershaft.

2. Install countershaft front bearing into case. Install bearing cover and gasket.
3. On 5 speed units, install tapered roller bearing cone, with large end of rollers facing gear, on rear end of countershaft. Install countershaft assembly, then the rear bearing retainer and gasket in case. Install countershaft rear bearing shim pack adjacent to bearing. Install countershaft tapered roller bearing cone, with small end of rollers next to shim pack and secure with clamp plate, then install rear bearing cover and gasket on case.
4. On 10 and 15 speed units, place countershaft assembly into case with small end of shaft extending out of rear bearing opening. Install rear bearing cup into case. Install tapered roller bearing cones back-to-back on rear end of countershaft, and secure with clamp plate. Position countershaft into front bearing. Push bearing cup in front end of compound countershaft constant-mesh gear and the bearing cone on rear end. Push countershaft rear bearing cup in compound case and install shim pack and bearing cover. Temporarily assemble compound countershaft case gasket and compound case assembly onto main case and countershaft assembly.

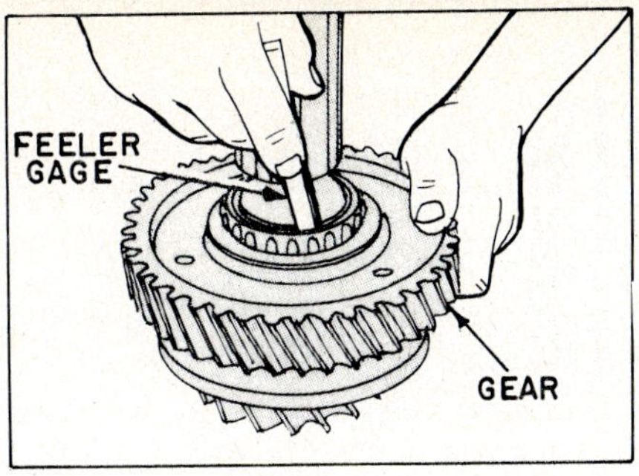

Fig. 11 Checking Gear Bushing To Mainshaft Clearance

5. Preload countershaft bearings as follows:
   Lubricate all bearings with an approved grade of oil. Wrap a piece of string around countershaft 2nd speed gear (largest integral gear) several times. Place spring scale in looped free end of string and with a steady pull, rotate assembly slowly. Repeat procedure and adjust thickness of rear bearing shim pack until recommended rolling drag is acquired on scale. On 10 and 15 speed units, when correct shim pack thickness is established, remove compound case and countershaft from main case.
6. Position reverse cluster gear shaft so that slot in end of shaft will permit shaft lock to engage it. Advance shaft into case, and assemble roller bearings and spacer into cluster gear. Position assembled gear into case, then using a soft faced mallet, tap shaft into case and onto the gear. Install shaft lock and bolt. Torque bolt to specifications. Using a piece of soft iron safety wire, wire up shaft lock bolt to prevent bolt from working loose during vehicle operation.
7. On 5 speed units, slide mainshaft rear ball bearing thrust washer on rear end of mainshaft, then install the bearing. Install third speed thrust washer Woodruff key in shaft.
8. On 10 and 15 speed units, position compound mainshaft on a suitable work stand and install compound mainshaft pilot bearing and snap ring onto mainshaft. Install rear bearing cone, bearing retainer with cups, bearing cone shims, front bearing cone and bearing split thrust washer. Install thrust washer Woodruff key into keyway provided in shaft.
9. Install mainshaft into case through rear bearing opening. As mainshaft is advanced into case, slide the following parts onto mainshaft:
   Install first speed gear and reverse gear splined sleeve. Install reverse gear, second speed gear, third speed splined sleeve clutches, third speed gear, thrust washer and Woodruff key into shaft. Insert sleeve onto fourth and fifth speed gear and install assembly onto mainshaft. Using suitable tool, rotate fourth and fifth speed sleeve teeth so inner teeth align with mainshaft splines. Install Woodruff key into keyway provided on mainshaft, then install fourth and fifth speed sliding clutches onto shaft splines.
10. On 10 and 15 speed units, install mainshaft rear bearing bolts and torque to specifications.

NOTE: To prevent bolts from working loose, it is recommended that soft iron safety wire be used to secure bolt after torquing.

11. Install pilot bearing inner race and snap ring onto front end of mainshaft.
12. Install mainshaft pilot bearing and snap ring onto front end of main drive pinion. Place assembly into a soft jaw-vise, then install pinion ball bearing and nut. Torque nut to specifications.

NOTE: Nut is left hand threaded. To tighten and torque nut, rotate nut counterclockwise.

13. Install oil tube in pinion gear end, then position outer race of ball bearing so filler notch is easily accessible. Install pinion assembly into transmission case. Install oil pump vane onto pinion. Also install bearing cover and bolts. Torque bolts to specifications.
14. Using hoist, position transmission main case vertically, and clamp compound mainshaft to a soft jaw-vise. Assemble compound mainshaft as follows: On 15 speed units, install pilot bearing inner race, high split gear and key onto mainshaft. Install low split fixed gear and key onto mainshaft. Install rear bearing cone onto compound mainshaft, and bearing spacer onto bearing. Install high split clutch onto fixed clutch on mainshaft, then install clutch nut and torque to specifications. On 10 speed units, install pilot bearing inner race on pilot end of shaft. Slide compound low split floating gear on shaft, then press inner rear bearing cone on shaft with large end of rollers facing low split gear. Install bearing spacer next to bearing, then sliding clutch on splines of compound mainshaft.
15. On 10 and 15 speed units, slide interconnecting oil tube into end half of mainshaft. Carefully engage pilot of assembled compound mainshaft with oil tube and pilot bearing in end of mainshaft. Position compound countershaft on rear bearing cone. Be sure gears are properly meshed. Apply approved sealer RTV732 or equivalent, onto case gasket. Carefully lower compound case down over assembled parts; ensure bearings are properly seated on their shafts. Install and torque case bolts to specifications.
16. On 15 speed units, through top of compound case install low split shifter fork, and engage fork into clutch fork groove. Install low split rail into case and onto fork. Install rocker arm pivot pin and low split rocker into slot provided in rail. On all units, install speedometer drive gear on shaft, also the fifth speed thrust washer, if equipped. Apply approved sealer RTV732 or equivalent, onto gasket and install onto rear bearing. Install rear bearing retainer and bolts. Torque bolts to specifications.
17. Lock up gear train by engaging both sliding clutches of the compound gearset. Torque drive flange nut to specifications, then install nut cotter pin. Return sliding clutches to their normal neutral positions.
18. Install main case cover in soft jaw vise, then install transmission case breather and toggle shifter levers.
19. Advance first and reverse speed shifter rail in cover and first and reverse speed shifter. Install spring and compress poppet ball into cover. Push rail over compressed ball and through first and reverse speed rocker shifter into neutral position. Install large interlock pin next to rail.

ALL FORKS IN SLIDING CLUTCHES

	MIN. NEW	MAX. NEW	MAX.* WEAR
SIDE CLEARANCE	.005	.020	.050

*NOTE - IF UNIT HAS EXPERIENCED DISENGAGEMENT, SIDE CLEARANCE MUST NOT EXCEED .030 MAX.

	BEARING SPACER PART NUMBER	THICKNESS
J	28-KD-139-P270 THRU 28-KD-139-P334 (USE (1) AS REQUIRED)	.270 THRU .334
K	28-KD-140-P306 THRU 28-KD-140-P361 (USE (1) AS REQUIRED)	.306 THRU .361

	BEARING SHIM PART NUMBER	THICKNESS
L	74-KC-229A-P1	.005
	" P2	.007
	" P3	.020
	" P4	1/32
	(10-SPEED DUPLEX MODELS)	
M	55-KH-115-P1	.003
	" P2	.005
	" P3	.020
	" P4	.062
	(5-SPEED MODELS)	

SCREW TORQUES IN LBS. FT.

SCREW		LUBRICATED
1	60	80
2	45	65
3	125	155
4	225	255
5	145	175

CAPSCREWS WHICH ARE COATED OR DIPPED WITH PERMATEX OR OTHER SEALANT ARE TO BE CONSIDERED LUBRICATED.

FLOATING GEAR	GEAR END-PLAY MIN. NEW	GEAR END-PLAY MAX. NEW	GEAR END-PLAY MAX. WEAR	BUSHING CLEARANCE MIN. NEW	BUSHING CLEARANCE MAX. NEW	BUSHING CLEARANCE MAX. WEAR
(A) 4TH/5TH SPEED	.003	.007	.015	.0036	.0046	.0076
(B) 3RD. SPEED	.004	.008	.015	.0043	.0053	.0080
(C) 2-ND. SPEED	.004	.008	.015	.0043	.0053	.0080
(D) IST. SPEED	.006	.011	.015	.0039	.0049	.0079
(E) COMPOUND	.005	.009	.015	.0036	.0046	.0076

THRUST WASHER	AVAILABLE THICKNESSES	
(X) 3RD SPEED GEAR USE (1) AS REQUIRED	.271	.291
(Y) 1 ST. SPEED GEAR SELECT AS REQUIRED FOR TIGHT FIT IN GROOVE (10-SPEED DUPLEX ONLY)	.365 .390 .400 .410 .420	.370 .395 .405 .415 .425

RECOMMENDED TORQUES FOR CAPSCREWS NOT ILLUSTRATED

BELL HOUSING-TO-MAIN CASE CAPSCREWS	4	SEE CHART
MAIN CASE-TO-COMPOUND CASE DOWEL BOLT	5	"
CAPSCREWS (8)	5	"

MAINSHAFT BEARING ADJUSTMENT

	MIN. NEW	MAX. NEW	MAX. WEAR
END-PLAY AT J	.001	.005	.015
END-PLAY AT K	.003	.007	.017

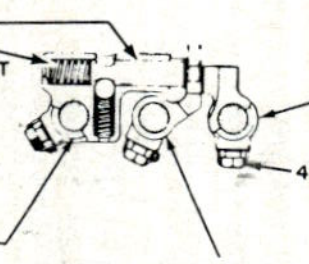

FRONT VIEW OF SHIFTER ARRANGEMENT

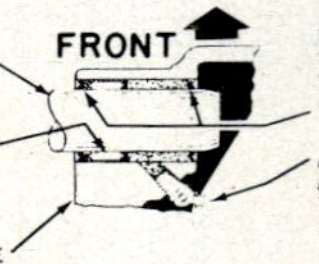

R.H. MOUNTING ARRGT. OF CLUTCH RELEASE SHAFT

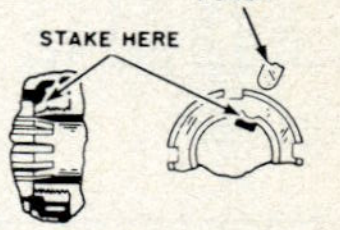

POSITIVE STAKING METHOD

USE ROUND-NOSED PUNCH TO STAKE METAL LIP OF NUT INTO MILLED SLOT IN SHAFT

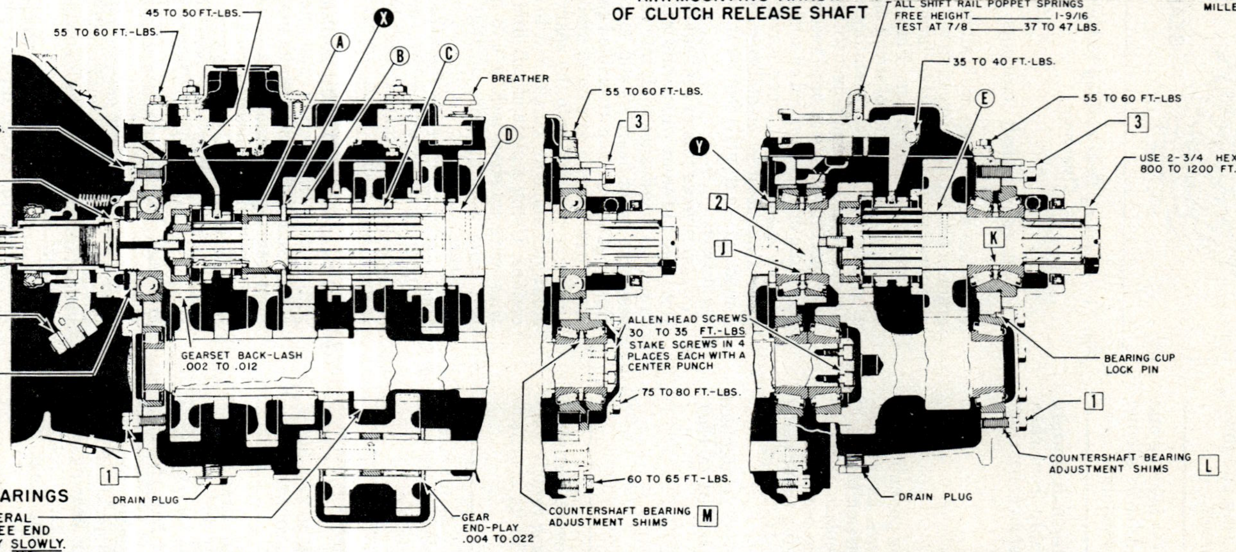

PRELOADING OF COUNTERSHAFT BEARINGS

WRAP STRING AROUND 2ND. SPEED GEAR SEVERAL TIMES. ATTACH SPRING SCALE TO LOOPED FREE END AND WITH A STEADY PULL, ROTATE ASSEMBLY SLOWLY. ADJUST REAR BEARING SHIM PACK UNTIL ROLLING—DRAG IS:

NEW TAPERED BEARINGS 8 TO 11 LBS. PULL.
USED OR RUN-IN TAPERED BEARINGS 4 TO 5-1/2 LBS. PULL.

SECTIONAL VIEW OF BASIC 5-SPEED TRANSMISSION

REAR BEARINGS OF 5-SPEED BOX

COMPOUND SECTION OF 10-SPEED DUPLEX TRANSMISSION

Fig. 12 Mack TR-72, 722 5 & 10 speed transmissions

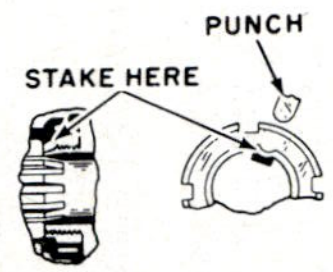

POSITIVE STAKING METHOD
USE ROUND-NOSED PUNCH TO STAKE METAL LIP OF NUT INTO MILLED SLOT IN SHAFT

	THRUST WASHER	AVAILABLE THICKNESSES	
X	3-RD SPEED GEAR USE (1) AS REQUIRED	.271	.291
Y	1-ST SPEED GEAR SELECT AS REQ'D. FOR TIGHT FIT IN GROOVE	.365 .390 .400 .410 .420	.370 .395 .405 .415 .425
Z	COMPOUND COUNTER-SHAFT LO-SPLIT GEAR	.363	.371

	BEARING SPACER PART NUMBER	THICKNESS
J	28 KD 139— P270 THRU 28 KD 139— P334 USE (1) AS REQUIRED	.270 THRU .334
K	28 KD 140— P306 THRU 28 KD 140— P361 USE (1) AS REQUIRED	.306 THRU .361

	BEARING SHIM PART NUMBER	THICKNESS
L	741 KC 229A-P1	.005
	" P2	.007
	" P3	.020
	" P4	1/32

	FLOATING GEAR	GEAR END-PLAY MIN. NEW	GEAR END-PLAY MAX. NEW	GEAR END-PLAY MAX. WEAR	BUSHING CLEARANCE MIN. NEW	BUSHING CLEARANCE MAX. NEW	BUSHING CLEARANCE MAX. WEAR
A	4-TH/5-TH SPEED	.003	.007	.015	.0036	.0046	.0076
B	3-RD SPEED	.004	.008	.015	.0043	.0053	.0080
C	2-ND SPEED	.004	.008	.015	.0043	.0053	.0080
D	1-ST SPEED	.006	.011	.015	.0039	.0049	.0079
E	COMPOUND HI-SPLIT	.005	.009	.015	.0048	.0058	.0080
F	COMPOUND LO-SPLIT	.005	.009	.015	.0031	.0041	.0071

TORQUES FOR FASTENERS NOT ILLUSTRATED	
MAIN CASE-TO-COMPOUND CASE, DOWEL BOLT & CAPSCREWS	2

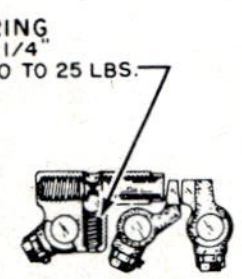

DETAIL OF SHIFTER & REVERSE LOCKOUT ARRANGEMENT AS VIEWED FROM THE FRONT

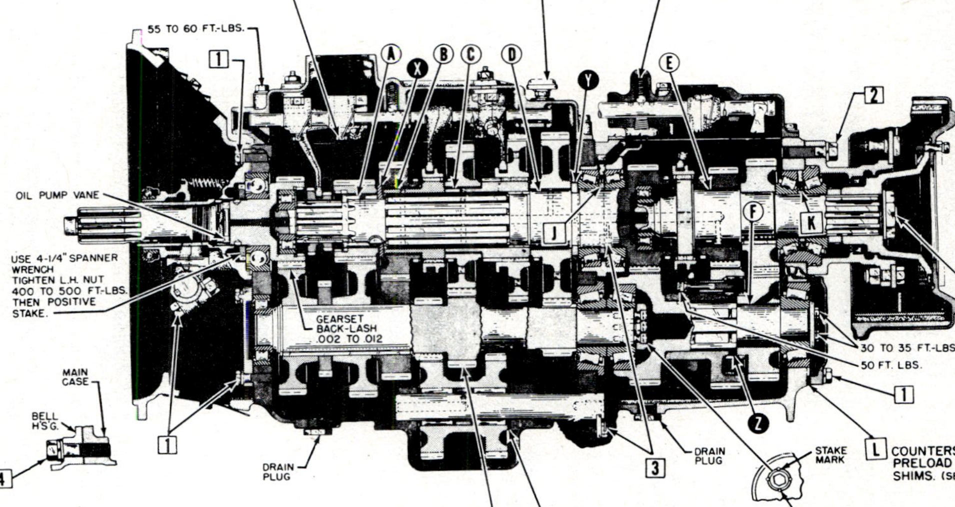

SCREW TORQUES IN FT. LBS.		
SCREW	LUBRICATED	
1	60	80
2	125	155
3	45	65
4	225	255

FRONT

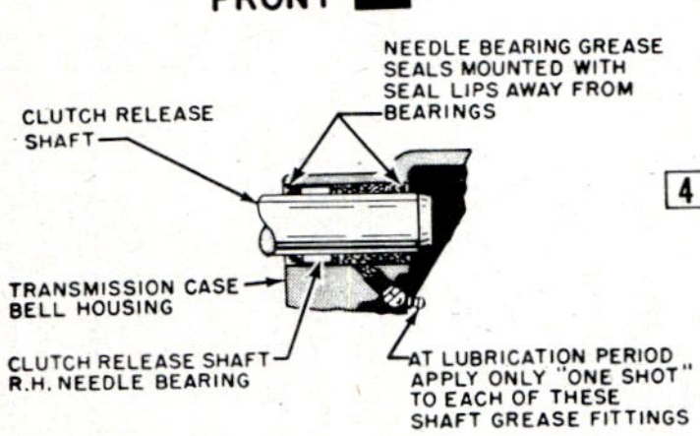

R.H. MOUNTING ARRANGEMENT OF CLUTCH RELEASE SHAFT

PRELOADING OF COUNTERSHAFT BEARINGS

WRAP STRING AROUND 2ND. SPEED GEAR SEVERAL TIMES. ATTACH SPRING SCALE TO LOOPED FREE END AND WITH A STEADY PULL, ROTATE ASSEMBLY SLOWLY. ADJUST REAR BEARING SHIM PACK UNTIL ROLLING—DRAG IS:

NEW TAPERED BEARINGS ... 8 TO 11 LBS. PULL.
USED OR RUN-IN TAPERED BEARINGS ... 4 TO 5-1/2 LBS. PULL

REVERSE IDLER GEAR
END-PLAY .... .004 TO .022

MAINSHAFT BEARING ADJUSTMENT	MIN. NEW	MAX. NEW	MAX. WEAR
END-PLAY AT J	.001	.005	.015
END-PLAY AT K	.003	.007	.017

ALL FORKS IN SLIDING CLUTCHES	MIN. NEW	MAX. NEW	MAX.* WEAR
SIDE CLEARANCE	.005	.020	.050

*NOTE - IF UNIT HAS EXPERIENCED DISENGAGEMENT, SIDE CLEARANCE MUST NOT EXCEED .030 MAX.

Fig. 13 Mack TRTL-72, 722 15 speed transmission

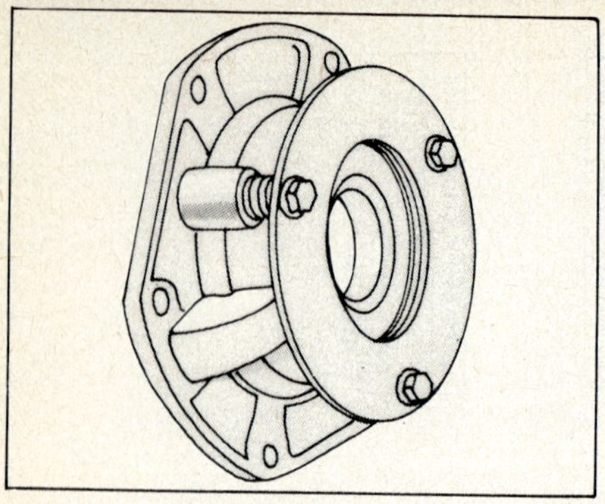

Fig. 14 Main Drive Pinion Bearing Cover, Early Model

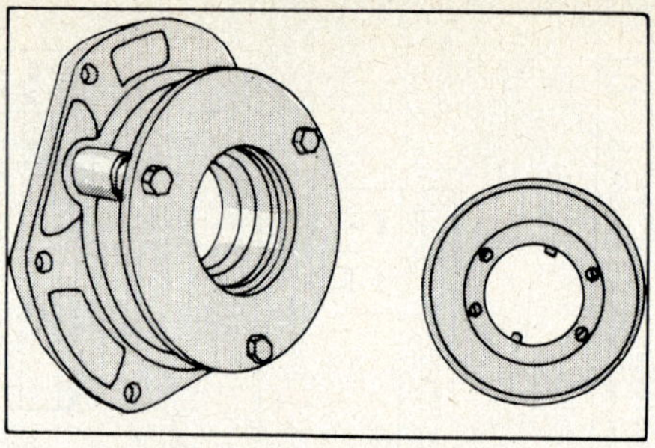

Fig. 15 Main Drive Pinion Bearing Cover, Field Conversion

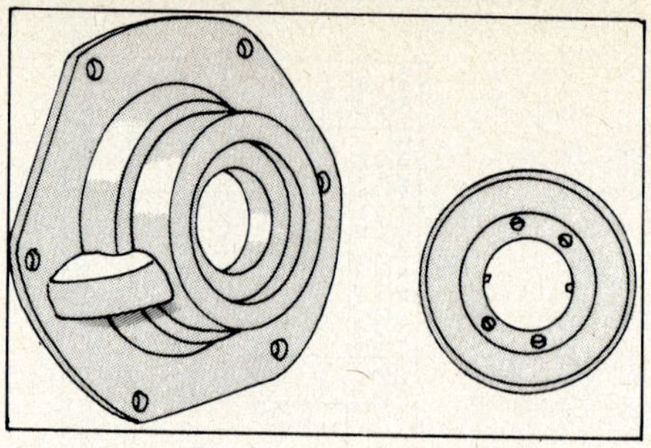

Fig. 16 Main Drive Pinion Bearing Cover, Later Model

20. For direct gear only, advance second and third speed rail into cover. Install second and third speed shifter onto rail. Install spring and compress poppet ball into cover, push rail over compressed ball and through second and third speed fork into neutral position. Install smaller interlock pin in second and third speed rail, then install large interlock pin next to rail. Install reverse speed rail with first and reverse speed fork into cover. Advance fifth speed rail into cover, then through fourth and fifth speed fork and shifter. Install and compress spring and detent ball. Pass shifter rail over compressed spring and detent ball and into its working position.
21. For over gear only, install second and third speed rail into cover. Push shifter rail through fourth and fifth speed fork, then second and third speed shifter. Install spring, poppet and ball into cover and shifter rail over compressed spring and ball assembly through second and third fork and into its neutral position. Install small interlock pin in second and third speed rail; install large interlock pin next to rail. Install first and reverse speed rail with first and reverse speed fork into cover. Install rail through fourth and fifth speed rocker shifter and over compressed spring assembly, then pass shifter rail over ball and into working position.
22. Install and safety wire to its rail, shifter and fork bolts. Install new welch plugs into shifter rail openings.
23. On 10 and 15 speed units, place compound case cover in a soft jaw-vise and install cover oil seals and cover onto compound. Install and compress detent ball and spring into cover, then pass low split rail over ball and spring assembly into working position. On 15 speed units, install high split and detent ball into cover and through fork, then into working position. Install pipe plug in pinhole. Install fork bolt and torque to specifications.

NOTE: Check interlock for proper function. Be sure not more than one rail of any group can be moved into engaged position simultaneously.

24. Apply approved sealer RTV732 or equivalent, onto compound case openings and cover gaskets on cases. Position sliding gear and all sliding clutches into neutral position. Install cover bolts and lockwashers. Torque bolts to specifications.
25. Apply sealer compound to all other cover gaskets, then install covers and bolts. Ensure proper operation of shifting lever and free rotation of gears. Torque bolts to specifications.

MACK AIR SHIFT MAXITORQUE TRTXL107 & TRTXL1070 TRANSMISSION

Disassembly

1. Tag and disconnect air lines from shift cover.
2. Drain lubricant from transmission.
3. Remove clutch release yoke bolts, and the yoke.
4. Remove Woodruff key, clutch release shaft and clutch release yoke.
5. Remove clutch brake from input shaft.
6. Remove cover bolts and the cover.
7. Remove front compound poppet spring and ball.
8. Remove main case top cover bolts and the top cover.
9. Remove air shift cover bolts and the shift cover.
10. Remove shifter rail poppet ball cover bolts and cover.

CAUTION: Cover is spring loaded. Alternately remove bolts directly opposite each other.

11. Remove poppet springs and balls.
12. From rearward controls, remove front compound shifter bolt, shifter, rail and fork.
13. From forward control, remove front compound, shifter bolts, shifter, reverse speed lockout spacer and first speed shifter.
14. Position transmission vertically and remove screws retaining compound case to main case.
15. Using a hoist, remove main and rear cases from front compound case.

NOTE: If cases do not separate, gently pry them apart with a pry bar. Do not pry on a mating surface.

16. Remove bolts and first, second and third speed shift forks.
17. Remove bolt, then fourth and fifth speed shift forks.
18. From forward controls, remove first and second shift rail from case.
19. From rear controls, remove first speed shifter bolt and shifter.
20. From rearward controls, first speed shifter, reverse lockout spacer and first speed shifter shift fork.
21. From rearward controls, remove second and third speed shifter belt, then second and third speed shift rail and shifter fork.
22. Position two sliding clutches into engaged position to lock up mainshaft for drive flange or yoke assembly removal.
23. Remove clamp plate bolt, clamp plate and yoke assembly.
24. Remove main case to rear case bolts and nuts.
25. Using a piece of wire, secure high-low shift fork. Position fork as not to interfere with removal of rear case from front case.
26. Using hoist, remove rear case from front case. Keep cases in alignment while separating to prevent binding which could damage bearings. If cases will not separate, use a pry bar and gently pry cases apart.

Sub-Assembly Disassembly

Main Case

1. Position main case horizontally, and remove main drive pinion bearing cover bolts, cover and main drive pinion assembly from main case.

NOTE: Tap end of pinion, if necessary, with a soft faced mallet to loosen pinion.

2. Remove countershaft front bearing snap ring, and bearing.
3. Position main case vertically and remove mainshaft oil tube. Remove tube only if replacement is necessary.

NOTE: Beginning with Serial Number 903151, TRTXL1070 transmissions are built without oil transfer tubes and drilled out oil passages in the main drive pinion and mainshaft.

4. Remove compound main drive gear snap ring and main drive gear.
5. Remove countershaft snap rings and countershaft.
6. Remove first speed gear sliding clutch, first speed gear and thrust washer.
7. Remove second speed gear and thrust washer.
8. Remove mainshaft and three countershaft assemblies.

Rear Compound Case

1. Remove rear countershaft rear bearing cover bolts and bearing covers.
2. Remove rear mainshaft rear bearing cover bolts and rear bearing cover.

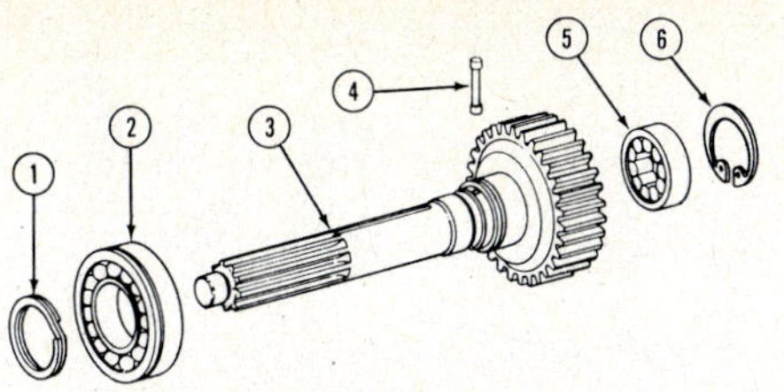

1. Spirolox Snap Ring
2. Bearing
3. Main Drive Pinion
4. Oil Pump Vane
5. Pilot Bearing
6. Snap Ring

Fig. 17 Disassembled View of Front Compound Main Drive Pinion

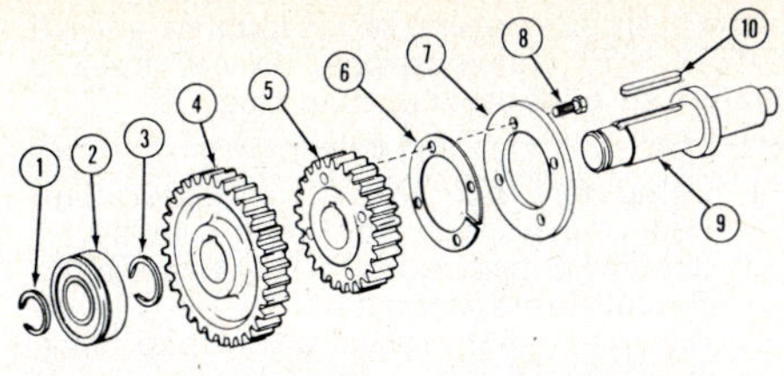

1.	Snap Ring	6.	Shim
2.	Bearing	7.	Washer
3.	Snap Ring	8.	Capscrew
4.	Direct Gear	9.	Countershaft
5.	Lo-Range Gear	10.	Key

Fig. 18 Disassembled View Of Front Compound Countershaft

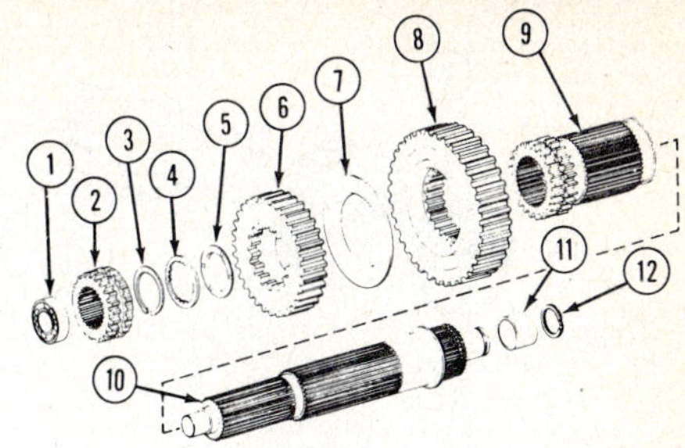

1. Pilot Bearing
2. 4th/5th Sliding Clutch
3. Snap Ring
4. Splined Thrust Washer
5. Flanged Thrust Washer
6. 4th Speed Gear
7. Flanged Thrust Washer
8. 3rd Speed Gear
9. 2nd/3rd Sliding Clutch
10. Mainshaft
11. Bearing Inner Race
12. Snap Ring

Fig. 19 Disassembled View Of Main Case Mainshaft

3. Remove high-low range snap ring, sliding clutch and shift fork.
4. Remove high-low range gear snap ring, rear compound mainshaft and high-low range gear front thrust washer.
5. Push reverse gear rearward to engage the three reverse idler gears, then withdraw rear mainshaft. At the same time, remove rear compound mainshaft high-low gear thrust washer.
6. Remove rear compound mainshaft high-low range gear from rear case.
7. Remove rear countershaft rear bearing snap ring, then drive countershaft rearward, to provide access to rear bearing snap ring.
8. Install tool No. S593C or equivalent, on rear bearing snap ring, then remove snap ring and rear bearing.
9. Remove reverse gear and shifter fork from rear case.
10. Remove main case countershaft rear bearing snap ring and bearings from front of rear case.
11. Remove reverse idler shaft expansion plugs from rear case.
12. Install tool No. J-28668 into the internal threads of the reverse idler shaft, and tighten threaded shaft of tool until reverse idler shaft is pulled through top of case.
13. Remove reverse idler gear, bearing and thrust washer.

Front Compound Case

1. Remove front compound main drive pinion bearing cover bolts and cover.
2. Remove oil pump vane from front compound main drive pinion.
3. Remove front compound countershaft front bearing cover bolts and cover.
4. Remove front bearing and front compound main drive pinion bearing snap rings.
5. Using a hoist, position front compound case vertically.
6. Place a soft iron wedge between gear teeth so gears cannot turn, and remove countershaft gear thrust washer bolts.
7. Remove front compound main drive pinion sliding clutch and low range gear from front compound case.
8. Remove front compound countershaft rear bearing snap ring, then using a soft faced mallet, tap on front end of front compound countershafts until shafts slide rearward and partially through rear bearing opening.
9. Using tool No. J-25031-2 or equivalent, remove countershaft rear bearings.
10. Remove three front compound countershafts from front compound case.

EARLY & LATE MODEL FRONT COMPOUND MAIN DRIVE PINION BEARING COVER, DESCRIPTION

Main drive pinion bearing covers may be found in three different styles, as shown in Figs. 14, 15, and 16. Fig. 14 shows an early production model used on all TRTXL107 and TRTXL1070 transmissions prior to Serial Number 8R2957. Springs and washers incorporate this unit. This arrangement was used with either three- or two-piece clutch brakes.

Fig. 15 shows a field conversion arrangement. The springs have been replaced by spacers, and the washers have been removed. This unit is used with a one-piece torque limiting clutch brake. Fig. 16 shows the later production model, used on TRTXL1070 transmissions with Serial Number 8R2957, and later. The cover is of one-piece design. Again, the one-piece torque limiting clutch brake is used.

Front Compound Main Drive Pinion Bearing Cover Early Models, Disassembly

NOTE: If the early model drive pinion bearing

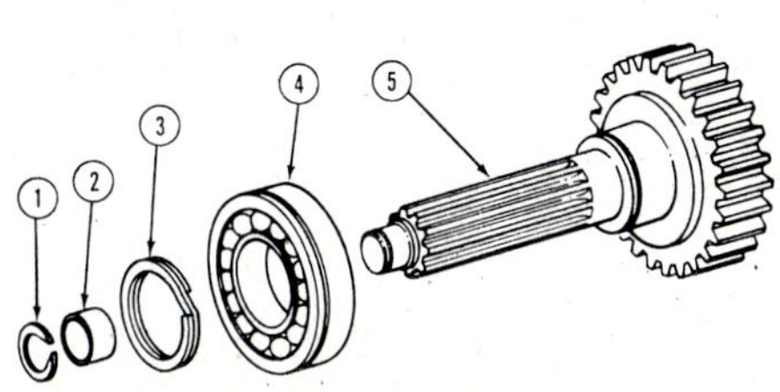

1. Snap Ring
2. Bearing Inner Race
3. Spirolox Snap Ring
4. Bearing
5. Main Drive Pinion

Fig. 20 Disassembled View Of Main Case Main Drive Pinion

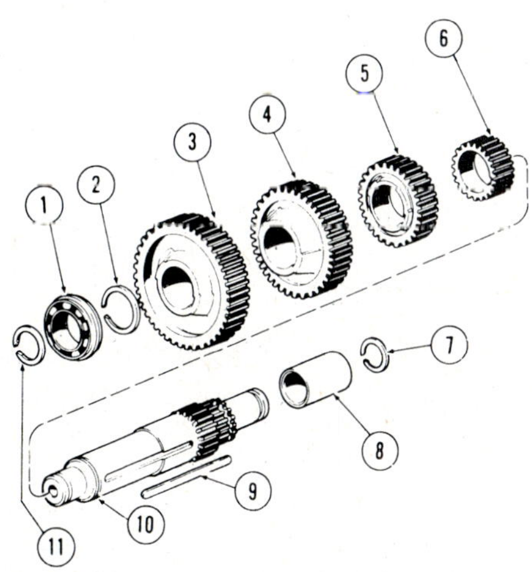

1.	Bearing	8.	Bearing Inner Race
2.	Gear Retaining Snap Ring	9.	Key
3.	Fifth Speed Gear	10.	Front Countershaft
4.	Fourth Speed Gear	11.	Bearing Retaining Snap Ring
5.	Third Speed Gear		
6.	Second Speed Gear		
7.	Snap Ring		

Fig. 21 Disassembled View Of Main Case Countershaft

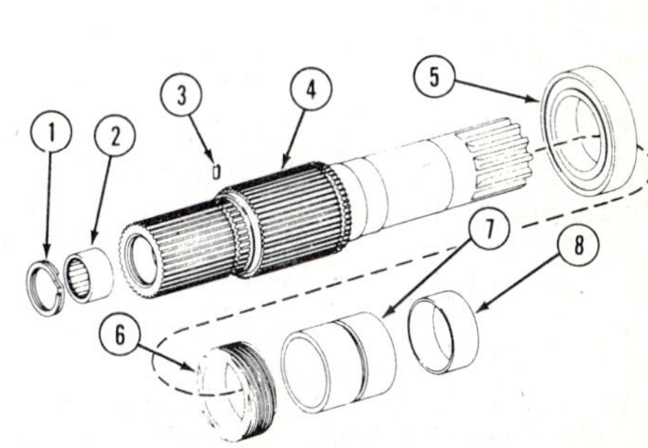

1.	Spirolox Snap Ring	6.	Speedometer Gear
2.	Front Bearing	7.	Spacer
3.	Pin	8.	Bearing Race
4.	Mainshaft		
5.	Rear Bearing		

Fig. 22 Disassembled View Of Rear Compound Mainshaft Rear Bearing Cover

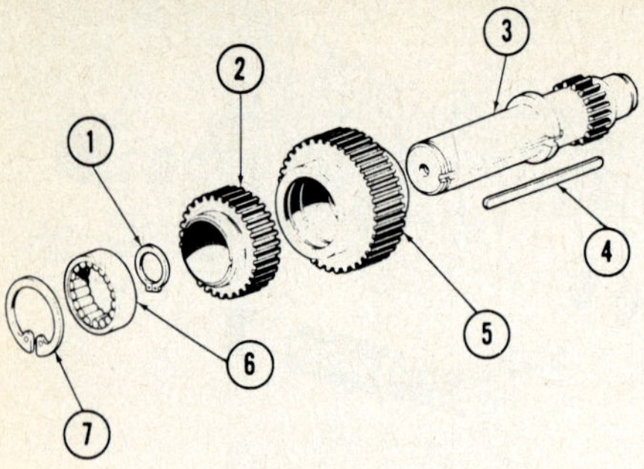

1. Gear Retaining Snap Ring
2. Direct Gear
3. Rear Countershaft
4. Key
5. Lo/Hi Range Gear
6. Bearing
7. Bearing Retaining Snap Ring

Fig. 23 Disassembled View Of Rear Compound Countershaft

cover is incorporated on transmission, it is recommended that the current model cover be installed in its place.

1. Install tool No. J-23387-01 or equivalent into bearing cover. Position spreader jaws behind oil seal and tighten bolt until jaws grip rear of seal.
2. Place assembly into a soft jaw-vise, and press seal and pump sleeve from house-ing.

Front Compound Main Drive Pinion Bearing Cover Late Models, Disassembly

NOTE: Oil seal is not installed in pump sleeve, and can be removed separately. Remove seal by driving it out of cover with a drift and hammer.

Front Compound Main Drive Pinion, Fig. 17

1. Remove bearing spirolock snap ring and bearing.
2. Using an arbor press, remove main drive pinion bearing from shaft.
3. Remove pilot bearing snap ring, then using tool No. 956 or equivalent, remove pilot bearing.

Front Compound Countershaft, Disassembly, Fig. 18

1. Remove front compound countershaft front bearing snap ring. Then using tool No. J-25031-2, or equivalent, remove bearing.
2. Remove gear snap ring. Then using an arbor press, remove front compound countershaft direct gear.
3. Using an arbor press, remove front compound countershaft low range gear.

Main Case Mainshaft, Disassembly, Fig. 19

NOTE: Prior to Serial Number 9X1342, TRTXL107 transmission model, a snap ring was used to retain mainshaft pilot bearing and must be removed. TRTXL107 transmissions built after above serial number and all TRTXL1070 transmissions, do not have a mainshaft pilot bearing snap ring.

1. Using tool No. CG-250 or equivalent, remove mainshaft front pilot bearing.
2. Remove fourth and fifth speed sliding clutches from mainshaft.
3. Remove fourth speed gear snap ring, fourth speed gear splined thrust washer and flanged thrust washer.
4. Remove fourth speed gear, third speed gear flanged thrust washer and third speed gear.
5. Remove second and third speed sliding clutches, and clutch.

Main Case Drive Pinion, Disassembly, Fig. 20

1. Remove bearing spirolock snap ring and bearing.
2. Using an arbor press, remove main drive pinion bearing from main drive pinion shaft.
3. If replacement of front compound main drive pinion pilot bearing is necessary, remove main case main drive pinion snap ring.

Main Case Countershaft, Disassembly, Fig. 21

1. Remove main case countershaft snap ring. Then using tool No. S593C, remove countershaft front bearing.
2. Remove countershaft gear snap ring, and countershaft gear.
3. Remove bearing inner race snap ring from rear of main case countershaft. Using tool No. S500 or equivalent, remove bearing inner race from rear main case countershaft.

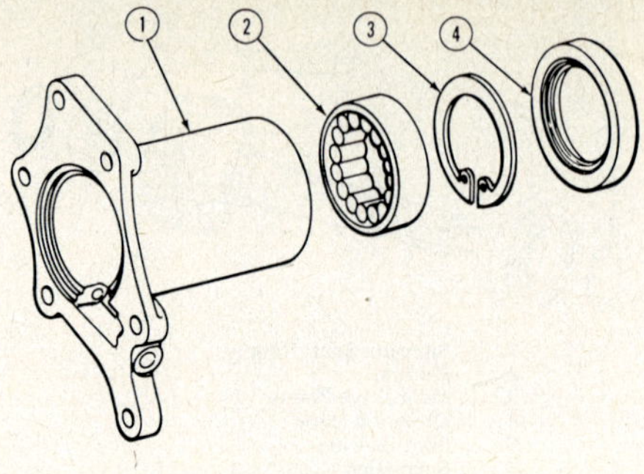

1. Rear Bearing Cover
2. Bearing
3. Snap Ring
4. Oil Seal

Fig. 24 Disassembled View Of Rear Compound Mainshaft

4. Position main case countershaft in a press arbor, and remove gears.

Rear Compound Mainshaft, Disassembly, Fig. 22

1. Remove rear compound mainshaft front snap ring and bearing.
2. Position mainshaft in an arbor press and remove rear bearing, speedometer gear, spacer and bearing race.

Rear Compound Countershaft, Disassembly, Fig. 23

1. Remove rear compound countershaft

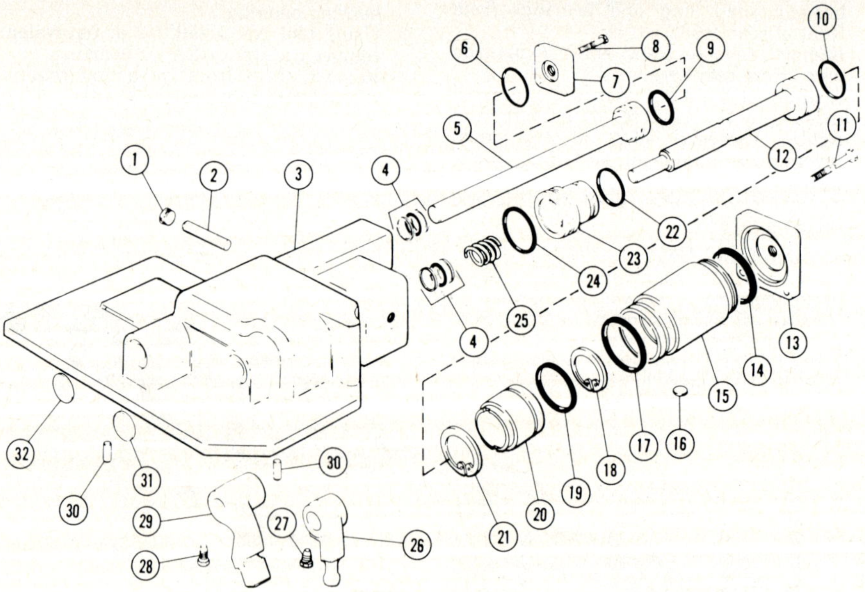

1. Plug
2. Interlock Plunger
3. Shift Cover
4. O-Ring W/Teflon Rings
5. Reverse Shift Rail
6. O-Ring
7. Reverse Shift Cover
8. Capscrew
9. O-Ring
10. O-Ring
11. Capscrew
12. Hi/Direct Shift Rail
13. Hi/Direct Shift Cover
14. O-Ring
15. Shift Cylinder
16. Breather
17. O-Ring
18. Snap Ring
19. O-Ring
20. Hi-Range Piston
21. Snap Ring
22. O-Ring
23. Direct Piston
24. O-Ring
25. Spring (See Note Below)
26. Hi/Direct Shifter
27. Setscrew
28. Setscrew
29. Reverse Shifter
30. Dowel Pin
31. Expansion Plug
32. Expansion Plug

Fig. 25 Disassembled View Of Air Shift Cover

front bearing snap ring and bearing.
2. Remove countershaft gear snap ring and countershaft gear.
3. Position gears in an arbor press, and remove gears from countershaft.

Rear Compound Mainshaft Rear Bearing Cover, Disassembly, Fig. 24

1. Using a drift and hammer, remove rear bearing cover oil seal from cover.
2. Remove rear bearing snap ring, and the bearing.

Air Shift Cover, Disassembly, Fig. 25

NOTE: TRTXL1070 transmissions Serial Number 7F0879, and later.

1. Position shifter rails into neutral, and remove reverse shifter bolt and shifter.
2. Remove reverse shift cover bolts, cover and "O"-ring.
3. Remove shift rail, shifter and "O"-ring from cover.
4. Remove interlock plunger plug from air shift cover.
5. Remove high direct shift cover bolts, cover and "O"-ring.
6. Remove snap ring from inside high direct shift cylinder and remove piston.
7. Remove "O"-ring from high direct shift cylinder.
8. Remove high direct shifter bolt, shifter rail and shifter.

NOTE: TRTXL1070 transmissions Serial Number 8S6798 and later, remove spring from high direct shift rail.

9. Remove direct piston and "O"-rings from high direct shift rail.
10. Remove "O"-rings and Teflon seals from reverse and high direct shift covers.

Selectair Valve Disassembly, Fig. 26

1. Tag and disconnect air lines from valve.
2. Remove valve from shift lever.
3. Remove valve cover bolts and cover.

CAUTION: Cover is spring loaded. Alternately remove bolts directly opposite each other.

Sub-Assembly, Assembly

Selectair Valve

1. Install detent plunger in cover.
2. Install rubber gaskets onto slots provided in valve body.
3. Install exhaust valve "O"-ring, valve spring and valve into body.
4. Secure valve in position against spring. Install rotor into valve body, then place the first detent wafer onto top of rotor.
5. Position second detent wafer on top of first installed, and align depression in the first detent wafer with tangs on the second wafer.
6. Align handle or rotor directly over reference extension of valve body, and install spring into cover.
7. Connect the two halves, aligning rotor handle with slot provided in cover.
8. Press halves together, compressing the detent spring. Rotate each half slightly, until tangs on first detent wafer seat correctly in slots provided in cover.
9. Install cover screws.
10. Adjust detent plunger so it provides added resistance to movement of shift lever between direct and neutral positions.

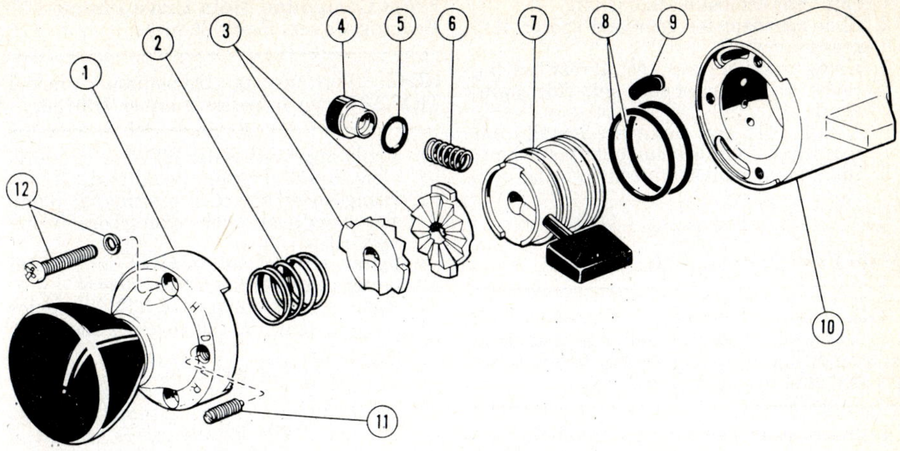

1. Cover
2. Detent Spring
3. Detent Wafers
4. Exhaust Valve
5. Exhaust Valve O-Ring
6. Exhaust Valve Spring
7. Rotor
8. O-Rings
9. Gasket
10. Valve Body
11. Detent Plunger
12. Screw

Fig. 26 Disassembled View Of Selectair Valve

Air Shift Cover

1. Install "O"-ring and Teflon seals into the reverse and high direct openings of shift cover.
2. Install "O"-rings onto direct piston and high direct shift rail.
3. Install direct piston onto high direct shift rail.
4. On transmission models, TRTXL1070 Serial Number 8S6798 and later, install spring onto high direct shift rail.
5. Install high direct shift rail in shift cover, then install high direct shifter onto shift rail.
6. Install high direct shifter bolt, and torque to 35–45 ft. lbs.
7. Install "O"-ring and snap ring onto high direct shift cylinder.
8. Install high range piston into the high direct shift cylinder, then install high direct shift cylinder onto shift cover.
9. Install "O"-ring onto end half of high direct shift cylinder, and install shift cover onto high direct shift cylinder. Torque shift cover bolts to 24–30 ft. lbs.
10. Position high direct shift rail into neutral, and install interlock plunger and plug in shift cover.
11. Install "O"-ring onto reverse shift rail, then slide reverse shift rail into shift cover, and install reverse shifter.
12. Install reverse shift cover "O"-ring, cover and bolts. Torque bolts to 81–101 inch. lbs.
13. Install reverse shifter bolt and torque to 35–45 ft. lbs.

Rear Compound Mainshaft Rear Bearing Cover

1. Install bearing and snap ring in rear mainshaft rear bearing cover.
2. Apply a bead of approved RTV732 sealer or equivalent, to bearing cover shoulder.
3. Using an arbor press, install oil seal in rear bearing cover.

Rear Compound Countershaft, Assembly

NOTE: The direct and low-high range gears incorporate a press fit with the rear countershaft. The gears should be heated, using a lamp or hot oil to 270–300°F, before being pressed onto the countershaft. The heating process should not last for more than a period of 30 minutes.

1. Insert gear key into countershaft keyway.
2. Apply a light coat of oil onto countershaft, then align low-high range gear with large hub facing opposite reverse gear. Using an arbor press, install gear onto shaft until it seats properly on shaft shoulder.

NOTE: Be sure gear keyway is properly aligned with key, before final seating of gear.

3. Align direct gear with larger hub facing opposite from low-high range gear, then install gear onto shaft until it seats properly against low-high range gear.
4. Install gear snap ring.
5. Install rear countershaft front bearing and snap ring.

Rear Compound Mainshaft

1. Apply lubricant to bearing, and install

onto rear compound mainshaft.
2. Slide speedometer gear and spacer onto rear mainshaft.
3. Using an arbor press, install rear bearing cover bearing inner race onto rear mainshaft.
4. Using drift and hammer, install rear mainshaft front bearing onto mainshaft.
5. Install rear mainshaft front bearing snap ring.

Main Case Countershaft

NOTE: The gears incorporate a press fit with the countershaft, and should be heated using a lamp or hot oil to 270–300°F, before being pressed onto the countershaft.

1. Insert gear key into countershaft keyway.
2. Apply a light coat of oil to the countershaft, then align keyway of second speed gear with key.
3. Using an arbor press, install gear onto countershaft.

NOTE: Be sure gear keyway is properly aligned with key, before final seating of gear.

4. Install fourth gear, fifth gear and snap ring onto countershaft.
5. Using a wooden drift and hammer, install front bearing and snap ring onto countershaft.
6. Using a wooden drift and hammer, install front bearing inner race and snap ring onto countershaft.

Main Case Main Drive Pinion

1. Install front compound main drive pinion pilot bearing inner race onto front main case main drive pinion, and secure with snap ring.
2. Using an arbor press, install main drive pinion bearing onto mainshaft. Be sure outer race snap ring is positioned opposite main drive pinion gear teeth.
3. Install bearing spirolock snap ring onto mainshaft.

Front Compound Countershaft

1. Install key into keyway of countershaft, and tap key into place, using soft faced mallet.

NOTE: The gears incorporate a press fit with the countershaft, and should be heated using a lamp or hot oil to 270–300°F, before being pressed onto the countershaft.

2. Apply a light coating of oil to the countershaft, then align key with keyway.
3. Using an arbor press, install countershaft into low range gear.
4. Oil shaft again, align key with keyway and press shaft into the direct gear.
5. Install gear snap ring, and front compound countershaft front bearing and snap ring onto countershaft.

Front Compound Main Drive Pinion

1. Install bearing and snap ring onto main drive pinion.
2. Install pilot bearing and snap ring onto main drive pinion.

Front Compound Main Drive Pinion Bearing Cover, Early Models

NOTE: Pertains to transmission model TRXL107 prior to Serial Number 7B0736.

1. Apply approved sealer RTV732 or equivalent, to shoulder on the oil seal.
2. Using an arbor press, position oil pump sleeve with oil seal counterbore facing upward.
3. Install oil seal into sleeve, until metal flange of seal is flush with sleeve.
4. Apply a bead of approved RTV732 sealer or equivalent, in the cover of the cover bore.
5. Align oil pump sleeve oil inlet slot with pump oil supply slot in bearing cover.
6. Install oil pump sleeve assembly in bearing cover. Wipe off any excess sealant.

Front Compound Main Drive Pinion Bearing Cover, Later Models

NOTE: Pertains to transmission models TRTXL107 Serial Number 7B0736 and TRTXL1070 Serial Number 8R2957, and later.

1. Apply a bead of approved RTV732 sealer or equivalent to the oil seal shoulder.
2. Position assembly into an arbor press and install seal onto cover.

Rear Compound Case

1. Insert expansion plug into reverse idler shaft, and partially install reverse idler shaft in case.
2. Install reverse idler gear, bearing and thrust washers in rear case.
3. Using a wooden drift and hammer, partially drive reverse idler shaft into rear case. Ensure proper alignment of gear, bearing and thrust washers before driving shaft into case.
4. Apply approved sealer RTV732 or equivalent, around the outer edge of the reverse idler shaft. Drive reverse idler shaft the rest of the way, until reverse idler shaft is flush with case.
5. Using feeler gauge, measure reverse idler gear end-play. End-play should be .003–.023 inch. If end-play is not within tolerance, check thrust washers, reverse idler gear and case for wear or damage. Replace any part which has been found to be damaged.
6. Apply sealer to end of reverse idler shaft, including area around the expansion plug.
7. Install bearing and snap ring into main case countershaft.
8. Install reverse gear and shift fork. Engage reverse gear teeth with teeth of the three reverse idler gears.
9. Locate timing marks on each of the rear countershaft low-high range gears. Mark gear tooth top land on direct gear that is in line with timing mark. Then install rear countershaft into rear case.
10. Position a block of metal at front end of rear countershaft, to provide a solid base. Using a brass drift and hammer, install bearing onto rear countershaft. Be sure drift only contacts bearing inner race when installing bearing.
11. Install snap ring onto rear countershaft.
12. Install low-high range gear into rear case and position it between countershaft direct gear and countershaft low-high range gear.
13. Install rear mainshaft into rear case. Then install mainshaft low-high range gear thrust washer onto shaft.

NOTE: Thrust washer must be installed so that notch in washer aligns with pin in mainshaft.

14. Push mainshaft through low-high range gear until rear bearing snap ring seats completely against rear case.
15. Install "O"-rings into countershaft rear bearing covers.
16. Install cover and bolts, torque bolts to 36–44 ft. lbs.
17. Apply approved sealer RTV732 or equivalent, onto rear mainshaft rear bearing cover and gasket.
18. Install cover and bolts. Torque bolts to 106–130 ft. lbs.
19. Disengage reverse gear from idler gears, and locate three timing marks on face of mainshaft low-high range gear. Also locate marks on the countershaft direct gear.
20. Align timing marks, and engage mainshaft low-high range gear with the countershaft low-high range gears.
21. Install low-high range gear front thrust washer, and snap ring into rear case.
22. Install low-high range sliding clutch and shift fork into rear case.
23. Using a piece of wire, secure high-low shift fork from moving forward.

Main Case

1. Position main case vertically, and install three countershafts.
2. Position main case horizontally, and install countershaft front bearing dished retaining ring.
3. Install main case mainshaft third speed gear, ensuring timing marks on gear line up with marks on countershaft third speed gears.
4. Install main drive pinion into main case.

NOTE: After main drive pinion is installed, countershafts are timed and held in position by the pinion.

5. Install main drive pinion bearing cover and bolts. Torque bolts to 36–44 ft. lbs.
6. Install second and third speed sliding clutches onto main case mainshaft.
7. Install third speed gear, thrust washer, fourth speed gear, thrust washer and snap ring onto mainshaft.
8. Install forth and fifth sliding clutches onto mainshaft.
9. Using a wooden drift and mallet, install pilot bearing onto front of mainshaft.
10. On transmission models TRTXL107 prior to Serial Number 9X1342, a snap ring was used retaining mainshaft pilot bearing and must now be installed.
11. Install mainshaft assembly into main case. Drive mainshaft in completely to ensure proper seating of the pilot bearing into main case main drive pinion.

NOTE: Beginning with Serial Number 903151, TRTXL1070 transmissions are built without oil transfer tubes and drilled oil passages in the main drive pinion and mainshafts.

12. Apply lubricant to the second speed gear to retain thrust washer, and install onto mainshaft. Be sure second speed gear teeth engage with countershaft second speed gears.

13. Apply lubricant to the first speed gear to retain thrust washer, and install onto mainshaft. First speed gear teeth should engage with countershaft first speed gears.
14. Install first speed gear sliding clutch onto mainshaft.
15. Install main case countershaft selective thrust washers and snap rings onto the three countershafts.

NOTE: Before measuring mainshaft end-play, the three countershafts must be secured in their proper operating positions. This may be done either by installing front compound case into main case, or by temporarily installing front compound countershaft front bearing covers onto front of main case countershafts.

16. Insert feeler gauge between mainshaft first speed gear and countershaft selective washer squarely against its snap ring, and measure end-play. End-play should be .015–.035 inch, and must not vary in excess of .005 inch between the countershafts. If end play is not within limits given, install thicker or thinner countershaft washers.
17. Install oil tube, if transmission is so equipped.
18. Install compound main drive gear, and snap ring.

Front Compound Case

1. Loosely install gear retaining thrust washers onto each of the front compound countershafts.
2. Install three countershaft assemblies into front compound.
3. Position a steel plate under front end of each countershaft to act as a base.
4. Using a wooden drift and hammer, install rear bearing onto countershaft.
5. Install snap ring onto rear bearing.
6. Install front compound main drive pinion into case.
7. Align timing marks on main drive pinion and countershaft gears.
8. Position transmission horizontally, and install front compound main drive pinion bearing and front compound countershaft bearing snap rings.
9. Position oil pump vane into its bore in the compound main drive pinion.

NOTE: Vane should be flush with the compound shaft, protruding slightly on the opposite end of shaft.

10. Install main drive pinion bearing cover gasket onto case.
11. On transmission models that use pull-type clutches only, install front compound main drive pinion bearing cover and bolts. Torque bolts to 36–44 ft. lbs.
12. Install "O"-ring in each front compound countershaft front bearing covers; install covers onto countershafts, and torque bolts to 24–30 ft. lbs.
13. Install the low range gear onto the front compound main drive pinion sliding clutch.
14. Position front compound vertically, and install sliding clutch assembly, engaging low range gear teeth of the three countershaft gears.
15. Install split adjusting shims between the three front compound countershaft low range gears and their gear retaining thrust washers.
16. Loosely install thrust washer bolt, then position a soft iron wedge between the gear teeth and torque bolt to 24–30 ft. lbs.
17. Install first, second and third speed shift fork bolts, torque bolts to 35–45 ft. lbs.
18. Install a new gasket onto rear compound mating surface. Using a hoist, install rear compound and main case to front compound.
19. Install front compound case to main case bolts. Torque bolts to 84–94 ft. lbs.
20. Install front compound case to main case bolts, located inside the front compound case, and torque to 84–94 ft. lbs.
21. On forward controls, install reverse lockout spacer and first speed shifter onto end of first speed shift rail.
22. Install second, third, fourth and fifth speed shifters onto their shift rails. Torque shifter bolts to 35–45 ft. lbs.
23. Install front compound shift fork into front compound.
24. Slide front compound shift rail through front of case, front compound shift fork and intermediate bore, until rail reaches neutral position.
25. Install front compound shift fork bolt, and torque to 35–45 ft. lbs.
26. On rearward controls, install front compound shifter onto front compound shifter rail.
27. Install front compound shifter bolt, and torque bolt to 35–45 ft. lbs.
28. Install shifter rail poppet balls, springs, cover and bolts. Torque bolts to 24–30 ft. lbs.
29. Install gasket and main case top cover. Torque main case top cover bolts to 24–30 ft. lbs.
30. Install front compound poppet ball and spring.
31. Install gasket and front compound top cover. Torque front compound top cover bolts to 24–30 ft. lbs.
32. Install clutch brake onto input shaft, and insert clutch release stub shaft into clutch release yoke. Position bell housing and tap stub into housing.
33. Install bolt and torque to 14–24 ft. lbs.
34. Using wooden drift and hammer, tap splined clutch release shaft into housing and install Woodruff key.
35. Install bolt and torque to 14–24 ft. lbs.

Transmission Assembly

1. Install a new gasket onto main case mating surface. Using a hoist, install rear case to main case.
2. Properly align cases, and install dowel type bolts. Torque bolts to 66–80 ft. lbs.
3. Install main to rear case bolts, and torque to 56–70 ft. lbs.
4. Remove wire previously installed securing low-high shift fork.
5. Install drive flange or yoke onto rear mainshaft.
6. Lubricate and install drive flange or yoke clamp plate bolt, then place sliding clutches into engaged position and torque clamp plate bolt to 474–574 ft. lbs.
7. Install second and third speed shift fork into case.
8. On rearward controls, install second and third speed shift rail and shifter into case.
9. On rearward controls, advance shifter rail through the intermediate bore of main case and through the second and third speed shift fork. Shifter rail at this point should be in the neutral position for installation of the second and third speed shifter bolts. Torque bolts to 35–45 ft. lbs.
10. Install first speed shift fork onto first speed sliding clutch.
11. On rearward controls, slide first speed shift rail through front of main case and install first speed shifter and reverse lockout spacer onto rail.
12. On rearward controls, advance first speed shift rail through intermediate bore of main case and through the first speed shift fork.
13. Install first speed shifter bolt, and torque bolt to 35–45 ft. lbs.
14. On forward controls, install first, second and third speed shift rails through front of main case. Advance them through the intermediate bore until shift forks reach their neutral positions.
15. Install fourth and fifth speed shift fork onto its sliding clutch, then slide fourth and fifth speed shift rail through front of case.
16. Install bolt onto fourth and fifth speed shift fork, and torque to 35–45 ft. lbs.
17. Position front compound horizontally, then using a feeler gauge, measure end-play of the compound main drive pinion and low range gears. End-play should be .010–.020 inch. If measurements are not within tolerance given, add or remove shims as required.

Air Shift Cover Assembly, Early Models

1. Temporarily install air shift cover on rear compound.
2. Connect shop air lines to the low-high and direct fittings of shift cover, and apply equal air pressure to both sides.
3. Insert screw driver into slot provided in forward end of low-high shift rail.
4. Turn shift rail clockwise, and at the same time rotate output flange or yoke until gear clash is heard.
5. Turn shift rail counterclockwise, counting number of turns until gear clash is again heard.
6. Turn shift rail clockwise again for half the number of turns counted from gear clash to gear clash. Shift rail will now be in neutral position.
7. Disconnect air lines and remove air shift cover from transmission.
8. Lock low-high shifter in neutral by tightening locknut.
9. Install expansion plug into shift cover at front end of low-high shift rail.
10. Install gasket, shift cover and bolts. Torque bolts to 24–30 ft. lbs.

Air Shift Cover Assembly, Late Models

1. Install a new gasket onto air shift cover.
2. Install air shift cover and bolts. Torque bolts to 24–30 ft. lbs.

TRXL107 & TRXL1071 6 SPEED TRANSMISSIONS

Transmission Disassembly

1. Drain lubricant from transmission.
2. Remove hand brake components, clutch release shaft and yoke.
3. Remove bolts and transmission case top covers.
4. Remove shifter rail poppet ball covers, spring and bolts from top covers.
5. Remove shifter fork bolts, fork and shifter rails from front case.

6. Shift transmission lever into two gears to lock up the assembly, then remove drive flange clamp plate bolt, clamp plate and drive flange from rear mainshaft splines.
7. Remove transmission front case to rear case bolts. Using a hoist, separate rear case from front case.

CAUTION: Front and rear cases must be in alignment during separation of cases.

8. From front case, remove mainshaft low-range main drive gear snap ring and the gear.
9. Remove front countershafts, selective thrust washer snap rings and washers from front countershafts.
10. Remove first speed gear sliding clutch, first speed gear, first speed gear thrust washer, second speed gear and thrust washer.
11. Remove front mainshaft assembly from transmission case.
12. Using a hoist, position transmission frontcase in a horizontal position and remove countershaft front bearing covers and main drive pinion bearing cover assembly from front case.
13. Remove oil pump vane from main driving pinion. If necessary, tap on gear end of main driving pinion and remove assembly from case.
14. Remove countershaft front ball bearing snap ring, then position front case vertically and remove three countershaft assemblies from case.
15. From rear case, remove counter and mainshaft rear bearing bolts, and cover.
16. Slide direct and low-range (low-low range sliding clutch for transmission models TRXL1071) sliding clutch from mainshaft.
17. Remove direct and low-range shift fork from case.
18. Remove low-range gear (low-low range for TRXL1071 transmissions) snap ring and thrust washer.
19. Withdraw rear mainshaft from case through rear bearing opening, and remove low-range gear rear thrust washer from mainshaft.
20. Remove low-range gear (low-low range gear for TRXL1071 transmissions) from rear case.

NOTE: Reverse gear must be engaged before removing low-range gear.

21. Slide the three rear countershafts rearward, and partially through the rear bearing opening.
22. Install puller tool on countershaft rear bearings, then remove bearings and countershaft assemblies from rear case.
23. Remove reverse gear and shifter fork.
24. Remove idler gear shafts, gears, bearings and thrust washers from rear case.
25. Remove front countershaft rear roller bearing snap ring and bearing from rear case.

Sub-Assembly Disassembly

Main Drive Pinion

Remove bearing snap ring and pinion ball bearing from main drive pinion shaft.

Main Drive Pinion Bearing Cover, Early Models

Using oil pump sleeve remover tool No. J-23387-01 or equivalent, remove oil pump sleeve and seal assembly as follows:

1. Install spreader jaws tool behind oil seal, and tighten bolt to expand jaws until jaws grip rear of oil seal.
2. Place assembly in an arbor press, and remove seal and pump sleeve from housing.

Main Drive Pinion Bearing Cover, Late Models

NOTE: Oil seal is not installed in oil pump sleeve, therefore during replacement of oil seal, do not remove sleeve unless replacement of sleeve is required.

Front Mainshaft

1. Remove front mainshaft front pilot ball bearing snap ring, and ball bearing.
2. Remove fourth and fifth speed gear sliding clutch.
3. Remove fourth speed gear snap ring, and remove the following parts:
   a. Fourth speed gear splined thrust washer, and the fourth speed gear.
   b. Third speed gear thrust washer, then second and third speed sliding clutch.
   c. Mainshaft rear bearing inner race snap ring, and bearing inner race.

Rear Mainshaft

1. Remove mainshaft front roller bearing snap ring, and the roller bearing.
2. Remove rear bearing cover bearing inner race, spacer, speedometer gear and rear ball bearing from mainshaft.
3. Remove reverse sliding gear stop snap ring.

Front Countershafts

1. Remove countershaft front bearing snap ring and bearing.
2. Remove countershaft gear snap ring.
3. Remove rear countershaft bearing inner race, spacer and front countershaft rear bearing inner race from shaft.

Rear Countershafts

1. Remove countershaft front snap ring, and bearing.
2. Remove countershaft gear snap ring, and gear.

Rear Mainshaft Rear Bearing Cover

1. Remove oil seal from bearing cover.
2. Remove roller bearing retainer snap ring and bearing.

Sub-Assembly Assembly

NOTE: Coat all components with a suitable grade of oil prior to assembly of transmission.

Main Drive Pinion

1. Install ball bearing and snap ring, with outer race facing toward pilot end of pinion.
2. Install bearing snap ring.

Main Drive Pinion Bearing Cover

1. Apply approved RTV732 gasket sealer to the press fit on the outer diameter on oil seal. Do not apply sealer on sealing lip.
2. Place oil pump sleeve in an arbor press, and install oil seal into counterbore.

Front Mainshaft

NOTE: The mainshaft third speed gear will be needed for timing the countershafts. Do not assemble front mainshaft at this time.

Rear Mainshaft

1. Using an arbor press, install rear ball bearing onto shaft.
2. Install snap ring onto outer race of ball bearing.
3. Install speedometer gear, spacer, rear bearing cover bearing inner race and reverse stop snap ring.
4. Install mainshaft front roller bearing, and snap ring.

Front Countershafts

NOTE: The gears incorporate a press fit with their countershafts and should be heated with a lamp or hot oil to 270–300°F. The heating process should not exceed a period of more than 30 minutes.

1. Insert gear key into countershaft keyway, and using an arbor press, install gears and snap rings onto the countershafts.
2. Using an arbor press, install ball bearings onto ends of countershafts.
3. Install ball bearing snap rings.
4. Turn countershafts over and install front countershaft rear bearing inner race, spacer, rear countershaft inner race and snap ring.

Rear Countershafts

NOTE: The gears incorporate a press fit with their countershafts and should be heated with a lamp or hot oil to 270–300°F. The heating process should not exceed a period of more than 30 minutes.

1. Insert gear key into countershaft keyway, and using an arbor press, install gears onto countershafts.
2. Install gear snap ring.
3. Install front bearing into countershaft gear, and then the snap ring.

Rear Mainshaft Rear Bearing Cover

1. Install roller bearing into cover and secure with snap ring.

Front Case

1. Using a hoist, position front case vertically and install three front countershafts.
2. Position front case horizontally, and install front countershaft ball bearing snap rings onto outer race of bearings. Using a soft faced mallet, tap countershaft components rearward until snap ring seats properly against case.
3. Temporarily install front mainshaft third speed gear into case, aligning timing marks on face of gear with marks on countershaft third speed gears.
4. Install main drive pinion assembly into case.
5. Install oil pump vane into pinion. Vane should be flush with right side of pinion shaft, protruding slightly on left side. Apply approved sealer compound onto pinion gasket and cover. Install cover and bolts. Torque bolts to 35–45 ft. lbs.
6. Install "O"-ring in each of the three countershaft front bearing covers. Install cover assemblies and bolts onto front case. Torque bolts to 35–45 ft. lbs.
7. Apply a light coat of suitable grease to front face of third speed gear hub to retain thrust washer, then install second and third speed sliding clutch, third speed gear, thrust washer and fourth speed gear onto front mainshaft.

8. Install fourth speed gear thrust washer, splined thrust washer and snap ring onto mainshaft.
9. Install fourth and fifth speed gear sliding clutch, pilot ball bearing and snap ring onto front of mainshaft.
10. Install mainshaft assembly in front case. Advance mainshaft until pilot bearing seats completely in case.
11. Apply a suitable grease onto front face of second speed gear hub and position second speed gear thrust washer over hub. Install second speed gear and thrust washer. Be sure gear teeth on second speed gear mesh with teeth on countershaft second speed gear.
12. Apply a suitable grease onto front face of first speed gear hub and position first speed gear thrust washer over hub. Install first speed gear and thrust washer. Be sure gear teeth on first speed gear mesh with teeth on countershaft first speed gear.
13. Slide first speed gear sliding clutch onto mainshaft.
14. Install front countershaft selective thrust washers onto the three countershafts and secure with snap rings.
15. Using feeler gauge, check end-play clearance between first speed gear and countershaft selective thrust washer. First speed gear end-play should be .025–.035 inch. End-play should not vary more than .005 inch between the three selective thrust washers.
16. Slide low-range main drive gear onto mainshaft and install snap ring.

Rear Case

1. Install reverse idler gear shafts, roller bearing and thrust washers in rear case.

NOTE: Apply approved RTV sealer around idler gear shafts before final seating of gear.

2. Engage reverse speed shifter fork with groove provided in reverse speed gear. Install reverse gear with shifter fork in rear case. Engage gear with the idlers, and at the same time allow the fork to enter its bore in rear case.

NOTE: Secure fork and gear assembly temporarily with bolt or wire.

3. Install front countershaft rear roller bearing rear snap ring, roller bearing and snap ring onto countershaft. Be sure snap ring seats properly in groove provided in rear case.
4. Align timing marks on the low-range countershaft gears with the yellow paint mark on the direct gear tooth directly opposite low-range gear timing marks. Install the three countershaft assemblies in the rear case.
5. Install bearings and snap rings onto the three countershaft assemblies.

NOTE: Transmission model TRXL-1071 incorporates a larger diameter low-low rear mainshaft gear. Therefore, permanently install the left hand and lower countershaft assemblies, but position the right countershaft in rear case without installing the rear bearing.

6. Through front opening of rear case, install mainshaft low-range gear.

NOTE: For transmission models TRXL-1071 only, install right side countershaft rear bearing and snap ring onto shaft.

7. Advance rear mainshaft assembly into rear case until rear bearing positioning snap ring seats properly against case. Position reverse gear to neutral before timing assembly.
8. Align three alignment marks on face of mainshaft low-range gear with countershaft direct gears previously marked with yellow paint. When timing marks align, engage rear mainshaft low-range gear.
9. Slide low-range gear front thrust washer onto countershaft and install snap ring.
10. Engage direct low-range shifter fork with groove provided in the direct low-sliding clutch. Slide assembly onto rear mainshaft and at the same time allow fork to enter its bore at rear of case.
11. Apply approved RTV sealer compound onto mainshaft rear bearing cover gasket and cover assembly. Install cover bolts and torque to 55–75 ft. lbs.
12. Install "O"-rings in each of the three countershaft rear bearing covers. Install cover and cover bolts onto rear case. Torque bolts to 55–75 ft. lbs.

Transmission Assembly

1. Install front case to rear case gasket. Using a hoist, install rear case to front case. Install rear case-to-front case bolts, and torque to 55–75 ft. lbs.
2. Install drive flange onto rear mainshaft splines, and position transmission into two gears to lock up the assembly. Install drive flange clamp plate and bolt, and torque bolt to 55–75 ft. lbs.
3. Engage second and third speed shifter fork with second and third speed sliding clutch. Engage first speed shifter fork with first speed sliding clutch.
4. Install second and third speed shifter rail and shifter in case. Advance rail through intermediate bore of case and second and third speed shifter fork, until rail reaches neutral position.
5. Install first speed shifter rail, first speed shifter and reverse speed lockout spacer in intermediate bore of case, until shifter rail reaches neutral position.
6. Install fork and shifter bolts, and torque bolts to 35–45 ft. lbs.
7. Install fourth and fifth speed shifter rail through front of case, and engage fourth and fifth speed shifter fork with fourth and fifth speed sliding clutch. Advance rail through hub of fork and intermediate bore of case, until rail reaches neutral position. Install fork bolt, and torque to 35–45 ft. lbs.
8. Install shift rail poppet balls and springs into front and rear cases.
9. Install poppet ball covers and bolts, and torque bolts to 55–75 ft. lbs.
10. Install transmission top and power take-off covers. Install and torque bolts to 55–75 ft. lbs.
11. Install clutch release shaft needle bearings and seals. Install clutch release shaft, yoke and any other external components removed during disassembly.

MACK MAXITORQUE TRL1076 & TRL1078 5-SPEED TRANSMISSION

Disassembly

NOTE: Unless a complete overhaul is necessary, remove only those parts required to gain access to the defective parts. Do not remove parts having heavy press fits unless replacement is necessary. If so, use proper tools so that usable parts are not damaged.

1. Drain lubricant and remove transmission from vehicle.
2. Remove hand brake components, clutch release shaft and yoke. Remove transmission case top covers. Remove shifter rail poppet ball cover, pins and springs. Poppet balls may be removed with a magnet.
3. Remove setscrews from shifters and shifter forks except second and third. Remove reverse speed shifter fork front snap ring. Remove shifter rails, shifters and shifter forks from case. Slide reverse shifter fork snap ring forward, then slide fork forward. While moving rail forward, alternately move front and rear snap rings rearward until rail is removed from case.
4. Place transmission in two gears which will lock up assembly to facilitate removal of drive flange clamp plate screw and clamp plate. Install suitable puller and remove drive flange from mainshaft splines. Remove mainshaft rear bearing cover, then slide speedometer gear off of mainshaft.
5. Remove main driving pinion bearing cover. Tap on gear end of main driving pinion with a soft faced mallet if necessary and remove assembly from case.
6. Remove fourth/fifth speed sliding clutch from mainshaft.
7. Remove fourth speed gear selective thrust washer retaining snap ring through main driving pinion bearing opening, then slide fourth speed gear selective thrust washer off of mainshaft.
8. Carefully remove mainshaft from case through rear bearing opening.
9. Remove fourth speed gear and thrust washer through transmission case top cover opening.
10. Remove countershaft rear bearing cover, then the power take-off covers.
11. Remove lower right-hand rear countershaft oil slinger front snap rings, or power take-off gear front snap ring, if used.
12. Remove the three rear countershafts from case through rear bearing cover openings. While withdrawing the lower and right-hand countershafts, slide oil slinger retaining spring and front snap ring off of shaft. If power take-off gear is used instead of oil slinger, slide gear and front snap ring off of shaft.
13. Place main case in a vertical position. Remove all case-to-case dowel bolts and capscrews. With a soft-faced mallet, tap rear case and separate it from front case. Remove rear case and place on bench.
14. Remove front countershaft thrust washer snap ring, then the thrust washer from countershafts.
15. Remove mainshaft sliding clutches and gears, with thrust washers, from case.
16. Place transmission in a horizontal position and remove countershaft front bearing covers. Remove snap rings from outer race of countershaft front bearings.
17. Position transmission vertically, then remove the three front countershaft assemblies from case.

SUB-ASSEMBLIES

Main Driving Pinion

Remove bearing snap ring. Remove ball bearing by pressing or tapping off. Invert

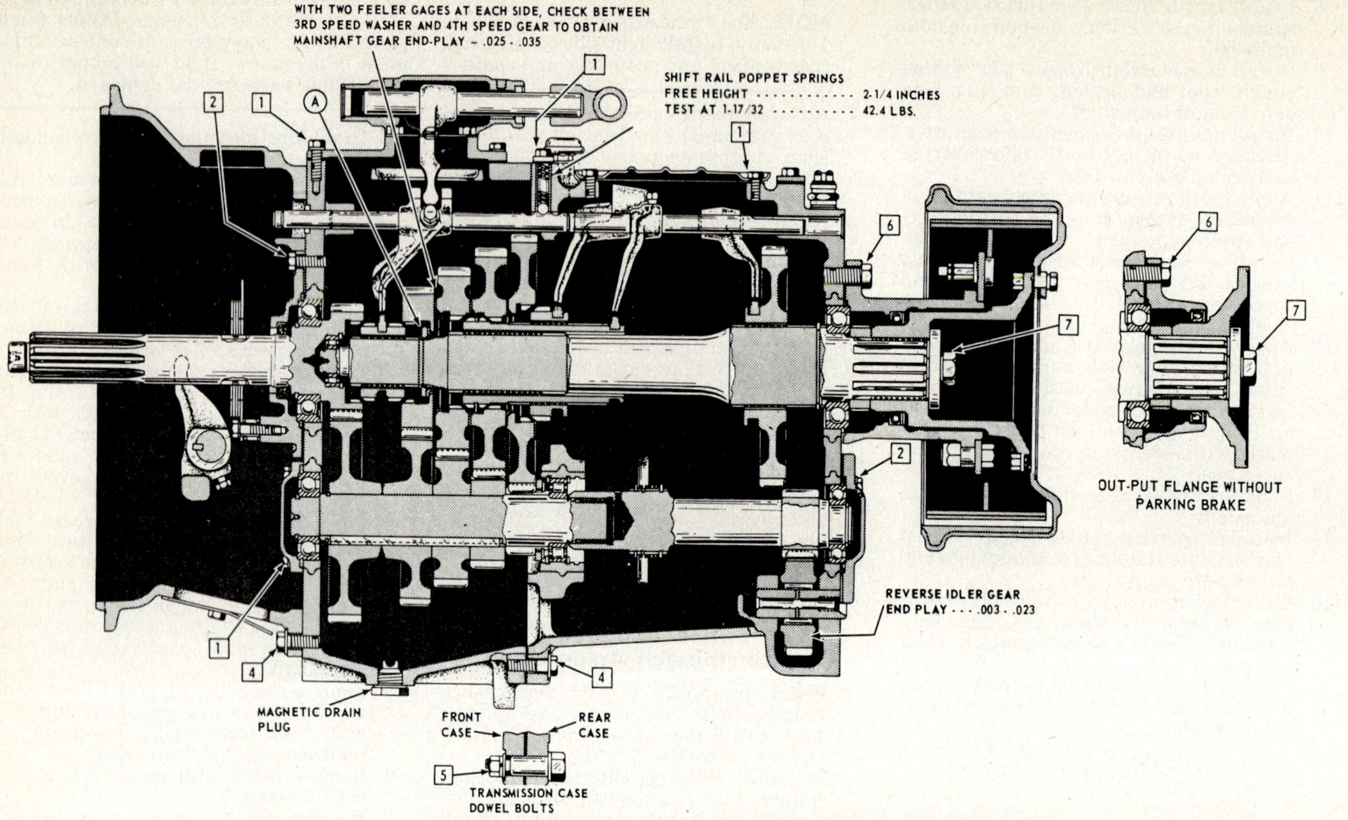

Fig. 27 Mack TRL-1076, 1078 5 speed triple countershaft transmission

pinion and remove snap ring and bearing.

Mainshaft

Remove mainshaft rear ball bearing snap ring, then press shaft out of bearing. Place mainshaft in soft-jawed vise and remove bearing snap ring and bearing inner race. Remove reverse stop snap ring from shaft.

Rear Case

Using a sharp tool, punch a hole in idler shaft expansion plug and pry plug out. Using a suitable puller, remove the three idler gear shafts from case. Remove idler gears, bearings, thrust washers and reverse speed gear from case. Remove countershaft roller bearings from case.

Rear Countershafts

Remove countershaft rear bearing snap rings, then press countershafts out of rear ball bearing.

Front Countershafts

Remove countershafts front bearing and gear retaining snap ring. Invert countershafts and remove roller bearing inner race from shafts.

NOTE: On TRL1076 only, remove rear roller bearing inner race snap ring from upper left hand countershaft before removing inner race.

Inspection

Clean transmission parts thoroughly using a suitable cleaning solvent to remove all grease and foreign matter. Dry parts with compressed air.

Soak bearings in suitable solvent to loosen grease and foreign matter. Strike bearings flat against block of wood several times then immerse in solvent while turning races slowly. Repeat until bearings are clean, then blow dry with compressed air. Do not spin bearing with compressed air as damage or personal injury may result.

Check bearings for wear or damage. Replace as necessary.

Replace gear if teeth show any sign of wear or damage. Gears may be checked by Magnaflux or similar method for cracks which otherwise would not be visible.

Replace forks and/or clutches if side clearance in groove exceeds specified limits. Replace shaft rails if cracked in either the poppet or stake screw holes. If the clearance exceeds .010 inch maximum, check to determine which member is worn before replacement. Shift wear may be checked by comparing shaft diameter at an unworn point as against the worn point.

Reassembly

NOTE: All working parts, should be coated with clean engine oil while the transmission is being assembled. This will insure lubrication when first starting and will prevent seizing of these parts. As moving parts are assembled, check frequently to see that they move freely. For torque recommendations refer to Figs. 27 and 28.

1. Position front case vertically and install the three front countershaft assemblies in case.
2. Position front case horizontally, then install front countershaft ball bearing snap rings in outer race of bearings. Tap countershaft assemblies rearward until snap ring seats against case.
3. Install O-ring in countershaft front bearing covers, then install the covers. Torque capscrews to specifications.
4. Place front case in a vertical position, then install mainshaft third speed gear so that alignment marks on face of gear mate with alignment marks on countershaft third speed gears.
5. Install second/third speed sliding clutch and engage with third speed gear clutch teeth.
6. Apply suitable grease to thrust surface of second and first speed gear, then place thrust washer on front face of gear hub.
7. Install second speed gear (clutch teeth to the rear) with thrust washer over second/third speed sliding clutch and engage with countershaft second speed gears.
8. Install first speed gear (clutch teeth to the rear) with thrust washer and engage with countershaft first speed gears. Position first sliding clutch on second/third speed sliding clutch.
9. Install thrust washer and snap ring on each front countershaft.
10. Apply suitable gasket sealer compound to front case mating surface, then install gasket. Apply suitable sealer to gasket and lower rear case assembly over assembled countershafts. Tap case into position so that bearings are seated properly, then install capscrews and dowel bolts and torque to specifications.

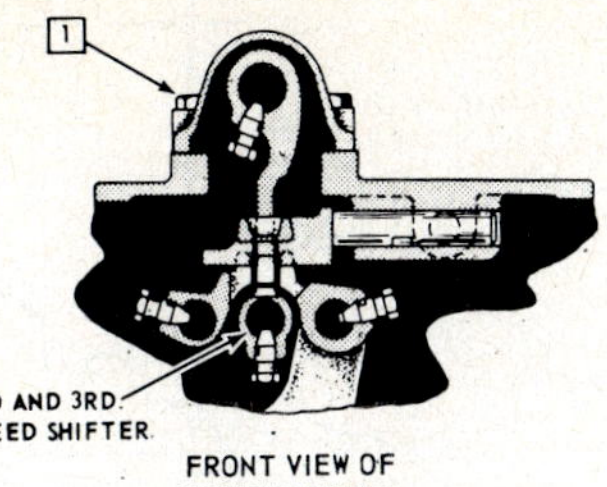

FRONT VIEW OF SHIFTER DETAIL

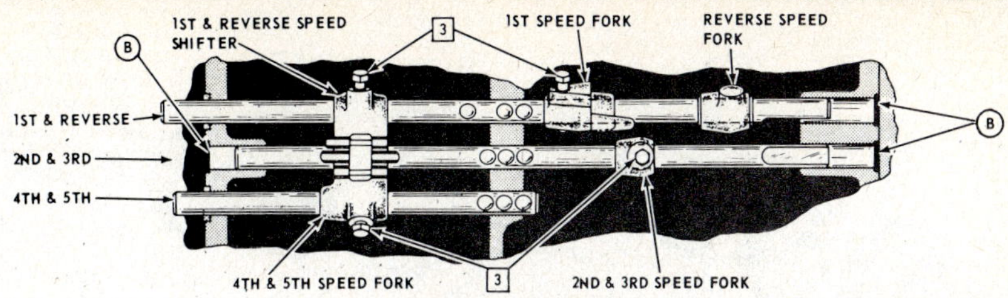

TOP SECTIONAL VIEW OF TRANSMISSION SHOWING SHIFTER RAIL ARRANGEMENT

FORK INTERLOCK BALL SPRING
FREE HEIGHT .... 1-1/4 INCHES
TEST AT 1-1/8 INCHES .... 38 TO 46 LBS.

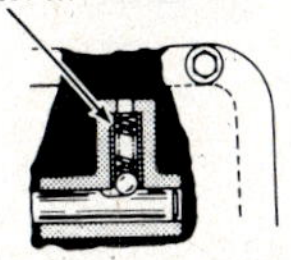

DETAIL OF SHIFTER FORK INTERLOCK SPRING-LOADED BALL AND DETENT ARRANGEMENT

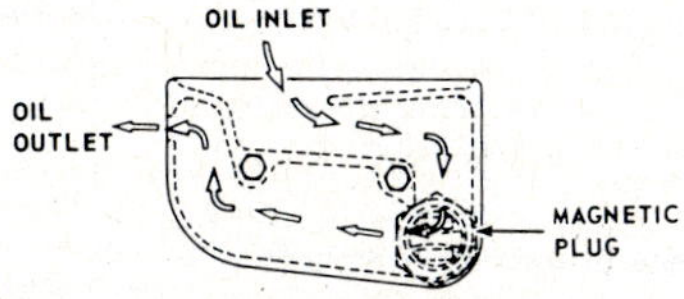

VIEW SHOWING MAGNETIC OIL FILTER ON INSIDE WALL OF TRANSMISSION CASE

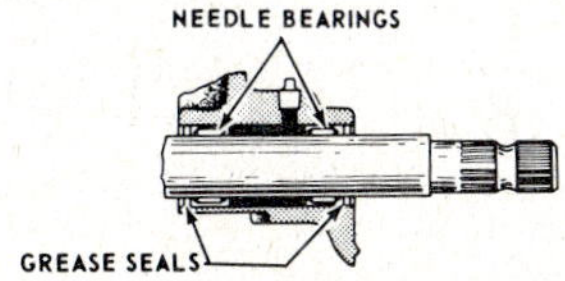

CLUTCH RELEASE SHAFT ARRANGEMENT AS VIEWED FROM FRONT OF TRANSMISSION

SCREW TORQUES IN LBS. FT.	
SCREW LOCATION	TORQUE (LUBRICATED)
1 -	18 TO 28
2 -	30 TO 40
3 -	35 TO 45
4 -	65 TO 75
5 -	55 TO 75
6 -	105 TO 115
7 -	480 TO 520

THRUST WASHER	AVAILABLE THICKNESS
(A) 4TH SPEED GEAR SELECT FOR TIGHTEST FIT IN GROOVE TO OBTAIN 0.025 TO 0.035 MAIN-SHAFT GEAR END-PLAY.	0.318 - 0.321 0.304 - 0.306 0.289 - 0.291 0.274 - 0.276 0.259 - 0.261 0.244 - 0.246 0.229 - 0.231 0.214 - 0.216 0.199 - 0.201 0.184 - 0.186

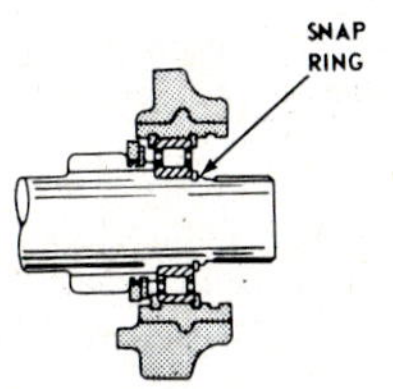

VIEW SHOWING LEFT HAND COUNTERSHAFT TRL1076

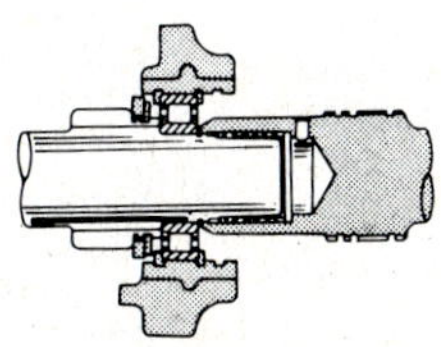

VIEW SHOWING LEFT HAND COUNTERSHAFT TRL1078

(B) IMPORTANT - BE SURE EXPANSION PLUGS ARE IN PLACE BEFORE INSTALLING UNIT IN CHASSIS (2 REAR - 1 FRONT)

ALL FORKS IN SLIDING CLUTCHES			
	MIN. NEW	MAX. NEW	MAX. * WEAR
SIDE CLEARANCE	0.005	0.020	0.050

* NOTE: IF UNIT HAS EXPERIENCED DISENGAGEMENT SIDE CLEARANCE MUST NOT EXCEED 0.030 MAX.

Fig. 28 TRL-1076, 1078 torque & adjustment chart

11. Position the transmission horizontally, then install rear countershaft through rear bearing opening in case.
12. While installing countershaft, install oil slinger retaining spring on shaft. With flared lip of slinger toward the rear, install slinger on countershaft splines. Compress retaining spring with slinger and install snap ring. If equipped with power take-off gear, position gear on countershaft splines, then install snap ring.
13. Align timing marks on rear countershaft with timing marks on front countershaft and continue advancing shaft until the countershaft rear bearing snap ring seats against case.
14. Install O-rings in countershaft rear bearing covers, then install the covers. Torque capscrews to specifications.
15. Install third speed gear thrust washer in third speed gear.
16. Install fourth speed gear in case (clutch teeth forward) with thrust washer and engage with countershaft gears. Engage reverse speed sliding gear with reverse idler gears.
17. Install mainshaft in case through gear bearing opening. Advance shaft through reverse speed gear, second and third speed sliding clutch and fourth speed gear until rear bearing snap ring seats against case. Install speedometer drive gear on shaft. Apply suitable sealant and install rear bearing cover gasket and cover. Torque capscrews to specifications. Oil lip of cover seal, then install clamp plate and screw.
18. Place transmission in two gears which will lock up assembly, then tighten clamp plate.
19. Install forth speed gear selective thrust washer and snap ring on mainshaft.
20. Using two feeler gages, check between third speed washer and fourth speed gear for recommended end-play.
21. Install fourth and fifth speed sliding clutch on splines of pinion, and engage with fourth speed gear.
22. Position and drive pinion assembly into place being careful not to damage bearing.
23. Install tool J23796 in main driving pinion bearing cover and through cover oil seal. This tool will eliminate damage to oil seal when installing cover over splined end of main driving pinion. Apply suitable sealant and install gasket and cover assembly. Remove tool from cover assembly. Install capscrews and torque to specifications.
24. Install reverse speed fork on reverse speed sliding gear then the first speed fork on first speed sliding clutch.
25. Install first and reverse speed shifter rail and shifter through front of case. Advance rail through intermediate bore of case, first speed fork, front snap ring, reverse speed fork and rear snap ring until rail reaches neutral position. Install fork and shifter setscrews and torque to specifications.

NOTE: Assemble snap rings with rounded edge of rings against shift fork. After assembling snap rings, rotate 30° to assure rings are properly seated.

---

26. Install second and third speed shifter rail through hole provided in rear of case. Engage second and third speed fork with second and third speed sliding clutch. Advance rail through hub of fork and intermediate bore of case.
27. Install second and third speed shifter on rail and continue advancing rail into front support until it reaches neutral position. Install fork and shifter setscrew and torque to specifications.
28. Install fourth and fifth speed shifter rail in hole provided in front of case. Engage fourth and fifth speed fork with sliding clutch. Advance rail through hub of fork and intermediate bore until rail reaches neutral position. Install fork setscrew and torque to specifications. Make sure that the two shift rail oil seals and the expansion plug are installed in front of case and the two shift rail expansion plugs are installed in rear of case.
29. Install shift rail poppet balls and springs

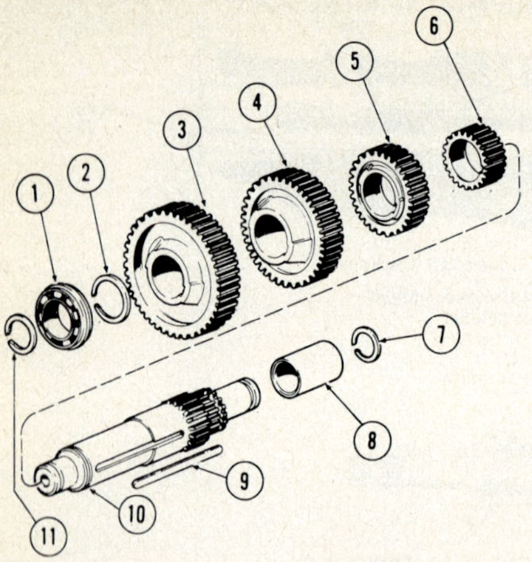

1. Bearing
2. Gear Retaining Snap Ring
3. Fifth Speed Gear
4. Fourth Speed Gear
5. Third Speed Gear
6. Second Speed Gear
7. Retaining Snap Ring
8. Bearing Inner Race
9. Key
10. Front Countershaft
11. Bearing Retaining Snap Ring

Fig. 29 Exploded view of front countershaft

in holes in top of case. Install cover and torque capscrews to specifications.

30. Apply suitable sealant to transmission case top cover gaskets and install on case, then install covers and torque capscrews to specifications. Apply suitable sealant to surface of power take-off covers, then install the covers.

MACK MAXITORQUE TRDL1070, TRDLG1070, TRXL107, 107A, 1070 & 1070A TRANSMISSIONS

Disassembly

1. On all except TRXL107 and 1070, tag and disconnect air lines from shift cover.
2. Drain lubricant, then remove transmission from vehicle. Clean exterior of transmission and install unit in suitable overhaul stand.
3. Remove clutch release yoke tapered bolts, then drive splined clutch release shaft inward and remove Woodruff key and the release shaft.
4. Remove clutch release stub shaft with clutch release yoke, then the clutch brake from input shaft.
5. Remove front case top cover bolts, then the cover.
6. On all units except TRXL107 and 1070, remove air shift cover and rear case top cover bolts, then the covers.
7. Remove shifter rail poppet ball cover bolts, then the cover, poppet spring and ball.

NOTE: The cover is spring loaded. Alternately remove bolts directly opposite each other.

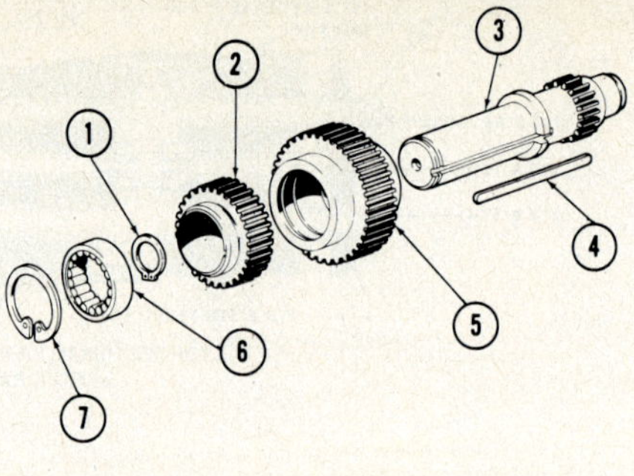

1. Gear Retaining Snap Ring
2. Direct Gear
3. Rear Countershaft
4. Key
5. Hi-range Gear
6. Bearing
7. Bearing Retaining Snap Ring

Fig. 30 Exploded view of rear countershaft

8. Remove bolts from first speed, second/third and fourth/fifth speed shift forks.
9. Slide fourth/fifth speed shift rail forward and remove fourth/fifth speed shift fork.
10. Remove first speed shifter bolt, then slide shift rail forward and remove first speed shifter, reverse lockout spacer and first speed shift fork.
11. Remove expansion plug from front of shaft by driving second/third speed shifter and shift rail forward.
12. Remove second/third speed shifter bolt, then the shifter and shift fork.
13. Position two sliding clutches into engaged position to lock up mainshaft for drive flange or yoke assembly removal.
14. Remove clamp plate bolt and clamp plate, then the drive flange or yoke.
15. Remove front case to rear case attaching bolts, then place transmission in a vertical position and loosen the case-to-case dowel bolt nuts. Drive dowel bolt out.
16. Using a piece of wire, secure direct-high range shift fork on TRD series units, or direct-low range shift fork on TRX series units, to prevent it from sliding forward.
17. Using suitable hoist, remove rear case from front case. Keep cases in alignment while separating to prevent binding which could damage bearings. If cases will not seperate, use a pry bar and gently pry cases apart.

Sub-Assembly Disassembly

Front Case

1. Position front case vertically and remove mainshaft oil tube, if necessary.
2. Remove mainshaft high range, on TRD series units, or low range on TRX series units, main drive gear snap ring, then the gear.
3. Remove front countershaft selective thrust washer snap rings, then the thrust washers.
4. Rmeove first gear sliding clutch, then the gear and thrust washer.
5. Remove second speed gear and thrust washer, then the front mainshaft assembly.
6. Position front case horizontally. Remove countershaft front covers, then the main drive pinion bearing cover.
7. Remove oil pump vane from main drive pinion, then the main drive pinion assembly from front case.

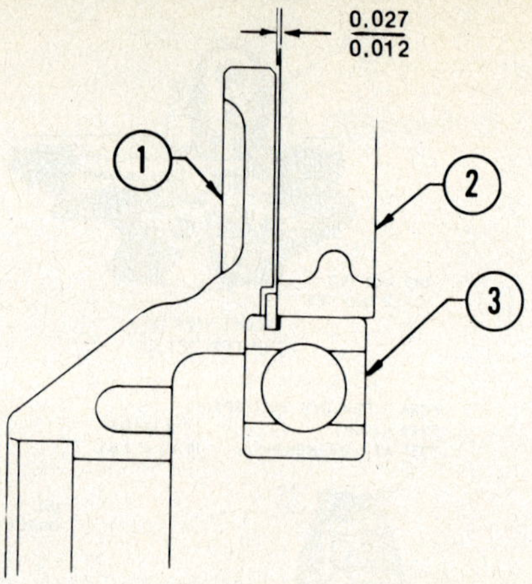

1. COVER 2. CASE 3. BEARING

Fig. 31 Measuring main drive pinion cover clearance

NOTE: If necessary, tap on end of pinion with a soft faced mallet to loosen pinion.

8. Remove countershaft front bearing snap rings, then position front case vertically and remove the three front countershaft assemblies.

Rear Case

1. Remove rear countershaft and mainshaft rear bearing covers. Note location of any hollow rear countershafts for rear mounted P.T.O., so unit can be assembled with shafts in same position.
2. On TRD series units, remove direct-high range sliding clutch and shift fork as an assembly, then remove high range gear snap ring through rear case front opening. Remove Hi range gear thrust washer.
3. On TRX series units, remove direct-low range sliding clutch and fork as an assembly. Remove low range, or low-low range gear snap ring through front opening in rear case, then remove the thrust washer.
4. Slide reverse sliding gear rearward to engage it with reverse idler gears, then start mainshaft out of case by sliding it rearward. Remove gear thrust washer from mainshaft, then the mainshaft from case.
5. On TRD series units, remove high range gear from front of case. On TRXL107 and 107A units, remove low range gear through front of rear case.
6. Remove rear countershaft rear bearing snap ring, then drive countershaft rearward to gain access to positioning snap ring.
7. Remove rear snap ring and bearing using tool S593C, or equivalent.
8. On TRXL1071 and 1071A units, remove low-low range gear through top opening in rear case after rear right countershaft has been moved rearward 1–2 inches.
9. Remove the three rear countershaft assemblies, then the reverse gear and shift fork from case.
10. Remove front countershaft rear bearing snap rings and bearings from front of rear case.

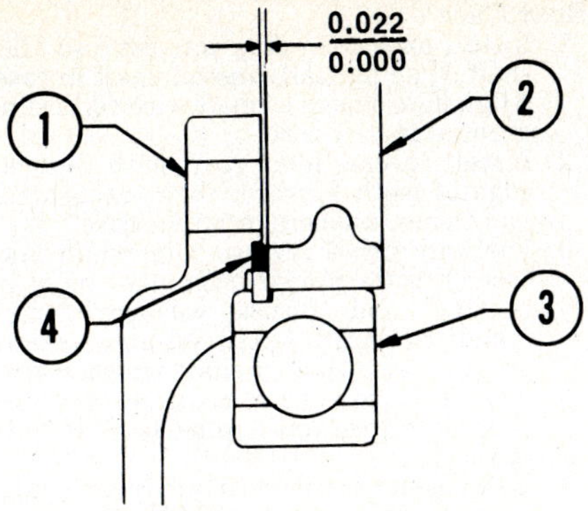

1. COVER 2. CASE
3. BEARING 4. O-RING (56AX 352)

Fig. 32 Measuring countershaft front cover clearance

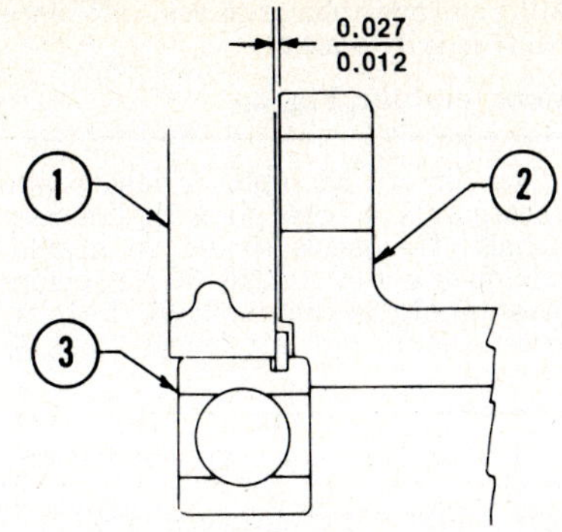

1. CASE 2. COVER 3. BEARING

Fig. 33 Measuring mainshaft rear cover clearance

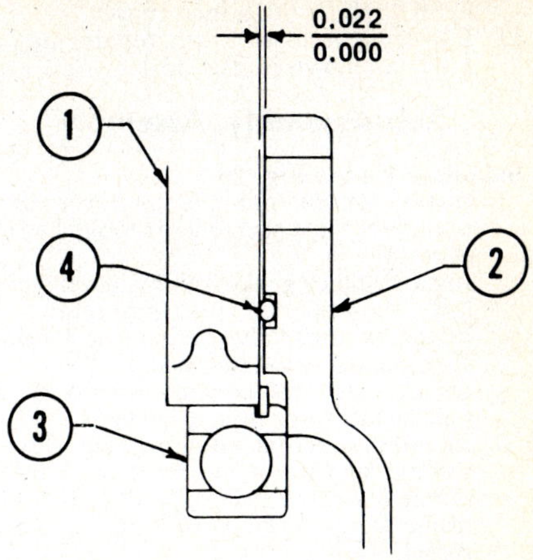

1. CASE 2. COVER
3. BEARING 4. O-RING (56AX 431)

Fig. 34 Measuring countershaft rear cover clearance

11. Remove reverse idler shaft expansion plugs from rear case.
12. Install tool J-28668 into internal threads of reverse idler shaft and tighten threaded shaft of tool until reverse idler shaft is pulled through top of case.
13. Remove reverse idler gear, bearing and thrust washers.

Main Drive Pinion

1. Remove bearing retaining spirolox snap ring.
2. Using an arbor press, remove main drive pinion bearing from shaft.

Main Drive Pinion Bearing Cover

1. On TRD series units, remove oil seal by driving it out of housing with a punch and hammer.
2. On TRX series units, remove seal using tool J-23387-01. Position jaws behind seal, then tighten screw to expand jaws. Install assembly in a press and press out seal.

Front Mainshaft

1. Remove front mainshaft bearing using suitable puller.
2. Remove fourth/fifth speed sliding clutch, then the fourth speed gear snap ring and thrust washers.
3. Remove fourth speed gear, then the third speed gear thrust washer and the gear.
4. Remove second/third speed sliding clutch.

Rear Mainshaft

1. Remove gear mainshaft front bearing spirolox snap ring, then the bearing using tool S-501, or equivalent.
2. Install rear mainshaft in an arbor press and simultaneously remove rear bearing, speedometer gear, rear spacer and rear bearing cover bearing inner race.

Front Countershaft, Fig. 29

1. Remove front countershaft front bearing snap ring, then the bearing using tool S593C.
2. Remove front countershaft gear snap ring.
3. Remove bearing inner race snap ring, then the inner race from rear of countershaft.
4. Position front countershaft in an arbor press and remove gears.

Rear Countershaft, Fig. 30

1. Remove rear countershaft front bearing snap ring, then the bearing using tool CG-270.
2. Remove countershaft gear snap ring, then press gears off shaft.

Rear Mainshaft Rear Bearing Cover

1. Using a drift and hammer, remove rear bearing cover oil seal from cover.
2. Remove rear bearing snap ring, then the bearing.

Air Shift Cover, Fig. 25

NOTE: TRXL107A and TRXL1071A units use a low-direct rather than a high-direct shift rail which is used on TRDL series units. Any reference to high-direct components will apply to the low-direct components on these units.

1. Position both shift rails in neutral position, then remove reverse shifter bolt.
2. Remove reverse shifter cover bolts, then the cover and "O" ring.
3. Remove shift rail and shifter from cover, then the "O" ring from shift rail.
4. Remove interlock plunger plug, then the plunger from air shift cover.
5. Remove high-direct shift cover bolts, then the cover. Remove "O" ring from cover.
6. Remove high-direct shift cylinder from air shift cover, then the high range piston from shift cylinder.
7. Remove snap rings and "O" rings from shift cylinder and high range piston. Remove high-direct shifter bolt.
8. Remove high-direct shift rail and shifter through bottom opening. Remove locknut and lockwasher, if equipped.

NOTE: TRDL1070 units beginning with Serial No. 8S6796, remove spring from high-direct shift rail.

9. Remove direct piston and "O" ring from high-direct rail, then the "O" rings from direct piston.
10. Remove "O" ring and two Teflon rings from reverse and high-direct openings of shift cover.
11. Clean air breather in high-direct shift cylinder.

Selectair Valve

Refer to Fig. 26., to disassemble valve. Use care when removing cover and exhaust valve as these components are spring loaded.

Cleaning & Inspection

1. Clean all components in suitable solvent and dry with compressed air.
2. Replace bearings which show any amount of wear or damage.
3. Inspect gears for wear or damage. Also, check gears for cracks, which may not be visible, by magnetic particle test.
4. Measure clearance of forks in sliding clutches. Allowable side clearance for new parts is .005–.020 inch. Wear must not exceed .050 inch, or .030 inch if the unit has been disengaged.
5. Inspect shift rail poppet and stake screw holes for cracks and replace as necessary. Measure clearance between shift rail and mating housing bore. If clearance exceeds .010 inch, determine which member is worn before replacing.
6. Replace any damaged oil pump components. Vane must not have more than .006 inch play in mating bore.
7. Check oil pump for proper operation by pouring oil into pump pickup trough while rotating main drive pinion clockwise. Proper operation will be indicated by oil appearing at various outlets along the mainshaft.
8. Inspect all Selectair valve components and replace as necessary. If any one rubber part is worn, all should be replaced.
9. Check poppet spring tension and replace as necessary. The spring should measure 42.4 lbs. of resistance when compressed to 1 17/32 inches. Spring free height should measure 2 1/4 inch.
10. Measure clearance between covers and case, Figs. 31 thru 34. Replace covers as necessary. If clearance between main driving pinion cover and case, or mainshaft rear cover and case, is .015 inch or more, use gasket 628KB310-P2 (.031 inch thick). If less than .015 inch, use gasket

628KB310-P1 (.016 inch thick).
11. If bearing bores are worn, repairs may be made as shown in Fig. 35.

Sub-Assembly, Assembly

Selectair Valve, Fig. 26

1. Install detent plunger in cover until plunger tip just protrudes through bottom of cover.
2. Install rubber gaskets into slots in valve body, then the "O" rings onto rotor.
3. Install exhaust valve spring and "O" ring, then install valve into body.
4. Secure valve into position against spring and install rotor into valve body.
5. Install first detent wafer onto top of rotor so that depression on back of wafer is aligned with raised portion of rotor and wafer tangs align with slots in top of rotor.
6. Install second wafer on top of first. Align depression in top wafer with tangs on bottom wafer.
7. Align rotor handle directly over reference extension of valve body, then install spring into cover.
8. Connect the two halves, aligning rotor handle with slot provided in cover.
9. Press halves together, compressing the detent spring. Rotate each half slightly until tangs on first detent wafer seat in slots in cover.
10. Install cover screws.
11. Adjust detent plunger so it provides added resistance to movement of shift lever between direct and neutral positions.
12. Connect air lines to valve, then install valve on shift lever.

Air Shift Cover, Fig. 25

1. Install "O" ring and Teflon seals into reverse and high-direct openings of shift cover.
2. Install "O" rings onto direct piston and high-direct shift rail.
3. Install direct piston onto high-direct shift rail.
4. If equipped, install spring onto high-direct shift rail.
5. Install high-direct shift rail into shift cover, then the high-direct shifter onto shift rail. Install locknut and lockwasher, if equipped.
6. On TRDL1070 units prior to Serial No. 7F1122, TRDLG1070 prior to 7F1260 and TRXL107A, 1071A prior to 700525, turn high-direct shift rail to screw it into locknut, lockwasher and shifter. On all other units, install high-direct shifter bolt, and torque to 41–49 ft. lbs.
7. Install "O" ring and snap ring onto high range piston.
8. Install "O" ring and snap ring into high-direct shift cylinder, then install piston into cylinder.
9. Install a high-direct shift cylinder, then the "O" ring onto end of cylinder.
10. Install high-direct shift cover onto cylinder. Torque bolts to 24–30 ft. lbs.
11. Position high-direct shift rail into neutral, then install interlock plunger and plug in shift cover.
12. Install "O" ring into reverse shift rail, then slide rail into shift cover and install reverse shifter.
13. Install reverse shift cover "O" ring, cover and bolts. Torque bolts to 81–101 inch lbs.
14. Install reverse shifter bolt and torque to 41–49 ft. lbs.

Rear Mainshaft Rear Bearing Cover

1. Install bearing and snap ring into cover.
2. Apply a bead of RTV sealant to cover seal mating surface, then press seal into place until it is fully seated.

Rear Countershaft, Fig. 30

NOTE: The direct and high, or low, range gears incorporate a press fit with the rear countershaft. The gears should be heated, using a lamp or hot oil, to 270–300° F, before being pressed onto the countershaft. The heating process should not last longer than 30 minutes.

1. Insert gear key into countershaft keyway.
2. Apply a light coat of clean engine oil onto countershaft, then align high, or low, range gear with large hub facing opposite reverse gear. Using an arbor press, install gear onto shaft until fully seated on shaft shoulder.

NOTE: Be sure keyway is properly aligned with key before final seating of gear.

3. Align direct gear with larger hub facing opposite high, or low, range gear, then install gear onto shaft until fully seated against high, or low, range gear.
4. Install gear snap ring.
5. Install rear countershaft front bearing and snap ring.

Front Countershaft, Fig. 29

NOTE: The gears incorporate a press fit with the countershaft and should be heated, using a lamp or hot oil, to 270–300° F before being pressed onto countershaft.

1. Install gear key to countershaft keyway.
2. Lubricate countershaft with clean engine oil, then align keyway of second speed gear with key and press gear onto shaft.

NOTE: Remove any burrs between gear and its mating surface before final seating.

3. Install remaining gears. Fourth speed gear hub must face away from third speed gear and fifth speed gear hub must be next to fourth speed gear hub.
4. Install gear retaining snap ring, then the front countershaft front bearing and snap ring.
5. Install bearing inner race and snap ring on rear of front countershaft.

Rear Mainshaft

1. Press rear bearing onto shaft, then slide speedometer gear and spacer onto shaft.
2. Press rear bearing cover bearing inner race onto shaft.
3. Install front bearing and snap ring onto mainshaft.

Main Drive Pinion Bearing Cover

1. Apply a bead of RTV sealant to cover seal mating surface.
2. Position assembly into an arbor press and install seal into cover until it is fully seated.

Main Drive Piston

1. Press bearing onto main drive pinion shaft. Make sure that outer race snap ring is positioned away from pinion gear teeth.
2. Install bearing retaining spirolox snap ring.

Rear Case

1. Install expansion plug into reverse idler shaft, then partially install shaft in case. Align flats on end of shaft with centerline of countershaft bore.
2. Install reverse idler gear, with rounded edge of teeth forward, then the bearing and thrust washers into rear case.
3. Partially drive reverse idler shaft into case. Ensure proper alignment of gear, bearing and thrust washers before installing shaft. Apply suitable sealant around outer edge of shaft where it contacts the case, then continue to drive shaft into case until relief is flush with case.
4. Using feeler gauge, measure reverse idler gear and play. Maximum end play is .023 inch. If end play exceeds specification, replace thrust washers. If still out of specification, replace reverse idler gear. If end play is still greater than .023 inch, the case must be replaced.
5. Apply suitable sealant to end of reverse idler shaft, including area around expansion plug.
6. Install front countershaft rear bearings and snap rings, then install reverse gear and shift fork.
7. Locate timing marks on high, or low, range gears. Mark direct gear tooth top land that is directly in line with each alignment mark, then install rear countershaft assemblies into case.
8. Position a block of metal at front end of rear countershaft to provide a solid base, then drive rear countershaft rear bearing onto shaft. Apply force only to the inner race.
9. On TRXL1071, 1071A units, install low-low range gear through opening in top of rear case.
10. Install snap ring onto rear countershafts.
11. Install high range gear (TRD series), or low range gear (TRXL107, 107A), and temporarily position it between countershaft direct gear and high, or low, range gear.
12. Install rear mainshaft into rear case. Slide shaft through reverse gear, then install rear thrust washer onto shaft.
13. Continue to advance shaft through high, low, or low-low gear until rear bearing snap ring seats against case.
14. Install "O" ring in countershaft rear bearing cover, then install the cover. Torque bolts to 36–44 ft. lbs.
15. Apply suitable sealant to mainshaft rear bearing cover and gasket, then install the cover. Torque bolts to 96–120 ft. lbs.
16. Disengage reverse gear from idler gears and locate three timing marks on face of mainshaft high, low, or low-low range gear and on the countershaft direct gears.
17. Align timing marks and engage mainshaft high, low, or low-low range gear. Install front thrust washer and snap ring, then the sliding clutch and shift fork into case.
18. Using a wire, secure direct-high, low, or low-low shift fork from moving forward.

Front Case

1. Position front case vertically and install three front countershafts.
2. Position case horizontally and install front countershaft bearing snap rings, then lightly tap countershafts rearward until snap ring seats against case.
3. Temporarily install front mainshaft third speed gear so that the three alignment marks on face of gear and countershaft third speed gear mate.

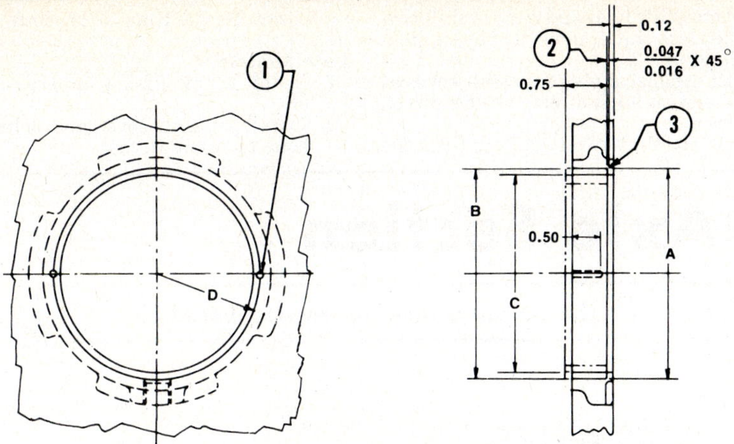

1. DRILL & REAM $\frac{0.1245}{0.1240}$ DIA.
   FOR TWO PINS, 31AX 289
2. ON SLEEVE
3. $\frac{0.016}{0.000}$ R. IN CASE

INSTRUCTIONS
1. APPLY LOCTITE TO SLEEVE AND PRESS INTO CASE.
2. FACE SLEEVE FLUSH WITH CASE.
3. DRILL AND REAM FOR 2 STRAIGHT PINS.
4. INSTALL 2 STRAIGHT PINS.
5. BORE TO 'C' DIMENSION.

NOTE:
SLEEVES TO BE MADE FROM LOW CARBON STEEL TUBING S.A.E. J403 NO. 1010 OR EQUAL (E.G. 1018, 1020, 1112, 1118)

CAUTION
SLEEVES MUST BE BORED CONCENTRIC WITHIN 0.002 IN. OF EACH OTHER.

PART NO.	BORE LOCATION	A CASE BORE	B SLEEVE O.D.	C BEARING BORE	D PIN HOLE R.
281KB 562A AUXILIARY CASE	COUNTERSHAFT	3.794/3.793	3.7950/3.7945	3.544/3.543	1.897
	MAINSHAFT	5.172/5.171	5.1730/5.1725	4.922/4.921	2.586
284KB 5159B REAR CASE	COUNTERSHAFT	3.400/3.399	3.4010/3.4005	3.150/3.149	1.700
	MAINSHAFT	5.172/5.171	5.1730/5.1725	4.922/4.921	2.586
284KB 5163 FRONT CASE	COUNTERSHAFT	3.794/3.793	3.7950/3.7945	3.544/3.543	1.897
	MAINSHAFT	5.172/5.171	5.1730/5.1725	4.922/4.921	2.586

Fig. 35 Repairing bearing bore

4. Install main drive pinion into case.

NOTE: After main drive pinion is installed, the pinion times the countershafts and holds them into place. Therefore, the third speed gear can now be removed from the case until the front mainshaft is assembled.

5. Install oil pump vane into main drive pinion bore. The vane should be flush with right side of pinion shaft and protrude slightly from left side of shaft, as viewed from front of transmission.
6. Install main drive pinion bearing cover gasket, then the cover. Torque bolts to 36–44 ft. lbs.

NOTE: On units equipped with pull type clutch, use tool J-23796 when installing bearing cover to protect lip of oil seal.

7. Install "O" rings in three front countershaft front bearing covers, then install the covers. Torque bolts to 24–30 ft. lbs.
8. Install second/third speed sliding clutch onto front mainshaft. Make sure oil passage is free of obstructions.
9. Install third speed gear and flanged thrust washer onto front mainshaft.
10. Install fourth speed gear, then the flanged and splined thrust washers and snap ring onto mainshaft.
11. Install bearing onto front of front mainshaft, then install mainshaft into case until bearing is fully seated in main drive pinion.
12. Position second speed gear thrust washer on front face of second speed gear and hold in place with a small amount of grease.
13. Install second speed gear, with beveled side of clutch teeth towards front of transmission, and engage with countershaft second speed gears.
14. Position first speed gear thrust washer on front face of first speed gear and hold in place with a small amount of grease. Install first speed gear, beveled side of clutch teeth toward rear of transmission, with thrust washer and engage with countershaft first speed gear.
15. Install first speed gear sliding clutch.
16. Install front countershaft thrust washers and snap rings on the three countershafts.
17. Check end play of each countershaft by inserting a feeler gauge between mainshaft first speed gear and countershaft thrust washer. Hold thrust washer squarely against snap ring. End play should be .025–.035 inch and must not vary more than .005 between the countershafts. If end play is not within specifications, install thicker or thinner washers as necessary.
18. Install mainshaft rear oil tube, if previously removed.
19. Install mainshaft high, or low, range main drive gear and snap ring.

Transmission Assembly

1. Install a new gasket on front case mating surface.
2. Using suitable hoist, install front case to rear case. Torque capscrews to 56–70 ft. lbs. and dowel bolts to 66–80 ft. lbs.
3. Remove wire which was previously installed to hold direct-high, or low, shift fork.
4. Install drive flange or yoke onto rear mainshaft splines. Lubricate clamp plate screw threads, then install the clamp plate and screw.
5. Place two sliding clutches into engaged position to lock up mainshaft, then torque drive flange or yoke clamp plate bolt to 474–574 ft. lbs.
6. Install first speed shift fork onto first speed sliding clutch.
7. Slide second/third speed shift rail through front of case, then install second/third speed shifter onto rail.
8. Advance rail through intermediate bore of case and second/third speed shift fork until it reaches the neutral position. Install shifter bolt and torque to 41–49 ft. lbs.
9. Install first speed shift rail through front of case, then install first speed shifter and reverse lockout spacer onto rail. Advance rail through intermediate bore of case and first speed shift fork until it reaches the neutral position. Install shifter bolt and torque to 41–49 ft. lbs.
10. Install second/third and first speed shift fork bolts and torque to 41–49 ft. lbs.
11. Install fourth/fifth speed shift fork onto fourth/fifth speed sliding clutch. Install shift rail through front of case, through shift fork, until it reaches the neutral position. Install shift fork bolt and torque to 41–49 ft. lbs.
12. Install expansion plug into second/third speed shift rail.
13. Install clutch brake onto input shaft.
14. Install clutch release shaft into clutch release yoke, then tap the shaft into bell housing. Install retaining bolt and torque to 13–21 ft. lbs.
15. Install poppet ball and springs, then the poppet ball cover. Torque bolts to 24–30 ft. lbs.
16. Install top case cover gasket, then the top cover. Torque bolts to 24–30 ft. lbs.

Air Shift Assembly

Early Models

1. Temporarily install air shift cover assembly onto rear case.
2. Connect air lines to high, or low, and direct fittings of shift cover. Apply equal air pressure to both sides.
3. Insert screwdriver into slot provided in forward end of high-direct, or low-direct, shift rail.
4. Turn shift rail clockwise with screwdriver, while simultaneously turning output flange or yoke, until gear clash is heard.

5. Turn shift rail counterclockwise and note number of turns until gear clash is heard again.
6. Turn shift rail clockwise half the number of turns counted from gear clash to gear clash. The shift rail will now be in neutral position.
7. Disconnect air lines and remove shift cover, then lock shifter in neutral by tightening locknut.
8. Install expansion plug into shift cover at front end of high-direct, or low-direct, shift rail.
9. Install air shift cover gasket, then the cover. Torque bolts to 24–30 ft. lbs.

Late Models

1. Install a new gasket onto air shift cover.
2. Install air shift cover and bolts. Torque bolts to 24–30 ft. lbs.

Rear Axle Section

NOTE: For Timken (Rockwell) driving axles, see the Timken chapter for service procedures

REAR AXLES

Mack Unit No.	Type	Oil Capacities In Pints	Mack Unit No.	Type	Oil Capacities In Pints	Mack Unit No.	Type	Oil Capacities In Pints
RAS23C	S.R.	28	SHB342C-W	D.R.	58	SS50C	D.R.	55
RAD23C	D.R.	31	SA38C-W	D.R.	59	SW50C	D.R.	55
RAD231C	D.R.	31	SC38C-W	D.R.	59	SS58C	D.R.	55
RAD29C	D.R.	31	SS38C-W	D.R.	59	SW58C	D.R.	55
RAD38C	D.R.	31	SW38C-W	D.R.	59	SS65C-W	D.R.	74
SS34C-W	D.R.	59	SH38C-W	D.R.	59	SW65C-W	D.R.	74
ST34C-W	D.R.	59	SSB38C-W	D.R.	59	SW80W	D.R.	74
SC342C-W	D.R.	58	SWB38C-W	D.R.	59	SWP800W	D.R.	74
SS342C-W	D.R.	58	SHB38C-W	D.R.	59	SWP801W	D.R.	74
SW342C-W	D.R.	58	SS440C	D.R.	59	EW48C-W②	D.R.	88
SH34C-W	D.R.	59	SW440C	D.R.	59	EW60C②	D.R.	74
SH342C-W	D.R.	58	SS441C	D.R.	49	ES60C	D.R.	74
SWB342C-W	D.R.	58	SW441C	D.R.	49			

①—Capacity of complete system.
②—Tri-axle bogies.

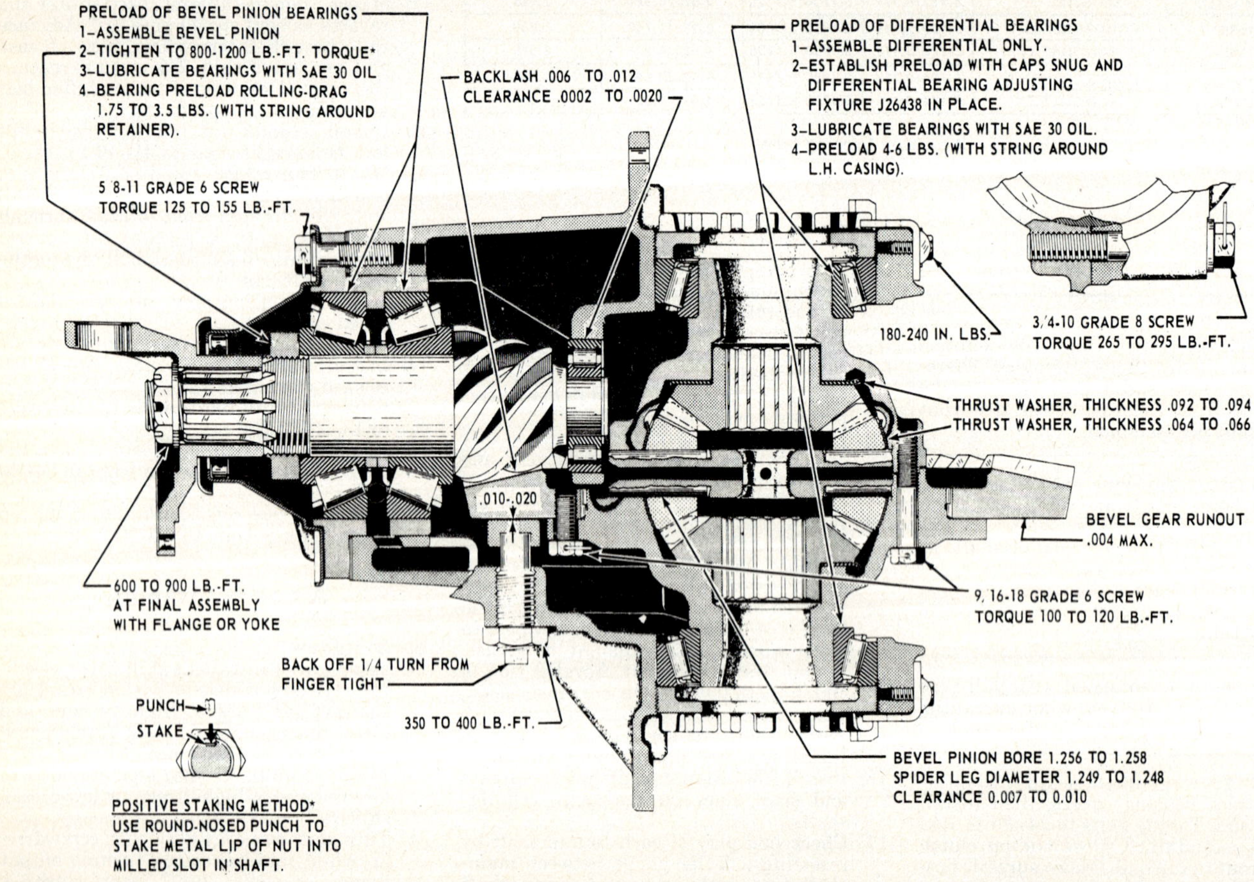

Fig. 1 Single reduction carrier. Models CRS-84 & CRS-840

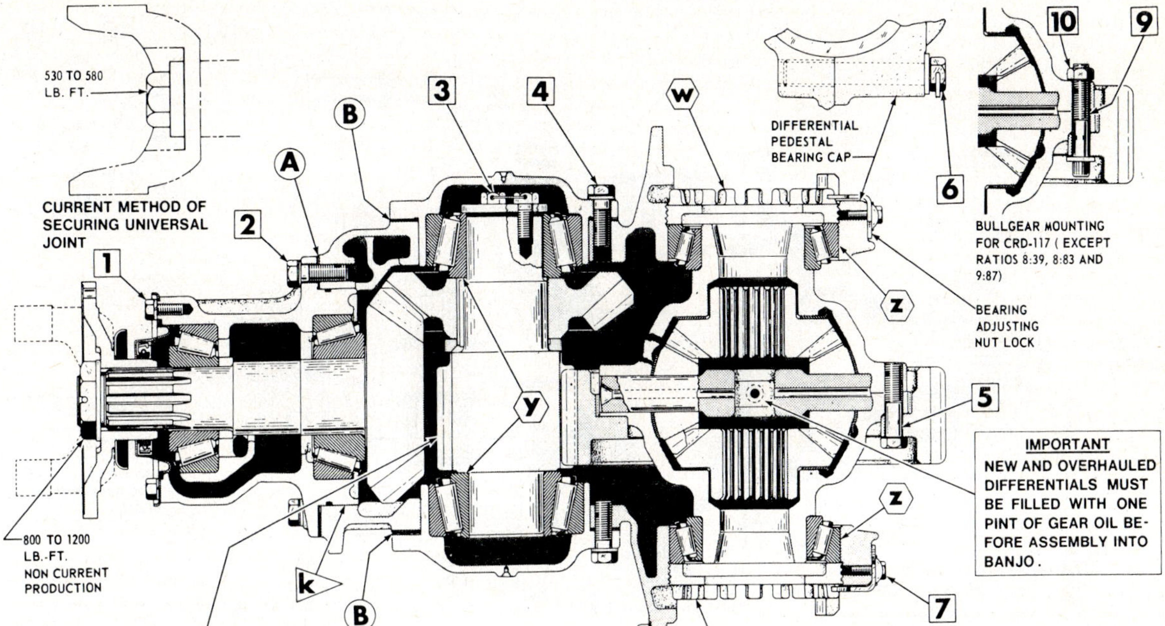

J WRAP STRING AROUND TEETH AND WITH A SLOW STEADY PULL MEASURE LBS.-PULL AS FOLLOWS:

NUMBER OF GEAR TEETH	POUNDS PULL REQ'D TO TURN ASSEMBLY
13	4.9 TO 9.8 LBS. PULL
15	4.5 TO 8.9 LBS. PULL
16	4.3 TO 8.5 LBS. PULL
17	4.1 TO 8.2 LBS. PULL

NOTE – BEVEL GEAR RATIOS 4:88, 6:36 & 7:48 HAVE MARKED TEETH AND MUST BE "TIMED" FOR BEST RESULTS.

CARRIER-TO-AXLE HOUSING CAPSCREW TORQUES

(11) 5/8-18 SCREWS -- 134 TO 148 LB.-FT.
(1) 3/4-16 SCREW -- 238 TO 264 LB.-FT.

W

TO LOCK CAPSCREW, BEND WASHER UP AGAINST HEAD ON ONE SIDE AND DOWN AGAINST LOCK ON OPPOSITE SIDE

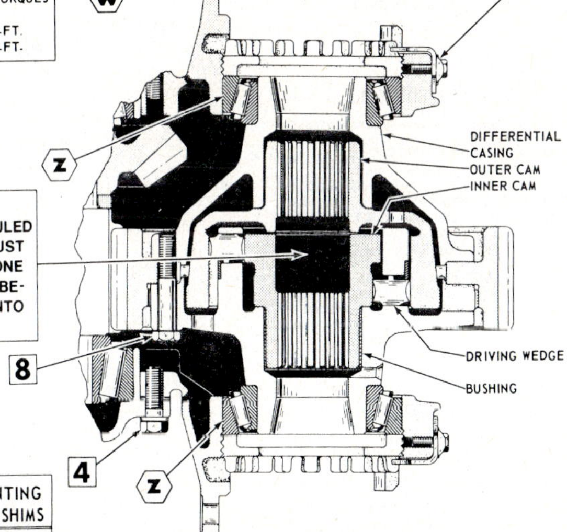

MACK INTER-WHEEL POWER DIVIDER DIFFERENTIAL PROVIDED IN CRD-1171 & CRD-1161

STEP NO. 1 PRELOAD PINION BEARINGS (X)

A - ASSEMBLE PINION AND BEARINGS IN HOUSING OMITTING BEARING COVER AND DRIVE YOKE OR FLANGE.
B - LUBRICATE BEARINGS WITH SAE-30 OIL AND PLACE ASSEMBLY ON PRESS TABLE.
C - APPLY 30,000 LB. LOAD TO BEARING CONES.
D - WRAP STRING AROUND HOUSING PILOT AT POINT (K). ATTACH SCALE TO LOOPED FREE END OF STRING, AND WITH A STEADY PULL, SLOWLY ROTATE HOUSING, AT THE SAME TIME, MEASURE THE ROLLING-DRAG WHICH MUST BE WITHIN 2.8 TO 5.6 LB. PULL. IF NECESSARY TO ADJUST BEARING SPACER (C), CONSULT SPACER CHART FOR AVAILABLE SIZES.
E - FINISH ASSEMBLY OF PINION AND ESTABLISH THICKNESS OF SHIM PACK (A) WITH PINION SETTING GAGE TOOLS. TEMPORARILY, REMOVE PINION HOUSING AND SHIMS THEN PROCEED WITH NEXT STEP.

STEP NO. 2 PRELOAD HELICAL PINION SHAFT BEARINGS (Y)

A - ASSEMBLE HELICAL PINION COMPLETELY AND INSTALL IN CARRIER HOUSING. LUBRICATE BEARINGS WITH SAE-30 OIL.
B - PRELOAD BEARINGS BY ADJUSTING SHIM PACKS (B). SEE CHART FOR ROLLING-DRAG REQUIRED PER RATIO AT (J).
C - ALSO ADJUST (B) TO OBTAIN CORRECT GEAR TOOTH CONTACT.
D - REMOVE BEVEL PINION ASSEMBLY FOR NEXT STEP.

STEP NO. 3 PRELOAD DIFFERENTIAL BEARINGS (Z)

A - LOOSEN HELICAL PINION BEARING COVER CAPSCREWS TO RELIEVE PRELOAD.
B - ASSEMBLE DIFFERENTIAL COMPLETELY AND INSTALL.
C - LUBRICATE BEARINGS WITH SAE-30 OIL AND INSTALL FIXTURE J26438 ON BEARING CAP LUGS.
D - WITH HOOK IN BULL GEAR, SLOWLY ROTATE ASSEMBLY WITH SCALE ATTACHED AT 12 IN. RADIUS. TIGHTEN NUTS (W) UNTIL SCALE READS 2 TO 3 LB. PULL.
E - REMOVE TOOLS AND COMPLETE ASSEMBLY OF UNIT.

(A) PINION MOUNTING ADJUSTMENT SHIMS

51 KH 267 P10 (.010)
51 KH 267 P12 (.012)
51 KH 267 P15 (.015)
51 KH 267 P31 (.031)

USE AS REQUIRED

(B) GEAR MOUNTING ADJUSTMENT SHIMS

51 KH 244 P10 (.010)
51 KH 244 P12 (.012)
51 KH 244 P15 (.015)
51 KH 244 P31 (.031)

INTERCHANGEABLE BETWEEN SIDES – USE AS REQUIRED

(C) TWO-IN-COMBINATION BEARING CONE SPACERS ARE AVAILABLE TO SERVICE SINGLE SPACER 52KH171. PLEASE REFER TO **MASTER MANUAL PAGE 13.7**

THREADED FASTENERS				TORQUE LB. FT.		
	LOCATION	QUANTITY	SIZE	GRADE	LUBRICATED	
	1	(6)	7/16-14	5	30	40
	2	(9)	9/16-12	8*	130	145
	3	(3)	1/2-20	6*	75	85
	4	(14)	9/16-12	8*	130	145
#	5	(12)	9/16-18	8*	140	150
	6	(4)	3/4-10	8	265	295
	7	(2)	3/8-16	5	20	25
	8	(12)	9/16-18	8*	145	165
#	9	(12)	9/16-18	8*	140	150
#	10	(12)	9/16-18		80	100

* SELF-LOCKING CAPSCREWS   # USE LOCTITE 262

Fig. 2 Single reduction carrier. Models CRD-116 & CRD-117

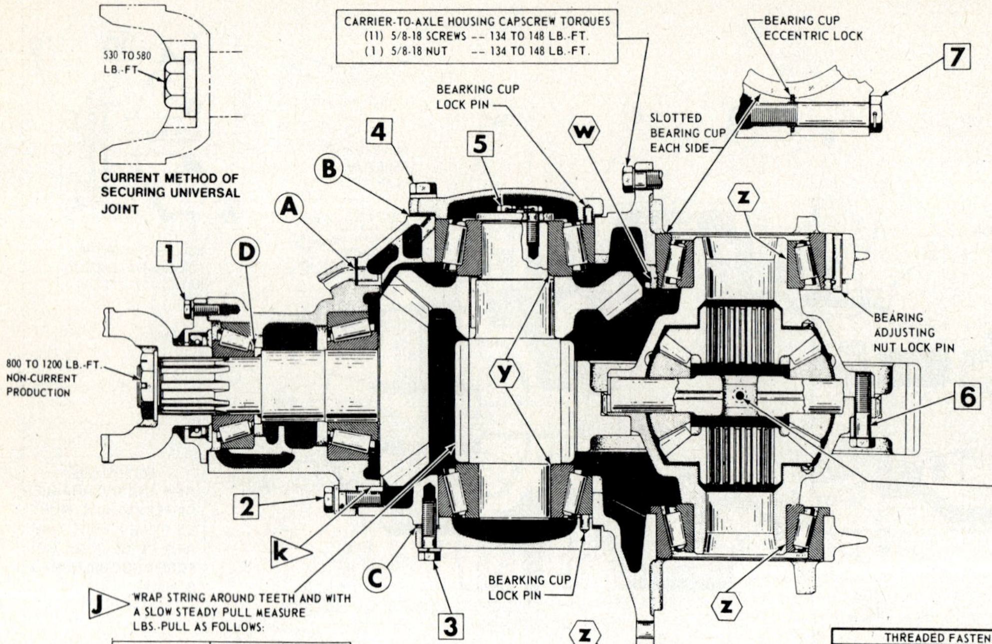

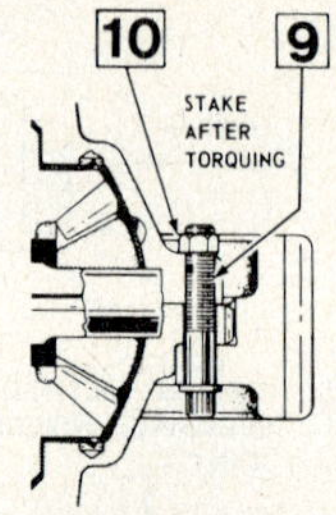

BULLGEAR MOUNTING FOR CRD-118 AND 119 CURRENT PRODUCTION

IMPORTANT
NEW AND OVERHAULED DIFFERENTIALS MUST BE FILLED WITH ONE PINT OF GEAR OIL BEFORE ASSEMBLY INTO BANJO.

J WRAP STRING AROUND TEETH AND WITH A SLOW STEADY PULL MEASURE LBS.-PULL AS FOLLOWS:

NUMBER OF GEAR TEETH	POUNDS PULL REQ'D TO TURN ASSEMBLY
12	4.8 TO 8.6 LBS. PULL
13	4.5 TO 8.2 LBS. PULL
15	4.0 TO 7.3 LBS. PULL

THREADED FASTENERS				TORQUE LB.-FT.	
LOCATION	QUANTITY	SI	GRADE		
1	(6)	7/16-14	5	30	40
2	(8)	5/8-11	8*	155	185
3	(4)	1/2-13	8*	70	90
	(1)	9/16-12	8*	100	120
4	(4)	5/8-11	8*	155	185
	(1)	5/8-11	8*	155	185
5	(3)	1/2-20	6	70	90
6	(12)	5/8-18	8*	165	195
7	(4)	7/8-9	6	380	420
8	(15)	1/2-20	5	105	115
# 9	(12)	5/8-18	8	200	210
10	(12)	5/8-18		80	100

*SELF LOCKING CAPSCREWS
#USE LOCTITE NO. 262

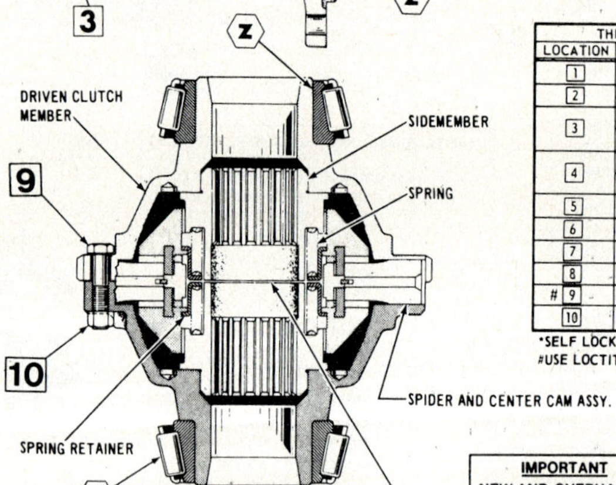

NO SPIN DIFFERENTIAL (WHEN SPECIFIED)

STEP NO. 1
PRELOAD PINION BEARINGS (X)

A - ASSEMBLE PINION AND BEARINGS IN HOUSING OMITTING BEARING COVER AND DRIVE YOKE OR FLANGE.
B - LUBRICATE BEARINGS WITH SAE-30 OIL AND PLACE ASSEMBLY ON PRESS TABLE.
C - APPLY 34,000 LB. LOAD TO BEARING CONES.
D - WRAP STRING AROUND HOUSING PILOT AT POINT K, ATTACH SCALE TO LOOPED FREE END OF STRING, AND WITH A STEADY PULL, SLOWLY ROTATE HOUSING. AT THE SAME TIME, MEASURE THE ROLLING-DRAG WHICH MUST BE WITHIN 2.8 TO 5.6 LB. PULL. IF NECESSARY TO ADJUST BEARING SPACER (D), CONSULT SPACER CHART FOR AVAILABLE SIZES. SEE (D) BELOW.
E - FINISH ASSEMBLY OF PINION AND ESTABLISH THICKNESS OF SHIM PACK (A) WITH PINION SETTING GAGE TOOLS. TEMPORARILY, REMOVE PINION HOUSING AND SHIMS; THEN PROCEED WITH NEXT STEP.

IMPORTANT
NEW AND OVERHAULED DIFFERENTIALS MUST BE FILLED WITH ONE PINT OF GEAR OIL BEFORE ASSEMBLY INTO BANJO.

(A) PINION MOUNTING ADJUSTMENT SHIMS
51 KH 36 P20 (.020)
51 KH 36 P22 (.022)
51 KH 36 P25 (.025)
51 KH 36 P31 (.031)
USE AS REQUIRED

STEP NO. 2
PRELOAD HELICAL PINION SHAFT BEARINGS (Y)

A - ASSEMBLE HELICAL PINION COMPLETELY AND INSTALL IN CARRIER HOUSING. LUBRICATE BEARINGS WITH SAE-30 OIL.
B - PRELOAD BEARINGS BY ADJUSTING SHIM PACKS (B) AND (C). SEE CHART FOR ROLLING-DRAG REQUIRED PER RATIO AT J.
C - ALSO ADJUST (B) AND (C) TO OBTAIN CORRECT GEAR TOOTH CONTACT.
D - REMOVE BEVEL PINION ASSEMBLY FOR NEXT STEP.

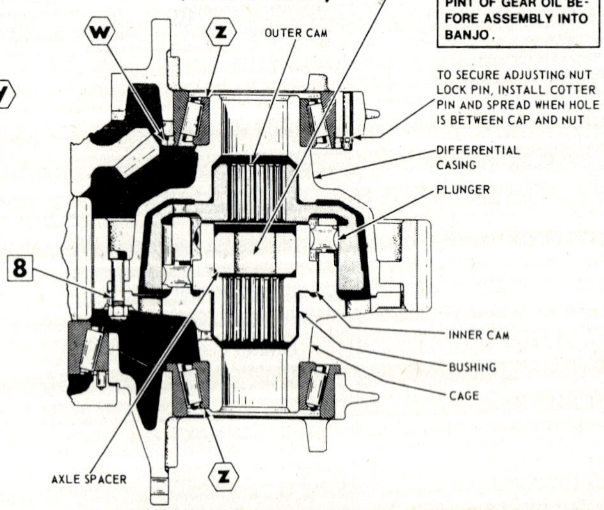

MACK INTER-WHEEL POWER DIVIDER DIFFERENTIAL PROVIDED IN CRD-1181 & CRD-1191 CARRIERS

(B) GEAR MOUNTING ADJUSTMENT SHIMS
51 KH 238 P10 (.010)
51 KH 238 P12 (.012)
51 KH 238 P15 (.015)
51 KH 238 P31 (.031)
USE AS REQUIRED

(C) GEAR MOUNTING ADJUSTMENT SHIMS
51 KH 254 P10 (.010)
51 KH 254 P12 (.012)
51 KH 254 P15 (.015)
51 KH 254 P31 (.031)
USE AS REQUIRED

STEP NO. 3
PRELOAD DIFFERENTIAL BEARINGS (Z)

A - LOOSEN HELICAL PINION BEARING COVER CAPSCREWS TO RELIEVE PRELOAD.
B - ASSEMBLE DIFFERENTIAL COMPLETELY AND INSTALL.
C - LUBRICATE BEARINGS WITH SAE-30 OIL AND INSTALL FIXTURE J26439 ON BEARING CAP LUGS.
D - WITH HOOK IN BULL GEAR, SLOWLY ROTATE ASSEMBLY WITH SCALE ATTACHED AT 12 IN. RADIUS. TIGHTEN NUT (W) UNTIL SCALE READS 2 TO 3 LB. PULL.
E - REMOVE TOOLS AND COMPLETE ASSEMBLY OF UNIT.

(D) TWO-IN-COMBINATION BEARING CONE SPACERS ARE AVAILABLE TO SERVICE SINGLE SPACER 52KH171. PLEASE REFER TO MASTER MANUAL PAGE 13.7

Fig. 3 Double reduction carrier. Models CRD-118 & CRD-119

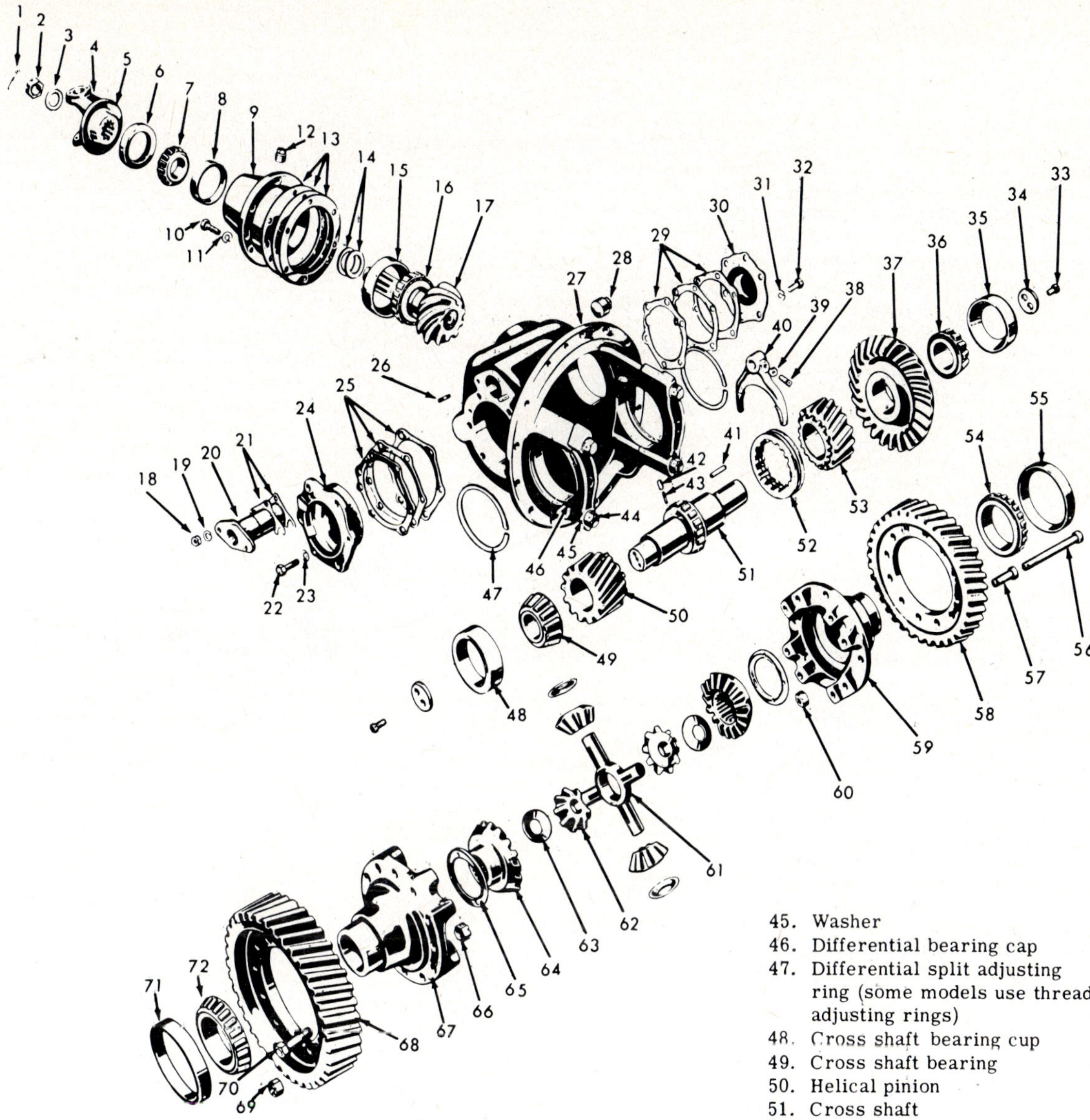

1. Cotter pin
2. Yoke nut
3. Washer
4. Yoke
5. Slinger
6. Pinion oil seal
7. Drive pinion outer bearing
8. Drive pinion outer bearing cup
9. Pinion housing
10. Capscrew
11. Lockwasher
12. Oil filler plug
13. Drive pinion shims
14. Drive pinion bearing spacer
15. Drive pinion inner bearing cup
16. Drive pinion inner bearing
17. Drive pinion
18. Nut
19. Lockwasher
20. Shift shaft sleeve
21. Sleeve shims
22. Capscrew
23. Lockwasher
24. Cross shaft bearing retainer
25. Shims
26. Sleeve stud
27. Differential carrier housing
28. Oil filler plug
29. Shims
30. Cross shaft bearing cover
31. Lockwasher
32. Capscrew
33. Bearing retainer capscrew
34. Bearing clamp plate
35. Cross shaft bearing cup
36. Cross shaft bearing
37. Drive gear
38. Shift fork setscrew
39. Shift fork setscrew lock nut
40. Shift fork
41. Hypoid gear key
42. Poppet
43. Spring
44. Differential bearing cap capscrew
45. Washer
46. Differential bearing cap
47. Differential split adjusting ring (some models use threaded adjusting rings)
48. Cross shaft bearing cup
49. Cross shaft bearing
50. Helical pinion
51. Cross shaft
52. Cross shaft shift collar
53. Helical pinion
54. Differential bearing
55. Differential bearing cup
56. Differential case bolt
57. Gear-to-case bolt
58. Helical gear
59. Differential case half
60. Gear-to-case bolt nut
61. Differential spider
62. Differential pinion
63. Differential pinion thrust washer
64. Differential pinion side gear
65. Differential pinion side gear thrust washer
66. Gear-to-case bolt nut
67. Differential case half
68. Helical gear
69. Differential case bolt nut
70. Gear-to-case bolt
71. Differential bearing cup
72. Differential bearing

Fig. 4 Exploded view of two speed, double reduction carrier

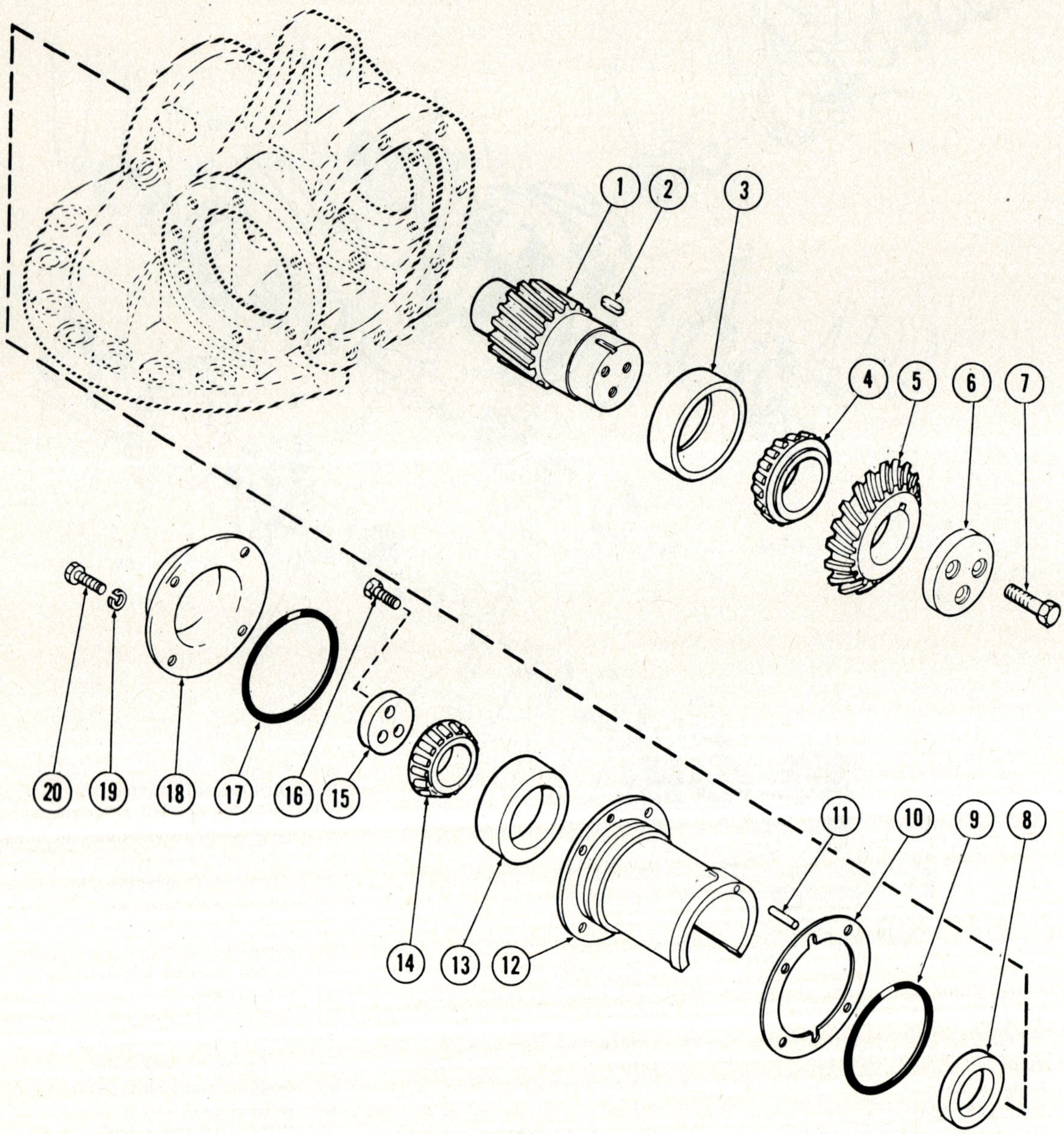

1. Pinion Shaft	6. Clamp Plate	11. Pin	16. Capscrew
2. Key	7. Capscrew	12. Housing	17. Gasket
3. Cup	8. Spacer	13. Cup	18. Cover
4. Cone	9. 'O' Ring	14. Cone	19. Lockwasher
5. Pinion Gear	10. Shims	15. Clamp Plate	20. Capscrew

Exploded view of helical or spur pinion shaft

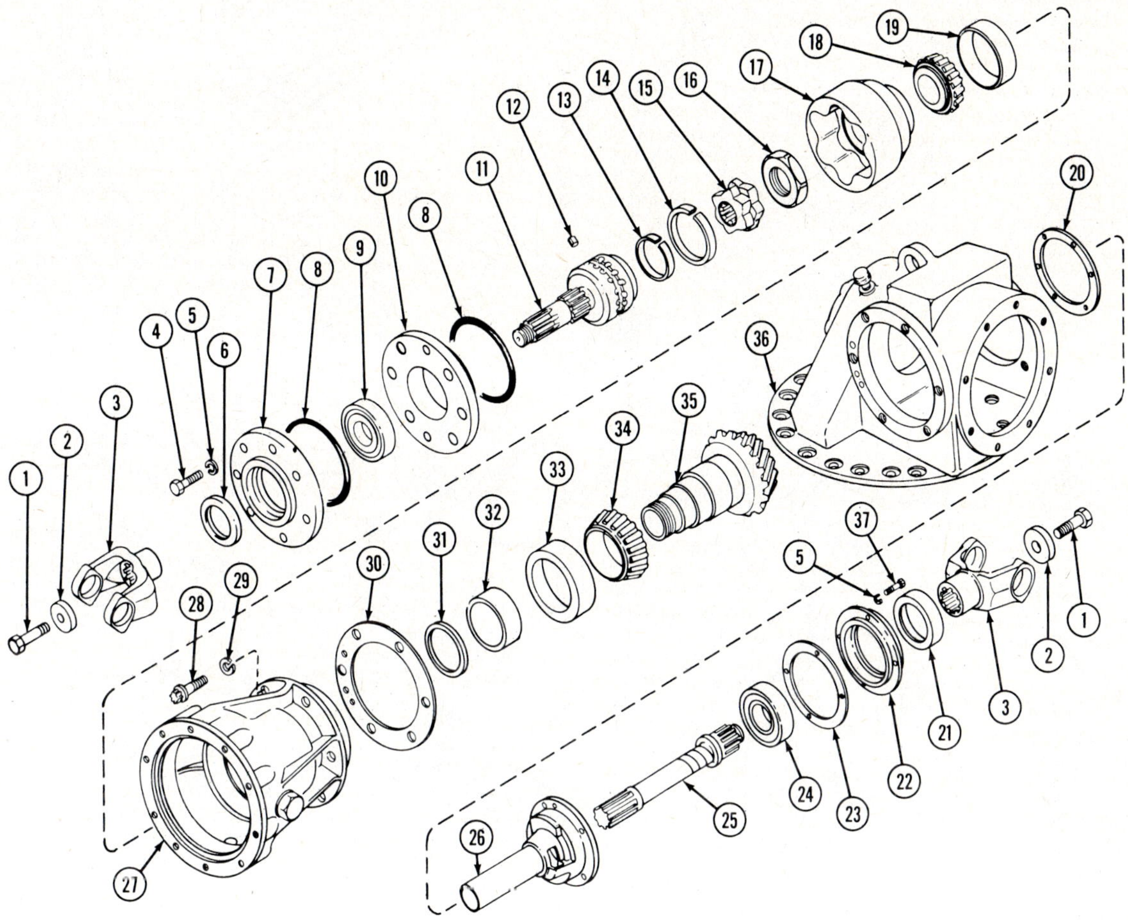

1. Capscrew
2. Clamp Plate
3. Yoke
4. Capscrew
5. Lockwasher
6. Oil Seal
7. Cover
8. "O" Ring
9. Bearing
10. Retainer
11. Cage
12. Wedge
13. Inner Ring
14. Outer Ring
15. Cam
16. Nut
17. Outer Cam
18. Cone
19. Cup
20. Gasket
21. Seal
22. Cover
23. Gasket
24. Bearing
25. Through Shaft
26. Tube
27. Housing
28. Capscrew
29. Lockwasher
30. Shim
31. Spacer
32. Spacer
33. Cup
34. Cone
35. Bevel Pinion
36. Carrier Housing
37. Capscrew

Exploded view of bevel pinion assembly with power divider/less air actuated control

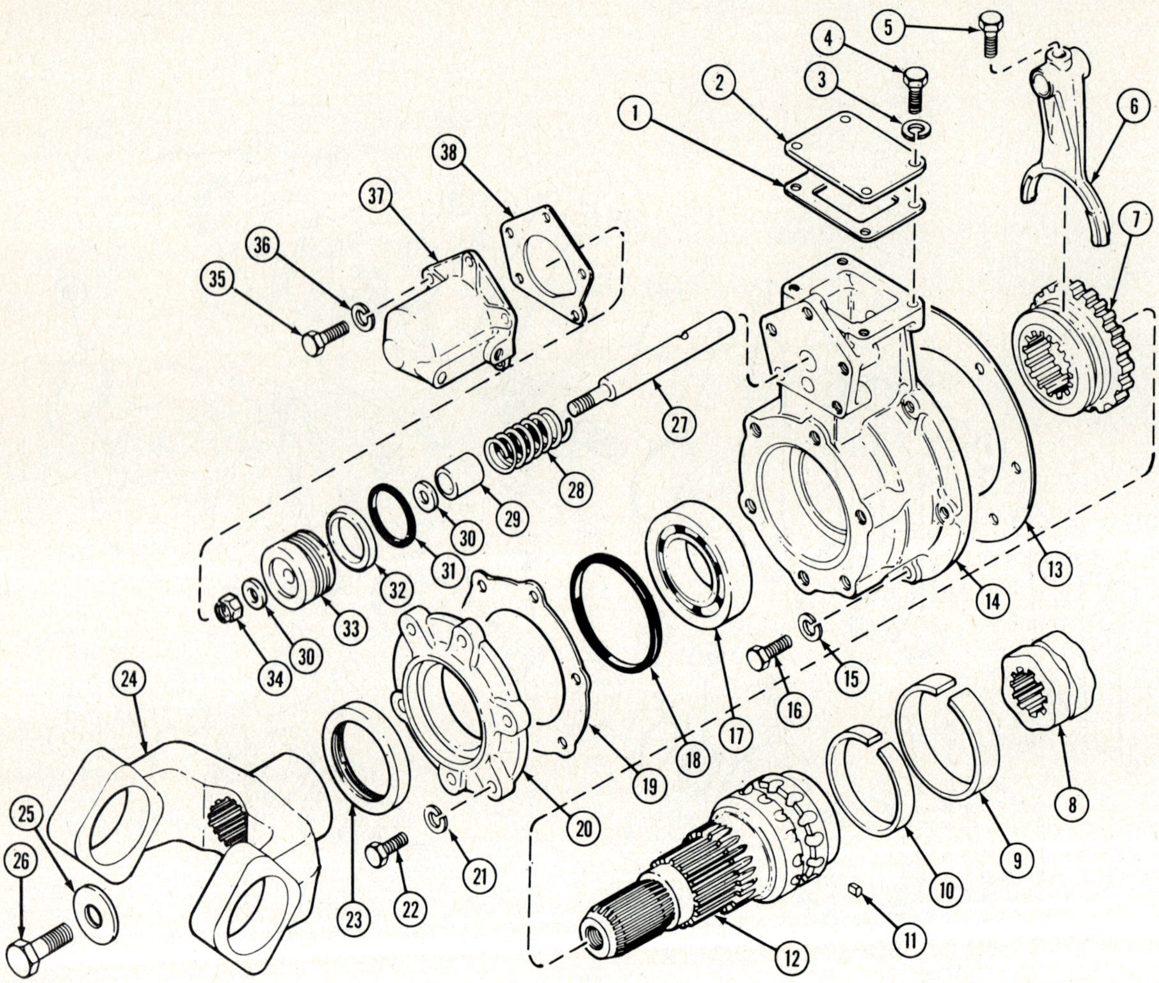

1. Gasket
2. Cover
3. Capscrew
4. Lockwasher
5. Capscrew
6. Fork
7. Clutch
8. Cam
9. Outer Ring
10. Inner Ring
11. Wedge
12. Cage
13. Gasket
14. Housing
15. Lockwasher
16. Capscrew
17. Bearing
18. "O" Ring
19. Shim
20. Cover
21. Lockwasher
22. Capscrew
23. Oil Seal
24. Yoke
25. Clamp Plate
26. Capscrew
27. Shaft
28. Spring
29. Sleeve
30. Spacer
31. Felt
32. "O" Ring
33. Piston
34. Nut
35. Capscrew
36. Lockwasher
37. Housing
38. Gasket

Exploded view of air actuated power divider lockout

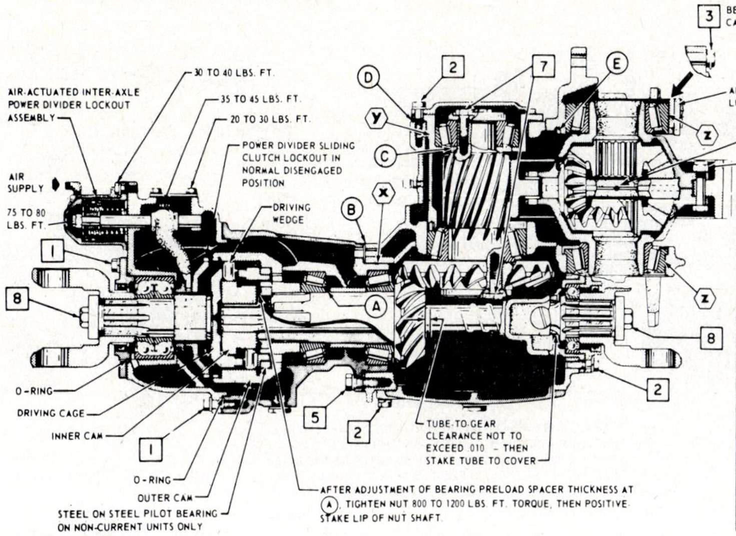

IMPORTANT

NEW AND OVERHAULED DIFFERENTIALS MUST BE FILLED WITH ONE PINT OF GEAR OIL BEFORE ASSEMBLY INTO BANJO.

TORQUE LIMITS			
LBS.-FT. FOR CRD-95/96 SERIES			
FASTENER TYPE	GRADE	TORQUE	
SCREW 1	5	65	75
SCREW 2	6	85	95
SCREW 3	6	440	485
SCREW 4	8	140	150
SCREW 5	6	165	185
BOLT 6	8	95	105
SCREW 7	8	190	210
SCREW 8		530	580

CRDP-95 FORWARD BOGIE CARRIER WITH INTER-AXLE POWER DIVIDER AIR-ACTUATED LOCKOUT

MACK INTER-WHEEL POWER DIVIDER DIFFERENTIAL (AVAILABLE WHEN SPECIFIED)

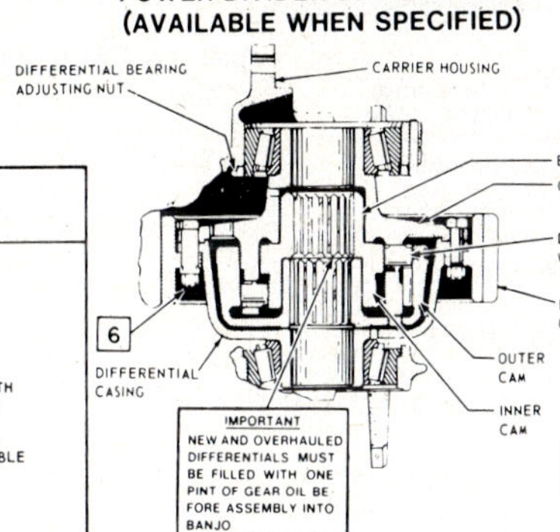

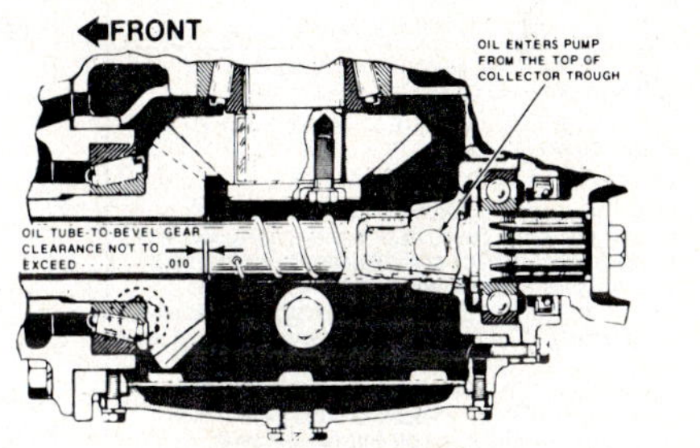

LUBRICATION SYSTEM

STEP NO. 1.
PRELOAD PINION BEARINGS (X)

A - ASSEMBLE PINION AND BEARINGS IN HOUSING OMITTING BEARING COVER AND DRIVE YOKE OR FLANGE.
B - LUBRICATE BEARINGS WITH SAE-30 OIL AND PLACE ASSEMBLY ON PRESS TABLE.
C - APPLY 34,000 LB. LOAD TO BEARING CONES.
D - WRAP STRING AROUND HOUSING PILOT AT POINT (X) ATTACH SCALE TO LOOPED FREE END OF STRING, AND WITH A STEADY PULL, SLOWLY ROTATE HOUSING. AT THE SAME TIME, MEASURE THE ROLLING-DRAG WHICH MUST BE WITHIN 3 TO 6 LB. PULL. IF NECESSARY TO ADJUST BEARING SPACER (A), CONSULT SPACER CHART FOR AVAILABLE SIZES.
E - FINISH ASSEMBLY OF PINION AND ESTABLISH THICKNESS OF SHIM PACK (B) WITH PINION SETTING GAGE TOOLS. TEMPORARILY, REMOVE PINION HOUSING AND SHIMS; THEN PROCEED WITH NEXT STEP.

AFTER ADJUSTMENT OF BEARING PRELOAD SPACER AT (A), TIGHTEN NUT 800 TO 1200 LBS. FT. TORQUE, THEN POSITIVE STAKE LIP OF NUT TO SHAFT

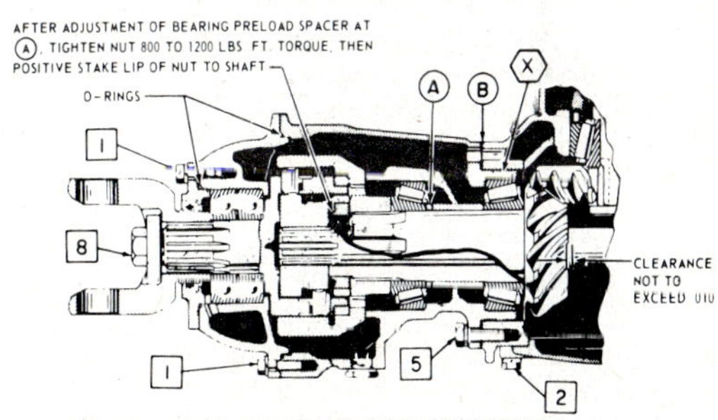

CRDP-95 FORWARD BOGIE CARRIER WITHOUT THE INTER-AXLE LOCKOUT ARRANGEMENT

STEP NO. 2
PRELOAD HELICAL PINION SHAFT BEARINGS (Y)

A - ASSEMBLE HELICAL PINION WITH BEARINGS IN BEARING RETAINER OMITTING CLAMP PLATES AND BEVEL GEAR.
B - LUBRICATE BEARINGS WITH SAE-30 OIL AND POSITION ASSEMBLY ON PRESS TABLE.
C - APPLY 34,000 LB. LOAD TO BEARING CONES.
D - WITH STRING WRAPPED AROUND RETAINER PILOT AT POINT (Y) ADJUST THICKNESS OF SPACER (C) UNTIL A ROLLING-DRAG OF 3 TO 6 LB. PULL IS OBTAINED.
E - PRESS BEVEL GEAR ON PINION SHAFT. INSTALL RETAINER AND PINION ASSEMBLY IN CARRIER HOUSING AND ADJUST SHIM PACKS (B) AND (D) AS REQUIRED TO OBTAIN CORRECT GEAR TOOTH CONTACT.
F - REMOVE BEVEL PINION FROM CARRIER HOUSING FOR NEXT STEP.

(A) TWO-IN-COMBINATION BEARING CONE SPACERS ARE AVAILABLE TO SERVICE SINGLE SPACER. PLEASE REFER TO CHART NO. 2 IN PINION BEARING PRELOAD SPACER WRITE-UP, MASTER MANUAL PAGE 13.7.

(B) PINION MOUNTING ADJUSTMENT SHIMS
51 KH 34 P20 (0.020)
51 KH 34 P22 (0.022)
51 KH 34 P25 (0.025)
51 KH 34 P31 (0.031)
USE AS REQUIRED

STEP NO. 3
PRELOAD DIFFERENTIAL BEARINGS (Z)

A - LOOSEN HELICAL PINION R.H. BEARING CLAMP PLATE CAPSCREWS AND WITH A PLASTIC MALLET, RAP ON CAPSCREWS LIGHTLY TO RELIEVE THE BEARING PRELOAD.
B - ASSEMBLE DIFFERENTIAL COMPLETELY AND INSTALL IN CARRIER HOUSING WITH PEDESTAL CAPSCREWS FINGER-TIGHT.
C - LUBRICATE BEARINGS (Z) WITH SAE-30 OIL, THEN INSTALL DIFFERENTIAL BEARING ADJUSTMENT FIXTURE J26441 ON BEARING CAP LUGS.
D - WITH CALIBRATED HOOK IN BULL GEAR, SLOWLY ROTATE ASSEMBLY WITH SCALE ATTACHED AT THE 12 IN. RADIUS MARK TIGHTEN ADJUSTING NUT (E) UNTIL A ROLLING-DRAG OF 2 TO 3 LB. PULL IS REQUIRED TO ROTATE THE ASSEMBLY.
E - TIGHTEN PEDESTAL BEARING CAPSCREWS (3) TO THE SPECIFIED TORQUE. THEN RECHECK ROLLING-DRAG.
F - IF CORRECT, REMOVE ADJUSTMENT FIXTURE AND COMPLETE THE ASSEMBLY OF THE CARRIER.

(C) TWO-IN-COMBINATION BEARING CONE SPACERS ARE AVAILABLE TO SERVICE SINGLE SPACER. PLEASE REFER TO CHART NO. 2 IN PINION BEARING PRELOAD SPACER WRITE-UP, MASTER MANUAL PAGE 13.7.

(D) GEAR MOUNTING ADJUSTMENT SHIMS
51 KH 232 P20 (0.020)
51 KH 232 P22 (0.022)
51 KH 232 P25 (0.025)
51 KH 232 P31 (0.031)
USE AS REQUIRED

(F) SPACER CHART
52 KH 163 - P583 (.583)
THROUGH
52 KH 163 - P683 (.683)
USE ONE AS REQUIRED

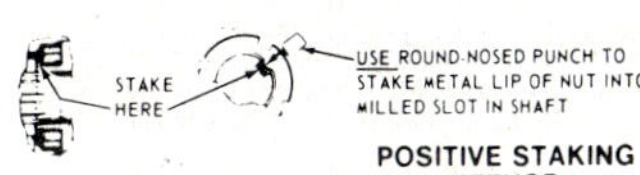

POSITIVE STAKING METHOD

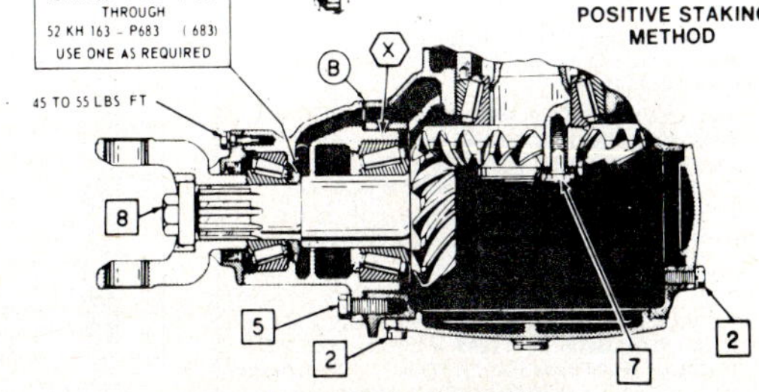

CRD-96 REARWARD BOGIE CARRIER

Bearing adjustment & screw torque chart. Current production CRD-95 & 96 top-mounted, dual-reduction bogie carriers

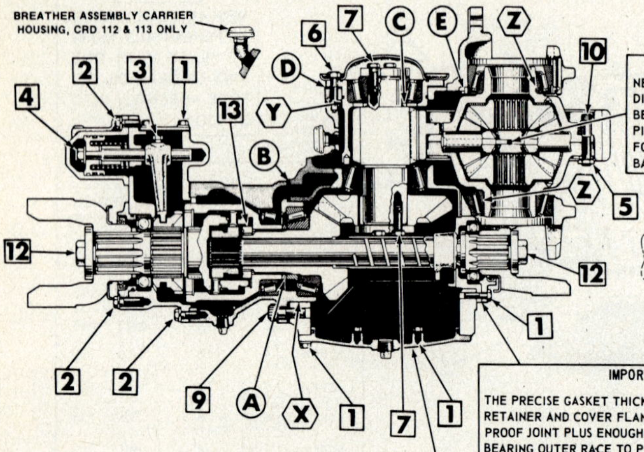

IMPORTANT
NEW AND OVERHAULED DIFFERENTIALS MUST BE FILLED WITH ONE PINT OF GEAR OIL BEFORE ASSEMBLY INTO BANJO.

IMPORTANT NOTE (F)
THE PRECISE GASKET THICKNESS REQUIRED BETWEEN THE RETAINER AND COVER FLANGES, MUST PROVIDE A LEAK-PROOF JOINT PLUS ENOUGH CLAMPING FORCE ON THE BALL BEARING OUTER RACE TO PREVENT ITS ROTATION OR CREEP IN THE BEARING RETAINER. SEE MANUAL TEXT FOR DETAILED INSTRUCTIONS.

CRDPC-92, 112 FORWARD BOGIE CARRIER WITH INTER-AXLE POWER DIVIDER AIR-ACTUATED LOCKOUT

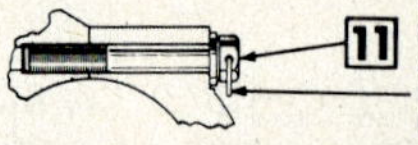

DIFFERENTIAL BEARING CAP

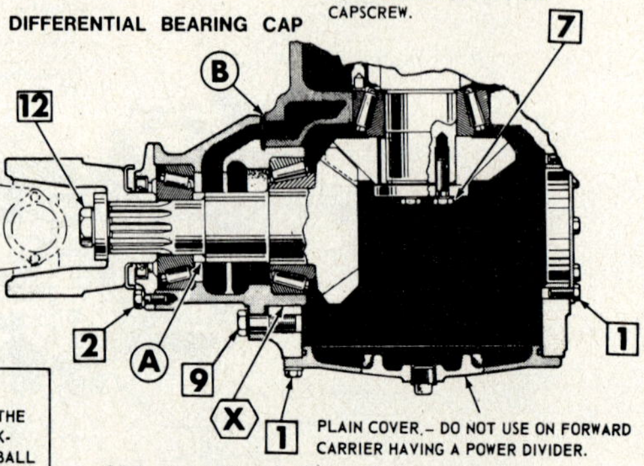

CRD - 93 OR 113 REAR BOGIE CARRIER

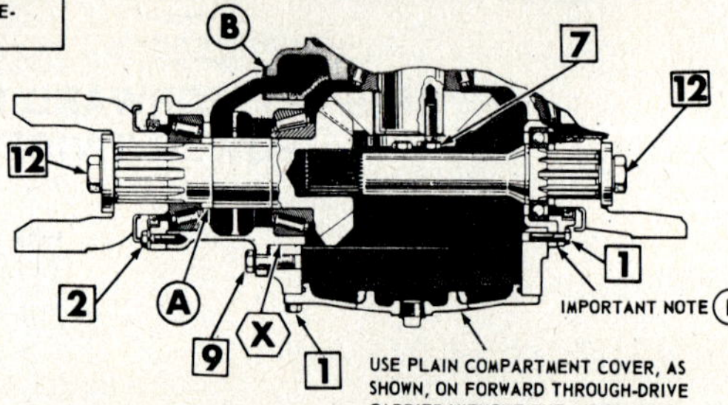

CRD - 92 OR 112 THROUGH-DRIVE FORWARD BOGIE CARRIER WITHOUT INTER-AXLE POWER DIVIDER

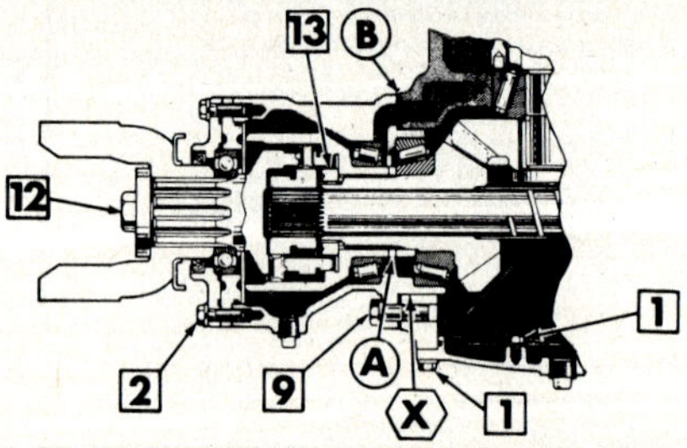

CRDPC-92 OR 112 FORWARD BOGIE CARRIER WITH INTER-AXLE POWER DIVIDER WITHOUT AIR-ACTUATED LOCKOUT

(X) STEP NO. 1 — PRELOAD PINION BEARINGS

A – ASSEMBLE PINION AND BEARINGS IN HOUSING OMITTING BEARING COVER AND DRIVE YOKE OR FLANGE.
B – LUBRICATE BEARINGS WITH SAE-30 OIL AND PLACE ASSEMBLY ON PRESS TABLE.
C – APPLY 34,000 LB. LOAD TO BEARING CONES.
D – WRAP STRING AROUND HOUSING PILOT AT POINT (X) ATTACH SCALE TO LOOPED FREE END OF STRING, AND WITH A STEADY PULL, SLOWLY ROTATE HOUSING. AT THE SAME TIME, MEASURE THE ROLLING - DRAG WHICH MUST BE WITHIN 3 TO 6 LB. PULL. IF NECESSARY TO ADJUST BEARING SPACER (A), CONSULT SPACER CHART FOR AVAILABLE SIZES.
E – FINISH ASSEMBLY OF PINION AND ESTABLISH THICKNESS OF SHIM PACK (B) WITH PINION SETTING GAGE TOOLS. TEMPORARILY, REMOVE PINION HOUSING AND SHIMS; THEN PROCEED WITH NEXT STEP.

(Y) STEP NO. 2 — PRELOAD HELICAL PINION SHAFT BEARINGS

A – ASSEMBLE HELICAL PINION WITH BEARINGS IN BEARING RETAINER OMITTING CLAMP PLATES AND BEVEL GEAR.
B – LUBRICATE BEARINGS WITH SAE-30 OIL AND POSITION ASSEMBLY ON PRESS TABLE.
C – APPLY 34,000 LB. LOAD TO BEARING CONES.
D – WITH STRING WRAPPED AROUND RETAINER PILOT AT POINT (Y) ADJUST THICKNESS OF SPACER (C) UNTIL A ROLLING-DRAG OF 3 TO 6 LB. PULL IS OBTAINED.
E – PRESS BEVEL GEAR ON PINION SHAFT. INSTALL RETAINER AND PINION ASSEMBLY IN CARRIER HOUSING. ALSO INSTALL BEVEL PINION IN CARRIER HOUSING AND ADJUST SHIM PACKS (B) AND (D) AS REQUIRED TO OBTAIN CORRECT GEAR TOOTH CONTACT.
F – REMOVE BEVEL PINION AND HELICAL PINION FROM CARRIER HOUSING FOR NEXT STEP.

(Z) STEP NO. 3 — PRELOAD DIFFERENTIAL BEARINGS

A – ASSEMBLE DIFFERENTIAL COMPLETELY AND INSTALL IN CARRIER HOUSING WITH PEDESTAL CAPSCREWS FINGER-TIGHT.
B – LUBRICATE BEARINGS (Z) WITH SAE-30 OIL, THEN INSTALL DIFFERENTIAL BEARING ADJUSTMENT FIXTURE J26440 ON BEARING CAP LUGS.
C – WITH CALIBRATED HOOK IN BULL GEAR, SLOWLY ROTATE ASSEMBLY WITH SCALE ATTACHED AT THE 12 IN. RADIUS MARK. TIGHTEN ADJUSTING NUT (E) UNTIL A ROLLING-DRAG OF 2 TO 3 LB. PULL IS REQUIRED TO ROTATE THE ASSEMBLY.
D – TIGHTEN PEDESTAL BEARING CAPSCREWS [11] TO THE SPECIFIED TORQUE. THEN RECHECK ROLLING-DRAG.
E – IF CORRECT, REMOVE ADJUSTMENT FIXTURE AND COMPLETE THE ASSEMBLY OF THE CARRIER.
F – TEMPORARILY INSTALL HELICAL OR SPUR PINION SHAFT. CHECK TOOTH CONTACT PATTERN. IF CORRECT COMPLETE HELICAL OR SPUR PINION SHAFT ASSEMBLY, INSTALL BEVEL PINION ASSEMBLY AND COMPLETE FINAL ASSEMBLY.

(A) TWO-IN-COMBINATION BEARING CONE SPACERS ARE AVAILABLE TO SERVICE SINGLE SPACER. PLEASE REFER TO TABLES IN BEVEL PINION REASSEMBLY.

(B) GEAR MOUNTING ADJUSTMENT SHIMS

51 KH 263-P10 (0.010)
51 KH 263-P12 (0.012)
51 KH 263-P15 (0.015)
51 KH 263-P31 (0.031)
USE AS REQUIRED

(C) TWO-IN-COMBINATION BEARING CONE SPACERS ARE AVAILABLE TO SERVICE SINGLE SPACER. PLEASE REFER TO TABLES IN HELICAL PINION REASSEMBLY.

(D) PINION MOUNTING ADJUSTMENT SHIMS

51 KH 264-P10 (0.010)
51 KH 264-P12 (0.012)
51 KH 264-P15 (0.015)
51 KH 264-P31 (0.031)
USE AS REQUIRED

TORQUE LIMIT CHART

FASTENER			TORQUE IN LB.FT.	
LOCATION	SIZE	GRADE	LUBRICATED	
1	3/8 - 16UNC	5	18	28
2	7/16 - 14UNC	5	30	40
3	7/16 - 20UNF		35	45
4	1/2 - 20UNF		75	80
5	1/2 - 20UNF		80	100
6	1/2 - 13UNC	5	45	65
7	1/2 - 20UNF	5	55	75
8	9/16 - 18UNF	6	100	120
9	5/8 - 11UNC	8	155	185
10	5/8 - 18UNF	8	200	210
11	3/4 - 10UNC	8	265	295
12	7/8 - 14UNF	8	480	520
13*	2-3/8 - 12UN		1000	1300

* AFTER ADJUSTMENT OF BEARING PRELOAD SPACER THICKNESS AT (A) TIGHTEN NUT 1000 TO 1300 LBS. FT. TORQUE. USE LOCTITE 262 ON THREADS OF BEVEL PINION GEAR.

IMPORTANT
NEW AND OVERHAULED DIFFERENTIALS MUST BE FILLED WITH ONE PINT OF GEAR OIL BEFORE ASSEMBLY INTO BANJO.

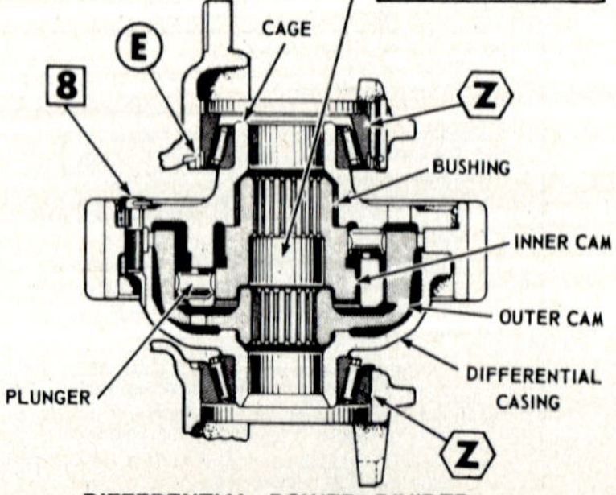

DIFFERENTIAL POWER DIVIDER
CRD - 921, 931, 1121, 1131

Bearing adjustment & screw torque chart. Current production CRD-92, 93, 112 & 113 top-mounted, dual-reduction bogie carriers

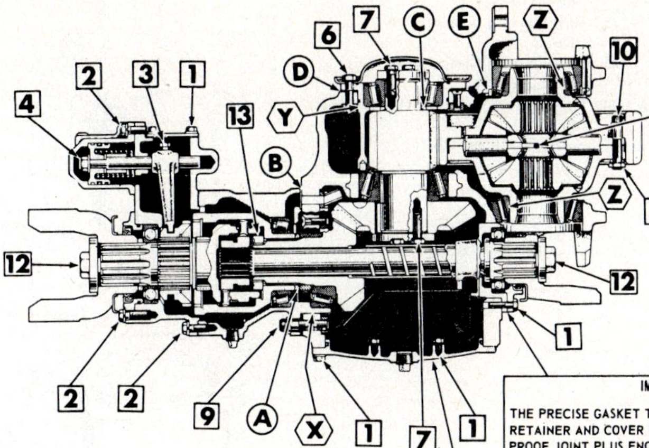

IMPORTANT
NEW AND OVERHAULED DIFFERENTIALS MUST BE FILLED WITH ONE PINT OF GEAR OIL BEFORE ASSEMBLY INTO BANJO.

IMPORTANT NOTE (F)
THE PRECISE GASKET THICKNESS REQUIRED BETWEEN THE RETAINER AND COVER FLANGES, MUST PROVIDE A LEAK-PROOF JOINT PLUS ENOUGH CLAMPING FORCE ON THE BALL BEARING OUTER RACE TO PREVENT ITS ROTATION OR CREEP IN THE BEARING RETAINER. SEE MANUAL TEXT FOR DETAILED INSTRUCTIONS.

CRDL P92,-112 FORWARD BOGIE CARRIER WITH INTER-AXLE POWER DIVIDER AIR-ACTUATED LOCKOUT

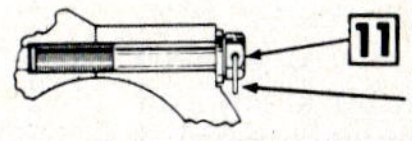

LOCKWIRE USED ON POWER DIVIDER DIFFERENTIAL ONLY. ALL OTHER ASSEMBLIES REQUIRE THE USE OF LOCTITE 262 ON THREADS OF DIFFERENTIAL BEARING CAP CAPSCREW.

DIFFERENTIAL BEARING CAP

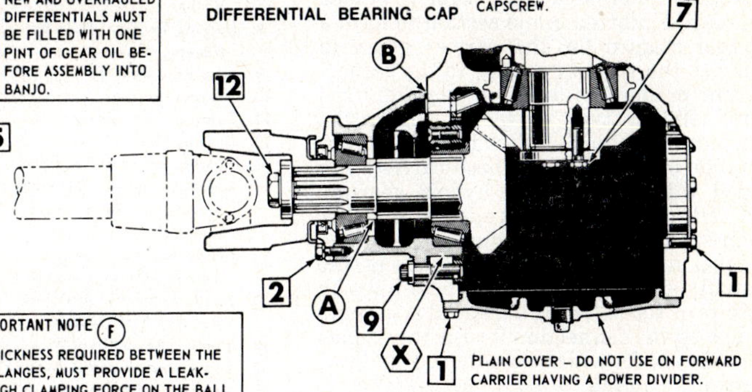

CRDL 93 OR 113 REAR BOGIE CARRIER

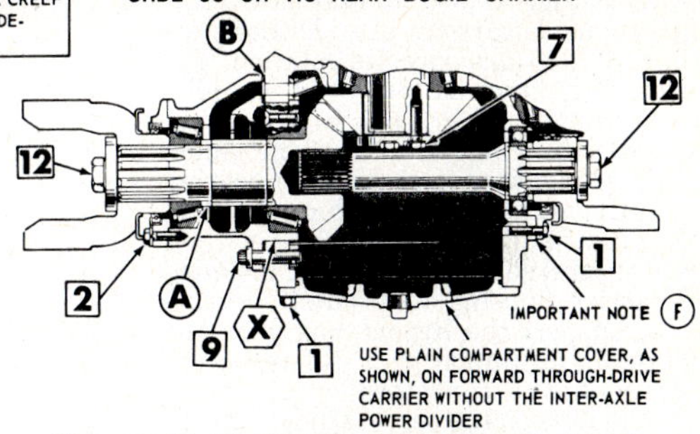

CRDL 92 OR 112 THROUGH-DRIVE FORWARD BOGIE CARRIER WITHOUT INTER-AXLE POWER DIVIDER

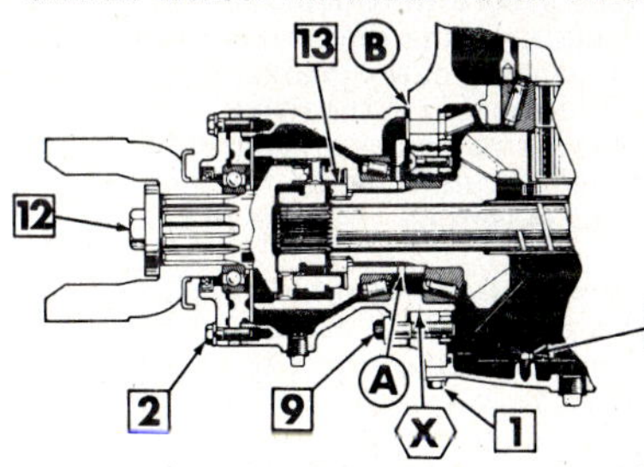

CRDLP 92 OR 112 FORWARD BOGIE CARRIER WITH INTER-AXLE POWER DIVIDER WITHOUT AIR-ACTUATED LOCKOUT

(X) STEP NO. 1 — PRELOAD PINION BEARINGS

A – ASSEMBLE PINION AND BEARINGS IN HOUSING OMITTING BEARING COVER AND DRIVE YOKE OR FLANGE.

B – LUBRICATE BEARINGS WITH SAE-30 OIL AND PLACE ASSEMBLY ON PRESS TABLE.

C – APPLY 34,000 LB. LOAD TO BEARING CONES.

D – WRAP STRING AROUND HOUSING PILOT AT POINT (X) ATTACH SCALE TO LOOPED FREE END OF STRING, AND WITH A STEADY PULL, SLOWLY ROTATE HOUSING. AT THE SAME TIME, MEASURE THE ROLLING - DRAG WHICH MUST BE WITHIN 3 TO 6 LB. PULL. IF NECESSARY TO ADJUST BEARING SPACER (A), CONSULT SPACER CHART FOR AVAILABLE SIZES.

E – FINISH ASSEMBLY OF PINION AND ESTABLISH THICKNESS OF SHIM PACK (B) WITH PINION SETTING GAGE TOOLS. TEMPORARILY, REMOVE PINION HOUSING AND SHIMS; THEN PROCEED WITH NEXT STEP.

(A) TWO-IN-COMBINATION BEARING CONE SPACERS ARE AVAILABLE TO SERVICE SINGLE SPACER. PLEASE REFER TO TABLES IN BEVEL PINION REASSEMBLY.

(B) GEAR MOUNTING ADJUSTMENT SHIMS

51 KH 263-P10 (0.010)
51 KH 263-P12 (0.012)
51 KH 263-P15 (0.015)
51 KH 263-P31 (0.031)
USE AS REQUIRED

(C) TWO-IN-COMBINATION BEARING CONE SPACERS ARE AVAILABLE TO SERVICE SINGLE SPACER. PLEASE REFER TO TABLES IN HELICAL PINION REASSEMBLY.

(Y) STEP NO. 2 — PRELOAD HELICAL PINION SHAFT BEARINGS

A – ASSEMBLE HELICAL PINION WITH BEARINGS IN BEARING RETAINER OMITTING CLAMP PLATES AND BEVEL GEAR.

B – LUBRICATE BEARINGS WITH SAE-30 OIL AND POSITION ASSEMBLY ON PRESS TABLE.

C – APPLY 34,000 LB. LOAD TO BEARING CONES.

D – WITH STRING WRAPPED AROUND RETAINER PILOT AT POINT (Y) ADJUST THICKNESS OF SPACER (C) UNTIL A ROLLING-DRAG OF 3 TO 6 LB. PULL IS OBTAINED.

E – PRESS BEVEL GEAR ON PINION SHAFT. INSTALL RETAINER AND PINION ASSEMBLY IN CARRIER HOUSING. ALSO INSTALL BEVEL PINION IN CARRIER HOUSING AND ADJUST SHIM PACKS (B) AND (D) AS REQUIRED TO OBTAIN CORRECT GEAR TOOTH CONTACT.

F – REMOVE BEVEL PINION AND HELICAL PINION FROM CARRIER HOUSING FOR NEXT STEP.

(D) PINION MOUNTING ADJUSTMENT SHIMS

51 KH 264-P10 (0.010)
51 KH 264-P12 (0.012)
51 KH 264-P15 (0.015)
51 KH 264-P31 (0.031)
USE AS REQUIRED

(Z) STEP NO. 3 — PRELOAD DIFFERENTIAL BEARINGS

A – ASSEMBLE DIFFERENTIAL COMPLETELY AND INSTALL IN CARRIER HOUSING WITH PEDESTAL CAPSCREWS FINGER-TIGHT.

B – LUBRICATE BEARINGS (Z) WITH SAE-30 OIL, THEN INSTALL DIFFERENTIAL BEARING ADJUSTMENT FIXTURE J26440 ON BEARING CAP LUGS.

C – WITH CALIBRATED HOOK IN BULL GEAR, SLOWLY ROTATE ASSEMBLY WITH SCALE ATTACHED AT THE 12 IN. RADIUS MARK. TIGHTEN ADJUSTING NUT (E) UNTIL A ROLLING-DRAG OF 2 TO 3 LB. PULL IS REQUIRED TO ROTATE THE ASSEMBLY.

D – TIGHTEN PEDESTAL BEARING CAPSCREWS [11] TO THE SPECIFIED TORQUE. THEN RECHECK ROLLING-DRAG.

E – IF CORRECT, REMOVE ADJUSTMENT FIXTURE AND COMPLETE THE ASSEMBLY OF THE CARRIER.

F – TEMPORARILY INSTALL HELICAL OR SPUR PINION SHAFT. CHECK TOOTH CONTACT PATTERN. IF CORRECT COMPLETE HELICAL OR SPUR PINION SHAFT ASSEMBLY, INSTALL BEVEL PINION ASSEMBLY AND COMPLETE FINAL ASSEMBLY.

TORQUE LIMIT CHART

FASTENER			TORQUE IN LB.FT.	
LOCATION	SIZE	GRADE	LUBRICATED	
1	3/8 - 16UNC	5	18	28
2	7/16 - 14UNC	5	30	40
3	7/16 - 20UNF		35	45
4	1/2 - 20UNF		75	80
5	1/2 - 20UNF		80	100
6	1/2 - 13UNC	5	45	65
7	1/2 - 20UNF	5	55	75
8	9/16 - 18UNF	6	100	120
9	5/8 - 11UNC	8	155	185
10	5/8 - 18UNF	8	200	210
11	3/4 - 10UNC	8	265	295
12	7/8 - 14UNF	8	480	520
13 *	2-3/8 - 12UN		1000	1300

* AFTER ADJUSTMENT OF BEARING PRELOAD SPACER THICKNESS AT (A). TIGHTEN NUT 1000 TO 1300 LBS. FT. TORQUE. USE LOCTITE 262 ON THREADS OF BEVEL PINION GEAR.

IMPORTANT
NEW AND OVERHAULED DIFFERENTIALS MUST BE FILLED WITH ONE PINT OF GEAR OIL BEFORE ASSEMBLY INTO BANJO.

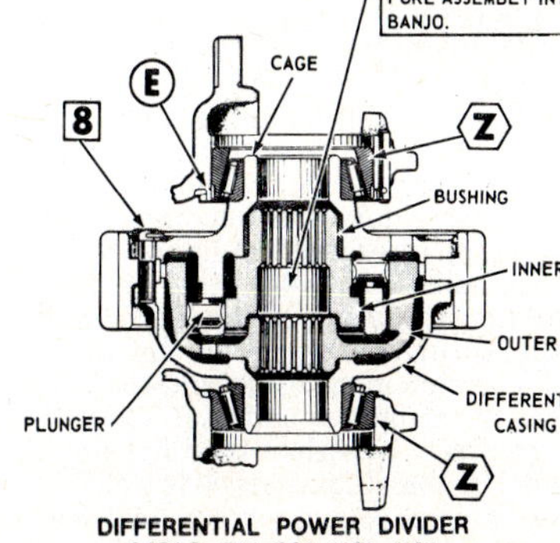

DIFFERENTIAL POWER DIVIDER CRDLP 921, 931, 1121, 1131

Bearing adjustment & screw torque chart. Current production CRDL-92, 93, 112 & 113 top-mounted, dual-reduction bogie carriers

SINGLE REDUCTION AXLE CRS-84, 840, FIG. 1

The axle housings are of the usual full-floating design. They consist of a one-piece steel housing, formed from tubular stock and heat-treated. A pressed steel cover is welded to the rear reinforced banjo section to form a leak-proof closure for the carrier gears in addition to the rigidity it adds to the banjo.

Present production axles employ the fixed eye and slipper type of spring suspension which does not require radius rods. To absorb the higher engine torque, these carriers are provided with coarse pitch hypoid gearing, closed self-lubricating differentials and four steel-on-steel differential pinions.

Integral flanges at the other ends of the axle shafts are secured to the wheel hubs by capscrews or studs. The differential side gears provide a spline connection for the inner ends of the axle shafts which employ the shallow-type involute spline.

The Mack single-reduction rear axle front-mounted (hypoid) carriers are identical, except the CRS-840 has opposite rotation. The hypoid driving pinion is carried in straddle-mounted arrangement by a pair of face-to-face tapered roller bearings in front of the pinion and by a case mounted straight roller bearing to support the rear end of the driving pinion in spigot arrangement.

A pair of tapered roller bearings, mounted in the carrier housing differential bearing pedestals, support the hypoid ring gear and conventional four-pinion gear differential in straddle-mounted arrangement.

Machined rests on the pedestal bearing caps fit against corresponding machined diameters in the axle housing bore to restrict pedestal deflection, thus maintaining the correct bearing preload and meshing of the gearset under load. In addition, deflection of the hypoid ring gear under load is restrained by an adjustable bronze thrust shoe, located behind the gear, opposite the point of contact with the hypoid pinion.

The hypoid pinion bearing preload is obtained by varying the length of the spacer between the front bearing cones. Gear tooth contact is adjusted by varying the shim pack thickness between the carrier housing and the driving pinion bearing retainer flange. Slotted nuts, which bear against the differential bearing cups, are threaded to the pedestals to preload the differential bearings and adjust the axial position of the hypoid ring gear.

LEGEND

1. Motor Cover Nylok Screw
2. Motor Cable Clip
3. Electric Motor
4. Motor Cover Gasket
5. Shift Motor Housing
6. Outside Terminal Jam Nut
7. Outside Terminal Fibre Washer
8. Outside Terminal Hole in Housing Bushing
9. Snap Ring Bearing
10. Bearing Locknut
11. Bearing Cover Gasket
12. Bearing Cover
13. Bearing Cover Nylok Screw
14. Automatic Switch, Terminal Screws, and Base Gasket Assembly
15. Inside Automatic Switch Terminal Elastic Stop Nut
16. Inside Automatic Switch-to-Housing Flat Head Screw
17. Winding Lever Pivot Pin
18. Drive Screw and Nut Assembly
19. Double-wound Helical Torsion Spring
20. Torsion Spring Lever Assembly
21. Shift Housing Compartment Cover Gasket
22. Shift Housing Compartment Cover
23. Shift Housing Compartment Oil Level Plug
24. Compartment Cover Nylok Screw
25. Seal (when required)

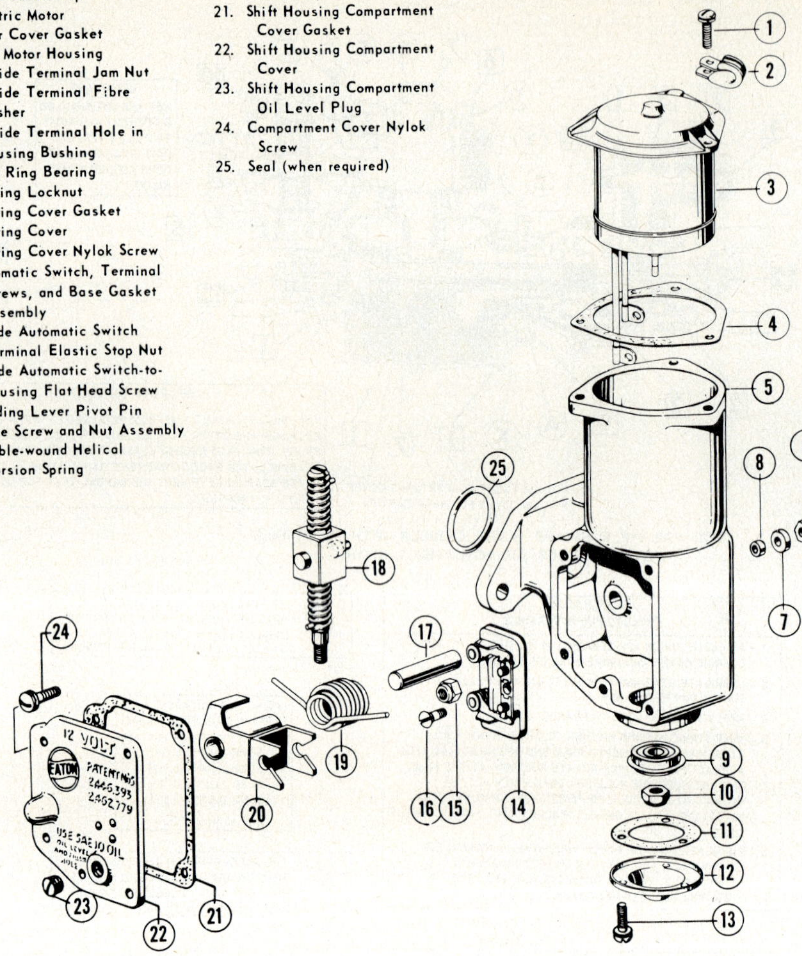

Fig. 5 Exploded view of electric shift unit

DUAL REDUCTION AXLE CRS-116, 117, 118 & 119 SERIES, FIGS. 2 & 3

Dual-reduction front-mounted rear axles provide a wide range of ratios, in the same design housing as a single reduction axle.

The extra heavy duty axle housings consist of top and bottom pressed steel halves welded together and spindles of alloy steel welded in place. The housing arms, extending slightly beyond the spring seats, are of full-corner rectangular section made of carbon steel. This design uses the strongest material where the stress is greatest. In either type housing, a pressed steel, dome-shaped cover is welded to the rear reinforced banjo section to form a leak proof closure for the carrier gears in addition to the rigidity it adds to the banjo. In all other respects, the suspensions, brakes, etc. are similar to the single reduction rear axles.

First-reduction in the carrier is provided by a durapoid bevel gear set that varies in ratio from 1.21 to 2.40 to 1.00. These gears are matched with, and drive, coarse-pitch, wide faced helical bull gear sets ranging in ratio from 3.00 to approximately 4.3; 1.00. A typical dual-reduction carrier ranges from a fast 4.25; 1.00 to a slow 9.87; 1.00.

TWO SPEED DOUBLE REDUCTION CARRIERS, FIG. 4

Each axle housing is a one-piece, high-carbon steel enclosure of rectangular shaped arms with a welded-on bowl cover at the banjo. Final drive reduction gearing for these axles is provided by two-speed, double reduction, front mounted carrier units to match the various axle housing capacities.

First reduction in the gear train is provided by a fixed-ratio hypoid gear set. Shims between the pinion and carrier housing provide for the hypoid pinion gear tooth contact adjustment with its mating hypoid ring gear which is keyed to the cross shaft.

A pair of tapered roller bearings, mounted in the carrier housing, carry the cross shaft assembly in straddle-mounted arrangement. Shims under the bearing covers provide a means of adjusting the bearing preload and hypoid gear set backlash.

Second reduction in the final gear train consists of two full-sized helical bull gear sets. One is a fast or high range ratio, the other a slow or low range ratio. The helical bull gears, rigidly bolted to the casing flanges on each side of the differential are in constant-mesh with their drive pinions carried in free-floating arrangement on the cross-shaft.

On the inner end of each helical pinion are integral, external spline clutching teeth. On the cross shaft is a row of integral splined clutching teeth. A shift fork and push rod assembly, actuated by an electric linear shift unit, moves the internally splined sliding shift collar carried on the splined portion of the cross shaft to engage the splined clutching teeth of either the high or low range helical pinion as selected by the driver.

The sliding shift collar has three parallel rows of clutching teeth. Each set or row of teeth perform only one function, either driving or locking. The middle row of narrow teeth perform only the locking function, thus elimi-

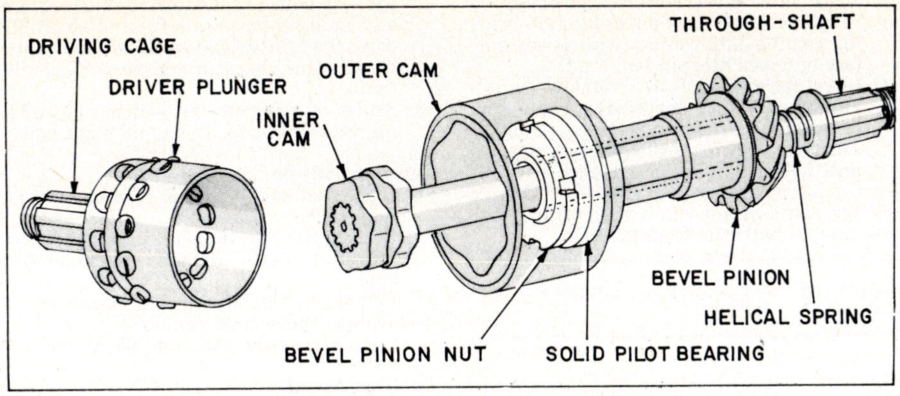

Fig. 6 Mack Inter-Axle Power Divider

nating the possibility of the collar slipping out of driving engagement during operation. The outer rows of wide-faced clutching teeth couple either the high or low range pinion to the cross shaft while the non-coupled pinion floats freely at its own speed on the cross shaft.

A third pair of tapered roller bearings, mounted in the integral pedestals, carry the entire differential and dual bull gear assembly in straddle-mounted arrangement. Machined lugs on the separable pedestal bearing caps bear against mating machined diameters in the inner bore of the banjo housing to prevent sidewise deflection of the bearing pedestals when under load. Thus, the present bearing preload is preserved and gear alignment maintained.

The engagement of either the high or low range pinion is made by the electric shifter unit which is mounted directly on the carrier housing. It is mechanically connected to the outer end of the shifter fork push rod. This provides a direct connection to the shift collar.

Two-Speed Axle Electric Shift Unit, Fig. 5

The electric shift unit is of the torsion spring type with a solid push rod and is mounted on the left side of the two-speed carrier parallel to the cross shaft. It consists of a reversible electric motor, a drive screw and nut assembly, an automatic travel-limiting switch, a bellcrank-type torsion spring winding lever and pivot pin, and a captive, double wound helical torsion spring all assembled in the shift motor housing.

Control of the electric motor and torsion spring winding mechanism is accomplished through an electrical two-way manual control switch mounted on the transmission shift lever.

TANDEM AXLES

These tandem double-reduction axle carriers each have a spiral bevel gearset for the first reduction, spur gears for second reduction and a three-pinion type differential. In addition, the forward carrier has an inter-axle power divider, Fig. 6, which consists of a driving cage, plungers, male and femal cams. The driving cage carries two rows of stagger-indexed plungers which reciprocate between the male and femal cam lobs when unequal resistance is offered by the cams and are stationary when equal resistance is offered. The femal cam drives the forward carrier and the male cam, connected through the inter-axle propeller shaft, drives the rearward carrier.

A separable housing, fastened to the front of the forward carrier by capscrews, contains the inter-axle power divider with the drive pinion and through shaft in a quill arrangement. The pinion is carried in a pair of tapered roller bearings and has a spigotted roller bearing at the forward end to maintain power divider cage and outer cam alignment.

In the rearward carrier, the drive pinion shaft is carried by a pair of tapered roller bearings in a separable bearing housing which is fastened to the front of the carrier housing.

Beyond the differences stated above, both carriers are the same. The spur pinion shaft and spiral bevel gear mounted at right angles to the bevel pinion shaft, are carried by a pair of tapered roller bearings, one mounted in the carrier housing and the other in a separable bearing retainer. Proper bevel gear mesh is obtained by varying the thickness of the shim packs between the carrier housings and the pinion gear housings or retainers.

Body-fitted bolts clamp the bull gear to the differential casing which also carries the differential pinion gears. Long body-fitted through bolts clamp the casing between the casing covers, thus forming an enclosure for the differential side and pinion gears. This assembly turns in a pair of tapered roller bearings at the outer ends of the casing covers, the bearings being mounted in the carrier pedestals. Pre-loading of these bearings is provided for by a slotted adjusting nut threaded to the inboard side of the right-hand bearing cup.

Front Driving Axles

ADJUSTMENTS

Fabco Axle

Wheel Bearings, Adjust

1. Raise and support front of vehicle.
2. Remove front wheels and brake drums.
3. Remove outer axle shaft stud nuts and lockwashers.
4. Remove bolt and washer from center of outer drive yoke, then remove the outer axle shaft.
5. Remove wheel bearing lock nut and washer from spindle, then loosen bearing adjusting nut.
6. Tighten adjusting nut while rotating hub to make sure bearings are properly seated.
7. Back nut off 1/2 turn, then torque to 50 ft. lbs. and back off 1/4 turn.
8. Install wheel bearing lockwasher over adjusting nut dowel with tang in keyway.
9. Install and tighten the wheel bearing lock nut. Spin hub and rotor assembly to make sure that bearings are not too loose or too tight.
10. Measure end play. End play should be .001–.010 inch.

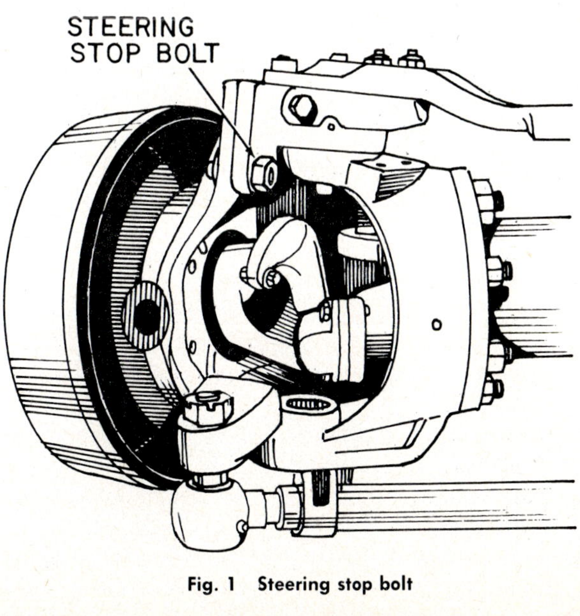

Fig. 1 Steering stop bolt

11. Install a new outer axle shaft flange gasket, then insert shaft into spindle bore. Align shaft splines with outer drive yoke splines and the shaft flange with drive studs.
12. Back up drive yoke with a suitable weight, then tap axle shaft flange end to drive tip of shaft into the split ring.
13. Apply suitable gasket compound to outer portion of one side of yoke retaining washer, then install yoke retaining bolt through sealing washer and the clean side of retaining washer. Insert assembly into center of drive yoke.
14. Apply pressure to outer drive yoke and measure clearance under axle shaft flange. Clearance should measure no less than .015 inch.
15. Install axle shaft retaining nuts and lock washers.
16. On oil lubricated wheel bearings, turn wheel hub to position oil plug in outer shaft horizontally, then fill hub with oil and install plug.

Kingpin, Adjust

1. Raise and support front of vehicle.
2. Loosen lower kingpin jam nut and fully back down the ball stud.
3. Raise wheel hub, spindle yoke and kingpin bracket assemblies with a suitable jack positioned under the torque plate. Raise unit until upward movement is resisted by bottom of upper kingpin bearing coming into contact with lower surface of upper kingpin ball stud.
4. Insert feeler gauges between lower kingpin bracket and suspension yoke. The feeler gauges should fill the gap snugly.
5. Turn lower kingpin ball stud upward until feeler gauges loosen, then back off the lower kingpin ball stud 1/4 turn.
6. Tighten jam nut while preventing lower kingpin ball stud from turning.

NOTE: Some free play in the kingpins is normal.

Steering Stop, Adjust

NOTE: The steering stop adjustment is pre-set and does not require checking unless a major component such as the suspension yoke, spindle yoke or kingpin bracket has been replaced. The purpose of this stop is to protect the universal joint and brake chamber from hitting another part in a full turn of the steering.

1. Adjust tie rod to give 0–1/8 inch toe-in.
2. Turn steering to put axle in a full left turn condition and check that the head of the stop bolt, Fig. 1, stops the turn before any rotating parts in left hand steerable drive end interfere with another part. Repeat this check in a right turn condition.
3. Adjust steering stop as necessary by adding washers under the adjustment bolt.

Camber, Adjust

1. Raise and support front of vehicle, then remove front wheels.
2. Loosen the four upper kingpin bracket retaining bolts. Remove the top two bolts.
3. Remove or install shims as necessary to achieve the desired camber.
4. Install bolts and torque to 65 ft. lbs.

Toe-In, Adjust

1. Raise and support front of vehicle.
2. Rotate tires and scribe a reference line around the tires.
3. Measure distance between scribed lines on front and back of tires.
4. Turn tie rod tube to adjust toe-in until front measurement is 0–1/8 inch less than rear measurement.
5. Torque tie rod clamp bolts to 60–80 ft. lbs. Roll vehicle forward for a distance equal to at least four full turns of the wheels, then recheck toe-in and correct as necessary.

WHITE

INDEX OF SERVICE OPERATIONS

MODEL INDEX & ENGINE APPLICATION

ENGINE IDENTIFICATION—White service parts identification plate is attached to the left hand cab door. Stamped on the plate is the model of the engine, transmission, auxiliary transmission, rear axle and front axle. This represents only the equipment on the vehicle when it was shipped from the factory.

Truck Model	Year	Standard Engine		Oil Pan Capacity Qts. Refill	Cooling System Capacity Qts.
		Make	Model		
1500 SERIES & COMPAC-VAN					
1550-01	1974	White	6-200A	—	—
1550-05	1974	Detroit	6V-53N	—	—
1550T-01	1974	White	6-200A	—	—
1550T-05	1974	Detroit	6V-53N	—	—
1564-01	1974	White	6-200A	—	—
1564-05	1974	Detroit	6V-53N	—	—
1564T-01	1974	White	6-200A	—	—
1564T-04	1974	Cummins	V8E-210	—	—
6000 SERIES					
6000-06	1974	Cummins	NH-230	—	—
6000T-06	1974	Cummins	NH-230	—	—
6064-06	1974	Cummins	NH-230	—	—
6064T-06	1974	Cummins	NH-230	—	—
9000 SERIES					
9064DS	1974–75	Cummins	CF-160	16	20
9364	1974–79	White	6-186A	—	—
C SERIES					
CM64	1974–76	Cummins	V-210	22	—
CL64	1974–76	Cummins	NH-230	30	—
R SERIES					
RBL42T	1974	Cummins	Super 250	—	—
RBL42T	1975–76	Cummins	NTC-250	—	—
RB2L42BT	1978	Cummins	Form. 290BC	—	—
RB2L42T	1977	Cummins	Form. 290	—	—
RB2L42T	1978–79	Cummins	Form. 290BC	—	—
RB2L42T	1980	Cummins	Form. 290	—	—

Continued

MODEL INDEX & ENGINE APPLICATION—Continued

Truck Model	Year	Standard Engine Make	Standard Engine Model	Oil Pan Capacity Qts. Refill	Cooling System Capacity Qts.
RB2L42T	1982–83	Cummins	Form. 300	—	—
RBL64	1974	Cummins	Super 250	—	—
RBL64	1975–76	Cummins	NTC-250	—	—
RBL64T	1974	Cummins	Super 250	—	—
RBL64T	1975–76	Cummins	NTC-250	—	—
RB2L64BT	1978	Cummins	Form. 290BC	—	—
RB2L64T	1977	Cummins	Form. 290	—	—
RB2L64T	1978–79	Cummins	Form. 290BC	—	—
RB2L64T	1980	Cummins	Form. 290	—	—
RB2L64T	1982–83	Cummins	Form. 300	—	—
RBM42T	1974	Cummins	Super 250	—	—
RBM42T	1975–76	Cummins	NTC-250	—	—
RB2M42BT	1978	Cummins	Form. 290BC	—	—
RB2M42T	1977	Cummins	Form. 290	—	—
RB2M42T	1978–79	Cummins	Form. 290BC	—	—
RB2M42T	1980–81	Cummins	Form. 290	—	—
RB2M42T	1982–83	Cummins	Form. 300	—	—
RBM64	1974	Cummins	Super 250	—	—
RBM64	1975–76	Cummins	NTC-250	—	—
RB2M64	1977	Caterpillar	3208	—	—
RBM2M64BT	1978	Cummins	Form. 290BC	—	—
RBM64T	1974	Cummins	Super 250	—	—
RBM64T	1975–76	Cummins	NTC-250	—	—
RB2M64T	1977	Cummins	Form. 290	—	—
RB2M64T	1978–79	Cummins	Form. 290BC	—	—
RB2M64T	1980–81	Cummins	Form. 290	—	—
RB2M64T	1982–83	Cummins	Form. 300	—	—
RBS42	1974	Caterpillar	1160	—	—
RBS42	1975–76	Caterpillar	3208	12	25
RB2S42	1977	Caterpillar	3208	—	—
RBS42T	1974	Caterpillar	Super 250	—	—
RBS42T	1975–76	Cummins	NTC-250	—	—
RB2S42T	1977	Cummins	Form. 290	—	—
RB2S42T	1978–79	Cummins	Form. 290BC	—	—
RB2S42T	1980–81	Cummins	Form. 290	—	—
RB2S42T	1982–83	Cummins	Form. 300	—	—
RBS64	1974	Cummins	Super 250	—	—
RBS64	1975–76	Cummins	NTC-250	—	—
RB2S64	1977	Caterpillar	3208	—	—
RBS64T	1974	Cummins	Super 250	—	—
RBS64T	1975–76	Cummins	NTC-250	—	—
RB2S64T	1977	Cummins	Form. 290	—	—
RB2S64T	1978–79	Cummins	Form. 290BC	—	—
RB2S64T	1980–81	Cummins	Form. 290	—	—
RB2S64T	1982–83	Cummins	Form. 300	—	—
RBWL42BT	1978	Cummins	Form. 290BC	—	—
RBWL42T	1978–79	Cummins	Form. 290BC	—	—
RBWL42T	1980	Cummins	Form. 290	—	—
RBWL64BT	1978	Cummins	Form. 290BC	—	—
RBWL64T	1978–79	Cummins	Form. 290BC	—	—
RBWL64T	1980	Cummins	Form. 290	—	—
RBWM42BT	1978	Cummins	Form. 290BC	—	—
RBWM42T	1978–79	Cummins	Form. 290BC	—	—
RBWM42T	1980–81	Cummins	Form. 290	—	—
RBWM64BT	1978	Cummins	Form. 290BC	—	—
RBWM64T	1978–79	Cummins	Form. 290BC	—	—
RBWM64T	1980–81	Cummins	Form. 290	—	—
RBWS42T	1979	Cummins	Form. 290BC	—	—
RBWS42T	1980	Cummins	Form. 290	—	—
RBWS64T	1979	Cummins	Form. 290BC	—	—
RBWS64T	1980	Cummins	Form. 290	—	—
RC42T	1974	Cummins	Super 250	—	—
RC42T	1975	Cummins	NTC-250	—	—
RC42TB	1974	Cummins	Super 250	—	—
RC42TB	1975	Cummins	NTC-250	—	—
RC64T	1974	Cummins	Super 250	—	—
RC64T	1975	Cummins	NTC-250	—	—
RC64TB	1974	Cummins	Super 250	—	—
RC64TB	1975	Cummins	NTC-250	—	—
RC2L42T	1976	Cummins	Form. 290	—	—
RC2L42T	1977	Cummins	Form. 290	—	—
RC2L42T	1978–79	Cummins	Form. 290BC	—	—
RC2L42T	1980–81	Cummins	Form. 290	—	—
RC2L42T	1982–83	Cummins	Form. 300	—	—
RC2L64T	1976–77	Cummins	Form. 290	—	—
RC2L64T	1978–79	Cummins	Form. 290BC	—	—
RC2L64T	1980–81	Cummins	Form. 290	—	—
RC2L64T	1982–83	Cummins	Form. 300	—	—
RC2M42T	1978–79	Cummins	Form. 290BC	—	—
RC2M42T	1980–81	Cummins	Form. 290	—	—
RC2M42T	1982–83	Cummins	Form. 300	—	—
RC2M64T	1978–79	Cummins	Form. 290BC	—	—
RC2M64T	1980–81	Cummins	Form. 290	—	—
RC2M64T	1982–83	Cummins	Form. 300	—	—
RC2S42T	1976–77	Cummins	Form. 290	—	—
RC2S42T	1978–79	Cummins	Form. 290BC	—	—
RC2S42T	1980–81	Cummins	Form. 290	—	—
RC2S42T	1982–83	Cummins	Form. 300	—	—
RC2S64T	1976–77	Cummins	Form. 290	—	—
RC2S64T	1978–79	Cummins	Form. 290BC	—	—
RC2S64T	1980–81	Cummins	Form. 290	—	—
RC2S64T	1982–83	Cummins	Form. 300	—	—
RCWL42T	1978–79	Cummins	Form. 290BC	—	—
RCWL42T	1980–81	Cummins	Form. 290	—	—
RCWL64T	1978–79	Cummins	Form. 290BC	—	—
RCWL64T	1980–81	Cummins	Form. 290	—	—
RCWS42T	1978–79	Cummins	Form. 290BC	—	—
RCWS42T	1980–81	Cummins	Form. 290	—	—
RCWS64T	1978–79	Cummins	Form. 290BC	—	—
RCWS64T	1980–81	Cummins	Form. 290	—	—
REWS64T	1981	Cummins	Form. 290	—	—
RX42	1974	Caterpillar	1160	—	—
RX42	1975–76	Caterpillar	3208	—	—
RX2-42	1977	Caterpillar	3208	—	—
RX2-42	1978–79	Cummins	NTC-230	—	—
RX2-42	1980–81	Cummins	Form. 290	—	—
RX2-42	1982–83	Cummins	Form. 300	—	—
RX42T	1974	Caterpillar	1160	—	—

Continued

MODEL INDEX & ENGINE APPLICATION—Continued

Truck Model	Year	Standard Engine Make	Standard Engine Model	Oil Pan Capacity Qts. Refill	Cooling System Capacity Qts.
RX42T	1975–76	Caterpillar	3208	—	—
RX2-42T	1977	Caterpillar	3208	—	—
RX2-42T	1978–79	Cummins	Form. 290BC	—	—
RX2-42T	1980–81	Cummins	Form. 290	—	—
RX2-42T	1982–83	Cummins	Form. 300	—	—
RX64	1974	Caterpillar	1160	—	—
RX64	1975–76	Caterpillar	3208	—	—
RX2-64	1977	Caterpillar	3208	—	—
RX2-64	1978–79	Cummins	NTC-230	—	—
RX2-64	1980–81	Cummins	Form. 290	—	—
RX2-64	1982–83	Cummins	Form. 300	—	—
RX64T	1974–76	Cummins	NH-230	30	—
RX2-64T	1977	Cummins	Form. 290	—	—
RX2-64T	1978–79	Cummins	Form. 290BC	—	—
RX2-64T	1980–81	Cummins	Form. 290	—	—
RX2-64T	1982–83	Cummins	Form. 300	—	—

GENERAL ENGINE SPECIFICATIONS

Engine Model	No. Cyls. & Valve Location	Bore & Stroke	Piston Displacement Cubic Inch	Compression Ratio	Horsepower @ R.P.M.	Torque Ft. Lb. @ R.P.M.	Engine Governed Speed	Normal Oil Pressure, Lbs.
6-110A	6 In Head	$3\frac{5}{8} \times 4\frac{1}{8}$	255	6.70	110 @ 3400	194 @ 1400	3400	40
120A	6 In Block	$3\frac{7}{8} \times 4\frac{1}{2}$	318	6.50	114 @ 2800	250 @ 1200	3000	30–35
130A	6 In Block	$4 \times 4\frac{1}{2}$	340	6.56	120 @ 3000	270 @ 1200	3000	30–35
6-130A	6 In Head	$3\frac{7}{8} \times 4\frac{1}{8}$	292	6.94	130 @ 3300	230 @ 1600	3300	40
140A	6 In Block	$3\frac{7}{8} \times 5\frac{1}{8}$	362	6.03	125 @ 3000	285 @ 1400	3000	30–35
6-145A	6 In Head	$4\frac{1}{8} \times 4\frac{1}{8}$	331	6.73	145 @ 3200	270 @ 1600	3200	40
150A	6 In Block	$4 \times 5\frac{1}{8}$	386	6.45	135 @ 3000	315 @ 1300	3000	30–35
6-170A	6 In Head	$4\frac{1}{8} \times 4\frac{1}{8}$	331	7.50	170 @ 3400	297 @ 1600	3400	40
6-185A	6 In Head	$4\frac{1}{4} \times 4\frac{1}{4}$	362	7.50	185 @ 3400	320 @ 1200	3400	40
6-186A	6 In Head	$4\frac{1}{4} \times 4\frac{1}{4}$	362	7.50	186 @ 3400	320 @ 1200	3400	40
6-200A	6 In Head	$4\frac{1}{4} \times 4.7$	400	7.50	200 @ 3400	365 @ 1600	3200	40
230A	6 In Block	$4 \times 4\frac{1}{2}$	340	6.56	130 @ 3000	275 @ 1200	3000	50–55
250A	6 In Block	$4 \times 5\frac{1}{8}$	386	6.45	145 @ 3000	328 @ 1250	3000	30–35
260A	6 In Block	$4\frac{3}{8} \times 5$	451	6.25	170 @ 3000	350 @ 1200	2800	40–45
280A	6 In Block	$4\frac{5}{8} \times 5$	504	6.00	184 @ 3000	405 @ 1200	2800	40–45
290A	6 In Block	$4\frac{5}{8} \times 5$	504	6.00	186 @ 3000	405 @ 1200	2800	40–45
380A	6 In Block	$4\frac{5}{8} \times 5$	504	6.00	185 @ 2900	405 @ 1400	2900	40–45
390A	6 In Block	$4\frac{3}{4} \times 5$	531	6.40	200 @ 2900	440 @ 1400	2900	40–45
450A	6 In Block	$4 \times 5\frac{1}{8}$	386	6.80	145 @ 3000	328 @ 1400	3000	50–55
460A	6 In Block	$4 \times 5\frac{1}{8}$	386	6.80	159 @ 3000	336 @ 1400	3000	50–55
462A	6 In Block	$4 \times 5\frac{1}{8}$	386	7.00	172 @ 3200	335 @ 1400	3200	50–55
470A	6 In Block	$4\frac{1}{2} \times 5$	477	6.60	200 @ 2900	395 @ 1400	2900	50–55
470G	8 In Head	$4\frac{5}{8} \times 3\frac{1}{2}$	470	8.10	250 @ 3400	312 @ 1800	3300	50–75
477A	6 In Block	$4\frac{1}{2} \times 5$	477	6.6	185 @ 2600	398 @ 1200	2600	50–55
490A	6 In Block	$4\frac{3}{4} \times 5$	531	6.40	215 @ 2900	440 @ 1200	2900	50–55
531A	6 In Block	$4\frac{3}{4} \times 5$	531	6.75	200 @ 2600	450 @ 1200	2600	50–55

TUNE UP & VALVE SPECIFICATIONS

Engine Model	Firing Order	Spark Plugs		Ignition Timing		Valve Seat Angle, Degrees	Valve Clearance H–Hot C–Cold		Valve Spring Pressure, Lb. @ In. Length
		Type⑥	Gap	Timing Mark	Location		Intake	Exhaust	
6-110A	153624	⑦	.025	DC 1-10°	Damper	30	.015H	.015H	183 @ 1 13/32
120A	153624	⑦	.025	①	Flywheel	45	Zero	Zero	103 @ 2 1/32
130A	153624	⑦	.025	①	Flywheel	45	Zero	Zero	103 @ 2 1/32
6-130A	153624	⑦	.025	DC 1-10°	Damper	30	.015H	.015H	183 @ 1 13/32
140A	153624	⑦	.025	①	Flywheel	45	Zero	Zero	103 @ 2 1/32
6-145A	153624	⑦	.025	DC 1-10°	Damper	30	.015H	.015H	183 @ 1 13/32
150A	153624	⑦	.025	①	Flywheel	45	Zero	Zero	145 @ 1.758
6-170A	153624	⑦	.025	DC 1-10°	Damper	30	.023C	.023C	183 @ 1 13/32
6-185A	153624	⑦	.025	DC 1-10°	Damper	30	.023C	.023C	183 @ 1 13/32
6-186A	153624	⑦	.025	DC 1-10°	Damper	30	.023C	.023C	183 @ 1 13/32
6-200A	153624	⑦	.025	DC 1-10°	Damper	30	.023C	.023C	183 @ 1 13/32
230A	153624	⑤	.025	Line	Flywheel	45	Zero	Zero	145 @ 1.758
250A	153624	⑤	.025	Line	Flywheel	45	Zero	Zero	145 @ 1.758
260A	153624	⑤	.025	Line	Flywheel	45	Zero	Zero	②
280A	153624	⑤	.025	Line	Flywheel	45	Zero	Zero	②
290A	153624	⑤	.025	Line	Flywheel	45	Zero	Zero	②
380A	153624	⑤	.025	Line	Flywheel	45	Zero	Zero	180 @ 1.827
390A	153624	⑤	.025	Line	Flywheel	45	Zero	Zero	180 @ 1.827
450A	153624	⑤	.025	Line	Flywheel	45	Zero	Zero	150 @ 1.758
460A	153624	⑤	.025	Line	Flywheel	45	Zero	Zero	150 @ 1.758
462A	153624	⑤	.025	Line	Flywheel	45	Zero	Zero	150 @ 1.758
470A	153624	⑤	.025	Line	Flywheel	45	Zero	Zero	182 @ 1.612
470G	15486372④	—	.025	6° BTC	Flywheel	30	.014H	.024H	213
477A	153624	⑤	.025	Line	Flywheel	45	Zero	Zero	182 @ 1.612
490A	153624	⑤	.025	Line	Flywheel	45	Zero	Zero	182 @ 1.612
531A	153624	⑤	.025	Line	Flywheel	45	Zero	Zero	182 @ 1.612

①—Line at bottom inspection hole or steel ball at side inspection hole.
②—Inner spring 95 @ 1.827", outer 104 @ 1.827".
③—Champion.
④—Cylinder numbering (front to rear): Right bank 1-2-3-4, left bank 5-6-7-8.
⑤—J6 for 14MM, D10 for 18MM.

WHEEL ALIGNMENT SPECIFICATIONS

Model	Caster, Deg.	Camber, Deg.	Toe-In, Inch	Kingpin Inclination, Deg.
WC MODELS				
14, 16 Series	2 5/6	1	1/8	8
18 Series	2 5/6	1	1/8	8
20, PLT, 20B, 20T	2 5/6	1	1/8	8
22 Series	2 5/6	1	1/8	8 1/2
4000, 4200, 4400, & 9000 MODELS				
4000T, 9000	2 1/8	1	1/8	6
4200T, 4400TD, 9000TD	2 1/8	1	1/8	6①

Model	Caster, Deg.	Camber, Deg.	Toe-In, Inch	Kingpin Inclination, Deg.
R SERIES				
F3W-100 thru F6W-140②③	2 5/6	④	3/32	⑤
F3W-100 thru F6W-140②⑥	5/6	④	3/32	⑤
FH, FL-901②	5/6	1	3/32	5 1/2

①—FG-900 axle 8°, FE-990 and FD-900 5 1/2°.
②—Axle models.
③—Exc. tandem axle models.
④—Right, 1/3°; left, 1/2°.
⑤—Right, 6 2/3°; left, 6 1/2°.
⑥—Tandem axle models.

PISTON, PIN, RING, CRANKSHAFT & BEARING SPECIFICATIONS

Engine Model	Wristpin Diameter	Piston Clearance, Inch	Ring End Gap, In. (Minimum) Comp.	Ring End Gap, In. (Minimum) Oil	Crankpin Diameter, In.	Rod Bearing Clearance, In.	Main Bearing Journal Diameter, In.	Main Bearing Clearance, Inch	Thrust On Bearing No.	Shaft End Play
6-110A	1.125	.003④	.018③	.012	2.1705–2.1712	.001–.0032	2.4975–2.4982	.001–.0035	1	.004–.007
120A	1.000	.0030–.0035①	.010	.010	2.187–2.188	.002–.003	2.9995–3.0005	.003–.004	Center	.003–.007
130A	1.187	.0040–.0045①	.013	.013	2.187–2.188	.002–.003	2.9995–3.0005	.003–.004	Center	.003–.007
6-130A	1.125	.003④	.020③	.012	2.1805–2.1812	.001–.0032	2.5075–2.5082	.001–.0035	1	.004–.007
140A	1.000	.0030–.0035①	.010	.010	2.187–2.188	.002–.003	2.9905–3.0005	.003–.004	Center	.003–.007
6-145A	1.125	.004④	.020③	.012	2.1805–2.1812	.001–.0032	2.5075–2.5082	.001–.0035	1	.004–.007
150A	1.187	.005①	.010	.013	2.187–2.188	.002–.003	2.9995–3.0005	.003–.004	Center	.003–.007
6-170A	1.0625	.003④	.020③	.012	2.3093–2.3100	.001–.0032	2.7485–2.7492	.001–.0035	1	.004–.007
6-185A	1.0625	.003④	.020③	.017	2.3093–2.3100	.001–.0032	2.7485–2.7492	.001–.0035	1	.004–.007
6-186A	1.0624	.003④	.020③	.017	2.3093–2.3100	.001–.0032	2.7485–2.7492	.001–.0035	1	.004–.007
6-200A	1.0624	.003④	.020③	.017	2.3093–2.3100	.001–.0032	3.1235–3.1242 ⑥	.001–.0035⑤	1	.004–.007
230A	1.187	.005	.013	.013	2.187–2.188	.002–.003	2.9995–3.0005	.003–.004	Center	.003–.007
250A	1.187	.005	.010	.013	2.187–2.188	.002–.003	2.9995–3.0005	.003–.004	Center	.003–.007
260A	1.249	.005①	.013	.013	2.4995–2.5005	.003–.004	3.2495–3.2505	.0035–.0045	Center	.003–.007
280A	1.249	.005①	.013	.013	2.4995–2.5005	.003–.004	3.2495–3.2505	.0035–.0045	Center	.003–.007
290A	1.249	.005②	.013	.013	2.4995–2.5005	.003–.004	3.2495–3.2505	.0035–.0045	Center	.003–.007
380A	1.249	.005②	.013	.013	2.4995–2.5005	.003–.004	3.2495–3.2505	.0035–.0045	Center	.003–.007
390A	1.249	.005②	.013	.013	2.4995–2.5005	.003–.004	3.2495–3.2505	.0035–.0045	Center	.003–.007
450A	1.187	.005②	.013	.013	2.1868–2.1875	.002–.003	2.9995–3.0005	.003–.004	Center	.003–.007
460A	1.187	.005②	.013	.013	2.1868–2.1875	.002–.003	2.9995–3.0005	.003–.004	Center	.003–.007
462A	1.187	.005②	.013	.013	2.1868–2.1875	.002–.003	2.9995–3.0005	.003–.004	Center	.003–.007
470A	1.249	.005②	.013	.013	2.4995–2.5005	.003–.004	3.2495–3.2505	.0035–.0045	Center	.003–.007
470G⑦	1.3739	.0085–.0110	.013	.013	2.499–2.500	.0015–.0045	3.249–3.250	.0015–.0045	—	.004–.014
470G⑧	1.3739	.0085–.0110	.013	.013	2.499–2.500	.0015–.0045	3.499–3.500	.0015–.005	—	.004–.014
477A	1.249	.004	.013	.013	2.4995–2.5005	.002–.003	2.2495–2.2505	.0035–.0045	Center	.006–.013
490A	1.249	.005②	.013	.013	2.4995–2.5005	.003–.004	3.2495–3.2505	.0035–.0045	Center	.003–.007
531A	1.249	.004	.013	.013	2.4995–2.5005	.002–.003	2.2495–2.2505	.0035–.0045	Center	.006–.013

①—When checked with a 1" wide feeler it should require a 7 to 10 lb. pull on a spring scale to withdraw the feeler past the piston.
②—Piston is properly fitted when it can be moved by applying pressure with one or two fingers.
③—For top ring, 2nd ring .012".
④—When checked with 1/2" wide feeler gauge, a pull of 3 to 8 lbs. is required.
⑤—Front bearing .0014–.0039".
⑥—No. 1: 3.1231–3.1238.
⑦—Engine Se. No. Prefix "D".
⑧—Engine Se. No. Prefix "F".

ENGINE TIGHTENING SPECIFICATIONS★

★ Torque Specifications are for clean and lightly lubricated threads only. Dry or dirty threads produce increased friction which prevents accurate measurement of tightness.

Engine	Spark Plugs Ft. Lbs.	Cylinder Head Bolts Ft. Lbs.	Intake Manifold Ft. Lbs.	Exhaust Manifold Ft. Lbs.	Rocker Arm Shaft Bracket Ft. Lbs.	Rocker Arm Cover Ft. Lbs.	Connecting Rod Cap Bolts Ft. Lbs.	Main Bearing Cap Bolts Ft. Lbs.	Flywheel to Crankshaft Ft. Lbs.	Vibration Damper or Pulley Ft. Lbs.
110A	①	100–105	—	—	—	—	65–70	85–90	—	—
120A	30	105–110	—	—	None	None	40–65	70–75	48–52	310–320
130A	30	105–110	—	—	None	None	40–65	70–75	48–52	310–320
6-130A	①	100–105	—	—	—	—	65–70	85–90	—	—
140A	30	105–110	—	—	None	None	40–65	70–75	48–52	310–320
6-145A	①	100–105	—	—	—	—	65–70	85–90	—	—
150A	30	105–110	—	—	None	None	50–65	80–90	48–52	200
6-170A	①	100–105	—	—	—	—	65–70	85–90	—	—
6-185A	①	100–105	—	—	—	—	65–70	85–90	—	—
6-186A	30	100–105	—	—	—	—	65–70	85–90	—	—
6-200A	30	100–105	—	—	—	—	65–70	85–90	—	—
230A	30	105–110	—	—	None	None	50–65	80–90	48–52	200

Continued

ENGINE TIGHTENING SPECIFICATIONS★—Continued

★ Torque Specifications are for clean and lightly lubricated threads only. Dry or dirty threads produce increased friction which prevents accurate measurement of tightness.

Engine	Spark Plugs Ft. Lbs.	Cylinder Head Bolts Ft. Lbs.	Intake Manifold Ft. Lbs.	Exhaust Manifold Ft. Lbs.	Rocker Arm Shaft Bracket Ft. Lbs.	Rocker Arm Cover Ft. Lbs.	Connecting Rod Cap Bolts Ft. Lbs.	Main Bearing Cap Bolts Ft. Lbs.	Flywheel to Crankshaft Ft. Lbs.	Vibration Damper or Pulley Ft. Lbs.
250A	30	105–110	—	—	None	None	50–65	70–75	48–52	200
260A	30	105–110	—	—	None	None	70–75	70–75	48–52	200
280A	30	105–110	—	—	None	None	70–75	70–75	48–52	200
290A	30	105–110	—	—	None	None	70–75	70–75	48–52	200
380A	28	105–110	—	—	None	None	80–120	80–90	48–52	200
390A	28	105–110	—	—	None	None	80–120	80–90	48–52	200
450A	28	105–110	—	—	None	None	50–65	80–90	40–44	200
460A	28	105–110	—	—	None	None	50–65	80–90	40–44	200
462A	28	105–110	—	—	None	None	50–65	80–90	40–44	200
470A, 477A	28	105–110	—	—	None	None	80–120	80–90	48–52	200
470G	25	100–105	25–30	30–32	100–105	—	②	175–185	115–120	95–105
490A, 531A	28	105–110	—	—	None	None	80–120	80–90	48–52	200

①—30 ft. lbs. for 14MM, 34 ft. lbs. for 18MM. ②—See text for template tightening method.

STARTING MOTOR SPECIFICATIONS

Starter Number	Pinion Clearance	Brush Spring Tension Ounces	No Load Test			Resistance Test②		Lock Test		
			Amperes	Volts	R.P.M.	Amperes	Volts	Amperes	Torque Ft-Lbs. Minimum	Volts
1107037	—	24–28	75	5.7	5000	—	—	525	12	3.4
1107224	.010–.140	35	55–80①	9	3500–6000	300–360	3.5	—	—	—
1107506	.010–.140	35	50–80①	9	5500–9000	520–590①	3.5	—	—	—
1107541	—	35	75–95	10.6	6400–9500	545–600	3.5	—	—	—
1107542	.010–.140	35	50–80①	9	5500–9000	520–590①	3.5	—	—	—
1107578	.010–.140	35	50–80①	9	5500–9000	520–590①	3.5	—	—	—
1107671	—	35	112①	10.6	3240	—	—	320–385①	—	3.5
1107884	.010–.140	35	50–80①	9	5500–10500	270–310①	4.3	—	—	—
1107920	—	24–28	80	5.67	5500	—	—	600	14	3
1107972	—	24–28	80	5.67	5500	—	—	600	14	3
1108111	—	24–28	70	10.0	7000	—	—	530	16	6.7
1108455	—	24–28	70	5	2500	—	—	600	28	6.7
1108533	—	24–28	70	10.0	2800	—	—	530	33	6.7
1108607	—	24–28	75	11.25	6000	—	—	615	29	5.85
1108621	1/64–9/64	24–28	75	11.25	6000	—	—	615	29	5.85
1108840	—	36–40	115	11.6	7000	—	—	570	20	2.3
1108863	3/16–1/4③	36–40	100	23	8000	—	—	525	32	3.0
1108872	3/16–1/4③	36–40	100	23.0	8000	—	—	525	32	3.0
1109173	—	36–40	65	11.4	6000	—	—	725	44	5
1109977	—	—	95–120①	20	5500–7500	—	—	—	—	—
1113328	—	48	65	11.4	6000	—	—	500	28	3.2
1113641	.010–.140	80	95	11.6	8000	—	—	600	20	2.2
1113658	—	—	130–160①	9	5000–7000	—	—	—	—	—
1113664	—	—	130–160①	9	5000–7000	—	—	—	—	—
1113670	—	—	130–160①	9	5000–7000	—	—	—	—	—
1113678	—	—	130–160①	9	5000–7000	—	—	—	—	—
1113708	3/16–1/4③	35–40	100	23.0	8000	—	—	500	28	3
1113800	3/16–1/4③	35–40	100	23.0	8000	—	—	500	28	3
1113842	23/64	80	60–90①	23	7000–10700	—	—	500	26	3.5

Continued

STARTING MOTOR SPECIFICATIONS—Continued

Starter Number	Pinion Clearance	Brush Spring Tension Ounces	No Load Test			Resistance Test②		Lock Test		
			Amperes	Volts	R.P.M.	Amperes	Volts	Amperes	Torque Ft-Lbs. Minimum	Volts
1113851	23/64	80	60–90①	23	7000–10700	—	—	500	26	3.5
1113855	23/64	80	60–90①	23	7000–10700	—	—	500	26	3.5
1113872	23/64	80	60–90①	23	7000–10700	—	—	500	26	3.5
1113887	—	—	75–95①	20	5500–7500	—	—	—	—	—
1113952	—	—	75–95①	20	5500–7500	—	—	—	—	—
1113968	—	—	75–95①	20	5500–7500	—	—	—	—	—
1114000	—	35	105	11.5	6000	—	—	600	19	2.3
1114052	—	—	120–150①	9	3000–4500	—	—	—	—	—
1114058	23/64	80	115–170①	11	6300–9500	—	—	700	15	1.5
1114070	—	—	140–190①	9	4000–7000	—	—	—	—	—
1114072	23/64	80	115–170①	11	6300–9500	—	—	700	15	1.5
1114077	—	—	120–150①	9	3000–4500	—	—	—	—	—
1114078	—	—	120–150①	9	3000–4500	—	—	—	—	—
1114086	23/64	80	115–170①	11	6300–9500	—	—	700	15	1.5
1114088	23/64	80	115–170①	11	6300–9500	—	—	700	15	1.5
1114096	—	—	120–150①	9	3000–4500	—	—	—	—	—
1114098	23/64	80	115–170①	11	6300–9500	—	—	700	15	1.5
1114110	—	—	140–190①	9	4000–7000	—	—	—	—	—
1114128	—	—	140–190①	9	4000–7000	—	—	—	—	—
1114135	—	—	120–150①	9	3000–4500	—	—	—	—	—
1114137	—	—	120–150①	9	3000–4500	—	—	—	—	—
1114143	—	—	140–190①	9	4000–7000	—	—	—	—	—
1114144	—	—	140–190①	9	4000–7000	—	—	—	—	—
1114145	—	—	140–190①	9	4000–7000	—	—	—	—	
1114147	—	—	140–190①	9	4000–7000	—	—	—	—	—
1114154	—	—	140–190①	9	4000–7000	—	—	—	—	—
1114157	—	—	140–190①	9	4000–7000	—	—	—	—	—
1114158	—	—	140–190①	9	4000–7000	—	—	—	—	—
1114159	—	—	140–190①	9	4000–7000	—	—	—	—	—
1114161	—	—	140–190①	9	4000–7000	—	—	—	—	—
1114163	—	—	140–190①	9	4000–7000	—	—	—	—	—
1114164	—	—	140–190①	9	4000–7000	—	—	—	—	—
1114170	—	—	140–190①	9	4000–7000	—	—	—	—	—

①—Includes solenoid.
②—Check capacity of motor by using a 500 ampere meter and a carbon pile rheostat to control voltage. Apply volts listed across motor with armature locked. Current should be as listed.
③—Pinion travel.

DISTRIBUTOR SPECIFICATIONS

Unit Part Number ①	Rotation ②	Cam Angle, Degrees	Breaker Gap, In.	Condenser Capacity, Mfds.	Breaker Arm Spring Tension, Oz. ③	Centrifugal Advance Deg. @ R.P.M. of Dist.		Vacuum Advance		
						Advance Starts	Full Advance	Inches of Vacuum to Start Plunger Movement	Inches of Vacuum for Full Plunger Travel	Maximum Vacuum Advance, Dist. Degrees
1110132	CC	31–37	.022	.20–.25	17–21	1 @ 425	9 @ 1600	5–7	16–18	7.5
1111790	C	31–37	.022	.20–.25	17–21	1 @ 375	9 @ 1300	4–6	13	7–8
1111866	C	38–45	.016	.20–.25	19–23	1 @ 275	15 @ 1150	None	None	None
1111905	C	38–45	.016	.20–.25	19–23	1 @ 450	15 @ 1700	None	None	None
1111922	C	38–45	.016	.18–.23	19–23	1 @ 450	13 @ 1875	None	None	None
1111923	C	28–35	.016	.18–.23	19–23	1 @ 425	12 @ 1700	None	None	None
1112254	C	35	.018–.024	.18–.23	19–23	1 @ 300	10 @ 1300	None	None	None
1112263	C	31–37	.022	.20–.25	17–21	1 @ 375	9 @ 1300	None	None	None
1112318	C	31–37	.022	.20–.25	17–21	1 @ 375	9 @ 1300	None	None	None
1112319	C	31–37	.022	.20–.25	17–21	1 @ 375	9 @ 1300	None	None	None
1112329	C	31–37	.022	.18–.23	19–23	2 @ 375	18 @ 1300	None	None	None
1112333	C	31–37	.022	.18–.23	19–23	2 @ 375	12 @ 1300	None	None	None

①—Stamped on plate riveted to housing.
②—As viewed from top; C—Clockwise. CC—Counter-clockwise.
③—Measurement taken from center of breaker point.

Engine Section

CYLINDER HEAD, REPLACE

Side Valve Engines

Cylinder heads should be tightened down by starting from the center and working outward from side to side and to the ends. Draw the nuts or bolts down evenly, repeating the operation until all are normally tight. After the engine has been run long enough to warm it up, a final tightening should be made in the sequence shown in Figs. 1 and 2 and to the torque given in the *Engine Tightening chart.*

Std. OHV Engines

When removing a cylinder head a chisel or screw driver should never be used to pry it loose from the cylinder block. Use an eyebolt in the third or fourth spark plug hole (6-cyl. engines), lifting up on the eyebolt while tapping lightly around the edges of the head with a lead or rawhide hammer or mallet. Be sure all the head stud nuts or bolts have been removed before starting to remove the head.

In replacing the head, the stud nuts or bolts at the center of the head should be tightened first and then work outward from side to side and to the ends. Draw the nuts or bolts down evenly until all are normally tight. After the engine has run long enough to warm it up, make a final tightening with a torque wrench to the torque loads given in the *Engine Tightening Chart* and in the sequence shown in Fig. 3.

470G Engine

Cylinder Head, Install

1. First make sure breather vent holes are plugged.
2. Place grommet retainers in water passages in block, small end up.
3. Lay head gasket on flat surface and place grommets in place by hand. Check grommet thickness; standard grommets are .065–.070" thick. *Resurfaced blocks require thicker head gaskets and grommets than new blocks.*
4. Install head gasket over ring dowels so that word TOP on gasket is visible. Be careful not to dislodge grommets as gasket is lowered over grommet retainers.
5. Install O-rings over dowels.
6. Install cylinder head over dowels.
7. Coat short capscrews with a suitable rust preventive and install. Snug tighten down in the sequence shown in Fig. 5. Final tightening is done later.

Valve Crossheads, Adjust

1. Insert crossheads over guides.
2. Hold crosshead firmly down on guide at stellite pad.
3. Run crosshead adjusting screw down until it just touches valve stem.
4. Torque adjusting screw lock nut to 20–30 ft-lbs.
5. Check clearance between crosshead and valve spring retainer with a wire feeler gauge; clearance should be .025".

Rocker Levers and Push Rods

1. Position rocker lever to cylinder head.
2. Coat rocker capscrews with a suitable rust preventive, and install in head.
3. Install push rods.
4. Check for a .010–.037" clearance between all rocker levers, spacers and lever brackets.

Tighten Cylinder Head Capscrews

1. Torque all capscrews to 100–105 ft-lbs in 25 ft-lb increments in the sequence shown in Fig. 4.

Push Rod Cavity Covers

1. Check for installation of pipe plugs in block.
2. Position cover with new gasket to block, and secure with flatwashers, lockwashers and capscrews.

Spark Plugs

1. Clean spark plug sleeve with a clean lint-free cloth.
2. Set spark plug gap at .025".
3. Install plugs and torque to 25 ft-lbs.

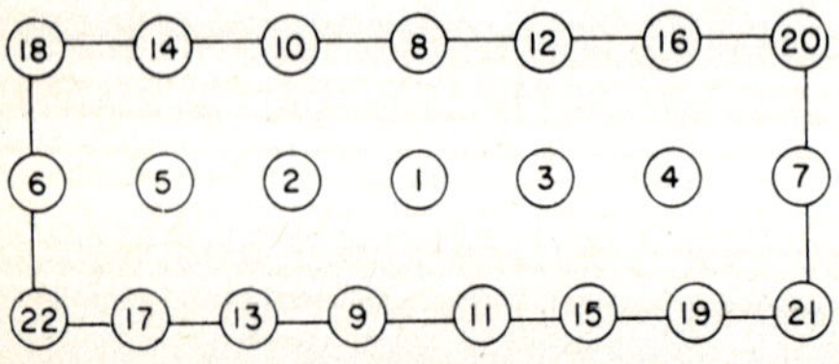

Fig. 1 Head tightening chart for side valve engine heads with 22 bolts

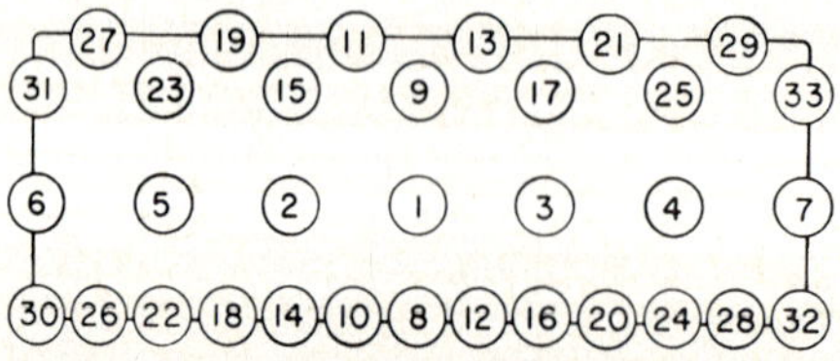

Fig. 2 Head tightening chart for side valve engine heads with 33 bolts

Fig. 3 Cylinder head tightening sequence for overhead valve sixes

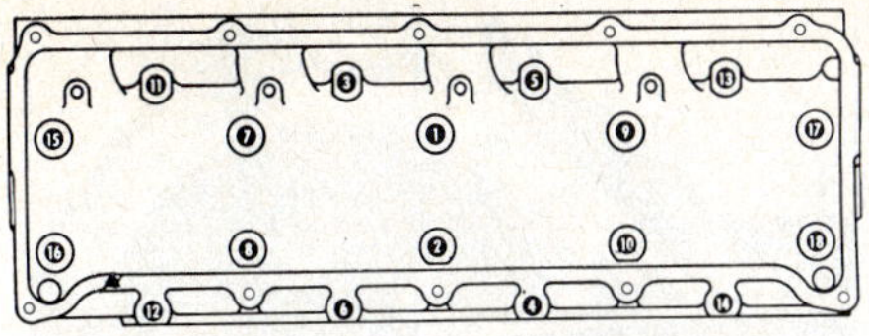

Fig. 4 Cylinder head tightening sequence. 470G engine

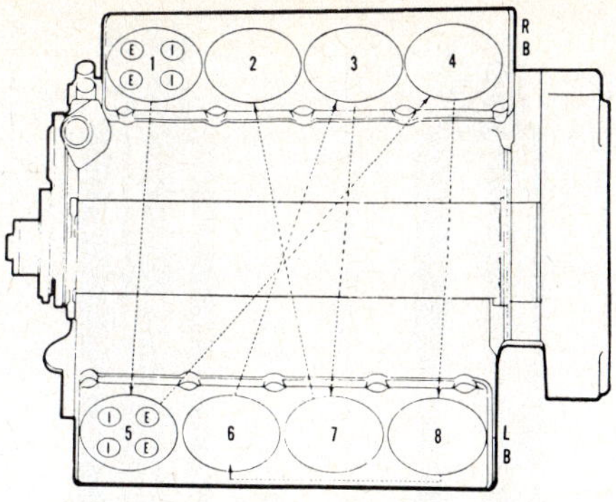

Fig. 5 Engine numbering and firing order diagram. 470G engine

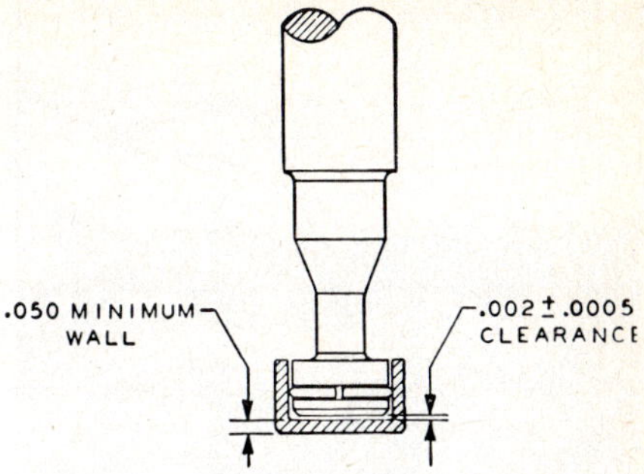

Fig. 6 Free valve stem-to-cap clearance. Side valve engines

Spark Plug Adapter Tubes

1. Lubricate O-rings with engine oil.
2. Install O-rings in grooves of spark plug adapter tube.
3. Install adapter tube in cylinder head, secure with hold-down clamp and socket-head capscrew.

Cylinder Head Cover

1. After adjusting valves as outlined below, position new gasket to cylinder head cover.
2. Position cover on cylinder head.
3. Insert spark plug adapter tubes in bores, and secure cover to head with flatwashers, lockwashers and capscrews.

VALVES, ADJUST

470G Engine

1. Remove spark plug from No. 1 cylinder.
2. Turn engine over in direction of rotation.
3. As piston comes up to top of stroke, stick a small diameter rod into spark plughole and continue to turn engine over until it is at the top of its stroke and just before it starts to back down. This is top center.
4. Observe TC mark on flywheel through hole in bottom of flywheel housing. Also, see "VS" marks on vibration damper.
5. Shake both valve rocker levers, which should be free. If not free, a complete revolution is required before adjusting procedure should begin.
6. Loosen lock nut and back off adjusting screw. Insert feeler gauge between rocker lever and top of crosshead. Turn screw down until lever just touches gauge. Run adjusting screw nut down. Valve clearances are .014" for intake and .020" for exhaust valves.
7. Tighten adjusting screw nut to 70–80 ft-lbs.
8. After crossheads and valves are adjusted in this cylinder, rotate crankshaft 90-deg. in operating direction to the next cylinder corresponding to the firing order of the engine (1-5-4-8-6-3-7-2), Fig. 5, and repeat the above procedure.

NOTE: Two complete revolutions of the crankshaft are needed to set all crossheads and valves. Crossheads and valves can be adjusted for only one cylinder at any one crankshaft position.

Engines Other Than 470G

Adjust the valves to the clearances given in the *Tune Up & Valve Specifications* table. If the valves are to be adjusted warm, run the engine for about 30 minutes to normalize the expansion of all parts.

VALVE ROTATORS

Some Side Valve Engines

The parts involved in the rotating mechanism are special spring seat retainer, a pair of flat halfmoon keys, a close fitting cap located on the valve stem, and a specially constructed valve stem, Fig. 6.

In order to accommodate valve expansion, the valve lash must be maintained. When camshaft rotation causes this lash or clearance to be taken up, the cap on the valve stem causes the valve keys to lower the spring retainer, removing the load of the valve springs from the valve *before* the valve is raised from its seat. The clearance shown in Fig. 6 should be maintained between the end face of the valve stem and cap. This is the distance the spring retainer is lowered *before* the valve is removed. The slow valve rotating motion is caused by vibration of the valve, the flowing of exhaust gases around the valve head, and a slight rotation motion imparted to the valve by the valve spring.

Whenever valve work is done, check the clearance, using a special gauge which is available for the purpose. If there is no clearance, grind off the end of the valve stem. If the clearance is too great, grind off the open end of the valve cap.

If the special gauge is not available, a check for clearance can be made by holding the assembled valve and rotating mechanism in the hands. With one hand press down on the valve cap and with the other hand turn the valve. If the valve turns readily at least it indicates that there is clearance present.

Std. OHV Engines

Increased exhaust valve lift is accomplished in these engines by the use of "Roto Valve" mechanism which is a special valve spring retainer and cap located on the end of the valve stem.

In operation, the rocker arm striking the cap lifts the retainer first, thus taking the spring load off the valve before lifting it from its seat. When the valve is in the raised position, it is free to rotate due to the vibration of the system.

When checking valve clearances, the feeler gauge is placed between the cap and socket as there is no other flat surface in the valve train.

VALVE LIFTERS

470G Engines

If the roller-type tappets (valve lifters) have been removed, install them as follows:

1. Install new spring clips on all tappets.
2. Install tappets in bores in block.
3. Hold tappet plate across tappet and tap lightly with a soft hammer to align tappets.
4. Place feeler gauge or shim between tappet plate and tappet, Fig. 7, to maintain .006–.014" clearance.
5. Secure tappet plate with lockwashers and capscrews, torquing capscrews to 18–20 ft-lbs.

Hydraulic Type

Fig. 8 illustrates the hydraulic lifter used in these engines. Because of the close fit between the cylinder and plunger, the plunger and cylinder should never be mismatched.

Failure of a hydraulic valve lifter is generally caused by an inadequate oil supply or dirt. An air leak at the intake side of the oil pump or too much oil in the engine will cause air bubbles in the oil supply to the lifters, causing them to collapse. This is a probable cause of trouble if several lifters fail to function, but air in the oil is an unlikely cause of failure of a single unit.

Remove the lock ring from the top of the lifter body and slide the complete unit out of the bottom of the lifter bracket. Take out the cylinder and plunger. Remove the plunger from the cylinder by twisting plunger counterclockwise to release the spring on the plunger.

Wash all parts in clean kerosene. *Never use gasoline or other fast drying agents.* Wash cylinder thoroughly until ball check rattles in cylinder.

To assemble, insert plunger in body. Plunger spring should be locked into cylinder body by turning clockwise with a twist of the plunger. Insert the lifter in the lifter bracket.

Before installing valve lifters in the engine, note the following items to make sure lifters are assembled properly. And be sure they are installed in the original position.

1. Ball check should rattle when cylinder unit is shaken.
2. Plunger should bounce back when pressed quickly into cylinder and released.
3. Plunger spring should be locked into cylinder body with a twist of the plunger.
4. Cylinder should slide smoothly into lifter body when free of oil.

Do not attempt to fill lifters with oil before assembly as this will be done automatically when the engine is running. After servicing and replacing the lifters, measure the clearance between the end of the plunger and valve stem. The valve must be fully seated. Use a feeler gauge to check the clearance, which should be from .040 to .080". If the clearance is more than .080", grind in the valve seat just

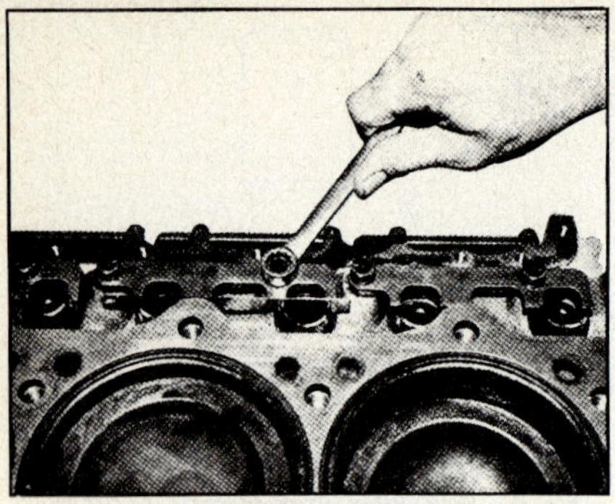

Fig. 7 Installing valve tapper plate. 470G engine

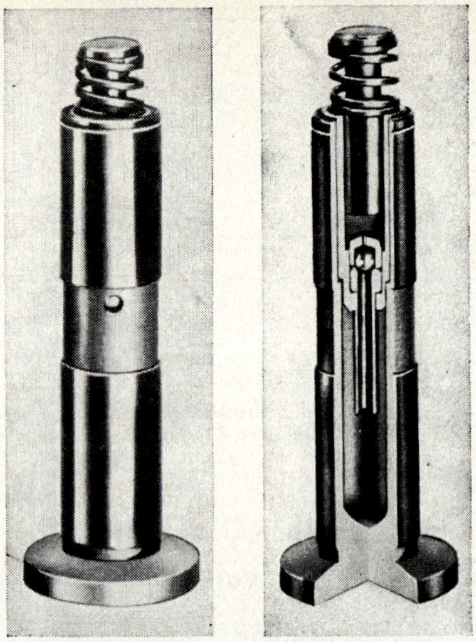

Fig. 8 Cutaway and exterior views of hydraulic valve lifter. Side valve engines

Fig. 9 Removing camshaft thrust plate capscrews. 470G engine

enough to obtain the proper clearance. If the clearance is less than .040", carefully grind the end of the valve stem.

Std. OHV Engines

These engines are equipped with barrel type lifters which can be lifted out from above after taking out the push rods. With this construction the oil pan and camshaft need not be disturbed.

CAMSHAFT & TIMING GEARS

470G Engine

The gear housing, located at the rear of the engine, contains a set of three gears; camshaft gear, crankshaft gear and lubricating oil pump gear. The camshaft gear is a press fit on the camshaft which, of course, requires the removal of the camshaft. In order to do this, the following procedure must be followed:

Removal Procedure

1. Remove engine from vehicle.
2. Remove capscrews and washers securing oil pan to block, flywheel housing and front cover. Then lift off oil pan.
3. Remove two flywheel capscrews on opposites sides of flywheel. Insert two threaded studs, six inches long, to provide support for flywheel during removal. *Insert a wood block between crankshaft and cylinder block to prevent crankshaft from turning while removing capscrews.*

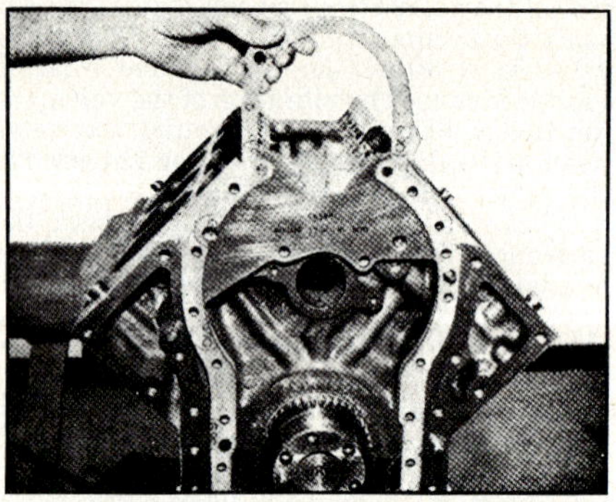

Fig. 10 Removing gear housing spacer plate. 470G engine

4. Remove remaining flywheel mounting capscrews.
5. In two holes provided, screw two flywheel mounting capscrews into flywheel. Turn in capscrews alternately to pull flywheel from crankshaft.
6. Remove capscrews and lift off flywheel, remove studs from crankshaft.
7. Remove pilot bearing from flywheel bore.
8. Unfasten flywheel housing from block and drive housing from dowels with block of wood or soft hammer.
9. Unfasten gear housing from block and, using a soft hammer, tap gear housing from dowels.
10. Remove lockplates and capscrews securing lubricating oil pump from block.
11. Remove capscrews and lockplates securing suction tube brackets and supports then lift off oil pump.
12. From behind camshaft gear, Fig. 9, unfasten camshaft thrust plate from spacer plate.
13. Rotate and lift camshaft slightly while pulling camshaft from block. *Do not remove gear from camshaft.*
14. Unfasten and lift off gear housing spacer plate, Fig. 10.

Installation Procedure

1. Position gear housing spacer plate with new gasket to rear of block. Gasket and plate must be flush with bottom of block.
2. Secure with lockplates and capscrews.
3. Install camshaft by lifting slightly and rotating it to assist in its installation.
4. Index "O" mark on camshaft gear with "O" mark on crankshaft gear to insure

Fig. 11 Valve timing marks aligned. 470G engine

Fig. 12 Checking camshaft end play 470G engine

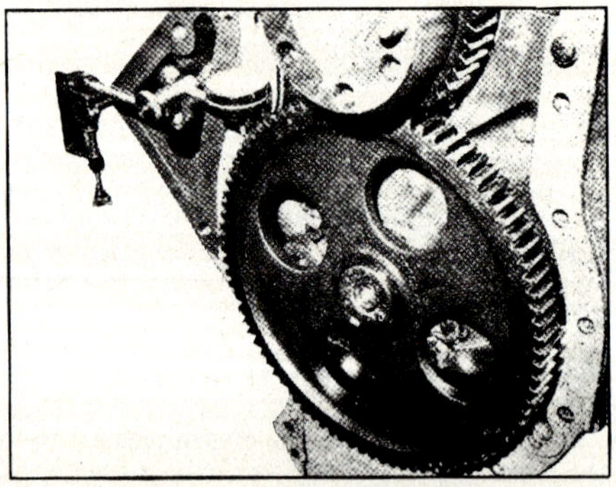

Fig. 13 Checking camshaft gear backlash. 470G engine

Fig. 14 Installing lubricating oil pump. 470G engine

Fig. 15 Checking oil pump drive gear backlash. 470G engine

correct timing, Fig. 11.

5. Secure camshaft thrust plate to block with new lockplates and capscrews.
6. With a dial indicator installed as shown, Fig. 12, check camshaft end play which must be .002–.005". If not within these limits, remove camshaft and change thrust washer.
7. Shift dial indicator to position shown in Fig. 13 and check gear train backlash which should be .003–.009". If backlash is not within limits, it is recommended that a complete new gear train be installed.

CAUTION: Insufficient gear lash or improperly mated gears will cause a noisy gear train and may cause gear failure.

---

8. Install lubricating oil pump, Fig. 14.
9. Attach dial indicator as shown in Fig. 15 and check oil pump drive gear backlash. Rotate gear by hand to advance position and "zero" the indicator. Back gear to retard position and note reading, which should be .002–.010". If not within limits, gear train must be replaced.
10. With new gasket in place, position gear housing to block over dowels and secure snugly with capscrews and washers.
11. Place dial indicator on end of crankshaft and check housing seal bore runout, which must not exceed .005". If run-out is excessive, remove housing and pull both dowels from block. Align housing to bore and ream for smallest permissible dowels. Install new pin dowels (use pin type only for replacement).
12. Remove gear housing and install new oil seal. Press seal into crankshaft bore from outside of housing until seal is flush with outer edge of bore.
13. Install flywheel housing and tighten capscrews in a criss-cross fashion gradually until all are tight; final tightening torque is 65–75 ft-lbs. Check housing face runout with a dial indicator. Run-out should not exceed .008". If dowels were removed, ream dowel holes to the smallest permissible oversize and install new pin dowels.
14. Install two 1/2" × 6" long guide studs in crankshaft flange and install flywheel. Insert capscrews and tighten in a criss-cross fashion gradually; final torque should be 100–105 lbs.
15. Install oil pan with new gasket. Tighten capscrews alternately and evenly. Torque 5/16" screws to 16–18 ft-lbs and 3/8" screws to 30–35 ft-lbs.

Overhead Valve Sixes

Before removing the timing gear cover it will be necessary to remove the oil pan and take out five fore-and-aft bolts between front oil pan arch and timing gear cover. Before removing cover bolts from front face, tap edge with soft hammer and lift off over two locating dowels.

TIMING GEARS

Side Valve Engines

Before removing timing gears be sure timing marks are legible. And when replaced valve timing will be correct if "X" marks on both gears are meshed as shown in Fig. 16.

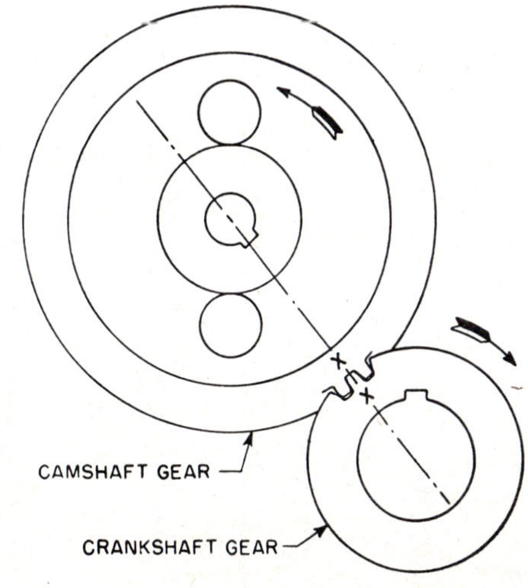

Fig. 16 Valve timing on all timing gear driven side valve engines

Std. OHV Engines

Both timing gears are punch-marked to automatically establish valve timing. On six-cylinder engines the punch-marks are aligned as shown in Fig. 17 with No. 6 piston on top of its firing stroke.

PISTONS & RODS, ASSEMBLE

470G Engine

Assemble pistons to rods so that when they are installed the numbered sides of rods will be toward outside of block.

NOTE: It is recommended that the connecting rod cap nuts be tightened as follows:

---

1. Tighten both nuts to 25–30 ft. lbs. Then tighten again to 50–55 ft-lbs.
2. Loosen both nuts completely to remove all tension. Then tighten them to 25 ft. lbs.
3. Advance nuts 60-deg. in 30-deg. increments.
4. When all assemblies are installed, push rods apart and check for .010–.018" clearance between rod on each crankpin, using feeler gauge at three locations around crankshaft.
5. Secure nuts by bending lockplate tangs against flat surface of nut.

All Except 470G

If the pistons and rods have factory markings to indicate their position in the cylinders, assemble and install them according to these marks, or according to the marks made during removal if they were not otherwise marked.

CYLINDER LINERS

470G Engine

Liners are the wet, replaceable sleeve type, precision machined to a uniform wall thickness to permit even expansion and contraction. The liners are a press fit in the top deck of the cylinder block and sealed at the lower bore with O-rings and crevice seals to prevent coolant leakage, Fig. 18.

All OHV Except 470G

Cylinder liners used in these engines are similar to those used in the Diesel engine but two O-ring seals are used at the lower bore to prevent coolant leakage.

When the liner is seated, check and make sure that top flange extends .002–.005" above block. If flange is below planed surface of block it will not be sealed and held in place by the cylinder head and gasket which may then allow water leakage into the bore. Shims are available to correct sleeve protrusion.

CRANKSHAFT END PLAY, ADJUST

470G Engine

End thrust of the crankshaft is controlled by thrust rings behind the crank gear. To check the end play, proceed as follows:

1. Attach a dial gauge to block with contact point of gauge resting on crankshaft flange face.
2. Using a pry bar, pry crankshaft toward front of engine. Remove bar and set gauge at "O". Now pry crankshaft toward rear of engine and again remove pry bar.
3. Total gauge reading should be .004–.014" with a new crankshaft and thrust rings.
4. Rotate crankshaft to see that it turns freely.
5. If end play is less than .004", loosen capscrews slightly and shift crankshaft toward front of engine, then toward rear of engine. Retighten capscrews as outlined below, and recheck end play.

Tightening Procedure

1. Lubricate bolt threads and start main bearing cap side bolts (tightening these bolts will be done later).
2. Tighten both capscrews on each cap in 55–65 ft. lbs. increments to a final torque of 175–185 ft. lbs.
3. Loosen all capscrews to remove all tension.

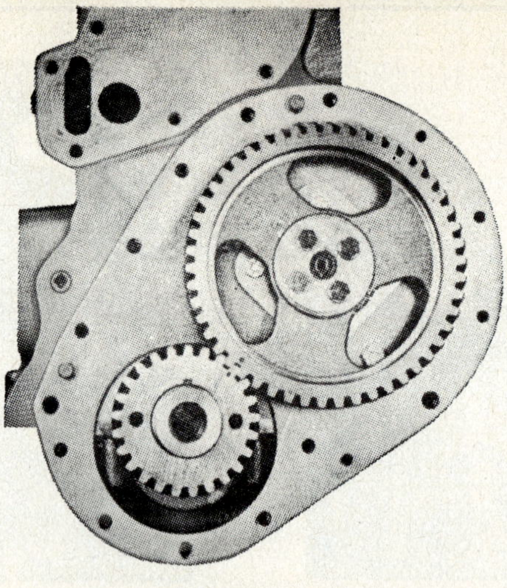

Fig. 17 Punch marks aligned for correct valve timing with No. 6 piston at top of its firing stroke. Overhead valve sixes

4. Repeat Step 2.
5. Tighten side bolts to 100–110 ft. lbs. in increments of 50–55 ft. lbs. using the sequence shown in Fig. 19.
6. Rotate crankshaft to see that it turns freely.

Side Valve Engines

Crankshaft end thrust is taken on the flanges of the center main bearing. Through normal wear end play will seldom become excessive but whenever new bearings are fitted, check the end play by forcing the crankshaft to the limit of its end play and check the clearance with a feeler gauge. If end play is greater than .007", install new thrust bearing shells. If clearance is less than .003", the bearing flanges may be dressed down to obtain the desired result.

Overhead Valve Engines

On six-cylinder engines crankshaft end thrust is taken up on the front main bearing. On early models it is controlled by flanges on the upper half only of this bearing. Later models have thrust flanges on upper and lower halves.

On V8 engines crankshaft end play is controlled by flanges on the rear main bearing.

A quick check of crankshaft end play can be made by anchoring a dial indicator to the chassis back of the radiator with the indicator on the front of the crankshaft pulley. Throwing the clutch in and out will cause a forward movement of the crankshaft to the extent of its end play. Using a piece of 2 × 4 wood ahead of the pulley, pry the crankshaft back (with foot off clutch pedal).

NOTE: *If check shows end play in excess of .012", immediate correction of the condition is absolutely necessary because if the engine is allowed to run with excessive end play it will destroy the highly polished face of the crankshaft against which the bearing thrust flanges ride. Ultimately, this polished surface will be scored to the extent that a new thrust bearing will wear out in a very few miles. When this condition is reached the only correction is a new crankshaft.*

OIL PUMP REPAIRS

470G Engine

To remove the pump, follow the procedure given under *Camshaft and Timing Gears*. Disassembly of pump is an obvious operation. Clean and inspect all parts and replace those that are worn or damaged. Then assemble pump as follows:

1. Press idler shaft into pump body so that shaft protrudes 1.847" above bottom surface of pocket.
2. Press driven gear on shaft flush to .035" below surface of gear, Fig. 20.
3. Install driven gear and shaft into pump body; support drive shaft and press drive gear on shaft to the dimension shown in Fig. 20 (concave side toward gear body).
4. Slide idler gear over idler shaft.
5. Install cover to body, using new gasket. Torque capscrews to 180–205 inch lbs.
6. With dial indicator, check for

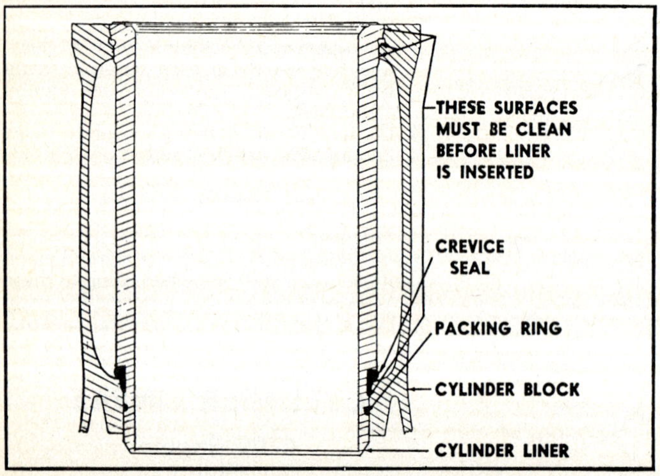

Fig. 18 Cylinder liner installed. 470G engine

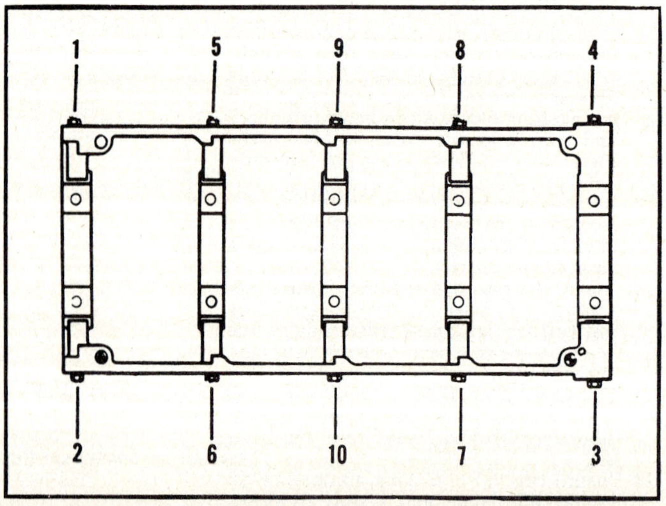

Fig. 19 Main bearing cap side bolt tightening sequence. 470G engine

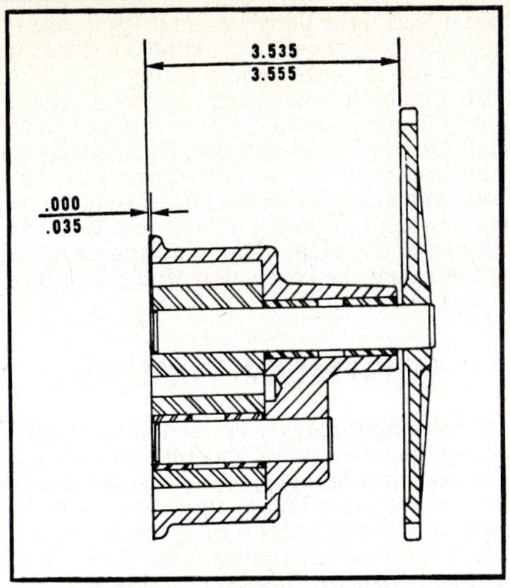

Fig. 20 Oil pump dimensions 470G engine

.0045–.0075" end clearance on pump drive shaft.

7. Install regulator plunger and spring.

Side Valve Engines

These pumps can easily be disassembled by taking off the upper drive gear and bottom cover, after which the internal parts can readily be removed. Be sure to examine the pump pressure ball check, spring and seat, and if these parts show wear, new parts must be installed.

NOTE: On some engines, the oil pump is the rotor type. It is easily disassembled but the following are several points which should be noted when the pump is serviced.

1. The end clearance should be held to .0015 to .0045" on the inner rotor and .001" to .004" on the outer rotor. These measurements are the difference between the depth of the counterbore in the body, including the gasket, and the length of the rotor. Always use a new gasket in assembling the pump.

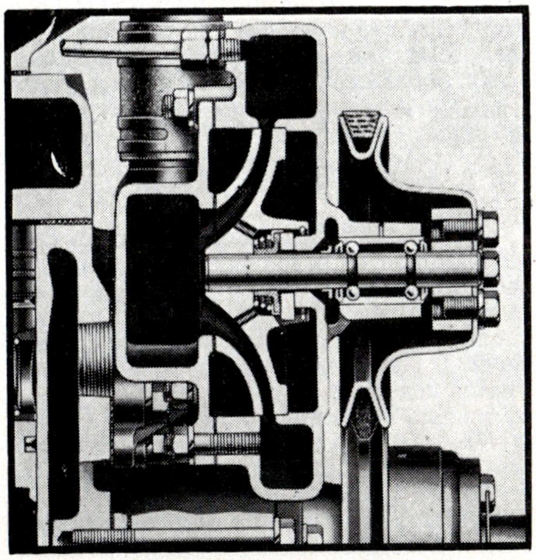

Fig. 23 Seated bearing type water pump

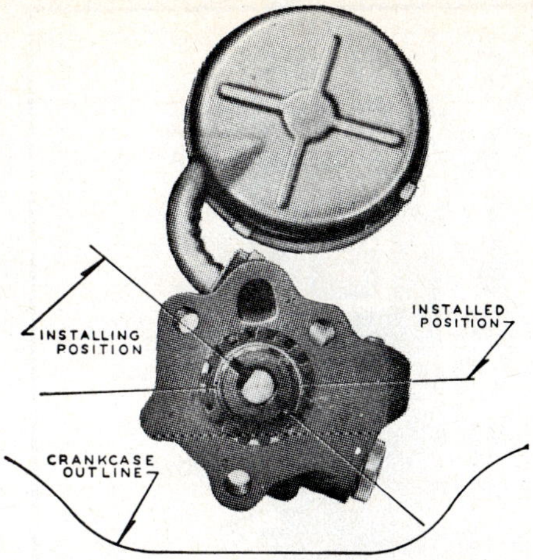

Fig. 21 Installing oil pump on overhead valve sixes

2. The inner and outer rotor should have a tooth clearance of .002 to .010".
3. The outer rotor outside diameter clearance should be from .003 to .010". The clearance is measured between the outside diameter of the outer rotor and the inside diameter of the pump body.
4. Always install a new snap ring when the rotor has been removed from the shaft.
5. When assembled, the pump should turn finger free before installing on the engine.

NOTE: On 110A, 120A, 140A, 150A, 250A, whenever the oil pump is removed, it is advisable to remove the ignition distributor as well, since it is driven directly from an extension of the oil pump shaft.

To install the pump, crank the engine until No. 1 piston is on its compression stroke and stop when it is on top dead center, as indicated by the timing mark on the flywheel. With the engine in this position, install the oil pump. Replace the oil pump extension shaft. With the tongue on the lower end of the shaft in mesh with the oil pump shaft groove, check the position of the groove in distributor end of the extension shaft. Location of the oil pump drive gears may not permit exact positioning of the groove (45 degree angle from centerline of crankshaft); however, make sure the pump gear is not meshing one or more teeth off the prescribed position.

Installation, 6 Cyl. Engines

When installing oil pump, distributor should be removed to prevent interference by distributor shaft tongue.

If engine has been completely disassembled, oil pump should not be installed until crankshaft, camshaft and timing gears have been installed. Install oil pump as follows:

1. Turn engine over to No. 6 firing position. Be sure marks on timing gears are aligned as shown in Fig. 17.
2. Place oil pump distributor drive slot in position shown in Fig. 21. Note that slot is offset and that narrow segment of drive end is at top and rotated to the right.
3. Install long studs in two bolt holes in crankcase as pilots. Place gasket on pilot studs. Slide pump assembly into position

Fig. 22 Checking water pump impeller to body clearance. 470G engine

in crankcase being careful not to turn gear before it meshes with drive gear on camshaft. As oil pump gear meshes with camshaft drive gear it rotates counterclockwise approximately two teeth and slot then faces as shown in Fig. 21, installed position.

4. Mark No. 6 terminal position on distributor body and install distributor. Rotor should face No. 6 mark when distributor cap snap springs are in a line approximately parallel to cylinder block.

CAUTION: Do not tighten oil pump cap screws until distributor is installed. Check carefully to determine that distributor and oil pump assemblies rotate freely together after tightening pump fastenings.

WATER PUMP REPAIRS

470G Engine

Disassemble

1. Pull impeller from shaft.
2. Pull drive pulley from shaft.
3. Remove snap ring from groove in pump body.
4. Press on impeller end of shaft to remove shaft and bearing assembly. If small bearing remains in pump body, press it out.
5. Remove carbon face seal from body.
6. Remove slinger from shaft and press shaft from large bearing.
7. Clean and inspect all parts and replace those that show wear or damage.

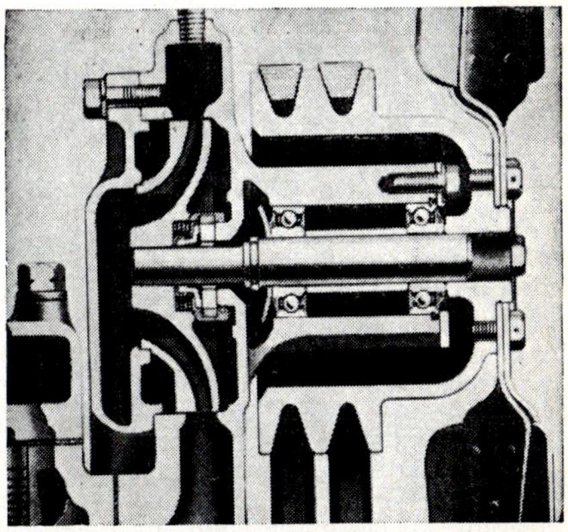

Fig. 24 Water pump with removable bearings

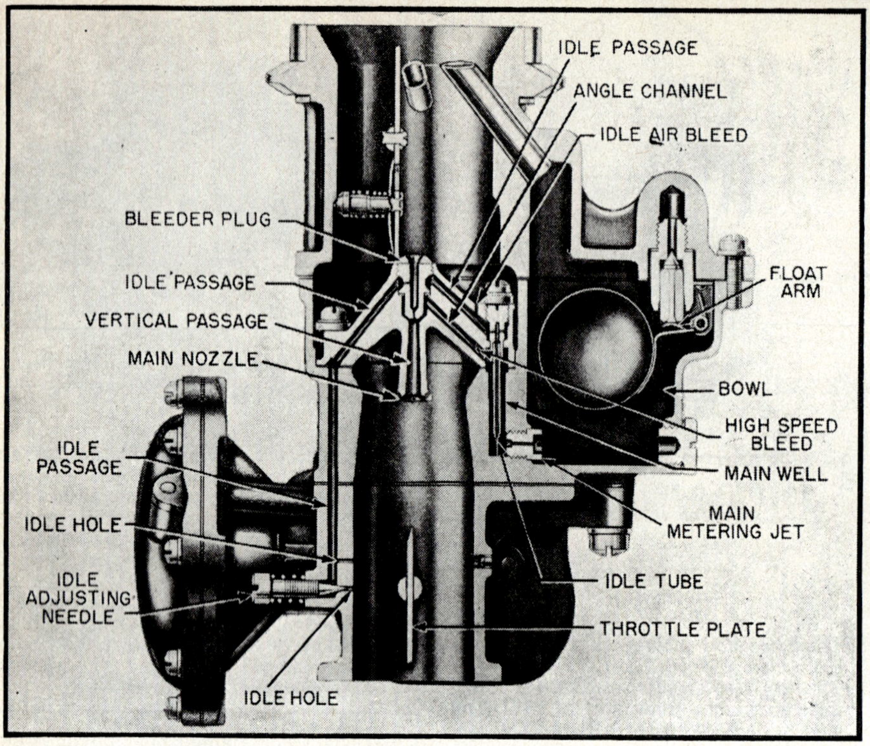

Fig. 25 Holley 847-FS carburetor. Typical of 852-FFC

Reassemble

1. Lubricate shaft and bearings with grease.
2. Pack bearings with grease.
3. Press small bearing on shaft with shielded side out toward impeller end of shaft until inner race seats against shoulder.
4. Press shaft into pump body bore.
5. Fill cavity of bore 1/2 to 2/3 full of grease.
6. Place large bearing over shaft and into bore of housing with shielded side out. Press bearing into its seat in housing bore.
7. Install large snap ring in pump body to secure bearings in place.
8. Support shaft on impeller end and press on drive pulley until pulley hub is tight against bearing inner race.
9. Apply suitable plastic lead sealer to bottom side or seat of carbon face seal driving flange.
10. Press in new carbon face seal. *Driving flange of seal case must seat on edge of bore in pump body.*
11. Support pump on pulley end of shaft; then press on impeller and maintain a .010-.020" clearance between impeller vanes and pump body, Fig. 22.

Sealed Bearing Type, Fig. 23

This type pump is packed with lubricant at the time of assembly and requires no further lubrication. When disassembling, carefully note the arrangement of the seal parts so that correct assembly may be assured.

Never soak the shaft and bearing assembly in cleaning solvent because it will dissolve the lubricant. To clean the shaft and bearing, merely moisten a cloth with cleaning fluid and wipe away any surplus lubricant or other foreign matter.

Be sure the face of the housing against which the thrust washer turns is smooth and true. If not, smooth it up with a special water pump facing cutter.

Removable Bearing Type, Fig. 24

To service the pump, remove the fan pulley. Release the snap ring from the front bearing. Take off the rear cover plate. Press the shaft out of the impeller from the rear and remove the shaft and bearings as a unit. Press the shaft from the bearings, sleeve and slinger. Remove the seal from the impeller, noting the arrangement of the seal parts so that the new seal may be installed correctly.

Before assembling the pump, check the thrust face of the housing and if the surface is not smooth or true it should be smoothed up with a special water pump facing cutter.

To assemble the pump, apply a liberal quantity of grease to the bearings and to the shaft between the bearings. Then install the shaft and bearings in the housing as a unit. Press the impeller and seal assembly down onto the shaft, pressing it down just flush with the end of the shaft. Complete the assembly by installing the cover plate and fan pulley.

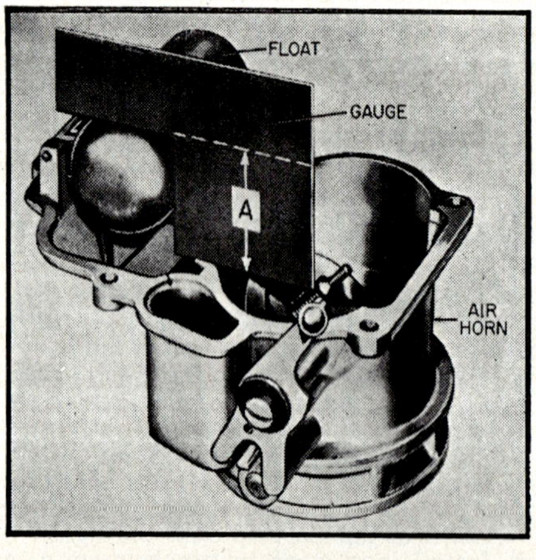

Fig. 26 Gauging float level. Holley carburetors

IGNITION TIMING

If a timing light is to be used to set ignition timing, disconnect the vacuum advance pipe to the carburetor and place a piece of tape over open end of pipe. *This is important as carburetor trouble can affect timing adjustments.*

Lacking a power timing light, an accurate method of setting ignition timing with the engine stopped is with the aid of a jumper light. Be sure to use a light bulb that corresponds with the system voltage of the vehicle.

1. Remove distributor cap and rotor and see that the breaker cap is set according to specifications.
2. Rotate engine until No. 1 cylinder is at the ignition timing point as indicated by the timing pointer and timing mark being lined up with each other.
3. Connect the jumper light between distributor ignition terminal and ground.
4. Turn on ignition switch.
5. Loosen distributor and turn it in the direction of normal rotation until the points just close (light out). Then slowly turn distributor in the opposite direction just to the exact point that the light goes on. Tighten distributor in this position.

HOLLEY CARBURETORS

Models 847-FS & 852-FFG

These carburetors are dual downdraft units incorporating an integral governor, Fig. 25. The governor throttle mechanism is part of the carburetor and has been designed to provide the needed power required for moving the throttle to governing speeds as controlled by a separately driven governor rotor. The combination of these two units gives instant response and accurate governing.

The carburetor and governor can be cleaned and serviced with ordinary tools and the governor rotor can be serviced independent of the carburetor. *Do not under any circumstances apply compressed air to an assembled carburetor as to do so may rupture the governor diaphragm.*

Fuel & Float Level

When checking the fuel level in an open vessel, under 4 lbs. (3 on 852) pump pressure the level should measure 11/16". The measurement is taken below the surface of the float chamber cover (gasket in place) to the level of the fuel. The level may be corrected by bending the float arm, Fig. 25.

The float level can be set when the air horn is off by gauging the distance between the bottom of the float and the flange surface of the air horn, Fig. 26. Holding the air horn upside down and the float in the closed position, measure the distance from the flange surface of the air horn to what would normally be the bottom side of the float (not the soldered seam). The correct distance should measure

$1^{9}/_{32}''$ on 847 units and $1^{11}/_{32}''$ on 852 carburetors. Bend the float lever arm as required to correct the float level.

Idle Mixture

The idle mixture is controlled by the idle adjustment needles. Turning these needles in makes a leaner mixture and turning them out makes the mixture richer. Turn the needles in until they just touch the seats and back off one full turn.

When the engine has warmed up to operating temperature, alter the adjustment as required to obtain smooth performance. The range should be between 1 and $1^{1}/_{2}$ turns.

Accelerating Pump

The quantity of the fuel discharged by the accelerating pump can be controlled by changing the position of the pump link in the throttle lever. Of the three different positions (marked 1, 2 and 3) number 2 is the average setting, number 1 is for summer and 3 is for winter operation.

Models 885-JJG & 885-JJSG

These units are dual concentric carburetors with all metering jets and passages grouped together and located in the center of the float bowl. The float bowl is a separate unit, being suspended inside the main body from the main body cover. With this design the fuel is kept cool by circulation of the main air supply around and under the float bowl.

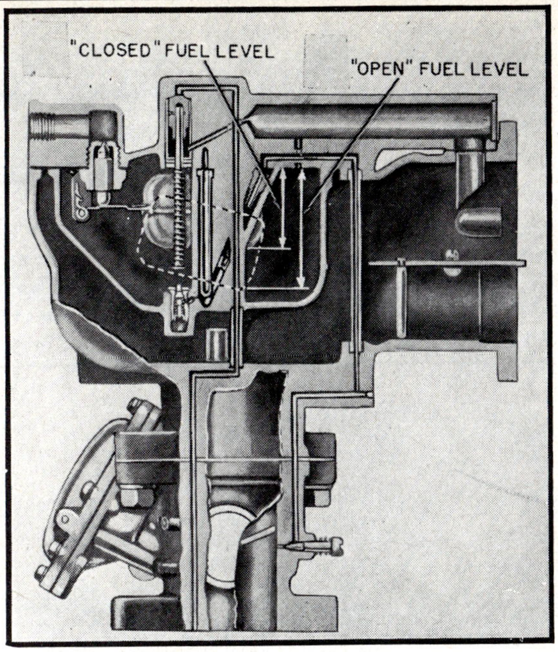

Fig. 27 Checking fuel level. Holley 885 carburetors

Fuel & Float Level

When checking the fuel level in an open vessel, under 5 lbs. pressure the fuel level should be $^{19}/_{32}''$ on 885-JJSG and $^{5}/_{8}''$ on 885-JJG. Measurement is taken from the float chamber cover with gasket in place to the level of the fuel in the bowl.

The float level can be set accurately by gauging the distance between the bottom of the floats and the chamber cover surface. When setting the float level, it is important that both the "Open" and "Closed" position of the floats be checked, Fig. 27.

To gauge the "Closed" position of the floats, hold the float chamber cover upside down. Then measure the distance from the float chamber gasket to what would normally be the bottom side of the free end of the floats. The distance should be $1^{5}/_{16}''$ on both models. Measurement is taken with fuel inlet valve in the closed position and the float chamber cover gasket in place. To correct the float setting for this position, the float lever arms should be bent up or down to bring the floats within the established limits and thus correct the fuel level for best operation. Pushing the floats up or toward the float cover, raises the fuel level, and down or away from the cover lowers the level.

The "Open" position is gauged by raising the floats up until the float stop arm comes to rest against the inlet needle housing. Then measure the float setting in the same manner as for the "Closed" position. The distance should measure $1^{15}/_{16}''$ on 885-JJSG ($1^{1}/_{4}''$ on 885-JJG) with the gasket in place and the float in the position illustrated in Fig. 27. Corrections for this setting can be made by bending the stop arm on the float lever.

Idle Mixture Setting

The idle needle valve setting from the closed position is from $^{3}/_{4}$ to $1^{1}/_{4}$ turns open. Care should be taken not to jam the needles against the seat tight enough to groove the points. If this occurs, the needles will have to be replaced before a satisfactory adjustment can be made.

When the engine has been warmed up, turn one idle needle in until the engine starts to slow down, then turn it out until it again slows down. A point midway between these two will be very close to the correct idle mixture. Repeat this procedure for the other needle. When this is done, cross check by repeating both of the above procedures. The idle speed should be reset, if necessary, between the above operations.

Transmission & Rear Axles

TRANSMISSIONS

Transmission Identification

NOTE: White model number stamped on plate attached to left-hand cab door. For service procedures and lubricant capacities on Clark, Fuller, New Process, Spicer and Warner transmissions, see the "name" chapters

White Model	Mfrs Model	Make	White Model	Mfrs Model	Make	White Model	Mfrs Model	Make
400B	T9A	Warner	524B	5A43	Fuller	574B	5A430	Fuller
418B	233F	Clark	525B	310V	Clark	581B	5AH650	Fuller
421B	5A650	Fuller	526B	318V	Clark	582B	305V	Clark
423B	8041	Spicer	527B	5AH65	Fuller	583B	315V	Clark
424B	185F	Clark	528B	308V	Clark	591B	5CW650	Fuller
426B	T98	Warner	529B	250V	Clark	800B	R45C	Fuller
427B	332F	Clark	530B	5CW65	Fuller	801B	R46	Fuller
501B	270V	Clark	531B	5AH65	Fuller	802B	R46S	Fuller
502B	205V	Clark	532B	5CW65	Fuller	1003B	303X	Clark
506B	5A65	Fuller	537B	5CW65	Fuller	1004B	403X	Clark
507B	290V	Clark	541B	5CW65	Fuller	1006B	10FB65	Fuller
509B	269V	Clark	551B	270 VO	Clark	1007B	R95C	Fuller
510B	5C72	Fuller	552B	205 VO	Clark	1026B	R63	Fuller
512B	4652	Spicer	556B	5A650	Fuller	1027B	R96	Fuller
513B	265V	Clark	557B	290 VO	Clark	1037B	R96S	Fuller
514B	300V	Clark	558B	8251	Spicer	1050B	10B1120	Fuller
515B	5A62	Fuller	559B	8051	Spicer	1056B	10FB650	Fuller
516B	5A65	Fuller	560B	5C720	Fuller	1057B	R950C	Fuller
517B	5A65	Fuller	561B	6453A	Spicer	1076B	R630D	Fuller
520B	295V	Clark	562B	4753	Spicer	1077B	R960	Fuller
521B	5A65	Fuller	565B	5A620	Fuller	1200B	8125	Spicer
522B	290V	Clark	566B	6455	Spicer			

REAR AXLES

NOTE: See "name" chapters for service procedures and lubricant capacities on Eaton and Rockwell (Timken axles)

White Model	Mfrs Model	Make	White Model	Mfrs Model	Make	White Model	Mfrs Model	Make
90C	SD3010	Rockwell	322C	R301	Rockwell	407C	32M	Eaton
94C	SFD460	Rockwell	323C	17800	Eaton	408C	SW3020	Rockwell
106C	H100	Rockwell	324C	22501	Eaton	410C	SFD75H	Rockwell
108C	L100	Rockwell	325C	Q300	Timken	411C	SLDD	Rockwell
109C	F145	Rockwell	327C	QT300	Timken	412C	SQD	Rockwell
116C	E100	Rockwell	328C	19500	Eaton	413C	SQDD	Rockwell
132C	R101	Rockwell	329C	18800	Eaton	414C	SQW	Rockwell
135C	R101	Rockwell	333C	QT360	Rockwell	415C	SFDD4600	Rockwell
136C	H140	Rockwell	333TC	RT340	Rockwell	417C	SRDD	Rockwell
138C	L140	Rockwell	335C	R301	Rockwell	418C	SLHD	Rockwell
206C	H200	Rockwell	336C	H340	Rockwell	420C	SQDD	Rockwell
208C	L200	Rockwell	338C	L340	Rockwell	421C	SRDD	Rockwell
232C	R201	Rockwell	342C	16800	Eaton	422C	SLDD	Rockwell
235C	R201	Rockwell	389C	QT360	Rockwell	423C	SLD	Rockwell
306C	H300	Rockwell	389TC	RT340	Rockwell	424C	SQD	Rockwell
308C	L300	Rockwell	400C	36M	Eaton	425C	SQHD	Rockwell
315C	R301	Rockwell	401C	SFD4600	Rockwell	426C	38DS	Eaton
317C	13600	Eaton	402C	28M	Eaton	427C	34DS	Eaton
318C	16600	Eaton	403C	22M	Eaton	429C	SRHD	Rockwell
319C	17600	Eaton	404C	SW456	Rockwell	430C	SJHD	Rockwell
320C	18500	Eaton	405C	SW3456	Rockwell	433C	38DP	Eaton
321C	20500	Eaton	406C	SLD	Rockwell			

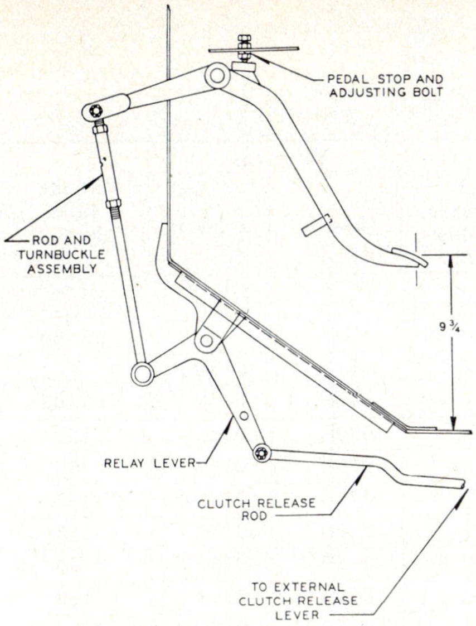

Fig. 1 Clutch linkage, 4000, 9000 and Road Boss series

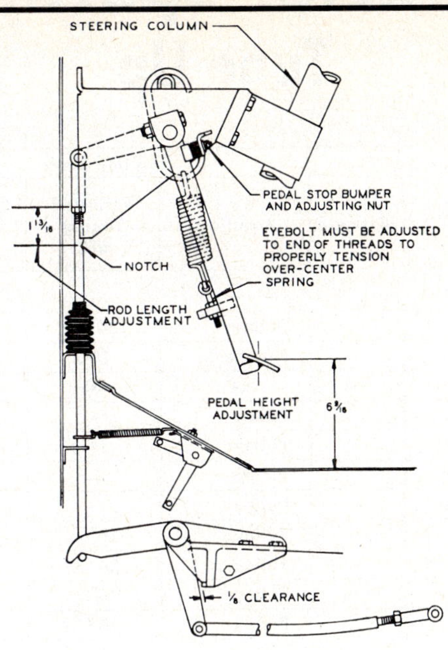

Fig. 2 Clutch linkage. 6000 series

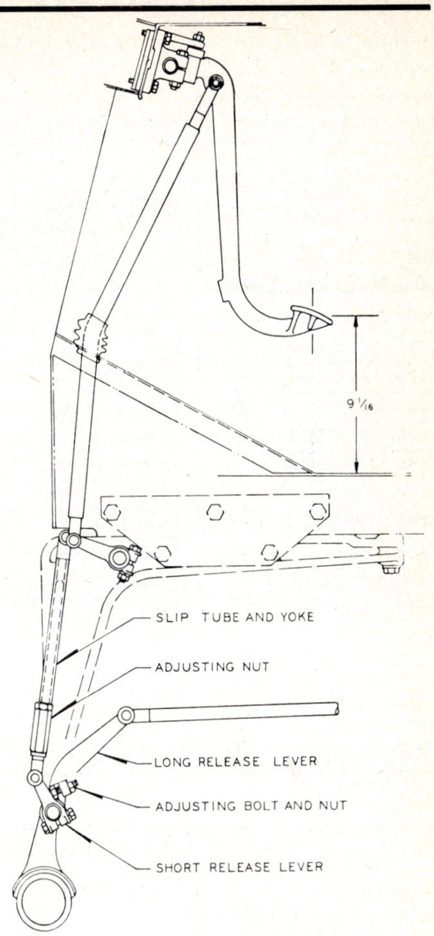

Fig. 3 Clutch linkage. 7400 and Road Commander series

CLUTCH LINKAGE ADJUSTMENT

4000, 9000 & Road Boss Series Fig. 1

1. Insert a 1/8 inch thick spacer between each yoke release lever finger and release bearing wear pad. Hold release yoke fingers firmly in contact with spacers by applying pressure to the external release lever while adjustments are being made.
2. Check clutch pedal height. If height is not 9 3/4 inches, turn pedal stop bolt as necessary to bring pedal height to specifications.
3. Loosen upper and lower jam nuts at turnbuckle in vertical linkage rod, then loosen turnbuckle until free play is obtained in linkage and retighten turnbuckle until all free play is just eliminated.

NOTE: Clutch pedal must be blocked in the "up" position or "clutch engaged" position while adjusting linkage.

---

4. Tighten jam nuts, remove spacers and check clutch pedal for free play.

6000 Series, Fig. 2

1. Insert a 1/8 inch thick spacer between each yoke release lever finger and release bearing wear pad. Hold release yoke fingers firmly in contact with spacers by applying pressure to the external release lever while adjustments are being made.
2. Check clutch pedal height. If height is not 6 9/16 inches, turn pedal stop adjusting nut as necessary to bring pedal height to specifications.
3. Loosen break-away clutch rod yoke jam nut.

NOTE: Jam nut is accessible from underneath instrument panel.

---

4. Tilt the cab and check clearance between relay lever and lever bracket. If clearance is not 1/8 inch, loosen jam nut, then disconnect clutch release rod and turn yoke as necessary to obtain desired clearance. After adjusting, reconnect release rod and tighten jam nut.
5. Rotate break-away clutch rod several turns to lengthen it, then return cab to driving position and check clearance between top of notch in relay rod and contact surface of break-away clutch rod.

NOTE: The 1 13/16 inch dimension shown in Fig. 2, is for assembly purposes only. Slight variations from this dimension are permissible when making final adjustments.

---

6. Tilt the cab and turn clutch rod as necessary until all clearance is just eliminated between the relay lever and clutch rod.
7. Return cab to driving position, then tighten clutch rod yoke jam nut, remove spacers from release bearing and check clutch pedal for free play.

7400 & Road Commander Series, Fig. 3

1. Insert a 1/8 inch thick spacer between each yoke release lever finger and release bearing wear pad. Hold release yoke fingers firmly in contact with spacers by applying pressure to the external release lever while adjustments are being made.
2. Check clutch pedal height. If height is not 9 1/16 inches, turn pedal stop adjusting bolt as necessary to bring pedal height to specifications.
3. Tilt the cab, then back off clutch rod jam nut to end of threads and turn adjusting nut to shorten slip tube end of adjustable clutch rod.
4. Loosen lock nut and turn adjusting bolt inward until there is clearance between bolt head and short release lever, then turn adjusting bolt outward until bolt head just contacts short release lever and tighten lock nut.

NOTE: Clutch pedal must be blocked in "up" position or "clutch engaged" position while adjustments are being made.

---

5. Return cab to driving position, then turn adjustable clutch rod adjusting nut until all clearance is just eliminated between the adjusting nut and slip yoke flange and tighten jam nut.
6. Remove spacers from release bearing and check clutch pedal for free play.

Farm Tractor Specifications
Tune Up

Tractor Model	Firing Order	Spark Plug Gap	Breaker Gap		Ignition Mark			Valve Lash		Cylinder Head Torque
			Dis-tributor	Magneto	Location	Retarded Timing BTC	Full Advance Timing BTC	Intake	Exhaust	
ALLIS CHALMERS										
160 Diesel	123	—	—	—	Pulley	24°	—	⑥	⑥	55–60
170, 175 Gasoline	1243	.025	.022⑦	—	Flywheel	TDC	"F-25"	.011H	.015H	②
170, 175 Diesel	1342	—	—	—	④	④	④	.010H	.010H	80–85
180, 185, 190 Gasoline	153624	.025	.016	—	Pulley	—	26°	.020H	.025H	150
180, 185, 190 Diesel	153624	—	—	—	Pulley	⑤	—	.015H	.015H	150
190 XT Gasoline	153624	.025	.016	—	Pulley	—	26°	.020H	.025H	150
190 XT LPG	153624	.020	.016	—	Pulley	--	27°	.020H	.025H	150
190 XT Diesel	153624	—	—	—	Pulley	26°③	③	.015H	.015H	150
200 Diesel	153624	—	—	—	Pulley	24°③	③	.015H	.015H	150
210, 220 Diesel	153624	—	—	—	Pulley	24°③	③	.015H	.015H	130–140
5020, 5030 Diesel	12	—	—	—	Pulley	20°①	—	.012C	.014C	65–80
7000 Diesel	153624	—	—	—	Pulley	18°③	③	.015H	.015H	150
7030 Diesel	153624	—	—	—	Pulley	24°③	③	.015H	.015H	150
7040 Diesel	153624	—	—	—	Pulley	16°③	③	.015H	.015H	150
7050 Diesel	153624	—	—	—	Pulley	26°③	③	.015H	.015H	150
7060 Diesel	153624	—	—	—	Pulley	18°③	③	.015H	.015H	150
7080 Diesel	153624	—	—	—	Pulley	22°③	③	.015H	.015H	150

①—Single notch on pulley located before double notches indicates 20° BTDC, No. 1 cylinder. Double notches indicate TDC, No. 1 cylinder.

②—1/2" capscrews 90–95 ft. lbs.; 7/16" capscrews 70–75 ft. lbs.

③—For injection timing, pointer on timing gear cover should be aligned with correct timing mark on pulley. Timing lines on governor plate and cam should then be aligned.

④—The "c" timing mark should align with straight edge of snap ring with No. 1 piston at 22° BTDC.

⑤—180, 185; 26°; 190; 28°.

⑥—.010 Hot; .012 Cold.

⑦—Dwell angle 31°–34°.

Tractor Model	Firing Order	Spark Plug Gap	Breaker Gap		Ignition Mark			Valve Lash		Cylinder Head Torque
			Dis-tributor	Magneto	Location	Retarded Timing BTC	Full Advance Timing BTC	Intake	Exhaust	
J. I. CASE										
430 Gasoline	1342	.025	.020	—	Flywheel	4° BTC	—	.014C	.014C	95–100
430 Diesel	1342	—	—	—	Flywheel③	8° BTC④	—	.014C	.014C	①
470 Gasoline	1342	.025	.020	—	Flywheel	4° BTC	—	.014C	.014C	95–100
470 Diesel	1342	—	—	—	Flywheel③	8° BTC④	—	.014C	.014C	①
530 Gasoline	1342	.025	.020	—	Flywheel	4° BTC	—	.014C	.014C	95–100
530 Diesel	1342	—	—	—	Flywheel③	8° BTC④	—	.014C	.014C	①
570 Gasoline	1342	.025	.020	—	Flywheel	4° BTC	—	.014C	.014C	95–100
570 Diesel	1342	—	—	—	Flywheel③	8° BTC④	—	.014C	.014C	①
630 Gasoline	1342	.025	.020	—	Flywheel	4° BTC	—	.014C	.014C	95–100
630 Diesel	1342	—	—	—	Flywheel③	8° BTC④	—	.014C	.014C	①
770 Gasoline	1342	.025	.020	—	Flywheel	5° ATC	28° BTC	.015C	.025C	210
770 Diesel	1342	—	—	—	Pulley②	29° BTC④	—	.015C	.025C	210
870 Gasoline	1342	.025	.020	—	Flywheel	5° ATC	28° BTC	.015C	.025C	210
870 Diesel	1342	—	—	—	Pulley②	29° BTC④	—	.015C	.025C	210
885 Diesel	123	—	—	—	Flywheel⑥	—	—	.010C	.010C	90
970 Gasoline	153624	.025	.020	—	Flywheel	—	27° BTC	.015C	.025C	210
970 Diesel	153624	—	—	—	Flywheel③	29° BTC④	—	.015C	.025C	210
995 Diesel	1243	—	—	—	⑦	⑦	—	.010C	.010C	90
1070 & 1090 Diesel	153624	—	—	—	Pulley②	30° BTC④	—	.015C	.025C	210
1170 Diesel	153624	—	—	—	Pulley②	30° BTC④	—	.015C	.025C	210
1175 Diesel	153624	—	—	—	Pulley②	30° BTC④	—	.015C	.025C	210
1210 & 1212 Diesel	1243	—	—	—	⑧	⑧	—	.010C	.010C	90
1270	153624	—	—	—	Pulley②	30° BTC④	—	.015C	⑤	210
1370	153624	—	—	—	Pulley②	30° BTC④	—	.015C	⑤	210
1410 & 1412 Diesel	1243	—	—	—	⑧	⑧	—	.010C	.010C	90
1570	153624	—	—	—	Pulley②	25° BTC④	—	.015C	⑤	210

Continued

J. I. CASE TUNE-UP NOTES—Continued

①—Head stud nut 95–100 ft. lbs., head cap screws 110–115 ft. lbs.
②—With timing correct mark on pulley & pointer on front cover aligned, align timing mark on injection pump housing with timing mark on drive gear hub.
③—With timing marks on flywheel & flywheel housing aligned, align timing mark on governor cage with timing mark on cam ring.
④—Injection timing.
⑤—Cold .025"; hot .020".
⑥—With timing pin inserted through housing TDC hole & into flywheel detent, align timing mark on pump housing with mark on drive flange.
⑦—With pump drive gear & engine timing gear marks aligned, align timing mark on pump housing with mark on engine front plate.
⑧—With pump drive gear & idler gear properly meshed, align timing mark on pump housing with mark on engine front plate.

Tractor Model	Firing Order	Spark Plug Gap	Breaker Gap		Ignition Mark			Valve Lash		Cylinder Head Torque
			Dis-tributor	Magneto	Location	Retarded Timing BTC	Full Advance Timing BTC	Intake	Exhaust	
JOHN DEERE										
820, 830 Diesel	123	—	—	—	Pulley③	②③	—	.014H	.018H	110
850 Diesel	—	—	—	—	Pulley	26° BTC③	—	.008C	.008C	41.3
950, 1050 Diesel	—	—	—	—	Pulley	25° BTC③	—	.006C	.006C	48.7
1020 Gasoline	123	.025	.020	—	Pulley	⑭	—	.014	.022	110
1020 Diesel	123	—	—	—	Pulley	⑮	—	.014	.018	110
1520 Gasoline	123	.025	.020	—	Pulley	⑭	—	.014	.022	110
1520 Diesel	123	—	—	—	Flywheel④	②④	—	.014H	.018H	110
1530 Diesel	123	—	—	—	Flywheel⑤	②⑤	—	.014C	.018C	110
2020, 2030 Gasoline	1342	.025	.020	—	Pulley	—	—	.014C	.022C	110
2020 Diesel	1342	—	—	—	Flywheel⑥	②⑥	—	.014C	.018C	110
2030 Diesel	1342	—	—	—	Flywheel⑤	②⑤	—	.014C	.018C	110
2040 Diesel	123	—	—	—	⑩	②⑩	—	.014	.018	110
2240 Diesel	123	—	—	—	Pulley⑪	—	—	.014	.018	110
2440 Diesel	1342	—	—	—	Pulley⑦	TDC	7°	.014	.018	95
2510, 2520 Gasoline	1342	.025	.020	—	Pulley	⑭	30°	.014C	.022C	110
2510, 2520 Diesel	1342	—	—	—	Flywheel⑦	TDC②	—	.014C	.018C	110
2630 Diesel	1342	—	—	—	Pulley⑦	TDC	7.5°	.014	.018	110
2640 Diesel	1342	—	—	—	Pulley⑦	TDC	6°	.014	.018	95
2840 Diesel	153624	—	—	—	⑩	②⑩	—	.014	.018	110
3010 Gasoline	1342	.025	.022	—	Flywheel	5° ATC	20°	.015H	.028H	115
3010 LP-Gas	1342	.015	.022	—	Flywheel	TDC	25°	.015H	.028H	115
3010 Diesel	1342	—	—	—	Flywheel⑧	10°②	10°	.018H	.018H	115
3020 Gasoline	1342	.025	.022	—	Flywheel	5° ATC	20°	.015H	.028H	115
3020 LP-Gas	1342	.015	.022	—	Flywheel	TDC	25°	.015H	.028H	115
3020 Diesel	1342	—	—	—	Flywheel⑧	TDC②	—	.018H	.018H	115
4010 Gasoline	153624	.025	.022	—	Flywheel	5° ATC	20°	.015H	.028H	115
4010 LP-Gas	153624	.015	.022	—	Flywheel	TDC	25°	.015H	.028H	115
4010 Diesel	153624	—	—	—	Flywheel⑧	14°②	14°	.018H	.018H	115
4020 Gasoline	153624	.025	.022	—	Flywheel	5° ATC	20°	.015H	.028H	115
4020 LP-Gas	153624	.025	.022	—	Flywheel	TDC	25°	.015H	.028H	115
4020 Diesel	153624	—	—	—	Flywheel⑧	TDC②	—	.018H	.018H	115
4030 Gasoline	153624	.025	.020	—	Pulley	—	24°BTC	.014	.022	110
4030 Diesel	153624	—	—	—	①⑫	TDC②	—	.014C	.018C	110
4040 Diesel	153624	—	—	—	①⑫	TDC②	—	.018	.028	150
4230 Gasoline	153624	.025	.022	—	Pulley	—	20°BTC	.015	.028	130
4230 Diesel	153624	—	—	—	①⑫	TDC②	—	.018C	.028C	130
4240 Diesel	153624	—	—	—	①⑫	TDC②	—	.018	.028	150
4320 Diesel	153624	—	—	—	Flywheel⑧	TDC②	—	.018	.022	130
4430 Diesel	153624	—	—	—	Flywheel⑬	TDC②	—	.018C	.028C	150
4440 Diesel	153624	—	—	—	Flywheel⑬	TDC②	—	.018	.028	150
4520 Diesel	153624	—	—	—	Flywheel	TDC②	—	.018	.022	130
4620 Diesel	153624	—	—	—	Flywheel	TDC②	—	.018	.022	130
4630 Diesel	153624	—	—	—	Flywheel⑬	TDC②	—	.018C	.028C	150
4640 Diesel	153624	—	—	—	Flywheel⑬	TDC②	—	.018	.028	150
4840 Diesel	153624	—	—	—	Flywheel⑬	TDC②	—	.018	.028	150
5010 Diesel	153624	—	—	—	Flywheel⑨	TDC②	—	.015H	.022H	⑯

Continued

FARM TRACTOR TUNE-UP—Continued

Tractor Model	Firing Order	Spark Plug Gap	Breaker Gap		Ignition Mark			Valve Lash		Cylinder Head Torque
			Dis-tributor	Magneto	Location	Retarded Timing BTC	Full Advance Timing BTC	Intake	Exhaust	
JOHN DEERE—Continued										
5020 Diesel	153624	—	—	—	Flywheel⑨	TDC③	—	.018H	.028H	⑭
6030 Diesel	153624	—	—	—	①⑨	TDC③	—	.018	.028	180
6030 Turbo Diesel	153624	—	—	—	Pulley⑮	TDC③	—	.018	.028	180

①—Pump window.
②—Injection pump timing.
③—Timing on 830 tractor models cannot be checked. On 820 models prior to serial #047753 and after #082143, after timing pin is set with flywheel, timing marks on governor weight retainer and pump cam should align. On 820 models between pump serial #047753 and #082142 use of flywheel and pump timing pins are necessary. If both pins index at same time, timing is correct.
④—On pump models between serial #83759 and #117499, use of flywheel and pump timing pins are necessary, if both index at same time, timing is correct. On pump models after serial #117500, after flywheel pin is inserted, governor weight retainer and pump cam timing lines should align.
⑤—On some models the CAV rotor pump was installed, in this case timing cannot be checked. In all other cases after flywheel pin has been inserted, governor weight retainer and pump cam timing lines should align.
⑥—On pump models before #50303 and after #117500, after flywheel pin has been installed, governor weight retainer and pump cam timing lines should align. On pump models between serial #50304 and #117499, use of flywheel and pump timing pins are necessary, if both pins index at same time, timing is correct.
⑦—On pump model DBG and JDB after flywheel pin has been inserted, governor weight retainer and pump cam timing lines should align. On pump model CDC and CBC use of flywheel and pump timing pins are necessary. If both pins index at same time, timing is correct.
⑧—With marks on flywheel and flywheel timing window aligned, align scribe lines on governor cage and cam ring.
⑨—With timing marks on flywheel and timing pointer aligned, align scribe marks on the cam and governor cage.
⑩—Injection pump timing gears are stamped with two timing marks, "3" or "4". Use timing mark "3" when timing a three or six cylinder engine. The injection pump is properly timed when timing mark "3" on pump timing gear is directly below timing tool, when tool is placed between center lines of crankshaft & injection pump shafts.
⑪—With No. 1 cylinder at TDC, align slot on pump hub flange with dowel pin in gear.
⑫—With No. 1 cylinder at TDC, mark on cam ring should be aligned with mark on governor weight retainer.
⑬—With timing pin inserted into flywheel hole, align pump drive hub mark with pointer.
⑭—Align "S" or "28" mark on crankshaft pulley with line on right hand side timing case cover at 2500 RPM.
⑮—Align "INJ" mark on pulley with notch on timing case cover, then remove timing hole plug & check to ensure that mark on pump drive hub is aligned with mark on pointer.
⑯—"D" grade cap screws, 170 ft. lbs.; "F" grade cap screws, 180 ft. lbs.

Tractor Model	Firing Order	Spark Plug Gap	Breaker Gap Dis-tributor	Breaker Gap Magneto	Ignition Mark Location	Retarded Timing BTC	Full Advance Timing BTC	Valve Lash Intake	Valve Lash Exhaust	Cylinder Head Torque
FORD										
TW-10, 20, 30	153624	—	—	—	Flywheel⑦	23°⑤	—	.015H	.018H	160
1000, 1600 Diesel	12	—	—	—	⑦	21°③⑦	—	.012C	.012C	108–111
2000 Gas	123	.025	.025	—	Flywheel	22°①	—	.015H	.018H	105–115
2000 Diesel	123	—	—	—	Flywheel⑥	19°③	—	.015H	.018H	105–115
2600 Gas	123	.025	.025	—	Flywheel	2°	—	.016H	.019H	110
2600 Diesel	123	—	—	—	Flywheel⑧	23°③	—	.016H	.019H	110
3000 Gas	123	.025	.025	—	Flywheel	①⑨	—	.015H	.018H	105–115
3000 Diesel	123	—	—	—	Flywheel⑥	19°③	—	.015H	.018H	105–115
3600, 4100, 4600, 4600 SU Gas	123	.025	.025	—	Flywheel	4°	—	.016H	.019H	110
3600, 4100, 4600, 4600 SU Diesel	123	—	—	—	Flywheel⑧	23°③	—	.016H	.019H	110
4000 Gas	123	.025	.025	—	Flywheel	①⑩	—	.015H	.018H	105–115
4000 Diesel	123	—	—	—	Flywheel⑥	19°⑤	—	.015H	.018H	105–115
5000, 7000 Gas	1342	.025	.025	—	Flywheel	2°①	—	.015H	.018H	105–115
5000, 7000 Diesel	1342	—	—	—	Flywheel④	19°③	—	.015H	.018H	105–115
5600 Diesel	1342	—	—	—	Flywheel④	19°③	—	.016H	.0194	105–115
6000 Gas	153624	.030	.025	—	Damper	4°	24°	.015H	.015H	95–105
6000 Diesel	153624	—	—	—	Damper⑤	26°③	—	.015H	.015H	95–105
6600, 6700 Gas	1342	.025	.025	—	Flywheel	2°	—	.015H	.018H	105–115
6600, 6700 Diesel	1342	—	—	—	Flywheel④	19°③	—	.015H	.018H	105–115
7600, 7700 Diesel	1342	—	—	—	Flywheel④	23°③	—	.015H	.018H	105–115
8000, 8600, 8700 Diesel	153624	—	—	—	Flywheel④	23°⑤	—	.015H	.018H	②
9000, 9600, 9700 Diesel	153624	—	—	—	Flywheel④	23°③	—	.015H	.018H	②

①—With vacuum advance line disconnected.
②—8000, 8600, 9000 & 9600 (1/2" bolts), 110 ft. lbs.; 8700 & 9700 (9/16" bolts), 160 ft. lbs.
③—Injection pump timing.
④—With timing mark on flywheel and and arrow on timing hole aligned, align timing mark on the drive gear adapter hub with timing mark on pump housing.
⑤—With timing mark on vibration damper on gear cover aligned, align timing mark on injection pump drive flange with pointer on injection pump front cover.
⑥—C.A.V. pump—align scribe line on pump with "O" mark on engine front plate. Simms pump—with timing mark on flywheel and arrow on timing hole aligned, align timing mark on pump flange with scribe line.
⑦—Injection pump is properly timed when timing marks on pump drive gear & idler gear are aligned & marks on pump coupling & pump drive gear are aligned.
⑧—Injection pump is properly timed when scribed line on pump body is aligned with "O" mark on engine front plate.
⑨—Dist. No. WC5NF-12127A, 18° BTDC; dist. No. WC7NF-12127A, 22° BTDC.
⑩—Models built prior to June 1968, 24° BTDC; models built from June 1968, 22° BTDC.

Continued

FARM TRACTOR TUNE-UP—Continued

Tractor Model	Firing Order	Spark Plug Gap	Breaker Gap		Ignition Mark			Valve Lash		Cylinder Head Torque
			Dis-tributor	Magneto	Location	Retarded Timing BTC	Full Advance Timing BTC	Intake	Exhaust	
INTERNATIONAL HARVESTER										
Hydro 84	1342	—	—	—	Pulley⑭	16°⑩	—	.012H	.012H	90
100	1342	.023	.020	.013	Pulley	TDC	30°①	.014H	.014H	㉑
Hydro 100 & 186	153624	—	—	—	Flywheel⑬	18°⑩	—	.020H	.025H	165
130	1342	.023	.020	.013	Pulley	TDC	26°①	.014H	.014H	㉑
140	1342	.023	.020	.013	Pulley	TDC	①㉒	.014H	.014H	㉑
154 Lo-Boy	1342	.023	.020	—	Pulley	TDC	16°	.015C	.015C	45
Farmall Cub	1342	.023	.020	.013	Pulley	TDC	16°	.013C	.013C	45
IH Cub, Cub Lo-Boy	1342	.023	.020	.013	Pulley	TDC	16°	.013C	.013C	45
200	1342	.023	.020	.013	Pulley	TDC	30°①	.014H	.014H	㉑
230	1342	.023	.020	.013	Pulley	TDC	30°①	.014H	.014H	㉑
240	1342	.023	.020	.013	Pulley	TDC	30°①	.014H	.014H	㉑
B275 Diesel	1342	—	—	—	Pulley⑯	20°⑩	—	.020H	.020H	70–75
354 Gas	1342	.024	.014	—	Pulley	5°	39°	.020H	.020H	75–80
354 Diesel	1342	—	—	—	Pulley⑯	20°⑩	—	.020H	.020H	75–80
364 Diesel	1342	—	—	—	Flywheel⑯	16°⑩	—	.020H	.020H	75–80
384 Diesel	1342	—	—	—	Flywheel⑲	16°⑩	—	.020H	.020H	75–80
404	1342	.023	.020	—	Pulley	TDC	20°	.014H	.020H	80–90
B414 Diesel	1342	—	—	—	Flywheel⑯	TDC⑩	—	.020H	.020H	75–80
B414 Gas	1342	.024	.014	—	Pulley	5°	39°	.020H	.020H	75–80
424, 444 Diesel	1342	—	—	—	Flywheel⑲	16°⑩	—	.020H	.020H	75–80
424, 444 Gas	1342	.023	.020	—	Pulley	TDC	17°	.014H	.020H	80–90
454 Gas	1342	.023	.020	—	Pulley	TDC	17°	.027H	.027H	90
454 Diesel	132	—	—	—	Pulley⑭	14°⑩	—	.010H	.012H	90
464 Gas	1342	.023	.020	—	Pulley	TDC	20°	.027H	.027H	90
464 Diesel	132	—	—	—	Pulley⑭	14°⑩	—	.010H	.012H	90
484 Diesel	132	—	—	—	Pulley⑭	14°⑩	—	.010H	.020H	90
544 Diesel	1342	—	—	—	Pulley⑭	14°⑩	—	.010H	.012H	90
544, 574, 674 Gas	1342	.023	.020	—	Pulley	TDC	20°	.027H	.027H	90
574, 674 Diesel	1342	—	—	—	Pulley⑭	14°⑩	—	.010H	.012H	90
584 Diesel	1342	—	—	—	Pulley⑭	14°⑩	—	.010H	.020H	90
656 Diesel	153624	—	—	—	Pulley⑰	TDC⑩	—	.027H	.027H	85–95
656 LPG	153624	.015	.020	—	Pulley	2°	24°	.027H	.027H	85–95
656 Gasoline	153624	.023	.020	—	Pulley	2°	19°⑳	.027H	.027H	85–95
666, Hydro 70 Gas	153624	.023	.020	—	Flywheel⑬	TDC	18°	.027H	.027H	90
666, Hydro 70 Diesel	153624	—	—	—	Flywheel⑬	18°⑩	—	⑳	.021H	165
684 Diesel	1342	—	—	—	Pulley⑭	16°⑩	—	.012H	.012H	90
686, Hydro 86 Gas	153624	.023	.020	—	Pulley	TDC	—	.027H	.027H	90
686, Hydro 86 Diesel Early	153624	—	—	—	Flywheel⑬	18°⑩	—	⑳	.021H	165
686, Hydro 86 Diesel Late	153624	—	—	—	Pulley⑭	8°⑩	—	.010H	.012H	90
706 Gasoline	153624	.023	.020	—	Pulley	④	⑤	.027H	.027H	90
706 LP-Gas	153624	.015	.020	—	Pulley	2° BTC	24°	.027H	.027H	90
706 Diesel	153624	—	—	—	Pulley⑪⑮⑱	②⑩	—	⑥	⑥	⑦
756 Gas	153624	.023	.020	—	Flywheel	1°	18°	.027H	.027H	90
756 LPG	153624	.015	.020	—	Flywheel	2°	24°	.027H	.027H	90
756 Diesel	153624	—	—	—	Flywheel⑪	14°⑩	—	.010H	.012H	90
766 Gas	153624	.023	.020	—	⑧	1°	18°	.027H	.027H	90
766 Diesel	153624	—	—	—	Flywheel⑬	18°⑩	—	.012H	.021H	165
784 Diesel	1342	—	—	—	Pulley⑭	16°⑩	—	.012H	.012H	90
806 Gasoline	153624	.023	.020	—	Flywheel	TDC	22°	.027H	.027H	90
806 LP-Gas	153624	.015	.020	—	Flywheel	2°BTC	24°	.027H	.027H	90
806 Diesel	153624	—	—	—	Flywheel⑮⑱	③⑩	—	.013H	.025H	135
826 Gas	153624	.023	.020	—	Pulley	TDC	22	.027H	.027H	135
826 LP-Gas	153624	.015	.020	—	Pulley	2°	24	.027H	.027H	135

Continued

FARM TRACTOR TUNE-UP—Continued

Tractor Model	Firing Order	Spark Plug Gap	Breaker Gap		Ignition Mark			Valve Lash		Cylinder Head Torque
			Distributor	Magneto	Location	Retarded Timing BTC	Full Advance Timing BTC	Intake	Exhaust	
INTERNATIONAL HARVESTER—Continued										
826 Diesel	153624	—	—	—	⑧⑪	16°⑩	—	.010H	.012H	135
856 Gas	153624	.023	.020	—	Flywheel	TDC	22°	.027H	.027H	90
856 LPG	153624	.015	.020	—	Flywheel	2°	24°	.027H	.027H	90
856 Diesel	153624	—	—	—	Flywheel⑮	6°⑩	—	.013H	.025H	135
884 Diesel	1342	—	—	—	Pulley⑭	16°⑩	—	.012H	.012H	90
886 Early	153624	—	—	—	Flywheel	⑩㉔	—	.010H	.012H	90
886 Late	153624	—	—	—	Flywheel	18°⑩	—	.012H	.021H	165
966 Diesel	153624	—	—	—	Flywheel⑬	18°⑩	—	.020H	.025H	165
986 Diesel	153624	—	—	—	Flywheel	18°⑩	—	㉕	.025	165
1026 Diesel	153624	—	—	—	Flywheel⑫	6°⑩	—	.013H	.025H	135
1066 Diesel	153624	—	—	—	Flywheel⑬	18°⑩	—	.020H	.025H	165
1086 Early	153624	—	—	—	Flywheel	18°⑩	—	.020	.025	165
1086 Late	153624	—	—	—	Flywheel	18°⑩	—	.025	.025	165
1206 Diesel	153624	—	—	—	Flywheel⑮	8°⑩	—	.013H	.025H	135
1256 Diesel	153624	—	—	—	Flywheel⑮	6°⑩	—	.013H	.025H	135
1456 Diesel	153624	—	—	—	Flywheel⑮	2° ATC⑩	—	.013H	.025H	135
1466 Diesel	153624	—	—	—	Flywheel⑬	18°⑩	—	.020H	.025H	165
1468, 1568 Diesel	18736542	—	—	—	Pulley⑨	30°⑩	—	.014H	.016H	110
1486, 1586 Diesel	153624	—	—	—	Flywheel⑬	⑩㉓	—	.020H	.025H	165
1566 Diesel	153624	—	—	—	Flywheel⑬	15°⑩	—	.020H	.025H	165
2404	1342	.023	.020	—	Pulley	TDC	20°	.014H	.020H	80–90
2424, 2444 Diesel	1342	—	—	—	Flywheel⑲	16°⑩	—	.020H	.020H	75–80
2424, 2444 Gasoline	1342	.023	.020	—	Pulley	TDC	17°	.014H	.020H	80–90
2706 Gasoline	153624	.023	.020	—	Pulley	④	⑤	.027H	.027H	90
2706 LP-Gas	153624	.015	.020	—	Pulley	2° BTC	24°	.027H	.027H	90
2706 Diesel	153624	—	—	—	Pulley⑪⑮⑱	②⑩	—	⑥	⑥	⑦
2756 Gas	153624	.023	.020	—	Flywheel	1°	18°	.027H	.027H	90
2756 LPG	153624	.015	.020	—	Flywheel	2°	24°	.027H	.027H	90
2756 Diesel	153624	—	—	—	Flywheel⑮	14°⑩	—	.010H	.012H	90
2806 Gasoline	153624	.023	.020	—	Flywheel	TDC	22°	.027H	.027H	90
2806 LP-Gas	153624	.015	.020	—	Flywheel	2° BTC	24°	.027H	.027H	90
2806 Diesel	153624	—	—	—	Flywheel⑮⑱	③⑩	—	.013H	.025H	135
2856 Gas	153624	.023	.020	—	Flywheel	TDC	22°	.027H	.027H	90
2856 LPG	153624	.015	.020	—	Flywheel	2°	24°	.027H	.027H	90
2856, 21256 Diesel	153624	—	—	—	Flywheel⑮	6°⑩	—	.013H	.025H	135
21206 Diesel	153624	—	—	—	Flywheel⑮	8°⑩	—	.013H	.025H	135
21456	153624	—	—	—	Flywheel⑮	2° ATC⑩	—	.013H	.025H	135

①—TDS, with Roosa-Master injection pump; 40° mark with 1H pump; 14° with Bosch pump.
②—8° BTDC with Roosa-Master injection pump; 50° mark with 1H pump.
③—Early TDC; late 1° BTDC.
④—Early, 22°; late, 18°.
⑤—Early, intake and exhaust .027H. Late, intake .010H, exhaust .012H.
⑥—Early, 115 ft. lbs.; late 90 ft. lbs.
⑦—Pulley or flywheel.
⑧—With timing marks on crankshaft pulley & pointer aligned, insert 1/4" diameter rod in front cover engaging hole in pump drive hub.
⑨—Injection timing.
⑩—Robert Bosch pump; with timing marks on flywheel & clutch housing aligned, align timing points on roller retainer ring.
⑪—With timing marks on flywheel & clutch housing or hydrostatic drive housing aligned, align timing mark on injection pump cam ring with timing mark on pointer.
⑫—With timing marks on flywheel & clutch housing or hydrostatic housing aligned, align timing mark on injection pump hub with timing pointer.
⑬—With timing marks on pulley & timing gear cover aligned, align timing line on pump face cam with timing pointer on roller retaining ring.
⑭—Roosa-Master pump; with timing marks on flywheel & clutch housing aligned, align injection pump cam ring with timing mark on pump mounting flange.
⑮—With mechanical governor; align timing marks on flywheel & left front flange of clutch housing, then align timing mark on front end plate with timing mark on pump mounting flange. With pneumatic governor; measure the distance from marker on pulley to timing pointer which should be 1 1/8" or 20° BTDC.
⑯—With timing marks on pulley & timing pointer aligned, align timing marks on governor drive plate & cam ring.
⑰—I H Model RD; with timing mark on flywheel aligned with timing pointer, align timing mark on governor weight shaft with scribe line on front side of pump timing hole opening.
⑱—With timing marks on flywheel & left front flange of the clutch housing aligned, align timing mark on front end plate with timing mark on pump mounting flange.
⑲—With Hyd. drive, 22° BTDC.
⑳—Prior to serial No. 2505, .012H; Serial No. 2505 & later, .016H.
㉑—Engines prior to serial No. 65001, 65 ft. lbs.; engines from serial No. 65001, 80 to 90 ft. lbs.
㉒—Engines prior to serial No. 73050, 26° BTDC; engines from serial No. 73050, 22° BTDC.
㉓—Engines prior to serial No. U063636, 18° BTDC; engines from serial No. U063636, 15° BTDC.
㉔—Injection pump No. 3218-238-R91, 16° BTDC; injection pump No. 3228-169-R91, 18° BTDC.
㉕—Engines prior to serial No. 83592, .020 inch; engines from serial No. 83592, .025 inch.

Continued

FARM TRACTOR TUNE-UP—Continued

Tractor Model	Firing Order	Spark Plug Gap	Breaker Gap		Ignition Mark			Valve Lash		Cylinder Head Torque
			Distributor	Magneto	Location	Retarded Timing BTC	Full Advance Timing BTC	Intake	Exhaust	
MASSEY-FERGUSON										
MF-25 Diesel	1342	—	—	—	Pulley②	18°①	—	.012C	.012C	40–42
MF-130 Diesel	1342	—	—	—	Pulley⑥	18°①	—	.012C	.012C	40–42
MF-135 Special Gas	1342	.025	.022	—	Flywheel	6°	30°	.013	.013	70–75
MF-135 Non Diesel	1342	.025	.022	—	Flywheel	6°	26°	.013H	.013H	70–75
MF-135 Diesel	123	—	—	—	Flywheel⑧	20°①	—	.010H	.010H	55–60
MF-150 Cont. Gas	1342	.025	.022	—	Flywheel	6°	26°	.013H	.013H	70–75
MF-150 Perkins Gas	123	.025	.021	—	Flywheel	12°	26°	.012	.015	55–60
MF-150 Diesel	123	—	—	—	Flywheel⑧	20°①	—	.010H	.010H	55–60
MF-165 Cont. Gas	1342	.025	.022	—	Flywheel	6°	26°	.016H	.018H	70–75
MF-165 Perkins Gas	1342	.025	.022	—	Pulley	12°	36°	.012	.015	85–90
MF-165 Diesel	1342	—	—	—	Flywheel⑧	24°①	—	.010H	.010H	55–60
MF-175, 180 Diesel	1342	—	—	—	Flywheel⑦	23°①	—	.010H	.010H	70–75
MF-175 Gas	1342	.025	.022	—	Pulley	6°	26°	.016H	.018H	70–75
MF-180 Perkins Gas	1342	.025	.022	—	Pulley	10°	32°	.012H	.012H	85–90
MF-180 Cont. Gas	1342	.025	.022	—	Pulley	6°	26°	.016H	.018H	85
MF-205, 210, 220	12	—	—	—	Pulley	20°①	—	.012C	.014C	65–80
MV-230 Gas	1342	.025	.020	—	Flywheel	8°	27°	.011H	.013H	70–75
MF-230 Diesel	123	—	—	—	Flywheel⑩	16°①	—	.010H	.010H	70–75
MF-235 Gas	1342	.025	.020	—	Flywheel	8°	28°	.011H	.013H	70–75
MF-235 Diesel	123	—	—	—	Flywheel⑩	24°①	—	.010H	.010H	70–75
MF-245 Gas	1342	.025	.020	—	Flywheel	8°	28°	.011H	.013H	70–75
MF-245 Diesel	123	—	—	—	Flywheel⑩	24°①	—	.010H	.010H	70–75
MF-255 Gas	1342	—	.022	—	Flywheel	12°	—	.012H	.015H	90–95
MF-255 Diesel	1342	—	—	—	Flywheel⑫	26°①	—	.010H	.010H	70–75
MF-265 Gas	1342	—	.022	—	Flywheel	11°	—	.012H	.015H	95–100
MF-265 Diesel	1342	—	—	—	Flywheel⑬	23°①	—	.010H	.010H	95–100
MF-275 Diesel	1342	—	—	—	Flywheel⑭	24°①	—	.010H	.010H	95–100
MF-285 Diesel	1342	—	—	—	Pulley②	20°①	—	.010H	.010H	85–90
MF-1080	1342	—	—	—	Pulley②⑤	28°①	—	.012C	.012C	⑨
MF-1085	1342	—	—	—	Pulley②⑤	20°①	—	.012C	.012C	⑨
MF-1100 Gas	153624	.025	.016	—	Flywheel	2°	—	.012C	.022C	95
MF-1100 LP Gas	153624	.025	.016	—	Flywheel	DC	—	.012C	.022C	95
MF-1100 Diesel	153624	—	—	—	Flywheel③	22°①	—	.012C	.012C	85
MF-1105 Diesel	153624	—	—	—	Flywheel⑪	26°①	—	.010H	.010H	85
MF-1130 Diesel	153624	—	—	—	Flywheel③	28°①	—	.012C	.012C	85
MF-1135 Diesel	153624	—	—	—	Flywheel⑪	30°①	—	.010H	.010H	85
MF-1150 Diesel	18754362	—	—	—	Flywheel④	30°①	—	.012C	.012C	120–125
MF-1155	18754362	—	—	—	Pulley	28°	—	.012C	.012C	120–125
MF-2675, 2705	153624	—	—	—	—	28°①	—	.008C	.018C	—

①—Injection pump timing.

②—With timing mark on crankshaft pulley and timing pointer aligned, align scribe mark on injection pump rotor with square end of snap ring.

③—When flywheel timing mark "STAT" is centered with pointer, "E" mark scribed on pump rotor should be aligned with square end of the snap ring.

④—With timing mark on flywheel and notch on inspection hole aligned, align slot in tab on injection pump camshaft with center of sight hole.

⑤—Early models use flywheel.

⑥—APE pump—with timing mark on flywheel and pointer on inspection port aligned, align timing line on pump drive gear hub with timing line on front flange of pump housing. PSB pump—with timing mark on flywheel and pointer on inspection port aligned, align line mark on apex of pump plunger drive gear teeth with mark on housing.

⑦—When flywheel timing mark "SPILL" is aligned with center of timing window, on regular diesel letter "C" on injection pump rotor should be aligned with lower hole of snap ring, on Direct injection diesel, letter "B" on injection pump rotor should be aligned with lower end of snap ring.

⑧—When flywheel timing mark "STAT" is centered with pointer, "E" mark scribed on pump rotor should be aligned with line scribed on snap ring.

⑨—Engines before serial #312VA1148, 105-110; Engines serial #318VA1148 and after, 85–90.

⑩—With no. 1 piston on compression, align the correct degree mark on flywheel with timing on housing, then align "E" mark on injection rotor with square end of snap ring.

⑪—With no. 1 piston on compression and "Static Timing" mark is aligned in timing window, align "F" mark on injection rotor with square end of snap ring.

⑫—With 26° mark on flywheel centered in timing hole, align "B" mark on injection rotor with square edge of snap ring.

⑬—With 23° mark on flywheel aligned with pointer, align "C" mark on injection rotor with square edge of snap ring.

⑭—With pointer halfway between 23° & 25° mark on flywheel, align "C" mark on injection rotor with square edge of snap ring.

Continued

FARM TRACTOR TUNE-UP—Continued

MINNEAPOLIS-MOLINE

Tractor Model	Firing Order	Spark Plug Gap	Breaker Gap		Ignition Mark			Valve Lash		Cylinder Head Torque
			Distributor	Magneto	Location	Retarded Timing BTC	Full Advance Timing BTC	Intake	Exhaust	
G550 Gasoline	153624	.025	.025	—	Flywheel	TDC	—	.010C	.016C	④
G550 LP-Gas	153624	.025	.025	—	Flywheel	TDC	—	.010C	.016C	④
G550 Diesel	153624	—	—	—	Flywheel	2°ATC	—	.010C	.020C	④
G750 Gasoline	153624	.025	.025	—	Flywheel	TDC	—	.020C	.030C	④
G750 LP-Gas	153624	.025	.025	—	Flywheel	TDC	—	.020C	.030C	④
G750 Diesel	153624	—	—	—	Flywheel	4°BTC	—	.030C	.030C	—
G850 Gasoline	153624	.025	.025	—	Flywheel	2°BTC	26°BTC	.020C	.030C	129–133
G850 Diesel	153624	—	—	—	Flywheel	4°BTC	—	.030C	.030C	129–133
G-955 LP-Gas	153624	.015	.021	—	Flywheel	—	16°	.012C	.028C	170–175
G-955 Diesel	153624	—	—	—	Flywheel①	2°②	—	.015C	.022C	③
G940 Gasoline	153624	.025	.025	—	Flywheel	TDC	24°BTC	.020C	.030C	129–133
G940 Diesel	153624	—	—	—	Flywheel	4°BTC	—	.030C	.030C	129–133
G-1355 LP-Gas	153624	.015	.021	—	Flywheel	—	17°	.010C	.028C	170–175
G-1355 Diesel	153624	—	—	—	Flywheel①	4°②	—	.010C	.020C	170–175

①—All models, align flywheel with correct timing mark. With APE pump, timing mark on gear flange should align with mark on pump body. With PSB pump, timing mark on pump coupling flange should align with timing pointer. With Roosa-Master pump, timing lines on governor cage and pump cam should align.

②—Injection timing.

③—Torque 9/16" nut to 130–135 ft. lbs. and 5/8" nuts to 170–175 ft. lbs.

④—Except oil cap screw, 112–117 ft. lbs.; oil cap screw, 96–100 ft. lbs.

OLIVER

Tractor Model	Firing Order	Spark Plug Gap	Distributor	Magneto	Location	Retarded Timing BTC	Full Advance Timing BTC	Intake	Exhaust	Cylinder Head Torque
1550, 1555 Gasoline	153624	.025	②	—	Flywheel	TDC	—	.010C	.016C	115①
1550, 1555 Diesel	153624	—	—	—	Flywheel⑧	2° ATC⑩	—	.010C	.020C	115①
1550, 1555 LPG	153624	.025	②	—	Flywheel	TDC	—	.010C	.016C	115①
1600 Gasoline	153624	.025	.016	—	Flywheel	③	—	.014C	.023C	115
1600 LPG	153624	.025	.016	—	Flywheel	5° BTC	—	.014C	.023C	115
1600 Diesel	153624	—	—	—	Flywheel⑧	7° BTC⑩	—	.010C	.020C	115
1650, 1655 Gasoline	153624	.025	.025	—	Flywheel	TDC	—	.020C	.030C	115
1650, 1655 LPG	153624	.025	.025	—	Flywheel	TDC	—	.020C	.030C	115
1650, 1655 Diesel	153624	—	—	—	Flywheel⑧	④⑩	—	⑤	.030C	⑥
1755 Gasoline	153624	.025	.025	—	Flywheel	2°	26°	.019C	.029C	129
1755 Diesel	153624	.025	—	—	Flywheel⑧	2°⑩	26°	.029C	.029C	⑦
1855 Gasoline	153624	.025	.025	—	Flywheel	TDC	24°	.019C	.029C	129–133
1855 Diesel	153624	.025	—	—	Flywheel⑧	TDC	24°	.029C	.029C	⑦
1955 Diesel	153624	.025	—	—	Flywheel⑧	2°⑩	—	.029C	.029C	⑦
2255 Diesel	12734568	—	—	—	⑨	16°	—	.015C	.025C	—

①—Drilled screw 96–100.

②—1550, .016; 1555, .025.

③—231 cu. in., TDC; 248 cu. in., 2°.

④—2° BTC; On series 1650 after tractor serial number 187 586-000 and all series 1655, 4° BTC.

⑤—Indirect injection .020C; direct injection .050C.

⑥—Indirect injection, 131 ft. lbs.; direct injection, tighten six center capscrews from 147–150 ft. lbs., all others to 131 ft. lbs.

⑦—Tighten 1 through 8, 11 & 12, 180 ft. lbs.; tighten 9, 10, 13 & 14, 133 ft. lbs.

⑧—With correct timing mark on flywheel aligned with pointer, timing lines in injection pump window should be exactly aligned.

⑨—With timing pin inserted into slot on injection pump camshaft, thread a timing gear cover cap screw into tapped hole on camshaft gear. If capscrew is centered in timing hole, pump is correctly timed.

⑩—Injection pump timing.

WHITE

Tractor Model	Firing Order	Spark Plug Gap	Distributor	Magneto	Location	Retarded Timing BTC	Full Advance Timing BTC	Intake	Exhaust	Cylinder Head Torque
2-70 Gasoline	153624	.025	.025	—	Flywheel	21° BTC	—	.020C	.030C	112–117③
2-70 Diesel	153624	—	—	—	Flywheel	4° BTC①	—	.030C	.030C	④
2-85, 2-105 Diesel	153624	—	—	—	—	30° BTC	—	.010H	.010H	100
2-135, 2-155 Diesel	153624	—	—	—	Pulley	20° BTC②	—	.010C	.025C	160
2-150	153624	—	—	—	Flywheel	4° ATC①	—	.010C	.020C	170–175⑤

①—With correct timing mark on flywheel aligned with pointer, timing lines in injection pump window should be exactly aligned.

②—With correct timing mark on crankshaft damper aligned with pointer, timing lines in injection pump window should be exactly aligned.

③—Torque oil screw to 96–100 ft. lbs.

④—Inner six bolts, 147–150 ft. lbs.; outer eight bolts, 129–133 ft. lbs.

⑤—Torque front nut securing water manifold flange to 100–105 ft. lbs.

Pistons, Pins, Rings, Bearings & Crankshaft

Tractor Model	Cyl. Bore Size In.	Piston Clearance	Pin Diam.	Ring End Gap Minimum		Rod Bearings			Main Bearings		
				Comp.	Oil	Crankpin Diameter	Bearing Clearance	Rod Bolt Torque	Journal Diameter	Bearing Clearance	Cap Bolt Torque
ALLIS-CHALMERS											
160 Diesel	3.6	.006–.008	1.2499	.009	.009	2.2485–2.2490	.0025–.004	45–50	2.7485–2.7490	.0025–.004	110–115
170, 175 Gasoline	4	.0025–.004	.9894	.009	.015	2.3745–2.375	.001–.003	45–55	2.9995–3.000	.001–.004	130–140
170, 175 Diesel	3 7/8	.007	1.375	.015	.011	2.499–2.4995	.0015–.003	65–70	2.9985–2.999	.0025–.0045	145–150
180, 185 Gas	3 3/4	.0015–.004	1.2516	.008	.013	2.373	.0009–.0034	40–45	2.7475	.0016–.0043	130–140
180, 185 Diesel	3 7/8	.0035–.006	1.2516	④	⑤	2.373	.0009–.0034	40–45	2.7475	.0016–.0043	130–140
190 Gasoline	3 3/4	.0015–.004	1.2516	.008	.013	2.373	.0009–.0034	40–45	2.7475	.0016–.0043	130–140
190 Diesel	3 7/8	.0035–.006	1.2516	④	⑤	2.373	.0009–.0034	40–45	2.7475	.0016–.0043	130–140
190XT Gasoline	3 7/8	.0015–.004	1.2516	②	.013	2.373	.0009–.0034	40–45	2.7475	.0016–.0043	130–140
190XT LPG	3 7/8	.0015–.004	1.2516	②	.013	2.373	.0009–.0034	40–45	2.7475	.0016–.0043	130–140
190XT Diesel	3 7/8	.0035–.006	1.2516	④	⑤	2.373	.0009–.0034	40–45	2.7475	.0016–.0043	130–140
200 Diesel	3 7/8	.0045–.007	1.2516	③	.010	2.373	.0009–.0034	40–45	2.7475	.0016–.0043	130–140
210, 220 Diesel	4 1/4	.0025–.005	—	.013	.008	2.7470–2.7485	.001–.004	①	3.247–3.248	.0019–.0051	170–190
5020 Diesel	3.6	.0035–.0047	1.1021	.008	.008	2.2819–2.2821	.0008–.0033	36–40	2.7543–2.7551	.0008–.0033	44–58
5030 Diesel	3.8	.0020–.0059	1.1021	.008	.008	2.2819–2.2821	.0008–.0033	36–40	2.7543–2.7551	.0008–.0033	44–58
7000 Diesel	3 7/8	.0045–.007	1.2516	③	.010	2.3720–2.3735	.0009–.0034	40–45	2.7465–2.7480	.0016–.0043	130–140
7030 Diesel	4 1/4	.0025–.005	1.5012	⑥	.009	2.7470–2.7485	.001–.004	⑦	3.2470–3.2480	.0019–.0051	170–190
7040 Diesel	4 1/4	.0025–.005	1.5012	⑥	.009	2.7470–2.7485	.001–.004	⑦	3.2470–3.2480	.0019–.0051	170–190
7050 Diesel	4 1/4	.0025–.005	1.5012	⑥	.009	2.7470–2.7485	.001–.004	⑦	3.2470–3.2480	.0019–.0051	170–190
7060 Diesel	4 1/4	.0025–.005	1.5012	⑥	.009	2.7470–2.7485	.001–.004	⑦	3.2470–3.2480	.0019–.0051	170–190
7080 Diesel	4 1/4	.0025–.005	1.5012	⑥	.009	2.7470–2.7485	.001–.004	⑦	3.2470–3.2480	.0019–.0051	170–190

①—Prior to engine serial #3D-09649, 65–70 ft. lbs.; later models, 80–85 ft. lbs.
②—Top ring .014″, others .017″.
③—Top ring, .019″; 2nd ring, .015″; 3rd ring, .015″.
④—Top ring, .017″; 2nd ring—early 190, .017″; others, .013″. 3rd ring—early 190, .017″; others, .012″.
⑤—3 piece ring, .013″; others, .015″.
⑥—Top ring, .013″; 2nd & 3rd rings, .009″.
⑦—Early models with six point cap screws, 60–70 ft. lbs.; late models with socket head screws, 80–85 ft. lbs.

Tractor Model	Cyl. Bore Size In.	Piston Clearance	Pin Diam.	Comp.	Oil	Crankpin Diameter	Bearing Clearance	Rod Bolt Torque	Journal Diameter	Bearing Clearance	Cap Bolt Torque
J. I. CASE											
430 Gasoline	3 3/8	.003	.8592	.010	.010	2.0605–2.0615	.0015–.002	45–50	2.623–2.624	.0015–.003	95
430 Diesel	3 13/16	.005	1.2498	.010	.010	2.0605–2.0615	.001–.0035	45–50	2.873–2.874	.0008–.0038	95
470 Gasoline	3 3/8	—	.8592	.010	.010	2.0605–2.0615	.0015–.002	45–50	2.623–2.624	.0015–.003	95
470 Diesel	3 13/16	.005	1.2498	.010	.010	2.0605–2.0615	.001–.0035	45–50	2.873–2874	.0008–.0038	95
530 Gasoline	3 1/2	.003	.9994	.010	.010	2.0605–2.0615	.0005–.0025	45–50	2.623–2.624	.0015–.003	95
530 Diesel	3 13/16	.005	1.2498	.010	.010	2.0605–2.0615	.001–.0035	45–50	2.873–2.874	.0008–.0038	95
570 Gasoline	3 1/2	—	.9994	.010	.010	2.0605–2.0615	.0005–.0025	45–50	2.623–2.624	.0015–.002	95
570 Diesel	3 13/16	.005	1.2498	.010	.010	2.0605–2.0615	.001–.0035	45–50	2.873–2.874	.0008–.0038	95
630	3 13/16	.005	1.2498	.010	.010	2.0605–2.0615	②	45–50	2.873–2.874	.0008–.0038	95
770 Gasoline	4	.0035	1.3584	.013	.015	2.748–2.749	.0013–.0038	95–105	2.998–2.999	.0016–.0046	145–155
770 Diesel	4 1/8	.005	1.4994	.013	.015	2.748–2.749	.0013–.0038	95–105	2.998–2.999	.0016–.0046	145–155
870 Gasoline	4 3/8	.004	1.3584	.013	.169	2.748–2.749	.0013–.0038	95–105	2.998–2.999	.0016–.0046	145–155
870 Diesel	4 5/8	.005	1.7994	①	.013	2.998–2.999	.0013–.0038	95–105	3.488–3.489	.0016–.0046	145–155
885 Diesel	3.939	.0065–.0080	1.250	.011	.011	2.3730–2.3735	.002–.004	50	2.4990–2.4995	.002–.004	120
970 Gasoline	4	.0035	1.3584	.013	.015	2.748–2.749	.0013–.0038	95–105	2.998–2.999	.0016–.0046	145–155
970 Diesel	4 1/8	.005	1.4994	.013	.015	2.748–2.749	.0013–.0038	95–105	2.998–2.999	.0016–.0046	145–155
995 Diesel	3.939	.0065–.0080	1.250	.011	.011	2.3730–2.3735	.002–.004	50	2.4990–2.4995	.002–.004	140
1070 & 1090 Diesel	4 3/8	.005	1.4993	①	.010	2.748–2.749	.0013–.0038	95–105	2.998–2.999	.0016–.0046	145–155
1170 Diesel	4 3/8	.005	1.6244	.013	.010	2.748–2.749	.0013–.0038	95–105	2.998–2.999	.0016–.0046	145–155
1175 Diesel	4 3/8	.004–.006	1.6244–1.6246	.013	.010	2.748–2.749	.0013–.0038	95–105	2.998–2.999	.0016–.0046	145–155
1210 & 1212 Diesel	3.939	.0065–.0080	1.250	.011	.011	2.3730–2.3735	.002–.004	50	2.4990–2.4995	.002–.004	140
1270	4 3/8	.004–.006	1.6244–1.6246	①	.010	2.748–2.749	.0013–.0038	95–105	2.998–2.999	.0016–.0046	③
1370	4 5/8	.0045–.0055	1.7994–1.7996	.015	.010	2.998–2.999	.0013–.0038	95–105	3.498–3.499	.0016–.0046	③

Continued

PISTONS, PINS, RINGS, BEARINGS & CRANKSHAFT—Continued

Tractor Model	Cyl. Bore Size In.	Piston Clearance	Pin Diam.	Ring End Gap Minimum		Rod Bearings			Main Bearings		
				Comp.	Oil	Crankpin Diameter	Bearing Clearance	Rod Bolt Torque	Journal Diameter	Bearing Clearance	Cap Bolt Torque
J.I. CASE—Continued											
1410 & 1412 Diesel	3.939	.0065–.0080	1.250	.011	.011	2.3730–2.3735	.002–.004	50	2.4990–2.4995	.002–.004	140
1570	4 5/8	.005	1.7795	.015	.016	2.998	.0013–.0038	100	3.498–3.499	.0016–.0046	③

①—Top compression .015, all others .013.

②—Diesel .001–.0035; non-Diesel .001–.002.

③—Capscrews with washers, 200–210 ft. lbs.; capscrew without washers, 145–155 ft. lbs.

Tractor Model	Cyl. Bore Size In.	Piston Clearance	Pin Diam.	Comp.	Oil	Crankpin Diameter	Bearing Clearance	Rod Bolt Torque	Journal Diameter	Bearing Clearance	Cap Bolt Torque
JOHN DEERE											
820, 830	6 1/8	.010	2.3547	①	①	3.9991–4.0005	.0015–.0049	210	4.4975–4.4985	③	150
820, 830 Diesel	3.86	.0028–.0052	1.1877	①	①	2.748–2.749	.0012–.0042	60–70	3.1235–3.1245	.0011–.0041	⑦
850 Diesel	3.15	.003–.005	1.0236	.011	.011	1.9665–1.9669	.001–.003	40	2.753–2.754	.0015–.0039	60
950 Diesel	3.54	.006–.009	1.1807	.011	.011	2.1240–2.1244	.001–.003	47	⑳	㉑	60
1020 Gasoline	3.86	.0034	1.1877	①	①	2.3085–2.3095	.0014–.0044	40–45	3.1235–3.1245	.0011–.0041	85
1020 Diesel	3.86	.0030–.0054	1.1891	①	①	2.7480–2.7490	.0012–.0042	60–70	3.1235–3.1245	.0011–.0041	85
1050 Diesel	3.54	.006–.009	1.1807	.011	.011	2.1240–2.1244	.001–.003	47	⑳	㉑	60
1520 Gasoline	4.02	.0024	1.1891	①	①	2.748–2.749	.0012–.0042	60–70	3.1235–3.1245	.0011–.0041	85
1520 Diesel	4.02	.0015–.0039	1.1891	①	①	2.748–2.749	.0012–.0042	60–70	3.1235–3.1245	.0011–.0041	85
1530 Diesel	4.02	.0015–.0039	⑧	①	①	2.748–2.749	.0012–.0042	60–70	3.1235–3.1245	.0011–.0041	110
2020 Gasoline	3.86	.0034	1.1891	①	①	2.3085–2.3100	.0014–.0044	40–45	3.1235–3.1245	.0011–.0041	85
2020 Diesel	3.86	.0030–.0054	1.1891	①	①	2.7480–2.7490	.0012–.0042	60–70	3.1235–3.1245	.0011–.0041	85
2030 Gasoline	4.02	.0024	⑧	①	①	2.3085–2.3100	.0014–.0044	40–45	3.1235–3.1245	.0011–.0041	85
2030 Diesel	4.02	.0015–.0039	⑧	①	①	2.7480–2.7490	.0012–.0042	60–70	3.1230–3.1240	.0011–.0041	85
2040 Diesel	4.02	.005	1.3750	①	①	2.748–2.749	.0012–.0040	60–70	3.1230–3.1240	.0012–.004	85
2240 Diesel	4.19	.004–.009	1.375	—	—	2.748–2.749	.0012–.0040	60–70	3.1230–3.1240	.0012–.0040	85
2440 Diesel	4.02	.0041–.0075	1.375	—	—	2.748–2.749	.0016–.0046	60–70	3.1230–3.1240	.0012–.0042	85
2510 Gasoline	3.86	.0023–.0047	1.1877	①	①	2.3085–2.3095	.0014–.0044	40–45	3.1235–3.1245	.001–.004	85
2510 Diesel	3.86	.0018–.0042	1.1877	①	①	2.7480–2.7490	.0012–.0042	60–70	3.1235–3.1245	.001–.004	85
2520 Gasoline	3.86	.0034	1.1877	—	—	2.308–2.309	.0014–.0044	40–45	3.123–3.124	.0016–.0046	85
2520 Diesel	3.86	.001–.0044	1.1877	—	—	2.748–2.749	.0016–.0046	60–70	3.123–3.124	.0016–.0046	85
2630 Diesel	4.19	.0018–.0053	1.375	—	—	3.063–3.064	.0011–.0041	90–100	3.1235–3.1245	.0011–.0041	85
2640 Diesel	4.19	.0018–.0053	1.375	—	—	3.063–3.064	.0011–.0041	90–100	3.1235–3.1245	.0011–.0041	85
2840 Diesel	4.02	.005	1.3750	①	①	2.748–2.749	.0012–.0040	60–70	3.1230–3.1240	.0012–.004	85
3010 Gasoline	4	.006④	1.2497	.013	.013	2.9985–2.9995	.0015–.0045	90	3.3720–3.3730	.0031–.0061	150
3010 Diesel	4 1/8	⑤	1.4997	.013	.013	2.9985–2.9995	.0015–.0045	125	3.3720–3.3730	.0031–.0061	150
3020 Gasoline	4 1/4	.0055	1.250	.013	.013	2.998–2.999	.0015–.0045	90	3.3715–3.3725	.0031–.0061	150
3020 Diesel	4 1/4	.0035	1.500	.013	.013	2.998–2.999	.0015–.0045	125	3.3715–3.3725	.0031–.0061	150
4010 Gasoline	4	.007④	1.2497	.015	.015	2.9980–2.9990	.0015–.0045	105	3.3730–3.3740	.0031–.0061	150
4010 Diesel	4 1/8	⑥	1.4997	.015	.015	2.9980–2.9990	.0015–.0045	125	3.3730–3.3740	.0031–.0061	150
4020 Gasoline	4 1/4	.0055	1.250	.013	.013	2.998–2.999	.0015–.0045	90	3.3715–3.3725	.0031–.0061	150
4020 Diesel	4 1/4	.0035	1.500	.013	.013	2.998–2.999	.0015–.0045	125	3.3715–3.3725	.0031–.0061	150
4030 Gasoline	3.86	.0000–.0034	1.1877	—	—	2.7480–2.7490	.0012–.0042	65	3.1230–3.1240	.0016–.0046	⑲
4030 Diesel	4.02	.0015–.0039	⑨	①	①	2.7480–2.7490	.0012–.0042	65	3.1230–3.1240	.0016–.0046	⑲
4040 Diesel	4.25	.0027–.0041	1.500	—	—	2.9980–2.9990	.0010–.0040	55⑮	3.3720–3.3730	.0010–.0040	150
4230 Gasoline	4.25	.0028–.0052	1.2499	—	—	2.9985–2.9995	.0015–.0045	90–100	⑩	⑪	150
4230 Diesel	4 1/4	.0027–.0053	1.5000	①	①	2.9982–2.9992	.0015–.0045	100–110	⑩	⑪	150
4240 Diesel	4.56	.0036–.0053	1.625	—	—	2.9980–2.9990	.0010–.0040	55⑮	3.3720–3.3730	.0010–.0040	150
4320 Diesel	4 1/4	.0023–.0047	1.625	①	①	2.9985	.0015–.0045	110	3.372	.0017–.0047	150
4430 Diesel	4 1/4	.0027–.0041	1.6250	①	①	2.9980–2.9990	.0015–.0045	110–140	⑬	⑭	150
4440 Diesel	4.56	.0036–.0053	1.8742	—	—	2.9980–2.9990	.0010–.0040	55⑮	3.3720–3.3730	.0012–.0042	150
4520 Diesel	4 1/4	.0023–.0047	1.625	①	①	2.9985	.0015–.0045	110	3.372	.0017–.0047	150
4620 Diesel	4 1/4	.0023–.0047	1.625	①	①	2.9985	.0015–.0045	110	3.372	.0017–.0047	150
4630 Diesel	4 1/4	.0027–.0041	1.6250	①	①	2.9980–2.9990	.0015–.0045	110–140	⑯	⑰	150
4640 Diesel	4.56	.0036–.0053	1.8742	—	—	2.9980–2.9990	.0010–.0040	55⑮	3.3720–3.3730	.0012–.0042	150

Continued

PISTONS, PINS, RINGS, BEARINGS & CRANKSHAFT—Continued

Tractor Model	Cyl. Bore Size In.	Piston Clearance	Pin Diam.	Ring End Gap Minimum		Rod Bearings			Main Bearings		
				Comp.	Oil	Crankpin Diameter	Bearing Clearance	Rod Bolt Torque	Journal Diameter	Bearing Clearance	Cap Bolt Torque
JOHN DEERE—Continued											
4840 Diesel	4.56	.0036–.0053	1.8742	—	—	2.9980–2.9990	.0010–.0040	55⑮	3.3720–3.3730	.0012–.0042	150
5010 Diesel	4¾	.0065	1.750	.013	.013	3.498–3.499	.0015–.0045	160	3.748–3.749	.0031–.0061	170
5020 Diesel	4¾	.0028–.0052	1.750	.013	.013	3.498–3.499	.0015–.0045	160	3.748–3.749	.0031–.0061	175
6030 Diesel	4¾	.0058–.0082	1.8742	—	—	3.498–3.499	.0024–.0054	160	3.748–3.749	.0019–.0049	210

①—Rings are provided with proper gap by supplier.
②—Right .005–.007, left .004–.006.
③—Right and left .006–.008, center .0021–.0055.
④—Thrust side.
⑤—Thrust side .006, anti-thrust side .020.
⑥—Thrust side .0065, anti-thrust side .017.
⑦—820: 85 ft. lbs.; 830: 110 ft. lbs.
⑧—Early models: 1.1886–1.1896; Late models: 1.376–1.377.
⑨—Early models, 1.871–1.1879 in.; late models, 1.3748–1.3752 in.
⑩—Up to Serial No. E395613, 3.3715–3.3725 in.; from Serial No. E395614 on, 3.3725–3.3735 in.
⑪—Up to Serial No. E395613, .0017–.0047 in.; from Serial No. E395614 on, .0007–.0037 in.
⑫—Up to Serial No. E410332, 3.3715–3.3725 in.; from Serial No. E410333 to E433251, 3.3725–3.3735 in.
⑬—Up to Serial No. E410332, .0017–.0047 in.; from Serial No. E410333 to E433251, .0007–.0037 in.
⑭—Up to Serial No. E410332, 100 to 110 ft. lbs.; from Serial No. E410333 on, 110 to 140 ft. lbs.
⑮—Plus ¼ turn.
⑯—Up to Serial No. E410264, 3.3715–3.3725 in.; from Serial No. E410265 to E432598, 3.3725–3.3735 in.
⑰—Up to Serial No. E410264, .0017–.0047 in.; from Serial No. E410265 to E432598, .0007–.0037 in.
⑱—Up to Serial No. 410264, 100 to 110 ft. lbs.; from Serial No. 410265 on, 110 to 140 ft. lbs.
⑲—Up to Serial No. 246640, 85 ft. lbs.; from Serial 246641 on, 110 ft. lbs.
⑳—Front & center journals, 2.7530–2.7540. Rear journal, 3.5413–3.5417.
㉑—Front & center bearing clearance, .0015–.0039. Rear bearing clearance .0027–.0051.

Tractor Model	Cyl. Bore Size In.	Piston Clearance	Pin Diam.	Comp.	Oil	Crankpin Diameter	Bearing Clearance	Rod Bolt Torque	Journal Diameter	Bearing Clearance	Cap Bolt Torque
FORD											
TW-10	4.4	⑦	1.4998	③	.013	⑤	①	60–65	⑥	.0022–.0045	115–125
TW-20, 30	4.4	⑦	1.6248	③	.013	⑤	①	60–65	⑥	.0022–.0045	115–125
1000	3.54	.006	1.26	.012	.012	2.3602–2.3610	.0023–.0510	51–54	2.6752–2.6760	.0016–.0004	—
1600	3.54	.006	1.26	.012	.012	2.3602–2.3610	.0023–.0510	51–54	2.6752–2.6760	.0016–.0004	—
2000 Gas	4.2	.0027	1.4998	.012	.013	2.749–2.750	①	60–65	3.3714–3.3722	.0022–.0045	115–125
2000 Diesel	4.2	.0075	1.4998	.012	.013	2.749–2.750	①	60–65	3.3714–3.3722	.0022–.0045	115–125
2600 Gas	4.2	.0032	1.4998	.015	.015	2.749–2.750	⑧	60–65	⑥	.0022–.0045	115–125
2600 Diesel	4.2	.0080	1.4998	.015	.015	2.749–2.750	⑧	60–65	⑥	.0022–.0045	115–125
3000 Gas	4.2	.0027	1.4998	.012	.013	2.749–2.750	①	60–65	3.3714–3.3722	.0022–.0045	115–125
3000 Diesel	4.2	.0075	1.4998	.012	.013	2.749–2.750	①	60–65	3.3714–3.3722	.0022–.0045	115–125
3600 Gas	4.2	.0032	1.4998	.015	.015	2.749–2.750	⑧	60–65	⑥	.0022–.0045	115–125
3600 Diesel	4.2	.0080	1.4998	.015	.015	2.749–2.750	⑧	60–65	⑥	.0022–.0045	115–125
4000 Gas	4.4	.0032	1.4998	.012	.013	2.749–2.750	①	60–65	3.3714–3.3722	.0022–.0045	115–125
4000 Diesel	4.4	.008	1.4998	.012	.013	2.749–2.750	①	60–65	3.3714–3.3722	.0022–.0045	115–125
4100 Diesel	4.2	.0080	1.4998	.015	.015	2.749–2.750	⑧	60–65	⑥	.0022–.0045	115–125
4600 Gas	4.4	.0034	1.4998	.015	.015	2.749–2.750	⑧	60–65	⑥	.0022–.0045	115–125
4600 Diesel	4.4	.0085	1.4998	.015	.015	2.749–2.750	⑧	60–65	⑥	.0022–.0045	115–125
4600 SU Gas	4.4	.0034	1.4998	.015	.015	2.749–2.750	⑧	60–65	⑥	.0022–.0045	115–125
4600 SU, Diesel	4.4	.0085	1.4998	.015	.015	2.749–2.750	⑧	60–65	⑥	.0022–.0045	115–125
5000, 7000 Gas	4.4	.0037	②	⑨	.015	⑤	①	60–65	⑥	.0022–.0045	115–125
5000, 7000 Diesel	4.4	.0085	②	⑨	.015	⑤	①	60–65	⑥	.0022–.0045	115–125
5600 Diesel	4.2	.0080	1.4999	⑨	.015	2.7496–2.7504	①	60–65	3.3713–3.3723	.0022–.0045	115–125
6000 Gas	3.62	.0015	.9122	.010	.010	2.2980–2.2988	.0024–.006	45–50	2.4963–2.4971	.0021–.0039	100–105
6000 Diesel	3.62	.005	1.1242	.010	.010	2.2980–2.2988	.0024–.006	45–50	2.4963–2.4971	.0021–.0039	100–105
6600 Gas	4.4	.0037	1.4999	⑨	.015	2.7496–2.7504	①	60–65	3.3713–3.3723	.0022–.0045	115–125
6600 Diesel	4.4	.0085	1.4999	⑨	.015	2.7496–2.7504	①	60–65	3.3713–3.3723	.0022–.0045	115–125
6700 Diesel	4.4	.0085	1.4999	⑨	.015	2.7496–2.7504	①	60–65	3.3713–3.3723	.0022–.0045	115–125
6700 Gas	4.4	.0037	1.4999	⑨	.015	2.7496–2.7504	①	60–65	3.3713–3.3723	.0022–.0045	115–125
7600 Diesel	4.4	.0085	1.6248	⑨	.015	2.7496–2.7504	①	60–65	3.3713–3.3723	.0022–.0045	115–125
7700 Diesel	4.4	.0085	1.6248	⑨	.015	2.7496–2.7504	①	60–65	3.3714–3.3723	.0022–.0045	115–125
8000, 9000 Diesel	4.4	⑦	1.4998④	③	.013	⑤	①	60–65	⑥	.0022–.0045	115–125
8600 Diesel	4.4	⑦	1.4998	③	.013	⑤	①	60–65	⑥	.0022–.0045	115–125
8700 Diesel	4.4	⑦	1.4998	③	.013	⑤	①	60–65	⑥	.0022–.0045	115–125
9600 Diesel	4.4	⑦	1.6248	③	.013	⑤	①	60–65	⑥	.0022–.0045	115–125
9700 Diesel	4.4	⑦	1.6248	③	.013	⑤	①	60–65	⑥	.0022–.0045	115–125

Continued

FORD PISTONS, PINS, RINGS, BEARINGS & CRANKSHAFT—Continued

FORD NOTES—Continued

①—Aluminum bearings .0025–.0046 inch, copper lead bearings .0017–.0038 inch.
②—5000 series, 1.4999; 7000 series, 1.6248.
③—Top ring .012; all others .010.
④—9000 Series, 1.6248.
⑤—Blue 2.7496–2.750; Red 2.750–2.7504.
⑥—Blue 3.3713–3.718; Red 3.3718–3.3723.
⑦—Cyl. # 1-5, .008–.009 inch; cyl. # 6, .010–.011 inch.
⑧—Aluminum bearing .0021–.0042 inch; copper lead bearing .0017–.0038 inch.
⑨—Top ring .015; all others .013.

Tractor Model	Cyl. Bore Size In.	Piston Clearance	Pin Diam.	Ring End Gap Minimum		Rod Bearings			Main Bearings		
				Comp.	Oil	Crankpin Diameter	Bearing Clearance	Rod Bolt Torque	Journal Diameter	Bearing Clearance	Cap Bolt Torque
INTERNATIONAL HARVESTER											
Hydro 70, 86 Non Diesel	3 3/4	.0027	.8748	.010	.010	2.373–2.374	.0009–.0034	50	2.748–2.749	.0012–.0042	80
Hydro 70, 86 Diesel	3.875	.0055	1.6249	.014	.010	2.5185–2.5193	.0029–.0055	63	3.1484–3.1492	.0029–.0055	⑰
Hydro 84	3 15/16	.0039–.0047	1.4172	.014	.010	2.5185–2.5193	.0023–.0048	61	3.1484–3.1492	.0028–.0055	③
Hydro 100 & 186	4.300	.045–.0065	1.6249	⑱	.010	2.9977–2.9990	.0018–.0051	130	3.3742–3.3755	.0018–.0051	115
100, 130, 140	3 1/8	.0015	.9193	.010	.010	1.749–1.750	.001–.0035	45	2.124–2.125	.001–.0035	75
154 LoBoy	2 5/8	.002	.6875	.007	.007	1.498–1.499	.002–.003	—	1.623–1.627	.002–.003	55
Farmall Cub	2 5/8	.002	.6875	.007	.007	1.4985	.002–.003	16	1.6235	.002–.003	55
IH Cub, Cub LoBoy	2 5/8	.002	.6875	.007	.007	1.4985	.002–.003	16	1.6235	.002–.003	55
200, 230, 240	3 1/8	.0015	.9193	.010	.010	1.749–1.750	.001–.0035	45	2.124–2.125	.001–.0035	75
B275 Diesel	3 3/8	.0048–.0056	1.1022	.012	.015	1.7495–1.7500	.0010–.0029	30–35	2.124–2.125	.002–.004	75
354 Diesel	3 3/8	.0031–.0039	1.1022	.010	.010	1.7495–1.7500	.0010–.0029	40–45	2.124–2.125	.002–.004	⑯
354 Gasoline	3 3/8	.0031–.0039	1.1022	.012	.012	1.7495–1.7500	.0010–.0029	40–45	2.124–2.125	.002–.004	⑯
364 Diesel	3 1/2	.0031–.0039	1.1022	.010	.010	1.7495–1.7500	.0010–.0029	40–45	2.124–2.125	.002–.004	⑯
384 Diesel	3 1/2	.0031–.0039	1.1022	.010	.010	1.7495–1.7500	.0010–.0029	40–45	2.124–2.125	.002–.004	⑯
404	3 1/4	.0015	.8592	.010	.010	1.7495	.0009–.0039	45	2.1245	.0009–.0039	75–80
B414 Diesel	3 1/2	.0031–.0039	1.1022	.010	.010	1.7495–1.7500	.0010–.0029	40–45	2.124–2.125	.002–.004	⑯
B414 Gasoline	3 3/8	.0031–.0039	1.1022	.012	.012	1.7495–1.7500	.0010–.0029	40–45	2.124–2.125	.002–.004	⑯
424, 444 Diesel	3 1/2	.0031–.0039	1.1022	.010	.010	1.7495–1.7500	.0010–.0029	40–45	2.124–2.125	.002–.004	⑯
424, 444 Gasoline	3 3/8	.001–.002	.8592	.010	.010	2.059–2.060	.0009–.0039	43–49	2.6235–2.6245	.0009–.0039	80–90
454 Gasoline	3 3/8	.002–.005	.8592	.010	.010	2.059–2.060	.0009–.0039	43–49	2.748–2.749	.0012–.0042	75–85
454, 464, 484 Diesel	3 7/8	.0039–.0047	1.4172	.014	.010	2.5185–2.5193	.0023–.0048	61	3.1484–3.1492	.0028–.0055	③
464 Gasoline	3 9/16	.002–.005	.8748	.010	.010	2.373–2.374	.0009–.0034	④	2.748–2.749	.0012–.0042	75–85
544 Gas	3 13/16	.0025	.87485	.010	.018	2.3735	.0009–.0034	④	2.7485	.0012–.0042	75–80
544 Diesel	3 7/8	.0042	1.41725	.014	.010	2.5185–2.5193	.0023–.0048	63	3.1484–3.1492	.0029–.0055	⑤
574, 674 Gasoline	3 13/16	.002–.005	.8748	.010	.010	2.373–2.374	.0009–.0034	④	2.748–2.749	.0012–.0042	75–85
574, 584, 674 Diesel	3 7/8	.0039–.0047	1.4172	.014	.010	2.5185–2.5193	.0023–.0048	61	3.1484–3.1492	.0028–.0055	③
656 Non-Diesel	3 9/16	.0025	.87485	.010	.010	2.3735	.0009–.0034	45–55	2.7485	.0012–.0042	75–85
656 Diesel	3 11/16	.0025	1.1248	.015	.010	2.3735	.0009–.0034	45–55	2.7485	.0012–.0042	75–85
666 Non-Diesel	3 3/4	.0025	.8748	.010	.010	2.3730–2.3740	.0009–.0034	45–50	2.748–2.749	.0012–.0042	80
666 Diesel	3.875	.0055	1.6249	.010	.010	2.9977–2.9990	.0018–.0051	130	3.3742–3.3755	.0018–.0051	115
684 Diesel	3 7/8	.0039–.0047	1.4172	.014	.010	2.5185–2.5193	.0023–.0048	61	3.1484–3.1492	.0028–.0055	③
686 Non-Diesel	3 3/4	.0025	.8748	.010	.010	2.373–2.374	.0009–.0034	50	2.748–2.749	.0012–.0042	80
686 Early Diesel	3.875	.0055	1.6249	.014	.010	2.5185–2.5193	.0018–.0051	⑰	3.3742–3.3755	.0018–.0051	115
686 Late Diesel	3.875	.0039–.0047	1.4172	.014	.010	2.5185–2.5193	.0023–.0048	63	3.1484–3.1492	.0029–.0055	115
706 Non-Diesel	⑥	.0025	.87485	.010	.010	2.373–2.374	.0009–.0034	④	2.748–2.749	.0012–.0042	80
706 Diesel	⑦	⑧	⑨	.015	.010	⑩	⑪	④	⑬	⑭	⑮
756 Non-Diesel	3 3/4	.0010	.87485	.010	.010	2.373–2.374	.0009–.0034	④	2.748–2.749	.0012–.0042	80
766 Non-Diesel	3 3/4	.0045	.8748	.010	.010	2.373–2.374	.0009–.0034	④	2.748–2.749	.0012–.0042	115
766 Diesel	3 7/8	.0065	1.6249	①	.010	2.9977–2.9990	.0018–.0051	130	3.3742–3.3755	.0018–.0051	115
784 Diesel	3 15/16	.0039–.0047	1.4172	.014	.010	2.5185–2.5193	.0023–.0048	61	3.1484–3.1492	.0028–.0055	③
806 Non-Diesel	3 13/16	.0025	.87485	.010	.010	2.373–2.374	.0009–.0034	④	2.748–2.749	.0012–.0042	80
806 Diesel	4 1/8	.004	1.4999	.017	.014	2.998–2.999	.0018–.0051	105	3.3742–3.3755	.0018–.0051	115
826 Non-Diesel	3 13/16	.0045	.8748	.010	.010	2.373–2.374	.0009–.0034	④	2.748–2.749	.0012–.0042	80
826 Diesel	3 7/8	.0047	1.4172	①	.010	2.5185–2.5193	.0023–.0048	45	3.1484–3.1492	.0029–.0055	⑰
856 Non-Diesel	3 13/16	.0010	.87485	.010	.010	2.373–2.374	.0009–.0034	④	2.748–2.749	.0012–.0042	80
856 Diesel	4 5/16	.0049	1.4999	.017	.014	2.998–2.999	.0018–.0051	105	3.3742–3.3755	.0018–.0051	115

Continued

PISTONS, PINS, RINGS, BEARINGS & CRANKSHAFT—Continued

Tractor Model	Cyl. Bore Size In.	Piston Clearance	Pin Diam.	Ring End Gap Minimum Comp.	Ring End Gap Minimum Oil	Rod Bearings Crankpin Diameter	Rod Bearings Bearing Clearance	Rod Bearings Rod Bolt Torque	Main Bearings Journal Diameter	Main Bearings Bearing Clearance	Main Bearings Cap Bolt Torque
INTERNATIONAL HARVESTER—Continued											
884 Diesel	3 15/16	.0039–.0047	1.4172	.014	.010	2.5185–2.5193	.0023–.0048	61	3.1484–3.1492	.0028–.0055	③
886 Early	3 7/8	.0039–.0047	1.4172	.014	.010	2.5185–2.5193	.0023–.0048	45	3.1484–3.1492	.0029–.0055	⑰
886 Late	3 7/8	.0045–.0065	1.6249	②	.010	2.9977–2.9990	.0018–.0051	130	3.3742–3.3755	.0018–.0051	115
966 Diesel	4.3	.0065	1.6249	①	.010	2.9977–2.9990	.0018–.0051	130	3.3742–3.3755	.0018–.0051	115
986 Diesel	4.300	.0045–.0065	1.6249	②	.010	2.9977–2.9990	.0018–.0051	130	3.3742–3.3755	.0018–.0051	115
1026 Diesel	4.321	.0069	1.6249	①	.013	2.998–2.999	.0018–.0051	105	3.3742–3.3755	.0018–.0051	115
1066 Diesel	4.3	.0065	1.6249	①	.010	2.9977–2.9990	.0018–.0051	130	3.3742–3.3755	.0018–.0051	115
1086 Early	4.300	.0045–.0065	1.6249	②	.010	2.9977–2.9990	.0018–.0051	130	3.3742–3.3755	.0018–.0051	115
1086 Late	4.300	.0045–.0065	1.6249	②	.010	2.9977–2.9990	.0018–.0051	130	3.3742–3.3755	.0018–.0051	115
1206 Diesel	4 1/8	.0043	1.4999	.017	.014	2.9983	.0018–.0051	105	3.3745	.0018–.0051	115
1256 Diesel	4 5/16	.0049	1.4999	.017	.014	2.998–2.999	.0018–.0051	105	3.3742–3.3755	.0018–.0051	115
1456 Diesel	4 5/16	.0049	1.6249	.017	.014	2.998–2.999	.0018–.0051	105	3.3742–3.3755	.0018–.0051	115
1466, 1566	4.300	.0045–.0065	1.6248–1.6250	⑱	.010–.020	2.9977–2.9990	.0018–.0051	130	3.3742–3.3755	.0018–.0051	115
1468, 1568	4.500	.006–.007	1.4998–1.5000	.013–.023	.013–.028	2.753–2.754	.0019–.0044	55⑲	3.123–3.124	.0018–.0048	130
1486 & 1586 Diesel	4.3	.0045–.0065	1.6249	⑱	.010	2.9977–2.9990	.0018–.0051	130	3.3742–3.3755	.0018–.0051	115
2404	3 1/4	.0015	.8592	.010	.010	1.7495	.0009–.0039	45	2.1245	.0009–.0039	75–80
2424, 2444 Diesel	3 1/2	.0031	1.1022	.010	.010	1.7495–1.750	.001–.0029	40–45	2.124–2.125	.002–.004	⑯
2424, 2444 Gas	3 3/8	.0015	.8592	.010	.018	2.059–2.060	.0009–.0039	43–49	2.6235–2.6245	.0009–.0039	⑯
2706 Non-Diesel	⑥	.0025	.87485	.010	.015	2.373–2.374	.0009–.0034	④	2.748–2.749	.0012–.0042	80
2706 Diesel	⑦	⑧	⑨	.015	.010	⑩	⑪	⑫	⑬	⑭	⑮
2756 Non-Diesel	3 3/4	.0010	.87485	.010	.010	2.373–2.374	.0009–.0034	④	2.748–2.749	.0012–.0042	80
2756 Diesel	3 7/8	.0039	1.4172	.014	.010	2.5185–2.5193	.0023–.0048	63	3.1484–3.1492	.0029–.0055	⑤
2806 Non-Diesel	3 13/16	.0015	.87485	.010	.010	2.373–2.374	.0009–.0034	④	2.748–2.749	.0012–.0042	80
2806 Diesel	4 1/8	.004	1.4999	.017	.014	2.998–2.999	.0018–.0051	105	3.3742–3.3755	.0018–.0051	115
2856 Non-Diesel	3 13/16	.0010	.87485	.010	.010	2.373–2.374	.0009–.0034	④	2.748–2.749	.0012–.0042	80
2856 Diesel	4 5/16	.0049	1.4999	.017	.014	2.9982.999	.0018–.0051	105	3.3742–3.3755	.0018–.0051	115
21206 Diesel	4 1/8	.0043	1.4999	.017	.014	2.9983	.0018–.0051	105	3.3745	.0018–.0051	115
21256 Diesel	4 5/16	.0049	1.4999	.017	.014	2.998–2.999	.0018–.0051	105	3.3742–3.3755	.0018–.0051	115
21456 Diesel	4 5/16	.0049	1.6249	.017	.014	2.998–2.999	.0018–.0051	105	3.3742–3.3755	.0018–.0051	115

①—Top ring, D360 .010 Top ring, D414 and DT414 .016. Second ring .020.
②—Top ring, D360, 012 All others .016 Second ring, all except DT-414B Serial No. 14001 and above .020. DT-414B serial No. 14001 and above .065.
③—Bolts with shank diameter of .41 inch, marked 10.9K, 80 ft. lbs.; marked 12.9K, 97 ft. lbs. Bolts with shank diameter of .5 inch, marked 10.9K, 105 ft. lbs.; marked 12.9K, 141 ft. lbs.
④—Place Bolt—50 ft. lbs. Pitch Bolt 45 ft. lbs.
⑤—Bolt Marked 10K-80 ft. lbs. Marked 12K-97 ft. lbs. with lubricated threads.
⑥—Early series 3 9/16; late, 3 3/4.
⑦—Early series 3 11/16; late, 3 7/8.
⑧—Early series, .005; late, .0039.
⑨—Early series, 1.1247; late, 1.4172.
⑩—Early series 2.373–2.374; late, 2.5185–2.5193.
⑪—Early series, .0009–.0034; late, .0023–.0048.
⑫—Early series, 45–50 ft. lbs.; late, 63 ft. lbs.
⑬—Early series, 2.748–2.749; late, 3.1484–3.1492
⑭—Early series, .0012–.0042; late, .0029–.0055.
⑮—Early series, 80; late, see note ⑤.
⑯—Pitch bolt 70–75 ft. lbs.; Place bolt 80–85 ft. lbs.
⑰—Bolt marked 10K or 10–9: 80 ft. lbs.; marked 12K or 12–9: 97 ft. lbs.; marked 10.9: 115 ft. lbs.; marked 12.9: 140 ft. lbs.
⑱—Top ring, .016"–.026"; Second ring, .020"–.030".
⑲—Tighten nut 1/6 turn after 55 ft. lbs. torque is reached.

Tractor Model	Cyl. Bore Size In.	Piston Clearance	Pin Diam.	Ring End Gap Minimum Comp.	Ring End Gap Minimum Oil	Rod Bearings Crankpin Diameter	Rod Bearings Bearing Clearance	Rod Bearings Rod Bolt Torque	Main Bearings Journal Diameter	Main Bearings Bearing Clearance	Main Bearings Cap Bolt Torque
MASSEY-FERGUSON											
MF-25	3 1/8	.005	.9376	.012	.009	1.9995–2.000	.0015–.003	30–35	2.248–2.2485	002–.0035	80–85
MF-130	3 1/8	.005	.9376	.012	.009	1.9995–2.000	.0015–.003	30–35	2.248–2.2485	.002–.0035	80–85
MF-135 Diesel	3.6	.006	1.250	.009	.009	2.2485–2.2490	.0025–.004	45–50	2.7485–2.749	.003–.005	④
MF-135 Gas	3.6	.003	.8592	.010	.010	1.9365–2.9375	.0003–.0023	40–45	2.249–2.250	.0005–.0027	85–95
MF-135 Spec. Gas	3.312	.0025	.8591	.010	.010	1.9365–1.0375	.0003–.0023	40–45	2.249–2.250	.0015–.0027	85–95
MF-150 Diesel	3.6	.006	1.250	.009	.009	2.2485–2.2490	.0025–.004	45–50	2.7485–2.7490	.003–.005	④
MF-150 Gas	3.6	.003	.8592	.010	.010	1.9365–1.9375	.0003–.0023	40–45	2.249–2.250	.0005–.0027	85–95
MF-165 Diesel	3 3/8	.006	1.250	.009	.009	2.2485–2.2490	.0025–.004	45–50	2.7485–2.7490	.004–.005	④

Continued

PISTONS, PINS, RINGS, BEARINGS & CRANKSHAFT—Continued

Tractor Model	Cyl. Bore Size In.	Piston Clearance	Pin Diam.	Ring End Gap Minimum		Rod Bearings			Main Bearings		
				Comp.	Oil	Crankpin Diameter	Bearing Clearance	Rod Bolt Torque	Journal Diameter	Bearing Clearance	Cap Bolt Torque
MASSEY-FERGUSON—Continued											
MF-165 Gas	3.6	.003	1.1251	.010	.010	2.0615–2.0625	.0007–.0027	40–45	2.374–2.375	.000–.0027	85–95
MF-165 Diesel	3.6	.005–.007	1.24975	.009	—	2.2485–2.249	.0025–.004	45–50	2.7485–2.749	.003–.005	④
MF-165 Gas	3 7/8	.007	1.3749	.018	.014	2.499–2.4995	.0015–.003	65–70	2.9985–2.999	.0025–.0045	145–150
MF-165 Gas	3.58	.0012	.8592	.10	.10	1.9365–1.9375	.0015–.0025	35–40	2.249–2.250	.0015–.0025	90
MF-175 Diesel	3 7/8	.007	1.3749	.018	.014	2.499–2.4995	.0015–.003	65–70	2.9985–2.999	.0025–.0045	145–150
MF-175 Cont. Gas	3 13/16	—	1.250	.010	.010	2.0615–2.0625	.0007–.0027	40–45	2.374–2.375	.0005–.0028	85–95
MF-175 Perkins Gas	3 7/8	—	.9998	.016	.012	2.499–2.4995	.0015–.003	65–70	2.9985–2.999	.0025–.0045	145–150
MF-180 Cont. Gas	3 13/16	.007	1.3749	.018	.014	2.499–2.4995	.0015–.003	65–70	2.9985–2.999	.0025–.0045	145–150
MF-180 Diesel	3 7/8	.007	1.3749	.018	.014	2.499–2.4995	.0015–.003	65–70	2.9985–2.999	.0025–.0045	145–150
MF-180 Perkins Gas	3 7/8	—	.9998	.016	.012	2.499–2.4995	.0015–.003	65–70	2.9985–2.999	.0025–.0045	145–150
MF-205	3.5	.0039–.0055	.9839	.008	.008	2.0850–2.0858	.0008–.0033	36–43	2.7543–2.7551	.0008–.0033	44–58
MF-210	3.6	.0039–.0055	1.1021	.008	.008	2.2819–2.2827	.0008–.0033	44–50	2.7543–2.7551	.0008–.0033	44–58
MF-220	3.8	.0039–.0055	1.1021	.008	.008	2.2819–2.2827	.0008–.0033	44–50	2.7543–2.7551	.0008–.0033	44–58
MF-230 Gas	3.375	—	.8592	.010	.010	1.9365–1.9367	.0006–.0031	40–45	2.249–2.250	.0005–.0032	85–95
MF-230 Diesel	3.6	—	1.24975	⑦	⑦	2.2485–2.2490	.0025–.004	40–45	2.7485–2.7490	.003–.005	110–115
MF-235 Gas	4.063	—	.8592	.010	.010	1.9365–1.9367	.0006–.0031	40–45	2.249–2.250	.0005–.0032	85–95
MF-235 Diesel	5	—	1.24975	⑦	⑦	2.2485–2.2490	.0025–.004	40–45	2.7485–2.7490	.003–.005	110–115
MF-245 Gas	4.063	—	.8592	.010	.010	1.9365–1.9367	.0006–.0031	40–45	2.249–2.250	.0005–.0032	85–95
MF-245 Diesel	5	—	1.24975	⑦	⑦	2.2485–2.2490	.0025–.004	40–45	2.7485–2.7490	.003–.005	110–115
MF-255 Gas	3 7/8	—	.9999	⑧	.016	2.4990–2.4995	.0015–.0030	65–70	2.9985–2.9990	.0025–.0045	170–180
MF-255 Diesel	3.602	.005–.007	1.24975	①	.011	2.2485–2.2490	.0025–.0040	40–45	2.7485–2.7490	.003–.005	110–115
MF-265 Gas	3 7/8	—	.9999	⑧	.015	2.4990–2.4995	.0015–.0030	65–70	2.9985–2.9990	.0025–.0045	170–180
MF-265 Diesel	3 7/8	—	1.3749	.016	.016	2.4990–2.4995	.0015–.0030	75–80	2.9985–2.9990	.0025–.0045	170–180
MF-275 Diesel	3.975	—	1.3749	.016	.016	2.4990–2.4995	.0015–.0030	75–80	2.9985–2.9990	.0025–.0045	170–180
MF-285 Diesel	4.5	—	1.4376	⑥	.018	2.749	.002–.0042	140–150	2.999	.0025–.0045	140–150
MF-1100 Gas	4 1/8	.003	1.3748	.013	—	2.624–2.625	.0005–.0030	85–90	2.624–2.625	.0015–.0045	110
MF-1100 Diesel	3 7/8	.007	⑤	.018	.014	2.499–2.4995	.0015–.0030	65–70	2.9985–2.999	.0025–.0045	145–150
MF-1105	3 7/8	—	1.4998–1.50	.016	—	2.499–2.4995	.0015–.003	③	2.9985–2.999	.0025–.0045	180
MF-1130 Diesel	3 7/8	.007	⑤	.021	.015	2.499–2.4995	.0015–.0030	65–70	2.9985–2.999	.0025–.0045	145–150
MF-1135	3 7/8	—	1.4998–1.50	.016	—	2.499–2.4995	.0015–.003	③	2.9985–2.999	.0025–.0045	180
MF-1080, 1085 Diesel	4.5	.006–.008	1.4375	⑥	.018	2.7483–2.749	.002–.0042	100–105	2.9985–2.999	.0025–.0045	140–150
MF-1150	4 1/4	.005–.007	1.6249	.017	.017	3.248–3.2485	.0023–.0042	90–95	3.9967–3.9972	.004–.0063	190–200
MF-1155	4 1/4	—	1.7498–1.7500	②	.017	2.9980–2.9985	.0018–.0038	100–105	3.9967–3.9972	.004–.0063	190–200
MF-2765, 2705	3 7/8	—	⑤	.016	.009	2.4988–2.4996	.0019–.0032	75	2.9984–2.9992	.0033–.0046	200

①—Top ring, .014. Second ring, .011. Third ring, .010. Fourth ring, .018 Bottom ring, .011.
②—Top ring .008. Second ring, .017.
③—Cadium plated nuts-70 ft. lbs., plain rod nuts-90 ft. lbs.
④—Bearings with shim washers and lock tabs 90–95; without washers and locks 110–115.
⑤—1.3748 on models without turbocharger and 1.4998 on models with turbocharger.
⑥—Top Comp. 018.—2nd & 3rd .014.
⑦—D and DL serial # suffix engines, top ring-.014, second ring-.011, third ring-.010, fourth ring-.018 and bottom ring-.011. DS and DLS serial # suffix engines, top ring-.014, second and third ring-.011 and fourth ring, .012.
⑧—Top ring, .016. Second and third ring, .012.

MINNEAPOLIS-MOLINE

Tractor Model	Cyl. Bore Size In.	Piston Clearance	Pin Diam.	Comp.	Oil	Crankpin Diameter	Bearing Clearance	Rod Bolt Torque	Journal Diameter	Bearing Clearance	Cap Bolt Torque
G550 Exc. Diesel	3 5/8	.006	1.2495	.010	.015	2.249–2.250	.0005–.0015	46–50	2.249–2.250	.0002–.0032	87–92
G550 Diesel	3 5/8	.006	1.2495	.010	.010	2.249–2.250	.0005–.0015	46–50	2.249–2.250	.0002–.0032	87–92
G750 Exc. Diesel	3 3/4	.006	1.2495	.010	.015	2.249–2.250	.0005–.0015	46–50	2.624–2.625	.0005–.0035	108–112
G750 Diesel	3 7/8	.006	1.2495	.010	.010	2.4365–2.4375	.0005–.0015	③	2.624–2.625	.0015–.0045	129–133
G850 Gasoline	3 7/8	.006	1.2495	.010	.015	2.4365–2.4375	.0005–.0015	44–46	2.624–2.625	.0015–.0045	129–133
G850 Diesel	3 7/8	.006	1.2495	.010	.010	2.4365–2.4375	.0005–.0015	44–46	2.624–2.625	.0015–.0045	129–133
G940 Gasoline	3 7/8	.006	1.2495	.010	.015	2.4365–2.4375	.0005–.0015	44–46	2.624–2.625	.0015–.0045	129–133
G940 Diesel	3 7/8	.006	1.2495	.010	.010	2.4365–2.4375	.0005–.0015	44–46	2.624–2.625	.0015–.0045	129–133

Continued

PISTONS, PINS, RINGS, BEARINGS & CRANKSHAFT—Continued

Tractor Model	Cyl. Bore Size In.	Piston Clearance	Pin Diam.	Ring End Gap Minimum		Rod Bearings			Main Bearings		
				Comp.	Oil	Crankpin Diameter	Bearing Clearance	Rod Bolt Torque	Journal Diameter	Bearing Clearance	Cap Bolt Torque
MINNEAPOLIS-MOLINE—Continued											
G-955 Gas	4 1/4	.0055	1.2495	.020	.015	2.9115–2.9120	.0014–.0039	70–75	2.911–2.912	.0014–.0044	①
G-955 Diesel	4 3/8	.006	1.8745	.015	.010	2.9115–2.9120	.0014–.0039	70–75	2.911–2.912	.0014–.0044	①
G-1355 Gas	4 5/8	.007	1.2495	.016	.013	2.9115–2.9120	.0014–.0039	70–75	2.911–2.912	.0014–.0044	①
G-1355 Diesel	4 3/4	.008	1.8745	.017	.015	3.498–3.499	.0023–.0053	80	3.498–3.499	.0033–.0053	②

①—Front 240, others 140.

②—Front 200, others 108–112.

③—Indirect injection models, 46–58 ft. lbs.; direct injection models, 46–50 ft. lbs.

Tractor Model	Cyl. Bore Size In.	Piston Clearance	Pin Diam.	Comp.	Oil	Crankpin Diameter	Bearing Clearance	Rod Bolt Torque	Journal Diameter	Bearing Clearance	Cap Bolt Torque
OLIVER											
1550, 1555 Gas	3 5/8	.002	1.2495	.010	.010	2.249–2.250	.0005–.0015	46–50	2.249–2.250	.0002–.0032	90
1550, 1555 Diesel	3 5/8	.002	1.2495	.010	.010	2.249–2.250	.0005–.0015	46–50	2.249–2.250	.0002–.0032	90
1600 Non-Diesel	①	.0025	1.2495	.010	.010	2.249–2.250	.0005–.0015	46–50	2.624–2.625	.0005–.0035	110
1600 Diesel	3 3/4	.003	1.2495	.010	.010	2.436–2.437	.0005–.0015	56–60	2.624–2.625	.0005–.0035	110
1650, 1655 Non-Diesel	3 3/4	.003	1.2495	.010	.010	2.249–2.250	.0005–.0015	46–50	2.624–2.625	.0005–.0035	110
1650, 1655 Diesel	3 7/8	②	1.2495	.010	.010	2.436–2.437	.0005–.0015	③	2.624–2.625	.0015–.0045	131
1755 Gasoline	3 7/8	.003	1.2495	.010	.015	2.4365–2.4375	.0005–.0015	44–46	2.624–2.625	.0015–.0045	127–133
1755 Diesel	3 7/8	.003	1.2495	—	.010	2.4365–2.4375	.0005–.0015	44–46	2.624–2.625	.0015–.0045	129–133
1855 Gasoline	3 7/8	.003	1.2495	.010	.015	2.4365–2.4375	.0005–.0015	44–46	2.624–2.625	.0015–.0045	129–133
1855 Diesel	3 7/8	.003	1.2495	.014	.010	2.4365–2.4375	.0005–.0015	44–46	2.624–2.625	.0015–.0045	129–133
2255 Diesel	4 1/2	—	1.4999	.015	.010	2.7496–2.7504	.0015–.0045	27–33	3.4995–3.500	.0015–.0045	165–185

①—Prior to Ser. No. 137613, 3 1/2"; after Ser. No. 137612, 3 5/8".

②—Direct injection (1650–1655), .002; indirect injection (1650), .004

③—Direct injection (1650–1655), 46–50; indirect injection (1650), 56–58.

Tractor Model	Cyl. Bore Size In.	Piston Clearance	Pin Diam.	Comp.	Oil	Crankpin Diameter	Bearing Clearance	Rod Bolt Torque	Journal Diameter	Bearing Clearance	Cap Bolt Torque
WHITE											
2-70 Gasoline	3 3/4	①	1.2496	.010	.015	2.249–2.250	.0005–.0015	46–50	2.624–2.625	.0005–.0035	87–92
2-70 Diesel	3 7/8	②	1.2496	.010	.010	2.4365–2.4375	.0005–.0015	46–50	2.624–2.625	.0015–.0045	87–92
2-85 Diesel	3 7/8	③	1.3749	④	.011	2.4990–2.4995	.0015–.0030	80	2.9985–2.9990	.0025–.0045	180
2-105 Diesel	3 7/8	③	1.4999	⑤	.012	2.4990–2.4995	.0015–.0030	80	2.9985–2.9990	.0025–.0045	180
2-135 Diesel	4 9/16	.006–.0085	1.6248	.015	.015	2.9970–2.9980	.0014–.0044	115–120	3.621–3.622	.0018–.0048	135–140
2-150 Diesel	4 3/4	.007–.009	1.8746	.017	.015	3.498–3.499	.0023–.0053	55–60	3.498–3.499	.0023–.0053	⑥
2-155 Diesel	4 9/16	.006–.0085	1.6248	.015	.015	2.9970–2.9980	.0014–.0044	115–120	3.621–3.622	.0018–.0048	135–140

①—Desired piston clearance in sleeve, requires a 3 to 6 pound pull on a spring scale to withdraw a 1/2 in. wide by .003 in. thick feeler gauge, positioned between sleeve & piston 90 degrees from piston pin bore.

②—Desired piston in sleeve clearance, requires a 3 to 6 pound pull on a spring scale to withdraw a 1/2 in. wide by .002 in. thick feeler gauge, positioned between piston & sleeve 90 degrees from piston pin bore.

③—Desired piston clearance in sleeve requires a 5 to 7 pound pull on a spring scale to withdraw a 1/2 in. wide by .006 in. thick feeler gauge, positioned between sleeve & piston 90 degrees from piston pin bore.

④—Top ring, .015 in.; others .011 in.

⑤—Top ring, .016 in.; others .012 in.

⑥—5/8 in. cap screws, 108–112 ft. lbs.; 3/4 in. cap screws, 196–204 ft. lbs.

STOCK ENGINE SPECIFICATIONS

AUTOMOTIVE TYPE

Continental

ENGINE IDENTIFICATION—Model stamped on plate attached to engine.

GENERAL ENGINE SPECIFICATIONS

Engine Model	Type	No. Cyls. & Valve Location	Bore and Stroke	Piston Displacement Cu. In.	Compression Ratio	Horsepower @ R.P.M.	Torque Lb. Ft. @ R.P.M.	Normal Oil Pressure, Lbs.	Engine Oil Capacity Qts.①
F-4140	Gasoline	4 In Block	3 3/16 × 4 3/8	140	6.99	52 @ 3200	108 @ 1600	35–40	4
F-4162	Gasoline	4 In Block	3 7/16 × 4 3/8	162	7.02	58 @ 3200	122 @ 1600	35–40	4
F-6186	Gasoline	6 In Block	3 × 4 3/8	186	6.70	77 @ 3500	142 @ 1600	35–40	5
F-6209	Gasoline	6 In Block	3 3/16 × 4 3/8	209	6.70	90 @ 3500	160 @ 1600	35–40	5
F-6226	Gasoline	6 In Block	3 5/16 × 4 3/8	226	6.70	99 @ 3500	180 @ 1600	35–40	5
K-6271	Gasoline	6 In Head	3 5/8 × 4 3/8	271	6.40	115 @ 3200	216 @ 1400	40–50	7
M-6271	Gasoline	6 In Block	3 5/8 × 4 3/8	271	7.70	97 @ 3000	209 @ 1400	40–50	7
K-6290	Gasoline	6 In Head	3 3/4 × 4 3/8	290	6.40	123 @ 3200	232 @ 1400	40–50	7
M-6290	Gasoline	6 In Block	8 3/4 × 4 3/8	290	6.70	108 @ 3000	226 @ 1400	40–50	7
K-6330	Gasoline	6 In Head	4 × 4 3/8	330	6.40	147 @ 3200	266 @ 1800	40–50	7
M-6330	Gasoline	6 In Block	4 × 4 3/8	330	6.70	125 @ 3000	258 @ 1400	40–50	7
M-6363	Gasoline	6 In Block	4 × 4 13/16	363	6.70	146 @ 3000	304 @ 1600	40–50	7
K-6363	Gasoline	6 In Head	4 × 4 13/16	363	6.50	162 @ 3200	300 @ 1800	40–50	7
B-6371	Gasoline	6 In Block	4 1/8 × 4 5/8	371	6.50	124 @ 3000	284 @ 1400	40–50	7
T-6371	Gasoline	6 In Head	4 1/8 × 4 5/8	371	6.40	144 @ 3000	297 @ 1400	40–60	7
B-6427	Gasoline	6 In Block	4 5/16 × 4 7/8	427	6.50	142 @ 3000	328 @ 1200	40–50	8
T-6427	Gasoline	6 In Head	4 5/16 × 4 7/8	427	6.40	170 @ 3000	350 @ 1400	40–60	8
U-6501	Gasoline	6 In Head	4 1/2 × 5 1/4	501	6.20	178 @ 2600	414 @ 1200	55–65	14
R-6513	Gasoline	6 In Head	4 1/2 × 5 3/8	513	6.00	192 @ 2800	422 @ 1400	55–65	14
R-6572	Gasoline	6 In Head	4 3/4 × 5 3/8	572	6.00	220 @ 2800	464 @ 1200	55–65	14
R-6602	Gasoline	6 In Head	4 7/8 × 5 3/8	602	6.00	232 @ 2800	484 @ 1200	55–65	14
S-6749	Gasoline	6 In Head	5 3/8 × 5 1/2	749	6.00	250 @ 2800	574 @ 1400	55–65	18
S-6820	Gasoline	6 In Head	5 5/8 × 5 1/2	820	6.83	275 @ 2800	628 @ 1400	55–65	12
V-8603	Gasoline	8 In Head	4 3/4 × 4 1/4	603	7.00	240 @ 3200	500 @ 1400	50	18
TD-6427	Diesel	6 In Head	4 5/16 × 4 7/8	427	14.50	116 @ 2400	307 @ 1300	40–50	18
RD-6572	Diesel	6 In Head	4 3/4 × 5 3/8	572	14.50	172 @ 2400	423 @ 1300	40–60	18
SD-6802	Diesel	6 In Head	5 9/16 × 5 1/2	802	14.70	225 @ 2200	620 @ 1300	55–65	20
VD-8603	Diesel	8 In Head	4 3/4 × 4 1/4	603	15.80	182 @ 2800	470 @ 1300	50	12

①—Capacity of crankcase only, less filters.

TUNE UP & VALVE SPECIFICATIONS

Engine Model	Firing Order	Spark Plug Gap	Breaker Gap	Ignition or Injection Timing		Cylinder Head Torque Ft. Lb.	Valve Seat Angle Degrees	Valve Clearance H-Hot C-Cold		Valve Spring Pressure (Minimum) Lbs. @ Inch Length	
				Mark	Location			Intake	Exhaust	Outer	Inner
F-4140	1342	.025	.022	9°BTDC	Flywheel	②	③	.014H	.014H	96 @ 1 27/64	⑤
F-4162	1342	.025	.022	9°BTDC	Flywheel	②	③	.014H	.014H	96 @ 1 27/64	⑤
F-6186	153624	.025	.020	①	Flywheel	②	③	.014H	.014H	103 @ 1⅜	⑤
F-6209	153624	.025	.020	①	Flywheel	②	③	.014H	.014H	96 @ 1 27/64	⑤
F-6226	153624	.025	.020	①	Flywheel	②	③	.014H	.014H	96 @ 1 27/64	⑤
K-6271	153624	.025	.020	½°BTDC	Flywheel	②	③	.018H	.022H	77 @ 1.01	⑤
M-6271	153624	.025	.020	¼°BTDC	Flywheel	②	③	.017H	.020H	115 @ 1.52	⑤
K-6290	153624	.025	.020	8°BTDC	Flywheel	②	③	.018H	.022H	77 @ 1.01	⑤
M-6290	153624	.025	.020	½°BTDC	Flywheel	②	③	.017H	.020H	115 @ 1.52	⑤
K-6330	153624	.027	.020	8°BTDC	Pulley	②	③	.018H	.022H	77 @ 1.01	⑤
M-6330	153624	.025	.020	½°BTDC	Flywheel	②	③	.017H	.020H	115 @ 1.52	⑤
M-6363	153624	.025	.020	½°BTDC	Flywheel	②	③	.017H	.020H	115 @ 1.52	⑤
K-6363	153624	.027	.020	5°BTDC	Pulley	②	③	.017H	.022H	77 @ 1.01	⑤
B-6371	153624	.025	.020	IGN	Flywheel	②	③	.017H	.022H	137 @ 1.31	⑤
T-6371	153624	.025	.020	①	①	②	③	.018H	.022H	125 @ 1.11	57 @ 1.01
B-6427	153624	.025	.020	3/16"BTDC	Flywheel	②	③	.017H	.022H	137 @ 1.31	⑤
T-6427	153624	.025	.020	3/16"BTDC	Flywheel	②	④	.018H	.022H	125 @ 1.11	57 @ 1.01
U-6501	153624	.025	.020	①	①	②	③	.017H	.022H	47 @ 1 25/32	⑤
R-6513	153624	.025	.020	⅝"BTDC	Flywheel	2	③	.018H	.024H	153 @ 1.61	86 @ 1.36
R-6572	153624	.025	.020	⅝"BTDC	Flywheel	②	③	.018H	.024H	153 @ 1.61	86 @ 1.36
R-6602	153624	.025	.020	IGN	Flywheel	②	③	.018H	.024H	153 @ 1.61	86 @ 1.36
S-6749	153624	.025	.020	8°BTDC	Flywheel	②	③	.020H	.024H	200 @ 2 3/16	100 @ 2 1/32
S-6820	153624	.025	.020	8°BTDC	Flywheel	②	③	.020H	.024H	200 @ 2 3/16	100 @ 2 1/32
V-8603	—	.025	.020	3°BTDC	Pulley	105	③	.020H	.028H	81 @ 2⅛	48 @ 1⅞
TD-6427	153624	None	None	32°BTDC	Flywheel	②	④	.018H	.022H	69 @ 1.54	29 @ 1.45
RD-6572	153624	None	None	30°BTDC	Flywheel	105	③	.020H	.024H	78 @ 2¼	⑤
SD-6802	153624	None	None	38°BTDC	Flywheel	②	③	.020H	.024H	93 @ 2¾	—
VD-8603	—	None	None	—	Pulley	105	45	.022H	.024H	78 @ 2¼	—

①—Timing varies with type of vehicle operation.
②—For ⅜" 35–40, 7/16" 70–75, ½" 95, 9/16" 135, ⅝" 150.
③—Intake 30, exhaust 45.
④—Intake 15, exhaust 45.
⑤—Single spring used.

PISTONS, PINS, RINGS, CRANKSHAFT & BEARINGS

Engine Model	Piston Clearance	Wrist Pin Diameter	Ring End Gap (Minimum)		Rod Bearings			Main Bearings		
			Comp.	Oil	Shaft Diameter	Bearing Clearance	Rod Bolt Torque Ft. Lb.	Shaft Diameter	Bearing Clearance	Main Bolt Torque Ft. Lb.
F-4140	.002	.859	.008	.007	1.9365–1.9375	.0002–.0022	①	2.249–2.250	.0002–.0024	①
F-4162	.002	.859	.007	.007	1.9365–1.9375	.0002–.0022	①	2.249–2.250	.0002–.0024	①
F-6186	.003	.859	.007	.007	1.9365–1.9375	.0002–.0022	①	2.249–2.250	.0002–.0024	①
F-6209	.002	.859	.007	.008	1.9365–1.9375	.0002–.0022	①	2.249–2.250	.0002–.0024	①
F-6226	.002	.859	.010	.008	2.0619–2.0627	.0000–.0018	①	2.3744–2.3752	.0000–.002	①
K-6271	.004	1.109	.013	.013	2.248–2.249	.0009–.0029	①	2.623–2.624	.0015–.0037	①
M-6271	.004	1.109	.008	.008	2.248–2.249	.0009–.0029	①	2.623–2.624	.0015–.0037	①
K-6290	.005	1.109	.013	.013	2.248–2.249	.0009–.0029	①	2.623–2.624	.0015–.0037	①
M-6290	.005	1.109	.008	.008	2.248–2.249	.0009–.0029	①	2.623–2.624	.0015–.0037	①
K-6330	.005	1.109	.013	.013	2.248–2.249	.0009–.0029	①	2.623–2.624	.0015–.0037	①

Continued

PISTONS, PINS, RINGS, CRANKSHAFT & BEARINGS—Continued

Engine Model	Piston Clearance	Wrist Pin Diameter	Ring End Gap (Minimum)		Rod Bearings			Main Bearings		
			Comp.	Oil	Shaft Diameter	Bearing Clearance	Rod Bolt Torque Ft. Lb.	Shaft Diameter	Bearing Clearance	Main Bolt Torque Ft. Lb.
M-6330	.005	1.109	.011	.010	2.248–2.249	.0009–.0029	①	2.623–2.624	.0015–.0037	①
M-6363	.005	1.109	.011	.010	2.248–2.249	.0009–.0029	①	2.623–2.624	.0015–.0037	①
K-6363	.005	1.109	.013	.013	2.248–2.249	.0009–.0029	①	2.623–2.624	.0015–.0037	①
B-6371	.005	1.250	.011	.007	2.498–2.499	.001–.003	①	2.873–2.874	.0015–.0037	①
T-6371	.005	1.250	.013	.013	2.498–2.499	.001–.003	①	2.873–2.874	.0015–.0037	①
B-6427	.005	1.250	.013	.013	2.498–2.499	.001–.003	①	2.873–2.874	.0015–.0037	①
T-6427	.005	1.250	.013	.013	2.498–2.499	.001–.003	①	2.873–2.874	.0015–.0037	①
U-6501	.005	1.500	.010	.010	2.7475–2.7485	.0011–.0036	①	2.7475–2.7485	.0011–.0041	①
R-6513	.005	1.500	.010	.010	2.999–3.000	.0012–.0028	①	3.249–3.250	.0022–.0046	①
R-6572	.005	1.500	.010	.010	2.999–3.000	.0012–.0028	①	3.249–3.250	.0022–.0046	①
R-6602	.006	1.500	.010	.010	2.999–3.000	.0012–.0028	①	3.249–3.250	.0022–.0046	①
S-6749	.008	1.750	.017	.022	3.499–3.500	.0015–.0045	①	3.749–3.750	.0016–.0046	①
S-6820	.008	1.750	.022	.022	3.499–3.500	.0015–.0045	①	3.749–3.750	.0016–.0046	①
V-8603	.002	1.500	.013	.015	3.124–3.125	.0016–.0042	①	3.499–3.500	.0018–.0044	①
TD-6427	.006	1.250	.013	.013	2.498–2.499	.001–.003	①	2.873–2.874	.0015–.0037	①
RD-6572	.006	1.500	.010	.010	2.999–3.000	.0012–.0028	①	3.249–3.250	.0022–.0046	①
SD-6802	.007	1.750	.013	.013	3.499–3.500	.0015–.0045	①	3.749–3.750	.0016–.0046	①
VD-8603	.005	1.500	.013	.013	3.124–3.125	.0016–.0042	①	3.499–3.500	.0018–.0044	①

①—5/16" 20–25, 3/8" 35–40, 7/16" 70–75, 1/2" 85–95, 9/16" 100–110.

Hall-Scott-Hercules

ENGINE IDENTIFICATION—Model stamped on plate attached to engine.

GENERAL ENGINE SPECIFICATIONS

Engine Model	Type	No. Cyls. & Valve Location	Bore and Stroke	Piston Displace-ment Cu. In.	Compression Ratio	Horsepower @ R.P.M.	Torque Ft. Lb. @ R.P.M.	Normal Oil Pressure, Lbs.	Engine Oil Capacity Qts.
190-2	Gasoline	6 In Head	5¼ × 6	779	6.00	240 @ 2200	625 @ 1300	80	20
190-5	Gasoline	6 In Head	5¼ × 6	779	6.00	240 @ 2200	625 @ 1300	80	20
590-GV-3	Gasoline	6 In Head	5 × 5	590	6.60	239 @ 2800	490 @ 1600	60	12
590-GV-4	Gasoline	6 In Head	5 × 5	590	6.60	239 @ 2800	490 @ 1600	60	12
590-BV-1	L. P. G.	6 In Head	5 × 5	590	8.70	256 @ 2800	530 @ 1600	60	12
590-GH-1	Gasoline	6 In Head	5 × 5	590	6.60	239 @ 2800	490 @ 1600	60	18
590-BH-1	L. P. G.	6 In Head	5 × 5	590	8.70	256 @ 2800	530 @ 1600	60	18
779	Gasoline	6 In Head	5¼ × 6	779	6.00	240 @ 2200	625 @ 1300	80	20
935-B-1	L. P. G.	6 In Head	5¾ × 6	935	8.10	330 @ 2400	880 @ 1400	60	22
935-G-1	Gasoline	6 In Head	5¾ × 6	935	6.40	294 @ 2400	820 @ 1400	60	22
1091-B-1	L. P. G.	6 In Head	5¾ × 7	1090	8.10	357 @ 2200	1070 @ 1200	60	22
1091-G-1	Gasoline	6 In Head	5¾ × 7	1090	6.40	318 @ 2200	945 @ 1400	60	22
6156-B-1	L. P. G.	6 In Head	5¾ × 6	935	8.10	342 @ 2400	890 @ 1400	60	22
6156-G-1	Gasoline	6 In Head	5¾ × 6	935	6.40	310 @ 2400	812 @ 1300	60	22
6182-B-1	L. P. G.	6 In Head	5¾ × 7	1091	8.10	368 @ 2200	1091 @ 1100	60	22
6182-G-1	Gasoline	6 In Head	5¾ × 7	1091	6.50	332 @ 2200	960 @ 1200	60	22

TUNE UP & VALVE SPECIFICATIONS

Engine Model	Firing Order	Spark Plug Gap	Breaker Gap	Ignition Timing		Cylinder Head Torque Ft. Lb.	Valve Seat Angle Degrees	Valve Clearance H-Hot C-Cold		Valve Spring Pressure (Minimum) Lbs. @ Inch Length	
				Mark	Location			Intake	Exhaust	Inner	Outer
190-2	153624	.025	.018	①	Flywheel	④	②	.022C	.022C	110 @ 1 15/16	143 @ 2
190-5	153624	.025	.018	①	③	④	②	.022C	.022C	110 @ 1 15/16	143 @ 2
590-GV-3	153624	.025	.020	"5"	③	⑤	45	.021C	.021C	80 @ 1 3/4	144 @ 1 3/16
590-GV-4	153624	.025	.020	"5"	③	⑤	45	.021C	.021C	80 @ 1 3/4	144 @ 1 3/16
590-BV-1	153624	.015	.020	"10"	③	⑤	45	.021C	.021C	80 @ 1 3/4	144 @ 1 3/16
590-GH-1	153624	.025	.020	"5"	③	⑤	45	.021C	.021C	80 @ 1 3/4	144 @ 1 3/16
590-BH-1	153624	.015	.020	"10"	③	⑤	45	.021C	.021C	80 @ 1 3/4	144 @ 1 3/16
779	153624	.025	.018	①	Flywheel	④	②	.022C	.022C	110 @ 1 15/16	143 @ 2
935-B-1	142635	.016	.021	3/4"BTDC	Flywheel	⑥	②	.021C	.031C	105 @ 1 15/16	138 @ 2
935-G-1	142635	.016	.021	3/4"BTDC	Flywheel	⑥	②	.021C	.031C	105 @ 1 15/16	138 @ 2
1091-B-1	142635	.016	.021	3/4"BTDC	Flywheel	⑥	②	.021C	.031C	105 @ 1 15/16	138 @ 2
1091-G-1	142635	.016	.021	3/4"BTDC	Flywheel	⑥	②	.021C	.031C	105 @ 1 15/16	138 @ 2
6156-B-1	142635	.016	.021	1 1/2"BTDC	Flywheel	⑥	②	.021C	.031C	110 @ 1 15/16	143 @ 2
6156-G-1	142635	.016	.021	1 1/2"BTDC	Flywheel	⑥	②	.021C	.031C	110 @ 1 15/16	143 @ 2
6182-B-1	142635	.016	.021	5/8"BTDC	Flywheel	⑥	②	.021C	.031C	110 @ 1 15/16	143 @ 2
6182-G-1	142635	.016	.021	5/16"BTDC	Flywheel	⑥	②	.021C	.031C	110 @ 1 15/16	143 @ 2

①—Three flywheel teeth before TDC mark on flywheel.
②—Intake 30°, exhaust 45°.
③—Flywheel and crankshaft pulley.
④—½" 35, ¾" 215.
⑤—Small 35, large 185.
⑥—Large 240, small nuts 40.

PISTONS, PINS, RINGS, CRANKSHAFT & BEARINGS

Engine Model	Piston Clearance	Wrist Pin Diameter	Ring End Gap (Minimum)		Rod Bearings			Main Bearings		
			Comp.	Oil	Shaft Diameter	Bearing Clearance	Rod Bolt Torque Ft. Lb.	Shaft Diameter	Bearing Clearance	Main Bolt Torque Ft. Lb.
190-2	.0105	1.400	.013	.025	2.7475–2.7480	.004	130–140	3.247–3.2475	.004	180–200
190-5	.0105	1.400	.013	.025	2.7475–2.7480	.004	130–140	3.247–3.2475	.004	180–200
590-GV-3	.003	1.500	.017	.013	2.9970–2.9975	.003	100–120	3.248–3.2485	.004	130–140
590-GV-4	.003	1.500	.017	.013	2.9970–2.9975	.003	100–120	3.248–3.2485	.004	130–140
590-BV-1	.003	1.500	.017	.013	2.9970–2.9975	.003	100–120	3.248–3.2485	.004	130–140
590-GH-1	.003	1.500	.017	.013	2.9970–2.9975	.003	100–120	3.248–3.2485	.004	130–140
590-BH-1	.003	1.500	.017	.013	2.9970–2.9975	.003	100–120	3.248–3.2485	.004	130–140
779	.0105	1.400	.013	.025	2.7475–2.7480	.004	130–140	3.274–3.2475	.004	180–200
935-B-1	.006	1.6247	.020	.013	3.2468–3.2473	.0035	130–140	3.4968–3.4973	.004	180–200
935-G-1	.006	1.6247	.020	.013	3.2468–3.2473	.0035	130–140	3.4968–3.4973	.004	180–200
1091-B-1	.006	1.6247	.020	.013	3.2468–3.2473	.0035	130–140	3.4968–3.4973	.004	180–200
1091-G-1	.006	1.6247	.020	.013	3.2468–3.2473	.0035	130–140	3.4968–3.4973	.004	180–200
6156-B-1	.006	1.6247	.020	.013	3.2468–3.2473	.0035	130–140	3.4968–3.4973	.004	180–200
6156-G-1	.006	1.6247	.020	.013	3.2468–3.2473	.0035	130–140	3.4968–3.4973	.004	180–200
6182-B-1	.006	1.6247	.020	.013	3.2468–3.2473	.0035	130–140	3.4968–3.4973	.004	180–200
6182-G-1	.006	1.6247	.020	.013	3.2468–3.2473	.0035	130–140	3.4968–3.4973	.004	180–200

Hercules

ENGINE IDENTIFICATION—Model stamped on plate attached to engine.

TUNE UP & VALVE SPECIFICATIONS

Engine Model	Firing Order	Spark Plug Gap	Breaker Gap	Ignition or Injection Timing		Cylinder Head Torque Ft. Lb.	Valve Seat Angle Degrees	Valve Clearance		Valve Spring Pressure (Minimum) Lbs. @ Inch Length	
				Mark	Location			H-Hot Intake	C-Cold Exhaust	Outer	Inner
DFXE	153624	None	None	27° BTDC	Flywheel	350	45	.010H	.016H	94 @ 3	57 @ 2 21/32
DFXH	153624	None	None	27° BTDC	Flywheel	350	45	.010H	.016H	124 @ 1 25/32	96 @ 1 21/32
DIX-4-D	153624	None	None	30° BTDC	Flywheel	158	45	①	①	49 @ 1.21	34 @ 1.09
DIX-6-272	153624	None	None	30° BTDC	Flywheel	158	45	①	①	49 @ 1.21	34 @ 1.09
DJXC	153624	None	None	26° BTDC	Flywheel	157	45	①	①	55 @ 1.40	37 @ 1.28
DJXH	153624	None	None	29° BTDC	Flywheel	157	45	①	①	55 @ 1.40	37 @ 1.28
DOOD	1243	None	None	25o BTDC	Flywheel	157	45	①	①	55 @ 1.40	37 @ 1.28
DRXC	153224	None	None	30° BTDC	Flywheel	②	45	①	①	48 @ 1.45	30 @ 1.33
DWXLD	153624	None	None	24° BTDC	Flywheel	158	45	①	①	84 @ 1 37/64	74 @ 1 31/64
HXE	153624	.025	.020	DC 1–6	Flywheel	105	30	.010H	.016H	84 @ 3 5/32	50 @ 2 13/16
IXB	1243	.025	.020	DC 1–4	Flywheel	40	30	.006H	.008H	42 @ 1 3/16	③
JXC, JXD	153624	.025	.020	DC 1–6	Flywheel	④	45	①	①	58 @ 1.60	③
JXLD	153624	.025	.020	DC 1–6	Flywheel	④	⑤	①	①	58 @ 1.60	③
QXD, QXLD	153624	.025	.020	DC 1–6	Flywheel	⑦	30	①	①	58 @ 1.60	③
RXC	153624	.025	.020	DC 1–6	Flywheel	⑧	45	①	①	40 @ 1 9/32	③
RXLDH	153624	.025	.020	DC 1–6	Flywheel	80	⑩	①	①	102 @ 2 5/32	③
TDC-895	153624	None	None	—	—	350	45	.010H	.016H	102 @ 2 5/32	③
WXLC-3	153624	.025	.020	DC 1–6	Flywheel	⑨	30	.012H	.016H	57 @ 2 21/32	94 @ 2 31/32
G-14, 1600	1342	.018	31–34	6° BTDC	Flywheel	60–85	—	.015	.015	102 @ 2 5/32	③
G-20, 2300	1243	.031	.022	TDC	Flywheel	—	—	.016H	.016H	—	—
G-3000	153624	.031	.022	6° BTDC	Flywheel	—	—	.016H	.016H	72 @ 1.292	—
G-3400	153624	.031	.022	TDC	Flywheel	—	—	.016H	.016H	72 @ 1.292	—
D-4800	153624	None	None	20° BTDC	Flywheel	140	—	.010C	.010C	—	—

①—See Instruction Plate on engine.
②—For ⅝" nuts 175, for 1" nuts 280.
③—Single spring used.
④—Nuts 65, screws on manifold side 75, screws marked "1041" 100, other screws 85.
⑥—Exhaust 45°, intake 30° or 45°.
⑦—Nuts 60, screws with washers 100.
⑧—Cap screws 75, nuts 85.
⑨—Cap screws 75, nuts 100.
⑩—Intake 30°, exhaust 45°.

PISTONS, PINS, RINGS, CRANKSHAFT & BEARINGS

Engine Model	Piston Clearance	Wrist Pin Diameter	Ring End Gap (Minimum)		Rod Bearings			Main Bearings		
			Comp.	Oil	Shaft Diameter	Bearing Clearance	Rod Bolt Torque Ft. Lb.	Shaft Diameter	Bearing Clearance	Main Bolt Torque Ft. Lb.
DFXE	.009	2.00	.018	.022	3.309–3.310	.003–.0055	263	4.499–4.500	.005–.006	260
DFXH	.0145	2.00	.018	.022	3.309–3.310	.003–.0055	263	4.499–4.500	.005–.006	260
DIX-4-D	.006	1.125	.012	.012	2.372–2.373	.003–.004	150	2.747–2.748	.003–.0045	①
DIX-6-272	.006	1.125	.012	.012	2.372–2.373	.003–.004	150	2.747–2.748	.003–.0045	①
DJXC, H	.0055	1.187	.018	.018	2.497–2.498	.003–.004	175	2.997–2.998	.003–.0045	②
DOOD	.0055	1.187	.018	.018	2.497–2.498	.003–.004	175	2.997–2.998	.003–.0045	②
DRXC	.0075	1.625	.018	.018	2.997–2.998	.003–.004	175	3.497–3.498	.003–.004	175
DWXLD	.0125	1.50	.015	.015	2.747–2.748	.0045–.0055	③	3.497–3.498	.0045–.0055	175
HXE	.006	1.50	.020	.020	2.997–2.998	.003–.004	263	3.497–3.498	.003–.004	④
IXB	.0035	.75	.015	.015	1.747–1.748	.001–.0015	42	1.987–1.988	.002–.0025	77
JXC, JXD	⑫	1.00	.015	.015	1.987–1.988	.0025–.003	56	2.497–2.498	.0035–.004	⑤
JXLD	.005	1.125	.015	.015	1.987–1.988	.0025–.003	56	2.497–2.498	.0035–.004	⑤
QXD, QXLD	⑬	.875	.015	.015	1.987–1.988	.001–.0015	39	2.497–2.498	.002–.003	⑤
RXC	.0045	1.25	.015	.015	2.622–2.623	.002–.003	⑦	2.997–2.998	.003–.004	⑧

Continued

PISTONS, PINS, RINGS, CRANKSHAFT & BEARINGS—Continued

Engine Model	Piston Clearance	Wrist Pin Diameter	Ring End Gap (Minimum)		Rod Bearings			Main Bearings		
			Comp.	Oil	Shaft Diameter	Bearing Clearance	Rod Bolt Torque Ft. Lb.	Shaft Diameter	Bearing Clearance	Main Bolt Torque Ft. Lb.
RXLDH	.0065	1.50	.015	.015	2.997–2.998	.004–.0045	158	3.497–3.498	.0045–.005	175
TCD-895	.008	2.00	.018	.018	3.309–3.310	.0035–.0045	263	4.449–4.500	.005–.006	260
WXLC-3	.0045	1.31	.015	.015	2.247–2.248	.0015–.0025	⑨	2.622–2.623	.0025–.0035	⑩
D-15, 17, 2000	.005	1.2498	.010	.010	2.373–2.374	.001–.003	—	2.8734–2.8744	.0009–.0034	—
D-21, 2300	.005	1.2498	.010	.010	2.373–2.374	.001–.003	—	2.8734–2.8744	.0009–.0034	—
D-30, 3400	.005	1.2498	.010	.010	2.373–2.374	.001–.003	—	2.8734–2.8744	.0009–.0034	—
G-1400	.002	—	.012⑪	.015	1.9979–1.9987	.001–.0028	30–35	2.2477–2.2485	.0004–.002	55–60
G-1600	.002	—	.010	.015	1.9979–1.9987	.0013–.0031	30–35	2.2477–2.2485	.0009–.0035	55–60
G-20, 2300	.003	1.2498	.010	.010	2.373–2.374	.001–.003	—	2.8734–2.8744	.0009–.0034	—
G-30, 3400	.003	1.2498	.010	.010	2.373–2.374	.001–.003	—	2.8734–2.8744	.0009–.0034	—

①—Front, center and rear 105, intermediate 85.
②—For 5/8" nuts 95, for 1/2" nuts 78.
③—For 1/2" nuts 158, for 9/16" nuts 225.
④—Front & inter. 210, center and rear 192.
⑤—Front and inter. 70, center and rear 60.
⑦—For 1/2" 105, for 9/16" 115.
⑧—Front and center 123, center and rear 105.
⑨—For 1/2" 105, for 7/16" 53.
⑩—Front & inter. 70, center and rear 75.
⑪—Top compression ring .013.
⑫—Model JXC, .0035; Models JXD, .0045.
⑬—Model QXD, .003; Models QXLD, .0035.

GENERAL ENGINE SPECIFICATIONS

Engine Model	Type	No. Cyls. & Valve Location	Bore and Stroke	Piston Displacement Cu. In.	Compression Ratio	Horsepower & R.P.M.	Torque Ft. Lb. & R.P.M.	Normal Oil Pressure, Lbs.	Engine Oil Capacity Qts.
DFXE	Diesel	6 In Head	5 5/8 × 6	895	14.80	227 @ 2100	685 @ 2100	50	24
DFXH	Diesel	6 In Head	5 3/4 × 6	935	14.80	260 @ 2100	750 @ 1200	50	24
DIX4D	Diesel	4 In Head	3 5/8 × 4	166	15.50	57 @ 3000	120 @ 1500	30	①
DIX6-272	Diesel	6 In Head	3 11/16 × 4 1/4	272	15.50	103 @ 3000	202 @ 1800	30	①
DJXB	Diesel	6 In Head	3 1/2 × 4 1/2	260	15.50	77 @ 2600	178 @ 1300	45	①
DJXC	Diesel	6 In Head	3 3/4 × 4 1/2	298	15.50	83 @ 2600	208 @ 1300	45	①
DJXH	Diesel	6 In Head	3 3/4 × 4 1/2	298	15.50	99 @ 2600	234 @ 1400	45	①
DOOD	Diesel	4 In Head	4 1/4 × 4 1/2	255	15.50	79 @ 2600	183 @ 1400	30	①
DRXC	Diesel	6 In Head	4 5/8 × 5 1/4	529	15.00	147 @ 2200	400 @ 1100	30	①
DWXLD	Diesel	6 In Head	4 1/4 × 5	426	15.50	142 @ 2600	328 @ 1600	40	①
D-1500	Diesel	3 In Head	3 3/4 × 4 1/2	149	17.50	46 @ 2600	113 @ 1500	40	①
D-1700	Diesel	3 In Head	4 × 4 1/2	169	17.50	50 @ 2600	128 @ 1500	40	①
D-2000	Diesel	4 In Head	3 3/4 × 4 1/2	198	17.50	70 @ 2600	156 @ 1500	40	①
D-2120	Diesel	4 In Head	3 7/8 × 4 1/2	—	—	—	—	40	①
D-2300	Diesel	4 In Head	4 × 4 1/2	226	17.50	73 @ 2600	182 @ 1400	40	①
D-3000	Diesel	6 In Head	3 3/4 × 4 1/2	298	17.50	114 @ 3000	244 @ 1400	40	①
D-3000T	Diesel	6 In Head	3 3/4 × 4 1/2	298	17	117 @ 2000	308 @ 2000	—	①
DH-3000	Diesel	6 In Head	3 3/4 × 4 1/2	298	17.50	101 @ 2400	229 @ 1800	—	①
D-3400	Diesel	6 In Head	4 × 4 1/2	339	17.50	115 @ 2600	276 @ 1400	40	①
D-4800	Diesel	6 In Head	4 9/16 × 4 7/8	478	18	143 @ 2400	350 @ 1800	45	①
G-1400	Gasoline	4 In Head	3 7/8 × 3	141	8.0	58 @ 2800	115 @ 2000	40	①
G-1600	Gasoline	4 In Head	4 × 3 1/4	163	8.0	60 @ 2400	140 @ 1400	40	①
G-2000	Gasoline	4 In Head	3 3/4 × 4 1/2	198	7.50	78 @ 2400	185 @ 1400	35	①
G-2300	Gasoline	4 In Head	4 × 4 1/4	226	7.50	81 @ 2200	207 @ 1400	35	①
G-3000	Gasoline	6 In Head	3 3/4 × 4 1/2	298	7.50	117 @ 2400	270 @ 1300	35	①
G-3400	Gasoline	6 In Head	4 × 4 1/2	339	7.50	132 @ 2400	310 @ 1300	35	①
HXE	Gasoline	6 In Block	5 3/4 × 6	935	6.20	227 @ 2000	750 @ 900	35	①
IXB	Gasoline	4 In Block	3 1/4 × 4	133	6.50	46 @ 3200	92 @ 1800	20	①

Continued

GENERAL ENGINE SPECIFICATIONS—Continued

Engine Model	Type	No. Cyls. & Valve Location	Bore and Stroke	Piston Displacement Cu. In.	Compression Ratio	Horsepower & R.P.M.	Torque Ft. Lb. & R.P.M.	Normal Oil Pressure, Lbs.	Engine Oil Capacity Qts.
JXC	Gasoline	6 In Block	$3\frac{3}{4} \times 4\frac{1}{4}$	282	6.50	103 @ 3200	207 @ 1400	26	①
JXD	Gasoline	6 In Block	$4 \times 4\frac{1}{4}$	320	6.50	112 @ 3200	240 @ 1400	26	①
JXLD	Gasoline	6 In Block	$4 \times 4\frac{1}{2}$	339	6.50	131 @ 3200	272 @ 1400	26	①
L-237	Gasoline	6 In Block	$3\frac{7}{16} \times 4\frac{1}{4}$	236.7	—	79 @ 2400	185 @ 1400	—	—
QXD	Gasoline	6 In Block	$3\frac{7}{16} \times 4\frac{1}{8}$	230	6.50	78 @ 3200	167 @ 1500	26	①
QXLD	Gasoline	6 In Block	$3\frac{7}{16} \times 4\frac{1}{4}$	237	6.50	92 @ 3200	190 @ 1400	26	①
RXC	Gasoline	6 In Block	$4\frac{5}{8} \times 5\frac{1}{4}$	529	6.30	143 @ 2400	372 @ 1100	26	①
RXLDH	Gasoline	6 In Block	$4\frac{3}{4} \times 5\frac{1}{4}$	558	6.50	181 @ 2600	443 @ 1400	36	①
TCD-895	Diesel	6 In Head	$5\frac{5}{8} \times 6$	895	14.80	318 @ 2000	846 @ 1800	50	①
WXLC-3	Gasoline	6 In Block	$4\frac{1}{4} \times 4\frac{3}{4}$	404	6.50	139 @ 2600	312 @ 1300	26	①

①—Depends upon type of oil pan used.

Waukesha

ENGINE IDENTIFICATION—Model stamped on plate attached to engine.

GENERAL ENGINE SPECIFICATIONS

Engine Model	Type	No. Cyls. & Valve Location	Bore and Stroke	Piston Displacement Cu. In.	Compression Ratio	Horsepower @ R.P.M.	Torque Ft. Lb. @ R.P.M.	Normal Oil Pressure, Lbs.	Engine Oil Capacity, Qts.①
135-DK	Diesel	6 In Head	$4\frac{1}{4} \times 5$	426	16.80	140 @ 2400	326 @ 1600	40	12
135-DKB	Diesel	6 In Head	$4\frac{1}{4} \times 5$	426	16.8	147 @ 2800	328 @ 1600	40	12
135-DKBS	Diesel	6 In Head	$4\frac{1}{4} \times 5$	426	16.8	185 @ 2800	400 @ 1800	40	12
135-GKB	Gasoline	6 In Head	$4\frac{1}{4} \times 5$	426	6.25	148 @ 3000	337 @ 1200	40	12
135-GZ	Gasoline	6 In Head	$4\frac{3}{8} \times 5$	451	6.30	153 @ 2800	351 @ 1300	40	12
135-GZB	Gasoline	6 In Head	$4\frac{3}{8} \times 5$	451	6.30	153 @ 2800	354 @ 1200	40	12
140-GK	Gasoline	6 In Head	$4\frac{1}{2} \times 5\frac{1}{2}$	525	6.40	155 @ 2250	420 @ 1000	40	10
140-GKB	Gasoline	6 In Head	$4\frac{1}{2} \times 5\frac{1}{2}$	525	6.40	220 @ 3200	466 @ 1800	40	10
140-GZ	Gasoline	6 In Head	$4\frac{5}{8} \times 5\frac{1}{2}$	554	6.40	170 @ 2250	468 @ 1000	40	10
140-GZB	Gasoline	6 In Head	$4\frac{5}{8} \times 5\frac{1}{2}$	554	6.40	226 @ 3200	490 @ 1800	40	10
145-GK	Gasoline	6 In Head	$5\frac{1}{4} \times 6$	779	6.20	240 @ 2400	594 @ 1000	40	18
145-GKB	Gasoline	6 In Head	$5\frac{1}{4} \times 6$	779	6.20	260 @ 2400	640 @ 1000	40	18
145-GZ	Gasoline	6 In Head	$5\frac{3}{8} \times 6$	817	6.30	250 @ 2400	630 @ 1200	40	18
145-GZB	Gasoline	6 In Head	$5\frac{3}{8} \times 6$	817	6.30	268 @ 2400	652 @ 1200	40	18
148-DK	Diesel	6 In Head	$5\frac{1}{4} \times 6$	779	16.50	197 @ 2000	584 @ 1000	40	18
148-DKB	Diesel	6 In Head	$5\frac{1}{4} \times 6$	779	16.50	200 @ 2100	584 @ 1000	40	18
148-DKBS	Diesel	6 In Head	$5\frac{1}{4} \times 6$	779	16.50	280 @ 2100	706 @ 1800	40	18
180-DLC	Diesel	4 In Head	$3\frac{1}{2} \times 3\frac{3}{4}$	144	17.10	45 @ 2400	102 @ 1800	25-30	5
180-GLB	Gasoline	4 In Head	$3\frac{1}{2} \times 3\frac{3}{4}$	144	7.00	45 @ 2400	118 @ 1600	25-30	5
185-DLC	Diesel	6 In Head	$3\frac{1}{2} \times 3\frac{3}{4}$	216	17.10	60 @ 2400	152 @ 1000	25-30	8
185-GLB	Gasoline	6 In Head	$3\frac{1}{2} \times 3\frac{3}{4}$	216	7.00	67 @ 2400	176 @ 1400	25-30	8
190-DLCA	Diesel	6 In Head	$3\frac{3}{4} \times 4$	265	17.00	85 @ 2600	191 @ 1400	25-30	7
190-GLB	Gasoline	6 In Head	$3\frac{3}{4} \times 4$	265	7.00	77 @ 2200	223 @ 1200	25-30	7
195-DLCA	Diesel	6 In Head	4×4	302	17.00	98 @ 2800	221 @ 1800	25-30	7
195-GKA	Gasoline	6 In Head	$4\frac{1}{8} \times 4$	320	6.20	125 @ 3200	263 @ 1600	25-30	7
195-GL	Gasoline	6 In Head	4×4	302	6.20	85 @ 2400	240 @ 1200	25-30	7
197-DLC	Diesel	6 In Head	4×4	302	18.50	91 @ 2800	216 @ 1600	40	7
197-DLCS	Diesel	6 In Head	4×4	302	18.50	131 @ 2800	275 @ 1800	40	7
MZA	Gasoline	6 In Block	$4\frac{1}{4} \times 4\frac{3}{4}$	404	5.60	130 @ 3000	289 @ 1000	40	8
WAKB	Gasoline	6 In Head	$6\frac{1}{4} \times 6\frac{1}{2}$	1197	7.00	280 @ 1800	997 @ 1000	40	32

Continued

GENERAL ENGINE SPECIFICATIONS—Continued

Engine Model	Type	No. Cyls. & Valve Location	Bore and Stroke	Piston Displacement Cu. In.	Compression Ratio	Horsepower @ R.P.M.	Torque Ft. Lb. @ R.P.M.	Normal Oil Pressure, Lbs.	Engine Oil Capacity, Qts.①
WAKC	Gasoline	6 In Head	6 1/4 × 6 1/2	1197	7.00	290 @ 1800	1020 @ 1000	40	32
WAKD	Diesel	6 In Head	6 1/4 × 6 1/2	1197	16.50	224 @ 1600	840 @ 1000	40	32
WAKDB	Diesel	6 In Head	6 1/4 × 6 1/2	1197	16.50	258 @ 1800	845 @ 1000	40	32
WAKDBS	Diesel	6 In Head	6 1/4 × 6 1/2	1197	16.50	400 @ 1800	1400 @ 1200	40	32
ROILINE									
H540	Gasoline	V8 In Head	4 1/2 × 4 1/4	540	6.70	170 @ 2400	430 @ 900	40	②
H570	Gasoline	V8 In Head	4 5/8 × 4 1/4	570	7.45	218 @ 2400	509 @ 1600	50	②
H844	Gasoline	V8 In Head	5 1/4 × 4 7/8	844	6.70	300 @ 2400	637 @ 1500	30-40	②
H884	Gasoline	V8 In Head	5 3/8 × 4 7/8	884	7.62	320 @ 2400	770 @ 1500	50	②

①—Sump capacity only. Add for filters, etc.
②—H540 16 qts., T-H540 9 qts.
③—H844 and H884 20 qts., T-H844 and T-H884 front sump 16 qts., center sump 20 qts.

TUNE UP & VALVE SPECIFICATIONS

Engine Model	Firing Order	Spark Plug Gap	Breaker Gap	Ignition or Injection Timing		Cylinder Head Torque (Oiled) Inch Lbs.	Valve Seat Angle Degrees	Valve Clearance⑨ C-Cold		Valve Spring Pressure (Minimum) Lbs. @ Inch Length	
				Mark	Location			Intake	Exhaust	Outer	Inner
135-DK	153624	None	None	⑧	Flywheel	2000–2100	45	.011C	.023C	82 @ 2 5/16	①
135-DKB	153624	None	None	FP⑧	Flywheel	2000–2100	45	.011C	.023C	82 @ 2 5/16	①
135-DKBS	153624	None	None	FP⑧	Flywheel	2000–2100	45	.011C	.023C	82 @ 2 5/16	①
135-GKB	153624	.025	.018	Dist.⑧	Flywheel	2000–2100	45	.011C	.023C	82 @ 2 5/32	①
135-GZ	153624	.025	.018	⑧	Flywheel	2000–2100	45	.011C	.023C	82 @ 2 5/32	①
135-GZB	153624	.025	.018	Dist.⑧	Flywheel	2000–2100	45	.011C	.023C	82 @ 2 5/32	①
140-GK	153624	.025	.018	Dist.⑧	Flywheel	2400–2500	②	.013C	.019C	86 @ 1 21/32	55 @ 1 7/16
140-GKB	153624	.025	.018	Dist.⑧	Flywheel	2400–2500	②	.013C	.025C	177 @ 1 41/64	①
140-GZ	153624	.025	.018	Dist.⑧	Flywheel	2400–2500	②	.013C	.019C	177 @ 1 41/64	①
140-GZB	153624	.025	.018	Dist.⑧	Flywheel	2400–2500	②	.013C	.025C	177 @ 1 41/64	①
145-GK	153624	.025	.018	Dist.⑧	Flywheel	③	②	.013C	.024C	118 @ 2 3/8	81 @ 2 1/16
145-GKB	153624	.025	.018	Dist.⑧	Flywheel	③	②	.013C④	.030C④	265 @ 1 63/64	①
145-GZ	153624	.025	.018	Dist.⑧	Flywheel	③	②	.013C	.024C	265 @ 1 63/64	①
145-GZB	153624	.025	.018	Dist.⑧	Flywheel	③	②	.013C④	.030C④	265 @ 1 63/64	①
148-DK	153624	None	None	⑧	Flywheel	3500–3600	45	.015C	.028C	158 @ 2 3/8	100 @ 2 1/16
148-DKB	153624	None	None	FP⑧	Flywheel	3500–3600	45	.015C	.028C	158 @ 2 3/8	100 @ 2 1/16
148-DKBS	153624	None	None	FP⑧	Flywheel	3500–3600	45	.015C	.028C	158 @ 2 3/8	100 @ 2 1/16
180-DLC	1243	None	None	FP⑧	Flywheel	1350–1400⑦	45	o010C	.020C	48 @ 1 5/16	①
180-GLB	1243	.025	.018	Dist.⑧	Flywheel	1100–1200	45	.010C	.020C	48 @ 1 15/16⑤	45 @ 1 47/64⑥
185-DLC	153624	None	None	FP⑧	Flywheel	1350–1400⑦	45	.010C	.020C	48 @ 1 15/16	①
185-GLB	153624	.025	.018	Dist.⑧	Flywheel	1100–1200	45	.010C	.020C	48 @ 1 15/16⑤	45 @ 1 47/64⑥
190-DLCA	153624	None	None	FP⑧	Flywheel	1350–1400⑦	45	.010C	.020C	71 @ 1 19/32	①
190-GLB	153624	.025	.018	Dist.⑧	Flywheel	1100–1200	45	.008C	.022C	48 @ 1 15/16⑤	45 @ 1 47/64⑥
195-DLCA	153624	None	None	FP⑧	Flywheel	1350–1400⑦	45	.010C	.020C	124 @ 1 15/16	①
195-GKA	153624	.025	.018	Dist.⑧	Flywheel	1100–1200	45	.015C	.023C	124 @ 1 15/16	①
195-GL	153624	.025	.018	Dist.⑧	Flywheel	1100–1200	45	.008C	.022C	124 @ 1 15/16	①
197-DLC	153624	None	None	FP⑧	Flywheel	1350–1400	45	.010C	.020C	124 @ 1 15/16	①
197-DLCS	153624	None	None	FP⑧	Flywheel	1350–1400	45	.008C	.020C	124 @ 1 15/16	①
MZA	153624	.025	.018	Dist.⑧	Flywheel	875–900	45	.099C	.020C	101 @ 1 21/32	①
WAKB	153624	.025	.018	Dist.⑧	Flywheel	2400–2500	②	.018C	.038C	50 @ 3 15/32	56 @ 3 5/32
WAKC	153624	None	None	FP⑧	Flywheel	3500–3600	②	.018C	.038C	50 @ 3 15/32	56 @ 3 5/32
WAKD	153624	None	None	FP⑧	Flywheel	3500–3600	②	.018C	.038C	50 @ 3 15/32	56 @ 3 5/32
WAKDB	153624	None	None	FP⑧	Flywheel	3500–3600	②	.014C	.024C	50 @ 3 15/32	56 @ 3 5/32
WAKDBS	153624	None	None	FP⑧	Flywheel	3500–3600	②	.014C	.024C	50 @ 3 15/32	56 @ 3 5/32
ROILINE											
H540	⑩	.023–.028	.018	TDC	Flywheel	1075–1100	45	.015C	.020C	106 @ 1.6	①
H570	⑩	.023–.028	.018	TDC	Flywheel	1075–1100	45	.015C	.020C	100 @ 1.8	①

Continued

TUNE UP & VALVE SPECIFICATIONS—Continued

Engine Model	Firing Order	Spark Plug Gap	Breaker Gap	Ignition or Injection Timing		Cylinder Head Torque (Oiled) Inch Lbs.	Valve Seat Angle Degrees	Valve Clearance⑨ C-Cold		Valve Spring Pressure (Minimum) Lbs. @ Inch Length	
				Mark	Location			Intake	Exhaust	Outer	Inner
H844	⑩	.023–.028	.018	TDC	Pulley	1075–1100	45	.015C	.020C	132 @ 1.8	①
H884	⑩	.023–.028	.018	TDC	Pulley	1075–1100	45	.015C	.020C	150 @ 1.9	①

①—Single spring used.
②—Intake 30°, exhaust 45°.
③—Long 2400–2500, short 1800–1850.
④—With sodium-cooled exhaust valves intake .011C, exhaust .020C.
⑤—Intake valve spring.
⑥—Exhaust valve spring.
⑦—Cap screws with former 3/16" hole drilled through to be torqued 950–100. Present hole (3/16 top—1/8 bottom) torque to 1150–1200.
⑧—Engines with degree marks instead of stamped marks on flywheel have timing data noted on engine.
⑨—Valve clearances stamped on WAUKESHA engine nameplate for cold settings only.
⑩—1L-4R-4L-2L-3R-3L-2R-1R.

PISTONS, PINS, RINGS, CRANKSHAFT & BEARINGS

Engine Model	Piston Clearance	Wrist Pin Diameter	Ring End Gap (Minimum)		Rod Bearings			Main Bearings		
			Comp.	Oil	Shaft Diameter	Bearing Clearance	Rod Bolt Torque (Oiled) Inch Lbs.	Shaft Diameter	Bearing Clearance	Main Bolt Torque Inch Lbs.
135-DK	.0065	1.50	.015	③	2.999–3.000	.001–.0035	1300–1350	3.249–3.250	.0015–.0045	1500–1600
135-DKB	.0065	1.50	.015	③	2.999–3.000	.001–.0035	1300–1350	3.249–3.250	.0015–.0045	1500–1600
135-DKBS	.0065	1.50	.015	③	2.999–3.000	.001–.0035	1300–1350	3.249–3.250	.0015–.0045	1500–1600
135-GKB	.003	1.50	.015	③	2.999–3.000	.001–.0035	1300–1350	3.249–3.250	.0015–.0045	1500–1600
135-GZ	.003	1.50	.015	③	2.999–3.000	.001–.0035	1300–1350	3.249–3.250	.0015–.0045	1500–1600
135-GZB	.003	1.50	.015	③	2.999–3.000	.001–.0035	1300–1350	3.249–3.250	.0015–.0045	1500–1600
140-GK	.006①	1.375	.030	.015	2.624–2.625	.001–.003	1150–1200	3.249–3.250	.002–.004	1550–1600
140-GKB	.006	1.375	.030	.015	2.624–2.625	.001–.003	1150–1200	3.249–3.250	.002–.004	1550–1600
140-GZ	.006	1.375	.030	.015	2.624–2.625	.001–.003	1150–1200	3.249–3.250	.002–.004	1550–1600
140-GZB	.006	1.375	.030	.015	2.624–2.625	.001–.003	1150–1200	3.249–3.250	.002–.004	1550–1600
145-GK	.007②	1.625	.030	.020	2.999–3.000	.001–.003	800–825	3.499–3.500	.002–.004	3200–3300
145-GKB	.0075	1.625	.030	.020	2.999–3.000	.001–.003	800–825	3.499–3.500	.002–.005	3200–3300
145-GZ	.007	1.625	.030	.020	2.999–3.000	.001–.003	800–825	3.499–3.500	.002–.004	3200–3300
145-GZB	.007	1.625	.030	.020	2.999–3.000	.001–.003	800–825	3.499–3.500	.002–.005	3200–3300
148-DK	.007	1.875	.035	.015	3.374–3.375	.0015–.004	800–825	4.249–4.250	.002–.0047	3500–3600
148-DKB	.007	1.875	.035	.015	3.374–3.375	.0015–.004	800–825	4.249–4.250	.002–.0047	3500–3600
148-DKBS	.007	1.875	.035	.015	3.374–3.375	.0015–.004	800–825	4.249–4.250	.002–.0047	3500–3600
180-DLC	.0045	1.25	.014	.007③	2.249–2.250	.001–.0015	550–600	2.249–2.250	.002–.003	1050–1100
180-GLB	.002	1.25	.010	.007	2.249–2.250	.0005–.0015	525–550	2.249–2.250	.0002–.0032	1050–1100
185-DLC	.0045	1.25	.014	.007③	2.249–2.250	.0005–.0015	550–600	2.249–2.250	.0002–.0032	1050–1100
185-GLB	.002	1.25	.010	.007	2.249–2.250	.0005–.0015	525–550	2.249–2.250	.0002–.0032	1050–1100
190-DLCA	.0045	1.25	.015	.010③	2.624–2.625	.0005–.003	1050–1100	2.624–2.625	.0015–.0045	1300–1350
190-GLB	.0035	1.25	.015	.010	2.4365–2.4375	.0005–.0015	675–700	2.624–2.625	.0005–.0035	1300–1350
195-DLCA	.005	1.25	.015	.010③	2.624–2.625	.0005–.0015	1050–1100	2.624–2.625	.0015–.0045	1300–1350
195-GKA	.0035	1.25	.013	③	2.624–2.625	.0005–.003	675–700	2.624–2.625	.0015–.0045	1300–1350
195-GL	.0045	1.25	.013	③	2.249–2.250	.0005–.0015	1050–1100	2.624–2.625	.0005–.003	1300–1350
197-DLC	.006	1.250	.015	.010③	2.624–2.625	.0005–.003	1050–1100	2.624–2.625	.0015–.0045	1500–1600
197-DLCS	.006	1.250	.015	.010③	2.624–2.625	.0005–.003	1050–1100	2.624–2.625	.0015–.0045	1500–1600
MZA	.004	1.00	.013	.015	2.249–2.250	.001–.003	800–825	2.624–2.625	.0015–.0035	1150–1200
WAKB	.009	1.875	.045	.020	3.374–3.375	.0015–.004	1025–1050	3.999–4.000	.0025–.005	3500–3600
WAKC	.009	1.875	.045	.020	3.374–3.375	.0015–.004	1025–1050	3.999–4.000	.0025–.005	3500–3600
WAKD	.009	2.187	.035	.020	3.749–3.750	.002–.0045	1300–1350	4.749–4.750	.0024–.0058	3500–3600
WAKDB	.009	2.187	.035	.020	3.749–3.750	.002–.0045	1300–1350	4.749–4.750	.0024–.0058	3500–3600
WAKDBS	.009	2.187	.035	.020	3.749–3.750	.002–.0045	1300–1350	4.749–4.750	.0024–.0058	3500–3600
ROILINE										
H540	.002–.0035	1.2499	.012 Min.	.012 Min.	2.748–2.749	.0011–.0036	800–825	3.2481–3.2490	.0014–.0044	1300–1350
H570	.002–.0035	1.2499	.012 Min.	.012 Min.	2.748–2.749	.0011–.0036	800–825	3.2481–3.2490	.0014–.0044	1300–1350
H844	.003–.005	1.4999	.015 Min.	.015 Min.	3.248–3.249	.0013–.0043	1075–1100	3.7480–3.7490	.002–.005	1775–1825
H884	.003–.005	1.4999	.015 Min.	.015 Min.	3.248–3.249	.0013–.0043	1075–1100	3.7480–3.7490	.002–.005	1775–1825

①—For aluminum; cast iron .0045".
②—For aluminum; cast iron .0055".
③—U-Flex ring—no gap.

LIGHT DUTY DIESEL ENGINE REPAIR SECTION

INDEX

CHEVROLET (GM V8-379, 6.2L) DIESEL ENGINE

INDEX OF SERVICE OPERATIONS

GENERAL ENGINE SPECIFICATIONS

Year	Engine	Carburetor	Bore and Stroke	Piston Displace-ment, Cubic Inches	Com-pression Ratio	Net Brake H.P. @ R.P.M.	Net Torque Lbs. Ft. @ R.P.M.	Normal Oil Pressure Pounds @ R.P.M.
1982–83	135 Horsepower① V8-379	Fuel Inj.	3.98 × 3.82	379	21.5	135 @ 3600	240 @ 2000	35 @ 3000
	130 Horsepower② V8-379	Fuel Inj.	3.98 × 3.82	379	21.5	130 @ 3600	240 @ 2000	35 @ 3000

①—Heavy duty emissions.

②—Light duty emissions

VALVE SPECIFICATIONS

Year	Model	Valve Lash		Valve Angles		Valve Spring Installed Height	Valve Spring Pressure Lbs. @ In.	Stem Clearance		Stem Diameter	
		Int.	Exh.	Seat	Face			Intake	Exhaust	Intake	Exhaust
1982–83	V8-379	Hydraulic①		46°	45°	1.811	230 @ 1.38	.0010–.0027	.0010–.0027	—	—

①—No adjustment

PISTONS, PINS, RINGS, CRANKSHAFT & BEARINGS

Year	Model	Piston Clearance	Ring End Gap			Wrist Pin Diam-eter	Rod Bearings		Main Bearings			
			Top	2nd	Oil		Shaft Diameter	Bearing Clearance	Shaft Diameter	Bearing Clearance	Thrust on Bearing No.	Shaft End Play
1982–83	V8-379	.004–.005	.011–.021	.029–.039	.009–.020	1.2203–1.2206	2.398–2.399	.004–.005	2.949–2.950	①	—	.0019–.0070

①—No. 1, 2, 3, 4: .0018–.0032; No. 5, 1982: .0022–.0037; No. 5, 1983: .0041–.0099.

ENGINE TIGHTENING SPECIFICATIONS

Year	Engine Model	Glow Plugs Ft. Lbs.	Cylinder Head Bolts Ft. Lbs.	Intake Manifold Ft. Lbs.	Exhaust Manifold Ft. Lbs.	Rocker Arm Shaft Bolts Ft. Lbs.	Rocker Arm Cover Ft. Lbs.	Connecting Rod Cap Bolts Ft. Lbs.	Main Bearing Cap Bolts Ft. Lbs.	Flywheel to Crankshaft Ft. Lbs.	Torsional Damper Ft. Lbs
1982–83	V8-379	10	96	31	22	34	17	48	①	—	151

①—Inner 111 ft. lbs. Outer 100 ft. lbs.

Engine Electrical Diagnosis Section

DIAGNOSIS

NOTE: On early production 1982 vehicles, the engine wiring harness connector at the glow plug controller, Fig. 1, has an open hole at the No. 2 pin connection. Accumulated moisture and dirt in this hole may cause a controller malfunction. If the engine is hard to start, or has burned glow plugs, remove the connector and clean the pin area on the controller and connector, then reconnect the connector. Apply a small amount of RTV sealant, or equivalent over the No. 2 pin hole to prevent further contamination. Check operation of the controller. If the controller does not cycle normally and/or the pins are excessively corroded, the controller must be replaced.

Also note that the wires from pins 5 and 6 are ground wires, and that the ground connection is at the rear of the right hand head on a stud which also ground the body ground strap. Ensure that this ground connection, Fig. 2, is secure.

Refer to Figs. 3 thru 10 for diagnostic procedures.

Fig. 1 Glow plug controller

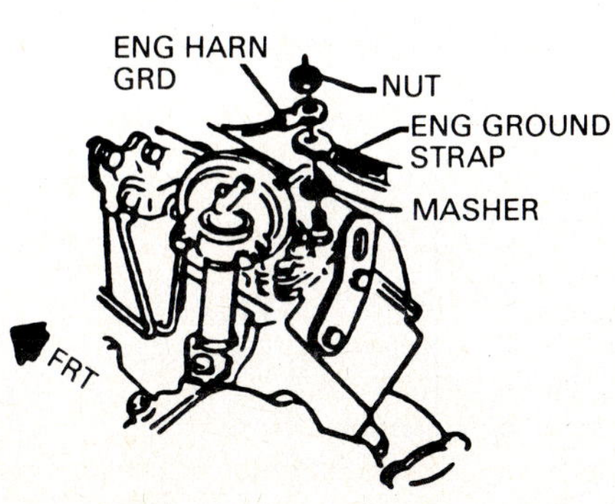

Fig. 2 Ground connections

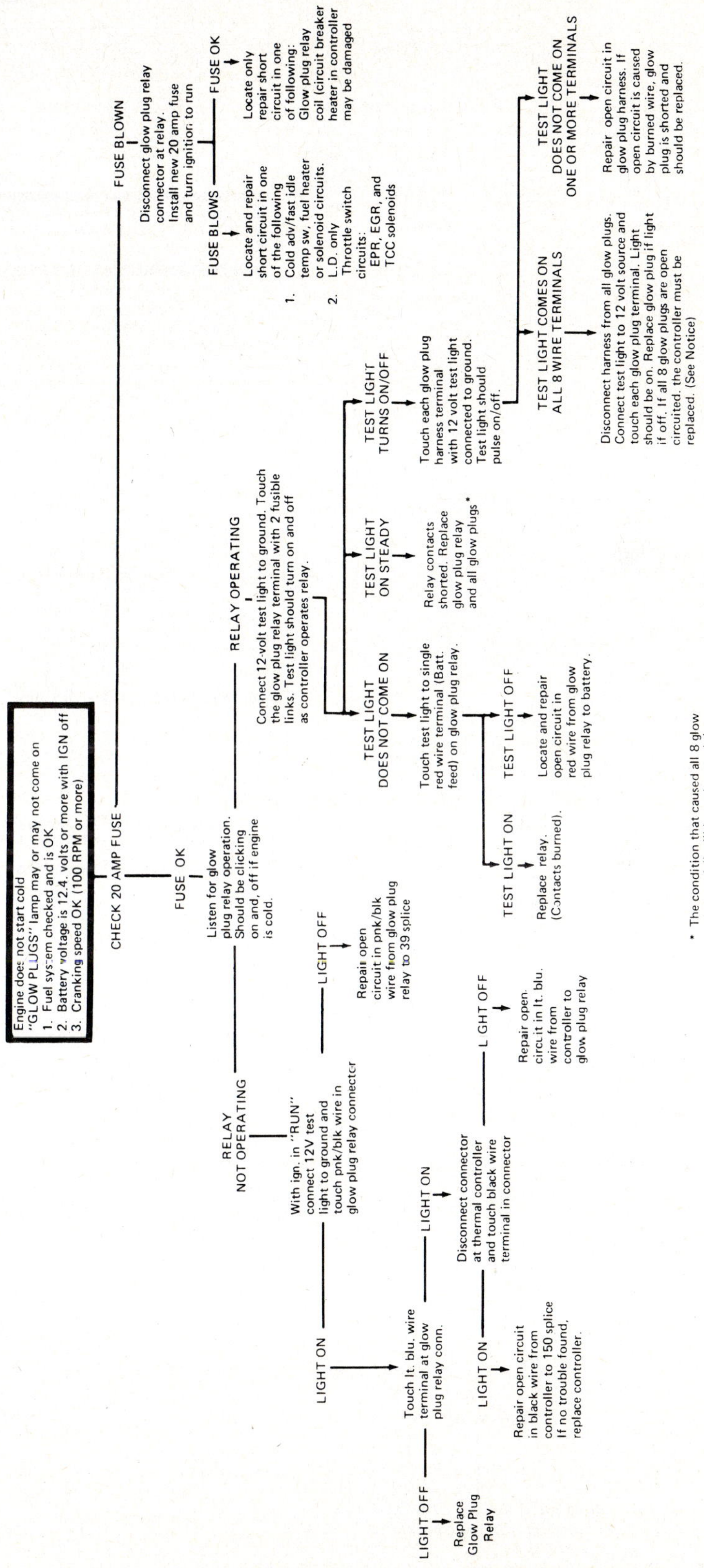

Fig. 3 Electrical system diagnosis

ENGINE CONTINUES TO RUN IGNITION KEY OFF

ENGINE CONTINUES TO RUN

↓

Disconnect pink wire at injection pump solenoid.

- ENGINE CONTINUES TO RUN → Remove injection pump for repair.
- ENGINE STOPS → Check ignition switch adjustment. If OK, replace ignition switch. Check ign. sw. assoc. wiring.

ENGINE STAYS ON FAST IDLE AT ALL TIMES

Turn A/C off, ignition on and disconnect fast idle solenoid – dark green wire.

- SOLENOID DOES NOT RETRACT → Check throttle linkage or solenoid plunger for binding. If linkage is OK, replace solenoid.
- SOLENOID RETRACTS → Connect solenoid wire. Open throttle slightly. Disconnect 2-wire connector at fast idle temperature sw.
  - SOLENOID RETRACTS → Check continuity of temperature switch with self powered test light (KD-125 or equiv): continuity below 90°F, no continuity above 122°F. Replace switch if bad.

ENGINE RUNS ROUGH ON COLD START GLOW PLUGS NOT CYCLING ON AND OFF AFTER ENGINE STARTS – 20 AMP FUSE OK

(All open circuit in the orn wire from controller to glow plug relay or an open heater element between pin 4 and 5 of the controller will cause this condition)

↓

WITH IGN OFF
Using self powered test lite check continuity of orn wire from pin 4 of controller, to the glow plug relay.

- NO CONTINUITY → Repair open circuit in orn wire from controller to relay.
- CONTINUITY → Disconnect controller connector and using a high impedance millimeters check pin 4 to pin 5 resistance (should be 27 ± Ω 3)
  - → INCORRECT READING OR OPEN CIRCUIT – Controller is defective.

Fig. 4 Electrical system diagnosis

NO GLOW PLUGS LAMP –

1. Check 20 Amp Fuse
   FUSE OK

↓

With ignition in "Run"
Check for glow plug relay operation by listening for click at relay.

- Relay clicking → With ignition switch OFF Check continuity of BLK wire circuit 150A, from bulb socket to fuse block bus bar grd.
  - Continuity → Check continuity of ORN/BLK wire circuit 503 from bulb socket to splice 503
  - No Continuity → Repair as necessary
- Not clicking → go to "Engine does not start cold" chart (relay not operating).

Fig. 5 Electrical system diagnosis

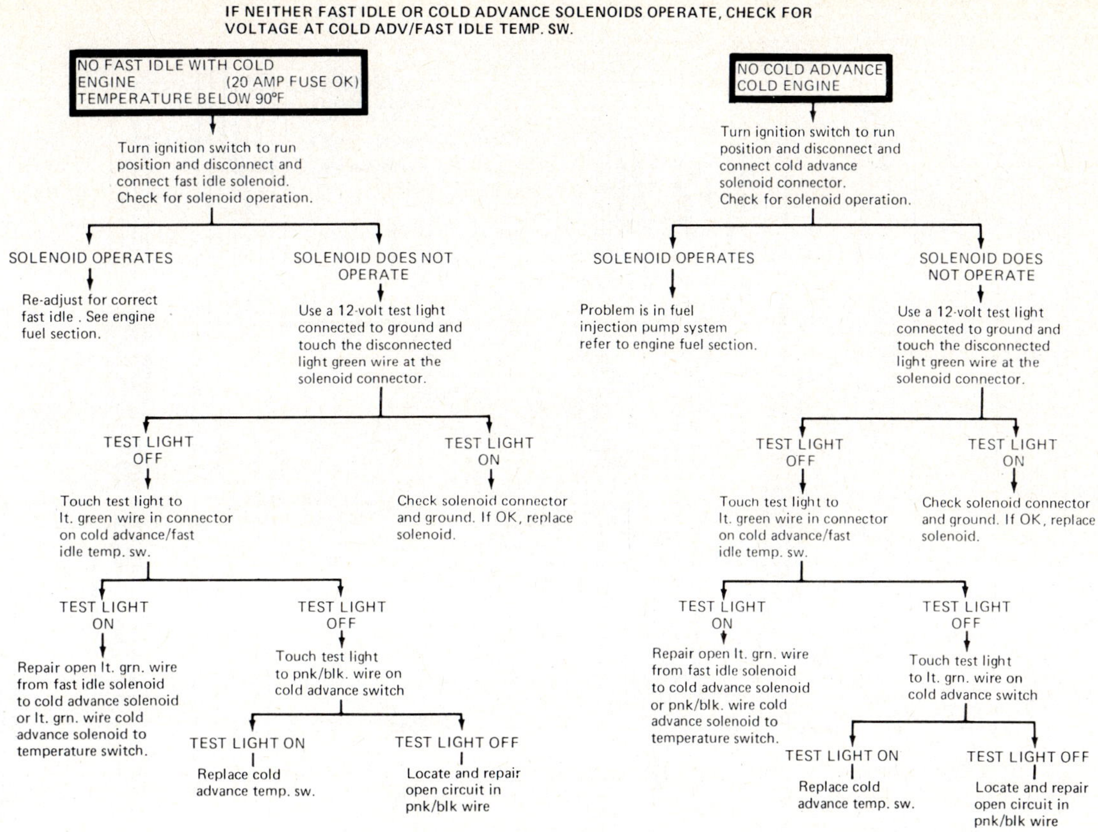

Fig. 6 Electrical system diagnosis

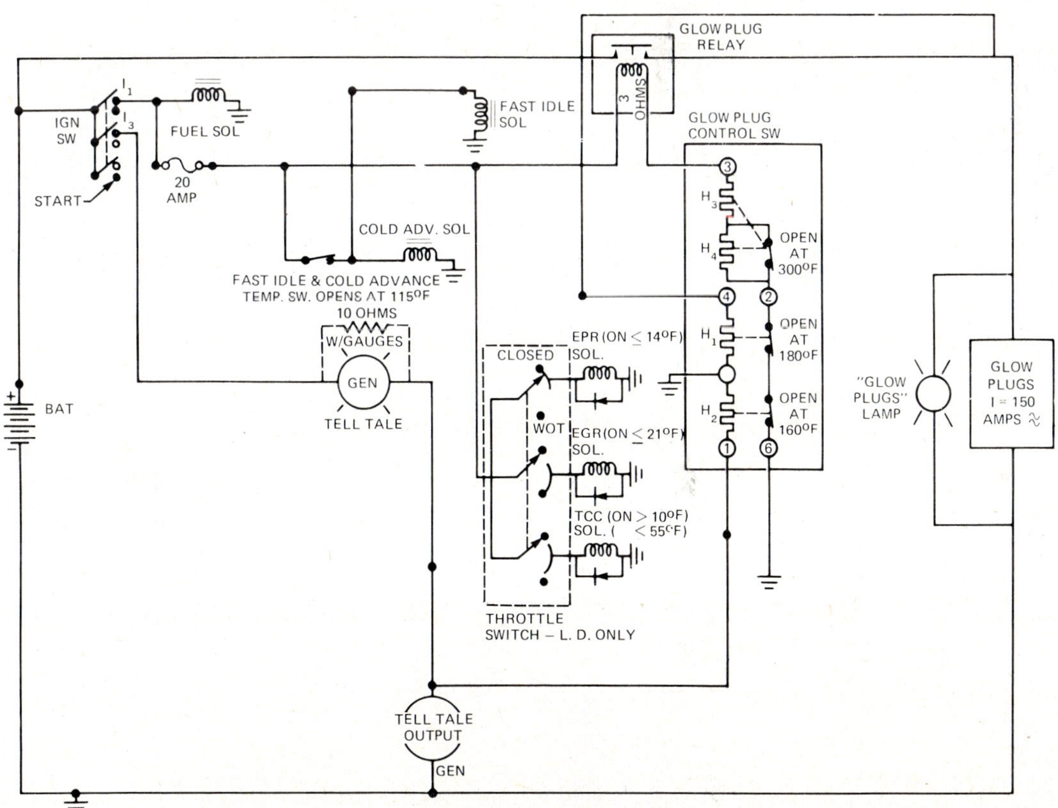

Fig. 7 Glow plug wiring schematic

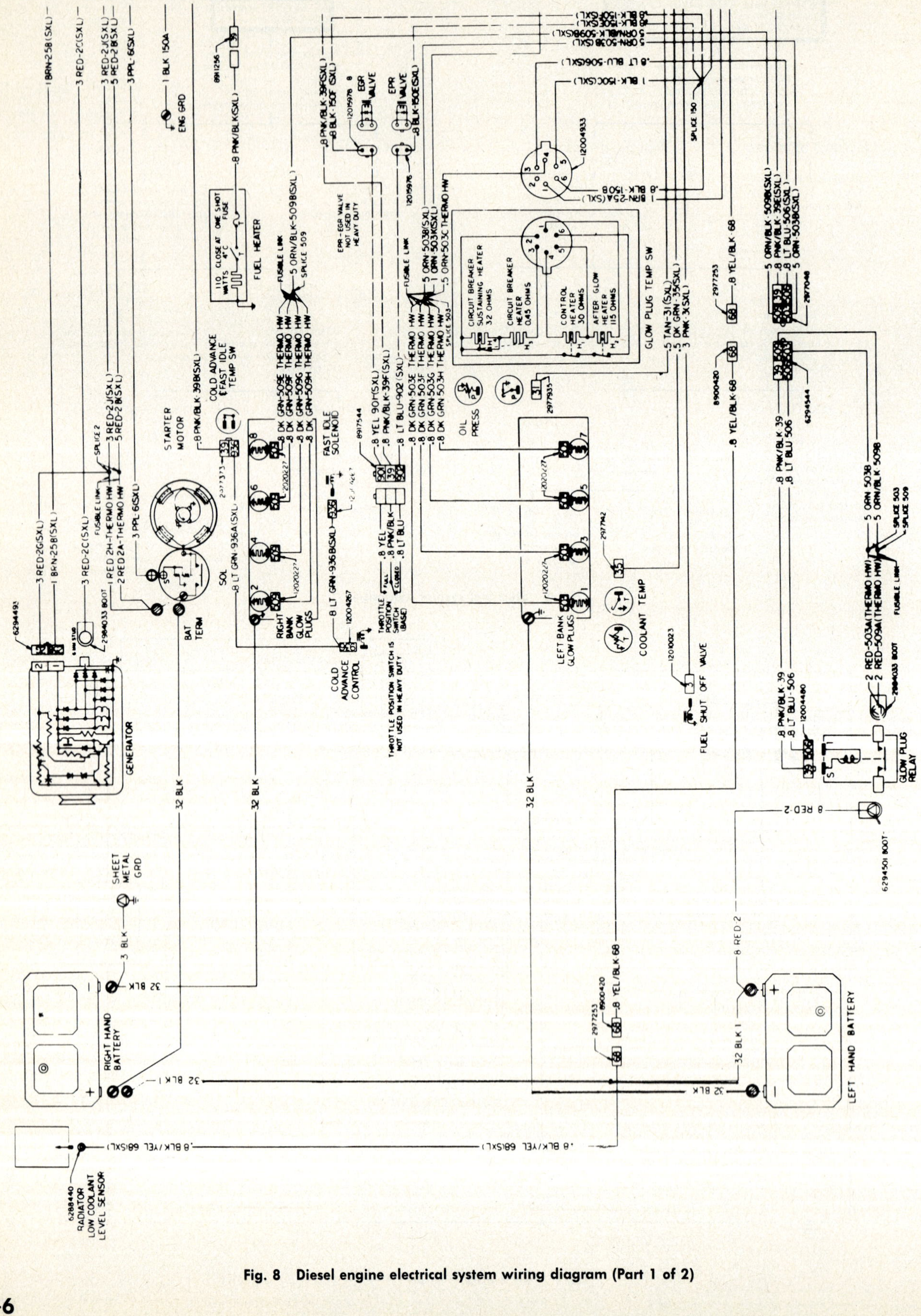

Fig. 8 Diesel engine electrical system wiring diagram (Part 1 of 2)

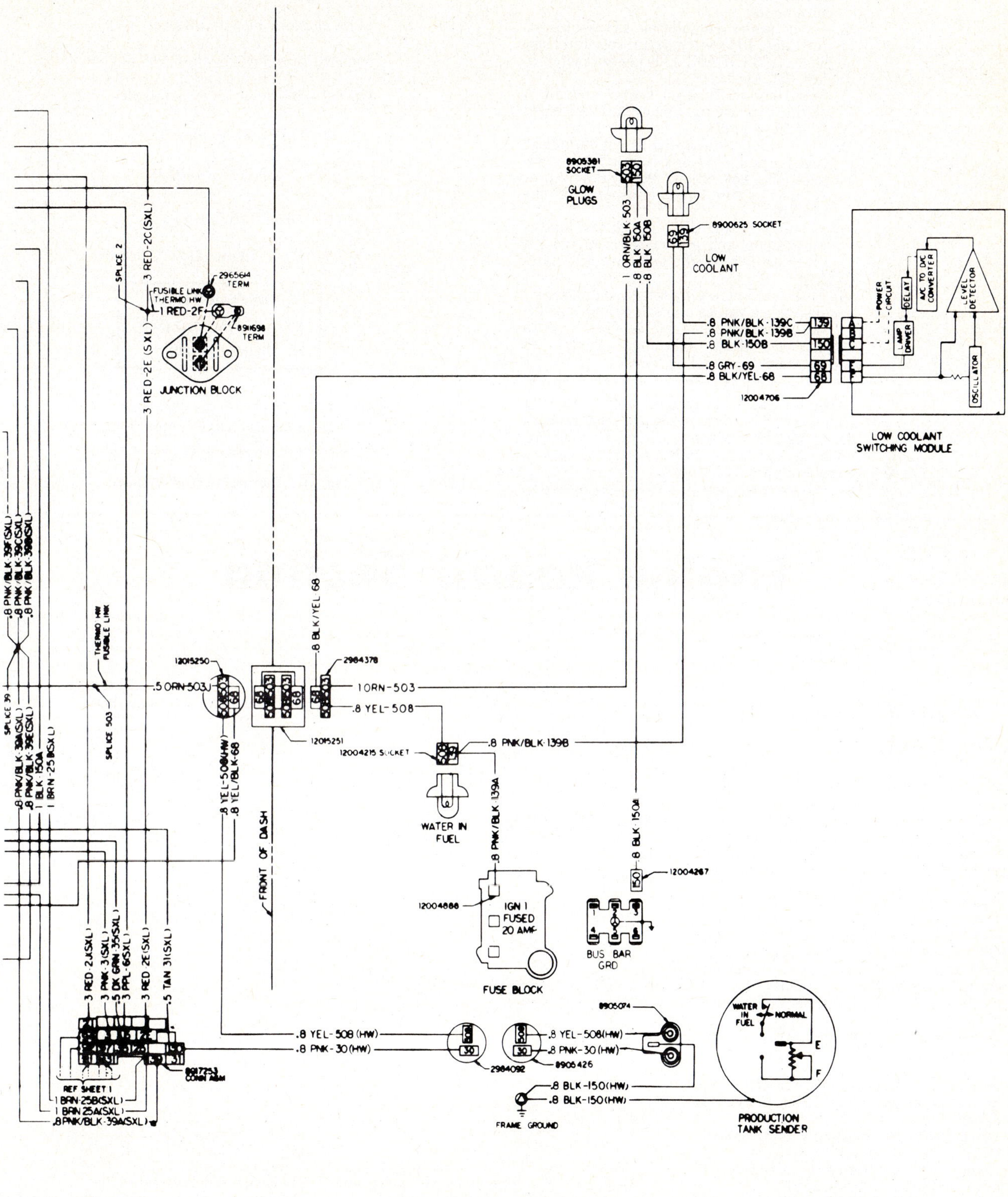

Fig. 9 Diesel engine electrical system wiring diagram (Part 2 of 2)

GLOW PLUG RESISTANCE PROCEDURE

1. Use Kent-Moore High Impedence Digital Multimeter (J-29125) for resistance measurements.
2. Select scales as follows: LH switch to "OHMS", RH switch to full counterclockwise, "200 OHMS", slide center switch to left "DC.LO."
3. Start engine, turn on heater and allow engine to warm up. Remove all feed wires from glow plugs.
4. Turn engine idle speed screw on side of injector pump to worst engine idle roughness. Do not exceed 900 RPM. Allow engine to run for a minimum of one minute.
5. Attach alligator clip to black test lead of multimeter and grounded to engine lift strap on left hand side of intake manifold.
6. Probe each glow plug terminal and record resistance values on each cylinder in firing sequence (1-8-7-2-6-5-4-3) with engine idling. Readings should be between 1.8 and 3.4 OHMS.

NOTE: If these readings are not obtained, turn engine off for several minutes and recheck glow plugs. Resistance should be .7–.8 OHMS. Readings of 1.2–1.3 OHMS would indicate an engine mechanical problem.

THERMAL CONTROLLER CHECK

With connector removed from controller the controller heater circuits may be checked using a high impedance ohmeter. However, this check will not determine shorted switches within the controller.

Pin 3 – Pin 2 .40 to .75 Ω

Pin 4 – Pin 5 27 Ω ± 3 Ω

Pin 5 – Pin 1 130 Ω ± 10%

Pin 2 – Pin 6 Continuity ("O" ohms)

GLOW PLUGS LAMP CYCLES ON AND OFF WARM ENGINE

This condition can be caused by an open circuit in 25 circuit from gen telltale output to pin 1 of the controller, or by generator output failure

Fig. 10 Thermal controller check

Engine Repair Section

ENGINE, REPLACE

Exc. Van

1. Disconnect batteries and raise vehicle.
2. Remove transmission dust cover and disconnect torque converter and exhaust.
3. Disconnect wires from starter and remove starter.
4. Remove transmission bell housing bolts and motor mount bolts.
5. Disconnect block heaters, then remove wiring harness, transmission cooler lines and front battery cable clamp at oil pan.
6. Disconnect fuel return lines and oil cooler lines from engine and remove lower fan shroud bolts and lower vehicle.
7. Scribe hood hinge locations and remove hood.
8. Drain cooling system, remove air cleaner with resonator attached and primary filter from cowling.
9. Disconnect ground cable at alternator bracket and remove alternator wires and clips.
10. Disconnect TPS, EGR-EPR and fuel cut-off at injection pump.
11. Remove harness from clips at rocker covers and disconnect glow plugs.
12. Disconnect EGR-EPR solenoids, glow plugs, controller and temperature sender. Move harness aside.
13. Disconnect ground strap from left side.
14. Remove fan, upper radiator hoses from engine, fan shroud, power steering pump and belt, and power steering reservoir.

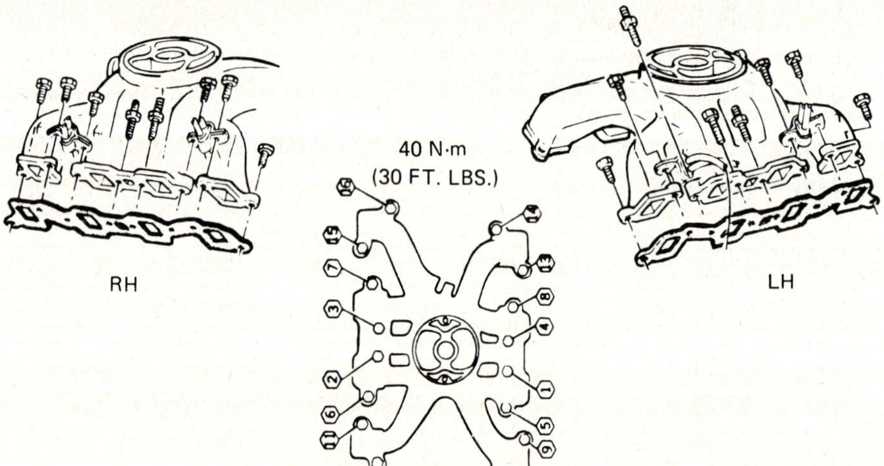

Fig. 1 Intake manifold tightening sequence

15. Disconnect vacuum at cruise servo and accelerator cable at injection pump.
16. Disconnect heater hose and lower radiator hose from engine and oil cooler lines from radiator.
17. Disconnect heater hose and overflow at radiator, automatic transmission cooler lines, upper radiator cover and radiator.
18. Remove detent cable.
19. Support transmission and remove engine.
20. Reverse procedure to install.

NOTE: If fuel filters were removed, they must be filled with clean diesel fuel.

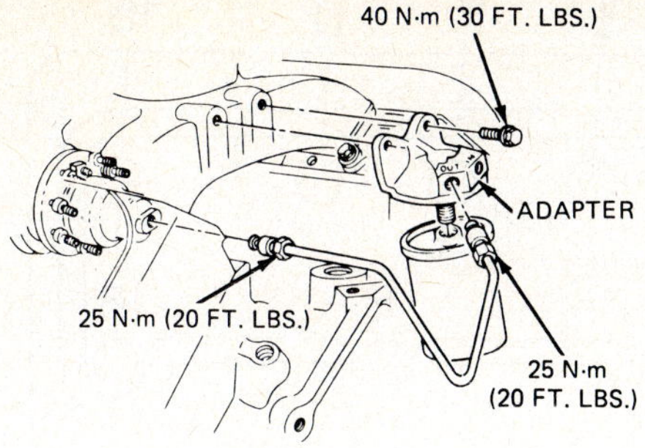

Fig. 2 Secondary fuel filter

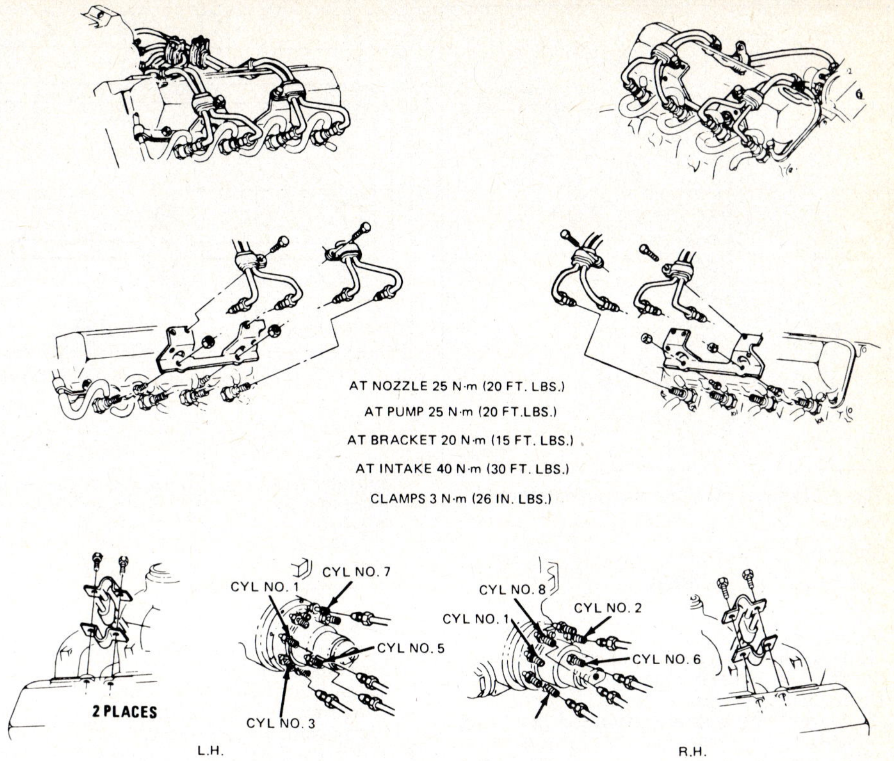

Fig. 3 Fuel injection lines

Van

1. Disconnect batteries.
2. Remove headlight bezels, then grille.
3. Remove bumper, then lower valence.
4. Remove hood latch assembly.
5. Remove coolant recovery body and upper fan shroud.
6. Remove upper tie bar, then engine cover.
7. If vehicle is equipped with A/C, discharge system, then disconnect A/C condenser inlet and outlet lines and remove condenser.
8. Disconnect low coolant sensor wire and drain radiator.
9. Disconnect transmission and engine oil cooler lines from radiator.
10. Remove radiator upper and lower hose, then remove radiator and fan assembly.
11. Remove fuel injection pump. Refer to "Injection Pump, Replace."
12. Raise and support vehicle.
13. Disconnect exhaust pipes from manifolds.
14. Remove inspection cover, then remove torque converter to flex plate bolts.
15. Remove motor mount through bolts.
16. Disconnect starter wiring, then remove starter.
17. Lower vehicle.
18. If equipped with cruise control, remove cruise control transducer.
19. If equipped with A/C, disconnect A/C compressor suction and discharge lines, then remove rear compressor brace, brackets and compressor.
20. Remove power steering pump and position aside.
21. Remove oil filler tube upper bracket.
22. Remove glow plug relay, then disconnect oil pressure sender and loom.
23. Remove air cleaner resonator and bracket.
24. Remove transmission filler tube bracket nut and position bracket aside.
25. Disconnect radiator, heater and bypass hoses at air crossover.
26. Remove generator upper bracket.
27. Remove coolant crossover.
28. Disconnect fuel lines from fuel pump.
29. Install engine lifting adapter J-33888 or equivalent.
30. Position suitable engine lifting device and connect to engine lifting adapter. Remove engine.
31. Reverse procedure to install.

INTAKE MANIFOLD, REPLACE

Exc. Van

1. Disconnect battery, remove air cleaner, crankcase ventilator tubes, secondary filter lines and secondary filter and adapter.
2. Loosen vacuum pump bolts and rotate pump for access to manifold bolts. Remove bolts.

NOTE: Injection line clips are retained by same bolts.

3. Remove intake manifold.
4. Reverse procedure to install. Refer to Fig. 1 for bolt torque and tightening sequence. Refer to Figs. 2 and 3 for fuel line installation.

NOTE: Gasket has opening for EGR on light duty application and insert to cover opening for heavy duty applications.

Van

1. Disconnect batteries, then remove engine cover.
2. Remove air cleaner assembly.
3. Remove EGR/EPR switches.
4. Remove crankcase depression regulator valve hoses from intake manifold.
5. If equipped with A/C, remove rear compressor bracket.
6. Remove fuel filter to intake manifold bracket.
7. Remove vacuum pump. After removal, place a rag or cover over hole to prevent foreign material from entering engine.
8. Remove intake manifold bolts and fuel line clips, then remove intake manifold.
9. Reverse procedure to install. Refer to Fig. 1 for bolt torque and tightening sequence. Refer to Figs. 2 and 3 for fuel line installation.

EXHAUST MANIFOLD, REPLACE

Exc. Van

Right Side

1. Disconnect battery and raise vehicle.
2. Disconnect exhaust pipe from manifold and lower vehicle.
3. Disconnect glow plug wires and remove air cleaner duct bracket.
4. Remove glow plugs and manifold bolts.
5. Reverse procedure to install. Refer to Fig. 4.

Left Side

1. Disconnect batteries.
2. Remove dipstick tube bracket nut, then remove dipstick.
3. Disconnect glow plug wires, then remove glow plugs.
4. Remove manifold bolts, then raise vehicle.
5. Disconnect exhaust pipe from manifold, then remove manifold from bottom of vehicle.
6. Reverse procedure to install. Refer to Fig. 4 for bolt location and torque.

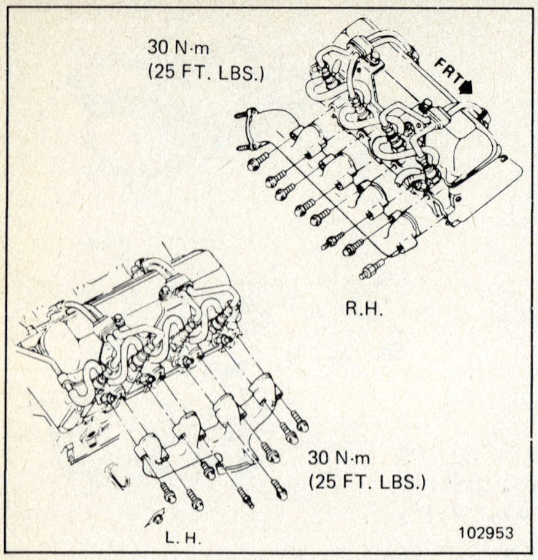

Fig. 4 Exhaust manifold tightening sequence

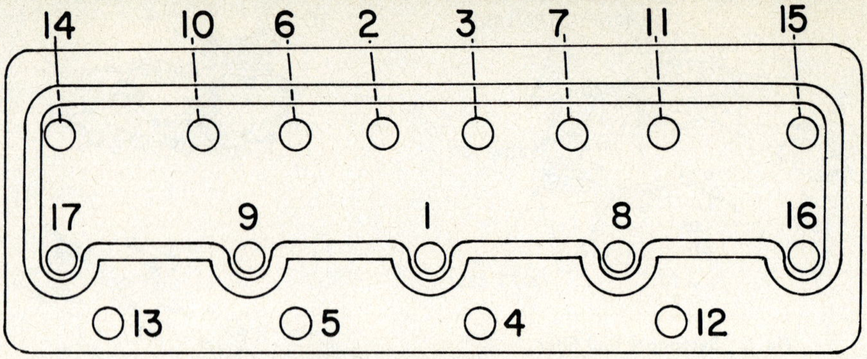

Fig. 5 Cylinder head bolt tightening sequence

Van

Right Side

1. Disconnect batteries and raise vehicle.
2. Disconnect exhaust pipe at manifold flange.
3. Lower vehicle, then disconnect glow plug wires.
4. Remove manifold bolts, then manifold.
5. Reverse procedure to install. Refer to Fig. 4 for bolt locations and torque.

Left Side

1. Disconnect batteries and raise vehicle.
2. Disconnect glow plug wires, then lower vehicle.
3. Disconnect exhaust pipe at manifold flange.
4. Remove manifold bolts, then manifold.
5. Reverse procedure to install. Refer to Fig. 4 for bolt locations and torque.

CYLINDER HEAD, REPLACE

Exc. Van

1. Remove injection lines, then the intake manifold and rocker arm covers.
2. Drain coolant, remove dipstick tube and disconnect ground wire at cowl.
3. Raise vehicle, disconnect exhaust manifold, and lower vehicle.
4. Remove A/C compressor (if equipped).
5. Remove generator and disconnect glow plug wires. Remove rocker arm assemblies.

NOTE: Mark rocker arms and push rods so they can be reinstalled in same location.

---

6. Disconnect radiator, bypass, heater hoses and ground strap.
7. Remove thermostat housing/crossover at cylinder head.
8. Remove cylinder head bolts and cylinder head.
9. Reverse procedure to install. Do not use any sealer on cylinder head gasket. Coat cylinder head bolts with sealing compound 1052080, or equivalent. Torque cylinder head bolts in sequence, Fig. 5, a little at a time, until 95 ft. lbs. of torque is reached.

Van

Right Side

1. Remove intake manifold. Refer to "Intake Manifold, Replace."
2. Remove injection lines.
3. If equipped with cruise control, remove cruise control transducer.
4. If equipped with A/C, remove upper fan shroud and A/C compressor belt.
5. Raise vehicle, then disconnect exhaust pipe at manifold flange.
6. If equipped with A/C, remove A/C compressor brace from exhaust manifold.
7. Disconnect glow plug wires and lower vehicle.
8. If equipped with A/C, discharge system, disconnect A/C compressor suction and discharge lines, then remove A/C compressor from brackets and remove.
9. Loosen dipstick tube front bracket and remove from stud.
10. Remove upper bracket from oil fill tube.
11. Remove wire loom, rocker cover bolts and rocker cover.
12. Remove rocker arm assemblies, then remove push rods.

NOTE: Prior to removing push rods, mark upper end as an aid during reassembly. Failure to identify push rods could result in incorrect installation, leading to premature wear or damage.

---

13. Drain cooling system, then remove air cleaner resonator and bracket.
14. Remove transmission fill tube bracket nut and position tube aside.
15. Disconnect heater, radiator and bypass hoses at coolant crossover, then remove crossover.
16. Remove generator upper bracket.
17. Remove cylinder head bolts.
18. Remove transmission dipstick at rear of head and remove tube.
19. Remove cylinder head.
20. Reverse procedure to install. Do not use any sealer on cylinder head gasket. Coat cylinder head bolts with sealing compound 1052080, or equivalent. Torque cylinder head bolts in sequence, Fig. 5, a little at a time, until 95 ft. lbs. of torque are reached.

Left Side

1. Remove intake manifold. Refer to "Intake Manifold, Replace."
2. Remove fuel injection lines from injection pump.
3. If equipped with cruise control, remove cruise control transducer.
4. If equipped with A/C, remove upper fan shroud and A/C compressor belt.
5. Raise vehicle, then disconnect exhaust pipe at manifold flange.
6. If equipped with A/C, remove A/C compressor brace from exhaust manifold.
7. Remove exhaust manifold.
8. Remove power steering pump lower adjusting bolts.
9. Disconnect glow plug wiring and temperature switch.
10. Remove injection lines from nozzles, then cap all openings. Lower vehicle.
11. If equipped with A/C, discharge system, disconnect A/C compressor suction and discharge lines, then remove A/C compressor from brackets and remove.
12. Remove power steering pump upper attaching bolt and position pump aside.
13. Remove stud from dipstick tube front bracket, then remove oil filler tube upper bracket.
14. Remove glow plug controller and bracket, then remove glow plug relay.
15. Disconnect T.V. cable.
16. Disconnect oil pressure switch and loom, then remove loom bracket.
17. Remove vacuum line clip attaching bolt from head.
18. Remove rocker cover bolts, then disconnect fuel return line bracket.
19. Remove rocker cover, then remove rocker arm assemblies and push rods.

NOTE: Prior to removing push rods, mark upper end as an aid during reassembly. Failure to identify push rods could result in incorrect installation, leading to premature wear or damage.

---

20. Drain cooling system, then remove air cleaner resonator and bracket.
21. Remove transmission fill tube bracket nut, then position tube aside.
22. Disconnect heater, radiator and bypass hoses at coolant crossover, then remove crossover.
23. Remove generator upper mounting bracket.
24. Remove cylinder head bolts, then cylinder head.
25. Reverse procedure to install. Do not use any sealer on cylinder head gasket. Coat cylinder head bolts with sealing compound 1052080 or equivalent. Torque cylinder head bolts in sequence, Fig. 5, a little at a time, until 95 ft. lbs. of torque is reached.

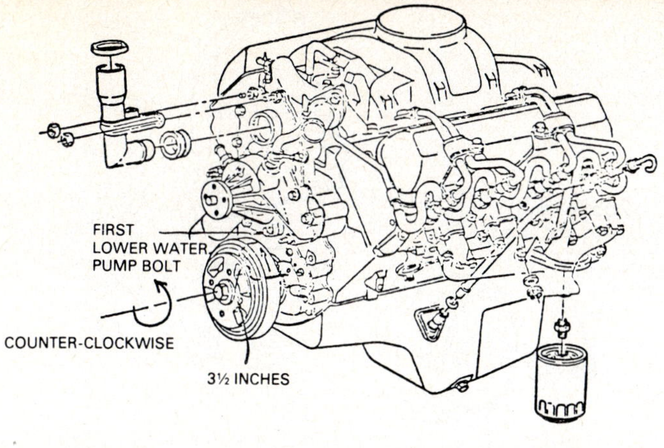

Fig. 6 Positioning engine for rocker arm shaft installation

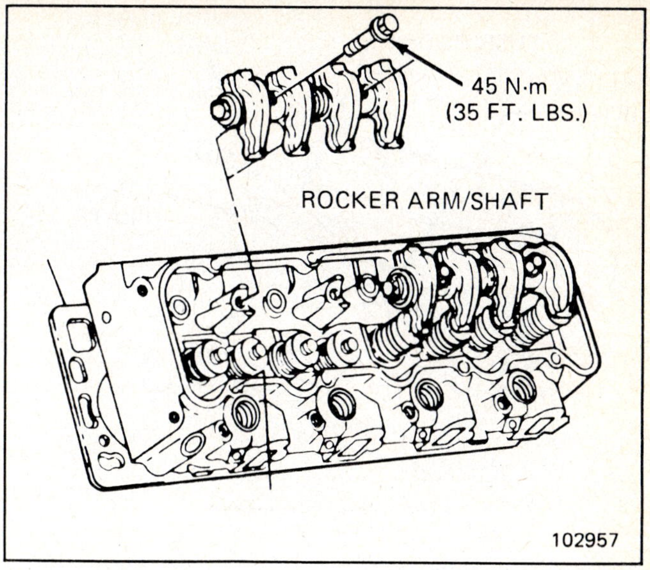

Fig. 6A Rocker arm shaft bolts

ROCKER ARMS & PUSH RODS, REPLACE

1. Remove valve cover.
2. Remove rocker arm and shaft. Remove cotter pin, and remove rocker arms from shaft.
3. Remove push rods and mark upper end for reinstallation.

NOTE: Failure to identify push rods could result in incorrect installation, leading to premature wear or damage.

---

4. Before installing rocker arm shafts, align TDC mark on engine with mark on engine balancer, then rotate crankshaft 3½ inches counterclockwise, or to first lower water pump bolt, Fig. 6. This will position engine so that all valves are closed. Torque rocker arm shaft bolts to 35 ft. lbs., Fig. 6A.

NOTE: The rocker arm shaft must be installed with the split in the ring around the shaft at bottom, Fig. 6B.

VALVE ARRANGEMENT

V8-379 Diesel .... I-E-I-E-I-E-I-E

VALVE TIMING

Intake Opens Before TDC

V8-379 Diesel 13 Degrees

VALVE GUIDES

Valves with oversize stems are available. To ream valve guide bores use Tool Set J-7049.

VALVE LIFTERS, REPLACE

Roller hydraulic lifters are used to reduce friction between valve lifter and camshaft lobe. Guides keep the lifters from rotating on the camshaft lobes, Figs. 7 and 8.

1. Remove rocker arm covers and rocker arms.
2. Remove guide clamps and guide plates.
3. Remove valve lifters through access hole in cylinder head using Tool J-29834 and a magnet.
4. Reverse procedure to install.

NOTE: Crankshaft must be manually rotated 720° after assembly of lifter guide plate clamp to insure free movement of lifters.

CRANKSHAFT PULLEY, REPLACE

1. Disconnect batteries.
2. On van models, remove upper fan shroud. On all models, remove generator belt.

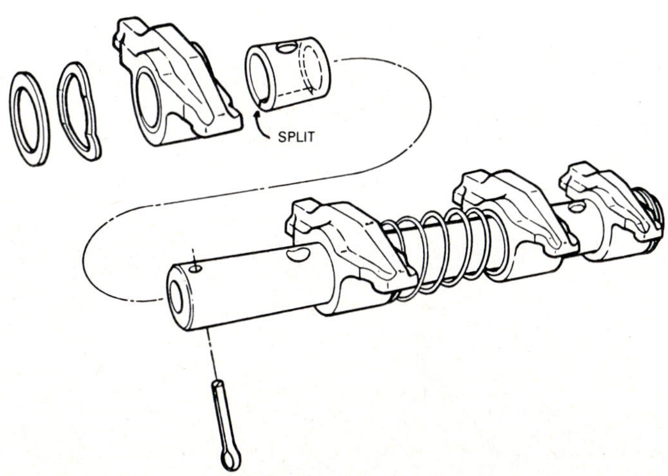

Fig. 6B Rocker arm shaft assembly

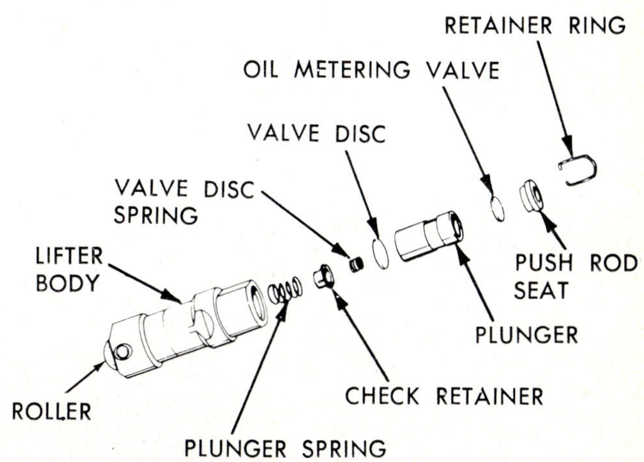

Fig. 7 Valve lifter

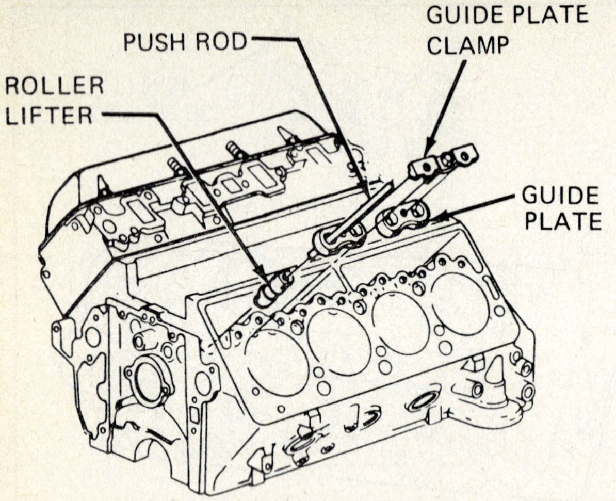

Fig. 8 Valve lifter removal

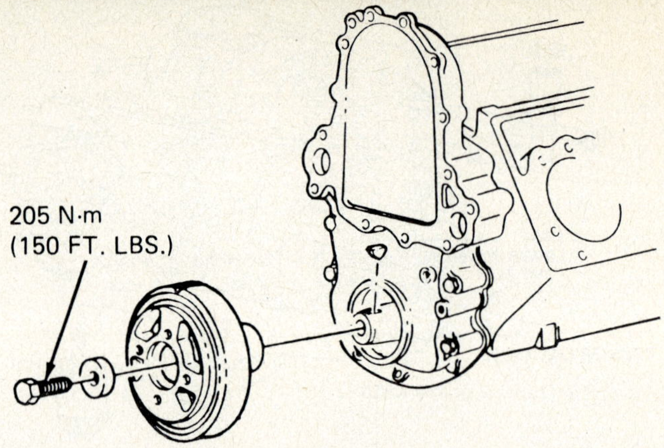

Fig. 9 Torsional damper

3. Remove power steering belt.
4. If equipped with A/C, remove A/C compressor belt.
5. On van models, raise vehicle.
6. Remove crankshaft pulley.
7. Reverse procedure to install.

TORSIONAL DAMPER, REPLACE

1. Disconnect battery ground cable, then remove crankshaft pulley as described under "Crankshaft Pulley, Replace".
2. Remove torsional damper using Tool J-23523 and suitable pilot.
3. Reverse procedure to install. Torque bolt to 150 ft. lbs., Fig. 9.

WATER PUMP, REPLACE

1. Disconnect battery.
2. Remove fan and fan shroud and drain radiator.
3. Remove A/C hose bracket nuts (if equipped). Remove oil fill tube, generator pivot bolt and generator belt.
4. Remove generator lower bracket, power steering pump and belt and A/C belt.
5. Disconnect by-pass hose and lower radiator hose. Remove water pump plate and water pump, Fig. 10.
6. Reverse procedure to install.

ENGINE FRONT COVER, REPLACE

1. Drain engine block and remove water pump.
2. Align marks on pump gear and camshaft gear, then scribe mark aligning injection pump flange and front cover.
3. Remove crank pulley, torsional damper, front cover to oil pan bolts, fuel return line clips and injection pump driven gear.
4. Remove injection pump retaining nuts from front cover.
5. Remove baffle, remaining cover bolts and front cover.
6. Reverse procedure to install. Apply a bead of sealant around sealing surface as shown in Fig. 11. Use sealant #1052357 or equivalent. Use RTV sealant around bottom portion of front cover which attaches to oil pan. Make sure scribe marks on injection pump and front cover are aligned and marks on cam gear and pump gear are aligned. Torque pump gear attaching bolts to 20 ft. lbs., Fig. 12.

FRONT OIL SEAL, REPLACE

1. Remove torsional damper and pry seal out of cover.
2. Install new seal using Tool J-22102.

TIMING CHAIN & GEARS, REPLACE

1. Remove front cover.
2. Remove cam shaft gear and injection pump gear.
3. Remove cam sprocket, chain and crank sprocket.
4. Reverse procedure to install. Install cam sprocket and crank sprocket so that timing marks are aligned. Next, rotate crankshaft 360° so that camshaft gear and injection pump gear timing marks

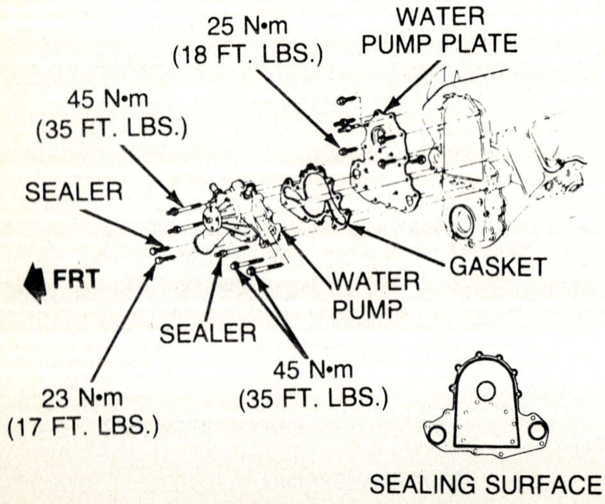

Fig. 10 Water pump installation

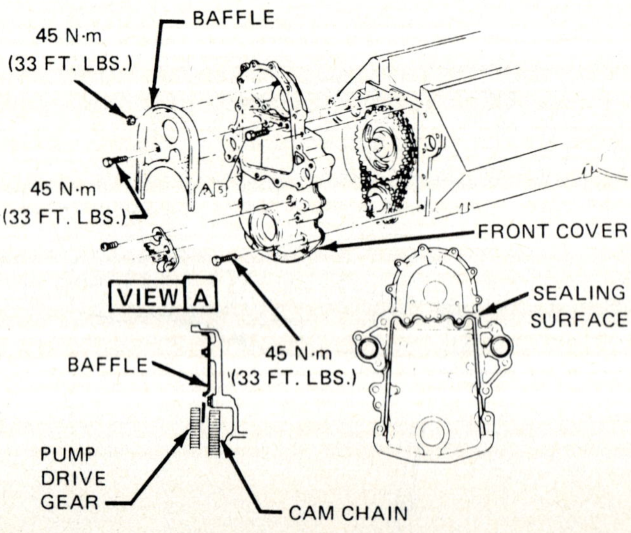

Fig. 11 Front cover installation

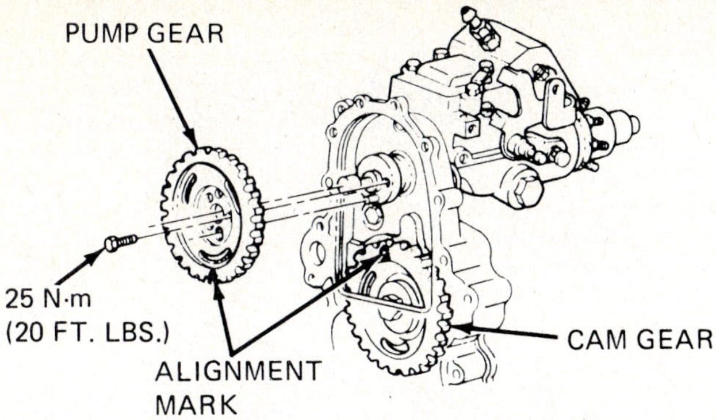

Fig. 12 Injection pump and cam gear alignment

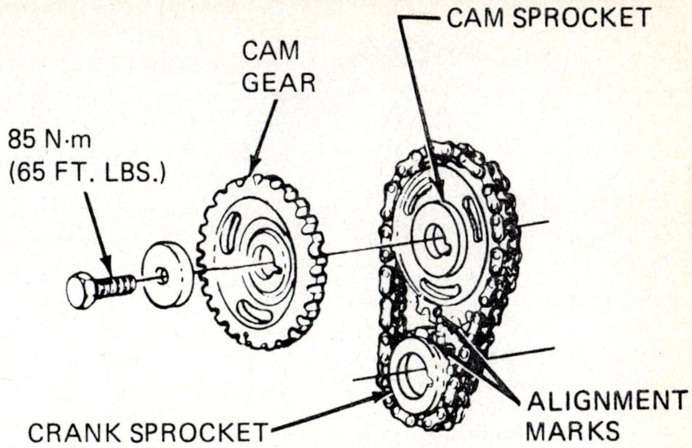

Fig. 13 Timing chain alignment marks

will align. Torque cam gear bolt to 65 ft. lbs., Fig. 13.

CAMSHAFT, REPLACE

Exc. Van

1. Disconnect battery, raise vehicle and drain radiator and block.
2. Disconnect exhaust pipe from engine and remove fan shroud-attaching bolts.
3. Lower vehicle and remove radiator, fan, vacuum pump and intake manifold.
4. Remove injector pump lines. Cap injector nozzles and identify lines for installation.
5. Remove water pump and injection pump gear.
6. Scribe a mark on front cover and injection pump flange for realignment. Remove injection pump from front cover.
7. Remove power steering pump, generator and (if equipped) A/C compressor.
8. Remove rocker arm covers, rocker arm shaft assemblies and push rods.

NOTE: Mark valve train location so parts may be reinstalled in original location.

9. Remove thermostat housing/crossover from cylinder head, and remove cylinder.
10. Remove valve lifter clamps, guide plates and valve lifters.

NOTE: Mark location of parts so they can be reinstalled in same position.

11. Remove front cover, timing chain, fuel pump and cam retainer plate. Remove A/C condenser (if equipped).
12. Remove camshaft.
13. Reverse procedure to install. Camshaft retainer plate should be torqued to 20 ft. lbs.

Van

1. Refer to steps 1 through 10 of "Engine, Replace" procedure to disassemble front of vehicle.
2. Remove cylinder heads. Refer to "Cylinder Head, Replace" procedure.
3. Remove generator lower bracket.
4. Remove water pump pulley, then crankshaft pulley.
5. Remove torsional damper. Refer to "Torsional Damper, Replace".
6. Remove water pump and water pump plate.
7. Rotate crankshaft and align timing marks.
8. Remove injection pump driven gear, then inner baffle.
9. Scribe a mark on front cover and injection pump flange to aid during reassembly. Remove front cover. Refer to "Engine Front Cover, Replace".
10. Remove fuel pump, valve lifters, injection pump drive gear, timing chain, crankshaft gear and cam retainer plate.
11. Remove camshaft.
12. Reverse procedure to install. Camshaft retainer plate should be torqued to 20 ft. lbs.

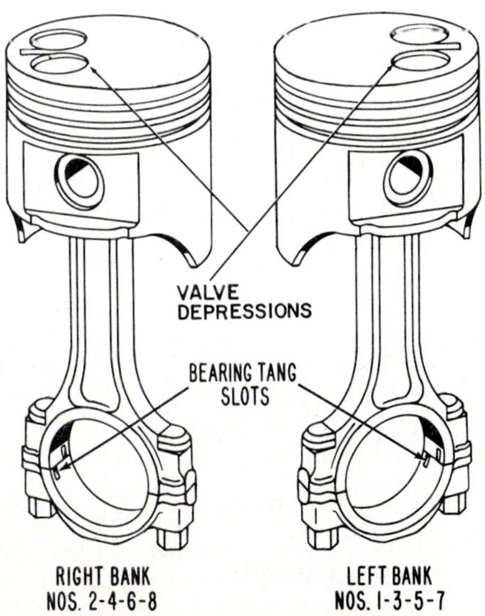

Fig. 14 Piston and rod assembly

PISTON & ROD ASSEMBLY

Assemble piston to rod and install into cylinder block. Install with depression on top of piston toward outside of engine, Fig. 14. Install connecting rod bearing with tang slots positioned on side opposite camshaft.

PISTON & RINGS

Pistons are available in standard sizes and oversize of .030 inch. Rings are available in standard sizes and oversize of .030 inch.

MAIN & ROD BEARINGS

Main bearings are available in standard sizes and undersizes of .005 and .010 inch. Rod bearings are available in standard sizes and undersizes of .010 inch and .020 inch.

CRANKSHAFT REAR SEAL SERVICE

Since the seal can be replaced only when the crankshaft is removed, the following repair procedure is recommended.

1. Remove oil pan, oil pump and bearing cap.
2. Drive end of old seal into groove using suitable packing tool approximately 1/4 inch. Do this on both sides of seal, Fig. 15.
3. Measure and note amount that seal was driven up on one side. Using old seal removed from bearing cap, cut a length of seal the length previously noted plus 1/16 inch. Do this for both sides of seal.
4. Pack cut lengths of seal into appropriate side of seal groove. Use packing tool J-33154-2 and guide tool J-33154-1 since these tools are machined with a built-in stop, Fig. 16.
5. Apply Loctite 496 or equivalent, and install new rope seal in lower bearing cap. Cut ends of seal flush with cap, using tool J-33153, Fig. 17.

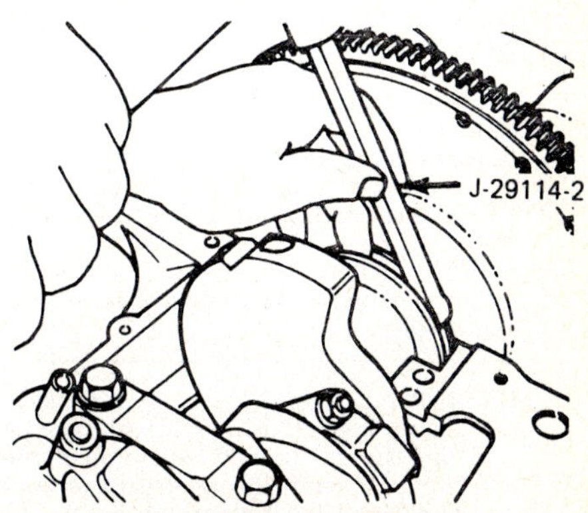

Fig. 15 Packing tool (upper half)

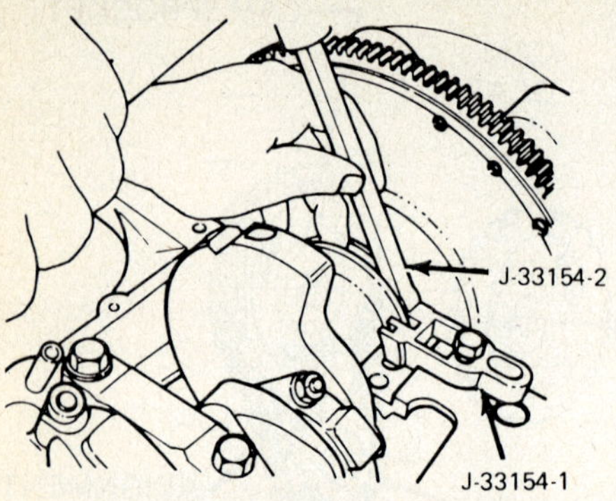

Fig. 16 Installation of guide tool

OIL PAN, REPLACE

Exc. Van

1. Disconnect battery and raise vehicle.
2. Drain oil and remove transmission dust cover, oil pan bolts and engine mount through bolt (left side).
3. Raise engine and remove oil pan.
4. Reverse procedure to install. Torque oil pan bolts to 6 ft. lbs., Fig. 18.

Van

1. Disconnect batteries and remove engine cover.
2. Remove engine and transmission oil dipsticks and tubes.
3. Disconnect T.V. cable at injection pump rod and remove upper routing clips if used.
4. Remove upper bellhousing to engine retaining bolts.
5. Remove vacuum pump assembly.
6. Raise and support vehicle.
7. Remove driveshaft, then disconnect speedometer cable from transmission.
8. Disconnect torque converter clutch connector from transmission.
9. Disconnect transmission shift linkage.
10. Disconnect transmission cooler lines from transmission.
11. Remove transmission flex plate inspection cover, then remove torque converter to flex plate retaining bolts.
12. Support transmission with a suitable jack.
13. Remove transmission mount, then remove crossmember.
14. Remove transmission to engine block retaining bolts, then remove transmission.
15. Remove flex plate.
16. Drain oil into a suitable container.
17. Disconnect engine oil cooler lines from engine block.
18. Remove starter.
19. Remove oil pan bolts, then lower pan from engine block.

NOTE: It may be necessary to gain additional clearance. This may be obtained by rotating the crankshaft so number one and two crankshaft throws are up, providing additional lower clearance.

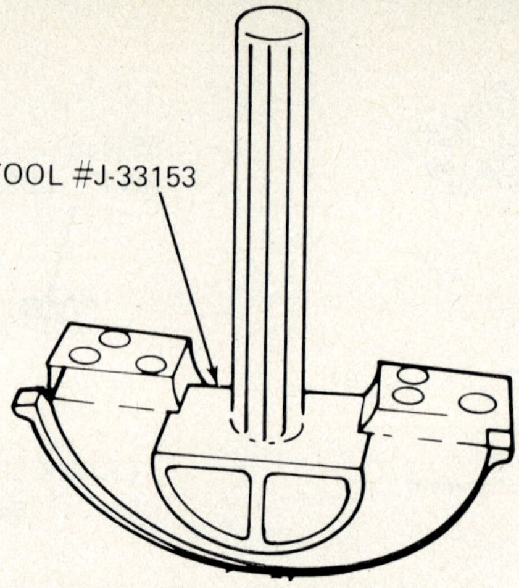

Fig. 17 Installing new seal (lower half)

20. Remove oil pump retaining bolt and allow oil pump assembly to fall into oil pan.
21. Remove oil pan.
22. Reverse procedure to install. Torque oil pan bolts to 6 ft. lbs., Fig. 18.

OIL PUMP, REPLACE & SERVICE

Removal

1. Remove oil pan as outlined previously.
2. Remove oil pump to rear main bearing cap attaching bolts and remove pump and extension shaft.

Service

Disassembly

1. Remove pump cover attaching screws and pump cover.
2. Remove drive gear and idler gear and shaft from pump body.

NOTE: Mark gear teeth so that they can be reassembled with same teeth indexing.

3. Remove retaining pin, pressure regulator valve and related parts.

Inspection

1. Inspect pump body and pump gears for damage or excessive wear. If either pump body or pump gears are damaged or worn, entire oil pump assembly must be replaced.
2. Check pressure regulator valve for fit and drive gear shaft for looseness in pump body.

Assembly

1. Install pressure regulator valve and related parts.

NOTE: Install idler gear with smooth side of gears toward pump cover opening.

2. Install gears and shaft in oil pump body.
3. Install pump cover.

Installation

1. Reverse removal procedures to install. Align hex on lower end of vacuum pump drive shaft with hex on top of pump drive extension shaft. Install oil pump to bearing cap bolts and torque to 67 ft. lbs.

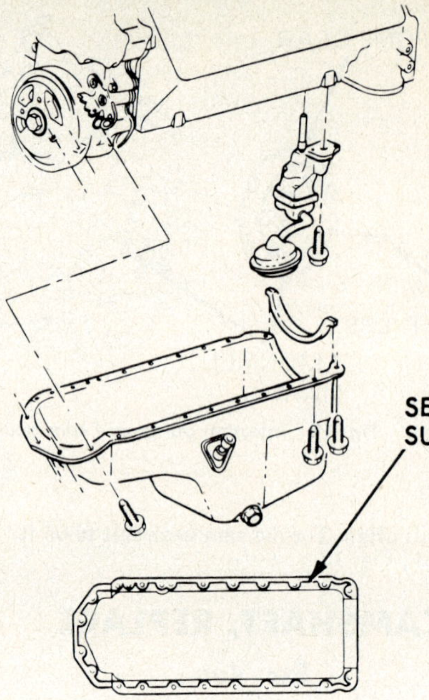

Fig. 18 Oil pan installation

INJECTION PUMP TIMING

1. Check marks on top of engine front cover and injection pump flange. These marks must be aligned for engine to be properly timed, Fig. 19.
2. To adjust:
   a. Loosen three pump retaining nuts.
   b. Align mark on front cover with mark on injection pump. Torque nuts to 30 ft. lbs.
   c. Adjust throttle rod.

INJECTION PUMP, REPLACE

Exc. Van

Removal

1. Disconnect batteries and remove fan and fan shroud.
2. Remove intake manifold and fuel lines.
3. Disconnect accelerator cable at injection pump. On models equipped with 700-4R auto. trans. also disconnect detent cable, Fig. 20.
4. Disconnect fuel return line and remove wires and hoses from injection pump.
5. Remove fuel line from pump.
6. Remove A/C hose retainer bracket, (if equipped).
7. Remove oil fill tube and PCV vent hose assembly.
8. Scribe a mark on front cover and pump for alignment at reinstallation.

NOTE: In order to gain access to injection

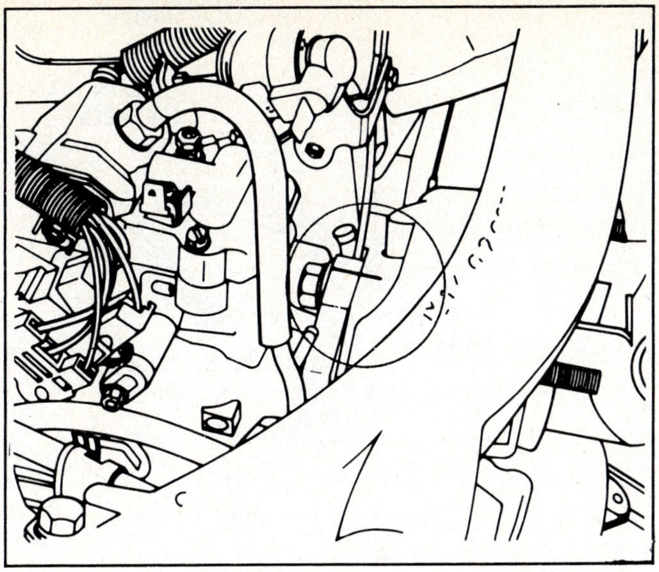

Fig. 19 Front cover to injection pump timing mark

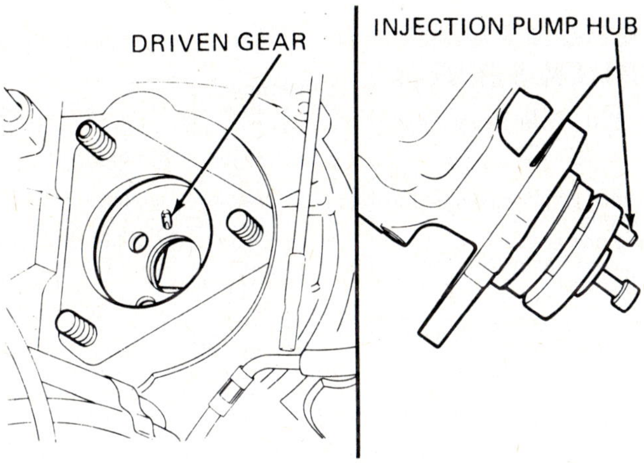

Fig. 22 Injection pump locating pin

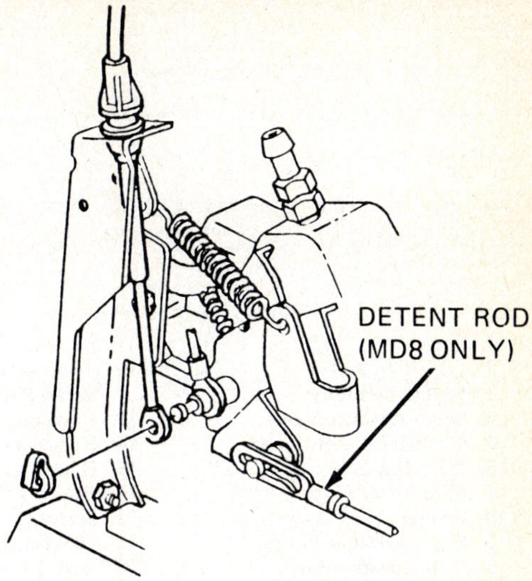

Fig. 20 Accelerator linkage

pump retaining bolts through oil filler neck, it will be necessary to rotate engine.

9. Remove front cover to injection pump attaching nuts and remove pump. Cap all open lines and nozzles.

Installation

1. Install new gasket. Align locating pin on pump hub with slot in injection pump gear. Align scribe mark on front cover and pump and torque attaching nuts to 30 ft. lbs., Figs. 21 and 22.
2. Install pump-driven gear to pump and torque bolts to 20 ft.lbs.
3. Install oil fill tube and PCV vent hose assembly. Install A/C hose retainer bracket (if equipped).
4. Install fuel line at pump and torque to 20 ft.lbs.
5. Connect fuel return line, accelerator cable, injection lines and necessary wires and hoses.
6. Install intake manifold, fan shroud and fan. Connect batteries.

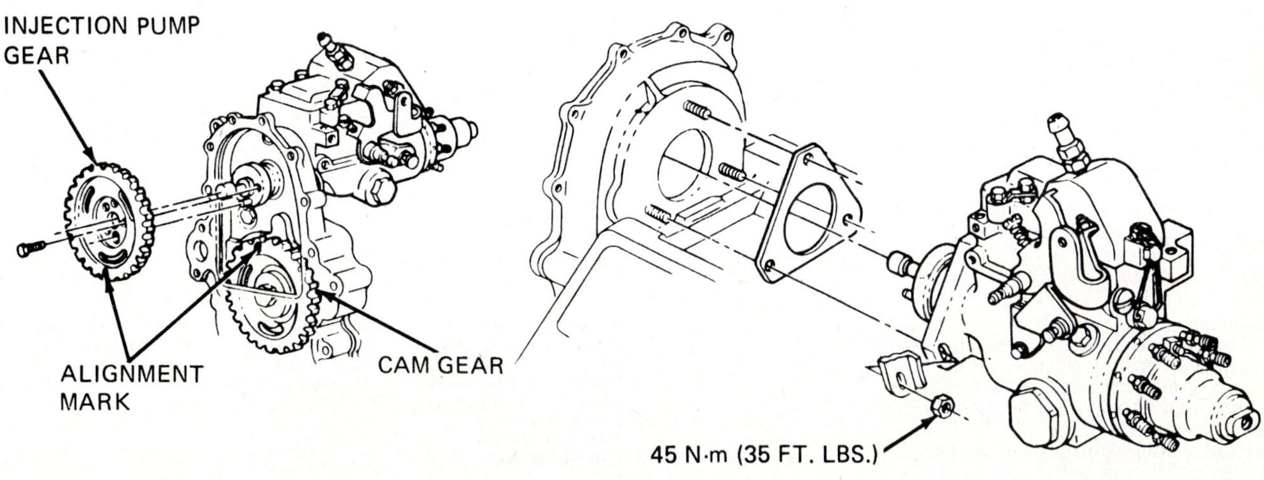

Fig. 21 Injection pump installation

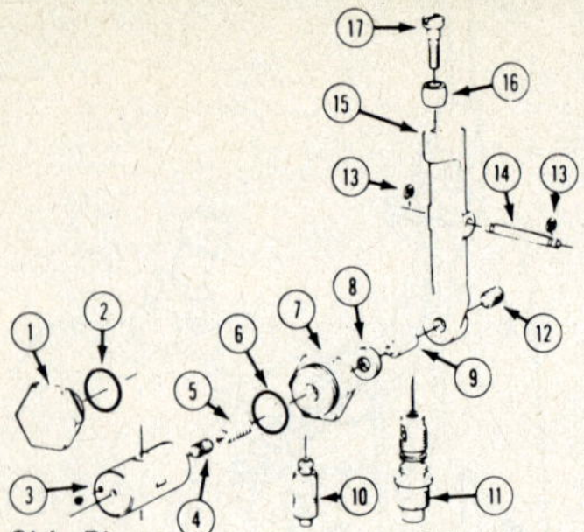

1. Power Side Plug
2. "O" Ring Seal
3. Piston Assembly
4. Servo Advance Valve
5. Mechanical Light Load
6. "O" Ring Seal
7. Spring Side Plug
8. Washer
9. Servo Advance Plunger
10. Cam Advance Pin
11. Head Locating Screw
12. Servo Advance Adjusting Screw (DO NOT ADJUST)
13. Retaining Ring
14. Rocker Lever Pin
15. Rocker Lever
16. Roller
17. Screw
18. "O" Ring
19. Cam Advance Pin Hole Plug

Fig. 23 Automatic advance group

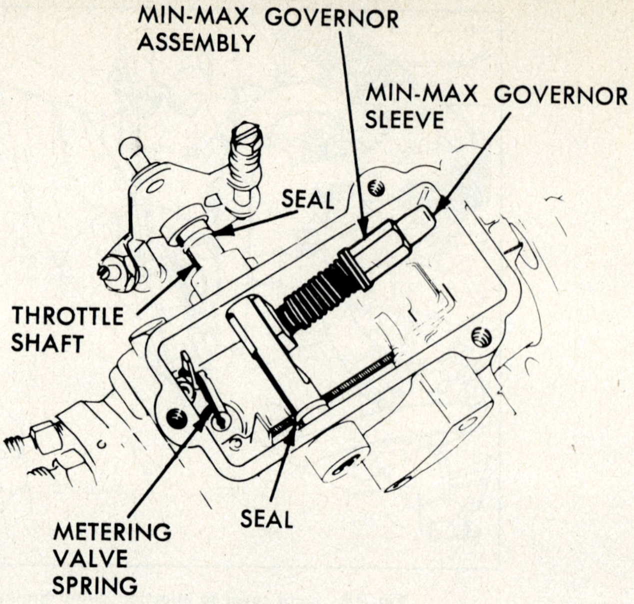

Fig. 24 Injection pump cover removed

Van

Removal

1. Remove intake manifold. Refer to "Intake Manifold, Replace".
2. Rotate air cleaner snorkel up to gain clearance.
3. Remove hood latch, then windshield washer bottle.
4. Remove upper fan shroud.
5. Remove oil fill tube and grommet.
6. Working through oil fill tube hole, remove injection pump drive gear bolts.

NOTE: It may be necessary to rotate engine to gain access to injection pump drive gear bolts.

7. Remove fuel filter, bracket and injection pump feed line.
8. Disconnect wire looms from injection lines.
9. Disconnect fuel injection lines from brackets, then remove engine oil dipstick.
10. Disconnect injection pump electrical connectors.
11. Disconnect T.V. cable (if equipped).
12. Disconnect throttle cable.
13. Disconnect injection lines 2, 4, 5, 6, 7 and 8.
14. Raise and support vehicle.
15. Disconnect injection lines 1 and 3.
16. Cap injectors and lines to prevent entry of dirt.
17. Identify injection lines at pump and remove. Cap line and pump openings.
18. Disconnect fuel return line.
19. Scribe a mark on front cover and pump flange to aid during reassembly.
20. Remove injection pump retaining bolts, then remove injection pump.

Installation

1. Install new gasket. Align locating pin on pump hub with slot in injection pump gear. Align scribe mark on front cover and pump flange, then torque attaching nuts to 30 ft. lbs., Figs. 21 and 22.
2. Install pump driven gear to injection pump and torque bolts to 20 ft. lbs.
3. Reverse remaining removal procedure to install.

INJECTION PUMP OFF-TRUCK SERVICE

Advance Pin Hole Plug Seal, Fig. 23

1. Loosen advance pin hole plug seal by tapping with hammer.
2. Remove plug and seal.
3. Reverse procedure to install. Torque plug to 75–100 inch lbs.

Auto Advance Seals, Fig. 23

1. Remove advance pin hole plug.
2. Remove spring side advance piston hold plug, plug, piston, spring and slide washer.
3. Remove power side advance piston hole plug, plug, piston, and slide washer.
4. Disassemble both plugs and pistons.
5. Reverse procedure to install. Torque plugs to 20 ft. lbs.

Hydraulic Head Seal-O-Ring

Removal

1. Remove throttle shaft and seals.
2. Remove metering valve, Fig. 24.
3. Remove housing vent screw assembly, Fig. 25.
4. Remove advance pin hole plug and advance pin, Fig. 23.
5. Remove head locating screws and seal.
6. Remove hydraulic head assembly and O-ring.

Installation

1. Install hydraulic head seal and place head assembly into pump housing.
2. Install two head locking screws finger-tight and turn pump upside down.
3. Install head-locating screw and torque to 15–18 ft. lbs., Fig. 23.
4. Torque head-locking screws to 15–18 ft.lbs.
5. Install advance pin and advance pin plug.

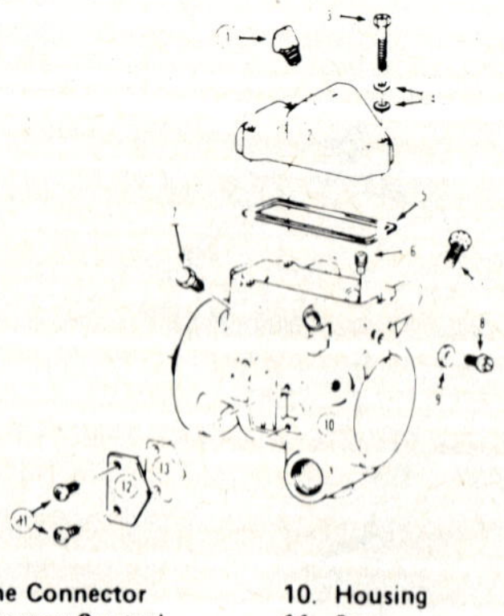

1. Return Line Connector
2. Cover, Governor Control
3. Screw
4. Lockwasher or Plain Washer
5. Governor Cover Gasket
6. Vent Screw Assembly
7. Head Locking Screw
8. Pressure Tap Screw
9. Seal
10. Housing
11. Screw
12. Timing Line Cover
13. Timing Line Cover Gasket

Fig. 25 Side housing and drive group

6. Turn pump so cover opening is up, and install metering valve.
7. Install throttle shaft, seals and pump cover.

INJECTION NOZZLE, REPLACE

NOTE: The fuel injection nozzles used on 1983 vehicles have a different thread pitch size than on 1982 vehicles. Also, the injection nozzles used on 1983 vans are shorter than those used on other models. If replacement becomes necessary, make certain the correct nozzle is used.

1. Disconnect batteries, fuel return clip, fuel return hose and fuel injection line.
2. Remove injection nozzle by applying torque to largest nozzle hex with tool J-29873, Fig. 26.
3. Reverse procedure to install. Torque nozzle to 50 ft. lbs. Torque fuel injection line to 20 ft. lbs.

TRANSMISSION VACUUM REGULATOR VALVE, ADJUST

1. Attach vacuum regulator valve to fuel injection pump.

NOTE: Vacuum regulator valve should be attached snugly. Do not tighten.

2. Attach vacuum gage to outboard vacuum nipple. Attach vacuum source to inboard vacuum nipple of 9.7 ±.7 PSI.
3. Open stop screw on throttle lever to wide open position and insert vacuum regulator gage bar between gage boss on injection pump and stop screw.
4. Rotate and hold throttle shaft against gage bar and slowly rotate vacuum regulator valve body clockwise until vacuum gage reads 5.6 ±.3 PSI.

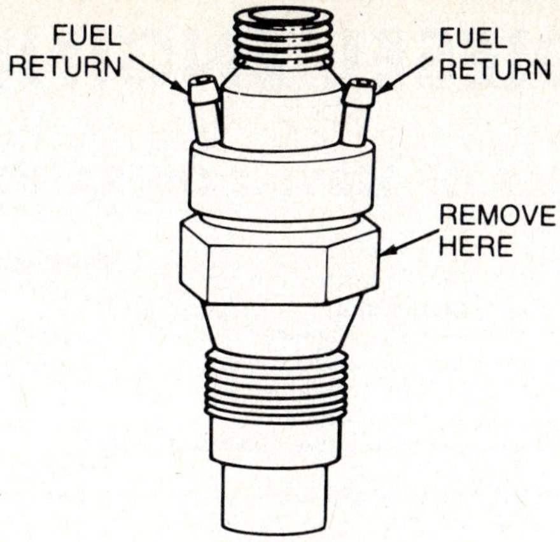

Fig. 26 Injection nozzle

NOTE: Clockwise direction is direction facing valve. Valve can be set while rotating valve body in this direction, only.

5. Check by allowing throttle shaft to return to idle stop position. Position throttle shaft back against gage bar. Vacuum gage should read 5.6 ±.3 PSI.

THROTTLE POSITION SWITCH, ADJUST, FIG. 27

1. Place throttle position switch on fuel injector pump with throttle lever in closed position.

NOTE: Throttle position switch should be attached snugly. Do not tighten.

2. Attach continuity meter across IGN (pink) and EGR (yellow) terminals or wires.
3. Turn stop screw on throttle shaft to wide open position and insert "switch closed" gage block between gage boss on injection pump and stop screw.
4. Rotate and hold throttle lever against gage block.
5. Rotate throttle switch clockwise (facing throttle switch) until continuity just occurs across IGN and EGR terminals or wires. Hold switch body in this position and tighten mounting screws.

NOTE: Switch point must be set only while rotating switch body in clockwise direction.

6. Check by allowing throttle lever to return to idle position. Remove "switch closed" gage bar and install "switch open" gage bar. Rotate throttle lever against it. There should be no continuity between IGN and EGR terminals or wires.

IDLE SPEED, ADJUST, FIG. 28

1. Install tachometer and adjust low idle speed screw to an engine RPM reading of 550 in drive, for automatic transmission and 575 RPM in neutral, for manual transmission.
2. Adjust fast idle speed:
   a. Remove connector from fast idle solenoid.

NOTE: To energize solenoid, use an insulated jumper wire from battery positive terminal to solenoid terminal.

   b. Energize fast idle solenoid plunger by momentarily opening throttle. Adjust by turning hex head to 700 RPM in neutral.

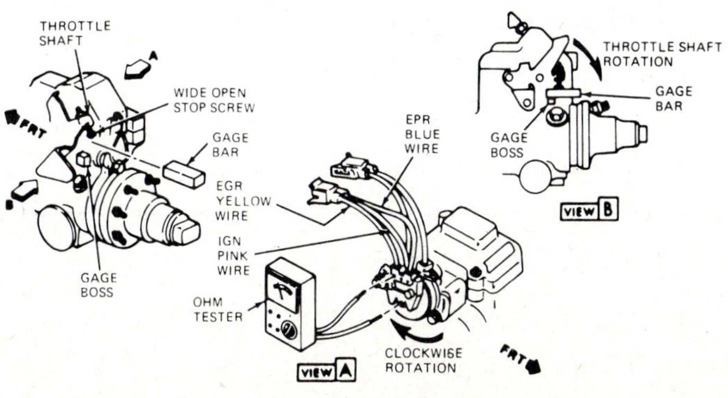

Fig. 27 Throttle position switch adjustment

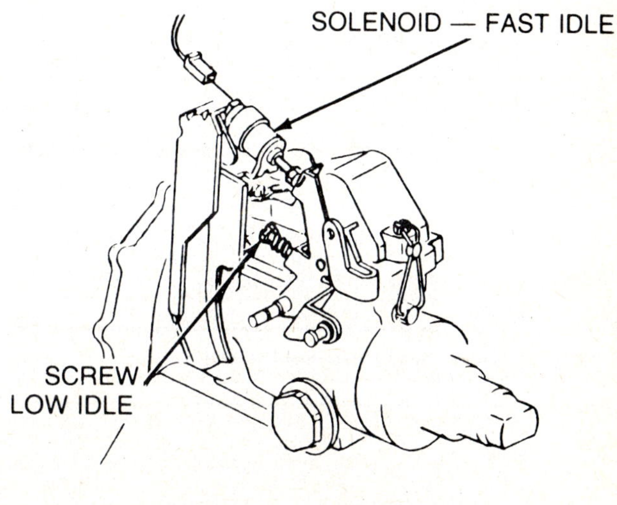

Fig. 28 Idle speed adjustment

INTERNATIONAL (FORD V8-420, 6.9L) DIESEL ENGINE

INDEX OF SERVICE OPERATIONS

GENERAL ENGINE SPECIFICATIONS

Year	Engine	Injection Timing	Idle Speed RPM	Bore and Stroke	Piston Displacement, Cubic Inches	Compression Ratio	Net Brake H.P. @ R.P.M.	Net Torque Lbs. Ft. @ R.P.M.	Normal Oil Pressure Pounds @ R.P.M.
1983	V8-420①	③	750	4.00 × 4.18	420	19.7	161 @ 3300	307 @ 1800	40-60 @ 2200
	V8-420②	③	750	4.00 × 4.18	420	19.7	146 @ 3300	278 @ 1800	40-60 @ 2200

①—Exc. high altitude.
②—High altitude.
③—Align marks on injection pump flange and adapter.

VALVE SPECIFICATIONS

Year	Engine	Valve Lash		Valve Angles Degree		Valve Spring Installed Height	Valve Spring Pressure Lbs @ In.	Stem Clearance		Stem Diameter	
		Int.	Exh.	Int.	Exh.			Intake	Exhaust	Intake	Exhaust
1983	V8-420	Hydraulic		30°	37.5°	—	60 @ 1.798	.0012-.0029	.0012-.0029	.37165-37235	.37165-.37235

PISTONS, PINS, RINGS, CRANKSHAFT & BEARINGS

Year	Engine	Piston Clearance	Ring End Gap			Wrist Pin Diameter	Rod Bearings		Main Bearings			
			Top	2nd	Oil		Shaft Diameter	Bearing Clearance	Shaft Diameter	Bearing Clearance	Thrust on Bearing No.	Shaft End Play
1983	V8-420	.0055-.0075	.014-.024	.010-.024	.060-.070	1.1099-1.1101	2.6905-2.6910	.0011-.0036	3.1228-3.1236	.0018-.0046	3	.0025-.0085

ENGINE TIGHTENING SPECIFICATIONS

Year	Engine	Glow Plugs Ft. Lbs.	Cylinder Head bolts Ft. Lbs.	Intake Manifold Ft. Lbs.	Exhaust Manifold Ft. Lbs.	Rocker Arm Bolts Ft. Lbs.	Rocker Arm Cover Ft. Lbs.	Connecting Rod Cap Bolts Ft. Lbs.	Main Bearing Cap Bolts Ft. Lbs.	Flywheel to Crankshaft Ft. Lbs.	Torsional Damper Ft. Lbs.
1983	V8-420	12	①	—	30	20	5½	45½-50½	95	52	90

①—Refer to cylinder head replacement section of text for procedure.

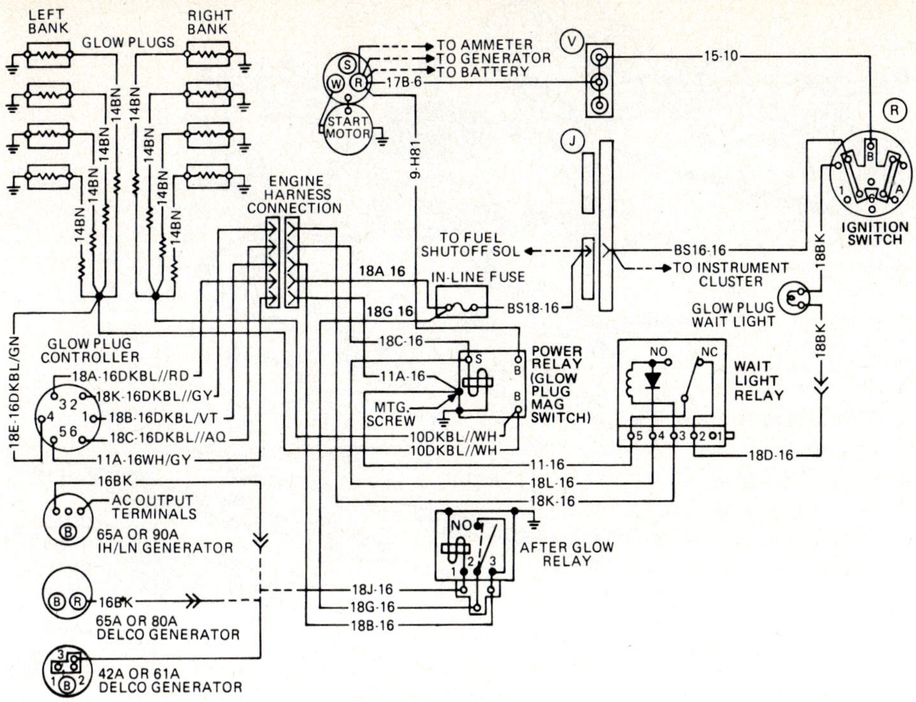

Fig. 1 Glow plug system wiring diagram

GLOW PLUG FAST START SYSTEM

The glow plug fast start system controls glow plug operation according to various operating conditions. When the ignition switch is turned on, the control switch is energized, which, in turn, energizes the power relay. The power relay contacts close and battery current is directed to the glow plugs. When the glow plugs reach a predetermined temperature, current to the plugs and the wait lamp is cut off. At this time, the engine can be started. After the engine is started, the glow plugs will cycle on and off for 40–90 seconds. This cycling of the glow plugs helps to eliminate start-up exhaust smoke.

A control switch, threaded into the cylinder head coolant jacket, senses engine coolant temperature. When restarting a warm engine, the glow plug system will not be activated unless the coolant temperature falls below normal operating temperature.

Refer to Fig. 1 for troubleshooting the glow plug system.

ENGINE, REPLACE

1. Disconnect battery ground cables and remove hood.
2. Drain cooling system, then remove air cleaner and intake duct assembly.
3. Place a suitable cover over intake manifold opening.
4. Remove fan shroud, then the fan and clutch assembly.

NOTE: The fan retaining nut has left hand threads.

5. Disconnect upper and lower hoses and the automatic transmission cooler lines from radiator.
6. Remove radiator.
7. If equipped, remove A/C compressor drive belt, then remove and position compressor on radiator upper support.
8. Remove power steering pump drive belt and remove and position pump aside.
9. Disconnect fuel supply line heater and alternator wiring from alternator.
10. Remove oil pressure sender from firewall and lay sender on engine.
11. Disconnect accelerator cable and speed control cable, if equipped, from injection pump.
12. Remove accelerator cable bracket, with cables, from intake manifold and position aside.
13. If equipped, disconnect transmission kickdown rod from injection pump.
14. Disconnect main wiring harness connector from right side of engine.
15. Disconnect ground strap from rear of engine.
16. Disconnect fuel return hose from left rear of engine.
17. Remove two upper engine-to-transmission attaching bolts.
18. Disconnect heater hoses from water pump and right hand cylinder head.
19. Disconnect water temperature sender wiring from left front of cylinder block.
20. Disconnect water temperature overheat light switch from top front of left hand cylinder head.
21. Raise and support vehicle.
22. Disconnect battery ground cables from lower front of engine and remove from vehicle.
23. Disconnect and plug fuel inlet line from supply pump.
24. Disconnect starter motor wiring.
25. Disconnect exhaust pipe from exhaust manifold.
26. Remove bolts attaching front engine mounts from front crossmember.
27. Remove flywheel cover, then the converter-to-flywheel attaching nuts.
28. Lower vehicle and support transmission with a suitable jack.
29. Attach suitable engine lifting equipment to engine and raise engine slightly to clear the front crossmember and pull engine forward. Move front of engine approximately 45 degrees toward the left side of engine compartment, then lift engine from vehicle.

NOTE: Use care not to damage the wiper motor when removing engine.

30. Reverse procedure to install.

INTAKE MANIFOLD, REPLACE

1. Disconnect battery ground cables, then remove air cleaner and install intake manifold cover, tool T83T-9424-A.
2. Remove injection pump as described under "Injection Pump, Replace".
3. Remove fuel return hose from rear injection nozzles, then remove return hose to fuel tank.
4. Remove engine wiring harness from engine. Disconnect engine harness ground cable from back of left cylinder head.
5. Remove intake manifold attaching bolts, then the intake manifold.
6. Reverse procedure to install. Refer to Fig. 2 for tightening sequence.

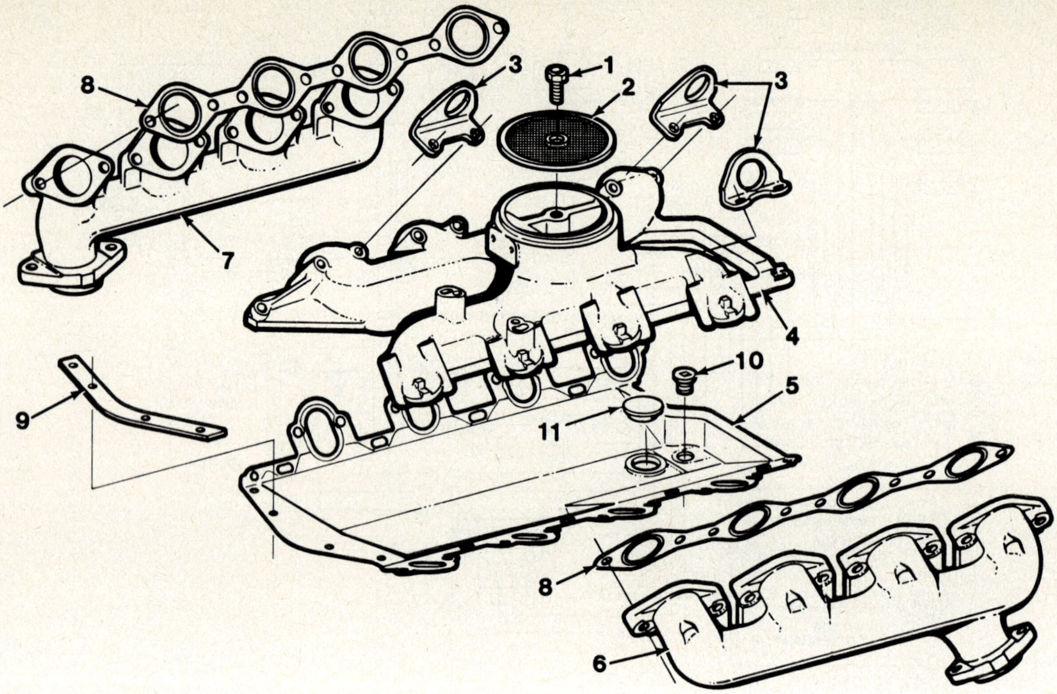

1. Stud
2. Guard
3. Lifting Bracket
4. Intake Manifold
5. Valley Pan and Gasket
6. L.H. Exhaust Manifold
7. R.H. Exhaust Manifold
8. Exhaust Manifold Gasket
9. Valley Pan Strap
10. Valley Pan Drain Plug
11. Valley Pan Closing Plug

Manifold assemblies

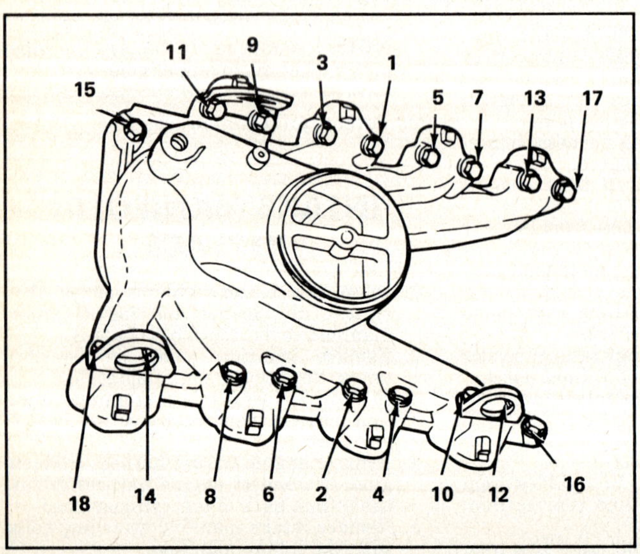

Fig. 2 Intake manifold tightening sequence

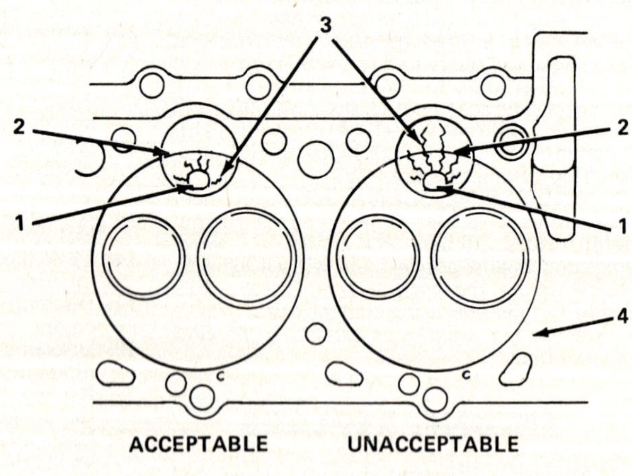

1. Throat of Pre-Combustion Chamber
2. Fire Ring Mark
3. Cracks
4. Cylinder Head (Bottom View)

Fig. 3 Pre-combustion chamber inspection

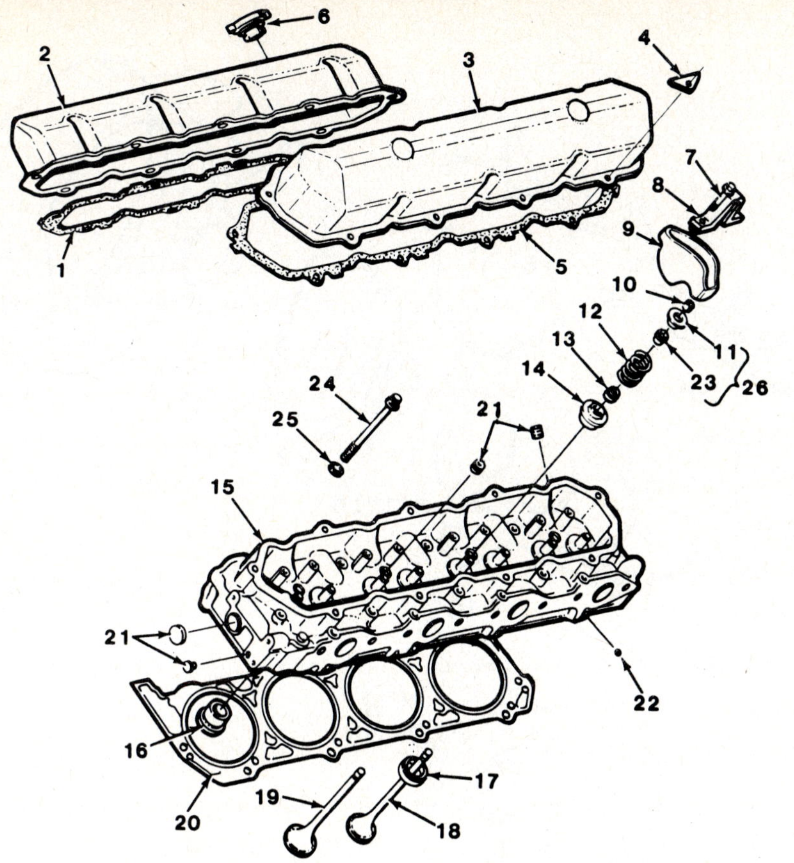

1. Valve Cover Gasket
2. Valve Cover (RH Side)
3. Valve Cover (LH Side)
4. Washer
5. Valve Cover Gasket
6. Closing Cap
7. Rocker Arm Bolt
8. Rocker Arm Post
9. Rocker Arm
10. Retainer Lock
11. Valve Spring Retainer
12. Valve Spring w/Damper
13. Valve Stem Seal (Intake Only)
14. Valve Rotator
15. Cylinder Head
16. Combustion Chamber Insert
17. Exhaust Valve Seat
18. Exhaust Valve
19. Intake Valve
20. Cylinder Head Gasket
21. Pipe Plug
22. Ball
23. Oil Shield
24. Cylinder Head Bolt
25. Washer
26. Spring Retainer Assembly

Cylinder head assembly

EXHAUST MANIFOLD, REPLACE

1. Disconnect battery ground cables.
2. Raise and support vehicle.
3. Disconnect exhaust pipe from exhaust manifold.
4. If removing left manifold, bend tabs for manifold attaching bolts, then remove bolts and the manifold.
5. If removing right manifold, lower vehicle, then bend tabs for manifold attaching bolts and remove the attaching bolts and manifold.
6. Reverse procedure to install.

CYLINDER HEAD, REPLACE

NOTE: The procedure outlined below is for the right bank cylinder head. The left bank cylinder head is removed in a similar manner.

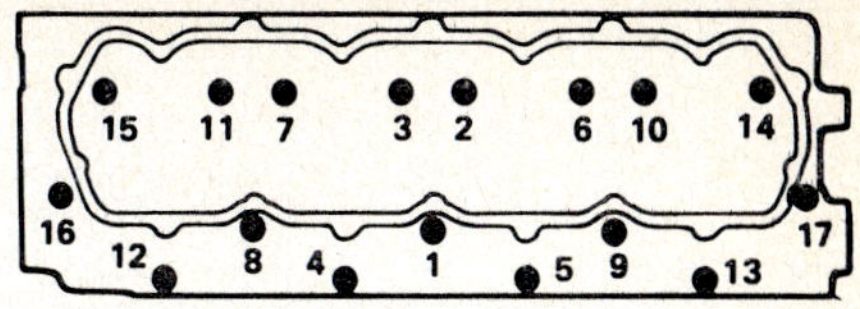

Fig. 4 Cylinder head tightening sequence

1. Disconnect battery ground cables, then drain cooling system.
2. Remove fan shroud, then the fan and clutch assembly.

NOTE: The fan retaining nut has left hand threads.

3. Disconnect fuel supply line heater and alternator wiring from alternator.
4. Remove alternator and vacuum pump.
5. Disconnect and plug fuel lines from fuel filter. Remove alternator and vacuum pump mounting bracket and the fuel filter with bracket attached.
6. Disconnect heater hose from cylinder head.
7. Remove injection pump and intake manifold.
8. Raise and support vehicle.
9. Disconnect exhaust pipe from exhaust manifold.
10. On right side, remove clamp or bolt securing engine oil and transmission oil dipstick tubes to engine.
11. Lower vehicle.
12. On right side, remove engine oil dipstick and tube.
13. Remove valve cover, rocker arms and push rods.
14. Remove injection nozzles and glow plugs.
15. Remove cylinder head bolts.
16. Attach suitable lifting equipment to cylinder head, then lift cylinder head from engine.

NOTE: Clean the pre-combustion chambers and inspect for cracks or burns. Some cracking and burning is acceptable, Fig. 3. When installed, the pre-combustion chambers may seat from .003 inch above to .001 inch below the cylinder head deck.

17. Reverse procedure to install. Torque cylinder head bolts in sequence, Fig. 4, and in three steps: first to 40 ft. lbs., then to 65 ft. lbs and finally to 75 ft. lbs. Torque intake manifold bolts in sequence, Fig. 2, to specifications.

ROCKER ARMS & PUSH RODS, REPLACE

1. Disconnect battery ground cables.
2. Remove valve cover.
3. Remove rocker arm posts and rocker arms and mark for assembly reference.
4. Remove push rods and mark for reinstallation.
5. Reverse procedure to install. Install rocker arm post attaching bolts with timing mark set at 11 o'clock position, as viewed from front of engine, and torque to 20 ft. lbs.

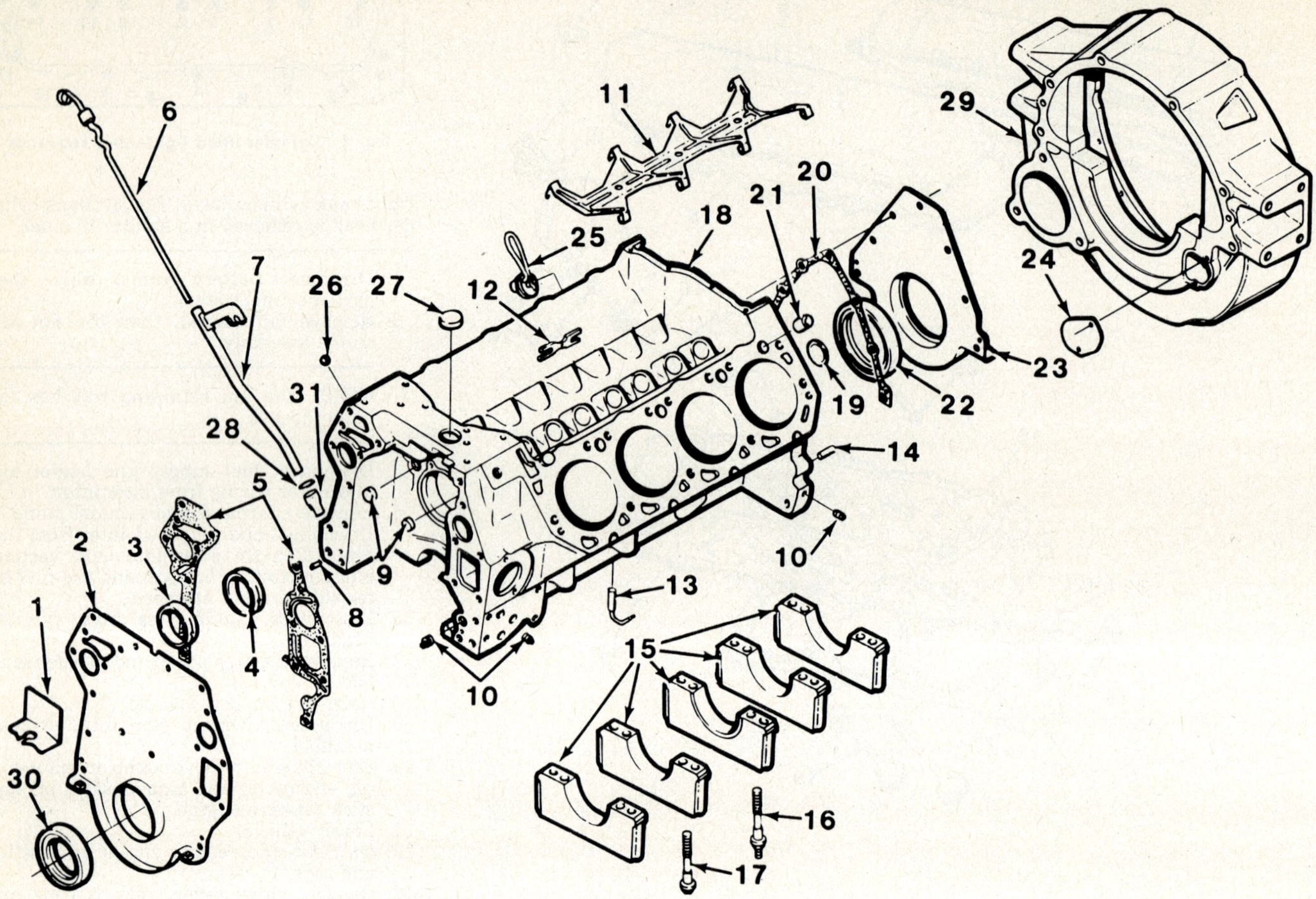

1. Timing Indicator
2. Front Cover
3. Bushing
4. Bushing
5. Front Cover Gaskets
6. Oil Level Gauge
7. Oil Level Gauge Tube, Upper
8. Dowel Pin
9. Cup Plug
10. Pipe Plug
11. Guide Retainer
12. Cam Follower Guide
13. Piston Cooling Tube
14. Dowel Pin
15. Bearing Cap
16. Bolt
17. Bolt
18. Crankcase with Plugs
19. Expansion Plugs
20. Rear Cover Gasket
21. Dowel Sleeve
22. Rear Oil Seal
23. Crankcase Rear Cover
24. Access Cover Plate
25. Block Heater
26. Ball 11/32''
27. Expansion Plug
28. ''O'' Ring
29. Flywheel Housing
30. Front Oil Seal
31. Oil Level Gauge Tube, Lower

Cylinder block assembly

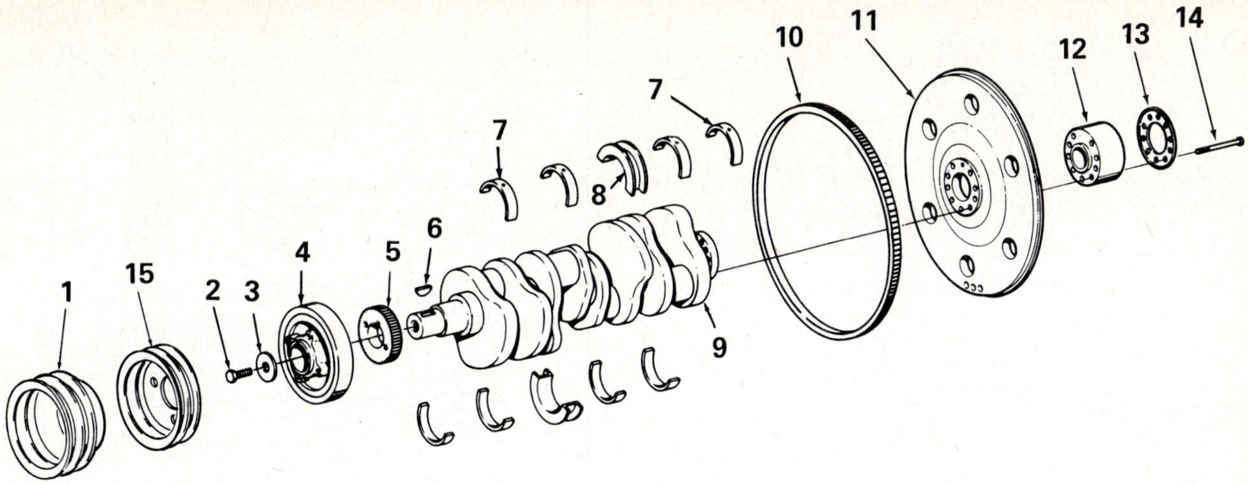

1. Pulley, Generator Drive
2. Screw
3. Crankshaft Washer
4. Vibration Damper
5. Crankshaft Gear
6. Woodruff Key
7. Bearing
8. Thrust Bearing
9. Crankshaft
10. Ring Gear
11. Flywheel
12. Flywheel Adapter
13. Reinforcement Ring
14. Bolt
15. Pulley, Crankshaft

Crankshaft assembly

VALVE ARRANGEMENT

V8-420 Diesel

Right .... I-E-I-E-I-E-I-E
Left .... E-I-E-I-E-I-E-I

VALVE GUIDES, REPLACE

If valve guides are damaged, or are larger than specifications allow, install repair insert as follows:

1. Drill out valve guide, then ream the drilled bore to correct size for insert sleeve.
2. Chill insert in dry ice, then carefully press insert in place.
3. Finish insert with reamer to specified valve guide diameter.

NOTE: Re-face valve seat after valve guide has been reamed, then, using a suitable scraping tool, break the sharp corner at top of valve guide.

VALVE LIFTERS, REPLACE

1. Remove intake manifold as described under "Intake Manifold, Replace".
2. Remove crankcase Depression Regulator (CDR) tube and grommet from valley pan.
3. Remove valley pan strap and drain plug, then the valley pan.
4. Remove rocker arms and push rods as described under "Rocker Arms & Pushrods, Replace".
5. Remove guide retainer, then the guides and valve lifters.
6. Reverse procedure to install.

VALVE LIFTERS, SERVICE

1. Disassemble valve lifter, Fig. 5, then clean all parts in solvent and wipe dry.
2. If any parts of lifter are worn or damaged, or plunger is not free in body, replace entire assembly.
3. Check that roller rotates freely and is not worn or damaged.
4. Lubricate all components with clean engine oil, then assemble as shown in Fig. 5.

TORSIONAL DAMPER, REPLACE

1. Disconnect battery ground cables.
2. Remove fan shroud, then the fan and clutch assembly.

NOTE: The fan retaining nut has left hand threads.

3. Remove drive belts.
4. Raise and support vehicle.
5. Remove crankshaft pulley.
6. Remove damper-to-crankshaft attaching bolt, then the damper.
7. Reverse procedure to install.

ENGINE FRONT COVER, REPLACE

1. Disconnect battery ground cables, then drain cooling system.
2. Remove air cleaner. Install intake air opening cap, tool T83T-9424-A.
3. Remove fan shroud, then the fan and clutch assembly.

NOTE: The fan retaining nut has left hand threads.

4. Remove fuel injection pump, injection pump adapter and water pump.
5. Raise and support vehicle.
6. Remove torsional damper, then disconnect ground cables from front of engine.
7. Remove front cover-to-engine block and oil pan attaching bolts.
8. Lower vehicle.
9. Remove front cover-to-engine block attaching bolts, then the front cover.
10. Using tool T80T-4000-W, remove seal from front cover, Fig. 6.
11. Reverse procedure to install. Lubricate new oil seal with DOAZ-19584-A grease, or equivalent. Apply a bead of sealant to mating surface as shown in Fig. 7.

FRONT OIL SEAL, REPLACE

1. Remove torsional damper, then pry seal out of cover.

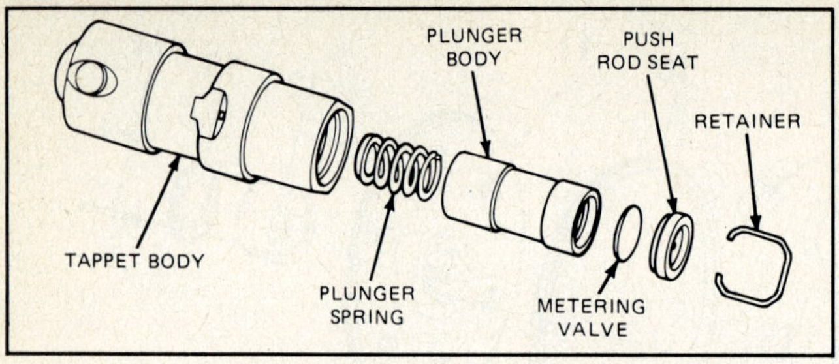

Fig. 5 Valve lifter, disassembled

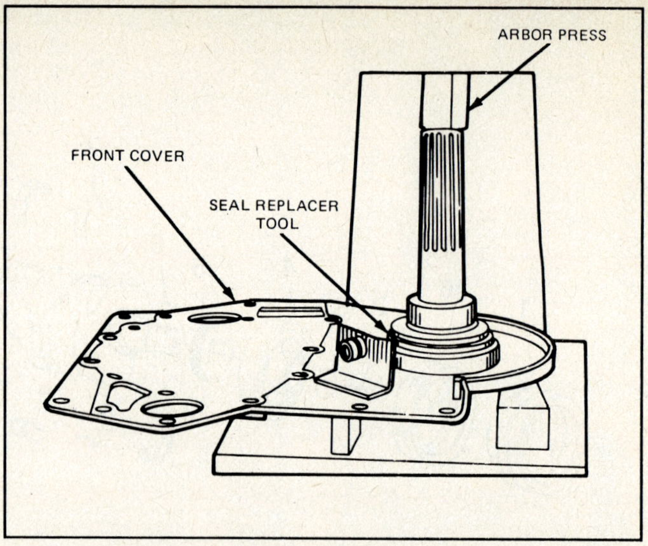

Fig. 6 Engine front cover seal replacement

2. To install new seal on engines without three weldnuts on front cover, position seal into tool T83T-6700 and install over end of crankshaft. Install tool T83T-6316-B over installation tool and tighten nut against tool and washer to seat seal.
3. To install new seal on engines with three weldnuts on front cover, position seal in tool T83T-6700-A and install over end of crankshaft. Attach bridge to weldnuts and draw seal into front cover by turning center screw clockwise.

NOTE: Prior to installation, lubricate oil seal with DOAZ-19584-A, or equivalent.

TIMING GEARS, REPLACE

1. Remove front cover.
2. Remove crankshaft drive gear, then the camshaft drive gear using tool T83T-6316-A.

NOTE: When removing crankshaft drive gear, use a breaker bar to prevent crankshaft rotation.

3. Reverse procedure to install. Align timing marks as shown in Fig. 8.

NOTE: The crankshaft drive gear may be heated in an oven to 300–350° F, to ease installation.

CAMSHAFT, REPLACE

1. Remove engine as described under "Engine, Replace".
2. Remove injection pump and adapter, intake manifold, valve lifters, engine front cover and fuel supply pump.
3. Remove camshaft drive gear, fuel supply pump cam, spacer and thrust plate from camshaft, Fig. 9.

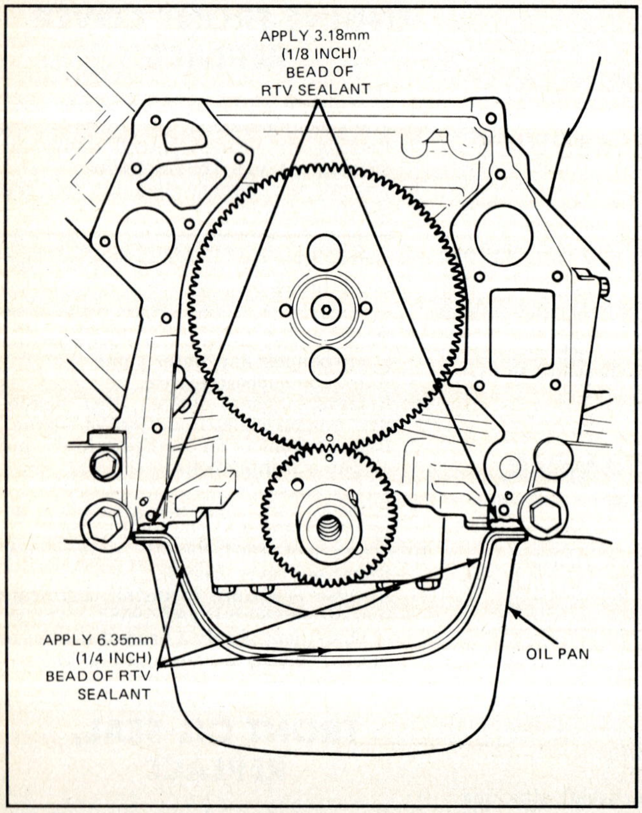

Fig. 7 Engine front cover installation

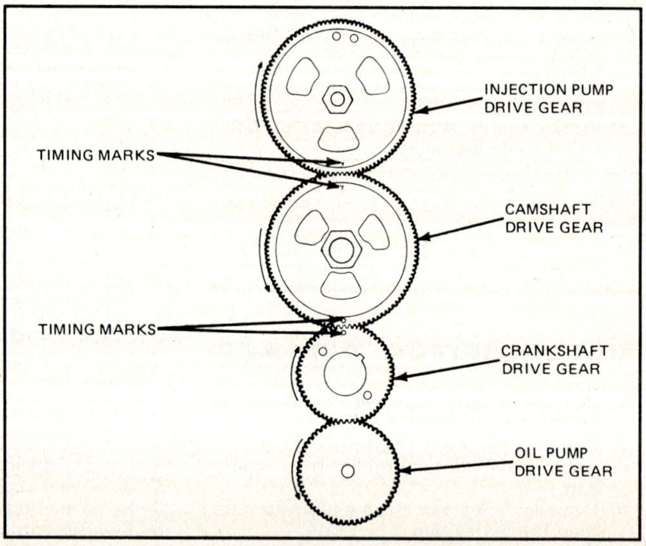

Fig. 8 Timing gear alignment

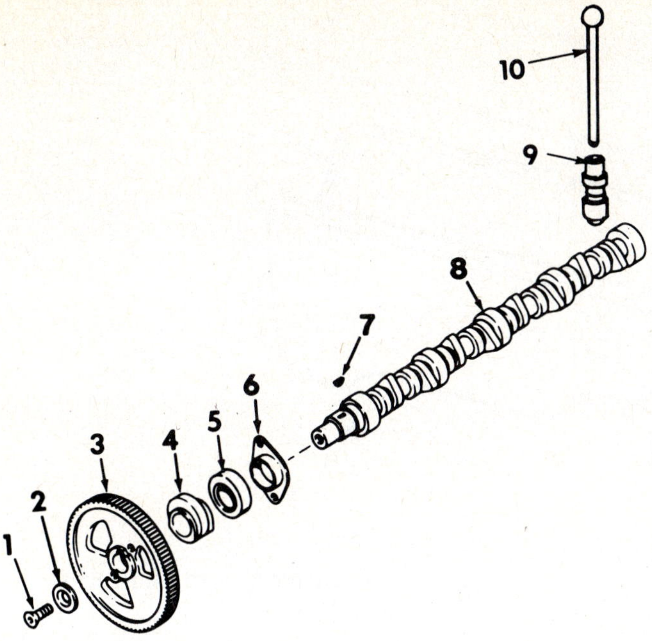

1. Camshaft Screw
2. Camshaft Washer
3. Camshaft Gear
4. Fuel Pump Cam
5. Thrust Flange Spacer
6. Thrust Flange
7. Woodruff Key
8. Camshaft
9. Cam Follower
10. Valve Push Rod

Fig. 9 Camshaft assembly

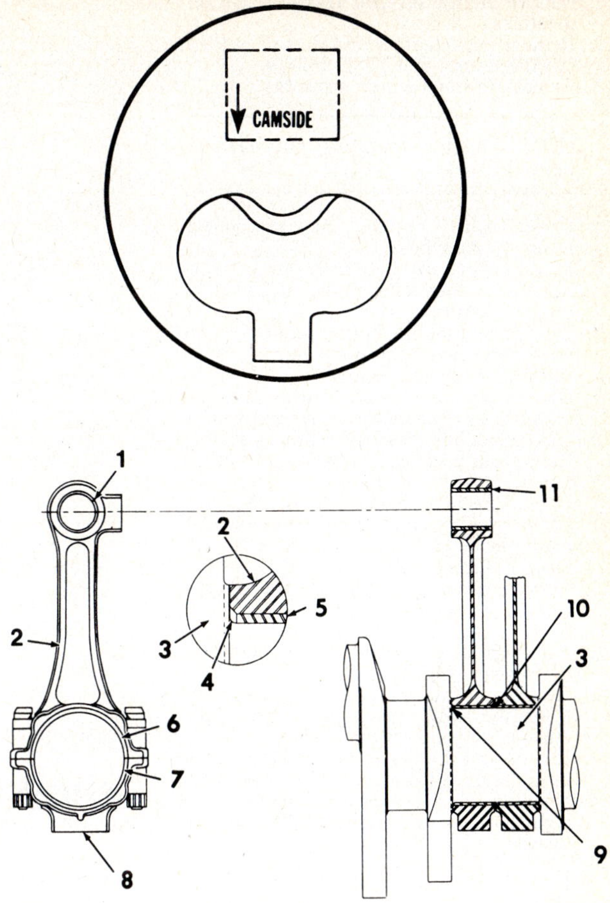

1. Split in Bushing
2. Connecting Rod
3. Crankshaft
4. Crankshaft Fillet
5. Bearing
6. Connecting Rod Bearing (Upper)
7. Connecting Rod Bearing (Lower)
8. Connecting Rod Bearing Cap
9. Large Chamfer Side
10. Small Chamfer Side
11. Bushing

Fig. 10 Piston & rod assembly

4. Remove camshaft using tools T65L-6250-A and Rotunda 14-0314, or equivalent.
5. Reverse procedure to install. Lubricate camshaft lobes with DOAZ-19584-A grease, or equivalent, and the journals with clean engine oil.

PISTON & ROD ASSEMBLY

Assemble piston to rod so that connecting rod weight pad is installed on the opposite side of the piston as the relief pocket, Fig. 10. Install the piston into the cylinder so the piston relief pocket is on the camshaft side of the engine and the arrow on the piston top also faces the camshaft side of the engine, Fig. 10.

PISTON & RINGS

Pistons are available in standard sizes and oversize of .003 inch.

FLYWHEEL, ENGINE REAR COVER & OIL SEAL, REPLACE

1. Remove transmission, then the clutch and clutch housing.
2. Remove flywheel, then the rear cover.
3. Using an arbor press and a 4⅛ inch spacer, remove oil seal.
4. Reverse procedure to install. Lubricate new oil seal with DOAZ-19584-A grease, or equivalent. Apply suitable sealant to sealing surfaces. Torque flywheel attaching bolts to specifications.

FLYWHEEL RING GEAR, REPLACE

1. Using a blow torch, heat ring gear on engine side of gear, then knock gear off of flywheel.
2. To install gear, heat it evenly until it expands enough to press onto flywheel.

NOTE: Do not allow temperature of ring gear to exceed 500° F.

OIL PAN & OIL PUMP, REPLACE

1. Disconnect battery ground cables, then

remove engine oil and transmission oil dipsticks.
2. Remove air cleaner and install intake opening cover, tool T83T-9424-A.
3. Remove fan and clutch assembly.

NOTE: The fan retaining nut has left hand threads.

4. Drain cooling system, then disconnect lower radiator hose.
5. Disconnect return hose from power steering pump, then plug pump and hose to prevent contamination.
6. Disconnect fuel line heater connector and wiring harness from alternator.
7. Raise and support vehicle.
8. Disconnect, then plug transmission oil cooler lines from radiator, if equipped.
9. Disconnect and plug fuel pump inlet line, then drain crankcase and remove oil filter.
10. Remove transmission oil filler tube.
11. Disconnect exhaust pipe from exhaust manifolds and muffler flange, then remove the inlet pipe.
12. Remove upper inlet pipe mounting stud from right exhaust manifold.
13. Remove nuts and washers securing engine insulators to No. 1 crossmember, then lower vehicle.
14. Attach suitable engine lifting equipment to engine and raise engine until transmission housing contacts body.
15. Install wood blocks between engine insulators and crossmember. Use a 2¾ inch block on left side and a 2 inch block on right side. Lower engine to rest on the blocks.
16. Raise and support vehicle.
17. Remove flywheel inspection plate.
18. Position fuel pump inlet line at rear of No. 1 crossmember and, if equipped, position transmission oil cooler lines aside.
19. Remove oil pan attaching bolts, Fig. 11, then remove oil pump and pick-up tube, Fig. 12, from engine and lay in oil pan. Remove oil pan.

NOTE: The crankshaft may need to be rotated to reposition counterweights in order to remove oil pan.

20. Reverse procedure to install. Apply sealant to oil pan mating surface as shown in Fig. 13.

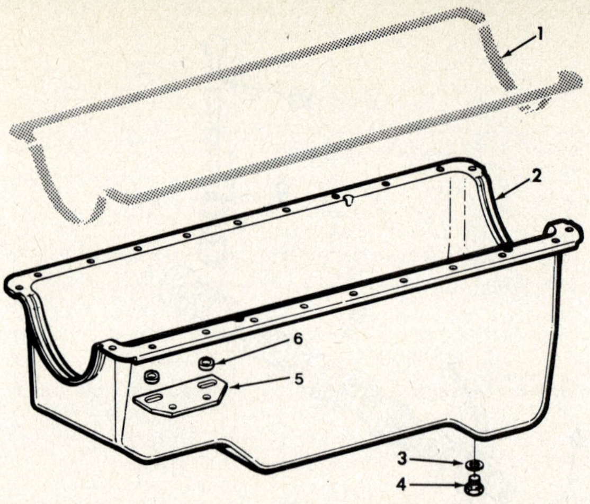

1. Oil Pan Gasket
2. Oil Pan
3. Washer
4. Drain Plug
5. Oil Pan Cover Mounting Bracket
6. Bracket Spacer

Fig. 11 Oil pan assembly

ENGINE OIL COOLER, REPLACE

1. Disconnect battery ground cables, then drain cooling system.
2. Remove fan shroud, then the fan and clutch assembly.

NOTE: The fan retaining nut has left hand threads.

3. Raise and support vehicle.
4. Drain engine oil and remove oil filter. Do not reinstall drain plug.
5. Remove left engine mount insulator-to-frame attaching nut.
6. Slightly raise left side of engine and install a one inch wood block between insulator and frame. Lower engine to rest on block.
7. Remove oil cooler attaching bolts, then the oil cooler, Fig. 14.
8. Reverse procedure to install.

WATER PUMP, REPLACE

1. Disconnect battery ground cables, then drain cooling system.
2. Remove fan shroud, then the fan and clutch assembly.

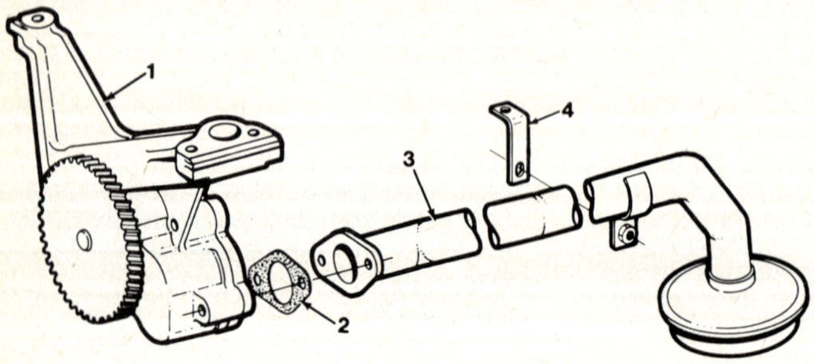

1. Oil Pump (With Gear)
2. Gasket
3. Oil Pick-Up Tube
4. Brace

Fig. 12 Oil pump, disassembled

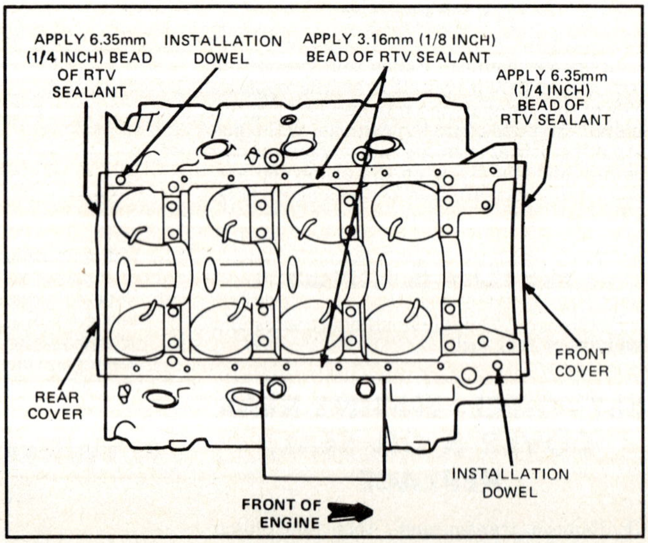

Fig. 13 Sealing oil pan mating surface

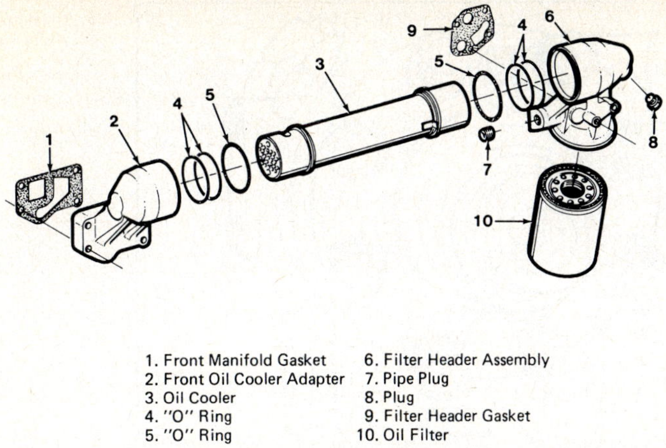

Fig. 14 Oil cooler, disassembled

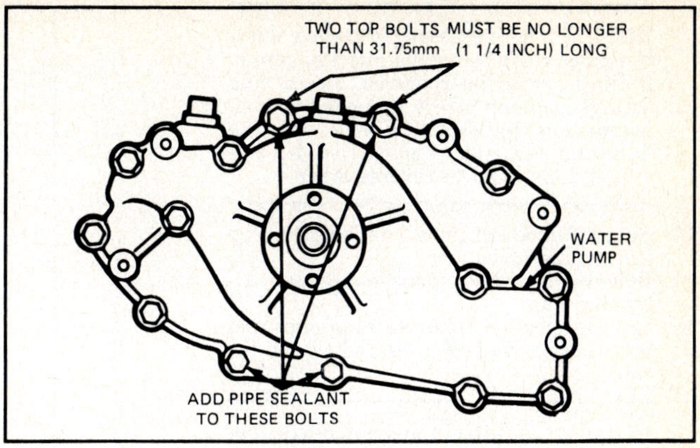

Fig. 15 Water pump installation

NOTE: The fan retaining nut has left hand threads.

3. Remove all drive belts, then the water pump pulley.
4. Disconnect heater hose from water pump, then remove heater hose fitting from pump.
5. Remove alternator adjusting arm and adjusting arm bracket.
6. Remove A/C compressor and brackets and position aside.
7. Remove power steering pump and bracket and position aside.
8. Remove water pump attaching bolts, then the pump.
9. Reverse procedure to install. Apply RTV sealer to two top and two bottom bolts, Fig. 15. Apply suitable pipe sealant to heater hose fitting. Torque water pump attaching bolts to 14 ft. lbs.

FUEL SUPPLY PUMP, REPLACE

1. Loosen fuel line connections. Do not remove fuel lines at this time.
2. Loosen fuel pump attaching bolts one to two turns. Using hand force only, loosen fuel pump to break gasket free.
3. Rotate engine by tapping starter until fuel pump cam lobe is at low position. This will reduce spring tension against pump attaching bolts.
4. Disconnect fuel pump inlet, outlet and return lines.
5. Remove fuel pump attaching bolts, then the fuel pump. Remove and discard gasket.
6. Reverse procedure to install. Run engine and check all connections for leaks.

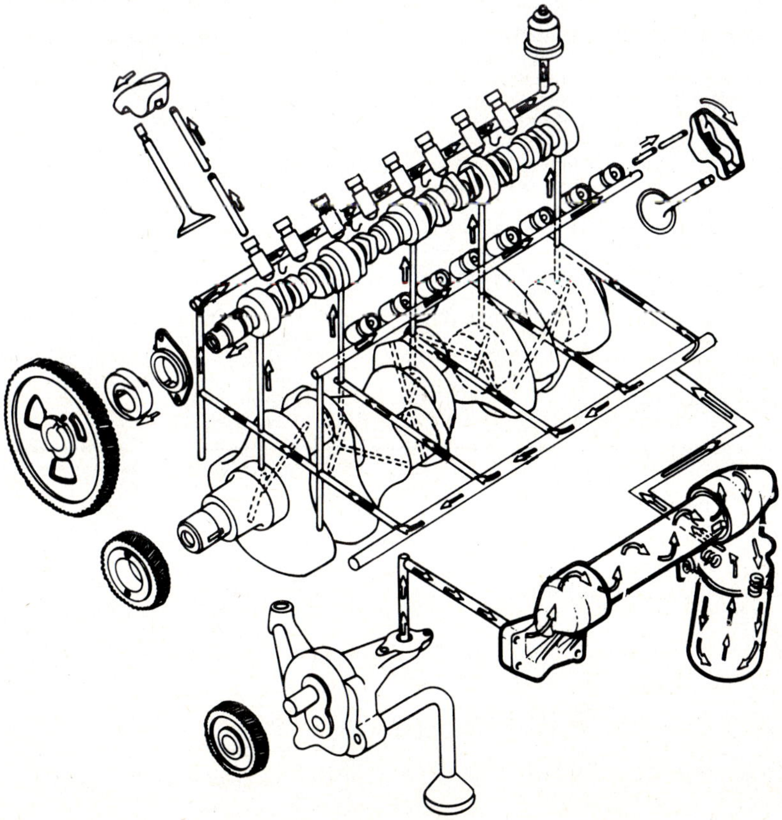

Engine oiling system

FUEL SUPPLY PUMP CAM, SPACER & THRUST PLATE, REPLACE

1. Remove camshaft drive gear, then the fuel supply pump as previously described.
2. Remove fuel pump cam and spacer using suitable puller.
3. Remove thrust plate attaching bolts, then the thrust plate.
4. Reverse procedure to install.

INJECTION PUMP, REPLACE

NOTE: Prior to removing any fuel lines, clean the exterior of lines with clean fuel oil or solvent to prevent contamination.

1. Disconnect battery ground cables, then remove engine oil filler neck.
2. Remove injection pump-to-drive gear attaching bolts.
3. Disconnect electrical connectors from injection pump.
4. Disconnect accelerator cable and, if equipped, speed control cable from throttle lever.
5. Remove air cleaner and install intake opening cover, tool T83T-9424-A.
6. Remove accelerator cable bracket, with cables attached, and position aside.
7. Remove fuel filter to injection pump fuel line and cap fittings.
8. Remove and cap pump inlet elbow and fitting adapter.
9. Remove fuel return line from pump and cap fittings.

NOTE: It is not necessary to remove injection lines when removing injection pump. If lines are to be removed, loosen line fittings at pump before removing it from engine. Fuel lines must be removed in the following sequence, Fig. 16, 5-6-4-8-3-1-7-2, and installed in reverse order.

10. Remove injection lines from nozzles. Cap lines and nozzles.
11. Remove injection pump-to-adapter attaching nuts.
12. Loosen injection line retaining clips and injection nozzle fuel lines and cap all fittings.

NOTE: Do not install injection nozzle fuel lines until pump is installed in engine.

13. Remove injection pump, with nozzle lines attached, from engine compartment.
14. Reverse procedure to install. Note the following:
    a. Install new "O" rings on drive gear end of injection pump and pump fitting adapter.
    b. Torque pump adapter nuts to 14 ft. lbs.
    c. Torque nozzle connector nuts and pump outlet fitting nut to 22 ft. lbs.
    d. Install elbow in pump adapter and torque to a minimum of 6 ft. lbs., then tighten further to align elbow with fuel inlet line, if necessary. Do not exceed one full turn or 10 ft. lbs.
    e. Apply suitable pipe sealant to pump elbow threads and a 1/8 inch bead of RTV sealant to adapter housing mating surface.

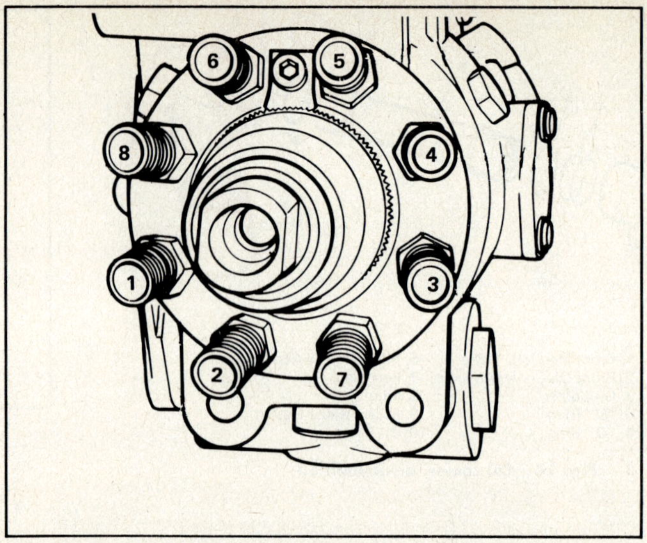

Fig. 16 Injection pump cylinder numbering sequence

INJECTION PUMP DRIVE GEAR & ADAPTER, REPLACE

1. Disconnect battery ground cables, then remove air cleaner and install intake opening cover, tool T83T-9424-A.
2. Remove injection pump as previously described.
3. Remove injection pump adapter housing.
4. Crank engine until No. 1 cylinder is at TDC of compression stroke using one of the following methods:
    a. Remove right valve cover and observe No. 1 cylinder valve lever.
    b. Remove No. 1 cylinder glow plug and listen for escaping air while turning engine in normal direction of rotation.
    c. Using the timing quadrant and vibration damper, scribe a line to indicate TDC.
5. With engine at TDC compression for No. 1 cylinder, gears with chamfers should be aligned and gears without chamfers should have the "Y" marks aligned.

NOTE: Injection pump and camshaft gears equipped with locating chamfers have one chamfer on the injection pump gear and two on the camshaft gear. Gears without chamfers are stamped with a "Y" on the face for proper alignment.

6. On gears with chamfers, scribe an alignment mark on front of injection pump gear where chamfer meshes with chamfer of camshaft gear, then remove gear.
7. On gears without chamfers, slide injection pump gear back so that top of camshaft gear is exposed when observed from top of front cover. Apply a small amount of paint to tip of a long screwdriver and mark top of camshaft teeth which mesh with injection pump gear. Also, scribe an alignment mark on front face of injection pump gear, then remove gear.
8. Reverse procedure to install. On gears with chamfers, position camshaft gear camfers in the 12 o'clock position, then install injection pump gear so that its chamfer meshes with those on camshaft gear. On gears without chamfers, mark new gear as old gear was marked in reference to "Y" mark on old gear, then install gear with marks aligned.

INJECTION NOZZLE, REPLACE

NOTE: Prior to removing nozzle, clean exterior of nozzle assembly with clean fuel oil or solvent to prevent contamination.

1. Remove fuel line retaining clamp from nozzle line.
2. Disconnect fuel inlet and leak-off lines from nozzle assembly. Cap all openings.
3. Remove nozzle from head by turning counterclockwise.
4. Reverse procedure to install. Note the following:
    a. Install a new copper gasket on each nozzle assembly.
    b. Apply suitable anti-seize compound to nozzle threads. Torque nozzle assembly to 33 ft. lbs.
    c. Install two new "O" rings in each fuel return tee.
    d. Run engine and check for leaks. If necessary, purge air from high pressure fuel lines by loosening connector one half to one full turn and cranking engine until fuel, free of air bubbles, flows from connector.

MITSUBISHI DIESEL ENGINES

INDEX OF SERVICE OPERATIONS

GENERAL SPECIFICATIONS

Engine Model	Bore & Stroke	Horsepower @ R.P.M.	Torque Ft. Lbs. @ R.P.M.	Firing Order	Compression Pressure @ 170 R.P.M.	Idle Speed	Governed Speed R.P.M.	Normal Oil Pressure @ R.P.M.
D-243-6	3.62 × 3.94 92mm × 100mm	100 @ 3700	163 @ 2200	1-5-3-6-2-4	425 PSI 2937 KPa	550–600	3700	42.7 PSI @ 1000 294 kPa @ 1000

VALVE SPECIFICATIONS

Engine Model	Valve Lash, Cold		Valve Face Angle	Valve Stem Diameter, Std.		Valve Stem Clearance		Valve Springs			
								Free Length		Lbs. @ In. Lgth.	
	Intake	Exhaust		Intake	Exhaust	Intake	Exhaust	Inner	Outer	Inner	Outer
D-243-6	0.012 0.3mm	0.012 0.3mm	45°	—	0.314 8mm	0.002–0.003 0.055–0.085mm	0.003–0.004 0.070–0.100mm	1.71 43.5mm	2.01 51.17mm	15.88 @ 1.530" 70.6N @ 39mm	37.26 @ 1.771" 165.74N 45mm

PISTONS, PINS, RINGS, CRANKSHAFT & BEARINGS

Engine Model	Piston Clearance	Piston Pin Diameter	Ring End Gap② Minimum			Rod Bearings		Main Bearings			
			Top	2nd 3rd	Oil	Shaft Diameter	Bearing Clearance	Shaft Diameter	Std. Bearing Clearance	Thrust Bearing	Shaft End Play
D-243-6	0.006- 0.008 0.145- 0.210mm	1.1021– 1.1023 27.994– 2800mm	0.001– 0.002 0.028– 0.050mm	0.001– 0.002 0.028– 0.050mm	0.012– 0.02 0.30– 0.50mm	2.281"– 2.282" 57.945– 57.965mm	0.0015"– 0.0044" 0.039– 0.113mm	2.754– 2.755 69.95– 69.97mm	0.0012– 0.0035 0.03– 0.089mm	7	0.004– 0.01 0.10– 0.25mm

ENGINE TIGHTENING SPECIFICATIONS

★ Torque Specifications are for clean and lightly lubricated threads only. Dry or dirty threads produce increased friction which prevents accurate measurement of tightness.

Year	Cyl. Head Ft. Lbs.	Exh. Man. Ft. Lbs.	Intake Man. Ft. Lbs.	Conn. Rod Ft. Lbs.	Main Bear. Cap. Ft. Lbs.	Flywheel to Crankshaft Ft. Lbs.	Damper to Crankshaft Bolt Ft. Lbs.	Camshaft Gear Bolt Ft. Lbs.	Oil Pan Ft. Lbs.
1978	90.4 122.6N	—	—	61.5–68.7 83.4–93.1N	①	75.9–83.2 102.9–112.8N	289.3 392.2N	21.7–28.9 29.4–39.1N	12.3–16.6 16.7–22.5N
1979	90.4 122.6N	—	—	57.9 78.5N	79.5 107.9N	65.1 88.3N	289.3 392.3N	25.3 34.3N	15.2 20.6N

①—65.1–72.3, 88.3–98.0N for all except w/"H" mark 76.7–85.4, 104.0–115.8N

SERVICE DIAGNOSIS

Condition	Possible Cause	Correction
HARD STARTING **Starter will not turn, or turns but fails to start engine**	1. Weak battery	1. Test battery specific gravity and recharge or replace as necessary
	2. Defective starter motor switch contacts	2. Replace switch
	3. Magnetic switch contact plate arced or defective	3. Polish or replace contact plate
	4. Magnetic switch pull-in coil open	4. Replace
	5. Bad brush contact	5. Correct brush contact
	6. Burned starter motor commutator	6. Repair commutator
	7. High commutator mica	7. Undercut commutator
	8. Starter motor field coil grounded	8. Replace coil
	9. Starter motor armature grounded	9. Replace armature
	10. Insufficient brush spring tension	10. Replace brush spring
	11. Improperly soldered field coil	11. Repair
	12. Worn starter motor bearing bushing	12. Replace bushing
	13. Open wiring between starter switch and magnet switch terminal	13. Repair or replace
	14. Faulty contact between battery and ground strap	14. Clean and correct
	15. Engine oil too thick	15. Change to specified oil
Starter turns but pinion fails to mesh with ring gear	1. Worn starter motor clutch pinion tip	1. Replace pinion tip
	2. Starter clutch races	2. Replace clutch
	3. Starter clutch drive spring defective	3. Replace drive spring
	4. Starter fails to rotate in overrun direction	4. Replace starter
	5. Splines meshing improperly	5. Lap, clean, and correct splines
	6. Worn starter bushing	6. Replace bushing
	7. Drive lever setting bolt missing	7. Replace bolt
	8. Magnetic switch pinion moves out to improper position	8. Adjust stud bolt
	9. Worn bearing bushing	9. Replace bushing
	10. Worn ring gear	10. Replace ring gear
Starter turns but fails to stop	1. Magnetic switch coil layer shorted	1. Replace switch
	2. Magnetic switch contact plate arced	2. Replace switch
	3. Starter switch fails to disengage	3. Replace switch
Engine	1. Burned valve	1. Lap or replace valve
	2. Intake manifold gasket loose	2. Tighten manifold bolts or replace gasket
	3. Worn piston, piston rings, and cylinder sleeve	3. Overhaul engine
	4. Defective head gasket	4. Replace head gasket
Fuel	1. Damaged air horn gasket or loose set-screw	1. Replace gasket or tighten set-screw
Ignition	1. Faulty glow plug operation	1. Check glow plug connections and glow plug operation
INADEQUATE IDLING **Valves**	1. Improper valve clearance	1. Adjust valve clearance
	2. Valve and valve seat not properly fitted	2. Lap valve
	3. Excessive clearance—valve stem to valve guide	3. Replace valve and valve guide
Cylinder Head	1. Head gasket leaking	1. Replace head gasket
Injection System	1. Improper injection timing	1. Adjust timing
	2. Dirty or worn fuel injection pump or injector nozzles	2. See "Injection System" group
	3. Leaking fuel lines	3. Tighten connections or replace line
ABNORMAL COMBUSTION **Fuel**	1. Dirty or clogged fuel pipe	1. Clean or replace fuel pipe
	2. Air leak into intake manifold	2. Tighten manifold bolts or replace gasket
Valves	1. Improper valve clearance	1. Adjust valve clearance
	2. Valve sticking	2. Disassemble and repair or replace valve
	3. Valve spring defective	3. Replace valve spring
Cylinder Head	1. Carbon deposits in combustion chamber	1. Remove carbon
	2. Clogged cylinder head water tube	2. Clean or replace water tube
	3. Gas leaking through head gasket	3. Replace head gasket
Injection System	1. Improper injection timing	1. Adjust injection timing
	2. Dirty or worn fuel injection pump or fuel injection nozzles	2. See "Injection System" group
Cooling System	1. Overcooling	1. Check cooling system components

Continued

Condition	Possible Cause	Correction
ENGINE NOISE	1. Excessive main bearing clearance	1. Replace bearing
	2. Main bearing scored	2. Replace bearing and check lubrication system
	3. Excessive connecting rod bearing clearance	3. Replace bearing
	4. Bent connecting rod	4. Replace rod
	5. Rod bearing scored	5. Replace bearing and check lubrication system
	6. Excessive piston clearance from worn cylinder sleeve	6. Rebore and hone cylinder sleeve and fit oversize piston
	7. Worn piston and piston pin	7. Replace piston and piston pin
	8. Piston seizure	8. Replace piston
	9. Damaged piston rings	9. Replace piston rings
	10. Excessive camshaft end play	10. Replace camshaft thrust plate
	11. Worn crankshaft thrust bearing	11. Replace bearing
	12. Worn timing gear	12. Replace timing gear
	13. Excessive valve clearance	13. Adjust valve clearance
	14. Defective water pump bearing	14. Replace bearing
	15. Loose or bent fan	15. Tighten or replace fan
	16. Defective fan belt	16. Replace fan belt
INSUFFICIENT ACCELERATION	1. Burnt or misadjusted valve	1. Replace or adjust valve
	2. Insufficient compression	2. Disassemble and repair engine
	3. Leaking head gasket	3. Replace head gasket
	4. Improper injection timing	4. Adjust injection timing
	5. Improper fuel injection adjustment	5. See "Injection System" group
	6. Dirty air filter element	6. Replace air filter element
INSUFFICIENT POWER Insufficient Compression	1. Improper valve clearance	1. Adjust valve clearance
	2. Compression leaks at valve seat	2. Remove cylinder head and lap valve
	3. Valve stem seizure	3. Correct or replace valve
	4. Defective or broken valve spring	4. Replace valve spring
	5. Leaking head gasket	5. Replace head gasket and tighten head bolts to specifications
	6. Sticking or worn piston rings	6. Replace piston rings
	7. Worn piston rings or cylinder	7. Disassemble and repair engine
Injection System	1. Improper injection timing	1. Adjust injection timing
Fuel System	1. Clogged fuel pipe	1. Clean pipe
	2. Air in fuel system	2. Check each pipe connection Tighten connections and bleed fuel system
	3. Loose fuel pipe connection	3. Tighten connection
	4. Fuel pipe cracked	4. Replace fuel pipe
	5. Clogged fuel filter	5. Clean or replace filter element
	6. Clogged air cleaner	6. Clean or replace air cleaner element
Overheating	1. Insufficient coolant	1. Add coolant and check for leaks
	2. Radiator leaking	2. Repair or replace radiator
	3. Loose or damaged radiator connection	3. Tighten clamp or replace hose
	4. Water pump leaking	4. Repair or replace pump
	5. Loose or damaged heater hose	5. Tighten or replace hose
	6. Head gasket leaking coolant	6. Tighten head bolts to specifications
	7. Cracked cylinder head or block	7. Replace or replace damaged component
	8. Loose fan belt	8. Adjust or replace fan belt
	9. Worn or broken fan belt	9. Replace fan belt
	10. Oil on fan belt	10. Replace fan belt
	11. Defective thermostat	11. Replace thermostat
	12. Defective water pump	12. Replace water pump
	13. Clogged radiator	13. Clean radiator and water jacket
	14. Improper engine oil	14. Change to specified oil
	15. Improper valve clearance	15. Adjust valve clearance
	16. Excessive back pressure in exhaust system	16. Clean or replace exhaust system
Overcooling	1. Faulty thermostat	1. Replace thermostat
	2. Extremely low outside temperature	2. Decrease radiator air passage area
EXCESSIVE FUEL CONSUMPTION Engine	1. Improper valve clearance	1. Adjust valve clearance
	2. Compression leaks at valve seat	2. Remove cylinder head and lap valve
	3. Seized valve stem	3. Correct or replace valve stem
	4. Defective or broken valve spring	4. Replace valve spring
	5. Leaking head gasket	5. Replace head gasket and tighten head bolts to specifications
	6. Sticking or worn piston rings	6. Replace piston rings
	7. Worn piston rings or cylinder sleeve	7. Disassemble and repair engine
Injection System	1. Improper injection timing	1. Adjust injection timing

Continued

Condition	Possible Cause	Correction
Injection System	2. Injection system leaks	2. Tighten connections and replace leaking pipes
	3. Injection nozzles dribbling	3. See "Injection System" group
EXCESSIVE ENGINE OIL CONSUMPTION Oil Leaks	1. Loose oil pan drain plug	1. Tighten drain plug
	2. Loose oil pan bolts	2. Tighten pan bolts to specifications
	3. Defective pan gasket	3. Replace pan gasket
	4. Loose timing gear cover or defective cover gasket	4. Tighten bolts or replace gasket
	5. Defective crankshaft front oil seal	5. Replace oil seal
	6. Defective crankshaft rear oil seal	6. Replace oil seal
	7. Defective cylinder head cover gasket	7. Replace gasket
	8. Loose oil filter or defective gasket	8. Replace oil filter
Oil in Combustion Chamber	1. Improper piston ring gap position	1. Adjust ring position
	2. Worn or sticking rings or ring grooves	2. Replace piston or rings
	3. Plugged oil return hole in oil ring	3. Replace oil ring
	4. Excessively worn piston and cylinder sleeve	4. Replace piston or rebore sleeve
	5. Excessive worn valve guide and/or valve stem	5. Replace valve guide insert and valve
	6. Clogged air filter element	6. Replace air filter element

DRIVEABILITY DIAGNOSIS

Condition	Cold Engine Symptom	Corrective Action
ENGINE CRANKS, BUT WILL NOT START (No fuel reaching injection pump)	1. Air trapped in fuel system	1. Bleed air from fuel filter and injection pump
	2. Fuel line drawing air	2. Replace cracked line or tighten loose connector
	3. No fuel in tank	3. Fill tank with specified fuel
	4. Gauze filter (strainer) at feed pump inlet clogged	4. Remove filter, clean, and replace
	5. Fuel line clogged	5. Disassemble and clean line
	6. Fuel filter clogged	6. Clean filter case and replace element
	7. Feed pump not delivering fuel	7. Valve inoperative or plunger stuck. Disassemble and repair
(Fuel reaches injection pump)	8. Injection pump seized	8. Replace injection pump
	9. Injection pump loose on mountings	9. Tighten injection pump mounting bolts
	10. Pump control rack seized Plunger stuck to barrel Control sleeve stuck	10. Replace injection pump
	11. Delivery valve stuck or spring broken	11. Replace injection pump
(Slight smoke in exhaust; insufficient fuel injected)	12. Air trapped in fuel system	12. See 1 above
	13. Gauge filter (strainer) at feed pump inlet clogged	13. See 4 above
	14. Fuel line clogged	14. See 5 above
	15. Fuel filter clogged	15. See 6 above
	16. Feed pump delivering insufficient fuel	16. See 7 above
	17. Pump control rack movement restricted	17. Adjust governor
	18. Accelerator linkage binding	18. Free and adjust accelerator linkage
	19. Injector line connector loose	19. Tighten connector
	20. Injection pump tappet sticking	20. Replace injection pump
	21. Injection pump tappet roller or cam excessively worn	21. Replace injection pump
	22. Injection pump tappet plunger stuck or spring broken	22. Replace injection pump
	23. Injection pump plunger excessively worn	23. Replace injection pump
	24. Injection pump control pinion set-screw loose	24. Align control pinion with line on control sleeve and tighten screw
	25. Fuel too viscous to flow freely	25. If weather is cold, heat fuel. If temperatures are normal, fuel is improper. Drain fuel and replace with specified grade
	26. Fuel leak or dribble at nozzle, insufficient fuel injection	26. Replace faulty nozzle. Check injection cycle following manual procedures
(Thick smoke in exhaust; incorrect injection timing)	27. Injection pump improperly installed and timed	27. Reinstall pump and time properly. Refer to "Injection Pump Timing"
	28. Injection pump tappet or cam worn excessively, throwing timing off	28. Replace injection pump and check timing

Continued

Condition	Cold Engine Symptom	Corrective Action
(Thick smoke in exhaust; incorrect injection timing)	29. Injection pump adjusting bolt loose, throwing timing off	29. Adjust injection timing and tighten adjusting bolt
(Nozzle Does Not Spray Fuel Properly)	30. Needle valve stuck in nozzle body—nozzle remains open	30. Disassemble nozzle. Replace if necessary
	31. Nozzle valve does not seat properly and allows fuel to leak	31. Disassemble nozzle. Replace if necessary
	32. Injection pressure too low	32. Use nozzle tester. Adjust injection pressure to 11 756 kPa (1705 psi) with adjusting screw
	33. Nozzle holder spring broken	33. Replace spring—adjust injection pressure
	34. Nozzle packing clogged	34. Remove packing and carbon deposits

Condition	Warm Engine Symptom	Corrective Action
ENGINE POWER OUTPUT INSUFFICIENT (Exhaust smoke thin; insufficient fuel supply)	35. Gauze filter (strainer) in feed pump inlet clogged	35. See 4 above
	36. Feed valve clogged; insufficient fuel feed	36. See 5 above
	37. Fuel feed pump delivering insufficient fuel	37. See 7 above
	38. Injector line connector leaking fuel	38. Tighten connector on injector line
	39. Injection pump tappet stuck	39. Replace injection pump
	40. Injection pump tappet roller or cam excessively worn	40. Replace injection pump
	41. Injection pump plunger stuck or spring broken	41. Replace injection pump
	42. Injection pump control pinion set-screw loose	42. See 24 above
	43. Injection pump delivery valve damaged or spring broken	43. Replace injection pump
	44. Nozzle leaking fuel	44. Replace defective nozzle
(Thick white smoke in exhaust; injection timing late)	45. Fuel too viscous to flow freely	45. See 25 above
	46. Injection interval improperly timed	46. Reset injection timing
	47. Injection pump tappet roller or cam excessively worn	47. Replace injection pump
	48. Injection pump tappet adjusting bolt loose throwing timing off	48. Adjust injection timing and tighten adjusting bolt
(Thick black smoke in exhaust; injection timing early)	49. Injection interval improperly timed	49. Reinstall pump and time properly Refer to "Injection Pump Timing"
	50. Air cleaner clogged	50. Replace air cleaner element
	51. Air intake restricted	51. Clean air intake path
(Thick black smoke in exhaust; fuel charge between cylinders non-uniform)	52. Worn tappet roller or cam timing	52. Replace injection pump and recheck
	53. Loose tappet adjusting bolts	53. Set timing and tighten adjusting bolts
	54. Stuck tappets; poor tappet return	54. Replace injection pump
	55. Worn plungers; stuck plungers; broken springs	55. Replace injection pump
	56. Control pinion set-screws loose	56. Align control pinion with line on control sleeve and tighten set-screws
	57. Delivery valves damaged or springs broken	57. Replace injection pump
(Improper nozzle spray pattern)	58. Needle valve stuck in nozzle body; nozzle stays open	58. Disassemble nozzle; replace if necessary
	59. Nozzle valve does not seat; fuel leaks	59. Disassemble nozzle; replace if necessary
	60. Injection pressure too low	60. Use nozzle tester. Adjust injection pressure to 11 756kPa (1705 psi) with adjusting screw
	61. Nozzle holder spring broken	61. Replace spring; adjust injection pressure to specifications
	62. Nozzle packing clogged	62. Remove packing and carbon deposits
ENGINE KNOCKS	63. Injection too early. Engine knocks hard and produces black smoke	63. Remove and reinstall injection pump Time properly. Refer to "Injection Pump Timing"
	64. Injection timing retarded. Engine knocks weakly and produces white smoke	64. Remove and reinstall injection pump. Time properly. Refer to "Injection Pump Timing"
	65. Cylinders into which too much fuel is injected are knocking	65. See (52) through (57) above
	66. Improper fuel spray from nozzles	66. See (58) through (62) above

Condition	Warm Engine Symptom	Corrective Action
EXCESSIVE SMOKE EMITTED DURING OPERATION (Thick white smoke in exhaust)	67. Injection timing retarded	67. Remove and reinstall injection pump. Time properly. Refer to "Injection Pump Timing"
	68. Water in fuel	68. Drain fuel filter. Drain fuel system. Replace with specified fuel
(Thick black smoke in exhaust)	69. Injection timing advanced	69. See (67) above
	70. Excessive fuel delivery from injection pump	70. Replace injection pump
	71. Non-uniform fuel delivery from injection pump	71. See (52) through (57) above
	72. Nozzle spray non-uniform	72. See (58) through (62) above
	73. Clogged air cleaner	73. Replace air cleaner element
ENGINE SPEED NOT SMOOTHLY CONTROLLED (Engine runs at high speed and will not stop)	74. Accelerator linkage binding between pedal and injection pump	74. Check accelerator linkage and ensure free operation
	75. Defective governor linkage	75. Replace injection pump
	76. Injection pump rack binding	76. Replace injection pump
	77. Injection pump control pinion set-screw loose	77. Align control pinion with line on control sleeve and tighten set-screw

Engine Repair Section

ENGINE, REPLACE

1. Disconnect battery cables and remove battery.
2. Remove air cleaner.
3. Scribe hood hinge locations on hood, then remove hood.
4. Drain cooling system and disconnect radiator hoses from engine.
5. Remove coolant reserve tank.
6. Raise and support vehicle.
7. Disconnect transmission oil cooler lines from radiator, if equipped.
8. Disconnect the lower radiator and fan shroud attaching screws.
9. Lower vehicle and remove upper radiator and fan shroud attaching screws. Remove radiator and fan shroud.
10. Disconnect heater hoses and position aside.
11. Disconnect speedometer cable housing from engine.
12. Disconnect all electrical wiring from engine that may interfere with engine removal. Position wiring harnesses aside.
13. Disconnect and cap fuel line at transfer pump inlet.
14. Disconnect and plug return line at injector lines bleed-back connection.
15. Disconnect and remove injector pump linkage. Disconnect and remove accelerator and throttle cable linkage.
16. Disconnect starter motor wiring and remove starter motor.
17. Disconnect battery ground cable from cylinder block. Remove engine ground strap.
18. Disconnect and plug power steering hoses from gear and position aside.
19. Raise and support vehicle.
20. Disconnect exhaust pipe from manifold.
21. Drain oil pan and remove dipstick tube.
22. Remove oil cooler lines, if equippped, and the road draft tube bracket from oil pan.
23. Remove oil pan bolts. Support engine under A/C compressor bracket.
24. Remove transmission inspection plate, then the oil pan.

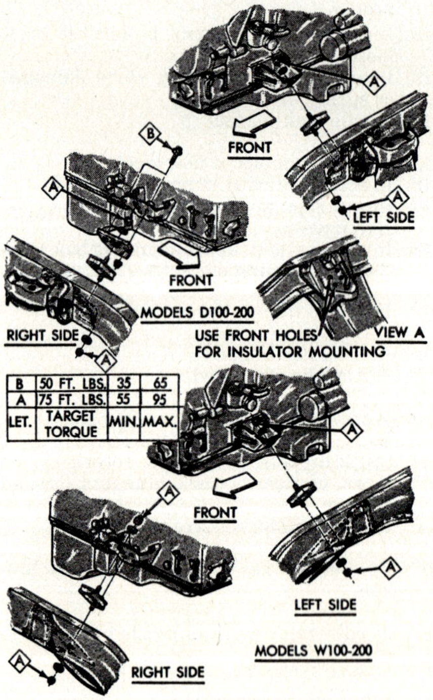

Fig. 1 Engine mounts

25. Remove torque converter to flywheel bolts.
26. Remove exhaust pipe bracket and lower bell housing bolts.
27. Remove engine support placed under A/C compressor bracket.
28. Support transmission with a suitable jack.
29. Remove rocker arm cover.
30. Attach suitable engine lifting equipment to engine lift points.
31. Remove bolts and nuts from engine mounts, Fig. 1.
32. Remove upper bell housing bolts.
33. Pull engine forward to separate from transmission, then lift engine from vehicle.
34. Reverse procedure to install.

CYLINDER HEAD, REPLACE

1. Disconnect battery ground cable and drain cooling system.
2. Remove air cleaner.
3. Disconnect hoses from fuel filter at transfer pump and injection pump. Drain fuel filter and remove from rear of air manifold.
4. Remove nuts securing air manifold and air cleaner mounting brackets.
5. Disconnect the No. 3 and No. 6 fuel injection lines from injection pump.
6. Remove air intake manifold and gaskets from cylinder head.
7. Push exhaust manifold shield aside.
8. Disconnect bypass hose and heater hoses from engine.
9. Remove thermostat housing and upper radiator hose from water manifold. Remove spray gasket.
10. Disconnect fuel line mounting brackets from cylinder head and push fuel lines aside.
11. Remove three exhaust manifold bridges, then the water manifold and gasket from cylinder head.
12. Raise and support vehicle.
13. Remove exhaust pipe from exhaust manifold.
14. Lower vehicle and remove exhaust manifold, heat shield and gasket.
15. Disconnect wiring from glow plug buss bar.
16. Disconnect injection lines from fuel injection pump.
17. Disconnect fuel injection line and fuel injection line bracket and ground strap

from cylinder head.
18. Disconnect alternator bracket and engine lifting fixture and position aside.
19. Loosen and remove cylinder head bolts in sequence, Fig. 2.
20. Remove rocker arm and shaft assembly.
21. Remove push rods. Mark location of push rods to ensure installation in original position.
22. Remove injector tubes, injector holders and injectors.
23. Remove glow plug buss bar and the glow plugs.
24. Remove cylinder head from engine.
25. Reverse procedure to install. Torque cylinder head bolts in sequence, Fig. 3, to specifications.

ROCKER ARMS, REPLACE

NOTE: Dimensions of the intake and exhaust rocker arms are different since a different dimension exists between the rocker shaft centerline and the intake and exhaust push rods, Fig. 4.

1. Remove cylinder head cover and gasket.
2. Remove nozzle holders and glow plugs.
3. Remove rocker shaft bracket retaining bolts.
4. Remove rocker arm and shaft assembly, Fig. 5.
5. Reverse procedure to install.

VALVE ARRANGEMENT

6-243 .... E-I-E-I-E-I-E-I-E-I-E-I

VALVE TIMING

Intake Opens Before TDC

Engine	Year	Degrees
6-243	1978	32

VALVES, ADJUST

The valves are adjusted with the engine cold. Adjust clearance between rocker arm and top of valve stem, Fig. 5A to .012 inch at top dead center, compression stroke of each cylinder.

VALVE GUIDES, REPLACE

1. Remove valve guide with tool No. 31691-10500, Fig. 6.
2. Press new valve guide into cylinder head with tool No. 31691-10600. The valve guide installed length is .7583–.791 inch, Fig. 6.

VALVE SEAT INSERTS, REPLACE

1. Grind the valve seat insert to a thickness of .02–.04 inch. Then remove the insert at room temperature.
2. Heat cylinder to a temperature of 167–212° F.
3. Immerse new insert in ether or alcohol with dry ice.
4. Press cooled insert into heated cylinder head, Fig. 7.
5. Allow cylinder head to cool to room temperature.
6. Caulk valve seat insert circumference with appropriate caulking tool, Fig. 8, and machine seat width to .079 inch.

TIMING GEAR CASE, REPLACE

1. Disconnect battery ground cable and drain cooling system.
2. Remove upper and lower radiator hoses. Disconnect overflow line from coolant reserve tank.
3. Raise and support vehicle.
4. Disconnect transmission oil cooler lines from radiator.

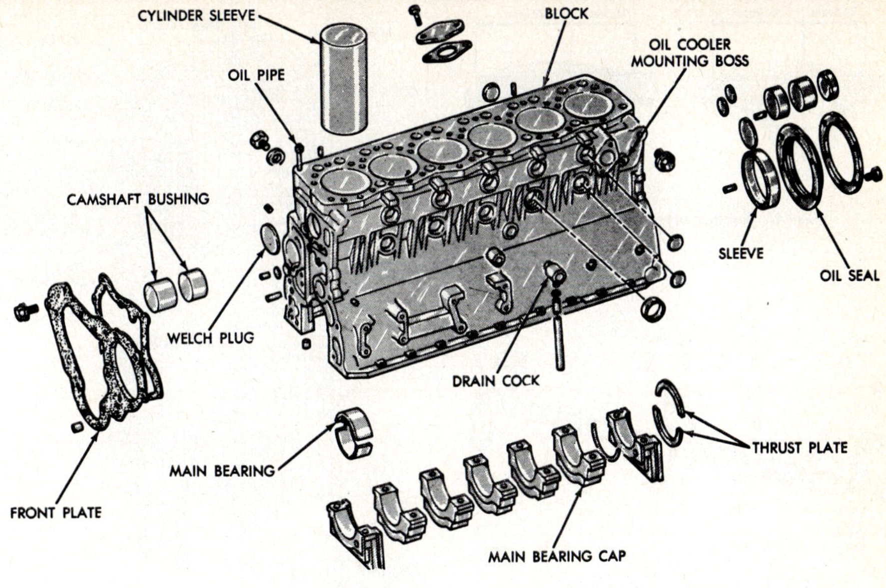

Cylinder block assembly

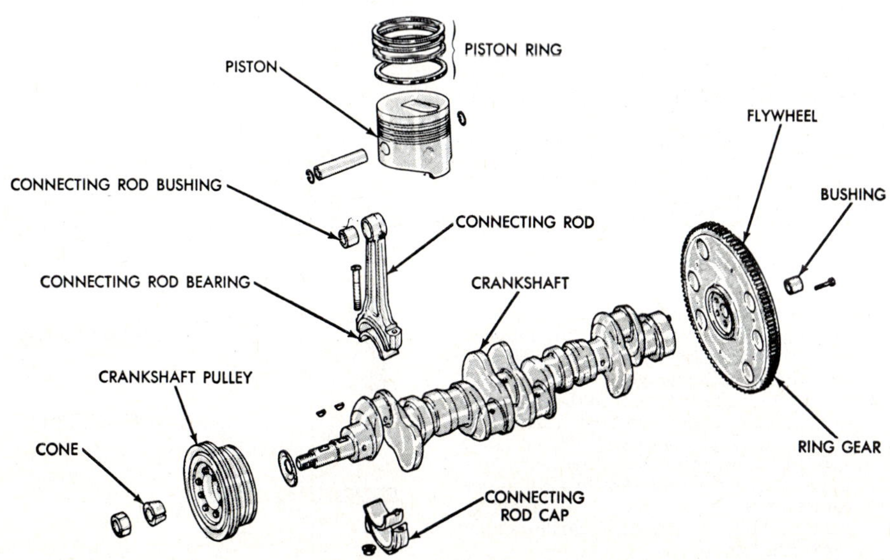

Crankshaft, piston & connecting rod assembly

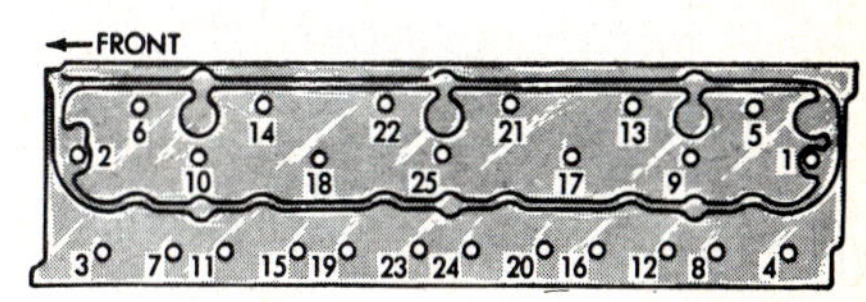

Fig. 2 Cylinder head bolt removal sequence

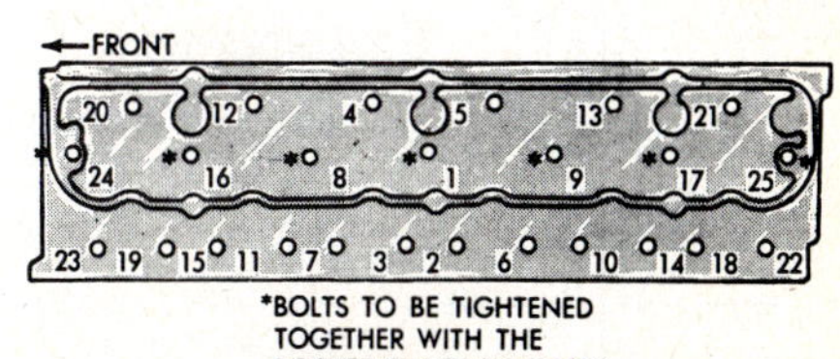

Fig. 3 Cylinder head bolt tightening sequence

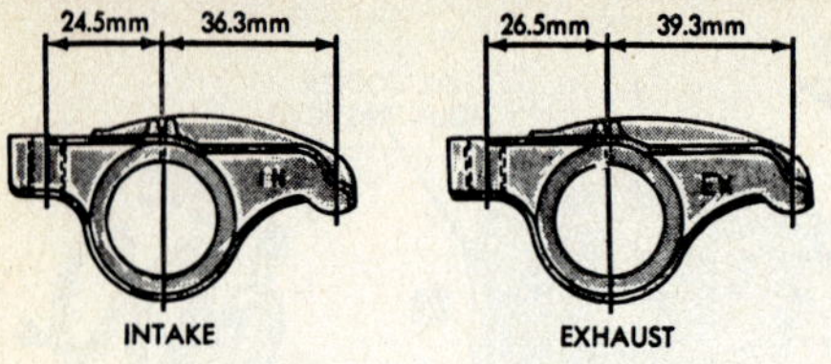

Fig. 4 Rocker arm identification

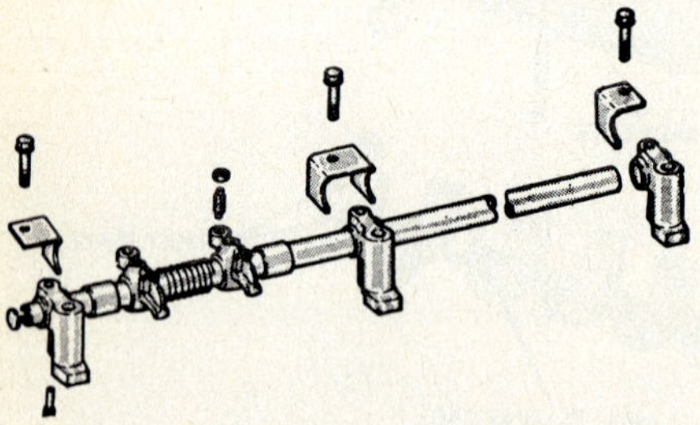

Fig. 5 Rocker shaft assembly

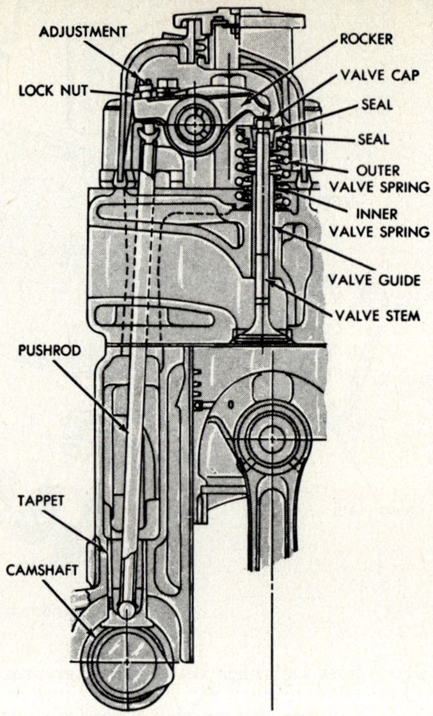

Fig. 5A Valve train

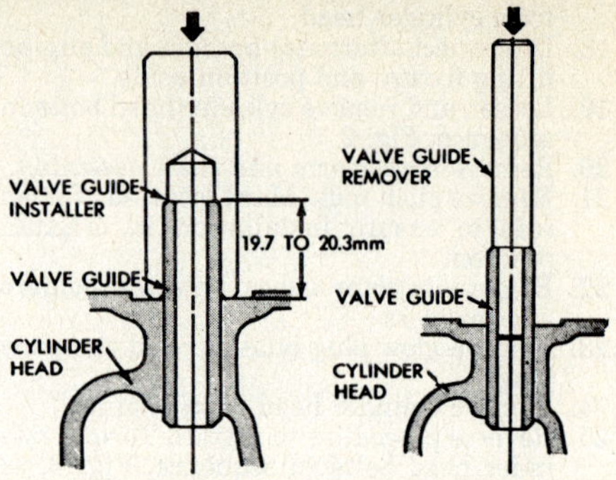

Fig. 6 Valve guide replacement

5. Lower vehicle.
6. Loosen radiator mounting bolts.
7. Remove fan.
8. Remove rubber shield between radiator and grille, then the radiator and fan shroud.
9. Remove drive belts and idler pulley.
10. Remove crankshaft pulley retaining nut. Rotate crankshaft until keyway is at 12 o'clock position. Remove crankshaft pulley and damper with a suitable puller.
11. Remove timing gear case, gasket and idler pulley bracket, Fig. 9.
12. Reverse procedure to install.

FRONT OIL SEAL, REPLACE

The front oil seat may be replaced after removing the timing gear case.

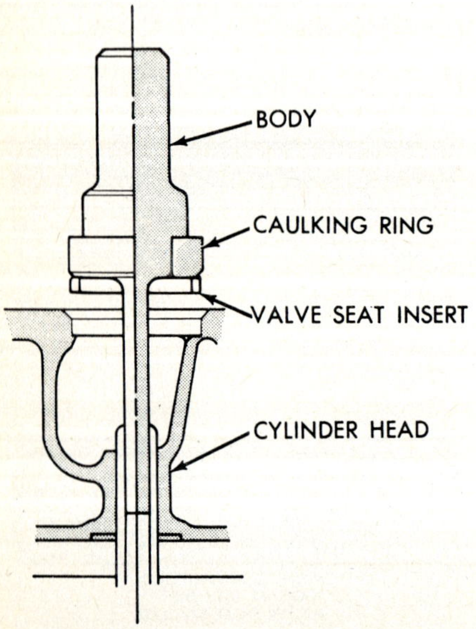

Fig. 7 Installing valve seat insert

TIMING GEAR ASSEMBLY, REPLACE

Removal

1. Remove timing gear case, gasket, front oil seal and idler pulley bracket.
2. Align timing marks on gearset, Fig. 10.
3. Remove camshaft drive gear with a suitable puller.
4. Rotate injection pump drive gear to allow notch in drive gear to pass idler gear teeth.
5. Loosen idler gear mounting bolt, then remove thrust plate and idler gear.
6. Disconnect injector pipes from injection pump.
7. Disconnect and cap fuel line from transfer pump.
8. Disconnect the transfer pump to fuel filter and the fuel filter to injection pump hoses.
9. Loosen the stay bolt in the rear of the injection pump. Remove the pump with the automatic timer gear and flange toward rear of engine.

Installation

1. Ensure engine is placed at top dead center, compression stroke, No. 1 cylinder.
2. Install idler shaft with a suitable drift and hammer, if removed.
3. Install idler gear on shaft so the marks on camshaft drive gear mate with marks on idler gear, Fig. 10. Install thrust plate and attaching bolt.
4. Install camshaft gear and thrust plate on camshaft with the marks on camshaft gear meshed properly with marks on idler gear, Fig. 10.
5. Place injection pump into position. Mesh the pump drive gear with idler gear, ensuring marks are properly aligned, Fig. 10. Also, ensure that the injection pump mounting flange scale is set at the injection timing point.
6. Install mounting bolts to timing gear case and flange mount.
7. Connect fuel feed pipe and filter hoses to pump assembly.
8. Operate primer pump and bleed air from fuel filter and injection pump.
9. Connect injection pipes to injection pump.
10. Check idler gear end play with a feeler gauge placed between gear and thrust plate. End play should be .002–006 inch. If end play exceeds repair limit of .014 inch, replace thrust plate and recheck end play.
11. Check gear camshaft backlash with a dial indicator. Backlash should be .004–.009 inch. If backlash exceeds repair limit of .01 inch, replace gear.
12. Install new front seal.
13. Install timing gear case and gasket.

CAMSHAFT, REPLACE

1. Remove engine from vehicle, and place in a suitable workstand.
2. Remove cylinder head.
3. Remove crankshaft pulley and damper.
4. Remove flywheel housing and flywheel.
5. Remove timing gear case.
6. Rotate crankshaft until No. 1 cylinder is at top dead center, compression stroke.
7. Remove injection pump.

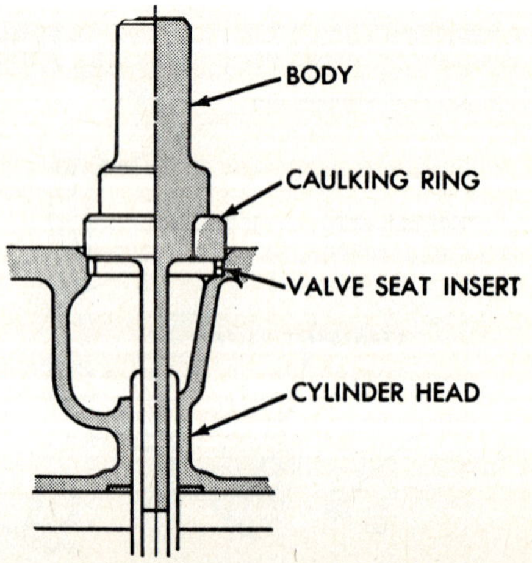

Fig. 8 Caulking valve seat insert

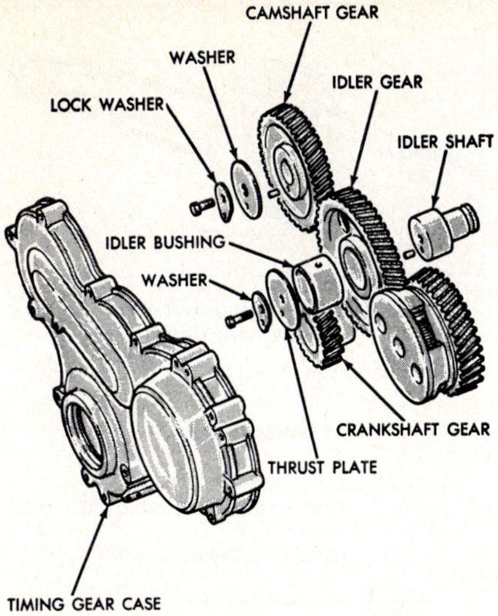

Fig. 9 Timing gear case assembly

8. Remove idler gear mounting bolt, thrust plate and idler gear.
9. Remove crankshaft gear and baffle.
10. Remove camshaft gear, then the oil pan.
11. Remove oil pickup tube and strainer. Loosen joint bolt and remove oil pump.
12. Remove thrust plate, then the camshaft from front of engine. Use care not to damage camshaft bearings.
13. Reverse procedure to install.

PISTON & ROD ASSEMBLE

The aluminum alloy piston has a top recess which forms a combustion chamber, Fig. 11. Assemble piston to rod with combustion chamber (recess) side of piston facing the side of the connecting rod having the weight stamping.

CYLINDER SLEEVES, REPLACE

Removal

1. Mount a portable boring bar to top of cylinder block.
2. Align boring bar with center of cylinder sleeve at bottom where eccentric wear is minimum.
3. Bore sleeve wall out to a thickness of .02 inch.
4. Extract sleeve from cylinder block with suitable equipment. Use caution not to damage cylinder block inner surface.
5. Check block bottom hole condition after sleeve has been removed. Bottom hole must be rebored before installing new sleeve if damaged or other defects are noted.

Installation

If Bottom Hole Is Not Rebored

1. Measure inside diameter of bottom hole and outside diameter of sleeve being installed. Sleeve to bottom hole clearance must be .003–.006 inch after installation.
2. Press cylinder sleeve into cylinder block until sleeve top surface is flush with cylinder block.
3. Bore the sleeve and finish to an inside diameter of 3.620–3.623 inch by honing.

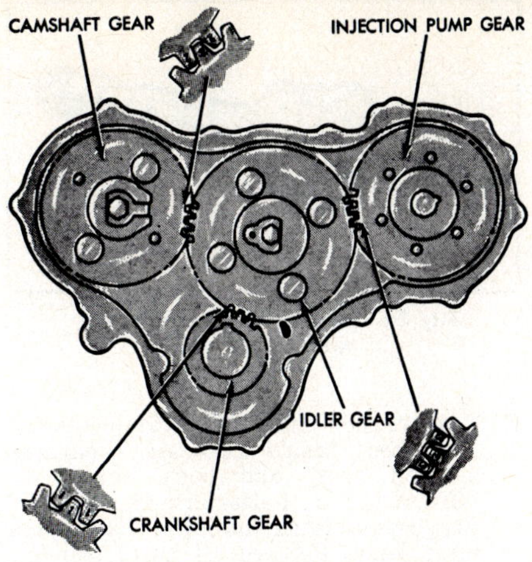

Fig. 10 Timing marks aligned

If Bottom Hole Must Be Rebored

1. The standard bottom hole in the block is 3.738–3.739 inch. Select an oversize sleeve with an outside diameter .02 inch larger.
2. Bore hole in bottom of block with a diameter of 3.738–3.739 inch if a standard sleeve is being used. If an oversize sleeve is being used, the bottom hole should have an inside diameter of 3.758–3.759 inch. This ensures clearance between bottom hole and sleeve outside diameter of .003–.006 inch.
3. Press sleeve into cylinder block until sleeve top surface is flush with cylinder block.
4. Bore the sleeve and finish to an inside diameter of 3.620–3.623 inch by honing.

OIL PAN, REPLACE

1. Disconnect battery ground cable and remove oil dipstick.
2. Raise and support vehicle. Drain oil pan.
3. Remove oil dipstick tube from oil pan.
4. Disconnect transmission oil cooler lines and road draft tube.
5. Remove oil pan attaching bolts.
6. Lower vehicle.
7. Remove nuts from engine mounts.
8. Raise and support vehicle. Support engine with a stand under compressor bracket, if equipped.
9. Remove transmission inspection plate.
10. Remove oil pan from vehicle.
11. Reverse procedure to install.

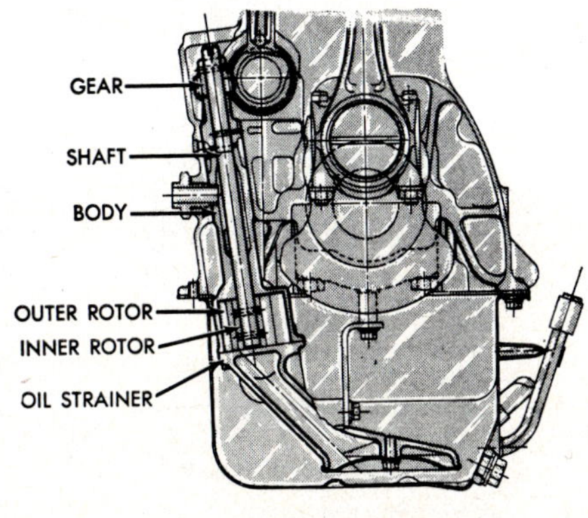

Fig. 12 Oil pump installation

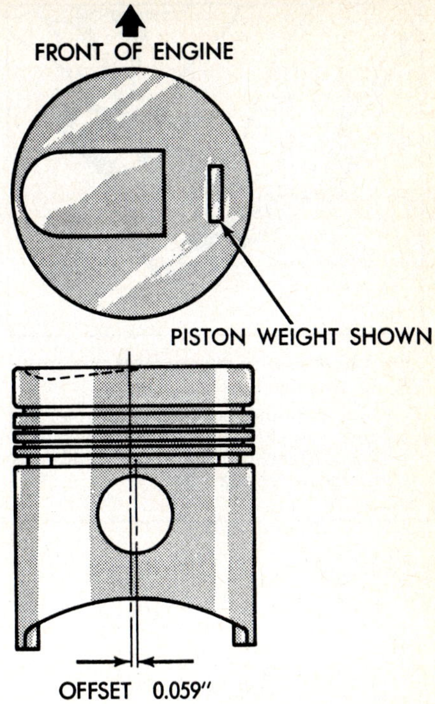

Fig. 11 Piston identification

OIL PUMP, REPLACE

1. Remove oil pan.
2. Remove oil pickup tube, strainer and gasket, Fig. 12.
3. Remove oil pump joint bolt.
4. Remove oil pump.
5. Reverse procedure to install.

OIL PUMP SERVICE

Disassembly

1. Loosen bolts and remove strainer (cover) from oil pump.
2. Remove inner rotor and shaft.
3. Lift out outer rotor from pump.

Inspection

1. Clean all parts of pump. The cover mating face should be smooth. Replace cover if scratched or grooved.
2. Install outer rotor into pump body. Install inner rotor into outer rotor and seat in pump body.
3. Place a straightedge across pump body face. Insert a feeler gauge between straightedge and rotors, Fig. 13. Clearance should be .0014–.004 inch. If clearance exceeds .006 inch, replace inner and outer rotors as an assembly.
4. Check clearance between inner and outer rotors, Fig. 14. Clearance should be .007 inch or less. If clearance exceeds .010 inch, replace inner and outer rotor as an assembly.

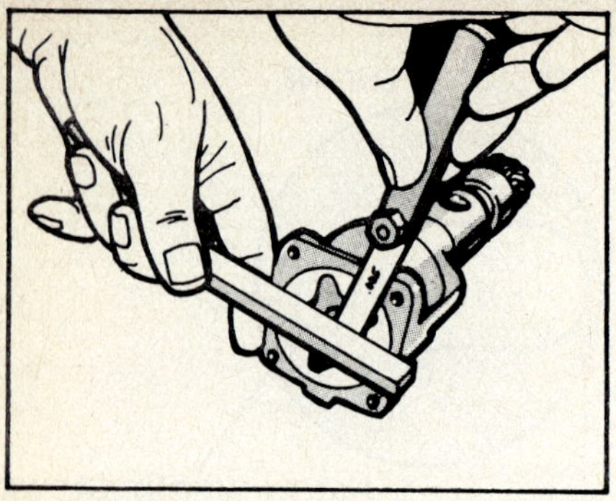

Fig. 13 **Measuring rotor to cover clearance**

5. Check clearance between outer rotor and pump body, Fig. 15. Clearance should be .008–.010 inch. If clearance exceeds .020 inch, replace outer rotor.

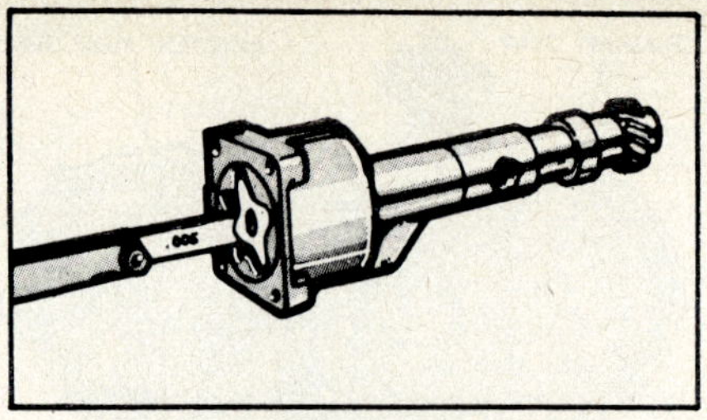

Fig. 14 **Measuring inner to outer rotor clearance**

6. Measure rotor shaft outside diameter and pump body inside diameter. Clearance between shaft and body should be .001–.003 inch. If clearance exceeds .006 inch, replace inner rotor and shaft and/or pump body. Replace all parts showing excessive wear.

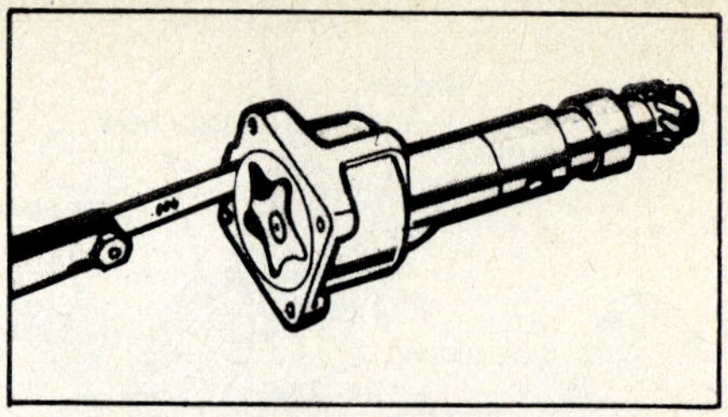

Fig. 15 **Measuring outer rotor to body clearance**

Assembly

1. Install outer rotor in pump body.
2. Install inner rotor and shaft in outer rotor.
3. Install strainer (cover) on pump body.

Fuel System

DESCRIPTION

The fuel system consists of the fuel tank, fuel line, feed pump, fuel filter, injection pump, injector lines, nozzle holders and nozzles, leak-off lines and the fuel return line, Fig. 1.

The feed pump draws fuel from the tank through the fuel line. The fuel then passes through the fuel filter, mounted at the rear of the intake manifold, and enters the injection pump. At the proper intervals, the injection pump forces the fuel under high pressure into the injector lines. This high pressure fuel is sprayed into the combustion chambers by the injector nozzles. Fuel not used during operation is drawn off by the leak-off lines and returned to the fuel tank by the fuel return line.

This fuel system utilizes three fuel filters. One filter is part of the gauge sending unit assembly immersed in the fuel tank at the end of the fuel line. This filter does not normally need service, however, can be replaced if necessary. The second filter (strainer) is a gauze type filter located at the inlet port of the feed pump, Fig. 2. It operates under suction and removes large size particles of dirt and foreign matter. This filter should be removed and cleaned every 12,000 miles. Clean the wire gauze in a suitable solvent. After cleaning, reinstall filter and fuel line connector and bleed air from fuel system. The third filter is a paper element type, Fig. 3, mounted at the rear of the intake manifold, located between the feed pump and injection pump in the system.

The injection pump is a fixed stroke, plunger type pump. Fuel is injected at high pressure into the injector lines by movement of the plungers in the plunger barrels. Delivery valves at the top of the plunger barrels prevent fuel injection until sufficient injector pressure has built up in the injection pump and "After-drip" from the nozzle tips after injection has ended.

The amount of fuel delivered to the engine is controlled by an RU type maximum-minimum mechanical governor, Fig. 4, mounted at the rear of the injection pump housing. The governor regulates the amount of fuel injected by controlling the movement of the injection pump control rack and sleeve proportional to engine speed. The governor operates only at engine idle and at high speed operation. It is inoperative at intermediate engine speeds.

A gear type automatic timer, Fig. 5, mounted to the front of the injection pump camshaft, advances injection timing as engine speed increases to ensure injection of fuel at the proper intervals.

The nozzle tip is installed in the nozzle holder, Fig. 6, and is secured by a retaining nut. Fuel from the injection pump enters the upper part of the nozzle holder and flows into the fuel reservoir in the nozzle through a fuel passageway. The nozzle tip atomizes the high pressure fuel and sprays the fuel into the precombustion chamber.

When no high pressure fuel is present at the nozzle holder input, the needle valve holds the nozzle tip orifice closed. Fuel delivered at the specified injection pressure overcomes spring pressure and the needle valve opens. Fuel is then injected into the precombustion chamber. When injection pressure drops, the spring closes the needle valve. Fuel remaining in the nozzle reservoir is bled off to the fuel return line through the leak off line.

OPERATION

Fuel system operation during starting, stopping and at high speed is controlled by the Electrical Diesel Injection Control (EDIC) system. The EDIC system consists of the key switch, fuel control relay operated by the key switch signal, fuel control motor, capsule rod and injection pump.

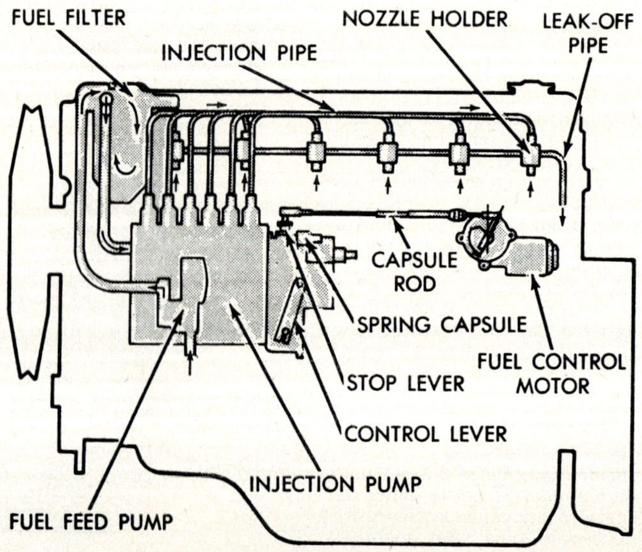

Fig. 1 **Fuel system**

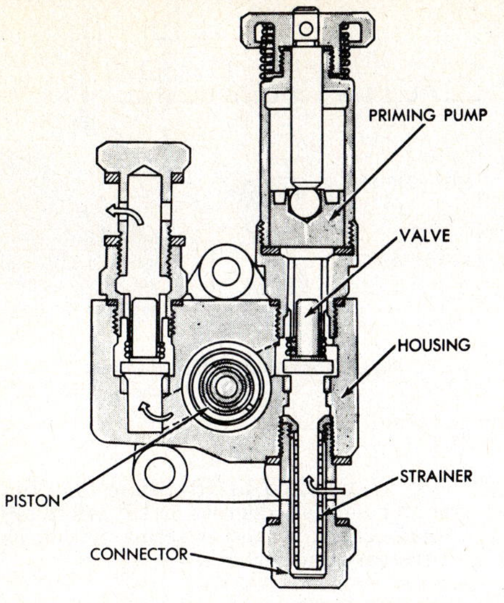

Fig. 2 Feed pump inlet fuel filter (strainer)

The key switch position determines fuel control relay input. The fuel control relay passes current to the fuel control motor in accordance with key switch position. This current actuates the control motor which operates the injection pump stop lever through the capsule rod.

When the key switch is turned from "Off" to "Start", the starter motor begins to crank the engine. The fuel control relay actuates the control motor which pulls the injection pump stop lever to the "Start" position, compressing the spring in the spring capsule located at the top of the governor. With the key switch at "Start", the control motor continues to revolve, transmitting this movement to the stop lever.

When the stop lever is moved to the "Start" position by compressing the spring capsule, it restricts movement of the capsule rod, drive lever and stopper. The motor continues to rotate through the control assembly clutch which prevents damage to the rotating parts.

When the engine starts, the key switch is returned to the "On" position. The "On" signal at the fuel control relay reverses the rotation of the control motor, returning the stop lever to the "Drive" position.

When the key switch is turned from "On" to "Off", the fuel control relay senses the "Off" signal and actuates the control motor to push the injection pump lever to the "Stop" position. When the control motor limiter plate reaches the "Stop" position, current to the motor is cut off and the motor stops.

The capsule rod with the enclosed spring permits the drive lever to push the rod after the stop lever reaches the "Stop" position. The motor "Stop" position may overrun the stop lever "Stop" position to ensure a positive stop.

EDIC OPERATIONAL CHECKS

Start Position Check

Turn key switch to the "Start" position. The control motor should move the injection pump stop lever to the "Start" position, Fig. 7.

Drive Position Check

After the engine starts, observe position of drive lever. The end of the drive lever at the motor should be positioned between the two machined marks, Fig. 8.

Stop Action Check

Turn the key switch to "Off" position. The injection pump lever should move to the "Stop" position and the engine should stop, Fig. 9.

FUEL CONTROL RELAY TEST

1. Disconnect electrical connector from fuel control relay.
2. Measure resistance between the following terminals on the relay connector socket, Fig. 10: 1 and 8, 4 and 6, 5 and 6, 6 and 7, 2 and 6, 3 and 6. All readings should be zero ohms.
3. If any readings indicate more than zero ohms in step 2, replace fuel control relay.
4. If readings are satisfactory in step 2, apply battery voltage (12 volts) between terminals 4 and 6, Fig. 10. Measure resistance between terminals 1 and 8. Reading should be infinite. If not, replace fuel control relay. If resistance is infinite, apply the battery voltage to terminals 5 and 6, Fig. 10.
5. Measure resistance between terminals 1 and 3, Fig. 10. Reading should be zero ohms. If not, replace fuel control relay. If resistance is zero ohms, connect the battery voltage to terminals 6 and 7, Fig. 10.
6. Measure resistance between terminals 3 and 6, Fig. 10. Reading should be zero ohms. If not, replace fuel control relay. If all readings are satisfactory, the fuel control relay is satisfactory.

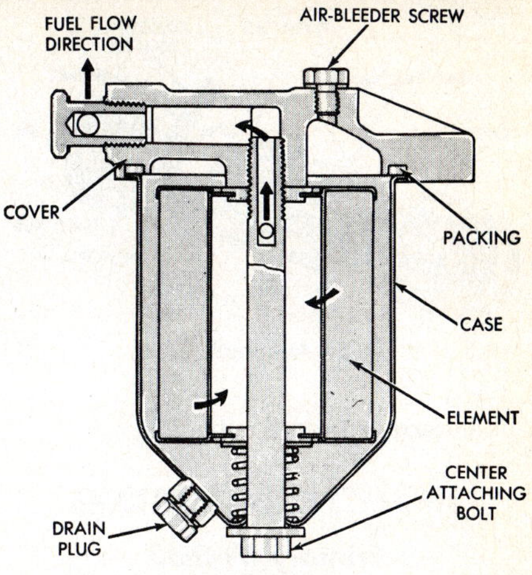

Fig. 3 Fuel filter

BLEEDING AIR FROM SYSTEM

1. Loosen fuel filter petcock or valve, Fig. 3, and operate the priming pump on the feed pump. If the filter is filled with fuel, fuel containing air bubbles will be discharged from the petcock or valve. Continue pumping until the discharged fuel contains no air bubbles. Tighten the petcock or valve securely.
2. Loosen the air bleeder screw at the top of the injection pump and operate the priming pump. Continue pumping until all air is bled from the fuel in the pump reservoir. Tighten bleeder screw securely.

NOTE: Approximately 1/4 pint of fuel should be drained from the fuel filter daily to remove dirt and water accumulation. Remove drain plug or open petcock, if equipped, at bottom of filter and drain fuel into container.

FUEL CONTROL MOTOR, REPLACE

1. Disconnect motor connector.
2. Disconnect capsule rod at control motor and injection pump stop lever and remove rod.
3. Remove the three motor mounting bolts.
4. Remove motor from side of block.

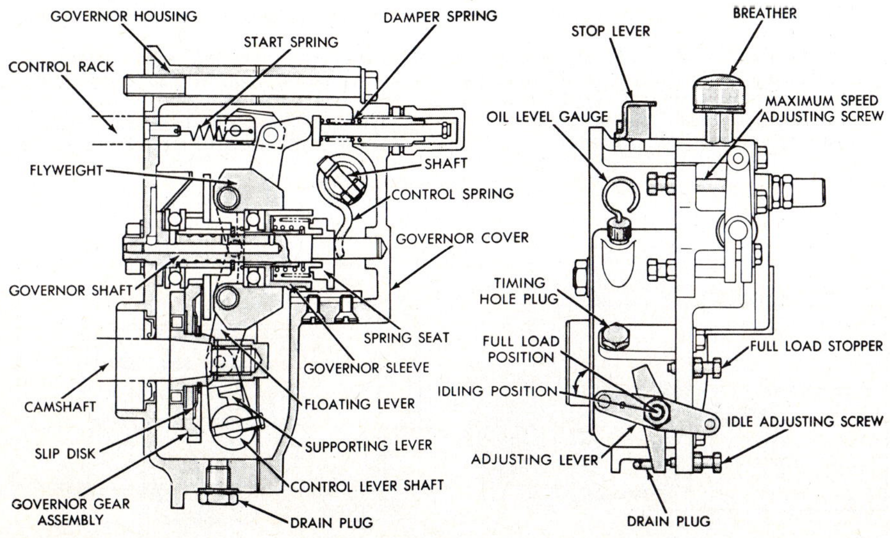

Fig. 4 Mechanical governor

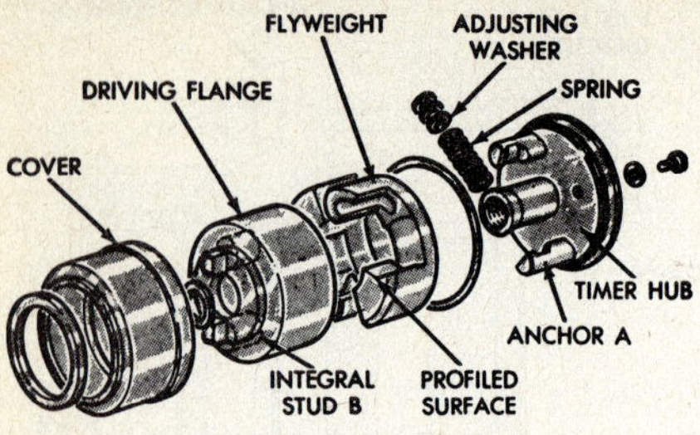

Fig. 5 Automatic timer

5. Reverse procedure to install.

INJECTION TIMING

Primary Method

1. Disconnect battery ground cable.
2. Disconnect fuel shutoff rod at injection pump stop lever.
3. Clean dirt and foreign material from No. 1 delivery valve, injection pipe and adjacent pump area.
4. Rotate crankshaft to place No. 1 cylinder at top dead center, compression stroke. Crankshaft position is identified by alignment of the degree marks on the crankshaft damper rear face with pointer on bottom of timing gear case.
5. Rotate crankshaft 1¾ revolutions.
6. Disconnect No. 1 injection pipe from delivery valve.
7. Remove lock plates from No. 1 delivery valve holder and unscrew delivery valve holder from pump carefully to prevent loss of delivery valve spring and any shims, Fig. 11.
8. Remove delivery valve springs and any shims. Reinstall delivery valve holder snugly.
9. Connect a clean plastic tube into open end of delivery valve holder and place the other end in a container.
10. Rotate crankshaft to approximately 30 degrees before top dead center. Unscrew handle of priming pump on fuel feed pump to unlock. Operate the pump until fuel flow is observed through the plastic tube.
11. Rotate crankshaft one degree and operate priming pump and observe fuel flow. Continue to rotate crankshaft in one degree increments and operate the priming pump until fuel no longer flows through plastic tube. This is the existing injection timing point.
12. If the existing timing point is not 18° before top dead center, loosen the four pump to flange plate nuts and rotate pump to adjust pump to specified timing point, Fig. 12. Turn the pump toward the cylinder block to advance timing or away from cylinder block to retard timing. The crankshaft angle varies by six degrees per division on pump flange plate scale, Fig. 12.
13. Tighten pump to flange plate nuts, Fig. 12.
14. Repeat step 5 and steps 10 through 12.
15. Remove plastic tube. Ensure that delivery valve spring and shims are clean and reinstall in holder. Tighten delivery valve holder and reinstall lock plates, Fig. 11.
16. Connect No. 1 injection pipe to delivery valve holder.
17. Install oil fill cap and lock the priming pump handle.
18. Connect fuel shutoff rod and the battery ground cable.
19. Start engine and check for leaks and proper operation.

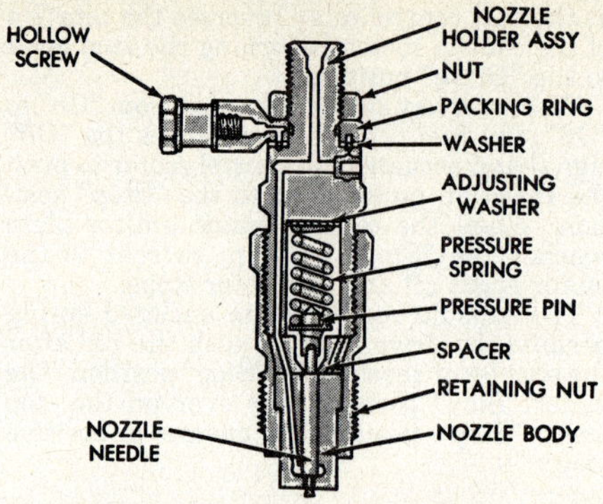

Fig. 6 Injection nozzle & holder

Alternate Method

1. Perform steps 1 through 6 as outlined under "Primary Method".
2. After removing the No. 1 injection pipe from the delivery valve, rotate the crankshaft in small increments. Stop when fuel begins to flow from the delivery valve holder. Fuel injection begins at this point.
3. Injection timing should be 18° before top dead center minus 2 degrees. When using the alternate timing method, 2 degrees of crankshaft rotation will be taken up in delivery valve delay. Therefore, the timing point should be 18° before top dead center minus 2 degrees or 16° before top dead center. Adjust injection timing as necessary and recheck.

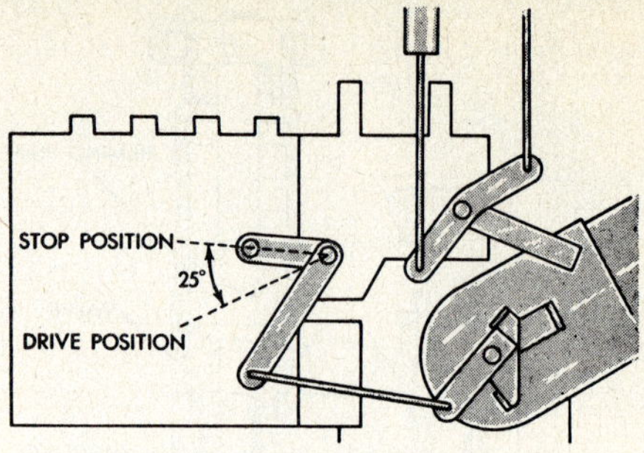

Fig. 7 Start operational check

INJECTION PUMP, REPLACE

Removal

1. Disconnect battery ground cable.
2. Disconnect fuel shutoff rod at stop lever.
3. Remove power steering pump and bracket assembly from engine and position aside.
4. Clean dirt and foreign material from fuel line, hose fittings and injection pipes at injection pump.
5. Drain oil pan and remove dipstick and tube.
6. Disconnect throttle cable and linkage from injection pump control lever.
7. Remove throttle control bracket assembly from cylinder block and the control motor bracket and position aside.
8. Disconnect fuel supply line to feed pump, loosening anchor clamps as necessary.
9. Disconnect filter hoses from feed pump and injection pump.
10. Rotate crankshaft until No. 1 piston is positioned between 7° before top dead center and top dead center, compression stroke. Check timing pointer. It should be located approximately midpoint between the TDC and 14 degree line on the damper.
11. Disconnect injection pipes from delivery valves and position away from cylinder block.
12. Disconnect injection pump lube line at block fitting near starter motor forward end.
13. The injection pump assembly is secured to the engine by five screws and one bolt.

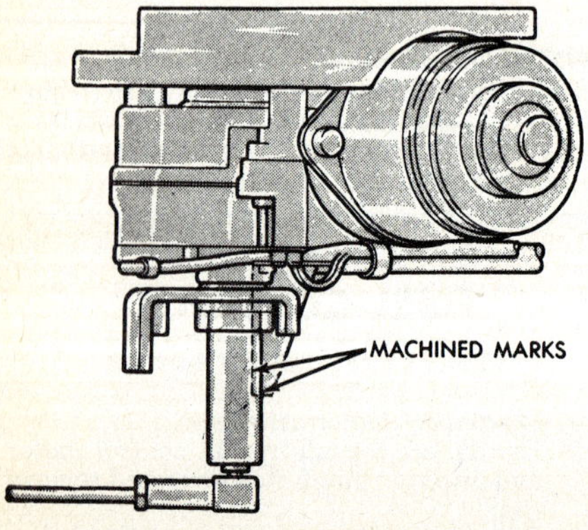

Fig. 8 Drive lever position check

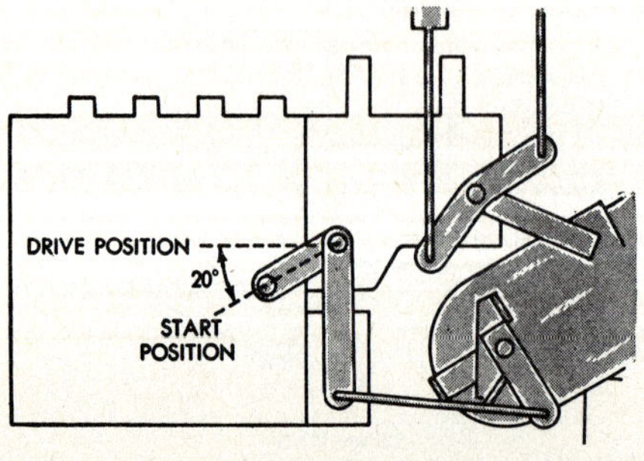

Fig. 9 Stop action check

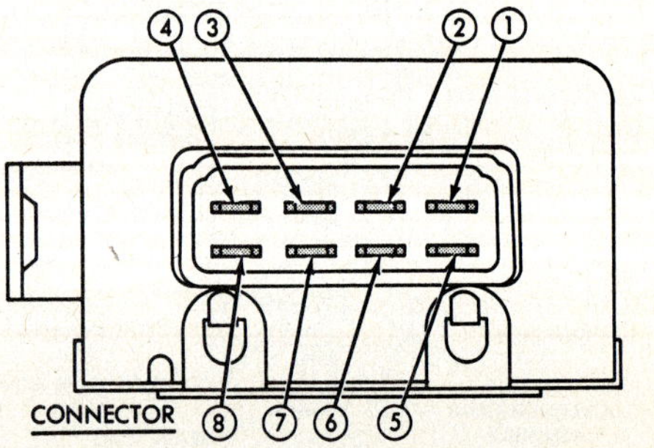

Fig. 10 Fuel control relay connector

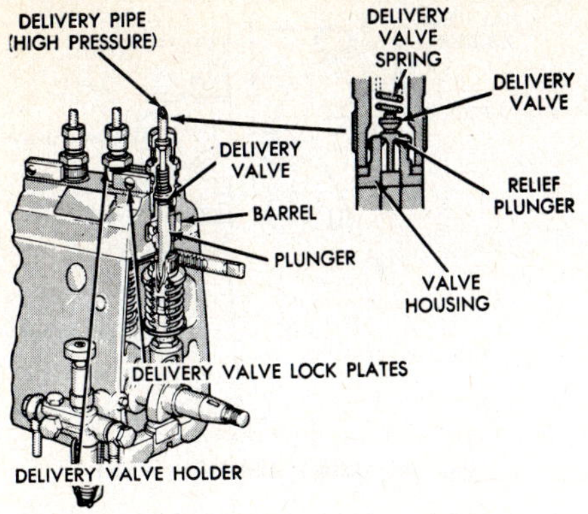

Fig. 11 Delivery valve assembly

The front screws extend through the timing case and engine front plate into the pump flange plate. The rear bolt secures the flange plate to the engine front plate, Fig. 12. Remove the screws and bolt.

14. Pull injection pump rearward to disengage from engine front plate and timing gear case. Twist pump toward cylinder block and pull rearward until automatic timer is free of case.

Installation

1. Loosen four nuts attaching injection pump to mounting flange plate. Align center timing mark on pump flange with pointer on plate.
2. Ensure that the "O" ring seal is in place on forward face of pump mounting flange.
3. Remove threaded timing port plug on governor housing behind control lever to expose pump camshaft bushing timing mark, Fig. 13. Turn pump drive gear to align timing mark on camshaft bushing with pointer on governor. The guide plate notch on drive gear will be approximately at the 8 o'clock position as viewed from the front.
4. Ensure that the crankshaft is still positioned as in step 10 in "Removal" procedure.
5. Insert automatic timer into timing gear case. Turn injection pump inward until against block. Then turn pump drive gear clockwise or counter-clockwise to mesh drive and idler gears. Push pump forward into timing gear case and rotate away from block to align mounting holes.

NOTE: Correct gear mesh is assured by drive gear guide plate. If pump cannot be pushed forward manually until flange plate seal diameter contacts engine front plate, the gear mesh is incorrect. Do not force pump into position. Retract the pump and turn drive gear as necessary to achieve correct gear mesh. Refer to Fig. 14 for correct gear mesh.

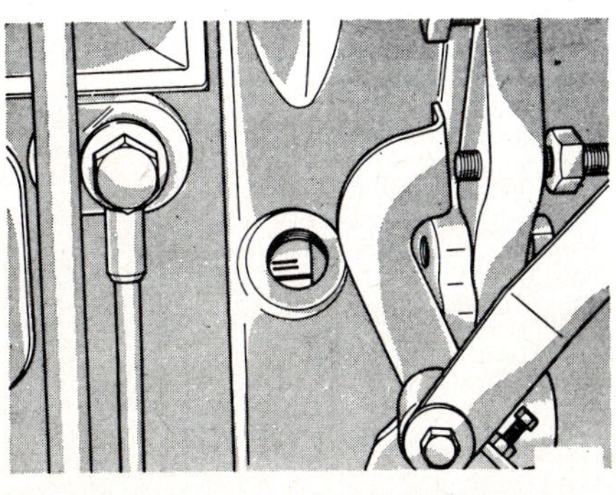

Fig. 13 Camshaft bushing timing mark

6. Attach pump to timing gear case. Turn crankshaft in opposite direction of normal rotation until it reaches the specified timing point. The governor pointer and pump camshaft bushing timing marks should be aligned. If not, the pump is installed improperly and must be removed and reinstalled.
7. When timing marks are aligned, install governor housing timing port plug.
8. Reverse removal procedure but do not connect the No. 1 injection pipe or the battery ground cable.
9. Fill oil pan with specified oil.
10. Bleed air from fuel filter and injection pump as outlined previously.
11. Connect No. 1 injection pipe and the battery ground cable.
12. Check injection timing point and adjust as necessary as previously outlined.
13. Start engine and check for leaks and proper operation.

INJECTION NOZZLE SERVICE

Removal

1. Disconnect injector line from nozzle holder at cylinder.
2. Remove nozzle holder and nozzle assembly.

Disassembly

1. Place nozzle holder in a soft jawed vise. Tighten the vise against the retaining nut.
2. Remove nozzle holder body from retaining nut, Fig. 6.
3. Lift pressure pin, spring, washer, spacer and nozzle tip from nozzle holder, Fig. 6.

Assembly

1. Secure nozzle holder retaining nut in a vise.
2. Immerse nozzle tip and needle valve assembly in clean fuel oil.
3. Assemble spacer, pressure pin and pressure spring onto nozzle tip, Fig. 6.
4. Install adjusting shim into nozzle holder body.
5. Assemble nozzle holder and tip to retaining nut and torque nozzle holder to 43.4–57.9 ft. lbs.

Installation

1. Install nozzle holder and nozzle assembly.
2. Connect injector line to nozzle holder.

CAPSULE ROD, ADJUST

1. Check capsule rod installation at motor drive lever. With injection pump lever in "Drive" position, the drive lever at the motor should be between the two machined marks.

Fig. 12 Injection pump timing marks & attachment

2. Turn drive lever with a screwdriver until it points to the 12 o'clock position.
3. Disconnect the capsule rod. Ensure that the injection pump stop lever is in the "Drive" position.
4. Loosen capsule rod lock nut and adjust rod length until the drive lever is positioned between the two machined marks.
5. Tighten the lock nut and connect capsule rod to drive lever. The nominal length of the capsule rod is 12.7 inch, Fig. 15.

IDLE SPEED, ADJUST

1. Remove cover and gasket from tachometer take-off on right side of engine in front of the oil filter assembly. Install mechanical tachometer adapter and attach mechanical tachometer and drive cable.
2. Turn hand throttle counter clockwise and pull fully outward. Depress accelerator pedal to floor and hold until engine starts. Permit engine to warm up and release accelerator slowly until engine runs smoothly. When engine begins to warm up, turn hand throttle clockwise to reduce engine speed to idle speed.

CAUTION: If a new accelerator pump has been installed, do not allow engine speed to rise above 1300 RPM. If engine over-

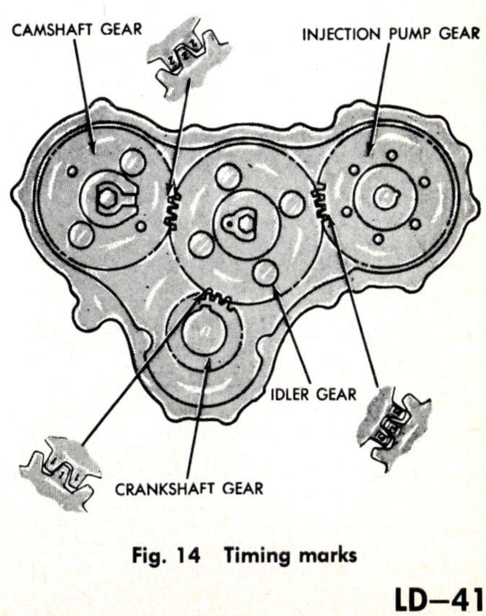

Fig. 14 Timing marks

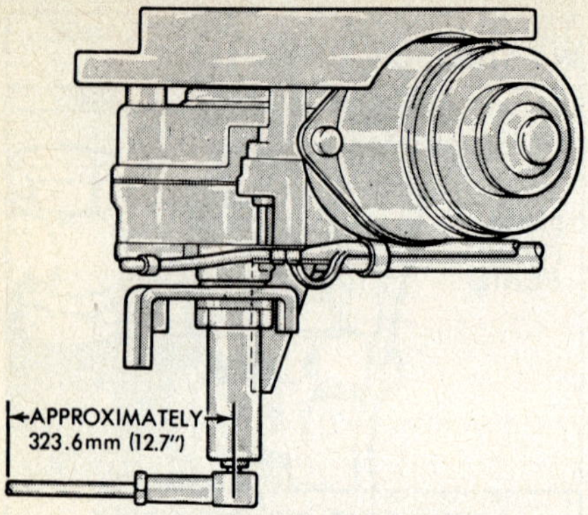

Fig. 15 Adjusting capsule rod

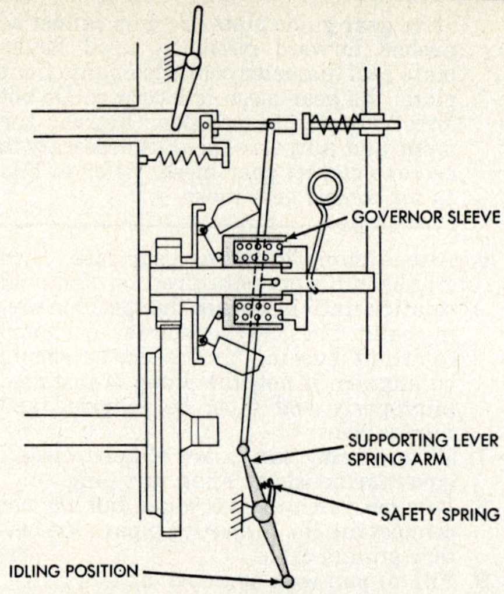

Fig. 16 Idling control

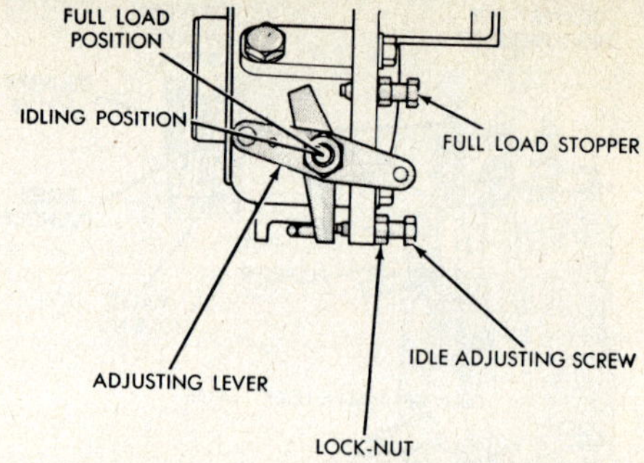

Fig. 17 Idle speed adjustment

speeds, it may run away and damage or destroy itself.

3. Ensure that governor control is at idling position before adjusting idle speed, Fig. 16.
4. Observe tachometer. Idle speed should be 550–600 RPM.
5. If idle speed is not within specification, loosen idle adjusting screw locknut. Adjust screws as necessary to obtain specified idle setting. Turn screw inward to raise idle speed or outward to lower idle speed, Fig. 17.
6. With idle speed at specified RPM, tighten idle adjusting screw lock nut. Recheck idle speed after tightening lock nut.
7. Remove tachometer, drive cable and adapter.
8. Install tachometer take-off gasket and cover.

NISSAN (INTERNATIONAL) DIESEL ENGINE

INDEX OF SERVICE OPERATIONS

GENERAL ENGINE SPECIFICATIONS

Engine Model	Bore & Stroke	Horsepower @ R.P.M.	Torque Ft. Lbs. @ R.P.M.	Firing Order	Compression Pressure P.S.I. @ R.P.M.	Idle Speed	Governed Speed R.P.M.	Normal Oil Pressure P.S.I. @ R.P.M.
SD-33								
6-198	3.27 × 3.94	88 @ 3800	141 @ 1800	1-4-2-6-3-5	426 @ 200	600	—	14.2 @ 600

VALVE SPECIFICATIONS

Engine Model	Valve Lash, Cold		Valve Face Angle Degrees	Valve Stem Diameter, Std.		Valve Stem to Guide Clearance		Valve Springs	
	Intake	Exhaust		Intake	Exhaust	Intake	Exhaust	Free Length Inch	Pressure Lbs. @ Inch Length
SD-33									
6-198	.0138	.0138	45	.314	.314	.0006–.0018	.0016–.0028	1.977	66 @ 1.535

PISTONS, PINS, RINGS, CRANKSHAFT & BEARINGS

Engine Model	Piston Clearance	Piston Pin Diameter	Ring End Gap Minimum		Rod Bearings		Main Bearings			Crankshaft End Play
			Compression	Oil	Shaft Diameter	Bearing Clearance	Shaft Diameter	Bearing Clearance	Thrust Bearing No.	
SD-33										
6-198	.0047–.0075	1.024	.0118	.0118	2.087	.0012–.0036	2.795	.0016–.0036	3	.0024–.0055

ENGINE TIGHTENING SPECIFICATIONS

★ Torque Specifications are for clean and lightly lubricated threads only.
Dry or dirty threads produce increased friction which prevents accurate measurement of tightness.

Engine Model	Cyl. Head Ft. Lbs.	Intake Manifold Ft. Lbs.	Exhaust Manifold Ft. Lbs.	Connecting Rod Ft. Lbs.	Main Bearing Ft. Lbs.	Flywheel to Crankshaft Ft. Lbs.	Damper to Crankshaft Ft. Lbs.	Camshaft Gear Ft. Lbs.	Oil Pan Ft. Lbs.
SD-33									
6-198	①	10.9–13	10.9–13	37.65–41.24	108.5–115.7	32.54–36.2	217–238.7	32.54–36.2	2.89–4.34

①—Final torque—main bolts, 94 ft. lbs.; sub bolts, 36.2 ft. lbs.

Engine Disassembly Section

DISASSEMBLY, FIGS. 1 & 2

1. Remove air cleaner.
2. Remove intake and exhaust manifolds from cylinder head.
3. Remove starter motor.
4. Disconnect rubber hose and oil tube of vacuum pump from cylinder
5. Remove generator adjusting bolt and the drive belts.
6. Remove generator mount bolts and the generator. Remove bracket from cylinder block.
7. On right side of engine:
   a. Remove bolts at oil cooler end of the oil inlet and outlet lines.
   b. Remove oil filter mount bolts and the oil filter from cylinder block with lines attached.
   c. Remove water hose between cylinder head and oil cooler.
   d. Remove oil cooler mounting bolts and the oil cooler.
   e. Remove fuel filter inlet, outlet and overflow connector bolts and the feed pump connector bolt. Disconnect fuel hoses from filter. Do not lose the connector bolt nylon gaskets.
   f. Remove fuel filter to bracket bolts and the fuel filter.
   g. Loosen the nozzle and injection pump cap nuts and disconnect the nozzle line assembly.

NOTE: After removing the nozzle line assembly, cover the delivery holders of the injection pump to prevent entry of dirt and foreign material.

8. On front of engine:
   a. Remove the fan, fan spacer and fan pulley from water pump.
   b. Remove thermostat bypass hose.
   c. Remove water pump attaching bolts and the water pump and gasket.
   d. Remove flywheel housing check cover and pointer. Lock the flywheel in place by inserting the flywheel locking tool between teeth of flywheel ring gear.
   e. Remove crankshaft pulley nut. Using a brass drift and hammer, tap evenly around the circumference of the crankshaft pulley. This will cause the cone to protrude beyond the pulley. Remove the cone, Fig. 3.
   f. Drive the crankshaft pulley from crankshaft with a suitable mallet.
   g. Remove timing cover bolts and the timing cover.
   h. Remove oil seal from timing gear case, Fig. 4. Use care as not to damage seal.
   i. Remove timing gear case mounting bolts, loosen the timing gear case from cylinder block with a mallet, then remove timing gear case.

NOTE: Do not stike the front end plate.

   j. Remove oil slinger from end of crankshaft, Fig. 5.
9. On rear of engine:
   a. Measure flywheel runout by mounting a dial indicator so the plunger contacts face of flywheel clutch side. Rotate crankshaft to obtain reading. Note reading for assembly.
   b. Flatten lock plate, Fig. 6.
   c. Remove flywheel attaching bolts.
   d. Remove flywheel using a suitable puller.

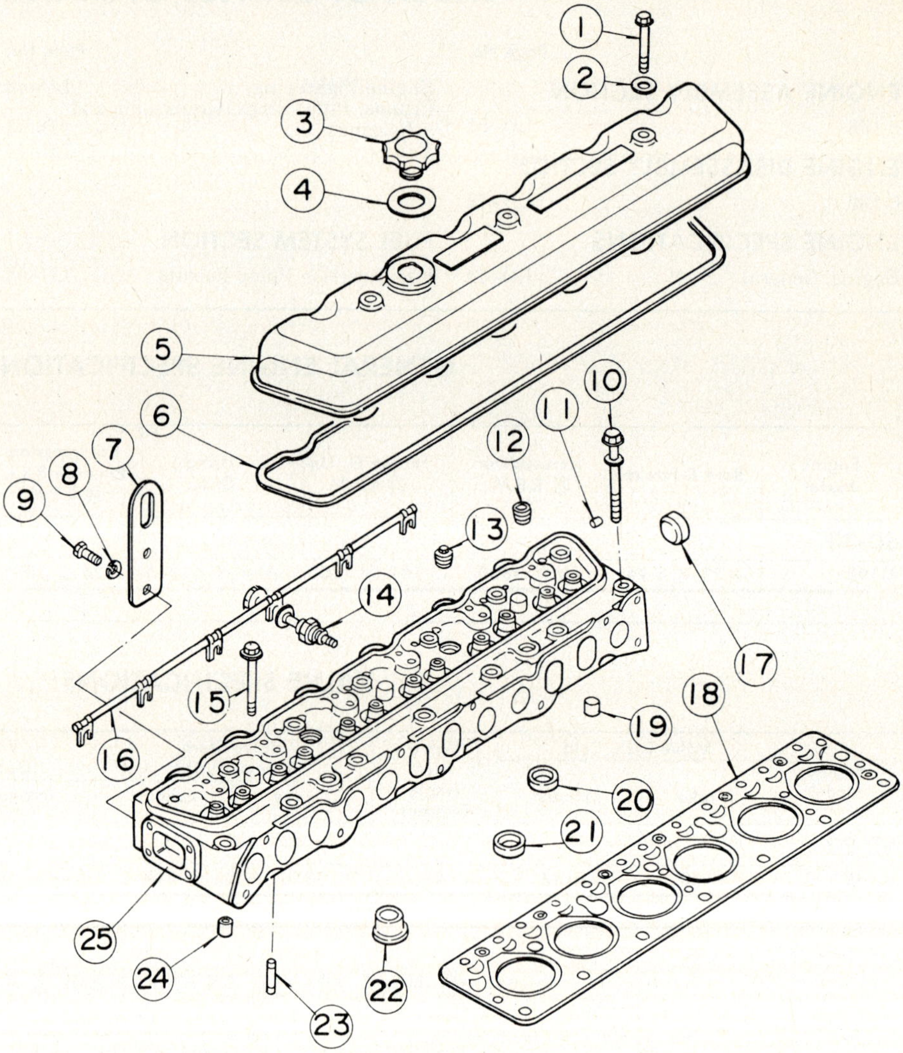

1	Rocker cover bolt	14	Glow plug
2	Gasket	15	Cylinder head bolt (sub)
3	Oil filler cap	16	Glow plug cable
4	Gasket	17	Plug
5	Rocker cover	18	Cylinder head gasket
6	Gasket	19	Plug
7	Slinger	20	Exhaust valve seat
8	Lock washer	21	Inlet valve seat
9	Bolt	22	Combustion chamber
10	Cylinder head bolt (Main)	23	Push rod pipe
11	Plug	24	Plug
12	Plug	25	Cylinder head
13	Plug		

Fig. 1 Cylinder head, disassembled

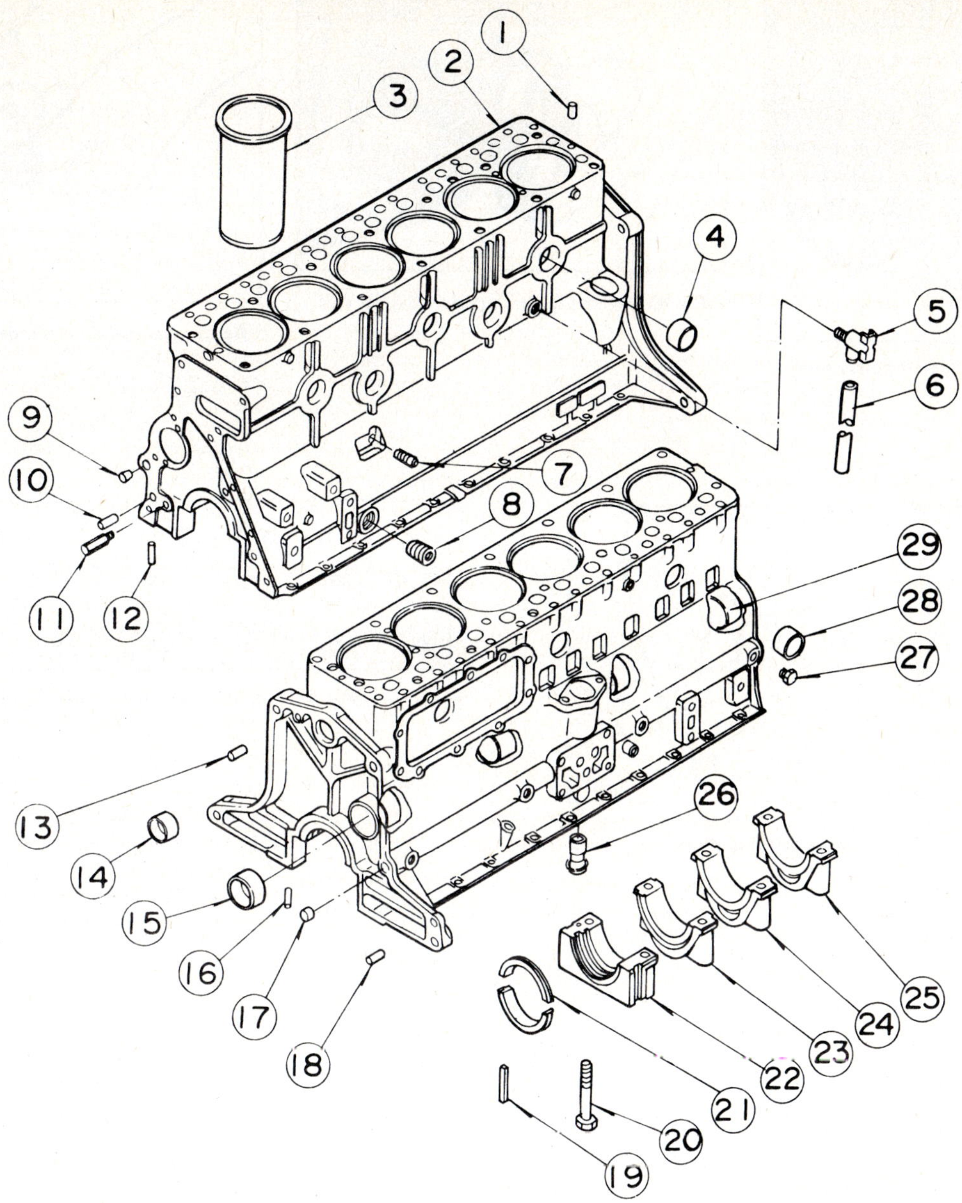

1	Dowel pin	11	Oil jet	21	Oil seal
2	Cylinder block	12	Plug	22	Main bearing cap (No. 4)
3	Cylinder liner	13	Straight pin	23	Main bearing cap (No. 3)
4	Plug	14	Plug	24	Main bearing cap (No. 2)
5	Water drain cock	15	Plug	25	Main bearing cap (No. 1)
6	Vinyl hose	16	Straight pin	26	Bushing
7	Plug	17	Plug	27	Plug
8	Plug	18	Straight pin	28	Plug
9	Plug	19	Oil seal	29	Camshaft bushing
10	Dowel pin	20	Main bearing cap bolt		

Fig. 2 Cylinder block disassembled

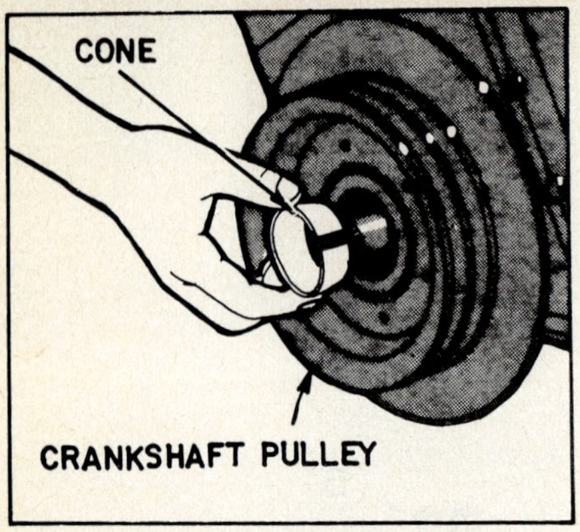

Fig. 3 Removing crankshaft pulley cone

Fig. 4 Removing timing gear case oil seal

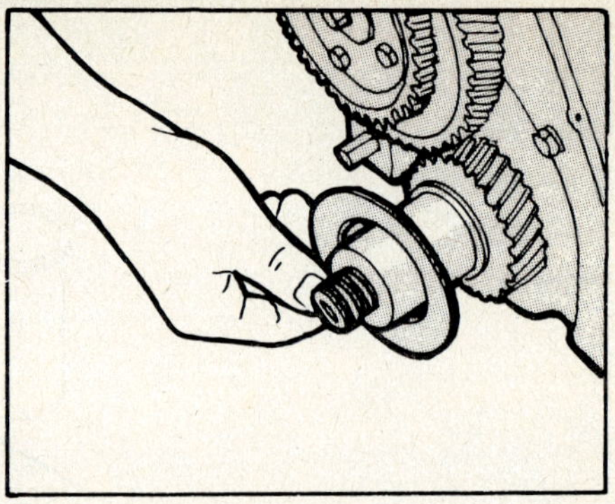

Fig. 5 Removing oil slinger from crankshaft

Fig. 6 Flattening flywheel bolt lock plate

10. Remove valve cover, Fig. 1.
11. Remove rocker arm shaft mounting bolts and the rocker arm shaft assembly.
12. Remove push rods.
13. Remove fuel return spill tube mounting bolts.
14. Remove nozzle holder assemblies from cylinder head.

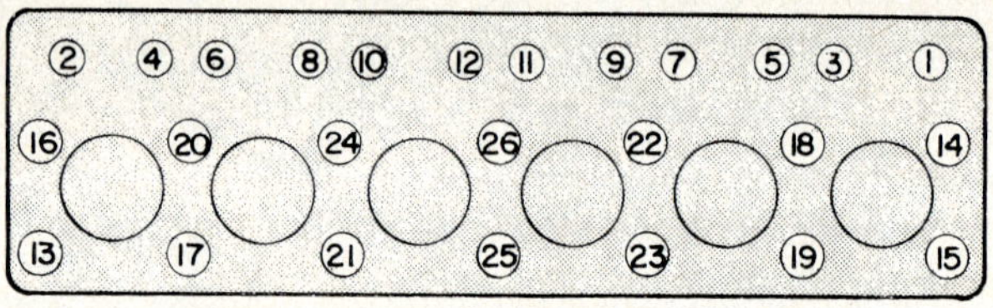

Fig. 7 Cylinder head bolt removal sequence

15. Remove the main and sub cylinder head bolts in sequence, Fig. 7.
16. Remove cylinder head from block.

NOTE: The combustion chambers may fall from cylinder head. If so, install the combustion chambers to the original positions.

17. Remove the cylinder head gasket and the rubber rings in the water and oil holes.
18. Remove flywheel housing mounting bolts and the flywheel housing. It may be necessary to loosen housing with a mallet.
19. Remove oil pan attaching bolts and the oil pan.
20. Measure gear train, Fig. 8, backlash as follows:
    a. Insert a length of solder between the respective gears, Fig. 9.
    b. Rotate crankshaft to crush the solder between gear teeth. Repeat procedure at 120 degree increments.
    c. Using a micrometer, measure thickness of the crushed solder. The gear

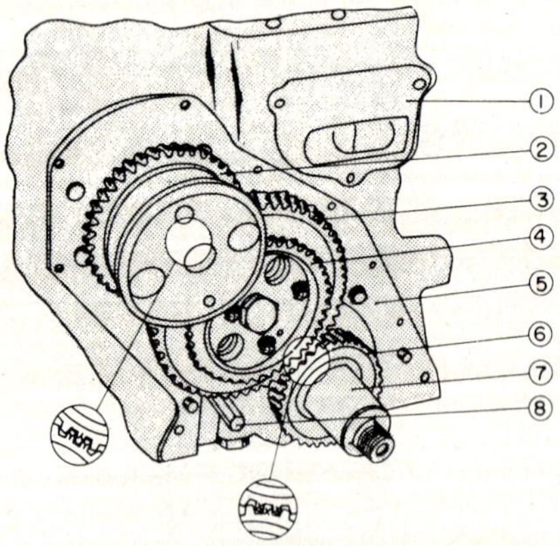

1	Cylinder block	5	Front end plate
2	Injection pump drive gear	6	Crankshaft gear
3	Camshaft gear	7	Crankshaft
4	Idle gear	8	Oil jet

Fig. 8 Gear train

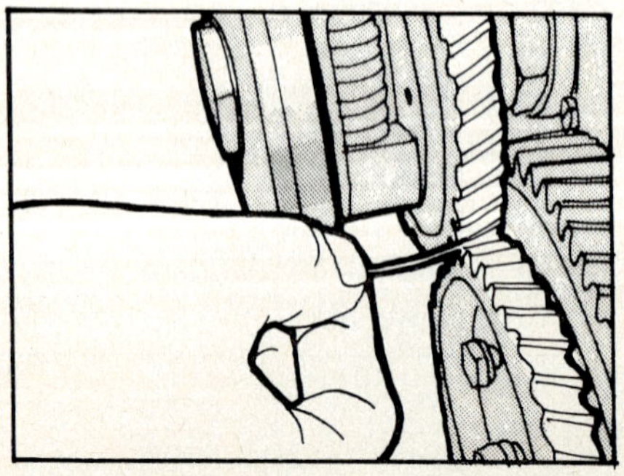

Fig. 9 Measuring gear train backlash

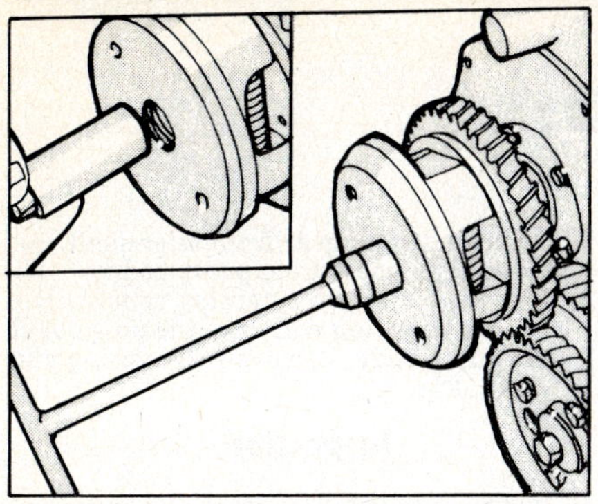

Fig. 10 Removing timer round nut

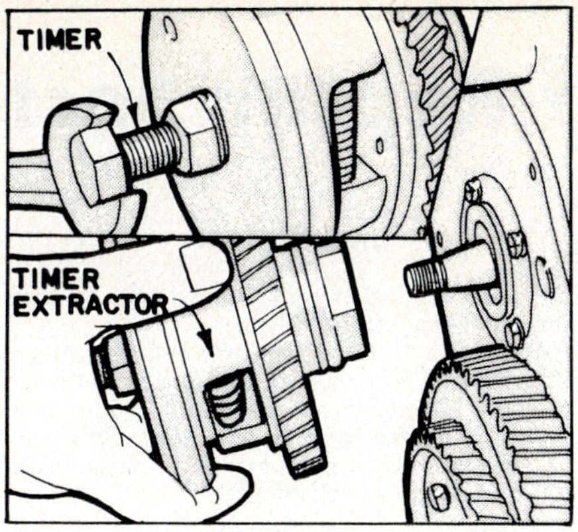

Fig. 11 Removing timing assembly

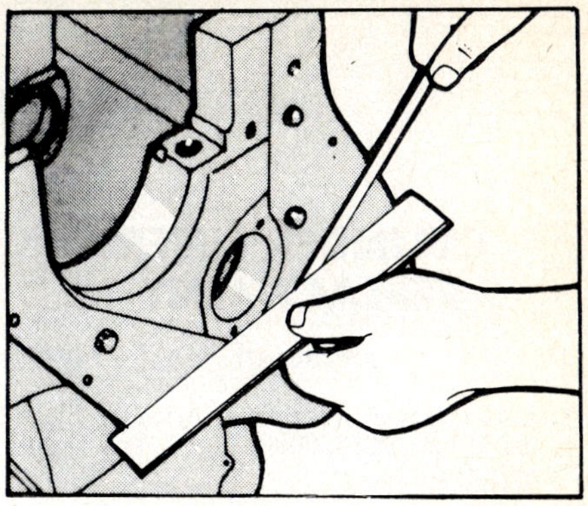

Fig. 12 Measuring front end plate warp

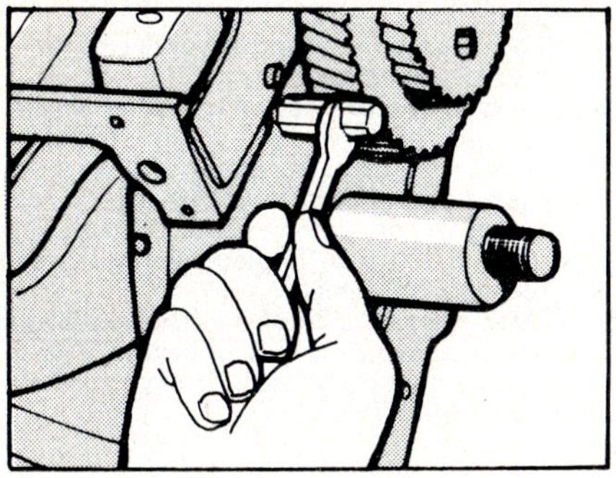

Fig. 13 Removing oil jet

backlash is the smallest of the three measurements.

d. Note the measurements for assembly.

21. Remove tachometer drive support nuts.
22. Remove the timer round nut, Fig. 10, with the flywheel locked. Remove timer assembly with tool 57926-581 threaded into timing weight holder, Fig. 11.
23. Remove injection pump attaching nuts and the injection pump.
24. Remove oil pump drive spindle bolts and the spindle support and drive spindle.
25. Remove camshaft locating plate bolts and the camshaft from engine.
26. Remove crankshaft gear from end of crankshaft with a suitable tool.
27. Measure front end plate warp as shown in Fig. 12. Note the results for assembly.
28. Remove front end plate mounting bolts, loosen the end plate from cylinder block by tapping with a suitable mallet and remove the end plate and gasket. Remove oil jet, Fig. 13.
29. Remove connecting rod nuts and cap bolts.
30. Remove connecting rod bearing caps and bearing. Identify each bearing cap with the proper cylinder number. To remove the upper bearing, rotate crankshaft in turn until each bearing can be removed. Keep the upper bearings with the respective bearing caps.
31. Remove piston and connecting rod assemblies by pushing upward on the connecting rod. It may be necessary to remove the carbon from the ridge at the top of the cylinder bore to facilitate piston removal.
32. Alternately loosen and remove main bearing cap bolts.
33. Remove bearing caps using tool ST16660000, Fig. 14.

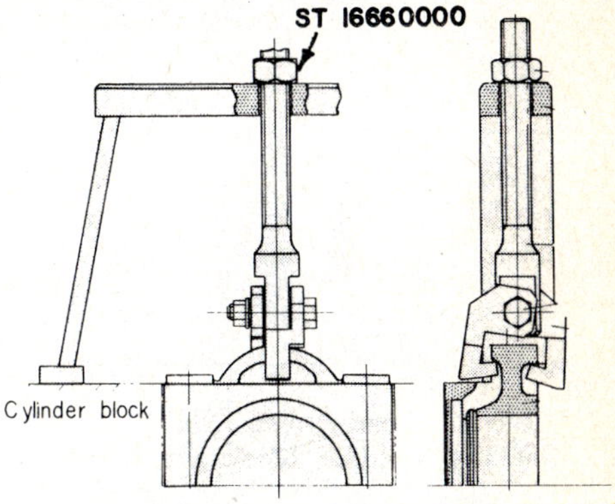

Fig. 14 Removing main bearing caps

NOTE: The No. 3 bearing cap contains a thrust washer. Retain washer with bearing cap.

34. Remove crankshaft from engine.
35. Remove upper main bearing shells. Identify shells for proper positioning.
36. Remove valve lifters from cylinder block.

Unit Repair Section

CYLINDER LINER REPLACEMENT

Removal

1. Pull cylinder liner from cylinder block using tool ST1030000, Fig. 1, and tightening the bolt.

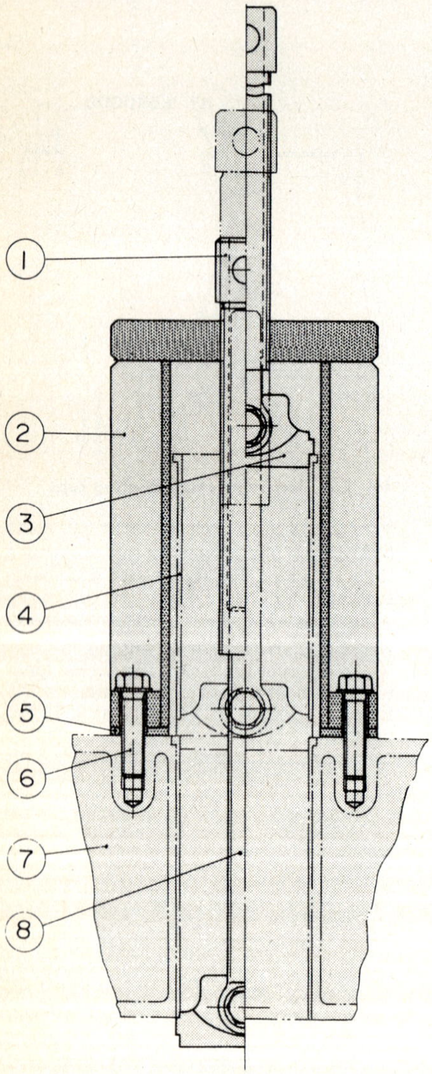

1	Bolt
2	Bracket
3	Patch plate
4	Cylinder liner
5	Packing
6	Bolt
7	Cylinder block
8	Bolt

Fig. 1 Replacing cylinder liner

Installation

1. Clean cylinder and liner surfaces and coat with engine oil.
2. Install cylinder liner into cylinder block with tool ST1030000, Fig. 1. The cylinder liner should protrude .0008–.0036 inch above surface of cylinder block. This protrusion should not vary more than .0020 inch between cylinders.

OIL PUMP SERVICE

Disassembly

1. Separate oil screen from cover, Fig. 2.
2. Remove four bolts and pump cover, Fig. 2.
3. Using a straight-edge and feeler gauge, measure clearance between side of gears and surface of pump body, Fig. 3.
4. Using a feeler gauge, measure clearance between gear teeth and pump bore, Fig. 4.
5. Measure gear backlash, Fig. 5.
6. Remove pin from drive shaft and using a suitable puller, remove drive gear from shaft, Fig. 6.
7. Remove driven gear from idler shaft.
8. Press idler shaft from pump body.
9. Remove cotter pin retaining relief valve. The valve spring will force the plug out of the pump body. Remove valve spring and valve, Fig. 7.

Inspection

1. If clearance between gears and surface of pump body is more than .0059 inch, replace both gears.
2. If clearance between gear teeth and pump bore is more than .0098 inch, replace both gears.
3. If gear backlash is more than .0197 inch, replace both gears.
4. Check pump body and cover for damage and replace as necessary.
5. Check diameter of drive and idler shafts. Diameter of drive shaft at tip should be .5943–.5950 inch. Diameter of drive shaft at gear end should be .5148–.5155 inch. Diameter of idler shaft should be .5178–.5182 inch. Replace shafts if wear exceeds .0039 inch.
6. Check clearance between drive shaft and body bore using a dial indicator, Fig. 8. Replace necessary parts if clearance exceeds .00492 inch.

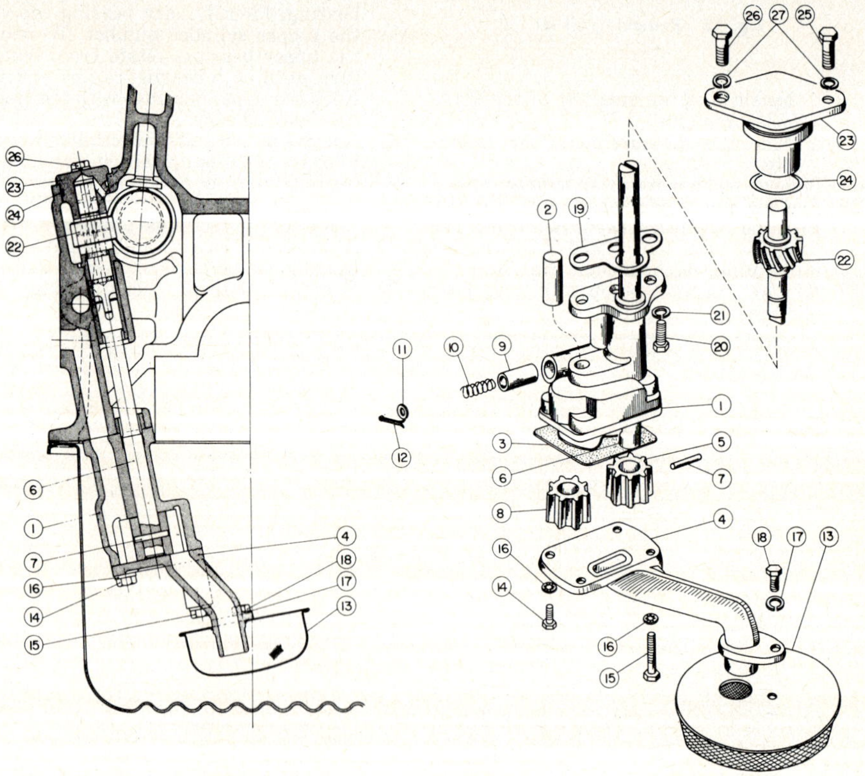

1	Oil pump body	8	Driven gear	15	Bolt	22	Driving spindle
2	Idler shaft	9	Relief valve	16	Washer	23	Driving spindle support
3	Packing (SD22)	10	Relief valve spring	17	Lock washer	24	O-ring
4	Oil pump cover	11	Washer	18	Bolt	25	Bolt
5	Drive gear	12	Cotter pin	19	Gasket	26	Bolt
6	Drive shaft	13	Oil screen	20	Bolt	27	Lock washer
7	Pin	14	Bolt	21	Lock washer		

Fig. 2 Oil pump assembly

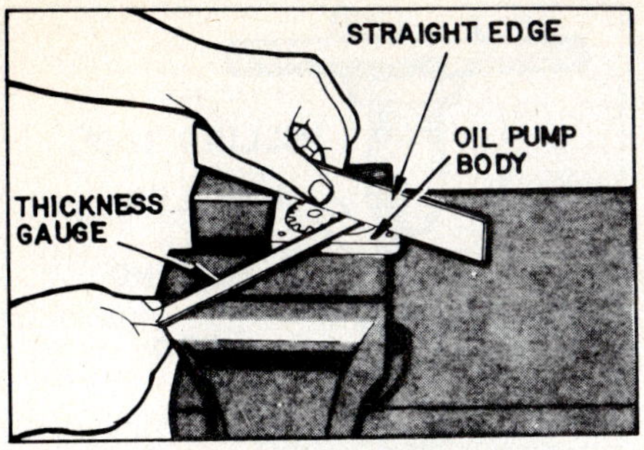

Fig. 3 Measuring clearance between side of gears & surface of pump body

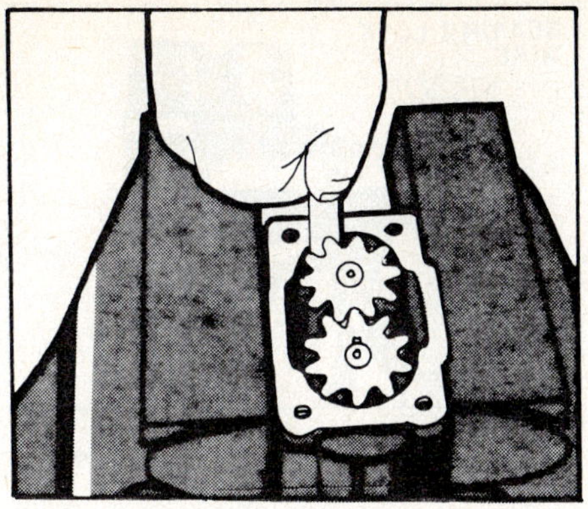

Fig. 4 Measuring clearance between gear teeth & pump body

Fig. 5 Measuring gear backlash

Fig. 6 Removing drive gear from shaft

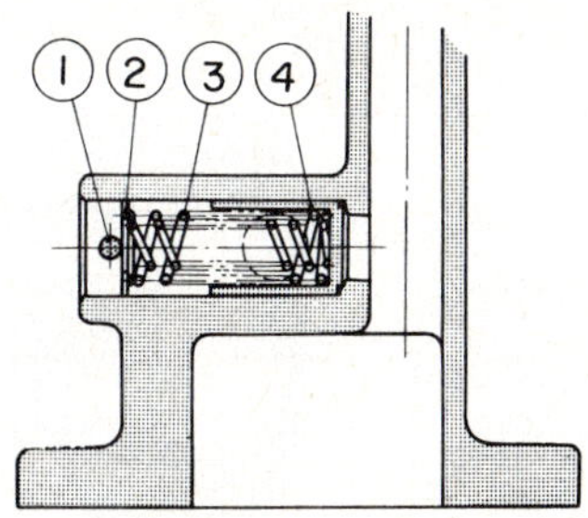

1	Cotter pin
2	Washer
3	Spring
4	Plunger

Fig. 7 Oil pump relief valve assembly

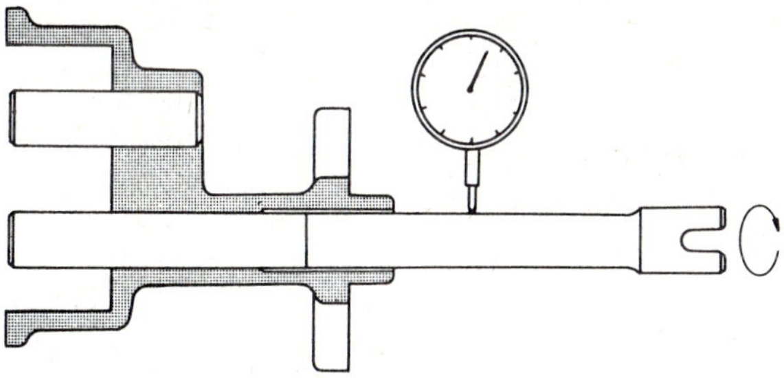

Fig. 8 Measuring clearance between drive shaft & body bore

7. If idler shaft replacement is necessary, ensure there is .0197 inch clearance between shaft and pump body cover when installing shaft, Fig. 9.

Assembly

1. Coat oil pump components with engine oil.
2. Reverse disassembly procedure to assemble pump. Install beveled side of gears facing toward pump body. Check clearances after gear installation.
3. Turn oil pump drive shaft by hand, ensuring pump rotates smoothly.

WATER PUMP SERVICE

Disassembly

1. Remove snap ring from front of pulley hub, Fig. 10.
2. Press pulley hub from water pump, Fig. 11. A suitable puller may also be used.
3. Remove bearing lock wire from slit of housing, Fig. 12.
4. Remove screws securing end plate to housing, then the end plate and gasket.
5. Press impeller and bearing from housing. Place a tube with a slightly larger diameter than the bearing over the bearing and contact with impeller. Support housing on press, then press impeller and bearing from housing, Fig. 13.
6. Drive water seal from housing using a drift and a suitable hammer, Fig. 14.

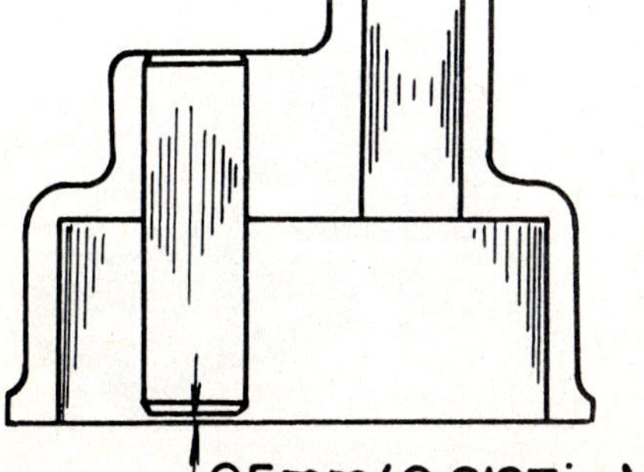

Fig. 9 Idler shaft installation

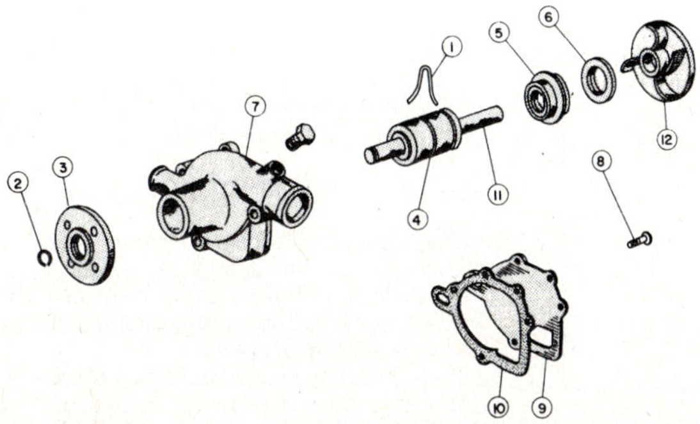

1	Lock wire
2	Snap ring
3	Pulley hub
4	Bearing
5	Water seal
6	Ceramic seat
7	Housing
8	Machine screw
9	Plate
10	Gasket
11	Shaft
12	Impeller

Fig. 10 Water pump disassembled

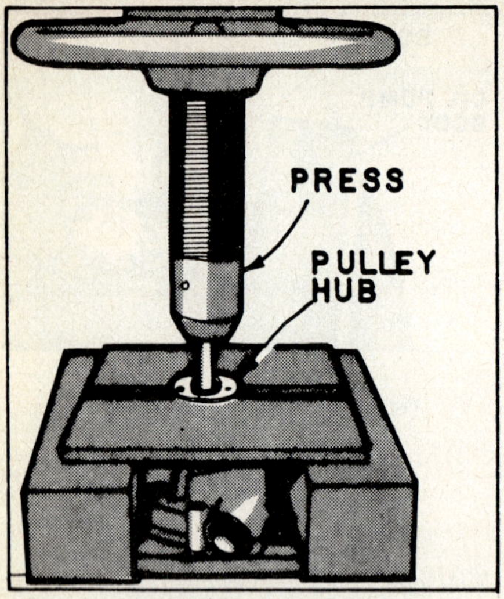

Fig. 11 Pressing pulley hub from water pump

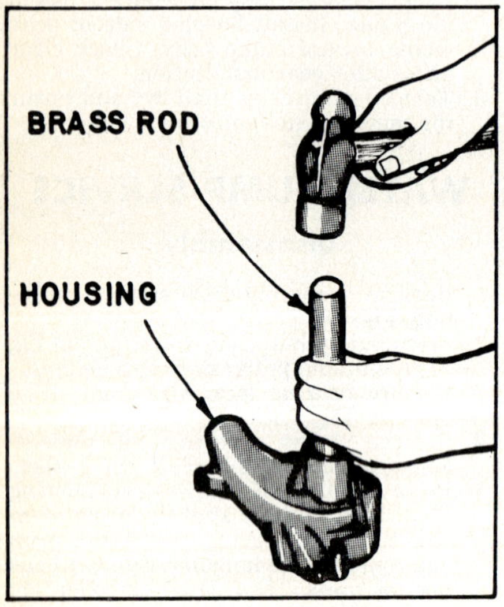

Fig. 14 Removing water seal from housing

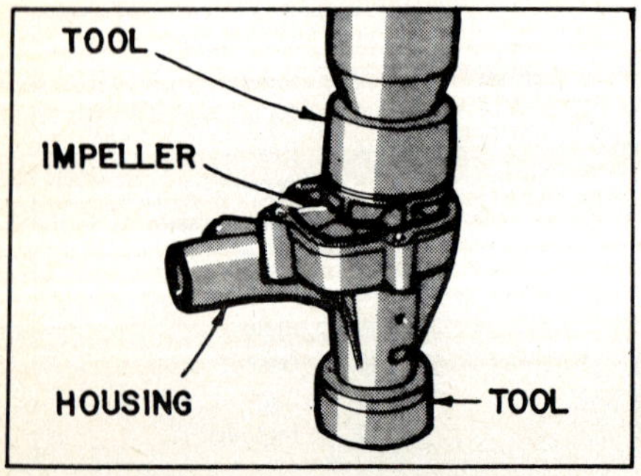

Fig. 17 Pressing impeller onto shaft

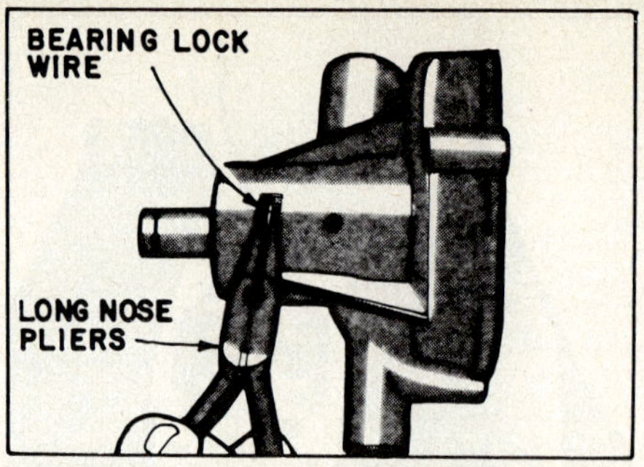

Fig. 12 Removing bearing lock wire

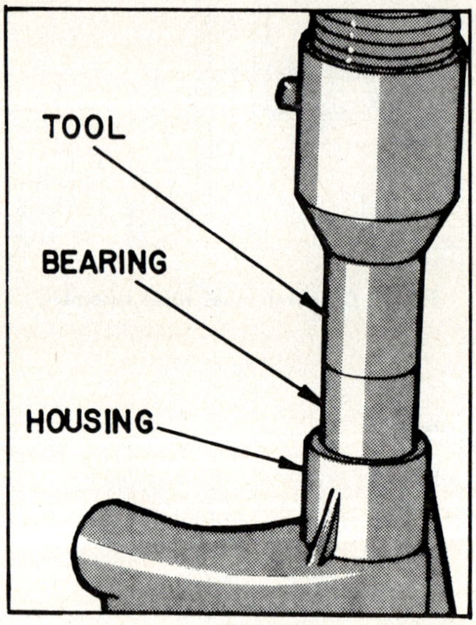

Fig. 15 Pressing bearing into housing

Inspection

1. Check for excessive play in water pump bearing. Listen for noise while turning bearing by hand. Replace shaft assembly if any defects are noted.
2. Check condition of impeller. If corrosion or other defects are noted, replace impeller.
3. Replace water seal, regardless of condition.
4. Check wear of surface of ceramic seat contacting water seal. If any defects are noted, replace ceramic seat.

Assembly

1. Coat housing side of water seal with Three-Bond No. 2 or equivalent. With a brass drift and a suitable hammer, drive water seal into housing. Ensure that the seal is seated properly.
2. Press bearing into housing by placing tool on outer race of bearing. Press into place from housing drive side, Fig. 15. Align bearing with housing lock wire groove.

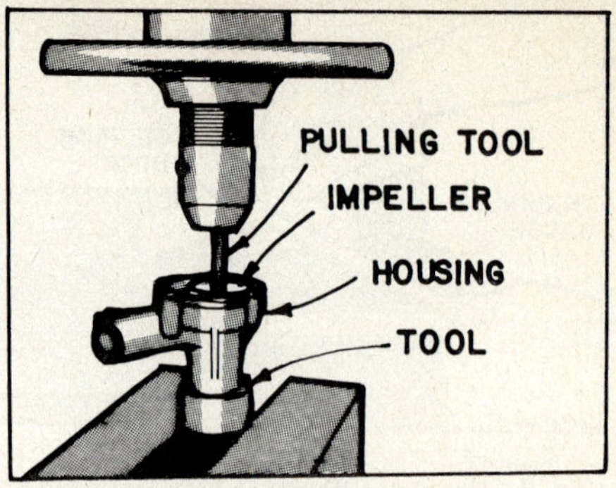

Fig. 13 Pressing impeller & bearing from housing

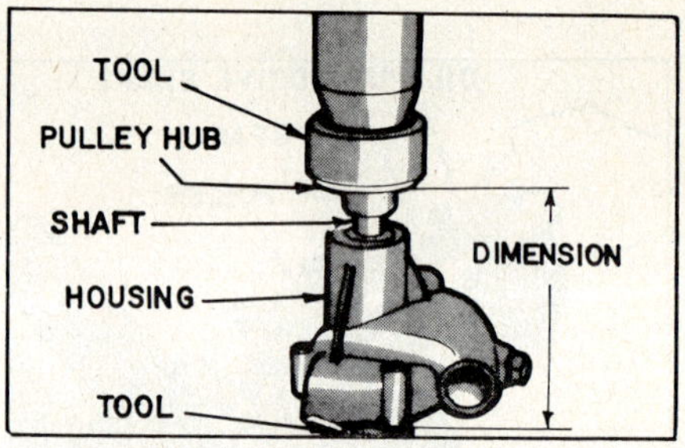

Fig. 16 Pressing pulley onto shaft

3. Drive lock wire into housing slit using a suitable hammer. Do not drive the lock wire into housing completely since any subsequent removal will become impossible. Let the end of the wire protrude slightly.
4. Press pulley onto shaft. When pressing pulley onto shaft, measure distance between end surface of housing (engine end) and the pulley. The distance should be 5.035–5.043 inch when pulley is properly installed, Fig. 16.
5. Install snap ring in shaft groove in front of pulley hub.
6. Press impeller onto shaft, Fig. 17. Place a .028 inch feeler gauge between impeller and housing, Fig. 18. Press impeller until impeller contacts feeler gauge. Remove feeler gauge. Proper clearance between impeller and housing is .016–.024 inch. Adjust if necessary.
7. Install gasket, end plate and attaching screws.

Fig. 18 Checking impeller to housing clearance

Engine Assembly Section

ASSEMBLY, FIGS. 1 & 2

NOTE: Refer to Figs. 1 and 2 in the "Engine Disassembly Section."

1. Coat the No. 4 (Rear) bearing area of cylinder block and main bearing cap with suitable sealer, then install the oil seal. Ensure that the oil seal is properly seated and that the seal end surface protrudes approximately .0197 inch from the surface of the main bearing cap.
2. Align the front end plate gasket and end plate holes, and the cylinder block dowel pins. Install front end plate and torque bolts to 7.23–9.4 ft. lbs.
3. Coat valve lifters with engine oil and install lifters into cylinder block.
4. Coat the camshaft bearings and camshaft engine oil. Install camshaft into engine, then the locating plate, Fig. 1.
5. Coat main bearing shells with engine oil and install upper bearing shells into cylinder block and the lower bearing shells into the bearing caps. Align the tongues and grooves.
6. Coat crankshaft main bearing journals and crankpins with engine oil and install crankshaft into cylinder block.
7. Check crankshaft end play, Fig. 2. Correct if necessary by replacing the thrust washer at No. 3 main bearing. Thrust washers are available in a standard thickness (.093 inch) and two oversizes, .0079 and .0157 inch. Face oil grooves toward the outside (crankshaft side), Fig. 3.
8. Install main bearing caps. Torque bolts to specifications.

NOTE: Ensure that the raised "F" mark faces toward front of engine, Fig. 4.

9. Install a side seal into rear bearing cap by forcing in with a screwdriver, Fig. 5.

NOTE: The longitudinal groove of the side seal should face toward the cap and block.

10. Assemble pistons to connecting rods, Fig. 6. The combustion at top of piston faces toward injection pump side of engine with the numbered side of connecting rod facing toward the exhaust manifold.
11. Install connecting and piston assemblies into engine. Install connecting rod bearing caps and torque bolts to specifications.
12. Measure piston top clearance, Fig. 7. Use a dial indicator to measure clearance between top of piston at top dead center and the top of cylinder block. The piston should be .0098 inch below top of cylinder block and can protrude .0039 inch above top of cylinder block.
13. Install oil pump assembly using a new gasket and torque bolts to 7.23–9.4 ft. lbs.
14. Install oil pump drive spindle, Fig. 8, by aligning spindle with oil pump drive shaft groove in cylinder block and camshaft oil pump drive gear. Install new O-ring on support and attach support to cylinder block with two bolts. When installing support, ensure that oil hole is located toward the block.
15. Install crankshaft gear, aligning the crankshaft gear and camshaft gear alignment marks, Fig. 9. Check gear backlash which should be .0028–.0079 inch.
16. Install oil slinger at front of crankshaft. Referring to Fig. 10, install oil jet so the jet faces the crankshaft, camshaft and idler gears.
17. Install timing gear case. Align dowel pins on cylinder block with holes in gear case. Torque small bolts to 3.6 ft lbs. and large bolts to 8.7 ft. lbs.
18. Install oil seal on crankshaft.
19. Install oil pan with new gasket and torque bolts to 2.89–4.34 ft. lbs.
20. Install flywheel housing. Align dowel pins on cylinder block with holes in flywheel housing. It may be necessary to use a mallet to drive the housing into position. Torque attaching bolts to 47–54.25 ft. lbs.
21. Install flywheel and torque bolts to 32.6–36.2 ft. lbs.
22. Insert key and install crankshaft pulley, cone and nut. Torque nut to 217–238.7 ft. lbs.
23. Check alignment of TDC mark on crank pulley and timing gear case mark, Fig.

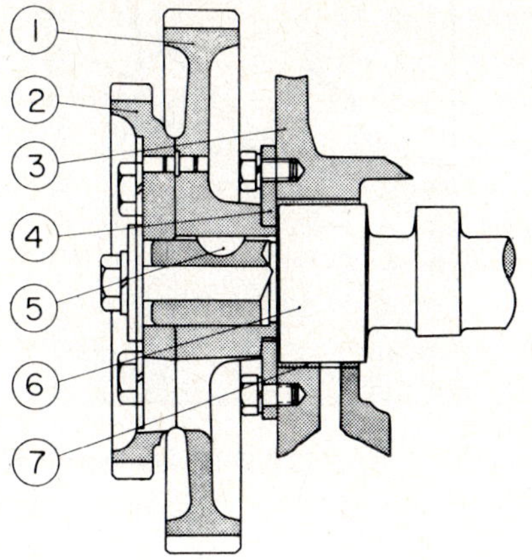

1	Camshaft gear	5	Key
2	Idle gear	6	Camshaft
3	Cylinder block	7	Camshaft bushing
4	Locating plate		

Fig. 1 Camshaft drive assembly

Fig. 2 Checking crankshaft end play

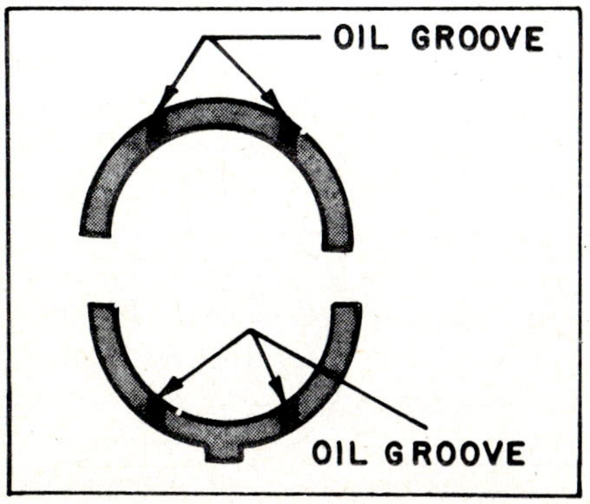

Fig. 3 Crankshaft thrust washers

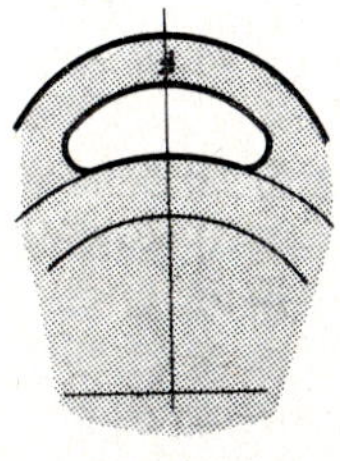

Fig. 4 Main bearing marking

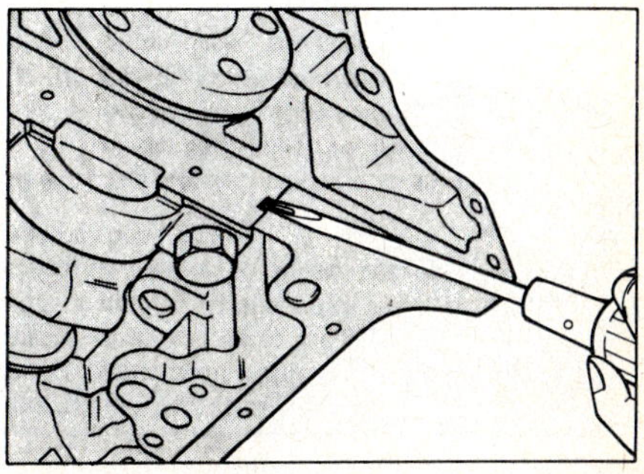

Fig. 5 Installing rear main bearing side seal

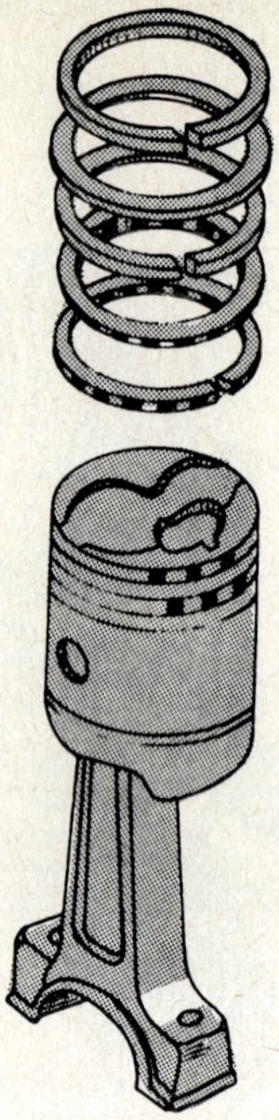

Fig. 6 Piston & rod assembly

Fig. 7 Measuring piston top clearance

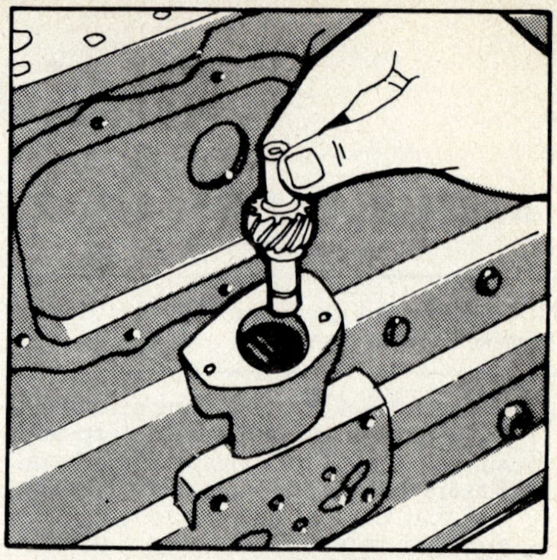

Fig. 8 Installing oil pump drive spindle

11. Use a dial indicator to determine top dead center of No. 1 piston. The TDC mark on crank pulley and timing gear case should be aligned. Also, the flywheel mark should be aligned.

24. Install cylinder head gasket with lips of largest stainless portions facing upward. Also, install rubber rings around water and oil holes.
25. Install cylinder head. Ensure that the pins are aligned and that the gasket and rubber rings do not slip during cylinder head installation.
26. Coat cylinder head main and sub bolts with engine oil and install bolts. Referring to Fig. 12, torque bolts in sequence as follows:
    a. Torque main bolts to 43.4 ft. lbs.
    b. Torque sub bolts to 21.7 ft. lbs.
    c. Torque main bolts to 94 ft. lbs.
    d. Torque sub bolts to 36.2 ft. lbs.
27. Install push rods with the rounded end facing downward. When installing push rods, turn push rods while applying downward pressure to properly seat push rods in valve lifters.
28. Install rocker arm shaft assembly on cylinder head and torque attaching bolts to 14.5–18.1 ft. lbs.
29. Thread nozzle holder assemblies into cylinder head, Fig. 13.
30. On right side of engine:
    a. Temporarily attach injection pump with gasket to front end plate.
    b. Rotate crankshaft to bring No. 1 piston to top dead center.
    c. Mesh injection pump drive gear and idler gear at "Y" marks, Fig. 14. After aligning keyway of injection pump, secure timing assembly with lock washer and round nut. Torque nut to 50.6–57.9 ft. lbs.
    d. Install tachometer drive coupling.
    e. Measure gear backlash between idler gear and injection pump drive gear. Backlash should be .0028–.0079 inch.
    f. Remove No. 1 cylinder holder clamp of injection pump, loosen delivery valve holder and remove delivery spring, Fig. 17. Then, torque valve holder to 21.7–25.3 ft. lbs.
    g. Connect fuel hoses so fuel can be supplied to priming pump.
    h. Rotate crankshaft to bring No. 1 piston 20° before top dead center. Align first mark on crank pulley in direction of normal rotation to gear case mark.
    i. Supply fuel with priming pump. With the injection pump pushed fully toward the engine, slowly return pump and note position at which fuel ceases to flow from valve holder. Lock injection pump in place since this is the beginning of injection for No. 1 cylinder. The injection pump and front end plate marks should now be aligned. If not, make a new mark on the front end plate.
    j. Remove injection pump delivery holder and install spring, Fig. 15. Torque holder to 21.7–25.3 ft. lbs.
    k. Connect nozzle lines between injection pump and nozzle holder assemblies.
    l. While meshing timing drive coupling of tachometer driver and the cover driven coupling, install timing gear case with gasket.
    m. Install fuel filter assembly to bracket and the attaching bolts.
    n. Connect the three filter hoses to injection pump using the connector bolts.

Fig. 9 Aligning crankshaft & camshaft gear marks

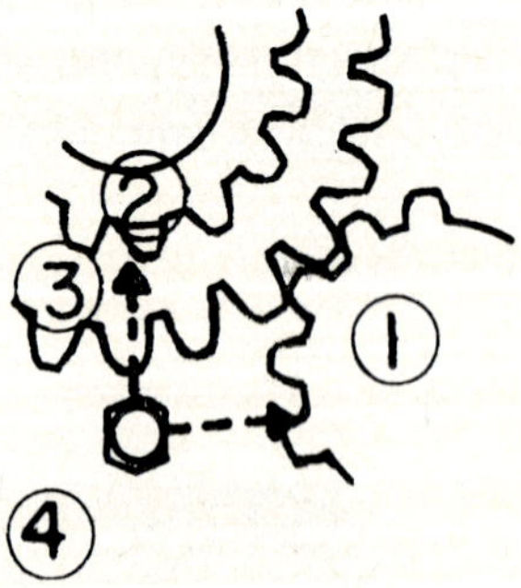

1 Crankshaft gear
2 Idle gear
3 Camshaft gear
4 Oil jet

Fig. 10 Oil jet installation

Fig. 11 TDG mark on crank pulley & timing gear case

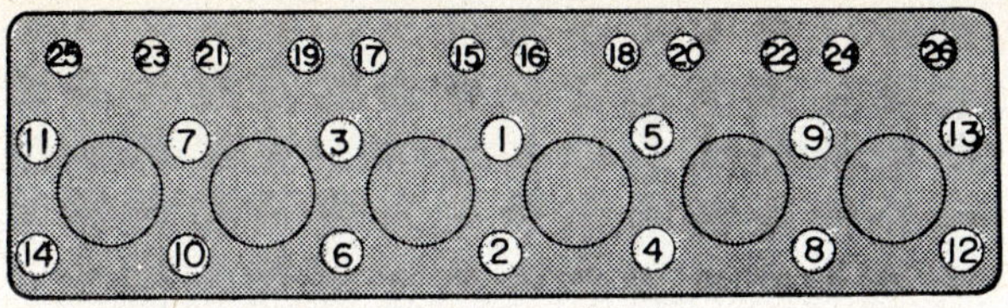

Fig. 12 Cylinder head bolt tightening sequence

Fig. 13 Nozzle holder installation

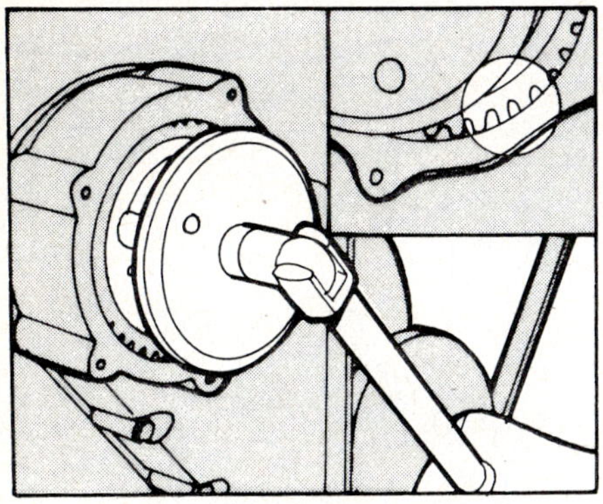

Fig. 14 Aligning injection pump drive gear & idler gear

Fig. 15 Removing or installing delivery spring

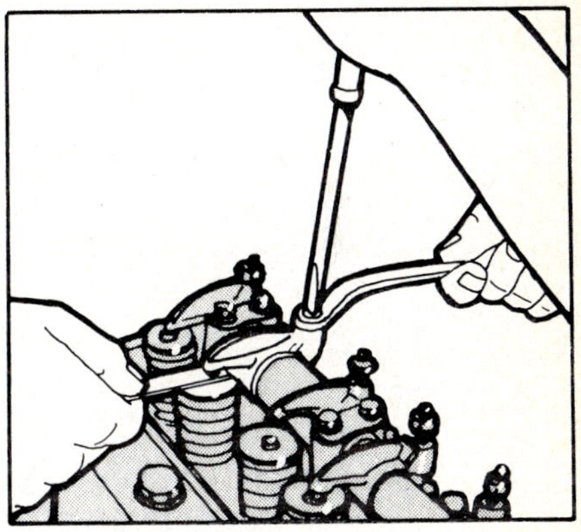

Fig. 16 Adjusting valve clearance

Use a nylon gasket at each connector bolt.

o. Attach oil cooler with gasket to cylinder block. Connect oil cooler lines to filter using rubber rings.

p. Attach oil filter assembly with gasket to cylinder block. Torque bolt to 14.5–18.1 ft. lbs. Tighten oil line bolts at cooler end.

q. Connect rubber water hose between ring connector of cylinder head and oil cooler.

r. Coat threaded portions of oil pressure and warning lamp switches with suitable sealer. Then, install switches into the proper cylinder block holes.

31. Attach water pump assembly with gasket to cylinder block with three bolts. Install belt adjuster. Do not secure water at this time. Final tightening will be at time of belt adjuster adjustment during generator installation.

32. Install spacer, pulley and fan.

33. On left side of engine:
   a. Install generator bracket.
   b. Temporarily secure generator to the belt adjuster and bracket.
   c. Connect rubber hose and oil tube of vacuum pump to cylinder block.
   d. Install drive belts. Move generator with a suitable tool until belt tension at midway between fan and generator pulleys is .314–.472 inch deflection when the belt is pushed with 22 lbs. of pressure. Tighten generator bolts.
   e. Torque water pump bolts to 7.2–9.4 ft. lbs. for .314 inch bolts and 14.5–18.1 ft. lbs. for .393 inch bolts.
   f. Install starter motor.
   g. Install intake and exhaust manifolds with gaskets to cylinder head. Use washers and yoke bolts. Torque bolts to 10.9–13 ft. lbs.
   h. Fill engine to proper level with engine oil. Pour oil over valve rocker arms.

34. To adjust valve clearance:
   a. Bring No. 1 or No. 6 cylinder to top dead center, compression stroke.
   b. With No. 1 cylinder at top dead center, check valve clearances on the following: Intake—Nos. 1, 3, 5. Exhaust—1, 2, 4.
   c. With No. 6 cylinder at top dead center, check valve clearances on the following: Intake—2, 4, 6. Exhaust—3, 5, 6.
   d. Adjust valve clearance to .0138 inch by loosening lock nut and turning adjusting screw to obtain proper clearance, Fig. 16.

Fuel Section

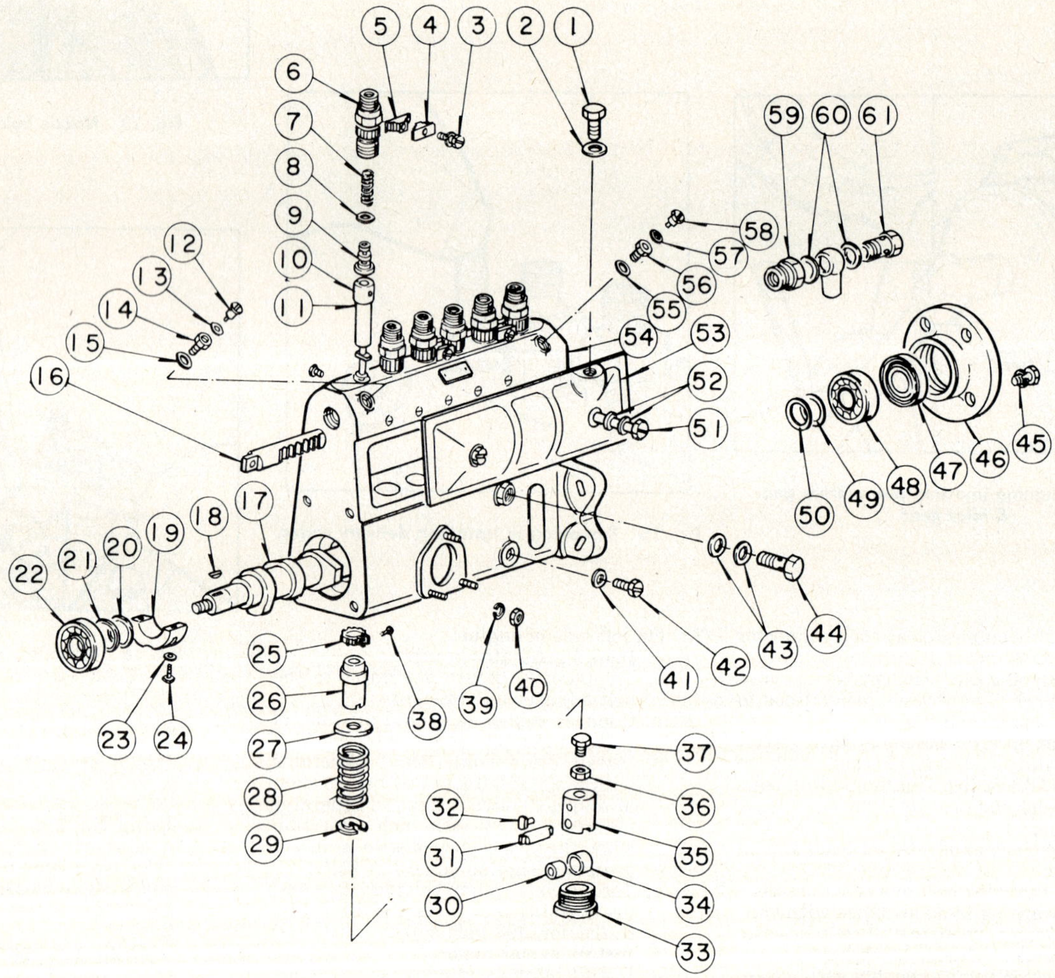

1	Screw plug	16	Control rack	31	Roller pin	46	Bearings cover
2	Gasket	17	Camshaft	32	Tappet guide	47	Oil seal
3	Bolt	18	Woodruff key	33	Screw plug	48	Tapper roller bearing
4	Plate	19	Center bearing	34	Tappet roller	48	Shim
5	Plate	20	Camshaft ring	35	Tappet	50	Camshaft ring
6	Delivery valve holder	21	Shim	36	Adjusting bolt nut	51	Cover screw
7	Delivery valve spring	22	Tapper roller bearing	37	Tappet adjust bolt	52	Gasket
8	Delivery valve gasket	23	Gasket	38	Pinion clamp screw	53	Cover assembly
9	Delivery valve assembly	24	Head bolt socket	39	Spring washer	54	Gasket
10	Plunger assembly	25	Pinion	40	Nut	55	Air bleeder adapter
11	Rack stop screw	26	Control sleeve	41	Gasket	56	Gasket
12	Air bleeder screw	27	Plunger spring upper seat	42	Plug screw	57	Air bleeder screw
13	Gasket	28	Plunger spring	43	Gasket	58	Adapter
14	Air bleeder adapter	29	Plunger spring lower seat	44	Pipe joint bolt	59	Gasket
15	Gasket	30	Tapper roller bush	45	Screw	60	Pipe joint bolt

Fig. 1 Fuel injection pump disassembled

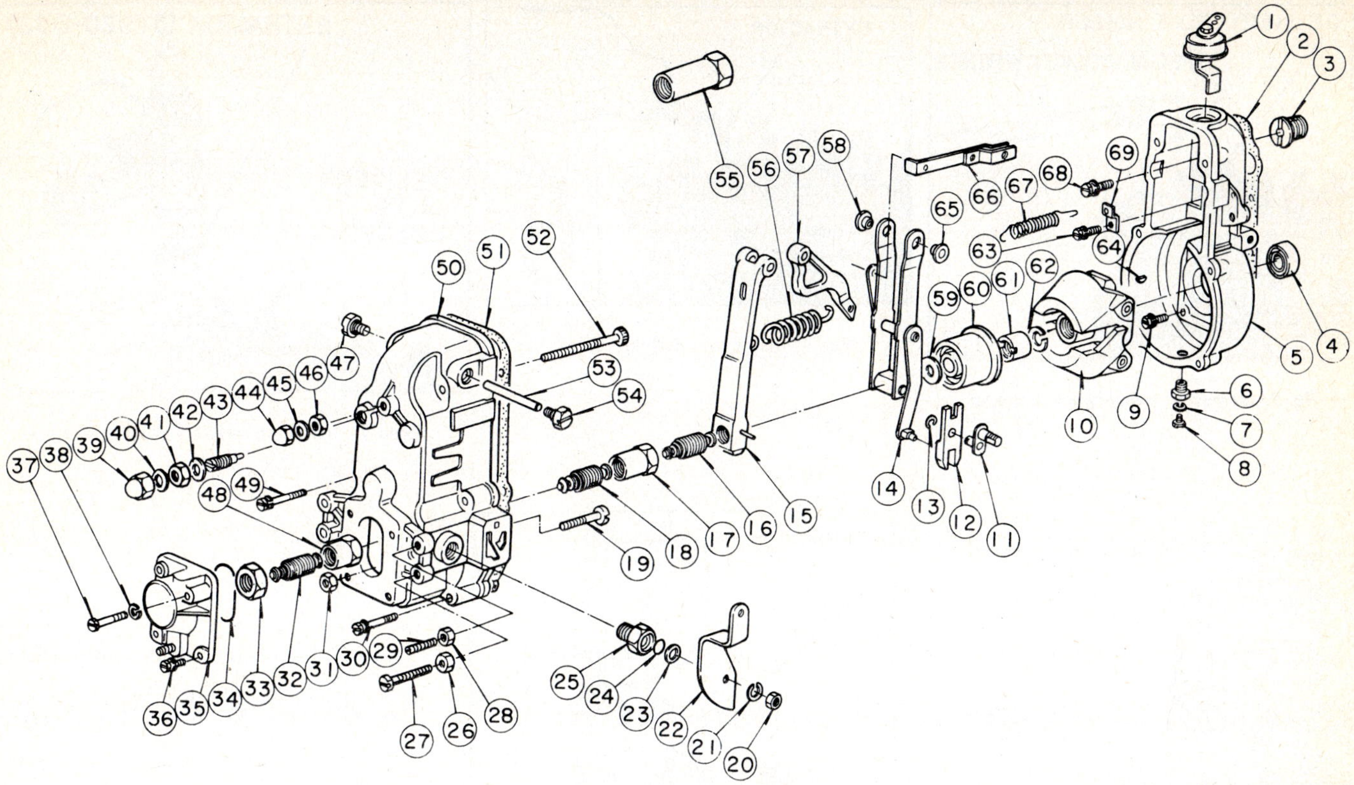

1	Stop lever	19	Stroke adjusting screw	37	Bolt	55	Limiter
2	Gasket	20	Nut	38	Lock washer	56	Governor spring
3	Adapter	21	Lock washer	39	Cup nut	57	Speed adjusting lever
4	Bearing	22	Control lever	40	Gasket	58	Collar
5	Governor housing	23	Shim	41	Nut	59	Shim
6	Adapter	24	O-ring	42	Gasket	60	Sleeve
7	Gasket	25	Bush	43	Damper spring	61	Round nut
8	Drain plug	26	Nut	44	Cap nut	62	Lock washer
9	Bolt	27	Full load stopper bolt	45	Gasket	63	Bolt
10	Flyweight	28	Nut	46	Nut	64	Woodruff key
11	Lever shaft	29	Idling stopper bolt	47	Plug	65	Collar
12	Supporting lever	30	Bolt	48	Nut	66	Floating lever link
13	Snap ring	31	Nut	49	Bolt	67	Start spring
14	Guide lever assembly	32	Angleich spring assembly	50	Governor cover	68	Bolt
15	Tension lever	33	Nut	51	Gasket	69	Spring eye
16	Idling spring assembly	34	O-ring	52	Speed adjusting bolt		
17	Nut	35	Closing cover	53	Tension lever shaft		
18	Angleich spring assembly	36	Bolt	54	Plug		

Fig. 2 Governor disassembled. RAD type

FUEL INJECTION PUMP TIMING

Refer to the "Engine Assembly Section," Step 30.

FUEL INJECTION PUMP REPAIRS, Figs. 1 & 2

Delivery Valve, Replace

Removal

1. Clean area around nozzle tube and delivery valve.
2. Remove nozzle tube.
3. Remove delivery valve holder lock plate.
4. Remove delivery valve holder, Fig. 3, and delivery valve spring.
5. Install delivery valve extractor tool (57930-132) and remove delivery valve, Fig. 4.

Installation

1. Install delivery valve in pump housing. Ensure that dirt does not enter between top of plunger barrel and the delivery valve.
2. Install delivery valve gasket with larger faced portion facing downward. Tap into position by placing extractor tool over gasket and tapping lightly with a hammer, Fig. 5.
3. Install delivery valve spring.
4. Torque delivery valve holder to 21.7–25.3 ft. lbs. Loosen holder and retorque.
5. Install lock plate, nozzle tube and nozzle tube clamp.

Plunger, Replace

Removal

1. Remove delivery valve as outlined previously.
2. Push plunger spring upward with two screwdrivers and remove lower spring seat from plunger, Fig. 6.
3. Insert a wire with a hooked end through

Fig. 3 Removing delivery valve holder

Fig. 4 Removing delivery valve

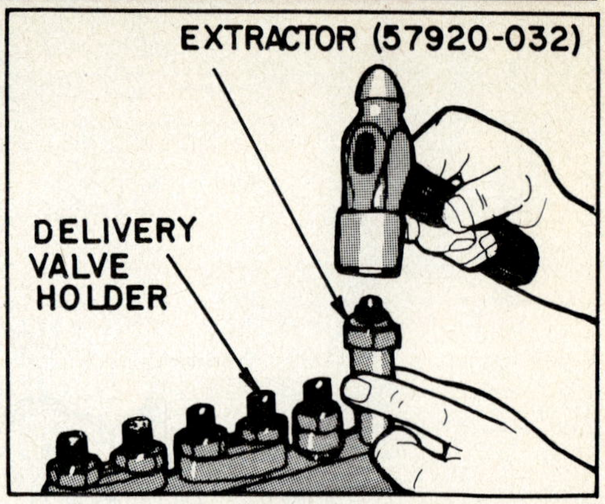

Fig. 5 Installing delivery valve gasket

Fig. 6 Removing lower spring seat

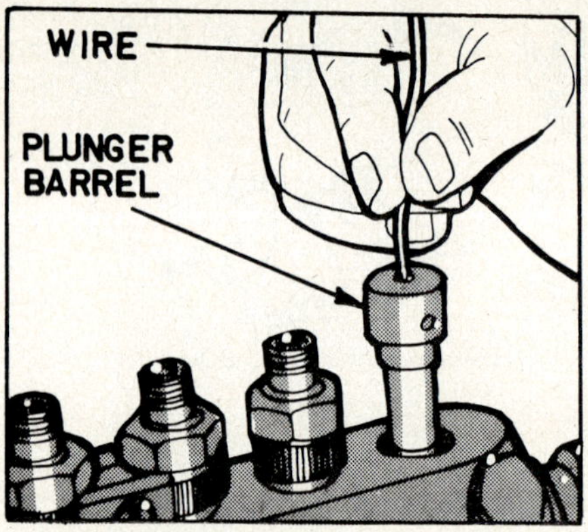

Fig. 7 Removing plunger & barrel

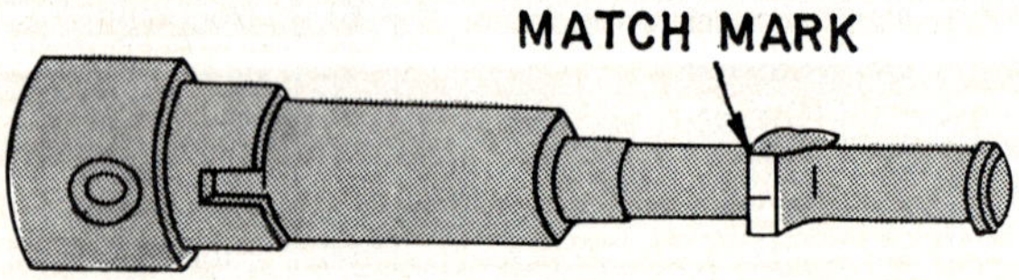

Fig. 8 Plunger & barrel identification marks

top of pump housing, through plunger hole and hook wire on the lead unit of the plunger. Pull wire upward to remove plunger and plunger barrel, Fig. 7.

Installation

1. Immerse new plunger in kerosene or solvent to remove rust preventative oil inside plunger.

NOTE: The plunger should be handled by the head or plunger barrel only.

---

2. Slowly insert plunger and plunger barrel into pump housing with the barrel groove and notch on plunger piston pin facing forward, Fig. 8.
3. Ensure that the plunger piston pin is installed properly in the control sleeve groove.
4. Push plunger spring upward with two screwdrivers and install lower spring seat.
5. Install delivery valve as outlined previously.

Plunger Spring, Replace

1. Remove delivery valve, plunger and plunger barrel as outlined previously.
2. Rotate camshaft to lower tappet to the bottom position.
3. Compress spring with a screwdriver, Fig. 9, pull control sleeve, upper spring seat and the plunger spring toward the front, then remove from pump housing.
4. Reverse procedure to install.

Tappet, Replace

1. Remove delivery valve, plunger, plunger barrel and plunger spring as outlined previously.
2. Remove tappet from pump housing, Fig. 10.
3. Reverse procedure to install.

Fig. 9 Removing plunger spring

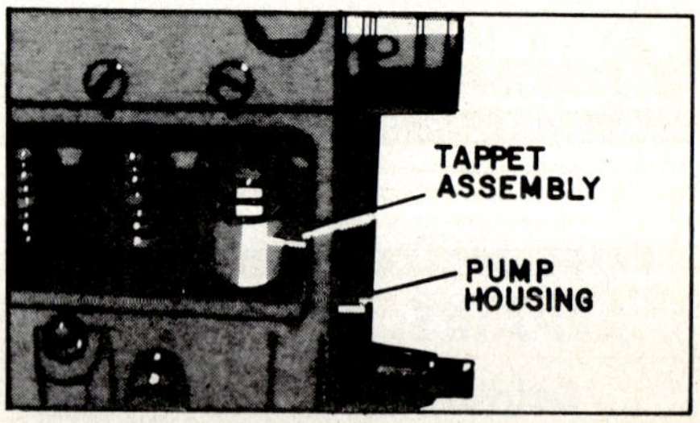

Fig. 10 Tappet assembly

OLDSMOBILE (GM V8-350) DIESEL

INDEX OF SERVICE OPERATIONS

GENERAL ENGINE SPECIFICATIONS

Year	Engine	Carburetor	Bore and Stroke	Piston Displacement, Cubic Inches	Compression Ratio	Net Brake H.P. @ R.P.M.	Net Torque Lbs. Ft. @ R.P.M.	Normal Oil Pressure Pounds @ R.P.M.
1978–80	125 Horsepower.... V8-350	Fuel Inj.	4.057 × 3.385	350	22.5	125 @ 3600	225 @ 1600	35 @ 3000
1981	125 Horsepower① ... V8-350	Fuel Inj.	4.057 × 3.385	350	22.5	125 @ 3600	225 @ 1600	35 @ 3000
	105 Horsepower② ... V8-350	Fuel Inj.	4.057 × 3.385	350	22.5	105 @ 3200	200 @ 1600	35 @ 3000
1983	105 Horsepower.... V8-350	Fuel Inj.	4.057 × 3.385	350	22.5	105 @ 3200	200 @ 1600	35 @ 3000

①—Exc. California.
②—California only.

DIESEL ENGINE PERFORMANCE SPECIFICATIONS

Year	Engine	Injection Timing	Curb Idle Speed	Fast Idle Speed
1978–80	V8-350①	4½°③④⑤	575D	650D
1978–80	V8-350②	5½°③④⑤	575D	650D
1981	V8-350①	4°③④⑤	⑥	⑦
1981	V8-350②	5°③④⑤	⑥	⑦
1983	V8-350	4°③⑤⑧	600D	750D

①—Except high altitude.
②—High altitude.
③—ATDC—After top dead center.
④—At 1200 RPM.
⑤—Using diesel timing meter J-33075 or equivalent.
⑥—Except California, 575D; California, 600D.
⑦—Except California, 650D; California, 750D.
⑧—At 1250 RPM.

VALVE SPECIFICATIONS

Year	Model	Valve Lash Int.	Valve Lash Exh.	Valve Angles Seat	Valve Angles Face	Valve Spring Installed Height	Valve Spring Pressure Lbs. & In.	Stem Clearance Intake	Stem Clearance Exhaust	Stem Diameter Intake	Stem Diameter Exhaust
1978–81	V8-350	Hydraulic①		②	③	1.670	151 @ 1.30	.0010–.0027	.0015–.0032	.3425–.3432	.3420–.3427
1983	V8-350	Hydraulic①		②	③	1.670	210 @ 1.22	.0010–.0027	.0015–.0032	.3425–.3432	.3420–.3427

①—No adjustment.
②—Intake 45°, exhaust 31°.
③—Intake 44°, exhaust 30°.

OLDSMOBILE (GM V8-350) DIESEL ENGINE

ENGINE TIGHTENING SPECIFICATIONS★

★ Torque specifications are for clean lubricated threads only. Dry or dirty threads produce increased friction which prevents accurate measurement of tightness.

Year	Engine Model	Glow Plugs Ft. Lbs.	Cylinder Head Bolts Ft. Lbs.	Intake Manifold Ft. Lbs.	Exhaust Manifold Ft. Lbs.	Rocker Arm Shaft Bracket Ft. Lbs.	Rocker Arm Cover In. Lbs.	Connecting Rod Cap Bolts Ft. Lbs.	Main Bearing Cap Bolts Ft. Lbs.	Flywheel to Crank-shaft Ft. Lbs.	Vibration Damper or Pulley Ft. Lbs.
1978–79	V8-350	12	130①	40①	25	25②	15	42	120	60	200–310
1980–81	V8-350	12	130①	40①	25	28②	15	42	120	60	200–310
1983	V8-350	12	130①	40①	25	28②	③	42	120	60	200–310

①—Clean and dip engine bolt in engine oil before tightening.

②—Rocker arm pivot bolt to head.

③—Fully driven, seated and not stripped.

PISTONS, PINS, RINGS, CRANKSHAFT & BEARINGS

Year	Model	Piston Clearance	Ring End Gap①		Wrist-pin Diam-eter	Rod Bearings		Main Bearings			
			Comp.	Oil		Shaft Diameter	Bearing Clearance	Shaft Diameter	Bearing Clearance	Thrust on Bear. No.	Shaft End Play
1978–81	V8-350	.005–.006	.015	.015	1.0951	2.1238–2.1248	.0005–.0026	2.9993–3.0003	②	3	.0035–.0135
1983	V8-350	.0035–.0045	③	.015	1.0946	2.1238–2.1248	.0005–.0026	2.9993–3.0003	④	3	.0035–.0135

①—Fit rings in tapered bores for clearance in tightest portion of ring travel.

②—No. 1, 2, 3, 4: .0005–.0021; No. 5: .0015–.0031.

③—Top compression ring, .019; 2nd compression ring, .013.

④—Nos. 1, 2, 3, 4: .0005–.0021; No. 5: .0020–.0034

Troubleshooting & Diagnosis

DIESEL ENGINE DIAGNOSIS

Condition	Possible Cause	Correction
ENGINE WILL NOT CRANK	1. Loose or corroded battery cables	1. Check connections at battery, engine block and starter solenoid.
	2. Discharged batteries	2. Check charging system.
	3. Starter inoperative	3. Check starting system.
ENGINE CRANKS SLOWLY—WILL NOT START (Minimum Engine Crank Speed—100 RPM)	1. Battery cable connections loose or corroded	1. Check connections at battery, engine block and starter.
	2. Batteries undercharged.	2. Check charging system.
	3. Wrong engine oil	3. Drain and refill with recommended oil.
ENGINE CRANKS NORMALLY—WILL NOT START	1. Incorrect starting procedure.	1. Use recommended starting procedure.
	2. Incorrect or contaminated fuel	2. Flush fuel system and install correct fuel.
	3. No fuel to nozzles	3. Loosen injection line at a nozzle. Do not disconnect. Use care to direct fuel away from sources of ignition. Wipe connection to be sure it is dry. Crank 5 seconds. Fuel should flow from injection line. Tighten connection. If fuel does not flow, check fuel solenoid operation as follows: Connect a 12 volt test lamp from wire at injection pump solenoid to ground. Turn ignition to "On". Lamp should light. If lamp does not light, check wiring to solenoid.
	4. No fuel to injection pump	4. Remove line at inlet to injection pump fuel filter. Connect hose from line to metal container. Crank engine. If no fuel is discharged, test the fuel supply pump. If lamp does not light, check wiring to solenoid. If the pump is OK, check the injection pump fuel filter and replace if plugged. If filter and inlet line to injection pump are OK, replace injection pump.

Continued

DIESEL ENGINE DIAGNOSIS—Continued

Condition	Possible Cause	Correction
ENGINE CRANKS NORMALLY—WILL NOT START (Continued)	5. Plugged fuel return system	5. Disconnect fuel return line at injection pump and route hose to a metal container. Connect a hose to the injection pump connection and route it to the metal container. Crank the engine; if engine starts and runs, correct restriction in fuel return system.
	6. Pump timing incorrect	6. Make certain that pump timing mark is aligned with mark on adapter.
	7. Glow plug control system inoperative	7. Refer to Glow Plug System Trouble-Shooting & Diagnosis.
	8. Glow plugs inoperative	8. Refer to Glow Plug System Trouble-Shooting & Diagnosis.
	9. Internal engine problems	9. Correct as necessary.
	10. No voltage to fuel solenoid	10. Connect a 12 volt test lamp from injection pump solenoid to ground. Turn ignition to "On"; lamp should light. If lamp lights, remove test lamp and connect and disconnect solenoid connector and listen for solenoid operation. If solenoid does not operate, remove injection pump for repairs. If lamp does not light, refer to Glow Plug System Trouble-Shooting & Diagnosis.
	11. Restricted fuel tank filter.	11. Remove fuel tank and check filter.
ENGINE STARTS BUT WILL NOT CONTINUE TO RUN AT IDLE	1. No fuel in tank	1. Install correct fuel in tank.
	2. Incorrect or contaminated fuel	2. Flush fuel system and install correct fuel.
	3. Limited fuel to injection pump	3. Test the fuel supply pump. Replace as necessary.
	4. Fuel solenoid disengaged with ignition switch in the "ON" position	4. Connect a 12 volt test lamp from wire at injection pump solenoid to ground. Turn ignition to "ON". Lamp should light. Turn ignition to "START". Lamp should light. If lamp does not light in both positions, check wiring to solenoid.
	5. Restricted fuel return system	5. Disconnect fuel return to a metal container. Connect a hose to injection pump connection and route to metal container. Crank engine; if engine starts and runs, correct restriction in fuel system.
	6. Fast idle solenoid inoperative	6. With engine cold, start car; solenoid should move to support injection pump lever in "Fast idle position" for about 5 seconds. If solenoid does not move, refer to Electrical System Diagnosis.
	7. Low idle incorrectly adjusted	7. Adjust idle screw to specification.
	8. Pump timing incorrect.	8. Make certain that timing mark, on injection pump, is aligned with mark on adapter.
	9. Glow plug control system malfunction	9. Refer to Glow Plug System Trouble-Shooting & Diagnosis.
	10. Injection pump malfunction	10. Install replacement pump.
	11. Internal engine problems	11. Correct as necessary.
ENGINE STARTS, IDLES ROUGH, WITHOUT ABNORMAL NOISE OR SMOKE	1. Low idle incorrectly adjusted	1. Adjust idle screw to specification
	2. Injection line leaks	2. Wipe off injection lines and connections. Run engine and check for leaks. Correct leaks.
	3. Restricted fuel return system	3. Disconnect fuel return line at injection pump and route hose to a metal container. Connect a hose to the injection pump container. Crank the engine; if engine starts and runs, correct restriction in fuel return system.
	4. Incorrect or contaminated fuel	4. Flush fuel system and install correct fuel.
	5. Nozzle(s) inoperative	5. With engine running, loosen injection line fitting at each nozzle in turn. Use care to direct fuel away from sources of ignition. Each nozzle should contribute to rough running. If nozzle is found that does not change idle quality, it should be replaced.

Continued

DIESEL ENGINE DIAGNOSIS—Continued

Condition	Possible Cause	Correction
ENGINE STARTS, IDLES ROUGH, WITHOUT ABNORMAL NOISE OR SMOKE (Continued)	6. Internal fuel leak at nozzle(s)	6. Disconnect fuel return system from nozzles on one bank at a time. With the engine running, observe the normal fuel seepage at the nozzles. Replace any nozzle with excessive fuel leakage.
	7. Fuel supply pump malfunctions	7. Test the fuel supply pump. Replace if necessary.
	8. Uneven fuel distribution to cylinders	8. Install new or reconditioned nozzles, one at a time, until condition is corrected as indicated by normal idle.
ENGINE STARTS AND IDLES ROUGH WITH EXCESSIVE NOISE AND/OR SMOKE	1. Injection pump timing incorrect	1. Be sure timing mark on injection pump is aligned with mark on adapter.
	2. Nozzle(s) inoperative	2. With engine running, crank injection line at each nozzle, one at a time. Use care to direct fuel away from sources of ignition. Each nozzle should contribute to rough running. If a nozzle is found that does not affect idle quality or changes noise and/or smoke, it should be replaced.
	3. High pressure lines incorrectly installed	3. Check routing of each line. Correct as required.
ENGINE MISFIRES BUT IDLES CORRECTLY	1. Plugged fuel filter	1. Replace filter.
	2. Incorrect injection pump timing	2. Be sure that timing mark on injection pump and adapter are aligned.
	3. Incorrect or contaminated fuel	3. Flush fuel system and install correct fuel.
ENGINE WILL NOT RETURN TO IDLE	1. External linkage misadjustment or failure	1. Reset linkage or replace as required.
	2. Internal injection pump malfunction	2. Install replacement injection pump.
FUEL LEAKS ON GROUND—NO ENGINE MALFUNCTION	1. Loose or broken fuel line or connection	1. Examine complete fuel system, including tank, supply, injection and return system. Determine source and cause of leak and repair.
	2. Internal injection pump failure	2. Install replacement injection pump.
SIGNIFICANT LOSS OF POWER	1. Incorrect or contaminated fuel	1. Flush fuel system and install correct fuel.
	2. Pinched or otherwise restricted return system	2. Examine system for restriction and correct as required.
	3. Plugged fuel tank vent	3. Remove fuel cap. If "hissing" noise is heard, vent is plugged and should be cleaned.
	4. Restricted fuel supply	4. Examine fuel supply system to determine cause of restriction. Repair as required.
	5. Plugged fuel filter	5. Remove and replace filter.
	6. External compression leaks	6. Check for compression leaks at all nozzles and glow plugs, using "Leak-Tec" or equivalent. If leak is found, tighten nozzle clamp or glow plug. If leak persists at a nozzle, remove it and reinstall with a new carbon stop seal and compression seal.
	7. Plugged nozzle(s)	7. Remove nozzles, check for plugging and have repaired or replaced.
	8. Internal engine problem	8. Correct as necessary.
	9. EGR malfunction	9. Correct as necessary
NOISE—"RAP" FROM ONE OR MORE CYLINDERS	1. Air in fuel system	1. Check for leaks and correct.
	2. Air in high pressure line(s)	2. Crack line at nozzle(s) and bleed air at each cylinder determined to be causing noise. Use care to direct fuel away from sources of ignition and be sure to carefully retighten lines.
	3. Nozzle(s) sticking open or with very low blowoff pressure	3. Replace the nozzle(s) causing the problem.
	4. Internal engine problem	4. Correct as necessary.

Continued

DIESEL ENGINE DIAGNOSIS—Continued

Condition	Possible Cause	Correction
NOISE—SIGNIFICANT OVERALL COMBUSTION NOISE INCREASE WITH EXCESSIVE BLACK SMOKE	1. Timing not set to specification 2. Internal engine problem 3. Injection pump housing pressure out of specifications. 4. Internal injection pump problem 5. EGR malfunction	1. Align timing marks on adapter and injection pump. 2. Check for presence of oil in the air crossover. If present, determine cause and correct. 3. Check housing pressure. If incorrect, replace fuel return line connector assembly. 4. Replace pump. 5. Correct as necessary.
NOISE—INTERNAL OR EXTERNAL	1. Fuel supply pump, alternator, water pump, valve train, vacuum pump, bearings, etc.	1. Inspect and correct as necessary.
ENGINE OVERHEATS	1. Coolant system leak or oil cooler system leak 2. Belt failure or slippage 3. Thermostat malfunction, head gasket failure or internal engine problem	1. Check for leaks and correct as required. 2. Replace or adjust as required. 3. Inspect and correct as necessary.
INSTRUMENT PANEL OIL WARNING LAMP "ON" AT IDLE	1. Oil cooler or oil cooler line restricted 2. Internal engine problem	1. Remove restriction in cooler or cooler line. 2. Correct as necessary.
ENGINE WILL NOT SHUT OFF WITH KEY **NOTE:** With engine at idle, pinch the fuel return line at the injection pump to shut off engine.	1. Injection pump solenoid does not drop out 2. Injection pump solenoid return spring failed	1. Refer to electrical diagnosis. If problem is determined to be internal with the injection pump, replace the injection pump. 2. Replace injection pump.

GLOW PLUG SYSTEM

1983

Refer to wiring diagram, Figs. 1 and 2 and Figs. 3 through 9, for troubleshooting and diagnosis procedures for the 1983 glow plug system.

1980–81

Refer to wiring diagram, Fig. 10, and Figs. 11 through 15, for troubleshooting and diagnosis procedures for the 1980–81 glow plug system.

1978–79

Refer to wiring diagram, Fig. 16, and use the following procedures for troubleshooting and diagnosis procedures for the 1978–79 glow plug system.

NOTE: Unless specified, testing is performed with a 12 volt voltmeter. Measurements should be made to a good engine ground. Grounding signals should be less than 1.5 volts. 12 volt signals should be 10.5–12.5 volts unless otherwise stated. Measurements are obtained with the ignition switch in the "On" position within a time period prior to emergency shut-down (2–5 minutes) after the engine is "On". Check battery voltage before testing. Charge batteries if voltage is less than 12.4 with ignition "Off".

Glow Plugs "Off" During Preglow & Afterglow

Possible Causes

1. Open 940 or 941 circuits from relays.
2. Open 2X or 2Y circuits from junction block relays.
3. Defective relay or relays.
4. Open 901 circuit between module and relays.
5. Open module ground at 150 circuit.
6. Open 3 circuit to relays or module.
7. Faulty 935 circuit to thermistor.
8. Open 925 circuit to alternator.
9. Defective module.

Diagnosis

1. Measure voltage of the following circuits at the relays: 2, 3, 150, 901, 940 and 941. Note the voltage readings. The voltage readings for the 2, 3, 940 and 941 circuits should be 12 volts. Voltage reading for the 150 circuit should be 9 volts. Voltage reading for the 901 circuit should be zero volts.
2. If circuits 2, 3, and 901 are satisfactory and circuits 940 and 941 are not, replace defective relay.
3. If circuits 2 and 3 are not satisfactory, repair open circuit or fusible link.
4. If circuit 901 is not satisfactory, measure voltage at module.
5. If circuit 150 is not satisfactory (right bank glow plugs not functioning), repair ground circuit.
6. Measure voltage of 940 and 941 circuits at glow plugs. If not satisfactory, repair open circuit.
7. Measure voltage of 901 circuit at module. If voltage is satisfactory, replace open relays. If voltage is not satisfactory, measure voltage of 3 and 150 circuits at module. Reading of circuit 3 should be 12 volts and reading of circuit 150 should be zero volts. If readings are not within specifications, repair open circuits.
8. Measure voltage of 25 circuit at module. Voltage should be less than 3 volts. If not, repair open circuit to alternator.
9. Replace the thermistor in the 935 circuit with a milliammeter placed on 0–1ma range. The reading should be .3–.5ma. If reading is within specifications, leave milliammeter in place and recheck 901 circuit. If 901 circuit is now satisfactory, check thermistor as follows: check resistance of the thermistor. Resistance should be 2000–5000 ohms at 60°–80° F.
10. If 901 circuit is unsatisfactory at this time, replace module.

Preglow Okay, Time Period Incorrect

Possible Causes

1. Open circuit to thermistor.
2. Faulty 935 circuit to thermistor.
3. Defective thermistor.
4. Defective module.

Diagnosis

1. Replace the thermistor in the 935 circuit with a milliammeter placed on the 0–1ma range. Reading should be .3–.5ma. If satisfactory, check thermistor resistance. Resistance should be 2000–5000 ohms at 60°–80° F. If satisfactory, replace module.

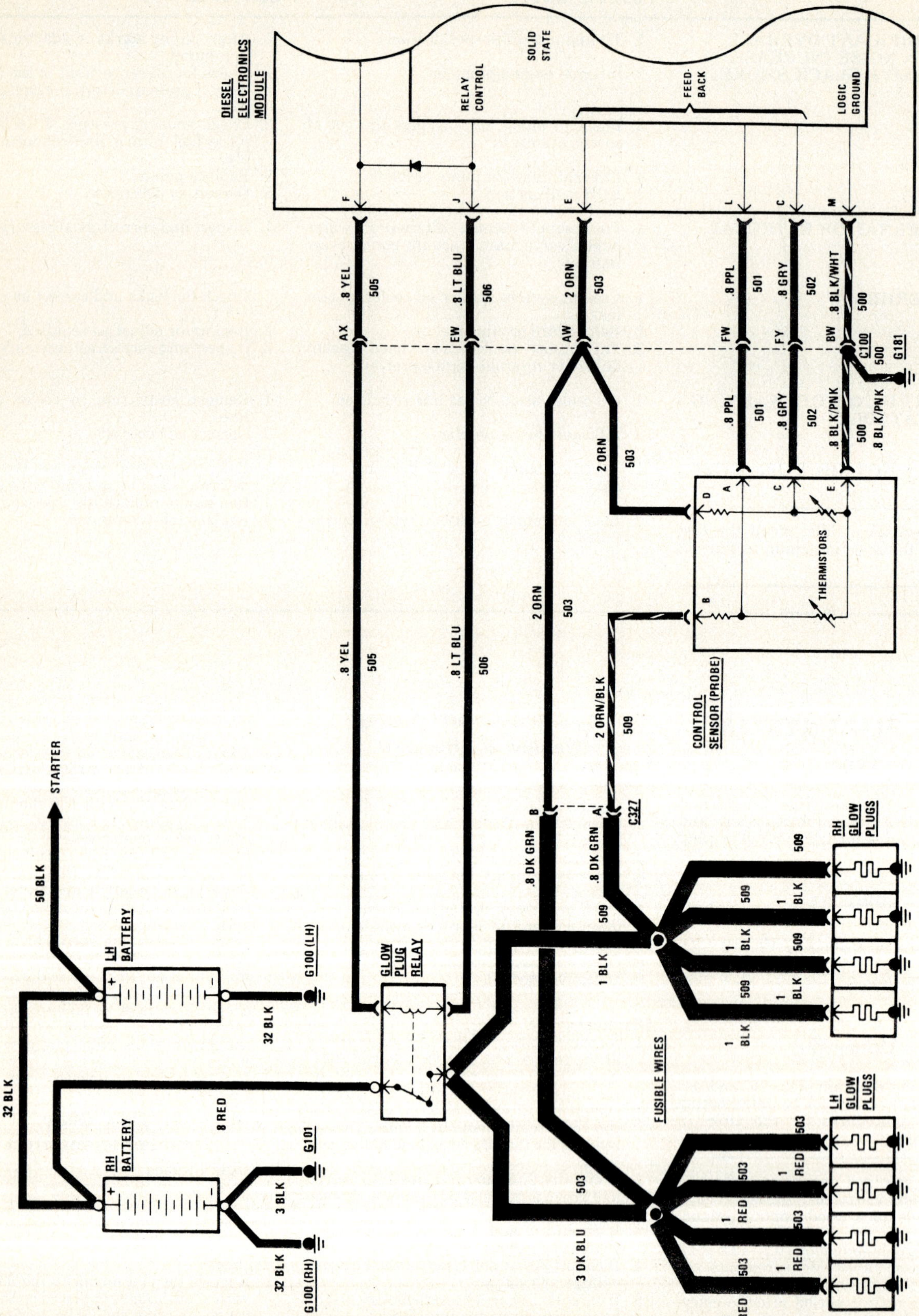

Fig. 1 (Part 1 of 2). Glow plug system wiring diagram. 1983

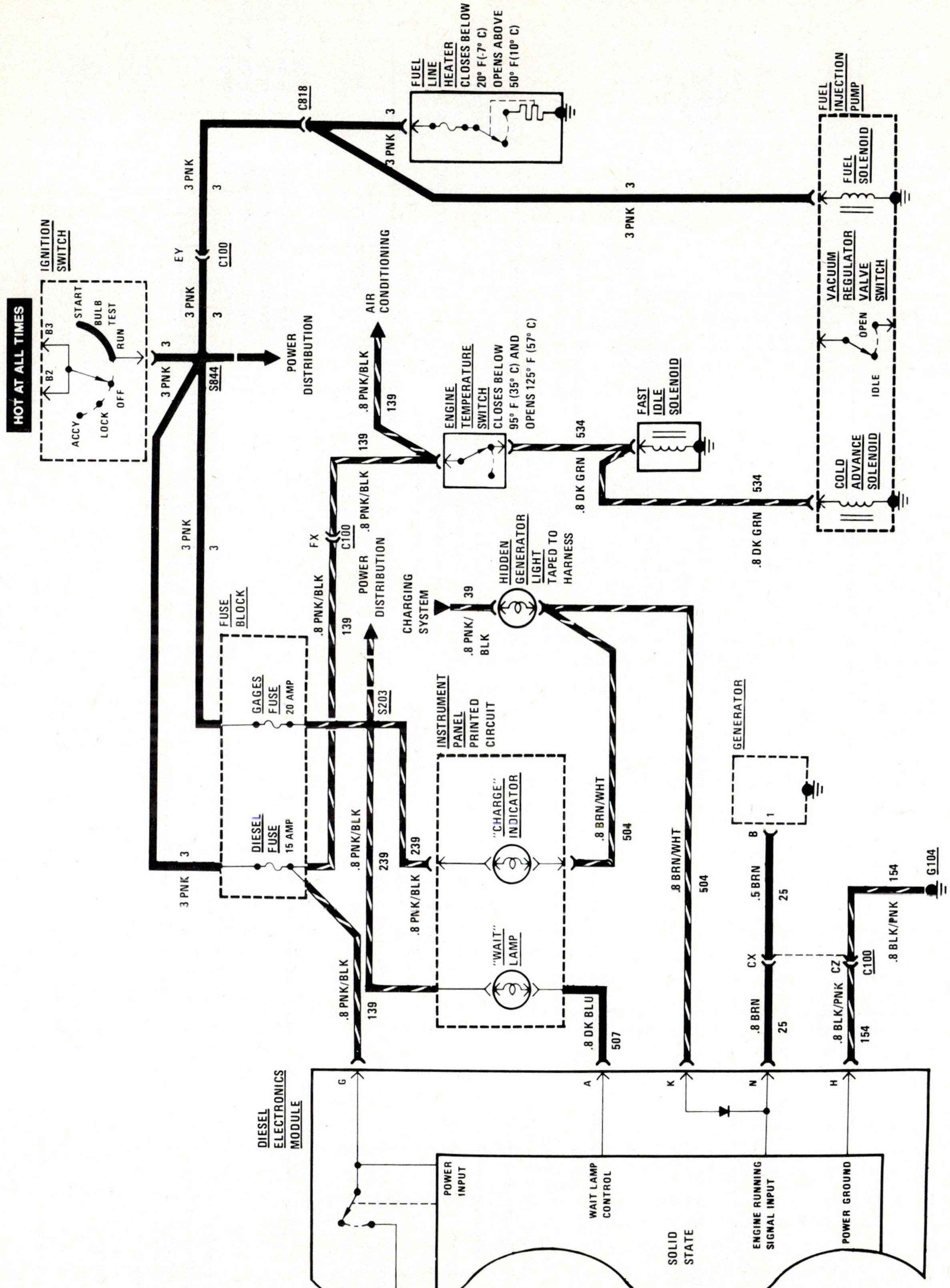

Fig. 2 (Part 2 of 2). Glow plug system wiring diagram. 1983

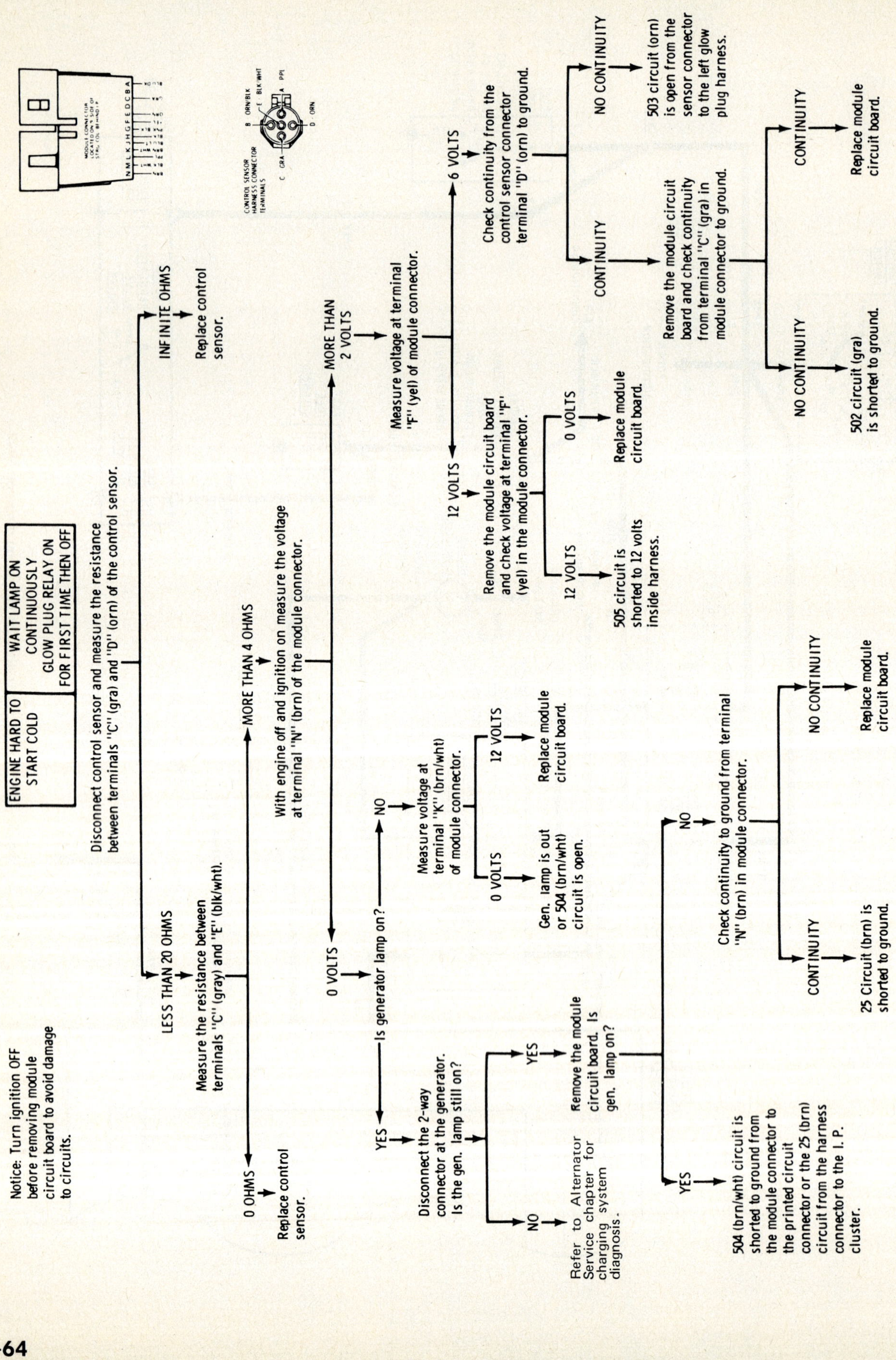

Fig. 3 Glow plug system diagnosis chart. 1983

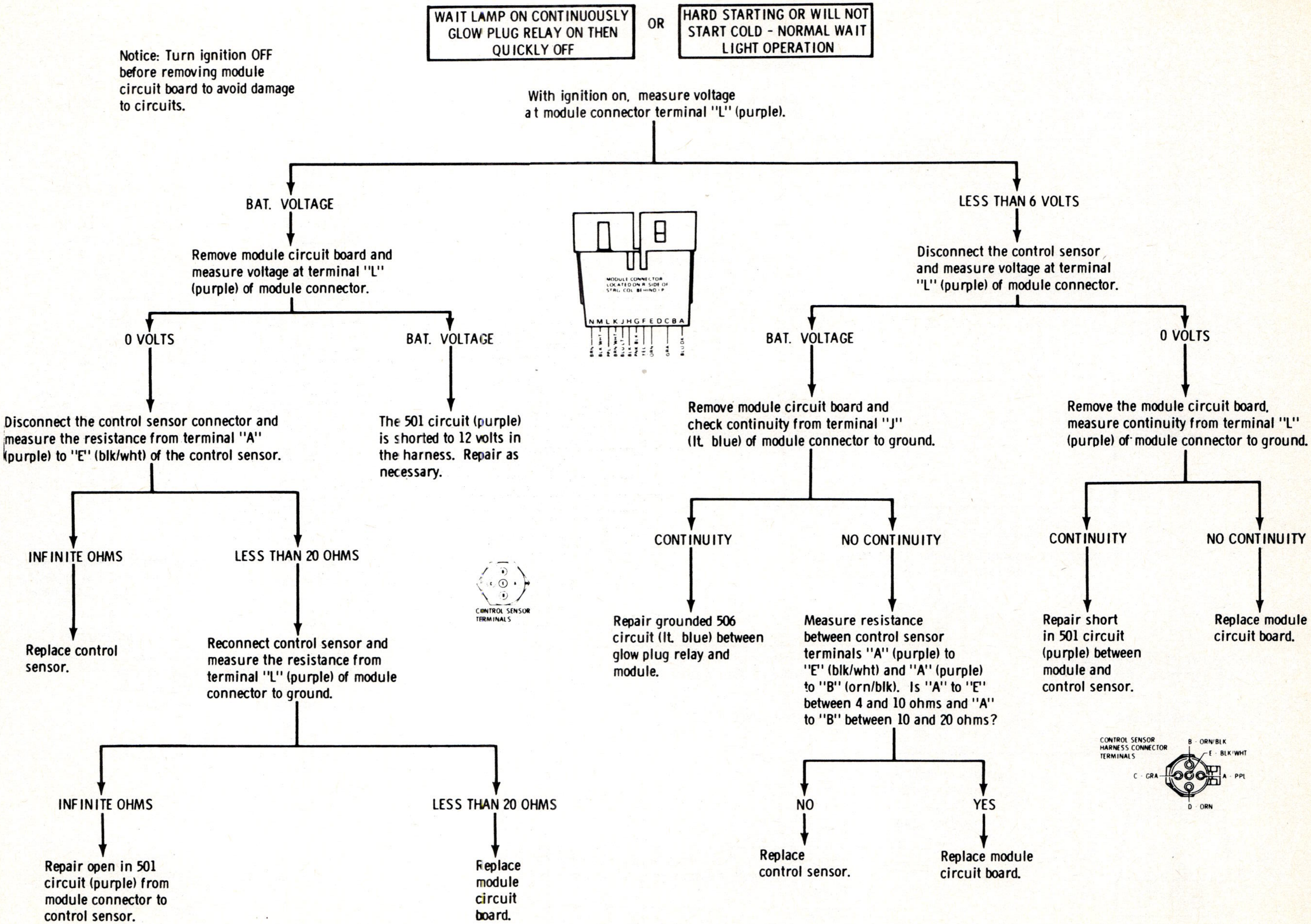

Fig. 4 Glow plug system diagnosis chart. 1983

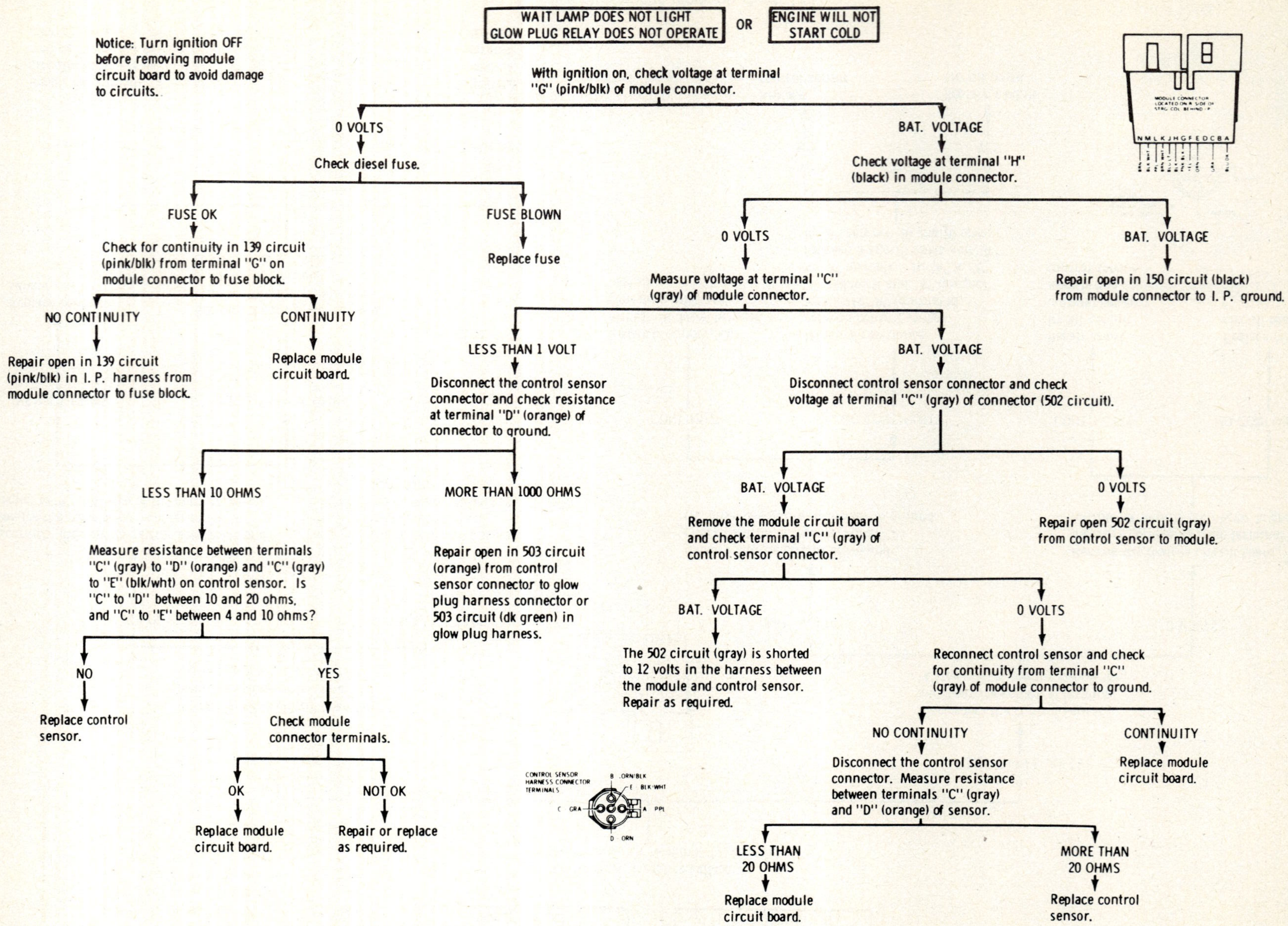

Fig. 5 Glow plug system diagnosis chart. 1983

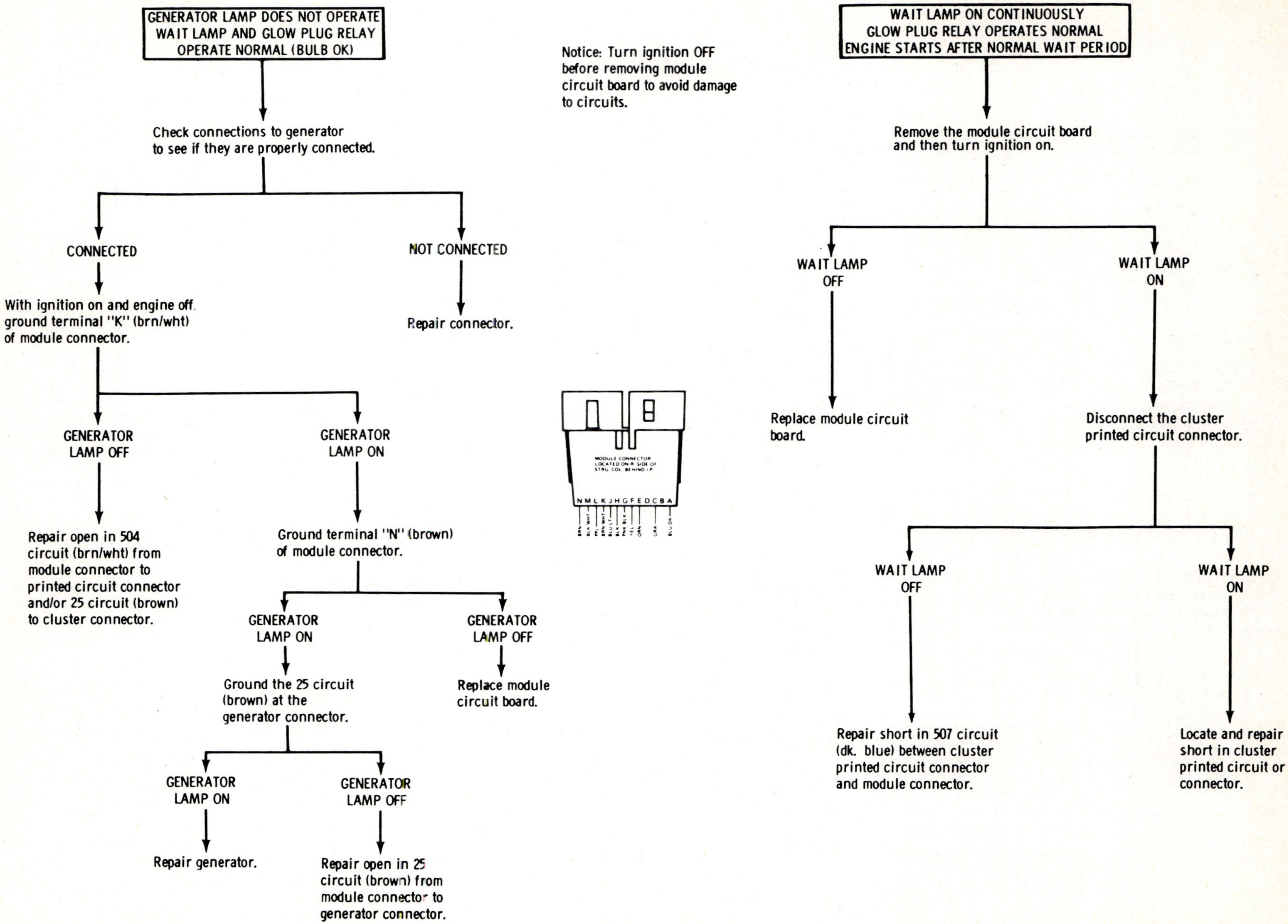

Fig. 6 Glow plug system diagnosis chart. 1983

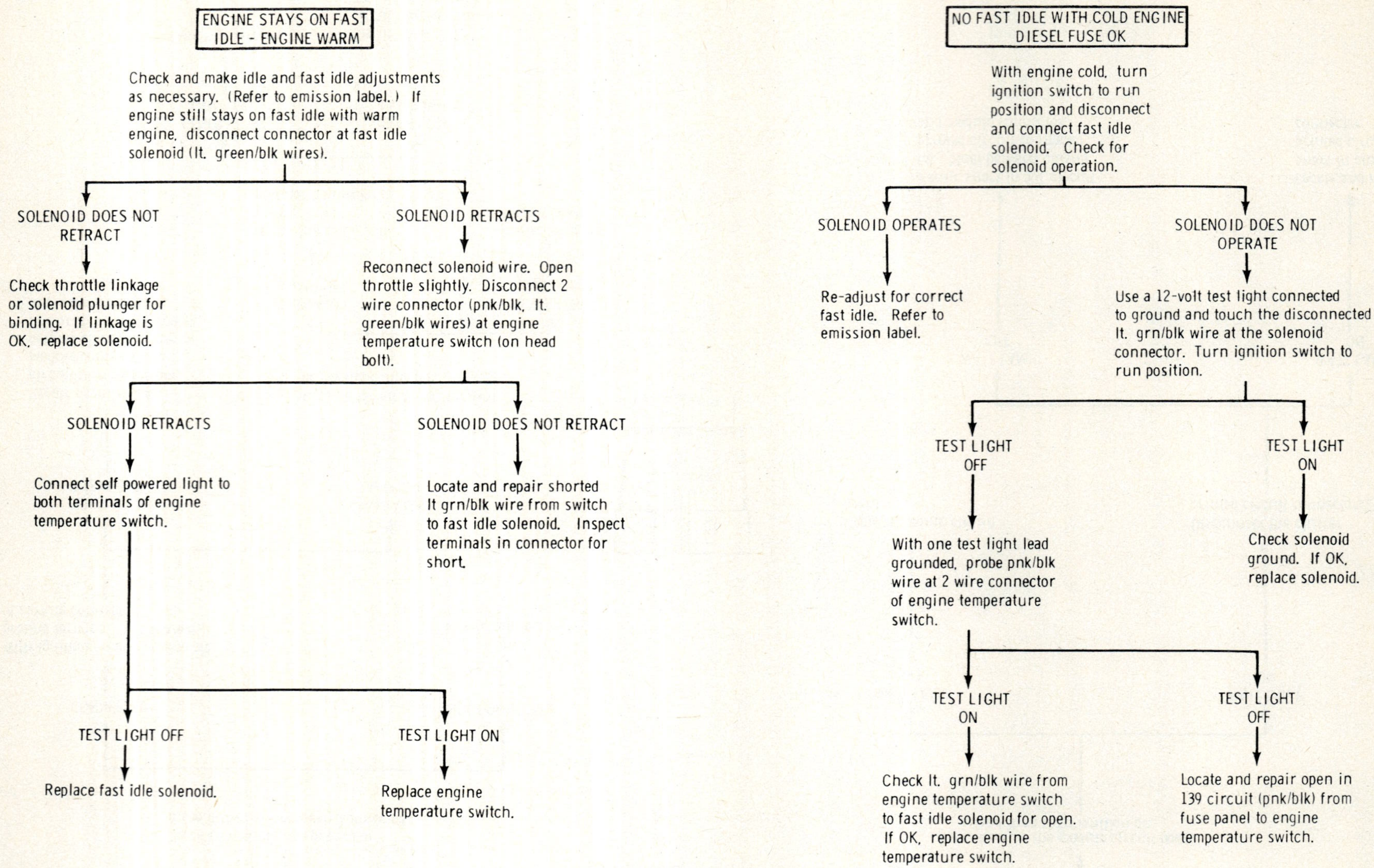

Fig. 7 Glow plug system diagnosis chart. 1983

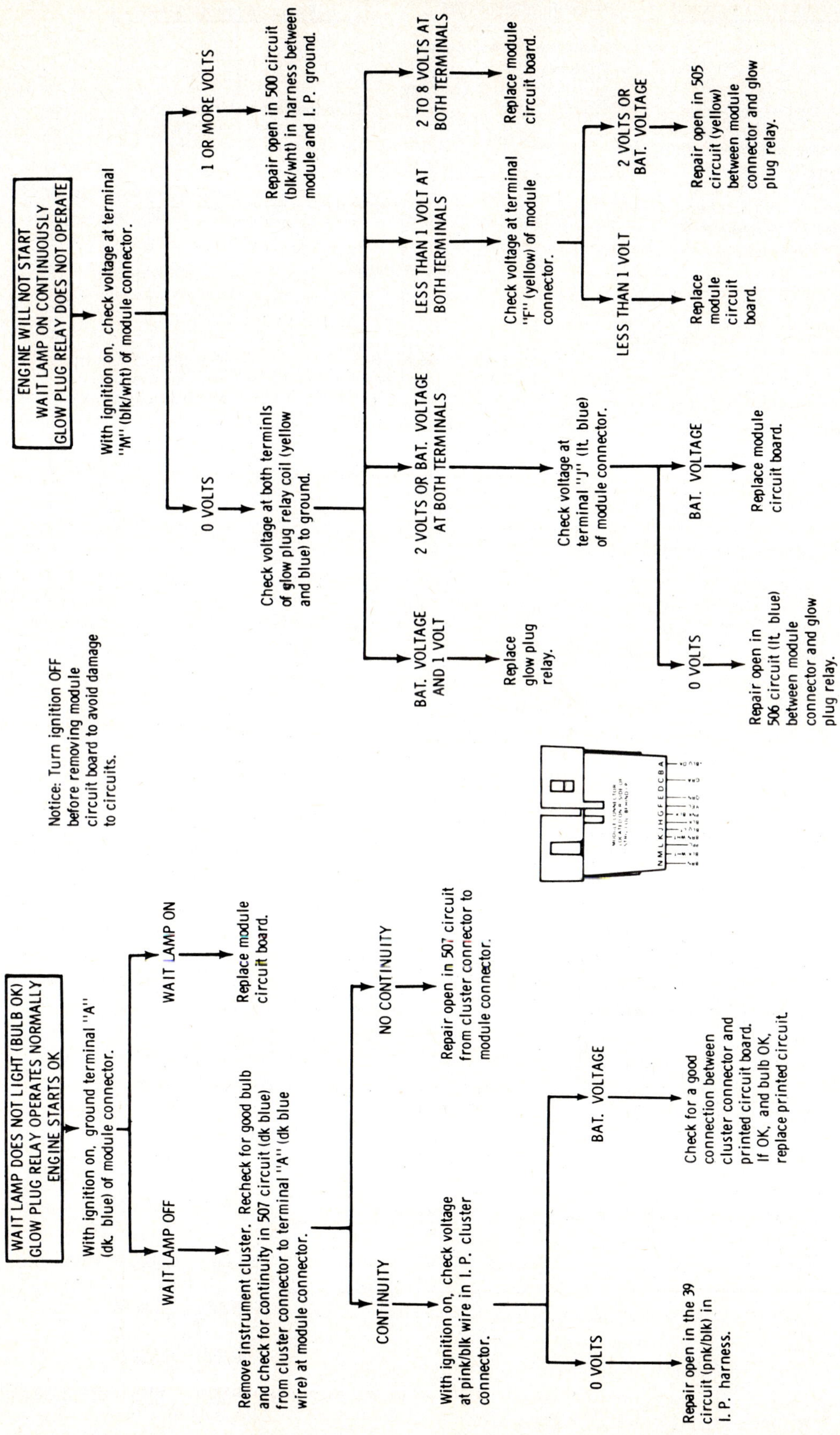

Fig. 8 Glow plug system diagnosis chart. 1983

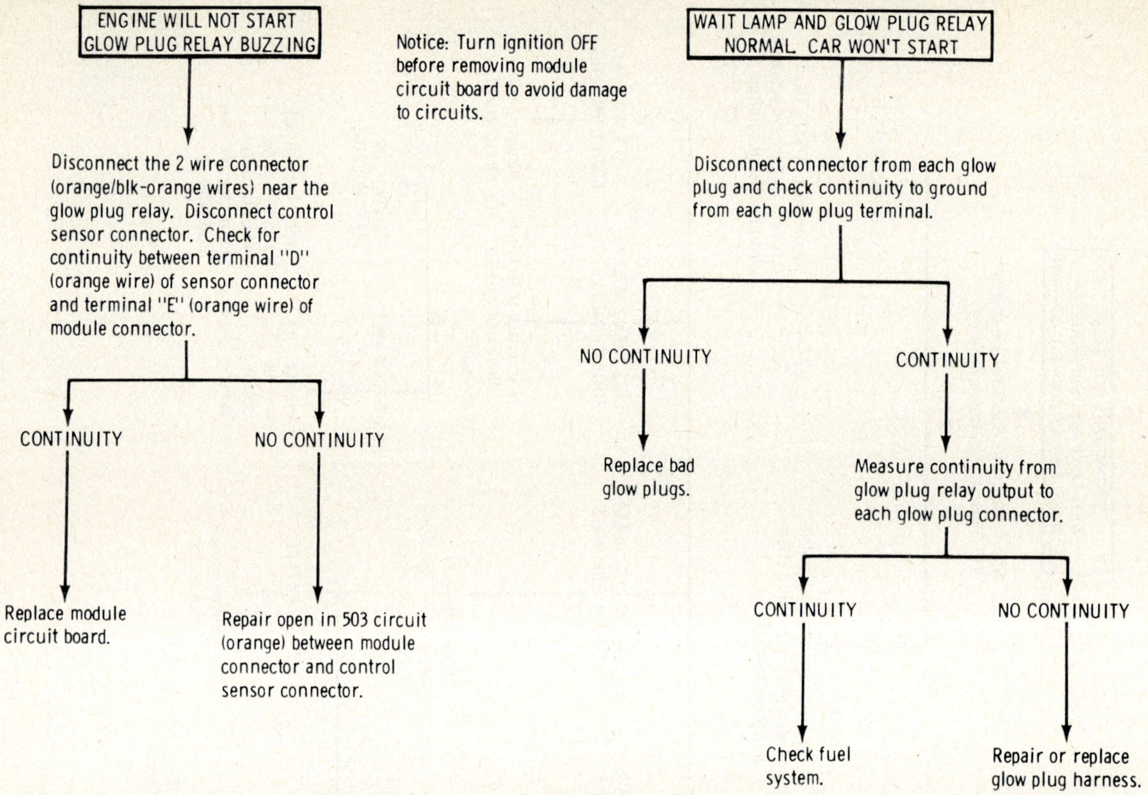

Fig. 9 Glow plug system diagnosis chart. 1983

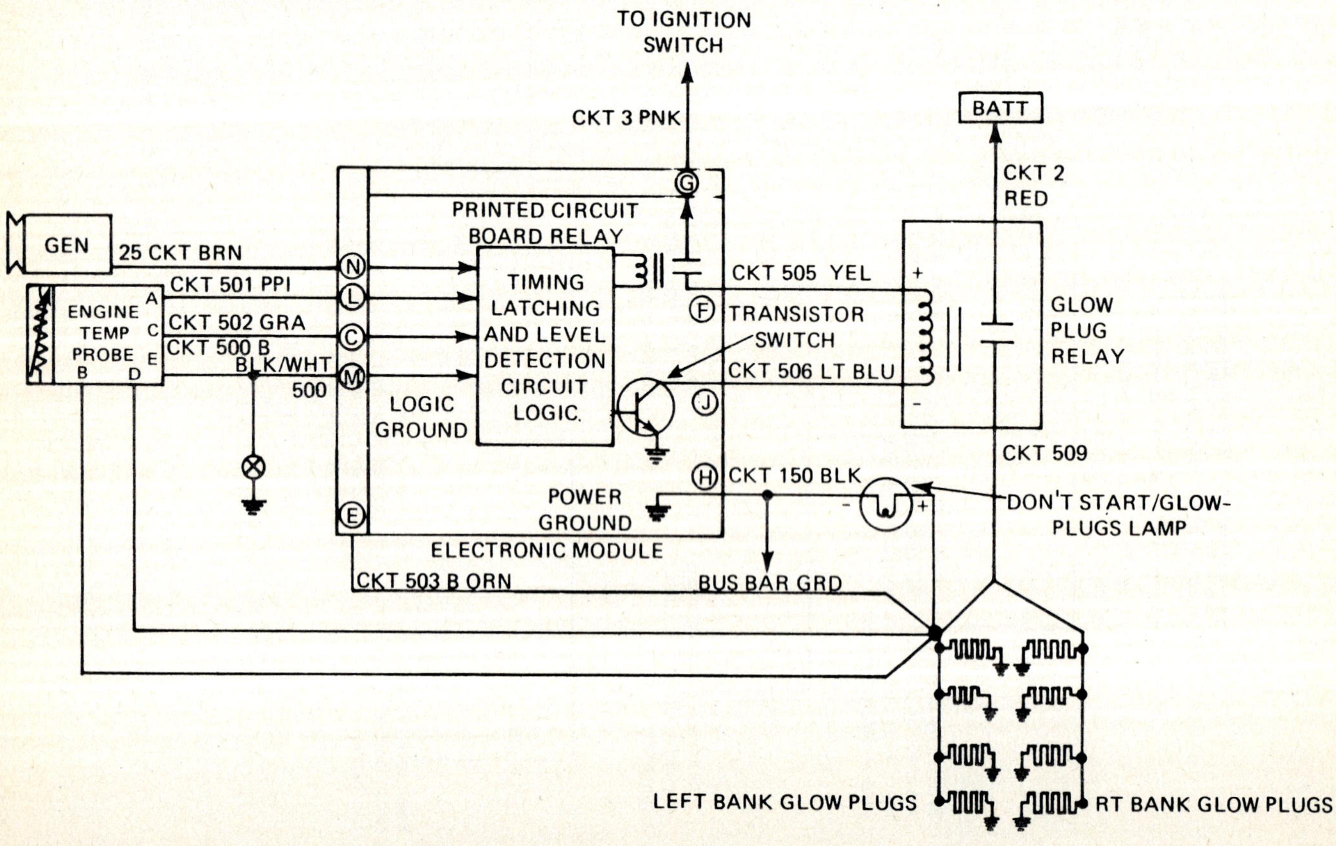

Fig. 10 Glow plug system wiring diagram. 1980–81

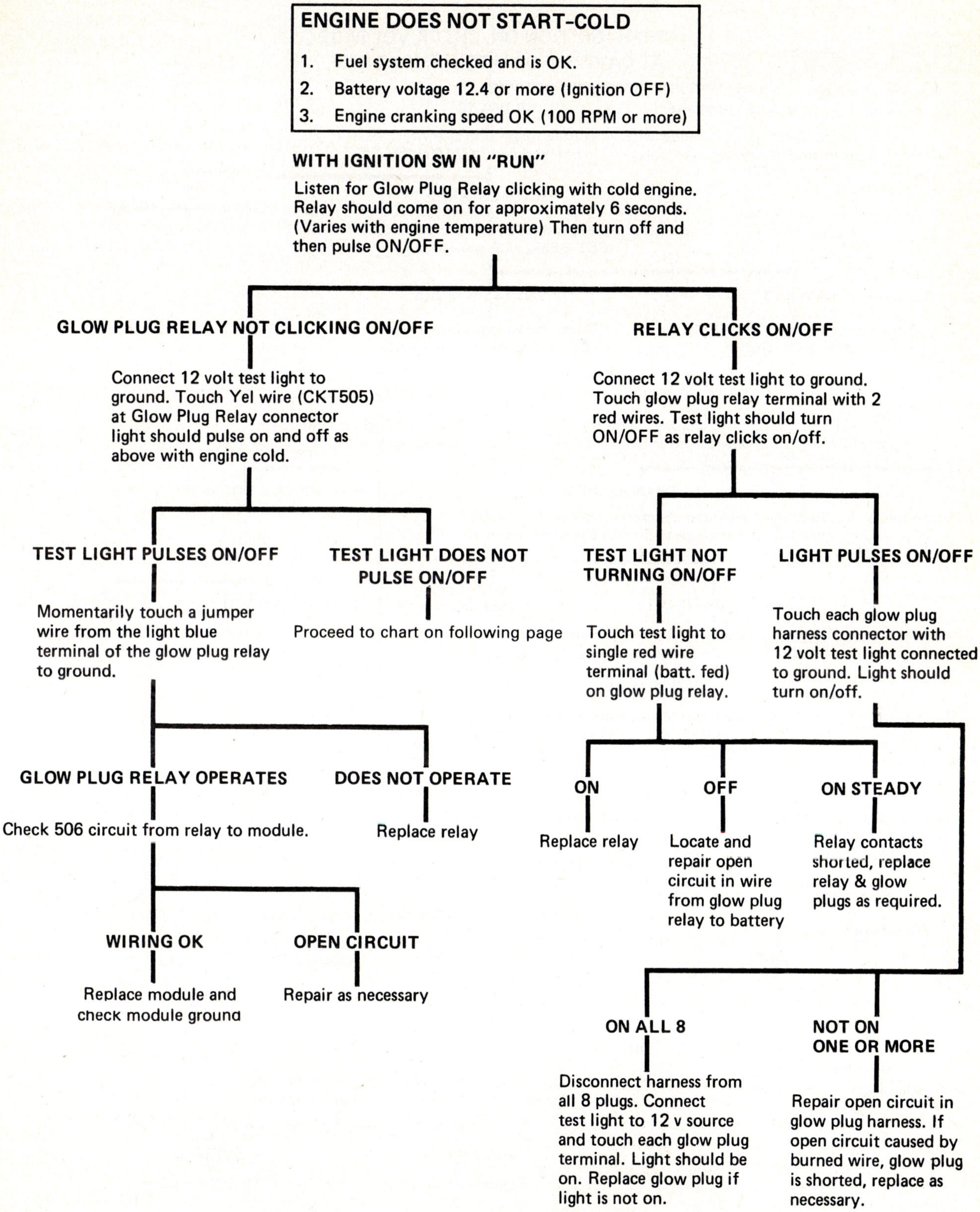

Fig. 11 Glow plug system diagnosis chart. 1980–81

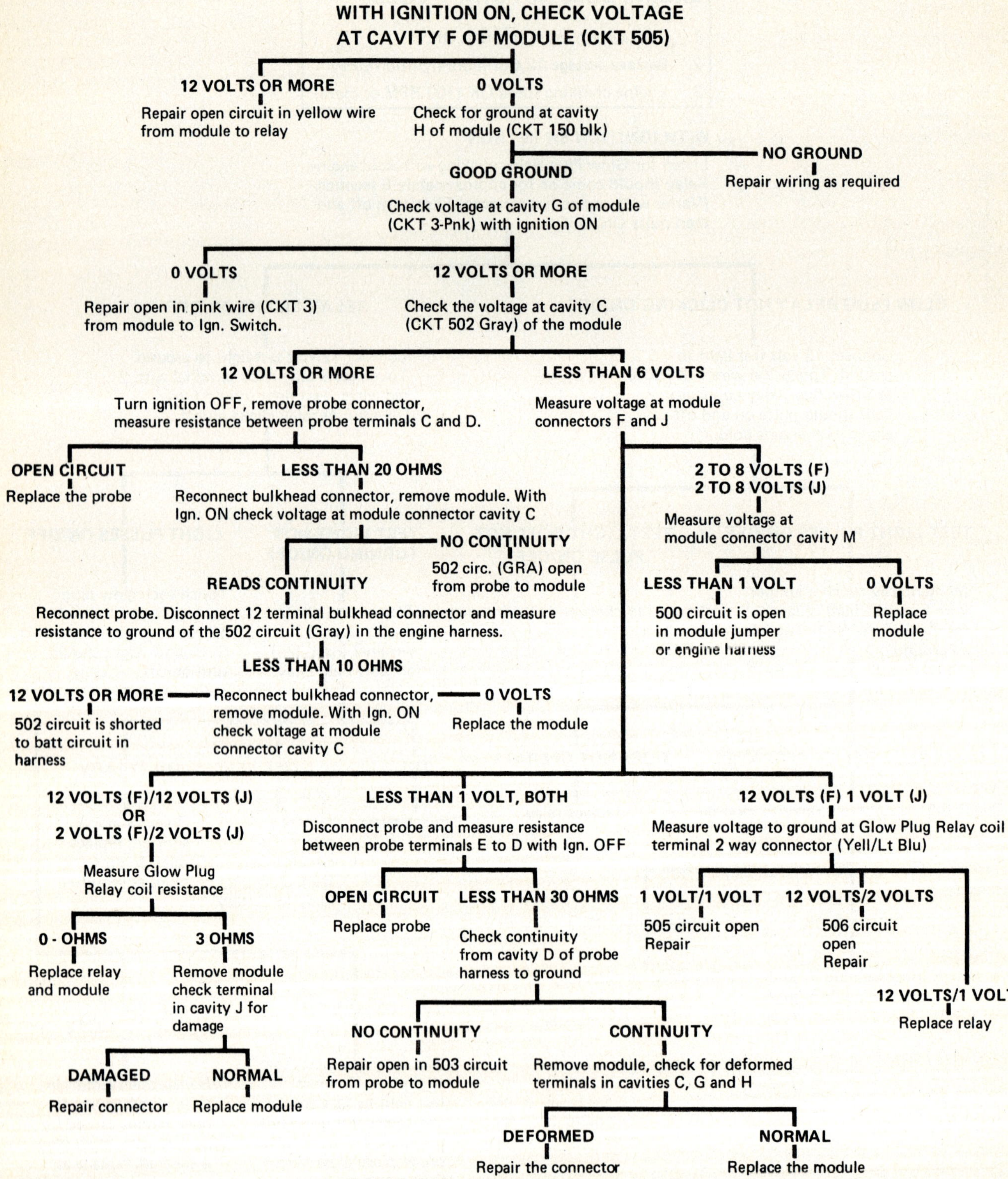

Fig. 12 Glow plug system diagnosis chart. 1980–81

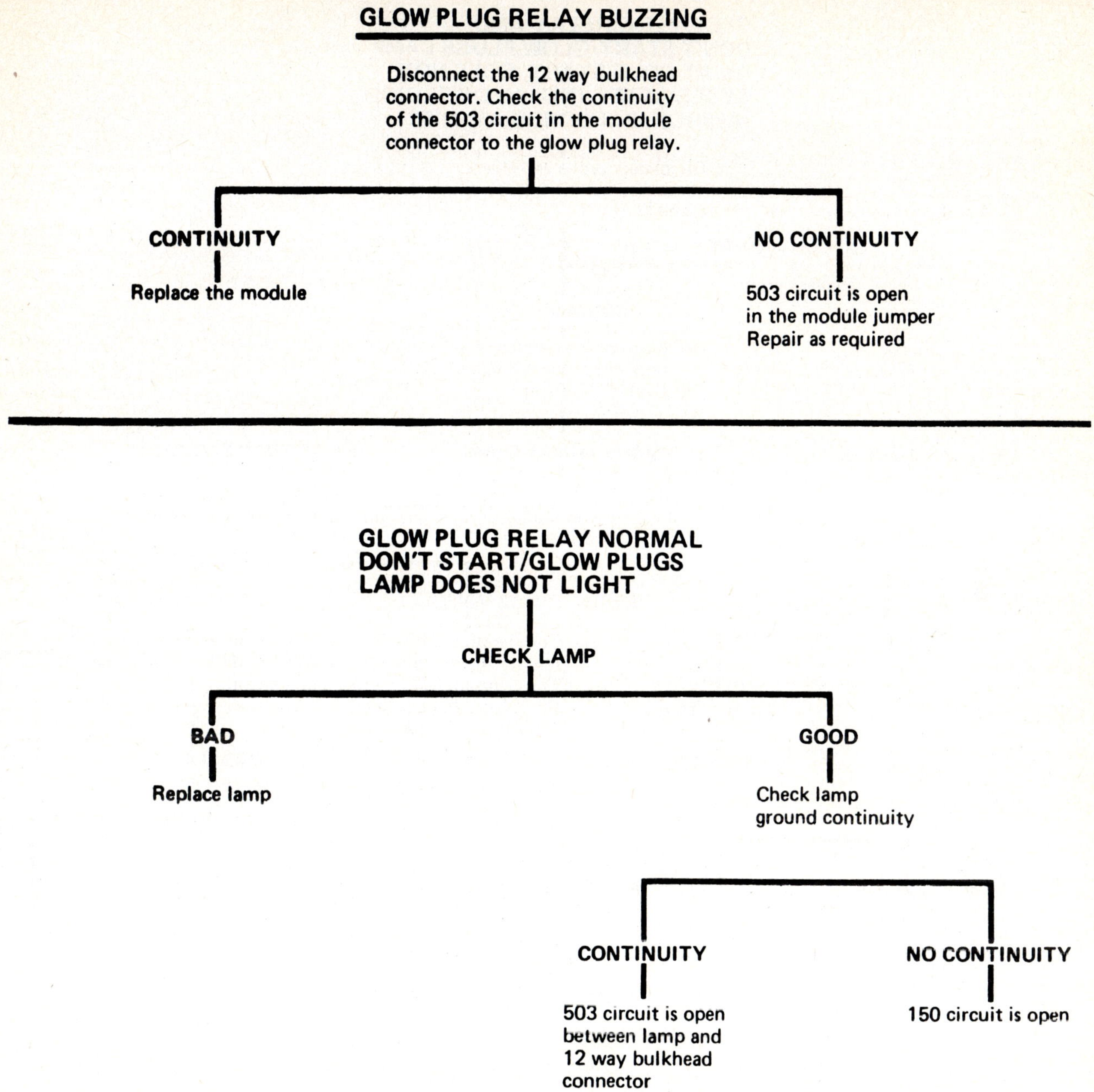

Fig. 13 Glow plug system diagnosis chart. 1980–81

Preglow Okay, Afterglow Time Period Incorrect

Possible Causes

1. Defective module.
2. Defective 906A circuit to module from starter solenoid.
3. Defective 925 circuit to module from alternator.

Diagnosis

1. Check voltage of 906A circuit at module while cranking engine. Voltage should be 8–12 volts. Then, check the voltage with ignition key in "Run" position. Voltage should be zero volts. If voltages are not within specifications, repair circuit.
2. On vehicles equipped with an alternator warning lamp, ensure that lamp is "On" before cranking engine and the lamp if "Off" after the engine starts. On vehicles equipped with a voltmeter or on vehicle equipped with an alternator warning lamp that checked satisfactory, check the voltage of the 925 circuit at the module with the engine running. Voltage should be 14–15 volts. If not, repair 25 circuit to alternator.

Glow Plugs Remain Activated After Required Time Period

Possible Causes

1. Defective relays.
2. Faulty 901 circuit.
3. Defective module.

Diagnosis

1. Check 901 circuit voltage with ignition "On." Voltage should be zero and after 2–5 minutes, voltage should be 12 volts without cranking engine. If voltages are satisfactory, replace relays if circuits 940 and 941 check at 12 volts.
2. If 901 circuit is not satisfactory, remove module and recheck 901 circuit for 12 volts. If no voltage is present, repair open 901 circuit. If a 12 volt reading is obtained, replace module.

Fast Idle "On" With High Engine Temperature &/Or A/C "Off"

Possible Causes

1. Defective fast idle solenoid.
2. Faulty 951 circuit to solenoid.
3. Defective module.

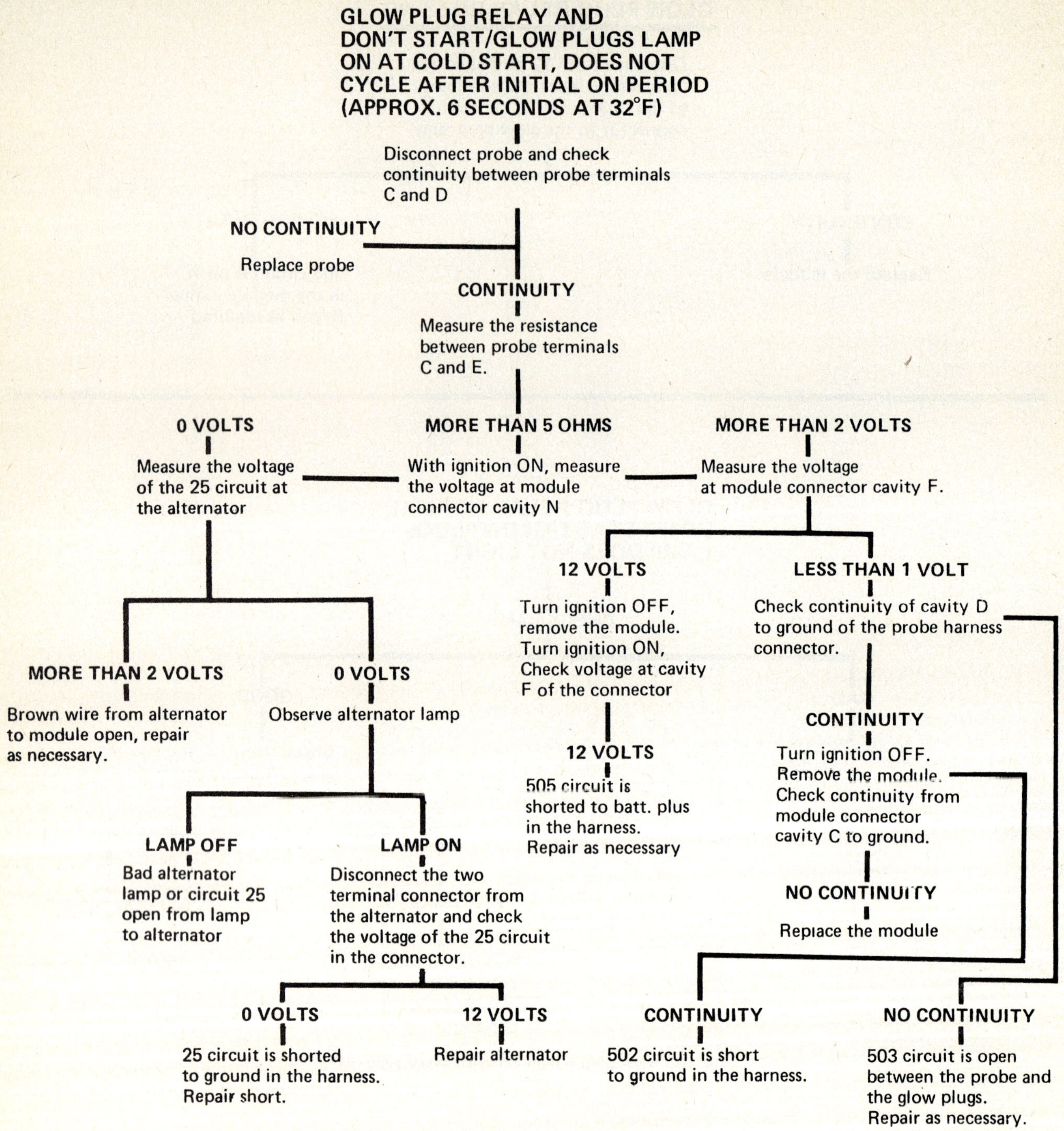

Fig. 14 Glow plug system diagnosis chart. 1980–81

4. Open 935 circuit to thermistor.
5. Defective thermistor.
6. Defective A/C controller.

Diagnosis
1. Remove solenoid connector, replace solenoid if it does not retract.
2. Check voltage at 951 circuits. Reading should be 12 volts. If not, disconnect module and recheck reading. If still not 12 volts, repair faulty 951 circuit.
3. If 951 circuit reads zero volts with module connected but 12 volts disconnected, apply a short to ground at 935 circuit at module. Recheck 951 with A/C "Off." If satisfactory, repair open 935 circuit to thermistor. If not, check voltage at 66 circuit. If zero volts, replace module. If no volts with A/C "Off", repair defective A/C controller switch.

Fast Idle "Off" With Cold Engine Or A/C "On"

Possible Causes
1. Defective fast idle solenoid.
2. Open 139 circuit feed to solenoid.
3. Defective module.
4. Open module ground 150 circuit.
5. Faulty 935 circuit to thermistor.
6. Defective thermistor.
7. No module feed, 3 circuit.
8. Open or faulty 66 circuit to A/C clutch.
9. No A/C voltage at feed point.

Diagnosis
1. Measure voltage at 139 circuit on solenoid. If not satisfactory, repair open 139 circuit. If voltage is satisfactory, measure 951 circuit. Reading should be zero volts. If within specifications, remove solenoid connector and apply 12 volts to one side of solenoid and ground the other side. If defective, replace solenoid.

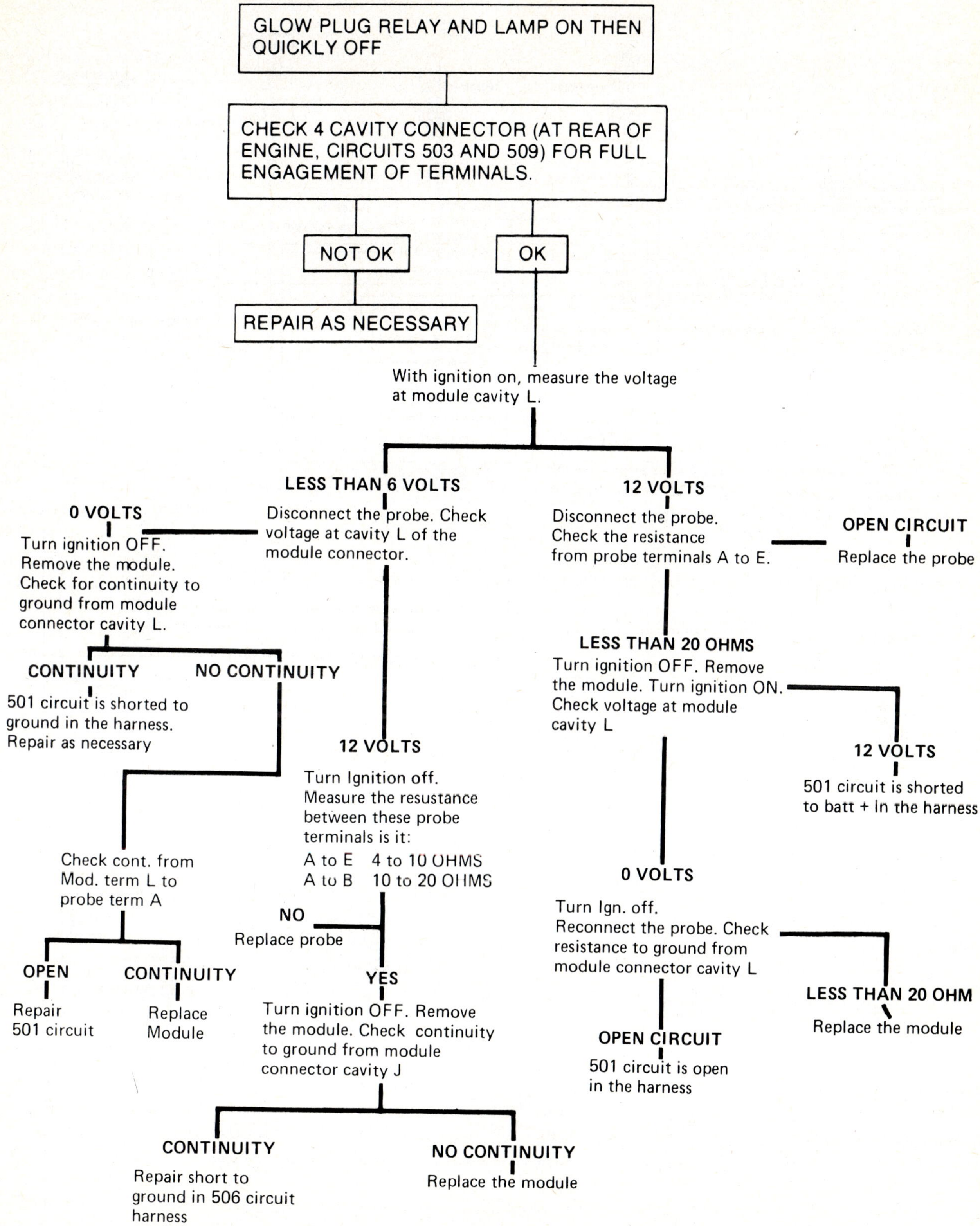

Fig. 15 Glow plug system diagnosis chart. 1980–81

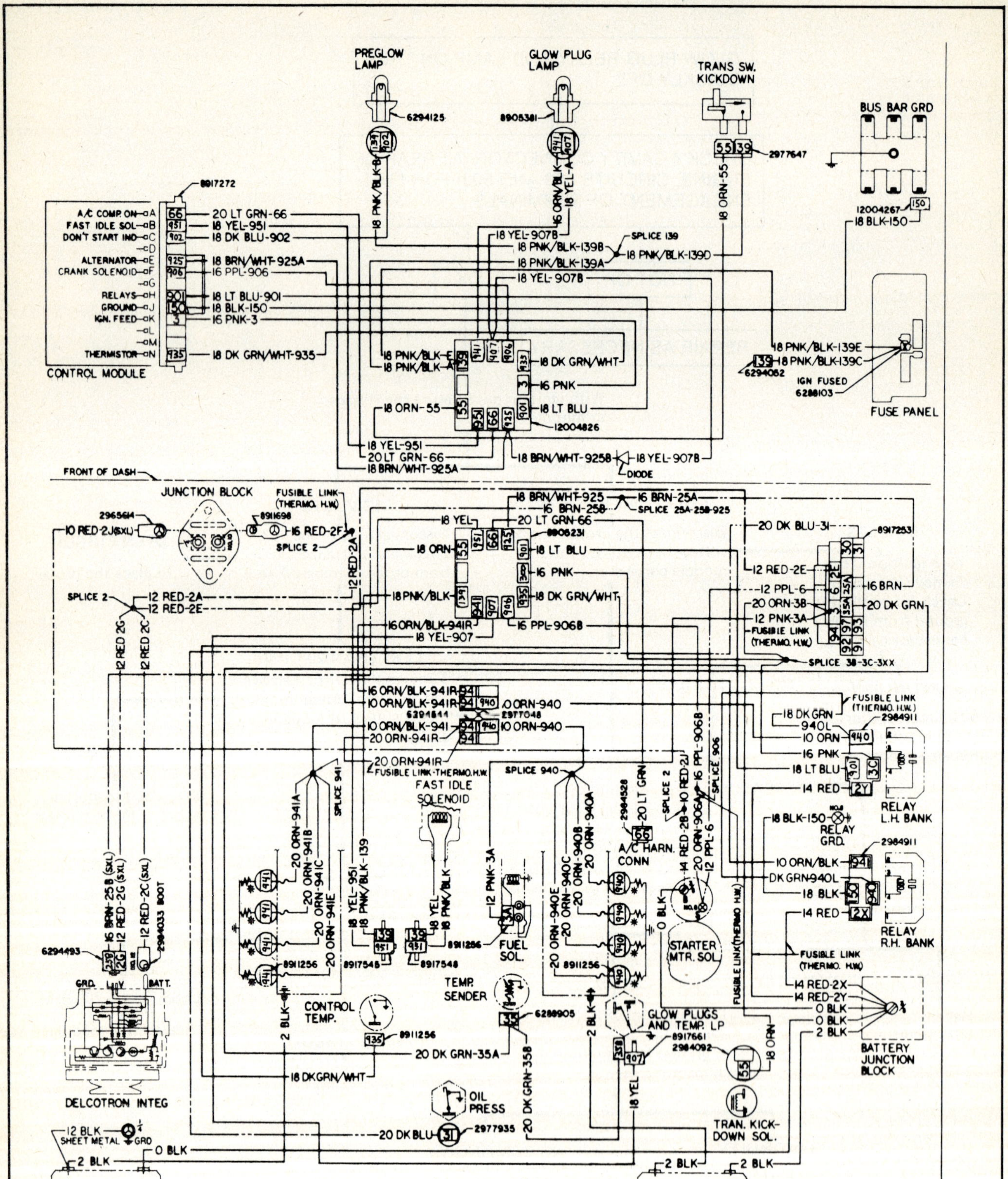

Fig. 16 Glow plug system wiring diagram. 1978–79

2. If circuit 951 is not zero volts, recheck at module. If satisfactory, repair open circuit.
3. If circuit 951 is not zero volts at module, check 3 circuit, 150 circuit, and 66 circuit at module. Repair if not as follows: 3 circuit, 12 volts; 150 circuit, zero volts; 66 circuit, 12 volts (A/C on). If satisfactory, replace module.
4. If solenoid is activated with A/C "On", but inactive with A/C off and engine cold, check 935 circuit to thermistor. If 935 circuit voltage is present (0.3–0.5 volts), check thermistor. If thermistor is satisfactory, replace module. Repair 935 circuit or thermistor for defects.
5. If 66 circuit voltage is zero volts with A/C "On", repair open circuit to A/C feed.

Fast Idle Not "On" For 5 Sec. After Each Engine Start With A/C "Off"

Possible Causes
1. Defective module.
2. Open 906 circuit to starter solenoid.

Diagnosis
1. If solenoid does operate when tested with jumper lead, but does not operate after each start with engine hot and A/C "Off", check 906A circuit at module with ignition switch in "Crank" position. If reading is 12 volts, replace module. If no reading is obtained, repair open 906A circuit.

No Voltage At Pump Solenoid With Ignition "On"

Possible Causes
1. Defective ignition switch.
2. No voltage at ignition switch.
3. Open circuit 3 from ignition switch to pump solenoid.

Diagnosis
1. Check 3A circuit voltage at solenoid. The reading should be 12 volts with ignition "On". If not, check 3C circuit at left bank glow plug relay. If satisfactory, repair fault in 3A circuit from bulkhead connector to solenoid. If not satisfactory at relay also, check instrument cluster gauges or tell-tale lamps for proper operation. If satisfactory, repair open circuit in 3 circuit from ignition switch to bulkhead connector.
2. If instrument cluster is not satisfactory, check headlamps, and if satisfactory, repair open circuit in 3 circuit from starter to ignition switch or replace ignition switch, if circuit 2 is satisfactory.
3. If circuit 2 and ignition are satisfactory, check 2 circuit at relays. If not satisfactory, check connections at junction block and battery voltage.

Voltage At Pump Solenoid With Ignition "Off"

Possible Causes
1. Defective switch.

Diagnosis
1. Check voltage at pump 3A circuit with ignition switch "Off". If zero volts is not obtained, check for maladjusted or defective ignition switch. Readjust or replace, as necessary.

"Don't Start" Lamp Not "On" During Preglow, Relays "On"

Possible Causes
1. Faulty "Don't Start" lamp.
2. Open 139A circuit feed to lamp.
3. Open 902 circuit feed to lamp.
4. Defective module.

Diagnosis
1. Check relays 940 and 941 circuits during preglow. If satisfactory but "Don't Start" lamp is not "On", check bulb and replace if necessary.
2. Check 902 circuit at module, if reading is zero volts, check at bulb. If zero volts at bulb, repair open circuit.
3. If relays are satisfactory during preglow and 902 circuit voltage is not zero volts during preglow, replace module.

"Don't Start" Lamp Remains "On"

Possible Causes
1. Defective 902 circuit.
2. Defective module.
3. Open 925 circuit to module from alternator.
4. Open alternator lamp circuit.
5. Defective alternator.

Diagnosis
1. Check 902 circuit voltage during start of afterglow. If reading is zero volts, disconnect module electrical connector and recheck bulb. If lamp remains "On", repair 902 circuit.
2. If lamp does "Out" with module disconnected, check 25 circuit during afterglow at module with engine running. If 14 volts are obtained, replace module.
3. If 925 circuit voltage is not approximately 14 volts with engine running, check 25 circuit at alternator. If 14 volts are obtained, repair 925 circuit open to module. If 14 volts are not obtained, check alternator lamp bulb for correct type.
4. Resistance of the 25 circuit should be approximately 10 ohms between the alternator and ignition switch with the ignition switch "Off".

"Don't Start" Lamp Not Recycled "On" After A No Start Crank

Possible Causes
1. Open 906A circuit to module.
2. Defective module.

Diagnosis
1. Check 906A circuit voltage at module when cranking engine. Voltage should be 8–12 volts. If voltage is within specifications, replace module. If zero volts are obtained, repair open 906A circuit between starter solenoid and module.

"Glow Plugs" Lamp Remains "On" After Engine Start, Glow Plugs "On"

Possible Causes
1. Faulty 907 circuit to "Glow Plugs" lamp.
2. Shorted glow plug temperature switch.
3. Glow plug temperature switch improperly calibrated.
4. Defective alternator.

Diagnosis
1. Disconnect the "Glow plugs-engine temperature switch", located at the front of the intake manifold. If "Glow Plugs" lamp goes "Out", with glow plugs still "On" in normal afterglow period, replace temperature switch. If lamp remains "On", check for defective 907 circuit and repair.
2. If lamp remains "On" with no defect indicated in the 907 circuit, check voltage of the 25 circuit at alternator. Reading should be 14 volts.

"Glow Plugs" Lamp Remains "On" After Engine Start, Glow Plugs "Off"

Possible Causes
1. Shorted diode.
2. Faulty 25 circuit.

Diagnosis
1. Replace diode in module harness.
2. Check 25 circuit for defect.

"Glow Plugs" Lamp "Off" Ignition Switch In "Run", Engine "Off"

Possible Causes
1. Glow plugs not "On".
2. Faulty diode.
3. Defective "Glow Plugs" bulb.

Diagnosis
1. Check for defective "Glow Plugs" bulb.
2. If bulb is satisfactory, check for open diode or open 907 or 25 circuit to alternator.

Engine Repair Section

DESCRIPTION

Engine Construction

The Oldsmobile four stroke cycle diesel V8-350 engine is basically the same in construction as the Oldsmobile gasoline V8-350 engine. The cylinders are numbered 1, 3, 5, 7 on the left bank and 2, 4, 6, 8 on the right bank. The firing order is 1-8-4-3-6-5-7-2. The major differences between the diesel and gasoline versions is in the cylinder heads, combustion chamber, fuel distribution system, air intake manifold and method of ignition. The cylinder block, crankshaft, main bearings, connecting rods, pistons and pins are of heavy construction due to the high compression ratio required to ignite the diesel fuel. The diesel fuel is ignited when the heat developed in the combustion chamber during the compression stroke reaches a certain temperature.

The valve train operates the same as in the gasoline engine, but are of special design and material for diesel operation. The stainless steel pre-chamber inserts in the cylinder head combustion chambers are serviced separately from the cylinder head. With the cylinder head removed, these pre-chamber inserts can be driven from the cylinder head after removing the glow plugs or injection nozzles.

On 1978–80 models and 1981 models except California, the glow plugs are threaded into the cylinder head and the injection nozzles are retained by a bolt and clamp. On 1981 California and all 1983 models, the glow plugs and the injection nozzles are both threaded into the cylinder head. The injection nozzles are spring loaded and calibrated to open at a specified fuel pressure.

Fuel System

The fuel injection pump is mounted on top of the engine and is gear driven by the camshaft and rotates at camshaft speed. This high pressure rotary pump injects a metered amount of fuel to each cylinder to the proper time. Eight high pressure fuel delivery pipes from the injection pump to the injection nozzles, Fig. 1 and 2, are the same length to prevent any difference in timing from cylinder to cylinder. The fuel injection pump provides the required timing advance under all operating conditions. Engine speed is controlled by a rotary fuel metering valve, Fig. 3. When the accelerator is depressed, the throttle cable opens the metering valve and allows more fuel to be delivered to the engine. The injection pump also incorporates a low pressure transfer pump to deliver fuel to the fuel line to the high pressure pump, Fig. 3.

The fuel filter is located between the mechanical fuel pump and the injection pump. The diaphragm type mechanical fuel pump is mounted on the right side of the engine and is driven by a cam on the crankshaft. The fuel tank at the rear of the vehicle is connected by fuel pipes to the mechanical fuel pump. Excess fuel returns from the fuel injection pump and injection nozzles to the fuel tank through pipes and hoses.

NOTE: Injection nozzles used on 1981 California models and all 1983 models do not use a fuel return line.

Water in Fuel System

This system is available on some 1980 and all 1981 and 1983 models. These vehicles have a "Water in Fuel" light mounted in the instrument panel. The "Water in Fuel" light has a bulb check feature and should light for 2 to 2½ seconds when the ignition is turned on. If not, the bulb is burned out or there is an open in the wiring circuit. When there is water in the fuel, the light will come back on and remain on after a 15 to 20 second delay.

A water sensing probe, mounted on the fuel sender, actuates the instrument panel light when it is partially covered with water. About 1 to 2½ gallons of water must be present in the fuel tank to cause the sensor light to activate.

If the "Water in Fuel" lamp goes on while the vehicle is being driven, the fuel system should be checked for water. If the lamp goes on immediately after refueling and before the vehicle is moved, there is a large quantity of water in the tank and should be removed immediately.

The water may be removed from the tank with a pump or by siphoning. The pump or siphon hose should be connected to the quarter inch fuel return hose (the smaller of the two hoses), located under the hood near the fuel pump. Refer to "Purging Water From Fuel Tank" procedure.

Fuel Heater

This cold weather device is installed on some 1981 and 1983 vehicles. This thermostatically controlled fuel heater reduces the possibility of wax plugging the filter when temperatures are below 20°F. Battery voltage is directed to the heater whenever the ignition is in the "Run" position.

The heater consists of a strip spiral wound around the fuel pipe and a bimetal thermal switch which closes and completes the circuit when fuel temperatures are approximately 20°F. When fuel temperatures reach 50°F. the thermal switch opens and deactivates the circuit.

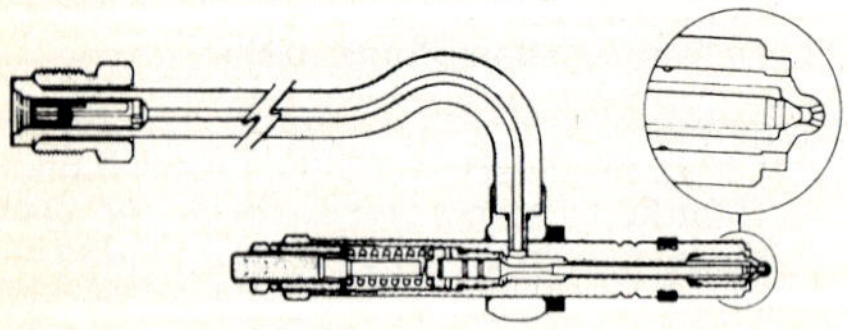

Fig. 1 Fuel injection nozzle. 1978–80 & 1981 exc. Calif.

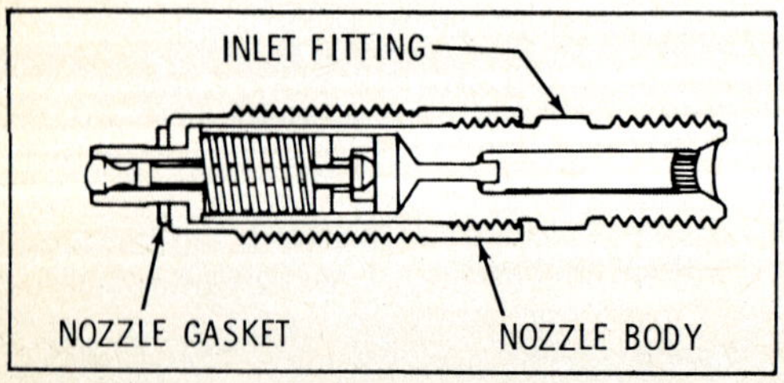

Fig. 2 Fuel injection nozzle. 1981 Calif. & all 1983

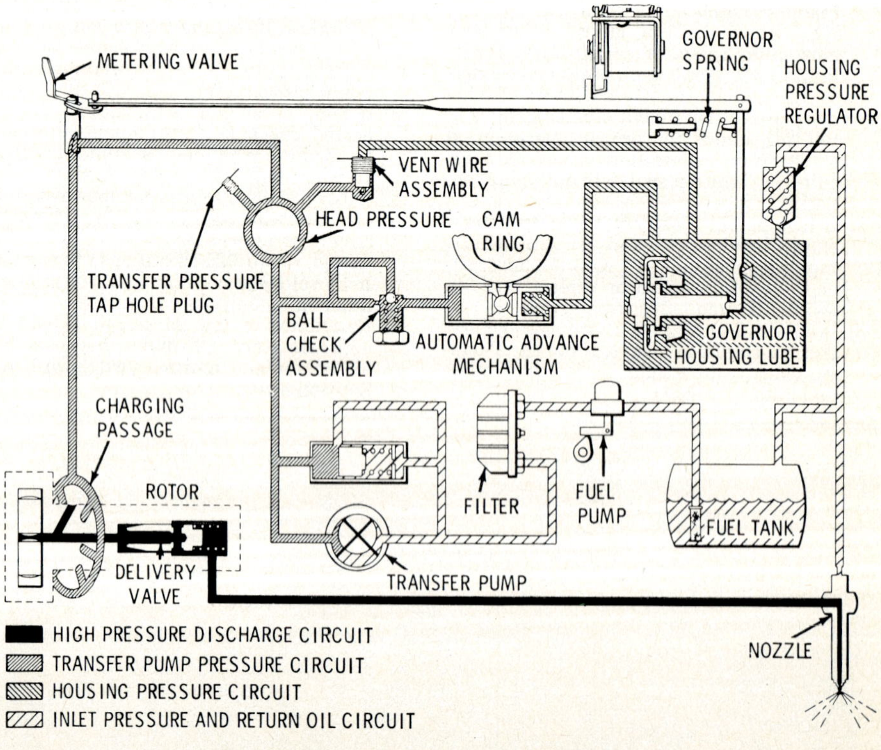

Fig. 3 Fuel injection pump circuit

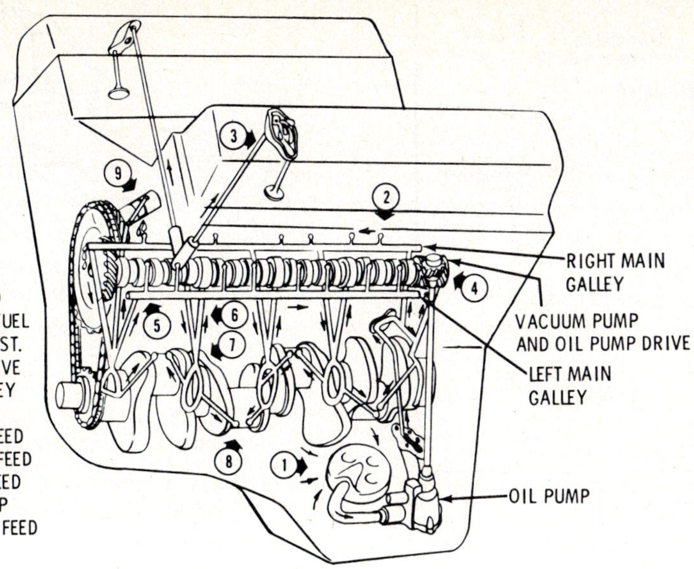

Fig. 4 Engine lubricating system

The fuel tank filter sock is equipped with a bypass valve which opens when the filter is covered with wax, allowing fuel to flow to the heater.

Housing Pressure Cold Advance (HPCA)

This feature is used on all 1981 and 1983 engines and advances the injection timing 3° during cold operation. This circuit is actuated by a temperature switch calibrated to open the circuit at 125° F. Below the switching point, housing pressure is decreased from 10 to 0 psi which advances the injection timing 3°. Above the switching point, the switch opens, de-energizing the solenoid and the housing pressure is returned to 10 psi. The fast idle solenoid is energized by the same switch and closes when the temperature falls below 95° F.

Engine Lubrication System

NOTE: On 1978 engines, the recommended diesel engine oil designation is SE/CD. On 1979–80 engines, the recommended diesel engine oil designation is SE/CC. On 1981 engines, the recommended diesel engine oil is SF/CC, SF/CD or SE/CC. On 1983 engines, the recommended diesel engine oil is SF/SC or SF/CD, it is also recommended that the oil be a fuel saving product.

The diesel engine lubrication system is basically the same as the gasoline engine. The fuel injection pump driven gear is lubricated by oil directed through a passage from the top of the camshaft bearing, Fig. 4. An angled passage in the shaft portion of the driven gear directs the oil to the rear driven gear bearing. At the front of the right oil gallery, a small orifice sprays oil to lubricate the fuel pump eccentric cam on the crankshaft and timing chain.

Engine Cooling System

The diesel engine cooling system is the same as the gasoline engine except the radiator incorporates two oil coolers. One cooler is used to cool the transmission fluid and the other cooler is used to cool the engine oil.

Engine Electrical System

Eight glow plugs are used to pre-heat the pre-chamber to aid in starting. The type 1 glow plugs are 12 volt heaters and are activated when the ignition switch is turned to the "Run" position. The type 1 system uses steady current applied to 12 volt glow plugs. The type 2 system uses 6 volt glow plugs with a controlled pulsating current applied to them for starting. The type 2 glow plug system uses an electromechanical controller for glow plug temperature, pre-glow time, wait/start lights and after glow time. The 6 and 12 volt plugs are not interchangeable and can be identified by the wire connector spade. The 6 volt glow plugs have a 5/16 inch wire connector spade, while the 12 volt plugs have a 1/4 inch wide spade. The glow plug remains activated for a short time after starting then is automatically turned "Off." Two 12 volt batteries, connected in parallel, are required for the higher electrical load due to the glow plugs and starting motor. The diesel starter motor is larger than the gasoline starter motor. It is designed to crank the engine at least the 100 RPM required for starting. An alternator supplies charging current to both batteries at the same time and there are no switches or relays in the charging circuit.

ENGINE, REPLACE

1. Disconnect ground cable from batteries and drain cooling system.
2. Remove air cleaner.
3. Scribe hood hinge locations and remove hood.
4. Disconnect ground wires at inner fender and the engine ground strap at right cylinder head.
5. Disconnect radiator hoses, oil cooler lines, heater hoses, vacuum hoses, power steering hoses from gear, A/C attached, fuel pump hose from fuel pump and the wiring.
6. Remove hairpin clip from bellcrank.
7. Remove throttle and throttle valve cables from intake manifold brackets and position cables aside.
8. Remove upper radiator support and the radiator.
9. Raise and support vehicle.
10. Disconnect exhaust pipes from exhaust manifold.
11. Remove torque converter cover and the three bolts securing torque converter to flywheel.
12. Remove engine mount bolts or nuts.
13. Remove three engine to transmission bolts on the right side.
14. Disconnect starter wiring and remove starter.
15. Lower vehicle.
16. Attach suitable engine lifting equipment to engine. Support transmission with a suitable jack.
17. Remove the three engine to transmission bolts on the left side.
18. Remove engine from vehicle.
19. Reverse procedure to install.

INTAKE MANIFOLD, REPLACE

1. Disconnect ground cables from batteries.
2. Remove air cleaner assembly.
3. Drain cooling system, then disconnect upper radiator hose and thermostat bypass hose from water pump outlet. Disconnect heater hose and vacuum hose from water control valve.
4. Remove breather pipes from valve covers and air crossover, Fig. 5.
5. Remove air crossover and cap intake manifold, Fig. 6.
6. Disconnect throttle rod and return spring. If equipped with Cruise Control, remove servo.
7. Remove hairpin clip from bellcrank and disconnect the cables. Remove throttle and throttle valve cables from intake manifold brackets and position cables aside.
8. Disconnect wiring as necessary.
9. Disconnect or remove alternator and A/C compressor as necessary.
10. Disconnect fuel line from fuel pump and filter and remove fuel filter and bracket.
11. Disconnect lines from injector nozzles and remove injection pump. Cap all open fuel lines and fittings.
12. Disconnect vacuum lines at vacuum pump. Remove vacuum pump, if equipped with A/C, or oil pump drive assembly, if less A/C, Fig. 7.
14. Remove intake manifold drain tube, Fig. 8.
15. Remove intake manifold bolts and the intake manifold.
16. Remove adapter seal and injection pump adapter.
17. Reverse procedure to install. Torque intake manifold bolts in sequence, Fig. 9, to specifications.

CYLINDER HEAD, REPLACE

NOTE: If a cylinder head is being removed to correct a head gasket problem on these engines, a few checks should be made to ensure the cause of the condition is corrected. The checks are as follows:

a. Cylinder head warpage. If any cylinder head is warped more than .010 inch, it should be replaced. Resurfacing is not recommended.

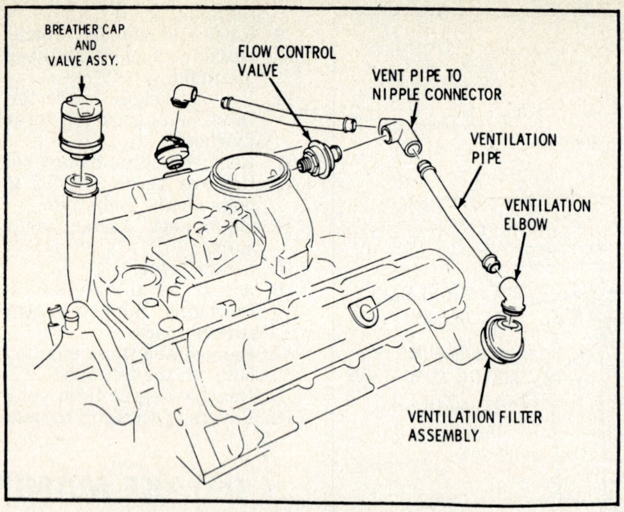

Fig. 5 Crankcase ventilation system

b. Cylinder head to cylinder block dowel pin interference. With the head properly positioned on the block, without the head gasket installed, run a .005 inch feeler gauge around the perimeter of the area where the head and block meet. There should be zero clearance, thus ensuring the head is flush with the block and no dowel pin interference exists.
c. Prechamber location. The cylinder head prechambers must not be recessed into the cylinder head or protrude out of the cylinder head by more than .004 inch as this will affect head gasket sealing.
d. Head gasket sealing bead. Ensure the head gasket sealing bead goes around all core holes in the block and head. If the sealing bead crosses a raised, depressed or stamped area, the raised area should be filed and the depressed or stamped area filled with RTV sealer and assembled before the RTV sealer cures.
e. Cylinder head cracks. Minor surface cracks in the cylinder head valve port area between the intake and exhaust ports are considered a normal condition and does not warrant replacement of the heads.

1. Remove intake manifold as outlined previously.
2. Remove valve cover. It may be necessary to remove any interfering accessory brackets.
3. Disconnect glow plug wiring.
4. If right side head is to be removed, remove ground strap from head.
5. Remove rocker arm bolts, pivots, rocker arms and push rod. Note locations of valve train components so they can be installed in original locations.
6. Remove fuel return lines from injection nozzles.
7. Remove exhaust manifold.
8. Remove engine block drain plug on side of block that cylinder head is being removed.
9. Remove cylinder head bolts, then cylinder head.
10. If necessary to remove pre-chamber, remove glow plug and injection nozzle, then tap out pre-chamber with a suitable drift, Fig. 10.
11. Reverse procedure to install. Do not use any sealing compound on cylinder head gasket. Torque cylinder head bolts in sequence, Fig. 11, to 100 ft. lbs., then 130 ft. lbs.

SERVICE NOTE

New head bolts with increased torque capacity were introduced into production during March 1980 and are now available, Fig. 12.

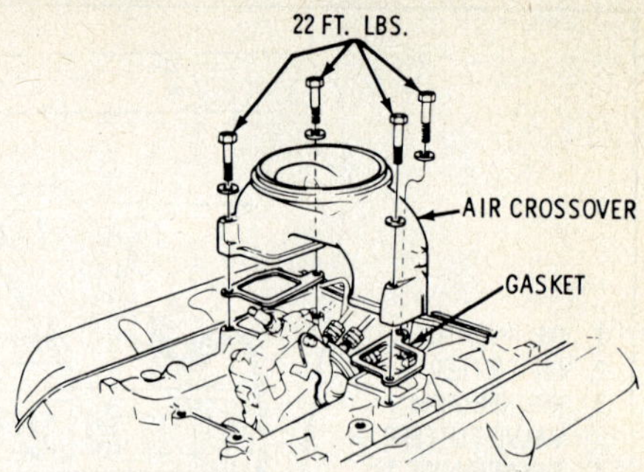

Fig. 6 Air crossover installation

When replacing a cylinder head, it is recommended new head bolts be used if they have not already been installed in the engine. When installing later production model head bolts, be sure to clean and oil the threads. Before installation of the cylinder head ensure bolt holes are tapped deep enough into the block. Blow out any chips or liquid in the bolt holes. Then position cylinder head on cylinder block without the cylinder head gasket. Install bolt and tighten by hand until bolt head contacts the cylinder head. This will indicate that the holes are tapped deep enough into the block, allowing for proper torque. For cylinder head bolt location, refer to Fig. 13.

A new designed head gasket went into production in March 1981, and is available for service replacement. It can be identified by the use of blue print O sealer instead of orange. This gasket is of the slotted and shim design.

No sealer should be used on the gasket during installation. Also the prechamber shield on the cylinder head gasket must be installed towards the cylinder head.

ROCKER ARMS

NOTE: This engine uses valve rotators, Fig. 14. The rotator operates on a sprag clutch principle utilizing the collapsing action of a coil spring to give rotation to the rotor body which turns the valve.

1. Remove valve cover.
2. Remove flanged bolts, rocker arm pivot

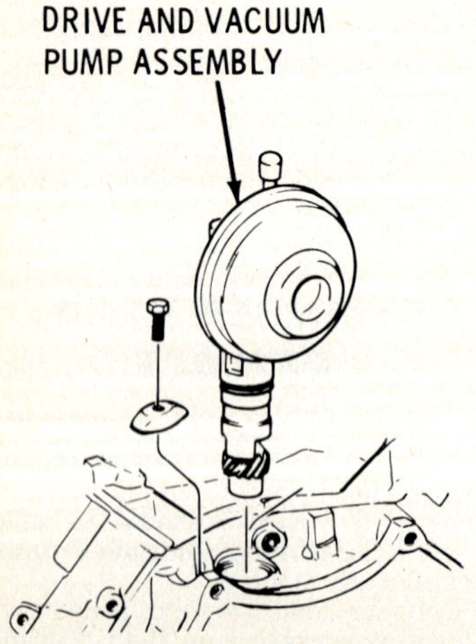

Fig. 7 Vacuum pump & oil pump drive assembly

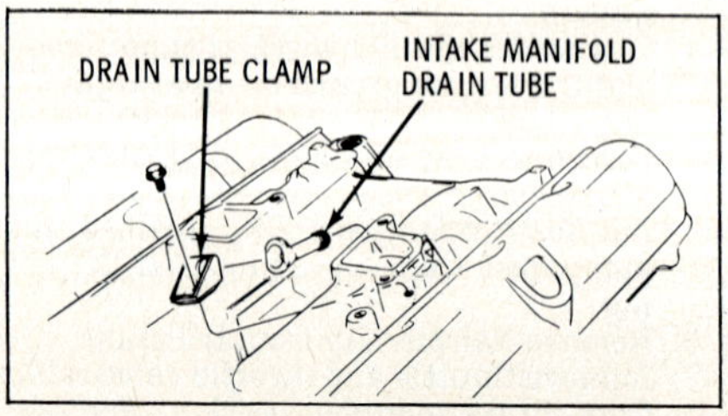

Fig. 8 Intake manifold drain tube installation

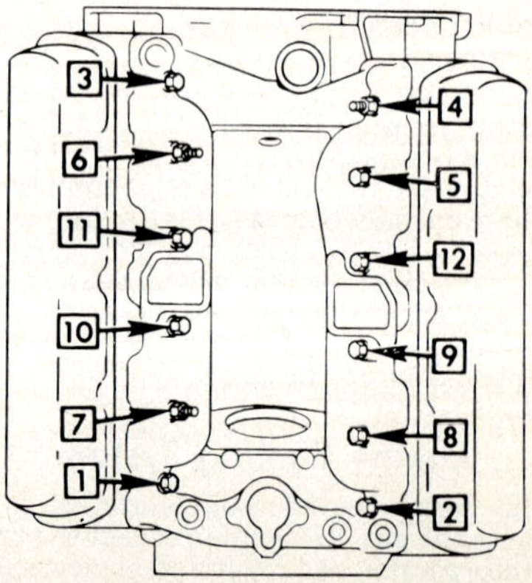

Fig. 9 Intake manifold tightening sequence

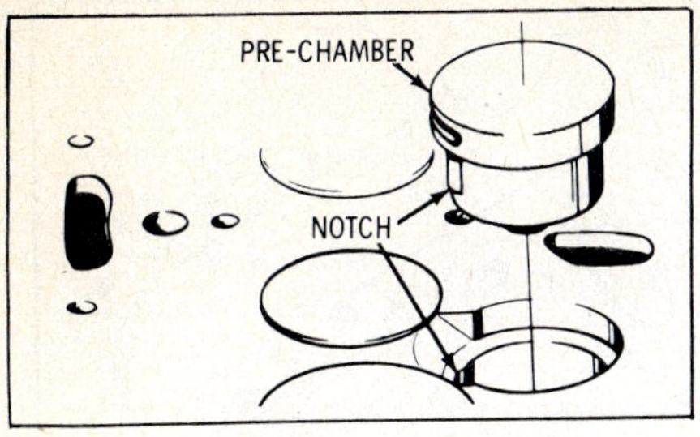

Fig. 10 Pre-chamber installation

and rocker arms.

3. To install rocker arm assemblies, lubricate wear surfaces with suitable lubricant. Refer to "Valve Lifter Bleed Down" for installation procedure. Final rocker arm flanged bolt tightening torque is 25 ft lbs. on 1978–80 models and 28 ft. lbs. on 1981 and 1983 models.

VALVE ROTATORS

The rotator operates on a Sprag clutch principle utilizing the collapsing action of coil spring to give rotation to the rotor body which turns the valve, Fig. 14.

To check rotator action, draw a line across rotator body and down the collar. Operate engine at 1500 rpm. Rotator body should move around collar. Rotator action can be in either direction. Replace rotator if no movement is noted.

When servicing valves, valve stem tips should be checked for improper wear pattern which could indicate a defective valve rotator, Fig. 15.

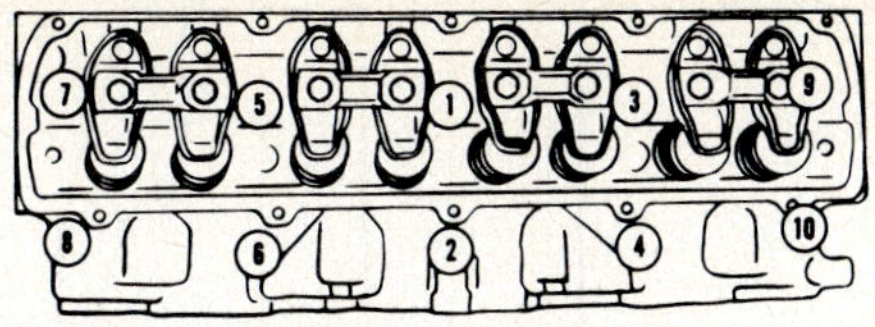

Fig. 11 Cylinder head tightening sequence

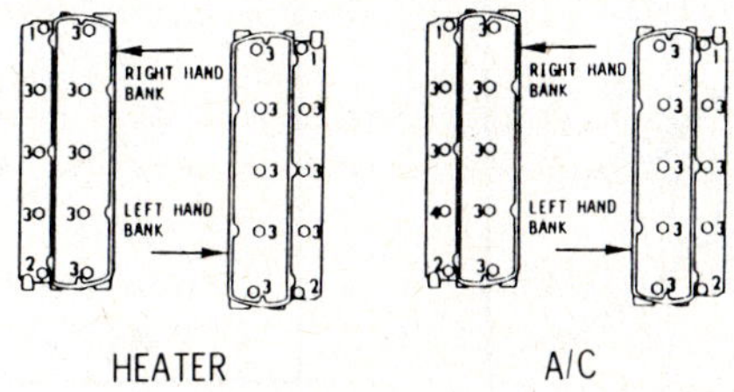

Fig. 13 Cylinder head bolt locations

VALVE ARRANGEMENT

V8-350 Diesel . . . . I-E-I-E-E-I-E-I

VALVE LIFT SPECS.

Engine	Year	Intake	Exhaust
V8-350 Diesel	1978–83	.375	.376

VALVE TIMING

Intake Opens Before TDC

Engine	Year	Degrees
V8-350 Diesel	1978–83	16

LOCATION NUMBER	PART NUMBER	SIZE
1	22510580	1/2 - 13 x 3.10
2	22510582	1/2-13 x 3.10 stud end
3	22510579	1/2 - 13 x 4.30
4	22510585	1/2 - 13 x 4.30 stud end

Fig. 12 Cylinder head bolt identification

VALVES, REPLACE

Whenever a new valve is installed or after grinding valves, it is necessary to measure the valve stem height with the special tool as shown in Fig. 16.

There should be at least .015 inch clearance between the gauge and end of valve stem. If clearance is less than .015 inch, remove valve and grind end of valve stem as required.

Check valve rotator height, Fig. 17. If valve stem end is less than .005 inch above rotator, the valve is too short and a new valve must be installed.

VALVE GUIDES

Valve stem guides are not replaceable, due to being cast in place. If valve guide bores are worn excessively, they can be reamed oversize.

If a standard valve guide bore is being reamed, use a .003" or .005" oversize reamer. For the .010" oversize valve guide bore, use a .013" oversize reamer. If too large a reamer is used and the spiraling is removed, it is possible that the valve will not receive the proper lubrication.

NOTE: Occasionally a valve guide will be oversize as manufactured. These are marked on the cylinder head as shown in Fig. 18. If no markings are present, the guide bores are standard. If oversize markings are present, any valve replacement will require an oversize valve. Service valves are available in standard diameters as well as .003", .005", .010" and .013" oversize.

VALVE LIFTERS, REPLACE

NOTE: Some engines have both standard and .010 inch oversize valve lifters. The .010 inch oversize valve lifters are etched with a "0" on the side of the lifter. Also, the cylinder block will be marked if an oversize lifter is used, Fig. 19.

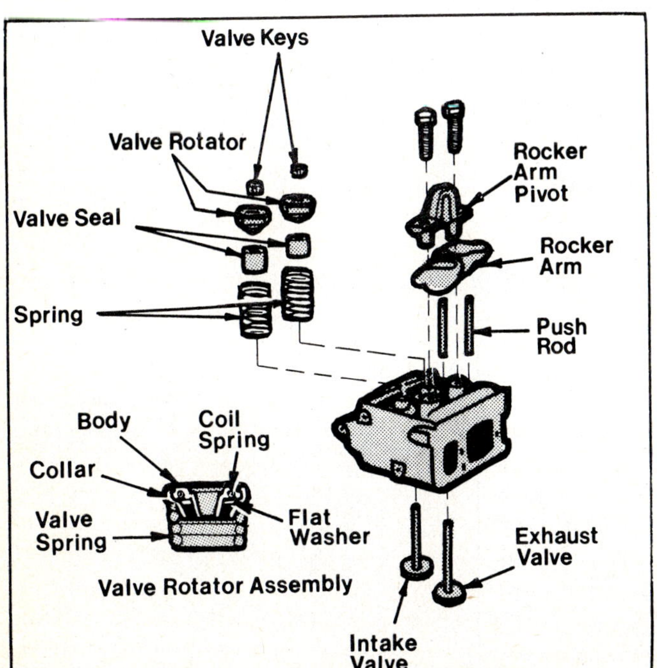

Fig. 14 Cylinder head exploded view

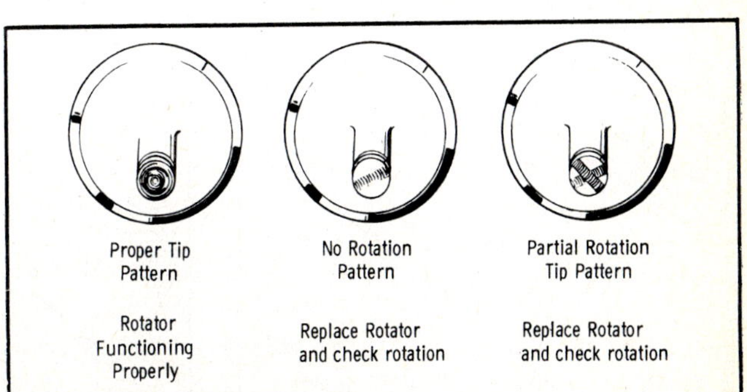

Fig. 15 Checking valve stem for rotator malfunction

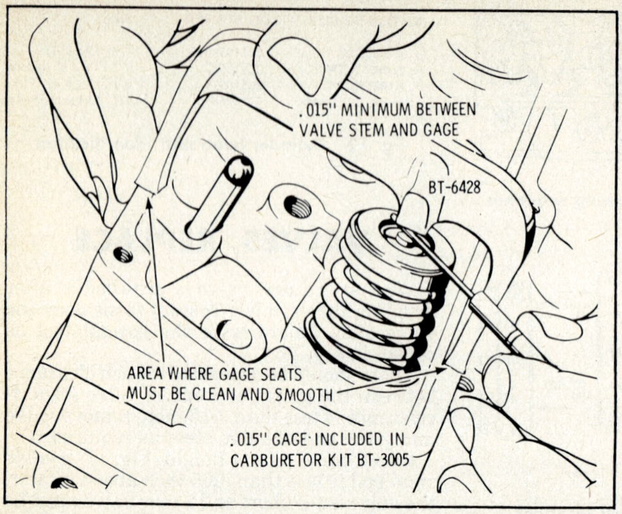

Fig. 16 Measuring valve stem height

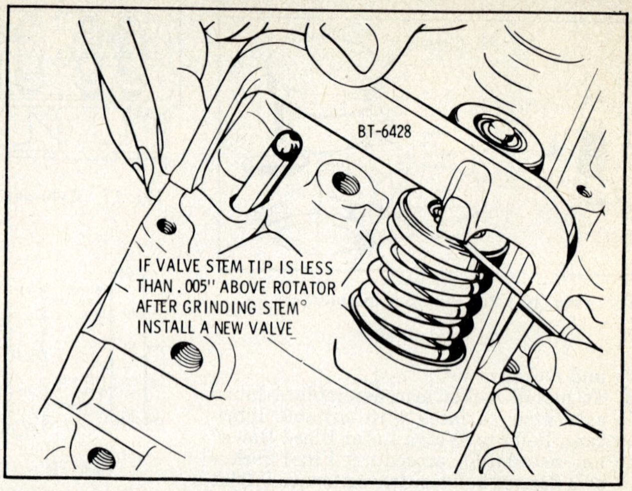

Fig. 17 Measuring valve rotator height

NOTE: 1981 and 1983 engines are equipped with roller lifters. These are the same internally as in the past. However, the lifter can not be "leak down" tested at this time as a tester adapter is not available. On this type lifter, a guide holds the lifters from rotating, Fig. 20. A spring steel retainer holds the guide in place.

1. Remove intake manifold as outlined previously.
2. Remove valve covers, rocker arm assemblies and push rods. On 1981 and 1983 models, remove the lifter retainer and lifter guides. Note location of valve train components so they can be installed in original position.
3. Remove valve lifters.
4. Reverse procedure to install.

VALVE LIFTERS, SERVICE

1. Remove valve lifters, refer to "Valve lifters, Replace".
2. Using a small screwdriver, remove retainer ring, Figs. 21 and 22.
3. Remove pushrod seat, oil metering valve, plunger and plunger spring.
4. Remove check ball retainer from plunger, then remove ball and spring.
5. Clean parts in a suitable solvent.

NOTE: Do not interchange parts between lifters. If any parts are worn, replace lifter.

6. Inspect all parts for nicks, burrs or scoring. If any parts are defective, replace lifter.
7. On 1981 and 1983 roller lifters, inspect roller. It should rotate freely, but without excessive play, also check for missing or broken needle bearing. If any parts are defective, replace lifter.
8. Apply a coating of light engine oil to all lifter surfaces.
9. Install ball check spring and retainer into plunger. Ensure retainer flange is pressed tight against bottom of recess in plunger.
10. Install plunger spring over check retainer.
11. Hold plunger with spring up and insert in lifter body. Hold plunger vertically to prevent cocking spring.
12. Submerge lifter assembly in clean diesel fuel or kerosene, then install oil metering valve and push rod seat into lifter and install retaining ring.

VALVE LIFTER BLEED DOWN

If the intake manifold has been removed and if any rocker arms have been removed or loosened, it will be necessary to remove those lifters, disassemble them, drain the oil from them and reassemble. Refer to "Valve Lifters, Service".

If the intake manifold has not been removed, but rocker arms have been loosened or removed, the valve lifters must be bled down to prevent possible valve to piston interference by using the following procedure;

1. Prior to installing rocker arms, rotate crankshaft pulley to a position 32° BTDC (before top dead center). This is approximately 2 inches counterclockwise from 0° pointer.
2. If the right side valve cover was removed only, remove cylinder No. 1 glow plug and determine if No. 1 piston is in the correct position. This can be determined by compression pressure.
3. If the left side valve cover was removed only, rotate crankshaft until No. 5 cylinder intake valve push rod is .28 inch above the No. 5 cylinder exhaust valve push rod.
4. If removed, install cylinder No. 5 pivot and rocker arms. Alternately torque the bolts until the intake valve begins to open and stop tightening.

NOTE: When torquing rocker arms, use

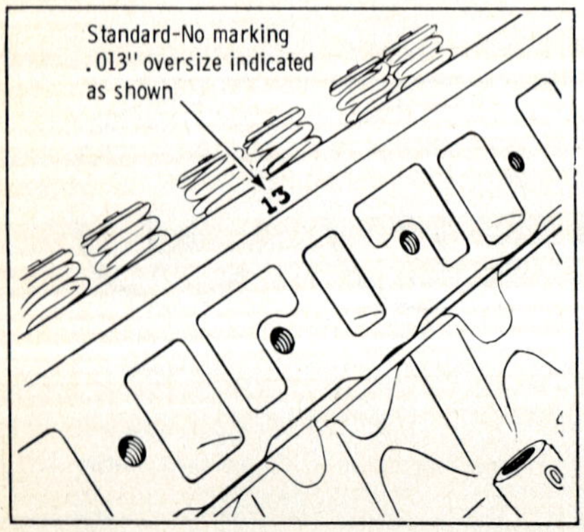

Fig. 18 Valve guide bore marking

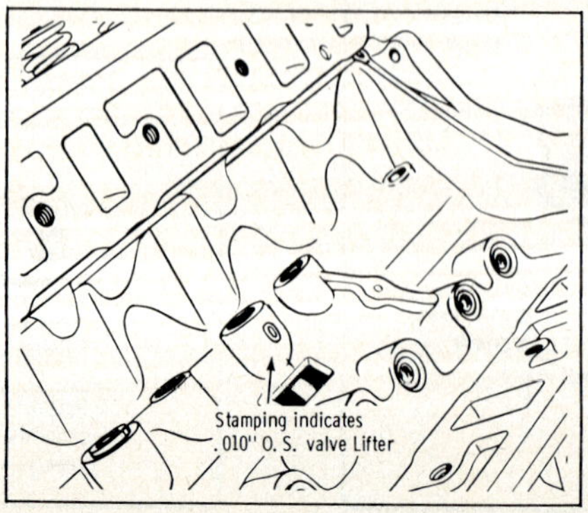

Fig. 19 Oversize valve lifter bore marking

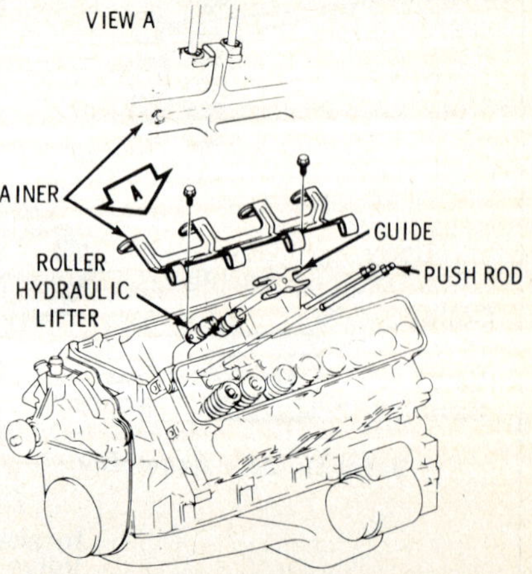

Fig. 20 Valve lifters & valve guide. 1981 & 1983

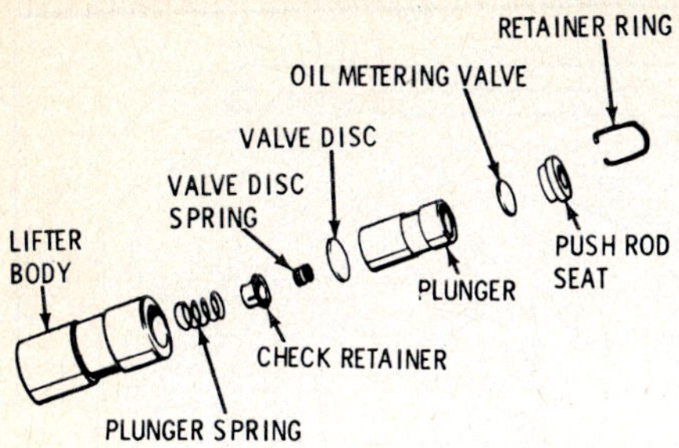

Fig. 21 Valve lifter exploded view. 1978–80

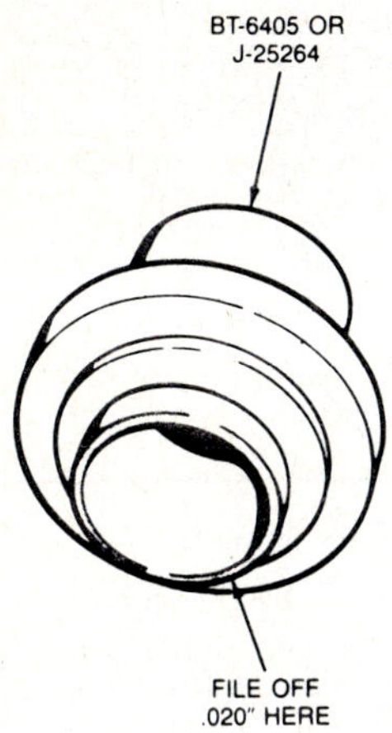

Fig. 24 Modifying tool for front seal installation

only hand wrenches to prevent engine damage.

---

5. Install remaining rocker arms except No. 3 exhaust valve. Torque bolt to 25 ft. lbs. on 1978–80 models and 28 ft. lbs on 1981 and 1983 models.
6. If removed, install No. 3 exhaust rocker arm and pivot, but do not torque beyond the point that the valve would be fully opened. This is indicated by a strong resistance while turning the pivot retaining bolt. Going beyond this point would bend the push rod.

NOTE: While performing step 6, torque bolt slowly allowing the lifter to bleed down.

---

7. Finish torquing No. 5 rocker arm pivot bolt slowly, allowing valve lifter to bleed down. Do not torque beyond the point that the valve would be fully opened. This is indicated by a strong resistance while turning the pivot retaining bolt. Going beyond this point would bend the push rod.

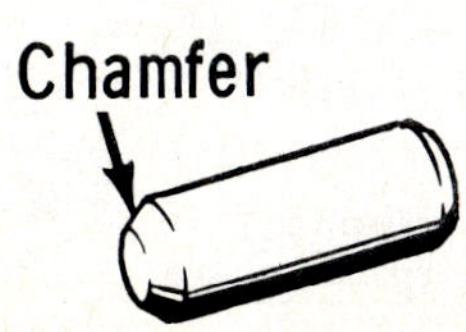

Fig. 26 Dowel pin chamfer

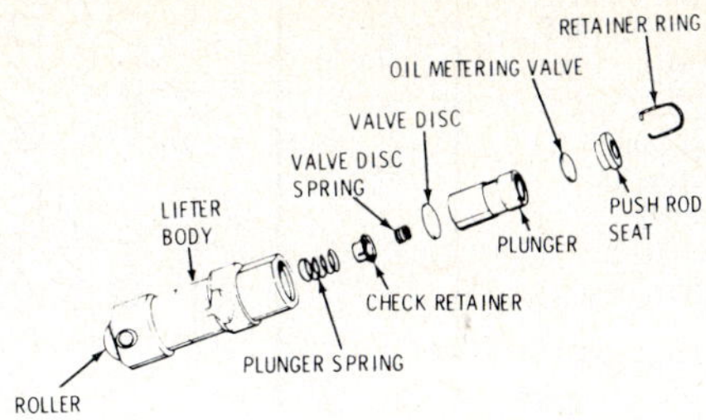

Fig. 22 Valve lifter exploded view. 1981 & 1983

8. Do not turn the crankshaft for at least 45 minutes while lifters bleed down.

NOTE: Do not rotate the engine until the valve lifters have bled down, otherwise engine damage might occur.

FRONT OIL SEAL, REPLACE

1. Disconnect ground cables from batteries.
2. Remove accessory drive belts.
3. Remove crankshaft pulley and harmonic balancer.
4. Using tool BT-6406, remove front oil seal, Fig. 23.
5. Apply suitable sealer to outside diameter of new oil seal.
6. File .020 inch from flange of tool No. BT-6405 Fig. 24, to prevent tool from contacting oil slinger before seal is properly seated in front cover.
7. Using tool BT-6611 and BT-6405, install new oil seal, Fig. 25.
8. Install harmonic balancer and crankshaft pulley.
9. Install and tension accessory drive belts.

ENGINE FRONT COVER, REPLACE

1. Disconnect ground cables from batteries.
2. Drain cooling system and disconnect radiator hoses and bypass hose.
3. Remove all drive belts, fan and pulley, crankshaft pulley and harmonic balancer, and accessory brackets.
4. Remove front cover to block attaching bolts, then remove front cover, timing indicator and water pump assembly.
5. Remove front cover alignment dowel pins. It may be necessary to grind a flat on the dowel pins to provide a rough surface for gripping.
6. Grind a chamfer on one end of each dowel pin, Fig. 26.
7. Cut excess material from front end of oil pan gasket on each side of cylinder block.
8. Trim approximately 1/8 inch from each end of new front pan seal, Fig. 27.
9. Install new front cover gasket and apply suitable sealer to gasket around coolant holes.

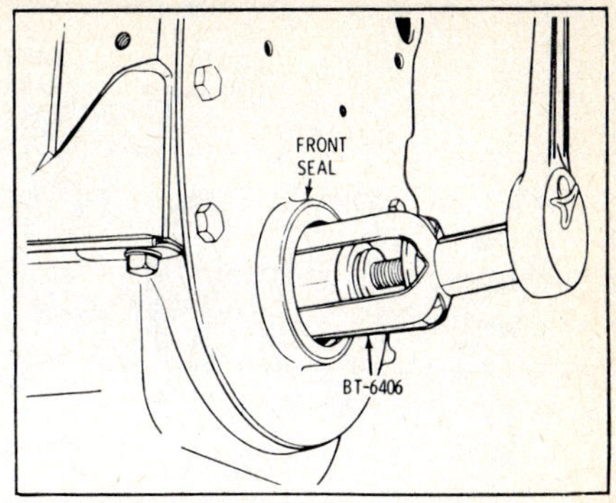

Fig. 23 Front oil seal removal

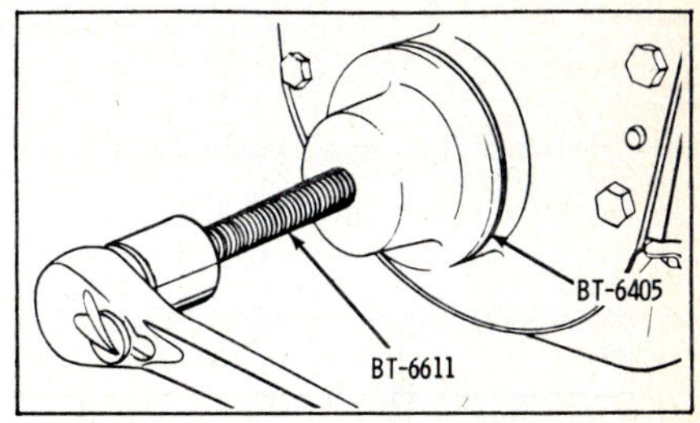

Fig. 25 Front oil seal installation

10. Apply RTV sealer to mating surfaces of cylinder block, oil pan and front cover.
11. Place front cover on cylinder block and press downward to compress seal. Rotate cover right and left and guide oil pan seal into cavity with a small screwdriver.
12. Apply engine oil to bolts.
13. Install two bolts finger tight to retain cover.
14. Install the two dowel pins, chamfered end first.
15. Install timing indicator and water pump and torque bolts as shown in Fig. 28.
16. Install harmonic balancer and crankshaft pulley.
17. Install accessory brackets.
18. Install fan and pulley and drive belts.
19. Connect radiator hoses and bypass hose.
20. Connect ground cables to batteries.

TIMING CHAIN & GEARS, REPLACE

1. Remove front cover as outlined previously.
2. Remove oil slinger, cam gear, crank gear and timing chain.

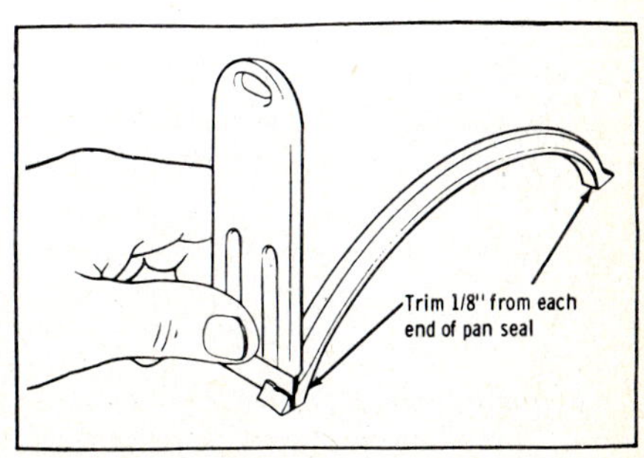

Fig. 27 Trimming oil pan seal

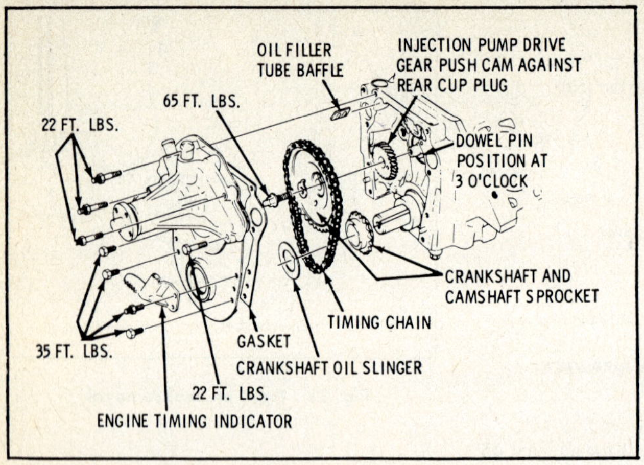

Fig. 28 Engine front cover installation

3. Remove fuel pump eccentric from crankshaft, if necessary.
4. Install key in crankshaft, if removed.
5. Install fuel pump eccentric, if removed.
6. Install cam gear, crank gear and timing chain with timing marks aligned, Fig. 29.

NOTE: With the timing marks aligned in Fig. 29. No. 6 cylinder is in the firing position. To place No. 1 cylinder in the firing position, rotate crankshaft one complete revolution. This will bring the camshaft gear mark to top and No. 1 cylinder will be in the firing position.

7. Install oil slinger.
8. Install front cover.

NOTE: Whenever timing chain and gears are replaced, it will be necessary to retime the engine. Refer to "Injection Pump Timing" procedure.

CAMSHAFT & INJECTION PUMP DRIVE & DRIVEN GEARS

1. Disconnect ground cables from batteries.
2. Drain cooling system.
3. Remove radiator upper baffle.
4. Disconnect upper radiator hose at water outlet and hose support clamp.
5. Disconnect cooler lines at radiator.
6. Remove fan shroud and radiator.
7. Remove intake manifold as outlined previously.
8. Remove engine front cover as outlined previously.
9. Remove valve covers.
10. Remove rocker arm bolts, pivots, rocker arms and push rod. Note valve train component locations to install components in original locations.
11. If equipped with A/C, discharge refrigerant system and remove condenser.
12. On all models, remove timing chain and gears as outlined previously.
13. Position camshaft dowel pin at 3 o'clock position.
14. While holding the camshaft rearward and rocking the injection pump driven gear slide, slide the injection pump drive gear from camshaft.
15. Remove injection pump adapter, snap ring, selective washer, injection pump driven gear and spring, Fig. 30.
16. Slide camshaft from front of engine.
17. Reverse procedure to install. Check injection pump driven gear end play. If end play is not .002–.006 inch, replace selective washer, Fig. 30. Selective washers are available from .080 to .115 inch in increments of .003 inch.

INJECTION PUMP
DRIVEN GEAR

SHIM

FUEL INJECTION PUMP
SHAFT SPRING

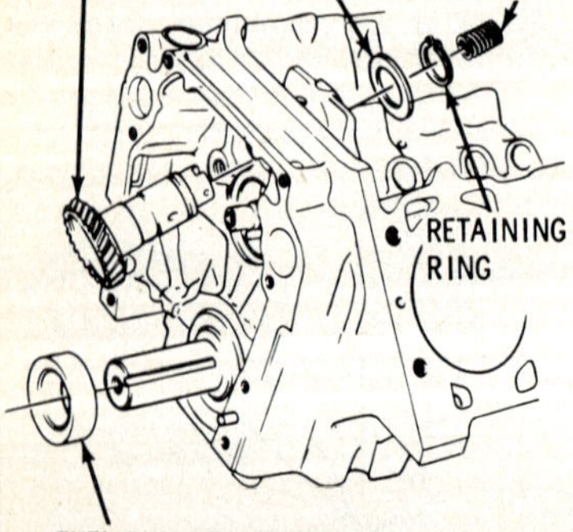

FUEL PUMP ECCENTRIC

Fig. 30 Fuel injection pump driven gear installation

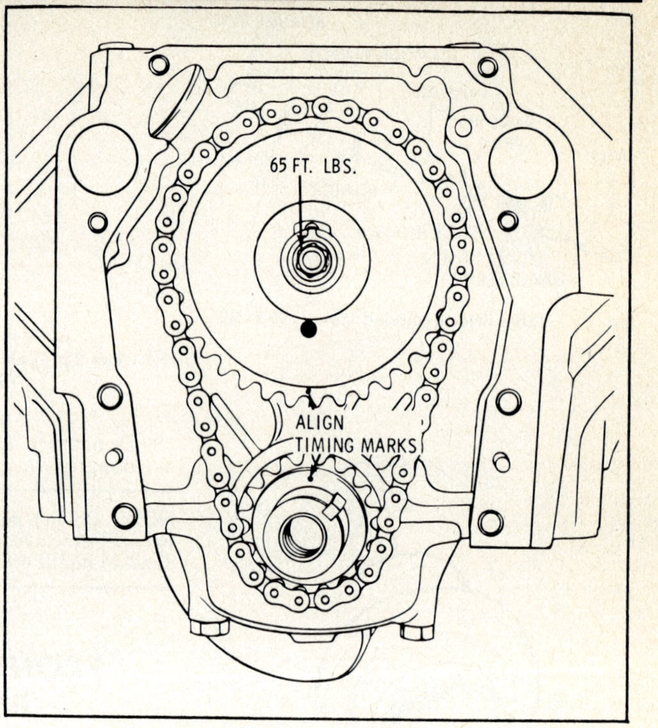

Fig. 29 Valve timing marks

PISTON & ROD ASSEMBLE

1983

Install piston and rod assembly so that notch at top of piston is facing toward front of engine, Fig. 31.

1978–81

Assemble piston to rod and install into cylinder block. The piston is installed with the valve depression facing toward the crankshaft. Also, there are two different pistons used in this engine. In cylinder numbers 1, 2, 3 and 4, the large valve depression faces the front of the engine, Fig. 32. In cylinder numbers 5, 6, 7 and 8 the large valve depression faces the rear of the engine, Fig. 32. The pistons are interchangeable between cylinder numbers 1, 3, 6 and 8 and 2, 4, 5 and 7.

PISTONS, RINGS & PINS

Pistons are available in standard sizes and oversizes of .010 inch.

Rings are available in standard sizes and oversizes of .010 and .030″.

MAIN & ROD BEARINGS

Main bearing inserts are available in standard sizes and undersizes of .0005, .0010 and .0015 inch. On 1983 models, a letter stamped on the bearing tang identifies the size of the bearing, Fig. 33. The undersize is also stamped on the bearing insert on the other side of the insert tang, Fig. 33. On 1978–81 models, the amount of undersize and part number is stamped on the bearing shell, Fig. 34.

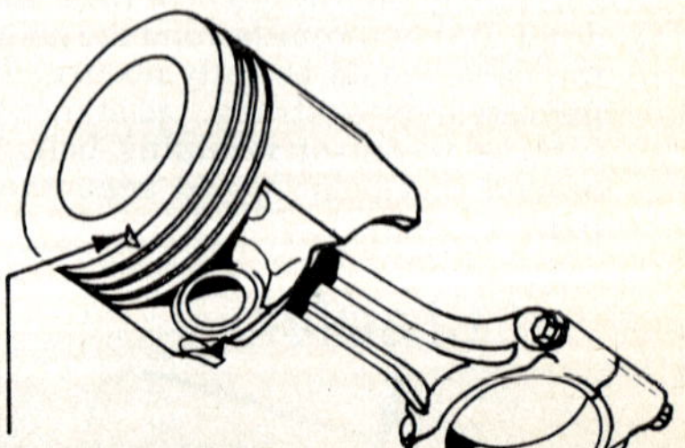

NOTCH TOWARD
FRONT OF ENGINE

Fig. 31 Piston & rod installation. 1983

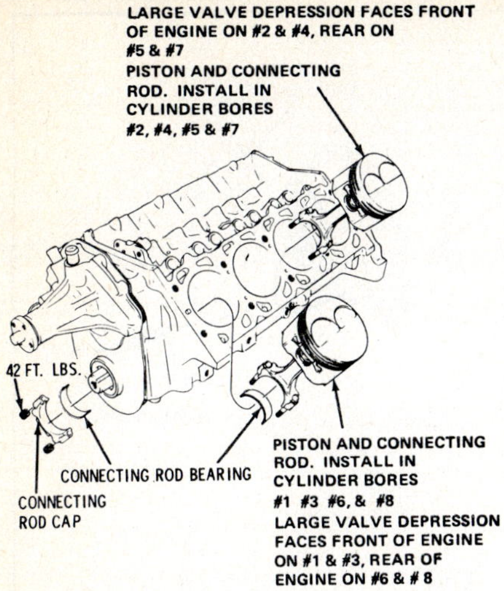

Fig. 32 Piston & rod installation. 1978–81

Rod bearings are available in standard sizes and an undersize of .010 inch.

SERVICE NOTE

Beginning in the 1981 model year, a 1/4 inch longer main bearing cap bolt entered production. The new part number is 22511818, replacing 392190. All 1981 and newer diesel engine main bearing bolt holes have a 1/4 inch deeper tap and 1/4 inch longer counter bore.

REAR CRANKSHAFT SEAL SERVICE

Since the braided fabric seal used on these engines can be replaced only when the crankshaft is removed, the following repair procedure is recommended.

1. Remove oil pan and bearing cap.
2. Drive end of old seal gently into groove, using a suitable tool, until packed tight. This may vary between 1/4 and 3/4 inch depending on amount of pack required.
3. Repeat previous step for other end of seal.

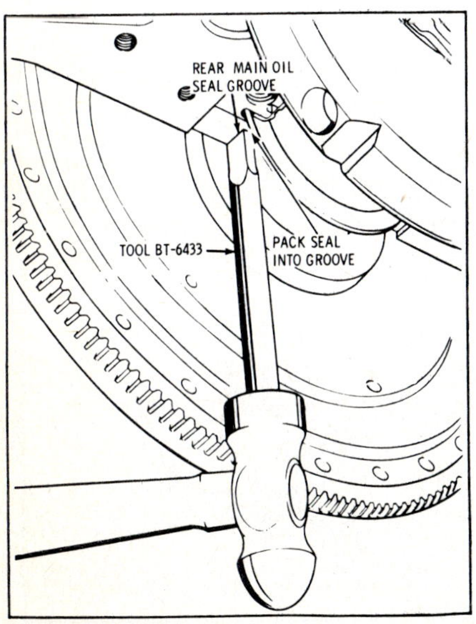

Fig. 35 Packing upper rear main bearing seal

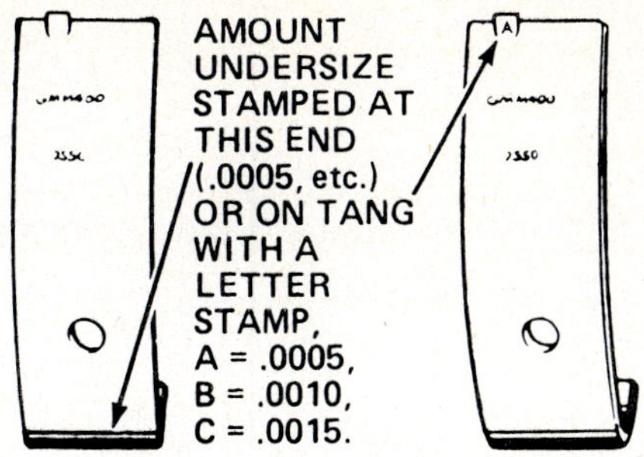

Fig. 33 Main bearing identification. 1983

4. Measure and note amount that seal was driven up on one side. Using the old seal removed from bearing cap, cut a length of seal the amount previously noted plus 1/16 inch.
5. Repeat previous step for other side of seal.
6. Pack cut lengths of seal into appropriate side of seal groove. A packing tool, BT-6433, Fig. 35, may be used since the tool has been machined to provide a built-in stop. Use tool BT-6436 to trim the seal flush with block, Fig. 36.
7. Install new seal in lower bearing cap.

OIL PAN, REPLACE

1. Disconnect battery ground cables.
2. Remove oil pump drive and vacuum pump.
3. Remove fan shroud attaching screws and pull upward from clips.
4. Raise and support vehicle. Drain oil pan.
5. Disconnect exhaust and crossover pipes, then remove oil cooler lines from filter base.
6. Remove flywheel cover and the starter motor.
7. Remove both engine mount through bolts and raise engine with suitable equipment.
8. Loosen the right-hand engine mount and remove the left-hand engine mount.
9. Remove oil pan attaching bolts and the oil pan.
10. Reverse procedure to install. Torque oil pan attaching bolts to 10 ft. lbs.

OIL PUMP, REPLACE & SERVICE

Replacement

1. Remove oil pan as outlined previously.
2. Remove oil pump to rear main bearing cap attaching bolts, Fig. 37.
3. Remove oil pump and drive shaft extension.
4. Reverse procedure to install. Torque attaching bolts to 35 ft. lbs.

Service

Disassembly

1. Remove oil pump drive shaft extension, Fig. 38. Do not attempt to remove washers from drive shaft extension. The drive shaft extension and washers is serviced

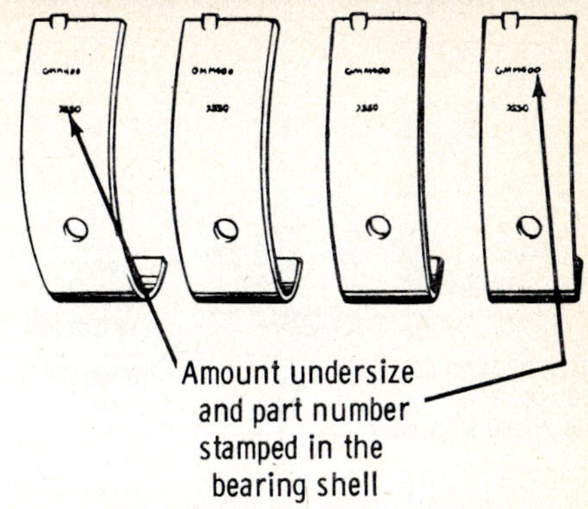

Fig. 34 Main bearing identification. 1978–81

as an assembly.

2. Remove cotter pin, spring and pressure regulator valve.

NOTE: Apply pressure on pressure regulator bore before removing cotter pin since the spring is under pressure.

3. Remove oil pump cover attaching screws and the oil pump cover and gasket.
4. Remove drive gear and idler gear from pump body.

Inspection

1. Check gears for scoring or other damage, replace if necessary.
2. Proper end clearance is .0005–.0075 inch.
3. Check pressure regulator valve, valve spring and bore for damage. Proper bore to valve clearance is .0025–.0050 inch.
4. Check extension shaft ends for wear, Fig. 39.

Assembly

1. Install gears and shaft in oil pump body.
2. Check gear end clearance by placing a straight edge over the gears and measure the clearance between the straight edge

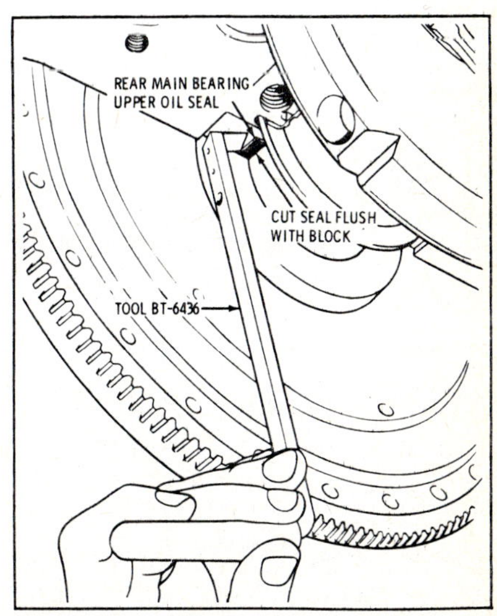

Fig. 36 Trimming upper rear main bearing seal

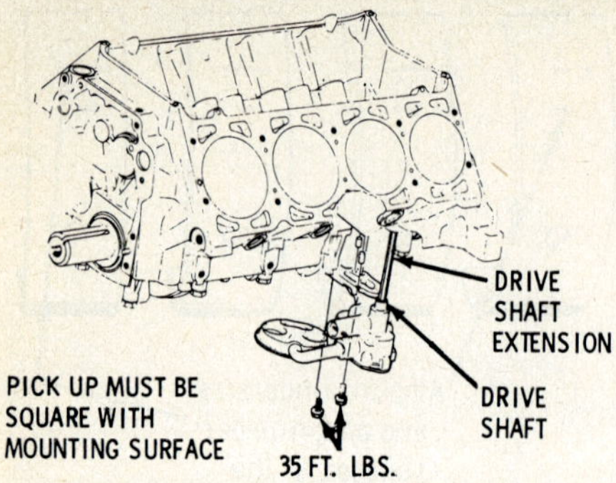

Fig. 37 Oil pump installation

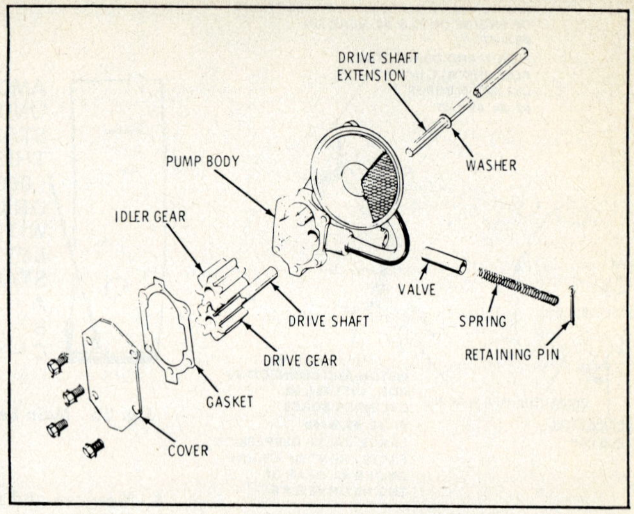

Fig. 38 Oil pump disassembled

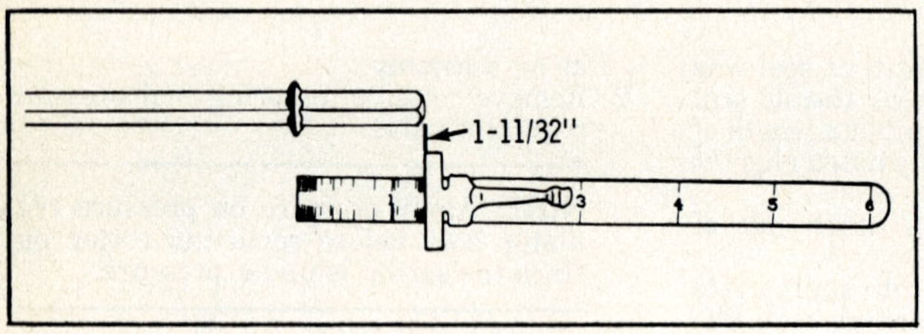

Fig. 39 Oil pump driveshaft extension

and gasket surface. If end clearance is excessive, check for scores in cover that would bring the clearance over specified limits.

3. Install cover and torque attaching screws to 8 ft. lbs.
4. Install pressure regulator valve, closed end first, into bore, then the valve spring and cotter pin.

WATER PUMP, REPLACE

1. Disconnect ground cables from batteries.
2. Drain cooling system.
3. Loosen drive belts and remove fan and pulley assembly.
4. Disconnect all hoses from water pump.
5. Remove alternator bracket, then the power steering pump and air conditioning compressor brackets, if equipped.
6. Remove water pump attaching screws and the water pump, Fig. 28.
7. Reverse procedure to install.

MECHANICAL FUEL PUMP, REPLACE

SERVICE NOTE

Some diesel engines may exhibit a condition of hard cold starts. If this condition exists, check the fuel pump housing for a crack in the area where the fuel line is connected to the fuel pump, which may allow air to enter the system. The cracked fuel pump housing could be caused by the fuel line vibrating.

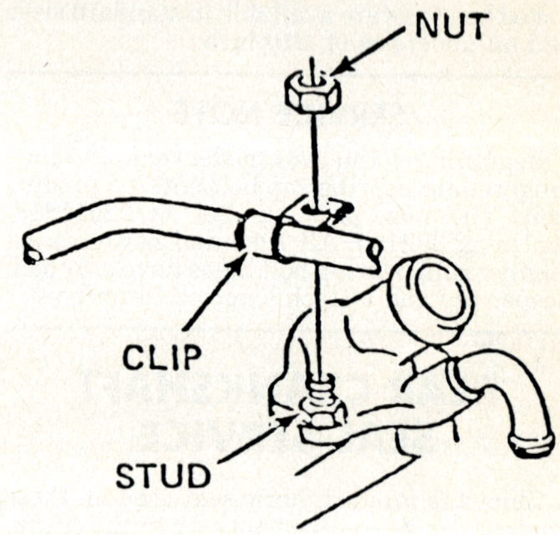

Fig. 39A Installing fuel line stud & clip assembly

To correct this condition, remove the right hand thermostat housing bolt and install stud, part No. 6270979. Disconnect fuel line at

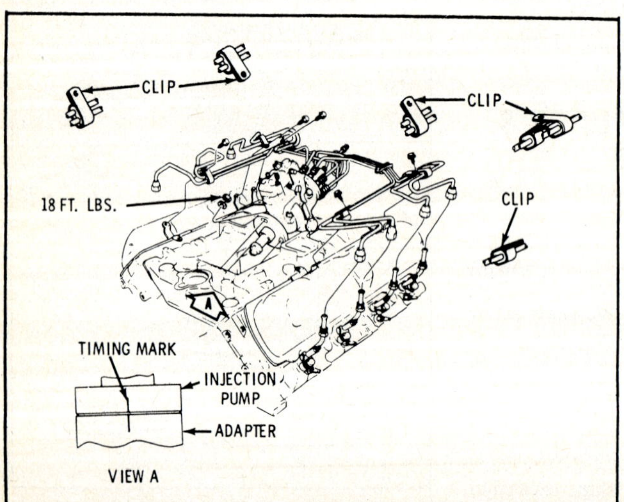

Fig. 40 Fuel injection pump timing marks

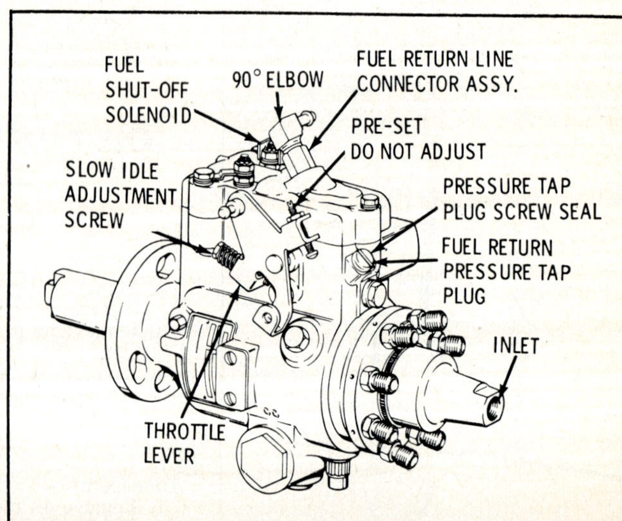

Fig. 41 Fuel injection pump connections. 1978–80

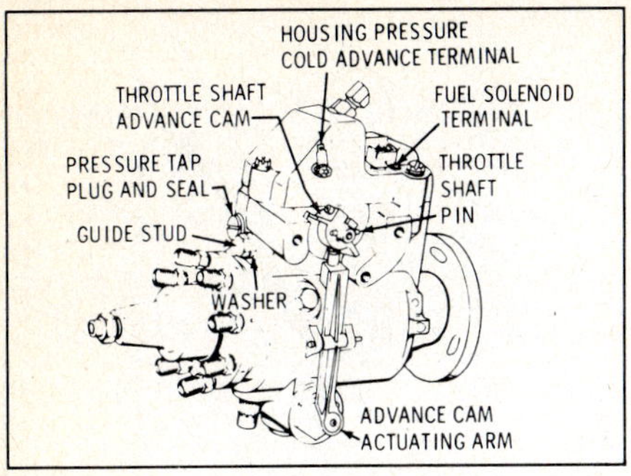

Fig. 42 Fuel injection pump housing right side view. 1981 & 1983

the fuel pump and install clip, part No. 343463 onto fuel line as shown in Fig. 39A. Install washer face nut, part No. 10008001. Replace the fuel pump and connect the fuel line.

1. Disconnect fuel lines from pump.
2. Remove fuel pump mounting bolts and the fuel pump.
3. Remove all gasket material from the pump and block gasket surfaces. Apply sealer to both sides of new gasket.
4. Position gasket on pump flange and hold pump in position against its mounting surface. Make sure rocker arm is riding on crankshaft eccentric.
5. Press pump tight against its mounting. Install retaining screws and tighten them alternately.
6. Connect fuel lines. Then operate engine and check for leaks.

SERVICE NOTE: Before installing the pump, it is good practice to crank the engine so that the nose of the crankshaft eccentric is out of the way of the fuel pump rocker arm when the pump is installed. In this way there will be the least amount of tension on the rocker arm, thereby easing the installation of the pump.

INJECTION PUMP TIMING

NOTE: Engine manufacturer recommends that diesel engine timing meter, J-33075 or equivalent, be used when setting injection timing to performance specifications. If timing meter is not available, use the following procedure.

1. The mark on the injection pump adapter must be aligned with the mark on the injection pump flange, Fig. 40.
2. To adjust:
   a. Loosen the three injection pump retaining nuts with tool J-26987.
   b. Align the mark on the injection pump flange with the mark on the injection pump adapter, Fig. 40.
   c. Torque injection pump retaining nuts to 18 ft. lbs.

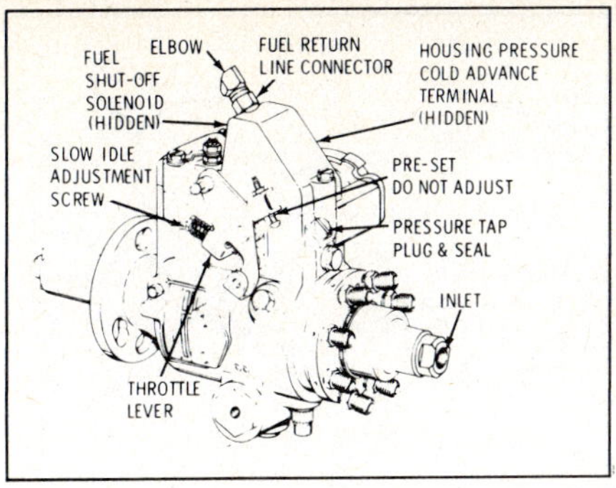

Fig. 43 Fuel injection pump housing left side view. 1981 & 1983

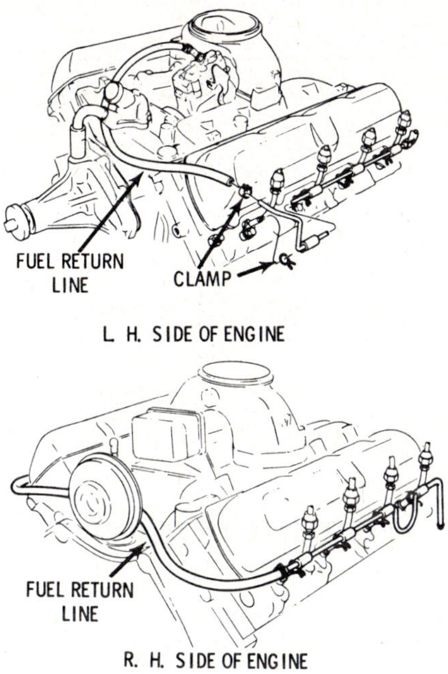

Fig. 45 Fuel return lines. 1978–80 & 1981 exc. Calif.

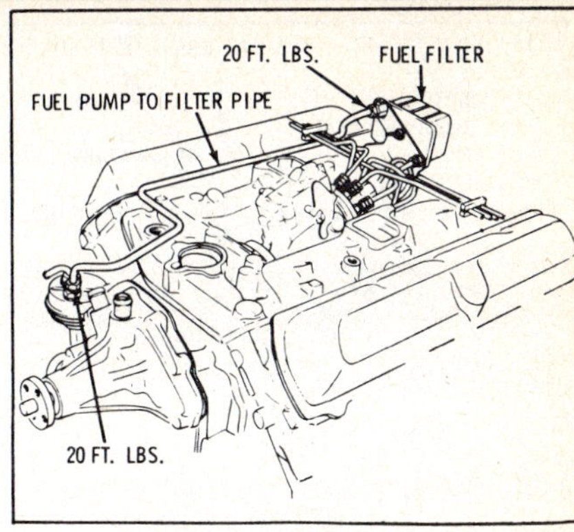

Fig. 44 Fuel filter & lines

FUEL INJECTION PUMP HOUSING FUEL PRESSURE CHECK

1. With engine thoroughly warmed up, remove air crossover and install screened covers, tool J-26996-2 or equivalent.
2. Remove fuel return pressure tap plug or torque screw, Figs. 41, 42 and 43. If equipped with torque screw, add a second nut to the lock nut and back out screw to avoid disturbing the adjustment.
3. Remove the seal from the pressure tap plug or torque screw and place seal on pressure tap adapter J-29382. Screw pressure tap adapter into pressure tap hole in pump, then screw adapter J-28526 into J-29382 adapter.
4. Attach a suitable low pressure gauge to J-28526 adapter.
5. Connect magnetic pick-up tachometer, tool J-26925, to engine.
6. With engine running at 1000 RPM in park, check fuel pressure. Pressure should be 8–12 PSI with no more than 2 PSI fluctuation.

NOTE: If vehicle is equipped with Housing Pressure Cold Advance (HPCA) and housing pressure is zero, disconnect electrical connector from housing pressure cold advance terminal, Figs. 42 and 43. If pressure remains zero, remove injection pump cover and check advance solenoid operation. If binding or not operating, free up or replace parts as necessary. If pressure is normal after electrical lead is disconnected, check for proper operation of temperature switch located on the cylinder head bolt.

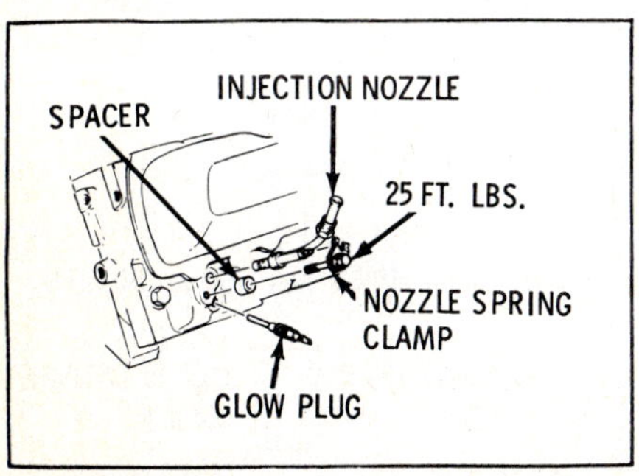

Fig. 46 Injection nozzle installation. 1978–80 & 1981 exc. Calif.

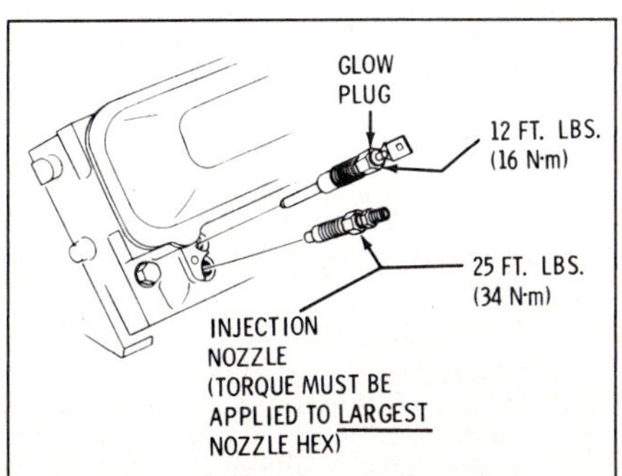

Fig. 47 Injection nozzle installation. 1981 Calif. & all 1983

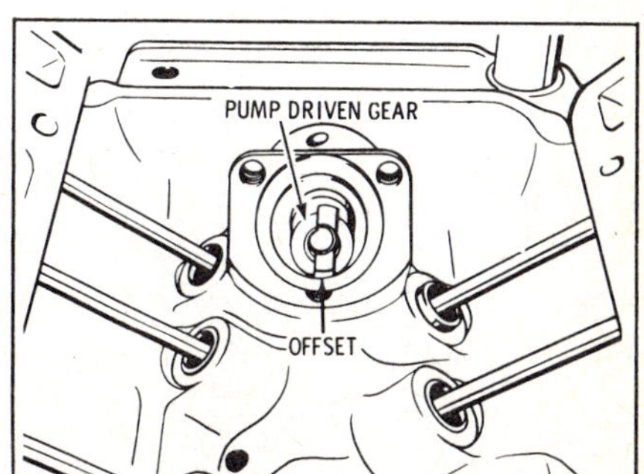

Fig. 48 Offset on fuel injection pump driven gear

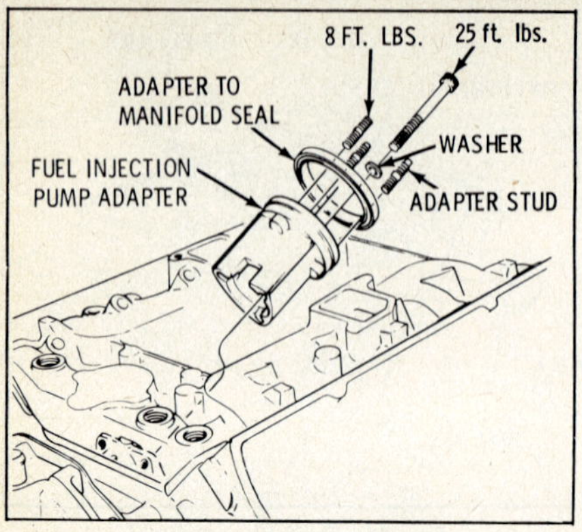

Fig. 49 Fuel injection pump adapter installation

7. If fuel pressure is too low, replace the fuel return line connector assembly. If fuel pressure is too high, the fuel return system may be restricted. Remove the fuel return line at the injection pump, then install a fitting and short piece of hose to discharge the return line fuel into a suitable container.
8. Recheck fuel injection pump housing fuel pressure. If fuel pressure is lower than in previous step, correct fuel restriction in fuel return line. If pressure is still too high, replace fuel return line connector assembly.

NOTE: Whenever the fuel return line connector assembly is replaced, the injection pump timing should be checked and reset as necessary. Refer to "Injection Pump Timing".

---

9. Recheck fuel injection pump housing fuel pressure. If still too high, remove injection pump for repair.
10. Remove tachometer, pressure gauge and adapters.
11. Install new pressure tap plug screw seal on the pressure tap plug or torque screw and replace in pump.
12. Remove screened covers and install crossover.

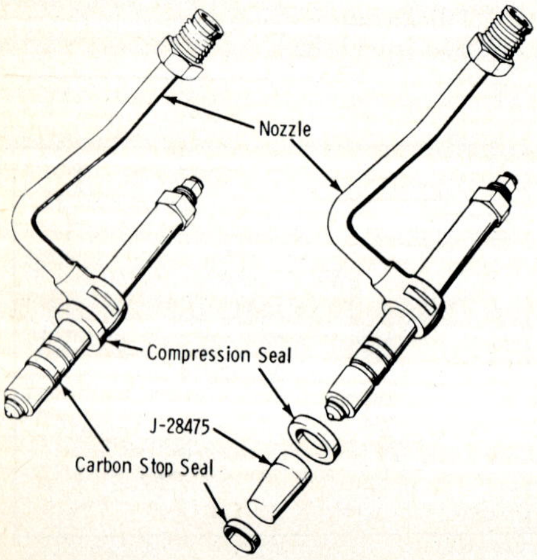

Fig. 53 Injection nozzle seal installation

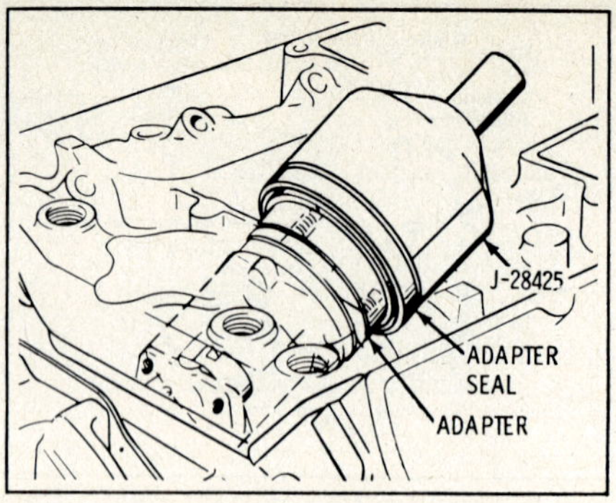

Fig. 50 Fuel injection pump adapter seal installation

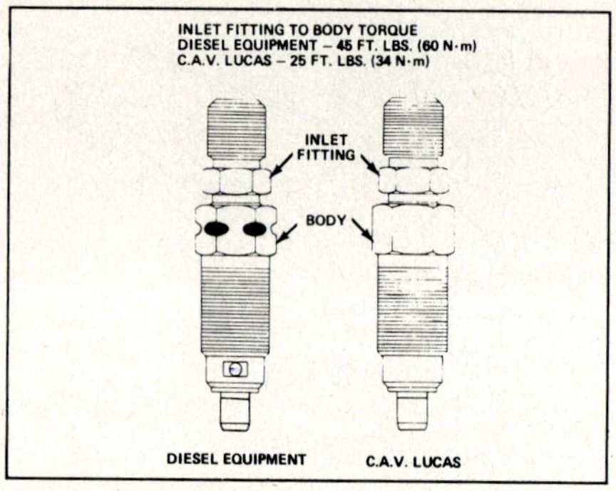

Fig. 52 Injection nozzle identification

INJECTION PUMP, REPLACE

Removal

1. Disconnect battery ground cables and remove air cleaner assembly.
2. Remove crankcase ventilation filters and pipes from valve covers and air crossover, Fig. 5.
3. Remove air crossover, then install screened covers J-26996-2 or equivalent.
4. Disconnect throttle rod and throttle return spring.
5. Remove bellcrank.
6. Remove throttle and T.V. detent cables from intake manifold brackets and position aside.
7. Remove fuel lines to fuel filter, then the fuel filter and bracket, Fig. 44. If equipped with A/C, remove rear compressor base.
8. Disconnect fuel return line from injection pump.
9. To remove injection pump only:
   a. Disconnect injection line clips,
   b. Disconnect injection lines from injection pump and plug all open lines.
   c. Reposition injection lines to gain access for pump removal.
   d. Using tool J-26987, remove 3 pump retaining nuts, then remove pump.
10. To remove injection pump with injection lines attached:
   a. On 1978–80 and 1981 models except California, slide clamp from fuel return lines at injector nozzles, Fig. 45, and remove fuel return lines from each bank.
   b. Disconnect injection pump lines at injector nozzles, Figs. 46 and 47. It is

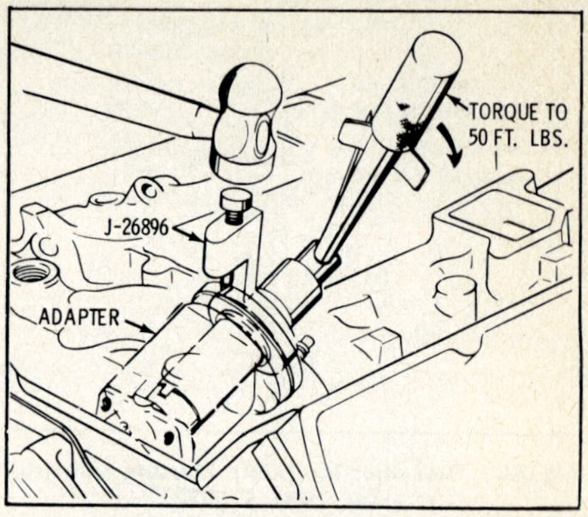

Fig. 51 Marking fuel injection pump adapter with new timing mark

necessary to use two wrenches.
   c. Remove three nuts retaining injection pump with tool J-26987.
   d. Remove injection pump and lines. Cap all lines and fittings.

Installation

1. With cylinder No. 1 at T.D.C., align offset tang on pump drive shaft with pump driven gear, Fig. 48.
2. Install new pump to adapter "O" ring, then install the pump or pump and lines, fully seating pump by hand. Loosely install 3 retaining nuts.
3. Align mark on injection pump with line on adapter, Fig. 40, then torque retaining nuts to 18 ft. lbs.
4. To install injection pump with injection lines attached:
   a. Connect injector lines at injector nozzles. Using a back up wrench on upper nozzle hex, torque line nuts to 25 ft. lbs., Figs. 46 and 47.
   b. On 1978–80 and 1981 except California, install fuel return lines to injectors on each bank, Fig. 45.
5. To install injection pump only:
   a. Position injector lines in proper position and connect injector lines to injection pump. Torque line nuts to 25 ft. lbs.
   b. Install injection line clips.
6. Connect fuel return line to injection pump.

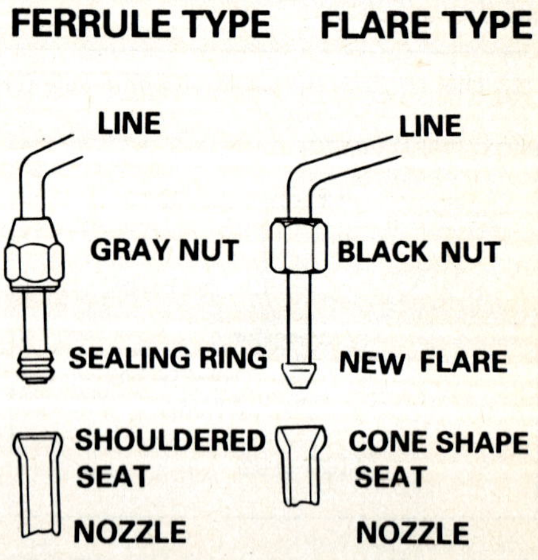

Fig. 54 Injection nozzle connections. 1979

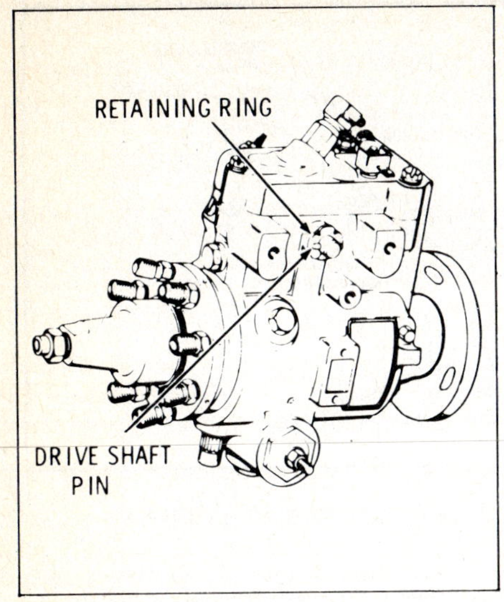

Fig. 55 Throttle shaft retention. 1978–79

7. Install fuel filter and bracket, then install fuel line. Return to Fig. 44, for torque specifications.
8. Install throttle and T.V. detent cables. Adjust T.V. cable.
9. Install bellcrank, then install throttle rod and throttle return spring. If equipped with A/C, install rear compressor brace.
10. Remove screened covers, then install air crossover.
11. Install crankcase ventilation filters and pipes to valve covers and air crossover, Fig. 5.
12. Install air cleaner assembly. Connect battery ground cables.
13. Start engine and allow to idle for several minutes. Turn ignition "Off", for approximately two minutes, then restart to allow air to bleed off within pump.

INJECTION PUMP ADAPTER, ADAPTER SEAL & NEW TIMING MARK

Adapter Seal, Replace

1. Remove injection pump. Refer to "Injection Pump, Replace".
2. Using a suitable probe, pry out adapter seal.

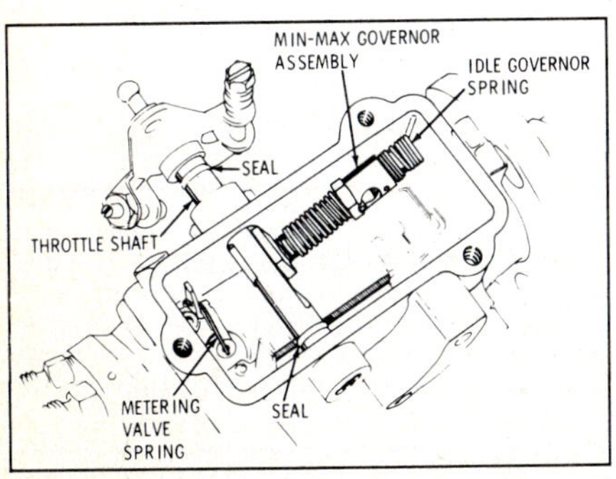

Fig. 57 Min-max governor removal

NOTE: When prying out adapter seal, use caution not to nick or gouge adapter.

3. Loosen bolts securing adapter to engine block, Fig. 49.
4. Lubricate new adapter seal inside and outside diameter with chassis lube.
5. Using seal installer J-28425, install seal on adapter, Fig. 50.
6. Remove seal installer tool and inspect seal for proper installation.
7. Torque adapter bolts to 25 ft. lbs.
8. Refer to "Marking Injection Pump Adapter" for remaining procedure.

Adapter, Replace

1. Remove injection pump. Refer to "Injection Pump, Replace".
2. Remove bolts securing adapter into engine block, Fig. 49. Then remove adapter.
3. Install replacement adapter in engine block and install bolts. Do not tighten.
4. Refer to "Adapter Seal, Replace" steps 4 through 8, for remaining procedure.

Marking Injection Pump Adapter

1. Using a suitable file, remove original timing mark from adapter. Do not remove timing mark from injection pump.
2. With cylinder No. 1 approaching TDC, align timing mark on balancer with zero mark on indicator. The injection pump driven gear should be slightly offset to the right when cylinder No. 1 is at TDC, Fig. 48.
3. Install timing tool J-26896, into pump adapter, Fig. 51. Using a suitable torque wrench, torque tool towards cylinder No. 1 to 50 ft. lbs. While holding torque steady, mark pump adapter, Fig. 51. Remove tool.
4. Reinstall injection pump. Refer to "Injection Pump, Replace".

INJECTION NOZZLE, REPLACE

1981 Calif. Models & 1983 All

SERVICE NOTE

Injection nozzle body leaks may be corrected by loosening the inlet fitting and retorquing to 45 ft. lbs. on DED type injection nozzles, and 25 ft. lbs. on CAV type injection nozzles, Fig. 52. In the event this does not correct the leak, remove the inlet fitting. Using a piece of crocus cloth, press and rotate the end of the inlet fitting against the crocus cloth back and forth about six times. After polishing has been performed, flush inlet fitting using diesel fuel and install fitting onto pump.

1. Remove fuel lines, using a backup wrench on upper injection nozzle hex.
2. Remove nozzle by applying torque to largest nozzle hex, Fig. 47.
3. Cap nozzle and lines to prevent entry of dirt. Also remove copper gasket from cylinder head if gasket did not remain with nozzle.
4. Reverse procedure to install. Torque nozzle to 25 ft. lbs. When tightening nozzle, torque must be applied to largest nozzle hex. Torque fuel line to 25 ft. lbs. using a backup wrench on upper injection nozzle hex.

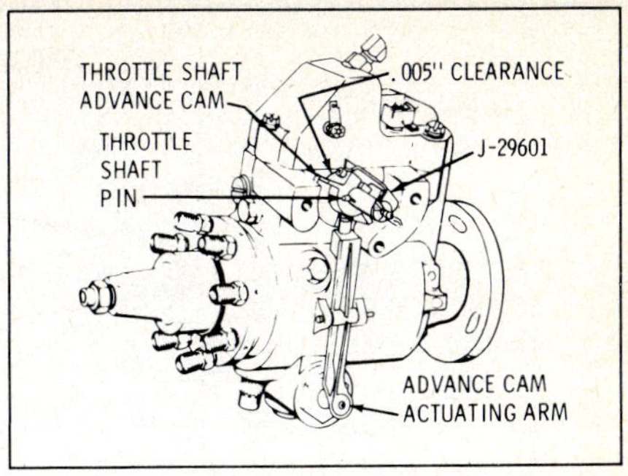

Fig. 56 Injection pump with tool J-29601 installed

1978–80 Models & 1981 Exc. Calif.

1. Remove fuel line from injector nozzle.
2. Remove fuel line clamps from all nozzles on bank where nozzle is being removed. Remove fuel return line from nozzle being replaced.
3. Remove nozzle hold down clamp and spacer, Fig. 46, then the nozzle with tool J-26952.
4. Cap nozzle inlet line and tip of nozzle.
5. Reverse procedure to install. Install new seals on injection nozzles, Fig. 53. Torque nozzle hold down clamp bolt to 25 ft. lbs.

NOTE: 1979 diesel engines use two different types of connections to attach the high pressure fuel lines to the injection nozzles. Before replacing an injection nozzle or high pressure fuel line, it is necessary to determine which type of connection is used, either the "Flare Type" or "Ferrule Type", Fig. 54. Also, check the replacement part to ensure that the proper connection is used.

THROTTLE SHAFT SEAL, REPLACE

1. Disconnect battery ground cables, then remove the air cleaner and crossover, and install screened covers, J-26996-2 or equivalent.
2. Disconnect injection pump fuel solenoid wire and fuel return line. On 1981 and 1983 models, disconnect housing pressure cold advance (HPCA) wires, Fig. 42.
3. On 1978–80 models, remove the trans-

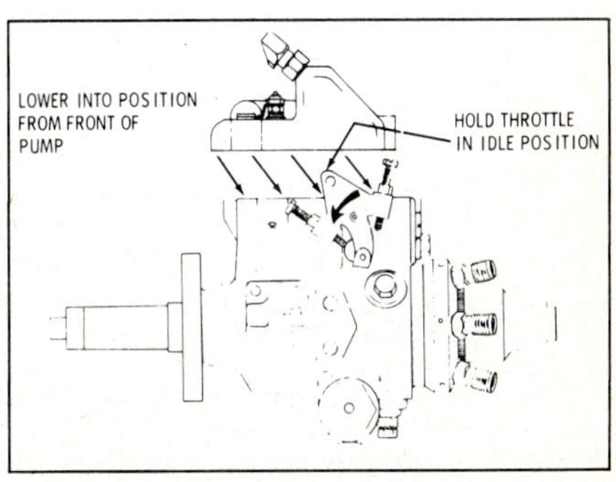

Fig. 58 Installing injection pump cover

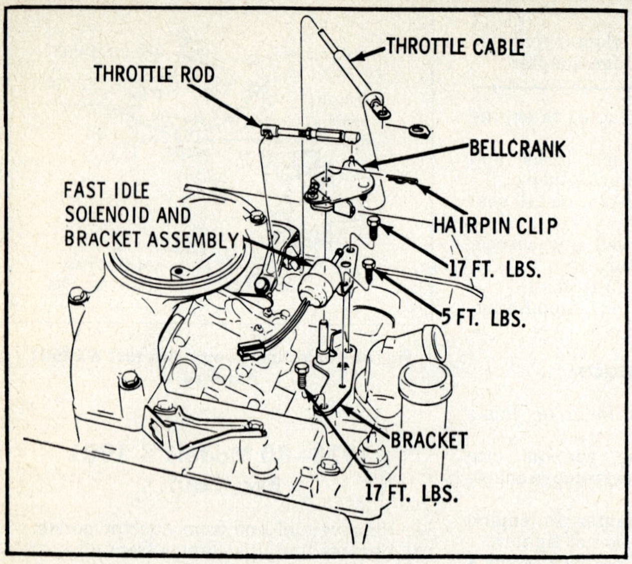

Fig. 59 Throttle linkage

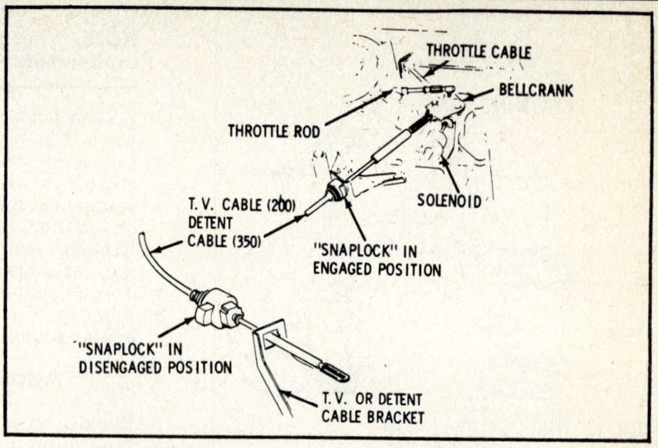

Fig. 60 Throttle valve or detent cable adjustment

mission vacuum regulator valve. On 1981 and 1983 models, scribe a line on the vacuum regulator valve and pump body to aid in reassembly, and remove the vacuum regulator. Refer to "Transmission Vacuum Regulator Valve, Replace".

4. Remove the throttle rod, return spring and throttle cable bracket.
5. On 1978–79 models, drive the throttle shaft drive pin out of shaft, then remove the shaft retaining ring and fiber washer, Fig. 55. On 1980–81 and 1983 models, position tool J-29601 over throttle shaft with tool slots engaging pin. Place the tool spring clip over the throttle shaft advance cam and tighten the wing nut, Fig. 56. Without loosening the wing nut, pull tool from throttle shaft. This will maintain proper alignment during reassembly. Drive pin from throttle shaft, loosen the clamp screw and remove the throttle shaft advance cam and fiber washer.
6. Remove any burrs on throttle shaft with shaft still installed in pump.
7. Thoroughly clean the injection pump cover, the upper portion of the pump and the throttle shaft and guide stud area. Place shop towels in the engine valley to catch spilled fuel.
8. Remove injection pump top cover, then remove screws from cover.

NOTE: When the injection pump top cover is removed, extreme care must be taken to keep foreign materials out of the pump. If any objects are dropped in the injection pump, they must be removed before the engine is started to prevent engine or injection pump damage.

9. Note position of metering valve spring over top of guide stud. This position must be duplicated exactly during reassembly.
10. Remove guide stud and washer, Fig. 42. Note part location prior to removal.
11. Rotate min-max governor assembly upwards to provide clearance, then remove from throttle shaft, Fig. 57.
12. Remove throttle shaft assembly and examine shaft for unusual wear or damage. Replace if necessary.

NOTE: When removing throttle shaft assembly, it may be necessary to loosen the injection pump mounting nuts and rotate pump slightly to provide clearance for shaft removal.

13. Examine pump housing bushings for unusual wear or damage. If wear is evident, the pump should be removed and sent to manufacturer for repairs.
14. Remove throttle shaft seals. Do not attempt to cut seals to remove, as nicks in the seat will cause fuel leakage.
15. Install replacement shaft seals taking care not to cut seals on shaft edges. Apply chassis lube lightly to the seals.
16. Slide the throttle shaft assembly carefully into the pump body to the point the min-max governor assembly will slide back onto the throttle shaft assembly, Fig. 57.
17. Rotate the min-max governor assembly downward into the pump body, then while holding in position, slide the throttle shaft and governor assembly into position.
18. On 1978–79 models, install new fiber washer, throttle shaft retaining ring and throttle shaft drive pin, Fig. 55. On 1980–81 and 1983 models, install new mylar washer, throttle shaft advance cam (do not tighten screw at this time) and throttle shaft drive pin. Align the throttle shaft advance cam so tool J-29601 can be reinstalled over the throttle shaft, pin in the slots and the spring clip over the advance cam, Fig. 56. Insert a .005 inch feeler gauge between the mylar washer on the throttle shaft and the pump housing. Squeeze the cam and throttle shaft together and tighten the cam screw. Torque to 30 in. lbs. Secure with suitable locking compound.
19. Install guide stud with new washer, Fig. 42. Ensure the metering valve spring upper extension rides on top of the guide stud. Torque guide stud to 85 in. lbs. Do not overtorque.
20. Hold throttle in idle position.
21. Install new pump cover seal. Ensure the securing screws are not in the cover and position the cover approximately 1/4 inch forward and 1/8 inch above the pump.
22. Move cover rearward and downward into proper position, using caution not to cut the seal, then install pump cover screws, Fig. 58. Torque screws to 35 in. lbs.

NOTE: Use caution not to drop and lose flat washer and internal lock washer with each screw. Flat washers must be against pump cover.

23. On 1978–80 models, install vacuum regulator valve. On 1981 and 1983 models, install vacuum regulator valve aligning marks made previously.
24. Connect battery ground cables.
25. Turn ignition to "Run" position. Touch pink solenoid wire to the injection pump solenoid terminal. A clicking noise should be heard indicating the fuel solenoid is operating. If clicking noise is heard, proceed to step 28. If no clicking noise is heard, the solenoid linkage may be jammed in a wide open position and the engine must not be started. Proceed to step 26.
26. Remove injection pump top cover, then ground the solenoid lead opposite the hot lead and connect the pink wire. With ignition switch in the "Run" position, the solenoid should move the linkage. If not, replace the solenoid.

NOTE: The minimum voltage across the solenoid terminals must be 12 volts.

27. Reinstall cover and repeat step 25.
28. Install throttle cable bracket and throttle rod, then install throttle cable and return spring. Connect solenoid wire and on 1981 and 1983 models housing pressure cold advance (HPCA) wire.
29. Install fuel return line. Ensure the timing mark on the pump is aligned with the adapter timing mark. Also ensure the injection pump retaining bolts are tight.
30. Start engine and check for leaks.

NOTE: If engine roughness is observed, it may be due to air in the pump. To purge air, let engine idle for several minutes. It may be necessary to shut engine down for several minutes to allow air bubbles to rise to top of pump where they will be purged.

31. Remove intake manifold screens, then reinstall air crossover and air cleaner.

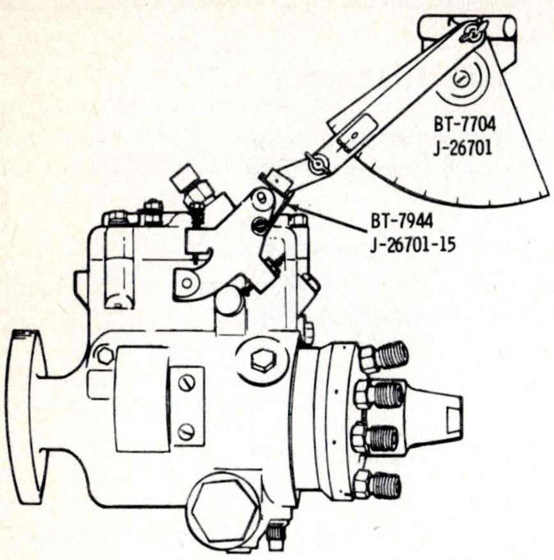

Fig. 61 Transmission vacuum regulator valve adjustment. 1979–83 Turbo Hydra-Matic 350

TRANSMISSION VACUUM REGULATOR VALVE, REPLACE

1. Note location of the valve vacuum hoses, then disconnect the vacuum hoses.
2. Remove the two valve attaching bolts and the valve.
3. Reverse procedure to install.
4. Adjust vacuum valve. Refer to "Transmission Vacuum Regulator Valve, Adjust".

THROTTLE ROD, ADJUST

1. If equipped with Cruise Control, remove clip from Cruise Control rod, then the rod from bellcrank.
2. Remove throttle valve cable from bellcrank, Fig. 59.
3. Loosen the throttle rod locknut and shorten the rod several turns.
4. Rotate the bellcrank to the full throttle stop, then lengthen the throttle rod until the injection pump lever contacts the injection pump full throttle stop. Release the bellcrank.
5. Tighten the throttle rod locknut.
6. Connect the throttle valve cable and Cruise Control rod, if equipped, to bellcrank.

THROTTLE VALVE OR DETENT CABLE, ADJUST

1. Remove throttle rod from bellcrank, Fig. 60.
2. Push snap lock to disengaged position.
3. Rotate bellcrank to full throttle stop position and push in snap lock until flush with cable end fitting. Release bellcrank.
4. Connect the throttle rod.

TRANSMISSION VACUUM REGULATOR VALVE, ADJUST

1979–83

1. Remove air crossover, then install screen covers J26996-2.
2. Remove throttle rod from throttle lever, then loosen transmission regulator vacuum valve injection pump bolts.
3. Install carburetor angle gauge adapter, J-26701-15 on injection pump throttle lever, then place angle gauge, J-26701 on adapter, Fig. 61.
4. Rotate throttle lever to wide open position and set angle gauge to zero degrees.
5. Center bubble in level, then set angle gauge to 49 degrees on 1979 units, 50 degrees on 1980 units and 58 degrees on 1981–83 units.
6. Rotate throttle lever so level bubble is centered.
7. Attach a suitable vacuum pump to center port of vacuum regulator valve, then install a vacuum gauge to outside port of vacuum regulator valve. Apply 18–22 inches of vacuum.
8. Rotate vacuum regulator valve clockwise to obtain 10.6 inches of vacuum on 1983 models, 8½ to 9 inches of vacuum on 1979 and 1981 models and 7 to 8 inches of vacuum on 1980 models.
9. Tighten vacuum regulator to injection pump retaining bolts, then remove vacuum gauge and pump.
10. Install throttle rod to bellcrank, then remove screened covers and install air crossover and air cleaner.

1978

1. Remove throttle rod from bellcrank.
2. Loosen transmission vacuum regulator valve attaching bolts to disengage valve from injection pump shaft.
3. Hold the injection pump lever against injection pump full throttle stop, rotate vacuum regulator valve to full throttle position and insert a .090 inch diameter pin to hold the valve in the full throttle position, Fig. 62.
4. Rotate assembly clockwise until the injection pump shaft lever is contacted, then tighten the valve attaching bolts.
5. Remove pin and release lever.
6. Connect throttle rod.

PURGING WATER FROM FUEL TANK

1980–83 Models With Water In Fuel Detector

Single Tank Models

Water in the fuel tank may be purged by siphoning or using a pump. The pump or siphon hose should be connected to the ¼ inch fuel return hose (smaller of the two hoses) under the hood near the fuel pump. Purging should continue until all water is removed from the fuel tank. Also, remove fuel filler cap while purging, and replace cap when completed.

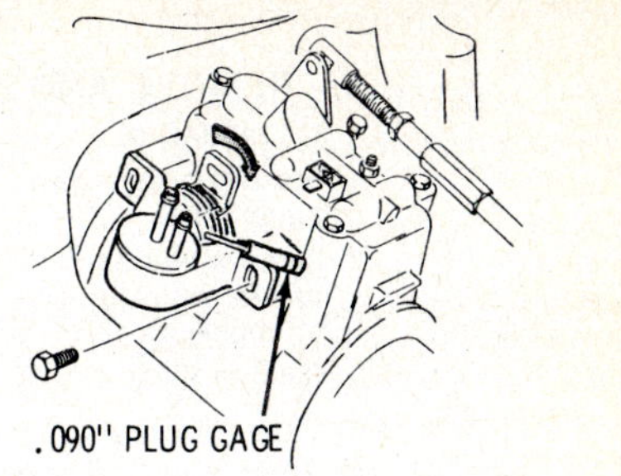

Fig. 62 Transmission vacuum valve adjustment. 1978 Turbo Hydra-Matic 350

CAUTION: Use all safety precautions while handling the fuel/water mixture.

Dual Tank Models

Water in the dual fuel tanks is purged in much the same way as single tank models, except that the fuel selector control should be switched to the main tank position. Siphoning should continue until all water is removed from the main tank. Place ignition switch in the "On" position, and move the fuel selector control to the auxiliary position. Turn ignition "Off" and continue to siphon water from auxiliary tank until all water is removed.

NOTE: Auxiliary tank must be siphoned within twenty minutes. If water cannot be siphoned within twenty minutes, cycle ignition switch "Run" and back to "Off" to obtain another twenty minute siphon period.

When siphoning water from tanks, fuel filler caps should be removed and reinstalled when siphoning is completed.

CAUTION: Use all safety precautions while handling the fuel/water mixture.

VACUUM PUMPS

NOTE: On engines equipped with gear driven type vacuum pump, do not operate engine unless vacuum pump and drive assembly are installed, as the drive gear of the vacuum pump also drives the engine lubricating oil pump.

On these engines, a vacuum pump is required to provide vacuum for accessory operation. The vacuum pump maybe either gear driven or belt driven, depending on vehicle application. The gear driven pump has a drive gear located at the lower end of the shaft which meshes with a gear on the engine camshaft. This vacuum pump drive gear also drives the engine lubricating oil pump. The belt driven pump has a pulley attached to the lower end of the shaft and is driven by an accessory belt.

When servicing the vacuum pump, refer to Fig. 63 for service procedures.

1. REMOVE AND INSTALL

REMOVE

1. Remove hose from pump inlet.
2. Remove bolt and bracket holding pump to engine block.
3. Remove pump.

INSTALL

1. Insert pump in engine, making sure the gears on the pump mesh with the gears on the engine cam shaft.
2. Rotate the pump into position so the bracket and bolt can be installed.
3. Install vacuum hose.

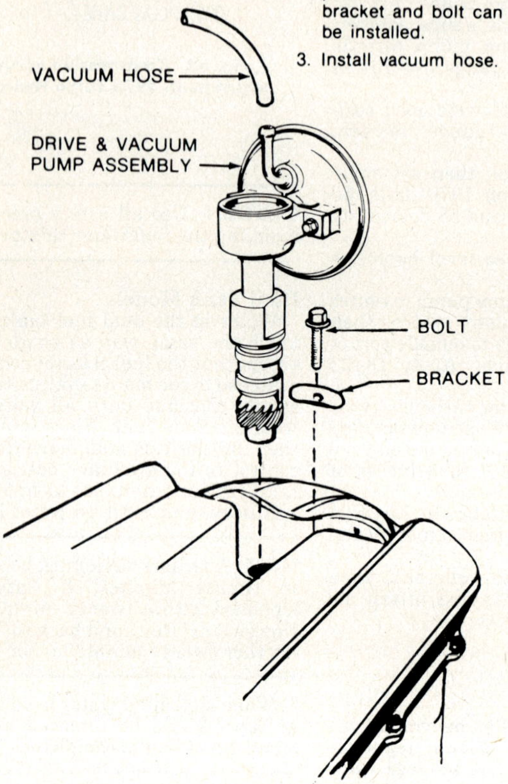

Removing assembly from engine.

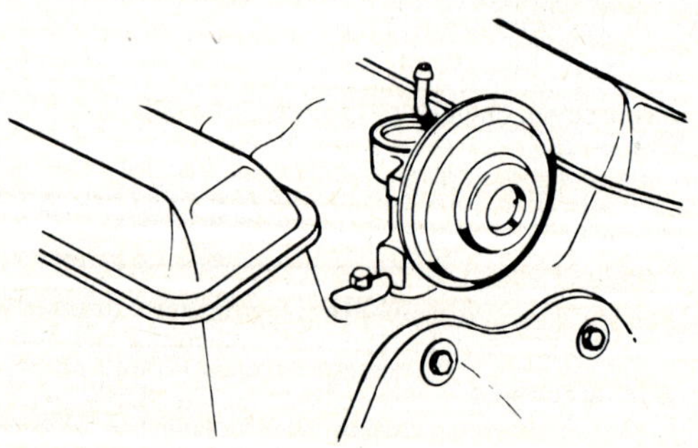

View showing assembly from rear of engine.

2. DISASSEMBLE AND ASSEMBLE

CAUTION: *Take extreme care when clamping the vacuum pump in vise.*

DISASSEMBLE

1. Remove hex head tapping screws.
2. Remove "O" Ring Seal.

ASSEMBLE

1. Install "O" Ring Seal in cavity of drive assembly.
2. Connect drive assembly to vacuum pump with hex head tapping screws.

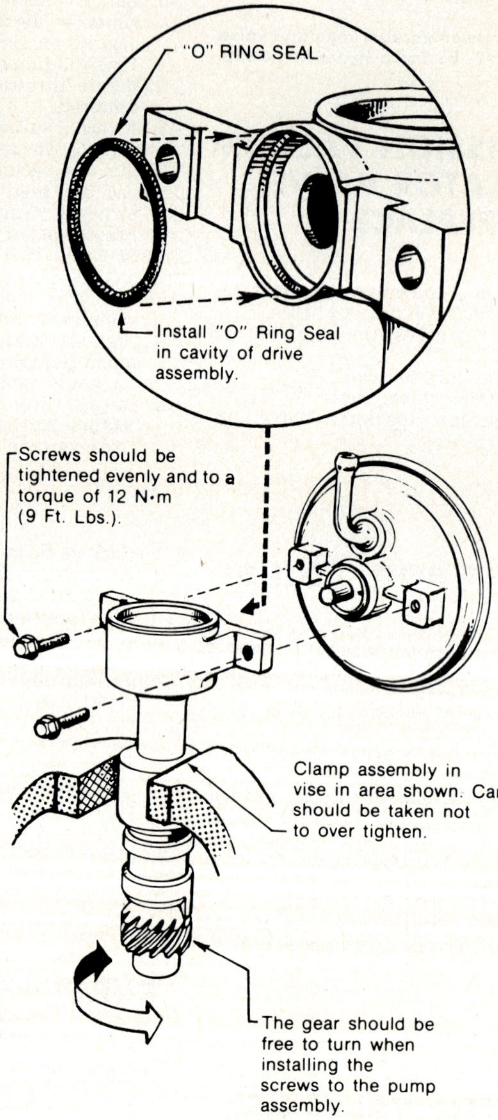

Fig. 63 Vacuum Pump service